Concise Encyclopedia Chemistry

Translated and revised by
Mary Eagleson

Walter de Gruyter Berlin · New York 1994

Title of the original, German language edition
ABC Chemie

Edited by
Dr. Hans-Dieter Jakubke and Dr. Hans Jeschkeit

Translated into English and revised by
Mary Eagleson, Ph. D.

♾ Printed on acid-free paper, which falls within the guidelines of the ANSI
to ensure permanence and durability.

Library of Congress Cataloging-in-Publication Data

ABC Chemie. English.
Concise encyclopedia chemistry / translated and revised by
Mary Eagleson.
ISBN 0-89925-457-8
1. Chemistry–Encyclopedias. I. Eagleson. Mary, 1945–
II. Title.
QD4.A2313 1994
540'.3–dc20

Die Deutsche Bibliothek – CIP-Einheitsaufnahme

Concise encyclopedia chemistry / transl. and rev. by
Mary Eagleson. – Berlin ; New York : de Gruyter, 1994
Dt. Ausg. u. d. T.: Fachlexikon ABC Chemie
ISBN 3-11-011451-8
NE: Eagleson, Mary [Übers.]

English language edition

Typesetting and Printing: Buch- und Offsetdruckerei Wagner GmbH, Nördlingen.
Binding: Lüderitz und Bauer, Berlin.
Cover Design: Hansbernd Lindemann, Berlin.

of the Elemen

radioactive elements the value in
entheses refers to the number of
cleons (mass number) of the isotope
h the longest half-life.

: gaseous
en: liquid
ck : solid
te : all isotopes radioactive

at STP – standard temperature and pressure ≙ 0° C and 1.0 bar

	Ib	IIb	IIIa	IVa	Va	VIa	VIIa	VIIIa
								2 4,003 – −268,6 – **He** Helium $1s^2$
			5 10,81 2,0 – 2300 **B** Boron $[He]2s^2p^1$	6 12,011 2,5 4827 3550 **C** Carbon $[He]2s^2p^2$	7 14,007 3,1 −195,8 −209,9 **N** Nitrogen $[He]2s^2p^3$	8 15,999 3,5 −183,0 −218,4 **O** Oxygen $[He]2s^2p^4$	9 18,998 4,1 −188,1 −219,6 **F** Fluorine $[He]2s^2p^5$	10 20,179 – −246,1 −248,7 **Ne** Neon $[He]2s^2p^6$
			13 26,982 1,5 2467 660,4 **Al** Aluminium $[Ne]3s^2p^1$	14 28,086 1,7 2355 1410 **Si** Silicon $[Ne]3s^2p^2$	15 30,974 2,1 280(P4) 44(P4) **P** Phosphorus $[Ne]3s^2p^3$	16 32,06 2,4 444 114,6 **S** Sulfur $[Ne]3s^2p^4$	17 35,453 2,8 −34,6 −101,0 **Cl** Chlorine $[Ne]3s^2p^5$	18 39,948 – −185,7 −189,2 **Ar** Argon $[Ne]3s^2p^6$
3 58,70 3 '32 453 **Ni** Nickel $[Ar]3d^84s^2$	29 63,55 1,8 2595 1083 **Cu** Copper $[Ar]3d^{10}4s^1$	30 65,38 1,7 907 419,6 **Zn** Zinc $[Ar]3d^{10}4s^2$	31 69,72 1,8 2403 29,8 **Ga** Gallium $[Ar]3d^{10}4s^2p^1$	32 72,59 2,0 2830 937,4 **Ge** Germanium $[Ar]3d^{10}4s^2p^2$	33 74,92 2,2 subl. **As** Arsenic $[Ar]3d^{10}4s^2p^3$	34 78,96 2,5 685 217 **Se** Selenium $[Ar]3d^{10}4s^2p^4$	35 79,90 2,7 58,8 −7,2 **Br** Bromine $[Ar]3d^{10}4s^2p^5$	36 83,80 – −152,3 −156,6 **Kr** Krypton $[Ar]3d^{10}4s^2p^6$
5 106,4 4 40 52 **Pd** Palladium $[Kr]4d^{10}$	47 107,87 1,4 2212 962 **Ag** Silver $[Kr]4d^{10}5s^1$	48 112,41 1,5 765 320,9 **Cd** Cadmium $[Kr]4d^{10}5s^2$	49 114,82 1,5 2080 156,6 **In** Indium $[Kr]4d^{10}5s^2p^1$	50 118,71 1,7 2270 231,9 **Sn** Tin $[Kr]4d^{10}5s^2p^2$	51 121,75 1,8 1750 630,7 **Sb** Antimony $[Kr]4d^{10}5s^2p^3$	52 127,60 2,0 890 449,5 **Te** Tellurium $[Kr]4d^{10}5s^2p^4$	53 126,90 2,2 184,4 113,5 **I** Iodine $[Kr]4d^{10}5s^2p^5$	54 131,30 – −107 −111,9 **Xe** Xenon $[Kr]4d^{10}5s^2p^6$
3 195,1 4 3830 72 **Pt** Platinum $[Xe]4f^{14}5d^96s^1$	79 196,97 – 2940 1064 **Au** Gold $[Xe]4f^{14}5d^{10}6s^1$	80 200,59 – 356,6 −38,9 **Hg** Mercury $[Xe]4f^{14}5d^{10}6s^2$	81 204,37 1,4 1457 303,5 **Tl** Thallium $[Xe]4f^{14}5d^{10}6s^2p^1$	82 207,2 1,6 1740 327,5 **Pb** Lead $[Xe]4f^{14}5d^{10}6s^2p^2$	83 208,98 1,7 1560 271,3 **Bi** Bismuth $[Xe]4f^{14}5d^{10}6s^2p^3$	84 (209) 1,8 962 254 **Po** Polonium $[Xe]4f^{14}5d^{10}6s^2p^4$	85 (210) 2,0 – – **At** Astatine $[Xe]4f^{14}5d^{10}6s^2p^5$	86 (222) – – – **Rn** Radon $[Xe]4f^{14}5d^{10}6s^2p^6$

Elements 107 – 109
Names and symbols have been suggested for the elements 107–109:
Nielsbohrium Ns (after Niels Bohr), Hassium Hs (named after Hessen; lat. Hassia)
and Meitnerium Mt (after Lise Meitner); these have yet to be approved by the IUPAC.

3 151,96) 97 22 **Eu** Europium $[Xe]4f^76s^2$	64 157,25 1,1 3233 1312 **Gd** Gadolinium $[Xe]4f^75d^16s^2$	65 158,93 1,1 3041 1360 **Tb** Terbium $[Xe]4f^96s^2$	66 162,50 1,1 2335 1409 **Dy** Dysprosium $[Xe]4f^{10}6s^2$	67 164,93 1,1 2720 1470 **Ho** Holmium $[Xe]4f^{11}6s^2$	68 167,26 1,1 2510 1522 **Er** Erbium $[Xe]4f^{12}6s^2$	69 168,93 1,1 1727 1545 **Tm** Thulium $[Xe]4f^{13}6s^2$	70 173,04 1,1 1193 824 **Yb** Ytterbium $[Xe]4f^{14}6s^2$	71 174,97 1,1 3315 1656 **Lu** Lutetium $[Xe]4f^{14}5d^16s^2$
5 (243) 1,2 – 1000 **Am** Americium $[Rn]5f^77s^2$	96 (247) ≈1,2 – ≈1340 **Cm** Curium $[Rn]5f^76d^17s^2$	97 (247) ≈1,2 – – **Bk** Berkelium $[Rn]5f^97s^2$	98 (251) ≈1,2 – – **Cf** Californium $[Rn]5f^{10}7s^2$	99 (254) ≈1,2 – – **Es** Einsteinium $[Rn]5f^{11}7s^2$	100 (257) ≈1,2 – – **Fm** Fermium $[Rn]5f^{12}7s^2$	101 (258) ≈1,2 – – **Md** Mendelevium $[Rn]5f^{13}7s^2$	102 (259) – – – **No** Nobelium $[Rn]5f^{14}7s^2$	103 (260) – – – **Lr** Lawrencium $[Rn]5f^{14}6d^17s^2$

Concise Encyclopedia Chemistry

From the preface of the original edition

This **Concise Encyclopedia of Chemistry** offers an informative overall view of the wide field of chemistry in a handy, compact volume.

Care has been taken to combine scientific precision with a clear presentation so that this handbook is not just a useful reference tool for chemists and laboratory technicians but also for lecturers, teachers and students of chemistry, biochemistry, pharmacy, biology or medicine.

The publishers and editors were presented with the difficult task of processing the vast increase of knowledge accumulated over the past twenty years and presenting it within the tried and tested framework of an alphabetical encyclopedia. It was at this point that the limits imposed on the volume of the book became clear.

Aspects of other related fields were kept to a minimum; please note that an encyclopedia of biochemistry has also been published*. A certain shift in emphasis can be seen towards the chemical substances and their structural formulas. However, apart from the chemical elements, their reactions and compounds, the most important natural substances, synthetics, pharmaceuticals, dyes etc. have been included, as well as the complex atomic and molecular structure of matter, stoichiometry, analysis, catalysis, chemical kinetic reactions and thermodynamics, electrochemistry, colloid chemistry, carbon chemistry and petrochemistry, etc.

An encyclopedia cannot – and should not – take the place of a textbook; rather, should give a concentrated overview of chemistry, such as a handbook does. However, a book such as this can provide a mine of information for the reader, who uses it not just as a reference work to be consulted in cases of sudden need, but will also often just browse through its pages.

The **Concise Encyclopedia of Chemistry** contains around 12,000 entries taken from the fields of general, inorganic, organic, physical and technical chemistry, complemented by some 1,600 figures and 300 tables. References have not been given due to the problem of the relatively long life of such encyclopedic works.

The internationally valid nomenclature has been used and a detailed entry "Nomenclature" informs the reader of the IUPAC regulations for the naming of chemical elements and compounds. The book contains a periodic table, together with tables of the elements, SI units and physical constants.

The publishers and editors are aware that, in spite of their efforts, it is impossible to include all aspects of chemistry in one manageable volume. Some readers may feel that the choice of entries, the weighting of the different fields of chemistry, and the method and depth of presentation are not always the best. Specialists in the various chemical disciplines may think that their particular area has not received its due attention. Everyone involved, scientists in research, teaching or industry, has done their utmost to ensure that this encyclopedia is of a high standard. Therefore, the publishers, editors and authors alike welcome your suggestions for the revision of any ensuing editions.

* **Concise Encyclopedia Biochemistry**, Second Edition.
Revised and expanded by Thomas Scott and Mary Eagleson.
1988. 650 pages. Hardcover. ISBN 3-11-011625-1
Published by Walter de Gruyter, Berlin · New York

How to use this book

Throughout this book, the chemical elements and compounds have been namend mainly according to the recommended rules of the IUPAC (International Union for Pure and Applied Chemistry).

Entries are listed in alphabetical order. Numbers, Greek letters and configurational numbers/ letters at the beginning of names are ignored in the allocations of alphabetical order, e.g. 2-Nitro-benzaldehyde is listed under N; α-Oximino ketones is listed under O.

Standard abbreviations, e.g. NMR, DNA, EDTA etc., are found as entries in the appropriate alphabetical positions. If the main entry title is repeated in the text, then the entry title is abbreviated to its first letter, e.g. "**Heme:** an iron (II) complex of a porphyrin. H. are found in nature". An exception to this rule can be observed under the entries of the elements where the entry title is either abbreviated to its first letter or its chemical symbol is given, e.g. "**Cadmium**, symbol *Cd* ... Properties: Cd is a silvery white, relatively soft metal". Other abbreviations – when not explained in the text – are listed below.

Cross referencing is indicated by the word "see" and the subject of the cross reference starts with a capital letter, e.g. "... is transported mainly in Pipelines (see) ...", or "... see Nickel sulfate".

The main entry title is printed in bold type, followed by synonyms in bold italics. Any further relevant terms within the text are stressed by means of italics.

Physical quantities are given in SI-units.

Some of the figures and formulas were taken directly from the original German book. Therefore, it is possible that, in some cases, the German abbreviations still remain in the figures e.g. AS = AA, DNS = DNA, Peptid = peptide.

Other abbreviations:

abb.	abbreviation
[α]	specific optical rotation
b.p.	boiling point
c	concentration
°C	degrees Celsius
crit.	critical
(d.)	with decomposition
d	density
IP	isoelectric point
M	molar mass
M	molar
m.p.	melting point
M_r	relative molecular mass
n	refractive index
syn.	synonym
T	temperature
V	volume
Z	atomic number

How to use this book

A

a: symbol for axial bonds; see Sterioisomerism, 2.2.

A: 1) formula symbol for absolute atomic mass. 2) *A*, symbol for activity of a radioactive substance. 3) A_E, see Electrophilic reactions. 4) A_N, see Nucleophilic reactions. 5) A_r, symbol for relative atomic mass. 6) A_R, see Radical reactions.

α: configuration symbol. In organic nomenclature, it denotes a substituent on the carbon adjacent to a principal group; for example,

$$CH_3{-}\underset{\displaystyle NH_2}{\underset{|}{CH}}{-}COOH$$

is α-aminopropanoic acid. In inorganic nomenclature, α is used to designate one of the modifications of a polymorphic element or compound (see Polymorphism).

AAS: abb. for Atomic absorption spectroscopy (see).

Abadol: same as 2-Aminothiazol (see).

Abietic acid: abieta-7,14-diene-19-carboxylic acid, a tricyclic diterpene acid; colorless crystals, m.p. 173 °C. It is practically insoluble in water, but soluble in alcohol and ether. The esters and salts are called abietates. A. is a resin acid, and the main component of conifer resins. Colophony consists of about 43% A. It is used industrially in the preparation of paints, soaps, varnishes and plastics.

CH_3 CH_3 H_3C H H HOOC CH_3

Abrasives: very hard materials used in sanding and polishing. The grains of the abrasive may either be fixed (as on sandpaper) or free. The main characteristics of A. are hardness and grain size. Both natural and synthetic A. are used.

1) ***Natural A.*** are hard minerals which are granulated and sorted by sieving. a) *Diamond* is the hardest naturally occuring mineral, with a Mohs hardness of 10. The larger crystals are used as grinding diamonds for preparation of other A. The small to fine grains are used for grinding disks and abrasive pastes for grinding the hardest substances, for example rubies for clock works, sintered corundum, sintered metals, etc. b) *Corundum* is chemically aluminum oxide, α-Al_2O_3. Its Mohs hardness is 9. c) *Emery* is a mixture of corundum, magnesite, specular iron and quartz. Because of their limited occurrence, corundum and emery are of limited significance.

2) ***Synthetic A.*** are produced in various ways (see Crystal growing). a) Synthetic *corundums* are used; their hardness depends on their content of Al_2O_3. Grindstones are made with normal corundum, containing 95 to 97% Al_2O_3 (medium hard), or white or pink fine corundum which contains about 99% Al_2O_3 (hard). Ruby corundum, consisting of Al_2O_3-Cr_2O_3 mixed crystals with up to 3% Cr_2O_3, is used mainly for fine grinding. b) *Silicon carbide*, α-SiC, has a Mohs hardness of 9.5. c) Synthetic *diamonds* are harder than natural diamonds. d) *Boron nitride*, (regular) BN, is as hard as synthetic diamonds.

Solid ***grindstones*** consist of the A. grains and binding phases consisting of hard (epoxide or phenol formaldehyde) and elastic resins, together with ceramic binding masses (mixtures of clay, feldspar, quartz and kaolin). A metallic binding matrix (Fe, Ti, V, Nb, Mo or W) is preferred for high-performance grindstones with diamond A. The properties of a grindstone depend on the nature of the binding matrix as well as on the type of A. and its grain size. Corundum disks are most commonly used for steels, while silicon carbide disks are used for chilled castings, aluminum, brass, copper and cast iron. Diamond stones are used for hard metals, glass, ceramics, concrete, stone and nonferrous metal alloys.

ABS: abb. for Alkylbenzene sulfonates (see).

Abscisic acid, obsolete name, ***dormin***: (S)-5-(1-hydroxy-2,6,6-trimethyl-4-oxo-1-cyclohexenyl)-3-methyl-*cis,trans*-penta-2,4-dienoic acid, a phytohormone; chemically, a monocyclic sesquiterpene. A. forms colorless crystals, m.p. 160-161 °C, which are readily soluble in organic solvents, but poorly soluble in water. It is very sensitive to light, which causes a rearrangement to the inactive *trans, trans*-isomer. A. suppresses the effects of other phytohormones, and thus is a natural growth inhibitor. Together with the growth-promoting phytohormones, e.g. the cytokinins, A. regulates ageing processes, shedding of leaves, formation of blossoms, fruit ripening and other important development processes in the plant.

H_3C CH_3 CH_3 6' 5' 1' OH 4' 3' 2' COOH O CH_3

Absorbance: same as Extinction (see).

Absorbate: the gas taken up by absorption.

Absorbent: a condensed phase which takes up the absorbate in the process of absorption.

Absorbt: the molecularly dispersed substance in the absorbent after absorption.

Absorptiometry: a term which encompasses Photometry (see) and Colorimetry (see).

Absorption:

1) A. of gases: uptake of a gas into the interior of a condensed phase in a molecular-disperse distribution. In reversible, physical A., the equilibrium partial pressure of the absorbate over the solution is significant, and depends on the concentration of gas already dissolved in the liquid. Complete uptake of the absorbate by the absorbent is not possible. However, if a chemical reaction occurs and the vapor pressure of the absorbed substance over the solution is low, a complete A. is theoretically possible. If the compound formed by the reaction is unstable, an equilibrium pressure of the absorbate over the solution will develop.

Henry's Law, which is analogous to the Nernst distribution, applies to ***physical A.***: at constant temperature, the saturation concentration c of a gas in a solution is proportional to the partial pressure p: $c = \alpha p = n/v$. In this formula, v is the volume of the liquid, c is the saturation concentration of the gas in the solution, n is the number of moles of the dissolved gas, and α is Bunsen's absorption coefficient.

In the A. of a mixture of gases, an equilibrium is established for each component which corresponds to its partial pressure and solubility coefficient. ***Exsorption*** is the loss of an absorbed gas by increase in temperature, decrease of pressure on the solution, or treatment with an inert gas. After *chemical A.*, exsorption is possible only if the compound formed by the reaction is unstable and a certain partial pressure of the absorbed gas can be established over the liquid.

A. is used in the laboratory to purify gases, and in the chemical industry for purification of gases or separation of gas mixtures (Table). Since the substance exchange in A. occurs via the contact surfaces of the two phases, apparatus is used to provide the largest possible surface area between the gas and liquid, such as columns packed with inert filler, spray towers, thin-layer absorbers, etc.

Industrial absorbtion processes

Gas to be removed	Absorption agent	Process
CO_2, COS and H_2S from natural, city and synthesis gas	Water	
	potassium carbonate solution	
	Methanol	Rectisol
	Sodium sarcosinate solution	Sulfosolvane
	Sulforlane/ diisopropanolamine	Sulfinol
	N-Methylcaprolactam	Gasolane
H_2O vapor	Glycols	
Ethylene	Acetone	
	Dimethylformamide	
	Methylpyrrolidone	
NH_3	Water	
SO_3	Conc. sulfuric acid	
CO	Cu salt/NH_3H_2O	Uhde

2) A. of radiation: Decrease in the intensity of electromagnetic (see Electromagnetic spectrum) or particulate radiation on passage through a solid, liquid or gas. A. involves transformation of some of the energy of the radiation into another form of energy, such as heat or energy of excitation. The ratio of absorbed (I_A) to incident (I_o) radiation is called the ***absorption***. The expression $I_A/I_o \cdot 100$ is called the percent A.

If monochromatic radiation passes through a homogeneous substance in a parallel beam, then equal fractions of the radiation are absorbed by layers of equal thickness. For example, if visible radiation of a given wavelength is reduced in intensity by one half by passing through 1 cm of the substance, then of the remaining half, half will be absorbed by the next 1-cm layer, so that 1/4 of the original intensity remains. The absorption law $I_D = I_o e^{-\alpha d}$ was first formulated by Lambert and Bouguer; here I_D is the intensity of the transmitted radiation, and I_o is that of the incident radiation. Usually, the absorption law is expressed in the form $\ln(I_o/I_D) = \alpha d$, or, if base-10 logarithms are used, $\log(I_o/I_D) = k \cdot d$, where $k = 0.43422\ \alpha$. The absorption coefficient depends greatly on the frequency of the radiation and on the nature of the sample. A. is the basis of radiation chemistry and photochemistry. The A. in solutions is discussed in the entry on the Lambert-Beer law (see).

Absorption spectrum: see Spectrum.

ABS plastics: see Polystyrenes.

Ac: symbol for Actinium.

Acaricides: compounds used to combat plant-damaging mites, especially spider mites. The use of insecticides often destroys the natural enemies of mites, without eradicating the mites themselves. Since mites have very short generation times, and quickly develop resistance to chemicals, their control has become a problem. Preparations are needed which are specific for mites, and which have different mechanisms of action so that they can be used in rotation to prevent resistance from developing. A. are classified as ovicides, larvicides and adulticides.

The most important specific A. are dinitrophenol derivatives, azo compounds, sulfides, sulfones, sulfonate esters, sulfonite esters, diphenylcarbinols, carbamates, formamidines and organotin compounds (Table). Many Organophosphate insecticides (see) and Fungicides (see) effective against true mildew also have acaricidic effects.

Main types of acaricides

Name (Mechanism of action)	Formula
Dinocap (Contact acaricide, fungicide for true mildew)	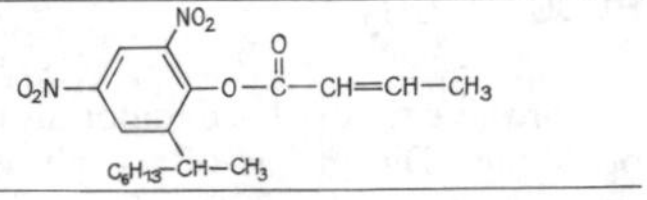 NO_2; O_2N; O–C(=O)–CH=CH–CH_3; C_6H_{13}–CH–CH_3
Fenazox (Fentoxan®) (Ovicide, larvicide, insecticide against white flies)	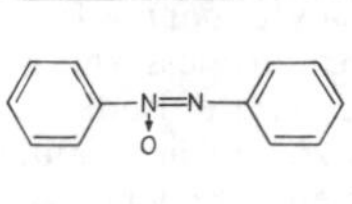 N=N; O
Chlorbenzide (Ovicide, larvicide)	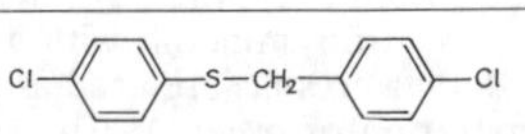 Cl; S–CH_2; Cl
Tetradifon (Ovicide, larvicide)	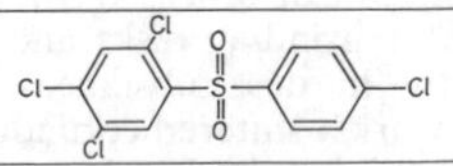 Cl; Cl; Cl; O; S; O; Cl
Chlorfenson (Ovicide)	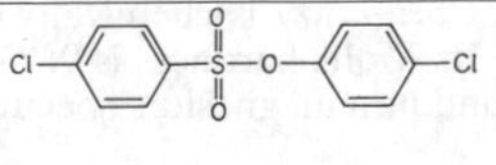 Cl; O; S; O; O; Cl

Name (Mechanism of action)	Formula
Propargite (Ovicide, larvicide, adulticide)	$(CH_3)_3C$–⟨benzene⟩–O–⟨cyclohexane⟩–O–S(=O)–O–CH_2–C≡CH
Dicofol (Contact acaricide, specific for soft-skinned mites)	Cl–⟨benzene⟩–C(OH)(CCl_3)–⟨benzene⟩–Cl
Chlorodimeform (Ovicide)	Cl–⟨benzene(CH_3)⟩–N=CH–N(CH_3)$_2$
Cyhexatin (Larvicide, adulticide)	(cyclohexyl)$_3$Sn–OH
Chlorbenzilate (Acts against all stages of spider mites)	Cl–⟨benzene⟩–C(OH)($COOC_2H_5$)–⟨benzene⟩–Cl

Acceptor number, abb. ***AN***: a measure of the electrophilicity of a solvent. The A. is equal to the ^{31}P chemical shift δ_{corr} of a complex of the solvent with triethylphosphine oxide, relative to the ^{31}P chemical shift of a 1:1 adduct of triethylphosphine oxide:antimony(V) chloride (δ_{corr} = 100) in 1,2-dichloroethane. The A. is related to other parameters which characterize solvents (Z, E_T, Y values, free solvation enthalpies of anions, redox potential of hexacyanoferrate(III)/hexacyanoferrate(II) in the solvent). Together with the Donicity (see), the A. is a useful criterion for selection of a solvent for a chemical reaction.

Acceptor substituent: atoms or groups of atoms which draw electron density to themselves from the molecule of which they are part. This can happen through an inductive effect (I-substituent) or a resonance effect (R-substituent).

ACE inhibitors: substances which reduce the activity of Angiotensin Converting Enzyme. This enzyme is present in blood plasma and converts the decapeptide angiotensin I into the octapeptide angiotensin II; the latter has a vasopressor effect. A. such as captopril and enalpril are used to reduce blood pressure.

N–C(=O)–CH(CH_3)–CH_2SH (pyrrolidine ring with COOH)

Captopril

Acenaphthene, ***1,2-Dihydroacenaphthylene***: a *peri*-condensed aromatic hydrocarbon. A. crystallizes in colorless needles: m.p. 96.2 °C, b.p. 279 °C, n_D^{95} 1.6048. It is insoluble in water and soluble in organic solvents. A. is present in coal tar and can be isolated from it. It can be synthesized from naphthalene and ethene or from 1-ethylnaphthalene by heating to high temperatures. A. is oxidized on an industrial scale to the yellow acenaphthene quinone, which is used for a dye. It is also used in the manufacture of plastics, insecticides and fungicides. Under extreme conditions naphthalene-1,8-dioic acid can be obtained from A.

Acenaphthene quinone: a quinone derived from acenaphthene. A. forms yellow crystals, m.p. 261 °C. It is soluble in alcohol and benzene, and can be synthesized by oxidation of acenaphthene with hydrogen peroxide in glacial acetic acid. A. has insecticidal and fungicidal properties; it can be used as an intermediate in the synthesis of dyes.

Acenaphthylene: a *peri*-condensed aromatic hydrocarbon. A. forms colorless crystals, m.p. 92 °C, b.p. 260 °C. A. is found in coal tar and can be isolated from it. Hydrogenation yields the industrially important acenaphthene.

Acenes: condensed aromatic hydrocarbons with linear arrangements of 2, 3, 4, 5 or 6 benzene rings. Naphthalene (see) and Anthracene (see) are followed by tetracene (naphthacene), a yellow-orange compound, pentacene (blue-violet) and hexacene (dark green). The aromatic character of the compounds decreases rapidly as the number of rings increases; they behave increasingly like conjugated polyenes instead. This results in the colors and sensitivity to oxygen and light.

Tetracene

Pentacene

Acephate: see Organophosphate insecticides.

Acetal: an organic compound with the general formula $R^1R^2C(OR^3)_2$. Formally, A. can be seen as diethers of the normally unstable geminal diols (see Erlenmeyer rule). A. are obtained by reaction of aldehydes with alcohols in the presence of acid catalysts. The reaction goes primarily via the semiacetals, which in rare cases can be isolated:

$$R^1\text{-CHO} \underset{}{\overset{+ROH(H^+)}{\rightleftharpoons}} R^1\text{-CH(OH)(OR)} \underset{}{\overset{+ROH(H^+)}{\rightleftharpoons}} R^1\text{-CH(OR)}_2 + H_2O$$

Semiacetal Acetal

A. are decomposed to the starting materials by aqueous acids. The A. of ketones generally cannot be obtained by direct addition of alcohols to ketones, because the addition equilibrium lies almost entirely on the side of the reactants. Ketone A. can be obtained by reaction of ketones with orthoformate esters:

$$R^1R^2C{=}O + HC(OR)_3 \xrightarrow{(H^+)} R^1R^2C(OR)_2 + HCOOR.$$

This method is also preferable to direct addition of the alcohol to α,β-unsaturated aldehydes.

Most A. are colorless, pleasant-smelling liquids. They are very stable to bases, and do not react with the common reagents for carbonyl detection under basic conditions. A. are thus a protected form of the aldehydes and ketones, and are often used instead of their parent compounds in organic reactions. In addition, they are used as solvents, softeners and perfumes.

Acetaldehyde, ***ethanal***: CH_3-CHO, a colorless, unstable liquid with a pungent smell. m.p. - 123.5 °C, b.p. 20.8 °C, n_D^{20} 1.3316. A. is infinitely soluble in water, alcohol and ether. It is easily ignited and burns with a pale flame. In higher concentrations, A. irritates the respiratory mucosae and has a narcotic effect on the central nervous system. In the presence of a small amount of conc. sulfuric acid it trimerizes spontaneously to Paraldehyde (see); at 0 °C the solid Metaldehyde (see) is formed.

Oxidizing agents convert A. to acetic acid; it can be reduced or hydrogenated to form ethanol. In the presence of dilute alkali or alkaline earth hydroxides, A. is converted by an Aldol reaction (see) to Acetaldol (see). In addition, it displays all the addition and condensation reactions typical for Aldehydes (see).

Occurrence. In nature, A. occurs mainly as an intermediate in the biological degradation of carbohydrates.

A. is obtained by oxidation or dehydrogenation of ethanol with oxygen or a silver contact, or with potassium dichromate and sulfuric acid. Industrially, the most common methods are water addition to acetylene in the presence of mercury salts, direct oxidation of ethene in air in the presence of palladium(II) and copper(II) chloride, and by hydrolysis of vinyl methyl ether, which is made industrially from acetylene and methanol.

Applications. A. is an important starting material for industrial production of an entire series of intermediates and end products, such as ethanol, acetic acid and its anhydride, acetaldol, crotonaldehyde, butadiene, acrolein, pentaerythritol, chloral hydrate and aldehyde resins.

Acetaldol, ***3-hydroxybutanal, 3-hydroxy-n-butaldehyde***, for short, ***aldol***: CH_3-CH(OH)-CH_2-CHO, a colorless, viscous liquid, b.p. 83 °C at $2.67 \cdot 10^3$ Pa, n_D^{20} 1.4238. A. is soluble in water and most organic solvents. Like all Aldehydes (see), A. is very reactive. In the presence of acids, it forms crotonaldehyde by splitting out water. A. is produced by aldol addition of acetaldehyde, usually under basic conditions. It is used as a hardener for gelatins and as an intermediate in the production of maleic acid and amino alcohols; it is also used in one method of butadiene synthesis. It is a sedative and narcotic.

Acetamide: CH_3-CO-NH_2 forms colorless and odorless deliquescent crystals, m.p. 82.3 °C, b.p. 221.2 °C. There is also a metastable form of A. which melts at 69 °C. A. is readily soluble in water and alcohol, and barely soluble in ether. It can react either as a weak base or as a very weak acid. It forms addition compounds with strong mineral acids, and ionic compounds of the general formula CH_3-CO-NHM^I with highly electropositive metals. Addition of water forms ammonium acetate, and elimination of water forms acetonitrile:

$$CH_3\text{–CO–}NH_2 + H_2O \longrightarrow CH_3\text{–COONH}_4$$

$$CH_3\text{–CO–}NH_2 \xrightarrow[-H_2O]{} CH_3\text{–C}{\equiv}N$$

A. is produced by heating of ammonium acetate, by reaction of ammonia with acetic anhydride, acetic acid or acetyl chloride, or by partial hydrolysis of acetonitrile. A. is used in the leather, textile and paper industries. In addition it is an intermediate for the preparation of pharmaceuticals and is used to accelerate vulcanization.

Acetanilide: $C_6H_5NHCOCH_3$, a derivative of aniline; colorless crystals, m.p. 114 °C. It is formed by acetylation of aniline. It was the first synthetic analgesic, introduced as "Antifebrin", but is no longer used in human medicine. It is used in the manufacture of other drugs and of dyes, as a stabilizer for H_2O_2 solutions and as an addition to cellulose ester varnishes.

Acetate fibers: synthetic fibers made by acetylation of cellulose with acetic acid/acetic anhydride and spinning of the acetone-soluble acetates (see Cellulose acetates, synthetic fibers, Table).

Acetates: salts and esters of acetic acid with the general formula CH_3-$COOM^I$ or CH_3-COOR. M^I symbolizes an ammonium residue or a monovalent metal, and R an aliphatic, aromatic or heterocyclic residue.

The salts are formed by reaction of ammonia, ammonium or metal hydroxides or carbonates with acetic acid; or by reaction of acetic acid with strongly electropositive metals, which evolves hydrogen. Basic and neutral A. can be formed with multivalent metals, e.g. CH_3-COO(OH)Mg and $(CH_3\text{-COO})_2$Mg. Ammonium, sodium and aluminum acetates are important compounds used as preservatives, antiseptics, buffers, caustics, etc.

The esters of acetic acid can be produced by reaction of the corresponding alcohols with acetic acid, acetyl chloride or ketene. Some esters are synthesized by special methods. They are used mainly as solvents for fats, oils, laquers, resins, cellulose nitrate, chlorinated rubber, celluloid and colophony, and also as perfumes.

Acetazolamide: see Diuretic.

Acetic acid, ***ethanoic acid***: CH_3-COOH, the most important monocarboxylic acid, a colorless, combustible, hygroscopic liquid with a pungent odor; m.p. 16.6 °C, b.p. 117.9 °C, n_D^{20} 1.3716. Anhydrous A. is also called ***glacial acetic acid***. It is readily soluble in water and most organic solvents. A.-air mixtures with 4 to 17 vol. % A. are explosive. Like many other Monocarboxylic acids (see), pure A. exists as hydrogen-bonded dimers in both the liquid and gas states. It is a weak acid; its salts and esters are called Acetates (see).

A. is rather stable to oxidation, and is often used as a solvent for oxidation reactions. It undergoes the reactions typical of aliphatic carboxylic acids, that is, it can be converted by suitable reagents into derivatives and substitution products. The reaction of perhydrol (30% aqueous solution of hydrogen peroxide) with A. forms ***peroxyacetic acid***, CH_3CO-OOH; this is used as a disinfectant. Pyrolysis of A. at 350 to 400 °C in the presence of alkaline catalysts generates ***acetone***, and Ketene (see) is formed on phosphorus-containing contacts at 700 °C: CH_3-COOH → CH_2=C=O + H_2O. Addition of A. to acetylenes yields the Vinyl acetates (see). A. can be detected qualitatively by the *cacodyl reaction*, in which it forms cacodyl oxide with arsenic(III) oxide; cacodyl oxide has a strong, unpleasant odor.

A. is found in nature in both free and bound form. It is a component of many plant juices, essential oils and animal secretions. Its esters are found in some essential oils. A. is formed as the stable end product of fermentation and rotting of plant and animal material. It has an important role in metabolism in the form of acetyl-coenzyme A. A. can be obtained by ***vinegar fermentation*** from alcohol or by pyrolysis of wood. It is produced industrially by oxidation of acetaldehyde or butane in the presence of manganese or cobalt salts, or by carbonylation of methanol in the presence of cobalt(II) iodide at 150 °C and about 2 to 5 MPa, or in the presence of rhodium complex catalysts (see Catalysis): CH_3-OH + CO → CH_3-COOH. More concentrated A. is produced by various dehydration methods, such as azeotrope distillation or liquid-liquid extraction; glacial acetic acid is obtained by the reaction of anhydrous sodium acetate with conc. sulfuric acid. A. is used as a solvent, to remove calcium from hides, as a reagent in dyeing and other textile treatments, and as a coagulant for latex. It is also an intermediate in the production of aromas, pharmaceuticals, dyes, acetates, acetic anhydride, acetyl chloride and table vinegar.

Acetic anhydride: $(CH_3\text{-}CO)_2O$, a colorless, lacrimatory liquid with a pungent odor; m.p. -73.1 °C, b.p. 139.55 °C, n_D^{20} 1.3906. A. is soluble in cold alcohol, ether and benzene. It is slowly solvolyzed in water or alcohol to acetic acid or acetate ester. A. is a suitable acetylation agent for alcohols, CH-acidic compounds, aromatic hydrocarbons and primary and secondary amines, e.g. R-NH_2 + $(CH_3\text{-}CO)_2O$ → R-NH-CO-CH_3 + CH_3-COOH. The methyl groups of A. are able to undergo condensation reactions due to the activating effects of the carbonyl groups. This reactivity is demonstrated, e.g. in the Perkin reaction (see) for preparation of cinnamic acid from benzaldehyde and A. A. is prepared industrially by two methods: in the *Wacher process* a ketene is prepared from acetic acid by intramolecular water elimination in the presence of phosphoric acid or phosphates at 700 to 750 °C; the ketene reacts with excess acetic acid to form A.: CH_3-COOH → CH_2=C=O + H_2O; CH_2=C=O + CH_3 → $(CH_3\text{-}CO)_2O$. In the second method, the *Knapsack process*, A. is made by oxidation of acetaldehyde in the presence of metal salts at 40 to 60 °C. In this synthesis the first product is peracetic acid, which reacts further with excess acetaldehyde to the end product: CH_3-CHO + O_2 → CH_3-CO-OOH; CH_3-CO-OOH + CH_3-CHO → $(CH_3\text{-}CO)_2O$ + H_2O. In the laboratory, A. can be synthesized in high yields by reaction of sodium acetate with inorganic acid chlorides, e.g. thionyl chloride or phosphorus oxide chloride. These reactions form first acetyl chloride, which reacts with unreacted sodium acetate to form A. A. is used mainly as an acetylation reagent, e.g. in the production of cellulose acetate, drugs, dyes, flavorings and perfumes. In addition, A. is used as a solvent and condensing agent.

Acetoacetate ester syntheses: see acetoacetate esters.

Acetoacetic acid, ***3-oxobutanoic acid, β-ketobutyric acid***: CH_3-CO-CH_2-COOH, an aliphatic β-ketocarboxylic acid. A. is a colorless, syrupy liquid which gives a strongly acidic reaction; it can be crystallized only with difficulty. m.p. 36 to 37 °C. On heating, A. decomposes around 100 °C to form acetone and carbon dioxide: CH_3-CO-CH_2-COOH → CH_3-CO-CH_3 + CO_2. A. is soluble in water, alcohol and ether. It is formed in the liver, starting from acetyl-coenzyme A, via acetoacetyl-coenzyme A, and in diabetics it is secreted into blood and urine as one of the ketone bodies (acetone is another). A. can be synthesized by hydrolysis of acetoacetate esters or by oxidation of butyric acid. Because of its low stability, it is used in organic syntheses only in special cases.

Acetoacetic acid ethyl ester: see Ethyl acetoacetate.

Acetoin, ***3-hydroxybutan-2-one***: CH_3-CH(OH)-CO-CH_3, one of the acyloins. A. is a colorless liquid; m.p. - 72 °C, b.p. 143 °C, n_D^{20} 1.4171. A. is soluble in water and acetone, insoluble in ether. It reduces Fehling's solution and dimerizes easily. Catalytic dehydration yields Methyl vinyl ketone (see). A. is made enzymatically from acetaldehyde or chemically by reduction of diacetyl.

Acetone, ***propanone, dimethyl ketone***: CH_3-CO-CH_3, a colorless, non-viscous, pleasant smelling, inflammable liquid. m.p. - 95.34 °C, b.p. 56.2 °C, n_D^{20} 1.3588. A. is miscible with water in all proportions, and miscible with most organic solvents. It is very stable to light and air. A. is found in volatile oils and in the urine of diabetics, where it can be detected with the Lieben iodoform test (see) or the Legal acetone test (see). The pathological accumulation of A. in urine is called *acetonuria*. A. is formed in the body via the ketone bodies. Like all Ketones (see), A. reacts with nucleophilic reagents, forming typical addition and condensation products. A. can be converted to isopropanol by reduction; and to bromoacetone by reaction with bromine.

Production. A. is formed in considerable amounts by dry distillation of wood, but its separation from other products, e.g. methanol, and purification is very tedious. A. can also be produced by pyrolysis of acetic acid on a manganese(II) oxide contact, or from acetylene and water vapor in the presence of zinc oxide. The more modern methods of producing A. include dehydrogenation of isopropanol, the Hock process (see), in which A. is obtained as a byproduct from cumene, and the direct oxidation of propene in the presence of copper(II) chloride and palladium(II) chloride.

Applications. A considerable fraction of the A. produced is used as a solvent and extraction agent for resins, fats, oils, colophony, cellulose acetate, acetylene and the like. It is also becoming steadily more important in the production of derivatives like diacetone alcohol, mesityloxide, chloroform, etc.

Acetone cyanohydrin, ***α-hydroxyisobutyronitrile***: $(CH_3)_2C(OH)CN$, a colorless, inflammable liquid which smells like hydrogen cyanide. m.p. - 19 °C, b.p. 82 °C at $3.07 \cdot 10^3$Pa; n_D^{20} 1.3996. A. is readily soluble in water, alcohol, chloroform and ether. It is obtained by addition of hydrogen cyanide to acetone in the presence of basic catalysts. Traces of alkali hydroxides cause it to split into the starting materials. This back reaction, which releases hydrogen cyanide, is responsible for the toxicity of A. A. is the starting material for methacrylate esters; it is also used as an insecticide.

α,α-Acetonedicarboxylic acid, ***3-oxopentanedioic acid, β-ketoglutaric acid***: $HOOC\text{-}CH_2\text{-}CO\text{-}CH_2\text{-}COOH$, a colorless, crystalline compound; m.p. 135 °C (dec.), which decomposes on distillation. It is readily soluble in water and alcohol, and slightly soluble in ether or chloroform. When stored, it slowly decomposes to acetone and carbon dioxide. When heated, or in the presence of mineral acids, metal salts or hydroxides, the cleavage occurs much faster. In every case, the unstable acetoacetic acid is formed as an intermediate:

$$HOOC\text{–}CH_2\text{–}CO\text{–}CH_2\text{–}COOH \xrightarrow[-CO_2]{}$$

$$CH_3\text{–}CO\text{–}CH_2\text{–}COOH \xrightarrow[-CO_2]{} CH_3\text{–}CO\text{–}CH_3$$

α,α-A. is formed by the action of fuming sulfuric acid on citric acid, by hydrolysis of 1,3-dicyanoacetone (obtained from 1,3-dichloroacetone and alkali cyanides), or from acetone and carbon dioxide. α,α-A. and its esters have several functional groups; the carboxyl or ester group, the keto group and the active methylene group can all undergo the reactions typical of these classes. α,α-A. is therefore useful for the syntheses of various heterocycles, e.g. some alkaloids. It is also used in the syntheses of amino acids, pharmaceuticals, insecticides, fungicides and disinfectants. Like all β-dicarbonyl compounds, α,α-A. forms chelate complexes with many metal ions, such as copper or iron, making it useful for the removal of traces of heavy metals from fats, oils and petroleum products.

Acetonitrile, ***methyl cyanide***: $CH_3\text{-}C{\equiv}N$, the simplest aliphatic nitrile. A. is a colorless, poisonous liquid with a pleasant odor; m.p. - 45.7 °C, b.p. 81.6 °C, n_D^{20} 1.3442. A. is infinitely soluble in water and most organic solvents. It burns with a bright pink flame. A. shows the reactions typical of aliphatic Nitriles (see). Because the nitrile group has an activating effect, the H-atoms of the methyl group undergo electrophilic substitution relatively easily. A. is contained in the crude benzene fraction of anthracite coal tar and in the liquid from low-temperature distillation of lignite coal. It can be synthesized according to Kolbe from methyl iodide and alkali-metal cyanides, by dehydration of acetamide or acetaldehyde oxime, and from acetylene and ammonia. Industrially, A. is obtained as a byproduct of acrylonitrile production through ammonoxidation of propylene. A. is used mainly as a solvent, but also as an intermediate in organic syntheses.

Acetonylacetone, ***hexane-2,5-dione***: $CH_3\text{-}CO\text{-}CH_2\text{-}CH_2\text{-}CO\text{-}CH_3$, the simplest 1,4-diketone. A. is a colorless liquid with an aromatic odor; m.p. - 5.5 °C, b.p. 194 °C, n_D^{20} 1.4421. It is soluble in water, alcohol, ether and acetone. It is produced by decarboxylation of diacetylsuccinic acid or by hydrolytic ring cleavage of 2,5-dimethylfuran. A. is used as a solvent for paints, laquers and cellulose acetate, and as a component in the synthesis of heterocyclic compounds.

Acetophenetidin, ***phenacetin***: 4-ethoxyacetacetanilide, a white, crystalline substance, m.p. 135 °C. A. can be made by the following steps: 1) nitration of chlorobenzene to 4-nitrochlorobenzene (and other products), 2) substitution of an ethoxy group for the activated Cl atom, forming 4-nitroethyoxybenzene, 3) reduction of the nitro group and 4) acetylation of the resulting amino group. It was introduced in 1887 as an analgesic. Kidney damage can result from long-term consumption of high doses. 4-Hydroxyacetanilide (white crystals, m.p. 168 °C) was introduced into pharmacy as *paracetamol* after it was found to be one of the biotransformation products of A., and the pain-killing effects were ascribed to it.

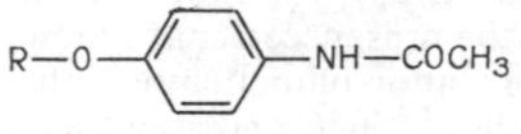

$$R\text{—}O\text{—}C_6H_4\text{—}NH\text{—}COCH_3$$

R = C_2H_5: Acetophenetidin
R = H: Paracetamol

Acetophenone, ***methyl phenyl ketone, acetylbenzene***: $C_6H_5\text{-}CO\text{-}CH_3$, a colorless compound with an aromatic odor. m.p. 20.5 °C, b.p. 202.6 °C, n_D^{20} 1.5372. A. is insoluble in water, but dissolves readily in most organic solvents. It is synthesized mainly through the Friedel Crafts reaction (see) or by catalytic oxidation of ethyl benzene. A. is used as a solvent, as a starting material for production of formaldehyde resins, pharmaceuticals, dyes and perfumes. It was formerly used as a hypnotic.

Acetyl-: term for the $CH_3\text{-}CO\text{-}$ group in a molecule, for the unstable radical $CH_3\text{-}CO^{\cdot}$ and for the cation $CH_3\text{-}CO^+$.

Acetylacetone, ***pentane-2,4-dione***: $CH_3\text{-}CO\text{-}CH_2\text{-}CO\text{-}CH_3$, the simplest aliphatic 1,3- or β-diketone. A. is a colorless, non-viscous liquid with a pleasant odor; m.p. - 23 °C, b.p. 140 °C, n_D^{20} 1.4494. A. is very soluble in water and most organic solvents. Like all 1,3-dicarbonyl compounds, it undergoes Keto-enol tautomerism (see).

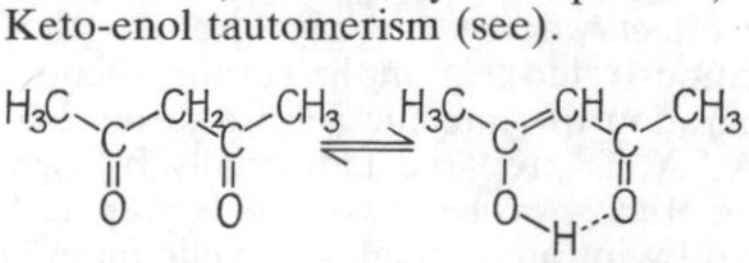

$$H_3C\text{–}C(=O)\text{–}CH_2\text{–}C(=O)\text{–}CH_3 \rightleftharpoons H_3C\text{–}C(O\cdots H)\text{=}CH\text{–}C(=O)\text{–}CH_3$$

Keto form Enol form

In the liquid state at 20 °C, 18% is in the keto form, and 82% in the enol form. In the vapor phase, it is nearly 100% in the enol form. Chelate acetylacetonates derived from the enol form are formed with metal salts. A. can be produced by acetylation of acetoacetate or acetone with acetic anhydride and boron trifluoride; or by pyrolysis of isopropylacetate. A. is used as a solvent, in the production of nitrogen-containing heterocycles in the pharmaceutical industry, and for the production of dyes and pesticides. It is also used in the plastics industry as a stabilizer, initiator of polymerization and hardener.

Acetyl acetic acid: same as Acetoacetic acid.

Acetylation: a preparative method for introduction of an acetyl group, CH_3-CO-, into an organic compound. A. is one of the most commonly used reactions for Acylation (see). Compounds with acidic H-atoms, such as alcohols, phenols, thiols, primary and secondary amines, are often acetylated to protect them from undesired side reactions. A. is used to determine the number of hydroxy, sulfhydryl or amino groups in organic compounds, the Acyl number (see). The A. of arenes in the Friedel-Crafts reaction (see) produces acetyl-substituted aromatics.

Acetylbenzene: see Acetophenone.

Acetylbromide: CH_3-CO-Br, a colorless liquid which fumes strongly in air; m.p. - 98 °C, b.p. 76 °C, n_D^{20} 1.4537. A. slowly turns yellow under long storage. It is readily soluble in ether, acetone, benzene and chloroform, but is solvolyzed by water or alcohol to form acetic acid or acetate esters. It is synthesized by reaction of phosphorus pentabromide with acetic acid. A. is used mainly as an acetylation agent in organic syntheses.

Acetylcellulose: see Cellulose acetate.

Acetylchloride, ***ethanoyl chloride***: CH_3-CO-Cl, a colorless liquid with a suffocating smell which is very irritating to the eyes and respiratory passages; m.p. - 112 °C, b.p. 50.9 °C, n_D^{20} 1.3897. On exposure to moist air, A. is hydrolysed with fuming, forming acetic acid and hydrogen chloride. It is soluble in most organic solvents. A. is a very reactive compound which forms acetamide with ammonia, acetate esters with alcohols, and *N*-substituted acetamides with primary and secondary amines. It is produced by reaction of acetic acid or its alkali salts with inorganic acid chlorides, e.g. sulfuryl chloride, thionyl chloride, phosphoryl chloride, phosphoryl oxochloride, etc. A. is produced industrially by reaction of hydrogen chloride with acetic anhydride; it is used mainly as an acetylation agent in organic synthesis. It is also used as a reagent in analytical organic chemistry for quantitative determination of hydroxy groups and to distinguish tertiary amines from primary and secondary amines.

Acetylcholine: 2-acetoxyethyl trimethylammonium hydroxide, the ester of choline with acetic acid. The neutral chloride, $[(CH_3)_3N^+\text{-}CH_2\text{-}CH_2\text{-}O\text{-}CO\text{-}CH_3]Cl^-$, is stable. A. is a neurotransmitter. It is administered parenterally in treatment of arterial blockage.

Acetylcholinesterase: an enzyme (see Esterases) which catalyses the cleavage of acetylcholine into choline and acetate ion. It is found in nervous tissue and is responsible for removal of acetylcholine released into the synapses during propagation of nervous impulses from one nerve cell to another.

A serine and a histidine residue are required for activity of A. The catalytically active serine hydroxyl is blocked by organophosphates, which inhibit the esterase activity. The enzyme can be reactivated by pralidoxime chloride (2-[(hydroxyimino)methyl]-1-methylpyridinium chloride), which is therefore used as an antidote to organophosphate poisoning.

Acetylcholinesterase inhibiters: see Parasympatheticomimetica.

Acetyl-coenzyme A: see Coenzyme A.

Acetylene: see Ethyne.

Acetylenecarboxylic acid: same as Propiolic acid.

Acetylene chemistry: a collective term for all chemical reactions starting from ethyne (acetylene). Ethyne is the parent compound of a very large number of organic compounds, so that it, together with petroleum and coal, is a basis for modern industrial organic chemistry. The development of A. occurred in two stages; the first was the work of Berthelot, Dupont, Nieuwland, Carothers, Hilebrone and Kutscherov, who discovered how to make acetaldehyde and its derivatives (acetic acid, ethyl acetate, acetic anhydride, acetone, ethanol, crotonaldehyde and butanol) from ethyne.

The second stage, which is extremely important to industry, is called ***Reppe chemistry*** after its founder, W. Reppe. Reppe opened a new field of A. by introducing pressure synthesis. He avoided the danger of explosion of the compressed ethyne by careful engineering. In addition to the high pressures, catalysts such as heavy metal acetylides, metal carbonyls and hydrogen metal carbonyls, are very important. The following reactions are especially significant:

1) *Hydration* of ethyne in the presence of 15% sulfuric acid with mercury(II) sulfate as catalyst leads via the unstable vinyl alcohol to acetaldehyde, the most important intermediate in A.: $HC{\equiv}CH + H_2O \rightarrow [H_2C{=}CHOH] \rightarrow CH_3\text{-}CHO$. When homologs of ethyne are used, ketones are formed; this hydration is a special case of vinylation.

2) *Catalytic hydrogenation* of ethyne and its homologs leads either to the corresponding alkane, or the C≡C triple bonds can be selectively converted to C=C double bonds with deactivated catalysts. The Lindlar catalyst, a palladium-calcium carbonate poisoned with lead, is used for this. The hydrogen transfer occurs by *cis*-addition; double bonds are not attacked by this method: R-C≡C-CH=CH-R + 2 H → R-CH=CH-CH=CH-R. Selective hydrogenation to the alkene is also possible with sodium in liquid ammonia, in which case *trans*-addition occurs:

R–C≡C–R + 2H ↗ (H)(R)C=C(H)(R) (cis)
R–C≡C–R + 2H ↘ (R)(H)C=C(R)(R) (trans)

3) *Addition of halogens*, e.g. chlorine or bromine, can be accelerated by Lewis acids or initiated by light. The reaction is a stereoselective *trans*-addition to 1,2-disubstituted ethenes, and further reaction produces

1,1,2,2-tetrahaloethane. This is converted to trihalooethene by elimination of hydrogen halide:

$$HC{\equiv}CH \xrightarrow{Cl_2} H(Cl)R{=}R(Cl)N \xrightarrow{Cl_2} HC(Cl)_2{-}C(Cl)_2H \xrightarrow{-HCl} HC(Cl){=}CCl_2$$

Trichloroethene is an important solvent.

4) *Dimerization* of ethyne with ammonium and copper(I) chlorides as catalysts produces vinylethyne, the starting material for chloroprene: $CH_2{=}CCl{-}CH{=}CH_2$.

Reppe syntheses include:

5) *Vinylation.* In addition to hydrogen halides and water, alcohols, phenols, thiols, amines, carboxylic acids and hydrogen cyanide add to ethyne to produce various types of vinyl compound. a) The addition of hydrogen chloride to ethyne at 140 to 200 °C in the presence of a mercury(II) chloride/activated charcoal catalyst produces vinyl chloride by the reaction: $HC{\equiv}CH + HCl \rightarrow CH_2{=}CHCl$; the product is used to make PVC. b) Addition of acetic acid to ethyne at 170 to 200 °C in the presence of a zinc acetate/activated charcoal or mercury(II) chloride catalyst produces vinyl acetate, which can be saponified to vinyl alcohol: $HC{\equiv}CH + CH_3COOH \rightarrow CH_2{=}CH{-}OOCCH_3$. c) Addition of hydrogen cyanide to A. yields the starting material for polyacrylonitrile fibers, acyl nitrile: $CH{\equiv}CH + HCN \rightarrow H_2C{=}CH{-}CN$. d) Vinyl ethers are formed by addition of alcohols to ethyne using alkaline catalysts: $HC{\equiv}CH + ROH \rightarrow H_2C{=}C{-}OR$.

Ethynylations are reactions of ethyne with aldehydes or ketones which maintain the triple bond. Copper(I) salts, including copper acetylide, are used as catalysts. Ethyne reacts with formaldehyde to form propargyl alcohol and but-2-yne-1,4-diol, which can be made into butadiene after hydrogenation to butane-1,4-diol, and with acetone to form 3-methylbut-1-yn-3-ol, from which isoprene can be made:

$$HC{\equiv}CH + HCHO \longrightarrow HC{\equiv}C{-}CH_2{-}OH + HCHO \longrightarrow HO{-}CH_2{-}C{\equiv}C{-}CH_2{-}OH;$$

$$HC{\equiv}CH + CH_3{-}CO{-}CH_3 \longrightarrow HC{\equiv}C{-}C(OH)(CH_3)_2.$$

The aminomethylation and the dimerization of ethyne to vinyl acetylene are also ethynylation reactions and are industrially significant.

7) In *carbonylations*, ethyne and carbon monoxide react in the presence of compounds with reactive hydrogen, e.g. water, alcohol or secondary amines, to acrylic acid or its derivatives: $HC{\equiv}CH + CO + H_2O \rightarrow CH_2{=}CH{-}COOH$ (acrylic acid), $HC{\equiv}CH + CO + ROH \rightarrow CH_2{=}CH{-}COOR$ (acrylic esters), $HC{\equiv}CH + CO + R_2NH \rightarrow CH_2{=}CH{-}CONR_2$ (acrylamide). Carbonylation occurs under pressure with carbonyl-forming metals, e.g. cobalt, nickel or iron, as catalysts.

8) *Cyclization* of ethyne in the presence of tricarbonyl(triphenylphosphine)nickel at 60 to 70 °C gives a mixture of benzene (88%) and styrene (12%). In the presence of nickel(II) cyanide and tetrahydrofuran, cyclooctatetraene is formed in 70% yield, along with other products.

Reactions 1 to 3 and 5c are uneconomical, and the corresponding products can be produced more cheaply from ethene, or in the case of acrylonitrile, through ammonoxidation of propene. A. has been partially displaced by ethylene chemistry. However, there has been a renaissance of carbochemistry, and ethyne and products of Reppe syntheses have again achieved great industrial importance. Vinyl ether is now produced exclusively from ethyne, and there is no competition for the synthesis of butyne-1,4-diol or butane-1,4-diol from ethyne.

Acetylene dicarboxylic acid, ***butynedioic acid***: $HOOC{-}C{\equiv}C{-}COOH$, the simplest dicarboxylic acid with a $C{\equiv}C$ triple bond. The crystals are colorless platelets, m.p. 179 °C. A. is very slightly soluble in water, alcohol and ether. It can be synthesized by dehydrobromination of dibromosuccinic acid. The diesters of A. are important as intermediates in the production of heterocycles and as dienophiles in Diels-Alder reactions.

Acetylene soot: soot produced by incomplete combustion of acetylene.

Acetylene tetrachloride: same as 1,1,2,2-Tetrachloroethane.

Acetylides: see Ethyne.

3-Acetyl-6-methyl-2H-pyran-2,4(3H)-dione: same as Dehydroacetic acid.

O-Acetylphenol: same as Phenyl acetate.

Acetyl salicylic acid, ***aspirin***: the crystals are colorless needles with a sour taste; m.p. 137 °C. A. is made by acetylation of salicylic acid. As a phenol ester, it is very sensitive to hydrolysis. It is very widely used as an analgesic and antipyretic. A. is less damaging to the gastric mucosa than salicylic acid. A. inhibits thrombocyte aggregation, by inhibiting the enzymes of prostaglandin synthesis. Since thrombocytes initiate blood coagulation, A. also inhibits clot formation, and this property has been utilized therapeutically in recent years. To prevent embolisms, A. is administered as tablets from which it is continuously released during passage through the intestinal tract (Micristin®, Colfarit®).

COOH
O—$COCH_3$

N'-Acetylsulfanilamide: see Sulfonamide.

Acheson process: see Graphite.

Achiral: see Stereoisomerism, 2.

Acid: see Acid-base concepts; Nomenclature, sect. II E.

Acid anhydride: see Carboxylic acid anhydride.

Acid-base concepts: systems based on definitions of acids and bases which permit classification of many compounds and interpretation of their reactions within a unified conceptual framework. The concept of acids and bases has undergone many changes in the course of its development. The definitions were generalized in several stages to include a broader range of compounds, and in addition, different definitions evolved from the application of the concept to different areas.

Arrhenius acids and bases: In 1887 Arrhenius defined acids as substances which dissociate to form H^+ ions in aqueous solutions, and bases as substances which dissociate to form OH^- ions in aqueous solutions. The reaction of an acid with a base is called

Important conjugate acid-base pairs with their pK_a and pK_b values

Acid strength	pK_a	Acid	$+ H_2O \rightleftharpoons H_3O^+ +$ Base	pK_b	Base strength
	− 10	$HClO_4$	$+ H_2O \rightleftharpoons H_3O^+ + ClO_4^-$	24	Very weak
Very strong					
	− 10	HI	$+ H_2O \rightleftharpoons H_3O^+ + I^-$	24	
	− 9	HBr	$+ H_2O \rightleftharpoons H_3O^+ + Br^-$	23	
	− 6	HCl	$+ H_2O \rightleftharpoons H_3O^+ + Cl^-$	20	
	− 3	H_2SO_4	$+ H_2O \rightleftharpoons H_3O^+ + HSO_4^-$	17	
	− 1.74	H_3O^+	$+ H_2O \rightleftharpoons H_3O^+ + H_2O$	15.74	
Strong	− 1.32	HNO_3	$+ H_2O \rightleftharpoons H_3O^+ + NO_3^-$	15.32	Weak
	1.92	HSO_4^-	$+ H_2O \rightleftharpoons H_3O^+ + SO_4^{2-}$	12.08	
	1.96	H_3PO_4	$+ H_2O \rightleftharpoons H_3O^+ + H_2PO_4^-$	12.04	
	3.14	HF	$+ H_2O \rightleftharpoons H_3O^+ + F^-$	10.86	
Medium	4.76	CH_3COOH	$+ H_2O \rightleftharpoons H_3O^+ + CH_3COO^-$	9.24	Medium
	5.52	H_2CO_3	$+ H_2O \rightleftharpoons H_3O^+ + HCO_3^-$	7.48	
	6.92	H_2S	$+ H_2O \rightleftharpoons H_3O^+ + HS^-$	7.08	
Weak	9.21	NH_4^+	$+ H_2O \rightleftharpoons H_3O^+ + NH_3$	4.79	Strong
	9.40	HCN	$+ H_2O \rightleftharpoons H_3O^+ + CN^-$	4.60	
	12.90	HS^-	$+ H_2O \rightleftharpoons H_3O^+ + S_2^{2-}$	1.10	
Very weak	15.74	H_2O	$+ H_2O \rightleftharpoons H_3O^+ + OH^-$	− 1.74	
	23	NH_3	$+ H_2O \rightleftharpoons H_3O^+ + NH_2^-$	− 9	
	34	CH_4	$+ H_2O \rightleftharpoons H_3O^+ + CH_3^-$	− 20	

Neutralization (see); the essential process is the combination of H^+ and OH^- ions to form water molecules. The neutral point (see Ion product) is a state in which the H^+ and OH^- concentrations are equal. The strength of an acid or base corresponds to the dissociation constant of the compound in water. The disadvantages of this definition are that it is limited to aqueous systems, and that there are compounds which do not contain OH^- groups, but have the ability to neutralize acids, e.g. ammonia, organoelement compounds, etc., which are not included in the definition of bases. The Arrhenius concept has now been replaced by the Brönsted system.

Brønsted acids and bases: Brønsted and Lowry independently and simultaneously (1923) defined acids as systems which can donate protons, and bases as systems which can accept protons. This essentially functional definition includes charged particles, and a distinction is therefore made between *neutral* acids (e.g. H_2SO_4, HNO_3) and bases (e.g. NH_3, H_2O), *anionic* acids (e.g. HCO_3^-, $H_2PO_4^-$) and bases (e.g. HCO_3^-, OH^-), and *cationic* acids (e.g. NH_4^+, OH_3^+) and bases (e.g. $[Al(OH)(H_2O)_5]^{2+}$). Acids which can donate one proton are called *monoprotic* (e.g. HNO_3, HCl, $HClO_4$); those which can donate two protons are *diprotic* (e.g. H_2SO_4, H_2CO_3), and those which can donate three, *triprotic* (e.g. H_3PO_4, H_3AsO_4). Di- and triprotic acids can form acidic salts after partial dissociation. Similarly, *monoacidic bases* (e.g. OH^-) can accept one proton, and *diacidic bases*, two (e.g. O^{2-}, SO_4^{2-}, HPO_4^{2-}).

The Brønsted concept is not limited to aqueous systems, but can be applied to reactions in non-aqueous solvents and the gas phase.

Since protons never exist in the free state, the dissociation of an acid (*proton donor*) must always occur in the presence of a base (*proton acceptor*). The proton transfer process is called *protolysis*. When an acid donates a proton, the remaining part of the compound is naturally able to accept a proton, and is thus a base; the original base, having accepted a proton, has been transformed into an acid: acid HA + base B $\rightleftharpoons$ Base A^- + acid HB^+. If the only difference between the acid and base is the presence of the proton on the acid, they constitute a *conjugate acid-base pair*. For example, the chloride ion is the base corresponding to the acid HCl, and the ammonium ion is the acid corresponding to the base NH_3 (Table).

Compounds which can act either as acids or as bases are called *Ampholytes* (see) (e.g. HCO_3^-, HPO_4^{2-}, H_2O, NH_3). Amphoteric behavior is a necessity for *autoprotolysis (acid-base disproportionation)*, i.e. the ability of a system to protonate itself, e.g. $HPO_4^{2-} + HPO_4^{2-} \rightleftharpoons H_2PO_4 + PO_4^{3-}$. The autoprotolysis equilibrium for water according to the equation $H_2O + H_2O \rightleftharpoons H_3O^+ + OH^-$ is of fundamental importance for an understanding of many processes which occur in water (see pH, Ion product).

If the strength of an acid or base is to be described in terms of the extent of its dissociation, this dissociation must be related to an interaction with a standard base or acid. Water, which is amphoteric, is suitable in most cases, and the *acid strength (acidity)* is described by the equilibrium constant for the reaction of the acid (HA) with water. This equilibrium constant is called the *acid constant* K_a, and its negative logarithm to the base ten is called its pK_a:

$$HA + H_2O \rightleftharpoons A^- + H_3O^+,$$

$$\frac{a_A \cdot a_{H_3O}}{a_{HA}} = K_a$$

$$pK_a = -\lg K_a;$$

Similarly, the measure of *base strength (basicity)* is the equilibrium constant of the reaction of the base (B) with water, the *base constant* K_b; the pK_b is the negative logarithm of K_b:

$$B + H \rightleftharpoons A + HO,$$

$$\frac{a_{BH^+} \cdot a_{OH^-}}{a_B} = K_B$$

$$pK_B = -\lg K_B.$$

Strong acids have large acid constants and small pK_a values, while weak acids have small acid constants and large pK_as; the relation between base constants and basicity is similar. The acid and base constants of a conjugate acid-base pair are inversely related to each other. Their product is equal to the ion product of water: $K_a \cdot K_b = K_w = 10^{-14}$; $pK_a = pK_b = pK_w = 14$ (see Ion product). In other words, a strong acid always has a weak conjugate base, while a weak acid always has a strong conjugate base. The pK values for some important conjugate acid-base pairs are listed in the table. This relationship makes it clear why, for example, strong acids are able to release weak acids from their salts, and why there is a considerable change in the pH when some salts are dissolved in water. For example, in the reaction of hydrochloric acid with sodium acetate, acetic acid is formed, because the acetate ion, as the strong conjugate base of weak acetic acid, binds the proton which the strong HCl is ready to donate. In other words, in the competition for protons, the CH_3COO^- is more successful than the Cl^- ion.

Similarly, with very strong acids the basicity of water is fully expressed, i.e. very strong acids completely transfer their protons to the water, and in so doing dissociate completely. This means that very strong acids cannot exist in water; they are always present in the form of H_3O^+ ions and their conjugate bases. Thus aqueous solutions of all very strong acids have the same acidity, due to the H_3O^+ ion, which is the strongest acid which can exist in water. This phenomenon is called the *leveling effect of water*. Likewise, stronger bases than the hydroxyl ion OH^- do not exist in water, but react immediately to form the conjugate acids and the OH^- ion. The protonation activity of extremely strong acids with respect to very weak bases can therefore be observed only in nonaqueous solutions (see Superacids).

If a salt is dissolved in water, its components, which always have acid-base properties, will always interact with the amphoteric water. For example, in aqueous sodium carbonate solutions, the strongly basic CO_3^{2-} ion binds protons from the water. Due to the constancy of the ion product of water, this causes a shift in pH toward the basic range. Similarly, ammonium salts, for example, give an acid reaction: the NH_3^+ ions, as the conjugate acid of the weak base ammonia, transfer some of their protons to the water.

Protolysis reactions of acids and bases with water determine the pH values of aqueous solutions. On the other hand, the protolysis equilibrium of the conjugate acid-base pair is determined by the pH of the solution (see Buffer solutions).

The Brønsted acid-base definition is extremely useful in treatment of proton-transfer reactions in water and nonaqueous solutions, especially in analytical chemistry, and is universally accepted. This is due in large part to the ease with which protolysis equilibria can be described in terms of this definition.

Lewis acids and bases: According to Lewis (1923), bases are molecules or ions which are able to donate an electron pair to a reaction partner to form a covalent bond, that is, they are *electron donors* and, depending on reaction kinetics, potential *nucleophiles*. Acids are molecules or ions which lack an electron pair and serve as *electron pair acceptors* and, seen kinetically, as *electrophiles*, which can form a covalent bond with a Lewis base. Thus any particle which has an unoccupied low-energy atomic or molecular orbital is a Lewis acid. The group includes cations in general (e.g. Ag^+, Cu^{2+}, Br^+, R_3C^+), molecules with incomplete electron octets (e.g. BF_3, $AlCl_3$, SO_3, $BeCl_2$), unsaturated coordinate compounds in which the central atom can undergo octet expansion (e.g. SiF_4, $SnCl_4$, PCl_3, SbF_5) and molecules with polar multiple bonds, in which the more positive atom carries the acidity (e.g. CO_2, carbonyl compounds). Correspondingly, *Lewis bases* are those particles which have occupied, relatively high-energy atomic or molecular orbitals. These include molecules with free electron pairs not used in bonding, anions (which act as bases with respect to metal cations or in complex formation) and molecules with polar multiple bonds in which the more negative atom acts as the center of basicity. Reactions between Lewis acids and bases yield neutralization products in which the components are joined by covalent bonds:

Acid	+ Base	⇌ Product
BF_3	$+ NH_3$	$\rightleftharpoons F_3\overset{-}{B}-\overset{+}{N}H_3$
$AlCl_3$	$+ Cl^-$	$\rightleftharpoons [AlCl_4]^-$
Ag^+	$+ 2\,NH_3$	$\rightleftharpoons [Ag(NH_3)_2]^+$
CO_2	$+ OH^-$	$\rightleftharpoons HCO_3^-$
H^+	$+ OH^-$	$\rightleftharpoons H_2O$

The Lewis definition of acids and bases has very wide applications, especially in discussions of reaction mechanisms in organic and coordination chemistry (HSAB concept, see below).

Although the Lewis and Brønsted base definitions include the same compounds, the definitions of acids are fundamentally different, and it is problematical that compounds which are usually classified as acids, e.g. HCl, H_2SO_4, H_3PO_4, etc., are not Lewis acids. To deal with this problem, Bjerrum proposed the following definition in 1951.

In the ***Bjerrum base-antibase system***, the term "acid" is reserved for proton donors, while Lewis acids are called *antibases*:

Base	+ Antibase	⇌ Product
NH_3	$+ H^+$	$\rightleftharpoons NH_4^+$
Cl^-	$+ AlCl_3$	$\rightleftharpoons [AlCl_4]^-$
S^2	$+ SnS_2$	$\rightleftharpoons SnS_3^{2-}$

This definition can be extended to include a suggestion made by Lux (1939) and Flood (1947), that the proton transfer which is the basis of the Brønsted definition is only a special case of ion transfer reactions. Many reactions which occur in molten salts can be formulated as oxide ion transfers (*oxidotropism*). *Oxide ion donors* are defined as bases, and *oxide ion acceptors* as acids, and on this basis, corresponding base-antibase systems can be formulated, e.g.:

Base	⇌ Antibase	+ O^{2-}
CO_3^{2-}	$\rightleftharpoons CO_2$	$+ O^{2-}$
$2\,SO_4^{2-}$	$\rightleftharpoons S_2O_7^{2-}$	$+ O^{2-}$

For example, the conversion of SiO_2 into nonrefractory form by fusion with a sodium-potassium carbonate mixture can thus be understood as a base-antibase reaction in which, as a result of an oxide ion transfer, the starting products are converted to the corresponding antagonists:

$$CO_3^{2-} + SiO_2 \rightleftharpoons SiO_3^{2-} + CO_2$$
Base + Antibase Base + Antibase

Ion-transfer reactions are also the basis of the ***solvent concept***, which was proposed by Gutmann and Lindquist (1939) and by Ebert and Konopik (1949), especially for reactions in aprotic solvents. According to this concept, acids are substances which increase the concentration of the cations typical of the solvent, while bases increase the concentration of solvent-typical anions. For example, in the phosphorus oxychloride system, which has the following dissociation reaction: $2\ POCl_3 \rightleftharpoons POCl_2 + POCl_4^-$, chloride ion donors would be bases, and chloride ion acceptors would be acids.

An even more inclusive generalization is the ***Ussanovitsch definition*** (1939), according to which acids are substances which react with bases, which cleave off protons or other cations (*cation donors*), or which can accept anions or electrons (*anion acceptors, electron acceptors, oxidation agents*). Bases are substances which react with acids, cleave off anions or electrons (*anion donors, electron donors, reduction agents*) or which can accept protons or other cations (*cation acceptors*). Thus this definition includes redox reactions as well as acid-base reactions, and in fact sees all chemical reactions, except for combination of radicals to form covalent compounds, as acid-base reactions.

In the ***HSAB concept*** (hard and soft acids and bases), which was published in 1963 by Pearson, an attempt is made to find criteria for the stability of the products of reactions of Lewis acids and basis. It turns out that the equilibrium position of the reaction
Lewis acid + Lewis base $\rightleftharpoons$ acid-base complex
depends primarily on the electronegativity and polarizability, i.e. on the deformability of the electron shells, of the species involved in the reaction. Lewis acids and bases can be classified on this basis as follows: *Hard acids* are Lewis acids which are not easily polarized, i.e. small, highly charged cations and molecules in which a high positive charge is induced on the central atom (e.g. H^+, Li^+, Mg^{2+}, Al^{3+}, Ti^{4+}, Fe^{3+}, CO_2, SO_3). *Soft acids* are Lewis acids which are highly polarizable, e.g. cations with large radius and low charge, or molecules with relatively high-energy occupied molecular orbitals (e.g. Ag^+, Cu^{2+}, Pd^{2+}, Pt^{2+}, carbenes, Br_2, I_2). *Hard bases* are Lewis bases with high electronegativity and thus low polarizability (e.g. NH_3, H_2O, OH^-, OR^-, F^-). *Soft bases* are Lewis bases with low electronegativity and thus higher polarizability (e.g. H^-, CN^-, R_3P, RSH, I^-). The terms hard and soft are relative, and gradual transitions are possible. Pearson found that hard acids react preferentially with hard bases, while soft acids react more readily with soft bases. This principle has been very useful for estimating equilibrium positions and the stability of the products, especially in organic and coordination chemistry. As an example, a complex of hard central ion and hard ligands or of soft central ion and soft ligands is much more stable than a hard-soft combination.

Acid-base disproportionation: see Acid-base concepts, section on Brønsted definition.

Acid-base titration: same as Neutralization analysis (see).

Acid carmoisin B: see Fast red.

Acid cleavage: the cleavage of 1,3-dicarbonyl compounds by heating in alkali solutions. A 1,3-diketone yields a ketone and a carboxylic acid:

$$R\text{–}CO\text{–}CH_2\text{–}CO\text{–}R \xrightarrow{OH^-} R\text{–}CO\text{–}CH_3 + R\text{–}COO^-.$$

Cyclic diketones yield δ-oxocarboxylic acids, and C-alkylated acetoacetates yield acetic acid and a higher carboxylic acid:

$$CH_3\ CO\text{–}CHR\text{–}COOC_2H_5 \xrightarrow{2HO^-} CH_3COO^- + R\text{–}CH_2\text{–}COO^- + 2C_2H_5OH.$$

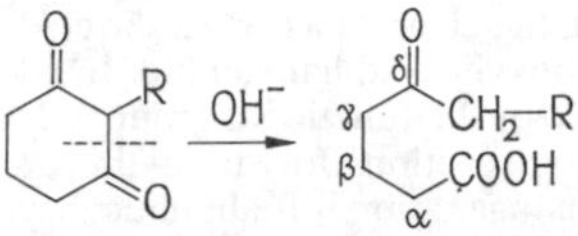

Acid constant: see Acid-base concepts, section on Brønsted definition.

Acid consumption: see Hardness, 2.

Acid dyes: a group of synthetic dyes which are directly adsorbed to animal fibers (wool, silk); the dye is precipitated on the fiber by addition of sulfuric, acetic or formic acid and sodium sulfate (formation of a water-insoluble "dye acid"). The A. are adsorbed to plant fibers (cotton) only after pretreatment of the fibers (see Mordant dyes). The most important A. are azo dyes, but the group also includes nitro, triphenylmethane, anthraquinone, pyrazolone, quinoline and azine dyes.

Acid halides: 1) derivatives of oxygen acids in which one or more OH groups are replaced by halogen atoms. Some important A. of inorganic oxygen acids are silicon tetrafluoride, SiF_4, sulfuryl chloride, SO_2Cl, thionyl chloride, $SOCl_2$, chlorosulfonic acid, HSO_3Cl, phosgene, $POCl_2$, phosphorus oxygen chloride, $POCl_3$, phosphorus(III) chloride, PCl_3, phosphorus(V) chloride, PCl_5, nitrosyl chloride, NOCl and chromyl chloride, CrO_2Cl_2.

2) ***Acyl halides***, ***carboxylic acid halogenides***: derivatives of carboxylic acids in which the hydroxy group in the carboxyl group is replaced by a halogen atom. The name is constructed either from the names of the skeleton hydrocarbon and the suffix -oyl halide, as in ethanoyl chloride, or by addition of the suffix -carbonyl halide to the name of the root hydrocarbon which is one C-atom shorter, as in methanecarbonyl chloride. Trivial names are also used, e.g. acetyl chloride. Acyl halides are reactive compounds, and they are used as acylating reagents, especially the chlorides. They are usually less water-soluble and have lower boiling or melting points than the parent carboxylic acid. They are made from the carboxylic acid or its salts or anhydride; the acid is reacted with an inorganic A. For example, the

chlorides are made by reaction of the acid, salt or anhydride with thionyl chloride, phosphorus(III) chloride, phosphorus oxygen chloride or phosphorus(V) chloride. Carboxylic acid chlorides can be converted to the bromides or iodides by reaction with hydrogen bromide or iodide. The carboxylic acid fluorides can be made by reaction of the chlorides with KHF_2. Acyl halides are excellent acylation reagents for compounds with active hydrogen atoms. On hydrolysis, they yield carboxylic acids, by an addition-elimination mechanism. Reaction with alcohols and phenols yields carboxylic acid esters; reaction with ammonia or amines yields acyl amides, with hydrazine or substituted hydrazines, hydrazides; with hydroxylamine, hydroxamic acids; with sodium azide, carboxylic acid azides; with sodium salts of carboxylic acids, carboxylic acid anhydrides. Acyl halides are used as the acylation reagents in the Friedel-Crafts acylation. They form ketones on reaction with Grignard compounds. Acyl chlorides can be reduced to aldehydes or primary alcohols (see Rosenmund reduction). The Arndt-Eistert synthesis (see) is a method for lengthening the chain of a carboxylic acid, and starts from the carboxylic acid halogenide. In addition to these reactions of the functional group, acyl halides display the usual reactions for the hydrocarbon part, including halogenation of aliphatic acyl halides. Acetyl and benzoyl chlorides are commonly used acyl halides.

Acidimetry: methods of Neutralization analysis (see) in which acids are used as standard solutions. Occasionally, the term is also used when bases are used as standard solutions.

Acidity: 1) acid strength; see Acid-base concepts, Brönsted definition. ***Acidic compounds***, e.g. acidic hydrocarbons, are those which are able to transfer protons to bases. 2) A measure of the H_3O^+ ion concentration of an aqueous solution (see pH).

Acid precipitation, often called ***acid rain***: precipitation (rain, snow, etc.) which has absorbed acidic or acid-forming pollutants (see Air pollution) in addition to carbon dioxide. Even in areas where the air is pure, the carbon dioxide in precipitation results in pH values as low as 5.6. A. is thus any precipitation in which the pH is lower than 5.5. If alkaline gases and particulates are present, higher concentrations of acidic pollutants are required to reduce the pH of the precipitation below 5.5.

Acid protection: a collective term for Corrosion protection (see) of metals and concrete subjected to corrosive media (not just acids) combined with thermal and mechanical stress. A. is achieved by application of ceramic or organic coatings (see Protective layers), or by massive construction using acid-resistant materials (hard polymers, plastics, ceramics).

A. is applied in the chemical and metal industries (pickling of metals), and to some degree also in coal and nuclear power plants, as well as in the production of cellulose, paper, textiles, synthetic fibers and leather products.

Acid reaction: a reaction, e.g. of a pH indicator, which shows that the pH (see) of an aqueous solution is less than 7 (see Ion product of water).

Acid violet, ***formyl violet***: various types of water-soluble, acid triphenylmethane pigments. They are used to dye paper and paints.

Aconitic acid, ***propene-1,2,3-tricarboxylic acid***: HOOC-CH_2-C(COOH)=CH-COOH, an unsaturated, water-soluble, triprotic carboxylic acid which can have either the *Z*- or the *E*-configuration. A. is synthesized by dehydration of citric acid at about 175 °C.

(Z)-A. is a colorless, crystalline compound which is barely soluble in ether; m.p. 130 °C. (*Z*)-A. is an important intermediate in the conversion of citric acid to isocitric acid in the presence of the enzyme aconitase; this is one of the steps of the citric acid cycle. When heated, (*Z*)-A. is readily converted to (*E*)-A.

(E)-A. forms colorless leaflets or needles which are soluble in alcohol; m.p. 198-199 °C. It is found mainly in the form of calcium or magnesium salts in horsetail, sugar cane and grain plants.

The name A. was coined by Peschier, who first isolated it in 1820 from aconite (*Aconitum napellus*). A. is used in the production of softeners and wetting agents for special purposes.

ACP: abb. for Acyl carrier protein (see).

Acridine, ***dibenzo[b,e]pyridine***, a heterocyclic compound which forms colorless needles with a characteristic odor; m.p. 111 °C (subl.), b.p. 346 °C. A. is nearly insoluble in water, but is soluble in organic solvents and can be steam distilled. The solutions have a blue fluorescence. The tertiary nitrogen atom causes A. to give a basic reaction and to form *acridinium salts* with acids and alkyl and aryl halides. Oxidation of A. with sodium dichromate in acetic acid yields Acridone (see). A. vapors irritate the skin and mucous membranes, and can lead to chronic diseases. A. is a component of coal tar, and can be synthesized by heating diphenylamines with formic acid in the presence of zinc chloride or by zinc powder distillation of acridone. A. is the skeleton of a numbers of dyes (see Acridine dyes) and pharmaceuticals, including antimalarials, bacteriocides and antiseptics. It was discovered in 1870 in coal tar by Graebe and Caro.

N

Acridine dyes: a group of synthetic dyes based on the acridine skeleton; the auxochromic groups are primary or dialkyl-substituted amino groups. The A. are made by condensation of 2,4-diaminotoluene with formaldehyde or benzaldehyde. The A. are basic mordant dyes used on cotton, leather and silk. The salts of acridine bases alkylated on the ring nitrogen, the acridinium salts, are used as antiseptics and occasionally as antimalarials.

R^7 R^2 R^6 N R^3

Acridine dye	R^2	R^3	R^6	R^7
Acridine orange	H	$N(CH_3)_2$	$N(CH_3)_2$	H
Acridine yellow	CH_3	$N(CH_3)_2$	$N(CH_3)_2$	CH_3

Some important A. are ***acridine orange***, 3,6-tetramethyldiaminoacridine, and ***acridine yellow***, 3,6-

tetramethyldiamino-2,7-dimethylacridine. The zinc chloride double salt of acridine orange gives a green fluorescence in aqueous solutions; the dye imparts an orange color to cotton.

Acridone: a heterocyclic compound derived from Acridine (see). A. is a stable compound which crystallizes from alcohol in yellow needles; m.p. 354 °C. It is rather insoluble in organic solvents, and displays neither ketonic nor phenolic properties; instead, it exists as an internal salt. Reduction with zinc powder produces acridine. A. is made by ring closure of diphenylamine-2-carboxylic acid (phenylanthranilic acid) with concentrated sulfuric acid, or by oxidation of acridine with sodium dichromate in acetic acid.

O

N

H

Acrilan®: see Synthetic fibers.

Acrolein, ***prop-2-enal, acrylaldehyde***: CH_2=CH-CHO, a colorless, mobile liquid, m.p. - 87.7 °C, b.p. 52.5 °C, n_D^{20} 1.3998. It is a lacrimator and has a pungent odor. A. is soluble in water and most organic solvents. It is the simplest unsaturated Aldehyde (see), and displays all the addition and condensation reactions typical of this class. As an α,β-unsaturated aldehyde with conjugated π-electrons, A. also has unusual reactions with nucleophilic reagents, which attack not only the carbonyl carbon atom, but also the β-C atom. This nucleophilic attack can be understood on the basis of the following canonical resonance formulas: $^+CH_2$-CH=CH-O^- ↔ CH_2=CH-CH=O ↔ CH_2=CH-$^+$CH-O^-.

> Acrolein is a strong poison which is extremely irritating to the mucous membranes of the respiratory passages and eyes, and can cause bronchitis and bronchopneumonia.

Because of its very high reactivity, A. tends to polymerize. It is easily converted to acrylic acid by oxidizing agents. Partial hydrogenation of A. on a nickel contact produces propionaldehyde; in the presence of copper-cadmium catalysts, allyl alcohol is formed. A. may be produced in the laboratory by heating glycerol with a dehydrating agent to about 200 °C. Industrially, A. is produced either by an aldol reaction of acetaldehyde and formaldehyde in the presence of silica gel or lithium phosphate, or by oxidation of propene or allyl alcohol on a metal oxide catalyst. A. can be used in the Diels-Alder reaction as the dienophile to synthesize formyl-substituted adducts. A. is also used in the synthesis of pharmaceutical products and many organic compounds.

Acrylaldehyde: same as Acrolein (see).

Acrylates: see Acrylic acid.

Acrylic acid, ***propenic acid, vinylcarboxylic acid, ethene carboxylic acid***: CH_2=CH-COOH, the simplest unsaturated monocarboxylic acid. A. is a colorless liquid with a pungent odor; m.p. 13 °C, b.p. 141.6 °C, n_D^{20} 1.4224. It is readily soluble in water and most organic solvents. It can be stored in monomeric form for a long time in the dark, if stabilizers, such as hydroquinone, are present. Under normal conditions, however, it polymerizes very readily to polyacrylic acid. The reactivity of A. is a result of the C=C double bond and the carboxyl group. These can undergo the reactions typical of their classes either singly or simultaneously, including additions to the C=C double bond, and various types of functionalizations of the carboxyl group. The products are substituted propanoic acids, Diels-Alder cyclization products, acrylic acid derivatives and heterocyclic systems. The salts and esters of A. are called *acrylates*. A. can be synthesized by oxidation of acrolein or allyl alcohol, hydrolysis of acrylonitrile, or dehydration of β-hydroxypropionic acid. Industrially, it is more effective to synthesize A. by gas phase oxidation of propene, by addition of water and carbon monoxide to acetylene in the presence of carbonylnickel(IV), or by catalytic ring cleavage of β-propiolactone. A. is used mainly to produce polyacrylic acid, polyacrylic esters and copolymers, and for organic synthesis.

Acrylic fibers: see Polyacrylonitrile fibers.

Acrylics: see Polymethacrylates, Polyacrylates.

Acrylonitrile, ***vinyl cyanide, acrylic nitrile***: CH_2=CH-C≡N, a colorless liquid with a pungent odor; m.p. - 83.5 °C, b.p. 77.5 °C, n_D^{20} 1.3911. A. is slightly soluble in water and readily soluble in most organic solvents. Because both an activated C=C double bond and the nitrile group, -C≡N, are present, A. is extremely reactive. It polymerizes very readily, forming polyacrylonitrile. In the presence of polymerization inhibitors, such as hydroquinone or pyrocatechol, A. can be stored for longer periods. Chemical reactions can occur on both functional groups, either separately or simultaneously. For example, CH-acidic compounds can add to the C=C double bond (Michael addition) to form cyanoethylated derivatives: R-H + CH_2=CH-C≡N → R-CH_2-CH_2-C≡N. A. can be hydrolysed by aqueous mineral acids to acrylamide or acrylic acid:

> Acrylonitrile is a strong poison which can enter the body through the lungs or the skin. It also irritates the skin and mucous membranes of the nose and eyes.

A. can be synthesized by addition of hydrocyanic acid to ethyne, or from ethylene oxide and hydrocyanic acid. Industrially, it is produced by ammoniooxidation of propylene:

A. is used mainly for the production of polyacrylonitrile, and also as an intermediate for organic syntheses.

ACTH: see Corticotropin.

Actin: A type of contractile protein found in many cell types. A. is an essential component of the contractile complex of Muscle proteins (see). Microvilli, microspikes (filopodia), and stereocilia (hair cells in the cochlea of the ear and other related organs) consist of A. associated with other proteins. Microfilaments in the cell cytoplasm consist of F- (polymerized) A. Monomeric, or G-A. (M_r 41 720) is an irregular mass, approximately ellipsoidal in shape. A consensus model of F-A. shows a helical filament with a diameter of 90-100 Å. The monomers lie with their long axes nearly perpendicular to the filament axis. The positions of the monomers within the filament are flexible, so that binding of proteins (e.g. tropomyosin) to the filament may impose a periodic but non-helical structure; the repeat distance is 7 monomers.

The structure of A. has been highly conserved in the course of evolution; this may be due to the large number of proteins with which it interacts specifically. In muscle fibrils, it forms a complex with regulator proteins, of which the best known are ***tropomyosin*** (M_r 68,000) and ***troponin*** (M_r 80,000). These are associated with the protein myosin in the *actomyosin complex*, which is responsible for the contraction of muscle. A. is present in all eukaryotic cells, as part of the *cytoskeleton*, where it makes up the microfilaments, and is involved in cell motions.

Actinium, symbol ***Ac***: a radioactive element from Group IIIb of the periodic system, the Scandium group (see). A. is a heavy metal, atomic number 89. The natural isotopes are members of the ^{235}U and ^{232}Th decay series, and have mass numbers 227 (β-emitter, $t_{1/2}$ 21.772 years) and 228 (β-emitter, $t_{1/2}$ 6.13 h). The atomic mass of Ac is 227.0278, its valence is III, and its density is calculated to be 10.07; m.p. 1050 °C, b.p. 3300 °C, standard electrode potential (Ac/Ac^{3+}) = - 2.6V.

A. is a regular, silvery-white metal forming cubic close-packed crystals; because of its radioactivity, it shines in the dark. Metallic A. is obtained by reduction of actinium(III) fluoride with lithium vapor at 1100 to 1300 °C. The chemistry of A. is very similar to that of lanthanum, its lighter homolog in the scandium group. Ac is very reactive, with surface oxidation occuring in air even at room temperature; in compounds, its valence is always +3. It is found naturally in uranium ores, but because of its short half-life, it is present only in very small concentrations. For example, 1 t pitchblende contains only 0.15 mg Ac. Ac makes up about $6.1 \cdot 10^{-14}$% of the earth's crust. It is produced in gram amounts by neutron irradiation of radium:

$$^{226}_{88}\mathrm{Ra} \xrightarrow{+n} {}^{227}_{88}\mathrm{Ra} \xrightarrow{-\beta^-} {}^{227}_{89}\mathrm{Ac}.$$

In addition to the naturally occurring isotopes, 24 synthetic isotopes with mass numbers from 209 to 226 and 229 to 232 are known; there are two nuclear isomers each with mass numbers 216 and 222.

Historical. A. was discovered in 1899 by André Debierné in pitchblende residues. The name (from the Greek "aktinoeis" = "shining") refers to the radioactivity of the element.

Actinium emanation: see Radon.

Actinoid: one of the 14 radioactive elements following actinium in the periodic system, with atomic numbers 90 to 103: thorium (Th), protactinium (Pa), uranium (U), neptunium (Np), plutonium (Pu), Americium (Am), curium (Cm), berkelium (Bk), californium (Cf), einsteinium (Es), fermium (Fm), mendelevium (Md), nobelium (No), and lawrencium (Lr). They are represented by the common symbol ***An***. A. with atomic numbers 93 to 103, the Transuranium elements (see), can only be obtained by nuclear reactions.

Table 1. Properties of the actinoids

	Th	Pa	U	Np	Pu	Am	Cm	Bk	Cr	Es	Fm	Md	No	Lr
Nuclear charge	90	91	92	93	94	95	96	97	98	99	100	101	102	103
Electron configuration*	$6d^2 7s^2$	$5f^2 6d^1 7s^2$	$5f^3 6d^1 7s^2$	$5f^5 7s^2$	$5f^6 7s^2$	$5f^7 7s^2$	$5f^7 6d^1 7s^2$	$5f^8 6d^1 7s^2$	$5f^{10} 7s^2$	$5f^{11} 7s^2$	$5f^{12} 7s^2$	$5f^{13} 7s^2$	$5f^{14} 7s^2$	$5f^{14} 6d^1 7s^2$
Atomic mass	232.0381	231.0359	238.029	237.0482	244	243	247	247	251	252	257	258	259	260
Atomic radius [pm]	179.8	160.6	139	131	151.3	173.0	174.4	176.7	186	186				
Density [g cm^{-3}]	11.724	15.37	18.97	20.48	19.737	13.671	13.51	13.25						
m.p. [°C]	1755	1568	1132	639	639.5	1173	1350	986	900					

* Electrons outside the Rn shell.

Table 2. Colors of some important actinoid ions

An^{3+}			U^{3+} Red-brown	Np^{3+} Faint purple	Pu^{3+} Blue violet	Am^{3+} Pink	Cm^{3+} Pale Yellow
An^{4+}	Th^{4+} Colorless	Pa^{4+} Yellow-green	U^{4+} Emerald green	Np^{4+} Yellow-green	Pu^{4+} Brown	Am^{4+} Reddish pink	Cm^{4+} Pale Yellow
AnO_2^+		PaO_2^+ Colorless	UO_2^+ Pale lilac	NpO_2^+ Green	PuO_2^+ Pale violet	AmO_2^+ Yellow-brown	
AnO_2^{2+}			UO_2^{2+} Yellow	NpO_2^{2+} Wine red	PuO_2^{2+} Fire red	AmO_2^{2+} Dark yellow	

A. have occupied outer orbitals ($6s^2$, $6p^6$, $7s^2$, and sometimes $6d^n$ where $n = 1$ or 2), and differ with respect to the occupation of the 5f shell.
Chemically, the A. differ more from one another than do the homologous lanthanoid group elements. With the exception of thorium, all A. form An^{3+} ions, the colors of which vary in a characteristic fashion (Table 2).

The absorption spectra of An^{3+} ions have narrow bands which are about ten times more intense than the bands of the corresponding Ln^{3+} ions; they are due to electron transitions within the 5f shell. Other similarities between the A. and lanthanoids are seen in the decrease in ionic radii of the An^{3+} ions with increasing nuclear charge number (see Actinoid contraction), the isomorphism of many actinoid and lanthanoid derivatives, such as the trichlorides, $AnCl_3/LnCl_3$, and dioxides, AnO_2/LnO_2, in the magnetic properties of An^{3+} and Ln^{3+} ions, and in their behavior on ion exchangers.

Thorium, protactinium and uranium have some similarities to the elements of Groups IVb, Vb and VIb. Plutonium, neptunium and americium are especially closely related to uranium, and the heavy A. curium to lawrencium are quite similar to the lanthanoids, with the +3 oxidation state the dominant one (Table 3).

Table 3. Valencies of the actinoids

Th	Pa	U	Np	Pu	Am	Cm	Bk	Cf	Es	Fm	Md	No	Lr
					2			2	2	2	2	2	
	3	3	3	3	3	3	3	3	3	3	3	3	3
4	4	4	4	4	4	4	4	4					
	5	5	5	5	5								
		6	6	6	6								
			7	7	7								

The A. are silvery-white, reactive metals which are tarnished by air and react at elevated temperatures with non-metals such as oxygen, hydrogen, nitrogen, carbon or halogens, to form the corresponding binary compounds (see Actinoid compounds).

Of the A., essentially only thorium, uranium and protactinium are found in nature; plutonium and neptunium are found only in exceedingly small traces. Those which are not found in nature can be made from uranium, or from pure transuranium elements obtainable from uranium, by nuclear reactions. Plutonium (^{239}Pu) and neptunium (^{237}Np) are generated in considerable quantity in nuclear reactors, and can be obtained on industrial scale by working up spent uranium fuel rods. A. with atomic numbers > 100 are obtained by bombardment of lighter A. with α-particles or accelerated carbon or boron nuclei:

$$^{253}_{99}Es + ^{4}_{2}He \longrightarrow ^{256}_{101}Md + ^{1}_{0}n$$

$$^{244}_{96}Cm + ^{12}_{6}C \longrightarrow ^{252}_{102}No + 4\,^{1}_{0}n$$

$$^{252}_{98}Cf + ^{11}_{5}B \longrightarrow ^{257}_{103}Lr + 6\,^{1}_{0}n$$

The lighter A. are separated using the considerable differences in stability of their various oxidation states ($UO_2^{2+} > NpO_2^{2+} > PuO_2^{2+} > AmO_2^{2+}$; U^{4+}, Np^{3+};≪ $Pu^{3+} < Am^{3+}$). The heavier A., like the lanthanoids, are separated by ion-exchange chromatography on the basis of the actinoid contraction. The smaller their atomic numbers, the more tightly A^{3+} ions are bound to cation exchangers, while extraction with complex formers, such as citrate, lactate or α-hydroxyisobutyrate, occurs more readily for the heavier An^{3+} ions with smaller ionic radii. Thus the A. ions appear in the eluate in the order Lr^{3+}, No^{3+} ... Bk^{3+}, Cm^{3+}.

Metallic A. can generally be obtained by reduction of the anhydrous fluorides MF_3 or MF_4 with magnesium or lithium-calcium vapor at 1100 to 1400 °C, or by melt electrolysis.

Actinoid compounds: The variety of A. is greater than that of the homologous lanthanoids, due to the ability of actinoids to exist in more oxidation states and to the differing stabilities of these states in different elements. For the elements curium to lawrencium, the +3 state is most important; the chemistry of thorium is dominated by the +4 state, which also is important for americium and plutonium. Protactinium and neptunium prefer the +5 state, while the +6 state is typical for uranium. With regard to the number of possible oxidation states, neptunium, plutonium and americium (+3 to +7) are comparable to uranium (+3 to +6), and these four elements tend to form dioxometal(VI) cations, AnO_2^{2+}, in aqueous solutions. The +2 oxidation state is limited to americium, as eka-europium, and the heavier actinoids californium to lawrencium. Actinoids form binary compounds with nonmetals such as hydrogen, carbon, fluorine and chlorine, including hydrides with the general formula AnH_3 (An = Pa, U, Np, Pu and Am), carbides AnC (cubic, NaCl-type crystals), An_2C_3 and AnC_2, and halides AnX_3 and AnX_4. The trichlorides, $AnCl_3$, of uranium to curium have ninefold coordination; while those of californium and einsteinium are dimorphic. The trifluorides AnF_3 of uranium to curium are isomorphic with the lanthanoid(III) fluorides, and also form polyhedra with a coordination number of 9. The trifluorides of berkelium and californium are dimorphic. The tetrafluorides AnF_4 are generally characterized by 8-fold coordination (distorted antiprisms). An^{3+} and An^{4+} ions, like the lanthanoid ions Ln^{3+}, can be precipitated by addition of fluoride or oxalate as hydrated actinoid(III) or (IV) fluorides or oxalates. Actinoid(IV) oxides, AnO_2, which crystallize in fluorite lattices, are known for An = thorium to californium; octahedral actinoid(VI) fluorides, AnF_6, have been obtained for An = uranium, neptunium and plutonium. The +5 and +6 valence states are represented by dioxoactinoid(V) and dioxoactinoid(VI) cations, which have been reported for protactinium to americium (AnO_2^+) and uranium to americium (AnO_2^{2+}), respectively. Actinoid(VII) derivatives are represented by lithium perneptunates and perplutonates of the type Li_5AnO_6 (An = Np, Pu).

Actinoid contraction: the continuous decrease in ionic radii with increasing nuclear charge among the actinoids: U^{3+}, 102.5 pm; Np^{3+}, 101.2 pm; Pu^{3+}, 100 pm; Am^{3+}, 97.5 pm; Cm^{3+}, 96.0 pm; Bk^{3+}, 95.5 pm; Cf^{3+}, 94.2 pm; Es^{3+}, 92.8 pm. Like the Lanthanoid contraction (see), the A. is caused by incomplete shielding of the increasing nuclear charge by the 5f electrons.

Actinometer: a device to determine the number of photons in a beam of ultraviolet or visible light, either as a time integral or per unit of time. Such information is needed to determine the quantum yields of photochemical reactions and photophysical processes. Bolometers and photodiodes are examples of physical A. In chemical A., the amount of radiation is determined by the amount of substance undergoing a photochemical reaction of known quantum yield. There are different A. for the various wave-length ranges. The ***ferrioxalate A.*** is very frequently used; it is based on the decomposition of $K_3[Fe(C_2O_4)_3]^{3-}$ into Fe^{2+} and oxalate ions and carbon dioxide. This reaction is useful because of its high quantum yield (1.25 to 0.9, depending on the wavelength), its wide range (250 to 480 nm) and the ease of measuring the concentration of the products (calorimetric determination of Fe^{2+} as the o-phenanthroline complex).

Actinomycins: a large group of peptide lactone antibiotics produced by various strains of *Streptomyces*. These highly toxic red compounds contain a chromophore, 2-amino-4,6-dimethyl-3-ketophenoxazine-1,9-dioic acid (actinocin), which is linked to two 5-membered peptide lactones by the amino groups of two threonine residues. The various A. differ only in the amino acid sequence of the lactone rings. In vivo, A. inhibit DNA-dependent RNA synthesis at the level of transcription by interacting with the DNA. The concentration required for inhibition depends on the base composition of the DNA; more is required for DNA with a low guanine content. A. are pharmacologically very important due to their bacteriostatic and cytostatic effect. Actinomycin D (Fig.) is one of the most widely occurring A. Its spatial structure has been elucidated by NMR studies, and the specificity of its interaction with deoxyguanosine was demonstrated by X-ray analysis. Actinomycin D is used as a cytostatic, e.g. in the treatment of Hodgkin's disease.

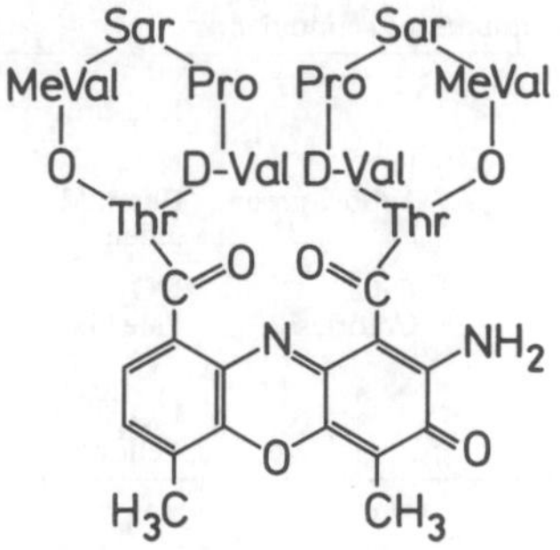

Actinomycin D

Actinon: see Radon.

Action constant: see Arrhenius equation.

Activated acetic acid: see Coenzyme A.

Activated complex: see Kinetics of reaction (theory).

Activated sludge: see Sewage treatment.

Activation analysis: a physical method of quantitative analysis based on the reaction of atomic nuclei with neutrons or charged particles, leading to formation of a radioactive isotope of the element to be analysed. The reaction utilized in neutron A. can be described by the following equations: 1) formation of the radioactive isotope:

$${}^{A}_{Z}X + {}^{1}_{0}n \longrightarrow {}^{A+1}_{Z}X + \gamma$$

2) Decay of the isotope:

$${}^{A+1}_{Z}X \longrightarrow {}^{A}_{Z+1}Y + \beta^-$$

The amount of radioactive isotope formed depends on the neutron flux, the number of reactive nuclei in the sample, and the capture cross section σ of these nuclei. σ is a measure of the probability of reaction 1), and is inversely proportional to the energy of the neutrons. Therefore A. is usually done with slow neutrons, which are present in high flux densities in nuclear reactors.

In practice, the unknown sample is usually irradiated along with a comparison sample of known composition, under identical conditions, and the activities generated after a sufficient time are compared. The simple equation $G_x/G_c = n_x/n_c$ is applied; here G_x is the amount of the substance being determined in the unknown sample, G_c is the amount in the comparison sample, n_x is the count rate of the unknown sample and n_c is the count rate of the comparison.

A. is a very sensitive method used mainly in Trace analysis (see), since the β-radiation emitted according to eq. 2) is readily measured. Some disadvantages are the very long time required for activation, and the fact that the process must be done in a nuclear reactor.

Activation energy: see Arrhenius equation.

Activation enthalpy: see Kinetics of reactions (theory).

Activation entropy: see Kinetics of reactions (theory).

Activator: see Catalysis, sect. III; Trace elements, 2.

Activity: 1) in solution theory, the corrected mass number for the composition of real solutions. The most common measures of Concentration (see) are mole fractions x_i, concentrations c_i and molalities m_i. In an ideal solution, the sum of the molecular interactions between the components is the same as in the pure substances. In real solutions, additional interaction energies appear, and lead to deviations in all the laws of solutions which were originally derived on the assumption of ideal behavior (e.g. the mass action law, the Nernst equation or the solubility product). In order to retain the equations for ideal solutions, Lewis introduced a corrected composition variable, the A. a, which is the product of the concentration and the ***activity coefficient*** f: $a_x = f_x x$; $a_c = f_c c$; $a_m = f_m m$.

The definition of A. is based on the chemical potential μ_i: $\mu_{real} = \mu_{ideal} + \Delta\mu_i = \mu_i^o\ RT \ln x_i + \Delta\mu_i$. Here μ_i^o is the standard chemical potential of substance i, x_i is the mole fraction of substance i in the solution, and $\Delta\mu_i$ is the difference between the chemical potential in the real solution and an ideal solution. $\Delta\mu_i$ corresponds to the reversible molar work of transferring substance i from the ideal into the real solution. According to Lewis, $\Delta\mu_i$ is set equal to $RT \ln f_i$; it follows that $\mu_{real} = \mu_i^o + RT \ln x_i + RT \ln f_{x,i}$ and $\mu_{real} = \mu_i^o + RT \ln a_{x,i}$ where $a_{x,i} = f_{x,i} x_i$. The mole fraction times the correction factor is the A., $a_{x,i}$, and the correction factor $f_{x,i}$ is the activity coefficient.

For practical purposes, chemical potentials can be formulated with concentrations c_i or molalities m_i and corresponding standard potentials μ_i^o. Deviations from ideal behavior require the definition of A. as $a_{c,i} = c_i f_{c,i}$ or $a_{m,i} = m_i f_{m,i}$. The subscripts c and m indicate that the A. are corrected concentrations and molalities, respectively. The numerical values of the activity coefficients defined in these three ways are not the same.

Activity coefficients depend on concentration, because the additional interactions vary with the composition of the solution. In the ideal solution, the activity coefficients are equal to 1. The ideal solution is defined in different ways. For mole fraction activities $a_{x,i}$, it is usually the pure substance, i.e. $f_{x,l} = 1$ for $x_i = 1$. For concentration and molality A., the ideal solution is an infinitely dilute one, i.e. $f_{x,i} = 1$ for $c_i \rightarrow 0$ and $f_{m,i} = 1$ for $m_i \rightarrow 0$.

Although solutions of neutral molecules have nearly ideal behavior at low concentrations, and the mass action or Raoult laws (for example) can be approximately formulated with concentrations instead of activities, in electrolyte solutions the long-range electrostatic interactions of the ions produce large deviations from ideal behavior even at concentrations $<10^{-3}$.

Activity coefficients must be determined experimentally. For dilute solutions of strong electrolytes, they can be calculated using the Debye-Hückel theory (see).

2) See Optical activity.

Activity coefficient: see Activity.

Acyclic: a term for organic compounds in which the carbon atoms are arranged in a chain, especially in contrast to cyclic compounds.

Acyl ...: a term for the atomic group R-CO- in a molecule, for the unstable radical R-CO· and for the cation $R\text{-}CO^+$.

Acylalanine fungicide: a systemic Fungicide (see) containing an acylated phenylalanine component. The substances effective against oomycetes specifically inhibit RNA synthesis. *Metalaxyl* (Ridomil®), which is effective in low concentrations against *Phytophthora, Pythium* and *Peronospora*, is used on potatoes to prevent tuber and leaf rot. It is usually combined with other fungicides to prevent development of resistance. A. is synthesized by reaction of dimethylaniline with 2-bromopropionic acid methyl ether; the product is then acylated to methoxyacetyl chloride. *Furalaxyl* is used on ornamentals.

Metalaxyl ($R = CO{-}CH_2{-}OCH_3$)
Furalaxyl ($R = CO$–2-furyl)

Acylals: organic compounds of the type $R^1R^2C(O\text{-}CO\text{-}R^3)_2$; they are esters of geminal diols with carboxylic acids. A. are formed, for example, by oxidation of methylarenes with chromic acid in acetic anhydride. They prevent further oxidation to the carboxylic acid, and thus give access to formyl arenes, which are readily obtained by hydrolysis of the isolated A.

Acylaniline herbicides: Herbicides (see) which contain an anilide structure.

Table 1. Anilide herbicides $R{-}C(=O){-}NH{-}Ar$

Name	R	Ar
Propanil	C_2H_5	3,4-dichlorophenyl
Pentanochlor	$H_3C{-}CH(C_3H_7)$	3-chloro-4-methylphenyl
Monalid	$H_7C_3{-}C(CH_3)_2$	4-chlorophenyl
Chloranocryl	$H_2C{=}C(CH_3)$	3,4-dichlorophenyl

1) Anilides derived from low-molecular-weight carboxylic acids (Table 1) act as photosynthesis inhibitors; in most cases they are applied as post-emergence, selective contact herbicides.

2) Chloroacetanilide herbicides, i.e. derivatives of chloroacetic acid (Table 2) usually inhibit protein synthesis. They are usually used as pre-emergence herbicides which are highly effective against grasses.

Table 2. Chloroacetanilide herbicides $Cl{-}CH_2{-}C(=O){-}N(R){-}Ar$

Name	R	Ar
Propachlor	$(H_3C)_2CH$	phenyl
Prynachlor	$HC{\equiv}C{-}CH(CH_3)$	phenyl
Alachlor	H_3COCH_2	2,6-diethylphenyl (H_5C_2, H_5C_2)
Butachlor	$H_9C_4OCH_2$	2,6-diethylphenyl (H_5C_2, H_5C_2)

Acylation: a preparative method for introduction of an acyl group, R-CO-, into an organic compound. It is usually the H atom of a hydroxyl group (*O-acylation*), sulfhydryl group (*S-acylation*), amino group (*N-acylation*) or arene or alkene (*C-acylation*) which is replaced by the acyl group. The most important acylation reagents are the halides, anhydrides and esters of carboxylic acids. O-, S- and N-acylation is frequently used to protect OH, SH and NH_2 groups from undesired side reactions; the acyl group can later be removed easily by hydrolysis. The products of O-, S- or N- acylation are esters, thioesters or amides. These compounds often crystallize well, and are therefore used to characterize alcohols, phenols, thiols and amines. An important A. method for preparative and analytical purposes is the Schotten-Baumann reaction (see) with benzoyl chloride. C-acylation of arenes by the Friedel-Crafts reaction (see) produces aromatic-aliphatic or pure aromatic ketones. The Vilsmeier-Haack reaction on arenes (see) is another A. A further application of A. is the determination of the number of OH, SH or NH_2 groups in an organic compound, the acyl number.

Acyl azides: derivatives of carboxylic acids with the general formula $R{-}CO{-}N_3$. The electronic structure can be described by canonical resonance structures:

$$R{-}C(=\overline{O})\!{-}\overline{N}{=}\overset{+}{N}{=}\overset{-}{\underline{N}} \longleftrightarrow R{-}C(=\overline{O})\!{-}\overset{-}{\underline{N}}{-}\overset{+}{N}{\equiv}N| \longleftrightarrow R{-}C(\overline{O}^{-}){=}N{-}\overset{+}{N}{\equiv}N|$$

C. are obtained by reaction of carboxylic acid chlorides with NaN_3, or by reaction of nitrous acid with carboxylic acid hydrazides. They are very reactive, and the smaller molecules tend to decompose explosively. When heated in a solvent, they undergo Curtius decomposition (see). In inert solvents (benzene, chloroform), isocyanates can be isolated. C. are important acylating reagents. N-Protected amino acids or peptide azides are very important in Peptide synthesis (see).

Acyl carrier protein, abb. ***ACP***: a globular protein with relative molecular mass up to 16,000. The A. is the carrier of the fatty acid chains produced in fatty-acid biosynthesis. They are part of a multienzyme complex, fatty acid synthetase. A serine residue on the ACP is linked to phosphopantetheine, which acts as the "swinging arm" and moves the acyl group from one active site to the next within the complex. The fatty acid is esterified to the thiol group of the phosphopantetheine.

Acyl halides: see Acid halides, 2.

Acyl hydrazines: see Hydrazides.

N-Acylneuraminic acid: same as Sialic acid (see).

Acyl number: the number of acyl groups in a substance. The A. is determined by hydrolysis in alcoholic potassium hydroxide and neutralization of the released acid. For example, 1 ml of a 0.5 M potassium hydroxide solution corresponds to 21.52 mg acetyl or 52.51 mg benzoyl.

Acyloin: a general term for an α-hydroxyketone, R-CH(OH)-CO-R. The R groups can be alkyl, aryl or heterocyclic substituents; mixed A. are also known. A. are in tautomeric equilibrium with enediols, and are therefore capable of reducing Fehling's solution.

$$R{-}CH(OH){-}C(=O){-}R \rightleftharpoons R{-}C(OH){=}C(OH){-}R$$

Aliphatic A. are formed when carboxylate esters are heated with sodium in inert solvents, in the absence of water, oxygen and alcohols. Cyclic A. are formed from dicarboxylic acid esters. Aromatic and heterocyclic A. are formed mainly through Acyloin condensation (see). Some important A. are Acetoin (see), Benzoin (see) and Furoin (see).

Acyloin addition: same as Acyloin condensation.

Acyloin condensation, ***acyloin addition***: 1) a dimerization of aromatic or heterocyclic aldehydes which do not have α-hydrogen atoms to form acyloins.

The classic example of A. is the ***benzoin condensation***, in which benzaldehyde forms benzoin in the presence of potassium cyanide or special thiazolium or imidazolium salts in a water/alcohol solution:

$$2C_6H_5{-}CHO \xrightarrow{(CN^-)} C_6H_5{-}CH(OH){-}CO{-}C_6H_5$$

The following reaction mechanism is considered definitive:

$$R{-}CHO \underset{}{\overset{+CN^-}{\rightleftharpoons}} R{-}CH(CN){-}\overline{\underline{O}}|^- \rightleftharpoons R{-}\overset{-}{C}(CN){-}OH$$

$$R{-}CH{=}O + {}^{-}C(CN)(OH){-}R \rightleftharpoons R{-}CH(\overline{\underline{O}}|^-){-}C(CN)(OH){-}R$$

$$R{-}CH(O^-){-}C(CN)(OH){-}R \underset{}{\overset{-CN^-}{\rightleftharpoons}} R{-}CH(OH){-}C(=O){-}R$$

Anisaldehyde and furfural, for example, react in the same way to form symmetric, substituted α-hydroxyketones, anisoin and furoin. A number of unsymmetric benzoins can also be synthesized under the conditions of the A. For example, a mixture of benzaldehyde and 4-dimethylaminobenzaldehyde in the presence of cyanide ions yields a benzoin with the following structure: $(CH_3)_2N\text{-}C_6H_4\text{-}CO\text{-}CH(OH)\text{-}C_6H_5$.

2) The formation of acyloins from aliphatic carboxylic acid esters and sodium in inert solvents.

N-Acylsphingoid: same as Ceramide (see).

"Adam and Eve" cycle: see Synthesis gas.

Adamantane, ***tricyclo[3.3.1.1^{3,7}]decane***: a saturated hydrocarbon forming colorless crystals with a camphor-like smell. It sublimes at room temperature; m.p. 270 °C. A. is found in petroleum and was first isolated from that source.

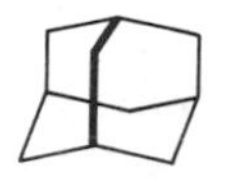

Adamantane

It is synthesized by isomerization of *endo*-tetrahydrodicyclopentadiene by aluminum chloride. A. is the parent compound of the "diamond-like" compounds. 10 carbon atoms are linked to form four completely equivalent, non-stressed cyclohexane rings which are fixed in the chair form. The highly symmetrical structure is an optimal arrangement and approximates a sphere (it is also similar to Urotopin). A. is therefore very stable and unreactive. It is impossible for steric reasons to form a double bond, i.e. Bredt's rule applies strictly to each C atom.

Adamsite: see Chemical weapons.

Addition: incorporation of atoms or molecules into unsaturated organic compounds. In the process, triple bonds are converted to double bonds, and double bonds, to single bonds. If cyclic compounds are formed by A., one speaks of Cycloaddition reactions (see). A. can occur as Electrophilic reactions (see) (A_E), Nucleophilic reactions (see) (A_N) or Radical reactions (see) (A_R). The A. of hydrogen is called Hydrogenation (see); A. of water is called Hydration (see). A. reactions are catalysed by acids. A. of bases to special carbonyl compounds with cumulative double bonds are important in the synthesis of carboxyl derivatives, for example A. to carbon dioxide to form carboxylic acid derivatives, A. to carbon disulfide to form xanthogenates, A. to mustard oils and isocyanic acid or its esters to form urea derivatives and urethanes, or A. to ketenes to form acetic acid derivatives. The reactions of carbonyl compounds with CH-acidic compounds are also A. in which the anion of the CH acidic compound is added to the carbonyl carbon atom by a nucleophilic mechanism. Seen in this way, the reaction of carbonyl compounds with certain bases occurs by an addition-elimination mechanism (see Substitution 1.3.2). A. can also be assumed to be steps in the nucleophilic substitutions of benzene derivatives, which occur by addition-elimination or elimination-addition mechanisms.

Addition complex: see Coordination chemistry.

Addition-elimination mechanism: see Substitution 1.3.2.

Additives: chemical substances added in small amounts to materials to achieve a desired effect or diminish undesired properties.

1) In the petroluem industry, the most notorious is tetraethyllead, $Pb(C_2H_5)_4$, which increases the octane rating of gasolines.

a) *Antioxidants* suppress autooxidation of lubricants; 2,6-di-*tert.*-butyl-p-cresol is widely used for this purpose.

b) *DD additives* (detergents-dispersants) are intended to prevent formation of deposits in combustion engines. Insoluble tars and solids which form in the lubricating oil or are picked up by it are kept in fine suspension and prevented from forming a sludge in the engine. Salts of alkylarylsulfonic acids are used for this purpose.

c) *HD additives* (heavy-duty A.) have a function similar to that of the DD-A., but they are used in more heavily stressed engines, particularly those burning sulfur-containing diesel fuels.

d) *Stock-point lowering A.* prevent formation of paraffin crystals in lubricating oils at low temperatures. If such crystals form, the oil components are adsorbed on them, causing the entire oil to solidify prematurely. Low-molecular-weight ethylene-vinyl acetate copolymers and highly alkylated naphthalene derivatives made from chloroalkanes and naphthalene are used for this purpose.

e) *Viscosity-temperature A.* are used to counteract the decrease in viscosity of lubricants at higher temperatures. These compounds are high polymers, such as polyisobutylene, copolymers of styrene with C_8 to C_{12} alkenes, and polymethacrylates with molecular weights between 10,000 and 30,000.

f) *EP additives* (extreme-pressure A.) are added to gear oils. They usually have high contents of sulfur or phosphorus. When they are used in heavily loaded gears, a sulfide or phosphide film forms on the metal surfaces and prevents binding.

2) In the food industry, A. are substances deliberately added to foods and ingested by the consumers. A. are intended to improve the appearance, consistency and flavor of the product or prolong its shelf life. A. are subject to regulations concerning their effects on health; they must be harmless and must not lead to formation of harmful substances in the foods. Some examples of A. are Food colorings (see), Flavorings (see), Antioxidants (see), Preservatives (see), sweeteners, acidifiers, leavenings and emulsifiers (see Emulsion).

Additivity principle: It is possible to calculate a number of molecular properties to a good approximation from the contributions (increments) of atoms, atomic groups and individual bonds in the molecule. The atomization energy E_A of a molecule can be determined by addition of mean Bond energies (see) E_B(X-Y). As an example, for methanol one obtains $E_A(CH_3OH) = 3\,E_B(C\text{-}H) + E_B(O\text{-}H) + E_B(C\text{-}O)$. In more exact increment systems, the hybridization states of the atoms involved in the bonds must be taken into consideration. The dipole moment of a molecule can be calculated by vectorial addition of bond dipole moments or group moments: $\mu(H_2O) = 2\,\mu_{O\text{-}H}\cos(HOH/2)$. The molar refraction (see Polarization) is the sum of the atomic refractions R_A with special increments for multiple bonds: $R(C_2H_5OH) = 6\ R_H + 2\ R_C + R_O$. Alternatively, it can be obtained from the bonding reactions $R_{X\text{-}Y}$: $R(H_3C\text{-}CH_2\text{-}OH) = 5\ R_{C\text{-}H} + R_{C\text{-}C} + R_{C\text{-}O} + R_{O\text{-}H}$. Conjugated multiple bonds increase the molar refraction (*exaltation*). The A. arises from the fact that the electron density of many molecules can be adequately described by bonding orbitals.

Ade: abb. for Adenine.

Adenine, ***6-aminopurine*** abb. ***Ade***: a component of adenosine and its nucleotides, of RNA and DNA, and of coenzymes such as NAD(P) and FAD. A. crystallizes with 3 mol water; at 110 °C it loses this water of crystallization. It sublimes around 200 °C. It is nearly insoluble in cold water, but readily soluble in mineral acids or alkali hydroxides, with which it forms salts. It is practically insoluble in ether and chloroform. A. derivatives have cytokinin activity. A. is degraded in the human body to uric acid. It is used to treat liver disease. ***Adenine sulfate*** prolongs the storage life of erythrocytes and is used in blood-stabilizing solutions.

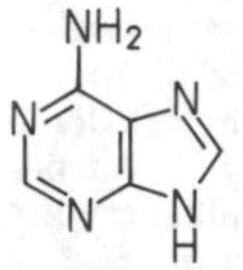

Adenosine, ***6-aminopurine-9-β-D-ribofuranoside***, abb. ***Ado***: a purine nucleoside. It crystallizes with 1.5 mol water of crystallization. A. is a white powder, m.p. 229 °C (dec.), $[\alpha]_D^{20}$ - 60° (water), which is soluble in hot water, but practically insoluble in ethanol. It can be isolated from yeast nucleic acid after alkaline hydrolysis. A. has a weak dialtory effect on blood vessels.

Adenosine 5′-diphosphate: see Adenosine phosphates.

Adenosine monophosphates: see Adenosine phosphates.

Adenosine 3′,5′-phosphate: see Cyclic nucleotides.

Adenosine 3-phosphate-5′-phosphosulfate: see Nucleotides.

Adenosine phosphates, ***adenylic acids***: nucleotides of adenosine. ***Adenosine 5'-monophosphate***, abb. ***AMP*** (***muscle adenylic acid***) is a white, crystalline powder; M_r 347.2, m.p. about 196 °C (dec.). It is soluble in hot water, and slightly soluble in ethanol. ***Adenosine 3'-monophosphate*** (***yeast adenylic acid***) forms white crystalline needles; m.p. 208 °C (dec.). It is soluble in hot water. ***Adenosine 5'-diphosphate***, abb. ***ADP***, M_r 427.2. ***Adenosine 5'-triphosphate***, abb. ***ATP***, M_r 507.2. In pure form, ATP is a white powder which is readily destroyed by acids or bases; however, as the dibarium salt, it precipitates readily in anhydrous form and is relatively stable. ATP is readily soluble in hot water; it forms insoluble salts with barium, lead, silver and mercury.

AMP, ADP and ATP form a metabolic system which is essential to life. It connects the energy-producing processes (photosynthesis, respiration and glycolysis) with energy-consuming synthetic processes. ATP is the most important intracellular energy transferring compound. When hydrolysed to ADP, it releases 32.66 kJ mol^{-1} energy.

ATP is synthesized in the course of energy-producing reactions by phosphorylation of ADP. In addition to their roles in intermediary metabolism, ATP-cleaving enzymes (ATPases) are active in the transport of various compounds and ions across biological membranes and in the contraction of actomyosin complexes, which power muscle contractions and also motility of non-muscle cells. ATP was first isolated by Lohmann in 1929. ADP and ATP have a dilatory effect, especially on coronary arteries; AMP and adenosine have a similar, but weaker effect. ATP can be isolated as the barium salt from meat extracts.

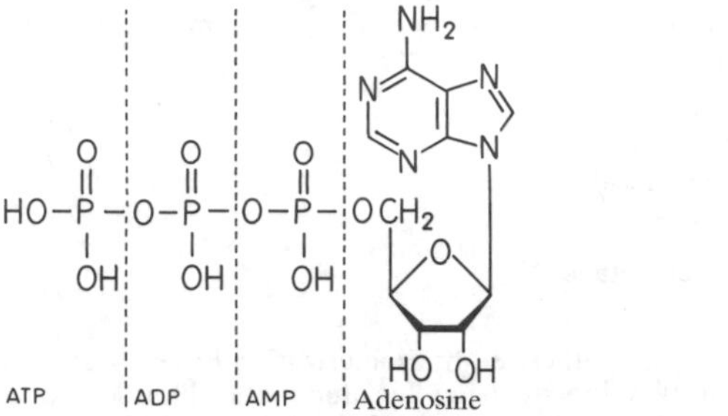

ATP | ADP | AMP | Adenosine

Adenosine 5′-triphosphate: see Adenosine phosphates.

Adenosine triphosphatases: see ATPases.

S-Adenosylmethionine, abb. ***AdoMet***, a coenzyme of many enzymes which transfer methyl groups (methyl transferases).

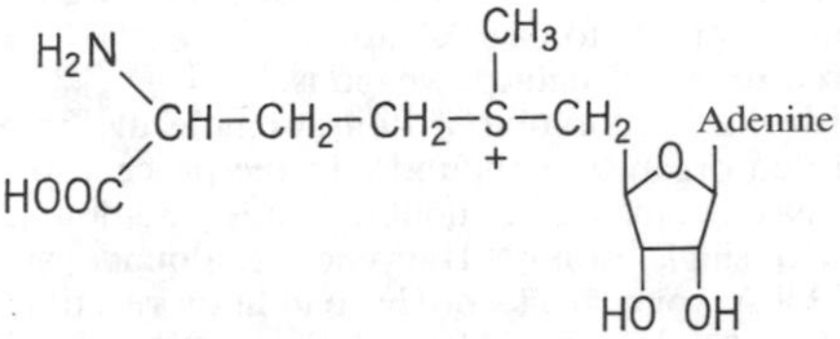

Adenine

Adenylate cyclase system: see Hormones.

Adenylic acids: same as Adenosine phosphates (see).

Adenylic acid system: see Adenosine phosphates.

Adermine: same as Vitamin B_6 (see).

ADH: see Vasopressin.

Adhesion: the sticking together of molecules and particles, or of particles or extended phases of different materials. A. is caused by electric attraction between the particles, due to permanent or induced dipoles (Debye-Keesom forces) or to dispersive forces (London forces, van der Waals forces). ***Adhesive forces*** play a part in adsorption, wetting of solids by liquids, gluing, adhesion of paints, coating of paper, textiles and metals, in the production of filled polymers, in construction materials and in pressure processes. For optimum A., both the properties of the substances and the number of atomic (or molecular) contact sites are important. If the surface is rough or very rigid, glues or higher temperatures are used to promote A.

Adiabat: a curve in a diagram joining all the points through which an adiabatic process passes. If this change in state is also reversible, the A. is also an ***Isentrope***.

Adiabatic: 1) in thermodynamics, a process which occurs without exchange of heat with the environment; see Adiabatic process. 2) In kinetics, a reaction which occurs without any changes in the electronic states of the reactants. The reaction can be described as motion along a potential surface. For thermal reactions at relatively low temperatures, this is the potential surface of the electronic ground state. Nonadiabatic reactions are characterized by a change in the total electron spin, that is, by a transition between two potential surfaces. They occur when the potential surfaces intersect, or at least come very close together.

Adiabatic process: a thermodynamic process which occurs without exchange of heat with the environment, that is, $dq = 0$ (q = heat). An A. is an ideal, limiting case which can only be approximated in practice, for example, by very efficient thermal insulation, or by a reaction which is so rapid that heat transport can be neglected (adiabatic compression and expansion in sound waves or detonation fronts). The opposite ideal case, in which the heat exchange with the environment is so rapid that thermal equilibrium is always maintained, is the ***isothermal process***. Between the two limiting cases lies the region of Polytropic processes (see). From the First Law of Thermodynamics, it follows that since $dq = 0$, $de = dw$, that is, the change in internal energy e of a system is equal to the work exchanged with the environment. When work is done by the system (expansion), it cools, and when work is done on the system (compression), it is heated. For an ideal gas (see State, equation of, 1.1), it follows that $c_v dT = -p dv = -(NRT/v) dv$. Bringing in the equations $c_p - c_v = nR$ and $c_p/c_v = \delta$, we obtain the differential equation $d \ln T = (1 - \varkappa) d \ln v$. The solution of the equation, $Tv^{\varkappa - 1}$ = const. or $pv^{\varkappa}$ = const. is called the ***Poisson equation***. Here c_p and c_v are the heat capacities at constant pressure and constant volume, respectively. It follows from the Poisson equation that the pressure of an ideal gas increases more rapidly during adiabatic compression than during isothermal compression, because $\varkappa > 1$ for all gases (see Molar heats).

Adipic acid, ***hexanedioic acid, butane-1,4-dicarboxylic acid***: $HOOC-(CH_2)_4-COOH$, a saturated, aliphatic dicarboxylic acid. A. crystallizes in colorless and odorless, monoclinic prisms; m.p. 153 °C, b.p. 265 °C at 13.3 kPa. It is slightly soluble in cold water, and more readily soluble in alcohol and hot water. A. is a stable compound which can be converted to cyclopentanone at high temperatures by loss of carbon dioxide and water. Like all dicarboxylic acids, A. forms mono- and diesters with alcohols. It is found in free form in the juice of sugar beets. It can be prepared by a malonic ester synthesis from sodium malonic esters and 1,2-dibromoethane. Industrially, A. is made by oxidation of cyclohexanone or cyclohexanol with dilute nitric acid or atmospheric oxygen in the presence of a catalyst. A. is synthesized in large quantities as a raw material for production of polyamides and polyesters. It is also used for synthesis of adipic nitrile, softeners and perfumes.

Adipiodone: same as iodipamide; see X-ray contrast agents.

Adiuretin: same as Vasopressin (see).

Ado: abb. for Adenosine (see).

AdoMet: abb. for *S*-Adenosylmethionine (see).

ADP: see Adenosine phosphates.

Adrenalin: see Noradrenalin.

Adrenergics: see Sympathicomimetics.

Adrenal corticosteroids: a group of steroid hormones formed in the adrenal cortex in response to adrenocorticotropic hormone (ACTH). The natural N. are derived from pregnane, and therefore contain 21 C atoms. Oxygen functions are present on C3 and C20; unlike the gestagens, the A. also have oxygen functions on C21 and/or C11 and C17. Over 30 steroids have been isolated from the adrenal cortex, but only a few of them have hormone activity. The biological precursor of the A. is ***progesterone***.

The A. regulate mineral and carbohydrate metabolism. The ***mineralocorticoids*** regulate the plasma concentrations of sodium and potassium ions, causing increased potassium excretion and retention of sodium and water. The first natural mineralocorticoid isolated was *cortexolone*, 17α,21-dihydroxy-4-pregnene-3,20-dione, prepared in 1937 by Reichstein. Its m.p. is 205 °C, $[\alpha]_D^{20}$ + 132° (in ethanol). *Deoxycorticosterone* (cortexone), 21-hydroxy-4-pregnene-3,20-dione, m.p. 142 °C, $[\alpha]_D^{20}$ + 178° (in ethanol), was the first mineralocorticoid used (in the form of its acetate, deoxycorticosterone acetate) for treatment of Addison's disease and shock. The compound was first partially synthesized by Reichstein and Steiger before it was isolated from the adrenal cortex in 1938. *Aldosterone* has about 30 times the mineralocorticoid potency of deoxycorticosterone; in this compound, the C18 is oxidized to a formyl group which forms a semiacetal with the β-OH group on C11. Since aldosterone is poorly absorbed and has an elimination half-life of only 30 minutes, the partially synthetic *fludrocortisone* is used instead, sometimes in the form of its C21 acetate. Fludrocortisone has a somewhat higher glucocorticoid effect than aldosterone, and a much higher mineralocorticoid effect.

Glucocorticoids promote the breakdown of proteins to make amino acids available for gluconeogenesis; this elevates the blood sugar level and leads to glycogen synthesis in the liver. These hormones are also used therapeutically in cases of adrenal insufficiency. However, they are therapeutically more important because of their anti-inflammatory, anti-allergenic, anti-exsudative and immunosuppressive effects. Glucocorticoids are used to treat rheumatism, asthma and various skin conditions. Although a mineralocorticoid effect is desirable for substitution therapy, this effect is undesirable in glucocorticoids for broad use as anti-inflammatory agents, and it has been possible to eliminate it by partial synthetic modification. The most important naturally occurring glucocorticoids are cortisone and hydroxycortisone (*cortisol*), both of which have considerable mineralocorticoid activity. *Cortisone*, 17α,21-dihydroxy-4-pregnene-3,11,20-trione, m.p. 215 °C, $[\alpha]_D^{20}$ +209° (in ethanol) has a weaker glucocorticoid effect than *hydrocortisone*, 11β,17α,21-trihydroxy-4-pregnene-3,20-dione, m.p. 220 °C, $[\alpha]_D^{20}$ +167° (in ethanol).

An important advance in the development of compounds with specific glucocorticoid effects was made by the introduction of *prednisone* and *prednisolone*. The glucocorticoid effect of these is 5 times greater than that of hydroxycortisone, while their mineralocorticoid effect is 3 times lower. An improvement in the anti-inflammatory effect and simultaneous elimination of mineralocorticoid activity was achieved by introduction of a fluorine atom in the 9α-position, and introduction of a methyl or hydroxy group on C16. Some examples are *triamcinolone* and *dexamethasone*, which are applied systemically and topically. For topical application only, highly lipophilic compounds are needed which penetrate the skin easily but have no systemic effects. Esters and ketals such as *flumethasone pivalate* and *fluocinolone acetonide* have proven useful in this context.

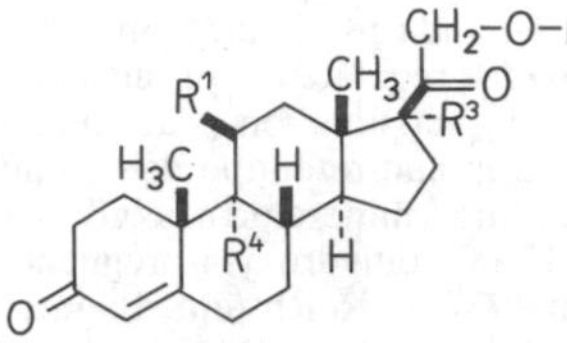

Name	R^1	R^2	R^3	R^4
Cortexolone	H	H	OH	H
Deoxycortone	H	H	H	H
Deoxycortone acetate	H	CH_3CO	H	H
Fludrocortisone acetate	OH	CH_3CO	OH	F

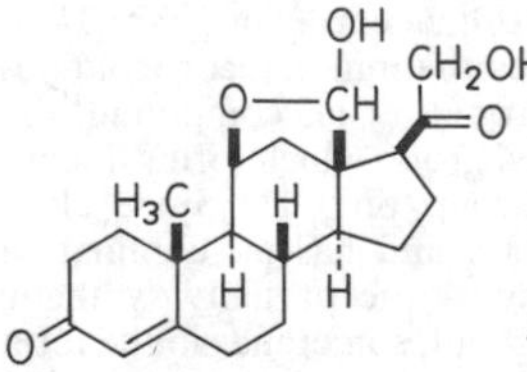

Aldosterone

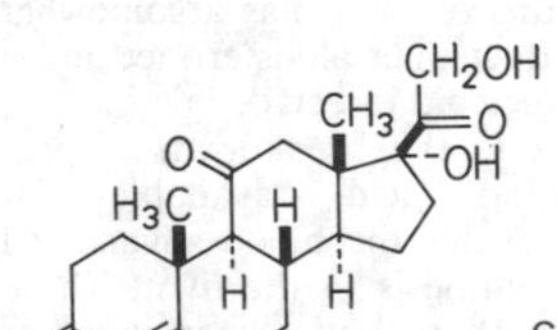

Cortisone

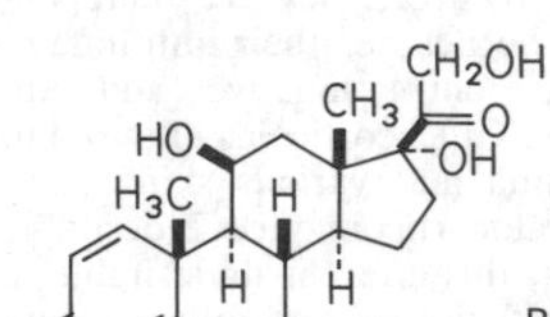

Prednisolone

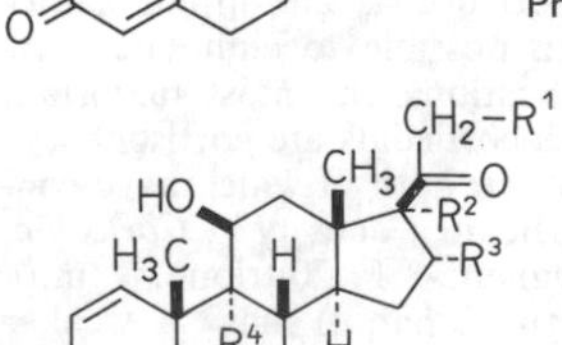

Name	R^1	R^2	R^3	R^4	R^5
Prednisolone	OH	OH	H	H	
Triameinolone	OH	OH	OH	F	H
Dexamethasone	OH	OH	CH_3	F	H
Fluocinolone acetonide	OH	$-O-C(CH_3)_2-O-$		F	F
Flumethasone pivalate	$(CH_3)_3C-COO-$	OH	CH_3	F	F

Adrenocorticotropin: same as Corticotropin (see).

Adrenolytics: see Sympathicolytics.

Adriamycin: see Anthracyclins.

Adsorbate: term for a system consisting of an adsorbed substance and an adsorbent.

Adsorbent: a condensed phase on the surface of which adsorption occurs. The most commonly used A. are activated charcoal, silica gel, aluminum oxide, zeolite, ion exchangers or polyamides.

Adsorber: an apparatus for carrying out adsorption and desorption, such as a column.

Adsorpt: a substance adsorbed to the surface of a condensed phase.

Adsorption: the enrichment of a substance at the surface of a condensed phase. The process: adsorbed substance + adsorbent ⇌ adsorbate can be treated in the same way as a chemical reaction. The heat released by the process is called ***adsorption energy*** (at constant volume) or ***adsorption enthalpy*** (at constant pressure). Depending on the magnitude of the molar adsorption enthalpy, a distinction is made between reversible ***physical A.*** (physisorption) and irreversible ***chemical A.*** (chemisorption). The former can be reversed by desorption, without chemical change of the adsorbed substance or adsorbent, while the latter cannot. However, a sharp distinction cannot be made. Physical A. is due mainly to van der Waals forces between the adsorpt and adsorbent, and the molar adsorption enthalpies $\Delta_A H$ are between - 5 and - 40 kJ mol^{-1}. In chemical A., chemical bonds form on the surfaces, and $\Delta_A H$ are on the order of - 40 to - 400 kJ mol^{-1}. The surfaces of real solids are not energetically or structurally homogeneous. At the beginning of an A., particles are adsorbed to energetically favored sites, with the result that the enthalpy of A. depends on the amount of substance adsorbed. Heats of A. measured when the amount of adsorbed substance are equal are called ***isosteric heats of adsorption***.

The amount of a substance i adsorbed to a surface is indicated by the surface concentration $\Gamma_i = n_i/S$, where n_i is the number of moles of i adsorbed to surface area S (in m^2 or cm^2). If the surface area of the adsorbent is not known, the amount of adsorbed substance per g of adsorbent is used. In the ideal case, the adsorbent has an energetically homogeneous surface which is saturated with a monomolecular layer of adsorpt. The ratio $\Theta = \Gamma_i/\Gamma_\infty$ is the degree of coverage, where Γ_∞ is the saturation concentration.

In physical A., a temperature- and pressure-sensitive equilibrium is established between the concentration c_i of the adsorpt in the homogeneous phase and on the surface concentration Γ_i. The dependence of Γ_i on the concentration c_i or partial pressure p_i of gaseous substances at constant temperature is called the ***adsorption isotherm***: $\Gamma_{i,T} = f(c_i \text{ or } p_i)$. The dependence of Γ_i on temperature at constant pressure p_i is the ***adsorption isobar*** $\Gamma_{i,p}D = f(T)$. The ***adsorption isoster*** shows the concentrations c_i which lead to the same surface concentration Γ_i at different temperatures: $c_i,\Gamma = f(T)$.

In practice, the most important of these curves are the A. isotherms. At very low coverage, the amount of adsorbed substance increases almost linearly with the concentration. At very high concentrations, it approaches a limiting value Γ_∞, which can be regarded as monomolecular coverage of the surface. However, for many adsorbents, there is then a renewed increase which can be caused by formation of several layers of the adsorbed substance, or by capillary condensation of the this substance in micropores of the adsorbent.

There are various mathematical approximations to describe the A. isotherms of gases on solids:

1) *Freundlich's A. isotherm*: $\Gamma_i = \alpha p_i^\beta$, where p_i is the partial pressure of the adsorpt, α and β are constants; 2) *Langmuir's A. isotherm*: $\Gamma_i = \Gamma_\infty \alpha p_i/(1 + \alpha p_i)$; 3) *Brunauer, Emmett and Teller's A. isotherm* (BET equation): $\Gamma_i = \Gamma_\infty \alpha p_i/(p_{i0} - p_i)[1 + (\alpha - 1)p_i/p_{i0}]$, where Γ_∞ is the maximum surface concentration in a monolayer and p_{i0} is the saturation vapor pressure of the pure liquid adsorpt. This equation applies to multi-layer adsorption. 4) the *Gibbs A. isotherm* gives the amount of a substance adsorbed to the surface of a liquid:

$$\Gamma_i = -\frac{C_i}{RT}\left(\frac{\partial 6}{\partial C_i}\right)_{p,T,O}$$

Here c_i is the concentration of substance i in the interior of the liquid phase, R is the general gas constant, T is the temperature in K, and $(\delta\sigma/\delta c_i)$ is the change in surface tension σ of the liquid when the concentration of surface-active substance i changes.

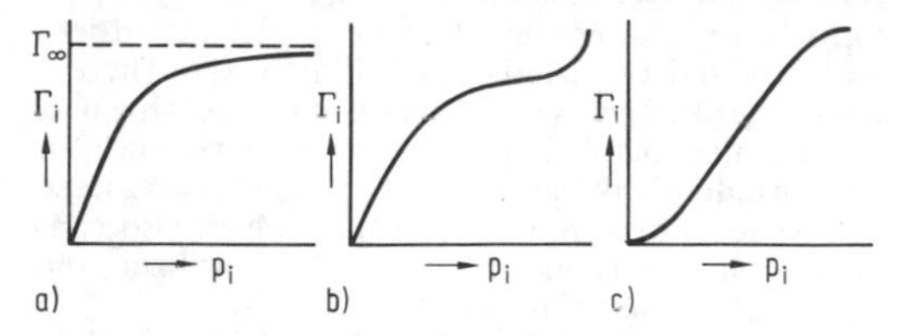

Various types of adsorption isotherms.

When gaseous compounds are adsorbed at high pressures by porous adsorbants, the gases condense in the pores (*capillary condensation*).

For practical applications, the most important A. processes are those on highly disperse solids, which are often porous as well; these have large specific surface areas. Examples are activated charcoal, molecular sieves and aluminum oxide. They are used for selective removal of certain impurities from gaseous and liquid mixtures (e.g. to decolor solutions, remove sulfur from smoke and to filter air in gas masks), to remove residual gases in high-vacuum systems (getter pumps) and for separations (see Parex process), etc. In heterogeneous Catalysis (see), the A. of reactants on the catalyst surface is essential. A. is also an integral part of the process of gas and adsorption chromatography. Some industrially important applications of A. are listed in the table.

Important industrial adsorption process

Adsorbed compound	Adsorbent
a) from the gas phase	
Water	8-Al_2O_3, molecular sieves, silica gel
Hydrocarbons	Activated charcoal
Solvents, e.g. halomethanes	Activated charcoal, silica gel
Sulfur compounds	Fe_2O_3, ZnO
Paraffins (parex process)	Molecular sieve
b) from the liquid phase	
Softening and deionizing of water	Ion exchangers
Removal of phenols from sewage	Activated charcoal,
Paraffins (molex process)	Molecular sieve
p-Xylene (parex/UOP process)	Molecular sieve
Olefins (olex process)	Molecular sieve

Adsorption chromatography: a separation method based on the difference in adsorption of molecules to a condensed phase (see Adsorbent) whose activity depends on the nature of the material and its surface characteristics (particle size, porosity). In A., the compounds are repeatedly adsorbed and desorbed, resulting in establishment of an equilibrium. The differences in the structures of the various adsorbates are reflected by differences in the strength of their interactions with the adsorbent, and these lead to differences in their migration rates and thus to separation. Solid adsorbants such as aluminum oxide, silica gel, activated charcoal, zeolites or starch, polyamide or dextran gels are generally used. The mobile phase can be selected from an eluotropic series of solvents (see High performance liquid chromatography). A. can be done as gas-solid or liquid-solid chromatography in columns or layers (see Thin-layer chromatography). Additional separation effects are obtained because of specific interactions (ion-exchange, biospecific affinity, molecular sieve effect, formation of inclusion compounds). Affinity chromatography (see) is a special form of A.

Adsorption isobar: see Adsorption.

Adsorption isoster: see Adsorption.

Adsorption isotherm: see Adsorption.

Adsorption charcoal: same as activated charcoal; see Charcoal.

Adsorptive: the substance enriched by adsorption, in the isolated state.

Adumbran: see Oxazepam.

A_E reaction: see Electrophilic reactions.

Aerogel: see Permeation chromatography.

Aeration element, *Evans element*: a concentration element formed by differential aeration of the electrolyte. Metal goes into solution at the poorly aerated local anode, and the electrolyte becomes acidic through hydrolysis. At the well aerated local cathode, oxygen is reduced to the hydroxide ion, which causes the pH value to rise. The oxygen corrosion of iron in aqueous solution usually occurs via many parallel A. The local corrosion in cracks and holes (see Crack corrosion; Pitting corrosion) can be considered aeration corrosion.

Aerosol: a dispersion of solid or liquid particles in a gas. Fine dusts, mists and smoke are examples of A. Clouds of fine ice particles (cirrus clouds) or water droplets occur naturally in the atmosphere; in addition natural A. occur as results of volcanic eruptions and forest fires. Anthropogenic A. are the undesirable byproducts of industrialization: exhaust, smoke, dust and smog. Natural A. act as nucleation centers for the condensation of water vapor; they also serve as catalysts of atmospheric chemical reactions between pollutant gases, such as chlorine, nitrogen oxides and sulfur oxides. Anthropogenic A. cause considerable damage to the environment, although they may also be applied deliberately as fertilizers and pesticides in agriculture. Nearly all chemical weapons would be used in the form of A.

A. particles acquire electric charge by absorption of ions (or electrons) from the atmosphere; these charges are usually small and vary from one particle to the next. If the particles are produced by dispersion, positive, negative and neutral particles are all generated by the process. The number of charge carriers per particle depends greatly on its nature; A. of nonpolar liquids are usually weakly charged, while water droplets are highly charged (waterfall electricity, for example). A. can be destroyed by application of a centrifugal force, an electrical field or filtration.

AES: 1) abb. for atomic emission spectroscopy. 2) abb. for Auger electron spectroscopy.

Aesculetin: see Coumarin, Table.

Affination: see Parting.

Affinity: an historical term, now obsolete, for the tendency of substances to react with each other. The heat of reaction was formerly considered a measure of the A. (see Berthelot-Thomsen principle). In Thermodynamics (see), the maximum useful work $\Delta_R G$ is equal to the A.: A = - $\Delta_R G$.

Affinity chromatography: a special form of Adsorption chromatography (see) in which the adsorbent is biospecific. For example, specific interactions between antibodies and antigens, enzymes and their inhibitors, nucleic acids of complementary sequences, lectins and polysaccharides, avidin and biotin or receptors and hormones can be utilized. The stationary phase consists of the bioselective *ligand* chemically bound (by esterification, azo-coupling, etc.) to an inert, porous *matrix* (agarose gel, glass beads, cellulose, polyacrylamide, cross-linked dextrans). Since the active center of a biological substance is often deep inside the molecule, the capacity of the adsorbent can often be increased by introducing *spacers* between the matrix and ligand. This reduces steric hindrance to binding of the two biological partners. Like the matrix, the ligand should be inert (fig.). In A., the components with selective affinity to the ligand are retained, while other substances are immediately eluted. The bound substance can then be eluted in pure form using a solvent which has been altered in some way (ionic strength, pH, addition of free ligand).

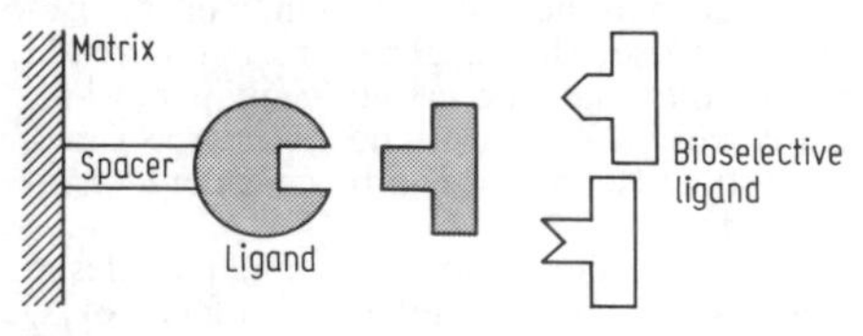

Affinity chromatography

A. is an important method in biochemistry. It is used routinely to isolate specific sequences of DNA, antibodies, enzymes, etc.; it can also be used to fractionate cell cultures.

Affinoelectrophoresis: see Electrophoresis.

Aflatoxins: metabolic products of *Aspergillus flavus* and other molds. The A. are derivatives of furocumarin, and are biosynthesized as Polyketides (see). The main product is A.B_1 (Fig.). The A. are colorless and lipophilic, and are highly carcinogenic (liver). A. diffuse into the food beyond the area occupied by the mold. Their effect is due to covalent binding to the DNA of the cell. The A. were discovered in 1960.

Aflatoxin B_1

AFS: abb. for Atomic fluorescence spectroscopy (see).

Ag: symbol for Silver (see).

Agar: a substance obtained from various red algae, especially those of the genuses *Gelidium* and *Gracilaria*. A. is more than 90% polysaccharides, and contains 0.3 to 0.7% sulfur in the form of sulfate monoesters. The component with the greatest tendency to swell is *agarose*, which makes up about 70% of A. Agarose is a linear polysaccharide consisting of alternating units of 3-O-substituted β-D-galactopyranose and 4-O-substituted 3,6-anhydro-α-L-galactopyranose. To obtain A., the algae are dried, bleached and extracted with boiling water. The extract is purified and solidified by cooling. A large part of the water-soluble impurities can be removed from the partially dehydrated gel by freezing it. Agarose can be separated from agaropectin, which also contains uronic acids and pyruvic acid, by utilizing the differences in solubilities of the acetates.

A. swells in water and when heated, forms a clear solution. When it is cooled, as little as 0.5 to 1.0% in solution gives a stable gel. A. is used in the food industry as a gelling agent and in pharmaceutical preparations. Because of its ability to swell, it is used as a mild laxative. Agar gels are used as carriers for electrophoresis and in microbiology, as nutrient media for microorganisms. Special agarose preparations, such as Sepharose, are used in gel chromatography.

Agarose: see Agar.

Agarose gel: see Permeation chromatography.

Agent Orange: a term for a herbicide consisting of equal parts of the butyl esters of 2,4-dichloro- and 2,4,5-trichlorophenoxyacetic acid. It was used by the US Army in the Vietnam War to defoliate forests. The severe consequences of use of this herbicide are due to the presence of approximately 40 g/t of the byproduct 2,3,7,8-Tetrachlorodibenzo-1,4-dioxin (see).

Aggregate state: a physical state of matter. The three classical A. are the solid, liquid and gas states;

plasma is often called the fourth A. The atoms, ions or molecules of a substance attract each other to a greater or lesser degree, and in the absence of thermal motion, these attractive forces would bind them rigidly together. The state of a given substance at any temperature is determined by the strength of the attractive forces relative to the tendency of thermal motions to disperse the particles into a gas.

Solids have the highest degree of order. In the limiting case of the *ideal solid*, only the forces of interaction affect the particles and they have no kinetic energy. They are arranged with complete regularity in a three-dimensional lattice; there is a strict *long-range order* (see Crystal). This state is most closely approximated at low temperatures and high pressures, but even under these conditions, it is never completely realized. Even at absolute zero, particles have a certain amount of kinetic energy, called the zero-point energy, which causes them to vibrate about their equilibrium positions. As the temperature increases, the vibrations become much more vigorous. In *real solids*, there are in addition imperfections in the regular arrangement of the particles. The model of the *perfect solid*, which has no such structural or chemical Crystal lattice defects (see), but in which the particles do vibrate thermally, has been closely approximated in a few cases by Crystal growing (see). Solids which do not have long-range structural order are called *Amorphous* (see); they are relatively rare, compared to crystalline solids. Because their constituent particles have fixed positions and because the interactions between them are strong, solids have fixed shapes and volumes.

A ***gas*** is the A. with the lowest degree of order. The average distance between gas particles is very large compared to their own size, and they move constantly in random directions which are frequently changed by collisions with other particles or the walls of the vessel (see Kinetic gas theory). In the limiting case of the *ideal gas*, which is approximated by real gases at high temperatures and low pressures, the particles do not exert any forces on each other, and their motion is completely random. In addition, the volume of the particles is so small compared to the total volume available to them that it can be ignored. All *real gases* deviate from this ideal state to a greater or lesser degree. A gas will fill the entire volume available to it, so it has no fixed shape. Gases are fully miscible with one another in any proportions.

Liquids have a position intermediate between solids and gases. They display a definite *short-range order*; there is a certain regularity of orientation and distance between immediate neighbors at any time, and this regularity approaches that of a crystal lattice. However, due to the constant thermal motion of the particles, this order is much weaker even for next-nearest neighbors, and at a distance of a few particle diameters, it cannot be recognized at all. The short-range order is dynamic; it is limited not only in spatial dimensions, but in time as well. Because of the irregular motions of the particles, the ordered microregions are constantly formed and dispersed, and each particle's position changes both in space and with respect to its neighbors. There is still no complete structural theory of liquids capable of accounting for all their macroscopic properties. Because of the ease with which their particles shift positions, liquids have no fixed shapes; they adopt the shape of the vessel and develop flat surfaces due to the effect of gravity. However, because the particles are relatively densely packed, they resist changes in volume.

The degree of order in the three A. can be visualized with the help of *radial density distribution functions W(R)* derived from diffraction experiments (Fig. 1). In a coordinate system with its origin on an arbitrarily chosen particle, *W(R)* is the probability that a vector of length *R* starting at the origin will end on another particle. The periodic spacing of the maxima of *W(R)* for a solid indicates the long-range order of the crystal. In liquids, the first maximum has the same position and height as the first maximum in the corresponding solid, but subsequent maxima decline rapidly, indicating the spatial limitation of the short-range order. For a gas, beyond a certain distance determined by the diameter of the particle, the probability of finding another particle at any given distance is equal to that at any other distance.

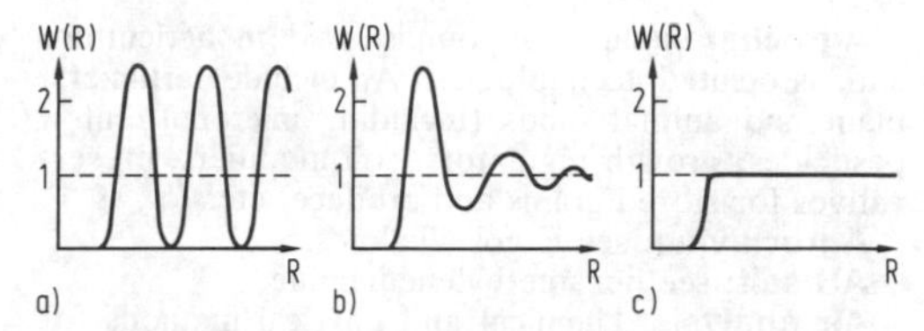

Fig. 1. Radial distribution function *W(R)* of a solid *(a)*, a liquid *(b)* or a gas *(c)*.

All elements and many compounds can exist in all three A. The transitions between them occur at characteristic temperatures, which depend on the pressure; decreases in order are accompanied by absorption of heat, and increases in order, by release of heat. These phase changes are described by the Clausius-Clapeyron equation. The three A. of a substance can be plotted in a diagram of state, e.g. a *p-T* diagram. The state surfaces of each A. are separated by curves, at which the phase changes occur. The phase-change curve between the liquid and gas states ends at a critical point, above which it is no longer possible to distinguish between liquid and gas. Each A. can occur in a very broad temperature range. The range of existence and the relationships for the transitions between A. can be seen in Fig. 2. The diagram includes the ***plasma*** into which a gas is converted at

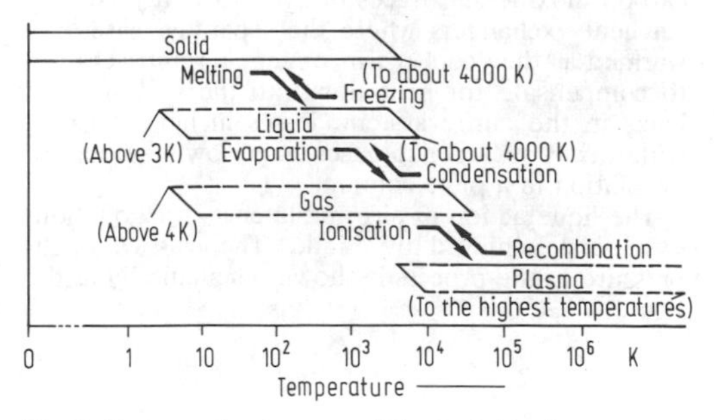

Fig. 2. Ranges of existence of the individual aggregate states.

sufficiently high temperatures. Since the plasma state is fundamentally different from the other A., it can be considered a fourth A.

The boundary between the A. is not always clear. Liquid crystals (see) and Amorphous solids (see) are intermediate states between solids and liquids.

The thermodynamic Phase (see) must not be confused with the concept of A. For example, two immiscible liquids form two phases, but are in a single A.

Aggregate system: a macroscopic system of aggregated particles. Its properties are determined by the nature and strength (range) of the interaction between the particles. A highly ordered A. with a periodic arrangement of particles extending over a long range is a crystal. Lower-order A., in which the order extends only to the immediate neighborhood of a given particle (short-range order) include amorphous solids and liquids. Liquid crystals occupy an intermediate position.

Aggregation number: see Micelle.

Aglycon: the non-sugar component of a Glycoside (see).

Agrochemicals: compounds used in agriculture and associated technologies. A. include fertilizers, plant and animal foods (including micronutrients), pesticides, growth regulators, soil modifiers, preservatives for stored grains and produce, etc.

Agroclavine: see Ergot alkaloids.

AH salt: see hexamethylenediamine.

Air analysis: chemical and physical methods for quantitative determination of the concentration of air pollutants, both indoors and outdoors. For indoor monitoring, semiquantitative analyses using standardized test kits are usually sufficient, but monitoring of outdoor pollutants requires highly sensitive and selective methods. Gaseous pollutants are usually sampled by impingers, and measured by colorimetric tests, titration, ion-sensitive electrodes, atomic absorption spectrophotometry or other methods. There are automatic, continuously recording devices which measure conductivity. To measure the amount of particulates, a certain volume of air is sucked through a special filter. Sedimentation particulates are collected for 30 days in vessels with known trapping surface and determined gravimetrically.

Air fractionation: low-temperature separation of air into its component gases, chiefly nitrogen and oxygen, but also the noble gases argon, krypton, xenon, neon and helium. As a rule, A. involves the following steps: compression, purification (removal of water, carbon dioxide and traces of hydrocarbons), cooling (in heat exchangers where the separated gases are warmed as they cool the incoming air), refrigeration to compensate for heat flow into the system (heat leaks in the entire system, losses in heat transfer, withdrawal of liquefied gases), and low-temperature distillation in a plate column.

The liquefaction of air, and later its fractionation, were first achieved by Linde. The classical high-pressure Linde process is shown schematically in the figure.

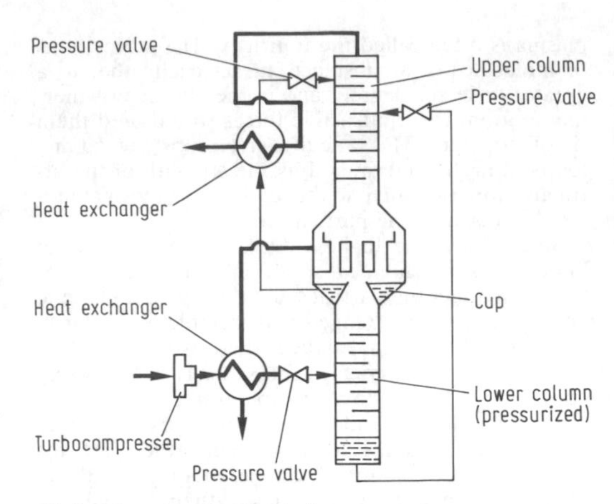

Air fractionation by the Linde process.

Air liquefaction: see Gas liquefaction, see Air fractionation.

Air pollution: in the widest sense, a term for solids, liquids and gases which were not originally present in the atmosphere. In a narrower sense, only those substances which have entered the air through human activity (***anthropogenic pollutants***) are included. Natural A. is caused, for example, by volcanoes, decomposition of organic materials, wildfires and natural dust. The main sources of anthropogenic A. are industry, transportation, space heating and agriculture. Of the industrial sources, heating and electric power plants are the major contributors; they are followed by the chemical industry, metallurgy, the cement industry and the ceramic industry. Combustion of fossil fuels (other than natural gas) converts the sulfur in the fuel to sulfur dioxide, and most of this escapes with the smoke.

A mixture of smoke and fog is called "smog", and the term has become widely synonymous with A. There are two major types of smog: industrial and photochemical. Industrial (London-type) smog consists mainly of sulfur trioxide (SO_3) and particulates (ash, soot and partially burned hydrocarbons) formed by combustion of fossil fuels. Sulfur trioxide is the anhydride of sulfuric acid; atmospheric moisture converts it to the acid and precipitation washes it out of the air as *acid rain*. In aerosol form, H_2SO_4 is damaging to human and animal lungs, metals and building materials such as marble. In the form of acid rain, H_2SO_4 has caused heavy losses of fish in lakes throughout northeastern North America and Europe.

Photochemical (Los Angeles type) smog is produced by the action of ultraviolet light on the products of internal combustion engines: nitrogen oxides and unburned hydrocarbons. It contains little or no sulfur trioxide. The major components of photochemical smog are ozone, peroxyacyl nitrogen, peroxides, and various hydrocarbons. This type of smog damages plant life and is thought by some to be responsible for the dying of forests in Europe in the past 25 years. Other significant pollutants, which do not necessarily create smog, are fluorine compounds, nitrogen oxides, hydrogen chloride and chlorides, ammonia, chlorofluorocarbons and heavy metal compounds. Hydrogen fluoride, hydrogen chloride and nitric acid contribute to acid rain (see above). The presence of heavy metals in the soil is damaging to both plants and the animals which feed on them. The presence of chlorine compounds in the atmosphere catalyses the destruction of stratospheric ozone, which protects the surface of the earth from solar ultraviolet radiation; the decrease has recently become statistically significant.

Because A. is noxious and even small amounts interfere with a sense of well being, the concentration of pollutants in air must be limited. Governments establish limits for the maximum acceptable concentrations of various pollutants in the workplace and in the atmosphere outside the workplace (see Indoor air quality). In both cases, both chronic and peak values are regulated to prevent damage by extremes.

A. can be prevented both by engineering in new plants and vehicles and by removal of pollutants by Waste gas purification (see) or by Flue gas desulfurization (see). The construction of very high smokestacks prevents concentrated A. near the plant producing the smoke, but is believed to be responsible for acid precipitation falling hundreds of kilometers away in nonindustrial areas. In the Vehicle exhaust (see) sector, both reduction of emissions and better organization of traffic can make significant contributions. A. is monitored by methods of Air analysis (see).

Ajmaline: see Rauwolfia alkaloids.

Al: symbol for Aluminum.

Ala: abb. for Alanine (see).

Alachlor: see Acylaniline herbicides.

Alamethicin: a membrane-active peptide antibiotic isolated from culture filtrates of the fungus *Trichoderma viride*: Ac-Aib-Pro-Aib-Ala-Aib-Ala-Gln-Aib-Val-Aib-Gly-Leu-Aib-Pro-Val-Aib-Aib-Glu-Gln-Phol. A. is an amphiphilic polypeptide which, like ***suzukacillin*** and ***trichotoxin***, can generate a fluctuating, voltage-dependent ion flux with action potentials. It is therefore of great interest as a model system for nerve conduction. Aggregation of these peptides forms pores with different conductive capacities, so that mechanisms of membrane penetration can be studied. A. was originally thought to have a cyclic octadecapeptide structure, but in the revised linear structure, the *N*-terminal α-aminoisobutyric acid residue (Aib) is acetylated, and the *C*-terminus is bound to a phenylalaniol group (Phol). Suzukacillin A and trichotoxin A40 are natural analogs of A. with similar membrane-penetrating properties; they have a high degree of homology with A.

Alanates: hydrido complexes of aluminum with the composition $M[AlH_4]_n$. The tetrahedral AlH_4^- anion is formally the product of addition of a hydride ion to aluminum hydride, $(AlH_3)_x$. The most important is Lithium alanate (see), which is widely used as a reducing agent. Other elements in Groups Ia to IIIa also form A.

Alane: same as aluminum hydride.

Alanine, ***aminopropionic acid***, abb. ***Ala***: 1) α-Alanine, CH_3-$CH(NH_2)$-COOH is a proteogenic Amino acid (see). It is one of the main components of silk fibroin. The chemical synthesis according to Strecker and Bucherer gives an 88% yield of DL-A. which, in the form of *N*-acetyl- or *N*-chloroacetyl-DL-A., is converted to L-A. by hydrolysis catalysed by immobilized acylase. A biotechnological pathway to L-A. starts from L-aspartic acid, which is converted to L-A. by microbial or enzymatic decarboxylation. In the organism, A. is formed by transamination of pyruvic acid. *N*-Methylalanine is used to dissolve hydrogen sulfide from industrial gases (sulfosolvane process).

2) β-Alanine, H_2N-CH_2CH_2-COOH, is an important, naturally occurring β-amino acid. It is formed by enzymatic decarboxylation of aspartic acid, and is a component of anserine, carnosine and coenzyme A.

Albumins: simple proteins consisting largely of glutamic and aspartic acids (20 to 25%), leucine and isoleucine (up to 16%) and relatively large amounts of cysteine and methionine, but little glycine (< 1%). The A. are very soluble in water and crystallize well; they can be precipitated by high concentrations of netural salts. Their isoelectric points are in the weakly acid range. The most important animal and plant A. are listed below. ***Ovalbumin*** from chicken eggs has M_r 44,000. It has a phosphate group esterified to serine and a carbohydrate component. The heat-resistant ***lactalbumin*** is present in all types of milk; it contains 123 amino acid residues (M_r 14,176) and four disulfide bridges. ***Serum albumin*** makes up to 60% of the dry mass of blood serum; its M_r is 67,500. It has a high binding capacity for calcium, sodium and potassium ions, fatty acids and drugs. The main function of the serum A. is maintenance of the osmotic pressure in the blood. Some important plant A. are the toxic ***ricin*** from *Ricinus* seeds (castor oil is produced from *Ricinus communis*), ***leucosin*** from wheat and other grains, and ***legumetin*** from legumes.

Alcohol: see Alcohols; in general usage, the same as ethanol.

Alcoholates, ***alkoxides***: ionic derivatives of alcohols formed by substitution of a metal of the first to third main group for the H of the OH group. The most common A. contain Na, K, Mg or Al. Sodium methylate, CH_3ONa, potassium *tert*.-butylate, $(CH_3)_3COK$, magnesium methylate, $(CH_3O)_2Mg$ and aluminum isopropylate $[(CH_3)_2CHO]_3Al$ are of practical importance. A. are moisture-sensitive solids which hydrolyse immediately on contact with water, forming the alcohol and metal hydroxide: $CH_3O^-Na^+ + H_2O \rightarrow CH_3OH + NaOH$. The aluminum A. are relatively stable, and can be distilled in vacuum without decomposition. A. are readily soluble in excess alcohol. They are used as nucleophiles for syntheses in anhydrous solutions.

A. are produced by reaction of the metals with anhydrous alcohols: CH_3-CH_2-OH + Na $\rightarrow$ CH_3-CH_2-O^-Na^+. Sodium and potassium react spontaneously and release so much heat that the reaction mixture may have to be cooled. Magnesium and aluminum have to be activated by treatment with iodine or mercury(II) chloride. The reaction releases atomic hydrogen, which can be used for hydrogenation. A. react with haloalkanes (see Williamson synthesis) to form ethers, which is especially useful as a method for producing unsymmetric ethers: R^1-Cl + $NaOR^2 \rightarrow R^1$-O-R^2 + NaCl.

A. are used as condensation reagents for the Claison condensation (see) and as catalysts for organic redox reactions (see Oppenauer oxidation,

Meerwein-Ponndorf-Verley reduction). The reaction of reactive metals with alcohols is important in the synthesis of anhydrous alcohols, e.g. with magnesium methylate, and for safe disposal of excess sodium or potassium.

Alcohol dehydrogenases: enzymes which catalyse the reversible oxidation of primary and secondary alcohols to the corresponding aldehydes or ketones; NAD^+ is the hydrogen acceptor. The most famous example is ***yeast A.***, a zinc-containing protein of M_r 145,000. It consists of four subunits, each with a binding site for $NAD^+/NADH$; as the last enzyme in alcoholic fermentation, it catalyses the conversion of acetaldehyde to ethanol. A. are found in all organisms. To determine the concentration of ethanol in the blood, the change in optical extinction of a solution of $NAD^+/NADH$ at 340 nm in the presence of the enzyme is measured.

Alcoholic fermentation: anaerobic conversion of glucose to ethanol and carbon dioxide, especially by yeast or enzymes from yeast. The process is used on an industrial scale to produce ethanol. The overall equation for A. is $C_6H_{12}O_6 + 2\ H_3PO_4 + 2\ ADP \rightarrow 2\ C_2H_5OH + 2\ CO_2 + 2\ H_2O + 2\ ATP$; the conversion of ADP to ATP supplies the cell with the energy needed to maintain its metabolism. The glucose is converted to pyruvic acid by the reactions of Glycolysis (see). Pyruvate decarboxylase converts pyruvic acid to acetaldehyde, which is then converted to ethanol by alcohol dehydrogenase; NAD^+ is simultaneously regenerated:

$$CH_2OH\text{-}CO\text{-}CH_2OPO_3H_2 \rightleftharpoons CHO\text{-}CHOH\text{-}CH_2OPO_3H_2 \underset{NADH}{\overset{NAD^+}{\rightleftharpoons}} COOH\text{-}CHOH\text{-}CH_2OPO_3H_2 \rightleftharpoons$$

$$COOH\text{-}CHOPO_3H_2\text{-}CH_2OH \xrightarrow{-HOH} COOH\text{-}C(OPO_3H_2)\text{=}CH_2 \longrightarrow COOH\text{-}CO\text{-}CH_3 \xrightarrow{-CO_2} CHO\text{-}CH_3 \underset{NAD^+}{\overset{NADH}{\rightleftharpoons}} CH_2OH\text{-}CH_3$$

A. depends on the presence of a sufficient concentration of phosphate, a temperature between 30 and 37 °C, a pH between 4.5 and 5.7, and a sugar concentration of 20 to 25%. Higher sugar concentrations and alcohol concentrations above 20% inhibit growth of the yeast cells.

The fermented liquids generally contain 10 to 18% ethanol; the maximum yield is 0.5 kg ethanol per kg glucose.

A byproduct of A. is glycerol, which may account for up to 3% of the sugar consumed. This yield can be considerably increased by addition of sodium hydrogensulfite to the fermentation liquid. This is "Neuberg's second type of fermentation". The acetaldehyde which normally acts as hydrogen acceptor is trapped by the hydrogensulfite, and the reducing equivalents of $NADH + H^+$ are transferred to dihydroxyacetone phosphate. The resulting glycerol phosphate is then dephosphorylated to glycerol. "Neuberg's third type of fermentation" occurs when an alkali added to the fermentation causes a Canizzaro reaction to take place, converting acetaldehyde to ethanol and acetic acid; dihydroxyacetone phosphate acts as the hydrogen acceptor and leads to production of glycerol. $2\ C_6H_{12}O_6 + H_2O \rightarrow CH_3\text{-}CH_2\text{-}OH + CH_3\text{-}COOH + 2\ CO_2 + 2\ CH_2OH\text{-}CHOH\text{-}CH_2\text{-}OH$.

Historical. A. was known several thousand years ago, and is considered the oldest biotechnology. The first scientific studies were made by Lavoisier (1789), who realized that carbohydrates are converted to alcohol and carbon dioxide. In 1815, Gay-Lussac established the net equation for A. Around 1857, Pasteur promoted the hypothesis that A. could only occur in the presence of living organisms ("vitalistic" fermentation theory). In 1897, Buchner showed that A. could be carried out by cell-free preparations from yeast. This discovery was the starting point for modern enzymology. Further important studies on the complicated course of fermentation reactions were done by Neuberg (1877-1956), Harden (1865-1940), Embden (1874-1933) and Meyerhof (1884-1951).

Alcoholometry: determination of the alcohol content of aqueous ethanol solutions using an alcoholometer (density gauge). The instrument is calibrated so that the ethanol content can be read directly in mass % or volume % (Table).

Density of solution	Mass %	Vol. %	g alcohol per liter
0.998	0.15	0.19	1.5
0.990	4.70	5.90	46.5
0.980	11.47	14.24	112.4
0.970	19.11	23.48	185.4
0.960	26.15	31.80	251.0
0.950	32.24	38.80	306.3
0.940	37.62	44.80	353.6
0.930	42.57	50.15	395.9
0.920	47.26	55.07	434.8
0.910	51.78	59.69	471.2
0.900	56.18	64.04	505.6
0.890	60.50	68.21	538.5
0.880	64.78	72.22	570.1
0.870	69.01	76.06	600.4
0.860	73.20	79.74	629.5
0.850	77.32	83.26	657.2
0.840	81.38	86.60	683.6
0.830	85.37	89.76	708.6
0.820	89.24	92.69	731.8
0.810	92.25	95.38	752.9
0.800	96.48	97.77	771.8

Alcohols: organic compounds containing hydroxyl (-OH) groups on sp^3-hybridized C atoms. Formally, the A. can be derived by substitution of H atoms in hydrocarbons: Alkanols (see) from alkanes, Cycloalkanols (see) from cycloalkanes, etc. The C atom to which the OH group is bound can also be part of a side chain of an aromatic compound, as in Benzyl alcohol (see), or of a heteroaromatic compound, as in Furfuryl alcohol (see).

Mono-, di-, tri- and poly A. contain the corresponding number of OH groups, each bound to a different C atom. Some important monoalcohols are Methanol (see) and Ethanol (see); Glycol (see) and Glycerol (see) are simple di- and trialcohols, respectively (Tables 1 and 2). Reduction of monosaccharides leads to hexaalcohols, such as D-Glucitol (see) and D-Mannitol (see). A primary A. is one in which the OH group is bound to a terminal C atom; a secondary A. is one in which the C to which the OH is bound has bonds to two other carbons, and a tertiary A. is one in which the C to which the OH is bound is linked to three other C atoms. Thus the general formulas are:

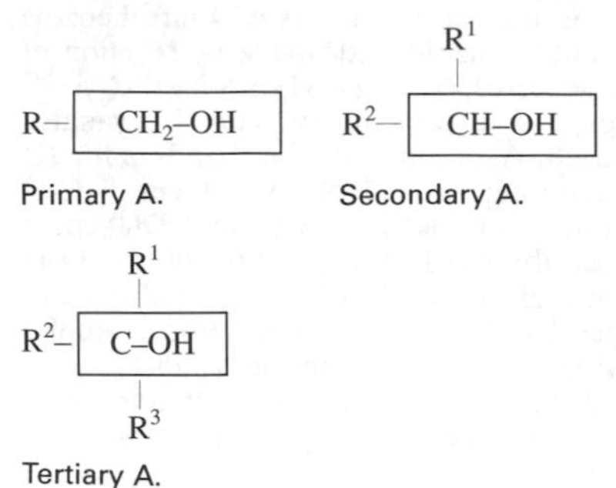

Table 1. Homologous series of monoalcohols

Methanol	CH_3-OH
Ethanol	CH_3-CH_2-OH
Propan-1-ol	$CH_3-CH_2-CH_2-OH$
Propan-2-ol	$CH_3-CH(OH)-CH_3$
Butan-1-ol	$CH_3-CH_2-CH_2-CH_2-OH$
Butan-2-ol	$CH_3-CH_2-CH(OH)-CH_3$
2-Methyl-propan-1-ol (Isobutanol)	$CH_3-CH(CH_3)-CH_2-OH$
2-Methyl-propan-2-ol (tert-Butanol)	$CH_3-C(CH_3)_2-OH$

Table 2. Some di- and polyalcohols

Glycol (Ethan-1,2-diol)	$HO-CH_2-CH_2-OH$
Propylene glycol (Propan-1,2-diol)	$HO-CH_2-CH(OH)-CH_3$
Propan-1,3-diol	$HO-CH_2-CH_2-CH_2-OH$
Glycerol (Propan-1,2,3-triol)	$HO-CH_2-CH(OH)-CH_2-OH$
Pentaerythrol	$HO-CH_2-C(CH_2-OH)_2-CH_2-OH$

The simplest primary A., methanol, is obtained by formal substitution of R by H.

The monoalcohols are named by combination of the root name for the hydrocarbon skeleton with the suffix -ol. If the root ends in -e, this is elided. Isomers are distinguished by indicating the position of the OH group by a numeral in front of the suffix, e.g. propan-2-ol. In radicofunctional nomenclature, the word "alcohol" is added to the name of the radical, as in methyl alcohol. Thus the simplest tertiary A. can be named in three ways: *tert.*-butanol, *tert.*-butyl alcohol and 2-methylpropan-2-ol. When two or more OH groups are present, the suffixes -diol, -triol, etc. are used, and the positions of the OH groups are indicated, using the smallest numbers possible, e.g. butan-1,4-diol. Some trivial names are still accepted in the IUPAC system.

Properties. Low-molecular-weight A. are liquids with a typical kind of odor and a burning taste; compared to hydrocarbons of similar size, they have relatively high boiling points, due to intermolecular hydrogen bonding. This property is even more pronounced in the viscous polyalcohols, such as glycols or glycerol. A. with higher molecular weight are solids with weak odors. Because of the interactions of the hydrophilic polar OH group with water molecules, the lowest A. are miscible with water in all proportions. The boundary of solubility is found at C_4 for monoalcohols, but is much higher for polyalcohols. *tert.*-Butanol is still miscible with water in any proportion, while the isomeric amyl A. are practically insoluble in water; here the hydrophobic character of the alkyl radicals predominates. D-Glucitol and mannitol are readily soluble in water, but they have lipophilic properties; they are not soluble in diethyl ether or ethanol. Many A. have physiological effects (see Methanol, Ethanol, Fusel oil). The lower monoalcohols are combustible and ignite readily. Methanol, which can be synthesized on a large industrial scale, and ethanol, which is also produced on a large scale, by fermentation, can be used as fuels for motor vehicles. A. give neutral reactions with acid-base indicators, but they can form salts with reactive metals of the first to third main groups of the periodic system. Substitution of a metal atom for the H atom of the OH group produces an Alcoholate (see): $CH_3OH + Na \rightarrow CH_3O^-Na^+ + H$. The reaction is also of practical interest for producing nascent hydrogen, which can react immediately, in the alohol solution, with a substrate (see, for example, Bouveault-Blanc reduction). The reaction can also be used to produce absolutely dry A. with magnesium methylate, or for the safe disposal of alkali metals. The latter two applications are based on the ease of hydrolysing the alcoholates.

In terms of the Brönsted acid-base theory, A. are ampholytes:

$$ROH \rightleftharpoons RO^- + H^+;\ ROH + H^+ \rightleftharpoons R\text{–}\overset{+}{O}H_2$$

However, A. are weaker acids than water, with pK values around 16, so that there is practically no spontaneous dissociation. The formation of alkoxonium ions, which can appear as intermediates in chemical reactions, occurs only with strong Brönsted acids or with Lewis acids, such as BF_3: $R\text{-}O\text{-}H + BF_3 \rightarrow$

$$R\text{–}\overset{+}{O}(H)\text{–}\overset{-}{B}F_3$$

Reactions. A. combine with inorganic acids to form alkyl halides, alkyl nitrites, alkyl nitrates, alkyl sulfates, dialkyl sulfates and trialkyl phosphates, and with carboxylic acids or their anhydrides to form esters: $R\text{-}CH_2\text{-}OH + HBr \rightarrow R\text{-}CH_2\text{-}Br + H_2O$; $C_2H_5OH + R\text{-}COOH \rightleftharpoons R\text{-}COOC_2H_5 + H_2O$. The reaction of A. with 3,5-dinitrobenzoyl chloride to form 3,5-dinitrobenzoates is used to identify them:

$$(O_2N)_2C_6H_3\text{–}C(=O)Cl + HO\text{–}R \xrightarrow{-HCl} R\text{–}O\text{–}C(=O)\text{–}C_6H_3(NO_2)_2$$

The hydrochloric acid released is neutralized by a base.

When heated with sulfuric, phosphoric or oxalic acid, A. undergo intramolecular dehydration to form alkenes or an intermolecular water loss to form ethers. The two reactions compete with one another. High temperatures promote alkene formation, while lower temperatures promote ether formation: $C_2H_5OH \rightarrow CH_2{=}CH_2 + H_2O$, $2\ C_2H_5OH \rightarrow C_2H_5\text{-}O\text{-}C_2H_5 + H_2O$.

The situation is similar in gas-phase hydrogenation on Al_2O_3. At 350 to 400 °C, alkenes are formed almost without byproducts, while at 260 °C, the products are mainly ethers. The ease of intramolecular water elimination (alkene formation) increases in the order primary A. < secondary A. < tertiary A. Ether formation is promoted by an excess of A.; primary A., as would be expected, are more likely to form symmetric ethers than are tertiary A.

In the formation of alkenes, the thermodynamically more stable product is formed (Zajcev orientation); for example, but-2-ene is formed from butan-2-ol, not but-1-ene:

$$CH_3CH_2\text{–}CH(OH)\text{–}CH_3 \xrightarrow[-H_2O]{} CH_3\text{–}CH{=}CH\text{–}CH_3.$$

A. can be oxidized or dehydrogenated to aldehydes, ketones and carboxylic acids. Primary A. produce aldehydes, which can then be further oxidized to carboxylic acids:

$$R\text{–}CH_2\text{–}OH \xrightarrow[-H_2O]{+O} R\text{–}CHO \xrightarrow{+O} R\text{–}COOH.$$

Secondary A. are oxidized to ketones:

$$R^1\text{–}CHOH\text{–}R^2 \xrightarrow[-H_2O]{+O} R^1\text{–}CO\text{–}R^2.$$

Tertiary A. are not changed in neutral or basic solution, but in acid media, they can be oxidatively degraded to their hydrocarbon skeletons. Identification of the various oxidation products is used in analytical chemistry to distinguish primary, secondary and tertiary A. Potassium dichromate in sulfuric acid, chromium(VI) oxide, *tert.*-butylchromate or a pyridine-chromic acid complex is used as oxidizing agent. Potassium permanganate or nitric acid can also be used for complete oxidation of primary A. to carboxylic acids. Activated manganese dioxide and selenium(IV) oxide are used to convert primary A. to aldehydes. Primary and secondary A. can also be oxidized to aldehydes and ketones with acetone in the presence of aluminum isopropylate (see Oppenauer oxidation). Industrially, the dehydrogenation to aldehydes or ketones is done with air oxygen on copper, silver or zinc oxide contacts at 400 to 600 °C.

Analytical. Qualitative tests are a neutral reaction and the reaction with a small amount of metallic sodium, generating hydrogen; for methanol, a test is formation of the boric acid methyl ester which burns with a green flame. The iodoform test is used for ethanol or compounds with the CH_3-CHOH- group. Primary, secondary and tertiary A. can be distinguished by their different oxidation products. These are identified in the form of esters of 4-nitrobenzoic acid or 3,5-dinitrobenzoic acid, made by reaction of the A. with the corresponding acyl chlorides. Crystalline urethanes made by reaction with aryl isocyanates may also be used. A. are identified by bands in the IR spectrum in the range from 1050 to 1300 cm^{-1} (C-O valence vibration) and between 3200 and 3700 cm^{-1} (O-H valence vibration). The position of the O-H vibration is strongly affected by the degree of association, and thus by the concentration. For example, free hydroxyl groups have absorption bands between 3590 and 3650 cm^{-1}; these bands are shifted to lower wavenumbers by intramolecular (3420 to 3590 cm^{-1}) or intermolecular (3200 to 3550 cm^{-1}) hydrogen bonds. In NMR spectra, the chemical shifts of the OH protons depend on the polarity of the O-H bond and the degree of association; in other words, these spectra are not very useful for identification of A. The mass spectra of A. often display only weak molecular peaks. The fragmentation pattern is determined by elimination of water and cleavage of the C-COH bond. The ion $[CH_2{=}OH]^+$ appears as a characteristic fragment.

Occurrence. A few A. occur in nature, in free or bound state. Glycerol, for example, is a component of lipids, and A. with higher molecular weights, such as cetyl, ceryl and myricyl A., are components of waxes. A few other biological A. of interest are geraniol and menthol, which are terpenes, farnesol, a sesquiterpene, retinol (vitamin A_1), a diterpene, and cholesterol, a steroid A.

A. can be synthesized chemically in many ways:

1) Hydration of alkenes in the presence of acids:

$$R\text{–}CH{=}CH_2 + H_2O \xrightarrow{H^+} R\text{–}CH(OH)\text{–}CH_3$$

2) Epoxidation of alkenes with subsequent hydrolysis of the epoxides (oxiranes):

$$R\text{–}CH{=}CH\text{–}R + CH_3\text{–}COOOH \longrightarrow$$

$$R\text{–}\underset{\diagdown O \diagup}{CH\text{–}CH}\text{–}R \xrightarrow{H_2O/H^+} R\text{–}CH(OH)\text{–}CH(OH)\text{–}R$$

This is done, for example, with peroxyacetic acid, or oxygen and silver oxide.

3) Addition of hypochlorous acid, HOCl, to an alkene followed by basic hydrolysis of the intermediate chlorohydrine and epoxide:

$$R\text{–}CH{=}CH_2 + HOCl \longrightarrow R\text{–}CH(OH)\text{–}CH_2\text{–}CL \xrightarrow{-HCl}$$

$$R\text{–}\underset{\diagdown O \diagup}{CH\text{–}CH_2} \xrightarrow{+H_2O} R\text{–}CH(OH)\text{–}CH_2\text{–}OH.$$

4) Reaction of ethyne with formaldehyde (see Acetylene chemistry), followed by catalytic hydrogenation of the resulting butyne-1,4-diol:

$$HC{\equiv}CH + 2\ CH_2 \longrightarrow HOCH_2{-}C{\equiv}C{-}CH_2OH \xrightarrow{H_2/Catalyst} HO{-}CH{-}CH_2{-}CH_2{-}CH_2{-}OH$$

5) Reaction of alkyl halides with alkali hydroxides:

$$R{-}CH_2{-}Cl + OH^- \to R{-}CH_2{-}OH + Cl^-.$$

6) Hydrolysis of carboxylic acid esters which are either natural products or can be synthesized from alkyl halides.

7) Reduction of carboxylic acid esters: a) Bouveault-Blanc reduction. b) Catalytic hydrogenation using copper(II)-chromium(III) oxide mixed catalysts: $R^1\text{-}COOR^2 + 4\,H \rightarrow R^1\text{-}CH_2OH + R^2\text{-}OH$.

8) Reduction of aldehydes and ketones: a) catalytic hydrogenation with Raney nickel: $R\text{-}CHO + H_2 \rightarrow R\text{-}CH_2\text{-}OH$; b) reduction with sodium and ethanol: $R\text{-}CHO + 2\,H \rightarrow R\text{-}CH_2\text{-}OH$; c) reduction with complex metal hydrides, such as lithium aluminum hydride or sodium borohydride: $R\text{-}CHO + 2\,H \rightarrow R\text{-}CH_2\text{-}OH$; d) reduction of ketones to pinacols; e) Meerwein-Ponndorf-Verley reduction (see); f) Cannizzaro reaction (see) and Claisen reaction (see).

9) Grignard reactions with aldehydes, ketones and carboxylic acid esters: a) reactions of formaldehyde with Grignard reagents to form primary A.:

$$HCHO + RMgX \rightarrow H{-}\underset{H}{\overset{OMgX}{C}}{-}R \xrightarrow{H_2O} R{-}CH_2{-}OH.$$

b) reaction of other aldehydes, ketones and carboxylic acid esters to secondary and tertiary A. (see Grignard reaction).

10) Complete catalytic hydrogenation of phenols to cyclohexanols under pressure and at high temperatures in autoclaves.

11) Reaction of primary aliphatic amines with nitric acid: $R\text{-}CH_2\text{-}NH_2 + HNO_2 \rightarrow [R\text{-}CH_2\text{-}N{=}N\text{-}OH] \rightarrow R\text{-}CH_2\text{-}OH + N_2$.

12) Oxidation of industrially available aluminum trialkyls with subsequent hydrolysis:
$[CH_3\text{-}(CH_2\text{-}CH_2)_n]_3Al \rightarrow [CH_3\text{-}(CH_2\text{-}CH_2)_nO]_3Al \rightarrow 3\ CH_3\text{-}(CH_2\text{-}CH_2)_n\text{-}OH + Al(OH)_3$
($n = 11, 13, 15, 17$).

This reaction is important for industrial production of long-chain alkan-1-ols as starting materials for detergents.

Applications. Monoalcohols of low molecular weight are used in large quantities as solvents and fuels, and in the synthesis of esters, ethers, alkenes and other organic compounds. Simple dialcohols and trialcohols are used in cosmetics, as antifreezes, and in the production of explosives, lubricants and textile conditioners.

Alcoholysis: same as Esterification (see Esters).

Aldaric acids, ***sugar dicarboxylic acids***: dicarboxylic acids with the general formula $HOOC\text{-}(CHOH)_n\text{-}COOH$. Formally, they are the products of oxidation of both terminal groups of aldoses. The names are based on the roots of the aldose names, with the suffix -aric acid. Trivial names are generally preferred, e.g. tartaric instead of threaric acid, mucic instead of galactaric acid, saccharic instead of glucaric acid. Salts and esters of glucaric acid are called ***saccharates***. Calcium saccharate is used as a concrete liquefier, and is added to calcium gluconate solutions as a stabilizer.

Aldazines: symmetric or unsymmetric biscondensation products of hydrazine with aldehydes. The class of Azines (see) consists of A. and ketazines.

Aldehyde: a carbonyl compound containing one or more aldehyde (-CHO) groups in the molecule. Depending on the nature of the organic residue to which the aldehyde group is bound, the A. can be classified as aliphatic, aromatic or heterocyclic. According to the IUPAC nomenclature, an aliphatic A. is named by adding the suffix "-al" to the root of the corresponding hydrocarbon skeleton name:

$H{-}C({=}O){-}H$ $CH_3{-}C({=}O){-}H$ $CH_3{-}CH_2{-}CH_2{-}C({=}O){-}H$

Methanal Ethanal Butanal

An older system of trivial names is also still in use; these names were derived from the Latin names of the carboxylic acids formed by oxidation of the A.:

$$H{-}C({=}O){-}H \xrightarrow{O} H{-}C({=}O){-}OH$$

Formaldehyde → Formic acid (acidum formicium)

$$CH_3{-}C({=}O){-}H \xrightarrow{O} CH_3{-}C({=}O){-}OH$$

Acetaldehyde → Acetic acid (acidum aceticum)

The aromatic and heterocyclic aldehydes, in which the aldehyde group is bound directly to a carbon atom of the ring, are usually given trivial names, e.g. benzaldehyde. Their systematic names are formed from the name of the root compound with the suffix "carbaldehyde", e.g. pyrrol-2-carbaldehyde.

Properties. Most of the lower aliphatic A., except for the gaseous formaldehyde, are non-viscous liquids with pungent odors. As the molecular mass increases, the A. become more viscous; those with the highest molecular masses are solids. Because they do not form intermolecular hydrogen bonds, the boiling points of the A. are lower than those of the corresponding alcohols. They are generally readily soluble in organic solvents. Certain A. have pleasant, characteristic odors, e.g. benzaldehyde and cinnamaldehyde.

Except for formaldehyde, the saturated, aliphatic aldehydes are relatively non-toxic, except for their irritation of the mucous membranes. Their mildly narcotic effects are only noticeable at high concentrations. The unsaturated aldehydes, e.g. acrolein and crotonaldehyde, are more irritating to the mucous membranes and to the external skin. The toxicity of the aldehydes generally decreases as molecular mass increases. Halogen-substituted aldehydes are considerably more irritating to the mucous membranes.

Reactions. The reactivity of the A. is due mainly to the presence of the Carbonyl group (see). In the addition and condensation reactions which are typical of the A., nucleophilic reaction partners add to the partially positive carbonyl carbon atom, while electrophilic reagents are bound to the partially negative oxygen atom (condensation). In contrast to the condensation products, most of which are stable, addition products have varying degrees of stability, depending largely on the nature of the substituents and the catalytic conditions.

When water is added to A., the products are aldehyde hydrates (1,1-diols), most of which are unstable (see Erlenmeyer rule); the equilibrium position of the reaction

$$R\text{–}CHO + H_2O \rightleftharpoons R\text{–}CH(OH)_2$$

usually lies on the side of the reactants. One stable compound of this type is chloral hydrate.

Alcohols or thioalcohols can be added in analogous fashion to A. in the presence of acidic catalysts like hydrogen halides or zinc chloride. The unstable semiacetals formed thus usually react further under the reaction conditions, adding another alcohol or thioalcohol molecule, with cleavage of water, to form the acetal or thioacetal:

$$R\text{–}C(H){=}O + HX\text{–}R^1 \overset{(H+)}{\rightleftharpoons}$$

$$R\text{–}CH(XR^1)(OH) \underset{-HX\text{–}R^1}{\overset{+HX\text{–}R^1}{\rightleftharpoons}} R\text{–}CH(XR^1)_2 + H_2O$$

where X = O or S.

Because acetals are very stable in the presence of bases, they are often used in syntheses instead of the free A. They can be easily converted back to the A. by dilute acids. Reducing agents, such as hydrogen, lithium aluminum hydride and sodium borohydride, convert A. into primary alcohols. The reduction can be achieved by the Meerwein-Ponndorf-Verley reduction (see), which uses aluminum isopropanolate in a reversal of the reaction of the Oppenauer oxidation (see) by isopropanol. Under the conditions of the Wolff-Kishren reduction (see), A. are reductively converted to the corresponding hydrocarbons. The action of oxidation agents on A. produces the corresponding carboxylic acids. This reaction occurs very easily, since A. are reducing agents. It is therefore often used to characterize this class of compounds and to distinguish them from Ketones (see). The Tollens, Nylanders or Fehling reagent is used as an oxidation agent for this purpose.

Another well-known addition reaction of A. is the addition of sodium hydrogensulfite to form insoluble, crystalline hydrogensulfite compounds:

$$R\text{–}C(H){=}O + NaHSO \longrightarrow R\text{–}CH(SO_3^-Na^+)(OH)$$

Since the bisulfide adducts can easily be removed again by dilute acids or sodium carbonate solution, this reaction is often used to purify the A.

A reaction of A. which is important for numerous syntheses is the formation of cyanohydrins by addition of hydrogen cyanide to the carbonyl group:

$$R\text{–}C(H){=}O + HCN \rightleftharpoons R\text{–}CH(OH)(CN)$$

When A. react with ammonia, the first products are the aldehyde-ammonia adducts, most of which are unstable, and readily eliminate water to form aldimines:

$$R\text{–}C(H){=}O + NH_3 \longrightarrow R\text{–}CH(NH_2)(OH)$$

$$\longrightarrow R\text{–}CH{=}NH + H_2O$$

Aldimines often trimerize to triazine derivatives. In the case of formaldehyde, a further reaction produces hexamethylene tetramine. The reaction of benzaldehyde with ammonia differs from the ordinary reactions of A. with ammonia; it forms hydrobenzamide under these conditions.

A. condense with primary amines to azomethines:

$$R\text{–}C(H){=}O + H_2N\text{–}R^1 \rightleftharpoons R\text{–}CH{=}N\text{–}R^1 + H_2O$$

In the reaction with secondary amines, A. react with α protons to form enamines. In the absence of α-H atoms, aminals are formed. A. can be converted into secondary alcohols by Grignard reactions (see).

The condensation of A. with hydrazine or substituted hydrazines, such as phenylhydrazine, 4-nitrophenylhydrazine or 2,4-dinitrophenylhydrazine to form the corresponding hydrazones:

$$R\text{–}C(H){=}O + NH_2\text{–}NH\text{–}R^1 \longrightarrow R\text{–}CH{=}N\text{–}NH\text{–}R^1 + H_2O$$

are very important, not least in the separation and characterization of A.

The condensation of A. with hydroxylamine, forming the oximes, are of similar importance:

$$R{-}C(H){=}O + H_2N{-}OH \longrightarrow R{-}CH{=}N{-}OH + H_2O$$

Semicarbazones and thiosemicarbazones are formed from A. and semicarbazide or thiosemicarbazide. They too can be used for the purification and characterization of the A.

A. with α hydrogen atoms undergo the aldol reaction in the presence of bases or acids to form aldols; when these are subjected to higher temperatures, they split out water to form α,β-unsaturated A.:

$$2CH_3{-}C(H){=}O \rightleftharpoons CH_3{-}CH(OH){-}CH_2{-}C(H){=}O \xrightarrow[-H_2O]{} CH_3{-}CH{=}CH{-}C(H){=}O$$

Aromatic A. disproportionate under these conditions to the corresponding carboxylic acids and alcohols, in the Cannizzaro reaction (see). In the presence of aluminum alcoholates, this disproportionation also occurs with aliphatic A.

In reactions similar to the aldol reaction, A. can also condense with other CH-acidic reactants, such as malonic acid and its derivatives in the Knoevenagel condensation (see), acetanhydride/sodium acetate in the Perkin reaction (see) and with α-halogencarboxylate esters in the Darzens-Erlenmeyer-Claisen condensation (see).

Many aromatic A. form benzoins in the presence of potassium cyanide or certain thiazolium or imidazolium salts through the benzoin condensation. The structurally analogous aliphatic compounds, the acyloins, can only be obtained directly from aliphatic A. by use of enzymes from certain species of yeast:

$$2CH_3{-}C(H){=}O \xrightarrow{\text{(Carbinolase)}} CH_3{-}CH(OH){-}C({=}O){-}CH_3$$

Analytical. A. and ketones can be characterized chemically in the form of derivatives, by IR and NMR spectroscopy, and by mass spectrometric fragmentation. Their IR spectra are characterized by intense bands for the C=O valence vibration in the range of 1680 to 1740 cm^{-1}. The C-H vibrations of A. appear between 2665 and 2880 cm^{-1}. The signals of the aldehyde protons in ^1H NMR spectra are to be expected in the range of δ = 9 to 10 ppm. In ^{13}C NMR, signals in the range of δ = 180 to 210 ppm are very probably due to the C=O groups of A. and ketones. In the mass spectra of aromatic A. and ketones, a characteristic key fragment is the benzoyl cation.

Occurrence and isolation. Various A. are found in low concentrations in plants, especially in numerous etheric oils, e.g. citral, citronellal. For the synthesis of A., there are special methods in addition to the numerous general methods which are also applicable to ketones. The best known and most important method for the synthesis of aliphatic A. is partial oxidation or dehydrogenation of primary alcohols:

$$R{-}CH_2{-}OH \xrightarrow{O} R{-}C(H){=}O + H_2O$$

Suitable oxidation agents are chromium(VI) oxide or potassium dichromate in sulfuric acid, oxygen in the presence of heated copper or silver, manganese dioxide and selenium dioxide. Another important industrial technique is pyrolysis of mixtures of a carboxylic acid and formic acid in the presence of manganese(II) oxide:

$$R{-}COOH + H{-}COOH \xrightarrow{(MnO)} R{-}C(H){=}O + CO_2 + H_2O$$

Aliphatic A. can also be produced by cleavage of glycols or hydroformylation of alkenes with carbon monoxide and water in the presence of dicobalto octacarbonyl around 150 °C and pressure of about $3 \cdot 10^4$ kPa. Some other synthetic methods for A. are the Rosenmund reduction (see) of carboxyl chlorides, the Grignard reaction (see) of ortho-formate esters with the Grignard reagent, the Sommelet reaction (see) of alkyl halides with urotropin, the Stephen reduction (see) of nitriles, the Nef reaction (see) of primary nitroalkanes with dilute mineral acids, the Krönke reaction (see) from nitrones, the Vilsmeier-Haack reaction (see) of arenes, the Gattermann synthesis (see) of arenes with hydrogen cyanide and hydrogen chloride in the presence of aluminum chloride and the Gattermann-Koch synthesis (see).

Applications. A. are used mainly as starting materials for numerous syntheses, including industrially important ones, and as perfumes and flavorings in cosmetics and foods. A much broader area is the use of A. as intermediates in the industrial synthesis of styryl and azomethine dyes and many classes of organic compounds, e.g. carboxylic acids, alcohols, nitriles and amines.

Aliphatic and aromatic aldehydes

Formaldehyde	$H{-}CHO$
Acetaldehyde	$CH_3{-}CHO$
Propionaldehyde	$CH_3{-}CH_2{-}CHO$
Butyraldehyde	$CH_2{-}(CH_2)_2{-}CHO$
Acrolein	$CH_2{=}CH{-}CHO$
Crotonaldehyde	$CH_3{-}CH{=}CH{-}CHO$
Benzaldehyde	$C_6H_5{-}CHO$
Cinnamaldehyde	$C_6H_5{-}CH{=}CH{-}CHO$

Aldehyde carboxylic acids: see Oxocarboxylic acids.

Aldehyde collidines: see Collidines.

Aldehyde group: the functional group, -CHO, of an Aldehyde (see).

Aldehyde hydrates: addition products of aldehydes and water with the general formula $R\text{-}CH(OH)_2$. Most A. are very unstable compounds, because the equilibrium of the addition reaction lies on the side of the reactants. A. are stabilized by strongly electron-withdrawing substituents, e.g. chloral hydrate.

Aldehyde resin: a synthetic resin produced by condensation of aldehydes, especially acetaldehyde, acrolein or furfural. Industrial syntheses in aqueous or alkaline media containing strong alkalies (concentrated sodium or potassium hydroxide) are most common. The first product is a water-sensitive, soft, crude resin; this is converted into an elastic, polishable, thermostable product which is insensitive to strong acids and bases by special treatments. These include exposure to steam or hot air, esterification with glycerol and fusion with other resins (colophony, cumarone-indene resin) or metal hydroxides. Because they are stable to light, have no odor and are not toxic, these products are used for paints and polishes. Acrolein A. have good electrical properties. Condensation of furfural in alkaline media produces dark, alcoholic varnishing resins; in acidic media, furfural condenses to hard rubber substitutes which are highly resistant to alkalies and acids. Rapidly drying varnishes are often based on A.

Aldicarb: see Carbamate insecticides (table); A. is also a nematocide.

Aldimorph: see Morpholine and piperazine fungicides.

Aldimines: Organic compounds with the general formula R-CH=NH. They are formed by the reaction of ammonia with aldehydes, via the unstable aldehyde ammine:

$$R\text{-}C(=O)H \xrightarrow{+NH_3} R\text{-}CH(NH_2)(OH) \xrightarrow{-H_2O} R\text{-}C(=NH)H$$

Aldehyde ammine

Alditols, ***sugar alcohols***: reduction products of monosaccharides. The name is formed from the root name for the aldose plus the suffix -itol. Reduction of the sugars is carried out in neutral media, e.g. by amalgams or catalytically. The reduction of a ketose produces a new center of chirality, and thus two isomeric A., for example, glucitol and mannitol from fructose. A. taste sweet and are used as sugar substitutes (see Sweeteners). The best-known A. are *D-glucitol (sorbitol), D-mannitol* and *xylitol*.

Aldolases: enzymes which catalyse the cleavage of hexose chains into two C_3 units, e.g. fructose 1,6-bisphosphate into D-glyceraldehyde 3-phosphate and dihydroxyacetone phosphate. Examples are ***liver A.*** (M_r 158,000) and ***muscle A.*** (M_r 160,000). The human skeletal musculature contains very high A. activity; the appearance of elevated levels of A. in the serum is diagnostic of certain muscle diseases.

Aldol condensation: same as aldol reaction.

Aldol reaction, ***aldol condensation, aldol addition***: originally, the acid- or base-catalysed dimerization of aliphatic aldehydes with α-hydrogen atoms to form β-hydroxyaldehydes (aldols). For example, under the conditions of A., acetaldehyde is converted to acetaldol:

$$CH_3\text{-}CHO + CH_3\text{-}CHO \xrightarrow{(OH)} CH_3\text{-}CH(OH)\text{-}CH_2\text{-}CHO.$$

In the general sense, all reactions of ketones and aldehydes with CH-acidic compounds can be considered A., since the reaction mechanism is the same in all cases. The mechanism of A. under basic conditions can be formulated as follows:

$$CH_3\text{-}CHO + OH^- \rightleftharpoons |\bar{C}H_2\text{-}CHO + H_2O$$

$$CH_3\text{-}CH(=O) + |\bar{C}H_2\text{-}CHO \rightleftharpoons CH_3\text{-}CH(\text{-}|\bar{O}|^-)\text{-}CH_2\text{-}CHO$$

$$CH_3\text{-}CH(|\bar{O}|^-)\text{-}CH_2\text{-}CHO + H_2O \rightleftharpoons CH_3\text{-}CH(OH)\text{-}CH_2\text{-}CHO + OH^-$$

The nucleophilic attack of the carbanion at the carbonyl carbon atom of the aldehyde is the actual aldol addition. The addition product of base-catalysed A. is in many cases stable enough to be isolated.

Acid-catalysed A. is initiated by formation of the enol form of the CN-acidic component:

$$CH_3\text{-}CH{=}O + H^+ \rightleftharpoons CH_3\text{-}\overset{+}{C}H\text{-}OH \rightleftharpoons CH_2{=}CH\text{-}OH + H^+$$

In the further course of the reaction, the basic properties of the C=C double bond of the enol cause the nucleophilic attack on the C atom of the carbonyl group:

$$H^+ + O{=}C(CH_3)H + CH_2{=}CH\text{-}OH \rightleftharpoons HO\text{-}CH(CH_3)\text{-}CH_2\text{-}\overset{+}{C}H\text{-}OH$$

$$\rightleftharpoons CH_3\text{-}CH(OH)\text{-}CH_2\text{-}CH{=}O + H^+$$

In acid-catalysed A., the addition product cannot be isolated, because under these conditions water is inevitably split off:

$$CH_2\text{-}CHO + H^+ \rightleftharpoons CH_3\text{-}CH(^+OH_2)\text{-}CH_2\text{-}CHO \rightleftharpoons CH_3\text{-}CH{=}CH\text{-}CHO + H_2O + H^+.$$

The reaction product is an α,β-unsaturated carbonyl compound, i.e. a product of aldol condensation in the narrow sense of the term. Base catalysis is much more widely used in synthesis than acid catalysis.

Aldehydes which do not have an α-hydrogen atom, e.g. benzaldehyde, cannot undergo A. However, they are able to serve as the carbonyl component in a reaction with other aldehydes or ketones to form the corresponding aldol reaction products:

$$\underset{\text{Carbonyl component}}{C_6H_5\text{-}CHO} + \underset{\text{Methylene component}}{CH_3\text{-}CO\text{-}C_6H_5} \xrightarrow[-H_2O]{} C_6H_5\text{-}CH{=}CH\text{-}CO\text{-}C_6H_5$$

When aromatic reaction components are used, water is usually split off the addition products, even under basic conditions, leading to α,β-unsaturated carbonyl compounds. The driving force for this reaction step is the formation of a system of conjugated double bonds.

Aliphatic ketones react analogously to the aldehydes, forming β-hydroxyketones, e.g. the A. of acetone produces diacetone alcohol:

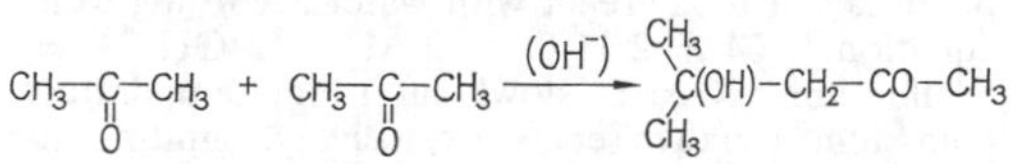

$$CH_3-\underset{\underset{O}{\|}}{C}-CH_3 + CH_3-\underset{\underset{O}{\|}}{C}-CH_3 \xrightarrow{(OH^-)} (CH_3)_2C(OH)-CH_2-CO-CH_3$$

In the acid-catalysed reaction, this is dehydrated to mesityl oxide.

If a mixture of an aldehyde and a ketone, each with an α-hydrogen atom and thus capable of A., is subjected to this reaction, the aldehyde always acts as the carbonyl component, due to its higher carbonyl reactivity, and the ketone acts as the methylene component. The reaction product in such cases is always a β-hydroxyketone. Instead of an aldehyde or ketone, other CH-acidic compounds such as malonate esters, cyanoacetate esters, malonic acid dinitrile, acetoacetate esters and aliphatic nitro-compounds, can be used as the methylene component in A. Thus reactions like the Knoevenagel condensation (see), the Perkin reaction (see) or the Darzens-Erlenmeyer-Claisen condensation (see) can be considered special cases of A. The reactivity of the carbonyl components increases in the series: ketones, branched aldehydes, straight-chain aldehydes. The A. is reversible at each step. The addition products can be cleaved into the starting components by acids or bases.

Aldonic acids: monocarboxylic acids formed by oxidation of aldehyde groups on aldoses. If the secondary hydroxyl of an A. is oxidized to a carbonyl group, the product is a ***ketoaldonic acid***. Depending on the number of C-atoms, an A. is classified as an aldotrionic, aldotetronic, aldopentonic, etc. acid. A. are formed by mild oxidation of monosaccharides, e.g. with bromine water. Bacteria and other microorganisms contain glucose oxidases which oxidize D-glucose to ***D-gluconic acid*** (abb. *GlcA*). In free A., nucleophilic attack on the carboxyl group by an hydroxyl group readily leads to formation of a 1,4- or 1,5-lactone. In aqueous solution, the equilibrium lies on the side of the lactone. A. form insoluble salts with alkaline earth metal cations. Both free and bound A. are rare in nature. D-Gluconic acid is used as an additive to soft drinks and as an etching reagent for metals. Its calcium salt is used for calcium therapy. D-Glucono-1,5-lactone is added to sausages. Ascorbic acid is a derivative of an A.

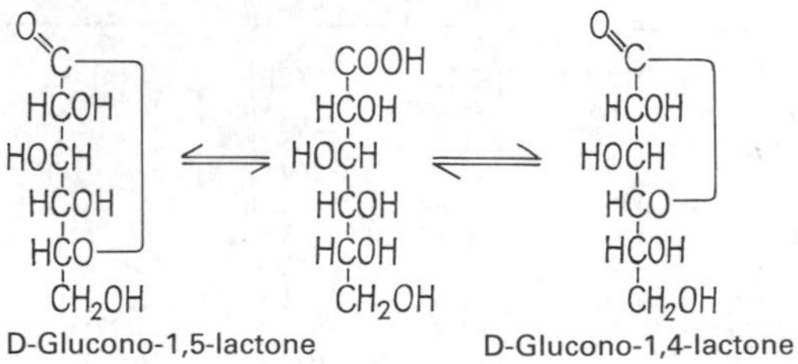

D-Glucono-1,5-lactone ⇌ D-Gluconic acid ⇌ D-Glucono-1,4-lactone

Aldoses: see Monosaccharides.

Aldosterone: see Adrenal cortex hormones.

Aldosterone antagonists: see Diuretics.

Aldoximes: see Oximes.

Aldrin: see Cyclodiene insectides.

Alginates: see Alginic acid.

Alginate fibers: synthetic fibers made by spinning a solution of sodium alginate into a precipitation bath. The sodium alginate solution is made by extraction of alginic acid from dried algae with a sodium carbonate solution. ***Calcium alginate fibers*** are made by spinning the sodium alginate solution into a weakly acidic calcium chloride solution. The A. do not "pill", are not inflammable and are insoluble in organic solvents; however, they are soluble in alkaline aqueous solutions.

In the textile industry, the A. are used as supplemental threads and to achieve special effects. They are used in dentistry as self-dissolving threads.

Alginic acid: a linear heteropolysaccharide found in the cell walls of algae, especially brown algae. A. forms a colorless, highly hygroscopic powder. It is only slightly soluble in cold water, but swells strongly; it is insoluble in organic solvents. Stable gels are formed by association of the polyuronide chains of the A. with divalent cations. Salts and esters of A. are called ***alginates***. A. consists of (1→4)-glycosidically linked β-D-mannuronic acid and α-L-guluronic acid esters. It is produced from algae growing on the coasts of the Atlantic and Pacific Oceans, especially *Laminaria* species, *Macrocystis pyrifera, Nereocystis luetkeana, Ascophyllum nodosum* and *Ecklonia maxima*. Up to 40% of the dry mass of the algae can be A. The A. is extracted with sodium carbonate solution and precipitated again with dilute hydrochloric acid. The applications of A. are based on its ability to form gels and to stabilize emulsions. Alginates are used in food, cosmetics and pharmaceuticals as thickeners and protective colloids, and also to make coatings and dental impressions.

Alginite: see Macerals.

Alicyclic: a term for hydrocarbon ring compounds of various sizes which can be visualized as having been formed from aliphatic compounds.

Aliphatic: an adjective applied to hydrocarbon compounds with their carbon atoms arranged in chains (but not containing systems of conjugated double bonds).

Alizarin pigments: a group of synthetic pigments based on the alizarin skeleton. This is modified by addition of reactive groups, such as hydroxyl, amino, nitro and sulfo groups (Table). A. are important mordant dyes; with aluminum or chromium mordants, they make very fast wool dyes, e.g. ***alizarin black S***. *Alizarin*, 1,2-dihydroxy-9,10-anthracenedione occurs in the root of the madder plant, *Rubia tinctorum*, in combination with 2 moles glucose. It was known and used in ancient Egypt, Persia and India.

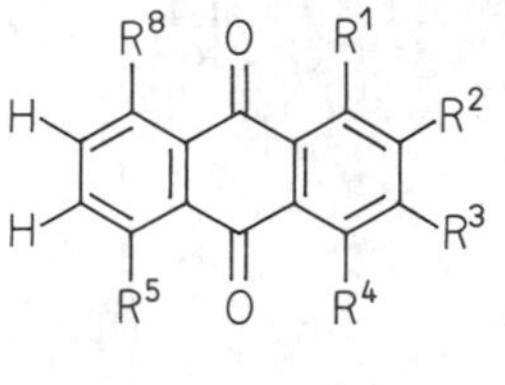

For key to substituents, see p. 36

Pigment	R^1	R^2	R^3	R^4	R^5	R^8
Alizarin	OH	OH	OH	H	H	H
Alizarinbordeaux	OH	OH	H	H	OH	OH
Purpurin	OH	OH	H	OH	H	H
Alizarin Orange	OH	OH	NO_2	H	H	H
Alizarin Red	OH	OH	SO_3H	H	H	H
Alizarin Saphirol B	NH_2	H	SO_3H	OH	NH_2	OH
	(+SO_3H in position 7)					

Alizarin violet: same as Gallein (see).

Alkali: a substance which gives alkaline aqueous solutions (see Basic reaction). The most important A. are the hydroxides of sodium and potassium, which are called ***caustic*** A. because of their caustic properties (see Caustification). They have low melting points and a characteristic flavor of lye. In the broad sense, the A. also include the alkaline earth hydroxides and aqueous ammonia. The alkali metal carbonates, such as sodium carbonate (soda), Na_2CO_3, and potassium carbonate (potash), K_2CO_3, are called ***mild A.*** because of their weaker basicity.

Alkali fusion: a method for production of phenols. A. is used mainly for replacement of aromatic sulfur groups by hydroxyl groups, using alkalies.

Alkali metals: the elements lithium (Li), sodium (Na), potassium (K), rubidium (Rb), cesium (Cs) and francium (Fr), which comprise the 1st main group of the periodic system. The chemical and physical properties of the A. are similar, although there are regular variations within the group. A. are light metals with low melting and boiling points; they can be cut with a knife. They form alloys with each other, some of which are liquid at room temperature. For example, the eutectic Na-K mixture containing 77.2% K melts at - 12.3 °C. The A. are produced by melt electrolysis of their chlorides.

The A. are highly electropositive, in agreement with their electron configuration, and thus their position in the periodic system. Their characteristic electron configuration is a single s-electron outside the filled noble gas shell of the next lower element. As can be seen from the 1st ionization potential, this electron is easily lost; as a result, the A. occur almost without exception in the positive monovalent state. For the same reason, they are also highly reactive, usually as reducing agents. Their standard electrode potentials are so high that the A. occupy the extreme negative end of the electrochemical potential series, and are thus among the strongest known reducing agents, as shown by their reactions with oxygen, hydrogen, water and halogens. The silver sheen of freshly cut surfaces of the A. are rapidly covered, when exposed to moist air, with a dull layer of the oxide or hydroxide. This makes it necessary to keep the A. under inert liquids, usually mineral oil, which prevent access of air. The reactivity of the A. increases markedly from lithium to cesium. This is due to the decreasing coulomb interaction between the nucleus and the outer electron as the atomic radius increases; it is reflected by the decreasing ionization potentials. The A. react with water according to the equation: $2\ M + 2\ H_2O \rightleftharpoons 2\ M^+ + 2\ OH^- + H_2$. While lithium reacts slowly at room temperature, potassium usually reacts explosively, igniting the evolved hydrogen. The A. burn in air, lithium to Li_2O, sodium primarily to the peroxide Na_2O_2 and both potassium and rubidium to mixtures of the peroxides and superoxides, K_2O_2 and KO_2, Rb_2O_2 and RbO_2. Since they are very electropositive, the A. react preferentially with the electronegative elements on the right side of the periodic table, forming ionic compounds. Many of these salts have industrial applications. Aqueous solutions of the hydroxides are widely used as strong bases (see Alkali).

The A. dissolve in liquid ammonia to form dark blue solutions which conduct electricity. Both phenomena are due to the presence of electrons solvated by the ammonia. Volatile A. compounds produce characteristic colors in a gas flame which are used for qualitative and quantitative analysis of these elements.

In accordance with the general diagonal relationship in the Periodic system (see), lithium resembles magnesium more closely than it does the other members of the A. The unusually small ionic radius of the Li^+ ion makes it highly polarizing. As a result, Li^+ has a very pronounced tendency to become solvated, and the lattice energies of unsolvated Li salts are high. These effects are responsible for the differences between the solubilities of Li salts and those of heavier homologous compounds, and they explain the similarity to mganesium salts. The polarizing effect of the Li cation is also the reason for a tendency of lithium to form covalent compounds, a tendency not shared by its heavier homologs. This gives lithium its special importance in organoelement chemistry.

The A. are found in nature as various salts. Sodium and potassium make up considerable fractions of the earth's crust, but lithium, rubidium and cesium are relatively rare elements. As a member of natural radioactive decay series, francium occurs in trace amounts in nature (Table).

Properties of the elements of the 1st main group of the periodic system

	Li	Na	K	Rb	Cs	Fr
Nuclear charge	3	11	19	37	55	87
Electronic configuration	[H] $2s^1$	[Ne] $3s^1$	[Ar] $4s^1$	[Kr] $5s^1$	[Xe] $6s^1$	[Rn] $7s^1$
Atomic mass	6.941	22.9897	39.0983	85.4678	132.905	(223)
Atomic radius [pm]	134	154	196	211	225	·
Ionic radius [pm]	60	95	133	148	169	·
Electronegativity	0.97	1.01	0.91	0.89	0.86	·
1st ionization potential [eV]	5.392	5.139	4.341	4.177	3.894	·
Standard electrode potential [V]	–3.045	–2.7109	–2.924	–2.925	–2.923	·
Density [g cm^{-3}]	0.534	0.968	0.86	1.532	1.878	·
m.p. [°C]	180.54	97.81	63.65	38.89	28.40	(27)
b.p. [°C]	1317	882.9	774	688	678.4	(677)

Properties of the elements of group IIa of the periodic system

	Be	Mg	Ca	Sr	Ba	Ra
Nuclear charge	4	12	20	38	56	88
Electronic configuration	[He] $2s^2$	[Ne] $3s^2$	[Ar] $4s^2$	[Kr] $5s^2$	[Xe] $6s^2$	[Rn] $7s^2$
Atomic mass	9.0122	24.312	40.08	87.62	137.34	226.0254
Atomic radius [pm]	112	160	197	215	222	.
Ionic radius [pm]	31	65	99	113	135	.
Electronegativity	1.47	1.23	1.04	0.99	0.97	(0.97)
1st ionization potential [eV]	9.322	7.646	6.113	5.695	5.212	5.279
2nd ionization potential [eV]	18.211	15.035	11.871	11.030	10.004	10.147
Standard electrode potential [V]	–1.70	–2.375	–2.76	–2.89	–2.90	.
Density [g cm^{-3}]	1.85	1.74	1.54	2.6	3.51	5.50)
m.p. [°C]	1278	648.8	839	769	725	(700)
b.p. [°C]	2970	1107	1484	1384	1640	(1140)

Alkalimetry: a method of Neutralization analysis (see) in which bases are used as standard solutions. However, titration with acid is still occasionally called A., without regard for the usual rules.

Alkaline earth metals: group IIa of the periodic system, including beryllium, magnesium, calcium, strontium, barium and radium. With the exception of radium, the A. are light metals. There is a distinct difference between the properties of calcium, strontium and barium, the A. in the narrower sense, and those of beryllium. The properties of magnesium are intermediate.

The A. are strongly electropositive elements, in accordance with their ns^2 electron configuration and their position in the periodic system. The ionization potentials, especially for the second electron, are considerably higher than those of the alkali metals. However, this energy, which must be expended to form compounds, is more than compensated by the high lattice energies of solid salts and by the high solvation enthalpies of the divalent cations in solutions. These considerations apply to magnesium, calcium, strontium and barium, and explain why these elements always have the +2 oxidation state in their stable compounds; they also explain the high reducing capacity of these metals, which is reflected by their high standard electrode potentials.

The unusually small atomic and ionic radii of beryllium, which lead to very high ionization potentials, make it impossible for Be to form divalent cations. Instead, it forms two covalent bonds which, however, only complete an electron quartet. This electron-deficient situation accounts for the Lewis acidity of many Be compounds, and is often compensated by polymerization. For example, solid beryllium chloride, $BeCl_2$, consists of long chains in which the individual Be atoms are linked by two chlorine bridges.

Cl Be Cl Be Cl Be
Cl Cl Cl

Thus with a coordination number of 4, each Be atom achieves a noble gas configuration. On the whole, beryllium resembles aluminum more than it does magnesium (diagonal relationship in the Periodic system (see)). However, magnesium also has a recognizable tendency to form covalent bonds; because it is able to form stable bonds to carbon, organomagnesium compounds (see Grignard compounds) are important in organic and organometallic syntheses.

As expected, the reactivity of the A. increases with increasing atomic mass. Although calcium, strontium and barium react vigorously with water to form the hydroxides and evolve hydrogen, magnesium is attacked only by hot water, and the reaction is slow. Beryllium is passivated by an oxide skin, and does not react with water. The stability in air also decreases with increasing atomic mass. At high temperatures, the A. react with oxygen to form the oxides (e.g. MgO, CaO). Barium forms the peroxide above 500 °C; this decomposes reversibly at 700 °C:

$$2\,BaO + O_2 \underset{700°C}{\overset{500°C}{\rightleftharpoons}} BaO_2.$$

The strong reducing effect of the A. is utilized in the production of other metals, such as titanium, uranium, and even potassium. The basic character of their hydroxides and oxides increases with increasing atomic mass. Although beryllium hydroxide, $Be(OH)_2$, is amphoteric, barium hydroxide, $Ba(OH)_2$, is a strong base. This trend is related to the increase in decomposition temperatures of A. carbonates (e.g. $CaCO_3 \rightarrow CaO + CO_2$) on going from magnesium to barium. Many A. salts are insoluble in water, and their precipitation is used to detect the presence of the elements. The solubilities of the hydroxides and oxides increase with increasing atomic mass, while those of the carbonates, sulfates and chromates decrease in the same order.

Magnesium (1.9%) and calcium (3.4%) make up significant fractions of the earth's crust. Beryllium is a rare element. The A. occur naturally as carbonates, sulfates, silicates and chlorides in many minerals. Radium occurs as a product of natural radioactive decay of uranium in uranium minerals. The A. are generally produced by melt electrolysis of their chlorides.

Alkaline zinc-manganese dioxide cell: an Electrochemical current source (see) which is an improved version of the Leclanché cell (see). The electrolyte is a 45 to 50% potassium hydroxide solution saturated with zinc oxide. The anode is formed by a zinc powder with a large surface area; this gives a higher energy density, higher capacity and better storage properties.

Alkaloids: biogenic, nitrogen-containing com-

Alkaloids

The most important groups of alkaloids

Chemical group	Parent compound	Examples	Occurrence
Pyrrolidine A.	Pyrrolidine	Hygrin	*Erythroxylaceae (Erythroxylon)*
Indole A.	Pyrrolidinoindole	Physostigmine	*Fabaceae (Physostigma)*
	β-Carboline	Rauwolfia A. (reserpine, ajmaline)	*Apocynaceae (Rauwolfia)*
		Yohimbine	*Rubiaceae (Pausinystalia)*
		Vinca A. (vinblastine, vincristine)	*Apicynaceae (Vinca, Catharanthus)*
	Carbozole	Strychnose A. (strychnine, brucine, calebash curare A.)	*Loganiaceae (Strychnos)*
		Aspidosperma A.	*Apocynaceae (Aspidosperma, Vinca Pleiocarpa)*
		Catharanthus A.	*Apocynaceae (Catharanthus)*
	Ergoline	Ergot A.	Fungus *(Claviceps)*
Pyrrolizidine A.	Pyrrolizidine	Senecio A.	*Asteraceae (Senecio)*
Piperidine A.	Piperidine	Piperine	*Piperaceae (Piper)*
		Coniine	*Apiaceae (Conium)*
		Lobelia A. (lobelin)	*Campanulaceae (Lobelia)*
		Areca A.	*Palmae (Areca)*
	Tropane	Hyoscyamine, scopolamine, atropine	*Solanaceae (Atropa, Datura, Hyoscyamus, Scopolia)*
		Cocaine	*Erythroxylaceae (Erythroxylon)*
Quinolizidine A.	Quinolizidine	Sparteine, cytisine, lupinine	*Fabaceae (Lupinus, Cytisius, Genista)*
Pyridine A.	Pyridine	Nicotine	*Solanaceae (Nicotiniana)*
Quinoline A.	Quinoline	Quina A. (quinine, quininidine)	*Rubiaceae (Cinchona)*
Isoquinoline A.	Isoquinoline	Anhalonium A.	*Cactaceae (Anhalonium)*
	Benzylisoquinoline	Opium A. (papaverine)	*Papaveraceae (Papaver)*
	Bisbenzylisoquinoline		*Annonaceae, Berberidaceae, Hernandiaceae, Lauraceae, Magnoliaceae, Monimaceae, Nymphaceae, Ranunculaceae*
		Tubocurarine	*Menispermaceae (Chondrodendron)*
	Phthalidisoquinoline	Opium A. (narcotine)	*Berberidaceae (Berberis)*
		Narcotine	*Papaveraceae (Papaver)*
	Aporphine	Apomorphine	
	Morphinane	Opium A. (morphine, codeine, thebaine)	*Papaveraceae (Papaver)*
	Benzoquinazoline	Ipecacuanha A. (emetine)	*Rubiaceae (Cephalis)*

Typical alkaloid reagents

Reagent	Composition	Reaction
Mayer's	$K_2[HgI_4]$	Yellowish-white precipitate
Dragendorff's	$K[BiI_4]$	Orange precipitate
Sonnenschein's	Phosphomolybdanic acid	Yellow precipitate which turns blue-green
Scheibler's	Phosphotungstic acid	Precipitate
Wagner's	KI_3	Brown precipitate
Erdmann's	HNO_3/H_2SO_4	Coloration
Fröhde's	Molybdanic acid/H_2SO_4	Coloration
Marquis'	Formaldehyde/H_2SO_4	Violet color (opium A.)

pounds. Most are N-heterocyclic. Amino acids, peptides, nucleosides, amino sugars and antibiotics are not considered A.

Classification. A. are classified according to their structures or their origins (Table). The largest groups are the indole and isoquinoline A., and these two groups also include numerous ***dimeric A. (bisalkaloids)***. The pseudo- and protoalkaloids are distinguished from these ***A. in the narrow sense***. ***Pseudoalkaloids*** are compounds whose basic carbon skeletons are not derived from amino acids, including, for example, the steroid alkaloids and coniine. ***Protoalkaloids*** are compounds in which the N atom derived from an amino acid is not part of the heterocycle; this group includes the Biogenic amines (see).

Occurrence. A. are most abundant in higher plants, especially the *Magnoliatae*, and less commonly in the *Liliatae* or *Pinidae*. The families *Apocynaceae, Buxaceae, Asteraceae, Euphorbiaceae, Loganiaceae, Menispermaceae, Papaveraceae, Rutaceae* and *Solanaceae* are especially rich in A. Usually a plant contains a primary A. and numerous secondary A., which differ with respect to methylation or hydrogenation. About 20% of higher plants contain A. They are found in some non-spermatophytes, including club mosses, horse-tails and fungi (e.g. ergot). A. are also found sporadically in animals, such as salamanders, toads (indolylalkylamines), frogs (batrachotoxin, pumiliotoxins), fish (tetrodotoxin) and millipedes (quinazoline derivatives).

Properties and determination. With a few exceptions, in which the N atom is acylated (e.g. colchicine), the A. are basic. The basicity depends on the heterocyclic skeleton and its substituents. Some A. bases, e.g. nicotine, hygrine and sparteine, are liquids, but most A. crystallize. The free bases, most of which are colorless, are only slightly soluble in water but are more readily soluble in organic solvents. The salts of A. are soluble in water, and they are used therapeutically in this form. The A. are determined by precipitation with picric acid or the reagents listed in the table, or by more or less group-specific color reactions. They are tested for purity mainly by thin-layer chromatography.

Isolation. A. are found in plants as salts of organic or inorganic acids. The A. is isolated from the plant material by treating it with alkali; the free alkaloid base is then extracted with organic solvents, or, more rarely, by steam distillation. The extracted A. are usually mixtures, which are then separated by fractional crystallization of suitable salts (e.g. hydrohalides, perchlorates, picrates or oxalates) or by chromatography.

Biosynthesis. Incorporation of radioactively labelled amino acids has demonstrated that nearly all A. are synthesized from the amino acids phenylalanine (isoquinoline A.), tryptophan (indole A.), lysine (piperidine A.) or proline or ornithine (pyrrolidine A.). A. can also contain isoprenoid components; examples are the steroid A. and some indole A.

Applications. Many A. are biologically active. Plants containing A. have therefore long been used in medicine. Today, pure A. are usually used instead of the plant preparations. Some of the most important are morphine and its derivatives (as analgesics), papaverine (as a spasmolytic), curare A. (as muscle relaxants), rauwolfia A. (as antihypertonics and neuroleptics), quinine (against malaria), catharanthus A. (as a cytostatic), quinidine and ajmaline (as antiarrhythmics) and cocaine (as a local anaesthetic, and as an illicit drug). Many pharmaceuticals were modelled on A. A few A. are used as pesticides (e.g., nicotine).

Alkanals: saturated aliphatic Aldehydes (see).

Alkanes, ***paraffins***: aliphatic hydrocarbons with the general formula C_nH_{2n+1}. A. contain exclusively sp^3-hybridized C atoms which are saturated with hydrogen.

The A. constitute a homologous series of compounds in which two neighboring members of the series differ by one $-CH_2-$ group (Table 1). The compounds are chemically similar, and their physical properties change in a continuous manner along the series.

Table 1. Homologous series of alkanes

Methane	CH_4	Decane	$C_{10}H_{22}$
Ethane	CH_3-CH_3	Undecane	$C_{11}H_{24}$
Propane	$CH_3-CH_2-CH_3$	Dodecane	$C_{12}H_{26}$
Butane	$CH_3-(CH_2)_2-CH_3$	Tridecane	$C_{13}H_{28}$
Pentane	$CH_3-(CH_2)_3-CH_3$	Tetradecane	$C_{14}H_{30}$
Hexane	$CH_3-(CH_2)_4-CH_3$	Pentadecane	$C_{15}H_{32}$
Heptane	$CH_3-(CH_2)_5-CH_3$	Eicosane	$C_{20}H_{42}$
Octane	$CH_3-(CH_2)_6-CH_3$	Triacontane	$C_{30}H_{62}$
Nonane	$CH_3-(CH_2)_7-CH_3$	Tetracontane	$C_{40}H_{82}$

Beginning with butane, C_4H_{10}, there are isomers of each compound; the number of isomers increases with the number of carbon atoms. A distinction is made between *linear* and *branched* A. From C_7H_{16} on, some of the isomers contain asymmetric carbon atoms, so there are stereoisomers in addition to structural isomers (Table 2). The rules for naming A. are given under Nomenclature (see), sect. III C.

Properties. The first four A. of the homologous series are gases, The next members of the series, up to C_{16}, are liquids, and the compounds from C_{17} on are solids. The boiling points of branched A. are lower than those of the isomeric *n*-A. The boiling points of the higher members of the series are closer together, making separation by distillation more difficult.

Table 2. Number of isomeric alkanes

No. of C atoms	Name of compound	No. of structural isomers	No. of structural and stereoisomers
1	Methane	1	1
2	Ethane	1	1
3	Propane	1	1
4	Butane	2	2
5	Pentane	3	3
6	Hexane	5	5
7	Heptane	9	11
8	Octane	18	24
9	Nonane	35	55
10	Decane	75	136
20	Eicosane	366 319	3 395 964

The gaseous and solid A. have no odor, and the liquids usually have a typical gasoline smell. The A. are not miscible with water; they dissolve in ether and ethanol, but their solubility decreases with increasing molar mass. A. are combustible, and the gaseous and liquid compounds are more readily ignited than their higher homologs. The lower compounds also form explosive mixtures with air. Complete combustion produces carbon dioxide and water, while incomplete combustion also produces carbon monoxide and/or soot. Although the reactivity of A. is low, relative to many other classes of compounds, most will undergo radical reactions under appropriate conditions, and these are utilized industrially. The branched A. are generally more reactive than *n*-A.

Natural occurrence. A. are found in natural gas and petroleum deposits, and their extraction and use are major contributors to the world economy. Petrochemical methods are used to produce both pure A. and various fractions of A. mixtures.

Production. A. are usually produced from natural gas, petroleum or, especially in countries with large coal deposits, by liquefaction of coal. The following methods of synthesis are suitable for laboratory purposes:

1) Catalytic hydrogenation of alkenes and alkynes:

$$R^1{-}C{\equiv}C{-}R^2 + H_2 \longrightarrow R^1{-}CH{=}CH{-}R^2 \xrightarrow{+H_2} R^1{-}CH_2{-}CH_2{-}R^2$$

Platinum, palladium or finely divided nickel can be used as catalyst.

2) Reduction of alcohols or alkyl iodides with hydrogen iodide: $R\text{-}OH + HI \rightarrow R\text{-}I + H_2O$; $R\text{-}I + HI \rightarrow R\text{-}H + I_2$. It is convenient to carry out the reaction with a mixture of hydrogen iodide and red phosphorus, so that the hydrogen iodide is regenerated: $2\,P + 3\,I_2 \rightarrow 2\,PI_3$; $2\,PI_3 + 6\,H_2O \rightarrow 2\,P(OH)_3 + 6\,HI$.

3) Hydrolysis of Grignard compounds: $R\text{-}X + Mg \rightarrow RMgX$ (X = Cl, Br, I); $RMgX + H_2O \rightarrow R\text{-}H + Mg(OH)X$.

4) Reaction of alkyl halides with sodium (see Wurtz reaction).

5) Electrolysis of the alkali salts of carboxylic acids in aqueous solutions (see Kolbe synthesis), which, like the Wurtz reaction, is suitable for synthesis of long-chain A.

6) Aldehydes and ketones can be converted to A. by the Wolff-Kishner reduction (see) of hydrazone, or by direct reduction with zinc amalgam (see Clemmensen reduction).

The typical reactions of A. are radical substitutions. For example, at high temperatures and in the presence of UV light, chlorine can react with methane in the gas phase to produce chloromethane, dichloromethane, chloroform and carbon tetrachloride:

$$CH_4 + Cl_2 \rightarrow CH_3Cl/CH_2Cl_2/CHCl_3/CCl_4.$$

By suitable choice of reaction conditions and reactant concentrations, the relative yields of the products can be manipulated as desired; they can also be separated by fractional distillation.

Under similar conditions, a mixture of chlorine and sulfur dioxide produces sulfonyl chlorides:

$$Cl_2 \rightarrow 2\,Cl\cdot$$
$$R\text{-}H + Cl\cdot \rightarrow R\cdot + H\text{-}Cl$$
$$R\cdot + SO_2 \rightarrow R\text{-}SO_2\cdot$$
$$R\text{-}CO_2\cdot + Cl_2 \rightarrow R\text{-}CO_2Cl + Cl\cdot$$

Alkane sulfonyl chloride

The end products of this radical chain reaction, which is used industrially with long-chain A., are valuable intermediates for production of detergents. Sulfoxidation with sulfur dioxide and oxygen produces mixtures of alkane sulfonic acids: $2\,R\text{-}H + SO_2 + O_2 \rightarrow 2\,R\text{-}SO_3H$ (alkane sulfonic acid). Nitration of A. in the gas phase at 400 °C with nitric acid yields mixtures of isomeric nitroalkanes; partial oxidative cleavage of these gives lower nitroalkanes as well:

$$CH_3{-}CH_2{-}CH_3 \longrightarrow CH_3{-}CH_2{-}CH_2{-}NO_2 / CH_3{-}\underset{\displaystyle NO_2}{\underset{|}{CH}}{-}CH_3$$

(1-nitropropane/2-nitropropane) and $CH_3\text{-}CH_2\text{-}NO_2$/$CH_3\text{-}NO_2$ (nitroethane/nitromethane). Nitration with nitrosyl chloride, NOCl, yields nitrosoalkanes, which can be rearranged to oximes:

$$R^1R^2CH + NOCl \xrightarrow{-HCl} R^1R^2CH\text{-}NO \longrightarrow R^1R^2C{=}NOH \text{ (Oxime)}$$

Such reactions are important in the production of polyamides by subsequent Beckmann rearrangements. The oxidation of A. in excess air or oxygen produces CO_2 and H_2O; incomplete combustion also produces CO or soot. The latter reaction is carried out under controlled conditions with methane, to produce soot used as filler material for tires. Controlled oxidations can also be carried out in such a way that C-C cleavage produces lower or medium-length carboxylic acids (see Paraffin oxidation).

Recently, ionic reactions of A. have become more significant; these include rearrangements and isomerizations such as the conversion of butane to isobutane when heated with aluminum(III) chloride and alkenes. Reactions which might permit a rapid identification of A. by formation of derivatives are not known. They must therefore be characterized on the basis of their physical constants and spectral data. Mixtures are usually separated, and the individual components are identified by combination of gas chromatography and mass spectroscopy.

A. are used as heating gases, e.g. natural gas (methane), propane and butane; the liquid A. C_5 to C_{16} are used as fuel in internal combustion engines, diesel engines, furnaces and as lubricants. Because they can form explosive mixtures with air or oxygen, the A. up to about C_{10} must be handled with caution. However, their use as fuels for engines depends precisely on this property. Branched A. are especially important for production of an even combustion without residues or "knocking" of the engine (see Octane rating). Pure A. of various chain lengths, e.g. pentane or hexane, are used as solvents. The semisolid and solid mixtures of compounds containing more than 16 or 17 C atoms are used as vaseline, soft and hard paraffins, pharmaceutical preparations, paraffin packings in medicine, and for candles.

Alkane sulfonates, also ***secondary A.***, abb. ***SAS***: an economically important group of anionic Surfactants (see); a mixture of isomers and homologs R^1-$HC(SO_3Na)$-R^2 (where R^1, R^2 = C_{11} to C_{17}) formed by sulfoxidation or sulfochlorination of alkanes and subsequent neutralization or saponification of the sulfochlorides by sodium hydroxide. A. are biodegradable, and are used mainly as detergents for washing dishes, clothes and household cleaning; they are also used to some extent in personal hygiene products, in emulsion polymerization reactions and in fire extinguishers. A. are very stable to hydrolysis, and can therefore also be used in acidic or alkaline cleansers. They can also be incorporated into powdered laundry products.

Alkane sulfonic acids: see Sulfonic acids.

Alkannin, ***alkanna, alkanna red***: a dark red to yellowish red natural pigment, m.p. 149 °C. A. can be extracted from the roots of *Lawsonia alba* or *L. inermis*, or from *Alcanna tinctoria*; it was formerly used in small amounts as a textile dye. ***Alkanna paper*** or ***Böttger's paper*** is an indicator impregnated with an alcoholic solution of A.; it is turned green to blue by bases.

Alkanols: monoalcohols derived from alkanes with the general formula $C_nH_{2n+1}OH$, for example, methanol and ethanol (see Alcohols).

Alkanones: saturated aliphatic Ketones (see).

Alkazide process, same as Sulfosolvane process (see).

Alkenes, ***olefins, ethylene hydrocarbons***: unsaturated hydrocarbons, that is, those containing C-C double bonds. In the broad sense, A. include compounds with one, two or several double bonds. When two or more double bonds are adjacent to each other, they are called cumulative double bonds (C=C=C), and the compound is a Cumulene (see); when two or more double bonds alternate with single bonds (C=C-C=C), they are conjugated (see Dienes or Polyenes). Isolated double bonds are separated by two or more single bonds (C=C-C-C=C).

A. in the narrow sense, simple A., contain only one double bond. These compounds form a homologous series with the general formula C_nH_{2n}, beginning with ethene (Table). Structural and configuration isomers are possible, starting with butene.

Homologous series of alkenes

Ethene	$CH_2{=}CH_2$
Propene	$CH_2{=}CH{-}CH_3$
But-1-ene	$CH_2{=}CH{-}CH_2{-}CH_3$
But-2-ene	$CH_2{-}CH{=}CH{-}CH_3$
Isobutene, 2-methylprop-1-ene	$(CH_3)_2C{=}CH_2$
Pent-1-ene	$CH_2{=}CH{-}CH_2{-}CH_2{-}CH_3$
Hex-1-ene	$CH_2{=}CH{-}CH_2{-}CH_2{-}CH_2{-}CH_3$

The rules for naming A. are found under Nomenclature (see), sect III C.

Properties. The physical properties of A. are very similar to those of the analogous alkanes. In the homologous series, there are gaseous (C_2 to C_4), liquid (C_5 to C_{17}) and solid compounds. In pairs of E,Z-isomers, the E-diastereomer usually has a higher melting point, a lower boiling point and a lower index of refraction than the Z-diastereomer.

A. are only slightly soluble in water, but dissolve in ether and ethanol. They are combustible, and generally burn to carbon dioxide and water; under oxygen-deficient conditions, carbon monoxide and soot are also produced.

Unlike the alkanes, A. are extremely reactive. The typical reactions are *additions*, because the chemical behavior of the A. is determined mainly by the π-bond with its relatively low π-bonding energy. Since the C=C double bond is nucleophilic, electrophilic reactions occur in addition to radical additions. 1) Catalytic hydrogenation to alkanes in the presence of platinum or palladium catalysts generally occurs at room temperature and normal pressure, or at a slight overpressure (in the laboratory); in industry, elevated temperatures and higher pressures are used:

$$R{-}CH{=}CH{-}R + H_2 \xrightarrow{\text{cat.}} R{-}CH_2{-}CH_2{-}R.$$

This reaction is used both for synthesis of saturated compounds from unsaturated compounds and for quantitative determination of the number of double bonds in the molecules.

2) Addition of halogens usually occurs in the sense of a *trans*-addition, at room temperature; 1,2-adducts are formed: $R{-}CH{=}CH{-}R + Br_2 \rightarrow R{-}CH(Br){-}CH(Br){-}R$. This reaction is especially significant as a qualitative test for a C=C double bond, e.g. by decoloration of a dilute bromine solution when it is added, dropwise, to a sample solution. It is also used for quantitative determination of the number of double bonds in the molecule, e.g. in determination of the iodine number of fats (addition of iodine to the double bonds in the fat molecules). The decoloration is due to the formation of nearly colorless 1,2-adducts, consuming the bromine or iodine. An excess of iodine can also be used, and the iodine not consumed can then be determined quantitatively by iodometric titration.

3) Addition of hydrogen halides produces halogenated alkanes: $CH_2{=}CH_2 + HBr \rightarrow CH_3{-}CH_2{-}Br$. The reactivity of the A. towards hydrogen halides increases with increasing acidity: HF < HCl < HBr < HI. In unsymmetrically substituted A., the electrophilic addition occurs regioselectively according to Markovnikov's rule: the electronegative halogen atom is added to the most hydrogen-poor C atom: $CH_3{-}CH{=}CH_2 + HBr \rightarrow CH_3{-}CH(Br){-}CH_3$ (2-bromopropane; Markovnikov product). If peroxides are added to the mixture, the reaction can occur as a radical addition according the Kharasch; here the anti-Markovnikov product is favored, e.g. 1-bromopropane from propene:

$$CH_2{-}CH{=}CH_2 + HBr \longrightarrow CH_2{-}CH_2{-}CH_2{-}Br$$

(1-Bromopropane; anti-Markovnikov product).

4) Hydration, that is, addition of water to form an alcohol, is the reverse of formation of A. from alcohols (dehydration). It is catalysed by acids, follows the Markovnikov rule, and is industrially important, for example, for production of *tert.*-butanol (see Butanols). When sulfuric acid is used, it is added first, forming acidic sulfate esters; the sodium salts of these esters are used as detergents (see Surfactants). Hydrolysis of the alkyl sulfates also produces alcohols, e.g.:

$$CH_3–CH=CH_2 + H_2SO_4 \rightarrow (CH_3)_2CH–O–SO_3H \xrightarrow{+H_2O} CH_3–CH(OH)–CH_3 + H_2SO_4.$$

On the other hand, alkyl sulfates can react readily with an excess of alcohol to form ethers.

5) Hydroxylation to *cis*-1,2-diols (glycols) can be done with a dilute alkaline solution of potassium permanganate, which is bleached and precipitates as manganese dioxide. This reaction can be used to detect the A. double bond (Baeyer's test):

$$R–CH=CH–R \xrightarrow{KMnO_4/H_2O} R–CH(OH)–CH(OH)–R + MnO_2.$$

A. can also be hydroxylated on laboratory scale using osmium(VIII) oxide or the less toxic ruthenium(VIII) oxide. This reaction is an addition in the presence of water and the oxidizing agent; glycol esters of manganic, osmic or ruthenic acids are formed as intermediates and then hydrolysed:

$$R_2C=CR_2 \xrightarrow{OsO_4} \text{[cyclic osmate ester]} \longrightarrow R_2C(OH)–C(OH)R_2$$

6) Epoxidizing is an addition of oxygen to a double bond which produces an Epoxide (see):

$$CH_2=CH_2 \xrightarrow{+O(Ag_2O)} H_2C–CH_2 \text{ (bridged by O)}$$

For example, the reaction of ethene with silver oxide yields ethylene oxide (oxirane). Some other possible reagents are the per-acids, such as perbenzoic and monoperphthalic acids, or hydroperoxides in the presence of molybdenum and tungsten catalysts. Oxiranes can be converted to *trans*-1,2-diols by acid-catalysed ring opening.

7) Carbonylation is the reaction of A. with carbon monoxide and water in the presence of tetracarbonylnickel at 250 °C and 20 MPa pressure to form carboxylic acids. Acid catalysts can also be used, under different reaction conditions:

$$CH_3–CH=CH_2 + CO + H_2O \xrightarrow{cat.} (CH_3)_2CH–COOH.$$

8) Ozonization is a reaction initiated by ozone addition; it leads via a primary ozonide (1,2,3-trioxolane) to a secondary ozonide (1,2,4-trioxolane):

$$R_2C=CR_2 + O_3 \longrightarrow \text{1,2,3-Trioxolane}$$

1,2,3-Trioxolane

$$\longrightarrow [R_2C=O + R_2C^+–O–O^-] \longrightarrow \text{1,2,4-Trioxolane}$$

1,2,4-Trioxolane

Both ozonides are formed by 1,3-dipolar cycloaddition. In general, the more stable secondary ozonides are cleaved by reduction with zinc and acetic acid, or by catalytic hydrogenation, forming aldehydes and/or ketones (depending on the substitution). In this way, the aldehydes/ketones can be obtained synthetically, or the reaction can be used to determine the position of a double bond in a molecule. It must be kept in mind that ozonides tend to decompose explosively.

9) Carbene transfer is an addition of the carbene to the A. double bond, forming a cyclopropane derivative, as in a Simmons-Smith reaction. 1,1-Dichlorocarbene is easily made from chloroform, and can be added, e.g. to cyclohexene:

$$\text{cyclohexene} + CCl_2 \longrightarrow \text{7,7-dichlorobicyclo[4.1.0]heptane}$$

In addition to the simple addition reactions, A. tend to *polymerize*, that is, to undergo *polyaddition* of many molecules (monomers) to form macromolecules (polymers):

$$n\,CH_2=CH_2 \rightarrow [CH_2–CH_2]_n$$

Ethylene Polyethylene

Polyethylene, polypropylene, polyvinyl chloride (PVC), polystyrene, polyacrylonitrile, polyacrylic esters, etc., are industrially very important. The wide variety of synthetic polymers of A. and substituted A. indicate the significance and applications of simple compounds with A. double bonds.

A. also undergo an extraordinary number of substitution reactions (*allyl substitutions*) under radical conditions. For example, at 400 to 600 °C, propene and chlorine form allyl chloride: $CH_2=CH\text{-}CH_3 + Cl_2 \rightarrow CH_2=CH\text{-}CH_2\text{-}Cl + HCl$. The reaction with *N*-bromosuccinimide in the presence of peroxides or UV light (see Wohl-Ziegler reaction) is an example of an allyl substitution.

Analytical. Double bonds are detected using alkaline permanganate solution (Baeyer's test) or bromine solution in chloroform; both reagents are bleached on reaction with a double bond. A. form yellow charge-transfer complexes with tetranitromethane. Hydroxylation of the double bond, followed by cleavage of the glycol and ozonolysis, is used to elucidate structure. IR spectra of A. are characterized by strong bands in the region of C=C valence vibrations between 1620 and 1680 cm^{-1}; the C-H valence vibrations are seen between 3010 and 3095 cm^{-1}. E and Z-isomers have different C-H deformation vibrations: E at 960 to 980 cm^{-1}, Z at 650 to 720 cm^{-1}. A. absorb in the UV between 180 and 200 nm. In ^1H NMR, the signals for olefinic proteins are found at δ = 4.3 to 6 ppm, and are unique to this group. E and Z configuration can often be easily determined from the vicinal coupling constants J: J_Z = 5 to 16 Hz, while J_E = 13 to 21 Hz. The mass spectroscopic fragmentation of A. is characterized by allyl cleavage of the molecular ion; the resulting allyl cations are resonance stabilized. The position of the C=C double bond cannot be determined by mass spectroscopy.

Synthesis. Lower A. (C_2 to C_5) are isolated on the industrial scale from the refinery gases from petroleum processing. They are also made by catalytic dehydrogenation of alkanes or pyrolysis of gasoline or kerosene at 800 to 900 °C. In addition, there are many characteristic methods of synthesis for the laboratory scale or for scientific purposes: 1) dehydration of alcohols by heating with sulfuric or phosphoric acid or zinc chloride:

$$CH_3\text{–}CH_2\text{–}CH(OH)\text{–}CH_3 \xrightarrow[-H_2O]{} CH_3\text{–}CH\text{=}CH\text{–}CH_3.$$

A more efficient method is catalytic dehydration on aluminum oxide or thorium oxide catalysts in the gas phase; the competing reactions, such as formation of ethers, are largely suppressed by this method.

2) Dehydrohalogenation of haloalkanes with bases: $CH_3\text{-}CH_2\text{-}Br + KOH \rightarrow CH_2\text{=}CH_2 + KBr + H_2O$.

3) Dehalogenation of 1,2-dihaloalkanes by zinc, sodium iodide in methanol, chromium(II) salts or sodium thiosulfate in dimethylsulfoxide:

$$R\text{–}CH(Br)\text{–}CH(Br)\text{–}R \xrightarrow[-ZnBr_2]{+ Zn} R\text{–}CH\text{=}CH\text{–}R.$$

4) Introduction of a double bond by Hoffmann degradation of quarternary ammonium hydroxides.

5) Introduction of a double bond by the Cope reaction, starting from *tert.*-amine oxides.

6) Reaction of aldehydes or ketones with alkylide phosphoranes in the sense of the Wittig reaction.

7) Reductive coupling of aldehydes or ketones by treatment with lithium aluminum hydride: $2\,(R)_2C\text{=}O \rightarrow (R)_2C\text{=}C(R)_2$.

8) Pyrolysis of carboxylic acid esters at temperatures around 500 °C, leading to synchronous *cis*-elimination:

[Cyclic ester transition state: $R^1\text{–}C(=O)\text{–}O\text{–}CH_2\text{–}CH(R^2)\text{–}H$] $\xrightarrow{\Delta T}$ $R^1\text{–}C(=O)OH + R^2\text{–}CH\text{=}CH_2$

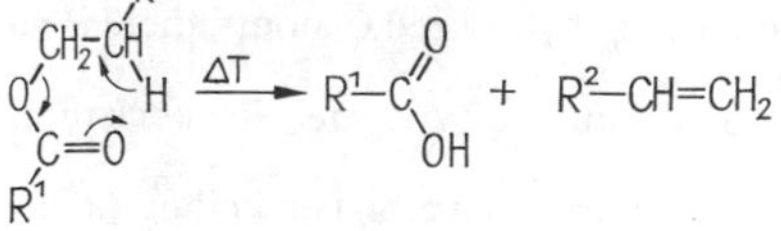

9) Pyrolysis of xanthogenic acid esters (Čugaev reaction) around 200 °C also leads to synchronous *cis*-elimination:

[Cyclic xanthate transition state: $R^1S\text{–}C(=S)\text{–}O\text{–}CH_2\text{–}CH(R^2)\text{–}H$] $\xrightarrow{\Delta T}$ $O\text{=}C(SH)(SR^1) + R^2\text{–}CH\text{=}CH_2$

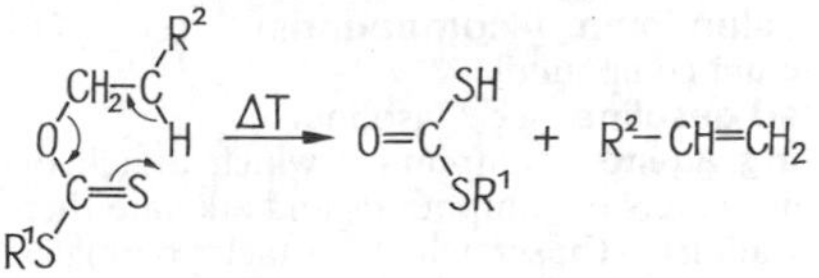

The applications of A. result from their numerous reactions. A. obtained on a large industrial scale from petroleum are used mainly to make highly branched alkanes, alcohols, ethers, epoxides, glycols, carboxylic acids, halogen alkanes, alkyl benzenes, alkyl sulfates and synthetic polymers.

Alkoxide: same as alcoholate (see).

Alkyde resins: synthetic polyester resins made by polymerization of polyvalent alcohols (e.g. glycerol, glycol or pentaerythritol) with dicarboxylic acids (e.g. phthalic, adipic, succinic, maleic acids or their anhydrides). A. with phthalic or maleic acids are called ***phthalate*** and ***malate resins***, respectively, while those consisting of glycerol and phthalic acid are called ***glyptal resins***.

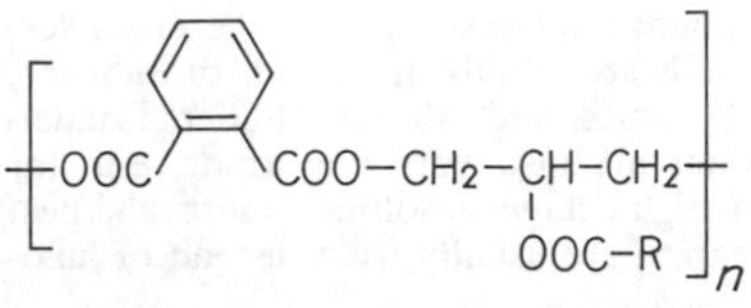

$$\left[\text{–OOC–}C_6H_4\text{–COO–CH}_2\text{–CH(OOC–R)–CH}_2\text{–}\right]_n$$

A. can be modified by incorporation of further molecular groups, for example, by esterification of the secondary hydroxyl group of glycerol; such modified A. form elastic films with good binding qualities. 1) Fatty-acid-modified A. are made from the first runnings from paraffin oxidation, the acids of tall oil or unsaponified fats which, when heated with glycerol and phthalic anhydride, are transesterified. 2) Oil-modified A. (oil alkydes) are obtained by heating phthalic anhydride and glycerol with linoleic or ricinoleic acid, or with oils such as wood or poppy-seed oil. 3) Resin-modified A. (resin alkydes) are made by incorporation of colophony or copal; abietic acid or copalic acid acts as the esterifying component. 4) Phenalkydes are A. combined with phenol derivatives, and 5) styrolized A. are combinations of styrene and A.

The large selection of polyvalent alcohols, saturated and unsaturated aliphatic and aromatic dicarboxylic acids as starting materials, and the various types of modification give a wide spectrum of variations. There are a few hardening A., but in general they are thermoplastic and soluble in organic solvents. They are used mainly as paint bases which combine well with other components; drying agents such as lead, manganese or cobalt salts of fatty or resin acids are added to the products containing A. based on unsaturated dicarboxylic acids. About 60% of the synthetic paint bases are related to A. A. are also used to make printer's inks, glues, insulation, plastics, floorings and textile treatments.

Alkyde resin varnishes: see Varnish.

Alkyl-: term indicating an atomic group C_nH_{2n+1}, either as a group R in the systematic terminology of compounds, or as an unstable free radical or carbenium ion. A. is derived from "alkane".

Alkyl aluminum compounds: see Organoaluminum compounds.

Alkylated gasoline: see Gasoline.

Alkylating agents: compounds which attack the nucleophilic centers of compounds and alkylate them (see Alkylation). Those which attack biological molecules under physiological conditions are particularly significant. The attack of N atoms in nucleotide bases in nucleic acids, especially the N7 atom of guanine, by di- and trifunctional A. leads to cross-linking of DNA. The A. include mustard gas derivatives, such as chlorambucil and cyclophosphamide, aziridine derivatives, (e.g. thio-tepa), aliphatic sulfonates (e.g. busulfane, CH_3-SO_2-O-$(CH_2)_4$-O-SO_2-CH_3) and certain oxiranes. A. are used as cytostatics and immunosuppressives; they are also mutagenic.

Alkylation: introduction of alkyl groups such as methyl (CH_3-), ethyl (C_2H_5-), etc., into an organic compound. Some well-known alkylating reagents are dialkyl sulfates and haloalkanes, which react preferentially with CH-, NH- and OH-acidic compounds in the presence of bases. For example, alkynes, 1,3-dicarbonyl compounds, amines, nitrogen heterocycles, phenols and enols are readily alkylated. In industry, alkylations of arenes and alkanes under Friedel-Crafts conditions are also very important, e.g. for production of high-octane gasolines. Here alkenes, which are cheaper, are usually used instead of haloalkanes.

Alkylbenzenes: aliphatic-aromatic hydrocarbons in which H atoms on the benzene ring are replaced by alkyl groups, as in toluene or xylene. A. with longer carbon chains are required for the production of alkylbenzene sulfonates, which are used as detergents, e.g. dodecylbenzene or tetradecylbenzene. These A. are produced by Friedel-Crafts reactions from benzene and haloalkanes with 8 to 18 carbon atoms. In technical syntheses, the alkene byproducts from petroleum refining can be used instead of the haloalkanes.

Alkylbenzene sulfonates, abb. ***ABS***: economically the most important group of anionic detergents (surface-active substances); a mixture of (R^1,R^2)-CH-C_6H_4-SO_3Na, where R^1 and R^2 = C_9 to C_{12}. They are made by sulfonation of alkylbenzenes with sulfur trioxide, followed by neutralization of the alkylbenzene sulfonic acids, usually with sodium hydroxide. A. are used mainly in powdered laundry products and liquid household cleansers.

Linear A., abb. ***LABS*** or ***LAS***, are biodegradable, in contrast to the highly branched, biologically "hard" ***tetrapropylenebenzene sulfonates*** used prior to the mid-1960's.

Alkyl halides: same as Haloalkanes (see).

Alkylidene-: a term for the atomic group R-CH= in a molecule, where R can be H or an alkyl group.

Alkylidine-: a term for the atomic group R-C≡C- in a molecule; with R = H (*ethylidine*), R = CH_3 (*propylidine*), etc.

Alkyl malonate esters: see Malonic ester syntheses.

Alkyl nitrates: R-O-NO_2, esters of alcohols and nitric acid. A. are pleasant-smelling liquids which explode when heated above their boiling points. They are obtained by reaction of alcohols with concentrated nitric or nitrating acid. Some important examples are Ethyl nitrate (see) and Glycerol trinitrate (see).

Alkyl nitrites: R-O-NO, esters of alcohols and nitrous acid. A. are made by reaction of dinitrogen trioxide with alcohols, or by reaction of alcohols with sodium nitrite as an equivalent amount of hydrochloric or sulfuric acid is added dropwise. A. are pleasant smelling liquids which are readily cleaved. Therefore, they are used instead of alkali nitrites in organic solvents to release nitrous acid (e.g. for diazotizations). The most important A. are Ethyl nitrite (see) and Isoamyl nitrite (see). A. with longer alkyl chains rearrange by a photochemical reaction into γ-nitrosoalcohols.

Alkyl sulfates: mono- and diesters of sulfuric acid.

1) ***Monoalkyl hydrogensulfates*** are most often used as the sodium salts, R-O-SO_3Na. These are white powders, readily soluble in water. A. of longer-chain alcohols, with 10 to 18 carbon atoms, are used as detergents, wetting agents and emulsifiers. Even small amounts of A. significantly reduce the surface tension of water and other solvents. They have the advantage over soaps that they form soluble compounds with Ca^{2+} and Mg^{2+} ions, and are not hydrolysed (neutral reaction). A. of primary alcohols (fatty alcohols) are made by sulfation with concentrated sulfuric acid, chlorosulfonic acid, sulfur trioxide, sulfur trioxide/pyridine or amidosulfonic acid. Secondary alcohols are sulfated with sulfuric acid and then neutralized. The surfactant properties of secondary A. are not as good as those of the primary A.

2) ***Dialkyl sulfates***, R-O-SO_2-O-R, are the neutral esters of sulfuric acid with alcohols. They are made by reaction of fuming sulfuric acid with alcohols in excess. A. are used as alkylating agents.

Alkynes: in the narrower sense, unsaturated hydrocarbons containing one C-C triple bond, with the general formula C_nH_{2n-2}. In the wider sense this class also includes hydrocarbons with several triple bonds, which can be isolated or conjugated; they are found widely in plants, e.g. in the Compositae. The A. form a homologous series of compounds (table) beginning with ethyne (acetylene).

Homologous series of alkynes

Ethyne (acetylene)	HC≡CH
Propyne	HC≡C−CH_3
But-1-yne	HC≡CH−CH_2−CH_3
But-2-yne	CH_3−C≡C−CH_3
Pent-1-yne	HC≡C−CH_2−CH_2−CH_3
Pent-2-yne	CH_3−C≡C−CH_2−CH_3
Hex-1-yne	HC≡C−CH_2−CH_2−CH_2−CH_3

Starting with butyne, structural isomerism with different positions for the triple bond become possible (but-1-yne and but-2-yne). However, since the triple bond is made up of sp-hybridized C atoms, there is no *cis-trans* isomerism.

For the nomenclature of A., see Nomenclature, sect. CIII.

Properties. Alk-1-ynes have higher boiling points than the corresponding alkanes. Ethyne, propyne and but-1-yne are gaseous at room temperature, and the higher homologs are liquid. The relatively higher boiling points are due to intramolecular attraction;

the A. have permanent dipole moments D, which in the alk-1-ynes range from 0.75 to 0.9. Compared to ethane and ethene, ethyne is more soluble in water and easier to liquefy, but liquid ethyne explodes violently when jarred or heated. Another property of ethyne and alk-1-ynes is that they are CH-acidic, i.e. that the H-atoms bound to the sp-hybridized C-atoms can be substituted by metal ions. The resulting metal compounds are called *acetylides*.

Reactions. *Addition reactions* are as typical of A. as of compounds with C-C double bonds (alkenes). Usually these reactions can be conducted in such a way that either substituted alkenes or the fully saturated alkane derivatives are formed. Triple bonds are more rapidly hydrogenated than double bonds, but in many other addition reactions, the reactivity of the A. is less than that of the alkenes. For further reactions, see Acetylene chemistry.

Analytical. The formation of copper(I) and silver(I) acetylides in ammoniacal solution can be used to detect A. with terminal triple bonds. In contrast to the alkenes, they do not react with tetranitromethane. The C≡C valence vibration appears in the IR spectrum in the range 2080 to 2280 cm^{-}. The UV absorption of non-conjugated C≡C bonds occurs below 200 nm. In ^1H NMR spectra, the signals of acetylenic protons appear between δ 2 and 3 ppm. There are often long-range couplings, that is, couplings over more than three bonds.

Production. Only the most important A., ethyne (see Ethyne, sect. production), is produced industrially. Homologous A. or cycloalkynes can be synthesized by the following methods: 1) dehydrohalogenation of 1,2- or 1,1-dihalogen compounds or halogen vinyl compounds with alcoholic potassium hydroxide solution or sodium amide:

$$R\text{-}CHCl\text{-}CHCl\text{-}R \xrightarrow{KOH} R\text{-}CHCl{=}CH\text{-}R \xrightarrow{KOH} R\text{-}C{\equiv}C\text{-}R \text{ or}$$

$$R\text{-}C(Cl)_2\text{-}CH_2\text{-}R \xrightarrow{KOH} R\text{-}C{\equiv}C\text{-}R.$$

2) Oxidation of bihydrazones of 1,2-diketones with mercury(II) oxide:

$$R\text{-}C({=}N\text{-}NH_2)\text{-}C({=}N\text{-}NH_2)\text{-}R \xrightarrow[-2H_2O/-2N_2]{+\,2O(HgO)} R\text{-}C{\equiv}C\text{-}R$$

Low-molecular-mass cycloalkynes like cyclopentyne or cyclohexyne can be formed in this way, although to date these highly strained compounds have only been demonstrated through subsequent reactions.

3) A. with terminal C-C triple bonds can be coupled to form alkadiynes by heating with copper(II) salts in pyridine (see Glaser reaction).

The applications of A. are based on their industrial availability and the many possible reactions (see Ethyne, Acetylene chemistry). Homologous A. are used for syntheses in scientific laboratories.

Allantoin, ***5-ureidohydantoin***: a degradation product of purine. It forms colorless crystals, m.p. 238.4 °C. A. is slightly soluble in water and alcohol, but dissolves readily in alkalies by forming salts. It is synthesized by oxidation of uric acid in a slightly alkaline medium, or by heating urea with dichloroacetic acid.

[Structure: H_2NCONH–CH in hydantoin ring with C=O, NH, C=O, NH (N-H)]

A. is found in the urine of many mammals (not including human beings and apes). It is formed from uric acid in the organism by the action of the enzyme uricase. The degradation of A. to ***allantoic acid*** is catalysed by the enzyme allantoinase. A. is also found in plants. It is used as an additive to cosmetics.

Allelochemicals: see Semiochemicals.

Allenes: unsaturated aliphatic hydrocarbons with two cumulative double bonds (see Dienes); the group is named for its simplest representative, ***allene (propadiene)***, $CH_2{=}C{=}CH_2$. Allene is a colorless, combustible gas; m.p. - 136 °C, b.p. - 34.5 °C, n_D^{20} 1.4168.

Typical reactions of the A. are base-catalysed isomerizations to alkynes and addition reactions similar to those of the alkenes. For example,

$$CH_2{=}C{=}CH_2 \xrightarrow{OH^-} HC{\equiv}C\text{-}CH_3$$

converts allene to propyne, while

$$CH_2{=}C{=}CH_2 + H_2O \xrightarrow{H+} CH_2{=}C(OH)\text{-}CH_3 \rightarrow CH_3\text{-}CO\text{-}CH_3$$

converts allene to acetone. Homologous A. and large carbon rings with cumulative double bonds react in the same way, so that it is possible to synthesize cycloalkynes and cycloketones of the same ring size. Of the many methods of synthesis, the following are important:

1) Dehalogenation, e.g. debromination, of 2,3-dihaloalk-1-enes with zinc:

$$R_2C{=}C(Br)\text{-}C(R)(Br)\text{-}R \xrightarrow[-ZnBr_2]{+\,Zn} R_2C{=}C{=}CR_2$$

2) Cyclopropylidene-allene rearrangement. Lithium alkylene reacts with 1,1-dichlorocyclopropane, causing the three-membered ring to open; the reaction probably has a carbene intermediate:

$$\text{1,1-dichloro-2,3-(H,R)-cyclopropane} \xrightarrow[-RCl/LiCl]{+RLi} [\text{cyclopropylidene (H,R)}] \rightarrow HRC{=}C{=}CRH$$

Since such cyclopropane derivatives are readily synthesized by dichlorocarbene addition to alkenes, this method is broadly applicable for synthesis of A.

Allethrin: see Pyrethroids.

Allomones: see Semiochemicals.

Allopurinol: a synthetic isomer of hypoxanthine which inhibits xanthine oxidase, the enzyme responsible for oxidative degradation of the purine skeleton to uric acid. In the presence of A., the danger of forming and depositing uric acid crystals (urate crystals) is reduced, because the purines are excreted as hypoxanthene, xanthine and uric acid. A. is used as a drug against gout.

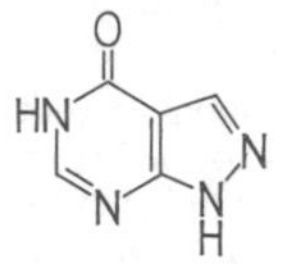

Allose: see Monosaccharides, fig.

Allotropism: see Polymorphism.

Alloxan, ***hexahydropyrimidine-2,4,5,6-tetrone***: the ureide of mesooxalic acid. A. forms yellow crystals which are readily soluble in water but insoluble in ether; m.p. 256 °C (dec.). Solutions of A. give the skin a purple-red color. A. forms a colorless hydrate, ***alloxan hydrate***. It damages the pancreas, and is therefore used in animal experiments to induce diabetes. A. is a degradation product of uric acid.

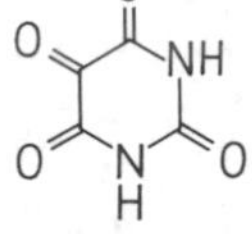

It is synthesized by oxidation of barbituric acid with chromium(V) oxide, or of uric acid with nitric acid or potassium perchlorate in hydrochloric acid solution. A. is used in organic syntheses, and in the cosmetic industry to make lipsticks and self-browning skin creams.

Alloxan 5-oxime: same as Violuric acid (see).

Alloxazine, ***2,4-dihydroxybenzopteridine***: a heterocyclic ring system which is a tautomer of ***isoalloxazine*** (flavin). The isoalloxazine structure as such can exist only if the central nitrogen in position 10 has a carbon-containing substituent instead of hydrogen. A. is a gray-green powder, and its disodium salt forms yellow crystals. It decomposes above 300 °C without melting; it is a weak acid. It is slightly soluble in alcohol, insoluble in water and ether, and readily soluble in alkali hydroxide solutions. It is made by condensation of o-phenylenediamine with alloxan. When heated with sodium hydroxide, it splits into urea and 2-hydroxyquinoline 3-carboxylic acid.

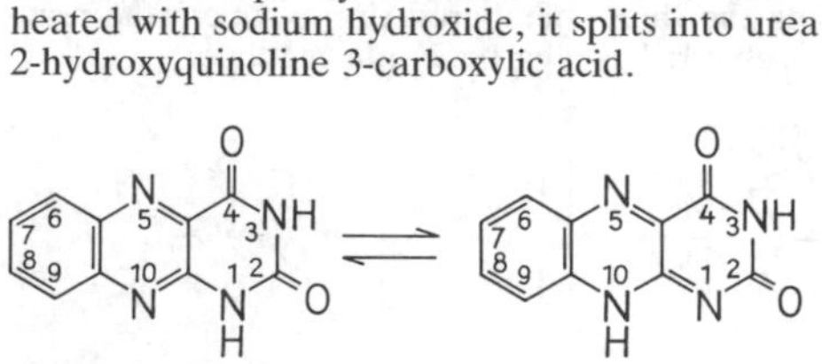

Alloxazine   Isoalloxazine

Alloy: a metallic substance consisting of at least two metallic elements, or of one metallic and one nonmetallic element. The physical and chemical properties of the A. depend on the purity, isotropism and lattice defects in the crystals. The elements which make up an A. are called components; depending on the number of components, an alloy is called a *binary, ternary,* or *quaternary* A. A distinction is made between the base metal, which is the major component of the A., and the additives. The A. is named for the base metal (e.g. Aluminum alloys (see) or Copper alloys (see)), although the additives are often named as well to characterize the A. more fully (e.g. Iron-carbon alloys (see)). Nearly all the metals used in industry are A.

There are three limiting cases of A. structure: 1) phases with the character of compounds (see Intermetallic compounds), 2) solid solutions with complete or limited miscibility of the components (substitution and inclusion mixed crystals; see Mixed crystals), and 3) mixtures of pure components.

An A. is *homogeneous (monophasic)* if the crystallites all have the same structure and composition, even if they differ in shape, size or spatial orientation. An A. is *heterogeneous* if it consists of at least two different phases, e.g. two mixed crystal types with different concentrations, or a mixture of pure components (see Eutectic mixture). If the A. is at or near thermodynamic equilibrium, the solubility relationships of its phases can be expressed as functions of the temperature and pressure in a diagram of state. All A., with the exception of congruently melting intermetallic phases or eutectic mixtures, have a freezing or melting interval.

If thermodynamic equilibrium is not demanded, but only sufficient stability for use (metastable structures), there are further possibilities for production of A. Some examples of A. with nonequilibrium structures are the amorphous metals, also called metallic glasses, and highly disperse heterogeneous structures which consist of supersaturated mixed crystals.

The physical and chemical properties of the base metal are changed by alloying; for example, its mechanical strength and resistance to corrosion can be increased, but its electrical and thermal conductivity reduced. Often even very small concentrations of alloying elements are sufficient to cause a large change in properties (e.g. microalloyed steel).

A. are usually produced by melting the components together. To accelerate the process, the elements can be combined in a ***prealloy***, which contains the element in a higher concentration. High-melting components, or those which are immiscible in the liquid state, can also be alloyed by mixing and pressing their powders together, after which they are heated and pressed (sintered) into compact objects. A sintered material made by mixing powdered components which are not miscible in the liquid state is called a ***pseudoalloy***.

The classical method of producing pieces of A. is casting of blocks, heat working the blocks by hammering or rolling, and casting the metal into a mold (***casting A.***) or stamping it (***stamping A.***). Very pure pieces which are isotropic and free of macroscopic structural defects are made by special processes, such as the production of granulates by high-speed cooling and sintering the granulate by isostatic hot presses; electric slag remelting and casting, and thermomechanical working to increase the strength of the

A. structure. Reactive, high-melting metals are remelted by means of electric arcs, electron beams or plasma beams in high vacuum.

Mixtures of polymers, that is multicomponent systems consisting of organic polymers, are sometimes called ***polymer alloys***.

All-purpose cleanser: see Household cleansers.

Alphabet acids: sulfonic acids of the naphthalene series; their trivial names are constructed from a letter and the word "acid" (Table). Some of these compounds which are important in the production of azo dyes are also named for the authors who discovered them. Different colors of azo dyes can be obtained by choosing appropriate sulfonic acids for use as coupling components.

Some important alphabet acids

Trivial name	Systematic name
Tobias acid	2-Aminonaphthalene-1-sulfonic acid
Naphthionic acid	4-Aminonaphthalene-1-sulfonic acid
Peri acid	8-Aminonaphthalene-1-sulfonic acid
Nevile-Winther-acid	4-Hydroxynaphthalene-1-sulfonic acid
L acid	5-Hydroxynaphthalene-1-sulfonic acid
Cleve acid	4-Aminonaphthalene-2-sulfonic acid
1,6-Cleve acid	5-Aminonaphthalene-2-sulfonic acid
Schäffer acid	6-Hydroxynaphthalene-2-sulfonic acid
F acid	7-Hydroxynaphthalene-2-sulfonic acid
I acid	7-Amino-4-hydroxynaphthalene-2-sulfonic acid
G acid	7-Hydroxynaphthalene-1,3-disulfonic acid
R acid	3-Hydroxynaphthalene-2,7-disulfonic acid
H acid	4-Amino-5-hydroxynaphthalene-2,7-sulfonic acid

Allyl bromide, ***3-bromopropene***: $Br\text{-}CH_2\text{-}CH{=}CH_2$, a colorless, strongly lacrimatory liquid with an unpleasant odor; m.p. 119.4 °C, b.p. 71 °C, n_D^{20} 1.4697. A. is slightly soluble in water but dissolves readily in alcohol and ether. Its reactions are similar to those of Allyl chloride (see); it can be synthesized from allyl alcohol and hydrogen bromide or by bromination of propene at 300 °C. A. is used for organic syntheses.

Allyl chloride, ***3-chloropropene***: $Cl\text{-}CH_2\text{-}CH{=}CH_2$, a colorless liquid with a pungent smell; m.p. - 134.5 °C, b.p. 45 °C, n_D^{20} 1.4157. A. is slightly soluble in water, and readily soluble in most organic solvents. Under normal conditions, it is rather stable. As a bifunctional compound, it undergoes both the reactions typical of a C=C double bond and the usual halogen substitution reactions (see Haloalkanes). Allyl alcohol, allyl isothiocyanate and allylamine can be obtained from A. by nucleophilic exchange of the chlorine. A. is synthesized by chlorination of propene at 500 °C. It is used mainly in the production of glycerol, and is also used for the synthesis of other industrially important allyl compounds.

Allyl isothiocyanate: $CH_2{=}CH\text{-}CH_2\text{-}N{=}C{=}S$, a colorless, lacrimatory liquid with an odor similar to mustard; m.p. - 80 °C, b.p. 152 °C, n_{20}^D 1.5306. A. is a strong respiratory poison; it produces blisters and oozing, poorly healing ulcers on the skin. It is slightly soluble in water, and readily soluble in alcohol, ether and benzene. When stored for long periods under normal conditions, it decomposes. A. is present in the seeds of black mustard and horseradish roots, in the form of the glycoside sinigrin. When the glucoside is cleaved by the enzyme myrosin, A., glucose and potassium hydrogensulfate can be isolated. A. can be synthesized from allyl halides and silver rhodanide, or by isomerization of allyl rhodanide; this compound can be made from allyl halides and alkali rhodanides. A. is used as a perfume, germination inhibitor and insecticide.

Allyl substitution: radical substitution of a hydrogen atom in an alkene or cycloalkene in the allyl position by another atom or group. With chlorine, A. produces allyl chlorides at high reaction temperatures. Allyl bromides are formed by reaction with *N*-bromosuccinimide in carbon tetrachloride solution; the reaction is initiated with peroxides or UV light (see Wohl-Ziegler reaction). In certain alkenes, in which a stabilized cation intermediate can form, A. occurs by an addition-elimination mechanism.

Allyl rearrangement: a rearrangement of the allyl system observed in nucleophilic substitution reactions (see Substitution, 1) of allyl compounds. An intermediate allyl cation is formed by an S_N1 reaction. Because of the partial positive charges on the terminal C atoms of the allyl system, the nucleophilic reagent can react with either of these two atoms. The reaction product is formed by the S_N1' mechanism as a mixture of two structural isomers:

$$R\text{-}CH{=}CH\text{-}CH_2\text{-}X \xrightarrow{-X^-} R\text{-}CH{=}CH\text{-}\overset{+}{C}H_2 \longleftrightarrow R\text{-}\overset{+}{C}H\text{-}CH{=}CH_2$$

$$\downarrow +Y^- \qquad\qquad \downarrow +Y^-$$

$$R\text{-}CH{=}CH\text{-}CH_2\text{-}Y \qquad R\text{-}\underset{\displaystyle Y}{CH}\text{-}CH{=}CH_2$$

Certain steric conditions on the substrate and highly nucleophilic reagents can make the allyl substitution occur by an S_N2 mechanism with A.:

$$Y^- + \underset{\displaystyle R}{CH}{=}CH\text{-}CH_2\text{-}X \longrightarrow [Y\cdots\underset{\displaystyle R}{CH}\text{=}CH\text{=}CH_2\cdots X] \xrightarrow{-X^-} Y\text{-}\underset{\displaystyle R}{CH}\text{-}CH{=}CH_2$$

The reaction then takes place as a vinyl or normal S_N2 substitution, and leads to the rearrangement product (S_N2' mechanism) and/or the normal substitution product.

Aloe emodin: see Anthraglycosides.

Aloin: see Anthraglycosides.

Aloxidation: see Anodic oxidation.

Alpaka®: see New silver.

Alternating solid-liquid phase peptide synthesis: a method for synthesis of peptides using insoluble solid and soluble carriers. The principle of the A. is that a polypeptide chain is changed in the course of the coupling reaction so that the starting material and product are easily separated. The amino acid to be coupled is linked via its amino group to an insoluble polymeric carrier by an anchor group which can be selectively cleaved; the amino component is coupled to a soluble carrier. In the course of the reaction, the polypeptide is transferred to the solid phase. The product is separated by filtration and washing; after cleavage of the amino-protective group, the peptide re-enters the liquid phase and, depending on the carboxyl protective group and the chain length of the peptide, it can be separated from excess polymer-linked amino acid by shaking, ultrafiltration, gel permeation chromatography or precipitation. The next synthetic cycle then begins with addition of the next N^a-polymer-bound amino acid. Although A. is flexible and has some methodological advantages, it has not yet been widely tested in practice. The method was used for synthesis of somatostatin on a scale of 100-500 g.

Alternative prohibition: a selection rule in vibration spectroscopy which says that for a molecule with a center of symmetry, a vibration can be active either in IR or in Raman spectroscopy, but not in both. See Infrared spectroscopy; Raman spectroscopy.

Altrose: see Monosaccharides, Fig.

Alum: a double salt of the type $M^IM^{III}(SO_4)_2 \cdot 12H_2O$, in which M^I = Na, K, Rb, Cs, NH_4 or Tl and M^{III} = Al, Ga, In, Sc, V, Cr, Mn, Fe or Co. The name A. is derived from that of the best-known representative of the group, potash alum (Potassium aluminum sulfate, see). The A. usually crystalize in well-formed octahedra or cubes. In the crystal lattice, both the monovalent and the trivalent metal ions are solvated by six water molecules each, yielding the dodecahydrate formula.

Aluminates: salts in which aluminum is part of the anion. ***Hydroxoaluminates***, with the composition $M^I[Al(OH)_4]$, $M^I_2[Al(OH)_5]$ or $M^I_3[Al(OH)_6]$, are obtained by dissolving aluminum hydroxide or aluminum in the corresponding metal hydroxide solutions. For example, $Al(OH)_3 + KOH \rightarrow K[Al(OH)_4]$. The tetrahydroxoaluminate ion can condense, via the dialuminate ion $[(OH)_3Al\text{-}O\text{-}Al(OH)_3]^{2-}$, to linear polymers; in this it resembles silicic acid. The polymers can be dehydrated at high temperatures to anhydrous A. with the composition M^IAlO_2, a compound hased on cross-linked tetrahedral AlO_2^- units. These anhydrous A. can also be obtained by fusion of metal oxides with aluminum oxides, for example, $Na_2O + Al_2O_3 \rightarrow 2\ NaAlO_2$. The corresponding derivatives of many divalent metals are found in nature as spinels, e.g. $MgAl_2O_4$ or $ZnAl_2O_4$.

Aluminizing: a process by which an aluminum protective layer is added to steel or cast iron; the layer is highly resistant to scale formation, especially in the presence of hot combustion gases. By long diffusion of aluminum into the iron at high temperature, a zone of iron-aluminum mixed crystals is formed in the iron material; above it there are brittle intermetallic layers (Al_3Fe, Al_2Fe). The outer zone of the layer is nearly pure aluminum with residues of Al_3Fe and Al_2Fe. In ***calorization***, the object is heated in an aluminum or iron-aluminum alloy powder to which ammonium chloride and clay powder have been added. At a temperature of 850 to 1050 °C, the aluminum diffuses into the surface of the iron object. In ***dip aluminizing***, the object is dipped into an aluminum melt at 700 to 800 °C. In ***spray aluminization*** or ***alumetization***, melted aluminum is sprayed onto the surface of the object (see Metal spraying). In this case, the alloy layer is formed when the object is heated to higher temperature (about 800 °C). A. can be used to protect automobile mufflers, palettes for annealing of metal parts, parts for electric heating devices and metal stovepipes.

Aluminothermal process, ***thermite process, Goldschmidt process***: a method introduced in 1894 for carbon-free production of metals which are difficult to reduce and melt at high temperatures (chromium, manganese, vanadium, cobalt, boron, silicon, iron, etc.). The metals are produced from their oxides by reduction with aluminum, e.g. $Cr_2O_3 + 2\ Al \rightarrow 2\ Cr + Al_2O_3$. The metal oxide is mixed with coarse aluminum powder, and ignited, usually with an ignition pellet of magnesium powder mixed with barium peroxide or potassium chlorate. The reaction is highly exothermic, and temperatures over 2000 °C are reached. The metal can therefore flow into a compact mass on which the slag floats. The reaction is driven by the high free energy of formation of aluminum oxide; this is a result of the high lattice energy of the oxide. The A. can also be used to weld iron parts, such as streetcar tracks by igniting a mixture of iron oxide and aluminum: $3\ Fe_3O_4 + 8\ Al \rightarrow 4\ Al_2O_3 + 9\ Fe$, $\Delta H = -2980$ kJ mol^{-1}.

The Al_2O_3 slag which is always formed is synthetic corundum, and can be used as an abrasive and for fire-resistant coatings.

Aluminum, symbol ***Al***: an element of Group IIIa of the periodic system, the Boron-aluminum group (see). Al is a light metal and an isotopically pure element, Z 13, with atomic mass 26.98154, valence III, very rarely I, density 2.702, m.p. 660.37 °C, b.p. 2467 °C, electrical conductivity 40 Sm mm^{-1} (at 0 °C), standard electrode potential (Al/Al^{3+}) - 1.706 V.

Properties. Al is a silvery white, relatively soft, ductile metal crystallizing in a cubic face-centered lattice. It is easily rolled into thin foils or drawn into wire. Its tensile strength is 70 to 120 MPa, depending on the degree of purity. A. is an excellent heat conductor, and its electrical conductivity is about 65% of that of copper. At 600 °C, it takes on a granular structure. Stirring of a supercooled melt produces coarse aluminum powder, and fine powder is produced by pounding.

Al tends to give up its three valence electrons to form colorless Al^{3+} cations. Because of their high positive charge, these are highly hydrated in water. Solid aluminum salts also often contain $[Al(H_2O)_6]^{3+}$ cations. The strong reductive activity which might be expected from the high standard electrode potential is often not seen, because a solid, firmly attached layer of oxide forms on the surface of the metal (passivation). This oxide layer can be thickened by treating the Al with an oxidizing agent, or by electrolysis (eloxal process), which significantly increases its resistance to corrosion. Al treated in this way is stable to atmospheric components, water, dilute acids and bases. It also resists concentrated nitric acid, because of the oxide layer. If the layer is destroyed, for example by amalgamation, the Al reacts rapidly with air to form aluminum oxide, or with water to form aluminum hydroxide and hydrogen. Al dissolves in strong acids or bases with evolution of hydrogen: $Al + 3\ H_3O^+ \rightarrow Al^{3+} + 3/2\ H_2 + 3\ H_2O$; $Al + OH^{-1} + 3\ H_2O \rightarrow [Al(OH)_4]^{-1} + 3/2\ H_2$. Finely divided Al burns to give aluminum oxide, and releases heat: $4\ Al + 3\ O_2 \rightarrow 2\ Al_2O_3$. The strong tendency of Al to oxidize is utilized in industry; for example, oxides dissolved in molten iron are removed by reduction with Al (deoxidation). Al is also used to prepare many metals from their oxides (see Aluminothermal process). Al combines with halogens in exothermic reactions to form the aluminum halides (AlF_3, $AlCl_3$, etc.). These compounds are interesting because of their structures, but also because of their Lewis acidity. Al reacts with sulfur and nitrogen at very high temperatures, forming aluminum sulfide, Al_2S_3 or aluminum nitride, AlN. Compounds of Al with oxidation number +1 are obtained by reduction of aluminum(III) compounds with Al. They are endothermic compounds (e.g. Al_2O, AlF), and are stable only at high temperatures.

Analytical. The qualitative identification of Al in the ammonium sulfide group is made by precipitation as aluminum oxide hydrate, or as a color lake with alizarin S. Another qualitative test is the formation of $CoAl_2O_4$ (Thénard's blue) by heating aluminum hydroxide, $Al(OH)_3$, in the presence of cobalt(II) nitrate, $Co(NO_3)_2$. Al can be determined quantitatively by precipitation of the oxide hydrate, conversion of the precipitate to aluminum oxide by heating to red heat, cooling and weighing. Complexometric determination with EDTA is less cumbersome; this is usually done as a back titration. Very small concentrations of aluminum are detected by photometric techniques with various complex formers, such as Chromazurol S.

Occurrence. A. makes up 8.1% of the earth's crust, and is thus the third most abundant element, after oxygen and silicon, and the most abundant metal, before iron. It is always found in compounds in the form of various feldspars, glimmers, clays and bauxite. The most important Al minerals are orthoclase (potassium feldspar), $K(AlSi_3O_8)$, albite (sodium feldspar), $Na(AlSi_3O_8)$, anorthite (calcium feldspar), $Ca(Al_2Si_2O_8)$, muscovite, $KAl_2(AlSi_3O_{10})(OH,F)_2$, and cryolite, Na_3AlF_6. Clays of various compositions are formed by weathering of rocks containing feldspars. Clays containing silicon dioxide and iron oxide are called loams, while clays containing calcium or magnesium carbonate are called clay marls. Kaolin, $Al_2O_3 \cdot 2SiO_2 \cdot 2H_2O$, is the starting material for mak-

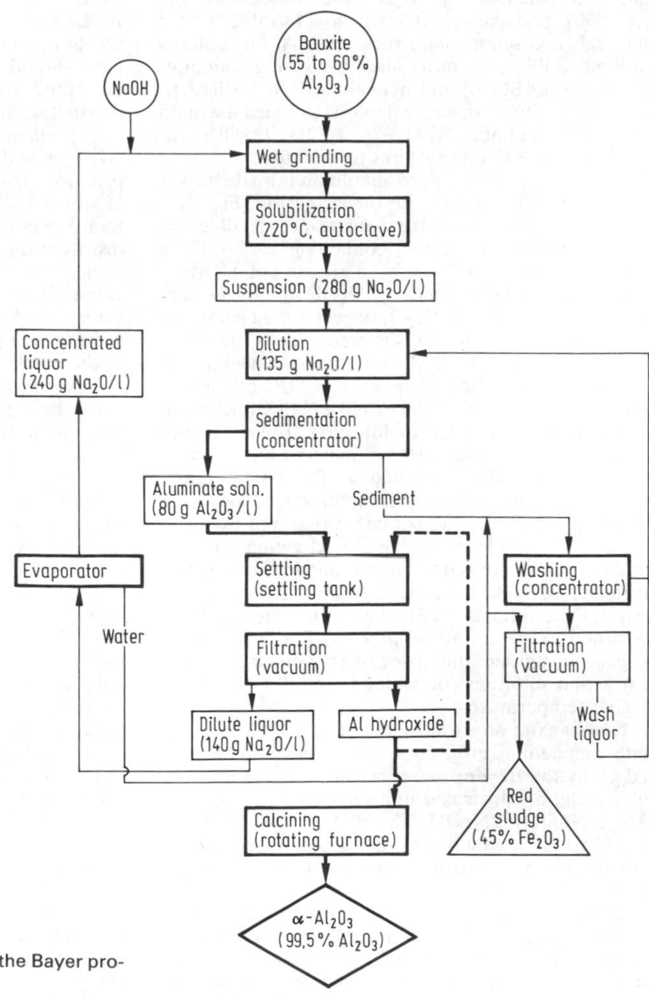

Fig. 1. Diagram of alumina production by the Bayer process.

ing porcelain. Pure aluminum oxide is found in the form of corundum; with chromium oxide impurities it is known as ruby, and when colored with titanium oxide, it is sapphire. Emery is also a form of aluminum oxide.

Al is a component of the soil and is also found in plant and animal tissues. However, it seems to be physiologically inert.

Al is extracted from oxidized ores. Aluminum oxide cannot be reduced carbothermically, because of the formation of Al_4C_3; however, Al-Si alloys can be produced in this way. Because of the high affinity of Al for oxygen, it is purified by electrolysis of a cryolite-alumina melt in which Al is the least reactive metal component. Reduction to the pure metal takes place in two steps: production of pure Al_2O_3 (aluminum oxide, alumina) and melt electrolysis of Al_2O_3 in molten cryolite (sodium fluoroaluminate), Na_3AlF_6.

1) Production of alumina. Nearly all alumina is produced from the sedimentary rock bauxite, which consists of aluminum and iron aquaoxides, calcium oxide and silicon dioxide. The Al_2O_3 content should be high (> 55%), and the SiO_2 content low (< 5%). Other minerals and sedimentary rocks with Al_2O_3 contents around 30% are now also used, e.g. alunite, $KAl_3[(OH)_6/(SO_4)_2]$ and nepheline, $KNa_3(AlSiO_4)_4$. It seems probable that clay deposits in which the main mineral is kaolinite, $Al_2O_3 \cdot 2SiO_2 \cdot 2H_2O$, will eventually become significant ores for alumina.

In most operations, pure aluminum is made by wet extraction of the ore by the *Bayer method* (Fig. 1). The Al compounds in the bauxite are dissolved in sodium hydroxide solution containing 200 to 350 g Na_2O/l at 160 to 240 °C and at a pressure of 0.8 MPa: $Al(OH)_3 + NaOH \rightleftharpoons Na^+ + [Al(OH)_4]^-$; the ratio of $Na_2O:Al(OH)_3$ is 1.7. Fe_2O_3 does not dissolve under these conditions, and SiO_2 is converted to insoluble sodium aluminosilicate (loss of NaOH and Al_2O_3). The alumina solution is cooled to 100 °C by decompression and diluted to 100 to 140 g Na_2O/l; the insoluble sludge is separated by filtration. The clear solution is further cooled to 60 °C, then hydrargillite seed crystals (γ-$Al(OH)_3$) are added. The crystallization occurs in 30 to 70 hours, in large, stirred vats. The $Al(OH)_3$ which crystallizes out is removed by filtration; 80% is used for seeding the extraction solution, and the remainder is dried in rotating or fluidized bed furnaces and calcined at 1200 to 1300 °C to obtain α-Al_2O_3. The filtered Na_2O solution is condensed and returned to the extraction process. The energetically expensive condensation process is avoided by use of extraction solutions containing only 140 g Na_2O/l, at higher temperatures.

Non-bauxite ores which are rich in SiO_2 (clays, alunite, nepheline, etc.) cannot be converted to pure Al_2O_3 by the Bayer process because of the formation of sodium hydroxyaluminosilicate, $3\ Na_2O \cdot 3Al_2O_3 \cdot 6SiO_2 \cdot 2NaOH$. The SiO_2 can be bound to CaO by basic extraction at high temperatures, or the aluminum can be extracted with mineral acids which do not dissolve the SiO_2.

In the basic *lime-sinter process*, nepheline or clay is thermally extracted with lime at 1300 °C in a rotating furnace; $(Na,K)AlO_2$ and Ca_2SiO_4 are formed. After leaching of the alkali metal aluminate from the porous sinter product with water or soda solution, the polymeric silicic acids are precipitated from the partially dissolved silicates in autoclaves at 150 to 175 °C and separated. CO_2 is then passed through the solution to precipitate the $Al(OH)_3$, which is calcined to alumina. Byproducts are alkali metal carbonates, obtained by evaporation of the mother liquor, and cement from the silicate residues.

Before it can be used for acidic extraction, kaolinite must be converted by thermal decomposition at 750 °C to acid-soluble metal kaolinite, $Al_2O_3 \cdot 2SiO_2$. The iron-containing impurities in the ore go into solution with the Al_2O_3. Although the recovery of the extraction acid and its corrosive effects on the equipment are problems, acid extraction will become more significant as time goes on, because of the shortage of bauxite. The advantages of the *sulfurous acid process* are the ready decomposition of aluminum sulfite and the circulation of SO_2; the crude aluminum hydroxide is worked up by the Bayer process. In the *sulfuric acid process*, pressurized extraction of clays does not dissolve the iron compounds; the subsequent pressure hydrolysis at 220 °C produces $H_3OAl_3(SO_4)_2(OH)_6$. In the *hydrochloric acid process*, the resulting $AlCl_3$ solution contains $FeCl_3$, which can be separated by liquid-liquid extraction. $AlCl_3 \cdot 6H_2O$ is crystallized out of the $AlCl_3$ solution and converted to alumina by thermal decomposition.

2) *Melt electrolysis*. Since melt electrolysis of pure Al_2O_3 would require a very high temperature, Al is produced from a melt of synthetic cryolite, Na_3AlF_6, in which 5% Al_2O_3 is dissolved. The electrolysis can then be done at 950 to 980 °C. The liquid cryolite dissolves the Al_2O_3 by forming a complex: $Al_2O_3 + [AlF_6]^{3-} \rightarrow 3\ [AlOF_2]^-$, and provides the current carriers (Na^+, $[AlF_6]^{3-}$, F^-). The fluoride is not consumed by the electrolysis, that is, the electrolytic decomposition produces Al at the cathode, and oxygen at the carbon anode, which reacts to form CO_2. The reaction mechanism is very complex and is still not completely elucidated; in simplified form the cathode and anode processes can be formulated as follows: $8\,AlF_3 + 12\,e \rightarrow 4\,Al + 4\,[AlF_6]^{3-}$; $2\,Al_2O_3 + 4\,[AlF_6]^{3-} + 3\,C - 12\,e \rightarrow 3\,CO_2 + 8\,AlF_3$.

Melt electrolysis is done in shallow carbon tubs. The anodes are suspended above the melt and dip into it; they consist of carbon blocks or self-combusting Söderberg electrodes (Fig. 2).

The practical decomposition voltage is 1.7 V; power consumption may be as high as 100 kW per cell. The production of 1 t Al requires 15 MWh of electricity, 1.9 t Al_2O_3 and 0.5 t carbon. The liquid Al collects on the bottom of the cell (cathode) and is sucked into a vacuum crucible. It is then either poured into ingots or remelted in large furnaces. Electrolysis Al is 99.9% pure Al.

Very pure Al, with a purity greater than 99.99%, is produced by three-layer electrolysis at 750 °C. The lowest layer is an alloy of 70% Al and 30% Cu, and serves as a liquid anode. It is covered by a layer of electrolyte melt, consisting of 60% $BaCl_2$, 24% AlF_3, 12% NaF and 4% NaCl. The top layer consists of liquid purified aluminum. The carbon cathodes are surrounded by the pure aluminum layer; they dip into the salt melt layer, where reduction to the metal occurs.

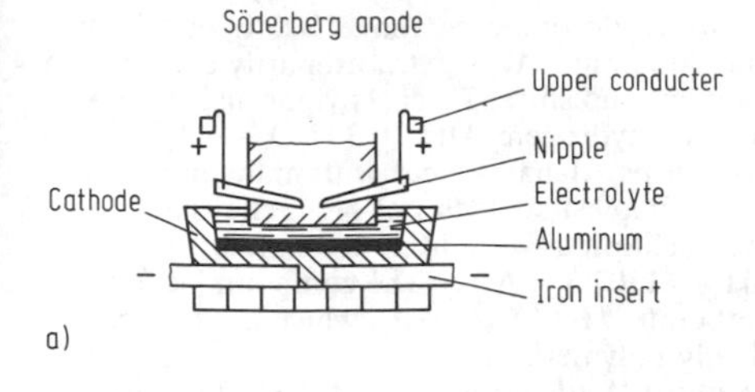

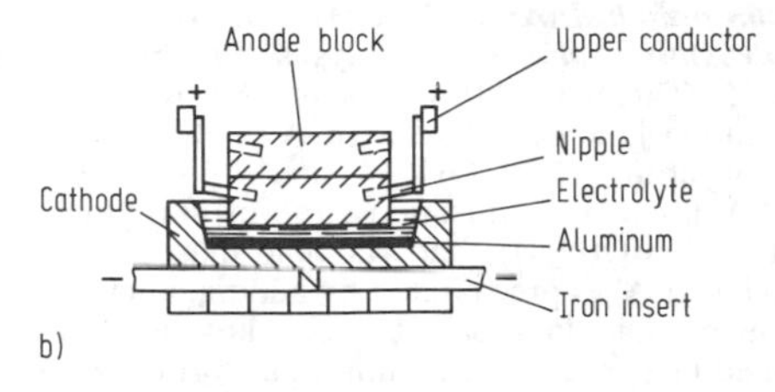

Fig. 2. Electrolysis cells for production of aluminum: *a* with Söderberg anodes; *b* with pre-fired continuous block anodes.

Recently, $AlCl_3$ melt electrolysis has been developed to the stage of industrial application. The electrolyte consists of 5% $AlCl_3$, 53% NaCl and 42% LiCl. The advantages are the low working temperature and thus the lower current consumption, and the advoidance of environmental contamination by release of fluorine-containing gases.

Applications. Its low density, favorable mechanical properties, good heat conductivity and ductility, and adequate resistance to corrosion make Al a valuable material for production of equipment, armatures, containers, etc. and for construction of vehicles and aircraft. Because of its high electrical conductivity, it is used in electrical cables and wires. Because it is not toxic, it is used to make tanks and containers for foods and beverages, cooking ware, aluminum foil, etc. Vacuum deposited layers of Al on glass make excellent reflective layers for light and heat; this is utilized, for example, in the construction of mirror telescopes. Aluminum powder is used to make rust-protective paints and for fireworks; coarse powder is used in metallurgy to produce various metals by the aluminothermal process. Al is also used as a filler and pigment. Most of the Al produced is alloyed (see Aluminum alloys).

Historical. The name A. is derived from the Latin "alumen" = "alum". This term appeared in the 5th century B.C. in the writings of Herodotus. Marggraf first produced alumina in 1754, and Oersted first produced the free metal in 1825 by reduction of the chloride with potassium amalgam. In 1827, Wöhler improved the process by reduction with potassium; he is considered the actual discoverer of Al. The first industrial process for Al production was developed by St. Claire Deville in 1854; he reduced $AlCl_3$ with the cheaper sodium. However, the high price prevented wide use of the new metal. Bunsen, Deville, Le Chatelier and others worked on methods for electrolytic production of Al, but the electrochemical process did not become practical before the invention of the dynamo by von Siemens in 1866/67. The year 1886 can be considered to have seen the birth of industrial Al electrolysis; in that year the Frenchman Heroult and the American Hall independently proposed electrolytic decomposition of aluminum oxide dissolved in cryolite. The process was developed in 1888 by Kiliani. The first hard Al alloy (duraluminum) was develped in 1906 by Wilm, and the three-layer process for very pure Al was introduced in 1919 by Hoopes.

Aluminum acetates: the aluminum salts of acetic acid. ***Aluminum triacetate***, $Al(OOCCH_3)_3$, is a neutral salt, while ***aluminum diacetate***, $HOAl(OOCCH_3)_2$, and ***aluminum monoacetate***, $(HO)_2AlOOCCH_3$, are basic. The A. are colorless solids which have not been well characterized. The triacetate is obtained by dissolving aluminum sulfate in lead or barium acetate solution, or from the reaction of aluminum hydroxide, glacial acetic acid and acetic anhydride. In aqueous solution, it forms basic aluminum acetates by hydrolysis. The diacetate can also be produced directly by reaction of sodium aluminate with acetic acid. The A. are used as mordants in dyeing. The diacetate can be used in treating cuts because of its astringent and antiseptic effect; it is called *acetic alumina*. Modern preparations contain tartaric acid as a stabilizer.

Aluminum alloys: alloys of aluminum, most often with copper, magnesium, manganese, silicon and zinc, but also containing small concentrations of nickel, lead, chromium, titanium and antimony. Iron is generally undesirable. The A. are harder than very pure aluminum, but they have lower electrical conductivity, and they are also often less resistant to corrosion. A. may be produced by casting or kneading. Thermal treatment of many A. increases their strength. This is true for ***duralumin*** (2.5 to 5.5% Cu, 0.2 to 0.5% Mg, 0.5 to 1.2% Mn and 0.2 to 1.0% Si), which has a density of 2.75 to 2.87 and is therefore used mainly for construction of vehicles and aircraft. Other important A. are ***hydronalium*** (3 to 12% Mg, 0.2 to 0.8% Mn and 0.2 to 1.0% Si), which is very resistant to seawater, and ***Silumin***® (up to 14% Si), which is used as a pressure-resistant alloy in the construction of motors and other apparatus.

Aluminum bromide:, $AlBr_3$, forms colorless rhombic crystals which fume in moist air; K_r 266.71, D. 2.64, m.p. 97.5 °C, b.p. 263.3 °C. A. is a molecular solid consisting of Al_2Br_6 molecules; it is soluble in benzene and many other organic solvents. Two Br^- ions act as bridge ligands, giving the Al atoms electron octets. A. reacts violently with water, hydrolysing the Al-Br bonds. The solution is therefore very acidic. A. crystallizes out of aqueous solutions as the hexahydrate $AlBr_3 \cdot 6H_2O$; below − 9 °C, $AlBr_3 \cdot 15H_2O$ is formed. A. is produced by passing bromine gas over a glowing mixture of aluminum oxide and carbon, or by the direct reaction of bromine and aluminum. It is used as catalyst in organic syntheses such as the Friedel-Crafts reactions, polymerizations and brominations.

Br, Br, Br / Al, Al / Br, Br, Br

Al_2Br_6 molecule

Aluminum bronze: copper-aluminum alloy containing no more than 9% aluminum. The homogeneous copper-aluminum mixed crystals have excellent resistance to heat, corrosion and scaling. Multi-component bronzes with iron and nickel in addition to aluminum are also produced. A. are used to make acid-resistant parts for the chemical and food industries, for hot-steam armatures, valve seats, sliding parts, tooth and spiral gears. Ships' screws are cast from an alloy of 9.5% aluminum, 5% nickel, 4% iron, 1.5% manganese and the rest copper.

Aluminum carbide: Al_4C_3, forms colorless, hexagonal crystals; M_r 143.96, density 2.36, dec. 1400 °C. As an ionic carbide, A. is decomposed by acids to aluminum salt solutions and methane: Al_4C_3 + 12 HCl → 4 Al^{3+} + 12 Cl^- + 3 CH_4. A. can be produced from the elements in an electric furnace.

Aluminum chloride: $AlCl_3$, forms a colorless hexagonal crystalline powder but, because of impurities, it usually appears light yellow. The crystals fume strongly in moist air and are hygroscopic. A. is soluble in many organic solvents; its M_r is 133.34, density 2.44, m.p. 190 °C at a pressure of 0.253 MPa, subl.p. 182.7 °C. In the liquid and vapor phases, and in some solvents, A. is dimeric, $Cl_2AlCl_2AlCl_2$, with four-fold coordinated aluminum in a structure analogous to that of Aluminum bromide (see). In solid A., the aluminum is 6-fold coordinated by Cl^-. Its solvation in water is strongly exothermic, involving hydrolysis of the largely covalent A. into chloride ions and hexaaquaaluminum ions: $AlCl_3$ + 6 H_2O → $[Al(H_2O)_6]^{3+}$ + 3 Cl^-. The conversion of these cations into the hydroxo compound is responsible for the very acidic reaction of A.: $[Al(H_2O)_6]^{3+}$ + H_2O → $[Al(H_2O)_5OH]^{2+}$ + H_3O^+. Hexaaquaaluminum ions are also present in the rhombic hexahydrate, $AlCl_3 \cdot 6H_2O$, which crystallizes out of aqueous solution. When heated, this does not lose water of crystallization, but eliminates HCl to become basic aluminum chloride and finally, aluminum oxide. Water-containing A. can also be produced by dissolving aluminum in hydrochloric acid. Anhydrous A. is prepared by passing chlorine over a mixture of aluminum oxide and carbon at 800 °C, according to the equation: Al_2O_3 + 3C + $3Cl_2$ → $2AlCl_3$ + 3 CO, or it can be made directly from the elements: 2 Al + 3 Cl_2 → 2 $AlCl_3$. Anhydrous A. is a strong Lewis acid used as a catalyst in organic syntheses, e.g. in Friedel-Crafts alkylations and acylations, dehydrogenations and polymerizations, and as a condensing agent and a halogen carrier. The hexahydrate is used in the soap and textile industries, in deodorants, as an antiseptic, to protect wood, etc.

Aluminum fluoride: AlF_3, a colorless, triclinic crystalline powder which is barely soluble in water and organic solvents; M_r 83.98, density 2.88, subl.p. 1291 °C. Several hydrated forms are also known. A. is an industrially important compound produced by passing hydrogen fluoride over red-hot aluminum or aluminum oxide: Al_2O_3 + 6 HF → 2 AlF_3 + 3 H_2O. A. forms complex salts of the type $M^I_3[AlF_6]$ with metal fluorides, for example the reaction of sodium fluoride produces sodium hexafluoroaluminate, $Na_3[AlF_6]$. A. is used in metallurgy as a flux and is also used in aluminum production.

Aluminum hydride, ***alane***: $(AlH_3)_x$, a colorless powder which decomposes above 100 °C into hydrogen and aluminum. A. is extraordinarily sensitive to oxidation and moisture. It reacts vigorously with water to release hydrogen: AlH_3 + 3 H_2O → $Al(OH)_3$ + 3 H_2. A. forms Alanates (see) with many metal hydrides. The simplest synthesis of A. is the reaction of aluminum chloride with lithium alanate: $AlCl_3$ + 3 $LiAlH_4$ → LiCl + 4 AlH_3. The compound is formed as the etherate, $H_3Al\text{-}C(C_2H_5)_2$, which is a monomer; it gradually polymerizes to A.

Aluminum hydroxide: There are two forms, ***aluminum orthohydroxide***, $Al(OH)_3$, and ***aluminum metahydroxide (aluminum oxygen hydroxide)***, AlO(OH). Three modifications of $Al(OH)_3$ are known, monoclinic hydrargillite (α-$Al(OH)_3$), hexagonal bayerite (β-$Al(OH)_3$) and γ-$Al(OH)_3$. AlO(OH) has two modifications, the orthorhombic boehmite (α-AlO(OH)), and rhombic diaspore (β-AlO(OH)). If A. is precipitated by adding ammonia to an aqueous aluminum salt solution, the compound is obtained in a voluminous, amorphous form known as ***aluminum oxide hydrate***. This very slowly changes, via boehmite and bayerite, to the thermodynamically stable hydrargillite. If carbon dioxide is passed through a sodium aluminate solution at 80 °C, crystalline α-$Al(OH)_3$ forms directly. However, if the precipitation is done below room temperature, or if the high-temperature precipitation is done too quickly, the first product is bayerite, which slowly converts to α-$Al(OH)_3$. If hydrargillite is heated in an autoclave to 300 °C, it is partially dehydrated to crystalline boehmite. Diaspore is made by heating boehmite in aqueous sodium hydroxide to 280 °C, under 50 MPa pressure. All forms of A. are dehydrated to aluminum oxide, Al_2O_3, by strong heating.

A. is amphoteric; it dissolves in acids to give the corresponding aluminum salt solutions, and it is converted by bases to aluminates: $Al(OH)_3$ + OH^- → $[Al(OH)_4]^-$. The reactivity depends on the form of the A.; for example, crystalline $Al(OH)_3$ dissolves very much more slowly in acids than the amorphous product. Hydrargillite and diaspore are components of natural bauxite. Bayerite and hydrargillite are important as intermediates in the production of aluminum.

Aluminum nitrate: $Al(NO_3)_3 \cdot 9H_2O$, colorless, hygroscopic, rhombic crystals, m.p. 73.5 °C, dec. 150 °C, M_r 375.13. A. is soluble in water, alcohol and acetone. It is made by dissolving aluminum hydroxide in nitric acid, and is used as a mordant in dyeing.

Aluminum nitride: AlN, blue-gray, hexagonal crystalline powder, M_r 40.99, density 3.26, m.p. in N_2 atmosphere, 2200 °C, subl. p. 2000 °C, Mohs hardness 9 to 10. Water very slowly and incompletely hydrolyses A. to aluminum hydroxide and ammonia; in sodium hydroxide solution, A. is decomposed to ammonia and aluminate solution: AlN + NaOH + 3 H_2O → NH_3 + $Na[Al(OH)_4]$. A. is made by heating aluminum oxide with carbon and nitrogen in an electric furnace: Al_2O_3 + 3 C + N_2 → 2 AlN + 3 CO.

Aluminum oxide, ***alumina***: Al_2O_3. The important modifications are the cubic γ-Al_2O_3 and the rhombic corundum (α-Al_2O_3). γ-Al_2O_3 is a colorless, loose, hygroscopic powder which does not dissolve in water, but does dissolve in strong acid or base. It is made by cautious dehydration of hydrargillite or boehmite (see Aluminum hydroxide). Above 1000 °C, it is converted into α-Al_2O_3, which is not soluble in acids or bases; M_r 101.96, density 3.97, m.p. 2015 °C, b.p. 2980 ± 60 °C.

α-Al_2O_3 is made in large amounts as an intermediate in the production of aluminum metal, and also occurs in nature as corundum. It has a Mohs hardness of 9, and is therefore used as an abrasive, in bearings for watches and analytical instruments, etc. Sintered α-Al_2O_3 (sintered corundum) is fire-resistant and is used to make laboratory crucibles and furnace cladding. Cutting tools are also made from sintered corundum. Rubies and sapphires are forms of corundum containing small amounts of Cr_2O_3 and TiO_2, respectively, as impurities. These gemstones, which are now important in laser technology, are prepared by melting the corresponding mixed metal oxides in an oxyhydrogen flame. A. forms Aluminates (see) in stoichiometric proportions with various metal oxides.

γ-Al_2O_3 is a very porous material. Its surface structure depends strongly on the temperature at which it is produced, and it is used widely as an adsorbent. It is also used as a catalyst and a carrier for other catalysts.

Aluminum oxide hydrate: see Aluminum hydroxide.

Aluminum oxygen hydroxide: see Aluminum hydroxide.

Aluminum phosphate: ***Aluminum orthophosphate***, $AlPO_4$, is a colorless, rhombic crystal powder, M_r 121.95, density 2.566, m.p. 1500 °C. It is insoluble in water, but dissolves in acids and bases. It is formed by adding sodium phosphate to an aluminum salt solution. In nature it occurs in the form of double salts, as wavellite, $Al_3(PO_4)_2(F,OH)_3 \cdot 5H_2O$, and turquoise, $Al(OH)_3 \cdot AlPO_4 \cdot H_2O$. It is used in the ceramics industry to make glazes. ***Aluminum metaphosphate***, $Al(PO_3)_3$, is a colorless, tetragonal crystalline powder; formula mass 263.90, density 2.779. It is insoluble in water and acids; its structure is based on long-chain and cyclic polyphosphate ions. It is formed by reaction of aluminum hydroxide with excess orthophosphoric acid and heating the mixture above 300 °C. It is used to prepare glazes, enamels, etc.

Aluminum silicates: crystalline silicates in which the aluminum has the cation function, in contrast to the Alumosilicates (see). The aluminum cation is octahedrally coordinated by oxygen atoms. In many clays, the aluminum is bifunctional (aluminum alumosilicates).

Aluminum sulfate: $Al_2(SO_4)_3$, a colorless powder; M_r 342.15, density 2.71. A. dissolves in water, giving an acidic reaction, and crystalizes out of this solution at room temperature as monoclinic $Al_2(SO_4)_3 \cdot 18H_2O$. It also forms hydrates with 6, 10 and 27 moles water. At 340 °C, the salt is completely dehydrated, and above 770 °C, it decomposes to aluminum oxide, Al_2O_3, and sulfur trioxide, SO_3. The hydrate is obtained by dissolving pure aluminum oxide in concentrated sulfuric acid. A. forms double salts with sulfates of monovalent metals; these have the composition $M^IAl(SO_4)_2 \cdot 12H_2O$ (see Alum). A. is used in the paper industry as sizing; it is also used as a mordant in dyeing, as a protective coating for seeds, to treat water and as a component of antifoaming agents.

Alumon: see AMO explosive.

Alumosilicates: silicates in which some of the SiO_4 tetrahedra are replaced by isomorphic AlO_4 tetrahedra. The additional positive charge required to neutralize the AlO_4 is supplied by incorporation of an equivalent number of cations (usually alkali or alkaline earth metal cations). Potassium feldspar, $KAlSi_3O_6$, is a typical A.; in it, every fourth silicon atom in the anionic network structure is replaced by an aluminum atom. The amount of substitution is limited; according to a rule of Loewenstein, AlO_4 tetrahedra can be linked only to SiO_4 tetrahedra. Other examples of A. are zeolites, ultramarine and clay minerals.

Alytensin: see Bombesin.

Am: symbol for Americium.

Amadori rearrangement: the rearrangement of glycosylamines (*N*-glycosides) into 1-amino-1-deoxy-2-ketoses. These compounds are made, for example, by reaction of aldoses with aromatic amines. The first products are glycosylamines, which rearrange on heating, in an acid- or base-catalysed step, into arylaminoketoses. A. corresponds to isomerization of the aldoses (see Monosaccharides).

Amalgamation: an enrichment step in the extraction of metals, expecially gold and silver, from their ores. The noble metal is dissolved in mercury, which has the capacity to form amalgams with many metals (see Mercury alloys). Gold or silver can be recovered from the amalgam by distilling off the mercury. A. can only be used with gold particles above a certain minimum size, because smaller particles are not sufficiently wetted by the mercury. A. has been largely replaced by Cyanide leaching (see).

Amanin: see Amatoxins.

Amanitins: see Amatoxins.

Amantadine: 1-aminoadamantane, a cycloalkylamine which was developed as a virostatic for prevention of flu. Other cycloalkylamines and 1-aminoadamantane derivatives with substituents on the N atom also have antiviral effects. The main use of A. at present is in the therapy of Parkinson's disease.

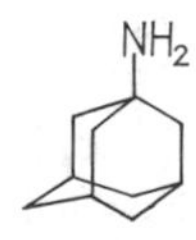

Amaranth, ***Red no. 2***: a dark, reddish-brown powder, soluble in water and very slightly soluble in alcohol. It is made by diazotization of naphthionic acid and development with naphth-2-ol-3,6-disulfonic acid. It is used to dye wool and silk bright bluish-red from an acid bath, as an indicator in hydrazine titrations, and in color photography. It has been banned by the FDA for use in foods, drugs or cosmetics. Formula, see Fast red (p. 400), where R, R_3 and R_6: $-SO_3Na$ and R_2: -OH.

Amatoxins: bridged, heterodetic, cyclic octapeptides which, together with the phallotoxins, are the most important toxins in the death cap mushroom *Amanita phalloides*. The A. include α-, β- and γ-amanitin and amanin (Fig.) and are found in high concentrations (0.2 to 0.4 mg/g fresh weight) in *Amanita phalloides*, *A. virosa* and in other mushrooms, including some in the genus *Galerina* and *Lepiota*. At a concentration of 10^{-8} M, A. inhibit the DNA-dependent RNA polymerase. By blocking protein biosynthesis at the level of transcription, they cause necrosis of a large fraction of the liver cells. Although the effects of A. probably begin within half an hour, the liver cells do not die until the second or third day after the poison has been consumed. Over 90% of the fatal mushroom poisonings are due to the A. The toxic effects of α-amanitin can be reversed, under special conditions, by antamanide, which is also found in the death cap fungus. The LD_{50} for the A. are about 0.5 mg/kg in the mouse. The structures and syntheses of the A. were reported by Th. Wieland et al.

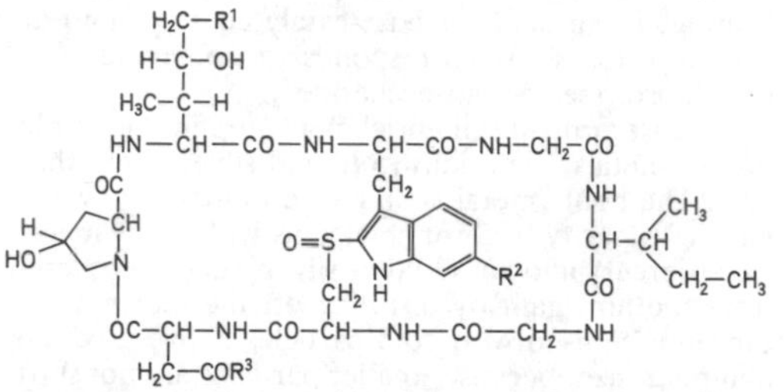

Amatoxin	R^1	R^2	R^3
α-Amanitin	–OH	–OH	$-NH_2$
β-Amanitin	–OH	–OH	–OH
γ-Amanitin	–H	–OH	$-NH_2$
Amanin	–OH	–H	–OH

Ambazone: 1-amidinohydrazono-4-thiosemicarbazono-2,5-cyclohexadiene; a compound used as an oral disinfectant.

Amber, ***succinite***: a fossil resin which is a mixture of different compounds. Its average composition is 78% carbon, 10% hydrogen, 11% oxygen, 0.4% sulfur, etc. A. is transparent to translucent, and its color ranges from light yellow to orange and yellowish red, brown, yellowish white, sometimes striped, and also skyblue to dark blue. Buried A. is covered by a weathered layer which is often lacking from A. which has been recovered from water. A. is amorphous and has conchoidal breaks; it is brittle and has a Mohs hardness of 2 to almost 3. Its density is 1.050 to 1.096. A. burns with an aromatic odor (incense odor) and it is an excellent insulator for electric current. Rubbed A. acquires a negative charge. It melts around 375 °C, and it dissolves only partially in alcohol, ether, chloroform, terpentine oil, etc. When it is dry distilled, oil of amber and succinic acid are released as volatiles; the residue is A. colophony, which has a green fluorescence. A. is not dissolved by hydrofluoric acid or alkalies.

A. is the resin of Lower Tertiary pines and firs (mainly *Pinus succinifera*) which was washed into the sea in the early Oligocene and deposited together with clay, sand and gravel as "blue earth". Insects, spiders, feathers and parts of plants are often preserved in the resin, which was originally soft. It is thrown up along the entire coastline of the Baltic and North Seas.

A. is used mainly to make ornaments and objets d'art. Oil of A., a steam-distillable oil obtained by distillation of A. scraps, is used in paints.

Ambident ligand: see Coordination chemistry.

Ambidence: the presence of two or more centers (atoms) in a reagent where reactions can take place. Some examples of ***ambident*** or ***ambifunctional reagents*** are, the nucleophilic nitrite, cyanide, phenolate and enolate ions. Nucleophilic substitution reactions with these reagents often produce mixtures of products. Primary and secondary alkyl halides react with ambident nucleophiles by an S_N2 mechanism (see Substitution, 1), preferentially at the center of higher polarizability or lower electron density (center of greatest nucleophilicity). Tertiary alkyl halides, however, react by an S_N1 mechanism. The intermediate alkyl cation attacks the nucleophilic reagent on the atom with the highest electron density or lowest polarizability (see Kornblum rule).

A1 mechanism: the rate-determining step in S_N1 reactions (see Substitution 1.1). A protonated intermediate is cleaved in a monomolecular step:

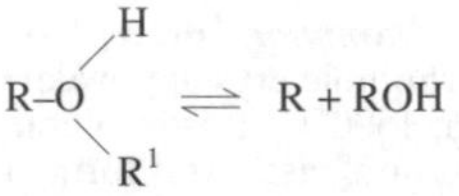

$$R{-}O(H)(R^1) \rightleftharpoons R + ROH$$

Because they are highly nucleophilic, it is not possible for hydroxy or alkoxy groups to be split off as the anions OH^- or RO^- in substitution and elimination reactions.

A2 mechanism: bimolecular cleavage of a protonated intermediate involving a nucleophilic reagent. Especially in S_N2 reactions (see Substitution 1.2), the leaving tendency of the charged group of the oxonium ion may be too low to permit a monomolecular reaction step; however, the reaction does occur with the help of a nucleophile.

Ameletin: see Scotophobin.

Ameliotex: see Synthetic fibers.

Americium, symbol ***Am***: a radioactive element which is available only through nuclear reactions; it is a member of the actinide series in the periodic table (see Actinides). Am is a heavy metal, atomic number 95. The known isotopes have the following mass numbers (in parentheses the decay type, half-life and nuclear isomers): 232 (spontaneous decay; 1.4 min), 234 (spontaneous decay, 2.6 min), 237 (K-capture; α; 75 min; 2), 238 (K, α; 1.6 h; 2), 239 (K, α; 11.9 h; 2), 240 (K, α; 50.8 h; 2), 241 (α; 433 a; 2), 242 ($β^-$, K; 16.01 h; 3), 243 (α; 7400 a; 2) 244 ($β^-$, K; 26 min; 3) 245 ($β^-$; 2.05 h; 2); 246 ($β^-$; 25 min; 3), 247 ($β^-$; 22 min). The atomic mass of the most stable isotope is 243. The valence is III, sometimes also II, IV, V, VI, VII; density 13.671, m.p. 1173 °C, b.p. 2610 °C, standard electrode potential (Am/Am^{3+}) - 2.320 V.

Am is a silvery white, very soft metal which occurs in two modifications. Its chemistry is very similar to that of plutonium; it can be obtained in the metallic state by reduction of americium (III) fluoride with barium at 1100 to 1200 °C or by reduction of americium(IV) oxide with lanthanum. During enrichment of Am, lanthanides are used as carrier substances; Am can be separated from them by fractional fluoride precipitation, because americium(III) fluoride is somewhat more soluble than the lanthanide fluorides.

Am was discovered in 1944 by the Americans Seaborg, James, Morgan and Ghiorso in the form of the isotope ^{241}Am; it was the product of neutron irradiation of plutonium 239 in a nuclear reactor:

$$^{239}_{94}Pu \xrightarrow[-\gamma]{+n} {}^{240}_{94}Pu \xrightarrow[-\gamma]{+n} {}^{241}_{94}Pu \xrightarrow[(14.89a)]{-\beta^-} {}^{241}_{95}Am$$

Americium 241 is now available on the 100-g scale. As a β-emitter, it is converted into the long-lived neptunium 241. The most stable americium isotope, ^{243}Am, is also obtained by neutron bombardment of plutonium 239; it is now available on the 10-g scale.

Am is determined radiometrically, spectrophotometrically or gravimetrically by precipitation as americium(III) hydroxide or oxalate, then calcination to the dioxide, AmO_2, above 450 °C.

Am was named as a parallel to the lanthanide with the comparable electron configuration, europium.

Americium compounds: compounds in which americium is most often in the +3 oxidation state, but is also frequently in the +2 state; in this respect it is similar to the homologous lanthanide, europium. Compounds in which americium is in the +4, +5, +6 or +7 state are similar to the corresponding plutonium compounds. Americium(II) compounds can be obtained from americium(III) compounds by reduction with sodium amalgam. ***Americium(II) sulfate***, $AmSO_4$, like $EuSO_4$, is barely soluble in water. When americium(III) nitrate or oxalate is heated in an oxygen atmosphere to 700 to 800 °C, ***americium(IV) oxide***, AmO_2, is formed. This is a black compound which crystallizes in a fluorite lattice; M_r 275.13, density 11.68. When it is heated with tetrachloromethane, at temperatures between 800 and 900 °C, ***americium(III) chloride***, $AmCl_3$ is formed; this forms pink, hexagonal crystals which sublime at 850 °C; M_r 349.49, density 5.78. ***Americium(III) bromide***, $AmBr_3$ (pink, orthorhombic crystals, M_r 482.86), and ***americium(III) iodide***, AmI_3 (yellow, orthorhombic crystals, M_r 623.84, density 6.9) can be made by reaction of AmO_2 with the corresponding aluminum halides. ***Americium(III) fluoride***, AmF_3 (light pink, hexagonal crystals, M_r 300.12, density 9.53), is obtained from the reaction of hydrofluoric acid with AmO_2 at 650 °C. Americium(III) halides are isotypes of the trihalides of plutonium and neptunium. ***Americium(III) oxide***, Am_2O_3 (red-orange cubic or hexagonal crystals, M_r 534.26), is formed by reduction of AmO_2 with hydrogen at 600 °C. High-temperature reactions of americium(III) compounds with fluorine lead to yellow-brown ***americium(IV) fluoride***, M_r 318.89. This compound is an isotype of other actinide(IV) fluorides AnF_4, where An = U, Np, Pu, Cm. AmF_4 reacts with alkali metal fluorides to form fluoro complexes of the type $[AmF_5]^-$, $[AmF_6]^{2-}$ and $[AmF_8]^{4-}$.

Strong oxidizing agents, such as peroxodisulfate, convert americium to the +5 and +6 oxidation states. The resulting linear cations, AmO_2^+ and AmO_2^{2+}, are yellow-brown and red-brown, respectively. Americium(V) and americium(VI) compounds tend to disproportionate. For example, in acidic solution, americium(V) is observed to disproportionate into americium(III) and americium(VI) according to the equation $3\,AmO_2^+ + 4\,H^+ \rightleftharpoons Am^{3+} + 2\,AmO_2^{2+}$. In strongly alkaline solution, americium(VI) compounds disproportionate to americium(V) and the very unstable americium(VII) compounds: $2\,AmO_2(OH)_2 + 3\,OH^- \rightleftharpoons AmO_2OH + AmO_5^{3-} + 3\,H_2O$.

The americium(VI) complex $Na[AmO_2(CH_3COO)_3]$ is obtained by oxidation of americium(III) with peroxodisulfate in acidic solution in the presence of sodium acetate.

Amiben: see Herbicides.

Amide: 1) a derivative of an inorganic (see Cyanamide) or organic acid in which an OH group is replaced by an amino group. The H atoms of the amino group can be replaced by other groups R^1 and R^2. Most A. have trivial names derived from the trivial names of the acyl group, in which the -yl ending is replaced by -amide, as in acetamide. The systematic name is based on that of the acid, with the ending "-ic acid" replaced by "-amide". With the exception of the liquid formamide, A. are crystalline solids at room temperature. They are hydrolysed by aqueous acids or bases, releasing the acid. A. are amphoteric, and form salts with strong acids and bases. They display Amide-iminol tautomerism (see) and can be reduced with strong reducing agents (Na/ethanol, $LiAlH_4$). Amines are also formed by catalytic hydrogenation of A. Hofmann degradation (see) or degradation with lead(IV) acetate produces the amine with one C atom fewer. Organic A. are synthesized by heating ammonium salts of the carboxylic acids, or the carboxylic acids themselves, with urea; by reaction of carboxylic acid anhydrides, halides or esters with ammonia or primary or secondary amines; by partial hydrolysis of nitriles, by Beckmann rearrangement (see), the Gattermann-Hopff reaction (see), the Willgerodt-Kindler reaction or by the reaction of arenes with isocyanates. When an A. is formed between the carboxyl group of an amino acid and the amino group of a second amino acid, the result is called a Peptide bond (see). Polyamides are made by reaction of di- or polycarboxylic acids with di- or polyamines. Thioamides are formed from thiocarboxylic acids, amidines from imidoesters and ammonia, and sulfonamides from sulfonic acids and amines. Reactions with aniline yield the corresponding anilides. Replacement of a hydrogen atom in the NH_3 molecule by a metal atom gives a metal amide (e.g. sodium amide). Formamide and *N,N*-dimethylformamide are common solvents; the latter is used as a formylating reagent in the Vilsmeier-Haack reaction (see). Urea is the diamide of carbonic acid.

2) A compound formed by substitution of a metal atom for an H atom in ammonia, with the general formula M^INH_2; an example is sodium amide.

Amide degradation: see Hoffmann degradation.

Amide hydrazone: see Amidrazone.

Amide-iminol tautomerism: a tautomeric equilibrium between an acid amide and a non-isolable iminol form:

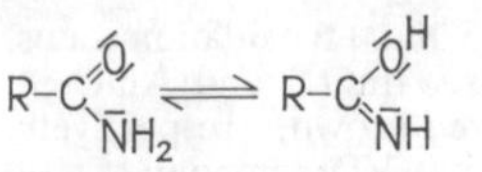

$$R{-}C(=O){-}NH_2 \rightleftharpoons R{-}C(-OH){=}NH$$

Amide form Iminol form

The equilibrium lies far on the amide side, but derivatives of the iminol form are known: imide halides and imidoesters. A special case of A. is Lactam-lactim tautomerism.

Amide resins, ***amino resins***, ***amino plastics***: polycondensation products made from amines and aldehydes, usually formaldehydes. The A. are classified according to the amine used as 1) protein-formaldehyde plastics, 2) aniline resins, 3) urea resins, 4) melamine resins and 5) dicyanodiamide resins. The production of protein-formaldehyde plastics, such as artificial horn, and of aniline resins has declined compared to other A. The urea and Melamine resins (see) are now the most important A., but the Dicyanodiamide resins (see), which were first developed in the 1960s, are growing in importance because they are more temperature-resistant than urea resins. A. are generally molded to make various articles; they are also used as molded sheets, glues and foams, textile additives, bonders for plastic parts and as paint bases.

Amidines: nitrogen-containing derivatives of carboxylic acids with the general formula R-C(=NH)-NH_2. The H atoms of the strongly basic amidine or guanyl group may be replaced by aliphatic, aromatic or heterocyclic groups. A. are readily hydrolysed in the presence of water, forming carboxylic acid amides and ammonia or amines. Their salts are much more stable. A. can be synthesized by the reaction of ammonia or amines and imino esters: $R{-}C(=NH){-}OR^1 + NH_3 \rightarrow R{-}C(=NH){-}NH_2 + R^1OH$. Unsubstituted A. are important starting materials for production of nitrogen-containing heterocycles, e.g. imidazoles and pyrimidines, which are used in the manufacture of drugs.

Amidol, ***2,4-diaminophenyldihydrochloride***: colorless to gray crystals, sinters at 227 °C, m.p. 230-240 °C (dec.). A. is very soluble in water. It is synthesized by reduction of 2,4-dinitrophenol to 2,4-diaminophenol, a compound with leaflet crystals, m.p. 78-80 °C (dec.). A. is a high-quality, non-streaking photographic developer. It is also used to dye furs and hair.

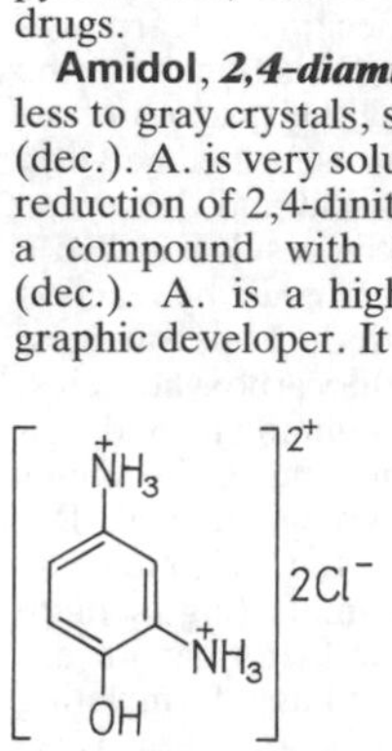

$$[C_6H_3(OH)(\overset{+}{N}H_3)_2]^{2+}\ 2Cl^-$$

Amidosulfonic acid, formerly ***sulfaminic acid***: $H_2N{-}SO_3H$, colorless, orthorhombic crystals, M_r 97.09, density 2.126, m.p. 200 °C (dec.). A. is soluble in water and gives an acidic reaction. It is synthesized by reaction of chlorosulfuric acid or sulfur trioxide and ammonia, or by reaction of urea with sulfuric acid or oleum. A. is used in the textile industry, in galvanizing, in pesticides and as a reagent to detect and destroy nitrites. Its acid properties are utilized in the removal of calcium deposits and as a component of bubbling bath salts, etc.

Amido trizoate: see X-ray contrast media.

Amido yellow E.: see Nitro dyes.

Amidrazone: a term for a nitrogen-containing carboxylic acid derivative with the general formula R-C(=N)-NH-NH_2 or R-C(=N-NH_2)-NH_2. Depending on the position of the double bond, these compounds are also called ***hydrazide imides*** or ***amide hydrazones***. A. are formed by reaction of hydrazine and imino esters: $R{-}C(=NH){-}OR^1 + H_2N{-}NH_2 \rightarrow R{-}C(=NH){-}NH{-}NH_2 + R^1OH$.

Amikacin: see Kanamycins.

Aminals: organic compounds with the general formula $R{-}CH(NR_2)_2$. They are formed by the reaction of an aldehyde, in which there is no α-hydrogen atom, with a secondary amine; water is split off:

Amination: a reaction which introduces an amine group, $-NH_2$, -NHR or $-NR_2$, into a compound. A. is essentially a substitution reaction in which compounds with reactive halogen atoms are converted to the corresponding substituted amino derivative by reaction with ammonia or an amine (see Aminolysis). In some cases, hydroxy or sulfonic acid groups can be replaced by amino groups. Some other possibilities for A. are offered by the Čičibabin reaction for pyridines and the addition of primary and secondary amines to alkynes.

Amine: an organic compound in which, formally, one or more of the hydrogen atoms in ammonia is replaced by a carbon atom. The substituents can be aliphatic or aromatic hydrocarbons or heterocycles. Primary A. have one substituted hydrogen atom, R-NH_2, secondary A. have two, R_2NH, and tertiary A. have three, R_3N. The substituents on the nitrogen atom can be identical or different. There are also compounds with quaternary nitrogen atoms, the quaternary ammonium hydroxides $(R_4N^+)OH^-$, and their salts, $(R_4N^+)X^-$, and simple ammonium salts, which can have one, two or three substituents (e.g. $(R_3HN^+)X^-$).

The number of amino groups on a molecule is indicated by a prefix such as mono-, di-, tri-, etc., as in "diamine".

The name of an aliphatic monoamine is derived from the name of the hydrocarbon group to which the N atom is bound, followed by the suffix "-amine". Examples are methylamine, $CH_3{-}NH_2$, ethylamine, $CH_3{-}CH_2{-}NH_2$, dimethylamine, $CH_3{-}NH{-}CH_3$, trimethylamine, $(CH_3)_3N$, or benzylamine, $C_6H_5{-}CH_2{-}NH_2$. With primary aliphatic amines, it is also possible to construct the name from the root for the hydrocarbon and the prefix "amino-", as in aminoethane, $CH_3{-}CH_2{-}NH_2$. Similarly, di- and triamines and arylamines are considered amino-substituted hydrocarbons, for example, 1,3-diaminopropane, $NH_2{-}(CH_2)_3{-}NH_2$, or aminobenzene. There are also a number of trivial names which are either permitted by IUPAC or are in common use, e.g. aniline, ethylenediamine, putrescine and cadaverine.

Properties. The first two members of the homologous series, methyl- and ethylamine are gaseous, combustible compounds with an ammonia-like odor. The next aliphatic A. are liquids with fishy odors, while those with larger molecules are solids and have no odors. Arylamines are liquid or solid, have unpleasant odors, and slowly turn brown in the air, due to oxidation. The aromatic A. are highly toxic. Because of the free electron pair on the N atom, the A. are basic, that is, they are able to add protons (they are proton acceptors in the Brönsted sense): R_3N: + $HX \rightarrow R_3N^+HX^-$.

Comparison of the pK_a values of aliphatic and aromatic A. reveals distinct differences; the aliphatic A. have higher pK_a values, and are thus more basic, than the aromatic A. The reason is that the free electron pair on the nitrogen atom of an aromatic A. is drawn into its π-electron system.

> Gaseous aliphatic amines irritate the mucous membranes of the eyes and respiratory passages. When the skin is wetted with liquid alkylamines, it is severely burned. Poisoning by inhalation of higher concentrations can lead to elevation of blood pressure and temporary cramps. On the toxicity of aromatic amines, see Aniline.

Reactions. The reactions of various A. with nitrous acid are very important for many syntheses and also for characterization. Primary aliphatic A. are deaminated, forming alcohols, but primary aromatic A. form diazonium salts which can be converted to azo compounds by azo coupling around 0 °C. Secondary A. form yellow to orange nitrosoamines with nitrous acid. Tertiary aliphatic A. give unstable ammonium nitrites, and tertiary A. are converted under the same conditions to green, crystalline *p*-nitrosoarylamines. Some other important reactions of primary A. include the isonitrile reaction with chloroform and alkali hydroxide solution and the formation of azomethines with some carbonyl compounds, including aldehydes and ketones. A. can be converted to carboxylic amides by acylating reagents, and with hydrogen halides, such as hydrogen chloride, they form ionic compounds, the *amine hydrohalides* (e.g. *amine hydrochlorides*). Depending on the hydrocarbon group, these can also be called alkyl or aryl ammonium halides.

Analysis. A. can be identified chemically or by IR, NMR, or mass spectroscopy. The IR spectra of amines have absorption bands for the C-N valence vibration between 1020 and 1220 cm^{-1} for aliphatic A., and between 1250 and 1360 cm^{-1} for aromatic compounds. For primary and secondary A., the N-H valence vibrations absorb between 3300 and 3500 cm^{-1}, and the N-H deformation vibrations are between 1550 and 1650 cm^{-1}. In ^{1}H-NMR spectra, the signals for aliphatic NH groups are in the range of $\delta = 1$ to 2 ppm, and for aromatic A., between 2.6 and 4.7 ppm. A. with odd numbers of N atoms always have odd mass numbers. The occurrence of odd-numbered molecular ion peaks always indicates the presence of nitrogen. Aliphatic A. very often yield alkylamine cations as key fragments. Aromatic A. readily cleave off hydrogen cyanide.

Synthesis. Aliphatic A. are usually synthesized technically by alkylation of ammonia. The resulting mixture of amines can be separated by fractional distillation or by the Hinsberg separation (see). Some other methods of synthesis are the Gabriel synthesis (see), the Curtius degradation of acid azides (see), the Hoffman degradation of acid amides (see), the Ritter reaction (see) and the reduction of nitro compounds; this last method is the most important industrial source of primary aromatic amines.

Applications. A. are used as starting materials for the synthesis of pigments (azo and azomethine dyes), drugs (e.g. sulfonamides) and polyamide fibers.

Amine oxides, *N-oxides*: organic compounds with the general formula $R_3N \rightarrow O$. The bond between the N and O atoms is called a semipolar bond, because it is formally a combination of a single N-O covalent bond and an N^+O^- ionic bond. This type of bond can be symbolized as follows:

$$R_3N \longrightarrow \bar{O}| \quad \text{or} \quad R_3\overset{+}{N}-\bar{\underline{O}}|^-$$

A. have very high dipole moments and high melting and boiling points, and can only be distilled in vacuum. They form salts with acids, with the general formula $(R_3N^+\text{-}OH)X^-$. Some of these compounds are used as chemotherapeutics.

A. are formed by oxidation of tertiary amines, e.g. by peroxides, or by oxidation of heterocycles with tertiary N atoms, e.g. pyridine and quinoline. A. can be converted back to the tertiary nitrogen bases by reaction with suitable reducing agents, such as triphenylphosphine. A. are used mainly for organic syntheses. For example, in the Cope reaction, alkenes are produced by thermolysis of aliphatically substituted A.

Amine resins: same as Amide resins (see).

Amine hydrochloride: see Amine.

Amino acids, *amino carboxylic acids*: organic acids in which one or more of the C-atoms in the hydrocarbon portion carries an amino group, $-NH_2$. Aliphatic A. are designated, α-, β-, γ-, ω-, etc., depending on the position of the NH_2 relative to the terminal COOH group.

$$R-\underset{NH_2}{\overset{H}{C}}-COOH \qquad R-\underset{NH_2}{\overset{H}{C}}-CH_2-COOH$$

α-Amino acid β-Amino acid

If the NH_2 group is bound to a ring, the relative positions of the functional groups are indicated by the numbering of the ring atoms, e.g. 2-aminobenzoic acid. γ- and δ-A. are formed by heating cyclic lactams, for which a Lactam-lactim tautomerism (see) can be formulated. ε-A. and A. with greater distances between the NH_2 and a terminal COOH group can readily be converted to polyamides by intermolecular condensations; this is important in the synthesis of synthetic fibers, e.g. from ε-aminocaproic acid.

More than 500 A. are known to occur in nature; of these, the ***proteogenic A.*** (Table 1), which are incorporated into proteins, are of special significance. All are α-A., and with the exception of glycine, all are optically active. Only the L-configuration is found in proteins. D-A. are found in the cell walls of bacteria and various antibiotics, however.

Amino acids

Table 1. Proteogenic amino acids

Trivial name	Formula	m. p. (dec.) in °C	$[\alpha]_{25}^{D}$	IEP
Alanine Ala, A	$CH_3-CH(NH_2)-COOH$	297	+ 14.47 c = 10.0 in 6N HCl	6.00
Arginine Arg, R	$HN-CH_2-CH_2-CH_2-CH(NH_2)-COOH$, with HN–C(=NH)–$NH_2$ on the N	238	+ 27.6 c = 2 in 6N HCl	11.15
Asparagine Asn, N	$H_2N-CO-CH_2-CH(NH_2)-COOH$	236	− 5.6 c = 2 in 5N HCl	5.41
Aspartic acid Asp, D	$HOOC-CH_2-CH(NH_2)-COOH$	271	+ 24.6 c = 2 in 6N HCl	2.77
Cysteine Cys, C	$HSH_2C-CH(NH_2)COOH$	240	+ 9.7 c = 8 in 1N HCl	5.02
Glutamine Gln, Q	$H_2N-OC-CH_2-CH_2-CH(NH_2)-COOH$	185	+ 31.8 c = 2 in 5N HCl	5.65
Glutamic acid Glu, E	$HOOC-CH_2-CH_2-CH(NH_2)-COOH$	202	+ 31.8 c = 2 in 5N HCl	3.22
Glycine Gly, G	H_2N-CH_2-COOH	292	–	5.97
Histidine His, H	imidazole ring (N, NH)$-CH_2-CH(NH_2)-COOH$	287	− 39.2 c = 3.8 in water	7.47
Isoleucine Ile, I	$(H_3C-CH_2)(H_3C)CH-CH(NH_2)-COOH$	284	+ 40.6 c = 2 in 6N HCl	5.94
Leucine Leu, L	$(H_3C)_2CH-CH_2-CH(NH_2)-COOH$	315	− 10.4 c = 2 in water	5.98
Lysine Lys, L	$H_2N-CH_2-CH_2-CH_2-CH_2-CH(NH_2)-COOH$	225	+ 25.9 c = 2 in 5N HCl	9.59
Methionine Met, M	$H_3C-S-CH_2-CH_2-CH(NH_2)-COOH$	283	+ 23.4 c = 5 in 3N HCl	5.74
Phenylalanine Phe, F	$C_6H_5-CH_2-CH(NH_2)-COOH$	284	− 35.1 c = 2 in water	5.48
Proline Pro, P	pyrrolidine ring (N–H)$-COOH$	222	− 85 c = 1 in water	6.30
Serine Ser, S	$HOH_2C-CH(NH_2)-COOH$	228	− 6.8 c = 10 in water	5.68
Threonine Thr, T	$H_3C-CHOH-CH(NH_2)-COOH$	253	− 28.6 c = 2 in water	5.64
Tryptophan Trp, W	indole ring (N–H)$-CH_2-CH(NH_2)-COOH$	281	− 32.1 c = 1 in water	5.89
Tyrosine Tyr, Y	$HO-C_6H_4-CH_2-CH(NH_2)-COOH$	290	− 7.3 c = 4 in 6N HCl	5.66
Valine Val, V	$(H_3C)_2CH-CH(NH_2)-COOH$	315	+ 28.8 c = 3.4 in 6N HCl	5.96

IEP = isoelectrical point, c = concentration, N = normality

The general formula for a proteogenic A. is NH_2-CHR-COOH. The side chain, R, may be polar or nonpolar; the polar A. have acidic, neutral or basic side chains, and all are hydrophilic. The neutral, polar A. include serine, threonine, cysteine, asparagine, glutamine and tryptophan; the acidic A. are aspartic acid, glutamic acid and tyrosine; and the basic A. are lysine, arginine and histidine. The nonpolar A. are glycine and alanine; the hydrophobic A. are valine, leucine, isoleucine, proline, methionine and phenylalanine. The configurations of proteins in aqueous or nonaqueous media (e.g. cell membranes) are determined by the interactions of the A. side chains with the solvent and with each other.

A. are classified according to their metabolic fates as ***glucoplastic*** or ***ketoplastic***. Glucoplastic A. are degraded to C_4 dicarboxylic acids or succinic acid, and can be converted to carboydrates, while ketoplastic A. are degraded to ketone bodies, especially acetoacetate. A different system of classification is based on the ability of human or animal bodies to synthesize the proteogenic A.; ***essential A.*** are those which cannot be synthesized and are therefore required in the diet in the form of suitable Proteins (see) (Table 2). The requirements for essential A. depend on age and physiological state; growing and pregnant animals require more than non-pregnant adults. Ruminants obtain A. from the bacteria in their rumens, and can exist without other dietary protein if they have an adequate source of inorganic nitrogen (ammonium salts or urea).

Table 2. Human minimum daily requirement for essential amino acids

Amino acid (in g)		Amino acid in mg/kg body mass		
		Adults	Infants	Children (10–12 years)
Arg	1.8	.	.	.
His	0.9	.	28	.
Ile	0.7	10	70	30
Leu	1.1	14	161	45
Lys	0.8	12	103	60
Met	1.1	13	58	27
Phe	1.1	14	125	27
Thr	0.5	7	87	35
Trp	0.25	3.5	17	4
Val	0.80	10	93	33

Properties and reactions. The A., having NH_2 and COOH groups, are amphoteric, and their solutions are ampholytes. In the solid state and in solution, the A. have dipolar zwitterion structures, H_3N^+-CHR-COO^-, which accounts for their good solubility in polar solvents such as water and ammonia, and for their high melting points (200-300 °C). A. react with acids or bases by uptake or loss of protons:

$$H_3N^+\text{–CHR–COOH} \underset{H^+}{\overset{OH^-}{\rightleftharpoons}} H_3N^+\text{–CHR–COO}^-$$

Cation Zwitterion

$$\underset{H+}{\overset{OH^-}{\rightleftharpoons}} H_2N\text{–CHR–COO}^-,$$

Anion

so the A. exist in acidic solutions as ammonium cations, and in alkaline solutions as carboxylate anions. In an electric field, the A. move toward the anode in an alkaline solution, and toward the cathode in acidic solution. At its isoelectric point (IEP), the pH at which it exists as a zwitterion, an A. does not migrate. At this point, the solubility of most A. in water is very low, because the hydrophilicity of the NH_2 and COOH groups is cancelled. This fact is useful in purification and recrystallization, and for electrophoretic separation of A. The acid-base behavior of all A. can be characterized by the pK values associated with their α-amino, α-carboxyl and side chain functions (K_1, pK_2, pK_3). The pK_1 values of the A. are lower than those of the comparable carboxylic acids; for example, the COOH group of glycine (pK_1 = 2.35) is much more acidic than that of acetic acid (pK = 4.76). Since the A. differ in their acidity, they can be conveniently separated by ion exchange chromatography. The titration curves of the A. have two regions of steep slope, in which they can be used as buffers. The acid-base behavior of the A. serves as a model for that of proteins and peptides, of which they are the components.

Chemically, the amino and carboxyl functions of the A. can react independently. Two reactions of practical significance are acylation to the very acidic acylamino acids, or esterification to amino acid esters, which act as strong organic bases.

Other characteristic reactions of the A. are N-alkylation, e.g. to N-methylamino acids, complex formation with metal ions, reaction with nitrous acid, which yields α-hydroxyacids upon N_2 cleavage, decarboxylations to Biogenic amines (see) and transaminations, in which the NH_2 group of one A. is reversibly transferred to the α-C of an α-ketoacid. The amide linkage of A. to form peptides and proteins is extremely important.

Analytical. Qualitative analysis of A. depends mainly on chromatographic separation and subsequent identification by a color reaction. The *ninhydrin reaction* produces a blue-violet pigment (absorption maximum at 570 nm) by reaction of ninhydrin (2,2-dihydroxy-IH-indene-1,3(2H)-dione) with the A. *Fluorescamine*, 4-phenyl[furan-2H(3H)-1'-phthalane]-3,3'-dione, reacts with the A. to produce highly fluorescent compounds which can be detected in the nanomolar range at 336 nm. Some other highly sensitive reagents for A. are 2,4,6-trinitrobenzenesulfonic acid, Folin's reagent (1,2-naphthoquinone-4-sulfonic acid) and 4,4'-tetramethyldiaminodiphenylmethane (TDM). Intensely fluorescent amino acid derivatives are formed with o-phthalaldehyde in the presence of reducing agents, or with dansyl chloride (5-dimethylaminonaphthalinesulfonyl chloride).

Mixtures of A. are separated by electrophoresis, paper and thin-layer chromatography; for quantitative determination, ion exchange chromatography is used. The first automated amino acid analyser developed by Spackman, Moore and Stein required 24 h for analysis of a protein hydrolysate; modern machines need only 45 min, and the limit of detection lies in the picomole range.

Production. A. can be obtained by isolation from protein hydrolysates, by biotechnological means, or by chemical synthesis. The classic method for isolation from proteins begins with hydrolysis of the protein in hydrochloric acid for 12 to 72 h; tryptophan is completely destroyed and serine/threonine are partially degraded (up to 10%) by this treatment. Cysteine and tyrosine are precipitated from the hydrolysate by neutralization, and the A. remaining in the mother liquor are separated by chromatography on synthetic resin ion exchangers or by extraction. The enzymatic hydrolysis of proteins by bacterial proteases is becoming more important.

All proteogenic A. can now be produced by biotechnological means, either by microbial production (fermentation), using mutants which produce more of a particular A. than they need, or biotransformation. The enzymes for biotransformations are usually attached to a solid carrier, and the reagents flow past them. They carry out specific steps in the synthesis, usually those for which enantiomeric selectivity is required. Glutamic acid (see) is now produced by fermentation, while the enzymatic synthesis of Lysine (see) is an example of biotransformation.

The most important methods of chemical synthesis of A. are 1) reaction of halocarboxylic acids with ammonia, e.g. for glycine; 2) the Strecker synthesis (see); and 3) ***malonic ester synthesis***: malonic ester reacts with nitrous acid to form oximinomalonic ester, which is converted to formyl- or acetaminomalonic ester by reduction and acylation of the free amino group. The side chain of the A. is introduced by reaction with aldehydes or alkyl halides. This complicated synthesis is used for only a few A., e.g. asparagine and tryptophan.

Since the chemical synthesis of A. always produces DL compounds, the enantiomers must be separated at the end of the synthesis. The industrially useful methods are: a) spontaneous crystallization of one enantiomer from the supersaturated solution of amino acid salts; b) reaction with optically active compounds to produce diasteriomeric salts, which can be fractionally crystallized; and c) the enantioselective hydrolysis of amino acid derivatives (acetyl- and chloroacetylamino acids or amino acid esters) by immobilized acylases or esterases.

The separation of enantiomers is not needed in asymmetric synthesis of A., e.g. by the Strecker synthesis with optically active amines, or by enantioselective hydrogenation of α,β-dehydroamino acids, $R\text{-}CH{=}C(NH_2)COOH$, in the presence of optically active Wilkinson catalysts. Such methods produce A. of relatively high optical purity. For example, an industrial synthesis of DOPA (3,4-dihydroxyphenylalanine) yields 97% L- and 3% D-DOPA.

The world production of A. in 1980 had passed the 500,000 ton level. The largest amount was of L-glutamic acid (340,000 t), followed by DL-methionine (120,000 t) and lysine (34,000).

Applications. A. are used in large amounts as flavorings in foods (monosodium glutamate, aspartic acid, cystine, alanine and glycine), as infusion solutions for intravenous feeding (all proteogenic A.), for supplementation of foods (e.g. rice, maize, wheat flour) and feeds (e.g. chicken feed), for supplementation of the diets of domestic animals (lysine, methionine, threonine), as additives to skin cremes, shampoos and other cosmetics (cysteine derivatives) and, in small amounts, as vulcanization promoters, additives for galvanic baths and corrosion inhibitors. In organic synthesis, A. are used as starting materials for synthesis of peptides, pharmaceuticals and pesticides, and, in increasing amounts, as chiral components and intermediates for synthesis of optically active compounds and nitrogen heterocycles.

Aminoanisoles: same as Ansidines (see).

Aminoanthraquinones: derivatives of anthraquinone in which one or more aromatic H atoms are replaced by amino groups, $-NH_2$. A. are colored compounds which are only slightly soluble in water, but dissolve readily in alcohol and ether. In A. with primary or secondary amino groups in positions 1, 4, 5 or 8, relatively strong hydrogen bonds are formed, as in hydroxyanthraquinones. A. are synthesized mainly by reduction of nitroanthraquinones or anthraquinone sulfonic acids with ammonia. They are used as starting materials for synthesis of dyes.

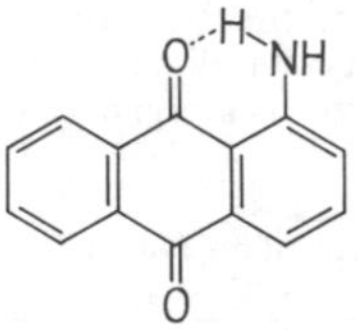

1-Aminoanthraquinone 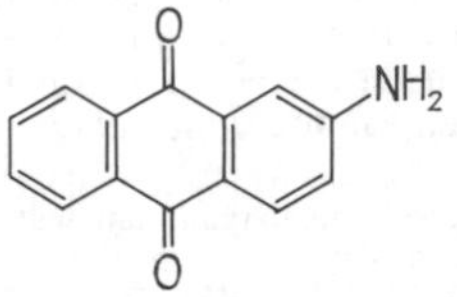   2-Aminoanthraquinone

4-Aminoazobenzene, ***aniline yellow***: yellow leaflets or needles; m.p. 127 °C, b.p. > 360 °C. A. is only slightly soluble in water, but dissolves readily in alcohol, ether, benzene or chloroform. It is synthesized by reaction of benzene diazonium salts with aniline hydrochloride and rearrangement of the primary product, diazoaminobenzene. It may also be synthesized by direct reaction of aniline hydrochloride with diazoaminobenzene. The ***diazoamino-aminoazo rearrangement*** is an intermolecular process which can be formulated as follows:

$$C_6H_5\text{–}NH\text{–}N{=}N\text{–}C_6H_5 \xrightarrow{+H^+} C_6H_5\text{–}\overset{+}{N}H_2\text{–}N{=}N\text{–}C_6H_5 \longrightarrow$$

$$C_6H_5\text{–}NH_2 + \overset{+}{N}{=}\overset{+}{N}\text{–}C_6H_5 \xrightarrow[-H^+]{} C_6H_5\text{–}N{=}N\text{–}C_6H_5\text{–}NH_2.$$

The diazonium ion can be detected by trapping reactions, and this confirms the intermolecular nature of the rearrangement.

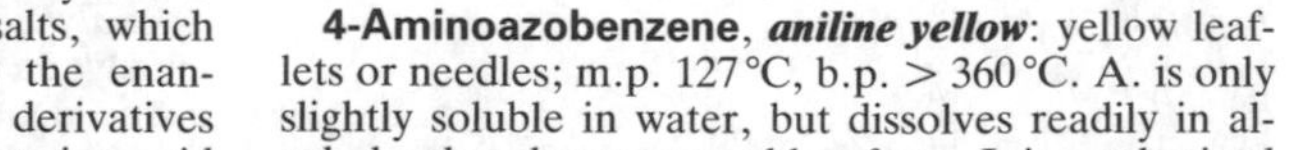

Diazoaminobenzene   4-Aminoazobenzene

In the presence of strong mineral acids, A. forms steel-blue, stable salts; there is also a violet form which is less stable. A. is used as an intermediate in the production of azo dyes, which are used mainly as dispersion dyes.

5-Aminobarbituric acid: same as Uramil (see).

Aminobenzene: same as Aniline (see).

4-Aminobenzenesulfonic acid: same as Sulfanilic acid (see).

Aminobenzoic acids: three structural isomers which are slightly soluble in cold water, and somewhat more soluble in hot water and alcohol. Because of their amphoteric nature, they form salts with either strong bases or mineral acids. A. can be sublimed in vacuum without decomposition. ***o-*** or ***2-aminobenzoic acid, anthranilic acid*** forms light yellow leaflets; m.p. 146-147 °C. The solutions give a blue fluorescence. 2-Aminobenzoic acid is found as the methyl ester in many plants, such as jasmine and bergamotte. It is synthesized industrially, mainly by Hofmann degradation of phthalimide. It can also be made by reduction of 2-nitrobenzoic acid, reaction of 2-chlorobenzoic acid with ammonia in the presence of copper salts, or by intramolecular redox reaction of 2-nitrotoluene. 2-Aminobenzoic acid is used in the synthesis of indigo, thioindigo, azo dyes, perfumes and salves for burns. It is also used as an indicator for many metals. ***m-*** or ***3-aminobenzoic acid*** forms pale yellow crystals, m.p. 174 °C. It is made technically by reduction of 3-nitrobenzoic acid. 3-Aminobenzoic acid is an important intermediate for the synthesis of azo dyes. ***p-*** or ***4-aminobenzoic acid***, abb. ***PABA*** is a colorless, crystalline compound; m.p. 186-187 °C. It is found in free or bound form in yeasts, enzymes and folic acid. It is essential for the growth of many microorganisms, including lactic acid bacteria, streptococci, staphylococci and pneumococci. It acts as an antagonist to sulfonamides used as antibiotics. In the human body, it is detoxified by formation of *N*-acetylaminobenzoic acid. 4-Aminobenzoic acid is made industrially by reduction or hydrogenation of 4-nitrobenzoic acid. It is an intermediate in the synthesis of azo dyes, folic acid and a number of esters used as local anesthetics; it is also the active ingredient in sunblocking skin creams used to prevent sunburn.

COOH COOH COOH
NH_2 NH_2 NH_2
2-A. 3-A. 4-A.

Aminobutanes: see Butylamines.

γ-Aminobutyric acid: see Biogenic amines, Table.

4-Aminobutyric acid lactam: same as Pyrrolidone (see).

ε-Aminocaproic acid: $H_2N\text{-}(CH_2)_5\text{-}COOH$, a white, crystalline substance; m.p. 202-203 °C. ε-A. is soluble in water and insoluble in ether. It is used as an antifibrinolytic and is an intermediate in the polymerization of ε-Caprolactam (see) to make polyamide fibers.

Aminocarboxylic acids: same as Amino acids (see).

Aminochromes: see Catecholamines.

Amino compounds: see Amines.

Aminocyclitol antibiotics: a group of antibiotics in which a base-substituted cyclitol is glycosidically bound to two or three monosaccharides, usually amino sugars. A. are found in actinomycetes (*Streptomyces, Micromonospora*). The therapeutically important A. are derived from streptomycin, neomycin and kanamycin. The streptomycin type A. have streptidin(1,3-deoxy-1,3-diguanidino-myo-inositol) as their cyclitol. The other two types of A. contain 2-deoxystreptamine (1,2-diamino-1,2,3-trideoxy-myo-inositol).

Aminocyclohexane: same as Cyclohexylamine (see).

p-Amino-N,N-diethylaniline: same as *N,N*-Diethyl-p-phenylenediamine.

5-Amino-2,3-dihydrophthalazine-1,4-dione: same as Luminol (see).

5-Amino-1,4-dihydroxyphthalazine: same as Luminol (see).

Aminodimethylbenzene: same as Xylidine (see).

Aminoacetic acid: same as Glycine (see).

Aminoethane: same as Ethanolamine (see).

2-Aminoethanethiol: same as Cysteamine (see).

Aminoethyl alcohols: same as Ethanolamines (see).

4-(2-Aminoethyl)pyrocatechol: same as Dopamine (see).

β-Aminoethylmercaptan: same as Cysteamine (see).

Aminoformic acid: same as Carbamic acid (see).

Aminoglycoside antibiotics, ***oligosaccharide antibiotics***: a group of antibiotics in which glycosidic aminosugars and monosaccharide components are linked to each other and/or to basic cyclohexane derivatives containing multiple hydroxyl groups. The A. have a broad action spectrum and are used mainly against infections by gram-negative bacteria. However, they are relatively toxic. Resistance can be developed by enzymatic acetylation of the amino groups or phosphorylation and adenylation of the hydroxyl groups. (Adenylation is the transfer of an adenosine 5'-phosphate group to the molecule.) The A. include streptomycin, neomycin, paromomycins, kanamycins and gentamycins.

Amino group: in the narrow sense, the group $-NH_2$, which is the functional group in all primary amines and amides. In the wider sense, the term includes mono- and disubstituted nitrogen derivatives of the type RHN- and R_2N-, such as the methylamino group $(CH_3)HN$- or the ethylmethylamino group, $(C_2H_5)(CH_3)N$-.

1-Aminohexane: same as Hexylamine (see).

6-Aminohexanoyl lactam: same as ε-Caprolactam (see).

Aminolysis: the reaction corresponding to hydrolysis in which the reactive substituent in a compound is replaced by an amine group. For example, the halogen atom in a haloalkalne can be replaced by an amino group. Similarly, carboxylate esters can be converted to amides. If A. is carried out in ammonia, the reaction is called ***ammonolysis***, e.g. C_2H_5Cl + 2 $NH_4 \rightarrow C_2H_5NH_2 + NH_4Cl$.

Aminomethane: see Methylamine.

Aminomethylation: see Mannich reaction.

Aminomethylpropanes: two structurally isomeric amino derivatives of 2-methylpropane; see Butylamines.

Aminonaphthalenes: same as Naphthylamines (see).

Aminonaphthols: 14 isomeric aminohydroxynaphthalenes. These compounds form colorless crystals; they are amphoteric and dissolve in acids and alkaline solutions. They are slightly soluble in water, and readily soluble in alcohol and ether. They are usually synthesized by reduction of the corresponding nitronaphthols, or by alkaline hydrolysis of aminonaphthalenesulfonic acids. A. are used as couplers for dyes.

OH (position 1 or 2)

NH_2 (any other position)

Aminonitrobenzenes: same as Nitroanilines.

Aminopeptidases: a class of exopeptidases (see Proteases) which catalyse the cleavage of the amino-terminal amino acid from a polypeptide chain. A. are large proteins consisting of several subunits (M_r up to 330,000), and most of them have divalent metal ions as effectors. The best known is ***leucine aminopeptidase***, which can be isolated in crystalline form from bovine eye lenses, kidneys or the mucous membrane of the small intestine. A. are used for the step-wide degradation of peptide chains (sequence analyses). Peptide bonds to proline are not cleaved by A.

Aminophenazone: see Pyrazolones.

Aminophenetols: same as Phenetidines (see).

Aminophenols: phenols with one or more amino groups on the aromatic ring. A. are amphoteric, but the basic characteristics dominate. The most important A. are 2-A., 3-A. and 4-A.

H_2N–

OH (position 2, 3 or 4)

2-Aminophenol (o-aminophenol) forms colorless leaflets or needles; m.p. 174 °C, sublimation at 153 °C and 1.47 kPa. It is nearly insoluble in cold water, but somewhat more soluble in alcohol and ether. It is synthesized by reduction of 2-nitrophenol with iron in weakly acidic solution, or by reduction of nitrobenzene via phenylhydroxylamine. This intermediate rearranges under the acidic reaction conditions to 2- and 4-A. 2-A. is used in the production of pigments, drugs and hair dyes.

3-Aminophenol (m-aminophenol), colorless, hexagonal prisms; m.p. 123 °C, b.p. 164 °C at 1.47 kPa. It is slightly soluble in water, and readily soluble in alcohol and ether. 3-A. is obtained mainly by the reaction of alkali hydroxides with 3-aminobenzenesulfonic acid or from resorcinol and ammonia in the presence of catalysts. It is used to synthesize the important drug, aminosalicylic acid, and rhodamine dyes.

4-Aminophenol (p-aminophenol), colorless leaflets; m.p. 186-187 °C, sublimation at 180 °C and 360 Pa. It is nearly insoluble in water, somewhat soluble in alcohol, and soluble in chloroform and benzene. The most important method of synthesis of 4-A. is the reduction of 4-nitrophenol with iron in weakly acidic solution or the reduction of nitrobenzene via phenylhydroxylamine. 4-A. is used as a photographic developer. It is also used in the synthesis of modified photodevelopers, sulfur and azo pigments, and drugs.

Aminophenyl ethyl ether: same as Phenetidine (see).

Aminophenyl methyl ether: same as Anisidine (see).

3-Aminophthaloyl hydrazide: same as Luminol (see).

Aminophyllin: see Theophyllin.

Amino plastics: see Amide resins.

Aminopropanols, ***propanolamines***: aliphatic compounds with an amino and a hydroxyl group in the molecule. They are colorless liquids with fishy odors and are soluble in water and alcohol. They display the chemistry of both amines and alcohols, and form salts with acids. Of the five possible isomeric A. with primary amino groups, two are used extensively in industry. ***3-Aminopropan-1-ol***, $H_2N\text{-}CH_2\text{-}CH_2\text{-}CH_2\text{-}OH$, b.p. 187-188 °C, n_D^{20} 1.4617., is synthesized by catalytic hydrogenation of ethylene cyanohydrin in the presence of ammonia. It is used to make drugs, dyes, synthetic resins and soaps. ***1-Aminopropan-2-ol (isopropanolamine)***, $CH_3\text{-}CH(OH)\text{-}CH_2\text{-}NH_2$, m.p. 1.9 °C, b.p. 159.2 °C, n_D^{20} 1.4478, is a degradation product of vitamin B_{12}. It is synthesized by reaction of propylene oxide with ammonia under pressure. 1-Aminopropan-2-ol is used to synthesize dyes, soaps, drugs and pesticides.

In addition to *N*-unsubstituted A., there are many *N*-mono- and *N*-dialkyl compounds of this type which are synthesized and utilized in industry.

Aminopropionic acid: same as Alanine (see).

6-Aminopurine: same as Adenine (see).

Aminopyridine: an amino derivative of pyridine. ***2-Aminopyridine*** has colorless, crystalline leaflets, m.p. 57-58 °C, b.p. 204 °C. It is soluble in water, alcohol and ether, and is a base. When heated with 2-chloropyridine in the presence of zinc chloride, it forms di-α-pyridylamine. 2-Aminopyridine is obtained by the Čičibabin reaction (see). It is used in industry in the production of pigments, sulfonamides and antihistamines. ***3-Aminopyridine*** forms yellowish crystalline leaflets, m.p. 65 °C, b.p. 252 °C. It is synthesized from nicotinamide by Hofmann degradation. ***4-Aminopyridine*** forms yellowish needles; m.p. 158-159 °C, b.p. 180 °C at 1.74 kPa. It is formed by decarboxylation of 4-aminopicolinic acid. ***2,6-Diaminopyridine*** forms colorless crystals; m.p. 121-122 °C, b.p. 148-150 °C at 0.67 kPa. It is obtained by heating pyridine with two moles sodium amide in cumene. Further amination of 2-aminopyridine also yields 2,6-diaminepyridine. It is very interesting that A., in contrast to pyridine itself, readily undergo electrophilic reactions. For example, nitration yields 2-amino-3-nitropyridine, which can be hydrogenated to form 2,3-diaminopyridine. A. are used as intermediates in the production of dyes.

NH_2

N

5-Aminopyrimidine-2,4,6-triol: same as Uramil (see).

p-Aminosalicylic acid, abb. ***PAS***: a compound used as a tuberculostatic. It forms white crystals which are soluble in ethanol but only slightly soluble in water or ether; m.p. 150-151 °C. It decomposes by decarboxylation to form 3-aminophenol, or by oxidation to colored products. It is synthesized by the Kolbe-Schmitt synthesis from the sodium salt of 3-aminophenol and CO_2. High doses of the sodium or potassium salt are used in therapy. A. acts as an antimetabolite of *p*-aminobenzoic acid.

α-Aminosuccinic acid: same as Aspartic acid (see).

Aminosugars: monosaccharides in which one or more alcoholic hydroxyl groups have been replaced by amino groups. The free A. are relatively strong bases, and are not very stable. In nature, A. are always bound glycosidically; they are especially plentiful in animals and microorganisms. With the exception of a few basic A.-glycosides in microorganisms, which include the Aminoglycoside antibiotics (see), the amino group of the A. is always amidated as N-acetyl or, less commonly, as N-sulfuryl. Neither group is basic. The most abundant A. in animals are derivatives of 2-amino-2-deoxyhexoses, such as D-Glucosamine (see), D-Galactosamine (see) and D-mannosamine. These A. are components of polysaccharides, e.g. of Chitin (see), Mucopolysaccharides (see), Blood group substances (see) and many glycoproteins. In addition to 2-amino-2-deoxyhexoses, microorganisms also contain methylated A. and 3-amino-2-deoxy-, 6-amino-6-deoxy- and 2,6-diamino-2,6-dideoxysugars. Among other functions, A. serve as components of bacterial cell walls (see Murein). Another important A. is Neuraminic acid (see).

2-Aminothiazole: a heterocyclic compound forming yellow crystals with m.p. 93 °C, b.p. 140 °C at 1.47 kPa. A. is readily soluble in hot water, but less soluble in alcohols and ethers. Like an aromatic amine, it can be diazotized; it also couples with diazonium salts to form azo dyes. It is obtained by reaction of chloroacetaldehyde with thiourea, or better, from vinyl acetate and thiourea in the presence of sulfuryl chloride. A. is used as a starting material for synthesis of the sulfonamide sulfathiazole.

Aminotoluene: same as toluidine.

Aminotransferases, ***transaminases***: enzymes which reversibly transfer the α-amino group of an amino acid to a 2-oxoacid. Aspartate aminotransferase (known in medicine as ***glutamic-oxaloacetic transaminase, GOT***) and alanine aminotransferase (known in medicine as ***glutamic-pyruvic transaminase, GPT***) are diagnostically useful because their serum concentrations rise in certain illness (cardiac infarction, hepatitis).

Aminoxylenes: same as Xylidines (see).

Amitriptyline: see Dibenzodihydrocycloheptadienes.

Amitrol: see Herbicides.

Ammine complexes: complexes in which ammonia, NH_3, is present as a ligand. Chromiacs (see) and Cobaltiacs (see) are important examples of this type of compound.

Ammonia: NH_3, a colorless gas with a characteristic and very strong odor; M_r 17.03, m.p. - 77.4 °C, b.p. - 33.35 °C, crit. temp. 132.5 °C, crit. pressure, 11.25 MPa, crit. density, 0.235, density of liquid A. at - 34.4 °C, 0.683, at 0 °C, 0.639; vapor pressure at 0 °C, 0.438 MPa, at 30.0 °C, 1.19 MPa; heat of evaporation of the liquid at - 33.35 °C, 1372.0 kJ kg^{-1}, at 0 °C, 1264.3 kJ kg^{-1}, and at 32.2 °C, 1137.3 kJ kg^{-1}. Solid A. consists of colorless, cubic crystals.

Properties. The ammonia molecule is pyramidal, with an H-N-H angle of 107.3°. From this one can conclude that the N atom is largely sp^3 hybridized. It sits at the center of a tetrahedron with protons on three vertices; the free electron pair is oriented toward the fourth vertex (Ψ-tetrahedral molecule). The NH_3 pyramid is not stable; even at low temperatures, it undergoes rapid inversion in which the N atom passes through the plane of the three H atoms. The energetic barrier to inversion is 24.8 kJ mol^{-1}. The high polarity of the NH_3 molecule and the formation of strong hydrogen bonds cause liquid A. to be highly associated (dielectric constant at - 50 °C is 22.7). This is the reason for the unusually high boiling point and the high enthalpy of evaporation of A.

NH_3-Pyramid

As a result of its molecular structure, A. is able to act either as a donor or acceptor molecule, that is, it can solvate either anions or cations. This gives liquid A. solvent properties similar to those of water. Many inorganic and organic compounds are soluble in A., including many halides, pseudohalides, nitrites and nitrates, sulfur, selenium, phosphorus, alcohols, aldehydes, phenols and esters. Liquid A., like water, undergoes autoprotolysis: $2\ NH_3 \rightleftharpoons NH_2 + NH_4^+$. Thus there is a neutralization in the ammono system, consisting of the combination of amide and ammonium ions to A., which is comparable to neutralization in water. An example is the reaction $NH_4I + NaNH_2 \rightarrow 2\ NH_3 + NaI$. Alkali metals and calcium, strontium and barium dissolve in liquid A., forming dark blue solutions. The electrical conductivity and paramagnetism of these solutions are due to the presence of solvated electrons:

$$M \xrightarrow{\text{liq. } NH_3} M^+ + [e(NH_3)_x]^-.$$

A. dissolves aggressively in water. At 0 °C, 90.7 g A. dissolve in 100 ml water; this is equivalent to 117.6 l gaseous A. At 100 °C, 7.4 g A. dissolves in 100 ml water. The enthalpy of solution of gaseous A. at 25 °C is 30.64 kJ mol^{-1}. The concentration of an aqueous NH_3 solution can be determined directly from its density:

D	0.880	0.900	0.920	0.940	0.960	0.980	0.990
% NH_3	34.35	27.33	20.88	14.88	9.34	4.27	1.89

In water, most dissolved A. is molecular. However, due to the ability of A. to accept protons, a slight amount of protolysis occurs: $NH_3 + H_2O \rightleftharpoons NH_4^+ + OH^-$ ($pK_B = 4.75$). The aqueous solution is thus basic. The basic character of A. is also expressed by its ability to form ammonium salts with strong acids. In the presence of strong bases, A. is released from ammonium salts. For example, $NH_4Cl + NaOH \rightarrow NH_3 + NaCl + H_2O$. A. can act as an acid with respect to very strong bases, such as organometal compounds or ionic hydrides, and is converted to its corresponding base, the amide ion NH_2^-. Amides are also formed by the action of electropositive metals on A. at high temperatures, e.g. $Na + NH_3 \rightarrow NaNH_2 + 1/2\ H_2$. The amides M^INH_2 can be converted to metal imides, M^I_2NH and metal nitrides, M^I_3N, by increasing the temperature.

A. is stable at normal temperatures, but decomposes into its elements when heated in the presence of certain catalysts: $2\ NH_3 \rightleftharpoons N_2 + 3\ H_2$, $\Delta H = 92.5$ kJ mol^{-1}. A. burns in pure oxygen, mainly to nitrogen and water: $2\ NH_3 + 3/2\ O_2 \rightarrow N_2 + 3\ H_2O$. Mixtures of A. and oxygen are explosive in the concentration range of 13.5 to 82 vol. % A., and A.-air mixtures in the range of 15.5 to 28 vol. % A. If the combustion takes place in the presence of platinum or platinum/rhodium catalysts, nitrogen monoxide is formed: $4\ NH_3 + 5\ O_2 \rightarrow 4\ NO + 6\ H_2O$, $\Delta H \approx -900$ kJ mol^{-1}. This reaction is the basis of the Ostwald process for synthesis of nitric acid. Due to its free electron pair, A. is a good complex ligand in coordination compounds, and a strong nucleophile in ammonolysis.

Both liquid ammonia and its aqueous solutions are irritating to the skin and mucous membranes. The eyes are especially at risk. If ammonia gets into the eyes, they must immediately be rinsed with large amounts of water.

1.5 to 2.5 g m^{-3} air causes death within 30 to 60 minutes. Inhalation for a short period can cause burns and lead to inflamation of the respiratory passages and lung edema. In this case, complete quiet and inhalation of water and glacial acetic acid vapors are recommended. When aqueous ammonia is swallowed, acid solutions (acetic, tartaric or citric acid) should be administered.

Analytical. A. is detected qualitatively by formation of the blue $[Cu(NH_3)_4]$ complex or by Nessler's reagent (see); it is determined quantitatively by distilling it out of an alkaline aqueous solution, absorbing it in acid and back titrating the excess acid. A. can also be determined with NH_3-sensitive electrodes.

Occurrence. A. is the biological degradation product of many organic nitrogen compounds, and occurs in nature, usually in the form of ammonium salts, as a result of decay of organic matter. A few minerals contain small amounts of A.

Production. Today, nearly all A. is synthesized by direct combination of nitrogen and hydrogen on the principle of the *Haber-Bosch process*. The processes based on this method differ only in the means of production and purification of the synthesis gas, the choice of reaction conditions (such as pressure and temperature), the catalysts and the technical realization of the individual steps of the process. The reaction $N_2 + 3\ H_2 \rightleftharpoons 2\ NH_3$, $\Delta H = -92.5$ kJ mol^{-1} requires the input of large amounts of activation energy, due to the high dissociation enthalpy of the N_2 molecule. However, because it is an exothermal reaction, the amount of heat which can be applied is limited; as the temperature rises, the equilibrium position is shifted to the left. For this reason, catalysts have been developed which permit economically useful rates at a reaction temperature of 400 °C. (The catalysts are iron oxides with small amounts of the oxides of aluminum, calcium, potassium, magnesium, titanium, etc. as promotors; after reduction by hydrogen, the iron oxide is converted to the active α-form.)

The figure is a diagram of an ammonia plant. There are two types of reactor: cylinders filled with catalyst are cooled over their entire length, while layered reactors contain 5 to 10 layers of catalyst and are cooled by input of cold gas. Before it enters the reactor, the synthesis gas is heated to the working temperature of the catalyst (400-500 °C) by heat exchange with the reacted gas. After the reacting mixture passes each layer of catalyst, cold gas is added to cool it to the appropriate temperature. After about the 5th to 10th catalytic layer has been passed, the gas has reached the equilibrium concentration. It leaves the reactor and passes through a cooling vat, a heat exchanger to preheat the synthesis gas, and a water cooler. At this point, part of the A. precipitates. The remainder is removed from the circulation by a low-temperature cooler filled with A. The unreacted gas is returned to the reactor with fresh synthesis gas (with a ratio of N_2 to H_2 of 1:3). A compressor in the circulation compensates for the loss of pressure. Part of the flow is continuously decompressed, to prevent accumulation of inert gases (argon and methane) in the circulating gas. Argon can be extracted from these decompression gases.

Other possibilities for A. synthesis are fixation with organometallic complexes, biochemical fixation and conversion of nitrogen to nitrogen monoxide in a plasma. A. can also be extracted from coking plant wash water. In future, the plasma methods may become significant in connection with nuclear energy. In the laboratory, A. is produced by the reaction of strong bases with ammonia salts.

A. is commercially available in steel bottles, or as a 25 to 35% solution in water (***ammonium hydroxide***).

Applications. A. is a raw material for nearly all industrial syntheses of nitrogen compounds. The greatest fraction of the world production is used to make nitrogen fertilizers, especially urea and ammonium sulfate; liquid A. is also applied directly. A. is used in the production of soda, nitric acid, various

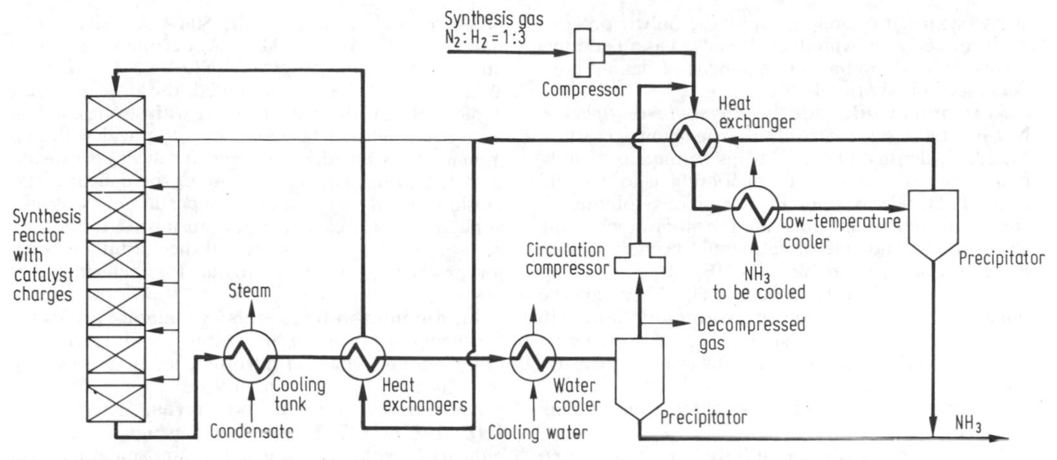

Diagram of an ammonia synthesis plant.

ammonium salts, sodium cyanide, hydrogen cyanide, hydrazine, nitriles, amines, amide resins, synthetic fibers, dyes, explosives, etc. It is used in refrigeration and in metallurgy (for nitro hardening). Ammonium hydroxide is used in bleaching and dyeing, and generally as a cheap base.

Ammonia resin: see Latex resins.

Ammonia-soda process: see Soda.

Ammonites: see Ammonium nitrate explosives.

Ammonium acetate: CH_3COONH_4, a colorless, hygroscopic mass, which is readily soluble in water and ethanol; M_r 77.08, density 1.17, m.p. 114 °C. A. is synthesized by reaction of glacial acetic acid with ammonia or ammonium carbonate. It is used to preserve meat and in dyeing.

Ammonium alum: same as ammonium aluminumsulfate.

Ammonium aluminumsulfate, ***ammonium alum***: $NH_4Al(SO_4)_2 \cdot 12H_2O$, colorless, water-soluble, cubic crystals, M_r 453.33, density 1.64, m.p. 93.5 °C. A. is used in water purification, paper-making, tanning, flame-retardants, caustics and in medicine as an astringent.

Ammonium carbamate: $H_2N\text{-}CO\text{-}O^- NH_4^+$, the ammonium salt of carbamic acid. A. forms colorless, water-soluble rhombic crystals which smell like ammonia. In aqueous solution it hydrolyses at 60 °C, forming ammonium carbonate, which readily decomposes to ammonia and carbon dioxide. A. is formed by heating dry ammonia with carbon dioxide. It is an intermediate in the synthesis of Urea (see), and is used as a cleanser, caustic and neutralizing reagent. One type of baking powder is a mixture of A. with ammonium hydrogencarbonate.

Ammonium carbonate: $(NH_4)_2CO_3 \cdot H_2O$, colorless, water-soluble, cubic crystals; M_r 114.10. A. smells of ammonia even at room temperature, and at 58 °C, it decomposes completely into NH_3, CO_2 and H_2O. It is synthesized by reaction of carbon dioxide with ammonia in aqueous solution: $2\ NH_3 + CO_2 + H_2O \rightarrow (NH_4)_2CO_3$, or by heating a mixture of ammonium sulfate and calcium carbonate: $(NH_4)_2SO_4 + CaCO_3 \rightarrow CaSO_4 + (NH_4)_2CO_3$. The A. sublimes off, together with the ammonium hydrogencarbonate and ammonium carbamate byproducts. It is used in dyeing and as a CO_2 source in fire extinguishers.

Ammonium carbonate group: see Analysis.

Ammonium chloride: NH_4Cl, colorless, cubic crystals with a bitter taste; M_r 53.49, density 1.527, soluble in water and ethanol. The aqueous solution is slightly acidic. As the temperature is increased, A. becomes volatile, due to thermal dissociation into ammonia, NH_3, and hydrogen chloride, HCl. At 340 °C, it sublimes rapidly. This process, like the combination of NH_3 and HCl to form A., depends on the presence of catalytic amounts of water. A. melts under pressure at 520 °C. It is obtained by adding hydrochloric acid to aqueous ammonia solutions, or to the gas scrub water from coking plants. A. is a byproduct of Soda (see) production by the Solvay process. It was formerly used as the nitrogen component of combination fertilizers. It is still used in dyeing, tanning leather and making galvanoplastics, iron-rust cement and cold mixtures. Its use in soldering, zinc plating and tinning is based on its ability to react with metal oxides and form volatile chlorides, and thus to clean the metal surface. A. is used as an electrolyte in dry batteries, and as an expectorant in medicine.

Ammonium cyanate: NH_4OCN, colorless crystals, soluble in ethanol and water, M_r 60.06, D. 1.342. The salt can be made by neutralization of cyanic acid with ammonia; above 60 °C it is converted to urea: $NH_4OCN \rightarrow H_2N\text{-}CO\text{-}NH_2$. This rearrangement was first carried out in 1828 by Wöhler. At the time, it was believed by many that natural products could not be synthesized outside living organisms; this synthesis was taken as proof to the contrary.

Ammonium hydrogencarbonate, ***primary ammonium carbonate***, formerly, ***ammonium bicarbonate***: NH_4HCO_3, colorless, water-soluble powder or rhombic or monoclinic crystals; M_r 79.06, density 1.58. Even at 40 to 60 °C, A. decomposes into ammonia, carbon dioxide and water. If the aqueous solution is heated, carbon dioxide escapes and forms neutral ammoniumcarbonate: $2\ NH_4HCO_3 \rightarrow (NH_4)_2CO_3 + CO_2 + H_2O$. It is made by passing carbon dioxide through aqueous ammonia solution.

In the synthesis of Soda (see) by the Solvay process, A. is formed as a byproduct. It is used as an additive for cattle feed, and as a component of baking powders (see Hartshorn salt).

Ammonium fluorides: ***Ammonium fluoride***, NH_4F, colorless, water-soluble, hexagonal crystals; M_r 37.04, density 1.009. NH_4F is formed as a sublimate when a mixture of ammonium chloride and sodium fluoride is heated. In aqueous solution, it splits off ammonium, and after evaporation of the solution, the rhombic or tetragonal crystals of ***ammonium hydrogenfluoride***, NH_4HF_2, M_r 57.04, density 1.50, m.p. 125.6 °C, are left. NH_4HF_2 can also be obtained by neutralization of hydrofluoric acid with ammonia. Among other things, it is used to etch glass and metal and as a disinfectant in breweries and distilleries.

Ammonium hexachlorostannate: same as Pink salt (see).

Ammonium hydrogensulfate, formerly ***ammonium bisulfate***, NH_4HSO_4, colorless, water-soluble, rhombic crystals, M_r 115.11, density 1.78, m.p. 146.9 °C. A. is used as an acidic condensation reagent in organic synthesis.

Ammonium hydrogensulfide: NH_4HS, colorless, poisonous rhombic crystals, readily soluble in water; M_r 51.11, density 1.17. A. decomposes even at room temperature into ammonia, NH_3, and hydrogen sulfide, H_2S. The aqueous solutions are alkaline; they are obtained by saturation of aqueous ammonia with hydrogen sulfide. Conversion of A. into a neutral ammonium sulfide, $(NH_4)_2S$, for example by addition of a second equivalent of NH_3, does not occur, because the ammonium ion, NH_4^+, is a stronger acid than hydrogen sulfide, HS^-, by a factor of about 10^4. The aqueous solution of A. is used mainly for separation of cations in qualitative analysis.

Ammonium hydroxide: a term used for aqueous ammonia solutions; it is a misnomer, because the compound NH_4OH does not exist (see Ammonia). The solid compound which can be made at low temperatures from NH_3 and H_2O (melts at - 77 °C) should be formulated as ammonia hydrate, $NH_3 \cdot H_2O$.

Ammonium iodide: NH_4I, colorless, hygroscopic, cubic crystals, M_r 144.94, density 2.514, subl.p. 551 °C. The salt is obtained by neutralization of hydroiodic acid with ammonia; it is used in medicine as an expectorant.

Ammoniumiron alum: same as ammoniumiron(III) sulfate.

Ammoniumiron(II) sulfate, ***Mohr's salt***: $(NH_4)_2Fe(SO_4)_2 \cdot 6H_2O$, light green, water-soluble monoclinic crystals; M_r 392.14, density 1.864, dec. 100 to 110 °C. The double salt is less sensitive to oxidation than iron(II) sulfate, and is therefore easier to purify. For this reason, it is used to calibrate $KMnO_4$ standard solutions.

Ammoniumiron(III) sulfate, ***ammoniumiron alum***: $NH_4Fe(SO_4)_2 \cdot 12H_2O$, pale violet, water-soluble, cubic crystals, M_r 482.19, density 1.71, m.p. 39 to 41 °C, conversion at 230 °C to the anhydrous form. A. is used as a mordant in dyeing and printing of textiles and as an astringent in medicine. It is used in quantitative analysis as an indicator in Volhard's titration.

Ammonium nitrate: NH_4NO_3, colorless, hygroscopic monoclinic crystals, rhombic at room temperatures below 32.1 °C; M_r 80.04, density 1.725, m.p. 169.6 °C. Above 200 °C, A. decomposes into dinitrogen oxide and water: $NH_4NO_3 \rightarrow N_2O + 2\,H_2O$. When suddenly heated or ignited, this process occurs explosively. A. dissolves in water with strong cooling. It is made by neutralization of nitric acid with ammonia. A. is an ideal nitrogen fertilizer, because it contains no ballast, but because of the danger of explosion and its hygroscopic properties, it is usually applied in mixtures with other substances. It is a component of safety explosives and the starting material for production of nitrous oxide for medicinal purposes.

Ammonium nitrate gels: gelatinous explosives consisting of nitroglycerin gelatinized with a mixture of collodium wool and nitrotoluene. A. are used to blast hard rocks, even under water.

Ammonium nitrate explosives: often called ***PAC, ANC*** or ***ANFO explosives***: a group of important industrial explosives which contain ammonium nitrate as the main component, a carbon carrier such as coal, oil or sawdust, and sometimes aluminum powder. A. which contain water are important as Slurry blasting agents (see). A. which contain 20 to 40% gelatinized nitroglycerin in addition to ammonium nitrate are called ***gelatin donarites***; they are used in hard rock. ***Donarites*** are A. containing 70 to 80% ammonium nitrate, 4 to 6% nitroglycerin, trinitrotoluene, sawdust, etc. A. which contain no nitroglycerin are called ***ammonites***; examples are powdered mixtures of 80 to 82% ammonium nitrate and 18 to 20% trinitrotoluene. AN-D explosives (see) are ammonites.

Ammonium nitrite: NH_4NO_2, colorless crystalline solid, readily soluble in water; M_r 64.04, density 1.69. In aqueous solution, A. decomposes into nitrogen and water when heated: $NH_4NO_2 \rightarrow N_2 + 2\,H_2O$. The solid explodes at 60 to 70 °C. As a product of various oxidation processes, A. is present in small amounts in the air and in rainwater. It is made by absorption of equal parts of nitrogen monoxide and dioxide in aqueous ammonia; it is occasionally used in the laboratory to produce pure nitrogen.

Ammonium oxalate: $(NH_4)_2C_2O_4 \cdot H_2O$, colorless rhombic crystals, soluble in water and ethanol; M_r 142.11, density 1.50. A. is made by neutralization of oxalic acid with ammonia solution, and is used mainly in analytical chemistry.

Ammonium oxalatoferrate(III), ***iron(III) ammoniumoxalate***: $(NH_4)_3[Fe(C_2O_4)_3] \cdot 3H_2O$, bright green crystals obtained by reaction of iron(III) chloride solutions with ammonium oxalate. Light causes A. to decompose to carbon dioxide and iron(II) compounds. This reaction can be used to measure light and also for photosensitive papers.

Ammonium peroxodisulfate, ***ammonium persulfate***: $(NH_4)_2 \cdot S_2O_8$, colorless monoclinic crystals, readily soluble in water; M_r 228.18, density 1.982. Dry, pure A. is stable up to about 120 °C, but when moist it decomposes rapidly, forming ozone-containing oxygen: $(NH_4)_2S_2O_8 + H_2O \rightarrow 2\,(NH_4)HSO_4 + 1/2\,O_2$. The aqueous solution is strongly oxidizing. A. is obtained by anodic oxidation of ammonium sulfate solution which contains sulfuric acid. It was formerly used mainly to produce hydrogen peroxide. Now it is used as an initiator in polymerization reactions, as a bleach, deodorant and disinfectant, and in photography and galvanization.

Ammonium persulfate: same as Ammonium peroxodisulfate.

Ammonium phosphates: ***Ammonium dihydrogenphosphate, primary ammonium phosphate***: $NH_4H_2PO_4$, colorless, water-soluble, tetragonal crystals; M_r 115.03, density 1.803, m.p. 190 °C, decomposition above 200 °C with release of ammonia and water. The salt is obtained by neutralization of phosphoric acid with ammonia to a pH of 4. If more ammonia is added, ***diammonium hydrogenphosphate, secondary ammonium phosphate***, $(NH_4)_2HPO_4$, is obtained. This compound forms colorless, water-soluble monoclinic prisms; M_r 132.05, density 1.619. $(NH_4)_2HPO_4$ is a valuable, ballast-free component of multiple-nutrient fertilizers, and is also used as a flame retardant for wood, paper, etc. ***Triammonium phosphate, tertiary ammonium phosphate***, $(NH_4)_4PO_4 \cdot 3H_2O$, is obtained from the reaction of gaseous ammonia with $(NH_4)_2HPO_4$. It forms colorless, water-soluble prisms, M_r 203.13. It is unstable in the solid form, releases ammonia even at room temperature, and, like the other A., decomposes on stronger heating into ammonia and polyphosphoric acid.

Ammonium phthalate: see Phthalamic acid.

Ammonium picrate: $NH_4O\text{-}C_6H_2(NO_2)_3$, the ammonium salt of picric acid, a crystalline compound with a bitter taste, which exists in a yellow and a red modification; m.p. 257 to 259 °C (dec.). A. is soluble in water and ethanol. It is made from ammonia and picric acid, and is a highly brisant explosive. The rate of detonation is 6950 s m^{-1}, which gives A. an explosiveness equivalent to that of 2,4,6-trinitrotoluene (TNT). Its disadvantages, however, are that the melting point is so high and that it cannot be poured into cartridges in molten state.

Ammonium rhodanide: same as Ammonium thiocyanate (see).

Ammonium sodium hydrogenphosphate, ***phosphor salt***: $NH_4NaHPO_4 \cdot 4H_2O$, colorless, water-soluble, monoclinic crystals; M_r 209.07, density 1.574. When heated, A. decomposes, forming sodium polyphosphate: $NH_4NaHPO_4 \rightarrow NaPO_3 + NH_3 + H_2O$. It is used in qualitative analysis to make Microcosmic salt beads (see).

Ammonium sulfate: $(NH_4)_2SO_4$, colorless, rhombic crystals, soluble in water with strong cooling; M_r 132.14, density 1.769. Above 235 °C it releases ammonia and is converted to ammonium hydrogensulfate. A. is made on laboratory scale by neutralization of sulfuric acid with ammonia. Industrially, carbon dioxide and ammonia are passed through an aqueous suspension of finely ground gypsum. The calcium carbonate precipitates, and A. is obtained by evaporation of the solution: $2\,NH_3 + CO_2 + CaSO_4 + H_2O \rightarrow CaCO_3 + (NH_4)_2SO_4$. A. was for a long time the most important ammonia fertilizer, but because of its relatively low nitrogen content and acidity, it has been replaced by ammonium nitrate and urea.

Ammonium sulfide: see Ammonium hydrogensulfide.

Ammonium sulfide group: see Analysis.

Ammonium thiocyanate, ***ammonium rhodanide***: NH_4SCN, colorless, deliquescent, monoclinic crystals, readily soluble in water and ethanol; M_r 76.12, density 1.305, m.p. 149.6 °C. At 170 °C, A. decomposes with partial formation of thiourea: $NH_4SCN \rightarrow H_2C(S)NH_2$. It is obtained by the reaction of ammonia with carbon disulfide under pressure. A. is used in analytical chemistry to detect iron(III) ions and in halide determination according to Volhard.

Ammonolysis: see Aminolysis.

AMO explosive: an explosive made of ammonium nitrate and hydrocarbons, such as diesel fuel. A. do not detonate easily and usually require an initial charge. For this reason they are preferred as safety explosives in mining; examples are Dekamon I (94% ammonium nitrate and 6% diesel fuel) or Alumon (addition of sawdust or aluminum powder).

Amonton's law: see State, equation of 1.1.

Amorphous (from the Greek "amorphos" = "without shape"): a term applied to solids in which the constituent particles (atoms, ions or molecules) do not display periodic long-range order. In A. substances, the particles are not completely randomly distributed, however. There is a certain regularity in their arrangement with respect to distance and orientation of nearest neighbors (short-range order). Thus A. substances correspond in their structure to liquids, but differ from them because the particles are not mobile. They are sometimes called *supercooled liquids* in which the friction between particles is very great. All A. have a tendency to convert to the crystalline state, but the process can occur extremely slowly. It can be considerably accelerated by heating.

Solid A. substances are isotropic (crystals are anisotropic). They have no defined melting point, but are converted to the liquid state by a process of slow softening. They can be distinguished from crystalline solids by x-ray diffraction because they do not give sharp interferences; instead they show only a very few diffuse interferences at small diffraction angles. Substances which give this type of x-ray diffraction diagrams are called ***x-ray amorphous***. Only a few naturally occurring substances are x-ray amorphous.

There are many methods of producing the A. state, but only a few of them are universally applicable. Most are limited to a few substances. The most important methods are evaporation and condensation onto a very cold surface (used mainly to produce thin A.layers), very rapid cooling of a melt below its freezing point (used especially for silicate, borate and phosphate glasses and A. semiconductors), precipitation from solutions, application of great mechanical stress to crystalline solids by grinding or polishing, and irradiation with high-energy particles such as ions or neutrons.

Glasses are among the most important A. Occasionally gases and liquids are called A. due to their lack of long-range order.

Amoxicillin: see Penicillins.

AMP: see Adenosine phosphates.

Amperometry: a method of electrochemical analysis based on the concentration dependence of the diffusion current. An electrochemical cell is used with a reference electrode with a constant electrode potential, an indicator electrode (dropping mercury or platinum electrode) and a suitable electrolyte. A sufficient voltage is applied to cause an electrochemical reaction of the substance to be determined at the reference electrode. The current flowing in the cell as a result of this reaction is proportional to the concentration of the reacting substance. A well known example of an amperometric determination is the oxygen determination with the Clark electrode. The principle of A. can also be used as a method for recognition of the endpoint of a titration (***amperometric titration***). In the course of the titration, the diffusion current of the substrate drops, reaching the level of the constant base current at the equivalence point. If the titrator rather than the titrand is electrochemically active, an inverted curve is obtained (Fig.).

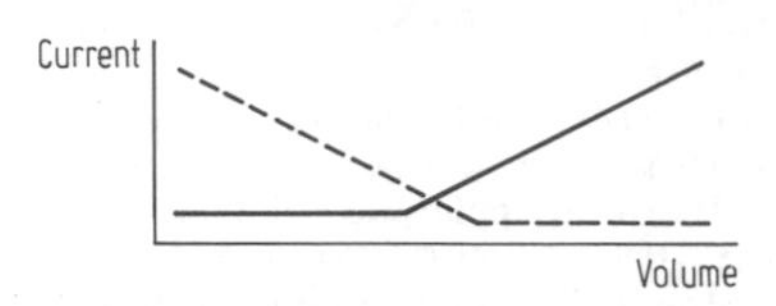

Signal curves of an amperometric titration. *dashed curve*: titrand electrochemically active; *black curve*: titrator electrochemically active.

Both in A. and in amperometric titration, a second indicator electrode can be used instead of the reference electrode. For A. with two indicator electrodes, the term ***biamperometry*** was formerly used.

"Dead stop" methods are those in which an amperometric titration is carried out with two polarizable electrodes, usually platinum electrodes. The endpoint of the titration is indicated by a sudden flow of current, or by cessation of current ("dead point"). The titrations are usually done at voltages of about 10 mV, and a titration curve is not recorded. The method requires only small amounts of equipment and is used mainly to determine small concentrations of substance.

Amphetamine: C_6H_5-CH_2-$CH(CH_3)$-NH_2, a phenylethylamine derivative which acts as a stimulant; it is addictive. The (S)-configured (+)-enantiomer is called ***dexamphetamine***. A. is the racemate.

Amphibian toxins: toxins sequestered in skin glands of various species of amphibians. Most are mixtures of simple biogenic amines, peptides, steroids and alkaloids. They affect the heart, muscles and nerves. An example of a central effect is the hallucinogenic activity of o-methylbufotenin from the Arizona toad *Bufo alvarius*. Some important A. are batrachatoxin from the Columbian arrow frog, *Phyllobaktes aurotaenia*, with a LD_{50} (mouse) of 2 μg kg^{-1}, and tarichatoxin, which is identical to Tetrodotoxin (see) from pufferfish. Tarichatoxin is present in the Californian salamander *Taricha torosa*, and its LD_{50} (mouse) is 8 μg kg^{-1}. Other A. are samandarin from the European fire salamander, *Salamandra salamandra*, with a LD_{50} (mouse) of 300 μg kg^{-1}, and bufotenin from the toad *Bufo bufo*, with a LD_{50} (mouse) of 400 μg kg^{-1}.

Amphiphilic: adjective describing a compound with both hydrophobic and hydrophilic behavior; surfactants are examples of A. substances.

Ampholyte: a term coined by Brönsted for a substance which acts as both an acid and a base, i.e. one which can either accept or donate a proton. Some typical ampholytes are:

Base effect		Ampholyte		Acid effect	
H_2SO_4	$+ OH^- \rightleftharpoons$	HSO_4^-	$+ H_2O \rightleftharpoons$	SO_4^{2-}	$+ H_3O^+$
H_2CO_3	$+ OH^- \rightleftharpoons$	HCO_3^-	$+ H_2O \rightleftharpoons$	CO_3^{2-}	$+ H_3O^+$
NH_4^+	$+ OH^- \rightleftharpoons$	NH_3	$+ H_2O \rightleftharpoons$	NH_2^-	$+ H_3O^+$
H_3O^+	$+ OH^- \rightleftharpoons$	H_2O	$+ H_2O \rightleftharpoons$	OH^-	$+ H_3O^+$

The amphoteric behavior of water is extremely important for the definition of acid and base strength (see Acid-base concepts, definition of Brønsted acids) and in connection with phenomena related to the Ion product (see) of water.

Amphotenside: amphoteric tensides (surfactants); see Surfactants.

Amphoteric: reacting with both sides. Substances which can act as both acids and bases in the Brønsted sense are A. (see Ampholytes).

Amphotericin B: a polyene antibiotic used as an antimycotic.

Ampicillin: see Penicillins.

AMS: see Herbicides.

Amygdalin: D(-)-mandelic nitrile-β-D-gentiobioside, a cyanogenic glycoside. A. is found in the seeds and leaves of members of the *Rosaceae* family. For example, bitter almonds contain 3 to 5%. A. is cleaved by emulsin, a mixture of enzymes, into glucose, benzaldehyde and hydrogen cyanide.

Amyl acetate: CH_3-CO-OC_5H_{11}, a colorless mixture of isomers with an odor like bananas or pears. The technical product has a boiling range of about 105-148 °C. A. is slightly soluble in water, but dissolves readily in most organic solvents. It is synthesized by esterification of the mixture of isomeric amyl alcohols with glacial acetic acid in the presence of sulfuric acid. A. is used mainly as a solvent for cellulose nitrate, alkyde resins, chlorinated latex, polyvinyl acetate, polystyrene, fats and oils, and as a flavoring and perfume.

Amyl alcohols, ***pentanols***: monovalent alcohols with the general formula $C_5H_{11}OH$; there are eight structural isomers (Table). Except for 2,2-dimethylpropan-1-ol, which is a solid, the A. are colorless liquids with a characteristic odor. They are only slightly soluble in water, but are largely miscible with alcohol and ether.

Amyl alcohols are injurious to the human organism. Inhaled vapors irritate the mucous membranes, and in higher concentrations act as anaesthetics. The consumption of 12.5 g amyl alcohol has approximately the same toxic effects as 100 g ethanol. Some isomers are highly irritating to the skin; an effect which can be perceived even at 10^4-fold dilution.

For the above reasons, liquors are not allowed to contain more than 0.1% fusel oils. Fusel oils contain 60 to 80% A., and were formerly the most important starting material for production of A. Today they are synthesized mainly by chlorination of pentanes, followed by hydrolysis of the amyl chlorides. As an alternative, the pentene-containing C_5 fractions formed as industrial byproducts can be used; water is added in the presence of sulfuric acid. A. are used widely as

(Amyl alcohols)
Physical properties of the isomeric amyl alcohols

Name	Formula	m.p. [°C]	b.p. [°C]	n_D^{20}
Pentan-1-ol (amyl alcohol)	CH_3-CH_2-CH_2-CH_2-CH_2OH	−79	137.3	1.4101
Pentan-2-ol* (*sec*-amyl alcohol)	CH_3-CH_2-CH_2-CHOH-CH_3	–	119	1.4053
Pentan-3-ol	CH_3-CH_2-CHOH-CH_2-CH_3	–	115.5	1.4087
2-Methylbutan-1-ol*	CH_3-CH_2-$CH(CH_3)$-CH_2OH	–	129	1.4098
3-Methylbutan-1-ol (*iso*-amyl alcohol)	CH_3-$CH(CH_3)$-CH_2-CH_2OH	–	131	1.4053
2-Methylbutan-2-ol (*tert*-amyl alcohol)	CH_3-CH_2-C(OH)(CH_3)2	–	102	1.4052
3-Methylbutan-2-ol*	CH_3-$CH(CH_3)$-CHOH-CH_3	–	112	1.3973
2,2-Dimethylpropan-1-ol	$(CH_3)_3C$-CH_2OH 52-53	52–53	113	

* Compounds with asymmetric carbon atoms.

solvents and to make esters. Amyl acetate (see) is particularly important. Pentan-3-ol is also used as a flotation agent in ore processing and as an intermediate for organic syntheses, for example, for production of diethyl ketone by dehydrogenation.

Amylases: enzymes which catalyse the cleavage of 1,4-glycosidic bonds in polysaccharides such as starch, glycogen and dextrins.

α-***Amylases*** are found in saliva, pancreatic secretion and plants, but are now produced industrially using microorganisms. They break down amylose and amylopectin into smaller components (liquification of starch), producing small amounts of maltose and glucose, but mainly the limit dextrins of higher molecular weight. α-A. attack the interior of the starch molecule. Microbial A. are metalloproteins and are stabilized by calcium ions. The pH optimum is 6.5 to 7; the α-A. from *Bacillus licheniformis* is catalytically active up to 110 °C, and is thus one of the most thermostable enzymes.

β-***Amylases*** have so far been found only in plant seeds, where they are active in the breakdown of storage carbohydrates. They are exoenzymes, attacking the molecules from the ends of the chains. They break every second 1,4-glycoside bond, releasing maltose.

Glucoamylase is an exoenzyme which degrades starches and dextrins from the non-reducing chain ends, releasing glucose. The enzyme is formed extracellularly by lower fungi, and is produced on large scale, mainly from *Aspergillus* species (300 t/year). Glucoamylase is thermostable up to 60 °C, and is used in combination with α-A. to make starch hydrolysates containing 94 to 96% glucose.

In the liquor industry, bacterial A. are increasingly replacing the plant enzymes present in sprouting grain (malt).

Amyl chlorides, ***monochloropentanes***: the eight isomeric monochloro derivatives of pentane. ***Amyl chloride (1-chloropentane)***, CH_3-$(CH_2)_4$-Cl, m.p. -99 °C, b.p. 107.8 °C, n_D^{20} 1.4127; ***2-chloropentane***, CH_3-CHCl-CH_2-CH_2-CH_3, as the racemate, m.p. -137 to -139 °C, b.p. 96.9 °C, n_D^{20} 1.4069; ***3-chloropentane***, CH_3-CH_2-CHCl-CH_2-CH_3, m.p. -105 °C, b.p. 97.8 °C, n_D^{20} 1.4082; ***1-chloro-2-methylbutane***, CH_3-CH_2-$CH(CH_3)$-CH_2-Cl, as the racemate, b.p. 99.9 °C, n_D^{20} 1.4102; ***isoamyl chloride (1-chloro-3-methylbutane)***, CH_3-$CH(CH_3)$-CH_2-CH_2-Cl, m.p. -104.4 °C, b.p. 98.5 °C, n_D^{20} 1.4084; ***tert-amylchloride (2-chloro-2-methylbutane)***, CH_3-$CCl(CH_3)$-CH_2-CH_3, m.p. -73.5 °C, b.p. 85.6 °C, n_D^{20} 1.4055; ***sec-isoamylchloride (2-chloro-3-methylbutane)***, CH_3-CHCl-$CH(CH_3)$-CH_3, as the racemic form, b.p. 92.8 °C, n_D^{20} 1.4020; ***neopentylchloride (1-chloro-2,2-dimethylpropane)***, m.p. -20 °C, b.p. 84.3 °C, n_D^{20} 1.4044.

A. are colorless liquids which are only slightly soluble in water, but dissolve readily in many organic solvents. They are synthesized by thermal chlorination of pentane-isopentane mixtures. The reaction mixture contains all isomeric A. except neopentyl chloride. This can be produced by chlorination of neopentane at 0 °C. A. are used to produce amyl alcohols, amyl amines, amyl mercaptans and aromatic amyl derivatives.

Amylenes: same as Pentenes (see).

Amylopectin: see Starch.

Amylose: see Starch.

Amylum: same as Starch (see).

Amyran: see Triterpenes.

Amyrin: see Triterpenes.

An: a frequently used general symbol for actinides.

Anabolics, ***anabolic steroids***: a group of synthetic steroids which have anabolic effects: they promote protein biosynthesis and retention of nitrogen in the body. This leads to an increase of muscle mass. In 1935, the anabolic effect of testosterone was discovered. In order to separate the androgenic and anabolic effects of the hormone, so that the anabolic effect could be utilized in therapy of women and children, partially and totally synthetic steroids were developed. Some examples of A. are *nandrolone* (Fig.) and its esters, such as the decanoate; others are *methenolone acetate (Primobolane®)* and *stanozolol (Stromba®)*.

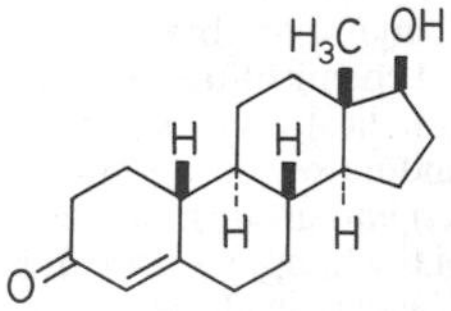

Nandrolone

Anabolic steroids: same as Anabolics (see).

Analeptics: compounds which enhance the excitability of the central nervous system. In therapeutic doses, they act as stimulants, especially of the respiratory and vasomotor centers. In higher doses, they cause cramps, which can be relieved by barbitals. Natural products such as strychnine, picrotoxin, camphor and lobelin have analeptic effects. They are now rarely used for this purpose, however. Caffeine has a weak analeptic effect. Today the most commonly used A. are synthetic compounds such as pentetrazol and bemegrid, and various *N,N*-dialkylsubstituted aromatic or heteroaromatic carboxylic acid amides. Phenylethylamine derivatives which stimulate the central nervous system, and thus counteract the need for sleep, are sometimes considered A. These compounds temporarily increase physical and mental capacities, and in some persons cause euphoria. They have only slight effects on the heart activity and blood pressure. A side effect is depression of the appetite, which has been utilized in Appetite suppressants (see). The best known phenylethylamines of this type are Amphetamine (see) and Methamphetamine (see).

Analgesics: compounds which, in therapeutic doses, are able to relieve pain through their effects on centers in the cerebral cortex. The do not affect the other functions of the central nervous system, and do not cause loss of consciousness. Strong A. are also called *narcoanalgesics*; weaker A. cannot suppress severe pain, but often also act as Antipyretics (see) and Antiphlogistics (see).

The strong A. include morphine and structurally similar compounds, pethidine and methadone. The weak A. include derivatives of salicylic acid, such as acetylsalicylic acid and salicylamide, pyrazolones such as phenazone, aminophenazone, analgin and propyphenazone, and aniline derivatives, such as phenacetin and paracetamol.

Various A. are combined with other drugs, such as barbitals, and are very widely used.

Analgin: see Pyrazolones.

Analysis: the determination of the composition of a substance or mixture of substances. The process is subdivided into qualitative, quantitative and structural A.

Qualitative A. is the determination of the nature of the components of a substance. In inorganic chemistry, qualitative A. is essentially elemental analysis, but in organic chemistry, where the elemental compositions of the molecules are very similar, a different type of information is also sought. Both chemical and physical methods are used. This section treats mainly the chemical methods, because the physical methods can also be used for quantitative and structural analysis, and are discussed in those sections.

Preliminary tests are used to determine whether certain substances are present, and/or to obtain information on the composition and chemical behavior of the sample. They are used to indicate the best strategy for the subsequent A., and thus precede but do not take the place of separation into groups. General information on the behavior of the sample is provided by solubility experiments, dry heating in closed tubes, Blowpipe analysis (see) and Microcosmic salt beads (see). There are also special tests used for certain elements or substances. Such tests are used especially to detect strong poisons, for example, the Berzelius test (see), the Bettendorf test (see), the Gutzeit test (see) and the Marsh test (see) indicate the presence of arsenic. In organic qualitative A., certain tests (see Beilstein test, Hepar test and Lassaigne test) indicate the presence of certain elements.

After the preliminary tests, solid samples are dissolved, which homogenizes them. If this is not possible, or if a solid residue remains, Solubilization (see) is required. The components of the resulting sample solution are then identified by adding reagents which cause them to precipitate or change color. The sensitivity of such reactions can be characterized by the Limit concentration (see) and the Limit of detection (see). Identification of a component is only positive if the reagent is specific for that component alone. However, most reagents are not selective for a single substance, but rather react with a number of substances, so the identification is not unique.

A systematic *separation process* is used to separate the components of a sample into groups. Usually this is done by adding "group reagents" in a sequence which separates possible components into groups by precipitation. Within each group, a positive identification of the components can be made with nonspecific reagents or reactions. Different separation processes are known for cations, anions and organic compounds. In the Fresenius system for cations (Table 1), the *hydrochloric acid group*, silver, mercury(I) and lead ions, is precipitated as chlorides, and thus separated from the other cations.

Table 1. The Fresenuis separation system for cations.

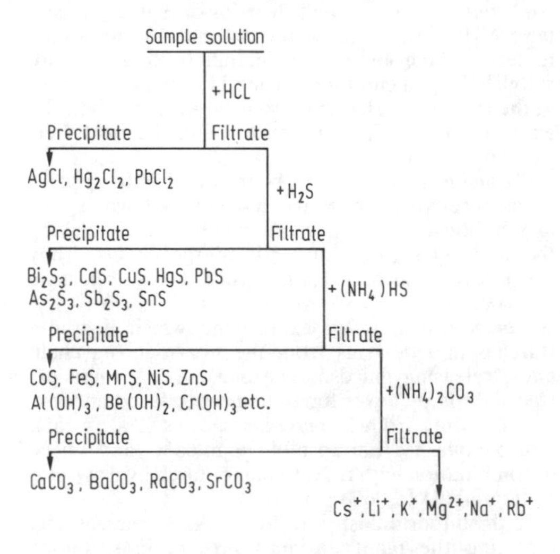

The *hydrogen sulfide group*, consisting of arsenic, antimony, lead, cadmium, copper, mercury(II), bismuth and tin ions, is separated as the insoluble sulfides. The *ammonium sulfide group* consists of iron, cobalt, nickel, manganese, indium, thallium and zinc ions, which are precipitated as their sulfides, and aluminum, beryllium, chromium, hafnium, lanthanum, niobium, tantalum, titanium, thorium, scandium, uranium, yttrium and zirconium ions, which are precipitated as hydroxides or oxide hydrates. The *ammonium carbonate group* consists of calcium, barium, radium and strontium ions, which precipitate as carbonates.

In this way it is possible to determine the qualitative composition of all inorganic substances using chemical methods. Methods of Microanalysis (see) are available for use with very small amounts of sample; Drop analysis (see) is a special type of microanalysis.

In qualitative organic A., the elements can be identified with simple preliminary tests, but it is difficult or imposible to identify definite compounds by chemical methods alone. Therefore, the first step is to identify the functional groups which are present, e.g. carboxyl, carbonyl, hydroxyl, amino, nitro, mercapto and sulfonic acid groups, using chemical methods. It is often necessary to carry out separations, such as distillation, extraction, crystallization, or especially chromatography, before identification is possible. The isolated compounds are then often identified by characteristic constants or by means of other physical methods. In all qualitative A., the course of the identification reaction or the intensity of the signal from a physical method gives an indication of the amount of the component present.

Quantitative A. is used to determine the amounts of the components of a substance. It requires knowledge of the qualitative composition of the sample. In many cases, the results of quantitative A. are economically important, as in the determination of metal contents of ores, control of the compositions of alloys, or quality control of chemical processes. In such cases, the sample is taken from a very large amount of material, and accurate Sampling (see) techniques are required. There are many different physical and chemical methods for quantitative A. Chemical methods require relatively little apparatus and do not need to be calibrated, because they are based on the laws of Stoichiometry (see). Their disadvantage is the amount of time they require, which is often considerable, and their unsuitability for series A. They are also of little use for continuous determinations for controlling and directing industrial processes.

For quantitative A., as for qualitative A., solid samples must first be dissolved using chemical methods. Before the actual determination, the component to be determined must often be separated from interfering substances which may be present. This can be done by extraction or chromatography. In some cases, such separations can be avoided by use of Masking reagents (see).

The chemical methods of quantitative A. can be classified as 1) Gravimetric A. (see), 2) Volumetric A. (see), electrochemical A., such as Amperometry (see), Coulometry (see), Conductivity measurements (see), Polarography (see) and Potentiometry (see), 4) optical methods such as Colorimetry (see), Nephelometry (see) and Photometry (see).

In contrast to chemical methods, the physical methods of quantitative A. are relative rather than absolute, and must therefore be calibrated. The composition of the calibration reagents must be determined by very precise chemical methods. The advantage of physical methods is that they can be done rapidly and are especially suitable for series analyses; the sample preparation is also often simpler. The disadvantage is that the equipment is often very expensive. These methods can be classified on the basis of the underlying physical processes, for example, 1) atomic spectroscopy (see Atomic emission spectroscopy, Atomic absorption spectroscopy and Atomic fluorescence spectroscopy) 2) x-ray spectroscopy (see X-ray fluorescence analysis), 3) radiochemical methods (see Activation analysis) and 4) Mass spectroscopy (see).

The methods used for structural A. can also be used in special cases for quantitative A.

The results of a quantitative A. are usually given in mass % of the components. If all the components of a substance are determined, one speaks of ***complete A.***

The methods of A. can be classified according to the amounts of sample used: ***macroanalyses*** require about 1 g sample, while ***microanalyses*** require only μg to mg of sample. Further subdivisions are ordinarily indicated by use of the prefixes "semi-", "ultra-" and "sub-".

Another classification can be based on the amounts of the components. Major component A. is concerned with the components which make up 10 to 100% by mass, minor component A. deals with the range of 1% to 10%, and Trace analysis (see) applies to the range below 1%. Traditionally, the methods of A. are also classified according to the nature of the substance under study, for example, Elemental A. (see), 2) Gas A. (see), 3) Clinical A. (see), 4) Metal A. (see), 5) Silicate A. (see) and Water A. (see).

The capacities of the various methods cannot be generalized. For example, the values for reproducibility and accuracy of an analytical method which are determined by statistical methods depend also on the nature and composition of the sample. On the other hand, the limits of detection of the various methods can be relatively accurately indicated (Table 2).

Table 2. Comparison of the limits of detection of a few selected analytical methods with those of natural sense organs

Method	Limit of detection in g
Chemical methods	to 10^{-9} g
Electrochemical methods	10^{-12}
Atom emission spectrometry	10^{-12}
Mass spectroscopy	10^{-13}
Human nose	10^{-13}
Activation analysis	10^{-14}
Electron microprobe	10^{-16}
Canine nose	10^{-18}
Bee sense of smell	10^{-20}

Structural A. is concerned with elucidation of the structure of a compound, and is of great importance in organic chemistry, but is also applied in inorganic chemistry, especially in complex chemistry. The sample must be a pure substance, not a mixture. The main methods of structural A. are those of molecular spectroscopy, e.g. Infrared spectroscopy (see), UV-VIS spectroscopy (see) and NMR spectroscopy (see), combined with Mass spectroscopy (see). These methods do not give direct information on the structure of a compound, but only indicate the ranges in which a molecule can absorb energy. The connection between the structure of the molecule and the signal of the method can only be interpreted on the basis of empirical comparisons. A combination of the results of different methods often provides a clear indication of the structure of the compound. However, it is not possible in this way to obtain information on the bond lengths and angles in large molecules. A complete structural elucidation of very complicated compounds is only possible with diffraction methods, such as x-ray and electron diffraction (see X-ray structure analysis). These methods can be used only with solids in the form of single crystals (to some extent, also with powders), and require extensive calculations, so they have only become widely applicable since the introduction of modern data analysis systems. The result is a complete picture of the structure of the compound studied, which, however, is exact only for its crystalline state.

Analytical chemistry: see Chemistry.

AN-D explosives: abb. for ammonium nitrate-diesel explosives used in industry. They consist of 92.5 to 96% fine-grained, free-flowing ammonium nitrate, NH_4NO_3 and 4 to 7.5% diesel fuel. Because they are not very sensitive to mechanical effects (they require a special ignition cap for detonation), these are relatively safe explosives. In some countries, they are mixed at the site of use, and then poured or blown into the boreholes with compressed air.

Androgens: a group of steroids which act as male sex hormones. The natural A. are derivatives of ***androstane***; they thus contain 19 C atoms and have oxygen functions on C atoms 3 and 17. The main natural A. is ***testosterone***, 17β-hydroxyandrost-4-ene-3-one. Other derivatives of androstane, such as the testosterone metabolite ***androsterone***, have an androgenic effect, although it is much weaker. Testosterone is synthesized in the interstitial cells of the testes, the *Leydig cells*, in response to gonadotropic hormone (see Gonadotropins). 17α-Hydroxyprogesterone is converted to 4-androstene-3,17-dione by removal of the C_2 chain on C-17; this is then reduced to testosterone, a white crystalline solid with m.p. 155 °C, $[\alpha]_D^{20}$ + 109° (ethanol).

Androsterone   Mesterolone

Name	R^1	R^2
Testosterone	H	H
Testosterone propionate	CH_3-CH_2-CO-	H
Testosterone enanthate	$CH_3-(CH_2)_5-CO-$	H
Methyl testosterone	H	CH_3

Testosterone is soluble in acetone, ethanol and chloroform. It is partially synthesized from androstenolone (5-androsten-3-ol-17-one), or is produced microbiologically from progesterone. The sexually specific effects of testosterone include induction of secondary male sexual characteristics and maintenance of accessory gland function in the genital tract, spermogenesis and libido. In addition, it has sexually unspecific, anabolic effects (see Anabolics). Testosterone is injected in patients who do not produce enough androgens; orally applied testosterone is too rapidly metabilized to be useful. Esters, such as testosterone propionate, are effective for longer periods and are used for that reason. Esters with longer-chain fatty acids, such as the enanthate (heptoate), have depot effects. ***Methyl testosterone***, with a 17α-methyl group, and ***mesterolone***, with a 1α-methyl group, are effective upon oral administration. The 17α-methyl group inhibits dehydrogenation to the 17-keto compound, and the 1α-group prevents binding to metabolizing enzymes.

Historical: The first A. isolated was androsterone, which was extracted from human urine in 1931 by Butenandt and Tscherning. Testosterone was isolated from bovine testes in 1935 by Laqueur. Its structure was determined and partial synthesis reported in the same year; the principal investigators were Butenandt and Ružcvka. Total synthesis was achieved in 1955 by Johnson.

Androstane: see Androgens.

Androsterone: see Androgens.

Andrusov process: see Natural gas.

Anergy: see Exergy.

Anesthesin®: see Anesthetics, local.

Anesthetics, local: compounds which reversibly inhibit the excitability and conductivity of nerves in a limited area. The effect depends on an inhibition of the flow of sodium ions into the nerve cells. A. are used for surface or mucous membrane anesthesia, in which primarily the nerve endings are affected, for conduction anesthesia, in which the transmission of impulses in a nerve trunk is interrupted, and for infiltration anesthesia, in which the nerve endings and smaller nerve fibers in the infiltration area are affected.

A. which are relatively stable to metabolism are also used in cases of cardiac arhythmia. The first A. was the alkaloid Cocaine (see), which was used in opthamology in 1884. Synthetic compounds replaced natural alkaloids many years ago. The first of these was ***benzocaine*** (***Anesthesin®***), 4-$(H_2N)C_6H_4COOC_2H_5$, the ethyl ester of 4-aminobenzoic acid. Because of the low basicity of the amino group, solutions of the hydrochloride are acidic and are not suitable for injection. The next development was of basic esters of 4-aminobenzoic acid; the first of this group was ***procaine*** (***Novocaine®***), 4-$(H_2N)C_6H_4COOCH_2CH_2N(C_2H_5)_2$, the 2-diethylaminoethanyl ester of 4-aminobenzoic acid. Aqueous solutions of procaine hydrochloride are nearly neutral and are very suitable for injection. Later, variations in the substituents on the benzene ring and in the alcohol component were developed. The most widely used of these compounds is ***tetracaine***, 4(n)C_4H_9-NH-C_6H_4-COO$(CH_2)_2$-N$(CH_3)_2$, a surface anesthetic. A. with ester structures are rather unstable. Base-substituted carboxyanilides such as ***lidocaine*** (***Xylocitrin®***), 2,6-$(CH_3)_2$-C_6H_3-NHCOCH$_2$-N$(C_2H_5)_2$, are more stable, due to the ortho-substituent and the anilide structure. The surface anesthetic ***propipocaine***

(*Falicaine*®) is an example of a β-aminoketone type A.

Anethol: 4-methoxypropenylbenzene, a phenol ether. The crystals are colorless, pleasant-smelling platelets; m.p. 22-23 °C, b.p. 235 °C, n_D^{20} 1.5615. A. is insoluble in water, soluble in ethanol, ether and other organic solvents. It is synthesized from 4-methoxyphenylmagnesium bromide and allyl bromide. The primary product is estragol (4-methoxyallylbenzene); this isomerizes in the presence of alkali hydroxide to A. Oxidation by nitric or chromic acid forms anisaldehyde. A. is the main component of anis oil, and is also present in tarragon and fennel oil. It can be extracted from anis oil and is used in perfumes and liquors. Because of its mucus-loosening effect, it is also used in cough medications.

OCH_3 (para) – benzene ring – $CH{=}CH{-}CH_3$

Anethol

Aneurin: same as Vitamin B_1 (see).

Angelicin: see Furocoumarins, Table.

Angeli-Rimino reaction: a method of synthesizing hydroxamic acids from aldehydes and benzenesulfonylhydroxylamine under alkaline conditions:

$$R{-}CHO + C_6H_5{-}SO_2{-}NH{-}OH \xrightarrow{(OH^-)} R{-}CO{-}NH{-}OH + C_6H_5{-}SO_2H$$

The A. is also a reductive conversion of the sulfonic acid derivative to a sulfinic acid.

Angiotensin, ***angiotonin, hypertensin***: a peptide tissue hormone. The inactive precursor of A., ***Angiotensin I*** is released from a plasma α_2-globulin (angiotensinogen) by the kidney protease renin. A converting enzyme, a chloride-dependent, EDTA-sensitive dipeptidase, converts A. I to a vasoactive octapeptide, ***angiotensin II***. The [Ile^5]A.II of human beings, pigs and horses has the same biological effects as [Val^5]A.II from cattle.

Angiotensinogen

Asp–Arg–Val–Tyr–Ile–His–Pro–Phe–His–Leu–Leu–Val–Tyr–Ser–
(1, 5, 10)

↓ Renin

Asp–Arg–Val–Tyr–Ile–His–Pro–Phe–His–Leu Angiotensin I
(1, 5, 10)

↓ Converting enzyme

Asp–Arg–Val–Tyr–Ile–His–Pro–Phe Angiotensin II
(1, 5, 8)

A. strongly elevates blood pressure and stimulates aldosterone production in the adrenal cortex. It also has an effect on the smooth muscles, which is not thought to have any physiological significance. The first total synthesis was reported in 1957. Many analogs have been prepared; the antagonists are of great interest, as are inhibitors of the converting enzyme, for reducing high blood pressure.

Angiotensinogen: see Angiotensin.

Angiotonin: same as Angiotensin (see).

Anhalonium alkaloids: see Alkaloids, Table.

Anharmonic oscillator: see Infrared spectroscopy.

Anhydrite: anhydrous calcium sulfate; see Gypsum.

Anhydro compound: see Purpurea glycosides.

Anhydrosugars: intramolecular ethers of monosaccharides. The oxygen ring usually contains three or five atoms. The A. are synthesized from halogen compounds, or sulfonate or sulfate esters. A. occur naturally, in the form of 3,6-anhydro-L-galactopyranose residues in various galactans.

Anid®: see Synthetic fibers.

Aniles: see Schiff's bases.

Anilides: N-phenyl-substituted carboxamides with the general formula $R{-}CO{-}NH{-}C_6H_5$. They can be hydrolysed by bases or mineral acids to aniline and the corresponding acids. A. are formed by reaction of anhydrous carboxylic acids, acyl halides or anhydrides with aniline. For example, acetanilide is formed from acetic anhydride and aniline: $(CH_3{-}CO)_2O + NH_2{-}C_6H_5 \rightarrow CH_3{-}CO{-}NH{-}C_6H_5 + CH_3{-}COOH$. A. are often used to characterize carboxylic acids and in the production of drugs and pesticides.

Aniline, ***aminobenzene, phenylamine***: $C_6H_5{-}NH_2$, the simplest aromatic primary amine. A. is a colorless, oily liquid with an unpleasant odor, which quickly turns brown in the air; m.p. - 6.3 °C, b.p. 184 °C, n_D^{20} 1.5863. A. is nearly insoluble in water but dissolves readily in most organic solvents; it can be steam distilled. It can be detected by the Runge lime chloride test; oxidation of A. leads to a characteristic red-violet, quinoid pigment. The free electron pair on the N atom makes A. basic, but resonance interaction of this pair with the ring electrons makes A. less basic than aliphatic primary amines. It forms stable salts with strong acids, for example, Aniline hydrochloride (see) with hydrochloric acid. Acylation of A. produces Anilides (see). Another important reaction of A. is condensation with aldehydes and ketones, forming Schiff's bases. The reaction of A. with nitric acid produces a phenyldiazonium salt which is essential for the production of azo dyes. A. reacts wtih carbon disulfide on heating in alcoholic solution, forming N,N'-diphenylthiourea, from which phenyl isothiocyanate can be formed. Under suitable conditions, A. can be oxidized specifically to phenylhydroxylamine, nitrosobenzene or nitrobenzene. The oxidation of A. to *p*-benzoquinone is important, and its condensation reactions with azobenzene derivatives are of interest for synthesis of symmetric or unsymmetric azobenzene derivatives. Under strongly basic conditions, A. forms the unpleasant-smelling benzoisonitrile (see Isonitrile reaction).

Aniline is a strong blood and nerve poison. It can enter the body through the lungs or skin. Very small amounts can be detoxified in the body by conversion to 4-aminophenol, and excreted in this form. Mild poisoning produces blue coloration of the lips, nose, ears and fingernails (cyanosis), dizziness and excitability. Oral intake leads to vomiting, cramps and diarrhea. In severe poisoning, aniline causes headaches, apathy, faintness, loss of awareness, irritation of the bladder and difficulty in breathing. Chronic symptoms are weakness, loss of appetite, neurasthenic symptoms, skin rashes and bladder growths which can become carcinomas. In the blood, aniline leads to formation of methemoglobin, and thus to loss of oxygen transport.

First aid: bring the patient into fresh air; remove clothing which is soaked in aniline, and wash the affected skin. Spray with cold water, give artificial respiration, preferably with oxygen, and call the doctor!

A. occurs naturally in coal tar. Industrially, it is made in large amounts by the Bechamp method, by reduction of nitrobenzene with iron and hydrochloric acid, or by catalytic hydrogenation of nitrobenzene in the presence of copper catalysts. It is also produced industrially by ammonolysis of chlorobenzene or phenol with suitable catalysts. A. is one of the most important starting materials for industrial synthesis of aromatic products, such as dyes, pharmaceuticals, photographic developers, pesticides, substituted A., A. resins, phenylhydrazine, etc.

Historical. A. was first isolated in 1826 by Unverdorben, by lime distillation of natural indigo. It was later detected in coal tar distillate by Runge (1834); in 1841, Fritsche discovered the same compound in alkali melts of indigo. Finally, it was synthesized in 1842 by Sinin, by reduction of nitrobenzene. In 1843, Hofmann demonstrated the identity of the four isolated substances. After Perkin had made the first synthetic aniline dye, industrial production of A. began in 1857.

Aniline black: an important black developing dye. A. is almost completely insoluble in water. It is nearly always synthesized directly on the fiber by oxidation of aniline with potassium chlorate (***chlorate black***) or potassium dichromate (***dichromate black***). A complicated series of reactions produces a chain of eight aromatic rings. Some intermediate products are the red and blue ***emeraldine bases***, which are indamines and are converted to light-fast triazine derivatives. A. was discovered in substance by F.F. Runge.

Aniline dyes: see Pigments.

Aniline hydrochloride, ***anilinium chloride***: C_6H_5-NH_3Cl, the most important salt of aniline. A. forms white crystals with m.p. 198 °C, b.p. 245 °C. It is soluble in hot water and alcohol, and insoluble in ether. In aqueous solution, A. is largely dissociated. It is often used in syntheses instead of aniline, for example in the production of phenylisocyanate from A. and phosgene. A. is produced from aniline and hydrochloric acid, and is used for synthesis of various aniline intermediates and dyes.

Aniline purple: same as Mauveine (see).

Aniline resin: a hard aminoplastic. There are three types. 1) Non-cross-linked, non-hardening, brittle, colophony-like, soluble and meltable A. are obtained by condensation in neutral or slightly alkaline medium.

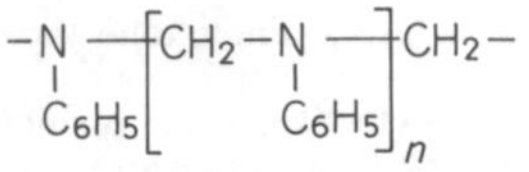

$$-N(C_6H_5)-[CH_2-N(C_6H_5)]_n-CH_2-$$

1

The starting materials for this type are formaldehyde and, usually, anilines, but toluidines, naphthylamines, diamines and secondary amines (e.g. benzylaniline) are also used. Other aliphatic or aromatic aldehydes can be used instead of formaldehyde. This type of A. is used for paints and textile conditioning.

2) Cross-linked, non-melting A. are made by condensation in highly acidic media. These materials are thermoplastic around 150 °C, but they do not become liquid. The aldehyde component is always formaldehyde, and only primary aromatic amines are used. The primary chains are cross-linked by addition of excess formaldehyde, creating resins which can be used with fillers to make plastics with good mechanical and electrical properties.

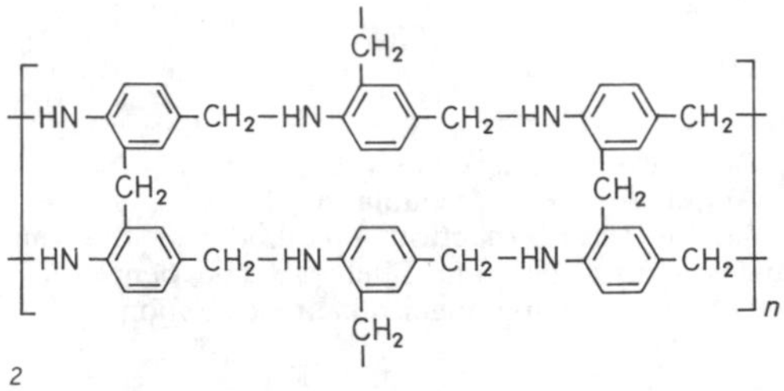

2

3) Completely insoluble A. which do not soften when heated are made by condensation of aniline and other amines with furfural.

Aniline yellow: same as 4-Aminoazobenzene (see).

Anilinium chloride: same as aniline hydrochloride.

Anilinoacetic acid: same as *N*-Phenylglycine (see).

Animal pigments: natural pigments from animals; see Pigments.

Animal starch: same as Glycogen (see).

Animal toxins: Poisons (see) regularly formed by many animals under normal living conditions; a high toxicity is necessary. The A. include chemically very diverse substances; most of them are mixtures of different components (e.g. the snake toxins). Their chemical nature has not been thoroughly explored, and the toxic effects cannot be ascribed to chemically defined substances in every case.

Many A. contain proteins of high molecular weight; these are related to the microbial toxins. Many of them have high sulfur contents, which often parallel their toxicity.

Toxic substances of low molecular weight include many which are probably protein degradation products, e.g. methylated amines in the foul-smelling secretions of many beetles and caterpillars; betaines such as stachydrin and trigonellin in sea urchins; adrenalin and tryptamin derivatives (e.g. bufotenin) in various toad poisons.

The effects of the individual A. and their components are various, and the mechanisms are often unknown. The heart is commonly affected (toad poisons), or the blood (e.g. agglutination, coagulation, inhibition of clotting), or central nervous system (e.g. depression of the respiratory center, excitation). The most important of the A. are found under the following separate entries: Ant toxins, Amphibian toxins, Bee toxins, Fish toxins, Insect toxins, Jellyfish toxins, Snake toxins, Scorpion toxins.

Anion: an Ion (see) with one or more negative charges. The term was originally defined by Faraday only in connection with electrolysis: the particles which move toward the anode (the positive electrode) in electrolysis are A.

Anion base: see Acid-base concepts, section on Brønsted definition.

Anion chromatography: see Ion chromatography.

Anion exchanger: see Ion exchanger.

Anionics: non-ionic Surfactants (see).

Anion acid: see Acid-base concepts, section on Brønsted definition.

Anisaldehyde, ***4-methoxybenzaldehyde***: a colorless, pleasant-smelling liquid; m.p. - 0.02 °C, b.p. 248 °C, n_D^{20} 1.5730. A. is slightly soluble in water and infinitely soluble in alcohol and ether. It is steam volatile and can be separated from reaction mixtures and purified by steam distillation. Like most Aldehydes (see), A. is very reactive and undergoes the typical addition and condensation reactions, with the exception of those in which α-H atoms are required. It is readily converted to anisic acid by oxidation in air. A. occurs in nature as a component of anis and fennel oils. It can be synthesized technically from anethol by oxidation with nitric acid, chromic acid or ozone. A. can also be synthesized by methylation of 4-hydroxybenzaldehyde or by a Vilsmeier-Haack reaction (see) from anisol. A. is used mainly as a perfume.

$CH_3O-C_6H_4-CHO$

Anisidines, ***aminoanisoles, methoxyanilines, aminophenyl methyl ethers***: the three isomeric methoxy deriviatives of aniline. ***o-Anisidine*** is a yellow, unstable liquid; m.p. 6.2 °C, b.p. 224 °C, n_D^{20} 1.5713. ***m-Anisidine*** is an oily liquid, b.p. 251 °C, n_D^{20} 1.5794. ***p-Anisidine***, crystal platelets (from water); m.p. 57.2 °C, b.p. 115 °C at 1.73 kPa. The A. are slightly soluble in water, but readily soluble in alcohol, acetone, ether and benzene. They are more basic than aniline, due to the electron-shifting effects of the methoxy group. The A. are synthesized by reduction of the corresponding isomeric nitroanisols. They are used in the production of drugs and azo dyes.

NH_2, OCH_3
(position 2, 3 or 4)

Anisol, ***methoxybenzene, methyl phenyl ether***: C_6H_5-OCH_3, a colorless, combustible liquid with a pleasant odor; m.p. - 37.5 °C, b.p. 155.4 °C, n_D^{20} 1.5179. A. is insoluble in water but readily soluble in ethanol and ether. It is obtained by reaction of phenol with dimethylsulfate in alkali hydroxide solution. It is used as a high-boiling solvent in organic syntheses, e.g. in Grignard reactions.

Anisotropy: a dependence of physical properties and behavior of substances on direction. A. is most common in crystals, but is also found in liquid crystals and, under certain conditions, even in liquids. Polycrystalline substances can be anisotropic if the crystallites are not randomly oriented (see Texture). The A. of a crystal is determined by its lattice structure (geometric A.); it is observed, e.g. in electrical and thermal conductivity, elasticity, hardness, cleavage directions, thermal expansion, rate of crystal growth and optical properties. All crystals are not equally anisotropic; e.g. cubic crystals are optically isotropic, although they are anisotropic in other respects, such as growth rates and cleavage planes. Antonym: Isotropy (see).

Anisic acid, ***4-methoxybenzoic acid***, 4-CH_3O-C_6H_4-COOH, colorless prisms or needles, m.p. 184 °C, b.p. 277 °C. A. is slightly soluble in water, but readily soluble in alcohol, ether and benzene. It dissolves in bases, forming salts. A. is present in anethol-containing essential oils. It can be synthesized by oxidation of anethol with potassium permanganate or by a Grignard reaction (see) from 4-methoxyphenylmagnesium bromide and carbon dioxide.

Annealing: treatment of a hardened or cold-worked metal object (usually steel) at a temperature below the first crystal-structure conversion temperature. A. at high enough temperature to improve the mechanical properties is called tempering. During A. thin layers of oxide form on the surface of the metal, and these display interference colors. If objects with clean surfaces are used, these colors permit certain determination of the annealing temperature in the range from 220 to 230 °C.

A. may also be used to dissolve alloy components which precipitate out at low temperatures, or, in a vacuum or under a protective gas, A. can be used to remove an oxide layer.

Annihilation: the interaction between two electronically excited species leading to formation of one species in the ground state and one in an electronically excited state. In triplet-triplet A., both of the interacting species are in excited triplet states; one of the resulting species is in the singlet ground state and the other is in a singlet excited state. The singlet excited species is usually deactivated by a delayed fluorescence.

Annulenes: cyclic, conjugated hydrocarbons with different ring sizes. The A. are named by placing the even number *n* of CH groups in the ring system in square brackets in front of the word "annulene"; *n* = 4, 6, 8, 10, etc.

In addition, the compounds can be given systematic names: cyclobutadiene, cyclooctatetraene, cyclodecapentaene, etc. The "hypothetical" cyclohexatriene does not exist as a cyclic conjugated ring system and is excluded (see Benzene). Cyclobutadiene

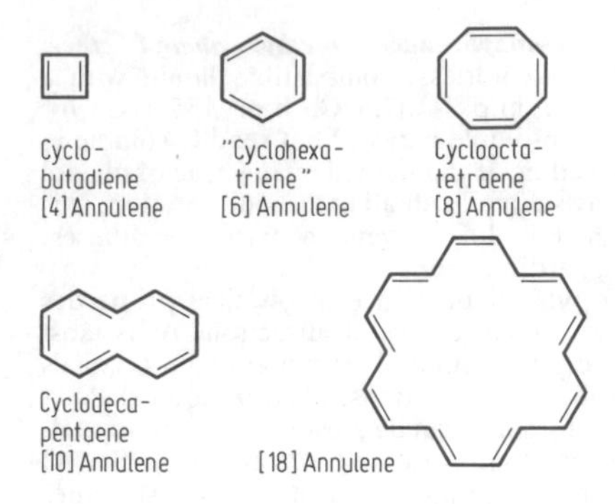

(see) and Cyclooctatetraene (see) are also often treated separately. The four- and eight-membered rings and other A. with $4n\pi$ electrons (n = 1, 2, 3, 4...) have higher energies than hydrocarbons with the corresponding numbers of isolated double bonds, and are called "*antiaromatic*". If the cyclic conjugated system contains $4n + 2$ electrons, as benzene does (n = 1), the system has lower energy than comparable systems with isolated double bonds, and it is *aromatic* (Hückel rule). [10]Annulene corresponds to the Hückel rule with respect to its electronic structure. It has 10 π electrons in a cyclic conjugated ring system, but it is not planar and is therefore not resonance stabilized; the two H atoms in positions 1 and 6 prevent flattening. Because of this steric hindrance, it does not generate a diamagnetic ring current in a magnetic field, as would be typical of an aromatic, and it is atropic. [14]Annulene, too, which should be aromatic by the Hückel rule, is not completely planar for steric reasons and does not display a diamagnetic ring current. Only in [18]annulene are all the conditions for an aromatic compound fulfilled. In addition to having $4n + 2\ \pi$ electrons (n = 4), it is completely planar and has a diamagnetic ring current. Therefore, in a nuclear magnetic spectrometer, all the peripheral protons give a single signal because all are equivalent (δ = 8.8 ppm). [18]Annulene is diatropic, and it is chemically very stable. It is one of the non-benzoid aromatics.

A. with 12, 16 or 20 ring members are antiaromatic and display the properties of polyenes. [16]Annulene has been particularly thoroughly studied; in it, a paramagnetic ring current is induced by an external magnetic field. This means that the signals from the external ring protons are shifted to higher field, while those from the internal ring protons are shifted to lower field.

Anode: see Electrode.

Anodic corrosion protection: see Electrochemical corrosion protection.

Anodic oxidation: oxidation which occurs at the anode of an electrolysis cell through which current is flowing. A. is the basis of various industrial processes, e.g. electrolytic production of chlorine.

Protective oxide coatings on metals, especially aluminum, magnesium and their alloys, can also be produced by A. In aluminum, A. is also called ***aloxidation*** or ***eloxation*** (***eloxal process***, abb. for electrolytically oxidized aluminum). This process is used to generate a corrosion- and wear-resistant layer, 10 to 30 μm thick, on the surface of aluminum and its alloys. The workpiece is connected to the cell as anode, and dipped in a bath of sulfuric or oxalic acid. The resulting colorless layer can also be dyed with an organic pigment, e.g. for materials used in construction of buildings. The porosity in the layer can be removed by boiling in water or treatment with hot chromate, acetate or silicate solution; this process is called ***sealing***. The layers are electrically insulating.

A special process of A. is ***hard eloxation***, which produces gray to gray-black, hard and abrasion-resistant layers up to about 150 μm thick.

With other metals, A. can be used to a certain extent to produce thin, non-porous oxide layers with special electrical properties, e.g. for electrolyte capacitors and rectifiers. A. is sometimes used to create an oxide layer with better lubricating qualities than the underlying metal, which makes it easier to shape the piece.

Organic substrates can also be subjected to A. (see Organic electrosynthesis). Finally, A. is also used in waste-water treatment.

Anolyte: see Electrolysis.

Anomers: see Stereoisomerism, 1.2.1; see Monosaccharides.

Anorexics: same as Appetite suppressants (see).

Ansa compounds: see Cyclophanes.

Ansamycins: antibiotic ansa compounds produced by microorganisms (*Streptomyces, Nocardia* and *Micromonospora* species). They contain an aromatic nucleus which is bridged by an aliphatic side chain. The A. are classified on the basis of the aromatic nucleus as naphthalene or naphthoquinone derivatives, including the streptovaricins, rifamycins, halomycins, tolypomycins and naphthomycins, and the benzene or benzoquinone derivatives, including, for example, geldamycin. The maytansinoids isolated from plants, which have cytotoxic effects, are also A. Only the Rifamycins (see) are therapeutically important.

Antabuse®: same as Disulfiram (see).

Antacids: compounds or preparations which are used to bind excess hydrochloric acid in the stomach. The pH of gastric juice should be raised to 3.0 to 4.0; in no case should the content of the stomach be neutralized, because in this case more gastric juice will be secreted. This danger is present when carbonates (e.g. magnesium carbonate), hydrogencarbonates (e.g. sodium hydrogencarbonate) or oxides (e.g. magnesium oxide) are used. Colloidal aluminum hydroxide and aluminum magnesium silicates are better for the purpose, because they have a good capacity to bind acid without increasing the pH above 4.0.

Antamanid: cyclo-(Pro-Phe-Phe-Val-Pro-Pro-Ala-Phe-Phe-Pro-), a homodetic, cyclic decapeptide isolated from the death's cap mushroom which can counteract the toxicity of phalloidin (see Phallotoxins) and α-amanitin (see Amatoxins). Reliable protection against the lethal effect of the toxins is only obtained, however, when the protective dose of A. (0.5 mg/kg mouse against 5 mg/kg phalloidin) is ingested before or at the same time as the poison. A. was discovered in 1968 by T. Wieland and coworkers, who also elucidated the structure and synthesized it.

Antarafacial: see Woodward-Hoffmann rules.

Antelepsin®: see Benzodiazepines.

Anthelminthics: compounds used to combat parasitic worms in human beings and animals. They are effective against worms in the gastrointestinal tract, and in some cases, against tissue parasites. Some compounds effective against nematodes (threadworms) and ascarides (eelworms) are piperazine, cyanine dyes (such as pyrvinium embonate), benzimidazol derivatives (e.g. mebendazol) and cyclic amidines (such as pyrantel). At present, the most important medication for cestodes (tapeworms) is the salicylanilide derivative *niclosamide*.

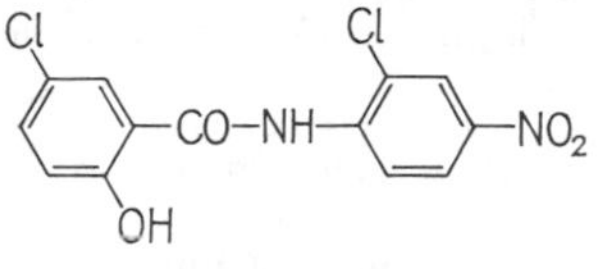

Cl Cl CO–NH NO_2 OH

Niclosamide

Anthocyanidins: see Anthocyans.

Anthocyanins: a group of glycosidic plant pigments. The most important of them are cyanin, pelargonin, delphin, idaein, malvin, petunin, keracyanin, micocyanin, fragarin, paeonin, oenin and chrysanthemin. A. are hydrolysed by acids or glycosidases to the corresponding aglycons, the ***anthocyanidins*** (Table), which are flavylium salts. The chlorides are usually dark violet, brown or dark red crystals with m.p. > 300 °C. The natural compounds differ with respect to the number of hydroxyl groups: pelargonidin, cyanidin and delphinidin have one, two and three, respectively. In other anthocyanidins, some of the hydroxyl groups are methylated, as in peonidin, petunidin, malvidin and hirsutidin. In the A., the sugar residues are bound to the 3- or 5-positions of the anthocyanidins: monosaccharides are found at the 3- or, more rarely, the 5-position, or at both positions, or a disaccharide group is present at the 3-position. The most common monosaccharides are glucose and galactose. In addition, the phenolic hydroxyl groups may be acylated with phenol carboxylic acids or other carboxylic acids. For example, the acylated A. of red cabbage contain p-coumaric, ferulaic and sinapic acids.

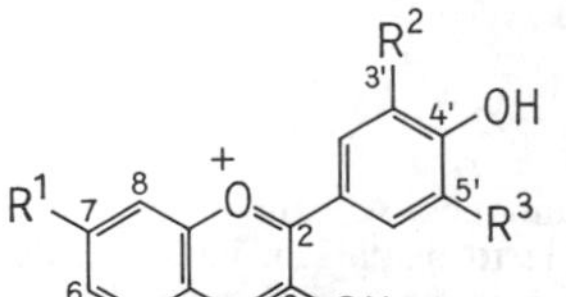

R^2 3' 4' OH R^1 7 8 O + 2 5' R^3 6 5 4 3 OH OH

Anthocyanidine

R^1	R^2	R^3	Anthocyanidine
OH	H	H	Pelargonin
OH	OH	H	Cyanidin
OH	OH	OH	Delphinidin
OH	OCH_3	H	Peonidin
OH	OCH_3	OH	Petunidin
OH	OCH_3	OCH_3	Malvidin
OCH_3	OCH_3	OCH_3	Hirsutidin

The A. are common red, violet or blue pigments in blossoms, berries and other parts of the plant. They are responsible for the colors of strawberries, raspberries, blueberries, cherries and plums. The color of the A. depends on pH. Blue A. are present in the plant, e.g. in cornflowers, as Al(III) and Fe(III) complexes which are also bound to polyuronides. A. are used as food colorings.

Anthracene: a condensed aromatic hydrocarbon, one of the acenes. It crystallizes in colorless, blue-fluorescing leaflets which sublime: m.p. 216.3 °C, b.p. 340 °C (subl.). It is insoluble in water, slightly soluble in ethanol and ether, somewhat soluble in boiling benzene. A. is found in coal tar, and is still isolated from it industrially.

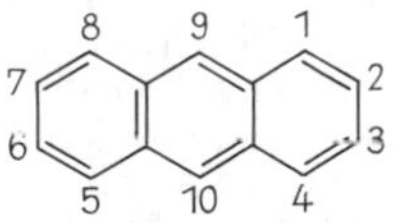

8 9 1 7 2 6 3 5 10 4

It can be synthesized by pyrolysis of 2-methylbenzophenone, by a Friedel-Crafts reaction of 2-bromobenzylbromide, or by reaction of benzene with phthalic anhydride in the presence of aluminum chloride. Only one of the rings in A. has six π-electrons. It is therefore rather reactive, especially at positions 9 and 10. It is easily oxidized here to anthraquinone; addition of nascent hydrogen produces 9,10-dihydroanthracene, and the Diels-Alder reaction with maleic anhydride also occurs readily. The oxidation:

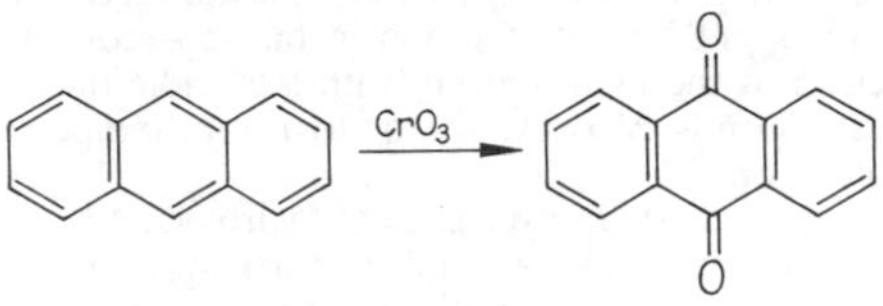

CrO_3 O O

is industrially important.

Electrophilic reagents like sulfuric acid can also attack the 1 or 2 position, so that both anthracene 1-sulfonic acid and anthracene 2-sulfonic acid are formed together. A. is used mainly to produce alizarin and indanthrene dyes; it also serves as starting material for agricultural chemicals and tanning agents.

Anthracene blue: 1,2,4,5,6,8-Hexahydroxyanthraquinone, a pure blue, synthetic pigment; one of the anthraquinone dyes. A. is produced by reaction of mildly fuming sulfuric acid with 1,5-dinitroanthraquinone. It is used to dye wool and cotton.

Anthracene-9,10-diol, ***anthrahydroquinone***: an anthracene diphenol compound. A. forms brown crystals which dissolve in ethanol with a green fluorescence; m.p. 180 °C (dec.).

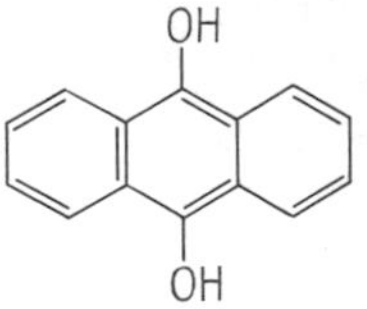

OH OH

Anthracene-9,10-diol

A. is synthesized by treating anthraquinone with zinc powder and alkali hydroxide or with sodium dithionite. The alkaline solution of A. is dark red, and the color disappears when the solution is shaken in air, due to oxidation to anthraquinone. The preparation of anthraquinone dyes for use (vatting) depends on the above reduction reactions. The salts of sulfate esters of A. are used for dyeing and printing; they are called anthrasols.

Anthracene oil: see Tar.

Anthracite: a mineral coal at the last stage of carbonization; its vegetable origin can be detected only by polarized-light microscopy. A. is very hard and shiny, and contains only 6 to 10% volatiles. It is difficult to ignite, burns without smoke, and has a low ash content. Its water content is only about 2%. A. is more than 91.5% carbon, less than 3.75% hydrogen and less than 2.5% oxygen. A. has the highest heating value of any coal: up to 35,000 kJ kg^{-1}. When gasified, it yields a powdery residue and an almost tar-free gas. A. is used mainly as a special fuel and reducing agent, and in production of water gas. It is used to produce large electrodes.

Anthracyclins: a group of antibiotics with anthraquinone derivatives as aglycons. They contain fused linear cyclohexane rings (anthracyclinones). An aminosugar is glycosidically linked to the aglycon. The orange to red A. are formed by various species of *Streptomyces*, e.g. *adriamycin* by *S. peuceticus* and *daunomycin* by *S. coeruleorubidus*. They are used as antineoplastics, for example, in the treatment of acute leukemia. Their activity is due to intercalation of the planar part of the molecule into the DNA of the dividing cell and its fixation in that position by interaction of the amino group with the acidic phosphodiester bonds. Most A. are too toxic for therapeutic applications.

Anthraglycosides: glycosides of hydroxylated derivatives of anthraquinone and its partially reduced relatives, such as anthranols, anthrones and bianthrones. The most important aglycons of the anthraquinone type, which are called *emodins*, are rhein, aloe emodin, chrysophanol and frangula emodin. Similarly substituted aglycons of the anthrone and bianthrone types are also known; in compounds of the dimeric bianthrone type, a single molecule can have two identical or two different substitution patterns on its aglycon rings.

Name	R^1	R^2
Rhein	COOH	H
Aloe emodin	CH_2OH	H
Chrysophanol	CH_3	H
Frangula emodin	CH_3	OH

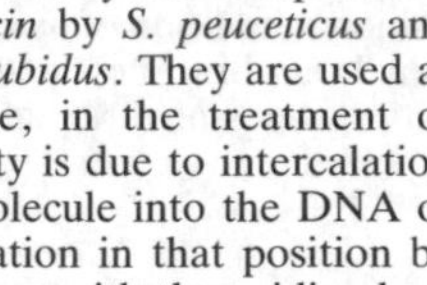

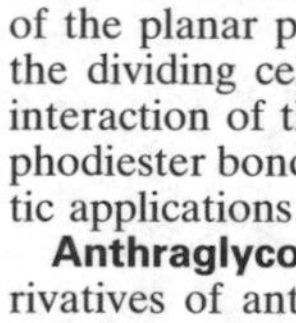

Bianthrone

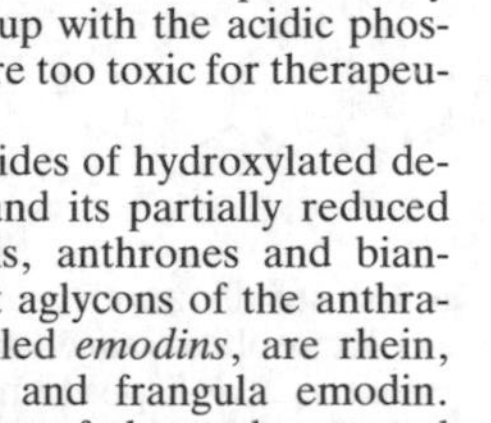

Aloin

The most important medicinal plants containing A. are the rhizomes of certain species of rhubarb, senna leaves, buckthorn (black alder, dogwood) bark and various species of aloe. Rhubarb contains anthraquinone, anthrone and bianthrone aglycons, mainly as glucosides. Buckthorn bark contains mainly bianthrone glycosides of the frangula emodin type; these are poorly tolerated, and within a year of storage they are converted by oxidation to anthraquinone compounds. The main active compounds in senna leaves are sennosides A and B. Sennoside A is the (+)-form, and sennoside B, the *meso*-form with a varying configuration at C-9 and C-9'. The most important component of aloe is aloin, the 10-C-glycopyranosyl compound of aloe emodin.

1,8-dihydroxyanthraquinone derivatives are detected by the red color they develop with hydroxide ions (Bornträger's reaction), or sometimes after cleavage of the glycoside bond and oxidation of the corresponding anthrones and bianthrones to anthraquinones.

A. are laxatives which irritate the mucous membrane of the large intestine and increase peristalsis. It is thought that only the anthrone and bianthrone compounds are effective.

Anthrahydroquinone: same as Anthracene-9,10-diol (see).

Anthralane dyes: a group of acid dyes which are very fast to light and give the fibers (mainly wool) an even color.

Anthranilic acid: see Aminobenzoic acids.

Anthranilic acid methyl ester: $2-H_2N-C_6H_4-CO-OCH_3$, a colorless, crystalline compound with an odor similar to orange blossoms; m.p. 24-25 °C, b.p. 256 °C. It is slightly soluble in water, and readily soluble in alcohol and ether. A. is found naturally in essential oils, for example jasmine and orange-blossom oil, oil of lemon peel, narcissa oil and bergamotte leaves. It is used in perfumes.

Anthraquinone, ***9,10-dihydroanthracene-9,10-dione***: the quinone derived from anthracene. A. crystallizes in light yellow needles, subl. 286 °C, b.p. 379.8 °C. A. is insoluble in water, and soluble in alcohol, ether and benzene.

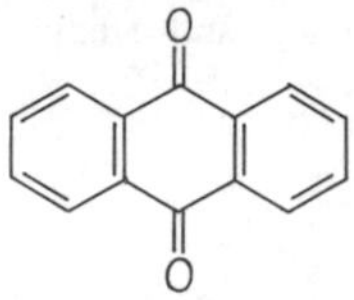

In contrast to most quinones, A. cannot be steam-distilled. It is very stable to oxidation, but is easily reduced to anthrahydroquinone, for example by sodium dithionite. This reaction is the basis of fixing of anthraquinone dyes. Reduction with tin and hydrochloric acid in glacial acetic acid produces anthrone. A. can undergo electrophilic substitution reactions, such as nitration or sulfonation, giving mono- or disubstitution products. For the dye industry, the most important of these are the Anthraquinone sulfonic acids (see). The CO groups of A. react only slightly with the typical ketone reagents. A. is produced by oxidation of anthracene with nitric or chromic acid, or by the Friedel-Crafts reaction of benzene and phthalic anhydride; the latter method is also suitable for synthesis of substituted A. A. is the key compound for production of Anthraquinone dyes (see); it is also used as a bird repellent.

Anthraquinone dyes: a group of synthetic dyes which are very colorfast, especially with respect to light fading. The ***anthraquinone acid dyes*** have sulfonic acid groups on the anthraquinone nucleus to make them water-soluble; in addition, they contain other ring systems such as acridone and thiazole. The A. are used as vat, mordant and chrome dyes. The anthraquinone vat dyes are a diverse group of compounds, some based on the simple anthraquinone skeleton, but others consisting of several anthraquinone molecules. This group includes some indanthrene dyes.

Anthraquinone acid dyes: see Anthraquinone dyes.

Anthraquinone sulfonic acids: sulfonation products of anthraquinone with one or two sulfonic acid groups in the molecule. Depending on the reaction conditions, sulfonation produces mostly anthraquinone 1-sulfonic acid or 2-sulfonic acid. Further sulfonation produces anthraquinone disulfonic acids, in which the sulfonic acid groups are located at the 1,5- and 1,8- or 2,6- and 2,7- positions. The A. can be converted by melting with alkali to hydroxyanthraquinones; both groups are important intermediates in the dye industry.

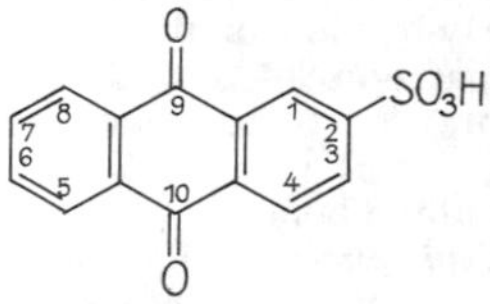

O, 9, 10, O, 1, 2, 3, 4, 5, 6, 7, 8, SO_3H

Anthraquinone 2-sulfonic acid

Anthranol: the enol form of Anthrone (see).

Anthra red B: same as thioindigo (see Indigo).

Anthrone: a carbonyl compound derived from anthracene. A. forms pale yellow needles, m.p. 155 °C. It is not soluble in water, but is soluble in alcohol, benzene and acetone. It dissolves in alkali hydroxide solutions when heated, and the tautomeric ***anthranol*** (anthr-9-ol) can be isolated from these solutions after acidification.

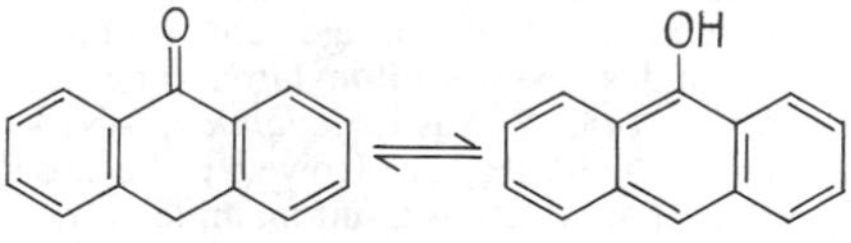

O ⇌ OH

Anthrone Anthranol

This type of tautomerism is called *transannular tautomerism*. Anthranol forms orange crystals which fluoresce strongly in solution; m.p. 120 °C. The stable A. is formed by reduction of anthraquinone with tin and hydrochloric acid. It is used in the production of vat dyes and in the determination of carbohydrates.

anti-: see Stereoisomerism 1.2.3.

Antiallergics: see Antihistamines.

Antiandrogens: antagonists of testosterone; the best known is ***cyproterone acetate***. Because of its gestagen effect, this is used as a component of contraceptives. Cyproterone acetate is to treat male hypersexuality and masculinization of women.

Antianginous drugs: see Coronary drugs.

Antiaromatic: see Aromaticity.

Antiarrhythmics: drugs used to relieve cardiac arrhythmia. Quinidine, ajmaline and its derivative detajmium bitartrate, β-receptor blockers, e.g. propanolol and talinolol, and local anesthetics such as lidocain and procainamide, p-H_2N-C_6H_4-CO-NH-CH_2-CH_2-$N(C_2H_5)_2$, are used as A.

Antiauxochromes, ***antiauxochromic groups***: groups such as -NO_2 and >C=O which form resonance structures with Auxochromes (see) in chromogens (see Chromophores) and thus deepen the color of the compound (see Bathochromicity).

Antibase: see Acid-base concepts, Bjerrum definition.

Antibiotics: compounds produced by microorganisms, and their partially or totally synthetic analogs, which at low concentrations can inhibit the growth of other microorganisms. The A. are chemically very heterogeneous. Some of the major classes are: amino acid derivatives (for example, the amino acid antagonists cycloserine and azaserine, the β-lactam and polypeptide A.), aminocyclitols and nucleoside A. Many A. are biosynthesized by the polyketide pathway, such as the macrolide antibiotics, the tetracyclins, griseofulvin and cycloheximide.

Most of the A.-producing organisms are bacteria of the genera *Bacillus* and *Streptomyces* or molds of the genera *Penicillium, Aspergillus* and *Cephalosporium*. Of approximately 5,500 A. which have been isolated from microorganisms, about 4000 are products of actinomycetes. About 300 new A. are reported anually.

For production of A., the microorganisms are grown in suitable nutrient media in surface or submersion culture (fermentation). Production strains with high productivities are used; these are obtained by selection, often after artificially induced mutation of the original strain. Productivity is affected not only by the genetic potential of the cultured organisms, but by the nutrient medium and, for aerobes in submersion culture, by the aeration. After a maximum concentration of A. has been reached, the fermentation is terminated, and the microorganisms are removed by filtration. The culture medium is then processed by methods which depend on the properties of the A. The most common are liquid-liquid multi-step extraction of relatively lipophilic A. with organic solvents (e.g. benzylpenicillin), ion-exchange chromatography of A. with acidic or basic groups (e.g. the aminocyclitols) and perhaps precipitation of insoluble derivatives (e.g. oxytetracycline). In some cases, new A. were obtained by addition of precursors to the nutrient medium (e.g. phenoxyacetic acid for production of phenoxymethylpenicillin). Partial synthesis plays an important role, especially in the production of the penicillins and cephalosporins. Only a few A. are made by total chemical synthesis (e.g. chloramphenicol).

The action spectra of individual A. vary. Most penicillins, erythromycin and the polypeptide antibiotics are only effective against gram-positive bacteria, while the polymyxins affect only gram-negative bacteria. ***Broad-spectrum A.*** are active against either gram-positive or gram-negative bacteria. This group includes mainly the tetracyclins, chloramphenicol, the cephalosporins and a few partially synthetic penicillins (e.g. ampicillin). Only a few A. affect fungi; griseofulvin and a few polyene A. are used. Some A., including mitomycin, rifamycin and the anthracyclins, inhibit the growth of cancer cells.

In addition to their use as chemotherapeutics for treatment of human and animal infections, and of malignant tumors, A. are now used in large quantities on plants and in industrial animal production, as additives to feeds. These additives produce higher slaughter weights of the animals.

A problem with the application of A. is that the microbes develop resistance, that is, an inherited insensitivity of certain strains to the A. The resistance can also be transferred from one species to the next on extrachromasomal DNA (plasmids). Biochemically, resistance arises in several ways. The membrane may change so that an otherwise effective A. is no longer absorbed; this type of resistance is observed after tetracyclin treatments. The resistant microbes often form enzymes, usually hydrolases or transferases, which inactivate the A. This is the case in resistance to the β-lactam A. (β-lactamase, penicillinase), chloramphenicol (chloramphenicol acetyltransferase) and aminocyclitol A. (acetyl, phosphate and adenylate transferases). Cross-resistance can develop between A. with related chemical structures.

The mechanisms of action of A. are varied. Some points of attack are DNA (mitomycin, actinomycin, anthracyclins), protein synthesis (chloramphenicol, aminocyclitols, tetracyclins, rifamycin), the membrane (ionophores, polyene A.) and the biosynthesis of the bacterial cell wall (β-lactam A., cycloserine, bacitracin).

Historical. The first hints of A. were observed in 1877 (Pasteur and Joubert) and 1889 (metabolic product of *Bacillus pyrocyaneus*). In 1929, Fleming discovered penicillin. It was not purified until 1940, however (Florey, Chain). The term A. was introduced by Waksman in 1942. In 1944, streptomycin, the first aminocyclitol A., was obtained (Waksman). The first broad-spectrum A. was chloramphenicol, isolated in 1947, and synthesized shortly thereafter (Rebstock). In 1980, the world production of A. was about 25,000 t, of which about 17,000 t were penicillins, and 5000 t were tetracyclins.

Antibiotic A-23187, ***calcimycin***: a natural ionophore. It is a crystalline powder which is soluble in ethyl acetate, chloroform, methanol and dimethylsulfoxide, but only slightly soluble in water. A. is commercially available as either the free acid or the calcium/magnesium salt. It can transport divalent cations through lipophilic membranes, and binds Mn^{2+} $> Ca^{2+} > Mg^{2+}$ with a relative binding affinity of 210:2.6:1. It is used as a calcium ionophore for studies in cell biology. In vitro, it has a weak effect against gram-positive bacteria.

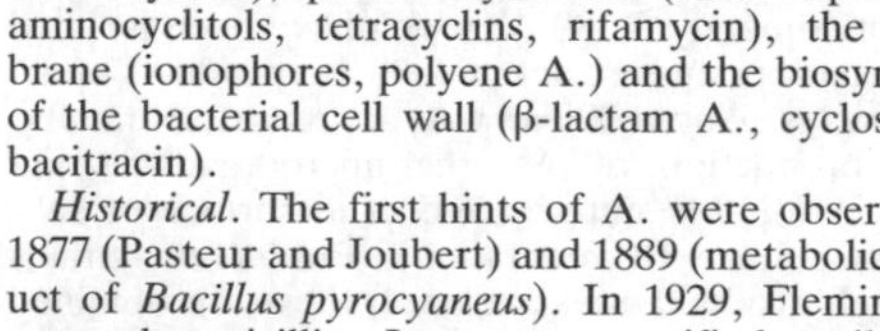

A-23187

Antibiotic X-537a, ***lasalocide***: an antiobiotic isolated from a streptomycete. It forms colorless crystals; m.p. 100-109 °C, which are soluble in organic solvents but practically insoluble in water. A. is a lipophilic salicylic acid derivative; it acts as a calcium ionophore (see Ionophore).

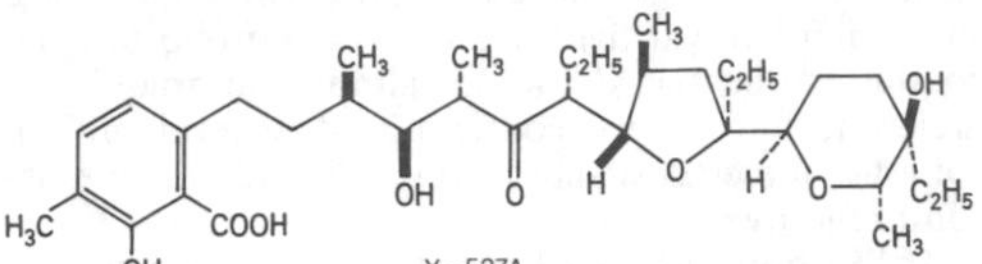

X-537A

Antibodies: proteins (immunoglobulins) synthesized by vertebrate lymphocytes in response to antigenic stimulation. They bind very specifically to the antigen, often causing it to precipitate (***antigen-antibody reaction***). The A. consist of two light and two heavy peptide chains linked by disulfide bridges. Part of the sequences of these chains are relatively invariant and part of them (the variable regions) are specific for the antigen. There are several different types of heavy chain which account for the different types of A. (Immunoglobulin = Ig): IgA, IgM, IgG, IgE, IgD. IgG is commonly known as γ-globulin because it migrates with the γ-fraction of the blood serum in electrophoretic separations.

Anti-clinal conformation: see Stereoisomerim, Fig. 14.

Anticoagulants: see Antithrombotics.

Anticonvulsants, ***antiepileptics***: compounds which prevent the incidence of central nervous system convulsions or reduce their intensities. An epileptic seizure can involve tonic and clonic convulsions (grand mal) or it may not produce generalized convulsions (petit mal). Important A. are found among the barbitals, the hydantions and the succinimides (see Ethosuximide). *Carbamazepin* is another important A. which is also used to treat trigeminus neuralgias.

Antidepressives: Psychopharmaceuticals.

Antidiabetics: drugs used to treat diabetes mellitus. The disease is due to a lack of insulin in the blood, and can be relieved by administration of insulin. Oral A. can be used to treat some forms of diabetes, particularly diabetes of age. This group of compounds includes *N*-arylsulfonylurea and biguanides. The first oral A. was *carbutamide*, $4\text{-}NH_2\text{-}C_6H_4\text{-}SO_2\text{-}NH\text{-}CO\text{-}NH\text{-}C_4H_9(n)$. However, because of its undesirable side effects as a sulfonamide derivative, it is no longer used. Other *N*-arylsulfonylureas with other substituents instead of the aromatic amino group act as A. *Tolbutamide*, $4\text{-}CH_3\text{-}C_6H_4\text{-}SO_2\text{-}NH\text{-}CO\text{-}NH\text{-}C_4H_9(n)$ was the first to become prominent. Replacement of the methyl group on the benzene ring by more hydrophobic groups created much stronger A. such as *glibenclamide*. The compounds listed so far stimulate insulin production in cases where it is inadequate. In rare cases, biguanides such as butylbiguanide (*buformin*) and *metformin* are used; they act by a different mechanism.

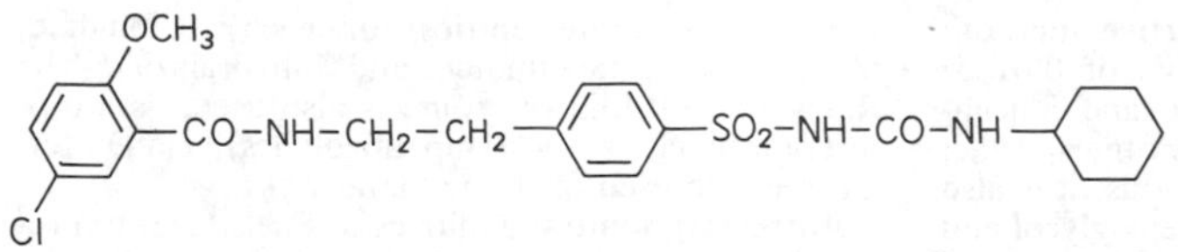

Glibenclamide

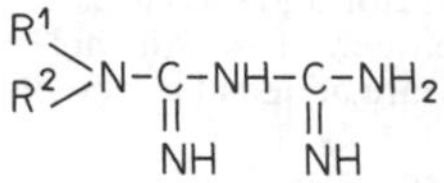

$R^1 = H$, $R^2 = C_4H_9(n)$: Buformin

$R^1, R^2 = CH_3$: Metformin

Antidiuretic hormone: same as Vasopressin (see).

Antidiuretin: same as Vasopressin (see).

Antidote: a substance which is administered to counteract the effects of a poison. The essential components of treatment for poisoning, in addition to an A., are symptomatic therapy, maintenance of vital life functions, measures to prevent further absorption of the poison and acceleration of its elimination. Occasionally substances which affect absorption or elimination of the poison (emetics, laxatives, adsorptive substances, oxidizing agents) are called unspecific A. However, A. in the narrow sense are specific drugs. Their mechanism of action may be a) chemical: they may convert the poison to an insoluble and therefore non-poisonous compound (e.g. precipitation of soluble barium and lead compounds with sodium sulfate); the soluble poison may be converted to a soluble but non-poisonous compound (e.g. conversion of cyanides into rhodanides by sodium thiosulfate, or binding of cyanide to methemoglobin, which can be formed in the organism by administration of compounds such as nitrites or methylene blue); or acids can be "neutralized" with dilute bases such as sodium hydrogencarbonate, soapy water or magnesium oxide. b) Pharmacological A. counteract the pharmacological effect of the poison, e.g. by competion for cellular receptors as in the case of acetylcholine and atropine. Poisoning by substances which inhibit the degradation of physiologically formed acetylcholine (see Poisons), is actually a form of endogenous acetylcholine poisoning, which can be treated by relatively high doses of atropine. A. can also reactivate enzymes which have been blocked, e.g. cholinesterase inhibited by phosphate esters can be reactivated by pralidoxime (PAM), or enzymes containing thiol groups blocked by metal ions can be rescued by 1,2-dithioglycerol, which binds the metal ions (e.g. mercury, arsenic) more tightly than the endangered enzymes do (Table).

Antidote	Application
Bemegride	Stimulation of the respiratory center after barbiturate poisoning
Biperiden	Neuroleptica poisonings
Atropine sulfate	Cholinesterase inhibitor poisoning
Dexamethasone isonicotinate	Poisoning with phosgene, nitrosulfuric gases and other lung irritants
Calcium thiosulfate	Fluoride, oxalate or citrate poisoning
Methylthionine chloride (methylene blue)	Poisoning by methemoglobin-forming compounds (nitrobenzene, aromatic nitro and amino compounds, nitrites
Deferoxamine mesylate	Iron poisoning
Dimethylaminophenol	Hydrocyanic acid and cyanide poisoning
Clomethiazole	Alcohol delirium
Sodium calcium EDTA	Heavy metal poisoning, esp. by Pb, Ni, Cu, Zn, Cr, Cd, Mn Prophylaxis against heavy metal poisoning
Polystyrolsulfonate resin, Ca form	Potassium poisoning
Dimeticon	Poisonings with tensidine and other foaming agents
Hexamethylenetetraamine	Phosgene poisoning
Isoamylnitrite	Hydrocyanic acid and cyanide poisonings
Pyrostigmine	Anticholinergic (atropine) poisoning
Phytomenadione	Poisoning with indirect coagulants
Nalorphine hydrobromide	Overdoses of morphine or its derivates
Sodium thiosulfate	Hydrocyanic acid or cyanide poisoning
Obidoxime	Alkylphosphate poisonings
Penicillamine	Heavy metal poisoning, esp. by Pb, Hg, Cu, Au, Co, Zn
Protamine sulfate	Heparin poisoning
Dimercaptrol	Heavy metal poisoning, esp. by As, Hg, Au, Ni, Cu, Sn

Antienzymes: polypeptides or proteins which act as enzyme inhibitors. The term includes those antibodies which bind to the enzyme in such a way as to inhibit it, but not those which bind without affecting the activity.

Antiepileptics: same as Anticonvulsants (see).

Antifeedants: see Insect repellents.

Antifibrinolytics: compounds used to treat pathologically activated fibrinolysin and to control therapeutically applied fibrinolysins (streptokinase treatments). They inhibit activation of plasminogen or plasmin, and thus of fibrinolysin, which breaks down polymeric fibrin. ε-Aminocaproic acid, 4-aminomethylbenzoic acid (PAMBA®) and the *trans*-isomer of 4-aminomethylcyclohexane carboxylic acid (tranexamic acid) are used as A.

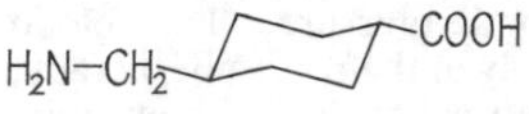

$H_2N-CH_2-C_6H_{10}-COOH$

Tranexamic acid

Antifluorite type: see Fluorite type.

Antifoaming agents, ***foam inhibitors, foam breakers***: substances used to prevent formation of foam. Because they do not mix readily with the foaming liquid, they tend to accumulate on the liquid-gas interface. Alkylpolysiloxanes are effective for both aqueous solutions and oils in amounts of 0.01 to 0.1%. In lubricating oils, a critical siloxane concentration must be exceeded, because they otherwise act as foam generaters. For aqueous solutions, it is also possible to use higher alcohols, propylene glycol and ethylene oxide-propylene oxide adducts as A.

Antifreeze: materials which are added to water to lower its freezing point, such as methanol, ethanol, ethylene glycol, glycerol, sodium, potassium, calcium or magnesium chloride, or potassium carbonate. They are used in engine coolants, to make cooling brines for refrigeration equipment, in hydraulic systems, to de-ice aircraft and motor vehicles, and in the liquids used to clean streets in winter. A. mixtures often contain corrosion inhibitors (such as borax, phosphates or chromates) and antifoaming agents (e.g. fatty acid esters or silicon oils). A. are used in construction so that concrete can be poured even at low temperatures. Some other special applications are Freezing point lowering (see) of diesel fuels (see Additives) and prevention of icing of carburetors (see Anti-icing agents).

Antigens: substances which induce the formation of Antibodies (see). They are usually foreign to the body; induction of antibodies to the body's own proteins is the cause of autoimmune diseases. A. are most often proteins or polysaccharides with molecular masses greater than 2000. Artificial A. can be synthesized by covalently linking a small molecule, such as a benzene derivative, to a protein.

Antihistamines: compounds which reverse or reduce the physiological effects of histamine. The reduction in blood pressure induced by histamine is mediated by H_1 and H_2 receptors; correspondingly, the A. are classified as H_1 and H_2 antagonists. The classic A., which have long been known, are also called *antiallergics*; these compounds are H_1 antagonists. Examples are *diphenylhydramine* (Diabenyl®), *etholoxamine* (AH 3®), *tripelannamine* (Dehistin®), *talastine* (Ahanon®) and *dioxopromethazine* (Prothanon®). The phenothiazine derivatives promazine and promethazine are used in the same way (see Phenothiazines). The close structural similarities of H_1 antagonists with other types of drugs explains the side effects of these compounds, such as depression of the central nervous system, spasmolysis, sympathicolytic and parasympathicolytic effects and local anesthesia.

The H_2 antagonists were discovered much later than the H_1 antagonists. H_2 antagonists are able to inhibit the histamine-induced secretion of gastric juice. The first compound of this type was *cimetidine*, which is used to treat gastric and duodenal ulcers.

Anti-Hückel system: see Aromaticity.

Antihypertensives: compounds which reduce pathologically high blood pressure. The blood pressure should drop slowly in the course of therapy, and remain within the normal range, without great deviations, during maintenance therapy. In the majority of cases, the cause of the hypertension is unknown (*essential* or *primary hypertension*). *Symptomatic* or *secondary hypertension* is caused by pathological changes in the organs, for example the kidneys (*renal hypertension*) or the endocrine glands. The following A. are used primarily for essential hypertension (see separate entries for each): Clonidine, Methyldopa, Guanethidine, and Dihydralazine. The Rauwolfia alkaloid reserpine is also used, as are β-receptor blockers (see Sympathicolytics), and Diuretics (see) are used for basic therapy.

Anti-icing agents: additives to Fuels (see) to prevent ice formation in the carburetor under conditions of high air humidity and low temperatures. Alcohols, ether, etc. are suitable A.; they are added at a ratio of about 100 ppm fuel.

Antiisotypism: see Isotypism.

Antiknock compounds, ***knock inhibitors***: compounds added to gasoline in small amounts to increase its resistance to Knocking (see). A. must be readily soluble in the fuel and have about the same volatility. They have no effect on the density, boiling, heating value or similar properties of the fuel. The resistance of a gasoline to knocking is expressed by its Octane rating (see). Although normal alkanes tend to cause knocking, iso-alkanes, naphthenes and aromatics are very knock-resistant components of gasoline.

The most effective A. are organometallic compounds, especially tetraethyl and tetramethyl lead, methylcyclopentadienyl tricarbonyl manganese, pentacarbonyl iron and carbonyl nickel. However, N-containing aromatic compounds, such as aniline and xylidine, are also effective. Relative to tetraethyl lead, the amounts of compound required to give the same knock resistance are as follows: Tetraethyl lead, 1; methylcyclopentadienyl tricarbonylmanganese, 1.2; pentacarbonyliron, 3; tetracarbonylnickel, 8; aniline, 90; xylidine, 150. The combustion products of tetraethyllead are poisonous, and in areas with dense traffic, their concentration becomes a health hazard. In addition, these products poison the catalysts used to reduce the other emissions of the engine (unburned hydrocarbons and CO) which are serious environmental contaminants (see Smog). For this reason, unleaded gasolines are required for cars equipped with catalytic converters, which are required by law in many places (see Fuels).

The effect of the metal-containing A. depends on the fact that they decompose thermally in the combustion space before the fuel-air mixture ignites, leading to a finely divided metal or metal compounds, which react with unstable intermediates of "cold combustion" and thus prevent premature ignition of the fuel.

Antimalarial agents: compounds used for therapy and prophylaxis of malaria, a disease which is transmitted by the *Anopheles* mosquito and is very widespread in the tropics. The developmental cycle of the malaria organisms within the human host is very complicated; the asexual schizonts and/or the sexual gametocytes inside and outside the erythrocytes can be the point of attack by A. Some A. can also be used successfully for prophylaxis of malaria. The classical A. is quinine. Some synthetic compounds now used are chloroquine, primaquine, pyrimethamine and proguanil.

Anti-Markovnikov orientation: see Markovnikov rule.

Antimetabolites: chemical compounds which are structurally similar to metabolites, and therefore compete with them for binding sites on enzymes in biochemical reaction chains. They thus block the metabolic pathways at specific sites. Some examples of A. are the synthetic nucleic acid bases fluorouracil, cytarabin and 6-mercaptopurine; the folic acid A. methotrexate; the *p*-aminobenzoic acid A. *p*-aminosalicylic acid; and the sulfonamides. The nucleic acid A. and methotrexate are used as cytostatics; the sulfonamides are antibiotics, and *p*-aminosalicylic acid is used against the tuberculosis bacillum.

Antimonates: ***Antimonates(III)***, same as Antimonites (see). ***Antimonates(V)***, the salts of antimonic(V) acid, which does not exist in pure, defined form. Antimonic acid and A. differ from phosphoric acid and phosphates in that the antimony is *sixfold* coordinated. Antimonic(V) acid should probably be formulated as $H[Sb(OH)_6]$, so that the A. are more correctly called ***hexahydroxoantimonates(V)***, $M^I[Sb(OH)_6]$. Most of these salts are relatively insoluble in water; they are obtained by melting the metal hydroxides with antimony(V) oxide. Potassium hexahydroxoantimonate, $K[Sb(OH)_6]$, is used as a precipitating reagent for sodium ions.

Antimonic acid: see Antimony oxides; Antimonates.

Antimonites, ***antimonates(III)***: the salts of antimonous acid, $HSbO_2$, which does not exist in the free state. The A. are made by dissolving antimony(III) oxide in alkali hydroxide solutions, e.g. $Sb_2O_3 + 2\,KOH \rightarrow 2\,KSbO_2 + H_2O$. They are strong reducing agents.

Antimonous acid: see Antimonites.

Antimony, ***stibium***, abb. ***Sb***: an element of the 5th main group of the periodic system, the Nitrogen-phosphorus group (see). Sb is a semimetal, *Z* 51. The natural isotopes have mass numbers 121 (57.25%) and 123 (42.75%); the atomic mass is 121.75. Sb has valences III and V; its Mohs hardness is 3, density 6.684, m.p. 630.5 °C, b.p. 1750 °C, electrical conductivity 2.56 Sm/mm² at 0 °C, standard electrode potential 0.1445 V ($2\,Sb + 3\,H_2O \rightleftharpoons Sb_2O_3 + 6\,H^+ + 6\,e$).

Properties. Gray A. is a silvery white, shiny, very brittle and readily pulverized metal. Its lattice corresponds to that of gray metallic arsenic. Other modifications described in the older literature, such as yellow A., black A. or the glassy-amorphous explosive A. are now considered uncertain or have been shown to be multi-component systems. A. vapor consists of Sb_4 molecules, which dissociate into Sb_2 units at higher temperatures. Above its melting point, A. burns in air to form antimony(III) oxide, Sb_2O_3. In finely divided form, it burns in chlorine to antimony(V) chloride, $SbCl_5$. Its position in the electrochemical potential series is such that it is not attacked by non-oxidizing acids; nitric acid oxidizes A. to Sb_2O_3 or Sb_2O_5, depending on its concentration. In melts with sulfur, A. forms the antimony sulfides Sb_2S_3 and Sb_2S_5.

The increasing stability of the +3 oxidation state observed in the progression from phosphorus to arsenic is continued to A., and antimonates(V) are strong oxidizing agents, especially in acid solution. In aqueous solution, antimony(III) salts typically form SbO^+ cations.

Analysis. In systematic qualitative analysis, A. is precipitated with the H_2S group in the form of a sulfide which is soluble in ammonium sulfide. After removal of other elements, it is identified as the red-orange Sb_2S_3. The Marsh test produces a metal mirror which is insoluble in hypochlorite solution. Iodometric or bromatometric methods can be used for quantitative analysis, or, for low concentrations, atomic absorption spectroscopy.

Occurrence. The fraction of A. in the earth's crust is 10^{-4}%. It is occasionally found in the elemental state in nature, but antimony sulfides and oxides and metal antimonides are much more common. The important minerals are antimonite (antimony glance, stibnite), Sb_2S_3 and its weathering product, valentinite (antimony bloom), Sb_2O_3, antimony blende, $2\,Sb_2S_3 \cdot Sb_2O_3$, and the antimonides breithauptite, NiSb, and discrasite, Ag_2Sb.

Production. The metal is usually produced from the sulfide, Sb_2S_3. This is either roasted to the oxides Sb_2O_3 or Sb_2O_4 and then reduced to A. with carbon, or it is only partially roasted and the oxide reacts directly with the remaining sulfide, corresponding to the equation $3\,Sb_2O_4 + 2\,Sb_2S_3 \rightarrow 10\,Sb + 6\,SO_2$.

Applications. Pure A. is of little commercial significance. It is used mainly as a component of alloys of tin and lead, which it makes harder (see Tin alloys, Lead alloys). A few organic A. compounds are used in medicine, in particular in the treatment of tropical diseases.

Historical. A. was known 3000 years ago in China, and later in Babylon. The Greeks and Romans used stibnite for makeup to darken their eyelids and lashes. Around 1600, the Frankenhäuser salt dealer Thölde wrote a book, *Triumphwagen des Antimonii*, in which the production of the metal, its uses in alloys and the synthesis of a few derivatives were described.

Antimony butter: see Antimony chlorides.

Antimony chlorides: ***Antimony(III) chloride, antimony trichloride, antimony butter***, $SbCl_3$, colorless, soft rhombic crystals which fume in moist air, M_r 228.11, density 3.140, m.p. 73.4 °C, b.p. 283 °C. $SbCl_3$ is soluble in ether. With a small amount of water, it forms a clear solution, but at higher dilutions, insoluble oxygen chlorides, SbOCl and $Sb_4O_5Cl_2$, precipitate. The salt is obtained by dissolving antimonite in concentrated hydrochloric acid, or by chlorination of antimony. It is used as a caustic in medicine and as a catalyst in organic synthesis.

Antimony(V) chloride, antimony pentachloride, $SbCl_5$, is a colorless liquid when pure; usually, however, it is pale yellow. M_r 299.02, density 2.336, m.p. 2.8 °C, b.p. 79 °C at about 3 kPa. In the gas and solid states, it consists of trigonal bipyramidal $SbCl_5$ molecules. When heated, $SbCl_5$ decomposes to $SbCl_3$ and chlorine. It is hydrolysed in water to antimony(V) oxide hydrates. $SbCl_5$ reacts with many metal chlorides to form hexachloroantimonates, $M^I[SbCl_6]$. It is a strong Lewis acid and forms addition compounds with many donor molecules (e.g. $SbCl_5 \cdot OPCl_3$). It is made by chlorination of $SbCl_3$. $SbCl_5$ is used as a strong chlorinating reagent and as a catalyst in organic syntheses.

Antimony electrode: an oxide electrode used as an indicator electrode in the measurement of pH. The electrode consists of metallic antimony, the surface of which is coated with a thin layer of antimony(III) oxide, Sb_2O_3. The potentiometric reaction is: $Sb + 3\,H_2O \rightarrow Sb(OH)_3 + 3\,H^+ + 3e$. The reversible electrode voltage is thus $E = E^o + (RT/3F)\ln a_{H^+}{}^3$, that is, $E = -0.059 \cdot \mathrm{pH}$. Here E^o is the standard electrode potential, R, the gas constant, T, the temperature in kelvin, F, Faraday's constant and a_{H+}, the activity of the hydrogen ions. A. can be used to measure pH from 3 to 11. Its advantage is its excellent mechanical stability (especially compared to gas electrodes), and its disadvantage is its slow response time and its sensitivity to redox systems present in the solution.

Antimony fluorides: ***Antimony(III) fluoride, antimony trifluoride***, SbF_3, colorless, rhombic crystals, M_r 178.75, density 4.379, m.p. 292 °C, sbl.p. 319 °C. SbF_3 is slowly hydrolysed in water. It is made by dissolving antimony(III) oxide in concentrated aqueous hydrofluoric acid.

Antimony(V) fluoride, antimony pentafluoride, SbF_5, is a colorless, oily compound with trigonal bipyramidal molecules; M_r 216.74, density 2.99, m.p. 7 °C, b.p. 149.5 °C. As a strong Lewis acid, SbF_5 forms stable complexes with many organic and inorganic donor compounds. It reacts with metal fluorides to form hexafluoroantimonates, $M^I[SbF_6]$. SbF_5 is most conveniently synthesized by the reaction of anhydrous hydrofluoric acid with antimony pentachloride. Both A. are used as fluorinating reagents in organic synthesis.

Antimony hydride: see Stibane.

Antimony oxides: ***Antimony(III) oxide, antimony trioxide***, Sb_2O_3, M_r 291.50, exists in two modifications which occur naturally as well. In the form which is stable at room temperature, the $[SbO_3]$ pyramids are linked in infinite double chains. This form is converted at 606 °C to a cubic form, which is built up of tetrahedral Sb_4O_6 units; these are also present in the vapor. The molecular structure is like that of P_4O_6; see Phosphorus oxides. The high-temperature modification melts at 656 °C, and sublimes 1550 °C. Sb_2O_3 is amphoteric, dissolving in alkali hydroxides with formation of antimonites: $Sb_2O_3 + 2\,M^IOH \rightarrow 2\,M^ISbO_2 + H_2O$. It dissolves in sulfuric or hydrochloric acid to form antimony(III) sulfate, $Sb_2(SO_4)_3$ or antimony(III) chloride, respectively. Sb_2O_3 is obtained by roasting antimonite, Sb_2S_3, or by burning antimony. It is used as an opacifier in the production of enamel.

Antimony(V) oxide, antimony pentoxide, Sb_2O_5, is a yellow, cubic, crystalline powder; M_r 323.50, density 3.80, Above 330 °C, it loses oxygen and is converted to Sb_2O_4. Aqueous suspensions of the rather insoluble Sb_2O_5 are acidic, but a defined antimony(V) acid is not known. The products which were formerly called antimonic acids, obtained, for example, by hydrolysis of antimony(V) chloride, are not exactly defined, especially with respect to their water contents, and are more correctly termed ***antimony(V) oxygen hydrates***. Sb_2O_5 reacts as the anhydride of this hypothetical acid, for example, by forming potassium hexahydroxoantimonate, $K[Sb(OH)_6]$ with potassium hydroxide (see Antimonates). Sb_2O_5 is made by dehydration of the oxide hydrate, which is formed by oxidation of antimony with concentrated nitric acid.

Antimony(III,V) oxide, Sb_2O_4, is a colorless rhombic or monoclinic crystal powder which is only slightly soluble in water; M_r 307.50. It is made by heating Sb_2O_5 or Sb_2O_3 in the air. Above 930 °C, it releases oxygen and is converted to Sb_2O_3.

Antimony sulfides: ***Antimony(III) sulfide, antimony trisulfide***, Sb_2S_3, red-orange, amorphous powder; M_r 339.69, density 4.12, m.p. 550 °C, b.p. about 1150 °C. When heated in the absence of air, Sb_2S_3 is converted to a rhombic, gray-black modification. It is only slightly soluble in water and dilute acids, but dissolves in ammonium sulfide, forming thioantimonites: $Sb_2S_3 + 3\,S^{2-} \rightarrow 2\,SbS_3^{3-}$; in ammonium polysulfide solution, it is oxidized to thioantimonates; $Sb_2S_3 + 2S + 3\,S^{2-} \rightarrow 2\,SbS_4^{3-}$. It is obtained by passing hydrogen sulfide through an acidified solution of Sb(III) or also by fusing the elements . SbS_3 is found in nature as antimonite (antimony glance, stibnite).

Antimony(V) sulfide, antimony pentasulfide, gold sulfide, Sb_2S_5, a yellow-orange powder; M_r 403.82, density 4.12, dec. 75 °C. Sb_2S_5 is only slightly soluble in water and dilute acids, and reacts with ammonium sulfide solution to give thioantimonates. It is made by fusing the elements .

The A. are used as pigments, in the vulcanization of rubber, in fireworks and in matches.

Antimycotics: compounds with more or less specific effects on fungal infections. Certain disinfectants, such as halogenated phenols, 8-hydroxyquinoline derivatives, invert soaps and triphenylmethane dyes are used as A. Thiocarbamide esters such as tolnaftate (Bocima®), dibenzthione [tetrahydro-3,5-bis(phenylmethyl)-2*H*-1,3,5-thiadiazine-2-thione], imidazole derivatives, such as clotrimazol, and compounds such as miconazol are also used. Certain antibiotics, such as the polyenes amphotericin B and nystatin, and griseofulvin, also have antimycotic effects. Most A. are suitable only for local application. There are no generally reliable systemic A. available at present. Under certain conditions, amphotericin B and miconazol can be used for generalized mycoses. Applied orally, griseofulvin is effective against some fungal diseases of the toes and fingernails. A. are called ***fungistatics*** when they only inhibit growth of the fungus; and ***fungicides*** if they kill the fungi.

Antineoplastics: same as Cytostatics (see).

Antineuritic vitamin: same as Vitamin B_1 (see).

Antioxidants: substances which inhibit oxidation of readily oxidized materials such as mineral, transformer and turbine oils, jet fuels, plastics, rubber, edible fats and soaps.

In food chemistry, A. are substances which improve the shelf life of certain foods, especially fats and fat-containing foods, by preventing their oxidation (which leads to rancidity). A. can be natural components of the foods (tocopherols) or synthetic additives (e.g. gallates, nordihydroguajaretic acid, butylhydroxyanisole).

See also Additive, 1 (petroleum chemistry).

Antiparkinson's agents: compounds which at least partially and temporarily relieve the symptoms of Parkinson's disease, such as the trembling of the limbs, slowing of the gait and stiffness. In some forms of Parkinson's disease, the amount of the inhibitory transmitter Dopamine (see) in the brain stem is reduced. To increase the dopamine content, its biosynthetic precursor, *levodopa (L-dopa, L-3,4-dihydroxyphenylalanine)* is administered; in contrast to dopamine, L-dopa is able to cross the blood-brain barrier. In the brain, it is decarboxylated to dopamine. L-Dopa is often administered together with a decarboxylase inhibitor, which inhibits its conversion to dopamine outside the brain. Amantadine (see), which was introduced as a virostatic, acts as an indirect dopamine agonist and increases its availablity.

Parasympathicolytics (anticholinergics) are used to suppress the excitatory transmitter acetylcholine. The tropa alkaloids, such as atropine and scopolamine and plant extracts containing them were formerly used. Today the drugs of choice are basic ethers of benzhydrol, such as *ethylbenzhydramine* (Antiparkin®), aminopropanol derivatives such as *trihexyphenidyl* (Parkopan®) and triperidene (Norakin®), and the thioxanthene derivative metixene.

Antiparticles: see Elementary particles.

Anti-periplanar conformation: see Stereoisomerism, Fig. 14.

Antipermeability factor: same as Rutin (see).

Antiphlogistics: compounds which reduce inflammation. They are used in rheumatic diseases, sometimes over long periods of time. Many weak Analgesics (see) are also used for their antiphlogistic activity, for example, derivatives of salicylic acid. More recently, pyrazolidindione derivatives, such as phenylbutazone and kebuzone, the indole derivative indometacin and aralkylcarboxylic acids, such as ibuprofen have been introduced as A. Glucocorticoids are also major A. Gold compounds and chloroquine (which was introduced as an antimalarial) are also used for long-term therapy.

Antiprotozoics: compounds used in the treatment and prophylaxis of protozoan infections. Protozoa which are pathogenic in human beings include, for example, the malaria plasmodia, the sleeping-sickness trypanosomes, the enteroamebas which cause diarrhea, and the trichomonads, which cause trichomoniasis and leishmanias such as kala azar. Quinine and various synthetic drugs are used as Antimalarial agents (see). Suramin was developed for use against sleeping sickness, but has now been replaced by other drugs, such as the bisamidines (pentamidine). The ipecuacuanha alkaloid (-)-emetin is effective against amebic infections. Trichomoniasis is treated with nitroimidazol derivatives, such as metronidazol. Leishmanias are treated chiefly with organic compounds of pentavalent antimony.

Antipsychotics: see Psychopharmaceuticals.

Antipyretics: compounds which reduce fever by their action on the heat center of the central nervous system. Many weak analgesics, such as derivatives of salicylic acid, pyrazolone and aniline analgesics act as A.

Antipyrin®: see Pyrazolone.

Antiseptics: see Disinfectants.

Antistatics: preparations which reduce the static charge on synthetic textile fibers. Most A. are surface active substances, which are applied to the fibers in aqueous solution and form a film on their surface with the hydrophobic groups adjacent to the surface of the fiber and the hydrophilic groups pointing toward the air. These hydrophilic groups are able to take up moisture from the air, so that as a result of ionic mobility, electric charges can be quickly discharged. Furthermore, the A. act as lubricants and reduce the formation of triboelectricity.

Anti-Stokes lines: see Raman spectroscopy.

Antique purple: an animal dye from the juice of the purple snail *Murex brandaris* which was highly prized in antiquity (it was the imperial color in Rome). Structurally, P. is 6,6'-dibromoindigo (see Indigo dyes).

Antisymmetry: A property of the wavefunction Ψ of a multi-electron system with respect to exchange of electron coordinates. The wavefunction must be antisymmetric, that is, its sign must change if the numbering of the coordinates of two electrons is reversed: $\Psi(1,2) = -\Psi(2,1)$. For a system containing two electrons, this is achieved by writing the wavefunction in terms of the orbital spin functions φ_1 and φ_2: $\Psi(1,2) = \varphi_1(1)\varphi_2(2) - \varphi_1(2)\varphi_2(1)$. In the general case, this requirement is fulfilled by writing the wavefunction as a matrix in which every row contains all the orbital spin functions of electron coordinates with the same number i:

$$\Psi(1,2,3) = N \begin{vmatrix} \varphi_1(1)\varphi_2(1)\varphi_3(1) \\ \varphi_1(2)\varphi_2(2)\varphi_3(2) \\ \varphi_1(3)\varphi_2(3)\varphi_3(3) \end{vmatrix}$$

N is the normalization factor (see Normalization conditions). Exchanging electron coordinates means exchanging two rows of the matrix, which changes its sign. The antisymmetry requirement is the basis of the Pauli principle.

Antithrombotics: compounds which prevent formation of fibrin, and thus of blood clots (***anticoagulants***), or are able to dissolve clots which have already formed (***fibrinolytics***). 4-Hydroxycoumarin derivatives, such as dicoumarol, ethyl biscoumacetate, phenprocoumon and warfarin; alkylindanediones, such as chlorindione; and heparin and heparinoids are used as anticoagulants. Hydroxycoumarins and arylindanediones inhibit the synthesis of coagulation factors, and are therefore called *indirect anticoagulants*. Heparin activates coagulation factors which inhibit thrombin, and is thus a *direct anticoagulant*. Streptokinase and urokinase are used as fibrinolytics.

Anticoagulants are used as Rodenticides (see).

Antonov's rule: the surface tension between nonmiscible liquids (σ_{12}) is equal to the difference between the surface tensions of the two with respect to air: $\sigma_{12} = \sigma_1 - \sigma_2$. This empirical rule applies only when the mutual interaction is due only to London forces (see Van der Waals bonding forces) and the contributions of the London forces to the surface tensions of the two liquids are equal. Complete spreading of the liquids is also a requirement.

Ant toxins: poisons produced by ants which have a special poison gland; they are used for attack or defense. Most A. are not dangerous to human beings. The longest known is highly concentrated formic acid, which is used by some Formicidae; in the red forest ant *Formica rufa*, for example, the formic acid concentration is 70%. It acts as a respiratory poison in lower animals. Some ants have a poisonous sting. An example is the use of a mixture of various piperidine derivatives which are highly insecticidal by the fire ant (*Solenopsis saevissima richteri*). Others (*Dolichoderiae*) produce poisons in separate anal glands; these have both antibiotic and insecticidal effects. Examples are iridomyrmecin or ketene-dialdehyde mixtures. The Poneridae have a highly developed poison apparatus with a sting for hunting prey. Little is known about the poisons used by the migratory and driver ants; they contain little or no formic acid and are similar in effect to bee and wasp toxins.

ANTU: a Rodenticide (see).

Antu: see α-Naphthylthiourea.

AO: abb. for atomic orbital.

Apamine: a heterodetic cyclic branched 18-peptide amide with two intrachain disulfide bonds. A. is the neurotoxic component of bee venom. It was isolated in 1965 and its structure was elucidated two years later by Habermann and coworkers.

Cys-Asn-Cys-Lys-Ala-Pro-Glu-Thr-Ala-Leu-Cys-Ala-Arg-Arg-Cys-Gln-Gln-His-NH_2

Apholates: see Chemical sterilizers.

Aphrodisiacs: substances which are supposed to increase the sex drive. Formerly various plant parts were recommended, and also yohimbine.

Apigenin: see Flavones.

Apoatropine: see Atropine.

Apoenzyme: the protein part of an enzyme which requires a covalently bound coenzyme (a low-molecular-weight organic compound) or cofactor (such as a metal ion) for activity. The A. and coenzyme or cofactor together form the holoenzyme.

Apomorphine: a product of the reaction of morphine with concentrated sulfuric acid at 150 °C; m.p. 195 °C. The compound is a pyrocatechol derivative which is oxidized in the air and becomes discolored. A. is administered parenterally as a rapidly acting emetic.

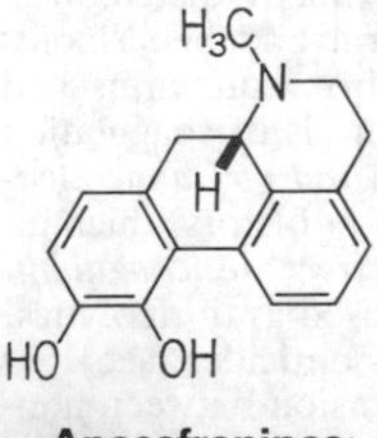

H_3C N H HO OH

Aposafranines: see Azine pigments.

Appearance voltage: the minimum voltage required for formation of fragment ions in Mass spectroscopy (see). It exceeds the ionization potential by an amount equivalent to the dissociation energy of the broken bond(s).

Appetite suppressants, ***anorexics***: compounds which reduce appetite and thus reduce food intake. Under medical supervision, they can therefore be used to reduce body weight. As a rule, A. also excite the central nervous system. Many are derivatives of β-phenylethylamine, and are related to the Sympathicomimetics (see). *Norpseudoephedrin (Exponcit®)*, C_6H_5-CH(OH)-CH(CH_3)-NH_2 and *phendimetrazine (Sedafamem®)* are examples of A. Norpseudophedrin has a *threo*-configuration and is applied as the racemate.

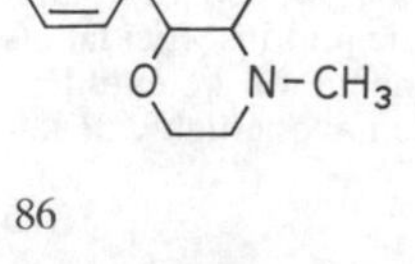

CH_3 O N-CH_3

Phendimetrazine

Aprobarbital: see Barbitals, table.

Aqua complexes (also commonly called ***aquo complexes***): metal complexes which contain water, H_2O, as a ligand. Di- and trivalent ions of the metals in the 3d series form A. with coordination numbers of six, $[M(H_2O)_6]^{n+}$ (n = 2 or 3). A. with higher coordination numbers are formed by heavier transition metals, for example the lanthanides. Due to hydrolysis, A. are acidic in aqueous solution: $[M(H_2O)_6]^{n+} + H_2O \rightleftharpoons [M(H_2O)_5OH]^{(n-1)} + [H_3O]^+$.

Aquametry: general term for methods of quantitative determination of water in solid and liquid, inorganic and organic substances. These determinations are now made almost exclusively with Karl-Fischer solution (see) as titrant.

Aqua regia: "water of kings", so called because it can dissolve gold, the royal metal. A mixture of 3 parts concentrated hydrochloric acid and 1 part concentrated nitric acid. The two acids react with each other to form chlorine, nitrosyl chloride and water: $3\ HCl + HNO_3 \rightarrow 2\ Cl + NOCl + 2\ H_2O$. Most metals, including gold, are dissolved by the nascent chlorine and the nitrosyl chloride: $Au + 2\ Cl + NOCl \rightarrow AuCl_3 + NO$.

Aquo complexes: see Aqua complexes.

Aqua oxides: same as Oxide hydrates (see).

Ar: 1) symbol for argon. 2) abb. for aryl-.

Ara: abb. for arabinose.

Arabans, ***arabinans***: polysaccharides made up of L-arabinose. A. are found, for example, in Pectin substances (see) and various plant Gums (see).

arabino: prefix indicating a certain configuration, especially in Monosaccharides (see).

Arabinose abb. ***Ara***: a pentose (monosaccharide with five carbon atoms). A. is a colorless powder with a slightly sweet taste which dissolves readily in water and is practically insoluble in organic solvents. ***L-A.***: m.p. 158-160 °C, $[\alpha]_D$ after mutarotation, +104.6°. A. is found naturally as the L- form and more rarely as the D-form. It is a component of plant heteropolysaccharides (arabans, arabinogalactans, arabinoxylans, etc.).

Arachidonic acid: all-*cis*-5,8,11,14-eicosatetraenoic acid, abb. Δ_4Ach or 20:4(5,8,11,14). $H_3C(CH_2)_3(CH_2CH{=}CH)_4(CH_2)_3COOH$ is a polyunsaturated, essential fatty acid. It is a pale yellow liquid, m.p. - 49.5 °C, which is very easily oxidized. Human fat tissue contains 0.3 to 0.9% A. It is enriched in the phospholipids of the brain and liver, where it is bound preferentially to phosphatidylethanolamine. It is the most effective fatty acid in correcting the symptoms of essential fatty acid deficiency. A. is released from glycerophospholipids in cell membranes by the enzyme phospholipase A_2. It is the starting material for biosynthesis of the eicosanoid hormones (thromboxanes), as it is a substrate for the cyclooxygengenases and lipoxygenases. A. is used to treat skin diseases.

Arachinic acid: same as Eicosanoic acid (see).

Arboricide: see Herbicides.

Arbutin, ***hydroquinone β-D-glucopyranoside***: a hydroquinone glucoside which occurs widely in plants, especially the Ericaceae. A. forms colorless, bitter-tasting crystals, m.p. 200 °C, $[\alpha]_D^{20}$ - 64.3°. A. and methylarbutin are the major active components of bearberry leaves, which as a dried medicinal herb should contain at least 6% A. The herb is used to treat infections of the urinary tract; after oral intake, the hydroquinone is present in the alkaline urine and kills the bacteria.

Arc spectra: see Atomic spectroscopy.

Ardein fibers: a Protein fiber (see).

Areca alkaloids: basic substances present at 0.2 to 0.5% in the seeds of *Areca catechu* (betel nut), a tropical palm. The main alkaloid is Arecolin (see). The betel nut is used in Southeast Asia and East Africa and a mild intoxicant.

Arecolin: an areca alkaloid; a very basic oil; b.p. 209 °C. A. is very toxic. It reacts with the acetylcholine receptors and acts as a parasympathicomimetic. A. is used in veterinary medicine as a wormer.

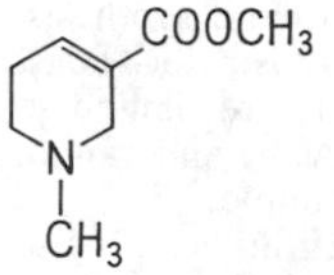

$COOCH_3$ N CH_3

Arene, ***aromatic hydrocarbon***: a cyclic, conjugated compound based on the benzene ring, generally with a relatively low energy content and high stability. The A. are classified on the basis of their structures into monocyclic and polycyclic A. The ***monocyclic A.*** include benzene and its homologs (alkyl benzenes, e.g. toluene); the ***polycyclic A.*** are further subdivided into hydrocarbons in which two or more rings are directly linked (e.g. biphenyl), di- and polyarylalkanes (e.g. diphenylmethane or triphenylmethane) and hydrocarbons with fused rings, one or more of them benzoid in nature (e.g. naphthalene or anthracene).

Fig. 1. Monocyclic arenes

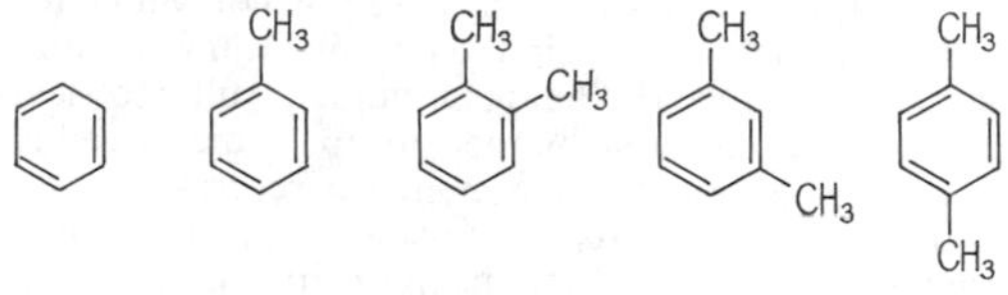

Benzene Toluene o-Xylene m-Xylene p-Xylene

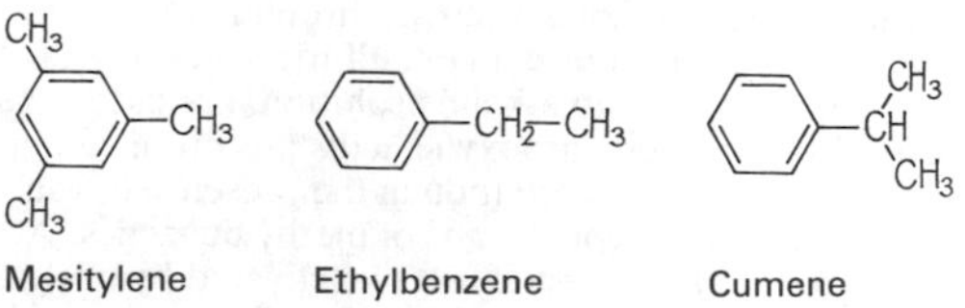

Mesitylene Ethylbenzene Cumene

Fig. 2. Polycyclic arenes

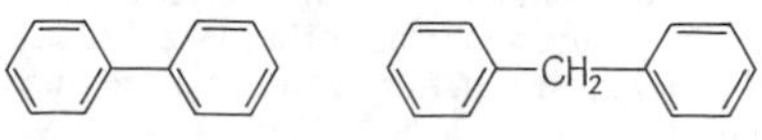

Biphenyl Diphenylmethane

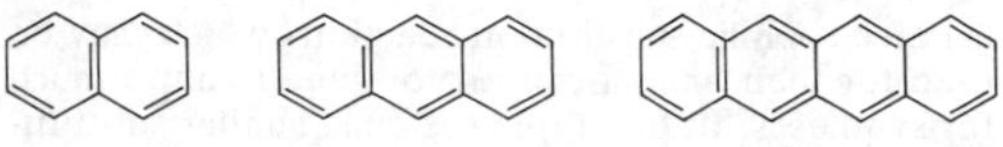

Naphthalene Anthracene Tetracene

Phenanthrene Pyrene

Most A. are designated by allowed trivial names. This applies in some cases even to homologs of benzene which might rationally be named as alkyl-, dialkyl-, trialkylbenzene, etc. If a ring system has more than one substituent, these are assigned the lowest possible numbers, as with 1,2-dimethylbenzene (*o*-xylene). For isomeric disubstituted molecules, the prefixes *ortho*- (*o*-) indicates 1,2-; *meta*- (*m*-) indicates 1,3-; and *para*- (*p*-) indicates 1,4-. Formal removal of an H atom from the arene ring produces an ***aryl*** group, such as phenyl-, *p*-tolyl-, 1-naphthyl-, etc.; these should not be confused with the benzyl group which is derived from toluene by removal of an H atom from the methyl group (Fig. 1). The phenyl group is important in nomenclature of polycyclic A.; the benzyl group is used in the names of many compounds which are formally derived from toluene, such as benzyl alcohol, benzyl cyanide, etc. The names of complicated fused rings are constructed in the same manner as those of comparable heterocycles (see Nomenclature).

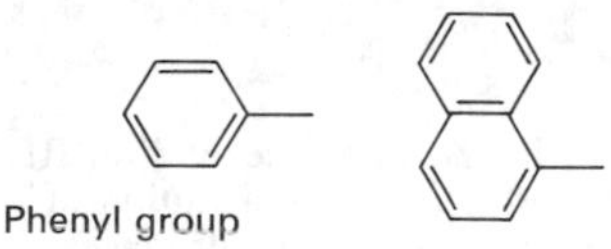

Phenyl group 1-Naphthyl group

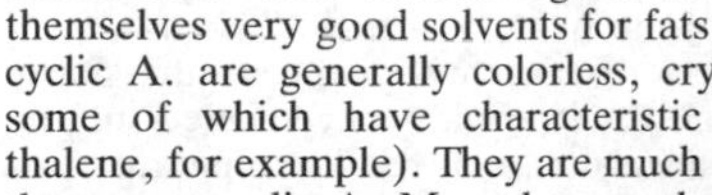

p-Tolyl group Benzyl group

Properties. Monocyclic A. are colorless, flammable liquids which produce large amounts of soot when they burn. They are essentially insoluble in water, but are miscible with various organic solvents, and are themselves very good solvents for fats and oils. Polycyclic A. are generally colorless, crystalline solids, some of which have characteristic odors (naphthalene, for example). They are much less flammable than monocyclic A. Many have a characteristic UV fluorescence, while others absorb visible light and are therefore colored (for example, tetracene, pentacene, coronene). They are insoluble in water. Because of their special bonding and electronic structures (see Aromaticity), the monocyclic A. have considerable delocalization energies, which makes the aromatic ring system very resistant to attack, even by aggresive reagents. For example, these compounds are not destroyed by concentrated sulfuric or nitric acid, unlike the alkanes. Instead, they undergo typical electrophilic substitution reactions which may be used for their analysis, or are of general importance for syntheses. In fused ring systems, similar substitution reactions are possible, with certain ring positions being favored. Only benzoid rings are inert to hydrogenation, that is, in naphthalene, for example, one of the two rings can be relatively easily hydrogenated (see Tetralene). In anthracene and phenanthrene, the middle rings are relatively easily hydrogenated or oxidized (see Anthraquinone and Phenanthroquinone).

Electrophilic substitutions are the typical reactions of the A.; they conserve the benzoid ring system (Fig. 3).

Fig. 3. Reactions of arenes

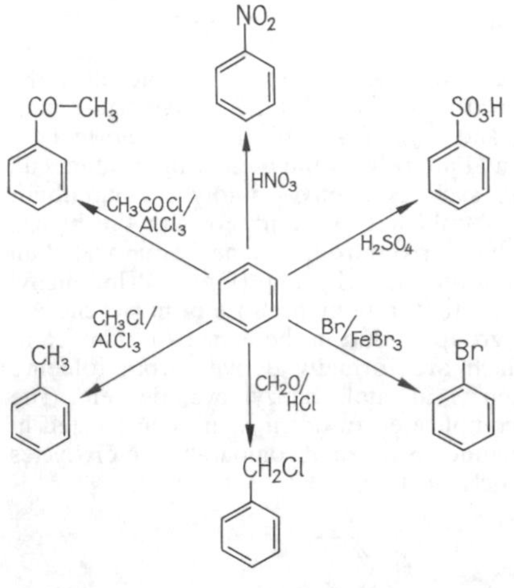

Halogenation occurs in the presence of iron(III) halide as catalyst on mild heating; substitution of a halogen atom for an H in the arene ring produces an ***aryl halide (haloarene)***. In the presence of excess halogen, and with strong heating, two or three halogen atoms can be introduced; these are always ortho or para to each other. The formation of structural isomers is governed by substitution rules which are based on the S_E mechanism. Only Cl or Br atoms can be introduced in this way; F and I atoms must be introduced in a series of steps, because their activities are too high and too low, respectively (see Sandmeyer reaction). The reaction conditions required for chlorination or bromination depend on the structure and reactivity of the A.

2) *Nitration*, a reaction with concentrated nitric acid or a mixture of nitric with concentrated sulfuric acid (nitrating acid) occurs at room temperature, or slightly above or below it, depending on the reactivity of the A. *Nitroarenes* are made by substitution of nitro groups for one or more H atoms in the ring system.

Toluene reacts much more readily than benzene, which can be seen from the fact that it can easily be converted to 2,4,6-trinitrotoluene. In contrast, only two nitro groups can easily be introduced into the benzene molecule (1,3-dinitrobenzene).

3) *Sulfonation* can be carried out with concentrated sulfuric acid at room temperature or a little above it. Substitution of an H atom produces the strongly acidic sulfonic acids, most of which are soluble in water. In general, sulfonic acid groups can be introduced to increase the water solubility of many derivatives of the A., such as dyes. In addition, the sulfonic acid group may later be changed to a derivative (see Sulfonamides).

4) In *hydroxymethylation*, the A. reacts with formaldehyde to form a hydroxymethylarene, C_6H_5-CH_2OH. With relatively inert A., such as benzene, the reaction requires an acidic catalyst, such as hydrogen chloride. However, the reaction then often produces a chloromethylarene, polysubstituted products or even condensation products. These reactions are industrially important for the production of plastics from phenols and formaldehyde. Homologous aliphatic or aromatic aldehydes can also be used: C_6H_6 + R-CHO $\rightarrow$ C_6H_5-CHOH-R.

5) *Formylation reactions* produce aromatic aldehydes in which an H atom of the arene ring is replaced by the aldehyde group (see Gattermann synthesis, Gattermann-Koch synthesis and Vilsmeier-Haack reaction). These reactions, however, are limited to especially reactive A. ("activated A."), and cannot be carried out with benzene, for example.

6) *Nitrosylation*, a reaction with nitrous acid, is also possible only with activated A. such as phenols or tertiary aromatic amines; it introduces a nitroso group para to the activating substituent:

$$R\text{–}C_6H_5 + HNO_2 \xrightarrow{-H_2O} R\text{–}C_6H_5\text{–}NO.$$

Similarly, azo coupling with diazonium salts, which gives azo dyes, and carboxylation with carbon dioxide (see Kolbe-Schmitt synthesis), which gives carboxylic acids, give practical yields only with activated A.

7) *Reactions with alkyl halides* (see Friedel-Crafts reactions).

8) *Chloromethylation* (see Blanc reaction). Substitution of a nucleophile for an H atom in the arene ring is extremely difficult, and is practically never done. However, SN reactions are possible with various substituted A., if suitable leaving groups are present. These include halogen atoms, sulfonic acid groups and the diazonium group, which readily cleaves off nitrogen (see Sandmeyer reaction).

Oxidation reactions are possible with many alkyl arenes and fused ring systems, for example the oxidation of toluene to benzoic acid or *p*-xylene to terephthalic acid.

The classic oxidizing agents, chromium(VI) compounds in glacial acetic acid or sulfuric acid, or potassium permanganate in alkaline solution, are used. Industrially, aromatic carboxylic acids are produced in large amounts by air oxidation in the presence of vanadium(V) oxide or cobalt salts of methylbenzenes. A. can be reduced, with catalytically activated hydrogen under pressure and at high temperatures, to saturated cyclic ring systems; for example, benzene to cyclohexane or naphthalene to decahydronaphthalene. This property is utilized in industry and in the laboratory to synthesize compounds which are otherwise difficult to produce. Usually such reactions require a pressure vessel (autoclave).

Analytical. A. are detected by nitration and reaction with aluminum chloride and chloroform (Friedel-Crafts reactions). They are identified either by sulfochlorination (reaction with chlorosulfonic acid) followed by aminolysis, or by formation of adducts with picric acid. A. with alkyl side chains can be oxidized with potassium permanganate in alkaline solution to carboxylic acids; a few A., such as anthracene or phenanthrene, are oxidized by chromium(VI) compounds to quinones. The IR spectra of benzene and its derivatives show C-H valence vibrations around 3030 cm^{-1} and benzoid C=C valence vibrations at about 1500 cm^{-1} and 1600 cm^{-1}. For identification of substituted compounds, the range from 650 to 850 cm^{-1} (out-of-plane vibrations) is important. The UV spectra of benzene and its alkyl derivatives have three absorption bands at 180 nm, 200 nm and 255 nm ($\pi \rightarrow \pi^*$ transition). For many fused rings (especially linear fusions), the longest-wave absorption maximum is shifted into the visible range. Introduction of certain substituents, such as nitro or nitroso groups, also produces bathochromic shifts. In the ^1H NMR spectrum, the proton signals appear between 6.5 and 8.5 ppm; the ring current withdraws shielding from the protons. The coupling constants give important clues to the positions of substituents: J_{ortho} = 6 to 10 Hz, J_{meta} = 1 to 3 Hz, and J_{para} = 0 to 1 Hz. The mass spectrum of benzene or one of its derivatives has a strong molecular peak. The basis peak of an alkyl benzene is formed from the tropylium ion $C_7H_7^+$ (M = 91); some other typical fragments appear at M = 77, 65, 53, 51, 50 and 39.

Occurrence and extraction. Benzene and many of its homologs are found in petroleum from various sources, coke gas and coal tar, and they are extracted from these sources on an industrial scale. To meet the increasing demand for these hydrocarbons, aliphatic and saturated cyclic components of petroleum are now converted to A. by dehydrocyclization, dehydroisomerization, dehydrogenation or high-temperature cracking. Many polycyclic A. are found, in considerable quantities, in coal tar or the tars resulting from pyrolysis of gasoline or kerosene, and they are extracted from these sources. In addition, there are many synthetic methods, some of which are carried out industrially or in the laboratory.

1) Alkylbenzenes are synthesized from halobenzenes and haloalkanes by the Wurzt-Fittig reaction, using sodium; from A., haloalkanes and aluminum chloride by the Friedel-Crafts reaction; or from arylmagnesium chloride and haloalkanes by the Wurzt-Grignard reaction. With the versatile Friedel-Crafts acylations, the industrially available alkenes ethene and propene are often used instead of the haloalkanes as alkylation reagents.

2) Diphenylmethane and triphenylmethane are also accessible through Friedel-Crafts reactions, for example by reaction of benzene with benzyl halide, or of benzene with chloroform.

3) Fused A. can be obtained on a preparative scale by Friedel-Crafts acylation, for example naphthalene from benzene and succinic anhydride, or anthracene from benzene and phthalic anhydride.

Applications. Benzene and its homologs, especially toluene and the xylenes, are used as organic solvents, although their toxicity must be kept in mind (see Benzene). Methylbenzenes are converted to aromatic carboxylic acids by oxidation; alkyl-substituted cyclohexanes are obtained by hydrogenation. Fused A. are oxidized to quinones or carboxylic acids, depending on their structures. Hydrogenation can be carried out on selected rings or it may be total, producing other solvents or intermediates. Many aromatic functional compounds, which serve as intermediates in the synthesis of dyes, drugs, optical brighteners, laboratory and fine chemicals and explosives, are produced by electrophilic substitution of A. and subsequent reactions. Thus nitro compounds can easily be reduced to amines (see Aniline), sulfonic acids can be converted to phenols, and haloarenes can be further reacted with Grignard compounds. Halogenation of alkyl groups on A. to mono-, di- or trichloro compounds is also important; these compounds are converted by hydrolysis to alcohols, aldehydes or carboxylic acids. A. thus provide access to almost all functional compounds.

Arene complexes: see Organo-element compounds.

Arene diazonium salts: aromatic diazonium salts with the general formula R-N̊≡N|X⁻, where R is an aromatic or heteroaromatic group and X^- is a negative ion. Unlike aliphatic diazonium salts, A. are relatively stable, crystalline compounds which can be isolated, due to resonance stabilization of the charge. They tend to detonate when heated; the nitrates and perchlorates are especially dangerous. With the exception of the tetrafluoroborates, A. are readily soluble in water. They can be synthesized by Diazotization (see). A distinction is made between reactions in which the diazo group is displaced from the molecule in the form of nitrogen (see Diazo cleavage) and those yielding other nitrogen compounds, such as those leading to phenylhydrazine, phenylazide and the industrially important azo dyes (see Azo coupling).

Arene oxides: see Benzene oxide.

Arene thiols: same as Thiophenols (see).

Arene sulfonic acids: see Sulfonic acids.

Arg: abb. for arginine.

Arge high capacity process: see Fischer-Tropsch synthesis.

Argentan®: see New silver.

Argentometry: the most important method of precipitation analysis in which silver ions are the titrator. These form insoluble precipitates with halide and pseudohalide ions. The standard solution can be easily prepared from silver nitrate, a titration standard, but they are sensitive to light and should be stored in brown bottles.

There are several methods for recognizing the endpoint in A. The oldest is the ***Gay-Lussac method***, in which there is no indicator. If the titration is done with silver nitrate solution, the silver halide at first precipitates as a colloid. At the equivalence point, the precipitate clumps together and the solution becomes completely clear. This method can also be used in the reverse process in which the halide ions are used as titrator to determine silver concentration.

The ***Mohr method*** uses chromate ions as indicator at a pH of 5.5 to 8.0. A dark brown precipitate of silver chromate forms only after complete precipitation of silver chloride or bromide, and thus indicates the endpoint of the titration.

The ***Fajans method*** makes use of absorption indicators. Such indicators, e.g. fluorescein, dichlorofluorescein or eosin, are adsorbed on the surface of the silver halide precipitate after the equivalence point has been reached, and change their color. High salt concentrations in the sample solution interfere with this method, because they affect the adsorption of the indicator.

The ***Volhard method*** is suitable only for indirect titrations. After addition of an excess of silver nitrate standard solution, a thiocyanate standard solution is used for back titration. Iron(III), which forms a dark red complex with thiocyanate, is used as indicator.

The endpoint in A. can also be determined by electrochemical methods, which are sensitive enough to permit simultaneous determination of several halide ions.

Argentum: see Silver.

Arginase: an enzyme which catalyses the hydrolysis of arginine to ornithine and urea. A. (*M*r 118,000) consists of four subunits, each of which contains an Mn^{2+} ion.

Arginine, abb. ***Arg***: α-amino-δ-guanidinovaleric acid, the most basic of the proteogenic amino acids (formula and physical properties, see Amino acids, Table 1). A. is particularly abundant in protamines and histones. In free form, it is found in red algae, cucurbits and conifers. Industrially, A. is obtained almost exclusively from hydrolysis of gelatins. It was first isolated in 1885 from lupine seedlings by Schulze and Steiger.

Argon, symbol ***Ar***: chemical element of the zeroth or eighth main group of the periodic system of the elements, the Noble gases (see); *Z* 18, natural isotopes with mass numbers 36 (0.337%), 38 (0.063%) and 40 (99.600%). Atomic mass 39.948, valence 0, density 1.784 g l^{-1} at 0 °C, m.p. - 189.2 °C, b.p. - 185.7 °C, crit. temp. - 122.3 °C, crit. pressure 4.8 MPa.

Properties. A. is a colorless, odorless, tasteless monoatomic gas. At 0 °C, 52 ml A. dissolves in 1 l water.

A. forms a very unstable hydrate and a clathrate with hydroquinone, but true "valence" compounds of A. are not known.

Analysis. A. is most conveniently separated by gas chromatography, and detected spectroscopically.

Occurrence. A suitable starting material for industrial production of A. is the residual gas from ammonia synthesis. Because fresh, A.-containing nitrogen is continuously introduced into the circulation, and the nitrogen is removed in the form of ammonia, A. is enriched to concentrations of more than 10%. The gas mixture is liquefied and fractionated by low-temperature distillation, yielding A. at very high purity. Air can also be used as a source of A., but the proximity of the boiling points of A. and oxygen (O_2 boils at - 182.97 °C) make it necessary to use a very efficient rectifying column and chemical treatment is still required after distillation.

Applications. As a cheap inert gas produced in large quantities industrially, A. is being used increasingly as a protective gas for reactions of readily oxidized substances and as a rinsing gas to remove gases from metal melts. A. is used as a protective gas in electrowelding, especially of readily oxidized light metals (aluminum, magnesium) and metals with very high melting points (titanium, zirconium, tantalum, molybdenum, tungsten; A. arc method). It is also used to fill light bulbs.

Historical. A. was discovered by Ramsay and Rayleigh in 1894 as a component of air.

Arndt-Eistert synthesis: a sequence of reactions described in 1935 for lengthing carboxylic acid chains by one CH_2 unit by reaction of carboxylic acid chlorides with diazomethane. The first product is an α-diazoketone, which releases nitrogen when heated with silver or silver oxide; a ketene is formed via a ketocarbene:

$$R{-}CO{-}Cl + |\bar{C}H_2{-}\overset{+}{N}{\equiv}N| \longrightarrow$$

$$R{-}CO{-}|\bar{C}{-}\overset{+}{N}{\equiv}N| \xrightarrow{(Ag_2O)}$$

$$R{-}CO{-}\bar{C}H \longrightarrow R{-}CH{=}C{=}O \xrightarrow{+H_2O} R{-}CH_2COOH$$

The shift of the residue R with the bonding electron pair to the carbene carbon atom is known as the Wolff rearrangement (see). The reaction of water with the resulting ketene produces the corresponding carboxylic acid. If the diazoketone is decomposed in the presence of an alcohol or ammonia, the carboxylic acid ester or amide is formed directly.

Arnel®: a Synthetic fiber (see).

Arogenic acid: an aminocarboxylic acid discovered in 1974 as an intermediate in the biosynthesis of phenylalanine and tyrosine in prokaryotes (bacteria and blue-green "algae"), yeasts and green plants. It is formed by transamination of prephenic acid.

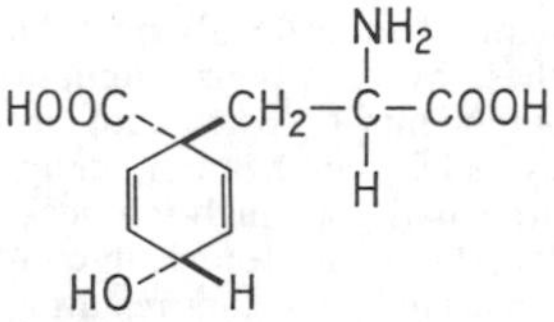

Aromatic: see Aromaticity.

Aromaticity: certain cyclic conjugated systems are more stable than the corresponding straight-chain compounds with the same number of π-electrons. The first theoretical explanation of A. was given by the Hückel method (see). From the HMO scheme for cyclic conjugated hydrocarbons, it can be seen that for systems with $4n+2$ π-electrons ($n = 0, 1, 2...$), there are very stable singlet states (*Hückel rule*). Conjugated ring hydrocarbons which obey the Hückel rule are called *aromatic*. Using the Hückel approximation, it can be calculated that cyclic conjugated systems are energetically more stable than linear conjugated systems. For example, the π-electron system of benzene ($n = 1$) has much lower energy than 1,3,5-hexatriene. In contrast to this, the π-electron systems of conjugated cyclic hydrocarbons with $4n$ π-electrons ($n = 1, 2...$) are less stable than the corresponding linear conjugated systems. This applies to cyclobutadiene, which has higher energy than 1,3-butadiene. Cyclic conjugated hydrocarbons with $4n$ π-electrons are therefore called *antiaromatic*.

If a $2p_z$ orbital is replaced by a 3d atomic orbital, for example in cyclobutadiene, the topology of the bonds changes; there is a phase shift associated with a change in sign of an orbital lobe (Fig.). Such a system is called a *Möbius system*; it can occur, for example, in π-complexes of transition metals. In Möbius systems, in contrast to Hückel systems, even the lowest-energy molecular orbital is doubly degenerate. As a result, the aromaticity properties are reversed. Möbius-type cyclic conjugated systems are aromatic if they have $4n$ π-electrons ($n = 1, 2 \ldots$), while those with $4n+2$ π-electrons ($n = 0, 1, 2\ldots$) are antiaromatic. Möbius systems are therefore also called *anti-Hückel systems*).

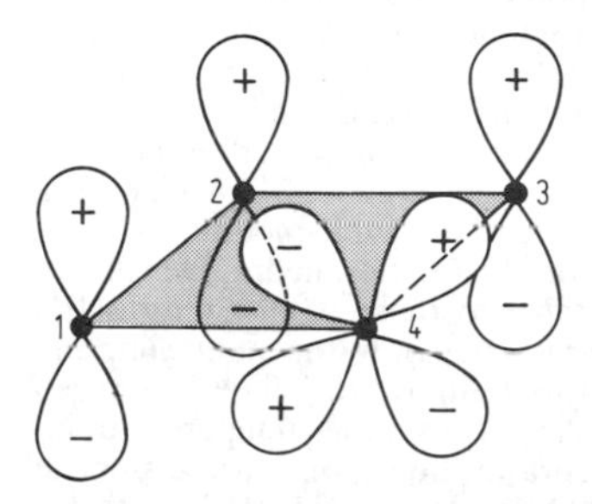

Möbius system (phase shift between atoms 1 and 4).

In general, the following rule applies to the topology of cyclic conjugated systems and their A.: if the π-atomic orbitals can be arranged in such a way that the number of minimal phase shifts is zero or even, the system is Hückel type (*Hückel topology*). If the number of phase shifts is odd, it is Möbius type (*Möbius topology*). Cyclic conjugated molecular ions (e.g. cyclopentadienyl anion) and heterocyclic conjugated systems (e.g. pyridine) can also be treated by this rule.

Arosolvane process: a process for obtaining aromatics by Extraction (see).

Aroxyls: term for aryl-substituted oxygen radicals of the phenoxyl type, C_6H_5O.

Arrhenius acid-base definition: see Acid-base concepts.

Arrhenius equation: in kinetics, an equation first suggested by S. Arrhenius (1869) which describes the temperature dependence of the rate constants of a very large number of reactions: $k = A \exp(-E_a/RT)$. k is the rate constant, R, the general gas constant and T the absolute temperature. The factor A (also called the *frequency factor* or *action constant*) and the *activation energy* E_0 are characteristic of each reaction, and must be determined experimentally. The A. can be derived theoretically from collision theory.

Arrow poisons: poisons (usually from plants) which are rapidly toxic or fatal when injected parenterally (through skin wounds). They are used for hunting and war by certain indigenous peoples. The best known P. are the African A. ***ouabain*** (chemically, g-strophanthin), a cardiac poison; the South American A. ***curare***, which paralyzes the voluntary muscles; and ***Javanese A.***, which contains either strychnine (a convulsant which acts primarily in the spinal cord) or antiarine (which has an effect similar to those of strophanthin and strichnine).

Arsane: same as Arsenic hydride (see).

Arsenates: salts of Arsenic acid (see).

Arsenic: (the poison) see Arsenic oxides.

Arsenic, symbol ***As***: chemical element from the fifth main group of the periodic system, the Nitrogen-phosphorus group (see). It is a semimetal, Z 33, with only one natural isotope, atomic mass 74.9216, valences III and V, standard electrode potential 0.2475 V ($As + 3\,H_2O \rightleftharpoons H_3AsO_3 + 3\,H^+ + 3e$).

Properties. A. exists in three modifications. The ***gray, metallic*** modification is thermodynamically stable at room temperature; its lattic consists of infinite layers in which the As atoms are linked in puckered, six-membered rings. It forms a steel-gray, brittle, rhombohedral crystal with a metallic sheen. Its Mohs hardness is 3 to 4, density 5.727, electrical conductivity 3.0 Sm mm^{-2} (at 20 °C). In a closed tube, it melts at 817 °C under its own vapor pressure of 2.8 MPa, and sublimes at 613 °C, forming a yellow vapor which consists of As_4 molecules. The proportion of As_2 molecules in the vapor increases as the temperature increases. If arsenic vapor is quenched, it condenses as waxy, ***yellow nonmetallic A.***, which is soluble in carbon disulfide. Its density is 2.064, and its cubic lattice is comparable to that of white phosphorus, in that it consists of As_4 molecules. Yellow A. is rapidly converted to the gray form, especially when exposed to light. If arsenic vapor is condensed on surfaces where the temperature is 100 to 200 °C, it forms the hard, brittle, black and shiny ***amorphous A.***. This form does not conduct electricity. In the presence of mercury, it is converted at 125-175 °C to the unstable orthorhombic ***black A.***, which has a structure analogous to that of black phosphorus.

A. burns in air > 180 °C to form arsenic(III) oxide, As_2O_3. It burns in chlorine to give arsenic(III) chloride, $AsCl_3$. Concentrated nitric acid or aqua regia oxidizes A. to arsenic acid: $As + 5\,HNO_3 \rightarrow 5\,NO_2 + H_3AsO_4 + H_2O$; in dilute nitric acid or concentrated sulfuric acid, A. is dissolved to form arsenous acid: $2\,As + 3\,H_2SO_4 \rightarrow 2\,H_2AsO_3 + 3\,SO_2$. A. forms alloys with many metals; most of these alloys are brittle.

> Pure arsenic is reported to be nontoxic. However, since it is readily oxidized in the air, and it is difficult to be sure no impurities are present, the element should be treated with caution.

Analysis. The Gutzeit test (see), Marsh test (see), Bettendorf test (see) or Berzelius test (see) can be used for qualitative analysis. In the systematic separation procedure, A. is precipitated as arsenic sulfide, separated by treatment with $(NH_4)_2S$, and its presence indicated by precipitation as ammonium magnesiumarsenate, $NH_4MgAsO_3 \cdot 6H_2O$. Iodometry or atomic absorption spectroscopy is used for quantitative determination of As^{3+}.

Occurrence. A. makes up $5.5 \cdot 10^{-4}$% of the earth's crust. It is occasionally found in the elemental state in nature, but it is more commonly found as arsenides, arsenic sulfides and arsenic oxides. Some important minerals are arsenopyrite, FeAsS, loellingite, $FeAs_2$, cobaltine (cobalt glance), CoAsS, gersdorffite (arsenic nickel glance), NiAsS, realgar, As_4S_4 and oripiment, As_2S_3. Claudetite or arsenolite, As_2O_3, are weathering products of arsenic ores. A. occurs very widely; it can always be detected in the soil and in plant and animal tissues.

Extraction. If arsenopyrite is heated in the absence of air, elemental As forms and sublimes off: $FeAsS \rightarrow FeS + As$. However, the element is usually produced industrially from arsenic-containing byproducts of various metallurgical processes. These are worked up to arsenic(III) oxide, As_2O_3, and reduced to As with carbon. For the production of very pure As, the oxide can be reduced with hydrogen.

Applications. Very pure A. is used to dope silicon and germanium semiconductors, and in the production of gallium arsenide. Most of the As produced is used in alloys, especially of lead and copper. For example, the lead-antimony-arsenic alloy used as the lattice in lead batteries contains 0.5% As.

Historical. A few arsenic compounds were known in antiquity. Production of the elemental form was first described by Albertus Magnus about 1250.

Arsenic acid: H_3AsO_4, has been isolated only as the semihydrate, $H_3AsO_4 \cdot 1/2H_2O$, colorless, hygroscopic crystals; M_r 150.95, m.p. 35.5 °C. In aqueous solution, A. is a medium strong triprotic acid: $pK_1 = 2.32$, $pK_2 = 7$, $pK_3 = 13$. There are three series of salts derived from A., the primary arsenates or dihydrogenarsenates, $M^IH_2AsO_4$, the secondary or hydrogenarsenates $M^I_2HAsO_4$ and the tertiary arsenates $M^I_3AsO_4$. Overall, A. and the arsenates are very similar in their chemistry to phosphoric acid and the phosphates. For example, A. forms a relatively insoluble, colorless ammonium magnesium arsenate, $NH_4MgAsO_4 \cdot 6H_2O$ and an insoluble, chocolate-brown silver arsenate, Ag_3AsO_4. Like the phosphoric acids, A. can be a component of heteropolyacids. For example, in nitric acid solution, ammonium molybdate and A. form the relatively insoluble yellow ammonium dodecamolybdatoarsenate $(NH_4)_3[As(Mo_{12}O_{40})]$. All the above precipitations can also be used as detection reactions for arsenic or arsenates. In contrast to phosphoric acid, A. is a strong oxidizing agent; for example, it can oxidize iodide to iodine. A. is obtained by oxidation of arsenic or arsenic(III) oxide with conc. nitric acid.

Arsenic(III) chloride, *arsenic trichloride*: $AsCl_3$, a colorless, oily and very poisonous liquid which fumes in moist air; M_r 181.28, density 2.163, m.p. - 8.5 °C, b.p. 130.2 °C. A. is obtained by burning arsenic in a chlorine atmosphere, or by reaction of arsenic(III) oxide, As_2O_3 with dry hydrogen chloride at 180-200 °C. The reaction of aqueous hydrochloric acid with As_2O_3 leads to an equilibrium: $H_3AsO_3 + 3\,HCl \rightleftharpoons AsCl_3 + 3\,H_2O$; in the presence of a large excess of HCl, this equilibrium is shifted to the right. By distilling off the water, it is possible to convert As_2O_3 completely to $AsCl_3$. A. is an important starting material in the preparation of various arsenic compounds, especially organic ones.

Arsenic hydride, *arsine, arsane*: AsH_3, a colorless and very poisonous gas with an unpleasant, garlic-like odor; M_r 77.95, m.p. - 116.3 °C, b.p. - 55 °C. A. burns in air with a pale blue flame to give arsenic(III) oxide and water: $2\,AsH_3 + 3\,O_2 \rightarrow As_2O_3 + 3\,H_2O$. However, if a cold object is held in the flame, the combustion is not complete and a black arsenic mirror precipitates: $4\,AsH_3 + 2\,O_2 \rightarrow 4\,As + 6\,H_2O$. When heated, A. decomposes into arsenic and hydrogen. The last two reactions are used in the Marsh test (see). A. is a strong reducing agent, which is able, for example, to reduce silver cations to metallic silver. It reacts with solid silver nitrate to form the double salt $Ag_3As \cdot 3AgNO_3$ (see Gutzeit test). A. is formed in the reaction of nascent hydrogen with arsenic or arsenic compounds. It is used as a dopant in the semiconductor industry.

Arsenic oxides: ***Arsenic(III) oxide, arsenic trioxide, arsenic***; M_r 197.84, exists in two modifications. The cubic modification is stable at room temperature, has density 3.865, sublimes 193 °C, and is converted at 221 °C to a monoclinic form, density 4.15, b.p. 457.2 °C. As_2O_3 is obtained by combustion of arsenic; it exists either as an amorphous, colorless powder, density 3.738, m.p. in a closed tube, 312.3 °C, or as a glassy mass. The lattice of the cubic modification and the vapor consist of As_4O_6 tetrahedra, which are structurally comparable to those of phosphorus(III) oxide (see Phosphorus oxides). In the monoclinic and glassy forms, $[AsO_3]$ pyramids are joined by oxygen atoms to infinite layers. As the temperature increases, the tetrahedra in the vapor phase dissociate to As_2O_3 units. As_2O_3 is not very soluble in water. It is amphoteric with respect to acids and bases. In alkali hydroxide solutions, it dissolves to form arsenites (see Arsenous acid), while it comes to an equilibrium with hydrochloric acid: $H_3AsO_3 + 3\,HCl \rightleftharpoons AsCl_3 + 3\,H_2O$. When heated with carbon, hydrogen or sodium cyanide, As_2O_3 is reduced completely to arsenic.

> Arsenic(III) oxide and arsenites are strong poisons. The lethal dose for a human being is approximately 0.1 g As_2O_3. It is absorbed from the gastrointestinal tract, and to a lesser extent through the skin as well. Dust is also absorbed in the lungs. The individual sensitivities to As_2O_3 vary widely. By consuming gradually increasing amounts, a person can become habituated (arsenic eaters), so that significantly larger doses can be tolerated without severe damage. The symptoms of acute poisoning are pains in the body, vomiting, diarrhea, dryness in the mouth and throat, convulsions, drop in blood pressure, etc., leading to paralysis and death.

As_2O_3 is found in nature as cubic arsenolite and monoclinic claudetite. It is obtained industrially by roasting of arsenopyrite according to the equation: $2\,FeAsS + 5\,O_2 \rightarrow Fe_2O_3 + 2\,SO_2 + As_2O_3$. The oxide sublimes off. As_2O_3 was formerly used in medicine. Now it is used as a rodenticide, to preserve furs, bird skins and hides, and is the starting material for synthesis of various arsenic compounds, especially organic ones.

Arsenic(V) oxide, arsenic pentoxide: As_2O_5, colorless amorphous powder or glassy mass; M_r 229.84, density 4.32, dec. 315 °C. As_2O_5 dissolves in water, forming arsenic acid, H_3AsO_4. It decomposes when heated, forming As_2O_3 and oxygen. It is obtained by dehydration of arsenic acid.

Arsenic sulfides: ***Arsenic monosulfide***, $(AsS)_4$, ruby-red crystals or an amorphous mass; M_r 427.94. The monoclinic lattice consists of As_4S_4 molecules, the structure of which is derived from that of As_4 tetrahedra by insertion of sulfur in four As-As bonds (Fig.). As_4S_4 occurs in nature as realgar; it is synthesized by fusion of equimolar amounts of elemental sulfur and arsenic, or by heating arsenopyrite with pyrite: $4\,FeAsS + 4\,FeS_2 \rightarrow 8\,FeS + As_4S_4$. It is used to remove hair from hides which are to be tanned.

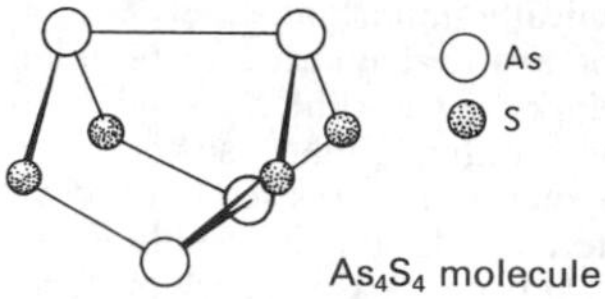

As
S

As_4S_4 molecule

Arsenic(III) sulfide, arsenic trisulfide, As_2S_3. A lemon-yellow, monoclinic crystalline powder; M_r 246.04, density 3.43, m.p. 300 °C, b.p. 707 °C. The compound is insoluble in water and acids (pK_s = 25.3) and is obtained by precipitation of As^{3+} ions with hydrogen sulfide. Crystalline As_2S_3 has a layered structure comparable to that of monoclinic arsenic(III) oxide, and the vapor phase consists of As_4S_6 molecules which correspond structurally to those of P_4O_6 or As_4O_6. As_2O_3 dissolves in ammonium sulfide solution to form thioarsenites according to the equation $As_2S_3 + 3\,S^{2-} \rightarrow 2\,AsS_4^{3-}$. As_2S_3 occurs in nature as oripiment. The pure compound is not poisonous, because of its insolubility, and is used as a pigment in paints.

Arsenic(V) sulfide, arsenic pentasulfide, As_2S_5, is a yellow, amorphous powder; M_r 310.16. As_2S_5 is insoluble in water and acids, but dissolves in ammonium sulfide to form thioarsenates. It is made by passing hydrogen sulfide through a very acidic solution of arsenic acid, or by fusion of As_2S_3 and sulfur.

Arsenides: compounds of arsene with metals. A. are formed by fusion of the elements. The alkali and alkaline earth arsenides, and zinc arsenides, are derived from arsenic hydride, AsH_3, and should be considered its salts. They are decomposed by water with generation of AsH_3. The heavy metal A. often have complicated stoichiometries and are considered intermetallic phases. Gallium arsenide, GaAs, and indium arsenide, InAs, are important in semiconductor technology.

Arsenites: the salts of Arsenous acid (see).

Arsenous acid, *arsenic(III) acid*: H_3AsO_3, a vcry weak, triprotic acid (pK_1 = 9.22) which cannot be isolated. The dilute aqueous solution of A. is obtained by dissolving its anhydride, arsenic(III) oxide, in water. Its salts, the poisonous *arsenites* or *arsenates(III)*, are derived either from the ortho form of A. or from the meta form, $HAsO_2$. There are three types of ortho-form derivatives: $M^IH_2AsO_3$, $M^I_2HAsO_3$ and $M^I_3AsO_3$. The meta-form derivatives have the formula M^IAsO_2, and contain polyanions. In strong hydrochloric acid, A. is in equilibrium with arsenic(III) chloride: $H_3AsO_3 + 3\,HCl \rightleftharpoons AsCl_3 + 3\,H_2O$. Elemental arsenic is precipitated from this solution by tin(II) chloride. On the other hand, arsenites or A. are oxidized by iodine in neutral solution to arsenates: $H_3AsO_3 + I_2 + H_2O \rightleftharpoons H_3AsO_4 + 2\,H^+ + 2\,I^-$. In acid solution, the reaction reverses.

Arsine: same as Arsenic hydride (see).

Arsines: see Organoarsenic compounds.

Arsinic acids: see Organoarsenic compounds.

Arsonium salts: see Organoarsenic compounds.

Arsonic acids: see Organoarsenic compounds.

Arsphenamine, ***Salvarsan®***: an organic arsenic compound with which it was possible, for the first time, to heal syphilis with a chemical pharmaceutical. A. is usually shown as an arseno compound (-As=As-), but it may be a trimer with a ring of 6 As atoms. The active form of A. is oxophenarsine.

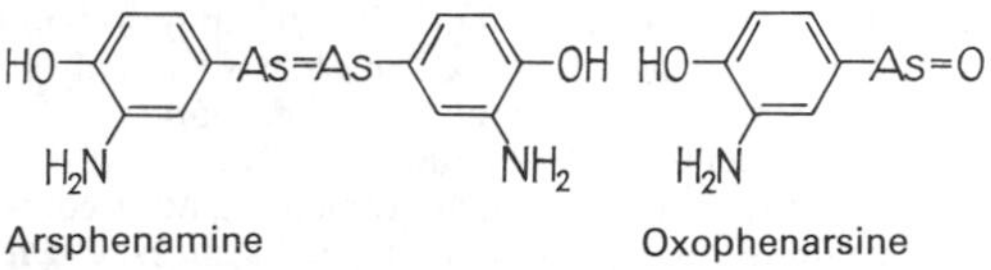

HO–C₆H₃(NH₂)–As=As–C₆H₃(NH₂)–OH   HO–C₆H₃(NH₂)–As=O

Arsphenamine   Oxophenarsine

Because of its high toxicity, A. is no longer used. It was developed in 1909 by P. Ehrlich.

Aryl-, abb. ***Ar-***: a term for a group derived from an arene by removal of one of the hydrogen atoms on the aromatic ring, e.g. phenyl- or naphthyl. Aryl groups are important in the systematic nomenclature of aromatic compounds. Aryl radicals are unstable and very short-lived, as are the corresponding carbenium ions.

Arylation: introduction of an aryl group into an organic compound. An important example is the addition of aryl radicals to unsaturated compounds. The radicals can be generated photochemically from halogen-substituted arenes or diazonium salts.

Aryl halides: same as Haloarenes (see).

Aryloxyalkanoic acid herbicides: same as Growth-hormone herbicides.

Arylsulfatases: see Esterases.

Arynes: aromatic ring systems with a formal triple bond in the aryne ring. The simplest compound of this type is *1,2-dehydrobenzene (benzyne)*. A. are short-lived, very reactive intermediates which can be generated and detected by UV spectroscopy at - 265 °C:

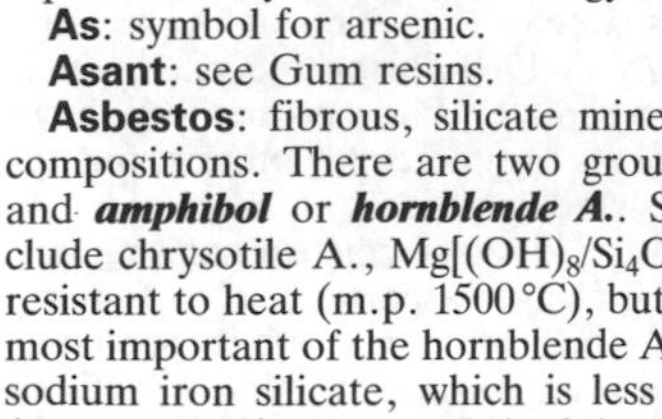

$\xrightarrow{-2\,CO_2}$

Actually, the structure of these compounds is best represented by the lowest-energy singlet state.

As: symbol for arsenic.

Asant: see Gum resins.

Asbestos: fibrous, silicate minerals with varying compositions. There are two groups, ***serpentine A.*** and ***amphibol*** or ***hornblende A.***. Serpentine A. include chrysotile A., $Mg[(OH)_8/Si_4O_{10}]$, which is very resistant to heat (m.p. 1500 °C), but not to acids. The most important of the hornblende A. is crocydolite, a sodium iron silicate, which is less resistant to heat (m.p. 1150 °C). About 90% of the A. used as a heat-resistant material is chrysotile A., and only about 5%

is crocydolite. Because of its carcinogenicity (bronchial carcinomas from asbestos dust), A. is being replaced in many applications by heat-resistant mineral and glass fibers. The world production reached a maximum in 1975 of about 5.5 million tons. Crocydolite is more carcinogenic than chrysotile A.. The mineral mixture talcum can contain up to 50% asbestos.

Ascaridol: 1,4-peroxo-*p*-menth-2-ene, a monoterpene found in the oil of American wormseed (*Chenopodium ambrosiodes* var. *anthelminticum*). The oil was formerly used as an drug against intestinal worms (ascarides, hookworms). However, it has severe side effects.

Ascarite: sodium hydroxide adsorbed to asbestos. A. is used to absorb carbon dioxide, forming sodium carbonate and water. It is used in elemental analysis and for gravimetric determination of carbon.

Ascorbic acid: same as Vitamin C (see).

Ashes: the inorganic residue remaining after complete combustion of plant or animal substances. ***Plant A.*** contain water-soluble potassium and sodium carbonates, sulfates and chlorides, as well as insoluble calcium, magnesium and iron carbonates, phosphates and silicates. ***Coal A.*** contain mainly clay, iron oxide and sulfates or silicates. They therefore are of no use as fertilizers. ***Bone A.*** consist essentially of calcium phosphate and can be used as fertilizer.

ASIS effect (for Aromatic Solvent Induced Shift), the shift in an NMR signal caused by transferring the substance from a nonpolar (e.g. deuterochloroform, $CDCl_3$) to an aromatic (e.g. benzene) solvent. The A. is often used to elucidate stereochemical problems. Benzene, for example, forms collision complexes with dissolved polar molecules; on the time average, these prefer certain spatial orientations. The chemical shifts (see NMR spectroscopy) of the individual nuclei of the dissolved molecule are affected differently, depending on their spatial relationship to the aromatic ring. The solvent shift $\delta = \delta_{CDCl_3} - \delta_{C_6H_6}$ for individual nuclei gives essential information on the structure of the dissolved compound (e.g. about its conformation). The ASIS has been particularly useful in the steric assignment of ketones.

Asn: abb. for Asparagine.

Asp: abb. for Aspartic acid.

Asparagine, abb. ***Asn***: the β-semiamide of aspartic acid, and a proteogenic amino acid (formula and physical properties, see Amino acids, Table 1). As a component of proteins, A. occurs widely in nature, and is found in free form in the sprouting seeds of many plants. It was the first proteogenic amino acid discovered; it was isolated from asparagus (*Asparagus officinalis*) in 1806 by Robiquet.

Aspartame, ***L-asparagyl-L-phenylalanine methyl ester***: H-Asp-Phe-OMe, a synthetic sweetener about two hundred times sweeter than sucrose. A. was discovered accidentally in 1969 during recrystallization of a peptide intermediate. Various methods have been developed for its commercial synthesis, including a protease-catalysed synthesis.

Aspartase: a lyase found in microorganisms and higher plants which cleaves aspartic acid to fumaric acid and ammonia. The reverse reaction is used commercially: A. or microorganisms with high A. activity are used to synthesize aspartic acid from fumaric acid and ammonia on an industrial scale.

Aspartic acid, abb. ***Asp, α-aminosuccinic acid***: $HOOC\text{-}CH_2\text{-}CH(NH_2)\text{-}COOH$, a proteogenic amino acid found in many plant and animal proteins. In the presence of aminotransferases, A. is reversibly coverted to oxaloacetic acid, an important member of the citric acid cycle. Thus protein and carbohydrate metabolisms are linked through A. DL-A. is synthesized industrially by addition of ammonia to maleic or fumaric acid under high pressure. However, the enzymatically catalysed addition of ammonia to fumaric acid, in which yields of 90% L-A. are obtained, is also economically important.

Asphalt: natural or artificial mixture of bitumen with minerals. A. is dark brown to black, solid, usually not very hard, with either a matte surface or a pitch-like shine. It sheds water and is resistant to alkalies and dilute mineral acids, but is soluble in organic solvents such as benzene, gasoline, chloroform and carbon disulfide. When heated, A. becomes soft or liquid. Asphaltenes, asphalt acids, etc. can be extracted from A., which have widely varying compositions. Natural A. cover the entire range from mineral-free bitumen to bituminous rocks containing relatively little bitumen.

1) ***Natural A.*** is probably an oxidation and polymerization product of petroleum produced by the action of microorganisms. It is found in large deposits. *Asphaltite* is a hard natural A. with less than 30% oil content, found in Utah and West Virginia in the USA. *Asphalt rocks* (less than 10% bitumen) are found on Trinidad, in the USA, the Andes and Albania. *A. limestones* are found in West Germany, France, Switzerland and Italy. *A. sand* (3 to 24% bitumen) is found in the USA, Canada, Spain, the USSR, Rumania and Bulgaria. *Lake A.*, found on Trinidad, contains 39% bitumen, 30% mineral components and 31% emulsified lake water. *A. shale* is found in West Germany.

2) ***Artificial A.*** *Petroleum A.* are residues left after distillation of crude oil or its A.-rich distillates; it is black and rigid at normal temperatures. *Chemical A.* (*acid A., blast A.*) are products which accumulate in many processes and from various organic raw materials, especially mineral and resin oils, tars, resins, waxes and natural A. through the action of various chemicals.

Applications. The greatest use for A. is in making roads. ***Cold A.*** is an emulsion of bitumen and water, stabilized by emulsifiers. Cold A. is used to spray onto gravel on roadbeds; the emulsion is broken and the gravel is stuck together. Cold A. is also used as a protective paint.

A solution of A. in benzene or turpentine oil is ***A. paint***, which is used to cover metal surfaces in galvanizing and etching processing, and as a rust protection.

Aspidosperma alkaloids: see Alkaloids, Table.

Aspirin: see Acetosalicylic acid.

Association colloids: see Colloids.

Astatine, symbol ***At***: a radioactive, very short-lived element from the 7th main group of the periodic system, the Halogens (see). At is a nonmetal, *Z* 85, with isotopes with mass numbers 196 to 219 (two nuclear isomers each for the mass numbers 198, 200, 202 and 212). The atomic mass of the most stable isotope is 210 ($t_{1/2}$), which decays by K-capture (99.83%) or α-emission (0.17%). The element has valences -1, +1, +5 and +7, m.p. ≈ 300 °C, b.p. ≈ 370 °C, standard electrode potential (At^-/At_2) is +0.25 V.

The isotopes ^{215}At, ^{218}At and ^{219}At have very short half-lives: ^{215}At, $t_{1/2}$ $1.64 \cdot 10^{-4}$ s, α-emission; ^{218}At, 1.3 s, 99.9% α-emission and 0.1% β-emission; and ^{219}At, 0.9 min, 97% α and 3% β emission. These are members of the natural decay series initiated by β-decay of polonium or α-decay of francium, and are therefore found in extremely small amounts in the earth's crust. (The At content of the lithosphere is estimated to be about 70 mg.)

The relatively stable isotopes with mass numbers 209 to 211 have half-lives between 5.5 and 8.1 h. They are formed by irradiation of bismuth 209 with 30-50 Mev α-particles, e.g. according to $^{209}Bi + {}^4He \rightarrow {}^{211}At + 2\,{}^1n$. They are also found as spallation products from irradiation of thorium or uranium with very energetic protons. These isotopes are separated from their mixtures by wet chemistry or gas thermochromatography. A. is less volatile than iodine. It is readily soluble in organic solvents and forms interhalogen compounds such as AtCl, AtBr and AtI with other halogens. At is reduced by strong reducing agents to At^-; ***astatine hydride***, HAt, is a stronger reducing agent than hydrogen iodide, HI. Mild oxidizing agents, such as Fe^{3+} or Br_2, convert At to the level of ***hypoastatous acid***, HAtO. Strong oxidizing agents like peroxodisulfate oxidize A. to ***astatate(V)***, $[AtO_3]^-$, while xenon difluoride in hot, alkaline solution oxidizes At^- to ***perastatate, astatate(VII)***, $[AtO_4]^-$. Many ***organoastatine compounds***, e.g. C_2H_5At, C_6H_5At, C_6H_4AtX (X = Cl, Br, I) and $AtCH_2COOH$, have been synthesized.

Historical. The possibility that iodine had a heavier homolog (eka-iodine) was predicted by Mendeleyev. Corson, McKenzie and Segrè discovered A. in bismuth which had been irradiated with 20-MeV α-particles. It was only later shown that extremely small amounts of At (≈ $3 \cdot 10^{-24}$% of the earth's crust) exist in nature, in uranium and thorium minerals. The name A. comes from the Greek "astatos" = "unstable".

Astatine emanation: see Radon.

Aston's isotope rule: see Isotopes.

Astraphloxin FF: see Polymethane pigments.

Astrazone red, GB: see Polymethane pigments.

asymm-: see Nomenclature, sect. III D.

Asymmetric: see Nomenclature, sect. III D.

Asymmetric carbon atom: see Stereoisomerism.

Asymmetric synthesis: synthesis of unequal amounts of the enantiomers of a chiral molecule (see Stereoisomerism, 1.1), starting from an achiral compound. Since not all chiral compounds are asymmetric, it is better to call this process ***stereoselective synthesis***. 1) ***Absolutely stereoselective syntheses*** are those in which the optically active compound is made from the non-active starting material without the help of an optically active reagent, simply through the effects of chiral physical processes, such as circularly polarized light or the enantiomorphic crystal structure of one of the reactants. The optical yield is only a few percent. 2) ***Diastereoselective syntheses (internally asymmetric syntheses)***. A new center of chirality is formed in the reaction of a molecule with a *diastereotypic group* or *side* (see Topic groups); the original chirality center influences the formation of the new one. This type of diastereoselective synthesis is seen, for example, in the reaction of carbonyl compounds which have chirality centers adjacent to the carbonyl with Grignard reagents. These obey *Cram's rule*: if the carbonyl compound tends to be in the conformation in which the C=O group is flanked by the two smaller groups of the adjacent chirality center, the reagent attacks preferentially from the side of the smallest ligand, and selectively forms the corresponding diastereomer (Fig.).

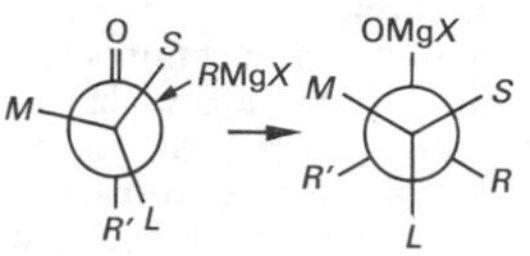

Reaction according to Cram's rule. *S*, small; *M*, medium; *L*, large.

3) ***Enantioselective syntheses (external asymmetric syntheses***. Optically active compounds are made from compounds containing enantiotopic groups or sides (true achiral substances) by means of a chiral reagent or catalyst.

Asymmetry effect: see Debye-Hückel theory.

At: symbol for astatine.

Atactic: see Polymers.

Ataractica: see Psychopharmaceuticals.

At. U.: abb. for Atomic units (obsolete).

Atebrin®: see Mepacrine.

Atmospherilics: air pollutants which stimulate Corrosion (see) through the atmosphere. Sulfur dioxide, chloride, particulates and atmospheric moisture all play a part in corrosion.

Atmungsferment: see Cytochrome oxidase.

Atom: the smallest electrically neutral particle of a chemical element, which cannot be further divided by chemical means. The properties of the A. determine the chemical and physical properties of the element. A. can be subdivided by physical means, and in this way they have been observed to consist of positively charged ***nuclei*** and negatively charged ***electrons*** which form a cloud around the nucleus (see Atom, models of). If electrons are removed or added, the particle is called a positive or negative ion, respectively.

The nucleus contains protons and, in all cases but one (^1H), neutrons. The ratio of neutron number to proton number increases from 1.0 in ^4He to higher values in heavier elements. The nucleons (protons and neutrons) make up nearly the total mass of the A., and the number of nucleons is equal to the *mass number M* of the A. (see Atomic mass). A. with the same atomic number (number of protons) can have different numbers of neutrons. These atoms have nearly identical chemical behavior, and are called Isotopes (see) of the element.

The relative size of an A. is the Atomic radius (see). The positions of A. in crystal lattices have been determined by x-ray, electron and neutron beam interference (see X-ray structure analysis). An atomic model is a conceptual image of the structure of an A.

Atomic absorption spectroscopy, abb. ***AAS***: methods of Atomic spectroscopy (see) for determination of single elements, especially of inorganic traces and minor components.

Principle. A. utilizes the absorption of single atoms for quantitative determination of their amounts. The free atoms are usually obtained from solutions by thermal evaporation and dissociation in an absorption chamber. This chamber is irradiated by light from a special source; the change in intensity of the chosen spectral line (the extinction) is proportional to the number of atoms present in the chamber, and thus to their concentration in the solution.

Fig. 1. is a diagram of an atomic absorption spectrometer. Free atoms are generated in the atomizer, a process which requires energy. This is provided by chemical processes (flames) or electric resistance heating (graphite or quartz atomizers). In the first case, the dissolved sample is sprayed into the flame as an aerosol. Combustion of ethyne/air mixtures yields temperatures up to 2500 K, while mixtures of ethyne/NO_2 produce flame temperatures of 3100 to 3200 K; these are available for samples which are less readily volatilized.

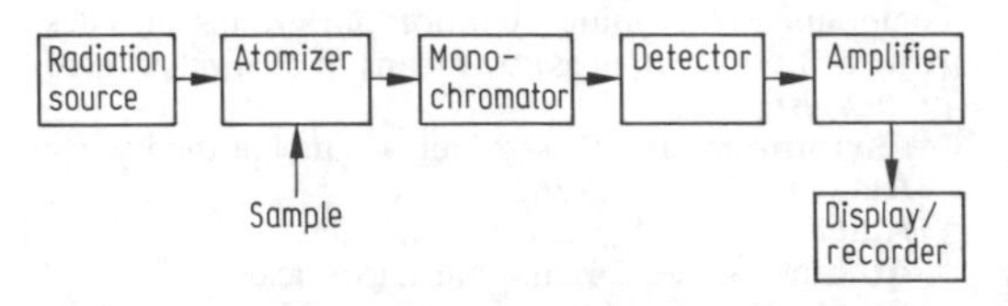

Fig. 1. Diagram of an atomic absorption spectrometer.

Flameless A. is also widely used, usually with electrothermal atomization of the sample in a graphite cuvette. A solution of the sample is usually injected into the cuvette by micropipette, and very rapidly heated to 2000 to 3000 K by means of a resistance heater. The sample is dried, incinerated and finally atomized.

In neither method is the temperature of the Plasma (see) high enough to excite the free atoms to electronic states above the ground state (the table shows the ratio of the number N of excited atoms to the number N_0 in the ground state for various elements and temperatures, calculated from the Maxwell-Boltzmann distribution (see). There are three advantages of A. which result from this fact. 1) Since the observed absorption processes start from the ground state, they are of high intensity, and this is the basis of the high analytical sensitivity of the method. 2) Since only those absorption lines are observed which start from the ground state, the spectra have relatively few lines, and there is thus little overlapping of spectral lines from different elements. As a result, the method is highly selective. 3) Since the fraction of atoms in the ground state at these temperatures is very high, it can be taken as approximately constant. As a result, the absorption process is largely independent of temperature, in contrast to the emission process.

Only some solids can be studied. Special methods, such as vaporization of microamounts of solids with laser beams, can be used to examine solids which cannot be readily vaporized from solution.

Ratio N/N_0 for the resonance lines of a few elements

Resonance line	Temperature in K			
	2000	3000	4000	5000
Na 589.0 nm	$9.86 \cdot 10^{-6}$	$5.88 \cdot 10^{-4}$	$4.44 \cdot 10^{-3}$	$1.51 \cdot 10^{-2}$
Ca 422.7 nm	$1.21 \cdot 10^{-7}$	$3.69 \cdot 10^{-5}$	$6.03 \cdot 10^{-4}$	$3.33 \cdot 10^{-3}$
Zn 213.9 nm	$7.29 \cdot 10^{-5}$	$5.58 \cdot 10^{-10}$	$1.48 \cdot 10^{-7}$	$4.32 \cdot 10^{-6}$

The *light sources* used in A. are chosen to have narrow linewidths and high spectral intensity. Hollow-cathode lamps fulfill these requirements. These lamps consist of a glass tube filled with a noble gas at 250 to 500 Pa; the cathode is made of the same element as is to be detected (Fig. 2). A separate lamp is required for each element. When a voltage is applied, there is a glow discharge in the lamp. Both the noble gas and the metal vapor emit their characteristic spectra; by reduction of the collision and Doppler spreading (see Linewidth), the emitted lines can be made very narrow, even narrower than the corresponding absorption lines (Fig. 3).

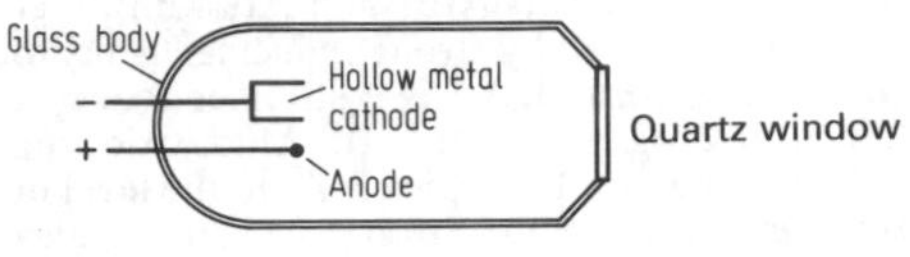

Fig. 2. Hollow cathode lamp

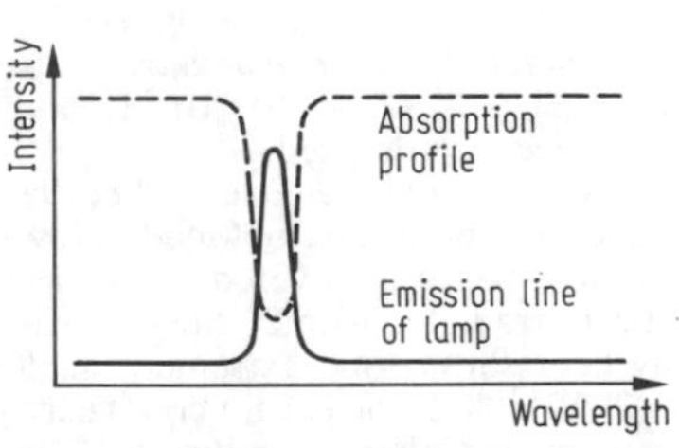

Fig. 3. Linewidths of emission and absorption lines.

This improvement was introduced by Walsh in 1955; it insures that all the incident radiation can be absorbed, and this leads to a great improvement in the sensitivity of the method.

Another type of light source is an electrodeless discharge lamp, in which excitation is achieved by inductive coupling with the high-frequency energy of a transmitter.

The absorption process is described by an equation analogous to the Lambert-Beer law (see): $E = \log I_0/I_D = k'N$, where E is the extinction, N is the number of absorbing particles, and k' is a constant.

Since the radiation from the radiation source contains several spectral lines, a *monochromator* is used to remove all but one line. As a rule, the resonance line corresponding to the transition from the ground state to the first excited state of the element is used for its determination. Since the most important absorption lines of most elements are in the UV, the optical elements must be made of quartz. The intensity of the incident and transmitted light is measured by a Secondary electron multiplier (see) used as the *detector*. It is placed at the exit slit of the monochromator, and converts the light signal into an electric signal which is amplified and displayed.

Atomic absorption spectrometers can be built either as single or as double-beam devices (see Spectral instruments). Single-beam instruments require stability in the emission from the light source, but intensity fluctuations do not affect double-beam measurements. To eliminate interfering emissions of flames and atomizers, the radiation from the hollow-cathode lamps is usually modulated.

The results are evaluated either by use of a calibration curve, in which the measured extinctions E of various standards are plotted against their concentrations c, or by an addition method. If properly prepared, the calibration curve gives more precise and more accurate results. The addition method is more rapid, but often produces systematic errors. The accuracy of the results can be affected by Matrix effects (see). In the case of A., these are most often factors which affect the vaporization and dissociation of the sample. Efforts should be made to eliminate them, for example by addition of substances which bind the interfering matrix, or changing the temperature of the flame.

Applications of A. In the past 25 years, A. has become the most important method of analysis for determination of traces of inorganic substances in solutions. Using A., about 70 of the elements can be determined, mostly metals and semimetals. Determination of nonmetals is often difficult, because the resonance lines of these elements are usually below 200 nm, and because some of them form stable molecules which do not dissociate readily. A. can be used to analyse both micro- and macrosamples. Electrothermal atomization in graphite cuvettes is especially useful for microsamples. This is because there are no losses of solution through ashing of the sample in the flame, and all of the sample is vaporized essentially simultaneously. Use of solutions requires that the sample be dissolved.

The applications of A. are extremely varied. It is used in any situation where inorganic components, especially metals, are analysed. This is the case in numerous areas of science, medicine and technology, including biochemistry, toxicology, environmental protection, food analysis, agriculture and metallurgy.

Atomic bond: see Chemical bond.

Atomic emission spectroscopy, abb. ***AES***: methods of atomic spectroscopy used for qualitative and quantitative determination of elements in solid and liquid samples. The sample is converted to the plasma state by an excitation source (e.g. flame, arc, spark or high frequency electromagnetic radiation). The light emitted by the plasma is focused on the entrance slit of a monochromator, spectrally resolved and allowed to impinge on a photographic plate (spectrograph) or photoelectric detector (spectrometer). The positions of the lines permit qualitative analysis of the sample. The relationship between line intensity and concentration permits quantitative analysis.

Spectral apparatus for AES consist of three main components: an excitation source, a monochromator and a detector.

Excitation sources have two main functions: conversion of the sample to the gas state, so that free atoms are present, and excitation of the sample so that it radiates light. The intensity of the spectral lines should be high and that of the background (see Plasma) should be low. Constant intensity of the lines is also desirable. The most important excitation sources used in A. are flames, electric arcs, high-voltage sparks and Inductively coupled plasma (see).

Flames are relatively low-energy sources, so they are used mainly for those elements which are readily excited. This technique is often considered a separate method, Flame spectrophotometry (see).

When excitation of an element to the point of light emission requires higher energy, *electric arcs* are used. The arc is created between two electrodes, one or both containing the sample. Either constant or alternating current, at 3 to 20 A, can be used to generate the arc plasma. Temperatures between 4000 and 7000 K are produced in the plasma. Because of the large amount of energy transfered, the electrodes become very hot. This permits vaporization and thus analysis of relatively non-volatile samples. However, for volatile substances (e.g. mercury or zinc), this is a disadvantage. For such materials, periods of cooling must be allowed. This is achieved by the use of intermittent arcs, which are periodically interrupted and spontaneously re-established. Because it consumes a large amount of material, the electric arc permits resolution of low concentrations, and is very suitable for trace analysis. For analysis of main components of samples, however, it is less suitable, because the measurements are not highly reproducible. The spectra generally contain only atom lines (fewer lines from ions). The arc spectra therefore contain relatively few lines and are good for qualitative analysis because there is little overlap.

A *high-voltage spark* is obtained by use of a transformer to produce about 10 kV direct current. This is used to charge variable-capacity capacitors. The high-voltage spark is produced by a brief discharge of the capacitor across a spark gap. The time course of this discharge is influenced by the capacity, self-induction and resistance of the oscillating circuit and the condition of the spark gap. In analysis, a series of individual sparks with identical time courses are discharged about every 10^{-2} s; each spark lasts about 10^{-4} to 10^{-5} s. The current density and thus the temperature in the spark plasma are very high ($\approx$ 100 A, $\approx 5 \cdot 10^4$ K). However, the average current density is much lower. The electrodes remain at a low temperature during the spark discharge. Because of the high temperature in the spark plasma, many ionic lines are seen in addition to the atomic lines, which makes the spectra less useful for qualitative analysis. Only a small amount of material is consumed in the high-voltage sparks, so that this method is not suitable for trace analysis. In addition, the low temperature of the electrodes makes excitation of non-volatile substances difficult. However, the reproducibility of a spark excitation is significantly better than that of an arc excitation, so the spark excitation is used mainly for quantitative analysis; both major and minor components can be studied.

Recently, *lasers* have also been applied for vaporization and excitation of samples; since the laser can be focused on a very small area of the surface, local analyses are possible.

The light coming from an excitation source is focused by a system of lenses on the entrance slit of a monochromator, where it is spectrally resolved by means of a prism or diffraction grating. The resolution required of the monochromator depends on the type of analysis being done. Much lower resolution is required for the relatively line-poor spectra of the alkali metals than for the spark spectra of samples containing numerous heavy metals, for example. For the latter purpose, high-resolution plane gratings are used.

The choice of optical material depends on the spectral region being measured. For the visible range, glass optics are suitable, but for the near UV (400 to 200 nm), quartz optics are needed. For studies in the far UV (below 200 nm), which are done mainly to determine elements such as sulfur, phosphorus or carbon, a vacuum grating apparatus is used.

The *detector* is either a photographic plate or a suitable photoelectric detector. The photographic plate has a number of advantages: it permits simultaneous measurement of wavelengths and intensities over the entire spectral range. For example, a single photograph is sufficient for analysis of an unknown sample containing many components. In addition, it is a permanent document of the analysis. The main disadvantage is that the analysis of the plates is cumbersome. Photoelectric detectors are increasingly being used for precise and rapid measurements. With secondary electron multipliers or photodiodes, for example, the process can be automated. Photoelectric detectors have the great advantage over photographic plates that there is a linear relation between the concentration and the voltage generated in the detector over a much broader range ($1:10^5$). In the simplest case, the photoelectric detector is applied to the exit slit of the monochromator. As the dispersive element is rotated (see Spectral apparatus), the various wavelengths pass through the slit and are registered by the detector (sequential analysis). Simultaneous determination of several elements requires a multichannel system which contains several photoelectric detectors, each placed to receive light of a certain wavelength. Devices with more than 30 secondary electron multipliers have been described for use in simultaneous multielemental analysis.

Sample preparation. A. is used mainly with solid and liquid samples. If a solid sample is an electrical conductor, it can be used directly as an electrode. The counter electrode can be made of the same material or of Fe, Cu or C. If the sample is nonconducting, graphite electrodes are used. A hole is bored in one electrode and filled with finely powdered sample. Nonconducting or inhomogeneous samples can be dissolved, which may be followed by concentration of trace elements or separation of interfering impurities. Solutions have the advantage over solids that the matrix effect is reduced, and the preparation of calibration samples is easier. There are various methods of applying the solution to the electrodes.

Applications of A.: *Qualitative analysis*. Since each type of atom emits a characteristic spectrum, the emission spectrum of a mixture (e.g. of an alloy) contains the characteristic spectra of all the individual components superimposed on each other. A. is therefore very well suited for multielement analysis. More than 70 elements can be determined with samples of only a few mg. An element can be detected simply by locating its characteristic lines in the complex spectrum; the lines used are the most intense, so that they are the last to disappear as the concentration decreases. The lack of a line is a rather reliable negative indicator. However, since the spectra can contain a large number of lines, the lines of different elements may coincide. Therefore a positive identification of an element requires the presence of several of its detection lines. The limits of detection of the method are summarized in the table.

Limits of detection of some elements with the direct current arc

Element	Analysis wavelength in nm	Limits of detection	
		percent	micrograms
Ag	328.1	0.0001	0.01
Ca	393.4	0.0001	0.01
Cd	228.8	0.001	0.1
Cu	324.7	0.00008	0.008
K	344.7	0.3	30
Mg	285.2	0.00004	0.0004
Na	589.6	0.0001	0.01
Zn	334.5	0.003	0.3

Quantitative analysis. The relation between light intensity and concentration is complicated, because the plasma contains atoms in both the ground and excited states and ions in an equilibrium which depends on the temperature. The fraction of ions is given by the Saha equation:

$$\frac{M_o}{N_o} = \frac{A}{N_e}(kT)^{5/2}\, e^{-E_i/kT}$$

Here M_0 is the concentration of the ions, N_0 the concentration of the atoms and N_e is the concentration of the electrons. E_i is the ionization energy, k the Boltzmann constant, T the absolute temperature and A is a constant. From this equation it can be seen that as the temperature increases, so do the numbers of ions, M_0, and electrons, N_e. At plasma temperatures up to 4000 K, the ion concentration can be ignored, however, and a simple Boltzmann distribution can be taken as the distribution of the atoms between the ground and excited states.

$$\frac{N_a}{N_g} = \frac{g_a}{g_g}\, e^{-E_a/kT}$$

Here N_a is the concentration of the atoms in the excited state, N_g is their concentration in the ground state, g_a and g_g are the statistical weights of the states, and E_a is the excitation energy. From equation (2), it can be seen that the concentration of atoms in excited states increases with increasing temperature. However, calculation shows that the fraction in excited states is small, even at high temperatures. For example, the ratio N_a/N_g for the resonance line of cesium at 2000 K is $4.44 \cdot 10^{-4}$; at 3000 K, $7.24 \cdot 10^{-3}$; at 4000 K, $2.98 \cdot 10^{-2}$, and at 5000 K, $6.82 \cdot 10^{-2}$. Thus even at 5000 K, there is only one atom in the excited state for every 682 in the ground state. Since the ratio N_a/N_g increases as the temperature rises, the intensity of the emitted light ought also to increase with temperature.

However, this is only partially correct, because the number of ions also increases, at the expense of the neutral atoms, and this leads to a loss of intensity of the atomic line. The increase in intensity to be expected from equation (2) occurs only so long as the concentration of ions is small. At high temperatures, the decrease in the number of neutral atoms outweighs the increase in intensity. Thus the line intensity passes through a temperature-dependent maximum, and the temperature of the maximum is different for each line of an element. This fact is very useful for identification of elements. The intensity I of the light emitted by the excited atoms is

$$I = h \cdot \nu \cdot N_a \cdot A \cdot V. \quad (3)$$

Here $h \cdot \nu$ is the energy of the photon, A the Einstein transition probability of the transition $N_a \rightarrow N_g$, and V is the volume. Substitution in equation (2) gives

$$I = h \cdot \nu = g_a/g_g \cdot N_g \cdot e^{-E_a/kT} \cdot A \cdot V \quad (4)$$

From this it can be seen that the intensity depends on the concentration of atoms in the plasma. Provided that the concentration of atoms in the plasma is proportional to their concentration C in the sample,

$$I = K \cdot C, \quad (5)$$

where K is a function of the temperature.

This equation is fulfilled only by very dilute plasmas. For real plasmas, there is no strict proportionality between the intensity and the concentration, because other factors affect the intensity. The most important of these are: 1) *self-absorption*, the reabsorption of primary emitted photons by atoms in the ground state; 2) *nonradiative transitions*, in which the excited atoms give up their energy of excitation by non-radiative processes, such as collisions; 3) *dissociation processes* which affect the formation of free atoms when the dissociation energies are high; and 4) *vaporization processes* of salts, which affect the number of atoms in the plasma. These processes are often affected by other elements in the sample (see Matrix effects).

Because of the factors discussed here, theoretical treatment of the relation between intensity and concentration is very difficult, and in general, quantitative analysis has an empirical basis. The empirical calibration function most commonly used today is

$$I = aC^b \quad (6)$$

where a reflects the effects of matrix and temperature, and b the effect of self absorption.

In most cases, however, the absolute intensities are not measured; instead, the signal is measured as the ratio of two intensities. The intensity of an analytical line is related to a reference line which can come from the sample itself or an additive, or to the spectral background.

$$I/I_R = a' \cdot c^{b'} \quad (7)$$

A typical *application* of quantitative A. is determination of many elements simultaneously in metal samples; for continuous quality control in production, the apparatus is automated. Compared to other methods of analysis, A. has an average lower limit of detection, average to good selectivity, a very wide range of application and the possibility of simultaneous multielement analysis. The disadvantages are the considerable expense, the time required when photographic plates are used, and the matrix-sensitive effects of temperature.

Atomic fluorescence spectrometry, ***AFS***: A technique using the fluorescence of free atoms for quantitative determination. The principle is shown in the block diagram of an atomic fluorescence spectrometer (Fig. 1).

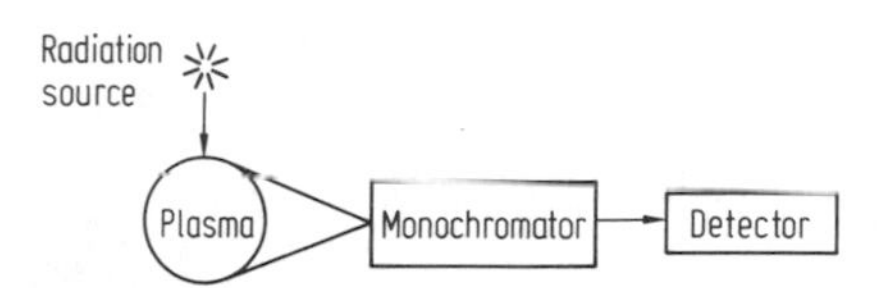

Fig. 1. Block diagram of an atomic fluorescence spectrometer.

An intense monochromatic beam of light from a dye laser or electrodeless discharge lamp is focused on a Plasma (see) which is generated by a low-background flame or electric heating. The sample is present in the plasma in the form of free atoms, most of which are in the ground state. The frequency ν of the excitation radiation is chosen so that its energy exactly corresponds to the excitation energy ΔE of the atom ($\Delta E = h\nu$; h is Planck's constant). Upon absorbing this light, the atoms enter an excited state, where they remain for a very short time ($\approx 10^{-8}$ s) before re-emitting all or part of the excitation energy in the form of fluorescence light as they return to the ground state or a lower excited state. The emitted fluorescence light is measured by a detector [e.g. a Photomultiplier (see)] at right angles to the exciting light, in order to avoid recording the latter. Since the fluorescence light can contain light of various wavelengths as well as scattered light, the fluorescence line used for analysis must be isolated by a monochromator. To eliminate interference by the emission of the flame, the excitation radiation is modulated so that only the fluorescence light resulting from it is registered by the detector.

There are 4 different types of fluorescence: 1) *Resonance fluorescence*, which occurs when the fluorescence light has the same wavelength as the absorbed radiation, permits the most sensitive detection. 2) *Direct line fluorescence* occurs at longer wavelengths than the absorbed radiation, because the fluorescing atom enters a lower excited state instead of returning to the ground state (Fig. 2). 3) *Stepped fluorescence* occurs when an atom in a higher excited state drops to the first excited state and then decays to the ground state by emission of fluorescence (Fig. 3). 4) *Sensitized fluorescence* occurs when the atom enters the excited state as a result of colliding with an excited atom of another element. For example, a plasma containing both Hg and Tl atoms, when irradiated with the Hg line at 253.7 nm, emits the two Tl lines at 377.6 and 535.0 nm.

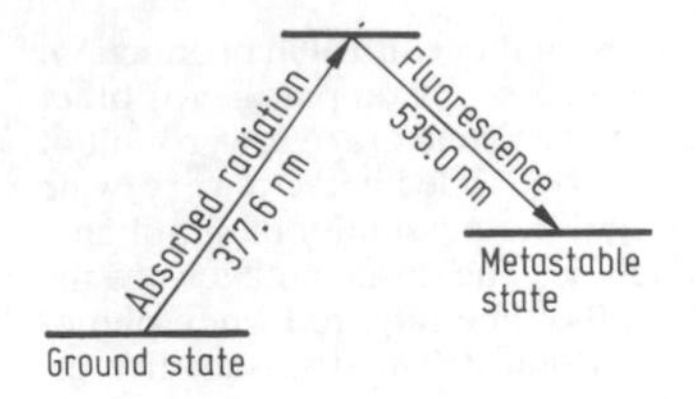

Fig. 2. Direct line fluorescence of thallium.

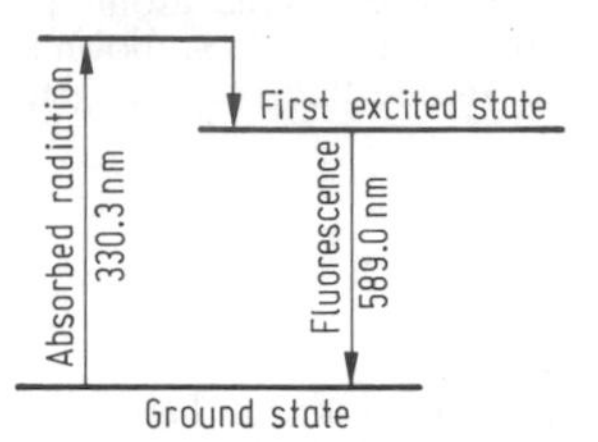

Fig. 3. Stepped fluorescence of the sodium ion.

The intensity of the fluorescence depends on the intensity of the exciting radiation, the fraction of the radiation absorbed by the atoms, the quantum yield of the fluorescence and the fraction of the fluorescent radiation which is reabsorbed by other atoms (self-quenching). For practical purposes, it is essential that the fluorescent radiation be proportional to the intensity of the exciting radiation and to the number of atoms in the absorbing plasma. If the quantum yield of the fluorescence is high, it can give an extremely sensitive method for detection of certain elements, which makes A. especially suitable for trace analyses.

The fluorescence measurements are evaluated using calibration curves, which are often linear with respect to the fluorescence intensity and concentration over a wide range. However, there are many factors which can cause errors and which have prevented wide use of A.

A. is used most often for detection of traces of such metals as Zn, Cd, Hg, Tl, Cu, Ag, Au, Pb and Mg, because at present the majority of instruments still use electrodeless discharge lamps for excitation. As dye lasers become more widely used, this selection will change.

Atomic heat: see Molar heat.

Atomic lattice: see Lattice type.

Atomic mass: 1) ***absolute A.***, the mass of an atom. It is too small (on the order of 10^{-24} to 10^{-22} g) to be measured directly, but it can be calculated for any element X as the quotient of its molar mass M_X divided by Avogadro's number N_A: $A_X = M_X/N_A$. Using this equation, the absolute A. of hydrogen, the mass of an H atom, is calculated as

$$A = \frac{1.008 \text{g mol}^{-1}}{6.023 \times 10^{23} \text{ mol}^{-1}} = 1.674 \times 10^{-24} \text{g}$$

The absolute A. is an inconvenient parameter for calculations, and is rarely used in chemistry.

2) ***Relative A.***, A_r, is the ratio of the mass of an atom to 1/12 of the mass of the carbon isotope ^{12}C. The term *atomic weight* for A_r is based on a confusion of the terms mass and weight, and should not be used. The relative A. of an element X is the ratio of its absolute A. to a uniform reference mass, the *atomic mass unit u*:

$$\underline{A}_{r,x} = \frac{A_x}{\underline{u}} \text{ with } u = \frac{1}{12}\ \underline{A}_{12C} = \frac{1}{12} \times 1.992 \times 10^{-23} \text{g}$$

$= 1.660 \cdot 10^{-24}$ g. The relative A. of the elements are in the same ratios as their absolute A. Most chemical elements are mixtures of isotopes, in which the abundances of the individual isotopes are almost always in a constant ratio. The A. of such mixtures is therefore a weighted average: $A_{r,X} = \sum_i x_i A_r, X_i$, where x_i is the relative molar abundance (see Composition parameters) and A_r, X_i is the relative mass of each isotope of the element X. Relative A. were originally determined by measurements of gas densities and by very precise quantitative analyses of suitable compounds. In this way, the average relative masses of isotopically mixed elements were obtained. Higher precision is now possible by use of mass spectroscopy, which gives the relative A. of the individual isotopes.

Other relative masses such as the relative Molecular mass (see), Formula mass (see) and Equivalent mass (see) are defined analogously.

Historical. The term A. was introduced by J. Dalton around 1805 in connection with his atomic hypothesis and used in establishing the basic laws of stoichiometry. He assigned the value of 1 to the mass of the lightest element, hydrogen. In the further development of chemistry, oxygen compounds were most often used in the determination of relative A. It therefore seemed reasonable to express the value directly in terms of this element. In 1815, J.J. Berzelius used $A_{r,O} = 100$ for his table of atomic masses; however, this was not widely accepted. In order to retain the approximate value of 1 for the relative A. of hydrogen, the relative A. of oxygen was set equal to 16 on the suggestion of the Belgian chemist J.S. Stas. After the establishment of $A_{r,O} = 16.0000$ in 1905, the ratio of $A_{r,H}:A_{r,O} = 1.0080:16.0000$. The choice of oxygen as the basis for the atomic mass scale proved problematic, however, when it was discovered that this element consists of small amounts of the heavier isotopes ^{17}O and ^{18}O in addition to the main isotope ^{16}O, and that the isotopic ratios can vary somewhat depending on the source of the oxygen. Therefore, in addition to the *chemical atomic mass scale* with the value $A_{r,O} = 16.0000$ for the natural isotope mixture, a *physical atomic mass scale* was introduced in which the mass of the isotope ^{16}O was the reference parameter. To convert the values for relative A. in the chemical scale to the physical scale, they had to be multiplied by the factor 1.000275. To remove this inconsistency, which caused considerable confusion, in 1961 a new international system was introduced: the carbon scale. In it, the isotope ^{12}C has an A_r of exactly 12. The values from the chemical oxygen scale had to be multiplied by 0.999957, and those from the physical scale by 0.999682, to convert them to the carbon scale.

The precise determination of relative A. by chemical methods owed much to the German chemist O. Honigschmid (1878-1945). The relative A. are still subject to constant improvement in precision, and their latest values are published at intervals by a special commission of the IUPAC.

Atomic mass unit: see Atomic mass.

Atomic moment: see Dipole moment.

Atomic number, abb. ***Z***: originally, the number indicating the position of an element in the periodic system. The A. is equal to the number of protons in the nucleus, and, therefore, to the number of electrons in the neutral atom of an element. For this reason, it is also called the ***nuclear charge number***. The A. of the elements can be determined experimentally from the K_α x-ray lines (Moseley's law).

Atomic orbital: see Atom, models of.

Atomic polarization: see Polarization.

Atomic radius: a relative measure of the size of an atom. Since an atom is not a sharply limited system (see Atom, models of: quantum mechanical model), it is also not possible to give its size precisely. The following A. are defined: the ***covalent A.*** is equal to half the distance between identical atoms in an atomic bond. Covalent radii for the atoms depend on the bond order of the bond. The bond lengths of single or multiple bonds can thus be given approximately as the sums of the corresponding covalent A. The ***theoretical A.*** is the maximum of the radial distribution (see Atom, models of) of the outermost occupied orbital. The ***metallic A.*** is defined as half the distance between adjacent atoms in the crystal. This is usually referred to lattice types with coordination numbers of 12 (metal crystals). To estimate the spatial extent of a molecule, Stuart and Briegleb introduced the ***van der Waals A.***, which is also called the ***effective radius***. It approximates the action range of an atom in terms of Van der Waals bonding forces (see) of a molecule, and therefore permits estimates of intermolecular distances in molecular crystals. Van der Waals A. are larger than covalent A., and when used in models they show interpenetrating atomic spheres. These are the basis of the space-filling models which are very useful for visualizing molecular configurations and conformations in organic chemistry. The different definitions of A. lead to different absolute values for the same type of atom. However, the changes in the different types of A. throughout the periodic system generally show the same tendencies (Table). The A. are greatly changed by formation of ions. Cation radii are smaller, and anion radii are larger than those of the corresponding neutral atoms (see Ionic radius).

Atomic radii of some representative elements on pm

	Covalent atomic radius	Theoretical atomic radius	Metallic atomic radius	van der Waals atomic radius
H	30	53	.	120
Li	135	159	150 ... 155	.
Na	155	171	185 ... 190	.
K	195	216	235 ... 255	.
Mg	140	128	155 ... 160	.
Ca	.	169	190 ... 195	.
Sr	.	184	210 ... 215	.
Ba	.	206	220 ... 225	.
B	80 ... 90	78	.	.
Al	125 ... 130	131	140 ... 145	.
C	75 ... 85 (67, 60)*	60	.	125 ... 135
Si	117 (107, 100)*	107	135	170 ... 200
Sn	140	124	160	.
Pb	145	122	160 ... 175	.
N	70 ... 80 (60, 55)*	49	.	150
P	110 (100, 93)*	92	.	190
As	115 ... 120 (111)*	100	135 ... 150	200
Q	65 (55)*	41	.	130 ... 140
S	105 (94)*	81	.	155 ... 185
Se	115 (107)*	92	160	200
Te	130 ... 135 (127)*	121	170	220
F	65	36	.	135
Cl	100	73	.	180
Br	115	85	.	195
I	135	104	.	215

* Numbers in parentheses indicate covalent A. for double and triple bonds.

Atomization: the distribution of a liquid into fine droplets by spraying. It creates a large surface area, which accelerates interactions between the atomized liquid phase and the surrounding gas phase (acceleration of chemical reactions; for example, in the synthesis of ammonium sulfate, liquid sulfuric acid is atomized in an ammonia atmosphere; rapid heat exchange is achieved by spraying molten solids into areas where the temperature is below their melting point, and a fine-grained powder is obtained). In other cases, the desired result is the finest possible distribution of the liquid medium on a large surface (liquid fertilizers, hair spray and perfume atomizers). The A. of a liquid can also be used to bind and precipitate fine dust or ash. A. is also frequently used for drying (see Spray drying).

In a pressure atomizer, the liquid is pressed through an opening of 0.3 to 4 mm diameter by a pressure of 3 to 7 MPa. In a compressed air (or gas) atomizer, the liquid is carried through the valve by the stream of propellant gas. In centripetal or centrifugal atomizers, the liquid is sprayed off a disk which is rotating at a high speed - up to 250 m s^{-1}. Centripetal atomizers can be used to atomize even highly viscous solutions, suspensions and other liquids which would corrode or plug valves.

A. can also be achieved electrostatically. A high voltage is applied between the valve and a counterelectrode. This method is used for paint sprayers for painting metal objects; the object to be painted acts as the counter electrode.

Atomization drying: same as Spray drying (see).

Atomic spectroscopy: analysis of the electromagnetic spectra of free atoms. Since free atoms absorb and emit at the same wavelengths, atomic spectra can be observed as emission or absorption spectra, or as a combination of the two, as fluorescence spectra. An individual line is characterized by its frequency ν, and corresponds to the energy difference ΔE between two defined energy states E_1 and E_2 of the atom in question:

$$\Delta E = E_2 - E_1 = h\nu \quad (1)$$

In optical A., which is limited to the UV, visible and IR portions of the electromagnetic spectrum, these energy states correspond to different arrangements of the valence electrons. Electronic transitions of the inner electrons produce X-ray spectra (see X-ray spectroscopy). Only optical A. will be discussed here.

In order for a sample to exist as free atoms, it must be converted to a plasma before the spectrum is observed.

Theoretical. Since the spectra are specific for the type of atom which absorbs or emits them, they must be closely related to the structure and properties of the atoms. They are therefore interpreted in terms of a suitable model, either the Bohr-Sommerfeld or the quantum mechanical model. Comparison of spectra of different atomic spectra shows that the arrangement of the spectral lines is related to the position of the element in the periodic system. Elements with a single valence electron, e.g. the alkali metals, have relatively few lines in their spectra, whereas elements with several valence electrons, e.g. the transition elements, have numerous lines in their spectra. The simplest spectrum is that of the neutral hydrogen atom. It contains a succession of sharp lines which are closer together at shorter wavelengths (Fig. 1); such a group of lines is called a series. At the short-wavelength end, the lines converge on a position called the series limit.

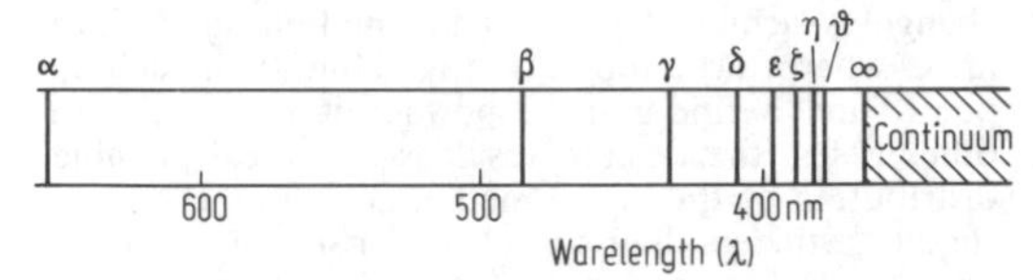

Fig. 1. Balmer series of the hydrogen spectrum.

On the other side of the series limit is a region of continuous absorption which arises from transitions of electrons from discrete energy states into energy states outside the atom, or conversely. The states outside the atom are not quantized, so the free electron can have any arbitrary amount of kinetic energy.

In the spectrum of atomic hydrogen, there are five series. The wavenumbers $\tilde{\nu}$ of all the lines of the hydrogen spectrum can be expressed by the equation

$$\tilde{\nu} = Z^2 R \left(\frac{1}{n^2} - \frac{1}{m^2} \right)$$

where m is the principal quantum number of the higher excited state, n is the principal quantum number of the lower excited state, R is the Rydberg constant (see) (109677.759 cm^{-1}) and Z is the nuclear charge number. The wavenumber of a spectral line is thus the difference between two expressions $T_n = R/n^2$ and $R_m = R/m^2$, which are called *terms*. They represent the energy states of the atom divided by hc (c is the velocity of light). A spectral series arises when a variable term R/m^2 is subtracted from a fixed term R/n^2; R/m varies as the quantum number m takes on a series of values. For hydrogen $Z = 1$; thus eq. 2 yields the following formulas:

Lyman series (in the UV)	$\tilde{\nu} = R\left(\frac{1}{1^2} - \frac{1}{m^2}\right)$	$m = 2, 3, 4 \ldots$	(3a)
Balmer series (in the visible)	$\tilde{\nu} = R\left(\frac{1}{2^2} - \frac{1}{m^2}\right)$	$m = 3, 4, 5 \ldots$	(3b)
Paschen series (in the IR)	$\tilde{\nu} = R\left(\frac{1}{3^2} - \frac{1}{m^2}\right)$	$m = 4, 5, 6 \ldots$	(3c)
Bracket series (in the IR)	$\tilde{\nu} = R\left(\frac{1}{4^2} - \frac{1}{m^2}\right)$	$m = 5, 6, 7 \ldots$	(3d)
Pfund series (in the IR)	$\tilde{\nu} = R\left(\frac{1}{5^2} - \frac{1}{m^2}\right)$	$m = 6, 7, 8 \ldots$	(3e)

It is conventional to represent these relationships in graphic form, in a term scheme, where the individual terms are represented as horizontal lines, so that the distance between them is proportional to the wavenumber of the corresponding spectral line (Fig. 2). It can be seen how the terms get closer together as the series limit is approached.

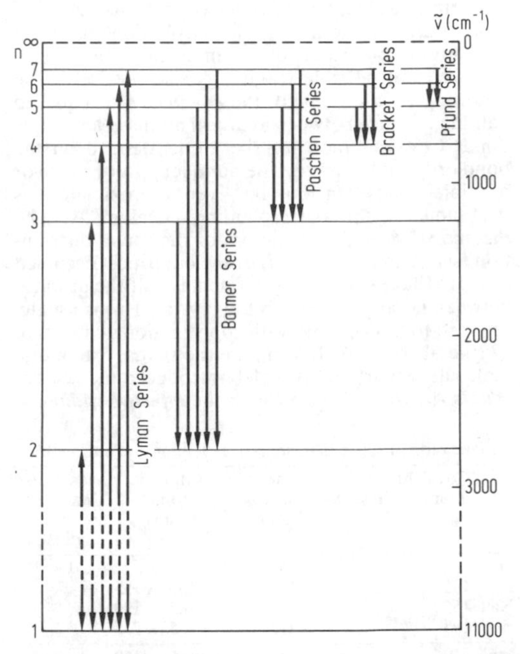

Fig. 2. Term scheme of the hydrogen atom showing spectral series.

The H atom should, in principle, be able both to emit and to absorb at all wavenumbers corresponding to transitions in Fig. 2. However, since H atoms are in the electronic ground state ($n = 1$) at normal temperatures, absorption is only possible for the Lyman series, in which the lower term is the ground state. The lines which end in the ground state are called resonance lines; they are the only ones which can be observed by both emission and absorption. All the other series are observed only as emissions, except when the atoms are already in an excited state.

The spectra of the helium ion He^+ and the lithium ion Li^{2+} are very similar to that of the hydrogen atom; each of these ions also has just one electron. In accordance with eq. 2, the spectral lines of the series of the He^+ ion ($Z = 2$) are at wavenumbers four times as large as those of the hydrogen spectrum, while the wavenumbers of the Li^{2+} lines ($Z = 3$) are nine times as large. These regularities are summarized in the

spectroscopic shift rule of Sommerfeld and Kossel, according to which the spectrum of any atom is similar to that of the singly charged positive ion of the following element in the periodic system, and of the doubly charged positive ion of the next element after that. Spectra of singly and multiply ionized atoms are called the first, second, etc. *spark spectra* because they are obtained from electric spark discharges. The spectrum of the neutral atom is called the *arc spectrum*, because it is usually observed in an electric arc. The spectra of alkali metals are relatively simple; the valence electron has a spectrum similar to the hydrogen spectrum. However, there are more energy states, and they cannot be described in terms of a single quantum number as the hydrogen states can. In addition to the principal quantum number, the following 3 quantum numbers are needed to describe these energy levels and the transitions between them:
- the secondary or orbital angular momentum quantum number l, which can have the values 0, 1, ..., n - 1; the corresponding electronic states are indicated by the letters *s, p, d, f*;
- the spin quantum number S, which can have the values +1/2 and - 1/2;
- the total angular momentum quantum number j, which is the vectorial sum of the orbital angular momentum and spin.

The various term series (Fig. 3) arise from different values of the orbital angular momentum (s, p, d, f). The observed series can be characterized as follows: principal series, s - np; sharp secondary series, p - ns; diffuse secondary series, p - nd; and fundamental (Bergmann) series, d - nf.

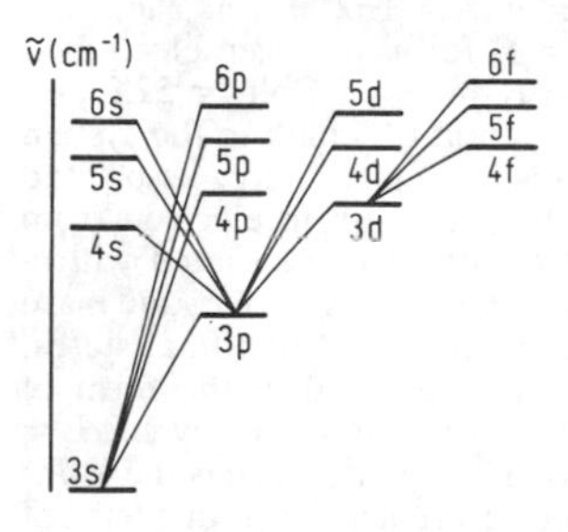

Fig. 3. Simplified term scheme for the sodium atom with the transitions corresponding to the various spectral series.

In the principal series the electron transitions occur between the p states of different principal quantum numbers n and the lowest s state (ground state); in the first secondary series they occur between the higher s states and the lowest p state, and so on. In contrast to the situation with the hydrogen atom, here the possible transitions between energy states do not all occur; there is a selection rule which requires that $\Delta l = \pm 1$. Closer inspection of the alkali metal spectra shows that all lines are actually double lines. This can be explained by the total angular momentum quantum number j. For $l = 1$, j can have the two values +1/2 and - 1/2 corresponding to the two possibilities for $s = \pm 1/2$. For $l = 0$, there is only a single possibility for s, so that $j = 1/2$. Therefore all terms except the s states are doublet terms, and the magnitude of the doublet splitting increases rapidly in the progression to alkali metals with higher atomic numbers. Only transitions in which $\Delta j = 0, \pm 1$ are allowed. This splitting of spectral lines into multiplets due to the quantized interaction between the orbital angular momentum and the spin is called fine structure.

In systems with several valence electrons, the complexity of the spectra increases further. The angular momenta l and s of the individual electrons add vectorially to give the total momenta L and S: $l_1, l_2, l_3, \ldots L$ and $s_1, s_2, s_3 \ldots S$. These then add vectorially according to the Russel-Saunders coupling to give the total angular momentum quantum number J.

There is a system of symbols denoting individual terms. For example, $3\ ^2P_{3/2}$ indicates a state in which the principal quantum number is 3 and $L = 1$ (P). The left-hand superscript 2 indicates the multiplicity of the term ($2s+1$), and the subscript 3/2, the value of J.

In the spectra of atoms with more than one valence electron, e.g. the alkaline earth metals, there are several term systems which in general do not combine with each other and which differ in their multiplicity. In the alkaline earth metals, which have 2 valence electrons, there are a singlet and a triplet system. Each system consists of a number of series similar to those of the alkaline earth metals. Another complication arises from the possibility of simultaneously exciting both valence electrons (double excitation). In general, even and odd multiplicities alternate through the periodic system (Table 1). Even numbers of electrons are associated with odd multiplicities, and vice versa (spectroscopic alternation). The facts mentioned above and others which have not been mentioned make the series character of most spectra, especially those of heavier elements, difficult to recognize, because the spectra are so complex.

Table 1. Spectroscopic alternation rule

Number of electrons	1	2	3	4	5	6	7
Multiplicity	2	1,3	2,4	1,3,5	2,4	1,3	2

At higher resolution, there is in addition to the fine structure a further, very slight splitting, called *hyperfine structure*. This is due to the interaction of electrons with the nucleus. It can occur because the element consists of several isotopes, or it can be due to an interaction with the nuclear spin (see NMR spectroscopy). The splitting of spectral lines in an external electric field is called the Stark effect (see), and the splitting in an external magnetic field is called the Zeeman effect (see).

Applications. A. can be used for qualitative and quantitative analysis. If an atom is elevated into an excited state, it can lose the energy of excitation by emission of a photon, which is the basis for Atomic emission spectroscopy (see) and Flame spectrophotometry (see). The absorption of photons is the basis of Atomic absorption spectroscopy (see). If the energy taken up by absorption of photons is re-emitted in the form of a photon, the phenomenon is atomic fluorescence. The atomic spectroscopic analysis methods based on the processes of light emission and absorption are shown in Table 2.

Table 2. Atomic spectroscopic analysis methods

Sample	Plasma Main analytical generation	Excitation	Measured process	Methods
Solid or solution	Arc, spark Qualitative inductively coupled plasma	Electric (heat)	Emission	Emission spectroscopy AES
Solution	Flame Quantitative	Chemical (heat)	Emission	Flame spectro-photometry FSP
Solution	Flame Quantitative	Chemical, special light	Absorption	Atomic absorp-tion spectroscopy

Atomic units: a system of measures based on atomic parameters. The unit of mass is the electron mass m_e, the unit of angular momentum is $\hbar = h/2\pi$ (h is Planck's constant), and the unit of charge is the charge of the electron e. Using the formula $e_\pi = e/\sqrt{4\pi\varepsilon_0}$ (ε_0 is the electric field constant), the energy is 1 Hartree = $e_\pi^4 m_e/h^2 = 4.3593 \cdot 10^{-18}$ J. This is twice the ionization energy of the hydrogen atom. The A. of length is the radius of the first Bohr radius in the hydrogen atom (see Atom, models of): 1 $a_0 = \hbar/(m_e e_\pi)$ = 1 Bohr = 52.917 pm. Other parameters can also be given in terms of m_e, $\hbar$ and e. The use of A. is very convenient in quantum theoretical calculations for atoms and molecules, because the value of 1 is used for e, $\hbar$ and m_e in the equations. The use of A. also has the advantage that the results in this mass system are independent of corrections in the numerical values of the natural constants.

Atomic weight: see Atomic mass.

Atomization energy: see Bond energy.

Atom, models of: The current model of the structure of the atom has been constructed and revised over a period of time on the basis of experimental results. The earliest model, proposed by Dalton, was a rigid, homogeneous *sphere*. This model was used successfully to derive the basic equations of the kinetic theory of gases. The discoveries of the electron, proton and neutron indicated that the spherical model was incomplete. Rutherford's scattering experiments, in which α-particles (see Radioactivity) were deflected by thin gold foils, indicated that the atom has *structure*; this was corroborated by Lenard's electron-scattering experiments. The very existence of sub-atomic particles, as well as natural fission, indicate that the atom is not indivisible.

The ***Rutherford model*** was based on the results of α-particle scattering; it posits a very small *nucleus* with a radius of approximately 10^{-14}m. The nucleus contains Z positive unit charges and nearly the entire mass of the atom; it is surrounded by a cloud of Z *electrons* (in the neutral atom). The number Z corresponds to the atomic number of the element. The electrostatic attraction of unlike charges should cause the electrons and protons to adhere to each other. It was suggested that they could be separated because the electrons have kinetic energy and move around the nucleus in circular or elliptical orbits; the centrifugal force is compensated by the electrostatic attraction of the nucleus.

According to classical physics, the motion of an electron around the nucleus would give rise to an oscillating dipole, which would continuously emit energy. As it did, the orbit would steadily be reduced in size until the electron spiraled into the nucleus. It was clear not only that atoms do not emit light continuously (i.e. electrons do not spiral into the nucleus), but also that the emission of light from energetically excited atoms does not yield a continuous spectrum; instead, the light is emitted in discrete bands at certain wavelengths.

To resolve this dilemma, Bohr suggested that atomic electrons can move only in certain allowed orbits, or stationary states, and that they do not radiate energy so long as they remain in these states. In the ***Bohr model***, the allowed orbits are those in which the orbital angular momentum is an integral multiple of the natural constant $\hbar = h/2\pi$ (h = Planck's constant). In this way, Bohr arrived at the quantization of the orbital radii r_n and the energies E_n of electrons with quantum numbers n (n = 1, 2, 3...). The smallest orbital radius (n = 1) for a hydrogen electron is the *Bohr atomic radius* $r_1 = a_0$ = 1 Bohr = 52.9 pm. This number is used as a unit of length in the system of atomic units. There is a definite energy associated with each orbit; and transition of an electron from one stationary state to another is associated with a change in its energy $\Delta E = E_n - E_m$, where m and n are the quantum numbers of the initial and final states. The energy is absorbed or emitted in the form of electromagnetic radiation with the frequency ν: $\Delta E = h\nu$. The electron energies calculated using the Bohr model can thus be used to predict the frequencies of light emitted or absorbed by atoms; for the hydrogen atom, the predicted frequencies agree very closely with those measured spectroscopically.

The Bohr model was refined by Sommerfeld, who included calculations for elliptical orbits; these require the introduction of another quantum number, and permit a theoretical explanation of the fine structure of the spectrum of the hydrogen atom. The great achievement of Bohr was the recognition that the electrons in an atom cannot have any arbitrary energy, but only discrete energy levels. Even though the Bohr postulates stand in contradiction to classical physics, and represent an arbitrary application of quantum conditions to the classical description of the electron, their heuristic value for the further development of atomic physics was immense. The difficulties of the Bohr model were first resolved by the development of a newer, more general quantum mechanical theory. This was the accomplishment of Schrödinger, Heisenberg, Born and de Broglie in the 1920s.

The ***quantum mechanical model*** is fundamentally a new conception of the atom. It is based on the dual nature of matter: photons, electrons and other submicroscopic particles have both particle and wave characteristics. This is expressed by the *de Broglie relation* $\lambda = h/p$, where the wavelength λ and momentum p of a particle are related by the Planck constant h. From this relationship follows a basic principle of the new theory: the *Heisenberg Uncertainty Principle*, which states that there is an unavoidable degree of uncertainty in the simultaneous determination of the location and momentum of a particle.

Mathematically, the uncertainty is expressed by the equation: $\Delta x \Delta p \geqq h$; the uncertainty in the location Δx and the momentum Δp must be at least as great as Planck's constant. For sub-microscopic objects like electrons and nucleons, therefore, an orbital curve cannot be determined. Thus the Bohr model violates the uncertainty principle by assigning definite orbits and velocities to the electrons. In the quantum mechanical model, orbits are replaced by *wave functions* which describe the behavior of the particle. For time-independent processes, the probability dW of finding a particle in the volume element $d\tau = dxdydz$ is given by the square of the wavefunction $|\Psi|^2 = \Psi^*(x,y,z) \cdot \Psi(x,y,z)d\tau$. $|\Psi|^2$ is the product of the wavefunction Ψ and its complex conjugate Ψ^*, and is called the *probability density*.

In addition to wavefunctions, *operators* $\hat{A}$, which are a form of mathematical prescription, are fundamental to quantum mechanics. It has been found that each physical parameter can be assigned an operator which, when applied to the wavefunction, yields the observable values in the form of *eigen values*. This is expressed by the equation $\hat{A}\Psi_n = A_n\Psi_n$. The functions Ψ_n are called the *eigen functions*, and the numbers A_n, the *eigen values* of the operator $\hat{A}$. The *eigen equation* for energy, called the *Schrödinger equation* after its inventor, is central to the quantum mechanical model of the atom. For a particle of mass m, it has the following form in cartesian coordinates:

$$\left[-\frac{h}{8\pi^2 m}\left(\frac{\delta^2}{\delta x^2}+\frac{\delta^2}{\delta y^2}+\frac{\delta^2}{\delta z^2}\right)+V_{(x,y,z)}\right]\Psi_{(x,y,z)}$$

$$= E\Psi_{(x,y,z)}.$$

The expression in the square brackets is called the *Hamiltonian operator* $\hat{H}$. It represents the classical potential energy $V(x,y,z)$ of the particle and the second-order partial derivatives of the variables. The Schrödinger equation for the hydrogen atom can be solved exactly, if polar coordinates (r, ϑ, φ) are used instead of cartesian coordinates; in this way a very simple expression for the potential energy of the electron is obtained, and the variables can be separated. With the product expression $\Psi(r,\vartheta,\varphi) = R(r)\Theta(\vartheta)\phi(\varphi)$, the solution of the second-order partial differential equation with three variables is reduced to the solution of three corresponding differential equations with one variable each. It follows from the mathematical character of the Schrödinger equation and the physical boundary conditions that solutions exist only for certain wave functions Ψ_n (eigen functions) with the corresponding eigen values E_n for energy. The Ψ-functions, which describe stationary states of the electron, are called "*orbitals*". If they apply to electrons in an atom, they are called "*atomic orbitals*" (abb. *AO*). The wave functions, i.e. the atomic orbitals $\Psi_{n,l,ml}(r,\vartheta,\varphi)$ depend on the three variables r, ϑ, and φ, and in the hydrogen atom, or in the independent-particle approximation of the multi-electron atom (see Central field model, Electron configuration), they are determined by the three quantum numbers n, l, m. The quantum numbers can assume only certain values, which are related as follows: $n = 1, 2, 3...$; $l = 0, 1, 2..., n - 1$; $m_l = -l, -l+1,... -1, 0, +1,...l$. n is the *principal quantum number*, l is the *orbital momentum quantum number*, and m_l is the *magnetic quantum number*. l is the absolute value of the orbital angular momentum L_l and m_l is the component of L_l in the z direction.

Atomic orbitals with the same principal quantum number comprise a *shell*. For $n = 1,2,3,...$, the letters K, L, M ... are used to designate the corresponding shells. Atomic orbitals with the same principal and orbital momentum quantum numbers comprise a *subshell*. The atomic orbitals with various l values are indicated by letters which are historical; for l = 0,1,2,3, the letters s, p, d and f are used (from the "sharp", "principal", "diffuse" and "fundamental" series of spectral lines of sodium). The symbols for atomic orbitals consist of a numeral indicating n and a letter indicating l, e.g. a state with $n = 3$ and $l = 1$ is a 3p atomic orbital.

In the hydrogen atom, the energy E_n of an electron depends only on the principal quantum number. Since for each value of n there are a total of n^2 states with the same energy, each energy level in the hydrogen atom is n^2-fold degenerate. Thus, for example, the 2s and 2p orbitals have the same energy. Graphic representation of the atomic orbitals $\Psi_{n,l,m}(r,\vartheta,\varphi)$ of the hydrogen atom, and especially of $|\Psi|^2$, which shows the electron densities around the nucleus, is very important for qualitative understanding of chemical bonding. However, for the sake of clarity, one is limited to partial representations. It is useful to write $\Psi_{n,l,m}(r,\vartheta,\varphi)$ as a product of a *radial function* $R_{n,l}(r)$ and an *angular function* $Y_{l,ml}(\vartheta,\varphi)$. The function $Y_{l,ml}(\vartheta,\varphi)$, normalized to 1, is called the spherical function. Real angular functions $Y_{real}(\vartheta,\varphi)$ are obtained by suitable linear combinations of the complex angular functions. The real functions have the advantage over the complex forms that their dependence on direction arises directly from the analytical expression. The $Y_{real}(\vartheta,\varphi)$ represent surfaces in space, and are usually represented as sections intersecting the xz or xy planes. They have different forms and orientations, depending on l and m_l (Fig. 1). The signs of the angular functions have no physical meaning, but they are useful for study of the symmetry properties of orbitals with regard to the electronic structure of molecules. The squares of the angular functions $Y_{real}^2(\varphi,\vartheta)$ are very similar in shape to the $Y_{real}(\varphi,\vartheta)$ and represent the distribution of the relative electron density over the surface of a sphere with a constant radius r. In general, there are l nodal surfaces ($Y_{real}^2 = 0$) in which the probability of electron occupation is 0. The parameter $4\pi r^2 R_{n,l}^2(r)$, called the *radial distribution*, is the most easily visualized. It gives the probability of finding the electron in a spherical shell with the radius r and thickness dr. The maximum in the probability distribution of an electron in the 1s orbital of the hydrogen atom is found at the Bohr radius a_0 (Fig. 2). In contrast to the Bohr model, however, the quantum mechanical electron has a non-zero probability of being at another distance from the nucleus. As n increases, the maxima of the radial distribution functions are shifted towards larger values of r. There are also $n - l - 1$ radial nodal surfaces between $0 < r < \infty$. By themselves, the representations of the squares of the angular and radial partial functions give an incomplete picture of the distribution of the electron density in space. For a complete characterization of an electron in an atom, a fourth quantum number, the *spin quantum number* m_s (see Electron spin) is required. It represents the angular momentum of the electron's spin, and can have the values +1/2, - 1/2.

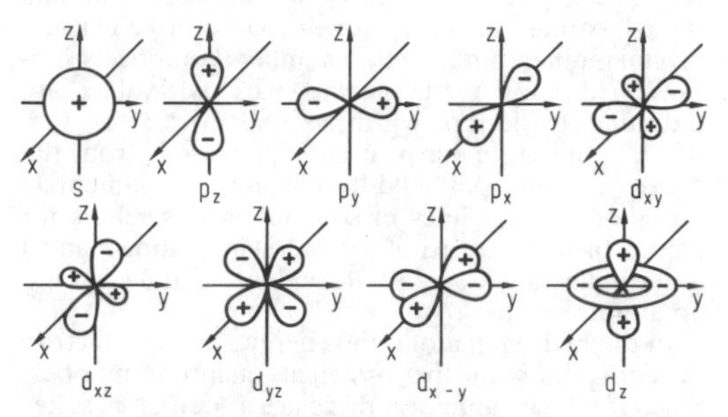

Fig. 1. Angular function *Y*real (ϑ, φ). s, p, d ... indicate atomic states with the quantum number l = 0, 1, 2, ...

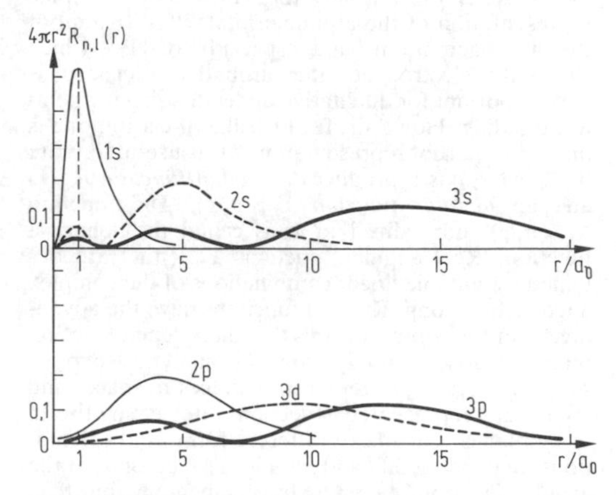

Fig. 2. Radial distribution of the electron in a hydrogen atom.

The insights into the electron structure of the hydrogen atom provided by the quantum mechanical model are used in the approximate treatment of multi-electron atoms (see Central field model). The advantage of the quantum mechanical model is its theoretical completeness, and the excellent agreement of its predictions with experimental results.

ATP: see Adenosine phosphates.

ATPases abb. for ***adenosine triphosphatases***: enzymes which catalyse hydrolysis of ATP to ADP and phosphoric acid. The energy released by the reaction (ΔH = - 30 kJ mol^{-1}) is used to drive most of the energy-requiring reactions in the cell.

Atramentization: see Phosphatization.

Atraton: see Triazine herbicides.

Atrazine: see Triazine herbicides.

Atropine: an ester alkaloid in which the OH group of Tropine (see) is esterified to the carboxyl group of R,S-tropaic acid. The tropaic acid component is a racemate of the naturally occurring S-hyoscyamine, in which the acid component has the S-configuration. ***S-Hyoscyamine*** forms crystalline needles; m.p. 109 °C, $[\alpha]_D^{20}$ - 22° (in 50% ethanol); ***atropine***, m.p. 118 °C, atropine sulfate, m.p. 194 °C. As ester alkaloids, A. and hyoscyamine are readily cleaved by hydrolysis. At higher temperatures, water is split off the tropaic acid component to form ***apoatropine***.

Atropine

R = $-CO-CH(CH_2OH)-C_6H_5$: Atropine

R = $-CO-C(=CH_2)-C_6H_5$: Apoatropine

R = $-CO-CHOH-C_6H_5$: Homatropine

A. is isolated from hyoscyamine-containing plants such as nightshade (*Atropa belladonna*), henbane (*Hyoscyamus niger*) and datura (*Datura stramonium*). The total alkaloid content of the leaves is 0.2 to 1% in nightshade, 0.04 to 0.08% in henbane, and 0.2 to 0.6% in datura. A. is a Parasympathicolytic (see). It is a neurotropic Spasmolytic (see), and the sulfate is used to relieve cramps in the abdominal and bronchial areas. It causes a long-lasting dilation of the pupils. In higher doses, it can cause undesired central effects. A. is the active ingredient of the belladonna extract. ***Homatropine*** is a partially synthetic derivative in which racemic mandelic acid is esterified to tropine. This has effects similar to those of A., but the pupil dilation does not last so long.

Historical. A. was independently isolated in 1833 by Main and by Geiger and Hesse. It was first synthesized by Wilstätter.

Atropine methobromide: a compound formed by quaternizing atropine with methyl bromide. A. is used as a spasmolytic. The central nervous effects of atropine are largely eliminated by the quaternization.

ATR technique: see Infrared spectroscopy.

Atropisomerism: a type of configurational isomerism which results when steric hindrance completely prevents rotation around a C-C single bond.

Enantiomers or diastereoisomers resulting from hindered rotation can be isolated if the rotation barrier at room temperature is above 85 kJ mol^{-1}. A. is observed in suitably substituted biphenyls (e.g. 6,6'-dinitrodiphenic acid), ansa compounds and cyclophanes.

Attractants: see Pheromones.

Attrinite: see Macerals.

Au: symbol for gold.

Auger electron spectroscopy, abb. ***AES***: a spectroscopic method for measuring the kinetic energy of secondary electrons (*Auger electrons*) emitted from a sample as a result of photoionization or electron collision ionization.

When atoms interact with X-rays or electron beams, electrons from inner shells are knocked out, leaving ionized atoms in excited states (see Photoelectron spectroscopy, X-ray spectroscopy). The resulting hole, e.g. in the K shell, is filled by an electron from a higher shell, such as the L_{II} shell. The energy released by this process,

$$E_K - E_{L_{II}},$$

can be emitted in the form of X-rays (X-ray fluorescence) or it may serve to release another electron (e.g. from the L_{III} shell), which takes with it the rest of the energy as kinetic energy. This produces an ion with a double positive charge. The kinetic energy of the Auger electron is approximately

$$E_{kin} \approx E_K - E_{L_{II}} - E_{L_{III}}.$$

Whether X-ray fluorescence or the Auger effect is more prominent in the deactivation of an excited state depends on the atomic number of the element. The Auger effect dominates in lighter elements.

Three electron states are involved in the process of Auger electron emission. The Auger electron is identified by adding the symbols of the electron states to that of the element. For example, in Mg-K $L_{II}L_{III}$, the first letter, K, indicates the level of the primary ionization of the magnesium atom, and the next two letters, L_{II} and L_{III}, the starting states of the electron which jumps into the inner shell and that of the Auger electron, respectively. The energy of the Auger electron corresponds to certain electron transitions, which are characteristic for the atom in question and are independent of the energy of excitation. It can therefore be used to identify a type of atom (qualitative analysis). Since the electron states are shifted slightly, depending on the bonding and the chemical environment of the atom, there is a chemical shift of the signal, just as in photoelectron spectroscopy. This is important for a detailed characterization of the position of the element in the sample molecules. Quantitative analyses may be made on the basis of the number of Auger electrons. Since Auger electrons can escape only from the uppermost layers of a solid sample, they can be used for analysis of very thin surface layers. This method is therefore especially well suited for surface studies, particularly for the study of such processes as catalysis, oxidation and corrosion.

August's formula: see Clausius-Clapeyron equation.

Auramine: $[(CH_3)_2N\text{-}C_6H_4]_2C{=}NH \cdot HCl \cdot H_2O$, a basic diphenylmethane dye. It is a bright yellow powder which is soluble in water and methanol; m.p. 136 °C. A. is used to dye paper, jute and coconut fibers; in medicine it is used to stain certain bacteria in sputum.

Aureolin: see Potassium hexanitritocobaltate(III).

Aureomycin®: see Tetracyclines.

Aurum: see Gold.

Austenite: a metallographic term for γ-iron, which is stable between 911 and 1392 °C, and for cubic face-centered γ-mixed crystals with a lattice constant of 356 pm of Iron-carbon alloys (see). A. with 0.8% carbon is stable above 723 °C, and at 1147 °C contains a maximum of 2.06% carbon. When cooled slowly, A. decomposes below 723 °C into Ferrite (see) and Cementite (see). Rapid cooling to room temperature leads in iron-carbon alloys to the formation of residual A. Addition of other alloy components, such as 18% chromium, 12% manganese and 8% nickel stabilizes the A. form in austenitic steels at low temperatures. A. was named for the English metallurgist Roberts-Austen.

Autocatalysis: a special case of Catalysis (see) in which a reaction product acts as a catalyst in the reaction. The reaction accelerates as it proceeds. Under certain conditions, A. can lead to bistability or to oscillations (see Oscillating reaction).

Autoclave: a vessel with pressure-resistant walls which can be closed air- and steam-tight and heated. The A. is equipped with a built-in thermometer and manometer, and sometimes also a stirring device. Substances can be heated to 400 °C and pressurized to 100 MPa in an A. The apparatus is used for reactions which do not give satisfactory yields under normal conditions. In medicine and biology, A. are used to sterilize equipment, growth media, drugs, etc. at 120 °C.

Autofining process: see Hydrorefining.

Autooxidation: oxidation of chemical compounds with atmospheric oxygen at room temperature or slightly above. In A., molecular oxygen is usually added to the oxidized molecules to form peroxide-like compounds; these then oxidize other molecules of the same or another substance. Substances which can be directly oxidized by molecular oxygen are called ***autooxidizers***; examples are sodium sulfite or hydrogen dissolved in palladium. If a substance can be oxidized only in the presence of an autooxidizer, it is called an acceptor, and it is oxidized by the peroxide-like intermediates of the autooxidizer. Some examples are ammonia, carbon monoxide, hydrogen iodide and oxalic acid. A. is initiated by a radical mechanism. Some industrially important A. are the oxidation of cumene (Hock's synthesis) and the drying of drying oils, such as linseed oil. Rusting of iron, ageing of rubber, resin formation in fuels and lubricants, rancidity in natural fats and oils and self-ignition of damp hay and straw are examples of undesired A. These processes can be inhibited or slowed by addition of antioxidants or stabilizers.

Autocide process: a method of exterminating insect pests by sterilizing them, e.g. by using Chemosterilizers (see).

Auwers-Skita rule: a rule for determining the configuration of *cis,trans*-isomers (see Stereoisomerism,

1.2.3) from physical properties (boiling point, index of refraction, density). The rule was established by Auwers and Skita between 1920 and 1923, and has been modified and limited since then. The modern form of the rule is that for alicyclic diastereomers with nearly identical dipole moments, the less stable isomer has the higher density, the higher index of refraction and the higher boiling point.

Auxins: a group of natural and synthetic growth regulators which stimulate cell division and cell extension in plants. The A. can either promote or inhibit growth, depending on the amount applied and the stage of development of the plant. The most important A. is Indol-3-ylacetic acid (see). Some important synthetic A. are indolyl-3-butyric acid, 1-naphthylacetic acid and 2,4-dichlorophenoxyacetic acid (2,4-D). Natural A. are formed mainly in rapidly growing tissues, especially at the tips of shoots and in young leaves and fruiting nodes. A. is used to promote the rooting of cuttings and setting of fruit, to thin fruits on trees, to inhibit growth of shoots in stored potatoes, and for defoliation. Some synthetic A. (2,4-D and 4-chlorophenoxyacetic acid) are also used as herbicides.

Auxochromes, ***auxochromic groups***: substituents with electron lone pairs which, when introduced into a colored compound (see Chromophores), increase the intensity and shift the wavelength of its absorption to longer wavelengths (see Bathochromicity). Some basic A. are $-NH_2$, $-N(CH_3)_2$, -NHR, $-NR_2$ (but not NH_3^+); some acidic A. are -OH, -OR and -OCOR. A pigment is not useful as a color-fast dye for fabric without such substituents; the basic or acidic properties of various A. give dyes the ability to bind to different fibers. However, the A. changes the color of the pigment by shifting its absorption. For example, quinone (yellow) → anilinoquinone (dark brown); dibenzoylethylene (colorless) → indigo (dark blue).

Avicides: poisons used to combat pest species of birds (e.g. crows or starlings). Some examples of A. set out in bait are strychnine (as poisoned wheat), yellow phosphorus (as poisoned eggs) and 3-halogen-*p*-toluidine. In some countries, thiophosphates of the same types as Organophosphate insecticides (see) are used either as bait or as area sprays.

A. are not so commonly used as Bird repellents (see). In addition to protecting newly sown crops and harvests from bird damage, they are used in cities, airports and rest areas used by migratory birds.

Avogadro's law: a basic law of stoichiometry: equal volumes of different gases at the same temperature and pressure contain the same number of particles (A. Avogadro, 1811). The particles postulated by Avogadro are molecules.

The discovery of A. was of great importance for the development of chemistry because, together with the Chemical volume law of Gay-Lussac (see), it made possible determination of the ratios of atoms in the molecules of gaseous compounds. It is a limiting law, and applies rigorously only to ideal gases.

Avogadro's number, N_A: the number of particles or elementary objects in 1 mol (see Mole). The best available numerical value is $(6.023045 \pm 0.000028) \cdot 10^{23}$. When used with the dimension mol^{-1}, A. becomes "Avogadro's constant".

A. is a proportionality factor linking molar with molecular parameters, $M = N_A \cdot m$, $V_M = N_A \cdot \varphi$, $R = N_A \cdot k$, $F = N_A \cdot e_0$, $E_\gamma = N_A \cdot h \cdot \nu$, etc. ($M$ is the molar mass, m the mass of a molecule, V_M the molar volume, φ the mean volume occupied by a molecule, R the general gas constant, k the Boltzmann constant, F the Faraday constant, e_0 the unit of electric charge, and E_ν the energy of 1 mol light quanta $h \cdot \nu$

Axerophthol: see Vitamin A.

Axial bond: see Stereoisomerism 2.2.

Aza[18]annulene pigments: a group of pigments in which the colored element is a cyclic system of conjugated double bonds with 18 π electrons. The most important examples are heme and chlorophyll; the synthetic phthalocyanin pigments are also A.

Azacyanine: see Cyanine pigments.

9-Azafluorene: same as Carbazole (see).

Azalene: see Pseudoazulenes.

Azamethonium bromide: see Ganglion blockers.

Azaserine: O-diazoacetyl-L-serine,

N=N-CH-CO-O-CH_2-CH(NH_2)-COOH,

an antibiotic synthesized by *Streptomyces* species; it was formerly used as an antineoplastic drug.

Azathioprine: see 6-Mercaptopurine.

Azeotropic distillation: a process of distillation used to separate azeotropes or mixtures of liquids with very similar boiling points which cannot be separated by rectification. The equilibrium curve of liquid mixtures which do not obey Raoult's and Dalton's laws has either an azeotropic point P at which x (molar fraction of the more volatile component in the liquid) is equal to y (molar fraction of this component in the vapor), or the curve approaches the diagonal asymptotically at the upper or lower end. Separation of these mixtures beyond this point is only possible under certain conditions.

There are two methods of A. 1) A change of the pressure of the system (vacuum or overpressure) influences the phase equilibrium, so that the azeotropic point is shifted (Fig. 1). The separation of the mixture is carried out at a succession of different pressures. 2) Another form of separation is based on addition of another component. In this case, an azeotrope between the added component and the component to be

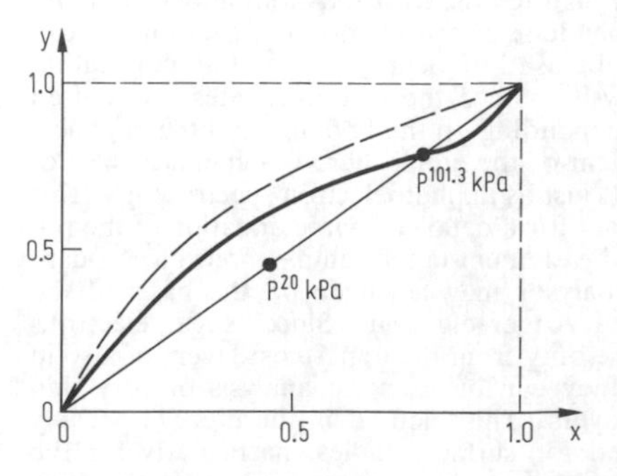

Fig. 1. Equilibrium diagram of an azeotropic mixture with the azeotropic point P. ---- Equilibrium line with the selective solvent; —— Equilibrium line without the selective solvent.

removed is deliberately created. The boiling point of this azeotrope is sufficiently different from that of the original mixture that it readily distills over; the added component must also be chosen, however, so that it can readily be removed from the azeotropic condensate.

An example of A. is the drying of ethanol by addition of benzene. At first a ternary azeotrope of 18.5% ethanol, 74.1% benzene and 7.4% water boils off at 64.9 °C, then a binary ethanol/benzene azeotrope (b.p. 68.2 °C). The benzene or toluene is removed from the condensed fraction by A. with acetone or methanol (Fig. 2).

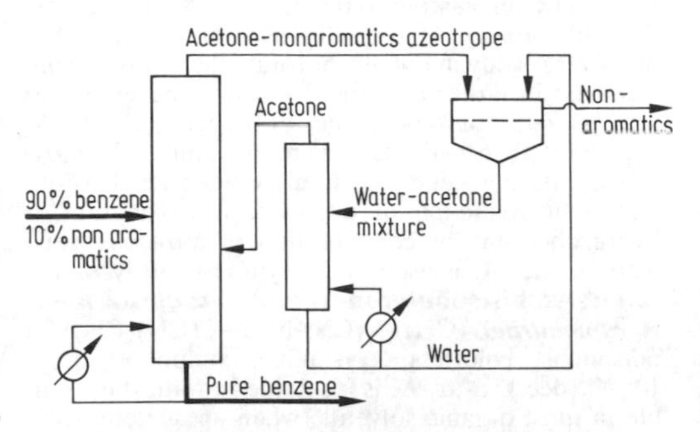

Fig. 2. Production of pure benzene by azeotropic distillation with acetone as selective solvent.

This A. is economical when small amounts (<10%) of nonaromatics are to be separated from the aromatic fraction, because in this case the major portion does not need to be vaporized. Extractive distillation is advantageous when the content of aromatics is between 90 and 60%. For aromatic contents < 60%, only liquid-liquid extraction is economical.

Azeotropic mixture: a mixture of two or more liquids which has a constant boiling point different from the boiling points of the individual components. A. cannot be separated by simple distillation. See Azeotropic distillation.

Azepines: seven-membered, heterocyclic compounds containing one nitrogen atom and three double bonds in the ring. They are antiaromatic and nonplanar, and therefore act as polyenes.

NH 1H-Azepine

Azetidine, ***trimethylenimine***: a saturated, four-membered heterocyclic compound with a nitrogen atom in the ring; b.p. 63 °C, n_D^{25} 1.4287. The doubly unsaturated analogs are called ***azetes***, and the monounsaturated compounds are ***azetines***. A. can be synthesized by cleavage of hydrogen halides out of γ-halogen amines. Their properties are very similar to those of the aziridines. Azetidin-2-ones are also β-lactams and can be obtained by removal of water from β-aminocarboxylic acids. Another possible method of synthesis is cycloaddition of ketenes to azomethines. Azetidin-2-ones can be reduced to A.

NH

Azidamphenicol: see Chloramphenicol.

Azides: derivatives of hydrazoic acid, HN_3.

1) Inorganic A. a) Salts with the general formula M^IN_3. *Ionic A.* are based on the azide ion,

$$\overset{-}{\underline{N}}=\overset{+}{N}=\overset{-}{\underline{N}} \longleftrightarrow |N{\equiv}\overset{+}{N}{-}\overset{2-}{\underline{N}}| \longleftrightarrow |\overset{2-}{\underline{N}}{-}\overset{+}{N}{\equiv}N|$$

which is a resonance-stabilized, linearly symmetrical structure. Because of its halide-like behavior, it is considered a Pseudohalide (see). The ionic A. formed with the most electropositive elements, e.g. Sodium azide (see), are relatively stable. However, because of a considerable covalent contribution and loss of symmetry of the anion, the heavy metal A. are highly explosive. The A. of lead and mercury explode on impact, and are therefore used as ignition explosives. The heavy metal A. are made by reaction of the corresponding metal salt solutions with sodium A. b) In nonmetal A., one or more A. groups are bound *covalently* to the nonmetal atom. Some well known examples are the unstable hydrazoic acid, HN_3, the A. of boron (Boron triazide, $B(N_3)_3$) and silicon (silicon tetraazide, $Si(N_3)_4$) and halogen A. of the type XN_3.

2) Organic A. a) Compounds with the general formula R-N=N=N. The names of these compounds are constructed from the names of the alkyl, aryl or heterocyclic groups R and the functional term "azide", as in phenyl azide, C_6H_5-N_3, or, if the molecule contains a higher-priority functional group, the name is constructed from the prefix "azido-" and the name of the root compound, e.g. azidobenzenesulfonic acid.

Low-molecular-weight A. are very unstable, and highly explosive. They are often processed only in the form of their dilute solutions. ***Aliphatic A. (alkyl A.)*** can be synthesized by reaction of haloalkanes with alkali azides, or from diazoalkanes and hydrazoic acid:

$$R{-}\overset{-}{C}H{-}\overset{+}{N}{\equiv}N + H_3N \longrightarrow R{-}CH_2{-}N_3 + N_2$$

Aromatic A. (aryl A.) are made by reaction of arene diazonium salts with alkali azides or by the reaction of nitrous acid with aryl hydrazines: Ar-NH-NH_2 + O=N-OH → Ar-N_3 + 2 H_2O. Aryl azides react with hydrogen chloride to form N-chloroamines, which are easily rearranged to form chloroanilines. Some other important reactions of the A. are 1,3-dipolar cycloaddition to unsaturated compounds, forming triazoles, and photolytic or thermolytic cleavage into nitrenes and nitrogen. b) Compounds with the general formula R-CO-N_3 are called ***Acyl A.*** (see).

Azine: 1) an incorrect group name for derivatives of phenazine; see Azine pigments.

2) The systematic name for biscondensation products of hydrazine with carbonyl compounds. Aldehydes form ***aldazines***, while ketones form ***ketazines***.

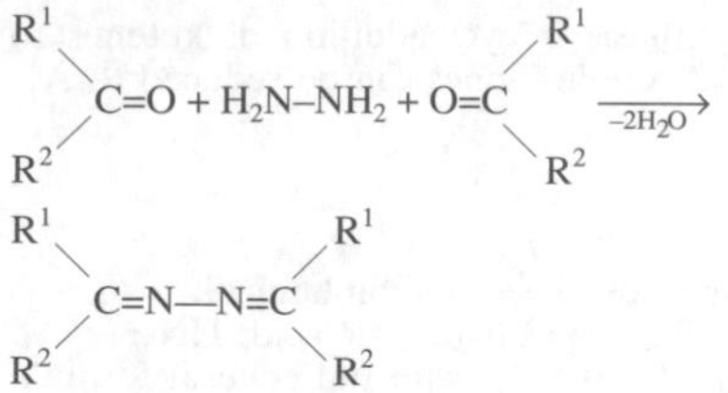

$$R^1R^2C{=}O + H_2N{-}NH_2 + O{=}CR^1R^2 \xrightarrow{-2H_2O} R^1R^2C{=}N{-}N{=}CR^1R^2$$

3) A collective term for six-membered, heterocyclic compounds with at least one nitrogen atom in the ring. If several nitrogen atoms are present in the same ring system, the compounds are called Diazines (see), Triazines (see) or Tetrazines (see). If the ring system contains oxygen in addition to nitrogen, the compound is an Oxazine (see), and if both N and S are present, it is a Thiazine (see).

Azine pigments, ***phenazine pigments***: a large group of synthetic pigments which have a phenazine ring as the common chromophore. The A. can be either acidic or basic. 1) ***Quinoxalines*** are synthesized from 2-quinones and 2-diamines. 2) ***Eurhodines*** and ***eurodols*** are produced by co-oxidation of 4-diamines or aminophenols and 3-diamines or 3-dihydroxy compounds, or by condensation of 4-nitroso compounds of secondary and tertiary aromatic amines with 3-diamines. 3) ***Aposafranines*** are made by condensation of aminoazo pigments with aromatic amines. 4) ***Safranines*** are produced by co-oxidation of 4-diamines and monoamines. 5) ***Induline*** (see Induline pigments) and ***nigrosines*** are produced by heating aminoazobenzene with amines.

Azinphos-methyl: see Organophosphate insecticides (Table 3).

Aziridine, ***ethyleneimine***: a saturated, three-membered heterocycle with a nitrogen atom in the ring. A. is a colorless liquid, b.p. 56 °C, with an odor like that of ammonia. It is easily ignited and fumes in the air. It is poisonous, very caustic and carcinogenic. Because of the high ring tension, it is very reactive and can polymerize explosively to polyethylenimine. A. can be alkylated and acylated. It is synthesized by treating ethanolamine hydrochloride with thionyl chloride, which yields β-chloroethylamine hydrochloride. When this compound is heated with sodium hydroxide, A. is produced. A. can be used to introduce the aminoethyl group into organic compounds. It is used in the synthesis of drugs and to modify synthetic polymers.

NH

Azirine: an unsaturated, three-membered heterocyclic compound with a nitrogen atom in the ring. If the double bond is between the two carbon atoms, the compound is called 1H-A.; otherwise, it is 2H-A.

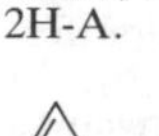

NH 1H-Azirine

Azlactones: see Oxazolinones.
Azlocillin: see Penicillins.

Azobenzene: the simplest aromatic azo compound. A. exists in two stereoisomeric forms; the E-form is stable.

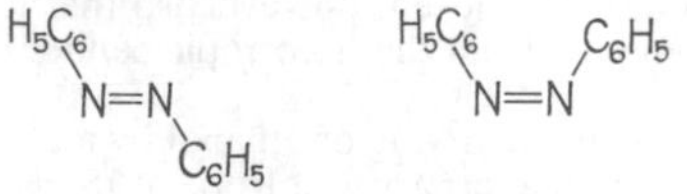

$H_5C_6{-}N{=}N{-}C_6H_5$ E- or trans-form; $H_5C_6{-}N{=}N{-}C_6H_5$ Z- or cis-form

It crystallizes in red leaflets; m.p. 68 °C. When a solution of (E)-A. is irradiated with UV light, an equilibrium between the E- and Z-forms is established. (Z)-A. can be separated from the mixture of isomers by chromatography; m.p. 71.4 °C, b.p. 297 °C [E- and Z-form]. A. is nearly insoluble in water, but dissolves readily in hot alcohol and ether. It is poisonous, and in larger amounts has the same effects as nitrobenzene. In the presence of reducing agents, A. is converted to hydrazobenzene or aniline. It is oxidized with peracetic acid to azoxybenzene. It is obtained by reduction of nitrobenzene, oxidation of hydrazobenzene or condensation of nitrosobenzene with aniline. A. is used in the synthesis of dyes.

α,α'-Azobisisobutyronitrile, ***2,2'-azobis(2-methylpropionitrile)***: $(CH_3)_2C(CN){-}N{=}N{-}C(CN)(CH_3)_2$, a poisonous, colorless, crystalline compound, m.p. 105 °C (dec.). α-,α'-A. is insoluble in water but soluble in most organic solvents; when an acetone solution of α,α'-A. is heated, it can explode. Above 35 °C, the compound decomposes, splitting out nitrogen and forming radicals. It is therefore used to initiate radical reactions. α,α'-A. is made by oxidation of the hydrazo compound.

Azo compounds: organic compounds containing the azo group, -N=N-, bound to the C atom of an aliphatic, aromatic or heterocyclic skeleton. A. can be oxidized to Azoxy compounds (see). The simplest representatives of the aliphatic and aromatic A. are Azomethane (see) and Azobenzene (see), respectively. The Azo pigments (see) are industrially important.

Azo coupling: the linking of arene diazonium salts with aromatic amines or phenols to form diazoamino or azo compounds.

Arene diazonium salts are electrophilic reagents which react with primary and secondary aromatic amines in weakly acidic solution to form diazoamino compounds or 1,3-diaryl triazenes:

$$R{-}\overset{+}{N}{\equiv}N| + H_2N{-}R \xrightarrow[-H^+]{} R{-}N{\equiv}N{-}NH{-}R.$$

These can undergo diazoamino-aminoazo rearrangement (see 4-Aminoazobenzene) to form aminoazo compounds.

With tertiary aromatic amines, diazonium salts couple directly to yield azo compounds:

$$R{-}\overset{+}{N}{\equiv}N| + C_6H_5{-}O^- \rightarrow R{-}N{=}N{-}C_6H_5{-}OH$$

Arene diazonium salts can also react with CH-acidic compounds such as β-ketoesters and β-diketones in an A. The azo compounds formed in such reactions rearrange to form hydrazones. A. is a very common reaction, especially in the chemistry of the azo pigments.

Azodicarboxylic acid esters: aliphatic azo compounds with the general formula ROOC-N=N-COOR. Most A. are yellow, unstable compounds which can decompose when distilled with a violent detonation. They are synthesized from chloroformate esters and hydrazine. The product of the first reaction step is a hydrazine *N,N'*-dicarboxylate which is converted to A. by oxidation with nitric acid or chlorine. A. are important in organic syntheses.

Azole: a five-membered heterocyclic compound containing 1 to 5 nitrogen atoms in the ring: Pyrazole (see) has two N atoms, Triazole (see) has three, and Tetrazole (see) has four. The ring may also contain other heteroatoms such as oxygen, sulfur or selenium.

Azole fungicides: Systemic Fungicides (see) based on imidazole or triazole. The most important imidazole type fungicide is *imazalil* [1-(β-allyloxy-2,4-dichlorophenylethyl)-imidazole], which is used against rose mildew and as a component in seed treatment for grains. The most important of the triazole types is *triadimefon*, m.p. 82.3 °C, p.o. $LD_{50} \approx 460$ mg/kg rat. This is used against mildews and rusts; it is synthesized from pinacolone. If the carbonyl group of triadimefon is reduced to the alcohol, the product is *triadimenole* (m.p. 112 °C, p.o. $LD_{50} \approx 950$ mg/kg rat). If the 4-terminal chlorine atom in this compound is replaced by a phenyl group, the systemic effect is weaker, but the action spectrum is broadened, and the protective and curative properties are still good: *bitertanole* (m.p. 123 to 129 °C; p.o. $LD_{50} > 5000$ mg/kg rat). *Propiconazole* and *etaconazole* are further modifications.

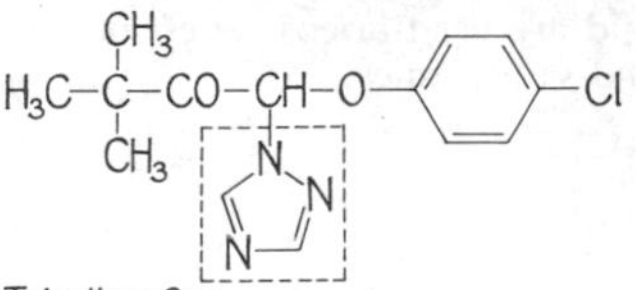

Triadimefon

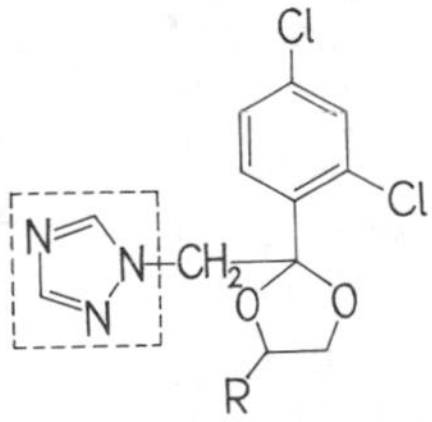

R = H_2: Bitertanole
R = C_2H_5: Etaconazole
R = C_3H_7: Propiconazole

Azolides: see Azomethines.

Azoimide: same as Hydrazoic acid (see).

Azomethane, *dimethyldiimide*: CH_3-N=N-CH_3, a colorless to slightly yellow gas; m.p. - 78 °C, b.p. 1.5 °C. A. is soluble in alcohol, ether and acetone. It can be made by oxidation of *N,N'*-dimethylhydrazine with copper(II) chloride.

Azomethines: same as Schiff's bases (see). A. acylated on the ring nitrogen atoms are called ***azolides***.

Azo pigments: a large class of synthetic pigments which contain azo groups bound to sp^2-hybridized carbon atoms. The azo groups are usually bound to benzene or naphthene rings, but in some A. they are bound to heterocycles or enolizable aliphatic compounds. The number of azo groups is indicated by the prefix mono-, di-, tri-, etc. The color depends on the structure of the A. Monoazo pigments, e.g. 4-aminoazobenzene (Fig.), absorb only in the short-wavelength range because their delocalized electron systems are small; therefore, they appear yellow or orange. The introduction of more azo groups produces red, green, blue and black pigments. The color-deepening effect (see Bathochromic shift) can be intensified by introduction of condensed ring systems. The great variety of A. is also due to the introduction of substituents into the rings to which the azo groups are attached; -OH, -COOH, -SO_3H, -NO_2, alkyl or alkoxy groups, halogens, etc. are used. The number of A. produced in the laboratory is probably well over 100,000, although only a fraction of these are produced on a large scale.

Most A. are synthesized in two steps. First a primary aromatic amine is diazotized to make a diazonium salt, and this is then "coupled" with a suitable second component. These coupling reactions (see Azo coupling) are sometimes carried out in acidic media, and sometimes in alkaline media. The second component can be an aromatic amine or phenol, or their sulfonic or carboxylic acid derivatives (e.g. the alphabet acids). A. can also be synthesized by oxidative coupling of hydrazones with activated aromatic compounds.

4-Aminoazobenzene

The A. are used as acid, mordant and direct dyes for wool, cotton, rayon, silk and synthetic fibers. Some are also used as food colorings. Diazo pigments usually have darker colors than the monoazo pigments, and are the most numerous group of A. Because of their greater molar masses, they are much better direct dyes than the monoazo pigments. The first and best known, although not the most significant, diazo pigment is congo red. Most triazo pigments are already rather complicated structurally; and are too insoluble in water to be used in traditional dyeing processes. Instead, the fibers are dyed with a diazo pigment which contains a free amino group; this is then diazotized and coupled to another component, such as pyrazolone or β-naphthol derivatives (development dyes) to produce the triazo pigment.

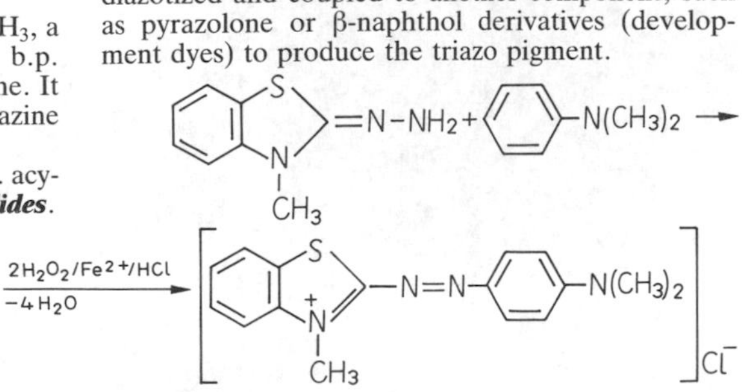

$2H_2O_2/Fe^{2+}/HCl$, $-4H_2O$

Synthesis of azo pigments by oxidative coupling.

Azote: French for nitrogen.

Azotometer: a gas buret for measurement of nitrogen volumes in elemental analysis. The A. consists of a calibrated measuring tube with a glass stopcock at the upper end. At the lower end of the tube there is a wide glass vessel with connections to a leveling vessel and the gas inlet. The A. is filled with very pure, 50% potassium hydroxide solution. The gas coming out of the combustion apparatus (CO_2 and N_2) is introduced into the A. CO_2 and H_2O are absorbed by the potassium hydroxide solution, while the nitrogen collects in the upper part of the tube. Its volume can be read, and after reduction to standard conditions, can be used to calculate the mass.

Azoxybenzene: the simplest aromatic azoxy compound. A. exists in two stereoisomeric forms, the Z or *cis*-form and the E or *trans*-form.

H_5C_6–$\overset{+}{N}$(–$\overline{\underline{O}}|^-$)=N–$C_6H_5$

Z-form E- form

The more stable Z-form of A. crystallizes in light yellow needles; m.p. 36 °C. The (E) isomer forms colorless crystalline needles; m.p. 87 °C. A. is insoluble in water, but soluble in alcohol, ether and ligroin. It is formed by mild reduction of nitrobenzene in alkaline solution with glucose. Reduction of the stereoisomeric forms of A. with lithium alanate produces (E)-A. A. is important in the dye industry, and is also used for organic syntheses.

Azoxy compounds: aromatic compounds containing the group

–N=$\overset{+}{N}$–$\overline{\underline{O}}|^-$.

A. are oxidation products of azo compounds. When heated in sulfuric acid, they undergo Wallach rearrangement (see), forming 4-hydroxyazo compounds. The simplest A. is Azoxybenzene (see).

Azulenes: bicyclic, non-benzoid, aromatic hydrocarbons. They are colored, usually blue. The structure consists of a fused cyclopentadiene and cycloheptatriene ring system in which two carbon atoms are shared by the two rings. This ring system contains ten C atoms and has 10 π-electrons in a planar arrangement, and thus has the typical properties of an aromatic compound. The parent compound, *azulene*, crystallizes in blue-violet leaflets, m.p. 99-100 °C, b.p. 115-135 °C at 1.33 kPa. Since 1955, its synthesis from Zincke aldehyde and cyclopenta-1,3-diene has been known. An intermediate of this reaction is fulvene, which is cyclized to A.

A. are termed non-alternating hydrocarbons, which means that the electron density distributions on individual C atoms are different. The electron density is particularly high at positions 1 and 3 (in the five-membered ring), so that electrophilic reactions occur there; examples are nitrations, sulfonations and Vilsmeier formylations. These different electron distributions are also reflected by the fact that A. has a dipole moment of about 1 D. Derivatives of the blue, crystalline parent compound are found in many essential oils, e.g. camomile oil, geranium oil and vetiver oil (see Guaiazulene and Vetiverazulene). A. have mild antiseptic effects, and are used in cosmetics such as toothpaste, soaps and skin cremes.

B

B: symbol for boron.

Ba: symbol for barium.

Bacitracins: peptide antibiotics produced by *Bacillus licheniformis*. The most important of them is ***bacitracin A*** (Fig.), a dodecapeptide with a thiazole structure as a heterocomponent. The thiazole is formed from the *N*-terminal isoleucine and the neighboring cysteine. The commercial preparation contains about 70% bacitracin A, and is used to treat superficial infections. B. are effective against gram-positive bacteria.

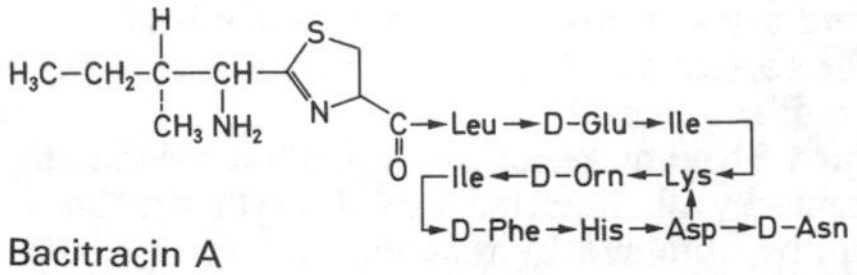

Bacitracin A

B. A forms a complex with undecaprenylpyrophosphate, a molecule which ferries hydrophilic cell-wall components across the bacterial plasmalemma, and prevents its enzymatic hydrolysis to the corresponding orthophosphate ester. This inhibits formation of the cell wall, and thus growth, of the bacteria. An intact thiazole ring and the histidine group are essential for the action of B.

Back bonding: a bonding model based on molecular orbital theory. In B., two parts of a molecule which are linked by a covalent bond display additional π-orbital interaction in which d orbitals are involved. It leads to charge equalization and thus to stabilization of the molecular system. The concept of B. is especially useful for explaining the bonding and stability of complex compounds of the transition metals, e.g. the metal carbonyls (Fig.).

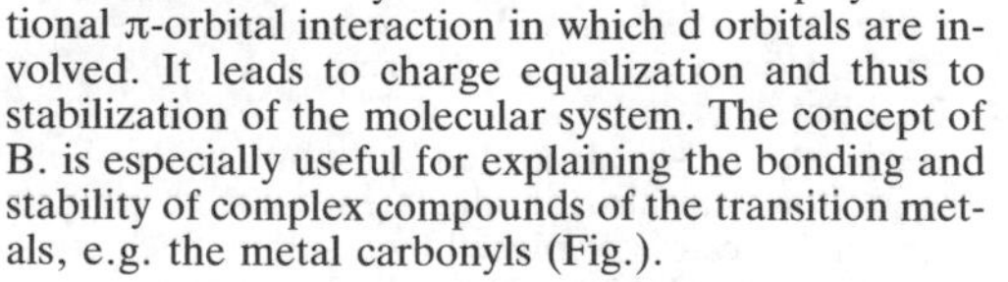

Bond model for metal carbonyls:
a, donor bond; *b*, back bonding

Overlap of an occupied σ-orbital of the CO molecule and an empty d-orbital of the metal atom produces a charge transfer from the ligand to the central atom (electron donor σ-bond, Fig. a). In addition, the π*-orbital (see Molecular orbital theory) of the CO molecule is a low-energy, unoccupied π-molecular orbital and can form an electron-acceptor π-bond (Fig. b) with an occupied π-type d orbital on the central atom (Fig. b). Thus there is a simultaneous transfer of an electron from the central atom to the ligand. This charge transfer from the central atom to the anti-bonding molecular orbital of the CO ligand weakens the C≡O bond in the carbonyl. This prediction is in agreement with the observed increase in the C-O distance and the decrease in frequency of C-O valence vibrations observed in these complexes. In addition, B. can explain why the M-CO bond is rather stable – the bond order is between 1 and 2 – and has a low bond dipole moment. Other ligands capable of B. are CN^-, RCN and PF_3. Since B. stabilizes a negative formal charge on the central ion through charge transfer to the ligand, it can explain why the oxidation state of the central atom in complexes with strong B. is usually very low. For example, it is 0 in $Ni(CO)_4$ and $[Ni(CN)_4]^{4-}$, and in $[Co(CO)_4]^-$, it is -1. The bonding in ethene complexes, e.g. $PtCl_2C_2H_4$, is similar to that in metal carbonyls. The ethyne derivatives R-C≡C-R and nitriles, R-C≡N, also form π-complexes with transition metal ions. In addition, R. can also be used in describing bonding in electron excess compounds, e.g. in noble gas compounds.

Bacteriochlorin: see Porphyrins.

Bacteriochlorophyll: see Chlorophylls.

Bacteriocide: a substance which kills bacteria and is used as a drug, disinfectant, preservative (also for protection of wood, leather and textiles) and pesticide. B. are used to combat the bacterial pathogens of plant diseases and to prevent rotting of harvested products. Many fungicides used to protect plants also have bacteriocidal or bacteriostatic effects, for example, organomercury compounds, organotin compounds, copper oxygen chloride, chloranil, 8-hydroxyquinoline, diphenyl, 2-phenylphenol, antibiotics (e.g. chloramphenicol), etc. *Technofthalan* (*N*-[2',3'-dichlorophenyl]-3,4,5,6-tetrachlorophthalamic acid), *bronopol* (2-bromo-2-nitropropane-1,3-diol) and various quaternary ammonium compounds are used specifically against bacterial pathogens.

Bacteriorhodopsin: see Vitamin A.

Badger-Bauer rule: an empirical relationship relating the association enthalpy of an H-bond and the frequency shift of the X-H valence vibration in the infrared (see Infrared spectroscopy); according to the B., the relationship is linear so long as solvation interactions and steric hindrance can be excluded. This rule is only true within limits. It is most applicable when structurally similar systems are studied, where either the proton donor in the H bridges is the same while the acceptor varies, or vice versa.

Baeyer's tension hypothesis: an hypothesis concerning spatial structure and reactivity of cycloalkanes. A. von Bayer assumed that the cycloalkanes were planar. From the magnitude of the deviation of the ring angles from the tetrahedral angle (109° 28'), he defined a *ring tension*, the *bond-angle* or *Bayer tension*, to explain the different thermodynamic stabilities of cycloalkanes with different ring sizes (the stabilities are derived from the heats of combustion). His relationship holds for cyclopropane, cyclobutane and cyclopentane, but not for cyclohexane or the larger cycloalkanes. The reason is that only cyclopropane is planar, as was predicted in 1890 by Sachse and 1918 by Mohr. The classical B. is therefore not applicable. The Bayer tension is small in all cycloalkanes, since even the angles in cyclobutane and cyclopentane are close to the tetrahedral value; however, together with the Pitzer tension (see Stereoisomerism, 2.1), it determines the stable conformations of small and normal rings. When transannular tension is added, these values determine the conformation of medium-sized rings also.

Baeyer tension: see Baeyer's tension hypothesis.

Baeyer test: a detection reaction for double bonds in organic compounds (see Alkenes); in the presence of double bonds, potassium permanganate is bleached.

Baeyer-Villinger oxidation: a method developed in 1899 for oxidation of ketones to carboxylic acid esters using peroxy acids, e.g. peracetic, perbenzoic or trifluoroperacetic acid:

$$R^1\text{–CO–}R^2 \xrightarrow{(R\text{–CO–}O_2H)} R^1\text{–CO–O}R^2$$

In the oxidation of cyclic ketones under these reaction conditions, lactones are formed:

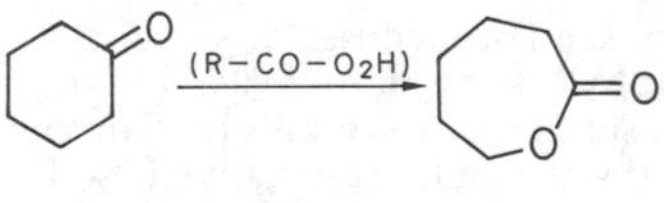

(R–CO–O₂H)

Bagasse: the lignocellulose residue remaining after extraction of the juice from the sugar cane (*Saccharum officinale*). B. is used as a source of cellulose (2.5 t B. yields 1 t paper) and furfural.

Bakelite: international term for Phenol resins (see).

Baker-Venkataraman rearrangement: conversion of 2-acyloxyacetophenones to 1,3-diketones in the presence of basic condensing agents, such as sodium, sodium amide or sodium hydroxide, in benzene, ether or toluene:

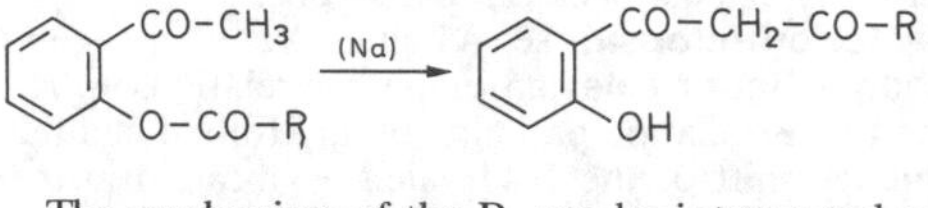

CO–CH₃ / O–CO–R —(Na)→ CO–CH₂–CO–R / OH

The mechanism of the B. can be interpreted as an intramolecular ester condensation.

Baking: the production of aminosulfonic acids by dry heating (baking) of the sulfates of various aromatic amines to about 260-280 °C. Sulfanilic acid, for example, is made from aniline sulfate via sulfaminic acid as an intermediate:

$$C_6H_5\text{–}NH_2 \cdot H_2SO_4 \xrightarrow[-H_2O]{} C_6H_5\text{–NH–}SO_3H$$

$$\rightarrow HO_3S\text{–}C_6H_4\text{–}NH_2.$$

The method can be used to sulfonate aromatic amines without protecting the amino group. Today, the reaction is often carried out in high-boiling, inert organic solvents, such as 1,2-dichlorobenzene. At the boiling temperature of the solvent, the water is cleaved off and then separated.

Bake-on paints: paints based on acrylic, epoxide, phenol, melamine, urea, polyurethane and other resins, which harden alone or in combination at temperatures up to 250 °C. They can be colored by suitable pigments. The silicone resin B. (see Polysiloxanes) are very important. B. are very hard and stretch well; they are resistant to scratching, the weather and chemical effects.

BAL: same as Dimercaprol (see).

Balata: a coagulated latex from the tree *Mimusops balata*, which is very similar to gutta percha. Unlike natural rubber, it is hard, tough and only slightly elastic. Like rubber, B. can be vulcanized, and it is sometimes added to rubber mixtures. B. was used for drive belts, and de-resined B. is used for the outer layer of golf balls.

Ball and stick models: see Stereochemistry.

Balmer series: see Atomic spectroscopy.

Balsams: see Resins.

Bamford-Stevens reaction: a method for making diazoalkanes by alkaline cleavage of tosylhydrazones. These can be obtained by reaction of carbonyl compounds with tosylhydrazine:

$$R^1R^2C\text{=O} + H_2N\text{–NH–}SO_2\text{–}C_6H_4\text{–}CH_3 \xrightarrow[-H_2O]{}$$

$$R^1R^2\text{=N–NH–}SO_2\text{–}C_6H_4\text{–}CH_3 \xrightarrow[-H_2O]{+OH^-}$$

$$R^1R^2\overset{-}{C}\text{–}\overset{+}{N}\equiv N| + CH_3\text{–}C_6H_4\text{–}SO_2^-.$$

Band model: same as Energy band model (see).

Band spectrum: see Spectrum.

Barastu process: see Water softening.

Barban: see Carbanilate herbicides.

Barbital: see Barbitals.

Barbitals: 5,5-disubstituted barbituric acids. A nitrogen atom can also be alkylated, and the oxygen atom on C-2 can be replaced by a sulfur. Such compounds are called ***thiobarbitals***. B. are synthesized by condensation of dialkylmalonate esters or dialkylcyanoacetate esters with urea or urea derivatives. B. are weak acids, so that aqueous solutions of their salts with alkalies give a basic reaction. B. decompose in alkaline media due to ring opening of the heterocyclic ring by the hydroxide ions; usually the reaction occurs at C-4 or C-6. The end product is a monoacylurea.

Unsubstituted barbituric acid and 5-monosubstituted barbituric acids are not hypnotic. 5,5-Dimethylbarbituric acid has a very weak effect. Increasing the hydrophobicity of the compounds, e.g. by lengthening the chains of the alkyl groups or introduction of cycloaliphatic or aromatic groups, increases their pharmaceutical effects. The greatest activity is observed in compounds in which the sum of the numbers of carbon atoms in the two groups bound to C-5 is 6 to 8. Branched-chain compounds are more effective and less toxic than isomeric compounds with straight-chain substituents.

B. are classified according to the length of time their effects persist. ***Long-term B.*** such as *barbital, phenobarbital* and *methylphenobarbital* are no longer widely used, although phenobarbital and methylphenobarbital are still used at low doses as anticonvulsants. ***Medium-term B.*** such as *crotylbarbital* are used as all-night sleep medicines, while ***short-term*** and ***very short-term B.*** such as *hexobarbital* are used as sleep inducers. Hexobarbital and thiobarbitals such as *ethylbutylthiobarbital* are used as injection Narcotics (see). A disadvantage of the B. is the depression of the respiratory control center which occurs at high doses.

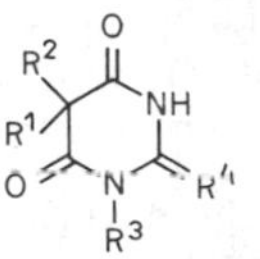

Name	R^1	R^2	R^3	R^4
Barbital (Ethylbarbital, Veronal®)	$-C_2H_5$	$-C_2H_5$	H	O
Phenobarbital (Lepinal®, Luminal®)	$-C_2H_5$	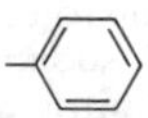 phenyl	H	O
Methylphenobarbital (Mephytal®)	$-C_2H_5$	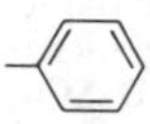 phenyl	$-CH_3$	O
Crotylbarbital (Kalypnon®)	$-C_2H_5$	$-CH_2-CH=CH-CH_3$	H	O
Aprobarbital (in Dormalon®, in Oramon®)	$-CH(CH_3)_2$	$-CH_2-CH=CH_2$	H	O
Cyclobarbital (Phanodorm®)	$-C_2H_5$	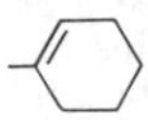 cyclohexenyl	H	O
Hexobarbital (Evipan®)	$-CH_3$	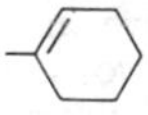 cyclohexenyl	$-CH_3$	O
Ethylbutylthiobarbital (Brevinarcon®)	$-C_2H_5$	$-CH(CH_3)-CH_2-CH_3$	H	S

Barbiturates: see Barbituric acid.

Barbituric acid, ***hexahydropyrimidine-2,4,6-trione***: the ureide of malonic acid arising by keto-enol tautomerism. B. forms white crystals, m.p. 248 °C (dec.). It is readily soluble in hot water, ether, alkalies and hydrochloric acid, but only slightly soluble in ethanol. Its aqueous solutions give a weakly acidic reaction. B. is synthesized by condensation of malonic acid with urea in the presence of phosphorus oxygen chloride or acetic anhydride, or from malonic acid diethyl ester and urea in the presence of sodium ethylate. B. is the basic structure of an extremely important group of sleep-inducers, sedatives, antiepileptics and narcotics called barbiturates or Barbitals (see).

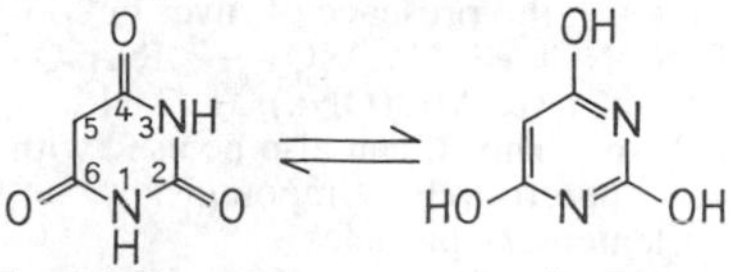

Lactam form ⇌ Lactim form

Substitution of a sulfur for the oxygen on C-2 (use of thiourea) leads to the pharmacologically active ***thiobarbiturates***. B. and its derivatives are generally quickly absorbed. They are removed from the body partly by excretion in the urine, and partly by degradation to malonic acid and urea.

Barite process: see Water softening.

Barium, ***Ba***: a chemical element from Group IIa of the periodic system, the Alkaline earth metals. Ba is a light metal, *Z* 56, and the mass numbers of its natural isotopes are 130 (0.101%), 132 (0.097%), 134 (2.42%), 135 (6.59%), 136 (7.81%), 137 (11.32%) and 138 (71.66%). The atomic mass is 137.34, valence II, Mohs hardness 2, density 3.51, m.p. 725 °C, b.p. 1640 °C, electrical conductivity 2.8 Sm mm^{-2} (at 0 °C), standard electrode potential (Ba/Ba^{2+}) - 2.90 V.

Properties. Ba is a soft metal with a silvery sheen on fresh surfaces. It resembles lead in appearance and crystallizes in a cubic body-centered lattice. Ba is the most reactive of the alkaline earth metals; its high negative standard electrode potential and low ionization potential (see Alkaline earth metals) characterize Ba as a strong reducing agent. Even at room temperature, it is attacked by air, which forms a black oxide coating on its surface. It must therefore be stored under mineral oil. Finely divided Ba is pyrophoric and burns to form barium oxide, BaO, and barium nitride, Ba_3N_2. At elevated temperatures, the compact metal will also burn. Ba reacts with water or alcohol, reducing the hydrogen to the elemental state: $Ba + 2\ H_2O \rightarrow Ba^{2+} + 2\ OH^- + H_2$. Ba dissolves in aqueous acids with the same reaction. An exception is sulfuric acid, which forms an insoluble, protective layer of barium sulfate, $BaSO_4$, on the surface of the metal, and therefore cannot dissolve it.

Barium and those of its compounds which are soluble in water or gastric juice are strong poisons. They inflame the mucous membranes and cause excitation or contraction of both the voluntary and involuntary muscles, nausea, vomiting, body pains, etc. Countermeasures: after oral ingestion of toxic amounts, vomiting should be induced immediately. Then sodium sulfate solution is administered, to convert the Ba^{2+} ions to nontoxic barium sulfate.

Analysis. In qualitative analysis, Ba is in the ammonium carbonate group and is detected by precipitation of barium sulfate or chromate. Ba also gives a green color to a flame, or can be recognized spectroscopically from its green and red emission lines. Quantitative analysis can be done by gravimetric determination of barium sulfate, complexometric titration with EDTA, or by various instrumental methods such as atomic absorption spectroscopy.

Occurrence. Ba makes up 0.04% of the earth's crust. The most important minerals are barite (heavy spar), $BaSO_4$, and witherite, $BaCO_3$.

Production. Ba is produced by reduction of barium oxide with aluminum or silicon in a vacuum furnace at 1200 °C, e.g. $3\ BaO + 2\ Al \rightarrow 3\ Ba + Al_2O_3$. The barium oxide is obtained by reduction of barium sulfate (barite) with carbon at 600 to 800 °C, forming barium sulfide: $BaSO_4 + 2\,C \rightarrow BaS + 2\,CO_2$. This is dissolved in water and precipitated as barium carbonate, $BaCO_3$, by passing CO_2 through the solution. Synthetic or natural $BaCO_3$ is finally converted to the oxide by roasting: $BaCO_3 \rightarrow BaO + CO_2$.

Applications. Elemental Ba is of minor industrial importance. It is used as a component of nickel and lead alloys, and serves as a getter metal.

Historical. Ba (from the Greek "barys" = "heavy") was first obtained in impure form in 1808 by H. Davy, who electrolysed barium hydroxide. Barium oxide had already been prepared in 1774 from barite by Scheele. Fairly pure Ba was first obtained in 1901 by Guntz, by reduction of barium oxide with aluminum.

Barium carbonate: $BaCO_3$, colorless, rhombic crystals, M_r 197.35, density 4.43, coversion to a cubic form at 982 °C. Above 1300 °C, B. decomposes to barium oxide, BaO, and carbon dioxide, CO_2. At a CO_2 pressure of about 9 MPa, it melts without decomposition at 1740 °C. B. is fairly insoluble in water (solubility product K_L at 25 °C is $8.1 \cdot 10^{-9}$), and it precipitates when solutions containing Ba^{2+} and CO_3^{2-} ions are mixed as colorless, fine crystals. Strong acids dissolve B., forming the corresponding salt solutions and CO_2. B. is poisonous (see Barium). It is found in nature in the form of witherite, which is isomorphic with aragonite and strontianite. B. is used to make low-melting, strongly refracting glasses, to soften water, as a filler in paint, paper and rubber, in fireworks (green fire), to make other barium compounds, and as a pesticide.

Barium chlorate: $Ba(ClO_3)_2 \cdot H_2O$, colorless, monoclinic crystals which dissolve readily in water; M_r 322.26, density 3.18, transition at 120 °C to the anhydrous form, m.p. 414 °C. B. is poisonous (see Barium). It is formed by the reaction of chlorine or chloric acid with barium hydroxide solutions. It fulminates with readily oxidized compounds, and it is therefore used mainly in fireworks (green flames).

Barium chloride: $BaCl_2$, colorless, very water-soluble, monoclinic crystals; M_r 208.25, density 3.856; conversion at 962 °C of α-B. to a cubic β-form, m.p. 963 °C, b.p. 1560 °C. B. crystallizes from water as the rhombic dihydrate, $BaCl_2 \cdot 2H_2O$. B. is poisonous (see Barium). It is made by the reaction of hydrochloric acid and barium sulfide (or barium carbonate): $BaS + 2\,HCl \rightarrow BaCl_2 + H_2S$. B. is used for the qualitative and quantitative analysis of sulfate ions (see Barium sulfate). In industry, it is used to harden steel, to impregnate wood, in glassmaking and in the production of other barium compounds.

Barium chromate: $BaCrO_4$, forms bright yellow, rhombic crystals; M_r 253.33, density 4.498. B. is insoluble in water (solubility product $K_s = 1.6 \cdot 10^{-10}$ at 18 °C), but it will dissolve in strong acids. In qualitative analysis schemes, the precipitation of B. serves to indicate the presence of and to separate barium. B. is formed by mixing solutions containing barium and chromate ions. It is used as a yellow pigment for paint, and to color glass and ceramics.

Barium hydride: BaH_2, a colorless powder; M_r 139.36, density 4.21, dec. above 675 °C (into its component elements). As a typical ionic hydride, B. reacts with water, alcohols and other proton donors, releasing elemental hydrogen, e.g. $BaH_2 + 2\,H_2O \rightarrow Ba(OH)_2 + 2\,H_2$. B. is poisonous (see Barium). It is made by the reaction of hydrogen with barium at about 200 °C, and is a strong reducing agent.

Barium hydroxide: $Ba(OH)_2 \cdot 8H_2O$, colorless, monoclinic crystals, M_r 315.48, density 3.18; the crystals melt in their own water of crystallization at 78 °C and are converted to the crystalline anhydrous form (density 4.5) above 95 °C. This compound is stable to about 600 °C, at which temperature it loses water to form barium oxide, BaO. B. is poisonous (see Barium). It dissolves in water more readily than any other alkaline earth metal hydroxide. The resulting solution, also called *baryta water*, is strongly basic and is used to determine carbon dioxide: $Ba(OH)_2 + CO_2 \rightarrow BaCO_3 + H_2O$. B. is obtained by dissolving barium oxide in water: $BaO + H_2O \rightarrow Ba(OH)_2$. It is used to synthesize other barium compounds, in glass and ceramics, and to soften water.

Barium nitrate: $Ba(NO_3)_2$, colorless, hygroscopic, moderately water-soluble, cubic crystals; M_r 261.35, density 3.24, m.p. 592 °C. B. is poisonous (see Barium). When heated, it loses oxygen, forming barium nitrite, $Ba(NO_2)_2$. On stronger heating, it is converted to barium oxide, BaO. B. occurs occasionally in nature as nitrobarite. It is made by reaction of barium chloride with sodium nitrite or by dissolving barium carbonate in nitric acid. B. is used mainly to produce pure barium oxide and barium peroxide and as green fire in fireworks.

Barium sulfate: $BaSO_4$, colorless, rhombic crystals; M_r 233.40, density 4.50. At 1148 °C, B. is converted to a monoclinic form, and above 1400 °C, it decomposes to barium oxide, sulfur dioxide and oxygen: $BaSO_4 \rightarrow BaO + SO_2 + 1/2\,O_2$. B. is insoluble in water ($K_s = 1.08 \cdot 10^{-10}$ at 25 °C, which corresponds to 0.24 mg B. in 100 g water). It therefore always precipitates out as fine crystals when solutions containing Ba^{2+} and SO_4^{2-} ions are mixed. This reaction is used for qualitative and quantitative analysis of both Ba^{2+} and SO_4^{2-} ions, and also for the industrial synthesis of B. B. dissolves to a significant extent in concentrated sulfuric acid, due to formation of the complex acid $H_2[Ba(SO_4)_2]$. B. occurs naturally as the mineral barite (heavy spar). Because of its chemical and thermal stability, B. is used as a pigment and whitener under the names permanent white, lithopone or blanc fixe. However, because it is not very opaque, it is usually mixed with zinc sulfide (see Lithopone). B. is used as a filler in the production of papers for artistic and photographic printing; it permits the paper to be very smooth and white. B. is also added as a filler to plastics and elastics. It is added to the cement used in the construction of nuclear reactors, because it absorbs radioactivity well. For the same reason, B. is used as an x-ray contrast material for the gastrointestinal tract.

Barton rule: see Elimination.

Bart reaction: a method for producing arene arsonic acids by reaction of arene diazonium salts with alkali metal arsenites in the presence of silver or copper powder: $R\text{-}N{\equiv}NCl + Na_3AsO_3 \rightarrow R\text{-}AsO \cdot (ONa)_2 + N_2 + NaCl$; $R\text{-}AsO(ONa)_2 + 2\,H^+ \rightarrow R\text{-}AsO(OH)_2 + 2\,Na^+$. The B. can also be used with anthraquinones and heterocyclic compounds.

Baryons: see Elementary particles.

Baryta: same as barium oxide.

Baryta water: see Barium hydroxide.

Basalt fibers: gray-brown fibers made from the rock basalt, which melts at 1400 °C. Limestone is sometimes added. The basalt melt is made into fibers by a centrifugal process (see Fiberglass). *Basalt silk* can be made by extrusion from fine nozzles. B. have an average chemical composition of 50% silicon dioxide, 12% aluminum oxide, 11% calcium oxide, 10% magnesium oxide, 7% iron(II) oxide, 5% alkali metal oxides Na_2O and K_2O, 3% titanium(IV) oxide, and 2% other oxides. They are used as insulating materials and for fiber-reinforced building materials.

Basalt silk: see Basalt fibers.

Base: see Acid-base concepts.

Base constant: see Acid-base concepts, sect. on Brønsted definition.

Base strength: see Acid-base concepts, sect. on Brønsted definition.

Basicity: see Acid-base concepts, sect. on Brønsted definition.

Basic reaction: a reaction, for example of a pH indicator, which shows that the pH (see) of an aqueous solution is greater than 7.

Basis peak: see Mass spectroscopy.

Basis vector: see Crystal.

Bassianolide: *cyclo*-(-D-Hyv-MeLeu-)$_4$, a peptide insecticide. The cyclic depsipeptide consists only of D-α-hydroxyisovaleric acid (D-Hyv) and *N*-methyl-L-leucine (MeLeu), and can be synthesized chemically. B. is a metabolic product of the entomopathogenic fungi *Beauveria bassiana* and *Verticillium lecanii*, and is very toxic to silk worm larvae.

Bathochromic shift: a shift toward longer-wavelength absorption caused by the presence of certain groups of atoms in a molecule. Such *bathochromic groups* are the same as Auxochromes (see). Blocking the electron lone pair, e.g. by salt formation, prevents the color shift due to the NH_2 group. For example, the nitroanilines are bright yellow, but their salts are colorless. When a salt is formed with a phenolic OH group, there is usually an additional bathochromic shift. For example, anthraquinone is yellowish brown, while its alkali salts are red-orange. Bathochromic shifts are also always observed when quinohydrones are formed. This is due to the simultaneous presence of a quinoid and a benzoid ring system in the same molecule. The opposite of B. is Hypsochromicity (see).

Batrachotoxin: a poisonous steroid alkaloid found in amphibians which irreversibly blocks nerve end plates and causes paralysis. B. is a basic pregnane derivative with a pyrrol-3-carboxylic acid group in position 20. It was isolated from the Columbian arrow-poison frog, *Phyllobates aurotaenia*.

Batu: see Resins.

Bayer process: see Aluminum.

Bay oil: a viscous, yellow oil which darkens when oxygen is present; it smells and tastes like cloves. It is readily oxidized and polymerized, and contains mainly eugenol, methyleugenol and citral. Myrcene is also present. B. is used in perfumes.

Be: symbol for Beryllium.

Bearing materials: materials used to make various kinds of bearings. B. must have little tendency to wear, the ability to be embedded in other materials and sufficient hardness, solidity and heat conductivity. White metals, copper alloys, aluminum alloys, and sintered metals used as B. are called ***bearing metals***. In addition, duromers (pressed plastics), plastomers (polyamides, polyurethanes, polyoxymethylene, polytetrafluoroethylene), cast iron, soft rubber, graphite, glass, ceramics, gemstones and semiprecious stones are used as B.

Bearing metal: see Bearing materials.

Beattie-Bridgman equation of state: see State, equations of, 1.2.3.

Bechamp reduction: a method for production of primary aromatic amines by reduction of nitroaryl compounds with iron or iron(II) salts and dilute mineral acids: $R\text{-}NO_2 + 2\,Fe + 6\,HCl \rightarrow R\text{-}NH_2 + 2\,FeCl + 2\,H_2O$. The method was studied by A.J. Bechamp in 1854, and is now frequently used for industrial syntheses because it is economical.

Beckmann rearrangement: rearrangement of ketoximes into *N*-substituted amides in the presence of phosphorus(V) chloride, sulfuric acid or polyphosphoric acid:

$$\mathrm{R(H)C{=}N{-}OH \longrightarrow H{-}C({=}O){-}NH{-}R}$$

In 1921, J. Meisenheimer demonstrated on unsymmetrically substituted ketoximes that the B. is an *anti*-rearrangement, that is, that the OH group changes places with the group *trans* to it.

$$\mathrm{R^1R^2C{=}N{-}OH \xrightarrow{+H^+} R^1R^2C{=}N{-}\overset{+}{O}H_2 \xrightarrow{-H_2O} R^1\overset{+}{C}{=}N{-}R^2 \xrightarrow{+H_2O} R^1(H_2\overset{+}{O})C{=}N{-}R^2 \xrightarrow{-H^+}}$$

$$\mathrm{R^1(HO)C{=}N{-}R^2 \rightleftharpoons R^1C({=}O){-}NH{-}R^2}$$

With cycloalkanoximes, the conditions of the B. produce lactams, e.g. cyclohexanone oxime is converted into the industrially important ε-caprolactam. In similar fashion as the ketoximes, (E)-aldoximes can be converted to *N*-substituted formamides in the presence of acetic anhydride:

Behenic acid: same as Docosanoic acid (see).

Beer's law: one of the basic laws of absorption spectroscopy which describes the effect of concentration on light absorption. According to B., the light absorbed by a sample is proportional to the number of particles in the layer through which the light passes (see Absorption, Lambert-Beer law).

Bee toxin: The toxin of the honeybee, *Apis mellifica* is similar in its composition to that of the wasp *Vespa vulgaris* and the hornet *Vespa crabro*. The toxins contain biogenic amines, peptides and enzymes. The only amine in B. is histamine; wasp toxin also contains serotonin, and hornet toxin contains histamine, serotonin and acetylcholine. Acetylcholine has an effect on the heart. The most significant peptides of B. are apamin, which has a strong central nervous effect, and mellitin, which is hemolytic. Mellitin is the main component of the toxin and is the main factor in its toxicity. Wasp and hornet toxins contain kinins, peptides which affect blood pressure and affect the smooth muscle system. All three types of toxin contain the enzymes phospholipase and hyaluronidase, which break down cell membranes of connective tissue. The toxic effect is the result of the complex interaction of all the components. B. is used therapeutically to treat rheumatic diseases.

Beilstein test: a qualitative test for the presence of halogens in organic compounds introduced by Beilstein. A copper wire coated with the substance to be examined is held in a non-glowing bunsen-burner flame. Halogens, if present, form volatile copper halides which color the flame green or blue. The B. is not specific, because volatile copper salts are also formed by other substances, including acetylacetone, urea and some pyridine derivatives.

Bell bronze: see Tin bronze.

Belousov-Žabotinski reaction: see Oscillating reaction.

Bemegride: 4-ethyl-4-methyl piperidine-2,6-dione used as an analeptic. It was developed as an antidote to barbital poisoning.

Benactyzine: the benzilic acid 2-diethylaminoethanylester. B. is used as a tranquilizer (see Psychopharmaceuticals).

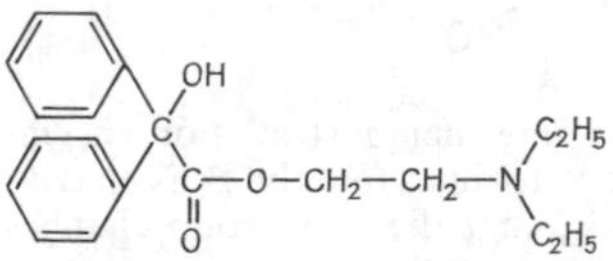

Benedict test: see Monosaccharides (Table 1).

Benfluralin: see Herbicides.

Bengal rose: see Eosine pigments.

Benodanil: see Carbamide fungicides, table.

Benomyl: see Benzimidazole fungicides.

Benquinox: see Fungicides.

Benzal acetone, ***4-phenylbut-3-en-2-one, benzylidene acetone***: C_6H_5-CH=CH-CO-CH_3, white platelets, m.p. 42 °C, b.p. 260-262 °C. B. is insoluble in water, but is soluble in most organic solvents. It is synthesized by an Aldol reaction (see) of benzaldehyde and acetone. B. is used mainly for organic syntheses.

Benzal acetophenone: same as Chalgone (see).

Benzal chloride, ***dichloromethylbenzene, benzylidene dichloride, α,α-dichlorotoluene***: C_6H_5-$CHCl_2$, a colorless, highly lacrimatory liquid, m.p. - 17.4 °C, b.p. 205 °C, n_D^{20} 1.5502. B. is insoluble in water, but is soluble in alcohol and ether. It is formed by side-chain chlorination of toluene in the presence of radical formers, benzyl chloride and benzotrichloride. Pure B. can be obtained by the reaction of thionyl chloride and benzaldehyde. Technical B. is used chiefly for production of benzaldehyde.

Benzaldehyde, ***formylbenzene, bitter almond oil***: C_6H_5-CHO, the simplest aromatic aldehyde; m.p. - 56 °C, b.p. 179.5 °C, n_D^{20} 1.5448. B. is a colorless liquid with a strong odor of bitter almonds; it is readily soluble in alcohol and ether, but only slightly in water. It can be steam distilled. Like all Aldehydes (see), B. is very reactive and undergoes the typical addition and condensation reactions. Because it is readily oxidized, B. is converted autocatalytically to benzoic acid merely on standing in air. Ammoniacal silver nitrate solutions are reduced by B., but Fehling's solution is not. In the presence of concentrated alkali hydroxide solutions, B. disproportionates in the sense of a Cannizarro reaction (see) to benzyl alcohol and benzoic acid. In the presence of cyanide ions or various thiazolium slats, B. is converted to benzoin by an Acyloin condensation (see). It reacts with ammonia to form hydrobenzamide. B. is found in free form in a few essential oils, but its principle natural source is the cyanogenic glycoside amygdalin, which is found in bitter almonds and the pits of many fruits.

B. is synthesized industrially, mainly from toluene, either by direct oxidation in the liquid phase with manganese dioxide or in the gas phase with atmospheric oxygen in the presence of vanadium pentoxide. It can also be made from benzal chloride by hydrolytic cleavage. Some other well-known syntheses for B. are the Rosenmund reduction (see) of benzoyl chloride and the Stephen reduction of benzonitrile.

B. is used as an odorant and flavoring in foods and as the starting material for many products in the pharmaceutical industry. It is used to make dyes and in cosmetics.

Benzaldehyde cyanohydrin: same as Mandelic nitrile (see).

Benzamide, ***benzene carboxamide***: C_6H_5-CO-NH_2, colorless leaflets, m.p. 132-133 °C, b.p. 290 °C. B. is slightly soluble in water, and readily soluble in alcohol and benzene. It is synthesized by the reaction of benzoyl chloride with ammonia or ammonium carbonate. B. is used as an intermediate in the synthesis of pigments.

Benzamine pigments: a group of synthetic pigments used to dye cotton. These pigments are made very fast to washing by diazotization on the fiber and coupling with β-naphthol.

Benzanthrone pigments: a group of synthetic vat dyes derived from benzanthrone or its dimer, violanthrone.

Benzazepine: a heterocyclic compound containing a benzene ring fused to an azepine ring system.

Benzene: C_6H_6, the basic unit of benzoid aromatic hydrocarbons (see Arenes). B. is a colorless, combustible liquid with a characteristic smell; m.p. 5.5 °C, b.p. 80.1 °C, n_D^{20} 1.501 l. B. is nearly immiscible with water, but is almost completely miscible with ethanol and ether. It is flammable and produces much soot when burned.

Benzene is very toxic. Its vapor is especially dangerous, even in low concentration. Short-term inhalation produces headache, tiredness, dizziness, nausea and a need for sleep; it can sometimes produce shortness of breath. Inhalation of relatively large amounts of benzene vapor in a short time causes acute poisoning, with cyanosis of the lips, ears and nose, a blue-green coloring of the skin and dark coloration of the urine. It can lead to loss of consciousness, vomiting and spasms; death may result due to respiratory paralysis. At a concentration of 2% benzene vapor in the air, death can occur in 5 to 10 minutes. Repeated inhalation of smaller amounts of benzene leads to chronic poisoning with damage to liver, kidneys and bone marrow, including a loss of red blood cells. The symptoms of poisoning often appear only after weeks or months. They include bleeding of the gums, mucous membranes and skin. The toxicity is increased by consumption of alcohol. Benzene is a resorptive poison, and is also taken up through the skin. To prevent chronic and acute poisoning, working conditions must be arranged so that no vapor can be inhaled.

Occurrence and production. B. is found in coal tar, petroleum and coke gases, from which it can be purified by extractive distillation (morphylane process), selective adsorption, washing with oil and similar processes. Technical grade B. contains thiophene, thiol alcohols, phenols, pyridine bases and unsaturated hydrocarbons. It can be purified by washing with concentrated sulfuric acid or hydrogenation under conditions which do not permit the B. to react. The pure B. obtained by hydrogenation contains only very small traces of sulfur compounds. In addition to the above sources, pyrolysis of alkanes to produce ethene is another industrially important source; B. is one of the pyrolysis products.

Applications. B. reacts with many electrophiles; it can be hydrogenated, chlorinated and sulfonated, or treated with Friedel-Crafts reagents, to form numerous products. It also forms metal complexes with sandwich structures, such as bis(benzene)chromium(O). It is no longer as important as a solvent as it once was, because of its toxicity and inflammability.

Historical. Faraday first isolated B. in 1825 from lighting gas. In 1834, Mitscherlich obtained B. by decarboxylation of benzoic acid, and in 1845, it was found in coal tar by Hofmann. The properties of this compound with the stoichiometric formula C_6H_6 could not be explained by the conventional ideas about bonding. In 1865, Kekulé von Stradonitz proposed his resonance formula. The Kekulé formula is still the most common representation of the B. molecule, expressing the experimental observation that all 6 C atoms and H atoms are equivalent, and that there can be only one monosubstitution product for each type of substituent. The fact that there are three, not four, structural isomers of a given disubstitution product, was explained by Kekulé in 1872 with his oscillation hypothesis. This states that there is a continuous rapid change between the C-C single and C=C double bonds, so that all the C-C bonds are equivalent (Fig. 1). Other B. formula were also suggested (Fig. 2). Except for the centric formula of Armstrong and Baeyer, and Thiele's formula, which are similar to the modern concept, all these proved inaccurate. Some of them are formulas for valence isomers of B. which were produced and found not to be planar (Dewar B., Ladenburg B., Hückel B.).

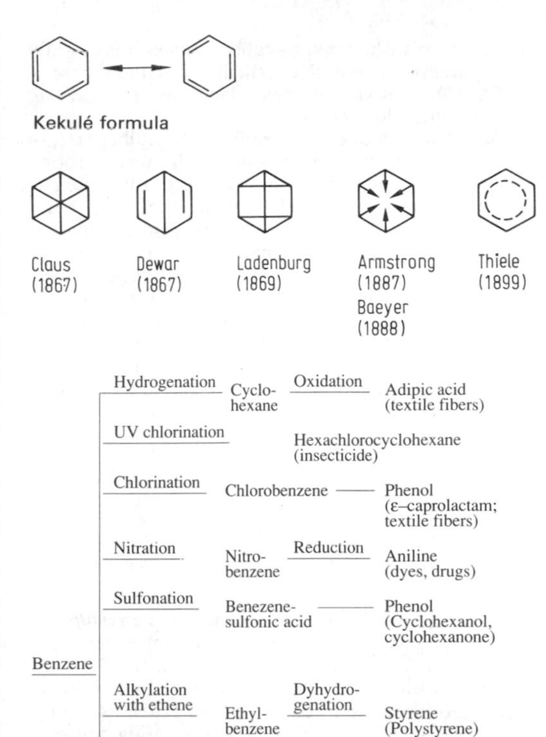

Benzene diazonium chloride: $C_6H_5\text{-}N^+{\equiv}NCl^-$, the simplest aromatic diazo compound. B. crystallizes out of the reaction solution as colorless needles, which deliquesce in the air with decomposition. They can be kept for a time in dry air and in the dark. Like all diazonium salts, B. tends to decompose explosively when heated or exposed to mechanical stress. B. is readily soluble in water, alcohol and acetone. An aqueous solution of B. can be made by diazotizing aniline with nitrous acid in a strongly acidic solution. B. can be obtained in crystalline form by reaction of aniline hydrochloride with alkyl nitrites in glacial acetic acid. B. is used to make phenylhydrazine and azo compounds.

Benzene dicarboxylic acids: the three structural isomers Phthalic acid (see), Isophthalic acid (see) and Terephthalic acid (see).

Benzene diols, ***dihydroxybenzenes***: see Pyrocatechol, Hydroquinone, Resorcinol.

Benzene disulfonic acids: three structural isomers with sulfonyl groups $-SO_3H$ at positions 1,2, 1,3 or 1,4 of the benzene ring. Benzene-1,3-disulfonic acid is readily made by heating benzene sulfonic acid with concentrated sulfuric acid to 200-230 °C. This isomer forms white, hygroscopic crystals. It is used in the synthesis of resorcinol (alkali hydroxide melts).

Benzene hexachlorides: same as 1,2,3,4,5,6-hexachlorocyclohexanes.

Benzene oxide: 7-oxabicyclo[4.1.0]hepta-2,4-diene, a heterocyclic compound which exists at room temperature only in a mixture with the valence tautomeric oxepin; formally it is the epoxide of benzene. B. is rearranged to phenol by acids. It is synthesized by dehydrobromination of 3,4-dibromo-7-oxabicyclo[4.1.0]heptane with sodium methanolate. Epoxides of aromatic compounds, the ***arene oxides***, are the primary products of oxidation of aromatic compounds to phenols. It is thought that arene oxides play a key role in the physiological and pharmaceutical effects of aromatic compounds, and in their toxicity and carcinogenicity.

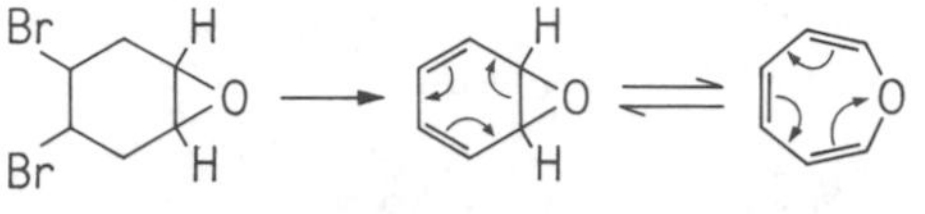

Benzene oxide Oxepin

Benzene sulfonic acid, ***benzene monosulfonic acid***: C_6H_5-SO_3H, colorless, hygroscopic needles; m.p. 50-51 °C, b.p. 135-136 °C at 2 kPa. B. is readily soluble in water and ethanol, slightly soluble in benzene and insoluble in ether. It is as acidic as strong inorganic acids. It is synthesized by treating benzene with conc. sulfuric acid at 80 °C. At higher temperatures, benzene-1,3-disulfonic and benzene-1,3,5-trisulfonic acids are formed. B. is used as a catalyst for esterification and condensation reactions.

Benzenesulfonyl chloride: C_6H_5-SO_2Cl, an oily liquid with an unpleasant odor; m.p. 14-15 °C, b.p. 251-252 °C. B. is insoluble in cold water, but dissolves in ethanol, ether and chloroform. It is synthesized by reaction of phosphorus pentachloride or phosphorus oxygen chloride with benzene sulfonic acid, or by reaction of benzene with chlorosulfuric acid. B. is used to make benzene sulfonamides; one of the uses of which is separation of primary and secondary amines (Hinsberg).

Benzenethiol: same as Thiophenol (see).

Benzenetriols, ***trihydroxybenzenes***: see Phloroglucinol, Pyrogallol, Hydroxyhydroquinone.

Benzhydrol, ***diphenylcarbinol***: $(C_6H_5)_2CHOH$, an aromatic secondary alcohol. B. forms colorless crystals which are insoluble in water; m.p. 69 °C. It is soluble in alcohol and ether. It is synthesized by catalytic hydrogenation of benzophenone.

Benzidine, ***4,4'-diaminobiphenyl***: colorless needles, often brownish or slightly reddish due to decomposition; m.p. 128 °C, b.p. 401 °C. B. is nearly insoluble in water and slightly soluble in alcohol. It is poisonous and gives rise to the same symptoms as Aniline (see); in addition, it is carcinogenic. With acids, B. forms salts of varying solubility; for example, benzidine dihydrochloride is readily soluble in water, while benzidine sulfate is nearly insoluble. Therefore B. is often used for quantitative determination of sulfuric acid. Oxidizing agents form red, green or blue products with B., depending on the reaction conditions. These are often quinone-like compounds. When B. reacts with nitrous acid, bisdiazonium salts are formed; these are very important in the synthesis of azo pigments. B. is synthesized by Benzidine rearrangement (see). It is used mainly in the dye industry, and also in analytical chemistry as a reagent for detection of a number of metals. It is used in forensic medicine to detect blood.

The isomeric 2,2'- and 3,3'-diaminobiphenyls are of little practical importance.

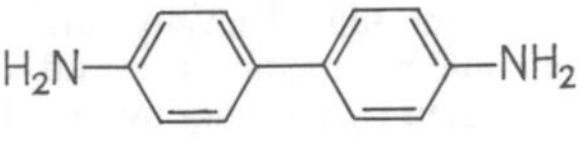

Benzidine dyes: a group of substantive diazo dyes formed by diazotization of both amino groups of benzidine and coupling with suitable compounds. The best known example is Congo red (see).

Benzidine rearrangement: an intramolecular rearrangement of hydrazobenzene into benzidine in the presence of strong mineral acids. The following mechanism can be considered verified:

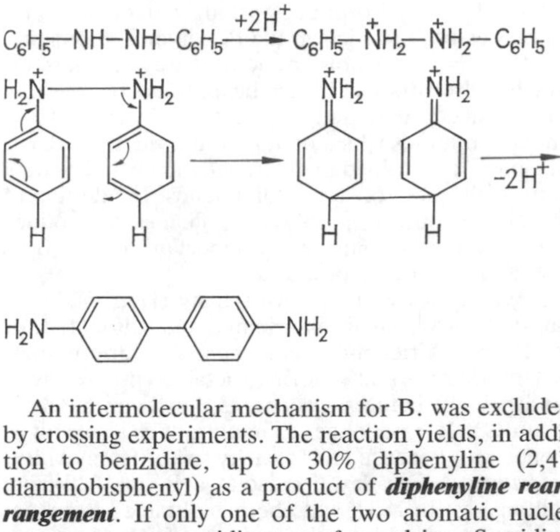

$$C_6H_5\text{—NH—NH—}C_6H_5 \xrightarrow{+2H^+} C_6H_5\text{—}\overset{+}{N}H_2\text{—}\overset{+}{N}H_2\text{—}C_6H_5$$

An intermolecular mechanism for B. was excluded by crossing experiments. The reaction yields, in addition to benzidine, up to 30% diphenyline (2,4'-diaminobisphenyl) as a product of ***diphenyline rearrangement***. If only one of the two aromatic nuclei moves, o- or p-semidines are formed in a Semidine rearrangement (see).

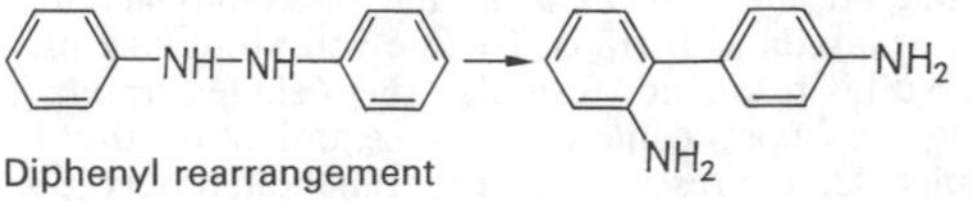

Diphenyl rearrangement

Benzil, ***dibenzoyl***: 1,2-diphenylethanedione, C_6H_5-CO-CO-C_6H_5, the simplest aromatic α-diketone. The crystals are light yellow prisms, m.p. 95 °C, b.p. 246-248 °C (dec.). It is insoluble in water, soluble in alcohol and ether. B. displays the typical ketone reactions, and all of them can occur on both carbonyl groups. B. is obtained by oxidation of benzoin with nitric acid or copper sulfate. It is used in the synthesis of benzilic acid and nitrogen-containing heterocycles.

Benzilic acid, ***2-hydroxy-2,2-diphenylethanoic acid, diphenylglycolic acid***, $(C_6H_5)_2C(OH)$-COOH, an α-hydroxycarboxylic acid. The crystals of B. are colorless needles, m.p. 151 °C. It is relatively insoluble in cold water, but readily dissolves in hot water, alcohol and ether. B. is formed in the Benzilic acid rearrangement (see) of benzil in the presence of alkalies. B. is used in the form of esters, chiefly in the pharmaceutical industry.

Benzilic acid rearrangement: a rearrangement first observed in benzil in the presence of strong alkalies:

$$C_6H_5\text{-CO-CO-}C_6H_5 \xrightarrow[2.\,+H^+]{1.\,+OH^-} (C_6H_5)_2\,C(OH)\text{-COOH}$$

The mechanism can be described as follows:

$$R-\underset{O}{\overset{\|}{C}}-\underset{O}{\overset{\|}{C}}-R \xrightarrow{+OH^-} R-\underset{O^-}{C}-\overset{R}{\underset{|\underline{O}|^-}{C}}-OH \longrightarrow$$

$$R-\underset{|\underline{O}|^-}{\overset{R}{C}}-\underset{O}{\overset{\|}{C}}-OH \rightleftharpoons R-\underset{OH}{\overset{R}{C}}-COO^- \xrightarrow{+H^+} R-\underset{OH}{\overset{R}{C}}-COOH$$

The B. has also been applied to other α-diketones; for example, furil gives furilic acid. Cyclic α-diketones react with a reduction in ring size; cyclohexane-1,2-dione becomes cyclopentane-1-ol-1-carboxylic acid under the conditions of the B.

Benzimidazole: a condensed ring system composed of fused benzene and imidazole rings. The crystals are colorless, m.p. 170.5 °C, b.p. over 360 °C. B. is readily soluble in ethanol, and slightly soluble in hot water; it is insoluble in benzene and light petroleum. It is less basic than imidazole. The benzene ring is oxidized by potassium permanganate, forming imidazole-4,5-dicarboxylic acid. B. is produced by the reaction of formic acid with o-phenylenediamine. It is used as the starting material for the synthesis of various products, such as viricides, anthelmintics, fungicides and sun creams.

Benzimidazole fungicides: Systemic Fungicides (see) derived from the benzimidazole structure; they are especially effective against *Ascomycetes*. All B. are metabolized to *carbendazim*, which interferes with oxidative phosphorylation and synthesis of DNA, RNA and proteins in the fungal organism. The B. are used as leaf and soil fungicides, and for treatment of seeds. A disadvantage is the appearance of resistance and cross resistance. B. is synthesized from o-phenylenediamine. The most commonly used are *benomyl* (p.o. LD_{50} > 9590 mg/kg rat), *carbendazim* (p.p. LD_{50} > 15,000 mg/kg rat; m.p. 307-312 °C (dec.)) and *thiabendazol*, 2-(thiazol-4-yl)-benzimidazole (p.o. LD_{50} = 3300 mg/kg rat; m.p. 305 °C). The ***phenylenethiourea fungicides*** are related systemic fungicides which are also converted to carbendazim-type compounds. Examples are *thiophanate* (m.p. 195 °C; p.o. LD_{50} > 15,000 mg/kg rat) and *methyl thiophanate* (m.p. 178 °C).

Carbendazim Benomyl

Thiophanate: R = C_2H_5
Methyl thiophanate: R = CH_3

Benzoates: the salts and esters of benzoic acid, with the general formulas C_6H_5-COOM^I or C_6H_5-COOR, where M^I is an ammonium ion or a monovalent metal, and R is an aliphatic, aromatic or heterocyclic group.

The salts of benzoic acid are formed by neutralization of the free acid with the corresponding hydroxide solutions, or by hydrolysis of benzoate esters. ***Sodium benzoate***, C_6H_5-COONa, is a colorless and odorless powder which is readily soluble in water and nearly insoluble in alcohol. Its aqueous solutions are weakly alkaline. Sodium benzoate is used as a preservative for foods, as an anticorrosive, and in the production of heavy metal benzoates.

The esters of benzoic acid are mostly colorless liquids with fruity odors; they are slightly soluble in water, and readily soluble in alcohol and ether. ***Methyl benzoate, niobe oil***, C_6H_5-$COOCH_3$; m.p. -12.3 °C, b.p. 199.6 °C, n_D^{20} 1.5164. It is used mainly as a solvent and as an aroma substance. ***Ethyl benzoate***, C_6H_5-$COOC_2H_5$; m.p. -34.6 °C, b.p. 213 °C, n_D^{20} 1.5051, is used in the aroma industry to make perfumes and fruit aromas. There are many other B. which are esters of longer-chain alcohols; these are also used in the synthesis of fruit essences.

Benzocaine: see Local anesthetics.

Benzodiazepines: bicyclic compounds with a seven-membered ring containing two N atoms (Fig. 1).

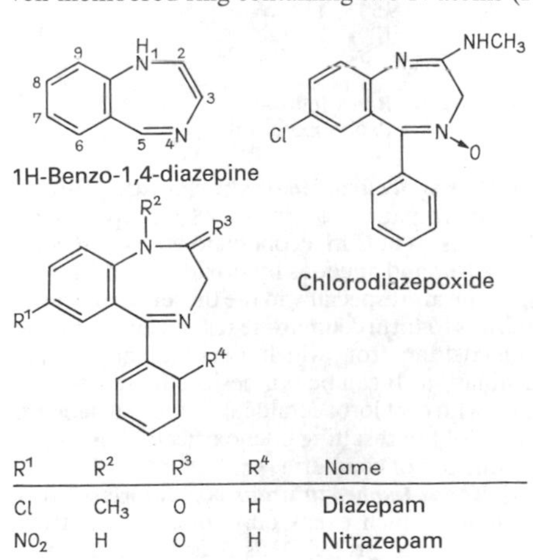

R^1	R^2	R^3	R^4	Name
Cl	CH_3	O	H	Diazepam
NO_2	H	O	H	Nitrazepam
Cl	CH_3	H_2	H	Medazepam
NO_2	H	O	Cl	Clonazepam

Fig. 1. Benzodiazepines

The most important group of tranquilizers are the 1H-benzo-1,4-diazepines. The first compound of this group to be widely used was *chlorodiazepoxide (Librium®, Radepur®)*. Later, compounds with lactam structures were introduced as drugs; of these, *diazepam (Valium®)* is the best known. Compounds in this group are also used as hypnotics, e.g. *nitrazepam (Radedorm®)*, anticonvulsants, e.g. *clonazepam (Antelepsin®)* and sedatives, especially daytime sedatives, e.g. *medazepam (Rudotel®, Nobrium®)*. The effects of the B. are explained by their interaction with benzodiazepine receptors which enhance the action of the inhibitory neurotransmitter γ-aminobutyric acid.

B. with lactam structures are synthesized from the corresponding substituted o-aminobenzophenones. Diazepam, for example, is made by cyclization of 5-chloro-2-methylaminobenzophenone with an ester of glycine (Fig. 2).

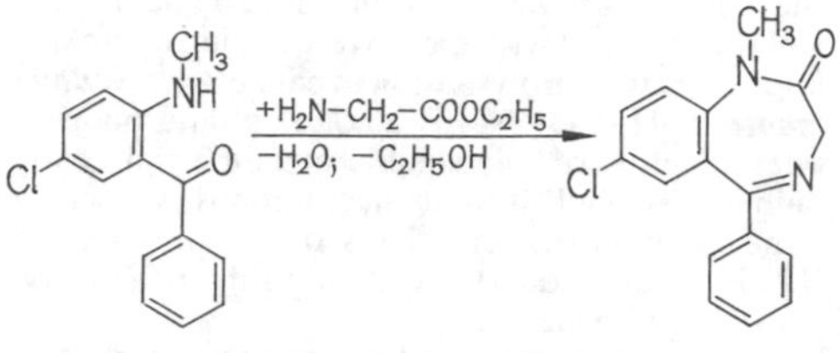

$+H_2N-CH_2-COOC_2H_5$ / $-H_2O; -C_2H_5OH$

5-Chloro-2-methylaminobenzophenone Diazepam

Fig. 2. Synthesis of diazepam

1,2-Benzodiazole: same as Indazole (see).

Benzo fast dyes: a group of synthetic di- and triazo dyes which are fast to light and chemically very stable, including benzo fast blue, benzo fast pink, benzo fast yellow.

Benzofurans: heterocyclic compounds containing a fused benzene and furan ring. The oxygen can be in either of two positions relative to the benzene ring.

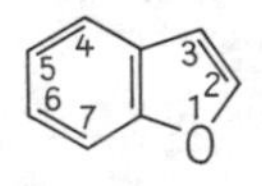

Benzo[b]furan (Coumarone)

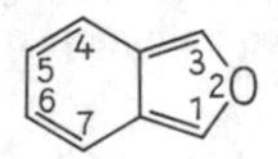

Benzo[c]furan (Isobenzofuran)

Benzo[b]furan (coumarone) is a colorless, oily liquid with an aromatic odor; m.p. - 28 °C, b.p. 174 °C, n_D^{17} 1.5615. It is soluble in alcohol and ether, but is insoluble in water and aqueous hydroxide solutions. On standing in the air (especially in the presence of acids), it polymerizes to form coumarone resin. Coumarone is present in coal tar, from which it is obtained by fractional distillation. It can be synthesized by reaction of phenolates with α-chloroacetaldehyde and subsequent dehydration of the resulting phenoxyacetaldehyde, or by ring reduction of coumarin.

Benzo[c]furan (isobenzofuran) is a colorless, very unstable liquid which exists only in solution; it has been characterized as a Diels-Alder adduct with maleic anhydride. An important derivative is ***1,3-diphenylisobenzofuran***, an intensely yellow compound; m.p. 127 °C. It is a very reactive diene and is used to capture unstable alkenes and alkynes.

Benzoic acid, ***benzenecarboxylic acid***: C_6H_5-COOH, the simplest aromatic monocarboxylic acid. B. forms colorless leaflets or needles, m.p. 122.13 °C, b.p. 249 °C, sublimation above 100 °C. It is nearly insoluble in cold water, and readily soluble in alcohol, ether, chloroform, benzene and hot water. B. is very resistant to oxidizing agents, and is converted to hexahydrobenzoic acid by hydrogenating reagents. It decomposes on heating in the presence of lime or alkali into benzene and carbon dioxide. In the presence of air and copper salts, the products are phenol and carbon dioxide. B. displays the reactions typical of aromatic compounds, that is, it is susceptible to electrophilic substitution. The entering group, e.g. $-NO_2$, halogen or $-SO_3H$, is directed toward the 3-position. In addition, the carboxyl group can form esters, acyl halides, amides, etc. The salts and esters of B. are called Benzoates (see). In the form of dust or vapor, B. has a weak irritating effect on the skin. When ingested, it is converted to Hippuric acid (see), and excreted in this form. B. is a strong poison for microorganisms. It is found both in free and bound form in many plants; the esters are very important. They are found in benzoin resins, blueberries, raspberries and cranberries. B. is present in amide linkage in leaves, bark, blossoms and fruits of many plants, and in the urine of plant-eating animals. B. is made by the hydrolysis of benzotrichloride, the oxidation of benzaldehyde or benzyl alcohol, the reaction of benzene with phosgene in the presence of aluminum chloride, or by decarboxylation of phthalic acid. Industrially, B. is produced primarily by direct oxidation of toluene with atmospheric oxygen in the presence of cobalt or manganese catalysts. B. is used in large amounts as a preservative for foods and to prevent the growth of molds in tobacco, glues, etc. It is also used as an intermediate for the production of pigments, paints, varnish and wetting agents, aromas and benzoyl chloride and benzotrichloride. Very pure B. is used in quantitative analysis and calorimetry as a reference substance.

Benzoic anhydride: $(C_6H_5\text{-}CO)_2O$, colorless prisms; m.p. 42-43 °C, b.p. 360 °C. B. is nearly insoluble in water, but dissolves readily in most organic solvents. It is hydrolysed by boiling water or aqueous hydroxide solutions, forming benzoic acid. B. is made by the reaction of benzoic acid with acetic anhydride. Industrially, it is synthesized from benzoic acid and benzoyl chloride or from sodium benzoate and benzoyl chloride. B. is used as a benzoylation reagent in organic chemistry and in the production of drugs.

Benzoid hydrocarbons: compounds of carbon and water derived from benzene; examples are toluene and xylene.

Benzoin: C_6H_5-CH(OH)-CO-C_6H_5, an aromatic, substituted Hydroxyketone (see). B. forms slightly yellow, columnar crystals; m.p. 137 °C (DL-form), b.p. 343-344 °C. B. is insoluble in water, but soluble in chloroform, glacial acetic acid and hot alcohol. Like all acyloins, it reduces Fehling's solution. B. is synthesized by benzoin condensation (see Acyloin condensation). Oxidation of B. produces benzil; reduction with tin and hydrochloric acid produces deoxybenzoin. B. is used in organic syntheses and as an initiator for photopolymerization.

Benzoin condensation: see Acyloin condensation.

Benzoin resin: a pleasant-smelling white gum from Southeast Asian and Indonesian trees. It dries in the air and turns dark. In addition to the coniferyl ester of benzoic acid, it contains various other esters of benzoic acid and other substances, depending on its source. B. is soluble in alcohol and alkalies, slightly soluble in ether and chloroform, and insoluble in water. When heated, it releases benzoic acid by sublimation. B. is obtained from the benzoin trees *Styrax tokinense* and *Styrax benzoides* by cutting into the trunk and branches.

The most valuable B. is ***Siam B.***, which contains milky white, opalescent pieces (m.p. 75°); the pieces can also be stuck together in blocks. ***Sumatra B.*** contains white grains in a colored base; m.p. 85 °C.

B. is used in perfumery, especially to fix soap perfumes, and to make pleasant-smelling varnishes and lacquers. It is also used to perfume tobacco. An alcoholic extract is used to improve the taste of medications.

1,2-Benzoisothiazol-3(2H)-one-1,1-dioxide: same as Saccharine.

Benzonitrile, ***phenylcyanide***: C_6H_5-C≡N, a colorless, poisonous liquid with an odor of bitter almonds; m.p. - 13 °C, b.p. 190.7 °C, n_D^{20} 1.5289. B. is only slightly soluble in water, but dissolves readily in alcohol, ether, acetone and benzene. It is stable to atmospheric oxygen and moisture. In the presence of mineral acids or bases, it is hydrolysed to form benzamide or benzoic acid. B. can be obtained by dry distillation of anthracite coals. It is synthesized by the Sandmeyer reaction from benzenediazonium salts and alkali cyanides in the presence of copper(I) cyanide, by dehydration of benzamide, or by heating the potassium salt of benzenesulfonic acid with potassium cyanide: $C_6H_5\text{-}SO_3K + KCN \rightarrow C_6H_5\text{-}C{\equiv}N + K_2SO_3$. B. is also synthesized directly from toluene and ammonia in the presence of air and manganese salts. B. is used as a solvent, as a swelling reagent for polyacrylonitrile, and as an intermediate in the production of pigments.

Benzophenone, ***diphenylketone***: C_6H_5-CO-C_6H_5, the simplest purely aromatic ketone. B. is found in various modifications. The stable α-B. crystallizes out of alcohol or ether as colorless, rhombic prisms; m.p. 48.1 °C, b.p. 305.9 °C. The other, unstable forms, have lower melting points. B. is insoluble in water, but is readily soluble in alcohol, benzene and ether. It is synthesized mainly by Friedel-Crafts reaction (see) from benzoyl chloride and benzene, or from benzene and carbon tetrachloride in the presence of aluminum chloride. B. is used as an intermediate in the pharmaceutical industry, as an aroma in the cosmetic industry, and as a photosensitizer.

Benzo[g]pteridine: a heterocyclic system made up of a fused benzene and pteridine ring system. Oxygen-containing B. are biologically important. B. is the parent compound of Alloxazine (see) and riboflavin (see Vitamin B_2).

[Structure: benzo[g]pteridine with ring numbering 1–10]

10-H-Benzo[g]pteridine-2,4-dione: same as Isoalloxazine (see).

Benzo[a]phenanthrene: same as Chrysene (see).

Benzopyrans, ***chromenes***: pyrans ortho-fused to a benzene ring. α-Chromene is derived from α-pyran, and γ-chromene from γ-pyran. Reduction of either chromene leads to Chromane (see).

2H-Chromene (α-Chromene) 4H-Chromene (γ-Chromene)

Coumarin (see) is derived from α-chromene, while Chromone (see) is derived from γ-chromene. Both α- and γ-chromenes and chromanes are components of a number of natural products (flavones, tocopherols).

4H-Benzo-1-pyran-4-one: same as γ-Chromone.

Benzo[c]pyrazole: same as Indazole (see).

Benzo[b]pyridine: same as Quinoline (see).

Benzo[c]pyridine: same as Isoquinoline (see).

Benzo[d]pyrimidine: same as Quinazoline (see).

Benzo[b]pyrrole: same as Indole (see).

1H,2H,3H-Benzo[b]pyrrole: same as Indoline (see).

Benzoquinones: quinones derived from benzene. ***1,4-(p-)Benzoquinone***, or ***quinone*** for short, forms golden yellow, volatile crystals with a pungent odor; they turn brown in air; subl.p. 115-117 °C. 1,4-B. is insoluble in water, and readily soluble in alcohol and ether. Its chemistry is similar to that of an α-β-unsaturated γ-diketone. It reacts with hydroxylamine to form a mono- or dioxime. The former is in a tautomeric equilibrium with the less stable 4-nitrosophenol.

1,2-Benzoquinone 1,4-Benzoquinone

Hydroquinone is formed by the action of reducing agents. Electrolytic reduction takes place via resonance-stabilized radical anions, the Semiquinones (see). 1,4-B. forms quinols with Grignard compounds. Addition of hydrochloric or hydrocyanic acid to 1,4-B. goes via a hydroquinone derivative. With alcohols or amines, it forms the corresponding 2,5-disubstituted 1,4-B. The addition of bromine to the C=C double bond of 1,4-B. leads via a dibromo adduct to quinone tetrabromide. 1,4-B. is synthesized by oxidation of hydroquinone, 4-aminophenol, *p*-anisidine or *p*-phenylenediamine, or by chromic acid oxidation of aniline in a solution of sulfuric acid. 1,4-B. and its derivatives are important intermediates in the production of pigments.

2) ***1,2-(o-)Benzoquinone*** forms unstable, bright red prisms; m.p. 60-70 °C (dec.). It is readily soluble in benzene, alcohol and ether. Unlike 1,4-B., 1,2-B. has no odor and is not steam volatile. It is formed by

oxidation of pyrocatechol with silver oxide in absolute ether, in the presence of a water-binding reagent, such as sodium sulfate.

Benzopyrazine: same as Quinoxaline (see).

Benzo-γ-pyrone: same as γ-Chromone (see).

Benzopyrylium salts: same as Chromylium salts (see).

Benzothiazole: a heterocyclic compound with a fused benzene and thiazole ring system; m.p. 2 °C, b.p. 231 °C, n_D^{20} 1.6379. B. is a weak base and is quaternized in the 3 position by alkyl halides. It is obtained by reaction of *N,N*-dimethylaniline with elemental sulfur.

B. substituted in position 2 can be made by condensation of 2-aminothiophenol with carboxylic acid derivatives. B. and 2-substituted derivatives, such as 2-mercaptobenzothiazole and 2-amidobenzothiazole, are intermediates in the synthesis of azo pigments and are also used as vulcanization accelerators.

Benzothiophenes: heterocyclic compounds containing fused benzene and thiophene rings.

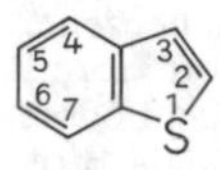

Benzo[b]thiophene Benzo[c]thiophene

Benzo[b]thiophene (thionaphthene) crystallizes in colorless leaflets which smell like naphthalene; m.p. 32 °C, b.p. 231 °C. It dissolves readily in organic solvents but is insoluble in water; however, it is steam volatile. Benzo[b]thiophene is a component of coal tar and is made on a large scale by addition of sulfur to styrene at 600 °C on an iron sulfide/aluminum contact. It is the parent compound of 3-hydroxybenzo-[b]thiophene (***thioindoxyl***) and the thioindigo pigments, and is the starting material for drugs. Unsubstituted ***benzo[c]thiophene (isothionaphthene)*** is known only in solution, and was characterized as a Diels-Alder adduct. An important derivative is ***1,3-diphenylisothionaphthene***, which crystallizes in yellow needles; m.p. 118-119 °C. It is a reactive diene and can be used to trap unstable alkenes and alkynes.

Benzotriazole, ***1H-benzotriazole***: a hetercyclic compound containing a fused benzene and 1,2,3-triazole ring system. B. forms colorless needles; m.p. 100 °C, b.p. 203 °C at 1.98 kPa. It is readily soluble in organic solvents. B. is made by reaction of *o*-phenylenediamine with nitrous acid. It is used as an additive to photographic emulsions; it stabilizes them and reduces streaking.

Benzotrichloride, ***trichloromethylbenzene***, ***α-α-α-trichlorotoluene***: C_6H_5-CCl_3, a colorless liquid with a pungent odor; m.p. - 4.5 °C, b.p. 220.9 °C, n_D^{20} 1.5580. B. is only slightly soluble in water, but dissolves in most organic solvents. The chlorine atoms can be easily removed by hydrolysis in acids, bases, or water upon heating; the product is benzoic acid. B. can be made by chlorination of toluene, as one of a mixture of products. It is used in the synthesis of triphenylmethane pigments, benzoic acid, benzoyl chloride and pesticides.

Benzoxazole: a heterocyclic fused ring system of benzene and oxazole. It forms colorless prisms with a smell like anthranilic acid; m.p. 31 °C, b.p. 182.5 °C. It is readily soluble in organic solvents and is highly steam volatile. B. is made by heating formic acid with 2-aminophenol or by dry distillation of 2-aminophenol with formamide. B. and its derivatives with substituents in position 2 are used mainly as optical brighteners.

Benzoxepine: a heterocyclic compound consisting of fused benzene and oxepine rings.

Benzoyl-: term for the atomic group C_6H_5-CO- in a molecule, for the unstable radical C_6H_5-CO · and for the cation C_6H_5-CO^+.

N-Benzoylaminoacetic acid: same as Hippuric acid (see).

Benzoyl chloride: C_6H_5-CO-Cl, the simplest aromatic acyl chloride. B. is a colorless, lacrimatory liquid with a pungent odor; m.p. - 1 °C, b.p. 197.2 °C, n_D^{20} 1.5537. B. is readily soluble in many aprotic solvents. It is hydrolysed in water or hydroxide solutions, forming benzoic acid or benzoates. It reacts under the conditions of the Schotten-Baumann reaction (see) with alcohols, phenols and amines, forming benzoylated derivatives. In the Friedel-Crafts reaction (see), B. can also be used to acylate aromatic hydrocarbons. It can be synthesized by partial hydrolysis of benzotrichloride, or by reaction of benzotrichloride with benzoic acid in the presence of zinc chloride:

$$C_6H_5\text{–}CCl_3 + H_2O \rightarrow C_6H_5\text{–}CO\text{–}Cl + 2\,HCl;$$

$$C_6H_5\text{–}CCl_3 + C_6H_5\text{–}COOH \xrightarrow{(ZnCl_2)} 2\,C_6H_5\text{–}CO\text{–}Cl + HCl.$$

In addition, B. can be obtained by the reaction of inorganic acyl halides, such as thionyl chloride, with benzoic acid in the presence of zinc chloride, or by chlorination of benzaldehyde at low temperatures: C_6H_5-CHO + Cl_2 → C_6H_5-CO-Cl + HCl. B. is used in the synthesis of benzoyl peroxide, drugs and pigments, and as a benzoylating reagent.

N-benzoylglycine: same as Hippuric acid (see).

Benzoylation: preparative methods for introducing a benzoyl group, C_6H_5-CO- into an organic compound. B. is one of the most important acylation reactions. Two well-known methods of B. are the Schotten-Baumann reaction for synthesis of benzoic acid esters and amides and the Friedel-Crafts reaction with benzoyl chloride, which goes via aromatic ketones.

Benzpinacol, ***tetraphenylethyleneglycol***: a diol with aromatic substituents; colorless crystals; m.p. 178-180 °C. B. is made by a photochemical redox reaction in which benzophenone is reduced by isopropanol in the presence of UV light. Ketyls are intermediates in the reaction.

$$C_6H_5-C(C_6H_5)(OH)-C(C_6H_5)(OH)-C_6H_5$$

3,4-Benzpyrene, ***benzo[a]pyrene***: an aromatic hydrocarbon consisting of five fused rings. 3,4-B. forms yellow crystalline needles; m.p. 176.6 °C, b.p. 312 °C at 1.33 kPa. It is insoluble in water, and soluble in organic solvents. 3,4-B. is present in coal tar at a concentration up to 1.5%, and because of its high carcinogenicity is considered the cause of the high frequency of skin cancers in workers in the coal tar industry. The compound has also been detected in tobacco smoke, so that it probably contributes to the induction of lung cancers in smokers.

Benzvalene, ***Hückel benzene, tricyclo [3.1.0.0^2,6]hex-3-ene***: a valence isomer of benzene. B. has a dipole moment of 0.88 D and a characteristic rotten odor; when isolated in pure form, it is very explosive. It is formed, along with fulvene and Dewar benzene, by UV irradiation of benzene in the liquid phase.

Benzvalene

At room temperature it converts to benzene with a half-life of 10 days.

Benzyl-: a term for the atomic group C_6H_5-CH_2- in a molecule, for the radical C_6H_5-$CH_2\cdot$ or the cation C_6H_5-CH_2^+.

Benzyl alcohol: C_6H_5-CH_2-OH, the simplest aromatically substituted alcohol, a colorless liquid with a weak, aromatic odor; m.p. - 15.3 °C, b.p. 205.4 °C, n_D^{20} 1.5369. B. is slightly miscible with water. It can be easily esterified or oxidized to benzaldehyde and then benzoic acid. B. occurs free or bound as an ester of acetic acid, benzoic acid or cinnamic acid in some essential oils, such as jasmine oil. It is found in maize, glucosidically bound. B. can be synthesized by a crossed Cannizzaro reaction from benzaldehyde and formaldehyde:

$$C_6H_5\text{–}CHO + HCHO \xrightarrow{OH^-} C_6H_5\text{–}CH_2\text{–}OH + HCOOH.$$

Industrially, it is synthesized by reaction of benzyl chloride with bases:

$$C_6H_5\text{–}CH_2\text{–}Cl + OH^- \rightarrow C_6H_5\text{–}CH_2\text{–}OH + Cl^-.$$

B. and its esters are used in perfumery to round out flower odors. Because of its antiseptic and local anesthetic effects, B. is used in medicaments. It is also an important solvent for pigments, cumarone resins, cellulose esters and ethers and sulfur.

Benzylamine: C_6H_5-CH_2-NH_2: an aliphatic primary amine with an aromatic substituent. B. is a colorless liquid with an ammonia-like odor; b.p. 185 °C, n_D^{20} 1.5401. It is readily soluble in water, alcohol, ether, benzene and acetone. Its reactivity is similar to that of other aliphatic Amines (see). It is synthesized by the reaction of benzylchloride and ammonia, catalytic hydrogenation of benzonitrile, or by catalytic hydrogenation of benzaldehyde in the presence of ammonia. B. is used in the synthesis of pigments and pharmaceutical products, and as a corrosion inhibitor.

Benzylcellulose: a cellulose ether.

Benzylchloride, ***chloromethylbenzene, α-chlorotoluene***: C_6H_5-CH_2-Cl, a colorless, lacrimatory liquid with a pungent odor; m.p. - 39 °C, b.p. 179.3 °C, n_D^{20} 1.5391. B. is only slightly soluble in water, but dissolves readily in alcohol, ether and chloroform. The chlorine atom in B. can be easily substituted by nucleophilic reagents, such as amines, alcoholates, phenolates and hydroxyl ions, forming the corresponding benzyl derivatives. B. is converted to benzaldehyde and benzoic acid by oxidizing agents. The Wurtz reaction (see) with B. yields 1,2-diphenylethane. B. is formed as one of several products of chlorination of toluene. Its main use is in the synthesis of benzyl alcohol, but it is also used as an intermediate in the production of aromas, pesticides, disinfectants, softeners and drugs, and as a benzylating reagent for organic syntheses.

Benzylcyanide, ***phenylacetonitrile***: C_6H_5-CH_2-C≡N, a colorless, poisonous liquid; m.p. - 23.8 °C, b.p. 234 °C, n_D^{20} 1.5320. B. is insoluble in water, but dissolves in alcohol, ether and acetone. It is obtained by reaction of benzyl halides with alkali cyanides and is used in the synthesis of drugs, insecticides and aromas.

Benzylidene- term for the atomic group C_6H_5-CH= in a molecule.

Benzylidine acetone: same as Benzal acetone (see).

Benzylidine dichloride: same as Benzal chloride (see).

Benzyl mandelate: see Mandelates.

Benzyloxycarbonyl chloride: see Chloroformates.

Benzylpenicillin: see Penicillins.

Benzylphenylketone: same as Deoxybenzoin (see).

Benzyne: see Arynes.

Bergamotte oil: a honey-yellow, essential oil with a pleasant fresh, sweet, fruity odor and a bitter taste. B. is often colored green by small amounts of copper or chlorophyll. It contains primarily linalyl acetate, which may account for up to 50% of the mass and is responsible for the typical odor. Other compounds in the oil are (+)-limonene, α- and β-pinene, nerol, citral, terpineol and β-caryophyllene. B. is produced chiefly in Italy, by pressing the rinds of the fruit of the bergamotte tree. It is used in perfumery (cologne) and Earl Grey tea.

Bergaptene: see Furocoumarins, table.

Bergaptol: see Furocoumarins, table.

Bergius process: see Coal hydrogenation.

Bergius-Rheinau process: see Wood hydrolysis.

Berkelium, symbol ***Bk***: a radioactive chemical element in the Actinide (see) group; it is not found on Earth naturally, but is the product of nuclear reactions. A heavy metal, *Z* 97, with the following known isotopes (decay type, half-life, number of nuclear isotopes in parentheses): 243 (K-capture, α; 4.5 h), 244 (K, α; 4.35 h; 2), 245 (K, α; 4.98 d; 2), 246 (K; 1.83 d), 247 (α; $1.38 \cdot 10^3$ a), 248 (β^-, K; 18 h; 2), 249 (β^-, α; 314 d), 250 (β^-; 3.22 h), 251 (β^-; 57 min). The most stable isotope has mass 247. The valence of Bk is III, less often IV. The density is 13.25, m.p. 986 °C, standard electrode potential (Bk^{3+}/Bk^{4+}) + 1.64 V.

B. is a silvery white metal which exists in two modifications (hexagonal and cubic). It is obtained by reduction of berkelium(III) fluoride, BkF_3, with lanthanum at 1025 °C. B. was first obtained in 1949 by Ghiorso, Thompson and Seaborg at the University of California at Berkeley (USA), by bombarding americium 241 with α-particles: $^{241}Am + ^4He \rightarrow ^{243}Bk + 2^1n$. B. can also be produced by the reaction of curium with neutrons, and is therefore available today on a mg scale:

$$^{248}_{96}Cm + ^1_0n \rightarrow ^{249}_{96}Cm \xrightarrow[62.2\ min]{\beta} ^{249}_{97}Bk.$$

Berkelium compounds: compounds in which berkelium occurs in the +3 and +4 oxidation states, as its lanthanide homolog, terbium, does. In contrast to americium(III), curium(III) and terbium(III), berkelium(III) can be oxidized to berkelium(IV) even in aqueous media, for example by bromate. It can be precipitated together with cerium(IV) or zirconium(IV) as the phosphate or iodate. If berkelium(III) compounds react with fluorine at high temperatures, the product is beige-colored ***berkelium(IV) fluoride***, BkF_4, which has the UF_4 structure. ***Berkelium(IV) oxide***, BkO_2, crystallizes in a fluorite lattice and can be reduced with hydrogen at 600 °C to yield pale green ***berkelium(III) oxide***, Bk_2O_3, which crystallizes in two modifications, hexagonal and cubic (m.p. 1920 °C). ***Berkelium(III) fluoride***, BkF_3, is dimorphous; the bright green ***berkelium(III) chloride***, $BkCl_3$, is isotypic with UCl_3.

Berry pseudorotation: a regular rearrangement process in penta-coordinated, trigonal-bipyramidal compounds in which a pair of equatorial ligands is shifted to apical positions, while the two apical ligands are shifted to equatorial positions (Fig.). A square pyramidal transition state is involved. B. is used to interpret the properties of penta-coordinated compounds and the stereochemical course of many reactions of derivatives of the heavy elements of groups IVa and Va, and of the transition metals.

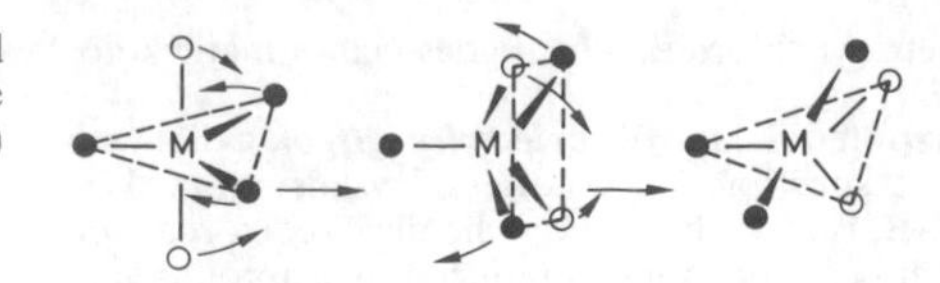

Berthelot-Thomson principle: a rule formulated in 1867 by M. Berthelot and J. Thomsen, that the affinity of a chemical reaction (its tendency to occur) is proportional to the heat released by it. According to the B., endothermic processes (with enthalpies of reaction $\Delta H > 0$) such as the carbide process or the dissolving of many salts in water should not occur spontaneously. However, it follows from the second law of thermodynamics that it is not the heat of reaction ($\Delta_R H$ or $\Delta_R E$), but the Gibbs free energy $\Delta_R G$, which is a measure of the spontaneity of a reaction. The relation between $\Delta_R G$ and $\Delta_R H$ is given by the equation $\Delta_R G = \Delta_R H - T\Delta_R S$. $\Delta_R S$ is the entropy of the reaction and T is the absolute temperature. Only in the special case that $T \rightarrow 0$ or $S \rightarrow 0$ is the B. approximately valid, because then $\Delta_R G \approx \Delta_R H$.

Berthollide, ***berthollide compound, non-stoichiometric compound***: (named after C.-L. Berthollet) a chemical compound whose composition can vary and which depends on the conditions of its production. The range of variation is called the *range of homogeneity* or *phase width*, and it is indicated in the chemical formula by ranges of the stoichiometry subscripts (e.g. $Fe_{0.89...1.00}S$, $TiO_{0.60...1.35}$). In general, a compound can be identified as a B. by the symbol ~, read as "approximately", which is placed before or over the ideal formula (e.g. ~FeS or $\widetilde{TiO}$). The B. include compounds of metals with metals (intermetallic compounds, e.g. the γ-phase of brass, $\sim Cu_5Zn_8$), of metals with semimetals (e.g. nickel arsenide, ~NiAs) and of metals with nonmetals (e.g. copper(I) oxide, $\sim Cu_2O$ and cadmium sulfide ~CdS). The variable composition of berthollide compounds can be explained by crystal lattice defects or by the formation of Mixed crystals (see). Antonym: Daltonides (see).

Beryllia: same as Beryllium oxide (see).

Beryllium, ***Be***: an element from group IIa of the periodic system, the Alkaline earth metals (see). Be is a light metal; *Z* 4, with natural isotopes of mass 9 (99.0098%) and 10 (0.0002%); atomic mass 9.0122, valence II, Mohs hardness 6 to 7, density 1.85, m.p. 1278 °C, b.p. 2970 °C, electrical conductivity 18.2 Sm mm^{-2} (at 0 °C), standard electrode potential (Be/Be^{2+}) - 1.70 V.

Properties. Be is a steel gray, hard metal, brittle at room temperature, which crystallizes in hexagonal close-packed array. Its chemical properties are determined largely by its very small atomic and ionic radii, and this explains the considerable difference between the chemistry of Be and that of its heavier homologs (see Alkaline earth metals). However, Be is very similar to aluminum, a group IIIa metal, because of a comparable charge to radius ratio (see Periodic system, diagonal relationships). The ionization potentials of Be are very high, and it exerts a strong polarizing effect on bonding partners, for the same reason. This is why its compounds are always covalent. However, the use of its two 2s electrons to form two covalent bonds provides the Be atom with only a quartet of electrons. It therefore attempts to achieve a coordination number of 4 to attain the noble gas configuration. This leads to polymerization, e.g. in solid beryllium fluoride, BeF_2, or beryllium chloride, $BeCl_2$. Be can also achieve the octet by coordination of electron donors, as in the formation of beryllium chloride etherate, $BeCl_2[O(C_2H_5)_2]_2$, the beryllium tetrafluoroberyllate anion $[BeF_4]^{2-}$, etc.

In spite of its high negative standard electrode potential, Be forms an impermeable oxide layer and therefore does not dissolve in water or concentrated nitric acid. It is not oxidized by air unless heated above 1000 °C, at which temperature it forms beryllium oxide, BeO. It reacts with dilute acids with evolution of hydrogen. Be also dissolves in aqueous alkalies, forming first the amphoteric beryllium hydroxide, $Be(OH)_2$, which is then converted to the hydroxoberyllate: $Be + 2\ NaOH + 2\ H_2O \rightarrow Na_2[Be(OH)_4] + H_2$. Be burns in gaseous chlorine, bromine or iodine, forming the beryllium halides. Beryllium powder is oxidized by gaseous sulfur, with flames, forming beryllium sulfide, BeS. Beryllium compounds taste sweet, and for this reason the element was formerly called glucinium.

> Beryllium and its compounds are extremely poisonous. The inspiration of dust or vapor of the metal or any beryllium compound leads to severe pneumonia. Skin contact with beryllium salts causes inflammation which is difficult to heal.

Analytical. Be forms a yellow-green fluorescing color lake with morin in alkaline solution; this serves as a qualitative indicator for Be and aluminum. For quantitative analysis, instrumental methods such as atomic absorption spectroscopy are used. In addition, it can be determined gravimetrically by precipitation with ammonia, NH_3; the precipitate is beryllium hydroxide, $Be(OH)_2$, which is roasted at 1000 °C to convert it to beryllium oxide, BeO. Interfering ions can be removed by a previous precipitation with oxine.

Occurrence. Be makes up $6 \cdot 10^{-4}\%$ of the earth's crust, and is thus one of the rare elements. Some important minerals are beryll, $Be_3Al_2[Si_6O_{18}]$. Colored forms of beryll are emerald and aquamarine, which are prized as gems. Other Be minerals are euclase, $BeAl[SiO_4](OH)$, gadolinite $Be_2Y_2Fe[Si_2O_{10}]$, chrysoberyll, $Al_2[BeO_4]$ and phenacite, $Be_2[SiO_4]$.

Isolation. The starting material is almost always beryll. This is either extracted by sintering with an extraction agent or converted to a soluble form by direct melting. In *sinter extraction*, beryll mixed with stoichiometric amounts of sodium fluorosilicate and soda is converted at 770 °C to sodium fluoroberyllate, aluminum oxide and silicon dioxide: $Be_2Al_2[Si_6O_{18}] + 2\ Na_2SiF_6 + Na_2CO_3 \rightarrow 3\ Na[BeF_4] + Al_2O_3 + 8\ SiO_2 + CO_2$. The fluoroberyllate is dissolved in water and sodium hydroxide is added to the solution, which causes beryllium hydroxide to precipitate. Most of the iron hydroxide present as an impurity is removed by dissolving the $Be(OH)_2$ in excess sodium hydroxide solution; the $Fe(OH)_3$ is left behind. The sodium hydroxoberyllate is decomposed by heating the solution under pressure, releasing beryllium hydroxide and sodium hydroxide: $Na_2[Be(OH)_4] \rightleftharpoons Be(OH)_2 + 2\ NaOH$. In *melt extraction*, the ground beryll is melted at 1650 °C and quenched with water; the glassy, grainy material is then heated to 250 or 300 °C with concentrated sulfuric acid. The resulting solution contains mainly beryllium sulfate, $BeSO_4$, and aluminum sulfate, $Al_2(SO_4)_3$. After addition of aqueous ammonia, ammonium aluminum sulfate, $NH_4Al(SO_4)_2 \cdot 12H_2O$, crystallizes out and is filtered off. Beryllium hydroxide is obtained by adding base to the filtrate and heating it.

The hydroxide obtained by one of the above methods is converted to the fluoride or chloride, depending on the process to be used to obtain the metal. To make beryllium fluoride, the hydroxide is treated with aqueous ammonium hydrogenfluoride solution, the precipitating ammonium tetrafluoroberyllate is separated, and decomposed into beryllium fluoride and ammonium fluoride by heating to 1000 °C: $(NH_4)_2BeF_4 \rightarrow BeF_2 + 2\ NH_3 + 2\ HF$. Beryllium fluoride is reduced to finely divided Be by heating it to 900 °C with magnesium; at 1300 °C, it melts to the compact metal. The chloride is obtained by converting the hydroxide to the oxide by heating, then mixing the oxide with carbon and chlorine: $BeO + C + Cl_2 \rightarrow BeCl_2 + CO$. The metal is produced by melt electrolysis of a mixture of $BeCl_2$ and NaCl.

Applications. As a light metal with a high melting point and good mechanical properties, Be is important in the construction of aircraft and rockets. Protective shields made of Be prevent the burning of space vehicles as they reenter the atmosphere. Be is transparent to x-rays, and is therefore used to make exit windows in x-ray apparatus. It is an important metal for improving the qualities of alloys of aluminum, iron, copper, cobalt and nickel; it increases their hardness, strength and resistance to corrosion. Non-sparking tools are made from Be-Ni alloys for use in areas where there is danger of explosions; Be-Ni and Be-Fe alloys are used for membranes and springs for high-temperature applications and for surgical instruments. Be alloys with magnesium to only a limited degree, but even 0.005% Be in a magnesium alloy significantly reduces its oxidation in the liquid state and its combustibility.

Historical. Beryllium oxide was recognized as a component of beryll in 1797 by Vauquelin. The impure metal was first produced in 1828 by Wöhler, by reduction of the chloride with potassium. Lebeau was able to produce the pure element in 1898 by electrolysis of sodium tetrafluoroberyllate.

Beryllium bronze: copper-beryllium alloy containing a maximum of 2% beryllium. B. can be hardened, because the solubility of beryllium in copper is decreased from 2.1% at 964 °C to 0% at room temperature. Springs which are highly resistant to corrosion and mechanical stress are made from B., as are tools which do not generate sparks.

Beryllium chloride: $BeCl_2$, colorless, very hygroscopic crystals; M_r 79.92, density 1.899, m.p. 405 °C, b.p. 520 °C. In the gas state, B. forms linear Cl-Be-Cl molecules. Solid B. consists of Cl-bridged chains (see Alkaline earth metals). It is soluble in ether, ethanol and water, and crystallizes out of water as the tetrahydrate, $BeCl_2 \cdot 4H_2O$. It is obtained by heating elemental beryllium in a stream of chlorine or hydrogen chloride. In industry, it is an intermediate in the production of Beryllium.

Beryllium hydroxide: $Be(OH)_2$, exists in two modifications. The α-form is a colorless, gel-like amphoteric substance obtained by addition of an alkali to an aqueous beryllium salt solution. It is only slightly soluble in water. The rhombic β-form is obtained from the α-form by allowing it to stand for a long period, or by boiling it. β-$Be(OH)_2$ is only slightly soluble in acids and bases. Freshly precipitated B. dissolves in bases, forming the beryllates, e.g. $Be(OH)_2 + 2\,NaOH \rightarrow Na_2[Be(OH)_4]$. $Be(OH)_2$ is dehydrated to beryllium oxide, BeO, by strong heating. $Be(OH)_2$ is the starting material for synthesis of other beryllium compounds.

Beryllium nitrate: $Be(NO_3)_2 \cdot 3H_2O$, colorless, deliquescent crystals, M_r 187.07, density 1.557, m.p. 60°C (in its own water of crystallization). B. is synthesized by dissolving beryllium hydroxide in nitric acid and evaporating the solution. It is used as a hardener for the mantles of gas lamps.

Beryllium oxide: BeO, a colorless, loose powder, or hexagonal crystals (Mohs hardness 9); M_r 25.01, density 3.01, m.p. 2530°C, b.p. about 3900°C. B. which has been calcined at high temperatures is insoluble in aqueous acids and bases. Calcination with carbon converts it to beryllium carbide, Be_2C. BeO is made by roasting $Be(OH)_2$. Because of its thermal stability, it is used as a lining for furnaces and to make melting crucibles. Because it brakes neutrons, but does not absorb them strongly, BeO is used as a moderator in nuclear reactors.

Berzelius test: a preliminary test for the presence of arsenic introduced by Berzelius. A small amount of the sample is heated with coal or calcium cyanide in a sealed test tube. Any arsenic compounds present are reduced to arsenic, which evaporates and is deposited on cooler parts of the tube as a black arsenic mirror. The B., in contrast to the Marsh test (see), is not affected by the presence of antimony.

Bessemer process: see Steel.

Betanal®: Carbanilate herbicides.

BET equation: see Adsorption.

Bettendorf test: a preliminary test for the presence of arsenic, introduced by Bettendorf in 1870. A small amount of the sample is dissolved in concentrated hydrochloric acid, and a few drops of a tin(II) chloride solution are added. In the presence of arsenic, dark brown flakes of elemental arsenic form. Noble metals or mercury interfere, as these are also reduced by the tin.

BHC: see Hexachlorocyclohexanes.

Bi: symbol for bismuth.

Biamperometry: see Amperometry.

Bianthrone: see Anthraglycosides.

Bicyclic: a term for carbon compounds in which two carbon rings are coupled.

Bicycloalkanes: saturated hydrocarbons in which two carbon rings are linked by one or several C atoms. The rings can be the same size or different. *Spiranes* are linked by a single common C atom, *condensed systems* by two, and *bridge ring systems* have more than two C atoms in common. Examples are spiro[3,3]heptane (1 common C atom), bicyclo [4,1,0]heptane (norcarane; 2 common C atoms) and bicyclo[2,2,1]heptane (norbornane; 4 common C atoms).

Spiro[3,3]heptane

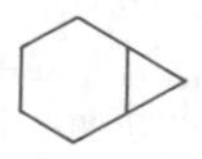

Bicyclo[4,1,0]heptane

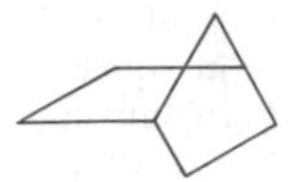

Bicyclo[2,2,1]heptane

The most common types of system in both natural products and industrially important compounds are the condensed ring system, e.g. decaline or indane, and bridged rings in terpenes, such as pinene or camphor. Most natural products are functional compounds with double bonds, hydroxyl or carbonyl groups, etc. The parent hydrocarbon B. are scientifically interesting because of their bonding and particular spatial structures. Many possibilities for linking two cycloalkane rings have been explored, including higher-order ring systems such as catenanes and toraxanes, which are also called topological isomers. These syntheses can also be used for tri- or tetracycloalkanes and other polycyclic saturated hydrocarbons. Some interesting systems of this type are prismane, adamantane, twistane, diamantane or cyclopentanoperhydrophenanthrene, all of which have been synthesized and studied.

Biladienes: see Bile pigments.

Bilane: see Bile pigments.

Bilatrienes: see Bile pigments.

Bile acids, ***steroid carboxylic acids***: steroids with a carboxyl group on the alkyl group attached to C17. G. are formed in the animal liver by oxidative degradation of the side chains of sterols, especially cholesterol, and excreted with the bile or directly into the intestine. Some of the B. are reabsorbed from the intestine (enterohepatic circulation). B. can contain 24 (*cholanoic acid derivatives*), 27 (*coprostanoic acid derivatives*) or, rarely, 28 carbon atoms. Other B. have been isolated from microorganisms. The B. are 5β-steroids, but in contrast to other steroids, the hydroxyl group on C3 is usually α in the B. In addition, B. may contain one or two more α-hydroxyl groups in positions 6, 7 or 12. Mammals synthesize almost exclusively C_{24} B. Human bile contains mainly chenodeoxycholanoic acid, cholanoic acid and deoxycholanoic acid. Bovine bile contains mainly cholanoic acid, and porcine bile, hyodeoxycholanoic acid. Amphibia and reptiles form C_{27} and C_{28} B., for example 3α,7α,12α-trihydroxycoprostanoic acid. B. are detected by the red color developed on treatment with sulfuric acid and sucrose or furfural (*Pettenkofer's reaction*). In the bile, the B. are present as the sodium salts of conjugates.

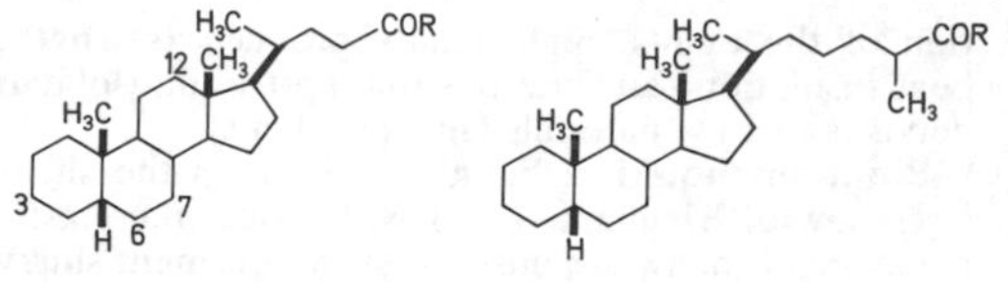

H3C, COR, 12 CH3, H3C, 3, 7, H 6

Cholanoic acid (R = OH)

H3C, COR, H3C, CH3, H3C, H

Coprostanoic acid (R = OH)

Taurine conjugates: R = $NHCH_2CH_2SO_3H$

Glycine conjugates: R = $NHCH_2COOH$

B. are conjugated by amide bonds to glycine or taurine. Taurine conjugates are more common; however, the ratio of glycine to taurine conjugates in human bile is 3:1. The sodium salts of the B. conjugates are amphiphilic, and therefore are surface active. They form micelles in aqueous solutions, and are important in the digestion and absorption of fats. Deoxycholanoic acid can form inclusion compounds with fatty acids and other substances; these are called *choleic acids*.

Bilenes: see Bile pigments.

Bile pigments: open-chain tetrapyrroles formed in vivo by oxidative cleavage of cyclic tetrapyrrols (see Porphyrins), usually with the loss of C-5. Most G. contain terminal pyrrolinone residues. An open-chain tetrapyrrole in which the 4 five-membered rings are linked by methylene groups, so that there is no conjugation, is called a ***bilane***. Compounds with one or more =CH- groups linking the rings are called ***bilenes, biladienes*** and ***bilatrienes***. Their absorptions in the visible differ markedly. In mammals, the B. are oxidation products of hemoglobins which are excreted with the bile. The primary degradation product is ***biliverdin*** (more exactly, biliverdin IXα); other B. are formed from it by secondary reactions. The most important B. is ***bilirubin***, which is partially conjugated to the glucuronide form. The bilenes ***mesobilin*** and ***stercobilin*** formed by spontaneous oxidation are responsible for the color of feces. Proteins bound to B. are called ***biliproteins***. Bilirubin, which is non-covalently bound to serum proteins, is a transport form. Covalently bound biliproteins are found in plants; some examples are the phycobiliproteins found in blue-green bacteria and the ***phytochromes*** of higher plants. These biliproteins serve as light-collecting pigments in photosynthesis and as light-sensing pigments. The ***phycocyanins*** (chromophore: phycocyanobilin) and ***phycoerythrins*** (phycoerythrobilin) are typical phycobiliproteins.

Skeleton	No. of =CH-groups	Color	Examples*
Bilane	0	Colorless	Mesobilinogen, stercobilinogen
Bilene	1	Yellow to orange	Bilenes (a): mesobilin, stercobilin
Biladiene	2	Red	Biladienes (a): bilirubin, mesobilirubin, phycoerythrobilin
Bilatriene	3	Green to blue	Biliverdin, phycocyanobilin

* The lower-case letters indicate the position of the =CH-group.

Biliproteins: see Bile pigments.

Bilirubin: see Bile pigments.

Biliverdin: see Bile pigments.

Bimetal, ***bimetallic strip***: a structure consisting of two different metals or alloys held together by plating, welding or riveting. Because the thermal expansion of the two components differs, the strip bends when the temperature changes, and it can therefore be used as a switch for controlling temperature. Examples are an iron-nickel alloy with 36% nickel (see Invar®) and an iron-nickel-manganese with 20% nickel, 6% manganese and the rest iron.

Binapacryl: see Fungicides.

Binders (construction materials): inorganic materials which bind mineral materials, such as sand, gravel, slag and ashes, into synthetic stone, mortars (mixtures of B. with sand) and concretes (mixtures of B. with coarser materials or porous synthetic stone). There are silicate (silicate cements, clays) and non-silicate B. (lime, gypsum, alumina melt cement). B. are divided into classes by the reaction mechanism of setting: air, hydraulic and hydrothermal B.

1) The most important ***air B.*** are slaked lime and burned gypsum. To make *slaked lime*, limestone ($CaCO_3$) is burned in a kiln with 10% coke as a fuel. The thermal decomposition reaction occurring at 1100-1300 °C is $CaCO_3 \rightarrow CaO + CO_2$. To make *mortar*, the burned lime is first slaked in an exothermal reaction with water: $CaO + H_2O \rightarrow Ca(OH)_2$; mixed with sand and water, this can be used directly for building. The process of setting involves uptake of CO_2 from the air according to the following equation: $Ca(OH)_2 + CO_2 \rightarrow CaCO_3 + H_2O$. Mixtures of water with gypsum burned at 120 °C, $CaSO_4 \cdot 1/2H_2O$, harden in the air by the formation of fibrous crystals of gypsum, $CaSO_4 \cdot 2H_2O$. Sand, sawdust or other filler may be incorporated into the mixture. Because $CaSO_4$ is relatively soluble in water, such plasters are only useful for interior walls and ceilings.

2) ***Hydraulic B.*** react with water to form insoluble compounds. They are thus more resistant to water than air B., and can harden under water. The prototype of hydraulic B. is *Portland cement*, the most common of the silicate cements. The raw materials are limestone, sandstone, clays, gypsum, silicate rocks, metallurgical cinders, slags and fly ash. Cement can be prepared directly from natural mixtures of limestone and clay (marls) of appropriate composition, but it is usually necessary to calculate the appropriate amounts of various natural materials and industrial waste materials to be mixed. For a cement with good hydraulic properties, the proper proportions of lime (CaO), silicon dioxide, SiO_2, aluminum oxide, Al_2O_3 and iron(III) oxide, Fe_2O_3 (hydraulic factors) must be achieved. In the course of the high-temperature treatment (burning), these form silicates, aluminates and ferrites. The quality of the Portland cement increases with increasing proportions of lime. The raw materials are first ground fine and homogenized. They can then be made into cement by the wet, semi-dry or dry process. The still widely used *wet process* has the disadvantage that a large amount of energy is required to remove the water from the crude slurry. The *dry process* has the disadvantage that large amounts of dust are generated. Therefore, the preferred technology today is

the *semi-dry process*, in which the powdered, dry raw materials are mixed in inclined granulaters with 10 to 15% water, forming marble-sized lumps. These are dried in a preheater with hot furnace gases, then burned in a rotating furnace at maximum temperatures of 1400-1500 °C to cement clinkers. These are cooled with air or water. Modern plants have capacities greater than 3000 t clinkers per day. During the burning process, sintering and solid-state reactions (changes in modifications, formation of compounds) occur at temperatures up to 1300 °C. Only above 1300 °C do the main reactions, solubilization and crystallization processes in the melt phase, take place. The melt phase should not exceed 20 to 30% of the raw material. Cement clinkers do not have hydraulic properties until they are finely ground; the increase in surface area produces the hydraulic properties. Grinding 3 to 4% gypsum with the cement prolongs the setting time (this is often desirable). Silicate waste materials, such as blast furnace slag, modify the type and properties of the cement. During the setting process, the cement mineral phases react with the water. The relatively soluble minerals dissolve and mineral phases with higher degrees of condensation (silicate anion complex) precipitate. Tri- and dicalcium silicate, $3\,CaO \cdot SiO_2$ and $2\,CaO \cdot SiO_2$, form gelatinous, semicrystalline calcium silicate hydrate phases with fibrous and layered structures (tobermorite), which fill the space between the grains (20% porosity remains) and solidify the mixture.

Special elements are used to achieve desired properties (chemical and thermal stability, less heat toning on setting) and to utilize waste materials (slags, ashes). The methods are often quite different from those used for Portland cement. ***Ferrocements*** are similar to Portland cements; they set slowly and are resistant to sulfate. In them, the aluminum oxide of Portland cement is largely replaced by iron(III) oxide. ***Blast furnace cements*** are made by adding ground blast furnace slag (up to 40% in iron Portland cement and up to 80% in blast furnace cement) to the normal Portland cement. Non-silicate clay melt cements are used as materials for heat-resistant claddings for industrial furnaces.

3) ***Hydrothermal B.*** are made in autoclaves at temperatures around 200 °C from sand, lime (sometimes with added gypsum or cement) and water. The binding reactions, as in cement hardening, involve calcium silicate hydrate phases. In the synthesis of ***lime sandstone***, quartz sand is mixed with 4 to 12% lime and water, and is pressed into molds. It is then allowed to react at 0.8 to 1.2 MPa and 175-200 °C for up to 20 hours. For ***silicate concrete***, sand and lime are ground together. Sand, gravel or filter ashes can be added to the dense B. In contrast, in the synthesis of ***gas silicate concrete***, a gas generator such as aluminum powder or paste, calcium carbonate/hydrochloric acid or hydrogen peroxide is added to the mixture to create a porous material which is used as a light-weight brick with good insulation qualities.

Hydrothermal B. synthesis has the following advantages: relatively low energy expenditure (no high-temperature reactions), much lower material costs (lower production costs for sand, smaller amounts of B. used, use of lime instead of cement, use of fine lime), complete mechanization and partial automation of the plants, high quality products (strength, heat insulation) and the possibility of using isolated deposits of raw materials (smaller plants).

Bingham liquid: a liquid which obeys the shear stress law of Bingham. This law describes the plastic behavior of many suspensions, such as cement slurry and drilling slurries: $\tau = \tau_0 + \eta_{pl}\Gamma$. Here τ is the shear stress, τ_0 is the flow limit, η_{pl} is the Bingham plasticity, and Γ is the deformation rate.

Bingham flow: same as Structural viscosity (see).

Binodal curve: see Mixing gap.

Bioassay: a method of analysis which makes use of the specific sensitivity of an organism. Until the structure of a biologically active compound has been elucidated, a B. may be the only available type of assay. For example, hormones may be assayed by measuring their effects on intact animals or plants.

Biochemistry: the study of the molecular basis of life. Chemical, physical and mathematical methods are used to examine the structures, functions and interactions of molecules found in organisms. The chemistry of living matter is highly complex, as organisms are capable of self-replication and self-regulation in a wide range of environmental conditions.

From a thermodynamic point of view, living organisms are "open systems" which exchange matter and energy with the environment. Their chemistry approximates a "steady state" in which the rates of influx and efflux of matter and energy are equal. In biochemical reactions, the reactants reach steady-state concentrations which are different from the equilibrium concentrations in the classical thermodynamic sense. A living cell can be considered an isothermal chemical machine, which absorbs energy from its environment and converts this into chemical energy, for example, in the form of adenosine triphosphate (ATP). The chemical energy of ATP is then utilized for nearly all the functions of the cell, such as synthesis of its component molecules, active transport of substances into and out of the cell, and osmotic or mechanical work (contraction and motility).

In living cells, essentially all reactions are catalysed by enzymes. These reactions are not independent of each other, but are organized in metabolic pathways, which may consist of 20 or more steps. Those metabolic pathways which serve to generate ATP from the breakdown of fuel substances are called *catabolic pathways*; two examples are the degradation of glucose to pyruvate or lactate (glycolysis) and the degradation of fatty acids to acetyl-coenzyme A by β-oxidation. Those pathways which serve to synthesize cell components are called *anabolic pathways*; examples are biosynthesis of glucose, fatty acids or proteins. There are also *amphibolic pathways*, which have both anabolic and catabolic functions, and *anaplerotic sequences*, which replace essential intermediates. Some pathways are cycles, for example, the citric acid (tricarboxylic acid) cycle or the glyoxalate cycle. Taken as a whole, the network of interconnected metabolic pathways channel the energy and some of the matter taken in from the environment into synthesis of cell components, and ultimately, into replication and growth.

The coupling of many reactions into a network is achieved by several levels of regulations, such as feedback inhibition (inhibition of a synthetic pathway by its end product), adaptation of the amount of enzyme synthesized to the metabolic situation (repression and derepression of the genes), and activation or inactivation of enzyme proteins by phosphorylation or other covalent modifications. Single-celled organisms regulate themselves in response to environmental conditions alone, while the cells of multicellular organisms are also responsive to intercellular signals such as hormones and nerve impulses.

The ability of living organisms to reproduce themselves is based on the complementary structures of the two strands of DNA molecules. These contain information in the form of a code for the synthesis of proteins - structural and enzyme proteins. The interactions of proteins with other classes of substances (fats, carbohydrates, etc.) create and maintain the structures of the organism.

Although organisms are very diverse, as are their component molecules, there are certain basic themes common to all. For example, the genetic material of all organisms consists of nucleic acids, DNA and RNA. All cells are enclosed by membranes which consist of lipid bilayers. Certain basic metabolic pathways are ubiquitous, such as glycolysis, the citric acid cycle, and β-oxidation of fats. The biosynthetic pathways for carbohydrates, amino acids and lipids are common to the the great majority of organisms, although there are some variations.

The discipline of B. is divided into descriptive, functional and applied B. ***Descriptive B.*** is essentially the same as natural products chemistry, and is concerned with the chemistry of biomolecules. ***Functional B.*** is the study of the biological functions as chemical or physicochemical processes. This division includes *cell biochemistry*, which is concerned with intermediary metabolism, *enzymology*, the study of biological catalysis, and *molecular biology*, which is concerned with the biosynthesis and functions of information-carrying biopolymers. B. is closely related to physiology or physiological chemistry. ***Applied B.*** is subdivided into specialties which are closely related to agriculture, medicine and industry.

Biocytin: see Biotin.

Bioelectrochemistry: a field of electrochemistry dealing with the electrochemical aspects of biological systems. Some essential problems in the field are the electrical phenomena associated with charge transport across cell membranes and synthetic model membranes, electrochemical oxidation and reduction of biological substances, electrochemical analysis in biological systems (especially in vivo) and the development of electrochemical sensors (see Ion-selective electrodes).

Biofilter: an installation for deodorizing waste gases using filtration through humus layers. The impurities are first trapped in the moist filter mass, then degraded by aerobic bacteria.

Biogas, ***marsh gas***: a mixture of gases produced by anaerobic microbial degradation of cellulose-containing wastes. It consists of 50 to 75% methane, 25 to 50% carbon dioxide and up to 1% hydrogen and hydrogen sulfide. The natural process leading to biogas has been known since the Middle Ages (Helmont 1639). Wherever organic materials are protected from air, they are decomposed by the cooperative action of various bacterial species. *Acidogenic bacteria* hydrolyse the biopolymers into smaller components and ferment them to fatty acids, acetic acid, hydrogen and carbon dioxide. *Acetogenic bacteria* degrade the higher fatty acids to acetic acid, hydrogen and carbon dioxide, and *methane bacteria* convert acetic acid or carbon dioxide to methane. For each kg organic dry matter, 200 to 600 l B. are formed.

These natural processes occur most readily at temperatures of 10 to 15 °C and 30 to 35 °C. Depending on the composition of the gas mixture, the heating value of B. is between 18 and 25 MJ m^{-3}.

Organic material	Gas (l) per kg dry mass	% Methane
Manure	250	60
Liquid manure	200–600	60–70
Kitchen wastes	200–400	65–70
Sludge	300–400	65–70
Potato vines	420	60

In large-scale agricultural installations and sewage treatment plants, B. is produced in biogas reactors. Liquid manure, organic wastes or sludge is pumped into the reactor, and is fermented within 8 to 20 days. For each m^3 reactor volume, the daily output is 0.5 to 3 m^3. In evaluating the economic significance of B., one must remember that the plant itself has a high energy consumption; usually part of the B. is used to run it. In addition, the distribution of B. is generally very limited; it is usually consumed on the site to produce heat and electricity. The production of B. in China and India has become very important, where in areas with low energy requirements, 7.5 million (China) and 70,000 (India) small fermenters have been built.

Species	Number of animals	Excrement l/day	Amount of biogas m^3/day	Electricity produced kWh
Milk cow (500 kg)	1	45	2.1	4.2
Beef steer (350 kg)	1	31	1.3	2.6
Pig (70 kg)	10	56	2.0	3.0
Hen (1.7 kg)	100	18	1.3	2.6

Generation of biogas from plant sources

Substrate	Amount	Biogas m^3/day	Electricity kWh
Beet foliage	1 kg	0.6	1.2
Straw	1 kg	0.5	1.0
Molasses mash*	1 m^3	12.0	24.3
Grain mash*	1 m^3	15.0	30.3

* Residual after distillation of liquor.

Biogenic amines: decarboxylation products of amino acids (Table). B. are found widely in microorganisms, plants and animals, either bound or in free form. B. or their derivatives act as neurotransmitters in animals. Some are components of coenzymes (cysteamine and β-alanine in coenzyme A, propanolamine in cobalamine), others, of phospholipids (colamine, choline). Spermidine is found in sperm. Putrescine and cadaverine are products of decay.

Precursor amino acid	Biogenic amine	Methylation product
Leucine	Isoamylamine	
Serine	Colamine	Choline
Threonine	Propanolamine	
Cysteine	Cysteamine	
Aspartic acid	β-Alanine	
Glutamine acid	γ-Aminobutyric acid	
Ornithine	Putrescine	
Lysine	Cadaverine	
Phenylalanine	Phenylethylamine	
Tyrosine	Tyramine	N-Methyltyramine, hordenine, mescaline
3,4-dihydroxyphenylalanine	Dopamine	Noradrenalin
	Adrenalin	
Tryptophan	Tryptamine	N-Methyltryptamine N,N-dimethyltryptamine
Histidine	Histamine	

Biogenetic drugs: see Pharmacognosy.

Biogenic growth regulators: same as Phytohormones (see).

Bioluminescence: a form of chemiluminescence displayed by a variety of organisms, including bacteria, fungi, coelenterates, worms, crustaceans and insects. In general, a *luciferin* is oxidized in the presence of oxygen, in a reaction which often consumes ATP. The enzyme catalysing the reaction is called a *luciferase*, and the product is an activated oxyluciferin. This compound spontaneously emits light in its conversion to oxyluciferin. The yield can be as high as 90%. The processes which have been studied to date produce light by decomposition of a 1,2-dioxetane.

1,2-Dioxetane

There are several luciferins with differing structures. Some known luciferins are benzthiazole derivatives, and there are also imidazolopyrazines.

Biopolymers: naturally occurring macromolecular compounds, including polypeptides and proteins, nucleic acids, polysaccharides, polyprenes and lignins. The term also applies to chemically synthesized compounds such as polyamino acids and polynucleotides. Polysaccharides and polypeptides can also be linked to other types of molecule. A distinction is made between proteoglycans and glycoproteins; the former have the characteristics of polysaccharides (glycans) and the latter, of proteins. ***Homopolymers***, such as homopolysaccharides and polyprenes, are less common than ***copolymers*** of different amino acids (proteins), monosaccharides (heteropolysaccharides) or nucleotides (nucleic acids). As a rule, polypeptides, proteins, nucleic acids and polyprenes are unbranched, while polysaccharides can be either branched or unbranched. Lignin is highly branched and cross-linked. ***Block polymers***, in which a short sequence of monomers is repeated, are found among the polysaccharides.

B. are the basis of organic life. Nucleic acids carry information (genes), proteins serve as catalysts and control elements, give mechanical support and provide motility. Some polysaccharides are reserve substances (starch and related polysaccharides), while others provide mechanical support in plants and some animals (cellulose and chitin). Wood, which is the main support element of land plants, consists mainly of lignin and cellulose. The bacterial cell wall consists of a peptidoglycan (see Murein). Proteins and polysaccharides are essential components of foods. Many B. are also important raw materials for the leather, textile and paper industries. Much of our present knowledge of macromolecular chemistry was gained from studies of B.

In proteins, nucleic acids and polysaccharides, there are different levels of structure. *Primary structure* is the sequence of the individual monomers in the polymeric chain, and it determines the properties of copolymers. *Secondary structure* is the conformation of individual parts of the chain. A chain can exist in a non-ordered (random-coil) structure, or it may take on an ordered structure maintained by non-covalent bonds, especially hydrogen bonds. The most important secondary structures are the Helix (see) and the pleated sheet (see Proteins). *Tertiary structure* is the spatial structure of the entire molecule, which includes the conformation of side chains and the interactions between different segments of the molecule. Interactions between different molecules can lead to self-assembly into higher-order structures; this level of organization is called *quaternary structure*.

Highly ordered structures of B. are changed by changes in the external conditions (temperature, pressure, pH, addition of organic solvents, etc.) into other, less ordered structures. The transition is called a *phase change*. *Denaturation* is the change of a native structure into one which is not biologically active. It can also be recognized by changes in optical or hydrodynamic properties. Denaturation can be either reversible or irreversible; in reversible denaturation, covalent bonds (except, in some cases, for disulfide bonds) are not broken. In irreversible denaturation, covalent bonds are broken.

Biopterin: see Pterin.

Bioreactors: see Biotechnology.

Bios II: same as Biotin (see).

Biotechnology: the production or modification of substances by enzymes (***enzyme technology***), microorganisms, especially bacteria and yeasts (***industrial microbiology***), or cultured plant or animal cells. Genetic engineering (see) has developed as an independent field of B.

The production of materials extends from the massive production of microorganisms for use as animal feeds to the production of *biochemicals*, such as antibiotics, vitamins, amino acids and, more recently, optically active bulk and fine chemicals. The production of energy carriers such as alcohols, hydrogen or biogas is increasingly important. Biotechnological processes are also used for purification of waste water and isolation of metals.

Biotechnological *partial syntheses* are used when chemical reactions cannot be used or are too expensive, usually for regioselective and stereoselective reactions. Other industrial processes are the formation of acetic acid (vinegar) and ethanol, production of sorbose from sorbitol (a step in the synthesis of ascorbic acid) or of 6-aminopenicillanic acid from benzylpenicillin (see Penicillins).

Enzyme technology has a central place in B. Its applications can be expanded by improvement of the technological properties of the enzymes, such as thermostability or pH optima. The use of enzymes in organic solvents is a new development which makes reactions with lipophilic compounds easier.

The production of materials by microorganisms requires high-yielding production strains, which must be maintained by selection for such qualities as productivity, stability, extracellular secretion of the product, substrate requirements, culture type, etc. Such strains can be produced by genetic manipulations including mutagenesis, recombination and genetic engineering. The selection of mutants is very laborious, involving product screening. Recombination is achieved by cell fusion (crossing), followed by selection.

Microbial production can be carried out on surfaces or in submersion culture in *fermenters* or *bioreactors*; these can be operated discontinuously (batch processes) or continuously.

The development of reusable biocatalysts and continuous production is very important for the economic viability of biotechnological processes. *Immobilization* of enzymes and microorganisms (*heterogeneous biocatalysis*) is an important concept for such development. The enzymes or microorganisms can be immobilized by physical enclosure in a polymeric matrix or microcapsule, by cross-linking with a bifunctional reagent such as glutaraldehyde or toluene diisocyanate, or by adsorptive or covalent bonding to a carrier. Loss of activity or impairment of its interaction with the substrate must be avoided. The kinetic and mechanical properties of the immobilized products are industrially important. Heterogeneous biocatalysis is used, for example, in the formation of 6-aminopenicillanic acid from benzylpenicillin, in partial syntheses of steroids, by racemate cleavage of amino acids and isomerization of glucose to fructose.

Animal (including human) cells are used to produce interferons and monoclonal antibodies. Efforts are being made to develop plant cell cultures to produce secondary plant products, especially those used as drugs.

Biotechnological processes have been used for thousands of years to produce foods and beverages: bread, cheese, vinegar, sour milk products, beer and wine. In the past few years, due to advances in genetics, the development of B. has been very rapid and is thought by many to be a key industry of the future.

Biotin, ***vitamin H, bios II***: a water-soluble vitamin which contains nitrogen and sulfur. B. forms crystalline needles; m.p. 232-233 °C, $[\alpha]_D^{21}$ + 91° (0.1M NaOH). It is practically insoluble in organic solvents. B. is present in all cells in small amounts. The richest source is the liver, which contains about 0.00025%. Determination of B. in biological materials must be done by microbiological tests, due to the very low concentrations in such sources. *Allescheria boydii* or *Lactobacillus arabinosus* is used as the test organism; these microbes require B. for growth. B. deficiency in human beings causes changes in the skin, including an increase in oil secretion. Because B. is ubiquitous in foods, essentially the only way to cause a deficiency is to consume large amounts of raw egg white, which contains the protein avidin. Avidin binds B. specifically and very tightly. B. is a coenzyme of enzymes which transfer carbon dioxide, such as carbon dioxide ligases and carboxytransferases; it is linked via its carboxyl group to the ε-amino group of a lysine residue in the apoenzyme; ε-*N*-biotinyllysine is called ***biocytin***. Carbon dioxide is bound to the N atom in position 1 of B. as carboxybiotin ("active CO_2").

Biotin

Biotransformation: the biochemical reactions of drugs and other foreign substances (xenobiotics) in an organism. B. occurs mainly in the liver, but to a smaller extent in other organs and body fluids as well. The liver enzymes of B. are localized in the microsomes. The products of B. are also called *metabolites*, and they may have weaker effects than the original compounds, or none at all (detoxication, inactivation), but in other cases they may be more active (activation). In the case of drugs, the inactive precursors are called Prodrugs (see). In many cases, the products of B. are more soluble in water, and thus more readily eliminated. In B., oxidation, reduction and hydrolysis are called Phase I reactions, while formation of conjugates with other compounds is called Phase II.

The most important ***Phase I reactions*** are oxidations such as C-hydroxylation, epoxidation, oxidative *O*-, *N*- and *S*-dealkylation, oxidative deamination, *N*- and *S*-oxidation and oxidation of alcohols. *C-Hydroxylation* can occur on aromatic and aliphatic C atoms. *Epoxidation* occurs at aliphatic double bonds; the resulting epoxide may then be hydrolysed to a vicinal diol. An epoxide intermediate (arene oxide) is also postulated for hydroxylation of aromatic compounds. In *oxidative dealkylation*, a C atom next to the heteroatom is first hydroxylated, then the group is split off as a carbonyl compound. *Oxidative deamination, N*- and *S-oxidation* occur by similar mechanisms, leading to *N*-oxides, sulfoxides and sulfones. *Oxidation of alcohols* produces aldehydes, ketones and carboxylic acids. Oxidation reactions are catalysed by various species of cytochrome P_{450}, iron-containing enzymes which use molecular oxygen. They are active in conversion of carbonyl compounds to alcohols, the nitro group to an amino group, the cleavage of an azo group into two amine fragments, and reductive dehalogenation. Hydrolases catalyse the cleavages of esters, amides and glycosides. Related hydrolases are also found in the plasma and tissues (fig.).

C-Hydroxylation	–C–H ⟶ –C–OH
Epoxidation	–C=C– ⟶ –C–C– (O) ⟶ –C–C– (OH OH)
Oxidative dealkylation	–X–C–H ⟶ [–X–C–OH] ⟶ –XH + –C(=O)–
	X = O,N,S
Oxidative deamination	–C(H)–NH ⟶ [–C(OH)–NH] ⟶ –C(=O)– + –NH_2
N-Oxidation	≥N ⟶ ≥N⟶O
S-Oxidation	–S– ⟶ –S(=O)– ⟶ –S(=O)(=O)–

Phase-I reactions

Phase II reactions depend on the presence of a reactive group (OH, NH_2, SH, COOH) in either the original compound or in a product of Phase I reactions. The most common of the Phase II reactions is *formation of β-D-glucuronides* through reactions of hydroxyl groups with UDP-glucuronic acid (active glucuronic acid). For example, phenol is converted to phenyl-β-D-glucuronide. NH_2, SH and COOH groups undergo similar reactions. OH groups can also be converted to sulfates by reaction with "active sulfate". Aromatic carboxylic acids are often conjugated with glycine, forming an amide compound. Polynuclear aromatics without functional groups often react with glutathione, leading via several intermediates to *formation of mercapturic acids*, e.g. (1-naphthyl)mercapturic acid, $C_{10}H_7$-S-CH_2-CH(COOH)NH-CO-CH_3, in which an *N*-acetylated cysteine residue is bound to the aromatic via its S atom. *O*-, *S*- and *N*-*methylations* are also considered Phase II reactions.

Biphenyl, ***diphenyl***: an aromatic hydrocarbon which crystallizes in colorless, shiny leaflets or monoclinic prismatic plates; m.p. 71 °C, b.p. 255.9 °C, n_D^{77} 1.5880. The leaflets can sublime. B. is insoluble in water but can be steam distilled. It is soluble in ethanol, ether and benzene. B. is present in coal tar, and is isolated from it.

It is synthesized industrially by thermal dehydrogenation of benzene around 800 °C. It can be made by a Wurtz-Fittig synthesis (see) from bromobenzene and sodium, or, more conveniently, by an Ullmann reaction (see) of iodobenzene with copper powder. B. is used as a heating bath liquid because of its high thermal stability. It is also used to preserve citrus fruits, and is an important starting material for organic intermediate products, such as benzidine for the dye industry.

Biphenyl

2,2'-Bipyridine: same as 2,2'-Dipyridyl.

Bird repellents: substances used to drive off birds (optical and acoustic methods are also used), and thus to prevent loss of seeds or fruits. Because the sense of smell is poorly developed in birds, seeds or other parts of plants are treated with anthraquinone. This causes vomiting in crows and other birds. To drive off flocks of starlings, sparrows, crows or other birds, bait is impregnated with 4-aminopyridine or 4-nitropyridine *N*-oxide.

Birkeland-Eyde process: see Nitrogen oxides.

Birch oil: a yellowish to brown essential oil with a pleasant, balsamic odor. It partially solidifies when cold, due to precipitation of paraffins. B. is obtained by steam distillation of various parts of the white birch. Bud oil comprises 3.5 to 8% of the buds, leaf oil makes up 0.04 to 0.1% of the leaves, and bark oil makes up about 0.05% of the bark. The main component of the bud oil is the sesquiterpene α-betulenol (α-betulol). B. is used in cosmetics, especially in hair tonics.

Bisacodyl: a compound made by condensation of pyridine-2-aldehyde (pyridine-2-carbaldehyde) with phenol in the presence of conc. sulfuric acid and subsequent acetylation of the hydroxy groups. It is used as a laxative.

Bisbenzenechromium, ***dibenzenechromium***: $Cr(C_6H_6)_2$, black crystals; M_r 208.22, m.p. 285 °C. B. is an important example of a metal aromatic complex. It is obtained by reaction of chromium(III) chloride, aluminum and aluminum chloride in benzene, and reduction of the resulting bisbenzenechromium(I) ion, $[Cr(C_6H_6)_2]^+$ with dithionite.

2,3;4,5-Bis-(butylene)tetrahydrofurfural: an Insect repellent (see).

Bischler-Napieralski synthesis: a synthesis of isoquinoline derivatives described in 1893. The molecule is formed by intramolecular cyclization of *N*-acylated 2-phenylethylamines in the presence of phosphorus oxygen chloride or zinc chloride at high temperature:

NH, O, R ⟶ ($POCl_3$), $-H_2O$ ⟶ N, R ⟶ (S), −2H ⟶ N, R

The primary 3,4-dihydroisoquinoline derivative can be catalytically dehydrogenated to the isoquinoline system.

Bischolinesuccinate dichloride: see Succinates.

Bis(2-chloroethyl) ether: Cl-CH_2-CH_2-O-CH_2-CH_2-Cl, a colorless liquid with an odor like chloroform; m.p. - 24.5 °C, b.p. 178 °C, n_D^{20} 1.4575. B. is insoluble in water, but is soluble in chloroform, acetone, dioxane and hydrocarbons. It is a strong respiratory poison and is carcinogenic. It is obtained from diethylene glycol and hydrogen chloride. It is used as a solvent for ethylcellulose, resins, fats, cleaners, textile-treating compounds and as a fumigant for pests.

Bis(2-chloroethyl)sulfide, ***2,2'-dichlorodiethylsulfide***: Cl-CH_2-CH_2-S-CH_2-CH_2-Cl, a very toxic liquid, which in pure form is nearly odorless; m.p. 13-14 °C, b.p. 217 °C, n_D^{20} 1.5312. The technical product contains colloidal sulfur as an impurity, and has a mustard-like or garlic-like odor. B. is not very soluble in water, but is readily soluble in the usual organic solvents. It is made from ethene and sulfur dichloride, SCl_2, or also from disulfur dichloride, S_2Cl_2. Another possibility is the reaction of ethylene chlorohydrin with sodium sulfide and further treatment with hydrogen halide.

B. is a strong cell poison, and is still considered the most dangerous skin poison. It penetrates textiles and leather, and acts on the skin, at first without visible irritation. Only after several hours does the skin redden, and then forms painful blisters which do not heal easily. B. is also a metabolic poison (liver damage) and causes mental impairment. Detoxication consists of oxidation to the sulfoxides or sulfones with chloramine or calcium chloride.

In the First World War, B. was used as a Chemical weapon (see).

Bis(chlormethyl) ether, ***1,1'-dichlorodimethyl ether***: $ClCH_2$-O-CH_2Cl, a colorless liquid with a suffocating smell; m.p. - 41.5 °C, b.p. 104 °C, n_D^{21} 1.4350. It is unstable and reacts with atmospheric moisture or water to form hydrogen chloride and formaldehyde. B. is very irritating to the eyes and respiratory passages, and is a highly carcinogenic compound. It is made by reaction of dimethyl ether with chlorine. It is used in the production of ion exchangers and as an alkylating reagent.

4,4'-Bis(dimethylamino)benzophenone: same as Michler's ketone (see).

Bis(2-hydroxyethyl)amine: same as diethanolamine; see Ethanolamines.

3,3-Bis(4-hydroxyphenyl)phthalide: same as Phenolphthalein (see).

Bismark brown R: same as Vesuvine (see).

Bismuth, symbol ***Bi***: an element of Group Va of the periodic system, the Nitrogen-phosphorus group (see). It is a heavy metal, $Z = 83$, a single isotope, atomic mass 208.9804, valences III and V, Mohs hardness 2.5, density 9.80, m.p. 271.3 °C, b.p. 1560±5 °C, electrical conductivity 25 Sm mm^{-2} at 20 °C, standard electrode potential 0.32 V ($Bi + H_2O \rightleftharpoons BiO^+ + 2 H^+ + 3 e$).

Properties. B. is a reddish-white, shiny, brittle metal which crystallizes in a rhombohedral lattice. It is isomorphic with gray metallic arsenic and gray antimony. Bi melts with a contraction of volume, and boils to form a vapor consisting of Bi_2 molecules. Bi is stable in dry air, and in moist air it is oxidized on the surface. At red heat, it burns to form yellow bismuth(III) oxide, Bi_2O_3. Its position in the electrochemical potential series is such that it is not attacked by non-oxidizing acids. In concentrated nitric acid or hot conc. sulfuric acid, it is dissolved to form Bi^{3+} salts, e.g. $Bi + 6 HNO_3 \rightarrow 2 Bi(NO_3)_3 + 3 NO + 3 H_2O$. Bi reacts with halogens to form the bismuth(III) halides BiS_3 (X = F, Cl, Br, I), and with sulfur when hot to form bismuth(III) sulfide, Bi_2S_3. It forms alloys with many metals, and some of these have very low melting points. The preferred oxidation state of Bi is +3. Bismuth(V) compounds are strong oxidizing agents. Unlike the other elements in Group Va, Bi can exist as the Bi^{3+} cation.

Analysis. In classical qualitative analysis schemes, Bi is an element of the H_2S group. It is separated as the black-brown bismuth sulfide, Bi_2S_3, and is detected as the colorless, flocculent bismuth hydroxide, $Bi(OH)_3$ or as insoluble, black bismuth iodide, BiI_3, which dissolves in excess potassium iodide to form the yellow potassium tetraiodobismuthate, $K[BiI_4]$. Quantitative determination depends on complexometry, or atomic absorption spectroscopy for low concentrations.

Occurrence. Bi is a very rare element, making up 10^{-5}% of the earth's crust. It is occasionally found in the elemental state, but it is more often found as the sulfide, bismuthinite, Bi_2S_3, or as its oxide weathering product, bismuth ochre, Bi_2O_3.

Extraction. The sulfide ores are usually enriched by flotation, then processed by the *roasting reduction process*. This consists of a roasting step followed by reduction with carbon. The metal is also produced by reduction with iron, $Bi_2S_3 + 3 Fe \rightarrow 2 Bi + 3 FeS$. The byproducts of smelting other metals, such as copper and lead, are important sources of Bi.

Applications. Most of the Bi is made into alloys, chiefly of lead and tin. Because of their low melting temperatures, these are used as soft solder, heating bath liquids, fuses, automatic fire alarms, etc. (see Wood's metal, Rose's meal, Lipowitz metal). Large amounts of Bi are used in pharmaceuticals; in the form of bismuth oxygen nitrate, carbonate, salicylate, etc., the element is present in astringents, antiseptics and remedies for intestinal illnesses.

Historical. The discovery of Bi is ascribed to Saxon miners. The metal was mentioned by T. Paracelsus (1493-1541), and had already been thoroughly discussed by G. Agricola (1490-1555).

Bismuth(V) acid: see Bismuthates.

Bismuthates: the salts of the hypothetical bismuth(V) acid; they have the compositions M^IBiO_3 and $M^I_3BiO_4$. The alkali bismuthates are made by reaction of a slurry of bismuth(III) hydroxide in alkali hydroxide solution with strong oxidizing agents, such as chlorine, peroxodisulfates or potassium permanganate. The compound formerly known as bismuthic acid, bismuth(V) oxide hydrate, $Bi_2O_5 \cdot xH_2O$, is obtained by acidifying an aqueous alkali bismuthate solution with nitric acid. The B. are extremely strong oxidizing agents.

Bismuth(III) chloride, ***bismuth trichloride***: $BiCl_3$, colorless, deliquescent crystals; M_r 315.34, density 4.75, m.p. 230-232 °C, b.p. 447 °C. B. dissolves in hydrochloric acid, and when the solution is highly dilute, it is hydrolysed to bismuth oxide chloride, BiOCl. It reacts with many metal chlorides to form chlorobismuthates(III), $M^I[BiCl_4]$, or $M^I_2[BiCl_5]$. B.

is obtained by dissolving bismuth in aqua regia, or bismuth oxide in hydrochloric acid and fractionating the mixture.

Bismuth electrode: an indicator electrode for electrochemical measurement of pH. Its construction, function and applications correspond to those of the Antimony electrode (see).

Bismuth hydride: BiH_3, colorless gas; M_r 212.00, b.p. 22 °C. B. is thermally labile, decomposing into its components even at room temperature. It is formed, along with much hydrogen, by the reaction of acids with a magnesium-bismuth alloy.

Bismuth(III) hydroxide: $Bi(OH)_3$, a colorless, amorphous powder which is barely soluble in water; M_r 260.00, density 4.36. Above 100 °C, B. is dehydrated to BiOOH, and > 400 °C, it is converted to bismuth(III) oxide, Bi_2O_3. It dissolves in acids to give the corresponding bismuth(III) salts. B. is produced by precipitation from Bi^{3+} salt solutions to which alkali hydroxides have been added.

Bismuth(III) nitrate: $Bi(NO_3)_5 \cdot 5H_2O$, colorless, hygroscopic, columnar, triclinic crystals; M_r 485.07, density 2.83. When heated, it is dehydrated, but nitric acid is also split off and the product is bismuth oxygen nitrate, $BiONO_3$. B. is made by dissolving bismuth in nitric acid. It is used in the production of many other bismuth derivatives, and as an astringent and antiseptic in medicine.

Bismuth oxides. ***Bismuth(III) oxide, bismuth trioxide***, Bi_2O_3, is a yellow, monoclinic crystalline powder which turns brown when heated. M_r 465.96, density 8.9, m.p. 825±3 °C. There is a tetragonal modification and probably also another cubic modification. Bi_2O_3 is insoluble in water and hydroxide solutions, but it dissolves in acids to form the corresponding bismuth(III) salts. It is obtained by combustion of bismuth or by heating of bismuth(III) hydroxide, carbonate or nitrate. Bi_2O_3 is used in the production of optical glasses with high indices of refraction, and is also the starting material for production of other bismuth compounds.

The dark red or dark brown ***bismuth(V) oxide, bismuth pentoxide***, Bi_2O_5, M_r 497.96, density 5.10, is extremely unstable and has not yet been accurately characterized. It decomposes at 100 °C, releasing oxygen and forming Bi_2O_3. It can be made, for example, by oxidation of Bi_2O_3 in a potassium chlorate melt.

Bismuth(III) sulfide: Bi_2S_3, brown-black, amorphous powder or rhombic crystals; M_r 514.14, density 7.39, dec. 685 °C. B. is barely soluble in water and dilute acids. It is made by fusion of the elements, or by precipitation from Bi^{3+} salt solutions with hydrogen sulfide. It precipitates first in the amorphous form, which gradually converts to the crystalline form. This is also found in nature as bismuthinite.

Bisphenol A, ***2,2-bis(4-hydroxyphenyl)propane***: a condensation product of phenol with acetone. B. forms colorless crystals; m.p. 156-157 °C. It is only slightly soluble in cold water, but is soluble in alkali hydroxide solutions and various organic solvents. Industrial production is coupled to the cumene-phenol process, from which both starting materials are obtained. B. is used in large amounts for the synthesis of thermoplastics, for example, its reaction with phosgene yields organic polycarbonates, which have thermoplastic properties. B. can also be cleaved into phenol and 4-isopropylphenol; it is used as an antioxidant, an inhibitor of ageing of natural and synthetic rubbers, and as a fungicide.

Bistability: see Oscillating reaction.

Bister: see Manganese oxide hydrates.

Bisulfite adducts: addition products of aldehydes or ketones with sodium hydrogensulfite, for example, $R\text{-}CHO + NaHSO_3 \rightleftharpoons R\text{-}CH(OH)SO_3^- Na^+$. B. are relatively insoluble compounds, and are often used to separate aldehydes and ketones from reaction mixtures. The carbonyl compounds can be retrieved from the B. with dilute acids or bases.

Bitertanol: see Azole fungicides.

Bitter almond oil: same as Benzaldehyde (see).

Bitter orange oil: a yellowish, bitter-tasting essential oil. It consists chiefly of terpenes (over 90%), mainly (+)-limonene, and also contains esters, such as linalyl, neryl, geranyl and citronellyl acetates, small amounts of alcohols (linalool, terpineol), phenols and free acids (formic, acetic and cinnamic). B. is obtained from the skins of bitter oranges by centrifugation or pressing; the yield is 0.3 to 0.4%. More oil can be extracted from the pressed peels by steam distillation, but it is less valuable than the pressed or centrifuged B. B. is used in perfumes, liquors and flavorings.

Bitumen: all naturally occuring liquid or solid hydrocarbon mixtures, or mixtures derived from them without decomposition. They are combustible, brownish yellow to black, highly complex mixtures of asphalts, resins, high-molecular-weight acids, solid paraffins, ceresins, ester waxes, hydrocarbon oils, kerogens, etc. Tars and pitches, which are obtained by destructive distillation, are not B., but are related substances.

1) The ***natural B.*** are products of microbiological decomposition and geological alteration of carbohydrates, proteins, resins, lignin, waxes and fats. They are subdivided as follows: a) mostly soluble in carbon disulfide and saponifiable natural B., such as sapropel wax, montane wax and fossil resins, such as amber; b) mostly soluble in carbon disulfide and unsaponifiable natural B., such as natural gas (however, natural gas is not considered a B. in all countries), paraffinic, paraffinic-asphaltic and asphaltic crude petroleum, petroleum residues, earth wax and the CS_2-soluble portions of asphaltite, natural asphalt and asphalt rocks; c) largely insoluble in carbon disulfide and nonsaponifiable natural B., also called ***pyrobitumen***, such as wurzilite, elaterite, albertite, impsonite, and the B. of kerogenic rocks, such as oil shale B., etc. The organic components of pyrobitumen do not melt and yield a distillate only at 400 to 600 °C.

2) ***Technical B.*** are products obtained from natural B. without decomposition. In the broad sense, these include the petroleum distillates, solid paraffins like ceresin and petrolatum, and in the narrow sense, the residues of gentle petroleum distillation, such as vaseline. In some countries, the residues of refining with acids and the products of treatment with selective solvents are included.

A ***bituminous substance***, e.g. oil shale, is a material impregnated with B., pitch and/or tar.

B. are used, depending on their properties and sources, as glues for floorings, floorings and roof coverings, waterproofing, and in mixtures with asbestos as ***bitumen mortar***. This is resistant to acids and does not swell. *Bitumen pulp* is used to impregnate, and combined with paint components (drying oils, chlorinated latex, resins) as a physically drying ***bitumen paint*** for insulation of buildings and metals.

Bitumen emulsions are made by stirring hot, liquid B. into water with about 1% emulsifying agent (soap, resin, clay) and are used in road construction, or are mixed with fillers such as asbestos or sawdust to make insulation and waterproofing materials. Vaseline is used as a lubricant and a base for salves, while ceresin, petrolatum, ozocerite, bleached montane wax, etc., are used in the production of waxes and polishes.

Bituminous coal: see Coal.

Bituminous shale: see Coal.

Biuret: $H_2N\text{-}CO\text{-}NH\text{-}CO\text{-}NH_2$, the amide of allophanic acid, which cannot be isolated under normal conditions. B. forms colorless, hygroscopic crystals, m.p. 192-193 °C. It is readily soluble in water and alcohol, but only slightly soluble in ether. When heated above the melting point, it decomposes to cyanuric acid and ammonia. In alkaline solution, B. and copper(II) salts form a violet complex, which can be used for qualitative and quantitative colorimetric determination of B. (see Biuret reaction). B. is formed by slow heating of urea, via isocyanic acid as an intermediate:

$$H_2N\text{–}CO\text{–}NH_2 \xrightarrow[-NH_3]{\Delta} O{=}CNH \xrightarrow{+O=C(NH_2)_2} H_2N\text{–}CO\text{–}NH\text{–}CO\text{–}NH_2$$

It is used to produce formaldehyde-biuret polymers and as a feed additive.

Biuret reaction: a method for determination of compounds with two or more peptide bonds. Biuret-positive substances form red-violet to blue-violet complexes with Cu^{2+} salts in alkaline solutions. The absorption due to the complex can be measured between 540 and 560 nm.

Bixin: an orange-red carotinoid pigment, m.p. 198 °C. B. is the monomethyl ester of norbixin, a C_{24} dicarboxylic acid with nine conjugated double bonds. B. is slightly soluble in cold glacial acetic acid, alcohol and ether; it is readily soluble in boiling glacial acetic acid or pyridine. B. is extracted from the red seed coats of the plant *Bixa orellana* (see Orlean) and is used as a food coloring.

Bjerrum's base-antibase system: see Acid-base concepts.

Bk: symbol for berkelium.

Black powder: an explosive mixture of potassium nitrate, charcoal and sulfur in an average ratio of 75:15:10. B. is the oldest known explosive. It was discovered around the 12th century in China and became known in Europe as the discovery of Berthold Schwarz. B. was originally used as gunpowder, and later also as an explosive for other purposes. It has now been almost completely replaced by Smokeless powder (see) and other modern Explosives (see), because it has a low reaction rate (it burns at about 400 to 500 m s^{-1}) and therefore low power. B. is extremely sensitive to sparks and fire. Coarse B. is used as a low-power blasting powder for obtaining valuable stones in quarries, and fine B. is used to fill ignition cords. B. is used in somewhat larger amounts in fireworks, with or without addition of coloring salts or metal powders. An analogous mixture based on sodium nitrate is called Saltpeter blasting powder (see).

Blaise-Guerin degradation: a method described in 1929 for the degradation of α-hydroxycarboxylic acids to aldehydes by loss of carbon monoxide and water around 200 °C:

$$R\text{–}CH_2\text{–}CH(OH)\text{–}COOH \xrightarrow[-CO,\,-H_2O]{} R\text{–}CH_2\text{–}CHO.$$

The B. is thus a means of converting aliphatic carboxylic acids to aldehydes with one CH_2 group fewer, via the acyl chloride and α-bromocarboxyl chloride.

Blanc fixe: see Barium sulfate.

Blanc reaction: a method described in 1923 for chloromethylation of aromatic hydrocarbons by formaldehyde and hydrogen chloride in the presence of aluminum chloride, zinc chloride, sulfuric acid or phosphoric acid:

$$R\text{–}H + CH_2O + HCl \xrightarrow{(AlCl_3)} R\text{–}CH_2\text{–}Cl + H_2O.$$

Monochloromethyl ether can be used instead of formaldehyde. In many cases, chloroethyl or chloropropyl groups can be introduced by using acetaldehyde or propionaldehyde. Hydrogen bromide or hydrogen iodide can be used instead of hydrogen chloride to produce the corresponding halogen derivatives.

Blast burner: a metal gas burner usually built on the principle of a Daniell valve. The actual burner is mounted on a heavy foot and can be rotated in any direction by means of a ball joint. It is suitable for generating temperatures up to about 2500 °C. The gas is burned with pressurized air or oxygen from a steel bottle. Temperatures of 3100 to 3300 °C can be obtained by burning hydrogen with oxygen; in this case the G. is called an ***oxyhydrogen burner***. The G. is used to melt high-melting glasses, metals, alloys and quartz.

Blast furnace process: see Steel, see Iron, crude.

Bleach activators: reactive organic *N*-acetyl or *O*-acetyl compounds which transfer the acetyl group to the bleach in the wash water, and thus improve its activity at 95 °C. At 60 °C and lower, the bleaching action does not occur at all in their absence. Some examples of B. are *N,N,N',N'*-tetraacetylethylene diamine (***TAED***) and *N,N'*-dimethyl-*N,N'*-diacetylurea.

Bleaches: substances used to remove stains or colored compounds from textiles or hard surfaces. Some examples of ***oxidative B.*** are sodium hypochlorite, sodium perborate, sodium percarbonate and percarbamide. Chlorine B. are not allowed as all-purpose laundry products in many European countries. ***Reductive B.*** are also used; sodium hydrogensulfite, for example, is used as a decolorizer in preparation of textiles.

Bleomycin: a complex misture of structurally related glycopeptide antibiotics used therapeutically as antineoplastics. They act by forming adducts to DNA.

Blood alcohol concentration: the ethanol concentration in the blood, measured in mg/g or pro mille, is the main criterion for determining nervous impairment after consumption of alcoholic beverages, for example, in judicial determination of guilt of driving under the influence of alcohol. The B. is usually determined first by a screening test (see Breath alcohol test), and if this is positive, by a Blood alcohol test (see).

Blood alcohol test: This test is usually undertaken after screening by the Breath alcohol test (see); venous blood is usually tested as follows: 1) evaporation of the ethanol and oxidation in the vapor phase over the blood sample, e.g. with chromium(VI) or vanadium(V) compounds. In the Widmark method, an excess of potassium dichromate is added, and the excess is back-titrated iodometrically. The amount of dichromate consumed corresponds to the ethanol content of the blood sample. In modern tests, the change of color of the reagent is determined photometrically. 2) Gas chromatography can also be used to determine the amount of ethanol over the blood sample (head-space technique), or the blood can be injected directly into a sample-preparation column attached to the gas chromatography column. 3) The ethanol is oxidized using alcohol dehydrogenase (ADH) and the amount of NAD^+ reduced in the reaction is measured photometrically.

Because the results of the tests may have serious consequences, (e.g. charges of drunk driving), most countries require the use of two independent analytical methods and agreement of several samples. Quality controls must also be regularly applied.

It often happens that several hours pass between the incident giving rise to suspicion and withdrawal of the blood sample, which means that the concentration of alcohol in the blood sample is lower than it was at the time of the incident. It is known that the degradation of ethanol in the organism is nearly independent of external conditions, and is a zero-order process. This permits extrapolation from the measured concentration to the probable blood concentration at the time of the incident.

Blood group substances: the terminal oligosaccharide residues of glycolipids or glycoproteins found on the surfaces of erythrocytes. They act as antigens and are responsible for blood group specificity. The most thoroughly investigated are the ***ABH antigens***, which are the basis for the classical ABO or Landsteiner blood group system. The terminal sugar groups can be detected with lectins. Several oligosaccharide groups active as B. can be present on a single protein.

The main types listed in the table can be subdivided (A^a to A^d, B_{Ia}, B_{II}, H_1, H_2, H_3) into groups on the basis of the structure further in from the terminal sugar group. The major part of blood group activity is due to oligosaccharide groups on glycoproteins; the sugars are usually linked to the protein via an arginine residue. In addition, glycolipids with long saccharide chains (polyglycosylceramides) contribute to the blood group activity. Short-chain glycolipids apparently make the smallest contribution to the activity. The bond between the galactose and N-acetylglucosamine groups can be either $\beta(1{\rightarrow}3)$- or, more commonly, $\beta(1,4)$-. In addition to the B. on erythrocyte membranes, secreted B. are present in the blood plasma; these include the ***Le substances*** (after the blood donor Lewis, in whose blood the group was discovered). Here again, subtypes are known. The Le substances display an $\alpha(1{\rightarrow}4)$-linked fucose on the GlcNAc residue shown in the table. The distribution of B. in different human populations varies.

Relationships among the ABH blood group substances (antigens) and the ABO blood group system.

Oligosaccharide protein or lipid

Blood group substance	R	Antibodies	Blood group
A(little H)	GalNAc	Anti-B	A
B (little H)	Gal	Anti-A	B
H (little A and B)		Anti-A and anti-B	0
A and B (little H)	Gal and GalNAc	neither anti-A nor anti-B	AB

Fuc = fucose; Gal = galactose; GlcNAc = N-acetylglucosamine.

Blood sugar: see D-Glucose.

Blossom pigments, ***flower pigments***: pigments responsible for the colors of blossoms. Yellow and orange blossoms usually contain Carotenoids (see), while red and blue colors are due chiefly to Anthocyans (see). The latter are present in the blossom bound to polysaccharides and as metal complexes (blue) or oxonium salts (red).

Blow-down system: a safety system in oil refineries for emergency removal of gas and liquid products from the entire system. The products are emptied into the B., which consists of a blow-down tank, a gas tank, a slop tank for liquids and a water tank (Fig.). The liquids and gases are separated in this B.

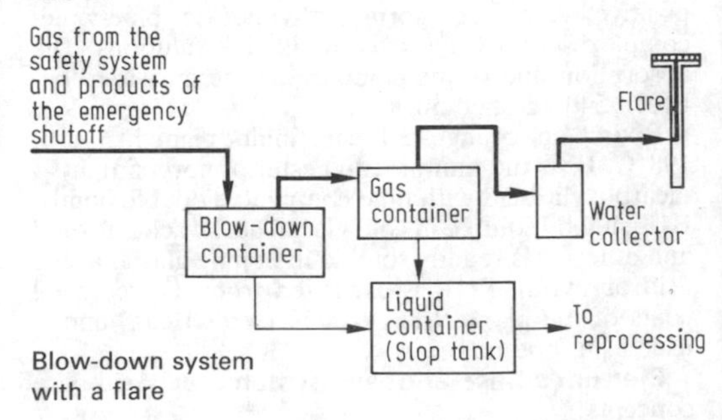

Blow-down system with a flare

The gases can be burned off in the flare. B. are arranged so that within a few minutes all combustible products are removed to a safe distance from the installation and can be "foamed" if necessary.

Blowpipe: a metal tube with a mouthpiece on one end. The other end has a 90° bend and ends in a jet. By blowing air into gas or alcohol flames with a B., one obtains very hot and pointed flames; depending on the position of the B., the flame can act reducing or oxidizing. The B. is used for Blowpipe analysis (see).

Blowpipe analysis: a very old method of qualitative analysis of ores and minerals. Today the B. is used mainly as a preliminary test for qualitative inorganic analysis. The sample is heated on charcoal using a Blowpipe (see), either by itself or with alkali metal salts, until it melts. Metals are formed by reduction, and oxidation forms deposits on the charcoal. The colors and appearance of the products give clues to the composition of the sample.

Blue gel: see Silica gels.

Blue water gas, ***blue gas***: see Water gas.

Bluing, ***new blue, permanent blue***: blue pigments added in small amounts to wash or rinse water to prevent yellowing of the laundry by iron oxide present in the water (see Optical brightener). Since blue is the complementary color to yellow, the combination gives a pure white. The pigments can be either inorganic (ultramarine, prussian blue) or organic (indigocarmine).

Boat form: see Stereoisomerism 2.2.

Bodenstein principle: same as Stationarity principle (see).

Boghead coal: see Coal.

Bohn-Schmidt reaction: a method for production of polyhydroxyanthraquinones from simple hydroxyanthraquinones by reaction of fuming sulfuric acid and boric acid in the presence of a small amount of mercury or selenium:

$\xrightarrow{(H_2SO_4/H_3BO_3)}$

The boric acid retards the oxidation process and promotes further hydroxylation.

Bohr model of the atom: see Atom, models of.

Bohr magneton: unit of magnetic moment. $\mu_B = eh/4\pi m_e = 9.274078 \cdot 10^{-24} JT^{-1}$, where e is the unit of electric charge, h is Planck's constant and m_e is the mass of the electron. The B. is the smallest magnetic moment possible for the electron in the hydrogen atom. All magnetic moments of atoms and molecules can be obtained by vectorial addition of integral multiples of the B.

Boiler coal: see Coal.

Boiler feedwater: same as Feedwater (see).

Boiler scale: a hard crust which precipitates out of hard water when it is heated and coats the inside of a boiler or other vessel. B. consists of the relatively insoluble salts of alkaline earth metals, e.g. calcium carbonate, $CaCO_3$, calcium sulfate, $CaSO_4$, calcium silicates, magnesium silicates, magnesium hydroxide, $Mg(OH)_2$, and magnesium carbonate, $MgCO_3$, and the oxides of other metals, e.g. iron(II) oxide, Fe_2O_3. B. deposited on the heating surfaces of boilers interferes with heat conduction; a far more dangerous consequence, however, is local overheating in steam systems which can lead to explosions. B. can be prevented by softening the water or adding phosphates to it. The phosphates form a loose sludge by reacting with the calcium and magnesium carbonate; this cannot be deposited as a hard scale on the insides of the pipes. Disodium phosphate, Na_2HPO_4, trisodium phosphate, Na_3PO_4, sodium hexametaphosphate, $(NaPO_3)_6$ and sodium polyphosphate are used for this purpose. It is not necessary to add a stoichiometric amount of phosphate. As a rule, a phosphate concentration of 1-2 g m^3 P_2O_5 is maintained in the boiler water.

Boiling: the transition of a liquid into the gas state, when the vapor pressure of the liquid is equal to the external pressure. Pure substances boil isothermally at their Boiling points (see), while mixtures boil over a temperature range. Since the vapor pressure is equal to the external pressure, vapor bubbles can form in the interior of the boiling liquid and pass into the gas phase. Below the boiling point, however, an evaporative equilibrium is established only at the surface of the liquid. If the gas phase is continuously removed, one speaks of an evaporation of the liquid; it occurs much more slowly than boiling.

The isothermal B. process requires a constant input of the heat of vaporization (see Phase-change heats), and its rate can thus be controlled by regulation of the flow of heat. Liquids which are free of dust and gas in carefully cleaned vessels can be heated (cautiously!) above their boiling points without beginning to boil (***boiling delay***). If the lack of nucleation centers for bubbles is corrected by impact or addition of a solid, evaporation occurs instantly and continues until the temperature is brought back to the boiling point. To avoid boiling delays, ***boiling stones*** are added to a liquid, or gas bubbles are introduced through ***boiling capillaries***.

Boiling diagram: a phase diagram in which the boiling point of a mixture at a fixed external pressure (e.g. atmospheric pressure) is plotted against the composition of the liquid phase. B. are closely related to Vapor pressure diagrams (see). In both types of diagram, the composition is plotted on the abcissa as mole fraction or mass fraction. Like vapor pressure diagrams, B. consist of two curves: the ***boiling curve*** gives the boiling point as a function of the composition of the liquid phase, and the ***condensation curve***, the equilibrium temperature of the gas phase as a function of composition. A difference between the two types of curve arises from the fact that vapor pressure and boiling point are inversely related; thus a mixture which shows a maximum in its vapor pressure curve will have a minimum in its B. at that composition, and vice versa. B. are the basis for separation of substances by distillation. Fig. 1-3 show the basic types of B. for binary systems. The two components A and B are miscible in all proportions (for analogous examples, see Vapor pressure diagram, Fig.). For example, in Fig. 1 a liquid mixture with the composition $s_B = x_1$ boils at T_1. The vapor is richer in component B, because it has the lower boiling point; its composition $x_B = x_2$ can be read from the condensation curve at the same temperature T_1. If this vapor is condensed, the resulting liquid with the composition x_2 boils at T_2, and produces a vapor which is even more enriched in B. Repetition of this process, e.g. in the individual plates of a distillation column, finally yields pure component B. At the same time, the starting mixture becomes depleted in B.

If the B. has a *minimum* (Fig. 2), the composition of the solution after repeated evaporation and condensation approaches that of the minimum, and it finally distills as a constant-boiling mixture with this composition (*azeotrope*). Some examples are the water/ethanol and water/dioxane systems. In a B. with a *maximum* (Fig. 3), the same process first yields a pure distilled component. The other component is enriched in the starting mixture, so that the composition of this mixture approaches that of the boiling maximum. Finally a mixture with the constant composition of the maximum boils. Some examples are the mixtures water/hydrogen chloride and chloroform/acetone.

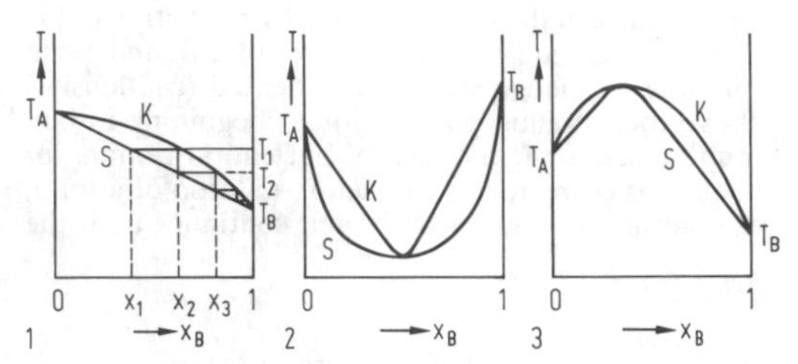

Types of boiling diagrams

B. for mixtures of more than two components, like the corresponding vapor pressure diagrams, cannot be represented in a plane. Liquids which are not completely miscible require a separate treatment (see Mixing gap).

Boiling point, ***boiling temperature***, abb. ***b.p.***: the temperature at which a liquid is converted to the gas phase (see Boiling); this is a highly pressure-dependent parameter. The B. is the same as the ***condensation point***, that is, the temperature at which a gas at the same pressure is converted to the liquid. Pure substances have constant B. The B. of a mixture depends on its composition, which usually changes during the process. Thus the B. of a mixture is not constant (see Boiling diagram). Solutions of a substance with a negligibly small vapor pressure have a higher B. than the pure solvent (see ***Boiling point elevation***). The B. is therefore an important criterion of purity.

The B. of a pure substance depends on the pressure. The B. at various pressures can be read off a Vapor pressure curve (see), or calculated using the Clausius-Clapeyron equation (see). B. are generally reported at standard pressure (101.3 kPa); when this convention is not followed, the pressure is given as a subscript to the abbreviation "b.p." In the older literature, the pressure was given in Torr (1 Torr = 133.3 Pa); thus "b.p.$_{20\ \mathrm{Torr}}$ 70.6 °C" means that the substance has a B. of 343.75 K (70.6 °C) at a pressure of 2.66 kPa (20 Torr). Many substances decompose before boiling at atmospheric pressure. In such cases, it is often possible to determine a B. at reduced pressure with the aid of a vacuum pump.

Boiling point elevation: the elevation of the boiling point of a solution compared to that of the pure solvent when the solute has a negligible vapor pressure. Raoult's second law (see Raoult's law) applies to ideal dilute solutions: $\Delta T = T - T_{0,1} = E_b \cdot m_2$, where $E_b = RT^2_{0,1}M_1/\Delta_pH$. Here T and $T_{0,1}$ are the boiling points of the solution and pure solvent, respectively, m_2 is the molality (see Composition parameters) of the solute, and E_b is the ***ebullioscopic constant***, which has a definite value for each solvent. This depends on the boiling point $T_{0,1}$, the molar mass M_1 and the molar enthalpy of vaporization Δ_pH (see Phase change heat) of the solvent. R is the gas constant.

The B. is often used to determine the molar masses of non-dissociating, non-volatile compounds (***ebullioscopy***). A given mass m_2 of the compound of interest is dissolved in a mass m_1 of the solvent, and the B., ΔT, is determined. Since the molality m_2 is defined as the amount n_2 of substance 2 which is dissolved in 1 kg of the solvent 1, it follows that $m_2 = n_2/m_1 = m_2/M_2 \cdot m_1 = \Delta T/E_b$, and the molar mass M_2 of the solute can be obtained from the equation $M_2 = (m_2/m_1) \cdot E_b/\Delta T$.

The B., like the freezing point lowering, is a colligative property, i.e. the size of the effect depends on the number of particles persent. In solutions of electrolytes, therefore, the increase in particle number resulting from electrolytic dissociation must be taken into account. In this case, Raoult's second law is $\Delta T = E_b m_2[1 + (\nu - 1)\alpha]$, where α is the extent of dissociation and ν is the number of ions into which the electrolyte dissociates. Complete dissociation of a binary electrolyte ($\nu = 2$) such as sodium chloride, thus produces a B. twice as great as a non-dissociating compound.

The above equations apply to ideal, dilute solutions. Significant deviations occur for non-electrolytes at concentrations above 0.1 molar, and for electrolytes at concentrations above 10^{-3} molar. These can be compensated by introduction of activity coefficients (see Activity).

Boiling temperature: same as Boiling point (see).

Bolathene®: see Synthetic fibers.

Boltzmann collision equation: see Kinetics of reactions (theory).

Boltzmann constant: symbol k_B or k, an important constant in statistical thermodynamics which relates absolute temperatures to energies: $k_B = R/N_A = (1.38054 \pm 0.00009) \cdot 10^{-23}$ JK^{-1}, where R is the general Gas constant (see) and N_A is Avogadro's number. The mean thermal energy per degree of freedom of a single particle is 1/2 k_BT.

Boltzmann distribution: see Maxwell-Boltzmann distribution.

Boltzmann factor: the numerical factor $\exp(-\varepsilon/k_BT)$, where ε is the energy, k_B is the Boltzmann constant and T is the absolute temperature. The B. is found in the equations of equilibrium statistics, such as the Maxwell-Boltzmann velocity distribution, the state sum, the Arrhenius equation, etc. It gives the fraction of particles with energies greater than or equal to ε at thermal equilibrium. For systems with discrete energy levels ε_i, $\exp(-\varepsilon_i/kT)$ is proportional to the probability of occupation, N_i/N, of the ith energy level; N_i is the number of particles in the ith level, and N is the total number of particles.

Bomb solubilization: see Solubilization.

Bomb tube, ***fusion tube***: a thick-walled tube made

of special glass and sealed on one end. After filling, the other end is melted shut. The B. is used for analytical solubilization of organic compounds (see Solubilization) or for reactions of small amounts of material to be carried out under pressure and at high temperatures in a closed space.

Bombesins: a group of peptides from amphibian skins which have effects similar to kinin. The compounds in this group have structures similar to that of bombesin from the skin of the European frogs *Bombina bombina* and *Bombina variegata*. Bombesin causes the smooth muscles to contract, reduces blood pressure and stimulates the kidneys and stomach. The thermoregulating activity of bombesin is also important; application of about 1 ng in the *cisterna cerebri* leads to a drop in the rectal temperature of 5° within 15 minutes. The effect lasts about 2 h.

Glu-Gln-Arg-Leu-Gly-Asn-Gln-Trp-Ala-Val-Gly-His-Leu-Met-NH_2
Bombesin

Glu-Gly-Arg-Leu-Gly-Thr-Gln-Trp-Ala-Val-Gly-His-Leu-Met-NH_2
Alytensin

Glu-Val-Pro-Gln-Trp-Ala-Val-Gly-His-Phe-Met-NH_2
Ranatensin

Glu-Gln-Trp-Ala-Val-Gly-His-Phe-Met-NH_2
Litorin

Alytensin from the skin of the obstetrical toad *Alytes obstetricans* differs from bombesin in only two amino acids; ***ranatensin*** (from the skin of *Rana pipiens*) and ***litorin*** (from the skin of *Litoria (Hyla) aurea*) are shorter sequences.

Bond: see Chemical bond, Molecular orbital theory.

Bond angle tension: see Baeyer's tension theory.

Bond dipole moment: see dipole moment.

Bond energy, ***mean bond energy***: in general, the energy E_B(X-Y) of the X-Y bond used to calculate the atomization energy E_A of a molecule (see Additivity principle). It is usually an empirically determined value. The *atomization energy* (total bond energy) is the amount of energy required to separate a molecule into its constituent atoms at 0 K. The values for E_B(X-Y) and E_A are usually given as molar energies. For molecules of the type AB_n, the mean B. is the atomization energy divided by *n*; this is also the *mean bond dissociation energy* E_D(A-B) of the A-B bond. It is not the same as the energy required to split an A-B bond in the molecule (see Dissociation energy). For example, the atomization energy of methane E_A = 1642 kJ mol^{-1}, so the mean bond dissociation energy E_D(C-H) = $E_A(CH_4)/4$ = 410.5 kJ mol^{-1}. The cleavage of the methane molecule into a methyl radical and a hydrogen atom requires the dissociation energy $E_D(H_3C\text{-}H)$ = 422.8 kJ mol^{-1}. In order to approximate the total bond energies of more complex molecules, an empirical system (incremental system) of energies E_B(X-Y) has been established; these are generally called B. The B. for some of the bonds which occur frequently in organic molecules are listed in the table.

Mean bond energy in kJ mol^{-1}

Bond	Energy	Bond	Energy
C—H	415	C—S	260
N—H	390	C—N	295
O—H	465	C≡N	780
S—H	340	N—N	160
C—C	350	N=N	420
C=C	620	C—F	440
C≡C	815	C—Cl	330
C—O	355	C—Br	275
C=O	715	C—I	240

Bond order: see Molecular orbital theory; see Hückel method.

Bonding orbital: see Molecular orbital theory.

Bone fat: a fat which is usually soft or liquid at room temperature; depending on the age and quality of the bones it can range in color from nearly white to dark brown. It is sometimes nearly odorless, but often has a bad smell. Its iodine number is 40 to 55, saponification number 190 to 200, acid number 40 to 80. B. usually consists of about 97% fatty components and at most 3% water. It can contain up to 50% free fatty acids. B. consists mainly of stearic, oleic, linoleic and linolenic acids. It is readily soluble in nearly all known organic solvents. B. is obtained as a byproduct of gelatin or glue production from bones, by solvent extraction. The yield of fat is 9 to 12%. B. can be purified by oxidation with nitric acid. It is used to produce soaps and candles, and as a lubricant for clocks and fine machines.

Bone meal: a product made of ground or grated fresh bones. Because it contains a large proportion of nitrogen compounds (collagen), fat, phosphorus, calcium, etc., B. is a valuable mineral additive to feeds. However, to avoid disease, it must be used immediately after it is prepared. B. can also be used to prepare compost and as a phosphate fertilizer for light soils.

Boosters: same as Foam stabilizers (see).

Boranates, ***hydridoborates***: complex metal boron hydrides. The boranate anions arise formally from boranes by uptake of hydride ions or elimination of protons. The B. based on tetrahedral BH_4^- anions are especially important; they are formed by reaction of diborane with ionic hydrides, e.g. $2\,NaH + B_2H_6 \rightarrow 2\,NaBH_4$. Most of the higher boranate anions can be described by the formula $B_nH_n^{2-}$. An interesting representative of this class is *closo*-dodecahydridododecaboranate(2 -), $B_{12}H_{12}^{2-}$, which is formed as its sodium salt from the reaction of sodium boranate with diborane in the presence of triethylamine: $2\,NaBH_4 + 5\,B_2H_6 \rightarrow Na_2B_{12}H_{12} + 13\,H_2$. The extremely stable $B_{12}H_{12}^{2-}$ anion has an icosahedral structure. The B_{12} cage is a resonance hybrid in which the boron atoms are linked by two- and three-center two-electron bonds. The 12 hydrogen atoms are bound to the boron atoms by normal covalent bonds.

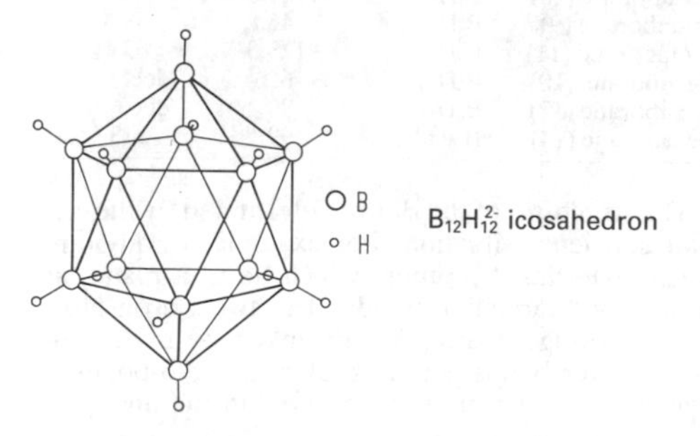

$B_{12}H_{12}^{2-}$ icosahedron

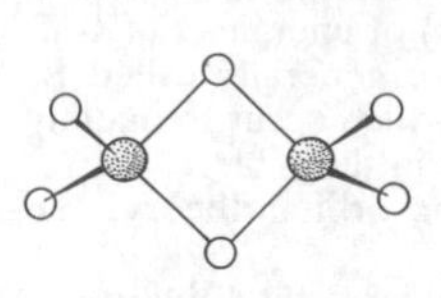

Diborane B_2H_6

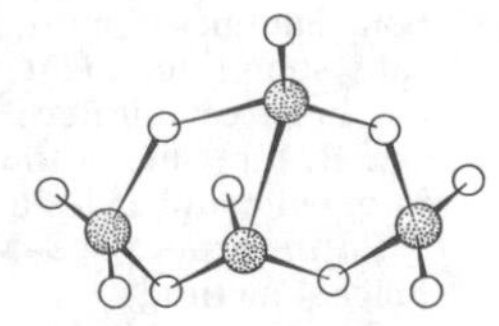

Tetraborane (10) B_4H_{10}

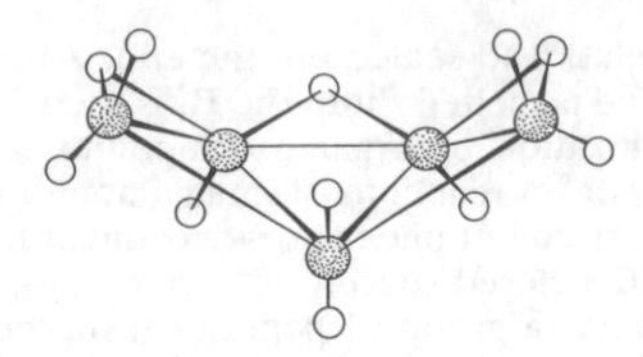

Pentaborane (11) B_5H_{11}

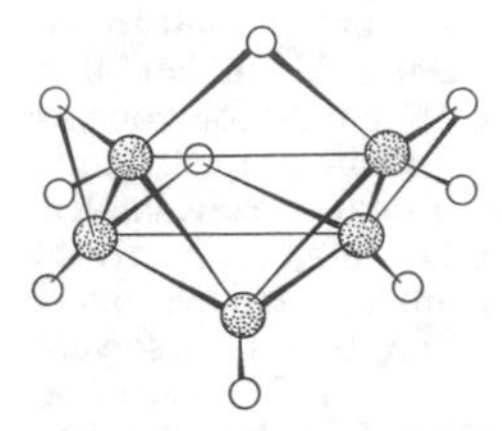

Pentaborane (9) B_5H_9

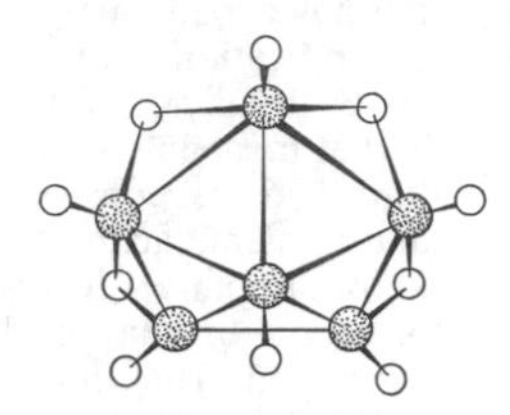

Hexaborane (10) B_6H_{10}

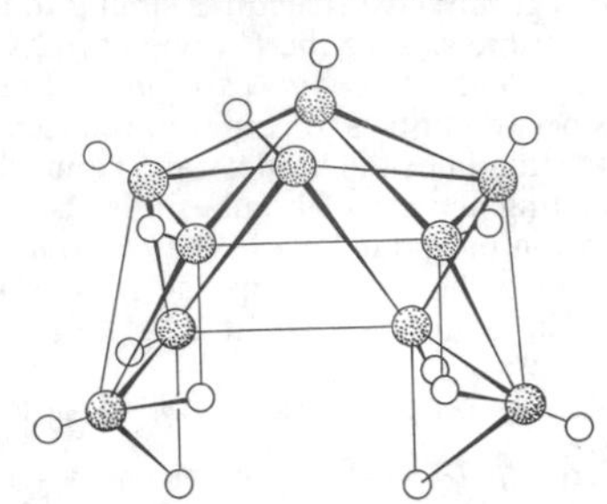

Decaborane (14) $B_{10}H_{14}$

Structures of some boranes.

Boranes, ***boron hydrides***: compounds of boron and hydrogen, of which about 25 examples are known. Their composition is reflected either by B_nH_{n+4} or B_nH_{n+6}. The names of these compounds are formed from a Greek prefix indicating the number of boron atoms, the root "borane" and a number in parentheses, which indicates the number of hydrogen atoms.

The lower members of these series are colorless gases or volatile liquids with unpleasant, nauseating odors. The higher B. are solid (Table). The simplest B. is diborane, B_2H_6; the compound BH_3 does not exist. Diborane is synthesized from boron trichloride and lithium alanate: $4\ BCl_3 + 3\ AlLiH_4 \rightarrow 2\ B_2H_6 + 3\ LiAlCl_4$, or from boron trifluoride and sodium boranate: $4\ BF_3 + 3\ NaBH_4 \rightarrow 3\ NaBF_4 + B_2H_6$. The higher B. are usually obtained by thermal cleavage of diborane at 150 to 250 °C; the composition of the reaction mixture depends mainly on the temperature and the hydrogen pressure. The reaction of acids with magnesium boride produces a mixture of boranes consisting chiefly of tetraborane(10).

Physical properties of some boranes

Name	Formula	m.p. in °C	b.p. in °C
Diborane (6)	B_2H_6	−165.5	−92.5
Tetraborane (10)	B_4H_{10}	−120.8	16
Pentaborane (9)	B_5H_9	− 46.8	58.4
Pentaborane (11)	B_5H_{11}	−123.3	63
Hexaborane (10)	B_6H_{10}	− 65	dec.
Hexaborane (12)	B_6H_{12}	− 90	dec.
Decaborane (14)	$B_{10}H_{14}$	99.5	213

The structures of the B. are determined by the electron deficiency situation. For example, in diborane, the four terminal protons, which lie in a plane, are bound by "normal" σ-bonds. The two boron atoms and the bridge hydrogens are linked by two B-H-B three-center bonds perpendicular to the σ-bonds. A similar bonding pattern is observed in the higher B., in which B-B-B three-center bonds are also present. The special structures of the B. are often indicated in their names by the prefixes "*nido*" (nest) or "*closo*" (closed), as in *nido*-decaborane(14) or *closo*-dodecahydrodecaboranate(2 -) (see Boranates).

The electron deficiency in the B. is also responsible for their reactivity. For example, in interactions with electron pair donors, such as tertiary amines, the three-center bonds are opened to form normal covalent bonds, e.g.

$$H_2B(\mu\text{-}H)_2BH_2 + 2\,NR_3 \rightarrow 2\,H_3B^{-}\text{-}N^{+}R_3.$$

The reaction with ammonia has a similar course, with formation of Borazine (see) as an intermediate. The reaction of diborane with water also follows this pattern, with subsequent elimination of hydrogen to give horic acid: $B_2H_6 + 6\ H_2O \rightarrow B_2O_3 + 3\ H_2O$. The reactions with ionic hydrides of the alkali or alkaline earth metals, or with various organometals to form metal boranates can also be understood on this basis (see Boranates).

The lower B. are pyrophoric in air, and burn to boron trioxide and water, e.g. $B_2H_6 + 3\ O_2 \rightarrow B_2O_3 + 3\ H_2O$. The B. are interconverted by thermolysis. Addition of diborane to C=C double and C≡C triple bond systems is used in the synthesis of organoboron compounds (see Hydroboronation).

The B. have played an important part in the development of an understanding of the structure of electron deficiency compounds. They are used practically as rocket fuels, as starting materials for production of boranates, and in hydroboronation.

Borates: 1) the salts of boric acids. The B. are based on anions with quite different types of structure. Salts of the monomeric orthoboric acid with the planar BO_3^{3-} anion are rare (e.g. scandium orthoborate, $ScBO_3$). Metaborates, with the composition

$[M^IBO_3]_n$, are much more common; these are derived from cyclic or linear polymeric boric acids. Anhydrous sodium metaborate, $Na_3B_3O_6$, for example, is based on a cyclic trimeric boric acid (1), while calcium metaborate, $Ca(BO_2)_2$, has a linear polyborate anion (2).

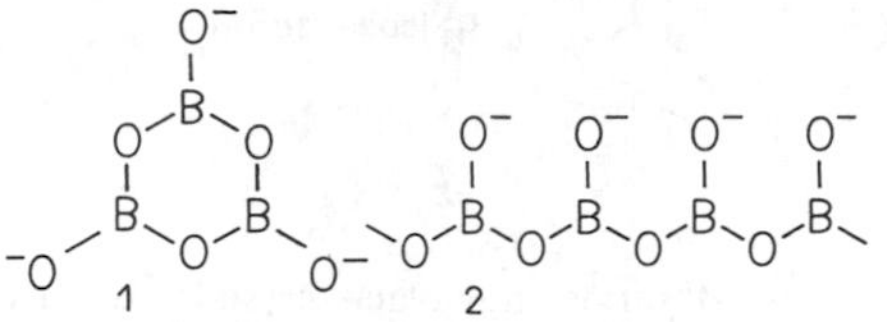

Metaborates

These compounds are made by fusing metal oxides with boron trioxide. The compounds containing more water, which are often considered hydrated B., are based on cyclic or linear polyborate anions built up of planar BO_3 or tetrahedral BO_4 units. Their stoichiometry usually does not indicate their structure. The best known example of this class of compounds is Borax (see). These B. are obtained by dissolving metal hydroxides in aqueous boric acid and evaporation of the solutions. B. are used as components of fertilizers and in making enamel and perborates.

2) The *esters* of boric acid, for example, Trimethylborate (see).

Borax, ***sodium tetraborate decahydrate***: $Na_2B_4O_7 \cdot 10H_2O$ or $Na_2[B_4O_5(OH)_4] \cdot 8H_2O$, colorless monoclinic crystals which weather in air; M_r 381.37, density 1.37. When heated rapidly, B. melts at 75 °C; it releases 8 molecules of crystal water at this temperature, and is converted at 320 °C to the anhydrous tetraborate, m.p. 741 °C. This dehydration process agrees well with the structure of B. Above 60 °C, a cubic or hexagonal pentahydrate, $Na_2B_4O_7 \cdot 5H_2O$ crystallizes from aqueous solution (***jeweler's borax***).

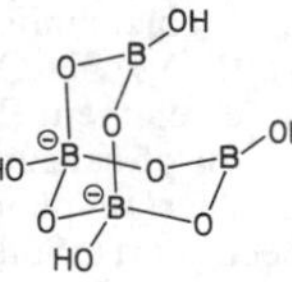

Anion of $Na_2[B_4O_5(OH)_4] \cdot 8H_2O$

B. is found as a mineral with the same name; it is also called tincal. Many salt lakes contain dissolved B. The compound is usually made technically by treatment of kernite with water under pressure, or by reaction of boric acid with soda. B. is used to make low-melting glazes for ceramic products, for enamels, in the production of temperature-resistant and optical glasses, as an additive for fertilizers and as a flux for hard soldering. In the laboratory, B. is used to make buffer solutions and for borax beads (see Microcosmic salt beads).

Borax beads: see Microcosmic salt beads.

Borazine: $B_3N_3H_6$, a colorless liquid with an aromatic smell; M_r 80.50, density 0.8614, m.p. - 58 °C, b.p. 55 °C. B. is a planar, cyclic conjugated 6π-electron system isosteric with benzene. Many of its physical properties are very close to those of benzene, and for this reason, it is often called ***inorganic benzene***. These analogies are particularly close for the derivatives of the two rings as well. However, the chemical behaviors are quite different. Because of the high bond polarities, the borazine system readily adds water, alcohols, hydrogen chloride, etc. In water it is hydrolysed to boric acid and ammonia. B. is formed by heating the diborane-ammonia adduct: $3\,(B_2H_6)(NH_3)_2 \rightarrow 2\,B_3N_3H_6 + 12\,H_2$. However, it is more conveniently made by the reaction of lithium boranate with ammonium chloride: $3\,NH_4Cl + 3\,LiBH_4 \rightarrow B_3N_3H_6 + 3\,LiCl + 9\,H_2$. N- and B-substituted products can be obtained by varying the starting materials.

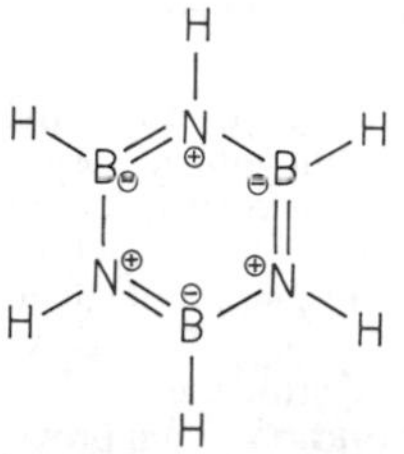

Bordeaux B: see Fast red.

Boric acid. ***Orthoboric acid***: H_3BO_3, colorless, scaly, water-soluble triclinic crystals, M_r 61.83, density 1.435, m.p. on rapid heating, 169 °C. However, B. dehydrates starting at 70 °C to ***metaboric acid*** $[HBO_3]_n$, which is converted around 300 °C to boroxide, B_2O_3. Metaboric acid is based on cyclic oligomeric or linear polymeric structures (see Borates).

Orthoboric acid is steam volatile. Its aqueous solution is weakly acidic. This is not the result of a simple dissociation of the B., but the result of a Lewis acid-base interaction with water: $B(OH)_3 + 2\,H_2O \rightarrow [B(OH)_4]^- + H_3O^+$. Orthoboric acid is found in free form in nature, for example, in various steam springs (fumaroles) and as ***sassolin*** in the boron springs in Toscana, Italy. The compound is produced industrially, usually by reaction of borax with sulfuric acid: $Na_2B_4O_7 + H_2SO_4 + 5\,H_2O \rightarrow 4\,B(OH)_3 + Na_2SO_4$.

B. is used chiefly in making glass, ceramics and enamels, in candles to stiffen the wicks, to preserve hides and glues, to impregnate wood and textiles for fire resistance, and as a boron fertilizer.

Borides: binary metal-boron compounds. In the ***high-boron B.***, with the compositions MB_4, MB_6 and MB_{12}, the metal atoms occupy free lattice places or holes in the lattice corresponding to linked boron polyhedra. They are obtained from the elements and are chemically and thermally very stable. Some are very hard, and some have semiconductor properties. High-boron B. are used as abrasives and in electronics. ***Low-boron B.*** have the formulae MB or MB_2, and especially those in which M is a transition metal have interesting materials properties. They are very hard, chemically and thermally very stable, and often display electrical conductivity. They are used in electronics.

Bornane: see Monoterpenes.

Borneol: a secondary alcohol derived from the bicyclic monoterpene bornane. There are 4 optically active forms: d-B. (also ***Borneo camphor***) and l-B. (also ***Ngai camphor***), and d- and l-isoborneol. B. forms colorless crystals with a camphor-like odor and a tendency to sublime. It is slightly soluble in water, and readily soluble in alcohol and ether; m.p. 203-210 °C (d-, l-, dl-B.) or 212 °C (d-, l-, dl-isoborneol). d- and l-B. and their esters are components of many essential oils. d-B. and d-bornyl acetate are found in the oils of many *Labiataea* and *Zingiberaceae*. l-B. is found in the oils of *Pinaceae* and *Asteraceae*.

CH_3 OH H_3C–CH_3 Borneol

CH_3 OH H_3C–CH_3 Isoborneol

l-B. can be obtained by saponification of l-bornyl acetate, which is present in large amounts in spruce needle oil. Most technical B. is obtained by hydration of pinene. Synthetic B. usually contain isoborneol. B. is used in the perfume indstry and is an intermediate in camphor synthesis.

Born-Haber cycle: see Haber-Born cycle.

Born-Oppenheimer approximation: an approximation going back to Born and Oppenheimer (1927): in the quantum mechanical treatment of atoms and molecules, electron motions are considered independent of the motions of the nuclei. This is based on the notion that the electrons move much faster than the nuclei, because they are far lighter, and their states instantaneously adapt to the changed nuclear position as the nucleus moves. It reduces the problem of quantum mechanical calculation for a molecule to the solution of the Schrœdinger equation for stationary states of the electrons in the fields of motionless atomic nuclei. If the nuclei are assumed to be at rest, the complete Hamiltonian operator $\hat{H}$ for a diatomic molecule can be split into a term for the kinetic energy of the nuclei and the electronic Hamiltonian $\hat{H}_R$. This operator and its eigenvalues E_R and eigenfunctions Ψ_R depend on the internuclear distance R, which plays the role of a parameter in the solution of the equation $\hat{H}_R\Psi_R = E_R\Psi_R$. The eigenvalues E_R are the energy levels of the molecule for a given internuclear distance. The eigenfunctions Ψ_R represent the electronic part of the total wave function. E_R is also called the electronic energy, and for a diatomic molecule, it can be represented in the form of a potential curve which is a function of R. The B. is an extremely good approximation in most cases, and is often the only possible way to make theoretical calculations for a molecular system.

Borohydrides: same as Boranes (see).

Boron, symbol ***B***: an element of Group IIIa of the periodic system, the Boron-aluminum group (see); a semimetal, Z 5, with natural isotopes of mass 10 (19.78%) and 11 (80.22%), atomic mass 10.811, valence III, standard electrode potential - 0.73 V ($B + 3\,H_2O \rightleftharpoons H_3BO_3 + 3\,H^+ + 3e$).

Properties. B exists in a tetragonal and two rhombic modifications. These crystalline forms contain B_{12} icosahedrons, and differ only with respect to the type of linkage between them. Both the bonding between the atoms within the B_{12} units and the bonding between units is best described as multi-center bonding.

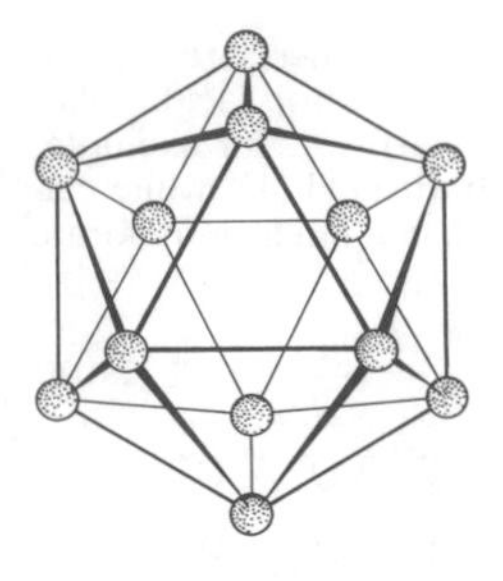

B_{12} icosahedron

Crystalline B forms gray-black crystals, density 2.34 to 2.37, m.p. about 2300 °C, b.p. 2550 °C, which are very hard (9.3 on the Mohs scale) and unreactive. There is also a brown, amorphous modification which has no odor or taste. This is very reactive; it ignites in air at 700 °C and burns with a reddish flame to boron trioxide, B_2O_3. Amorphous B forms boron trisulfide, B_2S_3, with sulfur. It reacts vigorously with fluorine, even at room temperature, to form boron trifluoride, BF_3; the other boron halides also form from the elements, but at higher temperatures. The reaction with nitrogen or ammonia around 900 °C produces boron nitride, BN. Alkaline fusion in the presence of air converts B to metal borates. At high temperatures, B is able to reduce many element oxides, sulfides or even halides to the corresponding elements, e.g. $3\,SiO_2 + 4\,B \rightarrow 3\,Si + 2\,B_2O_3$.

Analytical. The qualitative analysis of B is based on the green color which volatile boron compounds give to flames. After solubilization (often by a Na_2CO_3 fusion), boric acid is separated by distillation or extraction, and titrated in the presence of mannitol or another polyol.

Occurrence. B makes up about 0.005% of the earth's crust. It is found naturally as boric acid, e.g. as sassoline or dissolved in mineral springs, and in the form of its salts, the borates. Some important minerals are borax, $Na_2B_2O_9 \cdot 10H_2O$; kernite, $Na_2B_4O_4 \cdot 4H_2O$; borocalcite, $CaB_4O_7 \cdot 4H_2O$; boracite, 2 $Mg_3B_8O_{15} \cdot MgCl_2$; colemanite, $Ca_2B_6O_{11} \cdot 5H_2O$ and pandermite, $Ca_2B_6O_{11} \cdot 3H_2O$. B is found in soils and in plants as a trace element. It is involved in cell division and regulation of calcium metabolism in plants. A lack of B causes certain plant diseases, such as the decay of sugar beet hearts. Soils which are poor in B are therefore fertilized with the element in the form of mixed fertilizers.

Production. Crystalline B is produced by electrolysis of a molten mixture of boron trioxide, B_2O_3, potassium tetrafluoroborate, KBF_4 and potassium chloride, KCl, or by reduction of boron trichloride, BCl_3, with hydrogen in an electric arc. Very pure B is obtained from the crude material by the Vacuum deposition (see) method. Amorphous B is obtained by reaction of boron trioxide with sodium or magnesium.

Applications. B is used in metallurgy as a deoxidizing agent and a component of alloys. For example, the hardness of iron is significantly increased by even small additions of B. Because of its large neutron-capture cross section, the isotope ^{10}B is used as a neutron absorber in nuclear technology. "***Square B***", AlB_{12}, is obtained by reduction of boron trioxide with aluminum; it is used as an abrasive because of its great hardness.

Historical. Borax and boric acid have been known for a long time, but B (from the Arabic "bauraq" = "flux", a term for borax) was first obtained in 1808 by Gay-Lussac and Thénard by reduction of boron trioxide with potassium. A little later, Davy produced it by electrolysis. In 1909, Weintraub first obtained crystalline B by melting amorphous B in vacuum.

Boron-aluminum group, ***Boron group***: 3rd representative group of the periodic system, including the non-metallic boron and the metallic elements aluminum (Al), gallium (Ga), indium (In) and thallium (Tl). These metallic elements are also called ***earth metals*** because of the properties of their oxides. The elements of the B. have electron configuration ns^2p^1, and can achieve a stable noble-gas configuration by loss of three valence electrons and formation of the E^{3+} cations (E = element). As can be seen from the table, the ionization potentials for boron are so high, because of its very small radius, that no compounds containing the B^{3+} ion are known. Instead, boron forms covalent bonds. For the heavier elements, the E^{3+} ion exists in the oxides (Al_2O_3), fluorides (AlF_3) and hydrated ions (e.g. $[Al(H_2O)_6]^{3+}$).

In the progression from lower to higher atomic weights in the B., the monovalent E^+ state becomes increasingly important, and is dominant in thallium; thallium(III) salts are strong oxidizing agents. This phenomenon cannot be explained by the values of the ionization potentials, because the sum of the 2nd and 3rd ionization potentials of gallium is higher than that of thallium. Instead, the explanation is to be found in the decreasing ability of the s-electron pair either to ionize or to take part in covalent bonds (*inert pair effect*). This, in turn, is due to an increasingly unfavorable balance between the energies required to form bonds (promotion, ionization, etc.) and those released by bonding (covalent bond or lattice energies, solvation, etc.). The irregularities in the spacing of the ionization potentials, electronegativities and atomic radii which can be seen in the Table are due to the positions of the d- and f-elements.

The formation of three covalent bonds provides the elements of the B. with only an electron sextet. This electron deficiency situation has a decisive influence on the structure and chemical properties of EX_3 compounds. The element hydrides are characterized by electron-deficiency structures with 3-center, 2-electron bonds. In the planar boron halides, the electron deficiency of the central atom is compensated by intense $p_\pi p_\pi$ interactions with the ligands (BF_3), while the halides of the heavier elements achieve stable outer electron configurations by oligomerization, as found, for example, in dimeric Aluminum bromide (see). For the same reason, elements in this group have pronounced Lewis acidity. As might be expected, the electron pair acceptor capacity decreases markedly in the series $BX_3 > AlX_3 > GaX_3$.

The metallic character of the elements of the B. increases from top to bottom, and the acidity of the $E(OH)_3$ compounds decreases in the same direction. Although boric acid $B(OH)_3$ is decidedly acidic, aluminum orthohydroxide $Al(OH)_3$ is amphoteric, and the hydroxides of indium and gallium are basic.

Boron chemistry resembles aluminum chemistry less than it does silicon chemistry (diagonal relations in the Periodic system (see)). This is reflected, for example, by the acidity of boric H_3BO_3 and silicic H_4SiO_4 acids, the tendency of these acids to condense, the tendency of boroxide B_2O_3 and silicon dioxide SiO_2 to form glasses, and the hydrolysis behavior of the halides.

Boron carbide, B_4C, a black powder or shiny, black, rhombohedral crystal; M_R 55.29, density 2.52, m.p. 2350 °C, b.p. > 3500 °C. The hardness of B. is between that of silicon carbide and diamond. It is thermally and chemically extremely inert. It is synthesized by heating boron or boron oxide with carbon to 2500 °C. It is used as an abrasive and as a neutron absorber in nuclear reactors.

Boron chloride, ***boron trichloride***: BCl_3, a colorless liquid which fumes in air; M_r 117.71, density 1.349, m.p. - 107.3 °C, b.p. 12.5 °C. B. reacts with water to form boric acid and hydrochloric acid: $BCl_3 + 3H_2O \rightarrow H_3BO_3 + 3\,HCl$. The electron sextet on boron gives B. Lewis acidity. It reacts with many N-, P-, O- and S-nucleophiles, giving addition compounds. B. is produced by reaction of a mixture of boron oxide and coal with chlorine. It is used as a catalyst in Friedel-Crafts reactions, to remove carbides, nitrides and oxides from various alloys, and to synthesize many boron compounds.

Boron fluoride, ***boron trifluoride***: BF_3, a colorless gas with a suffocating odor; M_r 67.81, m.p. - 126.7 °C, b.p. - 99.9 °C. B. fumes in moist air due to hydrolysis. In the presence of water, the reaction goes to comple-

Properties of the elements of the 3rd representative group of the periodic system

	B	Al	Ga	In	Ti
Nuclear charge	5	13	31	49	81
Electron configuration	$[He]\,2s^2\,p^1$	$[Ne]\,3s^2\,p^1$	$[Ar]\,3d^{10}\,4s^2\,p^1$	$[Kr]\,4d^{10}\,5s^2\,p^1$	$[Xe]\,4f^{14}\,5d^{10}\,6s^2\,p^1$
Atomic mass	10.811	26.98154	69.72	114.82	204.37
Atomic radius [pm]	80	143	125	150	155
Ionic radius E^{3+} [pm]	20	39	47	80	88
Electronegativity	2.01	1.47	1.82	1.49	1.44
1st ionization potential [eV]	8.298	5.986	5.999	5.786	6.108
2nd ionization potential [eV]	25.154	18.828	20.51	18.869	20.428
3nd ionization potential [eV]	37.930	28.447	30.71	28.03	29.83
Standard electrode potential [V]	–0.73	–1.706	–0.560	–0.338	–0.3363*)
Density [g cm^{-3}]	2.34 ... 2.37	2.702	5.904	7.30	11.85
m.p. [°C]	ca. 2300	660.37	29.78	156.61	303.5
b.p. [°C]	2550	2467	2403	2080	1467 ± 10

*) for $Tl \rightleftharpoons Tl^+ + e$

tion: $BF_3 + 3 H_2O \rightarrow H_3BO_3 + 3 HF$. B. is a typical Lewis acid. For example, it reacts with HF to form tetrafluoroboric acid, HBF_4, and forms a boron trifluoride etherate, $F_3B\text{-}O(C_2H_5)_2$ (b.p. 126 °C) which is stable enough to be distilled. B. is synthesized by reaction of potassium borofluoride with boroxide and sulfuric acid. B. is used mainly as a catalyst in organic syntheses (e.g. in Friedel-Crafts reactions), for fluorinations and in the synthesis of other boron compounds. $^{10}BF_3$ is used to fill neutron counting tubes.

Boron group: same as Boron-aluminum group (see).

Boron nitride: BN, M_r 24.82, has three modifications: amorphous, cubic and hexagonal (density 2.25, subl. about 3000 °C). The colorless, hexagonal B. has a structure similar to that of graphite, with which it is isosteric; in this structure, the carbon atoms of graphite are replaced alternatingly by B and N atoms. It does not conduct electricity. Cubic B. has a diamond-like structure, in which B and N atoms alternate, and is nearly as hard as diamond. All three modifications are chemically and thermally extremely stable. They are synthesized either from the elements or by reaction of boron tribromide with ammonia. The hexagonal B. is formed first; it is converted at 1350 °C and 6000 MPa to the cubic form. Because of its great stability, B. is used as a cladding for the combustion chambers and jets of rockets, pumps, casting forms, etc. Hexagonal B. is used as a lubricant, and cubic B. as an abrasive. Boron nitride fibers have also been synthesized.

Boroxide, ***boron trioxide***: B_2O_3, colorless, glassy, hygroscopic mass or rhombic crystals; M_r 69.92, m.p. 450 °C, b.p. about 1860 °C. B. dissolves in water in a highly exothermal reaction, forming orthoboric acid: $B_2O_3 + 3 H_2O \rightarrow 2 H_3BO_3$. B. is formed by heating boric acid. It is used in the production of other boron compounds, e.g. boron carbide and boron fluoride, as a flux in metallurgy, and in the production of glass and enamel.

Bose-Einstein statistics: quantum statistics applicable to a system of bosons at equilibrium. ***Bosons*** are elementary particles with integral spins; they include photons and mesons. Their wavefunctions are symmetrical, that is, exchange of two bosons does not change the sign of the wavefunction of the system. Unlike the classical Maxwell-Boltzmann statistics (see Maxwell-Boltzmann distribution), quantum statistics treats individual particles as indistinguishable. The exchange of two bosons of the same type does not produce a different state. In contrast to Fermi-Dirac statistics (see) the number of particles which may occupy a given quantum state is unlimited. The ***distribution function*** in B. is $N_i = g_i/(e^{[E_i-\mu]/(k_BT}-1)$, where N_i is the mean number of particles i in the state with energy E_i, μ is the chemical potential of the whole system, k_B is the Boltzmann constant and T is the absolute temperature.

Bosons: see Bose-Einstein statistics.

Böttger's paper: see Alkannin.

Bottled gas: gases which are stored and transported in steel bottles. Gases which cannot be liquefied at normal temperature, e.g. oxygen, nitrogen, hydrogen and city gas, are condensed under pressure (up to 20 MPa) or, like ethylene, they are dissolved in a solvent under considerably lower pressure (up to 1.5 MPa). Gases which can be liquefied at normal temperatures (see Liquid gases) are stored at a pressure only slightly above that required to liquefy them.

The steel bottles in which gas is stored are slender cylinders of tough carbon steel which can withstand up to 800 MPa pressure. They are equipped with pressure valves through which the gas is removed for use.

Botulin toxin A: see Chemical weapons.

Boudouard equilibrium: an equilibrium among carbon, carbon monoxide and carbon dioxide: $C + CO_2 \rightleftharpoons 2 CO$, $\Delta H = +162.5$ kJ mol^{-1}. The position of the B. is shifted to the right with increasing temperature and decreasing pressure. When carbon is burned in insufficient oxygen, the product below 400 °C is therefore practically pure CO_2, and above 1000 °C, pure CO. The B. is important in all industrial processes in which carbon is incompletely burned (see Generator gas) or in which oxygen compounds are reduced with carbon (as in a blast furnace).

Bouguer-Lambert-Beer law: same as Lambert-Beer law (see).

Boundary layer photoeffect: see Photoeffect.

Bouveault-Blanc reduction: a method for conversion of aliphatic carboxylic acid esters to primary alcohols by reaction with sodium in ethanol:

$$R^1\text{–CO–OR}^3 \xrightarrow{+4H\ (Na,\ C_2H_5OH)} R^1\text{–CH}_2OH + R^2\text{–OH}$$

$$(CH_3)_2 = CH_2 + H_2O \xrightarrow{+H^+} H_3C\text{–}\overset{\displaystyle CH_3}{\underset{\displaystyle CH_3}{C}}\text{–OH.}$$

Isopropanol, butanol, cyclohexanol or methylcyclohexanols can be used instead of ethanol.

Boyle-Mariott law: see State, equation of, 1.1.

Br: symbol for bromine.

Bradykinin: see Plasmakinins.

Bragg's law: see X-ray structure analysis.

Brass: copper alloy with 18 to 50% zinc. Alloys which consist of homogeneous α-copper-zinc mixed crystals have the highest resistance to corrosion, which approaches that of copper. Heterogeneous alloys which contain β-copper-zinc mixed crystals in addition to α-copper-zinc mixed crystals are less resistant in a number of electrolytes. A unique corrosion form of B. is Zinc loss (see). In media containing ammonia, B. is subject to Stress-crack corrosion (see).

B. is made either as a cast or kneaded alloy. The most important kneaded B. contain 56, 58 and 65% copper. The last of these consists entirely of α-copper-zinc mixed crystals, and is used for deep-drawn sheets, condenser pipes, nuts and bolts. Brass sheets, strips and pipes are used in the electrical, chemical, optical, clock and metal-ware industries, in shipbuilding, musical instruments and fittings. Needles, sieves and springs are made from brass wire. If the zinc content is less than 18%, B. is reddish (***red brass, tombac***); if it is greater than 18%, the alloy is yellow to yellow-white (***yellow brass***). Muntz metal (see) is a B. with 40% zinc. B. which contain aluminum, manganese, nickel, silicon, tin and arsenic in addition to copper and zinc are called ***special B.***. The additions considerably improve the strength and resistance to corrosion of these alloys (see Delta metal®, Durana®). Aluminum and manganese improve the resistance to seawater. Special B. with the composition 61 to 66% copper, 2 to 5% manganese, 2.5 to 7.5% aluminum (***condenser B.***) is used to make condenser pipes for ships because of its resistance to seawater. ***Admirality B.***, made of 70% copper, 1% tin and 29% zinc, is not resistant to seawater.

Historical: B. was known around 1200 B.C., but was not widely used until the time of the Roman empire. Around 150 A.D., a center of B. production developed in the area of Aachen. In the 13th century, the Belgian city Dinant was the center of artistic B. production.

Bravais lattice: see Crystal.

Brazilin: $C_{16}H_{14}O_5$, a natural dye present in brazilwood. Pale yellow needles, m.p. 130 °C (dec.); soluble in water, alcohol and ether; turn orange in air. On oxidation, B. is converted to the actual dye, ***brazileine***. It is used to color inks, wood and textiles, as a histological stain and as an acid-base indicator.

Brazil-wood: the stem wood of *Caesalpinia echinata*. Like sappan wood from *Caesalpinia sappan*, it contains Brazilin (see), and aqueous extracts were formerly used to dye wool and print cotton. Today it is still used to dye leather.

Breakthrough reaction: see Electrode reaction.

Breath alcohol test: the concentration of ethanol in expired air is measured in pre-calibrated test tubes carried by police and amublances. It is assumed that the concentration of ethanol in the breath is proportional to the blood alcohol concentration. The test tubes usually contain a solution of potassium dichromate in silica gel soaked with concentrated sulfuric acid. A plastic bag connected to the tube is blown up by the subject; alcohol in the air reduces the potassium dichromate to green chromium(III) compounds, which are visible in the tube as a green band of color. The length of the colored zone is approximately proportional to the amount of alcohol in the air. The lower limit of detection is 0.3 pro mille, or 0.03% blood alcohol. If the result of this semiquantitative determination is positive, a Blood alcohol test (see) is required for confirmation; the "breatholyzer" serves only as a screening test.

Bremen blue: see Copper(II) hydroxide.

Brevinarcon®: see Barbitals.

Bridge carbon: see Nomenclature, sect. III.E.4.

Bridge ligand: see Coordination chemistry.

Bridgmen-Stockbarger process: see Crystal growing.

Brilliant black: various types of acidic, water-soluble secondary diazo dyes, such as brilliant black B, BL and BN. They are used to dye wool in an acid bath, and brilliant black BN is permitted as a food color. B. are synthesized by diazotization of 1-aminobenzene-4-sulfonic acid, coupling with 1-aminonaphthalene-7-sulfonic acid, renewed diazotization and coupling with the sodium salt of 1-acetyl-8-aminonaphthol-4,6-disulfonic acid.

Brilliant green, ***diamond green***: a basic triphenylmethane pigment which is soluble in water and alcohol. B. is made by condensation of benzaldehyde with diethylaniline and is used to color paper, straw, coconut fibers, etc., to make paints and as a skin disinfectant.

Brilliant indigo B: see Indigo dyes.

Brinell hardness: the hardness of a metal determined by a static sphere-impression experiment devised by the Swedish engineer Brinell. The test is usually conducting using a hardened steel sphere, 10 mm in diameter. It is pressed into the material being tested at a pressure of 29,420 N for 10 s; the surface area of the resulting impression is then measured. The test force in newtons, multiplied by the factor 0.102, and divided by the area in mm^2 of the impression, gives the B. in units of N/mm^2. If the tensile strength of the material is proportional to the B., it can be calculated from the B. For unalloyed, soft steels, a conversion factor of 0.36 is applied.

Britannia metal: see tin alloys.

Brockmann-Chen reaction: see Vitamin D.

Bromadiolone: a Rodenticide (see).

Bromates: the salts of bromic acid, $HBrO_3$, with the general formula $MBrO_3$. The bromate ion has a trigonal-pyramidal structure. Alkali and alkaline earth bromates are colorless, and with the exception of barium bromate, $Ba(BrO_3)_2$, they are readily soluble in water. B. are strong oxidizing agents; they detonate when heated with oxidizable substances. For example, ammonium bromate, Na_4BrO_3, tends to explode spontaneously even at room temperature.

> Bromates inhibit oxygen transport and also cause damage to the kidneys, spleen and liver, often fatal damage. The symptoms of poisoning are nausea, stomach pains and respiratory distress.
>
> As a countermeasure, the stomach must immediately be rinsed out with water, and a solution of Glauber's salt (sodium sulfate decahydrate) and a dose of sodium hydrogencarbonate should be administered.

B. are obtained by the reaction of bromine with warm alkali hydroxide solutions, or by electrolysis of alkali bromides.

Bromatometry: a method of redox analysis in which the bromate ion is used as titrator. This is a strong oxidizing agent in acidic solution, with a standard potential $E_0 = +1.42$ V. At the equivalence point of the titration, elemental bromine is formed by reaction of the bromate with the bromide ions. The very stable standard solutions are made from the titrimetric standard substance potassium bromate. The endpoint is usually recognized in B. with the aid of irreversible indicators, e.g. methyl orange or methyl red, which are destroyed by traces of elemental bromine and lose their color. More recently, reversible redox indicators, e.g. *p*-ethoxichrysoidin, have been discovered for B. The equivalence point can also be determined by electrochemical methods, e.g. potentiometry. B. is used mainly for determination of arsenic, antimony and tin, but can also be used for determination of phenols and other aromatic compounds.

Bromchlorophenol blue: see Bromphenol blue.

Bromcresol green: see Bromphenol blue.

Bromcresol purple: see Bromphenol blue.

Bromelain: an endopeptidase (see Proteases) isolated from pineapple fruits. It is used to cleave proteins and polypeptides into large fragments. B. is a thiol enzyme (M_r 33,000) with structure and activity similar to Papain (see). It is activated by cysteine and 2-mercaptoethanol, and inhibited by *N*-ethylmaleimide and monoiodoacetic acid.

Bromelia: same as 2-Ethoxynaphthalene (see).

Bromfemvinphos: see Organophosphate insecticides.

Bromic acid: $HBrO_3$, an oxidizing, strong acid which is stable only in aqueous solution. The salts of B. are called Bromates (see). Aqueous B. is made by oxidation of bromine with chlorine: $5\ Cl_2 + Br_2 + 6\ H_2O \rightarrow 2\ HBrO_3 + 10\ HCl$.

Bromides: compounds of bromine in which it is the more electronegative element. B. can be ionic, covalent or coordination-polymeric metal bromides, metalloid bromides and organic bromides such as alkyl or aryl bromides. Some B. occur in nature, e.g. silver bromide, AgBr (as bromite or bromargyrite). The B. are made, for example, by reaction of electropositive metals with hydrobromic acid, by reaction of metal oxides or hydroxides with hydrogen bromide, HBr, by bromination of elements with bromine liquid or vapor, or by bromination of hydrocarbons.

Bromination: see Halogenation.

Bromine, symbol ***Br***: a chemical element from group VIIa of the periodic system, the Halogen group (see). It is a nonmetal, *Z*35, with stable isotopes of masses 79 (50.537%) and 81 (49.463%), atomic mass 79.904, density 3.14, m.p. - 7.25°C, b.p. 58.78°C, valence - 1, +1, +3, +4, +5, +7; standard electrode potential (Br^-/Br_2) +1.0652 V.

Properties. Br is a dark brown, heavy liquid with a suffocating odor; it develops a corrosive, red-brown vapor. Aside from mercury, Br is the only element which is liquid at room temperature. In the solid state, Br has a faint metallic sheen, and its color lightens as the temperature drops. At 20K, it is orange. Br forms diatomic molecules. When heated, atomic Br forms; at 1500°C, about 30% of the molecules are dissociated. Br is only slightly soluble in water, but is more soluble than chlorine; 100 g H_2O at 20°C dissolves 3.55 g Br. The resulting solution, which is about 0.2 M, is called ***bromine water***; it is similar to chlorine water and decomposes in direct sunlight, forming oxygen and hydrogen bromide: $Br_2 + H_2O \rightarrow 2\ HBr + 1/2O_2$. Below 6.2°C, Br forms a hydrate with water: $Br_2 \cdot 8.5H_2O$. Br is relatively easily dissolved in solvents such as chloroform, carbon tetrachloride and carbon disulfide. Its chemistry is similar to that of chlorine, although Br is less reactive. It combines with most elements in either the liquid or vapor form, sometimes with the generation of flame. Although gold is readily oxidized by Br to gold(III) bromide, platinum is not attacked. Br readily undergoes substitution or addition reactions with organic compounds (see Halogenation).

Analytical. Br is recognized by its dark red-brown color and its odor. Inorganic bromides are oxidized to Br by heating them with conc. sulfuric acid; the Br_2 escapes as a red-brown vapor. Bromide ions are oxidized to Br_2 by chlorine water; they form a cheesy precipitate with silver ions. The slightly yellow AgBr is insoluble in nitric acid, but is dissolved by conc. ammonia, thiosulfate or cyanide solutions, in which it forms complexes. Br is quantitatively determined gravimetrically as silver bromide, or by argentometry.

Bromine causes deep, painful lesions in the skin. Bromine vapor attacks the eyes and causes oxidative burns in the respiratory tract.

The countermeasures are rest and fresh air, and inhalation of steam or sodium hydrogencarbonate vapor, and oxygen if necessary. Artificial respiration should be used only when absolutely necessary. The eyes must be rinsed with water. Liquid bromine which touches the skin must immediately be rinsed off with a rapid stream of water and then thoroughly washed with petroleum.

Habitual use of bromine preparations leads to a syndrome called bromism. The symptoms are a loss of ability to concentrate and sleeplessness, and progress to severe nerve and speech damage. Sodium chloride is an antidote; it replaces the bromide ions in the body.

Occurrence. Br makes up $7.8 \cdot 10^{-4}$% of the earth's crust. It is found only in compounds, nearly always in the form of bromides associated with chlorides. The ratio of Br to Cl in salt deposits is about 1:250. Bromine carnallite, $KBr \cdot MgBr_2 \cdot 6H_2O$ is an important mineral. Some rare minerals are bromargyrite (bromite), AgBr; embolite, Ag(Cl,Br); and iodoembolite, Ag(Cl, Br, I). Seawater contains about 0.0065% bromide, and large amounts of bromide (up to 1.5%) are present in the Dead Sea.

Production. The starting materials for the majority of the world production of Br are the end liquors from the potash industry or brines. The end liquors contain bromide salts, mainly bromine carnallite, which is enriched to 4%. This bromide-containing liquid trickles down through a tower, against a rising flow of chlorine, which oxidizes it to Br_2: $Cl_2 + 2\ Br^- \rightarrow 2\ Cl^- + Br_2$. The B. vapor is driven out, condensed, and purified by distillation over solid potassium bromide. In the USA, Br is obtained on an industrial scale from seawater. Here the bromide is oxidized with chlorine at pH 3.5, and the Br_2 is driven off by air. It is then converted back to bromide and bromate by sodium carbonate solution: $3\ Br_2 + 3\ Na_2CO_3 \rightarrow 5\ NaBr + NaBrO_3 + 3\ CO_2$. The element is then obtained by acidification with sulfuric acid, which leads to coproportionation: $5\ Br^- + BrO_3^- + 6\ H^+ \rightarrow 3\ Br_2 + 3\ H_2O$. In the laboratory, Br is obtained by the reaction of sulfuric acid, manganese dioxide and potassium bromide: $4\ HBr + MnO_2 \rightarrow MnBr_2 + 2\ H_2O + Br_2$.

Applications. About 90% of the Br produced industrially is used in the production of 1,2-dibromoethane, which in turn is added to fuels containing tetraethyllead. Br is also the starting material for production of numerous organic bromine compounds, which serve as sedatives, pesticides and pigments. The nuclide ^{82}Br, which is generated artificially in nuclear reactors, is a β-emitter with $t_{1/2}$ = 35.5 h; it is used as a radiolabel.

Historical. Balard discovered Br in 1826 in seawater and showed its relationship to chlorine and iodine.

The name is from the Greek "bromos" = "stink". Industrial production of Br began in the second half of the 19th century.

Bromine chloride: see Bromine halides, see interhalogen compounds.

Bromine fluoride: see Bromine halides, see Interhalogen compounds.

Bromine halides: very reactive Interhalogen compounds (see); BrX (X = Cl, F), BrF_3 and BrF_5 are formed by reaction of bromine with the lighter halogens. ***Bromine monofluoride***, BrF, is a very unstable, bright red gas; m.p. ≈ - 33 °C, b.p. ≈ +29 °C. ***Bromine monochloride***, BrCl, is a dark red-orange liquid; m.p. - 54 °C, b.p. (at 3.45 kPa) - 50 °C. ***Bromine trifluoride***, BrF_3, is a colorless liquid; m.p. +8.77 °C, b.p. 125.75 °C. ***Bromine pentafluoride***, BrF_5, is a colorless liquid; m.p. - 60.5 °C, b.p. +40.76 °C. BrF_5 is more reactive than BrF_3, which undergoes a certain amount of dissociation in the liquid phase: $2\ BrF_3 \rightleftharpoons BrF_2^+ + BrF_4^-$. ***Tetrafluorobromates*** can be made directly by reaction of BrF_3 with metal bromides, e.g. $BrF_3 + KF \rightarrow K[BrF_4]$.

Bromine oxides. ***Dibromine oxide***: Br_2O, a brown solid which is stable only at low temperature (up to - 40 °C), M_r 175.82. Even at the melting temperature, - 17.5 °C, it decomposes rapidly to bromine and oxygen. Br_2O dissolves in carbon tetrachloride in monomolecular form, giving a moss-green color. It reacts with sodium hydroxide to form hypobromite: $Br_2O + 2\ NaOH \rightarrow 2\ NaOBr + H_2O$.

Bromine dioxide, BrO_2, is an egg-yolk yellow solid which is stable only at low temperatures; M_r 111.91. It is decomposed on rapid heating to room temperature. Br_2O is obtained by the action of a glow discharge on a mixture of bromine and oxygen: $Br_2 + 2\ O_2 \rightarrow 2\ BrO_2$, or by ozonizing a solution of bromine in trichlorofluoromethane, $CFCl_3$ at - 50 °C. Bromine dioxide disproportionates in hydroxide solutions: $6\ BrO_2 + 6\ OH^- \rightarrow 5\ BrO_3^- + 3\ H_2O$.

Bromisovalum: see Sedatives.

Bromoacetic acid: $Br\text{-}CH_2\text{-}COOH$, a colorless, crystalline compound; m.p. 50 °C, b.p. 208 °C. B. is soluble in water, alcohol, ether, acetone and benzene. It is highly irritating to the skin. B. has the reactions typical of Carboxylic acids (see); the bromine atom can easily undergo nucleophilic substitution. B. is produced by bromination of acetic acid or acetic anhydride, or by the reaction of hydrogen bromide with chloroacetic acid. B. is used in organic syntheses.

Bromoacetone, ***bromopropanone***: $Br\text{-}CH_2\text{-}CO\text{-}CH_3$, a colorless to light brown liquid with a pungent odor; m.p. - 36.5 °C, b.p. 136.5 °C at $9.64 \cdot 10^4$ Pa, n_D^{15} 1.4697. B. is barely soluble in water, but dissolves readily in alcohol, ether and acetone. In pure state, B. can be kept for a long time, but in the presence of air, and especially light, it quickly turns dark brown. The vapor of B. is highly lacrimatory. B. is synthesized by bromination of acetone in glacial acetic acid. It is used as a chemical weapon in the form of grenades or as a spray; magnesium oxide is used as a stabilizer. B. is also used for many syntheses.

Bromoazide: see Halogen azides.

Bromobenzene, ***phenyl bromide***: $C_6H_5\text{-}Br$, a colorless liquid with an aromatic odor; m.p. - 30.8 °C, b.p. 156 °C, n_D^{20} 1.5597. B. is insoluble in water, but dissolves in alcohol, ether and benzene. Its reactions are typical of the Halogen arenes (see). B. can be made by bromination of benzene in the presence of iron(III) bromide, or by the Sandmeyer reaction (see). It is used as an intermediate for numerous organic syntheses, and as a solvent.

α-Bromobenzylcyanide, ***α-bromophenylacetonitrile***: $C_6H_5\text{-}CHBr\text{-}CN$, colorless, poisonous crystals; m.p. 29.5 °C (23.5 and 25.5 °C have also been reported), b.p. 132-134 °C at 1.6 kPa. α-B. is barely soluble in water, but more readily soluble in alcohol. When heated to 160 or 170 °C, it decomposes to hydrogen bromide and diphenylfumaric dinitrile. α-B. is one of the most aggressive eye irritants known. It can be made by the reaction of bromine and benzylcyanide at 105-110 °C; it is used for organic syntheses and as a chemical weapon.

Bromochloromethane: CH_2BrCl, a colorless liquid with a chloroform-like odor; m.p. - 86.5 °C, b.p. 68.1 °C, n_D^{20} 1.4838. B. is insoluble in water, but dissolves readily in most organic solvents. It is made, along with bromomethane, by the reaction of hydrogen bromide and chloromethane in the presence of aluminum chloride. B. is used as a fire-extinguishing liquid.

1-Bromo-1-chloro-2,2,2-trifluoroethane: same as Halothane (see).

Bromocriptin: see Ergot alkaloids.

Bromoethane: same as Ethyl bromide (see).

Bromoform, ***tribromomethane***: $CHBr_3$, a colorless, poisonous liquid with a sweetish odor; m.p. 8.3 °C, b.p. 149.5 °C, n_D^{20} 1.5976. B. is barely soluble in water, but is readily soluble in most organic solvents. It is synthesized by the Haloform reaction (see), or by reaction of chloroform with aluminum bromide. B. was formerly used as a medicament for whooping cough and as a sedative. It is now used as a sedimentation medium for ore separation.

Bromohexine: a bromine-containing aromatic amino compound which is contained in cough remedies in the form of its hydrochloride.

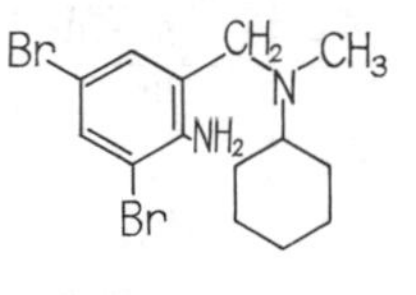

Bromomethane: same as Methyl bromide (see).

Bromometry: a seldom-used method of redox analysis in which elemental bromine is used as the titrator. B. has no advantages over Bromatometry.

4-Bromophenacyl bromide, ***4-α-dibromoacetophenone***: a crystalline compound; m.p. 110 °C. 4-B. is barely soluble in water and readily soluble in hot alcohol. It is formed from bromobenzene and α-bromoacetyl chloride in a Friedel-Crafts reaction, or by side-chain bromination of 4-bromoacetophenone in glacial acetic acid. It is used in organic syntheses and to characterize carboxylic acids as their 4-bromophenyacyl esters.

Bromophenols: isomeric mono-, di- and tribromophenols, the most important of which are 4-bromophenol and 2,4,6-tribromophenol. ***4-Bromophenol*** forms colorless crystals; m.p. 66-67 °C, b.p. 238 °C. It is soluble in water, alcohol and ether.

2,4,6-Tribromophenol consists of colorless to reddish-white, long needles; m.p. 96 °C, b.p. 244 °C. It is insoluble in water and readily soluble in ethanol and ethers. Both compounds can be made by the reaction of bromine and phenol, under different condition; 4-bromophenol at 0 °C in carbon disulfide. They are used as disinfectants, and 2,4,6-tribromophenol is also used as an antiseptic.

α-Bromophenylacetonitrile: same as α-Bromobenzylcyanide (see).

Bromophos: less toxic Organophosphate insecticides (see).

Bromopropanone: same as Bromoacetone (see).

3-Bromopropene: same as Allyl bromide (see).

1-Bromopyrrolidine-2,5-dione: same as *N*-Bromosuccinimide.

N-Bromosuccinimide, abb. ***NBS, 1-bromopyrrolidine-2,5-dione***: colorless crystals; m.p. 173-175 °C (dec.). N-B. is scarcely soluble in water or ether, but is readily soluble in acetone and ethyl acetate. It is synthesized by adding bromine to an ice-cold alkaline solution of succinimide. N-B. is an important bromination reagent used to substitute allyl hydrogen atoms (see Wohl-Ziegler reaction).

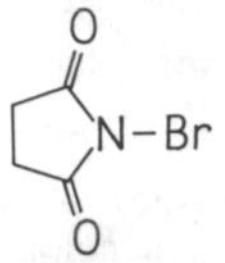

Bromosulfanes: see Sulfur bromides.

α-Bromoxylene: same as Xylyl bromide (see).

Bromoxynil: see Herbicides.

Bromphenol blue: 3,3',5,5'-tetrabromophenol sulfophthalein, a triphenylmethane pigment. B. forms colorless prisms which are insoluble in water but soluble in glacial acetic acid and acetone. It dissolves in alkalies, giving a blue color, and in acids, giving a yellow color; m.p. 279 °C (dec.).

Bromphenol blue and other important triarylmethane pigments

Pigment	$R^2 = R^{2'}$	$R^3 = R^{3'}$	$R^5 = R^{5'}$
Bromphenol blue	H	Br	Br
Cresol purple	CH_3	H	H
Cresol red	H	CH_3	H
Xylenol orange	H	CH_3	$C_5H_8NO_4$
Xylenol blue	CH_3	H	CH_3
Thymol blue	CH_3	H	C_3H_7
Methylthymol blue	CH_3	$C_5H_8NO_4$	C_3H_7
Chlorophenol red	H	Cl	H
Bromchlorphenol blue	H	Br	Cl
Bromphenol red	H	Br	H
Bromcresol purple	H	CH_3	Br
Bromcresol green	CH_3	Br	Br
Bromthymol blue	CH_3	Br	C_3H_7

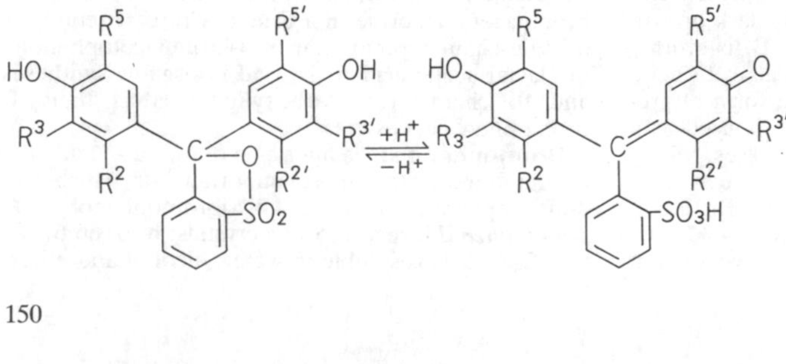

B. is made by bromination of phenolsulfophthalein in glacial acetic acid. It is used as an indicator in the pH range of 3.0 to 4.6 (color change from yellow to blue) and as a stain for microscopy. In paper electrophoresis, B. is used to develop (make visible) amino acids, peptides, etc.

Bromphenol red: see Bromphenol blue.

Bromthymol blue: see Bromphenol blue.

Bronopol: a Bacteriocide (see) which is also used in cosmetic preparations.

Brønsted acid-base definition: see Acid-base concepts.

Brønsted catalysis rule: for acid- or base-catalysed reactions, the rate constant k of the catalysed reaction depends as follows on the dissociation constant K_D of the catalyst: $k = bK_Da$. a and b are constants which are specific for the reaction being catalysed. The rule is approximate and is a special case of the LFE equations (see).

Bronzes: see Copper alloys; see Pigments.

Brownian motion: random motion of small particles, such as dust or smoke particles, suspended in a gas or liquid. B. was first observed in a light microscope by Brown, a botanist, in 1827. The smaller the particles are, the more pronounced the effect. It is caused by collisions of the particle with gas or solvent molecules, which transfer momentum to the particle and cause it to move. According to the kinetic gas theory of Maxwell and Boltzmann, all molecules are constantly in motion; their kinetic energy depends on their mass and the temperature. The kinetic energy of translation is $1/2m\ v^2 = 3/2k_BT$, where m is the mass of the molecule or particle, v is the mean velocity, k_B is the Boltzmann constant, and T is the temperature in kelvin.

B. leads to a random displacement of particles which can be understood as Diffusion (see). Einstein and Smoluchowski (1905/1906) derived the equation $x^2 = 2D\Delta t$ for the change in position in the x direction during time interval Δt. Here x^2 is the mean square displacement and D is the diffusion coefficient. D can be calculated from the observed displacement of the particle using the equation $x^2/2t = D = (k_BT)/(6\pi\eta\alpha)$, where η is the dynamic viscosity and a is the particle radius. However, it is not possible to determine particle radii from this equation, because the measurable displacement is not equal to the distance travelled by the particle in its random walk; it is only the microscopically observable net result of that walk.

Brucine: see Strychnine.

BTX aromatics: a term including benzene, toluene and xylene.

Bubble column reactor, ***bubble column***: a reactor for reactions between gases and liquids. The B. has a high thermal stability, a very simple heat removal system, and a small temperature gradient.

Bubble plate column: an apparatus for countercurrent exchange (e.g. in Distillation, see) in which the plates are perforated, and the perforations are covered by bell-shaped caps (see Columns).

Bucherer reaction: a method for producing naphthylamines from naphthols by reaction with ammonia in the presence of hydrogensulfites or sulfites. According to A. Rieche (1960), it occurs by the following mechanism:

OH +NH3 ⇌ NH2 +H2O

The reversal of this reaction, that is, the hydrolytic removal of the amino group to form the corresponding naphthol, is achieved by boiling aqueous alkalies in the presence of sodium sulfite.

The B. is especially suitable for production of primary amines of the naphthaline series. Phenols, with the exception of resorcinol, give unsatisfactory results.

Bufadienolides: see Cardiac glycosides.

Buffer solution: an aqueous solution in which the pH changes very little when small amounts of strong acid or base are added. A B. generally consists of a solution of a medium-strong acid and its salt with a strong base. The mass action expression for the equilibrium: acid + H_2O ⇌ base + H_3O^+ is $[HA][H_2O]/[A^-][H_3O^+] = K_a$; by taking logarithms and rearranging, the expression becomes the *Henderson-Hasselbach* equation: $pH = pK_a + \log([A^-]/[HA])$. Thus the pH depends on the pK_a and the logarithm of the ratio c_{base}/c_{acid}. Because the equation contains the logarithm of the concentration ratio, changes in the ratio itself have relatively little effect on the pH; that is, the pH remains nearly constant. To achicvc a high capacity of the B., the ratio c_{base}/c_{acid} should remain as close to 1 as possible. For each pH range, therefore, a suitable acid-base pair is needed for which the pK_a of the acid is as close as possible to the desired pH of the buffer; for $c_{base}/c_{acid} = 1$, the Henderson-Hasselbach equation simplifies to $pH = pK_a$.

B. are always used when it is necessary to maintain a definite pH in a solution. The B. system of human blood, which keeps the pH between 7.35 and 7.45, is based mainly on the CO_2/HCO_3 equilibrium.

Buformin: see Antidiabetics.

Bufotenin: see Amphibian toxins.

Builders: substances in laundry products which prevent mineral incrustations on textiles and deposits in washing machines, improve the action of the detergents and regulate the alkalinity of the wash water. They achieve these results by complex formation or ion exchange with the minerals in the water, particularly calcium and magnesium ions. Some typical B. are pentasodium triphosphate, the disodium salt of nitrilotriacetic acid, certain sodium aluminum silicates, for example zeolite 4A, and copolymers of acrylic acid and/or maleic acid. Because they contribute to eutrophication of natural bodies of water, the use of phosphate B. is discouraged but has not completely stopped.

Certain additives, in amounts greater than 5%, can stabilize amorphous and colloidal calcium carbonate by adsorption to the crystal nuclei, and thus delay precipitation (threshhold effect). An example of this type of B. is aminotrismethylene phosphonic acid.

Bullvalene, ***tricyclo[3.3.2.0^2,8^]deca-3,6,9-triene***: a stable, unstrained hydrocarbon with fluctuating bonds. The trivial name was suggested by Doering, who predicted the existence of the compound (1963), from the bull-like appearance of the molecule; the cyclopropane ring forms the head.

R R R=H

Bullvalene

B. forms colorless crystals; m.p. 96 °C. It is so stable that it does not decompose into naphthalene and hydrogen below 400 °C.

Molecules with fluctuating bonds undergo a continuous valence isomerization, which is called Cope rearrangement or topomerization. In B., this means that all 10 C atoms constantly exchange places at room temperature. About 1.2 million valence isomers can exist. The valence isomerization occurs so rapidly that at room reaction, all the H atoms have equivalent positions, on the average. Therefore, the nuclear resonance spectrum (above 13 °C) displays only a single signal for the protons. B. can be easily synthesized from cyclooctatetraene. In the first step, the cyclooctatetraene is heated to produce a dimer (m.p. 76 °C) which, when irradiated with UV light, cleaves off benzene to give B. in very good yield. B. undergoes the usual addition reactions, adding bromine and a total of 4 mol hydrogen for the three double bonds and the cyclopropane ring.

Buminafos: see Herbicides.

Bunsen valve: a simple pressure-release valve developed by Bunsen to prevent external air from entering a flask or buret. The B. consists of a rubber tube about 5 cm long in which there is a lengthwise slit about 1 cm long. One end of the tube is closed off, gas-tight, with a glass rod, and the other is pulled taut over a glass tube. The glass tube is passed through a rubber stopper which closes the apparatus. If excess pressure develops in the apparatus, the gas can escape through the slit in the rubber tube, but if the pressure inside the apparatus is lower than atmospheric, the edges of the slit are pressed together and air or moisture cannot enter.

Bupirimate: see Pyrimidine fungicides.

Burned lime: same as Calcium oxide (see).

Burnishing: a surface treatment used to create a thin black, bluish or brownish protective layer on steel objects. B. is intended either to improve the appearance of the object or to protect it from corrosion. It is usually done by dipping the object into boiling, highly concentrated aqueous solutions of sodium hydroxide, sodium nitrate and various additives; more rarely, it is achieved by burning off an applied layer of fat or oil.

Burn-up: 1) in metallurgy, the loss of metal by combustion, gasification, spattering and slag formation. Some forms of B. are the oxide layers which form on the surfaces of metal melts and the oxides which form on the surfaces of metals which are heated (see Scale).

2) In silicate technology, evaporation of components during heat treatments, for example, of relatively volatile oxides, such as boron trioxide and lead(II) oxide, or of fluorides.

Busulfane: see Alkylating agents.

Butacarb: a Carbamate insecticide (table).

Butachlor: see Acylaniline herbicides.

Butadiene, ***buta-1,3-diene***: $CH_2=CH-CH=CH_2$, the simplest 1,3-diene with conjugated double bonds; the parent compound of the dienes and polyenes. B. is a colorless gas which liquefies readily; m.p. - 108.9 °C, b.p. - 4.4 °C, n_D^{-25} 1.4292. It is barely soluble in water, but readily soluble in ethanol and ether. The significance of B. is that it can be polymerized to rubber, and is often used in mixtures with other polymerizable compounds. Many industrial syntheses for B. have been developed: 1) Catalytic dehydrogenation of the C_4 hydrocarbons from petroleum and natural gas, butane, but-1-ene and but-2-ene. Chromium oxide/aluminum oxide catalysts are used at temperatures of 600 to 650 °C. 2) Catalytic dehydration of butane-1,3-diol or butane-1,4-diol on sodium phosphate catalysts. The butane-1,3-diol can be made industrially from acetaldehyde via aldol, and butane-1,4-diol is obtained by reaction of ethyne with formaldehyde, via butyne-1,4-diol. 3) In the Lebedev process, ethanol vapor is passed over a zinc oxide/aluminum oxide catalyst at 400 °C. The intermediates in this process are acetaldehyde, acetaldol and butane-1,3-diol.

B. can form 1,2- and 1,4-adducts by addition and polymerization reactions.

Butadiene-1,4-dicarboxylic acid: same as Muconic acid (see).

Butadiyne, ***diacetylene***: $HC\equiv C-CH-C\equiv CH$, the simplest hydrocarbon with two triple bonds; a colorless gas with a very strong tendency to polymerize; b.p. - 78 °C. The copper and silver salts and the halogen adducts are highly explosive. B. is made from technical butynediol, which is first reacted with thionyl chloride to 1,4-dichlorobut-2-yne. In the presence of potassium hydroxide, this then eliminates hydrogen chloride to give B. The addition of alcohols to D. yields alkoxyvinylacetylenes, which are hydrogenated to alkyl butyl ethers.

Butanal: same as Butyraldehyde (see).

Butane: an alkane hydrocarbon with the general formula C_4H_{10}. There are two structural isomers, ***butane***, $CH_3-CH_2-CH_2-CH_3$, m.p. - 138 °C, b.p. - 0.5 °C, and ***isobutane (2-methylpropane)***, $CH_3-CH(CH_3)_2$, m.p. - 160 °C, b.p. - 11.7 °C. Both are colorless, combustible gases which are easily liquefied; they are insoluble in water but readily soluble in ethanol and ether. The B. are found in natural gas and petroleum, and are obtained by distillation or cracking processes. The two isomers are usually separated by adsorption and fractional desorption on activated charcoal or zeolites. Large amounts of isobutane are made from butane by catalytic isomerization with a mixture of aluminum chloride and hydrogen chloride.

Industrially, the most important reactions of B. are catalytic dehydrogenation to form unsaturated C_4 hydrocarbons, such as but-1-ene, but-2-ene, isobutene and buta-1,3-diene; reaction with sulfur at high temperature to form thiophene; alkylation reactions with isobutane and catalytic oxidation in the liquid phase. The reaction of isobutane with alkenes in the presence of sulfuric acid, hydrofluoric acid or boron trifluoride produces high-octane hydrocarbon mixtures (alkylate gasolines) which are used as additives to gasoline. For example, the reaction with isobutene leads to 2,2,4-trimethylpentane (isooctane). Oxidation in the liquid phase produces mainly acetic acid, while in the gas phase, the products are methanol, formaldehyde and acetaldehyde. Some other important reactions are the sulfochlorination of isobutane to 2-methylpropane-1-sulfochloride, nitration to 2-nitropropane, nitromethane and 1- and 2-nitroisobutane, chlorination to *tert.*-butylchloride, and oxidation in the presence of hydrogen bromide to form *tert.*-butylhydroperoxide and *tert.*-dibutylperoxide. The use of B. as a heating gas and coolant is not insignificant.

Butanedial: same as Succinaldehyde (see).

Butane-1,4-dicarboxylic acid: same as Adipic acid (see).

Butanediols, ***butylene glycols***: structurally isomeric dialcohols with the general formula $C_4H_{10}O_2$. B. are colorless, hygroscopic liquids which are soluble in water and lower alcohols, or crystalline solids with a slightly sweet taste. The following B. are important in industry:

Butane-1,4-diol, $HO-CH_2-CH_2-CH_2-CH_2-OH$, m.p. 20 °C, b.p. 235 °C, n_D^{20} 1.4660. It is obtained by catalytic hydrogenation of butyne-1,4-diol, which has a triple bond in its center and is easily synthesized by ethynylation. Butane 1,4-diol is an important intermediate in the production of tetrahydrofuran, butadiene, butyrolactone, polyurethanes, polyesters and polyamides. Its esters are used as softeners, textile treatments and synthetic resins. Butane-1,4-diol has hygroscopic and softening properties similar to those of glycol and glycerol, and it is used instead of these compounds in making paper, treating textiles and improving the quality of smoked products.

Butane-1,3-diol, $HO-CH_2-CH(OH)-CH_3$, b.p. 208 °C, n_D^{20} 1.4418 (racemate). It is made by catalytic hydrogenation of aldol or condensation of propene with formaldehyde. Butane-1,3-diol is an intermediate in the industrial synthesis of butadiene, and, like butane-1,4-diol, can be used as a softener, etc. The esters and ethers are good solvents for resins, waxes, cellulose esters and aromas.

Butane-2,3-diol, $CH_3-CH(OH)-CH(OH)-CH_3$, b.p. 179-182 °C, n_D^{25} 1.4364 (racemate). It is obtained by aerobic or anaerobic fermentation of carbohydrates. Depending on which organism is used for the fermentation, the product can be one of the optically active forms, the racemate or the *meso*-form. Butane-2,3-diol can be converted to butadiene by pyrolysis of its diacetate. All the B. are used as additives for cosmetics, paper, textiles, glue and gelatins. Butane-2,3-diol is also a good solvent for pigments, resins and certain types of rubber.

Butane-2,3-dione, ***diacetyl***: $CH_3-CO-CO-CH_3$, the simplest aliphatic diketone. B. is a steam volatile, yellow-green liquid with a quinone-like odor; at higher dilution the odor resembles that of butter. The m.p. is - 2.4 °C, b.p. 88 °C, n_D^{20} 1.3951. B. is soluble to a limited extent in water, and is infinitely miscible with most organic solvents. It is not stable to alkalies or light. It is present in coffee, cocoa, beer, honey and butter and contributes to the aromas of these products. It can be synthesized by oxidation of methyl ethyl ketone with selenium dioxide or nitrous acid. B. is used mainly as an aroma component in the food industry. The dioxime of B. (see Dimethylglyoxime) is used as a reagent in analytical chemistry.

Butanedioic acid: same as Succinic acid (see).

Butanols, ***butyl alcohols***: monoalcohols with the general formula C_4H_9OH; there are four structural isomers. The B. are readily soluble in most organic solvents. Their solubility in water varies; some are not very soluble in it. Their toxicity is low, and they are good solvents.

Butanol (butan-1-ol), CH_3-CH_2-CH_2-CH_2-OH, is a colorless liquid, m.p. - 89.8 °C, b.p. 117.3 °C, n_D^{20} 1.3993. It is synthesized by catalytic hydrogenation of crotonaldehyde, or by butanol-acetone fermentation.

Butan-2-ol, CH_3-CH_2-CH(OH)-CH_3, is a colorless liquid, m.p. - 114.7 °C, b.p. 99.5 °C, n_D^{20} 1.3978 (racemate), $[\alpha]_D^{20}$ 13.9° (optical antipodes). The racemate is synthesized by hydrogenation of but-1-ene.

Isobutanol (2-methylpropan-1-ol), CH_3-$CH(CH_3)$-CH_2OH, a colorless liquid, m.p. - 108 °C, b.p. 107.9 °C, n_D^{20} 1.3959.

tert.-Butanol (2-methylpropan-2-ol), rhombic, plate and prism crystals, m.p. 25.5 °C, b.p. 82.5 °C, n_D^{20} 1.3878. It is synthesized by hydration of isobutene:

$$(CH_3)_2C = CH_2 + H_2O \xrightarrow{+H^+} H_3C\text{-}\overset{\displaystyle CH_3}{\underset{\displaystyle CH_3}{C}}\text{-}OH.$$

The B. are used as solvents or intermediates in organic syntheses.

Butanolides: the systematic name for internal esters of hydroxybutyric acids. The name is derived from the root butan- and the suffix "-oic acid" is replaced by "-olide". The numbers indicating the carbon atoms to which the oxygen is bound are placed in front of the name, as in 1,4-butanolide (see γ-Butyrolactone).

Butan-2-one: same as Methyl ethyl ketone (see).

Butanoic acid: see Butyric acids.

Butanethiol, ***n-butylmercaptan***: CH_3-CH_2-CH_2-CH_2-SH, a colorless liquid with an unpleasant odor; in higher concentrations it is toxic. Its m.p. is - 115.7 °C, b.p. 98.4 °C, n_D^{20} 1.4440. B. dissolves in water and ethanol. It is synthesized industrially from *n*-butanol and hydrogen sulfide in the presence of thorium oxide, or from *n*-butene and hydrogen sulfide at 0 °C by UV irradiation. It is used in industry as an alarm gas added to various odorless but potentially explosive gases so that breaks or leaks in pipes will be noticed.

Butaperazine: see Phenothiazines.

But-2-enal: same as Crotonaldehyde (see).

Butene: an isomeric hydrocarbon with the general formula C_4H_8 (Table). All B. are colorless gases, which burn with very sooty flames and are readily liquefied. They are nearly insoluble in water, but dissolve readily in ethanol and ether. They are produced on a large scale from natural, refinery or cracking gases, or by dehydrogenation of butane and isobutane. Isobutene can be separated from the mixture of isomers using 65% sulfuric acid, and the other *n*-B. can be separated by extractive distillation. But-1-ene and the E,Z-isomeric but-2-enes are used in the synthesis of butadiene and butan-2-ol. *tert.*-Butanol and methylallyl chloride are synthesized from isobutene. These alkenes are also valuable alkylation reagents. For example, isobutene reacts with isobutane to form 2,2,4-trimethylpentane (isooctane), which is a very knock-resistant fuel. In addition, isobutene dimerizes to trimethylpentene isomers, which can also be hydrogenated to isooctane. These reactions are catalysed by boron trifluoride (Ipatiev, 1935) or concentrated sulfuric acid. Rubber-like polymers can also be obtained in this way.

Butenediol, ***but-2-ene-1,4-diol***: HO-CH_2-CH=CH-CH_2-OH, an unsaturated divalent alcohol, a colorless and odorless liquid; m.p. 4 °C, b.p. 238 °C. B. is infinitely miscible with water, alcohol and ether, and is very reactive. The double bond undergoes the reactions typical of alkenes; hydrogenation produces butanediol. B. can be catalytically oxidized with air to produce maleic acid, and removal of water by phosphoric acid leads to dihydrofuran. B. is synthesized industrially by partial catalytic hydrogenation of butynediol (see Acetylene chemistry).

Butenedioic acids: see Maleic acid, see Fumaric acid.

But-2-enoic acids: see Crotonic acid, see Isocrotonic acid.

But-3-en-2-one: same as Methyl vinyl ketone (see).

Butonate: see Trichlorphon.

Butter yellow: same as 4-(Dimethylamino) azobenzene.

Characteristics of the butenes

Name	Formula	m.p. [°C]	b.p. [°C]	n_D^{20}
But-1-ene	$CH_2 = CH-CH_2-CH_3$	−185.3	−6.3	1.3962
(Z)-But-2-ene	$(CH_3)(H)C = C(CH_3)(H)$ (cis)	−138.9	3.7	1.3931*
(E)-But-2-ene	$(CH_3)(H)C = C(H)(CH_3)$ (trans)	−105.5	0.9	1.3848*
Isobutene (2-methylpropene)	$CH_2 = C(CH_3)_3$	−140.4	−6.6	1.3811*

* at a temperature of −25 °C

Butyl acetates: the four isomeric butyl esters of acetic acid with the general formula CH_3-CO-OC_4H_9. B. are colorless liquids with a pleasant odor; they are nearly insoluble in water, but dissolve readily in alcohol and ether. They are synthesized by esterification of the corresponding butanols with acetic acid in the presence of sulfuric acid. ***Butyl acetate***, CH_3-CO-O-CH_2-CH_2-CH_2-CH_3; m.p. - 77.9 °C, b.p. 126.5 °C, n_D^{20} 1.3941. It is used in large amounts as a solvent for cellulose nitrate, celluloid and chlorinated rubbers, and as a thinner for paint. ***sec.-Butyl acetate***, CH_3-CO-O-CH(CH_3)-CH_2-CH_3; b.p. 112.2 °C (RS), n_D^{20} 1.3888. It is used in small amounts as a solvent. ***Isobutyl acetate***, CH_3-CO-O-CH_2-CH(CH_3)$_2$; m.p. - 98.58 °C, b.p. 117.2 °C, $n_D{}^{20}$ 1.3902. It is used as a solvent for wood lacquers. ***tert.-Butyl acetate***, CH_3-CO-O-C(CH_3)$_3$; b.p. 97-98 °C, n_D^{20} 1.3855. Its synthesis deviates from the usual pattern for esters; it is synthesized from *tert.*-butanol and acetic anhydride or ketene.

Butylamines: the four structurally isomeric monoamine derivatives of butane. ***Butylamine (1-aminobutane)***, CH_3-(CH_2)$_3$-NH_2, is a colorless liquid which smells like ammonia; m.p. - 49.1 °C, b.p. 77.8 °C, n_D^{20} 1.4031. It is produced by reduction of 1-nitrobutane, by the reaction of ammonia with butyl chloride, or by hydrogenating amination of butyraldehyde. ***sec-Butylamine*** [dl] (***2-aminobutane*** [dl]), CH_3-CH_2-CH(NH_2)-CH_3, is a colorless liquid; m.p. < -72 °C, b.p. 63.5 °C, n_D^{20} 1.3932. *sec*-B. can be synthesized by reduction of the oxime of methyl ethyl ketone. ***Isobutylamine*** (***1-amino-2-methylpropane***), (CH_3)$_2$ CH-CH_2-NH_2, is a colorless liquid; m.p. -85.5 °C, b.p. 68 °C. ***tert-Butylamine (2-amino-2-methylpropane***), (CH_3)$_3$C-NH_2, is a colorless liquid which smells like ammonia; m.p. -67.5 °C, b.p. 44.4 °C, n_D^{20} 1.3784. *tert*-B. can be made, for example, by reduction of 2-nitro-2-methylpropane. The B. are soluble in water and most organic solvents. Like all aliphatic amines, they are strong bases and irritate the mucous membranes and skin. They are used as intermediates in the production of drugs, pesticides, surfactants, resins and pigments.

Butylbiguanide: see Antidiabetics.

Butylene glycols: same as Butanediol (see).

Butyl mercaptan: same as Butanethiol (see).

tert.-Butyl methyl ether, ***methyl tert.-butyl ether***, abb. ***MTBE***: (CH_3)$_3$COCH_3, a colorless liquid, b.p. 53-56 °C, m.p. - 10 °C. *tert.*-B. is an important component of gasolines which is becoming more and more important as a substitute for lead-containing additives (see Fuels). Mixed with gasolines, it gives research octane ratings of 115 to 135. It is made industrially from methanol and isobutene in the presence of an acidic catalyst (e.g. an ion exchanger).

5-Butylpyridine-2-carboxylic acid: same as Fusaric acid (see).

Butyl rubber: see Rubber.

Butyne: a hydrocarbon with the general formula C_4H_6. There are two structural isomers, which differ markedly in their reactions (see Alkynes) due to the difference in position of the triple bond. ***But-1-yne***, HC≡C-CH_2-CH_3, is a colorless gas; m.p. - 125.7 °C, b.p. 8.1 °C, n_D^{20} 1.3962. ***But-2-yne***, CH_3-C≡C-CH_3, is a colorless liquid; m.p. - 32.3 °C, b.p. 27 °C, n_D^{20} 1.3921. The B. are nearly insoluble in water, but dissolve readily in ethanol and ether. They are easily ignited and burn with sooty flames. They are synthesized in the laboratory by alkylation of alkali acetylides, or by dehydrohalogenation of dihalogen alkanes. The B. have no particular industrial significance.

Butynediol, ***but-2-yne-1,4-diol***: HO-CH_2-C≡C-CH_2-OH, an unsaturated, dialcohol, which crystallizes in colorless leaflets; m.p. 58 °C, b.p. 126 °C. B. is readily soluble in water, but is nearly insoluble in ether; it recrystallizes well from acetoacetate. B. is slightly toxic and irritates the skin. It is synthesized by reaction of ethyne with formaldehyde (see Acetylene chemistry). B. is an important intermediate in organic syntheses. For example, it is hydrogenated to butenediol or butanediol, which then reacts further.

Butynedioic acid: same as Acetylene dicarboxylic acid (see).

Butyraldehyde, ***butanal***: CH_3-CH_2-CH_2-CHO, a colorless liquid with a pungent odor; m.p. - 99.0 °C, b.p. 74.7 °C, n_D^{20} 1.3843. B. is slightly soluble in water, and dissolves readily in most organic solvents. As a saturated aliphatic Aldehyde (see), it displays all the addition and condensation reactions typical of this class of compound. Small amounts of B. are found in free form in some essential oils, e.g. lavender oil. B. is synthesized by oxidation or dehydrogenation of *n*-butanol or by partial hydrogenation of crotonaldehyde. It is used to make paints and synthetic resins.

Butyrates: see Butyric acids.

Butyric acids: the two structurally isomeric, saturated monocarboxylic acids with the general formula C_3H_7-COOH. ***Butyric acid, butanoic acid, propane-1-carboxylic acid***, CH_3-CH_2-CH_2-COOH, is a colorless, oily liquid with an unpleasant, rancid odor; m.p. - 6.5 °C, b.p. 165.5 °C, n_D^{20} 1.3980. It is miscible in all proportions with water, alcohol and ether. Butyric acid can be steam distilled, and its reactions are typical of saturated monocarboxylic acids. Its salts and esters are called *butyrates*. *n*-Butyric acid is found as the glycerol ester in butter, and is released when the butter becomes rancid. It is also present in sweat and the liquid of meat. It can be produced from carbohydrates by certain fermentation processes, and is also made industrially by oxidation of butyraldehyde. B. is used to make cellulose esters, and its esters with lower alcohols are used as flavoring and perfume components, and as solvents.

Isobutyric acid, isobutanoic acid, 2-methylpropanoic acid, dimethylacetic acid, (CH_3)$_2$CH-COOH, is a colorless liquid with a very unpleasant smell; m.p. - 47.0 °C, b.p. 153.2 °C, n_D^{20} 1.3930. It is soluble in water, alcohol and ether. Its salts and esters are called *isobutyrates*. Iso-B. is present in sweat and carob fruits. Its esters are present in various essential oils, such as camomille, and also in excrements. Industrially, iso-B. is synthesized by oxidation of isobutyraldehyde. It is used as an intermediate in the production of perfumes, aromas, solvents and plastics. Some of its salts are used as preservatives and stabilizers.

γ-Butyrolactam: same as Pyrrolidone (see).

γ-Butyrolactone, ***1,4-butanolide, 4-hydroxybutyric acid lactone, tetrahydrofuran-2-one***: a colorless, hygroscopic, oily liquid which smells faintly like butyric acid; m.p. - 42 °C, b.p. 206 °C, n_D^{20} 1.4341. γ-B. is readily soluble in water and organic solvents, and is steam volatile. In aqueous solution there is an equilibrium between the lactone and hydroxybutyric

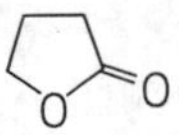

acid, which is shifted to the side of the acid by alkalies. γ-B. is obtained by catalytic dehydrogenation of butane-1,4-diol or by dehydration of 4-hydroxybutyric acid. It is used as a solvent for cellulose acetate, polyacrylonitrile, polystyrene, butadiene mixed polymers and especially for acetylene; it dissolves 15 times its own volume of acetylene at 27 °C. γ-B. is also used as a starting material for synthesis of pyrrolidone, *N*-methylpyrrolidone, softeners, synthetic resins, piperidine, etc.

Butyrophenones: basic compounds used as neuroleptics. A well-known B. is ***haloperidol*** (Fig.).

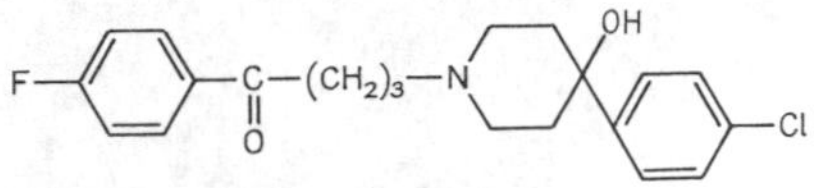

Haloperidol

BZ: see Chemical weapons.

C

c: 1) heat capacity; 2) concentration (molarity).

C: 1) symbol for carbon; 2) abb. for cytidine.

CA: abb. for cellulose acetate.

Ca: symbol for calcium.

CAB: abb. for cellulose acetobutyrate.

Cacodyl-: an obsolete term for the dimethylarsino group, as in cacodyl oxide, now bis(dimethylarsino)oxide.

Cacodyl reaction: see Acetic acid.

Cadaverine, ***pentamethylene diamine***, ***1,5-diaminopentane***: $H_2N-(CH_2)_5-NH_2$, a biogenic, aliphatic diamine. It is a syrupy, colorless, fuming liquid with an unpleasant odor; m.p. 9 °C, b.p. 178-180 °C, n_D^{25} 1.4561. P. is readily soluble in water and alcohol, but is nearly insoluble in ether. It undergoes the reactions typical of primary amines, and forms salts with acids. P. is formed by decarboxylation of lysine during bacterial decomposition of proteins. It can be synthesized via the phthalimido compound of 1,5-dihalopentane, or by hydrogenation of glutaric dinitrile.

Cadmium, symbol ***Cd***: a chemical element in Group IIb of the periodic system, the Zinc group (see). It is a heavy metal; *Z* 48, with isotopes with mass numbers 114 (28.86%), 112 (24.07%), 111 (12.75%), 110 (12.39%), 113 (12.26%), 116 (7.58%), 106 (1.15%) and 108 (0.875%). The atomic mass of Cd is 112.40 and its valence is II; its Mohs hardness is 2.0; density 8.642, m.p. 320.9 °C, b.p. 767.3 °C, electrical conductivity 13.2 Sm mm^{-2}, standard electrode potential (Cd/Cd^{2+}) - 0.4029 V.

Properties. Cd is a silvery white, relatively soft metal, which loses its shine upon exposure to air, due to formation of an oxide layer. It is similar to zinc in forming hexagonal close-packed crystals which are stretched in the direction of the six-fold lattice axis. Cd is ductile, can be rolled to a foil and drawn to wire; it is easy to cut shavings from a cadmium rod. When it bends, Cd, like tin, makes a crackling noise. It can be strengthened by alloying with zinc. Cd can be sublimed in a vacuum below 200 °C. When heated in the air, it burns with a red flame, forming a brown smoke of cadmium oxide, CdO. When heated, it readily combines with halogens. Oxidizing acids dissolve Cd, although pure Cd, like pure zinc, reacts very sluggishly with non-oxidizing acids. Unlike zinc, Cd is insoluble in alkali hydroxide solutions. In general, Cd is divalent in its compounds; there is only scattered evidence for the existence of cadmium(I) compounds. For example, when Cd is dissolved in molten cadmium(II) chloride in the presence of aluminum chloride, cadmium(I) tetrachloroaluminate, $Cd_2[AlCl_4]_2$ is formed; its Cd_2^{2+} cations react immediately with water, disproportionating to elemental Cd and Cd^{2+}. The coordination numbers 4 and 6 are most common in cadmium(II) complexes, which have tetrahedral and octahedral arrangements of ligands around the cations.

> Cadmium compounds are very toxic for many organisms, including human beings. The maximum allowable cadmium concentration in air is 0.1 mg m^{-3}. Inhalation of cadmium leads to damage of the respiratory passages. Ingestion of cadmium compounds poisons the gastrointestinal tract, and can also cause liver and kidney damage. Therefore, cadmium and its compounds should always be handled with caution.

Analytical. Cd associates with copper in qualitative analysis separation schemes. However, hydrogen sulfide precipitates only the yellow cadmium sulfide from a cyanide solution in which both elements are present as the cyanometallates $[Cd(CN)_4]^{2-}$ and $[Cu(CN)_4]^{3-}$. Classical gravimetric determination of Cd also depends on precipitation of CdS. Cd is determined complexometrically with EDTA at pH 10, using eriochrome T as an indicator, or at pH 4 with xylene orange as indicator. Some other effective methods are electrogravimetry, neutron activation analysis, polarography and flame photometry. Atomic absorption spectrometry permits detection of < 0.1 ppm Cd.

Occurrence. Cd is a rare element making up 10^{-5}% of the earth's crust. It accompanies zinc in sphalerite (zinc blende) and smithsonite (zinc spar), while pure cadmium minerals, e.g. greenockite (cadmium blende, CdS), monteponite, CdO, and otavite, $CdCO_3$, occur very sparsely. Cd is present in soils at about 0.1 ppm.

Production. Cd is obtained as a byproduct of zinc production and in the workup of many types of fly ash. In dry zinc production, the more volatile and more easily reduced Cd is considerably enriched in the first zinc dust fractions, and it is burned in the condenser to cadmium oxide, CdO. This first dust fraction is reduced with coke and subjected to repeated fractional distillation in a special muffle column until 99.5% Cd is finally obtained. In wet extraction, Cd is first precipitated by addition of metallic zinc to a zinc salt solution, according to the equation $Cd^{2+} + Zn \rightarrow Cd + Zn^{2+}$. The cadmium sponge which forms is oxidized and the oxide is dissolved in acid. It is then electrolysed with aluminum cathodes and lead anodes, yielding very pure electrolyte cadmium. However, Cd can be purified without previous enrichment by electrolysis of the original zinc sulfate solution, because the precipitation voltage of Cd is significantly lower than that of zinc.

Applications. Most Cd is used for Cadmium plating

(see). Cadmium alloys (see) are also used. Cd has a high neutron-capture cross section, especially the isotope 113, and this is therefore used to control and screen reactors. Cadmium compounds are used in small amounts as additives to polymers, polymerization catalysts, and as luminophors for television screens.

Historical. Cd was discovered in 1817 by Strohmeyer in Gœttingen, in the course of an investigation of a zinc carbonate from an apothecary which yielded a yellow oxide upon thermal decomposition. Almost simultaneously, Hermann in Magdeburg discovered the new element in a zinc oxide. The name "cadmium" is based on the Greek word "cadmeia", which was used in antiquity for minerals which yielded brass when worked with copper ores.

Cadmium alloys: metallic multi-component systems containing cadmium. C. are used as bearing metals for fast motors (maximum 97% cadmium, the remainder nickel, silver and copper). Special soft solders contain cadmium as a substitute for zinc. Aluminum is soldered with palladium-cadmium-tin solders. Silver containing 20% cadmium is a suitable contact alloy. Low-melting bismuth alloys which contain cadmium (see Wood's metal, Lipowitz' metal) are used in fire alarm systems. Cadmium amalgams are used in the production of standard electrolytic cells.

Cadmium bromide: $CdBr_2$, white, pearly leaflets; M_r 272.22, density 5.192, m.p. 567 °C, b.p. 863 °C. C. exists in four modifications, all of which form lamellar crystals. It crystallizes out of water as the tetrahydrate, $CdBr_2 \cdot 4H_2O$; M_r 344.28. C. is obtained by heating cadmium with bromine vapor or by reaction of the metal with bromine water, and is used in photography and lithography.

Cadmium chloride: $CdCl_2$, shiny rhombohedral crystals; M_r 183.32, density 4.47, m.p. 568 °C, b.p. 964 °C. C. is readily soluble in water, and somewhat soluble in methanol, ethanol and acetone. It crystallizes out of aqueous solution in the form of its hydrates, with 1, 2 or 4 moles water of crystallization. C. is obtained by heating cadmium or cadmium oxide in a stream of chlorine. It is used in analysis as an absorption reagent for hydrogen sulfide. C. is used in electroplating metals, dyeing and printing textiles, and in photography for production of copying emulsions.

Cadmium cinnabar: see Cadmium pigments.

Cadmium cyanide: $Cd(CN)_2$, a white compound soluble in strong acids; M_r 164.44, density 2.226. C. precipitates from cadmium salt solutions when alkali cyanides are added, but it dissolves in excess cyanide, forming cyanocadmates, $M[Cd(CN)_3]$ and $M_2[Cd(CN)_4]$. $H_2[Cd(CN)_4]$, for example, forms colorless, cubic crystals which are strongly refractive of visible light (M_r 294.68, density 1.846). It is used in electroplating with cadmium.

Cadmium hydroxide: $Cd(OH)_2$, a white powder; M_r 146.41, density 4.79, m.p. 300 °C (dec.). C. is insoluble in water or alkali hydroxide solutions, but dissolves in ammonia, forming a complex. It is made by precipitation of cadmium salt solutions with alkali hydroxide solution.

Cadmium iodide: CdI_2, colorless, shiny, trigonal plates; M_r 366.21, density 5.67, m.p. 388 °C, b.p. 790 °C. C. is formed from iodine and cadmium in the presence of water, or by dissolving cadmium oxide or cadmium carbonate in hydroiodic acid. It is used in photography and as a reagent, for example for detecting nitrous acid or alkaloids.

Cadmium iodide type, ***cadmium iodide structure***: a common Structure type (see) for compounds with the general composition AB_2 which crystallize in layered structures. In the crystal lattice of the prototype cadmium iodide, CdI_2, the I^- ions form a hexagonal close packing. The octahedral holes between the I^- layers are alternatingly occupied by a Cd^{2+} ions or vacant, so that I-Cd-I layer packets are formed (Fig.).

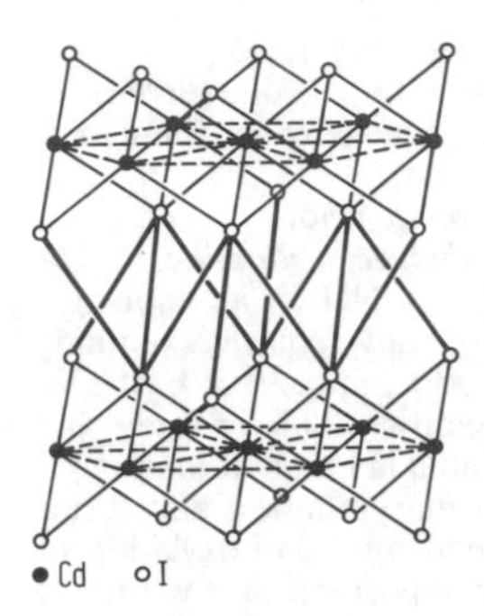

Layered lattice of the cadmium iodide structure

Each Cd^{2+} ion is surrounded in a distorted octahedral pattern by 3 Cd^{2+} ions; each I^- ion is coordinated by 3 Cd^{2+} ions. The bonding within a layer packet is intermediate between ionic and covalent; the layer packets are held together by van der Waals forces. The layer structure is responsible for a marked anisotropy of many properties.

Many diiodides (e.g. of the elements Mg, Ca, Cd, Pb, Fe, Zn), dibromides (e.g. of Mg, Fe and Co) and a number of dichalconides of tetravalent elements (e.g. TiS_2, $TeSe_2$, $TiTe_2$ and SnS_2) crystallize in the C. MoS_2, which is used as a lubricant, forms a layered lattice very similar to the C.

Cadmium nitrate: $Cd(NO_3)_2 \cdot 4H_2O$, hygroscopic columns and needles joined by rays; M_r 308.47, density 2.455, m.p. 59.4 °C, b.p. 132 °C. C. is extremely soluble in water. Other hydration levels which crystallize out of saturated solutions are $Cd(NO_3)_2 \cdot 9H_2O$ (- 16 to 3.5 °C), $Cd(NO_3)_2 \cdot 2H_2O$ (48.7-56.8 °C) and $Cd(NO_3)_2$ (>56.8 °C). C. is obtained by reaction of cadmium carbonate with dilute nitric acid. It is used to give a luster to glass and porcelain, and to make cadmium-nickel sinter plates for batteries.

Cadmium oxide: CdO, usually brown, cubic crystals; M_r 128.40, density 6.95, subl.p. 1559 °C. The color of C. depends greatly on the temperature at which it forms, and, depending on the size of the particles, it can vary from greenish yellow to brown to dark blue-black. C. is readily soluble in acids. It is found naturally in the form of black, shiny, octahedral crystals of monteponite. It is formed by combustion of cadmium. Single crystals can be obtained by hydrothermal techniques. C. is used for glazes and electroplating cadmium, to make metallic layers on plastics, and as a catalyst in redox reactions, dehydrogenations and polymerizations. It is also used in the synthesis of phosphors and semiconductors. It has a nematocidal effect and is used to protect plants from these pests.

Cadmium pigments: cadmium chalcogenides which can be used as pigments. 1) ***Cadmium yellow*** is cadmium sulfide, CdS; it is a bright yellow, nontoxic pigment which can be used with a wide variety of binders. 2) ***Cadmium red*** is cadmium selenide, CdSe; it is most often used in a mixture with CdS, and gives very bright colors, from orange to bordeaux red. It is very fast to light and heat, and is used in artists' colors, ceramics and enamels. 3) ***Cadmium cinnabar*** is an orange to red mixed pigment of CdS with 10 to 20% mercury sulfide, HgS (cinnabar). 4) see ***Cadmopone***.

Cadmium plating: the application of a layer of cadmium to an iron or steel object to protect it from corrosion. The corrosion properties of cadmium are similar to those of zinc. Cadmium forms water-soluble compounds which are very poisonous, so foods must not be allowed to come in contact with it. Electrolytically applied cadmium layers 6 to 12 μm thick are used to protect small steel parts from atmospheric corrosion. Cadmium layers are also applied by vacuum vaporization techniques.

Cadmium red: see Cadmium pigments.

Cadmium sulfate: $CdSO_4$, colorless crystals; M_r 208.46, density 4.691, m.p. 1000 °C. C. is readily soluble in water, but is insoluble in ethanol. It crystallizes out of aqueous solution above 41.5 °C as the monohydrate, $CdSO_4 \cdot H_2O$; M_r 226.48, density 3.79. C. is obtained by dissolving cadmium or cadmium carbonate in dilute sulfuric acid. It is used to detect hydrogen sulfide and as a starting material for synthesis of pigments and phosphors; it is also used in the Weston standard electrolysis cell.

Cadmium sulfide: CdS, lemon to reddish yellow, cubic crystals; M_r 144.46, density 4.82. When heated to 700-800 °C in the presence of sulfur vapor, cubic C. is converted into a cinnabar-red, hexagonal modification, which can also be obtained in impure form by passing hydrogen sulfide through cadmium halide solutions. C. can also be made by roasting cadmium oxide or carbonate with sulfur. It occurs sporadically in nature as greenockite (cadmium blende). C. is used to make photoresistors and photocells, in automatic aperture controls for cameras, and, doped with heavy metals, as a television phosphor. It is used as a pigment under the name cadmium yellow (see Cadmium pigments).

Cadmium yellow: see Cadmium pigments.

Cadmopone: a yellow pigment made in a similar way to Lithopone (see), by precipitation from cadmium sulfate and barium sulfate as a mixture of cadmium sulfide and barium sulfate.

Caffeic acid: see Hydroxycinnamic acids.

Caffeine: 1,3,7-trimethylxanthine, a very important purine alkaloid. C. is a weak base which crystallizes out of ethanol in white, bitter-tasting needles; m.p. 235-237 °C. C. is readily soluble in hot water, but only slightly soluble in ethanol and ether. It has a stimulating effect on the central nervous system, and thus suppresses tiredness and increases the ability to concentrate. It also increases the heart activity, dilates the coronary arteries and has a weak diuretic effect. When combined with analgesics, it increases their effects.

Coffee beans contain 0.6 to 3.0% C. It is also present in Chinese tea leaves (0.8 to 5.0%), mate leaves (1.0 to 2.0%) and cocoa beans (0.1 to 0.4%). It is extracted from natural products with benzene. C. can also be synthesized by methylation of theophylline.

To increase its solubility in water, C. is used in a mixture with alkali salts of organic acids, for example, as caffeine-sodium salicylate.

Cage effect: an effect on the rate of a reaction in solution caused by diffusion (see Diffusion control). One can visualize the dissolved particles as surrounded by a cage of solvent molecules. If such a particle undergoes a chemical reaction, e.g. a thermal or photochemical dissociation, or an electron exchange with another molecule in the same cage, the products must escape from the cage by diffusion before they can react with other substances present in the solution. Within the cage, there is therefore an increased probability that the products will collide with each other and react before they escape from the cage. This leads to a decrease in the yield of reactions with reactants outside the cage. For example, radical polymerizations are started by initiators; thermal cleavage of an initiator molecule produces two radicals in a common solvent cage. The probability of combination of these two radicals is increased by the C., and this means that a fraction of the radicals is unable to react with the monomers which are to polymerize. In photochemical reactions, C. reduces the quantum yield.

Cahn-Ingold-Prelog convention: see Stereoisomerism 1.1.

Cailletet-Mathias rule: a relationship established by L. Cailletet and E. Mathias: the arithmetic mean $(d_l + d_g)/2$ of the densities of a liquid d_l and a gas d_g phase in equilibrium with each other depends linearly on temperature. It can be used to determine exactly the critical molar volume $V_{crit} = M/d_{crit}$ from the molar mass M and the critical density d_{crit} (see Critical point). The densities of the liquid and gas phases are determined up to points near the critical point, and then $(d_l + d_g)/2$ is extrapolated to the critical temperature T_{crit}.

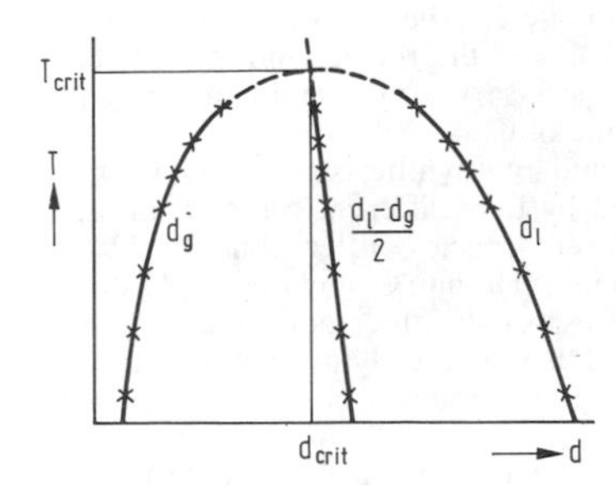

Determination of the critical density using the Cailletet-Mathias rule

cal: see Calorie.

Calamus oil: a viscous, thick, yellowish brown, bitter-tasting essential oil with a camphor-like odor. It is obtained from the roots of calamus plants by steam distillation. Fresh roots yield 0.5 to 0.8% C. by steam distillation. C. contains the following compounds: camphor, pinene, camphene, asarone, asarylaldehyde, eugenol, methyleugenol, palmitic acid, enanthic acid, various hydrocarbons and others. C. is used to perfume soaps, to make liquors, as a medication for gout and to improve digestion.

Calcaroni: see Sulfur.

Calciferol: same as Vitamin D (see).

Calcimycin: same as Antibiotic A-23187 (see).

Calcination: the process of heating a solid to a temperature at which it undergoes a certain degree of decomposition. For example, sodium hydrogencarbonate is converted by C. to anhydrous (calcined) soda, or sodium carbonate.

Calcitonin, ***thyrocalcitonin***: a peptide hormone consisting of 32 amino acids (Fig. 1). It reduces the levels of calcium and phosphate in the blood, and thus counteracts the effects of parathyrin. In mammals, C. is formed in the C cells of the thyroid in response to an increasing blood calcium level. In lower vertebrates and birds, it is formed in the ultimobranchial body, so that in these animals the term "thyrocalcitonin" is not appropriate. The C. formed in the ultimobranchial body has a 30 to 40-fold higher hypocalcemic effect than mammalian C. C. is biosynthesized via ***pre-procalcitonin*** (Fig. 2), a precursor of 136 amino acids. This is converted by removal of a 24-amino-acid signal sequence to ***procalcitonin***.

Cys–Gly–Asn–Leu–Ser–Thr–Cys–Met–Leu–Gly–Thr–
Tyr–Thr–Gln–Asp–Phe-Asn–Lys–Phe–His–Thr–Phe–
Pro–Gln–Thr–Ala–Ile–Gly–Val–Gly–Ala–Pro–NH_2

Fig. 1. Calcitonin

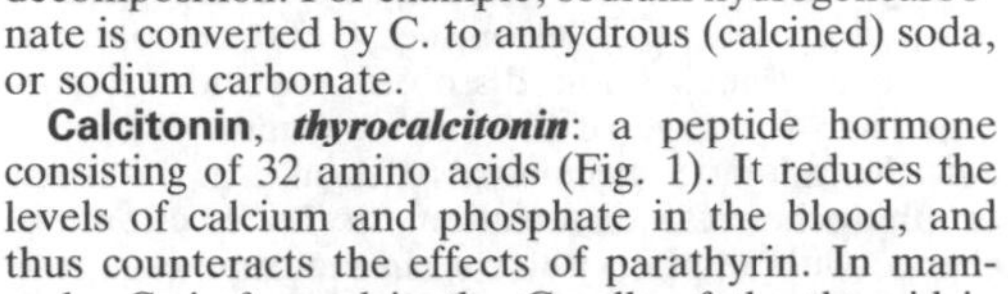

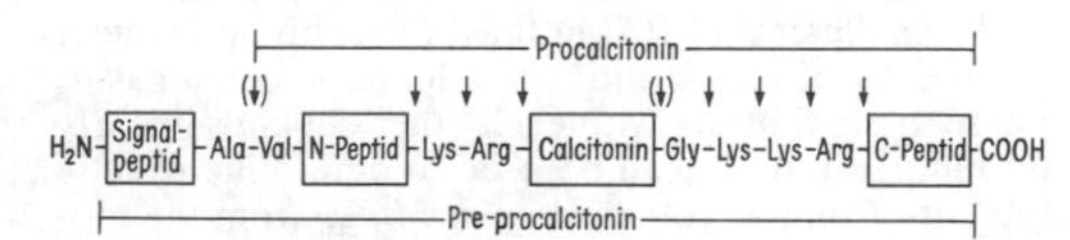

Fig. 2. Pre-procalcitonin

The basic amino acids flanking the C. sequence lead to its release by enzymes similar to trypsin and and carboxypeptidase B. The C-terminal part of the C. sequence is followed in procalcitonin by a glycine residue which is necessary for the amidination of the C-terminal proline of C.

The structure and total synthesis of porcine C. were reported in 1968 by four different research groups. The C. of different species differ considerably in amino acid sequence. Human C. and the C. of salmon and eels, and especially the more stable carba analogs, are used therapeutically in treatment of certain bone diseases (osteodystrophy deformans, osteoporosis, etc.).

Calcium, symbol ***Ca***: a chemical element of group IIa of the periodic system, the Alkaline earth metals (see); a light metal, *Z* 20, with isotopes of mass numbers 40 (96.947%), 42 (0.646%), 43 (0.135%), 44 (2.083%), 46 (0.186%) and 48 (0.18%), Mohs hardness 1.5, density 1.54, m.p. 839 °C, b.p. 1484 °C, electrical conductivity 25.0 Sm mm^{-2} (at 0 °C) and standard electrode potential (Ca/Ca^{2+}) - 2.76 V.

Properties. C. is a light metal with a silvery white shine on a freshly cut surface; however, in the air it rapidly acquires a layer of nitride and oxide. It exists in three allotropic modifications. Below 300 °C, the cubic body-centered α-C. is most stable; at 300 °C it converts to a β-form with lower symmetry. Above 450 °C, the hexagonal γ-C. is stable.

As a group IIa element, Ca tends to lose both 4s electrons and form the colorless Ca^{2+} cation. This is reflected by the high negative standard electrode potential, which indicates that Ca is a strong reducing agent. It is a decidedly electropositive element with little tendency to form covalent bonds, that is, the chemistry of Ca is essentially the chemistry of ionic compounds. Even at room temperature, Ca is attacked on the surface by air, forming chiefly the nitride. It should therefore be stored under petroleum. At higher temperatures, Ca burns to calcium oxide, CaO, and calcium nitride, Ca_3N_2. Water or alcohol reacts slowly with Ca when cold, but when heated the reaction is rapid and leads to generation of hydrogen: $Ca + 2\ H_2O \rightarrow Ca(OH)_2 + H_2$. In dilute acids, Ca dissolves with violent evolution of hydrogen. The reductive capacity is also expressed by its ability to release other metals from their oxides, for example, in the formation of titanium: $TiO_2 + 2\ Ca \rightarrow Ti + 2\ CaO$. Ca is only slowly attacked by halogens when cold, but when heated, the formation of calcium halides is rapid. Ca dissolves in liquid ammonia, NH_3, giving a blue-black color. When hot, it reacts with NH_3 to form calcium nitride, Ca_3N_2, and calcium hydride, CaH_2. The latter can also be made by direct reaction of elemental hydrogen with Ca.

Analysis. Volatile calcium salts give flames an intense red color. The red line at 622.0 nm and the green line at 553.3 nm are characteristic of Ca and serve for qualitative determination. In the qualitative analysis separation scheme, Ca is in the $(NH_4)_2CO_3$ group. It is separated by the chromate-sulfate or the alcohol-ether method, and its presence is demonstrated by precipitation as calcium oxalate, CaC_2O_4 or calcium ammonium hexacyanoferrate(II), $Ca(NH_4)_2[Fe(CN)_6]$. Quantitative determination can be done with instrumental methods, such as atomic absorption spectroscopy and flame photometry, or by gravimetric methods, such as precipitation of calcium oxalate, which is either weighed directly or first converted to calcium oxide. Ca can also be determined volumetrically by complexometry with EDTA.

Occurrence. Ca is the 5th most abundant element in the earth's crust, comprising 3.4%. It is always found in compounds, usually in the form of minerals. Calcium carbonate, $CaCO_3$, in the form of calcite, chalk, marble, aragonite and the double carbonate $CaCO_3 \cdot MgCO_3$ (dolomite), makes up entire mountain ranges. Gypsum, $CaSO_4 \cdot 2H_2O$, and anhydrite, $CaSO_4$, make up extensive deposits. In salt deposits, various double salts of Ca are also present, including tachhydrite, $CaCl_2 \cdot 2MgCl_2 \cdot 12H_2O$, and glauberite, $Na_2SO_4 \cdot CaSO_4$. Calcium fluoride, CaF_2, is found as fluorite (fluorspar). Phosphorite, $Ca_3(PO_4)_2$, and apatite, $Ca_5(PO_4)_3(OH,F,Cl)$ are very important as the basis of phosphate fertilizers. Of the calcium silicates, which form the great mass of silicate rocks, calcium feldspar, $CaAl_2Si_2O_8$ and wollastonite, $Ca_2Si_2O_6$ are very abundant. Various calcium salts are present in all natural bodies of water, as weathering products of calcium-containing minerals. Together with dissolved magnesium salts, they make the water hard. Seawater contains enough Ca to make about a 0.16% $CaSO_4$ solution.

The biological roles of Ca^{2+} are numerous. Hydroxylapatite, $Ca_5(PO_4)_3 \cdot OH$ is the main component of bone and tooth minerals. Egg shells and sea shells consist of $CaCO_3$. Ca^{2+} is essential for the functions of many enzymes, for nerve impulses, cell division, blood coagulation, muscle contraction, formation of cell walls, etc. Ca is an essential growth factor for plants as well as animals. A deficiency of Ca in the soil, which also causes degeneration of soil structure, must be compensated by fertilizer.

Production. Ca is made by melt electrolysis of calcium chloride and calcium fluoride or potassium chloride around 800 °C. Graphite anodes and iron cathodes are used. The iron cathodes are arranged in such a way that they just touch the melt. The high current density at the cathode causes molten Ca to precipitate. The iron rods are gradually raised during the electrolysis, thus producing long rods of metallic Ca. The use of these "contact electrodes" prevents the molten metal from diffusing into the cooler salt melt, and thus prevents loss of Ca. Very pure Ca is made by aluminothermal reduction followed by vacuum sublimation.

Applications. As a strong reducing agent, Ca is used in metallurgy to reduce other elements to the metallic state (e.g. titanium, zirconium, chromium and uranium), and as a deoxdizing agent for nickel and copper melts. Lead-calcium alloys are used as slide bearings. Ca is used as a getter metal and in the refining and desulfurizing of petroleum. In the laboratory, it is used to dry alcohols, as an auxiliary in organic syntheses, and to purify noble gases.

Historical. Ca (from the Latin "calx" = "stone, limestone") was first isolated in 1808 by Davy, from an amalgam obtained by electrolysis. The pure metal was first prepared in 1898 by Moissan, by reduction of anhydrous calcium iodide with sodium. Calcium compounds obtained by burning of limestone and quenching of the resulting lime, as well as gypsum, were known in antiquity as construction materials.

Calcium acetate: $Ca(CH_3COO)_2$, a colorless, hygroscopic mass which dissolves easily in water; M_r 158.17. C. crystallizes out of cold water as the dihydrate, or out of water at room temperature as the monohydrate. It is made by dissolving calcium carbonate in aqueous acetic acid and evaporating the solution. Above 160 °C, C. decomposes into calcium carbonate, $CaCO_3$, and acetone, CH_3-CO-CH_3, for which it is thus a source. C. is also used in printing with alizarin and in tanning hides.

Calcium acetylide: same as Calcium carbide (see).

Calcium bi-: see Calcium hydrogen-.

Calcium bromide: $CaBr_2$, forms colorless rhombic crystals, M_r 199.90, density 3.353, m.p. 730 °C, b.p. 806-812 °C. C. crystallizes out of aqueous solution as the hexahydrate, $CaBr_2 \cdot 6H_2O$ (m.p. 38.2 °C). It is formed by the reaction of elemental bromine with calcium oxide or calcium carbonate at glowing heat. C. is used in the photographic industry, and in medicine as a nerve tranquilizer.

Calcium carbide, ***calcium acetylide***: often called simply ***carbide***, CaC_2, forms colorless, tetragonal crystals; M_r 64.10, density, 2.22. The technical product is usually gray to black, due to impurities. C. is insoluble in organic solvents. It reacts with water and dilute acids, even when cold, evolving ethyne: $CaC_2 + 2H_2O \rightarrow Ca(OH)_2 + HC{\equiv}CH$. In oxygen, C. burns at red heat to calcium carbonate, $CaCO_3$. Elemental nitrogen converts hot C. to calcium cyanamide. Many metal oxides react with C. to form the corresponding metals or metal carbides.

C. is produced in large amounts industrially by reaction of burned lime with coke, in the ratio 60:40, in an electric furnace (Fig.), according to the following equilibrium equation: $CaO + 3\,C \rightleftharpoons CaC_2 + CO$, $\Delta H \approx 460$ kJ mol^{-1}. Since the reaction is highly endothermic, the equilibrium is shifted to the right only at very high temperatures. Below 1630 °C, in a pure CO atmosphere and at a pressure of 0.1 MPa, the reverse reaction is observed. On the other hand, at very high temperatures, C. reacts with calcium oxide, CaO, to form calcium and carbon monoxide, CO. Therefore a working temperature of 2100 °C is usually chosen, and the product is about 80% CaC_2, 12 to 15% calcium oxide and 1 to 2% carbon. Carbide furnaces consist of large iron structures, lined with firebricks, and with stamped coal floors and walls. These form one electrode. The counter electrode is provided by several mobile coal electrodes (usually Sœderberg electrodes). The entire furnace is surrounded by a water-cooled steel layer.

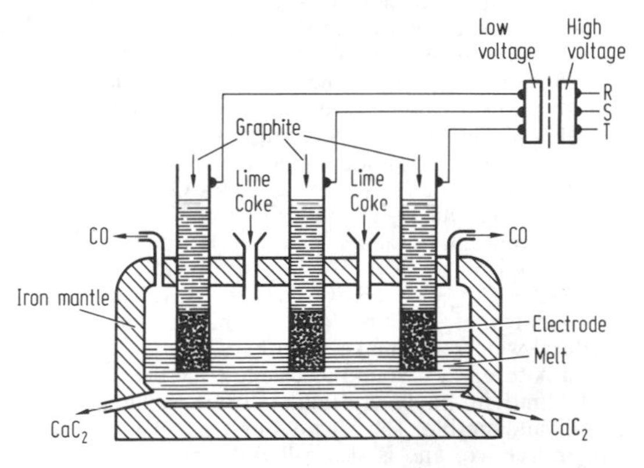

Carbide furnace

In modern installations, hollow electrodes are used; a stream of carbon monoxide loaded with coke and finely powdered lime is blown into the furnace through the electrodes. The C. accumulates as a liquid which is tapped periodically. After cooling, it is broken, sometimes milled, and then processed further or packed in air-tight containers and sold as 82% CaC_2 (technical CaC_2). For each ton of C., 950 kg lime, 550 kg coke, 20 kg electrode mass and 2100 KWh electrical energy are consumed. To reduce the amount of particulate contamination of the environment, completely closed furnaces are now used. Since the energy consumption is very high, there have been efforts to develop a synthesis which does not require electricity. For example, if part of the coke is burned with pure oxygen, the temperature required for formation of C. can be obtained. C. can also be produced by reduction of calcium phosphate, sulfate or sulfide with coal in an electric furnace.

C. is an important intermediate in the chemical industry which, as petroleum becomes less abundant, will become even more important. Most of it is converted the ethyne, a basic raw material for industrial chemistry. Many consumers generate ethyne for welding by adding water to C. Internationally, C. is processed mainly to calcium cyanamide, and is used in metallurgy for removal of sulfur.

Calcium carbonate: $CaCO_3$, M_r 100.09, exists in three modifications. The trigonal calcite (calcspar) is stable at room temperature; density 2.710. At a pressure of about 10.25 MPa and 1139 °C, it melts without decomposition. The other two modifications are the rhombic aragonite, density 2.930, and the hexagonal vaterite. At low temperatures, C. crystallizes out of water as the rhombohedral hexahydrate, $CaCO_3 \cdot 6H_2O$. However, this weathers rapidly in air, converting to the anhydrous form. C. is nearly insoluble in water, with a solubility product $K_s = 0.87 \cdot 10^{-8}$ at 25 °C (this corresponds to a solubility of about 0.0009 g $CaCO_3$ in 100 g water). Therefore, when Ca^{2+} and CO_3^{2-} ions are combined in solution, they form a colorless precipitate. C. is also formed by the reaction of carbon dioxide and calicum oxide or hydroxide. If C. is heated, the CO_2 pressure at 898 °C reaches atmospheric pressure. Above this temperature, C. therefore decomposes in air to form calcium oxide and carbon dioxide: $CaCO_3 \rightarrow CaO + CO_2$. This process, known as lime burning, is carried out on a very large scale to produce lime and CO_2 (e.g. for the Solvay process). As the calcium salt of the weak and volatile carbonic acid, C. decomposes even in weakly acidic solution, generating CO_2. The acidity of an aqueous ammonium chloride solution is sufficient: $CaCO_3 + 2\,NH_4Cl \rightarrow CaCl_2 + CO_2 + 2\,NH_3 + H_2O$. Strong acids release CO_2 from C. If CO_2 is passed through an aqueous suspension of C., it dissolves, forming calcium hydrogencarbonate: $CaCO_3 + H_2O + CO_2 \rightarrow Ca(HCO_3)_2$. In this way, hydrogencarbonates of calcium (and of magnesium) enter natural waters and make them hard.

In nature, C. is found as calcite (calcspar), aragonite, and in finely crystalline form, as limestone, chalk and marble. The calcite found in Iceland is doubly refractive, and is also called double spar. Limestone containing clays is called marl. C. is also found in dolomite, $CaCO_3 \cdot MgCO_3$. C. is found in the biosphere, e.g. in the shells of various marine animals (molluscs, corals), egg shells, and so on.

The thermal decomposition of C. to CO_2 and CaO is very important, as these are the basic materials for production of soda, potash, calcium carbide and slaked lime. Lime burned with clay forms cement. C. is used in the glass and ceramics industries, as a fertilizer and as whitewash. Pure, finely powdered C. is used to deacidify wine.

Calcium chlorate: $Ca(ClO_3)_2$, colorless, hygroscopic crystals; M_r 206.99, m.p. 340 °C. C. is readily soluble in water, alcohol and acetone. It crystallizes out of the aqueous solution as the dihydrate, $Ca(ClO_3)_2 \cdot 2H_2O$. C. is made by passing elemental chlorine through boiling calcium hydroxide suspension, or by anodic oxidation of calcium chloride. It is used in photography, pyrotechnics and as a herbicide.

Calcium chloride: $CaCl_2$, in anhydrous form, a colorless, grainy and extremely hygroscopic, cubic crystalline mass; M_r 110.99, density 2.15, m.p. 782 °C, b.p. 1600 °C. Molten C. conducts electricity well. Anhydrous C. dissolves in water with the generation of a great deal of heat, that is, the solvation enthalpy of the Ca^{2+} ions is much greater than the lattice energy.

C. crystallizes out of aqueous solution as the hexahydrate, $CaCl_2 \cdot 6H_2O$. When the hydrated cations go into solution, there is no significant enthalpy of solvation, so that the hexahydrate dissolves in water with marked cooling. It is therefore used to make cooling mixtures. When $CaCl_2 \cdot 6H_2O$ is mixed with ice in the ratio 1.44:1, it is possible to reach a temperature of -54.9 °C, the cryohydratic point.

The monohydrate, $CaCl_2 \cdot H_2O$, crystallizes out of $CaCl_2$ solutions above 176 °C. The dihydrate can be obtained by passing hydrogen chloride over the hexahydrate, or by mixing hydrochloric acid, HCl, with a concentrated, aqueous solution of C at low temperature. It is possible completely to dehydrate the hydrates by slow heating to 200 °C. However, when the substance is heated too rapidly, it may partially eliminate HCl and form calcium oxide. These preparations give an alkaline reaction when dissolved in water.

C. dissolves very readily in water (100 g water dissolves 60 g $CaCl_2$ at 0 °C, 159 g at 100 °C, and 347 g at 260 °C) and often forms supersaturated solutions. It is also soluble in ethanol and acetone. C. is synthesized by dissolving calcium carbonate (or oxide) in hydrochloric acid: $CaCO_3 + 2\,HCl \rightarrow CaCl_2 + H_2O + CO_2$. $CaCl_2$ solutions are made as byproducts of many chemical processes, for example, in the Solvay process for sodium carbonate production. C. is found in nature as tachhydrite, $2\,MgCl_2 \cdot CaCl \cdot 12H_2O$.

Anhydrous C. is a good and cheap drying agent for gases and organic liquids. It must be remembered that ammonia forms complexes with C., and alcohols also solvate it. Solutions which contain these compounds therefore cannot be dried with C. Solid C. is used to make cooling mixtures, to thaw ice on roads, and as a frost protective and setting accelerator for concrete. $CaCl_2$ solutions are used as heating bath liquids and cooling solutions. Pure C. is used in medicine, e.g. to stop bleeding, against caries and frostbite, and to treat various calcium deficiency diseases.

Calcium fluoride: CaF_2, colorless, cubic crystals: M_r 78.08, density 3.180, m.p. 1423 °C, b.p. about 2500 °C. C. is nearly insoluble in water (solubility product $K_s = 3.4 \cdot 10^{-11}$) and it is therefore nontoxic. When heated with concentrated sulfuric acid, it generates hydrogen fluoride: $CaF_2 + H_2SO_4 \rightarrow CaSO_4 + 2\,HF$. Dilute acids and even boiling alkali hydroxide solutions do not attack C. Calcium hydrogenfluoride $Ca(HF)_2 \cdot 6H_2O$, is obtained by evaporation of a solution of calcium hydroxide in excess hydrofluoric acid. C. exists in nature as fluorite (fluorspar), which is often colored, due to the presence of impurities. C. is formed as a flocculent, colorless precipitate when Ca^{2+} and F^- solutions are mixed. It is also obtained by adding calcium carbonate or oxide to aqueous hydrofluoric acid. C. is the starting material for industrial production of hydrofluoric acid and fluorine. It is used as a flux in many metallurgical processes, and is used to etch glass and as a precipitating agent in the enamel industry. Fluorspar crystals are transparent to infrared and ultraviolet rays, and are therefore used as lenses and prisms for optical devices, in astronomical telescopes and for lasers.

Calcium fluorosilicate: same as Calcium hexafluorosilicate (see).

Calcium hexafluorosilicate, ***calcium fluorosilicate***: $CaSiF_6$, colorless, tetragonal crystals; M_r 182.16, density 2.66; it is nearly insoluble in water. C. also crystallizes as a monoclinic dihydrate, $CaSiF_6 \cdot 2H_2O$. C. is made by concentration of a solution of calcium carbonate or calcium oxide in aqueous hexafluorosilicic acid. It is used as a flotation reagent, to protect wood and in the textile industry.

Calcium hexafluorosilicate, ***calcium fluorosilicate***: $CaSiF_6$, colorless, tetragonal crystals; M_r 182.16, density 2.66; it is nearly insoluble in water. C. also crystallizes as a monoclinic dihydrate, $CaSiF_6 \cdot 2H_2O$. C. is made by concentration of a solution of calcium carbonate or calcium oxide in aqueous hexafluorosilicic acid. It is used as a flotation reagent, to protect wood and in the textile industry.

Calcium hydride: CaH_2, colorless, rhombic crystalline mass, M_r 42.10, density 1.9. The commercial product, which can contain up to 10% calcium oxide and calcium nitride impurities, usually consists of gray, irregular lumps. As a typical ionic hydride, C. reacts with water with violent evolution of hydrogen: $CaH_2 + 2\ H_2O \rightarrow Ca(OH)_2 + 2\ H_2$. It is insoluble in the usual organic solvents, and only burns when strongly heated, to form calcium oxide and water. At elevated temperatures, it decomposes to hydrogen and calcium. C. is made by passing elemental hydrogen over metallic calcium at approximately 400 °C: $Ca + H_2 \rightarrow CaH_2$. C. is an excellent drying reagent for gases and organic solvents; it is also used as a source of hydrogen and as a reducing agent, especially to prepare numerous metals from their oxides.

Calcium hydrogencarbonate, formerly called ***calcium carbonate***: $Ca(HCO_3)_2$, a compound which is not known in pure, solid form. It forms as a solution when carbon dioxide is passed through an aqueous suspension of calcium carbonate: $CaCO_3 + CO_2 + H_2O \rightleftharpoons Ca(HCO_3)_2$. Together with the dissolved hydrogencarbonates of the other alkaline earth metals, it causes temporary hardness of water. When the solution is evaporated or boiled, the above equilibrium shifts to the left, causing the release of CO_2 and precipitation of neutral calcium carbonate (boiler scale). The formation of stalagmites and stalagtites in limestone caves is also due to this process.

Calcium hydrogensulfide, formerly ***calcium bisulfide***: $Ca(HS)_2 \cdot 6H_2O$, colorless, deliquescent crystals, usually colored gray by impurities. It melts at a low temperature in its own water of crystallization; M_r 214.32. Solutions of C. are made by passing hydrogen sulfide through suspensions of calcium hydroxide: $Ca(OH)_2 + 2\ H_2S \rightarrow Ca(HC)_2 + 2\ H_2O$. It can also be made by dissolving calcium sulfide in water. C. softens hairs and makes them swell, so that they can easily be removed. It is used as a cosmetic depilatory and to remove hair from hides for tanning.

Calcium hydrogensulfite, formerly ***calcium bisulfite***: $Ca(HSO_3)_2$, a compound known only in solution; M_r 202.22. The colorless aqueous solution smells like sulfur dioxide; it is made by dissolving calcium sulfite in aqueous SO_2 solution, or by passing SO_2 through calcium hydroxide suspensions. The latter method is used to make C. in industry, in that SO_2-containing roasting gases are absorbed by a $Ca(OH)_2$ suspension. C. is used in large amounts to make cellulose by the sulfite process. It is also used to disinfect barns, as an antiseptic in medicine, and as a preservative.

Calcium hydroxide, ***slaked lime, lime hydrate***: $Ca(OH)_2$, colorless, amorphous powder (density 2.08) or transparent, hexagonal crystals; M_r 74.09, density 2.24. At 580 °C, C. is converted to calcium oxide, CaO, by loss of water. C. is only slightly soluble in water: at 20 °C, 100 g water dissolves 0.165 g, and 0.077 g at 100 °C. The clear, aqueous solution (lime water) is strongly basic, has a taste of lye, and reacts with the CO_2 of the air to precipitate calcium carbonate. The suspension of C. in a saturated solution of C. is called lime milk. C. dissolves somewhat better in glycerol or sugar solutions, due to complex formation. C. is obtained in amorphous form by treating burned lime with water: $CaO + H_2O \rightarrow Ca(OH)_2$. This process, called ***slaking of lime***, releases a large amount of heat ($\Delta H = -67 kJ\ mol^{-1}$). Even when the suspension is cooled with ice, it may boil, so a large excess of water is normally used. The rate of slaking depends greatly on the grain size of the caustic lime and its purity. The resulting aqueous slurry is called ***slaked lime***.

C. is the cheapest industrial base, and as such is used widely. Slaked lime is used in the preparation of lime mortars. These are made by mixing C. or burned lime, CaO, with sand and water. The C. in the mortar reacts with the CO_2 in the air to form crystalline calcium carbonate (setting of the mortar) and this forms a solid mass with the added sand: $Ca(OH)_2 + CO_2 \rightarrow CaCO_3 + H_2O$. The sand is not involved in the setting process, but it increases the porosity of the material and thus permits the CO_2 to penetrate it more easily and the water to evaporate more quickly. The process can be accelerated by firing coke furnaces in new buildings. In principle, the setting of this material depends on the CO_2 in the air, and lime mortars are therefore also called air mortars.

Lime milk is used as whitewash; in the sugar industry, it is used to purify the crude juice and to remove the sugar from molasses. It is used in sewage treatment and as an industrial base for releasing ammonia from the ammonium chloride solutions resulting from the Solvay process. It was formerly used to make sodium hydroxide solution in the lime soda process.

Lime water is used in water treatment to remove hydrogencarbonate hardness. In medicine, it is used to treat burns and sulfuric acid and oxalic acid poisonings (by formation of insoluble calcium sulfate or oxalate). Lime water is used in qualitative analysis to detect CO_2.

Calcium hypochlorite: $Ca(ClO)_2$, colorless, highly corrosive crystalline powder which smells like chlorine; M_r 142.98, density 2.35. Technical C. with an effective chlorine content of 70 to 80% is made by passing chlorine through an aqueous suspension of calcium hydroxide. It is used as a bleach and disinfectant.

Calcium hypophosphite: $Ca(H_2PO_2)_2$, colorless, water-soluble, monoclinic crystals; M_r 170.06, density 2.51. When it is heated to about 300 °C, C. disproportionates, forming calcium phosphate and phosphine, PH_3, which ignites immediately. C. and phosphine are made by dissolving white phosphorus in boiling calcium hydroxide suspension. It is used in electroplating and in protecting metals from corrosion.

Calcium iodate: $Ca(IO_3)_2 \cdot 6H_2O$, colorless rhombic crystals, moderately soluble in water; M_r 497.98, with dehydration beginning above 35 °C. C. is formed by oxidation of calcium iodide, for example, with lime chloride. It is found in nature as lautarite.

Calcium iodide: CaI_2, colorless, deliquescent hexagonal crystals; M_r 293.89, density 3.956, m.p. 784 °C, b.p. 1100 °C. It is soluble in water, acetone and alcohol, and also forms a hexahydrate, $CaI_2 \cdot 6H_2O$ (m.p. 42 °C). In order to prevent oxidation of the iodide, C. must be stored away from light and air. It is obtained by dissolving calcium carbonate or calcium oxide in aqueous hydroiodic acid. C. is used for iodine therapy and as an expectorant.

Calcium molybdate: see Molybdate pigments.

Calcium nitrate: $Ca(NO_3)_2$, colorless, hygroscopic, cubic crystals; M_r 164.09, density 2.504, m.p. 561 °C. C. dissolves very readily in water and alcohol and crystallizes out of water below 132 °C as a colorless, monoclinic tetrahydrate, $Ca(NO_3)_2 \cdot 4H_2O$, which deliquesces in air. Its density is 1.82. If C. is heated above its melting point, it first evolves oxygen, and at higher temperature, nitrogen oxides are also evolved. C. occurs in nature as nitrocalcite (lime saltpeter). It often forms when nitrogen-containing organic materials decay in the presence of lime, for example as a bloom on the walls of stalls. At one time, C. was obtained in this way in saltpeter plants. It is now made industrially by dissolving limestone (or calcium hydroxide) in nitric acid and evaporation of the solution: $CaCO_3 + 2\ HNO_3 \rightarrow Ca(NO_3)_2 + CO_2 + H_2O$. C. is also made in the course of dissolving crude phosphates in nitric acid. It is precipitated from the solution either as $Ca(NO_3)_2 \cdot 4H_2O$ and worked up to solid C., or it is applied along with the dissolved phosphate as a nitrogen-phosphate mixed fertilizer. Industrially prepared C., often called ***Norway saltpeter*** because it was formerly prepared in this fashion only in Norway, is used in large amounts as a basic and rapidly acting nitrogen fertilizer. Its tendency to absorb water is a problem in this application, but it can be almost completely overcome by adding lime (CaO).

Calcium oxalate: CaC_2O_4, colorless, cubic crystals; M_r 128.10, density 2.2. C. also crystallizes as the monoclinic monohydrate, the tetragonal trihydrate and in a triclinic form with 2.5 mol crystal water. When heated, it decomposes to calcium carbonate and carbon monoxide: $CaC_2O_4 \rightarrow CaCO_3 + CO$. It is only slightly soluble in water (solubility product $K_s = 2.57 \cdot 10^{-9}$ at 25 °C), and is therefore used for qualitative and quantitative analysis of calcium. It is soluble in strong acids. C. is formed by mixing solutions of calcium and oxalic acid or alkali oxalates. It is found in nature as whewellite, and in the cells of many plants.

Calcium oxide, ***caustic lime, burned lime***: CaO, M_r 56.08; a colorless, soft, amporphous substance obtained by burning pure calcium carbonate. It can be converted by sublimation to colorless, cubic crystals; density 3.25-3.38, m.p. 2416 °C, b.p. 2850 °C. If C. is heated with an oxyhydrogen torch, it glows with a very bright, white light (Drummond's limelight). C. is converted exothermally to calcium hydroxide by water: $CaO + H_2O \rightarrow Ca(OH)_2$ (*slaking of lime*). C. reacts with carbon at about 2000 °C to form calcium carbide, CaC_2; when heated, magnesium reduces C. to elemental calcium. Above 500 °C, C. binds CO_2 to form calcium carbonate. C. is made by thermal decomposition of calcium carbonate (*lime burning*) at 900-1000 °C in shaft, ring or rotating furnaces: $CaCO_3 \rightarrow CaO + CO_2$; $\Delta H = 181$ kJ mol^{-1}. C. is used to make slaked lime, lime milk, lime water and mortar. It is also used to release ammonia from gas water or ammonium chloride solution (Solvay process) and to make lime chloride. It is a starting material for the production of calcium carbide and is used as a basic additive in smelting of ores. C. is used as a furnace lining, in glassmaking, in the sugar and cellulose industries and as a fertilizer.

Calcium peroxide: CaO_2, colorless, tetragonal crystalline powder which is nearly insoluble in water; M_r 72.08, density 2.92. C. dissolves in acids, such as sulfuric, with the formation of the corresponding calcium salt and hydrogen peroxide: $CaO_2 + H_2SO_4 \rightarrow CaSO_4 + H_2O_2$. The compound can be made by dehydration of its octahydrate, $CaO_2 \cdot 8H_2O$, which is made by the reaction of hydrogen or sodium peroxide with calcium hydroxide solution. It is used as an antiseptic.

Calcium phosphates: 1) in the narrower sense, calcium salts of orthophosphoric acid, H_3PO_4. a) ***Calcium dihydrogenphosphate, primary calcium phosphate, monocalcium phosphate***: $Ca(H_2PO_4)_2 \cdot H_2O$, colorless triclinic crystals which are readily soluble in water; M_r 252.07, density 2.220. At 109 °C the monohydrate converts to the anhydrous form. If it is heated above 200 °C, more water is lost, leading to mixtures of calcium metaphosphate, $Ca(PO_3)_2$, and neutral calcium diphosphate, $Ca_2P_2O_7$. The aqueous solution of the salt gives an acidic reaction. The compound can be produced by mixing stoichiometric amounts of calcium carbonate or oxide with pure phosphoric acid, followed by evaporation of the solution, or precipitation of tertiary calcium phosphate or hydroxylapatite with strong acids, e.g. $2\ Ca_5(PO_4)_3OH + 7\ H_2SO_4 + 12\ H_2O \rightarrow 3\ Ca(H_2PO_4)_2 + 7\ CaSO_4 \cdot 2H_2O$. The mixture of primary calcium phosphate and gypsum formed according to this equation is called superphosphate. It is widely used as a phosphate fertilizer, as is the pure $Ca(H_2PO_4)_2$ (double superphosphate) obtained by dissolving tertiary calcium phosphate or apatite in phosphoric acid. Mixed with alkali carbonates, primary calcium phosphate is used as baking powder.

b) ***Calcium hydrogenphosphate, secondary calcium phosphate, dicalcium phosphate***, $CaHPO_4 \cdot 2H_2O$ forms colorless, triclinic crystals which are nearly insoluble in water; M_r 172.09, density 2.306. The salt is soluble in acid. When it is in contact with water for a long time, it is gradually converted to hydroxylapatite, $Ca_5(PO_4)_3OH$ and calcium dihydrogenphosphate, $Ca(H_2PO_4)_2$. When it is roasted, it is converted by dehydration to calcium diphosphate, $Ca_2P_2O_7$. The compound is made by adding disodium hydrogenphosphate to a neutral calcium salt solution. This phosphate is found in nature in the form of the minerals bruskite and monetite; along with calcium metaphosphate, it is a component of guano. It is used as a feed additive, a polish in toothpaste, and in the glass and porcelain industries.

c) ***Tricalcium phosphate, tertiary calcium phosphate, calcium phosphate***, $Ca_3(PO_4)_3$, is a colorless, amorphous mass; M_r 310.18, density 3.14, m.p. 1670 °C. The compound is extremely insoluble in water, but it is slowly converted by water to hydroxylapatite, $Ca_5(PO_4)_3OH$, which is also insoluble. Tricalcium phosphate is dissolved by strong acids, such as sulfuric, nitric or phosphoric, which convert it to soluble calcium dihydrogenphosphate; this is soluble and therefore available for plants. For example $Ca_3(PO_4)_2 + 2\ H_2SO_4 + 4\ H_2O \rightarrow Ca(H_2PO_4)_2 + 2\ CaSO_4 \cdot 2H_2O$. Depending on the stoichiometric ratios, the reaction of sulfuric acid with apatite or phosphorite can also produce phosphoric acid. $Ca_3(PO_4)_2$ is dissolved to double superphosphate. At 1300 to 1450 °C in an electric furnace, $Ca_3(PO_4)_2$ reacts with carbon and silicon dioxide, SiO_2, to form elemental phosphorus, P_4. The compound exists in nature, as phosphorite and apatite, that is, double salts of $Ca_3(PO_4)_2$ with CaF_2, $CaCl_2$ or $Ca(OH)_2$. In the form of apatite, it is also the main component of bone and tooth minerals. Pure tricalcium phosphate is made by melting calcium oxide with phosphorus(V) oxide at 1100 °C. Tricalcium phosphate in the form of phosphorite or apatite is the starting material for production of phosphoric acid fertilizers, phosphoric acid and elemental phosphorus. It is used in comparatively small amounts as a polish in toothpaste and to make enamels and opaque glasses.

2) In the broader sense, calcium salts of condensed phosphoric acids. a) ***Calcium diphosphate, dicalcium diphosphate, calcium pyrophosphate***, $Ca_2P_2O_7$, is a colorless, amorphous powder, which is relatively insoluble in water; M_r 254.10, density 3.09, m.p. 1230 °C. It is made by roasting calcium hydrogenphosphate or by fusing calcium metaphosphate and calcium oxide. b) ***Calcium metaphosphate***, $Ca(PO_3)_2$, is a colorless amorphous substance, formula mass 198.02, density 2.82, m.p. 975 °C, which is nearly insoluble in water and aqueous acids. The salt consists of cyclic and linear condensed polyanions. The use of the term "metaphosphate" is not strictly correct in this context. The compound is made by severe roasting of primary calcium diphosphate; calcium dihydrogenphosphate is formed as an intermediate:

$$Ca(H_2PO_4)_2 \xrightarrow{-H_2O} CaH_2P_2O_7 \xrightarrow{-H_2O} Ca(PO_3)_2.$$

Calcium phosphide: Ca_3P_2, a red-brown, lumpy, amorphous, poisonous mass; M_r 182.19, density 2.51, m.p. about 1600 °C. It is slowly decomposed by water and rapidly by aqueous acids, evolving phosphine: $Ca_3P_2 + 6\ H_2O \rightarrow 3\ Ca(OH)_3 + 2\ PH_3$. C. is made by fusing phosphorus and calcium. It is used to combat voles and other pests.

Calcium silicates: the calcium salts of silicic acid. There are two types: ***calcium orthosilicates***, e.g. Ca_2SiO_4 ($2\ CaO \cdot SiO_2$) and $Ca_2Si_2O_7$ ($3\ CaO \cdot 2SiO_2$), are derived from orthosilicic or orthodisilicic acid, while ***calcium metasilicates***, $CaSiO_3$ are based on linear and cyclic condensed polysilicate anions. As mixed silicates (with aluminum, magnesium, iron and sodium), the C. are a major component of silicate rocks. For example, the mineral wollastonite, $Ca_2Si_2O_6$, is a pure calcium metasilicate. C. are made industrially by melting stoichiometric amounts of calcium oxide or carbonate with silicon dioxide. These compounds are essential components of cement and many glasses. The pure compounds are used, among other things, as adsorbants for gases and liquids, as pigment stretchers and filter aids.

Calcium silicides: gray intermetallic compounds with varying stoichiometric Ca:Si ratios. The best known is ***calcium disilicide***, $CaSi_2$; M_r 96.25, density 2.5. It is made by melting calcium and silicon or silicon dioxide in an electric furnace. The C. react with hydrochloric acid to form silicon hydrides. They are used in the steel industry as deoxidizing reagents.

Calcium silicofluoride: see Calcium hexafluorosilicate.

Calcium sulfate: $CaSO_4$, colorless, orthorhombic crystals found in nature as anhydrite; M_r 136.14, density 2.61. Below 66 °C, C. crystallizes out of water as a monoclinic dihydrate, $CaSO_4 \cdot 2H_2O$. This occurs naturally as gypsum, of which an especially beautiful form is alabaster. Both forms are only slightly soluble in water (solubility product $K_s = 2.45 \cdot 10^{-5}$ at 25 °C, which corresponds to 0.085 g in 100 g water). C. is synthesized by the reaction of sulfur trioxide and calcium oxide, or by mixing solutions containing calcium and sulfate ions. The thermal behavior and uses of the compound are discussed under Gypsum (see).

Calcium sulfide: CaS, a cubic, crystalline mass, which in pure form is colorless, but is usually yellowish and opaque; M_r 72.14, density 2.5. C. dissolves slowly in cold water, forming calcium hydrogensulfide, $Ca(HS)_2$; when heated, it evolves hydrogen sulfide, H_2S, and calcium hydroxide, $Ca(OH)_2$. C. is formed when hydrogen sulfide is passed over glowing calcium carbonate, or by reduction of calcium sulfate with carbon. It was an intermediate in the now obsolete Le Blanc process for sodium carbonate production. C. is phosphorescent, and is therefore used to make television screens. It is used as a hair remover in tanneries and as a lubricant.

Calcium sulfite: $CaSO_3 \cdot 0.5H_2O$, colorless, hexagonal crystals which are nearly insoluble in water; M_r 129.15. C. is a reducing agent, and is slowly converted to calcium sulfate, $CaSO_4$, by oxygen. It is formed when sulfur dioxide is passed through a solution of calcium hydroxide, or when solutions of calcium salts and alkali sulfites are mixed. C. is used as a disinfectant in the sugar industry and breweries, to preserve fruit juices and as an antichlorine in textile bleaches.

Calcium thiosulfate: $CaS_2O_3 \cdot 6H_2O$, colorless, triclinic crystals which are readily soluble in water; M_r 260.30, density 1.872. C. is obtained by boiling calcium hydroxide solution with sulfur, then passing sulfur dioxide through the solution. It is used as an antidote for metal poisonings.

Caldo process: see Steel.

Californium, symbol *Cf*: a radioactive element which does not occur naturally on earth; it belongs to the Actinides (see) of the periodic system; *Z* 98, with isotopes with the following mass numbers (decay type and half-life in parentheses): 240 (α; 1.06 s), 241 (K capture, α; 3.78 min), 252 (α; 3.68 min), 246 (α; 35.7 h), 247 (K; 2.45 h), 248 (α; 333.5 d), 249 (α; 350.6 a), 250 (α; 13.08 a), 251 (α; 898 a), 252 (α; 262 a), 253 (β^-, α; 17.8 d), 254 (spontaneous fission; 60.5 d). The stablest isotope has atomic mass 251, valences are II, II and IV; m.p. 986 °C.

C. is a silvery white metal which exists in three modifications; it is obtained by reduction of californium(III) oxide or fluoride with lanthanum. Some important californium(III) compounds are ***californium(III) oxide***, Cf_2O_3, a yellow-green compound which exists in three modifications, m.p. 1750 °C; bright green ***californium(III) fluoride***, CfF_3; emerald green ***californium(III) chloride***, $CfCl_3$; and lemon yellow ***californium(III) iodide***, CfI_3. The other two oxidation states are represented by the black-brown ***californium(IV) oxide***, CfO_2, the green ***californium(IV) fluoride***, CfF_4, yellow ***californium(II) bromide***, $CfBr_2$, and dark violet ***californium(II) iodide***, CfI_2. Californium(IV) compounds are strong oxidizing agents, and californium(II) compounds are strong reducing agents. Cf^{4+} ions (brown) are not stable in aqueous solution, in contrast to the green Cf^{3+} ions.

C. was first synthesized in 1950 at the University of California at Berkeley, by Thompson, Street, Jr., Ghiorso and Seaborg, by bombarding curium 242 with α-particles: $^{242}Cm + {}^{4}He \rightarrow {}^{245}Cf + {}^{1}n$. In 1958, Cunningham and Thompson isolated the Cf nuclides 249 to 252 from plutonium 239 which had been irradiated with neutrons for a period of 5 years in a nuclear reactor. Today these nuclides are available on a mg scale. Other syntheses are based on irradiation of uranium 238 with accelerated carbon or nitrogen nuclei: $^{238}U + {}^{12}C \rightarrow {}^{244}Cf + 6\,{}^{1}n$; $^{238}U + {}^{14}N \rightarrow {}^{247}Cf + 4\,{}^{1}n + {}^{1}H$.

Callendar's equation: see State, equation of, 1.2.1.

Calomel: see Mercury chlorides.

Calomel electrode: a type-two electrode which is very frequently used as a reference electrode. It consists of very pure mercury with a surface layer of mercury(I) chloride (calomel), Hg_2Cl_2. The electrolyte is a potassium chloride solution of defined activity which is saturated with calomel. A platinum wire dipped in the mercury, but not in contact with the electrolyte, is used for electrical contact with the remainder of the circuit. The potential-determining process is $Hg_2^{2+} + 2e \rightleftharpoons 2\,Hg$. The activity of the mercury(I) ions is determined by the chloride ions, via the solubility product of calomel. The following electrode potentials were determined for several different C.:

$Hg/Hg_2Cl(s)$, KCl(aq, saturated), +0.241 V;
$Hg/HgCl_2(s)$, KCl(aq, c = 1 mol/l), +0.280 V;
$Hg/HgCl_2(s)$, KCl(aq, c = 0.1 mol/l), +0.334 V
(s = solid, aq = aqueous solution, c = concentration).

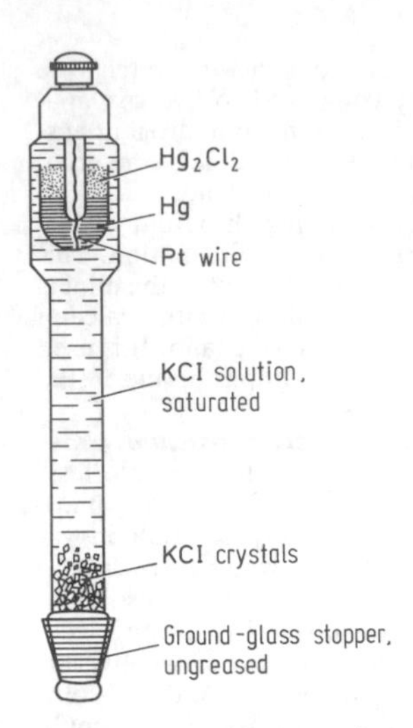

Saturated calomel electrode

Caloric equation of state: see State, equation of, 2.

Calorization: see Aluminization.

Calorie: a unit of heat, now obsolete. One calorie (lower-case "c") = 4.1868 joules. The Calorie (capital "C") is still popularly used as a unit for the energy in foods; one Calorie (1 kcal) = 1000 calories.

Calorimetry: the measurement of heat capacities of substances and of the heat released or absorbed by various processes, such as heats of reaction, mixing, solution and melting. A ***calorimeter*** is used to make such measurements; the heats are determined by means of temperature changes in the calorimeter liquid, or by comparison with other, known heats. There are three basic types of calorimeter: anisothermic, isothermic and adiabatic. In anisothermic calorimeters, the heat q generated by a process is calculated from the heat capacity c of the calorimeter and the temperature change ΔT: $q = c\Delta T$. In isothermic calorimeters, the heat produced is absorbed by another process, such as a phase change, so that the temperature remains constant. An example is the Bunsen ice calorimeter, in which the quantity of melted ice is measured. Adiabatic calorimeters work without heat exchange; this is achieved by heating the mantle so that its temperature remains equal to that in the interior.

There are different types of calorimeter for different applications: mixing calorimeters for heats of mixing and reaction in liquids; flow calorimeters, metal block calorimeters, calorimetric bombs for heats of combustion, and differential calorimeters.

Calvin cycle: see Photosynthesis.

cAMP: see Cyclic nucleotides.

Camphane: see Monoterpenes.

Camphechlor, ***polychlorocamphene, Toxaphen®, Melipax®***, a chloroterpene insecticide. C. is a mixture of chlorinated C_{10} hydrocarbons with the average composition $C_{10}H_{10}Cl_8$. The wax-like, yellowish mass has a melting range of 65-95 °C, density 1.65, and vapor pressure ≈ $6.6 \cdot 10^{-4}$Pa. Gas chromatography and mass spectrometry have revealed about 180 different components in the technical product; only a few of their structures have been elucidated. The mixture has a faint terpene odor. It dissolves readily in aromatic hydrocarbons, somewhat less readily in aliphatic hydrocarbons, and to about 6 ppm in water. In the presence of alkalies and when heated to 150 °C, it loses HCl and activity. C. is synthesized by photochlorination of camphene in tetrachloromethane to a chlorine content of ≈ 67%. The p.o. LD_{50} values (acute oral toxicity) are between 40 and 300 mg/kg rat. There is disagreement over its persistence and ability to accumulate in the fatty tissues of warm-blooded animals. In some countries, its application is limited or forbidden.

C. has a long-term insecticidal effect. It is primarily an ingestion poison, but has some contact effect on biting insect pests in agriculture and silviculture. C. is the only chloroterpene insecticide which is not harmful to bees, and can be used on blooming plants, e.g. rape. C. is used on non-edible crops, especially cotton. Its side effects as a rodenticide are utilized to a limited extent to combat field mice. Because of its high toxicity to fish, C. is used under certain conditions as a piscicide.

Camphor: a bicyclic monoterpene ketone. C. is optically active and exists in d- and l-forms. It forms colorless crystals with a characteristic odor; they are nearly insoluble in water, but dissolve very easily in ethanol; m.p. 178-179 °C (independent of optical activity). Most natural C. is in the d-form, and is synthesized mainly by *Lauraceae* and *Labiatae*.

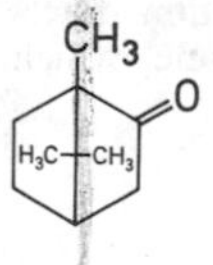

d-C., also known as ***Japanese camphor***, is obtained by steam distillation from the wood of *Cinnamomum camphora*. l-C., also called ***matriacaria camphor***, is found in many of the *Asteraceae*. C. is obtained industrially by partial synthesis from pinene, which is present in terpentine oil. The most commonly used synthetic pathway is a rearrangement of α-pinene to camphene, formylation to isoborneol formate, saponification to isoborneol and catalytic hydrogenation to C. This synthetic C. is optically inactive. It is used as a gelatinizing agent for cellulose acetate and nitrate. C. has a weak excitatory effect on the central nervous system; in higher doses, it causes paralysis and is an antiseptic. It is used externally to promote blood flow in the skin.

Camphoric acid: 1,2,2-trimethylcyclopentane-1,3-dicarboxylic acid. C. exists in a *cis* and *trans* form (also *iso*), and each of these structural isomers has a d-, l- and dl-form. Oxidation of camphor yields mainly *cis*-C. C. readily forms an anhydride.

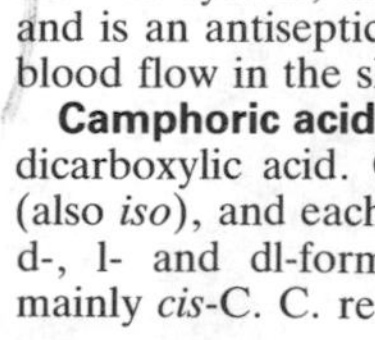

Camphor leaf oil: an essential oil from the leaves of the East Asian camphor tree; it is extracted from the leaves, which contain up to 1% oil, by steam distillation. The following compounds are usually present: camphor (up to 50%), caryophyllene (up to 20%) and small amounts of α-pinene, dipentene and camphene. C. is used in the perfume industry.

Camptothecin: an alkaloid isolated from the bark and wood of the Chinese tree *Camptotheca acuminata*. It is a pyrrolidinoquinoline derivative with antitumor activity, but has so far not been used in therapy.

Canada balsam: a liquid, tough natural resin which is colorless when fresh, but turns yellowish to light brown upon standing for a long time, and hardens. C. has a pleasant, balsamic odor and a bitter taste. Its main components are about 50% α- and β-canadinolic acid, $C_{16}H_{29}COOH$, about 15% canadinic acid, $C_{19}H_{37}COOH$ and about 5% canadoresene, $C_{21}H_{40}O$.

C. is obtained from the bark of certain conifers which grow in Canada and the northernmost parts of the USA, such as the firs *Abies canadensis* and *Abies balsamae*. C. is used as a glass glue to cement optical lenses and as an embedding and preserving material for various preparations.

Candelilla wax: a plant Wax (see) in which a high-molecular-weight fatty acid is esterified to a polyalcohol. It also contains free fatty acids, resinoic acids and lactones. C. is obtained from the Mexican euphorbia *Pedilanthus pavonis*. It is used as a substitute for carnauba wax in making floor and shoe polishes, and as an additive to soap emulsions.

Cannabinoids: dibenzopyranes present in the blossoms and upper leaves (*marijuana*) or the resins (*hashish*) of the female hemp plant (*Cannabis sativa*). C. make up 2 to 20% of the dried herb or resins.

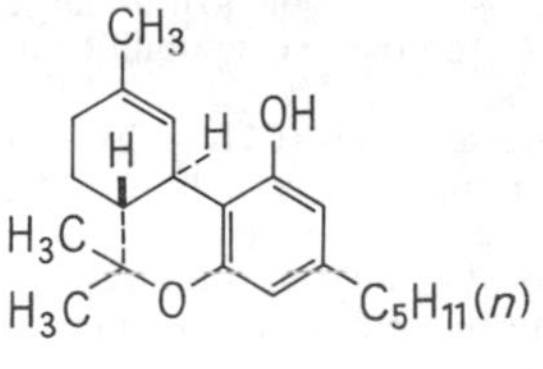

(−)-Δ^9-*trans*-tetrahydrocannabinol

The nature of the C. depends on the type of hemp. The most active compound is (-)-Δ^9-*trans*-tetrahydrocannabinol, which is present at an average concentration of 3% in hashish. C. are psychoactive; small doses lead to euphoria, and larger doses to visual and auditory impairments and hallucinations.

Cannel coal: see Coal.

Cannizzaro reaction: a disproportionation of aldehydes, especially aromatic and heterocyclic aldehydes and formaldehyde, in the presence of bases, to form the corresponding carboxylic acids and alcohols:

$$2\ R\text{–}CHO \xrightarrow{+KOH} R\text{–}COOK + R\text{–}CH_2\text{–}OH$$

It was reported in 1853 by S. Cannizzaro. The mechanism of the C. can be described as follows:

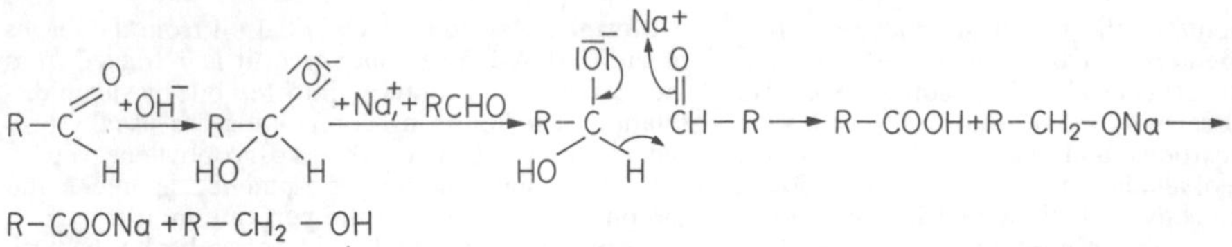

$$R-C(=O)H \xrightarrow{+OH^-} R-C(O^-)(HO)H \xrightarrow{+Na^+ + RCHO} R-C(O^-)(HO)H \cdot O=CH-R \rightarrow R-COOH + R-CH_2-ONa \rightarrow$$

$$R-COONa + R-CH_2-OH$$

Various aldehydes which cannot be enolized undergo the ***crossed C*** in the presence of alkalies. When formaldehyde is added, it always acts as the hydride ion donor, forming formic acid and the corresponding alcohol:

$$H{-}CHO{-} + R{-}CHO \xrightarrow{+NaOH} H{-}COONa + R{-}CH_2{-}OH.$$

Another variant of this reaction is the conversion of α-oxoaldehydes into α-hydroxycarboxylic acids through intramolecular C. in the presence of alkalies. For example, glycolic acid is formed from glyoxal in the presence of aqueous hydroxides:

$$OHC{-}CHO \xrightarrow{+H_2O(OH^-)} HO{-}CH_2{-}COOH$$

Canonical equation of state: see State, equation of, 3.

Caphnetin: see Coumarin.

Capillary condensation: see Adsorption.

Capillary gas chromatography: see Gas chromatography.

Capillary pressure: a pressure which arises at any curved phase interface: $P = 2\sigma/r$, where P is the C., σ is the surface tension and r is the radius of curvature of the interface. The C. is larger for small droplets than for larger ones, which leads, for example, to isothermal distillation of small droplets into large ones. The same applies to the relative solubilities of smaller and larger crystals.

Capric aldehyde, ***decanal***: CH_3-$(CH_2)_8$-CHO, a pleasant-smelling liquid; m.p. - 5 °C, b.p. 207-209 °C, n_D^{20} 1.4287. C. is present in free form in many essential oils, for example, the oils of mandarins, orange peel, lavender, camphor and terpentine. C. is used mainly as a flavoring and aroma.

Capri blue effect: see Spectral sensitization.

Capric acid: same as Decanoic acid (see).

Caproic acid: same as Hexanoic acid (see).

ε-Caprolactam, ***6-aminohexanoic acid lactam, hexahydro-IH-azepin-2-one***: the internal anhydride of ε-aminocaproic acid. It cyrstallizes in colorless, slightly hygroscopic leaflets, m.p. 70 °C, b.p. 268.5 °C, and is readily soluble in water, alcohol and ether. ε-C. is the product of Beckmann rearrangement (see) of cyclohexane oxime. When heated in the presence of traces of water, it polymerizes via the intermediate ε-aminocaproic acid to form poly-ε-aminocaproic acid, the starting material for the polyamides (perlon, nylon 6).

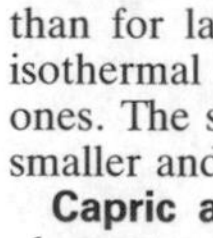

Fig. 1. ε-Caprolactam

Because of its great commercial importance, a number of syntheses for ε-C. have been developed. Most start from one of three substances: cyclohexane, phenol or toluene. Cyclohexanone oxime, the immediate precursor of ε-C., can be produced by the following methods:

1) *Hydrogenation of phenol* on a nickel contact (100 °C, 1.47 MPa) to form cyclohexanol, and subsequent dehydrogenation at 400 °C to produce cyclohexanone. Reaction with hydroxylamine produces cyclohexanone oxime, which is rearranged to ε-C.

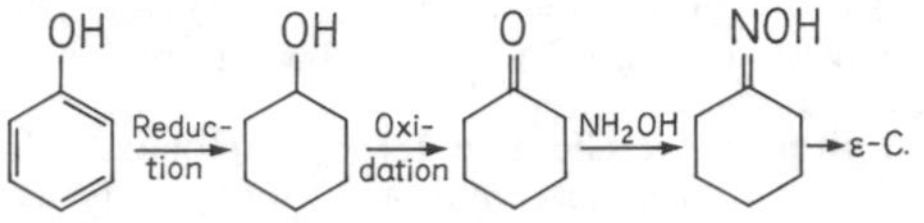

Fig. 2. Hydrogenation of phenol

2) *Catalytic oxidation of cyclohexane in air*. Cyclohexane is obtained by hydrogenation of benzene, and the cyclohexanone produced by its oxidation reacts with hydroxylamine to cyclohexanone oxime, which is rearranged to ε-C.

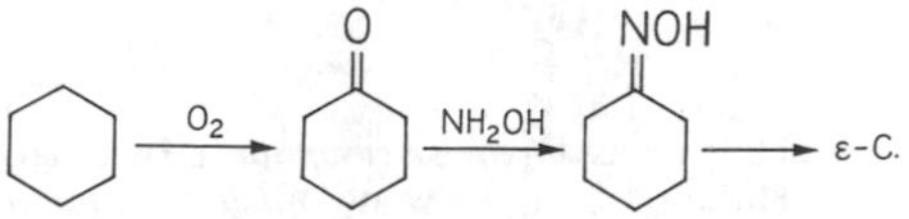

Fig. 3. Catalytic oxidation of cyclohexane

Nixan process: Cyclohexane is nitrated to nitrocyclohexane (130 °C, 0.29 MPa), and this is reduced on a copper catalyst (200 °C, 9.81 MPa) to nitrosocyclohexane. The nitrosocyclohexane rearranges in the presence of acid to cyclohexanone oxime; Beckmann rearrangement produces ε-C.

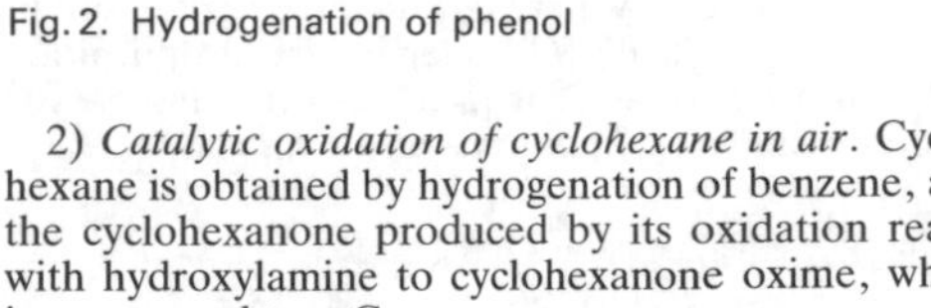

Fig. 4. Nixan process

4) *Photonitrosylation (PNC process)*: In this process, cyclohexane reacts with nitrosylchloride or a mixture of Cl and NO in the presence of UV light, producing nitrosocyclohexane. This rearranges to cyclohexanone oxime in the presence of acid, and Beckmann rearrangement yields ε-C.

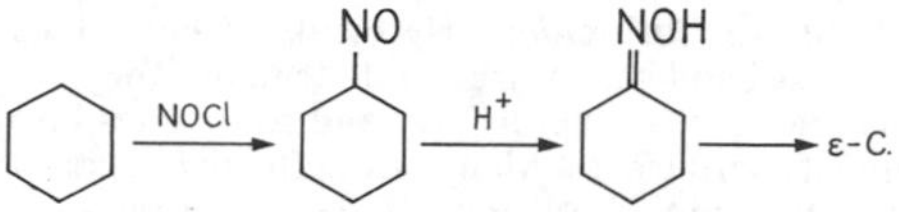

Fig. 5. Photonitrosylation (PNC process)

5) *SNIA-Viscose process*: Toluene is catalytically (cobalt stearate, 160 to 180 °C, 0.78 MPa) oxidized to benzoic acid, and this is hydrogenated, in the form of its methyl ester (180 °C, nickel-aluminum oxide catalyst), to hexahydrobenzoic acid. Hexahydrobenzoic acid reacts directly with nitrosylsulfuric acid, in a mixture of oleum and sulfuric acid, to yield ε-C.

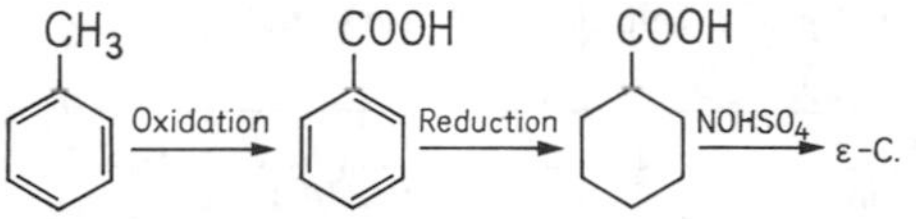

Fig. 6. SNIA-Viscose process

6) Starting from benzene, cyclohexylamine is produced via nitrobenzene and aniline. The cyclohexylamine is oxidized with hydrogen peroxide to cyclohexanone oxime, which is converted to ε-C. by Beckmann rearrangement.

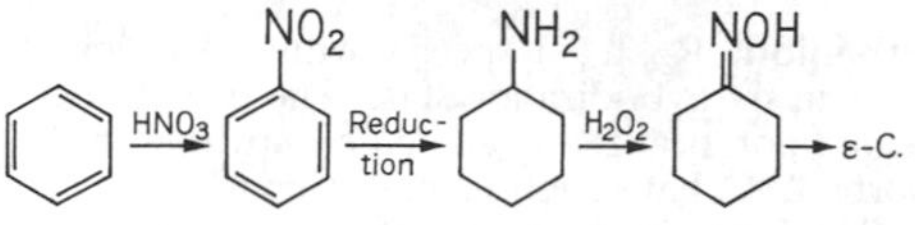

Capronaldehyde, ***hexanal, hexylaldehyde***: CH_3-$(CH_2)_4$-CHO, a colorless liquid soluble in many organic solvents; b.p. 128 °C. C. is found free in a few essential oils, such as templin and camphor oils.

Caprylaldehyde, ***octanal, octylaldehyde***: CH_3-$(CH_2)_6$-CHO, a pleasant smelling liquid; b.p. 170-173 °C, n_D^{20} 1.4217. C. is present in citron, orange, mandarin and lavender oils. It is used as an aroma and flavor in cosmetics and foods.

Caprylic acid: same as Octanoic acid (see).

Caps: small amounts of an explosive mixture, such as potassium chlorate, $KClO_3$, and red phosphorus, placed between two strips of paper, usually red. They are used in toy pistols and the like.

Capsanthin: see Xanthophylls.

Captan: see Fungicides.

Captopril: see ACE inhibitor.

Carane: see Monoterpenes.

Carbachol: $[(CH_3)_3N^+$-CH_2-CH_2O-CO-$NH_2]Cl^-$, the carbamate ester of choline in chloride form. C. is a direct parasympathicomimetic. It is more stable than acetylcholine, and can therefore be administered orally for intestinal or bladder atonia.

Carbamates: see Carbamic acid, see Urethanes.

Carbamate insecticides: a group of biologically active compounds which are *N*-methylcarbamic acid esters (Table).

Table. Important insecticidal *N*-methylcarbamates, R-O-CO-NH-CH_3

Name	R	m.p. in °C
Carbaryl	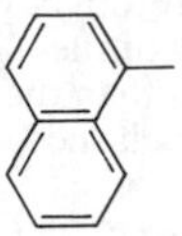	142
Carbofuran	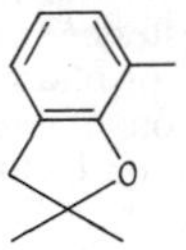	153–154
Dioxacarb	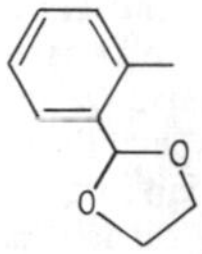	114–115
Propoxun	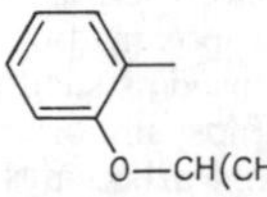 O—CH$(CH_3)_2$	91.5
Mercaptodimethur (Methiocarb)	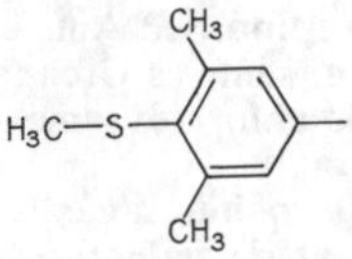 CH_3, H_3C—S—, CH_3	121.5
Ethiofencarb	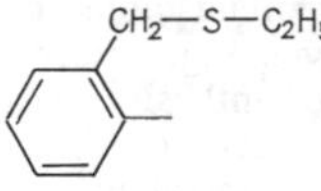 CH_2—S—C_2H_5	33.4
Mexacarbot	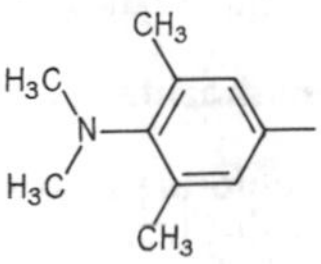 CH_3, $(H_3C)_2$N—, CH_3	85
Butacarb	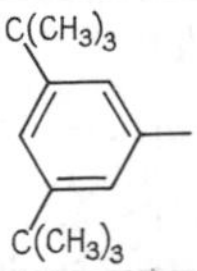 $C(CH_3)_3$, $C(CH_3)_3$	102–103
Aldicarb	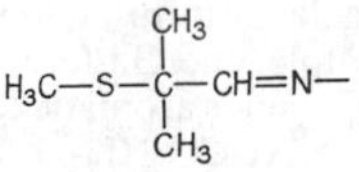 H_3C—S—C$(CH_3)_2$—CH=N—	100
Oxamyl	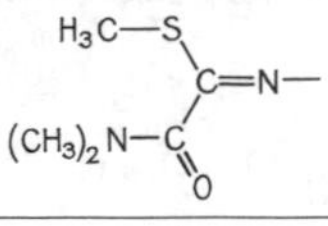 H_3C—S—C(=N—)—C(=O)—N$(CH_3)_2$	100–102
Methomyl	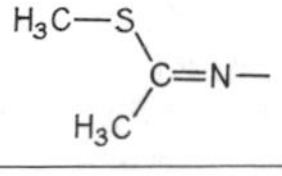 H_3C—S—C(H_3C)=N—	78–79

Since the introduction of carbaryl in 1958, the fraction of the total production of insecticides represented by C. has increased continuously. The C., especially carbaryl, are effective against DDT-resistant insects, and many of them are much less toxic to warm-blooded animals than the Organophosphate insecticides (see), although the C. are also cholinesterase inhibitors.

The technical synthesis of C. begins with substituted phenols or oximes, which are converted to the *N*-methylcarbamate esters. The reaction of phenol with methylisocyanate or phosgene to chloroformate esters, and the reaction of the latter with methylamine is another common method.

The main applications of C. are on cotton, fruit, vegetable and feed plants. Many C. are systemic (see Systemic pesticides), and can thus protect the plant against pests which are otherwise difficult to reach; however, some of these C. are also very toxic. The systemic action is especially valuable for the use of these compounds as soil insecticides and Nematocides (see); examples are aldicarb, carbofuran and oxamyl. The less toxic C. are used in hygiene and veterinary medicine, e.g. propoxur, butacarb and dioxacarb. Some C. are also used as snail poisons, e.g. mercaptodimethur. The most important subdivisions of the C. are the phenylcarbamates and the oximecarbamates.

Carbamazepine: see Anticonvulsants.

Carbamide: same as Urea (see).

Carbamide chloride: same as Carbamoyl chloride (see).

Carbamide resins: same as Urea resins (see).

Carbamic acid, ***aminoformic acid***: $H_2N\text{-}CO\text{-}OH$, in free form, an unstable compound. It has some stable derivatives, e.g. urea, which are used as intermediates in the synthesis of pesticides, drugs, synthetic fibers and special organic ocmpounds. The salts and esters of C. are called ***carbamates***; for example, see Ammonium carbamate. The esters are also called Urethanes (see).

Carbamic hydrazide: same as Semicarbazide (see).

Carbamoyl chloride, ***carbamide chloride, urea chloride***: $H_2N\text{-}CO\text{-}Cl$, an unstable, crystalline compound; melting is not sharp, around 50 °C; b.p. 61-62 °C. In aqueous solution, C. is decomposed to the unstable carbamic acid and hydrochloric acid. The carbamic acid immediately decomposes to carbon dioxide and ammonia. The reactions of C. are similar to those of the carboxylic acyl chlorides. It reacts with compounds with reactive H atoms, cleaving off hydrogen chloride to form amides, carbamates, ureas, etc. For example, the Gattermann-Hopff reaction of arenes with C. yields aromatic carboxamides: $Ar\text{-}H + H_2N\text{-}CO\text{-}Cl \rightarrow Ar\text{-}CO\text{-}NH_2 + HCl$. C. is synthesized by reaction of phosgene with molten ammonium chloride:

$$O{=}C(Cl)_2 + NH_4Cl \longrightarrow O{=}C(NH_2)(Cl) + 2\,HCl$$

C. and *N*-substituted derivatives are important intermediates for the synthesis of pesticides and pharmaceuticals.

Carbanilate herbicides: Herbicides (see) which contain a carbanilic acid structural element. They are used as selective soil herbicides, and sometimes also as contact herbicides. Many are effective against grasses and inhibit both mitosis (cell division) and photosynthesis. They are produced either by reaction of arylisocyanate with an alcohol or by reaction of an alkyl chloroformate with arylamine. *IPC* (propham: isopropyl-*N*-phenylcarbamate) and *CIPC* (chloropropham: 3-chloroisopropyl-*N*-phenylcarbamate) are used in large amounts both to combat weeds and to prevent sprouting of potatoes. Proximpham is a component in pre-emergence combination herbicides used on sugar beet fields. The "biscarbamate" phenmedipham is used as a post-emergence control in sugar beet fields.

Table. Carbanilate herbicides (X-substituted phenyl)–NH–COOR

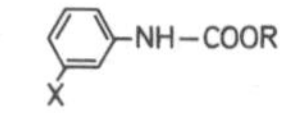

Name	X	R
Propham	H	$CH(CH_3)_2$
Chlorpropham	Cl	$CH(CH_3)_2$
Barban	Cl	$CH_2{-}C{\equiv}C{-}CH_2Cl$
Proximpham	H	$N{=}C(CH_3)_2$

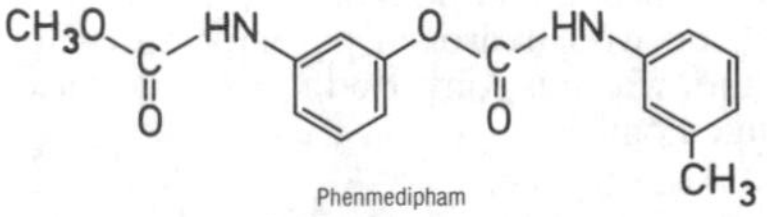

Phenmedipham

Carbanion: $R_3C|^-$, a species with a trivalent C atom in an sp^3 hybridization state. The two electrons which are not localized in a bond occupy one sp^3 hybrid orbital. C. have a tetrahedral structure. They are formed by heterolytic cleavage of the bond between a carbon atom and an atom of lower electronegativity, as in an organometal compound, or by proton loss from a CH acidic compound: $R_3C\text{-}X \rightarrow R_3C|^- + X^+$.

Carbanionoids: compounds in which a negative charge on a carbon atom is stabilized by a neighboring positively charged heteroatom. The compounds have a high degree of charge separation, but they are electrically neutral. Some examples are phosphorus, sulfur and nitrogen ylides.

Carbaryl: a Carbamate insecticide (see).

Carbazole, ***dibenzopyrrole, 9-azafluorene***: a heterocyclic compound which crystallizes in colorless leaflets or platelets, m.p. 247-248 °C, b.p. 355 °C; the solid sublimes readily. C. is readily soluble in pyridine, acetone and concentrated sulfuric acid, slightly soluble in benzene, ether and chloroform, and insoluble in water. Sulfonation, nitration and halogenation reactions introduce substituents in the 3 and 6 positions. C. can be synthesized from 2-aminobiphenyl by oxidation, from 2-nitrobiphenyl by reduction, from 2,2'-diaminobiphenyl in the presence of phosphoric acid or from diphenylamine. It is extracted industrially from the crude anthracene fraction of coal tar distillates, which contain about 20%, by treatment with potassium.

C. was discovered in 1872 by Graebe and Glaser in coal tar. It is a starting material for vinylcarbazole, pesticides, pigments and plastics.

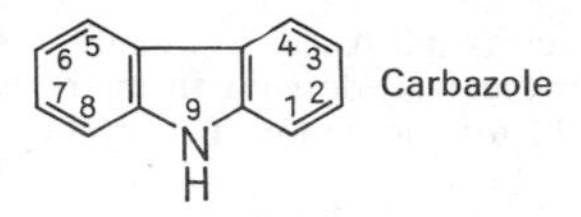

Carbendazim: see Benzimidazole fungicides.

Carbene: R_2C: an uncharged species in which a C atom is covalently bound to two substituents, and has two non-bonding orbitals occupied by its other two valence electrons. The geometric and electronic structure of C. depends on the nature of the substituents R. The simplest C., ***methylene*** (H_2C:) exists in a triplet state and has an angular structure (C_{2v} symmetry). In other cases, triplet carbenes have an sp-hybridized C atom, which dictates a linear structure. Formally, triplet carbenes are diradicals, but because of the localization of both unpaired electrons on the same C atom, they have some peculiarities. The ***dihalogen carbenes*** are relatively easy to prepare and are important in syntheses; they are singlet C. with angular structures. Although they have a nonbonding electron pair, they are generally electrophilic, due to the electron deficiency. C. are formed by photolysis or thermolysis of diazo compounds according to

$$R_2\overset{-}{C}-\overset{+}{N}\equiv N| \rightarrow R_2C: + N_2$$

or by α-elimination of HX from haloforms in the presence of strong bases (e.g. RO^-): $CHX_3 + RO^- \rightleftharpoons CX_3^- + ROH$, $CX_3^- \rightarrow X_2C: + X^-$. This can be achieved by phase-transfer catalysis of a two-phase reaction between haloform and sodium hydroxide in aqueous solution in the presence of a quaternary ammonium salt.

C. with different multiplicities have different reactivities. Some typical *reactions of the singlet carbenes* are 1) 1,2-sigmatropic shift with formation of ethylenes, e.g. 1,2-rearrangement of ketocarbenes to ketenes (see Wolff rearrangement); 2) stereospecific insertion into σ bonds, e.g.

$$H_2\ddot{C} + H-CH_2-Ph \rightarrow CH_3-CH_2-Ph$$

or 3) stereospecific [2+1]-cycloaddition to π-bonds:

$$R_2C: + \text{alkene} \rightarrow \text{cyclopropane } (R\,R)$$

(e.g. synthesis of tropones, 1,6-methano-[1-]-annulenes). If C. are added to alkenes or cycloalkenes in [2+1]-cycloadditions, forming cyclopropanes, the reaction is called a ***C. transfer***. 4) Addition of a nucleophilic reagent (in rare cases, even an electrophilic reagent), e.g. the Reimer-Tiemann reaction.

Some typical *reactions of triplet carbenes* are 1) abstraction of an atom to form other radicals, e.g.

$$C_6H_5\dot{C}\cdot + H-C(CH_3)_2-OH \rightarrow C_6H_5\dot{C}H + \cdot C(CH_3)_2-OH \rightarrow C_6H_5CH_2 + (CH_3)_2C{=}O$$

2) nonstereospecific addition to π-bonds, e.g.

$$R_2\dot{C}\cdot + \text{alkene} \rightarrow \text{cyclopropane } (R\,R) + \text{cyclopropane } (R\,R)$$

and 3) addition of radicals or radical-like substrates.

Carbene complexes: organometallic complexes of electron-rich transition metals with carbene (CR_2-) ligands. The metal-ligand bond in C. has considerable double-bond character ($M{=}CR_2$). First obtained in 1964 by Fischer and Maasbœl, C. can be made by alkylation of metal hexacarbonyls with lithium alkyls followed by methylation of the intermediate anion:

$$M(CO)_6 + CH_3^- \rightarrow [(OC)_5MC(O)CH_3]^-$$
$$[(OC)_5MC(O)CH_3]^- + CH_2N_2 + H^+ \rightarrow [(OC)_5MC(OCH_3)CH_3] + N_2$$

C. may also be obtained by cleavage of electron-rich alkenes.

Carbenicillin: see Penicillins.

Carbenium immonium ion: an unstable, reactive intermediate in the reaction of carbonyl compounds with ammonia, primary and secondary amines, hydroxylamine, hydrazines or semicarbazide:

$$R_2C{=}O + HNR^1R^2 \rightarrow R_2C(OH)-NR^1R^2 \xrightarrow[-H_2O]{+H^+} R_2\overset{+}{C}-NR^1R^2 \longleftrightarrow R_2C{=}\overset{+}{N}R^1R^2$$

The C. are stabilized by loss of a proton, by reaction with ammonia to give aldimines (R^1, R^2 = H), with primary amines to form Schiff's bases (R^1 = H), with secondary amines to form enamines, with hydroxylamine to form oximes (R^1 = H, R^2 = OH), with hydrazines to form hydrazones (R^1 = H, R^2 = NH_2) or with semicarbazide to make semicarbazones (R^1 = H, R^2 = $NHCONH_2$). In the reaction of benzaldehyde and formaldehyde with secondary amines, the C. is stabilized by addition of a second molecule of amine to form aminals. C. are formed as intermediates in the Mannich and Ugi reactions.

Carbenium ions: R_3C^+, classical ions with sp^2-hybridized, positively charged carbon atoms. C. have a planar configuration, with the unoccupied 2p atomic orbital arranged perpendicular to the σ-bonds. It is stabilized by electron-pair donor substituents, or by groups in which the positive charge can be delocalized (e.g. triarylcarbenium ions). On the other hand, bulky substituents, which do not fit in a planar arrangement, reduce the stability. As in the case of triaryl carbenium ions, some cyclic conjugated compounds which obey Hückel's rule have delocalized positive charges (cyclopropenylium and cycloheptatrienylium ions). C. appear as intermediates in substitutions, additions, eliminations and rearrangements. They are formed by heterolysis of C-X bonds (X is more electronegative than carbon), for example, in S_N1 reactions or when diazonium ions decay.

Carbenium oxonium ion: an unstable, reactive

intermediate in the reaction of a carbonyl compound with an alcohol in the presence of a strong acid:

$$>C=O + ROH \rightleftharpoons >C(OH)(OR) \xrightleftharpoons{+H^+, -H_2O} >\overset{+}{C}-OR \longleftrightarrow >C=\overset{+}{O}R.$$

The C. is stabilized by addition of a second alcohol molecule, which forms an acetal or ketal (analogous to the formation of the aminals).

Carbenoids: α-metal-α-halogen-hydrocarbons. They are formed as intermediates, for example in the reaction of butyllithium with 1,1-dihalogen-hydrocarbons:

$$R(R^1)CX_2 \xrightarrow[-R_2X]{R^2Li} R(R^1)C(X)Li$$

C. are formed in the Simmons-Smith reaction.

$$R(R^1)CX_2 \xrightarrow[-R^2X]{+R^2Li} R(R^1)C(X)Li$$

Carbide: see Calcium carbide.

Carbides: binary compounds of carbon with electropositive elements, particularly with numerous metals, boron and silicon. There are ionic, covalent and interstitial (metallic) C.

1) ***Ionic C.*** are derived from elements of the first three main groups of the periodic system. They have ionic lattices of negatively charged carbon species and metal cations. The carbon anions react with water as strong bases, forming the corresponding acids. Therefore, ionic C. are decomposed to hydrocarbons and metal hydrides by water. Beryllium and aluminum form methanides, Be_2C and Al_4C_3 (see Aluminum carbonide), which react with water to form metal hydroxides and ethyne (acetylene). The most important of this group is Calcium carbide (see), CaC_2, which is produced on an industrial scale to synthesize ethyne: $CaC_2 + H_2O \rightarrow CaO + HC{\equiv}CH$. Allenide ions, C_3^{2-}, containing ionic C. are also known.

2) ***Covalent C.*** are stable in water and acids, and crystallize readily. Typical examples are Silicon carbide (see), SiC, and Boron carbide (see), B_4C. In SiC, silicon and carbon atoms occupy the positions of a diamond lattice; in B_4C, the carbon is incorporated into the lattice of α-rhombohedral boron. This explains the chemical inertness and great hardness of covalent C.

3) ***Interstitial C.***, or ***metallic C.***, are non-stoichiometric compounds in which the carbon occupies some of the interlattice positions in the metallic lattice, usually of elements of Groups IVb to VIIIb. These compounds have the properties of alloys; they are often harder and melt at higher temperatures than the pure metals. When they are hydrolysed, mixtures of hydrocarbons are formed. Metallic C. are made by heating the metals or metal oxides with coke, or the metal is reacted with ethyne.

Carbinol: see Methanol.

Carbinolamines: a term still used in alkaloid chemistry for cyclic iminoacetals, derived from the obsolete name "carbinol" for methanol. The cyclic carbinolamine form

$$>NH + O{=}CH-.$$

is in equilibrium with the open aldehyde form

$$>N-C(OH)<$$

An example of a C. is the alkaloid berberin.

Carboanhydrase: a zinc-containing, single-chain enzyme which catalyses the reversible reaction of water and carbon dioxide to carbonic acid. It is important in respiration, CO_2 transport and acid-base regulation. High concentrations of C. are present in erythrocytes, gastric mucosa, kidneys and eye lens. The zinc atom, which is covalently bound to the polypeptide chain (M_r 28,000 to 30,000) is essential for activity. Sulfide, cyanide and cyanate ions inhibit C.

Carboanhydrase inhibitors: see Diuretics.

Carbocations: term for all positively charged carbon ions; they are divided into two groups, the Carbenium ions (see) and the Carbonium ions (see).

Carbochromes: see Coronary pharmaceuticals.

Carbocyclic: term for a compound containing a ring consisting entirely of carbon atoms.

Carbofuran: a Carbamate insecticide (see).

Carbohydrate, ***saccharide***: one of the natural products derived from Monosaccharides (see); the typical representatives have the net formula $C_x(H_2O)_y$. Monosaccharides can be linked by glycosidic bonds to form Oligosaccharides (see) and Polysaccharides (see). Mono- and oligosaccharides are also called *sugars*. The properties of oligosaccharides are similar to those of monosaccharides, while polysaccharides are polymers with quite distinctive properties.

Carbohydrate resins: synthetic resins made either by heating carbohydrates with added mineral acids, or by reaction of carbohydrates with phenols and phenol derivatives. If carbohydrates (e.g. cellulose, dextrin, starch) are heated with mineral acids, resins are obtained which are used mainly in paints. Pure carbohydrates, or also lignin-containing materials (such as wood, straw or reeds) react with phenols, cresols, naphthalene or aromatic amines upon heating in acidic media to form hardenable resins, from which cheap pressed materials can be made. The degree to which the material can be hardened can be regulated by addition of aldehydes, such as formaldehyde or furfural, which are incorporated during the condensation reaction.

β-Carboline: see Rauwolfia alkaloids.

Carbolineum: a red-brown mixture of coal-tar components, including phenols, cresols, anthracene, phenanthrene, naphthalene, chrysene, etc. It is insoluble in water, combustible, very irritating to the skin and carcinogenic if applied over a long period. C. vapors irritate the respiratory passages. C. is a disinfectant and inhibits spoiling. It is used to combat pests in fruit orchards and as a preservative paint for railroad ties, posts, walls, etc.

Carbolated oil: see Tar.

Carbolic acid: see Phenol.

Carbon: symbol ***C***, chemical element from the fourth main group of the periodic system, the Carbon-silicon group (see). C is a a nonmetal, atomic number 6, with stable isotopes of mass numbers 12 (98.89%) and 13 (1.11%). Its atomic mass is 12.01115, and its valence is IV, or, rarely, II.

In addition to the stable isotopes ^{12}C and ^{13}C, C exists as the radioactive isotope ^{14}C, which is formed in the high layers of the atmosphere through neutron capture by nitrogen nuclei according to the scheme: $^{14}N(n;p)^{14}C$. Its half-life is 5570 years, and it decays by β-emission to nitrogen. Organisms exchange CO_2 with the atmosphere as long as they live, and thus contain the same level of ^{14}C as the atmosphere. When they die, the exchange stops and the amount of ^{14}C in their tissues decays. Therefore the ^{14}C content of archaeological objects is an indicator of their age, and ^{14}C analysis is an important method of dating. The mass of an atom of the carbon isotope ^{12}C is the basis of the relative atomic mass scale in use since 1961.

Properties. Elemental C exists in three modifications, the colorless, cubic Diamond (see), the dark gray, hexagonal Graphite (see), and the black, powdery Fullerenes (see). The latter is the thermodynamically stable form at room temperature. Amorphous C (see Soot, Charcoal, Animal charcoal, Active charcoal) consists mostly of microcrystalline, highly distorted graphite. A colorless, transparent material called "white C" was obtained in 1969 by vacuum sublimation of graphite at 2250 K. However, it has not yet been fully characterized. Diamond converts to graphite at room temperature, but extremely slowly. Above 1500 °C, the process occurs rapidly. The reverse process, conversion of graphite to diamond, occurs under extreme conditions.

In accordance with its position in the periodic system, C occurs mainly in covalent, tetravalent form. Its ability to form stable C-H and C-C bonds, to form long chains and rings, and to form stable bonds with other elements of the first period gives C its special position among the elements (see Carbon-silicon group). The Tetrahedral model (see) of C is used to explain the isomerism of tetravalent C. Regardless of the modification, C is relatively unreactive at room temperature. It burns at higher temperatures, forming carbon monoxide (CO) or dioxide (CO_2), depending on the amount of oxygen available. When heated, C also reacts with sulfur to form carbon disulfide, CS_2, with hydrogen to form hydrocarbons, and with boron, silicon and various metals to form Carbides (see). C reacts even at room temperature with fluorine, forming carbon tetrafluoride, but it must be heated to react with chlorine; it then forms hexachloroethane and hexachlorobenzene.

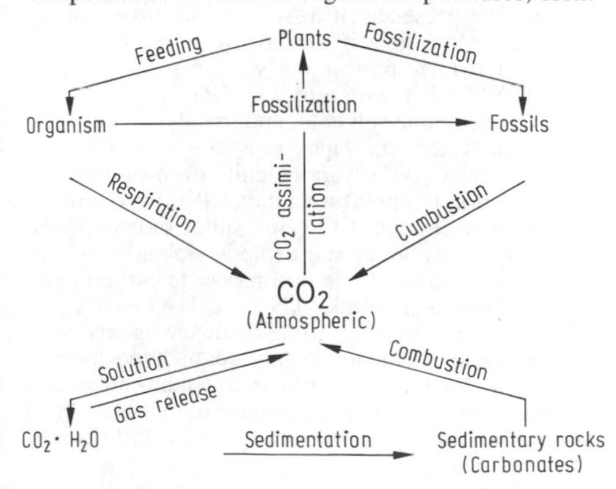

Analytical. Qualitative analysis for C is done by oxidizing the element or its compounds to CO_2 or carbonates (e.g. by combustion or in a KNO_3 melt). The CO_2 is identified, either directly or after release from the carbonate, by reaction with barium hydroxide solution to form barium carbonate, $BaCO_3$. For quantitative analysis (see Elemental analysis), C is burned in an oxygen stream in the presence of copper(II) oxide; the CO_2 formed is determined gravimetrically by absorption on sodium hydroxide or by gas chromatography.

Occurrence. Of the total carbon content of the earth (about 0.12% of the crust), more than 99.8% exists in minerals, including coal. About 0.1% is in the form of CO_2 in the atmosphere or seawater, and about 0.001% is in the biosphere. The most important minerals are calcium carbonate, $CaCO_3$ (limestone, chalk, marble), magnesium carbonate, $MgCO_3$ (magnesite), calcium magnesium carbonate, $CaCO_3 \cdot MgCO_3$ (dolomite) and iron carbonate, $FeCO_3$ (iron spar). C is the main component of coal, petroleum and natural gas, and of other fossil materials such as asphalt and mineral wax. The atmosphere contains about 0.03% CO_2. The natural C reservoirs are interconnected by the CO_2 cycle.

Details on the structure and behavior, isolation and applications of C are given in the entries on Graphite (see) and Diamond (see).

Carbonate: 1) a *salt* of carbonic acid, H_2CO_3. As a dibasic acid, H_2CO_3 forms two series of salts: the ***hydrogencarbonates (primary C.**, formerly **bicarbonates)*** which contain the HCO_3^- ion, and the ***neutral C. (secondary C.)***, which contain the CO_3^{2-} ion.

Neutral C. with the composition $M^I_2CO_3$ or $M^{II}CO_3$ are formed by the reaction of carbon dioxide with metal hydroxide solutions, or by addition of soluble C. to metal salt solutions. With the exception of alkali metal, ammonium and thallium(I) carbonate, the neutral C. are relatively insoluble. The CO_3^{2-} ion is the corresponding base of the relatively weak carbonic acid, and it therefore reacts with water to form its corresponding acid and release OH^- ions. This is the reason that aqueous solutions of neutral C. are basic. Strong acids release carbonic acid, which immediately decomposes to carbon dioxide and water. This reaction, coupled with the precipitation of barium carbonate from barium hydroxide solution, is also used to detect C.

In the presence of water and excess carbon dioxide, the neutral C. are converted to hydrogencarbonates: $M^{II}CO_3 + CO_2 + H_2O \rightleftharpoons M^{II}(HCO_3)_2$. When the temperature is raised, these decompose, in a reversal of their equation of formation. With the exception of sodium hydrogencarbonate, they are readily soluble in water. This is very important for the weathering of C.-containing rocks. The interaction of calcium and magnesium hydrogencarbonates with water and the

CO_2 of the air introduces calcium and magnesium carbonates into all natural waters, and accounts for their temporary hardness. The neutral C. precipitate as boiler scale (see Hardness, 2) when they are boiled.

Many C. are found in nature, including $CaCO_3$ (limestone, chalk, marble), $MgCO_3$ (magnesite), $MgCO_3 \cdot CaCO_3$ (dolomite), $FeCO_3$ (siderite), $MnCO_3$ (rhodochrosite), $ZnCO_3$ (zinc spar) and Na_2CO_3 (soda). Some C. are important industrial products, e.g. soda and potassium carbonate.

2) An ester of carbonic acid whith the general formula RO-CO-OH (carbonic semi-ester) or R^1O-CO-OR^2. ***Carbonic semi-esters***, RO-CO-OH, are very unstable, and most are known only in the form of their salts or their internal anhydrides. The esters R^1O-CO-OR^2 can be symmetrical or unsymmetrical, depending on the groups R^1 and R^2. They are stable compounds with a pleasant odor. They are insoluble in water, but dissolve readily in most organic solvents. They have the typical reactions of Esters (see). Symmetric C. can be synthesized from phosgene and alcohols. C. esters are used as solvents for polymeric compounds and as synthesis reagents for introducing alkoxycarbonyl groups into CH acidic compounds and for making polycarbonates.

Carbonate hardness: see Hardness 2.

Carbonate white: same as Lead white (see).

Carbon dioxide: CO_2, a colorless gas with a slightly acidic odor and taste; M_r 44.01, density of the solid, 1.53, m.p. - 51.6 °C at 0.52 MPa, subl.p. - 78.5 °C, crit. temp. 31 °C, crit. pressure 7.29 MPa. The CO_2 molecule is linear. Dry ice (see) is obtained by rapid expansion of compressed C. C. is rather stable and does not decompose to carbon and oxygen until it reaches very high temperatures. It is a very weak oxidizing agent, and does not support combustion or respiration. Hydrogen and a few electropositive metals such as sodium, potassium, magnesium and zinc are able to reduce C. at high temperatures to carbon monoxide. C. dissolves in water: at 15 °C and atmospheric pressure, 1 l water will dissolve about 1 l CO_2. About 0.1% of the dissolved C. reacts with the water to form carbonic acid: $CO_2 + H_2O \rightleftharpoons H_2CO_3$; the dissociation of carbonic acid is responsible for the weakly acidic reaction of the solution. Ammonia adds to C., forming ammonium carbonate, which is an intermediate in the industrial synthesis of urea. C. is detected by precipitation of barium carbonate out of barium hydroxide solution.

> Carbon dioxide cannot support respiration and therefore, if breathed for a long time at high concentrations, it causes death by suffocation.

C. is a component of the atmosphere (0.03%) and in this form it is assimilated by plants. Many natural springs contain C., and large amounts are dissolved in the oceans. In the form of carbonates, C. makes up a significant part of the earth's crust. It has a central position in the natural interconversions of various carbon compounds (CO_2 cycle; see Carbon).

C. is produced industrially by a lime-burning process: $CaCO_3 \rightarrow CaO + CO_2$, and by alcoholic fermentation. Large amounts of C. are produced by combustion of fossil fuels and in the conversion of carbon monoxide. To separate C. from gas mixtures, these are subjected to pressure scrubbing with water (after decompression, the C. escapes from solution), or it is absorbed in carbonate or amine solutions, from which it is released by heating. In the laboratory, C. is made by the reaction of carbonates with strong acids. C. is commercially available in liquid form.

C. is a raw material for industrial production of soda, urea, lead white, barium carbonate and ammonium sulfate. Large amounts are used in the beverage industry. It is also used as a fire extinguishing material, as a protective gas for welding and some chemical processes, for carbonic acid fertilization in greenhouses, and, in the form of dry ice, as a cooling material. C. is used in natural and artificial medicinal baths.

Carbon dioxide complexes: coordination compounds of electron-rich d-transition metals containing CO_2 as a ligand. Stable C. were only discovered in the last few years. They are synthesized by fixing CO_2 on transition metals with high electron densities by addition or substitution of labile ligands, e.g. $[Ni(P(C_6H_{11})_3)_2] + CO_2 \rightarrow [((C_6H_{11})_3P)_2NiCO_2]$ (Fig. 1). They can also be obtained by reduction of transition metal complexes in the presence of CO_2, as in the reaction $(h^5\text{-}CH_3C_5H_4)_2Nb(CH_2Si(CH_3)_3(Cl) + Na + CO_2 \rightarrow NaCl + [h^5\text{-}CH_3C_5H_4)_2Nb(CH_2Si(CH_3)_3(CO_2)]$, or by generating complex-fixed CO_2 within the coordination sphere, e.g. 4"$Re(CO)_5OH$" + 2 $CO_2 \rightarrow [\{(OC)_5Re_2\}(\mu\text{-}CO_2)_2\{Re(CO)_4\}_2] + 2\ H_2O$ (Fig. 2). The figures give an impression of the bonding in C. with known structures. At present, there is particular interest in C., because CO_2 fixation to transition metal centers would open the possibility of activating the potential C_1 synthetic unit CO_2.

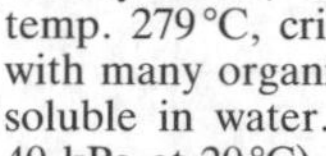

1 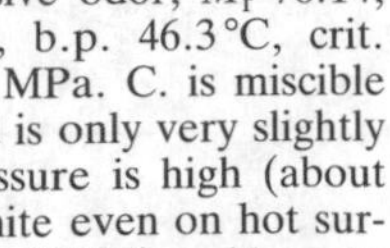 2

Carbon disulfide: CS_2, colorless liquid with a high index of refraction; due to impurities it may be light yellow. The presence of dissolved, sulfur-containing impurities often give it a repulsive odor; M_r 76.14, density 1.261, m.p. - 110.8 °C, b.p. 46.3 °C, crit. temp. 279 °C, crit. pressure 7.8 MPa. C. is miscible with many organic solvents, and is only very slightly soluble in water. Its vapor pressure is high (about 40 kPa at 20 °C). The vapors ignite even on hot surfaces (ignition temperature about 100 °C) and burn to form carbon dioxide, CO_2, and sulfur dioxide, SO_2. Mixtures of C. and air can explode violently. Above 150 °C, water causes C. to decompose to carbon dioxide and hydrogen sulfide: $CS_2 + 2\ H_2O \rightarrow CO_2 + 2\ H_2S$. It is oxidized by strong oxidizing agents, such as potassium permanganate and sodium hypochlorite, to carbon dioxide and sulfur or sodium sulfate. It reacts with chlorine in the presence of manganese(II) chloride to carbon tetrachloride, CCl_4, and disulfur dichloride, S_2Cl_2.

Many nucleophiles are able to add to the C=S bond of C. For example, C. reacts with sulfides to give trithiocarbonates: $CS_2 + SH^- + OH^- \rightarrow CS_3^{2-} + H_2O$. Addition of OH^- or O^{2-} ions leads to mixed O,S carbonates.

The dangers of working with carbon disulfide arise both from its exceptionally high toxicity and from the danger of fire or explosion. Carbon disulfide should be used only when it is really essential, and then with the greatest caution (hood, gas mask). It is absorbed through both the lungs and the skin. Inhalation leads to excitation, sometimes to convulsions; higher doses cause unconsciousness and possibly respiratory paralysis. Chronic poisoning leads to headache, tiredness, stomach complaints and visual impairment, later to severe damage to the central and peripheral nervous systems. Carbon disulfide is extremely inflammable. The flame point is at - 30 °C, and the ignition temperature is about 100 °C. Its vapors form explosive mixtures with air in the concentration range from 0.8 to 52.6%.

Alkoxides form xanthogenates with C., for example, $CS_2 + C_2H_5O^- \rightarrow C_2H_5OCSS^- Na^+$. This reaction type corresponds to the reaction of cellulose with C. and sodium hydroxide to form cellulose xanthogenate in the preparation of viscose. Secondary amines add to C. according to $CS_2 + HNR_2 + OH^- \rightarrow R_2N\text{-}CSS^- + H_2O$, forming dithiocarbamates, which are excellent chelating ligands. They form stable complexes with transition metals, some of which are used in agriculture as fungicides.

C. is made industrially by the reaction of sulfur vapor with coal around 900 °C. It is used mainly in the production of viscose and is an intermediate in the synthesis of carbon tetrachloride and dithiocarbamates. C. is an excellent solvent for fats, oils, rubber, various polymers and many other organic and inorganic compounds, but because of its high toxicity, its applications are limited. It is used as a reagent for organic syntheses.

Carbonic acid: H_2CO_3, a diprotic acid which is known only in water. It is made by dissolving carbon dioxide, CO_2, in water. However, according to the following equilibrium, only 0.1% of the dissolved CO_2 is actually C.: $CO_2 + H_2O \rightleftharpoons H_2CO_3$. Therefore, C. released from carbonates is largely decomposed into CO_2 and water. Depending on how the acid dissociation constant,

$$K_a = \frac{[HCO_3^-]\,[H_3O^+]}{[H_2CO_3]},$$

is calculated, that is, whether the total amount of dissolved CO_2 or only the actual amount of H_2CO_3 is used in the denominator, $K_a = 4.45 \cdot 10^{-7}$ (apparent dissociation constant) or $K_a = 2.5 \cdot 10^{-4}$ (true dissociation constant). Thus C. is actually a medium strong acid. In basic solutions, the dissociation equilibrium, and therefore the formation equilibrium for C., is shifted to the right. That is, CO_2 dissolves completely and reacts to form the Carbonates (see) $M^I_2CO_3$, the neutral salts of C. With excess CO_2, the carbonates form the hydrogencarbonates, M^IHCO_3.

Carbonic acid diamide: same as Urea (see).

Carbonic acid dichloride: same as Phosgene (see).

Carbonification: see Coal.

Carbonitration, ***cyanization***: a process for surface treatment of steels. A gaseous, solid or liquid reagent is applied to the steel at 800-950 °C; it consists of carbon and nitrogen compounds and increases the amount of those elements in the surface layers of the steel. Subsequent quenching (rapid cooling) increases the hardness and resistance to wear of the steel surface.

Carbonium ions: carbocations with pentacoordinated carbon atoms. However, the name is also applied to nonclassical ions, in which electrons from σ-bonds are involved in the delocalization of the positive charge. C. with pentacoordinated C atoms are formed in the reactions in a mass spectrometer; they are intermediates in the reaction of alkanes with superacids, such as the "magic acid" antimony(V) fluoride and fluorosulfonic acid, or with strong Lewis acids. Such reactions include electrophilic substitutions, protolyses and rearrangements.

The prototype of a nonclassical ion is the norbornyl cation with three normal two-electron bonds and one three-center, two-electron bond:

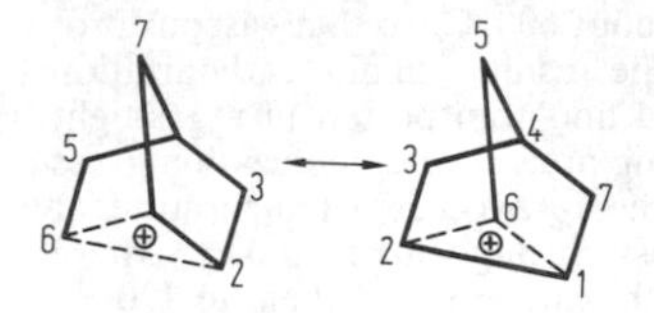

Carbonization: a form of Coal refining (see) in which coal, oil shale, peat and wood are decomposed in the absence of air at temperatures around 600 °C (low-temperature de-gassing, *dry distillation*). The volatile components of the material are driven off, and it is partially cracked.

1) C. of ***coal***.

a) The first plant for C. of lignite coals, chiefly those with at least 12% tar and at most 20% ash content, was opened in 1846 in Aschersleben.

At the beginning of this century, the most common process was ***hot surface C.***. The heat was transported through the ceramic walls of a cylindrical furnace. There were no particular grain size or hardness limitations on the material carbonized, and furnace production capacities were between 3 and 6 t coal and 200 kg tar per day. In addition, the long times spent in the furnaces led to cracking and thus to lower tar yields. Later, briquets were used, which increased the tar production to 500 kg per day and the daily consumption of briquets to 13 tons. Currently most lignite is processed in the form of briquets, in Lurgi rinsing gas furnaces (Fig.). The advantages of this method are that residual drying is separated from C. with relatively small amounts of foul water, that the tar produced is of high quality, and the coke is in small pieces. A modern rinse-gas furnace processes about 500 t lignite briquets daily. A loader distributes

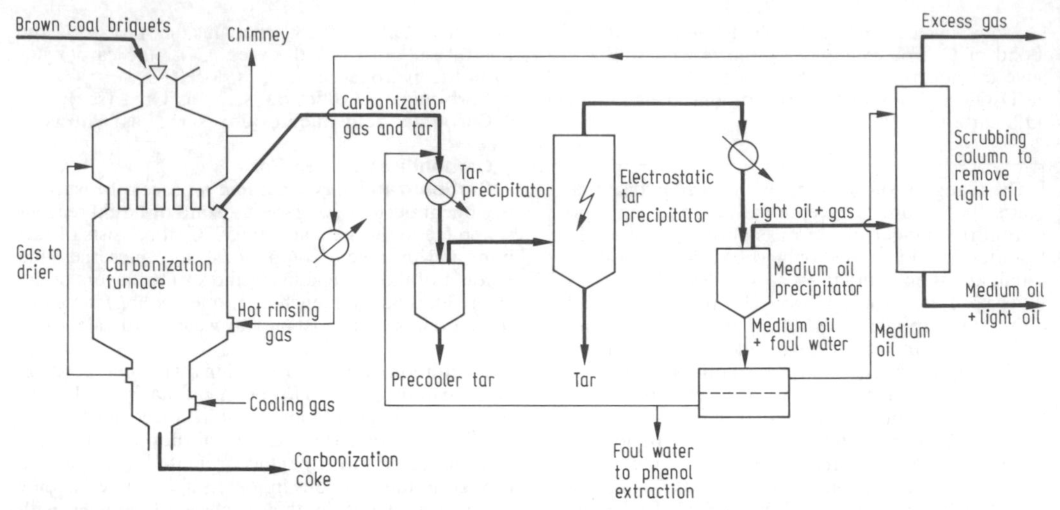

Diagram of a rinsing gas furnace

the briquets evenly over the entire furnace cross section. In the drier, which makes up the upper part of the furnace, the briquets are first warmed to 100-120 °C, so that the water content drops from 15 to 0.5%. In the next area, the carbonization zone, the dried briquets are rinsed by a hot gas (700 °C), and are thus heated to about 600 °C, so that gaseous products are formed. The residue, lignite carbonization briquets, is removed and transported in a gas-tight container to an ageing area. The carbonization gases are sucked out of the furnace at a temperature of 240 °C, and first pass through a pre-cooler, where they are sprayed with water to cool them to 120 °C. This causes part of the tar to condense, and washes out the coal dust which has been carried along. The gas then passes into an electrical purifier, where the remaining tar is precipitated. The gas from the furnace enters a cooler, where its temperature drops to 20 °C, and the middle oil and foul water condense. The remaining gasoline-like hydrocarbons are then washed out with middle oil. The purified gas is then led back to the furnace, where it is used to heat the drier and carbonizer. Since more carbonization gas is generated than is needed to heat the reactor, a large fraction of it can be used for other purposes (excess gas).

Balance of a Lurgi rinse-gas carbonization furnace processing 500 t lignite briquets per day

Product	Yield in t day^{-1}
Excess gas	37
Light oil	13
Tar	58
Coke	235

In *Collin-Humboldt cyclone C.*, non-baking fuels with diameters of 0 to 3 mm are carbonized by hot gases; the carbonization time is less than one minute, and the capacity of the furnace is more than 500 t per day. The heating gas enters the furnace at 900 to 1000 °C, and leaves it at 180-220 °C. b) The C. of anthracite is very much less important than that of lignite. Only those types of anthracite are used which are not suitable for coking. Here, too, the rinse gas method is very important.

Products of C. The main product of coal C. is *carbonization coke*, which, after tar, returns the most money. It is a matte black, porous, hygroscopic, grainy substance, with an ash content of about 25% and a heat of combustion of about 22,000 kJ kg^{-1}. C. coke is very reactive. In order to reduce its considerable tendency to ignite spontaneously, it is aged with steam, and must, in addition, have at least 20% moisture content when transported. It is used to produce synthesis gas (in Winkler generators), energy (dust firing in power plants) and, in small amounts, to heat houses. It is not suitable for metallurgical purposes. Anthracite carbonization coke has a heating value of about 28,000 kJ kg^{-1} and does not ignite spontaneously. It is used for heating houses, and as a starting material for synthesis gas. In addition, it is used in the chemical industry to produce ferrosilicon and calcium carbide, as a reducing agent in zinc and tin smelting, and as a fuel for furnaces.

The *foul water* is an undesired byproduct, from which phenol, cresols and xylenols are extracted by alkaline washing or by the phenosolvane process. *Carbonization tar* from some types of lignite has a high alkane content, while the C. tar from anthracite generally contains relatively large amounts of phenolic compounds, but small amounts of alkanes (see Tar). The composition of *C. gas* depends on the method. A lignite C. gas from the rinse gas method consists of 10 to 15% methane, 5 to 10% carbon monoxide, 15 to 20% carbon dioxide, about 10% hydrogen and about 50% nitrogen. Because of its low heating value, it is used mostly as a drying and heating gas in the C. plant itself, or it is mixed with city gas.

2) The C. of ***oil shales*** produces residues which are nearly worthless. It can also be carried out in a Lurgi rinse-gas furnace, or in a Lurgi tunnel carbonization furnace, Lurgi-Schweitzer furnace, etc. In a C. pro-

cess in situ, the oil shale is heated electrically, and enough air is added so that the heat of combustion from part of the oil shale produces the necessary C. temperature. The products of this process are shale tar, shale oil and gas. Shale tar and oil are processed mainly to make fuels, heating oils and lubricants. The gas is of little chemical use and is used for heating; the residue can be used for construction materials, e.g. cement.

3) In the C. of ***peat***, the high water content is a great disadvantage. The peat must be dried to at most 30% water content. The C. is carried out in a chamber or shaft furnace with direct or indirect rinse gas heating. The C. coke, which makes up 30 or 40% of the product, is very reactive, has a low ash content, and a heating value of about 30,000 kJ kg^{-1}. The C. coke contains about 20% coke dust, and is used as a substitute for charcoal, as a starting material for activated charcoal, in copper and zinc smelting, and as a starting material for gasification. Peat tar contains ammonia, acetic acid, wood alcohol, and other substances. The peat C. gases consist of about 22% hydrogen, 1% oxygen, 29% nitrogen, 8.5% carbon monoxide, 24% carbon dioxide and 14.5% methane.

4) The C. of ***wood*** is discussed under Charcoal (see).

Carbon monoxide: CO, a colorless and odorless gas; M_r 28.01, m.p. - 199 °C, b.p. - 191.5 °C, crit. temp. - 140 °C, crit. pressure 3.45 MPa. C. is only slightly soluble in water (2.32 ml C per 100 ml H_2O at 20 °C and atmospheric pressure). It exists in equilibrium with carbon and carbon dioxide: $CO_2 + C \rightleftharpoons 2\,CO$, $\Delta H = +162.5$ kJ mol^{-1} (see Boudouard equilibrium). However, this equilibrium is established so slowly at low temperatures that C. appears stable. In the presence of sufficient oxygen and temperatures above 700 °C, it burns completely, forming carbon dioxide: $CO + 1/2\,O_2 \rightarrow CO_2$, $\Delta H = -283$ kJ mol^{-1}. This property is utilized when C. is used as the energy carrier in city gas. Many metal oxides are reduced by C. to the metals (e.g. in the blast furnace process). C. is an excellent complex ligand and forms Metal carbonyls (see) with many transition metals. It reacts with chlorine to form phosgene, $COCl_2$, with bromine to form carbon oxygen bromide, $COBr_2$, and with sulfur to form carbon oxygen sulfide, COS. The hydrogenation of C. to methanol, higher alcohols or saturated or unsaturated hydrocarbons (Fischer-Tropsch synthesis) is of great technological importance. The products obtained depend on the catalyst and reaction conditions. Catalytic reduction of steam by C. to hydrogen and carbon dioxide is utilized in the production of synthesis gas.

Carbon monoxide is a strong poison, which forms a stable complex with hemoglobin and thus prevents oxygen transport in the body. The symptoms of CO poisoning are headaches, visual impairment, dizziness, excitation, nausea, etc., leading through various stages to unconsciousness and death. The antidote is immediate oxygen respiration.

C. is produced industrially by incomplete combustion of coke (see Generator gas), reaction of glowing coke with steam (water gas) or steam reforming of natural gas (see Synthesis gas). In the laboratory, C. is made by dehydration of formic acid with sulfuric acid:

$$HCOOH \xrightarrow{H_2SO_4} CO + H_2O.$$

C. is the starting material for the technical syntheses of methanol, phosgene, formamide, formates and hydrogen cyanide. It is hydrogenated to hydrocarbons by the Fischer-Tropsch process. As a component of generator gas, water gas, blast furnace gas, etc., it serves as an energy carrier. It is also used to make metal carbonyls and very pure metals, and in the synthesis of aldehydes (see Oxo synthesis), acetic acid, etc.

Carbon oxygen sulfide: COS, a colorless, odorless, readily ignited and extremely poisonous gas; M_r 60.07, m.p. - 138.3 °C, b.p. - 50.2 °C, crit. temp. 104.8 °C, crit. pressure 6.5 MPa. C. reacts slowly with water to form carbon dioxide, CO_2, and hydrogen sulfide, H_2S. It is formed at high temperatures from carbon monoxide and sulfur vapor.

Carbon-silicon group: 4th representative group of the periodic system, including the non-metal carbon (C), the semimetals silicon (Si) and germanium (Ge) and the metals tin (Sn) and lead (Pb). These elements have in common the ns^2p^2 electron configuration. The changes in physical properties from one member of the group to the next are fairly regular (Table); the irregularities in the series of electronegativities are due to the intervening transition series.

Corresponding to their position in the middle of the periodic system, the elements of the C. have little tendency to form ionic compounds. Only carbon can form the 4 - anion with highly electropositive metals (for example, in beryllium carbide); +4 cations are formed only by the heavier elements tin and lead (SnF_4, PbF_4). The formation of four covalent bonds is most common, with the central atom surrounded tetrahedrally by the ligands (for example, CH_4, $SiCl_4$, SiO_2, $SnCl_4$, etc.).

In the lighter elements, the +4 oxidation state predominates, but with increasing atomic number, the +2 oxidation state becomes more important. While divalent compounds of carbon and silicon exist only as labile intermediates or at very high temperatures, Sn^{2+} and Pb^{2+} salts are stable. Sn^{2+} is still a reducing agent, while Pb^{4+} compounds are strong oxidizing agents. This aspect of C. chemistry can be interpreted as a result of the inert pair effect (see Boron-aluminum group). As the atomic number increases, the metallic properties of elements within the group increase, as does the basicity of their oxides or hydroxides. This trend is even more pronounced in the shift from the +4 to the +2 oxidation state.

Carbon differs from its homologs in its ability to form stable C-H and C-C bonds, and thus to form long chains and rings of carbon atoms. The complexity of carbon chemistry is due to its ability to form $p_\pi p_\pi$ multiple bonds with itself and with nitrogen and oxygen; the variety and number of carbon compounds have led to the special position of carbon chemistry (*organic chemistry*) as a branch of chemistry. Silicon and the heavier elements do not form

Properties of the elements of the 4th representative group of the periodic system

	C	Si	Ge	Sn	Pb
Nuclear charge	6	14	32	50	82
Electron configuration	[He] $2s^2p^2$	[Ne] $3s^2p^2$	[Ar] $3d^{10}4s^2p^2$	[Kr] $4d^{10}5s^2p^2$	[Xe] $4f^{14}5d^{10}6s^2p^2$
Atomic mass	12.01115	28.0855	72.59	118.69	207.19
Atomic radius [pm]	77	118	122	140	
Electronegativity	2.50	1.74	2.02	1.72 (Sn^{IV})	1.61 (Pb^{II})
Density [g cm^{-3}]	2.25 (Graphite) 3.51 (Diamond)	2.32–2.34	5.35	5.75 (α) 7.28 (β)	11.3437
m.p. [°C]	> 3550 (Diamond)	1410	937.4	231.88	327.50
b.p. [°C]	4827 (Graphite)	2355	2830	2260	1740

$p_\pi p_\pi$. double bonds, but in contrast to carbon, they do have energetically favorable d-levels, which increase their coordination numbers to 5 and 6 and permit π-interactions with suitable ligands. Therefore 4-coordinated derivatives of these elements are coordinatively unsaturated; they are Lewis acids and readily form complexes with electron-pair donors. They also react readily with nucleophiles, as in the formation of complexes between $SnCl_4$ and pyridine, or the reactions of $SiCl_4$ and $SnCl_4$ with water.

Carbon suboxide: C_3O_2, a colorless, very poisonous gas with a suffocating odor; M_r 68.03, density (liq.) 1.114, m.p. - 111.3°C, b.p. 7°C. The molecule has a linear structure, O=C=C=C=O. C. is made by dehydration of malonic acid with phosphorus(V) oxide. It is extremely reactive, and even at room temperature tends to polymerize. It reacts with water to form malonic acid, and with ethanol to form diethyl malonate.

Carbon tetrachloride, ***tetrachloromethane***: CCl_4, a colorless, non-combustible, highly refractive liquid with a sweetish odor; m.p. - 23°C, b.p. 76.5°C, n_D^{20} 1.4601. C. is scarcely soluble in water, but dissolves readily in alcohol, ether, acetone and benzene. In the presence of water and light, it is slowly decomposed, forming hydrogen chloride and carbon dioxide. When heated in the presence of oxygen, C. is readily oxidized to phosgene, and the reaction is catalysed by metals. Mixtures of C. with aluminum powder or alkali metals are highly explosive, so C. must not be dried with elemental sodium!

C. is made by chlorination of methane or carbon disulfide. Either chlorine or disulfur dichloride can be used as a chlorination reagent for carbon disulfide: $CS_2 + 3\,Cl_2 \rightarrow CCl_4 + S_2Cl_2$, or $CS_2 + 2\,S_2Cl_2 \rightarrow CCl_6 + 6\,S$. C. is a very good solvent for fats, oils, resins, rubber, waxes and lacquers. It is therefore used in large amounts in industry. Because of the danger of poisoning by phosgene formation, C. is no longer used in fire extinguishers. It is used in the synthesis of fluorinated hydrocarbons.

> Carbon tetrachloride can be absorbed by inhalation or through the skin. It has a narcotic effect and can produce headaches, nausea, temporary deafness and reduced visual acuity, as well as severe damage to liver and kidneys. Antidotes: immediate fresh air, and, if necessary, artificial respiration. Severe intoxication can be fatal.

Carbonyl: see Metal carbonyls.

Carbonylation: introduction of a carbonyl group into alkenes, alkynes or arenes by reaction of alkenes or alkynes with carbon monoxide and water or compounds with mobile hydrogen atoms in the presence of special catalysts. Another method is the reaction of phosgene or oxalyl chloride with arenes under the conditions of Friedel-Crafts acylation. With nickel carbonyls as catalysts, C. of alkenes yields monocarboxylic acids. In the presence of metal carbonyls, alkynes give rise to carboxylic acids or their esters or amides, depending on the compound with the mobile hydrogen atom. Arenes form carboxylic acids.

Carbonyl chloride: same as Phosgene (see).

Carbonyl compounds: organic compounds with a Carbonyl group (see) in the molecule: aldehydes, ketones, quinones, carboxylic acids, etc. C. are among the most important organic compounds, due to their reactivity. They are essential starting materials for the synthesis of plastics, synthetic fibers, pigments, pharmaceuticals, cosmetics, flavorings, etc.

Carbonyl group, ***oxo group***: functional group

$$>C{=}O$$

found in aldehydes, ketones, quinones, carboxylic acids, carbon dioxide and most derivatives of carbonic acid. If the C-atom of the C. is bound to two C-atoms of aliphatic, aromatic or heterocyclic residues – as in the ketones – it is called a ***keto group***. Because of the difference in electronegativity of carbon and oxygen, the C=O double bond is polar:

$$>C{=}\bar{O} \longleftrightarrow >\overset{+}{C}{-}\bar{O}|^{-}$$

The reactivity of the C. depends greatly on the residues bound to the C-atom. Electron-withdrawing substances, e.g. halogen atoms, enhance the partial positive charge on the C-atom and thus increase its reactivity with respect to nucleophiles. Electron-donating substitutents reduce the reactivity of the C., e.g. the OH group in the carboxyl group. Thus carbonyl halides and aldehydes are very reactive carbonyl compounds, while carboxylic acids and the carboxylate anion react sluggishly or not at all with nucleophilic reagents.

Carbonyl iron: very pure iron which is obtained by thermal decomposition of iron pentacarbonyl,

$Fe(CO)_5$ (see Iron carbonyls). Iron pentacarbonyl, which is a liquid under normal conditions, decomposes above 150 °C into carbon monoxide and iron powder of a particle size of about $0.5 \cdot 10^{-4}$. This is worked by pressing at high temperatures.

C. is used mainly for electromagnetic components, such as mass nuclei for high-frequency technology, amplifier coils for cables, and in radio and television technology. It is also used in pharmacy (ferrum reductum).

Carbonylolefination: see Wittig reaction.

Carbonyl pigments: a class of pigments with at least two conjugated carbonyl groups (Fig. 1). The C. include indigo and its derivatives, anthraquinone substitution products, polycyclic carbonyl compounds and special carbonyl compounds.

1 2

Chromophore structures of the carbonyl pigments

1) *Indigo and its derivatives*: see Indigo, Indigosols.

2) *Anthraquinone substitution products* see Anthraquinone pigments.

3) *Polycyclic carbonyl compounds*. The characteristic trait of this group of pigments is that even the skeleton structures (Fig. 2) absorb in the visible. They are synthesized by ring closure in a Friedel-Crafts reaction of an acylium ion, or by nucleophilic attack on an activated aromatic carbon atom followed by cleavage of a substituent R^- or R^+ and two electrons. Useful dyes based on polycyclic carbonyl compounds are obtained by introduction of halogen atoms. The parent compound undergoes electrophilic substitution with the halogen in the last step of synthesis. Pigments in this group are used exclusively as vat dyes for textiles.

4) *Special C.* In addition to the C. listed above, there are several smaller groups of conjugated carbonyl compounds which are used as pigments on an industrial scale, for example, ***linear quinacridones*** (Fig. 3) and ***anthrone derivatives***, which are very fast pigments.

reddish violet pale yellow

Quinacridones

The ***perylenetetracarboxydiimide*** and ***naphthalenetetracarboxydiimide series*** are synthesized by reaction of perylene-3,4,9,10-tetracarboxylic acid or naphthalene-1,4,5,8-tetracarboxylic dianhydride with the corresponding amines. Examples are ***indanthrene Bordeaux RR*** and *indanthrene brilliant orange GR* (Fig. 4).

cis and trans

Indanthrene bordeaux RR — Indanthrene brilliant orange GR

Carbophanes: same as Cyclophanes (see).

Carbophosphonate: see Fungicides.

Carboranes: a group of organoboron compounds which are formally derived from the Boranates (see) by substitution of CH groups for one or more BH^- units. The best characterized is the $C_2B_{10}H_{12}$ system, which is isosteric with the $B_{12}H_{12}^{-2}$ anion. 1,2-Dicarba-*closo*-dodecaborane is formed from decaborane(14) and ethyne in diethyl sulfide as solvent: $B_{10}H_{14} + HC{\equiv}CH \rightarrow C_2B_{10}H_{12} + 2\ H_2$. This compound, also called ortho-carborane, isomerizes at 450 °C to 1,7- and at 620 °C to 1,12-dicarba-*closo*-dodecaborane (meta and para isomers).

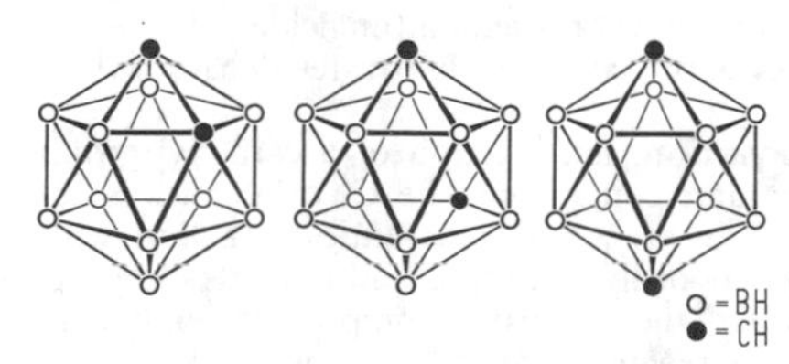

Structures of isomeric dicarba-*closo*-dodecaboranes

The C. are chemically and thermally very stable, and do not react with acids, bases or oxidizing agents. There are many possibilities for substitutions on the C atoms which bring the C. system into various classes of organic compounds.

Carborundum: see Silicon carbide.

Carbowax: same as Polyethylene oxide.

Carboxamide fungicides: fungicidal carboxamides, usually cyclic, α,β-unsaturated compounds with a *cis*-methyl group or a group of similar size on the β-C. The C. were developed in the 1960s; they are Systemic pesticides (see) selective for *Basidiomyces*, giving good protection against smut and rust fungi. *Carboxine* and related compounds with oxathiine skeletons are also called ***oxathiine fungicides***; an example is the sulfone, *oxycarboxine*. Some important C. are shown below.

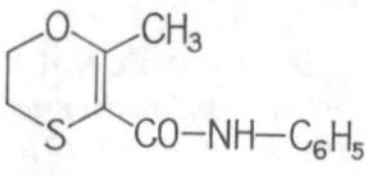

Carboxin

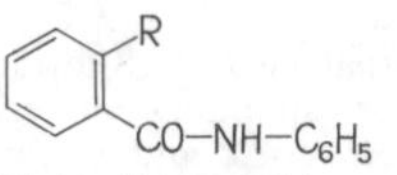

Mebenil (R= CH_3)
Benodanil (R= I)

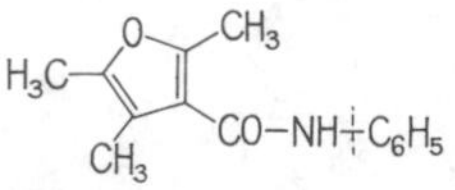

Methfuroxam

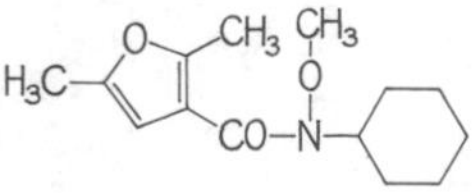

Furmethamide

Important carboxamide fungicides

Carboximide fungicides: fungicidal carboximides, most of which are *N*-(3,5-dichlorophenyl)-heterocycles. They are systemic fungicides with high activity, especially against *Botrytis cinerea*. Some important examples are shown below.

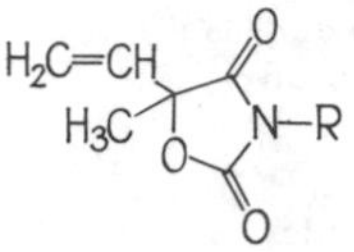

Vinclozoline

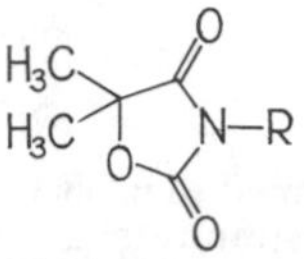

Dichlozoline

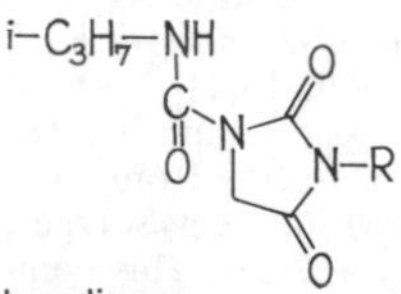

Iprodione

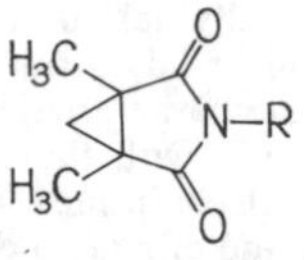

Proxymidone

Important carboximide fungicides

Carboxin: see Carboxamide fungicides.

Carboxylase: same as Pyruvate decarboxylase (see).

Carboxylation: introduction of a carboxyl group into an organic compound. The C. of alkanes is accomplished by a radical substitution reaction with phosgene or oxalyl chloride, and yields an acyl chloride. Of the aromatic compounds, only the phenols are reactive enough to react with carbon dioxide to give phenol carboxyic acids (see Kolbe-Schmitt synthesis).

Carboxyl group: the functional group -COOH, which is present in all carboxylic acids. Formally, the C. can be considered a combination of a carbonyl and a hydroxyl group. However, it has qualitatively new properties, such as the ready loss of the proton, and thus the acidic character of the group.

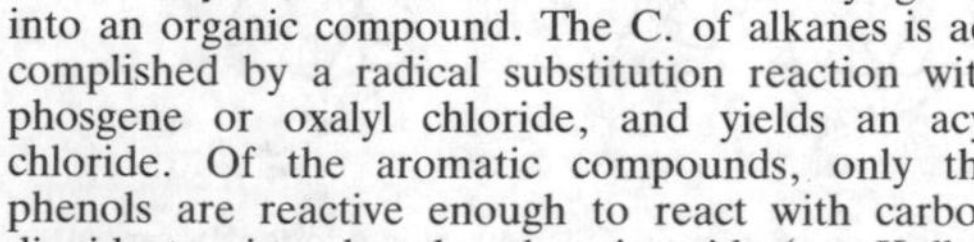

$$R{-}C(=O)OH + H_2O \rightleftharpoons \left[R{-}C(=O)O^- \longleftrightarrow R{-}C(O^-){=}O\right] H_3O^+$$

Carboxylic acids: organic compounds with one or more Carboxyl groups (see) in the molecule. They are classified as Monocarboxylic acids (see), Dicarboxylic acids (see), tricarboxylic acids, etc. They are classified as ***aliphatic, aromatic*** or ***heterocyclic C.*** on the basis of the organic structure to which the carboxyl group is attached.

The simplest aliphatic monocarboxylic acid is Formic acid (see), the simplest dicarboxylic acid is Oxalic acid (see), and the best known tricarboxylic acid is Citric acid (see). The parent compound of the aromatic C. is Benzoic acid (see). The simplest aromatic dicarboxylic acids include the structural isomers Phthalic acid (see), Isophthalic acid (see) and Terephthalic acid (see). One of the best known heterocyclic C. is Nicotinic acid (see). If a C. also contains a double bond, it is an ***unsaturated C.***, such as Acrylic acid (see), Cinnamic acid (see) or Oleic acid (see).

The trivial names of many C. are taken from their natural sources, where they are present in free form or in compounds, often esters. The systematic name is based on the name of the hydrocarbon with the same number of C atoms, to which the suffix "-oic acid" is added, e.g. propanoic acid, CH_3-CH_2-COOH. On the other hand, if the carboxyl group is considered the substituent, the name of the C. is taken from the root name of the hydrocarbon (with one C atom fewer) plus the suffix "carboxylic acid"; this method is preferred for cycloalkanes, heterocycles and arenes:

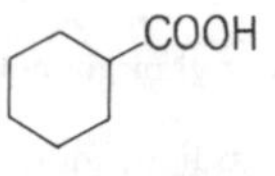

Cyclohexanecarboxylic acid 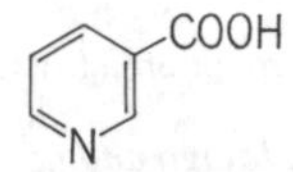 Pyridine-3-carboxylic acid

The chemical properties of the C. are determined chiefly by the functional Carboxyl group (see). They thus posess a readily cleaved, acidic H atom. In the presence of a base or strongly electropositive metal, the salt is formed; most of these salts are soluble in water. The acidity of the C. is distinctly lower than that of the strong mineral acids, however. The substitution of a hydrogen atom in the aliphatic, aromatic or heterocyclic skeleton of the C. produces a derivative such as a halocarboxylic acid, an amino acid, keto or hydroxycarboxylic acid. Functional derivatives of C. are obtained by conversion of the carboxyl group; some important types are acyl halides, anhydrides, amides, esters, azides, hydrazides and salts of C.

Analytical. C. dissolve in aqueous sodium hydroxide and sodium hydrogen carbonate solution with generation of carbon dioxide. For identification, the C. is converted to the amide or anilide by reaction with thionyl chloride and ammonia or aniline, respectively. Since carboxylic anhydrides and esters usually also react with ammonia or aniline, these derivatives are also characterized as amides. Some other important derivatives of the C. which can be used for identification are the 4-bromophenyl esters, which can be synthesized via the sodium salts. All carboxylic acid derivatives can be converted by hydrolysis to the parent C. The IR spectra of the C. display absorption bands for the C=O valence vibration between 3500 and 3560 cm^{-1} (free) and between 2500 and 2700 cm^{-1} (associated). The ^{1}H-NMR signal for the carboxyl group appears at δ values between 9.5 and 13 ppm. The mass spectra of aliphatic C. have weak molecular ion peaks; aromatic C. give stronger ones. The fragmentation of the monocarboxylic acids produces a characteristic acylium ion, R-CO$^+$ (mass no. = M_r - OH) and the fragment COOH$^+$ (mass no. = 45).

Some important methods for synthesis of C. are the oxidation of alkanes and alkylarenes, carbonylation of alkenes and alkynes, hydrolysis of trichloromethylarenes, esters, amides or nitriles, and the addition of carbon dioxide to Grignard compounds (see). C. can also be made by acylation of arenes with oxalyl chloride in the presence of aluminum chloride, by Malonic ester syntheses (see) and by Acid cleavage (see) of β-oxocarboxylic acid esters. There are also special methods for substituted or unsaturated C., such as the Perkin reaction (see) and Stetter synthesis (see). C. are used in many ways, as reaction media, as acidic reactants and to make derivatives for industry and organic synthesis.

Carboxylic anhydrides: derivatives of carboxylic acids formed from monocarboxylic acids by intermolecular elimination of water, and from di- and polycarboxylic acids by intramolecular elimination. They have the general formula:

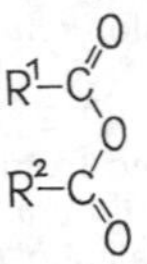

R^1–C(=O)–O–C(=O)–R^2

R^1 and R^2 can be aliphatic, aromatic or heterocyclic groups. If $R^1 = R^2$, the compound is a ***symmetric anhydride.***; if not, then it is an ***asymmetric (mixed) anhydride***. The names of these compounds are formed by substituting the word "anhydride" for "acid" in the name of the acid; for example, acetic anhydride. The anhydride of formic acid does not exist, but a mixed anhydride with acetic acid can be prepared.

The C. are liquids with pungent odors and higher boiling points than the related carboxylic acids. The C. of higher carboxylic acids and dicarboxylic acids are solids. C. of higher carboxylic acids can be made by heating the acids, but for lower acids, acetic anhydride must be used as a dehydrating agent. The reaction of carboxylic acid salts with acyl chlorides is another generally available method of producing C. There are special methods for a few C., such as Acetic anhydride (see). The C. are somewhat less reactive acylating agents than the acyl halides. They are used to make carboxylates (reaction with alcohols or phenols), amides (reaction with ammonia or amines) or Friedel-Crafts acylation (see Friedel-Crafts reactions). Hydrolysis of the C. produces carboxylic acids.

The C. of aliphatic carboxylic acids are used in the Perkin reaction for synthesis of α-substituted α,β-unsaturated arene carboxylic acids. Dicarboxylic anhydrides of industrial importance include maleic, succinic and phthalic anhydrides.

Carboxypeptidases: exopeptidases (see Proteases) which catalyse cleavage of the carboxy-terminal amino acid from a peptide chain. The most important of the animal C. are ***carboxypeptidase A*** (307 amino acids, M_r 34,409) and ***carboxypeptidase B*** (300 amino acids, M_r 34,000). C. A is specific for amino acids with aromatic or branched aliphatic side chains, while C. B acts only on peptide bonds in which an α-amino group of lysine or arginine participates. Both enzymes are synthesized in the pancreas as inactive precursors (pro-C. A, M_r 87,000; pro-C. B, M_r 57,400). They are secreted into the duodenum, where they are converted by trypsin into their active forms. The C. are important for digestion of proteins in the small intestine.

3-Carboxypyridine: same as Nicotinic acid (see).

Carburation: improving the heating value of gases by evaporation or thermal decomposition of liquid hydrocarbons into the gas, or by addition of gaseous hydrocarbons. For example, when air is passed through gasoline vapor, it produces a flammable gas with a high heating value.

Carburization: the formation of alloys of steels with carbon, forming mixed crystals and carbides. If C. is carried out as a surface treatment to make low-carbon steel (less than 0.2% carbon) easier to weld, it is also called ***casehardening*** or ***cementation***. The process can be carried out with solid, liquid or gaseous reagents. The most economical and best controlled is C. with mixtures of carbon monoxide and hydrogen; the process is carried out around 900 °C. The boundary layer of the piece should have a C content of 0.7 to 0.8%. If C. is carried out simultaneously with the uptake of nitrogen into the surface of the piece, the process is called Carbonitration (see).

Uptake of carbon into the metal lattice reduces the resistance of the steel to corrosion. Thus C. of non-rusting and acid-resistant chromium-nickel steels increases their susceptibility to intercrystalline corrosion. C. of heat-resistant steels reduces their resistance to scale formation.

Carbutamide: see Antidiabetics.

Carbyne: a form of elemental carbon, not yet completely confirmed, which is represented as a new modification. In it, the C atoms are linked in linear chains with alternating single and triple bonds.

Carbyne complexes: organometallic complexes of electron-rich transition metals with carbine (CR-) ligands; they are characterized by M-C triple bonds (M≡C). C. can be obtained, e.g. from carbene complexes and boron trihalides.

Carcinostatin: see Neocarcinostatin.

Cardenolides: see Cardiac glycosides.

Cardiac glycosides: steroid natural products and their partially synthetic derivatives which affect the heart. They contain a cyclopentanoperhydrophenanthrene skeleton, in which rings A/B are *cis*, B/C are *trans*, and C/D are *cis*. The ***cardenolide type*** has a 17β unsaturated, five-membered lactone ring, while the ***bufadienolide type*** has a doubly unsaturated, six-membered lactone ring at that site.

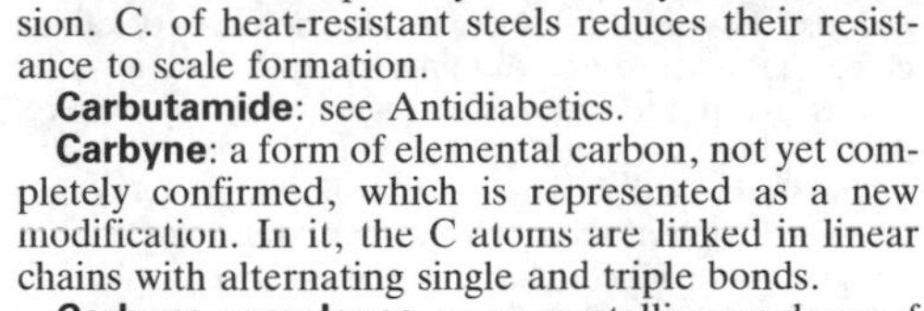

Cardenolide

Bufadienolide

The aglycones of the highly effective compounds have 3β- and 14β-OH groups and may have additional oxygen functions, most in the form of hydroxy groups. One to five monosaccharide groups are glycosidically bound to the O atom on C3. In addition to D-glucose and L-rhamnose, these monosaccharides include rare sugars such as 2,6-bisdeoxyhexoses (e.g. D-digitoxose), methyl ethers of 6-deoxyhexoses (e.g. D-digitalose) and methyl ethers of 2,6-bisdeoxyhexoses (e.g. D-cymarose and D-oleandrose). More than 400 glycosides have been isolated, mainly from the *Scrophulariaceae* (*Digitalis* species), *Liliaceae* (*Urginea maritima, Convallaria majalis*), *Ranunculaceae* (*Adonis* species) and *Apocyanaceae* (*Strophanthus* species, *Nerium oleander*). Non-glycosidic compounds of the bufadienolide type are found in the skin secretions of toads (*Bufo* species). The following C. are therapeutically important: digitalis glycosides (see Purpurea glycosides, Lanata glycosides), Strophanthus glycosides (see), Convallaria glycosides (see) and Scilla glycosides (see).

Historical. The use of medicinal herbs containing C. began in 1785 with the recommendation of the English physician William Withering that foxglove be used for dropsy. Because of the difficulty of isolating and determining the structures of pure, active components, aqueous extracts of the powdered leaves were used until after World War II. The leaves were biologically standardized, with the effectiveness being given in frog units. One gram of leaves which could kill 2000 g frogs by cardiac arrest had an activity of 2000 frog units. Because the leaves contain a varying spectrum of glycosides with different pharmacokinetic parameters, and because of the instability of the preparations, they are no longer used. The first crystalline glycoside mixture was obtained in 1872 by Nativelle; it was given the name digitoxin by Schmiedeberg. Pure digitoxin was not prepared until decades later.

Cardiolipin: 1,3-bis-(3-*sn*-phosphatidyl)glycerol, a glycerophospholipid in which the two primary hydroxyl groups of glycerol are esterified to phosphatidic acid. M. is present in the membranes of bacteria and mitochondria.

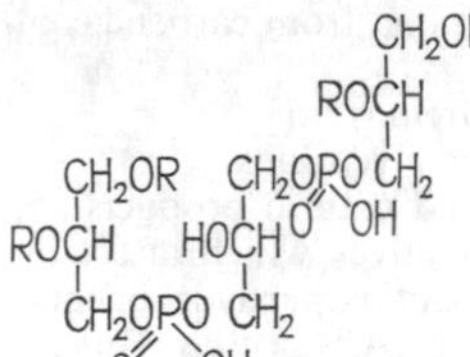

Cardiolipin (R = fatty acid)

C. from bovine heart makes up about 10% of the total phospholipid content. It is obtained as a viscous oil which is not stable at room temperature. It is readily soluble in benzene and chloroform, but is insoluble in methanol.

Carius method: a method introduced by G. L. Carius for quantitative determination of chlorine, bromine, iodine and sulfur in organic substances. The sample is placed in a sealed bomb tube with conc. nitric acid and silver nitrate, and is heated to 200-300 °C for several hours. The halides are converted to silver halides, and the sulfur to sulfuric acid. The final determination can be done by various methods.

Carmine lake: see Carminic acid.

Carminic acid: a red anthraquinone pigment found in various species of scale insects (see Cochenille). C. is extracted with water from the dried female insects. It is a tetrahydroxylated methylanthraquinone carboxylic acid bound glycosidically to D-glucose (m.p. 130 °C). It forms colored complexes with metal salts, for example tin salts; these are called carmine lakes. C. was formerly one of the most beautiful and costly mordant dyes for wool and silk.

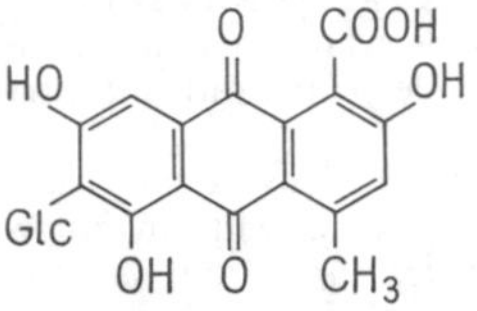

It is now used in the production of cosmetics, artists' colors and food colorlings, and as a color reagent in histology.

Carmoisin: see Fast red.

Carnauba wax: a light yellow wax obtained from the leaves of the carnauba palm *Copernicia cerifera*. It melts at 80 to 90 °C and is readily soluble in alcohol, ether and other organic solvents. C. is the hardest natural wax. In addition to cerotinic acid melissyl ester, it contains free cerotinic acid, carbaubic acid, higher monoalcohols, and hydrocarbons. Mixed with other waxes, C. increases their melting points. It is used as a polish and floor wax, as a dispersion medium for pigments, in the production of oilcloth and candles, and to impregnate textiles and paper. It was formerly used almost exclusively in the production of records (sound recordings).

Carnot cycle: an imaginary, reversible cyclic process invented by S. Carnot in 1824 which provides the theoretical basis for calculation of the efficiency of all periodically working cyclic heat machines. An idealized heat machine (Carnot machine) consisting of a cylinder with a frictionless piston, filled with an ideal gas as the working fluid, absorbs the heat $q_{1,2}$ at temperature T_u and transfers the amount $q_{3,4}$ to a heat reservoir at the lower temperature T_l. The difference in the two heat energies is the work (w) done by the machine, according to the first law of thermodynamics (Fig. 1).

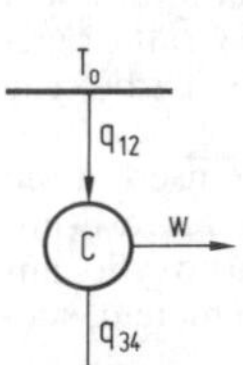

Fig. 1. Carnot cycle (schematic). C, Carnot machine; q, exchanged heats, T_u and T_l, temperatures of the upper and lower heat reservoirs

The C. consists of four steps (Fig. 2): 1) isothermal expansion in thermal contact with the heat reservoir at temperature T_u (heat uptake, production of work):

$$q_{1,2} = -w_{1,2} = \int_{v_1}^{v_2} p\mathrm{d}v = nRT_o \ln\frac{v_2}{v_1},$$

since from the equation of state of an ideal gas, $p = nRT/v$. 2) Adiabatic expansion, in which the gas cools from T_u to T_l, $q_{2,3} = 0$ and $w_{2,3} = \int(T_u, T_l).c_v\mathrm{d}T = -c_v(T_u - T_l)$. 3) Isothermal compression in thermal contact with the heat reservoir at temperature T_l (heat release, input of work) $-q_{3,4} = w_{3,4} = \int(v_3,v_4)p\,\mathrm{d}v = -nRT_l \ln v_4/v_3$. 4) Adiabatic compression back to the initial state $q_{4,1} = 0$, $w_{4,1} = \int(T_l,T_u).c_v\mathrm{d}T = c_v(T_u - T_l)$.

The total work of the C. is $w = w_{1,2} + w_{2,3} + w_{3,4} + w_{4,1} = -nR(T_u \ln v_2/v_1 + T_l \ln v_4/v_3)$. Application of the Poisson equation, $Tv^{\varkappa-1} = $ const., to the two adiabatic changes gives $v_2/v_1 = v_3/v_4$. Therefore, $w = -nRT(T_u - T_l) \ln v_2/v_1$.

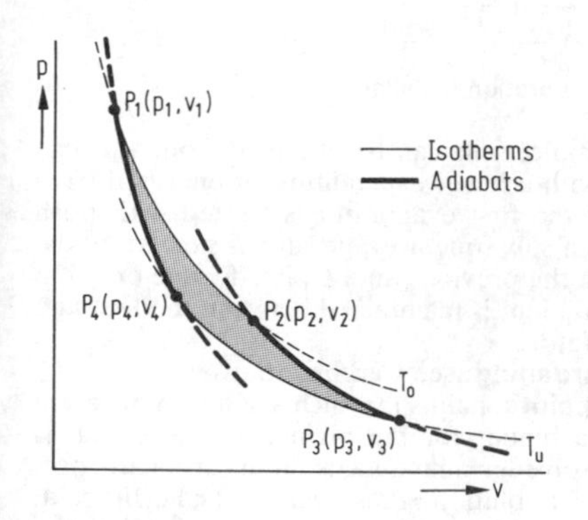

Fig. 2. Carnot cycle in a *p, v* diagram. gray, isotherms; black, adiabats. P_i (p_i, v_i) = points between which the changes of state occur

It can also be shown that the total change in the internal energy Δe over the course of the cycle is zero. The heat balance, on the other hand, is not zero: $q = q_{1,2} + q_{3,4} = -w \neq 0$. However, if the ratios of the heats to the temperatures at which they occur are taken, $q_{1,2}/T_u + q_{3,4}/T_l = 0$. The ratio q_{red}/T is called the *reduced heat*, after Clausius.

The ***efficiency*** η of a Carnot heat machine is defined as the ratio of the work done $|w|$ to the heat input at the upper temperature T_u: $\eta = |w|/|q_{1,2}| = (T_u - T_l)/T_u$. It is always less than one, and depends only on the temperatures of the two heat reservoirs between which heat is transferred. In real heat machines, these are the temperatures of the working fluid and the environment.

As a reversible cyclic process, the C. can be run in the opposite direction. Then an amount of work W is put into the machine, which draws the heat $q_{3,4}$ from the reservoir at the lower temperature T_l, and transfers to the reservoir at T_u the larger amount of heat $q_{1,2}$. It thus acts to cool the lower reservoir and heat the upper one. The efficiency of the heat pump is defined as $\eta' = |q_{1,2}|/|w| = T_u(T_u - T_l)$; it is greater than one.

Any reversibly functioning heat machine, regardless of the nature of the changes of state and of the working fluid, must have the same efficiency as the C. Otherwise, it would be possible to combine several different cyclic processes and create a perpetual motion machine of the second type, which would violate the second law of thermodynamics. Real heat machines always have lower efficiencies, because they do not work reversibly. In addition to friction losses, a reversible process is infinitely slow, while a real one proceeds at a finite rate.

Caro's acid: same as peroxomonosulfuric acid.

Carotenes: tetraterpenes with conjugated double bonds. The various C. differ in the structures of their end groups, which can be either acyclic (ψ-C.), cyclohexene rings (β-C. = β-ionone; ε-C. = α-ionone), benzene rings (φ- and χ-C.) or cyclopentane rings (κ-C.). A C. can have two different end groups. The C. isolated from carrots in 1831 by Wackenroder was later separated into three components by adsorption chromatography. Its average composition is 15% α-C. (new name, β,ε-C.), 85% β-C. (β,β-C.) and 0.1% γ-C. (β,ψ-C). α-C. and β-C. form purple crystals, m.p. of α-C. is 187.5 °C, m.p. of β-C. is 183 °C (in vacuum). γ-C. forms red leaflets, m.p. 152-154 °C. The C. are practically insoluble in water, barely soluble in alcohol, and readily soluble in nonpolar solvents. They are sensitive to light, air and heat. The C. can be synthesized via ylides by the Wittig synthesis. They are used as antioxidants and as food colorings, for example, in margarine. C. with β-ionone structures act as provitamins A.

Carotenoids: linear polyene tetraterpenes with conjugated double bonds, which have a characteristic tail-to-tail coupling in the middle of the molecule. The terminal groups may be cyclized. The C. include the Carotenes (see) and Xanthophylls (see). They are yellow to purple, crystalline, highly lipophilic compounds. With antimony(III) chloride (Carr-Price reaction) or sulfuric acid, the C. form highly colored complexes, which can be used for identification. C. are found in insects and birds, for example flamingoes, but they are most abundant in green plants; they are associated with chlorophyll in all green parts of the plants, but are also found in blossoms and ripe fruits. The C. serve as protective and light-collecting pigments for photosynthesis and phototropism in higher plants. They serve as phototaxis pigments in some algae, and are also produced by prokaryotes. C. and lipids are extracted from biological materials with nonpolar solvents; the C. can be separated from lipids by distribution between solvents (e.g. petroleum ether/90% methanol) or by chromatography.

Cleavage of the middle double bond produces ***apocarotenoids***, such as vitamin A and abscissic acid.

Carrageenan: a polysaccharide obtained from various red algae, such as *Chondrus crispus* or *Gigartina mamillosa*. The dried and bleached thalli of these algae are called ***carrageen*** or "***Irish moss***". C. forms very viscous solutions which become gels when they cool. It is a mixture of various galactans which vary in their structures and properties. ***κ-(k)-C.*** contains mainly 3-O-substituted β-D-galactopyranose 4-sulfate and 4-O-substituted 3,6-anhydro-α-D-galactopyranose, while **λ-(*l*)-*C.*** contains 3-O-substituted β-D-galactopyranose 2-sulfate and 4-O-substituted α-D-galactopyranose-2,6-bisphosphate. C. is used as a gelling agent and thickener for ice cream, pudding, sauces, jellies and jams, and in pharmaceuticals.

Carrier: 1) a molecule which transports substances into and within a cell. An imporant type is the Ionophore (see), which transports ions through the cell membrane; the best-known example of this group is Valinomycin (see).

2) An accelerator of the dyeing process used with synthetic fibers of high packing density (polyesters, triacetates and polyvinylchlorides). It causes the material to swell, and thus facilitates the diffusion of the dye into it. C. are used, generally in combination with tensides, in dyeing with dispersion dyes. C. are usually aromatic compounds, such as phenol, phenylphenol, salicylic acid and its derivatives, xylenes, etc.

Carrier gas: see Gas chromatography.

Carrier dyes: same as Substrate dyes (see).

Carrier vapor distillation: see Distillation.

Carr-Price reaction: see Vitamin A.

Cartap: an insecticide used as a contact and ingested poison against beetles and caterpillars.

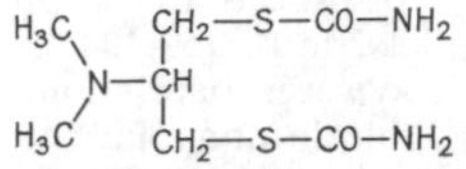

$(H_3C)_2N-CH(CH_2-S-CO-NH_2)_2$

It was obtained by modifying the structure of the animal product nereistoxin (4-dimethylamino-1,2-dithiolane), and commerciallized as a plant protectant.

Carthamine: see Safflower.

Carvacrol, ***3-isopropyl-6-methylphenol***: a colorless liquid with an odor like thyme; it is autooxidizable and gradually turns yellow; m.p. 1 °C, b.p. 237.7 °C, n_D^{20} 1.5230. C. is slightly soluble in water, and soluble in ethanol and ether. It occurs along with thymol in various essential oils, e.g. thyme, caraway and marjoram oils. It can be synthesized by sulfonation of *p*-cymene, followed by alkali fusion. C. is less toxic, but more strongly antiseptic, than phenol, and it is therefore often used as a disinfectant.

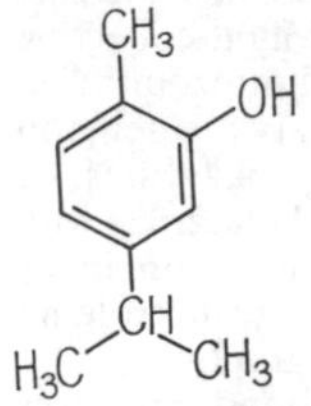

Carvacrol

Carvone: *p*-metha-6,8(9)-dien-2-one, a monocyclic monoterpene ketone. C. is a colorless liquid with an odor like caraway; it is nearly insoluble in water, but is soluble in ether and ethanol. D-C. from caraway oil has b.p. 230 °C at 101 kPa, $[\alpha]_D$ +59.6°; l-C. from spearmint oil has b.p. 230-231 °C at 102 kPa, $[\alpha]_D$ - 59.4°. C. makes up 45 to 60% of caraway oil. It is used to flavor liquors, mouthwashes, toothpastes and skin-care products.

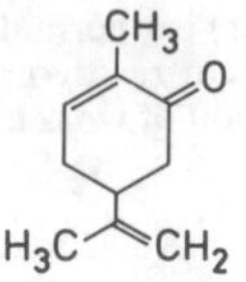

Cascade evaporation, ***multistep distillation***: a process for evaporation of aqueous solutions in a series of evaporators (Fig.).

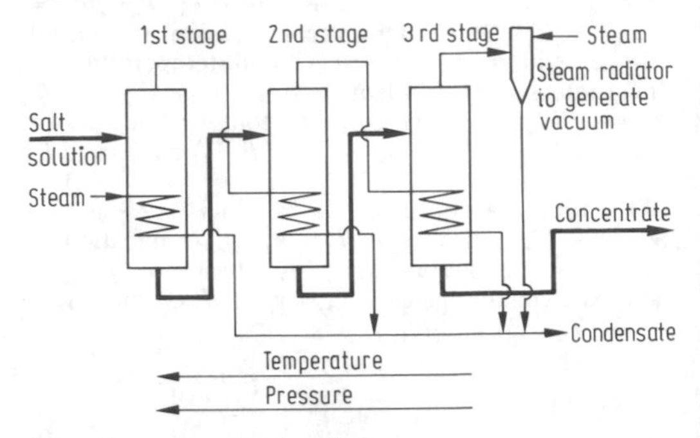

Cascade evaporation installation

For example, salts can be obtained from aqueous solutions with a lower expenditure of energy if C. is used; only the first evaporator is heated with fresh steam. Each subsequent evaporator is heated by the vapor from the previous one. C. is effective only if a pressure gradient is maintained between the first and last evaporator.

Case hardening: see Carburization.

Casein paints: paints in which a solution of casein serves as a binder for the pigments. The casein is usually dissolved in ammonia, sodium carbonate, potash or borax solution. Preservatives (salicylic acid, phenols, etc.) are used to prevent spoiling. If the freshly painted surface is sprayed with a solution of formaldehyde (formalin), it accelerates hardening of the paint.

CASING process: (abb. for Cross-linking by Activated Species of INert Gases) a process for surface cross-linking of polytetrafluoroethylene, polyethylene or other polymers. The plastic parts are placed in a helium or neon atmosphere and exposed to a glow discharge. The very reactive noble gas radicals formed by the discharge attack the surface and initiate cross-linking. Polymers which have been treated by C. can be printed and glued with the usual glues.

β-Casomorphin: Tyr-Pro-Phe-Pro-Gly-Pro-Ile, a heptapeptide with an opiate-like effect, made from an enzymatic degradation product of casein. Cleavage of the heptapeptide (**β-*casomorphin-7***) with carboxypeptidase Y yields the pentapeptide Tyr-Pro-Phe-Pro-Gly (**β-*casomorphin-5***), which has a stronger opiate effect. The name is derived from the presence of the heptapeptide sequence in β-casein. Unlike the endogenous opiate peptides, the endorphins, β-C. is an exorphin. Of many synthetic analogs which have been tested, Tyr-D-Ala-Phe-Pro-Tyr-NH_2 is the most active. **β-*Casomorphin-4*** (Tyr-Pro-Phe-Pro) has a strong effect on insulin release, but only a slight opiate effect.

Cassel brown, ***Cologne earth***: a decomposition product of wood found in peat and soft coal deposits. It consists mainly of huminic acids and its consistency is that of an earthy powder or soft blocks. C. is used as a brown pigment for oil paints. It decomposes in acids with precipitation of carbon; in bases it dissolves to give a brown color. ***Van-Dyck brown*** is a type with especially intense color which has been purified by washing. If C. is treated with alkali, it forms water-soluble huminic acid salts. This mass can be dried to give a scaly or grainy ***nut stain*** which, when dissolved in water, is used to stain wood.

Cassiopeium: see Lutetium.

Castable resins: synthetic resins, such as phenol, epoxide or polyester resins, which can be poured in the liquid state into molds where they slowly harden. They are transparent or colored, and are usually chemically resistant. They can be worked with cutting tools, and are used to make parts for electrical equipment. With added fillers such as fiberglass, they can be used for a wide variety of objects, including boats and automobile bodies.

Castner process: see Sodium.

Catadyn process: see Disinfection.

Catalase: an iron-porphyrin enzyme (M_r 245,000) which catalyses the degradation of hydrogen peroxide to water and oxygen. C. is found in animal and plant tissues and microorganisms; large amounts are present in the liver and erythrocytes. C. has a molecular activity of $5 \cdot 10^6$ molecules H_2O_2/minute, and is one of the most active enzymes known. Its function is to protect the organism from the toxic effects of H_2O_2.

Catalymetry: a term for methods of quantitative analysis in which the substance to be determined acts as a catalyst or inhibitor of a chemical reaction which is easy to follow, usually a redox reaction. Since the reaction rate is proportional to the amount of the catalytic substance present, its determination provides a very sensitive test for that substance.

C. is especially suitable for trace analysis, but so far has not been widely adopted into the practice of inorganic analysis. In clinical analysis, however, catalytic methods using enzymes are extremely important.

Catalysis: a process in which addition of a substance to a reaction increases the rate of the reaction without changing the added substance. The added substance is called the ***catalyst*** (also called ***contact*** in the case of solid catalysts used in industry); it can be solid, liquid (dissolved) or gaseous. A distinction is made between homogeneous and heterogeneous C., depending on whether the catalyst is in the same phase as the reaction (see below). In ***photocatalysis*** or ***electrocatalysis***, the catalyst is generated photo- or electrochemically. In ***biocatalysis***, Enzymes (see) act as catalysts. If addition of a substance reduces the reaction rate, the phenomenon is called *inhibition*, sometimes also "negative C." This effect can occur either by blocking or changing the catalyst of a catalysed reaction (*catalyst poisoning*), or by an *inhibition* (see Inhibitor).

I. *General*. There is a difference between catalysts and initiators: an initiator does not provide a new reaction pathway or mechanism, but a catalyst does. The new pathway allows the reaction to proceed faster, and includes regeneration of the catalyst. For a simple bimolecular reaction $A + B \rightarrow AB$, the principle of C. can be described as $A + B + C \rightarrow AB + C$. Since the catalyst C is not included in the matter balance, it also does not change the equilibrium position of the reaction. In order to open a new reaction pathway, the catalyst must form a more or less stable intermediate with one of the reactants: $A + C \rightleftharpoons AC$, $AC + B \rightarrow AB + C$, which provides a new transition state for the rate-determining step of the reaction, and this new transition state must have a lower free enthalpy of activation ΔG_{cat} (Fig. 1) (Eyring's theory; see Reaction kinetics).

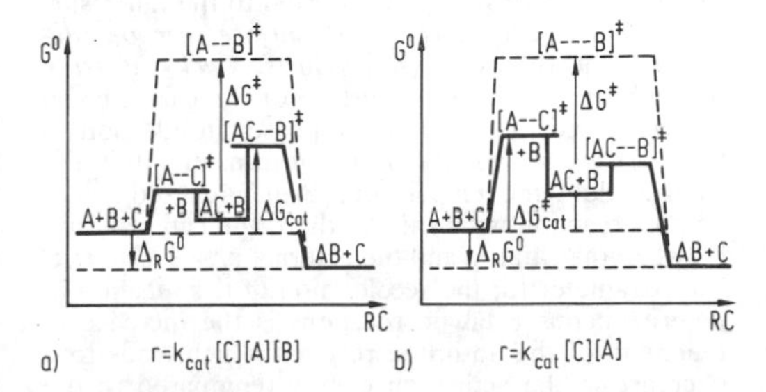

Fig. 1. Reaction diagram and rate law for a catalysed bimolecular reaction

$$A + B \xrightarrow{C} AB$$

a) with an Arrhenius intermediate AC, b) with a Van't Hoff intermediate AC. $\Delta_R G^0$ = standard free enthalpy of the reaction; $\Delta G_{cat}^{\#}$ = free enthalpy of activation of the catalysed reaction, and $\Delta G^{\#}$ = free enthalpy of activation of the non-catalysed reaction. Transition states are indicated by $^{\#}$; RC = reaction coordinate.

The greater the reduction of the free enthalpy of activation, the higher the rate of the catalysed reaction, or the activity of the catalyst. According to the Gibbs-Helmholtz equation $\Delta G = \Delta H - T\Delta S$, ΔG_{cat} depends on the enthalpy ΔH_{cat} and entropy ΔS_{cat} of activation, i.e. on the energy and probability of formation of the transition state with the catalyst in the rate-determining step of the reaction. Therefore, the nature of the interaction of the catalyst with the reaction is critical for its function. The reactivity of the reactants may be increased (activation) and/or their optimum spatial orientation may be promoted.

In the course of homogeneous catalytic reactions, there are two limiting cases: 1) the formation of the intermediate compound AC occurs more rapidly than the subsequent reaction, which is rate-determining. In this case, an Arrhenius intermediate (Fig. 1a) is said to form. This can in principle be detected, and the catalytic activity depends on the concentration and reactivity of this intermediate. 2) The formation of AC is rate-determining, because it occurs more slowly than the subsequent reaction. This type is called a Van't Hoff intermediate (Fig. 1b); it cannot be directly detected, and the catalytic activity is determined by the rate at which it forms. In each case, the catalyst is involved in the formation of the transition state in the rate-determining reaction step, and its concentration enters the rate equation of the reaction. Since a reaction product can also act as a catalyst (*autocatalysis*), a catalyst can be defined kinetically as a substance added to or formed by a reaction which appears in the rate equation of the reaction raised to a higher power than appears possible from the reaction equation. C. can be demonstrated kinetically by measurement of the dependence of the reaction rate on the catalyst concentration; this also gives a first indication for the mechanism of the catalytic reaction.

The course of a heterogeneous catalytic reaction includes the following steps: *external diffusion* of the reactants to the surface of the catalyst, *internal diffusion* or pore diffusion, i.e. diffusion to the inner surfaces of the pores; *adsorption*, *surface reaction, desorption* of the reaction products, *back diffusion* through the pores to the surface of the catalyst and diffusion into the surrounding phase. In addition to the mass transport, there is a heat transfer within the catalyst along the temperature gradient caused by the heat of reaction within the body of the catalyst, and between the catalyst and the external phase. The critical parameter for the acceleration of the reaction in heterogeneous catalytic reactions is the increase in reactivity of the adsorbed reactant which leads to a decrease in the activation energy (compared to the non-catalysed reaction) in the Arrhenius equation. In principle, it is possible that the energy of activation of the surface reaction is lower than the energy of adsorption or desorption. In this case, the adsorption or desorption will be rate-determining for the catalysed reaction.

The rate r of a heterogeneous catalytic reaction is given by the equation $r = (1/Q)\mathrm{d}\xi/\mathrm{d}t$, where ξ is the reaction progress, measured by a suitable component concentration, and Q is the volume of catalyst, or the total or a partial surface parameter. The rate laws of heterogeneous catalytic gas reactions have been found, purely empirically, to have the form $r = kp^n$ in many cases. However, the rate is often derived from a mechanism in which the surface concentration of the reactants is given by the Langmuir adsorption isotherms (see Adsorption). For the simplest case of a unimolecular, irreversible surface reaction, $A \rightarrow B$, for example, the following rate law of the Langmuir type is obtained; $K_{A(B)}$ is the adsorption equilibrium constant for A(B) and k is the rate constant for the surface reaction: $r = k\, K_A\, P_A/(1 + K_A\, p_A + K_B\, p_B)$. In bimolecular reactions, either two adsorbed molecules react with each other (*Langmuir-Hinshelwood mechanism*), or the reaction can occur between an adsorbed molecule and a molecule from the gas phase (*Rideal-Eley mechanism*). According to Hougen and Watson, rate laws for reaction with any arbitrary rate-determining step, which can be in the adsorption, desorption or surface reaction, can be described in the form r = kinetic term · potential term/ (adsorption term)^n; the individual terms are constructed from appropriate mathematical components. For kinetically more complex reactions, rate laws can be derived from the stationary state principle. If the rates of external and internal diffusion and heat transfer are too small, they reduce the activity of the catalyst and the rate of the reaction by reducing the relative penetration depth of the reactants into the catalyst. In more complex reaction, they can also affect the selectivity.

The following parameters of a catalyst are important: the *turnover number*, which gives the number of reactions per unit of time in moles product per mole catalyst or active center, characterizes the power of the catalyst and can be a measure of its catalytic activity. The lifetime of a catalyst is always limited by side reactions under the conditions of the main reaction, and this determines the *catalyst yield* or *productivity*, the amount of product which can be made per mole of catalyst. The *catalyst load* is the possible rate of reactant addition per amount of catalyst. It is given, e.g. as LHSV (liquid hourly space velocity) in volumes of liquid reactant per catalyst volume and time unit. The *selectivity* of a catalyst is expressed in the promotion of one reaction of several which are possible between the reactants; the *specificity* of a catalyst is its property of catalysing a certain type of reaction only in a substrate of a certain structure.

It is essential for high activity that a solid catalyst have a large, readily accessible surface, i.e. that it be finely divided and/or highly porous. (Its porosity is the ratio of the void volume to the total volume of the catalyst. The specific surface area (m^2 per g catalyst) can be calculated from the gas volume which is required to form a *molecular layer* over the surface of the solid. This volume of gas can be calculated from the low-temperature isotherms of physically adsorbed molecules, e.g. nitrogen, using the BET equation (see Adsorption). The surface area of the active component in carrier catalysts, the active (partial) surface area, is determined by specific chemisorption methods using hydrogen, oxygen or carbon monoxide as the adsorbed gas. The geometric parameters which characterize the void volume, such as shape, size and volume of the pores, are summarized by the concept of *texture*. Pores are classified on the basis of their diameters as micropores (< 2 nm), mesopores (2 to 50 nm) and macropores. Micro- and mesopores provide the major part of the internal surface; macropores make possible matter transport during the reaction. The surfaces of solid catalysts can be characterized by a number of methods, including adsorption and chemisorption methods monitored by calorimetry, IR and UV spectroscopy; ESCA and Auger electron spectroscopy.

Significance and applications. Kinetic inhibition of a reaction, which is generally the result of too low an attractive interaction between the reactants, can be circumvented by C. and the reaction can be steered in a certain direction. About 80% of the present chemical industrial processes depend on catalysts. The essential economic aspects of C. are increased reaction speed, avoidance of process steps, higher yields of the desired product and lower energy expenditures made possible by lower reaction temperatures and pressures. In many cases, materials which could not be made at all, or only with great difficulty, have become available through C. C. is also essential for biochemical processes. There are special applications of C. in analytical chemistry; the dependence of the reaction rate of a catalysed reaction on the nature and concentration of the catalyst can be used for qualitative analysis, end-point determination in titration (catalytic titration) and quantitative analysis (catalymetry); in principle a high sensitivity can be attained (trace analysis).

1) ***Brønsted acid-base C.*** The catalysts are proton donors and/or acceptors. If the rate of the catalysed reaction depends only on the H_3O^+ or OH^- concentration, one speaks of *special (specific) acid or base C.* In *general acid or base C.*, other acids or bases present in the system also act as catalysts. Many organic chemical reactions can be catalysed by protonation or deprotonation of the reactants, including additions to C=O and C=C double bonds, hydrolysis of esters (Fig. 2), condensation and polymerization reactions, eliminations and isomerizations.

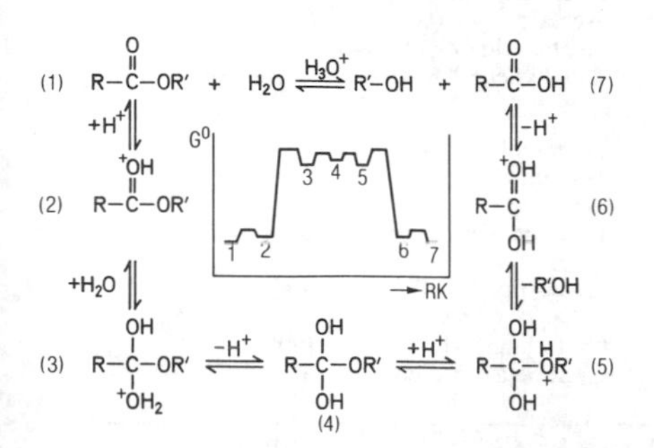

Fig. 2. Mechanism and reaction diagram of acid-catalysed ester hydrolysis

Proton transfers can occur extremely rapidly via hydrogen bonds, and in general a considerable increase in reactivity is associated with addition or loss of a proton. This is a result of the extremely strong polarizating effect of the proton. For example, with the strongest Brønsted acids, such as the combination FSO_3H/SbF_5 (magic acid), even saturated hydrocarbons can be protonated in liquid sulfur dioxide, forming highly reactive carbonium ions. In other cases, the formation of hydrogen bonds with the catalytic acid or base produces the necessary polarization of the substrate. The catalytic effect is often proportional to the pK value of the catalytic acid or base (*Brønsted catalysis relation*). An even greater activation may be achieved by combined effects of acid and base (*bifunctional C.*).

2) ***Electrophilic and nucleophilic C.*** is based on the activation of a substrate which is an electron pair donor or acceptor by a coordinative interaction with another acceptor or donor molecule. Some typical examples of electrophilic C. are the Friedel-Crafts alkylations (see Friedel-Crafts reactions) and the Meerwein-Ponndorf-Verley reduction. Nucleophilic C. occurs, for example, in the pyridine-catalysed acylation of amines, alcohols and phenols with acetyl chloride, and in cyanide-catalysed benzoin condensation.

3) ***Metal complex catalysis*** is based on the chemical reactivity of complex-bound metal ions and atoms in solution. Metal complexes can act as electrophilic reagents due to unoccupied valence orbitals on the central atom, or nucleophilic reagents due to the occupied inner d orbitals. In addition, especially with transition metal complexes, redox reactions are relatively easy. Lewis basic molecules can be activated for heterolytic bond cleavage or nucleophilic addition by complex formation, in a manner analogous to proton transfer. The polarizing effect of a metal ion is much less than that of a proton, but in neutral or weakly basic aqueous solution, it can be used catalytically. For this reason, the metal ions like Mg^{2+}, Zn^{2+} and Mn^{2+} are important activators in biocatalysis. Some examples for metal complex C. by coordinate interactions with metal(II) ions are the cleavage of oxaloacetic acid to pyruvic acid and CO_2 in aqueous solution, and the reaction of dimethylterephthalate with ethylene glycol to form polyethylene glycol terephthalate; here Cu(II) or Zn(II) ions have the highest activity. The C. of redox reactions can occur either via a redox reaction or a coordinative interaction of the reactant with the catalyst. Some examples for this are the catalytic decomposition of alkylhydroperoxides by Co(II) or Mn(II) compounds, which is important for the autooxidation of hydrocarbons, and the epoxidation of alkenes by alkylhydroperoxides catalysed by Mo(VI) compounds (Halcon process). However, substitution reactions can also be accelerated by a redox reaction with the catalyst. An example is the C. of solution of anhydrous chromium(III) chloride in coordinating solvents like water, tetrahydrofuran or pyridine, by Cr(II) ions, which are generated by addition of metallic zinc. Complex catalysts make possible considerable modification in the nature of the central atom, i.e. its valence orbital characteristics and its oxidation state, and its coordination sphere, which is determined by the donor-acceptor properties, geometric structure, volume, arrangement and mutual interactions of the ligands. Such modifications can be used to optimize the catalyst and in some cases to direct the catalytic reaction.

4) ***Organometallic complex C.*** is the C. of organic chemical reactions which involve formation of metal-carbon bonds. In 1953, K. Ziegler discovered the C. of ethylene polymerization under normal conditions by an organometallic mixed catalyst of $TiCl_4$ and $Al_2(C_2H_5)_6$ (see Organoaluminum compounds). A little later, G. Natta achieved stereoregular polymerization of propene and butadiene, using similar catalysts. In general, organometallic catalysts with a main-group metal are called *Ziegler-Natta catalysts*. The actual catalysts in these systems are very reactive σ- or π-organometal complexes with transition metals which are formed by reaction with the organyl component, sometimes involving the reactants. The organoaluminum compounds, which are readily soluble in hydrocarbons, have proved especially useful as the organometallic components (*cocatalysts*). Used in excess, they can protect the catalytically active complexes which form against poisoning by traces of O_2 and H_2O.

The profile of modern petroleum and high-polymer chemistry is essentially determined by organometallic complex catalysis (Table 1). Organometallic complex catalysts frequently are highly active, selective and specific. For example, the turnover number for nickel-complex-catalysed propylene dimerization was found to be $6 \cdot 10^5$ mol C_3H_6/mol Ni and second, which corresponds to the activity of highly active enzymes.

Table 1. Industrially important methods of generating organic intermediates and high polymers from simple precursors such as acetylene, ethylene, propylene, butadiene, carbon monoxide and hydrogen using organometallic catalysis (Ph = $-C_6H_5$, Ar = Aryl, R = acac

Reaction	Catalyst	Product (Selectivity)
Reppe syntheses with acetylene		
Vinylation with acetic acid	Hg^{2+}	Vinyl acetate (70%)
Ethynylation with formaldehyde	CuC_2	But-2-in-1,4-diol (90%)
Carbonylation with CO/H_2O	$Ci(CO)_4/HBr$	Acrylic acid (90%)
Cyclooligomerization	$Ni(CN)_2$	Cyclooctatetraene (70%)
Functionalization reactions		
Hydrosilylation of olefins	$Pt(C_2H_4)(PPh_3)_2$	Alkylsilane
Hydrocyanation of butadiene	$Ni(P(OPh)_3)_4$	Adipodinitrile
Ethylene oxidation	$O_2/PdCl_2/CuCl_2$	Acetaldehyde (95%)
Carbonylation of methanol	$Co_2(CO)_8/HI$	Acetic acid (90%)
Oxo-synthesis of l-olefins	$HCo(CO)_4/PR_3$	Aldehyde (80%) or Alcohol (80%)
C-C linking reactions		
Polymerization of ethylene	$TiCl_4/Al_2(C_2H_5)_6$	Polyethelene (97%)
Oligomerization of ethylene (Schell process)	$PhNi(PhC(O)$ - $CHPPh_2)PPh_3$	1-olefin $C_{10} \ldots C_{30}$ (98%)
Polymerization of propylene	$TiCl_3/MgCl_2/$ $Al(C_2H_5)_2Cl$	Isotactic polypropylene (95%)
Polymerization of butadiene	$Ni(O_2CR)_2/Al_2(C_2H_5)_6$ $/BF_3 \cdot (C_2H_5)_2$	1,4-cis-polybutadiene (96%)
Cyclotrimerization of butadiene	$TiCl_4/Al_2Cl_3(C_2H_5)_3$	Cyclododeca-1,5,9-triene (90%)
Cyclodimerization of butadiene	$Ni(acac)_2/Al_2(C_2H_5)_6/$ $P(OAr)_3$	Cycloocta-1,5-diene (96%)

Second-generation Ziegler-Natta catalysts have been improved by fixing them to carriers and treatment with Lewis bases; in ethylene polymerization they now yield over 10,000 kg polyethylene per mole of titanium, and the process step in which catalyst is removed is no longer necessary. The use of much more active rhodium complex catalysts in the synthesis of acetic acid from methanol and carbon monoxide and the oxo-synthesis of olefins permits increased selectivity and the high-pressure process has been converted to an energetically far less costly medium-pressure one. An example of high catalyst specificity is the Wilkinson complex $Rh(PPh_3)_3Cl$ (Ph = phenyl) used to hydrogenate olefinic double bonds, which does not attack other unsaturated groups such as -CHO, -CN, $-NO_2$, $-CO_2H$ and $-C_6H_5$. By introducing chiral phosphine ligands, it was possible to hydrogenate asymmetrically α-acetylaminocinnamic acids with an enantiomeric selectivity up to 99%; this made possible direct synthesis of the drug L-dopa (3,4-dihydroxyphenylalanine) on an industrial scale. The mechanism of action of an entire series of organometallic complex catalysts has been elucidated in broad outline.

The insight into the mechanisms of organometallic complex catalysts obtained by kinetic, spectroscopic and preparative studies has been of great importance in the development of the chemical theory of C. In general, the mechanism of homogeneous catalytic reactions is easier to explain, and in many cases provides models for the course of comparable heterogeneous catalytic reactions. A significant methodological disadvantage of organometallic complex catalysts is that they often do not have very great thermal or chemical stability, and they are difficult to separate. This problem has been approached by fixation to carriers, that is, binding the catalytically active complex chemically to an organic or inorganic solid.

III) *Heterogeneous C. (contact C., surface C.).* The catalyst here is usually an amorphous or crystalline solid on the surface of which gaseous or liquid reactants react. The interaction between catalyst and reactant can vary greatly, corresponding to the nature of the chemical bond. On this basis, the solid catalysts can be classified according to type of substance.

1) ***Metal and semiconductor catalysts***. The semiconductors (semiconductor oxides and sulfides) and metals which can transfer single electrons can act as electron donors and acceptors, and catalyse redox reactions. In *hydrogenations* and *dehydrogenations* metals, e.g. platinum, palladium, nickel, cobalt and iron, and oxides and sulfides of zinc, nickel, iron, chromium, molybdenum, tungsten and vanadium, are catalytically active; in practice, for the hydrogenation reactions at low temperatures, the more active, but more expensive metals, are preferred. For *oxidations*, vanadium(V) oxide, silver, copper and platinum are used. In the electronic theory of C. on metals and semiconductors, the catalytic properties of these solids are explained in terms of their electronic properties, such as conductivity type, position of the Fermi level and occupation density of electronic levels.

The preferred sites of reactions are, according to Tayler, active centers on the surface, e.g. atoms or ions which are not coordinatively saturated, especially at defect sites, crystal edges and corners, or whole groups of atoms (ensembles) on the surface. Such energy-rich deficit sites are generated in high concentration in alloy skeleton catalysts, the ***Raney catalysts***, e.g. aluminum from an aluminum-nickel alloy. The crystallographic orientation of the surface regions can also be important; a simultaneous action of several surface atoms requires a suitable geometric arrangement, according to the multiplet theory of Balandin. For example, the synthesis of ammonia from the elements occurs preferentially on the Fe(111) surface.

2) ***Acid-base catalysts***. Solids with acidic OH surface groups (Brönsted centers) are suitable as catalysts for reactions which occur by an acid-base mechanism, such as alkylation, isomerization, oligomerization, hydration, dehydration and cracking of hydrocarbons. The high thermal stability of some solid acids makes possible much higher reaction temperatures than are generally possible with homogeneous acid systems. This group of catalysts (acting as solid acids) includes, e.g. γ-Al_2O_3, Al_2O_3-SiO_2, MgO-SiO_2 and modified zeolite; all act as electron-pair acceptors on their surfaces. Type Y zeolites, doped with rare-earth metal cations, are used on large scale in industry for catalytic cracking of high-boiling petroleum fractions. In reactions in small-pore zeolites, molecular sieve effects can occur: reactant molecules which are larger than the micropore openings cannot enter them and are thus excluded from the reaction. An example is the type-selective cleavage of unbranched alkenes on small-pore zeolites. The differential catalytic action of various solids can also be coupled in bifunctional (or polyfunctional) catalysts: the platinum/γ-Al_2O_3 catalyst used to crack heavy gasoline simultaneously catalyses isomerization reactions (e.g. methylcyclopentane to cyclohexane) on the γ-Al_2O_3 and dehydrogenation reactions (cyclohexane to benzene) on the platinum.

3) ***Ionic and complex-like catalysts***. Ionic catalysts are used for addition and elimination of HX (X = halogen, acetate); the anion of the catalyst is the same as the group being introduced. Examples are the hydrochlorination of acetylene on $HgCl_2$/active charcoal, dehydrochlorination of tetrachloroethane on $BaCl_2$ and the Reppe synthesis (see Acetylene chemistry) with CuC_2 as catalysts. These and several other heterogeneous catalytic reactions, e.g. the disproportionation of olefins on MoO_3 and WO_3 and the medium-pressure polymerization of ethene on Cr_2O_3 are similar in their mechanisms to analogous homogeneous complex-catalysed reactions.

Structure, properties and applications of industrial solid catalysts. The heterogeneous catalysts used industrially are, as a rule, composite ***mixed*** or ***multi-substance catalysts***. Table 2 lists some industrially important heterogeneous catalytic reactions.

The effectiveness (activity, selectivity, lifetime) of some catalysts can be increased by addition of small amounts of other substances, *promoters* (*activators*). A promoter which suppresses the tendency of the active phase to sinter and thus lose surface area, is called a texture promoter (structure preserver). The effect of such promoters was first studied by Mittasch in the course of development of the catalyst for ammonia synthesis. For this reaction, a potassium and aluminum double promoted iron catalyst is used; the aluminum oxide acts as a texture promoter. In ***bimetallic*** and ***multimetallic catalysts***, both possible functions of a promoter – increasing the surface area and increasing the catalytic activity per unit of surface area – are provided by additional metal components. For example, rhenium acts mainly as a texture promoter in the Pt-Re/Al_2O_3 cracking catalyst. According to Rienäcker, the catalytic properties of ***alloy catalysts*** are often greatly modified by alloy formation. The incorporation into a semiconductor catalyst of a foreign substance which modifies its catalytic properties is called "doping".

Table 2. Heterogeneous catalysed industrial processes

Process	Catalyst
Ammonia synthesis (Haber-Bosch)	$Fe/Al_2O_3/K_2O$
Oxidation of sulfur dioxide to sulfuric acid (contact process)	V_2O_5/SiO_2
Oxidation of ammonia to nitrogen oxides for nitric acid synthesis	Pt/Rh
Fischer-Tropsch synthesis	$Fe,Co/ThO_2/MgO$
Synthesis of methanol	
High-pressure process	ZnO/Cr_2O_3
Low-pressure process	CuO/Cr_2O_3
Conversion of carbon monoxide	
High-temperature conversion	Fe_2O_3/Cr_2O_3
Low-temperature conversion	$ZnO/CuO/Cr_2O_3$
Steam conversion of hydrocarbons	$Ni/CaCO_3$
Catalytic cracking of hydrocarbons	Al_2O_3/SiO_2 (zeolite)
Hydrogen cracking of hydrocarbons	$Ni/Pd/Al_2O_3/SiO_2$ (zeolite)
Conversion of heavy gasoline	Pt/γ-Al_2O_3
Isomerization of the C_8 aromatics	$Pt/Al_2O_3/SiO_2$ (zeolite)
Hydrogen refining	$NiS/WS_2/Al_2O_3$ or $CoS/MoS_2/Al_2O_3$
Selective hydrogenation of phenol to cyclohexanone	Pd/Al_2O_3
Dehydrogenation of ethylbenzene to styrene	$Fe_3O_4/Cr_2O_3/K_2O$
Oxidation of ethylene to ethylene oxide	Ag/α-Al_2O_3
Oxidation of benzene to maleic anhydride and o-xylene	V_2O_5/SiO_2

In carrier catalysts, the catalytically active component is applied to a solid carrier with a high surface area (γ-aluminum oxide, silica gel, silicate, pumice, active charcoal). In the simplest case, such catalysts are obtained by soaking the carrier in a solution of a compound of the active substance, which is then activated by oxidation, reduction or thermal treatment. In addition to providing a uniform distribution of the active component, the carrier provides a high degree of dispersion and gives the catalyst the necessary solidity, as well as providing the pore system needed for transport of the reactants and products to and away from the active sites. In addition, the carrier can affect the catalytic properties of the active component by electronic interactions; the orientation of metal crystallites is also possible. In the extreme case, in bifunctional catalysts, the carrier itself is active in the reaction.

A high internal surface area is not always an advantage for a catalyst. For selective oxidations, compact catalysts with low porosity are used, because in very porous catalysts, the transport of reaction heat and reactants (products) is not rapid enough, and can lead to undesired secondary reactions (total oxidation!).

The *form of an industrial solid catalyst* is adapted to its use. The shape of the catalyst affects the pressure drop and flow conditions in the reactor, as well as particle density and mechanical stability (stability to pressure and abrasion) of the catalyst itself. The most important shapes of formed catalysts (which are often made by adding special binders) are cubes or tablets, rods, hollow rods and spheres. For catalytic cyclone processes, such as the ammoxidation of propane to acrylonitrile on $Bi_3O_2/MoO_3/P_2O_5$ and cracking of high-boiling petroleum fractions on acidic aluminosilicates, microspheroidal catalysts are produced by spray-drying. In most processes using heterogeneous catalysis, the catalyst is placed in a continuously operated solid-bed reactor. For special purposes, e.g. for hydrogenation of fats, discontinuously operated stirred vats are used.

Inactivation, regeneration, reactivation. The effectiveness of most catalysts decreases with time of use. The causes of this inactivation are poisoning, aging and coking.

Catalyst poisons (***contact poisons***). For metal catalysts, even traces of compounds from the Vth and VIth representative groups of the periodic system (sulfur, phosphorus, arsenic) act as poisons, as do carbon monoxide, halogen compounds and water. The *poisoning* can be either irreversible or reversible. Undesired side reactions can be suppressed by selective poisoning, e.g. in selective hydrogenation, addition of carbon monoxide suppresses the hydrogenation of olefins more strongly than that of acetylenes and dienes. A catalyst may also be poisoned by one of the reaction products.

The *aging* of solid catalysts is associated with changes in the structure and/or texture, and can be caused by sintering, recrystallization processes (e.g. crystal growth, collective crystallization of the active components in carrier catalysts, annealing of lattice defects) and by chemical changes in the active components, e.g. changes in the oxidation state or formation of compounds by solid reactions.

Coking is the deposition of highly condensed byproducts ("coke") on the catalyst in organic reactions. In regeneration, the coke is removed by burning it off.

The restoration of activity to an inactivated catalyst is called reactivation; it is advantageous to carry out the process in situ. It can include measures in addition to regeneration, such as redispersing the active components, as is done with the platinum in cracking catalysts.

Historical. In 1782, Scheele observed a catalytic acceleration of ester formation and saponification by mineral acids (H^+ ions). In 1823, Dœbereiner observed that hydrogen in contact with sponge platinum ignited in air at room temperature. Berzelius first defined the concept of C. in 1834, and W. Ostwald introduced the kinetic definition of C. in 1900. At that time, contact sulfuric acid was first made on a large industrial scale. In 1909, Mittasch developed mixed catalysts. In 1913 in Ludwigshafen, and in 1916 in Lena bei Merseburg, the first large high-pressure installations for ammonia synthesis by the Haber-Bosch process began production with iron mixed catalysts. In 1940, catalytic cracking in petroleum processing and petroleum chemistry was introduced by Houdry. In 1952, organometallic mixed catalysts for low-pressure polymerization of ethene were discovered by Ziegler, and in 1954, the stereoregular polymerization of propene and butadiene was introduced by Natta.

Catalyst: see Catalysis, see Fuels, see Exhaust, motor vehicle.

Catalyst poison: see Catalysis, sect. III.

Catanols: melting products of phenol-like substances with sulfur, and sometimes alkalies. The C. are used to treat cotton before dyeing with basic dyes, and in the paint and paper industries.

Catarol process: a process for obtaining aromatics from the aliphatic hydrocarbons in petroleum. A petroleum distillate is passed at 630-680 °C and atmospheric pressure through a pipe packed with copper contact materials. Gaseous byproducts are also formed, especially methane, ethene and ethylene.

Catcracking: see Cracking.

Catechol: see Catechol tannins.

Catecholamines: Alkylamino derivatives of pyrocatechol. The C. are readily oxidized to 2-quinones, which can cyclize to indole derivatives by an intramolecular nucleophilic attack of the amino group. The *leukochromes* formed in this way are readily further oxidized; anaerobic oxidation produces the colored *aminochromes*, while aerobic oxidation in the presence of polyphenol oxidases leads to the Melanins (see). Melanin-like products are also formed by oxidation of C. with chromate, a reaction which is used to detect C. in animal tissues (chromaffin tissues). The C. act as neurotransmitters in animals; this group includes dopamine, noradrenalin and adrenalin. Their biosynthesis begins with the amino acid tyrosine, which is first oxidized to 3,4-dihydroxyphenylalanine, then decarboxylated. The C. are thus biogenic amines.

Catechol tannins, ***condensed tannins***: a group of tannins in which the monomeric units are flavan-3-ol (catechols) or flavan-3,4-diol (leukoanthocyanidins) with phenolic hydroxyl groups preferentially in the 5,7,3',4' and 5' positions. The widely occurring parent compound is ***catechol***, a 5,7,3',4'-tetrahydroxyflavan-3-ol, of which there are 4 optical isomers (D- and L-catechol, D- and L-epicatechol). The C. are built up from these units by enzymatic dehydrogenation with polyphenol oxidases or by acid-catalysed autocondensation. The bark tannins from oak and horse chestnut and the wood tannin catechu are C.

Catechu: a dark brown, hard, brittle and opaque natural product with a bitter taste. C. is obtained by extraction of the twigs and leaves of *Uncaria gambier*, or of the red heartwood of the Indian acacias *Acacia catechu* and *A. suma*. The twigs and leaves or heartwood produce a reddish-brown extract in boiling water; the alcohol extract is a clear brown solution. C. contains the pigment ***catechol***, a pentahydroxy derivative of flavin. The aqueous solution turns bright red in the presence of alkalies, and upon boiling in the open air, it turns brown. C. also contains ***catechu tannic acid***, a non-crystalline, reddish-brown mass with an astringent taste. It is readily soluble in water and alcohol, but is insoluble in ether. C. was formerly used for tanning and as an aid to dyeing of cotton or silk. On account of its tannic acid content, it was also used to preserve fishnets, sails and leather.

Catena: see Nomenclature, sect. II.D.1.

Catenanes: see Topological isomerism.

Catergols: see Fuels.

Catforming process: see Reforming.

Catharanthus alkaloids: see Alkaloids.

Cathode: see Electrode.

Cathodic corrosion protection: see Electrochemical corrosion protection.

Cathodic reduction: reduction taking place at the cathode as electric current flows through an electrolysis cell. C. is used in the production of alkali metals and in galvanic technology.

Catholyte: see Electrolysis.

Cation: a positively charged Ion (see). The concept of C. was originally defined by Faraday only in connection with electrolysis: C. are those particles which move toward the cathode, that is, the negative electrode, in an electrolysis.

Cation base: see Acid-base concepts, sect. on Brønsted definition.

Cation exchanger: see Ion exchanger.

Cation chromatography: see Ion chromatography.

Cationic acid: see Acid-base concepts, sect. on Brønsted definition.

Cationics: see Surfactants.

CAT scan: see NMR tomography.

Caustic lime: same as Calcium oxide (see).

Caustic magnesia: same as Magnesium oxide (see).

Caustic potash: same as Potassium hydroxide (see).

Caustic soda: same as Sodium hydroxide (see).

Caustification: the conversion of alkali carbonates, e.g. sodium carbonate (soda), Na_3CO_3, or potassium carbonate (potash), K_2CO_3, into the corresponding hydroxides, NaOH and KOH, which are known as *caustic alkalies*. The process involved reaction of carbonates with slaked lime. As an industrial process, C. is now obsolete.

Cavitation corrosion: see Corrosion.

CCC: 1) see Countercurrent chromatography. 2) abb. for Chlorocholine chloride.

CCK: abb. for Cholecystokinin (see).

CCK-PZ: see Cholecystokinin.

CD: see Chirooptical methods.

Cd: symbol for cadmium.

Ce: symbol for cerium.

Cedar wood oil: a viscous, nearly colorless essential oil with a mild but long-lasting odor, containing up to 80% α- and β-cedrene. C. is obtained from chipped cedar wood by steam distillation and is used in making soaps and other cosmetic preparations. Condensed C. is used to enclose microscopic preparations.

Cefalexin: see Cephalosporins.

Cefaloridin: see Cephalosporins.

Cefalotin: see Cephalosporins.

Cefamandol: see Cephalosporins.

Cefotaxim: see Cephalosporins.

Cefoxitin: see Cephalosporins.

Ceiling temperature: the temperature above which a polymerization reaction no longer occurs, and the previously formed polymer depolymerizes to monomers or is completely decomposed. Equilibria between polymerization and depolymerization reactions are rare, being known only for a few mixed polymerizations with sulfur dioxide. Only in these cases is there depolymerization above the C.

Celestine blue: same as Cobalt blue (2) (see).

Cell current: the electrical current flowing through an electrochemical cell.

Cellitone fast dyes: a group of suspension or dispersion dyes which, with a certain dyeing technique, give very fast and even colors to acetate fibers. Several azo and anthraquinone dyes are used as C.

Cellitone fast yellow 7G: see Polymethine pigments.

Cellobiose: 4-(β-D-glucopyranosyl)-D-glycopyranose, a reducing disaccharide. C. exists in nature only in bound form. It can be obtained from cellulose as the octaacetate.

Cellophane: thin, solid but somewhat stretchable, transparent films used mainly as a packaging material. It is resistant to water, fats and oil, but is only slightly resistant to steam. It consists of regenerated cellulose.

Cellosolve: 1) in the narrower sense, a term for ethyl glycol (ethyleneglycol ethyl ether). 2) In the broad sense, low-molecular-weight polyalkylene glycols which are given the names of the alcohol group, e.g. methyl, butyl or hexyl cellosolve; the esters are named analogously, e.g. methyl cellosolve acetate.

The C. are used as solvents for paints, paint removers, textile cleaners and solvents for organic syntheses.

Cell reaction: the overall reaction of an electrochemical cell, the sum of the partial processes. Example: the C. of hydrogen chloride electrolysis:

$$\begin{array}{ll} \text{Anode:} & 2\,Cl^- \rightleftharpoons Cl_2 + 2\,e \\ \text{Cathode:} & 2\,H^+ + 2e \rightleftharpoons H_2 \\ \hline \text{Cell reaction:} & 2H^+ + 2\,Cl^- \rightleftharpoons H_2 + Cl_2 \end{array}$$

Cellulases: hydrolases from plants, microbes or fungi which degrade cellulose to cellobiose and glucose. The C. are often a mixture of different endo- and exoenzymes; they are more active against cellulose derivatives than against natural cellulose. The C. from *Penicillium notatum* (M_r 35,000) consists of 324 amino acids and contains one disulfide bridge. C. are used in the production of special foods, such as baby foods, and special diets, and they are increasingly used to produce glucose from cellulose-containing wastes.

Celluloid: a plastic based on cellulose dinitrate, and the first thermoplastic produced on a large scale (1869 by the Hyalt brothers). Pure C. is transparent, but it is easily colored with pigments, and it is readily soluble in many organic solvents. It softens about 80 °C, and its density is 1.38 g cm^{-3}. The disadvantage of C. is that it ignites and burns easily. It is made by mixing cellulose dinitrate (collodion cotton) with 25-30% softener (chiefly camphor), 5-10% vasoline or ricinus oil as a lubricant, and alcohol. It is kneaded into a plastic mass. Pigments or soluble organic dyes may be added. The doughy mass is further worked in a roller machine and pressed at about 100 °C and 120-200 kg pressure into homogeneous blocks. Plates or foils are cut or shaved off these blocks. These are usually worked further by warm pressing at about 80 °C or by planing. C. can be blown into hollow objects in a manner similar to glass blowing.

C. is used to make drawing stencils, combs, toys, glasses frames, ping pong balls, handles and medical instruments. It was used in the early film industry, but is no longer, because of the danger of fire.

Cellulose: β-D(1→4)-glucan, an unbranched polysaccharide made up of glucose. C. is a colorless substance and insoluble in ordinary solvents. In certain solvents, such as $[Cu(NH_3)_4](OH)_2$ solution (Schweitzer's reagent), C. dissolves by forming copper-chelate complexes. Regenerated (mercerized) C. can be recovered from the solution by precipitation. C. dissolved in concentrated alkali metal hydroxide solution is called ***alkali cellulose***. ***α-C.*** is the native C. which is insoluble in 17.5% NaOH or 24% KOH, ***β-C.*** is the part which can be precipitated from such solutions with methanol. The C. which remains in solution after methanol treatment is called ***γ-C.*** Treatment of C. with acetic anhydride and sulfuric acid produces octaacetylcellobiose. C. is degraded by acid-catalysed or enzymatic hydrolysis by cellulases to D-glucose; under gentle conditions, it is degraded to the disaccharide cellobiose. Cellulases are not present in mammals. Ruminants are able to utilize C. because microorganisms which do possess cellulases are present in their stomachs.

C. I is the term applied to the crystalline modification of native C., with parallel bands. In regenerated C. and alkali cellulose, the crystal lattice is changed; this modification is called ***C. II***. It contains antiparallel glucan chains held together by more intermolecular hydrogen bridges, which are responsible for the greater stability of this modification. C. fibers are partially broken down by wet milling. Dry milling and treatment with 1 N hydrochloric acid also break down larger structures and shorten the chains. The latter method is used to prepare microcrystalline C.

Occurrence. C. is the most important skeletal polysaccharide in plant cell walls. The primary cell wall contains, in addition to C., large amounts of hemicelluloses and pectin substances. The secondary, woody cell wall usually contains a larger proportion of C. The main components of wood, aside from C., are the hemicelluloses and lignin. The annual production of C. on the earth is between 100 and 150 · 10^9 t. C. is thus the most abundant organic substance. α-C. is present in unusually pure state in the cell walls of the algae *Valonia* and *Cladophora*. Other native C. usually contain small amounts of other monosaccharides; cellulose from cotton, for example, contains about 1.5% xylose and smaller amounts of other monosaccharides. The natural fibers cotton, linen, hemp, ramie and jute contain high percentages of C.; about 98% of the dry mass of cotton is C.

Cellulose: R = OH

Chitin: R = $NHCOCH_3$

Chitosan: R = NH_2

The average number of glucose units in native C. is about 20,000; the number detected depends greatly on the treatment of the sample. Chemical end-group determination gives a value of only about 1,000 units. The glucan chains of C. are connected into fibrils by intermolecular hydrogen bridges; in native C. the glucan chains are arranged in parallel bands. Native C. is 60 to 70% crystalline. The less ordered, amorphous sites are relatively easily attacked by enzymes and hydrolytic reagents. Thus during treatment of C., partial degradation begins at these sites.

Structure of a cellulose fibril showing regions of greater and lesser molecular order

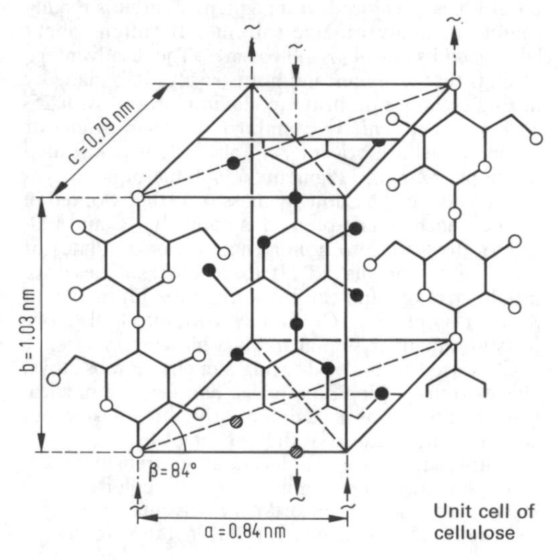

Unit cell of cellulose

Production. Before C. is used in the production of synthetic fibers and plastics, all the matrix compounds which accompany it in the cell walls (resins, fats, lignin, etc.) are removed chemically, leaving practically pure α-cellulose. Ordinary C. for use in paper production contains 85 to 88% β-C., while that for use in rayon production must be 90 to 92% α-C.; and C. for production of C. acetate and cuprammonia rayon must be 98% α-C.

The raw material for industrial production of C. is usually conifer wood, although for synthetic fibers, certain deciduous trees are used (beech or poplar). There are several chemical processes for extraction of C. 1) In the *sulfite process*, the wood is chipped, then cooked in an aqueous calcium hydrogensulfite solution containing excess sulfurous acid in 400 m^3 pressure vessels. The lignin of the wood is converted to soluble lignin sulfonic acid. After the cooking process, the softened pieces of wood are pulped, then washed and dried. The end product is raw C. pulp containing about 90% dry matter. The management of the huge amounts of sulfite liquor and washwater is a serious environmental problem (see Sulfite liquor).

2) In the *sodium hydroxide process* (***alkali C.***), the wood chips are boiled for a few hours in 6 to 8% sodium hydroxide solution, at 0.4 to 0.8 MPa. Because the sodium hydroxide is expensive and can be used only once, this process is used only to produce high-quality paper which does not become yellow or

brittle with age. For ordinary paper, it has been replaced by the 3) *sulfate process* (***sulfate C., kraft C.***), which is used mainly with resinous woods (pine, larch) or to produce high-quality C. from spruce or beech. Straw, which contains too much silicic acid to be worked up by sulfite, can be used with this process. One liter of the cooking liquor contains about 60 g sodium hydroxide, 22 g sodium sulfide, 15 g sodium carbonate and 4 g sodium sulfate. The pulp is cooked 3 to 6 hours at 170 to 180 °C under a pressure of 0.6 to 1.0 MPa. After the cooker is emptied, most of the liquor is removed on rotating filters; the C. is washed in diffusers by counter-current, and then dried to 50% dry mass. The cooking liquor is recovered. The process produces about 10 kg terpentine per ton of C., and 30 kg of a "resin soap" which, when acidified, yields tall oil.

The *nitric acid process* is used mainly with beech wood. In the impregnation phase, at a temperature below 50 °C, exactly enough 14% nitric acid is added to oxidize the lignin. After the acid has been pumped off, the wood chips are slowly heated to 100 °C in water; the C. is not attacked. After removal of the acidic lignin oxidation products, mainly oxalic acid, the pulp is cooked with hot, dilute sodium hydroxide. Finally it is treated with chlorine and 10% sodium hydroxide at a low temperature; its α-C. content is raised in this way to 98%. This product can be used directly for synthesis of cuprammonium rayon, as a substitute for linters.

5) The *Pomilio process* is used for straw, which is first treated with dilute sodium hydroxide and then with chlorine. Thereafter, it is extracted with strong sodium hydroxide solution.

The process of extraction and purification usually includes a bleaching step, usually with hypochlorite or the gentler chlorine dioxide.

Applications. C. is an essential raw material for production of paper, cardboard, textiles (both the natural fibers and synthetic ones such as rayon and viscose, etc.). Very pure C. products are used as bandage materials and absorbants. The sugar obtained by acid hydrolysis of C. (*wood sugar*) can be used to produce ethanol. There are many chemically modified C. products, especially the ethers (ethyl, benzyl, methyl and carboxymethyl) and esters (acetate, propionate, acetobutyrate, nitrate and xanthogenate), which have numerous applications.

Cellulose acetates, abb. ***CA, acetylcellulose***: cellulose esters formed by esterification of the free hydroxide groups of the cellulose with acetic anhydride. In the industrial production of C., the crude cellulose is first activated by treatment with acetic acid, then acylated with acetic anhydride in the presence of zinc chloride or sulfuric acid as catalyst. The ***cellulose triacetate*** thus formed is soluble only in a few solvents (methylene chloride, ethylene chloride, chloroform) and is incompatible with softeners. It is usually converted to ***cellulose 2 ½ acetate*** by heating with sulfuric acid to 50 to 70 °C or by treatment with hydrogen chloride bubbled through a zinc chloride-acetic acid solution containing it. In the 2 ½ acetate, every other glucose unit of the cellulose has a free hydroxyl group, and the polymer is soluble in acetone. The 2 ½ acetate is precipitated with water, washed and dried. It is used mainly to produce plastics and acetate fibers; a smaller amount is used as packaging and electrical insulation sheets. Plastics made of C. contain 10 to 30% softeners and are used for purposes where stability to light, surface shine and heat resistance are important (parts of tools, toys, combs, brushes, buckles, buttons, glasses frames, drinking straws, radio housings, etc.). The triacetate is used to produce electrical insulation and as the basis of photographic films; in contrast to cellulose nitrate films, these are not very flammable. The film bases are made from solutions of C. in methylene chloride/alcohol mixtures to which small amounts of softeners are added; they are formed on drums or ribbon casting machines. Both triacetate and 2 ½ acetate are starting materials for textiles.

Cellulose acetobutyrate, abb. ***CAB***: a cellulose mixed ester which is made by simultaneous reaction of acetic anhydride and butyric acid or butyric anhydride with cellulose. C. is better soluble in organic solvents than cellulose acetate. However, its solubility depends on the content of butyryl groups, C_3H_7CO-. C. is very resistant to oil, and is used mainly in the production of housings for machines and batteries, steering wheels, window coatings, and both lenses and frames for sun glasses.

Cellulose esters: Cellulose derivatives such as Cellulose acetate (see), Cellulose acetobutyrate (see), Cellulose propionate (see) and Cellulose nitrate (see), in which the hydroxyl groups in the cellulose molecule are partially or completely esterified to acids. The treatment with the esterification reagent causes partial degradation of the macromolecular cellulose molecule, so that the resulting C. is soluble in the esterification mixture.

Solid plastics made of C. are thermostable, give high acoustical damping, have good antistatic behavior and do not ignite readily. These properties can be further improved by addition of glass fibers, so that glass-fiber-reinforced thermoplastics based on C. are used increasingly to make electrical devices, toys, office machine and auto parts.

Cellulose ethers: cellulose derivatives in which some or all of the OH groups of the cellulose are in ether linkages. The starting material is alkali cellulose, which reacts with ether-forming reagents, such as haloalkyls, haloaralkyls, haloalkylcarboxylic acids or ethylene oxide. The degree of ether formation and solubility properties of the product depend on the reaction conditions. Incomplete ether formation yields a water-soluble product, while C. with a high percentage of bound OH groups are soluble only in nonpolar solvents.

Of the many possible C., only ***ethyl*** and ***benzyl cellulose*** are major industrial products. These are thermoplastics and are generally formed by injection molding. They are used to some extent for insulation, pipes and electrical insulators. They are also used in making paints, especially for underwater paints. Benzylcellulose binds well to light metals, and this property is utilized in making aluminum paints. ***Methyl cellulose*** and ***carboxymethyl cellulose*** are soluble in water and are used as binders for water-base paints, as glues, emulsifiers, textile conditioners and binders for printing cloth. Carboxymethyl cellulose is used as an additive to laundry products.

Cellulose nitrate, abb. ***CN, nitrocellulose***: a cellulose ester in which the hydroxyl groups of the cellulose are replaced by NO_3 groups. C. exists as white fibers which are odorless and tasteless, insoluble in water, but soluble in various organic solvents. Because it ignites easily, C. must always be kept moist with water or alcohol.

C. is prepared from thoroughly purified cellulose, which is submersed in nitrating acid (a mixture of nitric and sulfuric acids in water). The C. obtained from nitration is centrifuged out and the residual acid is removed by water rinsing. Because acid residues greatly reduce the stability of the C., and under some conditions can lead to an explosion, the material is stabilized by boiling with several changes of water. Industrial C. is a mixture of mono-, di- and triesters, of varying chain lengths. It is characterized by its nitrogen content. Fully nitrated cellulose (***cellulose trinitrate***) would theoretically have a nitrogen content of 14.14%, but the industrial product obtained by the usual nitration methods has at most 13.4% nitrogen. C. with a nitrogen content of 12.5 to 13.5% is called Gun cotton (see), and that with 10.6 to 12.5% nitrogen is called Collodion cotton (see). C. with 11.5 to 12.5% nitrogen is used in the production of crude films and paints.

The industrial production of C. in Germany began in 1845, but it was not produced in very large amounts until the First World War, when it was recognized that undesired explosions were due to traces of residual acids.

Cellulose nitrate lacquers: see Lacquers.

Cellulose nitrate powder: see Smokeless powder.

Cellulose propionate, abb. ***CAP***: a mixed cellulose ester obtained by simultaneous reaction of acetone hydride and propionic acid with cellulose. It is used in the same way as cellulose acetobutyrate.

Cellulose triacetate: see Cellulose acetate.

Cellulose xanthogenate: see Viscose.

Cell voltage: the electrical voltage between the poles of an electrochemical cell.

Cement: see Binders (building materials).

Cementation: the precipitation of a metal from a salt solution by a finely divided, active metal which has a more negative standard electrode potential and goes into solution in place of the more noble metal. Examples are the production of cement copper (see Copper) according to $Cu^{2+} + Fe \rightarrow Cu + Fe^{2+}$, purification of zinc sulfate solutions according to the equation: $Cd^{2+} + Zn \rightarrow Zn^{2+} + Cd$, and the recovery of gold according to the equation: $2\ [Au(CN)_2)]^- + Zn \rightarrow 2\ Au + [Zn(CN)_4]^{2-}$.

Cement black: see Manganese oxides.

Cementite, ***iron carbide***: Fe_3C, rhombic crystalline, metastable compound in which the C atoms occupy holes in the iron lattice. It is very hard and brittle. C. is a component of steel and there are three types which precipitate in Iron-carbon alloys (see): 1) ***primary C.*** forms from melts containing more than 4.3% carbon; 2) ***secondary C.*** forms at 723-1147 °C through decomposition of austenitic mixed crystals with more than 0.8% carbon; and 3) ***tertiary C.*** forms when α-mixed crystals are cooled below 723 °C (pearlite temperature).

Cement pigments: pigments suitable for coloring cement and concrete. Inorganic pigments are generally used; they must be stable to alkalies, very light fast, and free of water-soluble salts. The following pigments are suitable: titanium dioxide, lithopone, iron oxide yellow, Neapolitan yellow, mineral yellow, nickel titanium yellow, chromium oxide green, chromium oxide hydrate green, ultramarine, manganese blue, ultramarine violet, iron oxide red and iron oxide black. Organic phthalocyanin pigments are also used as C.

Centered lattice: see Crystal.

Central atom: see Nomenclature, sect. II G; see Coordination chemistry.

Central field model, ***independent particle model, shell model***: an approximation to the quantum mechanical description of the electron structure of multi-electron atoms. If the interactions between the electrons in a multi-electron system are taken into account in constructing the operator expression for potential energy, two-particle operators $l/rd2i_j$ are required. These depend on the distance r_{ij} between the electrons i and j, and thus simultaneously on the coordinates of both electrons; for this reason, the presence of two-particle operators makes it impossible to separate the Schrœdinger equation, or to solve it exactly (see Atom, models of). In the C., this difficulty is overcome by assuming that each electron moves independently of the others in an average effective potential E_{eff} generated by the nucleus and the charge cloud of all other electrons (see Shielding). The C. retains the concept of an electron state described by a wavefunction Ψ and determined by a Schrœding equation similar to that for the hydrogen atom. The average potential E_{eff}, which must be calculated from the wavefunction, itself depends on the single electron function, so that the only way to obtain an approximate numerical solution is by a process of iteration. The usual procedure is to start with a plausible approximation of Ψ for the occupied orbital, then to derive the potential associated with it. From this potential, a new electron function is calculated, and so on, until "self-consistency" is reached, that is, until the results obtained in successive rounds of calculation are approximately the same (*self-consistent field method*). Because the potential E_{eff} is not entirely coulombic, the single-electron functions differ from those of the hydrogen atom in the form of the radial function $R_{n,l}(r)$, and the energy of the state depends on the orbital angular momentum quantum number l. Thus the l-degeneracy is lifted in the multi-electron atom, although the spherical symmetry of the atom maintains the m_l degeneracy. The radial electron distribution of the electron density in the C. shows a typical shell structure for multi-electron atoms. For an atom in the ground state, the occupation of the energy levels by electrons follows certain rules (see Electron configuration). Although the C. gives a greatly simplified description of the electron interactions, it is sufficient to explain the structure of atoms and permits interpretation of their spectra. The C. is based on the work of Bohr, Slater, Pauli, Hartree and Fock in the years 1922 to 1930.

Centrifugation: the separation of mixtures by means of centrifugal force in a ***centrifuge***. The device consists of conical or cylindrical, rotating drums. Substances with different densities are separated in complete mantle centrifuges, while solids are separated from liquids in filtering centrifuges. In the latter type, differences in density are of subordinate importance.

In all centrifugation processes, the magnitude of the centrifugal acceleration $a = r\omega^2$ (r = drum radius, ω = angular velocity) is very important. The ratio of the centrifugal acceleration a to the gravitational acceleration $a/g = C$ is the ***separation factor*** or ***centrifuge number***. C therefore indicates how much greater the separation effect is in the centrifugal field than in the gravitational field. In calculating the effects of C., the equivalent clearing surface, S_{eq} is important. This indicates how large the clearing surface of a clearing basin would have to be to achieve the same separation (other conditions being the same) under the effect of gravity: $S_{eq} = S_c \cdot T$; here S_c is the centrifuge surface.

C. is used in the chemical industry mainly to separate solids from liquids, or liquids from liquids. Bottle or tube centrifuges are used mainly for discontinuous separations in the laboratory.

Cephalin: an old term for a mixture of glycerophospholipids which can be fractionated into phosphatidylethanolamine (***colamine cephalin***), phosphatidylserine (***serine cephalin***) and phosphatidylinositol.

Cephalosporins: a group of β-lactam antibiotics containing the bicyclic skeleton *cepham*. Most of the C. used therapeutically are derivatives of 7-aminocephalosporanic acid, in which the amino group is acylated and in some cases the acetoxymethyl group is modified. ***Cephalosporin C*** was isolated from *Cephalosporium sp.*; it has very little antibiotic activity. The biosynthesis of this compound starts in the same way as that of the penicillins, with L-2-aminoadipic acid, L-cysteine and D-valine; the L-2-aminoadipic acid is epimerized to the D-form, one methyl group of cysteine is incorporated into a 1,3-thiazine ring, and the second is converted to the acetoxymethyl group. The configurations at the two chiral C atoms are the same as those of the analogous chirality centers in penicillin. Cephalosporin C is degraded chemically to 7-aminocephalosporanic acid, from which the partially synthetic compounds are made.

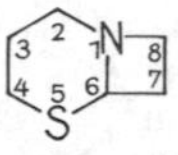

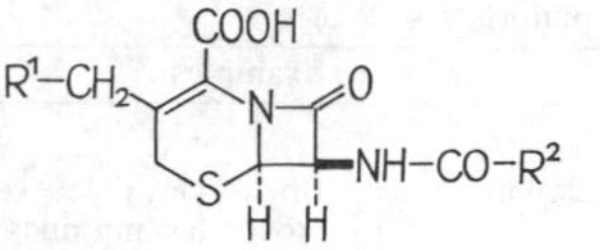

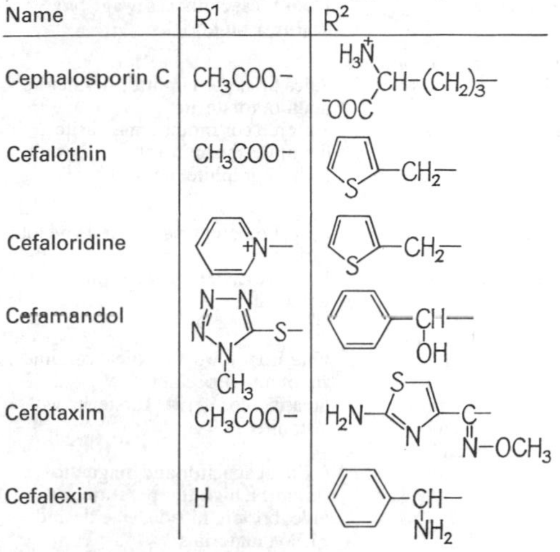

The C. have an antibiotic activity comparable to that of the penicillins. Their β-lactam rings are more stable, and they are more stable to acid. However, this could not be utilized with the older C., e.g. ***cefalothin*** and ***cefaloridine***, because they were poorly absorbed from the gastrointestinal tract. The newer C., such as ***cefotaxim*** and ***cefamandol***, have even greater lactam stability, a broader action spectrum and more biological activity. Oral C., such as ***cefalexin***, like ampicillin, contain a phenylglycine group. The cephamycin ***cefoxitin*** is related to the C.; it has an oxazine ring instead of the thiazine ring. It is also used therapeutically.

Cepham: see Cephalosporins.

Ceramic: a nonmetallic, inorganic material which is shaped and dried, then fired at high temperature to produce its characteristic properties. C. consist largely or completely of crystalline phases. During firing, both sintering processes and solid-state reactions occur. Because sintering processes are also a significant part of powder metallurgy and production of other high-temperature materials, the term C. has

Cepham (ring positions 1–8); general structure: R^1-CH_2–, COOH, =O, –NH–CO–R^2, H H

Derivatives of 7-aminocephalosporanic acid

Name	R^1	R^2
Cephalosporin C	CH_3COO-	$H_3\overset{+}{N}$–CH(–^-OOC)–$(CH_2)_3$–
Cefalothin	CH_3COO-	2-thienyl–CH_2–
Cefaloridine	pyridinium–$\overset{+}{N}$–	2-thienyl–CH_2–
Cefamandol	1-methyltetrazol-5-yl–S– (CH_3)	phenyl–CH(OH)–
Cefotaxim	CH_3COO-	H_2N–thiazolyl–C(=N–OCH_3)–
Cefalexin	H	phenyl–CH(NH_2)–

been transferred to these materials as well (e.g. metal, carbon or carbide ceramics).

The characteristic properties of ceramic products are high compressional strength, relatively low tensile strength, brittleness, high temperature resistance, low resistance to temperature changes, poor thermal and electrical conductivity (except for semiconductor, carbon, carbide and special ceramics), insolubility in water, insolubility in acids, bases and salt solutions. Aside from these common properties, there are other characteristics, such as structure, phase components, porosity, color and surface quality, which vary widely.

The applications of C. are very diverse, which makes classification difficult.

The ceramic technologies are determined by the composition and quality of the raw materials, especially the particle distribution and mineral contents of the particles. Clay mineral components are important for the silicate C., due to their plasticity and shrinkage on drying; some other main components are feldspar and quartz. Most of the raw materials are quarried by machinery which not only grinds the material (wet or dry grinding) but also mixes it. The raw materials are often sorted on the basis of the quality of the products to be made. The water added to the materials during wet grinding and sorting is largely removed by filtration, pressing and pulverizing driers. The products are shaped by turning, pressing and casting; the isostatic presses used for coarse ceramics, and more recently also for fine ceramics, make products which are highly uniform in size. The adsorbed water and pore water is lost from the clay as it dries, and leads to shrinkage without cracking. The chemically bound water (OH groups) is lost during firing. Firing takes place in round, tunnel, ring and oscillating dome kilns, at temperatures between 800 and

Ceramic materials

Ceramic	Examples
Coarse ceramics	
Construction materials	Bricks, clay pipes, terra cotta
Pottery	Stove tiles, majolica, faience, flower pots
Soneware	Floor tiles, clinker, sewage pipes, acid-resistant bricks
Fire-resistant ceramics	
Silicate ceramics	Silica bricks, chamotte, sillimanite mullite, forsterite
Nonmetallic ceramics	Sintered corundum, magnesite, dolomite chromite, silicon carbide (silite), graphite, coal
Fine ceramics	
Stoneware	Clay, lime, feldspar, mixed and talc stoneware
Porcelain	Hard, soft, dental and sanitary porcelain
Industrial ceramics	
Silicate ceramic on clay basis	Fine stoneware, electrical ceramics, zirconium porcelain
Silicate ceramics without clay component	Steatite, cordierite, forsterite wollastonite
Nonsilicate ceramics	Oxide, capacitor and magnetic ceramics; high-temperature carbide, boride, nitride, silicide and carbon materials

1600 °C. In addition to physical processes (sintering and melting), compounds are formed (e.g. silicate and aluminates) and decomposed (dehydration and decarboxylation reactions); polymorphic conversions also occur as solid state reactions. The surface of C. can be coated with a Glaze (see) and/or decorated (see Porcelain), sometimes in the course of a second firing.

Ceramic fibers: synthetic fibers which are resistant to high temperatures. They consist of 54% silicon dioxide, 45% aluminum trioxide and traces of iron, titanium, sodium, potassium and boron oxides. C. are resistant to most chemicals, especially to acids, and retain their properties to about 1435 °C. Spinnable C. are made from a mixture of quartz and alumina. Their density is 2.56 g cm^{-3}, and m.p. is 1700 °C. The fiber shrinkage at 900 °C is 0. The crude fibers have a very low conductivity, and because of enclosed air, the textiles made from them have excellent thermal insulating properties. It is possible to make C. with 85% SiO_2 by spinning a silk out of glass or chromium steel spinerets. Textiles made of C. can have long-term temperature resistance between 600 and 1000 °C, and are thus very superior to the old asbestos materials.

C. based on aluminosilicate are used together with inorganic fillers and organic binders to make *asbestos-free ceramic pulp*. This fire-resistant material can withstand temperatures up to 950 °C and is used to make insulation of all kinds, heat containment materials for welding, conveyor belts, etc.; it can be applied wherever asbestos was formerly used.

Ceramic pigments: inorganic pigments used for decoration of ceramics and enamels.

1) ***Subglaze pigments***. Kaolin, quartz, white ceramic powder and other silicate components are ground finely together with thermally stable pigments, such as vanadium(III) oxide, V_2O_3, chromium(III) oxide, Cr_2O_3, uranium(IV) oxide, UO_2, manganese(IV) oxide, MnO_2, iron(III) oxide, Fe_2O_3, cobalt(II) oxide, CoO, nickel(II) oxide, NiO and mixed oxides (spinels), or they are soaked in metal salt solutions (gold chloride, rare-earth metal chlorides) and fired at 900-1400 °C. The color is developed by reaction with the components of the glaze.

2) ***Glaze pigments (enamel colors)***. For low-melting glaze decoration, the colored oxides are ground with alkali metal and lead borosilicate glasses, which melt into the ceramic and give a hard glaze.

C. are applied by painting, screening or spraying onto the objects.

Ceramide, abb. ***Cer; N-Acylsphingoid***: the basic structure of the Sphingolipids (see). C. also occur in free form. They are formed by alkali-catalysed hydrolysis of sphingomyelins or acid-catalysed hydrolysis of glycosphingolipids.

Cerasin: see Cerebrosides.

Cerebron: see Cerebrosides.

Cerebrosides: 1-β-D-mono- or oligoglycosylceramides; members of the class of sphingolipids. C. isolated from bovine brain are obtained as a colorless, crystalline powder. The fraction containing non-hydroxylated fatty acids has m.p. 126-130 °C, while that containing hydroxyfatty acids melts between 153 and 157 °C. C. are soluble in hot alcohol, chloroform and pyridine. The C. of animals are usually galactosylceramides. Between 20 and 25% of the C. from brain are present as strongly acidic sulfate esters ("sulfatides"). Glucosylceramides are found in plants. Animal C. usually contain C_{24}-fatty acids, e.g. lignoceric, cerebronic, nervonic or oxynervonic acids. Some C. have special trivial names relating to these N-acyl groups, such as *cerasin* (lignoceric acid), phrenosin or *cerebron* (cerebronic acid), *nervon* (nervonic acid) or *oxynervon* (oxynervonic acid).

Ceresine: see Mineral wax.

Cerevitinov method: a method for quantitative determination of active hydrogen in organic compounds such as alcohols, phenols, enols, carboxylic acids, primary and secondary amines, using methylmagnesium iodide. The methane formed by the reaction is determined volumetrically: $R\text{-}OH + CH_3MgI \rightarrow Mg(OR)I + CH_4$. With primary amines, only one H-atom reacts at room temperature; elevated temperatures are needed to detect the second.

Cerimetry: a method of redox analysis in which cerium(IV) ion is used as titrator. This is as strong an oxidizing agent as permanganate in acidic solution. A cerium(IV) sulfate solution is often used; it must be calibrated, because cerium(IV) salts are not titrimetric standards. The same determinations are possible with C. as with Permanganometry (see), but C. has several advantages. The standard solutions are very stable and can be heated to the boiling point without decomposition. Chloride ions do not interfere with C. The reaction is unequivocal, because only cerium(III) ions can be formed. The endpoint can usually be readily recognized with the reversible redox indicator ferroin, but it can also be determined electrochemically.

Cerite earths: oxides of the rare earth metals lanthanum, cerium, praseodymium, neodymium and samarium. Because their ionic radii are nearly identical, the C. are always found together in nature. Cerium is the main component of C. minerals, such as monacite, $CePO_4$, bastnesite, $CeCO_3F$, and cerite, $Ce_3M^{II}H_3Si_3O_{13}$ (M^{II} = Ca, Fe).

Cerium, symbol ***Ce***: a rare-earth element of the Lanthanide (see) series of the periodic system; a heavy metal, *Z* 58, with natural isotopes of masses 140 (88.48%), 142 (11.08%, α-emitter, $t_{1/2}$ $5 \cdot 10^{16}$ a), 138 (0.25%), and 136 (0.19%), atomic mass 140.12, valencies II and IV, density 6.773, m.p. 798°C, b.p. 3257°C, standard electrode potential (Ce/Ce^{3+}) - 2.483 V.

C. is a silvery white, soft and very plastic metal which occurs in four allotropic modifications. Below - 150°C, it crystallizes in a cubic face-centered lattice (density 8.24), between - 158 and - 23°C, in a hexagonal face-centered form; between - 23 and +726°C, in a cubic face-centered form with different lattice spacing from the low-temperature form (density 6.771), and above 726°C, it adops a cubic body-centered form. The affinity of C. for oxygen is great. When it is heated in a stream of oxygen, C. ignites about 150°C and burns with a blinding white flame to cerium(IV) oxide, CeO_2. In general, C. can be rather easily converted to the +IV oxidation state, and in this respect it is distinctly different from most other lanthanides.

C. makes up $4.3 \cdot 10^{-3}$% of the earth's crust, and is the most abundant of the lanthanides. However, it always occurs together with these and other rare-earth metals. The most important cerium-containing minerals are cerite, bastnesite, monacite (turnerite) and orthite. See Lanthanides for more on the properties, analysis and production of C. C. is used in pure form or as cerium mixed metals, as an alloying additive to cast iron and steels and non-iron alloys. It improves their elasticity, casting properties, strength and ductility. C. is also used to make ignition stones. The synthetic radionuclide ^{144}Ce is used as a radiolabel.

Historical. C. was discovered in 1803 by Klaproth and, independently from him, by Berzelius and Hisinger. It was named after the planetoid Ceres, which had been discovered in 1801.

Cerium alloys: Alloys of rare earth metals consisting of 45 to 52% cerium, 20 to 27% lanthanum, 15 to 18% neodymium, 3 to 5% praseodymium, 1 to 3% samarium, about 3% terbium and yttrium about 5% other rare earth metals, up to 5% iron, and traces of aluminum, calcium and silicon. C. are very similar in their physical and chemical properties to cerium and the other lanthanides. Finely divided C. is pyrophoric. If it is rubbed on a hard, sharp-edged object, the metal filings rubbed off it ignite from the heat of friction and burn with a hot, bright flame. C. is the form in which the rare-earth metals are most often used. Although it was at first used almost exclusively for production of fire-starters, it was later discovered that it has a positive effect on the mechanical, and sometimes on the chemical properties of metals and alloys. Addition of C. to molten steel removes oxygen, sulfur and gases, and improves its ductility, flow, hot stamping and casting properties. C. decreases the porosity of the steel and increases its resistance to shock; it also improves its welding behavior. C. is also used in light metal alloys for aircraft, rockets and cars.

Cerium compounds: compounds in which cerium is in the +3 and +4 oxidation states. The colorless cerium(III) compounds can be converted by strong oxidizing agents to the yellow to brown cerium(IV) compounds, which are in turn strong oxidizing agents. Direct reaction of cerium with oxygen or fluorine yields cerium(IV) oxide or fluoride, while typical cerium(III) compounds, such as the chloride, fluoride and oxalate, are obtained by chlorination of the metal at elevated temperature, or by precipitation of the fluoride or oxalate from Ce^{3+}-containing aqueous solutions. ***Cerium(IV) oxide***, CeO_2, forms yellow-brown, cubic crystals, M_r 172.12, density 7.132; ***cerium(IV) fluoride***, CeF_4, forms colorless crystals, density 4.77, m.p. 977°C (dec.); ***cerium(III) fluoride***, CeF_3, forms colorless, hexagonal crystals, M_r 197.12, density 6.16, m.p. 1460°C; ***cerium(III) oxalate***, $Ce_2(C_2O_4)_3 \cdot 9H_2O$, yellow-white crystals, M_r 706.44; ***cerium(III) nitrate***, $Ce(NO_3)_3 \cdot 6H_2O$, colorless, triclinic crystals, M_r 434.23. The most commonly used C. is cerium(IV) oxide, which is used to polish optical glasses, as a coating for infrared filters, to color and decolorize glass, and as a catalyst in organic syntheses. Cerium glass is used in nuclear technology, because it does not turn dark due to the effects of ionizing radiation. Ce^{3+} salts are used as activators for cathode-ray luminophores, especially for those based on calcium magnesium silicate, yttrium silicate and yttrium aluminate. Cerium(III) nitrate is used to make filaments for gas lamps.

Cermets (abbreviation for ceramic + metal): mixed materials formed by Powder metallurgy (see) from metals and heat-resistant metal compounds. The metal component can be, for example, chromium, nickel, iron, molybdenum, tungsten, titanium or silicon. It is mixed with oxides, carbides, nitrides, borides or silicides and sintered. C. are heat-resistant, hard, non-scaling and heat-conducting high-temperature materials, used, for example, to make cutting tools (see Cutting ceramics).

Cerotinic acid: same as Hexacosanoic acid (see).

Cerulein: see Gallein.

Ceruleine: a decapeptide amide isolated from the skin of the Australian tree frog *Hyla caerula*; it has a kinin-like effect. Its blood-pressure reducing effect lasts longer than that of bradykinin (see Plasma kinins), while its effect on the smooth muscles is not as strong. The C-terminal sequence, and thus the effects, of C. are also similar to those of Cholecystokinin (see). The nonapeptide amide ***phylloceruleine*** from the skin of the South American frog *Phyllomedusa sauvagi* has a similar primary structure, and also displays the same effects as C., but is less potent.

```
            SO3H
            |
⌐Glu-Gln-Asp-Tyr-Thr-Gly-Trp-Met-Asp-Phe-NH2     Ceruleine

        SO3H
        |
⌐Glu-Glu-Tyr-Thr-Gly-Trp-Met-Asp-Phe-NH2

                                                 Phylloceruleine
```

Ceruleum: same as Cobalt blue (2) (see).

Ceruloplasmin: a mammalian copper storage and transport protein. Human C. is a four-chain glycoprotein (M_r 160,000) with 16% carbohydrate. The protein contains 8 Cu^{2+} ions per molecule. Wilson's disease is caused by a mutant C. gene; the absence of active C. results in life-threatening deposition of metallic copper in the body. In addition to its storage and transport functions, C. has specific oxidase activity.

Ceryl alcohol, ***hexacosan-1-ol***: $CH_3(CH_2)_{24}CH_2OH$, forms colorless crystals, m.p. 79.5 °C. C. is insoluble in water, but is soluble in alcohol and ether. It is found in esterified form in many waxes, for example, in Chinese insect wax. It is also found, free or esterified, in the bark of some trees, including spruces and red and white beeches. It has also been detected in spinach leaves, wool grease, scale insect wax and the waxy coat of green poppy capsules. It is obtained by alkaline hydrolysis of natural waxes.

Cesium, symbol ***Cs***: a chemical element from the first representative group of the periodic system (see Alkali metals); a light metal; *Z* 55, one isotope, atomic mass 132.905, valence I, density 1.878, m.p. 28.40 °C, b.p. 678.4 °C, electrical conductivity, 5.5 Sm mm^{-2} at 0 °C, standard electrode potential (Cs/Cs^+) - 2.923 V.

Properties. C. is a wax-soft metal with a golden yellow sheen on a freshly cut surface. It crystallizes in a cubic body-centered lattice. Its chemical behavior is very similar to that of potassium and rubidium. It is a strong reducing agent and a decidedly electropositive bonding partner. It reacts explosively with water, forming cesium hydroxide, CsOH, and hydrogen. In the presence of oxygen, it ignites spontaneously and burns to cesium hyperoxide, CsO_2.

Analysis. The reactions used to precipitate K^+ ions can also be used to detect Cs. For quantitative analysis, it can be precipitated as cesium tetraphenylboranate, $CsB(C_6H_5)_4$. Cs and its salts give a flame a violet-rose color. The Cs spectrum is characterized chiefly by the two blue lines at 455.5 nm and 469.3 nm. Atomic absorption spectroscopy can be used for quantitative determination.

Occurrence. Cs makes up 0.003% of the earth's crust. It is found in low concentrations as a companion of the other alkali metals. Its most important mineral is pollucite (pollux), $Cs[AlSi_2O_6] \cdot 0.5H_2O$.

Isolation. Cs is produced by reduction of cesium hydroxide, CsOH, with magnesium or calcium, or by reaction of cesium dichromate with zirconium (this is similar to the production of rubidium). It can also be produced by melt electrolysis.

Applications. Cs is used mainly in the production of photocells, but it is also used to fill metal vapor lamps and as a getter metal.

Historical. Cs was discovered in 1860 by Bunsen and Kirchhoff, who observed the blue spectral lines in Dürkheimer mineral water. The name is from the Latin word "cesius" = "skyblue". The metal was isolated in 1882 by Setterberg by melt electrolysis of a mixture of cesium and barium cyanide.

Cesium chloride type, ***cesium chloride structure***: a common Structure type (see) for compounds with the composition AB. The crystal lattice of the prototype CsCl consists of two primitive cubic lattices of the two types of ions, Cs^+ and Cl^-, which are displaced with respect to each other along the diagonals of the unit cells by an amount equal to half their length. Each ion is surrounded by 8 ions of the other type in a cubic arrangement (Fig.). For compounds with high ratios of ionic radii ($r_A/r_B = 0.732$), the C. is energetically the most favorable structure type for AB compounds, due to the high coordination numbers. For lower ratios, it is unstable with respect to the Sodium chloride type (see).

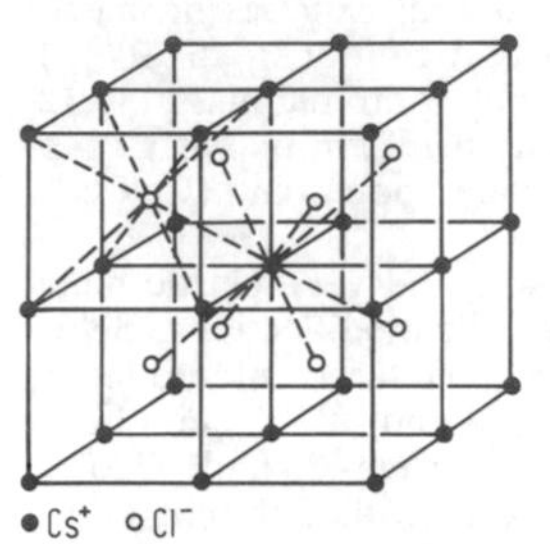

Cesium chloride structure (segment of the crystal lattice)

The cesium and thallium halides, with the exception of the fluorides, zintl phases of the MgTl type (see Intermetallic compounds) and many other alloys with the corresponding stoichiometric compositions crystallize in the C.

Cesium compounds: these compounds have the same stoichiometries as potassium compounds and are also similar in their properties. Most of them are water-soluble, and the Cs^+ ions are colorless. ***Cesium hydride***, CsH can be obtained at 300 °C from the elements; it is extremely reactive and ignites spontaneously in air. ***Cesium hydroxide***, CsOH, is made by reaction of ***cesium sulfate***, Cs_2SO_4, with barium hydroxide, $Ba(OH)_2$. It is used to some extent as an electrolyte for galvanic cells. ***Cesium hyperoxide***, CsO_2, is made by burning cesium. It reacts with water to form cesium hydroxide, hydrogen peroxide and oxygen: $2\ CsO_2 + 2\ H_2O \rightarrow 2\ CsOH + H_2O_2 + O_2$. Thermal decomposition of CsO_2 yields the ***cesium oxides*** Cs_2O, Cs_2O_2 and Cs_2O_3. The ***cesium halides*** are obtained by dissolving cesium carbonate, Cs_2CO_3, in the corresponding hydrohalic acids; ***cesium iodide*** is used to make prisms for IR spectrophotometers. ***Cesium bromide*** and ***iodide*** are used in scintillation counters.

Cetane rating: a measure of the ignitability of diesel fuels. The shorter the time between the entry of a fuel into the cylinder and its ignition, the greater its ignitability. The C. is established by measuring either the ignition of the fuel when the engine is started, or the time between injection of the fuel and the start of combustion. The measurements are based on comparison between the fuel being tested and a standard mixture of cetane and α-methylnaphthalene. The readily ignited cetane, $C_{16}H_{24}$, has a C. of 100, and α-methylnaphthalene, $C_{11}H_{10}$ has a C. of 0. Thus a fuel with a C. of 50 ignites as readily as a mixture of equal parts cetane and α-methylnaphthalene. The C. requirement of an engine depends on its construction (compression, shape of the combustion chamber, arrangement of the injection aggregate). Slow-running diesel engines, usually stationary machines or ship engines, require C. of 10 to 40. Rapidly running engines, such as those in vehicles, require C. above 40.

The C. of a fuel is determined either in a test engine or with an areometer, because the ignitability decreases as the density increases. Some diesel oils made by the Fischer-Tropsch process have C. above 100. Addition of tetranitromethane increases the C. of a mixture.

The ***cetene rating*** was formerly used; here cetene, $C_{16}H_{32}$, was used instead of cetane. A cetene rating of 100 = a C. of 116.

Cetene rating: see Cetane rating.

Cetyl alcohol: same as Palmitic alcohol (see).

Cetylic acid: same as Palmitic acid (see).

Cevanine: see Steroid alkaloids.

Cf: symbol for californium.

cGMP: see Cyclic nucleotides.

CH acid compounds: pseudoacids which lose protons only in the presence of strong bases, releasing anions. These compounds contain electron-withdrawing substituents which increase the polarity of the CH bond, and also become conjugated with the free electron pair of the resulting anion. Examples are aldehydes, ketones, acids, esters, nitriles and nitro compounds with a hydrogen atom in the α-position relative to the functional group, and hydrocyanic acid and alkynes with terminal triple bonds. The position of the equilibrium

$$\overset{\ominus}{B}| + H{-}\overset{|}{\underset{|}{C}}{-}X \rightleftharpoons BH + {}^{\ominus}|\overset{|}{\underset{|}{C}}{-}X \longleftrightarrow \;\rangle C{=}X^{\ominus}$$

depends on the basicity of the base B and of the anion. The acidity of pseudoacids increases with increasing inductive and resonance effects of the substituent X (Cl < COOR < C=N < $COCH_3$ < CHO < NO_2). The pK_a values of the CH-acid compounds range from about 40 (very weak) to about 5 (strong). Under suitable conditions, these compounds undergo addition reactions which can be considered, in the broadest sense, Aldol reactions (see).

Chain lattice: see Lattice type.

Chain reaction: a Complex reaction (see).

Chair form: see Stereoisomerism 2.2.

Chalcogran: see Pheromones.

Chalgon, ***benzal acetophenone, 1,3-diphenylprop-2-enone***: C_6H_5-CH=CH-CO-C_6H_5, an aryl-substituted vinyl ketone. C. is a bright yellow, crystalline compound, m.p. 57 °C, b.p. 345-348 °C. It is insoluble in water, but is soluble in ether, benzene and chloroform. C. is formed by an aldol reaction of benzaldehyde and acetophenone. Substituted C. can be made in similar fashion from substituted benzaldehyde and acetophenone derivatives. They are used for organic syntheses.

Chalcogens: (from the Greek for "ore formers"), the chemical elements oxygen, sulfur, selenium and tellurium from group VIa of the periodic system, the Oxygen-sulfur group (see).

Chalcogenides: binary compounds of the chalcogens in which they are in the -2 oxidation state, that is, oxides, sulfides, selenides and tellurides.

Chalcopyrite type: see Zinc blende type.

Chamazulene: 1,4-dimethyl-7-ethylazulene, a blue liquid, b.p. 160 °C at 1.5 kPa. C. is formed from Matricin (see) during steam distillation, and is responsible for the antiphlogistic effect of camomile extracts.

Chanoclavine: see Ergot alkaloids.

Chapman-Enskog model: see Kinetic gas theory.

Characteristic group: see Functional group.

Charcoal: the product of pyrolysis of wood, animal products (blood, bones) or other organic materials in the absence of oxygen. The composition depends on the source and conditions of pyrolysis.

Wood C. usually has a carbon content between 75 and 90%, and an ash content between 1 and 4%. In contrast to coal, C. contains only insignificant amounts of sulfur. When wood is pyrolysed, its volume decreases by 35 to 45%, and its mass by 65 to 75%. A cubic meter of birch wood, for example, yields between 55 and 185 kg C. The porosity of the C. depends on the type of wood and on the rate and final temperature of the pyrolysis, but lies between 77 and 87%. The crude density of the C. is between 0.15 and 0.40 g ml^{-1}, while the density without pores is between 1.38 and 1.46 g ml^{-1}. The higher the pyrolysis temperature, the higher the heating value; values between 28 and 35 MJ kg^{-1} are obtained. Elevation of the pyrolysis temperature from 400 to 1200 °C increases the specific heat of the C. from 1.02 to 1.60 kJ kg^{-1}. Criteria for the quality of C. are the size of the lumps and its content of water, ash and volatile components.

Fresh, dry C. taken from the pyrolysis oven tends to ignite spontaneously; therefore an ageing process under controlled conditions is necessary in the first 48 hours after pyrolysis.

Production. There are three groups of production processes. In the oldest, the heat for the process is supplied by burning part of the wood in the charcoal pile or furnace. In retort processes, the heat is supplied indirectly, through the wall of the retort. In the third group, the heat is supplied directly, e.g. by blowing heating gases into the reactor.

The yields of solid, liquid and gaseous pyrolysis products depends on the method, rate of heating, final pyrolysis temperature, the size of the wood chips, the time, and so on. On the average, the products are 30 to 40% wood vinegar, 25 to 30% charcoal, 10 to 15% wood tar and 10 to 15% wood gas.

Activated C. is a porous C. with an extremely large surface area (up to 800 $m^2\,g^{-1}$) and a high absorption capacity. A. is microcrystalline, highly distorted graphite obtained by pyrolysis of wood or animal materials (bones, blood) in the presence of an inorganic substance such as zinc chloride, which prevents sintering of the particles. After the pyrolysis, the inorganic salt is washed out with water. Activated C. from bones contains 80 to 90% of a ground substance consisting of calcium phosphate and calcium carbonate.

C. is used as a fuel in much of the world, and it is becoming more important in industry because of its high porosity, and low contents of ash, phosphorus and sulfur. Activated C. is used as a filtering and absorbing material in many chemical processes, and medicinally, in the gastrointestinal tract. It is used in some processes as a catalyst, e.g. in the synthesis of phosgene, cyanuric acid and sulfuryl chloride. It is also used to absorb sulfur from synthesis gas.

Charcoal exchanger: see Ion exchanger.

Chargaff's rules: see Nucleic acids.

Charge number: see Valence.

Charge order: see Hückel method.

Charge transfer, ***CT***: an electron transfer in which a charge is completely or nearly completely transferred from a donor to an acceptor group. If the two groups are attached to the same molecule, the process is an intramolecular C. If two different molecules or ions are involved, which may also be loosely bound, it is an intermolecular C. An example of intramolecular C. is seen in aniline, in which the UV absorption at 280 nm is due to an electron transfer from the NH_2 group (donor) to the phenyl ring (acceptor) (Fig.).

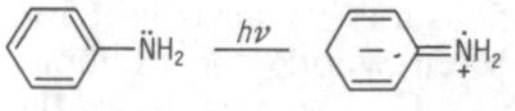

Intramolecular charge transfer in aniline

Intermolecular C. occurs, for example, in benzene/iodine complexes, in which the absorption maximum at 290 nm is due to an electron transfer from I_2 to benzene, or in complexes of the type $[Co(NH_3)_5X]^{3+}$, in which the strong UV absorption is due to an electron transfer from the halogen X to the metal ion.

Charles' law: see State, equation of, 1.1.

Chaulmoogric acid: (S)-cyclopent-2-en-1-tridecanoic acid, M_r 280.5, a branched fatty acid found as its glycerol ester in chaulmoogra oil, the seed oil of *Flacourtiaceae hydnocarpus kurzii*. The oil contains about 23% C. as well as 35% (S)-cyclopent-2-en-1-undacanoic acid (hydoncarpic acid), and other cyclopentenoic acids. C. forms crystalline leaflets, m.p. 68-69 °C, $[\alpha]_D^{25}$ +62° chloroform), which are practically insoluble in water, but dissolve readily in ether and chloroform. Like the homogous ***hydnocarpic acid***, C. is toxic for the tuberculosis and lepra pathogens, and is used for treating leprosy. However, C. is also rather toxic to mammals.

n = 10: Hydnocarpic acid
n = 12: Chaulmoogric acid

Cheddite: (named for Chedde, the place in southern France where it was first produced) a chlorate explosive. Its sensitivity to friction is reduced by addition of nitronaphthalene, viscous oils, etc.

Chelate complex: see Coordination chemistry.

Chelate effect: see Coordination chemistry.

Chelatometry: same as Complexometry (see).

Cheletropic reactions: synchronous opening or formation of two σ-bonds ending on the same atom. See Woodward-Hoffmann rules.

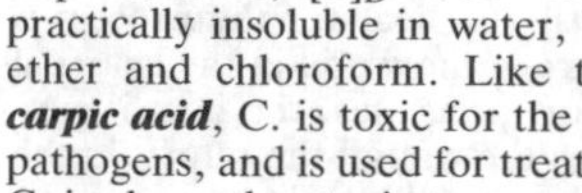

Chelidonic acid, ***4H-pyran-4-one-2,6-dicarboxylic acid, jervic acid***: a heterocyclic dicarboxylic acid. C. forms colorless crystals; m.p. 262 °C (dec.). It is slightly soluble in alcohol and cold water, and soluble in boiling water. Upon distillation, it yields 4*H*-pyran-4-one. C. is an acid component in *Veratrum* alkaloids and is also found in asparagus and chelidonium. It is synthesized in a several-step reaction from acetone and oxalic acid diethyl ester.

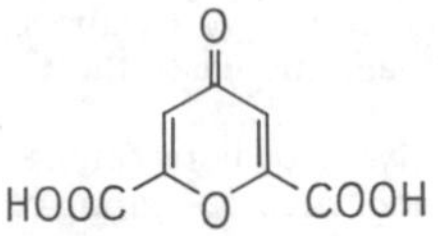

Chemical bonding: the binding of atoms and groups of atoms in molecules, crystals and surfaces. The preclassical theories and models of C. were reached inductively, and could not explain the forces which hold atoms in molecules. The concept of Valence (see) was developed to describe the possibilities for bonding of an element in molecules. A physical interpretation of C. was not possible before the development of theories of atomic structure (see Atom, models of). The classical theories of C. were based on the *octet principle*, the tendency of an atom to surround itself with 8 valence electrons, that is, with 8 electrons in its outermost shell. This electron configuration is called the noble-gas configuration, and it is very stable.

1) In ***covalent bonding (atomic, homeopolar, nonpolar, electron-pair bonding)***, it is believed that two atoms share one or more electron pairs (*bonding electron pairs*); this idea was suggested by Lewis. Sharing of electron pairs frequently fills the electron octets of both atoms. The number of shared electron pairs determines the *bond order*; a pair of atoms can be linked by one, two or three electron pairs, in which cases the bonds are *single*, *double* or *triple*, respectively. In *electron formulas*, the valence electrons are represented by dots:

:C̤̈l· + ·C̤̈l: → :C̤̈l:C̤̈l:, N̈· + ·N̈ → :N:::N:

In Langmuir *valence-dash formulas*, dashes between atoms represent bonding electron pairs, and dashes above or beside atoms indicate *free electron pairs*. Bonding and free electron pairs are arranged so that each atom has an octet, if possible, as in

| C̲̅l–C̲̅l |, | N≡N |.

The *formal charge* on an atom is determined by dividing the bonding electron pairs equally between the two atoms which share them, then subtracting the number of valence electrons on the unbonded atom from the total number of electrons on the bonded atom, including the shared electrons. For example, in carbon monoxide the valence-dash formula indicates a triple bond, which yields the following formal charges:

$|\overset{-}{C}\equiv\overset{+}{O}|$

From the nature of the covalent bond, it becomes clear that the bonding forces are directional and lead to small coordination numbers. The *coordination number* of an atom or ion indicates the number of closest neighbors. For example, in N_2, each atom has a coordination number of 1, in S_8 (ring), the coordination number is 2, in P_4 (tetrahedron), 3; and in {C} (diamond), 4. The two electrons shared in a covalent bond do not necessarily have to be contributed by different atoms. It is possible that one of the bonded atoms provides both of them. This type of bond is called a *coordinate* or *dative covalent bond*. Dative covalence is seen, for example, in the reaction of ammonia and boron trifluoride to form the compound $NH_3 \cdot BF_3$. It can be represented as $H_3N{\rightarrow}BF_3$ or $H_3N\text{-}BF_3$. It is essentially identical to a normal atomic bond.

Pure covalent bonds are found in diatomic, homonuclear molecules, where the bonding electron pair belongs to both atoms equivalently, and no dipole moment arises. When atoms of different elements form bonds, one of them will attract the bonding electron pair more strongly, and the bond becomes polar. The polarity of a bond is indicated qualitatively by the *partial charge* δ:

$$\overset{\delta+}{H}-\overset{\delta-}{F}$$

Dipole moments in diatomic, heteronuclear molecules are the result of polar bonds. In the limiting case, the bonding electron pair resides entirely on one of the atoms; this corresponds to an ionic bond (see below). The polarity of a bond can be estimated from the Electronegativity (see) of the bonded atoms. The bond strength is indicated by the Bond energy (see). In classical bonding theory, predictions of molecular geometry are based mainly on models in which the valence electron pairs are arranged so that their mutual electrostatic repulsion is as low as possible (see Valence shell electron pair repulsion model). Quantum mechanical theory is able to provide a quantitative explanation of the nature of the C. and the structure of molecules (see Molecular orbital theory, Valence structure theory).

2) The ***ionic (heteropolar) bond*** is due to the electrostatic attraction between oppositely charged ions. Ions are formed from atoms by loss or acquisition of electrons, which in either case leads to a stable noble-gas configuration (Kossel, 1916). Elements with small numbers of valence electrons, namely metals, are most likely to form cations. Anions are formed most readily by elements with high electronegativities, that is, members of the 6th and 7th main groups of the periodic system. In complex cations and anions, the central atoms can be either metals or nonmetals: SO_4^{2-}, MnO_4^-, NH_4^+, $Cu(NH_3)_4^{2+}$ (see Coordination chemistry). The number of positive or negative unit charges on an ion or in a charged group of atoms is called the *ionic valence* (*ionic charge, charge number*). For example, by losing an electron, sodium achieves the configuration of neon, while chlorine achieves the argon configuration by acquiring an electron:

$$Na\cdot + \cdot\overline{\underline{Cl}}| \longrightarrow Na^+ + |\overline{\underline{Cl}}|^-$$

The interactions between spherical ions can be modelled approximately on the Coulomb law. The approach of ions with opposite charges is limited, however, because there are repulsive interactions in addition to Coulomb attraction E_{att} (see Repulsion energy). A formula for approximate calculation of the repulsion energy (E_{rep}) was suggested by Born. For the total interaction energy E_{tot} between a cation and an anion with charge numbers z_+ and z_- at a distance R, $E_{tot} = E_{att} + E_{rep} =$

$$-\frac{z_+z_-e^2}{4\pi\varepsilon_o R}+\frac{b}{R^n};$$

where e is the unit charge, ε_0 is the electric field constant, b is a substance-specific constant and n is the Born repulsion exponent. Since the interaction forces of ions are isotropic, ionic bonding is not directional, in contrast to covalent bonding. Therefore this type of bonding is characterized by high coordination numbers, and in the solid state, by formation of giant molecules consisting of unlimited numbers of anions and cations: *ionic crystals*. The arrangement of ions in the ionic crystal is such that the total interaction energy is a minimum. The coordination numbers and distances between ions in the lattice, and thus the lattice type, depend on the size of the ions (ratio of radii) and on their charges. The stability of the ionic lattice is characterized by the molar lattice energy.

3) ***Metallic bonding*** is seen in metals and alloys in the solid or liquid states. Metals are elements with few s and p valence electrons, and most crystallize with high coordination numbers of 8 or 12. Their cohesion cannot be explained either by covalent or by ionic bonding. The bonding in metals is explained by the assumption that the positively charged atomic cores form a lattice which is held together by a freely mobile gas of valence electrons (*electron gas*). This model was developed around the turn of the century by Drude and Lorenz; it explains the characteristic properties of metals, such as their metallic lustre, their high thermal and electrical conductivities, and their malleability. The quantum mechanical description of metallic bonding is given by the Energy band model (see).

4) ***Intermolecular binding (molecular interactions)*** depends on interaction between two units with saturated valencies. In contrast to the bonding types discussed so far, these bonds are very weak, and are generally not regarded as chemical bonds. They are mediated by orientation, inductive and dispersion forces (see Van der Waals binding forces). At sufficiently low temperatures or high pressures, they cause atoms (noble gases) and molecules to liquefy or solidify, and are responsible for the deviations of real gases from ideal behavior. A somewhat stronger type of intermolecular binding occurs in molecules with certain functional groups (e.g. -OH or -NH) through formation of Hydrogen bonds (see); these have a major influence on the properties of such substances.

The types of bonds discussed here are limiting types; real chemical substances display a continuum of intermediate types which are combinations of the limiting types.

Chemical equation: a representation of a chemical process using empirical, constitutional or structural formulas for the reactants and products in the stoichiometrically correct proportions. The *net equation* gives the net turnover of a reaction in terms of the empirical formulas of the reactants and products. Both substances (sums of the same nuclides) and charges (sums of the charges) must be balanced on the two sides of a C. Those reactions which run essentially to completion (irreversible reactions) are indicated by an arrow which shows the direction of the reaction (e.g. from the reactants to the products). Reversible reactions are indicated by two arrows, one over the other, pointing in opposite directions. The conditions of the reaction (presence of a catalyst, elevated temperature, irradiation with light) can be indicated in parentheses above the arrow. A short-hand notation is often used in which certain reagents are written above the arrow, with a positive sign, if they add to the reactants, or below the arrow and with a negative sign if they are released from the reactants. For example:

$$CH_3{-}CH_2{-}OH \xrightarrow[-H_2O]{+H^+} CH_2{=}CH_2$$

Chemical equilibrium: see Equilibrium, see Mass action law.

Chemical equivalent: see Equivalent.

Chemical ionization: an ionization technique used in Mass spectroscopy (see) (*CI mass spectroscopy*), in which ion-molecule reactions are used for ionization. In the ion source of the mass spectrometer, ions are generated in an auxiliary gas, e.g. methane, CH_4, at a relatively high pressure ($\approx 10^3$ Pa) by electron collisions. The following ion-molecule reactions then occur: $CH_4^{+\cdot} + CH_4 \rightarrow CH_5^+ + CH_3\cdot$ and $CH_3^+ + CH_4 \rightarrow C_2H_5^+ + H_2$. The CH_5^+ and $C_2H_5^+$ ions ionize the molecules (M) of the sample, usually by transfer or cleavage of hydrogen, by addition or by charge exchange.

$$RH^+ + M \rightarrow (M + H)^+ + R$$
$$RH^+ + M \rightarrow (M - H)^+ + R + H_2$$
$$R^+ + M \rightarrow RM^+$$
$$R^+ + M \rightarrow R + M^+$$

This method of ionization is gentler than electron collision ionization, so that the mass spectra resulting from C.i. have fewer fragments. The "quasi-molecular ions" $(M + H)^+$, $(M - H)^+$ and RM^+, however, have high intensities.

Chemically induced dynamic electron polarization, ***CIDEP***: a special technique in electron spin resonance spectroscopy. The electron occupation of the two spin states of a radical is brought out of equilibrium, for example by a chemical reaction; this leads to emission of HF radiation, which causes anomalous intensities in the ESR spectrum.

Chemically induced dynamic nuclear polarization, ***CIDNP***: intensified appearance of emission or absorption lines in the NMR spectrum (see NMR spectroscopy) during a radical reaction. For example, when dipropionyl peroxide is thermally decomposed in an NMR sample tube containing thiophenol, the resulting ethyl radicals react with the thiophenol, forming thiophenetol: C_6H_5-CH + $C_2H_5\cdot \rightarrow C_6H_5$-S-$CH_2$-$CH_3$. If the NMR spectrum of thiophenetol is measured during the reaction, a CIDNP spectrum (Fig. 1) is observed. This is compared to the normal NMR spectrum (Fig. 2). The mechanism of this phenomenon is complicated, and depends on how certain nuclear spins interact with the electronic spin of the radical. The appearance of the CIDNP effect is taken as proof that a radical reaction is occurring.

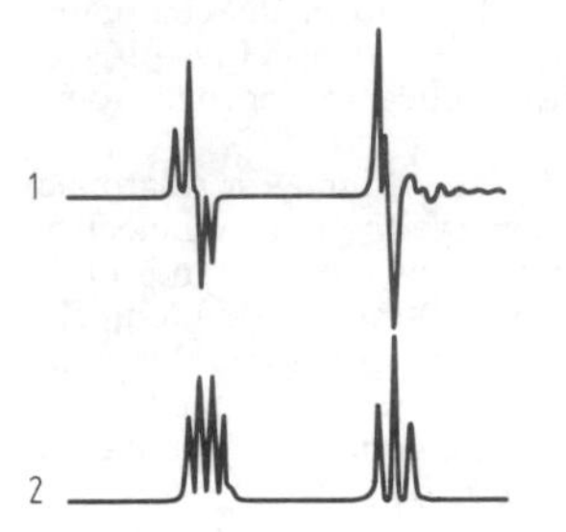

NMR spectra of the CH_2-CH_3 group in thiophenetol: *1* CIDNP spectrum. *2* Normal spectrum

Chemical oxygen demand, abb. ***COD***: the mass of oxygen required to oxidize substances present in water. It corresponds to the fraction of an added oxidizing agent added to the water being tested. The COD_{Cr} is measured using potassium chromate, K_2CrO_4, while the COD_{Mn} is measured using potassium permanganate, $KMnO_4$. The COD_{Mn} for drinking water, for example, should be $\leqq$ 12 mg/l $KMnO_4$.

Chemical potential μ_i: the partial molar free energy (g) of substance i in a mixture, defined by the equation $\mu_i = (\delta g/\delta n_i)p, T, n_j$ (see Partial molar parameters). The C. is the change in the capacity of a mixture to do work when one mole of substance i is added, assuming that pressure, temperature and composition (all other mole numbers n_j) are constant. The molar free entropy $G_i = g_i/n_i$ of the pure substance can also be taken as the C. In this case, the C. in the standard state is used, $\mu_i^0 = g_i/n_j$. In the mixture, $\mu_i = \mu_i^0 + RT\ln a_i$, where a_i is the activity of 1 mole of pure substance. $RT\ln a_i$ is an excess or additional parameter which gives the change in free energy when 1 mol of pure substance enters the mixture with activity a_i.

The C. is very important in determining the equilibrium of a process. If substance i has a C. in the initial condition μ_i', and μ_i'' in the final condition, matter will be transferred between the two conditions until the potential difference is reduced to 0: $\mu_i' - \mu_i'' = 0$, or $\mu_i' = \mu_i''$. This equilibrium criterion is the starting point for derivation of all equilibrium phase laws, and it also applies to each component of a chemical reaction.

Chemical reaction: the chemical conversion of elements or compounds (reactants) into other compounds or the elements (products). C. can be described by Chemical equations (see). ***Ionic reactions*** occur between ions, especially when these are in solution. They occur very rapidly, and in the direction which leads to an undissociated or insoluble compound (see Acid-base concepts), and are most important in inorganic chemistry. ***Electrochemical reactions*** in the broadest sense are those C. in which charges are exchanged, and in the narrow sense, are Electrode reactions (see). The time course of C. is described by kinetics, which distinguish between elementary and complex reactions (see Reaction, elementary and Reaction, complex). Endothermal reactions (see) are those which consume energy, and Exothermal reactions (see) are those in which energy is released.

The mechanism of a reaction can be used to classify it as an Electrophilic reaction (see), nucleophilic reaction (see) or Radical reaction (see).

Chemicals: products of the chemical industry used for chemical processes. The ***reagents*** used in the laboratory are inorganic or organic products with a certain degree of purity. ***Technical C.*** are usually inorganic compounds used for industrial purposes, such as mineral acids, alkalies, salts, chlorine, calcium carbide, etc.

Biochemicals are products made by methods of Biotechnology (see); their importance is steadily increasing.

Chemical shift: see Photoelectron spectroscopy, see Auger electron spectroscopy, see NMR spectroscopy, see Mœssbauer spectroscopy.

Chemical sterilants: compounds used to sterilize harmful insects. In addition to substances which affect their metabolism (see Antimetabolites), the most significant of these compounds are alkylating reagents with cytostatic effects. Most of them are compounds containing aziridine groups. Some of these alkylating agents, e.g. thiotepa, are also used in cancer therapy (Table).

Table. Chemical sterilizers

Name	Formel
Tepa X = O Thiotepa X = S	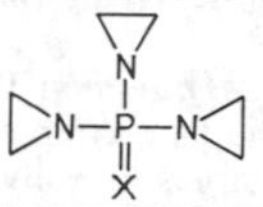
HMPA	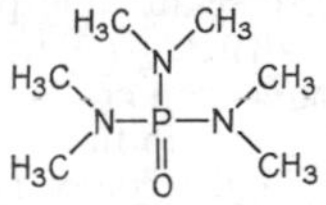
Apholate	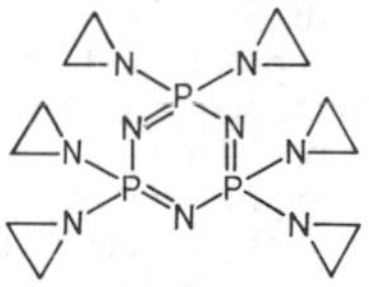
Tretamin	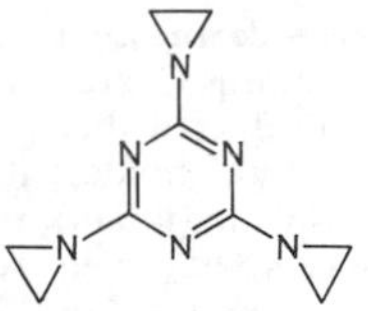

Not only high-energy radiation, but also chemicals can cause mutations and sterility in insects. The usual technique is to sterilize the males and to release very large numbers of them into the environment. They compete with the normal males for copulation with normal females and cause them to lay sterile eggs (because they were not fertilized). Under favorable conditions, this *autocide* method can eliminate entire local populations of insects. A less expensive approach not involving the mass breeding of insects is the sterilization of the wild population itself. However, most of the available C. are toxic to mammals and birds as well. A combination of attraction and sterilization seems more promising; a large fraction of the insect population is attracted to a small area by Pheromones (see), where they consume bait containing C. or are sterilized by contact with C.

Chemical volume law of Gay-Lussac: a basic law of stoichiometry: volumes of gases reacting in or produced by chemical reactions stand in the ratios of small integers (J.L. Gay-Lussac, 1808). For example, 1 liter hydrogen combines with 1 liter chlorine to produce 2 liters hydrogen chloride gas. In general, the volume ratios for gaseous reactants and/or products are the same as the ratios of the stoichiometric coefficients of the corresponding compounds in the reaction equation. The C. is a limiting case, which applies rigorously only to ideal gases. It is explained by Avogadro's law (see).

Chemical warfare agents: a general term for all industrially prepared chemical substances used to destroy, damage or hinder human beings in war, to destroy or damage animals and plants, installations, technical weapons, etc., and the accessories required for their military use. The C. include Chemical weapons (see), Sabotage poisons (see), Herbicides (see), fogging agents and Incendiaries (see).

Chemical weapons: substances or mixtures of substances produced industrially which, because of their high acute toxicity, have fatal or temporarily incapacitating effects, and which are used for military purposes. In contrast to other chemicals used in warfare, such as napalm or smokescreens, C. are used because of their toxicity. Some characteristics of interest in C. are their Toxicity (see), the dose required, their persistence, i.e. the average length of time the air or surface is poisoned after application under given environmental conditions, their chemical and thermal stability during storage or use, and the length of temporary incapacitation caused by those agents which are not meant to be lethal.

C. were first deployed using blowers, but later artillery was used. Modern deployment for C. includes artillery grenades, bombs, rocket warheads, mines, spray devices on planes and special equipment for police use: chemical guns and hand grenades, water throwers, etc.

Classification. The "classic" military categories "white cross", "green cross" and "yellow cross" are now obsolete and are simply of historical interest. Classification according to physical properties (physical state, vapor pressure) or chemical properties is possible, but inadequate. A modern military classification distinguishes lethal and non-lethal C., fast- or slow-acting C., C. suitable for air or surface (volatile or non-volatile) C. This classification scheme is based both on toxicological and physical chemical properties. The main parameters are the toxic (lethal or damaging) dose, latence time, persistence and also the intended mode of deployment.

C. can be divided into the following groups on the basis of their toxicology, taking into account other properties as well:

1) ***Irritants***: These include various chemical types, especially the halogenated or organoarsenic com-

pounds. The most important of the ***eye irritants*** are o-chloroacetophenone (CN), 2-chlorobenzylidene malonodinitrile (CS) and dibenz-(b,f)-1,4-oxazepin (CR). Whatever their chemical structure, eye irritants act on the sensitive nerve endings of the eye mucosae, probably by blocking sulfhydryl groups on enzymes and thus inhibiting them. The main symptoms are burning or stinging of the eyes, tearing, the sense of a foreign body in the eye, lid closing, temporary blindness and connective tissue inflammation. In high concentrations, permanent eye damage is possible. The upper respiratory passages can also be damaged by high doses, and in severe cases, toxic lung edema can lead to death. In addition, CS acting over long periods on the skin have been reported to cause reddening and blistering.

C. which act primarily on the upper respiratory passages are called ***nose and throat irritants***. The most important representatives of this group are organoarsenic compounds such as diphenylchloroarsine (Clark I), diphenylcyanoarsine (Clark II) and diphenylaminochloroarsine (Adamsite). Other representatives are trialkyllead compounds. These compounds irritate the mucosae in the nose and throat area, but also in the eyes. It has been suggested that the mechanism of action is blockage of sulfhydryl groups. The latence time can be as long as 30 minutes. The main symptoms are coughing and sneezing, increased secretion by the nose mucosae and salivary glands. Other effects are respiratory distress, headache and pain in the breast area. In high doses, the poisoning can be fatal (lung edema).

2) ***Psychotoxic C.***: substances which can cause temporary psychological anomalies (model psychoses). They are effective even in extremely low doses: milligrams or micrograms may suffice to poison a human being. There are two types of substance, the psychotomimetics which induce model psychoses, and psychotropic substances which reduce psychological ability to cope. The phenylglycollic esters or benzilic esters of heterocyclic iminoalcohols are of military interest. Derivatives of lysergic acid, tryptamine and cannabinol, and phenylalkylamines, were also tested. 3-Quinuclidinylbenzilate (BZ) seemed most suitable. The main symptoms of a BZ poisoning are pronounced confusion, disorientation and psychomotor excitation, as well as hallucinations of various kinds. Because of its cholinergic effect, BZ can lead to fatal poisoning in high doses.

3) ***Skin damaging C.***: These are generally slow-acting, thus persistent C., which cause temporary or long-term disability, or even death. The damage to the skin and mucosal membranes dominates the clinical picture. Chemically, these compounds are halogenated derivatives of thioethers, tertiary amines, primary arsines and oximes. Some compounds of military interest are bis-(2-chloroethyl)sulfide (mustard gas), which was used in WWI, its derivative 2-chloroethenyldichloroarsine (Lewisite), and some halogenated oximes, e.g. dichloroformoxin. The toxic properties of the skin-damaging C. are ascribed to their ability to form heterocyclic onium ions, which in turn react in numerous ways with proteins. The halogenated thioethers and amines have a symptomless latent period, but halogenated arsines and oximes act immediately. Depending on the dose, they cause reddening of the skin, blistering or necrotic tissue damage. There are also systemic toxic effects due to the relatively high degree of absorption into body fluids, especially with the mustard derivatives. Halogenalkylamines are very potent inhibitors of enzymatic processes. Mustard derivatives can damage various organs (liver, kidneys, spleen, brain, gastrointestinal tract, heart, lungs). The skin damage has a very poor tendency to heal, and the general resistance is greatly reduced (danger of secondary infections). The toxic effects of Lewisite are similar to those of the mustards, but if the acute poisoning is survived, the chances of healing is better. Halogenated oximes cause severe burning and pain in the skin with no latency. In addition, there is inflammation and weals form. Mustard derivatives have a high potential for alkylation, which explains the observed secondary damage after an acute or subacute intoxication, or after long-term exposure to very small doses. Mustard gas itself is one of the most potent carcinogens known.

4) ***Lung-damaging C.*** These had military significance only during the first world war. The main representatives are chlorine and phosgene; in addition, diphosgene and triphosgene were used, which split thermally into two or three moles of phosgene. Chloropicrin had some significance. The effectiveness of the lung-damaging C. is short, and there is a latency period up to several hours long. The resulting edema leads to increasing respiratory distress. At the same time, the increased struggle for air promotes the spread of the edema. At very high doses, the edema does not appear, and the victim dies in a short time of cramps.

5) ***General poisons***: These are mostly rapidly acting C. which had a certain role in the first world war. The main examples are hydrocyanic acid, halogen cyanides, arsenic and phosphorus hydrides and carbon monoxide. The toxic effects of these C. depend on specific effects on enzymatic or metabolic processes, especially those in the blood (see Poison). Poisoning occurs rapidly, accompanied by various symptoms, of which increasing respiratory distress, central nervous system symptoms and cramps are the most prominent. Death occurs within 15 to 30 minutes with cessation of breathing.

Other important types of compound are the derivatives of fluoroacetic acid. See Poison, Sabotage poisons.

6) ***Nerve-damaging C.***: The nerve-damaging organophosphorus C. are at present the most significant group. The first of them was synthesized in 1936 in Germany, by Schrader; it was dimethylaminoethylcyanophosphate (tabun), which now is of less military importance. The main representatives are O-isopropylfluoromethylphosphonate (sarin), O-pinacolylcuoromethylphosphate (soman) and the V substances, of which the most important is O-ethyl-S-(N,N-diisopropylaminoethyl)-methylphosphonate (VX). Because of its relatively high volatility, sarin is used mainly to poison the air. VX, by contrast, is a persistent substance used to poison land. Organophosphorus nerve poisons are both inhaled and, in the case of the V substances, absorbed through the skin. The clinical picture is determined primarily by blockage of acetylcholinesterase, an enzyme required for

the function of synapses in cholinergic nerves and motor endplates of the musculature (see Poison). The main symptoms are loss of heartbeat frequency and strength, contraction of the pupils, and the blood vessels of the heart and bronchia; those of the lungs, skin and musculature dilate. There is loss of bladder and bowel control. Direct cholinolytics (atropine) are used for therapy; these counteract the excess endogenous acetylcholine by competitive inhibition of its receptors. Certain reactivators which split the poison-enzyme complex are also used. Otherwise, therapy consists of continuing the oxygen supply (artificial respiration) and maintaining life functions.

In addition to these acetylcholinesterase-inhibiting nerve poisons, there are some bicyclic phosphoric esters which attracted military interest in the 1970's; they presumably act as antagonists of γ-aminobutyric acid (also a neurotransmitter). They are considered potential nerve weapons.

In addition to synthetic products, some natural toxins, especially microbial and animal products, are of military interest (see Sabotage poisons). The most important representative was botulin toxin A, an extremely poisonous metabolic product of the bacterium *Clostridium botulinum* which sometimes causes food poisoning. Regardless of the means of their production, toxins are included in the treaty banning biological weapons (concluded in 1972, in force since 1975), i.e. their development, production and storage for purposes of use in warfare is illegal under international law.

Identification and determination of C. The rapid and certain identification of C. and their quantitative determination are the first necessity in providing protection from them. Chemical, physical-chemical, physical and biochemical methods have been developed for rapid detection and identification. Color reactions are the most important of the chemical methods (test tubes, droplet reactions, test papers, etc.). The biochemical tests make use of inhibition of specific enzymes (e.g. cholinesterases) and are extremely sensitive, especially for the organophosphorus compounds. However, they are sometimes too slow for immediate detection. Optical, gas-chromatographic and mass spectroscopic methods have been developed both for use in rapid analysis and for identification and quantitative determination of the C. in the laboratory. More recently, distance measurement processes, such as Lidar, are being developed.

Material	Structure	Name	Physical	Toxicity
1) Eye irritants				
CN	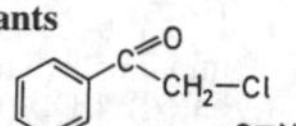	ω-Chloroacetophenone	m.p. 56.5 °C, b.p.(760) 247 °C $p(20)$ 1.7 Pa, $C_{max}(20)$ 0.105 mg l^{-1}	ICt_{50} 80 mg min^{-1} m^{-3}, LCt_{50} 10000 ... 11000 mg min^{-1} m^{-3}, cancerous
CS	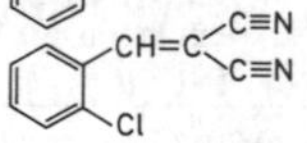	2-Chlorobenzylidene malonitrile	m.p. 95–96 °C, b.p.(760) 310–360 °C, no measurable vapor pressure	ICt_{50} 20 mg min^{-1} m^{-3}, LCt_{50} > 10000 mg min^{-1} m^{-3}
CR	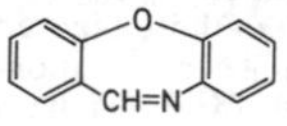	Dibenz-(b,f)-1.4-oxazepine	m.p. 72 °C. Use as aerosol or solution	More irritable than CS
2) Nose and throat irritants				
Clark I	$(C_6H_5)_2AsCl$	Diphenylchloroarsine	m.p. 44 °C, b.p.(760) 333 °C (dec.) $p(20)$ 0.007 Pa, mg $C_{max}(20)$ $6.8 \cdot 10^{-4}$ mg l^{-1}	ICt_{50} 15 mg min^{-1} m^{-3}, LCt_{50} 15000 mg min^{-1} m^{-3}
Clark II	$(C_6H_5)_2AsCN$	Diphenylcyanoarsine	m.p. 31.5 °C, b.p.(760) 346 °C (dec.),	ICt_{50} 25 mg min^{-1} m^{-3}, LCt_{50} 10000 mg min^{-1} m^{-3}
Adamsite	$NH(C_6H_4)_2AsCl$	Diphenylaminochloroarsine	$p(20)$ 0.03 Pa $C_{max}(20)$ $1.5 \cdot 10^{-4}$ mg l^{-1} m.p. 195 °C, b.p.(760) 410 °C, $p(20)$ $3 \cdot 10^{-11}$ Pa $C_{max}(20)$ $2 \cdot 10^{-15}$ mg l^{-1}	ICt_{50} 10 ... 20 mg min^{-1} m^{-3}, LCt_{50} 15000 mg min^{-1} m^{-3}
3) Psychotoxins				
BZ	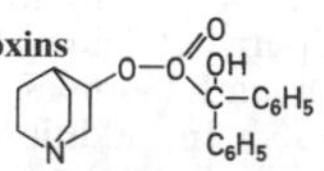	3-Quinuclidinylbenzilate	m.p. 189–190 °C, b.p.(760) 322 °C	ICt_{50} 110 mg min^{-1} m^{-3}, LCt_{50} 200000 mg min^{-1} m^{-3}
4) Skin-damaging compounds				
S-Mustard	$(Cl-CH_2-CH_2)_2S$	Bis-(2-chloroethyl) disulfide	m.p. 14.4 °C, b.p.(760) 216–218 °C, LCT_{50} $p(20)$ 15.3 Pa, $C_{max}(20)$ 0.625 mg l^{-1}	ICt_{50} 150 mg min^{-1} m^{-3}, LCt_{50} 1500 mg min^{-1} m^{-3} (when inhaled), 10000 mg min^{-1} m^{-3} (percutaneously)
N-Mustard	$(Cl-CH_2-CH_2)_3N$	Tris-(2-chloroethyl) amine	m.p. −4 °C, b.p.(760) 230–235 °C (dec.) $p(20)$ 92 Pa, $C_{max}(20)$ 92 Pa, $C_{max}(20)$ 0.07 mg l^{-1}	ICt_{50} 200 mg min^{-1} m^{-3}, LCt_{50} 1500 mg min^{-1} m^{-3} (when inhaled)
Lewisite	$Cl_2As-CH=CHCl$	2-Chloroethylenedichloroarsine	*cis*-form / *trans*-form m.p. −44.7 / −2.4 °C b.p.(760) 169.8 / 196.6 °C $p(20)$ 208.2 / 53.3 Pa	ICt_{50} 300 mg min^{-1} m^{-3}, LCt_{50} 1300 mg min^{-1} m^{-3}

Material	Structure	Name	Physical	Toxicity
Dichloro-formoxime	$Cl_2C=N-OH$	Dichloroformox-ime	$C_{max}(20)$ 2.3 m.p. 29–40 °C, b.p.(760) 129 °C, $C_{max}(20)$ 20–25 mg l^{-1}	Eye irritant 25 mg min^{-1} m^{-3}, Damage to skin 1000 to 25 000 mg m^{-3}, LD_{50} 30 mg kg^{-1}
5) Lung-damaging compounds Phosgene	$Cl_2C=O$	Carbonyl dichloride	m.p. −118 °C, b.p.(760) 8.2 °C $p(20)$ 156.4 kPa, $C_{max}(20)$ 6370 mg l^{-1}	ICt_{50} 2000 mg min^{-1} m^{-3}, LCt_{50} 3200 mg min^{-1} m^{-3}
6) General poisons Hydrogen cyanide	$H-C\equiv N$	Hydrogen cyanide	m.p. −13– −13.4 °C, b.p.(760) 25.6–26.5 °C $p(20)$ 81.6 kPa, $C_{max}(20)$ 873 mg l^{-1}	200 mg min^{-1} m^{-3}, LCt_{50} = 2000 mg min^{-1} m^{-3}, 5000 mg min^{-1} m^{-3}, lethal within one minute
7) Nerve poisons Sarin	$CH_3P(=O)(F)-O-C_3H_{7,iso}$	Methylfluorophos-phonic isopropyl ester	m.p. −57 °C, b.p.(760) 151.5 °C $p(20)$ 197 Pa, $C_{max}(20)$ 11.3 mg l^{-1}	At rest: ICt_{50} 40 ... 55 mg min^{-1} m^{-3}, LCt_{50} 70 ... 100 mg min^{-1} m^{-3}, LCt_{100} 150 ... 180 mg min^{-1} m^{-3} When active: ICt_{50} 8 ... 25 mg min^{-1} m^{-3}, LCt_{50} 15 ... 50 mg min^{-1} m^{-3}, LCt_{100} 30 ... 90 mg min^{-1} m^{-3}
ICt_{50} 8–25, Soman	$CH_3P(=O)(F)-O-CH(CH_3)-C(CH_3)_2-CH_3$	Methylfluorophos-phonic pinacolyl ester	hardens at −70– −80 °C, less volatile than GB	ICt_{50} 25 mg min^{-1} m^{-3}, LCt_{50} 70 mg min^{-1} m^{-3} (when inhaled), LCt_{50} 7500 ... 10 000 mg min^{-1} m^{-3} (percutane-ously)
	$C_2H_5P(=O)(C_2H_5O)-S-CH_2-CH_2-N(C_3H_{7,iso})_2$	O-Ethyl-S(N,N-diisopropyl-aminoethyl)-methylphos-phonothiolate	m.p. < −30 °C, b.p.(760) > 300 °C, $p(20) < 10^{-2}$ Pa C_{max} 10^{-3} – 10^{-4} mg l^{-1}	ICt_{50} 5 mg min^{-1} m^{-3}, LCt_{50} 36 ... 45 mg min^{-1} m^{-3} (when in-haled), LD_{50} (per-cutaneously) 15 mg / person

m.p., melting point; b.p.(760) boiling point at standard pressure, $p(20)$, saturation vapor pressure at 20 °C; C_{max}, saturation concentration in air at 20 °C, ICT_{50}, product of concentration and time which will incapacitate 50% of those exposed; LCt_{50}, median lethal product of concentration and time; LD_{50}, median lethal dose; LCT_{100}, lethal product of concentration and time for 100% of persons exposed. The lethal and toxic doses refer to human beings, but are extrapolated from animal data

There are devices to protect groups, such as isolated rooms and filtered air supplies. Individual protection includes gas masks and protective suits. The effectiveness of these measures is limited, and depends on such factors as availability and adequacy of the equipment, the training of soldiers and time. Medical care of those who are poisoned is also a protection factor, but in mass attacks is extremely difficult to provide. There are no universal antidotes, but specific antidotes are known for most C.

Detoxification can be achieved by physical or chemical methods. Persons are detoxified primarily by such physical methods as adsorption and washing, sometimes in combination with chemical detoxificants such as mild oxidizing agents. Internal poisoning can be countered by antidotes; an example is oxime therapy for nerve gas poisoning. Surfaces and interior spaces are detoxified by physical methods such as washing, adsorption or thermal treatment, or by chemical methods such as treatment with active chlorine compounds (lime chloride or chloramine), amine alcoholate mixtures, various oxidizing agents, or strong nucleophiles. Detoxification of land is usually limited to key areas, due to limited supplies of materials and personnel; it is done by physical methods, such as covering with uncontaminated material, removal of poisoned layers of soil, thermal treatment, or by chemical methods (lime chloride treatment). The effectiveness of these measures depends on such factors as the type and formula of the C., type and available amount of detoxifying agents, meterological factors, and time at which detoxification is begun. Detoxification requires trained personnel who are supplied with protective devices and take precautions to prevent secondary poisonings.

Historical. C. were first used in militarily significant amounts during World War I. The first massive deployment of C. is generally agreed to have been the chlorine gas attack on 22 April 1915 by the German troups in the Battle of Ypern (Belgium), although earlier the French-English and German sides had used less effective irritants. In addition to lung poisons (chlorine, phosgene, di- and triphosgene), new types of irritant (Clark I and II) and later skin poisons (mustard gas) were developed. Other new C. (Adamsite, Lewisite) were developed to the point of deployment. In the following years, C. were only occasionally used: during the Italian invasion of Abyssinia in the mid-1930's, by the Japanese in limited amounts in Manchuria during World War II. Although there was no militarily significant use of C. during the second world war, the organophosphorus nerve poisons (tabun, sarin and soman) were developed in Germany during that time, and some were produced. After the end of that war, the center of development and production of C. shifted to the

USA, where the organophosphorus V weapons were developed and went into production at the beginning of the 1960's. In the meantime, temporary, psychoactive C. were developed (e.g. BZ). Beginning in the mid 1950's, a new method of delivery for C. was developed, the binary weapon. These greatly reduce the danger of poisoning the troops who use them, because the last two chemical precursors of the C. are placed separately in the grenade, bomb or rocket, and they are not combined until the missile is on its way to the target area. Then they react to release the actual C., usually an organophosphorus nerve poison.

The use of C. during World War I was a breach of the Haague Convention of 1907. The principle was reaffirmed in the Geneva Protocol on Chemical and Bacteriological Warfare of 1925, which extended the ban to bacteriological (biological) weapons. Negotiations on a comprehensive ban of C. have been carried on since the end of the 1960's in the framework of the Geneva Disarmament Commission (now Conference).

Chemisorption: see Adsorption.

Chemistry: the science of substances: their properties, structures, compositions, syntheses, reactions and interactions. All chemical reactions are based on the release, uptake and distribution of electrons among atoms, groups of atoms and molecules, and the interactions between the energy levels of the electrons. The number of elements in existence on earth is 104, and the number of compounds so far synthesized is about 8 million. On the average, about a thousand new compounds are added to that number daily.

C. is classically subdivided into four main fields: inorganic, organic, physical and industrial; these differ in their specific goals and in the methods used, but there are many combinations and transitions between these areas.

Inorganic C. is concerned with the chemical elements and their compounds, with the exception of the carbon compounds, which are the province of organic chemistry. However, simple carbon compounds, such as the oxides of carbon, carbonic acid and its salts, carbides, metal carbonyls, etc., are included in inorganic C. Metals, inorganic polymers, semiconductors, silicates, coordination compounds and minerals are the main areas of study. The fields of geochemistry, complex chemistry and crystal chemistry are branches of inorganic C.

Organic C. is concerned with the compounds of carbon; the multitude of organoelement compounds and the carbon derivatives named above are shared with inorganic C. With the exception of the noble gases, nearly all the elements of the periodic system can be incorporated into organic compounds. The distinction between organic and inorganic C. is justified by the peculiar structure and reactivity of the carbon compounds. The original field of organic C. was the isolation, structure elucidation and synthesis of natural products from plants, animals and fossils. This "natural product C." is still an important field; the structures and biological functions of compounds, and the determination of conformation by means of highly developed techniques are at the center of interest. The isolation, sequencing and synthesis of genetic material are part of the modern field of natural products C. Using modern automated synthesis techniques, it is now possible to synthesize biologically active peptides, small proteins, oligonucleotides and polynucleotides; the latter are used in gene technology to manipulate genetic information. The close relationships to biochemistry are as apparent as those to cytochemistry, phytochemistry, food, forensic, macromolecular and pharmaceutical C. *Biochemistry* developed out of organic C.; in this field chemical, physical and mathematical methods are used to explore the chemical structures of living systems. *Pharmaceutical C.* is concerned with the chemical behavior, mechanisms of action, biotransformations, synthesis and analyses of drugs.

Physical C. is the study of all the physical phenomena which are related to reactions, and the effects of physical conditions on reactions. The field is subdivided into *electrochemistry, photochemistry, thermochemistry, mechanochemistry, magnetochemistry,* etc. *Radiation C.* explores those reactions which are induced by high-energy radiation and which lead to permanent changes in materials. *Nuclear C.* is the study, by chemical methods, of atomic nuclei and their reactions. A subdiscipline of nuclear chemistry, *radiochemistry*, is concerned with the properties and reactions of radioelements and their compounds, and with the radionuclides.

The application of chemical methods and knowledge to industrial processes and the development of the appropriate technology is the task of ***industrial C. (chemical technology)***. Some important subdisciplines are *high-polymer C., carbochemistry, petroleum C., agricultural C., food C., water C., textile C.* and *pigment C.*.

A division of C. into ***pure*** and ***applied C.*** is no longer customary, and there is little reason for it. The fields of ***quantum, cosmological, radiation, laser*** and ***plasma C.*** are significant in their own rights; ***analytical, preparative*** and ***ecological C.*** are related to all the major fields.

C. is involved in other natural sciences, especially physics and biology. At the end of the 19th century, the interactions between C. and physics led to the development of the theoretical basis of modern C. During this period, the concepts of atoms, ions, molecules and bonds were developed.

The main task of C. is the production and interconversions of substances which are not present in nature, or not present in sufficient quantity or in a useable form. C. thus contributes very significantly to human health, nutrition, clothing, housing, lengthened life expectancy, and transportation. Chemical industries have also contributed very greatly to the pollution of the environment, but it is hoped that further developments in C. can at least partially repair the damage. There is no aspect of life in an industrial society which is not affected by C.

The main problems to be solved by industrial C. in the near future are an adequate supply of raw materials and energy, and protection of the environment. At present, the primary raw materials are petroleum, natural gas and carbon; by the beginning or middle of the 21st century, the supplies of petroleum and natural gas will become critically depleted. The reserves of coal can be expected to last much longer, but its use as a fuel will be limited by the increasing

CO_2 levels in the atmosphere. Biological polymers formed by photosynthesis (especially cellulose) can serve as carbon reserves. Nuclear and solar energy, biogas and biotechnologically produced alcohol and hydrogen will probably contribute larger fractions of the total energy supply.

The production of the raw materials required by the chemical industry requires penetration into ever deeper and less accessible regions of the earth's crust, which raises the costs of extraction. In order for society to take full advantage of modern C., all reactions must be carried out under the mildest possible conditions, without byproducts, and with technologies which are far less damaging to the environment.

Historical. The word "chemistry" has its roots in "chemeia" or "chymeia". Etymologists formerly believed it was derived from the Egyptian "chemi" = "black", but it is now believed to stem from the Greek "chyma" = "metal casting". The term "alchemy" comes from Arabic. Chemical knowledge has always been closely linked with crafts. More than 1000 years before the Christian era, the Egyptians, Babylonians, Phoenicians, Chinese and Indians, and later the Greeks and Romans, had a large store of chemical knowledge related to the extraction of metals, making of alloys, colored glass, artificial gems and pearls, enamels, pigments, cosmetics, drugs and poisons. The first known natural philosophical theories of matter were developed in the fifth to third centuries BC, by the Greeks. The atomic theories of Leucippus of Milet and Democrites supposed that the world consists of an infinite number of tiny particles which cannot be further divided (a-tomos) and move through empty space. Empedocles demonstrated by experiment that air is a form of matter, and proposed the ordering of the elements earth, water, air and fire on the basis of the dynamic principle of constant change. In the 7th century, the Arabs absorbed the chemical knowledge of antiquity. They perfected practical methods and chemical apparatus, and developed the sulfur-mercury theory on the basis of Aristotle's speculations (sulfur as the principle of the combustible and volatile, mercury as the principle of the liquid and earth-like). Around this time, the first textbooks of C. were written [Abu Ali Ibn Sina, latinized to Avicenna (980-1037), Al-Razi (865-925), Djabir Ibn Haiian (721-815)]. In these books, transmutation was justified. This knowlege reached Europe in the 12th and 13th centuries, where it became fixed as the writings of Geber (latinized form of Djabir) and expanded by the discovery of the mineral acids.

Starting in the Renaissance, efforts were made to develop chemical knowledge for practical purposes, especially in mining and medicine. This period saw the discovery of gunpowder and the distillation of alcohol in the 13th century. G. Agricola (1494-1555) founded mineralogy, and Paracelsus (1493-1541) founded ***iatrochemistry***. This teaching was based on the idea that all processes in the body are chemical in nature, and involve the three elements sulfur, mercury and salt as the principles of the combustible, volatile and fire-resistant. An imbalance of these elements in the body would cause disease. The task of iatrochemistry, therefore, was to search for chemical substances which would reestablish the balance of the elements. It generated interest in the discovery of chemical processes, and opened the way for C. and medicine to enter the universities. The first chair for C. was founded in 1609 in Marburg. R. Boyle (1627-1691) showed that the theory of elements was untenable, introduced the idea that matter is corpuscular, and began to interpret chemical processes on this basis. He developed new methods of analysis (indicators) and made great progress in preparative C.

The phlogiston theory was developed by G.E. Stahls (1659-1734), who interpreted reduction as a combination of matter with the hypothetical substance phlogiston. This theory was the first to show that oxidation and reduction are related and opposing processes, and made possible a first systematization of chemical processes. It also provided a new concept of the tasks and goals of C. The work of J. Priestley (1733-1804) and C.W. Scheele (1742-1786) on gases, which led to the discovery of oxygen and other gases, and of M.W. Lomonossov (1711-1765), which showed that in the course of combustion something combines with the substance, prepared the way for the overthrow of the phlogiston theory by A.L. Lavoisier (1743-1794). He showed that every combustion is a combining of substance with oxygen. With this insight, he could also give a correct definition of a chemical element as a substance which cannot be further decomposed into other substances by chemical methods. Together with the French chemists de Morveau, Berthollet and Fourcroy, Lavoisier was responsible for developing the first scientific chemical nomenclature. Quantitative methods of research led to the discovery of the laws of constant and multiple proportions, and the foundation of stoichiometry. On the basis of these insights, Dalton proposed a new atomic theory in 1804 which took into account qualitative and quantitative relationships, assigned to each element a certain atomic weight, and explained chemical reactions as the result of combination and separation of these atoms.

Further major developments in inorganic C. were made by J.H. Berzelius (1779-1848), with his dualistic electrochemical theory, H. Davy (1778-1829), with the discovery of new elements (alkali and alkaline earth metals), and Gay-Lussac (1778-1850), who discovered the chemical volume law. The atomic theory was elegantly confirmed by the development of the periodic system by D.I. Mendeleyev (1834-1907) and, somewhat later, by L. Meyer (1830-1895).

Inorganic C. was given an important renewal in 1900 by the application of the coordination theory of A. Werner (1866-1919) to complex C. In 1915, the area of solid-state reactions was opened.

At the beginning of the 19th century, organic C. branched off as a field of C.; the development was started by the synthesis of urea in 1828 by F. Wœhler. This synthesis was a decisive defeat for the vitalistic theory that the substances of animal and plant organisms could not be synthesized in the laboratory; nothing more stood in the way of structural elucidation of organic substances and the synthesis of natural and artificial products such as alizarin, indigo, quinine, sugars, fats, alkaloids, terpenes, camphor, rubber and plastics.

Physical C. was developed in the last decades of the 19th century as efforts were made to give theoretical explanations of many empirical facts known at the time. Carnot, Kirchhoff, Helmholtz, Gibbs, Nernst and Boltzmann founded chemical thermodynamics. At the same times, the extended fields of atomistics, reaction kinetics and colloid C. were developing. The application of quantum mechanical methods to chemical problems led to better understanding and predictability of organic reactions (Ingold, Pauling). Other scientists involved in this effort were J. von Liebig, F. Wœhler, J.B. Dumas, M.F. Chevreul, E. Mitscherlich, F.F. Runge, C. Gräbe, A.W. von Hofmann, H. Kolbe, E. Fischer, H. Fischer and W.J. Reppe. The theory of organic C. was further developed by major contributions from A. Butlerow, A. Kekulé, van'tHoff, S. Cannizzaro, J. Wislicenus and A. von Baeyer.

The industrial production of synthetic organic (coal hydrogenation, fuels, synthetic rubber, plastics, acetylene C., etc.) and inorganic chemicals (sulfuric acid, ammonia, nitric acid) was given an enormous impetus by the discovery of the mechanisms of catalysis.

The great scientific achievements of the 20th century in the area of C. have been honored by bestowal of the Nobel Prize (see) for C. The scientific activities summarized by the list of those prizes are evidence for the tremendous progress in C. in the last 90 years.

Chemotherapy: a systemic therapy with drugs which are intended to destroy pathogens or cancer cells with the least possible damage to the host. The term was coined by P. Ehrlich, and it rests on the *concept of selective toxicity*. The selectivity of chemotherapeutics is based on the presence of different structural elements or different biochemical reactions in the host and pathogen or cancer cells. ***Chemotherapeutics*** are used to combat infections by bacteria, protozoa, fungi, worms or viruses, or tumors. These compounds can be either synthetic or biogenic. The first important synthetic chemotherapeutic was Salvarsan® (see Arsphenamine), which was developed by P. Ehrlich in 1909 and introduced in 1910. The treatment of bacterial infections first became possible with the introduction of the sulfonamide antibiotics, and now there are many others available as well. Antiprotozoics, e.g. the antimalaria drugs, anthelminthics, antimycotics and virostatics are also chemotherapeutics. The use of these compounds is complicated in many cases by the development of a resistance on the part of the pathogens to frequently used compounds. Development of resistance is an important reason for the continuous development of new chemotherapeutics.

Chemotronics: a subdivision of applied electrochemistry concerned with the production and function of electrochemical elements for regulation and measurement. (See Electrochemical components.)

Chenodeoxycholic acid, also ***anthropodeoxycholic acid***: 3α,7α-dihydroxycholic acid, a bile acid. C. forms colorless crystals (from acetate); m.p. 140 °C, $[\alpha]_D$ +11.1° (ethanol). It is scarcely soluble in water, but dissolves readily in organic solvents. C. is found in human bile. It forms soluble complexes with cholesterol, and is therefore used to remove cholesterol-containing gallstones.

Chiral: see Stereoisomerism, 1.1.

Chirality: see Stereoisomerism, 1.1.

Chirality element: see Stereoisomerism, 1.1.

Chiroptic methods: methods for study of the chirality of substances, including *circular dichroism (CD), optical rotatory dispersion (ORD), magnetic circular dichroism (MCD)* and *magnetic optical rotatory dispersion (MORD)*. C. measure the interactions of polarized light with a chiral medium, and are of great importance for structural determination of optically active compounds. As shown in Fig. 1, a linearly polarized beam of light can be thought of as the sum of two coherent, circularly polarized beams rotating in opposite directions.

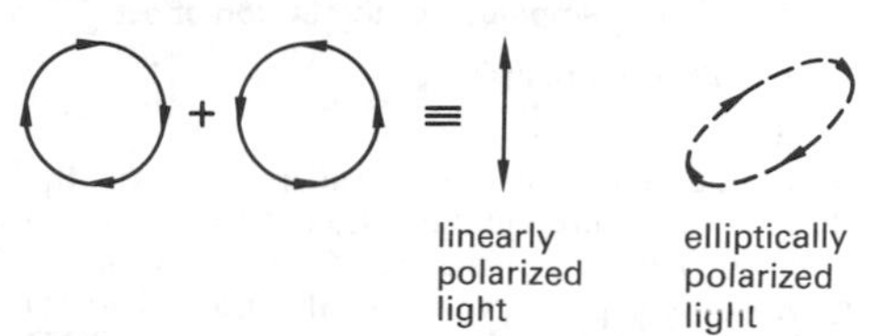

Fig. 1. circularly polarized light

ORD and CD arise as follows:

1) The indices of refraction, and thus also the velocity in the medium, differ for the left- and right-polarized beams which constitute the linearly polarized light. Therefore, when the light leaves the medium, there is a phase difference between the two circularly polarized beams, and the resultant linearly polarized beam oscillates in a different plane, at an angle α to the plane of the original beam. This phenomenon is known as optical rotation; it is given as the specific rotation [α] or the molar rotation $[\Phi] = M \cdot 10^{-2}[\alpha]$ (see Polarimetry for the definition of [α]. ORD is the dependence of optical rotation on the wavelength of the polarized light; it is measured in spectral polarimeters. ORD curves are plots of the molar rotation vs. the wavelength (Fig. 2).

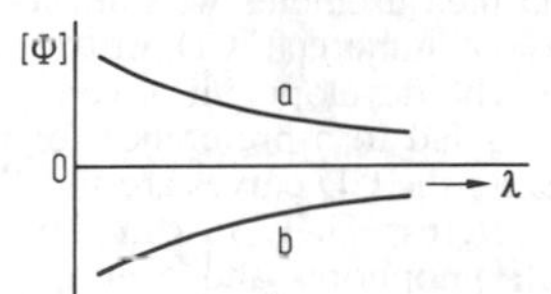

Fig. 2. Normal positive (a) and negative (b) ORD curves

2) The molar extinction coefficients for circularly polarized light rotating clockwise and counterclockwise are different. As a result, the two beams have different intensities as they leave the optically active medium, and their superposition no longer produces linearly polarized, but ellipsoidally polarized light (Fig. 1). This phenomenon is called circular dichroism (CD). CD is measured with a dichrometer; its magnitude is given by the difference in the molar extinction coefficients for left- and right-polarized light, $\Delta\varepsilon = \varepsilon_l - \varepsilon_r$, or by the molar ellipticity [Θ]. The two quantities are related by the equation $[\Theta] = 3300\,\Delta\varepsilon$. CD curves are plots of Δε or [Θ] against the wavelength (Fig. 3). CD has been observed only in absorption bands resulting from electronic transitions, although there are reports of CD in vibrational bands.

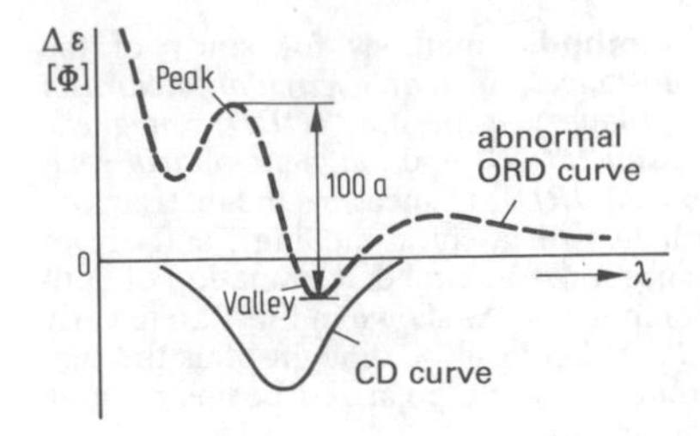

P, peak; V, valley, *a* amplitude of the Cotton effect

Fig. 3. Negative Cotton effect

If the dependence of the optical rotation on the wavelength is measured in the region of an absorption band, the CD in this region produces an S-shaped curve which is superimposed on the normal ORD curve. This type of curve is called the abnormal ORD curve. The sum of the abnormal ORD and the CD (Fig. 3) is the *Cotton effect*. If one of the two curves is known, the other can be calculated. If the ORD shows fine structure in the region of an electronic excitation, this is called a *multiple Cotton effect*. If the CD is positive, the abnormal ORD curve displays a peak followed by a valley (going from higher to lower wavelength). With negative CD, the valley comes before the peak (positive and negative Cotton effects, respectively). The magnitude of the Cotton effect is given by the amplitude $a = ([\varphi_1] - [\varphi_2]) \cdot 10^{-2}$.

The magnitude and sometimes the sign of the Cotton effect depend on the solvent. The Cotton effect appears when a chiral compound has a Chromophore (see) near its center of chirality, provided that the absorption of this chromophore is not so strong that all the light is absorbed; the carbonyl group is an ideal chromophore.

Until about 1965, ORD measurements were most commonly made, because commercial CD instruments were not available. The development of commercial CD instruments has led to a preference for these measurements, because the CD curves are simpler and easier to interpret, especially in complex molecules with several chromophores and overlapping absorption bands.

Faraday discovered that optical activity can be induced by a magnetic field (see *Faraday effect*), so that ORD and CD can be observed even in achiral compounds; here they are called MCD and MORD.

There are empirical and semi-empirical rules relating the sign and magnitude of the Cotton effect to the geometry of the molecule. These are used mainly to determine absolute and relative configurations of substances, chiefly natural products. Fig. 4 shows

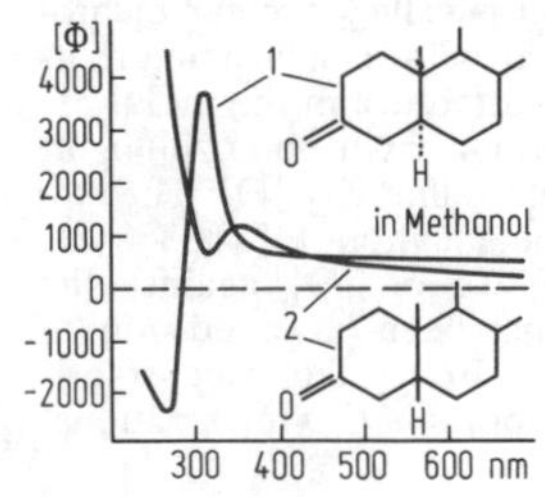

Fig. 4. ORD curves of stereoisomers

how sensitive the method is even to slight steric differences.

Chitin: a homopolysaccharide consisting of $\beta(1\rightarrow4)$ linked D-*N*-acetylglucosamine (formula, see Cellulose). C. is not soluble in ordinary solvents. It is degraded in strong acid or base to soluble components. C. is a structural polysaccharide which is part of the cell walls of fungi and green algae, and is the main component of the exoskeletons of arthropods. C. can be isolated in relatively pure form from the shells of lobsters or crabs. There are strong intermolecular hydrogen bridges between the glycan chains, formed between hydroxyl and amide groups. These strong intermolecular forces are also the reason for the insolubility of C. in water. X-ray studies have revealed several crystalline modifications of C., called α-, β- and γ-C. The most common is α-C., in which the glycan chains are antiparallel. Treatment of C. with strong base removes some of the acetyl groups; the resulting polymer is called *chitosan*. It dissolves in dilute acid to give a viscous solution. Inhibitors of C. biosynthesis have been developed recently for use as selective insecticides.

Chitosamine: see D-Glucosamine.

Chitosan: see Chitin.

Chloral hydrate: CCl_3-$CH(OH)_2$, the hydrate of chloral (CCl_3-CHO). C. is a crystalline compound; m.p. 51.7 °C. C. does not follow the Erlenmeyer rule; in spite of the fact that it is an aldehyde hydrate, it is stabilized by the inductive effect of the three chlorine atoms and intramolecular hydrogen bonds, and is thus a stable compound. In alkaline media, it is split into chloroform and formate. C. is used as a hypnotic. The active form is 2,2,2-trichloroethanol, CCl_3CH_2OH, which is formed by metabolism of C.

Chloralkali electrolysis: the most important large-scale industrial process for producing chlorine by electrolysis of an aqueous alkali chloride solution. Alkali hydroxide solution (lye) and hydrogen (electrolyte hydrogen) are byproducts. The basic equations of the process are:

at the cathode: $H_2O + e \rightarrow 1/2\ H_2 + OH^-$,

at the anode: $Cl^- \rightarrow 1/2\ Cl_2 + e$.

There are three basic methods:

The *diaphragm method* (Fig. 1) was developed by A. Breuer in 1884 and significantly improved by Billiter. The electrolysis cells are steel-reinforced concrete vats which are sealed gas-tight by their covers. They contain many graphite or magnetite anodes, and the cathode is a screen of iron or copper wire, or several bundles of wire net. As the graphite electrodes are oxidized, the distance between the electrodes increases, and therefore, so does the voltage drop in the electrolytes. The graphite anodes must therefore be changed frequently. In modern installations, dimensionally stable anodes made of titanium coated with titanium dioxide and platinum oxides are used instead. This increases the operation time of the cell several-fold; the specific consumption of electricity can be reduced and the volume-time yield is increased. The cathodes are either horizontal, as in the Billiter, Wellrost and Pestalozza cells, or vertical, as in the Bitterfelder, Griesheimer, Gauss, Gibbs-

Vorce, Nelson, Hooker and Krebs cells. Cells with vertical cathodes have larger electrode surfaces, and permit a higher current densities in the same volume, so they have higher capacities than cells with horizontal cathodes. The anode and cathode spaces are separated by diaphragms which prevent mixing of the anode and cathode solutions, but are porous and permit current to flow. In cells with horizontal cathodes, the diaphragm consists of asbestos paper, covered with a paste of barite and asbestos fibers. This lies on top of the wire net. Vertical cathodes are coated with a layer of asbestos fibers. The asbestos diaphragms change their properties during operation of the cell, and are gradually consumed. Therefore, dimensionally stable diaphragms are used in addition to dimensionally stable anodes; the diaphragms consist of a mixture of asbestos and fluorine-containing plastic fibers which are applied to a metal net cathode and then tempered.

The sodium chloride solution flows into the cell from above. Electric current flow generates chlorine at the anodes, and hydrogen is released at the cathode. The sodium ions remain in solution, forming sodium hydroxide. The sodium hydroxide solution flows out the bottom, and the hydrogen is drawn off at the side; the chlorine escapes through pipes in the lid of the vat.

The cells are usually operated at temperatures between 50 and 90 °C, and at currents of 25 to 150 kA; the vat voltages are 3.3 to 4.5 V, and the current yields are 93-97%. The energy expenditure is about 3.0 kWh per kg chlorine. The chlorine obtained by the diaphragm method is especially pure, but the sodium hydroxide solution is only 12 to 16% and still contains about 100 g NaCl per liter. Diaphragm cells with bipolar electrodes (e.g. Glanor cells), which permit a current as high as 800 kA, are especially efficient.

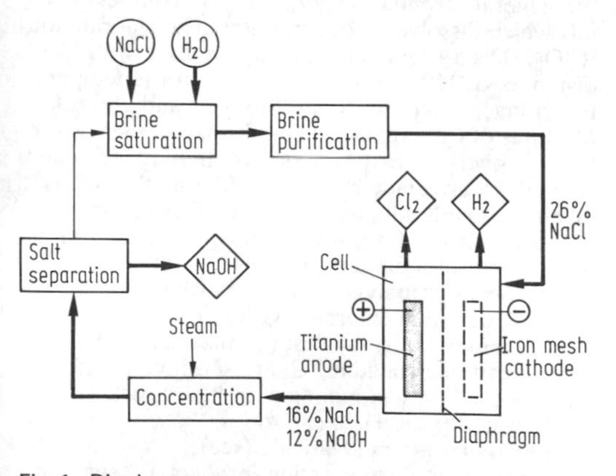

Fig. 1. Diaphragm process

2) The *amalgam* or *mercury process* (Fig. 2) was developed in 1892 by Castner and Kellner. Various types of cell are used. Concrete cells can handle up to about 9 kA current, while carrier cells, which consist of a double T carrier with gum-coated interior walls, and sometimes a gum-coated floor, can handle up to 20 kA. Frame cells, in which gummed, U-shaped frames are screwed into an iron floor, can carry up to 200 kA. Vertical cells have not been successful. The cells are sealed gas-tight with covers. The amalgam process is carried out in two cells, the actual electrolysis cell (primary cell) and the decomposition cell (secondary cell, pile).

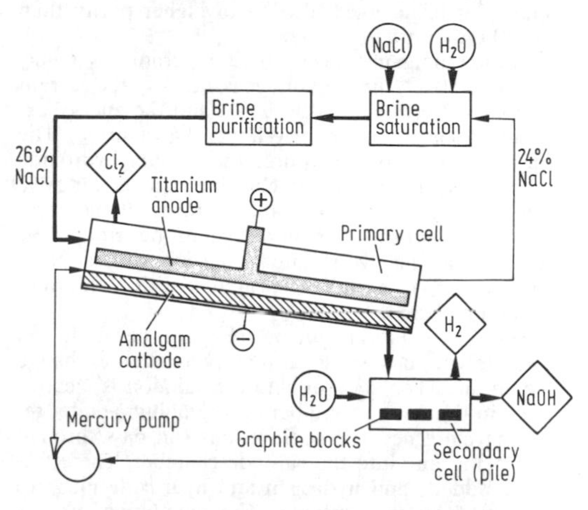

Fig. 2. Amalgam process

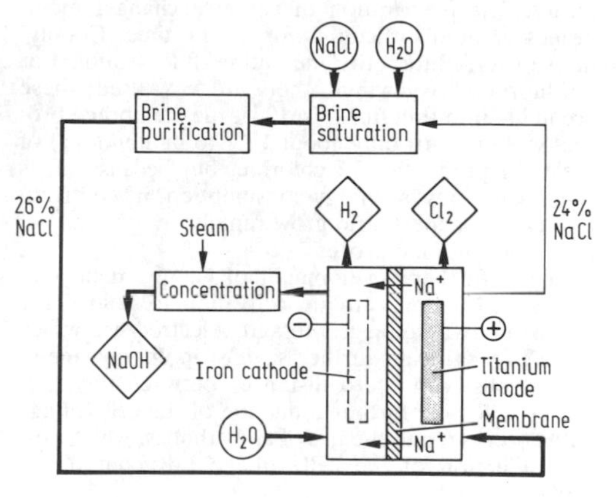

Fig. 3. Membrane process

A stream of mercury pumped through the primary cell acts as the cathode; graphite plates above the metal act as anodes. The sodium ions are discharged on the mercury cathode and form a sodium amalgam with it even before they can react with water to sodium hydroxide. Chlorine is developed at the anode and leaves the cell with the brine. In contrast to the diaphragm method, the brine is circulated, that is, it does not lose its entire salt content in one pass through the cell. The amalgam is transported to the pile, where it is separated into sodium hydroxide, hydrogen and mercury by reaction with water. The mercury is returned to the electrolysis cell. The concentration of the sodium hydroxide solution depends on the amount of water added, and any desired concentration up to 75% can be obtained. With dimensionally stabilized metal anodes, very high current densities can be tolerated; in modern installations, the distance between electrodes is controlled by a computer. A serious problem with the method is that the loss of mercury is still very high, more than 100 g per ton of chlorine; this creates a serious environmental load. The brine must also be of higher purity than with the diaphragm method.

In the amalgam process, the temperature is generally 70 to 80 °C, the vat voltage is 4 to 5 V, the current yields are between 94 and 95%, and the energy expenditure is about 3.5 kWh per kg chlorine. The chlorine is 98 to 99% pure, the sodium hydroxide solution is nearly free of chlorine, and about 50% NaOH. The hydrogen contains traces of mercury, and for certain uses, the mercury must be removed by cooling the gas with water, freezing with a brine (simultaneous drying) or sprinkling with a chloride-containing brine.

3) In the *membrane process* (Fig. 3), the anode and cathode volumes are separated by a cation exchange membrane. The sodium chloride solution is electrolysed in the anode chamber on a titanium anode, releasing chlorine. The sodium ions can pass through the membrane into the cathode chamber. There water is added, and hydrogen and hydroxide ions are produced on the cathode. The membrane process yields a 10 to 33% sodium hydroxide solution of high purity. The preparation of cation exchanger membranes created difficulties for a long time. Copolymers of tetrafluoroethylene and perfluorosulfonyl or perfluorocarboxyl vinyl ethers are now used; these are made into thin films. In 1985, the membrane process was used for only about 10% (3 million tons) of the total production of chlorine, but because of its significantly lower energy consumption, its contribution can be expected to grow rapidly.

Fig. 3. Membrane process.

In the further development of C., the main emphasis will be on cathode activation to reduce the hydrogen overvoltage, pressure electrolysis, which leads to significant further savings in energy, membrane cells with "zero distance" between the electrodes, allowing further reduction of the cell voltage (SPE process) and the use of air cathodes, which permit reduction of the cell voltage by about 0.8 V (20%).

Chlorambucin: see Mustard gas derivatives.

Chloramine T., ***sodium tosylchloramide***: a disinfectant synthesized from 4-toluenesulfonic acid via 4-toluenesulfonyl chloride and 4-toluenesulfonamide by reaction with sodium hypochlorite. The Cl atom in C. is in the +1 oxidation state. Its disinfectant effect is due to the formation of hypochlorite or chlorine.

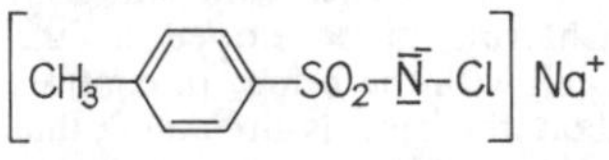

$$[CH_3-C_6H_4-SO_2-\bar{\underline{N}}-Cl]\ Na^+$$

Chloramphenicol, ***chloromycetin***: a broad-spectrum antibiotic synthesized by *Streptomyces venezuelae*. Only the D-*threo*-form with the 1R,2R configuration is biologically active. C. is a white, crystalline powder which is moderately soluble in water; m.p. 150 °C, $[\alpha]_D^{25}$ - 25.5° (in ethyl acetate), +19° (in ethanol). C. is now synthesized chemically.

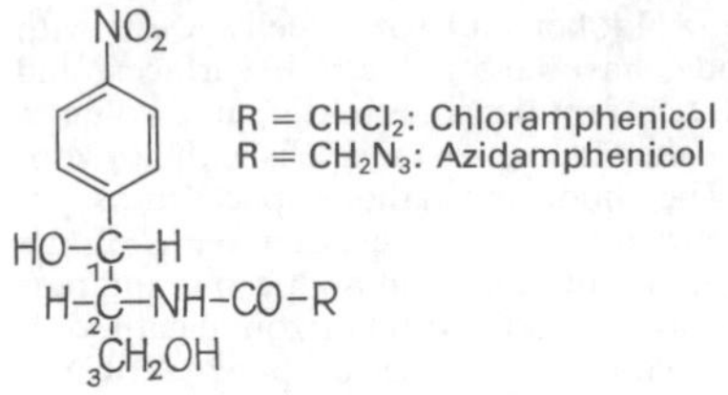

$NO_2-C_6H_4-$ (1) $HO-C-H$; (2) $H-C-NH-CO-R$; (3) CH_2OH

R = $CHCl_2$: Chloramphenicol
R = CH_2N_3: Azidamphenicol

Related compounds in which the primary OH group is esterified are also used therapeutically, for example the insoluble chloramphenicol palmitate and chloramphenicol hemisuccinate, which forms water-soluble salts. ***Azidamphenicol***, which is somewhat more soluble in water than C., is used in eye drops.

Chloranil, ***tetrachloro-1,4-benzoquinone***: golden yellow leaflets with an unpleasant, clinging odor; m.p. 290 °C (in a sealed tube), subl.p. 80 °C. C. is only slightly soluble in water, but dissolves readily in ether. It is formed by heating 1,4-benzoquinone with hydrochloric acid and potassium chlorate, or by the reaction of chlorine and phenol in anhydrous sulfuric acid. C. is used as a dehydrogenating reagent in the dye industry and in preparative chemistry; it is also used as a fungicide on plants.

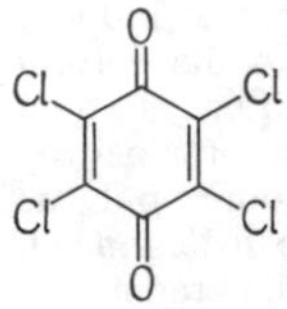

[Structure: 2,3,5,6-tetrachloro-1,4-benzoquinone, O=C with four Cl]

Chloranocryl: see Acylaniline herbicides.

Chlorates: the salts of chloric acid, $HClO_3$, with the general formula M^IClO_3. C. form colorless crystals which dissolve readily in water. The chlorate ion $[ClO_3]^-$ has a trigonal-pyramidal structure with Cl-O distances of 157 pm. C. decompose in the temperature range of 300-400 °C into oxygen and chlorides. Mixtures of C. and oxidizable substances, such as sulfur or phosphorus, explode on impact or when rubbed. C. are made by passing chlorine through hot alkali hydroxide solutions; direct synthesis by chloralkali electrolysis of a warm alkali chloride solution in the absence of a diaphragm is also possible.

Chlorate explosives: explosives which contain 85 to 90% potassium chlorate, $KClO_3$, with added non-viscous mineral oils, sawdust or other materials, such as chloratite or cheddite. C. are sensitive to friction and are used only in mining potash and rock salts.

Chlorazide: see Halogen azides.

Chlorbenzide: an Acaricide (see).

Chlorbuna: a chlorination product of synthetic Rubber (see).

Chlordane: a Cyclodiene insecticide (see).

Chlordimeform: an Acaricide (see).

Chlorfenprop: see Herbicides

Chlorfenson: an Acaricide (see).

Chlorfenvinphos: an Organophosphate insecticide (see).

Chloric acid, ***chloric(V) acid***: $HClO_3$, a strong acid (p*K* - 2.7) which is stable only in aqueous solution; its salts are the Chlorates (see). C. is a strong oxidizing reagent made by reaction of barium chlorate with dilute sulfuric acid: $Ba(ClO_3)_2 + H_2SO_4 \rightarrow 2\ HClO_3 + BaSO_4$. The resulting colorless solution can be concentrated in vacuum to an acid content of about 40%. C. is used in the laboratory as a strong oxidizing agent, and as such, it can be used in a mixture with fuming hydrochloric acid (see Euchlorin).

Chloridazon: see Herbicides.

Chlorides: compounds of chlorine in which Cl is the more electronegative bond partner. The C. include ionic, covalent or coordination polymeric metal C., covalent metal dichlorides and organic C., such as alkyl, aryl or acyl chlorides. Most electropositive metals, such as the alkali and alkaline earth metals, form ionic C., such as sodium and magnesium chlorides, NaCl and $MgCl_2$. Other main group metals, such as aluminum, germanium and tin, form covalent, readily hydrolysed C. such as aluminum(III) chloride, $AlCl_3$, germanium(IV) chloride, $GeCl_4$ or tin(IV) chloride, $SnCl_4$. Transition metals, especially in the +2 and +3 oxidation states, form coordination polymeric C., such as iron(II) chloride, $FeCl_2$, cobalt(II) chloride, $CoCl_2$ and chromium(III) chloride, $CrCl_3$. In the higher oxidation states, they form covalent, hydrolysable C., such as titanium(IV) chloride, $TiCl_4$, niobium(V) chloride, $NbCl_5$, or tungsten(VI) chloride, WCl_6. Some examples of covalent, volatile metalloid chlorides are disulfur dichloride, S_2Cl_2, phosphorus(III) chloride, PCl_3 and boron trichloride, BCl_3. Ionic C. are found in nature in huge salt deposits, e.g. rock salt, NaCl, or sylvin, KCl.

The C. are produced by the following methods: chlorination of the elements with chlorine at elevated temperature; chlorination of metal oxides at high temperatures with chlorination reagents such as disulfur dichloride or carbon tetrachloride; reductive chlorination of metal oxides with chlorine in the presence of carbon; reaction of metal oxides, hydroxides or carbonates with hydrochloric acid; reaction of reactive metals with hydrochloric acid; and chlorination of hydrocarbons. The C. are important starting materials, intermediates and products of large-scale chemical industry.

Chlorinated hydrocarbon insecticides: most insecticides in this group are characterized by: 1) high insecticidal effectiveness; 2) low acute toxicity for warm-blooded animals; 3) broad spectrum of action; 4) simple production and application; 5) favorable economics and 6) long persistence in the field. C. have had a leading role in the recent history of chemical crop protection. The leading representative is DDT (see), which introduced the era of modern synthetic contact insecticides. However, the long-term effectiveness of these compounds, which is an advantage in application, is also coupled to a high persistence in the environment. Since most C.h.i. are very slowly degraded by chemical, physical or microbial pathways, they have been forbidden or subjected to severe restrictions in many industrialized countries.

The C.h.i. include DDT and structurally related compounds, lindane (γ-HCH) and its homologs, chloroterpene and cyclodiene insecticides, and HCCP dimers or their products.

Chlorination: see Halogenation; see Disinfection.

The symptoms of C.h.i. poisoning can include restlessness, irritability, headaches and impairment of equilibrium and speech. Intake of higher doses may cause vomiting, diarrhea, gastro-intestinal cramps, delierium and loss of consciousness. Therapy. Induce vomiting, rinse stomach with medicinal charcoal added to a saline laxative. Under no circumstances should milk or castor oil be used! Paraffin oil (100 to 200 ml orally) is recommended to trap the poison in non-absorbable form. Short-term hypnotica for sedation! Intravenous administration of calciumthiosulfate (10 ml), repeated if necessary. Liver-protecting therapy! Infusion of isotonic NaCl solution. In case of unconsciousness, oxygen respiration! Caution! No adranalin or ephedrin preparations due to danger of ventricular fibrillation!

Chlorin: see Porphyrins.

Chlorin®: see Synthetic fibers.

Chlorine, *Cl*: an element from group VIIa of the periodic system, the Halogen group (see). Cl is a nonmetal, *Z* 17, with natural isotopes with mass numbers 35 (75.529%) and 37 (24.47%); atomic mass 35.453, valence - 1, +1, +3 to +7; density (at the b.p.) 1.565, m.p. - 101.00 °C, b.p. - 34.06 °C, crit. temp. 143.5 °C, crit. pressure 7.61 MPa, crit. density 0.57, standard electrode potential (Cl^-/Cl_2) +1.3595 V.

Properties. C. is a yellow-green, highly reactive and very poisonous gas with a suffocating odor. It exists as Cl_2 molecules; at - 34 °C, or at room temperature and a pressure of 0.66 MPa (20 °C), it condenses to a yellow liquid. Solid C. has an orthorhombic molecular lattice. It is readily soluble in water: 1 l water at 20 °C dissolves 2.3 l C. An approximately 0.1 M solution is called ***chlorine water***. When a saturated, aqueous solution of chlorine is cooled, hydrate, $Cl_2 \cdot 7.3H_2O$, crystallizes out. Chlorine water should be stored in brown bottles, because in the sunlight it rapidly decomposes to hydrochloric acid and oxygen, via hypochlorous acid: $Cl_2 + H_2O \rightleftharpoons HCl + HOCl$; $HOCl \rightarrow HCl + 1/2O_2$. The nascent oxygen is especially reactive, and this is the basis for use of moist Cl as a bleach. Cl combines directly with all elements except carbon, oxygen, nitrogen and the noble gases. Reactive metals, especially the alkali and alkaline earth metals, release a large amount of heat as they combine with Cl, and often develop flames. Even relatively unreactive metals such as bismuth, arsenic or copper, when finely divided, react with C. at room temperature or slightly above. For example, sodium heated to 100 °C burns in a stream of C. with an intense yellow light, forming sodium chloride. The noble metals and those metals which are easily passivated, such as chromium, tantalum or tungsten, are chlorinated only when heated to red heat, especially when they are in a compact state. At room temperature, C. reacts at least as vigorously with nonmetals, such as phosphorus, sulfur, iodine and hydrogen, as it does with the alkali metals, and at higher temperatures, the reactions are often rather violent. For example, an equimolar mixture of C. and hydrogen can be stored unchanged in the dark at room temperature, although it explodes in direct sunlight or upon local heating (see Chlorine detonating gas). Moist C. is much more reactive than dry. For example, dry C. attacks iron only above 270 °C, and thus it can be conducted through steel pipes or stored in steel bottles. Removal of tin from plated steel is also possible because dry C. reacts only with the tin, forming liquid tin(IV) chloride, but does not attack the iron. Because of its affinity for hydrogen, C. reacts with many hydrocarbon compounds, water, hydrogen sulfide, hydrogen bromide or hydrogen iodide, forming HCl. The reactions of C. with more complicated organic compounds are mainly oxidations and substitutions.

Chlorine is a lung poison which burns and inflames the mucous membranes of the respiratory organs. Severe poisonings lead to convulsive coughing and a feeling of suffocation, and later to respiratory distress, lung inflamation and bleeding. A concentration of 2.5 mg/l is immediately fatal, and a concentration of 0.15 mg/l is lethal after a longer period of inhalation.

Antidotes in chlorine poisonings are inhalation of steam, hydrogen carbonate solution or dilute ammonia. In cases of respiratory distress, oxygen should be administered or - in cases of extreme emergency only - artificial respiration should be attempted.

Analysis. Free C. can be recognized from its odor and, in high concentrations, from its yellow-green color. It colors potassium iodide/starch paper blue. Chloride ions are determined qualitatively and quantitatively by the reaction with silver nitrate, which produces an insoluble, white, curdy precipitate of AgCl. This is insoluble in nitric acid, but dissolves in conc. ammonia solution, forming the complex $[Ag(NH_3)_2]Cl$. The presence of bromide, iodide or pseudohalides interferes with this test, however. Another method is conversion to the volatile chromyl chloride, CrO_2Cl_2; however, fluoride interferes with this test.

Occurrence. C. makes up 0.19% of the earth's crust; it is the 12th most abundant element. In nature, it is always found in compounds, most frequently in the form of alkali and alkaline earth chlorides, in salt deposits and in seawater. The most important C. minerals are rock salt, NaCl, silvin, KCl, carnallite, $KCl \cdot MgCl_2 \cdot 6H_2O$, mainite, $KCl \cdot MgSO_4 \cdot 3H_2O$ and bischofite, $MgCl_2 \cdot 6H_2O$. Some heavy-metal minerals also contain C.: chlorargyrite, AgCl, atakamite, $Cu_2(OH)_3Cl$ and pyrmorphite, $Pb_5Cl \cdot (PO_4)_3$, for example. The human organism contains about 0.12% Cl, mainly in the form of sodium chloride, and also as hydrochloric acid in gastric juice. Cl is a micronutrient contained in soil at about 100 ppm, and in plants at about 0.01%.

Production. C. is usually produced in the laboratory by treating manganese(IV) oxide with conc. hydrochloric acid or with a mixture of sodium chloride or moderately conc. sulfuric acid. Manganese(IV) chloride is formed as an intermediate: $MnO_2 + 4\,HCl \rightarrow MnCl_4 + 2\,H_2O$, and rapidly decomposes to manganese(II) chloride and C.: $MnCl_4 \rightarrow MnCl_2 + Cl_2$. C. is produced on a large scale industrially, mainly by Chloralkali electrolysis (see). The world production in 1980 was more than 30 million tons. The C. from electrolysis is still moist and therefore aggressive; it is cooled with water and dried with sulfuric acid in protected towers. It is then purified in precipitators or filters. C. must be liquefied before transport to distant consumers; this is done at low temperature (- 40 to - 50 °C and atmospheric pressure), high pressure (0.8 to 1.3 MPa; liquefaction temperature about 25 °C at 0.8 MPa) or 0.3-0.4 MPa and - 20 °C. The amount of hydrogen in the gas may not exceed 8%, as the mixture is explosive above this limit.

Before the chloralkali electrolysis process was developed, the *Deacon process* was used to produce C. industrially; it is still used to some extent in a modified form (*Shell-Deacon process*). In the Deacon process, hydrogen chloride is oxidized to chlorine with air in the presence of copper(II) chloride:

$$4\,HCl + O_2 \xrightleftharpoons{CuCl_2} 2\,Cl_2 + 2\,H_2O$$

The *Weldon process*, which used to be common, now has no economic significance. In it, as in the laboratory process, hydrochloric acid is oxidized with manganese(IV) oxide.

Applications. The demand for C. is very large and growing. The main consumer is large-scale chemical industry, which uses it in the synthesis of inorganic and especially organic chlorine compounds (see Halogenization). Some important inorganic chlorine compounds produced on industrial scale are hypochlorite, chlorate, chlorine dioxide, disulfur dichloride, phosphorus chloride, silicon tetrachloride and germanium tetrachloride. The most common organic products include chlorine-containing solvents, such as carbon tetrachloride, chloroform and methylene chloride, and chlorinated plastics, such as polyvinyl chloride and chlorinated rubber. In addition, chlorinated pesticides and intermediates for the dye industry are of great importance. C. is used for chlorinating solubilization of low-quality ores, to remove tin from tin plate wastes, to disinfect ground water and swimming pools, and to treat sewage.

Historical. Free C. was first produced in 1774 by Scheele, who oxidized hydrochloric acid with manganese dioxide. Scheele called C. "dephlogisticated hydrochloric acid". Since the aqueous solution develops oxygen when exposed to sunlight, C. was thought to be an oxygen-containing compound. Only when all atempts to split oxygen out of C. had failed did Davy recognize the elemental nature of the gas, in 1810.

Chlorine bleaching: oxidative removal of colored impurities from textiles by sodium hypochlorite, or more rarely, by organic chlorine carriers, such as sodium dichloroisocyanurate. See Bleaches.

Chlorine detonating gas: an equimolar mixture of chlorine and hydrogen which, when irradiated with UV light or heated, reacts explosively to form hydrogen chloride, HCl. The chain reaction which occurs here is initiated by the cleavage of Cl_2 molecules: $Cl_2 \rightarrow 2\,Cl$ ($\Delta H = +\,243.3$ kJ). Other steps in the reaction are: $Cl + H_2 \rightarrow HCl + H$ ($\Delta H = -\,3.97$ kJ); $H + Cl_2 \rightarrow HCl + Cl$ ($\Delta H = -\,188.6$ kJ). The overall reaction is exothermal: $H_2 + Cl_2 \rightarrow 2\,HCl$; $\Delta H = -\,184.7$ kJ. Therefore the reaction temperature increases rapidly, and this causes the explosion. The temperature of a chlorine detonating gas flame is about 2200 °C.

Chlorine fluorides: ***Chlorine monofluoride***, ClF, a nearly colorless, extremely aggressive gas; K_r 54.45, m.p. - 155.6 °C, b.p. - 101.1 °C. ClF reacts so vigorously with many metals and with organic substances that flames appear. It attacks glass, forming explosive chlorine oxides. It is made by combining chlorine and fluorine at 250 °C. ***Chlorine trifluoride***, ClF_3, is a colorless gas which can easily be condensed to a bright green liquid; M_r 92.45, m.p. - 76.32 °C, b.p. 11.75 °C; the molecules are T-shaped. ClF_3 is synthesized by reaction of the elements at 200 to 300 °C, and is purified by reaction with potassium fluoride to form $KClF_4$, followed by thermal decomposition of this salt at 130-150 °C. The compound is extremely reactive and is commercially available; it is used as a fluorination reagent and as an additive to welding gases to increase their combustion temperature. ***Chlorine pentafluoride*** ClF_5, is a gas; its molecules have a quadratic pyramidal structure; M_r 130.44, m.p. - 103 °C, b.p. - 13.1 °C. ClF_5 is even more reactive than ClF_3. Above 165 °C, it is in equilibrium with ClF_3 and F_2: $ClF_5 \rightleftharpoons ClF_3 + F_2$. It is obtained by reaction of F_2 with ClF_3 above 200 °C, or by reaction of KCl and fluorine at 200 °C under pressure: $KCl + 3\,F_2 \rightarrow KF + ClF_5$.

Chlorine oxides: ***Dichlorine oxide***, Cl_2O, is a yellowish red, heavy gas with an unpleasant odor; M_r 86.90, m.p. - 120.6 °C, b.p. 2.0 °C. The Cl_2O molecule is bent (bond angle 111.2°), and the Cl-O distance is 169.3 pm. Cl_2O is the anhydride of hypochlorous acid. It is made by passing chlorine over mercury(II) oxide at 0 °C: $2\,Cl_2 + HgO \rightarrow Cl_2O + HgCl_2$. When it is heated slightly, or comes into contact with combustible substances, it explodes, decomposing to the elements.

> Dichlorine oxide attacks the respiratory organs. Countermeasures: see Chlorine.

Chlorine dioxide, chlorine(IV) oxide, ClO_2, is a greenish-yellow, extremely explosive gas with a penetrating odor; M_r 67.45, m.p. - 59 °C, b.p. 9.7 °C. The ClO_2 molecule has a bond angle of 117.7°, and the Cl-O distances are 147.5 pm. ClO_2 is one of the few molecules with an odd number of electrons which has no tendency to dimerize. Aqueous solutions of ClO_2 can be kept in the dark for long periods. However, the compound gradually reacts with water according to $2\,ClO_2 + H_2O \rightarrow HClO_2 + HClO_3$. ClO_2 explodes when heated even slightly, in the presence of oxidizable substances or upon impact. It is therefore stored in the cold in the presence of inert gas. ClO_2 is obtained by the reaction of dilute sulfuric acid with potassium chlorate in the presence of oxalic acid: $2\,HClO_3 + H_2C_2O_4 \rightarrow 2\,ClO_2 + 2\,CO_2 + 2\,H_2O$. In the industrial synthesis, sodium chlorate is reduced with sulfurous acid: $2\,NaClO_3 + H_2SO_3 \rightarrow 2\,ClO_2 + Na_2SO_4 + H_2O$. ClO_2 is used to make sodium chlorite and, stabilized as the pyridine adduct $ClO_2 \cdot C_5H_5N$, it is used as a bleach and disinfectant.

Dichlorine hexoxide, chlorine(VI) oxide, Cl_2O_6, is a dark red liquid which fumes in air; M_r 166.90, density 1.65, m.p. 3.5 °C, b.p. 203 °C (extrapolated). Solid Cl_2O_6 has the structure $[ClO_2]^+[ClO_4]^-$. The compound, which is the mixed oxide of chloric and perchloric acids, is diamagnetic and a strong oxidizing agent. Pure Cl_2O_6 is rather stable at room temperature, but it explodes violently upon contact with organic compounds. It is made by the reaction of ozone and chlorine dioxide diluted with CO_2 at 0 °C: $2\,ClO_2 + 2\,O_3 \rightarrow Cl_2O_6 + 2\,O_2$.

Dichlorine heptoxide, chlorine(VII) oxide, Cl_2O_7, is a colorless, volatile, oily liquid; M_r 182.90, density (0 °C) 1.86, m.p. - 91.5 °C, b.p. 81.5 °C. Cl_2O_7 has the structure $O_3ClOClO_3$; the terminal Cl-O distances are 140.5 pm and the bridge Cl-O distances, 170.9 pm; the Cl-O-Cl angle is 118.6°. As the anhydride of perchloric acid, Cl_2O_7 reacts with cold water: $Cl_2O_7 + H_2O \rightarrow 2\,HClO_4$. It explodes very violently upon contact with a flame or upon impact. Cl_2O_7 can be made by dehydration of perchloric acid with phosphorus pentoxide and subsequent cautious vacuum distillation.

> All chlorine oxides are highly endothermal compounds and explode under certain conditions, e.g. upon heating, pouring or on contact with organic substances.

Chlorites: the salts of chlorous acid, $HClO_2$, with the general formula $MClO_2$. The chlorite ion, $[ClO_2]^-$, is an angular anion with Cl-O distances of 164 pm. C. are very unstable; $NaClO_2$ decomposes violently when heated above 200 °C; other C. explode when heated or upon impact. C. also explode in mixtures with combustible substances; heavy metal C. are especially likely to explode. C. undergo partial hydrolysis in aqueous solution: $ClO^{+\,-} + H_2O \rightleftharpoons HClO_2 + OH^-$. Addition of acid leads to formation of chlorine dioxide: $5\,ClO_2^- + 5\,H^+ \rightarrow 4\,ClO_2 + HCl + 2\,H_2O$. This is the basis for use of Sodium chlorite (see).

Chlormadinone: see Gestagens.

Chloroacetanilide herbicides: see Acylaniline herbicides.

Chloroacetic acid, ***monochloroacetic acid***: $Cl\text{-}CH_2\text{-}COOH$, colorless, hygroscopic, prismatic monoclinic crystals. C. can exist in three different modifications: α-C., m.p. 63 °C; β-C., m.p. 56.2 °C; γ-C., m.p. 52.5 °C. The existence of a δ-modification has been reported but not confirmed. C. boils at 187.9 °C. It is readily soluble in water, ethanol, ether and acetone, but only barely soluble in hydrocarbons. The aqueous solutions of C. are very acidic. C. undergoes the reactions typical of Carboxylic acids (see). In addition, the chlorine atom can readily undergo nucleophilic substitution. C. is most often synthesized by chlorination of acetic acid in the presence of acetic anhydride and sulfuric acid, with phosphorus or sulfur catalysts, or by hydrolysis of trichloroethylene with approximately 75% sulfuric acid:

$$CHCl{=}CCl_2 + 2\,H_2O \xrightarrow{(H^+)} Cl{-}CH_2{-}COOH + 2\,HCl.$$

Because it is bifunctional, C. is used for the synthesis of many chain and ring compounds. It is used to make carboxymethylcellulose, glycine, indigo, glycolic acid, thioglycolic acid, barbiturates and vitamin B_6. It is also used as a caustic to remove corns and warts.

Chloroacetone, ***chloropropanone***: Cl-CH_2-CO-CH_3, a colorless, lacrimatory liquid with an unpleasant odor; m.p. - 44.3 °C, b.p. 119 °C. C. is soluble in water, alcohol, ether and chloroform. It is synthesized by chlorination of acetone in the presence of water. C. was used as a chemical weapon in the First World War. At present, it is used as a starting material for numerous syntheses in preparative chemistry.

o-Chloroacetophenone, ***phenacyl chloride***: C_6H_5-CO-CH_2-Cl, a stable, colorless, crystalline compound; m.p. 56.5 °C, b.p. 247 °C. o-C. is insoluble in water, but dissolves easily in alcohol, ether, acetone or benzene. It is a Chemical weapon (see) and an eye irritant. Since it is stable to detonation and high temperatures, it is suitable for use in grenades and smoke bombs. It can also be used as an aerosol.

Chloroacetyl chloride, ***monochloroacetyl chloride***: Cl-CH_2-CO-Cl, a colorless, very caustic liquid with a pungent odor; b.p. 107 °C, n_D^{20} 1.4541. C. is soluble in most organic solvents and is hydrolysed by water. It is made by reaction of chloroacetic acid with phosphorus(III) chloride, thionyl chloride or phosgene, or by chlorination of acetyl chloride. Because of its high reactivity, it is suitable for many syntheses, for example, that of adrenalin. It is often used instead of the less reactive chloroacetic acid.

Chloroaniline: an aniline with a chlorine substituent on the aromatic ring. There are three isomeric monochloroanilines, ***2-*** or ***o-chloroaniline*** (m.p. - 14 °C, b.p. 208.8 °C), ***3-*** or ***m-chloroaniline*** (m.p. - 10.3 °C, b.p. 229.9 °C) and ***4-*** or ***p-chloroaniline*** (m.p. 72.5 °C, b.p. 232 °C). There are also various dichloroanilines, e.g. ***2,3-dichloroaniline*** (m.p. 24 °C, b.p. 252 °C) and ***2,4-dichloroaniline*** (m.p. 63-64 °C, b.p. 245 °C), and a few trichloroanilines which are of industrial importance. Depending on the position of the halogen atoms, the C. are synthesized by chlorination of acetanilide and subsequent removal of the acetyl group, or by catalytic reduction of the corresponding chloronitrobenzenes. The C. are important intermediates in the synthesis of pigments, pesticides and drugs.

Chloroazide: see Halogen azides.

Chlorobenzene: C_6H_5-Cl, the simplest aromatic chlorohydrocarbon. C. is a colorless liquid with a pleasant odor; m.p. - 45.6 °C, b.p. 132 °C, n_D^{20} 1.5241. It is volatile with steam. It dissolves only slightly in water, but easily in alcohol, ether, benzene and chloroform. As in all nonactivated Halogen arenes (see), the chlorine atom in C. is relatively strongly bound to the aromatic ring. It can be replaced by other substituents only under special reaction conditions. The substitution of a hydroxyl group by reaction with aqueous alkali hydroxide solution does not occur below 250 °C, and produces phenol. The exchange of chlorine for ammonia in the presence of copper(I) oxide requires a temperature around 200 °C. C. is synthesized by chlorination of benzene in the presence of catalysts at 40 to 50 °C in the liquid phase, a reaction which also produces more highly chlorinated benzene derivatives. C. can be separated from the mixture by fractional distillation. It is used as a solvent for fats, resins and rubber, and as an intermediate in the synthesis of phenol, pesticides, pigments, pharmaceuticals, aromas and textile conditioners.

Chlorobenzilate: an Acaricide (see).

Chlorobutanol: CCl_3-C$(CH_3)_2$OH, a halogenated tertiary alcohol which is used as a preservative due to its disinfectant effect. C. is also a sedative and hypnotic.

Chlorocholine chloride, abb. ***CCC***: 2-chloroethyltrimethylammonium chloride, Cl-CH_2CH_2N$(CH_3)_3$Cl, a synthetic plant growth regulator used widely to stabilize the straw in winter wheat. Use of C. reduces lodging of the grain, particularly when it has been heavily fertilized. C. is a colorless, crystalline compound with m.p. 239-243 °C.

2-Chloro-1,3-diene: same as Chloroprene (see).

Chlorodifluoromethane: $HCCClF_2$, a colorless, non-combustible gas with a sweetish odor; m.p. - 160 °C, b.p. - 40.8 °C. C. is only slightly soluble in water, but is readily soluble in organic solvents. It is made by reaction of chloroform with hydrogen fluoride in the presence of antimony(V) chloride. C. is used to make Tetrafluoroethene (see), as a propellant for aerosols, and as a refrigerant (R 22).

1-Chloro-2,2-dimethylpropane: see Amyl chlorides.

1-Chloro-2,4-dinitrobenzene: one of six possible isomeric chlorodinitrobenzenes. 1-C. is a yellow compound which irritates the skin; it exists in one stable (α-form) and two unstable modifications; m.p. 53 °C (stable), 43 °C and 27 °C (unstable); b.p. 315 °C (dec.). It is insoluble in water, but dissolves readily in ether, benzene and hot alcohol. The Cl atom of 1-C. can readily be exchanged for nucleophilic subsubstituents, such as HS-, RO-, NH_2- or HO-, due to the electron-withdrawing effect of the nitro group. The reaction of 1-C. with pyridine derivatives is nearly quantitative, forming pyridine hydrochlorides, and is thus used for quantitative analysis of pyridines. 1-C. is made by nitration of chlorobenzene. It is important as an intermediate in the synthesis of pigments and explosives.

1-Chloro-2,3-epoxipropane: same as Epichlorohydrin (see).

Chloroethane: same as Ethyl chloride (see).

2-Chloroethanol: same as Ethylene chlorohydrin (see).

Chloroethene: same as Vinyl chloride (see).

Chlorofluorohydrocarbons: see Fluorohydrocarbons.

Chloroform, ***trichloromethane***: $CHCl_3$, a colorless, volatile liquid with a sweetish odor; m.p. - 63.5 °C, b.p. 61.7 °C, n_D^{20} 1.4459. C. is only slightly soluble in water, but dissolves readily in most organic solvents. Its vapors have a narcotic effect. In liquid form, C. is not combustible, but its vapors burn in an open flame, forming phosgene. In the presence of oxygen and light, C. is decomposed to phosgene and hydrogen chloride. It is therefore stored in brown bottles and stabilized with alcohol to remove any phosgene which forms. The halogen atoms in C. can be replaced by nucleophilic substitution in aqueous hydroxide solutions or by alkali alcoholates, forming formic acid or orthoformate esters. C. can be synthesized by the classic methods, from ethanol and lime chloride, via acetaldehyde and chloral. The last step of this process is the alkaline cleavage of chloral with calcium hydroxide, forming C. and calcium formate: $2\ CCl_3\text{-}CHO + Ca(OH)_2 \rightarrow 2\ CHCl_3 + (HCOO)_2Ca$. Industrially, C. is usually synthesized by chlorination of methane. This produces a mixture of chlorine derivatives which are separated by fractional distillation.

C. was formerly the most important inhaled anesthetic. However, it caused complications in some cases, and damages the liver, and it was therefore later replaced by diethyl ether. C. is very important as a solvent for resins, rubber, lacquers, fats, oils, paraffins, iodine, sulfur and penicillin. It is also an intermediate in the production of chlorofluoromethanes, orthoformic acid esters and other organic products.

Chloroformates, ***esters of chloroformic acid***: Cl-CO-OR. Chloroformic acid cannot be isolated in the free state. C. are lacrimatory liquids with a suffocating odor, which are soluble in most organic solvents. They are solvolysed by protic solvents; in alcohols, for example, they form esters. Some important C. are ***methyl chloroformate*** (b.p. 70-71 °C, n_D^{20} 1.3868), ***phenyl chloroformate*** (b.p. 189 °C, n_D^{20} 1.5160), ***benzyloxycarbonyl chloride*** (b.p. 103 °C at 2.66 kPa, n_D^{20} 1.5150) and ***isobutyl chloroformate***. The last two C. are used in peptide syntheses. C. are converted to urethanes by reaction with ammonia, primary or secondary amines, with loss of hydrogen chloride. Isothiocyanates are formed by the reaction of *N*-substituted dithiocarbamates with C. C. are synthesized by the reaction of phosgene with alcohols or phenols at about 0 °C: Cl-CO-Cl + ROH → Cl-CO-OR + HCl. Because of their high reactivity, C. are used for many special syntheses. In addition, they are used in the production of catalysts and pesticides. Some C. are very toxic, e.g. trichloromethyl chloroformate, which was used in the First World War as a chemical weapon.

Chloroformic acid trichloromethyl ester: same as perchloroformic acid methyl ester.

Chlorogenic acid: 3-caffeoylquinic acid. C. forms colorless crystals; m.p. 207-208 °C, $[\alpha]_D$ - 33 to - 35°. It is found in coffee beans and many temperate-zone fruits and vegetables.

Chlorohemin: see Hemins.

Chlorohemoglobin: see Hemoglobins.

Chlorohexidine: a bis-biguanide, which is usually used in the form of its diacetate as a disinfectant and preservative in pharmaceuticals, such as eye medications.

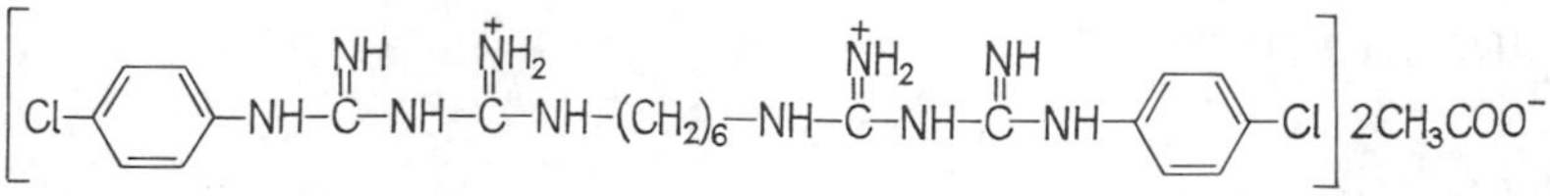

Chloromycetin: same as Chloramphenicol (see).

Chloropentanes: same as Amyl chlorides (see).

Chlorophene: see Disinfectants.

Chlorophenethiazine: see Phenothiazines.

Chlorophenols: I) ***Monochlorophenols*** three isomeric phenol derivatives:

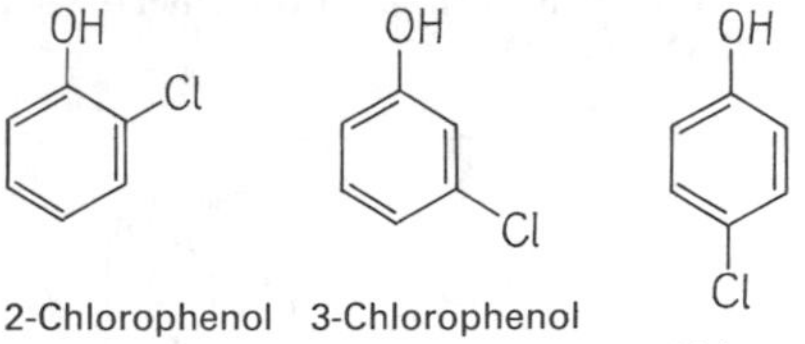

2-Chlorophenol 3-Chlorophenol 4-Chlorophenol

2-Chlorophenol is a colorless liquid with an unpleasant odor like that of iodoform; m.p. 9 °C, b.p. 174.9 °C, n_D^{20} 1.5524. It is formed, along with 4-chlorophenol, by the reaction of phenol with chlorine or sulfuryl chloride. The two isomers are separated by vacuum distillation. In the presence of excess chlorine, 2,4-dichlorophenol and 2,4,6-trichlorophenol are the major products. 2-Chlorophenol is used in the synthesis of pyrocatechol. ***3-chlorophenol*** forms crystalline needles; m.p. 33 °C, b.p. 214 °C, n_D^{40} 1.5565. It is obtained from 3-chloroaniline by diazotization followed by boiling of the diazonium salt. ***4-Chlorophenol*** forms colorless crystalline needles with a characteristic, phenol-like odor; m.p. 43.7 °C, b.p. 219.8 °C, n_D^{55} 1.5419, n_D^{40} 1.5579. It is obtained in relatively high yield by the reaction of phenol with sulfuryl chloride at 25 °C, and can be further processed to yield hydroquinone.

The C. are barely soluble in water, but are readily soluble in ethanol, ether and aqueous alkali hydroxide solutions. They are intermediates in the production of dyes and are used as disinfectants. They are stronger acids and stronger antiseptics than the parent compound, phenol.

II) ***Dichlorophenols***: six isomeric phenol derivatives, of which the readily synthesized 2,4-dichlorophenol is important in the production of the herbicide 2,4-dichlorophenoxyacetic acid (2,4-D). ***2,4-Dichlorophenol*** forms colorless needles; m.p. 45 °C, b.p. 210 °C. It is made by reaction of phenol with excess chlorine or sulfuryl chloride. A byproduct is ***2,6-dichlorophenol*** (m.p. 68-69 °C, b.p. 219 °C). The reaction of dichlorophenols with chloroacetic acid in alkaline solutions is used in the synthesis of herbicides. The C. are also used as moth repellants.

$$\left[Cl\text{-}C_6H_4\text{-}NH\text{-}\overset{NH}{\overset{\|}{C}}\text{-}NH\text{-}\overset{\overset{+}{N}H_2}{\overset{\|}{C}}\text{-}NH\text{-}(CH_2)_6\text{-}NH\text{-}\overset{\overset{+}{N}H_2}{\overset{\|}{C}}\text{-}NH\text{-}\overset{NH}{\overset{\|}{C}}\text{-}NH\text{-}C_6H_4\text{-}Cl\right] 2CH_3COO^-$$

Chlorohexidine diacetate

Chloromethane: same as Methyl chloride (see).

Chloromethylbenzene: same as Benzyl chloride (see).

Chloromethylation: see Blanc reaction.

3-Chloro-2-methylprop-1-ene: same as methyl allyl chloride.

Chlorophyllinite: see Macerals.

Chlorophylls those Mg-porphyrin complexes which act as primary photoreceptors in photosynthesizing plants and bacteria. Commercially available

C. is a mixture of C. a and b, and can be obtained by extraction of dried plant material with acetone. C. is soluble in alcohol and ether, but practically insoluble in water and aliphatic hydrocarbons. ***C. a*** forms blue-green crystalline leaflets, m.p. 117-120 °C (from acetone), λ_{max} (in ether): 660, 613, 577, 531, 498, 429, 409 nm. ***C. b*** forms green needles, m.p. 120 to 130 °C (from chloroform/methanol), λ_{max} (in ether): 642, 593, 565, 545, 453, 427 nm.

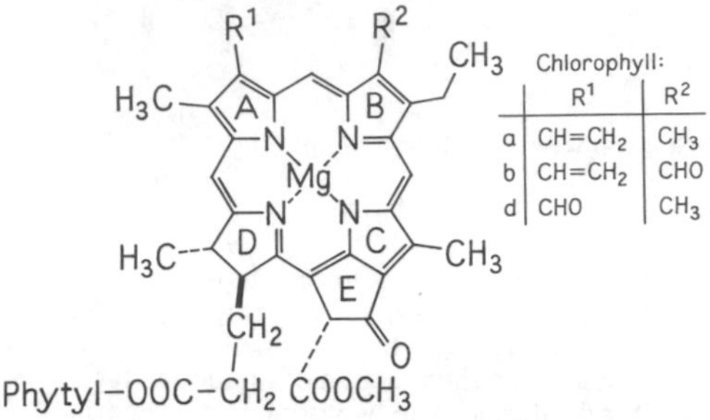

Chlorophyll:

	R^1	R^2
a	$CH=CH_2$	CH_3
b	$CH=CH_2$	CHO
d	CHO	CH_3

Phytyl-OOC-CH_2 $COOCH_3$

The most widely occurring C. is C. a, which is accompanied in higher plants and green algae by C. b. Algae contain other C., including c_1, c_2 and d. Chlorobium bacteria contain ***chlorobium chlorophylls***, and red sulfur bacteria, ***bacteriochlorophyll***. In contrast to other porphyrins, the C. have their acidic side chains esterified with methanol or phytol. All C. contain an additional alicyclic ring E. Careful acid hydrolysis of the C. produces magnesium-free ***pheophytins***, from which the ***pheophorbides*** are obtained by cleavage of the two ester bonds. The C. are optically active. The presence of the phytyl group makes them lipophilic and surface active.

C. play a crucial role in photosynthesis, converting light energy to redox potential.

Chloropicrin: same as Trichloronitromethane (see).

Chloroprene, ***2-chlorobuta-1,3-diene***: $CH_2=CCl-CH=CH_2$, a colorless liquid; m.p. - 130 °C, b.p. 59.4 °C, n_D^{20} 1.4583. C. is nearly insoluble in water, but dissolves easily in most organic solvents. It undergoes the addition reactions typical of alkenes. C. polymerizes very easily, forming high polymers. It is synthesized by addition of hydrogen chloride to vinyl acetone, or from butadiene and chlorine via 3,4-dichlorobut-1-ene, which can be converted into C. by splitting off hydrogen chloride:

$$CH_2{=}CH{-}CH{=}CH_2 \xrightarrow{+Cl_2} CH_2Cl{-}CHCl{-}CH{=}CH_2 \xrightarrow[-HCl]{} CH_2{=}CCl{-}CH{=}CH_2.$$

C. is used to synthesize polychlorobutadiene, a synthetic rubber.

Chloropropanes: the two isomeric monochloro derivatives of propane. ***Propyl chloride (1-chloropropane)***, $CH_3-CH_2-CH_2-Cl$, is a colorless liquid; m.p. - 122.8 °C, b.p. 46.6 °C, n_D^{20} 1.3879. It is barely soluble in water, but dissolves readily in alcohol, ether and benzene. Propyl chloride is synthesized by reaction of propan-1-ol with concentrated hydrochloric acid in the presence of zinc(II) chloride, and is used as a solvent and in organic syntheses. ***Isopropyl chloride (2-chloropropane)***, $CH_3-CHCl-CH_3$, is a colorless, combustible liquid; m.p. - 117.2 °C, b.p. 35.7 °C, n_D^{20} 1.3777. It is barely soluble in water, and dissolves readily in most organic solvents. It is synthesized by the reaction of hydrochloric acid with isopropanol, or by addition of hydrogen chloride to propene in the presence of aluminum(III) chloride. Isopropyl chloride is used mainly as a solvent, and as an intermediate in the production of isopropyl derivatives.

Chloropropanone: same as Chloroacetone (see).

3-Chloropropene: same as Allyl chloride (see).

1-Chloropyrrolidine-2,5-dione: same as *N*-Chlorosuccinimide (see).

Chloroquine: a base-substituted 4-aminoquinoline derivative used as an Antimalaria agent (see) and in the long-term therapy of certain rheumatic diseases. C. acts preferentially against the schizonts, and is suitable for prophylaxis of malaria. Certain strains of malaria are developing resistance to C.

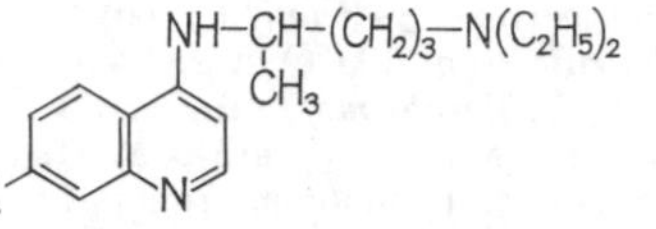

$NH-CH(CH_3)-(CH_2)_3-N(C_2H_5)_2$

Chlorosulfuric acid, ***chlorosulfonic acid***: $ClSO_3H$, the monochloride of sulfuric acid; a colorless liquid with a pungent odor which fumes copiously in moist air; M_r 116.52, density 1.766, m.p. - 80 °C, b.p. 158 °C. It reacts extremely vigorously with water to form sulfuric acid and hydrogen chloride: $ClSO_3H + H_2O \rightarrow H_2SO_4 + HCl$. Aromatic hydrocarbons are sulfonated by C., for example, $C_6H_6 + ClSO_3H \rightarrow C_6H_5SO_3H + HCl$. C. is made by combining sulfur trioxide dissolved in sulfuric acid with dry hydrogen chloride, or by the reaction of phosphorus pentachloride with concentrated sulfuric acid. C. is used in organic syntheses for sulfonation, sulfochlorination and chlorination. It is also used to generate artificial fog.

N-Chlorosuccinimide, abb. ***CNCS, 1-chloropyrrolidine-2,5-dione***: colorless platelets, m.p. 150 °C (dec.). It is soluble in acetone and glacial acetic acid. N-C. is synthesized by passing chlorine through an aqueous alcoholic solution of succinimide. It is used as a selective chlorinating and dehydrogenating reagent, analogously to N-Bromosuccinimide (see).

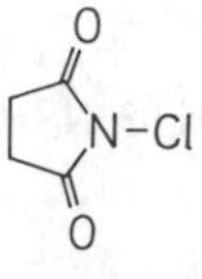

Chlorosulfanes: see Sulfur chlorides.

Chlorosulfonic acid: same as Chlorosulfuric acid (see).

Chloroterpene insecticides: a subgroup of chlorinated hydrocarbon insecticides, made by polychlorination of terpenes. Various bicyclic terpene hydrocarbons undergo additions, substitutions and rearrangements upon chlorination, leading to polychloroterpene mixtures which have insecticidal properties. The structures of these products are mostly unknown; the most important C. is Camphechlor (see).

Chlorotetracycline: see Tetracyclines.

1-Chloro-2,4,6-trinitrobenzene: same as Picryl chloride (see).

α-Chlorotoluene: same as Benzyl chloride (see).

Chloroxuron: see Urea herbicides.

Chlorphacinon: a Rodenticide (see).

Chlorphenol red: see Bromphenol blue.

p-Chlorphenylsilatran: a Rodenticide (see).

Chlorpromazine: see Phenothiazines.

Chlorpropham: see Carbanilate herbicides.

Chlorsulfone: see Urea herbicides.

Chlorthiamide: see Herbicides.

Chlorthiazide: see Diuretics.

Chlortoluron: see Urea herbicides.

Cholane: see Steroids.

Cholanoic acid: see Bile acids.

Cholecalciferol: see Vitamin D.

Cholecystokinin, abb. ***CCK, cholecystokinin-pancreozymin,*** abb. ***CCK-PZ***: a tissue hormone which stimulates contraction of the gall bladder and secretion of the pancreatic enzymes. C. is a 33-peptide amide, Lys-Ala-Pro-Ser-Gly-Arg-Val-Ser-Met-Ile-Lys-Asn-Leu-Gln-Ser-Leu-Asp-Pro-Ser-His-Arg-Ile-Ser-Asp-Arg-Asp-Tyr(SO_3H)-Met-Gly-Trp-Met-Asp-Phe-NH_2. It was isolated from crude extracts from intestinal mucosa, and it was only in 1964 that both hormone effects were shown to be due to the same peptide. Although the use of the double name is completely justifiable, in recent years the term C. has become more common. The C-terminal sequence of C. has a high degree of homology with gastrin and ceruleine. Interestingly, the C-terminal dodeca and octapeptide amides have higher biological activity than the native C. Partial sequences of C. act as neuropeptides.

Cholecystokinin-pancreozymin: see Cholecystokinin.

Choleic acids: see Bile acids.

Cholestane: see Steroids.

Cholesterinic: see Liquid crystals.

Cholesterol: cholest-5-ene-3β-ol forms colorless crystals, m.p. 147-150 °C (anhydrous), $[\alpha]_D$ - 31 °C, which are insoluble in water, moderately soluble in ethanol and very soluble in chlorform. C. is the main sterol in vertebrates, and is also present in invertebrates. It occurs in free form in human gallstones and biological membranes. Human serum contains about 200 mg C/100 ml; about 70% of this is esterified to fatty acids. In arteriosclerosis, C. is deposited on the inner walls of blood vessels. The serum level of C. is positively correlated with the development of arteriosclerosis and heart disease; this level can be reduced by certain drugs and diet. C. is obtained by extraction of bovine brain with dichloroethane; about 10 to 17% of dried brain mass is C. Other sources of C. are gallstones, wool grease and spinal cord. C. can be purified by adding bromine across the double bond to form the dibromide, followed by removal of the bromine with zinc. Another purification method is precipitation as the digitonide (see Digitonin). C. is used as an emulsifying agent in the preparation of salves, as a starting material for partial synthesis of steroid hormones, and for purification of cholecalciferol, with which it forms an addition compound.

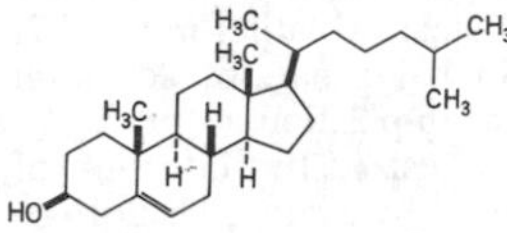

Cholic acid: 3α,7α,12α-trihydroxycholanoic acid, the most widely occurring bile acid. It forms colorless crystals (from ethanol); m.p. 196-197 °C (anhydrous). $[\alpha]_D$ + 35° (ethanol). The compound is scarcely soluble in water, but is readily soluble in aqueous ethanol or ether. The alkali salts are readily soluble in water. C. is present in human and bovine bile as glyco- and tauro-conjugates; it can be released from these conjugates by alkaline hydrolysis.

Choline: 2-hydroxyethyltrimethylammonium hydroxide, $[(CH_3)_3N^+\text{-}CH_2\text{-}CH_2\text{-}OH]OH^-$, the basic component of phosphatidylcholine (lecithin), a compound found in all organisms. C. is a crystalline, hygroscopic substance which binds carbon dioxide from the air, forming the carbonate. As a quaternary ammonium base, it is strongly basic. It can be synthesized by reaction of oxirane with trimethylamine.

Cholinergics: see Parasympathicomimetics.

Chondroitin: an acid mucopolysaccharide consisting of β(1→3)D-glucuronic acid units alternating with β(1→4)*N*-acetyl-D-galactosamine units. ***Chondroitin 4-sulfate*** and ***chondroitin 6-sulfate*** are major components of the connective tissue ground substance in cartilage and bone.

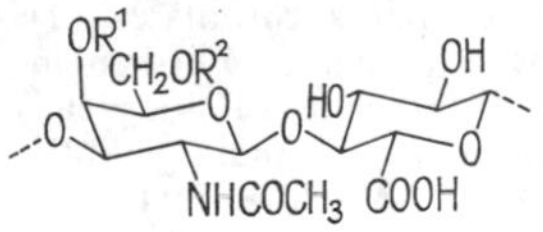

Chondroitin (R^1, R^2 = H)
Chondroitin 4-sulfate ($R^1 = SO_3H$; R^2 = H)
Chondroitin 6-sulfate (R^1 = H; $R^2 = SO_3H$)

The chondroitin sulfates are very hydrophilic and strongly acidic, due to their sulfate groups. Both the free acids and sodium salts of C. have been shown by x-ray analysis to be helical.

Chondroitin sulfate B: see Dermatan sulfate.

Chondrosamine: same as Galactosamine (see).

Choriogonadotropin, ***human chorionic gonadotropin***, abb. ***HCG***: a protein hormone formed by the placenta in the first weeks of pregnancy. It is a glycoprotein of two polypeptide chains, α and β The β-chain can combine with the α-chains of either lutropin or thyrotropin to form a fully active HCG. HCG has approximately the same effects as lutropin and prolactin. Its detection in urine is the basis of Aschheim and Zondek's pregnancy diagnostic test.

Choriomammotropin, ***human chorionic somatomammotropin***, abb. ***HCS***: a protein hormone formed in the placenta. C. is a linear polypeptide with 190 amino acids and two intrachain disulfide bridges. It has structural similarities to somatotropin. It is produced in greatest quantities in the last part of pregnancy, and has approximately the same effects as somatotropin and prolactin.

Chromane, ***dihydrobenzo[b]pyran***: the dihydro derivative of chromene. C. is an oil with an odor similar to peppermint; it is steam volatile; m.p. 5 °C, b.p. 215 °C, n_D^{20} 1.5444. In halogenation, acylation and nitration reactions, the 6-position is preferentially substituted. C. is synthesized by reduction of 2H-chromene or chromone, or by heating 1-chloro-3-phenoxypropane with tin(IV) chloride. Its basic structure is part of a series of natural products, such as the tocopherols and rotenoids.

Chromates: anionic complexes of chromium. C. in the narrow sense are ***oxochromates(VI)***, which are subdivided into ***monochromates***, $[CrO_4]^{n-}$ (n-2, 3, 4), ***dichromates(VI)***, $[Cr_2O_7]^{2-}$ and ***polychromates(VI)***, $[Cr_nO_{3n+1}]^{2-}$. Monochromates(VI) are yellow, unless the cation affects the color. Isolated tetrahedral CrO_4 units are also present in the black or blue-black oxochromates(V) $M_3[CrO_4]$ and in blue-black oxochromates(VI), $M_4[CrO_4]$. C.(VI) are strong oxidizing agents, which react in alkaline medium with hydrogen peroxide, forming red peroxochromates(V), $[Cr(O_2)_4]^{3-}$. When aqueous solutions of monochromates are acidified, dichromates, $M_2Cr_2O_7$, are formed by condensation reactions: $2\ CrO_4^{2-} + 2\ H^+ \rightleftharpoons Cr_2O_7^{2-} + H_2O$. The dichromates are readily soluble in water; they are orange to red. The dark red silver chromate $Ag_2Cr_2O_7$, for example, is relatively insoluble. Dichromates are also strong oxidizing agents; they form blue peroxochromates(VI), $[HCrO_6]^-$, with hydrogen peroxide. Further increase in the acidity of the dichromate solutions leads via tri- and tetrachromates to polychromates with the general formula $M_2Cr_nO_{3n+1}$, while the final product of this condensation reaction in very acidic medium is chromium(VI) oxide. All chromates(VI) are very poisonous (see Chromium). The most important industrially are sodium mono- and dichromates, which are often used as starting materials for further C.

In the general sense, many other anionic complexes of chromium are C. The chromates(III) were formerly also known as ***chromites***; an important group are the hexahydroxochromates(III), $M_3[Cr(OH)_6]$, which are formed by reactions of chromium oxide hydrates with strong bases. The chromates(III) also include acido complexes, such as hexaisothiocyanato and hexacyanochromate(III), $[Cr(NSC)_6]^{3-}$ and $[Cr(CN)_6]^{3-}$, and anionic Chromiacs (see), e.g. $[CrAc_4(NH_3)_2]^-$.

Chromatin: see Nucleic acids.

Chromatizing: a process for protecting metals (especially zinc and cadmium) from corrosion. It consists of treatment of the metal with chromates in sulfuric acid solutions; protective layers containing chromates form on the metal surfaces.

Chromatofocusing: an isoelectric focusing method analogous to column chromatography; it is used to separate proteins on the basis of their Isoelectric points (see). C. does not use an electric field; it works instead with a linear pH gradient formed by titration of an ion exchanger in a column with a low pH buffer. Applied proteins are focused to sharp bands and eluted from the column. The method can be used to separate proteins with isoelectric points between 11 and 4.

Chromatogram: the result of Chromatography (see) in which the separated products are fixed in the column or on a layer of paper or gel (***internal C.***), or a graphic representation of the detector signal during the elution of the components (***external C.***).

Chromatograph: an apparatus for chromatographic separations. See Chromatography.

Chromatography: a physical-chemical method of separating mixtures of substances by distribution between two non-miscible phases. There are many forms of C., including Adsorption C. (see), Distribution C. (see), Ion-exchange C. (see), Permeation C. (see) and Affinity C. (see). The separation occurs because different types of molecules or ions have different distribution coefficients between the *mobile* and *stationary* phases of the chromatographic system. Those which are relatively more attracted to the stationary state migrate more slowly than those with little affinity for it. Systems of C. may be categorized on the basis of the physical states of their components as ***liquid-solid*** (abb. LSC), ***liquid-liquid*** (LLC), ***gas-solid*** (GSC) and ***gas-liquid*** (GLC). On the basis of their geometry, separation methods may be classified as ***column, layer*** or ***flat-bed*** C. The chromatogram is developed by the Elution technique (see), the Displacement technique (see) or the Frontal technique (see). C. can be applied in the molecular mass range from 10 to 10^{15}.

Theory. C. consists of a repetitive *separation process* in which a substance is distributed between a stationary and a mobile phase. The *separation function* can be an adsorption, distribution or exchange equilibrium. Mathematical models used to treat the separation processes are somewhat simplified by abstraction.

The theoretical-plate approach describes the separation distance as consisting of a series of sections, called *theoretical plates*. The length or height is abbreviated HETP (height equivalent to one theoretical plate). The total separation distance, divided by HETP, yields the number n of theoretical plates. n is directly proportional to the retention time, and inversely proportional to the width of the analog signal (peak).

The dynamic theory is based on the *Van Deemter equation*: $HETP = A + B/u + C\,u$, where A is the scattering diffusion, B is the diffusion of the mobile phase, C is the diffusion and exchange of substance between the mobile and stationary phases, and u is the linear flow rate of the mobile phase (Fig. 1).

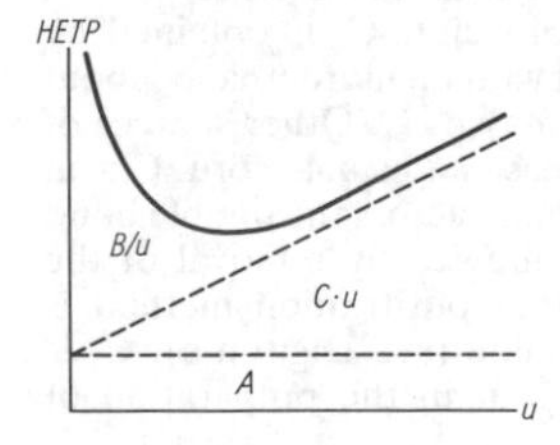

Chromatography: van Deemter curve

The kinetic theory treats the passage of a substance in analogy to the motion of boats with the current of a stream.

Historical. In 1903, the Russian botanist Michael Tsvett first separated leaf pigments by adsorption C. on calcium carbonate (chalk). In 1941, Martin and Synge developed liquid-distribution C., and for this achievement, received the Nobel Prize in chemistry in 1952.

Chromatometry: methods of redox analysis in which chromate or dichromate ions are used as titrators. These have a standard potential $E_0 = +1.36$ V in acidic solution, and are thus somehwat weaker oxidizing agents than bromate, cerium(IV) or permanganate ions. However, C. has some advantages. The standard solutions can be easily prepared from the primary titrant substance, potassium dichromate, and they are stable indefinitely. Chloride ions do not interfere with C. The end point of the titration can be determined using the reversible redox indicators diphenylamine and diphenylamine 4-sulfonic acid. It can also be readily determined by electrochemical methods, such as Potentiometry (see).

C. is used mainly to determine iron in a large number of industrial products. However, it can be used to determine many other inorganic reducing agents. In organic analysis, C. can be used to determine hydroquinones and alcohols, in some cases using back titration with iron(II) ions.

Chrome: see Chromium.

Chrome alum: same as Potassium chromium(III) sulfate.

Chrome dyes: see Mordant dyes.

Chrome green: a pigment which is a mixture of certain types of chrome yellow with Prussian blue. C. has good covering qualities. It is used to make printing inks, paints, colored papers and wall papers.

Chrome mordant dyes: a group of Mordant dyes (see) which includes the fast chromes, bichromin, anthracyl chrome, azidole chrome, chromoxane, metachrome, eriochrome and acid chrome dyes.

Chrome plating: the production of chromium protective layers on metallic objects. ***Hard chrome plating*** is used to create wear-resistant surfaces on tools; a thin layer of hard chromium is electroplated from a chromic acid solution onto steel, with no intermediate layer. Objects of unalloyed steel can be protected from corrosion and decorated by ***shiny chrome plating***; the chromium is electroprecipitated onto the object. For economic reasons, the thickness of the chromium layer is only 0.3 to 1 µm. Because of its porosity, it is applied to a nickel or nickel-copper intermediate layer. Chromizing (see) is a method of creating chromium-enriched steel surface layers by a diffusion process. These are resistant to corrosion and wear.

Chromenes: see Benzopyrans.

2H-Chromene-2-one: same as Coumarin (see).

Chrome orange: a pigment, lead oxide chromate, $PbO \cdot PbCrO_4$, orange, tetragonal crystals; density 6.9. C. has the same type of structure as ***chrome red***, which is chemically lead hydroxide chromate, $Pb(OH)_2 \cdot PbCrO_4$. Both pigments are insoluble in water, but dissolve readily in strong acids. They are very fast to light and resistant to alkalies. Hydrogen sulfide gradually turns them black, due to formation of black lead sulfide, PbS. C. and chrome red are synthesized industrially by boiling solutions of basic lead salts, potassium chromate and sodium hydroxide. Both pigments are used for oil paints, printer's inks, and to color linoleum, rubber and paper.

Chrome oxide green: see Chromium oxides.

Chrome oxide hydrate green: see Chromium(III) oxide hydrate.

Chrome red: see Chrome orange.

Chrome yellow: a pigment consisting of mixed crystals of lead chromate, $PbCrO_4$, and lead sulfate, $PbSO_4$. C. exists in three modifications, a light yellow rhombic form, a dark yellow monoclinic form and a red tetragonal form. The higher the proportion of lead sulfate, the lighter the color. The light yellow rhombic C. contains more than 50% lead sulfate. This also gives it a high intensity of color, good spreading ability and drying ability. If the lead sulfate is coprecipitated and not mixed in, it significantly increases the stability of the lead chromate and prevents thickening of the pigment. Rhombic C. is made by adding a sulfuric acid solution of potassium dichromate, $K_2Cr_2O_7$, to a very dilute lead(II) salt solution at a temperature below 40 °C. The precipitate is dried in vacuum at 35 to 50 °C. The color of the C. can be varied from very light yellow to orange by using different lead salts, variation of the pH and the use of additives to the precipitation solution. The light fastness of C. can be improved by precipitation of titanium dioxide hydrate, $TiO_2 \cdot xH_2O$ and/or cerium oxide hydrate, $Ce_2O_3 \cdot xH_2O$ onto the pigment. C. is used as a pigment for oil paints, printing inks, linoleum, rubber and paper, and in the synthesis of chrome green.

Chromiacs: a group of octahedral chromium(III) ammine complexes. The basic structure is the hexaamminechromium(III) cation, $[Cr(NH_3)_6]^{3+}$. Other C. are derived from it by exchange of some of the ammonia groups for other neutral or acido (Ac) ligands. Some examples are: hexaamminechromium(III) complexes $[Cr(NH_3)_6]X_3$, ammineaquochromium(III) complexes $[Cr(NH_3)_n(H_2O)_m]X_3$ $(n + m = 6)$, acidoamminechromium(III) complexes, $[Cr(Ac)_n(NH_3)_{6-n}]X_{3-n}$, acidoamminechromates(III), $[Cr(Ac)_n(NH_3)_{6-n}]^{(n-3)}$, and acidoammineaquochromium(III) complexes, $[Cr(Ac)_l(H_2O)_m(NH_3)_n]X_{3-l}$, $(l + m + n = 6)$. An important example of the tetraacidoamminechromates(III) is *Reinecke salt*, $NH_4 \cdot [Cr(NCS)_4(NH_3)_2]$, which is used in analytical chemistry for detection and gravimetric analysis of metals. Polynuclear C. contain two or more chromium atoms in the complex, usually joined by a hydroxo or amido bridge group. Some examples of dinuclear C. are the rhodo- and erythrochromium salts $(NH_3)_5Cr\text{-}OH\text{-}Cr(NH_3)_5]X_5$ and $[(NH_3)_5Cr\text{-}NH_2\text{-}Cr(NH_3)_4(OH)_2)]X_5$. All C. are colored, usually red.

Chromic acid, ***monochromic acid***: H_2CrO_4, an acid which exists only in aqueous solution and in equilibrium with dichromic acid, $H_2Cr_2O_7$: $2\ H_2CrO_4 \rightleftharpoons H_2Cr_2O_7 + H_2O$. If one tries to concentrate the solution, the equilibrium is shifted to the side of the dichromic acid, as can be seen from the change in color from the yellow of C. to the orange of dichromic acid. Further condensation reactions lead via tri- and tetrachromic acids, $H_2Cr_3O_{10}$ and $H_2Cr_4O_{13}$, and further polychromic acids, $H_2Cr_nO_{3n+1}$, to chromium(VI) oxide, the anhydride of C. Like dichromic acid, C. is a strong, dibasic acid; its salts are called Chromates (see). C. is highly caustic (see Chromium). An aqueous solution of C. can be made by dissolving chromium(VI) oxide in a large amount of water.

Chromites: see Chromates.

Chromium, symbol ***Cr***, a chemical element of group VIb, the Chromium group (see). It is a heavy metal, *Z* 24, with natural isotopes of mass numbers 52 (83.75%), 53 (9.55%), 50 (4.31%) and 54 (2.38%); atomic mass 51.996. Cr takes on a number of valence states, commonly II, III and VI, but also V, IV, I and 0. Density 7.14, m.p. 1875 °C, b.p. 2672 °C, electrical conductivity 7.75 Sm mm^{-2}, standard electrode potential (Cr/Cr^{3+}) - 0.744 V.

Properties. C. is a silver-white metal which in pure form is tough, ductile and malleable; it crystallizes in a cubic body-centered lattice. Even small amounts of impurities, e.g. oxygen, increase its hardness and make it brittle. Very finely divided Cr is pyrophoric. The compact metal, however, is chemically rather inert, and is not noticeably oxidized by either air or water. At high temperatures, Cr combines with nonmetals such as oxygen, halogens, nitrogen, boron or phosphorus. It is solubilized by alkali hydroxide fusions above 600 °C. The behavior of Cr with respect to acids depends on its pretreatment. If it is dipped into solutions of strong oxidizing agents, such as nitric or chromic acid, or if it is oxidized anodically, it is passivated by formation of a thin oxide layer. This is expressed by an increase of the standard potential to +1.3 V. Passivated Cr is thus not soluble in dilute acids. The passivation is removed if the metal is used as the cathode or if it is dipped into a reducing solution. Oxidizing acids do not attack Cr at low temperatures, due to the passivation process, and react only slightly with the metal when it is warmed.

In most of its compounds, Cr is in the +2, +3 or +6 oxidation state. The states +4 and +5 also occur, and strong complex formers, such as dipyridyl or carbon monoxide, can also stabilize lower oxidation states, +1, 0, - 1 and - 2. Chromium(VI) compounds are strong oxidizing agents, while chromium(II) compounds are reducing agents.

Chromium is extremely poisonous, especially in the +6 oxidation state. Chromium(VI) oxide, chromic acid, chromyl chloride, chromates, dichromates and chromium dust cause deep, chronic burns in the skin and mucous membranes. Frequent exposure to chromate solutions produces recalcitrant skin ulcers (chromium ulcers or eczema). Inhaled chromium or chromate dust causes ulcers in the nasal mucous membranes, catarrh of the lower respiratory passages and lung ulcers. Swallowing a chromate solution causes severe burns of the gastrointestinal tract and kidney damage. 0.6 g of chromium(VI) oxide absorbed through the stomach is fatal.

Rubber gloves and breathing masks are recommended to prevent chronic poisoning; good ventilation is also needed. In cases of acute internal burns, the stomach is rinsed, then milk and magnesia mixture are administered.

Analysis. As preliminary tests for Cr, oxidative fusion (yellow color in the presence of Cr due to formation of chromate) and phosphorus salt or borax beads (green color in the presence of Cr) may be used. Alkali hydroxide or ammonia solutions cause gray-green chromium oxide hydrate to precipitate from chromium(III) salt solutions; this can be oxidized to the yellow chromates(VI) in alkaline medium with strong oxidizing agents. Chromates are characterized qualitatively by converting them to blue chromium peroxide, CrO_5, by shaking with hydrogen peroxide and ether. Cr is determined gravimetrically as chromium(III) oxide or barium chromate(VI). Chromates(VI) can also be determined by iodometric titration.

Occurrence. Cr makes up $6.4 \cdot 10^{-3}$% of the earth's crust. The most important chromium ore is chromite, $FeCr_2O_4$. Crocoite, $PbCrO_4$, is a rare C. mineral. Cr is present in varying trace amounts in soils and plants, between 2 and 200 ppm.

Production. Chromite is enriched by flotation and finely ground; it is then oxidized in a furnace with added sodium carbonate and calcium oxide, in the presence of air, at 1000 to 1300 °C, forming sodium chromate: $2\ FeCr_2O_4 + 4\ Na_2CO_3 + 7/2\ O_2 \rightarrow 4\ Na_2CrO_4 + Fe_2O_3 + 4\ CO_2$. The product is cooled and extracted with a sodium carbonate solution. Addition of sulfuric acid or CO_2 converts sodium chromate to sodium dichromate, which precipitates as the dihydrate, $Na_2Cr_2O_7 \cdot 2H_2O$, when the solution is evaporated. Reduction with sulfur or coal yields chromium(III) oxide, from which Cr is obtained by an aluminothermal process: $Cr_2O_3 + 2\ Al \rightarrow Al_2O_3 + 2\ Cr$. The molten aluminum oxide produced by this process is "synthetic corundum", an excellent abrasive; its production is partly responsible for the fact that the process is economical. Very pure Cr is made by electrolytic reduction of chromic acid or chromium(III) salt solutions; the resulting Cr, which at room temperature contains up to 60 volume parts hydrogen, is degassed by heating in high vacuum. The very purest Cr is produced by a deposition process, in which the electrolyte Cr is purified by formation and thermal decomposition of chromium(III) iodide. The industrially important chromium-iron alloy *ferrochrome* is made from chromite by an electrothermal process with addition of coke.

Applications. C. is used to chrome plate other metals, protecting them from oxidation. It is used extensively for production of alloys; it is the main component of Chromium alloys (see) and is also added, for example, to Nickel alloys (see) and Super alloys (see). Addition of small amounts of C. to copper yields hardenable alloys with good electrical conductivity. C. is also used as an alloy component for cutting metals.

Historical. C. was discovered in 1797 by Vauquelin during analysis of crocoite. In 1799, it was found in chromite by Tassaert. The metal was prepared in 1844 by Berzelius, by reduction of chromium(III) chloride with potassium, and by Wœhler, in 1859, by reduction of chromium oxide with aluminum.

Chromium acetates: ***Chromium(II) acetate***, $Cr_2(CH_3COO)_4 \cdot 2H_2O$, red, monoclinic, diamagnetic crystals; M_r 494.30. The dinuclear complex has bidentate terminal acetate ligands and acetate bridges, and a direct Cr-Cr interaction. The compound is only slightly soluble in cold water and alcohol. It is obtained by combining chromium(II) chloride with sodium acetate in water; it is used in the laboratory as a strong reducing agent.

Chromium(III) acetate, more exactly, ***hexaaquachromium(III) acetate***, $[Cr(H_2O)_6](CH_3COO)_3$, blue-violet crystalline needles. The acetate is obtained by dissolving freshly precipitated chromium(III) oxide hydrate in glacial acetic acid; it is used as a mordant in textile printing, and to fix Vigoreux dyes.

Chromium alloys: alloys in which chromium is the main element. C. contain various carbide and intermetallic phases, which are finely divided and embedded in a mixed crystal matrix. Chromium brings to the alloys its ability to be passivated. In iron-chromium alloys, at least 12% chromium is required for the property to appear. Because of the difficulty of working them, alloys with more than 30% chromium have few industrial applications. Chromium is used mainly to alloy with iron, forming steels resistant to rust and acid, special high-strength steels and electrical resistance wires, and with nickel to form oxidation-resistant alloys. Even in concentrations of 1 to 3%, chromium increases the hardness of steels. In addition to the resulting increase in strength in cold and heat, the resistance to pressurized hydrogen is also increased (see Metal corrosion, chemical). Formation of special carbides increases the resistance to abrasion compared to the base metals, iron or nickel, and gives a higher resistance to scale. Because chromium has a higher affinity for oxygen than iron or nickel, the chromium is enriched in the scale layer. Such oxide layers resist ion diffusion, and thus growth of the layer, more effectively than oxide layers consisting only of iron or nickel oxides. ***Ferrochrome***, containing 52 to 75% chromium, is used as a pre-alloy for production of chrome steels. Chromium is also used in cobalt and nickel-based alloys (see Super alloys) and in composite materials (see Cermets, Stellite®). Chromium-nickel casting alloys (50 to 60% chromium remainder nickel) are resistant to high temperatures.

Chromium carbides: ***Trichromium dicarbide***, Cr_3C_2, gray, rhombic, hard and brittle compound; M_r 180.02, density 6.68, m.p. 1890 °C, b.p. 3800 °C. Cr_3C_2 is very resistant to corrosion, and is scarcely attacked even by highly concentrated and strongly oxidizing acids, or by alkali hydroxide solutions. Alkali fusions, however, convert it in the presence of oxygen to chromate and carbonate. Cr_2C_2 is made by heating chromium oxide with coal. It is one of the most abrasion-resistant materials, and is used to make sintered hard metals. Chrome steels contain ***heptachromium tricarbide***, Cr_7C_3; it forms silvery hexagonal crystals; M_r 400.00, density 6.92, m.p. 1650 °C.

Chromium carbonyl, ***chromium hexacarbonyl***: $Cr(CO)_6$, colorless, highly refractive, orthorhombic crystals; M_r 220.06, density 1.77. C. sublimes slowly even at room temperature; when heated in a melting tube it melts at 140-150 °C. It is obtained by reaction of chromium(III) chloride, aluminum and aluminum chloride in benzene, under a CO pressure of 30 MPa. It is used to improve metal surfaces, as an additive to fuels and as a catalyst in oxosyntheses and polymerizations.

Chromium chlorides: ***Chromium(II) chloride, chromium dichloride***, $CrCl_2$, colorless needles with a silky sheen; M_r 122.90, density 2.878, m.p. 824 °C. $CrCl_2$ is a strong reducing agent which readily takes up gaseous ammonia to form $[Cr(NH_3)_6]Cl_2$ and reacts with chloride ions to form tetrachlorochromate(II), $[CrCl_4]^{2-}$. $CrCl_2$ is obtained by reduction of chromium(III) chloride with hydrogen at 600 °C, or by reaction of chromium with hydrogen chloride at 700 °C. In aqueous solution, it is used for adsorption of oxygen, e.g. in gas analysis.

Chromium(III) chloride, chromium trichloride, $CrCl_3$, forms red-violet, shiny hexagonal leaflets; M_r 158.35, density 2.76, m.p. ≈ 1150 °C. In a stream of chlorine, $CrCl_3$ sublimes at 600 °C and decomposes in the absence of chlorine in this temperature range, forming $CrCl_2$ and chlorine. In the crystal lattice of $CrCl_3$, chloride ions are present in cubic close packing, with 2/3 of the octahedral holes between each second Cl double layer occupied with Cr^{3+} ions. When heated in the air, the compound is converted to chromium(III) oxide, while reaction with hydrogen sulfide, phosphine or ammonia at high temperature leads to chromium(III) sulfide, phosphide or nitride. Pure $CrCl_3$ is insoluble in water. In the presence of traces of chromium(II) salt, however, it dissolves with an exothermal reaction, forming a dark green hydrate complex. The solution slowly turns a lighter blue green, and finally turns violet. This change in color is related to the formation of hydrate isomers of chromium(III) complexes; $[CrCl_2(H_2O)_4]Cl \cdot 2H_2O$ (dark green), $[CrCl(H_2O)_5]Cl_2 \cdot H_2O$ (light green) and $[Cr(H_2O)_6]Cl_3$ can be isolated. $CrCl_3$ can be synthesized by chlorinating solubilization of chromite in the presence of coal at 900 to 1050 °C, followed by fractional condensation of $CrCl_3$ and $FeCl_3$. For synthesis on a laboratory scale, chromium can be made to react with chlorine at red heat. $CrCl_3$ is used to make very pure chromium and other chromium compounds, as a mordant in dyeing cotton, and in chromizing metals. ***Chromium(IV) chloride***, $CrCl_4$, is not capable of existence as the solid, but it can be detected in the vapor phase in equilibrium with $CrCl_3$: $2\, CrCl_3 + Cl_2 \rightleftharpoons 2\, CrCl_4$.

Chromium complexes: coordination compounds of chromium in which the central atom can be in an oxidation state between - 2 and +6; the most common coordination number is 6. The octahedral chromium(III) complexes are especially numerous and have been thoroughly studied; they are characterized by high kinetic stability. A typical representative is the violet hexaaquachromium ion $[Cr(H_2O)_6]^{3+}$; the numerous chromium(III)ammine complexes (see Chromiacs), are also typical. The sky-blue hexaaquachromium(II) ion $[Cr(H_2O)_6]^{2+}$ and the dimeric chromium(II) salts of fatty acids, e.g. $Cr(OOCCH_3)_2$, are typical chromium(II) complexes. The fatty acid salts are characterized by strong metal-metal interactions. The oxochromates of the type $[CrO_4]^{n-}$ (n = 2-4) have a tetrahedral configuration, while strong ligands such as dipyridine or carbon monoxide are able to stabilize lower oxidation states of chromium (-2 to +1): $[Crdipy_3]^+$, $[Crdipy_3]$,

$[Cr \cdot (CO)_6]$, $[Cr_2(CO)_{10}]^{2-}$, $[Cr(CO)_5]^{2-}$. Bisbenzenechromium (see) is an important representative of the chromium(0) complexes.

Chromium fluorides: ***Chromium(III) fluoride, chromium trifluoride***, CrF_3, green, rhombic crystalline needles; M_r 108.99, density 3.78, m.p. above 1000 °C, subl.p. 1100 to 1200 °C. CrF_3 synthesized in aqueous solution crystallizes with 3 to 9 mol water of crystallization. These hydrates exist in various isomeric forms, depending on the conditions of production, for example, as the violet hexaaquachromium(III) fluoride hydrates $[Cr(H_2O)_6]F_3 \cdot nH_2O$, as the green fluoroaquachromium(III) fluoride hydrates $[CrF_n(H_2O)_{6-n}]F_{3-n} \cdot mH_2O$, or as the green fluoroaquachromium(III), $[CrF_3(H_2O)_3]$. CrF_3 is obtained by heating chromium(III) chloride in a stream of hydrogen fluoride at 600 °C, and the hydrates are made by addition of alkali fluorides to chromium(III) salt solutions. Hydrated CrF_3 is used as a mordant for dyeing and printing wool, to protect wool from moths and to dye marble. ***Chromium(V) fluoride***, CrF_4, is a green-black solid; M_r 127.98, m.p. 277 °C, made when fluorine reacts with chromium or chromium(III) chloride at 300 to 350 °C. The reaction of chromium with fluorine under pressure and elevated temperature produces the fire-red ***chromium(V) fluoride***, CrF_5; M_r 146.99, m.p. 30 °C, b.p. 117 °C, and the very unstable, lemon yellow ***chromium(VI) fluoride***, CrF_6; M_r 165.99, which is converted to CrF_5 in vacuum above - 100 °C.

Chromium group: Group VIb of the periodic system, including the metals chromium (Cr), molybdenum (Mo) and tungsten (W). These elements are heavy metals with high melting and boiling points; because of their similar atomic and ionic radii, molybdenum and tungsten are very similar to each other. Chromium is more similar to the 3d elements which are its neighbors in the periodic system. The +6 oxidation state is typical for molybdenum and tungsten; however, chromium(VI) compounds are strong oxidizing agents. The metal(VI) oxides MO_3 of this group are acidic. The hexa- and pentahalides MX_6 and MX_5 are covalent, volatile and readily hydrolysed. Chromium(III) complexes are especially stable, and numerous. The di- and trihalides of molybdenum and tungsten have cluster structures. Metals of the C. form hexacarbonyls $M(CO)_6$ with CO.

Properties of Group VIb elements

	Cr	Mo	W
Atomic number	24	42	74
Electron configuration	$[Ar]3d^5 4s^1$	$[Kr]4d^5 5s^1$	$[Xe]4f^{14} 5d^4 6s^2$
Atomic mass	51.996	95.94	183.85
Atomic radius [pm]	117.6	129.6	130.4
Electronegativity	1.56	1.30	1.40
Standard electrode potential (M/M^{3+}) [V]	−0.744	−0.20	−0.11
Density [g cm^{-3}]	7.14	10.28	19.26
m.p. [°C]	1875	2620	3410
b.p. [°C]	2672	4825	5660

Chromium hydroxide: see Chromium(III) oxide hydrate.

Chromium(III) nitrate: $Cr(NO_3)_3 \cdot nH_2O$, violet crystals with a variable water content which depends on the method of preparation. The most important C. is the purple monoclinic hexaaquachromium(III) nitrate trihydrate, $[Cr(H_2O)_6](NO_3)_3 \cdot 3H_2O$; M_r 400.15, m.p. 60 °C. It is readily soluble in water. The violet solution turns green on heating, due to formation of isomeric hydrate species. C. is made by dissolving chromium(III) oxide hydrate in nitric acid; it is used as a mordant in textile printing and in dyeing, as a corrosion protection and in the production of alkali-free catalysts.

Chromium(III) oxide hydrate: $Cr_2O_3 \cdot xH_2O$, a light blue-green to dark green powder, depending on the conditions of synthesis; it contains varying amounts of water. If violet chromium(III) salt solutions are precipitated by addition of alkali, then air dried, the composition is $Cr(OH)_3(H_2O)_3$; therefore C. is widely, but incorrectly known as ***chromium hydroxide***. These preparations are light blue-green powders with limited stability; above 50 °C they lose water and are converted to oxygen-bridged polynuclear complexes. C. is relatively insoluble in water. As an amphoteric compound, it dissolves in acids, forming chromium(III) salts, and with strong bases, it forms hydroxochromates(III), $[Cr(OH)_6]^{3-}$. If freshly precipitated C. is boiled or stored for a long time under water, it ages and is then only slightly soluble in acids or bases. When it is heated, C. first loses water and is converted to amorphous Cr_2O_3. When this is heated to glowing to remove the last traces of water, it is converted to the lower-energy, crystalline oxide. In the laboratory, C. is made by adding alkali salts to chromium(III) salt solutions. In industry, it is made by reduction of aqueous sodium dichromate solutions with sodium sulfide, or by hydrolysis of chromium tetraborate, $Cr_2(B_4O_7)_3$. (The latter is obtained by fusing potassium dichromate and boric acid at red heat.)

C. is used as a treatment for textiles and to make other chromium compounds. Because it is very resistant to water, bases, light and weathering, it is used as a pigment for paints under the names ***chrome oxide hydrate green, Guignet green, Victoria green*** or ***emerald green***. Mixed with heavy spar to increase its covering capacity, it is known as ***permanent green***.

Chromium oxides: ***Chromium(II) oxide, chromium monoxide***, CrO, a black compound which crystallizes in a NaCl lattice; M_r 68.00. CrO is formed by thermal decomposition of hexacarbonylchromium at 200 to 250 °C in vacuum. If it is heated to a higher temperature, it disproportionates: $3\ CrO \rightarrow Cr + Cr_2O_3$.

Chromium(III) oxide, Cr_2O_3, is a green compound which is insoluble in water. It also forms hard, black, hexagonal crystals with a metallic sheen; M_r 151.99, density 5.21, m.p. 2435 °C, b.p. about 4000 °C. Roasted Cr_2O_3 is only slightly soluble in acids and bases. When it is treated with conc. sulfuric acid, chromium(III) sulfate forms; fused alkali hydroxides react with Cr_2O_3 in the presence of oxidizing agents to form chromates(VI). Fusion reactions with the oxides of divalent metals lead to double oxides of the spinel type, $MeO \cdot Cr_2O_3$, of which the most important is the mineral chromite, $FeCr_2O_4$. In the laboratory, Cr_2O_3 can be obtained by heating ammonium dichromate: $(NH_4)_2Cr_2O_7 \rightarrow Cr_2O_3 + 4\,H_2O + N_2$. In industry, Cr_2O_3 is made by reduction of sodium dichromate with sulfur: $Na_2Cr_2O_7 + S \rightarrow Cr_2O_3 + Na_2SO_4$, or with coal: $Na_2Cr_2O_7 + 2\,C \rightarrow Cr_2O_3 + Na_2CO_3 + CO$. Crystalline Cr_2O_3 is made from amorphous preparations by heating to 900 °C. Amorphous Cr_2O_3 is widely used as a pigment, ***chrome oxide green***; it is characterized by exceptional stability, intense color, good covering and miscibility. It is used for paints for buildings, machine parts, heaters, fences, etc., for printing chemical-resistant papers such as those used for money and stamps, in porcelain and enamel painting, etc.

Chromium(IV) oxide, chromium dioxide, CrO_2, forms black, tetragonal crystalline needles; M_r 83.99, density 4.8. CrO_2 is made by thermal decomposition of CrO_3. It is ferromagnetic and is used together with iron(III) oxide as the magnetic material in audio tapes and information storage.

Chromium(VI) oxide, chromium trioxide, CrO_3, forms long, dark red, rhombic crystalline needles; M_r 99.99, density 2.70, m.p. 196 °C. CrO_3 is somewhat volatile even at its melting temperature. It is extremely poisonous (see Chromium). CrO_3 is polymeric: each Cr atom is surrounded by four O atoms, and the CrO_4 tetrahedra are linked by common vertices. The Cr-O distances within the chain are 174.8 pm and 159.9 pm for terminal positions. Above 50 °C, CrO_3 readily splits off oxygen, and is converted via the intermediate forms Cr_3O_8, Cr_2O_5 and CrO_2 to Cr_2O_3. Because it so readily releases oxygen, it is a strong oxidizing agent, and it may react explosively with oxidizable substances. Methanol ignites spontaneously when dropped onto CrO_3. If dry ammonia is passed over CrO_3, it produces a flame as it is oxidized to nitrogen: $2\,NH_3 + 2\,CrO_3 \rightarrow N_2 + Cr_2O_3 + 3\,H_2O$. CrO_3 reacts with a large excess of water to form yellow chromic acid, H_2CrO_4; if less water is added, it forms yellowish red to red polychromic acids, $H_2Cr_nO_{3n+1}$.
CrO_3 combines with basic oxides to give mono- or polychromates. It is made by adding conc. sulfuric acid to conc. alkali chromate or dichromate solutions: $Na_2Cr_2O_7 + 2\,H_2SO_4 \rightarrow 2\,CrO_3 + 2\,NaHSO_4 + H_2O$. The precipitated CrO_3 crystals are washed with conc. nitric acid, and dried on clay at 70 °C. Very pure preparations can be obtained by recrystallization from chromyl chloride. CrO_3 is used mainly to treat the surfaces of metals, and also for electrolysis baths for galvanizing, as an oxidizing agent in preparative chemistry, as a caustic in medicine, to harden microscopic preparations, as a mordant in dyeing and to bleach mineral wax.

Chromium pigments: inorganic chromium compounds which can be used as pigments. Some examples are chrome oxide green and chrome oxide hydrate green (chromium oxide pigments). All chromate pigments (chromium yellow, chrome orange, chrome red, chrome green, zinc yellow) and the zinc oxide chromates, lead silicochromate, barium chromate, strontium chromate and barium-potassium chromate are also C.

Chromium sulfates. ***Chromium(II) sulfate***, $CrSO_4 \cdot 7H_2O$, blue crystals, M_r 274.17. The heptahydrate is obtained by reduction of a solution of chromium(III) sulfate in sulfuric acid with zinc. Chromium(III) sulfate solutions in sulfuric acid are used in gas analysis to absorb oxygen.

Chromium(III) sulfate, $Cr_2(CO_4)_3$, a red-violet powder; M_r 392.18, density 3.012; it dissolves in water only in the presence of reducing agents. $Cr_2(SO_4)_3$ forms a series of violet and green hydrates. The violet hydrates have the general formula $[Cr(H_2O)_6]_2(SO_4)_3 \cdot nH_2O$ (n = 1-9, 12) and when their solutions are heated, they form less soluble green hydrate complexes, which are sulfato-chromium(III) derivatives, such as $[Cr(SO_4)(H_2O)_4]_2 \cdot 2H_2O$. In some cases, hydrolysis leads to basic chromium(III) sulfates, e.g. hydroxopentaaquachromium(III) sulfate. Hydrated chromium(III) sulfate is made by dissolving chromium(III) oxide hydrate in sulfuric acid; it is commercially available as gelatinous, dark green, shiny leaflets. The basic chromium(III) sulfates, which are widely used in tanning leather, are obtained as a byproduct of industrial oxidation of organic compounds with sodium dichromate, or by reduction of sodium dichromate with sulfur dioxide or molasses. Neutral chromium(III) sulfate is used to treat cotton, produce green glazes, in paints and inks and in galvanizing.

Chromium sulfides: ***Chromium(II) sulfide***, CrS, black, hexagonal crystals; M_r 84.06, density 4.85, m.p. 1550 °C. CrS is formed from the elements, or from chromium and hydrogen sulfide at high temperatures.

Chromium(III) sulfide, Cr_2S_3, forms brown-black, hexagonal leaflets; M_r 200.18, density 3.77, m.p. 1350 °C, with loss of sulfur. Cr_2S_3 is insoluble in water, and is rather resistant to acids, but it is converted by oxidizing alkali fusions to chromate(VI). If alkali sulfides are melted with Cr_2S_3, alkali thiochromates(III), $M[CrS_2]$, are formed. Cr_2S_3 must be made in the absence of water, either from the elements or by the reaction of hydrogen sulfide with chromium(III) chloride at red heat; $2\,CrCl_3 + 3\,H_2S \rightleftharpoons Cr_2S_3 + 2\,HCl$.

Chromiumsulfuric acid, ***cleaning solution***: a strongly oxidizing mixture of potassium dichromate and conc. sulfuric acid, used in the laboratory to clean glassware. C. is a dark red-brown, oily liquid made by dissolving finely powdered, anhydrous alkali dichromate in conc. sulfuric acid; it contains both alkali hydrogensulfate and the strongly oxidizing chromium(VI) oxide.

Protective glasses must be worn when using chromiumsulfuric acid; because of the toxic chromium(VI) oxide vapors, it should be kept under a hood. In the presence of chlorides, chromyl chloride is formed; it is also very poisonous. Further material on toxicity, see Chromium.

Chromizing: the production of surface layers of steel which have high concentrations of chromium (see Protective coatings). C. is accomplished by diffusion of chromium into the surface of the steel, which must have a carbon content below 0.1%; the purpose is to improve the hardness of the steel and its resistance to wear and corrosion. Halogen compounds are used as chromium donors at temperatures above 950 °C. C. is done in the gas phase or in salt melts. There are several methods: hydrogen saturated with hydrogen chloride is passed at 1050 °C over a mixture of grainy ferrochrome and porous ceramic material; or the parts are imbedded in a mixture of ferrochrome, kaolin and ammonium iodide. Provided that the chromium-enriched layer is non-porous, chromized steel is as corrosion-resistant as chromium steel, and resists scale formation up to about 800 °C.

Chromocene: see Organochromium compounds.

Chromodiopsin: see Scotophobin.

Chromogen: see Chromophoric groups.

α-Chromone: same as Coumarin (see).

γ-Chromone, ***benzo-γ-pyrone, 4H-benzo-l-pyran-4-one***: a heterocyclic comopund which is a constitutional isomer of coumarin. γ-C. forms colorless needles, m.p. 59 °C. It is soluble in alcohol, ether, chloroform and benzene, but insoluble in water. γ-C. is the precursor of the Flavones (see) and isoflavones. γ-C. is synthesized by condensation of 2-hydroxyacetophenone with formic esters and cyclizing the resulting 3-(2-hydrophenyl)-3-oxopropionaldehyde with dilute sulfuric acid.

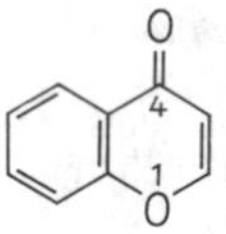

Chromophore, ***chromophoric group***: an unsaturated substituent which makes relatively simple molecules colored. If one or more C. is introduced into an organic compound, the absorption bands of the new compound are shifted towards the visible range due to promotion of resonance states by the π electrons of the C. The compound thus becomes a *chromogen*, or colored compound. Some typical C. are the nitroso group -N=O, the azo group -N=N-, the carbonyl group >C=O, the thiocarbonyl group >C=S and the azomethene group >C-N-.

Chromoproteins: conjugated proteins which contain chromophores as the non-protein components. They include hemoproteins, biliproteins, flavoproteins, visual pigments, etc.

Chromylium salts, ***benzopyrylium salts***: colored compounds made by condensation of salicylaldehyde with aldehydes or ketones in the presence of hydrochloric acid and iron(III) chloride.

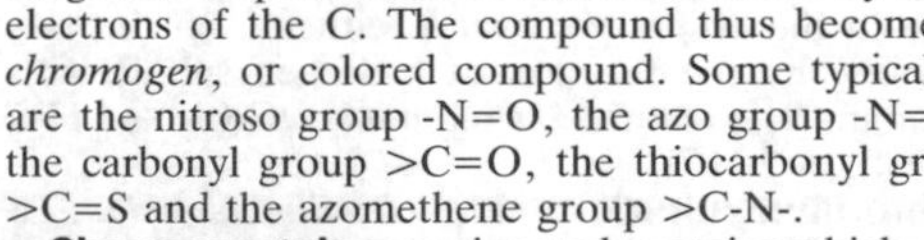

When hydroxide ions are added, they are converted to the colorless 2-hydroxy-α-chromenes, which revert to the C. in the presence of acid. C. with a phenyl substituent in position 2 are called ***flavylium salts***.

Chrysanthemin: see Cyanidin.

Chrysene, ***benzo[a]phenanthrene***: a fused cyclic aromatic hydrocarbon. It forms colorless, rhombic crystals which fluoresce in the red-violet; m.p. 256 °C, b.p. 448 °C. It is insoluble in water, slightly soluble in ethanol and ether, and soluble in boiling benzene. C. is formed by dry distillation from coal or the pyrolysis of fats and oils. It is isolated from coal tars and purified by chromatography. It can be synthesized by heating a mixture of ethyne and hydrogen to 800 °C. C. and its derivatives are carcinogenic.

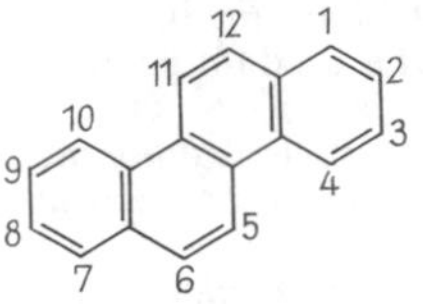

Chrysin: see Flavones.

Chrysolin: same as Fast yellow (see).

Chrysophanol: see Anthraglycosides.

Chrysophenin®: a yellow diazo dye. It is produced by coupling diazotized 4,4'-diaminostilbene-3,3'-disulfonic acid with 2 mol phenol; the phenolic hydroxyl groups are then ethylated with ethyl bromide. C. is resistant to alkalies, and binds directly to cotton, wool and silk.

Chymotrypsins: a group of serine proteases which have homologous structures and activities; they are synthesized and stored in the pancreas as inactive precursors (***chymotrypsinogens***). The most abundant is ***C. A (α-C.)***, which consists of 241 amino acid residues (M_r 25170). It is produced by a complicated enzymatic release process from chymotrypsinogen A (245 amino acid residues, M_r 25670). It hydrolyses proteins, peptides, amino acid esters and amides; it is most active at the carboxyl groups of aromatic amino acids. Fig., see Enzymes.

Çibaviolet 3B: see Indigo pigments.

Čičibabin reaction: amination of pyridine with sodium amide in toluene. The reaction is a nucleophilic substitution of the pyridine ring in position 2, which is typical for π-deficiency heterocycles. If the positions 2 and 6 are occupied, 4-aminopyridine can be obtained. The yields are between 75 and 90%.

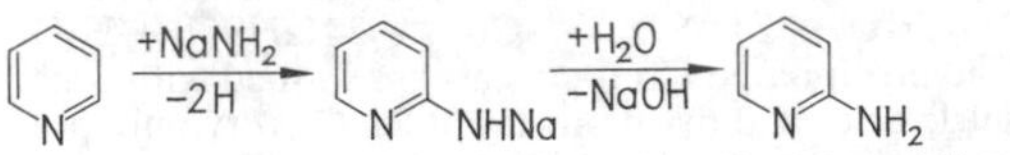

Quinolines react in the same manner as pyridines. It must be assumed that the reaction begins with the attack of an amide anion. The hydride ion eliminated by it immediately reacts with a proton of the NH_2 group, producing molecular hydrogen, which escapes, and the sodium salt of the aminopyridine, which is hydrolysed to the free base. The reaction has become important because aminopyridines are very reactive.

CIDEP: abb. for Chemically Induced Dynamic Electron Polarization (see).

CIDNP: see Chemically induced dynamic nuclear polarization.

Ciguteratoxin: a toxin produced by dinoflagellates (*Gambierdiscus toxicus*) which is taken up via the food chain into certain ocean fish (barracudas, bass, snappers, parrot fish) in the Carribean, Pacific and Indian Oceans. The toxin accumulates in the tissues of the fish; it is heat-resistant and is not destroyed by cooking. Symptoms of poisoning include gastrointestinal and neurotropic disorders, anxiety and sleeplessness. LD_{50} (mouse, intraperitoneal) is 0.08 mg kg^{-1}.

CI mass spectroscopy: abb. for Chemical Ionization mass spectroscopy; see Chemical ionization.

Cimetidine: see Antihistamines.

Cinchomeronic acid, ***pyridine-3,4-dicarboxylic acid***: colorless prisms or needles; m.p. 262 °C (dec.). C. is soluble in alcohol and acetone. It is formed by oxidation of 3,4-dimethylpyridine with potassium permanganate. It is also formed, along with phthalic acid, by oxidation of isoquinoline with potassium permanganate. When heated above its melting point, C. is decarboxylated, forming nicotinic and isonicotinic acids.

Cinchona alkaloids: a group of alkaloids with a quinoline ring system, found in the bark of various cinchona species. In the industrially treated bark of special domestic races, the total alkaloid content can be as high as 17%. Over 25 C. have been isolated from the bark. The main alkaloids are ***quinine, quinidine, cinchonine*** and ***cinchonidine***. The C. are biosynthesized from tryptophan and a monoterpene. They consist of a heteroaromatic quinoline ring system and a heteroaliphatic quinuclidine ring system, which are linked by a hydroxymethylene group (ruban). The C. are diacidic bases, which can form approximately neutral-reacting basic salts by protonation of the more basic N atom of the quinuclidine ring. Protonation of the quinoline N atom as well yields acid-reacting neutral salts.

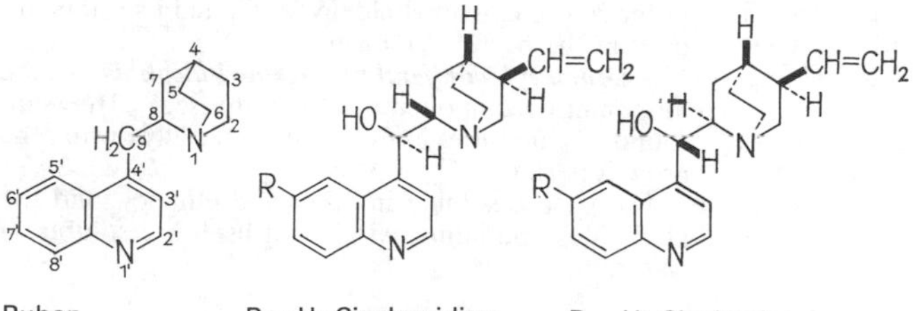

Ruban

R = H: Cinchonidine
R = CH_3O: Quinine

R = H: Cinchonine
R = CH_3O: Quinidine

The C. contain 4 chiral C atoms. The absolute configuration of quinine and cinchonidine is 3R, 4S, 8S, 9R, and of quinidine and cinchonine, 3R, 4S, 8R, 9S. In addition to the above C., there are dihydro alkaloids with an ethyl group instead of the vinyl group at C3, and the epibases with the opposite configuration at C9. The therapeutically important quinine and quinidine are difficult to separate from the corresponding dihydro alkaloids; therefore, only a certain percentage of dihydro compounds are permitted to be present in quinine and quinidine preparations.

The bases form colorless crystals which are slightly soluble in water, but are soluble in ethanol and chloroform. C. have an intensely bitter taste. Quinine base, m.p. 57 °C (trihydrate), m.p. 177 °C (anhydrous), $[\alpha]_D^{20}$ - 284.5° (in 0.1M H_2SO_4); quinine · HCl, m.p. 158-160 °C, $[\alpha]_D^{17}$ - 134° (in water); quinidine base, m.p. 173 °C (anhydrous), $[\alpha]_D^{15}$ +334° (in 0.1M H_2SO_4); quinidine · HCl, m.p. 259 °C, $[\alpha]_D^{20}$ +200° (in water); cinchonidine base, m.p. 211 °C, $[\alpha]_D^{15}$ - 108° (in ethanol); cinchonidine · HCl, m.p. 242 °C (dec.), $[\alpha]_D^{20}$ - 118 °C (in water); cinchonine base, m.p. 255 °C, $[\alpha]_D^{20}$ +229° (in ethanol).

Quinine and quinidine give an intense blue fluorescence in sulfuric acid solution. This and the thalleioquine reaction, which gives a green color when bromine water and ammonia solution are added in excess to an aqueous solution, are used in analysis.

Quinine is used in the form of its hydrochloride as an Antimalarial agent; it acts against the schizonts. This effect is due to intercalation of the planar aromatic group between the base pairs of the organism's DNA; the alkaloid molecule is then held in this position by the interaction between the N atom of the quinuclidine ring and the acidic phosphodiester groups. The basic-substituted quinoline derivatives, such as chloroquine and primaquine, act against the malaria organisms by the same mechanism. The antipyretic effect of quinine is relatively weak, and it is toxic to cells. Quinidine has the same basic effect as quinine, but it is less active as a chemotherapeutic. It has a stronger effect on the heart, however, and it is used as an Antiarrhythmic (see).

Historical. Cinchona bark was used as a remedy for fever in Europe as early as the first half of the 17th century. The first C. isolated was cinchonine, in 1810, by Gomes. Quinine was discovered in 1820 by Pelletier and Caventou. Skraup, Königs and Rabe contributed significantly to the elucidation of its structure; the configuration was determined in 1944 by Prelog. Woodward and von Doering achieved the total synthesis in 1945.

Cinchonidine: see Cinchona alkaloids.

Cinchonine: see Cinchona alkaloids.

Cinerins: see Pyrethrins.

Cine-substitution: nucleophilic substitution on aromatics and heteroaromatics in which the entering nucleophile not only assumes the ring position of the leaving group, but also enters the adjacent positions (Bunnett). C. occurs by an elimination-addition mechanism in which short-lived, non-isolable arynes or heteroarynes are formed as intermediates. These have been demonstrated as Diels-Alder adducts after trapping reactions.

Cinnamates: the salts and esters of Cinnamic acid (see), with the general formula C_6H_5-CH=CH-CO-OR, where R is a monovalent metal ion or ammonium ion, or an alkyl or aryl group. Although the salts are of little practical importance, many esters of cinnamic acid are used in perfumes and cosmetics, and some are found in essential oils. They are synthesized by esterification of cinnamic acid with alcohols in the usual manner, or by condensation of cinnamic aldehyde with acetic acid esters (see Claisen reaction). ***Methyl cinnamate*** is insoluble in water, but soluble in alcohol and ether. It forms colorless crystals with a strawberry-like odor; m.p. 36 °C, b.p. 262 °C. It is found in the root oil of *Alpinia officinarum* (Chinese ginger) and in camphor basil oil. ***Ethyl cinnamate*** is a colorless liquid which is insoluble in water but soluble in alcohol and ether; its odor is similar to cinnamon; m.p. 12 °C, b.p. 271 °C; it is also found in the essential oils of the ginger family. ***Cinnamyl cinnamate***, the cinnamate of cinnamic alcohol, forms colorless needles which are insoluble in water but soluble in alcohol and ether; it is found in Peru balsam and various essential oils.

Cinnamic acid, ***(E)-3-phenylpropenic acid, β-phenylacrylic acid***: a phenyl-substituted, unsaturated aliphatic carboxylic acid. In common usage, C. is understood to be both the E- or *trans*-form and the Z- or *cis*-form, although strictly speaking, the term should be reserved for the E-form. The Z-form is ***allocinnamic acid*** (m.p. 68 °C).

C_6H_5(H)C=C(H)(COOH)

Cinnamic acid

C_6H_5(H)C=C(COOH)(H)

Allocinnamic acid

There are in addition two further, less stable modifications of the Z-form, ***isocinnamic acids***, which melt at 42° and 58 °C. These interconvert readily. The C. found in nature is almost exclusively the E-(*trans*-)form, which may be free or bound. It crystallizes in colorless, stable leaflets; m.p. 135-136 °C, b.p. 173 °C at 1.33 kPa.

C. is barely soluble in water, but is readily soluble in alcohol, ether, benzene and acetone. It can be converted to styrene by decarboxylation; stronger oxidizing agents convert it to benzaldehyde or benzoic acid. The reactions of C. are typical of carboxylic acids and alkenes. Its salts and esters are called Cinnamates (see). Hydrogenation produces Hydrocinnamic acid (see). C. is readily converted by photochemical dimerization to Truxillic acid (see) and Truxic acid (see). It is synthesized by the Perkin reaction (see) from benzaldehyde, acetic anhydride and alkali acetate, or by Knoevenagel condensation (see) from benzaldehyde and malonic acid. It can also be made by oxidation of cinnamic aldehyde. The Z-stereoisomeric forms can be made by UV irradiation of C. Allocinnamic acid may also be synthesized by partial hydrogenation of phenylpropiolic acid, C_6H_5-C≡C-COOH.

C. is used mainly for synthesis of Cinnamates (see), which are used as perfumes and medicaments. It is also used as a preservative.

Cinnamon oil: a bright yellow essential oil with a pleasant spicy odor and taste. The main component is cinnamyl aldehyde (up to 75%), and it also contains eugenol, (-)-α-pinene, (-)-α- and (-)-β-phellandrene, *p*-cymol, nonyl aldehyde, hydrocinnamyl aldehyde, couminyl aldehyde, benzaldehyde, furfural, amyl methyl ketone, and sometimes camphor. C. is obtained by steam distillation from the bark and stems of the cinnamon tree, *Cinnamomum celanicum*, with a yield of 0.5 to 1.0%. Distillation of the leaves of the tree yields ***cinnamon leaf oil***. C. is used mainly in the preparation of candies, pharmaceuticals and cosmetics.

Cinnamyl alcohols: a group of unsaturated aromatic alcohols (table). Cinnamyl alcohol itself forms colorless needles with an odor similar to that of hyacinths. It is found in resins and balsams, esterified to cinammic acid. It is obtained by hydrolysis of natural esters, or by the Meerwein-Ponndorf-Verley reduction of cinnamyl aldehyde. C. and its esters are used in the perfume industry.

p-Coumaryl, coniferyl and ***sinapyl alcohols*** are the monomeric components of Lignin (see). They are found as glucosides in the plant, especially during the growth period.

The C. are soluble in ether and ethanol, and the glucosides and cinnamyl alcohol itself are soluble in water.

Cinnamyl alcohols

R^1	R^2	R^3	Name	m.p. in °C
H	H	H	Cinnamyl alcohol	33-35
H	OH	H	p-Coumaryl alcohol	·
H	O-β-D-Glc	H	p-Coumaryl alcohol glucoside	·
OCH_3	OH	H	Coniferyl alcohol	74-75
OCH_3	O-β-D-Glc	H	Coniferol	185
OCH_3	OH	OCH_3	Sinapyl alcohol (Syringenin)	63-65
OCH_3	O-β-D-Glc	OCH_3	Syringin	191-192

R^1, R^2, R^3-substituted benzene ring bearing CH=CH-CH_2OH

Cinnamyl alcohols

Cinnamyl aldehyde, ***3-phenylpropenal***: C_6H_5-CH=CH-CHO, an unsaturated, aromatic aldehyde. C. is a bright yellow liquid with the odor of cinnamon; m.p. - 7.5 °C, b.p. 252 °C, n_D^{20} 1.62. It is insoluble in water, but dissolves readily in many organic solvents, such as ether, ethanol and chloroform. C. is very reactive, with the formyl group undergoing the addition and condensation reactions typical of the Aldehydes (see). C. is converted by mild oxidants, e.g.

atmospheric oxygen, to cinnamic acid. Strong oxidizing agents also cleave the C=C double bond, forming benzoic acid. Reduction of C. produces cinnamyl alcohol.

C. is the main component (up to 70%) of cinnamon oil, and can be obtained from the oil via its hydrogensulfite. It is used as a perfume in cosmetics and as an additive to spices, aromas and pharmaceuticals.

Cinnamyl cinnamate: see Cinnamates.

CIPC: see Carbanilate herbicides.

Circular dichroism: see Chiroptic methods.

Circular filter methods: see Paper chromatography.

Circulin A.: see Polymyxins.

Cisplatin: *cis*-diamminedichloroplatinum(II), a planar complex of divalent platinum, which acts as a bifunctional alkylating agent because both Cl atoms are substituted by nucleophilic centers of nucleic acids. C. is used as a cytostatic.

cis-trans isomerism: see Stereoisomerism, 1.2.3.

Citral: an acyclic monoterpene aldehyde; a colorless, non-viscous, optically inactive liquid with a characteristic odor reminiscent of lemon; b.p. 117-119 °C at 2.7 kPa. C. makes up at least 75% of lemon grass oil, from which it is isolated industrially by fractional distillation. C.a (*geranial*) is the main component. C.a can be separated from C.b (*neral*) via the hydrogen sulfite adduct. C.a is formed by oxidation of geraniol, and C.b by oxidation of nerol. C. is synthesized from the acyclic monoterpene alcohols geraniol and nerol or linalool. It is used in the aroma industry to perfume soaps and other cosmetics, and as a starting material for the synthesis of ionone and vitamin A.

CH_3 CHO H_3C CH_3 Citral a

CH_3 CHO H_3C CH_3 Citral b

Citrates: the salts and esters of citric acid, with the general formula $RO\text{-}CO\text{-}C(OH)(CH_2\text{-}CO\text{-}OR)_2$, where R symbolizes monovalent metal or ammoniun ions, or an organic group. As a tribasic acid, citric acid can form both acidic and neutral salts and esters. The water-soluble alkali salts are especially important. The sodium salts, especially disodium hydrogencitrate, are used to make physiological buffers. They also inhibit blood clotting and are used for storing blood and blood products. They are used as buffers for cosmetics, foods and boiler water. The citric acid salts of alkaline earth and some other metals are insoluble in water.

Citric acid, ***2-hydroxypropane-1,2,3-tricarboxylic acid***: colorless, prismatic crystals; m.p. 153 °C (anhydrous). C. is only slightly soluble in ether, but is soluble in ethanol and water. It crystallizes out of aqueous solution as the monohydrate; m.p. 100 °C. Heating above 175 °C or conc. sulfuric acid dehydrates C. to aconitic acid; with fuming sulfuric acid, the product is acetone dicarboxylic acid. As a tribasic acid, C. forms three series of salts and esters, the Citrates (see), as well as double and complex salts. The insoluble calcium citrate is used for analysis.

Consumed in large amounts, C. is poisonous; 25 g can be fatal. Aqueous magnesium oxide, aluminum oxide suspensions or an egg stirred into milk are antidotes. C. makes up 6 to 8% of lemon juice, and is present in smaller amounts in other fruits (raspberries, currants). It is present in low concentration in blood and milk (≈ 0.2%). C. is obtained from fruit juices or from citric acid fermentation of mono- and disaccharides.

The chemical synthesis starts from 1,3-dichloroacetone, with formation of the cyanohydrin. Nucleophilic substitution of nitrile groups for the chlorine atoms and subsequent hydrolysis lead to C.

C. has a central position in the citric acid cycle, which is a major metabolic pathway. It is used in foods and drinks, as a treatment for textiles, and as a buffer for vitamin D preparations used to treat rickets.

$$\begin{array}{c} CH_2-COOH \\ | \\ HO-C-COOH \\ | \\ CH_2-COOH \end{array}$$

Citric acid cycle, ***tricarboxylic acid cycle, Krebs cycle***: a metabolic cycle in which citric acid is converted, in several steps catalysed by specific enzymes, to oxaloacetic acid. An aldol addition of the acetyl group from coenzyme A to oxaloacetic regenerates the citric acid.

$$H_3C\text{-}COSCoA + \begin{array}{l} CO\text{-}COOH \\ | \\ CH_2\text{-}COOH \end{array} \xrightarrow{H_2O} \begin{array}{l} CH_2\text{-}COOH \\ | \\ C(OH)\text{-}COOH + CoA\text{-}SH \\ | \\ CH_2\text{-}COOH \end{array}$$

The total balance of the C. is $H_3C\text{-}COSCoA + 3\,H_2O \rightarrow 2\,CO_2 + 8\,H + HS\text{-}CoA$; in other words, the cycle serves to convert the acetyl group of acetyl-CoA to CO_2. The hydrogen released from the acetyl is converted to reducing equivalents which are transported to the respiratory chain; the operation of this chain generates ATP from ADP in the course of transferring the reducing equivalents to molecular oxygen to make water. Certain intermediates of the C. are precursors for the biosynthesis of various compounds. The scheme below shows the linkage of the C. to carbohydrate, protein and fat metabolism.

Carbohydrates ⇅ ; Peptides ⇅

Fats ⇌ $H_3C\text{-}CO\text{-}COOH$ ⇌ Amino acids

⇅ Acetyl-coenzyme A ; ⇅ Proteins

↓ C_2 → C_6 → ($-CO_2$) C_5 → ($-CO_2$) C_4 → C_2

C_4: Oxaloacetic acid (Malonic, fumaric, succinic acids)
C_5: α-Ketoglutaric acid
C_6: Citric acid (Isocitric, aconitic acids)

Citronellal: $(CH_3)_2CH{=}CH(CH_2)_2CH(CH_3)CH_2CHO$, acyclic monoterpene aldehyde. C. is a colorless liquid which smells like balm mint; b.p. 205-208 °C. It is found in essential oils (usually the d-form), and in large amounts in citronella oil. C. is very sensitive to oxidizing agents, and easily cyclizes to *iso*-pulegol. It is an important starting material in the perfume industry.

Citronella oil: a colorless essential oil which smells somewhat like lemons and balm. Chemically, C. consists mainly of geraniol (25-45%), citronellal (25-45%), citral, eugenol and vanillin. It is made by steam distillations of certain types of grasses. It is produced in relatively large amounts, and is used mainly in the perfume industry, and also as a source of the main components (geraniol and citronellal), which are used as aroma components.

β-Citronellol: $(CH_3)_2CH{=}CH(CH_2)_2CH(CH_3)CH_2CH_2OH$, an acyclic monoterpene alcohol. C. is a colorless liquid which smells like roses; b.p. 103 °C at 1.3 kPa. It is practically insoluble in water, and is soluble in alcohol. C. is a component of many essential oils; large amounts of the l-form are present in roses, and both l- and d-forms are present in geranium oils. C. is used in perfumes. It is the main component of rose combinations.

Citronin A: same as Naphthol yellow S (see).

Citrus oil: same as Lemon oil (see).

City gas, formerly also called ***lighting gas***, after its main use at the time: a combustible gas with a heat of combustion greater than 15.9 MJ m^{-3} (s.t.p.), obtained by degassing or gasification of fossil carbon sources. The most important requirements for a C. are an oxygen content < 0.7 vol. %; hydrogen sulfide content < 0.5 or 2 g per 100 m^3 (s.t.p.), depending on the gas pressure; ammonia content < 0.5 g per 100 m^3 (s.t.p.) etc. For consumer safety, an even gas pressure of 600 to 1500 Pa in the supply network is required; in addition, an odorant is often added to make detection of leaks easier (see Gas odorizing). The composition of a C. depends on the method of production; it is (by volume) about 18% carbon monoxide, 5% carbon dioxide, 50% hydrogen, 19% methane, 6% nitrogen and 2% hydrocarbons.

C. is produced by 1) degassing hard coals in gas works and coking plants (see Coking); 2) gasification of solid fuels such as peat, coal or coke with a mixture of oxygen or of air and steam under pressure (see Synthesis gas); 3) partial oxidation and cracking of liquid fuels, such as crude oil, gasoline, heating oil or refinery residues, using oxygen, air or oxygen-enriched air and steam under pressure (oil pressure gasification, see Synthesis gas); 4) catalytic cracking of natural gas or refinery gases with steam (steam reforming, see Synthesis gas).

The individual methods of production are followed by methods of Gas processing (see). The quality required of a C. is much lower than that of a synthesis gas.

Civetone, ***cycloheptadec-9-en-1-one***: an unsaturated, macrocyclic, aliphatic ketone. C. forms colorless crystals with a musk-like odor; m.p. 32.5 °C, b.p. 158-160 °C at 266 Pa. C. is steam volatile and dissolves readily in alcohol and benzene. The Z-form of C. is the most important aroma substance in civet, a glandular secretion of the civet cat. It is used as a fixative in perfumes.

HC=CH ring: $HC{-}(CH_2)_7{-}C{=}O$, $HC{-}(CH_2)_7{-}$ (HC=HC joined, both $(CH_2)_7$ chains closing on C=O)

Cl: 1) symbol for chlorine; 2) unit symbol for the Clausius (see).

Claisen condensation: a condensation reaction described in 1887 by L. Claisen between carboxylate esters and CH-acidic methyl or methylene compounds in the presence of strong bases, e.g. sodium ethanolate, sodium amide or sodium hydride. Depending on the structure of the ester or methylene compound used, the following can be produced by the C.: 1) β-diketones by R^1-CO-OR^2 + CH_3-CO-R^3 → R^1-CO-CH_2-CO-R^3 + R^2OH; 2) β-ketoaldehydes by H-CO-OR^1 + CH_3-CO-R^2 → OHC-CH_2-CO-R^2 + R^1OH; 3) β-ketocarboxylate esters by R^1-CH_2-CO-OR^2 + R^1-CH_2-CO-OR^2 → R^1-CH_2-CO-CHR^1-CO-OR^2 + R^2OH; and 4) β-ketocarbonylnitriles by R^1-CO-OR^2 + R^3-CH_2-CN → R-CO-CHR^3-CN. The reaction mechanism of the C. is demonstrated by the synthesis of acetylacetone from ethyl acetate and acetone:

$$CH_3{-}\underset{\|}{\overset{}{C}}(O){-}CH_3 \xrightarrow[-ROH]{+RO^{\ominus}} {}^{\ominus}|CH_2{-}C(O){-}CH_3$$

$$CH_3{-}C(OC_2H_5)(O) + {}^{\ominus}|CH_2{-}C(O){-}CH_3 \longrightarrow CH_3{-}C(OC_2H_5)(O^{\ominus}){-}CH_2{-}C(O){-}CH_3$$

$$\xrightarrow{-C_2H_5O^{\ominus}} CH_3{-}C(O){-}CH_2{-}C(O){-}CH_3$$

In addition to monocarboxylate esters, oxalate esters and carbonate esters can be used as the carbonyl components. Longer-chain dicarboxylate esters can undergo a Dieckmann condensation (see), an intramolecular condensation analogous to the C.; in this case both the methylene group and the carbonyl component are contained in a single molecule.

Claisen reaction: a condensation of aromatic aldehydes and aliphatic carboxylates in the presence of alkali alcoholates or sodium at low temperatures. It was first investigated in 1890. An aldol addition intermediate is immediately dehydrated to form the corresponding α,β-unsaturated carboxylate ester, e.g. cinnamate esters from benzaldehyde and acetate esters according to the equation

$$C_6H_5{-}CHO + CH_3{-}COOR \xrightarrow{(Na)} C_6H_5{-}CH{=}CH{-}COOR.$$

Claisen rearrangement: thermal rearrangement of phenylallyl ethers into 2-allylphenols. The mechanism of the C. was proven by crossing experiments, isotope labelling and structural variations. The reaction proceeds via a cyclic transition state, and is thus sigmatropic.

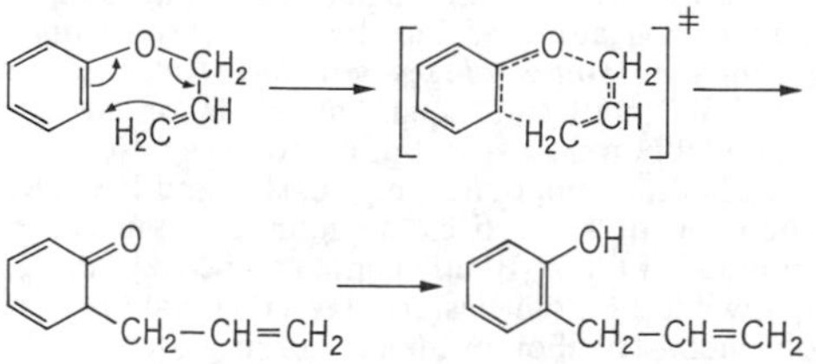

Claisen rearrangement

The C. of phenylallyl ethers which are blocked in the 2- and 6-positions by substituents produces 4-allylphenols. In this case, isotopic labelling showed that the reaction consists of two consecutive C.

Clarification: 1) In general, the removal of solid impurities from liquids by filtration, sedimentation or centrifugation. 2) The removal of substances from water which cause cloudiness, coloration, odor or taste, or of hygienically questionable micro-impurities. Even very small amounts of phenols, for example, react during chlorination of the water to form chlorophenols which have strong odors and tastes, giving the water a "drug-store" smell and taste. Water which contains such impurities can be clarified, after certain pretreatment steps, by filtration through activated charcoal or by treatment with ozone, chlorine dioxide or chloramines.

Clarite: see Coal.

Clark: see Organoarsenic compounds; see Chemical weapons.

Clathrates: inclusion or cage compounds, in which atoms or molecules occupy the hollows in a suitable molecular lattice. These enclosed molecules are not fixed in place by chemical bonds, but by the hollow spatial structure of the host lattice. Some typical examples are the solid gas hydrates, e.g. $6X \cdot 46H_2O$ (X = Ar, Kr, Xe, Cl_2, CH_4, etc.) in which X is enclosed in the hollows made by hydrogen bonding of the water in $(H_2O)_{20}$ dodecahedra. Urea, cyclodextrins, hydroquinone, phenol, toluene and other organic compounds, including some coordination polymers, form host lattices suitable for C.

Clausius, symbol ***Cl***: an obsolete unit of Entropy (see) named for R.J.E. Clausius. 1 Cl = 1 cal deg^{-1}. The standard SI unit is 1 J K^{-1}; the conversion is 1 Cl = 4.1868 J K^{-1}.

Clausius-Clapeyron equation: an equation describing the temperature dependence of the vapor and sublimation pressures of pure liquids and solids. It was established in 1834 by E. Clapeyron, and later given a theoretical foundation by R.J.E. Clausius. The C. also applies to the equilibrium pressure of melts and to the interconversions of various modifications; thus it applies to all equilibrium curves in the Phase diagram (see) of a pure substance. The C. is $dp/dT = \Delta_P H/T(V_2 - V_1)$, where $\Delta_P H$ is the molar phase-change enthalpy for the transition from phase 1 to phase 2, and V_1 and V_2 are the respective molar volumes.

Applications. 1) Evaporation and sublimation: the enthalpies of evaporation and sublimation of all substances are highly positive, and $V_2 \gg V_1$, i.e. the differential dp/dT is positive. The vapor and sublimation pressures increase with temperature. If the molar volume V_1 in the condensed phase is ignored, because it is much smaller than that in the gas phase, V_2, and if the ideal gas law is taken as an approximation, $V_2 = RT/p$, then the C. is $dp/dT = p\Delta_P HRT^2$, or $d \ln p/dT = \Delta_P H/RT^2$. Over small temperature ranges, $\Delta_P H$ can be taken as independent of temperature. Integration of the above equation gives $\lg p = \Delta_P H/2.303RT + \text{const.} = -A/T + B$, a relationship which was discovered empirically in 1828 by E.F. August (***August's formula***).

For larger temperature intervals, the temperature dependence of $\Delta_P H$, given by the Kirchhoff law (see), must be included in the integration. The result is $\lg p = -\Delta_P H/2.303\ RT + (\Delta_P C/R)\lg T + \text{const.}$, where $\Delta_P C$ is the difference in molar heats of the vapor and liquid (see Dupré-Rankin equation). As the critical point is approached, the above approximations are no longer justified, so that the general form of the C. must be used.

2) Melting equilibria and modification changes. Since these are equilibria between two condensed phases, the molar volumes V_1 and V_2 are similar in magnitude. The difference $V_2 - V_1$ is small, and therefore the temperature dependence of the equilibrium pressure is very large. In general, $V_{liq} > V_{solid}$, i.e. dp/dT and its reciprocal, dT/dp, are positive, and the melting temperature increases with pressure. However, the reverse can also occur. For example, for water near 273 K, $V_{liq} < V_{solid}$, and the melting temperature decreases with increasing pressure. Both possibilities are also observed with transitions between modifications.

The C. can be used not only to described phase-change equilibria quantitatively, but to calculate phase-change enthalpies from experimentally determined equilibrium pressures.

Claus process: see Sulfur.

Clavin alkaloids: see Ergot alkaloids.

Clavulanic acid: see β-Lactam antibiotics.

Cleaning solution: see Chromiumsulfuric acid.

Clear point: see Liquid crystals.

Clemmensen reduction: method for reduction of aldehydes and ketones to hydrocarbons using amalgamated zinc and hydrochloric acid: $R^1R^2C{=}O + 2\ Zn + 4\ HCl \rightarrow R^1R^2CH_2 + 2\ ZnCl_2 + H_2O$. The C. works with aliphatic and unsaturated ketones and with many aldehydes with good yields. Water-insoluble aldehydes and diaryl ketones react only with great difficulty. In addition, the method produces by-products in many cases, e.g. alcohols, alkenes and higher-molecular mass hydrocarbons. Various modifications of the C., e.g. use of solvation agents or different types of amalgamation of the zinc have considerably extended the range of compounds with which the method can be used.

Clenbuterol, ***contraspasmin***®: a selective β_2-sympathicomimetic, which also has secreolytic and antiallergic effects. It is used for long-term treatment of bronchial asthma and chronic bronchitis.

Clevic acid: an Alphabet acid (see).

Clindamycin: see Lincomycin.

Clinical analysis: analysis of human body fluids, mainly blood and urine, but also peritoneal fluid, pancreatic secretion, amniotic, brain, joint and cardiac fluids, gastric juice, pleural exudate, sweat, saliva and stools. C. is used for diagnosis and monitoring of therapy, for judging the prognosis of a disease and prophylaxis.

The usual analytical methods are used for the electrolytes, usually sodium, potassium, calcium, magnesium chloride, hydrogencarbonate, phosphate and sulfate ions; in addition trace elements and toxic heavy metals may be determined. Special biochemical methods are used for biological molecules (see Enzymes, Hormones and Vitamins).

Clofibrinic acid: $Cl\text{-}C_6H_4\text{-}O\text{-}C(CH_3)_2\text{-}COOH$, a 4-chlorophenoxyacetic acid derivative which reduces the serum lipid level. C. is used for therapy and prophylaxis of arteriosclerosis, since it reduces the cholesterol level of the blood. The ethyl ester is used under the name *Clofibrate*.

Clomipramine: see Dibenzodihydroazepines.

Clonazepam: see Benzodiazepines.

Clonidine, ***Haemiton***®: a cyclic guanidine derivative in which two nitrogen atoms of the guanidine group are part of an imidazoline ring. C. is an antihypertensive; it reduces blood pressure by interacting with α-receptors in the central nervous system. C. reduces the internal pressure in the eyes, and is therefore also used for glaucoma.

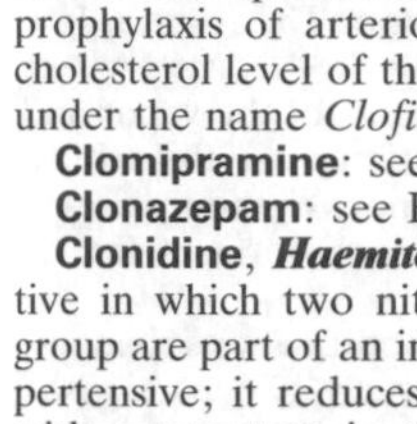

Clonitranilide: a Molluscicide (see) and veterinary medicine.

Clorindione, ***chlorathrombon***®: 2-(4-chlorophenyl)indane-1,3-dione, a compound used as an anticoagulant. Compounds which have an H atom or a methoxy group in the place of the chlorine atom have the same effect.

Clotrimazol: an Antimycotic (see).

Clove oil: a yellowish essential oil with a spicy odor; it turns dark in the presence of atmospheric oxygen. The main components of C. are eugenol, acetyleugenol, caryophyllene and furfural; the methyl esters of benzoic and salicylic acids are also present. C. is obtained from the dried flower buds and stems (cloves) of a member of the myrtle family, *Syzygium aromaticum*, by steam distillation. Similar oils are obtained from the leaves and stems. C. is used in perfumes, and sometimes in medicine as a local analgesic.

Cloxacillin: see Penicillins.

Clusius-Dickel process: see Isotope separation.

Cluster compounds: homogeneous or heterogeneous associations with finite numbers (2 to about 10^6) of atoms or molecules or structural units with homonuclear bonding, which are held together by weak (van der Waals or hydrogen) or strong (ionic, covalent or metallic) bonds. The C. can be classified by size as well as by bond type: microcluster compounds have 2 to about 15 structure units, small C. have from 10 to 100 units, and large C. have more than 100 units.

Clusters are usually polyhedra, in which the vertices are occupied by nonmetal atoms or transition metal atoms surrounded by ligands; these are linked to neighboring atoms within the C. There are ***nonmetal clusters (nonmetal cage structures)*** such as the tetrahedral P_4 molecule, square pyramidal or trigonal bipyramidal boranes (B_5H_9) or carboranes ($B_3C_2H_5$), and octahedral compounds, e.g. $B_6H_6^{2-}$ and $B_4C_2H_6$. Polyhedra with 10 vertices are found in many cage compounds of the A_4B_6 adamantane type (e.g. P_4O_6, P_4O_{10}), while 12-atom clusters play a dominant role in boron chemistry. For example, boron icosahedrons are present in all forms of elemental boron, in $B_{12}H_{12}^{2-}$ and in carboranes of the $B_{10}C_2H_{12}$ type.

Transition metal clusters, which are usually considered to include trigonal compounds, can be classified in two groups: halides and oxides of lower-valent 4d and 5d metals, and polynuclear carbonyls and their derivatives. Examples of trinuclear (trigonal) clusters are rhenium(III) chloride, Re_3Cl_9, and its derivatives. Tetrahedral $Co_4(CO)_{12}$, square pyramidal $Fe_5(CO)_{15}C$ and $Fe_3(CO)_9S_2$ and trigonal bipyramidal $Os_5(CO)_{15}$ are examples of clusters with four and five vertices. The octahedral transition metal clusters are very numerous; examples are $Rh_6(CO)_{16}$ (metal carbonyl) and metal halides of the types $[M_6X_{12}]X_2$ (M = Nb, Ta) and $[M_6X_8]X_4$ (M = Mo, W). Transition metal clusters can be used as catalysts, for example, in CO oxidation on $Rh_6(CO)_{16}$ and the conversion of nitrobenzene to aniline on $Ru_3(CO)_{12}$.

Cm: symbol for curium.

CMC: see Colloids.

CN: 1) abb. for cellulose nitrate. 2) see Chemical weapons.

Co: symbol for cobalt.

CoA: abb. for coenzyme A.

Coacervation: the separation of highly solvated, lyophilic colloids, such as protein solutions, into two liquid phases; the colloid phase is enriched in one of these phases.

Coagulation: the aggregation of colloidal particles. The first step is the approach of particles, either through Brownian motion (see) (***perikinetic C.***) or flow processes (***orthokinetic C.***), to an equilibrium distance. This can lead, in a second step, either to a loose aggregation (Flocculation, see) or to a compact phase (see Coalescence). The approach of colloidal particles toward one another is subject to a number of forces: electrostatic repulsion due to interpenetration of their diffuse electrochemical double layers, van der Waals attractive forces, and steric repulsion which arises when the adsorption layers penetrate each other. The latter are caused by oriented adsorption of solvent molecules, or by adsorption layers of tensides or macromolecules (see Protective colloids). Since the electrostatic repulsion of spherical particles is proportional to $\exp(-\varkappa d)$, where d is the distance between them, while the van der Waals attraction energy is proportional to d^{-1}, the attraction energy can be higher at large distances, and is always higher at short distances. $\varkappa$ is the Debye parameter. The interplay of forces leads to important properties of coagulates, such as structure, thixotropy, dilatance, flow limit and creep deformation.

Coagulation inhibitors: see Antithrombotics.

Coagulation structures: see Gel.

Coal: 1) in the widest sense, a decomposition product of organic substances formed either by geological processes or artificially (see Charcoal). C. is a brown to black substance with a high carbon content.

2) ***Natural C.*** are combustible, organogenic sedimentary rocks (castobioliths) which were formed by accumulation and decomposition of vegetable matter. Natural C. are, for the most part, the remains of plants in large moors (humus C.) which formed under certain biological, climatic and geological conditions. The vegetable matter has undergone a complicated process of chemical and structural changes, called *carbonization*. The biochemical phase of carbonization (diagenesis) produces peats and lignites (soft and dull lignite C.), and geochemical metamorphosis transforms these to glance lignite, hard coals and Anthracite (see).

C. is found in *seams*, which are layers or lenses of varying thickness and wide horizontal extent. One or more seams which fulfill certain conditions (thickness, geological or industrially utilizable amounts, quality) and which can be economically mined, make up a *coal deposit*.

Classification. C. has a special place in geology, partly due to its organic origin and the special conditions for its formation, but also because of its economic importance and the resulting differentiated requirements on C. as raw materials. Accordingly, there are numerous classification schemes for C. Bode (1932) classified the C. according to their geological and petrographic characteristics (Table).

Most of the lignite and mineral C. on the earth are humus C.

C. can be classified as lignite or mineral on the

Lignite coals	Soft	Clay Shale
	Hard	Dull Glance
Mineral coals	Humus (banded)	Flame Gas flame Gas Fat Forge Lean Anthracite
		Channel (sporophyte)
	Sapropel	Bog head (algal)

basis of their heating values (of air-dry, ash-free substance); the dividing line is at 24 MJ kg^{-1} (= 5,700 kcal kg^{-1}). Lignite C. have lower heating values, and mineral C. have higher ones. For comparison of various types of C., their heating values are converted to tce (Tons of coal equivalent; see), 29.3 MJ kg^{-1} (= 7,000 kcal kg^{-1}).

C. are generally classified as ***coking*** or ***boiler*** C.; internationally, this applies mainly to mineral C. Lignites are divided into the following classes: ***coking*** (for lignite high-temperature coking after Rammler and Bilkenroth), ***briquette, carbonization*** and ***extraction C.***, or ***bituminous*** and ***boiler*** C. Special technological requirements lead to further classification of various types of C. (see Coal chemistry).

Salt C. occupy a somewhat unusual position, due to the conditions of their formation and the present possibilities for their utilization. These are C. with more than 0.5% Na_2O (in the dry C.).

World coal reserves

Continent	Fraction of world coal reserves (%)	Distribution among geological periods (%)					
		Carbon-iferous	Permian	Triassic	Jurassic	Creta-ceous	Tertiary
Europe	19.3	77	2	–	–	–	21
Asia	27.9	43	23	–	28	–	5
North and Central America	45.5	30	3	–	2	33	32
South America	1.1	28	–	–	–	21	51
Africa	2.6	–	88	5	5	1	–
Australia and Oceania	3.5	–	18	2	–	–	79

American system of coal classification

Class	Group	Fixed carbon, % (Dry, free of mineral matter)	Volatile matter, % (Dry, free of mineral matter)	Heat content Btu/pound (Moist, free of mineral matter)
I Anthracite	1. Meta-anthracite	>= 98	<= 2	
	2. Anthracite	92–98	2– 8	
	3. Semianthracite	86–92	8–14	
II Bituminous	1. Low volatile bituminous	78–86	14–22	
	2. Medium volatile bituminous	69–78	22–31	
	3. High volatile A bituminous	<69	>31	≥ 14,000
	4. High volatile B bituminous	–	–	13,000–14,000
	5. High volatile C bituminous	–	–	11,500–13,000
III Subbituminous	1. Subbituminous A	–	–	10,500–11,500
	2. Subbituminous B	–	–	9,500–10,500
	3. Subbituminous C	–	–	8,300– 9,500
IV Lignite	1. Lignite A	–	–	6,300– 8,300
	2. Lignite B	–	–	≤ 6,300

The American Society for Testing and Materials uses the classification system on p. 233, which is based on the carbon content and heating values.

Conditions for C. formation. The basic requirement for formation of humus coals is the existence of land plants, which was first met at the boundary between the Silurian and Devonian periods, about 400 million years ago. The evolution of plants has led, in the course of geological time, to increasingly diversified and more highly specialized groups. The associated changes in anatomy, biological functions and composition of the vegetable starting materials is to some degree still visible in the different characteristics of C. from different geological periods.

The C. seams of the present are the bogs of past geological ages. Bogs form only in certain geographical locations and change in a predictable way during the course of geological development. The sinking of the ground, the height of the ground-water level, the soils, climate, organic rate of growth and the evolution of characteristic plant communities in the paleobog are closely related to the processes of C. formation.

A necessary condition for the preservation, and thus the accumulation of large amounts of vegetable matter, is that it sink below the ground-water level. Otherwise, the plant material is completely decomposed. Thus the sinking of the ground and the upward growth rate of the plants must be balanced. The causes of the different characteristics of C. are manifold and complex. Lignite C. especially, as products of biochemical carbonization only, are characterized by the biological and chemical variety of the vegetation zones from which they arose. The process of coal formation is still not completely understood, and there are a number of different theoretical models of it. For that reason, only a few aspects of C. geology will be presented here.

The *biochemical phase* of carbonization (diagenesis) begins immediately after the plant matter dies. It extends through the formation of peat and its conversion to lignite C. The biochemical conversion of the organic substance by microbes (bacteria, fungi, etc.) begins in an aerobic (oxidation) zone, then continues in an anaerobic (reducing) zone. Most of the plant substances, the cellulose and lignins, are converted to humus (huminic acids, humins, humates). Resins and waxes are not greatly changed. Under distinctly aerobic conditions, the relative amounts of these compounds in the substrate increase, forming the protobitumina of the C. The intensity of the decomposition processes in the relatively narrow aerobic zone has a decisive influence on the structural characteristics of the soft lignite C. This intensity of decomposition and humification of the plant matter depends mainly on climate, geochemical milieu, the degree of aerobicity and the geological facies in the paleobog.

As the plant matter sinks and is covered, it enters the anaerobic conditions of the reducing zone. This latent state of biochemical carbonization is reached in the peat stage, and it continues over a very long period of geological development until the stage of lignite is reached. As the material sinks deeper and is pressed by the weight of the material above, it becomes more solid and loses water, until the peat is converted to lignite, the end stage of biochemical carbonization.

If geological processes lead to higher temperatures, *geochemical carbonization* (metamorphosis) occurs. This is an epigenetic process, and begins with the dull lignite stage. The results are glance lignite, mineral C. and anthracite. As geochemical carbonization progresses, the phytogenic differences in the C. are lost.

The main factor in the metamorphosis of C. is the length of time it is exposed to high temperatures, which in turn depends on geological processes, such as the sinking of the C. seam to greater depth (geothermal depth stage), large-scale magma motions (plutonism, volcanic covering), local magma motions (galleries, sills, pipes) or hydrothermal effects. These factors determine the regional and local variations in the carbonization in the seams and deposits of C., and they are of great importance to the value of C. as raw materials. In the present view, pressure has only a modifying effect on the geochemical process.

It is thought that the geochemical conversion of organogenic substances begins in the temperature range of 100 °C to 150 °C. The marked petrographic and material changes which occur with the vitrinization of dull lignite to glance lignite are characteristic of the *1st carbonization jump*.

In the course of further geochemical carbonization, there are characteristic physical-chemical changes in the C., such as an increase in its carbon content, a decrease in its contents of oxygen, hydrogen and volatile components, and an increase in its density. These changes do not occur evenly over the stages of carbonization.

In the carbonization stage between gas and fat C.,

	Carbon (dry and ash-free) %	Hydrogen (dry and ash-free) %	Volatiles (dry and ash-free) %	Water %
Wood	50	6	–	–
Peat	59	6	63...70	90...75
Soft lignites	60...65	5 ...6	56...63	35...75
Dull lignites	65...71	5 ...6	49...56	25...35
Glance lignites	71...77	5 ...6	43...49	8...10
Flame coals	76...80	4.5...5.5	40...43	5... 6
Gas flame coals	80...83	5.0...5.8	34...40	4... 5
Gas coals	83...86.5	5.0...5.8	28...34	2... 4
Fat coals	86.5...89	5.0...5.5	19...28	1... 2
Forge coals	89...91	4.5...5.5	14...19	1
Lean coals	91...92	4.0...4.5	10...12	1
Anthracite	92...96	2.0...4.0	10	0.5... 1

the bituminous components (wax, resins, spores, pollen, etc.) of the C. are decomposed, and there is an increased release of CH_4 (*2nd carbonization jump*). This is the reason for the increased danger of mine gas in the mining of seams in this carbonization range, and also for the formation of natural gas. The C. which are most suitable for coking are found in the range of fat to forge C.

The processes of C. formation are an aspect of geological development which fits well into the current concepts of mountain formation and plate tectonics. The interaction between the tectonic regime of geological structures, such as sinking and magma flows, and C. formation are recorded in the structures of C. deposits. There are three major groups of C. deposits: 1) geosynclinal, 2) plateau and 3) intermediate types. These can be further subdivided into many groups and individual types on the basis of their characteristics.

1) *Geosynclinal* deposits account for about 40% of the world reserves of C. The most important types are:

a) Deposits formed in subducting regions. The productive (seam-bearing) series of layers in this type can be 10 to 18 km thick. In general, there is a large number of relatively thin (0.7 to 2.5 m) seams (sometimes over 100). Extensive folding and breaking of the layers lead to rather complicated deposits in which all stages of carbonization are found. Some important examples of this type are the deposits in the Pechora (Russia) and Donets (Ukraine) Basins, Górny Śląsk (Poland), the Ruhr Basin (Germany), Liege Basin (Belgium), Valenciennes (France), South Wales Basin (United Kingdom), Appalachian Basin (USA) and the Sydney Basin (Australia).

b) C. deposits which form in intramontane basins (molasses basins). The productive series are generally between 300 and 900 m thick. These deposits are generally small in area, and often have only a few, somewhat thicker seams. The seam-bearing series is not folded, or only slightly, but it is more severely broken. The C. are mainly mineral C. Typical examples include the Saar-Lothring Basin (Germany, France), the basin in the French central plateau, the basin of Kladno (Czechoslovakia) and many intramontain C. basins in the young ranges of the Alps, Andes, etc.

2) *Plateau-type deposits*. a) About 34% of the world C. reserves are in deposits in sinking interiors of plateaus. They generally have very large areas (some over 100,000 km^2). Only a few seams are formed (1 to 6), and their extent and thickness (10 to 15 m) vary considerably. The seams are horizontal to somewhat undulating, but they are not tectonically distorted. The C. in this type of deposit is generally lignite. In a few deposits, lava layers above the C. have caused thermal metamorphosis (glance lignite to mineral C.). Some typical deposits of this type are the central and southern Chinese basins, the Michigan and western intercontinental basins in the USA, and the Moscow Basins in Russia.

b) C. deposits on sinking edges of plateaus. This type of deposit covers a large area (up to 50,000 km^2), and has 1 to 3 seams of considerable thickness (50 to 60 m, sometimes as much as 200 m). The deposits are characterized by gently undulating to horizontal positions of the seams, with few if any faults. The C. are lignites to mineral C. with low degrees of carbonization. This type of deposit has so far been found only at the outer edge of the Siberian plateau (Kansk-Atschinsk, Irkutsk, South Yakut). These deposits alone account for about 12% of the world reserves.

c) C. deposits in regions of salt structures and reactivated structures of the underlying foundation. These deposits are relatively small in area, contain only a few seams (1 to 3), some of them thick (10 to 60 m). Just under 9% of the world reserves are in this type of deposit, which contains lignites.

The first formation of humus C. could not have begun before the Silurian/Devonian boundary (emergence of land plants). Worldwide, the formation of C. was not associated with any one geological period. The tectonic processes which occured on the present continents at various times insured that C. and C. deposits were formed in essentially all periods.

Coal production in the major producing countries (1979)

	Mineral coals [Mt]	Lignite coals [Mt]	Fraction of world coal production [%]
USA	667	37	19.1
China	663	n. a.	17.9
USSR	496	161	17.7
East Germany	–	256	6.9
Poland	201	38	6.5
West Germany	93	131	6.1
United Kingdom	120	–	3.2
Australia	84	32	3.1
India	105	3	2.9
South Africa	96	–	2.6

Classification, characteristics. As products of biochemical and/or geochemical carbonization, C. are complicated mixtures of substances, characterized by extraordinarily varied petrographical, chemical and physical properties. This variety in the composition and texture of C. determines the possibilities for use of C.

The main classification parameters are used mainly for statistical comparisons; among lignite C., they are water content, tar content, sometimes heating value, and sometimes ash content. The parameters for mineral C. are content of volatile components, carbon content, various parameters of coking behavior (Roga, swelling indices, etc.), sometimes heating value and sometimes ash content.

Macro- and micropetrographic studies and other special petrological and technological methods are now used in addition to the physical-chemical parameters.

World C. reserves and production. The exact size of the C. reserves of the world and individual countries cannot be stated. There are differences between countries in the principles on which the reserves are estimated, and in the geological knowledge of the territories. At the 10th World Energy Conference, the world resources were given as $10{,}125 \cdot 10^6$ tce. Only $636 \cdot 10^6$ tce, that is, 6.3% of the world resources, were considered economically and technologically usable reserves. In spite of this, the world C. reserves represent 5 times as much energy potential as the presently known petroleum reserves of about $135 \cdot 10^6$ tce.

World coal reserves (in tce). From the World Energy Conference 1977.

	Geological resources		Reserves	
	Anthracite	Lignite	Anthracite	Lignite
Europe (without the USSR)	535 664	55 241	94 210	33 762
Asia (incl. USSR)	5 494 025	887 127	219 226	29 626
America	1 306 541	1 408 838	126 839	71 081
Africa	172 714	190	34 033	90
Australia and South Pacific	213 890	49 034	18 164	9 333

Of the geological resources and reserves, about 75% are mineral C. About 25% of the reserves are estimated as coking C. About one third of the reserves can be reached by strip mining. About 50% of the mineral C. reserves are in seams deeper than 650 m and about 80% are in seams less than 1 m thick.

The world C. reserves are geographically widely distributed. Deposits are known in more than 80 countries. About 88% of them, however, are concentrated in the former USSR, USA and China; the USSR had 48% of the total, and of these more than half, i.e. 25% of the world reserves, are in Siberia. 91% of the world C. reserves are distributed in the 10 countries with the highest C. production. About 95% of the world reserves are found in the northern hemisphere. It is possible that this ratio will change as exploration of Latin America, Africa and Southeast Asia continues.

Extraction. a) Lignite C. seams are usually found at modest depths under thin or non-consolidated layers of rock. This makes possible the extraction of most (88%) lignite C. by strip mining. At present, depths of 300 m are reached by strip mining, and in the near future, it is expected that depths of 500 m will be reached. A strip mine can produce about 60 Mt per year.

Mineral C. have traditionally been mined underground, and about 70% is still obtained from deep mines. The greatest depth is now about 1,500 m. Methane, C. dust explosions, the mine climate and problems of mining mechanics usually make deep mining very difficult. Therefore, there is an increasing tendency to strip mine mineral C. as well, and new technologies to handle the more consolidated overburden and the more compact C. are being developed.

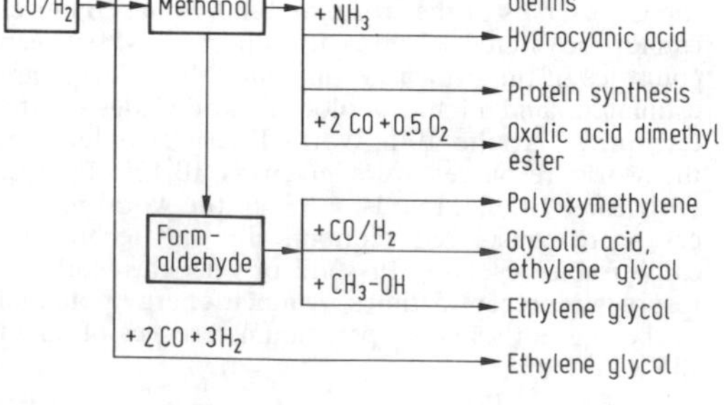

Fischer-Tropsch synthesis
Paraffins
Modified Fischer-Tropsch synth.
Ethene / propene
Alcohol synthesis
C_2 to C_4 alcohols
$+CO/H_2$
Ethanol and higher alcohols
+ CO
Acetic acid
MTG process
Fuels, aromatics olefins
CO/H_2
Methanol
$+NH_3$
Hydrocyanic acid
Protein synthesis
$+2\,CO + 0.5\,O_2$
Oxalic acid dimethyl ester
Polyoxymethylene
Form-aldehyde
$+CO/H_2$
Glycolic acid, ethylene glycol
$+CH_3\text{-}OH$
Ethylene glycol
$+2\,CO + 3\,H_2$
Ethylene glycol

Coal chemistry: a sub-field of Coal refining (see) dealing with applications of coal or coal products. Gaseous compounds such as carbon monoxide, hydrogen and methane, liquids such as light and medium oils, phenols and pyridine bases, and solid coke are obtained from coal refining processes such as gasification, carbonization and extraction. Before the Second World War, most products of the chemical industry were derived from coal. With the introduction of the cheaper and more easily handled petroleum, the significance of C. declined in the 1950s, but with the increases in petroleum prices in the 1970s, interest in C. was renewed. In addition to the classic methods of C., such as the Fischer-Tropsch synthesis and methanol synthesis, syntheses starting from carbon monoxide and hydrogen or methanol are of greatest current interest.

Coalescence: the fusion of colloidal particles into larger particles or a compact phase. C. can be observed directly in foam bubbles and emulsion droplets. With solid particles, direct contact cannot always be recognized immediately, because the fusion proceeds by a crystallization process, and it is slow.

Coal extraction: a method of Coal refining (see) for obtaining desirable compounds by extraction of coal with a pressurized solvent. In the *Pott-Broche process*, a tetralene-cresol mixture is used at 400 °C and a pressure around 10 MPa to extract most of the organic substances from hard coal. The solution is depressurized and filtered, and the solvent is distilled off in vacuum. The residual coal extract (montane wax) is a very good starting material for coking to electrode coke.

In 1967, the Consolidation Coal Co. (West Virginia, USA) built a pilot plant with a capacity of 20 t per day based on the *CSF process*. A hydrogen-generating solvent is used at a pressure of 10 to 30 MPa to extract the hard coal. At present, C. is not economically important; processes for Coal hydrogenation (see) with limited amounts of hydrogen are more important (e.g. the SRC method). However, these are closely related to C.

Coal gas: a fuel gas made by degassing hard coal.

Coal gasification: a method of Coal refining (see) in which anthracite, lignite, peat, coke or lignite briquets are converted to heating or Synthesis gas (see) The process takes place in a Generator (see) with the addition of air or oxygen. If the process is done with air, the gas (see Generator gas) has a heat value of only about 5000 kJ m^{-3} (s.t.p.), due to the large amount of nitrogen in air. A heating gas with a heat value of about 10,000 kJ m^{-3} (s.t.p.) is obtained if steam is passed through the hot coal or coke layer.

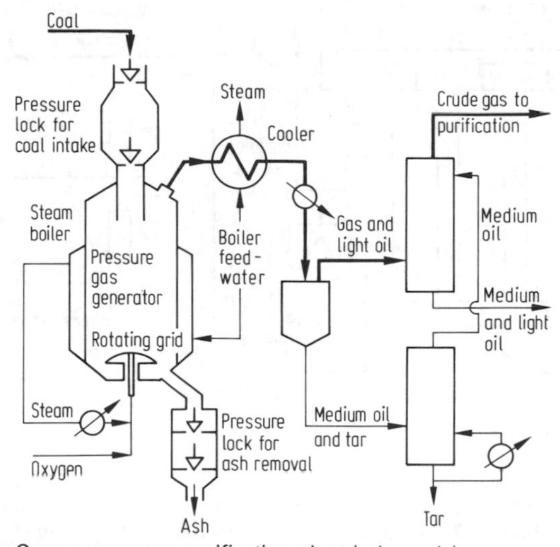

Oxygen pressure gasification plant (schematic)

This type of gas is called Water gas (see). Since the steam process is endothermic, the coke layer cools and must therefore be reheated by passing air through it. Therefore, water-gas generators are driven by passing air and steam through them in alternation. In modern plants for heating and synthesis gas, the process is run in a rotating grid generator with a mixture of steam and oxygen under pressure (***pressure gasification***) (Fig.).

The exothermal reaction with oxygen, $2\,C + O_2 \rightarrow 2\,CO$; $\Delta_R H = -218$ kJ mol^{-1}, runs simultaneously with the endothermal reaction with steam, $C + H_2O \rightarrow CO + H_2$; $\Delta_R H = +130$ kJ mol^{-1} at approximately 1000 °C. When anthracite or lignite coal is used, the coal in the upper part of the generator undergoes Carbonization (see). As coal is added to the top of the generator, it passes first through the heating zone (50 to 500 °C), the carbonization zone (500 to 800 °C), and then the gasification zone (800 to 1000 °C). Finally, the ashes collect under the rotating grid, and are removed by a pressure hose. The steam generated in the water jacket is used as process steam for the gasification. The gas leaving the generator passes through a cooling unit where it generates steam, and is then further cooled with water. In the process, tar and middle oil are precipitated. Light oil is removed from the gas stream by washing with middle oil. The crude gas passes into a purifier, where it is freed of sulfur-containing impurities (hydrogen sulfide, carbon oxygen sulfide) and carbon dioxide. As natural gas supplies are depleted, the pressure gasification of coal will become more important. Natural gas has a much greater heating value; therefore the synthesis gas from the pressure process is methanized: $CO + 3H_2 \rightleftharpoons CH_4 + H_2O$, $\Delta_R H = -205$ kJ mol^{-1} (synthetic natural gas).

Pressure gasification of coal can also be used to generate synthesis gas.

Coals with high ash contents and those which are difficult to mine can be gasified in situ. (The same type of process can be used with oil shales). Holes are drilled into the seam at several sites, and the coal is

Gas	Composition in %					Heating value
	H_2	CH_4	CO	CO_2	N_2	kJ m^{-3}*
Coke gas from anthracite	53	25	6	2	12	19000
Coke gas from lignite	35	15	19	19	12	12500
Carbonization gas from lignite	8	12	9	18	53	7000
Generator gas	–	–	32	3	65	5000
Water gas	50	–	40	5	5	10000
Pressure gas from lignite	50	25	20	3	2	19000

* (s. t. p.)

ignited; after this, air is pumped in. The gas, a mixture of carbon monoxide, hydrocarbons and nitrogen, escapes through the other shafts, and is refined above ground. From time to time, the circulation of air and gas through the various shafts is changed.

Coal hydrogenation: a process of coal refining (coal liquefaction) to obtain liquid products similar to petroleum. It is done by high-pressure hydrogenation of lignite or hard coal. C. was developed at the end of the 1920s in Germany, to meet the rising demand for liquid fuels. In the classic ***Bergius process***, finely ground coal is mixed with iron oxide powder and a heavy oil to form a slurry. This is hydrogenated at 400 to 490 °C and 23 to 70 MPa pressure. The products are then distilled into gasoline, medium oil, heavy oil and a solid residue. To increase the yield of gasoline, the Bergius process is carried out in two steps. In the first, the coal slurry is hydrogenated in the solid-containing liquid phase, forming some gasoline, but mainly medium and heavy oil. In the second step, which occurs in the gas phase, the medium oil is prehydrogenated and cracked to gasoline.

Table 1. Composition and octane ratingr of the gasoline fractions from liquid-phase hydrogenation of hard coals (in M %)

Phenols in crude gasoline	18
(Boiling point up to 100 °C)	23
Paraffins	32.5
Naphthenes	35
Aromatics	22.5
Alkenes	10
Octane rating (research)	80

1) Liquid phase hydrogenation (Fig. 1). The starting materials are coal, tar, petroleum, tar and petroleum residues and pitch. When tar is used, the fraction boiling up to 325 °C is removed (top distillation); when petroleum is used, top distillation goes up to 340 °C. The distillate is hydrogenated in the gas phase, and the residue is returned to the liquid phase. The coal must be dried down to a water content of 5%, and it must be ground to a grain size of 1 mm. It is then mixed with an oil to produce a product which can be pumped; this consists of 3 to 5% water, 48 to 50% oil and 45 to 48% solids. The oil is usually the fraction from liquid-phase hydrogenation which boils above 325 °C. A catalyst of bauxite residue, or brown iron ore (45% Fe_2O_3, 55% water) is used; the amount of catalyst (contact) used is 6% of the coal mass. The starting material, including the contact, is brought to the reaction pressure, and compressed hydrogen is added. The coal-gas or oil-gas mixtuer is passed over heat exchangers into the preheater, which brings it to

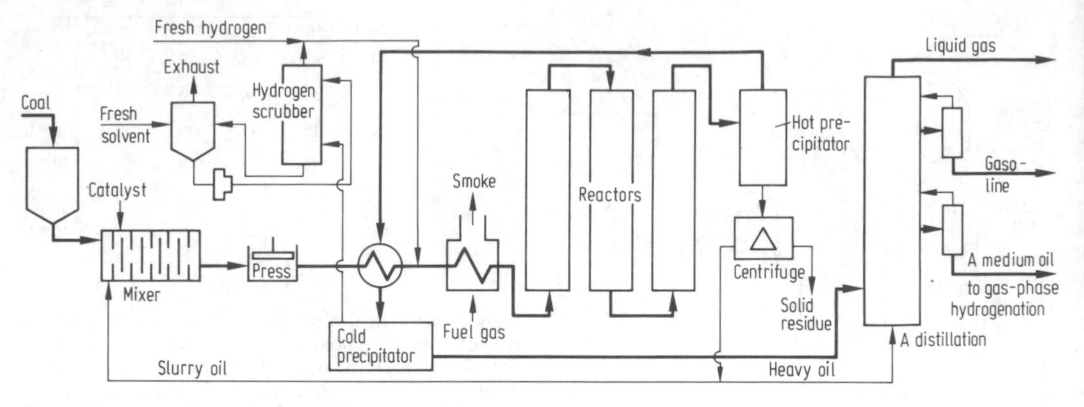

Fig. 1. Diagram of liquid-phase hydrogenation in the Bergius process

the reaction temperature of 470 to 490 °C. Hydrogenation of lignite coal occurs at 20 MPa pressure; hydrogenation of hard coals, at 70 MPa.

The reactors consist of chromium-nickel steel pipes, 12 to 18 m long and with interior diameter 0.8 to 1.2 m; the volume is thus 9 to 13 m^3. The slurry stays in the reactor for about 30 minutes. The product then leaves the reactor, passes through a heat exchanger, and enters a hot precipitator, in which the temperature is 10 to 40 °C lower than the reaction temperature. The lighter volatile oils and the hydrogen leave the precipitator as gases; the residue, which contains liquid, asphalt-like substances and solid contact, is drawn off as a sludge of 20 to 38% solid. Part of this is returned to the circulation, but the larger fraction is exchanged for new contact slurry.

The gaseous hydrocarbons are cooled further, first in heat exchangers and then in coolers, until they liquefy. They are then depressurized in slanted vessels, the strippers, in two steps. This releases a large amount of gas, which was dissolved at the high pressure. The first depressurization releases *poor gas*, which contains mainly hydrogen, and the second, *rich gas*, which contains large amounts of hydrocarbons, especially propane and butane. The condensate remaining after depressurization is drawn off; it consists of an oil layer, the stripper product, and an aqueous layer. The stripper product is distilled into gasoline, A-medium oil boiling between 180 and 325 °C, and heavy oil. The heavy oil, which makes up 40% of the product, is returned to the liquid phase hydrogenation. The A-medium oil goes into the gas phase hydrogenation, unless it is used directly as diesel oil. The gaseous hydrocarbons, such as propane and butane, are compressed and are used as fuels.

After the depressurization of the sludge, that part which is to be exchanged for new contact is separated by centrifugation. The solid residue is carbonized, and the oil supernatant, along with the carbonization oil, is added to the coal powder.

The heat of reaction from C. can be removed by adding cold hydrogen. As a rule, 3 to 5 reactors form a unit, which is housed in a chamber of reinforced concrete, for safety reasons.

2) Gas-phase hydrogenation (Fig. 2). In this step, cracking hydrogenation produces gasoline, B-medium oil and lubricating oil. Gas-phase hydrogenation usually is done in two steps, pre-hydrogenation and gasoline production.

a) Pre-hydrogenation. The A-medium oil from liquid-phase hydrogenation or corresponding fractions from distillation of tar or crude oil and hydrogen are

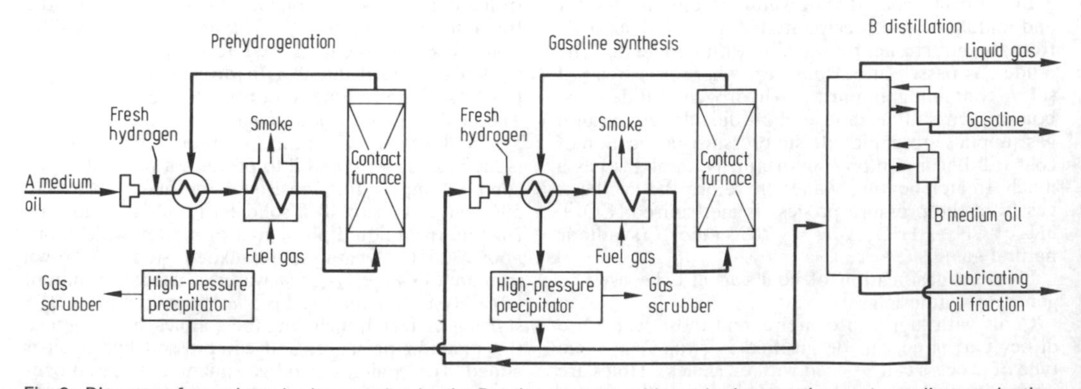

Fig. 2. Diagram of gas-phase hydrogenation by the Bergius process, with pre-hydrogenation and gasoline production

preheated to 300 °C in heat exchangers; the heat is supplied by the reaction products. The temperature is then brought to 360 to 445 °C by a heater. The reaction is catalysed by a mixture of 70% aluminum oxide, 27% tungsten sulfide and 3% nickel sulfide. The effect of the mixture is mainly a refining one; it does little cracking. The pre-hydrogenation converts harmful compounds of oxygen, sulfur and nitrogen into water, hydrogen sulfide and ammonia. The process and equipment for pre-hydrogenation are similar to those for liquid phase hydrogenation, that is, the medium oil is pressurized and led into the reactor with compressed hydrogen, via the heat exchanger and preheater. The basic difference is that that the solid contact mass is arranged on grids in the reactor. The hydrogen and hydrocarbons are passed through the reactor from top to bottom, to avoid stirring up the contact, which remains active for 1 to 2 years. The reaction product is separated, after cooling and depressurization in a stripper, into liquid and gaseous fractions. The liquid fraction is distilled into pre-hydrogenation gasoline and unreacted medium oil, which is either used directly as high-quality diesel oil, or for gasoline production. The gasoline has a low octane rating, and must be reformed.

Table 2. Composition and octane rating of the gasoline fractions from gas-phase hydrogenation

	Lignite	Hard coal
Boiling begins (°C)	45	44
Boiling ends (°C)	139	156
Parafins	51.5	37.5
Alkenes (in M %)	1	1
Naphthalenes	38	53
Aromatics	8.5	8.5
Octane	≈ 83	≈ 86

b) Gasoline production. Medium oil is cracked around 400 °C and at hydrogen pressure of 23 to 30 MPa. The fixed contact consists of 10% tungsten sulfide on fuller's earth or aluminum silicate as a carrier. The gasoline produced in the process has a relatively high octane rating. The system is similar to the one used for pre-hydrogenation, although the gasoline is not produced in a single step. After distillation of the product, the non-gasoline fraction is returned to the reactor for further cracking.

Attempts have been made to combine the pre-hydrogenation and gasoline production into a single step, using a combination contact of tungsten sulfide on fuller's earth activated by hydrofluoric acid. However, only medium oils with low contents of oxygen and nitrogen are suitable for this process. With the Bergius process, 14 t crude lignite coal are needed to produce 1 t gasoline; of this amount, only 37% is used in the actual hydrogenation, and the rest is used to produce hydrogen (40%) and to provide the energy for the process. The energy efficiency, relative to the yield of gasoline and liquefied gas, is 36%. The energy requirements could be reduced by 25 to 30% in modern plants, which is equivalent to increasing the energy efficiency from 36 to 50%.

Because of the high consumption of hydrogen and the expensive high-pressure installations, C. became uneconomical by the end of the 1930s. Most plants therefore no longer use coal, but instead process tars from carbonization or residues of petroleum refining. The ***low-temperature, high-pressure hydrogenation process*** (see Hydrorefining) was developed from this type of processing; it produces long-chain n-paraffins in addition to gasoline and medium oil. After World War II, the hydrogenation plants were converted to the more economical refining of petroleum.

As the prices of petroleum have risen, C. has again become interesting (see Coal chemistry). International attention is focused on the improvement of the Bergius process by reducing the pressure from 70 MPa to 30 MPa, with a simultaneous reduction in hydrogen consumption and changes in the processing of residues. The separation and processing of the hydrogenation sludge from liquid-phase hydrogenation present a key problem. The centrifugation and carbonization of the centrifugation residues can no longer be carried out in modern large plants. The centrifuge supernatant oil, which in the Bergius process was added to fresh coal to make the slurry, contains large concentrations of asphalt, which was always returned to the reactor and could only be degraded by a high hydrogen pressure. In newer methods, the hydrogenation sludge is withdrawn from the hot precipitator at a temperature of about 400 °C, and

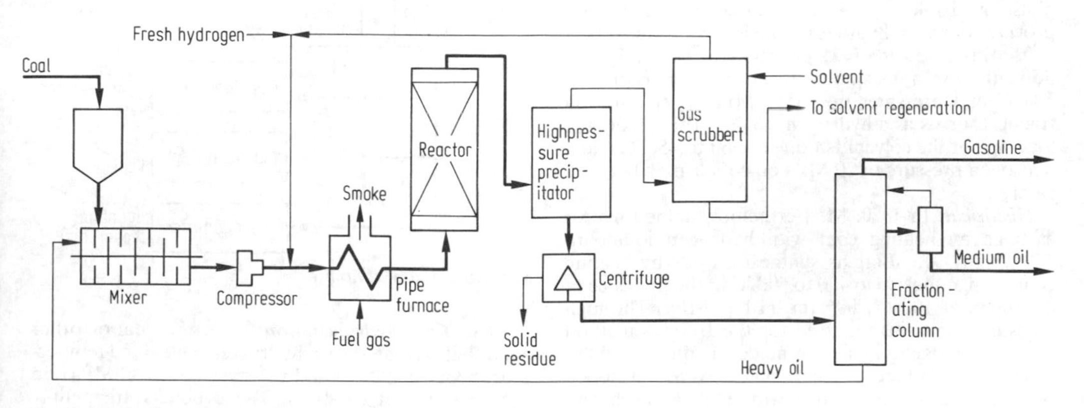

Fig. 3. Synthoil process

is distilled in vacuum. The vacuum distillate does not contain asphalt, and is used to make the coal slurry. The asphalt-containing residue is passed into a pressure gassification process to produce hydrogen. In the ***synthoil process*** (Fig. 3), a vigorously stirred slurry of coal in heavy oil is led past a cobalt-molybdenum carrier catalyst at 450 °C, hydrogen pressure of 12 MPa. The light gasoline is separated in a precipitator, and the residue goes to a centrifuge. The liquid products are distilled, and the filter is burned in a power plant or used to generate synthesis gas. The heavy oil from the distillation is returned to the process to form the coal slurry.

In the ***H-coal process***, an oil-coal suspension is passed through a reactor at high velocity. The coal slurry and hydrogen are led into the reactor from below, and bubble up through the ebuliated-bed catalyst of cobalt-molybdenum cylinders. The products leave from the top of the reactor; the catalyst is continuously exchanged to maintain its activity. The reaction temperature is 450 °C, and the pressure, 20 MPa. The circulating oil is separated by a hydrocyclone, and the solids are removed by precipitation with a solvent. The stripper product has the following composition:

Boiling range, °C	Mass %
< 50	11
50–200	20
200–340	11
340–520	12
Residue	38
H_2S, NH_3, CO_2, H_2O	8

The ***COED process*** is a combination of carbonization and hydrogenation. The main product is coke. The carbonization tar is refined in a hydrogen pressure process.

In addition to C., Coal extraction (see) under pressure with a solvent is becoming more significant. Catalysts are not generally used for this. The ***SRC process*** (abb. for Solvent Refined Coal process) is a combination of C. and coal extraction. The coal is made into a slurry with a tar oil distillate, then cracked at a hydrogen pressure of 10 MPa, without a catalyst. About 90% of the coal is dissolved. The solids are removed by special filters, and the liquid phase is distilled. Another indirect hydrogenation process, in which hydrogen transfer from the solvent is used, is called the ***EDS process*** (for Exxon Donor Solvent). At a maximum reaction temperature of 480 °C and pressures up to 17 MPa, a medium oil fraction serves as a hydrogen carrier. The "hydrogen loading" of the solvent is done at 360 to 450 °C and a hydrogen pressure of 20 MPa on cobalt-molybdenum catalysts.

Historical. In 1870, M. Berthelot obtained hydrocarbons by heating coal with hydrogen iodide to 270 °C. In 1913, Bergius synthesized oils by heating coal in an autoclave to 400 to 500 °C in the presence of hydrogen at 100 to 200 atm. (10-20 MPa) The high pressure apparatus required for the Bergius method of coal liquefaction must be made of sulfur- and hydrogen-resistant steels. The reactors were developed on the basis of experience with Haber-Bosch ammonia synthesis. In 1924, M. Pier discovered the first catalysts which were not poisoned by the sulfur in the raw materials, and divided the process into the liquid and gas phases. In 1927, the first large plant was built in Leuna; production reached 100,000 t fuel per year. Other plants were later built in Zeitz and Bœhlen. Between 1927 and 1943, 12 C. plants with a total annual capacity of about 4 million tons liquid products were built. In the USA, C. did not become significant, since petroleum was abundant there. In England (Billingham), a hydrogenation plant with a capacity of 150,000 tons gasoline per year was built in 1923.

Coal liquefaction: a method of coal refining in which liquid hydrocarbons are produced from coal or synthesis gas (carbon monoxide/hydrogen). Coal hydrogenation (see) and the Fischer-Tropsch synthesis (see) are the large-scale processes used for C.

Coal refining: a term for all the processes for treatment of coal and its products.

1) Physical refining of coal includes preparation for various processes, including making briquets (see Coal).

Thermal refining. When substances of vegetable origin, e.g. coal, oil shale, peat or wood, are heated in the absence of air (dry distillation), the product decomposes into gas, liquid products and coke. The yield and chemical nature of the decomposition products depend on the nature of the raw material and the conditions under which the dry distillation is done. Decomposition within a temperature range of 450 to about 600 °C is called ***Carbonization*** (see) or low-temperature degassing. At temperatures above 900 °C, the process is called ***Coking*** (see) or high-temperature degassing. Degassing at intermediate temperatures is becoming more important.

2) Chemical refining.

a) Degassing includes chemical reactions of combustible substances, e.g. coal, peat or coke, with gas-forming reagents (see Coal gasification).

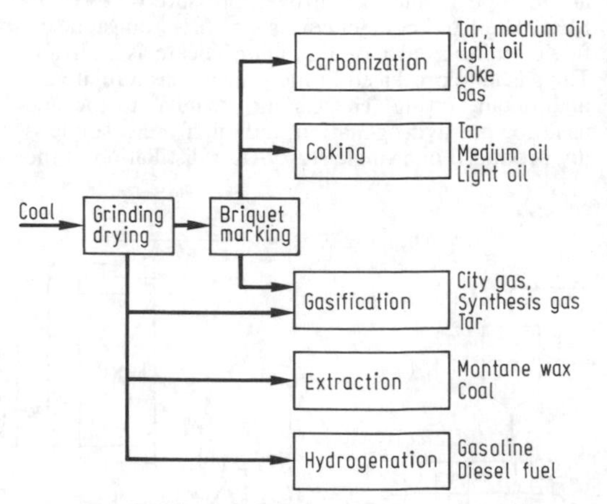

Methods of coal refining

b) In ***Coal hydrogenation*** (see), the coal or other solid fuel is treated with hydrogen at elevated temperature (350 to 490 °C) and pressure (10 to 40 MPa), in the presence of catalysts. The products are mainly mixtures with low boiling points, such as gasoline.

c) In addition to the above refining processes, ***Coal extraction*** (see) can be used with suitable coals. The coals are extracted with gasoline, medium oil and oxygen-containing solvents at high pressures (about 10 MPa) and temperatures (about 400 °C) to obtain liquid fuels and other products, such as montane wax.

Historical. It has long been known that coal, when heated in the absence of air, releases a combustible gas. Complete extraction and utilization of the valuable substances present in coal did not become possible until the technology developed. The first small-technological-scale use of coke instead of charcoal for reduction of ore was in 1585. The first large-scale use of coke as a reducing agent was in 1708, in England, where it was discovered how to smelt a useful crude iron in this way. Starting around 1880, modern hard-coal coking technology was developed in Belgium and Germany. In the 1930s, the first pilot plant for lignite coal coking was started, and in 1954, the first large-scale plant, developed by G. Bilkenroth and E. Rammler, went into production. The byproducts of coking have become the basis for a modern chemical industry, producing tar, dyes, explosives, pharmaceuticals and plastics.

Coal water gas: see Water gas.

Cobalamin: see Vitamin B_{12}.

Cobalt, symbol ***Co***: an element in group VIIIb of the periodic system, the Iron group (see). Co is a heavy metal with a single natural isotope, *Z* 27, atomic mass 58.9332, valence primarily II, III, but also - 1, 0, IV, V; Mohs hardness 5.5, density 8.89, m.p. 1492 °C, b.p. 3100 °C, electrical conductivity 16.3 Sm mm $^{-2}$, standard electrode potential (Co/Co^{2+}) - 0.277 V.

Properties. C. is a steel-gray, shiny metal which is hard, strong and ductile. It exists in two modifications: a hexagonal α-form stable at room temperature and a cubic face-centered β-form to which the α-form is converted slowly above 417 °C. Co is ferromagnetic; its Curie temperature is 1121 °C. Compact Co is stable to air and water at room temperature, while finely divided Co is pyrophoric. Non-oxidizing acids attack Co only slowly, but dilute nitric and sulfuric acid dissolve it with evolution of hydrogen. Concentrated nitric acid passivates Co. At elevated temperatures, Co reacts with oxygen to yield the cobalt oxides CoO and Co_3O_4, and it also reacts with other nonmetals, such as the halogens, phosphorus, sulfur, arsenic, silicon and boron, to form binary compounds. Co forms dicobalt octacarbonyl with carbon monoxide at 200 °C and 10 MPa.

In its compounds, Co is generally in the +2 or +3 oxidation state. Both complex and non-complex Co(II) compounds are stable, and Co(III) forms very stable octahedral complexes. However, cobalt(III) salts are very unstable. Most cobalt(III) complexes are diamagnetic, and are characterized by high kinetic stability. Cobalt(II) forms both octahedral and tetrahedral complexes; π-acceptor ligands stabilize the lower oxidation states of C. (see Cobalt carbonyls), while π-donor ligands, such as fluoride or oxide, make possible the higher oxidation states (+IV, +V). C. is a vital bioelement, found, for example, as the central atom in the corrin ring system of vitamin B_{12}.

Occurrence. C. makes up about $3.7 \cdot 10^{-3}$% of the earth's crust. C.-containing copper ores and magnetic pyrites are essential sources of C. production. The most important C. minerals are smaltite, $CoAs_{2-3}$, cobaltite (cobalt glance), CoAsS, linneite (cobalt pyrite), Co_3S_4, and erythrite (cobalt blossom), $Co_3(AsO_4)_2 \cdot 8H_2O$. C. is present in the soil as a trace element, at about 8 ppm, and in plants as about 0.2 ppm of the dry matter.

Production. C. production begins with the prepared ores from lead, copper or nickel production, or with C. ores. These contain C., nickel and copper as sulfides or arsenides. They are roasted with soda and nitrates to form the metal oxides, which are called *zaffer*, or sulfates and arsenates. The latter are leached out with water, and the metal oxides, which are insoluble in water, are dissolved in hot hydrochloric or sulfuric acid. The solution is fractionally precipitated with lime milk and lime chloride. The final product is cobalt(II,III) oxide, which is reduced to the metal, usually with coke: $Co_3O_4 + 4\,C \rightarrow 3\,Co + 4\,CO$.

Applications. Most of the C. produced is used as a component of alloys (see Cobalt alloys). C. plays an important role as a metal binder in the production of hard metals. Because of its hardness and resistance to corrosion, C. is also used in electroplating. The most widely used cobalt compounds are smaltes, which are used as pigments in ceramics and glasses; other important compounds are used as catalysts in oxo synthesis and the Fischer-Tropsch process. The radioactive nuclide ^{60}Co, which is obtained by irradiating ^{59}Co with thermal neutrons in a nuclear reactor, is used as a γ-source ($t_{1/2}$ = 5.26 a) in medicine; it is used for radiation therapy of cancer and in industry for materials testing.

Historical. Medieval miners gave the name "Kobalt" (a mischievous or evil sprite) to ores which resembled copper or silver ores but, when smelted, did not yield the desired metal. Later, C.-containing ores which were difficult to smelt but gave a blue color to glass fluxes were given this name. The metal was discovered in 1735 by Brandt.

Cobalt(II) acetate: $Co(CH_3COOH)_2 \cdot 4H_2O$, red, monoclinic, water-soluble crystals; M_r 249.08, density 1.705. C. is obtained, for example, by reaction of cobalt(II) carbonate with acetic acid. It is used as a bleach and drying agent in paints and lacquers, and to produce cobalt protective layers by electroplating. It is also used in the synthesis of catalysts and as a pigment for anodizing aluminum.

Cobalt alloys: alloys in which cobalt is the major element. 1) The Superalloys (see) are resistant to high temperatures. 2) Stellite® (see) and Triballoy® are very resistant to abrasion. 3) Certain C. have high thermal dimensional stability (see Super Invar®); some have a constant modulus of elasticity over a wide temperature range (see Co-Elinvar®, Velinvar®), others have high resistance to metal fatigue (see Elgiloy®), heat (see Cobanik®, Cochrome®) or corrosion in the human body (see Vitallium®).

Cobalt alums: see Cobalt sulfates.

Cobalt blue: a term applied to two cobalt pigments. 1) ***cobalt blue (I), cobalt ultramarine, Thénard's blue*** is cobalt aluminate, $CoAl_2O_4$, a blue powder which is insoluble in water. C. (I) is very fast to light and resistant to high temperatures, alkalies and most acids. It dissolves gradually in hot hydrochloric acid. It is used to make oil, glass and porcelain paints, and in printing money, because it can be identified unequivocally by photochemical methods. It is also used to detect aluminum. A related pigment, ***Leithner's blue***, is obtained by roasting a mixture of cobalt arsenate or phosphate with alumina hydrate. 2) ***Cobalt blue (II), celestine blue, ceruleum*** is cobalt(II) stannate, $CoSnO_3$, a sky-blue powder. C. (II) is used in oil and aquarell paints and for porcelain, due to its stability to heat.

Cobalt(II) carbonate: a compound resulting from addition of alkali carbonates to cobalt(II) salt solutions saturated with carbon dioxide; it precipitates as the hexahydrate, $CoCO_3 \cdot 6H_2O$, M_r 227.03. Commercial C. is usually a partially hydrolysed precipitation product $xCoCO_3 \cdot yCo(OH)_2 \cdot zH_2O$, which has a cobalt content of about 46%. C. is used to synthesize other cobalt(II) compounds.

Cobalt carbonyls: polynuclear cobalt(0) complexes. ***Octacarbonyl dicobalt***, $Co_2(CO)_8$, orange crystals, M_r 341.95, density 1.73, m.p. 51 °C (dec.). $Co_2(CO_8)$ is obtained by heating finely divided cobalt with carbon monoxide to 150-200 °C under pressure (10 MPa); addition of energy produces larger C.: $Co_4(CO)_{12}$, black crystals, M_r 571.68 °C, m.p. 60 °C (dec.), and $Co_6(CO)_{16}$, black crystals, M_r 801.76, m.p. 105 °C (dec.). Reduction of $Co_2(CO)_8$ yields ***carbonyl hydrogen cobalt***, $HCo(CO)_4$, a trigonal bipyramidal, bright yellow compound; M_r 171.84, m.p. - 26 °C. $Co_2(CO)_8$ is used as a catalyst in oxosynthesis; the carbonyl hydrogen cobalt formed under the process conditions is the actual complex-active species.

Cobalt(II) chloride: $CoCl_2$, a blue, sublimable compound which crystallizes in a cadmium chloride lattice; M_r 129.84, density 3.356, m.p. (in HCl atmosphere) 724 °C. C. is readily soluble in water and polar organic solvents, such as alcohol and acetone; it reacts with chlorides in nonaqueous media to form tetrahedral tetrachlorocobaltates(II), $M_2[CoCl_4]$. In moist air, C. slowly turns pink. The process involves formation of various colored hydrates, finally leading to the pink hexahydrate, $CoCl_2 \cdot 6H_2O$, M_r 273.93, density 1.924. This gives up its hydrate water readily, even at 35 °C. The color change is used to indicate the degree of atmospheric humidity, to indicate the moisture content of silica gels used for drying, and to make secret inks. $CoCl_2$ is made by dehydrating its hydrates, or by heating cobalt in a stream of chlorine. It is used in electroplating, as a trace element source for animal feeds, in painting glass and porcelain and for toning in photography.

Cobalt complexes: coordination compounds of cobalt in the - 1 to +5 oxidation states. The most stable C. are octahedral cobalt(III) complexes, most of which are diamagnetic and have high kinetic stability. These complexes are colored (see Cobaltiacs). Both octahedral and tetrahedral types are characteristic of cobalt(II), for example, the pale red cobalt(II) hexaaqua ion $[Co(H_2O)_6]^{2+}$ and the dark blue tetracyanatocobaltate(II), $[Co(NCO)_4]^{2-}$. π-acceptor ligands, such as CO, isocyanides or CN^-, stabilize lower oxidation states of cobalt, as in $Co_2(CO)_8$, $M[Co(CO)_4]$ (see Cobalt carbonyls), $[Co(CNR)_5]^+$ and $[Co(CN)_8)]^{8-}$. The oxidation states +4 and +5, which are unusual for cobalt, are observed in the fluoro- and oxocobaltates $Cs_2[CoF_6]$, $Ba_2[CoO_4]$ and $K_3[CoO_4]$.

Cobalt cyanides. ***Cobalt(II) cyanide***: $Co(CN)_2 \cdot 2H_2O$, red-brown needles; M_r 147.01, density 1.87. Cobalt(II) cyanide is obtained by combining aqueous cobalt(II) salt solutions with alkali cyanides. When it is heated to 200 °C, it is converted to the blue anhydride, $Co(CN)_2$. $Co(CN)_2$ reacts with potassium cyanide to form ***potassium cyanocobaltate(II)***, which is present in solution as the olive-green, paramagnetic $K_3[Co(CN)_5]$ complex. The crystalline form is the violet, diamagnetic dinuclear complex $K_6[Co_2(CN)_{10}] \cdot 4H_2O$. Pentacyanocobaltate(II) is able to take up hydrogen according to the equation $2\,[Co(CN)_5]^{3-} + H_2 \rightarrow 2\,[Co(CN)_5H]^{3-}$; it is therefore able to act as a catalyst for homogeneous hydrogenations.

Potassium hexacyanocobaltate(III): $K_3[Co(CN)_6]$ forms pale yellow, monoclinic prisms, M_r 332.35, density 1.88. It is formed by reaction of potassium cyanide with cobalt(II) salts in the presence of oxidizing agents, and can be reduced with potassium in liquid ammonia to $K_4[Co(CN)_4]$, a brown-violet, pyrophoric compound.

Cobalt fluoride. ***Cobalt(II) fluoride***, CoF_2, rose-red, monoclinic crystals; M_r 96.93, density 4.46, m.p. about 1200 °C. CoF_2 is obtained by reaction of cobalt carbonate with aqueous hydrofluoric acid. Anodic oxidation of a solution of CoF_2 in 40% hydrofluoric acid or the reaction of fluorine with cobalt(II) chloride or cobalt at 250 °C produces ***cobalt(III) fluoride***, CoF_3, which forms light brown, hexagonal crystals; M_r 115.93. CoF_3 is a good fluorination agent.

Cobalt glass: 1) a blue glass colored with cobalt oxide. Because it absorbs yellow sodium light and permits red potassium light to pass, it is used as a filter for detection of potassium by the flame test.

2) a blue pigment; see Smalt (1).

Cobalt green, ***Rinmann's green, turquoise green***: a cobalt pigment produced in various ways by roasting zinc and cobalt salts. It is used as an oil and aquarell pigment for artists' colors and in painting porcelain. C. was first produced in 1780 by Rinmann.

Cobalt group: same as Cobalt triad (see).

Cobaltiacs: a term for amminecobalt(III) complexes. ***Mononuclear C.*** are derived from the hexaamminecobalt(III) cation $[Co(NH_3)_6]^{3+}$ by substitution of another neutral or acidic ligand for one or more of the ammine ligands. The resulting complexes are of the types $[Co(NH_3)_{6-n}L_n]^{3+}$ and $[Co(NH_3)_{6-n}Ac_n]^{3-n}$, (L = neutral, Ac = acidic ligand). In ***polynuclear C.***, the cobalt(III) central atoms are joined by ligand bridges, such as oxide, sulfide, imide, peroxide, hydroxide or amide groups. C. are intensely colored, stable, diamagnetic complexes. Research on these complexes played an important part in the development of complex theory by Alfred Werner. For example, he discovered mirror-image isomerism in C. C. were formerly often named for their colors; for example, derivatives of $[Co(NH_3)_6]^{3+}$ were called luteosalts (from the Latin "luteus" = "yellow orange"); derivatives of $[CoCl(NH_3)_5]^{2+}$, purpureosalts (Latin "purpureus" = "purple"); derivatives of $[Co(NH_3)_5OH_2]^{3+}$, roseosalts (Latin. "roseus" = "rose"); derivatives of $[Co(NO_2)(NH_3)_5]^{2+}$, xanthosalts (Greek "xanthos" = "yellow"); and compounds of the types [CoCl-

$(H_2O)(NH_3)_4]^{2+}$ and $[CoCl_2(NH_3)_4]^+$ in their *trans*-forms were called praseosalts (Latin "prasinus" = "blue-green"), and in their *cis*-forms, flavosalts (Latin "flavus" = "yellow"). Nitrocobaltiacs such as $[Co(NO_2)_2(NH_3)_4]^+$ were called flavosalts in the *cis*-form and croceosalts (Latin "croceus" = "safron yellow") in the *trans*-form.

Cobalt(II) nitrate: $Co(NO_3)_2 \cdot 6H_2O$, red, monoclinic, water-soluble platelets, M_r 291.04, density 1.87, obtained by reaction of cobalt or cobalt(II) carbonate with dilute nitric acid. C. is used as a starting material for synthesis of pigments. If C. is roasted together with aluminum(III) oxide or zinc oxide, blue or green spinels, $CoAl_2O_4$ (see Cobalt blue) or $ZnCo_2O_4$ (see Cobalt green). These spinels are used as a means of detecting aluminum or zinc.

Cobalt oxides. ***Cobalt(II) oxide***, CoO, green-brown, cubic crystals; M_r 74.93, density 6.45, m.p. 1935 °C. CoO is obtained by heating cobalt(II) carbonate in the absence of air. It is used to make blue cobalt pigments. When heated above 400 °C in a stream of oxygen, CoO is converted to the blue-black, cubic ***cobalt(II,III) oxide***, Co_3O_4, M_r 240.80, density 6.07. This compound crystallizes in a spinel lattice, with Co^{2+} in the tetrahedral lattice sites and Co^{3+} in the octahedral sites. It is used in enamels, in the production of ferrites and thermistors, and as a catalyst for complete combustion of exhaust gases.

Cobalt pigments: inorganic cobalt compounds used as pigments. They include Smalt (see), Cobalt blue (see), Cobalt green (see) and cobalt yellow (see Potassium hexanitrocobaltate(III).

Cobalt sulfates. ***Cobalt(II) sulfate***: $CoSO_4 \cdot 7H_2O$, carmine red, monoclinic, water-soluble crystals; M_r 281.10, density 1.948, m.p. 96.8 °C. Cobalt(II) sulfate is obtained by reaction of cobalt, cobalt(II) oxide or carbonate with dilute sulfuric acid. It is used in electroplating, as a fertilizer for pasture land, and as a mineral additive for cattle feed.

Cobalt(III) sulfate, $Co_2(CO_4)_3 \cdot 18H_2O$, forms bluc, crystalline needles; M_r 730.33. Cobalt(III) sulfate is obtained by anodic oxidation of cobalt(II) sulfate, and the compound is stable in dilute sulfuric acid. It reacts with water, generating oxygen. ***Cobalt alums***, $MCo(SO_4)_3 \cdot 12H_2O$, arc obtained with alkali sulfates.

Cobalt triad, ***cobalt group***: the elements cobalt, rhodium and iridium which are part of group VIIIb of the periodic table. The elements of the C. are chemically very similar, and all form a great variety of stable octahedral $M^{III}L_6$ complexes, such as the homologous compounds of the types $M_2(CO)_8$ (M = Co, Rh), $M_4(CO)_{12}$, $M_6 \cdot (CO)_{16}$.

Cobalt ultramarine: same as Cobalt blue (1) (see).

Cobalt yellow: see Potassium hexanitrocobaltate(III).

Cobanik®: Cobalt-nickel alloy consisting of 45% cobalt, 55% nickel and 0.1% carbon. Because of its high resistance to heat, C. is used for heating elements, metal threads and cathodes.

Cobenium®: same as Elgiloy (see).

Cobyrinic acid: see Vitamin B_{12}.

Cobyric acid: see Vitamin B_{12}.

Coca alkaloids: see Tropane alkaloids.

Cocaine: an ester alkaloid in which the OH group of ecgonine is esterified with benzoic acid, and the carboxyl group with methanol. It forms colorless prisms, m.p. 98 °C, $[\alpha]^{20}_D$ - 16° (in chloroform); cocaine hydrochloride forms prisms, m.p. 197 °C, $[\alpha]^{20}_D$ - 72° (in water). As an ester alkaloid, C. is readily hydrolysed. It is the primary alkaloid in the coca bush *Erythroxylon coca*. The total alkaloid content is 0.5 to 1.0%. In South American coca leaves, C. makes up about 90% of the total alkaloids, while in Southeast Asian coca, it is only about 25%. The secondary alkaloids contain other carboxylic acids instead of benzoic acid, such as cinnamic acid and its dimers, truxinic and truxillic acids. In these dimers, the addition of two molecules of cinnamic acid at the olefinic double bonds forms a cyclobutane ring. Another secondary alkaloid is ***tropacocaine***, Ψ-tropine benzoate. To obtain very pure C. in high yield, a crude alkaloid mixture is first extracted, then hydrolysed. Ecgonine is purified from the mixture, then esterified with methanol and benzoic acid.

C. is used as its hydrochloride as a surface anesthetic for certain conditions in eye medicine. It is not widely used in medicine because of its toxicity and above all, because it is addictive. It is, however, one of the most widely consumed "recreational" drugs.

Historical. C. was discovered in 1860 by Niemann, and was introduced as a local anesthetic in eye medicine by Koller in 1884. The structure was proven in 1898 by Willstätter, by synthesis.

H H COOCH₃ N-CH₃ O-CO- H H

Cochenille: a term applied to various species of scale insects which contain a red pigment, Carminic acid (see), in their body fluids. The most important is the true cochenille insect, *Coccus cacti*, which is native to Mexico.

Cochrome®: a cobalt-chromium alloy consisting of 60% cobalt, 24% iron, 12% chromium and 2% manganese. Its uses are similar to those of the very heat-stable alloy Cobanik® (see).

COD: abb. for Chemical oxygen demand (see).

Code, ***genetic code***: During protein synthesis, the

First letter	Second letter: U	C	A	G	Third letter
U	Phe	Ser	Tyr	Cys	U
	Phe	Ser	Tyr	Cys	C
	Leu	Ser	ocker	opal	A
	Leu	Ser	amber	Trp	G
C	Leu	Pro	His	Arg	U
	Leu	Pro	His	Arg	C
	Leu	Pro	Gln	Arg	A
	Leu	Pro	Gln	Arg	G
A	Ile	Thr	Asn	Ser	U
	Ile	Thr	Asn	Ser	C
	Ile	Thr	Lys	Arg	A
	Met	Thr	Lys	Arg	G
G	Val	Ala	Asp	Gly	U
	Val	Ala	Asp	Gly	C
	Val	Ala	Glu	Gly	A
	Val	Ala	Glu	Gly	G

sequence of bases in a nucleic acid is converted into a sequence of amino acids in a protein. Each three bases (*triplet*) denotes a single amino acid. Since there are 4 bases in DNA or RNA, the number of possible triplets is $4^3 = 64$; however, there are only 20 proteogenic amino acids. Therefore the code is *degenerate*; each amino acid except tryptophan and methionine is denoted by more than one triplet. Several of the triplets do not direct incorporation of an amino acid; instead, they indicate the beginning (start) and end (stop) of a protein.

Genetic code. The triplets are shown as they appear in the mRNA, reading in the 5'→3' direction.

must be kept cool and protected from moisture. Weakly acidic solutions can be stored as long as one week. C. is sensitive to oxidation; the first reaction is an oxidation to the disulfide: 2 CoA-SH → CoA-S-S-CoA. C. is found in every cell, where it has a vital biochemical role in both the tricarboxylic acid cycle and fatty acid metabolism. Fatty acids are activated by formation of thioesters with the mercapto group of the cysteamine portion of C; these thioesters are energy-rich compounds. ***Acetyl-coenzyme A*** (***acetyl-CoA***) is a key intermediate in metabolism required for synthesis and degradation of carbohydrates and lipids.

HS–CH₂–CH₂–NH–CO–CH₂–CH₂–NH–CO–C(H)(OH)–C(CH₃)₂–CH₂–O–P(O)(OH)–O–P(O)(OH)–O–CH₂Ade (ribose 3'-phosphate: O–P(=O)(OH)–OH)

Cysteamine | β-Alanine | Pantoic acid | Pantothenic acid | Pantotheine

Codeine: see Morphine.

Codon-anticodon interaction: see Nucleic acids.

COED process: see Coal hydrogenation.

Co-Elinvar®: a cobalt alloy with an elasticity modulus which is nearly constant over a wide range of temperature. It consists of 57-60% cobalt, 25-35% iron and 8-15% chromium. It is used to make capillary springs for measuring instruments and chronometers.

Coenzyme: a non-protein cofactor of an enzyme; a low-molecular-weight organic compound which takes a direct part in the catalysed reaction. It is changed by the reaction and is regenerated in the subsequent enzyme reactions. In the narrow sense, the term is applied to a realtively loosely bound cofactor, which is not covalently bound to the enzyme and can be removed by dialysis. Covalently bound cofactors are called ***prosthetic groups***. The enzyme from which a C. has been removed is the ***apoenzyme***. C. can also have a catalytic effect in the absence of the apoenzyme, although they are usually less specific.

Enzyme class	Coenzyme
Oxidoreductases	Nicotinamide nucleotides Flavin nucleotides Metalloporphyrins: cytochromes Quinones: ubiquinone, menaquinone
Transferases	Coenzyme A Pyridoxal phosphate Tetrahydropteroylglutamic acid, folic acid Adenosine triphosphate *S*-Adenosylmethionine
Lyases	Thiamin pyrophosphate Lipoic acid
Ligases	Biotin Corrinoids, Vitamin B_{12}

Coenzyme A, abb. CoA: a white, amorphous powder, M_r 767.6, λ_{max} = 260 nm (pH 7), which is soluble in water. C. is commercially available as the free acid, the trisodium salt or the trilithium salt. It

Coenzyme M: 2-mercaptoethanesulfonic acid. C. plays a role in methyl transfer in bacterial methane production.

Coenzyme Q: same as Ubiquinone (see).

Coexisting phases: see Gibbs phase law.

Cofactor: a non-protein structure required by an enzyme for catalytic activity; it may be a metal ion or a low-molecular-weight organic compound (see Coenzyme).

Cogasine: (from COke-GAs-gasolINE), a mixture of liquid, aliphatic hydrocarbons (paraffins) similar to diesel oil, but made by the Fischer-Tropsch synthesis. ***Cogasine I*** has a boiling range from 180 to 230 °C, and an average chain length of 10 to 13 carbon atoms; ***cogasine II*** has a boiling range of 230-320 °C and chains of 14 to 18 carbon atoms. The alkene content of C. after normal-pressure synthesis with cobalt catalysts is about 6 to 7%; C. prepared by medium-pressure synthesis is about 40% alkenes.

Cogasine process: same as Fischer-Tropsch synthesis.

Cohesion pressure: see Van der Waals equation.

Coinage alloys: alloys of Group Ib metals (coinage elements) with aluminum, nickel and zinc. The latter elements extend the noble metal and increase resistance to wear. Silver coins are still in circulation in several countries, but gold coins in only a few. New silver, pure nickel, white nickel alloys, red copper, yellow aluminum bronze and reddish bronzes are used for low-value coins (small change). Aluminum is now also used, and if metal is scarce, iron or zinc.

Coinage gold: see Gold alloys, see Coinage alloys.

Coinage metal: see Copper group.

Coinage silver: see Silver alloys, see Coinage alloys.

Coke: the black, solid residue produced by Coking (see) or Carbonization (see). See also Petroleum coke.

Coke water gas: see Water gas.

Coking: the oldest process of Coal refining (see). Hard coal, lignite coal, pitch, oil shale, peat or wood is heated in the absence of air to about 1000 °C (high-temperature C.) or about 800 °C (medium-temperature C.); the process is also called *dry distillation*. The volatile components of the starting material are driven out, and partially decomposed, by the heat.

1) Not all hard coals are suitable for C.; the best are fat coals (which have high contents of volatile materials). Coke from a coking plant comes in larger lumps and is harder than ***gas coke***, which is a byproduct of city gas generation. C. is done in byproduct ovens about 12 m long, 5 m high and 0.45 m wide. In large plants, up to 500 ovens are arranged in groups, and produce up to 10,000 t coke each day. After the C. is complete, the glowing coke is pushed out of the ovens and quenched with water in washing towers, or cooling gases are circulated around it. After this, the coke is broken into pieces and seived to give the various commercial types. Lump coke is gray to black, dense and resistant to pressure. It is 96 to 97% carbon, and has a heating value of more than 28,000 kJ kg^{-1}. It is used primarily as a fuel, as a reducing agent in iron smelters, and in lime burning. It is also used as a starting material for the production of calcium carbide and synthesis gas.

The crude gases sucked out of the oven compartments are separated, through several cooling steps, into tar, crude gasoline, ammonia and coke-oven gas. The latter consists, on average, of 55% hydrogen, 6% carbon monoxide, 2% carbon dioxide, 10% nitrogen, 20% methane and 7% other gaseous hydrocarbons. Methane, ethane, ethene and higher hydrocarbons can be isolated from it. The remaining gas is used as a fuel in the coking plant to heat the ovens and to generate steam. Some of the gas is also used, after purification and mixing with water gas, as city gas.

2) For C. of lignite coal, only prepared coal with low ash and sulfur contents is suitable. Since lignite coal has no coking quality, it must first be ground and dried, then pressed into hard briquets with even grain size. In the high temperature process (Fig.) of G. Bilkenroth and E. Rammler, the briquets are dried in the upper part of the vertical C. oven with hot rinse gas, leaving only about 1% water content in them. They then slide into the C. chamber, which is heated from the outside to a temperature of 1000 to 1200 °C. The resulting lignite coal high temperature coke is carried out of the oven on rollers, cooled with inert gases, then sprayed with water and sorted. The process is semicontinuous; the briquets remain in the drier and C. chamber for 12 hours. The exhaust gas, which carries the moisture from the briquets, makes the white plumes which are characteristic of a lignite coal C. plant. Although there is some melting in hard coal C., the briquets of lignite shrink during the production of coke. Lignite coke is not as solid as hard coal coke, but it is more reactive. It is used in iron smelting, lead ovens, smelting of nonferrous metals, lime kilns, phosphorus furnaces, calcium carbide production, gas works, etc. The gaseous products sucked out of the C. chambers are condensed to tar and medium oil. The latter is separated by distillation into heavy gasoline and light oil. Part of the residual gas is used in the C. plant itself, and part is processed to a high-quality city gas and to synthesis gas.

3) The C. of pitch can be regarded as both a final distillation of coal tar and a degassing process. First the components which can be distilled without decomposition are driven off, after which the residue is decomposed to form gas. The C. is ended at temperatures around 1350 °C. ***Pitch coke*** is made as lumps with more than 98% carbon content; it has a low ash content and is used as electrode material.

Balance of the coking of 100 t hard coal or 100 t lignite coal briquets

Product	Yield in t	
	Hard coal	Lignite briquets
Coke	80	45
Tar	5	3
Crude gasoline or light oil	2	3
Ammonia	0.3	–
Excess gas	8	20

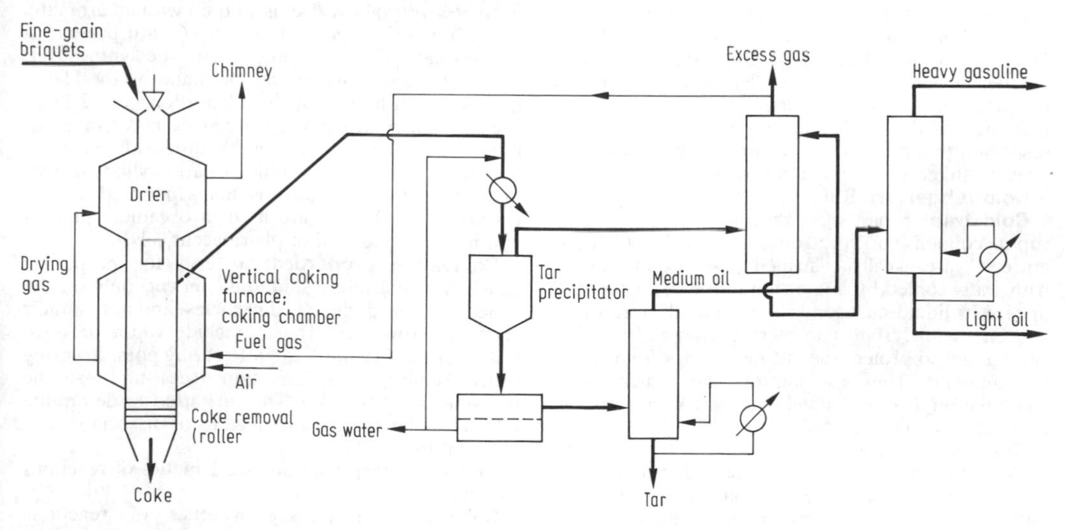

Diagram of a coking plant for lignite coal

Coking coal: see Coal.

Colamine: see Ethanolamine.

Colation: the removal of coarse components of a suspension by filtration with relatively coarse filter textiles made of linen, wool or a similar fiber. The filter cloth is usually stretched on a frame, but it can be made into a bag. The filtrate is usually cloudy, and is called the colate.

Colchicine: a tricyclic, non-basic alkaloid with an amide structure and a tropolone ring. C. forms white crystals which are soluble in water and readily soluble in ethanol and chloroform; m.p. 160-164 °C. The optical rotation is very sensitive to the solvent: $[\alpha]_D^{20}$ -430° (in water), - 245° (in ethanol), - 121° (in chloroform). C. is the main alkaloid in the autumn primrose, *Colchicum autumnale*.

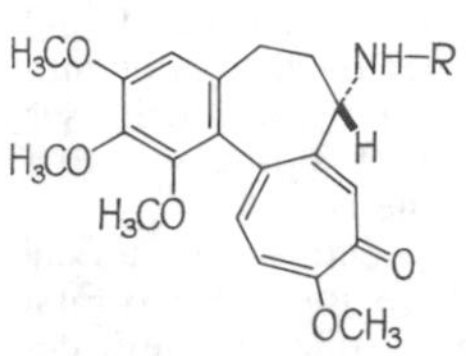

H_3CO H_3CO H_3CO NH–R H O OCH_3

R = $COCH_3$: Colchicine
R = CH_3: Demecolcine

The seeds of the plant contain 0.2 to 0.6% C. It is biosynthesized from phenylalanine, which is converted to cinnamic acid, and tyrosine, which is first converted to tyramine. C. is a mitosis inhibitor. It is used in plant breeding to induce polyploidy, which can give desirable properties in plant strains. It is also used to treat acute attacks of gout. It is of minor importance as a cytostatic.

Demecolcine (deacetyl-*N*-methylcolchicine) is a partially synthetic derivative of C. which is less toxic than C.; it is a basic compound, m.p. 185 °C, $[\alpha]_D^{20}$ - 129° (in chloroform).

Cold flame: the appearance of one or more rhythmic light emissions at the beginning of oxidation of hydrocarbon vapors at relatively low temperatures (500 to 700 K). The temperature range and the number of flames which occur depend on the type of hydrocarbon, the mixing ratio with oxygen, the total pressure and the dimensions of the reactor. It is now suspected that knocking in reciprocating gasoline engines is linked to the phenomenon of C.

Cold rubber: see Rubber.

Cold trap: a part of a vacuum system in which vapors condense or freeze out due to the low temperature. C. are usually U-shaped glass or metal tubes with walls cooled by immersion in a coolant such as dry ice or liquid nitrogen. They are used to trap water, oil, hydrocarbon and mercury vapors from the system and to protect the vacuum pumps from these contaminants. They also improve the vacuum. The combination of sorption and cooling (cooled sorption traps) is also very effective.

Colistine: see Polymixins.

Coli titer: an important parameter in water quality criteria. The C. is the average volume of water containing one coliform bacterium. For example, a C. of 100 means that 100 ml water contains 1 bacterial cell; a C. of 0.01 means that the average volume per cell is 0.01 ml, or that 1 ml contains 100 bacterial cells.

Collagens: the most abundant proteins in animals; they constitute 25 to 30% of the total proteins. They are found in the fibrils and ground substance in connective tissue, and they are components of ligaments, tendons, skin, cartilage, scales and blood vessel walls. They most abundant amino acids in C. are glycine (35%), alanine (11%), proline (12%) and hydroxyproline (9%). The presence of hydroxylysine is important for cross-linking of individual protein chains; C. also contain 1 to 2% carbohydrate. Sulfur-containing amino acids are found only in the C. of invertebrates. The basic structure of C. is a rod-shaped ***tropocollagen molecule***, 3000 nm long, 1.5 nm in diameter and with a relative molecular mass of 300,000. The tropocollagens of cartilage consist of three identical polypeptide chains, which form stretched left-handed helices with three amino acid residues per turn. The repetitive sequences (-Gly-X-Pro-)$_n$, (-Gly-X-Hypro-)$_n$ and (-Gly-Pro-Hypro-)$_n$ cause the individual chains to twist around each other in a right-handed *super* or *triple helix*. The helix structures line up to form fibrils with diameters up to 500 nm. Within the fibrils, the individual molecules are parallel, but are displaced by 1/4 their length with respect to each other. This gives rise to the typical cross banded pattern of the C. fibril. The distance between stripes is between 60 and 70 nm, depending on the type of C.

The triple helical structure of C. is destroyed when the protein is heated in an aqueous solution. The cooled solution forms Gelatin (see).

Collidines: alkyl derivatives of pyridine with the formula $C_8H_{11}N$, including trimethylpyridines and ethylmethylpyridines. ***α-C., 4-ethyl-2-methylpyridine***, is a liquid with a characteristic odor; b.p. 178 °C. ***β-C., 3-ethyl-4-methylpyridine***, is a hygroscopic liquid with a characteristic odor; b.p. 198 °C. β-C. is volatile in steam. ***Symm. C., 2,4,6-trimethylpyridine***, is a liquid with a characteristic odor; m.p. - 44.5 °C, b.p. 177 °C, n_D^{20} 1.5014. ***Aldehyde collidine, 5-ethyl-2-methylpyridine***, is a liquid with an aromatic odor, b.p. 177 °C, n_D^{20} 1.4974. The C. are present in the distillate of coal tar and can also be synthesized. Symm. C., for example, can be made by the Hantz pyridine synthesis, by condensation of 2 mol acetoacetate with 1 mol acetaldehyde and 1 mol ammonia. The first product is a dihydropyridine derivative which is oxidized to collidine carboxylic acid ester under mild conditions; this is then saponified and decarboxylated. The C. are used in organic syntheses and in the production of pharmaceuticals.

Colligative properties: a term for properties which, in ideal dilute solutions, depend only on the concentrations of dissolved particles, and not on their chemical properties. The C. include vapor pressure reduction (see Raoult's law), Freezing point lowering (see), Boiling point elevation (see) and osmotic pressure (see Osmosis). They are used to determine relative molar masses and degrees of dissociation.

Collinite: see Macerals.

Collision cross section: see Kinetics of reactions (theory).

Collision theory: see Kinetics of reactions (theory).

Collision tube method: see Kinetics of reactions (experimental methods).

Collodion cotton, ***colloxylin***: a nitrated cellulose with a nitrogen content of 10.6 to 12.5% (see Cellulose nitrate). C. is used to make nitrocellulose lacquers, celluloid and artificial leather. It is used as a gelatinization substance for glycerol trinitrate and ethylene glycol dinitrate to make gelatinous explosives. It is mixed with gun cotton to make smokeless powder.

A solution of C. in alcohol and ether is called ***collodion***. It is a clear, very viscous liquid which is used as a color binder. A solution of C. in acetone and acetates is called varnish.

Collodium: a solution of Collodion cotton (see).

Colloid, ***colloidally disperse system***: a dispersion of molecules or aggregates (consisting of about 10^3 to 10^9 atoms) in a solvent. The aggregates may be particles (diameter 10^{-7} to 10^{-4} cm) or thin films (less than 100 nm thick). C. occupy an intermediate position between molecular and macrodisperse systems (see Disperse system). In contrast to molecular disperse systems, the state of a C. cannot be characterized by its pressure, temperature and composition; the properties also depend on the size, shape and structure of the particles. The reason is that the atoms or molecules at or near the phase boundary, which have higher energy than those in the interior, constitute a non-negligible fraction of the total number. Small particles and thin films are not thermodynamically stable; increases in particle size release free energy. However, strongly lyophilic dispersions, such as solutions of macromolecules or tenside solutions above the critical micellar concentration (CMC) are exceptions to this generalization. Such macromolecular or associational (micellar) C. are thermodynamically stable; the macromolecules have defined diameters and the micelles display a definte degree of association.

C. can be classified as dispersion, associational (micellar) or molecular.

1) ***Dispersion C.***. can be classified on the basis of the physical states of the disperse phase and the dispersion medium: a) solid particles in a gas (dust), b) droplets of liquid in a gas (mist), c) liquids and solids in gas (smoke), d) gas bubbles in a liquid (foam), e) liquid droplets in a liquid (emulsion), f) solid particles in a liquid (sol, solid dispersion), g) gas bubbles in a solid (foam), h) liquid droplets in a solid and i) solid particles in solids (glasses, ceramics). Highly concentrated, structured solid dispersions are called Gels (see).

Production: 1) Compact matter must first be divided by external forces. Solids can be ground or broken, and liquids or gases can be dispersed by supersonic vibrations or centrifugal force. Disadvantages of dispersion methods are low energy yields (a large part of the applied energy is converted to heat) and the wide range of particle sizes obtained. 2) Condensation processes are much more important for production of C. for such applications as photographic films, magnetic tapes for data storage, catalysts, etc., because in principle any desired particle size, homodisperse systems and particles of a desired shape can be made from molecular disperse systems. Colloidal particles can be obtained either from highly concentrated (saturated or supersaturated) solutions by homogeneous seed formation, or from very dilute solutions by heterogeneous seed formation. It is important to interrupt the process of particle growth when the desired size range is reached. The molecular disperse system from which the C. has formed is removed by dialysis or ultrafiltration. 3) Peptization is the spontaneous dispersion of aggregates of colloidal particles, for example by adsorption of ions. The particles thus acquire a similar electric charge and repel each other.

Stability. Since atoms or molecules at or near the phase boundary have higher free energies than those in the interior of the phase (see Surface tension), it follows that dispersion C. have a large excess of free energy: the ratio of surface area to volume increases as particle size decreases. Therefore, dispersion C. tend to form coarser aggregates (see Coagulation), with a loss of free energy in the process. They can be stabilized a) if the particles have a sufficiently strong electric charge; this is achieved by dissociation of surface groups or by preferential adsorption of ions with the same charge. The particles then attract a shell of counter ions, which compensate for the electrical charge on their surfaces. Electrostatic repulsion occurs if the double layers of ions around two particles interpenetrate. This energy must compensate for the kinetic energy of Brownian motion and the van der Waals energy. In practice, all dispersion C. carry electric charges; metal C. are usually negatively charged, while oxides are either negative or positive, depending on the pH. b) Stabilization also occurs if the colloidal particles are highly solvated; in this case, their free energy decreases because of the solvation energy. Highly solvated C. are often called ***lyophilic C.***. c) Finally, addition of surface-active compounds, such as tensides or macromolecules, stabilizes C. The tensides are called *dispersives, emulsifiers* or *foam stabilizers*, and the macromolecules, *protective colloids*.

The destruction of C. is just as important in purification of waste water or gas, and removal of water from petroleum as stabilizing C. in pharmaceuticals, paints, etc. C. which are stabilized by electric repulsion can be made to coagulate by addition of electrolytes. These work either by causing the particles to adsorb ions of the opposite charge, thus neutralizing their surface charges, or by compressing the diffuse electrical double layer and causing the van der Waals attraction between particles to outweigh the electrostatic effects. The higher the valence of its counter ions, the more effective an electrolyte will be. The threshhold of coagulation concentrations for mono-, di- and trivalent counter ions stand in the ratio 10,000:50:1 (*Schulz-Hardy rule*).

Determination of particle size: Dispersion C. have a uniform particle size only in exceptional cases; as a rule, they are *polydisperse*. Their properties are affected by their degree of dispersion and polydispersity. The mean particle size can be determined by measurement of osmotic pressure, diffusion rate, light scattering or turbidity, and by electron microscopy. The last method and sedimentation analysis in a centrifuge also permit the size distribution of the particles to be determined.

Properties. 1) Electrical properties: colloidal particles migrate in an electric field. The potential which develops on the shear surface between the particle and electrolyte solution is called the *zeta potential*. The potential on the shear surface is smaller than the surface potential which is calculated from the number of charge carriers per unit of surface area; it corresponds roughly to the Stern potential (see Electrochemical double layer). In porous systems, application of an electric field causes a migration of the liquid phase, called electroosmosis. Here too, an electrokinetic potential develops on the shear surface. On the other hand, migration of particles in the same direction as the liquid flow through the porous system leads to potential differences such as the sedimentation potential and flow potential. 2) Optical properties: colloidal solutions scatter light strongly (see Tyndall effect). This was formerly considered a specific difference between colloidal and molecular disperse systems. Rod-shaped and leaflet-shaped anisometric particles (e.g. vanadium pentoxide) are oriented by the flow, causing streaming birefringence, which may be detected by crossed nicol prisms. 3) Rheological properties: the viscosity of a colloidal solution depends on its particle concentration, temperature, the shape of the particles, their size if they are anisometric, and on interactions between the disperse particles. When the interactions are strong, the viscosity is very high, as in gels. The viscosity does not obey the Newtonian friction law, but depends on the shear stress applied to it (see Viscosity, structural; Thixotropy).

II) ***Associational colloids*** (***micellar colloids***) are C. which form spontaneously when a molecular disperse substance is dissolved in a suitable solvent at a sufficiently high concentration. These substances include low-molecular weight surface active substances (tensides), of which the best known is soap. The ability to form associational colloids arises from the structure of these molecules, which consist of a hydrophobic part and a hydrophilic part. The hydrophobic part is made up of hydrocarbon groups such as alkyls, alkenyls, aryls or alkylaryl groups. The hydrophilic groups can be carboxyl, sulfate, sulfonate, phosphate, polyalcohol and polyether groups. The formation of micelles can occur in solvents other than water; for example, they can be detected in benzene or octane.

The physical properties of an aqueous solution of a surface-active substance change more or less abruptly when its concentration is increased. The concentration at which this change occurs is called the *critical micelle concentration*, abb. *CMC*. It depends on the chain length of the hydrocarbon group, and, to a lesser extent, on the nature of the hydrophilic group, the electrolyte content of the solution and the temperature. It is independent of the path by which it is reached, that is, it is the same whether a dilute solution is concentrated or a concentrated solution is diluted. Therefore the phenomenon is a true thermodynamic equilibrium between individual molecules and micelles. The dependence of tenside solubility on the temperature is quite different from that of other small molecules. At low temperatures, the solubility at first changes little as the temperature is raised. However, at a characteristic temperature for the substance, the *force point*, the amount of tenside dissolved increases rapidly. Conversely, when the solution is cooled, dissolved tensides precipitate suddenly at this temperature. The force points of a homologous series of compounds increase regularly with the length of the alkyl group. The appearance of the force point is linked to the formation of micelles in the solution. At low temperatures, the solubility of the tenside is low, so that the CMC is not reached. If the CMC is exceeded as the temperature rises, the solubility increases rapidly.

The CMC can be determined in a number of ways: 1) measurement of the surface tension, which is constant after the CMC is reached; 2) measurement of the electrical conductivity, which is greatly reduced when the CMC is reached; 3) measurement of the refractive index, which gives different slopes above and below the CMC; 5) measurement of the absorption spectrum, since this is different for single molecules and micelles of some substances; 6) measurement of the electromotive force (emf); a number of tensides form insoluble mercury salts, so that a mercury electrode records a continuous change in emf until the CMC is reached; 7) polarographic methods, both direct current polarographic determination of the maxima damping and alternating current polarographic measurement of absorption. Because of the added electrolyte, however, this method is suitable only for nonionic tensides; 8) solubilization of tensides. The light absorption of some water-soluble dyes changes when they are incorporated into micelles. Above the CMC, the micellar size is constant, that is, the system is *isodisperse* or *monodisperse*. The micelle is probably spherical, with the polar groups on the periphery and the alkyl groups in the interior, in aqueous media. At high tenside concentrations, *laminar micelles* can also form.

III) ***Molecular colloids*** consist of molecules with colloidal dimensions. Unlike dispersion and association C., which are held together by non-covalent bonds, these colloidal particles (molecules) are held together by chemical bonds. Solutions of proteins, cellulose, latex and synthetic macromolecules are examples of molecular C. Individual macromolecules can also be collected into larger aggregates held together by non-covalent bonds, as in biocolloids. Most macromolecular solutions consist of particles of varying size, that is, they are polydisperse. The mean particle size can be determined by the same methods listed for dispersion C. and, in addition, the mean size of macromolecules with thread-like structures (***linear C.***) can be determined by viscosity measurements.

C. are important both in nature and in chemical technology. For example, blood, protoplasm, milk, butter, skin creams, glues and pharmaceuticals are C. Some colloid-chemical processes are laundry, dyeing, lubrication, flotation, emulsion polymerization and dust removal. About 1/10 of all minerals, including hematite, manganate, opals and zeolites, are produced by colloidal-chemical processes. In the soil, colloidal processes are important for plant growth. Colloidal structures are also found in interstellar space, where they induce cloud formation. Gas dispersions are also industrially important. For example, oxygen hydrosols are used in oxidation of linseed and paraffin oils, and hydrogen hydrosols are used in industrial hydrogenations.

Historical. T. Graham is considered the founder of colloid chemistry (1861); he studied the special properties of silicic acids and proteins, introduced separation by dialysis, and coined the term "colloid". Discrete particles were discovered in gold suspensions in 1903 by Siedentopf and Zsigmondy, who observed the scattering of light by the particles in the high-power microscope. Microscopic studies by Einstein (1905) and Smoluchowski (1906) led to the proof of the theory of Brownian motion and diffusion of colloidal particles. The formation of micelles in tenside solutions at concentrations above a certain boundary (CMC) was observed in 1911 by McBain. The development of colloid chemistry in macromolecular solutions was greatly influenced by Staudinger (1950). Derjaguin and Landau (1939) and Vervey and Oberbeek (1948) gave quantitative descriptions of the stability of lyophobic C.

Colloidally disperse system: same as Colloid (see).

Colloid chemistry: a branch of physical chemistry dealing with the physics and chemistry of disperse systems and the properties of extended phase boundaries, insofar as the asymmetry of the intermolecular interactions at the phase boundary is responsible for its properties. The asymmetry of intermolecular interactions does not depend on the degree of dispersion. The effect of the degree of dispersion is important when the number of atoms or molecules near the surface is no longer negligible compared to the total number in a particle or film. As a result, there are two separate areas of study, one dealing with the physical and chemical properties of small particles and thin films, and the other with interactions between dispersed particles.

Colloxylin: same as Collodion cotton (see).

Colophony: a natural Resin (see) obtained as a distillation residue from the balsam of conifers or by extraction of pine roots. It consists primarily of abietic acid and related resin acids; it is hard and brittle and is soluble in many organic solvents. C. is produced in various degrees of lightness. It is used in the production of lacquers, paints, printing inks, soaps, glues, solder flux, to paint violins, as brewer's pitch, and, after boiling with soda, it is used together with alum to make paper resistant to ink. Heavy metal salts of C. are used as siccatives.

Color: a quality of light perception normally induced in the eye by electromagnetic radiation of certain wavelengths; it can also be induced by other stimuli such as an impact or electric current applied to sensitive neurons. Substances which are perceived to have C. are called Pigments (see) or Dyes (see).

Color coupler: a compound which gives one of the three basic colors of color photography upon reaction with the developer oxidation product. Purple couplers are usually pyrrazolone derivatives, blue-green couplers are substituted α-naphthols, and yellow couplers are C-H acidic compounds with β-diketone structure elements (Fig. 1).

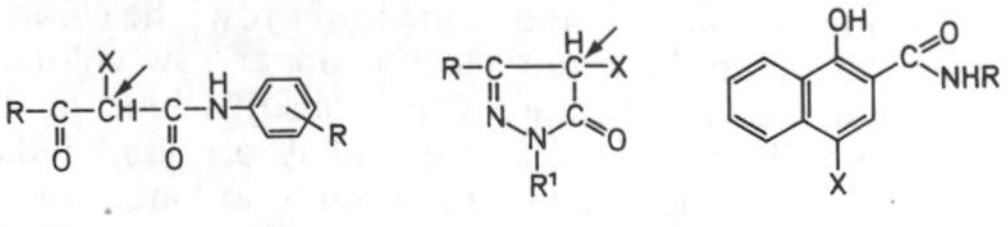

Yellow coupler | purple coupler | blue-green coupler

Examples of color couplers in color photography. The groups R or R' are usually long-chain alkyl groups, which prevent diffusion of the C. The substituents X are groups which have various structures and functions:

A) X = H: *four-equivalent coupler*. Four equivalent silver ions are required for oxidation and coupling (two for the oxidation of the color developer to its oxidation product, quinine diimine, and two for oxidation of the two hydrogen atoms at the coupling site).

b) X = halogen, sulfonic acid, thiocyano, imide, alkoxy, etc. group: *two-equivalent couplers*: only two equivalents of silver are required for oxidation of the color developer, because group X is eliminated with a hydrogen atom during the development (Fig. 2).

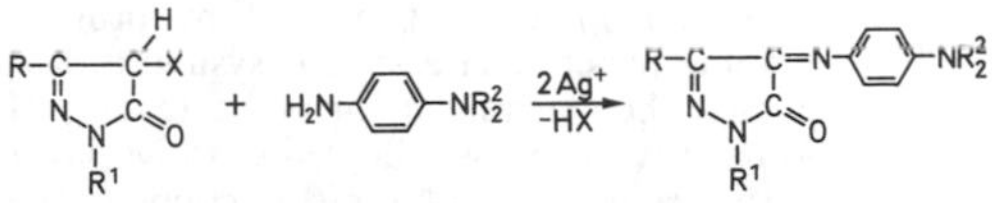

Color coupler + developer $\xrightarrow[-HX]{2Ag^+}$ pigment (red)

c) X = N=N-aryl: *masking coupler*: These are used in the red- and green-sensitized layer to prevent undesired absorption, e.g. in the blue region of the spectrum. The arylazo group is cleaved off as nitrogen and benzene derivative during color development.

d) X = SR''': *DIR coupler* (Development Inhibitor, Realizing coupler): On development, these split off a development inhibitor, HSR''', which improves the sharpness and color reproduction.

In diazotizing reproduction, polyhydroxybenzene is used to generate brown and red pigments, 3-substituted (COHNaryl) 2-hydroxynaphthalene for blue pigments, α-ketomethylene compounds or substituted resorcinol and m-aminophenols as yellow couplers, and 1-arylpyrrazolone as red couplers.

Colorimetry: a semiquantitative method of analysis based on visual comparison of color intensity of solutions. C. can be done without any apparatus, and does not require monochromatic light. After addition of a reagent to a sample solution, the resulting color is compared with a series of standard solutions under the same conditions.

Colorimetric methods have now been replaced in nearly all situations by Photometry (see), but the two terms are not always distinguished in everyday speech.

Color photography: the total process of creating true-color images of objects on light-sensitive paper or film. In the wavelength range of visible light (400 to 800 nm), the human eye distinguishes three primary colors of light: blue, green and red. Addition of the three gives white. All natural colors are generated by mixing of the primary colors. The color remaining after a color is subtrtacted from white is its complementary color. A color film usually consists of three light-sensitive layers, each made specific by sensitizers for one of the three primary colors. In *negative-positive processes*, the *negative film* has a top layer sensitive to blue light which contains a yellow coupler; a middle layer sensitive to green and containing a purple coupler, and a bottom layer sensitive to red and containing a blue-green coupler. The silver halide in the exposed areas is reduced to silver and the developer, an *N,N*-dialkylphenylenediamine, is simultaneously oxidized to a quinone diimine derivative. This reacts with the color couplers in the three layers to form the corresponding pigments; the amount of pigment is proportional to the exposure of each layer. A yellow pigment is formed in the blue layer, a purple pigment in the green-sensitive layer, and a blue-green pigment in the red-sensitive layer. The reduced silver is then oxidized to silver cations with potassium hexacyanoferrate(III); all the silver, including the unexposed silver halide, is is now complexed with thiosulfate and washed out of the emulsion. The colors of the resulting image are complementary to those in the original; it is a color negative. A *positive image* is produced by printing on photographic paper with a similar system of layers, except that their arrangement is the reverse of that in the negative process: the red-sensitive layer is on top and contains a blue-green coupler; the green-sensitive middle layer has a purple coupler, and the bottom, blue-sensitive layer has a yellow coupler.

Color reversal materials are used to make transparencies. The structure of the layers is in principle the same as in color-negative materials, except that there is a yellow filter layer between the blue-sensitive layer and the other two, to prevent blue light from penetrating and developing the green- and red-sensitive layers. Development begins with a black-and-white-type process which produces reduced silver in the exposed areas. The film is then exposed to diffuse light or a reducing bath (e.g. in $SnCl_2$); either process generates development nuclei in the previously unexposed grains. The color developer is added and oxidized by the silver halides in the non-exposed regions, after which it forms pigments with the color couplers. This step is followed by removal of the silver and fixing. The image is a positive because the complementary colors are formed where light did not strike the film in the original exposure.

In silver-bleaching methods, non-bleaching, non-diffusing yellow, purple and blue-green pigments (azo dyes) are incorporated into the layers sensitized for those colors. Following exposure, the silver produced in the developing step reduces the azo dyes to the corresponding amines, and thus destroys them. Although this material is not very sensitive, it has the great advantage of producing very light-stable and color-true copies, with excellent reproduction of details, from color slides.

A third principle of C., pigment transfer, is utilized in films for immediate development (e.g. Polaroid films). With these films, the colored picture is ready within minutes after exposure, without complicated processing. The material consists of a light-sensitive part and an image reception part. The light-sensitive film has three layers of silver halide emulsions (each sensitized for one of the three primary colors) and between these, layers containing developer compounds. Each developer molecule consists of two parts, the pigment and the developer, joined by a bridge structure (DBP structure).

Developer Bridge Pigment

Just under the blue-sensitive layer is a layer containing yellow pigment; under the green-sensitive is a layer containing purple pigment; and under the red-sensitive layer is a blue-green pigment layer. The image reception part consists of an image reception layer and several other layers which fix the picture or improve the quality of the development process. After the film is exposed, it is pulled out of the camera between rollers which break a capsule of developer paste, spread the paste between the two parts of the material, and press these two parts tightly together. The developer paste consists of alkali and special additives which ionize the DBP molecules and allow them to diffuse into the layers above them. Each type of DBP molecule can be oxidized, by the reduction of silver in the exposed grains, to a quinone which is not capable of further diffusion. If it is not trapped in the layer immediately above it, it diffuses into into the image reception layer, where it is fixed. The entire process lasts about 1 minute.

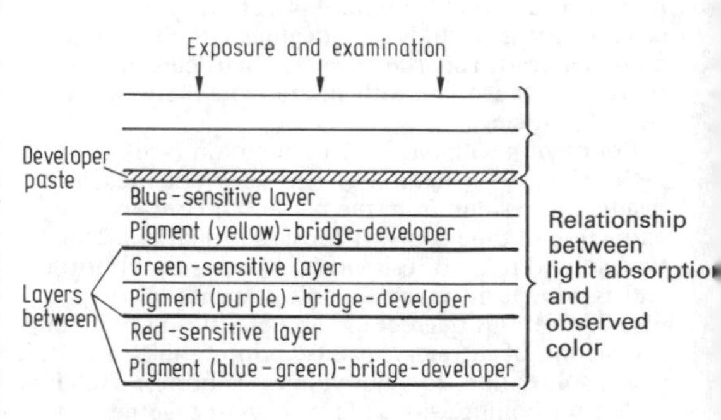

Layer structure of Polaroid film

Color theory: explains the colors of substances, especially Dyes (see). The color of a substance is due to its ability to absorb part of the light from the visible spectrum. The human eye is not able to resolve polychromatic light into its components; a color is perceived only as the sum of all the light entering the eye. If the spectral composition of the light is the same as that of daylight, white is perceived. Color is perceived if part of the spectrum is removed by absorption; the perceived color is complementary to the spectral color which has been absorbed. In the upper row of the table, the wavelength ranges of the spectral colors are shown, and in the lower row, the colors which are complementary to the upper row colors. For example, a substance which absorbs blue light appears yellow. If there are several absorption bands in the visible range, the relationship is no longer simple, and the perceived color cannot be easily predicted from its absorption spectrum.

λ in nm	400	440	480	490	500	560	580	595	605	750 →
absorbed light	UV	violet	blue	green-blue	blue-green	green	yellow-green	yellow	orange	red	IR
perceived color		yellow-green	yellow	orange	red	purple	violet	blue	green-blue	blue-green	

The color of organic compounds can be explained by the existence of three idealized basic structures, the aromatic, the polyenic and the polymethylene structures (*triad theory*; Fig. 1).

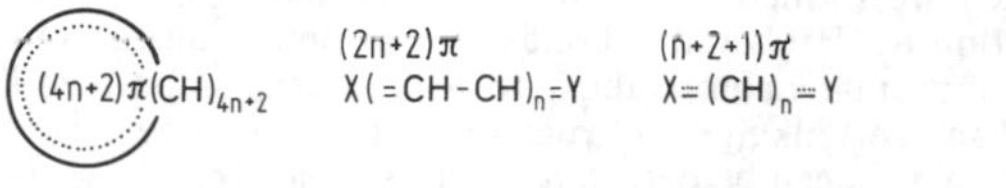

$(4n+2)\pi(CH)_{4n+2}$ $\qquad (2n+2)\pi$ $X(=CH-CH)_n=Y$ $\qquad (n+2+1)\pi$ $X=(CH)_n=Y$

Fig. 1. Aromatic Polyene Polymethylene

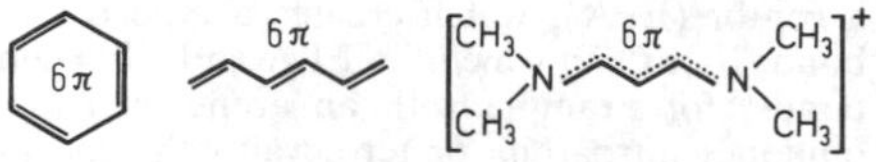

		Polyene	Polymethylene
λ_{max} in nm	250	268	312
ε_{max} in $mol^{-1} \cdot l \cdot cm^{-1}$	250	34700	64600

Combinations and intermediates between these structures account for the multitude of colored organic compounds. The following rules for the colors of ideal polymethylene structures have been developed on the basis of the triad theory: 1) For the same molecular size or the same number of π electrons, the polymethylene structures absorb at the longest wavelength and have the most intense color (Fig. 2). 2) If a modification, such as a branching, maintains the charge density, then the compound absorbs at a shorter wavelength than the longest polymethylene fragment in the molecule would. Compounds of this type are called branched or alternating polymethylenes. 3) If the characteristic charge distribution is changed by a modification of the basic structure, the compound absorbs at a longer wavelength than the longest polymethylene fragment in the molecule would. Such compounds are called non-alternating polymethylenes. 4) If the polymethylene structure in a non-alternating polymethylene compound is interrupted by competition from an aromatic structure, the compound absorbs at shorter wavelengths than the longest polymethylene fragment. The compounds then behave as substituted aromatics. 5) If the symmetry of the π-electron distribution along the polymethylene chain is reduced, the color becomes less intense. Aside from the nitrosoalkanes, nearly all organic compounds which contain only single or isolated multiple bonds are colorless. A single multiple bond leads to one or more absorption bands between 150 and 300 nm.

Even before these relationships had been worked out, it was determined empirically that atomic groups with multiple bonds, called Chromophores (see), lead to color. Other groups, the Auxochromes (see), do not themselves lead to color, but they intensify the colors of colored compounds. Certain other substituents, which have the opposite effect, are called Antiauxochromes (see). Most pigments contain both chromophores and auxochromes. Other groups cause shifts to longer (see Bathochromicity) or shorter wavelengths (see Hypsochromicity). In some cases, e.g. in the triphenylmethane pigments, colorless compounds become colored when a central C-atom either loses an electron or gains a free electron pair through salt formation. This facilitates resonance among the aromatic nuclei (*halochromaticity*).

Structural elucidation on the basis of the P. is based on UV-VIS spectroscopy (see). Pigments can have different colors in solvents with different polarities (solvatochromaticity). The difference in polarity of the solvents accounts for transitions between polymethylene and polyene structures, or between aromatic and polymethylene structures (Fig. 3).

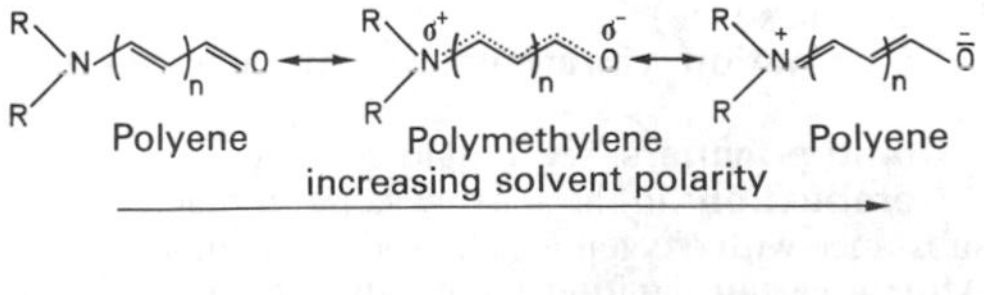

Polyene ⟷ Polymethylene ⟷ Polyene

increasing solvent polarity →

Column: an apparatus for distillation, extraction, absorption or adsorption of mixtures. In industry, C. are usually made of metal, or, for special purposes, of industrial glass. Laboratory C. are made of glass. Most C. consist of hollow, vertical cylinders in which there are several horizontal plates, or they may be filled with a porous material. The other parts of the C. depend on their application. Distillation C. have a container at the lower end, called the pot or sump. The upper end (head) of the C. is topped by a cooler. The built-in plates come in various forms, leading to designation of the C. as a *bubble-plate, sieve-plate, turbogrid*, etc. C.

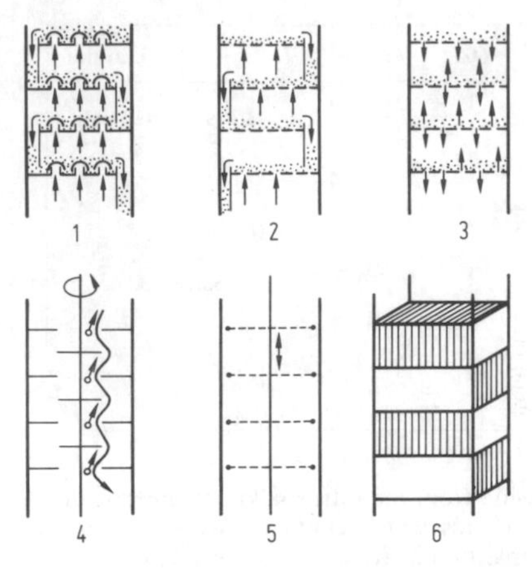

Types of columns: *1*, Bubble plate column; *2*, sieve-plate column; *3*, turbogrid column; *4*, rotating disc column; *5*, pulsator column; *6*, plate packet for a trickle-plate column

In a *filler column*, Filler (see) is used instead of plates. In *trickle-plate columns*, vertical plates of metal or plastic (trickle plates) are built in. Extraction columns are often provided with a vertical, rotating axis which has disks at certain intervals (*rotating disk column*). During the exchange process, these rotating disks provide better liquid distribution. The same effect is achieved with sieve plates, which are shaken up and down along the column axis. Such C. are called *pulsator C.*. This type can also be constructed so that the sieves are fixed as plates, while the entire liquid column is caused to vibrate vertically. Absorption columns have the same types of interiors as C. for distillation and extraction. Adsorption C. filled with an adsorbant, such as activated charcoal, are often called *adsorption towers* (adsorbers).

Colthrup tables: collections of the most important characteristic frequencies in infrared spectroscopy.

Combination principle: a principle established by Ritz, that the wavenumber of a spectral line is obtained from the difference of two terms (see Multiplet structure), provided that it is not forbidden by Selection rules (see).

Combination vibrations: see Infrared spectroscopy.

Comb polymers: see Liquid crystals.

Combustion: in the narrow sense, a reaction of a substance with oxygen which produces heat and light. After a certain ignition temperature is reached, the reaction occurs very rapidly. C. in pure oxygen is much more vigorous than in air. The final products of complete C. of organic substances are carbon dioxide and water. The C. of other substances, e.g. sulfur, phosphorus, sodium, magnesium, etc., produces the respective oxides. In the broader sense, C. is an oxidation process which occurs without flames; it is then called ***silent C.*** (see Autooxidation). This definition includes rusting of iron, all decay processes and respiration in organisms.

Combustion heat: the heat of reaction for complete combustion of a chemically uniform substance to the stable oxides of its elements. It is generally given for reaction of 1 mole of the substance. There is a difference between the ***molar combustion energy*** $\Delta_c E$ at constant volume, and the ***molar combustion enthalpy*** $\Delta_c H$ at constant pressure. The C. of an element is equal to the energy or enthalpy of formation of the oxide (see Formation reaction). $\Delta_C E$ is determined in a calorimetric bomb, and is used in thermochemistry as the basis for calculating the energies and enthalpies of formation. In industry, the negative C. $\Delta_c H$ is used as the upper limit of the Heating value (see).

Common name: an internationally standardized trivial name for a substance which is free for general use, for example, for pesticides or drugs.

Competing reaction: see Reaction, complex.

Complementary colors: colors which, when mixed, give white (light); for example, red and green, or yellow and blue (see Color theory).

Complex: see Nomenclature, sect. II G.

π-Complex: a donor-acceptor complex between an electrophile and an alkene or aromatic compound. π-C. are formed as intermediates in electrophilic addition to C=C double bonds, electrophilic substitution of aromatic compounds, or between transition metal ions and alkenes, alkynes or arenes. In some cases, the π-C. can be detected spectroscopically. A π-C. is formed by overlap of an occupied π-molecular orbital in the electron pair donor (EPD) with an unoccupied, antibonding σ-molecular orbital of the electron pair acceptor (EPA), which creates a type of diffuse σ-bond. In π-C. in which the EPA and EPD contain π-bonds, for example with an arene as EPD and a quinone, nitroarene or tetracyanoethylene as EPA, additional π-π overlap is possible. In organometallic π-C., the simultaneous overlap of an occupied pd hybrid orbital on the transition metal cation and an unoccupied, antibonding π-molecular orbital of the EPA leads to a complex π-molecular orbital (*π-back donation*). The bonding in organometallic π-C. is comparable to the C=C double bond; it consists of a σ- and π-bond combination. Organometallic π-C. are important in certain syntheses, for example the homogeneous hydrogenation, dimerization and oligomerization of alkenes, dienes and alkynes, carbonylation, hydroformylation and direct oxidation of alkenes (ethene to acetaldehyde).

σ-Complex: an intermediate formed in electrophilic or nucleophilic substitution of an arene. In *electrophilic aromatic substitution*, the σ-C. is an arenium ion with a coordinated trivalent carbon, in which n π-electrons are delocalized over $n+1$ C atoms (Fig. 1). The formation of a σ-C. is usually the rate-determining step in electrophilic aromatic substitutions. A σ-C. is always stabilized by substituents which are electron pair donors, and destabilized by those which are electron pair acceptors. Under special conditions, relatively stable σ-C. can be isolated, e.g. in the alkylation of mesitylene with fluoroethane/boron trifluoride at - 80 °C (Fig. 2).

In *bimolecular nucleophilic substitution* of benzoid compounds, a σ-C. with the nucleophilic reagent is formed by an addition-elimination mechanism; this decomposes into the reaction products. In a few cases, the σ-C. can be isolated, e.g. in the reaction of 2,4,6-trinitroanisol with potassium methanolate (Meisenheimer complex, Fig. 3).

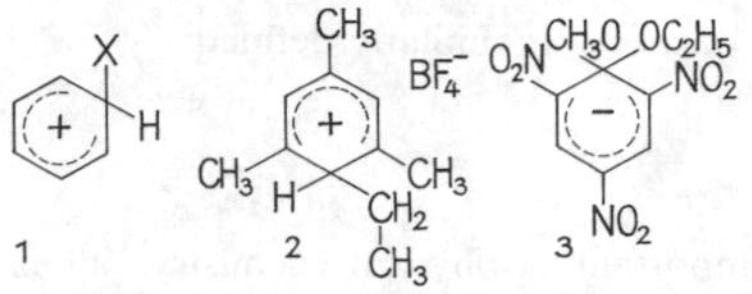

Complex isomerism: isomerism observed in coordination compounds. The most important C. are ***stereoisomerism***, such as the ***cis-trans-isomerism*** observed in square planar or octahedral mixed-ligand complexes of the type MA_2B_2, e.g. $PtCl_2(NH_3)_2$, (Fig. 1) or MA_2B_4, e.g. $[CoCl_2(NH_3)_4]^+$, (Fig. 2) and ***optical isomerism*** (mirror image isomerism) in octahedral complexes with bidentate ligands, e.g. oxalato complexes $[M(C_2O_4)_3]^{3-}$, (Fig. 3).

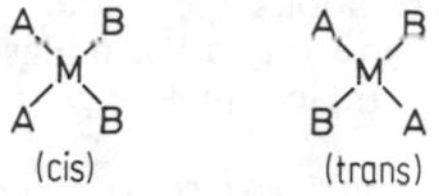

Fig. 1. *cis-trans*-Isomerism in square planar complexes of the MA_2B_2 type

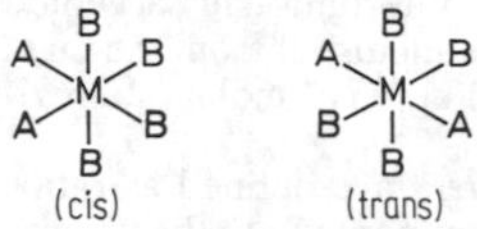

Fig. 2. *cis-trans*-Isomerism in octahedral complexes of the MA_2B_4 type

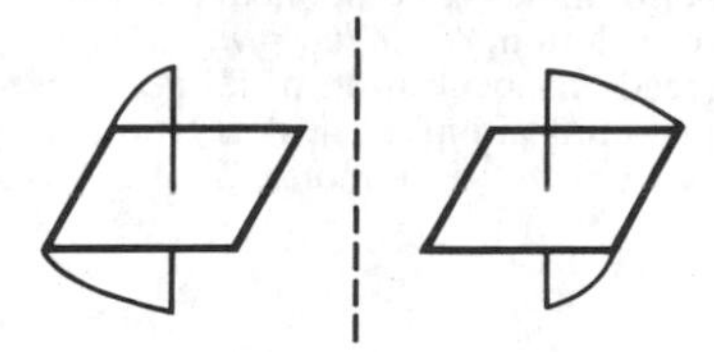

Fig. 3. Optical isomers of octahedral tris chelate complexes

In addition, complexes with the same composition can sometimes form different polyhedra. For example, the diphosphine complex $[NiCl_2(P(C_6H_5)_2(CH_2)_2P(C_6H_5)_2)_2]$ exists in both planar and tetrahedral forms. In ***hydrate isomerism***, there is a difference in the bonding of water molecules, e.g. $[Cr(H_2O)_6]Cl_3$ and $[Cr(H_2O)_5Cl]Cl_2 \cdot H_2O$. In ***ionization isomerism***, ionic ligands are either coordinatively or ionically bound, e.g. $[CoCl(NO_2)en_2]SCN$ and $[CoCl(SCN)en_2]NO_2$. ***Bonding isomerism*** is important in complexes with ambivalent ligands, such as nitrite, NO_2^-, or thiocyanate, NCS^-. One may compare, for example, the pairs $[Co(NO_2)_2en_2]^+$ and $[Co(ONO)_2en_2]^+$, $[Pd(NCS)_2(P(C_6H_5)_3)_2]$ and $[Pd(SCN)_2(P(C_6H_5)_3)_2]$. ***Coordination isomerism*** is limited to cation-anion complexes, and indicates a differential ligand distribution between complex cations and anions, e.g. $[Cu(NH_3)_4][PtCl_4]$ and $[Pt(NH_3)_4][CuCl_4]$. ***Complex polymerism*** describes complexes which have the same summary formulas but differ in molecular size, e.g. $[PtCl_2(NH_3)_2]$ and $[Pt(NH_3)_4 \cdot [PtCl_4]$

Complexometry, ***chelatometry***: the most important volumetric method for quantitative analysis of nearly all multivalent metal ions. It is based on the formation of very stable, water-soluble complexes with Complexons (see). The most important complexon used in C. as titrator is the disodium salt of ethylenediamine tetraacetic acid (EDTA). Other complexons, e.g. ethylene glycol bis(2-aminoethylether)tetraacetic acid (EGTA) or 1,2-diaminocyclohexane-tetraacetic acid (DCTA), are used only in special cases. If a metal ion with charge n is denoted by M^{n+}, and the doubly deprotonated anion of the complexon by H_2Y^{2-}, the formation of a complex in C. can be described by the general reaction equation $M^{n+} + H_2Y^{2-} \rightleftharpoons MY^{4-n} + 2\,H^+$. According to this equation, the reaction always has the stoichiometric ratio of 1:1, regardless of the charge on the metal ion. This means that, in contrast to all other methods of volumetric analysis, molar solutions of the titrator are used in C. Since hydrogen ions are released by complex formation, a buffered solution must be used. Otherwise, the pH drops during the titration, and complex formation is incomplete.

The endpoint of the titration is usually determined using metal indicators (see Indicators), but it can also be determined by electrochemical methods (see Potentiometry). Since suitable indicators are now available for almost all metal ions, direct titration, i.e. the titration of the sample solution with a standard solution of a complex, is the most common in C.

C. is not a selective method. Usually the metal ion to be determined must be separated from others. However, the selectivity of the determination of a metal ion can be considerably improved by use of masking reagents. For example, the alkaline earth metal ions can be selectively determined in the presence of nearly all transition metal ions by using cyanide ions as masking reagent. Another possibility is titration at different pH values, since the formation of complexes depends on the pH, as indicated by the reaction equation. For example, metal ions which form very stable complexes, e.g. bismuth, iron(III), thorium and zirconium, can be determined at pH values of 0 to 3. Metal ions for which the stability constants of the complexes are between 10^{16} and 10^{18}, e.g. lead, copper and zinc, can be titrated in solutions at pH 5 to 6. For those metal ions for which the stability constants of the complexes are between 10^8 and 10^{10}, on the other hand, the pH of the titration must be at least 10. This dependence of complex formation on pH also makes possible simultaneous determinations, e.g. in mixtures of bismuth and lead, the bismuth can be determined by titration at pH 1, and the lead by subsequent titration at pH 5.5. Such simultaneous determinations are in general possible only when the stability constants of the complexes differ by more than 5 orders of magnitude.

Complexon: a collective term coined by G. Schwarzenbach for aminopolycarbonic acids which form stable chelate complexes with multivalent metal ions. Some important C. are, e.g. nitrilotriacetic acid and ethylenediamine tetraacetic acid (EDTA).

C. are used in Complexometry (see). They are also used to mask multivalent metal ions in titrations, precipitations, electrolyses, polarography and chromatography. They are used industrially to purify water. In dyeing, C. are used to keep hardeners in solution. Metal ions which interfere in the production of drugs, laundry products and foods, can be removed by C. In pharmaceutical preparations, addition of C. can increase the stability of solutions which decompose easily in the presence of metal ions (e.g. ascorbic acid, adrenalin and hydrogen peroxide). C. are also used in treatment of heavy-metal poisoning, especially lead poisoning, and in removal of radioactive isotopes which have been incorporated into the body. Calcium deposits in the cornea are also treated with C. In some cases, C. can dissolve kidney stones.

Composition parameter, ***composition variable***: the quantitative amount of a component of a solution or mixture. A distinction is made between extensive and intensive C. (see State, parameters of).

A) ***Extensive C.***. The composition of a solution can be uniquely determined by the mass m_i or the amount n_i of each component i. Extensive C. are of little practical importance; they are used primarily in equations for extensive functions of state.

B) ***Intensive C.*** are parameters in which the extensive parameters of component i (mass m_i, molar amount n_i, volume v_i) are given as ratios to extensive parameters of the entire solution. An intensive C. has the same value for any arbitrary aliquot of the solution. The most important intensive C. are:

1) the ***fraction*** of component i in a system of several components is the ratio of an extensive parameter of that component to the same parameter for the entire system. The C. may be a mass, molar or volume fraction. The *mass fraction* w_i or ξ_i of component i is the ratio of the mass of i to that of the entire mixture m_M; m_M is the sum of all the masses of the components j:

$$w_i = m_i / m_M = m_i / \sum_j m_j$$

The *molar fraction* x_i is similarly defined:

$$x_i = n_i / \sum_j n_j.$$

It is very important in physical chemistry. When the *volume fraction* $\Phi_i = v_i/v_M$ is used, it must be taken into account that the total volume v_M is the sum of the component volumes v_j only in an ideal solution. For real solutions, it can be different. The volume fraction is used mainly in liquid mixtures for commercial purposes (e.g. alcohol content of liquor). The fractions are always ratios and dimensionless. Their values are always between 0 and 1. Sometimes the values are multiplied by 100 and given as percents; in this case one has the *mass percent* (*mass %* or *M%*, sometimes simply %), *mole %* or *volume percent* (*vol.%*). A C. for use with gas solutions is the Partial pressure (see) p_i. For ideal gas solutions, the relationship between p_i, the total pressure p and the molar fraction x_i is described by the simple equation $p_i = x_i p$.

For solutions the following C. are preferred:

2) the ***molality*** $\bar{m}$ is the ratio of the molar amount $n_i = m_i/M_i$ (M_i is the molar mass of i) and the mass of the solvent m_S: $\bar{m}_i = n_i/m_S = m_i/(M_i m_S)$. (The symbol $\bar{m}$ is used here instead of the internationally accepted m for molality in order to distinguish it from the mass m.) A solution of $\bar{m} = 1$ mol kg^{-1} is 1 molal (abbreviated 1 m).

3) ***Concentration parameters*** are defined as ratios of extensive parameters of component i to the volume v_M of the solution. The *mass concentration* $\varrho_i = m_i/m_M$; commonly used units are g cm^{-3} and g l^{-1}. The *molar concentration* (*concentration* or *molarity*) c_i is equal to the number of moles of component i divided by the volume of the solution: $c_i = n_i/v_M = m_i/(M_i v_M)$. The notation [i] (read "concentration of i") is often used for c_i; the usual units are mol l^{-1}. A solution of concentration c = 1 mol l^{-1} is 1 molar (1 M). All

Equations for conversion of composition parameters.

given → sought ↓	n_i	w_i	x_i	c_i	$\bar{m}_i$
$n_i =$	n_i	$= w_i \frac{m_M}{M_i}$	$= x_i \Sigma n_j$	$= c_i v_M$	$= \bar{m}_i m_s$
$w_i =$	$n_i \frac{M_i}{\Sigma n_j M_j}$	$= w_i$	$= x_i \frac{M_i}{\Sigma x_j M_j}$	$= c_i \frac{M_i}{\rho_M}$	$= \bar{m}_i \frac{M_i\ ^{*)}}{1 + \Sigma \bar{m}_j M_j}$
$x_i =$	$n_i \frac{1}{\Sigma n_j}$	$= w_i \frac{1}{M_i \Sigma (w_j/M_j)}$	$= x_i$	$= c_i \frac{M_s\ ^{*)}}{\rho_M + \Sigma c_j (M_s - M_j)}$	$= \bar{m}_i \frac{1\ ^{*)}}{\frac{1}{M_s} + \Sigma \bar{m}_j}$
$c_i =$	$n_i \frac{1}{v_M}$	$= w_i \frac{\rho_M}{M_i}$	$= x_i \frac{\rho_M}{\Sigma x_j M_j}$	$= c_i$	$= \bar{m}_i \frac{\rho_M\ ^{*)}}{1 + \Sigma \bar{m}_j M_j}$
$\bar{m}_i =$	$n_i \frac{1}{m_S}$	$= w_i \frac{1}{w_S M_i}$	$= x_i \frac{1}{x_S M_S}$	$= c_i \frac{1\ ^{*)}}{\rho_M - \Sigma c_j M_j}$	$= \bar{m}_i$

Meaning of a) symbols for parameters: c, (molar) concentration (molarity); m, mass; $\bar{m}$, molality; M, molar mass; n, number of moles; v, volume; w, mass fraction; x, molar fraction; ρ, density; b) subscripts: i, arbitrary component; j, running index for summation; S, solvent; M, total mixture.

*) The sum includes only the parameters of the dissolved components, but not those of the solvent.

concentration parameters have the disadvantage (with respect to molality) that they are temperature-dependent.

A special case of molarity is the *equivalent concentration (normality)*, $c_{ev,i} = n_{ev,i}/v_M$. Using the equation $n_{ev,i} = n_i z_i$ (z_i is the stoichiometric valency of i (see Mole)), it follows that $c_{ev,i} = n_i z_i / v_M$. In words, normality = molarity · valency. A solution of equivalent concentration c_{ev} = 1 mol l^{-1} is 1 normal (1 N). *Normal solutions* are solutions of standard normality (e.g. 1 N, 0.1 N...) used in Volumetric analysis (see) as standard solutions, because they simplify the stoichiometric analysis of titrations (see Stoichiometry). However, the stoichiometric valence of a compound, and thus the chemical equivalent, are in many cases not constant parameters, but depend both on the reaction conditions and the nature of the reaction partner. Therefore the use of normality as a C. and work with normal solutions may lead to uncertainty as to the equivalent. For example, 0.1 M $KMnO_4$ for a standard solution is correct and unequivocal, while the same solution can react as 0.5 N $KMnO_4$ (in acid solution) or as 0.3 N $KMnO_4$ (in neutral solution). Therefore normality should not be used as a C., or only with caution (with an indication of the equivalent).

A conversion of various C. is often necessary; this can be done using the equations given in the table. For very dilute solutions, the following approximation can be used: $x_i \approx n_i/n_S = \bar{m}_i M_S \approx c_i M_S/\varrho_S$.

Composition variable: same as Composition parameter (see).

Compound: a substance in which atoms of more than one element are chemically bonded, usually in fixed stoichiometric ratios.

Compressibility coefficient: see State, equation of, 1.

Computer assisted tomography: see NMR tomography.

Concanavalin A: see Lectins.

Concentration: a measure of the amount of a component of a solution or mixture. See Composition parameters.

Concentration cell: see Electrochemical metal corrosion.

Concentration parameters: see Composition parameters.

Concrete: see Binders (construction materials).

Concurrent distillation: see Distillation.

Concurrent process: a principle for carrying out heat or material exchanges. The two media flow past each other in the same direction, in as intimate contact as possible. In pure heat exchange, they are separated by walls which do not permit passage of material, but are good heat conductors.

When the concurrent principle is used in Drying (see), the exchanging media are brought into direct contact. The opposite of C. is a Countercurrent process (see), which is more efficient.

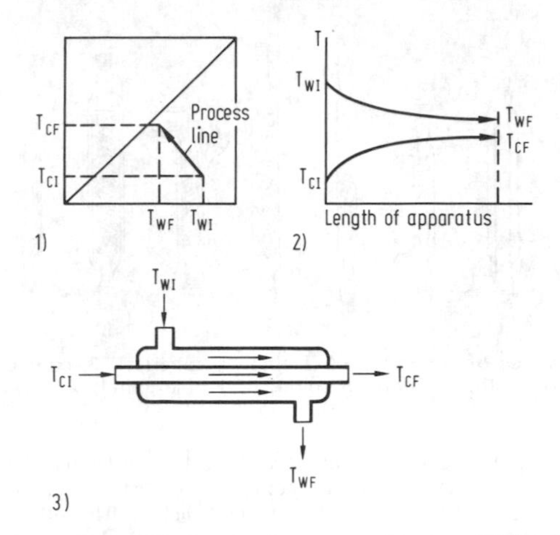

Concurrent process: *1,* Heat exchange on an equilibrium diagram. *2,* Temperatures of the media in a heat exchanger. *3,* Double walled heat exchanger. *W* = warmer medium, *C* = cooler medium, *S* = start of process, *E* = end of process, *T* = temperature.

Condensation: 1) in chemistry, the joining of molecules (intermolecular) or atomic groups within a molecule (intramolecular) with a concomitant loss of a simple molecule such as H_2O, NH_3, H_2S, HX or an alcohol. The reactions of carbonyl groups with bases and CH-acidic compounds are often C. The C. of oxo and hydroxo complexes with loss of water can produce polycations (e.g. hydroxo-oxo complexes of aluminum), while the C. of the oxygen acids of boron, silicon, phosphorus, chromium, molybdenum and tungsten leads to isopolyacids. In intramolecular C., a small molecule is split out of two groups within a molecule (e.g. ring closure of triphosphoric acid to form trimetaphosphoric acid). Polycondensation (see) is exceptionally important.

2) In physics, the conversion of a vapor to the liquid or solid state. C. is the opposite of Evaporation (see) of liquids or Sublimation (see) of solids (see Phase diagram, Phase change energy).

Condensation point: see Boiling point.

Condensation resins: synthetic resins formed by polycondensation of bi- and trifunctional starting materials; they are hardened by further progress of the polycondensation reaction. This group includes the Phenol resins (see) and Aminoplastics (see).

Condenser: an apparatus for Cooling (see), a Heat transferrer (see). In a C., the temperature of a gas or liquid is reduced by removal of heat, leading to liquefaction of vapors or solidification of liquids or gases. Water and air at ambient temperature are most often used as heat-absorbing media. However, in the chemical industry and for many other purposes, it is often necessary to reduce the temperature below the ambient, using a Refrigerator (see). In the C., the coolant and the medium to be cooled flow past each other, separated by a wall. The separating surface is at the same time the heat-transferring surface, to which the cooling (heat removal) is proportional.

1) In the laboratory, C. are usually made of glass. When vapors are liquefied, the resulting condensate is collected in a special vessel. For cooling substances with a boiling temperature > 150 °C, air cooling is used; between 120 and 150 °C, still water; < 120 °C, flowing water. The flowing water (normally tap water) should flow through the C. from bottom to top, or in the opposite direction from the vapor (which enters at the top). Some commonly used (1-3) and special (4-7) types of C. are shown in the figure.

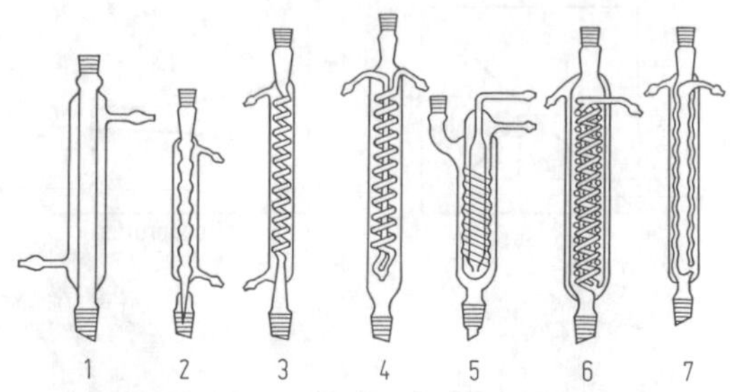

Types of condenser: 1, Liebig; 2, spherical; 3, coil; 4, Dimroth; 5, Friedrich spiral; 6, double spiral; 7, wavy tube

Only the Liebig and wavy tube C. can be used in the conventional arrangement for distillation, in which the C. is placed at a slight incline. The others must be built into the apparatus in a vertical position, above the boiling flask. They are often used as reflux C. (dephlegmatizers) and are then set vertically onto the apparatus so that the rising vapors are condensed and flow back into the flask.

2) A number of types of C. are used in the chemical industry. These are generally built of metal, and they can be closed apparatus or containers, such as bundled pipe C. (Fig. 8), submerged C., mantle C. (Fig. 9) or sprinkling C. The Kratz and cracking gas C. are special types for certain chemical processes.

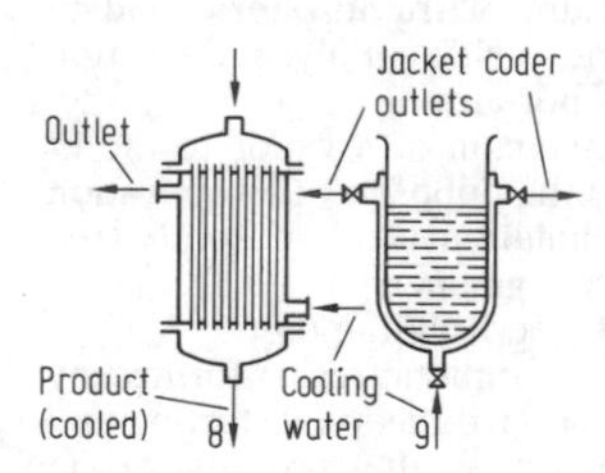

In C. with air cooling, the heat-transfer surface is enlarged (ribs, lamellae), and the heat transfer can be intensified by forced ventilation. The hot water from water-cooled C. is cooled in a cooling tower built of wood, steel or reinforced concrete. These contain structures of wood or plastic over which the water is distributed and trickles down. Because of the chimney effect of the tower, the air flows past the water from below. The degree to which the water is cooled depends on the temperature difference between the water and air and on the evaporation effect. In addition to such natural draft towers, there are cooling towers with forced ventilation, which can be built much lower (Fig. 10).

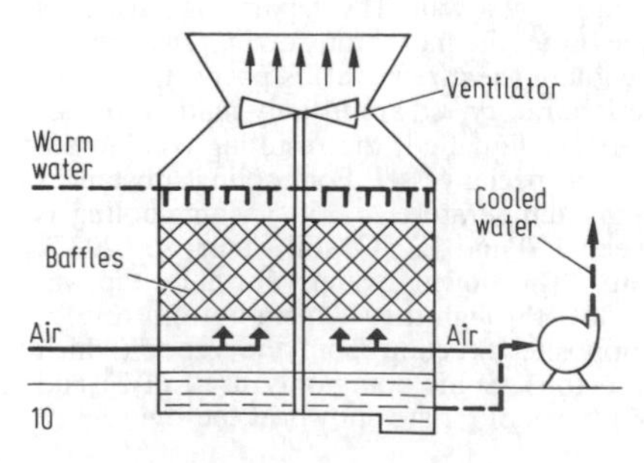

C. for low temperatures are made with an evaporating coolant which is recycled through the refrigerator. The term C. is thus also used in refrigeration technology to indicate the evaporater in a refrigerator.

Conducting electrolyte: same as Conducting salt (see).

Conducting salt, ***conducting electrolyte***: a salt which carries current in an electrolyte solution subjected to electrolysis. The C. reduces the ohmic resistance of the solution and transports the depolarization to the electrodes by diffusion. It does not take part in the electrode reactions. L. play an essential role in polarography and in organic electrosynthesis.

Conduction band: see Energy band model.

Conductivity: see Electrical conductivity.

Conductivity measurement: the electrical conductivity of electrolyte solutions is measured by determining their resistances, usually between two electrodes which dip into an electrolyte solution. The experimental values depend not only on the resistance of the electrolyte, but also on the voltage- or current-dependent polarization resistance which arises at the phase boundary between the electron-conducting electrodes and the ion-conducting solution. When the measurement is made with direct current, the polarization resistance can be very high, but if alternating current at high frequency (up to 50 kHz) is used, along with electrodes with high double-layer capacity (e.g. platinum-plated platinum), the polarization resistance is small enough to be ignored. Therefore measurements of electrolytic conductivity are normally made with alternating current using a bridge circuit.

Conductivity measurements can be used in two ways in chemical analysis: 1) for aqueous solutions of an electrolyte, there is a relationship between the specific conductivity and the concentration, so that concentration can be determined directly by measurement of conductivity. This type of measurement is made routinely for quality control in chemical manufacturing, and to test the purity of water (e.g. distilled water or water for boilers). 2) The end of a titration can be recognized, in suitable systems, by the conductivity. ***Conductometric titrations*** are used in acid-base and precipitation titrations. In the titration of HCl with NaOH, for example, H_3O^+ ions are first replaced by Na^+ ions, which have a lower limiting conductivity (see Hydronium ion conductivity). Therefore the conductivity of the solution decreases and reaches a minimum at the equivalence point. Thereafter, the addition of more NaOH causes the conductivity to rise because there are more ions in solution (Fig.).

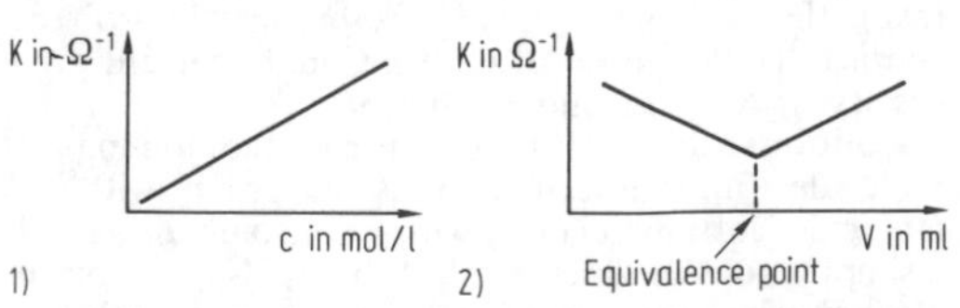

Signal curves in conductivity measurement (1) and conductometric titration (2). *C*, specific conductivity; *c*, concentration; *V*, volume

Conductor: a substance which, because of its particular structural properties, is able to conduct electric current, heat or sound. See Electron conductor, Ion conductor, Electrical conductivity, Energy band model.

Configuration: see Stereochemistry 1.

D,L-Configuration: see Stereoisomerism 1.1.

Configurational isomerism: see Stereoisomerism 1.

Conformation: see Stereochemistry 2.

Conformational equilibrium: see Stereoisomerism 2.1.

Conformational isomerism: see Stereoisomerism 2.

Conformation analysis: see Stereoisomerism 2.1.

Conformation anchor: see Stereoisomerism 2.2.

Congo paper: see Congo red.

Congo red: the sodium salt of benzidinediazobis-1-naphthylamine-4-sulfonic acid. C. is a red-brown powder (m.p. > 360 °C) which gives a colloidal suspension in water. It is formed by coupling of bis-diazotized benzidine with naphthionic acid. C. is a secondary, symmetric diazo dye (benzidine dye) used as a substantive cotton dye.

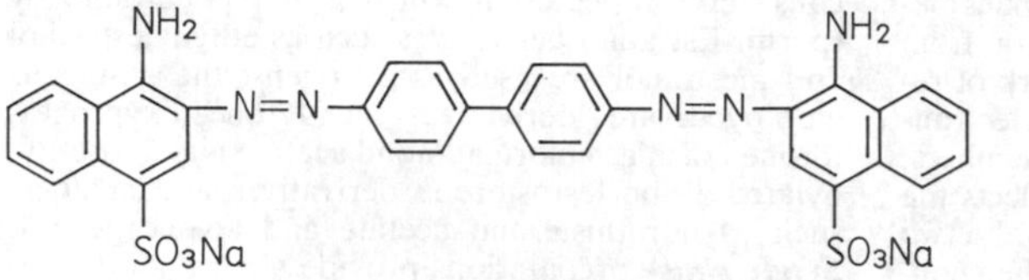

From a neutral or weakly alkaline solution, C. attaches directly to the cotton, while ordinary azo dyes bind only to wool and silk in acidic media. A disadvantage of C. is its lack of fastness to light and sensitivity to acid. C. is also used as a stain for microscopic preparations (e.g. for staining cytoplasm and erythrocytes) and as an acid-base indicator; at a pH between 3.0 and 5.2, its color changes from blue-violet to red. C. can be used to distinguish mineral and organic acids. It was discovered in 1884 by P. Bœttiger.

Congressane, ***diamantane***, ***pentacyclo[7.3.1.1^{4,12}0.^{2,7}.0^{6,11}]tetradecane***: an exceptionally stable hydrocarbon which, like adamantane, is a "diamantoid" compound. D. forms colorless crystals, m.p. 236 to 237 °C.

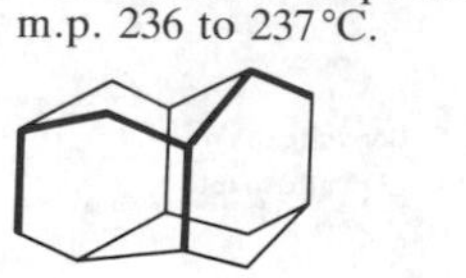

In D. all the H atoms are equivalent, i.e. there is only a single NMR signal for all the protons. From the structural formula, it can be seen that this molecule is a "section" of a diamond crystal lattice. It can be synthesized by photochemical dimerization of bicyclo[2.2.1]heptane or by $AlCl_3$-catalysed rearrangement of photochemical norbornene dimers:

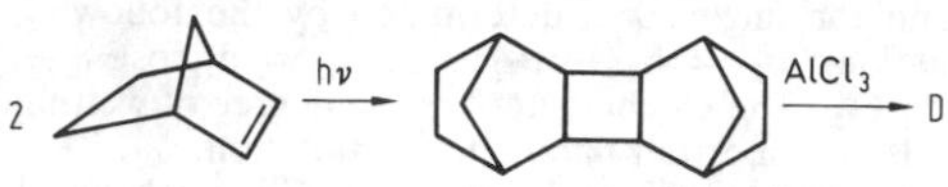

The symbol for C. is the congress emblem of the International Union of Pure and Applied Chemistry.

Coniferin: see Cinnamyl alcohols.

Coniferyl alcohol: see Cinnamyl alcohols.

Coniine, ***2n-propylpiperidine***: the major alkaloid of poison hemlock (*Conium maculatum*). C. is an oil which turns brown in air; it has a sharp taste and smells like mouse urine. The plant contains 1.5 to 2.0% piperidine alkaloids. C. is very poisonous, causing respiratory paralysis. C. was the first alkaloid to be synthesized (1886, Ladenburg).

Conjugated double bonds: two or more carbon-carbon double bonds in a linear or cyclic compound which are separated by single bonds: −C=C−C=C−... (see Dienes, Polyenes). C. display a characteristic interaction.

Connode: in a phase diagram, the line joining two points which represent the compositions of two phases in equilibrium with each other. For example, C. can join two points on the boiling and condensation curves of a boiling diagram, the solidus and liquidus curve of a melting diagram, or the two sides of a diagram of a mixing gap.

Conrad-Limpach synthesis: a method for producing hydroxyquinoline from β-ketocarboxylate esters and aryl amines:

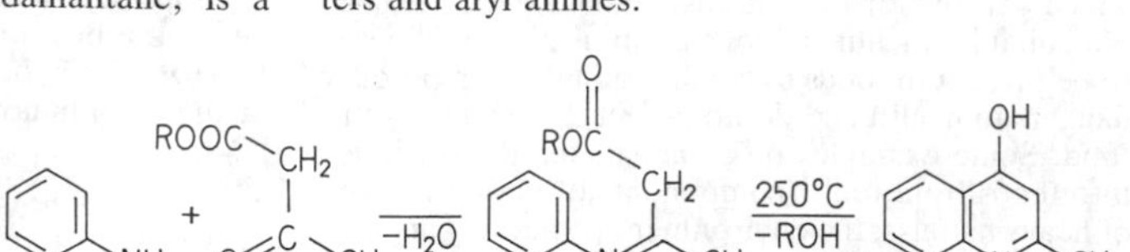

The 2-hydroxyquinoline can be formed in addition to the substituted 4-hydroxyquinoline, if the primary reaction between the amine and the ester group forms the anilide. Since the first reaction step in either case is reversible, the ring closure is determined by various factors, such as the relative volatilities of water and the alcohol which is cleaved off.

Conrotatory: see Woodward-Hoffmann rules.

Constant proportions, law of: see Stoichiometric proportions, law of.

Contact: same as solid catalyst. See Catalysis.

Contact angle: an angle established at the phase boundary between three mutually non-miscible phases, for example, in limited wetting of a solid by a liquid or of one liquid by another (see Wetting, Fig. a and b). The relationship between the surface tension σ and the angle Θ is determined by the following equations: $\sigma_{13} = \sigma_{23} + \sigma_{12}\cos\Theta$; $\sigma_{13} = \sigma_{12}\cos\Theta_{12} + \sigma_{23}\cos\Theta_{23}$. The C. characterizes the degree of wetting and is an important parameter in flotation.

Contact corrosion element: see Electrochemical metal corrosion.

Contact poison: 1) the active material in a pesticide which enters the body of a pest organism through its integument (skin, cuticle, etc.). Many modern insecticides are C. (*contact insecticides*). A few chemical weapons could also be considered C., in a wide sense, including all skin poisons and the organophosphate materials such as sarin and soman.

2) ***Catalysis poison***: see Catalysis, sect. III.

Contact process: see Sulfuric acid.

Contact voltage: a voltage which arises when two chemically different substances are in contact. Their surface areas acquire very small, opposite electrical charges, because one of the substances donates electrons and the other accepts them. This happens because the work required to remove an electron from the electron acceptor is greater than the work of removing an electron from the donor. The electrons move constantly back and forth, in small numbers, and an equilibrium is established which reflects the relative binding strengths of the electrons to the two materials. The result is a *contact potential*. When nonconductors are involved, the substance with the lower dielectric constant acquires the negative charge. The necessary close contact is achieved, with poor conductors, by rubbing them together. If metallic conductors are arranged according to their contact potentials in such a way that the more negative metal of any pair is on the right, the result is the *electrical potential series*: (+) K, Na, Al, Zn, Pb, Sn, Sb, Bi, Fe, Cu, Ag, Au, Pt (-).

Contaminants: in a broad sense, impurities such as chemical substances, microorganisms or their metabolic products (mycotoxins, biogenic amines, etc.), or radioactive substances. In a narrower sense, a term used in food chemistry for foreign materials which get into or onto foods, remain there and are not suitable for human consumption. C. are allowed to be present in foods only in amounts which pose no dangers to health and do not affect the quality of the food. Some examples of C. are materials which get into foods from tools (components of plastics, traces of heavy metals), the environment (especially air, water and soil pollutants such as industrial emissions, traffic exhuast, water-soluble chemicals), residues of pesticides used to protect growing and stored crops and materials used to treat animals (antibiotics, hormones). Foods which exceed a legally established maximum concentration of C. are considered unfit for consumption.

Contamination: 1) the enrichment of harmful substances in soil and ground water, animals, human beings, plants and foods. Radioactive C. is of special significance; it is the distribution of radioactive substances in sufficient concentration to endanger health or to make certain installations unusable. Radioactive C. occurs as a result of bombs, reactor accidents and careless handling of radioactive preparations. These substances can be incorporated via the lungs, skin (especially through wounds or sores) and food.

C. with chemical toxins can be a result of excessive application, accidental spills and releases or migration of the toxin to other environmental compartments (such as ground water). Chemical C. can also be the result of use of chemical weapons.

The removal of the C. is called Detoxication (see).

2) In analysis, the enrichment of a substance (either the substance being analysed or one which affects analysis) during sample work-up, that is, between the time the sample is taken and the actual determination. Sample C. is a particularly serious problem in trace analysis.

Continuous wave method: see NMR spectroscopy.

Contraceptives: means of preventing conception. Oral C. generally consist of combinations of an estrogen and a gestagen. They may act by inhibiting lutropin release and thus preventing ovulation, by preventing implantation of the fertilized ovum in the uterine mucosa, or by promoting formation of a viscous cervical secretion which is not penetrable by sperm. Estradiol derivatives such as ethynylestradiol and mestranol are used as estrogens; the gestagens employed are derivatives of 17α-acetoxyprogesterone, such as chlormadinone acetate and 17α-ethynylated 19-nortestosterone derivatives and analogs, such as norethisterone acetate and levonorgestrel. ***Single phase*** preparations provide a constant combination of a gestagen with a small amount of estrogen; ***sequential*** preparations consist first of estrogen alone (possibly containing a small amount of gestagen) and later a mixture of gestagen/estrogen. Hormonal contraception is also possible by a continuous release from a depot gestagen.

Contracid: a proprietary name for an alloy of 58-61% nickel, 12-19.5% iron, 15% chromium, 2% manganese and the remainder molybdenym, tungsten, cobalt and beryllium. C. is resistant to acids, and is used in dentistry and to make surgical instruments.

Convallaria glycosides: cardiac glycosides which make up 0.2 to 0.6% of dried lily-of-the-valley, *Convallaria majalis*. About 50 glycosides have been isolated; the best known are ***convallatoxin*** and ***convallatoxol***. Their therapeutic effects are similar to those of the Strophanthus glycosides (see).

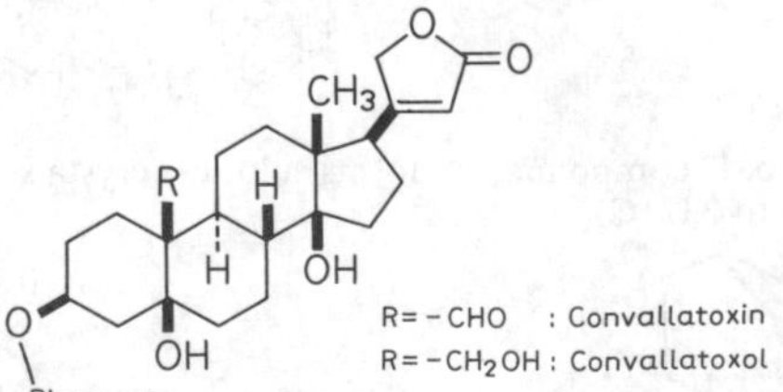

R = -CHO : Convallatoxin
R = $-CH_2OH$: Convallatoxol
Rhamnose

Convallatoxin: see Convallaria glycosides.

Convallatoxol: see Convallaria glycosides.

Conversion, ***1) CO conversion***: the reaction of carbon monoxide, CO, with steam to produce hydrogen, H_2, and carbon dioxide, CO_2.

2) COS conversion: the reaction of carbon oxygen sulfide, COS with steam to produce hydrogen sulfide, H_2S, and carbon dioxide, CO_2. C. is applied in production of synthesis gas for synthesis of ammonia and methanol, and to obtain hydrogen for hydrogenation. C. is also important for detoxication of city gas. Organic sulfur compounds, which are preesnt as contact poisons in synthesis gas, are also cleaved during C. to form hydrogen sulfide and sulfur.

C. is carried out in *conversion furnaces*, in which catalysts are present. ***Low temperature C.*** occurs around 250 °C, and ***high temperature C.***, around 350 °C. In modern plants, C. is done under a pressure of 3 to 10 MPa. Since the C. reaction is exothermic, the catalyst temperature must be kept low to allow the reaction to go as far as possible towards completion, and to make the operation efficient. Therefore multi-layered contact furnaces are used. The gases pass through heat exchangers before entering and after leaving the furnace. Hot water circulating through the heat exchangers transports the heat energy to a site where it can be utilized.

Conversion electron spectroscopy: spectroscopy of the kinetic energy of electrons which are ejected from the electronic shell of an atom by transfer of excitation energy from the nucleus (see Gamma ray spectroscopy). The kinetic energy E_{kin} of the conversion electron is $E_{kin} = E_{\Delta} - E_B$, where E_{Δ} is the energy difference between the nuclear states and E_B is the bond energy of the electron. When the resulting electron hole is refilled by other electrons, there is a characteristic emission of x-rays or Auger electrons (see X-ray spectroscopy, Auger electron spectroscopy).

Converter: in metallurgy, a furnace without external heating used to purify metals. The C. is a pear-shaped, or, rarely, a cylindrical vessel which can be rotated around the horizontal axis. It is lined with firebrick. A stream of oxygen or air is blown through the furnace and burns off impurities in the crude melt. The heat released by the combustion is sufficient to keep the metal liquid or even to increase its temperature. C. have capacities up to 370 t crude iron. There are various types (see Steel).

Cooling: a basic thermal operation used to reduce the temperature of a substance or keep it constant, to liquefy vapors or to convert a vapor or liquid to the solid state (crystallization, solidification, desublimation) by removal of heat. In the chemical industry, the main uses are to remove heats of reaction, solution and phase change. The methods used are determined mainly by the amount of heat to be removed, the properties of the substance to be cooled and the working temperature range. The most frequently used are submersion, trickle, and mantle C., and C. in a closed, separate appratus. The apparatus, which is a Heat transferrer (see) by function, is generally called a Condenser (see).

C. can be carried out continuously or periodically. C. with ice is a special case, in which small pieces of water ice or dry ice are added directly to the material to be cooled. In every case, C. involves a transfer of heat to a coolant at a lower temperature. There is a fundamental difference between C. above and below the ambient temperature. In the first case, water or air is commonly used as a coolant. When a material must be cooled to a temperature below the ambient temperature, refrigeration is required.

The removal of heat from systems and processes at very high temperatures is also C.; the heat absorbed by the coolant in such cases can be reused in some fashion.

Cooling brine: an aqueous salt solution with a freezing point far below that of water. C. are used as refrigerant liquids which are circulated between the refrigerator and the area to be cooled. The temperature at which crystallization begins, which depends on the nature and concentration of the dissolved salt, is adjusted according to the requirements of the application (the water concentration is as high as possible). Chloride brines can be used to about - 45 °C; carbonate brines are also used.

Theoretically, the minimum temperature is that of the eutectic mixture. Eutectic brines (see), which freeze in the same manner as pure substances, can be used to store cold. Commercial brines contain added buffers and corrosion inhibitors.

Cooling curve: a curve from which a phase diagram can be constructed; see Thermal analysis.

Cooling mixture: a mixture of salt and ice used for cooling. Under adiabatic conditions, the enthalpy of melting and of mixing must be obtained from the thermal energy of the system. The salt and ice melt and dissolve, respectively, so the mixture can only be used once. The lowest temperatures are obtained when the mixing ratio corresponds to the eutectic mixture. The following table gives examples.

C. were once sold commercially for cooling beverages and making ice cream, as well as for use in the laboratory. In the laboratory, a mixture of Dry ice (see) and methanol or acetone is sometimes used for the same purpose (cooling a reaction mixture), although in this case no melting occurs. With dry ice and acetone or methanol, temperatures as low as - 70 to - 90 °C can be obtained.

Salt	Mixture (g salt per 100 g ice)	Temperature (°C)
NaCl	31 (E)	−21.2
$MgCl_2 \cdot 6H_2O$	84 (E)	−33.6
$CaCl_2$	40	−35
$CaCl_2 \cdot 6H_2O$	125	−40
$CaCl_2 \cdot 6H_2O$	143 (E)	−55

(E) = eutectic mixture

Coordination chemistry: chemistry of complexes, which are higher-order compounds in which a *central atom* is surrounded by *ligands* (atoms, ions or small molecules); the number of ligands bound to the central atom is the *coordination number* (abb. n). For each coordination number, there is at least one characteristic *coordination polyhedron*, and there is often a correlation between the type, electron occupation of the central atom and the preferred spatial ligand orientation. The coordination number and geometry offer one important basis on which complex compounds can be classified; other bases are the type of ligands, thermodynamic and kinetic stabilities of the complex, magnetic and optical properties, and the predominant type of bonding.

Coordination numbers and geometries (Fig. 1). Complexes with n = 2 are linear and are formed mainly by d^{10} ions such as copper(I), silver(I) and gold(I). Examples: $[CuCl_2]^-$, $[Ag(NH_3)_2]^+$, $[AuCl_2]^-$. Complexes with n = 3 are relatively rare, and are usually trigonal planar (a) structures. Examples: $[HgI_3]^-$, $Ni(C_2H_4)_3$.

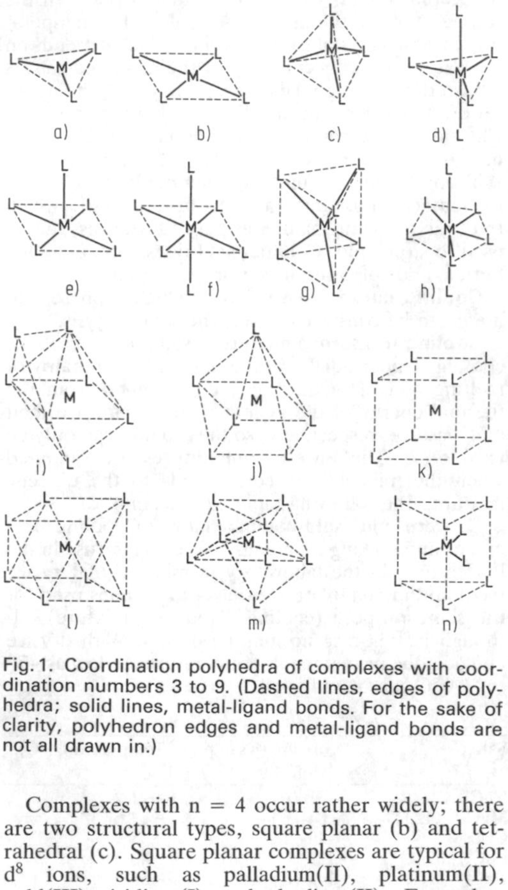

Fig. 1. Coordination polyhedra of complexes with coordination numbers 3 to 9. (Dashed lines, edges of polyhedra; solid lines, metal-ligand bonds. For the sake of clarity, polyhedron edges and metal-ligand bonds are not all drawn in.)

Complexes with n = 4 occur rather widely; there are two structural types, square planar (b) and tetrahedral (c). Square planar complexes are typical for d^8 ions, such as palladium(II), platinum(II), gold(III), iridium(I) and rhodium(II). Examples: $[PdCl_4]^{2-}$, $PtCl_2(NH_3)_2$, Au_2Cl_6, $[IrClCO(P(C_6H_5)_3)_2]$, $[RhCl(CO)_2]_2$. They are also observed for nickel(III), e.g. $[Ni(CN)_4]^{2-}$. Tetrahedral complexes occur with d^{10} and d^5 ions like zinc(II), cadmium(II), mercury(II), copper(I), manganese(II) and iron(III). Examples: $[Zn(MH_3)_4]^{2+}$, $[HgCl_4]^{2-}$, $[Cu(CN_4)]^{3-}$, $[FeCl_4]^-$. *cis-* and *trans-* stereoisomerism exists for square planar complexes of the type MA_2B_2 (see Complex isomerism). The coordination number 5 occurs more commonly than was at first believed; it is associated with trigonal-bipyramidal (d), e.g. $Fe(CO)_5$, or square-planar complexes (e), e.g. $NiBr_3[P(C_2H_5)_3]_2$. Complexes with n = 5 are characterized by non-rigid, fluctuating molecular skeletons. The two coordination polyhedra easily interconvert (see Berry pseudorotation), and intermediate forms are possible. n = 6 is very common, and is almost exclusively associated with an octahedral arrangement of ligands, either a regular (f) or a tetragonally, trigonally or rhombically distorted arrangement. Complexes with a trigonal prismatic ligand arrangement (g) are extremely rare, e.g. MoS_2. Complexes of the 3d metals usually have n = 2 to 6; those of transition metals with 4d, 5d, 4f and 5f electrons frequently have n > 6. Like n = 5 complexes, n = 7 coordination compounds are not rigid: the three possible polyhedra, the pentagonal bipyramid (h), e.g. $[UF_7]^{3-}$, the singly capped octahedron (i), e.g. $[NbOF_6]^{3-}$, and the singly capped trigonal prism (j), e.g. $[TaF_7]^{2-}$, are easily interconverted. The possible polyhedra for complexes with n = 8 are the cube (k), the square antiprism (l) and the dodecahedron (m). The cube is extremely rare in discrete complexes and the other two are relatively easily interconverted, as can be shown, for example, with the octacyanometallates $[M(CN)_8]^{n-}$ (M = Mo, W, n = 3,4). In complexes with n = 9, the triply capped trigonal prism arrangement (n), e.g. $[ReH_9]^{2-}$, is most common.

Nomenclature. In the names of coordination compounds, the name of the anion follows that of the cation. The name of the complex (whether it is neutral, cationic or anionic) gives first the number of ligands, then the names of the ligands (in mixed complexes, the acidic ligands before the neutral ligands), then the central atom (including the charge on the central atom as a Roman numeral). If the central atom is part of a complex anion, it is given the ending -ate with an oxidation number appended. Anionic ligands are given the ending -o, e.g. Cl^-, chloro, OH^-, hydroxo, CN^-, cyano, O^{2-}, oxo. Special terms are used for important neutral ligands, such as aqua (aquo) for water and ammine for ammonia. Otherwise, the names of neutral molecules are unchanged. Bridge ligands are indicated by a μ prefix. For the nomenclature of organometallic complexes, see Organometallic compounds. Some examples of the rational nomenclature of classic coordination compounds are: potasium hexacyanoferrate(II), $K_4[Fe(CN)_6]$; hexaamminenickel(II) sulfate, $[Ni(NH_3)_6]SO_4$; dichlorodiammineplatinum(II), $PtCl_2(NH_3)_2$; μ-hydroxobis(pentaamminechromium(III)) chloride, $[(NH_3)_5Cr(OH)Cr(NH_3)_5]Cl_5$.

Ligands. Complex ligands can be divided into two types. Classical ligands, the σ-donor ligands, usually have a free electron pair. In multiatomic ligands, the atom on which this free pair is localized is called the *donor* or *ligator atom*. σ-donor ligands can form complexes with all Lewis acids. Non-classical ligands, the π-bonding ligands, form complexes only with transition metals by simultaneous donor and acceptor interaction between occupied and free orbitals of the π ligand and the d orbitals of the metal. The π ligands include carbon monoxide, CO, cyanide, CN^-, phosphines, PR_3, unsaturated (alkenes, alkynes) and aromatic hydrocarbons, and other molecules which have multiple bond systems, such as dinitrogen, N_2, carbon dioxide, CO_2, or carbon disulfide, CS_2. π ligands play an important role in the organic chemistry of the transition metals (see Organometallic compounds) and in homogeneous catalysis.

If a ligand coordinates via a donor atom, it is called *monodentate*; if it has several potential ligator atoms, it is *multidentate*. Coordinate interactions between multidentate ligands and a central atom lead to ring-shaped compounds, the *chelate complexes*; five- and six-membered ring systems are most common. A typical bidentate chelator is ethylenediamine, $NH_2(CH_2)_2NH_2$, while diethylenetriamine, $HN(CH_2CH_2NH_2)_3$, the *tridentate ligand* triethylenetetramine, $N(CH_2CH_2NH_2)_3$ and ethylenediaminetetraacetate, $[(OOCH_2)_2N(CH_2)_2N(CH_2COO)_2]^{4-}$, are examples of tri-, tetra- and hexadentate ligands which can form multiple chelate rings because of their flexible molecular skeletons. Multidentate cyclic ligands, e.g. the phthalocyanins or porphyrins (heme or chlorophyll), are called *macrocyclic ligands*. The macrocyclic polyethers (see Crown ethers) and bicyclic aminopolyethers (see Cryptates) are multidentate ligands.

Monoatomic ligands which, like the halides X, have several free electron pairs, multiatomic ligands in which the donor atom has several free electron pairs, e.g. OH, and multiatomic ligands with several potential donor atoms, e.g. CN^-, NCS^-, N_3^- or NO_2^-, can act as *bridge ligands* binding several metal atoms together.

Ligands which have more than one kind of donor atom, e.g. NO_2^- or NCS^-, are called *ambivalent* or *ambidentate ligands*. The bonding in complexes containing such ligands is indicated by the name. A distinction is made between nitro (M-NO_2) and nitrito (M-ONO), isothiocyanate (M-NCS) and thiocyanate (M-SCN) complexes.

Complex formation in solution. Formation of a complex in solution, where the metal ion is always solvated, is necessarily a ligand exchange reaction. The exchange occurs in steps, as indicated below for a reaction in aqueous solution:

$$M(H_2O)_6 + L \rightleftharpoons M(H_2O)_5L + H_2O$$
$$M(H_2O)_5L + L \rightleftharpoons M(H_2O)_4L_2 + H_2O$$
$$M(H_2O)L_5 + L \rightleftharpoons ML_6 + H_2O$$

The concentrations of the complex species simultaneously present in solution are determined by the individual formation constants $k_1 = [M(H_2O)_5L]/[M(H_2O)_6]\cdot[L]$, $k_2 = [M(H_2O)_4L_2]/[M(H_2O)_5L]\cdot[L]$, ... $k_6 = [ML_6]/[M(H_2O)L5]\cdot[L]$; the formation constant of the end complex ML_6, which is a measure for the thermodynamic stability of this complex, is $K = k_1 \cdot k_2 ... k_6 = [ML_6]/[M(H_2O)_6]\cdot[L]^6$. As a rule, the individual formation constants decrease in order $k_1 . k_2 > ... k_6$, for statistical, electrostatic and steric reasons. In general, chelate complexes are significantly more stable thermodynamically than complexes with homologous, monodentate ligands, a phenomenon known as the *chelate effect*. It is caused by an increase in entropy upon formation of a chelate complex. The chelate effect can be illustrated by comparison of the stability constants *K* for the nickel(II) complexes $[Ni(NH_3)_6]^{2+}$ ($K = 10^{8.61}$) and $[Ni(NH_2(CH_2)_2NH_2)_3]^{2+}$ ($K = 10^{18.28}$). For homologous octahedral 3d metal complexes of the type ML_6, the thermodynamic stability follows the *Irving Williams series* $Mn^{2+} < Fe^{2+} < Co^{2+} < Ni^{2+} < Cu^{2+} < Zn^{2+}$. In a given metal-ligand system, the M^{3+} complex is more stable than the M^{2+} complex. Comparison of the thermodynamic stabilities of complexes with a wide variety of ligands and central atoms indicates that the HSAB concept (see Acid Base concepts) can be used as a basis of classification. Stable complexes result from combination of "soft" metals, e.g. Hg(II), Pd(II), Pt(II), with "soft" ligands (I^-, B^-; P- or S-donor ligands) or "hard" metals, e.g. d-metals of the third to sixth transition groups, f-metals, with "hard" ligands (F^-, O- or N-donor ligands).

Kinetic stability. Complex compounds have a wide range of kinetic stabilities. The kinetically stable complexes include primarily octahedral chromium(III), cobalt(III) and planar platinum(II) complexes. In general, substitution reactions on coordination complexes proceed either by a dissociative S_N1 or an associative S_N2 mechanism; the intermediate complexes formed have coordination numbers 1 lower (S_N1) or 1 higher (S_N2) than the reactants and products. Thus in a substitution reaction on an octahedral complex, the first, rate-limiting step is the formation of a square-pyramidal intermediate, while the formation of a pentagonal bipyramidal intermediate is the rate-limiting step in the associative substitution mechanism. In substitution reactions on the square-planar platinum(II) complex, which are believed to proceed by an associative S_N2 mechanism, a pronounced *trans*-directing effect of the ligands is observed. This *trans*-effect is a kinetic effect which describes the influence of the ligands on the rate of substitution in the *trans* position. The strength of the *trans* effect exercised by common ligands increases in the following order: H_2O, OH^-, NH_3, C_5H_5N < Cl, Br, SCN^-, I, NO_2^- < CH_3 < H^-, PR_3 < C_2H_4, CN^-, CO. The *trans* effect appears to be determined by the polarizability of the ligands and by their ability to form π bonds, which both weakens the coordinate bond in the position *trans* to the ligand in question, and promotes the 5-fold coordinate intermediate. The *trans* effect is often used in planning the synthesis of platinum(II) complexes, as is shown in Fig. 2 for the synthetic pathway for *cis*- and *trans*-dichlorodiammineplatinum(II).

$$[PtCl_4] \xrightarrow[-Cl^-]{+NH_3} [PtCl_3(NH_3)] \xrightarrow[-Cl^-]{+NH_3} cis\text{-}[PtCl_2(NH_3)_2]$$

$$[Pt(NH_3)_4] \xrightarrow[-NH_3]{+Cl^-} [PtCl(NH_3)_3] \xrightarrow[-NH_3]{+Cl^-} trans\text{-}[PtCl_2(NH_3)_2]$$

Magnetic properties. Magnetic studies are an important method for classification of transition metal complexes. If the magnetic moments of the complexes are nearly the same as those of the free ions, the complexes are called *high-spin complexes*; but if the paramagnetism is greatly reduced or the complex is diamagnetic, they are called *low-spin complexes*. For high-spin complexes, the magnetic moments are approximately given by the spin moment $\mu_S = g\sqrt{S(S+1)}\,\mu_B$; $g = 2.0003$ is the Landé factor, *S* is the spin quantum number, μ_B is the Bohr magneton. The total momentum, which includes the orbital angular momentum, $\mu_{S+L} = \sqrt{4S(S+1)+L(L+1)}\ \mu_B$ (L is the orbital angular momentum quantum number), is not usually calculated. The spin moments for the d^n systems are 1.73 (d^1), 2.83 (d^2), 3.87 $(d)^3$, 4.90 (d^4), 5.92 (d^5), 6.93 (d^6) and 7.94 (d^8) μ_B. The 3d transition metals in the +2 and higher oxidation states preferentially form high-spin complexes; low-spin complexes are formed mainly by cobalt(III) in octahedral and nickel(II) in planar ligand environments, and by octahedral cyano complexes of the $3d^4$, d^5 and d^6 ions. Low spin complexes are also common among the 4d and 5d transition metals; for example, square planar palladium(II) and platinum(II) complexes are all diamagnetic. The magnetic behavior of the transition metal complexes can be predicted using ligand field theory and comparing the spin pairing energy P, that is, the energy required to pair two electrons in the same orbital, with the ligand field strength parameter $10\ Dq$: if $10\ Dq > P$, the complex is low-spin; if $10\ Dq < P$, it is high-spin. If the spin pairing energy and the ligand field strength are approximately the same, and complex isomers exist, high- and low-spin complexes can coexist in a temperature-dependent equilibrium (*spin crossover*).

Bonding. The coordinate bonds in coordinate complexes can often be interpreted either in terms of the electrostatic or the covalent model. In most complexes, the bonding is intermediate between electrostatic and covalent, although the compounds of electron-rich d metals with π ligands are typical covalent compounds. For theoretical treatment both of electrostatic complexes and the broad field of complexes characterized by superposition of electrostatic and covalent components of the bonds, Ligand field theory (see) is used; it is capable, on the basis of UV/VIS spectra, of providing quantitative information on the magnetic, thermodynamic and kinetic behavior of the complexes.

Covalent bonding model. A first covalent bonding model was developed in the 1930's on the basis of the valence structure theory. According to this model, coordinate metal-ligand bonds form by overlapping of occupied ligand orbitals and empty metal orbitals of suitable symmetry. The magnetically normal *addition complexes* were distinguished from *penetration complexes* with reduced paramagnetism or diamagnetism; addition complexes were assumed to have an essentially electrostatic interaction between the central ion and the ligands. For penetration complexes, complex formation was proposed to cause electron pairing within the partially occupied d-shell, formation of $d^ks^mp^n$ hybrid orbitals and overlap of the resulting empty metal hybrid orbitals with occupied ligand orbitals. Diamagnetism and octahedral, trigonal-bipyramidal or planar structures, e.g. of low-spin d^6 or low-spin d^8 complexes, can be readily interpreted by the assumption that the t_{2g} orbitals are occupied by metal electron pairs, and bonding occurs by formation of d^2sp^3, dsp^3 or dsp^2 hybrid orbitals.

The VB model of coordination compounds is limited to qualitative predictions, while the MO theory of complexes also provides quantitative information; it is used primarily with covalent complexes of electron-rich transition metals, because here the electrostatic bonding model assumed by the ligand field theory naturally breaks down. The MO theory of complexes assumes a combination of central atom orbitals with suitable ligand orbital functions to form bonding or anti-bonding molecular orbitals and calculates their positions on the energy scale. The following metal orbitals have maximum electron densities in the directions of octahedral ligands, and are therefore suitable for formation of σ-bonds in octahedral complexes: d_z, d_{x2-y2} (E_g), s(A_{1g}),p_x, p_y and p_z(T_{1u}). The other d-orbitals, d_{xy}, d_{xz} and d_{yz}(T_{2g}), are capable only of π-bonding, and therefore in octahedral complexes with pure σ-donor ligands, they form non-binding molecular orbitals. The six bonding σ-molecular states have a stronger ligand orbital character, while the six T_{2g} electrons represent pure metal electrons (Fig. 3).

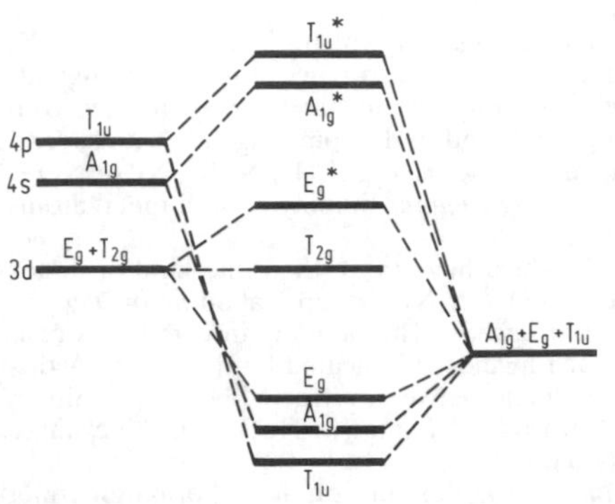

Fig. 3. MO diagram of octahedral complexes ML_6 (with pure Δ interaction between ligand and metal)

The MO diagram of an octahedral complex with π-bonding ligands is much more complicated. A qualitative description of the d-transition metal interaction with π ligands like carbon monoxide (or isoelectronic ligands such as CN^- or N_2) follows: The coordinate metal-carbon bond in carbonyls or cyanocomplexes, or the metal-nitrogen bond in dinitrogen complexes, has partial double bond character arising from a superposition of σ- and π-components. An overlap of occupied ligand orbitals and empty metal orbitals of suitable symmetry (σ-bonding) is supplemented by "back donation" of electrons from occupied metal d-orbitals into empty, anti-bonding π^* orbitals of the ligand (π-bonding) (Fig. 4).

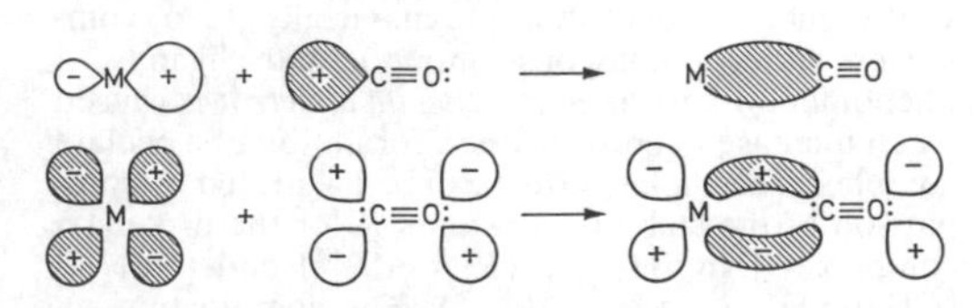

Fig. 4. Synergistic bonding mechanism: overlap of carbon-metal δ- and metal-ligand π-bonds in metal carbonyls

Historical. The groundwork of modern C. was laid by Alfred Werner (1866-1919). Important contributions to the development of the theory were made by Magnus (1922) (classical electrostatic model), Sidgwick and Pauling (1923/1931) (covalent bonding model), Bethe (1929), and Hartmann, Orgel et al. (1951) (Crystal field-ligand field theory).

Coordination compounds: see Nomenclature, sect. II G.

Coordination lattice: see Lattice type.

Coordination number: see Valence, see Coordination chemistry.

Coordination polyhedra: see Coordination chemistry.

Coordinative bonding: see Chemical bond.

Copals: a term for a large group of very hard, natural resins of recent or fossil origin from various botanical and geographical sources. Chemically, the C. are terpenoids, and probably consist of complicated mixtures of congocopalic acid, $C_{36}H_{58}(COOH)_2$ and congocopalolic acid, $C_{21}H_{32}(OH)COOH$. The less valuable ***soft C.*** with melting points of 180-200 °C are made either from the roots or from the bark of living copal trees. The more valuable ***hard C.*** with melting points from 200 to 360 °C are dug from the sites where copal trees stood formerly. After these trees decay, only the chemically very resistant resin remains. Depending on the origin, the size of the fossil C. lumps ranges from 30 cm diameter to pea size. The most important is ***Zanzibar C.***, which is found on the coast of southeast Africa. It is bright red to yellowish red, very hard, readily soluble in alcohol, and slightly soluble in glacial acetic acid, benzene and chloroform. Some other C. are ***Mozambique C.***, ***kiesel C.***, ***Kauri C.*** and ***Manilla C.***. Most C. is produced in East and West Africa, Madagascar and South America.

C. are used mainly as paint materials (now usually after esterification), to make oil lacquers and putty, and as additives to synthetic resins. The C. are often esterified with alcohols, especially glycerol. Very hard C. are used as substitutes for amber in making jewelry.

Cope reaction: pyrolytic cleavage of an amine oxide in which a synchronous *cis*-elimination forms an alkene and a hydroxylamine.

$$R^1_2\overset{+}{N}(O^-)-CH_2-CH(H)-R^2 \longrightarrow CH_2{=}CH-R^2 + R^1_2N-OH$$

The C. is especially important for production of *trans*-cycloalkenes.

Copper, ***cuprum***, symbol ***Cu***: a chemical element from group Ib of the periodic system, the Copper group (see). C. is a heavy metal, atomic number 29. The masses of the natural isotopes are 63 (69.17%) and 65 (30.83%); the atomic mass is 63.546, valence I and II, rarely III, IV; hardness on the Mohs scale 3; density 8.92; m.p. 1083 °C, b.p. 2595 °C; electrical conductivity 59.5 Sm/mm^2; standard electrode potential (Cu/Cu^{2+}) +0.337 V.

Properties. C. is a reddish, soft but tough and very malleable metal which can easily be drawn into fine wires and beaten into extremely thin, translucent leaflets which appear green in transmitted light. The crystals are regular, cubic close-packed arrays. After silver, C. has the highest electrical and thermal conductivity of any metal; it is the only metallic element other than gold with a characteristic color. C. is a typical transition metal which displays alternate oxidation levels, forms colored compounds, and has a distinct tendency to form complexes. The occupation of the outer $3d^{10}4s^1$ electron shell means that copper(I) ions have a completed $3d^{10}$ shell, while copper(II) ions have a $3d^9$ configuration.

In accordance with its position in the electromotive force series and the relatively low heats of formation of its oxides, C. is a relatively unreactive metal. It is only very slowly oxidized in air to copper(I) oxide. In the presence of carbon dioxide or sulfur dioxide (e.g. from the atmosphere), layers of the green basic carbonate, $CuCO_3 \cdot Cu(OH)_2$, or the basic sulfate, $CuSO_4 \cdot Cu(OH)_2$ (see Patina), form, protecting the underlying metal from further reaction. C. is attacked only by oxidizing acids, e.g. nitric; reactive metals like iron, zinc or magnesium cause C. to precipitate out of its salt solutions. At elevated temperatures C. is oxidized by oxygen and halogens. Sulfur and sulfur compounds attack C. in a characteristic manner, forming surface layers of copper sulfides.

Analytical. Copper compounds give a bunsen flame a characteristic green color. Cu^{2+} ions are precipitated from a weakly acidic solution as black copper(II) sulfide, CuS. Copper(II) is identified by means of the dark blue complex formed when ammonia is added to aqueous copper(II) salts, $[Cu(NH_3)_4(H_2O)_2]^{2+}$; this complex is also used for colorimetric determination of Cu(II). Potassium ferrocyanide gives a characteristic red-brown precipitate with Cu^{2+}: copper(II) hexacyanoferrate(II), $Cu_2[Fe(CN)_6]$. Electrogravimetry (precipitation of C. from sulfuric or nitric acid solution at 1.7 to 2.5 V on a platinum cathode) is used for quantitative determination; copper-ion-sensitive electrodes are now more commonly used. The reaction with iodide ions to copper(I) iodide, CuI, and iodine is used for volumetric determination of Cu^{2+}; the iodine is back-titrated with thiosulfate (see Iodometry).

> Copper in the form of Cu^{2+} ions is very poisonous to lower organisms. For example, bacteria and other decay microorganisms die in water in a copper vessel, and copper compounds prevent growth of algae.
>
> Copper compounds induce vomiting in human beings, so that a 5% copper sulfate solution can be used as an emetic for general laboratory poisonings.

Occurrence. C. makes up 0.01% of the earth's crust. Elemental deposits have become rare; most copper ores are sulfides, such as chalcopyrite (copper pyrite), $CuFeS_2$; bornite, Cu_5FeS_4/Cu_3FeS_3; and copper glance, Cu_2S. Oxide ores such as cuprite, Cu_2O, malachite, $CuCO_3 \cdot Cu(OH)_2$, and azurite ($Cu_3(OH)_2(CO_3)_2$ are also economically important. The average copper content in the copper ores now mined is 1.5 to 2%.

C. is an essential trace element for human beings, animals and plants. It is found in varying amounts in soils, from about 0.5 ppm to 80 ppm. Many soils, including agricultural soils, are deficient in C., and the resulting deficiency diseases are healed by intake of copper salts. The optimum amount of C. varies through several orders of magnitude for different types of organism. For lower organisms, e.g. algae, it is very low. Human beings consume 1 to 5 mg C. per day in their food, and the human body contains 100 to 150 mg. C. is involved mainly in electron transfer reactions in the biosphere. It is required for chlorophyll synthesis of plants and is a component of several enzymes involved in the synthesis of hemoglobin. In the blood of molluscs and marine crustaceans, the oxygen transport protein is the C.-containing hemocyanin. There are a number of C. proteins in human blood also; many of them are blue. Superoxide dismutase, which protects against oxygen radicals, is also a C. protein.

C. is usually produced industrially from sulfide concentrates which are obtained by flotation ore enrichment. Another metallurgical concentration method is smelting of a copper matte, followed by a roast reduction to crude copper which is electrolytically purified.

The sulfide concentrates obtained by flotation contain up to 30% C. Oxide ores, e.g. $CuCO_3 \cdot Cu(OH)_2$, are usually treated by a hydrometallurgical process.

1) Pyrometallurgical production of crude copper (Fig. 1). In the first step, the sulfide concentrate is converted to the copper matte for smelting and a slag containing up to 0.5% Cu. The matte contains 38 to 50% Cu, 22 to 33% iron and 24 to 26% sulfur. First the desired Cu : S ratio is established, then the matte is melted in a flame furnace at 1200 to 1400 °C with addition of limestone and sand as slag-producers. The copper matte and the copper slag above it are removed separately; the slag contains about 35% SiO_2, 40% FeO and 5% CaO.

Roasting and smelting methods in which the heat from the sulfide roasting is used for the melting in a suspension-melting furnace are more energy efficient. The C. is removed from the copper-rich slags produced in this way in electric furnaces. Sulfide ores which cannot be prepared by flotation are worked up by melting an ore-coke mixture in a blast furnace. In addition to the copper matte, a large amount of slag is formed.

When the copper matte is smelted, zinc and lead compounds are volatilized, and precipitate with the fly ash; noble metals and nickel collect in the matte.

Crude copper is produced by a roasting reaction of the molten copper matte in a drum converter. Air blasted into the melt through the bottom of the vessel partially converts the copper(I) sulfide into copper(I) oxide, which reacts with more sulfide according to the equation: $2\ Cu_2O + Cu_2S \rightarrow 6\ Cu + SO_2$. SiO_2 is added to the mixture so that the iron sulfides are converted into copper-rich silicate slags, which are added to the copper-matte melt. The copper content of crude copper is 97 to 99%. The sulfur dioxide is used to make sulfuric acid.

2) Refining of crude copper. Some of the impurities in crude copper, such as iron, zinc, tin, lead, arsenic and antimony, are removed by selective oxidation in flame furnaces with air, causing slag formation and

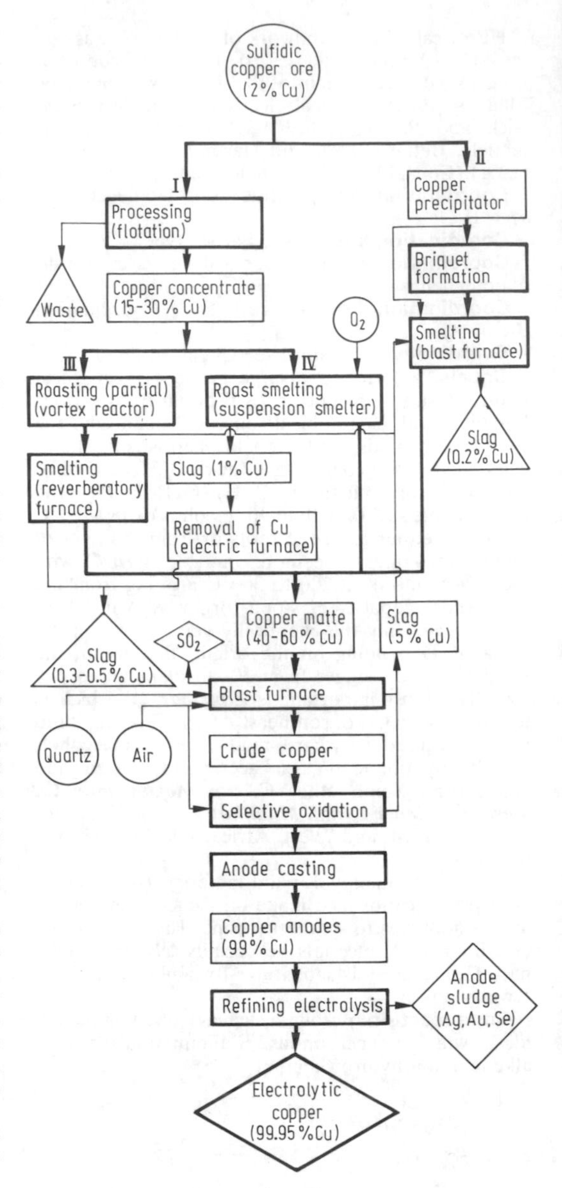

Fig. 1. Pyrometallurgical processes for sulfide copper ores and concentrates. I, flotable copper ores; II, non-flotable copper ores; III, standard process; IV, roasting and smelting process

volatilization. This step is necessary because these impurities interfere with electrolytic refining. Electrolysis is carried out at 50 °C in a sulfuric acid/copper sulfate electrolyte. Of the impurities remaining after pre-refining, most of the gold, silver, selenium, tellurium, lead and antimony collect in the anode sludge, while most of the nickel and arsenic remain in the electrolyte. This must therefore be regenerated (electrolytic removal of copper). The noble metals and selenium are recovered from the anode sludge. The cathodic copper is 99.95% Cu.

3) Hydrometallurgical process (Fig. 2). Oxide and oxide-sulfide ores and waste products with copper contents of 0.4 to 1.5% Cu are preferentially extracted with sulfuric acid or iron(III) sulfate solutions. The C. is precipitated from the dilute solutions by Cementation (see) with scrap iron, and is separated from the solution as cement C. It is then smelted to crude copper. Solutions with higher concentrations of Cu (30 to 40 g/l) can be electrolysed directly. Hydrometallurgical production can also be done by liquid-liquid extraction with hydroxide oximes in benzene, which concentrates and purifies the dilute primary solution so that it can be electrolysed.

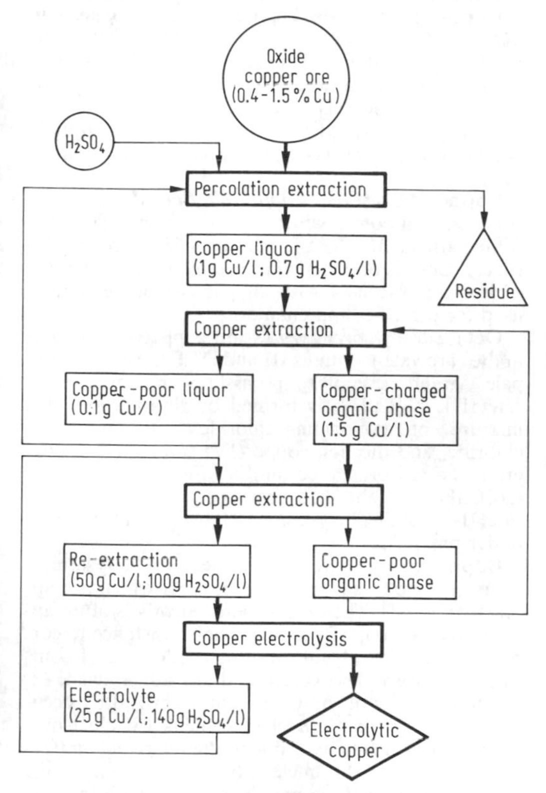

Fig. 2. Hydrometallurgical copper production from oxide ores

Applications. Because of its high electrical and thermal conductivity, malleability and ability to form alloys, its high resistance to corrosion, ease of galvanic precipitation and attractive color, more C. is used than any other non-ferrous heavy metal. More than half the world production of metallic C. goes to the electric industry, mainly for production of transmission wires, but also in switches, generators, motors and transformers. Pure C. is also used to make containers and pipes, heat exchangers, and apparatus for the chemical, food and beverage industries. C. is used to make household appliances, for coins and for roofing, mainly of historical buildings, and alloys. About 2% of the world copper production is made into copper compounds which have many applications.

Historical. C. was the first metal which was used for producing tools and weapons. The occurrence of natural elemental copper, and of easily smelted ores, was a major factor. By 4000 B.C., the Egyptians were mining C. on the Sinai peninsula. C. was also used very early in India (3500 B.C.), the Aegean (about 3000 B.C.), in Spain and China (2500 B.C.), and in Central Europe and the British Isles (ca. 2000 B.C.). C. deposits on Cyprus were important in antiquity, from which the name "aes cyprium" was taken; this became "cuprum" in medieval Latin. Copper alloys, especially bronze (about 3000 years ago), had an extremely important effect on the development of human culture. The oldest discoveries of bronze are from the Indus valley, Sumerian Ur (modern southern Iraq), Egypt and China. Bronze smelting sites were abundant in Central Europe, e.g. in Thüringen and Hessia, the Hartz and Vogt regions. After a decline in the Dark Ages, copper mining was reestablished in Central Europe: in 922 in Frankenberg/Sachsen, 928 in Rammelsberg bei Goslar, 1165 in Freiberg/Sachsen, 1199 in Hettstedt/Mansfeld.

Copper(II) acetate: $Cu(CH_3COO)_2 \cdot H_2O$, dark green, monoclinic prisms; M_r 1.99.65, density 1.882, m.p. 115 °C. C. is soluble in water, and dimeric in the solid state. It is used to make copper pigments, as an astringent and mild caustic in medicine, and for electroplating copper. Verdigris (see) is a mixture of basic C.

Copper(II) acetate arsenite: $Cu(CH_3COO)_2 \cdot 3Cu(AsO_2)_2$, an intensely green crystalline powder; M_r 1013.71. C. is very resistant to light and air, but is decomposed by acids or bases. It is synthesized by reaction of copper(II) acetate with arsenous acid, or of copper arsenite with acetic acid. C. is no longer used as a pigment (Schweinfurt green) because of its high toxicity.

Copper acetylide: same as Copper(I) ethynide.

Copper alloys: alloys of copper, mainly with zinc, tin, nickel, aluminum, manganese and silicon. C. are stronger than pure copper, can be worked cold, and are more easily welded than copper. However, their electrical and thermal conductivities are lower. C. are named and classified according to the main non-copper component.

1) ***Copper-zinc alloys*** are called Brass (see). If part of the zinc is replaced by nickel, the alloy is called New silver (see) or German silver; it is similar to brass.

2) ***Bronzes*** are all C. which do not contain zinc as the main non-copper component. Originally, the term applied only to copper-tin alloys (see Tin bronze). Today aluminum, beryllium, manganese, lead, silicon and phosphorus bronzes are produced (see separate entry on each of these). Copper-nickel alloys (see Nickel bronze) have many applications, e.g. for electrical resistors (see Konstantan®, Nickelin®). Kuprodur® (see) is a conductor.

Small amounts of copper are alloyed with the noble metals (see Gold alloys) to harden them, and with aluminum (see Aluminum alloys). Copper-based alloys are also used for jewelry (see Talmi gold, Nürnberger gold).

Copper(II) arsenite: $Cu_3(AsO_3)_2$, a yellow-green powder which is insoluble in water and alcohol. ***Basic copper arsenite*** is prepared by combining copper sulfate and an alkaline arsenite solution; it was formerly used as a paint pigment under the name Scheel's green or Swedish green. It is no longer legal for use as a pigment, because of its toxicity. However, it can still be used to paint ships or as a wood protectant. Scheel's green reacts with hot acetic acid to form copper(II) acetate arsenite (Schweinfurt green).

Copper carbide: same as copper(I) ethynide (see).

Copper carbonate: copper(II) carbonate has not yet been obtained in pure form. When aqueous copper salt solutions are mixed with alkali carbonate solutions, ***basic copper(II) carbonates*** of varying composition are obtained. Basic C. are found in nature, as malachite, $CuCO_3 \cdot Cu(OH)_2$, M_r 221.10, density 4.0, and as azurite, 2 $CuCO_3 \cdot Cu(OH)_2$, M_r 344.65, density 3.88. Malachite is used as a pigment and is a semiprecious stone; azurite is blue, and is also used as a pigment. Patina (see) is also a basic C.

Copper chlorides: ***Copper(I) chloride***, CuCl, is a colorless compound which crystallizes in a zinc blende structure analogous to that of diamond; M_r 98.99, density 4.14, m.p. 430 °C, b.p. 1490 °C. CuCl is obtained by reduction of copper(II) chloride with copper. It is barely soluble in water, but dissolves readily in hydrochloric acid or ammonia, forming linear complexes $[CuCl_2]^-$ or $[Cu(NH_3)_2]Cl$ and the tetrahedral tetraammine copper(I) chloride, $[Cu(NH_3)_4]Cl$. The latter complex absorbs carbon monoxide in aqueous solution, forming $[CuCl(CO)(H_2O)_2]$, an ability which is utilized both in industrial gas purification and in gas analysis.

Copper(II) chloride, $CuCl_2$, is a yellow-brown compound; M_r 134.44, density 3.386, m.p. 620 °C, b.p. 993 °C. $CuCl_2$ is formed from the green dihydrate $CuCl_2 \cdot 2H_2O$; M_r 170.74, density 2.54, by heating it in a stream of hydrogen chloride at 150 °C. It forms a layered lattice with square-planar $CuCl_4$ units and weaker Cu-Cl interactions in the *z* direction. $CuCl_2$ is used in the Deacon process, as an oxygen carrier in pigment syntheses and pyrotechnology, to etch copper and to bleach photographic negatives.

Copper complexes: coordination compounds of copper (see Coordination chemistry) in which copper(I) characteristically has a coordination number of 2, and copper(II) has 6 or 4. C. of the Cu^IL_2 type (L = ligand) are usually linear, while most complexes of copper(II) are distorted octahedral complexes CuL_6. As a result of the Jahn-Teller effect, these complexes have longer CuL distances for axial ligands than for equatorial ones. Because there are large variations in the degree and direction of the distortions, the coordination sphere of copper appears to be "plastic"; in addition to C. with coordination number 6, there are also some with 4 or 5, which have planar tetrahedral or trigonal-bipyramidal structures. C. are used in cyanide leaching and in galvanic technology. They are also used to determine copper or glucose (see Fehling's solution), and they have many important roles in the biosphere.

Copper cyanides: ***Copper(I) cyanide***: CuCN, white, monoclinic crystals; M_r 89.56, density 2.92, m.p. 473 °C. CuCN is a coordination polymer with a linear chain structure: ...-Cu-C≡N-Cu-C≡N-... It is insoluble in water and dilute acids, but dissolves in alkali cyanide solutions, forming complex cyanocuprate(I), $[Cu(CN)_2]^-$, and tetrahedral $[Cu(CN)_4]^{3-}$. Dicyanocuprate(I), in contrast to the linear complexes dicyanoargentate(I) and dicyanoaurate(I), has a spiral-like polymer structure with terminal and bridging CN groups (Fig.). Aqueous solutions of the alkali cyanocuprates(I) are very important in galvanic technology.

Copper(II) cyanide, $Cu(CN)_2$, is a brownish yellow compound which crystallizes in a layered lattice; M_r 115.58. $Cu(CN)_2$ decomposes even at room temperature to dicyanogen and CuCN.

Copper(II) dicyanamide: see Dicyanamides.

Copper(I) ethynide, ***copper acetylide, copper carbide***: Cu_2C_2, a red-brown powder, M_r 151.10. C. is insoluble in water and most organic solvents. It is obtained by passing ethyne through an ammoniacal solution of copper(I) chloride. C. is very sensitive when dry, and explodes even when rubbed slightly; it is therefore used as a component of electrical ignition caps.

Copper fluorides. ***Copper(II) fluoride***: CuF_2, a colorless compound which crystallizes in a distorted rutile lattice; M_r 101.54, density 4.23, m.p. 950 °C, (dec.). CuF_2 is obtained via the dihydrate by reaction of hydrofluoric acid with copper carbonate, and is used for ceramics and enamels.

Octahedral ***fluorocuprates*** are copper compounds in the rare valence states III and IV. For example, the pale green complex potassium hexafluorocuprate(III), $K_3[CuF_6]$, is formed by fluorination of a mixture of potassium chloride and copper(II) chloride; and the first copper(IV) compound discovered, the red-orange cesium hexafluorocuprate(IV), $Cs_2[CuF_6]$, is obtained from cesium trichlorocuprate(II), $Cs[CuCl_3]$, cesium chloride and fluorine under pressure.

Copper group: Group Ib in the periodic system; it includes copper, silver and gold, all of which are unreactive metals. The term ***coinage metals*** is often applied to this group. Because of their resistance to corrosion, they have been used for production of coins since they were discovered, and are still so used.

The elements of the C. are characterized by occupation of the nd^{10}, $(n+1)s^1$ orbitals by electrons. Their lack of reactivity is due to the relatively ineffective shielding of the nucleus by the outer electrons. The inactivity of these metals increases with increas-

Properties of the Group Ib elements

	Cu	Ag	Au
Nuclear charge	29	47	79
Electron configuration	$[Ar]3d^{10}4s^1$	$[Kr]4d^{10}5s^1$	$[Xe]5d^{10}6s^1$
Atomic mass	63.546	107.868	196.9665
Atomic radius in pm	117.3	133.9	133.6
Electro-negativity	1.75	1.42	1.42
Density in $g\ cm^{-2}$	8.92	10.491	19.32
m.p. in °C	1083	960.8	1063.0
b.p. in °C	2595	2212	2600

ing nuclear charge. All C. metals have high melting and boiling points. They can exist in the +1 oxidation state; copper and gold can also display the +2 and +3 states, respectively. The small ionic radii of these metals gives their compounds with electronegative partners, such as the halogens and chalcogens, a covalent character. They also have a decided tendency to form complexes. For the +1 oxidation state, linear complexes with coordination number 2 are typical, while copper(II) and gold(III) complexes tend to be distorted octahedral or square planar structures.

Copper(II) hexacyanoferrate(II): $Cu_2[Fe(CN)_6]$, a reddish brown compound; M_r 339.04. C. is obtained by combining a copper(II) salt with potassium hexacyanoferrate, $K_4[Fe(CN)_6]$. It was widely used as a red-brown to brown-violet oil paint.

Copper hydride: CuH, a reddish compound; M_r 64.55, density 6.38. C. is formed by the reaction of phosphoric(I) acid and copper sulfate solution, or directly by the reaction of copper(I) iodide with lithium alanate.

Copper(II) hydroxide: $Cu(OH)_2$, a bright blue powder; M_r 97.56, density 3.368. It is made by treating a boiling copper sulfate solution with a 25% ammonia solution. The precipitate is washed with water and digested at 20 to 30 °C with moderately conc. sodium hydroxide solution. Gelled C. loses water when heated, and is rapidly converted to black copper(II) oxide, CuO. C. dissolves in ammonia to give the dark blue tetramminecopper(II) hydroxide (*Schweitzer's reagent*). C. was formerly used as a pigment called Bremen blue. It is still used as a pesticide and to treat seeds. It is a component of Bordeaux mixture.

Copper(I) iodide: CuI, a colorless, crystalline powder; M_r 190.44, density 5.62, m.p. 605 °C, b.p. 1290 °C. C. is nearly insoluble in water, but dissolves readily in potassium cyanide, potassium iodide or ammonia solutions, in which it forms complexes. It is formed by the reaction of iodide with copper(II) salts. It is used as a temperature indicator, especially in the form of the copper iodomercurate complex.

Copper(II) nitrate: $Cu(NO_3)_2$, a blue compound; M_r 187.56. In solid phase, C. can exist in two coordination polymer forms with a three-dimensional linking of the copper atoms by the anionic ligands. In solution and the gas phase, there are discrete $Cu(NO_3)_2$ molecules. Anhydrous C. is made by reaction of dinitrogen tetroxide, N_2O_4, with copper in acetic acid, via the adduct $Cu(NO_3)_2 \cdot N_2O_4$ ($NO^+[Cu(NO_3)_3]^-$), which is heated to 90 °C in vacuum to yield $Cu(NO_3)_2$. ***Copper(II) nitrate hydrates***, $Cu(NO_3)_2 \cdot 3H_2O$, M_r 241.60, density 2.32, and $Cu(NO_3)_2 \cdot 6H_2O$, M_r 295.64, density 2.074, are obtained from the reaction of copper with nitric acid; after the solution has been highly concentrated, the blue hydrates crystallize out. They are used as inks, to color copper black and to burnish iron.

Copper oxides. ***Copper(I) oxide***: Cu_2O, forms yellow to red, regular crystals; M_r 143.08, density 6.0, m.p. 1235 °C. In the lattice of Cu_2O, each copper atom forms a linear unit with two oxygen atoms, while each oxygen atom is surrounded tetrahedrally by 4 copper atoms. Cu_2O is found in nature as cuprite. It is formed by the reaction of copper(I) salts with alkali hydroxide solutions, or by reduction of alkaline copper(II) salt solutions (see Fehling's solution). Electrochemical processes for making Cu_2O make use of anodic oxidation of copper in an alkali or alkaline earth chloride electrolyte in which the concentration of foreign ions is very low. It is used as a pigment for painting ships below the water line (antifouling paint), as a fungicide, and, at concentrations of only about 0.25%, to color glass, enamel and ceramic glazes.

Copper(II) oxide, CuO, is a black compound which is insoluble in water; M_r 79.54, density 6.48, m.p. 1326 °C. CuO is soluble in acids and potassium cyanide or ammonium carbonate solutions. It is reduced by hydrogen or carbon monoxide to copper, even below 250 °C. When heated to higher temperatures (> 1000 °C), it is converted to copper(I) oxide and oxygen. CuO is formed when copper(II) hydroxide is heated, or when copper is heated with oxygen to temperatures above 850 °C. It is synthesized industrially by precipitation methods in which water-soluble copper(II) salts are converted to copper hydroxide by addition of lye, and then, upon boiling, to CuO. CuO is used as a pigment for giving glass and enamel black, blue or green color. The ancient Egyptians knew how to prepare a blue glass melt by adding CuO. CuO is also used as an additive in optical glasses containing silver halide, in elemental analysis, in the petroleum industry to desulfurize petroleum, and in the production of thermistors.

Copper(II) oxygen chloride: $3\ Cu(OH)_2 \cdot CuCl_2$, green crystalline needles; M_r 427.10. C. is one of the most important copper salts in industry. It has four modifications, of which the rhombic bipyramidal atacamite and the rhombohedral paratacamite are found in nature; technical C. is a transition form. C. is obtained mainly by oxidation of copper(I) chloride with atmospheric oxygen, or by reaction of copper(II) chloride with copper(II) hydroxide; for this reaction, secondary raw materials such as scrap copper, fly ash or copper etching solutions can be used. In the most common process, the copper(I) chloride formed as an intermediate is dissolved by adding potassium chloride, and the oxidation is carried out at 60-90 °C. C. is widely used in fungicides, which are carefully formulated and include adhesive and dispersive agents, fillers and other active ingredients.

Copper(II) phosphate: $Cu_3(PO_4)_2 \cdot 3H_2O$, blue-green, rhombic crystals; M_r 434.61. C. is insoluble in water, but dissolves in acids and ammonia. It is used as a fungicide and in ceramics production.

Copper phthalocyanin: see Phthalocyanin pigments.

Copper pigments: copper compounds used as pigments, such as malachite and azurite (see Copper carbonates), Schweinfurt green (see Copper(II) acetate arsenite), Scheel's green (see Copper(II) arsenite), Bremen blue (see Copper(II) hydroxide) and Verdigris (see).

Copper plating: the production of thin copper layers on other metals. Copper layers to serve as intermediate layers for nickel and chromium plating are applied by ***galvanic C.***, i.e. by electrolysis in a bath of copper sulfate in sulfuric acid solution, or in an alkaline cyanide-containing copper bath. Copper can also be applied chemically, by dipping in a bath containing copper sulfate and sulfuric acid, or copper sulfate and copper tartrate, or by ***contact C.***, in which zinc or aluminum is the sacrifice metal. Copper layers can also be applied by Metal spraying (see) and Plating (see).

Copper(II) sulfate, ***copper sulfate*** (obsolete, ***copper vitriol***): $CuSO_4 \cdot 5H_2O$, transparent, blue, triclinic crystals; M_r 249.68, density 2.286. C. contains copper in a distorted octahedral coordination with four water molecules in equatorial positions, while oxygen atoms from the sulfate group occupy axial positions. The fifth water molecule is linked via hydrogen bonds to sulfate and hydrate oxygens. C. is readily soluble in water. When heated above 100 °C it loses four water molecules, while the fifth is not released until the temperature is over 150 °C. The resulting colorless, anhydrous C., $CuSO_4$, M_r 159.60, density 3.603, forms rhombic crystals; it readily takes up water and turns blue. It can therefore be used to indicate the presence of water. C. is made in large amounts industrially by reaction of granulated copper with 70% hot sulfuric acid in the presence of air. It is used as a pesticide, as a fertilizer additive to correct for copper deficiency in the soil, to preserve wood and animal hides, to treat seeds, to color metals, in copper engraving and copper plating, to make mineral pigments and as a flotation reagent for zinc blende ores. A 5% solution of copper sulfate can be used as a rapidly acting emetic in the laboratory in cases of poisoning.

Copper sulfides: ***Copper(I) sulfide***, Cu_2S, rhombic, blue-gray crystals which are converted at 150 °C to a hexagonal, high-temperature form; M_r 159.14, density 5.6, m.p. 1100 °C. Cu_2S is found in nature as chalcosite (copper glance). It can be synthesized by heating the elements in a hydrogen atmosphere, and is an essential intermediate in pyrometallurgical copper production. It is used as an additive in the production of luminophores.

Copper(II) sulfide: CuS, a black, insoluble compound; M_r 95.60, density 4.6. CuS is converted at high temperature to Cu_2S and sulfur vapor, and is also reduced to Cu_2S by hydrogen. Like Cu_2S, CuS has a relatively high electrical conductivity. It occurs in nature as a hexagonal mineral, covellite (indigo copper), which has a complicated layered lattice structure. CuS is used to make oxidation pigments, such as aniline black, and as an additive to antifouling paints.

Copper type, ***copper structure***: a Structure type (see) in which the atoms are in a cubic close packing (see Packing of spheres); that is, they occupy the vertices and face centers of a cubic unit cell. The coordination number of this arrangement is 12, and the packing density is 0.74. About 15 metallic element modifications crystallize in C.; in addition to the prototype copper, e.g. silver, gold, nickel, palladium, platinum, aluminum, lead and iron (as γ-Fe between 1178 and 1677 K) and all noble gases except helium. The metals of the C. are very ductile, which is due to the larger number of preferred slippage planes in the lattice, compared to the Tungsten type (see) or Magnesium type (see).

Copper vitriol: see Copper(II) sulfate.

Coproporphyrin: see Porphyrins.

Coproportionation, ***synproportionation***: a redox reaction in which two compounds of the same element, one with a higher and one with a lower oxidation state of the element, are converted to a compound with an intermediate oxidation state of the element. Typical examples of C. are the reaction of permanganate with manganese(II) ions to form manganese(IV) oxide:

$$3\,\overset{+2}{Mn}{}^{2+} + 2\,\overset{+7}{Mn}O_4{}^- + 4\,OH^- \rightleftharpoons 5\,\overset{+4}{Mn}O_2 + 2\,H_2O$$

or the combination of sulfur dioxide and hydrogen sulfide to give sulfur:

$$2\,H_2\overset{-2}{S} + \overset{+4}{S}O_2 \rightarrow 3/8\,\overset{\pm}{S}_n + 2\,H_2O.$$

Coprostanic acid: see Bile acids.

Cord factor: trehalose 6,6'-dimycolate, a toxic glycolipid found in the cell walls of various bacteria, including *Mycobacteria, Corynebacteria* and *Nocardia*. C. can be extracted from the lipids of *Mycobacterium paraffinicum* by column chromatography. It is a colorless, waxy substance which is readily soluble in chloroform and slightly soluble in alcohols, ether and acetone. C. modulates the immune response and attacks mitochondrial membranes.

Coronary drugs: compounds used to treat diseases of the coronary arteries and the surrounding tissues. C. increase the circulation in the heart and reduce flow resistance, and above all, counteract oxygen deficiency. They are also called *antianginous drugs* because oxygen deficiency leads to angina pectoris. β-Receptor blockers such as talinolol and propranolol are used as C., as are some nitrates, such as nitroglycerol, isosorbide dinitrate and pentaerythritol tetranitrate. Other C. include araliphatic amines, which act primarily as calcium antagonists; examples are Nifedipine (see), verapamil and prenylamine, $(C_6H_5)_2CH\text{-}CH_2\text{-}CH_2\text{-}NH\text{-}CH(CH_3)\text{-}CH_2\text{-}C_6H_5$. Heteroaromatic compounds used as *coronary dilators* are not effective in an acute attack; these include carbochromene, a synthetic derivative of coumarin, dipyridamol and trapidil. Some classes in this group, for example dipyridamil, are thought to act by inhibiting aggregation of thrombocytes.

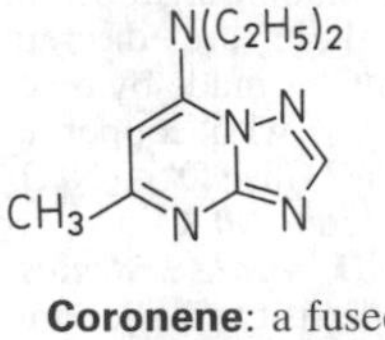

$N(C_2H_5)_2$ ... CH_3 ... N N N N

Trapidil

Coronene: a fused-ring aromatic hydrocarbon. C. forms yellow and blue fluorescing crystalline needles; m.p. 431 °C, b.p. 535 °C. C. is insoluble in water and soluble in benzene. It occurs in coal tar pitch and has also been synthesized, mainly for scientific reasons.

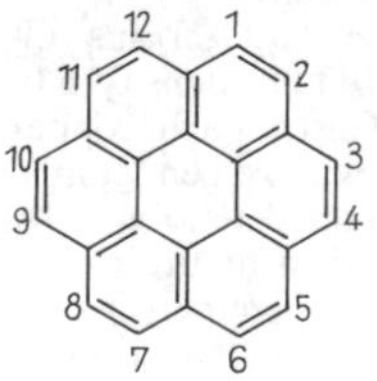

12 1 2 3 4 5 6 7 8 9 10 11

Corpohuminite: see Macerals.

Corpus luteum hormones: same as Gestagens (see).

Correspondence principle, ***theorem of corresponding states***: a hypothesis studied by van der Waals, Nernst and Kammerlingh-Onnes on the existence of a thermal equation of state valid for all substances. It is assumed that all substances at the critical point are in a comparable, "corresponding" state. If the thermal equation of state is expressed in terms of the dimensionless, reduced state parameters $p_r = p/p_{crit}$, $V_r = V/V_{crit}$ and $T_r = T/T_{crit}$, it should apply universally to all substances. One only needs to know the critical data p_{crit}, V_{crit} and T_{crit} in order to calculate the thermal behavior. For example, the van der Waals equation in reduced form is $(p_r + 3/V_r^2)(3V_r - 1) = 8\ T_r$.

The T. cannot be strictly true because of differences in molecular structure and polarity, and because of the resulting differences in interactions between molecules. A reduced equation of state yields useful results only for limited groups of substances which are chemically very similar to each other.

Corresponding acid-base pair: see Acid-base concepts, section on Brœnsted definition.

Corresponding redox pair: see Redox reactions.

Corrin: an approximately planar, macrocyclic ring system containing four pyrrole rings which, in contrast to the pyrrole rings in porphin, are hydrogenated. Two of them are connected by a direct C-C bond. The derivatives of C. are called ***corrinoids***.

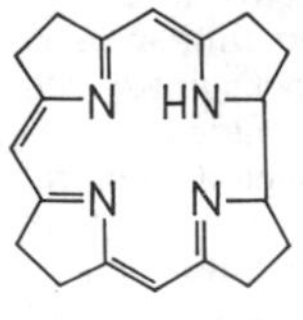

Metal-free corrinoids have been isolated from phototrophic bacteria. The cobalt-containing corrinoids occur widely; the most important group are Vitamin B_{12} (see) compounds and the coenzyme B_{12} derived from them. Corrinoids are biosynthesized only by microorganisms.

Corrinoids: see Corrin.

Corronel®: acid-resistant nickel-chromium alloy (see Nickel alloy) containing 35 to 37% chromium; the rest is nickel. C. is very resistant to highly concentrated nitric acid and mixtures of nitric and hydrofluoric acid.

Corrosion: in the broad sense, damage to a material (metal, alloy, glass, plastic, construction material) caused by reaction with a substance in the environment. In the narrow sense, changes resulting from chemical and electrochemical attack on a metal surface. The metal is oxidized and forms a compound (*corrosion product*).

The *corrosive medium* which causes the C., together with the corroded material, forms the corrosion system. Some corrosion media are, e.g. hot dry gases, components of the atmosphere, electrolytes, oxide and salt melts, metal melts and non-electrolytic liquids.

Corrosion products are, e.g. Rust (see) on iron-containing materials, Zinc rust (see), Patina (see) on copper, Forge scale (see) formed on unalloyed steels in the air at high temperature, Silicic acids (see) formed when aqueous solutions act on silicate materials, Ettringite (see) formed by action of soluble sulfates on concrete, and acid and amine chain fragments of polyurethane resins caused by action of strong acids or bases.

Mechanism of C. Physical and biological processes, as well as chemical and electrochemical ones, can take part in C. If metals are corroded by chemical reactions, the process is called Chemical metal corrosion (see); and if electrochemical reactions occur in conducting electrolytes, it is called Electrochemical metal corrosion (see). Depending on the zone of attack, C. can occur as volume C. (three-dimensional), surface C. (two-dimensional), line C. (one-dimensional) or pitting (zero dimensional).

The C. medium reaches the phase boundary of the material by transport processes (diffusion and convection) and adsorption. In three-dimensional processes, transport, adsorption and solubilization occur in the capillaries of the porous material. Physical factors in the process are mechanical, thermal, electrical and radiation effects. The combination of C. with tensile, vibrational and impact stress leads to Stress crack C. (see), Vibrational crack C. (see) and forms of Wear C. (see), respectively. Wear C. includes cavitational and erosional C. and tribocorrosion. Erosional C. can be caused, e.g. by a rapid flow of aggressive liquid past the surface of the material. In erosional C., plastic and elastic deformation of the material, together with the C. medium, can also induce cracks in the surface region. The mechanical energy expended on the surface of the material leads to formation of layers of C. product. Tribocorrosion occurs when, for example, a machine runs without lubrication. If the corrosion medium contains microorganisms, these can produce Microbiological C. (see).

Metal can be corroded either by even removal of the material parallel to the surface, or unevenly. Uneven C. leads to holes in the material; it can be caused by selective attack on one type of domain in an alloy, along the grain boundaries of a metal (see Metal corrosion, electrochemical) or along cracks formed by stress.

For inorganic, non-metallic materials, C. can occur by surface removal (scaling), removal of components by selective leaching, crystallization of voluminous salts, cracking (also due to the pressure exerted by newly formed salts) and blooms of soluble salts.

The characteristic signs of C. or polymeric organic materials are discoloration, swelling and disintegration, increasing brittleness and cracking.

Corrosion testing (see) is used in development and production of materials, finding substitutes and protecting materials from C.

Losses due to C. The main reason for C. testing is to reduce the economic losses caused by it. Direct losses are the replacement costs for corroded parts and maintenance costs for C. protection, including labor costs. Indirect losses include reduced production, losses of products and energy, efficiency losses, environmental damage and over-proportioning of construction components.

Corrosion protection (see) is necessary to reduce C.-caused losses.

Corrosion cell: see Electrochemical metal corrosion.

Corrosion current density: see Electrochemical metal corrosion.

Corrosion fatigue: see Corrosion.

Corrosion inhibitors: substances added to a corrosion medium which reduce the rate of corrosion of a metal (see Corrosion). The mechanism of action can be physical (***adsorption inhibitors***) or chemical. Depending on the type of reaction, chemical C. can be classified as ***passivators*** (see Passive state), ***formers of protective layers*** (e.g. phosphates, borates) and ***destimulators***, which reduce the effects of corrosive substances in the corrosion medium, e.g. sodium sulfite or hydrazine, which bind oxygen. The adsorption inhibitors include amines, imidazolines and quinolines which are used in petroleum refining to protect against corrosion by hydrochloric acid and hydrogen sulfide. Others are used to prevent iron from dissolving or being made brittle when rust and scale are removed from construction steels by dilute acids. Additives to oils (see Corrosion protection) used as temporary corrosion protection are also in this class, as are long-chain aliphatic and alicyclic amine nitrites and amine carbonates used as vapor-phase inhibitors. Dicyclohexylammonium nitrite used as a vapor-phase inhibitor also acts as a passivator.

Cathodic C. inhibit the cathodic part of the corrosion reaction; reduction inhibitors are examples. Anodic C. inhibit the anodic part of the reaction; examples are the passivators.

A certain minimum concentration of the C. is required for their protective effect. In the case of passivators, if this minimum is not present, holes may be formed in the material.

Corrosion medium: see Corrosion.

Corrosion potential: see Electrochemical metal corrosion.

Corrosion product: see Corrosion.

Corrosion protection: the protection of materials, buildings and machines from Corrosion (see).

1) ***Active C.*** is the use of methods which interfere with the corrosion process by changing the conditions in the corroding system. Active C. is achieved by design, inhibition, electrochemistry, change in the corrosion medium and formation of protective layers.

Design includes a) use of sites where the aggressiveness of the environment is least, b) use of more resistant materials, e.g. steels resistant to rust and acids, noble metals, titanium, porcellan, glasses and resistant plastics, c) architectural measures which prevent accumulation of corrosion-promoting materials, and d) use of combinations of materials which do not permit contact corrosion.

Inhibition is achieved by use of Corrosion inhibitors (see).

Electrochemical corrosion protection (see) is the use of cathodic or anodic polarization of the metallic material to be protected.

Changes in the corrosion medium are used to reduce the concentrations of corrosion stimulators, e.g. drying of air, reduction of the SO_2 content of the atmosphere, neutralization of acids, or their complete removal (e.g. water softening, removal of oxygen and chloride ions). Protective layers (see) are formed by reaction of some metal surfaces with the corrosion medium; a continuous and non-porous layer of corrosion product considerably reduces the corrosion rate.

2) ***Passive C.*** is the application of a protective layer of a more resistant material, the *corrosion protector*, to all surfaces of the material to be protected, so that it does not come into direct contact with the corrosion medium. Such *protective layers* can be produced by application of a different material, or by a chemical reaction at the surface of the material. A protective layer can act as a cathode or anode with respect to the material to be protected, it can interrupt transport of the corrosion medium to the surface, and/or it can inhibit the corrosion flow by resistance polarization. A protective layer must be firmly attached to the surface of the material to be protected.

Protective coatings (see) include the following: rust-preventing paints, organic coatings, and metallic and non-metallic inorganic protective layers.

The effectiveness of C. is expressed by the *corrosion protection value*. Active C. can be applied over the entire lifetime of the material, without interruption, but passive C. layers usually do not last as long as the material they protect. Thus they must be repeatedly maintained or renewed. Nonetheless, passive C. is more common.

If materials are subjected to combined stresses by chemicals, temperature, mechanical forces and high-energy radiation, special C. measures are required; see Acid protection.

Temporary C. is used during transport, storage, working or periods of inactivity of materials or machinery. Such measures include protective layers on the surface of a product, surrounding it with an inert atmosphere, and use of materials in packaging which inhibit corrosion. Protective layers are made of materials which can readily be removed, such as mineral oil, wax, resins and corrosion inhibitors, and they contain various inhibitory additives. The atmosphere around a product can be made more inert by use of drying agents, vapor-phase inhibitors (see Corrosion inhibitors), or replacement with ammonia, nitrogen or argon; such measures are usually applied for internal protection in electrical cables, apparatus and containers. Packaging can be used which prevents contact of the product with the atmosphere, and it can also include drying agents and Vapor-phase inhibitors (see). Protection in the form of packaging has the advantage that only the packaging has to be removed, while protective layers on the product itself must be removed before use by washing or stripping.

Corrosion testing: tests to determine the behavior of a material with respect to corrosion. The goals are the development and application of materials, determination of C. in machines during operation, checking and control of corrosion prevention measures, and study of corrosion damage and its prevention (see Corrosion protection). C. can be done in the laboratory with controlled stresses (dipping, stirring, flow, storage, spraying, boiling, steam exposure and pressure). In operational testing, the samples are exposed to the conditions of actual use, and in field experiments, they are exposed to the natural environment. The latter include experiments in soil corrosion and weathering.

The progress of corrosion may be observed by measurement of hydrogen production, reduction in strength, mass or thickness, increases in brittleness and changes in shape.

Some corrosion parameters independent of the length of the experiment are: *corrosion loss* $C = \Delta G/S$ [g m^{-2}] where ΔG is the loss of mass and S the area of the corroded surface; and *linear losses* $L = \Delta h = \Delta G/S\varrho$ [mm]; here Δ h is the loss of thickness and ϱ is the density of the material.

Some parameters which depend on the time are the *corrosion rate* $v = K/t$, where t is the time of corrosion; $v = \Delta G/S \cdot t$ [g m^{-2} d^{-1}], and the *linear corrosion rate* $v_L = v/\varrho$ [mm a^{-1}].

Cortexolone: see Adrenal cortex hormones.

Cortexone: see Adrenal cortex hormones.

Corticoids: same as Adrenal cortex hormones (see).

Corticoliberin: see Liberins.

Corticosteroids: same as Adrenal cortex hormones (see).

Corticotropin, ***adrenocorticotropin, adrenocorticotropic hormone***, abb. ***ACTH***: a polypeptide hormone with 39 amino acid residues. The human sequence is Ser-Tyr-Ser-Met-Glu-His-Phe-Arg-Trp-Gly-Lys-Pro-Val-Gly-Lys-Lys-Arg-Arg-Pro-Val-Lys-Val - Tyr - Pro - Asn - Gly - Ala - Glu - Asp - Glu - Ser - Ala-Glu-Ala-Phe-Pro-Leu-Glu-Phe. C. is produced in the anterior pituitary and regulates the growth of the adrenal cortex and the formation of steroid hormones in it: cortisol, cortisone and corticosterone. When the blood levels of these hormones are optimum, the secretion of C. by the pituitary is inhibited by feedback mechanisms. When the blood level of the adrenal cortex hormones is too low, or when the organism is under stress, the hypothalamus releases corticoliberin, a hormone which stimulates secretion of C. The relation between the amino acid sequence of this hormone and its biological activities has been extensively studied. The N-terminal sequence, amino acids 1-18, contains practically all the information required by the adrenal cortex, fat cells (lipolytic effect) and melanophore cells. Amino acids 19-24 specifically enhance the steroidogenic acitivity. The sequence of the C-terminus, amino acids 25-39, is species specific. For example, porcine C. contains Leu instead of Ser at position 31, and bovine and ovine C. has Gln instead of Glu at position 33. The C-terminus determines antigenicity and some transport properties. The active center is located at amino acids 5-10, and the receptor-binding region comprises amino acids 11-18. Amino acids 1-13 of the human hormone have the same primary structure as α-melanotropin. C. is biosynthesized by cleavage of the precursor glycoprotein *pro-opiomelanocortin*, which, in addition to C., contains the sequences of β-lipotropin, and thus also of the endorphins and melanotropins.

D. has been synthesized chemically, and ACTH (1-24) is used as a pharmaceutical preparation for special forms of pituitary insufficiency, allergies, arthritis, etc. Certain partial sequences of C., ACTH (4-10) and some of its analogs, affect the strength and the length of the extinction phase of various conditioned responses.

The first total synthesis of pork ACTH with the sequence then attributed to the hormone was reported in 1963 by R. Schwyzer.

Corticotropin releasing hormone: see Liberins.

Cortisol: see Adrenal cortex hormones.

Cortisone: see Adrenal cortex hormones.

Corundum type, ***corundum structure***: a Structure type (see) for certain compounds with the general composition A_2B_3. In the crystal lattice of corundum (α-Al_2O_3), the oxygen atoms form a hexagonal close packing, and the aluminum atoms occupy 2/3 of the octahedral holes between the layers of oxygen atoms. Each Al atom is surrounded by 6 O atoms in a distorted octahedral arrangement, and each O atom has 4 Al atoms as nearest neighbors in an approximately tetrahedral coordination. Other compounds which crystallize in the C. are Ti_2O_3, V_2O_3, Cr_2O_3, α-Fe_2O_3, etc.

Cotton effect: see Chiroptic methods.

Cottrell-Möller process: see Electrostatic gas purification.

Coulomb integral: see Hückel method.

Coulomb's law: a quantitative expression of the force between electrical charges, discovered in 1785 by Coulomb. The force F between two point charges q_1 and q_2 is directly proportional to the product of the two charges and inversely proportional to the square of the distance between them:

$$F = \frac{1}{\varepsilon 4\pi\varepsilon_o} \frac{q_1 q_2}{r^2},$$

where $1/4\pi\varepsilon_0 = 8.9876 \cdot 10^9$Vm/As and $\varepsilon_0 = 8.8542 \cdot 10^{-12}$ As/Vm. ε_0 is the *electric field constant* and ε is the *relative dielectric constant*. The latter must be taken into account when the volume between the point charges contains matter rather than a vacuum. For the vacuum, $\varepsilon = 1$. Opposite charges attract each other, and like charges repel. The repulsive forces are given a positive sign, and attractive forces, a negative sign. If the charge q_1 is brought from distance r_1 to r_2 in the field of charge q_2, and if $\varepsilon = 1$, the work done is equal to the change in potential energy

$$W = -\int_{r_1}^{r_2} F\,\mathrm{d}r = V_2 - V_1 = \frac{1}{4\pi\varepsilon_o} q_1 q_2 \left(\frac{1}{r_2} - \frac{1}{r_1}\right)$$

With the reference point $V = 0$ for $r \to \infty$, the potential energy of charge q_1 at the distance r from q_2 is given by the equation: $V = q_1q_2/4\pi\varepsilon_0 r$. For opposite charges, $V < 0$, and for like charges, $V > 0$.

Coulometer: a device for measuring electric current. According to Faraday's laws, the current-time integral can be determined from the amount of product formed in an electrolytic reaction. The ***copper coulometer*** consists of a copper anode and a copper cathode dipping into a copper sulfate electrolyte. When current flows through this cell, the copper dissolves from the anode and is deposited on the cathode. If the difference in mass of the electrodes before and after electrolysis is determined by weighing, the amount of current which has flowed can be calculated from Faraday's laws. Other C. are the silver coulometer, in which two silver or platinum electrodes dip into a silver nitrate solution, the oxygen-hydrogen C., in which nickel electrodes dip into a 15% potassium hydroxide solution, and the mercury

C., in which the current measurement is based on the volume of mercury deposited at the cathode. This type of C. was formerly used in households to measure the consumption of direct current.

Today, electrochemical C. have largely been replaced by electronic C.; they are mainly used in Coulometry (see).

Coulometry: a method of electrochemical analysis based on Faraday's laws. The amount of a substance is determined by the amount of current flowing in an electrolysis cell. A prerequisite is a selective, quantitative electrode reaction, that is, at the electrode at which the reaction of interest is occurring, no other reaction can occur, and the reaction of interest must give 100% current yield.

There are various techniques for coulometric determinations; the most important are potentiostatic C. and coulometric titration.

In ***potentiostatic C.***, the voltage at the working electrode is kept constant while the electrolysis takes place. The current decays exponentially with time until it reaches the value of the base current (which is subtracted). When the base current has been reached, the reaction is stopped, and the amount of electricity used is measured by a coulometer. From this measurement, the amount of substance is calculated, using Faraday's laws.

Coulometric titration is more common. Here the electrolysis takes place at constant current, generating a reagent which reacts quantitatively with the substance being analysed. When the equivalence point is passed, excess reagent appears; its presence is signaled in suitable fashion, for example, colorimetrically or potentiometrically. Some typical coulometric reagents are hydroxide ions, chlorine, bromine, iodine or silver ions.

Coulter counter: a device for determining the number and sizes of non-conductive particles in an electrically conductive dispersion medium. The particles pass through a narrow, exchangeable capillary with a diameter between 20 and 400 μm. Electrodes are placed on opposite sides of the capillary opening. The voltage or current pulses caused by changes in resistance are measured. The lower limit of particle size (radius) is about 0.2 μm, and the upper limit is about 100 μm.

Coumachlor: a Rodenticide (see).

4-Coumaric acid: see Hydroxycinnamic acids.

Coumarin, *α-**chromone, 2H-chromen-2-one, 2-hydroxycinnamic acid lactone***: the internal ester of 2-hydroxycinnamic acid. C. forms colorless crystals, m.p. 70 °C, b.p. 301 °C. It is readily soluble in alcohol and ether, but only slightly soluble in water. C. and its derivatives (Table) are found widely in plants, usually as the glycosides. C. is found in woodruff and many types of grass and clover. Some C. and various Furocoumarins (see) are mutagens.

C. are synthesized by condensation of salicylaldehyde with acetic anhydride in the presence of sodium acetate (see Perkin reaction). C. are used as perfumes in soaps and tobacco. C. and woodruff were formerly used to flavor drinks (May wine) and foods, but this is no longer allowed. A number of C. derivatives are used as optical brighteners and as laser pigments. C. is a constitutional isomer of γ-Chromone (see).

Trivial name	R^1	R^2	R^3	m.p in °C
Coumarin	H	H	H	67– 71
Umbelliferone	H	OH	H	225–228
Aesculetin	OH	OH	H	268–272
Scopoletin	OH	OCH_3	H	204–205
Daphnetin	H	OH	OH	255–256

Coumarone: see Benzofurans.

Coumarone-indene resin: a synthetic hydrocarbon resin made by polymerization of coumarone, indene and a few related compounds. The C. are light-colored, brown or black and have varying hardness; they soften between 30 and 160 °C. They are produced by treating the light oil fraction of coal tar distillation (anthracite) with concentrated sulfuric acid or aluminum chloride. This causes polymerization of the unsaturated fractions, such as coumarone, indene and cyclopentadiene. The resulting crude resin is purified and used in paints. However, because it is not fast to light, it is used mainly for relatively cheap materials such as paints for concrete or ship hulls. C. are also used as glue for linoleum, as softeners and vulcanization promoters for rubber, as electrical insulation, in phonograph records, etc.

p-Coumaryl alcohol: see Cinnamyl alcohols.

Countelle®: see Synthetic fibers.

Counter-current chromatography, abb. ***CCC***: a combination of liquid-liquid distribution chromatography with counter-current distribution in the absence of a solid carrier; either the hydrostatic or hydrodynamic equilibrium of a two-phase solvent system is used (Fig. 1).

A column is filled with one solvent (*stationary phase*) and another solvent which is immiscible with the first is introduced at one end (*mobile phase*). The mobile phase travels through the column under the influence of gravity. The mobile phase is constantly renewed by continuous elution, while the stationary

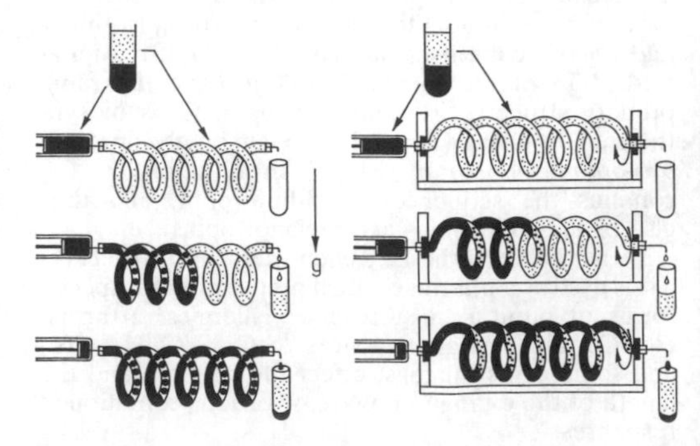

Fig. 1. Basic model of counter-current chromatography using hydrostatic (left) or hydrodynamic (right) equilibrium. g = gravity

phase remains in the column. A sample mixture applied to the column is distributed between the phases and separated in accordance with the distribution coefficients of its components.

Several variants of CCC have been developed by modification of the geometry of the column. In ***droplet counter-current chromatography*** (abb. ***DCCC***), the column is straight and vertical, and the mobile phase is a stream of droplets, which must of course be smaller than the inner diameter of the column (Fig. 2).

Rotation locular CCC (***RLCCC***) makes use of a system of glass columns which are subdivided into segments by centrally perforated Teflon disks and are connected with each other by Teflon capillaries. By rotation of the columns into an inclined position, a good exchange between the mobile and stationary phases is achieved. For some separations, RLCCC can also be carried out as reversed phase chromatography (see High-performance liquid chromatography).

Fig. 2. Droplet counter-current chromatography. Ascending (left) and descending (right) methods

C. makes possible effective separations of small samples in short times, with excellent reproducibility, high fraction purity and good sample recovery. It is used to separate biological samples such as natural and synthetic peptides, prostaglandins, cells and macromolecules, and for continuous extraction and enrichment of medicinal plants and their metabolites.

Counter current distillation: same as Rectification.

Counter current process: a method of carrying out exchange of heat or material. The two media are brought into intimate contact as they flow past one another. In pure heat exchange, the media are separated by walls which do not permit passage of material, but are good conductors of heat. This process is carried out technically in heat exchangers, coolers, condensers, evaporators, etc. In pure material exchange (e.g. in extraction), or in coupled material and heat exchange (e.g. in distillation, sorption, drying, and in some cases extraction), the two media are brought into direct contact. The same is true for those processes in which the densities of the two media are so different that they separate spontaneously, as in decanting.

The opposite of C. is a Concurrent process (see). Both processes are ***driving force processes***, that is, they depend on the presence of a driving force.

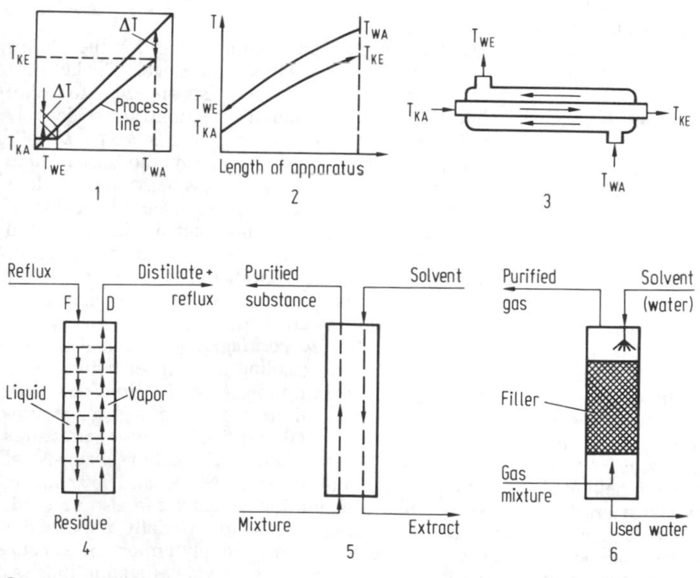

Countercurrent process: *1*, Heat exchange on an equilibrium diagram. *2*, Temperature curve during heat exchange in the apparatus. *3*, Double-walled heat exchanger. *4*, Countercurrent between a liquid and a vapor in rectification. *5*, Countercurrent in a liquid-liquid extraction. *6*, Gas-liquid counterflow during absorption. *w* = warmer medium, *C* = cooler medium, *S* = start of the process, *E* = end of the process, *T* = temperature, Δ *D* = driving force

Courlene®: see Synthetic fibers.

Covalent bond: see Chemical bond.

Covolume: see Van der Waals equation.

Cp: see Lutetium.

Cr: symbol for chromium.

CR: see Chemical weapons.

Crack corrosion: an uneven corrosion in cracks, either cracks in the material itself or between two parts of a machine or other structure which are made of the same material. The cause is restricted diffusion in cracks, which leads to formation of concentration elements. For example, a lack of oxygen transport to the surface of chromium-nickel steels under insulating material and local loss of passivity produces corrosion scars and corrosion in cracks between welded segments of heat-transfer pipes.

Cracking: a process of converting hydrocarbon fractions to mixtures with lower average molecular masses. C. allows the yields of verious fractions from distillation of crude petroleum to be matched to the demand. At temperatures between 450 and 650 °C, the large hydrocarbon molecules of these fractions are cleaved in endothermic reactions (that is, energy is supplied). The products are low-boiling hydrocarbons (gasoline) and cracking gases such as hydrogen, methane, ethane, propane, butane, ethene, propene and butene. These are the starting materials for further chemical processes (see petroleum chemistry). In the broad sense, C. includes pyrolysis processes, which occur at temperatures above 600 °C (up to electric arc temperatures); these cause extensive cracking of the starting material, so that most of the product is gaseous (see Pyrolysis.).

The gasoline produced by simple distillation of petroleum, called straight-run gasoline, is not obtained in sufficient quantity to meet the demand for gasoline.

C. can be accomplished by heat alone, or catalytically. Since catalytic gasolines contain more branched alkanes, and these produce less knocking, catalytic gasolines are qualitatively better than thermal gasolines.

1) ***Thermal C. (low-temperature cracking)*** is essentially obsolete. In ***liquid-phase C.*** (see Visbreaking) the temperature is around 500 °C and the pressure between 1 and 7 MPa, while in ***gas*** or ***mixed-phase C.***, the temperature can be as high as 600 °C and the starting material is largely vaporized; the pressure in this case is much lower. Thermal C. occurs by a radical chain mechanism (see Pyrolysis).

The oldest liquid-phase C. was carried out discontinuously in large vessels heated between 500 and 600 °C. The technology was improved by the introduction of flow-through oil-heated furnaces in which the material being cracked flows continuously, under pressure, through heated stainless-steel pipes. Coke formation is kept within limits by suitable technology. In contrast, in the *coking method*, the material heated in the pipe furnace is held for a long period in special coking chambers at cracking temperature, until the coke formation is finished. The petroleum coke precipitated in the chambers is mechanically removed; because of its low ash content, it is a valuable starting material, for example, for carbon electrodes. This is the principle of the *delayed coking process*.

The *fluid coking process* (Fig. 1) is a further development. Here the hot vapors of heavy residual products are blown into a fluidized bed reactor at a temperature of 500 to 570 °C. The fluidized bed consists of hot coke grains. The cracking reaction occurs mainly on the surface of the coke, on which the petroleum coke also precipitates. The coke is circulated through a regenerator, in which a small part is burned by adding air. The heat released is used to maintain the endothermal C. reaction. The excess coke is continuously removed from the installation. The fly ash is removed by cyclones, and the product is separated into gas-gasoline and cocker gasoil in a fractionating column.

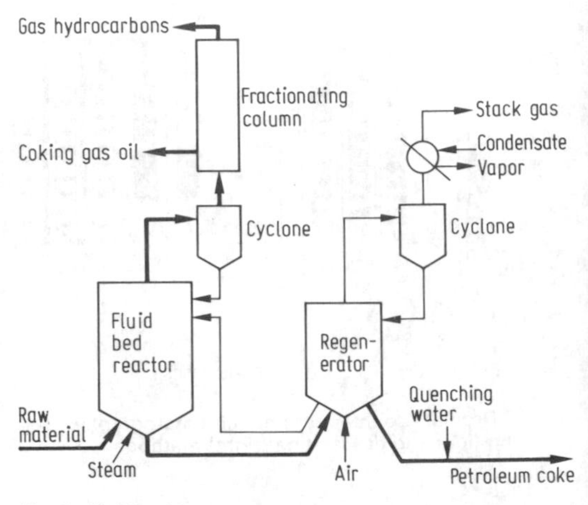

Fig. 1. Fluid coking process

Modern thermal installations, e.g. for the *Dubbs process*, usually consist of several stages. The heaviest part of the starting material is converted to coke, while the light and medium fractions are selectively cracked to heating oil and gasoline (***selective C.***).

In the last few years, thermal methods of residue C., which serve mainly to process heavy top residues into products of lower viscosity (see Visbreaking), have become more important. Such products are used either directly as heating oils or as starting materials for catalytic C. The conditions used are relatively mild, such that only the high-molecular-weight components are slightly cracked.

2) ***Catalytic C. (Catcracking)***. Like thermal C., this process yields gas, gasoline and higher-boiling fractions. With a few exceptions (e.g. the *Houdresid process*), catalytic C. is done in the vapor phase at temperatures between 460 and 500 °C, and at pressures between 0.1 and 0.15 MPa. The catalysts consist of about 10% aluminum oxide, Al_2O_3, and 90% silicon dioxide, SiO_2; magnesium silicates can also be used. Ion-exchange zeolites (X- and especially Y-molecular sieves) are becoming increasingly important as catalysts for C. The reaction goes via carbenium ions, so that in addition to the actual C. reaction, isomerizations and polymerizations can occur. In addition, dehydrogenation of naphthenes produces aromatics.

Mechanism of catalytic cracking:

Initiation:

$$R_1{-}CH_2{-}CH_2{-}CH_2{-}CH_2{-}R_2 + H^+ \rightarrow$$
$$R_1{-}CH_2{-}CH_2{-}\overset{+}{C}H{-}CH_2{-}R_2 + H_2$$

Chain reaction:

$$R_1{-}CH_2{-}CH_2{-}\overset{+}{C}H{-}CH_2{-}R_2 \rightarrow$$
$$R_1{-}\overset{+}{C}H_2 + CH_2 = CH{-}CH_2{-}R_2$$
$$R_1{-}\overset{+}{C}H_2 + R_1{-}CH_2{-}CH_2{-}CH_2{-}CH_2{-}R_2 \rightarrow$$
$$R_1{-}CH_3 + R_1{-}CH_2{-}CH_2{-}\overset{+}{C}H{-}CH_2{-}R_2$$

Chain terminaton:

$$R_1{-}CH_2{-}CH_2{-}\overset{+}{C}H{-}CH_2{-}R_2 \begin{cases} \nearrow R_1{-}CH_2{-}CH_2{-}CH{=}CH{-}R_2 + H^+ \\ \searrow R_1{-}CH_2{-}CH_2{-}CH_2{-}CH_2{-}R_2 + H^+ + H_2 \end{cases}$$

Isomerization on an acidic carrier catalyst:

$$R_1{-}CH_2{-}CH_2{-}\overset{+}{C}H{-}CH_2{-}R_2 \rightleftharpoons$$
$$R_1{-}\underset{}{\overset{CH_3}{\overset{|}{C}H}}{-}\overset{+}{C}H{-}CH_2{-}R_2$$
$$R_1{-}CH\overset{CH_2}{\underset{H^+}{\text{---}}}CH{-}CH_2{-}R_2 \rightleftharpoons$$
$$R_1{-}\overset{+}{C}H{-}\overset{CH_3}{\overset{|}{C}H}{-}CH_2{-}R_2$$

During the C. process, coke precipitates on the catalyst and reduces its activity. The contact is therefore regenerated periodically by burning off the coke in air at temperatures between 600 and 680 °C. The amount of coke and gas formed by catalytic C. is less than that produced by thermal C.; therefore the amount of material reacting in a single pass through the C. reactor can be much higher than in a thermal C. The gasoline yields depend on the starting material and the conditions in the reactor; because the catalytic process is more readily controlled and the conditions are milder, they are again higher than in the thermal process. With gas oils as starting material, up to 60% gasoline, 30 to 40% cracking gas oil and cracking heating oil and 10 to 15% gases are produced. The quality of the gasoline is higher than with the thermal process.

There are three types of process: fixed-bed, moving-bed and fluid bed.

a) The *fixed-bed process* was introduced by Houdry in 1936, and is now obsolete.

b) *Moving-bed processes*, also developed by Houdry, were introduced in 1940. The most important of these are the TCC (thermofor catalytic cracking) and the Houdry-flow processes. In the *TCC process*, the catalyst, usually a powder, drifts down through the reactor under the influence of gravity; the reaction temperatures are 450 to 510 °C. The coked catalyst falls into the regenerator below the reactor and there the coke is burned off in air. After cooling, the regenerated catalyst is returned by a pneumatic transport to the top of the reactor. The cracking vapors are processed in a fractionation column. The *Houdry-flow process* is basically the same as the TCC process, but the regenerated catalyst is returned to the reactor with a pneumatic column of exhaust gas rather than air.

c) Of the *fluid-bed processes*, the most important is the *FCC* (fluid catalytic cracking) process. The first FCC installation began operating in 1942 (Standard Oil Co. of New Jersey). Alumosilicates or modified zeolites, in the form of a very fine dust, with grains 0.05 to 0.2 mm in diameter, are used as catalysts. The mechanical properties of this dust are similar to those of a liquid. It permits easy transport, even distribution of heat and good heat transfer to the oil. The contact is heated to 500 to 550 °C, and evaporates the oil. The oil vapors pick up the catalytic dust and carry it through the reaction zone, and the intimate contact of catalyst and oil leads to rapid completion of the C. The pressure in the reactor is 0.07 MPa.

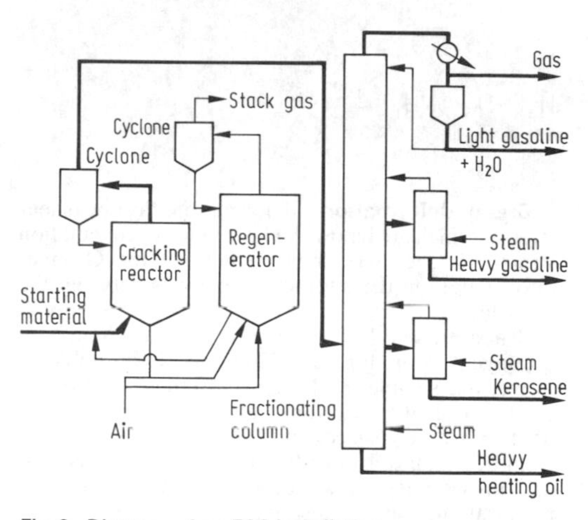

Fig. 2. Diagram of an FCC installation

The mixture of oil vapor and catalyst leaving the reactor is separated in a cyclone, and the coke-loaded catalyst is carried to the regenerator in a stream of air. The heat remaining in the exhaust gases is used to heat the reactor. The oil vapors are fractionated into gasoline, gas oil, heating oil, circulating oil and residue in a fractionation column. The fractions must be hydrorefined to remove unsaturated compounds, especially diolefins.

The capacity of an FCC installation can be up to 3 million tons/year.

3) ***Catalytic C. in the presence of hydrogen.***. Fractions with low hydrogen contents (residual oils) are not suitable for catalytic C. The low hydrogen content and the heterocompounds present in the residue rapidly deactivate the catalyst. Residual oils are

therefore cracked in the presence of hydrogen (see Hydrocracking).

4) ***Pyrolysis processes***. These have been developed from vapor-phase C.; they are used to produce mainly gaseous alkenes and certain amounts of aromatic hydrocarbons from lower- or higher-molecular-weight starting materials. These are only byproducts of the C. processes described above. All pyrolysis processes are carried out in vacuum at temperatures above 600 °C. There are two basic pyrolysis reactions: C., in which a long-chain alkane is split into a shorter-chain alkane and an alkene, and dehydrogenation processes, in which an alkane loses hydrogen and becomes an alkene. The longer the carbon chain, the more C. predominates over dehydrogenation (see Pyrolysis).

Cracking gas: see Synthesis gas.

Cram's rule: see Asymmetric synthesis.

Creatinine: 2-imino-1-methylimidazolidin-4-one, colorless crystals; M_r 113.1, m.p. 212-217 °C (dec.). It is soluble in water, but is practically insoluble in alcohol and ether. With acids it forms crystalline salts. C. is synthesized in muscle tissue from creatine. A relatively constant amount is excreted in the urine; C. clearance by the kidneys is used in clinical chemistry as a test of their function. C. is present in bouillon cubes and powders.

Creep deformation: deformation at low shear stresses which are below the flow limit. In coagulation structures (see Gel), measurement of the C. gives information on the interaction energies between the particles.

Creosote: a yellowish, oily, poisonous liquid with a pungent odor; b.p. 200-220 °C. C. is insoluble in water, but soluble in alcohol and ether. It also dissolves in alkalies, forming salts. The main components of C. are guaiacol and cresol, and it is obtained from beech tar or by distillation of coal tars. It is used as a preservative for wood, a flotation agent for ore preparation, and as a disinfectant in medicine.

Cresols, ***methylphenols, hydroxytoluenes***: three isomeric compounds which smell like phenol. They are slightly soluble in water, and readily soluble in ethanol and ether.

o-Cresol *m*-Cresol *p*-Cresol

o-Cresol (2-methylphenol) is a colorless crystalline mass or liquid which turns dark upon standing in air; m.p. 30.9 °C, b.p. 190.9 °C, n_D^{20} 1.5361. ***m-Cresol*** (3-methylphenol) is a colorless liquid; m.p. 11.5 °C, b.p. 202.2 °C, n_D^{20} 1.5438. ***p-Cresol*** (4-methylphenol) forms colorless crystals; m.p. 34.8 °C, b.p. 201.9 °C, n_D^{20} 1.5312.

The C. are found in coal tar fractions, from which the parent compound is also obtained. Because of their solubility in alkali hydroxide solutions, they can be extracted with sodium hydroxide, or they can be obtained by fractional distillation. The *m*- and *p*-compounds, which have similar boiling points, can be separated by forming the sulfates or by special precipitation methods. In industry, the mixture of the C. is often used as crude cresol. C. are very reactive (see Phenols). They have bacteriocidal effects and can be used as disinfectants. *Lysol* is a solution of C. and soap in water. C. are also used to preserve wood and to make phenoplastics, aromas, indicator dyes and antioxidants. *p*-C. is formed by biological degradation of tyrosine. In the human body, it is esterified with sulfuric acid to detoxify it, and excreted in the urine.

Cresol purple: see Bromphenol blue.

Cresol red: see Bromphenol blue.

CRH: see Liberins.

Criegee cleavage: a quantitative method for cleavage of glycols using lead(IV) acetate in dilute glacial acetic acid or benzene solution. The products are aldehydes and ketones: $R_2C(OH)\text{-}C(OH)R_2 + Pb(OOCCH_3)_4 \rightarrow 2\ R_2C(OH) + Pb(OOCCH_3)_2 + 2\ CH_3COOH$. The C. can be used to cleave 1,2-diols which are not split with periodates (see Malaprade reaction).

Crimidine: a Rodenticide (see).

Cristobalite type, ***cristobalite structure***: a Structure type (see) for compounds with the general composition AB_2 in which the ratio of atomic radii is low ($r_A/r_B \leqq 0.414$). In the crystal lattice of the SiO_2 modification cristobalite, the Si atoms form a Diamond structure (see); each is surrounded tetrahedrally by 4 O atoms. Each O atom lies between two neighboring Si atoms (Fig.).

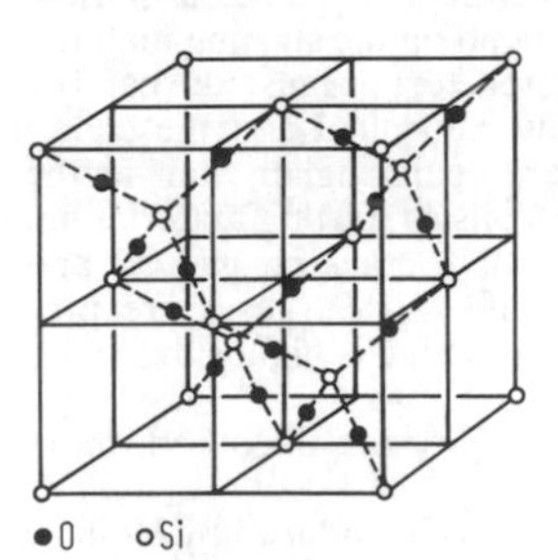

Cristobalite structure (unit cell of the crystal lattice)

Critical data: see Critical point.

Critical micellar concentration: see Colloids.

Critical point, ***critical state***: in general, the point in a two-phase system at which the two coesixting phases have identical properties (e.g. density, composition, index of refraction), and therefore represent a single phase.

1) The C. in the narrower sense is the point on the equilibrium curve between the liquid and gas phases of *pure substances* at which the differences between the two phases disappear. The vapor-pressure curve ends at this point. Above the C., there is only a uniform supercritical phase, and it is no longer useful to describe it as a gas or a liquid. There is so far no conclusive evidence for the existence of an analogous C. for the liquid-solid phase equilibrium at very high pressures.

The state parameters which determine the C. of a substance are called its ***critical data***: *critical pressure* p_{crit}, *critical temperature* T_{crit} and *critical molar volume* V_{crit}. T_{crit} and p_{crit} can be determined experimentally, by heating a closed two-phase system until the phase boundary disappears, or by cooling the supercritical phase until it falls into two phases. The latter is easier to recognize, because at first extremely fine droples of liquid appear, and these cause the previously clear substance to appear cloudy. Determination of the critical volume follows from the critical density d_{crit} using the Cailletet-Mathiassen rule (see). According to *Guldberg's rule*, the boiling points of non-associative liquids at normal pressures are about 2/3 their critical temperatures. Critical data can also be calculated from the thermal State equations (see) of a real substance, because the critical isotherm in the pv diagram has a horizontal turning point. From the van der Waals equation, for example, $V_{crit} = 3b$, $p_{crit} = a/27b^2$, $T_{crit} = 8a - 27\,b \cdot R$, where a and b are the van der Waals substance constants. The critical data calculated in this way, like the van der Waals equation itself, are only approximate (Table).

Critical data of some substances

	T_{crit} in K	p_{crit} in MPa	V_{crit} in dm^3	d_{crit} in kg m^{-3}
Helium	4.22	0.226	0.061	69
Hydrogen	33.25	1.29	0.061	31
Nitrogen	126.15	3.39	0.087	311
Oxygen	154.75	5.06	0.074	430
Carbon dioxide	304	7.29	0.096	460
Ammonia	405.6	11.3		235
Chlorine	417	7.61		573
Pentane	470	3.3	0.310	232
Methanol	513	10.0	0.117	358
Benzene	561.8	4.85	0.256	305
Chloro-benzene	632.4	4.52	0.308	365
Water	647.05	22.04	0.056	329

2) In binary systems which contain two substances which are not miscible in all proportions (see Mixing gaps), there is an *upper critical mixing point* if there is complete miscibility above this temperature, and a *lower critical mixing point* if there is complete miscibility below this temperature. C. for miscibility are observed both in liquid and solid mixtures. In general they are determined by the composition and temperature, and also depend on pressure.

Crocin: $C_{44}H_{64}O_{24}$, the digentiobiose ester of α-crocetin. C. is the orange pigment of saffron, m.p. 186 °C. The gentiobiose groups are bound β-glucosidically to the two carboxyl groups of the crocetin. C. and some of its degradation products are of biological interest. They increase the motility of the gametes of single-celled algae (*Chlamydomonas eugametos*) as they move toward each other.

Cromoglic acid: a γ-chromone derivative used for prophylaxis of asthmatic attacks. It acts to inhibit release of histamine and other mediators.

Crossing experiment: a chemical reaction with two or more chemically similar reactants are used to decide whether a mechanism is intramolecular or intermolecular. For example, it was found by a C. with two similarly substituted hydrazobenzene molecules that the benzidine rearrangement is a synchronous intramolecular reaction.

Cross-linking: the linking of long polymeric molecules along their length by covalent cross links. This produces a three-dimensionally linked macromolecule, which can be rubber elastic, soft, hard or brittle, depending on the number of cross links (see Plastics, hard). In plastics, the formation of cross-links is called *hardening*. C. can be induced either by chemically reactive cross-linking reagents, high-energy radiation (e.g. with polyethylene) or mechanically. ***Cross-linkers*** are used to bind two or more reactive sites in adjacent molecules; they are then incorporated into the three-dimensional network. Some examples are sulfur (for rubber), styrene, divinylbenzene and triallylcyanurate (for unsaturated polyester resins) and diisocyanates (for macromolecular substances with free OH, NH_2 and COOH groups).

Crotonaldehyde, ***but-2-enal***: CH_3-CH=CH-CHO, a colorless liquid with a pungent odor; m.p. - 74 °C, b.p. 104 °C, n_D^{20} 1.4366. C. is readily soluble in nearly all organic solvents and water, and is steam volatile. C. can exist in either of two configurations, the Z and the E form; the latter is very important industrially. The separation of the two isomers is relatively complicated.

C. contains a conjugated π-electron bond system, which causes its reactions to differ somewhat from those of saturated aldehydes. For example, the terminal methyl group is so highly activated that it is able to undergo aldol addition. C. also undergoes the typical Aldehyde (see) reactions. It occurs naturally in some plants which contain croton oil. It is most readily synthesized by distillation of acetaldol in the presence of acetic acid. C. is used industrially in the synthesis of important intermediates and end products, such as butanol, butyraldehyde, croton alcohol, maleic acid, quinaldine and a series of synthetic resins.

Z-Form E-Form

Crotonic acid, ***trans-but-2-enoic acid, (E)-but-2-enoic acid***: an unsaturated, aliphatic monocarboxylic acid. C. forms colorless needles or leaflets, m.p. 71.5 °C, b.p. 185 °C, which smell like butyric acid. It is soluble in hot water, alcohol, acetone and ether. When heated above 125 °C, C. isomerizes to the less stable Z-form, Isocrotonic acid (see), which converts back to C. around 180 °C. C. is formed by condensation of acetaldehyde with malonic acid in the presence of pyridine:

$$CH_3\text{–}CHO + CH_2(COOH)_2 \xrightarrow[-H_2O]{} CH_3\text{–}CH{=}C(COOH)_2$$

$$\xrightarrow[-CO_2]{} CH_3\text{–}CH{=}CH\text{–}COOH.$$

Industrially, C. is synthesized by oxidation of crotonic aldehyde with oxygen. It is used in the production of mixed polymers, and its esters are used as softeners.

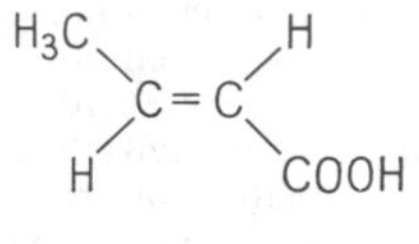

H_3C / H / C=C / H / COOH

Crotyl barbital: see Barbitals.

Crown ether: a macrocyclic polyether, which is enabled by its structure to bind metal ions in the center of the ring. The term C. was chosen because all the oxygen atoms surround the metal ion like a wreath or crown, and form coordinate bonds to the metal ion. In addition to C. with oxygen atoms, which immediately aroused great interest because of their ability to complex alkali metal ions, sulfur compounds have also been synthesized in increasing numbers. A prototype for these compounds is 18-crown-6, an 18-membered macrocycle with six oxygen atoms; its systematic name is 1,4,7,10,13,16-hexaoxacyclooctadecane (Fig.).

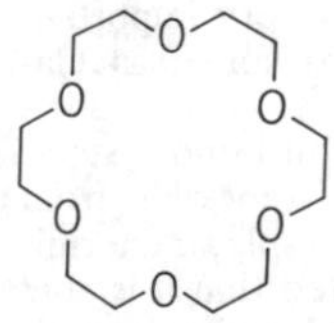

18-Crown-6

This compound can be synthesized in greater than 90% yield from potassium *tert*-butanolate and triethyleneglycol ditosylate. C. are used as model systems for carriers of alkali metal ions in membrane chemistry and, because of their solubility in organic solvents, for mild oxidations of alkenes, alkylbenzenes and aldehydes with potassium permanganate in benzene. They are also used for phase transfer catalysis. There is growing interest in their effects on reactions of other metal ions, such as silver.

Crum-Brown-Walker synthesis: a method for preparation of long-chain dicarboxylic acid esters by electrolysis of the potassium salts of the dicarboxylic acid monoesters: 2 $ROOC\text{-}(CH_2)_n\text{-}COO^-\ K^+$ + 2 $H_2O \rightarrow ROOC\text{-}(CH_2)_{2n}\text{-}COOR$ + 2 KOH + H_2 + 2 CO_2. For example, the potassium salt of monoethyl malonate can be converted by this method to diethyl succinate.

Crylor®: see Synthetic fibers.

Cryochemistry, often also called ***low-temperature chemistry***: the chemistry of reactions and processes at very low temperatures, in the range between approximately - 196 °C and - 270 °C. For example, reaction products which would normally decompose immediately are precipitated onto a very low-temperature surface. In this way it is possible to synthesize compounds which cannot be made in any other way. It is also possible to generate significant concentrations of free radicals and to store them. Because of the very low reaction rates at low temperatures, it is possible to study in detail reactions which occur explosively at room temperature.

Cryohydrate: see Melting diagram.

Cryolite process: see Soda.

Cryopump: a device for pumping gases out of a vacuum system by condensation and adsorption to surfaces at very low temperatures. The pumping effect of cooled surfaces is utilized in Cold traps (see) to remove water, oil and other vapors (down to about - 190 °C). Binding of nitrogen, oxygen or argon requires a temperature below - 235 °C. Effective removal of neon, hydrogen and helium is difficult with C. However, they make possible the production of very high vacuums.

Cryoscopy: see Freezing point lowering.

Cryotechnology: a part of Refrigeration technology (see) dealing with very low temperatures between about - 160 °C and absolute zero, - 273.15 °C. There are essential differences between C. and the methods used down to about - 100 °C. In C., gas refrigerators (see Refrigerators) are used. With helium, which has the lowest boiling point of any substance, temperatures of about - 270 °C can be reached. There are special processes (such as adiabatic demagnetization) for reaching still lower temperatures.

The importance of C. is growing. Liquefied gases are needed for many applications (CH_4, O_2, N_2, H_2, He) in energy, electronics, nuclear and vacuum technology, medicine and space travel.

Cryptand: a bicyclic compound of carbon, nitrogen and oxygen, sometimes also sulfur, which, like the Crown ethers (see), can complex alkali and alkaline earth metal ions by sequestering the metal ions in "hollows" in the macrocycles. A prototype for such compounds is diazahexaoxabicyclo[8.8.8]hexacosane (Fig.), which has various configurations at the bridgehead atoms.

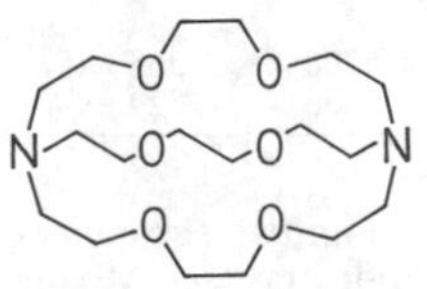

C. are soluble in organic solvents and are used for separations from cation mixtures, development of ion exchangers, selective extraction of elements from salt solutions and detoxification in the human body.

Cryptic ions: "hidden" ions. The ions which participate in organic chemical ionic reactions are often not originally present, but are formed during the course of the reaction.

Cryptobases: see Hydride ion donors.

Crystal: a solid in which the component atoms, ions or molecules are periodically arranged in a three-dimensional array. A C. forms through growth on a seed C. (see Crystallization, Crystal growing).

An *real C.* is a strictly periodic mathematical abstraction of a natural C. In a real C., there are deviations from strict periodicity because of 1) its finite size, 2) C. lattice defects and 3) the oscillations of the atoms or ions around their equilibrium positions, which are not completely "frozen out", even at absolute zero, and which have the effect that the periodicity exists at best as a time average, but not at any given point in time. The study of the ideal structure is the object of X-ray structure analysis (see). Many typical properties of C. can be explained by the ideal structure, including symmetry properties (see below), the plane surfaces of C. which grow undisturbed, the angles between these surfaces and the Anisotropy (see) of many properties. Some properties are not affected, or only slightly affected, by large deviations from the ideal structure; these are called *defect-insensitive properties*, or, less aptly, *structure-insensitive properties*. On the other hand, *defect-sensitive properties*, such as hardness, ion and electron conductivity in ionic or semiconductor C. and light absorption are seriously affected, and sometimes determined, by the deviations of the real C. from its ideal structure. The model of the *real C.* or the *real structure of the C.* is helpful in understanding such properties. The model describes Crystal lattice defects (see) such as point defects, displacements and foreign inclusions. There are several forms in which the deviations from ideal structure are extreme. *Disordered C.* lack periodicity in one or even two directions. For example, layers which are close to ideal structure can be stacked in random sequence, and thus form the C. This and similar phenomena are called *positional disorder*; under certain precisely defined geometrical conditions, one may speak of "order-disorder" C. (OD-C.). *Liquid crystals* (see) have so little order in many of their modifications that use of the term "crystal" in the sense defined at the beginning seems scarcely justified.

A *single crystal* is a body consisting of a single, uniform C. in which all parts are in a definite orientation with respect to the others. In many cases, e.g. in metals and ceramics, however, there are numerous small C. (*crystallites*) with different orientations within the solid (see Polycrystalline material), which is thus *polycrystalline*. Two single C. which have fused through regular growth are called *twins* (there are also triplets, and in general, multiplets), and can be described by certain "twin rules".

The examination and description of the structure of a C. is the goal of *C. structure theory*. It starts from a consideration of Symmetry (see) properties as characteristics of the (ideal) C. The basic symmetry element of every C. is *translation*, the motion of a point along a certain vector **a** (which may be repeated as often as desired). This generates an (infinite) series of points (one-dimensional lattice). The absolute value of the translation vector, $|\mathbf{a}| = a_0$, is called the *period* (identity period or distance). A three-dimensional point lattice is generated by periodic translation of a point along the vectors **a**, **b** and **c** (*space lattice*, Fig. 1). The vectors **a**, **b** and **c** are called the *basis vectors* of the lattice; they form a coordinate system. Their lengths a_0, b_0 and c_0 and the angles between them, α, β and γ are called *lattice constants*. The parallelopiped formed by the three basis vectors is called the *unit cell*. Since the lattice is formed by periodic repetition of (infinitely many) unit cells in all three directions, it can be completely described by the 6 lattice constants. The volume V of a unit cell can be calculated as follows from the lattice constants:

$$V = a_0 b_0 c_0 \sqrt{1-\cos^2\alpha-\cos^2\beta-\cos^2\gamma+2\cos\alpha\cos\beta\cos\gamma}.$$

The mathematical abstraction of the point lattice is translated into a C. by placing identical atoms, ions or molecules at each lattice point. These are called the *basis of the C. structure*; thus Lattice + Basis = C. structure. Two of the most important properties of a C. follow from its lattice structure: it is *homogeneous*, since it consists of identical unit cells, and it is anisotropic (see Anisotropy), because the distances between lattice points in different directions are usually different.

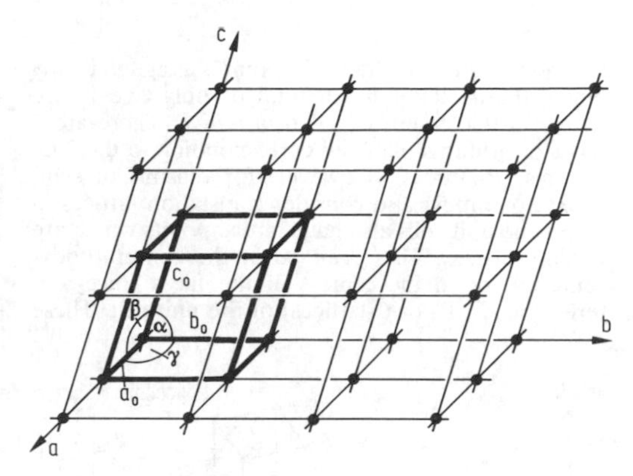

Fig. 1. Lattice with unit cell, Lattice constant and basis vector

The choice of the unit cell is not immediately obvious. In principle, any number of different unit cells, and thus triplets of basis vectors or sets of lattice constants could be found with which the lattice could be described. In limiting the choice of the unit cell, not only the translational symmetry, but the total symmetry of the C. is taken into account, and the following rules are applied: 1) The origin of the unit cell must lie on a geometrically determined point. 2) The unit cell must have the highest degree of symmetry which is possible within the framework of the total symmetry of the C. 3) The angle between basis vectors must be 90° if possible. 4) The volume of the unit cell must be as small as possible. Application of these rules, in the order given, produces 7 types of unit cell or crystallographic axis systems, which are called the 7 *crystal systems*. They are listed in Table 1 with the relations between lattice constants which characterize them.

Sometimes the hexagonal and rhombohedral C. systems are considered as one system. The trigonal system, which is also frequently used for historical and practical reasons, can be classified as hexagonal. By vector addition of the basis vectors, one obtains translation lattices in which the unit cells have lattice points only at their vertices. Since such a lattice point belongs simultaneously to 8 unit cells which meet at

Table 1. The 7 crystal systems

Crystal system	Unit cell (Relations among basis vectors)	Required (minimum) Symmetry element
Triclinic	$a_0 \neq b_0 \neq c_0 \neq a_0$ $\alpha \neq \beta \neq \gamma \neq \alpha$	None
Monoclinic	$a_0 \neq b_0 \neq c_0 \neq a_0$ $\alpha = \gamma = 90°, \beta \neq 90°$	2-fold rotation axis or mirror plane
Orthorhombic	$a_0 \neq b_0 \neq c_0 \neq a_0$ $\alpha = \beta = \gamma = 90°$	Any combination of three mutually perpendicular rotation axes or mirror
Hexagonal	$a_0 = b_0 \neq c_0$ $\alpha = \beta = 90°, \gamma = 120°$	6-fold rotation axis or 6-fold rotary inversion
Rhombohedral	$a_0 = b_0 = c_0$ $\alpha = \beta = \gamma \neq 90°$	3-fold rotation axis or 6-fold rotary inversion
Tetragonal	$a_0 = b_0 \neq c_0$ $\alpha = \beta = \gamma = 90°$	4-fold rotation axis or 4-fold rotary inversion
Cubic	$a_0 = b_0 = c_0$ $\alpha = \beta = \gamma = 90°$	4 3-fold rotation axes at an angle of 109° 28' to each other

this vertex, and only 1/8 of the point is assigned to any unit cell, the latter has a total of only one lattice point. In this case it is called *primitive*. There are 7 types of primitive unit cell corresponding to the 7 C. systems (P-lattices). However, for reasons of symmetry, one must also consider translation lattices in which the unit cells also have lattice points which are not at vertices. These can be in the center (body-centered) or in the centers of the faces (face-centered, or A-, B- or C-lattices) of the unit cell. These are called *centered* lattices. In addition to the 7 primitive lattices, there are 7 centered lattices, so that there are in all 14 types of translation lattice for C., which are called *Bravais lattices* (after A. Bravais, 1848) (Fig. 2, Table 2).

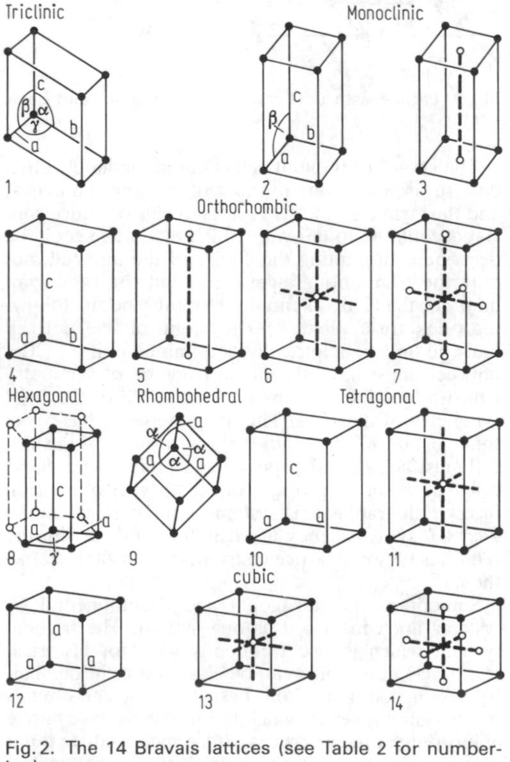

Fig. 2. The 14 Bravais lattices (see Table 2 for numbering)

All C. structures are based on the 14 Bravais lattices. In substances which consist of only one kind of atom the unit cell can be directly represented by one of the types shown in Fig. 2. For example, metallic copper crystallizes in a cubic face-centered lattice, and tungsten in a cubic body-centered lattice. The structures of compounds can be described by interpenetrating Bravais lattices (see Sodium chloride type, Cesium chloride type).

In addition to translation, other crystallographic symmetry operations are needed to describe the structures of C. These are listed, together with the corresponding symmetry elements and the internationally accepted symbols introduced by Hermann-Mauguin, in Table 3. The limitation of crystallographic rotation axes to 2-, 3-, 4- and 6-fold necessarily follows from the lattice structure of C. C. structures may contain no symmetry elements, or several. Combination of certain symmetry elements often yields others, for example, the combination of a 2-fold rotation axis with a mirror plane perpendicular to it automatically creates an inversion center; these three symmetry elements are thus coupled.

The possible combinations for the elements of C. structure symmetry are called *space groups*. Systematic studies have revealed the existence of 230 different space groups (E.S. Fedorov, 1885; A. Schoenflies, 1889). Of these, 2 are triclinic, 13 are monoclinic, 59 are orthorhombic, 45 are hexagonal, 7 are rhombohedral, 68 are tetragonal and 36 are cubic. Any C. structure can be assigned to one of the 230 space groups on the basis of its symmetry, and knowledge of the space groups is necessary for elucidation of the structure of a C. by C. structural analysis.

An obvious, though not essential characteristic of many C. is well formed limiting surfaces. *C. morphology* is the study of the shapes of C., their regularity and formation. There is a close correspondence between the form and structure of a C., which also applies to its symmetry. The symmetry of the C. form (macrosymmetry) is equivalent to the structural (mi-

Table 2. The 14 Bravais lattices

Crystal system	No. in Fig. 2	Type	Symbol	No. of lattice points per unit cell
Triclinic	1	Primitive	P	1
Monoclinic	2	Primitive	P	1
	3	Side-centered	C	2
Orthorhombic	4	Primitive	P	1
	5	End-centered	C (or A, B)	2
	6	Body-centered	I	2
	7	Face-centered	F	4
Hexagonal	8	Primitive	P	1
Rhombohedral	9	Primitive	P	1
Tetragonal	10	Primitive	P	1
	11	Body-centered	I	2
Cubic	12	Primitive	P	1
	13	Body-centered	I	2
	14	Face-centered	F	4

Table 3. Crystallographic symmetry operations

	Symmetry operation	Symmetry element	Hermann-Mauguin symbol
Point symmetry operations	2-fold rotation	2-fold rotation axis	2
	3-fold rotation	3-fold rotation axis	3
	4-fold rotation	4-fold rotation axis	4
	6-fold rotation	6-fold rotation axis	6
	Inversion (reflection through a point)	Inversion center	1
	Reflection	Mirror plane	*m*
	3-fold rotary inversion (6-fold rotary reflection)	3-fold rotary inversion axis (6-fold rotary reflection axis)	$\bar{3}$
	4-fold rotary inversion (4-fold rotary reflection)	4-fold rotary inversion axis (4-fold rotary reflection axis)	$\bar{4}$
	6-fold rotary inversion (3-fold rotary reflection)	6-fold rotary inversion axis (3-fold rotary reflection axis)	$\bar{6}$
	Translation		
	2-fold helix	2-fold helical axis	2_1
	3-fold helix	3-fold helical axis	$3_1, 3_2$
	4-fold helix	4-fold helical axis	$4_1, 4_2, 4_3$
	6-fold helix	6-fold helical axis	$6_1, 6_3, 6_5, 6_2, 6_4$
	Translational reflection	Translational reflection plane with translation component $a/2$	*a*
		$b/2$	*b*
		$c/2$	*c*
		$(a + b)/2$ or $(b + c)/2$ or $(c + a)/2$ or $(a + b)/2$	*n*
		$(a + b)/4$ or $(b + c)/4$ or $(c + a)/4$ or $(a + b + c)/4$	*d*

cro-) symmetry. There is an important difference between the two types of symmetry, however. Because of the small size of translations in the crystal lattice (about 100 pm), translation is a specific structural symmetry operation. Like the symmetry elements which include translation components (twist axes and translational mirror planes), translation is not expressed in the morphology of the C. Systematic examination of the combinatorial possibilities of the remaining symmetry elements revealed 32 crystallographic point groups or *C. classes* (J.F.C. Hessel, 1830). A C. is always placed in the most symmetric class of which it has the symmetry elements.

A plane in the C. lattice which passes through at least three non-colinear lattice points, is called a *lattice plane*. The shortest distance between two adjacent lattice planes is the lattice plane distance. The correspondence between C. form and structure is expressed by the fact that the external C. surfaces are parallel to the lattice planes in the C. The spatial position of a C. surface, or also of a lattice plane, is characterized as follows (Fig. 3):

A lattice point is chosen as the origin, and the basis vectors **a**, **b** and **c** are used as axes of a coordinate system. The inverses of the intercepts of the lattice plane with the axes (in units of the lattice constants a_0, b_0 and c_0) are taken, and these fractions are multiplied by their lowest common denominator. The resulting integers h,k,and l are called the *Miller indices* of the lattice plane or surface, and the index triplet enclosed in parentheses (hkl) is the *surface symbol*. If the lattice plane is parallel to an axis, the corresponding index is equal to zero (e.g. 100). If the intercept is on the negative side of an axis, the minus sign is set above the corresponding index (e.g. 2í2). The morphologically important surfaces, i.e. those which occur frequently in a given type of C. and display well formed surfaces, usually have small Miller indices. These correspond to lattice planes which pass through relatively large numbers of lattice points.

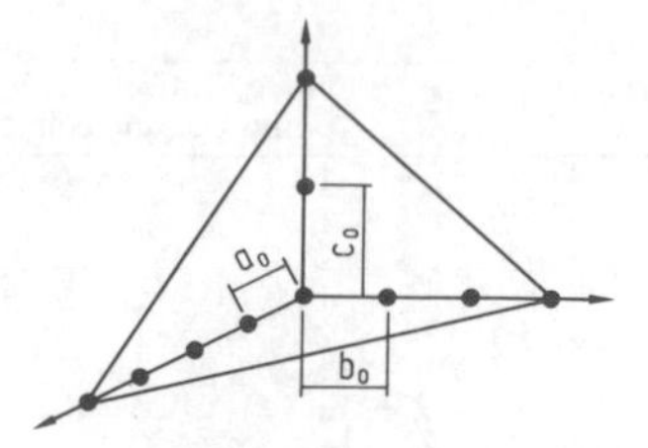

Fig. 3. Position of a surface (lattice plane) with the Miller indices (346)

The symmetrically equivalent surfaces of a C. are summarized in a complex called the *C. form*. For example, a 4-fold rotary axis gives rise to a form with 4 surfaces (tetragonal pyramid). The number of C. surfaces is called the tenacity of the C. form. The form of a C. is indicated by the symbol {hkl}, where the indices hkl apply to one of the surfaces of the habit (e.g. the 6 surfaces of the cube are indicated by the symbol {100}). There are 48 different C. forms. The *C. Tracht* is the totality of the forms of the C., independent of their relative importance, while the C. habit is the most frequently occurring form. For example, if a cubic C. has the forms {100} (cube) and {111} (octahedron), and the cube form is more pronounced, it has a cubic habit; if the octahedron predominates, it has an octahedral habit.

For general characterization of the habit, expressions such as "tabular", "prismatic", and "isometric" are used. The Tracht and habit depend on the internal structure of the C., but also to a great degree on the growth conditions.

Historical. The content and definition of the concept of C. have changed in the course of time. The ancient Greeks used the word C. for ice. The concept was then transferred to the mineral rock crystal, which was thought to be an extremely supercooled form of ice which was no longer capable of melting. Later the term was used for all natural substances which display regular, plane surfaces. Homogeneity and anisotropy were found to be essential attributes of a C., which gave rise to the classic definition of a C. as a solid, homogeneous and anisotropic body with naturally plane surfaces. However, by the latest at the time that X-ray diffraction by C. was discovered by von Laue, Friedrich and Knipping in 1912, it was proven that the essential aspect of a C. is its regular internal structure, which leads to the characteristics on which the original definition was based. The characterization by a certain external form is due to the lattice structure of the C., but it is non-essential, and in many cases such characterization is difficult or impossible. X-ray studies have shown that nearly all solids consist of C., and that the crystalline form is normal for the solid state.

Crystal chemistry, ***chemical crystallography***: the study of the relationships between the chemical composition, structure and properties of crystalline substances. C. is a subfield within modern crystallography, and is closely related to chemistry and physics. The basis of modern C. is knowledge of the geometric arrangements of atoms in crystals, and the forces between them. Crystal structure analysis can now provide very precise experimental information on the structure of crystalline materials, and chemical bond theory provides substantiated information on the *lattice forces* which hold the crystal together.

Crystals may be composed of atoms, ions or molecules. The four different types of lattice forces, covalent, ionic and metallic bonding, and intermolecular interactions (van der Waals forces) determine the properties of the crystals and therefore provide a suitable basis for their classification as ionic, metallic, covalent and molecular crystals (see Lattice type). However, the distinctions between these types are not always sharp, because the above lattice forces correspond to the limiting types of chemical bonding, and the real forces are almost always due to combinations of these idealized types. Thus the bonding in most crystals is complicated. For example, in the alkali halide series, as the atomic number of the alkali metal decreases and that of the halogen increases, the ionic nature of the bond decreases in favor of a more covalent nature. On the other hand, the covalent Si-O bonds in silicates and the P-O bonds in phosphates have considerable ionic character, while in elemental bismuth, the bond type is intermediate between metallic and covalent.

If all the bonds in a crystal are of the same type, then according to R.C. Evans, the structure is called *homodesmic*; if there are different types of bonds, it is *heterodesmic*. Some examples of homodesmic structures are diamond, sodium chloride, copper and crystallized noble gases. The most common form of heterodesmic structures are molecular crystals, which have intramolecular covalent bonding and intermolecular van der Waals forces, e.g. naphthalene, sulfur, oxides of phosphorus such as P_4O_6 and P_4O_{10}. The salts of inorganic oxygen acids are heterodesmic; for example, sodium sulfate, Na_2SO_4, has both polar covalent bonds between the S and and O atoms and ionic bonding between the Na^+ and SO_4^{2-} ions. The physical properties of heterodesmic substances are determined mainly by the weakest type of bonds in the structure. Some examples are the low melting temperatures of molecular substances caused by the weak intermolecular van der Waals forces, the use of graphite as a lubricant, and the ionic character of substances made up of organic ions such as tetraphenylammonium.

Modern C. has provided structural chemical explanations for many classical concepts, such as polymorphism, isomorphism, diodochism and isotypism, which were originally based purely on morphology. This area, which was originally a purely descriptive science, has grown through the discovery of general principles into a discipline which is able to systematize and correlate the structures and physical-chemical properties of substances and to apply this knowledge in practical areas.

In many cases, it is appropriate to regard the atoms and ions in a crystal structure as rigid spheres, and to assign their radii on the basis of the experimentally determined particle distances. This is the basis for use of close packing models (see Spherical packing) for interpretation of the structures of most metals. Similarly, the structures of many ionic crystals can be visualized as a packing of atomic ions. The resulting coordination numbers and geometries are determined here by the stoichiometric composition of the compound and the relative sizes of the ions. For ionic substances with polyatomic ions, such as CO_3^{2-}, SO_4^{2-}, BF_4^- or SiF_6^{2-}, it has been shown that these ions are complex units with the same structures in different compounds. The silicates are a thoroughly studied example for the linking of coordination polyhedra into rings, chains and layers, and finally into spatially linked three-dimensional structures. The C. of complex compounds, with its data on coordination geometries and symmetries in transition elements, has contributed greatly to the development of ligand field theory.

At present, organic C. is still developing. The crystalline structure of organic compounds is determined largely by the structures of the individual molecules. However, all types of molecules, whether nearly spherical or long and flat, prefer a crystalline arrangement which permits the greatest possible number of interatomic interactions and the highest possible Packing density (see). The crystal structures of both inorganic (e.g. hydrated salts) and organic compounds (especially when certain functional groups, such as -OH and -NH are present) are greatly affected by the formation of hydrogen bonds; these are therefore an especially interesting topic of study in C. The study of inclusion compounds is a separate area of C. The explanation of their structural principles is not only of theoretical interest, but is also of direct practical importance, because zeolites and similar materials with hollows and channels of defined sizes are widely used in catalysis and for separation of materials (see Molecular sieves, see Ion exchangers). Finally, it should also be mentioned that there is a close relationship between C. and solid state chemistry, because a deeper understanding of Solid state reactions (see) requires a knowledge of the structure of the solid reactants and products.

Historical. The first steps in the development of a C. were taken quite early. These older studies first focused on the relationship between chemical composition and external form of the crystals. As early as 1669, N. Steno formulated the law of constant angles which (in its modern version) says that the surface angles of a crystal are constants intrinsic to the substance. From this, the notion developed that there is a regular relationship between the chemical composition of a substance and its crystal shape, and thus a comparison of the morphology of the crystals of two substances would permit determination of their identity or non-identity (J.R. Haüy, around 1800). Apparent exceptions to this basic principle of classical C. arose with the discovery of isomorphism (E. Mitscherlich, 1819) and polymorphism (E. Mitscherlich, 1822), but these phenomena later led to a deeper understanding of the principle. A closely related phenomenon is mixed crystal formation (see Mixed crystals), first studied in depth with iron and zinc sulfates (F.S. Beudant, 1818); a special case of polymorphism is Polytypism (see) (H. Baumbauer, 1911). The discovery that molecular optical isomerism of the two optically active forms of tartaric acid is related to the formation of mirror-image crystals (L. Pasteur, 1860) provided concrete evidence for the correlation between chemical structure and morphology. A decisive chapter in the history of C. followed the discovery of x-ray diffraction in crystals (M. von Laue, W. Friedrich, P. Knipping, 1912) and its subsequent application to crystal structure analysis. The structural results which have accumulated since then are the actual basis of modern C., and insure its continued progress. Great contributions in this development were made by W.H. and W.L. Bragg, V.M. Goldschmidt, J.D. Bernal, L. Pauling, N.W. Below, M.A. Porai-Koschits, J.M. Robertson, D.C. Hodgkin and A.N. Kitai-Gorodski.

Crystal field theory: see Ligand field theory.

Crystal growing: the production of high-quality single crystals (see Crystal) for experimental, industrial and other purposes. The methods of C. depend on the properties of the substance and the requirements for the finished crystal. The methods of C. can be classified according to the number of components used. Single-component methods involve crystallization from a melt, or through sublimation, while multicomponent methods include crystallization from solution or as a result of a chemical reaction. In some methods, one starts with a small crystal, called the *seed crystal*, which is enlarged by controlled growth (see Crystallization). In other methods, only a single seed crystal or very few are allowed to form or grow. The properties of the crystals can be affected by addition of foreign substances to the starting material (see Doping).

The numerous and in some cases highly specialized processes of C. can be classified according to the physical state of the starting material.

1) C. from melts. This is the most common method, both in the laboratory and on an industrial scale, for obtaining large single crystals. The methods described below are based on variants of "controlled freezing" of a melt. The crystals are either drawn out of a melt which is in a crucible, or they can be grown outside the crucible. Because the necessary temperatures for crucible methods are usually high, there is a risk of contamination of the melt with crucible materials such as quartz, graphite, platinum or special ceramics. The main problem with C. from melts is the maintenance of suitable temperatures.

Verneuil or *flame melt process*. This non-crucible method was developed in 1892 by A. Verneuil, and has since found wide practical application. The finely crystalline starting material is melted in an oxyhydrogen flame or an electric arc into tiny droplets, which drop onto a chamotte rod. As this is gradually lowered, a pear-shaped single crystal of considerable mass (50 to 100 g in 3 to 5 hours) forms. The method is used for corundums (pure Al_2O_3 with small amounts of Cr_2O_3 for rubies, or TiO_2 and Fe_2O_3 for sapphires), spinels, titanates and tungstates, which are used mainly for bearings in watches and other precision instruments, for abrasives, sound pickup crystals and jewelry. The single crystals made by this process contain many lattice defects and impurities, and cannot be used in modern electronics or optics.

b) *Bridgman-Stockbarger process*. Here the crucible with the polycrystalline starting material is slowly moved vertically through a furnace in which there is a temperature gradient. The material is first melted, then slowly cooled according to a precise program; as it moves into the cooler regions of the furnace it gradually forms a single crystal. At pass-through rates of 1 to 100 mm h^{-1}, high-quality single crystals of most metals and many organic and inorganic substances can be obtained.

c) *Nacken-Kyropoulos process*. The crucible containing the melt is immobile in a furnace. A seed crystal on a cooled support is dipped into the melt; under suitable conditions, crystal growth occurs on the seed; it can be continued by increasing the cooling or by reducing the temperature of the melt until a single crystal of several kilograms mass is formed. This method is used, for example, to make single crystals of alkali halides or calcium fluoride for prisms for IR and UV spectroscopy.

d) *Czochralski process*. This is a further development of the previous method. The growing crystal is pulled out of the melt at a rate of 1 to 100 cm h^{-1}; at the same time, it is rotated to equalize an incompletely symmetrical temperature distribution. With this method, which also permits C. in an inert gas atmosphere, cylindrical single crystals of high perfection and purity are made; their diameters can be up to 100 mm, and lengths up to 600 mm; they are used in the semiconductor industry (silicon, germanium, gallium phosphide, gallium arsenide, etc.) and in lasers (ruby crystals).

e) *Zone melt process*. This is now used mainly in its modern, non-crucible form as a *suspended zone process*. A cylindrical, polycrystalline rod supported only at its ends is held in a vertical position and passed through an induction coil. The coil causes melting in a region a few millimeters long; it is held together by surface tension and is slowly moved through the rod. In the course of several passes, the rod can first be purified in its middle region, then converted to a single crystal. C. by this process is carried out automatically in highly sophisticated apparatus, under high vacuum or in a protective gas atmosphere. It is used to produce silicon and germanium single crystals on a technical scale.

2) C. from solution is used widely in laboratories to produce single crystals for structure analysis and other physical studies, and it is also widely used in industry. Most commonly, the crystal is grown from a supercooled (supersaturated) solution (see Crystallization). If the volume of solution is large compared to that of the crystal, the degree of supersaturation can be kept constant by a defined program of cooling. The same effect can be achieved with a circulation system between a saturation vessel, where the substance is continuously dissolved, and the somewhat cooler crystallization vessel, where it is incorporated into the growing crystal. By these methods, crystals of considerable size and excellent optical properties can be obtained from aqueous solutions within a period of several months. They are used for alums, Seignette salt, ammonium dihydrogenphosphate, etc. In vaporization crystallization, a stream of dry gas is passed over the surface of the solution. A special method of C. from aqueous solution under extreme conditions is Hydrothermal synthesis (see). In *melt solution processes*, molten metals or salts are used as solvents in which crystals of metals, metal oxides, titanates, tungstates, etc. are grown.

3) C. from the gas phase. This type of C. is used mainly for producing monocrystalline thin layers by vapor deposition (see Epitaxis), and also when other methods fail or are too laborious, and small crystals are adequate. The crystal material or its components are sublimed and redeposited; under favorable conditions single crystals with well-formed growth surfaces are obtained in sizes up to several centimeters. The use of chemical transport reactions is important for substances which do not sublime, or do so only at very high temperatures. Chemical reactions between gaseous components can also be used for C. For example, a precisely measured mixture of gallium(III) chloride and hydrogen arsenide vapors react in the presence of hydrogen according to the equation: $GaCl_3 + AsH_3 \rightarrow GaAs + 3\ HCl$. At a temperature of about 1000 K, gallium arsenide, GaAs, formed in this reaction precipitates as single crystals. Single crystals of Fe_3O_4, WO_3, GaAs, GaP, ZnS, CdS, etc. can be obtained by C. from the gas phase.

4) C. from solid phases includes increasing of crystal size by recrystallization of a polycrystalline material, which is also called *collective crystallization*. It is used for growing metal crystals which do not contain displacements. Diamond synthesis represents C. under extreme conditions; at high pressures ($\geqq 500$ MPa) and temperatures ($\geqq 1520$ K), graphite is converted into diamond.

Applications. Crystals grown by these methods are used as semiconductor components in microelectronics (this is the area with the highest economic value and largest production), in optics, electroacoustics, high-frequency applications and as hard substances. Solid-state physical studies are increasingly carried out on monocrystalline samples with defined properties. Most jewelry is now made from artificial gems.

Crystal habit: see Crystal.

Crystal lattice defect: Irregularity in a real crystal where its structure deviates from the ideal (see Crystal). C. do not include the thermal oscillations of the lattice components or the termination of the periodic arrangement at the crystal surface.

C. are usually classified according to their size. A distinction is made between macroscopic and microscopic C.

1) Macroscopic C. have linear extent greater than 10^{-5} cm, and they can be seen with the naked eye or the light microscope. Macroscopic C. include gross defects such as scratches, cracks, internal hollow areas and inclusions.

2) Microscopic C. (more correctly, submicroscopic C.) have linear extent less than 10^{-5} cm, and may be no larger than atomic dimensions. Microscopic C. are further subdivided into structural and chemical C.

a) ***Structural C.*** are usually classified according to their geometry, that is, the number of dimensions in which their extent is greater than atomic dimensions.

There are two types of ***zero-dimensional C.*** (*point defects*), *vacancies* and *interlattice occupation*. Vacancies are sites in the regular lattice which should be occupied but are not, while interlattice occupation occurs when a component of the crystal is in the "wrong" place between the normal lattice sites. Both types of defect cause an elastic deformation of the lattice in their immediate neighborhood. *Frenkel defects* arise when components leave their normal positions and occupy interlattice sites. Frenkel defects thus consist of an equal number of vacancies and interlattice occupations, and in ionic crystals, they are usually possible only for the cations, which are normally smaller than the anions (Fig. 1a). *Schottky defects* occur when components migrate from the interior of the crystal to its surface; they therefore consist only of vacancies. In ionic crystals, these occur in pairs in the cation and anion parts of the lattice (Fig. 1b). *Anti-Frenkel defects*, in which anions are in interlattice positions and there are vacancies in anion sites, and *anti-Schottky defects*, in which cations and anions are in interlattice sites, are possible, but less common than the first two types.

Point defects are distributed completely randomly in the lattice, and are in thermal equilibrium. Their equilibrium concentrations at low temperatures are very low, but increase rapidly with increasing temperatures. In 1 cm^3 of a crystal with about 10^{24} lattice sites, about 10^{10} defects are in equilibrium at room temperature, about 10^{16} at 575 K, and about 10^{19} at 1275 K.

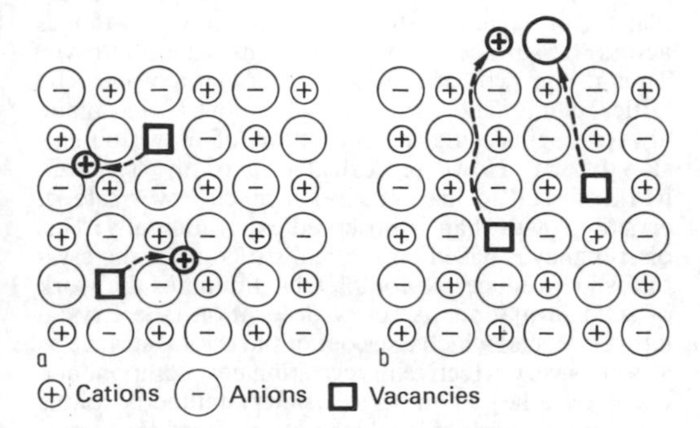

Fig. 1. Frenkel (a) and Schottky (b) defects in an ionic crystal

One-dimensional C. (***linear defects***) extend along open or closed lines in the crystal. They include *dislocations*, which occur through non-equilibrium processes (e.g. through mechanical stress). In an *edge dislocation*, two parts of the crystal are displaced by a distance corresponding to one lattice translation (insertion of a lattice plane; in the symbol ⊥, the vertical line indicates the inserted lattice plane, and the horizontal line, the direction of dislocation). This defect can migrate through the crystal under a dislocation pressure and appear at the surface as a step (Fig. 2).

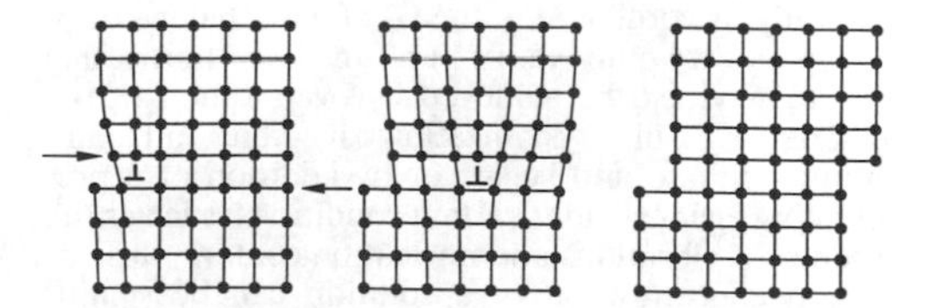

Fig. 2. Formation and migration of an edge dislocation

In a *screw dislocation*, one part of the lattice is rotated with respect to the adjacent part, and as a result, the lattice planes form a type of spiral around the axis of the defect (Fig. 3).

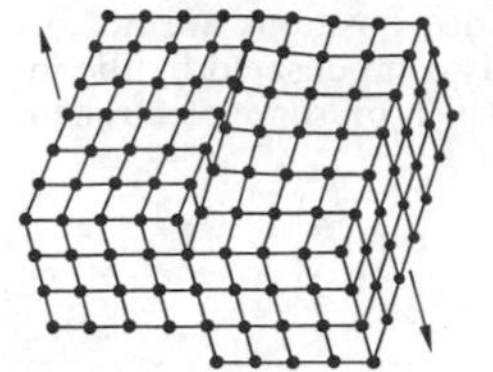

Fig. 3. Screw dislocation

The most important ***two-dimensional C.*** (*surface defects*) are stacking defects, grain boundaries and twin planes. *Stacking defects* are defects in the normal layering sequence in the crystal structure (see Polytypism). A *grain boundary* is the boundary between two regions of the crystal which have different orientations. Nearly all crystals have a *mosaic texture*, i.e. they consist of a large number of blocks which are aligned at slight angles (a few seconds of a degree) to one another; their linear dimensions are about 10^{-5} cm. The area of transition from one block to the next is the *small-angle grain boundary*, which is a series of dislocations (Fig. 4). By contrast, *large-angle grain boundaries* within a polycrystalline structure (see Polycrystalline material) do not have a dislocation structure, due to the random orientations of the crystallites with respect to each other. *Twin planes* occur where two regular crystals are fused. They can arise either through crystal growth (growth twins) or mechanical stress (deformation twins).

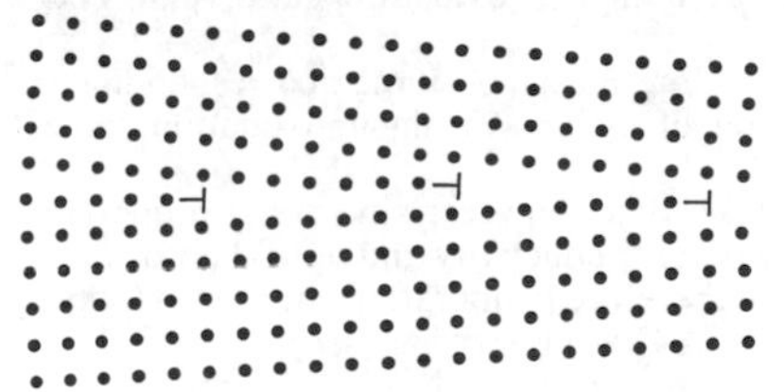

Fig. 4. Small-angle grain boundary consisting of edge dislocation

Structural C. have a decisive effect on those crystal properties which are sensitive to defects (see Crystal), and affect such phenomena as crystal growth, catalytic activity, diffusion, the course of solid-state reactions, etc.

b) ***Chemical C.*** are based on the presence of foreign substances in the crystal. Since the amount of impurities can scarcely be reduced below 10^{-8} mass %, even with very great effort, chemical C. are always present. In semiconductor crystals, foreign substances are intentionally added (see Doping). The foreign components occupy normal lattice positions, and in this way create *substitution defects*. When large amounts of the foreign component are included, the lattice becomes a Mixed crystal (see). The existence of non-stoichiometry (see Berthollides) in ionic or ionogenic covalent bonds is due in many cases to a chemical defect. For example, in the lattice of iron (II) oxide with the composition $Fe_{0.90...0.95}O$ (*wustite* *phase*), individual Fe^{2+} lattice positions are not occupied. This cation deficit is compensated by the incorporation of Fe^{3+} ions, which preserve electroneutrality (Fig 5).

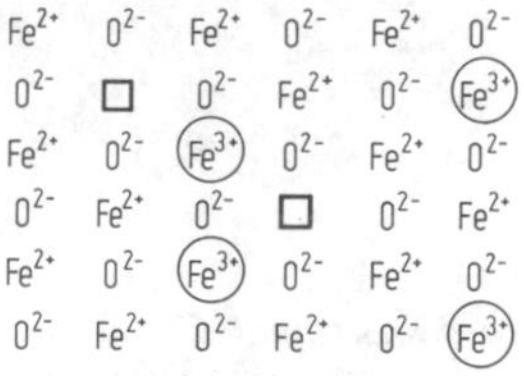

Fe^{2+} O^{2-} Fe^{2+} O^{2-} Fe^{2+} O^{2-}
O^{2-} □ O^{2-} Fe^{2+} O^{2-} Fe^{3+}
Fe^{2+} O^{2-} Fe^{3+} O^{2-} Fe^{2+} O^{2-}
O^{2-} Fe^{2+} O^{2-} □ O^{2-} Fe^{2+}
Fe^{2+} O^{2-} Fe^{3+} O^{2-} Fe^{2+} O^{2-}
O^{2-} Fe^{2+} O^{2-} Fe^{2+} O^{2-} Fe^{3+}

Fig. 5. Chemical defects in iron(II) oxide

Experience has shown that either cations or anions in a crystal can have stoichiometric excesses and deficits. Since ions can occupy interlattice positions and electrons can occupy anion positions under certain conditions, the following possibilities exist for chemical C. in nonstoichiometric compounds (the ideal formulas are given for the example compounds below):

Cation excess:
1.– Electrons in anion positions, e.g. alkali halides with color centers:
1.– Cations and electrons in interlattice positions: e.g. ZnO, CdS, BeO.

Anion excess:
1.– Vacancies in the cation lattice, with incorporation of cations at a higher oxidation state, e.g. FeO, Cu_2O.
1.– Anions in interlattice positions, corresponding to incorporation of cations in higher oxidation states, e.g. UO_2.

Chemical C. determine certain crystal properties, such as electrical conductivity and optical absorption, and they are therefore deliberately introduced by special methods of Crystal growing (see).

Crystallite: see Polycrystal.

Crystallization: the process by which a substance enters the crystalline state from the gaseous, liquid or solid phase. C. can take place in a liquid cooled below its melting point, from a supersaturated solution, as condensation from the vapor phase, as a precipitation from the vapor phase as a result of a Transport reaction (see), during phase changes in a solid (see Polymorphism), as a result of a Solid-state reaction (see) and from an amorphous solid phase.

Regardless of the phase from which C. occurs, it always consists of two stages, formation of a seed crystal and crystal growth. The process of *seed formation* is the formation of a submicroscopic crystal which is capable of growth, the *seed crystal*. It can form, e.g. in a supercooled liquid, or a supersaturated solution or vapor, through the random collision of crystal components. Such an aggregate, with ten to one hundred components, has a very large surface in relation to its volume, and is therefore in a very high-energy state. Since seed formation is energetically unfavorable, the *work of seed formation* must be expended in the process; this corresponds to the activation energy of a chemical reaction. If a critical seed size (10^{-7} to 10^{-5} cm^3) is attained, the seed spontaneously grows into a larger crystal. Homogeneous seed formation occurs in a system which consists of a single phase, that is, where there are no limiting surfaces; when several phases are present, e.g. in the form of foreign particles or solid surfaces, which serve as nucleation centers, the process is called heterogeneous seed formation.

The second process in C. is *crystal growth*, the increase in size from a submicroscopic seed crystal to a macroscopic crystal. The theory of crystal growth, which has developed since about 1920 (M. Volmer, W. Kossel, J.N. Strangski and R. Kaischew) proceeds from the assumption that different amounts of energy are released by addition of a crystal component to different sites on the crystal surface. The energetically most favorable positions are always occupied by incoming components, either immediately or through processes of surface migration by exchange of positions. The most favorable sites are in the *semicrystalline positions* on the edge of an incomplete lattice plane. This edge is a step in the surface which moves across the surface as components are added in rows. Thus the growth of a crystal occurs lattice plane by lattice plane. The formation of a new lattice plane always begins with the formation of a two-dimensional seed. However, calculation of the required work of seed formation gives a much lower rate of crystal growth than is observed experimentally. This discrepancy is based on the real structure of the crystal, which reduces or completely eliminates the work of seed formation. A screw dislocation (see Crystal lattice defects) which rises out of the crystal surface as a step, is very effective in recruiting new components. It creates a large number of semicrystalline positions without any work of seed formation. Since the step is retained during the spiral growth process, it acts as a continuous surface seed (spiral growth theory of C.F. Frank).

C. is one of the most important methods for obtaining pure solids. In the laboratory, it is used both as the last stage of a chemical synthesis, in which the product is isolated and purified, and for quantitative separation and determination of components of solutions (see Gravimetric analysis). In the chemical industry, C. in the form of *mass crystallization* is an important stage in production of fertilizers, soda, sugar, etc. The above examples generally involve C. out of a solution. The necessary supersaturation is often achieved by temperature changes in substances which have large temperature coefficients of solution, generally by cooling, e.g. the C. of naphthalene or p-xylene and the removal of paraffins from lubricating oils. It may also be achieved by evaporation of the solvent, e.g. in crystallization of table salt, ammonium sulfate and sugar, or by addition of a precipitating agent. Under certain conditions, such as the absence of vibrations and nucleation centers, the saturation concentration of a solution can be significantly exceeded without initiation of C. (C. delay). If the solubility limit is rapidly passed, and many seed crystals are present, the product consists of small crystals. It often is of higher purity than larger crystals of the same substance which arise during slow growth and can easily incorporate the mother liquor (*occlusion*). If small crystals stand for a long time in contact with a saturated solution, they will gradually grow, because large crystals are less soluble than small ones and grow at their expense.

In *electrocrystallization*, the crystal growth is associated with electrolysis of a solution or melt. Some important applications are the production of galvanic coatings and production or refining of metals.

There are special methods of C. for production of single crystals of high quality and large size; see Crystal growing.

Crystallization heats: see Phase-change heats.

Crystallizers: 1) ***Crystallization apparatus***: devices for producing crystals. There are cooling, vacuum and evaporation C. in which the supersaturated solution is made, seed crystal formation or seeding occurs, and the crystals are grown. Fractional crystallization on an industrial scale can be done in a crystallization column.

2) ***Mineralizers***: a term for certain ions which are added to natural or synthetic melts to initiate crystal precipitation. Lithium, boron, tungsten, chloride or other ions can be used as C.

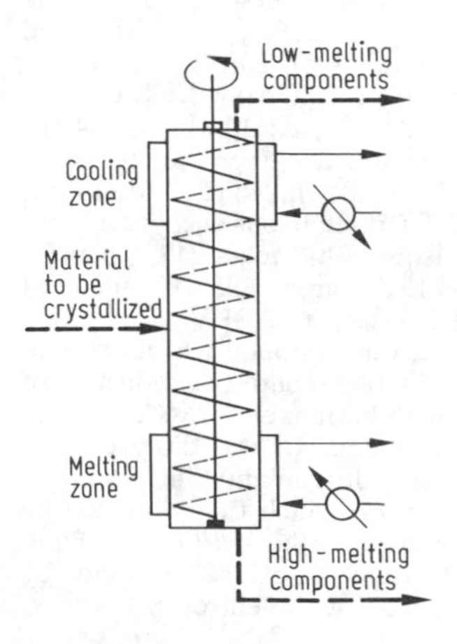

Crystallography: the study of structure and properties of crystalline substances. C. began as an area of mineralogy, and until the beginning of the 20th century, it was limited mainly to the phenomenological description of crystal properites. Modern C. began with the discovery of x-ray diffraction in crystals (M. von Laue, W. Friedrich, P. Knipping, 1912), which made possible the analysis of internal crystal structure at atomic dimensions. C. makes use of the experimental methods of physics, chemistry and physical chemistry to study crystals, and has become integrated into these fields as well.

Crystallose: see Sucrose.

Crystal morphology: see Crystal, see Crystallography.

Crystal structure analysis: see X-ray structure analysis.

Crystal violet: a triphenylmethane pigment; chemically, hexamethylparafuchsin. C. forms bronze-colored crystals. It is made by condensation of Michler's ketone with dimethylaniline. It dissolves in water, giving a violet color; if hydrochloric acid is added continuously, the color lightens from blue and green to yellow. This is due to the fact that the nitrogen atoms are sequentially quaternized, after which they are no longer able to participate in resonance structures and their bathochromic effect is lost. C. is used in making typewriter ribbons.

$(CH_3)_2N-C_6H_4-C(C_6H_4N(CH_3)_2)=C_6H_4=\overset{+}{N}(CH_3)_2\ Cl^-$

Cs: symbol for Cesium.

CS: see Chemical weapons.

CSF process: see Coal extraction.

CTP: see Cytidine.

CT transition: abb. for Charge Transfer transition.

Cu: symbol for copper.

Cubane, ***pentacyclo[4.2.0.0^{2,5}.0^{3,8}.0^{4,7}]octane***: a hydrocarbon which forms a highly strained cubic system; a cage compound. C. forms colorless, shimmering rhombic crystals; m.p. 130-131 °C. It decomposes above 200 °C. It is usually synthesized by photochemical cycloisomerization in which four-membered rings are formed from opposing double bonds. Hydrogenation leads to addition of hydrogen with ring opening, as with cyclobutane derivatives. C. first adds one mole hydrogen to a partially opened cube, and further uptake of hydrogen converts it to tricyclo and bicyclooctane.

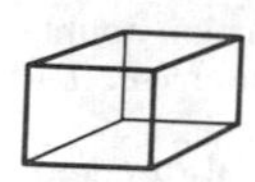

Čugaev's reagent: see Dimethylglyoxime.

Cumaline: see Pyrones.

Cumasina process: see Disinfection.

Cumene, older names, ***cumol, isopropylbenzene***: an aliphatic-aromatic hydrocarbon which is a colorless, combustible liquid; m.p. - 96 °C, b.p. 152.4 °C, n_D^{20} 1.4915. C. is insoluble in water and soluble in ethanol and ether. It is produced on a large industrial scale by reaction of benzene with propene in the presence of phosphoric acid at about 250 °C and about 40 MPa pressure. It is used mainly for production of acetone and phenol by the cumene-phenol process (see Hock process). It is also used as an additive for aviation gasoline to improve its anti-knock quality.

$C_6H_5-CH(CH_3)_2$

Cumene hydroperoxide: a colorless to pale yellow liquid; b.p. 53 °C at 13 to 14 Pa. C. is slightly soluble in water, and more soluble in alcohols and other organic solvents. It is made by the reaction of cumene with atmospheric oxygen at 90-130 °C. The reaction can be run industrially in a heterogeneous system, an aqueous alkaline emulsion with sodium stearate as emulsifying agent, or in a homogeneous system to which some C. has been added. In either process, oxidation proceeds up to a content of 25-30% C. Thereafter, the mixture is concentrated by vacuum distillation to a content of 75-90% C. Acid cleavage by the cumene-phenol process starts with this concentrate. This reaction is the most important application of C., but the compound is also used as a polymerization catalyst or hardener.

CH_3
–C–O–OH
CH_3

Cumene-phenol process: same as Hock process (see).

Cuminal: same as Cuminaldehyde.

Cuminaldehyde, ***cuminal, 4-isopropylbenzaldehyde***: 4-$(CH_3)_2CH$-C_6H_4-CHO, an oily liquid which smells like carroway; b.p. 235.5 °C, n_D^{20} 1.5301. C. is insoluble in water but dissolves readily in ordinary organic solvents. It is a component in the oils of carroway, eucalyptus, Celon cinnamon, rue and templin, and is used as a perfume.

Cumol: same as Cumene (see).

Cumulenes: unsaturated aliphatic hydrocarbons with three or more cumulative double bonds, such as $CH_2=C=C=CH_2$ (butatriene) or $CH_2=C=C=C=CH_2$ (pentatetraene).

C. are homologs of the more important and more thoroughly studied Allenes (see). They are produced by the same synthetic methods as allenes and have comparable properties and reactions.

Cumulative double bonds: carbon-carbon double bonds immediately adjacent to one another (see Dienes, Allenes).

Cupellation: 1) In general, the removal of a substance by a chemical reaction. For example, in the production of bromine, the final liquor is placed in a C. tower and a stream of chlorine passes through it from the bottom. This drives out the bromine.

2) In metallurgy, the isolation of noble metals by oxidation of the less noble companion metals, for example, in the production of silver.

Cuprammonia rayon: synthetic fibers and silks consisting of cellulose produced from a copper oxide-ammonia solution. Cotton linters or cellulose is dissolved in a solution of copper(II) hydroxide in ammonia, forming a cellulose tetraammine copper complex. The purified spinning solution is extruded into hot water as a precipitating bath, and the freshly extruded fibers are rapidly picked up, being stretched in the process to more than 300 times their original length. Some of the copper is dissolved in the precipitation bath as the tetraammine complex, and is hydrolysed. The copper hydroxide remaining in the fibers is dissolved with dilute sulfuric acid, and is removed as copper sulfate.

C. are especially fine and have a soft, silky texture. Their wet strength is low, and their chemical properties are similar to those of viscose fibers. They are used for woven textiles for clothing, linings, underwear and decoration fabrics. It is very important for economical production of the C. that the copper and ammonia be completely recovered. The most important trade names of C. are given under Synthetic fibers (see).

Cuprene: a polymerization product of ethyne (see Polyacetylenes).

Cuprum: see Copper.

Curare alkaloids: quaternized alkaloids isolated as the active components of curare. They are bound to the acetylcholine receptors of striated muscles and cause paralysis. Orally, the C. have little effect. ***Curare*** is an arrow poison used by South American Indians. *Calabash curare* is obtained from the bark of strychnose species and is stored in calabash gourds. *Tube* or *pot curare* is an extract from chondoden species. The active substances in the two types of curare are quite different. Calabash C. are similar in structure to Strychnine (see). The active compounds (C-curarin, C-toxiferin) are dimers and contain two quaternary N atoms. The tube C. on the other hand are bis-benzylisoquinoline alkaloids. The most important of them is tubocurarine chloride, which is used as a Muscle relaxant (see).

Curium, ***Cm***: a radioactive element which can be made by nuclear reactions; it does not occur on earth. Cm belongs to group IIIb of the periodic system, the Actinides (see); it is a heavy metal, *Z* 96 with the following known isotopes (decay type, half-life and nuclear isomers in parentheses): 238 (K-capture, α; 2.4 h); 239 (K; 3 h); 240 (α; 26.8 d); 241 (K, α; 36 d; 2); 242 (α; 163 d; 2); 243 (α; K; 30 a); 2); 244 (α, 18.099 a; 2); 245 (α; $4.8532 \cdot 10^3$ a; 2); 246 (α; $4.82 \cdot 10^3$ a); 247 (α; $1.56 \cdot 10^7$ a); 248 (α; $3.61 \cdot 10^5$ a); 249 (β^-; 64.2 min); 250 (spontaneous decay; $1.13 \cdot 10^4$ a). The most stable isotope has mass 247, valencies III and IV, density 13.51, m.p. 1350 °C, standard electrode potential (Cm/Cm^{3+}) - 2.31 V.

C. is a silvery white, ductile metal which exists in two modifications; it can be prepared by reduction of curium(III) fluoride with barium at 1275 °C. It is less stable in air than plutonium. C. was discovered in 1944 by Seaborg, James, Morgan and Ghiorso in the form of the isotope ^{242}Cm, which they obtained by irradiation of plutonium 239 with α-particles: $^{239}Pu + {}^4He \rightarrow {}^{242}Cm + {}^1n$. Curium 242 can also be made by bombarding americium 241 with neutrons:

$$241_{Am} \xrightarrow[-\gamma]{+n} 242_{Am} \xrightarrow[(16.01\ h)]{-\beta^-} 242_{Cm}.$$

Neutron irradiation of plutonium 239 yields curium 244, which is now available on the 100-g scale:

$$239_{Pu} \xrightarrow{+4n} 243_{Pu} \xrightarrow{-\beta^-} 243_{Am};$$

$$243_{Am} + 1_n \rightarrow 244_{Am} \xrightarrow[(16\ min)]{-\beta^-} 244_{Cm}.$$

C. isotopes are used in atomic batteries (for generation of current in satellites and heart pacemakers) and as a radiation source for materials testing. The element was named for Marie and Pierre Curie.

Curium compounds: compounds in which curium is most commonly in the +3 oxidation state, and less commonly in the +4 state. The unusual stability of curium(III) compounds is due to the fact that in this oxidation state, the 5f orbitals are half filled. However, curium(III) compounds can be converted by very strong oxidizing agents into curium(IV) compounds. Curium is the first actinide element which has chemical properties comparable to that of its homolog in the lanthanide series, gadolinium. Some examples of C.: ***curium(III) hydroxide***, $Cm(OH)_3$, is a yellowish compound which can be obtained by ammonia precipitation from curium(III) salt solutions. Upon heating, it is converted to ***curium(III) oxide***, Cm_2O_3, a colorless compound which can exist in five enantiotropic modifications, m.p. 2270 °C, b.p. 3700 °C. ***Curium(III) fluoride***, CmF_3, is a yellowish white compound, m.p. 1679 °C; it can be obtained by adding hydrofluoric acid to curium(III) solutions. Fluorination of CmF_3 yields ***curium(IV) fluoride***, CmF_4, a greenish-beige compound with coordination number 8. ***Curium(IV) oxide*** is a black compound which can be made by reaction of curium(III) oxide with oxygen at about 300 °C.

Current density: the ratio of current strength to effective electrode surface in an electrolysis.

Current yield: a parameter which characterizes the effectiveness of an electrolysis. $Y = (Q_p/Q) \cdot 100\%$. Q_p is the amount of current which can be calculated using the Faraday law (see) from the amount of electrolysis product, and Q is the amount of current actually used in the electrolysis.

Curtius degradation: a reaction of acyl azides leading to primary amines. When the azides are heated in alcoholic solution, they first form isocyanates by nitrogen cleavage and nucleophilic 1,2-shift of the alkyl or aryl residue. The isocyanate reacts with the alcohol to form an N-substitued urethane which, on hydrolysis, produces primary amines, alcohol and carbon dioxide. The isocyanate intermediate can be isolated if an inert solvent, such as benzene or chloroform, is used.

$$R-C(=O)-\bar{N}^{-}-N^{+}\equiv N \xrightarrow{-N_2} \left[R-\bar{N}^{-}-C^{+}=O \leftrightarrow R-\bar{N}=C=O\right] \xrightarrow{+R^1OH} R-NH-C(=O)-OR^1$$

$$\xrightarrow[-R^1OH,\,-CO_2]{+H_2O} R-NH_2$$

Cusein fiber: a Protein fiber material (see).

Cutin: a plant biopolymer found in cuticles. Waxes are embedded in the C. of land plants. C. is a polyester containing hydroxy- and epoxyfatty acids with 16 and 18 C atoms. The main components are dihydroxypalmitic, o-hydroxyoleic, o-hydroxy-9,10-epoxystearic and trihydroxy stearic acids. ***Suberin*** has a structure similar to that of C.; in addition to hydroxyfatty acids, it contains phenylpropane derivatives. Suberin is a component of the secondary cell wall and the secondary endodermis cells of the underground organs of the plant. C. and suberin are insoluble in ordinary solvents. They can be depolymerized by alkaline hydrolysis, transesterification with methoxide, or treatment with lithium alanate.

Cutinite: see Macerals.

Cutting ceramics: a term for ceramic sintered materials which are very hard and resist high temperatures, and which are relatively resistant to changes in temperature. They are used to cut metallic and nonmetallic materials. C. often consist of oxides, e.g. aluminum oxide, or carbides, nitrides, borides or silicides, and metals. To improve the cutting properties of these components, they are often made into mixed sintered materials (see Cermets).

Cutting metals: see Hard metals.

CVD diamond synthesis: (see p. 300).

CW method: see NMR spectroscopy.

Cyanamide: NH_2CN, colorless, hygroscopic, orthorhombic crystals; M_r 42.043, m.p. 45 °C. C. is obtained by hydrolysis of calcium cyanamide: $CaNCN + H_2O + CO_2 \rightarrow NH_2CN + CaCO_3$. Calcium cyanamide is obtained on an industrial scale by the Frank-Caro process. Above 80 °C, C. dimerizes to Dicyanodiamide (see), $NCN{=}C(NH_2)_2$, which is also colorless and crystalline, M_r 84.08. C. also adds to a series of H-acidic compounds HZ (where Z is, e.g. OH, SH, OR, SR, NR_2) according to the equation

$$NH_2CN + HZ \rightarrow NH_2-C(=NH)-Z \;.$$

C. is soluble in water, acetone, ether, ethanol and dioxane, slightly soluble in benzene and chloroform, and insoluble in hexane.

C. is a weak, dibasic acid (pK_1 10.26, pK_2 14.60); the ***cyanamide anion*** $[NCN]^{2-}$ is isosteric with cyanate, $[NCO]^-$. It is a Pseudochalcogenide (see). Dialkali cyanamides such as ***disodium cyanamide***, Na_2NCN (colorless, water-soluble compound, M_r 86.00, density 1.96, m.p. 550 °C (dec.)), react with nonmetal oxides to form cyanamide acylates. ***Lead cyanamide***, PbNCN, is a lemon-yellow powder, M_r 247.23, used as a pigment in rust-protective paint.

C. are used to produce intermediates for pharmaceuticals or agrochemicals; and are used as herbicides and defoliants.

Cyanates: derivatives of hydrocyanic acid, H-OCN. ***Ionic C.*** are Pseudohalides (see) based on the linear, resonance-stabilized anion $[NCO]^-$. Alkali cyanates such as ***sodium*** or ***potassium cyanate*** are colorless salts obtained by reaction of urea with alkali hydroxides or carbonates at elevated temperature. Protonation of C. produces ***isocyanic acid***, N-NCO, a compound which is stable only at low temperatures; M_r 43.2, m.p. - 86 °C. In aqueous solution, it is a weak acid (K_a $2 \cdot 10^{-4}$). The isocyanate bonding type is also present in heavy-metal and covalent non-metal cyanates, such as ***silicon tetraisocyanate***, $Si(NCO)_4$, a colorless liquid, m.p. 26 °C, b.p. 186 °C; ***phosphorus triisocyanate***, $P(NCO)_3$, a colorless liquid, b.p. 169.3 °C; ***iodoisocyanate***, INCO, m.p. - 50 °C. Bonding isomeric nonmetal cyanates are represented by ***phosphate isocyanates***, $(RO)_2P(O)$-NCO, and ***phosphate cyanates***, $(RO)_2P(O)$-OCN.

The cyanate anion is a good complex ligand, forming coordinate bonds primarily via the nitrogen, so that nearly all cyanate complexes so far reported are of the isocyanate type, M-NCO. In the bridge function, the anion can be coordinatively multidentate. For example, water-insoluble silver(I) cyanate contains cyanate bridges of the type

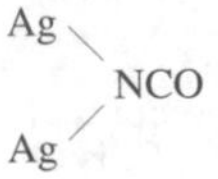

Ag
NCO
Ag

Cyanide: a derivative of hydrogen cyanide, HCN. Ionic C. contain the anion $[CN]^-$, which is a Pseudohalide. ***Alkali cyanides*** are colorless, crystalline compounds, MCN, which are synthesized as the salts of hydrocyanic acid by addition of HCN to aqueous alkali hydroxide or carbonate solutions. The anion is an excellent complex former. In monodentate complexes, the coordinate bonding occurs exclusively via the carbon, M-CN; the resulting ***cyano complexes*** are generally very stable compounds. Because of the high ligand field strength of the anion, most are of the low-spin type. In bridge functions, the anion acts as a bidentate ligand, and the resulting bridge type M-CN-M' is present in many transition metal cyanometallates, for example iron(III) hexacyanoferrate(II) (see Prussian Blue). Nonmetal C., like hydrocyanic acid HCN, have covalent bonds. Examples: ***tricyanophosphate***, $P(CN)_3$, colorless crystals, M_r 109.03; ***dicyanosulfane***, $S(CN)_2$, colorless, rhombic crystals which tend to polymerize, M_r 84.10; ***tetracyanomethane***, $C(CN)_4$, hexagonal, colorless crystals, M_r 116.09, m.p. 160 °C (dec.); there are also ***Cyanogen halides*** (see), XCN. ***Alkyl*** and ***aryl C.***, RCN, are also called Nitriles (see); they are to be distinguished from the very reactive ***alkyl*** and ***aryl isocyanides*** (see Isonitriles).

> Cyanides are extremely poisonous. See Hydrocyanic acid. They must be handled with extreme caution.

C. are used in extraction of gold and silver (see Cyanide leaching) and in electroplating. They are frequently used reagents in the chemical laboratory.

Cyanide leaching: a process of extraction of silver or gold by treating the ores or residues from the amalgam process with 0.1 to 0.25% sodium or potassium cyanide solution. The solution reaction can be described, for example, by the equations $2\,M + 1/2\,O_2 + 4\,NaCN + H_2O \rightarrow 2\,Na[M(CN)_2] + 2\,NaOH$ (M = Ag, Au) and $Ag_2S + 4\,NaCN \rightleftharpoons 2\,Na[Ag(CN)_2] + Na_2S$. The process utilizes the exceptional thermodynamic stability of the cyanometallates(I). The noble metal is precipitated from the cyanide leachate by addition of zinc powder: $2\,Na[M(CN)_2] + Zn \rightarrow Na_2 \cdot [Zn(CN)_4] + 2\,M$.

Cyanidin: 3,5,7,3',4'-pentahydroxyflavylium cation, the aglycon of many Anthocyanins (see); M_r of the chloride, 322.7, red-brown needles, m.p. > 200 °C (dec.). Glycosides of C. and a few acylated derivatives are found widely in plants; their oxonium salts are responsible for the dark red color of many blossoms and fruits, e.g. red roses, geraniums, and tulips. Chelate formation with iron(III) or aluminum ions gives dark blue complexes, which are bound to a polysaccharide carrier and thus in the native state are chromosaccharides. An example is *protocyanin*, the blue pigment of cornflowers.

The structures of more than 20 natural glycosides of C. have been determined. Examples are *chrysanthemin* (3-β-glucoside), chloride m.p. 205 °C (dec.) from the red autumn leaves of some types of maple, blackberries, etc.; *idaeine* (3-β-galactoside), chloride m.p. 210 °C (dec.), from the leaves of the copper beech, apples, etc.; *mecocyanin* (3-gentiobioside) from hibiscus blossoms and morello cherries; *keracyanin* (3-rhamnoglucoside) from the blossoms of snapdragons, tulips, etc., and cherries; and *cyanin* (3,5-di-β-glucoside), chloride m.p. 204 °C, from the blue cornflower, violets, red roses, etc.

Cyanin: see Cyanidin.

Cyanin pigments: a group of cationic Polymethine pigments (see). The C. consist of a conjugated polymethine chain with an odd number of carbon atoms between two nitrogen heterocycles which act as donor or acceptor groups in the π-electron system. The C. are subdivided into ***true C.***, ***streptocyanins*** and ***hemicyanins*** (Fig.).

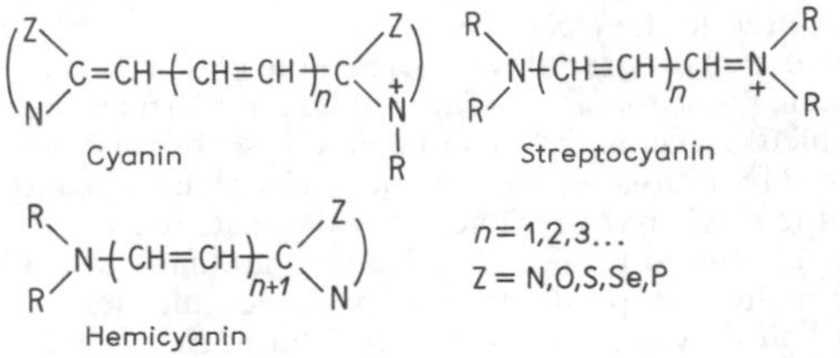

Cyanin

Streptocyanin

Hemicyanin

$n = 1,2,3\ldots$

Z = N,O,S,Se,P

The heterocycles can be, for example, pyrrole, tetrazene, pyridine, thiazene or quinoline. In the ***azacyanins***, one or more methine groups are replaced by azamethine groups -N=. Some of these are very fast dyes, which are used especially to dye polyacrylonitrile fibers. The true C. are not widely used as dyes, because they are not fast to light, but they are used in photochemistry as sensitizers for films and photographic plates. For example, quinoline blue is a sensitizer for the colors red, yellow and orange. Quinoline yellow is used to make colored inks, and astrazone red GB, a hemicyanin, is a dye for polyacrynonitriles.

Cyanizing: same as Carbonitration (see).

Cyanoacetate: same as Ethyl cyanoacetate (see).

Cyanoacetic acid, ***malonic mononitrile***: $NC\text{-}CH_2\text{-}COOH$, colorless, deliquescent crystals; m.p. 70-71 °C, b.p. 108 °C at 20 Pa. When heated to about 165 °C, C. decomposes into acetonitrile and carbon dioxide. It is readily soluble in water, ether and alcohol, but nearly insoluble in benzene and chloroform. It is synthesized by reaction of the alkali salts of chloroacetic acid with potassium cyanide. C. is used in preparative chemistry for such purposes as synthesis of amino acids, malonic acid and malonate esters.

Cyanoaurates: see Gold cyanides.

Cyanocobalamin: see Vitamin B_{12}.

Cyanoferrates: anionic cyanocomplexes of iron. ***Hexacyanoferrates(II)*** are well defined, yellow, diamagnetic compounds which are readily soluble in water. Some important examples are *potassium hexacyanoferrate(II), potassium ferrocyanide*, $K_4[Fe(CN)_6] \cdot 3H_2O$, M_r 422.39; *sodium hexacyanoferrate(II)*, $Na_4[Fe(CN)_6] \cdot 10H_2O$, M_r 484.04 and *ammonium hexacyanoferrate(II)*, $(NH_4)_4[Fe(CN)_6] \cdot 3H_2O$, M_r 338.15. With heavy metal ions, hexacyanoferrates(II) form characteristic colored coordination polymers which precipitate out of aqueous solutions. These include iron(III)hexacyanoferrate(II) (Prussian blue; see) and copper(II)hexacyanoferrate(II), $Cu_2[Fe(CN)_6] \cdot nH_2O$. The latter is used for production of semipermeable membranes in osmotic cells. Protolysis of hexacyanoferrates(II) yields hexacyanoiron(II) acid, $H_4Fe(CN)_6$ in the form of white, rhombic crystals; M_r 215.99, m.p. 190 °C (dec). Industrial production of potassium hexacyanoferrate(II) usually goes via calcium hexacyanoferrate(II) (calcium ferrocyanide); hydrogen cyanide reacts with an iron(II) chloride solution in the presence of calcium hydroxide: $FeCl_2 + 6\,HCN + 3\,Ca(OH)_2 \rightarrow Ca_2[Fe(CN)_6] + CaCl_2 + 6\,H_2O$. Addition of potassium chloride causes precipitation of the mixed salt $K_2Ca[Fe(CN)_6]$, which reacts with potassium carbonate to form potassium hexacyanoferrate(II). This is an important starting material for synthesis of blue pigments and potassium ferricyanide. Sodium hexacyanoferrate(II) is made by direct reaction from iron(II) chloride-hydrogen cyanide (or sodium cyanide) and sodium hydroxide; it is used in vat dyeing and as a flotation agent.

Hexacyanoferrates(III). Alkali hexacyanoferrates(III) are red, poisonous, water-soluble complexes of the low-spin type; the potassium salt is *potassium hexacyanoferrate(III)*, $K_3[Fe(CN)_6]$, M_r 329.24; it is also called *potassium ferricyanide*. Its protolysis produces the yellow-brown hexacyanoiron(III) acid, $H_3Fe(CN)_6$, M_r 214.98, which forms crystalline needles. $K_3[Fe(CN)_6]$ is obtained by anodic oxidation of $K_4[Fe(CN)_6]$; it is used, e.g. as an oxidizing component in development baths for color film development.

Pentacyanoferrates are compounds of the type $[Fe(CN)_5X]^{n-}$ obtained by substitution of one of the CN ligands of hexacyanoferrate. They are also called ***prussiates*** or ***prussides***. The most important of them is *sodium pentacyanonitrosylferrate*, or *sodium nitroprusside*, $Na_2[Fe(CN)_5NO] \cdot 3H_2O$, which forms ruby-red, rhombic crystals, M_r 297.95. This compound is obtained by reaction of potassium hexacyanoferrate(II) with nitric acid, followed by addition of sodium carbonate; it is used, e.g. to detect sulfide and sulfite.

Cyanoform: see Tricyanomethanide.

Cyanogen bromide: see Cyanogen halides.

Cyanogen chloride: see Cyanogen halides.

Cyanogen fluoride: see Cyanogen halides.

Cyanogen halides: colorless, volatile, poisonous compounds with the general formula XCN. C. are stable at room temperature, but in the presence of acids or iron salts, for example, they may be explosively converted to cyanuric halides, $(XCN)_3$. ***Cyanogen fluoride***, FCN, is a colorless gas, M_r 45.02, m.p. - 82 °C, b.p. - 46.2 °C; ***cyanogen chloride***, colorless, readily condensed gas, M_r 61.47, m.p. - 6.5 °C, b.p. +12.5 °C; ***cyanogen bromide***, BrCN, colorless crystals, M_r 105.92, m.p. 52 °C, b.p. 61 °C; ***cyanogen iodide***, ICN, colorless crystals, M_r 152.92, m.p. 146 °C (triple point, p = 0.13 MPa).

The C. are hydrolysed in different ways, due to the differences in electronegativity of the halogens and the resulting differences in the polarity of the X-C bond: $XCN + 2\,NaOH \rightarrow NaOCN + NaX + H_2O$ (X = F, Cl), or $XCN + 2\,NaOH \rightarrow NaOX + NaCN + H_2O$ (X = Br, I). Above 150 °C, cyanogen iodide splits off iodine: $2\,ICN \rightarrow (CN)_2 + I_2$. While cyanogen fluoride is obtained by pyrolysis of cyanuric fluoride, the remaining C. are obtained by reaction of sodium cyanide with the halogens. Cyanogen chloride is synthesized on an industrial scale from hydrochloric acid, chlorine and hydrogen cyanide, and is processed further to cyanuric chloride.

Cyanogen iodide: see Cyanogen halides.

1-Cyanoguanidine: same as Dicyanodiamide (see).

Cyanohydrins, ***α-hydroxynitriles***: general formula $R^1R^2C(OH)\text{-}CN$, addition products of aldehydes or ketones and hydrocyanic acid, e.g. $(CH_3)_2C{=}O + H\text{-}CN \rightleftharpoons (CH_3)_2C(OH)\text{-}CN$ (acetone cyanohydrin). C. are poisonous, colorless liquids or solids with low melting points; they dissolve readily in many organic solvents. Most C. are thermally unstable and are therefore stabilized before distillation with phosphoric, sulfuric or chloroacetic acid. C. are readily decomposed to their starting materials by strong bases. They are synthesized by direct addition of hydrocyanic acid to the aldehyde or aliphatic ketone in the presence of a basic catalyst, or by reaction of bisulfite adducts with sodium cyanide. Purely aromatic ketones scarcely react with hydrocyanic acid.

An entire series of organic syntheses are based on the possibility of making compounds with carbon chains extended by one C atom via C. (***cyanohydrin syntheses***). This includes the hydrolysis of C. to α-hydroxycarboxylic acids, e.g. mandelic acid from benzaldehyde cyanohydrin, citric acid from the C. of 1,3-dichloroacetone, or chain extension of aldoses via the corresponding C.

If C. are synthesized in the presence of ammonia, the products are α-aminonitriles, which can be hydrolysed to α-amino acids. Examples are the synthesis of serine from glycolaldehyde, hydrocyanic acid and ammonia. Acetocyanohydrin (see) is industrially important.

Cyanonickelates: see Nickel cyanides.

Cyanoplatinates(II): square planar anionic complex compounds of platinum with the general formula $M_2[Pt(CN)_4]$. The C. are formed by reaction of platinum(II) salts with cyanides. Some examples: ***potassium tetracyanoplatinate(II)***, $K_2[Pt(CN)_4] \cdot 3H_2O$, pale yellow, rhombic needles, M_r 431.41, density 2.45; ***barium tetracyanoplatinate(II)***, $Ba[Pt(CN)_4] \cdot 4H_2O$, is dimorphic, forming either yellow monoclinic or green rhombic crystals; M_r 508.56. Its fluorescence makes it suitable for detection of X-rays, gamma rays and cathode rays. The crystal lattice of the tetracyanoplatinates(II) is made up of parallel stacking of the complex anions; the Pt-Pt distances within the resulting central atom chain are affected by the counter ions. For the potassium salt, the distance is 348 pm, and for the strontium salt, 309 pm. Partial oxidation of the tetracyanoplatinates(II), e.g. with chlorine or bromine, yields bronze-colored complexes of the type $M_2Pt(CN)_4X_{0.3} \cdot 3H_2O$. These are characterized by an increase in the Pt-Pt interaction and a corresponding decrease in the Pt distance to about 290 pm. These compounds have the properties of one-dimensional conductors.

Cyanosin: see Eosin pigments.

Cyanuric acid, ***2,4,6-trihydroxy-1,3,5-triazine***: the trimerization product of isocyanic acid. C. forms colorless crystals; m.p. 320-330 °C. It is soluble in hot water and alcohol, but only slightly soluble in ether and other organic solvents. C. exists in two tautomeric forms.

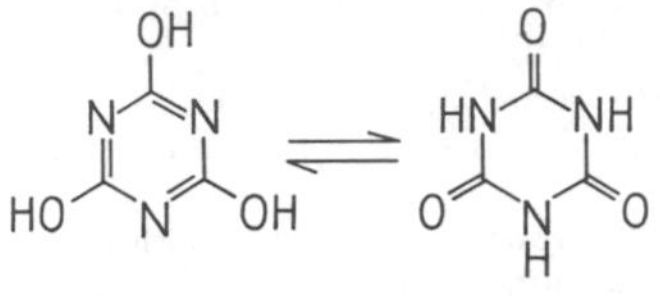

Cyanuric acid   Isocyanuric acid

It is synthesized by hydrolysis of cyanuric chloride or by heating urea in the presence of zinc chloride to 170-220 °C. C. is used in the form of its *N*-chlorinated derivatives as a bleach and disinfectant in industry and households.

Cyanuryl chloride, ***2,4,6-trichloro-1,3,5-triazine***: monoclinic crystals with a pungent odor; M_r 184.41, m.p. 146 °C, b.p. 190 °C. It is soluble in alcohol, acetone, chloroform and carbon tetrachloride. In water, C. is hydrolysed in steps to cyanuric acid. C. can be considered the trimerization product of cyanogen chloride (see Cyanogen halides). As an acidic chloride, it is reactive, and forms esters with alcohols and phenols, and amides with ammonia and amines. In the laboratory synthesis of C., hydrocyanic acid and chlorine are simultaneously bubbled through chloroform. Cyanogen chloride formed according to $HCN + Cl_2 \rightarrow ClCN + HCl$ is trimerized to C. by the hydrogen chloride which forms. Industrially, cyanogen chloride is induced to trimerize by passing it over activated charcoal at 200 to 500 °C. C. is an important intermediate for the synthesis of pigments, optical brighteners, drugs and pesticides (see Triazine herbicides). Its esters with alcohols are used as softeners and as cross-linkers for synthetic resins.

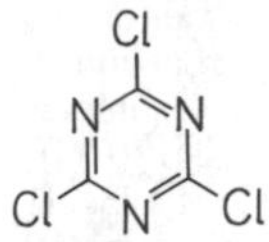

Cyanuryl triamide: same as Melamine (see).

Cyclamate: the sodium salt of cyclohexylsulfaminic acid. C. is used as a sweetener, especially for dietetic purposes, e.g. to sweeten baked products. It is synthesized by reaction of cyclohexylamine with chlorosulfonic acid.

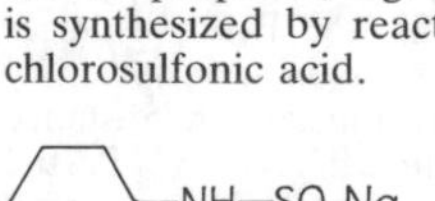

Cyclic: a term indicating a ring-shaped compound; these are extremely numerous among carbon compounds.

Cyclic adenosine monophosphate: see Cyclic nucleotides.

Cyclic guanosine monophosphate: see Cyclic nucleotides.

Cyclic nucleotides: nucleotides in which a single phosphate group is linked to two hydroxyl groups on the sugar moiety. ***Adenosine 3',5'-phosphate (cyclic adenosine monophosphate***, abb. ***cAMP)*** and ***guanosine 3',5'-phosphate (cyclic guanosine monophosphate***, abb. ***cGMP***) act as Second messengers (see) which are generated when a peptide hormone has bound to its receptor in the cell membrane; they act as signal transducers to the interior of the cell (see Hormones). C. are hydrophilic and cannot pass the cell membrane.

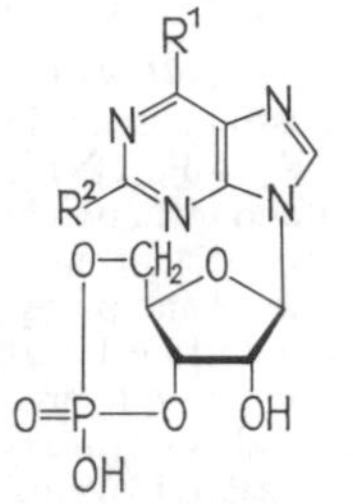

	R^1	R^2
cAMP	NH_2	H
cGMP	OH	NH_2

Cyclic triangular wave voltammetry: an electrochemical analytical technique based on analysis of current-voltage curves. The working electrode is polarized with a triangular voltage. If the voltage function is first negative, reduction can occur. After passing through an intermediate range, the voltage then becomes positive. If the reaction is reversible, the product of the reduction is reoxidized; thus for a reversible reaction, both a reduction and an oxidation peak are observed. If the reaction is completely irreversible, only one peak appears. The position of the peak potential and the magnitude of the peak current as functions of the rate of voltage change give clues as to the mechanism of the electrochemical reaction.

Cyclic process: a series of changes of state of a thermodynamic system which lead through a number of intermediate states S_1, S_2,...back to the initial state S_i. Although the system returns to its initial state, there may be permanent changes in its environment as a result of the C.

If the initial and intermediate states of the system are characterized by a function of state X, such as the internal energy, and the changes in this parameter are denoted by ΔX_{ij}, then according to the definition of a function of state, the sum over the entire cycle is $\sum_\circ \Delta X_{ij} = 0$,

or, for differential changes, $\oint dX = 0$.

In thermodynamics, C. are often used to derive laws; some important examples are the Carnot cycle (see) and the Haber-Born cycle (see).

Cyclitols: cyclohexanes with more than three hydroxyl groups. The hexahydroxycyclohexanes (see Inositols) are of particular significance in biochemistry. An example of a pentahydroxycyclohexane is quercitol, which is found in oak trees, and quinic acid is an example of a tetrahydroxycyclohexane carboxylic acid.

Cyclization: in general, the formation of ring compounds from one (electrocyclic, ring-closure reactions) or more open chained compounds (cycloadditions). The C. of ethyne (acetylene) is important industrially. Under certain conditions and with specific catalysts, ethyne and other alkynes can be cyclized to benzene, styrene, cycloocta-1,3,5,7-tetraene or other 4-, 5-, 6-, 7- and 8-membered rings.

Cyclo: see Nomenclature, section II D 1.

Cycloaddition: see Woodward-Hoffmann rules.

Cycloaddition reactions: synchronous or asynchronous addition reactions which form ring compounds. C. can be initiated thermally or photochemically. The selection rules governing synchronous C. are the Woodward-Hoffmann rules. C. are characterized by numerals in square brackets, which give the number of C atoms in the chains of the reactants forming the ring; for example, the Diels-Alder reaction is a [4+2] cycloaddition. Some important types of C. are *cyclodimerization* (e.g. [2+2] cycloaddition to form a four-membered ring), *cyclopolymerization* (e.g. of ethyne to form benzene or cyclooctatetraene), the Diels-Alder reaction (see) and 1,3-Dipolar cycloaddition (see).

Cycloalkanes, *cycloparaffins*: in the narrow sense, saturated cyclic hydrocarbons with the general formula C_nH_{2n}. Like straight-chain alkanes, C. contain sp^3-hybridized C atoms. They form a homologous series of compounds with different ring sizes, beginning with cyclopropane. Cyclopentane and cyclohexane are significant as the parent compounds of many naturally occurring ring systems, such as terpenes or steroids, and in industrially important products, such as the naphthenes.

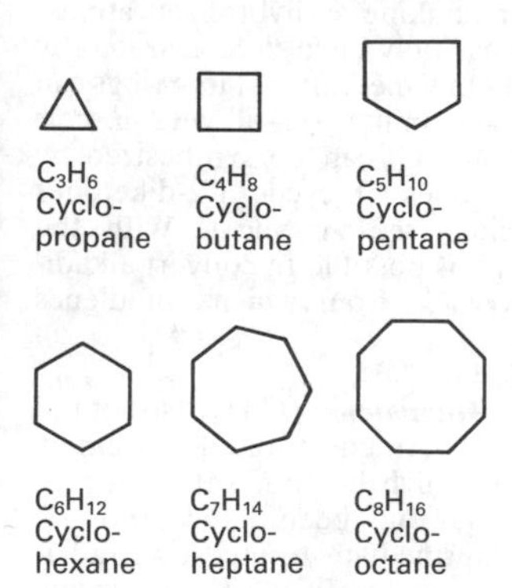

The names are constructed by combining the prefix cyclo- with the name of the n-alkane with the same number of C atoms.

The properties of the C. are similar to those of the n-alkanes, but they differ considerably within the series with regard to their stability; this is a function of ring size. Cyclopropane and cyclobutane are under considerable ring tension (115.5 and 109.6 kJ mol^{-1}, respectively). They do not tend to form easily during chemical reactions, and have a strong tendency to open. This is due to the fact that their bond angles are rather different from the tetrahedral angle. The difference in ring tension between the three- and four-membered rings is called bond-angle or Baeyer tension (see Baeyer's tension theory). With cyclopentane, cyclohexane and cycloheptane, the ring tension energy decreases sharply (C_5H_{10}, 27.2; C_6H_{12}, 0; C_7H_{14}, 25.9 kJ mol^{-1}) because the angles in these rings are close to the tetrahedral angles. The remaining small amount of tension is due to torsional or Pitzer tension. The distinctly higher ring tension energies of about 42 to 53 kJ mol^{-1} in the homologous C_8 to C_{11} compounds are due to bond-angle and torsional tension, and also to transannular interactions. Large rings with more than 12 C atoms are nearly free of tension, and are very similar in their properties to the long-chain n-alkanes. Because of these trends, the C. are classified as *small rings* (3 or 4 C atoms) with low stability and low tendency to form, *normal rings* (5 to 7 C atoms) with relatively high stability and tendency to form, *medium rings* (8 to 11 C atoms), comparable to the small rings, and *large rings* (12 or more C atoms) with high stability and tendency to form.

Chemical reactions of the C. depend on the ring size. Cyclopropane, like the alkenes, can be catalytically hydrogenated under mild conditions. This leads to ring opening and formation of propane:

$$C_3H_6 + H_2 \xrightarrow{(Pt)} CH_3\text{–}CH_2\text{–}CH_3.$$

Cyclobutane reacts similarly at higher temperatures, yielding n-butane. Cyclopropane also reacts with hydrogen bromide and bromine with ring opening, yielding the bromopropane derivatives: C_3H_6 + HBr → CH_3-CH_2-CH_2-Br (1-bromopropane); C_3H_6 + Br_2 → Br-CH_2-CH_2-CH_2-Br (1,3-dibromopropane). Under the same conditions, cyclobutane does not react with these reagents. Rearrangements leading to ring expansion or ring contraction are typical of many C. For example, ethylcyclobutane is converted to methylcyclopentane, or cycloheptane is converted to methylcyclohexane, when heated with aluminum chloride.

Special synthetic methods are required for small and medium rings, such as the Ruggli-Ziegler dilution principle. Intermolecular rearrangements which lead to polymerization are observed to compete with ring closure reactions. The yields therefore vary and depend on the differing tendencies to form rings. The following *cyclization reactions* (equations 1 to 5) are suitable for the synthesis of C. and their derivatives: 1) Intramolecular Wurtz reaction (1) of α,-o-dihaloalkanes for C. of ring sizes n = 3 to 6. 2) Dieckmann condensation (2) of dicarboxylate esters to form cycloketones C_5 to C_7; these can be reduced to C. In the first step of the reaction, the sodium salt of the cycloalkane carboxylate is formed; it is hydrolysed with water and acid. This is followed by a thermal decarboxylation to the ketones with n = 3 to 5. 3) Acyloin reduction (3) of dicarboxylate esters for large and medium rings; the dilution principle is not required. The acyloins can also be reduced as required. 4) Pyrolysis (4) of calcium, barium or thorium salts of dicarboxylic acids to form cycloketones with ring sizes n = 5, 6 or 7, and n = 12 or more. 5) Intramolecular

Thorpe reaction (5) of dinitriles to form cycloketones using the dilution principle for rings with 5 to 8 and 14 to 33 carbon atoms.

In addition to these methods for synthesis, Ring expansion reactions (see) Ring contraction reaction (see) and Cycloaddition reactions (see) of various types are important. Starting from cyclic compounds, the above methods can also be used to synthesize multiple-ring cycloalkanes.

In the broad sense, C. are considered to include ring systems with more than one ring; these are linked to each other in various ways.

$$(CH_2)_{n-2}(CH_2X)_2 + 2\,Na \xrightarrow{-2\,NaX} (CH_2)_n$$

1

$$(CH_2)_n(COOR)(CH_2\text{–}COOR) + NaOR \longrightarrow (CH_2)_n(CO)(CH\text{–}COOR) \longrightarrow (CH_2)_n(CO)(CH\text{–}COOH) \longrightarrow (CH_2)_n(CO)(CH_2)$$

2

$$(CH_2)_n(COOR)_2 \xrightarrow[b)\ +2H_2O/-2NaOH]{a)\ +4Na/-2RONa} (CH_2)_n(CO)(CHOH)$$

3

$$(CH_2)_{n-1}(COO^-)_2\ Ca^{2+} \xrightarrow{-CaCO_3} (CH_2)_{n-1}CO$$

4

$$(CH_2)_{n-2}(CN)_2 \xrightarrow{C_6H_5N(C_2H_5)Li} (CH_2)_{n-2}(C{=}NH)(CH\text{–}CN) \xrightarrow[b)\ T/-CO_2]{a)\ +3H_2O/H^+} (CH_2)_{n-1}CO$$

5

Cyclization reactions

Cycloalkanols: cyclic monoalcohols with the general formula $C_nH_{2n-1}OH$; for example, see Cyclohexanol.

Cycloalkanones: ketones derived from cycloalkanes.

Cycloalkenes, ***cycloolefins***: unsaturated cyclic hydrocarbons containing C=C double bonds. In the broad sense, C. are compounds with one or more double bonds; the largest possible number of double bonds, which are usually conjugated, depends on the size of the ring. For example, see Cyclopentadiene or Cyclooctatetraene. C. in the narrow sense (with one double bond) form a homologous series of compounds of varying ring size, with the general formula C_nH_{2n-2}, beginning with cyclopropene.

The nomenclature of the C. is analogous to that of the alkenes; when more than one double bond is present, the system is analogous to that of the dienes, trienes, etc. (see Dienes, Polyenes), with the prefix "cyclo-" placed in front of the name. The properties of the C. are similar to those of the straight-chain alkenes, dienes or polyenes, although peculiarities can arise with multiple unsaturation. Valence isomerization, e.g. in bullvalene or cyclooctatetraene, and the ease of conversion of cyclopentadiene into the aromatic cyclopentadienyl anion, or of cycloheptatriene into the aromatic tropylium cation, are examples.

Cycloalkene ring systems with one or more double bonds are found in many natural products, including terpenes, steroids, vitamins and some fatty acids (hydnocarpic and chaulmoogric acids). Many C. can be synthesized by the Diels-Alder diene synthesis or by dehydration of cycloalkanols. C. are very important in organic syntheses. Catalytic hydrogenation yields saturated ring systems, and by functionalization of the double bonds, many intermediates can be created which are useful in industrial chemistry, such as polyamides. Oxidative degradation yields dicarboxylic acids, and derivatives of ethylbenzene and *p*-xylene can be made by aromatization reactions.

Cycloalkynes: unsaturated cyclic hydrocarbons containing C≡C triple bonds; the general formula is C_nH_{2n-4}. The smallest C. which can exist at room temperature is *cyclooctyne*. Normal rings containing five, six or seven C atoms and a triple bond are under great tension at their flanking sp-hybridized atoms, and their production has only been demonstrated by subsequent reactions. Only medium or large rings can form stable C. These are similar to n-alkynes in their properties and reactions. C. can be synthesized by oxidation of bishydrazones of cyclic 1,2-diketones with mercury(II) oxide (see Alkynes). With the Glaser reaction (see), it is possible to convert alkadiynes to cycloalkapolyynes, from which Annulenes (see) can be synthesized.

Cyclobarbital: see Barbitals.

Cyclobutadiene, ***[4]annulene***: C_4H_4 (structural formula, see Annulenes). An unsaturated, extremely unstable hydrocarbon, which has not yet been prepared in free form. It has only been detected, by IR spectroscopy, as an intermediate formed by irradiation of pyridine in an argon matrix at 8 K. According to the Hückel rule, C. is antiaromatic (see Annulenes). Derivatives with push-pull substituents, such as NR_2 and COOR groups, are more stable and have been synthesized, as have various metal complexes.

Cyclobutane, ***tetramethylene***: C_4H_8 (structural formula, see Cycloalkanes), a cyclic hydrocarbon and a colorless, combustible gas which is insoluble in water; m.p. - 90°C, b.p. 12°C, n_D^{20} 1.4260. C. is made by reaction of sodium with 1,4-dibromobutane, or from cyclobutanone by the Wolff-Kishner reduction. It is more stable than cyclopropane; for example, the ring is not opened at room temperature by sulfuric acid, bromine or hydrogen bromide. Catalytic hydrogenation to n-butane requires a temperature of 120 °C (see Cycloaklanes). Cyclobutane rings are components of the terpene hydrocarbons α- and β-pinene, truxillic acid (2,4-diphenylcyclobutane-1,3-dicarboxylic acid) and its isomer truxic acid (3,4-diphenylcyclobutane-1,2-dicarboxylic acid). (The last two acids are obtained by hydrolysis of coca secondary alkaloids.)

Cyclodextrins: see Dextrins.

Cyclodiastereoisomerism: see Cyclostereoisomerism.

Cyclodiene insecticides: a subgroup of chlorohydrocarbon insecticides in which the skeleton is constructed on the principle of the Diels-Alder reaction.

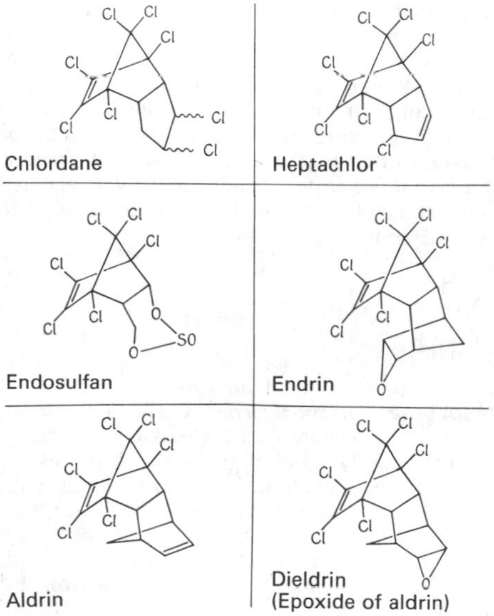

Chlordane | Heptachlor

Endosulfan | Endrin

Aldrin | Dieldrin (Epoxide of aldrin)

With the exception of endrin, an *endo*-isomer of dieldrin, C. are synthesized using hexachlorocyclopentadiene (HCCP) as the diene component. The reaction partners are acyclic (*endosulfan*), monocyclic (*chlordane, heptachlor*) or bicyclic (*aldin, dieldrin*) dienophiles.

The C. are extremely effective contact and ingestion poisons, and some are also respiratory poisons. They act primarily as nerve poisons. With a few exceptions (see Endosulfan), they are highly persistent, toxic to warm-blooded animals (causing liver damage among other things) and have a high tendency to accumulate in fatty tissues. Their once widespread use has been greatly reduced because of the problem of residues, and in many countries they have been banned.

Cyclododecane: $C_{12}H_{24}$, a cycloalkane hydrocarbon; a colorless solid, m.p. 61 °C. It is readily made from technical cyclododeca-1,5,9-triene by catalytic hydrogenation, and is used as a starting material for polyamide syntheses.

Cyclododecanone: $C_{12}H_{22}O$, a cyclic aliphatic ketone. C. forms colorless crystals with a camphor-like smell; m.p. 61 °C, b.p. 277.5 °C. C. is insoluble in water, but is soluble in alcohol, benzene and chloroform. C. can be made by dehydrogenation of cyclododecanol. It is used as an intermediate in the synthesis of perfumes for cosmetics.

Cyclododecatriene, ***cyclododeca-1,5-9-triene***: an unsaturated hydrocarbon. There are several configurational isomers. The most important is ***(Z,E,E)-cyclododeca-1,5,9-triene***, a colorless liquid; m.p. - 18 °C, b.p. 237.5 °C or 104 °C at 1.33 kPa, n_D^{20} 1.5081.

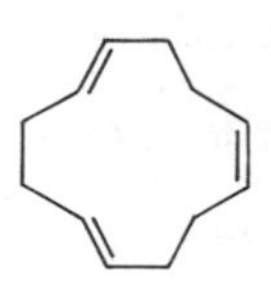

(Z, E, E)-Cyclododeca-1,5,9-triene

This isomer is obtained in good yields as the main product of cyclotrimerization of buta-1,3-diene in the presence of mixed Ziegler catalysts. ***(E,E,E)-Cyclododeca-1,5,9-triene***, a colorless, crystalline compound (m.p. 34-36 °C) can be obtained as a by-product. This industrial synthesis of C. made various medium and large ring compounds available. Selective hydrogenation with suitable catalysts yields cyclododecadiene or (Z)-cyclododecene, and from this, cyclododecanol and cyclododecanone can be obtained. This makes possible the synthesis of lauric lactam and special polyamides. In addition, oxidative ring opening yields long-chain dicarboxylic acids.

Cycloenantiomerism: see Cyclostereoisomerism.

Cycloheptadec-9-en-1-one: same as Civetone.

Cycloheptane, ***heptamethylene***: C_7H_{14} (structural formula, see Cycloalkanes), a cycloalkane hydrocarbon, a colorless, combustible liquid; m.p. - 12 °C, b.p. 118.5 °C, n_D^{20} 1.4436. C. is insoluble in water, but is readily miscible with ethanol and ether. The lowest-energy conformational isomer is a twisted chair form, which is very mobile (see Stereoisomerism). C. is found in petroleum and can be isolated from it. It can be synthesized by the Clemmensen reduction of cycloheptanone (suberone). The cycloheptane ring is a component of a few alkaloids, such as atropine and cocaine.

Cycloheptanone, ***suberone***: $C_7H_{12}O$, a cycloaliphatic ketone. C. is a colorless liquid, b.p. 178-179 °C, n_D^{20} 1.4608. It is insoluble in water, but dissolves readily in alcohol and ether. C. can be synthesized from cyclohexanone and diazomethane. It is used mainly for organic syntheses.

Cycloheptatriene: same as Tropilidene (see).

Cyclohepta-2,4,6-trienone: same as Tropone (see).

Cyclohexane, ***hexamethylene, hexahydrobenzene***: C_6H_{12} (structural formula see Cycloalkanes), a cycloalkane hydrocarbon and a colorless, combustible liquid which smells like petroleum; m.p. 6.5 °C, b.p. 80.7 °C, n_D^{20} 1.4266. C. is insoluble in water, but easily miscible with ethanol and ether. The two conformational isomers of C., the chair and boat forms, are separated by an energy barrier of about 25 kJ mol^{-1}. They can be interconverted when energy is available; the more stable is the chair form (see Stereoisomerism). C. is an important component of certain petroleums, and can be isolated from them. It can also be made industrially by catalytic hydrogenation of benzene on nickel or platinum catalysts around 200 °C. C. is used as a solvent, but its main use is as a starting material for polyamide syntheses, because it can be oxidized to cyclohexanol, cyclohexanone or adipic acid as desired. Cyclohexanone oxime is obtained by using nitrosyl chloride. The cyclohexane ring occurs widely in natural products, such as terpenes, steroids, vitamins and hormones.

Cyclohexanol: a cyclic, saturated monoalcohol. It is an oily, hygroscopic liquid with a camphor-like odor; m.p. 25.2 °C, b.p. 161.5 °C, n_D^{20} 1.4641. C. is partially miscible with water, and readily miscible with organic solvents. It irritates the mucous membranes, and its vapors have a weak narcotic effect. It is obtained by complete hydrogenation of phenol, air oxidation of cyclohexane, or addition of water to cyclohexene in the presence of sulfuric acid. When heated with acids, such as sulfuric or phosphoric acid, C. splits out water and forms cyclohexene. Oxidation with chromic acid produces cyclohexanone; with potassium permanganate, the ring is opened, and adipic acid is formed. C. is a commonly used solvent, e.g. for waxes, resins, caoutchouc and cellulose acetate. The esters can be used as softeners; the acetate is a versatile solvent and is used in perfumery. C. is an intermediate in the industrial production of adipic acid, caprolactam and polyamides.

H OH

Cyclohexanone: a cyclic, aliphatic ketone. C. is a colorless liquid which smells like peppermint; m.p. - 16.4 °C (also reported as - 45 °C), b.p. 155.6 °C, n_D^{20} 1.4507. C. is partially soluble in water, and is readily soluble in most organic solvents. It is synthesized by catalytic dehydrogenation of cyclohexanol, by liquid-phase oxidation of cyclohexane or by catalytic hydrogenation of phenol in the presence of palladium or platinum. C. is used mainly for the synthesis of ε-caprolactam and adipic acid, and also as a solvent for polyvinyl chloride, nitrocellulose and other synthetic resins.

=O

Cyclohexene: an unsaturated cycloalkene hydrocarbon, a colorless liquid; m.p. - 103.5 °C, b.p. 83 °C, n_D^{20} 1.4464. C. can be synthesized by acid or gas-phase dehydration of cyclohexanol. It undergoes the typical addition reactions of the alkenes, for example with bromine or potassium permanganate and water, and is useful for classroom demonstrations. It is also an intermediate in the synthesis of many functional cyclohexane derivatives.

Cycloheximide: an antibiotic; chemically it is a derivative of glutamine. C. was isolated from the culture medium of a strain of *Streptomyces griseus*. It is cytotoxic for eukaryotic cells, but is too toxic for use in cancer chemotherapy. C. is used in biochemical studies to inhibit protein synthesis on the ribosomes.

O OH
H_3C
H
CH_3
O N O
H

Cyclohexylamine, ***aminocyclohexane***: a cycloaliphatic primary amine. C. is a colorless liquid with a fishy smell; m.p. - 17.7 °C, b.p. 134.5 °C, n_D^{20} 1.4592. It is soluble in water and most organic solvents. The chemical behavior of C. is similar to that of the acyclic aliphatic primary amines. It is made industrially by catalytic high-pressure hydrogenation of aniline, or by amination of cyclohexanol. It is used for organic syntheses, to make softeners and emulsifying agents, pesticides and vulcanization accelerators. It is also used as a corrosion protection.

H
NH_2

Cyclooctane, ***octamethylene***: C_8H_{16} (structural formula, see Cycloalkanes), a cycloalkane hydrocarbon, a colorless, oily liquid; m.p. 14.3 °C, b.p. 149 °C, n_D^{20} 1.4586. C. is a medium hydrocarbon ring with relatively high ring tension (see Cycloalkanes); there are several conformational isomers with the same energy levels (see Stereoisomerism). C. is obtained by catalytic hydrogenation of cyclooctatetraene, which is industrially available.

Cyclooctatetraene, ***cycloocta-1,3,5,7-tetraene, [8]annulene***: an unsaturated cyclic hydrocarbon; a gold-colored liquid with m.p. - 4.7 °C, b.p. 140.5 °C, n_D^{20} 1.5381. C. is insoluble in water, but dissolves in organic solvents. It is made by cyclotetramerization of ethyne in tetrahydrofuran in the presence of nickel salts, according to Reppe. C. is very reactive. Suberic acid can be made from it by partial oxidation to cyclooctene and oxidation. Addition reactions usually produce bicyclic compounds; for example, addition of chlorine yields dichlorobicyclo[4,2,0]octadiene, since C. undergoes a valence isomerization to (Z)-bicyclo[4,2,0]octa-2,4,7-triene, and this compound adds chlorine more rapidly than C.

H
H

Reaction with water in the presence of mercury(II) sulfate produces phenylacetaldehyde on heating; reaction with hypochlorous acid leads to terephthalic dialdehyde, which can be further oxidized to the industrially important terephthalic acid. Thus C. has become an important intermediate in organic syntheses.

Cycloolefins: same as Cycloalkenes (see).

Cycloparaffins: same as Cycloalkanes (see).

Cyclopentadiene, ***cyclopenta-1,3-diene***: an unsaturated, cyclic hydrocarbon, and a colorless liquid; m.p. - 85 °C, b.p. 42 °C, n_D^{20} 1.4440. C. is insoluble in water, but dissolves in ethanol, ethene and other organic solvents. On standing, it readily polymerizes to ***dicyclopentadiene*** (m.p. 32.5 °C). C. is present in the light oil fraction of coal tar (anthracite) and is obtained from this source industrially. It is formed in large amounts during thermal cracking of petroleum fractions. C. is used as the diene component in the Diels-Alder reaction, and is used in this way in the synthesis of various insecticides. Fulvenes can be made by condensing C. with aldehydes and ketones, and C. reacts with sodium to form cyclopentadienylsodium, with the aromatic cyclopentadienyl anion. Cyclopentadienyl anions are also present in the metallocenes, such as ferrocene (bis(π-cyclopentadienyl)iron), where they are bound to iron in a sandwich structure.

Cyclopentadienyl-: 1) a term for the group C_5H_5-, which is derived from cyclopentadiene by removal of an H atom, in systematic names of compounds. 2) A term for the aromatic carbanion $C_5H_5^-$, which is found in the form of salts, such as cyclopentadienylsodium. The carbanion is stabilized by a π-electron sextet.

Cyclopentane, ***pentamethylene***: C_5H_{10} (structural formula, see Cycloalkanes), a cycloalkane hydrocarbon, a colorless, combustible liquid which is not miscible with water; m.p. -93.9 °C, b.p. 49.2 °C, n_D^{20} 1.4065. In the laboratory, C. is most easily made by reduction of cyclopentanone, which is easily synthesized. It is also found in large amounts in petroleum (see Naphthenes). C. is chemically similar to the normal alkanes. Its reactions are discussed under Cycloalkanes (see).

Cyclopentanone: a cyclic, aliphatic ketone. C. is a colorless liquid which smells faintly like peppermint; m.p. - 51.3, b.p. 130.6 °C, n_D^{20} 1.4366.

C. is insoluble in water, but dissolves readily in alcohol, ether and acetone. It can be synthesized by heating adipic acid salts, and is used for organic syntheses.

Cyclopenthiazid: see Diuretics.

Cyclophanes, ***carbophanes***: hydrocarbons in which arene rings are bridged by alkyl chains. C. are also called ***ansa compounds***; they are compounds with two or more rings, one of which can be considered the "handle" (Latin "ansa" = "handle").

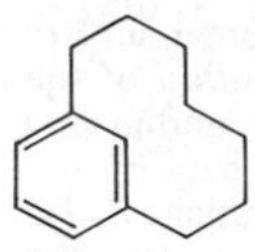

[7](1,3)Cyclophane or [7] Metacyclophane

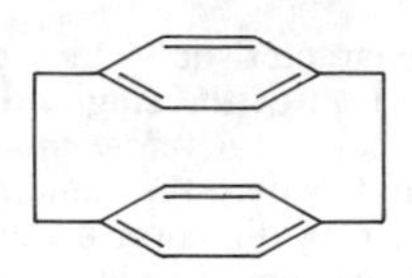

[2.2](1,4)Cyclophane or [2.2]Paracyclophane

Metacyclophanes are those in which the handle is anchored at the meta positions of the aromatic ring; the bridgeheads in ***paracyclophanes*** are in the para positions. The properties of C. depend on their structures. For example, if the number of methylene groups in [7]metacyclophane is reduced to six or five, the ring system is stressed, and this also reduces the aromaticity of the benzene ring. In [10]paracyclophane, on the other hand, the benzene ring can rotate freely around the para axis. However, if bulky substituents are added to the benzene ring, the rotation is hindered and optical antipodes are generated. [2.2]Paracyclophane can be polymerized to thermoplastics.

Cyclophosphamide: see Mustard gas derivatives.

Cyclopropane, ***trimethylene***: C_3H_6 (structural formula see Cycloalkanes), the simplest hydrocarbon of the cycloalkane series, a colorless, combustible gas which is insoluble in water; m.p. - 127.6 °C, b.p. -32.7 °C, n_D^{-42} 1.3799. It is similar in its chemical behavior to ethene. The ring is opened by conc. sulfuric acid, bromine, hydrogen bromide or catalytically activated hydrogen (see Cycloalkanes). Unlike ethene, C. is not attacked by cold, aqueous potassium permanganate solution or ozone. Thermal or catalytic isomerization yields propene. Many natural products contain C. rings, for example chrysanthemic acid, lactobacillic acid and some terpenes. C. is synthesized by dehalogenation of 1,3-dibromopropane with zinc, and is widely used as a harmless inhalation anaesthetic which has no after-effects. Substituted C. can be synthesized by reactions of alkenes with carbenes or by the Simmons-Smith reaction (see).

Cycloreversion: see Woodward-Hoffmann rules.

Cycloserine: D-4-aminoisoxazolin-3-one, an antibiotic formed by *Streptomyces orchidaceus*. The compound is also available synthetically. It was formerly used against tuberculosis.

Cyclosporin A: a cyclic peptide consisting of 11 amino acids, of which 7 have methyl groups on their N atoms. C. is a metabolic product of the fungus *Trichoderma polysporum* and has a strong immunosuppressive effect.

Cyclostereoisomerism: isomerism occurring in cyclic, directional, planar compounds in which there are $2n$ chirality centers (see Stereoisomerism 1.1) in the ring and one half is enantiomeric to the other.

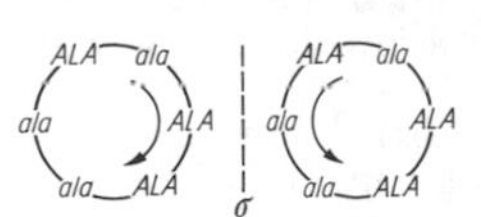

Cycloenantiomers of cyclohexaalanines. ALA = (R)-configuration; ala = (S)-configuration

The characteristics which generate isomerism are the direction of the ring and the distribution of the chirality centers. *Cycloenantiomerism* is found in ring compounds where the same distribution pattern and differing ring direction produce mirror images. *Cyclodiastereoisomerism* is the condition in which ring compounds have the same distribution pattern and different ring directions, but the two isomers are not mirror images.

Cyclotrimethylene trinitramine: same as 1,3,5-Trinitro-1,3,5-triazacyclohexane (see).

Cyd: abb. for cytidine.

Cyhexatin: an Acaracide (see).

p-Cymene: ***1-isopropyl-4-methylbenzene***: an aliphatic aromatic hydrocarbon, a colorless liquid with a characteristic odor; m.p. - 68 °C, b.p. 177.1 °C, n_D^{20} 1.4909. *p*-C. is insoluble in water, but dissolves in ether and ethanol. It is a component of many essential oils, such as eucalyptus or carroway oil, and can be extracted from them. It is structurally similar to the monoterpenes and can readily be synthesized from α-pinene or limonene. *p*-C. is used as a starting material for synthesis of the antiseptics carvacrol and thymol and as a thinner for paints, lacquers and varnishes. It is also added to industrial products as a covering perfume.

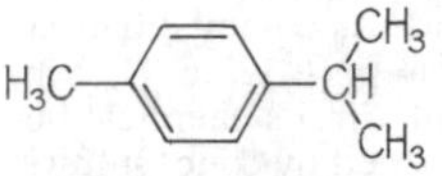

$H_3C-C_6H_4-CH(CH_3)_2$

Cypermethrin: see Pyrethroids (see).

Cyproterone acetate: see Antiandrogens.

Cys: abb. for cysteine.

Cystamine: see Cysteamine.

Cysteamine, ***2-aminoethanethiol***, ***β-aminoethylmercaptan***: $H_2N-CH_2-CH_2-SH$, a decarboxylation product of cysteine, and a component of coenzyme A. It forms colorless crystals with an unpleasant odor, m.p. 99-100 °C, which dissolve easily in water and ethanol. The m.p. of the hydrochloride is 70-71 °C. C. is easily oxidized in air to the disulfide, ***cystamine***. It is used for radiation protection, because it is a radical trapper.

Cysteine, abb. ***Cys***: α-amino-β-mercaptopropionic acid, $HSCH_2-CH(NH_2)-COOH$, a sulfur-containing, proteogenic amino acid, which is especially abundant in keratin of hair (14.4%), feathers (8.2%) and wool (11.9%). C. is obtained by reduction of its disulfide, ***cystine***, $HOOC-CH(NH_2)-CH_2-S-S-CH_2-CH(NH_2)-COOH$, which is obtained by acid hydrolysis of keratins. Keratin is hydrolysed for 18 h with a mixture of hydrochloric acid and formic acid at 110 °C. The hydrolysate is brought to a weakly alkaline pH, and the precipitated crude cystine is purified by repeated precipitation in the presence of activated charcoal (which adsorbs the accompanying tyrosine). 100 g L-cystine can be obtained from 2 kg hair. It is reduced to C. cathodically in hydrochloric acid solution, or with nascent hydrogen.

C. is very important in redox reactions in the organism. In proteins, free SH groups of C. are necessary for the activity of many enzymes, and the disulfide bridges formed between C. residues in proteins are responsible for stabilization of tertiary protein structure. In the oxidative metabolic degradation of C., the first product is ***cysteic acid***, $HOOC-CH(NH_2)-CH_2SO_3H$; the decarboxylation of this compound yields ***taurine***, $H_2N-CH_2CH_2-SO_3H$. In cystinuria, C. and cystine are found in the urine, and may lead to formation of cystine stones (bladder stones).

C. is used in medicine for prophylaxis and therapy of liver and metabolic disorders. The protective effect of C. and peptides with free SH-groups (glutathione) against radiation is due to their capture of reactive radicals generated within the cell by the radiation (e.g. gamma rays). C. is used in the food industry as a flavor intensifier for dried onion and garlic products, to stabilize aromas and in the production of artificial meat aromas. A typical artificial chicken aroma, for example, contains 13.0 g C., 6.7 g glycine, 10.8 g dextrose, 8.0 g arabinose and 60 ml water in a special preparation.

Cystine: see Cysteine.

Cyt: abb. for cytosine.

Cytarabin: 1-β-D-arabinofuranosylcytosine, a pyrimidine nucleoside antimetabolite used as a virostatic and antineoplastic. C. is a stereoisomer of cytosine with D-arabinose as its sugar component.

Cytidine, abb. ***Cyd*** or ***C***: 3-β-D-ribofuranosylcytosine, a nucleoside which forms colorless needles; m.p. 231 °C. C. is soluble in water. It is a component of ribonucleic acid synthesized as ***cytidine-5'-triphosphate (CTP)*** by amination of uridine triphosphate.

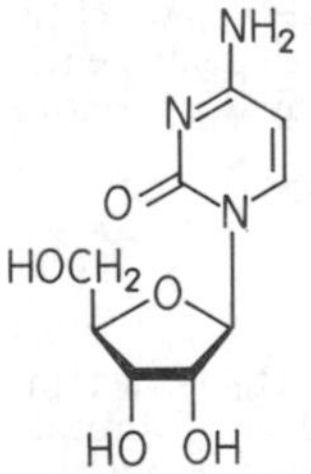

NH_2 N O N $HOCH_2$ O HO OH

Cytisine: an Alkaloid (see).

Cytochrome: an electron-transferring hemoprotein. The transfer of electrons depends on the valence change of the central atom, $Fe^{2+} \rightleftharpoons Fe^{3+}$, of the prosthetic group, i.e. of the non-protein part. At present, more than 30 C. are known, and are classified in four main groups according to their prosthetic groups: ***cytochromes a, b, c*** and ***d*** (table). C. a contain formylporphin iron, C. b, like the hemoglobins, contain ***protoheme***. The other prosthetic groups are derived from protoheme. ***Cytohemes*** a, b and d can be separated from the proteins with acetone/HCl and extracted with ether. In vertebrate cytochrome c, the heme is covalently bound to the 104-amino-acid polypeptide chain by 2 thioether bonds between the vinyl groups of the porphyrin and cysteine residues of the protein. The prosthetic group of C. d is derived from a dihydroporphyrin. The α-bands of the electronic spectrum of the pyridine-hemochromes are used to distinguish the prosthetic groups of the individual C. The C. occur widely in all organisms; the C. in the respiratory chain and photosynthetic apparatus play an essential role in electron transport. These are among the oldest functional proteins; their basic structure has remained essentially unchanged for 2 billion years. The different C. differ in the substituents on the porphyrin rings and the carrier proteins have somewhat different redox potentials.

Cytochrome	Prosthetic group	Examples
a	Cytoheme a (Heme a)	Cytochrome oxidase
b	Cytoheme b (Protoheme)	Respiratory chain cytochrome b
c	Cytoheme c (Heme c)	Respiratory chain cytochromes C and C_1
d	Cytoheme d	Cytochromes of chloroplasts and microorganisms

Cytochrome P-450, one of the C. b complex, is a monooxygenase involved particularly in steroid hydroxlations; its absorption maximum is at 450 nm. The enzyme (M_r 850,000) consists of 16 identical subunits with 8 heme groups. It is involved in the metabolism of pharmaceuticals in the liver through hydroxylation, demethylation and N-oxidations.

Cytochrome *a/a*$_3$ complex: see Cytochrome oxidase.

Cytochrome oxidase, ***Warburg's respiratory enzyme***: an oxidoreductase discovered by Otto Warburg which catalyses the last step of respiratory electron transfer: the transfer of electrons to molecular oxygen. The catalytic reaction is: $4\,H^+ + O_2 + 4\,Cyt\,c^{2+} \rightarrow 2\,H_2O + 4\,Cyt\,c^{3+}$, where C. acts as a coupler between the one-electron donor, cytochrome c (Cyt c), and the four-electron acceptor, O_2. The protein component of C., the ***cytochrome a/a*$_3$ *complex*** (M_r 200,000), consists of 12 subunits of varying sizes. It is bound to two hemes and two copper ions which, like the central iron ion, undergo a valence change ($Cu^{2+} + e \rightleftharpoons Cu^+$) and take part in the electron transfer process.

Another important function of C., which is found in all eukaryotic cells and some bacteria, is its participation in proton transport through the inner mitochondrial membrane. This transport takes place through an ion channel formed by one of the four subunits of the C.

Cytoheme: see Cytochrome.

Cytokinins: a group of phytohormones which are derivatives of adenine or adenosine substituted on the primary amino group. A natural C. is ***trans-zeatin***, which was isolated from maize. ***Kinetin*** (5-furfurylaminopurine), which is formed by hydrolysis of yeast material containing nucleic acids, has C. activity, as does 6-(isopent-2-enylamino)purine, which is formed from tRNA. The C. regulate growth, differentiation and development in plants. They have an effect on cell division and promote growth of shoots and side buds. C. are formed in young roots. In addition to the adenine derivatives, synthetic compounds such as benzimidazole or N,N'-diarylurea have C. activity, and are used as growth regulators.

NHR

Kinetin: R = CH_2–(2-furyl)

Zeatin: R = CH_2–CH=C(CH_3)–CH_2OH

6-(Isopent-2-enylamino)-purine: R = CH_2–CH=C(CH_3)–CH_3

Cytosine, abb. ***Cyt***: 4-amino-2-hydroxypyrimidine, a nucleic acid base. The monohydrate forms colorless crystals which release water of crystallization about 100 °C; m.p. 320-325 °C (dec.). C. is slightly soluble in water and alcohol, and practically insoluble in ether. C. is a component of all nucleic acids.

Cytostatics, ***antineoplastics***: compounds used in chemotherapy of cancers, especially inoperable and metastasized tumors and generalized malignant diseases such as leukemias. C. are also used prophylactically after surgical intervention, to prevent metastasis and relapse. They are usually used in combinations of compounds with different points of attack. Chemotherapy with C. has many side effects, and requires strict medical supervision. It must also be considered that many C. are themselves carcinogens, and some also act as immunosuppressants.

C. can be classified as 1) *Antimetabolites* (see), 2) *alkylating agents* (see), 3) *mitosis inhibitors* such as the vinca alkaloids vinblastine and vincristine, 4) *antibiotics*, such as the anthracyclins daunorubicin and doxirubicin and the actinomycin dactinomycin, and 5) *other compounds*, for example the methylhydrazine derivative procarbazine, the triazene derivative dacarbazine, *N*-nitrosoureas such as lomustin (with a β-chloroethyl group), and platinum complexes such as cisplatin. Finally, synthetic sex hormones of the opposite sex can be used on tumors which are dependent on sex hormones, e.g. fosfestrol, or antihormones like tamoxifen can be applied.

Czochralski process: 1) a general method for Crystal growing (see); 2) a special method for Silicate synthesis (see).

CVD diamond synthesis: a method of diamond synthesis by chemical vapor deposition, developed in the 1980s. In contrast to the high-pressure synthesis of diamond, this is a low pressure method which can be used to form thin films and free membranes of diamond. These are used to form new compounding powders and to coat cutting edges of tools. In C., mixtures of gaseous hydrocarbons are decomposed into reactive radicals and molecular fragments, which precipitate onto a hot surface as diamond. The H atoms generated in the gas phase react more rapidly with the graphite and amorphous carbon which form than with diamond and therefore under these conditions metastable diamond is formed.

The four most commonly used methods are as follows: A) hot wire method, in which a mixture of CH_4 and H_2 is decomposed by an electrically heated tungsten, molybdenum or tantalum wire. B) In the microwave-plasma method, the reacting gases are converted into a plasma in a partially evacuated chamber by microwaves. C) Direct-current arc discharge: a direct-current plasma at about 5000 °C is generated in a mixture of Ar, H_2 and CH_4. D) Flame CVD: acetylene is burned with oxygen in a welding torch.

D

D: 1) Symbol for deuterium. 2) Configurational symbol; see Stereoisomerism, 1.1. 3) Abb. for debye.

2,4-D: see Growth hormone herbicides.

Dacron®: see Synthetic fibers.

DADI: see Mass spectroscopy.

Dakin reaction: method for synthesizing polyvalent phenols by oxidation of 1,2- and 1,4-phenol aldehydes with hydrogen peroxide in alkaline solution.

$$\text{Salicylaldehyde (OH, CHO)} \xrightarrow[-\text{HCOOH}]{+\text{H}_2\text{O}_2} \text{Pyrocatechol (OH, OH)}$$

Salicylaldehyde   Pyrocatechol

1,3-Phenolaldehydes cannot be converted to the corresponding resorcinols under the conditions of the D.

Dakin-West reaction: a method for synthesizing *N*-acetylated α-aminoketones by reaction of α-aminocarboxylic acids with acetic anhydride in the presence of pyridine:

$$\underset{\text{NH}_2}{\text{R-CH-COOH}} \xrightarrow[-\text{CH}_3\text{-COOH}]{+(\text{CH}_3\text{-CO})_2\text{O}} \underset{\text{NH-CO-CH}_3}{\text{R-CH-COOH}} \xrightarrow[-\text{CH}_3\text{-COOH},\ -\text{CO}_2]{+(\text{CH}_3\text{-CO})_2\text{O}} \underset{\text{NH-CO-CH}_3}{\text{R-CH-CO-CH}_3}$$

Dalapon: see Herbicides.

Daltonide, ***daltonide compound, stoichiometric compound***: (named for J. Dalton) a chemical compound with a constant composition which obeys the basic stoichiometric rules. Nearly all molecular substances, e.g. water, methane, and most salts, are D. Antonym: Berthollide (see).

Dalton's law: 1) a law proposed by J. Dalton in 1801 which states that the total pressure p of a mixture of ideal gases is equal to the sum of their partial pressures p_i: $p = p_1 + p_2 + \ldots = \Sigma p_i$. If all the components are under the same pressure before they are mixed, and the mixing occurs isobarically, the resulting total volume is equal to the sum of the volumes of the components. The partial pressure p_i is related to the total pressure by the equation $p_i = x_i p$, where x_i is the mole fraction. It follows that $p = \Sigma p_i = \Sigma x_i p$.

2) See Stoichiometric proportions, law of.

3) See Henry-Dalton law.

Dammar: a natural resin from the Dipterocarpaceae (yang teak woods) of Indonesia. It softens at 90 °C and at 150 °C becomes liquid and clear. There are white, brown and black types of D.; the resin is readily soluble in chloroform, benzene and carbon disulfide, and partially soluble in alcohol, ether, etc. It is used in pharmacy as a component of sticking plaster, and is also used to make paints which are resistant to acids and alkalies. D. is used to seal microscope slides.

Danaidon: see Pheromones.

Danamid®: see Synthetic fibers.

Daniell cell: one of the oldest and best-known galvanic cells (see Electrochemical current sources), discovered in 1835 by J.F. Daniell. The cell shown in the figure is constructed so that it can produce electric current.

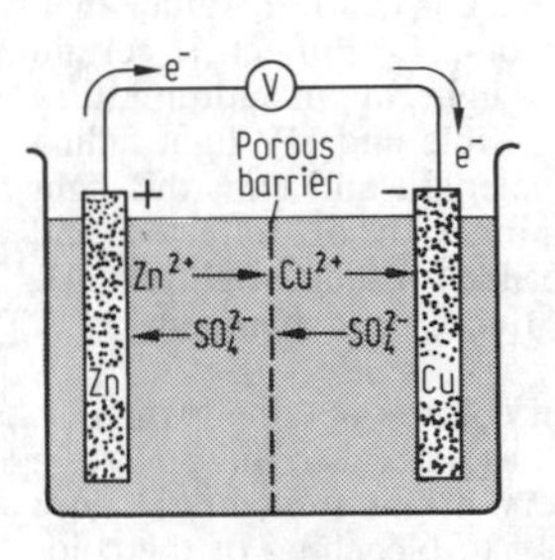

Daniell cell
Anode: $Zn \rightarrow Zn^{2+} + 2e$
Cathode: $Cu^{2+} + 2e \rightarrow Cu$

The left half-cell has a metallic zinc electrode and a zinc sulfate solution. The right half-cell has a copper electrode and a copper sulfate solution. To prevent mixing of the solutions in the half-cells, they are separated by a porous wall which prevents the passage of zinc and copper ions. If the zinc and copper electrodes are connected by a wire, electrons flow from the zinc to the copper electrode. At the same time, the metallic zinc is oxidized to zinc ions, while at the copper electrode, copper ions are reduced to metallic copper. The zinc electrode (oxidation) is thus the anode, and the copper electrode (reduction), the cathode. The cell potential of the D. is approximately 1.10 V.

Danger symbols: pictographs used as warnings; as a rule, the warning is repeated in print (e.g. "poison" or "radioactivity"). The shapes of D. vary from one country to the next; the most commonly used are the skull and crossbones for poison and glasses for danger of splashing.

Daphnin: see Eosine pigments.

Dapsone: bis-(4-aminophenyl)sulfone, an antibacterial compound used mainly for treatment of leprosy.

Darzens-Erlenmeyer-Claisen condensation, ***glycide ester synthesis***: a method for preparation of glycide esters through condensation of aldehydes or ketones with α-chlorocarboxylate esters in the presence of sodium methanolate or sodium amide in ether, benzene or xylene:

$R^1–CO–R^2 + R^3–CHCl–COOC_2H_5 \xrightarrow[-HCl]{} R^1R^2C – CR^3–COOC_2H_5$ (epoxide, bridged by O).

Hydrolysis of the glycide ester and subsequent thermal decarboxylation leads to the formation of an aldehyde or ketone, depending on R^3. The reaction of an aldehyde with a chloroacetate ester under the conditions of D. thus offers the possibility of a preparative lengthening of the aldehyde by one CH_2 group.

Daunomycin: see Anthracyclines.

Dayglow colors: paints and textile brighteners which are unusually bright. They are important for warning and safety marking, because even at a distance they still are perceived as colored, rather than gray. Like all other colors, D. reflect a certain wavelength range of visible light, but in addition, they convert the rays of both visible and UV light falling on them into longer wavelength and emit this light (fluorescence). The combination of reflected and fluorescent light is responsible for their brightness. D. consist of transparent, fluorescent pigments in a transparent binder.

The greatest amount of color is obtained when D. are painted onto a white base, because this absorbs a minimum of light. In very bright sunlight, D. lose their fluorescence, leading to bleaching of the color and loss of brightness. To prolong the fluorescence, the paint is coated with an optically non-absorbing layer of lacquer which contains substituted benzophenone as a UV absorber. The lacquer is supposed to absorb the higher-energy rays in the UV range.

Dazomet: 3,5-dimethyltetrahydro-1,3,5-thiadiazine-2-thione, a nematocide synthesized from carbon disulfide, formaldehyde and methylamine. m.p. 102-103 °C, solubility in water, 0.12% at 30 °C.

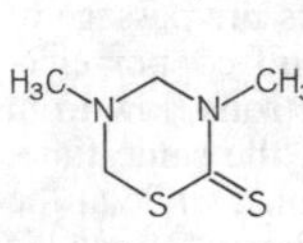

H_3C–N, N–CH_3, S, =S (structure)

DCCC: see Counter-current chromatography.

DD®: see Nematocides.

DDA: see DDT.

DDE: see DDT.

DDNP: see 4,6-dinitro-2-Diazohydroxybenzene.

DDT: abb. for ***D***ichloro***d***iphenyl***t***richloroethane, one of the chlorinated hydrocarbon insecticides which acts as a contact poison. DDT is a colorless, crystalline substance, D. 1.556, m.p. 108.5 °C, b.p. 185-187 °C at 7 Pa, vapor pressure at 20 °C $4.5 \cdot 10^{-5}$ Pa. DDT is readily soluble in cyclohexanone, dioxane, trichloroethane, benzene, xylene, acetone; soluble in chloroform and ether; and has low solubility in ethanol, methanol; ≈ 0.001 mg/l in water.

DDT is largely stable to light, air and acid. Under alkaline conditions, however, it easily eliminates HCl to form the biologically inactive *D*ichloro*d*iphenyldichloro*e*thene (*DDE*). This reaction is catalysed by iron and aluminum compounds. DDT is synthesized by condensation of chloral with chlorobenzene in the presence of sulfuric acid or oleum:

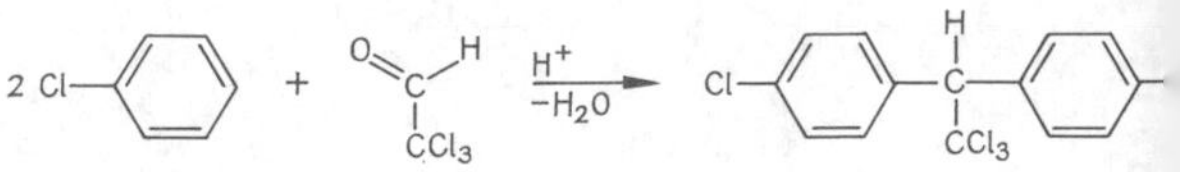

$$2\ Cl–C_6H_5 + OHC–CCl_3 \xrightarrow[-H_2O]{H^+} Cl–C_6H_4–CH(CCl_3)–C_6H_4–Cl$$

The industrial product is a solid, yellowish mass with a softening point ≈ 90 °C. It contains about 70% p,p'-DDT and, as the most important byproduct, up to 20% o,p-DDT, which is only weakly effective as an insecticide.

DDT is a very strong contact and oral poison for many insect species. Diptera (flies and mosquitoes) and biting insects (caterpillars, beetles) are very sensitive. Aphids are less sensitive, and spider mites are insensitive. The substance is highly lipophilic, which makes it penetrate the insect cuticle easily, and thus act as a contact poison. It is actually a nerve poison. The acute toxicity to warm-blooded animals is relatively low: oral LD_{50} = 250 to 500 mg/kg rat. The accumulation of DDT and its main metabolite, DDE, in fatty tissue is a problem. Warm-blooded animals excrete water-soluble dichlorodiphenylacetic acid (*DDA*) in their urine as the end product of DDT metabolism. The fatal oral dose for an adult human heing is estimated at 20 to 30 g. Cases of chronic poisoning are not known.

Of the insecticides structurally related to DDT, Methoxychlor (see) is the most important.

Historical. The insecticidal properties of DDT were discovered in 1939 by Paul Müller in Switzerland. It was used as the preferred broad-spectrum insecticide against chewing insects in horticulture, against flies and mosquitoes (especially in areas where these are disease vectors) and other insects. Due to its high persistence, and its accumulation in the environment, DDT (along with other chlorinated hydrocarbons) was banned in most industrial countries during the 1970s.

DDVP: same as Dichlorvos (see).

Deacidification: in water treatment, the removal of acidic components from drinking or industrial water. Acidic water is generally corrosive and causes damage to pumps and plumbing (see Calcium-carbon-dioxide equilibrium). The acids can be removed mechanically or chemically, depending on their volatility (carbon dioxide and hydrogen sulfide are volatile, sulfuric acid, for example, is not). The most important methods of *mechanical* or *physical* D. are cascade, pressure, spray or roller aeration (see Gas exchange). There are special forms of aeration, for example, trickling through aerated beds, but these are generally less effective than the above processes and are only used in special circumstances. *Chemical D.* consists almost exclusively of addition of lime or sodium hydroxide. Small amounts of acid in the water can also be removed by filtration over marble or other alkaline materials, such as decarbolith. However, it must be remembered (especially during removal of free excess carbon dioxide) that the calcium-carbon-dioxide equilibrium will be established in the water.

Deacon process: see Chlorine.

Deactivation: see Decontamination.

"Dead-stop" method: see Amperometry.

Dealuminization: see Demetalization.

Deamination: the removal of an amino group from an organic compound. The most important synthetic method of D. is diazotization of primary amines or amides, followed by nitrogen cleavage. The resulting unstable carbenium ion reacts to form a stable product, the nature of which depends on the reaction conditions. In the presence of reducing agents, such as alcohol and copper(I) oxide, hypophosphorous acid, formaldehyde or tin(II) compounds, aromatic primary amines can be converted via the diazonium salts to the deaminated, unsubstituted hydrocarbons. Decomposition of diazonium salts in water produces alcohols. Loss of a proton from the carbenium ion forms alkenes in a competing reaction:

$$R{-}CH_2{-}CH_2{-}\overset{+}{N}{\equiv}N| \xrightarrow[-N_2]{}$$

$$R{-}CH_2{-}\overset{+}{C}H_2 \xrightarrow[-H^+]{+H_2O} R{-}CH_2{-}CH_2{-}OH$$

$$R{-}CH_2{-}\overset{+}{C}H_2 \xrightarrow{-H^+} R{-}CH{=}CH_2$$

Amides are deaminated under these conditions to form carboxylic acids. Arene diazonium salts can be similarly converted to phenols. In the presence of copper(I) salts or copper powder, the diazonium group can also be replaced by other anions, e.g. chloride, bromide, cyanide, rhodanide or nitrate, in the sense of the Sandmeyer reaction (see). In the reaction of α-amino acid esters and α-amino ketones with nitric acid at low temperatures, resonance-stabilized α-diazo esters and α-diazoketones are formed. D. of these compounds does not occur until they are heated, or exposed to dilute acids. The oxidative D. of α-amino acids is biologically important. A flavin enzyme first catalyses formation of an α-iminocarboxylic acid, which is hydrolysed to an α-ketoacid by loss of ammonia: $R\text{-}CH(NH_2)\text{-}COOH \rightarrow R\text{-}C(=NH)\text{-}COOH \rightarrow R\text{-}CO\text{-}COOH$. The transfer of an amino group from an α-amino acid to an α-ketoacid is also a D. from the point of view of the amino acid.

The nitrogen released by the D. of primary amines or amides with nitrous acid is used for quantitative determination of these compounds by the method of van Slyke.

De Broglie equation: see Atom, models of.

Debye, *D*: unit of Dipole moment (see).

Debye forces: see Van-der-Waals forces.

Debye-Hückel theory: a theory which explains the deviation of dilute solutions of strong electrolytes from the behavior of ideal solutions in terms of ionic interaction. The D. is based on the following assumptions: Dissolved ions which are solvated can move freely past each other. The relative positions of the ions are determined by attractive and repulsive electrostatic forces and by thermal motion. The electrostatic forces are responsible for short-range order, while the thermal motion counteracts this order. On the average over time, a single ion will be surrounded by a spherically symmetrical cloud of counterions. This order is distorted by application of an electric field, since the positive and negative ions move in opposite directions in the field. The ions therefore migrate out of their charge clouds, and new ones are formed. However, since this process takes a certain amount of time, the central ion will always move a little ahead of its charge cloud. As a result, it feels a certain counter force, the size of which depends on the distance of the ions from each other, i.e. on the concentration. Higher concentrations mean that the interionic distances are smaller, and thus the interactions are stronger (*relaxation effect, asymmetry effect*). Higher concentrations of ions also lead to a greater resistance to the ion clouds moving in the opposite direction (*electrophoretic effect*), with the result that all properties which depend on the particle number (colligative properties such as conductivity or osmotic pressure) are proportional to an "effective concentration", the *activity a*, rather than the actual concentration. The activity is obtained by multiplication of the concentration c by an activity coefficient f: $a = f \cdot c$. The activity coefficients of cations and anions cannot be determined independently; instead, a mean activity coefficient $f_\pm$ is measured. The D. now permits calculation of $f_\pm$ from the equation: $\log f_\pm = 0.0591 \cdot z_+ \cdot z_- \cdot I$ (***Debye-Hückel limit law***), where z_+ and z_- are the charges on the ions and I is the ionic strength according to Lewis and Randall ($I = \Sigma c_i \cdot z_i^2$, where c = concentration). The application of the D. to the concentration dependence of electrolytic conductivity Λ_{eq} gives the equation: $\Lambda_{eq} = \Lambda_\infty - (B_1\Lambda_\infty + B_2)\sqrt{c}$; this is identical to the square root law determined empirically by Kohlrausch (see Electrical conductivity). B_1 and B_2 are constants specific for each substance.

Debyeogramm: see X-ray structure analysis.

Debye-Scherrer method: see X-ray structure analysis.

Debye's T³ law: see Molar heats.

Decahydronaphthalene: same as Decalin® (see).

Decalin®, ***decahydronaphthalene, bicyclo[4,4,0]decane***: a bicyclic, saturated hydrocarbon, which is also called a hydroaromatic because it can be synthesized from aromatic compounds (arenes). D. is a colorless liquid with a camphor-like odor; its *cis*-form has m.p. - 43.3 °C, b.p. 195.7 °C, n_D^{20} 1.4810; *trans*-form, m.p. ≈ 30.4 °C, b.p. 187.3 °C, n_D^{20} 1.4695. D. is insoluble in water, but is soluble in ethanol and ether. It is synthesized industrially by hydrogenation of naphthalene or tetralin® in liquid phase at 200-260 °C and at pressures of 2.5 to 4 MPa in the presence of nickel catalysts. D. is used as a solvent for fats, plant oils and lacquers, and in the production of shoe polish and polishing wax.

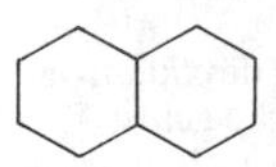

Decamethonium: see Muscle relaxants.

Decamethrin: see Pyrethroids.

Decametry: same as Dielectrometry.

Decanal: same as Capric aldehyde (see).

Decane: $CH_3(CH_2)_8CH_3$, an alkane hydrocarbon, a combustible, colorless liquid with no odor; m.p. -30 °C, b.p. 174 °C, n_D^{20} 1.4119. D. is insoluble in water, but dissolves very readily in ethanol and ether.

Decanedioic acid: same as Sebacic acid (see).

Decanoic acid, ***capric acid***: CH_3-$(CH_2)_8$-COOH, a higher, saturated, monocarboxylic acid. The IUPAC nomenclature no longer permits use of the trivial name. D. forms colorless crystals with a rancid odor; m.p. 31.5 °C, b.p. 270 °C. It is insoluble in water, but is soluble in ether, hot alcohol and benzene. It is found in free or esterified form in palm heart, coffee bean, camomile and fusel oils. The glycerol esters of D. are present in cow's and goat's butter and in coconut oil. D. can be obtained by saponification of palm heart or coconut oil. Some of its esters are used to make fruit aromas and other aromas.

Decanol, ***decan-1-ol***: $CH_3(CH_2)_8CH_2OH$, a monoalcohol, a colorless liquid which is insoluble in water; m.p. 7 °C, b.p. 229 °C, n_D^{20} 1.4366. D. is found in musk seed oil. It is made industrially by Oxosynthesis (see) or by catalytic hydrogenation of the fatty acid mixture which results from saponification of natural fats. D. is a very versatile solvent; it is also used as a stabilizer of drying solutions and as an additive to hydraulic fluids. It is very important in the production of wetting agents and laundry products. The esters are used as softeners in lacquers and plastics, and in perfumes as aromas.

Decantation: cautious pouring off of a liquid above a sediment, such as a precipitate, which is to be kept. In the laboratory, there are special decanting vessels for separation of the two phases. See Sedimentation.

Decarbonation: 1) the removal of the carbonate hardness of water; see Water softening.

Decarbonization: In steelmaking, the removal of carbon from a) the iron melt, by oxidation (refining), via the iron(II) oxide and other oxides in the slag, which dissolve in the melt and react with the carbon ($FeO + C \rightleftharpoons Fe + CO$); or b) from the surface layer of the steel by chemical-thermal treatment (oxidative diffusion removal), to achieve a change in the composition and structure.

Decarbonylation: cleavage of carbon monoxide from an organic compound. Ketones, α-ketocarboxylic acids and lactides can be decarbonylated by thermal or photochemical activation. Example: D. of acetone to ethane:

$$CH_3\text{–}CO\text{–}CH_3 \xrightarrow{h\nu} CH_3\text{–}CH_3 + CO.$$

Decarboxylases: enzymes which catalyse the cleavage of carbon dioxide from organic acids. The ***amino acid decarboxylases***, which convert amino acids to amines, and Pyruvate decarboxylase (see) are important examples.

Decarboxylation: cleavage of carbon dioxide from carboxylic acids and substituted carboxylic acids. The D. of aliphatic monocarboxylic acids or their salts requires high temperatures, and alkanes are the products. Oxalic and malonic acids and substituted malonic acids (see Malonate ester syntheses) undergo D. relatively easily, forming formic or acetic acid or substituted acetic acids. α-Ketocarboxylic acids are very easily decarboxylated upon treatment with dilute sulfuric acid and heating. β-Ketoacids form ketones upon D. (see Ketone cleavage). The D. of α,β-unsaturated carboxylic acids occurs via a cyclic transition state with formation of alkenes. α-Amino acids are decarboxylated when heated with soda lime or barium hydroxide, forming primary amines. This reaction is also catalysed by enzymes. Carbamic acids, which are formed as intermediates in the Hoffmann degradation, are so unstable that they are immediately decarboxylated.

Decay series: see Radioactivity.

Decolorizing: in the textile industry, the removal of color by stripping off or destroying the pigment, in the processing of rags, preparation of cloth for dyeing, etc. Sodium dithionite, $Na_2S_2O_4$ (hydrosulfite) is widely used. Azo dyes (R_1-N=N-R_2) are cleaved by reduction, producing primary amino groups (R-NH_2); the color of the azo dye is converted to that of the products. These are usually colorless or water soluble, so the textile is decolorized. Azo dyes on cellulose fibers, such as substantive dyes, diazotization, reactive or development pigments can be decolorized by treatment with 2 to 3 g/l sodium dithionite in alkaline medium (0.5 to 1 g/l soda). Stabilized hydrosulfite preparations are used to decolorize protein fibers (wool, silk) and acetate synthetics. In contrast to azo dyes, anthraquinone dyes are only temporarily decolorized by sodium dithionite reduction; reoxidation in the presence of air causes their color to return. For decolorizing textiles dyed with mordant or sulfur pigments, sodium dithionite must be supplemented with a reagent which has an affinity to the pigment and prevents it from being reabsorbed by the fibers. Polyvinylpyrrolidone is particularly effective in this manner. Pigments which are particularly difficult to remove can be oxidized by sodium hypochlorite in weakly acidic medium.

Decomposition point: see Melting point.

Decomposition voltage: see Electrolysis.

Decontamination: the removal of radioactive, biological or chemical contamination (see Detoxification). See Contamination.

Decylaldehyde: same as Capric aldehyde (see).

Dederon®: see Synthetic fibers.

De-emulsification: breaking down (destruction, coalescence) of emulsions. Either the stabilizing surfactant is displaced from the phase boundary (desorption) or holes are temporarily created in the adsorption layers. Desorption is achieved by changes in temperature (for example, with nonionic surfactants), or by addition of deemulsifiers (as in dehydration of petroleum) which have stronger surface activity but are not stabilizers. Water-in-oil (W/O) emulsions can be broken by strong electric fields; voltage breaks through the oil films between coagulated water droplets. Oil-in-water (O/W) emulsions which have been stabilized by alkali salts of alkanoic acids are broken by acidification, and W/O emulsions which have been stabilized by alkaline earth salts of alkanoic acids are deemulsified by alkalinization. Filtration through hydrophilic filters destroys W/O emulsions, and filtration through hydrophobic filters destroys O/W emulsions.

Deer repellents: preparations based on animal oils or petroleum residues which are applied to plants to protect them from foraging by deer. The sticky consistency or unpleasant smell makes the plants less attractive to the animals.

DEET: an Insect repellent (see).

Deferrization: 1) in water treatment, the reduction of the iron content of the water to a level consistent with the intended use; for example, in drinking water, to 0.1 mg/l. Iron forms deposits in the plumbing and has an unpleasant taste.

Since the main source of drinking water is ground water, there is little dissolved oxygen in the untreated water, and the iron is mostly in the divalent form. It is thus in true solution.

There are two different means of removing iron: a) direct oxidation with air oxygen or other oxidation agents (for iron contents up to about 15 mg/l) and b) precipitation of Fe^{2+} oxide hydrate by addition of OH ions and subsequent oxidation (for iron contents above about 15 mg/l). The reactions can be represented by the following simplified equations: for a), $4\ Fe^{2+} + O_2 + 2(x + 2)H_2O \rightarrow 2\ Fe_2O_3 \cdot xH_2O + 8\ H^+$; for b), $4\ Fe^{2+} + O_2 + 8\ OH^- + 2(x - 2)H_2O \rightarrow 2\ Fe_2O_3 \cdot xH_2O$. The most widely used processes of D. are, for case a), aeration and filtration, and for case b), aeration, rough work-up and filtration.

2) In metallurgy, see Demetallization.

Defect electron conductor: see Semiconductors.

Deflagration: a very rapid combustion of an explosive substance in which the expansion of the reaction through the material occurs at less than the speed of sound, that is, less than 2000 m s^{-1}. It is self-sustaining, due to the release of the reaction enthalpy; flames and glowing particles are formed. In contrast to the situation in a Detonation (see), in a D. the reaction products flow in the opposite direction from the expansion of the reaction. See Explosion. The initiation of a D. is called ***ignition***.

Deficiency conductor: see Semiconductor.

π-Deficiency heterocycles: heteroaromatic compounds characterized by low electron densities on the carbon atoms of the heterocyclic ring, especially in positions 2 and 4 (relative to the heteroatom). Compounds of this type, especially Six-membered heterocycles (see), react preferentially with nucleophilic reagents (*S_N-reactions*). The prototype of this group is pyridine, which, for example, is readily converted to aminopyridine by the Čičibabin reaction.

Defoliants: compounds used to kill the leaves of cultivated plants, e.g. of potato plants, seed plantations and cotton, to accelerate ripening and to simplify the harvest. Herbicides, such as diquat, are usually used as D.

Deformational vibrations: vibrations in a molecule which involve deformation of the valence angle. There are various possible D. which are shown using the CH_2 group as an example (Fig.).

Motions of various deformational vibrations: a) scissors; b) swing; c) bending; d) torsional; vibration. → indicates motion in the plane of the paper, + and – indicate motions perpendicular to the paper.

Degasification: 1) the removal of volatile components from organic raw materials (coal, wood, etc.) by heating them in the absence of air to produce charcoal or coal products. D. at lower temperatures is called carbonization; D. at medium and high temperatures is called coking (see Coal refining).

2) The removal of oxygen and carbon dioxide from boiler feedwater is a type of Water treatment (see). D. is achieved by heating, boiling or adding reagents. Oxygen can be removed chemically in one of several ways. a) In sulfite D., sodium sulfite, Na_2SO_3, is added; it is converted to sodium sulfate: $2\ Na_2SO_3 + O_2 \rightarrow 2\ Na_2SO_4$. However, this naturally increases the salt content of the water. Sodium hydrogensulfite, $NaHSO_3$, or sulfurous acid, H_2SO_3, can also be used to bind the oxygen. b) In hydrazine D., hydrazine is added and converted to nitrogen and water: $N_2H_6(OH)_2 + O_2 \rightarrow N_2 + 4\ H_2O$. There is no increase in the salt content of the water. c) Oxygen removal by iron filters has proven satisfactory for large operations and relatively low oxygen contents. Steel wool in a pressure filter reacts with oxygen in the water: $4\ Fe + 3\ O_2 + x\ H_2O \rightarrow 2\ Fe_2O_3 \cdot xH_2O$. The filter must be back flushed from time to time, and at longer intervals, the steel wool must be replaced.

Chemical binding of carbon dioxide is achieved by addition of ammonia; the hydrolysis of the resulting ammonium carbonate in the boiler does not increase the salt content: $CO_2 + 2\ NH_3 + H_2O \rightarrow (NH_4)_2CO_3$. Instead of ammonia, amines such as morpholine or cyclohexylamine can be used.

Degeneracy: property of a quantum mechanical operator which yields several independent linear eigenfunctions with the same eigenvalue. The number of eigenfunctions is called the *degree of degeneracy*. For example, in the hydrogen atom, all atomic orbitals with the same principal quantum number n are energetically degenerate, and the degree of degeneracy of the energy level E_n is n^2 (see Atoms, models of).

Degradation: the cleavage of a compound into smaller molecular fragments, as in the D. of higher alkanes by cleavage of C-C bonds in cracking, the thermal D. of carbonates to form CO_2 or the hydrolytic D. of biolopymers to monomeric units. The stepwise D. of proteins and nucleic acids is used to determine their primary structures.

Degrees of freedom: 1) the number of independent coordinates which may be arbitrarily chosen to determine uniquely the position of a system of mass points (e.g. atoms) in space. Each equation which fixes a spatial relationship between the mass points, e.g. in the form of a fixed distance due to a bond, reduces the number of external D.

The number of D. plays an important role in statistical Thermodynamics (see), the calculation of Molar heats (see) or State sums (see). An atom has 3 translational D., so a system of N atoms therefore has $3N$ translational D. If the N atoms combine to form a molecule, it will have only 6 external D., three translational and three rotational. The remaining $3N - 6$ D. are internal D. of the vibrational motions of the atoms with respect to each other.

2) In equilibrium thermodynamics, the number of independent parameters, e.g. temperature, pressure

or composition, through which the state of a macroscopic system is established. If these parameters are varied within certain limits, the structure of the system does not change qualitatively. The number of these D. is determined by the Gibbs phase rule.

Dehalogenation: cleavage of halogens from 1,2-dihalogen alkanes or 1,2-dihalogen cycloalkanes, forming alkenes or cycloalkenes. D. occurs most readily in 1,2-dibromo compounds. D. of 2,3-dihalogenpropenes with zinc produces allenes in good yields. Ketenes can be made by D. of α-haloacylhalides with zinc in diethyl ether.

Dehydration: 1) loss of water from an organic compound. *Intramolecular D.* of alcohols is a β-elimination and yields alkenes or cycloalkenes; it is catalysed by sulfuric or phosphoric acid or zinc chloride. In industry, the D. of alcohols is carried out in the gas phase on aluminum oxide or thorium catalysts. *Intermolecular D.* of alcohols with concentrated sulfuric acid, anhydrous phosphoric acid, aluminum oxide or aluminum phosphate catalysts produces ethers. D. of diols produces dienes, and intramolecular D. of 1,4-diols gives tetrahydrofurans. Hydroxycarboxylic acids give characteristic dehydration products which depend on the relative positions of the hydroxyl and carboxyl groups. Carboxamides and aldoximes are dehydrated to nitriles, 1,4-dicarboxylic acid compounds to furans, and carboxylic acids to their anhydrides.

2) Removal of water from a material. Solids can sometimes be dehydrated even at room temperature, but usually the process requires heat, up to 200 °C, in a drying oven, or at higher temperatures in a muffle furnace. Materials containing small amounts of water can be dried in a desiccator. D. can be accelerated by applying a vacuum or by passing dry air across the material (see Drying). Aqueous solutions are treated in the same way as solids. However, they can also be dehydrated by freezing. Nonaqueous solutions or solvents are dried by putting hygroscopic drying agents, such as phosphorus pentoxide, roasted sodium sulfate, calcium chloride or sodium wire, into them. More on this subject is given under Solvents (see). The substance which remains behind after D. is the Dry mass (see).

In chemical technology, there are also a number of mechanical methods for dehydration of solid-liquid mixtures, e.g. vacuum and pressure filtration and D. on sieve, dish and jet centrifuges, on D. sieves or drying drums.

Dehydroacetic acid, ***3-acetyl-6-methyl-2H-pyran-2,4(3H)-dione***: colorless crystals, m.p. 112 °C, b.p. 270 °C. D. is soluble in ether, acetone and hot water, and slightly soluble in alcohol. It is synthesized by heating acetoacetic acid ethyl ester with sodium hydrogencarbonate or barium oxide, or by heating diketene in the presence of a basic catalyst such as triethylamine or pyridine. When D. is boiled with dilute mineral acid, hydrolysis of the lactone ring

O COCH3 5 4 3 6 1 2 H3C O O

produces a β-ketoacid, which is decarboxylated and dehydrated to form 2,6-dimethyl-4H-pyran-4-one. This is used in organic syntheses, as a softener, additive to toothpastes and skin cremes, and as a fungicide and herbicide.

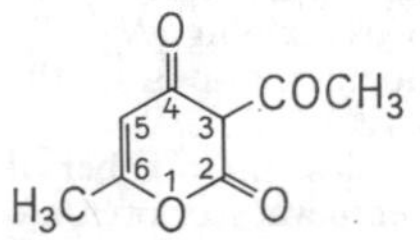

1,2-Dehydrobenzene: see Arynes.

Dehydrogenases: enzymes which catalyse oxidoreduction reactions by transferring hydrogen from one substrate (hydrogen donor) to another (hydrogen acceptor). The most important D. are Flavin enzymes (see) and pyridine-nucleotide-dependent D., in which NAD^+ or $NADP^+$ is the primary H acceptor.

Dehydrohalogenation: elimination of hydrogen halide from organic compounds. D. of haloalkanes and halocycloalkanes produces alkenes or cycloalkenes. Alkynes can be produced from 1,2- or 1,1-dihalo- or halovinyl compounds by D. D. of acyl halides can be used to produce ketenes. D. is important for certain heterocycle syntheses, e.g. those of oxiranes and oxetanes from β-halogen alcohols or γ-halogen alcohols, or of benzene oxide or oxepin by dehydrobromation of 3,4-dibromo-7-oxabicyclo[4.1.0]heptane:

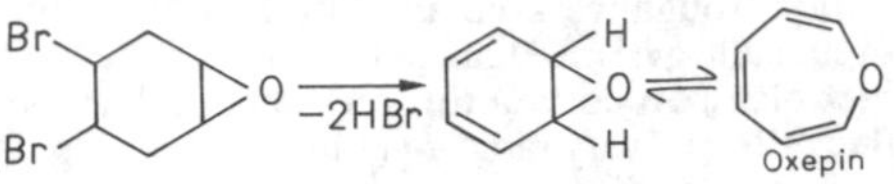

Br Br O —2HBr H O H ⇌ O Oxepin

Dekamon I: see AMO explosives.

Delayed-coking process: see Cracking.

d-Elements: see Elements.

Delocalization energy: the difference between the energy of a conjugated π-electron system and the additive energy of a reference state with strictly localized two-center bonds. In the Hückel approximation, this difference is ascribed to the delocalization of the π-electrons and is expressed in units of the resonance integral (see Hückel method). The D. can be determined experimentally by thermochemical measurements, for example of hydrogenation and combustion energies. In the calculation of the D. of benzene, the reference system is the hypothetical cyclohexatriene with three isolated double bonds. From this point, the ground state of benzene is calculated in two steps. First the compression of the σ-bonded skeleton requires 77.3 kJ mol^{-1}. The delocalization of the six localized π-electrons yields the *vertical D.* of -227.9 kJ mol^{-1}. Only the residual energy ΔE = -150.5 kJ mol^{-1} is detected by thermochemical measurements. Thus it is clear that the D. depends greatly on the definition of the reference state.

Delphin: see Delphinidin.

Delphinidin, ***delphinidin chloride***: 3,5,7,3',4',5'-hexahydroxyflavylium chloride, the aglycon of numerous Anthocyans (see). In the form of its glycosides, D. occurs widely in higher plants, for example as *delphin* (3,5-di-β-glucoside) in the blossoms of delphiniums, and as *violanin* in violets.

Delta metal®: a special brass (see Brass) which can be worked and cast. It consists of 41% zinc, 1% lead, 1% iron, 1% manganese and the rest copper; it is used in machines.

Demanganization: in water treatment, reduction of the manganese content to a permissible value; e.g.

in drinking water, to 0.05 mg/l. Manganese is found in iron-containing waters at concentrations from 0.5 to 5 mg/l. It gives an unpleasant taste to the water and forms deposits in plumbing. It is usually removed by aeration and filtration.

Demecolcine: see Colchicine.

Demelverin: see Spasmolytics.

Demetallization: 1) the partial or complete removal of metal coatings without attack on the underlying metal. Depending on the coating metal removed, the process is called, e.g. decopperizing, dezinking, detinning, etc. D. is done either chemically (dipping process) or electrochemically (electrolytic D.). In the *dipping processes*, the objects are treated with solutions, e.g. hydrochloric acid for zinc and cadmium coatings, potassium cyanide for gold and copper, and sodium hydroxide for tin. In *electrolytic D.*, the object is used as the anode. The composition of the bath is not essentially different from that used in the dipping process, but electrolytic D. is faster. Voltages of 2 to 6 V are used.

Detinning with dry chlorine gas is an exception; here liquid tin tetrachloride is the product. This process is used to recover the tin from tinned steel scraps.

2) Corrosion of alloys: the more reactive metal is preferentially dissolved; for example, in the corrosion of brass, Zinc loss (see) occurs. A comparable process occurs with aluminum bronzes in some corrosive media, e.g. concentrated sulfuric acid. Copper-nickel alloys sometimes lose nickel in air-containing hydrofluoric acid. The form of corrosion known as graphitization in gray cast iron is due to a loss of iron.

3) Surface depletion of the metal concentration in an alloy by sublimation of one of the components; for example, zinc loss from brass objects occurs by sublimation of the zinc at high temperatures.

Demeton: see Organophosphate insecticides.

Demineralized water: water from which all salts have been removed.

Demister: a device for forcing precipitation of the finest droplets of liquids from gases, vapors or mists. They are made of wire-knit packing and have a very large internal surface area.

Demulsification: the separation of emulsions by physical or chemical methods. Mechanical, electrical or thermal forces or chemical additives (demulsifiers) are applied to oppose the stabilizing emulsifiers. ***Demulsifiers*** are substances which change the charge and surface tension of the emulsified particles. Mixtures containing sodium salts of sulfonic acids, starch or sodium oleate are often used.

D. is very important for separation of the oil-water emulsions which are frequently pumped out of oil wells.

Denaturation: see Biopolymers.

Denatured alcohol: ethanol to which another substance has been added to make it unfit for drinking. Methanol, acetone or pyridine bases can be used as denaturants. D. is used as a fuel, for example, under chafing dishes.

Denaverin: see Spasmolytics.

Dendrites: branching growth forms of crystals which resemble trees. D. are single crystals, and their branches are crystallographically oriented. They can arise during crystallization from solutions or melts when the growth rate is high, due to a high degree of supersaturation or supercooling. Under these conditions, the edges and vertices of the crystal grow at the expense of the flat surfaces, due to inadequate surface diffusion. Such incomplete crystal forms are generally called *skeletons*; when the phenomenon is mild, it generates *hollow forms*, and D. are an extreme case. Dendritic growth is generally undesirable for crystal growing, because the products are impure and the form is irregular.

Denickelization: see Demetalization.

Denitrification: see Nitrification.

Densinite: see Macerals.

Density: symbols d, D, ϱ; the mass of a substance per unit of volume. The term *specific gravity* is now obsolete. For homogeneous bodies, $d = m/v$, where m is the mass and v is the volume. The *density number* is the ratio of the mass of a body to the mass of the same volume of water at 4 °C. The symbol D_4^{20} indicates the density number for a body at 20 °C relative to water at 4 °C.

Density sorting: a process for separation of the fractions of a solid mixture on the basis of their differing densities. The solid mixture is added to a liquid with a density intermediate between the densities of the fractions to be separated. The lighter material floats, and the heavier sinks. With ores, for example, the ore particles may sink while the matrix rock floats. Very dense liquids are needed as separation media. In practice, sludges of fine-grained solids are used; for example, quartz sand, ferrosilicon, galenite, magnetite or arsenopyrite.

Dental alloys: alloys which resist corrosion and wear, used in dentistry. They are based on platinum, gold, silver, mercury, iron and cobalt. Rods for anchoring crowns are platinum-iridium alloys with 10 to 20% iridium. Fillings, crowns, bridges and plates can be made of gold alloys containing at least 50 mol-% gold plus platinum (see Goplat). Although silver quickly tarnishes in the mouth, 20 to 30% palladium and 0 to 10% gold makes a resistant silver alloy (see Sipal). In Germany during the Second World War, the shortage of gold and palladium led to development of a silver alloy with cadmium and tin (see Sident). Plastic fillings are made of silver amalgam. The dentist takes 1 part by weight of a "filling", a silver alloy with 50 to 74% silver and the rest tin and copper, and rubs it together with 0.8 to 1.5 parts mercury. The hardened alloy consists mainly of tin-mercury mixed crystals, the Ag_3Hg_4 phase and unreacted Ag_3Sn. Crowns, plates and braces are made of stainless steels. Cobalt-chromium alloys (see Vitallium®) are used to cast fillings, bridges and prostheses.

Dental silver: see Silver alloys, see Dental alloys.

Deoxo process: a purification process for synthesis gases. It was developed for gas flows between 0.15 and 600 $m^3 h^{-1}$, and uses noble metal catalysts under pressures of 0.3 to 1.0 MPa. With the D., it is possible to remove small amounts (< 2 vol. %) of the following gases: oxygen from hydrogen, oxygen or hydrogen from nitrogen, carbon dioxide, hydrocarbons or noble gases; oxygen or hydrogen in the presence of carbon monoxide; hydrogen from oxygen or gases containing large amounts of oxygen; unsaturated hydrocarbons from various gases; carbon monoxide from gases which do or do not contain hydrogen.

Deoxyaldoses: see Deoxysugar.

Deoxybenzoin, ***benzyl phenyl ketone***: C_6H_5-CH_2-CO-C_6H_5, an aliphatic-aromatic ketone. D. crystallizes out of alcohol as white platelets, m.p. 60 °C, b.p. 320 °C. It is insoluble in water, but is soluble in alcohol, ether and chloroform. It undergoes the reactions typical of Ketones (see). D. can be synthesized by reduction of benzoin with tin and hydrochloric acid, or by a Friedel-Crafts reaction of benzene with phenylacetyl chloride. It is used for organic syntheses.

Deoxycholic acid: 3α,12α-dihydroxycholanoic acid, a bile acid. D. forms colorless crystals which are only slightly soluble in water, but dissolve easily in alcohol; m.p. 176-177 °C, $[\alpha]_D$ +57° (ethanol). The residual solvent is difficult to remove from the crystals. D. forms choleic acids (see Bile acids) with fatty acids. It is found in conjugated form in bovine and human bile.

Deoxycortone: see Adrenal cortex hormones.

Deoxycortone acetate: see Adrenal cortex hormones.

6-Deoxygalactose: same as Fucose (see).

Deoxyribonucleic acids: see Nucleic acids.

2-Deoxyribose: 2-deoxy-D-*erythro*-pentose, colorless crystals; m.p. 90 °C, $[\alpha]_D$ - 58° (in water). 2-D. is soluble in water and slightly soluble in alcohols. It is present as the *N*-glycoside in deoxyribonucleic acids and their components, the deoxyribonucleosides and deoxyribonucleotides.

Deoxysugar: a monosaccharide in which one or more hydroxyl groups has been replaced by a hydrogen atom. The naturally occurring D. are almost all *deoxyaldoses*, and they are always found in glycosidic linkage. Either a primary or a secondary hydroxyl group can be missing.

The ***6-D.*** (***methyloses, methylpentoses***) include rhamnose and fucose. The most important of the 2-D. is 2-deoxyribose, the sugar component of the genetic material DNA. Glycosides of 2-D. are more readily cleaved by acid catalysis than those of normal sugars with hydroxyl groups in the 2-position. The cardiac glycosides (lanata and purpurea glycosides) are especially rich in D., e.g. the 2,6-dideoxyaldohexoses and their methyl ethers. 3,6-dideoxyaldohexoses are found as components of lipopolysaccharides of microorganisms.

Deparaffinization: see Paraffin.

Dephlegmator: a reflux condenser with only partial condensation.

Dephosphorization: in steel production, the removal of phosphorus from the iron melt by oxidation (refining) by the low-temperature reaction 10 FeO + 4 P → 10 Fe + P_4O_{10}. Melts which are rich in phosphorus form calcium phosphates with basic, lime-rich slags; the reaction is exothermic.

Depolarizer: in the broadest sense, a substance which counteracts the polarization of an electrode; in the narrow sense, a substance which can be polarographically reduced or oxidized (see Polarography).

Depolymerization: a process in which a polymeric substance is broken into smaller pieces by mechanical or thermal effects. For example, caoutchouc is partially depolymerized by mastication, and polystyrene is almost completely depolymerized by heating to 300 °C (see Ceiling temperature). The D. of old tires and plastic wastes is becoming increasingly important.

Deposition: the settling of pollutants (see Air pollution) on the soil, vegetation, buildings and waters. ***Wet D.*** occurs as Rainout (see) and Washout (see); precipitation (rain, snow, etc.) can thus have a certain cleansing effect on the air. In certain cases, Acid precipitation (see) results. In ***dry D.***, gases and dust affect vegetation, soil, water, etc. without the help of rain.

Depsides: condensation products of aromatic hydroxycarboxylic acids. Depending on the number of carboxylic acid units, one speaks of di-, tri-, ... oligodepsides. The gallic acid derivatives which are components of Gallotannins (see) are D.

Depsipeptides: heterodetic peptides which contain ester as well as peptide bonds. Thus the D. include homeomeric O-peptides and peptide lactones of the hydroxyamino acids (serine, threonine, etc.) and of heteromeric peptides with hydroxyacid components in the sequence; these are also called ***peptolides***. Most peptolides are cyclic compounds. D. are metabolic products of microorganisms, and usually have strong antibiotic effects; they are therefore of great interest. Some important peptide lactones are the actinomycins and etamycin; the peptolides include the enniatins, sporodesmolides, valinomycin, espirin, etc.

Derivative: a compound which can be derived formally or actually synthesized from a parent compound. D. are prepared, for example, for identification and characterization of organic compounds (see Structural elucidation).

Dermatan sulfate, formerly ***chondroitin sulfate B*** or ***B-heparin***: an acidic mucopolysaccharide. D. is a copolymer of β(1→3)-*N*-acetyl-D-galactosamine 4-sulfate and β(1→4)-L-iduronic acid. It occurs bound to proteins, especially in the skin, and also in the tendons and the aorta.

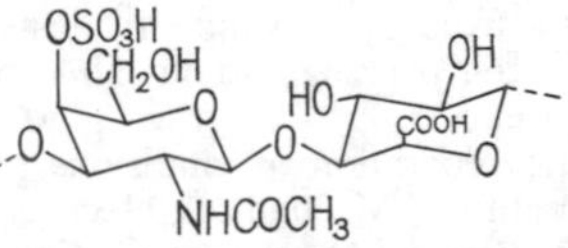

Dermorphins: a new class of opiate-like peptides found in amphibian skins. D., Tyr-D-Ala-Phe-Gly-Tyr-Pro-Ser-NH_2 and [6-hydroxyproline]D. were isolated from methanol extracts of the skin of the South American frogs *Phyllomedusa sauvagai* and *Phyllomedusa rhodei*. The presence of D-alanine in both D. is surprising and completely novel for peptides which are not of microbial origin. The D. and their analogs can be synthesized chemically. They have a very intense and long-lasting peripheral and central opiate-like activity. The analgesic effect of D. is a thousand times stronger than that of morphine, when injected intracerebroventricularly.

Desalination: the removal of salts from water. Complete desalination, the removal of all ions from the water, is done on an industrial scale to prepare water for use in steam boilers. Partial desalination is necessary in many countries to produce water of suitable quality for industrial or even agricultural use; it may be done to provide potable water from seawater on ships and in dry coastal areas where economical energy sources (e.g., solar) are available. A few plants are combined with nuclear power plants (e.g. in the former USSR).

There are a number of methods for D.

1) In the ***thermal method***, the water to be desalted is either heated to produce steam or cooled to produce ice. The residual liquid is greatly enriched in salts. The vapor, which contains only a few salt particles and gases such as oxygen, carbon dioxide and nitrogen, is condensed. The ice is mechanically separated from the salty water and warmed to provide fresh water. To increase the efficiency of energy use, and thus make such plants more economical, the heat of condensation or melting is recycled.

a) In *multistep distillation (cascade distillation)*, the heat of condensation of the water vapor is utilized in successive distillation kettles. To compensate for the unavoidable heat losses, the successive kettles are operated at increasing levels of vacuum. b) In *thermocompression*), the heat of compression removed from the water vapor being condensed is used to preheat the salty water (heat pump principle). c) *Vacuum distillation* works at temperatures of 50 °C with periodic pressure reductions, which cause the amount of water vapor to increase. d) in *solar distillation*, the solar irradiation of 28,000 kJ m^{-2} per day in desert regions is sufficient for about 14 l per m^2 of evaporation surface and day. e) *Freezing methods*, in which sea water is frozen to produce fresh water, have been tested. However, their use will remain limited to a few special cases, because these methods usually require a very large input of energy.

2) ***Semipermeable membrane methods*** (see Osmosis, Hyperfiltration) include *electroosmosis*, which is suitable for water with a sodium chloride content up to 5000 mg/l. The water is separated into several cells by electrolyte-permeable membranes of polystyrene or cellulose acetate and direct current passes through it. Water flowing between a pair of membranes, one anion-selective and the other cation-selective, is depleted in ions (desalted; Fig.).

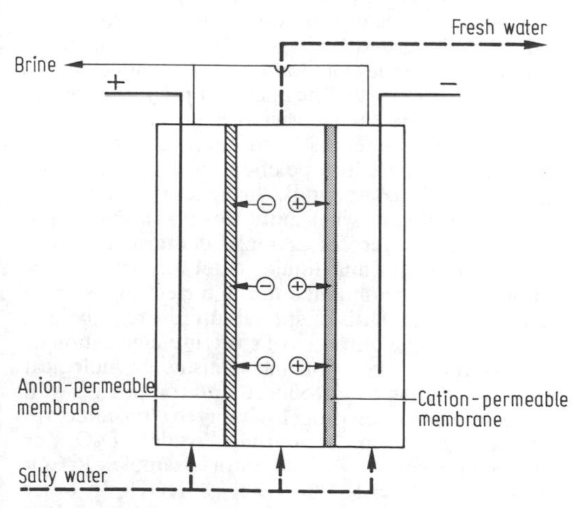

Scheme for desalination of brackish water or seawater by electroosmosis

3) ***Chemical methods***. a) *Liquid-liquid extraction*. This method is based on the abilities of various substances, some of them liquid, to adsorb electrolytes. Other substances, such as certain aliphatic amines, are able to absorb water. Depending on the nature of the extraction agent, the salt components can be separated from the solvent water. b) D. by removal of the carbonate hardness: see Water softening.

Descaling: the removal of thin layers of corrosion and oxide, usually from steel, by pickling with dilute hydrochloric or sulfuric acid. D. is a process of Surface pretreatment (see).

Desensitization: see Spectral sensitization.

Desiccants: substances used to dry plants to accelerate and facilitate the harvest, for example, bipyridylium herbicides (see Herbicides, see Drying agents).

Desiliconizing: in steel production, the oxidative removal of silicon from the iron melt (refining) according to the equation $2\ FeO + Si \rightarrow 2\ Fe + SiO_2$. Calcium silicates form in the lime-containing slag.

Desipramine: see Dibenzodihydroazepines.

Desizing agent: see Sizings.

Desmetryn: see Triazine herbicides.

Desmoids: see Saponins.

Desmosine: see Elastin.

Desorption: 1) the removal of dissolved gases from a liquid. This may be done by use of an inert gas, evaporation of the solution, application of a vacuum or by a combination of these methods. D. is used to obtain an absorbed gas component and to regenerate the solvent. 2) removal of adsorbed gases from an adsorbant by heating or displacement.

Destimulator: see Corrosion inhibitors.

Desulfonation: hydrolysis of arene sulfonic acids. Sulfonation of arenes is a readily reversible process. Benzenesulfonic acid undergoes D. when treated with dilute hydrochloric acid at temperatures between 150 and 200 °C: $C_6H_5\text{-}SO_3H + H_2O \rightarrow C_6H_5 + H_2SO_4$. Other arenesulfonic acids are desulfonated even by steam distillation.

Desulfuration: cleavage of sulfur as hydrogen sulfide (in the form of the nickel salt NiS) from thiols and certain five-membered sulfur heterocycles, through the reaction with hydrogen and Raney nickel, for example:

$$\text{(thiophene)} \xrightarrow[-NiS]{+6H,\ +Ni} CH_3\text{-}CH_2\text{-}CH_2\text{-}CH_3$$

Desulfurizing: 1) in metallurgy, the removal of sulfur from an iron melt by formation of calcium sulfide with lime, CaO, or calcium carbide, CaC_2, under reducing conditions: $FeS + CaO \rightarrow CaS + FeO$; $FeS + CaC_2 \rightarrow CaS + Fe + 2C$. Some D. is accomplished by manganese present in the melt: $FeS + Mn \rightarrow MnS + Fe$. Calcium sulfide and manganese sulfide are absorbed by the slag. Addition of soda, Na_2CO_3, to the melt produces volatile sodium sulfide, Na_2S. During refining, the sulfur is oxidized: $2\ FeS + 3\ O_2 \rightarrow 2\ FeO + 2\ SO_2$.

2) In fuel gases: see Gas purification mass.

Detajmium bitartrate: see Rauwolfia alkaloids.

Detectors: see Gas chromatography.

Detergent: a collective term for cleaning products based on surfactants. The term D. is not identical with the term surfactant.

Detinning: see Demetalization.

Detonating cap, ***blasting cap***: a copper or aluminum shell filled with an explosive and used to ignite an explosive. D. generally have a basic charge of a powerful explosive (usually pentaerythritol tetranitrate or 2,4,6-trinitrophenyl methylnitramide) and an ignition charge which is sensitive to sparks or flame; for copper caps this consists of mercury fulminate, and for aluminum caps, lead azide with added lead trinitroresorcinate. S. are connected to an ignition fuse or electric wire and placed in the explosive charge. Delay caps have a delaying element between the ignition charge and the electric fuse.

Detonating oils: a term applied to Glycerol trinitrate (see) (Nitroglycerol) and Ethyleneglycol dinitrate (see) and mixtures of these explosives.

Detonation: a particularly violent kind of Explosion (see), in which the explosive substance usually reacts in a branched chain reaction (see Reaction, complex). A shock wave, the ***detonation wave***, is formed which travels faster than the speed of sound (sometimes up to 9000 m s^{-1}). Behind the shock wave, adiabatic compression leads to a large increase in temperature, which immediately causes the explosive in that region to react. In addition, the adiabatic compression causes the speed of sound to increase, so that the gaseous products formed as the reaction expands flow out toward the original detonation wave with very high velocity and reinforce it. This creates a very steep shock wave with pressures of 10^9 to 10^{10} Pa, which can cause enormous damage to the environment. See also Initiation.

Detoxication: 1) the conversion of a toxic substance, especially a Chemical weapon (see), to a harmless one by physical, chemical and/or biochemical processes. The poison is either converted to a form which is no longer active (immobilized), diluted to the point that acute intoxication is no longer a danger (washing processes) or converted to a less toxic form by chemical and/or enzymatic treatment. The most important D. reagents which are relatively universal in their effects are lime chloride, chloramines, alcoholates and alcohol-amine mixtures, and nearly all alkalies, especially sodium hydroxide, potassium hydroxide and ammonia solutions. Immobilization can be achieved with adsorption agents such as activated charcoal, certain layer silicates which can swell, synthetic resins or zeolites. In this case, immobilization is partly a process of ion exchange.

Both organic solvents (including gasoline or diesel fuel) and water (if possible, hot water or steam) can be used for washing.

The methods of D. depend on the object. a) ***Persons***. If the person has absorbed the poison, medicinal detoxicants (see Antidotes) are administered, if possible under medical supervision. For example, phosphoric acid ester damage is treated by injection of atropine and pralidoxime (PAM), and organoarsenic compounds are counteracted by treatment with dimercaprol (BAL). If the poison is stuck to the skin, it should be washed thoroughly and treated with mild alkalies, chloramine or lime chloride slurry. There are also detoxicant salves which quickly transport the active substance into the deeper skin layers.

b) ***Clothing*** is treated by special laundry methods or by treatment with hot steam or steam/ammonia. The method and reagent depend on the type of fiber.

c) ***Equipment*** is treated by a variety of methods, depending on whether it is a motor vehicle, a fine-mechanical or optical device or a corrosion-resistant object of wood, ceramics, glass, plastic, stainless steel, etc.

d) D. of ***land, streets and buildings*** requires very large amounts of detoxicants and must therefore be reserved for vital centers. Otherwise, the decomposition of the poison must be left to the effects of weather. In some cases, affected areas must be closed for long periods, and can only be visited by persons wearing protective clothing.

The personnel who carry out D. require special protection measures, such as protective clothing and continual medical supervision.

2) The biochemical binding of toxic products of metabolism or ingested toxins by physiological mechanisms. This ability of the body to detoxify certain substances is limited to relatively low concentrations, and therefore plays no significant role in acute poisoning. The endogenous detoxication capacities of individuals vary; there are people who can process relatively high concentrations of poisons (e.g. potasium cyanide, KCN) and show no symptoms, while others who are more sensitive would be in acute danger from the same concentrations.

Detrite: see Coal.

Deuterium, ***heavy hydrogen, D, ^2H***: a hydrogen isotope with mass number 2; the nucleus (***deuteron***, symbol d) contains a proton and a neutron; atomic mass 2.0140. The density of liquid molecular D_2 at the b.p. is 0.169, m.p. - 254.0 °C, b.p. - 249.7 °C. D. is present in natural hydrogen at a concentration of 0.015%. Like dihydrogen, it can exist in two forms, in which the relative orientations of the nuclear spins are different, ***ortho*** and ***para D.*** (see Hydrogen). The mass difference between a ^1H and a D atom is about 100%, which is the greatest relative difference found between any pair of isotopes. Therefore, protium and D. differ significantly in some physical properties, such as melting and boiling points, densities, vapor pressures, heats of evaporation, heat conductivities, rates of diffusion, etc. The chemical properties of the two isotopes and their derivatives are nearly identical, although differences in reaction rates can often be observed. These isotope effects are used to enrich and isolate D. compounds. In general, the D. derivatives react somewhat more slowly, so that in the electrolysis of water, for example, deuterium oxide is enriched in the residual liquid. After multiple fractionation, D. can be obtained through electrolysis or by chemical means. D. is of special interest for labelling molecules for the purposes of structure elucidation or for the study of reaction mechanisms or biological processes. D. can be introduced into compounds with sufficiently acidic hydrogen simply by repeated recrystallization from deuterium oxide, D_2O, or deuterium alcohol, C_2H_5OD; for example, R-CO-NH-R + D_2O $\rightleftharpoons$ R-CO-ND-R + HOD. In other cases, special synthetic pathways must be used. It is hoped that industrial realization of the fusion of D.

nuclei to helium will make a decisive contribution to the solution of the world energy problem. D. was discovered in 1933 by Urey.

Deuterium oxide, ***heavy water***: D_2O, a liquid which is comparable to water; M_r 20.031, density (20 °C) 1.105, m.p. 3.82 °C, b.p. 101.42 °C. D. makes up 0.015% of natural water, and in a few saline lakes and in deep seawater, it is somewhat more concentrated. In mixtures of D. and water, the D and H exchange rapidly, so that most of the deuterium is present as HOD. When the water is electrolysed, H_2 is generated preferentially, so that D. becomes enriched in the residual liquid. By repeated electrolysis, it is possible to obtain pure D. D. is used in nuclear reactors as a coolant and neutron modulator; it is the starting material for production of deuterized compounds. At high concentrations in water, D. is toxic to organisms.

Deuterium hydride, ***HD***: a compound of deuterium and hydrogen.

Deuteron: see Deuterium.

Deuteroporphyrin: see Porphyrins.

Devard's alloy: an alloy consisting of 50% copper, 45% aluminum and 5% zinc. It is very brittle and can therefore be pulverized easily. In acidic and alkaline solutions, the aluminum and zinc react with the solvent, forming hydrogen, which is such a strong reducing agent in statu nascendi that, for example, it can reduce nitrates to ammonia. D. is therefore used in the laboratory as a reducing agent.

Development dyes: a group of water-insoluble azo dyes which are generated directly on the fibers. The fibers to be dyed are first impregnated with either a diazonium salt solution or a solution of the coupling component. The color is then developed with a coupling component (if the fibers were impregnated with the diazonium salt) or a diazonium salt (if the fibers were impregnated with the coupling component). The D. include the ice dyes, such as naphthol AS dyes and many other azo dyes, aniline black, naphthogen dyes and Zambesi dyes.

Dewar benzene, ***bicyclo[2.2.0]hexa-2,5-diene***: a valence isomer of benzene. D. is a bent molecule and is not aromatic, in contrast to the planar benzene. The trivial name was given in connection with the formula for Benzene (see) suggested by J. Dewar, which was not correct. D. can be made by UV irradiation of benzene. It is very unstable, and when heat is added, it is reversibly converted to benzene. Alkyl and especially *tert.*-butyl substituted derivatives are much more stable and have a significant tendency to form. For example, tri-*tert.*-butylbenzene is very readily converted by UV irradiation to the stable D. with *tert.*-butyl substituents. A substituted D. is formed by cycloaddition of tetra-*tert.*-butylcyclobutadiene with acetylene dicarboxylate.

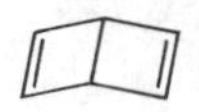

Dew point: the temperature at which water vapor present in a mixture of gases begins to condense. At the D., the vapor pressure is equal to the saturation pressure. Below the D., clouds, dew or frost can form. The moisture content of the air can be determined by measuring the D.

Dexamethasone: see Adrenal cortex hormones.

Dexamphetamine: see Amphetamine.

Dexpanthenol: see Pantothenic acid.

Dextran: an extracellular microbial polysaccharide, formed by bacteria of the *Leuconostoc* genus, e.g. *L. mesenteroides*. It is a highly branched glucan with an α(1→6)-linked backbone and α(1→3), α(1→4) and α(1→2) branches. The α(1→2) branches are less frequent than the others. The molecular mass of native D. can be very high, and the viscosity of its solutions increases with increasing molecular mass. D. sometimes causes clogging of filters and pipes in sugar processing plants. The film of D. which forms on teeth is an important factor in the development of caries.

Solutions of partially hydrolysed D. with molecular masses between 50,000 and 100,000 are used as plasma expanders to replace plasma in patients who have lost large quantities of blood. *Sephadex* is a D. cross-linked with epichlorohydrin to form a molecular sieve for gel chromatography.

Dextrin: an oligosaccharide obtained by partial degradation of starch. ***Heat D.*** are dark-colored products of heating starch to 100-200 °C; they form on the surface of foods during baking. ***Acid D.*** are formed by acid-catalysed hydrolysis of S. and subsequent heating to temperatures above 140 °C. The yellow D. produced by hydrolysis alone can be obtained as a colorless product by reprecipitation from water/ethanol. D. is soluble in water, and the solution rotates polarized light to the right. D. which are not too highly degraded give a red color with iodine solution. ***α-D.*** contains 6 glucose units; ***β-D.*** 7, and ***γD.***, 8. ***Limit D.*** are the products of amylopectin degradation with β-amylase, which removes only α(1→4)-linked glucose units at chain termini. It therefore stops when it reaches an α(1→6) branch point. D. are used as glues, thickeners and sizings, and are also used in the pharmaceutical industry.

Schardinger D. or ***cyclodextrins*** are ring-shaped oligosaccharides consisting of 5 to 8 glucose units. They are formed by degradation of amylose (see Starch) by an amylase from *Bacillus macerans*; they arise by enzymatic linkage of a helical turn of the amylose. α-cyclodextrin (cyclohexoamylose) consists of 5 to 6 glucose units, and readily forms inclusion complexes in aqueous solution.

Dextromethorphan: see Levorphanol.

Dextrose: same as D-Glucose (see).

DFP: see Fluostigmine.

DHD process: see Reforming.

Diabetic foods: foods intended primarily for patients with diabetes mellitus. They differ from ordinary foods in that they contain little or no digestible carbohydrate (e.g. glucose, maltose, lactose or sucrose), and contain artificial sweeteners such as saccharin or cyclamate, or sugar substitutes such as D-glucitol (sorbitol), xylitol or fructose instead. The fat and carbohydrate contents are reduced, and they are enriched with fiber.

Diacetonamine: 4-amino-4-methylpentan-2-one, (CH_3)-$C(NH_2)$-CH_2-CO-CH_3, a reaction product of acetone and ammonia. The first step of the reaction forms diacetone alcohol, which then reacts further with ammonia to form D. ***Triacetonamine*** (2,2,6,6-tetramethylpiperidin-4-one) is a byproduct which is the main product at elevated temperature. Its formation can be explained by reaction of D. with acetone.

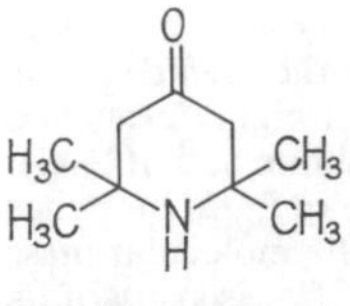

Triacetonamine

Diacetone alcohol, ***4-hydroxy-4-methylpentan-2-one***: $(CH_3)_2C(OH)\text{-}CH_2\text{-}CO\text{-}CH_3$, a colorless, combustible liquid; m.p. - 44 °C, b.p. 164 °C, n_D^{20} 1.4213. D. is readily soluble in water and most organic solvents. It has a weak narcotic effect and irritates the mucous membranes. Water is easily eliminated, forming mesityl oxide. D. is synthesized from acetone in the presence of alkaline earth hydroxides. It is used mainly as a solvent for resins, fats, cellulose ethers, chlorinated rubber and colophony. It is also used as a starting material for organic syntheses.

Diacetyl: same as Butane-2,3-dione (see).

Diacetyldioxime: same as Dimethylglyoxime (see).

Diacetylene: same as Butadiyne (see).

Diacetyltannin-protein-silver, ***Targesin***®: a brown substance which forms platelets with a metallic sheen. It is readily soluble in water. The silver content is 6.0-6.5%. The compound is used as an antiseptic, chiefly in the eye, but also for gastritis.

Diadochism: in crystal chemistry, a term for the ability of individual components of a crystal structure to replace isomorphic components. The requirement for D. are similar atomic or ionic radii, similar bonding character and preservation of the electrostatic equilibrium in ionic lattices. A D. between two elements is not a fundamental relationship; it can occur in certain crystalline substances but not in others. Potassium and rubidium, for example, are diadochic in most cases, as are magnesium and iron. "Coupled D." is a common phenomenon among the silicates; for example, in the feldspars a coupled substitution of calcium and aluminum for sodium and silicon are necessary to preserve the charge equilibrium.

D. explains the variable but often very stable compositions of many minerals.

Diagonal relationships: see Periodic system.

Dialdehydes: organic compounds containing two aldehyde groups, -CHO, in the molecule. There are aliphatic, aromatic and heterocyclic D. The aliphatic D. are very important in chemical syntheses. Because of the reactivity of the two aldehyde groups, they are used, for example, in the synthesis of carbocyclic and heterocyclic compounds. D. are usually synthesized by special methods. Some important examples are glyoxal, malonic dialdehyde, succinaldehyde, glutaraldehyde, glutaconic dialdehyde and the aromatic isomers phthalaldehyde, isophthalaldehyde and terephthalaldehyde.

Diallate: see Herbicides.

Dialuramide: same as Uramil (see).

Dialysis: a physical process used to separate low-molecular weight substances from solutions of macromolecular or colloidal substances. It is used mainly to purify macromolecular substances, especially to free biopolymers from salts bound to them. D. is used to detoxify blood (***hemodialysis***: artificial kidney). D. is often used to determine the colloidal structure of the dissolved substance. D. is based on the ability of low-molecular substances to pass through certain types of membranes, such as parchment, cellulose acetate or cellulose nitrate, animal skin, etc., which do not permit passage of macromolecular or colloidal substances. D. is accomplished in a dialysis cell or simply a bag made of a suitable membrane. The ***dialysis cell*** consists of two chambers separated by a membrane. One is filled with the solvent, e.g. water, and the other with the solution. The low-molecular-weight substances then diffuse through the membrane into the solvent, while the macromolecular or colloidal substance remains behind. The ***dialysis rate*** depends on the surface area of the membrane relative to the amount of liquid being dialysed, and on the difference in concentration between the two sides of the membrane. It is useful to allow the solvent to flow through the cell continuously, so that the low-molecular-weight substance cannot come to equilibrium. The rate can be increased by increasing the temperature and applying an electric field (see Electrodialysis). A special case of D. is Diasolysis (see). To separate colloids of different sizes from each other, Ultrafiltration (see) can be used.

D. is very important for removing Hemicelluloses (see) from spent leachate in viscose fiber plants. The leachate (18% sodium hydroxide solution) becomes enriched in hemicelluloses, so that a portion of it must be constantly renewed. Using D., the spent leachate is separated into an 8% sodium hydroxide solution free of hemicelluloses and a hemicellulose solution (Fig.). After the sodium hydroxide solution has been reconcentrated, it can be reused.

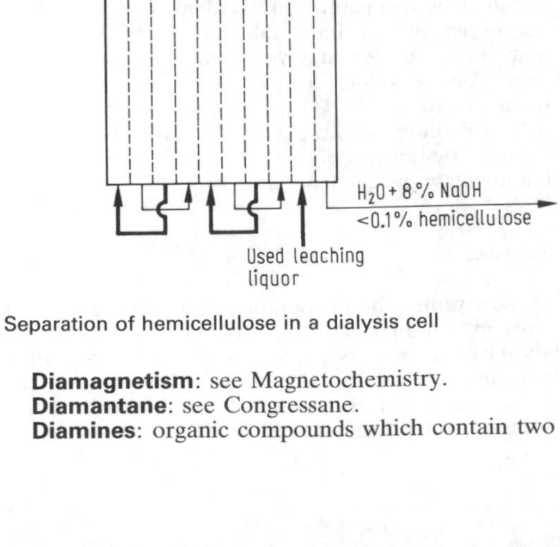

Separation of hemicellulose in a dialysis cell

Diamagnetism: see Magnetochemistry.

Diamantane: see Congressane.

Diamines: organic compounds which contain two

amino groups in the molecule, e.g. ethylenediamine or putrescine (diaminobutane). D. are basic and undergo the reactions typical of amines. They form mono- and diacidic salts, which are more or less stable, depending on the substituents. D. are very rare in nature. Only a few aliphatic D., such as putrescine and cadaverine, are found as bacterial degradation products of the amino acids ornithine and lysine. These two D. belong to the class of ptomaines (corpse poisons). The D. are synthesized by the same methods as Amines (see), and are used as bifunctional reagents in the production of polyamides and other polycondensates.

Diaminobenzenes, ***phenylenediamines***: three isomeric aromatic diamines with the general formula $C_6H_4(NH_2)_2$: ***1,2-D. (o-phenylenediamine)***, m.p. 103 °C, b.p. 256-258 °C; ***1,3-D. (m-phenylenediamine)***, m.p. 63-64 °C, b.p. 282-284 °C; and ***1,4-D. (p-phenylenediamine)***, m.p. 140 °C, b.p. 267 °C.

D. are colorless, crystalline compounds which turn dark in the presence of air. They are partially soluble in water, and dissolve readily in alcohol and ether. They undergo the reactions typical of primary aromatic amines, and form defined salts with mineral acids.

> Diaminobenzenes are poisonous. Like aniline, they act as blood poisons and can cause cyanosis and liver damage. They can also cause allergic skin reactions.

The D. are synthesized by reduction of the corresponding nitranilines. They are used mainly for synthesis of pigments and dyes. In addition, 1,3-D. is an important intermediate in the synthesis of heterocyclic compounds with at least two nitrogen atoms in the ring. 1,2- and 1,4-D. and some of their *N*-alkylated derivatives are used as developers in photography.

4,4'-Diaminobiphenyl: same as Benzidine (see).

1,4-Diaminobutane: same as Putrescine (see).

4,4'-Diamino-3,3'-dimethoxybiphenyl: same as Dianisidine (see).

4,4'-Diamino-2,2'-dimethylbiphenyl: see Tolidines.

4,4'-Diamino-3,3'-dimethylbiphenyl: see Tolidines.

1,2-Diaminoethane: same as Ethylenediamine.

1,6-Diaminohexane, ***hexamethylenediamine***: $H_2N\text{-}(CH_2)_6\text{-}NH_2$, an aliphatic diamine which forms colorless, fuming crystals with a silky sheen; m.p. 41-42 °C, b.p. 204-205 °C. H. is soluble in water, alcohols and hydrocarbons, and gives the reactions typical of a primary amine. It forms stable mono- and diacid salts with inorganic acids. H. is synthesized mainly by catalytic hydrogenation of adipic dinitrile; it is used in the form of the bisadipic acid salt (*AH salt*) for production of polyamides and polyurethanes.

1,5-Diaminopentane: same as Cadaverine (see).

2,4-Diaminophenol dihydrochloride: same as Amidol (see).

Diamond: the cubic modification of carbon, and the hardest mineral, with a Mohs hardness of 10. Its density is 3.50 to 3.52. The properties of D. are the result of the strong covalent bonds between the tetrahedrally coordinated carbon atoms with C-C distances of 154 pm. D. is a brittle mineral, and can readily be pulverized by cleavage along octahedral cleavage planes. Pure D. is colorless and transparent, but impurities such as calcium, aluminum, silicon, titanium, chromium and iron give colors of yellow, blue-green, gray and black. The dark ***carbonados*** are polycrystalline D. with up to 4% impurities, some of which are gas and liquid inclusions. Because of the covalent localization of its electrons, D. is an electrical insulator, and because of its short C-C distances, a relatively good conductor of heat. D. is chemically relatively inert; however, it is attacked by oxidizing liquids, and burns in pure oxygen at 720 °C, or in air at 800 °C, to carbon dioxide, CO_2. Under an inert atmosphere, D. converts at 1500 °C into graphite, in an exothermal reaction ($\Delta_R H$ = 2.69 kJ mol^{-1}.) The mass of D. is given in carats: 1 carat = 200 mg.

Synthesis of D. from graphite requires pressures of more than 12,500 MPa and temperatures around 3000 K. (See also CVD diamond Synthesis, p. 300.) The conversion rate is increased by catalysts such as chromium, iron or platinum. Industrial diamonds obtained in this way are very small (0.1 carat). Like natural D., they are used for drilling, sanding and cutting hard materials; in dies for drawing very fine wires, and as bearings for precision instruments. Large, polished D. crystals are the most expensive gemstones. The largest D. is Cullinan, which weighs 3,025 carats.

Diamond black: 1) a diazo pigment, a black-brown powder which dissolves in water with a blue-violet color. D. dyes wool black by the chromizing process and is very fast to light. It is made by diazotizing 4-aminosalicylic acid and coupling the product with α-naphthylamine in an acidic medium. The resulting product is again diazotized and coupled with naphth-1-ol-4-sulfonic acid or naphth-1-ol-5-sulfonic acid.

2) a deep black type of gas soot, which becomes shiny when oil is stirred into it; it is used as an artist's oil paint.

Diamond green: same as Brilliant green (see).

Diamond type, ***diamond structure***: a Structure type (see) which is characterized by a regular arrangement of tetrahedrally coordinated atoms of the same type in a crystal lattice. In the lattice of diamond, each carbon atom lies at the center of a tetrahedron consisting of 4 other C atoms; each C atom is both a central and a vertex atom of coordination tetrahedrons. The unit cell of the D. (Fig. 1) contains 8 atoms, half in positions of a cubic face-centered lattice and the other half centered in the cubes made by dividing the unit cell into 8 parts. Diamond is the classic example of a crystal with pure covalent bonding forces. The strong, symmetrical σ-bonds based on sp^3 hybrid orbitals of the carbon atom form a network through the entire crystal, so that it can be considered a single giant molecule. The shortest interatomic distance in the lattice (154 pm) agrees with the C-C distance in aliphatic compounds.

The structure of the D. explains the unusual properties of substances which crystallize in it, especially their high melting points and the extreme hardness of diamond. Its preferred cleavage parallel to morphological octahedral surfaces ([111] plane of the lattice) is due to the fact that in this direction relatively fewer bonds must be broken. In the puckered layers,

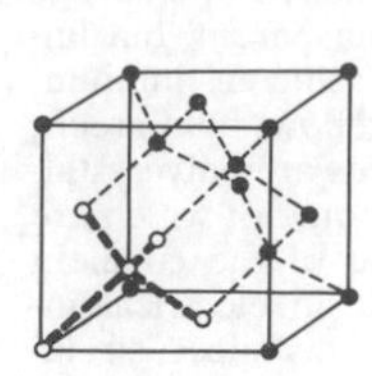

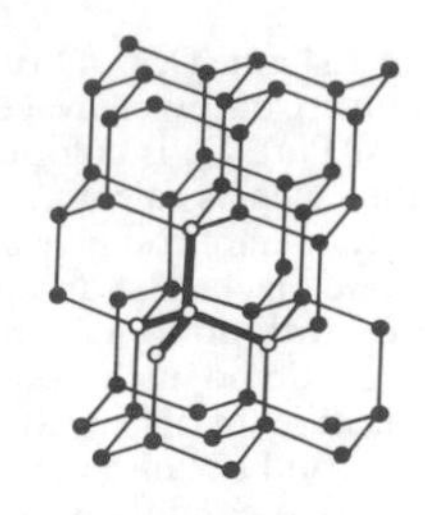

Diamond lattice: 1) unit cell, 2) lattice section with the [111] direction vertical (= direction of the space diagonals in 1)

3/4 of the bonds are parallel to the [111] plane, and only 1/4 are perpendicular to it (Fig. 2). Carbon, silicon, germanium and α-tin (gray tin) crystallize in the D. If the lattice positions in the D. are occupied in alternation by two different types of atoms, the structure is the Zinc blende type (see).

Dian: same as Bisphenol A.

Dianisidine, ***4,4'-diamino-3,3'-dimethoxybiphenyl***: colorless crystalline leaflets which turn violet in the air; m.p. 137 °C. D. is barely soluble in water, but dissolves easily in most organic solvents. It is synthesized by a Benzidine rearrangement (see) of 2,2'-dimethoxyhydrazobenzene. D. is used mainly to synthesize azo pigments.

Diaphragm: in electrochemistry, a porous partition wbich separates the cathode and anode spaces in electrolysis cells (e.g. in chloralkali electrolysis) without preventing a flow of current. Ceramic plates, asbestos layers, plastic foils and other materials are used as D.

Diarylmethane pigments: a group of pigments in which a central carbon atom is bound to two aryl groups (usually phenyl groups); see Triarylmethane pigments.

Diasolysis: a special case of Dialysis, in which a substance dissolves through a membrane into another liquid. Unlike the membranes used in dialysis, the membranes used for D. act as a solvent for the substance which passes through them. A process in which dialysis and D. occur simultaneously is conceivable, if the membrane is porous and its material is also able to dissolve and diasolyse the substance passing through it. The membrane materials used consist either of cellulose derivatives or polyvinyl chloride, polystyrene or polyethylene. D. is used to separate hydrophilic and hydrophobic compounds.

Diastereoisomerism: see Stereochemistry 1.2.1.

Diastereoselective synthesis: see Asymmetric synthesis.

Diastereotopic groups: see Topic groups.

Diatomaceous earth: same as Kieselguhr (see).

Diazepam: see Benzodiazepines.

Diazepines: a term for triply unsaturated, seven-membered heterocyclic compounds containing two ring nitrogens.

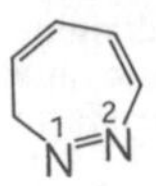

1,2-Diazepine

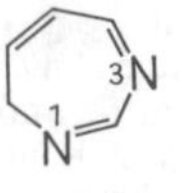

1,3-Diazepine

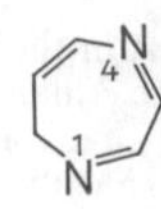

1,4-Diazepine

1,2-D., ***1,3-D.*** and ***1,4-D.*** differ in the positions of their nitrogen atoms. 1,4-benzodiazepine is the parent compound of a series of tranquilizers (see Benzodiazepines).

Diazines: the systematic name for six-membered heterocycles containing two nitrogen atoms. Three different ring systems are possible; see Pyridazine (1,2-diazine), Pyrimidine (1,3-diazine) and Pyrazine (1,4-diazine).

Diazinone: an Organophosphate insecticide (see) which is less toxic than most.

Diaziridine: a saturated, three-membered heterocyclic compound with two ring nitrogen atoms. It is synthesized by reaction of ammonia and chlorine with an aldehyde or ketone, or by the reaction of a Schiff's base with chloramine. The D. are strong reducing agents, and can be dehydrogenated to Diazirines (see).

Diazirine: an unsaturated, three-membered heterocyclic compound with two nitrogen atoms in the ring. D. is a colorless gas; b.p. - 14 °C. It is synthesized from difluoroamine and *tert.*-butylazomethine. The two reactants are mixed in carbon tetrachloride and the resulting D. is trapped by application of a vacuum in a cold trap at - 128 °C. ***1H-D.*** and ***3H-D.*** differ with respect to the position of the double bond. When heated or exposed to light, the D. form carbenes by splitting off nitrogen.

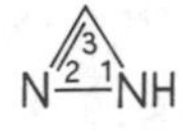

Diazoacetate: see Ethyl diazoacetate.

Diazoalkanes: very reactive, poisonous, aliphatic diazo compounds with the general formula $R^1R^2CN_2$. Two examples are diazomethane and diazocyclopentadiene. D. can be synthesized by alkali cleavage of substituted *N*-nitrosoalkylamides, dehydrogenation of ketohydrazones with mercury(II) oxide, the Bamford-Stevens reaction or Diazo group transfer (see). They are used for alkylation reactions, synthesis of α-diazoketones, ring expansion of cycloalkanones, 1,3-dipolar cyclo additions, production of carbenes, etc.

Diazoamino-aminoazo rearrangement: see 4-Aminoazobenzene.

Diazoaminobenzene, ***1,3-Diphenyltriazene***: C_6H_5-N=N-NH-C_6H_5, an aromatic diazo compound; yellow crystalline leaflets; m.p. 98 °C. D. is nearly insoluble in water but dissolves readily in hot alcohol, ether, benzene and pyridine. In a weakly acidic medium, it rearranges to form 4-aminoazobenzene. D. can be synthesized by partial diazotization of aniline and subsequent neutralization of the acidic solution with sodium acetate. It is used for synthesis of 4-aminoazobenzenes.

Diazo cleavage: a reaction in which the diazo group

$$-\overset{+}{N}\equiv N$$

of a diazonium salt is replaced by another atom or group, releasing nitrogen. Using D., hydrogen atoms, halogen atoms, hydroxyl, mercapto, azido or aromatic residues can be introduced in a specific position in a molecule. D. includes hydrolysis to phenols, the Sandmeyer reaction (see), the Bart reaction (see), the Schiemann reaction (see) and the Gomberg-Bachmann reaction (see).

Diazo compounds: organic compounds in which two N atoms are bonded to each other by a double or triple bond, and one of them is bonded to a hydrocarbon skeleton. The other N atom can be bound to a heteroatom or a CN group. D. can be synthesized by Diazo group transfer (see). Some typical D. are Diazomethane (see), Benzene diazonium chloride (see), Diazoaminobenzene (see) and benzene diazohydroxide.

Diazocyanides: organic diazo compounds with the general formula R-N=N-CN. D. can exist in either a stable E- or an unstable Z-form:

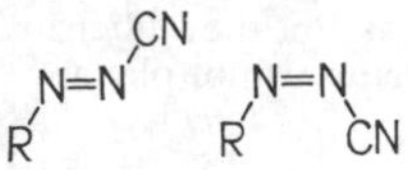

CN / N=N / R (E)-form; N=N with R and CN on same side (Z)-form

(E)-form (Z)-form

The Z isomers have lower melting points and decompose explosively upon moderate heating. D. are readily soluble in organic solvents. They are synthesized by reaction of arene diazonium salts with alkali cyanides.

Diazocyclopentadiene: an aliphatic diazo compound formed by Diazo group transfer (see) from arenesulfonyl azides and cyclopentadiene in the presence of a base. The structure can be described by the following resonance structures:

$$C_5H_4^{-}-\overset{+}{N}\equiv N| \longleftrightarrow C_5H_4=\overset{+}{N}=\overset{-}{\underline{N}}$$

D. contains the aromatic cyclopentadienyl anion.

Diazodinitrophenol: same as 4,6-Dinitro-2-diazohydroxybenzene (see).

Diazo group transfer: a method of synthesizing aliphatic diazo compounds by reaction of arenesulfonyl azides with CH-acidic reactants, such as cyclopentadiene.

In D., the N_2 structural element from the azide group is transferred to the compound with the active methylene group.

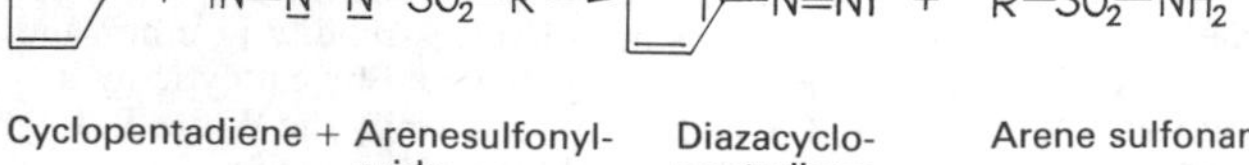

Diazoketones: with the general formula R-CO-$\bar{C}$H-$\overset{+}{N}\equiv$NI are relatively stable aliphatic diazo compounds. Their stability is due to the contribution of the keto group to the resonance system, and the associated release of additional delocalization energy:

$$R-\underset{\underset{|O|}{\|}}{C}-\bar{C}H-\overset{+}{N}\equiv N| \longleftrightarrow R-\underset{\underset{|O|_{-}}{|}}{C}=CH-\overset{+}{N}\equiv N| \longleftrightarrow \ldots$$

D. can be synthesized by reaction of acyl halides with diazomethane. When they are heated, especially in the presence of silver or silver oxide, D. cleave off nitrogen, forming a reactive ketocarbene; migration of the anion R^- converts the compound to a ketene. This reaction is known as the Wolff rearrangement (see), and makes possible chain lengthening of a carboxylic acid by one C atom. The method has been thoroughly studied in the context of the Arndt-Eisert synthesis (see).

1,2-Diazole: same as Pyrazole (see).

1,3-Diazole: same as Imidazole (see).

Diazomethane: CH_2N_2, the simplest aliphatic diazo compound. D. is a yellow gas; m.p. - 145 °C, b.p. - 24 °C. It is highly explosive as a gas, in condensed form and also in concentrated solutions. The ground state of D. is best described by the following zwitterionic limit structures:

$$|\bar{C}H_2-\overset{+}{N}\equiv N| \longleftrightarrow |\bar{C}H_2-N=\overset{+}{N}| \longleftrightarrow CH_2=\overset{+}{N}=\overset{-}{\underline{N}}|$$

Diazomethane is very poisonous. Inhalation of small amounts causes respiratory distress, coughing and confusion. Ether solutions of diazomethane are very irritating to the skin. Like phosgene poisoning, severe diazomethane poisoning can produce fatal lung edema with a latency time of a few hours or days. In experimental animals, a short exposure by inhalation led after a longer period to lung carcinomas. Artificial respiration and immediate medical help are necessary for acute diazomethane poisoning.

D. can be synthesized by alkaline cleavage of *N*-methyl-*N*-nitrosourethane or of the more stable *N*-methyl-*N*-nitrosotoluene-4-sulfonamide:

$$H_2N-CO-N-(CH_3)-N=O \xrightarrow{+OH^-} CH_2-N\equiv N + 2\,H_2O.$$

In addition, D. can be made by reaction of formaldoxime with chloramine or of chloroform with hydrazine in the presence of alkali hydroxides. D. is used as a reactive methylation reagent for H-acidic compounds, such as carboxylic acids, phenols, enols and hydrogen halides. Phenol, for example, is methylated according to: C_6H_5-OH + CH_2-N≡N →

$$C_5H_6 + |\overset{+}{N}=\underline{N}-\overset{-}{\underline{N}}-SO_2-R \longrightarrow C_5H_4^{-}-\overset{+}{N}\equiv N| + R-SO_2-NH_2$$

Cyclopentadiene + Arenesulfonyl-azide; Diazacyclo-pentadiene; Arene sulfonamide

$[CH_3\text{-}N{\equiv}N]^+[C_6H_5O]^- \rightarrow C_6H_5\text{-}O\text{-}CH_3 + N_2$. D. is also used as a reagent for 1,3-dipolar cycloadditions, in the synthesis of diazoketones, and for ring expansion of cyclic ketones.

Diazonium salts: nitrogen-containing, organic compounds with the general formula $[R\text{-}N{\equiv}N]^+X^-$. Aliphatic D. are so unstable that they are short-lived intermediates. The aromatic compounds of this type, the Arene diazonium salts (see) are more stable.

Diazo pigments: see Azo pigments.

Diazotates: salts of arene diazohydroxides with the general formula $R\text{-}N{=}N\text{-}O^-M^+$. They are formed by the reaction of alkali hydroxides with arene diazonium salts via the intermediate stages of arene diazonium hydroxide and arene diazo hydroxide:

$$[R\text{-}\overset{+}{N}{\equiv}N]Cl \xrightarrow[-NaCl]{+NaOH} [R\text{-}\overset{+}{N}{\equiv}N]OH^- \rightleftharpoons$$

$$R\text{-}N{=}N\text{-}OH \xrightarrow[-NaCl]{+NaOH} R\text{-}N{=}N\text{-}O^-Na^+.$$

In this reaction, the primary product is the *Z*- or *n*-isomer, which then rearranges to the more stable *E*- or iso-D.

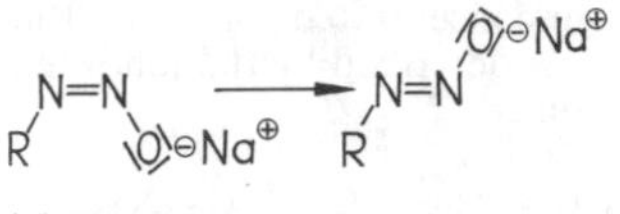

(*Z*)-Diazotate (*E*)-Diazotate

Strong mineral acids can reconvert D. to diazonium salts.

Diazotization: a process for making arene diazonium salts by reaction of primary aromatic amines with nitrous acid or nitrite esters in a strongly acidic medium: $R\text{-}NH_2 + 2\ HX + NaNO_2 \rightarrow R\text{-}N^+{\equiv}N|X^- + 2\ H_2O + NaX$. The reaction must be carried out under cooling, because the diazonium salts are unstable. D. is very important for the synthesis of azo pigments.

Diazotype copying: a method of copying which utilizes the light sensitivity of diazonium salts. The light-sensitive material (paper or plastic film) is prepared by coating with an aryl diazonium salt (e.g. *p*-dialkylaminobenzene diazonium salt), a phenol (e.g. resorcinol) as coupler and an organic acid (tartaric acid, aryl sulfonic acid) as a coupling inhibitor. When it is exposed to light through a mask, the diazonium salt is decomposed in the exposed areas, and remains unreacted only at the nonexposed sites. In the presence of ammonium vapor, it can bond to the coupler to form an azo pigment. In this application, D. is a positive process with a steep gradient, and is suitable for copying documents and line drawings. With suitable spectral sensitization, the process can be used to duplicate movie films. When the *o*-quinone diazide of naphthalene is used, D. is a positive method for production of printing sheets. In the exposed areas, the *o*-quinone diazides cleave off nitrogen, then rearrange to form indenecarboxylic acids; these can be washed out with lye. This generates a relief with raised parts in the unexposed sites, which can be used as a printing matrix. The D. can also be used as a negative process when ketenes formed by exposure to light polymerize and the *o*-quinone diazides in the unexposed sites are dissolved away with organic compounds.

Dibenzodihydroazepines, ***iminodibenzyls***: tricyclic compounds with one seven-membered ring containing a nitrogen atom (Fig.). In contrast to the Phenothiazines (see), these compounds are not planar in their tricyclic portions. The side chains are basic. These compounds are used as thymoleptics. The Diazodihydrocycloheptadienes (see) have a related structure.

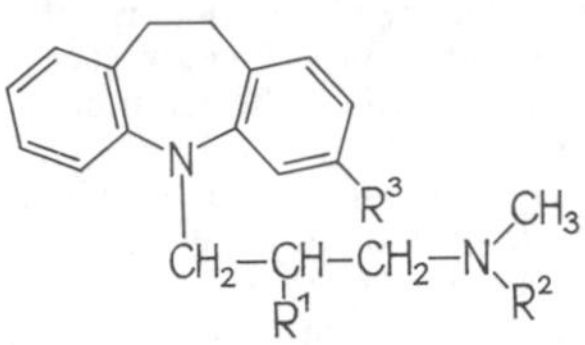

R^1	R^2	R^3	Name
H	CH_3	H	Imipramine
H	H	H	Desipramine
CH_3	CH_3	H	Trimipramine
H	CH_3	Cl	Clomipramine

Dibenzodihydrocycloheptadienes: tricyclic compounds containing a seven-membered ring, such as amitriptyline and noxiptilin. As in the dibenzohydroazepines, the tricyclic structure is not planar.

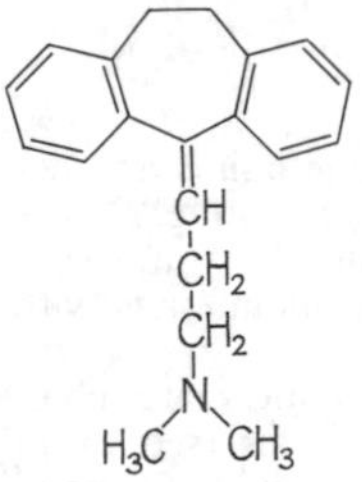

Amitriptyline 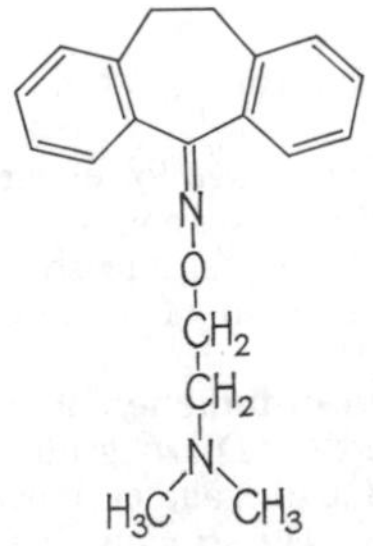 Noxiptilin

These compounds are used as thymoleptics. The basic side chain can be bound to the ring by a C=C double bond or an =N-O bridge. Compounds in which the seven-membered ring of amitriptyline has an O or S atom instead of a CH_2 group have the same effects.

Dibenzo-4H-1,4-oxazine: same as Phenoxazine (see).

Dibenzoyl: same as Benzil (see).

Dibenzo-γ-pyran: same as Xanthene (see).

Dibenzopyrazine: same as Phenazine (see).

Dibenzo[b,e]pyridine: same as Acridine (see).

Dibenzo-γ-pyrone: same as Xanthone (see).

Dibenzopyrrole: same as Carbazole (see).

Dibenzyl, ***1,2-diphenylethane***: $C_6H_5\text{-}CH_2\text{-}CH_2\text{-}C_6H_5$, an aliphatic-aromatic hydrocarbon; colorless crystals, m.p. 52.2°C, b.p. 285°C, n_D^{20} 1.5476. D. is insoluble in water but dissolves readily in ether and carbon disulfide. It is synthesized by catalytic hydrogenation of benzil or stilbene, or by the Wurtz-Fitting synthesis from benzyl chloride and sodium.

Dibenzyl ether: C_6H_5-CH_2-O-CH_2-C_6H_5, a liquid with a slightly fruity, rose-like odor; m.p. 3.6 °C, b.p. 298 °C (dec.), n_D^{20} 1.5168. D. is insoluble in water and readily soluble in ethanol and ether. It is obtained by heating benzyl alcohol with conc. sulfuric acid; industrially, it is a byproduct of benzyl alcohol production. D. is used as a softener in plastics and a solvent for musk and other aroma substances in perfumery.

Diborane: see Boranes.

4,ω-Dibromoacetophenone: same as 4-Bromophenacyl bromide (see).

1,2-Dibromoethane: same as Ethylene dibromide (see).

Dibromomethane: same as Methylene bromide.

Dibromosulfane: see Sulfur bromides.

2,6-Di-tert.-butyl-4-methylphenol: colorless crystals, m.p. 70 °C, b.p. 265 °C. The compound is insoluble in water, but is soluble in alcohols, hydrocarbons, fats and oils. It must be protected from light and kept cool. 2,6-D. is a radical trapper and is used as an antioxidant in food wrappings, rubber, wax paper and petroleum chemistry.

Dicamba: see Herbicides.

Dicarba-*closo*-dodecaborane: see Carboranes.

Dicarbonyl compounds: in the broad sense, organic compounds with two carbonyl groups >C=O in the molecule; in the narrow sense, compounds in which the carbonyl groups are linked directly or via an aliphatic chain. Ketone, aldehyde and ester groups can act as carbonyl groups. If the carbonyl groups are in positions 1 and 2 relative to each other, it is an α-D; if they are 1,3-, it is a β-D., and so on. Some well known 1,2-D. are glyoxal, methylglyoxal, diacetyl and benzil. The most important class of D. are the 1,3-D., which include malonic dialdehyde, acetylacetone and acetoacetate. 1,3-D. undergo keto-enol tautomerism. Some important 1,4-D. are succinic dialdehyde and acetonylacetone.

D. are essential starting materials for the synthesis of heterocyclic compounds. For example, quinoxaline is synthesized from 1,2-D. and *o*-phenylenediamine; pyrazole is made from 1,3-D. and phenyl hydrazine; and pyrrole from 1,4-D. and ammonia. These reactions are also used to distinguish different D.

Dicarboxylic acids: organic compounds containing two carboxyl groups in the molecule. ***Aliphatic D.*** occur widely in nature. They are named from the root for the aliphatic hydrocarbon (which may be unsaturated) plus the suffix "-dioic acid", for example, ethandioic acid, HOOC-COOH, propandioic acid, HOOC-CH_2-COOH and (Z)- or (E)-butendioic acid, HOOC-CH=CH-COOH. These compounds have better-known trivial names derived from their plant or animal sources, for example, Oxalic acid (see), Malonic acid (see), Maleic acid (see) and Fumaric acid (see). The systematic names of ***aromatic*** and ***heterocyclic D.*** are formed from the appropriate roots and the suffix "-dicarboxylic acid"

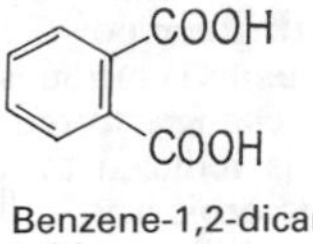

COOH
COOH

Benzene-1,2-dicarboxylic acid
Phthalic acid

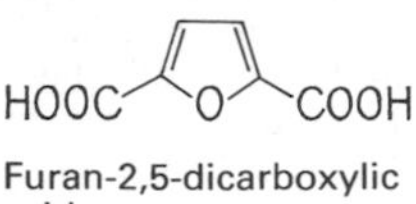

HOOC O COOH

Furan-2,5-dicarboxylic acid
Dehydrofuroic acid

This type of nomenclature is sometimes also used for aliphatic compounds, e.g. nonane-1,9-dicarboxylic acid, HOOC-$(CH_2)_9$-COOH.

D. are crystalline compounds, usually colorless. The smaller molecules may dissolve readily in water, but the solubility declines rapidly with increasing molecular mass. As is the case for the *n*-alkanes and Monocarboxylic acids (see), a saturated aliphatic D. with an even number of C atoms will always melt at a higher temperature than the acid with the next higher odd number of C atoms.

D. are able to release two protons, one at a time:

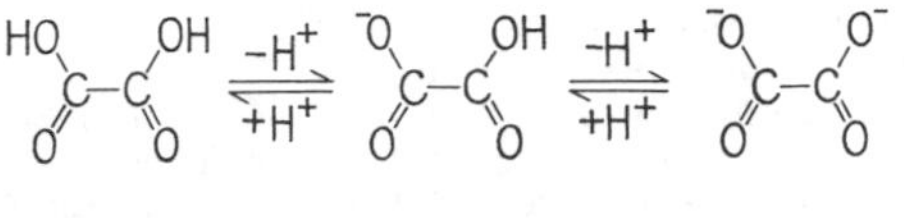

$$\mathrm{HOOC{-}COOH} \underset{+H^+}{\overset{-H^+}{\rightleftharpoons}} \mathrm{^-OOC{-}COOH} \underset{+H^+}{\overset{-H^+}{\rightleftharpoons}} \mathrm{^-OOC{-}COO^-}$$

The first proton is more readily lost than the second, because the carboxylate anion inhibits the dissociation of the second proton. The D. lose their first proton more readily than the unsubstituted monocarboxylic acids (are stronger acids), except for formic acid, which in turn is a stronger acid than any other carboxylic acid except oxalic and malonic acids. In the longer-chain D., the acidities of the two carboxyl groups do not significantly affect each other.

The chemical reactions of the D. are essentially the same as those of the Monocarboxylic acids (see). The carboxyl groups can react simultaneously or sequentially. A series of acidic or neutral functional derivatives can thus be obtained.

The following regularity is observed in the thermolytic reactions of the D.: oxalic and malonic acid are decarboxylated by heating to about 200 °C, or 150 °C, forming formic and acetic acids, respectively. The next members of the series, succinic and glutaric acids, form cyclic anhydrides by loss of water:

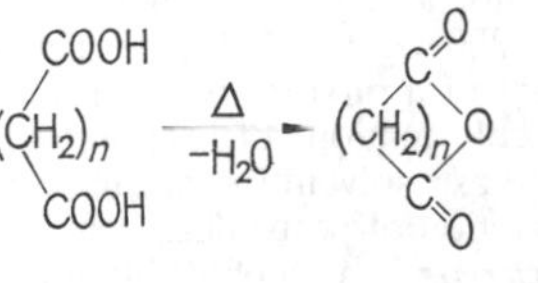

$$\mathrm{HOOC{-}(CH_2)_n{-}COOH} \xrightarrow[-H_2O]{\Delta} \mathrm{(CH_2)_n(CO)_2O}$$

Maleic and phthalic acids react similarly. When heated to about 300 °C, adipic, pimelic and suberic acids lose both water and carbon dioxide to form the corresponding cycloalkanones, with one C atom fewer. The higher D. can be converted to cyclic ketones by thermolysis of the alkaline earth or thorium salts.

D. are synthesized mainly by hydrolysis of dinitriles, oxidation of diols, or the Malonic ester synthesis (see). Aromatic D. can be synthesized, for example, by oxidation of dimethyl substituted aromatic compounds, e.g. terephthalic acid from *p*-xylene by air oxidation in the presence of cobalt(II) or manganese(II) bromide.

D. and their substitution products and derivatives are used mainly as monomers for formation of polymers, e.g. polyester or polyamide fibers.

Aliphatic saturated dicarboxylic acids

Trivial name	Systematic name	Formula	m.p. in °C
Oxalic acid	Ethanedioic acid	HOOC-COOH	101.5
Malonic acid	Propanedioic acid	HOOC-CH_2-COOH	135.6
Succinic acid	Butanedioic acid	HOOC-$(CH_2)_2$-COOH	188
Glutaric acid	Pentanedioic acid	HOOC-$(CH_2)_3$-COOH	99
Adipic acid	Hexanedioic acid	HOOC-$(CH_2)_4$-COOH	153
Pimelic acid	Heptanedioic acid	HOOC-$(CH_2)_5$-COOH	106
Suberic acid	Octanedioic acid	HOOC-$(CH_2)_6$-COOH	144
Azelaic acid	Nonanedioic acid	HOOC-$(CH_2)_7$-COOH	106.5
Sebacic acid	Decanedioic acid	HOOC-$(CH_2)_8$-COOH	134.5
–	Undecanedioic acid	HOOC-$(CH_2)_9$-COOH	111

Dichlone: see Fungicides.

Dichlorfluanide: see Fungicides.

1,3-Dichloroacetone: Cl-CH_2-CO-CH_2-Cl, colorless, volatile crystalline leaflets with an intense odor; m.p. 45 °C, b.p. 173.4 °C. 1,3-D. is soluble in water, alcohol and ether. It is very irritating to the mucous membranes. 1,3-D. can be synthesized by chlorination of acetone in aqueous solution. It is used mainly for organic syntheses, for example, in the production of citric acid.

Dichlorobenzene: three structural isomers which are stable, colorless compounds. They are very slightly soluble in water, but dissolve easily in many organic solvents. As in other non-activated Haloarenes (see), the chlorine atoms are relatively tightly bound to the benzene ring, and can undergo nucleophilic substitution only under special conditions. ***1,2-*** or ***o-dichlorobenzene*** is a combustible liquid; m.p. -17 °C, b.p. 180. 5 °C, n_D^{20} 1.551.

Cl

Cl (1,2, 1,3 or 1,4 position)

It forms 2-chlorophenol and pyrocatechol in the presence of alkalies at 200 °C. 1,2-Dichlorobenzene can be produced as part of a mixture of chlorobenzene derivatives by chlorination of benzene or chlorobenzene. It is used as a solvent for lacquers and resins, and in the production of pesticides.

1,3- or ***m-dichlorobenzene*** is a combustible liquid with a strong odor; m.p. - 24.7 °C, b.p. 173 °C, n_D^{20} 1.5459. 1,3-Dichlorobenzene is also produced by chlorination of benzene, and its yield as a percentage of the product mixture can be greatly increased by using a reaction temperature above 400 °C. Pure 1,3-dichlorobenzene is obtained from the Sandmeyer reaction with 3-chloroaniline or by chlorination of 1,3-dinitrobenzene. It is an important intermediate in the production of pigments.

1,4- or ***p-dichlorobenzene***, crystalline leaflets with an intense odor; m.p. 53.1 °C, b.p. 174 °C. It sublimes even at room temperature. 1,4-Dichlorobenzene is the main product of chlorination of benzene. It is used as an intermediate in the production of pigments, pesticides, pharmaceuticals and a room deodorizer.

2,2′-Dichlorodiethyl sulfide: same as Bis(2-chloroethyl) sulfide.

Dichlorodifluoromethane: CCl_2F_2, a colorless, non-combustible gas; m.p. - 158 °C, b.p. - 29.8 °C. D. is soluble in alcohol, ether and glacial acetic acid, and practically insoluble in water. It is formed by the reaction of hydrogen fluoride with carbon tetrachloride, or by reaction of methane with chlorine and hydrogen fluoride: $CH_4 + 4\,Cl_2 + 2\,HF \rightarrow CCl_2F_2 + 6\,HCl$. D. is used in a mixture with other chlorofluoroalkanes in the production of foam products and aerosols, and as a coolant (R 12).

Dichloroacetic acid: Cl_2CH-COOH, a colorless, caustic liquid; m.p. 13.5 °C, b.p. 194 °C, n_D^{20} 1.4658. D. is readily soluble in water, ether, methanol, ethanol, chloroform and benzene. Its aqueous solutions are strongly acidic. D. undergoes the reactions typical of carboxylic acids. The chlorine atoms can be replaced by nucleophiles, but they are more resistant to hydrolysis than those in Chloroacetic acid (see). D. is formed by the reaction of chloral hydrate with potassium cyanide: $Cl_3C\text{-}CHO + KCN + H_2O \rightarrow Cl_2CH\text{-}COOH + HCN + KCl$. It can be made industrially by chlorination of acetic acid; it is then obtained in a mixture with mono- and trichloroacetic acids. D. can also be synthesized by hydrolysis of tetrachloroethylene with steam, or by hydrolysis of dichloroacetyl chloride (which can be obtained by oxidation of trichloroethylene):

$$Cl_2C{=}CHCl \xrightarrow{+O} Cl_2CH\text{–}CO\text{–}Cl \xrightarrow[-HCl]{+H_2O} Cl_2CH\text{–}CO\text{–}OH.$$

D. is used for organic syntheses, for production of glyoxalic acid and to remove corns and warts.

1,2-Dichloroethane: same as Ethylene dichloride (see).

1,1-Dichloroethylene: same as Vinylidene chloride (see).

Dichloroformoxime: see Chemical weapons.

Dichloromethane: same as Methylene chloride.

Dichloromethylbenzene: same as Benzal chloride (see).

2,6-Dichlorophenol-indophenolsodium, ***2,6-dichloro-N-(4-hydroxyphenyl)-1,4-benzoquinonimine sodium salt***, ***Tillman's Reagent***: a dark green powder soluble in water and alcohol. The aqueous solution is dark blue, and when acid is added, it changes to red. D. is used as a redox indicator. It is reduced to a colorless leukoform by iodide ions, ascorbic acid and cholinesterase. This is the basis of an important reaction for detection of vitamin C. Formula p. 319.

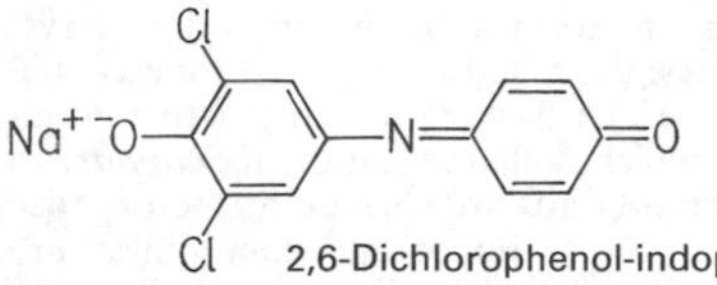

Cl, Na⁺ ⁻O, N, O, Cl — 2,6-Dichlorophenol-indophenolsodium

Dichlorophosphazenes: same as Phosphorus nitride dichlorides (see).

Dichloroprop: see Growth hormone herbicides.

Dichlorosulfane: see Sulfur chlorides.

α,α-Dichlorotoluene: same as Benzal chloride (see).

Dichlorvos, DDVP: abb. for O,O-*d*imethyl-O-(2,2-*d*ichloro*v*inyl)-*p*hosphate, an Organophosphate insecticide (see; Table 1) which is a colorless liquid, D_4^{25} 1.415; m.p. at 130 Pa, 74 °C; vapor pressure at 20 °C, 1.6 Pa; solubility in water ≈ 1 g/100 ml. D. is miscible with most organic solvents and is rapidly hydrolysed by bases.

It is synthesized by reaction of trimethylphosphite (produced from phosphorus trichloride and methanol in the presence of base) with anhydrous chloral, or by removal of HCl from Trichlorfon (see):

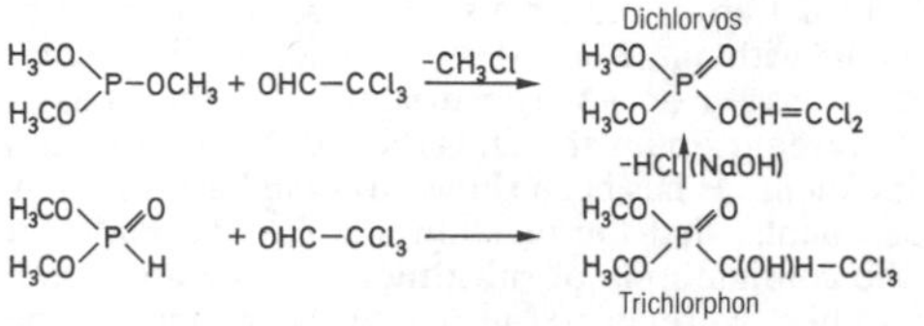

$$(H_3CO)_2P{-}OCH_3 + OHC{-}CCl_3 \xrightarrow{-CH_3Cl} (H_3CO)_2P(=O){-}OCH{=}CCl_2 \text{ (Dichlorvos)}$$

$$(H_3CO)_2P(=O)H + OHC{-}CCl_3 \longrightarrow (H_3CO)_2P(=O){-}C(OH)H{-}CCl_3 \text{ (Trichlorphon)} \xrightarrow{-HCl\ (NaOH)} \text{Dichlorvos}$$

D. is a contact and respiratory poison used to combat biting and sucking insects. Because of its high effectiveness in the vapor phase, it is used to kill flies and other pests in rooms and to protect stored foodstuffs and feeds.

Dichlozoline: see Carboximide fungicides.

Dichromate black: see Aniline black.

Dichrome process: see Chromizing.

Diclofop: see Growth hormone herbicides.

Dicloran: see Fungicides.

Dicloxacillin: see Penicillins.

Dicofol: an Acaricide (see).

Dicumarol see Ethyl biscoumacetate.

Dicyanamides: derivatives of cyanocarbodiimide, H-N=C=N-CN, which has so far not been reported to exist as such. Ionic D. contain the bent (∢ CNC 120.3°), resonance-stabilized anion $[N(CN)_2]^-$, which is a Pseudohalide (see). Alkali dicyanamides such as ***sodium*** or ***potassium dicyanamide***, $NaN(CN)_2$ or $KN(CN)_2$, are obtained as colorless salts by reaction of aqueous cyanamide solutions with cyanogen halides in the presence of alkali hydroxide. The transition metal dicyanamides are only slightly soluble in water; some examples are the green ***copper(II) dicyanamide***, $Cu[N(CN)_2]_2$ and the bright blue ***nickel(II) dicyanamide***. Like the water-insoluble, colorless ***silver(I) dicyanamide***, $AgN(CN)_2$, these are coordination polymers. In its complex-forming behavior, the dicyanamide anion resembles the homologous pseudohalides tricyanomethanide and cyanates, preferentially forming coordinate bonds via the cyano nitrogen. In bridge functions, it can be polydentate. Some covalent dicyanamide derivatives are ***aryl cyanamides***, $R{-}N(CN)_2$ and the isomeric ***aryl cyanocarbodiimides***, R-NCNCN, and ***phosphate ester dicyanamides*** and their isomers, the ***phosphate ester cyanocarbodiimides***, $(RO)_2P(O){-}N(CN)_2$ or $(RP)_2P(O)$-NCNCN.

1,2-Dicyanobenzene: same as Phthalic dinitrile (see).

Dicyanodiamide, ***1-cyanoguanidine***: the dimer of cyanamide. D. can exist in two tautomeric forms:

$$H_2N{-}C({=}NH){-}NH{-}CN \rightleftharpoons H_2N{-}C(NH_2){=}N{-}CN$$

It forms colorless and odorless cyrstals, which cannot be distilled; m.p. 211-212 °C. D. is only slightly soluble in cold water, alcohol or acetone, but dissolves readily in dimethylformamide. D. is synthesized industrially by the reaction of carbon dioxide and water with calcium cyanamide. The cyanamide which is initially formed by this reaction dimerizes in aqueous solution, forming D.: $CaN{-}CN + H_2O + CO_2 \rightarrow H_2N{-}CN + CaCO_3$; $2\ H_2N{-}CN \rightarrow H_2N{-}C(NH_2){=}N{-}CN$. D. is used to make dicyanamide-formaldehyde resins, melamine resins and guanidine salts. It is used as a hardener for epoxide resins.

Dicyanodiamide resins: durable amide resins made by polycondensation of diacyanodiamide with formaldehyde. Like the Melamine resins (see), the D. are very hard and insoluble in water, and in this respect are superior to the urea resins. D. are pressed together with fillers, such as short-fiber cellulose and textile shreds, at 150-180 °C to make thermally resistant dishes, etc. D. are also used in wood glues, and alcohol-modified D. are used as paint bases.

Dicyanogen: $(CN)_2$, a colorless gas with a pungent odor; M_r 52.02, m.p. - 27.98 °C, b.p. - 21.15 °C. D. can be made by oxidation of cyanide with copper(II) in aqueous solution: $Cu^{2+} + 4\ CN^- \rightarrow 2\ CuCN + (CN)_2$; other methods are based on the oxidation of hydrogen cyanide with hydrogen peroxide, or on reaction of cyanogen chloride with silicon or aluminum: $4\ ClCN + Si \rightarrow 2\ NC{-}CN + SiCl_4$. D. is a pseudohalogen (see Pseudohalides). At present there is no industrial application, but its use in welding has been suggested, because of its high temperature of combustion in oxygen (≈ 4600 K).

Dicyanoketenimine: see Tricyanomethanide.

Dicyanosulfane: see Cyanides.

Dicyclopentadiene, ***tricyclo[5.2.1.0^{2,6}]deca-3,8-diene***: a dimerization product of cyclopenta-1,3-diene, which forms readily at room temperature and can be cleaved reversibly at higher temperatures.

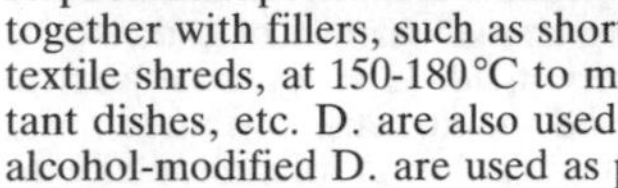

$$2\ \text{cyclopentadiene} \underset{180\,°C}{\overset{\approx 20\,°C}{\rightleftharpoons}} \text{dicyclopentadiene}$$

D. exists in the *endo*-form; it is a colorless solid, m.p. 32-33 °C, b.p. 64-65 °C at 1.86 kPa, n_D^{35} 1.5050. The synthesis of D. corresponds to a diene synthesis (Diels-Alder reaction), and the reverse reaction corresponds to a retro-Diels-Alder reaction.

Didropyridine: see Hypnotics.

Didymium metal: see Neodymium.

Didymium oxide: a mixture of rare earths which was first obtained in 1839 by Mosander in the course of separating the rare earths. After separation of samarium (europium) and gadolinium, the main component (also called didyme) was later separated into neodymium and praseodymium by Auer von Welsbach in 1885. See Lanthanides.

Dieckmann condensation: intramolecular condensation of aliphatic dicarboxylate esters to cyclic β-ketoacylate esters in the presence of sodium or sodium methanolate:

$$\begin{matrix} CH_2-CH_2-COOR \\ | \\ CH_2-CH_2-COOR \end{matrix} \xrightarrow[-\,ROH]{(Na\,OC_2H_5)} \begin{matrix} CH_2-CH-COOR \\ | \quad\quad \backslash \\ | \quad\quad\quad CO \\ | \quad\quad / \\ CH_2-CH_2 \end{matrix}$$

The β-ketoacylate ester can be converted into the corresponding alicyclic ketone by subsequent ketone cleavage with dilute acid or base. For steric reasons, the formation of cyclopentanone or cyclohexanone carboxylate esters is especially favored in the D. The cycloheptanone derivative forms considerably less readily. Esters of pentanedioic (glutaric), nonanedioic (azelaic) or decanedioic (sebacic) acids cannot be cyclized with sodium methanolate, or only in very low yields. The reaction mechanism of the D. is the same as that of the Claisen condensation (see).

Dieldrin: a Cyclodiene insecticide (see).

Dielectric constant, abb. ***DC***: the measure of the dielectric shift of the charge carriers of a dielectric substance in an electric field. In an external electric field, the electrons and the nucleus are shifted away from their rest positions; in the case of molecules with permanent dipole moments, the entire molecule can be rotated in the electric field (see Polarization, 2).

In gases, the D. decreases with decreasing density until the D. of a vacuum is reached. The relative D. of the vacuum is equal to 1, and it is usually a suitable approximation to set the D. of a gas equal to 1. Dipole effects in gases increase the D. only very slightly. On the other hand, dipolar liquids have very high D., because here the molecule is oriented in the field by rotation. This orientation is continuously disrupted by the thermal motion of the molecules, so that an equilibrium is reached at any temperature. Therefore, the D. is highly temperature dependent, and this fact is used to determine the dipole moment. When liquids solidify, the large D. is lost, because in the solid crystal lattice, the dipoles are fixed. The D. depends on the frequency of the applied external field as well as temperature. The frequency dependence is due mainly to the inertia of molecules with dipole moments. The magnitude of the D. and its dependence on temperature and frequency give important information on the structure of a substance.

Dielectrometry, ***Decametry***: an electrochemical analytical method which measures the dielectric constant as an indicator of concentration. The dielectric constant of a solution is proportional to the concentration of a solute.

Diels-Alder reaction, ***diene synthesis***: a method first described in 1928 by O. Diels and K. Alder for 1,4-addition of an activated alkene or alkyne to a conjugated diene to form a six-membered ring system. In most cases, the reaction is a synchronous (4 + 2) cycloaddition of a 1,3-diene and the activated alkene or alkyne, which is also called the *dienophile*. In the reaction, two σ-bonds are formed at the expense of two π-bonds. Some important dienophiles are maleic anhydride, acrolein, tetracyanoethylene, acrylic acid ester, p-benzoquinone, acetylene dicarboxylic acid esters and azodicarboxylic acid esters. The diene component can be an aliphatic or alicyclic 1,3-diene, or an aromatic or heterocyclic system, such as s-(Z)-buta-1,3-diene,

1,3-Diene Dienophile

cyclopenta-1,3-diene, cyclohexa-1,3-diene, anthracene or a furan. Analogs of the 1,3-dienes in which one C=C double bond is replaced by carbon double bonded to a heteroatom, e.g. C=O, C=S or C=N, can also enter the D. D. occur particularly readily if the alkene or alkyne is activated by substituents with an - I or - M effect, and the diene component has +I or +M substituents. Another important prerequisite for the D. is the s-(Z)-conformation of the diene. It has been shown, on the basis of many experiments, that the reaction is a *cis*-addition, that is, the configuration of substituents in the diene and dienophile is retained in the cyclic product. The cleavage of Diels-Alder adducts into the starting components, which is often observed even under very mild conditions, is called the *retro-Diels-Alder reaction*. For example, *endo*-dicyclopentadiene decays around 180 °C into cyclopentadiene. D. has been used in the elucidation of numerous structures of unsaturated natural products, and in the syntheses of a series of insecticides, e.g. chlordane, aldrin and dieldrin.

Dienes: unsaturated aliphatic hydrocarbons with two double bonds in the molecule; the general formula is C_nH_{2n-2}. There are three possible types, depending on the relative positions of the double bonds: 1) allenes, $CH_2=C=CH_2$, with cumulative double bonds (see Allenes, Cumulenes); 2) dienes, $CH_2=CH-CH=CH_2$, with conjugated double bonds (see 1,3-Dienes, Polyenes); and 3) diolefins, $CH_2=CH-CH_2-CH=CH_2$, with isolated double bonds (***diolefins***).

While the reactivities of allenes and diolefins are nearly the same as those of the alkenes, D. of the diene type have some unusual properties, such as 1,4-additions (see 1,3-Dienes).

1,3-Dienes: unsaturated aliphatic hydrocarbons with two conjugated double bonds (see Dienes). The simplest member of this class is butadiene.

1,3-D. are more reactive in polar and radical additions than are hydrocarbons with isolated double bonds, and they polymerize very readily. They are characterized by the ability to undergo 1,4-additions in addition to 1,2-additions; in some cases the 1,4-addition is exclusive (see Diels-Alder reaction). This reactivity is due to the conjugated system of double bonds; because of the resonance of the π-electrons of the four sp^2-hybridized C atoms, this system has lower energy than a compound with two isolated π-bonds.

The following reactions are of particular interest: 1) *Reduction* with sodium in liquid ammonia. This reaction is possible only with hydrocarbons with conjugated double bonds. Butadiene, for example, is converted to but-2-ene:

$$CH_2{=}CH{-}CH{=}CH_2 \xrightarrow[-2Na^+]{2Na} \overset{-}{C}H_2{-}CH{=}CH{-}\overset{-}{C}H_2 \xrightarrow[-2NH_2^-]{2NH_3} CH_3{-}CH{=}CH{-}CH_3.$$

2) *Addition of halogens*. When equimolar amounts of diene and chlorine or bromine are used, mixtures of 1,2- and 1,4-adducts are formed; the 1,2-adduct is kinetically favored, but the thermodynamically more stable 1,4-dihalogenalk-2-ene eventually predominates, because it is formed by a rearrangement of the 1,2-adduct, via a resonance-stabilized cation:

$CH_2{=}CH{-}CH{=}CH_2 \xrightarrow{Cl_2}$ $CH_2{=}CH{-}CHCl{-}CH_2CL$ 1,2-Adduct

$ClCH_2{-}CH{=}CH{-}CH_2CL$ 1,4-Adduct

3) *Diene synthesis (Diels-Alder reaction)*. 1,3-D. combine with activated alkenes (dienophiles) in a synchronous [4+2]-cycloaddition under mild conditions to form cyclohexene derivatives, usually in high yields.

[Diene + alkene with R^1, R^2 → cyclohexene with R^1, R^2]

This reaction is a *cis*-addition, in which the configurations of the substituents on the diene and dienophile are retained in the adduct. Adducts with maleic anhydride are used in analytical characterization of 1,3-D.

4) *Electrocyclizations*. The ends of an open conjugated system can form a σ-bond at the expense of one π-bond in a synchronous reaction.

1,3-D. cyclize to cyclobutene derivatives, and 1,3,5-trienes to cyclohexenes (see Woodward-Hoffman rules, sect. on electrocyclic reactions).

5) *Polymerization* of 1,3-D. can occur as a series of 1,2- and 1,4-additions. In general, mixed structures are formed in the presence of various catalysts; a high proportion of 1,4-polybutadiene is desirable for the properties it confers on the synthetic rubber.

$[-CH_2-CH(-CH{=}CH_2)-]_n$ $[-CH_2-CH{=}CH-CH_2-]_n$

1,2-Polybutadiene 1,4-Polybutadiene

There are many ways to synthesize 1,3-D., such as the dehydration of 1,4- and 1,3-diols or of unsaturated alcohols, pyrolysis of esters of the 1,4- and 1,3-diols, dehydrogenation of alkanes or alkenes, and the Wittig reaction.

Diene synthesis: same as Diels-Alder reaction (see).

Dienophile: see Diels-Alder reaction.

Diesel fuel: see Fuels.

Diesel oil: see Fuels.

Diet foods: foods which are suitable for special diets required by individuals with nutritional or metabolic disorders or by those whose age or physiological condition (infancy, pregnancy, old age, exceptional working conditions, etc.) place special demands on their nutrition. There are various types of D., such as Diabetic foods (see), Low-calorie foods (see), low-protein foods with a maximum total protein content of 3% for patients with chronic kidney failure; foods which are free of gluten (including the proteins gliadin, hordein and avenine found in wheat, rye, barley and oats) for patients with celiac disease or sprue; low-salt (or salt-free) foods containing little or no sodium for patients with coronary disease and certain skin diseases; foods rich in polyunsaturated fatty acids (linoleic, arachidonic) for patients with coronary-circulatory diseases; high-fiber foods with a minumum of 10% fibers (cellulose, hemicellulose, pectin, lignin, etc.) for patients with chronic constipation; and infant foods which are specially formulated to provide for the nutritional needs of infants and small children.

Diethylene dioxide: see Dioxane.

Diethyleneglycol dinitrate, ***diglycol dinitrate, dinitrodiglycol***: $O(CH_2\text{-}CH_2\text{-}O\text{-}NO_2)_2$, an oily liquid which tends to explode; b.p. ≈ 160 °C (explosion). The explosiveness of D. is only about ⅔ that of glycerol trinitrate, and it is also less sensitive to impact. D. can be gelatinized with collodium wool and is a component of smokeless gunpowder.

Diethyl ether, ***ethyl ether***, often called ***ether*** for short: $CH_3\text{-}CH_2\text{-}O\text{-}CH_2\text{-}CH_3$, a water-clear, colorless, non-viscous liquid with a characteristic, ether-like odor; m.p. - 116.2 °C, b.p. 34.5 °C, n_D^{20} 1.3526, flame point - 40 °C, self-ignition temperature 186 °C.

D. is only slightly soluble in water, but is miscible with organic solvents. It evaporates very easily, is inflammable and burns with a pale flame. The vapor is denser than air, but forms explosive mixtures with air at concentrations above 1.8 vol. %. D. readily picks up electric charge when it is shaken. Transfer of larger amounts from one container to another should therefore be done only with electrically grounded equipment. Special protective measures should be observed in working with D., such as working with amounts over 50 ml only in separate ether rooms. It should be heated only in a waterbath without electric or gas heating (steam heat only for the waterbath), and electric charge buildup or discharge, sparks and flames must be avoided. To prevent formation of peroxide, D. is to be stored airtight in brown bottles (more on safety, see Ethers). D. can contain small amounts of dissolved water. It can be dried with metallic sodium, with which it does not react: after a preliminary drying, sodium wire is dipped into the ether.

D. is an inert solvent, and dissolves many organic and inorganic substances, such as fats, oils, resins, neutral organic compounds, alkaloids, phosphorus, sulfur, bromine and iodine. It is also miscible with strong aqueous mineral acids, such as conc. hydrochloric acid, forming soluble oxonium salts (see Ethers).

High concentrations of D. in the air can lead to paralysis of the respiratory center and circulation, with fatal results. Chronic intoxication leads to the same symptoms as found in chronic alcoholism.

D. is synthesized for industrial purposes from ethene and sulfuric acid, or from ethanol and sulfuric acid. D. for anaesthetic use is made by passing ethanol vapor over aluminum oxide at 240-260 °C. Inhalation of ether vapor at low concentrations leads to intoxication, and higher doses to paralysis and unconsciousness (narcosis).

The use of D. as an inhalation anesthetic in medicine has declined, because of the relatively long recovery period and unpleasant after-effects, such as nausea and unease. It is rarely used as a single anesthetic. A mixture of 3 parts ethanol and 1 part D. is known as Hoffmann's drops and has an animating effect. In industry, D. has been largely replaced by solvents which are less likely to burn or explode.

Historical. The use of D. in medical anesthesia was recognized in 1842 by Long, and used by him in surgery. The first publication, hwoever, was in 1846, by Morton and Jackson.

Diethylketone, ***pentan-3-one***: CH_3-CH_2-CO-CH_2-CH_3, a colorless, easily combustible liquid which smells like acetone; m.p. - 39.8 °C, b.p. 101.7 °C, n_D^{20} 1.3924, flame point 12 °C. D. is slightly soluble in water, but dissolves easily in most organic solvents. It forms an azeotropic mixture with water which boils at 82.9 °C and contains 86% D. D. is synthesized by hydroformylation of ethene, or by condensation of propionic acid in the presence of various metal oxide catalysts. D. is used mainly as a solvent for lacquers, fats and waxes, as a deparaffinization reagent for petroleum fractions, and as a reagent for organic syntheses.

Diethylmalonate, ***malonic acid diethyl ester***: H_5C_2O-CO-CH_2-CO-OC_2H_5, a colorless liquid with a pleasant odor; m.p. - 48.9 °C, b.p. 199.3 °C, n_D^{20} 1.4139. M. is nearly insoluble in water, but dissolves readily in most organic solvents. Under normal conditions, it is almost completely in the keto form. M. is a very reactive compound, due in part to activation of the CH_2 group by the neighboring ester groups. This accounts, for example, for the Malonate ester syntheses (see), Michael addition (see) and the Knoevenagel condensation (see). In addition, M. undergoes the reactions typical of a carboxylic acid ester. For example, because it has the two ester groups in the 1,3-positions, it is useful for synthesis of heterocyclic compounds. Oxidation of M. by selenium dioxide converts it to Mesoxalic acid (see). M. is synthesized industrially by esterification of malonic acid with ethanol in the presence of concentrated sulfuric or hydrochloric acid. M. is used as an intermediate for synthesis of barbituric acid and its derivatives, pigments, aromas, pharmaceuticals and amino acids.

N,N-Diethyl-p-phenylenediamine, ***p-amino-N,N-diethylaniline***: a yellow, oily compound; b.p. 260-262 °C.

H_2N–C_6H_4–$N(CH_2$–$CH_3)_2$

D. is readily soluble in benzene. It is made by reduction of *p*-nitroso-*N,N*-diethylaniline with zinc and hydrochloric acid. It is used as a developer in color photography and in the synthesis of a number of pigments, such as indophenols, indamines and azine and thiazine dyes.

Diethylstilbestrol: see Estrogens.

Diethylsulfate, ***sulfuric acid diethyl ester***: $(C_2H_5O)_2SO_2$, a poisonous oil which smells like peppermint; m.p. - 24.5 °C, b.p. 208 °C (dec.), n_D^{20} 1.4004. D. is insoluble in water. It is mutagenic and a carcinogen in animal tests. It is synthesized by reaction of ethanol with concentrated sulfuric acid, or by passing ethene through cold, conc. sulfuric acid. It is used as an ethylating reagent.

Diethylsulfide, ***diethylthioether***: a colorless liquid with a characteristic, ether-like odor; m.p. - 103.8 °C, b.p. 92.1 °C, n_D^{20} 1.4430. D. is practically insoluble in water, but is readily soluble in ethanol and ether. It is readily oxidized to the sulfoxide or sulfone, and can be converted to diethylalkylsulfonium salts by alkyl halides. D. is made by reaction of potassium sulfide with ethyl halides. It is used as a solvent for cellulose esters, rubber and resins.

Difex process: a process for obtaining butadiene by Urea separation.

Differential pulse polarography: see Polarography.

Differential reactor, ***differential circulation reactor***: primarily a research reactor in which only partial consumption of the reactants is desired. Using differential turnover, it is easier to remove or add reaction heat and thus to run the reactor isothermally. D. are used mainly for study of heterogeneous catalysis of reactions.

Differential thermoanalysis, abb. ***DTA***: a method of thermal analysis in which the heat absorbed or released by a sample is measured as a difference between its temperature and that of a thermally inert reference substance (e.g. Al_2O_3, SiC) in the same furnace. The difference in temperature ΔT between the sample and reference substance is measured with thermoelements and is plotted against the temperature of the reference substance. The Fig. shows the differential thermogram obtained by heating calcium oxalate monohydrate.

The maxima correspond to exothermal processes, and the minima to endothermal ones. These can be the result of chemical change, as in the example, or of physical processes, such as evaporation, sublimation,

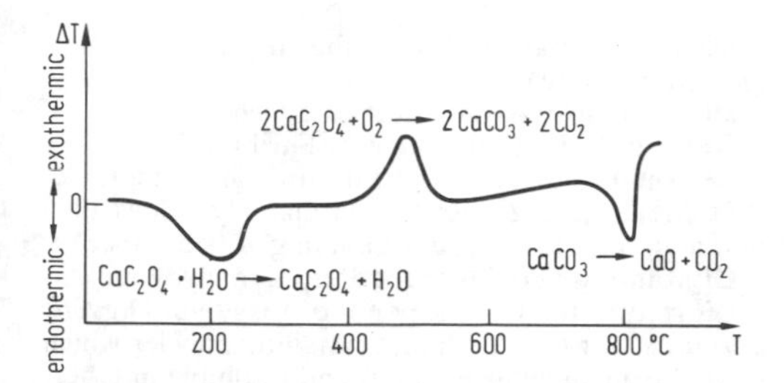

Differential thermogram of $CaC_2O_4 \cdot H_2O$ in the presence of O_2

adsorption, desorption, or phase changes. The surface arcas under the peaks are a measure of the amount of heat absorbed or released, and therefore, with the appropriate calibration, they can be used to determine the amount of the substance. Since the surfaces are affected by many factors, such as the heat conductivity of the sample, the geometry of the apparatus or the atmosphere in the furnace, the quantitative accuracy is limited.

D. can be used not only to study phase changes, but also recrystallization processes in glasses, the healing of lattice defects, desorption of gases and chemical reactions in condensed substances. D. has also been used to study and characterize polymers, which form characteristic peaks even in mixtures.

D. may be coupled with ***differential thermogravimetry***, abb. ***DTG***, in which the mass difference between two samples is determined. In this way, reactions in which gas is taken up (oxidation) or lost can be readily studied.

Differential thermogravimetry: see Differential thermoanalysis.

Difform system: in physical chemistry, a system of particles which extends much further in one or two dimensions than in the third. The type of system represented, for example, by cellulose or asbestos can be made by drawing a material (as a wire), and is called ***fibrillar***. The second type, which is found in nature in the form of mica or talcum, can be made by rolling (thin foils) and is called ***laminar***.

Diffraction methods: methods for determining crystal structures from the diffraction patterns of x-rays, neutrons and electrons. The most important of these methods are X-ray structure analysis (see), Neutron diffraction (see) and Electron diffraction (see).

Diffusion: the transport of gases, liquids or solids driven by differences in concentration. D. occurs spontaneously, due to microscopic particle motion, and in a direction which tends to equalize concentration differences. D. is an irreversible process coupled to production of entropy; it can be reversed only by investment of work.

In gases and liquids, the random thermal motion of particles leads on the statistical average to matter transport in the direction of the lower concentration. In crystalline solids, there are various D. mechanisms: 1) direct exchange of two crystal components (exchange of positions), 2) migration through interlattice spaces, 3) migration through empty lattice positions (hole mechanism), 4) migration along lattice imperfections or on crystal boundaries, 5) exchange of positions on crystal surfaces (surface diffusion) or in the pores of porous solids (pore diffusion).

The ***diffusion rate*** generally increases with temperature, because the particles are moving faster at higher temperature, and decreases as the density increases. In gases, the concentration equalization due to D. occurs within minutes in volumes of a few liters. In liquids, the process takes days or weeks. In solids, D. does not occur at a perceptible rate except at temperatures close to the melting point, where the number of lattice imperfections is greatly increased.

The mathematical treatment of D. began with Fick's laws (1855). *Fick's first law* applies to a stationary, i.e. temporally and spatially constant, concentration gradient dc/dx: $dn/dt = -DA(dc/dx)$. Here dn/dt is the D. rate, i.e. the number of molecules dn which in the time dt pass through a surface element with area A in the direction x. D is the ***diffusion coefficient***, which is usually given in units of $cm^2\ s^{-1}$. It is assumed that D is independent of concentration, an assumption which is fulfilled only by gases and very dilute solutions. Constant concentration gradients are found, e.g. as a solid dissolves, because at its surface there is a constant replenishment of the diffusing substance. If the concentration gradient is not spatially and temporally constant (***non-stationary D.***), *Fick's second law* applies. For the one-dimensional case of D. in the x-direction, this is $(\delta c/\delta t)_x = D(\delta^2 c/\delta x^2)_t$. This differential equation has many solutions, which depend on the boundary conditions of the D. process in question. The solutions describe how the concentration changes as a function of position and time.

If the diffusion coefficient depends on concentration and position, the general transport equation

$$(\delta c/\delta t)t = \frac{\delta}{\delta x}\left(D\frac{\delta c}{\delta x}\right)$$

must be used to describe D.

The D. coefficient in gases, liquids and solids differ by many orders of magnitude (table). There is a distinction between self D. and counter D.: ***self diffusion*** is the exchange of positions of one type of particle with others which are chemically identical; it is usually determined by means of isotopic labelling. ***Counter diffusion*** is the D. of different substances into each other; it always occurs when two substances are mixed.

Selected diffusion coefficients at standard pressure

	D in $cm^2\ s^{-1}$	T in K
H_2 in O_2	0.78	298
N_2 inO_2	0.19	298
H_2O vapor in air	0.23	298
CH_3OH in H_2O (0.78 mol dm^{-3})	1.2×10^{-5}	298
Sucrose in H_2O (0.73 mol dm^{-3})	0.5×10^{-5}	298
Au in Cu	2.1×10^{-11}	1023
Fe in Al	6.2×10^{-14}	632
$^{110}Ag^+$ in AgCl	2.5×10^{-14}	423
	7.1×10^{-8}	623
^{36}Cl in AgCl	3.2×10^{-16}	623

The self-D. coefficient of a gas calculated from the kinetic theory of gases is $D = v\ l/3$, where v is the mean particle speed and $l = 1/(4\sqrt{2}\ r^2\ \pi\ ^1N)$ is the mean free path of the molecules (k is Boltzmann's constant, r is the molecular radius, ^1N is the particle density). For counter diffusion between two substances 1 and 2, $D_{12} = (N_1v_1l_1 + N_2v_2l_2)/3(N_1 + N_2)$. For light gases with approximately the same mean speeds and free paths, $D_{12} \approx D$.

Diffusion coefficients in the liquid phase can be estimated by the *Einstein Stokes equation*: $D = kT/6\pi\eta r$. Here η is the viscosity and r the radius of spherical particles. In solids, the temperature dependence of

$$D = D_0 e^{-E_D/RT}$$

Here E_D is an activation energy of D., which is between 150 and 500 kJ mol^{-1}. D_0 is a constant. The equation can be applied to liquids; the values of E_D are between 5 and 20 kJ mol^{-1}.

Like heat transport, D. plays a central role in industrial-scale reactions. In most reactions of solids and in rapid reactions in solution, the reaction rate is determined by the D. rate (see Diffusion control). D. is utilized in the production of semiconductors, electronic components and surface layers of metals with special properties. Various methods of separation and vacuum diffusion pumps also utilize D.

A special form of D. occurs in very dilute gases, in which the mean free path of the particles is large compared to the dimensions of the vessel. In this case, there are practically no collisions between molecules; only with the walls. The equalization of differences in concentration and pressure occurs much more slowly than in normal D.; this phenomenon is called ***Knudsen D.*** (***Knudsen flow***). It plays a role in high-vacuum physics, and also in the transport of molecules through very narrow pores in adsorbents and catalysts. Another special case of D. is Thermodiffusion (see).

Diffusion coefficient: see Diffusion.

Diffusion control: a term for the observation that the rate of a chemical reaction is not controlled by the actual chemical process, but by the transport of the reacting particles to each other by diffusion. Since the rate of diffusion is highest in gases and lowest in solids, D. plays a role in gas reaction only in exceptional cases. Most solid reactions, on the other hand, are diffusion controlled. In solution, diffusion controlled reactions such as combination of radicals or oppositely charged ions (e.g. neutralization) have the highest rate constants, in the range of 10^{10} to 10^{11} $dm^3\ mol^{-1}\ s^{-1}$. Their temperature dependence is the same as that of the diffusion coefficent: $D = D_0 \exp(-E_D/RT)$. Here T is the absolute temperature, R the general gas constant, D_0 and E_D are constants characteristic of the substance (see Diffusion). A special case of D. is the Cage effect (see).

Diffusion current, ***diffusion boundary current***: an electrical current in an electrolysis cell, the magnitude of which is determined by diffusion to or from the electrode (see Polarography).

Diffusion potential: a Galvanic voltage (see) which appears on the surface between two electrolyte solutions.

Diffusion thermoeffect: the reverse of Thermodiffusion (see).

Diflubenzurone: an Insecticide (see).

Di-α-furyl diketone: same as Furil (see).

Digitalinum verum: see Purpurea glycosides.

Digitalis glycosides: the cardiac glycosides Purpurea glycosides (see) and Lanata glycosides (see).

Digitanol glycosides: see Purpurea glycosides.

Digitonin: the main saponin of foxglove, *Digitalis purpurea*. It forms a white, crystalline powder which is soluble in aqueous ethanol, and insoluble in ether; m.p. 235 °C, $[\alpha]_D$ -54.3° (methanol). The saponin mixture which makes up about 2% of the seeds of digitalis contains only about 40% pure D. The aglycone is digitogenin; it has a branched pentasaccharide glycosidically bound to the 3-position. D. is used for isolation and quantitative determination of sterols, especially cholesterol. Sterols with 3β-hydroxy groups form insoluble complexes (*digitonides*) with D.; in contrast to D., these are not hemolytic.

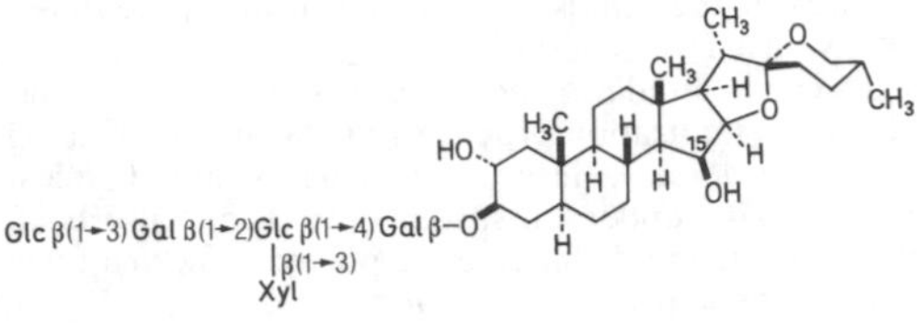

Glc β(1→3) Gal β(1→2)Glc β(1→4) Gal β–O (Xyl β(1→3))

Digitoxin: see Purpurea glycosides, see Lanata glycosides.

Digitoxigenin: see Lanata glycosides, see Purpurea glycosides.

Diglycol: same as Diethylene glycol (see).

Diglycol dinitrate: see Diethylene glycol dinitrate.

Diglycol distearate: $(C_{17}H_{35}\text{-CO-O-}C_2H_4)_2O$, a white, wax-like ester which is soluble in alcohol and hydrocarbons, and can be emulsified with water. D. is used as an emulsifier in polishes, cleansers and dispersing agents.

Diglycol monolaurate, ***diglycol monolauric acid ester***: $HO\text{-}CH_2\text{-}CH_2\text{-O-}CH_2\text{-}CH_2\text{-O-CO-}C_{11}H_{23}$, an oily, bright yellow liquid which is insoluble in water. D. can be produced by esterification of lauric acid with an equimolar amount of diglycol. It is used as an emulsifier and for synthesis of textile conditioners, soaps, drilling lubricants and cleansers.

Diglycol monolauric acid ester: same as Diglycol monolaurate (see).

Diglycol powder: see Smokeless powder.

Diglycolic acid, ***2,2'-oxydiacetic acid***: $HOOC\text{-}CH_2\text{-O-}CH_2\text{-COOH}$. As the monohydrate, it forms colorless and odorless, monoclinic prisms; m.p. 148 °C. D. cannot be distilled without decomposition. It is readily soluble in water and alcohol, and less soluble in ether. D. is more acidic than acetic acid, and forms a series of defined salts. It is synthesized by reaction of chloroacetic acid with barium hydroxide. D. is used as a coagulating agent and for organic syntheses. Its esters are used as softeners.

Digoxigenin: see Lanata glycosides.

Digoxin: see Lanata glycosides.

Dihydralazine, ***Depressan®***: a bishydrazino derivative of phthalazine. It acts centrally and peripherally as an antihypertensive.

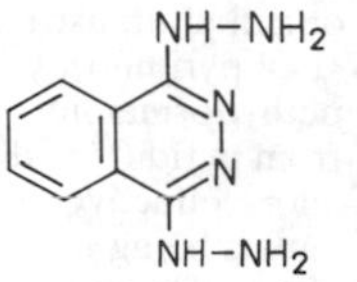

NH–NH2 / N / N / NH–NH2 Dihydralazine

9,10-Dihydroanthracene-9,10-dione: same as Anthraquinone (see).

Dihydrobenzo[b]pyran: same as Chromane (see).

Dihydro-1,3-dioxol: same as 1,3-Dioxolane (see).

2,3-Dihydroflavone: see Flavones.

2,5-Dihydrofuran-2,5-dione: same as Maleic anhydride (see).

2,3-Dihydroindole: same as Indoline (see).

9,10-Dihydrophenanthrene-9,10-dione: same as Phenanthrene quinone (see).

Dihydrotachysterol: a compound formed by partial hydrogenation of tachysterol, a byproduct of vitamin D_2 synthesis from ergosterol. D. increases the blood calcium level, and thus has an effect similar to that of parathormone. It can be used in replacement therapy for the hormone.

Dihydroxyacetone: 1,3-dihydroxypropan-2-one, $HOCH_2$-CO-CH_2OH, the simplest ketose; m.p. about 80 °C. It is readily soluble in water. D. is formed by enzymatic dehydrogenation of glycerol. Non-enzymatic oxidation of glycerol produces a mixture of glyceraldehyde and D. The monophosphate of D. is an intermediate in carbohydrate metabolism. D. produces a brown color on the skin, due to reaction with the free amino groups in the skin.

3,4-Dihydroxybenzaldehyde: same as Protocatechualdehyde (see).

Dihydroxybenzenes: same as Benzene diols (see).

2,5-Dihydroxybenzoic acid: same as Gentisic acid (see).

2,4-Dihydroxybenzopteridine: same as Alloxazine (see).

2,5-Dihydroxybenzyl alcohol: same as Gentisyl alcohol (see).

3,4-Dihydroxycyclobut-3-ene-1,2-dione: same as Quadratic acid (see).

5,8-Dihydroxy-1,4-naphthoquinone: a quinone pigment formerly called ***naphthazarine***; it crystallizes in brown needles with a greenish sheen; m.p. 237 °C. D. is only slightly soluble in water and ether, but better soluble in alcohol. It is an excellent mordant dye used on wool and in printing cloth, usually as the hydrogensulfite compound.

3,4-Dihydroxyphenylalanine: see Tyrosine.

2,3-Dihydroxypropanol: same as Glyceraldehyde (see).

Diiodomethane: same as Methylene iodide (see).

Diisoamyl ether, ***isoamyl ether***: $(CH_3)_2CH$-CH_2-CH_2-O-CH_2-CH_2-$CH(CH_3)_2$, a colorless, combustible liquid with a pleasant, fruity odor; b.p. 172-173 °C, n_D^{20} 1.4085. It is obtained by heating isoamyl alcohol with concentrated sulfuric acid. The solvent properties of D. are similar to those of diethyl ether, but it is less volatile. It is therefore used as a solvent for perfumes, fats, metal oxides and lacquers.

Diisopropyl ether, ***isopropyl ether***: $(CH_3)_2CH$-O-$CH(CH_3)_2$, a colorless, combustible liquid with an ether-like odor; m.p. - 86 °C, b.p. 68 °C. D. is slightly soluble in water, and is miscible with ethanol, ether, benzene and chloroform. It irritates the eyes and respiratory passages and has a narcotic effect. It tends very easily to form highly explosive peroxides. It is synthesized from isopropanol and sulfuric acid, or by catalytic addition of isopropanol to propene. Because of its solvent properties, it is used as a solvent for extraction of fats, oils, resins, waxes and cellulose acetate; mixed with ethanol, it is also used for cellulose nitrate. D. is also used to remove the sheen from acetate silks and, mixed with isooctane, as an antiknock additive for gasoline.

Diisopropylidene acetone: same as Phoron (see).

Diketene, ***vinylacetic acid β-lactone, 1,3-but-3-enolide***: a dimerization product of ketene, a colorless liquid with a pungent odor; m.p. - 7.5, b.p. 127 °C. D. is very reactive and must be stabilized as a liquid by addition of hydroquinone. Addition of ethanol to D. yields ethyl acetate, which decomposes back to ketene on heating to 500 °C. D. is synthesized by cautious heating of liquefied ketene in the presence of aluminum chloride.

HC=C–O
HC–CO

Diketones: organic compounds with two ketone groups, >C=O, in the molecule. D. are dicarbonyl compounds. Depending on the relative positions of the keto groups, they are 1,2- or α-, 1,3- or β-, 1,4- or γ- , and so forth. D. are synthesized by various methods, the choice depending mainly on the relative positions of the keto groups. Like all bifunctional carbonyl compounds, D. are important reagents for the synthesis of heterocycles and carbocyclic compounds.

Dilatance: Some colloids become rigid under mechanical stress and recover the capacity to flow when the stress is relieved. A thick suspension of uncooked cornstarch displays D., as does wet sand.

Dimedone: 5,5-dimethylcyclohexane-1,3-dione, a substituted alicyclic 1,3-diketone. D. is steam volatile; it crystallizes out of water in yellow needles, and out of alcohol in colorless prisms; m.p. 148-149 °C. D. is slightly soluble in water, and readily soluble in alcohol and chloroform. It is synthesized from mesityl oxide and diethyl malonate.

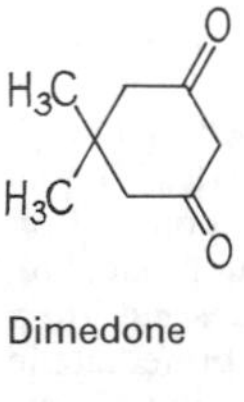

O / H3C / H3C / O — O O / H3C CH3 / CH– / R / H3C CH3 / O O

Dimedone 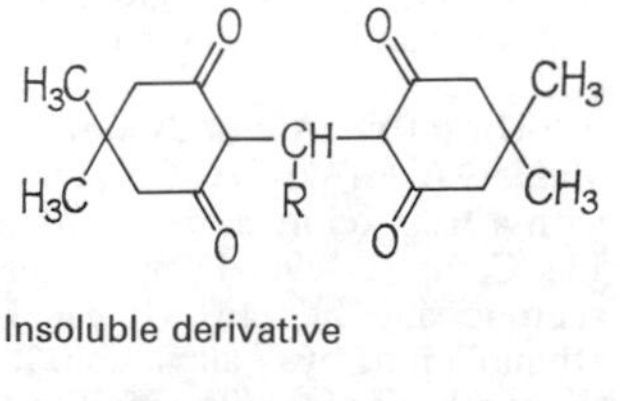 Insoluble derivative

The first two steps are a Michael addition and subsequent Claisen condensation to give the cyclohexane carboxylic acid ethyl ester, which can be converted to D. by ketone cleavage. D. is a sensitive detection reagent for aldehydes, with which it forms insoluble, crystalline derivatives.

Dimefox: see Organophosphate insecticides.

Dimercaprol, ***BAL***: $CH_2OH\text{-}CHSH\text{-}CH_2SH$, a yellow liquid with a mercaptan-like odor, which is soluble in water to a concentration of 5%. It is synthesized by addition of bromine to allyl alcohol; afterwards, the bromine atom is replaced by halogen sulfide. D. is an antidote to mercury and arsenic poisoning, because it can release the blocked SH groups of enzymes. D. was developed as an antidote to the arsenic-containing chemical weapon Lewisite; BAL is an acronym for British Anti-Lewisite.

Dimers: see Polymers.

Dimethirimol: see Pyrimidine fungicides.

Dimethoate: O,O-dimethyl *S*-methylcarbamoylmethyl phosphorodithioate, a systemic and contact Organophosphate insecticide (see; Table 3). It is a solid, D^{65} 1.277, m.p. 51 °C, b.p. 117 °C at 13 Pa, vapor pressure at 20 °C, $1.13 \cdot 10^{-3}$ Pa. D. is readily soluble in most organic solvents, except the saturated hydrocarbons. In water it is 2.9% soluble at 20 °C. D. is slowly hydrolysed in acidic medium, and quickly in alkaline medium. It is manufactured by reaction of alkali dimethyldithiophosphate with *N*-methylchloroacetamide, and is used to combat biting and sucking insect pests.

3,4-Dimethoxybenzaldehyde: same as Veratrum aldehyde (see).

1,2-Dimethoxybenzene: same as Veratrol (see).

Dimethylacetaldehyde: same as Isobutyraldehyde (see).

Dimethylamine: see Methylamines.

4-(Dimethylamino)azobenzene, ***dimethyl yellow, butter yellow***: yellow crystals which are insoluble in water, but soluble in alcohol, ether, glacial acetic acid, benzene and pyridine:

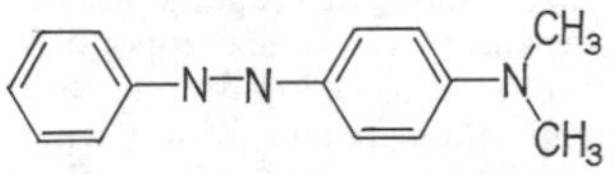

m.p. 117 °C. 4-D. can be synthesized by azo coupling of benzene diazonium chloride and *N,N*-dimethylaniline in weakly acidic solution. It is used as an indicator in analytical chemistry. It changes color in the pH range between 2.9 and 4.0, from red to yellow. 4-D. was formerly used to color butter yellow. Since rats fed large quantities of it developed bladder cancer, however, its use was forbidden. More recent studies on its carcinogenicity in higher animals, such as rabbits and monkeys, have failed to demonstrate ill effects.

2-Dimethylaminoethanol, ***dimethylethanolamine***: $(CH_3)_2N\text{-}CH_2\text{-}CH_2\text{-}OH$, a colorless liquid with a fishy odor; m.p. < - 70 °C (Stock point), b.p. 134 °C, n_D^{20} 1.4296. D. is soluble in water, alcohol, acetone and benzene. It can be synthesized from ethanolamine by *N*-alkylation. It is an intermediate in the synthesis of textile conditioners, corrosion protection preparations, emulsifiers, pigments and pharmaceuticals.

3-(Dimethylaminomethyl)indole: same as Gramine (see).

4-Dimethylaminopyridine: a colorless, crystalline compound; m.p. 112-113 °C. It is soluble in methanol, acetone, acetate and chloroform, and slightly soluble in cyclohexane and water. 4-D. is obtained industrially in good yields by reaction of 1-(4-pyridino)pyridinium dichloride with *N,N*-dimethylformamide; the first of these is readily obtained from pyridine and thionyl chloride. 4-D. is about 10^4 more effective as an acylation catalyst than pyridine, and is being used increasingly in acylations which are not adequately catalysed by pyridine. The extraordinary catalytic effet is due to the fact that even in nonpolar solvents, a high concentration of the *N*-acylpyridium salt is achieved, and the salt is present as a loosely bound ion pair.

N,N-Dimethylaniline: see *N*-Methylaniline.

Dimethylanilines: same as Xylidines (see).

Dimethylbenzenes: same as Xylenes (see).

Dimethylbenzidines: same as Tolidines (see).

3,3-Dimethylbutan-2-one: same as Pinacolone (see).

Dimethylcarbate: an Insect repellent (see).

Dimethyldiimide: same as Azomethane (see).

Dimethylacetic acid: see Butyric acids.

Dimethylethanolamine: same as 2-Dimethylaminoethanol (see).

Dimethyl ether, ***methyl ether, methoxymethane***: $CH_3\text{-}O\text{-}CH_3$, a colorless, inflammable and narcotic gas with a typical ether-like odor; m.p. - 138.5 °C, b.p. - 23 °C. At 18 °C, 1 l water dissolves 37 l D. The gas forms explosive mixtures with air. D. is used for methylation reactions, and sometimes as a propellant gas for aerosols.

Dimethylformamide, abb. ***DMF***: $O{=}CH\text{-}N(CH_3)_2$, a colorless, slightly hygroscopic, polar liquid; m.p. - 60.5 °C, b.p. 153 °C, n_D^{20} 1.4305. D. is readily soluble in water and most organic solvents, but is barely soluble in aliphatic hydrocarbons. In the absence of acids or bases, D. is very resistant to hydrolysis. Mixtures of D. with a number of halogenated hydrocarbons, such as carbon tetrachloride, can react explosively at high temperatures; the reaction is catalysed by iron salts. D. is irritating to the skin and especially the mucous membranes. It can be taken up through the skin and causes liver damage. D. is synthesized industrially by reaction of methyl formate with dimethylamine, or from carbon monoxide and dimethylamine in the presence of catalysts. D. is an excellent dipolar, aprotic solvent, which is used in large amounts in the chemical industry and research. It is a good solvent for polymers and many inorganic salts, for gases such as ethene, ethyne, hydrogen sulfide, sulfur dioxide, hydrogen chloride, etc., and for a number of pigments. D. is used in the Vilsmeier-Haack reaction (see) as a reagent for introducing the formyl group into organic compounds.

Dimethylglyoxime, ***diacetyldioxime***: $CH_3\text{-}C({=}N\text{-}OH)\text{-}C({=}N\text{-}OH)\text{-}CH_3$, the dioxime of a 1,2-diketone. It forms colorless, crystalline needles; m.p. 245-246 °C, subl. about 234 °C. D. is insoluble in water but readily soluble in alcohol, ether and acetone. It is most commonly synthesized from butane-2,3-dione and hydroxylamine or from methyl ethyl ketone. An alcoholic solution of D. is known as *Čugaev's reagent*, and is used for qualitative and quantitative determination of nickel.

N,N-Dimethylurea: $CH_3\text{-}NH\text{-}CO\text{-}NH\text{-}CH_3$, a colorless, hygroscopic crystalline compound; m.p. 108 °C, b.p. 268-270 °C. *N,N'*-D. is soluble in water and alcohol, but barely soluble in ether. It can be synthesized by transamidination of urea, or from phosgene or methylisocyanate and methylamine: CH_3-N=C=O + CH_3-NH_2 → CH_3-NH-CO-NH-CH_3. *N,N'*-D. is used in the synthesis of purine derivatives, such as caffeine and theophyllin, and in urea-formaldehyde resins.

Dimethyl yellow: same as 4-(Dimethylamino)azobenzene (see).

2,6-Dimethylhepta-2,5-dien-4-one: same as Phoron (see).

Dimethyl ketone: same as Acetone (see).

N,N-Dimethyl-4-nitrosoaniline: same as *p*-Nitrosodimethylaniline.

Dimethylphenols: same as Xylenols (see).

Dimethylphthalate: an Insect repellent (see).

2,2-Dimethylpropanoic acid: see Valeric acid.

Dimethylpyridine: see Lytidine.

2,6-Dimethylpyrone, ***2,6-dimethyl-γ-pyrone, 2,6-dimethyl-4H-pyran-4-one***: a readily synthesized γ-pyrone derivative. It forms colorless crystals; m.p. 135 °C, b.p. 248-250 °C. 2,6-D. is readily soluble in water, alcohol and ether. It is synthesized by heating Dehydracetic acid (see) with dilute mineral acids. It was the first compound with an ether oxygen which was shown to have basic properties. 2,6-D. forms crystalline salts with acids, for example, a hydrochloride, m.p. 154 °C. It forms addition compounds (prototypical oxonium salts) with inorganic salts, e.g. copper or zinc chloride. While the oxonium salt of diethyl ether and hydrochloric acid is hydrolysed, the oxonium salts of 2,6-D. are stable; see Pyrylium salts.

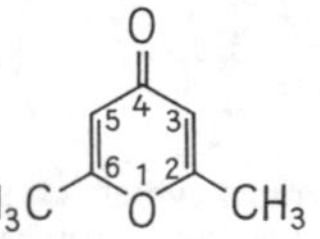

Dimethyl sulfate, ***sulfuric acid dimethyl ester***: $(CH_3O)_2SO_2$, a colorless, combustible liquid, m.p. -31.7 °C, b.p. 188.5 °C, n_D^{20} 1.3874. D. is very poisonous. For external detoxication, use 3% ammonia solution in water. When D. is distilled under normal pressure, it decomposes. It is slightly soluble in water and aliphatic hydrocarbons, and readily soluble in alcohol, acetone, dioxane, ether, esters, aromatic hydrocarbons and chlorinated hydrocarbons. D. is obtained by reaction of methanol with fuming sulfuric acid, or of dimethyl ether with sulfur trioxide. It is often used as a methylation reagent. In the presence of alkali hydroxide, it replaces the acidic hydrogen atoms of hydroxy, mercapto, amino or imino groups. As a rule, only one methyl group per molecule of D. is transferred, but at temperatures above 100 °C, the second methyl group is also transferred. Alcohols can be methylated by phase transfer catalysis, and aromatic hydrocarbons with aluminum chloride as catalyst.

Dimethyl sulfide, ***dimethyl thioether***: CH_3-S-CH_3, a colorless, combustible and poisonous liquid with a penetrating odor; m.p. - 98 °C, b.p. 37 °C. D. is insoluble in water, but soluble in ethanol and ether. It occurs naturally in coffee, cocoa and other aromatic plant products, and has also been shown to be a natural alarm substance. D. can be synthesized from the leachates from cellulose production, by reaction of dimethyl ether with hydrogen sulfide, or by the general synthetic methods for sulfides (see Thioethers). D. is used to synthesize dimethylsulfoxide, methionine, surfactants and other sulfur compounds. It is also used to odorize gas.

Dimethylsulfone, IUPAC name, ***methylsulfonylmethane***: CH_3-SO_2-CH_3, forms colorless and odorless crystals; m.p. 110 °C, b.p. 238 °C. D. is soluble in water. It is found naturally in cows' milk and human urine. Is is synthesized by oxidation of dimethylsulfide with hydrogen peroxide in excess, or with concentrated nitric acid. It is used as a solvent for polymers and other inorganic and organic compounds, and as a softener and ignition accelerator in fuels.

Dimethylsulfoxide, IUPAC name, ***methylsulfinylmethane***: CH_2-SO-CH_3, a colorless and odorless, hygroscopic liquid; m.p. 18.4 °C, b.p. 189 °C, n_D^{20} 1.4770. D. is soluble in water and many organic solvents, but is insoluble in alkanes. It is obtained by oxidizing dimethylsulfide with atmospheric oxygen in the presence of dinitrogen tetroxide, or in the laboratory with hydrogen peroxide, nitric acid or chromic acid. Because of its high dielectric constant (ε = 46.7) and dipole moment (4.3 D), D. is an excellent solvent. It is simultaneously an extraction solvent, oxidizing agent and base component (forming CH_3-CO-CH_2^-, the dimsyl anion) in many reactions. The dimsyl anion is a strong base and a strong nucleophile. As a typical oxidizing agent, D. reacts with halomethylarenes to form aromatic aldehydes.

D. is absorbed through the skin and acts as a carrier for toxic substances or for drugs. Its use for medical or cosmetic purposes is a controversial subject.

Dimilin®: an Insecticide.

Dimorphism: see Polymorphism.

Dimroth cooler: see Coolers.

Dinitrobenzene: the three structurally isomeric dinitro derivatives of benzene. ***1,2-(o-)-Dinitrobenzene***: colorless needles or platelets, m.p. 118.5 °C, b.p. 319 °C. 1,2-D. is practically insoluble in water, but is soluble in alcohol, ether and chloroform. It is steam volatile. Because of the electronic effects of the nitro group, it is possible to replace one of the two nitro groups by other substituents, such as -OH or -NH_2. 1,2-Dinitrobenzene can be reduced to 2-nitroaniline or *o*-phenylenediamine. It is synthesized by oxidation of 2-nitroaniline or by the reaction of sodium nitrite with 2-nitrobenzene diazonium salts. Direct nitration of benzene or nitrobenzene yields 1,2-D. in only small yields, and mixed with the other isomers.

1,3-(m-)Dinitrobenzene: bright yellow platelets, m.p. 90 °C, b.p. 302 °C. It is slightly soluble in water, and readily soluble in chloroform, acetone and benzene. Neither nitro group can be substituted under the same conditions which permit substitution of the 1,2- and 1,4-compounds. 1,3-D. can be easily synthesized by direct nitration of benzene or nitrobenzene. It is used to synthesize 3-nitroaniline.

1,4-(p-)Dinitrobenzene, a colorless, crystalline compound which can be steam distilled; m.p. 174 °C, b.p. 299 °C. 1,4-D. is slightly soluble in water, and readily soluble in acetone, benzene and glacial acetic acid. Its chemical behavior and synthesis are similar to those of the 1,2-isomer. The toxicity of D. is similar to those of Nitrobenzene (see) and Aniline (see).

NO_2 / NO_2 (1,2, 1,3 or 1,4 position)

Dinitrobenzene

Dinitrocresol, ***2-methyl-4,6-dinitrophenol***: yellow, strongly staining prisms which are steam volatile (m.p. 87.5 °C) and very poisonous. D. is slightly soluble in water, and is soluble in ethanol, ether and benzene. It is made by nitration of *o*-cresol with concentrated nitric acid in glacial acetic acid, and is used in the form of its water-soluble salts to combat pests and weeds, and to impregnate wood.

OH, O_2N, CH_3, NO_2

4,6-Dinitro-2-diazohydroxybenzene, ***diazodinitrophenol***, abb. ***DDNP***: yellow, light crystals which decompose at 177 °C without melting. D. is synthesized by mixing sodium nitrite and sodium picramate in a solution of hydrochloric acid. It is an initiating explosive used in some ignition caps.

Dinitrodiglycol: same as Diethylene glycol dinitrate.

Dinitrogen complexes: coordination complexes of the transition metals with molecular nitrogen, N_2, as a ligand. The first D., $[Ru(NH_3)_5N_2]^{2+}$, was discovered in 1965 by Allen and Senoff. Since then, many have been synthesized. The methods include direct substitution of labile ligands by N_2 according to:

$$[Ru(NH_3)_5(H_2O)]^{2+} + N_2 \xrightarrow{H_2O} [Ru(NH_3)_5N_2]^{2+} + H_2O,$$

reduction of transition metal complexes in the presence of suitable coligands in a N_2 atmosphere according to:

$$2\ Ni\ (acetylacetonate)_2 + 4\ P(C_6H_{11})_3 + N_2 \xrightarrow{Al(CH_3)_3} [Ni(N_2)\ (P\ (C_6H_{11})_3)_2]_2$$

and conversion of suitable, already coordinated N ligands into coordinated N_2 according to: $[Mn(h^5\text{-}C_5H_5)(CO)_2(N_2H_4)] + 2\ H_2O \rightarrow [Mn(h^5\text{-}C_5H_5)(CO)_2(N_2)] + 4\ H_2O$. Most D. contain the N_2 ligands bound "end-on" (M-N≡N), but N_2-bridged complexes are known, e.g. $[(H_3N)_5Ru\text{-}NN\text{-}Ru(NH_3)_5]^{4+}$. A few cases of "side-on" binding have been obtained; these complexes are of complicated composition. The bonding in the D. with "end-on" coordination is similar to that in metal carbonyls (see Coordination chemistry); the coordination of the dinitrogen ligand weakens the N-N bond.

D. play an important role in nature, for example in nitrogen fixation by the root-nodule bacteria which live in symbiosis with the legumes. The metalloenzyme responsible for this reaction, nitrogenase, consists of two different proteins, one of which is a molybdenum-iron protein ($M_r \approx 220{,}000$; contains 2 Mo, 24 Fe and 24 S atoms) and binds the N_2. The second is an iron protein ($M_r \approx 56{,}000$) containing a Fe_4S_4 ferredoxin unit; it functions as an electron carrier in N_2 reduction.

Dinitrophenols: six isomeric dinitro derivatives of phenol; only ***2,4-dinitrophenol*** is industrially significant. It crystallizes in bright yellow, steam-volatile needles; m.p. 114 °C.

OH, NO_2, NO_2

It is slightly soluble in water, and readily soluble in ether, acetone and benzene. It is synthesized by nitration of 2- or 4-nitrophenol with concentrated nitric acid. 2,4-Dinitrophenol is a relatively strong acid (pK_a 4.03) and is used as an indicator (colorless at pH 2.6, yellow at pH 4.4); it is also used to make pesticides and wood protectants, azo dyes, sulfur dyes and explosives.

2,4-Dinitrophenylhydrazine: poisonous, red crystals; m.p. 198-200 °C (dec.). 2,4-D. is slightly soluble in water and alcohol, but dissolves well in glacial acetic and dilute mineral acids.

O_2N—, —NH—NH_2, NO_2

It forms 2,4-dinitrophenylhydrazones which crystallize well with aldehydes and ketones, and is therefore highly suitable for characterization of these carbonyl compounds. 2,4-D. is used in the synthesis of 1-chloro-2,4-dinitrobenzene with hydrazine.

Dinocap: an Acaricide (see) which is also effective against true mildew mold.

Dinoseb: see Herbicides.

1,2-Diols: same as Glycols (see).

Diolefins: see Dienes.

Diolen®: see Synthetic fibers.

Diosgenin: (25S)-spirost-5-en-3β-ol, the aglycon of saponins. D. is isolated mainly from the roots of *Dioscorea* species, where it is present at 5 to 6% of the dry mass. The roots are collected from wild plants in Mexico, China, India and South Africa. D. is an important starting material for partial syntheses of steroids.

Dioxacarb: a Carbamate insecticide (see).

Dioxane: a six-membered, heterocyclic compound with two oxygen atoms in the ring. It exists in three isomers, but only ***1,4-D., tetrahydro-1,4-dioxin, diethylene dioxide***, is of industrial importance. This is a colorless, combustible liquid with a pleasant odor; m.p. 11.8 °C, b.p. 101 °C, n_D^{20} 1.4224. 1,4-D. is miscible with water and organic solvents in any proportions. As a cyclic ether, it is chemically inert, but because of its low melting point and the ease with which it undergoes autooxidation, forming peroxides, it should be handled with caution. 1,4-D. should be kept in the dark and away from air, or stabilized with reducing agents such as iron(II) sulfate. 1,4-D. is manufactured by dehydration of ethylene glycol or dimerization of ethylene oxide. It forms crystalline adducts with bromine and sulfur trioxide, and these can be used as gentle bromination or sulfonating reagents. 1,4-D. is used in large quantities as a solvent for cellulose esters, chlorinated rubbers, fats, waxes, resins and lacquers, and as an extraction solvent for pharmaceutical preparations. Chlorination products of 1,4-D. are also used as pesticides.

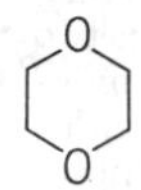

1,2-Dioxetane: a saturated, four-membered heterocyclic ring system with two oxygen atoms in the ring; the unsaturated compounds are called ***dioxetes***. The 1,2-D. are highly endothermal compounds, but because of their kinetic stability, they can be synthesized by photooxygenation of olefins. Oxygen is passed through a solution of an olefin which contains a sensitizing pigment, and simultaneously, light shines on the solution. Singlet oxygen is formed and adds to the C=C double bond.

Dioxete: see 1,2-Dioxetane.

"Dioxin": see 2,3,7,8-Tetrachlorodibenzo-1,4-dioxin.

1,4-Dioxin: an unsaturated, six-membered heterocyclic compound with two oxygen atoms in the ring. 1,4-D. is a colorless liquid, b.p. 75 °C, n_D^{20} 1.4350. It is miscible with ether, acetone and benzene. 1,4-D. undergoes the addition reactions typical of alkenes. 2,3,7,8-Tetrachlorodibenzo-1,4-dioxin (see) is an important derivative.

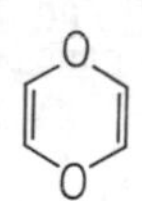

1,3-Dioxolane, ***dihydro-1,3-dioxol***: a saturated, five-membered heterocyclic compound with two oxygen atoms; a colorless liquid; m.p. - 95 °C, b.p. 78 °C, n_D^{20} 1.3974. It is soluble in water and organic solvents. 1,3-D. is stable to bases, but is hydrolysed by dilute acids to carbonyl compounds and diols. It is formed by heating of aldehydes or ketones with 1,2-glycols. For example, 1,3-D. is formed in the reaction of ethylene glycol with formaldehyde in benzene in the presence of toluene-4-sulfonic acid as catalyst. 1,3-D. and its derivatives are used as solvents.

Dioxopromethazine: see Antihistamines.

2,4-Dioxotetrahydroimidazole: same as Hydantoin (see).

Dioxygenases: see Oxygenases.

Dioxygen complexes: coordination compounds of the transition metals with dioxygen (O_2-) ligands. D. are either 1:1 or 2:1 complexes in terms of the metal:oxygen ratios.

Coordination types of the dioxygen ligand

A metal-ligand charge transfer always occurs during O_2 complex formation, leading to the four types of complex shown in the figure. Types I (e.g. $[Co(CN)_5O_2]^{3-}$) and II (e.g. $[(NC)_5Co\text{-}O_2\text{-}Co(CN)_5]^{5-}$ are generally known as superoxo compounds, while types III (e.g. $[IrCl(CO)(P(C_6H_5)_2)_2(O_2)]$) and IV (e.g. $[(P(C_6H_5)_3)_2RhCl(\mu O_2)]_2$ can be classified as peroxo complexes with terminal or bridging O_2 ligands. D. of the superoxo and peroxo types have distinctly longer O-O distances (125-130 pm and 130-155 pm, respectively) than the bond length in molecular oxygen (120.7 pm).

D. are generally obtained through reversible addition reactions with coordinatively unsaturated complexes, for example, with the Vaska complex: *trans*-$[IrCl(CO)(P(C_6H_5)_3)_2]$ + O_2 $\rightleftharpoons$ $[Ir(CO)Cl(O_2)(P(C_6H_5)_3)_2]$. They can also be obtained through O_2 ligand substitution, which is often irreversible, for example: $[Pt(P(C_6H_5)_3)_4]$ + $O_2 \rightarrow$ $[Pt(P(C_6H_5)_3)_2(O_2)]$ + 2 $P(C_6H_5)_3$, or $2[Co(NH_3)_6]^{2+}$ + $O_2 \rightarrow$ $[(H_3N)_5Co\text{-}O_2\text{-}Co\text{-}(NH_3)_5]^{4+}$. D. are of special interest as key compounds for understanding the biochemistry of oxygen transport and storage by metalloproteins such as hemoglobin and myoglobin, and they are also model substances for homogeneous catalytic oxidation reactions.

Dipentene: see Limonene.

Dipeptides: see Peptides.

Diphacinone: a Rodenticide (see).

Diphenhydramine: see Antihistamines.

Diphenyl: same as Biphenyl.

Diphenylacetylene: same as Tolane (see).

Diphenylamine, ***N-phenylaniline***, ***N-phenylaminobenzene***: $C_6H_5\text{-}NH\text{-}C_6H_5$, a secondary aromatic amine. D. forms colorless crystals which smell like flowers; m.p. 54-55 °C, b.p. 302 °C. It is slightly soluble in water, but dissolves easily in organic solvents and mineral acids. The salts of D. are largely hydrolysed in aqueous solution, because D. is a very weak base. It is an important reagent for nitric and nitrous acids and their salts, with which it develops an intense blue color in acidic solution. D. is synthesized chiefly from aniline in the presence of various deaminating catalysts. Its synthesis from aniline and chlorobenzene or phenol is less important. D. is used as an intermediate in the dye industry, as a stabilizer for glycerol nitrate and cellulose nitrate, and as a compound to protect rubber from ageing.

Diphenylcarbinol: same as Benzhydrol (see).

3,4-Diphenylcyclobutane-1,2-dicarboxylic acids: same as Truxic acids (see).

2,4-Diphenylcyclobutane-1,3-dicarboxylic acids: same as Truxillic acids (see).

1,2-Diphenylethane: same as Dibenzyl (see).

1,2-Diphenylethene: same as Stilbene (see).

Diphenyl ether, ***phenoxybenzene***: C_6H_5-O-C_6H_5, colorless, monoclinic crystals which smell like geraniums; m.p. 26.8 °C, b.p. 257.9 °C, n_D^{25} 1.5787. D. is insoluble in water but is soluble in ethanol, ether and benzene. It is synthesized by heating phenol with zinc chloride or aluminum chloride, or by reaction of potassium phenolate with bromobenzene in the presence of finely divided copper. D. is used as a heat-transfer material and in perfumes, e.g. as a rose perfume for soaps.

Diphenylethyne: same as Tolane (see).

Diphenylglycolic acid: same as Benzilic acid (see).

Diphenyline rearrangement: see Benzidine rearrangement.

1,3-Diphenylisobenzofuran: see Benzofurans.

1,3-Diphenylisothionaphthene: see Benzothiophenes.

Diphenylketene: $(C_6H_5)_2C{=}C{=}O$, an orange liquid; b.p. 265-270 °C (dec.). In a vacuum, D. can be distilled without decomposition. It is soluble in ether and benzene. In contrast to unsubstituted ketene, D. does not spontaneously dimerize to the corresponding diketene. It can be synthesized by a Wolff rearrangement (see) from azobenzil.

Diphenyl ketone: same as Benzophenone (see).

Diphenylmethane: C_6H_5-CH_2-C_6H_5, an aliphatic-aromatic hydrocarbon, colorless needles with a pleasant, orange-like odor; m.p. 25.4 °C, b.p. 264.3 °C, n_D^{20} 1.5753. D. is insoluble in water, but is soluble in ethanol and ether. It can be synthesized from benzene, dichloromethane and aluminum chloride, or from benzyl chloride, benzene and aluminum chloride. When D. is heated to a high temperature, fluorene is formed. D. is the parent compound of the ***diphenylmethane pigments***.

Diphenylmethane pigments: see Triarylmethane pigments.

Diphenyloids: see Nomenclature, sect. III E 1.

1,3-Diphenylprop-2-enone: same as Chalcone.

N,N'-Diphenylthiourea, ***thiocarbanilide***: S=C(NH-C_6H_5)$_2$, colorless, crystalline leaflets; m.p. 154-155 °C. *N,N'*-D. is insoluble in water, but is soluble in alcohol, ether and chloroform. In the presence of mineral acids, *N,N'*-D. is split into phenylisothiocyanate and aniline. It is synthesized by reaction of carbon disulfide with aniline in the presence of basic catalysts: $S{=}C{=}S + 2\,H_2N\text{-}C_6H_5 \rightarrow S{=}C(NH\text{-}C_6H_5)_2 + H_2S$. *N,N'*-D. is used as a vulcanization accelerator, a PVC stabilizer, a flotation agent in ore preparation and an intermediate for the production of pharmaceuticals and pigments.

1,3-Diphenyltriazene: same as Diazoaminobenzene (see).

Diphosgene: same as Methyl perchloroformate (see).

Diphosphates, ***pyrophosphates***: the salts of diphosphoric acid. ***Acid D.***, $M^I_2H_2P_2O_7$, are obtained by cautious heating of primary phosphates: $2\,M^IH_2PO_4 \rightarrow M^I_2H_2P_2O_7 + H_2O$; they are converted by stronger heating to polyphosphates (see Phosphates). ***Neutral D.***, $M^I_4P_2O_7$, are formed by heating of secondary phosphates: $2\,M^I_2HPO_4 \rightarrow M^I_4P_2O_7 + H_2O$.

Diphosphine: see Phosphane.

Diphosphopyridine nucleotide: see Nicotinamide nucleotides.

Diphosphoric acid, ***pyrophosphoric acid***: $H_4P_2O_7$, colorless, hygroscopic, water-soluble crystals; M_r 177.98, m.p. 61 °C. D. is the first of the series of Polyphosphoric acids (see). The molten compound reacts to form an equilibrium mixture of orthophosphoric acid, D. and more highly condensed polyphosphoric acids. D. is a tetrabasic acid with two readily ionized protons and two which are only weakly ionizable (pK_a values of 1.0, 2.0, 6,6 and 9.6). It forms two series of salts, the acidic and the neutral Diphosphates (see). D. is conveniently synthesized from its alkali salts by ion exchange or reaction of lead diphosphate with hydrogen sulfide.

Dipicrylamine, ***2,4,6,2',6'-hexanitrodiphenylamine***, trivial names ***hexyl, hexamine, hexite***: yellow, poisonous, explosive crystals; m.p. 243 °C (dec.), detonation rate 7150 m/s at density 1.67. D. is insoluble in water, ether or chloroform, but is soluble in acetone, glacial acetic acid, nitric acid and alkalies. It can be made by nitration of diphenylamine in sulfuric acid, or by nitration of 2,4-dinitrodiphenylamine, which can be made from 2,4-dinitrochlorobenzene and aniline; the second synthesis is better. In the second world war, D. was used as an underwater explosive. It is used as a reagent for gravimetric and colorimetric determination of potassium, rubidium, mercury, lead, etc., and as a precipitation reagent for alkaloids.

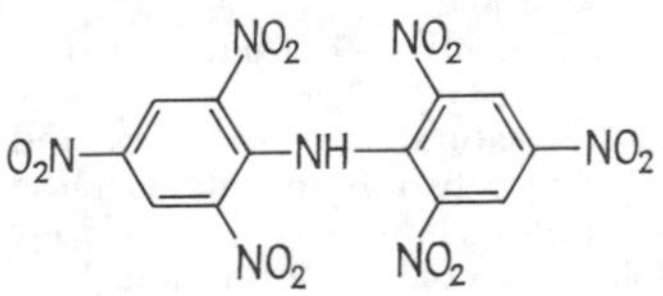

NO_2 NO_2
O_2N—NH—NO_2
NO_2 NO_2

Di-pi (π) methane rearrangement: a sigmatropic 1,2-rearrangement of acyclic, cyclic or aryl-alkenyl dienes, in which the double bonds are separated by an sp^3-hybridized C atom; the products are cyclopropane derivatives.

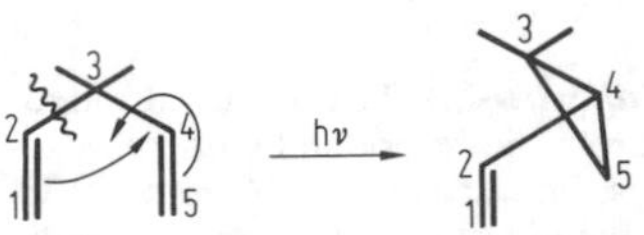

3 2 4 1 5 hν 3 4 2 5 1

Dipiproverin: see Spasmolytics.

1,3-Dipolar cycloaddition: a synchronous *cis*-addition of a 1,3-dipole to the multiple bond of a dipolarophile, forming a heterocyclic five-membered ring (1). Ozonization is a reaction consisting of two successive 1,3-D. (2). 1,3-D. are especially significant for synthesis of five-membered heterocycles. For example, diazoalkanes act as 1,3-dipoles for addition to alkynes to form pyrazolene (3); azide acts as a 1,3-dipole for addition to alkynes to form 1,2,3-triazolene, or for addition to nitrile to form tetrazolene.

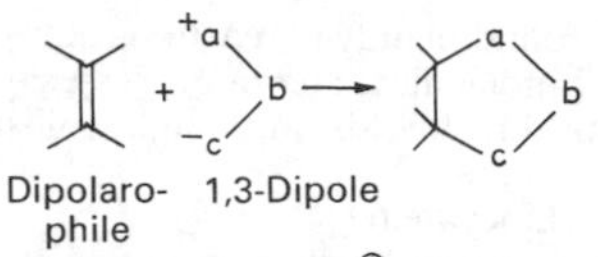

Dipolaro-phile 1,3-Dipole

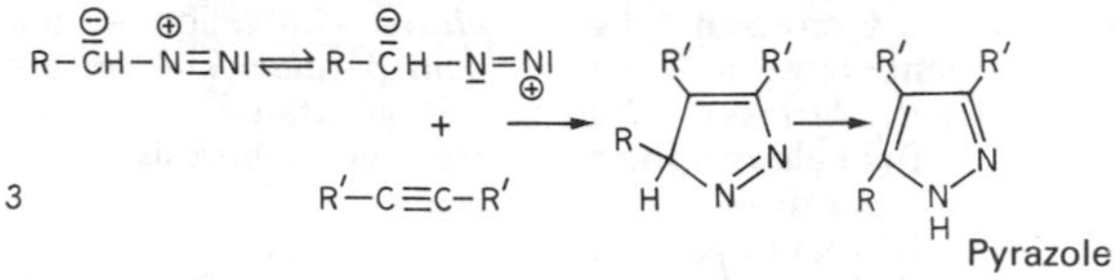

1

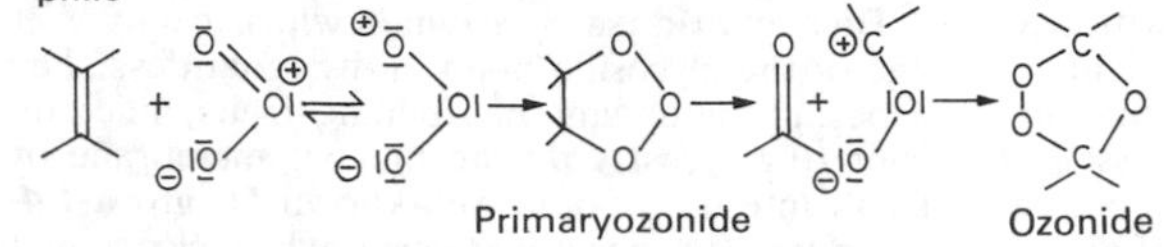

2 Primaryozonide Ozonide

3 Pyrazole

Dipole moment: product of the charge q of two point charges with the same absolute value but opposite signs and the vector $\mathbf{r}$ pointing from the positive to the negative pole. The length l of $\mathbf{r}$ is equal to the distance between the two point charges: $\boldsymbol{\mu} = q\mathbf{r}$. Such an arrangement of two point electric charges of the same magnitude and opposite signs at a distance l is called a ***dipole***. Since experimental determination of the direction is difficult, usually only the absolute value of the D. is given: $|\mu| = ql$. The unit of the D. is Cm (coulomb meter). Molecular D. were formerly given in Debye units (D), because their numerical values in this system lie between 0 and 10: 1 D = $0.336 \cdot 10^{-29}$ Cm. In a molecule, only the positively charged nuclei can be regarded as point charges. The contribution resulting from the continuous charge distribution of the electrons can only be approached through the wavefunction Ψ. Single atoms have equal charges in their nuclei (positive) and electrons (negative), and are therefore electrically nonpolar. Only in an electric field do they develop a slight D. By contrast, many molecules have *permanent D.* due to the difference in electronegativities of the individual atoms; this is independent of an external field (Table 1).

Table 1. Dipole moments of molecules in Cm

Molecule	$\mu \times 10^{29}$
Hydrogen H_2	0
Chlorine Cl_2	0
Hydrogen chloride HCl	0.36
Hydrogen bromide HBr	0.26
Hydrogen iodide HI	0.13
Carbon monoxide CO	0.04
Carbon dioxide CO_2	0
Nitrogen monoxide NO	0.05
Nitrogen dioxide NO_2	0.06
Methane CH_4	0
Water H_2O	0.61
Hydrogen sulfide H_2S	0.61
Methanol CH_3OH	0.56
Ethanol C_2H_5OH	0.53
Benzene C_6H_6	0
Chlorobenzene C_6H_5Cl	0.51
Benzonitrile C_6H_5CN	1.38

As an approximation, the permanent D. of a molecule can be calculated by vectorial addition (see Additivity principle) of the individual *bond dipole moments* (Table 2.)

Bond and group dipole moments $\mu_{A-B} \cdot 10^{29}$ in Cm (orientation A → B)

H–C	0.13	C–F	0.61
H–N	0.55	C–Cl (alkyl)	0.68
H–O	0.53	C–Cl (aryl)	0.52
H–S	0.22	C–Br (alkyl)	0.68
C–O	0.97	C–Br (aryl)	0.51
C=O	0.9	C–I	0.6
C–N	0.2	H_3C	0.13
C≡N	1.31	NO_2	1.32

For molecules with non-bonded electron pairs, the *atomic moments* must also be considered. The direction of the bond D. is determined by the electronegativities of the bonded atoms.

Measurements of D. provide important information on molecular structure. For example, it can be concluded from the observed $\mu(CO_2) = 0$ that the structure of the CO_2 molecule is linear. The *cis-* and *trans-*isomers of 1,2-dichloroethene can be distinguished by measurement of the D., because only the *cis-*form has a D. ($\mu = 0.43 \cdot 10^{-29}$ Cm); the D. of the *trans-*form is 0. Deviations between the D. calculated by addition for molecules of known structure and the experimental values indicate unusual bonding, e.g. strong interactions between neighboring bonds or delocalized bonds.

If a molecule is placed in an external electric field., the electrons are displaced toward the positive pole, and the nuclei toward the negative pole, which produces an *induced D.* (see Polarization 2).

Dippel's oil: a yellow, evil-smelling oil which turns brown in the presence of light. It is readily soluble in alcohol, ether and fatty oils, and consists mainly of nitrogen cmpounds, such as pyridine, aniline, quinoline, nitriles, methylamines and ammonium salts. It also contains low-boiling fatty acids and alkanes. D. is a product of distillation of bone tar (see Tar). It is used to denature ethanol (25 ml D. in 100 l ethanol).

Dipropyl ether, ***propyl ether***: CH_3-CH_2-CH_2-O-CH_2-CH_2-CH_3, a colorless, combustible liquid with an ether-like odor; m.p. - 122 °C, b.p. 91 °C, n_D^{20} 1.3809. D. is slightly soluble in water, and soluble in ethanol and ether. Its peroxides are highly explosive.

D. is synthesized by reaction of propanol with sulfuric acid. It forms azeotropic mixtures with many solvents, and is used widely for separation processes. It is also used as a stabilizer for chlorine-containing solvents, especially trichloroethylene.

Dipyridamol, ***Curantyl***®: a homopurine with several basic substituents. In homopurine, the five-membered ring of purine is expanded to a pyrimidine ring. D. is used as a coronary drug. It dilates the vessels of a healthy heart, and thus allows better oxygenation. It is not suitable for treatment of an attack of angina pectoris. D. also inhibits aggregation of thrombocytes.

$N(CH_2-CH_2OH)_2$

$(HOH_2C-H_2C)_2N$

2,2'-Dipyridyl, ***2,2'-bipyridine***: colorless crystals with an aromatic odor; m.p. 70 °C, b.p. 273 °C. It is readily soluble in alcohol, ether, chloroform or benzene, but only very slightly soluble in water. 2,2'-D. is obtained by heating pyridine with iron(III) chloride or from 2-bromopyridine and copper powder. It reacts with iron(II) salt solutions to give a dark red color. The corresponding salts dye untreated silk and wool rose-red. 2,2'-D. is used as a reagent to detect silver, cadmium and molybdenum. It inhibits carboxypeptidases, and some derivatives are effective herbicides.

Dipyrrolylmethane, ***dipyrrylmethane, dipyrrol-2-ylmethane***: colorless leaflets or needles; m.p. 73 °C, b.p. 163-167 °C at 1.6 kPa. It is readily soluble in ether and benzene, insoluble in water, acids and bases.

D. is made by condensation of pyrrole with formaldehyde. It can be readily oxidized to the orange ***dipyrrolylmethene***.

Diquat: see Herbicides.

Direct dyes: see Substantive dyes.

Direct reaction: a reaction in which the lifespan of the collision complex is so short ($< 10^{-12}$s) that no statistical energy redistribution can occur within the collision complex. The theory of the activated complex (see Kinetics of reactions, theory) is not applicable to a D. D. are found mainly in reactions between atoms or very small molecules, and in processes with very low potential threshholds and high collision energies.

Dirhodane: see Thiocyanates.

Disaccharidases: enzymes which catalyse cleavage of the glycoside bond in disaccharides. They are especially abundant in ripening fruits, microorganisms (e.g. yeast) and the mucous membrane of the small intestine. Some well-known D. are ***α-1,4-glycosidases*** (formerly ***maltases***), which cleave maltose and sucrose; ***β-1,4-glucosidases*** with cellobiose and gentiobiose as substrates and ***β-galactosidase*** (formerly ***lactase***) with lactose as substrate.

Disc electrophoresis: see Electrophoresis.

Discotic: see Liquid crystals.

Diselenocyanogen: see Selenocyanates.

Dishwashing detergents: for manual dishwashing, D. are generally liquid, highly foaming products. The main ingredients are anionic surfactants in neutral, aqueous solution. Alkane sulfonates and/or alkylbenzene sulfonates and ether sulfates are most common, but nonionogenic surfactants, such as fatty alcohol oxethylates or long-chain amine oxides are used; urea may also be present. To improve the clarity of the dishwater, low-molecular weight alcohols or short-chain alkylarylsulfonates are added; perfumes complete the recipe.

Products for dishwashing machines form very little foam; they consist mainly of sodium triphosphate and sodium silicates; small amounts of nonionic surfactants are added as wetting agents. To prevent formation of salt residues on the dishes, low-foam surfactants are added as clear rinses; these can also be acidic.

Disinfectants: compounds which are able to render microorganisms harmless or to kill them. They are used chiefly against pathogenic microorganisms, to prevent infection. The various microorganisms have widely varying sensitivities to different D. Bacterial spores are particularly resistant. Compounds used to treat the surface of the body and body cavities (nose, mouth, throat) are also called *antiseptics*. D. are not suitable for systemic use, either because they are too toxic, or because they lose their effect in the presence of serum.

The following are used as D.: 1) oxidizing inorganic and organic compounds such as chlorine, tincture of iodine, lime chloride, hydrogen peroxide, chloramine T, peroxyacetic acid and potassium permanganate; 2) mercury and silver compounds such as sublimate (mercury(II) chloride), thiomersal, phenylmercury borate and diacetyltannin-protein-silver; 3) alcohols, such as ethanol, propan-1-ol, propan-2-ol and chlorobutanol, and aldehydes, such as formaldehyde and glutaraldehyde; 4) alkyl-, aryl- and halophenols, such as thymol, cresol, chlorophene (2-benzyl-4-chlorophenol), hexachlorophene and hexylresorcinol; 5) quinoline and acridine derivatives, such as 8-hydroxyquinoline, ethacridine and acriflavinium chloride; 6) biguanide and thiosemicarbazone derivatives, such as chlorohexidine and ambazone; 7) invert soaps of the quaternary ammonium types, such as benzododecinium bromide; and 8) triphenylmethane dyes such as malachite green, brilliant green and crystal violet.

Disinfection: 1) In general, removal of nearly all microorganisms and prevention of their multiplication. D. prevents infections, but does not lead to complete sterility. It is achieved using chemical substances such as chlorine, iodine, potassium permanganate, alcohols, acids, soaps, etc., or by physical measures such as exposure to hot air, steam or UV light. Objects, parts of the body and rooms can be disinfected.

2) In water treatment, an important process to reduce the number of both pathogenic and harmless microorganisms (see Germ removal). The chemical and physical methods of water treatment are not capable of reducing the number of microorganisms to the legally prescribed concentration. For this, special oxidizing agents are required.

a) ***Chlorination*** of drinking water is achieved by addition of concentrated aqueous chlorine solutions, gaseous chlorine, lime chloride or chlorine bleach. The effect depends on the temperature, pH, time of action and chlorine concentration in the water. The chlorine content of drinking water at the periphery of the supply network should be at least 0.1 mg/l. When phenol-containing water is chlorinated, bad smelling and tasting chlorination products are formed; these can be removed by Clarification (see).

b) ***D. by chlorine dioxide***, ClO_2. Chlorine dioxide is added to drinking water which still contains slight amounts of phenols. This avoids the smell and taste which arise by chlorination. Chlorine dioxide is synthesized in the water treatment plant, usually from sodium chlorate, $NaClO_3$, or sodium chlorite, $NaClO_2$.

c) ***D. by chloramine***. Since chlorine is lost by oxidation processes in the supply network, the prescribed chlorine content of 0.1 mg/l may not be maintained at some of the peripheral points. However, the original chlorine dose at the treatment plant cannot exceed 0.5 mg/l, because the odor is too unpleasant for the consumers. For this reason, D. is often achieved by the chloramine process. First ammonia is added to the water, then chlorine; the effect of chlorine is considerably delayed in ammonia-containing water. However, the necessary reaction time is increased to about 1 hour.

d) ***Ozonization*** is accomplished by open or closed gasification, emulsification or combined methods. A reaction time of about 1.5 min is sufficient, regardless of the pH, to achieve D. The ozone concentration must be higher than 0.1 mg/l. In spite of its high activity, ozone has been largely replaced by chlorine, which is cheaper.

e) In the ***cumasina process***, the oligodynamic disinfectant effect (see Oligodynamism) of heavy metal salts, e.g. of silver (see Silvering), is utilized. Silvering can be done with colloidal silver solutions, solid preparations with oligodynamic silver in a state from which it can be reactivated, or with activated combination preparations. This method is used mainly to disinfect wells and parts of large water treatment plants. The *catadyne process* is essentially analagous to the cumasina process.

f) ***D. with iodine*** (for emergencies). This is recommended for use when the chlorination machinery fails, or when there is a shortage of chlorine due to irregular delivery of chlorine to the plant. Addition of 5 g iodine per m^3 water kills all germs; the time required is 20 min.

g) ***D. by ultraviolet radiation*** is a purely physical effect. It requires clear water, and the rays penetrate only a few centimeters deep into the water. Quarz lamps are used as light sources.

Dismutation: same as Disproportionation (see).

Disperse system, ***dispersion***: a two-phase system in which one phase, the ***disperse phase***, is distributed in the second, the ***dispersion medium***. According to the size of the particles, D. are classified as coarse (particle size, 10^{-4} cm), colloidal (10^{-4}-10^{-7} cm) or molecular ($<10^{-7}$ cm). Since there is no sudden change of properties with change of particle size, this system of classification is of limited value. The important D. are those in which the properties are determined by the thermal motion of the particles (see Brownian motion) and the forces between particles (see Colloids). If the particles of a D. are all the same size, the system is ***mono-*** or ***isodisperse***; if they vary in size, the system is ***poly-*** or ***heterodisperse***. Depending on the shape of the particles, the system may be ***corpuscular, laminar*** or ***fibrillar***. ***Lyophobic*** and ***lyophilic*** D. differ in their degrees of solvation. The dispersion medium can be a gas, liquid or solid; correspondingly, the system is an Aerosol (see), lysosol or xerosol.

Dispersing agent, ***emulsifier***: an organic or inorganic substance used to facilitate the dispersion of particles in a suspension. It reduces the surface tension of the closed phase and thus makes possible a fine distribution of the disperse phase. Most D. are Surfactants (see).

Dispersion: same as Disperse system (see).

Dispersion colloids: see Colloids.

Dispersion dyes: a group of synthetic dyes, especially water-insoluble monoazo and anthraquinone dyes and a few naphthol-AS dyes. D. are used mainly to dye acetate and polyamide fibers. They are dispersed finely in water, using a dispersing agent (emulsifier), and taken up by the fiber, which acts as a solid solvent. The uptake is governed by the Nernst distribution law. The dye not only is adsorbed to the surface, it diffuses into the fiber, giving a very fast and long-lasting color.

The azo dyes among the D. usually contain 2-amino-5-nitrobenzonitrile, 2-amino-3,5-dinitrobenzophenone or 2-amino-5-nitrothiazene as the diazo component. The most frequently used coupling components are *N*-ethyl-*N*-(3-oxobutanyl)aniline and *N*-β-cyanoethyl-*N*-β-hydroxyethylaniline.

Dispersion forces: see Van der Waals forces.

Displacement: see Crystal lattice defect.

Displacement technique: a method of chromatography in which the eluent is more effectively retained on the stationary phase than the sample is.

Disproportionation, ***dismutation***: a redox reaction in which an element in an intermediate oxidation state is simultaneously converted into derivatives with higher and lower oxidation states. Typical examples of D. are the reaction of chlorine in sodium hydroxide to chloride and hypochlorite:

$$\overset{0}{Cl_2} + OH^- \rightarrow \overset{-1}{Cl}{}^- + HO\overset{+1}{Cl}$$

and the formation of phosphane and phosphoric acid by heating of phosphorous acid:

$$4\,\overset{+3}{H_3P}O_3 \rightarrow \overset{-3}{P}H_3 + 3\,H_3\overset{+5}{P}O_4.$$

Disrotatory: see Woodward-Hoffmann rules.

Dissipative structures: a term coined in 1969 by Prigogine for spatially or temporally ordered states which arise in originally non-structured, homogeneous molecular systems which are far from thermodynamic equilibrium. They are stationary, non-equilibrium states which can arise in open systems by constant input of material and free energy; the energy is dissipated by the system. The transition from the disordered to the ordered state occurs suddenly. Not every system which is far from thermodynamic equilibrium forms D.; essential critera are highly nonlinear equations describing the system's approach to equilibrium and the stability of the spontaneously arising structure towards Fluctuations (see). D. are observed in numerous physical and chemical processes. Some examples are the formation of cell structures at the beginning of convection in a liquid which is heated from below (Benard effect), or at phase boundaries in flow processes. In chemical reactions, D. occur in the form of temporally or spatially periodic changes in the reaction rate and in the concentrations of intermediates (see Oscillating reactions). They also occur in the form of *bistabilities*. In biology and biochemistry, D. are of fundamental significance, e.g. for the conduction of nerve impulses and for the self-organization of matter at the beginning of evolution.

Dissociation: the cleavage of a molecule into two or more simpler molecules, atoms or ions.

1) *Electrolytic D.* is the reversible decomposition of an electrolyte into ions. Electrolytic D. occurs when the electrolyte dissolves in a solvent. In a *true electrolyte*, the ions exist in the solid in an ionic crystal. In this case, electrolytic D. consists of a solvation of the previously existing ions. In a *potential electrolyte*, on the other hand, the ions do not exist in the solid or gas, but are formed in the course of D. by reaction with the solvent.

For a *binary electrolyte*, the D. equilibrium is AB $\rightleftharpoons A^+ + B^-$. *Strong electrolytes* are almost completely dissociated (see Debye-Hückel theory), while *weak electrolytes* are only partly dissociated. The ***degree of dissociation*** α is given by the equation $\alpha = c_{A^+}/c_{B^-} = c_{B^-}/c_{AB}$, where c is the concentration of A, B or AB. Applying the mass action law, we obtain the ***dissociation constant K_D***: $K_D = (c_{A^+} \cdot c_{B^-})/c_{AB}$. Since $c_{A^+} = \alpha c_0$, $c_{B^-} = \alpha c_0$ and $c_{AB} = (1 - \alpha) \cdot c_0$ (c_0 is the initial concentration of AB), the following equation relates the D. constant K_D and the degree of D. α:

$$K_D = \frac{c_{A^+}\, c_{B^-}}{c_{AB}}.$$

K_D can be determined experimentally, for example from conductivity measurements (see Ostwald dilution law).

2) ***Thermal D.*** is the reversible decomposition of a compound upon absorption of an amount of heat called the Dissociation energy (see).

Dissociation energy: the energy required to cleave a molecule R_1-R_2 into the fragments R_1 and R_2 at 0K. The D. is different from the mean bond dissociation energy (see Bond energy).

Dissociation constant: the product of the activities (or concentrations) of the ions resulting from a dissociation reaction as a special case of the mass action law which applies when the activity of the undissociated component is constant (e.g. Solubility product (see)). The ***D. of water*** is of special importance. For the autoprotolysis equilibrium of water, corresponding to $H_2O + H_2O \rightleftharpoons H_3O^+ + OH^-$, K_w, the D. of water, depends on the temperature; it can be determined by conductivity measurements. If the activities of the hydrogen or hydroxyl ions in water are changed by addition of acid or base, dissociation of H_2O molecules or combination of OH^- and H_3O^+ ions will lead to a re-establishment of the equilibrium. K_w is always constant. For convenience the negative logarithm of K_w, pK_w, is often used. For the following temperatures, this has the values:

at 10 °C: $K_w = 0.36 \cdot 10^{-14}$; p$K_w$ = 14.45
22 °C: $K_w = 1.00 \cdot 10^{-14}$; p$K_w$ = 14.00
100 °C: $K_w = 74 \cdot 10^{-14}$; pK_w = 12.13

The activity of the hydronium ion is an extremely important parameter for characterization of an aqueous solution. This is seldom given in direct form, but rather in the form of the negative decadic logarithm, the pH (see): pH = $-\log a_{H_3O^+}$.

The pOH value is defined similarly: pOH = $-\log a_{OH^-}$, and thus pH + pOH = pK_w. The above relationship makes it possible, if the activity of either the H_3O^+ or the OH^- ion is known, to calculated the other. For dilute solutions, the molar concentrations may be used instead of the activities.

For a neutral aqueous solution, in which $c_{H_3O^+} = c_{OH^-}$, it follows that $c_{H_3O^+} = \sqrt{K_w} = 10^{-7}$ (at 22 °C); pH = 7 (***neutral point***). If the H_3O^+ and OH^- concentrations are different, the solution is defined as acidic if $c_{H_3O^+} > 10^{-7}$ mol l^- (pH < 7); and basic if $c_{H_3O^+} < 10^{-7}$ mol l^- (pH > 7).

Distapex process: a process for producing benzene by Extractive distillation (see).

Distex process: a process for producing benzene by Extractive distillation (see).

Distillation: the most important thermal method for separation of liquids by evaporation and subsequent condensation of the vapor. Since only the vapor phase moves in this case, one speaks of ***continuous flow D.***. If part of the condensed vapor, the *reflux*, flows back past the rising vapor into the distillation flask, one speaks of ***countercurrent distillation*** (see Rectification).

For a theoretical treatment of D., it is important to know the phase equilibrium between the vapor and the liquid to be separated. For partial pressures p_A and p_B of two components A and B in the vapor space, Raoult's law (ideal behavior) says that:

$$p_A = P_A \cdot x_A, \text{ and } p_B = P_B \cdot x_B. \quad (1)$$

Here P_A and P_B are the vapor pressures of the pure components A and B, and x_A and x_B are the mole fractions of the components in the liquid phase.

Since for a binary mixture, $x_B = 1 - x_A$, it follows

that the ratio of partial pressures in the vapor space is:

$$\frac{p_A}{p_B}=\frac{P_A}{P_B}\cdot\frac{x_A}{1-x_A}. \quad (2)$$

According to Dalton's law, the partial pressures p_A and p_B are related to the total pressure P in the vapor space by the equation

$$p_A=p\cdot y_A \text{ or } p_B=p\cdot y_B=p(1-y_A) \quad (3)$$

where y_A and y_B are the mole fractions of the two components in the vapor space. By substitution in (2), we obtain

$$\frac{y_A}{1-y_A}=\frac{P_A}{P_B}\cdot\frac{x_A}{1-x_A}. \quad (4)$$

It is a convention that x and y without subscripts always stand for the more volatile component; thus when the *relative volatility* α for an ideal mixture ($\alpha = P_A/P_B$, the ratio of the vapor pressures of the pure components) is introduced, (4) becomes:

$$\frac{y}{1-y}=\alpha\frac{x}{1-x} \text{ or } y=\frac{x\cdot\alpha}{1+x(\alpha-1)}. \quad (5)$$

One can see that the compositions of the vapor and liquid phases will differ only if $\alpha > 1$. In other words, separation by D. is only possible if $\alpha > 1$. The y/x diagram (equilibrium diagram, Fig. 1) can be calculated using (5). If the relative volatility α is large, a ***simple D.*** is sufficient for separation. This is done in the laboratory in a glass distillation apparatus (Fig. 2), and in a metal distillation column (see Column) in industry.

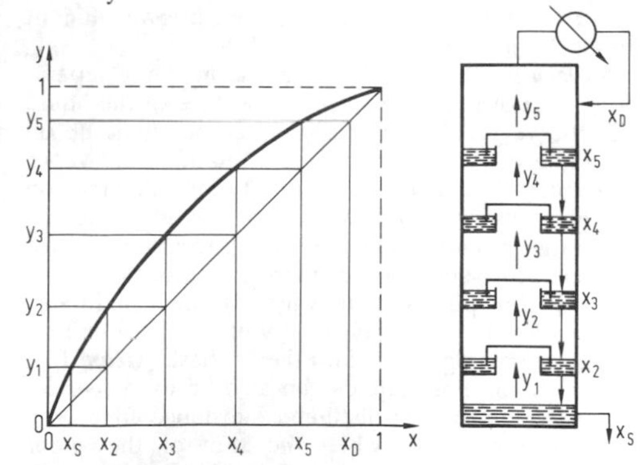

Fig. 1. Equilibrium diagram for an ideal binary mixture and determination of the theoretical plate number

Techniques. The mixture to be separated is placed in a heated distillation flask, or in a kettle below the distillation column. The vapor is condensed in a cooled condenser and collected in another flask. If the volatile components of the mixture are to be more highly enriched, the D. is carried out in several steps.

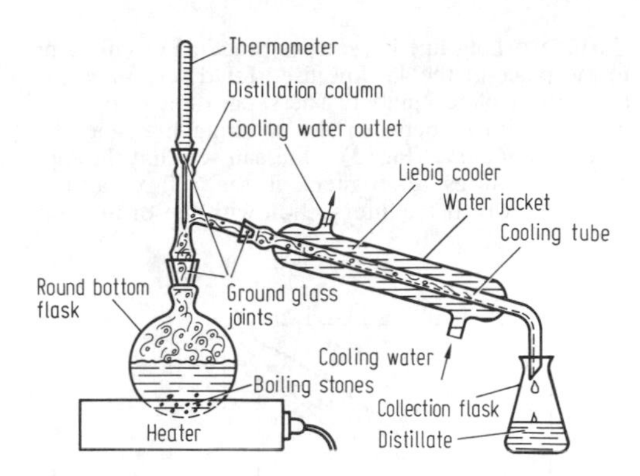

Fig. 2. Laboratory distillation apparatus

Fractional D. is used when a single-stage simple D. does not provide adequate separation. If a binary mixture with an initial concentration x_1 is evaporated, the vapor has a composition y_1. If this vapor is condensed, it becomes a liquid with composition x_2, which is in equilibrium with a vapor y_2; the condensate of this vapor has composition x_3, and so on. One proceeds along a stair-step curve between the equilibrium curve and a 45° line, until the desired distillate concentration x_D is reached. The number of steps is equal to the number of theoretical plates required. It is easy to see that the greater the distance between the equilibrium curve and the 45° line, the smaller the number of theoretical plates. The effectiveness of a plate depends on the degree to which the exchange process approaches the equilibrium. The effectiveness of a distillation column is designated by the number of theoretical plates (see Theoretical plate number). Step-wise separation can be done in a single column; the practical plate number is increased by special components (plates) which permit rapid mixing and a large contact area between the phases. The predictions made above hold only in the case that no distillate is removed from the rectification, and the entire condensate flows back through the column (total reflux). Under practical conditions, this equilibrium is interrupted by removal of part of the distillate from the process; this leads to the following balance of materials:
Total evaporated liquid (T) = reflux (R) + distillate (D). (6)

Substitution of the concentration factors y = concentration of the vapor, x = concentration of the liquid and x_D = concentration of the distillate in equation (6) and rearrangement gives

$$y=\frac{R\cdot x}{R+D}+\frac{D\cdot x_D}{R+D}. \quad (7)$$

Further rearrangement and introduction of the reflux equation $\nu = R/D$ gives a linear equation (8) with the slope $\nu/(1+\nu)$ and intersection of the ordinate $x_D(1+\nu)$:

$$y=\frac{\nu\cdot x}{1+\nu}+\frac{x_D}{1+\nu}. \quad (8)$$

This working line is used for practical calculations in the place of the 45° line used for determining the theoretical plate number. The separation step curve must be drawn between the working line and the equilibrium curve (Fig. 3). One can see that the separation requires fewer steps if the reflux ratio is larger, that is, if the intersection with the ordinate is lower.

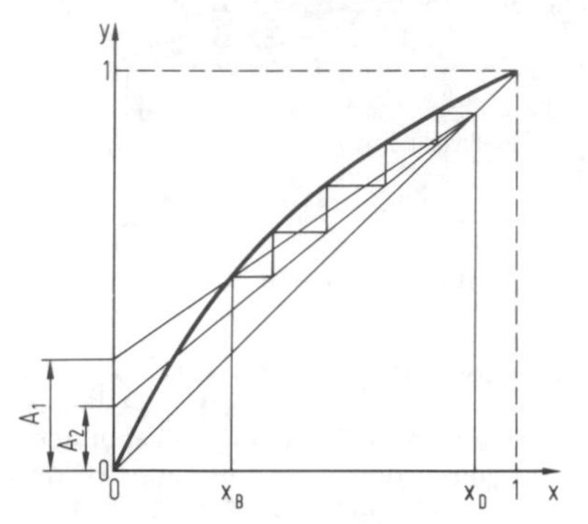

Fig. 3. Determination of the theoretical plate number by the Thiele-McCabe method.

From Fig. 3, the following limiting cases can be recognized:

1) The working line which intersects the ordinate at A_1 passes through the point on the equilibrium curve at abcissa value x_B. In this case, an infinite number of steps can be drawn between the working line and the equilibrium curve. For this reflux ratio, therefore, an infinite number of plates is required for separation. It is called the minimum reflux ratio, because the reflux ratio for the separation must exceed it.

The working line which intersects the ordinate at A_2 is a practical case. Between the two limits of minimum plate number and minimum reflux ratio, a lower plate number can be compensated by a greater reflux ratio, and conversely.

A D. can be either discontinuous or continuous. a) In ***discontinuous D.***, the amount of the liquid to be separated must be appropriate for the capacity of the apparatus. The individual components appear in the condensate in more or less pure state, in the order corresponding to their boiling points or ranges, and can be recovered separately. b) In ***continuous D.***, the mixture to be separated is fed into the column continuously, and the separated products are removed from the top or bottom. The portion of the column between the injection port and condenser is called the *amplifier column*, and the portion between the bottom drain and injection port, the *drain column* (Fig. 4).

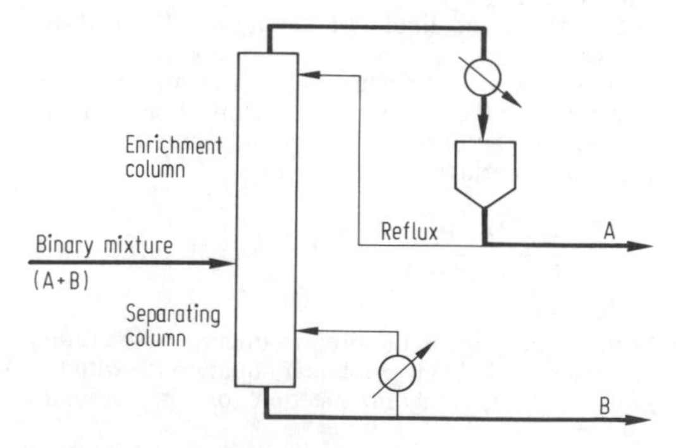

Fig. 4. Continuous distillation of a binary mixture; *A* is the more volatile component (distillate) and *B* is the bottom product

A binary mixture (*A* and *B* can be separated in a column into a top product *A* and a bottom product *B*. For a three-component mixture of *A*, *B* and *C*, at least two columns are required (Fig. 5) to separate all three into pure form.

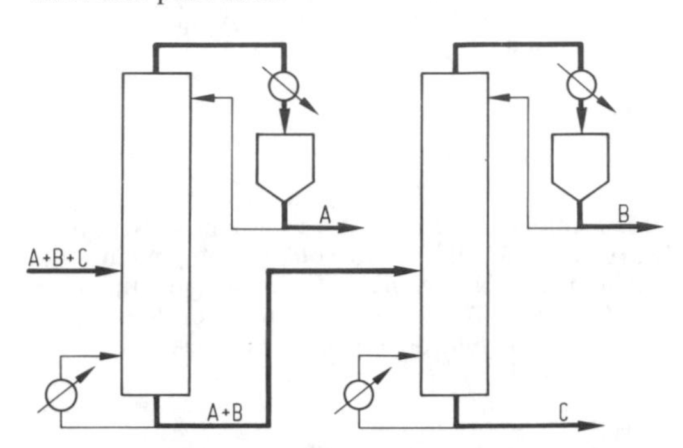

Fig. 5. Distillation of a three-component mixture (*A, B, C*) with two distillation columns

A mixture of several to many components can also be separated on a column with ports on the sides (see Petroleum refining).

Liquids with high boiling points or temperature-sensitive liquids can be separated by ***vacuum distillation*** at pressures of 0.1 to 2.5 kPa. ***High vacuum distillation*** is done at pressures of about 0.01 Pa. At these pressures, the boiling points are much lower than at atmospheric pressure (101.3 kPa).

Molecular D. is used for very demanding separations. The distance between the surface of the liquid and the very cold condensation surface must be on the same order as the mean free path of the vapor molecules, so that the evaporated molecule lands on the condenser surface without previously colliding with other gas molecules. This method is suitable, for example, for separation of mercury isotopes.

The boiling points of many liquids which are immiscible with water, or only slightly miscible, can be reduced by passing steam into the D. flask (***steam D.***). Other nonmiscible liquids can also be used (***carrier vapor D.***). The steam-distillable substance appears in the collection flask when the sum of the vapor pressures of the pure components (water and the substance being distilled) reaches atmospheric pressure. Steam D. can also be combined with vacuum D.

The equilibrium curve for nonideal mixtures of liquids may include a point which intersects the 45° line: the azeotropic point. Azeotropic mixtures, or mixtures of liquids with very similar boiling points, can be separated by Azeotrope distillation (see) or Extractive distillation (see); or the D. can be done at various system pressures.

In ***dry D.*** or ***pyrolysis*** of organic substances (in the absence of air), gaseous, liquid or solid decomposition products are formed. For example, dry D. of coal produces crude gas, tar and coke; dry D. of wood yields methanol, acetic acid, wood tar and charcoal. The applications of D. in the chemical industry and laboratory are extremely varied. It is used, for example, to separate purified crude oil into gasoline, kerosene, gas oils, lubricating oils and asphalts and to separate coal tar into light, medium and heavy oils, anthracene oil and pitch, to separate benzene from the wash oil of a coking plant, to obtain nitrogen, oxygen and noble gases from liquefied air, to produce 96% ethanol from mixtures with water, and so on.

Since D. requires large amounts of energy, the economics of its application depends heavily on the consumption of heat and cooling water. Cascade evaporation (see) and Vapor compression (see) offer possibilities for conservation of energy.

Distribution: see Extraction.

Distribution chromatography: a separation method based on the principle of continuous multiplicative distribution (see Extraction). A mixture of substances is distributed between a flowing (mobile) phase and a stationary phase fixed to a carrier. D. is carried out in columns or on surfaces. The materials used for the stationary phase are of uniform grain size and have inert surfaces, e.g. cellulose, silica gel, kieselguhr or starch.

Distribution equilibrium: a term for the equilibrium which forms when a substance is dissolved in two non-miscible phases which are in contact with each other. It is the basis for extraction and chromatographic separations. D. is described by the Nernst distribution law (see).

Distribution function, ***probability density***: 1) In statistics, a function describing the distribution of particles in an aggregate, such as the molecules of a gas, with respect to their locations, velocities and energies. The D. gives the probability that a particle will be found at a given location, with a given velocity or internal excitation state. For large aggregates, the D. can be formulated as a continuous function. In equilibrium statistics (see Thermodynamics, statistical), spatial distributions are isotropic. If the internal degrees of freedom are ignored, the classical D. depends only on the absolute value of the velocity (see Maxwell-Boltzman distribution).

Non-classical equilibrium distributions are the Fermi-Dirac distribution for fermions and the Bose-Einstein distribution for bosons.

2) In macromolecular chemistry, the function which describes the proportions of certain molar masses or degrees of polymerization in a polymeric compound.

3) In error calculations, a function which characterizes the distribution of random error in observation. The ***Gaussian error distribution*** is often applied to the distribution of random errors x about the probable value of the measurement:

$$\varphi(x) = \frac{1}{\sigma\sqrt{2\pi}} e^{-x^2/2\sigma^2}$$

σ is called the scatter. The smaller σ is, the more tightly the data are bunched around the mean.

Disulfates, ***pyrosulfates***: the salts of disulfuric acid, $H_2S_2O_7$, with the general formula $M^I_2S_2O_7$. In the disulfate ion, $S_2O_7^{2-}$, the two tetrahedral sulfur atoms are linked by an oxygen bridge (Fig.). The D. dissolve in water to form the corresponding hydrogensulfates. The latter are the starting materials for production of D., in that they lose water when heated above their melting points, e.g. 2 $NaHSO_4 \rightleftharpoons Na_2S_2O_7 + H_2O$.

Disulfate

Disulfiram, ***Antabuse®***: tetraethylthiuram disulfide, $(C_2H_5)N\text{-}C(S)\text{-}S\text{-}S\text{-}C(S)\text{-}N(C_2H_5)$, a compound used in the treatment of alcoholism. D. is synthesized by oxidation of the potassium salt of *N,N*-diethyldithiocarbamic acid, which is made by reaction of diethylamine, carbon disulfide and potassium hydroxide. m.p. 70-72 °C. After intake of D., consumption of an alcoholic beverage leads to nausea, sweating and other symptoms caused by an elevated blood level of acetaldehyde. D. inhibits oxidative degradation of the acetaldehyde.

Disulfites, ***pyrosulfites***, formerly ***metabisulfites***: the salts of disulfurous (pyrosulfurous) acid, $H_2S_2O_5$. The acid itself does not exist in free form; its salts have the general formula $M^I_2S_2O_5$. They are based on the structurally interesting disulfite ion $S_2O_5^{2-}$, which contains an S-S bond (Fig.). D. dissolve in water with the formation of hydrogensulfites: $S_2O_5^{2-} + H_2O \rightleftharpoons 2\ HSO_3^-$. The solutions act as reducing agents. D. are made by the reaction of excess sulfur dioxide with sulfite solutions: $M^I_2SO_3 + SO_2 \rightarrow M^I_2S_2O_5$. ***Sodium disulfite***, $Na_2S_2O_5$ is used as a reducing agent in photography and textile dyeing, as an antichlorine agent in the paper industry and as a food preservative.

Disulfite

Disulfoton: an Organophosphate insecticide (see) with systemic properties (see Systemic pesticides).

Disulfuric acid, ***pyrosulfuric acid***: $H_2S_2O_7$, colorless, hygroscopic crystals; M_r 178.14, density 1.9, m.p. 35 °C. When heated, D. decomposes to sulfuric acid, H_2SO_4, and sulfur trioxide, SO_3. It has very strong dehydrating and oxidizing effects. When added to water, it forms sulfuric acid with vigorous hissing: $H_2S_2O_7 + H_2O \rightarrow 2\ H_2SO_4$. D. is a strong diprotic acid, and its salts are Disulfates (see). D. is miscible with SO_3 and H_2SO_4 in all proportions. This mixture, known as Oleum (see) is an industrial byproduct from which D. is obtained by freezing out.

Disulfurous acid: see Disulfites.

Ditactic: see Polymers.

Ditalimfos: see Fungicides.

Diterpenes: terpenes consisting of 4 isoprene units, or 20 C atoms. The most important of the acyclic D. is the alcohol Phytol (see). Most D. are polycyclic; many are have a perhydrophenanthrene skeleton. Examples are the abietane, pimarane and kaurane groups. Methyl group migrations, ring openings and reclosings often change the pattern of methyl groups. Very few naturally-occurring D. are hydrocarbons; acids (see Resin acids, Gibberellins) or highly substituted alcohols, such as phorbol esters, are more common. A few plant alkaloids are derivatives of the D., such as those of the *Ranunculus* and *Delphinium* genuses. Most D. are highly viscous oils or solids which are not volatilized by steam. They are only rarely present in essential oils.

Dithiocarbamate: see Carbon disulfide.

Dithiocarbamate fungicides: derivatives of dithiocarbamic acid with fungicidal properties (see Fungicides). Metal salts of dimethyldithiocarbamic acid $[(CH_3)_2N\text{-}CS\text{-}S\text{-})]_n$ Me are synthesized from dimethylamine, carbon disulfide and alkali hydroxides, followed by precipitation of the insoluble zinc or iron salts. *Ziram* ($n = 2$, Me = Zn) and *ferbam* ($n = 3$, Me = Fe) are used as protective fungicides in orchards, vineyards and truck farms. The alkylene-bisdithiocarbamic acid derivatives are synthesized in similar fashion from ethylene or propylene diamine; they are polymers, more stable and more widely applicable. Examples are *zineb, maneb, mancozeb* and *propineb*, which are used on potatoes, root crops, tomatoes and hops.

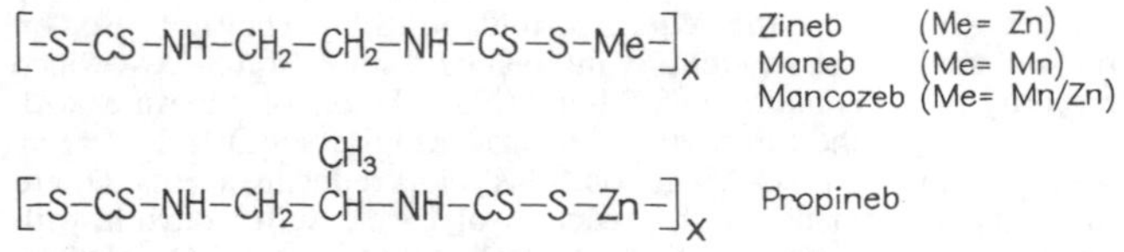

$[\text{-S-CS-NH-CH}_2\text{-CH}_2\text{-NH-CS-S-Me-}]_x$ Zineb (Me= Zn), Maneb (Me= Mn), Mancozeb (Me= Mn/Zn)

$[\text{-S-CS-NH-CH}_2\text{-CH(CH}_3\text{)-NH-CS-S-Zn-}]_x$ Propineb

Fungicidal alkylene-bisdithiocarbamates

The toxic principles of the D. are isothiocyanates produced by decomposition of the parent compounds; the isothiocyanates react with thiol groups of the proteins and enzymes of the fungi. Although the D. were formerly believed to by non-toxic to mammals, it has been found that their decomposition sometimes releases ethylene thiourea (*ETU*), which is carcinogenic. This has led to limitations on the use of D. *Thiram* (tetramethylthiuram disulfide, abb. *TMTD*), $(H_3C)_2N\text{-}CS\text{-}S\text{-}S\text{-}CS\text{-}N(CH_3)_2$, m.p. 156 °C), is a chemically similar compound which can be obtained by gentle oxidation of an alkali salt of dimethyldithiocarbamic acid; it was the first leaf fungicide reported which did not contain metal. It is also used to treat seeds and fumigate soils.

Dithiocyanogen: see Thiocyanates.

Dithioglycol: same as Ethane-1,2-dithiol.

1,3-Dithiolanes: saturated, five-membered heterocyclic compounds with two sulfur atoms in the ring. They can be considered cyclic dithioacetals. The unsaturated analogs are called ***dithiols***. The 1,3-D. are synthesized by reaction of aldehydes or ketones with ethane-1,2-dithiol. In the presence of Raney nickel, they are cleaved by hydrogenation. Thus the 1,3-D. provide a pathway for reduction of the carbonyl compounds to the corresponding hydrocarbons.

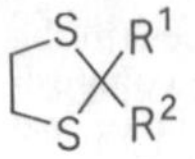

S R1 / S R2

Dithionates, formerly ***hypodisulfates***: the salts of dithionic acid, $H_2S_2O_6$, with the general formula $M^I{}_2S_2O_6$. The dithionate ion $S_2O_6{}^{2-}$ has both sulfur atoms in tetrahedral configurations. It is relatively stable to oxidizng and reducing reagents. When heated, D. disproportionates into sulfates and sulfur dioxide. It is made from hydrogen sulfates by anodic oxidation, or by reaction with manganese(IV) oxide: $MnO_2 + 2\,HSO_3^- + 2\,H^+ \rightarrow Mn^{2+} + S_2O_6^{2-} + 2\,H_2O$.

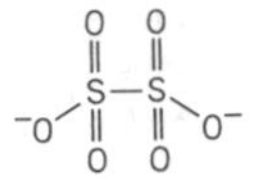

Dithionate 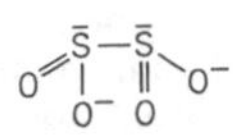　　Dithionite

Dithionites, formerly ***hypodisulfites***: the salts of the dibasic dithionous (formerly hypodisulfurous) acid, $H_2S_2O_4$; the acid does not exist in free form. The salts have the general formula $M^I{}_2S_2O_4$. The dithionite ion has a relatively long S-S bond (239 pm), which is relatively easily broken. When heated, dithionite solutions decompose to thiosulfate and hydrogen sulfite: $2\,S_2O_4^{2-} + H_2O \rightarrow S_2O_3^{2-} + 2\,HSO_3^-$.

D. are strong reducing agents which are oxidized to sulfite or sulfate, depending on the oxdizing agent. They are made by reduction of hydrogen sulfites, for example with zinc. Their properties, production and applications are discussed under Sodium dithionite (see).

Dithionous acid: see Dithionites.

Dithizone: 1,5-diphenylthiocarbazone, $C_6H_5\text{-}N{=}N\text{-}CS\text{-}NH\text{-}NH\text{-}C_6H_5$, a black or brownish black powder; m.p. 165-169 °C (dec.). D. is soluble in alcohol, chloroform and carbon tetrachloride, giving a green solution. It forms water-soluble internal complexes with various metals, such as lead, silver, copper and zinc. For this reason, D. is used in analytical chemistry for the qualitative and quantitative determination of metals. The complex compounds dissolve in carbon tetrachloride with characteristic colors which go from violet to red with increasing atomic radius of the metal. The colored complexes are determined by colorimetry.

Dithranol, ***Cignolin***®: a phenolic anthracene derivative with antiseptic, keratolytic and itch-soothing effects; it is used in treatment of psoriasis. D. can be synthesized chemically. Chrysarobin, an exudate of the native Brazilian tree *Andira araroba*, has the same effects and applications.

Diuretics: compounds which increase the production of urine by a direct effect on the kidneys. Excess salts leading to edema are excreted with the increased amount of urine. There are several types of substance which act as D.: 1) ***Osmotic D.*** are injected in hypertonic solution and, when excreted, bring a larger volume of water along. The electrolyte balance is scarcely affected. These are used for forced diuresis, in poisonings and kidney failure. Some important representatives of this group are *D-glucitol* and *D-mannitol*. 2) ***Mercury D.*** inhibit the resorption of sodium ions in the kidney tubules. The increased sodium excretion is accompanied by increased excretion of chloride ions and water. Due to the side effects, which are also observed with organic mercury compounds, this group has been largely abandoned for therapy. The most important example is *mersalyl*. 3) ***Carboanhydrase inhibitors*** inhibit the kidney enzyme (carboanhydrase) which converts carbon dioxide and water to hydrogencarbonate and hydrogen ions. The hydrogen ions are exchanged for sodium ions in the process of resorption of sodium; therefore, inhibition of the enzyme inhibits resorption of sodium and leads to increased excretion of sodium (and potassium) and water. Excretion of hydrogencarbonate reduces the reserve of alkali metal cations and leads to blood acidosis, which counteracts the effect. The most important example of this group is *acetazolamide*, which is also used for treatment of glaucoma. 4)***Saluretics*** cause excretion of alkali metal and chloride ions, together with water. Many of these compounds are 7-sulfamoyl-1,2,4-benzothiadiazine-1,1-dioxides, which are called *thiazides*. Some examples are *chlorothiazide, hydrochlorothiazide* and *cyclopenthiazide*. Compounds which do not contain an heterocyclic ring may also have saluretic effects, for example, *mefruside* and *furosemide*. 5) ***Aldosterone antagonists*** competitively displace the hormone aldosterone, and lead to increased sodium elimination and potassium retention, combined with excretion of water. They are used when the body is forming too much aldosterone and the alkali metal cation equilibrium is disturbed. An example is *spironolactone*.

Some other heterocyclic compounds are D., including purines or pteridines, like *theophyllin* and *triamterene*. With triamterene, less potassium than sodium is excreted. Compounds with this type of effect are called potassium-saving D.

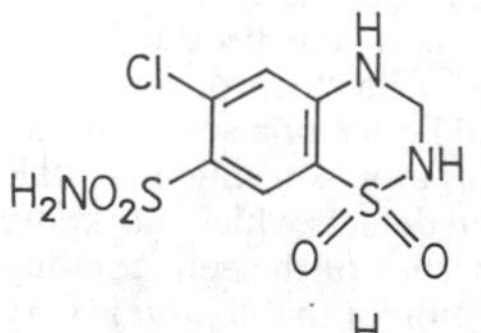

Hydrochlorothiazide

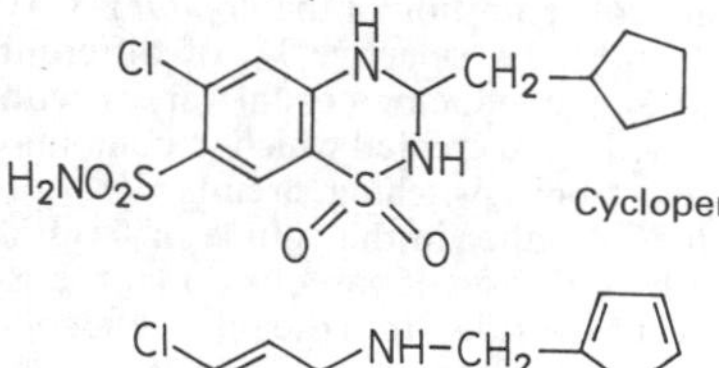

Cyclopenthiazide

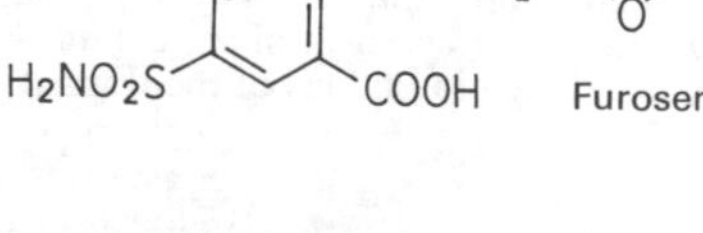

Furosemide

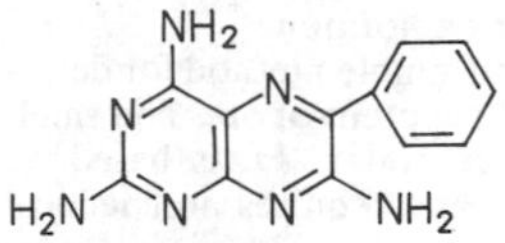

Triamterene

Diuron: see Urea herbicides.

Diva: see Vinylacetylene.

Divinylacetylene: see Vinylacetylene.

Divinylmethane rhythm: see Fatty acids.

Dixanthogen: an Insecticide (see).

DNA: see Nucleic acids.

DNA ligases, ***polynucleotide ligases***: enzymes which link DNA segments by forming internucleotide ester bonds between phosphoric acid and deoxyribose groups.

D. are important for the repair of DNA damage in the cell and are used in synthesis of genes in vitro.

DNA ligase method: see Gene synthesis.

DNA polymerases: enzymes which catalyse the construction of DNA polynucleotides from 3'-deoxyribonucleotide triphosphates on DNA templates (DNA replication). The enzymes can also be used for synthesis of DNA in vitro. The Kornberg enzyme (see) is an important representative of the group.

DNA polymerase method: see Gene synthesis.

DNBP: see Herbicides.

DNOC: abb. for 4,6-***Di**n*itro-***o**-c*resol, the dinitro derivative of 2-methoxyphenol. It forms yellow prisms; m.p. 85.8 °C, vapor pressure at 25 °C, $1.4 \cdot 10^{-2}$ Pa, water solublity at 15 °C, 130 ppm. DNOC is synthesized by nitration of *o*-cresol with a mixture of sulfuric and nitric acids at low temperatures. DNOC and its homologs and derivatives are used as ovicidal insecticides and acaricides, and also as herbicides and dessicants.

DNOC is fairly toxic to mammals, with a p.o. LD_{50} of 20 to 40 mg/kg rat. It is rapidly absorbed through the intact skin, especially in an oily solution. Its toxicity increases with increasing exterior temperature.

Poisonings with DNOC and other nitrophenol derivatives cause headaches, dizziness, nausea, fever, sweating and convulsions which may lead to collapse and coma. Delayed damage to heart, liver and kidneys are possible.

Therapy: After oral intake, immediate pumping and rinsing of the stomach with 5% sodium carbonate solution to which activated charcoal has been added. Use sodium sulfuricum, not ricinus oil, as laxative. After inhalation, fresh air, artificial respiration and oxygen. After contact with the skin, wash affected areas repeatedly with soap and water.

The patient should be cooled in an ice bath, with ice-water inflow, ice packings, artificial hypotonia with prophenin, protective therapy for the liver required.

Caution! Avoid central stimulants and ricinus oil, milk or alcohol.

DOCA: see Adrenal cortex hormones.

Docimasy: a very old but simple method for determination of the noble metal content of ores and smelter products. It is still used today. D. is based on pyrometallurgical methods and requires no chemical reagents.

Docosanoic acid, ***behenic acid***: CH_3-$(CH_2)_{20}$-COOH, a higher, saturated fatty acid. It crystallizes as colorless needles; m.p. 80 °C, b.p. 306 °C at $8 \cdot 10^3$ Pa. D. is only slightly soluble in water, but is more soluble in alcohol. It is found in many plant oils, for example peanut and rapeseed oils.

Dodecan-1-ol: same as Lauryl alcohol (see).

Dodecane: $CH_3(CH_2)_{10}CH_2$, an alkane hydrocarbon, a colorless, combustible liquid; m.p. - 9.6 °C, b.p. 216.3 °C, n_D^{20} 1.4216. It is soluble in ethanol, ether and chloroform, but it is not miscible with water. It is found in American petroleum and coal tar. D. can be synthesized according to Wurtz from hexyl bromide, or by hydrogenation of lauryl alcohol in the presence of a sulfidic catalyst.

Dodecanal: same as Lauraldehyde (see).

Dodecanoic acid: same as Lauric acid (see).

Dodemorph: see Morpholine and Piperazine fungicides.

Dodine: see Fungicides.

Doebner-von Miller synthesis: a synthetic method for substituted quinolines in which primary aryl amines react with α/β-unsaturated aldehydes or ketones in the presence of conc. hydrochloric acid or zinc chloride:

$C_6H_5NH_2$ + O=CH–CH=CH–CH_3 → 2-methyl-1,2-dihydroquinoline (N–H, C(H)CH_3)

$\xrightarrow[-H_2O]{+O}$ 2-methylquinoline (CH_3)

In the D., the unsaturated carbonyl compound is first formed by condensation of an aldehyde or/and a ketone. Addition of the arylamine to the C=C double bond of the α/β-unsaturated carbonyl compound and acid-catalysed ring closure yields the dihydroquinoline derivative. The compound is dehydrogenated to the quinoline system by the azomethine formed from the carbonyl compound and the aniline. The reaction steps and products of the D. are very similar to those of the Skraup synthesis (see).

Donarites: see Ammonium nitrate explosives.

Donicity, ***donor number***: a measure of the relative Lewis basicity of an electron-pair donor (EPD). The D. of an EPD is the negative of the standard reaction enthalpy $\Delta_R H_{298}$ for the reaction of this EPD with antimony(V) chloride, the standard electron pair acceptor (EPA), in 1,2-dichloroethane solution. The D. is one means of characterizing dipolar-aprotic solvents.

Donnan equilibrium: a membrane equilibrium which is established when two electrolyte solutions are separated by a semipermeable membrane which not all the components of the solutions can pass. A D. is established, for example, when an aqueous solution of sodium chloride is separated by a parchment membrane from an aqueous solution which contains, in addition to NaCl, the salt NaR, with a colloidal anion R^-. The nature of the membrane permits no diffusion of R^-, but water, sodium and chloride ions can pass. The resulting diffusion leads to an equilibrium subject to the conditions $a^I_{Na^+} \cdot a^I_{Cl^-} = a^{II}_{Na^+} \cdot a^{II}_{Cl^-}$, where a is the activity. Because electroneutrality must also be preserved, that is, $a^I_{Na^+} = a^I_{Cl^-}$ and $a^{II}_{Na^+} + a^{II}_{R^-}$, it follows that $a^{II}_{NaCl} < a^I_{NaCl}$. The resulting differences in the osmotic pressure and electric potential of the two solutions are called the Donnan osmatic pressure (Donnan potential). D. are very important for the chemistry of cells, because they arise at the semipermeable cell membrane and also at the membranes of cell organelles, such as chloroplasts and mitochondria.

Donor atom: see Coordination chemistry.

Donor substituent: an atom or group which gives up electron density to the molecule of which it is a part. D. generate an electron push through an inductive effect (+I substituent) or through a resonance effect (+R substituent).

Donor number: same as Donicity (see).

Dopa: see Tyrosine.

Dopamine, ***4-(2-aminoethyl)pyrocatechol*** a catecholamine which acts as a neurotransmitter in the sympathico-adrenal system. It is thus related to Noradrenalin (see). D. is biosynthesized by decarboxylation of 3,4-dihydroxyphenylalanine (***dopa***). It is used therapeutically for treatment of shock and refractory cases of coronary insufficiency.

3,4-(HO)$_2C_6H_3$–CH_2–CH(R)–NH_2

R = H: Dopamine
R = COOH: Dopa

Dopant: see Doping.

Doping: the addition of a defined amount of a foreign substance to a highly pure material, in order to achieve desired properties in the latter. The most important application of D. is in the production of semiconductor components. The incorporation of foreign atoms into the crystal lattice of a semiconductor, e.g. boron or phosphorus atoms in single crystals of silicon, creates substitution defects which markedly change the conduction properties of the semiconductor. The necessary concentration of the *dopant* is very low (10^{-8} to 10^{-5}%). By selective D. of different areas of a single semiconductor crystal, areas with different conductivities are created which in combination form microelectronic switching circuits.

D. can be achieved either in the course of Crystal growing (see) or by diffusion of gaseous dopant, e.g. boron or phosphorus hydrides, into the pure semiconductor crystal at high temperature. In *ion implantation*, the semiconductor surface in a high vacuum is exposed to a beam of accelerated ions of the dopant.

Dorlastan®: see Synthetic fibers.
Dorr process: see Sodium hydroxide.
Dose: see Toxicity.
Double contact process: see Sulfuric acid.
Double gas: see Water gas.
Double helix: see Nucleic acids.
Double layer: see Electrochemical double layer.
Double layer capacity: see Electrochemical double layer.
Double reaction: see Solid state reactions.
Double resonance: see NMR spectroscopy.
Double resonance methods: methods of High frequency spectroscopy (see) in which two resonance transitions of different frequencies are excited simultaneously. The D. extend the range of application of high-frequency spectroscopy by permitting simplification of spectra or increase in intensity; it can also yield additional information on the sample. Some typical applications of D. are the uncoupling techniques in NMR spectroscopy (see) or the ENDOR technique (see) in ESR spectroscopy.
Doublet: see Multiplet structure.
Downs process: see Sodium.
Doxycyclin: see Tetracyclines.
2,4-DP: see Growth hormone herbicides.
DPN: see Nicotinamide nucleotides.
Dragon's blood: a natural resin from the fruits of *Daemonorops sp.* It is a dark red-brown mass which is soluble in alcohol, chloroform, glacial acetic acid, benzene and carbon disulfide, but is insoluble in ether, petroleum ether and terpentine oil. D. is used as a colorant in plaster mixtures and glues for teeth; it is also used to make lacquers and wood stains.
Dragendorff's reagent: see Alkaloids.
Dralon®: see Synthetic fibers.
Dreiding model: see Stereochemistry.
Drilling fats: see Drilling oils.
Drilling flushing materials: water and suspensions of clay, with or without added inorganic or organic substances. D. are used in deep drilling (e.g. for oil) to bring the drilled rock to the surface. They lubricate the drill bit and drill rods and seal the borehole and the sides of the bit, in addition to bringing the fractured material to the surface. Bentonite, carboxymethylcellulose, hydrolysed polyacrylonitrile, tall oil fractions, protective colloids, defoaming agents and heavy materials to increase the density are used as additives. Examples of heavy materials are finely ground barium sulfate (heavy spar), hematite, etc.
Drilling oils, ***coolants***: are colloidal or true solutions of soaps in low-viscosity mineral oils, with or without solubilizers. D. are among the oils used in metal working. When thinned with water, they should dissolve or emulsify, with no separation of the oil (or only very slight separation). This degree of emulsification is achieved by addition of soaps. The soaps or emulsifiers used in common D. are alkali or ammonium salts of naphthenic, naphthensulfonic or stearin-free fatty acids, sulfonated or phenolated fatty oils, and resins. Free fatty acids, which attack steel when heated, must not be present. To increase and stabilize the emulsifying activity of the soaps in mineral oils, a small amount of water or alcohol is thought to be helpful.
Drilling fats contain 25 to 50% water, in addition to the emulsifiers and mineral oils. They have a soft consistency, similar to fats or salves.
Drop analysis: a method of qualitative Analysis (see) in which only one drop of sample solution is used for the detection reaction. In the test, a drop of sample solution is mixed with a drop of reagent solution on the surface of a porous or non-porous material, such as paper, glass or porcelain, or the drop of sample solution is applied to a filter paper impregnated with reagent. In the presence of the substance of interest, there is either a change in color, or a precipitate forms. In many cases, solid materials can be identified by application of a drop of reagent directly to the surface of the sample and observing the resulting change in color. Although efforts have been made to make the reactions as specific as possible, by use of organic reagents and Masking reagents (see), interfering ions must be removed in many cases. D. can also be used in qualitative organic analysis to detect certain elements in organic compounds, to determine functional groups and, in many cases, to detect individual organic compounds. Reagent papers have been developed which even permit semiquantitative detection of many ions and compounds. The advantages of the D. are their speed, the small amounts of material required, and the lack of apparatus requirements. However, unlike physical processes, D. requires very extensive knowledge of chemistry.
Drugs: originally, dried plants or parts of plants used for medicinal purposes, or for preparation of medicines (extracts, etc.). The term is now used to include synthetic pharmaceutical products as well as, loosely, psychoactive substances used or abused for nonmedical purposes (see Psychopharmaceuticals).
Addictive D. are substances which, when used repeatedly, lead to dependence due to adaptive changes in the central nervous system. Originally, the concept of addiction was relatively narrow. The criteria were development of tolerance (that is, increasing doses are required to achieve a certain effect) and the appearance of withdrawal systems (that is, acute symptoms occurring within hours after the last dose). In modern pharmacology, these phenomena are now considered physical dependence, and the term "addiction" has been expanded to include psychological dependence on certain psychopharmaceuticals.

The classical addictive compounds are opiates, barbituates and alcohol. Nicotine has more recently been recognized as addictive. Most of the "recreational D.", including marijuana, hashish, cocaine, mescaline, and LSD, are also addictive. Chronic use of these substances usually leads to changes in personality and socialization, as well as to pathological changes in the internal organs (often the liver).

Psychoactive D. lead to marked changes in consciousness, in some cases associated with hallucinations (***hallucinogens***). They can have either excitatory or inhibitory effects on the central nervous system, and usually strongly affect mood. The opiates and tetrahydrocannibinol (marijuana or hashish) depress the CNS, while cocaine, mescaline and LSD are excitatory.

The widespread misuse of addictive D. is a major national and international concern, and has led to governmental efforts to stop the international illegal trade in the substances. There is an international treaty of 1971 on psychotropic substances which, in its 1972 version, incorporates individual treaties on narcotic drugs from 1961. These treaties establish controls on international trade and limitations on production of the drugs.

Dry cell: same as Leclanché cell (see).

Dry ice: solid carbon dioxide, CO_2. At normal atmospheric pressures, D. is converted directly from the solid to the gas state (sublimation), and unlike water ice (see Ice), it does not leave a liquid residue. The reason for this behavior is the high triple point of CO_2, 0.52 MPa. At 0.101 MPa (atmospheric pressure), the sublimation temperature is - 78.5 °C. D. is opaque and white, and has a density between 1.4 and 1.57 kg dm^{-3}, depending on the method of production. Its heat of sublimation is 573 kJ kg^{-1}. D. is used to keep perishable items cold during transport. The usable heat-absorbing capacity is increased by the heat capacity of the gas, which in turn depends on the temperature at which cooling is required. At -30 °C, the heat capacity of D. is about 600 kJ kg^{-1}. Because of its high density, D. thus has an extraordinary heat capacity per unit of volume. Its advantages over water ice are its much lower temperature, its greater specific heat capacity and its direct conversion to the gas phase.

To make D., gaseous CO_2 is compressed, liquefied and then allowed to expand several times, with the adiabatic heat of expansion being used to cool the residual CO_2 until a fraction of it condenses in the form of ***carbon dioxide frost***. This is pressed into blocks. In some plants, a mixture of carbon dioxide frost and cold gas is used to freeze foods.

Drying: The removal of moisture (solvent residues) from a substance or mixture by evaporation, sorption, sublimation, selective distillation, freezing out or chemical reactions. In the laboratory, gases are dried by freezing out the moisture in a cold trap, absorption by a liquid drying agent (such as sulfuric acid), adsorption by solid drying agents (such as molar sieves or silica gel) or by hygroscopic compounds which form hydrates (e.g. calcium chloride) during the D. The drying process is carried out in wash bottles, drying tubes and drying towers (Fig. 1).

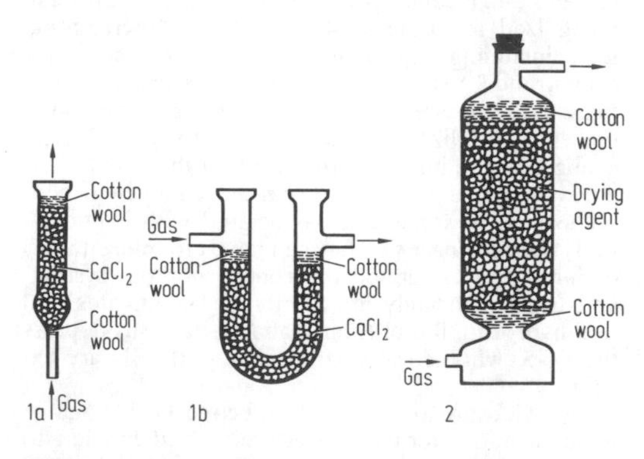

Fig. 1. Drying: *1*, Calcium chloride tube: *a* straight, *b* U-tube. *2* Drying tower

Liquids are dried by addition of drying agents which undergo chemical reactions (e.g. metals or oxides such as sodium, potassium or phosphorus pentoxide), physically adsorb the water or form hydrates. After a long, intensive contact with the drying agent, it is removed by decantation or filtration, or the dried liquid is distilled off. It is also possible to remove moisture by Azeotropic distillation (see).

Solids are usually dried by evaporation; the liquid is converted to the gas state by heating and is removed as steam or a mixture of gases. However, evaporation may also be allowed to occur at low temperatures and low vapor pressures. Heat may be supplied by convection (***convection D.***), radiation (***radiation D.***) or conduction (***contact D.***). Solids may also be dried by vacuum, freeze and high-frequency D.

Vacuum drying is an especially gentle method of D. It can be applied whenever the apparatus can be sealed airtight. Because of the low pressure in the apparatus, the boiling temperature of the moisture is reduced and it can evaporate very rapidly. Heat cannot be applied by convection with a heated gas, of course, but can be applied by radiation, conduction or previous heating of the material. Vacuum D. is particularly important in the production of drugs, dry, sterile sera and penicillin. It is also used for preservation of vitamin-rich foods, meat, fish and egg powder.

Freeze D. is very suitable for D. of readily decomposed substances, such as foods and drugs. The material to be dried is placed in a vacuum-tight chamber and very rapidly cooled, so that it freezes. A high vacuum is then applied. Because of the difference in vapor pressure between the frozen material and the condenser, the ice sublimes. The resulting vapors precipitate in the condenser.

In ***high-frequency D.*** or ***dielectric D.***, the material to be dried (wood, textiles, ceramic parts, molds, food and drugs) is placed in the alternating field of an electrical capacitator. The heat is controlled by the high-frequency voltage from the generator, or by the air gap between the material and the electrodes. Petroleum can also be dried by a strong alternating electric field (field strength 2 kV cm^{-1}). In this case, the asphalt capsules which form around the water are broken down by the alternating field.

A special type of D. is Atomization drying (see). D. by chemical reactions is discussed under Siccatives (see).

The D. process is characterized by the *D. rate*. The parameter which drives the process arises from the difference between the non-bound moisture and the equilibrium moisture. D. of solids occurs in two phases: 1) removal of the water sticking to the surface of the solid, and 2) removal of the water bound by capillary action, adsorption and chemical bonds.

To determine the achievable final moisture level, the sorption isotherms must be determined experimentally, and to establish which is the most suitable D. method, the D. curves must be measured. An enthalpy-moisture diagram (the Mollier *ix* diagram) is used to determine the specific requirements for air and heat.

Channel driers and *belt driers* (Fig. 2) are used to dry solids with warm air. *Roller driers* (Fig. 3) and *screw driers* are used for contact D.

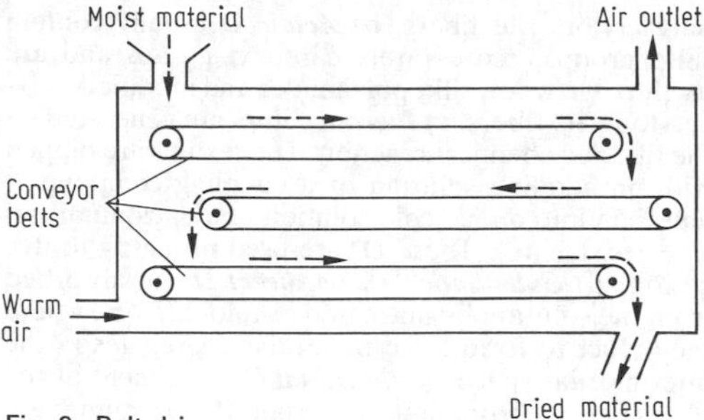

Fig. 2. Belt drier

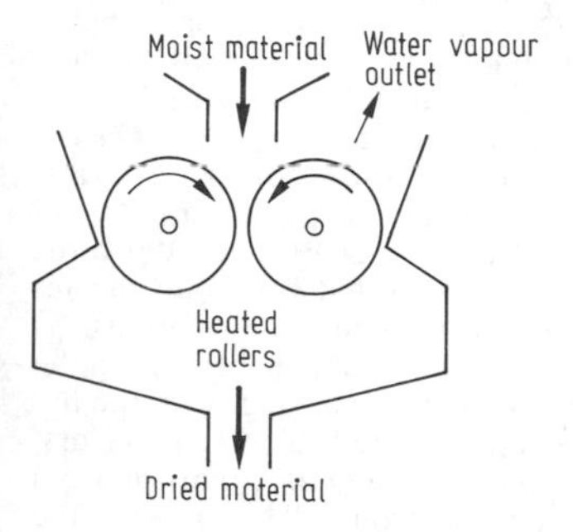

Fig. 3. Roller drier

In the laboratory, solids are dried in desiccators or heated drying apparatus (e.g. a drying pistol, Fig. 4).

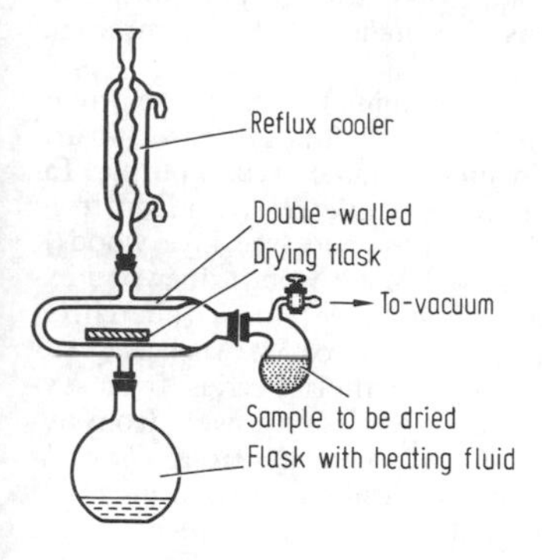

Fig. 4. Drying pistol

Drying agents: 1) substances which withdraw water from other materials (gases or liquids). The drying depends on chemical reaction, such as the formation of hydrates, or on adsorption and absorption.

a) D. for gases. Calcium chloride is most often used, as it is suitable for many gases, such as chlorine, hydrogen sulfide, ethyne and hydrogen chloride. Another D. is silica gel which, like calcium chloride, can dry gases down to about 0.02 mg water/l. Blue silica gel is most often used; its color changes to a light violet around 650 Pa (about 25% mass increase). Neutral gases can also be dried by molecular sieves; the degree of drying is about 0.02 mg water/l. Sulfuric acid works exceptionally rapidly and has a very high affinity for water; with it about 0.003 mg water/l can be reached. Magnesium perchlorate is very effective and can be used for nearly all gases which have no organic impurities. With it, less than 0.001 mg water/l can be achieved. Phosphorus pentoxide is the most effective of the generally used D., and it can be used with nearly all gases. It can remove all but 10^{-5} mg water/l from gases passed over it. Potassium hydroxide, calcium oxide and barium oxide can be used to dry all neutral and basic gases and amines and ethers. Potassium hydroxide and calcium oxide dry down to 0.002 mg water/l, while barium oxide goes to 0.004 mg water/l. Anhydrite and γ-aluminum oxide can also be used to dry gases.

b) D. for liquids. For some liquids, the same compounds can be used as for gases, as long as they do not dissolve in the liquid or react with it. Anhydrite and sodium sulfate are very handy, strong D. which work rapidly. Calcium chloride is usually used for preliminary drying of ether, carbon disulfide, carbon tetrachloride, etc. Basic D., such as potassium hydroxide, barium oxide, calcium oxide and potassium carbonate are used to dry all bases. Phosphorus pentoxide is used to dry carbon disulfide, carbon tetrachloride, chloroform, acids and hydrocarbons. Sodium wire or sodium-potassium alloy is used to dry ethers and hydrocarbons. For more information on methods of drying, see Solvents.

2) Same as Siccatives (see).

Dry matter: the solvent-free portion of a substance which is obtained by drying.

Dry metallurgy: same as Pyrometallurgy (see).

DTA: abb. for Differential thermoanalysis.

DTG: see Differential thermoanalysis.

dThd: abb. for Thymidine.

Dubbs process: see Cracking.

Duhem-Margules equation, ***Gibbs-Duhem-Margules equation***: an equation first derived in 1886 by P. Duhem, and thoroughly explored by M. Margules. It expresses a relation between the partial pressures (actually, the fugacities) in the vapor phase above a condensed mixture and the composition of the mixture. The D. follows from application of the Gibbs-Duhem equation (see) to the chemical potentials of the components of the mixture. For a binary mixture, it is:

$$x_1 \frac{d \ln p_1}{dx_1} + x_2 \frac{d \ln p_2}{dx_2} = 0.$$

Here x_1 and x_2 are the mole fractions of the two components in the liquid phase, and p_1 and p_2 are the partial pressures (or fugacities) in the vapor. Since $x_2 = 1 - x_1$, the equation can be transformed into:

$$x_1 \frac{d \ln p_1}{dx_1} - x_2 \frac{d \ln p_2}{dx_2} = 0, \text{ or}$$

$$x_1 \frac{d \ln p_1}{d \ln x_1} = x_2 \frac{d \ln p_2}{d \ln x_2}.$$

If the partial pressures of one component in an evaporation equilibrium have been determined experimentally as a function of its mole fraction, the partial pressure curve of the other components can be determined using the D.

Dulong and Petit, rule of: see Molar heats.

Dumas method: in elemental analysis, a method for quantitative determination of the nitrogen content of organic compounds. The substance is thermally decomposed in a stream of carbon dioxide, and the released nitrogen is determined in an Azotometer (see).

Dupré-Rankin equation: an equation describing the temperature dependence of the vapor pressure of a liquid: $\lg p = A - B/T + C \lg T$, where p is the vapor pressure, T the absolute temperature and A, B and C are empirical constants (see Clausius-Clapeyron equation). The D. is a better description of the temperature dependence of the vapor pressure than August's formula.

Durana®: a special brass (see Brass) of 29% zinc, 2% aluminum, 2% tin + antimony used in ship construction because of its resistance to seawater and its strength.

Duraluminum: an Aluminum alloy (see).

Durene, ***1,2,4,5-tetramethylbenzene***: an aliphatic-aromatic hydrocarbon, a colorless, crystalline compound with a camphor-like odor; m.p. 79.2°C, b.p. 196.8°C, n_D^{20} 1.5116. Even at room temperature, D. sublimes gradually. It is insoluble in water, moderately soluble in ethanol, and readily soluble in ether, benzene and acetone. D. is present in coal tar, some petroleum fractions and in the reaction mixture from catalytic hydrocarbon hydrogenation, and can be isolated from these sources. It is used for organic syntheses.

H_3C CH_3
H_3C CH_3

Durite: see Coal.

Duroscope hardness: see Hardness.

Dust removal: the precipitation of fine solid particulates from gases. *Collision precipitators* utilize the directional motion of dust particles; *cyclone precipitators* work on the centrifugal principle; and in *Venturi scrubbers*, water is sprayed into the accelerated gas stream. Woven or ceramic *filters* retain the particles which are larger than the pores of the filter, while the gas flows through them. In *metal filters*, the adhesion of the particles to oil-coated surfaces is utilized. Electrostatic precipitation (see) is especially suitable for removal of a high percentage of the particulates. Partial D. is achieved simply by slowing the flow of the gas stream. Thorough D. generally requires the used of several methods.

Dy: symbol for dysprosium.

Dye: coloring agent for textiles, leather, furs, paper, etc. In contrast to Pigments (see), which are always insoluble, most D. are soluble in solvent or binders. Fastness to light and laundering are the most important criteria of quality in D.

Classification. 1) D. can be classified according to their applications and the methods used to apply them (*coloristic classification*). a) the ***basic D.*** contain amino or alkylamino groups, or their salts with acids, and are used to dye wool, silk, treated cotton and polyacrylonitrile fibers. b) ***Acidic D*** usually contain sulfo groups, more rarely carboxyl groups and are used to dye wool, silk, polyamides and modified polyacrylonitrile fibers. c) ***Ingrain colors*** are generated on the fiber by chemical reaction. The textiles are dipped with an alkaline solution of a coupling component, and then into an ice-cold solution of a diazonium salt (see Ice dyeing). These D. are used primarily to dye cotton. d) ***Substantive D.*** or ***direct D*** are absorbed from the solution by untreated cellulose fibers, where they react to form large molecular aggregates in the intermicellar spaces. e) ***Mordant D.*** Treatment of the fibers with chromium(III), iron(III) or aluminum salts and subsequent steam treatment leaves finely divided metal hydroxides and metal oxide hydrates on the fibers. The D., which contains groups capable of forming metal complexes, e.g. -OH, -COOH, then forms insoluble, colored metal complexes on the treated wool, silk or cotton: the D. lake. f) ***Vat D.***: The insoluble D. is converted to a soluble, colorless leuko-compound by reduction. The fibers are dipped in the solution (substantive dyeing), and afterwards the leuko-compound is converted back into the insoluble D. by oxidation (usually air oxidation). g) ***Reactive D.*** react with certain functional groups in the fibers to form covalent compounds, e.g. with the hydroxy groups of the cellulose molecule to form esters, with primary and secondary amino groups in wool and silk, or with the numerous NH groups in the polyamides.

2) D. can be classified according to their chemical structure as nitro, nitroso, azo, di- and triarylmethane, acridine, xanthene, quinonimine, azine, oxazine, methylene, aza[18]annulene, carbonyl (anthraquinones, indigos, vat dyes), stilbene and sulfur D.

3) A distinction is also made between natural and synthetic D.

a) ***Natural D.*** may be of animal or vegetable origin. ***Animal D.*** are made from body fluids or parts of various animals; they include carmine, Tyrian purple, India yellow and sepia. ***Vegetable D.*** are made from various parts of plants: the wood (see Dye woods), e.g. redwood extract (see Brazil-wood), logwood extract (see Logwood), Fustic (see); bark (quercitron bark comes from several species of oak, such as *Quercus tinctoria*); fruit (e.g. buckthorn berries, from several species of *Rhamnus*); leaves (e.g. weld from dyer's weed, *Reseda luteola*, catechu from *Unicaria gambier*, henna from the shrubs *Lawsonia alba* and *L. inermis*, chlorophyll from various green leaves, such as spinach or nettles); roots (e.g. berberine from barberry, *Berberis vulgaris*, roots, stems and bark; curcumin (turmeric) from the curcuma plant, madder dyes, especially alizarin and purpurin from some Rubiaceae, alkanna red from *Lawsonia alba*, *L. inermis* and *Alkanna tinctoria*); from lichens (e.g. archil and litmus from various races of *Roccella lecanora*); from seeds (e.g. orlean from *Bixa orellana*); and from flowers (e.g. saffron from the pistils of *Crocus sativus*, safflower from *Carthamus tinctorius*, anthocyanins, pelargonidin, cyanidin, delphinidin, paeonidin, oenidin, malvidin, etc.).

Natural D. have largely been replaced by synthetic D. Some are still used in small amounts to color foods and cosmetics, e.g. carmine, buckthorn, alkanna, orlean, saffron, safflower, chlorophyll. Logwood extract is still used in large amounts to dye hides, silk and leather. Chlorophyll is used to color volatile oils, waxes, salves and cosmetics.

b) ***Synthetic D. (Coal tar D.***, also called ***aniline D.*** consist of complex organic compounds. Their use began with the synthesis of mauve by W.H. Perkin in 1856. In the subsequent years numerous D. were synthesized, usually starting from the aromatic compounds of coal tar. The term "coal tar D." was coined by F.F. Runge (1795-1867), who was the first to synthesize D. from the previously worthless coal tar, via aniline and phenol (hence the term "aniline D."). The synthetic D. can be classified on the basis of their chemical structure as shown in the following table.

The most important synthetic dyes

Class	Examples	Uses
Nitro and nitroso D.	Picric acid	Explosive, picraminic acid
	Naphthol yellow S	Wool, silk
	Amido yellow E	Wool, silk
	Naphthol green B	Wool
Azo D.	True red E	Acid dye
	Sudan red BB	Fat stain
	Permanent red	Print color
	Parared	Development dye
	Chrysoidine	Basic dye
	Tartrazine	Wool dye
	Congo red	Direct D., indicator
	Diamond black PV	Chrome dye
	Eriochrome blue-black B	Chrome dye Complexometric titration
Di- and triarylmethane dyes		
Diphenylmethanes	Auramine O	Basic dye
Triphenylmethanes	Malachite green	Basic dye
	Crystal violet	
	Phenolphthalein	Indicator
Acridine D.	Acridine orange	Basic dye
	Acriflavin	anti-trypanosome
	Atebrin	Therapy and prophylaxis of malaria
Xanthene D.	Chromoxan brilliant red	Chrome dye
	Fluoresceine	
	Eosine	
	Rhodamine B	
Quinonimines	Indophenols	Artists' colors
	Indanilines	
	Insamines	
Azine dyes	Mauve	First synthetic D.
	Safranine T	Basic dye
	Aniline black	Pigment
Oxazine dyes	Capri blue	Basic dye
	Gallocyanin	Chrome dye
	Sirius light blue	Substantive dye
Methylene dyes	Astraphloxin FF	Basic dye
	Astrazone yellow	Dyes for PAN
	Astrazone orange	fibers
	Pinaverdol	Green sensitizer
	Pinacyanol	Red sensitizer
Carotinoids	β-Carotene	Food dye
Aza[18]annulene D.		
Phthalocyanins	Heliogen blue	Pigment, reactive and vat dye
	Phthalocyanin developer	Formation of the complex on the fibers in aqueous solution
Carbonyl dyes		
Indigo dyes	Indigo	
	6,6-Dibromindigo	
	Thioindigo	Vat dyes
	Indanthrene brilliant pink R	
Anthraquinone D.	Alizarin	Mordant dye
	Alizarinsaphirol A	Wool
	Alizarin direct blue	Wool
	Indanthrone	
	Flavanthrone	Vat dyes
	Pyranthrone	
Higher annulated carbonyl dyes	Anthrapyrimidines	
	Anthrathrones	Vat dyes
	Violanthrone	
	Anthrimides	
Stilbene dyes	4,4-diaminostilbene 2,2'-disulfonic acid	Optical lightener
Sulfur dyes		
Baked dyes	Immedial orange	Vat dyes
Boiled dyes	Immedial black	
	Pyrogen indigo	

Historical. The use of D. goes back into prehistory; animal and vegetable products were used. The Phoenicians traded in Tyrian purple, and the Egyptians had developed the technique of vat dyeing with indigo by 5000 years ago. In royal Egyptian tombs the remains of textiles have been found which were dyed yellow with saffron, and others with alizarin (Turkish red). In addition, kermes, litmus and luteolin were used. These natural D. became far less important after it was discovered how to produce D. synthetically. F.F. Runge (1834) and Fritzsche and Zinin (1841/42) isolated aniline, the starting material for many D., in quantity. Mauve was synthesized by W.H. Perkin in 1856, and fuschin by Verguin in 1859. These discoveries led to a tremendous development of organic chemistry and, especially in Germany, to a major chemical industry. D., pharmaceuticals, fertilizers, explosives and so on were produced and led in a few decades to the founding of the IG-Farben concern. The basic research which led to dye chemistry included the studies of coal tar by A.W. von Hofmann (starting in 1843), the synthesis of madder red and alizarin by Graebe and Liebermann (1868), eosin by Baeyer and Caro (1874), anthracene dyes by Emil and Otto Fischer (1877), benzidine dyes by Bœttiger (1884), indigo by A. v. Baeyer (1881), the rational indigo synthesis of Heumann (1890), idanthrene dye (1901), naphthol AS dye (1913) and phthalocyanins by Linstead and de Diesbach (1928/29). Phthalocyanin development dyes have been commercially available since 1953. They are developed on the fibers in an aqueous bath of 3-amino-iminoisoindolenines in the presence of metal salts and have a high degree of fastness. Reactive dyes have been used since 1958.

Dye affinity: the tendency of a textile fiber to bind to a dye.

Dyeing aids: substances which improve the dyeing process by promoting evenness of color and thorough penetration of the fiber. D. include eveners, wetting agents, dispersing agents, penetrating agents, dye solvents, mordants, secondary treatments for improving fastness, etc.

Dye lake: a pigment made by precipitation of a water-soluble synthetic dye onto the fibers of a textile. The precipitating reagents most often used for acid and direct dyes are barium chloride and water-soluble salts of calcium, aluminum, manganese, zinc and lead; the corresponding metal salts of the dyes are insoluble. Basic D. are usually precipitated with tannin and tartar emetic. The colors are particularly fast if reactive groups on the fibers are involved in the formation of the D.

Dye woods: woods of various trees, most of them tropical, which contain dyes (see Brazil-wood, Logwood, Fustic). The dyes can be extracted from the chipped wood with water, and they were formerly widely used in the textile industry.

Dynamic equilibrium: see Equilibrium.

Dynamite: a group of explosives of which the most important is glycerol trinitrate. ***Gur-dynamite***, was first produced in 1867 by A. Nobel, and consisted of 75% glycerol trinitrate, 25% kieselguhr and some soda. Gur-D. burned without exploding and was difficult to make explode by impact. Today gur-D. has been replaced by mixtures in which the kieselguhr has been replaced by explosive substances which increase the power. The detonation rate of D. is around 6000 $m\,s^{-1}$. Instead of pure glycerol trinitrate, a mixture of glycerol trinitrate and ethylene glycol dinitrate is now most commonly used. The ***gelatin D.*** contain Explosives (see) instead of kieselguhr. D. is used to blast very hard rock and for underwater blasting. Its use has decreased in favor of Ammonium gelites (see), however.

Dynel®: see Synthetic fibers.

Dynorphin: a 32-peptide discovered in 1982; it has an opiate-like effect. The sequence is Tyr-Gly-Gly-Phe-Leu-Arg-Arg-Ile-Arg-Pro-Lys-Leu-Lys-Trp-Asp-Asn-Gln-Lys-Arg-Tyr-Gly-Gly-Phe-Leu-Arg-Arg-Gln-Phe-Lys-Val-Val-Thr. It contains the sequence of Leu-enkephalin both in the N-terminus and in amino acids 20-24, and the residues 1-17 constitute the D.-17 which was partially characterized in 1979 and was given its name because of its very high activity on the guinea pig ileum. The sequence D.-17 is called ***dynorphin A***, while the peptide with residues 20-32 is ***D. B***. Another fragment, ***D.-8***, contains residues 1-8, and was found in amounts similar to D. A in the hypothalamus. The mechanism of enzymatic cleavage of the Ile^{8}-Arg^{9} bond is not yet clear. D.-8 was classified as a highly active ligand of the kappa opiate receptor, and is probably a neurotransmitter. D. A, which is metabolically more stable, may act as a hormone. In 1982, Numa et al. deduced the sequence of pre-prodynorphin from cDNA; because of the similarity of this 256 amino-acid sequence to the enkephalin precursors, it was first called ***pre-proenkephalin B***. However, since Leu-enkephalin may not be a physiological product of this precursor, it has been renamed ***pre-prodynorphin***. Pre-prodynorphin also contains the sequence of neoendorphin.

Dysprosium, symbol ***Dy***: an element of the Lanthanide (see) series of the periodic system; a rare-earth and heavy metal; *Z* 66, with naturally occurring isotopes with mass numbers 164 (28.1%), 162 (25.5%), 163 (24.9%), 161 (19.0%), 160 (2.34%), 158 (0.10%) and 156 (0.06%). Its atomic mass is 162.50, valence is III or sometimes IV, m.p. 1407 °C, b.p. 2335 °C, standard electrode potential (Dy/Dy^{3+}) - 2.353 V.

D. is a silvery white, ductile metal which exists in two modifications. Below the conversion temperature of 1360 °C, it is hexagonal (density 8.559), and above this temperature, cubic. D. is found in the earth's crust at a concentration of about $4.2 \cdot 10^{-4}$%, and is associated with the other rare-earth metals. Monacite and bastnesite are the most important ores for its production. D. is used in the form of Cerium mixed metal (see); as an alloy with lead, it is used as a shielding material in nuclear technology. Its other properties, analysis, production and history are discussed under Lanthanides (see).

Dysprosium compounds: compounds in which dysprosium is usually in the +3 oxidation state, although scattered dysprosium(IV) compounds are known, such as $Cs_3[DyF_7]$. These can only be produced in nonaqueous media. The general properties and production are discussed under Lanthanide compounds (see). The most important D. are ***dysprosium(III) chloride***, $DyCl_3$, yellow leaflets, M_r 268.85, density 3.67, m.p. 718 °C, b.p. 1500 °C; ***dysprosium(III) fluoride***, DyF_3, colorless crystals, M_r 219.50, m.p. 1360 °C; ***dysprosium(III) oxide***, Dy_2O_3, colorless, M_r 373.00, density 7.81, m.p. 2340 °C; ***dysprosium(III) nitrate***, $Dy(NO_3)_3 \cdot 5H_2O$, yellow crystals, M_r 438.58, m.p. 88.6 °C.

E

e: 1) Symbol for electron. 2) Symbol for exergy. 3) Symbol for equatorial bonds; see Stereoisomerism 2.2.

E: configuration symbol; see Stereoisomerism 1.2.3.

E 600: see Paraoxon.

E 605: see Parathion.

EAA: abb. for essential amino acid (see Amino acids).

Earth metals: term for the metals of group IIIa of the periodic system (see Boron-aluminum group), aluminum, gallium, indium and thallium; it was formerly applied to the group IIIb rare-earth metals scandium, yttrium and lanthanum as well.

Eau de Javelle: see Potassium hypochlorite.

Eau de Labarraque: see Sodium hypochlorite.

Ebullioscopy: see Boiling point elevation.

Ecdysone: see Ecdysteroids.

Ecdysteroids: steroids which serve as molting or pupation hormones in certain invertebrates. The processes of molting and pupation are induced by action of the E. on epidermis cells. In insects, they are produced in the prothoracic glands.

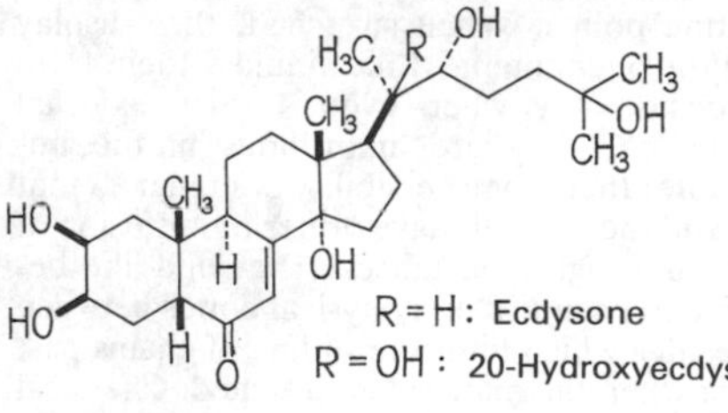

R = H: Ecdysone
R = OH: 20-Hydroxyecdysone

Zooecdysteroids are synthesized by insects, crustaceans and various worms; generally they are present at concentrations in the range of 10^{-5} to 10^{-9}%. The skeleton of the E. is ***ecdysone***, 2β,3β,14α,22R,25S-pentahydroxy-5β-cholest-7-en-6-one, also called α-ecdysone. Ecdysone and 20-hydroxyecdysone, also known as β-ecdysone or crustecdysone, are the most widely occurring E. Ecdysone was isolated in 1954 by Butenandt from silkworms. The compound is biosynthesized from cholesterol.

Phytoecdysteroids were first isolated from the Japanese conifer *Podocarpus nakai*, which contains relatively large amounts (about 1% of the dry mass), and later from plants of other families. Some of them still have alkyl groups in the side chain on C17. E. are characterized chemically by the 14α-hydroxy group and the 2-enone structure in ring B. As polyhydroxysteroids, they are readily soluble in polar solvents, but not in nonpolar solvents.

Ecgonin: see Tropane.

E-coke: see Petroleum coke.

Ecological chemistry: an area of chemistry dealing with the material processes in the ecosphere. E. studies the distribution, persistance and conversions of a chemical compound in the environment. *Xenobiotics* (substances not found in nature, such as anthropogenic biocides) which are intentionally or unintentionally released into the biosphere by human activities are of primary concern. The ecological significance of a substance depends on the amount produced, the pattern of application, its persistance, tendency to disperse and potential effects on the environment, its biotic and abiotic conversions and its ecotoxicology. The degree of pollution and ecotoxicological risks can be estimated by analysis of samples of air, water or soil for residues, e.g. of pesticides.

Edman degradation: see Proteins.

EDS process: see Coal hydrogenation.

EDTA: abb. for ethylenediaminetetraacetic acid.

Edwards equation: an LFE (see) equation established in 1956 by Edwards for substitution reactions $X^- + RY \rightarrow RX + Y^-$. The equation is $\lg(k/k_o) = aE + bH$. Here k is the rate constant for the substitution reaction with any reagent X^-, k_o is the rate constant for water and the same substrate RY, and a and b are empirical constants. H is a measure of the relative basicity of X^- compared to water: $H = pK_{HX} - pK_{H_3O^+}$. E is linked to the polarizability of X^- and represents the standard potential of the oxidation reaction $2\,X^- \rightleftharpoons X_2 + 2\,e$. The values of H and E are available in tabellar form for many nucleophilic reactions.

Effusion: the flow of gases out of very narrow tubes, which follows the *Graham-Bunsen flow law*, $v_1/v_2 = \sqrt{M_2/M_1}$. Here v_1 and v_2 are the rates of flow of two gases with molar masses M_1 and M_2. This law can also be applied to the penetration of porous membranes by gases.

Egg albumin: see Albumins.

Ehrlich's reagent: 1) a reagent for pyrrole derivatives, with which it develops a characteristic red color. The reagent is a solution of 1 g 4-dimethylaminobenzaldehyde in 100 ml 1N hydrochloric acid. E. is used in clinical chemistry to determine degradation products of hemoglobin in the urine, which are excreted in large amounts in certain diseases.

2) Reagent for phenols, phenol derivatives and some heterocycles which are easily coupled to red azo pigments. The reagent consists of a solution of sulfanilic acid and sodium nitrite in hydrochloric acid solution; these form a diazonium salt. Adrenalin, tyrosine, histidine, bile pigments and proteins give color reactions, so that the reagent is also used for clinical chemistry (*Ehrlich's diazo reaction*).

Eicosane: $CH_3[CH_2]_{18}CH_3$, an alkane forming colorless crystals; m.p. 36.8 °C, b.p. 343 °C or 150 °C at 0.13 kPa. It is insoluble in water, but soluble in alcohol and ether. It is a component of commercial paraffin and is found in the leaf waxes of certain rose species.

Eicosanoic acid, ***arachidic acid***: CH_3-$[CH_2]_{19}$-COOH, a higher, saturated fatty acid. E. crystallizes as shiny leaflets; m.p. 75.3 °C, b.p. 328 °C (dec.). It is insoluble in water, and soluble in ether and chloroform. E. is found as a component of triacylglycerols in many plant fats and oils, such as peanut, maize, rapeseed, olive and sunflower oils.

Eicosanoids: compounds with hormone effects, formed by an enzymatic peroxidation of multiply unsaturated C_{20} fatty acids, particularly arachidonic acid (20:4) and dihomo-γ-linolenic acid. The name E. is derived from eicosanoic acid, the saturated C_{20} fatty acid (since the IUPAC 1975 recommendations, actually icosanoic acid). The E. include products of cyclooxygenases and lipoxygenases. The products of cyclooxygenases are the cycloendoperoxides, Prostaglandins (see), thromboxanes and prostacyclin. The products of lipoxygenases are 5- and 12-hydroxyperoxyeicosatetraenoic acids (5- and 12-HPETE), which are formed as intermediates, various mono- and oligohydroxyeicosatetraenoic acids with chemotactic effects (e.g. 5- and 12-HETE) and the Leukotrienes (see).

Eigenfunction: see Atom, models of.

Eigenvalue: see Atom, models of.

Einstein: an obsolete, non-official stoichiometric unit in photochemistry. 1 einstein is 1 mol light quanta with the energy $h\nu$ (h = Planck's constant, ν = frequency of the electromagnetic radiation).

Einsteinium, symbol ***Es***: a radioactive element which is available only through nuclear reactions. It is an Actinide (see); Z 99, known isotopes with the following mass numbers (in parentheses, decay type; half-life; nuclear isomers): 243 (α; 20 s); 244 (K-capture; 40 s); 245 (K, α; 1.3 min); 246 (K, α; 7.3 min); 247 (K, α; 4.7 min); 248 (K, α; 28 min); 249 (K, α; 1.7 h); 250 (K; 8.3 h; 2); 251 (K, α; 33 h); 252 (α, K; 401 d); 253 (α; 20.47 d); 254 (β^-; 39.3 h; 2); 255 (β^-; 39.8 d). The most stable isotope has mass 252; valencies III and II, standard electrode potential (Es^{2+}/Es^{3+}) - 1.60 V.

E. is a silvery white metal which forms cubic crystals; it is obtained by reduction of einsteinium(III) fluoride with lithium. E. was first discovered in 1952 among the products of a thermonuclear explosion, and is now available on a µg scale by irradiation of plutonium 239:

$$^{239}Pu \xrightarrow[-5\beta^-]{+14\,n} {}^{253}Es$$

Some einsteinium compounds have been prepared on a nanogram scale and characterized; these include ***einsteinium(III) oxide***, Es_2O_3, cubic, white; ***einsteinium(III) chloride***, $EsCl_3$, hexagonal, UCl_3 type crystals, bright pink; ***einsteinium(III) bromide***, $EsBr_3$, $AlCl_3$ type crystals, amber yellow. Einsteinium(II) compounds are strong reducing agents.

Einstein-Stokes equation: see Diffusion.

Elaidic acid: *trans*-octadec-9-enoic acid, a monounsaturated fatty acid. E. is the *trans*-isomer of oleic acid. Its m.p. is 44.5-45.5 °C, which is significantly higher than that of oleic acid (13-16 °C). Oleic acid can be converted to the more stable E. by nitrogen oxide, sulfur or selenium. The rearrangement with nitrous acid, which is called the ***elaidic acid test***, is used in analysis; it causes fats rich in oleic acid to harden. Under the conditions of the E. test, an equilibrium is established with 66% E. E. is also formed from oleic acid in the presence of hydrogenating catalysts and enzymes. It is present in small amounts in milk and depot fats, and in margarine.

Elaidic acid test: see Elaidic acid.

Elana®: see Synthetic fibers.

Elastase: a serine protease which catalyses the cleavage of peptide bonds, especially in the neighborhood of neutral, hydrophobic amino acids. E. consists of 240 amino acid residues (M_r 25,700); the active center is formed by serine 196, aspartic acid 102 and histidine 57. It is important in the degradation of elastin, which is insoluble in water. E. is formed as a zymogen (***proelastase***, 251 amino acid residues, M_r 27,000) in the pancreas, and is secreted into the duodenum.

Elastic: a polymer which displays Rubber elasticity (see). Because their chains are highly mobile, the glass temperature of these compounds lies far below the temperature at which they are used (≈ - 80 °C). The E. include all natural and synthetic Rubber (see) products and many other polymers with similar properties (Table).

The E. occupy a position intermediate between solids and liquids. Like solids, they have a definite shape in their unstressed state, and they return to this state even after extreme deformation (except for a slight, permanent deformation). They also have relatively sharp melting points. When stretched, they display crystallization phenomena. Like liquids, their form changes considerably when even a small external force is applied; they are amorphous in the unstretched state, their compressibility is similar to that of liquids, and they can dissolve or permit diffusion of low-molecular weight substances. The solid-like behavior is due to chemical and physical bonds between chain molecules which limit the sliding of chains past one another when the material is stretched. Chemical bonds are created primarily by addition of cross-linking compounds during the polymerization process, and secondarily by vulcanization of the polymer. Examples are the introduction of sulfur bridges during sulfur vulcanization of R. or introduction of ether bridges in vulcanization of polychlorobutadiene with magnesium oxide. Physical bonds arise through formation of crystalline regions, by addition of substances such as carbon black, silicon dioxide, or aluminum oxide, or because of the tangling which occurs with very high relative molar masses. The liquid-like properties occur because the parts of the molecule are mobile, as in a liquid. The mobility can be increased by mixed polymerization (internal softening) or by mechanical working in of other substances (external softening).

In addition to the above conditions, which define a substance as an E., it must fulfill others to be of practical use. It must be able to absorb fillers such as carbon black, silicon dioxide, aluminum oxide, or zinc oxide; it must also be able to react with vulcanizing reagents and accelerators. The material must be resistant to light and heat in the presence of oxygen and moisture, and cannot swell more than slightly in water or organic solvents. In addition to the long-developed applications for natural and synthetic rubbers, E. with special properties, such as silicone rubber and fluorine rubber, have special applications.

General or chemical name	Abb.	Basic unit	Properties Applications
1) Natural rubber	NR	$[CH_2-C(CH_3)=CH-CH_2-CH_2-C(CH_3)=CH-CH_2]_n$	Good mechanical properties, not oil-resistant. After vulcanization raw material for tire and rubber industries.
2) Conversion products of natural rubber Chlorine rubber			Hard, non-inflammable, resistant to chemicals and heat to 80 °C. Uses: paint, protective agent against corrosion.
Rubber hydrochloride			Resists chemicals, penetration by water vapor. For impregnation and glue
Cyclo-rubber			Resists fats, dilute acids and bases. For paints, printer's inks and paper coatings.
3) Synthetic rubber Polybutadiene	BR	$[CH_2-CH=CH-CH_2]_n$ 1,4-cis-	After vulcanization, raw material for the tire and rubber industries.
Styrene-butadiene copolymers	SBR	$[CH_2-CH=CH-CH_2]_{n_1}\cdots[CH_2-CH(C_6H_5)]_{n_2}\cdots$	Very cold-resistant raw material for the tire and rubber industries
Acrylonitrile-butadiene copolymers	NBR	$[CH_2-CH=CH-CH_2]_{n_1}\cdots[CH_2-CH(C{\equiv}N)]_{n_2}\cdots$	Resist solvents, oils, fats and gasoline: for gasoline-resistant hoses, seals, sleeves, cable sheathing; in mixtures with phenol resins, for forms.
Polychloroprene (Chloroprene rubber)	CR	$[CH_2-CH=C(Cl)-CH_2]_n$	Heat- and oil-resistant; not easily inflamed: for conveyor belts, diver suits, coating of vessels and pipes, glue.
Polyisoprene	IR	Like natural rubber	Like natural rubber
Isobutene-isoprene copolymers (Butyl rubber)	IIR	$[CH_2-C(CH_3)_2]_{n_1}\cdots[CH_2-CH=C(CH_3)-CH_2]_{n_2}\cdots$	Resist weathering and oxidation, gas-tight: for automobile hoses, linings of vessels.
Ethylene-propylene copolymers	EPM	$[CH_2-CH_2]_{n_1}\cdots[CH_2-CH(CH_3)]_{n_2}\cdots$	Resist aging, weathering: for industrial rubber articles, seals, forms
Ethylene-propylene terpolymers (e. g. with ethylidene norbornene)	EPDM	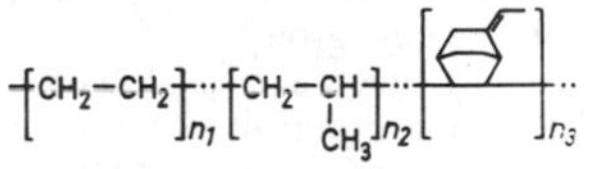 $[CH_2-CH_2]_{n_1}\cdots[CH_2-CH(CH_3)]_{n_2}\cdots[\text{ethylidene norbornene}]_{n_3}\cdots$	
Chlorosulfinated polyethylene	CSM	$[CH_2-CH_2]_{n_1}\cdots[CH_2-CH(Cl)]_{n_2}\cdots[CH_2-CH(SO_2Cl)]_{n_3}\cdots$	Resist oxidation, aging and discoloration. For industrial rubber articles.
Fluorine rubber	FKM	$[CF_2-CF(CF_3)]_{n_1}\cdots[CH_2-CF_2]_{n_2}\cdots$	Resist chemicals, but not heat; not easily inflamed. Resist aging. For fuel and oil lines, seals coatings, special rubber articles.
Nitroso rubber		$[CF_2-CF_2-N(CF_3)-O]_n$	
Triazine rubber		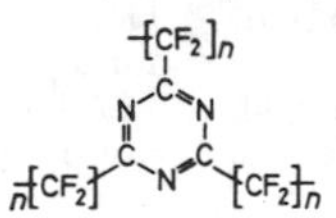 triazine ring (C_3N_3) with three $[CF_2]_n$ chains	
Fluorophosphazene rubber		$[N=P(OCH_2CF_3)_2]_n$	

General or chemical name	Abb.	Basic unit	Properties Applications
Polyurethane	AU	$\left[\mathrm{CONH{-}R^1{-}NHCO{-}O{-}R^2{-}O}\right]_n$	Gas-tight, resists oil and aging. For technical rubber articles
Silicone rubber	Q,MQ	$\left[\mathrm{Si(CH_3)_2{-}O{-}Si(CH_3)_2{-}O}\right]_n\cdots$	High temperature resistance, low resistance to chemicals
Polyalkene polysulfides (polysulfide rubber, thiokol)	TM	$\left[\mathrm{CH_2{-}CH_2{-}S_x}\right]_n$	Resistant to gasoline, heat and lubricants, low abrasion resistance. For container linings, gasoline and oil hoses, caulk for buildings
Epoxide rubber Polypropene oxide rubber with 5 to 10% allylglycide ether	PO PO, GPO	$\left[\mathrm{CH_2{-}CH(CH_3){-}O}\right]_n\cdots\left[\mathrm{CH_2{-}CH(CH_2{-}O{-}CH_2{-}CH{=}CH_2){-}O}\right]$	Cold resistant, high elasticity, low swelling, for special hoses, textile coatings
Epichlorohydrin rubber	CO	$\left[\mathrm{CH_2{-}CH(CH_2Cl){-}O}\right]_n$	
Fluorosilicone rubber		$\left[\mathrm{Si(CH_3)(CH_2{-}CH_2{-}CF_2){-}O}\right]_n$	Cold resistant, resistant to solvents. For seals in freezers, blood transfusion tubes

Elastic fiber: a fiber which has rubber elasticity. Such fibers are made from natural or synthetic rubbers by spining the latex and vulcanizing the threads. They made into the "elastic" used in underwear, waistbands, etc.

Elastin: a structural protein found in elastic fibers of connective tissue, the tendons, ligaments and artery walls. E. consists mainly of glycine, alanine, proline and aliphatic hydrophobic amino acids, but in contrast to collagen, it contains no hydroxylysine and only a small amount of hydroxyproline. Two sequences which repeat in E. are Lys-Ala-Ala-Lys and Lys-Ala-Ala. The highly elastic properties of the protein are due to a specific cross-linking of the polypeptide chains. Some of the lysine residues are oxidized to aldehydes, and then react with the free o-amino groups of other lysines, forming Schiff's bases. The resulting C=N double bonds are reduced, so that the actual bridge compound is a lysino-norleucine group. Another type of cross link is provided by reaction of three lysinaldehyde groups with one lysine group to form ***desmosine***. E. is exceptionally resistant to denaturing reagents and is attacked only by elastase.

Elastodiene: same as Elastic fiber.

Elaston®: see Synthetic fibers.

Elbatan®: see Herbicides.

Electrical conductivity: a measure of the ability of a substance to transport electric charge, either on electrons (class I conductors) or on ions (class II conductors). Mixed conductors can conduct electric current by either mechanism.

The E. depends on temperature. In electron conductors, it decreases as the temperature increases, while in ion conductors, the opposite is true. When the E. of a substance is reported, therefore, the temperature at which the measurement was made is always indicated.

By definition, the E. is the reciprocal of the electric resistance; its SI unit is the Siemens ($S = \Omega^{-1}$). Similarly, the reciprocal of the specific resistance is called the *specific conductivity* $\varkappa$; it is calculated from the general formula for electrical resistance R as follows:

$$R = \rho\frac{l}{q} = \frac{1}{\kappa}\cdot\frac{l}{q};\ \kappa = \frac{1}{R}\cdot\frac{l}{q}$$

where ϱ is the specific resistance, l the length and q the cross-section of the conductor.

For ionic conductors, the dimensions of specific conductivity are $\Omega^{-1}\ \mathrm{cm}^{-1}$; for metallic conductors, the dimensions of $\Omega^{-1}\ \mathrm{m\ mm}^{-2}$ are also used.

The specific conductivity of metallic conductors is several orders of magnitude higher than that of ionic conductors. For example, the values at 20 °C for some metals are:

Silver	$61\cdot10^4\ \Omega^{-1}\ \mathrm{cm}^{-1}$
Copper	$58\cdot10^4\ \Omega^{-1}\ \mathrm{cm}^{-1}$
Aluminum	$36\cdot10^4\ \Omega^{-1}\ \mathrm{cm}^{-1}$
Iron	$10\cdot10^4\ \Omega^{-1}\ \mathrm{cm}^{-1}$
Constantan	$2\cdot10^4\ \Omega^{-1}\ \mathrm{cm}^{-1}$
Mercury	$1\cdot10^4\ \Omega^{-1}\ \mathrm{cm}^{-1}$

If ionic conductors are in solution, their conductivity is called ***electrolyte conductivity***. In this case, the specific conductivity is referred to the resistance of an electrolyte solution between two electrodes at a distance of $l = 1$ cm, and with a cross section $q = 1\ \mathrm{cm}^2$. For 1 molar solutions at 18 °C, the following values are obtained:

Potassium chloride	$0.0982\ \Omega^{-1}\ \mathrm{cm}^{-1}$
Hydrochloric acid	$0.3010\ \Omega^{-1}\ \mathrm{cm}^{-1}$
Acetic acid	$0.0013\ \Omega^{-1}\ \mathrm{cm}^{-1}$
Silver nitrate	$0.0676\ \Omega^{-1}\ \mathrm{cm}^{-1}$

The specific conductivity is a function of the concentration. In dilute solutions, it is approximately directly proportional to concentration; as the concentration increases, the conductivity first reaches a maximum, and then decreases continuously. The reason for this decrease in strong electrolytes is a mutual inhibition of the motion of the ions by the ions of the opposite charge (see Debye-Hückel theory); in weak electrolytes it is due to a decrease in the degree of dissociation.

The concentration dependence of E. is more obvious when the molar conductivity is measured instead of the specific conductivity (variable amounts of electrolyte in 1 cm^3 solution). The *molar conductivity* Λ_m is defined as the conductivity due to a mole of electrolyte, dissolved in varying amounts of a solvent, measured between two electrodes of suitable size at a distance of 1 cm. The molar conductivity Λ_m is related to the specific conductivity K by the equation:

$$\Lambda_m = \frac{\kappa \cdot 1000}{c} \ \Omega^{-1}\ \mathrm{mol}^{-1}\ \mathrm{cm}^2$$

where the concentration is given in mol l^{-1}. If Λ_m is divided by the Equivalent number (see) n_e, the result is the equivalent conductivity Λ_{eq}:

$$\Lambda_{eq} = \frac{\Lambda_m}{n_e} = \frac{\kappa \cdot 1000}{n_e \cdot c} \ \Omega^{-1}\ \mathrm{eq}^{-1}\ \mathrm{cm}^2$$

Because the dependence of the specific conductivity on concentration is not linear, the molar and equivalent conductivities are also concentration dependent. This dependence is reflected in the empirical relationship for strong electrolytes discovered by Kohlrausch (*Kohlrausch square root rule*): $\Lambda_{eq} = \Lambda_\infty - k\sqrt{c}$, where Λ_∞ is the equivalent conductivity at infinite dilution (*limiting conductivity*) and k is a constant which depends on the viscosity and dielectric constant of the solvent.

E. is measured similarly to resistance, using a bridge circuit. To avoid polarization of the electrodes, alternating current is used. Concentrations of electrolytes can be determined by measuring conductivities of solutions (see Conductivity measurement).

Electrical conductor: a substance capable of conducting electric current (see Electron conductor, scc Ion conductor).

Electrical gas purification: see Electrostatic gas purification.

Electric double layer: same as Electrochemical double layer (see).

Electric field constant: see Coulomb's law.

Electric smelting process: see Steel.

Electroanalytical methods: same as Electrochemical analysis (see).

Electrocapillarity: an electrochemical phenomenon which describes the voltage dependence of the surface tension at the phase boundary between an electron conductor and electrolyte solution. Most studies on E. have been done with mercury electrodes. In 1875, Lippmann recognized that the surface tension of the mercury depends on the applied voltage, or on the charge on the surface. If there are excess electric charges on the surface, they repel each other and reduce the surface tension. If the surface tension is plotted against the voltage, the resulting *electrocapillary curve* is parabolic. The voltage at the maximum of the parabola is called the *zero charge voltage*.

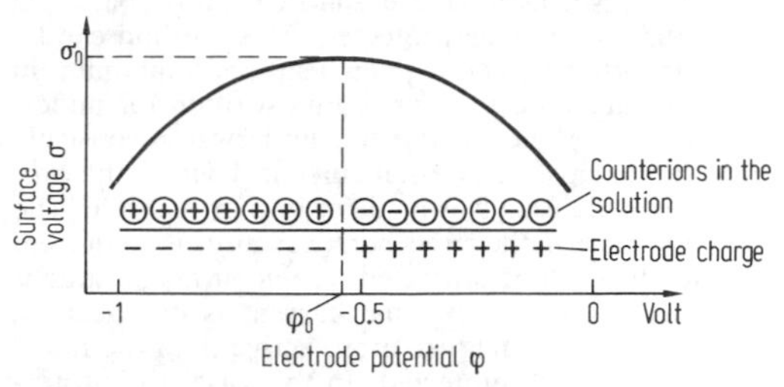

Electrocapillary curve. φ_{zero}, zero charge voltage; σ_{zero}, surface tension at the zero charge point

The change in the surface tension as the voltage changes can be calculated from the differential double layer capacitance (electrochemical double layer).

Electrocapillary curve: see Electrocapillarity.

Electrochemical analysis: Methods in which a signal is generated by an electrochemical reaction and is used for quantitative determination of a substance. There are three classes of E.: 1) methods in which neither the electrochemical double layer nor any electrode reaction needs to be considered, e.g. Conductometry (see), conductometric titration, High frequency conductometry (see) and high frequency conductometric titration; 2) methods which depend on double layer phenomena, but in which electrode reactions can be ignored, e.g. non-faraday admittance measurements (formerly tensametry); and 3) methods which depend on electrode reactions, e.g. Potentiometry (see), potentiometric titration, chronopotentiometry, Coulometry (see), Amperometry (see), amperometric titration, Electrogravimetric analysis (see), Polarography (see), Inverse voltammetry (see) and Cyclic triangular wave voltammetry (see).

Electrochemical cell: an arrangement consisting of at least two electrodes and an electrolyte. It can serve either as a source of current or as an electrolysis cell.

Electrochemical components: components for control and measurement based on the principles of electrochemistry. There are two groups: 1) *electrolytic elements*, in which the actual process occurs in an electrolyte or a phase boundary between an electrode and electrolyte (see Coulometer, Electrolytic timed switching cell, Electrolytic information storage) and 2) *electrokinetic elements*, which are based on processes taking place at the phase boundary between a polar liquid and a solid dielectric (see Electrolytic analog transistor).

Electrochemical corrosion protection: a method of active Corrosion protection (see) of metals, in which the object to be protected is made either the cathode (***cathodic corrosion protection***) or anode (***anodic corrosion protection***) in a direct current circuit. When the protective voltage has been reached, the corrosion current is smaller than at the rest voltage. In the *external current method*, the direct current is supplied by voltage-controlled rectifiers. In the *galvanic method* the protective current is provided by galvanic elements; in cathodic protection, these are created by an anode which is sacrificed, and in anodic protection, by local cathodes. In cathodic protection, the metal used as anode has a more negative potential than the metal being protected. This method can be used to protect steel, copper, lead and aluminum in all soils and waters. It is used most often for underground pipes and containers, underwater constructions and ships. In external current devices, the protective anode usually is made of iron or graphite. The negative pole of the direct current source is connected to the object to be protected. If zinc or magnesium is used as anode, an external current is not needed, because these metals are more active than the metal (e.g. iron) being protected. In this case, the protective anode becomes an active anode. To reduce the requirement for protective current, and to increase the effectiveness of the process, the objects to be protected are given protective coatings which are electrically insulating, such as polyethylene, coal tar or bituminous substances. If the metal to be protected is polarized as the anode by the external current, it is passivated (see Passivity).

Electrochemical current source, ***galvanic cell***: a collective term for devices which convert chemical energy into electrical energy. It includes Primary cells (see), Secondary cells (see) and Fuel cells (see). For the low and medium power range, E. are technically and economically the best available autonomous sources of electric energy.

Historical. It is thought that the Parthians used E. for gold plating of metal objects several centuries B.C. The historically documented (re)discovery of E. was made by Galvani in 1789. He suspended disected frog legs on copper hooks, and hung these on the iron lattice of his balcony; he observed that the legs began to twitch as soon as they touched the iron lattice. This phenomenon he ascribed to an "animal electricity". A few years later, Volta realized that the frog legs simply acted as the electrolytes; by replacing them with a salt solution, he made the first E., the "galvanic element". He made the first practical E., the *voltaic column*, which was named for him. This was a stack of silver, zinc and paper disks, in repeating sequence. The paper disks were soaked in a salt solution. The current produced by this column was orders of magnitude higher than that from an electrifying machine, and it stimulated fundamental research in electrochemistry and related areas. One of the oldest galvanic cells is the Daniell cell (see).

Electrochemical double layer, ***electric double layer***: a double layer of charge carriers formed at the boundary between two phases by transfer of charges. If two phases (such as an aqueous electrolyte solution and a metal) come into contact, charge carriers (electrons or ions) are often transferred between them. The charge carriers can either enter the aqueous phase or be absorbed from it, leading to a *potential difference* between the two phases. The charges at the phase boundary are compensated by the same number of counterions in the aqueous phase. Some of them may be adsorbed on the phase boundary, as a plate capacitor; these cause a decrease in the surface potential: (Ψ_0) drops to Ψ_δ, the ***Stern potential*** (Fig.). The rest of the counterions are found in the diffuse part of the double layer, in which the potential decreases exponentially with the distance. Relative motion of the two phases causes a part of the diffuse double layer to be sheared off; the resulting potential jump at the shear plane is called the Zeta potential (see). The zeta potential is less than or equal to the Stern potential. To date, no exact mathematical description of the E. has been discovered, but a well-known approximation was derived by Gouy and Chapman.

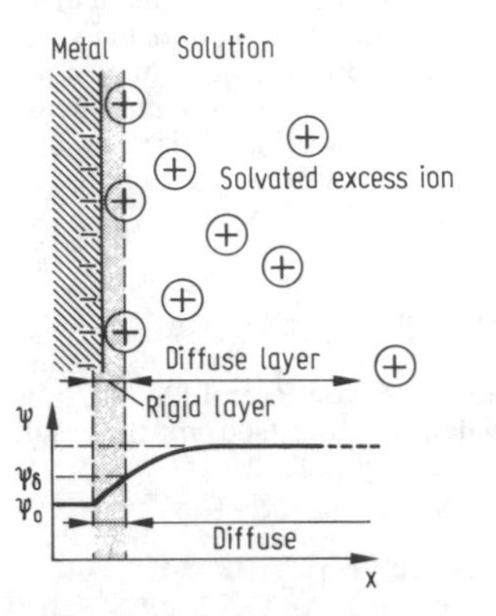

Structure of the electrochemical double layer according to Gouy, Chapman and Stern

Electrochemical equilibrium: an equilibrium which is established in heterogeneous systems and involves both charge carriers and transport of matter. It is characterized by differences in the chemical potentials of charge carriers in different phases which are in contact, and by different electrical potentials in these phases. If two phases are in chemical equilibrium with each other, the equilibrium condition $\mu_i(I) = \mu_i(II)$ holds for all components i, where $\mu_i(I)$ and $\mu_i(II)$ are the chemical potentials of the component i in phase I and II, respectively. For E., however, this equation is not sufficient to describe the position of the equilibrium. For example, if a metal such as silver dips into a solution of its ions (metal electrode), the equilibrium condition is generally not met at the instant the electrode enters the solution; μ_{Ag^+} (solution) $\neq \mu_{Ag^+}$ (metal). Therefore a reaction occurs, and, depending on the conditions, metal is dissolved or precipitated. Concomitantly, charge carriers (in this example, silver ions) are transported across the phase boundary, so that an electrical potential difference builds between the phases. This potential difference affects the position of equilibrium, and must be taken into account when electrochemical equilibrium conditions are set up. The equilibrium condition for E. is then $\mu_i(I) + z_iF\Phi(I) = \mu_i(II) + z_iF\Phi(II)$ where z_i is the electrochemical valence, F is Faraday's constant, $\Phi(I)$ and $\Phi(II)$ are the galvanic potentials. The expression $\mu_i + z_iF\Phi$ is called the *electrochemical potential* μ_i^*: $\mu_i^* = \mu_i + z_iF\Phi = \mu_i^o + RT \ln a_i + z_iF\Phi$ Here μ_i^o is the chemical standard potential of component i, R is the gas constant, T the temperature in Kelvin. If an E. has been established, $\mu_I^* = \mu_{II}^*$ and $\mu_I^\circ = RT \ln a_I + z_r F\Phi_I = \mu_{II}^\circ = RT \ln a_{II} + z_r F\Phi_{II}$ It follows from this that:

$$\varphi_I - \varphi_{II} = \frac{\mu_I^0 - \mu_{II}^0}{z_r F} + \frac{RT}{z_r F} \ln \frac{a_{II}}{a_I}$$

and

$$g_{eq}^{I,II} = g^{0I,II} + \frac{RT}{z_r F} \ln \frac{a_{II}}{a_I}.$$

Here $g_{eq}^{I,II}$ is the equilibrium galvanic potential and $g^{0I,II}$ is the standard galvanic potential. By its nature, the E. is a dynamic equilibrium. Even in the equilibrium state, charges are continuously exchanged between the electrode and electrolyte at the phase boundary.

Galvanic potentials of electrodes cannot be directly measured, but relative electrode potentials can be compared. For this purpose a reference electrode must be chosen, and the standard Hydrogen electrode (see) is used. Its standard potential has been arbitrarily set equal to zero. Thus the *relative electrode potentials E* are the voltages developed by galvanic cells consisting of the electrode to be measured and a reference electrode. In the literature, relative electrode potentials are also called *reference cell potentials, reference potentials* and, with reversed sign and limitation to equilibrium, *reference EMF values*. Relative electrode potentials referred to the standard hydrogen electrode are usually called *electrode potentials*. The electrode potential under standard conditions, i.e. at 25°C, 101.325 kPa (1 atmosphere) and activity 1, is the *standard electrode potential*. A potential series is an arrangement of electrodes in order of their standard electrode potentials.

Electrochemical potential: see Electrochemical equilibrium.

Electrochemistry: the area of physical chemistry concerned with processes and equilibria involving electrically charged particles. Such processes occur both in bulk phases (especially liquid solutions) and on the boundaries of these phases. In a narrow sense, E. is concerned with the interconversions of chemical and electrical energy, that is, with chemical reactions in which charges migrate or electrical potentials appear. The basic laws of E. were established in the first half of the 19th century by M. Faraday. Other important contributions were made by J.W. Hittorf (theory of ion migration), S. Arrhenius (theory of electrolytic dissociation), W. Ostwald (Ostwald's dilution law) and W. Nernst (theory of galvanic cells). E. is very widely applied. Many industrial processes make use of Electrolysis (see), and storage batteries of many types are in use (see Electrochemical current sources).

Electrochemical analysis methods (see) are very important in analytical chemistry. Bioelectrochemistry (see) and Chemotronics (see) are relatively new, rapidly developing fields.

Electrochromatography: see Electrophoresis.

Electrocrystallization: see Crystallization.

Electrocyclic reactions: see Woodward-Hoffmann rules.

Electrode: that part of a solid conductor which introduces electric current into a liquid, gas, vacuum or another solid, e.g. in electrolysis, gas discharge tubes and electron tubes.

An ***electrochemical E.*** is a multi-phase system in which an Electrochemical equilibrium (see) can be established between two electrically conducting phases linked in a series circuit. One phase of such a system consists of an electron conductor (or electron-defect conductor), and the other(s) of ion conductors. The term E. is sometimes used just for the electron (or electron-defect) conductor of such a system.

E. can be classified in different ways: 1) according to the type of process occurring between the anode and cathode, 2) according to the type of charge carrier which passes through the phase boundary, as ion electrodes and redox electrodes, 3) according to the type of electrode process and the speed at which equilibrium is reached, as reversible and irreversible E., 4) according to the number of phases involved in the electrode process, as type one and type two E., 5) according to the application, as reference and indicator E., and 6) according to the type of polarizability (see Polarization), as polarizable and nonpolarizable E.

Oxidations occur at the ***anode***; reductions at the ***cathode***. Whether an E. functions as anode or cathode depends on the direction of current flow. The anode of a galvanic cell becomes the cathode if this cell is operated as an electrolysis cell.

Ion electrodes are those in which ions cross the phase boundary in the potential-determining step. They can be divided into cation and anion E. *Cation E.* are, e.g., metal E. which dip into a solution of ions of that metal. The electrode potential of a metal E. is given by the Nernst equation: $E = E^\circ + (RT/zF) \ln A_{cation}$, where E is the E. voltage, E° is the E. voltage under standard conditions, z is the reaction charge number, R the gas constant, T the temperature in kelvin, F Faraday's constant and a_{cation} the activity of the cation. The Hydrogen electrode (see) is a type of cation E. *Anion E.* are those in which there is an electrochemical equilibrium between a substance which produces negative ions (usually a nonmetal) and a solution of these ions. The E. voltage can again be described by the Nernst equation (a_{cation} is replaced by a_{anion}). Oxygen electrodes (see) and halogen E. are anion E.

In ***redox electrodes***, it is electrons which cross the phase boundary in the potential-determining step. An inert metal (usually platinum) is dipped into a solution which contains two different oxidation states of a single substance. The EMF of a redox E. is described by: $E = E^\circ + (RT/zF) \ln (a_{ox}/a_{red})$, where a_{ox} is the activity of the oxidized form and a_{red} that of the reduced form. The quinohydrone E. is an example of a redox E.

In a ***reversible E.***, the E. process occurs without inhibition in either direction, and electrochemical equilibrium is rapidly established; this is in contrast to an ***irreversible E.***.

In ***type one E.*** two phases are involved in the potential-determining process. Typical examples are cation and anion E. ***Type two E.*** consist of three phases: a metal dips into a solution, which contains ions of that metal; these ions are in equilibrium with a precipitate of an insoluble salt. Type two E. include, e.g., the Oxide electrode (see) and the Mercury sulfate electrode (see). The EMF of a type two E. is calculated from equations of the following sort (those for the silver/silver chloride E. are shown as an example): $E_{Ag/AgCl} = E^{O}{}_{Ag/Ag^+} + (RT/F) \ln a_{Ag^+}$. With $K_S = a_{Ag^+} \cdot a_{Cl^-}$, we have

$$E_{Ag/AgCl} = E^{0}_{Ag/Ag+} + \frac{RT}{F} \ln \frac{K_S}{a_{Cl^-}}.$$

It also follows that

$$E_{Ag/AgCl} = \left[E^{0}_{Ag/Ag+} + \frac{RT}{F} \ln K_S \right] - \frac{RT}{F} \ln a_{Cl^-};$$

$$E_{Ag/AgCl} = E^{0}_{Ag/AgCl} - \frac{RT}{F} \ln a_{Cl^-}.$$

Here K_S is the solubility constant.

The relative potential of a type two E. thus depends on the corresponding anion. When the activity of this anion is constant, type two E. have a constant potential; they are therefore good as reference E.

Some common ***reference E.*** are the Calomel electrode (see) and the Silver/silver chloride electrode (see). With the ***indicator E.***, the EMF is directly proportional to the activity of the ion to be determined; these E. are therefore used in Potentiometry (see) and potentiometric titration. Some typical indicator E. are the glass E. (for protons), the silver E. (for silver or halide ions) and various ion-selective E.

Electrode coke: see Petroleum coke.

Electrode potential: see Electrochemical equilibrium.

Electrode reaction: an electrochemical reaction at an electrode which represents the sum of the *transit reaction* and the reactions before and after this step. The transit reaction is the process by which the electrons or ions cross the phase boundary between the electrode and the electrolyte.

Electrodialysis: a method for separation of dissolved substances, or of substances of different degrees of dispersion, by applying an electric field. The E. is based on the principle of dialysis, in which molecularly disperse substances or ions diffuse through a membrane, while colloidally disperse substances are held back. E. is used mainly to obtain drinking water by desalination of seawater. Cation and anion exchange membranes (see Ion exchanger) are arranged in an electrodialysis cell (Fig.) at a distance of 1.5 mm and parallel to each other. Application of a direct current voltage produces a potential drop. Salt water flows into the cell and desalinated water can be drawn off. E. is also used to remove salts from whey and tartar from wine.

Electrofax process: see Electrophotography.

Electrofilter: see Electrostatic gas purification.

Electrogravimetric analysis: an electrochemical method in which the substance to be determined is precipitated on an inert electrode (usually platinum) by electrolysis. The difference in mass of the electrode before and after electrolysis is equal to the amount of the substance to be determined. If several components of a sample could be precipitated, the analysis must be done at a constant voltage. Otherwise, a constant current can be maintained. For example, copper can be precipitated from acid solution onto a platinum cathode. If the electrolyte is depleted of copper, hydrogen is generated at the cathode. Since this does not change the mass of the cathode, it is not necessary to keep the voltage constant; the analysis can be done at constant current.

Electrokinetic phenomena: electric phenomena which arise from the relative motion of two mutually immiscible phases. The E. include the motion of particles in an electric field (see Electrophoresis), the motion of the fluid phase in an electric field (see Electroosmosis) and the generation of a potential difference due to relative motion (see Zeta potential).

Electrokinetic potential: same as Zeta potential (see).

Electrolysis, ***electrosynthesis***: chemical reactions which consume electric current. An E. involves the oxidation and reduction processes which occur in an electrolytic cell. A simple ***electrolysis cell*** consists of an anode (positive pole), where oxidation occurs, a cathode (negative pole), where reduction occurs, and an electrolyte (solution or melt). An external direct current voltage is applied to the cell. It must be greater than the characteristic *decomposition voltage*, at which oxidation (see Anodic oxidation) and reduction (see Cathodic reduction) are forced to occur (Fig.). In many cases, the decomposition voltage is higher than the difference between the electrode potentials, due to polarization or overvoltage.

Cl_2 Fresh water
Anode Cathode
Na^+ Na^+ Na^+ Na^+
Cl^- Cl^- Cl^- Cl^-
Saltwater inflow
Saltwater outflow
------- Cation-permeable membrane
·········· Anion-permeable membrane

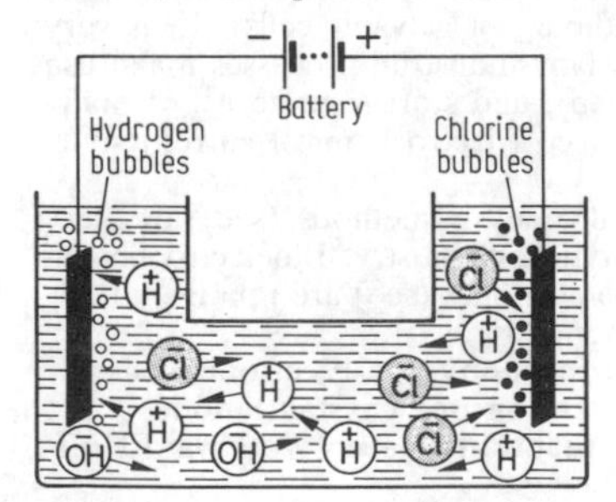

Electrolysis of an aqueous solution of hydrogen chloride

The anode and cathode spaces are often separated to prevent mixing of the products, and to prevent possible further reaction. The separation is achieved using a diaphragm made, for example, of a porous material or an ion-exchange membrane. In a divided E. cell, the electrolyte around the anode is called the *anolyte*, and the one at the cathode is the *catholyte*.

E. is used in chemical technology, e.g. to produce inorganic basic chemicals (see Chloralkali electrolysis), in electrometallurgy and galvanotechnology. Various organic compounds can also be produced by E. (see Organic electrosynthesis).

Electrolysis cell: see Electrolysis.

Electrolyte: a chemical compound which is dissociated into Ions (see) in the solid, liquid or dissolved state (see Dissociation, 1).

Electrolyte copper: see Electrolyte metal.

Electrolyte lead: see Electrolyte metal.

Electrolyte metal: a metal obtained or purified by an electrochemical method. Some important E. are aluminum, copper, iron, lead, magnesium, nickel, tin and zinc. Electrolytically purified metals are very pure. The purity of ***electrolyte copper*** (E-Cu) is characterized by its electrical resistance, and the maximum allowable level is $1.75 \cdot 10^{-8}$ $\Omega \cdot$m. Copper which conducts electric current as well as E-Cu but was purified by another method is still called E-Cu in industry. ***Electrolyte zinc***, E-Zn, is 99.99 to 99.999% pure. ***Electrolyte lead***, E-Pb, is 99.97 to 99.99% lead.

Electrolyte zinc: see Electrolyte metal.

Electrolytic analog transistor: an electrochemical component in which ions take over the function of electrons or holes, by changing their valence through an electrochemical process. If the reduced ionic form is the normal state of the electrolyte, the oxidized ions correspond to holes and the reduced ions to electrons. Thus a pnp transistor can be built on this basis. If the ions are oxidized in the normal state of the electrolyte, an npn transistor can be made.

The E. consists of a polymethacrylate casing containing an electrolyte and three electrodes. The collector and emitter consist of two highly polarized platinum foils, 1 to 2 cm in diameter. The basis electrode is an electrochemical half cell which cannot be polarized (e.g. mercury/mercury(I) chloride).

The function of the E. depends on the electrochemical oxidation or reduction of ions at the noble metal electrodes. These processes are possible when the emitter and collector are at different voltages with respect to the basis electrode.

Electrolytic conductivity: see Electrical conductivity.

Electrolytic corrosion cell: see Electrochemical metal corrosion.

Electrolytic information storage: an electrochemical component which can serve as either an analog or a binary storage element. The development E. was begun at the end of the 1950s with the construction of the memistor (memory resistor). This is a component in which the resistance of a copper conductor, supported by a relatively poorly conducting substrate material, is increased by anodic dissolution. The increase in resistance is a continuous function, and this is the basis for use of the element as an analog storage unit.

Electrolytic timed switching cell: an electrochemical component in which an electrochemical process occurs in a precisely defined period of time and causes a change in composition, geometry or internal resistance, or leads to a sudden change in voltage. Changes of this type are used as switching elements under otherwise constant conditions.

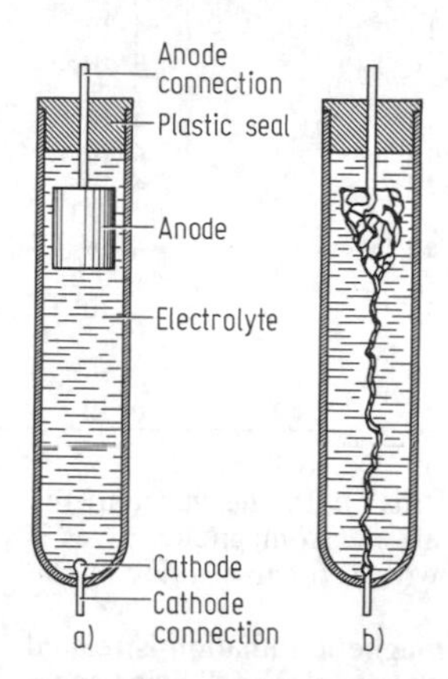

Timed switching cell with a dendritically growing cathode: *a* before use; *b* after use, or at the moment the circuit is closed

Electromagnetic spectrum: the entire range of electromagnetic waves. The wave nature of light was recognized early, on the basis of interference and refraction. For a long time it was not clear what oscillates in a light wave. Maxwell explained this in 1871 when he recognized light as electromagnetic radiation. According to this hypothesis, light is an electromagnetic wave consisting of electric and magnetic fields which are perpendicular to each other and to the direction of propagation of the wave. The oscillation is a periodic change in the strengths of the electric and magnetic fields. The two types of change occur in phase with each other (Fig.); thus electromagnetic waves are transverse waves. The entire range of electromagnetic waves is a continuum which includes much more than visible light. The subdivision shown in the Table is based on different methods of generating and applying various parts of the E., and the boundaries between the parts are often not sharp. Some of the ranges overlap.

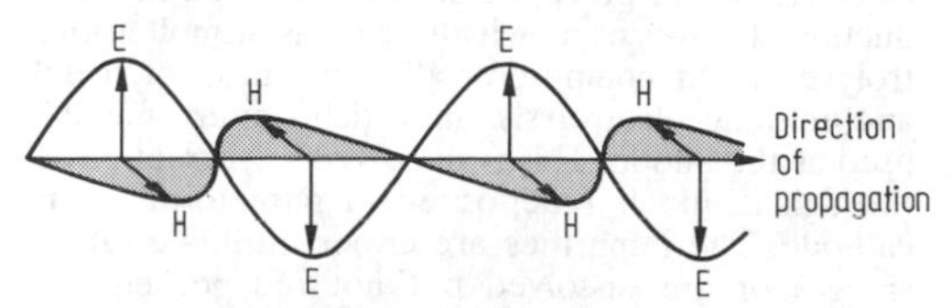

Electromagnetic wave. *E*, electric field vector; *H*, magnetic field vector

Electromagnetic radiation is described by the following parameters:

Wavelength λ (m) = length of a wave
Wavenumber $\tilde{\nu}$ (cm^{-1}) = number of waves per cm
Frequency ν (hz) = number of oscillations per second.

These parameters are related to each other by the equation $\tilde{\nu} = 1/\lambda = \nu/c$, where c is the velocity of light

The electromagnetic spectrum

Frequency ν [Hz]	Wavelength λ [m]	Energy E [J mol^{-1}]	Name	Interactions
10^1	$3 \cdot 10^7$	$4 \cdot 10^{-9}$	Alternating current	
10^3	$3 \cdot 10^5$	$4 \cdot 10^{-7}$	Long waves (radio)	
10^5	$3 \cdot 10^3$	$4 \cdot 10^{-5}$	Radio waves	Nuclear magnetic resonance
10^7	$3 \cdot 10^1$	$4 \cdot 10^{-3}$	Short waves	
10^9	$3 \cdot 10^{-1}$	$4 \cdot 10^{-1}$	Ultrashort waves	Electron spin resonance
10^{11}	$3 \cdot 10^{-3}$	$4 \cdot 10^{-1}$	Microwaves	Molecular rotations
10^{12}	$3 \cdot 10^{-4}$	$4 \cdot 10^1$	Infrared	Molecular vibrations
10^{14}	$3 \cdot 10^{-6}$	$4 \cdot 10^2$	Visible	Outer electron transitions
10^{16}	$3 \cdot 10^{-8}$	$4 \cdot 10^6$	Ultraviolet	Inner electron transitions
10^{18}	$3 \cdot 10^{-10}$	$4 \cdot 10^8$	X-rays	Excitation of atomic nuclei
10^{20}	$3 \cdot 10^{-12}$	$4 \cdot 10^{10}$	Gamma rays	
10^{23}	$3 \cdot 10^{-15}$	$4 \cdot 10^{13}$	Cosmic rays	

in a vacuum ($2.988 \cdot 10^8$ ms^{-1}). In the vacuum, the propagation velocity of all electromagnetic waves is constant. In a medium with refractive index n, the velocity is c/n.

The energy of electromagnetic radiation is related to the wavelength or frequency by the following equation: $E = h\nu = hc/\lambda$, where E is the energy of the radiation in joules (J) and h is Planck's constant ($6.626 \cdot 10^{-24}$ J s). The higher the frequency of the radiation, the higher its energy. To relate the energy to 1 mol, as is often done in chemistry, the expression $E = h\nu$ must be multiplied by Avogadro's number ($6.02 \cdot 10^{23}$ mol^{-1}. Light with the wavelength 100 nm $= 10^{-7}$ m has the energy $E \approx 2 \cdot 10^{-18}$ J. Multiplying by Avogadro's number gives $1.2 \cdot 10^6$ J mol^{-1} or 1200 kJ mol^{-1}. Electromagnetic radiation interacts with matter by the processes listed in the fifth column of the table; these can therefore be examined by spectroscopy.

Electromagnetic radiation: see Electromagnetic spectrum.

Electrometallurgy: a comprehensive term for all metallurgical processes for production and purification of metals using electric energy.

1) In the *electrochemical methods*, the electrical energy does chemical work (see Electrolysis). The electrochemical process can be subdivided into production electrolysis, in which aqueous or molten electrolytes are decomposed with deposition of metal, and refining electrolysis, in which impure metal is used as the anode. This is dissolved as current passes through it and is redeposited in pure form on the cathode. The impurities are either insoluble (anode sludge) or are dissolved but not redeposited. The electrochemically produced or refined metals are called ***electrolyte metals***.

2) In the *electrothermal process*, the electrical energy is converted into heat in an electric furnace.

In some cases, the two methods overlap.

Electromotive force, abb. ***EMF***: a term for the potential of a galvanic cell (see Electrochemical equilibrium).

Electron: 1) an elementary particle (see Atom, models of).

2) An old proprietary name for a group of magnesium alloys containing up to 10% aluminum, 5% zinc, 2% silicon and 0.5% manganese. There are a number of types of E. which differ in their resistance to corrosion and mechanical strength. E. are used as kneaded and cast alloys in aircraft construction, and to make suitcases and housings for optical devices.

Electron affinity: the energy required to separate an electron from an anion A^{n-} or the energy gained or lost when an electron is added to an atom, ion or molecule: $A^{n-} \rightleftharpoons A^{(n-1)^-} + e^-$. Since different signs can result, depending on the way the energy is measured, it is important to set values for the E. in relationship to the above equation. The E. is a measure for the tendency of elements to form anions. The halogens have the largest E.; by accepting an electron they can achieve a stable noble-gas configuration (octet principle) (Table). The values of E. for most other atoms are distinctly smaller or almost zero. The loss of an electron from a multiply charged anion with a noble gas configuration releases energy, even in highly electronegative elements (e.g. $O^{2-} \rightarrow O^- + e^-$), because the electrostatic repulsion of the electrons dominates. Experimental determination of E. is difficult and is usually done by indirect methods, e.g. the Born-Haber cycle. Therefore, exact values are known for only a few atoms. According to Mulliken, the E. and the 1st ionization energy of the atom can be used to define the Electronegativity (see) of the elements.

Table. Electron affinities of some atoms in kJ mol^{-1} ($A^- \rightarrow A + e^-$)

H	F	Cl	Br	I	O	S
72	338	367	338	299	222	251

Electron beam microanalysis, ***EMSA***, ***EMA***: a special method of X-ray spectroscopy (see) for local analysis in the µm range.

A beam of electrons is precisely focused on the surface of the sample, causing it to emit X-rays. These are analysed in an X-ray spectrometer. The chemical composition of the volume reached by the electron probe can be determined. All elements heavier than beryllium can be determined, at concentrations down to 10^{-15}g. Modern instruments are equipped so that the probe can be moved in a gridwork pattern across the surface. The method can be used for simultaneous analysis and microscopy, because the electrons falling on the surface of the sample are partly reflected and partly absorbed, followed by re-emission of lower-energy secondary electrons. These are collected and used to drive an oscilloscope, as in electron microscopy. The screen thus shows a picture of the surface which is characteristic of its chemical composition.

E. can be combined with Auger electron spectroscopy (see), because the two techniques make use of the same excitation conditions. They are used for surface studies, especially study of diffusion processes in catalysts, metals, and glasses, or to determine the thickness of metallic layers or films.

Electron beam microprobe: same as Electron beam microanalysis (see).

Electron beam paint hardening: see Radiation chemistry.

Electron conductor: a substance which conducts electric current by means of mobile electrons. See Electrical conductivity.

Electron configuration: the distribution of electrons in an atom among the atomic orbitals. In the independent-particle model (see Central field model), each electron is described by a single-electron function Ψ_{n,l,m_l,m_s} which, as in the hydrogen atom (see Atom, models of) is characterized by the four quantum numbers n, l, m_l and m_s. In this approximation, the energy of an electron in the multi-electron atom is determined by the quantum numbers n and l. The occupation of the energy levels follows the *Pauli exclusion principle*, which says that in a given multi-electron system, no two electrons can have an identical set of quantum numbers. An atomic orbital, which is characterized by the three quantum numbers n, l and m_l, can be occupied by at most two electrons which differ in their spin quantum number m_s. In the ground state, the electrons fill the lowest-energy orbitals available to them in accordance with the Pauli principle (*aufbau principle*). The occupation of energetically degenerate orbitals (see Degeneracy) follows *Hund's rule* (see Multiplet structure). Description of the E. of a given atom requires knowledge of the relative energies of the atomic orbitals. As an approximation, the qualitatively valid order, according to which the subshells (n l) are filled as the nuclear charge (atomic number) increases, is as follows: $1s < 2s < 2p < 3s < 3p < 4s \approx 3d < 4p < 5s \approx 4d < 5p < 6s \approx 5d \approx 4f < 6p < 7s \approx 6d \approx 5f < 7p$. $6s \approx 5d \approx 4f$, for example, means that filling the 5d and 4f orbitals competes with filling the 6s orbital. The E. of an atom is given symbolically by indicating each occupied subshell (n l) by a number and letter, with the number of electrons in the subshell denoted by a

Electron configurations of the elements in the ground state

Element		K	L		M			N				O				P			Q	
		1s	2s	2p	3s	3p	3d	4s	4p	4d	4f	5s	5p	5d	5f	6s	6p	6d	6f	7s
1	H	1																		
2	He	2																		
3	Li	2	1																	
4	Be	2	2																	
5	B	2	2	1																
6	C	2	2	2																
7	N	2	2	3																
8	O	2	2	4																
9	F	2	2	5																
10	Ne	2	2	6																
11	Na	2	2	6	1															
12	Mg	2	2	6	2															
13	Al	2	2	6	2	1														
14	Si	2	2	6	2	2														
15	P	2	2	6	2	3														
16	S	2	2	6	2	4														
17	Cl	2	2	6	2	5														
18	Ar	2	2	6	2	6														
19	K	2	2	6	2	6		1												
20	Ca	2	2	6	2	6		2												
21	Sc	2	2	6	2	6	1	2												
22	Ti	2	2	6	2	6	2	2												
23	V	2	2	6	2	6	3	2												
24	Cr	2	2	6	2	6	5	1												
25	Mn	2	2	6	2	6	5	2												
26	Fe	2	2	6	2	6	6	2												
27	Co	2	2	6	2	6	7	2												
28	Ni	2	2	6	2	6	8	2												
29	Cu	2	2	6	2	6	10	1												
30	Zn	2	2	6	2	6	10	2												
31	Ga	2	2	6	2	6	10	2	1											
32	Ge	2	2	6	2	6	10	2	2											
33	As	2	2	6	2	6	10	2	3											
34	Se	2	2	6	2	6	10	2	4											
35	Br	2	2	6	2	6	10	2	5											
36	Kr	2	2	6	2	6	10	2	6											

Electron configuration

Element		K	L		M			N				O				P			Q	
		1s	2s	2p	3s	3p	3d	4s	4p	4d	4f	5s	5p	5d	5f	6s	6p	6d	6f	7s
37	Rb	2	2	6	2	6	10	2	6		1									
38	Sr	2	2	6	2	6	10	2	6		2									
39	Y	2	2	6	2	6	10	2	6	1		2								
40	Zr	2	2	6	2	6	10	2	6	2		2								
41	Nb	2	2	6	2	6	10	2	6	4		1								
42	Mo	2	2	6	2	6	10	2	6	5		1								
43	Tc	2	2	6	2	6	10	2	6	6		1								
44	Ru	2	2	6	2	6	10	2	6	7		1								
45	Rh	2	2	6	2	6	10	2	6	8		1								
46	Pd	2	2	6	2	6	10	2	6	10		0								
47	Ag	2	2	6	2	6	10	2	6	10		1								
48	Cd	2	2	6	2	6	10	2	6	10		2								
49	In	2	2	6	2	6	10	2	6	10		2	1							
50	Sn	2	2	6	2	6	10	2	6	10		2	2							
51	Sb	2	2	6	2	6	10	2	6	10		2	3							
52	Te	2	2	6	2	6	10	2	6	10		2	4							
53	I	2	2	6	2	6	10	2	6	10		2	5							
54	Xe	2	2	6	2	6	10	2	6	10		2	6							
55	Cs	2	8		18			2	6	10		2	6			1				
56	Ba	2	8		18			2	6	10		2	6			2				
57	La	2	8		18			2	6	10		2	6	1		2				
58	Ce	2	8		18			2	6	10	2	2	6			2				
59	Pr	2	8		18			2	6	10	3	2	6			2				
60	Nd	2	8		18			2	6	10	4	2	6			2				
61	Pm	2	8		18			2	6	10	5	2	6			2				
62	Sm	2	8		18			2	6	10	6	2	6			2				
63	Eu	2	8		18			2	6	10	7	2	6			2				
64	Gd	2	8		18			2	6	10	7	2	6	1		2				
65	Tb	2	8		18			2	6	10	9	2	6			2				
66	Dy	2	8		18			2	6	10	10	2	6			2				
67	Ho	2	8		18			2	6	10	11	2	6			2				
68	Er	2	8		18			2	6	10	12	2	6			2				
69	Tm	2	8		18			2	6	10	13	2	6			2				
70	Yb	2	8		18			2	6	10	14	2	6			2				
71	Lu	2	8		18			2	6	10	14	2	6	1		2				
72	Hf	2	8		18			2	6	10	14	2	6	2		2				
73	Ta	2	8		18			2	6	10	14	2	6	3		2				
74	W	2	8		18			2	6	10	14	2	6	4		2				
75	Re	2	8		18			2	6	10	14	2	6	5		2				
76	Os	2	8		18			2	6	10	14	2	6	6		2				
77	Ir	2	8		18			2	6	10	14	2	6	7		2				
78	Pt	2	8		18			2	6	10	14	2	6	9		1				
79	Au	2	8		18			2	6	10	14	2	6	10		1				
80	Hg	2	8		18			2	6	10	14	2	6	10		2				
81	Tl	2	8		18			2	6	10	14	2	6	10		2	1			
82	Pb	2	8		18			2	6	10	14	2	6	10		2	2			
83	Bi	2	8		18			2	6	10	14	2	6	10		2	3			
84	Po	2	8		18			2	6	10	14	2	6	10		2	4			
85	At	2	8		18			2	6	10	14	2	6	10		2	5			
86	Rn	2	8		18			2	6	10	14	2	6	10		2	6			
87	Fr	2	8		18			2	6	10	14	2	6	10		2	6			1
88	Ra	2	8		18			2	6	10	14	2	6	10		2	6			2
89	Ac	2	8		18			2	6	10	14	2	6	10		2	6	1		2
90	Th	2	8		18			2	6	10	14	2	6	10		2	6	2		2
91	Pa	2	8		18			2	6	10	14	2	6	10	2	2	6	1		2
92	U	2	8		18			2	6	10	14	2	6	10	3	2	6	1		2
93	Np	2	8		18			2	6	10	14	2	6	10	5	2	6			2
94	Pu	2	8		18			2	6	10	14	2	6	10	6	2	6			2
95	Am	2	8		18			2	6	10	14	2	6	10	7	2	6			2
96	Cm	2	8		18			2	6	10	14	2	6	10	7	2	6	1		2
97	Bk	2	8		18			2	6	10	14	2	6	10	8	2	6	1		2
98	Cf	2	8		18			2	6	10	14	2	6	10	10	2	6			2
99	Es	2	8		18			2	6	10	14	2	6	10	11	2	6			2
100	Fm	2	8		18			2	6	10	14	2	6	10	12	2	6			2
101	Md	2	8		18			2	6	10	14	2	6	10	13	2	6			2
102	No	2	8		18			2	6	10	14	2	6	10	14	2	6			2
103	Lw	2	8		18			2	6	10	14	2	6	10	14	2	6	1		2
104	Ku	2	8		18			2	6	10	14	2	6	10	14	2	6	1		2

The solid horizontal lines separate the periods; the dotted lines enclose inner and outer transition elements.

superscript on the right, e.g. $1s^22s^22p^63s^23p^3$ for the ground state of the phosphorus atom. For elements with high atomic numbers, this becomes unwieldy, so the filled electron shells are indicated by writing the symbol of the corresponding noble gas. For example, the ground state of iron can be indicated by (Ar) $3d^64s^2$; (Ar) is an abbreviation for the E. of the argon atom. The E. for the ground states of the free atoms are shown in the table. Some of the chemical properties of an element can be deduced from the occupation of its outer electron orbitals. Atoms with filled electron shells, i.e. those in which the outer shell is occupied by 8 electrons (***noble gas configuration***), are unusually stable. E. with half-filled or completely filled subshells are also relatively stable.

In the central field model, the energy of an atom is calculated as the sum of the orbital energies indicated by the E. More exact treatment shows that if the electron interactions and the spin-orbital coupling are considered, the energy of the E. is split into a Multiplet structure (see).

Electron deficiency compound: see Three-center bond.

Electron diffraction: a method based on the diffraction of an electron beam passing through or reflected from a crystalline substance. In 1927, Davison and Germer made the first electron diffraction patterns with slow electrons (30 to 300 eV energy), and thus obtained proof for the wave nature of the electron. Bragg's Law (see X-ray structure analysis) determines the geometry of E. by a crystal, and establishes the useful energy range for the electrons; it lies between 10 eV and 1 keV. The electrons are scattered from nuclei, and the scattered beams reach high intensities. Electron beams penetrate only a few lattice planes, so that only very thin preparations can be observed by passing the electrons through them.

Electronegativity: a relative measure of the tendency of an atom in a molecule to attract bonding electron pairs. The E. is used in the estimating the polarities a bond; it was developed for this purpose by L. Pauling (1932). It was assumed that the bond energy E_{AB} for a heteronuclear compound AB is always greater than the arithmetic or geometric mean of the bond energies AA and BB: $E_{AB} > 1/2(E_{AA} + E_{BB})$ and $E_{AB} > \sqrt{E_{AA} \cdot E_{BB}}$. This suggests that the heteronuclear molecule is stabilized by a supplemental electrostatic term. The additional stabilization energy Δ_{AB} (in kJ mol^{-1}) should therefore be a measure of the polarity of the bond. Using the geometric mean, one obtains $\Delta_{AB} = E_{AB} - \sqrt{E_{AA} \cdot E_{BB}}$. Using Δ_{AB}, the difference in the E. of atoms A and B can be defined as $|X_A - X_B| = 0.088 \sqrt{\Delta_{AB}}$ kJ mol^{-1}. Using the arbitrary value of $X_F = 4$ for fluorine, a complete E. scale for the elements can be established from the differences in E. of individual elements (Table). It can be seen that the E. is a periodic property. It cannot be directly measured and is given as a dimensionless number. Mulliken (1934) showed that the sums of the first ionization energies E_I and the electron affinity E_E (both in kJ mol^{-1}) are proportional, to a fairly close approximation, to the E. values of Pauling. A numerically similar E. scale can therefore be established on this basis: $X = (E_A + E_I)/523.3$ kJ mol^{-1}. Other E. scales with similar values have been established using other parameters.

Table 1. Electronegativity values of some representative group elements (according to Pauling)

			H 2.1			
Li 1.0	Be 1.5	B 2.0	C 2.5	N 3.0	O 3.5	F 4.0
Na 0.9	Mg 1.2	Al 1.5	Si 1.8	P 2.1	S 2.5	Cl 3.0
K 0.8	Ca 1.0	Ga 1.6	Ge 1.8	As 2.0	Se 2.4	Br 2.8
Rb 0.8	Sr 1.0	In 1.7	Sn 1.8	Sb 1.9	Te 2.1	I 2.5

The E. is not an atomic constant. It depends greatly on the bonding state of the atom and its neighbors in the molecule. Based on this, and on several different E. scales, Bažanov introduced a system of E. values in 1962. It is true for all E. scales that the *polarity*, and thus the *ionic character* of the bond increases with increasing difference between the E. of the atoms in the bond.

Electroneutrality principle: a rule formulated in 1948 by Pauling: the greatest stability is achieved when the electrical charges in a molecule or ion are distributed in such a way that the effective charge on each atom is as small as possible. The E. can be used to explain the stability of compounds, especially complex compounds. The nature of the bonding between the central ion and the ligands is such that the charge of the central ion is distributed over the entire complex. For example, in the hexaamminecobaltate(III) ion, $[Co(NH_3)_6]^{3+}$, if it were assumed that the Co-N bonds were completely ionic, the total charge of +3 would be localized on the cobalt atom. If the bonds were purely covalent, the cobalt would have a charge of - 3 and each nitrogen would have a charge of +1. The real ionic nature of the Co-N and N-H bonds is such that the charges on the central atom and the nitrogen atoms are close to zero, and each hydrogen atom has a positive charge of approximately +1/6. The distribution of the charge over the surface of the nearly spherical complex ion is also the condition for the greatest electrostatic stability. Thus it is easy to understand why stable complex cations have hydrogen atoms on the outside (e.g. hydrates and ammines), while stable complex anions have electronegative exterior ligand atoms, as in $[Co(NO_3)_6]^{3-}$ or $[Fe(CN)_6]^{4-}$. In addition to the ionic character of a bond, its multiple bond character is an important factor leading to charge equalization and thus stabilization of the system. The relatively small dipole moments of carbon monoxide and nitrogen dioxide can be interpreted in terms of the E.

The model of back binding in the molecular orbital theory of complexes is a quantum mechanical interpretation of the E.

Electron excess compounds: see Three-center bond.

Electron-exchange chromatography: see Ion-exchange chromatography.

Electron excitation spectra: see UV-VIS spectroscopy.

Electron gas model: see Chemical bond.

Electronic formula: see Formula.

Electron pair bond: see Chemical bond.

Electron pair repulsion: see Valence shell electron pair repulsion.

π-Electron sextet: a term for the electronic structure of the p-electrons of benzene and benzenoid compounds which correspond to the Hückel rule (see Aromaticity).

Electron spin: An angular momentum of the electron which cannot be attributed to an orbital motion. The first indication of E., $\mathbf{L}_s$, was provided by the Stern-Gerlach experiment (1922) in which a beam of silver atoms in an inhomogeneous magnetic field was split into two bunches which were symmetric with respect to the primary beam. The experiment showed that the E. is quantized. Its z component has one of the two values $+1/2\ h/2\pi$ and $-\ 1/2\ h/2\pi$, where h is Planck's constant, i.e. there are two possible orientations for the electron, which are indicated by the spin quantum numbers $m_s = +1/2$ and $-\ 1/2$. The corresponding spin eigenfunctions are called α and β. The value of the E. is $L_s = \sqrt{s(s+1)}h/2\pi$, where s is the spin quantum number $s = 1/2$. Thus the state of an electron is characterized by four quantum numbers, n, l, m_l and m_s (see Atom, models of).

Electron spin resonance spectroscopy *ESR spectroscopy, electron paramagnetic resonance spectroscopy, EPR spectroscopy*: a method of High frequency spectroscopy (see) applied to paramagnetic substances. E. is used to study the effect of internal magnetic and electrostatic fields on the unpaired electron. These interactions have specific spectra with characteristic line numbers and line widths. Information on the structure, symmetry, nature of chemical bonding, chemical equilibria and reaction mechanisms can be derived from these spectra.

E. is closely related to NMR spectroscopy. Like the nuclei studied by NMR, the unpaired electron in a paramagnetic substance has a spin ($S = 1/2$) and therefore a magnetic moment which is split into two energy states by an external magnetic field. Electromagnetic radiation of suitable frequency induces transitions between the two energy states. Therefore the theoretical basis of E. is analogous in many respects to that of NMR spectroscopy. The magnetic moment of an electron is much larger than that of the nucleus, however, so that higher energy is required for "flipping" the electron spin [corresponding to frequencies in the microwave range of the Electromagnetic spectrum (see)]. The frequency of electromagnetic radiation which is in resonance with the paramagnetic substance at an external field strength H_0 is

$$\nu = (g\mu_B H_0)/h, \qquad (1)$$

where μ_B is the Bohr magneton, h is Planck's constant and g is the g factor whose magnitude depends on the contributions of the electron's spin and orbital angular momentum to the magnetic moment (see below).

Sample preparation and measurement. Fig. 1 is a diagram of an ESR spectrometer. The sample is placed in a resonator, which is in a strong, homogeneous magnetic field H_0. The strength of the field can be varied. A high-frequency electromagnetic alternating field perpendicular to the magnetic field is applied to the sample. If the resonance condition (1) is met, energy is absorbed from the applied alternating field,

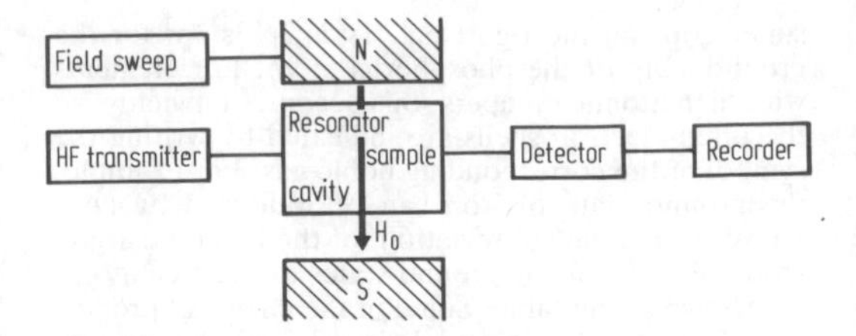

Fig. 1. Diagram of an ESR spectrometer

and this absorption is registered by a sensitive detector. After appropriate amplification it is displayed either on an oscilloscope or with a pen. In E., the differentiated signal (first derivative of the absorption signal) is usually displayed (Fig. 2).

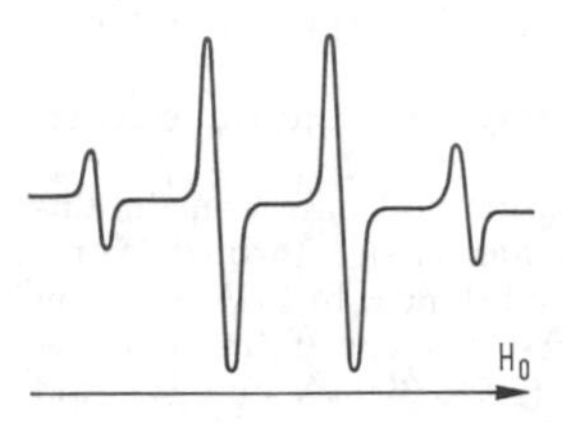

Fig. 2. ESR spectrum (derivative) of the methyl radical

Substances in any physical state can be examined by E., provided that they contain unpaired electrons. Such substances include atoms or ions with unfilled inner electron shells (e.g. ions of the transition metals), atoms or molecules with odd numbers of electrons (e.g. NO or NO_2), free radicals, metals, semiconductors and imperfections in solids. Since most substances are diamagnetic (see Magnetochemistry), it is important that many of them can be made tractable to ESR measurements by special methods of generating radicals (e.g. by irradiation or electrolysis).

Spectral parameters and their information content. The position of the signal is determined by the g factor in addition to the instrumental parameters ν and H_0. For a "solvated electron", such as is present, for example, in the dark blue solution of an alkali metal in liquid ammonia, the g factor is 2.00232. In any chemical compound, the unpaired electron is subject to certain internal fields which are superimposed on the H_0 field. Therefore the g factor of a chemical compound differs from that of the free electron and is a substance-specific parameter which is formally equivalent to the chemical shift in NMR spectroscopy. g factors can thus provide information on structure. The g factor of a symmetric line is measured at the position in the spectrum at which the absorption curve has its maximum, i.e. at the point at which the first derivative is 0. If the resonance frequency is known, the g factor is obtained by measurement of the magnetic field strength H_0. However, absolute measurements are usually dispensed with and relative measurements are made against a substance with a known g factor, e.g. the stable diphenylpicrylhydrazyl

(DPPH) radical with a g factor of 2.0036. The g factor, which is a dimensionless parameter, is then determined using the following relation: $g(\text{sample}) = g(\text{DPPH})\ H(\text{DPPH})/H(\text{sample})$, where H(DPPH) and H(sample) are the resonance absorptions of DPPH and the sample on the magnetic field scale. In organic radicals, the g factor is usually only slightly different from that of the free electron (Table). This indicates that in organic compounds there is practically complete suppression of the orbital component of the magnetic moment. If the unpaired electron is completely or partially located on a heteroatom, the g factor differs somewhat more from that of a free electron. As the table shows, this permits carbon, oxygen and nitrogen radicals to be distinguished. Paramagnetic metal ions can have g factors which differ considerably from that of the free electron. For lanthanoid ions, for example, g factors up to 18 have been found. This indicates a strong spin-orbital coupling in these atoms. g factors measured in liquids are isotropic, but if single crystals are examined, the g factors depend on the orientation of the crystal with respect to the field direction.

g factors of organic radicals

$CH_3\cdot$	2.0026	$CH_2{=}CHCH_2\cdot$	2.0025
$C_2H_5\cdot$	2.0026	$CH_2{=}C{-}O$	2.0045
$CH_2{=}CH\cdot$	2.0022	R–N(–O)–R	2.005...2.009

Hyperfine structure. The greatest analytical significance of E. is derived from the interactions of unpaired electrons with the magnetic moments of nuclei, which result in hyperfine structure (hfs). The electron may interact with the nucleus to which it belongs or with those of neighboring atoms. The distances between these hyperfine lines, their number and intensities give important information on the structures of radicals and paramagnetic centers.

Nuclei with magnetic moments which give rise to electron-nuclear spin interactions include ^{1}H, ^{14}N, ^{31}P and ^{77}Se. In analogy with the spin or orbital moments of the electron, they can orient in one of $2I+1$ directions relative to an external magnetic field. In the simplest case, an unpaired electron ($S = 1/2$) interacts with a nucleus with nuclear spin quantum number $I = 1/2$, e.g. the ^{1}H nucleus. Hydrogen nuclei can take one of two different orientations in an external field, with nearly equal probabilities. This means that about half of the hydrogen nuclei point in the same direction as the external field, and the other half point in the opposite direction. An unpaired electron is thus subject not only to the external field, H_0, but to the additional field from the nucleus, so that the actual field felt by the electron is H_0+H_{nucleus} or $H_0 - H_{\text{nucleus}}$, and the energy of the Zeeman level is split into two levels which depend on the nuclear orientation. Transitions can occur between these energy levels in the presence of radiation of the appropriate frequency (eq. 1). In contrast to the ESR spectrum of the free electron, which consists of a single signal, there is now a line splitting. Only those transitions are allowed which correspond to the selection rules $\Delta m_s = \pm 1$ and $\Delta m_I = 0$.

The simplest case of an hfs is found in the hydrogen atom, for which the term scheme is shown in Fig. 3. It can be seen that two transitions fulfill the selection rules, and therefore two hfs lines appear; their separation is given either in mT (milli-Tesla) or Hz and is called a coupling constant. The coupling constant is a measure for the strength of the interaction between the electron and nucleus, and is independent of the external magnetic field.

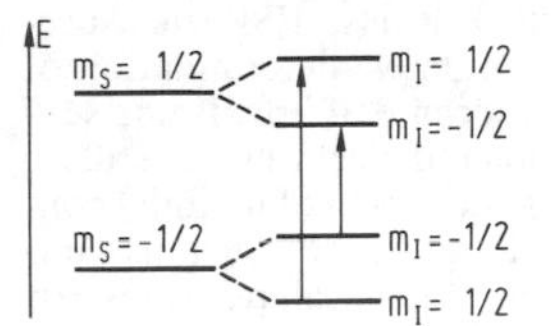

Fig. 3. Hfs scheme for $S = ½$ and $I = ½$

In general, the following rules apply to hfs splitting: 1) if the electron spin S interacts with a nucleus with spin I, the resulting multiplet has $2I+1$ lines. 2) If S interacts with n equivalent nuclei of type A and m nuclei of type B which are not equivalent to the first, the total number of lines is $(2nI_A+1)(2nI_B+1)$. 4) For nuclei with $I = 1/2$, the relative intensities within the multiplet are in the ratio of the binomial coefficients. *Example*: the methyl radical has an especially simple ESR spectrum consisting of 4 lines. According to rule 2, these would be expected from the interaction of S with 3 equivalent H nuclei ($I = 1/2$), and their relative intensities, in accordance with rule 4, are 1:3:3:1. The coupling constant derived from the distance between two adjacent lines is 2 mT.

In single crystals, the hfs factor, like the g factor, is anisotropic.

Signal intensities. The surface of the absorption signal is proportional to the number of unpaired electrons in the sample. However, since many instrumental factors affect the observed intensities, absolute measurements are essentially never made. Instead, relative measurements are made by measuring the surface under the absorption signal of the sample and comparing it with that of a standard (e.g. DPPH) with a known spin. Since the first derivative of the signal is normally shown in the ESR spectrum, the recorded curve must be integrated twice to obtain the surface area.

Applications. Determination of the structure of free radicals. The number of hfs lines and their relative intensities give an indication of the number and equivalence of the coupled nuclei. More information on the structure of radicals is obtained if the hfs coupling constant is also taken into account. The coupling constant is proportional to the probability of finding the electron on the nucleus in question. Since this probability can be calculated by quantum mechanical methods, there is a direct relation between E. and theory. With hfs, it is possible to demonstrate experimentally the delocalization of the unpaired electron often observed in free radicals. The unusual stability of the DPPH radical, for example, is ascribed to *resonance stabilization* in which the unpaired electron is located with equal probability on either of the two nitrogen atoms. If only one of these structures were present, electron-nuclear interaction with this N atom (I = 1) would lead to $2I+1$ = 3 hyperfine lines. If the unpaired electron were unevenly distributed between the two N atoms, $(2I_A+1)(2I_B+1)$ = 9 lines would be expected. The observed 5 equidistant lines with relative intensities of 1:2:3:2:1 can only be explained by an equal distribution over both N atoms, because then the rules predict $2nI+1$ = 5 lines.

Detection and quantitative determination of radicals. Because the signal intensity is proportional to the number of unpaired electrons, ESR measurements can be used for quantitative determination of paramagnetic substances. Because of its extreme sensitivity – as few as 10^{11} unpaired electrons can still be detected – E. is suitable for detection of minimal concentrations of paramagnetic substances, such as radical intermediates in certain chemical reactions, or also in enzyme systems or in living tissues. Short-lived radicals can be studied by flow techniques. Here the reactants which produce the radicals are mixed immediately before the solution flows into the sample cell, which is built as a flow-through cell. In this way, radicals with lifetimes in the ms range can be detected. The "*spin trapping*" technique is also useful for identification of the unstable radicals produced in reactions. Here the reaction is carried out in the presence of radical traps (e.g. aromatic or aliphatic nitroso compounds) which yield longer-lived nitroxide radicals. The ESR spectra of these radicals give information on the unstable radical, and the kinetics of the reaction can be elucidated by following the course of concentration changes in radical reactions.

Applications in inorganic chemistry. Ions and complexes of the transition metals and rare-earth metals contain unpaired d and f electrons, so they can be studied by E. Their spectra are more difficult to interpret, however, than those of free radicals. Coulombic repulsion between electrons and spin-orbital coupling produce g factors which are very different from that of the free electron. Measurements are usually made in a suitable diamagnetically diluted single crystal to avoid undesired interactions between the paramagnetic compounds. The g factor varies as a function of the orientation of the complex in the external field; for Fe^{3+} in hemoglobin, for example, it is 6 if the external field lies in the plane of the porphyrin ring, and 2 if the external field is perpendicular to the ring. The most important information which can be obtained concerns 1) the symmetry and strength of the ligand field, 2) the type of electronic ground state and 3) the delocalization of the unpaired electrons in the ground and excited states. Numerous studies have also been published on the valence states of the foreign atoms in semiconductors and on ferromagnetic compounds.

Studies of solid surfaces. E. is being increasingly used to study the surfaces of solids and reactions which occur on these surfaces, e.g. after absorption of certain atoms or molecules. Valuable information can be obtained on the presence of paramagnetic surface centers, the structure of radicals or transition metal complexes which may be formed, electron transfer between adsorbate and adsorbant, etc. These are especially significant in relation to the problems of adsorption and heterogeneous catalysis.

Electroosmosis: an electrokinetic phenomenon which occurs during electrolysis through a porous system, such as a diaphragm. It is caused by an Electrochemical double layer (see) which forms in the pores of the system. The water migrates in a direction determined by the charge excess on the liquid side of the double layer; usually into the cathode area, but sometimes into the anode area. E. is used to desalinate seawater, to dry wet mortar, etc.

Electroosmotic process: see Desalinization.

Electropherography: see Electrophoresis.

Electrophilic reactions: reactions of electrophilic reagents with nucleophilic substrates. There are two basic mechanisms, substitution by an S_E mechanism (see Substitution 2.) and electrophilic addition (A_E). The mechanism of the ***A_E* reaction** is analogous to electrophilic substitution, except for the stabilization of the carbenium ion. The electrophilic reagent forms a π-complex with the substrate which rearranges, in the rate-determining step, into a carbenium ion. Addition of a nucleophilic partner stabilizes the carbenium ion, and forms the product:

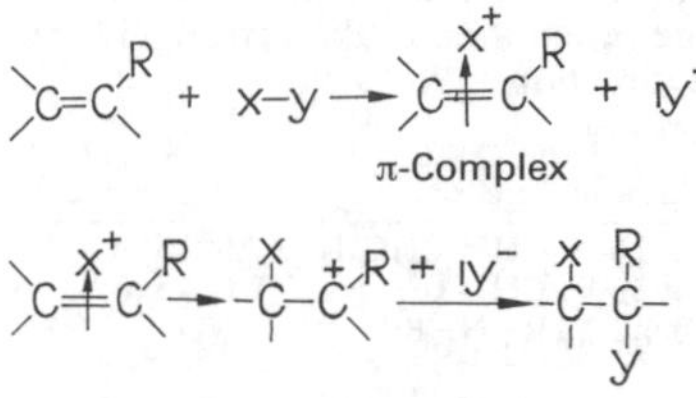

π-Complex

Electrophilic addition of protonic acids to unsymmetric alkenes is regioselective, and subject to the Markovnikov rule (see). A_E reactions are often stereoselective, giving *trans*- or *cis*-addition, although the carbenium ion postulated as an intermediate normally does not react stereoselectively. This may be explained by formulation of bridging symmetric or asymmetric cations:

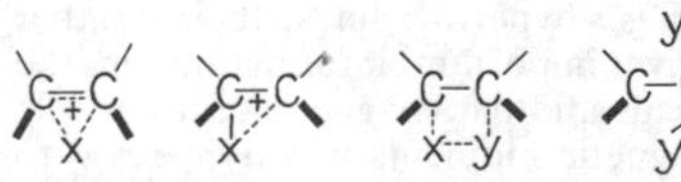

If the bridged cation is more stable than the carbenium ion, nucleophile Y can only attack from the back, as in the S_N-2 mechanism, and the product is the *trans*-adduct. If the carbenium ion is more stable (e.g. in addition of halogens to the alkene), there are two possible cases: 1) The carbenium ion forms a four-membered transition state with the reagent, which is transformed by heterolysis to a short-lived internal ion pair. The reaction is stereospecific as a *cis*-addition (e.g. hydroboronation of alkenes). 2) The carbenium ion has a sufficient lifetime so that the nucleophile can add to either side, giving a thermodynamically controlled product mixture (e.g. addition of water, alcohols or carboxylic acids to alkenes).

Which variant of the mechanism will occur depends on the structure of the substrate and the nature of the reagent. In the addition of halogens to cycloalkenes of medium ring size, 1,3- and 1,5-hydride shifts can lead to transannular addition. *cis*-Alkenes are normally more reactive in A_E reactions than the *trans*-isomers, due to their higher energy contents. However, steric effects of neighboring groups can affect the mechanism. Some important A_E reactions are addition of hydrogen halides, sulfuric acid, water, alcohols, halogens, interhalogens, hypohalogenous acids, oxygen (see Epoxidation and Hydroxylation), ozone (see Ozonization), and carbenes and cationic polymerization. Under certain conditions, if more than one nucleophilic reagent is present in the reaction mixture (e.g. reactions with chlorine and water), mixed adducts are formed. In the Ritter reaction, nitriles act as nucleophiles in acidic solution, and acylamines are formed. Friedel-Crafts alkylation is an electrophilic addition of aromatic compounds to alkenes, and the Prins reaction is the acid-catalysed electrophilic addition of formaldehyde to alkenes.

Electrophilic addition to dienes occurs by a similar mechanism, forming 1,2- and 1,4-addition products. The 1,4-adduct is normally more stable than the 1,2-adduct. However, the 1,2-adduct is formed preferentially in kinetically controlled reactions (e.g. addition of hydrochloric and hypochlorous acids). Some typical electrophilic 1,4-additions are addition of bromine or chlorine and the Diels-Alder reaction (see). Electrophilic 1,4-additions are also observed in quinones, e.g. HCl addition to p-benzoquinone to form 2-chlorohydroquinone. Alkynes are less reactive in A_E-reactions, due to the greater electronegativity of the sp-hybridized C atoms. The mechanism is comparable to that of electrophilic addition to alkenes, but in some cases cannot be clearly classified. Bromine, hydrogen halides and carboxylic acids, for example, add to alkynes.

Electrophilic reagent: an electron-pair acceptor (EPA) which can react with nucleophilic substrates. Acids and Lewis acids, positive ions (e.g. H^+, NO_2^+, SO_3H^+, X^+ where X is a halogen, R^+ and RCO^+), compounds with electron deficiencies (e.g. ozone, free oxygen or oxygen in peroxo acids or peroxides), ethyne and carbonyl compounds are examples of E.

Electrophilic substitution: see Substitution, 2.

Electrophilicity: the electron-seeking property of an electrophilic reagent; the ability to accept an electron pair from a nucleophilic partner, thus forming a bond with this nucleophile. E. is directly related to electron affinity.

Electrophoresis: in the broadest sense, the separation of charged particles of a solution on the basis of their different rates of migration in an electric field. The migration velocity v is the product of the net mobility m of the ions and the field strength E: $v = m \cdot E$. If two electrodes are connected by a continuous, conductive electrolyte (***electrophoretic system***), the field strength is constant within the system, and the migration rate is proportional to the net mobility of the particles. Since this parameter depends, under defined electrophoretic conditions, on the size, shape and charge of the particles, different types of particles can be separated from each other. In the narrow sense, E. is the migration of colloidal particles or macromolecules, while ***ionophoresis*** is the migration of smaller, inorganic ions. The migration of the particles in the electric field is countered by diffusion; its effect is inversely proportional to the particle size and electric field. As a result, the separation of small ions requires high voltages (1000 to 10,000 V), while large particles can be separated by ***low voltage E.*** (300 to 400 V). The migration rate of the ions is also highly dependent on the degree of dissociation, and thus the pH, of the solution. When the total charge on a particle is reversed due to a change in pH, the direction of its motion also reverses. For every substance, there is an isoelectric point at which its mobility is zero; many macromolecules precipitate at their isoelectric points. By a suitable choice of buffer, the mobility and solubility of the substances to be separated can be adjusted for optimum results.

E. can be carried out in free solution or in carrier materials such as starch, PVC powder, cellulose (paper), polyamide-agarose or dextran gels (***electropherography, electrochromatography***); here the electrophoresis is coupled with a chromatographic separation. In ***gel electrophoresis***, the charged molecules are driven through a porous gel by an electric field in a buffer solution. By varying the gel composition and the buffer, it is possible to separate the molecules on the basis of their charges, their isoelectric points, their net mobility (see Isotachophoresis) or their biospecific affinity (***affinoelectrophoresis***). In ***pore gradient electrophoresis***, the pore size in the gel decreases continuously in the direction of electrophoretic migration, so that molecules (such as natural proteins) are separated mainly on the basis of their sizes. The gel used in ***disc electrophoresis*** is discontinuous with regard to pH and pore size. A large-pore gel is used at the beginning of the gel to concentrate the sample in a narrow starting zone, and a small-pore gel is used for the separation (*permeation effect*). This affords a high separation capacity. In ***SDS electrophoresis***, the detergent sodium dodecylsulfate (SDS) is added to the sample. The negative charges of the SDS molecules completely mask the natural charge of the molecule, so the separation of the molecules is determined solely by their relative sizes.

Methods of E. Paper E. is very common; the E. chamber consists of two electrode chambers and a horizontal raised support for the paper strip. The sample is applied to the carrier at a point or along a strip, with the position determined by its expected direction of migration (Fig.).

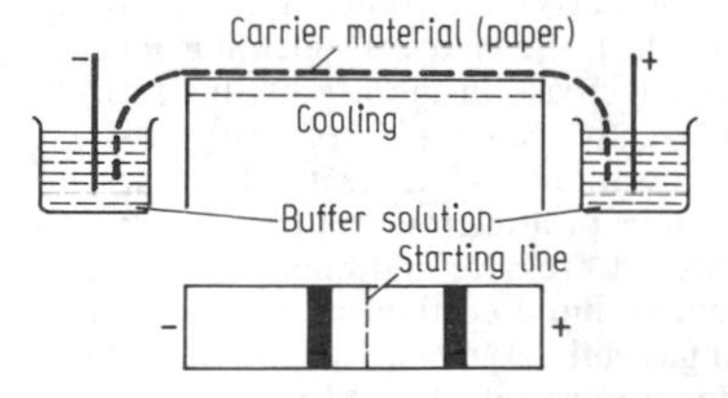

Electrophoresis apparatus and electropherogram

Qualitative and quantitative analysis of the paper strip is done by the same methods used in thin-layer or paper chromatography.

Applications. E. is used to separate charged particles which are stable in aqueous solutions, e.g. in the routine examination of biological fluids (urine, serum), for fractionation of nucleic acids, for characterization of enzymes and other proteins, and to determine molecular properties (isoelectric point, electrophoretic mobility, approximate molecular mass).

Electrophoretic effect: see Debye-Hückel theory.

Electrophotography: a widely used method of reproduction. The light-sensitive material consists of an electrically conducting base with a photoconductor layer above it. (In *xerography*, the photoconductor is a vapor-deposited layer of selenium; in the *electrofax process*, it is a layer of zinc oxide.) The photoconductive layer is electrostatically charged in the dark, leading to accumulation of the opposite charges on the boundary with the base. When light shines on the layer, the resistance of the photoconductor is reduced by 3 to 4 orders of magnitude, and a latent image is formed as a charge density profile in the photoconductor. The visible image is created by applying electrically charged particles (toner), which are distributed according to the charge density profile. This toner image is then transferred to a dieletric substrate (foil, normal paper) and fixed to the substrate, usually by means of heated rollers. Depending on the polarity of the toner charge, the image can be positive or negative.

Electroplating: a galvanic method of producing metallic Protective coatings (see) by electrolytic precipitation of the metal on the surface of the object to be coated, which serves as the cathode. A salt of the metal to be precipitated is used as electrolyte. If the plating metal is used as anode, the metal ion concentration in the bath remains approximately constant. Direct current is used, usually with a voltage of a few volts. Galvanic protective layers must be firmly bound to the underlying metal, which can often be achieved only by plating an intermediate layer of a third metal. In addition, the protective layer must be non-porous.

Electrostatic gas purification, *Cottrell-Müller process*: a method for removal of particulates and tar from gases. E. is used to clean waste gases (smoke and poison gases, cement industry waste gases), to remove tar from coking and synthetic gases, and to remove liquid droplets from waste gases (e.g. acid mists). E. works with a high-voltage constant current at 20 to 70 kV. The spray electrode is used to give the liquid or solid particles a negative charge, so that they migrate to the positive flat or cylindrical electrode, where they are discharged and precipitate. This precipitation electrode is cleaned by a beating mechanism. The steady increase in the use of this process (often, but incorrectly, called *electrofiltration*) is explained by its versatility and capacity. Particles no larger than fractions of a micrometer can be precipitated at little loss of pressure and high pass-through rates (up to several hundred thousand cubic meters per hour) from gas with large amounts of particulates and at high temperatures. 99.9% or more of the particulates can be precipitated.

Electrostatic separation: the separation of mineral mixtures into their components in an electric field. Differences in the surface conductivity, grain boundary resistance or dielectric constants causes particles which are introduced into an electric field to be deflected from their original direction of motion by different degrees. In 1) ***high-field separation***, ionization occurs in the spray field due to corona effects, and separation is based on the differences in surface conductivity. In 2) ***low-field separation***, separation occurs in the field of a capacitor on the basis of different dielectric constants. E. can be done in a *chamber* or *rotating drum separator* (Fig.). In the rotating drum separator, the particles are polarized in an electric field. As they leave it, the conductive particles lose their dipole moments immediately, and fall off the drum immediately, due to gravity. The less conductive particles retain their dipole moment and cling to the drum longer. E. requires that the material have a very low water content, because surface moisture practically equalizes any differences in surface conductivity.

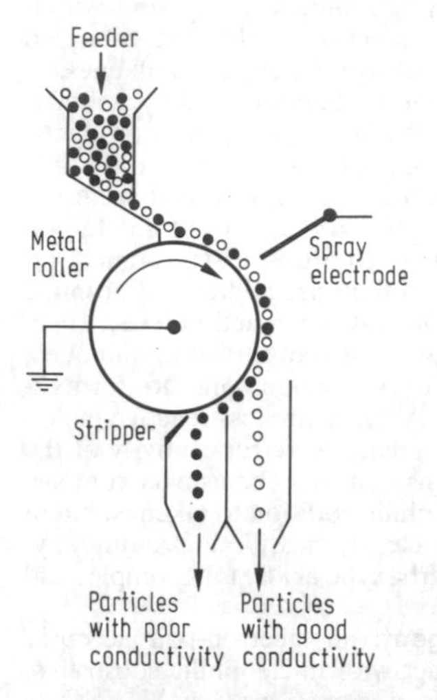

Electrostatic rotating drum separator

E. is used in processing potassium chloride, sodium chloride, iron ores and heavy mineral sands which contain diamonds, and to separate plastics.

Electrosynthesis: same as Electrolysis (see).

Eledoisin: see Tachykinins.

Element: a chemical substance in which all the atoms have the same nuclear charge. This definition has supplanted the classical definition of an E. as a substance which cannot be decomposed into simpler substances by chemical or physical means. The E. are characterized by the atomic number Z and the mass number M. They are arranged in order of increasing Z in the Periodic system (see); depending on their position in this system they are classified as ***main group*** or ***transition*** E.

The abundances of the E. in the earth's crust, the outer stony layer of about 16 km thickness, vary enormously. The E. with even atomic numbers are generally more abundant than those with odd atomic numbers (*Harkins' Rule*). Oxygen (50.5%) and silicon (27.7%) are the most abundant, followed by aluminum (8.1%, iron (3.38%), calcium (3.4%), sodium (2.63%) potassium (2.59%), magnesium (1.93%), hydrogen (0.81%), titanium (0.43%), chlorine (0.19%) and carbon (0.12%). The other E. are present in the following abundance ranges:

10^{-1} to 10^{-2}%: P, F, Mn, S, Ba, Sr, Zr, Rb;
10^{-3} to 10^{-3}%: V, Ce, Cr, Ar, Zn, La, Ni, Li, Yb, Y, Nd, B, Nb, Ga, Pb, Sc, Th, Co;
10^{-3} to 10^{-4}%: Br, Sm, Gd, Pr, Dy, U, Er, Sn, Cu,

Ta, Hf, Cs, As, Be, Ho, W, Ge, Tl, Eu, Tb, Mo, Lu;
10^{-4} to 10^{-5}%: I, Sb, Tm, Se, Hg, Cd;
10^{-5} to 10^{-6}%: Bi, In, Ne, Ag, Kr;
10^{-6} to 10^{-7}%: Pt, Pd, He, Au, Te;
10^{-7} to 10^{-8}%: Re, Ir, Os, Rh, Ru, Xe;
10^{-10} to 10^{-11}%: Ra, Pa;
10^{-14} to 10^{-16}%: Ac, Po, Np, Pu;
10^{-19} to 10^{-21}%: Rn, Fr.

Of the 109 known E., 19 are found ***isotopically pure*** in nature, i.e. all the atoms have the same proton and neutron numbers. This group consists of sodium, cesium, aluminum, phosphorus, arsenic, bismuth, fluorine, iodine, scandium, yttrium, niobium, manganese, cobalt, rhodium, gold, praseodymium, terbium, holmium and thulium. All other E. are found as ***isotopic mixtures***; the atoms all have the same number of protons, but varying numbers of neutrons (see Isotopes). The atomic masses of these E. are determined by the relative abundances of the naturally occurring isotopes. For each E. there is a chemical Symbol (see). A nucleus of a particular type (see Nuclides) is indicated by a mass number written as a superscript to the left of the chemical symbol. The heavy E. with Z > 83 exist only as unstable, radioactive isotopes (see Radionuclides). The E. with Z = 43 (technetium) and 61 (promethium), as well as the transuranium E. (Z > 92) are obtained only by nuclear reactions; likewise the E. astatine and francium were first obtained by nuclear reactions. Such nuclear reactions are now an important and widely used method for *transmutation* of E. Synthesis of E. by nuclear fusion is the source of stellar energy, e.g. the fusion of hydrogen to helium in the sun, which is catalysed by carbon and nitrogen. In addition to the above highly radioactive E., nuclides of lower activity and longer half-lives are known for many E.

The discovery of more new E. by nuclear reactions with highly accelerated heavy ions is expected; it is thought that superheavy E. with atomic numbers from about 110 to 114 will be relatively stable.

Elemental analysis: process of determining the elemental formula of an organic compound. E. may be qualitative or quantitative.

1) ***Qualitative E.*** is sometimes done as a preliminary to quantitative E. Carbon and hydrogen are detected by heating the sample with powdered copper oxide; the resulting carbon dioxide is passed through barium hydroxide solution and precipitates as barium carbonate. Hydrogen is recognized by the formation of water droplets in cooler parts of the apparatus. Nitrogen, sulfur and halogens are the elements other than carbon, hydrogen and oxygen which are most likely to be present in an organic molecule. They are detected by heating a small sample of the unknown substance with a pea-sized piece of sodium (or potassium) in a melt-sealed reagent tube to red heat. This is then placed in a beaker of cold, distilled water; the reagent tube breaks and the reaction products dissolve. Part of this solution is tested for nitrogen in the form of sodium cyanide by means of the Lassaigne test (see). Other portions of the solution are tested for sulfur by addition of a soluble lead salt and for halogens by addition of silver nitrate. Another test for halogens is the Beilstein test (see).

2) ***Quantitative E.*** is of much greater importance; it goes back to Liebig. In the course of it, all covalent bonds of the organic molecule are cleaved, and the elements which comprise it are quantitatively converted to inorganic substances which can be measured (mineralization of the organic substance), usually by combustion of the sample. The amounts of combustion products are converted to the masses of the elements in the sample and related to the weighed amount of sample, yielding the percent mass fractions of the corresponding elements. The results of an E. could be, for example, 52.1% C, 4.42% H, 20.32% N, 23.17% O. These values are divided by the corresponding atomic masses:

$$\frac{52.15}{12.011} : \frac{4.42}{1.008} : \frac{20.32}{14.007} : \frac{23.17}{15.9994}$$

= 4.34:4.39:1.45:1.45. Allowing for the limits of error of the determination, these values indicate the integral ratio of atoms and the relative formula $(C_3H_3NO)_x$. The value of x must be determined by an independent experiment; if in the above example the result were 138.3 g mol^{-1}, the summary formula of the sample would be $C_6H_6N_2O_2$. Quantitative E. is now usually carried out on a micro-scale. The most important methods are the following:

a) Determination of carbon and hydrogen. 2 to 4 mg of the sample is burned in a stream of oxygen in a quartz tube. To insure complete oxidation, catalysts (copper(II) oxide, cobalt(II,IV) oxide, etc.) are placed in the front part of the tube. Interfering combustion products, such as sulfur dioxide, halogens or nitrogen oxides, are removed by suitable absorption materials. The remaining products, carbon dioxide and water, are usually determined gravimetrically by absorption on magnesium perchlorate or sodium hydroxide (see Ascarite) on a carrier. The absorption agents are in special absorption tubes, which are weighed before and after the combustion. Gas chromatographic separation of the combustion products has largely replaced the above method; the products are detected by heat conductivity, IR photometry or other physical chemical methods. This greatly reduces the time required, and a number of sources of error are removed; in addition, the process can be automated.

b) Determination of oxygen is an underdeveloped method of E., and the content of oxygen is often inferred by difference from the values determined for all the other elements. Direct determination is possible by "cracking" the organic substance in a stream of nitrogen; the oxygen is quantitatively converted to carbon monoxide on a carbon contact at a temperature of 900 to 1100 °C.

Determination of nitrogen. The organic substance is thermally decomposed by the Dumas method (see); the Kjeldahl nitrogen determination (see), in which the substance is decomposed with sulfuric acid and catalysts, can also be used for some substances.

d) Halogens are determined after oxidative or reductive removal of the halogen from the organic molecule. Oxidative methods are combustion in an oxygen stream with a platinum catalyst, combustion in a hydrogen-oxygen flame, treatment with sodium peroxide, combustion in an oxygen-filled flask or a wet method using nitric acid (see Carius method) or chromosulfuric acid. Some reductive methods are combustion in a hydrogen stream on a catalyst or solubilization with sodium, potassium or magnesium. Both oxidative and reductive methods are carried out in bombs (see Solubilization). There are many methods available for determination of the halogens: gravimetric, acidimetric, argentometric, iodometric, mercurimetric, potentiometric or amperometric.

e) Determination of sulfur. The solubilization methods are similar to those used for halogen determination. Sulfur is either present as sulfate after the solubilization, or it is converted to sulfate. It is usually determined gravimetrically (precipitation as barium sulfate) or volumetrically. More recently, commercial multi-element analyzers are used, which in some cases are completely automated, from injection of the sample to print-out of the analysis. By electronic control of all analysis parameters, they achieve a high degree of consistency in the working conditions, and thus of reproducibility. Subjective sources of error are largely removed, and the time required is considerably reduced. The organic sample is mineralized by dry pyrolysis, and the combustion products are separated by chromatography or selective absorption. Carbon, hydrogen, nitrogen, etc. are usually detected by heat conductivity (static or dynamic).

Elementary process: in kinetics, a term for the interaction between single particles in defined excitation states. An E. can be an elastic (exchange of translational energy), inelastic (excitation of internal degrees of freedom) or reactive collision. In contrast to E., an elementary reaction (see Reactions, elementary) is a macroscopic single-step reaction.

Elemi: term for various resins, most of them from Burseraceae. The most commonly used is ***Manilla elemi***, a cloudy, soft, doughy mass with a greenish or yellowish white color which is soluble in many organic solvents. E. is used to make salves for stimulating the skin. It was formerly used in making varnishes and paints, especially as a whitener for spirit and nitro lacquers.

Elgiloy®, ***Cobenium®***: a cobalt alloy which is resistant to corrosion, has a high elasticity modulus and is resistant to metal fatigue. It is used for the springs of watches.

Elicitors: see Phytoalexins.

Elimination: splitting off of two atoms or groups from an organic compound, forming an unsaturated compound. E. is the reverse of an addition reaction. In an α-E., the two leaving atoms or groups leave from the same carbon atom, while in β-E., they leave from adjacent carbon atoms. ***α-E.*** occurs in the presence of very strong bases, which first abstract a proton. The remaining carbanion then eliminates an anion X^-, forming a carbene:

$$\text{H-}\overset{|}{\underset{|}{\text{C}}}\text{-}\overset{|}{\underset{\text{H}}{\text{C}}}\text{-X} + |\text{B}^{\ominus} \underset{}{\overset{-\text{HB}}{\rightleftharpoons}} \text{H-}\overset{|}{\underset{|}{\text{C}}}\text{-}\overset{|}{\underset{\ominus}{\text{C}}}\text{-X} \xrightarrow{-\text{X}^{\ominus}} \text{H-}\overset{|}{\underset{|}{\text{C}}}\text{-}\overset{|}{\underline{\text{C}}} \xrightarrow{\sim\text{H}} \text{>C=C<}_{\text{H}}$$

Carbanion Carbene

If there is an adjacent C-H bond in the molecule, the carbene can rearrange into an alkene, so that in this case the product is the same as in a β-E.

β-E. is an ionic 1,2-elimination (see Fragmentation). It can occur by various mechanisms, the individual steps of which are analogous to the S_N mechanisms (see Substitution 1). In *monomolecular* or *E1 E.*, which occurs by the *E1 mechanism*, the rate-determining step is the formation of a carbenium ion by heterolysis of a C-X bond. The carbenium ion can be stabilized either by formation of the substitution product (S_N1 mechanism) with the base Y^-, or by abstraction of a proton from the β-C atom by the base Y^-:

$$\text{H-C-C-X} \rightleftharpoons |\text{X}^- + \text{H-C-C}^+ \begin{cases} \xrightarrow{+\text{Y}|^-} \text{H-C-C-Y } (S_N1) \\ \xrightarrow{+\text{Y}|^-} \text{>C=C<} + \text{HY (E1)} \end{cases}$$

The E1 mechanism is promoted by special steric and electronic conditions (steric pressure from substituents promoting the formation of a planar carbenium ion and stabilization of the carbenium ion by electronic effects of substituents) and by high temperatures. It is observed, for example, in the dehydration of *sec.* and *tert.* alcohols in solution and in the gas phase, in the dehydrohalogenation of *sec.* and *tert.* alkyl halides, in the solvolysis of sulfate and sulfonate esters and in the deamination of amines with nitric acid. The E1 mechanism is less suitable for syntheses, because it is an equilibrium reaction, while the competing substitution reactions are usually irreversible.

Bimolecular or *E2 E.* occurs by the *E2 mechanism*, in which a proton is abstracted and a substituent on the adjacent C atom is simultaneously forced out of the molecule. High concentrations of strong bases (e.g. R_3N, $C_6H_5\text{-}O^-$, HO^-, RO^-, H_2N^-) promote the E2 mechanism:

$$\text{Y}|^- + \text{H-C-C-X} \begin{cases} \rightarrow \text{Y}\cdots\text{C}\cdots\text{X} \rightarrow \text{Y-C} + \text{X}|^- \; (S_N2) \\ \rightarrow \text{Y}\cdots\text{H}\cdots\text{C-C}\cdots\text{X} \rightarrow \text{Y-H} + \text{>C=C<} + \text{X}|^- \; (\text{E2}) \end{cases}$$

The competing reaction is an S_N2 reaction. Which of the two is favored depends on the basicity and nucleophilicity of the reagent Y^-; basicity promotes the E2 E. and nucleophilicity promotes the substitution. In contrast to the E1/S_N1 pair of mechanisms, the E2/S_N2 pair has different transition states. The E2 E. is irreversible. The leaving groups are

$-\overset{+}{N}R_3$, $-\overset{+}{P}R_3$ and $-\overset{+}{S}R_2$

(e.g. in the Hofmann degradation of ammonium, phosphonium and sulfonium hydroxides), -Cl, -Br, -I (e.g. in the dehydrohalogenation of alkyl halides) and $-OSO_2R$ (e.g. in alkene formation from esters of sulfuric and sulfonic acids). Which of the possible competing reactions (E1, E2, S_N1, S_N2) occurs depends on structural, steric and electronic factors. The formation of alkenes by E. generally increases on going from primary to secondary to tertiary alkyl compounds. Steric factors in the base also play a role. Bases with voluminous groups, e.g. ethyldicyclohexylamine, are more inclined to E. than to substitution, because the abstraction of a proton from the periphery of a molecule is not greatly influenced by steric factors. In primary alkyl compounds, E. leads exclusively to the alkene with the lowest number of alkyl groups (Δ^1 alkene; *Hofmann elimination product*), but in E. of secondary and tertiary alkyl compounds, several isomeric alkenes are often produced. In such cases, the most stable alkene with the largest number of alkyl groups (Δ^2 alkene) is the preferred product formed by cleavage of the proton from the C atom with the fewest H atoms (*Zajcev product*):

β α β
-HX
or
(Hofmann product)
(Zajcev product)

E1 E. yield mainly Zajcev products. In E2 E., the ratio of products depends on various factors; the Zajcev product is preferred in compounds with groups with high leaving tendency, and the Hofmann product formation is facilitated by the difficulty of cleaving a proton out of the middle of the molecule, which is determined by steric effects of the base.

E2 E. are generally stereospecific ***trans-E. (anti-E.)***, while the ionic E1 E. are not sterically uniform. In *trans*-E., atoms or groups *trans* to one another are stereospecifically split off. According to the Ingold rule (see), E2 E. are especially favored when the leaving substituents are completely *trans*, that is, when they have a staggered conformation. In E1 E., the planar carbenium intermediate makes cleavage from either a *cis* or a *trans* position possible. The application of these predictions to alicyclic compounds is governed by the *Barton rule*, which states that E2 E. from alicyclic compounds are especially favorable when both leaving atoms or groups are in an axial (*trans*) position, that is, are in a staggered conformation. On the other hand, E2 E. from equatorial *trans* positions are generally not possible. Compounds in which the leaving substituents are in a *cis* position react poorly or not at all. Leaving atoms or groups in an e,e-arrangement can be shifted to the alternative conformation, the a,a-arrangement, by an inversion of the ring:

In *trans*-E., the molecule passes through a transition state in which all four atoms involved in the reaction are in a plane (four-center principle). The reaction occurs in the way which requires the smallest changes in the positions of the substituents (*principle of least motion*).

cis-E. is a sterically uniform β-E. which can be ionic or nonionic. The two leaving atoms or groups depart from a *cis* position.

Nonionic cis-E. occurs under pyrolytic conditions and has a cyclic transition state. The breaking and formation of bonds are simultaneous:

The more strongly basic the atom Y which acts on the *cis*-H atom, the more easily this reaction occurs. These E. occur in xanthogenates (X = O, Y = S, Z = SR), urethanes (X = O, Y = O, Z = OR), carboxylic acid esters (X = O, Y = O, Z = R) and amine oxides (X-C = NR, Y = O^-, Z = R).

Ionic cis-E. occur when a nucleofugic substituent and the β H atom are fixed in an ecliptic arrangement (e.g. in rigid bicyclo[2.2.1]heptane or cyclopentane derivatives), or when the nucleofugic group has a low leaving tendency (trialkylammonium or dialkyl sulfonium groups). Steric factors such as bulky nucleofugic groups and bulky attacking bases and solvents which are only weakly dissociating promote *cis*-E.

Nucleophilic substitution of aromatic compounds occurs by various mechanisms in which E. appear as partial reactions (addition-elimination mechanism, elimination-addition mechanism; see Substitution 1.3). The condensation step which follows the addition step in the reaction of carbonyl compounds with bases and CH acidic compounds is an E.

Elimination-addition mechanism: see Substitution 1.3.3.

Ellipsometry: a method for study of extremely thin layers, preferably on solid carriers. Linear polarized light becomes elliptical polarized after reflection from a thin layer (thickness in the nanometer range). Elliptical polarized light arises from superposition of two mutually perpendicular, linear polarized, coherent beams which are out of phase with one another. The thickness or refractive index of the layer can be calculated from the ellipticity of the polarization. The lower limit of sensitivity for measurements of layer thickness is about 0.2 nm, and the upper limit is about 100 nm.

Eloxal process: see Anodic oxidation.

Eloxation: see Anodic oxidation.

Eluotropic series: see High performance liquid chromatography.

Elution method: a chromatographic method in which the elutant moves through the chromatographic separation material.

Elutriation: the separation of very fine solid particles from a granular mixture by sedimentation of the heavier components in water. The desired product can be either the sediment or the remaining suspended particles. In flotation, the finest particles are removed because they are not desired, while in kaolin preparation, it is precisely the fine kaolin particles which are valuable, and the coarser quartz and glimmer portions must be removed. The process of E. depends on the different times required for sedimentation of different sized particles. If several solids with different densities are present in the granular material, a sorting process also takes place. E. is carried out in concentraters, hydroseparators, screeners and other apparatus.

Elymoclavin: see Ergot alkaloids.

EMA: abb. for Electron beam microanalysis.

Emanation: see Radon.

Emeraldine base: see Aniline black.

Emetics: substances which induce vomiting. E. can act on the nervous system (e.g. apomorphine) or on the gastric mucosa (e.g. concentrated sodium chloride solution).

Emetine: an isoquinoline alkaloid. E. is the primary alkaloid in the Brazilian ipecac root (*Uragoga ipecacuanha*). It is present in the root at a concentration of 0.6 to 1.2%, and is accompanied by cephaeline and other ipecacuanha alkaloids. E. is an emetic and is used to treat amebic dysentery. The usual commercial form is the dihydrochloride, a colorless, crystalline powder.

EMF: abb. for Electromotive force; see Electrochemical equilibrium.

Emission: in the general sense, that which is sent out or given off. In a specific sense, 1) pollutants coming from a given source (see Air pollution). The source is usually a point source, such as a smokestack, but it can also be a surface source, such as a storage area. A quantitative estimate of an E. is often possible only for point sources. The E. can be given as *E. concentration* in g m^{-3} or as the *E. rate* in kg h^{-1}. The allowable E. for a factory or installation are prescribed by law as the MEC, or maximum allowable emission concentrations. The *E. conditions* are also regulated, such as the height of the source, the volume of waste gas flow, the presence of scrubbers, etc. The emittors are generally required to monitor themselves.

2) The spontaneous or induced production of wave or particulate radiation, such as light from a shining object, electrons from surfaces in thermal E. or field E., or α-, β- and γ-radiation from radioactive materials.

Emission damage: changes or damage to plants, animals and soil due to Air pollution (see), usually industrial air pollution, but also community smoke, exhaust and particulate emissions. The most damaging substances are sulfur dioxide and trioxide, hydrogen fluoride, hydrogen chloride, nitrous gases, lower hydrocarbons, ammonia, heavy metals and various particulates. Symptoms of E. in plants are reduced intensity of photosynthesis, and thus slower growth, discoloration of leaves or needles, accumulation of pollutants, and finally dying off of parts of the plant or the entire plant. Conifers are particularly susceptible. Animals react to these air pollutants with slower growth and lower productivity, e.g. in milk production. Damage to skeletal structure and infertility can also result. The effects on the soil are, for example, shifts in pH, enrichment of heavy metals and salt accumulation.

Emission spectrum: see Spectrum.

Emulsifier: see Emulsion.

Emulsin: a mixture of enzymes which can be extracted from almonds; it consists mainly of β-glucosidase. It splits the amygdalin of bitter almonds into glucose and mandelic nitrile.

Emulsion: a disperse system in which both the disperse phase and the dispersion medium are liquids. The most common combination is water with an organic liquid which has limited solubility in water; such E. are called ***oil in water*** (O/W) or ***water in oil*** (W/O) E. The relative lifetimes of the oil droplets in water and of the water droplets in oil determine which type of E. will form. If the lifetimes differ by an order of magnitude, the type of E. will form in which the longer-lived droplets are present. Since the lifetime of the droplets can be changed by addition of surface-active substances (***emulsifiers, surfactants***), the choice of a suitable compound is the most important step in creating an E. A simple criterion for deciding whether a particular surfactant will produce O/W or W/O E. is its relative solubility in the two phases. Surfactants can be empirically classified on the basis of their hydrophilic-lipid balance (HLB) values. ***Emulsification*** generally requires application of mechanical energy in the form of stirring, vibration or ultrasonic vibration; only in rare cases does it occur spontaneously. It requires the transport of a surface active material through a phase boundary, and turbulent flows result. F. can also be formed by polymerization and polycondensation processes.

Some ***natural E.*** are milk and latex; examples of ***artificial E.*** are foods (butter, margarine, mayonnaise), cosmetics, pharmaceuticals and pesticides.

Emulsion breaker: see De-emulsification.

Enamel: an oxide material on a metallic base. It serves to protect the metal from chemical attack and improves its appearance. According to Dietzel, an E. is defined as an inorganic, mainly glassy mass of chiefly oxidic composition applied to a metal base by complete or partial melting (fritting). It is applied in one or more layers, sometimes with additives. E. are multicomponent systems, consisting, for example, of silicon dioxide, SiO_2, borax, $Na_2B_4O_7 \cdot 10H_2O$, aluminum hydroxide, $Al(OH)_3$, calcium carbonate, $CaCO_3$, calcium fluoride, CaF_2, lithium carbonate, Li_2CO_3 and sodium carbonate, Na_2CO_3. It also contains coloring, binding and opacifying agents, such as iron(III) oxide, Fe_2O_3, cobalt oxide, CoO, nickel oxide, NiO, tin(IV) oxide, SnO_2, titanium(IV) oxide, TiO_2, and zirconium(IV) oxide, ZrO_2. These are melted together at temperatures between 1200 and 1400 °C. The glass frit is quenched in water or between cooled steel rollers and is then ground wet to a slip (***wet enamel***) or dry to a powder (***powder enamel***) which is melted onto the hot cast iron or steel object at 800-900 °C. The wet enamel is applied by spraying or dipping, or by electrophoresis or electrostatic methods. To mediate between the different thermal expansion of the metallic base and the surface layer, and to achieve bonding, a layer of ***base enamel*** is usually applied first to the metal, then covered with one or more layers of a ***coating enamel*** (multilayer enamelling). Nickeled steels can be covered with a single elastic coating layer (single-layer direct enamelling), which is resistant to bending and impact.

The top layer of coating enamel, which can be subjected to tempering to permit controlled, partial crystallization, is responsible for the quality of the mechanical, thermal and chemical resistance of the material.

E. are used to make household wares, automobile parts and large tanks for the chemical, food and beverage industries. Enamelled aluminum sheeting is used for decorative exterior siding and interior paneling in buildings. Enamelled objects for decoration are usually made with copper as the base metal (low-melting, borate-containing E.).

Enamines: α,β-unsaturated amines with the general formula $R_2C{=}CR\text{-}NH_2$. E. are formed by addition of amines to ethynes, or by the reaction of secondary amines with ketones or aldehydes with α-H atoms: $R\text{-}CH_2\text{-}CHO + R_2NH \rightarrow R\text{-}CH{=}CH\text{-}NR_2 + H_2O$. E. are very reactive and are used or obtained as intermediates in many syntheses, for example, in the synthesis of a number of heterocyclic compounds.

Enanthaldehyde, ***heptanal, n-heptyl aldehyde***: $CH_3\text{-}(CH_2)_5\text{-}CHO$, a colorless highly refractive liquid with an aromatic odor; m.p. - 43.7 °C, b.p. 152.2 °C. E. is barely soluble in water but dissolves very readily in most organic solvents, such as ethers, ethanol or chloroform. E. is very reactive, undergoing the reactions typical of Aldehydes (see). In the presence of hydrogen chloride, it polymerizes (when cooled) to *paraenanthal*. E. can be prepared by pyrolytic cleavage of ricinolic acid esters or by hydroformylation of hex-1-ene. E. is used, for example, to make *n*-heptanol or enanthic acid.

Enanthic acid: same as Heptanoic acid (see).

Enanthyl alcohol: see Heptanol.

Enantiomerism: see Stereoisomerism 1.1.

Enantioselective syntheses: see Asymmetric syntheses.

Enantiotopic groups: see Topic groups.

Enantiotropism: see Polymorphism.

Encron: see Synthetic fibers.

Endo: see Stereoisomerism 1.2.2.

Endogenous ligands: see Hormones.

Endopeptidases: see Proteases.

ENDOR (abb. for ***e***lectron ***n***uclear ***do***uble ***r***esonance): a double resonance method in which the sample is excited simultaneously with resonance frequencies of Electron-spin resonance spectroscopy (see) and NMR spectroscopy (see). The ENDOR methods are used with paramagnetic substances (see Magnetochemistry) to determine hyperfine structure coupling constants, when these cannot be clearly observed in the ESR spectrum. Strong microwave irradiation permits an ESR transition and reduces the difference in occupation numbers of the two levels involved, thus partially suppressing the ESR signal. While the magnetic field is left at this frequency, the sample is irradiated with a second frequency from the NMR range. This frequency is varied until a transition occurs between NMR levels. This partially relieves the saturation of the ESR signal, so that its intensity (ENDOR signal) increases again. ENDOR spectra are less complicated and better resolved than ESR spectra, so that coupling components can be more easily determined.

Endorphins, ***opiate peptides, opiate-like peptides***: peptides which are the natural ligands of the opiate receptors; their effects are similar to those of morphine (*endo*genous mo*rphine*). Although the term E. was intended to apply to all peptides with opiate-like effects, including the enkephalins, it is used in the narrower sense to mean higher-molecular-weight polypeptides, especially fragments of β-lipotropin (Fig.). Leu-enkephalin, α-neo-E. and dynorphin would not be included by this narrower definition. The relationships among the primary structures of the methionine-containing E. are shown in the diagram of the C-terminal region of porcine β-lipotropin, residues 61-91. β-Lipotropin itself has no opiate-like effect.

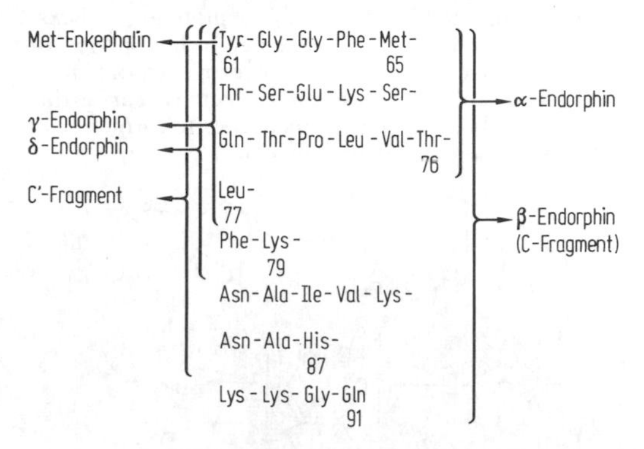

With the exception of met-enkephalin, all of these E. are derived from pro-opiomelanocortin, which contains the sequence of lipotropin and corticotropin.

E. have been found in the central nervous system, cerebrospinal fluid, kidneys, ganglia of the gastrointestinal tract, blood, placenta and pituitary. They act as opiate antagonists and cause a dose-specific reduction in the contractile waves of the mouse vas deferens and the guinea pig ileum. The inhibition of contraction caused by E. can be released by the opiate antagonist naloxon, which implies a specific interaction between E. and the opiate receptors. β-Endorphin has the highest analgesic effect. Unfortunately, the E. have not proven to be "non-addictive" analgesics. They play a part in acupuncture analgesia, and are thought to have a part in the pathogenesis of mental disorders (schizophrenia, hallucinations, etc.). A connection between E. and stress has been discussed. Although the physiological functions of the E. are only partly known, it is certain that the neuromodulating function of E. in the control of sensitivity to pain is only one aspect of their function. They also interact with the mechanisms of the autonomic nervous system (e.g. circulation, body temperature, sleep, appetite). Therapeutic applications may eventually be found for synthetic analogs.

Endosulfan: a Cyclodiene insecticide (see); m.p. of the α-form, 108-109 °C, vapor pressure at 80 °C, 1.2 Pa. To synthesize it, the Diels-Alder adduct of hexachlorocyclopentadiene and but-2-ene-1,4-diol is reacted with thionyl chloride to give the cyclic sulfite. In contrast to the other cyclodiene insecticides, E. does not leave detrimental residues, has low persistence, is mild in its effects on useful insects, especially honey bees, and is used on ornamental plants, field and fruit crops and in silviculture, especially to combat June beetles. It is very poisonous to fish (see Piscicide).

Endothermal reaction: a reaction in which energy, usually in the form of heat, is consumed. According to the thermodynamic sign convention, the reaction enthalpy $\Delta_R H$ and energy $\Delta_R U$ are positive.

Endotoxins: same as Lipopolysaccharides (see).

Endrin: a Cyclodiene insecticide (see).

Ene reaction: indirect substitution/addition of alkenes to philodienes. The result is an allylic substitution, which in many cases goes via a cyclic transition state (1). Carbonyl compounds and thioketones with electron deficits at the C=O or C=S function (e.g. chloral, perfluorocyclobutanone, hexafluorothioacetone) are suitable heterodienophiles for E. Alkenes with at least one hydrogen atom in an allyl position react with singlet oxygen by rearranging the double bond in an E., which leads to formation of allyl-hydroperoxy compounds (2).

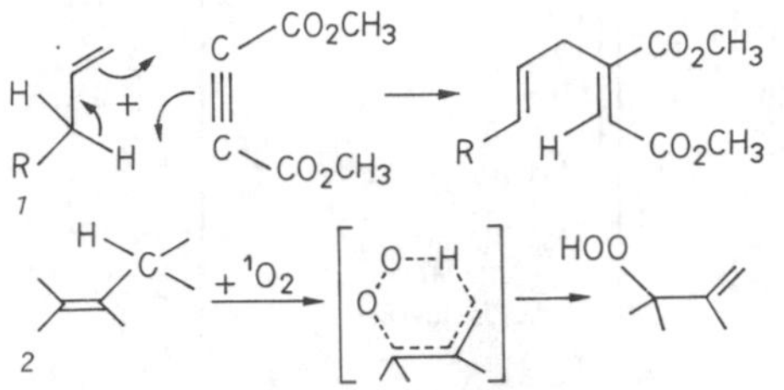

Energy band: see Energy band model.

Energy band model: quantum mechanical description of the electronic states of molecular (crystal) orbitals, especially those in metallic solids. The molecular orbitals are formed by combination of the large numbers ($N \approx 10^{23}$) of available equivalent atomic orbitals and they are delocalized over the entire metal crystal. The energy levels associated with the molecular orbitals are so close to each other that they cannot be experimentally distinguished, and they are therefore referred to as an ***energy band***. The strength of the interaction between the equivalent atomic orbitals, and thus the width of the energy band, depends mainly on the internuclear distance of the atoms in the lattice. The atomic orbitals combine to generate a *valence band*, which is partially or completely filled in the metal, and a *conduction band* (Fig.), which is empty. (The occupation of the levels in the energy bands obeys the Pauli principle.) Metals are characterized by overlapping of the valence and conduction bands, while in insulators and semiconductors, there is a gap between them. In semiconductors, the forbidden gap is much narrower than in insulators, so that thermal excitation of electrons in the valence band can transfer them to the conduction band. By addition (doping) of certain foreign atoms, the conductivity of a semiconductor can be increased. If silicon, for example, is doped with atoms of the 3rd or 5th main group, they create acceptor or donor levels within the gap which increase the conductivity of the material.

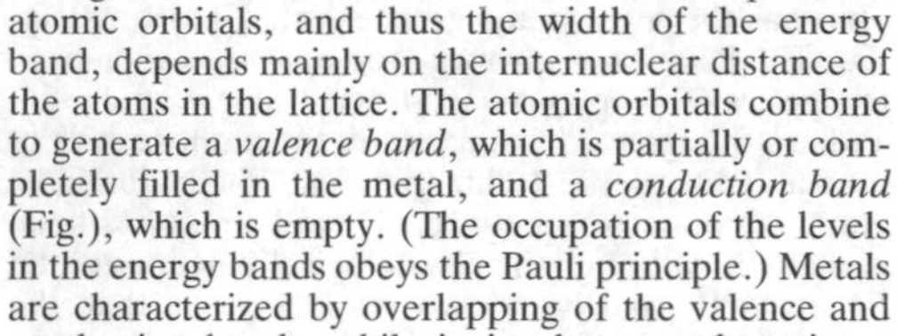

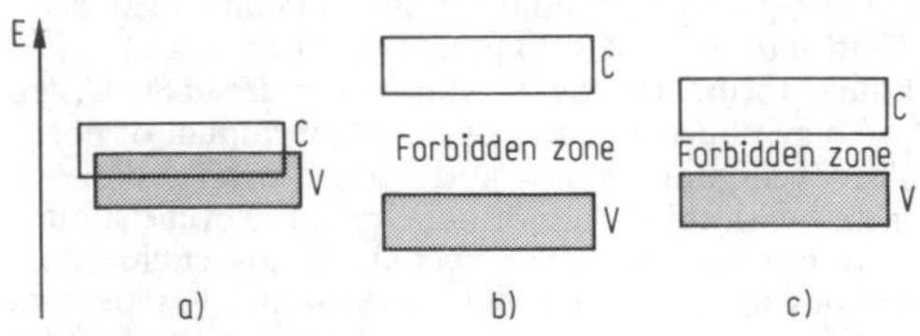

Energy band model of: (a) a metal with an occupied valence band; (b) an insulator; and (c) a semiconductor. *V*, valence band. *C*, conduction band.

Energy balance graphics: see Technological diagrams.

Energy distribution: in statistical theory, the way in which energy is distributed among the particles in a statistical unit. In classical statistics, the E. at equilibrium is given by the Maxwell-Boltzmann distribution (see). Quantum statistical distribution requires other methods, due to the indistinguishability of particles. For particles with integral spins, e.g. mesons, Bose-Einstein statistics (see) apply; for those with semi-integral spins, Fermi-Dirac statistics (see). The Pauli exclusion principle in the occupation of electronic states in atoms and molecules, the existence of the Fermi limit for electronic energies in metals and semiconductors and the phenomenon of superfluidity in liquid ^{4}He at temperatures below 2 K are results of quantum statistical E.

Energy dose: see Radiation chemistry.

Energy elasticity: the elasticity of hard solids, such as steel, which is due primarily to a change in the internal energy of the body. Its opposite is Rubber elasticity (see).

Energy flow chart: see Technological diagram.

Energy profile: the course of the potential energy along the reaction coordinates of a chemical reaction (see Potential surface).

Energy transfer: a photophysical process in which a species in an excited electronic state (donor) transfers excitation energy to another species (acceptor). The donor thus passes to a lower state (usually its ground state) while the acceptor enters a higher state. If the donor and acceptor are parts of the same molecule, the process is an intramolecular E. E. is one of the mechanisms of photosensitization, and can be classified as singlet-singlet or triplet-triplet E. (see Triplet generator). An effective E. requires that the energy difference between the two electronic states of the donor be greater than the difference between the highest occupied and the lowest unoccupied molecular orbital of the acceptor.

E. can occur by several mechanisms. 1) In the radiation mechanism, the light emitted by the excited donor is absorbed by the acceptor. 2) Resonance E. can occur only in singlet-singlet transfer over a distance of no more than 100 nm between the donor and acceptor (emitter-receiver mechanism). A condition for resonance is that the energy differences between the ground and excited states of donor and acceptor are equal. 3) Electron exchange mechanism: during a collision of an excited donor with an acceptor in the ground state, an electron can be exchanged, and as a result, the donor goes to the ground state while the acceptor enters the excited state. 4) Energy migration: this process is an exciton migration which occurs in solids (e.g. molecular complexes) over 10^4 to 10^6 formula units (singlet excitons) or up to 10^{11} formula units (triplet excitons) during the lifetime of the excited state. The energy migration mechanism is important for spectral sensitization of photographic films and in photosynthesis.

Engel-Precht process: see Potassium carbonate.

Engobe, ***slip***: a white or colored coating on a Ceramic (see) object. The E. is applied to the unbaked ware as a viscous slurry, which usually contains clay; it improves the appearance of the object. The rough surface can be smoothed by applying a glaze. The E. can be painted on with a brush to make a design.

Enin: see Malvidin.

Enkephalin: a mixture of endogenous peptides with morphine-like effects. ***Methionine-E.*** (abb. ***Met-E.***) is Tyr-Gly-Gly-Phe-Met, and ***leucine E.*** (abb. ***Leu-E.***) is Tyr-Gly-Gly-Phe-Leu; they differ only at the C-terminal amino acid and are found naturally in varying relative amounts. E. was discovered in 1975 by Hughes and Kosterlitz in porcine brain, and was the first endogenous opiate peptide (see Endorphins) to be found. It is present in various parts of the brain and in the substantia gelatinosa of the spinal cord and neuroplexus, but it is also present in the exocrine cells of the gastrointestinal tract. When applied intravenously, E. has little analgesic effect, because it is quickly degraded by enzymes. By replacing the glycine in position 2 by D-amino acids, and other strategies, analogs were obtained which are more stable to degradation and are more readily transported. Direct administration of E. and its synthetic analogs to the brain leads to short-term analgesia, but is not non-addictive, as was hoped. The E. is thought to act as a neurotransmitter or neuromodulator. In 1982, the structure of its biological precursor was elucidated; this protein consists of 267 amino acids. Pre-proenkephalin contains six Met-E. sequences and one Leu-E., but does not contain the sequences of other opiate peptides, such as dynorphin, α-neoendorphin or endorphin, which were previously thought to be precursors of E.

Enniatins: cyclic hexadepsipeptides with an antibiotic effect against mycobacteria. The E. are Ionophores (see) and are produced by strains of *Fusarium*. Some important examples are ***enniatin A***, cyclo-(D-Hyv-MeIle-)$_3$, and ***enniatin B***, cyclo-(-D-Hyv-MeVal-)$_3$. Both contain D-α-hydroxyisovaleric acid (D-Hyv) and L-*N*-methylisoleucine (MeIle) or L-*N*-methylvaline (MeVal). E. form complexes with K^+ ions and are involved in their transport across membranes. They are practically insoluble in water, but are soluble in organic solvents. They are sensitive to alkalies.

Enolase: an enzyme which catalyses the conversion of 2-phosphoglyceric acid to 2-phosphoenolpyruvic acid. The enzyme, which is found in yeasts, bacteria, muscle and liver, occurs in several forms (see Isoenzymes) and requires magnesium ions for its catalytic activity. It is specifically inhibited by fluoride ions, causing glycolysis and alcoholic fermentation to stop at the level of glyceric acid.

Enolization: conversion of the keto form of a compound to the enol form in Keto-enol tautomerism (see).

Enskog formula: see Kinetic gas theory.

Enterokinase: an exopeptidase (see Proteases) found in the duodenum which acts specifically on trypsinogen. E. consists of 1100 amino acid residues (M_r 196,000) and is a glycoprotein; the carbohydrate makes up 37% of its mass. It catalyses the cleavage of the *N*-terminal hexapeptide Val-(Asp)$_4$-Lys from the inactive trypsinogen to release trypsin.

Enthalpy, ***Gibbs heat function***; in older literature, also ***heat content***, symbol ***h***: a parameter of state in thermodynamics which is used in addition to the Internal energy (see) e as a matter of convenience. It is defined as $h = e + pv$, where pv is the volume work which can be calculated from the pressure p and the volume v. The E. has an important role in all situations in which material systems exchange heat q and volume work with their environment, but do no other kind of work. From the first law of thermodynamics, it follows for a differential change in E., $\mathrm{d}h = v\,\mathrm{d}p + \delta q$. For a process carried out at constant pressure, $\mathrm{d}p = 0$, and $\mathrm{d}h = \delta q$. The heat exchanged between the system and its environment is then equal to the change in E.

In practice, many processes are carried out isobarically. The heat changes observed in these processes are changes in the E., e.g. Reaction heats (see), Mixing heats (see) and Phase change heats (see).

Enthalpy function: the relatively seldom used thermodynamic function $h/T = s - \Phi$, where h is the enthalpy, s is the entropy, Φ is Plank's function and T is the absolute temperature. An analogous function is the Free enthalpy function (see).

Entropy: a parameter of state in thermodynamics, symbol s, which gives information on the reversibility (see Reversible) and direction of processes, and on the position of thermodynamic equilibrium. E. was defined in 1850 by Clausius as follows: if, in a reversible process, there is a differential heat exchange $\mathrm{d}q_{\mathrm{rev}}$ with the environment at temperature T, the ***reduced heat*** $\mathrm{d}q_{\mathrm{rev}}/T$ is equal to the change in a parameter of state $\mathrm{d}s$, i.e. $\mathrm{d}s = \mathrm{d}q_{\mathrm{rev}}/T$. The unit of E. is $\mathrm{J\,K^{-1}}$, or in the older literature, the clausius Cl ($\mathrm{cal\,K^{-1}}$). For a cyclic reversible process, it can be shown that $\overset{\Sigma}{_0} q_{\mathrm{rev}}/T = 0$, or for all the differential steps in the process, $\oint \mathrm{d}q_{\mathrm{rev}}/T = 0$ (see Carnot cycle). Here $\overset{\Sigma}{_0}$ and $\oint$ indicate that the sum or integral is over all the partial steps of the cycle. The expression $\mathrm{d}s = \mathrm{d}q_{\mathrm{rev}}/T$ thus fulfills the requirement for a parameter of state, that its change is independent of the path and is zero in a cyclic process. However, if there are irreversible steps in the cyclic process, the sum of the reduced heats $\overset{\Sigma}{_0} q/T < 0$. For example, in the Carnot machine the cooling might occur by direct heat transfer rather than by reversible adiabatic expansion of the gas. Then the reduced heat no longer fulfills the definition of a state parameter; for irreversible processes, it is not a measure of the E.

On this basis, it can be proved that for closed systems in which there is no exchange of matter or energy with the environment, $\mathrm{d}s > 0$ for irreversible changes of state, and for reversible changes of state, $\mathrm{d}s = 0$. This means that for spontaneous, and thus irreversible processes, the total E. of the system always increases. E. is a parameter which can change in only one direction, and thus gives information on the direction of macroscopic processes. In nature, only those macroscopic processes can occur which, taken as a whole, lead to an increase in E. This makes it possible to establish directionality in time. The state of higher E. comes later in time than that with lower E. If E. reaches a maximum, $\mathrm{d}s = 0$, and equilibrium has been reached. The process comes to a stop or proceeds reversibly via equilibrium states.

A closed system can be constructed which consists of closed subsystems which exchange energy among themselves. Examples are the two heat reservoirs and the working gas in the Carnot cycle, or a closed reactor in the heat bath of a thermostat. In such a subsystem, the E. can be changed by heat exchange with the environment or by entropy production in an irreversible process within the system. The change in E. ds_i in the subsystem can therefore be positive or negative. The spontaneity of the process is determined only by the E. balance for the total system, i.e. the sum of the E. of all subsystems, for which $ds > 0$ as long as equilibrium has not been reached.

E. is defined as a parameter of state by the second law of thermdynamics (entropy law). It is a function of state variables, such as pressure or volume, temperature and chemical composition. The state function $s = f(T, p, x_1...x_k)$ makes it possible to calculate the E. changes associated with state changes and, using the third law of thermodynamics, to calculate the absolute values of E.

A statistical definition of E. was given by Boltzmann and Planck. It is $s = k_B \ln P$. Here k_B is the Boltzmann constant and P is the thermodynamic probability of the macrostate with the E. s. A macrostate is conceptualized as consisting of a large number of microstates, which are characterized by type of particle, definite lattice positions, energies and orientations. The thermodynamic probability is the number of possible distinguishable (in principle) arrangements of system components among the various microstates. The fewer ways there are to create the macrostate, the lower are its probability and its E. An increase in E. indicates the conversion of a state into one with higher probability, i.e. less "order" (see Thermodynamics).

Entropy elasticity: same as Rubber elasticity (see).

Entropy law: second law of Thermodynamics (see).

Entropy production: see Thermodynamics II.

Envelope form: see Stereoisomerism 2.2.

Enzymatic peptide synthesis: see Peptide synthesis.

Enzyme: a protein catalyst. About 90% of cellular proteins are E., and some structural proteins (e.g. actin and myosin) are also E. Some species of RNA, notably the self-splicing introns of eukaryote messenger RNA, are also capable of highly specific catalysis and thus can be considered RNA E. The Nomenclature Committee of the IUB lists well over 2000 E. which have been well characterized; however, these are classified according to the reaction catalysed, not the identity of the proteins. The same reaction may be catalysed by non-homologous proteins in different organisms, and thus the number of E. actually characterized is far greater than 2000.

The mechanisms by which E. catalyse reactions are usually fairly simple. Like all catalysts, E. reduce the activation energies of the reactions they catalyse. The thermodynamic equilibrium of the reaction is not changed. The most striking difference between E. and chemical catalysts is the specificity of E, with regard both to their substrates and to the reactions catalysed. E. distinguish stereoisomers absolutely, due to the three-dimensional nature of the binding between substrate and enzyme.

The first step of catalysis is formation of an enzyme-substrate complex (ES) from the enzyme (E) and substrate (S) or substrates. Next, the actual catalytic step converts the ES to EP, the enzyme-product complex, which then dissociates to release the product and regenerate the E. Catalysis is accomplished by an E. in several ways: by bringing the reactants together in the appropriate orientations; by providing acidic and/or basic groups at the right positions to promote acid or base catalysis; by providing electrophilic or nucleophilic groups which form a temporary covalent bond with the substrate; and/or by providing for concerted mechanisms in which two catalytic groups on the E. attack the substrate molecule. Any of these forms of catalysis may be accompanied by conformational changes in the E. molecule which strain the substrate molecule and facilitate the breaking of bonds within it. The figure shows an example of a concerted catalysis in the cleavage of a peptide bond by chymotrypsin.

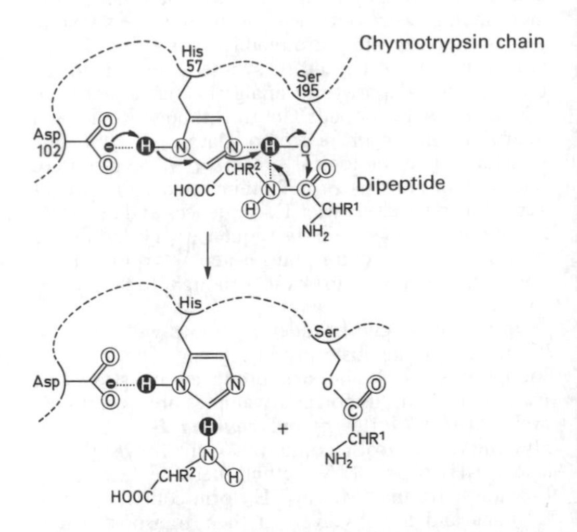

Simplified model of the cleavage of a peptide bond by chymotrypsin. R[1] and R[2] are side chains of amino acids 1 and 2.

Kinetics. The generalized equation for an E. reaction may be written:

$$E + S \underset{k_{-1}}{\overset{k_1}{\rightleftharpoons}} ES \xrightarrow{k_2} E + P$$

where k_1 and k_2 are rate constants. (Although the ES $\rightarrow$ E + P step is reversible, the reaction usually occurs so far from equilibrium that the back reaction can safely be ignored; for example, P will usually be removed immediately in a subsequent metabolic reaction.) In the *Michaelis-Menten* kinetic equation, it is assumed that the rate of change in [ES] is small compared to the change in [S] (the *steady state approximation*) because ES breaks down at the same rate at which it is formed. It follows that

$$[E]\,[S] = \frac{(k_2 + k_{-1})}{k_1}[ES] = K_m\,[ES]$$

where K_m is known as the Michaelis constant. If v_0 is the rate of an E. reaction and V_{max} is the maximum velocity when the E. is saturated with substrate, the *Michaelis-Menten equation* is

$$v_0 = \frac{V_{max}\,[S]}{K_m + [S]}$$

By setting $v_0 = 1/2\ V_{max}$, it can be seen that K_m is equal to the substrate concentration at which the initial rate is half maximal. High values of K_m indicate a low affinity of the E. for its substrate, and conversely, low K_m indicates high affinity. A plot of initial velocity vs. substrate concentration yields a hyperbolic curve if the E. follows Michaelis-Menten kinetics. A sigmoid curve is typical of allosteric E. for which the substrate acts as an activator.

Classification. The Nomenclature Committee publishes an enzyme catalog (abb. EC) in which each E. receives a four-part number. The first part indicates the general category of reaction (Table), the second and third, the subgroup and sub-subgroup, and the fourth part of the number is assigned arbitrarily, in consecutive order.

Structure. Like all proteins, E. consist of polypeptide chains with specific sequences of amino acids. ***Monomeric*** E. consist of single chains; ***oligomeric*** E. consist of several, usually 2 to 4. E. which catalyse subsequent steps in a reaction sequence may form ***multienzyme complexes*** which are held together by non-covalent bonds. In such complexes, substrates can be passed from one E. to the next, with no dilution of intermediates by the surrounding medium. Finally, two or more functional sites may occur on a single polypeptide chain, which is then a ***multifunctional E.*** The precise folding of a long polypeptide into a functional E. is determined by interactions among the amino acid residues in the chain, such as ionic, hydrogen and hydrophobic bonding, formation (and interruption) of helical segments, "pleated sheet" structures or loops, and by the nature and extent of modification of the protein by covalent bonding to sugar or phosphate groups. E. with the same catalytic function but slight variations in amino acid sequence are often formed in different tissues of the same organism; these are called Isoenzymes (see).

The amino acid residues which take part in catalysis and their immediate neighbors constitute the *active center* of the E. The *substrate binding site* is close to or includes the active center, but if the substrate is large, the binding site may extend beyond the active center. *Allosteric E.*, in which the rate of catalysis is affected by binding of a molecule other than the substrate, also have *effector binding sites* which are distinct from the substrate binding site. (The effector is often an end-product of the synthetic chain of which the allosteric E. is a part; inhibition of the E. by the effector permits feed-back regulation of the entire synthesis.)

Table 1. Classification of enzymes according to their specificity

1.	Oxidoreductases: catalysis of redox reactions between a pair of substrates	
1.1	Acting on CHOH	Alcohol dehydrogenase 1.1.1.1
1.2	Acting on C=O	Formate dehydrogenase 1.2.1.2
1.3	Acting on HC-CH-	Succinate dehydrogenase 1.3.99.3
1.4	Acting on C-NH_2	L-Aminoacid oxidase 1.4.3.2
2.	Transferases: catalysis of intermolecular group transfers	
2.1	C_1 group transfer	Aspartate-carbamoyl transferase 2.1.2.3
2.2	Carbonyl groups	Transketolase 2.2.1.1
2.3	Acyl groups	Choline acetyltransferase 2.3.1.6
2.4	Glycosyl groups	Glycosyl transferases 2.4
2.5	Alkyl, aryl groups	
2.6	Amino groups	Aminotransferases 2.6.1
2.7	Groups containing phosphorus	Nucleotidyl transferases 2.7.7
3.	Hydrolases: catalysis of hydrolytic cleavage of	
3.1	Ester bonds	Lipase 3.1.1.3
3.2	Glycoside bonds	β-Amylase 3.2.1.1
3.3	Ether bonds	Thioether hydrolases 3.3.1
3.4	Peptide bonds	Leucine aminopeptidase 3.4.11.1
3.5	Other C-N bonds	Urease 3.5.1.5
4.	Lyases: catalysis of elimination reactions with formation of double bonds, or, if named as *synthases,* addition to double bonds	
4.1	C=C	Pyruvate decarboxylase 4.1.1.1
4.2	C=O	Carboanhydrase 4.2.1.1
4.3	C=N-	Ammonia lyases 4.3
5.	Isomerases	
5.1	Racemases-epimerases	Alanine racemase 5.1.1.1
5.2	*cis-trans*-Isomerases	Retinal isomerase 5.2.1.3
5.3	Intramolecular oxidoreductases	Triose phosphate isomerase 5.3.1.1
5.4	Intramolecular transferases	Phosphotransferases 5.4.2
6.	Ligases (synthetases): catalysis of the linking of two molecules with consumption of ATP	
6.1	C-O bonding	Tyrosyl-tRNA synthetase 6.1.1.1
6.2	C-S bonding	Acetyl-CoA synthetase 6.2.1.1
6.3	C-N bonding	NAD synthetase 6.3.1.5
6.4	C-C bonding	Carboxylases 6.4.1

In many E., a non-protein group is required for catalysis. This may be either a Coenzyme (see), which is not covalently bound, a *prosthetic group*, which is bound to the E., or a metal ion. ***Metalloenzymes*** contain metal ions as a bound component. Examples are Fe^{2+} and Fe^{3+} in oxidoreductases, Cu^{2+} in oxidases, Zn^{2+} in dehydrogenases, Mo^{2+} in nitrogenase and Ca^{2+} in α-amylase. Many other E. require the presence of specific metal ions for activity.

Biosynthesis. Like other proteins, E. are synthesized on subcellular particles called ribosomes. The amino acid sequence is determined by the sequence of nucleotide bases in the molecule of messenger RNA which is transcribed in the synthesis of the protein. The rate of protein synthesis is controlled both at the level of mRNA synthesis from DNA (the gene) and at the level of mRNA translation on the ribosome. E. which are continuously synthesized in the cell are called ***constitutive***; those which are made only under certain conditions (the vast majority) are ***adaptive***. The latter category includes both ***inductive*** E. which are synthesized in response to an environmental stimulus (e.g. a particular carbohydrate nutrient or a hormone) and ***repressible*** E., the synthesis of which is inhibited by some environmental stimulus (such as the end-product of a biosynthetic pathway).

Most E. are strictly intracellular, but some are excreted into the body fluids or the medium of unicellular organisms. Proteolytic E. are secreted in the form of inactive precursors which are only activated after they are safely out of the cell; those which are not secreted are sequestered into a special organelle, the lysosome. Some E. are specific for particular organs or tissues, and their presence in the blood can be used as a diagnostic test for damage to the tissue of origin. Within the eukaryotic cell there is a fair amount of compartmentation of the E. (and reactions) within organelles. When the cell is fractionated, these E. can be used as *markers* for the various fractions (mitochondria, cytosol, liposomes, etc.).

Inhibition of E. The exact conformation of a polypeptide chain can be changed by *effectors* which may either activate or inactivate the E., but do not bind to the same site as the substrate. *Inhibitors* are molecules which bind reversibly to the substrate-binding site or the catalytic center. *Competitive inhibitors* compete with substrate for the binding site but are not covalently bound to it. The degree of inhibition depends on the relative concentrations of substrate and inhibitor. *Non-competitive inhibitors* bind to the active center and block it, either temporarily or permanently. The inhibition is not reversible by high concentrations of substrate. Examples are heavy metal ions which form complexes with catalytically active SH-groups and chelate formers which sequester metal ions required for E. activity. E. can be inactivated by substances which react irreversibly, forming covalent bonds with catalytically active amino acids (e.g. p-mercuribenzoate with -SH groups) or simply blocking the substrate-binding site permanently.

Enzyme units. *Enzyme activity* is indicated by the amount of substrate converted to product in a unit of time by a given amount of E. The course of the reaction is most conveniently followed spectroscopically, if possible. The current international standard catalytic unit is the katal (abb. kat), which is the amount of activity which can convert 1 mole substrate per second. This unit is inconveniently large, so that the microkatal (μkat), nanokatal (nkat) or picokatal (pkat) is more likely to be encountered. Many authors prefer the *enzyme unit* established prior to the katal; a unit is the activity which converts 1 micromole substrate per minute under standard conditions. For conversion between the two systems, $1\ \text{kat} = 6 \cdot 10^7\ \text{U}$; $1\ \text{U} = 16.67\ \text{nkat}$.

The *molecular activity* (formerly *turnover number*) is the number of substrate molecules which are converted per minute by a single E. molecule or active center.

Production. E. can be isolated from animal, plant or microbial material. After homogenization of the material, the E. are either extracted directly, or the material may first be dehydrated to a powder at low temperature, using an organic solvent (usually acetone). Purification is achieved by differential precipitation, chromatography (column, paper, ion-exchange, affinity, etc.), ultrafiltration and other techniques.

The most important method of industrial production is fermentation by microorganisms, especially highly productive mutant strains. Fermenters may contain up to 100,000 l, and the fermentation time can be 50 to 150 h. If the E. is excreted into the culture medium, extraction is simple, but if not, the cell walls of the microorganisms must be mechanically ruptured, e.g. by high-pressure homogenizers, before extraction can proceed.

Applications. In addition to their use as symptoms of tissue damage, E. are applied as reagents for the specific detection of other metabolites, such as glucose, ATP, lactate, etc., which are diagnostically significant. The proteases, in particular, may be applied to relieve poor digestion, problems of blood coagulation and fibrinolysis. Pathologically increased levels of proteases, on the other hand, may be treated with protease inhibitors from bovine organs, which do not provoke an immune response in humans.

With the development of appropriate methods for large-scale cultivation of microorganisms and isolation of E. in very large quantities, it has become economically feasible to use them in a number of industries, including food processing, textiles, paper, laundry products and pharmaceuticals. With the advent of genetic engineering, there is interest in developing artificial E. to catalyse various reactions in organic and pharmaceutical chemistry. A major advance in E. technology was the development of *immobilized E.*; the soluble E. is bound covalently to an inert carrier material, such as porous glass, ion-exchangers, Sephadex, etc., and the reactant solution is passed over a column containing the now insoluble E. Alternatively, the E. may be embedded in a matrix (e.g. polyacrylamide gel) or enclosed in cellulose fibers or microcapsules which greatly improve their chemical and thermal stability. All the above are ***carrier-fixed E.***; another method involves ***enzyme membrane reactors*** in which the native E. is freely mobile in the interior of the reactor, but is prevented from leaving it by a polymeric membrane through which substrates and products can diffuse freely. An advantage is that E. which are dependent on soluble coenzymes can be used; the coenzymes are bound to a water-soluble polymer, such as polyethylene glycol, which prevents them from diffusing through the barrier membrane.

Enzyme immunoassays make use of E. as "markers" for antibodies to hormones, antigens, drugs, etc. The amount of E.-labelled antibody bound to a sample is readily measured by an assay for E. activity. ***Enzyme electrodes*** are used for electrochemical process control. Among the best-known are glucose-sensitive electrodes with immobilized glucose oxidase as E. In this case the decrease in oxygen or the increase in hydrogen peroxide concentration is measured. If the glucose oxidase is coupled to a suitable hydrolase,

Table 2. Technical applications of selected enzymes

Enzyme (source)	Reaction catalysed	Applications
Bacillus proteases (*Bacillus* sp.)	Endohydrolysis of proteins	Laundry products (0.01–0.025%): degradation of protein stains
Fungal proteases (*Aspergillus* sp.)	Endohydrolysis of proteins	Hydrolysis of soy proteins (production of soy sauce) Additive to baked goods (partial hydrolysis of gluten) Preparation of protein hydrolysates from meat scraps and fish proteins
Bacterial β-amylase (*Bacillus* sp.)	Endohydrolysis of starch	Liquefaction of starch; preparation of starch paste; removal of starch from textiles; ethanol production from starchy raw materials
Glucoamylase (*Aspergillus* sp.)	Exohydrolysis of starch	Breakdown of starch to glucose in combination with β-amylase (high yields and purity of products)
Glucose isomerase (*Streptomyces* sp., *Bacillus* sp.)	Isomerization of D-glucose to D-fructose	Production of fructose syrup
Glucose oxidase (*Aspergillus* sp., *Penicillium* sp.)	Oxidation of glucose to gluconic acid	Preservation of foods and drinks by removal of O_2
Invertase (*Saccharomyces* sp.)	Cleavage of sucrose into glucose and fructose	Production of invert sugar
Catalase *Aspergillus* sp.)	Decomposition of H_2O_2	Used in combination with glucose oxidase for preservation of goods
Lipase (pancreas, *Aspergillus* sp.)	Fat cleavage	Production of unstable fatty acids, acceleration of cheese ripening
Cellulase (*Trichoderma reesei*)	Hydrolysis of cellulose	Production of glucose from cellulose-containing materials
Lactase (*Aspergillus* sp.)	Cleavage of lactose into galactose and glucose	Production of lactose-free milk
Pectinase (*Aspergillus* sp.)	Hydrolysis of polygalacturonic acid and its methyl ester	Removal of cloudiness from fruit juices
Penicillinamidase *Bacillus* sp.)	Formation of 6-aminopenicillinic acid from penicillin G	Production of semisynthetic penicillins
L-Amino acid acylases (kidneys, *Aspergillus oryzae*)	Enantioselective hydrolysis of DL-acylamino acids into L-amino acids and D-acyl amino acids	Production of essential amino acids for animal and human nutrition

the electrode can be used for rapid determination of glucose-containing oligosaccharides. ***Enzyme thermistors*** make use of the heat of reaction released by the E.-catalysed reaction as a signal.

The use of E. can be expected to increase both in analytical and selective organic synthetic chemistry. This will also be true of artificial *synzymes* created by incorporating "enzyme-specific" structural elements in synthetic macromolecules.

Enzyme immunoassay: see Immunoassay.

Enzyme technology: see Biotechnology.

Eosine: see Eosine pigments.

Eosine pigments: a group of xanthene pigments derived from fluorescein by introduction of halogens, nitro groups, etc. The most important of these is ***eosine (tetrabromofluorescein)***, a red, crystalline powder which is slightly soluble in water, readily soluble in alcohol and insoluble in ether. An alkaline solution of eosine is red. Eosine is made from fluorescein and elemental bromine in glacial acetic acid. It dyes wool and silk bright red, but because it is not fast to light, it is no longer used for this purpose. Today eosine is used, often in the form of its aluminum, lead or tin lake, to make red inks, lipsticks, spirit lacquers and poisons (as a warning color); it is used as a vital stain and, to a limited extent, to sensitize photographic emulsions. The sodium salt, which is often also called eosine, forms red needles and dissolves readily in water.

Other examples of E. are *erythrosine*, which has 4 iodine atoms instead of bromine, *rose bengale* (*Bengal rose*, chemically 3',4',5',6-tetrachloro-2,4,5,7-tetraiodofluorescein), *phloxine* (*cyanosine*, chemically 3',6'-dichloro-2,4,5,7-tetrabromofluorescein) and *daphnine*.

EP: abb. for polyepoxide; see Plastic.

Ephedrin: a phenylethylamine derivative which has a weak sympathicomimetic effect. It is the main alkaloid in various East Asian species of *Ephedra*, in which it is found in concentrations of 0.1 to 3%. E. was isolated from *Ephedra sinica* in 1887 by Nagai. It is biosynthesized from *R*-phenylalanine. E. has two centers of chirality; it has the *erythro* configuration and is levorotatory. The absolute configuration is 1*R*,2*S*; the configuration at C1 is thus the same as that at C1 of Noradrenalin (see) or adrenalin; m.p. 40 °C, $[\alpha]_D^{20}$ - 6.3° (in ethanol); ephedrin hydrochloride has m.p. 218 °C, $[\alpha]_D^{20}$ - 36.6° (in water); the racemate has m.p. 76 °C, racemate hydrochloride, m.p. 189 °C. E. undergoes *hydramine cleavage*; depending on how the reaction is carried out, it may yield a ketone as well as methylamine. Dry distillation produces propiophenone, $C_6H_5COCH_2CH_3$; heating E. with concentrated phosphoric acid produces phenylacetone, $C_6H_5CH_2COCH_3$. E. is used as a medication for asthma, weak circulation and bronchitis. It also has a weak excitatory effect on the central nervous system.

HO–C¹–H
CH₃HN–C²–H
CH₃

L(–)-Ephedrin (1R,2S)

epi-: (Greek for "on" or "to") a prefix used to denote 1) one of two diastereomers, usually the one discovered later; 2) the 1,6-position in substituted naphthalene compounds; or 3) the stereoisomers of the hexitols and aldoses.

Epichlorohydrin, ***1-chloro-2,3-epoxipropane***: a colorless, mobile, poisonous liquid with a pungent odor similar to that of chloroform; mp. - 48 °C (elsewhere reported as - 25.6 °C), b.p. 117 °C. It is soluble in ether, alcohol, carbon tetrachloride and benzene, and slightly soluble in water.

$Cl-CH_2-CH-CH_2$ (O bridging)

E. contains a reactive oxirane ring, but it is also able to act as an alkyl halide. In the presence of dilute sodium hydroxide solution, it is converted to ***glycerol***. In the laboratory, E. is synthesized by reaction of glycerol with phosphorus pentachloride. Its industrial synthesis is based on high-temperature chlorination of propylene to allyl chloride, which is then reacted with aqueous hypochlorous acid to form the isomeric glycerol-1,2- and glycerol-1,3-dichlorohydrins. These are converted to E. with calcium hydroxide or sodium hydroxide. E. is used in the production of epoxide resins and insecticides, and as a solvent for resins.

Epimerization: see Stereoisomerism 1.2.1.

Epinephrin: see Adrenalin.

Episulfides, ***thiiranes***: saturated, three-membered heterocyclic compounds with a sulfur atom in the ring. They are synthesized by the reaction of potassium thiocyanate with an oxirane or by splitting hydrogen halide out of a β-halothiol. The thiirane ring is readily cleaved by nucleophiles. E. are used as drugs, insecticides, fungicides and additives to synthetic polymers.

Epitaxis: oriented crystal growth on a crystalline substrate. Epitactic growth produces a second crystalline phase (deposit crystal) on a previously existing crystal (host, carrier crystal, substrate); there is a defined orientation of the deposited lattice with respect to the host lattice. E. is possible for a single substance (***homoepitaxis***) but is also very often observed in different materials (***heteroepitaxis***), and between quite different types of crystals (e.g. metallic, ionic and molecular crystals in all possible combinations). E. can occur from solutions or the vapor phase (the deposited component is sublimed in high vacuum). The properties of both the substrate and epitactic substances, as well as the growth conditions, determine the occurrence of E. Some important parameters are the geometries of the two lattices, the nature of the lattice forces and the distances over which they act, the temperature of the substrate, the rate of deposition and the real structure of the substrate. Heteroepitactic deposition requires the presence of certain structural analogies between the two substances, but it can also occur when there are large differences in the parameters of the lattice planes which fuse (misfit), if special lattice defects (adaptation displacements) are formed. In chemical reactions on crystal surfaces, the products can also be oriented with respect to the substrate lattice (see Topotaxis).

E. is of great practical significance, e.g. for heterogeneous catalysis, inhibitor effects and corrosion processes. A modern and extremely important application of E. is the production of semiconductor elements for optics and microelectronics. Here thin layers of various semiconductors are grown epitactically on suitable crystalline substrates.

Epithelium-protecting vitamin: see Vitamin A.

EPN: abb. for *O-E*thyl phenylthio*p*hosphoric acid *O*-4-*n*itrophenyl ester, an organophosphate insecticide (thiophosphate ester). EPN is a colorless, crystalline substance, D^{25} 1.27, m.p. 36 °C, oral LD_{50} = 14 to 42 mg/kg rat. It is both a contact and oral poison.

S, P, O, OC_2H_5, NO_2

Epoxide, ***oxirane***: a saturated, three-membered heterocyclic compound with an oxygen atom in the ring. E. are found sporadically in essential oils and antibiotics. They are formed by reaction of hydrogen peroxide or peracids on olefins (see Priležaev reaction) or by base-catalysed cleavage of hydrogen halides out of β-haloalcohols. The condensation of α-halogen carboxylic acid esters also produces E. (see Darzens-Erlenmeyer-Claisen condensation). On heating, or in the presence of Lewis acids, E. rearrange to form carbonyl compounds. The products are often mixtures of aldehydes and ketones. Another typical reaction of E. is opening of the ring by nucleophiles; for example, acid-catalysed hydrolysis gives 1,2-diols. Ethylene oxide and propylene oxide are the starting materials for many industrial products.

Epoxide resin: a synthetic resin which hardens; depending on the purity of the starting materials, an E. may be light yellow to dark brown. E. in the narrower sense are polyethers with terminal epoxide groups formed by reaction of epichlorhydrin with aromatic hydroxy compounds in the presence of alkali hydroxides. The component containing hydroxyl groups is usually 2,2-bis(p-hydroxyphenyl)propane (bisphenol A, diane), a low molecular weight phenol resin, or a heterocyclic nitrogen compound such as cyanuric acid. Diane is relatively easily synthesized from acetone and phenol. It reacts with epichlorhydrin in the presence of alkali, splitting off hydrogen chloride and forming dianediglycide ether, which polymerizes to an epoxide resin precondensate (Fig.) with a molecular mass between 100 and 5000. Cycloaliphatic di- and polyepoxides, formed by epoxidation of the corresponding olefinic compounds, can also be used as hardening resins; examples are fatty acid esters with two or more double bonds.

Further hardening of the E. occurs through reaction of the terminal ethylene oxide rings and hydroxyl groups in the macromolecule; the reaction occurs without pressure and without cleavage of low-molecular-weight products. The reaction is initiated by addition of acidic hardeners, such as dicarboxylic acid anhydrides or dicyanodiamide, at 120 to 200 °C. With basic (amine) hardeners, such as polyamines, especially diethylenetriamine, hardening occurs at room temperature. E. display very little shrinkage on hardening.

The industrial uses of E. as glue, casting or laminating resins, and as chemically resistant coatings are many and varied. E. have excellent adhesion to wood, metal and ceramics, and are widely used as glues. They are used as casting resins for embedding electrical and electronic components (microelectronics) and to repair cast metal parts. Epoxide laminating resins are used with fiberglass reinforcement for parts (pipes, containers) used in the chemical and electric industries (switching chambers for high-voltage switches); and as components of laminated building materials of various kinds. Epoxide resins are used as components in cold-hardening, air-drying, corrosion-resistant coatings, or as bake-on coatings.

E. were first produced on a laboratory scale in 1938, and have been produced industrially since about 1948.

1,2-Epoxypropane: same as Propylene oxide (see).

1,3-Epoxypropane: same as Oxetane (see).

EPR spectroscopy: same as Electron spin resonance spectroscopy (see).

Epsom salt: see Magnesium sulfate.

Eptam: see Herbicides.

Equation of state: see State, equation of.

Equatorial bond: see Stereoisomerism 2.2.

Equilibration vessel: a spherical or cylindrical vessel (***leveling sphere, tube***) connected by a tube to a fluid reservoir (e.g. gas buret). By raising or lowering the E., the desired level of the liquid can be obtained.

Equilibrium: 1) in mechanics, the state in which a body is at a minimum of potential energy. 2) In electrochemistry: see Electrochemical equilibrium. 3) ***Thermodynamic E.***: the state of a closed thermodynamic system in which no macroscopic changes occur with time, so long as the external conditions, e.g. pressure and temperature, remain constant. A system in thermodynamic E. is stable with respect to disturbances; small (differential) disturbances in the external conditions produce a change in the E. state. If the disturbance is removed, the system returns to the original E. state. The approach to the E. state usually occurs asymptotically, but it can also occur by damped oscillation (see Oscillating reaction). An E. is not instantaneously established; it requires time, and the amount of time can vary over orders of magnitude for different types of system (see Diffusion, Reaction kinetics). If variables of state are changed very rapidly in comparison to the time required for E. to be established, the system cannot follow these changes. The original state is "frozen in" at a different temperature or pressure. This situation is a *metastable state*, often (wrongly) called "metastable E.". Examples of such conditions are glasses, crystals with an unstable arrangement, supersaturated vapors or

$$\underset{Cl}{CH_2}-\overset{\diagdown O \diagup}{CH-CH_2} + HO-C_6H_4-C(CH_3)_2-C_6H_4-OH + \overset{\diagdown O \diagup}{CH_2-CH}-\underset{Cl}{CH_2}$$

Epichlorohydrin Diane

$$\downarrow (OH^-)$$

$$\underset{Cl}{CH_2}-\underset{OH}{CH}-CH_2-O-C_6H_4-C(CH_3)_2-C_6H_4-O-CH_2-\underset{OH}{CH}-\underset{Cl}{CH_2}$$

$$\downarrow \begin{matrix}-2H_2O \\ -2Cl^-\end{matrix} \; +2OH^-$$

$$\overset{\diagdown O \diagup}{CH_2-CH}-CH_2-O-C_6H_4-C(CH_3)_2-C_6H_4-O-CH_2-\overset{\diagdown O \diagup}{CH-CH_2}$$ Diane diglycide ether

$$\downarrow +H_2N-R-NH-R-NH_2$$

$$...-O-C_6H_4-C(CH_3)_2-C_6H_4-O-CH_2-\underset{OH}{CH}-CH_2-NH \leftarrow$$ Diane diglycide ether

$$\qquad\qquad\qquad\qquad | \; R$$

$$N-CH_2-\underset{OH}{CH}-CH_2-O-C_6H_4-C(CH_3)_2-C_6H_4-O-...$$

$$\qquad\qquad\qquad\qquad | \; R$$

$$...-O-C_6H_4-C(CH_3)_2-C_6H_4-O-CH_2-\underset{OH}{CH}-CH_2-NH \leftarrow$$ Diane diglycide ether

Epoxide resin lacquers: see Lacquer.

Epoxidation: oxidation of olefins with hydrogen peroxide or peracids to form Epoxides (see). E. is also a form of Biotransformation (see).

α,β-Epoxycarboxylates: same as Glycide esters (see).

solutions, and many chemical compounds. In the tool and glass industries, metastable states which have special properties, e.g. high surface hardness with high elasticity, are deliberately created by sudden cooling (quenching); in chemistry, high concentrations of labile substances (see Matrix isolation technique) are achieved. Metastable states do not react to small changes in external conditions in the same way as E. states; most such changes leave them unchanged, but they can also suddenly overcome their internal inhibitions and undergo an instantaneous conversion to E. (e.g. boiling of superheated liquids, crystallization after formation of seed crystals).

Some important thermodynamic E. are Phase E. (see), which are established between the phases of systems, and ***chemical E.*** which occur with chemical reactions. These states are unchanging only from a macroscopic point of view. Microscopically, there is a constant exchange of matter. For example, in a vapor pressure E., there is a constant exchange between the gas and liquid phases, but the quantitative flow of matter is the same in both directions, so that the vapor pressure over the liquid is constant. This is a ***dynamic E.***. Similarly, in chemical E., the rates of the forward and back reactions are equal, so that macroscopically the rate of change of substance is zero. This fact was the basis of Guldberg and Waage's kinetic derivation of the Mass action law (see).

The various criteria of E. (see Thermodynamics) can be used in the derivation of laws: a) in closed systems, the entropy s is at a maximum, i.e. $ds = 0$; b) in closed systems at constant temperature, the Gibbs free energy g is minimal, i.e. $dg = 0$; c) in multisubstance systems, each individual component i has a constant chemical potential, i.e. $d\mu_i = 0$.

A ***coupled E.*** is a process in thermodynamic E. which consists of several partial steps. These could be several chemical reactions, but they could also be physical processes such as phase transitions or adsorption processes. Some examples are the dissociation of multibasic acids or the coupling of matter transport, adsorption and surface reaction in heterogeneous catalysis. The *principle of detailled E.* or *microscopic reversibility* applies to coupled E.: a process consisting of several partial steps is at E. if and only if each partial step is at E.

4) ***Steady state*** or ***stationary state*** is a state which is constant in time; it is established if a steady stream of matter or energy flows through an open system. The energy or matter input, the process within the system and the energy or matter exported compensate each other in such fashion that there is no change in the system.

Stationary states are established, e.g. in industrial flow-through reactors or in biological systems. ***Radioactive E.*** within a radioactive decay series as also a stationary state, in which the amount of a radioactive nuclide remains constant because it is formed from the nuclide which preceeds it in the series at exactly the same rate as it decays.

Equilibrium cell voltage: see Terminal voltage.

Equilibrium constant: see Mass action law.

Equilibrium galvanic voltage, formerly ***equilibrium potential***: the molar chemical work of an electrode reaction divided by the molar reaction charge:

$$g_{eq}^{I,II} = (\varphi^{I} - \varphi^{II})_{eq} = \frac{\Sigma v_i \mu_i}{zF};$$

here $g_{eq}^{I,II}$ is the equilibrium galvanic voltage, φ^{I} and φ^{II} are the galvanic potentials at equilibrium, μ_i is the chemical potential of component i, z is the electrochemical valence and F is Faraday's constant.

The E. under standard conditions, i.e. at 298 K and 101.3 kPa for activity $a = 1$ is called the ***standard galvanic voltage***.

Equilibrium potential: older term for Equilibrium galvanic voltage (see).

Equilibrium reaction: a Complex reaction (see).

Equivalence point: see Volumetric analysis.

Equivalent, ***chemical equivalent***: the zth part of a particle (atom, ion, molecule, atomic group) where z is an integer equal to the stoichiometric valence of the particle. The concept of E. is always associated with a definite reaction in which the stoichiometric valence of the particle is established. For example, 1 E. sulfuric acid, H_2SO_4, is 1/2 in an acid-base reaction, because in such reactions, sulfuric acid is divalent. Chemical E. are always equivalent to a stoichiometric valence of 1. The symbol val is used for a molar E.; 1 mol E. = 1 val. When the ratios of reactants are equal to their stoichiometric ratios, is is said that the *chemically equivalent* amounts of the reactants are present. Because the E. of an atom or molecule can be different in different reactions, the use of this concept can be confusing and is discouraged; the use of moles is preferable.

Equivalent concentration: see Composition parameters.

Equivalent conductivity: see Electrical conductivity.

Equivalent groups: see Topic groups.

Equivalent mass: the relative equivalent mass, formerly called the ***equivalent weight***, is symbolized by E_r; it is the quotient of the relative formula mass, F_r, divided by the stoichiometric valence z of a compound: $E_r = F_r/z$. For an element, F_r in this definition is replaced by the relative atomic mass A_r; for a molecular substance, by the relative molecular mass M_r. The relative E. is the relative mass of a stoichiometric equivalent; it is a dimensionless quantity. The reference mass, as for all relative masses, is the atomic mass unit (see Atomic mass). The relative E. is generally not a constant for an element or compound; it depends on the reaction in question. For example, in the oxide FeO, iron has the relative E.

$$E_{r,Fe} = \frac{E_{r,Fe}}{2} = 27.93,$$

while in Fe_2O_3,

$$E_{r,Fe} = \frac{E_{r,Fe}}{3} = 18.62.$$

For sulfuric acid acting as a dibasic acid,

$$E_{r,H_2SO_4} = \frac{E_{r,H_2SO_4}}{2} = 49.04.$$

Equivalent number: absolute number of positive or negative charges released, per electrolyte particle, upon its electrolytic dissociation: $n_e = z^+n^+ = z^-n^-$. Here z^+ and z^- are the valences of the positive and negative ions, and n^+ and n^- are the numbers of ions released per electrolyte. For example, n_e of H_2SO_4 = 2.

Equivalent weight: see Equivalent mass, relative.

Er: symbol for erbium.

Erbine earths: see Ytter earths.

Erbium, symbol ***Er***: a chemical element in the Lanthanoid (see) group; Z 68, with natural isotopes of mass numbers 166 (33.41%), 168 (27.07%), 167 (22.94%), 170 (14.88%), 164 (1.56%), 162 (0.136%); atomic mass, 167.26, valence III, density 9.045, m.p. 1522 °C, b.p. 2510 °C, standard electrode potential (Er/Er^{3+}) - 2.296 V.

E. is a white, silvery metal which forms hexagonal crystals. It makes up about $2.3 \cdot 10^{-4}$% of the earth's crust, and is found in association with the other heavy lanthanoids primarily in yttrium minerals. E. is used occasionally in nuclear technology and to a certain extent in the form of Cerium mixed metal (see) in metallurgy. Its further properties, analysis, production and history are discussed under Lanthanoids (see).

Erbium compounds: compounds in which erbium is always present in the +3 oxidation state. Their general properties and production are discussed under Lanthanoid compounds (see). Some important E. are: ***erbium(III) oxide***, Er_2O_3, pink M_r 382.56, density 8.64; ***erbium(III) fluoride***, ErF_3, pink crystals, M_r 224.28, m.p. 1350 °C, b.p. 220 °C; ***erbium(III) chloride hydrate***, $ErCl_3 \cdot 6H_2O$, pink crystals, m.p. (anhydrous) 774 °C; and ***erbium(III) oxalate***, $Er_2(C_2O_4)_3 \cdot 10H_2O$, reddish, M_r 778.77, density 2.64, m.p. 575 °C.

Erdmann's reagent: see Alkaloids.

Ergobasine: see Ergot alkaloids.

Ergocalciferol: see Vitamin D.

Ergocornine: see Ergot alkaloids.

Ergocriptine: see Ergot alkaloids.

Ergoline: see Ergot alkaloids.

Ergometrine: see Ergot alkaloids.

Ergostane: see Steroids.

Ergosterol: 5,7,22-ergostatrien-3β-ol, colorless crystals, m.p. 165 °C (from 95% ethanol), $[\alpha]_D$ - 135°. E. is soluble in ether, chloroform and boiling ethanol, but practically insoluble in water. It is the main sterol of mushrooms, but is also found in higher plants. It was first isolated from ergot; it is now extracted from yeast residues. E. is a $\Delta_{5,7}$-diene, and therefore acts as a provitamin D; UV radiation converts it to ergocalciferol (see Vitamin D).

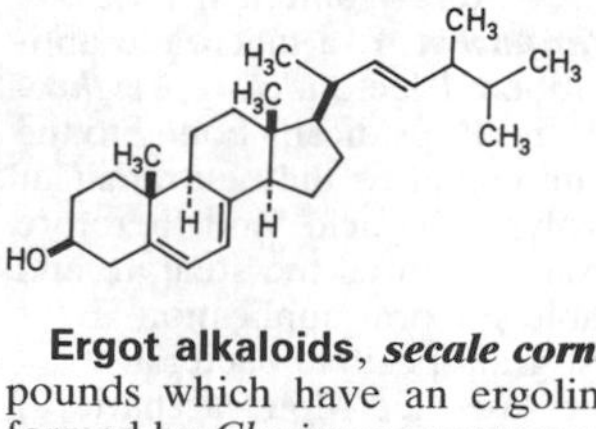

Ergot alkaloids, ***secale cornutum alkaloids***: compounds which have an ergoline skeleton; they are formed by *Claviceps purpurea* and a few other fungi. ***Spurred rye*** (*secale cornutum*) is the dried sclerotium of the fungus, which is a parasite of rye grains and other grasses. To date, more than 30 E. are known.

These compounds are biosynthesized from tryptophan and an activated derivative of isoprene. The first products are *clavine alkaloids*, which have no carboxyl group on the piperidine ring; instead they have a methyl group (*agroclavine*) or hydroxymethyl group (*elymoclavine*), or the piperidine ring is not yet closed (*chanoclavine*). These compounds have no pharmaceutical uses. Oxidation of the methyl or hydroxymethyl group on C-8 produces *5R,8R-lysergic acid (D-lysergic acid)*, which is a characteristic component of the pharmaceutically active E. preparations. Those E. in which the carboxyl group of lysergic acid is linked by an amide bond to a low-molecular-weight aminoalcohol are classified in the *ergometrine group*, while those in which a cyclic peptide group is bound in similar fashion are classified in the *ergotamine* or *ergotoxin groups*. Members of the ergometrine group are also termed ***water-soluble E.***, in distinction to the ***water-insoluble E.*** of the peptide alkaloids. In their peptide component, the ergotamine alkaloids all contain *R*-alanine, while members of the ergotoxin group contain *R*-valine. The cyclic peptides of the E. also contain *R*-proline, one other *R*-amino acid, and a ketocarboxylic acid.

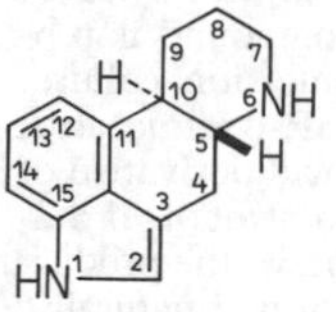

Ergoline

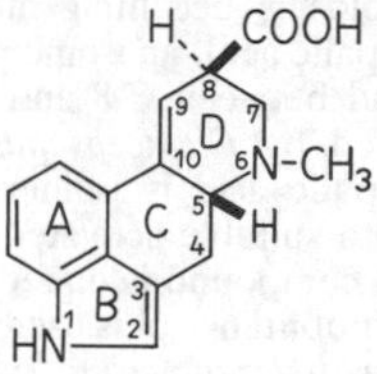

5R,8R-Lysergic acid

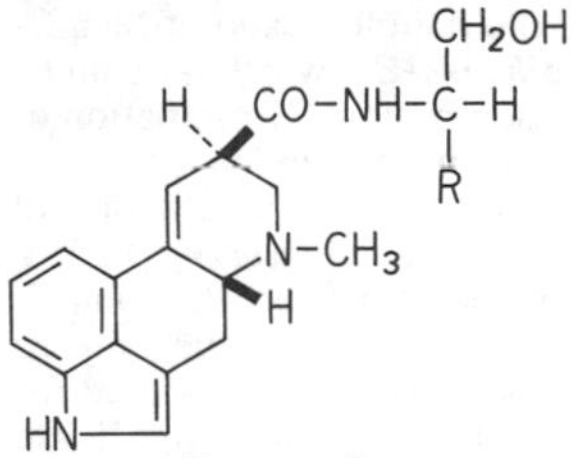

R = CH_3 : Ergometrine
R = C_2H_5 : Methylergometrine

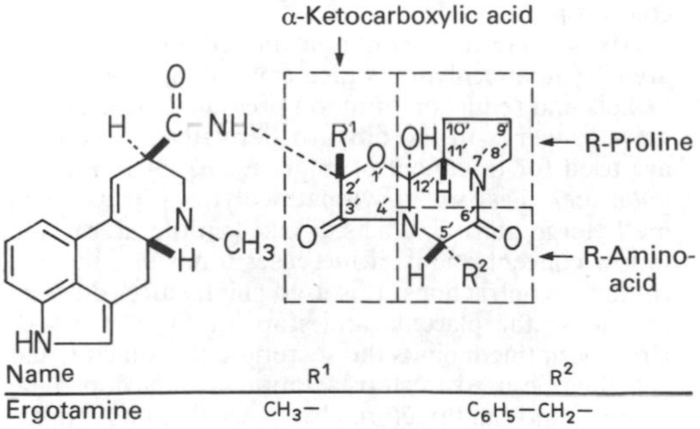

α-Ketocarboxylic acid ← R-Proline ← R-Amino-acid

Name	R¹	R²
Ergotamine	CH_3-	$C_6H_5-CH_2-$
Ergocristine	$(CH_3)_2CH-$	$C_6H_5-CH_2-$
α-Ergocriptine	$(CH_3)_2CH-$	$(CH_3)_2CH-CH_2-$
β-Ergocriptine	$(CH_3)_2CH-$	$CH_3-CH_2-CH(CH_3)-$
Ergocornine	$(CH_3)_2CH-$	$(CH_3)_2CH-$

Most natural E. from the ergometrine, ergotamine and ergotoxin groups are strongly levorotatory. Isomerization at C-8 of the lysergic acid portion of the molecule gives the 5R,8S-isolysergic acid (D-isolysergic acid) configuration; compounds with this configuration are highly dextrorotatory. They are named by adding the syllable "in" to the name of the corresponding natural E., as in ergometrinine and ergotaminine. Derivatives of 5R,8S-isolysergic acid are not physiologically active. The isomerization occurs easily, because the compounds have a double bond in the piperidine ring adjacent to the carboxyl group of the lysergic acid.

In acidic media, isomerization at C-2' of the peptide alkaloids leads to formation of the physiologically inactive aci-alkaloids. The action of light in acidic solution causes water to add to the double bond between C-9 and C-10, yielding lumi-alkaloids. The OH group in these compounds is bound to C-10.

The alkaloid content of wild spurred rye is between 0 and 1%, and is usually below 0.2%. E. are isolated by extraction of the sclerotia of cultivated high-yield strains, which usually form an alkaloid or group of alkaloids in higher concentration. Rye is artificially infected with these strains, or the fungi can be cultured on synthetic media. Partial synthesis of the alkaloids is becoming more important, starting with paspalic acid, an isomer of lysergic acid with a double bond between C-8 and C-9. This compound is produced by *Claviceps paspali* in submersion culture. Paspalic acid is isomerized to 5R,8R-lysergic acid, and a suitable activated carboxylic acid derivative of this compound is linked to an amino alcohol. If S-2-aminobutan-1-ol is used in this synthesis, the product is *methylergometrine*, which is not formed naturally. The analogous reaction with diethylamine produces LSD. Hydrogenation of the double bond between C-9 and C-10 produces ***dihydro-E.***, which are much more stable to isomerization on C-8. Bromination of α-ergocriptine at C-2 yields *bromocriptine*.

The E. are characterized by their specific optical rotation and by thin-layer chromatography. In UV light, the E. with a double bond between C-9 and C-10 have a bright blue fluorescence, but the dihydro compounds do not. The *van Urk reaction* is used for detection and photometric determination. The E. are condensed with 4-dimethylaminobenzaldehyde in acid in the presence of iron(III) compounds, forming colored products.

Alkaloids of the ergotamine and ergotoxine groups are α-sympathicolytics, which cause dilation of blood vessels and reduction of blood pressure; these effects are stronger with the dihydro derivatives. These E. are used for treatment of migrane. *Ergometrine (ergobasine)* has no α-sympathicolytic effects; like methylergometrine, it acts directly on the uterus and in low concentrations, it increases tonus and induces rhythmic contractions. It is used chiefly after delivery to release the placenta and stop the flow of blood. Bromocriptine inhibits the secretion of prolactin, and can thus be used to stop lactation. It is a dopamine agonist, and can therefore also be used as an Antiparkinson agent (see).

Historical In early times, mass poisonings were caused by consumption of ergot-containing grains; the major symptom was gangrene. In 1808, ergot was introduced as an oxytocic by Stearn. In 1875, Tancret began experiments on isolation of the E. The first pure E., ergotamine, was obtained in 1918 by Stoll. Ergometrine was first obtained in pure form in 1935, by several groups. The first total synthesis of ergometrine was achieved by Woodward and Kornfeld.

Ergot alkaloids

Name	m.p. [°C]	$[\alpha]_D^{20}$	Medium
Ergometrine	162	+ 90°	(H_2O)
– Ergometrinine	196	+414°	
Ergotamine	212	−160°	(Chloroform)
– Ergotaminine	242	+369°	
Ergocristine	156	−180°	(Chloroform)
– Ergocristinine	226	+366°	
Ergocornine	183	−186°	(Chloroform)
– Ergocorninine	228	+440°	

Ergotamine: see Ergot alkaloids.

Ergotoxin: see Ergot alkaloids.

Eriochrome dyes: a group of fast wool dyes which are not faded by light, washing or sweat, for example, eriochrome red (see Pyrazolone pigments). E. are applied to the wool in a solution of acetic or sulfuric acid, and developed with potassium dichromate.

Eriochrome black T, the sodium salt of 1-(1-hydroxy-2-naphthylazo)-5-nitro-2-naphthol-4-sulfonic acid is used as an indicator in complexometry; its color change interval is around pH 10, where the color changes from red to blue.

Erlenmeyer rule: a rule formulated by E. Erlenmeyer (1825-1909) which states that compounds with two or three hydroxy groups on a single C atom are not stable and cleave off water. The following equilibria therefore usually lie far to the right:

$$R{-}CH(OH)_2 \underset{+H_2O}{\overset{-H_2O}{\rightleftharpoons}} R{-}CHO \qquad R{-}C(OH)_3 \underset{+H_2O}{\overset{-H_2O}{\rightleftharpoons}} R{-}C(=O){-}OH$$

Exceptions to this rule are the hydrates of chloral, glyoxylic and mesoxalic acids and ninhydrin.

Erlenmeyer synthesis: a method of producing Amino acids (see).

Erosive corrosion: see Corrosion.

Erythritol: *erythro*-pentane-1,2,3,4-tetrol, a tetraalcohol which can exist as an optically active (D- or L-) or as a *meso*-form. It forms colorless crystals which taste sweet; m.p. 120 °C. It is readily soluble in water, slightly soluble in alcohol and insoluble in ether.

erythro-: a prefix indicating a certain configuration (see Monosaccharide, see Stereoisomerism 1.1.2).

Erythromycin, ***erythromycin A***: a macrolide antibiotic. It has a 14-membered lactone ring, *erythronolide*, as an aglycon. E. is glycosidically bound to the amino sugar D-desosamine and to the neutral sugar L-cladinose. It is hydrolysed by acid, and therefore only rather insoluble salts, such as the stearate and laurylsulfate, are suitable for oral application. E. is effective mainly against gram-positive bacteria.

Erythrose: an aldotetrose (see Monosaccharides) which does not occur naturally.

Erythrosine: see Eosine pigments.

Es: symbol for einsteinium.

ESCA: see Photoelectron spectroscopy.

Eschweiler-Clarke methylation: a reaction of primary and secondary amines with formaldehyde and formic acid, forming tertiary amines that contain methyl groups: $R_2NH + CH_2O + HCOOH \rightarrow R_2NCH_3 + CO_2 + H_2O$. The E. is a special variant of amine alkylation.

Eserine: same as Physostigmine (see).

ESR spectroscopy: abb. for electron spin resonance spectroscopy.

Essences: see Aroma substances.

Essential: required in the diet. See Amino acids, Fatty acids.

Essential oils: volatile, oily products with strong odors obtained by steam distillation of plants or plant parts, or by pressing the outer fruit rinds of some citrus fruits. In contrast to the fatty oils (e.g. linseed or rapeseed oil) which are also obtained from plants, E. evaporate completely and leave no oily spot on a piece of paper. The E. are mixtures, usually of 5 to 20 components. Some E. consist primarily of a single compound. Wintergreen oil, for example, is 99% methyl salicylate and lemon grass oil is 80% citral. Other E. consist of more than 100 components; mandarin orange peel oil, for example, consists of 148 compounds. The percent composition of the E. varies with the location of the plant, year and season, and also with the methods of extraction and storage.

The components of E. are mostly terpenes and sesquiterpenes. The fragrant components are oxygen derivatives, such as alcohols, ethers, aldehydes, ketones, esters, lactones and epoxides, but a few contain nitrogen or sulfur. The latter are usually present only in very small amounts. The terpene hydrocarbons generally have no odors and make the E. less soluble in ethanol.

The names of the E. come from the plants or plant parts from which they are extracted, for example, angelica root oil, cinnamon leaf oil, orange blossom oil, garlic oil, juniper berry oil, sandlewood oil and nutmeg oil. In a few cases, the E. from different parts of a plant are distinguished; for example the orange tree yields oil of neroli from the blossoms, oil of petitgrain from the leaves, twigs and unripe fruits, and oil of bitter orange from the fresh fruit peel.

Properties. The E. can be either colorless or colored, viscous or nonviscous, rarely of a semisolid, salve-like consistency. Most are lighter than water. Only a few can be distilled in vacuum without decomposition, and many are very sensitive to air, light and heat. Quality is now usually monitored by chromatographic methods (GC, HPLC, thin-layer) coupled with spectroscopy; these can also reveal adulteration.

Occurrence. E. are present in all plants with odoriferous parts. Of more than 3000 known E., however, only about 150 are utilized commercially. The most valuable and/or most abundant of these are anis, bay leaf, bergamotte, camphor, cinnamon, citronella, citrus, clove, eucalyptus, geranium, lavender, lemon grass, orange blossom, patchouli, peppermint, petitgrain, rose, rosemary, sandlewood, sassafras, spike lavender, spruce needle and ylang-ylang oils. The E. are present in the plants as droplets in the cells, or in larger vacuoles. In a few cases they are present as odorless glycosides (e.g. bitter almond oil and wintergreen oil) from which they are released by enzymatic hydrolysis when the plant is soaked in water. Interestingly, plants which are rich in alkaloids contain little or no E., and vice versa. The secretion of E. in the plant is irreversible, that is, they are not further metabolized. In the blossoms, they serve as insect attractants, and in the vegetative parts and roots, they serve as defense substances against insects and microorganisms. Many E. have fungicidal and bacteriocidal effects.

Preparation. The E. are obtained by steam distillation, extraction or pressing. *Steam distillation* yields only the volatile components of the plant, but it can be used only for relatively stable E. which are present in large amounts and are soluble in water. *Extraction* makes use of a wide variety of solvents. Enfleurage, the extraction of E. with highly purified thin layers of beef or mutton fat, and maceration, the extraction with fatty oils, are now of minor significance. The fats or oils obtained by these methods are used directly for salves, or are extracted with ethanol. ***Concrete oils*** are obtained by extraction with low-boiling solvents; they contain large amounts of waxes and paraffins. ***Absolute oils*** are obtained from concrete oils by extraction with ethanol, after which the alcohol is distilled off in vacuum. They are suitable for use in fine perfumes. The mildest method is extraction with liquefied gases, such as carbon dioxide, ammonia or ether, in the supercritical range. In this way, unchanged blossom oils can be obtained. The E. obtained from citrus fruit rinds by *pressing* have the same composition as in the rinds, and contain only small amounts of nonvolatile components.

Applications. After removal of the terpene hydrocarbons, which are odorless and insoluble in ethanol, the E. can be used to make perfumes, to perfume industrial products, to make aromas and essences, and as a source for purification of Scents (see).

Esterases: a group of Enzymes (see) which catalyse hydrolysis of ester bonds. There are several types of E.: ***carboxylases***, which cleave carboxylic esters to the corresponding carboxylic acids and alcohols (see Lipases, Acetylcholinesterase); ***Phosphatases*** (see) (including ***Phospholipases***, see) which cleave phosphate monoesters; ***phosphodiesterases*** (see Ribonucleases), which cleave the bonds between adjacent nucleotides in nucleic acids; the ***thiolesterases*** and the ***sulfatases***. Thiolesterases cleave the bond between acetyl-coenzyme A and an acyl group bound to it; formation of such an acyl-coenzyme A bond is tantamount to metabolic activation of the carboxylic acid. The sulfatases hydrolyse organic sulfuric esters of the type $R\text{-}O\text{-}SO_3H$; the group includes the ***arylsulfatases***, which act mainly on esters of nitrophenols.

Ester: a condensation product of an acid with an alcohol or phenol. E. are obtained by reaction of inorganic acids, carboxylic acids or their derivatives (acyl chlorides, acid anhydrides) with alcohols or phenols. Multibasic acids can form ***neutral E.*** such as diethyl sulfate, or ***acidic E.***, such as oxalic acid monomethyl ester. Hydroxycarboxylic acids form ***internal E.***, called lactones. The E. of inorganic acids are formed by reaction of these acids, their anhydrides or halides with alcohols. The alkyl halides can be considered E. of the hydrohalic acids, e.g. methyl chloride or benzyl chloride. Alkyl nitrates are E. of nitric acid, e.g. glycerol trinitrate (trinitroglycerol); alkyl nitrites are E. of nitrous acid. The trialkyl phosphates are neutral E. of phosphoric acid, trialkyl borates are neutral boric acid esters, e.g. trimethylborate, and tetraalkylsilicates are neutral E. of silicic acid. The carboxylic acid esters are formed from an aliphatic, aromatic or heterocyclic carboxylic acid and an alcohol, phenol or heterocyclic hydroxyl compound.

$$CH_3-C(=O)OH + HOC_2H_5 \rightleftharpoons CH_3-C(=O)OC_2H_5 + H_2O$$

$$SO_3H\text{-}OH + H\text{-}OC_2H_5 \rightleftharpoons SO_3H-O-C_2H_5 + H_2O$$

Acid + Alcohol Ester + Water

The name of an E. consists of the name of the alcohol radical and the root of the acid name, to which the suffix -ate is added, e.g. ethyl acetate. Alternatively, the name of the acid is used, then the name of the alcohol radical and the word "ester", e.g. acetic acid ethyl ester.

The E. of monocarboxylic acids are usually pleasant-smelling liquids or solids which occur, for example, in essential oils. The E. of long-chain monocarboxylic acids with long-chain monoalcohols are Waxes (see). Other naturally occurring E. are the triacylglycerols (see Fats), and complex lipids.

Some general methods for synthesis of carboxylic acid esters: a) Esterification of carboxylic acids with alcohols; the reactions usually occur by $A_{Ac}2$ mechanisms, but in special cases, for example, with tertiary alcohols, by an $A_{Ac}1$ mechanism. b) Acylation of alcohols and phenols with carboxylic anhydrides, acyl halides or monothiocarboxylic acids; the reaction is accelerated by bases, e.g. pyridine. c) Reaction of carboxylate ions, especially silver salts of carboxylic acids, with alkyl halides. d) The Claisen reaction (see). e) The Baeyer-Villinger oxidation (see). f) The Favorskii rearrangement (see). g) The addition of carboxylic acids to alkenes or alkynes (e.g. production of *tert*-butyl esters by acid-catalysed addition of carboxylic acids to isobutene, synthesis of vinyl esters). h) Reaction of carboxylic acids with diazoalkanes, e.g. the gentle synthesis of methyl esters by reaction of carboxylic acids with diazomethane in ether solution. i) Reaction of alkenes with carbon monoxide and alcohols in the presence of conc. sulfuric acid, e.g. the synthesis of formate esters by catalytic reaction of CO with alcohols under pressure. k) Hydrolysis of imidoesters. l) The Claisen condensation (see).

Some important reactions of the E. are acid-catalysed hydrolysis, which, as the reverse of esterification, occurs by the same mechanisms; and base-catalysed hydrolysis (saponification) to the alcohol and the alkali salt of the carboxylic acid using a stoichiometric amount of base. This reaction usually occurs by a $B_{Ac}2$ mechanism. Further reactions are transesterification, or alcoholysis, in which one alcohol group is replaced by another; acidolysis to exchange the carboxylic acid group of the E. for another; aminolysis to form an amide from the E. and ammonia or a primary or secondary amine; synthesis of hydrazines or hydroxamic acids from the E. and hydrazine or hydroxylamine. Primary alcohols are obtained from E. by catalytic hydrogenation with copper(II) oxide/chromium(III) oxide mixed catalysts, by the Bouveault-Blanc reduction (see) or by reduction with $LiAlH_4$. The reaction of E. with Grignard compounds produces ketones or tertiary alcohols (formate esters give aldehydes or secondary alcohols with Grignard compounds). Acyloins are obtained by Acyloin condensation (see) of carboxylates under inert conditions with sodium. β-Ketoesters also undergo Ketone cleavage (see) or Acid cleavage (see) after alkylation.

Applications. Acetoacetate is the starting material for synthesis of heterocycles, pigments and pharmaceuticals. Pyrolysis of carboxylate esters at 500 °C produces alkenes or cycloalkenes by *cis*-elimination. Xanthogenate esters react similarly, forming an alkene, a thiol and carbon oxide sulfide when heated to about 200 °C. Dicarboxylates are used for special syntheses, for example, malonate esters for Malonate ester syntheses (see), and the Dieckmann condensation (see) and acyloin reduction of dicarboxylate esters to produce medium and large alicyclic compounds. In addition, E. serve as alkylating agents (e.g. alkyl halides, dimethyl sulfate), solvents for varnish, resins and nitrocellulose, extraction solvents (e.g. for penicillin and for phenols in water treatment), as plastics and varnish bases (polyesters, cellulose esters), as starting materials for syntheses (malonates, acetates) and as flavorings and perfumes (fruit esters). Nitrites are used as ignition promoters in diesel fuels (ethyl nitrite) and in medicine as a drug for angina pectoris (iso-amyl nitrite). Nitrates are important as explosives (trinitroglycerol), silicates are used in the production of silicones, and phosphates are used as insecticides.

Esterification: reaction of an alcohol and an acid to form an Ester (see).

Ester pyrolysis: a synchronous *cis* elimination of carboxylic acid alkyl and cycloalkyl esters at high temperatures; the products are alkenes or cycloalkenes:

$$R-C(=O)-O-CH_2-CH(R')-H \longrightarrow RCOOH + CH_2{=}CH-R'$$

E. occurs via a cyclic transition state with negative activation entropy. Xanthogenates can also undergo this reaction.

Estradiol: see Estrogens.

Estrane: see Estrogens.

Estriol: see Estrogens.

Estrogens: a group of female sex hormones. Natural E. are derivatives of ***estrane***; they contain 18 carbon atoms and have an aromatic ring A with a phenolic hydroxy group on C-17. The most important E. is estradiol, 1,3,5-(10)-estratriene-3,17β-diol. It has two less active metabolites, estrone and estriol. The E. are produced in the Graafian follicles of the ovary and in the corpus luteum; during pregnancy they are also produced by the placenta.

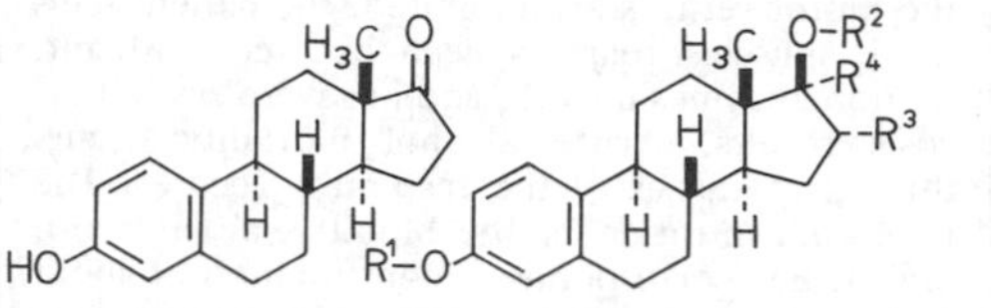

Estrone Estradiol derivatives

Name	R^1	R^2	R^3	R^4
Estradiol	H	H	H	H
Estriol	H	H	OH	H
Estradiol benzoate	C_6H_5–CO–	H	H	H
Estradiol valerate	H	CH_3–$(CH_2)_3$–CO–	H	H
Ethynyl estradiol	H	H	H	–C≡CH
Mestranol	CH_3	H	H	–C≡CH

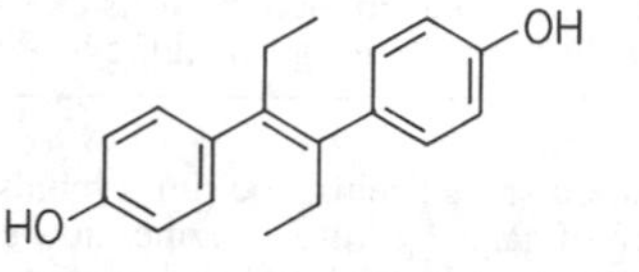

Diethylstilbestrol

The E. are also present in small amounts in the testes. They are biosynthesized via testosterone; the CH_3 group on C-10 of testosterone is eliminated as formaldehyde, and then ring A is aromatized. ***Estradiol, estrone*** and ***estriol*** are colorless, crystalline solids which are nearly insoluble in water, but are soluble in alcohol and chloroform. Estradiol: m.p. 187 °C, $[\alpha]_D^{18}$ +78° (ethanol). Estrone is polymorphic, displaying three forms: m.p. 254 °C, 256 °C and 259 °C, $[\alpha]_D^{20}$ +165° (chloroform). Estriol: m.p. 283 °C, $[\alpha]_D^{20}$ +34.4° (pyridine). The E. are partially synthesized from other steroids, and, increasingly, they are also totally synthesized. They induce secondary female sexual characteristics, and, together with the gestagens progesterone and the gonadotropins, they regulate the menstrual cycle.

E. are used therapeutically, usually in combinations with gestagens, for estrogen deficiency symptoms such as threatened abortion, amenorrhoea, dysmenorrhoea, climacteric problems and as Contraceptives (see). Of the natural E., only estriol is effective when administered orally. Therefore, esters of estradiol are used (e.g. the 3-benzoate for parenteral and the 17-valerate for oral application). Partially synthetic E. with 17α-ethynyl groups, such as ***ethynylestradiol*** and ***mestranol*** are highly effective when given orally; they are therefore used in contraceptives. The 17α-ethynyl group is introduced by ethynylation of compounds with keto functions on C-17. Synthetic compounds of the ***diethylstibestrol*** (E.-α,α'-diethyl-4,4'-dihydroxystilbene) type are also effective orally. In these compounds, the two hydroxyl groups are the same distance apart as in estradiol. The O,O-dimethyl ethers of these compounds are used as depot compounds. However, because they have many undesirable side effects, they are now only rarely used in human medicine. The tetrasodium salt of diethylstilbestrol bis-O-monophosphate (see Fosfestrol) is used to treat prostate carcinoma.

Historical: estrone was first isolated in 1920, from the urine of pregnant women, by Butenandt and Doisy, independently. It was the first sex hormone isolated. In 1930, Marrian obtained estriol in the same way. Estradiol was partially synthesized from estrone in 1932 by Schwenk and Hildebrand. In 1935 it was isolated from ovaries.

Estrone: see Estrogens.

Etaconazole: see Azole fungicides.

Etamycin, ***viridogrisein***: a heptapeptide lactone antibiotic produced by streptomycetes:

Hypic–Thr–D-Leu–D-aHyp–Sar–DiMeLeu–Ala–C-PhSar

E. is effective against gram-positive bacteria and *Mycobacterium tuberculosis*. The amino function of the threonine is linked to a hydroxypipecolic acid group (Hypic); other unusual amino acids are L-β-*N*-dimethylleucine (DiMeLeu), sarcosine (Sar) and L-α-phenylsarcosine (C-PhSar).

Etard reaction: methods of producing aromatic aldehydes by partial side chain oxidation of aromatic hydrocarbons with chromyl chloride in carbon tetrachloride or carbon disulfide. The primary product is a brown chromium complex, which precipitates out of the reaction solution, and thus protects the aldehyde from further oxidation:

$$C_6H_5{-}CH_3 \xrightarrow{+2\,CrO_2Cl_2} C_6H_5{-}CH(OCrCl_2OH)_2 \xrightarrow{(H_2O)} C_6H_5{-}CHO.$$

Decomposition of the complex with water releases the aldehyde, which can then be extracted from the solution. The E. can also be used to make ketones, by using suitable aromatic hydrocarbons.

Ethacridine: a derivative of acridine which is used in the form of its lactate as a disinfectant for wounds. E. is an intensely yellow compound.

Ethambutol: a (+)-*N,N'*-bis(hydroxyalkyl)ethylenediamine used as a tuberculostatic. The (+)-form has the R configuration at both chirality centers, and is considerably more active than the *meso*- or (-)-forms. The toxicity of all three stereoisomers, on the other hand, is approximately the same.

Ethanal: same as Acetaldehyde (see).

Ethane: CH_3-CH_3, a colorless, odorless and combustible hydrocarbon gas; m.p. - 172 °C, b.p. - 88.6 °C. E. is only slightly soluble in water, but it is readily soluble in absolute alcohol. Mixtures with air or oxygen are explosive. E. is produced from natural gas, vent gases from petroleum distillation, and hydrogenation of petroleum, tar, or coal. In the laboratory, E. is produced by electrolysis of a concentrated sodium acetate solution according to Kolbe, by the Wurtz reaction of methyl iodide, or by hydrolysis of diethyl zinc according to Frankland. E. is converted to ethyl chloride by thermal or photochemical chlorination; catalytic dehydrogenation yields ethylene. E. can react with steam in the presence of catalysts to form synthesis gas, or it can be converted to acetic acid by gas-phase oxidation in the presence of hydrogen bromide. E., like methane, is also used as a fuel.

Ethanedial: same as Glyoxal (see).

Ethane-1,2-diol: same as Ethylene glycol (see).

Ethanedioic acid: same as Oxalac acid (see).

Ethane-1,2-dithiol, ***dithioglycol***: HS-CH_2-CH_2-SH, an evil-smelling, poisonous, colorless liquid; m.p. - 41.2 °C, b.p. 146 °C, n_D^{20} 1.5590. It is insoluble in water, but soluble in ethanol and ether. It is synthesized by the usual methods for Thioalcohols (see). E. is used for the synthesis of cyclic thioacetals and thioketals, and for wetting natural rubber.

Ethanethiol, ***ethylmercaptan***: CH_3-CH_2-SH, an evil-smelling, poisonous liquid; m.p. - 144.4 °C, b.p. 35 °C, n_D^{20} 1.4310. E. can be smelled at very low concentrations; the limit of perception is at 1 mg/t. It is nearly insoluble in water, but is soluble in the usual organic solvents. It is found in coal tar, petroleum and gasoline, from which it can be removed by passage across platinum together with some hydrogen. E. is used as an intermediate in organic syntheses, and is added to other gases, such as natural gas, as an alarm gas which can easily be smelled if there is a gas leak.

Ethanol, ***ethyl alcohol***, often simply ***alcohol***: CH_3-CH_2-OH, the best known of the alcohols, a colorless, spicy-smelling liquid with a burning taste; m.p. - 114.5 °C, b.p. 78.3 °C, n_D^{20} 1.3614. E. is combustible and ignites easily. It is miscible in any proportion with water, ether, benzene and petroleum distillates. When E. is mixed with water, its temperature rises and the volume of the solution is less than the sum of the volumes of the two pure liquids. E. forms azeotropic mixtures with water and various organic substances; for example, a mixture of 95.6% E. and 4.4% water cannot be further separated by fractional distillation. Absolute E. is therefore obtained in the laboratory by distillation over calcium oxide, with subsequent further drying with calcium, sodium or sodium methylate.

E. is very reactive. Oxidation produces acetaldehyde or acetic acid; dehydration leads to ethene or diethyl ether. E. forms esters with carboxylic or mineral acids and their derivatives. With aldehydes and ketones it forms acetals or ketals. The 4-nitrobenzoate ester (m.p. 57 °C) and the 3,5-dinitrobenzoate (m.p. 94 °C) are used for analytic identification. E. gives the Lieben's iodoform test (see) with potassium hydroxide and iodine. The water content is determined quantitatively by measurement of the density (see alcoholometry).

In nature, E. is found widely, but in small amounts. The normal physiological E. content of the blood is 0.02 to 0.03%. The Breath alcohol test (see) and Blood alcohol determination (see) are widely used for law enforcement.

When consumed, small amounts of ethanol have a stimulant effect, but in larger concentrations it releases inhibitions, impairs judgement and is toxic. Even the inhaled vapors can cause organ damage. E. is absorbed from the mucous membranes or the skin. Inorganic and organic toxins and some medications reduce alcohol tolerance; or consumption of alcohol increases the sensitivity of the organism to these substances. Symptoms of acute alcohol intoxication are at first stimulation, elevated reflexes and increased motor drive; later the motor centers are inhibited, the patient feels tired, muscles relax, speech becomes difficult, motion becomes unsure, and in severe cases, narcosis results. Acute alcohol mortality occurs through paralysis of the respiratory center. The fatal concentration in the blood is about 5 pro mill, which corresponds to consumption of about 250 g pure alcohol. When the outside temperature is very cold, the intoxicated person is in danger of hypothermia, because the dilated skin blood vessels radiate heat and the body temperature may drop below 30 °C. Chronic alcohol intoxication (alcoholism) is characterized by irritation of the gastrointestinal tract with reduced absorption of nutrients, impaired digestion, damage to liver and cardiac muscle, and central nervous system defects. Alcohol hallucinations or delirium tremens are specific symptoms. In the body, ethanol is degraded mainly by oxidation to carbon dioxide and water. Some is oxidized to acetaldehyde or is excreted unchanged through the lungs or kidneys.

E. has been obtained since prehistoric times by alcoholic fermentation of carbohydrates. Fermentation is now carried out on an industrial scale to produce alcoholic beverages, such as beer, wine and distilled spirits, and to produce E. as a solvent or as a starting material for organic syntheses (see Alcohols). The sources of carbohydrate are fruits, starch from potatoes or grains, molasses, sugar cane and residues from cellulose processing. Hydration of ethene is also used for industrial synthesis of E. The addition of water occurs on phosphoric acid carrier catalysts at 300 to 400 °C, under 2 to 4 MPa pressure, or with sulfuric acid at 75-80 °C and pressures of 1.7 to 3.5 MPa.

Applications. E. is used on a large scale as a beverage and as a solvent. Mixed with gasoline, it can be used as a motor fuel. E. is also an important starting material for synthetic chemistry (see Alcohols).

Ethanolamines, ***aminoethanols, aminoethyl alcohols***: compounds which combine the properties of nitrogen bases and alcohols. Formally, E. can be derived by stepwise exchange of 2-hydroxyethyl groups for H atoms in ammonia. ***Monoethanolamine, 2-aminoethanol, 2-hydroxyethyl amine, colamine***, NH_2-CH_2-CH_2-OH, is a colorless, viscous liquid, m.p. 10.5 °C, b.p. 170-172 °C, n_D^{20} 1.4539. ***Diethanolamine, bis(2-hydroxyethyl)amine***, HN(CH_2-CH_2-$OH)_2$ forms colorless crystals, m.p. 28 °C, b.p. 271 °C, n_D^{20} 1.4776. ***Triethanolamine, tris(2-hydroxyethyl)amine***, N(CH_2-CH_2-$OH)_3$, is a very viscous, colorless oil, m.p. 21 °C, b.p. 360 °C, n_D^{20} 1.4852.

E. have weak, amine-like odors. They are highly hygroscopic and absorb carbon dioxide. They are readily soluble in water, alcohol, acetone and glycols, but are barely soluble or insoluble in aliphatic and aromatic hydrocarbons and ether. The aqueous solutions of the E. are basic. E. form salts with acids, and with fatty acids, they form the neutral to weakly basic ethanolamine soaps, which are soluble in water and some organic solvents. The sulfuric acid esters of E.

are suitable for synthesis of ethylenimine or morpholine. The nitric acid esters are used as explosives. E. are produced mainly by transesterification of ethylene oxide with aqueous ammonia. The resulting mixture can be separated by fractional distillation. E. are used mainly in the form of their fatty acid salts as cleansers, emulsifiers, salve bases, etc. They are also used in the production of synthetic resins, for corrosion protection and as softeners. They are used to separate hydrogen sulfide and carbon dioxide from gases and, mixed with Cu(I) chloride, to remove alkenes from hydrocarbon mixtures. Monoethanolamine is a biogenic amine formed by decarboxylation of serine.

Ethanoic acid: same as Acetic acid (see).

Ethanoyl chloride: same as Acetyl chloride (see).

Ethene, ***ethylene***: $CH_2{=}CH_2$, the first of the alkene series. E. is a colorless, combustible gas with a narcotic effect; m.p. - 169.1 °C, b.p. - 103.7 °C. It is slightly soluble in water, but readily soluble in ethanol. E. burns with a bright, sooty flame and forms explosive mixtures with air. It is present in coking gas and refinery gases from petroleum processing. It is now produced exclusively by pyrolysis of liquid gases or gasoline hydrocarbons, followed by low-temperature pressure distillation. After drying, all condensates formed by pyrolysis and cracking gas are led into the ethane column for distillation. All C_3 hydrocarbons and higher-boiling compounds are withdrawn from the bottom of this column, while E., ethane and ethyne are distilled over. The ethyne is removed by selective hydrogenation or by use of a selective solvent (see Gas scrubbing, see Extractive distillation). Because the difference between the boiling points of E. and ethane is slight, a very efficient column (80 to 100 theoretical plates) with a high reflux ratio (2.5 to 3) is needed to separate them. Very cold, liquid E. is used as coolant for the condenser.

In the USA and the former USSR, E. is produced mainly by pyrolysis of ethane, propane and butane, while in most European countries and Japan, gasoline hydrocarbons are pyrolysed. The most useful laboratory method of producing E. is dehydration of ethanol with sulfuric acid, phosphoric acid or potassium hydrogen sulfate. Other possibilities are the cleavage of hydrogen halides out of ethyl halides or the passage of ethanol vapor over aluminum oxide catalysts at 300 to 400 °C. The selective hydrogenation of ethyne on a palladium catalyst is no longer technologically significant.

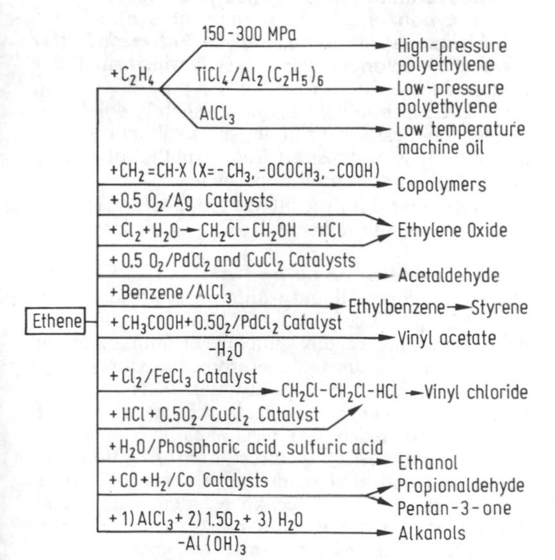

Overview of the technologically important reactions of ethene

E. is very reactive and undergoes numerous addition reactions. It is a key compound in the synthesis of many important compounds (table), and in many cases, it has replaced ethyne as a starting material for large-scale syntheses. A special application of E. is a result of the fact that it promotes ripening of fruits (see Phytohormones).

Ethenecarboxylic acid: same as Acrylic acid (see).

Ethephon: see Phytohormones.

Ether: an organic compound with the general formula R-O-R', where R and R' can be any alkyl or aryl groups. If it is not otherwise specified, "ether" used as the name of a compound means "diethyl ether". E. can be considered anhydrides of alcohols and/or phenols, or also as substitution products of water in which both H atoms are replaced by carbon groups.

A distinction is made between ***symmetric (simple)*** E., in which R = R', and ***unymmetric (mixed)*** E., in which R and R' are different hydrocarbon groups. Depending on the nature of the groups, the E. can be further classified as dialkyl, diaryl, enol or alkyl aryl (phenol) E.

Nomenclature. Symmetric E. are named as dialkyl, dicycloalkyl or diaryl ethers, for example, diethyl or diphenyl ether. Unsymmetric E. are named as alkoxy, cycloalkoxy or arylalkoxy derivatives of hydrocarbons. The hydrocarbon with the greatest number of C atoms is the parent compound; for example, methoxyethane or methoxybenzene. Unsymmetric E. are often given names such as ethyl methyl ether (or, incorrectly, methyl ethyl ether). Trivial names are allowed, especially for phenol E., for example, anisol and nerolin. Diethyl ether (see) is often called "ether" for short.

Properties. Dimethyl E. and methoxyethane are gaseous at room temperature. The next homologs are mobile, volatile and low-boiling liquids. Because they cannot form hydrogen bonds, and therefore are not associated, they are more volatile than alcohols or phenols with the same number of C atoms. They are less dense than water, and most are not soluble in water. Many E. have a pleasant ("etheric") odor and are therefore used in the composition of perfumes. The lower E. are easily inflammable and burn with bright flames in atmospheric oxygen.

In work with E., all precautions must be carefully followed, because the vapors form explosive mixtures with air. In addition, when E. have stood in air for a longer period, they form ether peroxides which explode when warmed. The first products are hydroperoxides, which are converted to polymeric peroxides. In the presence of metallic copper, diphenylamine, naphthalene or benzidine, peroxides do not form.

Analytical. E. can be identified by cleavage with hydrogen iodide. The resulting iodoalkanes are then converted to *S*-alkylisothiuronium salts by reaction with thiourea and characterized. Symmetric E. can also be cleaved with 3,5-dinitrobenzoyl chloride in the presence of zinc chloride, producing 3,5-dinitrobenzoic esters and the corresponding alcohols. The infrared spectra of dialkyl E. have typical bands in the region from 1060 to 1150 cm^{-1} and can be readily distinguished from aryl alkyl and diaryl E., which absorb between 1230 and 1270 cm^{-1}. The ^1H NMR spectra of methoxyalkanes have singlets at δ values of 3.2 to 3.4 ppm, while methoxyarenes have singlets around 3.7 ppm. The mass spectra of dialkyl ethers show only weak molecular ion peaks; fragmentation produces key fragments with the composition $RCH{=}OH^+$ (m = 31, 45, 59, ...). Alkyl aryl and diaryl ethers have strong molecular ion peaks.

Reactions. E. are very stable compounds and are therefore suitable inert solvents for chemical reactions. Since most are not attacked by sodium, they can be dried by this metal. When E. dissolve in concentrated strong acids, such as sulfuric, perchloric, hydrochloric or hydroiodic acids, they form *oxonium salts* which are hydrolysed by water:

$$R\text{–}\bar{\underline{O}}\text{–}R' + H^+X^- \longrightarrow R\text{–}\overset{+}{\bar{O}}(H)\text{–}R'X^-$$

E. form complexes with Lewis acids. The adduct of diethyl ether and boron trifluoride (boron trifluoride etherate) can even be distilled. Further reaction of such complexes with fluoroalkanes produces tertiary oxonium salts (Meerwein), which are excellent alkylating reagents:

$$C_2H_5\text{–}O\text{–}C_2H_5 + BF_3 \rightarrow (C_2H_5)_2\overset{+}{O}\text{–}\overset{-}{B}F_3 \xrightarrow{C_2H_5F} (C_2H_5)_3\overset{+}{O}\text{–}BF_4^-$$

(triethyloxonium tetrafluoroborate).

Diaryl ethers do not undergo all of these reactions. Dialkyl and aryl alkyl ethers are cleaved by concentrated hydroiodic or hydrobromic acid into a mixture of haloalkanes and alcohol, or phenol (*ether cleavage*): R-O-R + HI → R-I + R-OH. The reaction also occurs via formation of oxonium salts. E. cleavage is used for quantitative determination of methoxy or ethoxy groups according to Zeisel. *Autooxidation* of E. upon long standing in the air leads to hydroperoxides and peroxides:

$$CH_3\text{–}CH_2\text{–}O\text{–}CH_2\text{–}CH_3 + O_2 \longrightarrow CH_3\text{–}CH(O\text{–}OH)\text{–}O\text{–}CH_2\text{–}CH_3 \text{ (hydroperoxide);}$$

$$n\,CH_3\text{–}CH(O\text{–}OH)\text{–}O\text{–}C_2H_5 \longrightarrow n\,C_2H_5OH + \left[CH_3\text{–}CH(O\text{–}O\text{–})\right]_n \text{ („polymeric" ether peroxide).}$$

The ether peroxides are polymeric, non-volatile, oily liquids with a pungent odor; they are enriched in the residue after distillation of E., and they explode readily on heating. Therefore, E. must be tested before use with titanium(IV) oxygen sulfate or iron(II) salts and rhodanide ions to determine whether peroxides are present. They can be reductively destroyed with sodium, iron(II) or manganese(II) salts if necessary.

Synthesis of E. usually starts from alcohols or phenols. 1) Intermolecular dehydration of alcohols with concentrated sulfuric acid, anhydrous phosphoric acid or aluminum oxide catalysts produces symmetric E.: 2 R-OH → R-O-R + H_2O. 2) Both symmetric and unsymmetric E. can be made by acid-catalysed addition of alcohols to alkenes:

$$(CH_3)_2C{=}CH_2 + BOC_2H_5 \xrightarrow{H^+} (CH_3)_3C\text{–}OC_2H_5.$$

3) The Williamson synthesis (see) from alcoholates or phenolates and haloalkanes is particularly well suited for synthesis of dialkyl and aryl alkyl E. 4) Phenol E. can also be synthesized by alkylation of phenols with dialkyl sulfates in alkaline solution, or by methylation with diazomethane.

Applications. E. are used as solvents for many organic compounds. They are also used as scents in perfumes, softeners, disinfectants and anesthetics.

Ether cleavage: see Ethers.

Ether peroxides: non-volatile, high-molecular-weight, highly explosive compounds formed by Ethers (see) upon long standing in air.

Ethiofencarb: a Carbamate insecticide (see).

Ethirimol: see Pyrimidine fungicides.

Etholoxamine: see Antihistamines.

Ethosuximide: a Succinimide (see) with two alkyl substituents in the 3 position. E. is used as an anticonvulsant, as are other related compounds substituted on the N atom and the C3 atom.

Ethoxyaniline: same as Phenetidine (see).

Ethoxybenzene: same as Phenetol (see).

2-Ethoxynaphthalene, ***naphth-2-ol methyl ether (bromelia)***: colorless, shiny crystals which smell like orange blossoms; m.p. 37-38 °C, b.p. 282 °C, n_D^{26} 1.5975. E. is insoluble in water, slightly soluble in benzene and carbon disulfide, and soluble in oils and alcohols. It is synthesized from naphth-2-ol and diethylsulfate, and is used in perfumes.

Ethyl-: a term for the atomic group CH_3-CH_2- in a molecule, for the radical CH_3-$CH_2\cdot$ and for the cation CH_3-CH_2^+.

Ethyl acetate, ***acetic acid ethyl ester***: CH_3-CO-OC_2H_5 a colorless liquid with a pleasant odor; m.p. -83.6 °C, b.p. 77.06 °C, n_D^{20} 1.3723. E. is slightly soluble in water, and readily soluble in alcohol, ether and chloroform. It forms azeotropic mixtures with ethanol, methanol, ether and water. E. is slowly hydrolysed by water to form acetic acid and ethanol. It undergoes the reactions typical of carboxylic acid esters. E. is synthesized by direct ester formation from acetic acid and ethanol in the presence of sulfuric acid, by the Claisen-Tiščenko reaction from acetaldehyde, or by addition of ethanol to ketene. It is used mainly as a solvent for resins, fats, oils, celluloid, colophony, chlorinated rubber and cumarone resins, and for synthesis of acetoacetate, glues and scents.

Ethyl acetoacetate, ***acetoacetic acid ethyl ester***: CH_3-CO-CH_2-$COOC_2H_5$, a colorless liquid with a pleasant odor; m.p. -45 °C (also given as < - 80 °C), b.p. 180.4 °C, n_D^{20} 1.4194. E. is slightly soluble in water and miscible with most organic solvents. Due to Keto-enol tautomerism (see), E. under normal conditions is about 92.5% in the keto form and about 7.5% in the enol form: CH_3-CO-CH_2-$COOC_2H_5$ ⇌ CH_3-C(OH)=CH-$COOC_2H_5$. The mixture of the two components can be separated by fractional distillation in a quartz apparatus; the more volatile enol form (n_D^{20} 1.4432) goes over first. Under special conditions of separation, the keto form (m.p. - 39 °C, n_D^{20} 1.4171) can also be obtained in nearly pure form. In the presence of acids or bases, or when the compound is stored under normal conditions, the original equilibrium is rapidly re-established. Because the molecule contains several functional groups, E. is useful in numerous syntheses of chain and cyclic products. The acetoacetate syntheses and a number of heterocycle syntheses are of particular importance. E. is a weak acid which forms the corresponding salts with alkali hydroxides:

$$CH_3\text{–CO–}CH_2\text{–}COOC_2H_5 + NaOH \xrightarrow[-H_2O]{} [CH_3\text{–CO–CH–}COOC_2H_5]^- Na^+$$

The sodium salt thus formed serves as the starting reagent for ***acetoacetate syntheses***. In reactions with alkyl halides, the electrophilic attack on the anion usually occurs at the C-atom; O-alkylations are observed less often:

$$\left[CH_3\text{–}\underset{|\overline{\underline{O}}|^-}{C}\text{=CH–}COOC_2H_5 \longleftrightarrow CH_3\text{–}\underset{\overset{\|}{O}}{C}\text{–}\overline{C}H^-\text{–}COOC_2H_5\right] Na^+ \xrightarrow[-NaX]{+R-X} CH_3\text{–}\underset{\overset{\|}{O}}{C}\text{–CHR–}COOC_2H_5$$

Acid cleavage (see) can convert the alkylated E. into acetic acid, ethanol and the corresponding lengthened monocarboxylic acid R-CH_2-COOH, in a reversal of the Claisen condensation (see). In Ketone cleavage (see), the products are the ketone extended by CH_3-CO-CH_2-R, carbon dioxide and ethanol. Depending on the reaction conditions, C-acyl or O-acyl derivatives can be obtained by acylation of E.:

$$CH_3\text{–}\underset{\overset{\|}{O}}{C}\text{–}\overset{\overset{CO-R}{|}}{C}H\text{–}COOC_2H_5$$

C-acylacetoacetate ester

$$CH_3\text{–}\underset{\underset{O-CO-R}{|}}{C}\text{=CH–}COOC_2H_5$$

O-acylacetoacetate ester

Subsequent ketone cleavage of the C-acyl derivatives produces β-diketones with the general formula CH_3-CO-CH_2-CO-R. Thus with suitable alkylation and acylation agents, E. becomes the starting material for simple preparative synthesis of numerous ketones, β-diketones and carboxylic acids. E. is synthesized by Claisen condensation (see) of ethyl acetate or by reaction of diketene with ethanol.

E. is used as an intermediate in the preparation of drugs, dyes and pesticides, and as a solvent and aroma additive.

Ethyl alcohol: same as Ethanol (see).

Ethylamines: the primary, secondary and tertiary ethyl derivatives of ammonia. Along with the Methylamines (see), the E. are the most important industrial aliphatic amines. ***Monoethylamine*** (aminoethane), C_2H_5, is a combustible gas with an ammonia-like odor; m.p. - 81 °C, b.p. 16.6 °C, $n_D^{8.4}$ 1.3763. It is readily soluble in water, alcohol and ether. ***Diethylamine***, $(C_2H_5)_2NH$, is a liquid with an ammonia-like odor; m.p. - 48 °C, b.p. 56.3 °C, n_D^{20} 1.3864. It is readily soluble in water, alcohol and ether. ***Triethylamine***, $(C_2H_5)_3N$, is a liquid with an ammonia-like odor; m.p. - 114.7 °C, b.p. 89.3 °C, n_D^{20} 1.4010. It is readily soluble in water below 18.7 °C; above that temperature it dissolves in most organic solvents.

E. are stronger bases than ammonia. With mineral acids, they form salts which dissolve readily in water and alcohol. They also display all the typical reactions for primary, secondary and tertiary amines. E. can be made by reaction of ethyl halides with ammonia. Some newer catalytic methods of synthesis of E. include aminating hydrogenation of acetaldehyde, aminolysis of ethanol and hydrogenation of acetonitrile. All methods of synthesis produce mixtures of the three E., which must be separated by fractional distillation. E., especially diethylamine, are used as intermediates in chemical industry for production of synthetic resins, pharmaceuticals, vulcanization accelerators and insecticides. Triethylamine is also used as a solvent and acid-binding component in organic syntheses.

N-Ethylaminobenzene: same as *N*-Ethylaniline (see).

N-Ethylaniline, ***N-ethylaminobenzene***: an *N*-substituted aniline derivative. ***N-Ethylaniline (N-ethylaminobenzene)***, C_6H_5-NH-C_2H_5, is a secondary amine; it is a colorless, poisonous, highly refractive liquid; m.p. - 63.5 °C, b.p. 205 °C, n_D^{20} 1.5559. ***N,N-Diethylaniline (N,N-diethylaminobenzene)***, C_6H_5-$N(C_2H_5)_2$, is a tertiary amine; it is a colorless to bright yellow, poisonous liquid; m.p. - 38 °C, b.p. 216 °C, n_D^{20} 1.5421. The *N*-E. are insoluble in water, but are soluble in most organic solvents and mineral acids. They are synthesized by the reaction of aniline with ethanol or diethyl ether in the presence of a catalyst. They are used to synthesize triphenylmethane, azo or thiazine pigments.

Ethylbarbital: see Barbitals, table.

Ethylbenzene: an aliphatic-aromatic hydrocarbon and a colorless, combustible liquid; m.p. - 95 °C, b.p. 136.2 °C, n_D^{20} 1.4959. E. is nearly insoluble in water, but is soluble in most organic solvents and mineral acids. It can be synthesized by catalytic dehydrogenation and cyclization of certain petroleum fractions, by reaction of benzene with ethyl bromide in the presence of aluminum chloride according to Friedel and Crafts, or by reaction of benzene with ethene in the presence of a Friedel-Crafts catalyst. E. is used as a solvent and, in very large amounts, as a starting material for the synthesis of styrene by dehydrogenation.

$C_6H_5-CH_2-CH_3$ Ethyl benzene

Ethyl benzoate: see Benzoates.

Ethylbenzhydramine: see Antiparkinson's compounds.

Ethylbiscoumacetate: a bis-4-hydroxycumarol derivative used as an anticoagulant. The maximum effect is achieved after about 24 h, and the effect lasts for about 48 h altogether. The impetus for development of this class of compounds came from the observation that animals which eat spoiled sweet clover hay develop severe bleeding. The spoiled hay contains dicumarol, which is responsible for the effect.

Ethyl bromide, ***bromoethane***: CH_3-CH_2-Br, a poisonous, colorless liquid which smells like ether; m.p. - 118.6 °C, b.p. 38.4 °C, n_D^{20} 1.4239. E. has a narcotic effect. It is barely soluble in water, but dissolves readily in alcohol and ether. E. can be synthesized from ethanol and hydrobromic acid, or by addition of hydrogen bromide to ethene. It is used mainly as an ethylating reagent in organic chemistry. It is no longer used as an anaesthetic because of its toxicity.

Ethylbutylthiobarbital: see Barbitals, table.

Ethyl cellulose: a Cellulose ether (see).

Ethyl chloride, ***chloroethane***: CH_3-CH_2-Cl, a colorless gas with a sweetish odor; m.p. - 136.4 °C, b.p. 12.3 °C, n_D^{20} 1.3676. E. is only slightly soluble in water, but dissolves readily in alcohol, ether, acetone and chloroform. It ignites easily and burns with a green-edged flame. Under normal conditions, it is very stable and is only slowly hydrolysed or oxidized. The chlorine atom in E. can very easily be replaced by a nucleophilic reagent. E. is synthesized chiefly by chlorination of ethane around 400 °C, or by addition of hydrogen chloride to ethylene in the liquid phase in the presence of aluminum chloride, or in the gas phase, or by reaction of ethanol with hydrogen chloride. E. is used for the synthesis of many ethane derivatives, such as amines, thiols, ethers, thiocyanates or ethylbenzenes. A large fraction of the E. made today is still used in the production of tetraethyl lead, which is required as an anti-knock agent for gasoline. E. is also used as a coolant, local anesthetic and solvent.

Ethyl cinnamate: see Cinnamates.

Ethyl cyanoacetate: NC-CH_2-CO-OC_2H_5, a colorless liquid with a pleasant odor; m.p. - 22.5 °C, b.p. 207 °C, n_D^{20} 1.4176. C. is insoluble in water, but soluble in alcohol and ether. It is a very reactive compound, containing three reactive centers in the molecule: the ester, nitrile and CH-acidic methylene groups. Because of this high functionality, C. is used for the synthesis of many chain and ring compounds. It is synthesized by esterification of cyanoacetic acid with ethanol in the presence of mineral acids, or by reaction of ethyl chloroacetate with alkali cyanides. C. is used as an intermediate in the synthesis of β-alanine, B vitamins, barbituric acid derivatives, uric acid, cyanoacrylate esters, etc.

Ethyl diazoacetate, often shortened to ***diazoacetate***: N≡N-CH-$COOC_2H_5$, a relatively stable aliphatic diazo compound. E. is a non-viscous, yellow liquid with a pungent odor; m.p. -22 °C, b.p. 45 °C at $1.0 \cdot 10^3$ Pa, n_D^{20} 1.4605. It is barely soluble in water, but dissolves easily in most organic solvents. Its stability under neutral conditions is due to a high degree of resonance stabilization. In the presence of protons, it decomposes with the release of nitrogen and formation of substituted carboxylic acids:

$$N{\equiv}N{-}CH{-}COOC_2H_5 \xrightarrow{+HX} [N{\equiv}N{-}CH_2{-}COOC_2H_5]^+X^- \xrightarrow[-N_2]{} X{-}CH_2{-}COOC_2H_5$$

Since the rate of decay of E. is directly proportional to the hydrogen ion concentration, this reaction can be used for pH determination. E. is synthesized by the reaction of nitrous acid with ethyl aminoacetate in the cold. The compound is used for organic syntheses.

Ethylene: same as Ethene.

Ethylene bromide: same as Ethylene dibromide (see).

Ethylene chloride: same as Ethylene dichloride (see).

Ethylene chlorohydrin, ***2-chloroethanol***: Cl-CH_2-CH_2-OH, a chlorine substituted monoalcohol, a colorless liquid with a faint, ether-like odor; m.p. - 68 °C, b.p. 128-129 °C. E. is readily soluble in water, alcohol and ether.

> Ethylene chlorohydrin is a strong poison. Inhalation of the vapors can be fatal, because hydrolysis of the compound in the lungs produces hydrochloric acid. Ethylene chlorohydrin can also enter the body through the skin, sometimes with fatal results.

E. is synthesized in the laboratory by passing hydrogen chloride through ethylene glycol. In the industrial synthesis, ethylene oxide reacts with hydrogen chloride. E. is a good solvent for cellulose acetate, resins and pigments, and is an important intermediate in organic syntheses. Its phosphate esters are used as softeners.

Ethylene copolymers: mixed polymers formed from ethene and other vinyl monomers. ***Ethylene-propylene copolymers (polyallomers)*** are very important because of their rigidity, dimensional stability and low tendency to crack under tension. They are used for the synthesis of synthetic Rubber (see). The crystallinity of E. with α-olefins is determined essentially by the number of C atoms in the α-olefin, for example, ethylene-propylene copolymers (50% ethylene) or ethylene-hexene copolymers (60% ethylene). ***Ethylene vinyl acetate copolymers (EVA copolymers)*** are very resistant to impact and have a good resistance to tension cracking. They are used to make adhesives, to coat paper and to improve the impact resistance of polyvinyl chloride.

Ethylenediamine, ***1,2-diaminoethane***: H_2N-CH_2-CH_2-NH_2, an aliphatic diamine. E. is a colorless liquid which smells faintly like ammonia; m.p. 11 °C, b.p. 117 °C, n_D^{20} 1.4571. It is readily soluble in water and most organic solvents, but it does not dissolve so well in lower hydrocarbons. E. undergoes the reactions typical of primary amines and forms stable salts with acids. It forms metal complexes with transition metal salts. E. is synthesized by reaction of ethylene chloride with ammonia, ethanolamine with ammonia, or hydrocyanic acid, ammonia and hydrogen with formaldehyde in the presence of catalysts. The industrial significance of E. is growing. It is used in the production of pesticides, metal complexing agents, softeners, synthetic resins, drugs, textile conditioners and so on.

Ethylenediamine tetraacetic acid, ***ethylenedinitrilotetraacetic acid***, abb. ***EDTA***: $(HOOC\text{-}CH_2)_2N\text{-}CH_2\text{-}CH_2\text{-}N(CH_2\text{-}COOH)_2$, forms colorless and odorless crystalline needles; m.p. 242 °C (dec.). When slowly heated, it decomposes above 150 °C with release of CO_2. E. is only slightly soluble in water, but it dissolves in bases by forming salts. Only two of the four protons on thc carboxyl groups give acidic reactions, so E. is believed to have the structure of a double betaine. E. and its salts form soluble, stable complexes with polyvalent metal ions, such as calcium, magnesium, iron, copper, zinc, etc. These complexes are called *chelates*.

E. is synthesized by reaction of ethylene diamine with chloroacetic acid, or from ethylene diamine, formaldehyde and hydrocyanic acid or sodium cyanide. The E. and its alkali metal salts are used in nearly all areas of analysis, including polarography and complexometry, and as a calorimetric reagent, as a masking reagent in precipitations and extractions, in the determination and correction of water hardness, as an additive to shampoos, soaps and detergents, to remove iron salts from textiles, to remove heavy metal ions from foods, and as a means of detoxifying heavy metal ions and radioactive metal ions in cases of poisoning.

Ethylene dibromide, ***1,2-dibromoethane***, obsolete trivial name, ***ethylene bromide***: $Br\text{-}CH_2\text{-}CH_2\text{-}Br$, a colorless liquid which smells like chloroform; m.p. 9.8 °C, b.p. 131.3 °C, n_D^{20} 1.5387. It is scarcely soluble in water, but dissolves easily in most organic solvents. E. is synthesized by addition of bromine to ethene. It is used mainly as an additive for leaded gasolines and as a pesticide.

Ethylene dichloride, ***1,2-dichloroethane***, obsolete trivial name, ***ethylene chloride***: $Cl\text{-}CH_2\text{-}CH_2\text{-}Cl$, a colorless liquid which does not burn readily; m.p. -35.3 °C, b.p. 83.5 °C, n_D^{20} 1.4448. E. is only slightly soluble in water, but dissolves readily in alcohol, ether, acetone and benzene. It forms constant-boiling azeotropic mixtures with many solvents, including water, ethanol, methanol, carbon tetrachloride, and trichloroethylene. As in other Haloalkenes (see), the chlorine atoms can easily be substituted by nucleophilic reagents, forming the corresponding 1,2-disubstituted ethane derivatives, such as ethylene glycol and succinic dinitrile. E. is synthesized by addition of chlorine to ethene in the presence of iron(III) chloride or by catalytic oxichlorination of ethene: $CH_2{=}CH_2 + 2\ HCl + 1/2O_2 \rightarrow Cl\text{-}CH_2\text{-}CH_2\text{-}Cl + H_2O$. E. is used chiefly as an intermediate in the production of vinyl chloride. It is also used as a solvent for fats, oils, resins and rubber, and for the synthesis of many other ethane derivatives.

Ethylene glycol, ***ethane-1,2-diol***, sometimes simply ***glycol***: $HO\text{-}CH_2\text{-}CH_2\text{-}OH$, the simplest glycol; a viscous, sweet-tasting, hygroscopic liquid with no color or odor; m.p. 13 °C, b.p. 197 °C, n_D^{20} 1.4313. E. is miscible with water and ethanol, but is insoluble in ether. It is toxic and therefore may not be used in cosmetics or foods. E. is very reactive; depending on the conditions, oxidation produces one of several products (Fig.).

$$CH_2OH{-}CH_2OH \rightarrow \underset{\text{Glycol aldehyde}}{CHO{-}CH_2OH} \rightarrow \left[\underset{\text{Glycolic acid}}{COOH{-}CH_2OH},\ \underset{\text{Glyoxal}}{CHO{-}CHO}\right] \rightarrow \underset{\text{Glyoxylic acid}}{COOH{-}CHO} \rightarrow \underset{\text{Oxalic acid}}{COOH{-}COOH}$$

Glycol cleavage occurs with reagents such as lead(IV) acetate or periodic acid. When E. is heated with sulfuric or toluenesulfonic acid, it forms 1,4-dioxane, but when it is heated with phosphoric acid, it forms diglycol or polyethylene glycols. With hydrogen chloride, E. forms first ethylene chlorohydrin, and only at higher temperatures is 1,2-dichloroethane formed.

Production. In the laboratory, E. is obtained by hydrolysis of 1,2-dichloroethane or ethylene chlorohydrin, but on an industrial scale, it is made by reaction of ethylene oxide with water at 160 to 200 °C and 2 MPa pressure. Di-, tri- and polyethylene glycols are formed as byproducts of the ethylene oxide reaction. Reaction of ethylene oxide and alcohols or carboxylic acid produces the industrially important glycol monoethers and monoesters.

Applications. E., like glycerol, is used as an antifreeze and as a brake fluid in motor vehicles. It is also used as a solvent for cellulose esters and dyes, and for production of inks. Highly purified E. is used in the production of electrolyte capacitors. Its nitrate ester is an important safety explosive used in mining. The ester of G. with terephthalic acid is used to produce polyester fibers. The esters and ethers of E. are softeners and starting materials, or solvents, for nitrocellulose lacquers. The monocarboxylic acid esters are used as starting materials for detergents and emulsifiers.

Ethylene dinitrilotetraacetic acid: same as Ethylene diamine tetraacetic acid (see).

Ethylene glycol diacetate: see Glycol acetates.

Ethylene glycol dinitrate, ***glycoldinitrate***; in the industry, also *nitroglycol*: $C_2H_4(ONO_2)_2$, an oily, nonviscous, colorless and highly explosive liquid; b.p. >215 °C.

$$\begin{array}{ccc} CH_2 & \text{———} & CH_2 \\ | & & | \\ O{-}NO_2 & & O{-}NO_2 \end{array}$$

E. is made industrially by reaction of ethylene glycol with nitrating acid, a mixture of conc. nitric and conc. sulfuric acids. The detonation velocity in E. is 7400 m s^{-1}. Its ability to be gelatinized with gun cotton is very important for the synthesis of explosives. E. is very similar in its explosive properties to glycerol trinitrate. A important advantage is its very low freezing point, below - 20 °C. It is somewhat less sensitive to impact than glycerol trinitrate, and somewhat more stable. Because ethylene glycol is synthesized on a large scale and is readily available, and especially because of its low freezing point, E. has partially replaced glycerol trinitrate.

Ethylene glycol monoacetate: see Glycol acetates.

Ethylenimine: same as Aziridine (see).

Ethylene hydrocarbons: same as Alkenes (see).

Ethylene oxide, ***oxirane***: a colorless gas with a sweetish odor; m.p. - 111 °C, b.p. 10.7 °C, n_D^7 1.3597. It is readily soluble in water and many organic solvents. E. forms explosive mixtures with air. In the presence of catalytic substances such as acids or alkalies, it is polymerized in an exothermal reaction.

E. is poisonous. When larger amounts are inhaled, it leads to nausea and then rapid vomiting. If liquid E. is in contact with the skin over a long period, it causes freezing, due to its rapid evaporation. Exposure to an aqueous solution leads to poorly healing necroses on the skin.

E. can be made from ethylene chlorohydrin by cleavage of hydrogen chloride in the presence of calcium hydroxide. Most E., however, is made by direct oxidation of ethylene with atmospheric oxygen on a silver-aluminum oxide catalyst. E. is an important intermediate in the production of ethylene glycol, and di-, tri- and polyethylene glycols. Cyclization of two molecules of E. produces Dioxane (see). It forms ethylene glycol ethers with alcohols; these are used as solvents for lacquers and textiles. Polyethylene glycol ethers are used as emulsifiers, wetting agents and flotation agents. E. is also used in the synthesis of ethanolamines, acrylonitrile and phenylethyl alcohol.

Ethylene sulfide, ***thiirane***: a colorless liquid; b.p. 55 to 56 °C, n_D^{20} 1.4935. E. is readily soluble in acetone and chloroform. It can be made by reaction of ethylene oxide with potassium thiocyanate or thiourea, and it is an intermediate in the synthesis of rubber additives and insecticides.

Ethyl ether: same as Diethylether (see).

2-Ethylhexane-1,3-diol: an Insect repellent (see).

Ethyl mercaptan: same as Ethane thiol (see).

Ethylmethylacetic acid: see Valeric acids.

Ethyl methyl pyridine: see Collidine.

Ethylmorphine: see Morphine.

Ethyl nitrate: C_2H_5-O-NO_2, a colorless, combustible liquid with a pleasant odor; m.p. - 96.6 °C, b.p. 87.2 °C, n_D^{20} 1.3852. E. is barely soluble in water, but dissolves in organic solvents. It is made by reaction of ethanol with concentrated nitric acid. To prevent the formation of nitrous acid, which would oxidize the alcohol in an explosive reaction, some urea is added to the reaction mixture. E. explodes violently when heated above its boiling point.

Ethyl nitrite: C_2H_5-O-NO, a colorless, combustible liquid; b.p. 16 to 17 °C at 96 kPa, n_D^{10} 1.3418. E. is insoluble in water, but dissolves in organic solvents. It is obtained by reaction of ethanol with sodium nitrite and mineral acid, or with dinitrogen trioxide under cooling. It is readily hydrolysed, especially in the presence of acids and moisture. E. is used instead of alkali nitrite in reactions with nitrous acids in organic solvents.

Ethyl-N-phenylcarbamate: same as *N*-Phenylurethane.

Ethyl phenyl ether: same as Phenetol (see).

Ethyl phenyl ketone: same as Propiophenone (see).

Ethyl red: see Isocyanins.

Ethyl sulfate, ***ethyl hydrogensulfate, sulfuric acid monoethyl ester***: C_2H_5-O-SO_3H, colorless liquid; b.p. 280 °C, n_D^{20} 1.4105. E. can be obtained by reaction of ethanol with conc. sulfuric acid, by addition of one mole ethene per mole concentrated sulfuric acid, or in pure form, from ethanol and sulfur trioxide at 0 °C. The barium salt, in contrast to barium sulfate, is soluble in water.

Ethyne ***acetylene***: HC≡CH, the first member of the series of alkynes. E. is a colorless, inflamable gas; in the purest state it has almost no odor and a narcotic effect. m.p. - 81.5 °C; b.p. - 83.8 °C, heat content 57,000 kJ/M^3. Technical acetylene gas is produced from carbide, and has a characteristic unpleasant smell due to impurities, such as phosphane, hydrogen sulfide and thiol alcohols. E. is dissolves readily in water and even more readily in acetone. At 0 °C and 2.10 MPa it can be condensed to a liquid. E. burns with a very bright flame which has a temperature of about 1900 °C when fed with air, or 3000 °C when fed with oxygen. It produces a large amount of soot.

The unstability of E. is due to its high energy content; almost 220 kJ/mol is required for synthesis from the elements.

E. has the largest explosive range of all combustible gases. Explosive decomposition occurs even in the cold at a pressure above 0.2 MPa; under normal pressure, it occurs above 150 °C. Ethyne-air mixtures are explosive in the range from 2.3 to 82 vol % ethyne.

E. and homologous alkynes must not be allowed to come in contact with copper or silver (vessels or pipes!), because if forms highly explosive copper or silver acetylide; an exception is brass with a copper content up to 65%. This warning does not apply to situations in which these metals are used as catalysts.

To reduce the danger of explosion, acetylene gas is kept in steel bottles under pressures up to 1.5 MPa dissolved in acetone, which is absorbed by a porous substance, usually kieselgur. The hydrogen in E. and monoalkylethynes is acidic and can be replaced by metal, e.g. $HC{\equiv}CH + 2\,Cu^+ \rightarrow CuC{\equiv}CCu + 2\,H^+$ or $HC{\equiv}CH + Na \rightarrow CH{\equiv}C^-\,Na^+$. These salts are called *acetylides*; alkali metal and alkaline earth acetylides are hydrolysed by water; heavy metal acetylides by dilute hydrochloric acid.

The highly unsaturated condition and the high energy content of E. cause it to be extraordinarily reactive. Its reactions are discussed under Acetylene chemistry (see). For analysis, see Alkynes.

Production. E. is produced on a large scale by two methods: 1) reaction of calcium carbide with water: $CaC_2 + 2\ H_2O \rightarrow C_2H_2 + Ca(OH)_2$. The reaction is carried out with excess water in a wet generator, or with an approximately stoichiometric amount of water in a dry generator (Fig.), so that a friable calcium hydroxide powder is produced (carbide lime). Carbide lime is used in construction, the chemical industry and in agriculture. In modern plants, part of the carbide lime is recycled. Depending on the purity, 230 to 300 l technical E. is produced per kg carbide; it is purified in wash towers with chlorine water, sulfuric acid and sodium hydroxide to remove phosphorus hydrides, organic sulfur compounds and traces of hydrochloric acid.

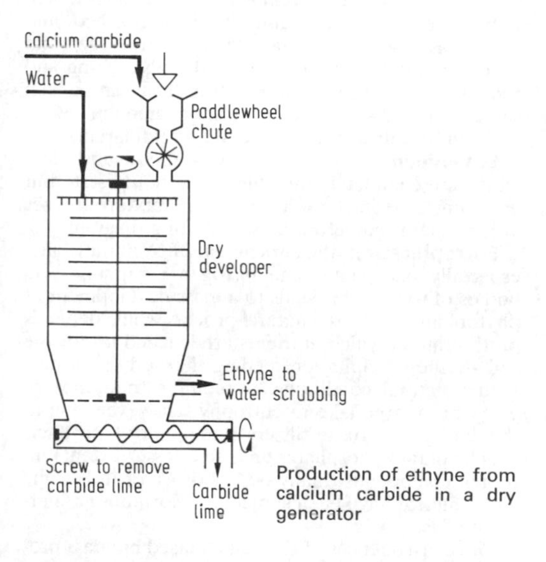

Production of ethyne from calcium carbide in a dry generator

2) high-temperature pyrolysis of hydrocarbons, especially methane (see Pyrolysis). The basic principle of this process is a rapid heating to a temperature above 1500 °C, an extremely short passage of the reactants through the reactor (10^{-2} to 10^{-3} s in the reactor), and a rapid quenching of the pyrolysis gas. The ethyne content of the pyrolysis gas is 5 to 18%; it is washed out with a selective solvent, such as dimethylformamide, *N*-methyl-pyrrolidone or acetone. It is then worked up in subsequent steps. There are three basic variants of this process: a) the allothermic process with direct heat transfer, e.g. the *sheath arc* and *hydrogen arc* pyrolysis methods (*plasma methods*); b) allothermic processes with indirect heat transfer, including the *Wulff process*, which works by the regenerative principle with alternating heating and cleavage, and the *Kureha process*) (see Pyrolysis); and c) autothermic processes, in which the required thermal energy is obtained by partial combustion of the starting material. The most commonly used is the *Sachse process*. This group also includes the *dipping flame process* and the *high temperature pyrolysis process (HTP process)*.

Most of the E. used by the chemical industry is obtained from petroleum. A recent aspect of its production is the international trend toward higher production temperatures, almost up to 1000 °C. This increases the E. content of the C_2 fraction, making isolation more economical.

Application. E. is one of the most important starting products in the organic chemical industry (see Acetylene chemistry). In the past few years, E. has been replaced in many syntheses by the much cheaper ethene, but some compounds, such as butane-1,4-diol and the vinyl esters, can at present be economically produced only from E. Due to its high heat contents, E. is used in large amounts as a combustion gas for welding and cutting metals; an oxyacetylene torch generates a 3000 °Cflame. The soot obtained by low-oxygen combustion of E. (acetylene black) is used as a filler material and to make water colors and oil paints. Polymers of E. are gaining importance in microelectronics.

Historical. E. was discovered in 1836 by Davy and was produced by Wohler in 1862 from carbide. The oxyacetylene torch for welding was introduced in 1906. The work of Reppe made E. an important starting material for large-scale organic chemical syntheses.

Ethynyl: a term for the atomic group HC≡C- in a molecule, the radical HC≡C· and the cation $HC{\equiv}C^+$.

Ethynylestradiol: see Estrogens.

Etioporphyrin: see Porphyrins.

Etofyllin: see Theophyllin.

Etomidate: see Narcotics.

Etridiazole: see Fungicides.

Ettringite: tricalcium aluminate sulfate hydrate; it is formed by the action of sulfate-containing water on the calcium-containing silicates and aluminates of concrete. The pressure of crystallization of E. damages the concrete.

ETU: see Dithiocarbamate fungicides.

Eu: symbol for europium.

Eucalyptus oils: a large group of essential oils, most of which are made in Australia by steam distillation of the leaves of eucalyptus species. The best known of the E. is ***globulus oil***, a colorless, slightly viscous liquid which smells like camphor. Its main components include cineol, camphene, pinene, butyraldehyde, valeraldehyde and capronaldehyde; it also contains terpenes and sesquiterpenes. E. are used medicinally for respiratory illnesses and as salves for rheumatic diseases.

Euchlorin: a mixture of 40% chloric acid, $HClO_3$, and 40% hydrochloric acid, HCl. Because it generates chlorine and chlorine dioxide, ClO_2, E. has a strong oxidizing effect. It is therefore used in chemical analysis to release inorganic components from organic substances.

Eugenol, ***1-allyl-3-methoxy-4-hydroxybenzene***: a colorless liquid which turns brown in air and has a typical clove-like odor; m.p. - 7.5 °C, b.p. 253.2 °C, n_D^{20} 1.5405. E. is slightly soluble in water, and readily soluble in ether and ethanol. It gives a blue color reaction with iron(III) chloride. E. is found in various essential oils, including clove, basil, pimento and cinnamon bark oils. It is extracted from clove oil by shaking with dilute potassium hydroxide. E. can be degraded to vanillin by oxidation with ozone or potassium permanganate. It is used in stomatology as an an anaesthetic and bacterial inhibitor. It is also used as such or in the form of its ester (***eugenol acetate***, m.p. 31 °C, b.p. 127 °C) or ether (***eugenol methyl ether***, b.p. 249 °C) in perfumes.

Eurhodines: see Azine pigments.

Eurhodols: see Azine pigments.

Europium, symbol ***Eu***: an element in the Lanthanoid group (see) of the periodic table; a rare-earth metal; *Z* 63, with natural isotopes with mass numbers 153 (52.18%) and 151 (47.8%); its atomic mass is 151.96. The valence of Eu is III or II, density 5.245, m.p. 826 °C, b.p. 1439 °C, standard electrode potential (Eu/Eu^{3+}) - 2.407 V.

E. is a bright, silvery, soft metal which crystallizes in a body-centered cubic lattice. It makes up only about $9.9 \cdot 10^{-6}$% of the earth's crust, and is thus the rarest stable lanthanoid. It is always found in association with other rare-earth metals, most often in the ytter earths. Monacite and bastnesite are the most important minerals bearing E., but even monacite sands contain only about 0.002% E. Because of its unusually high neutron-capture cross section, E. is used in control devices in nuclear reactors. Its other properties, analysis, production and history are discussed under Lanthanoids (see).

Europium compounds: compounds in which europium can occur in the +3 or +2 oxidation state. The yellow to red europium(III) compounds are relatively easily reduced to colorless europium(II) compounds. Europium(II) compounds are similar to the corresponding alkaline earth compounds; for example, europium(II) sulfate is scarcely soluble in water, while europium(II) hydroxide dissolves readily and gives an alkaline reaction. Eu^{2+} is obtained by reduction of Eu^{3+} with zinc amalgam or electrochemically, even in aqueous electrolytes. Some important E. are: ***europium(II) chloride***, $EuCl_3$, m_r 258.33, density 4.89, m.p. 850 °C, yellow needles; ***europium(III) oxide***, Eu_2O_3, M_r 351.92, density 7.42, pink; ***europium(III) fluoride***, EuF_3, M_r 208.96, m.p. 1390 °C; ***europium(III) nitrate***, $Eu(NO_3)_3 \cdot 6H_2O$, M_r 446.07, colorless; ***europium(II) chloride***, $EuCl_2$, M_r 228.87, colorless, m.p. 727 °C; and ***europium(II) sulfate***, $EuSO_4$, colorless, scarcely soluble in water, M_r 248.02, density 4.99.

An important modern application of E. is as activators for luminophores. Yttrium(III) oxide or vanadate activated with europium(III) make up the red components of color television screens.

Eutectic brine: an aqueous salt solution with the eutectic composition (see Eutecticum). The E. freezes or melts at a constant temperature, just as a pure substance does, and this temperature is the lowest one observed for any mixture of the components. Some eutectica are shown in the following table:

Dissolved salt	Mass fraction (rounded)	Eutetic temperature
NaCl	23%	−21.2 °C
$MgCl_2$	21%	−33.6 °C
$CaCl_2$	30%	−55.0 °C
K_2CO_3	40%	−36.8 °C

When a salt solution (brine) containing a lower concentration of salt than the eutecticum freezes, it first precipitates pure water ice. The fraction of ice increases and the temperature drops as more heat is removed, until the residual solution has reached the eutectic composition. This remainder then freezes as the eutecticum. An E. remains liquid as it is cooled until the eutectic temperature has been reached, and then freezes at this constant temperature. The solid eutecticum can also absorb heat during melting and remain at a constant temperature. Thus an E. can serve as a cold storage and transport medium.

Eutectic temperature: see Melting diagram.

Eutecticum: a mixture of two or more substances which are completely miscible in the liquid state but are immiscible in the solid state; the mixture freezes at a constant temperature. See Melting diagram.

Eutrophication: the enrichment of plant nutrients, especially phosphates and nitrogen compounds, in bodies of water. This leads to a mass development of phytoplankton. E. is a natural process which depends on the rate at which nutrients are leached out of the soil by the precipitation feeding the body of water. Under natural conditions, it requires thousands of years for a large lake to eutrophy. However, due to the heavy use of fertilizers and laundry products which contain phosphates and (to a lesser extent) nitrogen compounds, many bodies of water in Europe and America have been subject to enormous acceleration of E.

The consequences of E. are increased biomass production, discoloration and cloudiness of the water, consumption of dissolved oxygen and formation of hydrogen sulfides in the depths of standing waters. The habitats of fish and their prey are constricted, and anoxia may lead to fish kills. Similarly, the intensive photosynthesis during long summer days can increase the pH of the water and lead to ammonia poisoning. All these conditions also make treatment of the water for drinking or industrial purposes more difficult. Both salt and fresh waters can be affected, but usually freshwater lakes are of greatest concern.

E. can be reversed only with expensive and large-scale intervention. It is far more sensible and cheaper to undertake preventative methods, such as eliminating the nutrients from treated sewage (see Sewage treatment) and using less fertilizer.

EVA copolymers: see Ethylene copolymers.

Evans element: same as Aeration element (see).

Evans principle: see Woodward-Hoffman rules.

Evaporation: a change of state in which a liquid (sometimes a solid; see Sublimation) is converted to the gas phase by absorbing heat. The molecules leaving the liquid must overcome both the cohesive forces of the liquid and the external pressure. The kinetic energy required to overcome these forces is the ***heat of vaporization***. Since the external pressure is reduced when E. occurs in a vacuum, the molecules require less additional kinetic energy, so less heat must be supplied.

If the vapor pressure of the liquid is equal to the system pressure, the liquid boils. However, even if the vapor pressure is less than the atmospheric pressure, E. still occurs. In industrial E. processes, the solution is heated to boiling, and the vapor is either allowed to escape into the atmosphere or is collected for further use (see Distillation).

Enrichment of a dissolved substance in a solution by partial E. is called ***concentration***. In the laboratory, E. dishes (large, flat pans) are used; E. is also carried out in apparatus in which a reduction of pressure is possible. In the production of salt from brine, E. is carried out in lagoons.

The E. of a solvent from a solution can be carried out in a single apparatus or in a several-step process (see Cascade evaporation). There are many types of apparatus for E. (***evaporators***). The solution is usually carried by a bundle of pipes, while the vapor condenses in the area between the pipes. Thin-layer evaporators are used for gentle treatment of solutions which are to be concentrated. These consist of a vertical, sometimes conical part, on the inner surface of which the liquid trickles down. The surface of the liquid film is continuously renewed through the rigid or flexible arms of a rotor on a vertical axis. Evaporation towers are also used to concentrate solutions.

Evaporation entropy: see Phase change heats.

Evaporation heat: see Phase change heats.

Evaporation number: a measure for the time which a certain volume of liquid requires to evaporate at room temperature. The evaporation time of diethyl ether is generally set equal to 1 as a reference.

Evaporator: see Evaporation.

Evipan®: see Barbitals, table.

Ewens-Bassett system: see Nomenclature, II B 2b.

EXAFS: abb. for Extended X-ray Absorption Fine Structure; see X-ray spectroscopy.

Exaltation: see Additivity principle.

Exaltolide, ***1,15-pentadecanolide***: a musky-smelling substance. E. forms colorless crystals, m.p. 37-38 °C, b.p. 137 °C at 266 Pa. It can be made by heating 15-bromopentadecanoic acid with potassium carbonate, or by heating 15-hydroxypentadecanoic acid with benzene sulfochloride. E. is a component of *Angelica* oil, and is also a main component of musk. It was first synthesized in 1928 by Ruzicka.

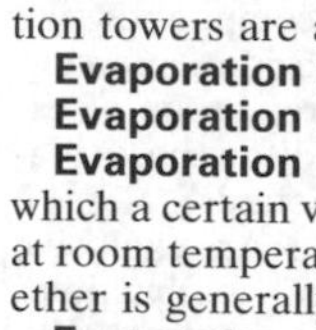

O O

Excelsior brown: same as Vesuvin (see).

Excess function: see Mixture.

π-Excess heterocycles: heteroaromatic compounds characterized by high electron densities on the carbon atoms of the heterocyclic ring. Compounds of this type, especially Five-membered heterocycles (see) react preferentially with electrophilic reagents (S_E *reactions*). The prototype of this group is pyrrole, which, for example, is readily converted to pyrrole 2-carbaldehyde by the Vilsmeier-Haack reaction (see).

Exchange reactions: see Solid-state reactions.

Excimer: (contraction of "excited dimer") an excited dimer which is non-binding in its electronic ground state. A few fused aromatic hydrocarbons, such as pyrene, form E. between an excited molecule and a molecule in the ground state. The molecular complex has properties which differ from those of the single molecules, such as difference fluorescence lines. The bond energies are roughly between 10 and 40 kJ mol^{-1}. E. may be in either singlet or triplet states.

Exciplex: (contraction of "excited complex") an electronically excited molecular complex which has a non-binding ground state. E. are formed as collision complexes between donor and acceptor molecules. One of the partners is in an excited singlet or triplet state. E. are formed only in nonpolar or weakly polar solvents; in a polar solvent, ion pairs are formed. The emission of E. is different from that of the excited single molecules from which it forms. The electron-pair donor might be, for example, an aromatic amine, while the electron acceptor could be a fused aromatic hydrocarbon (e.g. an E. from dialkylalanine + anthracene).

Excitation: transfer of a nucleus, atom, ion or molecule from its ground state into a higher-energy state by provision of energy. In general, excited states are not stable, and the excited system returns to the ground state after a very short time, releasing energy (e.g. in the form of light). It may return in steps or all at once. The frequency of light emitted in the return to the ground state, ν, is related to the difference in energy between the two states, ΔE by the equation $\Delta E = h\nu$, where h is Planck's constant.

Exciton: a coupled charge carrier pair consisting of an electron and a positive "hole" which can arise in a semiconductor or insulator (molecular complex). Such an E. is formed when an electron is elevated to an excited state by absorption of a photon or through collision. It diffuses through the solid until it either has the opportunity to enter the conduction band by adding to a charge carrier, or it recombines with the hole. In the first case, the conductivity of the material is increased, and in the second case, the released energy is converted to light or thermal energy. In the E. state, the motion of the pair of oppositely charged carriers does not carry current and makes no contribution to the conductivity of the semiconductor. The migration of E. is one of the mechanisms by which energy is conducted.

Exclusion chromatography: same as Permeation chromatography (see).

Exergy, abb. e: that portion of the total energy w which can be converted into economically useful form, such as electrical or mechanical work. The non-convertible portion is also called ***anergy*** b. According to the First Law of thermodynamics, $w = e + b$.

Electrical and mechanical energy is usable, and at least in the ideal case can be completely converted into other forms of energy, that is, they are pure E. The Second Law of Thermodynamics sets limits to the degree to which heat and chemical energy can be converted to other forms. In addition, a reference state, i.e. a zero for the E. scale, must be established. For purely thermal processes, this is the state of the environment. Heat at the temperature of the environment cannot be converted into other forms of energy without input of energy; it thus has zero E., and is pure anergy. For chemical processes, the stable end products, which cannot produce more energy, define the zero state.

Spontaneous processes (see Irreversible) such as friction, heat conduction or equilibration of concentration by diffusion are always associated with a loss of energy which can only be recovered by the expenditure of work. It follows that the E. of a closed system decreases as a result of any natural process which occurs within it. It remains constant only in the limiting case of a reversible reaction (principle of reduction of E.). The E. of an amount of heat q_1 in a heat reservoir at temperature T_1 can be derived from the Carnot cycle (see): $e = q_1(T_1 - T_e)/T_1$, where $(T_1 - T_e)/T_1 = \eta$, the thermal efficiency, and T_e is the temperature of the environment. For $T = 600$ K and $T_e = 300$ K, $\eta = 0.5$, that is, 50% of the heat q can be converted into work in an ideal heat engine. This is the quantity of E. For $T_1 < T_e$ (for example, in a refrigerator), η, and therefore e, is negative. Work must be done on the system to transfer heat to the environment. If the heat exchange between different temperatures T_i is taken into account, the E. balance for the limiting case of a reversible process is $\sum_i q_i(T_i - T_e)/T_i + a = 0$ (a is work), and if exergy Δe is lost due to irreversible steps, $\sum_i q(T_i - T_e)/T_i - \Delta e + a = 0$.

The balances of E. in chemical processes are much more complicated, because they contain, in addition to matter and heat exchanges, the reaction enthalpy, the reaction entropy and corresponding loss parameters. Such balances are the basis for economic evaluation of industrial processes.

Exhaust, motor vehicle: the gaseous combustion residues of gasoline and diesel engine fuels. The main components of E. are nitrogen, carbon dioxide and water, but it also contains nitrogen oxides and carbon monoxide, hydrocarbons, aldehydes and other products of incomplete combustion. The nitrogen oxides are formed at high combustion temperatures: $N_2 + O_2 \rightarrow 2\ NO$; $2\ NO + O_2 \rightarrow 2\ NO_2$, while the hydrocarbons are a result of incomplete combustion. E. also contains aerosols of unburned fuel and motor lubricants. Poorly maintained diesel engines also exhaust large amounts of particulates. The E. from gasoline engines may contain lead compounds from the gasoline additives which prevent Knocking (see). The most important air pollutants, in order of amount produced, are carbon monoxide, nitrogen oxides and hydrocarbons. The oxygen-containing compounds, mostly aldehydes (especially formaldehyde) irritate the mucous membranes. At high concentrations of E., the effects of light, UV radiation and ozone in the atmosphere lead to formation of smog.

Reduction of the air-polluting components of E. requires improved design of engines and carburetors, optimum maintenance of the vehicles, and minimizing unfavorable operating conditions (idling, acceleration and deceleration) by use of microelectronic control systems. The use of lead-free gasolines (see Fuels) and catalytic afterburners is also essential.

Exinite: see Macerals.

exo: see Stereoisomerism 1.2.2.

Exon: see Nucleic acids.

Exopeptidases: see Proteases.

Exorphins: exogenous neuroactive peptides with opiate-like effects. E. were first discovered in peptide hydrolysates of wheat gluten and casein. A well-characterized example is β-Casomorphine (see).

Exothermal reaction: a reaction in which energy (usually heat) is released. According to the thermodynamic sign convention, the reaction enthalpy $\Delta_R H$ and the reaction energy $\Delta_R U$ are negative.

Expansion: the increase in volume of a gas when the pressure is decreased or the temperature is increased.

Expansion coefficient: see State, equations of, 1.

Explosion: 1) in the broad sense, a sudden expansion of gases. It is insignificant whether the gases were present before the E. or are formed in it. For example, if the compressed gas in a damaged pressure vessel suddenly expands, it is called an E.

2) In the narrow sense, a very rapid chemical reaction of an explosive substance in which large amounts of gas and heat are released. Some highly exothermal reactions which tend to produce E. are combustion processes, some chlorination and fluorination reactions (e.g. the chlorine detonating gas reaction), and the decomposition of very unstable compounds like ethyne, heavy metal acetylides and azides, mercury fulminate and chlorine dioxide. A large group of explosives contains the oxygen required for the oxidation within the molecule (nitro compounds).

A special group of E. are ***gas E.***. An explosive mixture is defined as a mixture of combustibles (gases, aerosols of combustible liquids or dust) with an oxidizing agent, usually oxygen, in which a combustion process propagates independently after it has been initiated. ***Dust E.***, in which dust (coal dust, flour, etc.) burns explosively in atmospheric oxygen, are especially dangerous. Gas mixtures react explosively only if the ratio of components is within certain limits (upper and lower ***explosion limits***). Mixtures of gases or gasified liquids with air or oxygen are used as fuels in internal combustion engines and rocket or jet engines.

Exothermal reactions, especially chain reactions, often explode if the heat of reaction cannot be removed quickly enough. The temperature of the mixture increases and this further increases the reaction rate, finally causing an E. (***thermal E.***). In branched chain reactions (see Reaction, complex), the exponential growth in number of chain carriers is the actual cause of the sudden increase in reaction rate (***chain E.***). The gas formed as a result of the reaction and its expansion due to the high temperature creates a pressure wave which can travel at a rate between several hundred and 2000 m s^{-1}, and causes considerable damage (see Deflagration). An E. in which this rate is exceeded is called a Detonation (see).

The ***E. velocity*** is the rate at which the reaction zone in the explosive substance expands. This must be distinguished from the expansion velocity of the pressure wave.

E. can be initiated by heating the entire explosive mixture to a certain temperature, the ignition temperature, by local heating, sparks or UV light. Some substances explode when shaken or on impact. Small amounts of a highly sensitive explosive can be used to initiate an E. in a less easily ignited explosive (***ignition***).

Explosion limits: the lower and upper concentration limits for a mixture of combustible gas and air (or oxygen) between which the mixture can explode when ignited. E. depend on temperature and pressure.

Explosive: a technologically useful substance or mixture which is able to explode, releasing heat and large amounts of gas within a very short time (see Explosion). The reactions of most E. are oxidations of carbon with oxygen, where the oxygen is originally bound to nitrogen in the form of the nitro group, $-NO_2$, or it is present as an $-O-NO_2$ group. In explosive mixtures of components which are not explosive alone, the oxygen carrier is usually alkali nitrate, chlorate or perchlorate.

1) ***Explosive compounds*** are E. which require an initial shock to detonate them. They can be either single compounds or mixtures.

a) The single compounds are classified according to chemical structure: *nitrates* such as glycerol trinitrate (nitroglycerine, NG), ethylene glycol dinitrate (nitroglycol), pentaerythritol tetranitrate (PETN), cellulose nitrate (nitrocellulose), etc.; *nitro compounds* such as 2,4,6-trinitrophenol (picric acid), 2,4,6-trinitrotoluene (TNT), nitroguanidine, 2,4,6-trinitroresorcinol (styphnic acid), etc.; *nitramines*, such as 1,3,5-trinitro-1,3,5-triazacyclohexane (hexogen), 1,3,5,7-tetranitro-1,3,5,7-tetraazooctane, 2,4,6-trinitrophenylmethylnitramide (tetryl), etc.; *nitrosamines*, such as trimethylene trinitrosamine, and *salts* such as ammonium picrate. These groups include most explosive compounds.

b) Mixtures include black powder, dynamite, ammonium nitrate E., AMO E., gel E. (slurries), etc. Safety explosives (see) are used in mining.

2) ***Ignition E.*** are E. which detonate on impact, friction, flame or sparks. They are used to initiate explosion of more powerful E. Mercury fulminate, lead azide, lead trinitroresorcinol, 4,6-dinitro-2-diazohydroxybenzene (diazodinitrophenol), etc. are used as ignition E.

3) ***Smokeless powder*** (see) is an E. consisting chiefly of cellulose nitrate in the form of a gel.

4) ***Pyrotechnic charges*** are used in fireworks and for a variety of other purposes.

Applications. Powdered E. (black powder and similar mixtures) are used in quarries, where they are ignited by means of an ignition cable. The gelatinous E. are used in mining for ores and to quarry hard rocks. The less powerful, powdered E. (donarites, AMO E.) are used for softer rocks, such as limestone. The E. used for military purposes are nearly all the highly brisant explosives: 2,4,6-trinitrotoluene (TNT), 1,3,5-trinitro-1,3,5-triazacyclohexane (hexogen) and pentaerythritol tetranitrite (PETN). Smokeless powder is used for shells and cartridges for guns; either simple cellulose nitrate powder for single cartridges or the higher-energy glycerol trinitrate and ethylene glycol dinitrate powder for machine guns.

Explosive gel, ***slurry***: an explosive suspension consisting of a highly concentrated aqueous solution of nitrate (usually ammonium nitrate) in which more nitrates are suspended, along with a combustible substance such as aluminum powder or glycol. Other components (e.g. cross-linkers) can be used to adjust the power of the explosion. The consistency of the E. is adjusted by addition of a swelling component of a macromolecular compound, such as starch, guar, agar-agar or polyacrylates. Because these G. contain large amounts of water, they are not sensitive to heat or impact. They can therefore be pumped into large boreholes. E. are used in the production of all-weather explosives.

Explosive gelatin: a yellowish, translucent, gelatinous substance which can be cut or bent; density 1.6. E. is the strongest commercial explosive. It is made by slightly warming 92-93% glycerol trinitrate (nitroglycerol) with 7-8% collodium cotton (nitrogen content about 12%). The collodium cotton should have a high degree of polymerization, in order to gelatinize as slowly and evenly as possible, thus giving a readily shaped E. with the smallest possible amount of collodium. Enough collodium must be used to prevent sweating out of glycerol trinitrate. To make non-freezing E., the glycerol trinitrate is replaced by a mixture of glycerol trinitrate with 20 to 25% ethylene glycol dinitrate. E. is more explosive than dynamite. Because of the power of its explosiveness and its stability to water, it is used in the construction of tunnels. E. was first made in 1875 by A. Nobel.

Addition of fillers such as sawdust, flour or sodium nitrate gives the gelatin dynamites. See Dynamite.

Exsorption: see Absorption.

Extensive parameter: see Parameters of state.

Extinction, ***optical density***, ***absorbance***: a measure for the amount of light absorbed by a sample: $E = \log I_0/I_T = \log 1/T$, where I_0 is the intensity of the incident radiation, I_T is the intensity of the transmitted radiation, and T is the transparency I_T/I_0. See Lambert-Beer law.

Extra conductivity: the extremely high ionic limit conductivity of hydronium and hydroxide ions which is due to a special mechanism. While other ions in solution, such as metal ions, transport electrical charge by motion through the medium in response to an electric field, charge transport in a solution of H_3O^+ is due to the transfer of protons from one water molecule to the next. The individual protons do not travel further than the distance between water molecules, yet the overall effect is a rapid transfer of charge across the hydrogen bonds in the liquid medium. In the process, the electrons on the molecules rearrange (quantum mechanical tunneling effect). An analogous model has been proposed for hydroxide ions: a tunneling transfer of a proton H^+ from one water molecule to the next leaves a hydroxide ion OH^- behind. The positive charge now "migrates" in the direction of the cathode until it neutralizes an OH^- ion in the solution. As a result, an OH^- ion "migrates" in the direction of the anode.

Extraction: a method of separation which depends on selective solubility. E. can be carried out as solid-liquid E. or liquid-liquid E. In ***solid-liquid E.***, selected components are dissolved out of a solid mixture. If water is used as the solvent, the process is called *leaching*. This is used, for example, to obtain natural products or in the work-up of ores. In the laboratory, it is carried out in *Soxhlet extractors* (Fig. 1). In ***liquid-liquid E.***, a liquid or dissolved solid is separated from a solution. If a substance is distributed between two more or less immiscible liquid phases A and B, a solution equilibrium is established, in which the ratio of concentrations of the substance in the two phases is described by the *Nernst equation*: $c_A/c_B = K$; here c_A is the concentration of the substance in phase (solvent) A, c_B is its concentration in B and K is the distribution coefficient. If the equilibrium concentration of the substance in phase A is plotted against its concentration in phase B, the graph is its distribution isotherm. This is a straight line if the molecules of the distributed substance are the same in both phases. The slope of the line gives the distribution coefficient.

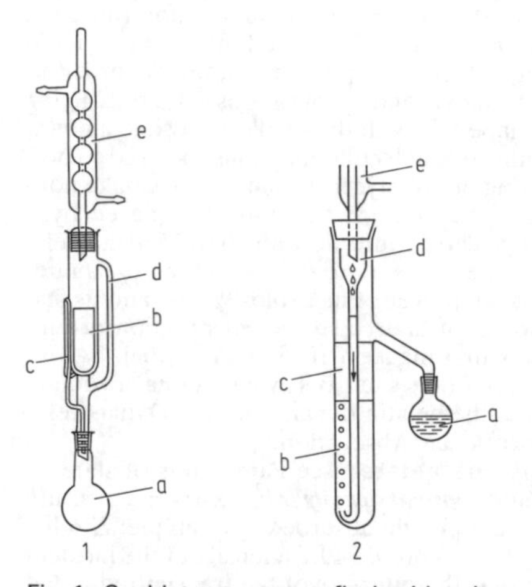

Fig. 1. Soxhlet extractor. *a*, flask with solvent, *b*, extractor, *c* rise tube, *d* side tube, *e* reflux cooler.
Fig. 2. Perforator. *a*, flask for evaporation of the mobile phase and collection of the extracted components; *b*, stationary phase being extracted; *c*, mobile phase, which runs back into the evaporation flask; *d*, funnel leading the mobile phase into the stationary phase; *e*, reflux condenser

There are two types of technique, involving step-wise and continuous distribution. In each type, the distribution can be either simple or multiplicative. In step-wise distribution, the individual operations, such as addition of fresh phase, achievement of equilibrium and phase separation are carried out discontinuously (in the laboratory, this is done in a separatory funnel), whereas in continuous techniques, the apparatus allows these steps to be done continuously. Simple E. (simple distribution) consists of establishing equilibrium between the substance to be extracted in solvent A and the extraction solvent B a single time; the resulting liquids are called ***extract*** and ***raffinate***. In the extract, the components to be separated are enriched according to their distribution coefficients. The large distribution coefficients of many organic compounds permit them to be almost completely transfered from an aqueous to an organic solvent. However, where the Nernst distribution equation applies, it can be proven that it is more efficient to do several extractions using fractional portions of a given volume of solvent, than to use the same total volume in a single step. A *perforator* (Fig. 2) is an apparatus for repeated E. The solvent is constantly evaporated from a flask, condensed in a reflux cooler, and allowed to trickle back through the solution to be extracted into the evaporation flask.

A better separation effect is achieved if extraction solvent B, after a simple distribution, is shaken with pure solvent A (***retrograde extraction***). Multiplicative distribution (countercurrent extraction) is based on this principle. In ***countercurrent extraction***, the two liquid phases are moved past each other in a countercurrent, and continuously re-equilibrated, so that the extract which is partially enriched with the extracted substance is equilibrated with fresh solution, and simultaneously, the partially extracted solution is equilibrated with fresh extraction solvent. Methods of multiplicative E. differ with respect to the motion of the two phases past one another; this can be either discontinuous or continuous. The material to be extracted can all be added at the beginning of the distribution, or gradually with each step; the extraction solvent can be added at the start or in the middle of the apparatus. Although in the laboratory, single-stream methods are more common (Craig, Martin-Synge distributions), in industry uniform countercurrent techniques are usual.

Multi-step countercurrent reactions are usually carried out in cylindrical columns (e.g. rotating disk columns), in which projections or pulsations are used to increase the area of the phase boundary and promote the diffusion of the substance between the two phases. The countercurrent and phase separation are achieved by gravity. Other apparatuses (e.g. mixer-settlers) consist of a series of mixing and settling chambers where first intense mixing and then phase separation can occur. The phase transport is accomplished by pumps or similar means. E. is used in a wide variety of industrial applications, e.g. for separation of aromatic hydrocarbons from the aromatic fractions resulting from petroleum refining. E. is also used in wet metallurgy, in production of essential oils, fats and pharmaceuticals, etc. See the table.

Methods of liquid-liquid extraction

Separation	Solvent	Process
Extraction of aromatics	*N*-Methylcaprolactam	Arex
	Diethyleneglycol	Udex
	N-Methylpyrrolidone	Arosolvane
	Tetramethylsulfone	Sulfolan
Extraction of butadiene	Copper acetate/ ammonia/water	CAA
Extraction of phosphoric acid	Butanol	Apex
Extraction of uranyl nitrate	Tributylphosphate	
Extraction of caprolactam	Benzene	
	Trichlorethylene	
Phenol removal from waste water	Butyl acetate	Phenosolvane
	Diisopropyl ether	Phenosolvane
	Tricresyl phosphate	
Extraction of m-xylene	HF/BF_3	Mex
	Hydrogen fluoride/ boron trifluoride	

Extraction coals: see Coal.

Extractive distillation: a process of Distillation (see) used to separate mixtures of liquids with similar boiling points. E. is based on the fact that the relative volatility of the components to be separated can be changed by dissolving them in a high-boiling solvent. The relative volatility of a real mixture of two substances is $\alpha = (f_A/f_B)(p_A/p_B)$. Here f_A and f_B are the activity coefficients of components A and B, and p_A and p_B are the vapor pressures of the pure components. Addition of the selective solvent changes f_A and f_B so that α becomes significantly larger, and distillation can separate the two components. If the selective solvent is volatile and goes over the top of the distillation column with the more volatile component A, one speaks of Azeotropic distillation (see); if the solvent is not volatile and remains behind, one speaks of E. The thermodynamic basis of E. and azeotropic distillation is the same. The vapor pressure of the selective solvent must be lower than that of the components of the mixture to be separated.

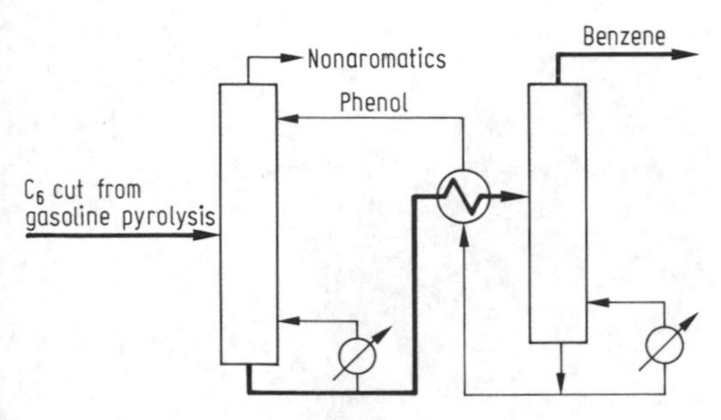

Production of benzene by extractive distillation in the distex process

E. has the advantage over azeotropic distillation that many more suitable solvents are available with volatilities which are significantly different from those of the components to be separated. In addition, the concentration of the solvent in E. can vary over a wide range. The solvent must not form an azeotrope with the components, and must be readily removed from them later. Since in E. the solvent is not vaporized, the energy consumption is smaller than in azeotropic distillation.

Technologically important extractive distillation processes

Mixture	Extracted compound	Selective solvent	Process
C_4 fraction of gasoline pyrolysis	Butadiene	Dimethylformamide *N*-Methylpyrrolidone Acetonitrile Furfural Dimethylsulfoxide	Difex
C_5 fraction of gasoline pyrolysis	Isoprene	Dimethylformamide *N*-Methylpyrrolidone	
C_6 fraction of gasoline pyrolysis	Benzene	Phenol *N*-Methylpyrrolidone *N*-Formylmorpholine	Distex Distapex Morphylan
Isopropanol/ water	Water	Calcium chloride/ water	
Hydrochloric acid/water azeotrope	Water	Sulfuric acid	

Eye irritants: see Chemical weapons.

Eyring equation: an equation for the temperature dependence of the rate constants k of a chemical reaction (see Kinetics of reactions, theory).

Eyring theory: see Kinetics of reactions, theory.

E,Z-isomerism: see Stereoisomerism 1.2.3.

F

f: symbol for free energy.

F: 1) symbol for fluorine. 2) F_r, symbol for formula mass.

Fabric finishes: a group of substances used to treat textiles to give them desired appearance, softness, smoothness and resistance to wrinkling, shrinkage, bacteria, molds and moths. F. are also used to create water-repellent, water-tight and fire-resistant cloth.

Permanent F. are those which are not lost when the cloth is washed or dry cleaned. They include 1) compounds which react with the fibers, 2) compounds which condense to resin-like deposits as they dry, and 3) compounds which are applied as polymers in the form of aqueous suspensions or solutions in organic solvents; these bind tightly to the textile surface after drying.

F. from the first two groups are used most often to make cellulose fibers resistant to wrinkling and shrinking. Most of them are water-soluble compounds of urea, cyclic ethyleneurea and its dihydroxy compound, of melamine, guanidine, etc. The resistance to wrinkling is due to absorption of the polymer of the F. in the amorphous regions of the fibers or to cross-linkage of the cellulose chains. F. in groups 1 (e.g. octadecylethyleneurea) or 3 (e.g. polyethylene suspensions) are used permanently to soften and smoothe the fabric and to make it easier to sew. Chromium(III) chlorostearate and pyridinium compounds of octadecyl chloromethyl ether and stearoyl chloromethylamide are used as permanent water repellents; they belong in group 1. Silicon compounds are examples of group 2 permanent water repellents. High-molecular-weight fluorocarbon compounds (group 2) are used as finishes to repel dirt and oil. Permanent shine or a matte finish is given by methoxy compounds, sometimes in the form of their methyl or butyl ethers. Products for permanent flame resistance belong in group 1, e.g. tetrakishydroxymethylphosphonium chloride, or 3, e.g. suspensions of antimony oxide and highly chlorinated or brominated hydrocarbons. Permanent texture finishes (stiffeners, fillers, sizings) and coatings ("oilcloth") belong in group 3, for example, suspensions of polymers or mixed polymers of vinyl chloride and vinyl acetate, acrylate and methacrylate esters, buta-1,3-diene and acrylonitrile, etc. The end effect depends on the type of F., the method of application and the amount.

Factice: a rubber substitute. ***Sulfur F.*** is obtained by the reaction of sulfur with unsaturated fatty oils or fats at temperatures of 130 to 160 °C. The sulfur adds to the double bonds, producing a soft, voluminous, elastic F. which is light to dark brown. ***Chlorosulfur F.*** is formed by the reaction of disulfur dichloride with the unsaturated compounds at temperatures as low as 20 to 50 °C. If refined plant oils are used in this reaction, the product is white F.

F. is usually made from natural raw materials such as linseed, ricinus, soy, rape or fish-liver oil; however, synthetic unsaturated compounds may also be used. F. is important as an additive to rubber mixtures, which it makes easier to work. In addition, F. is used to make upholstery materials, erasers, floorings and paints, and to impregnate fibers.

FAD: see Flavin nucleotides.

Faience: a ceramic product with a colored, porous body, usually yellowish. It is made from clays which contain limestone (marls). Modern F. also include ceramics similar to stoneware which have nearly white bodies. F. was formerly given a transparent glaze, but today an opaque white glaze consisting of clay, lead(II) oxide, alkali metal compounds and tin dioxide is commonly used. F. was formerly used to make dishes, but in recent times it has been used mainly for tiles.

Fanal dyes: a group of synthetic triarylmethane dyes. They form complexes with phosphoric-molybdic-tungstic acids which are very bright, colorfast and resistant to heat. ***Fanal blue B***, for example, is a complex salt of Victoria blue B with phosphoric-molybdic-tungstic acid. The F. are used for wallpaper, leather and printing inks.

Faraday current: the current in an electrolysis cell which is associated with the electrode reactions. The opposite is ***non-Faraday current***, which acts to charge the electrochemical double layer.

Faraday effect: rotation of the plane of polarization of light passing through a substance which is magnetized in the direction of the propagation of the light. The angle of rotation is usually proportional to the thickness of the medium and to the magnetic field (see Chiroptic effects).

Faraday's laws: the quantitative laws of electrolysis established by M. Faraday in 1833/1834. ***First F.***: the masses which react at the electrodes are proportional to the amount of electricity applied: $m = K \cdot Q$. Here m is the mass of the product, K is a proportionality factor and Q is the amount of electricity. The proportionality constant is $K = M(z \cdot F)$, where M is the molar mass, z the number of electrons exchanged and F the ***Faraday constant***, which has the numerical value 96,485.44 C. The Faraday constant is the product of the unit charge and Avogadro's number. ***Second F.***: For different electrode reactions involving the same amount of electricity, the ratio of the reacting masses is equal to the ratio of equivalent masses:

Faraday-Tyndall effect: same as Tyndall phenomenon (see).

Far infrared, *FIR*: see Infrared spectroscopy.

Farnesol: an acyclic sesquiterpene alcohol; a colorless liquid with a flowery odor. F. is a component of essential oils, e.g. lemon-grass, orange blossoms, citronella and rose oils, and is used in the perfume industry. Natural F. is the 6-*trans* isomer. The The F. isolated from the larvae of the flour beetle *Tenebrio molitor* has juvenile hormone activity. Some insect pheromones are derivatives of F.

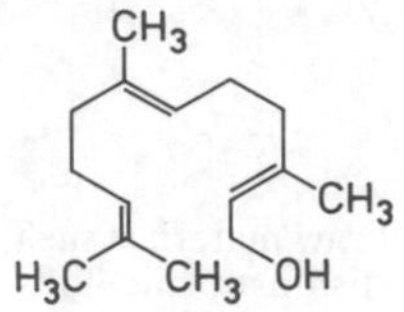

trans, trans-Farnesol

Farnoquinone: see Vitamin K.

Fast dye salts: double salts of diazonium and aromatic sulfonic acid, zinc chloride or boronhydrofluoric acid. They are not dangerous to handle and, in the completely dry state, they are stable. F. are used for development dyes. They contain diazetates instead of diazonium salts, but these are converted to diazonium salts in the presence of mineral acids.

Fast light yellow: see Flavazine, see Tartrazine.

Fast red: a term for a large number of pigments which vary widely in structure. The oldest red azo dye, fast red A, is made by diazotization of naphthionic acid and coupling with naphth-2-ol in alkaline medium. It is a very fast dye used to dye wool and silk in an acid bath, and for leather and paper. A related F. is Amaranth (see).

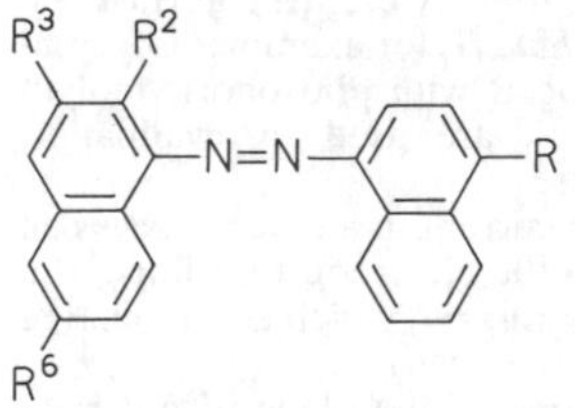

Fast red	R	R^2	R^3	R^6
A	SO_3Na	OH	H	H
B	H	OH	SO_3Na	SO_3Na
D	SO_3Na	OH	SO_3Na	SO_3Na
E	SO_3Na	OH	H	SO_3Na

Fast yellow, ***new yellow, chrysolin***: an azo dye. It is made by sulfation of 4-aminoazobenzene hydrochloride with fuming sulfuric acid. F. is licensed for use as a food color. It is also used to dye wool and silk, and is an intermediate in the production of diazo dyes such as Biebrich scarlet.

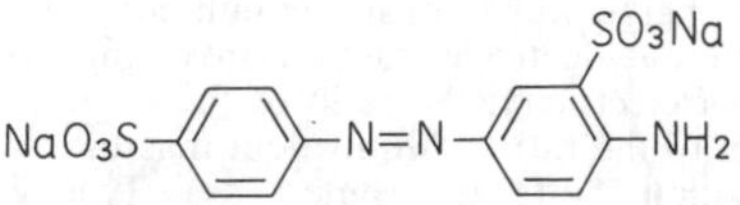

Fat coal: anthracite coal with a heating value of about 33,000 kJ kg^{-1}. It contains 87.5 to 89.5% carbon and 4.5 to 5% hydrogen. When coked, F. produces a gas which burns with a bright and sooty flame.

Fat pigments: fat-soluble pigments used to identify and label various kinds of fuel (gasoline, diesel, turbine and jet fuels, heating oil). F. are also used to color gear, lubricating and hydraulic oils, lubricants, waxes, shoe polish and plastics.

Fats and oils, ***acylglycerols, glycerides***: the mono- di- and triacyl esters of glycerol. Natural F. are about 98% mixed triacylglycerols. Monoacylglycerols are present only in traces (1%), and diacylglycerols only in small amounts (3%). The most common fatty acids in F. are palmitic ($C_{15}H_{31}COOH$) and stearic ($C_{17}H_{35}COOH$) acids in solid F., and oleic ($C_{17}H_{33}COOH$), linoleic ($C_{17}H_{31}COOH$) and linolenic ($C_{17}H_{29}COOH$) in oils. Animal F. also contain phospholipids, vitamins and small amounts of cholesterol. ***Milkfat***, for example, consists of more than 60 different fatty acids. Per kg, ***butterfat*** contains, in addition to the fatty acids, 4 to 14 mg vitamin A, 3 to 10 mg vitamin D, 25 to 30 mg vitamin E, 4 to 9 mg carotene, 2 to 3 g cholesterol and 2 to 5 g phospholipids.

Plant F. (oils) do not contain cholesterol; instead they contain up to 65% linoleic acid and 0.5 to 1.0 g vitamin E per kg. Certain F. contain substituted fatty acids in addition to the normal ones; for example, ricinolic acid (12-hydroxy-9-octadecenoic acid) makes up as much as 90% of ricinus oil; β-ketofatty acids comprise up to 0.03% of milkfat; alkyl branched fatty acids make up as much as 75% of microbial F. and as much as 3% of the total metabolic production of rumen bacteria in ruminants; cyclopentenecarboxylic acids, for example chaulmoogric acid (13-cyclopentene-2-tridecanoic acid), are found in the F. of certain tropical plants, and cyclopropanoic and cyclopropenoic acids may constitute up to 30% of bacterial fats.

Properties. The only difference between a fat and an oil is that the latter is liquid at room temperature. The m.p. of fats and oils are between about - 20 °C and +40 °C. The F. are lighter than water, with a density of about 0.9 g cm^{-1}, and insoluble in it. When finely dispersed, they form emulsions in water which can be stabilized by surface active compounds. F. are readily soluble in organic solvents other than alcohols. When they are boiled with alkalies, they are saponified. The yellow color of natural F. comes from their content of carotenoids, and their aroma and taste from the presence of other compounds. For example, diacetyl butane-2,3-dione, which is present at 0.5 mg/kg in butterfat, is responsible for the typical aroma of butter. Acetoin, butanol, butan-2-one, methylsulfide, ethyl formate and a number of aliphatic lactones, which are present in smaller amounts, are also involved in the aroma. F. become rancid in the presence of oxygen through autooxidation, and enzymes of microbial origin catalyse hydrolysis and decarboxylation, increasing the content of mono- and diacylglycerols and forming alkan-2-ones.

Analytical. The F. are characterized by their ***saponification number*** (abb. ***SN***), which indicates the number of milligrams of potassium hydroxide needed to saponify 1 g fat, and the ***iodination number*** (abb. ***IN***), which indicates how many grams iodine add to 100 g; it is thus a measure for the content of unsaturated compounds. Determination of fatty acid compositions is now done by gas chromatography (Table).

Characteristics of some important fats and oils

	m.p. [°C]	SN	IN
Lard	27–40	190–200	45– 70
Suet	40–50	190–200	35– 50
Goose fat	25–35	190–200	60– 75
Butterfat	15–25	220–235	20– 35
Coconut fat	20–25	250–265	7– 10
Palm heart fat	20–25	245–255	12– 17
Cocoa butter	22–26	190–200	35– 40
Cotton seed oil	0– 5	190–200	100–120
Peanut oil	−4 to −2	185–195	170–195
Linseed oil	<−15	190–195	120–140
Rapeseed oil	−3–0	170–180	95–105
Soy oil	<−15 to −10	190–195	120–140
Sunflower oil	−20 to −10	185–195	115–140
Sesame oil	−6 to −3	185–195	100–120

Occurrence and isolation. F. are found in all cells. In animals, F. are stored as energy reserves in fatty tissue under the skin, in the abdomen and around internal organs. In plants, F. are found mainly in fruits and seeds. The most important oil fruits are olives, coconuts, hemp and flax seeds, peanuts, maize, cotton seeds, soybeans, palm kernels and sunflower seeds. To isolate F., the fatty tissue is pressed out, melted (tried) or extracted with solvents such as benzene, carbon disulfide or trichloroethene; recently, extraction with supercritical gases has been introduced. Crude F. are usually refined by treatment with dilute sulfuric or phosphoric acid, then neutralized with alkalies (e.g. 8% sodium hydroxide) and freed of colors by adsorbents (fuller's earth or charcoal) and of undesired tastes by evaporation.

Synthetic fats. Oxidation of paraffin produces a mixture of fatty acids, which are esterified with glycerol at temperatures above 100 °C, with simultaneous removal of the resulting water. Unlike natural fatty acids, which always have even numbers of C atoms, synthetic fatty acids can have either odd or even numbers of C atoms. Although there are no physiological problems with the odd-numbered fatty acids, they are not at present used for human nutrition.

Emulsified fats. The most important of these is margarine, which is made by emulsification of plant oils and fats (some hydrogenated) by addition of about 0.5% emulsifiers, flavoring, color, preservatives, corn sirup, salt, citric acid and about 0.2% starch in water or sour skim milk. The fat content is 80%, or 40% in low-calorie margarines. Indigestible ***pseudofats***, e.g. ethers of glycerol and esters of propane-1,2,3-tricarboxylic acid or butane-1,2,3,4-tetracarboxylic acid, are not widely used, mainly for economic reasons.

Applications. 80% of the F. produced are used as food (1 g fat yields 38 kJ). F. are also used to produce fatty acids, glycerol, soaps, salves, candles, varnish, lacquer and paints.

Fatty acids: saturated and unsaturated aliphatic Monocarboxylic acids (see). The term F. is used because many of these acids, especially aliphatic acids with high molecular weights, are found in fats. Those present in lipids are usually unbranched and have an even number of carbon atoms. Both saturated and unsaturated F. are present in fats. Naturally occurring ***unsaturated F.*** have the Z configuration. In multiply unsaturated F., the ethene groups are separated by CH_2 groups, that is, they are not conjugated. This structural peculiarity of the natural F. is called the *divinyl rhythm*. ***Short-chain F.***, from C_4 to C_{10}, are present mainly in the milk fats of mammals. Palmitic and stearic acids are found in nearlly all animal and plant fats. ***Long-chain F.*** are present in brain lipids and in waxes. The most abundant unsaturated F. in nature is Oleic acid (see). ***Polyunsaturated F.*** are the main components of various plant fatty oils and fish liver oils. Linoleic and linolenic acids are present in large amounts in linseed oil. These polyunsaturated acids are among the ***essential F.***, which are required in the diets of human beings and higher animals to maintain normal body functions; they cannot be synthesized in the body.

The solubility of the F. depends on the length of the alkyl chain. The solubility in water decreases, and the solubility in nonpolar solvents increases, as the length of the chain increases. For example, the solubility of stearic acid in water at 22 °C is 0.29 g/100 ml, and it is readily soluble in ether. The higher F. can be obtained by alkaline hydrolysis of lipids. Other methods of synthesis are listed under Carboxylic acids (see).

Favorski rearrangement: conversion of α-haloketones into carboxylic acids or their esters by reaction with alkali hydroxides or alcoholates: $R\text{-}CO\text{-}CH_2\text{-}Cl + NaOH \rightarrow R\text{-}CH_3\text{-}COOH + NaCl$ or $R\text{-}CO\text{-}CH_2\text{-}Cl + NaOR \rightarrow R\text{-}CH_2\text{-}COOR + NaCl$. Cyclic haloketones undergo ring contraction to form cycloalkane carboxylic acids with one C-atom fewer, or their esters. Thus, for example, 2-chlorocyclohexanone is converted by the F. in the presence of an alcoholate to the cyclopentanecarboxylate of the alcohol:

2-chlorocyclohexanone $\xrightarrow[-Cl^-]{+RO^-}$ cyclopentyl–COOR

Two different reaction mechanisms are assumed for the F., depending on the structure of the α-haloketone used. In the rearrangement of an α-haloketone with an H atom in the α'-position, a cyclopropanone intermediate is formed; this was demonstrated by isotope labelling and trapping reactions. In α-haloketones which do not have an H atom in the α-position, the F. occurs by an anionotropic mechanism:

$$R\text{-}\overset{+}{C}(\text{-}\overline{\underline{O}}|^-)\text{-}CH_2\text{-}Cl \xrightarrow{+OR^-} R\text{-}C(OR)(\text{-}\overline{\underline{O}}|^-)\text{-}CH_2\text{-}Cl \xrightarrow{-Cl^-} R\text{-}CH_2\text{-}COOR$$

The F. of 2-bromocyclobutanone also goes by this reaction mechanism, which is essentially analogous to the benzilic acid rearrangement.

FCC process: see Cracking.

FDMS: abb. for Field Desorption Mass Spectroscopy.

Fe: symbol for iron.

Feedwater, ***boiler feedwater***: water which meets specific quality criteria for use in boilers. F. is treated to prevent corrosion and formation of boiler scale, to reduce the salt content and permit formation of pure steam. The first stage of treatment is mechanical separation of solids using screens and filters. Chemical treatment consists of adding lime, soda, sodium hydroxide, sodium phosphate, etc., to make the F. alkaline or neutral. In some cases, the F. is distilled before use in the boiler, or it may be completely deionized on ion-exchangers. The F. must be degassed (see Degasification) before it enters the boiler to remove the corrosive gases (O_2, CO_2). In some cases, F. must also be freed of fats, oils and other substances.

Fehling test: see Monosaccharides.

Fehling's solution: the alkaline solution of a copper(II) tartrate complex in water. F. is made by mixing copper sulfate (20 g $CuSO_4 \cdot 5H_2O$ in 500 ml water: Fehling I) and potassium sodium tartrate in sodium hydroxide (100 g $KNaC_4H_4O_6 \cdot 4H_2O$, 75 g NaOH in 500 ml water: Fehling II). F. is used to detect reducing substances, especially glucose in urine in clinical chemistry. For an analysis, equal volumes of Fehling I and II are mixed, then the sample is added, and the mixture is heated to boiling for a few minutes. If a reducing substance is present, the dark blue color of the copper(II) tartrate complex is bleached, and a red precipitate of copper(I) oxide forms. F. was introduced in 1850 by H. Fehling (1812-1885).

Feigl's reaction: a reaction for detection of silver ions using 1,4-dimethylaminobenzylidene thiocyanate.

Fenarimol: see Pyrimidine fungicides.

Fenazox: see Acaricides.

Fenchane: a Monoterpene (see).

Fenchone: a bicyclic monoterpene ketone, and a slightly oily liquid; m.p. 5-6°C. F. has an intense, camphor-like odor and a bitter taste. It is a component of some essential oils, such as fennel oil.

CH_3 O CH_3 CH_3

Fenitrothione: a relatively less toxic Organophosphate insecticide (see).

Fennel oil: an essential oil, usually pale yellow in color. The main constituent is anethol (up to 60%); it also contains camphene, pinene, anis ketone and fenchone. F. is obtained by steam distillation of the fruits of fennel, a member of the Umbelliferae which is native to southeast Europe and north Africa. The yield is 2 to 3%. F. is used as an aroma in sweets and liquors, and is used in medicine as a cough remedy and to prevent flatulence.

Fenpropimorph: a Morpholine and piperazine fungicide.

Fentin acetate: see Fungicides.

Fentin hydroxide: see Fungicides.

Fenton's reagent: a mixture of sodium hypobromite or hydrogen peroxide with iron(II) salts used to detect 1,2-diols, which form α-hydroxyaldehydes or ketones with it.

Fentoxan®: an Acaricide (see) which also kills white flies.

Fenuron: see Urea herbicides.

Fenvalerate: see Pyrethroid.

Ferbam: see Dithiocarbamate fungicides.

Ferment: see Enzymes.

Fermentation: microbial processes occurring in the absence of oxygen. They serve to generate metabolic energy by substrate phosphorylation; the hydrogen removed from the substrate is transferred to an organic H-acceptor. In contrast to Respiration (see), in which organic substrates are degraded to inorganic compounds, F. yields reduced organic compounds as end products. These are excreted and become enriched in the nutrient solution. The fermentation products are not further utilized by an anaerobically growing cell; they are metabolically equivalent to the CO_2 and H_2O produced by respiration. There are several types of F., classified according to the end products as Alcoholic F. (see), lactic acid, butyric acid, propionic acid, formic acid, homoacetate and methane F. The substrates of F. are organic compounds which can be partially oxidized by exergonic intramolecular cleavages, such as carbohydrates or organic acids. Some microorganisms produce acetic, citric, gluconic and keto acids, etc., by "incomplete oxidation" in the presence of oxygen; the term "aerobic F." for such processes is incorrect.

Lactic acid F. can be classified as homofermentative or heterofermentative, depending on the products; in the former, glucose is converted solely to lactic acid, and in the latter, there are other products as well. *Homofermentative lactic acid F.* yields pure or nearly pure (90%) lactic acid. The stereospecificity of the lactate dehydrogenase and the presence of a lactate racemase determine the relative amounts D(-)-, L(+)- and DL-lactic acids formed. Only a small fraction of the pyruvic acid is decarboxylated and converted to acetic acid, ethanol, carbon dioxide and acetoin. The mechanism of lactic acid F. corresponds to the scheme of Glycolysis (see). In *heterofermentative lactic acid F.*, the absence of aldolase and triosephosphate isomerase in the heterofermenting lactic acid bacteria forces the degradation of glucose by the pentose phosphate cycle. Glucose is converted to ribulose 5-phosphate, which is epimerized to xylulose 5-phosphate, and finally converted to glyceraldehyde phosphate and acetylphosphate. Depending on the species of bacteria, the products are lactic acid, ethanol and carbon dioxide, and also acetic acid and other products.

Other types of F.. Butyric acid, butanol, isopropanol, acetone and other organic acids and alcohols are products of F. of carbohydrates by Clostridia (anaerobic spore-formers). Propionic bacteria form propionic acid from glucose, sucrose, lactose or pentoses, and also from lactic acid, malic acid, glycerol and other compounds. In ***formic acid F.*** ("mixed acid F."), the most abundant product is formic acid; here the fermenting organisms are *Enterobacteriaceae*. Methane bacteria convert alcohols, organic acids, carbon dioxide and hydrogen to methane; various organic acids are only incompletely oxidized by them. In the production of vinegar, acetic acid bacteria (*Acetobacter sp.*) convert ethanol to acetic acid in the presence of atmospheric oxygen.

Fermi-Dirac stistics, ***Fermi statistics***: Quantum statistics for an equilibrium system of Fermions (see). In classic Maxwell-Boltzman statistics (see Maxwell-Boltzmann distribution), the particles are considered to be distinguishable, so that exchange of two particles of the same kind produces a new microstate (see Thermodynamics, III). In quantum statistics, the particles cannot be distinguished. According to the Pauli principle, which applies to fermions, each state can be occupied by only one particle. By contrast, in Bose-Einstein statistics (see), each state can be occupied by any number of particles. For a system of fermions, which do not interact with each other (ideal Fermi gas), the distribution function is

$$N_i = g_i/(e^{[E_i-\mu]/kT} + 1)$$

Here N_i is the mean number of particles in state i with energy E_i, k is the Boltzmann constant, T the absolute temperature, and μ is the chemical potential of the system. g_i is the statistical weight of state i, i.e. the number of states with the same energy E_i.

For high energies, $E_i - \mu \gg kT$, so that the 1 in the denominator can be ignored, and the F.-D. is reduced to Maxwell-Boltzmann statistics. The system behaves as a classical ideal gas. If the energies are not high, there are deviations from the ideal gas behavior; this is called *gas degeneracy*. It occurs especially at very low temperature. In the limiting case at absolute zero, the particles occupy only the lowest energy levels, and those above the chemical potential remain unoccupied (***Fermi limit, Fermi edge***).

The treatment of electrons in metals and semiconductors as a Fermi gas is very important, because this gas is already highly degenerate at room temperature. F. permits the prediction of specific heats of metals, electron emission of heated metal wires, the band structures of semiconductors, etc.

Fermion, ***Fermi particle***: a subatomic particle with a half-integral spin. These include electrons, positrons, protons, neutrons, neutrinos and muons. In quantum theory, F. are described by antisymmetric wavefunctions which change sign when two particles are exchanged. The Pauli principle, which states that two F. can never occupy exactly the same quantum state, is a consequence of the antisymmetric wavefunctions. It has important consequences for the structure of atoms, chemical bonding and the statistical treatment of large ensembles of such particles (see Fermi-Dirac statistics).

Fermi particle: same as Fermion (see).

Fermi resonance: a special type of band splitting in a vibration spectrum. F. can occur when a fundamental vibration has approximately the same frequency as a harmonic vibration. An interaction between the two increases the intensity of the harmonic vibration and reduces the intensity of the fundamental vibration. In the IR spectrum of cyclopentanone, for example, there are two intense absorption bands in the range expected for C=O vibrations, one at 1728 cm^{-1} and one at 1746 cm^{-1}. They arise through F. between the normal C=O vibration and a harmonic of a skeletal vibration which has a fundamental frequency of 889 cm^{-1}.

Fermi statistic: same as Fermi-Dirac statistics (see).

Fermium, symbol ***Fm***: a short-lived, radioactive element in the Actinide (see) series; it does not occur naturally on earth. Z 100, isotopes with the following mass numbers (in parentheses, decay type; half-life; nuclear isomers): 244 (spontaneous decay; $3.3 \cdot 10^{-3}$ s), 245 (α; 4.2 s), 246 (α; 1.2 s), 247 (α; 9.2 s; 2), 248 (α; 37 s), 249 (α, K-capture; 2.6 min), 250 (α; 30 min; 2), 251 (K, α; 5.30 h), 252 (α; 22.8 h), 253 (K, α; 3.0 d), 254 (α; 3.24 h), 255 (α; 20.1 h), 256 (spontaneous decay; 2.63 h), 257 (α, 100.5 d), 258 (spontaneous decay; $0.38 \cdot 10^{-3}$ s); the stablest isotope has atomic mass 257; valence III, II, standard electrode potential (Fm^{2+}/Fm^{3+}) -1.3 V.

F. has so far not been produced in weighable amounts. It forms stable Fm^{3+} ions in aqueous solution. Fm(II) compounds are strong reducing agents. F. was discovered in 1953 among the products of a thermonuclear reaction, and was later produced by reaction of plutonium 239 with neutrons:

$$^{239}Pu + 16\ ^{1}n \xrightarrow[-6\beta]{} {}^{255}Fm$$

or by bombarding uranium 238 with highly accelerated oxygen nuclei: $^{238}U + {}^{16}O \rightarrow {}^{249}Fm + 5\ {}^{1}n$; $^{238}U + {}^{16}O \rightarrow {}^{250}Fm + 4\ {}^{1}n$.

Fernichrome: a proprietary name for an alloy of 30% nickel, 37% iron, 25% cobalt and 8% chromium. F. has the same thermal expansion coefficient as glass and is used as a melt-in alloy.

Ferrates: in the broad sense, anionic complexes of iron. In the narrow sense, ***oxoferrates***, anionic complexes of iron in the +5 and +6 valence state with isolated FeO_4 tetrahedra, $M^I_3[FeO_4]$ or $M^I_2[FeO_3]$. Double oxides, such as $M^IFe^{III}O_2$, $M^I_2Fe^{IV}_2O_3$ and $M^{II}Fe^{III}_2O_4$ are also called F. The ferrates(III), which are usually called Ferrites (see), are of special interest.

There are many known F. which are anionic complexes of iron (see Iron complexes), especially in the +II and +III oxidation states. These include hydroxyferrates(II), $M_2[Fe(OH)_4]$ and the industrially important Cyanoferrates (see).

Ferredoxin: a low-molecular-weight iron-sulfur protein which is involved as an electron carrier in electron-transport processes. F. contain equal numbers of inorganic sulfur and iron atoms bound together covalently in the *iron-sulfur cluster*. In the ***8Fe-ferredoxins***, the protein consists of about 55 amino acids, including 8 cysteines, and contains two identical 4Fe-4S centers. Each of these forms a cube and is covalently linked with 4 cysteine side chains. Each 4Fe-4S center can transfer one electron. There are also ***4Fe ferredoxins***, each with a single 4Fe-4S cluster. Here the iron atoms are bound to the 4 cysteine residues in the molecule. This type of F. has been isolated mainly from bacteria. The ***high-potential iron-sulfur protein (HiPIP)*** is a special type of F. which has been isolated from photosynthetic bacteria. It also has only a single 4Fe-4S cluster, but unlike the other F., it has a positive standard potential of about 350 mV. The ***2Fe-ferredoxins*** isolated from blue-green bacteria, green algae and higher plants have 4 to 6 cysteine residues, four of which are covalently linked to the Fe-S center.

Ferri-: an obsolete term for iron(III)-, as in ferrisulfate for iron(III) sulfate.

Ferrite: a) ***α-ferrite***, a cubic body-centered iron mixed crystal formed by phase conversion in the solid state. α-F. is the plastic component of steel and determines its cold malleability (see Iron-carbon alloys). α-F. is ferromagnetic up to 768 °C and at 723 °C can dissolve a maximum of 0.02% C.

δ-Ferrite is a cubic body-centered iron mixed crystal which often appears in austenitic steels. It is undesirable because it reduces the corrosion resistance of the steel.

Ferrites: mixed oxide materials consisting of iron(III) oxide and one or more oxides of di- or trivalent metals. F. are magnetic, but in contrast to other magnetic materials (such as metals), they have high electrical resistance of 10^4 to 10^{12} ohm · cm. As a result, no eddy currents can arise and the materials can be used in high-frequency applications (10^3 to 10^{10} Hz).

Soft magnetic F. with the general formula $MO \cdot Fe_2O_3$ (M = Mn, Fe, Co, Ni, Cu, Zn, Cd, Mg) have low remanence and low hysteresis losses and crystallize mainly in cubic lattices; they have spinell structures. This group includes, for example, magnesium ferrite, $MgFe_2O_4$, cobalt ferrite, $CoFe_2O_4$, nickel ferrite, $NiFe_2O_4$, and a number of mixed F., such as manganese-zinc ferrite, $(Mn,Mg)Fe_2O_4$ for magnetic cores for use up to 10 MHz, and nickel zinc ferrite $(Ni,Zn)Fe_2O_4$ or manganese magnesium ferrite $(Mn,Mg)Fe_2O_4$ for use in magnetic data storage. The F. are generally produced by high-temperature reactions of the finely ground oxides, using ceramic techniques, e.g. $NiO + Fe_2O_3 \rightarrow NiFe_2O_4$. Soft magnetic F. can also be produced from oxide melts, e.g. cobalt ferrite with good magnetic properties precipitates below 1050 °C from a bismuth(III) oxide melt in which the starting materials, CoO and Fe_2O_3, have been dissolved. After it has solidified, the melt matrix is dissolved in dilute nitric acid, leaving well formed ferrite crystals. The thermal decomposition of double salts of the schoenite type, e.g. $(1/3Co \cdot 2/3Fe)SO_4 \cdot (NH_4)_2SO_4 \cdot 6H_2O$, can also produce soft F.

Hard magnetic F. with the general formula $MFe_{12}O_{19}$ (M = Sr, Ba, Pb) with high remanence crystallize mainly in hexagonal systems and have magnetoplumbite lattices. The main examples are the hexaferrites of strontium and barium. The hard F. are made by wet grinding the alkaline earth metal carbonates with highly reactive Fe_2O_3. These are filtered (also spray granulated) and presintered around 1000 °C. After the cooled material has been extremely finely ground in ball mills and organic binders have been added, the product is pressed (extrusion and magnetic field presses) and sintered at temperatures from 1100 to 1400 °C. The ***granate F.*** (granates) with the general formula $3\ M_2O_3 \cdot 5Fe_2O_3$ (M = rare earth metal) are used for special components in very high frequency devices.

Ferritin: the most important iron storage protein in mammals; it is found primarily in the spleen, liver and bone marrow. F. can carry up to 4300 iron(III) atoms per molecule of protein. The protein component of F., ***apoferritin***, has a relative molar mass of 445,000 and consists of 24 identical subunits which form a shell around the iron hydroxide micelle. The iron released to the blood by the intestinal mucosa cells is adsorbed to a carrier protein, transferrin, and transferred to the iron depots, especially those in the red blood marrow where red cells are formed.

Ferro-: in the broad sense, the same as iron-, and in the narrow sense, an obsolete designation for iron(II), as in ferrosulfate for iron(II) sulfate.

Ferrocene: bis(h^5-cyclopentadienyl)iron(II), $Fe(C_5H_5)_2$, a yellow-orange compound which is readily soluble in organic solvents but is insoluble in water. M_r 186.04, m.p. 174 °C, subl.p. 230 °C. F. has a sandwich structure in which the 5-membered ligand rings are parallel, and in a "staggered" configuration; the iron is between them (Fig.). F. can be synthesized, for example, by reaction of iron(II) chloride with cyclopentadienyl sodium; it is used as a combustion catalyst for smokeless combustion or as a catalyst for hardening polyester resins.

Ferrochrome: see Chromium; see Chromium alloys.

Ferroalloys: industrial alloys of iron with aluminum, boron, chromium, cobalt, manganese, molybdenum, nickel, niobium, phosphorus, silicon, tantalum, titanium, tungsten, vanadium and zirconium. F. are used as preliminary alloys for formation of other alloys, and as reactants in reaction treatment of steels.

Ferrocement: see Binders (construction materials).

Ferromanganese: see Manganese alloys.

Ferromolybdenum: see Molybdenum, see Molybdenum alloys.

Ferronickel: a technical alloy of iron with 25 to 50% Nickel (see). F. is used as a prealloy to make steels which can be cold-drawn and are resistant to corrosion and heat.

Ferroniobium: see Niobium.

Ferrophosphorus: an industrial compound or alloy of iron with 18 to 25% phosphorus. F. is a byproduct of phosphorus production in electric hearths. It is added to cast iron to improve the liquidity of the melt.

Ferrosilicon: industrial alloys of iron with about 15, 45, 75 or 90% silicon, and the remainder iron. F. are used in steel production as deoxidation and degassing agents, and as preliminary alloys for production of high-silicon steels and cast iron. F. with up to about 15% silicon are produced in blast furnaces. F. with higher silicon contents are made in electric hearths.

Ferrotantalum: see Tantallum alloys.

Ferrotitanium: industrial alloys of iron with about 30, 40 or 70% titanium. They are produced in electric hearths or by an aluminothermal process. F. are used as preliminary alloys for the production of chrome-nickel steels which are stabilized for welding and against corrosion.

Ferrotungsten: see Tungsten alloys.

Ferrovanadium: industrial alloys of iron with 60 to 80% vanadium, 0.1 to 1% carbon, and the rest iron. F. has a deoxidizing effect on steel, and the presence of vanadium increases the hardness, tensile strength and impact resistance of steels. F. is made either by aluminothermal reduction of a mixture of vanadium(V) oxide and scrap iron, or by reduction of a vanadium-iron mixture with coal. F. is added as an alloy component in amounts $< 0.2\%$ for construction steels, up to 0.5% for tool steels and up to 5% for rapid rotation steels.

Ferrum: the Latin name for Iron (see).

Fertilizer: a substance added to the soil to promote plant growth and crop yields. The main nutrients supplied by F. are nitrogen, phosphorus, potassium and calcium. Elements required in small amounts, such as iron, magnesium and sulfur, are usually present in the soil, but they may also be supplemented as trace elements in the F.

A distinction is made between organic and inorganic F. ***Organic F.*** consist mainly of plant residues and animal excrement, and they are usually applied to the same land which produces them. ***Inorganic F.*** are produced as salts (***mineral F.***) and urea and ammonia (***liquid F.***). The world production of F. is growing rapidly, and is at present about 30 megatons nitrogen, 40 megatons phosphorus pentoxide, P_2O_5, and 30 megatons potassium oxide, K_2O.

The most important inorganic fertilizers

Fertilizer	Nutrient content	Production and use
Nitrogen fertilizers		
Ammonia, liquid	82% N	from the elements, complicated application equipment
Urea	46% N	from ammonia and carbon dioxide; preferred nitrogen fertilizer
Ammonium nitrate	35% N	from ammonia and nitric acid; part of complex F.
Ammonium sulfate	21% N	from ammonia and sulfuric acid or as an industrial byproduct (caprolactam synthesis); use has leveled off
Ammonium chloride	24% N	from ammonia and hydrochloric acid; for rice
Phosphate fertilizers		
Double superphosphate	38% P_2O_5	Solubilization of crude phosphates with phosphoric acid
Superphosphate	16% P_2O_5	Solubilization of crude phosphates with sulfuric acid
Roasted phosphate	25% P_2O_5	Roasting of crude phosphates with soda, sodium sulfate/sand or lime/sand
Dicalcium phosphate	38% P_2O_5	Neutralization of phosphoric acid with calcium hydroxide
Thomas phosphate	13% P_2O_5	Smelting of phosphate-containing iron ores, slag as fertilizer
Soft earth crude phosphate	26% P_2O_5	mechanical preparation of special crude phosphates
Potassium fertilizers		
Potassium chloride	40–60% K_2O	from crude salt by recrystallization process; contains sodium chloride, calcium and magnesium sulfates
Potassium sulfate	54% K_2O	from potassium chloride and magnesium sulfate
Potassium nitrate	47% K_2O 14% N	from potassium chloride and sodium nitrate; high quality component
Lime fertilizers		
Slaked lime, lime hydrate	76% CaO	from burned lime, also byproduct of ethyne synthesis from calcium carbide
Lime nitrogen	47% CaO, 23% N	from calcium carbide, limited importance
Calcium ammonium nitrate	17–22% CaO, 21% N	from limestone or slaked lime, most important single fertilizer
Calcium nitrate	24% CaO, 12% N	as tetrahydrate from the ODDA process of nitric acid phosphate solubilization

In addition to the F. which contain only a single component, there is a wide variety of ***mixed F.***, which supply all the necessary nutrients, for example, NP (nitrogen + phosphate), NK (nitrogen + potassium) and NPK (nitrogen + phosphate + potassium). As agricultural technology intensifies, there is a growing tendency to deal in single-component F., so that agricultural concerns can mix their F. in accordance with the requirements of their own crops and soil.

Ferulic acid: see Hydroxycinnamic acids.

FFF: abb. for Field flux fractionation.

Fiberglass: fibrous glass products. ***Glass silk*** is a product consisting of several hundred strands of indefinite length, made by nozzle extrusion. Fibers 5 to 10 μm in diameter are drawn from 1 mm holes in the bottom of a V-shaped platinum/rhodium pan with about 1 liter capacity at a temperature of 1100 to 1300 °C; the threads are drawn at rates up to several thousand meters per minute. They are glued to form a thread with an organic material and wound on a spool. Glass silk can be processed as other fibers are, and when cut, is used as staple glass silk to reinforce plastic. The method of *rod drawing*, in which a large number of prepared glass rods are heated and drawn on their ends, is now used only to make silica glass fibers.

Glass fibers (glass wool) are 5 to 40 cm in length, and are 1 to 4 μm in diameter. They are produced by blowing, centrifugation or drawing. In *nozzle blowing* techniques, the raw materials (glass raw materials, slag, rock) are melted in a blast furnace and the glass flowing out of it is blown into fibers by high-pressure steam or air. However, these fibers contain a large proportion of glass beads. The fibers can be collected on a wire net drum as a fiberglass felt. *Centrifugal methods* produce higher quality fibers. In the *plate method*, a stream of glass is fed onto a rotating, hard porcelain plate with grooves; it is converted to fibers by centrifugal force. A newer and more productive method is the *Stillite process*. The fibers are created by impact with a rotating spherical ball, then stretched by spinning them against rotating disks. In the *nozzle-spinning process (Tel process)*, the glass stream is extruded through nozzles along the sides of a rotating, double-walled head which is provided with induction heating. The G. are additionally stretched by a tangential stream of hot air. Fibers of finite length are produced by the *rod-blowing technique*, which is now only rarely used. It differs from the rod-drawing method of making glass silk in that the molten glass is blown away from the end of the rod.

Superfine G. 3 to <1 μm diameter can be made by blowing primary glass fibers in a hot gas flow.

Composition and properties. F. consist of a number of oxidic components, chiefly silicon dioxide. Aluminum trioxide and boron trioxide improve the homogeneity and chemical resistance of the glass. Glasses for spinning are classified according to their alkali content (sodium and potassium oxides) as alkali-free (up to 1%), low-alkali (up to 5%) and high-alkali (> 17%). The electrical insulating capacity is improved with decreasing alkali content, and the leachability becomes lower.

F. are noncombustible and, in comparison to organic fibers, they are mechanically strong and more resistant to thermal, chemical and biological effects. Their strength increases with decreasing alkali content and fiber diameter, and is greater than that of metals. Their applications are limited by the softening temperature, which is between 500 and 800 °C. The strength decreases rapidly at higher temperatures.

Application. F. are used mainly as mats or loose stuffing for heat, sound and electrical insulation. Noncombustible, chemical-resistant fabrics of F. are used for industrial and decoration purposes, e.g. as filters, protective clothing, curtains and wallpaper. F. are very important as reinforcing materials for building materials, for example, F.-reinforced concrete. Organic polymeric materials are also reinforced with F., especially for autobodies, airplane and rocket parts, and cords for tires. Special papers are made from F. A special application of glass fibers is made in optics, where they serve in light conduits.

Fiberglass resins: Polyester resins (see) reinforced with fiberglass.

Fibrin: a protein formed as the endproduct of blood coagulation. In contrast to its precursor Fibinogen (see), F. is insoluble in neutral salt solutions, dilute acids or bases. It is produced from fibrinogen by the action of thrombin in the presence of calcium(II) ions; two pairs of peptides (fibrinopeptides) are split off, and the molecules then polymerize. F. is a fibrous protein with a highly regular structure; it forms a dense web of cross-banded fibrils. The diameter of the fibers is about 60 nm at pH 7.6. In the course of blood coagualtion, the precipitating F. traps blood cells, and thus forms a clot. Hemophiliacs lack one of the blood factors (clotting factors) which are activated in a cascade mechanism leading finally to activation of thrombin. They are thus unable to form F., or to form it as rapidly as non-hemophiliacs.

Fibrinogen: the coagulation protein present in the blood plasma of all vertebrates. F. is partially cleaved and then polymerizes to form Fibrin (see). The plasma concentration of F. in human beings is between 200 and 450 mg/100 ml. It is a very long, soluble glycoprotein (2% carbohydrate) with M_r 340,000, and consists of three pairs of polypeptide chains. These form a symmetrical molecule of two identical F. monomers; it is held together by several disulfide bridges between the chains. The disulfide bridges are located at the *N*-terminal end and form the disulfide knot. The most abundant amino acids in human F. are glutamic acid (14.5%), aspartic acid (10.01%), lysine (9.2%) and arginine (7.8%). In thromboses, the F. coagulates inside the blood vessels and blocks them.

Fibrinolysin: same as Plasmin (see).

Fibrinolytics: see Antithrombotics.

Fibroin: the protein of natural silk (silk fibroin, see Keratins).

Fibroin silk: a Protein fiber (see).

Fick's law: see Diffusion.

Ficoll®: a synthetic, polysaccharide-like polymer made by reaction of sucrose with epichlorohydrin. F. is a colorless, non-crystalline powder, which is readily soluble in water. Its molecular mass is about 400,000. F. is used in biochemical laboratories to concentrate aqueous solutions by dialysis (it has a low osmotic pressure), and in density centrifugation for separation and isolation of cells and organelles.

FID: abb. for Free Induction Decay. See NMR spectroscopy.

Field effect: see Inductive effect.

Field flux fractionation, abb. ***FFF***: a method of separating and characterizing macromolecules and particulate materials. The materials are passed through a narrow (50 to 500 μm) channel to which an external field is applied along the surface of the channel. The different interactions of the molecules or particles with this field leads to the formation of layers of different fractions along the wall of the channel. The rapidly flowing carrier liquid moves the bands along the wall, but their velocities are greatly diminished because of strong frictional interaction with the channel wall (Fig.). The separation is classified according to the nature of the external field, as ***electrical, sedimentation*** or ***thermal*** FFF. In each case, one specific property (charge, mass or size of molecule) is utilized for the interaction, and the elution time for the particles can be related to this property.

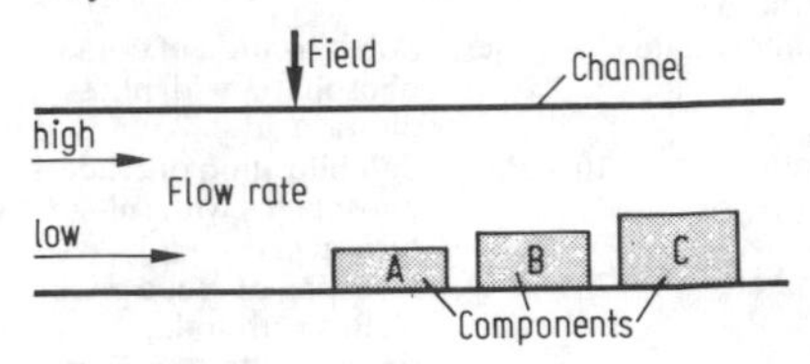

Field flow fractionation

FFF can be carried out continuously and can be used with either aqueous or nonaqueous systems. It can be used for either high-resolution analysis or preparative separations of inorganic particles, polystyrene latexes, viruses or biopolymers in the molecular mass range of 10^3 to 10^{17}, and the analysis times are short.

Field ionization: an ionization technique used in Mass spectroscopy (see) (*FI mass spectroscopy*). Electric fields with high field strengths ($\approx 10^8$ V/cm) are used. Metal points, thin wires or knife edges are used as anodes (field ion emitters); the gaseous sample is ionized close to the anode. The molecular ions formed in this way have lower energies than those formed by electron collision ionization, and FI mass spectra therefore have intensive molecule ion peaks and, usually, only a few fragment peaks.

Field strength parameter: see Ligand field theory.

Field sweep: see NMR spectroscopy.

Filipin, more exactly, ***filipin III*** (the original substance isolated and termed F. was heterogeneous): a polyene antibiotic with a pentaene structure. F. forms yellow crystals; m.p. 195-205 °C. It is soluble in most organic solvents, and is practically insoluble in water.

Filler: a material added to an amorphous matrix material during processing to give it certain properties. For example, barium sulfate, titanium white, kaolin, gypsum, etc. (up to 25%) are added to the pulp slurry for paper to make it white and opaque and to make print adhere better. The above substances are sometimes worked into a paste with casein, water, etc. and spread onto the crude paper in this form. Sawdust and powdered rock, cellulose or rags (e.g. linters), glass fibers and powder, carbon, metal fibers, etc. are added to plastics in solid or emulsified form. In the preparation of veneers, sawdust, powdered synthetic resins, etc. are added to the glue to prevent its being visible on the veneer surface.

Filter: a porous material used in the laboratory or industry to separate solids from liquids or gases (see Filtration).

Laboratory filters (Fig. 1). ***Filter papers*** for use in filter funnels are produced in various grades, both as circular sheets and fluted cones. Extraction shells made of filter paper are used for separation of substances which are soluble in organic solvents; used together with an extraction apparatus, these are quite versatile.

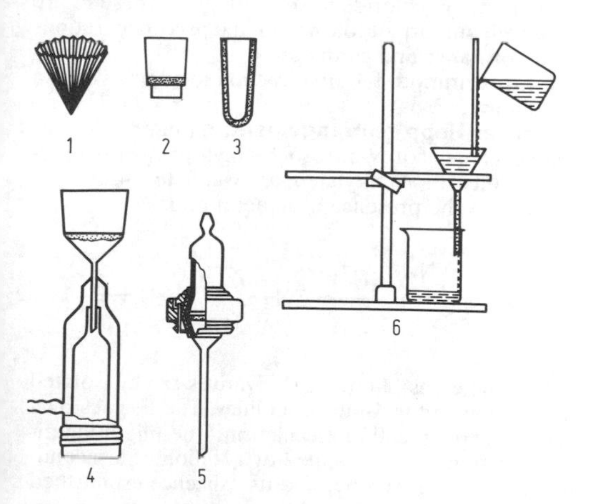

Fig. 1. *1* Fluted filter; *2* Filter crucible; *3* Filter candle; *4* Suction filter (bacterial filter); *5* Grabar pressure filter (membrane filter); *6* Filtration set-up

Filter plates (frits) are made either of sintered glass or ceramics. *Sintered glass filters* are usually round and consist of plane, glass filter plates polished on both sides. They come with different porosities, and thus permit different filtration speeds. The edge can be either fused, which prevents an escape of gases from the sides, or not. For special purposes, sintered glass filters are used as glass filter candles or extraction inserts, but they are usually used as the filters in glass filter crucibles or Büchner funnels; these devices are made with the glass filter fused to the glass sides of the crucible or funnel. *Ceramic F.* are usually made of fired ceramic materials of varying porosity, and thus different filtration rates are possible. Porcelain filters are most common. Colloidal suspensions are filtered through *membrane F. (ultrafiltration)*. These are made of animal or plant membranes, or simple paper filters impregnated, for example, with glacial acetic acid/collodium or hardened gelatin. Synthetic semipermeable materials are also used.

2) Industrial F. are made of a variety of materials. Layers of granular materials, such as sand and gravel, half-burned dolomites or anthracite, and more recently, polystyrol, are used for water in water- and sewage-treatment plants. Coke or anthracite is used for hot alkali liquors, oil-wetted Raschig rings are used for air. Cloth filters or filter cloths woven of natural or synthetic fibers or mineral fibers (asbestos or glass) are used to filter air, gases, aqueous suspensions, acids and alkalies. Sintered ceramics with pore diameters of 1 to 250 μm are resistant to high temperatures and are used for acids and bases. Rubber and plastic F. are used for corrosive and radioactive substances. F. plates made of sintered metal powders are used to filter liquid fuels, rayon, concentrated hot alkali liquors and sulfuric or nitric acids. Membrane F. are used as in the laboratory.

In many filtrations, the hydrostatic pressure of the liquid column above the F. is sufficient, as with sand or gravel filters used in water treatment. In some cases, overpressure is used, as in the filtration of rayon through porous steel plates with up to 70 MPa. In others cases, vacuum is used to increase the rate of filtration; depending on the nature of the solid, the pressure below the F. can be reduced to 10 to 100 kPa.

Some examples of F. used in discontinuous fashion are flow-through, one-layer and multi-layer F. used to purify air and water, F. presses for liquids with relatively large amounts of solids, bag, strainer or tube F. for air or liquids, F. candles, settling or Berkefeld F. for disinfection and purification of water, and sieve F. for purification of solvents, lubricants and fuels. Examples of continuous F. are rotating or ribbon F. for purification of air, etc. and rotating cellular F. for filtration of cellulose, petroleum and pigments, etc. (Fig. 2).

The removal of dust from gases can be achieved by an electrofilter (see Electrostatic gas purification).

Filter lime: a granular limestone used as a filter material.

Filter candle, ***candle filter***: a filtration device consisting of a porous hollow cylinder (usually of a ceramic material). Its purification effect is based on both sieving and adsorption. The purification (filtration) occurs as the liquid passes through the porous

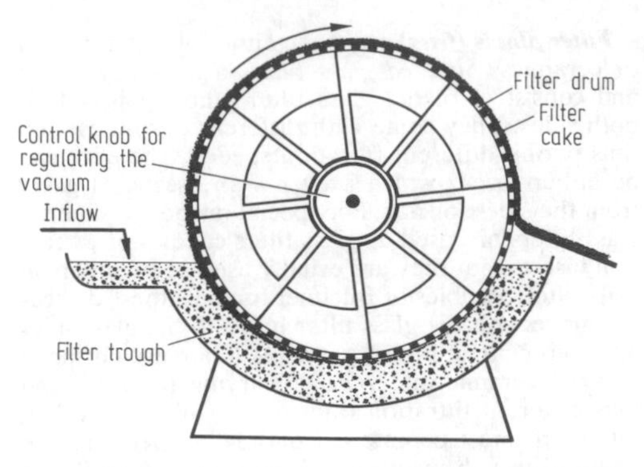

Fig. 2. Section through a vacuum rotating cellular filter

wall. Using F., it is possible to remove undissolved impurities of all kinds. F. are used in water treatment, for example, in the purification of small amounts of water.

Filtration: 1) the separation of a solid substance from a liquid using a Filter (see). If the pure liquid, the filtrate, is desired, the process is used for *clarification*; if the solid is the desired product, it is recovered as the *filter cake*. In some cases, both components are utilized.

F. of water and sewage is the most important step in treatment, which removes primarily the undissolved impurities. In Iron removal (see), the water is vigorously oxidized to cause dissolved iron to precipitate on an activated silica filter. F. of sewage over gravel or similar material is often sufficient to remove most of the impurities, and the clarified water can be reused directly, e.g. as cooling water.

2) The F. of gases is discussed under Dust removal (see).

FI-mass spectroscopy: abb. for Field Ionization mass spectroscopy; see Field ionization.

Fine dust: aerosols with solid disperse phases and particle sizes from 10 to 63 μm; it settles at great distances from the emitting sources. F. is involved in the formation of smog above population centers.

Fine structure: the splitting of spectral lines into multiplets due to interaction between the orbital motion and spins of electrons (see Atomic spectroscopy; see Multiplet structure).

Finkelstein reaction: a halogen exchange reaction between alkyl halides and suitable metal halides: $R\text{-}Cl + I^- \rightleftharpoons R\text{-}I + Cl^-$. In this reaction, the halogen with the lower nucleophilicity is replaced by another with a higher nucleophilicity. The F. makes possible the synthesis of iodoalkanes, which are difficult to obtain by other methods. In secondary and tertiary alkyl halides, there is usually no halogen exchange.

FIR: abb. for far infrared (see Infrared spectroscopy).

Fire point: see Ignition.

Fire retardants: substances which make combustible materials ignite less readily, delay their burning and suppress flame formation.

1) For wood, solutions of F. are applied by impregnation (long soaking under high pressure), spraying or painting. The active ingredients are a) ammonium salts, all of which release ammonia when heated (smothering the fire). Some also leave an acidic melt at the site which inhibits further burning. These ammonium salts include the chloride, NH_4Cl, bromide, NH_4Br, sulfate, $(NH_4)_2SO_4$, and phosphates, such as diammonium phosphate, $(NH_4)_2HPO_4$. Amidosulfonic acid, $NH_2(SO_3H)$, is also used. b) Borax, $Na_2B_4O_7 \cdot 10H_2O$, aluminum sulfate, $Al_2(SO_4)_3$, mixtures of borax and boric acid, diammonium sulfate and urea compounds form protective layers of foam when heated. c) Compounds which cleave off acids when heated, and which inhibit the glowing of coals by formation of a layer of charcoal on the wood. These include phosphoric acid, H_3PO_4, monoammonium phosphate, $NH_4H_2PO_4$, ammonium sulfate, $(NH_4)_2SO_4$, zinc chloride, $ZnCl_2$, antimony oxygen chloride, SbOCl, ammonium sulfonate, $NH_2(SO_3NH_4)$.

Sprayed-on F. penetrate the wood better when a wetting agent is added to the aqueous solution. Waterglass, especially sodium waterglass, Na_2SiO_3, is a useful paint. Paints based on polyvinyl chloride, chlorinated latex or polycondensates of phenol or urea and formaldehyde give very good protection.

2) Plastics can be protected in several ways: a) by addition of antimony trioxide, or by combination with halogen compounds during processing; b) by adding F. to the polymerization mixture, or flame-retarding monomers are included in the macromolecules (e.g. tetrabromodiane in epoxides); c) relatively non-inflammable plastics, such as polycarbonates, polyvinyl chloride and polytetrafluoroethylene, can be added to the inflammable plastics such as polyethylene and polypropylene.

3) For textiles, dipping in ammonium salt solutions (especially ammonium bromide) or disodium phosphate is very effective. Tin and tungsten salts are especially good. Borax is often also used, for example in a mixture with ammonium carbonate and gelatins for impregnating cotton batting to be placed under Christmas trees. To keep the F. in the textile, a glue is added to the mixture.

4) Paper and cardboard can be protected by spraying with ammonium sulfate, borax, alkali phosphates or ammonium chloride mixed with glue. These F. are especially important for theater and party decorations made of paper and cardboard.

First runnings: the first fraction to come over in a distillation.

Fischer-Hepp rearrangement: an intermolecular rearrangement of *N*-nitroso-*N*-alkyl- or *N*-nitroso-*N*-arylanilines to *N*-alkylated or *N*-arylated 4-nitrosoanilines in the presence of mineral acids:

$$C_6H_5\text{-}N(NO)\text{-}R \xrightarrow{(H^+)} ON\text{-}C_6H_4\text{-}NH\text{-}R$$

In some cases, 2-nitroso derivatives can be isolated in addition to the 4-nitrosoanilines. The F. goes by a cationotropic reaction mechanism. The migrating nitrosyl cation can be trapped by addition of a second, non-nitrosylated amine, and its existence established in this way.

Fischer indole synthesis: synthesis of indole derivatives by heating the phenylhydrazones of aliphatic aldehydes, ketones, ketoacids, etc. with zinc chloride, copper(I) chloride, sulfuric acid or polyphosphoric acid:

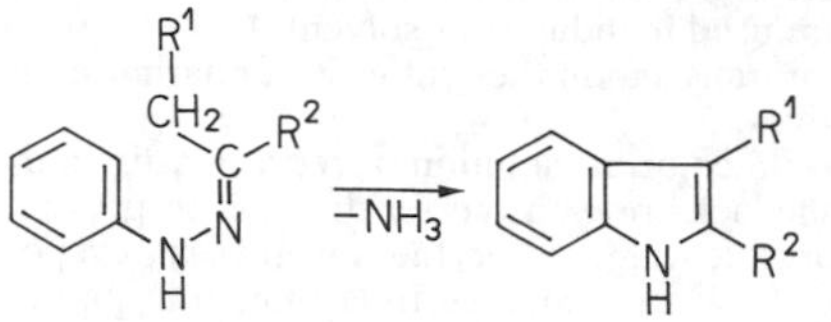

The formation of a bond between the C2 of the phenyl ring and the C of the CH_2 group of the aliphatic group is interpreted as the result of a sigmatropic rearrangement of the tautomeric enehydrazine:

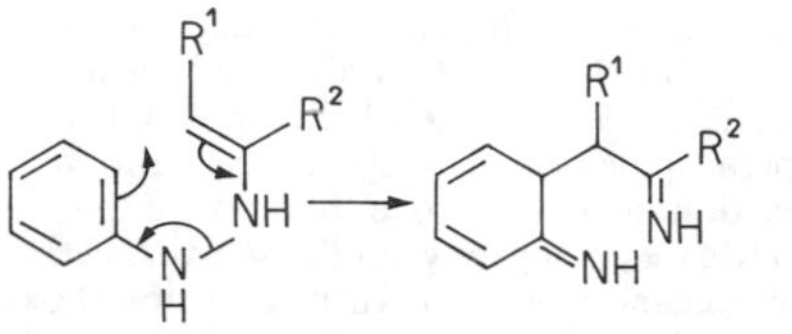

The rearranged product then undergoes ring closure with ammonia loss to give the end product. It was shown by the use of ^{15}N-labelled phenylhydrazine that the nitrogen atom lost in the form of ammonia was originally part of the azomethine group, and therefore must have been the β-nitrogen of the hydrazine group.

Fischer projection: see Stereochemistry.

Fischer's salt: same as Potassium hexanitrocobaltate(III).

Fischer-Tropsch synthesis: a method of coal refining (coal liquefaction) for production of hydrocarbons from synthesis gas. The synthesis was developed in 1926 by F. Fischer and H. Tropsch with the goal of producing aliphatic hydrocarbons in the boiling point ranges of gasoline and diesel fuel: $n\mathrm{CO} + (2n{+}1)\mathrm{H_2} \rightarrow \mathrm{C}_n\mathrm{H}_{2n+2} + n\mathrm{H_2O}$; $\Delta_R H = -\,n\ 160\ \mathrm{kJ\ mol^{-1}}$. The method has not been able to compete economically with petroleum. The most important development has taken place at the Sasol Works in South Africa; when Sasol III went into production in 1983, the annual production was about 4.5 million t liquid fuel from coal. With the prospect of depletion of the world oil reserves, the F. has again become of interest. The emphasis of recent developments has been the production of raw materials for the chemical industry, such as short- and long-chain olefins, especially ethene and propene, oxygen-containing compounds and polymethylene, by means of modified F. The range of products depends mainly on the catalyst used; the process conditions (pressure, temperature), on the other hand, have relatively little effect. Iron and cobalt with admixtures of thorium, magnesium, calcium and aluminum oxides are the most commonly used catalyst. Rhodium catalysts produce mainly oxygen-containing products, and polymethylene is generated with ruthenium catalysts.

The F. can be conducted in a fixed bed with a fixed catalyst, as a fly-ash synthesis with a mobile catalyst, or as a liquid-phase synthesis with either fixed or mobile catalyst.

Product range of the Fischer-Tropsch synthesis using a cobalt or an iron catalyst

Reaction products	Cobalt catalyst (%)	Iron catalyst (%)
C_2 to C_4 hydrocarbons	13	36
Crude gasoline	27	41.5
Diesel fuel	8	3
Heating oil	5	3
Paraffins	43	8
Alcohols	4	8.5

1) In the *fixed bed synthesis* with an alkalinized iron catalyst, a synthesis gas with a ratio of hydrogen to carbon monoxide of 1.3:1 to 2:1 is used at a pressure of 2 to 3 MPa and a temperature around 200 °C. Variants of this process are the *Arge high-load process* (Ruhrchemie, Lurgi, Fed. Rep. Germany), the *hot gas recycle process* (Bureau of Mines, USA) and the *stage-furnace process* (Lurgi, Germany).

2) *Fly-ash synthesis* is used in the ***synthol process*** (Kellogg synthesis) in South Africa. Iron-melt catalysts are used. The synthesis takes place at 210 to 240 °C and 2 to 2.5 MPa; the hydrogen to carbon monoxide ratio is 2.8:1 to 3.2:1. The main products of this process are the C_2 to C_4 hydrocarbons and crude gasoline.

3) *Liquid phase synthesis* is used in the *Rheinpreußen-Koppers process*, which is done in blast column reactors. The preferred catalyst is iron suspended in oil; the synthesis gas passes through the oil in extremely small bubbles. The heat of reaction is removed by water flowing through pipes. The reactor has a capacity 6 to 8 times that of a fixed-bed reactor of the same size. The product range can vary from 85% gasoline and liquefied gas to 65% soft and hard paraffin, and the alkene content can be as high as 95%.

The *Kölbel-Engelhard process* is similar, but uses carbon monoxide and steam according to the equation: $3n\ \mathrm{CO} + n\ \mathrm{H_2O} \rightarrow (\mathrm{CH_2})_n + 2n\ \mathrm{CO_2}$. Low quality gases can be used, such as blast-furnace gas, carbide oven gases, etc.

The products of the F. do not contain sulfur, aromatics or naphthenes.

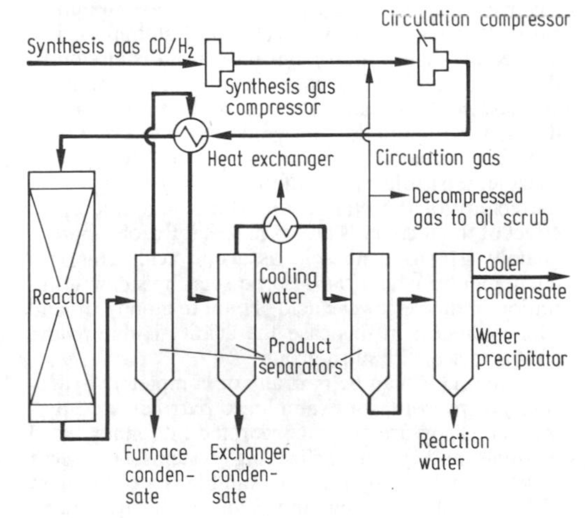

Diagram of a Fischer-Tropsch plant

Fisetin: see Flavones.

Fish toxins: 1) substances produced by living fish for offensive or defensive purposes; also toxins found in the tissues of the fish which are either produced by the fish itself (passively poisonous fish) or are taken up from its food (secondarily poisonous fish). a) Ichthyosarcotoxic species are subdivided into the cignateratoxic, secondarily poisonous species (see Ciguteratoxin) and tetrodotoxic, primary, passively poisonous species (see Tetrodotoxin). b) Icthyotoxic species are primary, passively poisonous, producing toxins in their gonads; c) ichthyohemotoxic species are primary, passively poisonous fish in which the blood serum contains a hemolytic protein. d) Actively poisonous species (about 250 such species are known) have a poison apparatus which is usually used to defend the animals against enemies. An example is the widely distributed scorpion fish, which has a poison gland at the base of the dorsal fin. The red fire fish *Pterosis volitans* of tropical oceans has 13 dorsal, 3 anal and 2 pectoral spines. This fish secretes a protein toxin with LD_{50} of 1.1 mg kg^{-1} mouse, intravenously.

2) Substances which are toxic to fish. Because of their sensitivity to toxins, a few fish species are used as bioindicators to monitor waters which are potentially contaminated with toxins.

Fittig synthesis: synthesis of bi- and polyaryls by reaction of aromatic halohydrocarbons with elemental sodium: 2 C_6H_5-Br + 2 Na → C_6H_5-C_6H_5 + 2 NaBr. The yields of the F. are considerably less than those of the Wurtz-Fittig synthesis (see).

Five-membered heterocycles: cyclic hydrocarbons containing one to four heteroatoms (see Heterocycles). The correct orientation of the structural formula is the one in which the highest-ranked heteroatom is shown at the top vertex of the pentagonal ring (position 1; see Nomenclature). In general, however, other orientations of the ring are shown, for example, with the heteroatom at the bottom. This is permitted and customary, since the ring can be rotated as desired.

Most aromatic F. are π-excess heterocycles, that is, they react extremely easily with electrophilic reagents (pyrrole, furan, thiophene). Therefore, their derivatives, such as halogenated and nitro compounds, carbaldehydes, carboxylic acids and sulfonic acids, are readily synthesized from the parent compounds. As the number of nitrogen atoms in the ring increases, this tendency decreases, and tetrazoles are π-deficient compounds. The gradation in properties is especially clear in the series pyrrole, imidazole, triazole, tetrazole, where there is a distinct increase in the acidity of the NH group. The electron acceptor effect of the tertiary N atoms makes tetrazole approximately as strong an acid as acetic acid. This is a phenomenon which can also be seen by stepwise nitration of the very weak acid phenol to mono-, di- and trinitrophenol; in this case the acidity is due to the (-)M effect of the nitro group.

Aromatic F. can be partially or completely hydrogenated; pyrrole, for example, to pyrroline and pyrrolidine. They are also incorporated into many fused benzene and heterocyclic ring systems. F. occur widely in nature; pyrrol, for example is a component of heme, bile pigments and chlorophyll; pyrrolidine makes up the side chains of the amino acids proline and hydroxyproline, indole is the side chain of tryptophan, and imidazole of histidine. Thiazolidine is part of the basic structure of the penicillins; F. structures are also present in thiamin, biotin and ascorbic acid. F. are used in industry as solvents (for example, tetrahydrofuran) and in the synthesis of pharmaceuticals and dyes.

Fixative: a colorless solution of resin, mastix, paraffin or shellack sprayed onto a drawing to prevent erasure or smearing. After the solvent has evaporated, the drawing is covered by a thin, transparent film.

Fixed bed: a term used for processes in which the catalyst is solid and does not move.

Fixed-bed process: see Cracking.

Fixed-bed reactor: a reactor in which the catalyst is firmly attached; it is used for heterogeneous catalyses of liquid-liquid, gas-liquid or gas-gas reactions. The reactor can filled with the catalyst bed, or it can be a pipe or shelf furnace in which the catalyst is packed into the pipes or on the shelves. In a shelf furnace, reaction heat can be either added or removed from the reactants between the shelves. F. are used in ammonia and methanol synthesis, sulfur dioxide oxidation and hydrocracking.

Flame coal: see Coal.

Flame melt process: 1) general process for Crystal growing (see); 2) special process for Silicate synthesis (see).

Flame point: see Ignition.

Flame spectrophotometry, ***FSP***: method of atomic emission spectrometry used for quantitative determination of those elements which can be excited by the thermal energy of the flame.

Detection of certain elements by the characteristic color they give flames is simple, rapid and sensitive. Although flame colors were formerly observed visually and used for qualitative determination of elements, modern analysis makes use of spectral instruments which disperse the light spectrally and thus permit recognition of characteristic emission lines. The intensities of these lines are related to the concentration of the element, so that quantitative analysis is possible. Instruments in which filters are used to isolate the lines of interest are called ***flame photometers***, while instruments in which monochromators (prism or diffraction grating monochromators) are called ***flame spectrophotometers***. For quantitative analysis, the intensities can be measured either by use of a photographic plate or a photoelectric device; today the latter are generally used. The intensity of the emitted radiation depends on the number of particles in the flame which are in the excited state. The sample to be examined must be in solution and is either sprayed into the flame or is mixed with the fuel before it is burned. The types of flame most commonly used for analysis are shown in the table. For quantitative analysis, the conditions must be kept constant, e.g. steady gas flow and constant flame temperature. The higher the temperature, the greater the number of excited atoms and the intensity of the emitted light.

Flames can be either laminar or turbulent. In *laminar flames*, the fuel and oxidant are mixed, and the sample solution is vaporized into the gas mixture, before they enter the combustion area. The flames burn quietly and are pale blue. The color of the flame, the *flame background*, is due to radicals, e.g. OH radicals) which form in the flame and emit a corresponding band system. Laminar burners are applicable for gases which burn slowly (e.g. acetylene-air mixtures). The low flow rate of the gases leads to a long residence time for the atoms in the flame. *Turbulent flames*, in which the fuel and oxidant are mixed immediately before they enter the combustion area, are used for mixtures which burn more rapidly, e.g. H_2/O_2 or acetylene/O_2. The sample solution is sprayed directly into the flame and causes large changes in the form and temperature of the flame. The flames in turbulence burners burn violently and very loudly.

Commonly used combustion mixtures

Fuel	Oxidant	Flame temp. in K	Flame type	Excitable elements
City gas	Air	2000	Laminar	Alkali metals
Acetylene	Air	2500	Laminar	Alkali and alkaline earth metals
H_2	O_2	3000	Turbulent	Alkali, alkaline earth and heavy metals
Acetylene	O_2	3350	Turbulent	Ag, Cu, Mn, etc.

In the flame, the ions present in the solution are converted to the free atoms, and their emission spectra are observed. The temperatures of the flames are relatively low compared to those of electric arcs and spark discharges, so that the higher energy states of the atoms are scarcely excited; for the most part only resonance lines, which are the strongest in the spectrum, are seen. They are often called the "last" lines, because when they are used to determine the concentration of the elements, they are the last which remain detectable as the concentration decreases.

F. is not an absolute method. For quantitative results, calibration curves must be used. Internal standards are often used; a constant amount of a similar ion is added to a series of solutions containing known amounts of the element to be determined. The calibration curve is made by plotting the emission intensities of the element being analysed, divided by the intensity of the reference emission, against the known concentration of the reference element. This method serves mainly to eliminate inconstancies of the flame and matrix effects, which can interfere with quantitative analysis.

Advantages of F. are 1) rapidity (if a calibration curve is available, the analysis can be made in seconds), 2) high sensitivity and 3) high selectivity, especially for elements with relatively few lines in their spectra. Some examples of limits of detection: Sr (acetylene/air), 0.01 mg/l; Li (acetylene/air), 0.0002 mg/l.

The disadvantages are that only a limited number of elements can be determined, and that solids must be dissolved. Aside from the alkali metals, Ca, Sr, Cu, Ga, In and Tl are suitable for flame spectrophotometric determination.

Flame spraying: a method of applying melted metals (see Metal spraying) or plastics (see Plastic spraying) by a spray gun, especially for corrosion protection. In F., the material to be applied is in the form of wire or powder in the spray gun. It is held in a flame in the gun, where it melts and is then converted to an aerosol by pressurized air. The flame is generated from gas and oxygen.

Flaring: the burning off of environment-polluting waste gases and vapors to dispose of them. The burners are usually 10 to 20 meters high, and consist of steel pipes in which combustion is essentially complete. Formation of soot is reduced by pumping steam into the gas stream.

Flash photolysis: see Kinetics of reactions.

Flash spectroscopy: spectroscopic methods for study of short-lived intermediate products of rapid reactions. The intermediate can be in either the ground state or an excited state, so long as its spectrum differs from those of the starting materials. An intense pulse of light is used to generate a high concentration of the short-lived intermediate in a cuvette; its absorption spectrum is then measured, using a second flash from the light source. The transmitted light is passed through a monochromator and is then registered by the detector (photographic plate or oscillograph). It is also possible to record the decrease in extinction at a single wavelength which corresponds to the disappearance of the intermediate. When this is done, a series of measurement pulses must follow the excitation pulse in an extremely short interval of time. The decrease in extinction with time is called the decay curve; it gives information on the lifetime of the intermediate of interest. It is essential for F. that the light pulse be shorter than this lifetime.

Light pulses are generated from 1) photoflash lamps with a pulse time of 10^{-5} s (to determine lifetimes in the μs range), 2) lasers with a pulse length of 10^{-8} to 10^{-12} s (to determine lifetimes in the ns range) or 3) special types of lasers, such as neodymium lasers, with a pulse length of 10^{-11} to 10^{-12} s (to determine lifetimes in the ps range).

F. is used to study the short-lived intermediates of photochemical reactions (see Photochemistry) such as radicals or triplet states.

Flat-bed chromatography: see Chromatography.

Flat-bed method: see Thin-layer chromatography.

Flavan, ***2-phenylchromane***: parent compound of the Flavanoids (see). F. forms colorless crystals, m.p. 44-45 °C, b.p. 332 °C. It is readily soluble in organic solvents. F. can be synthesized by heating 2-hydroxybenzyl alcohol with styrene to 230 °C, or by catalytic hydrogenation of 2-phenylchromene. It gives a pale violet color with iron(III) chloride in concentrated sulfuric acid, and when it is heated with sodium ethanolate in alcohol, 2-(3-phenylpropyl)phenol is formed.

Flavanoids: a large group of secondary plant metabolites derived from Flavan (see). The basic structure of the F. is ***flavanone*** (flavan-4-one), from which the derivatives of flavanol (flavan-3-ol), flavone and flavonol are derived. The Anthocyans (see) and catechols (see Catechol tannins) are also derivatives of flavan. The natural F. have hydroxy, or sometimes, methoxy groups at positions 5,7,3',4' and/ or 5'. F. form yellow salts with alkalies, via their phenolic hydroxyl groups. With strong acids, they form *oxonium salts (flavylium salts)*. F. are usually found as O-glycosides, but sometimes also as "C-glycosides". The most widely occuring F. are the Flavones (see). Phloretin (see) is a dihydrochalcone which is formed as an intermediate in the biosynthesis of F. Some F. are used therapeutically (see Rutin).

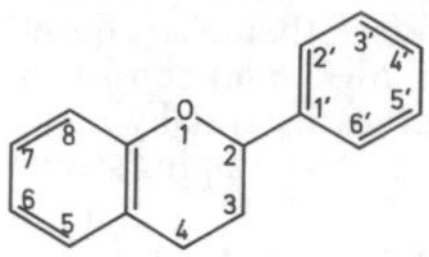

Flavan 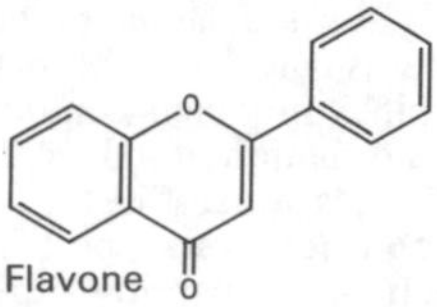 Flavone

Flavanone: see Flavones.

Flavazine: 1) ***Fast light yellow G, hydrazine yellow L***: a yellow, acidic pyrazolone dye which is readily soluble in water. It is made by diazotization of aniline and coupling with 1-*p*-sulfophenyl-3-methyl-5-pyrazolone, and it is used to dye animal fibers and in paints.

2) ***Flavazine S, hydrazine yellow S*** and ***SO***, a dark yellow, water-soluble acidic pyrazolone dye. It is produced by diazotization of aniline and coupling with 1-*p*-sulfophenyl-3-pyrazolone-3-carboxylic acid; it is used to dye animal fibers and to make paints.

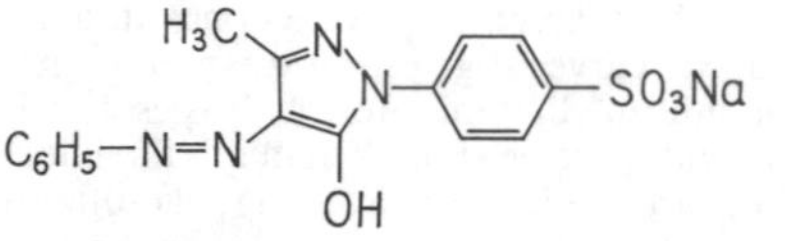

Fast light yellow G

3) ***Flavazine T***, same as Tartrazine (see).

Flavin: 7,8-dimethylisoalloxazine or one of its derivatives. F. can be made by condensation of the corresponding monosubstituted *o*-phenylenediamines with alloxan; for example, ***lumiflavin*** (7,8,10-trimethylisoalloxazine), a degradation product of vitamin B_2, can be synthesized from 4,5-dimethyl-2-methylaminoaniline and alloxan. The most important of the F. is ***riboflavin***, a member of the vitamin B_2 complex. Synthetic F. are of interest as biological antagonists of riboflavin, UV light absorbers in color photography and antioxidants.

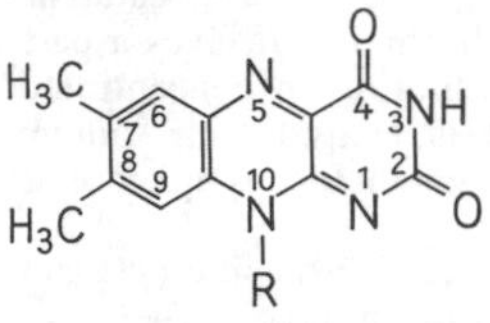

Flavin

Flavin-adenine dinucleotide: see Flavin nucleotides.

Flavin enzymes: a widely occurring group of oxidoreductases which contain flavin adenine dinucleotide (FAD) or, less frequently, flavin mononucleotide (FMN), as coenzyme. The F. catalyse hydrogen transfer reactions in which the hydrogen is not taken from NADH or NADPH (see Nicotinamide nucleotides), but directly from the substrate. Because the oxidized riboflavin part of the coenzyme is yellow, the F. were originally called ***yellow ferments***. A number of F. also contain metal ions (Fe, Mg, Cu, Mo) as cofactors. There are several types of F. 1) ***Oxidases*** react with oxygen as the hydrogen acceptor, and either form H_2O_2 as an intermediate electron carrier (e.g. Peroxidase (see), D(+)-glucose oxidase, amino acid oxidases, xanthine oxidase), or they act as four-electron transfer carriers and form water (e.g. Laccases (see), ascorbic acid oxidase). 2) ***Reductases*** usually react with cytochromes, e.g. Nitrate reductase (see). There are a few ***dehydrogenases***, of which Succinate dehydrogenase (see) is a well-known representative.

Flavin mononucleotide: see Flavin nucleotides.

Flavin nucleotides: a group of coenzymes with riboflavin as the biochemically active group. The riboflavin part reacts as a redox system. The oxidized form, ***flavoquinone*** is reduced to the flavin radical, ***flavin semiquinone***, and then to ***flavohydroquinone***. There are two coenzyme F., flavin mononucleotide and flavin adenine dinucleotide. The oxidized forms absorb in the visible between 445 and 450 nm.

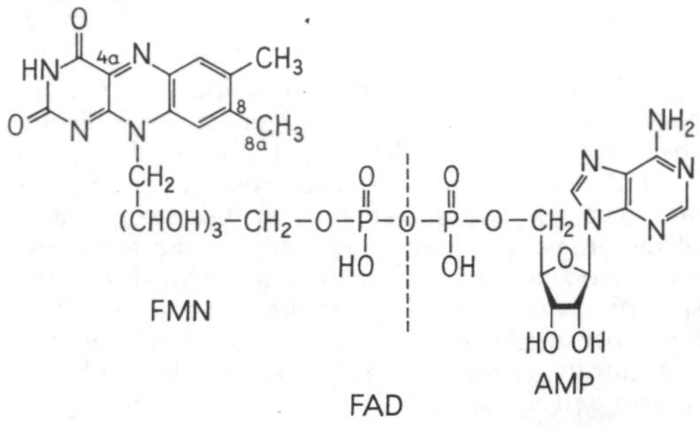

FMN FAD AMP

1) ***Flavin mononucleotide***, abb. ***FMN, riboflavin 5'-phosphate*** has M_r 456.4; the sodium salt has M_r 478.4, λ_{max} of the oxidized form is at 450 nm ($\varepsilon = 12.2 \cdot 10^3$ mol^{-1} cm^{-1}); E_0 (with enzyme protein) = -.06 V. 2) ***Flavin adenine dinucleotide***, abb. ***FAD***, M_r 785.6, as sodium salt, 807.6, λ_{max} of the oxidized form at 260, 366 ($\varepsilon = 46.2 \cdot 10^3$ mol^{-1} cm^{-1}) and 445 nm ($\varepsilon = 11.3 \cdot 10^3$ mol^{-1} cm^{-1}).

The F. are very sensitive to light. They are noncovalently attached to proteins (***flavoenzymes***), or sometimes they are covalently bound. There are several types of covalent bonds: from C-8a of the riboflavin to an imidazol N of histidine

$\underset{8a}{-CH_2}-N<$

or the thio group of a cysteine

$\underset{8a}{-CH_2}-S-$

or

$\underset{8a}{-CHOH}-S-$

The binding of the F. to the apoprotein causes a bathochromic shift in the absorption maximum of the flavin. Many oxidoreductases, including some of the components of the respiratory chain, are flavoenzymes. Most flavoenzymes contain FAD, but FMN is present in NADH reductase of the respiratory chain, and in some D-amino acid oxidases.

Flavoenzymes: see Flavin nucleotides.

Flavohydroquinone: see Flavin nucleotides.

Flavone: one of a group of yellow plant pigments of the flavanoid group; it has a flavone, isoflavone or flavonone skeleton. ***Flavone, 2-phenylchromane, 2-phenyl-4H-1-benzopyran-4-one*** (formula, see Flavanoids) forms colorless crystals; m.p. 99-100 °C. Flavone is soluble in most organic solvents, but insoluble in water. It is synthesized by heating phenol and benzoyl acetate in the presence of phosphorus pentoxide. Flavone forms white deposits in the aboveground parts of various primrose species, and is thought to be responsible for the allergenic effect of these plants. ***Isoflavone, 3-phenylchromone, 3-phenyl-4H-1-benzopyran-4-one***, forms colorless needles; m.p. 138 °C. It is soluble in most organic solvents, but insoluble in water. Isoflavone can be synthesized by condensation of 2-hydroxydeoxybenzoin with ethyl formate in the presence of sodium, followed by hydrolysis with concentrated sulfuric acid, or by heating 2-hydroxydeoxybenzoin with formamide. ***Flavanone, 2,3-dihydroflavone, 2-phenylchroman-4-one***, forms colorless needles; m.p. 75-76 °C. It exists in various stereoisomeric forms. Flavanone can be synthesized by treating 2'-hydroxychalcone with aqueous alcoholic sodium hydroxide solution. It can be converted to flavone by bromination followed by cleavage of the hydrogen bromide using concentrated potassium hydroxide solution. A subgroup of the F. is the ***flavanols***, which are hydroxylated in the 3 position.

F. occur widely in nature, for example, in blossoms, woods and roots, usually as glycosides or esters of tannic acid. They accompany the structurally closely related red, violet and blue anthocyans; this is why the same plant can have yellow and red blossoms, or red, blue and yellow colors in the same blossom. More than 20 F. have been isolated to date, including ***apigenin***, 5,7,4'-trihydroxyflavone (from dahlia, dandelion and camomile blossoms), ***chrysin***, 5,7-dihydroxyflavone (from poplar buds and conifers), ***fisetin***, 3,7,3',4'-tetrahydroxyflavone (from yellow cedar), ***genistein***, 7,4'-dihydroxyisoflavone (from gorse), ***hesperitin***, 5,7,3'-trihydroxy-4'-methoxyflavone (from lemon peel), ***kaempherol***, 3,4',5,7-tetrahydroxyflavone (from black alder berries, delphinium and grapefruit), ***luteolin***, 5,7,3',4'-tetrahydroxyflavone (in safflower, *Reseda luteola*, yellow digitalis and dahlias), ***morin***, 3,5,7,2',4'-pentahydroxyflavone (in the yellow wood extract of the black alder), ***myricetin***, 3,5,7,3',4',5'-hexahydroxyflavone (in black currents and potato blossoms) and ***quercetin***, 3,5,7,3',4'-pentahydroxyflavone (found in more than half of all angiosperms).

Flavone pigment: see Flavone.

Flavonol: see Flavanoid.

Flavosemiquinone: see Flavin nucleotides.

Flavoquinone: see Flavin nucleotides.

Flavorings: compounds or mixtures which have a particular odor or taste and are added to foods. Except for the four basic tastes, sweet, sour, salty and bitter, what is perceived as taste in food is the result of olfactory sensation. Thus in general, F. are substances which lend a particular aroma to food. They may be natural, nature-identical or synthetic. ***Natural F.*** are found in plant or animal materials, and are extracted by physical or chemical methods; examples are vanilla extract or peppermint oil. This type of product is an *essence*. ***Nature-identical F.*** are compounds identical to those found in natural F. but synthesized chemically or isolated chemically from raw materials. Vanillin is an example. ***Artificial F.*** are synthetic compounds not found in nature. Some examples are allylhexanoate, α-amylcinnamaldehyde, benzyl butyrate and propionate, butylbutyryl lactate, butyl cinnamate, glyceryl tributyrate, hydroxycitronellal, isoamylsalicylate, melonal, 6-methylcoumarin, methylheptyne carbonate, propenylguaiathol (2-ethoxy-5-(prop-1-enyl)phenol) and resorcine dimethyl ester.

Flavylium salt: see Flavanoid.

Flint glass; incorrectly called ***quartz glass***: a glass consisting almost completely of silicon dioxide, SiO_2. It has very good qualities, such as a low coefficient of thermal expansion, high transformation and softening temperatures, low electrical conductivity and high transparency to UV.

1) ***Transparent F.*** is produced by a variety of different methods, all of which require large amounts of energy and very pure raw materials. The methods which start with ground quartz begin with melting the granulate at about 2000 °C in a flintware tube in an acetylene/oxygen or oxyhydrogen flame, alternating with rolling into a raw material powder. This is followed by a drawing process. Another method is to spray powdered quartz onto a rotating flint glass core in a flame. In the *Hänlein process*, a F. is premelted under vacuum, then melted without bubbles in a molybdenum or tungsten crucible by resistance or induction heating; an outlet valve in the floor of the crucible makes possible the production of F. tubes. To make ***vycor glasses***, objects are made of a sodium borosilicate glass with a composition, for example, of 65% SiO_2, 25% B_2O_3 and 10% Na_2O; by tempering the objects, Subliquidus separation (see) occurs and forms a penetration structure. The sodium borate phase is then leached out with 3 N sulfuric acid, leaving a porous SiO_2 glass (pore diameter 2 to 5 nm) with less than 1% Na_2O and 2 to 3% B_2O_3. This is washed, carefully dried and sintered free of bubbles at 1100 °C; a 30% volume contraction occurs. The advantage of this method over the quartz method is that it requires much less energy. A very homogeneous F. is made by oxyhydrogen flame hydrolysis (gas-phase process) of silicon tetrachloride, $SiCl_4$.

Transparent F. is used in making vacuum systems and apparatus for use with UV light (lenses, prisms, cuvets, light sources, optical fibers), in electrotechnology for high-voltage insulators and for industrial and laboratory apparatus.

2) ***Flintware*** is a translucent F. which is clouded by the presence of many tiny air bubbles. It is made by layering quartz sand around a carbon heat conductor; passage of current through the apparatus melts the sand. Chemical reactions occurring in the regions of

contact with the carbon produce CO, which separates the glass from the heat conductor. Flintware is blown into iron molds to shape it. Blocks and plates are made by rolling. If several heat conductors are set up in parallel, flintware blocks can be made. The surface of flintware is polished by partial melting with an electric arc. Flintware is used mainly to make laboratory equipment, tubes and basins.

Flint glass fibers, often incorrectly called ***quartz glass fibers***: fibers of nearly pure silicon dioxide, SiO_2, which are similar to fiberglass. A distinction is made between ***flint glass silk*** and ***flint glass fibers***, on the basis of fiber diameter. They are made by drawing rods, valve drawing or valve blowing of a silica melt at 1800 °C. The starting materials are quartz and quartz sand. F. can also be made from alkali metal borosilicate glass fibers; they are tempered at a temperature which permits phase separation, then leached out with sulfuric acid. The remaining silica fraction is finally sintered. F. are very strong, and the specific strength increases with decreasing fiber diameter; they are thermally resistant up to 1200 °C and are resistant to most chemicals. F. are used as an electrical insulation material and, in the chemical industry, in the form of batting, felt and textiles. Temperature-resistant construction materials can be made by combining F. with organic polymers. Threads of F. are used in precision physical instruments.

Flintware: see Flint glass.

Flocculation: 1) The approach of colloidal particles to one another in the presence of a flocculating agent which provides countercharges for the colloidal particles. There are two main stages of F.: destabilization of the colloid and transport of the particles to one another. F. in this sense is equivalent to Coagulation (see).

2) In water treatment, the removal of undesired substances by formation of compounds which are insoluble in water. F. in aqueous suspensions is usually achieved by addition of suitable inorganic electrolytes with polyvalent cations (e.g. calcium hydroxide, iron and aluminum salts) or macromolecular organic flocculants (e.g. polyacrylamides, polyacrylic acids). In the first case, the potential barriers which prevent approach of the colloidal particles to each other are decreased by adsorption of the cations to the surfaces. In the second case, the particles are cross-linked by the thread-like molecules; binding groups on the flocculant molecules interact with the surfaces of the particles. Synthetic flocculants have become increasingly important in recent years.

3) In metallurgy, an undesired separation of materials in alloys, usually steels (e.g. chromium-nickel, chromium and manganese steels), by precipitation of hydrogen. The separation occurs when the solubility of hydrogen in the steel and its capacity to diffuse are reduced. It can be prevented by preheating of the metal before hammering, reducing the amount of hydrogen in the steel melt, or by vacuum melting or casting.

Floor waxes: products which leave a protective wax film on the treated floor. They contain a film-forming wax mixture and a solvent. The wax film must a) have a certain hardness, because this determines the stability of the shine; b) have a high melting point (about 70 °C), to prevent melting even in strong sunlight (melted wax is slippery); and 3) be easy to polish.

The F. should protect the floor from wear, penetration of moisture (swelling) and should serve to clean it and make it shiny. There are oil-base and emulsion products.

1) the ***oil-base products*** are either liquids or solids.

a) Solid oil-base products consist of about ⅓ wax mixture and ⅔ organic solvent, together with small amounts of perfume and soluble coloring. The wax mixture is about ⅓ hard wax and ⅔ hard paraffin; a certain amount of iso- or cycloparaffin should be present, because these compounds improve the retention of the solvent in the paste. Plant waxes, crude montane wax derivatives, macroparaffins, ozocerite and precipitation waxes are used as hard waxes. The hard paraffins are distillates of pyrolysis tars from soft coal or petroleum. The quality of the wax film can be improved by polyethylene waxes, or colloidal silicic acids. Terpentine oil and gasoline fractions are used as organic solvents.

b) Liquid oil-base products differ from solid ones only in the proportions of the wax and solvent; here the proportion of wax is about ⅙, while the solvent makes up ⅚ of the product. Because the solid content is lower, the resulting protective film is somewhat thinner, and wears off more rapidly.

2) ***Emulsion products*** can also be solids or liquids.

a) In solid emulsion products, some of the organic solvent is replaced by water. The wax mixtures and the solvents are the same as found in solid oil-base products. Partially saponified waxes, monoacylglycerols, etc. are used as emulsifiers. These F. are used mainly on very fragile floor coverings. The reduction in the amount of solvent reduces the cleaning power of the product, but also prevents the solvent from dissolving the floor surface.

b) Liquid emulsion products are made without organic solvents. They contain 12 to 15 parts of a wax mixture, 82 to 85 parts water and 3 to 4 parts emulsifier. The wax mixture is the same as in the oil-base products. The emulsifiers can be, for example, resin soaps, styrene emulsions, etc. The advantage of these liquid emulsions over the oil-base products is that they are easier to apply and leave a self-shining film of wax; they do not have to be polished. The disadvantage is that they do not clean the floor, because they do not dissolve the old wax film.

Self-polishing emulsions (see) are also F.

Flotation: separation process used to enrich ores and other minerals. The solid mixture is first ground to a fine powder, so that the individual components are in separate particles. The mixture is then dispersed in an aqueous suspension and a turbulent gas stream is blown through it. The hydrophobic particles are enriched on the phase boundary between the gas bubbles and water; in other words, the difference in wettability is the basis for the separation. The particles adhering to the gas bubbles rise to the surface and are skimmed off. The process is thus a ***density separation*** in which the different degrees of adsorption to the gas bubbles is the decisive step. Differential adsorption is achieved by specific adsorption of surface-active substances to the components to be separated; these are called *collectors*. In ***foam flotation***, air is blown into the aqueous suspension (Fig.), and a stirrer is used to increase the foaming and to prevent sedimentation of the dispersed particles. The mineral particles which have been made hydrophobic are enriched on the rising air bubbles and are deposited in a foam layer, which is made by addition of special tensides (*foaming agents*). The foam is mechanically separated from the aqueous suspension. The slag particles are not affected by the collectors, and sediment on the bottom of the F. cell.

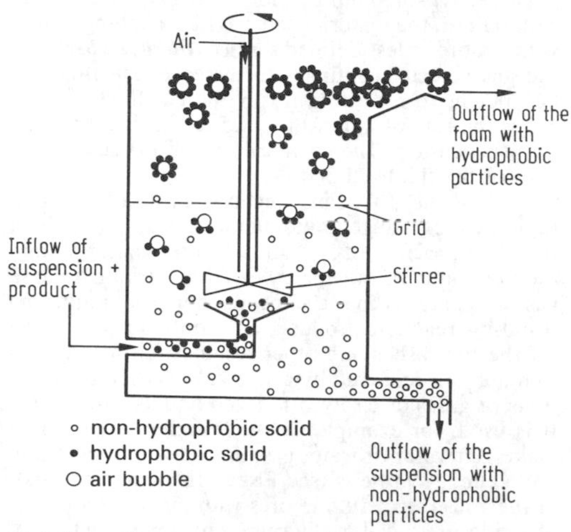

Flotation cell

Wetting agents are added to optimize the hydrophilicity of the slag. In ***oil F.***, the suspended particles are adsorbed by oil droplets instead of gas bubbles. ***Selective F.*** is the separation of several mineral components in successive stages using surface-active compounds which adsorb specifically. In ***ionic F.***, the binding of the desired ions to the surface-active substances is used to bring them into a foam.

F. agents. The surface-active substances used as *collectors* may contain anionic or cationic polar groups; they are responsible for the selective adsorption. The nonpolar part of the molecule is essential for the hydrophobicity of the compound. Some commonly used anionic collectors are alkyl xanthogenates such as methyl, ethyl and isopropylcyclohexyl xanthogenates, oleic acid, sodium palmitate, alkane sulfonates and phosphonates. Alkyl ammonium chlorides and pyridine compounds with similar structures are used as cationic collectors. The collector effect often increases as the length of the alkyl chain increases. Commonly used *foaming agents* include xylenols, cresols, pine oil and flotol (technical fennel alcohol). The foaming agent and collector can be identical. The wetting agents include *suppressers*, which make it possible for only one component to float, and *reactivators*, which make it possible for the "suppressed" component to be flotated.

Recently, the use of F. in sewage treatment has increased, especially to remove particulates which are difficult to sediment.

Flower pigments: see Blossom pigments.

Flow limit: the limiting shear below which no flow occurs. Above the F., the system may show Newtonian, elastic or plastic behavior.

Flow method: see Kinetics of reactions (experimental methods).

Flox®: see Synthetic fibers.

Fluates: see Fluorosilicates.

Fluazifop: see Growth hormone herbicides.

Fluazifprop: see Growth hormone herbicides.

Fluctuation: a random deviation of a parameter from an average value. In homogeneous material systems, F. are understood as local deviations, on the submicroscopic level, from the macroscopic average. F. can occur, for example, in the particle density, the index of refraction or the energy. Brownian motion is also a result of F. in the density of molecules in the space around a small particle, e.g. of dust or a macromolecule, which leads to an unequal number of molecular impacts on two sides of the particle. As a result, the particle moves about in a random fashion which is described by the Einstein-Stokes equation (see Diffusion). There is a theory of F. by means of which the size and probability of F. can be calculated.

Fludrocortisone: see Adrenal cortex hormones.

Flue gases: the gaseous products of combustion of solid fuels (see alse Waste gases). F. contain mainly carbon dioxide, CO_2 (as the actual product of combustion), nitrogen (from the combustion air), steam, H_2O (from moisture and oxidation of oxygen in the fuel) and unconsumed air. When the combustion occurs in insufficient air, the F. also contain methane, CH_4, hydrogen, H_2 and carbon monoxide, CO. The reduction of SO_2 emissions in F. is discussed under Flue gas desulfurization (see).

F. are gray or dark colored, due to solids (soot, fly ash) or tar vapors. The heat content of F. depends on the heating value and water content of the fuel, and on the degree of air excess; for hard coal lumps, there is about 40% more air, and for coal dust, about 15% more than is required stoichiometrically.

The conditions of the fire can be checked by physical (see Flue gas testers) and chemical methods (e.g. with the Orsat apparatus (see)) which automatically analyse the F. for its content of CO_2, CO, H_2 or O_2 (*Flue gas analysis*). For a given elemental analysis of the fuel, the maximum CO_2 content can be calculated; however, the practically attainable CO_2 values are lower than CO_{2max}. Since the same CO_2 values can occur in the range either of air deficiency or air excess, the CO_2 measurement must be supplemented by a second measurement of CO, H_2 or O_2. The practice of measuring O_2 alone is becoming increasingly widespread.

The F. are used for heating, in that they transfer their heat to the firebrick walls of furnaces or steam generators, and in the latter case, the heat is transferred to the steam. F. are also used to superheat the steam and to preheat the feed water. Sometimes they are added to the combustion air. They must not be cooled below a temperature of 180 °C, because this is required to make them rise in the smokestack.

Because of their suffocating and toxic components, F. pollute the air, inhibit the growth of plants, reduce the amount of sunlight reaching the ground, and cause corrosion.

Flue gas desulfurization: methods for reducing SO_2 emissions in Flue gases (see). F. is particularly important for removal of SO_2 from flue gases produced by coal-fired power plants. Sulfur is best removed from heating oils by hydrorefining the oil. Although a number of methods for F. have been proposed, many of them are too expensive for use. The practical methods available are absorption, adsorption, addition of additives to the fuel, oxidation and reduction.

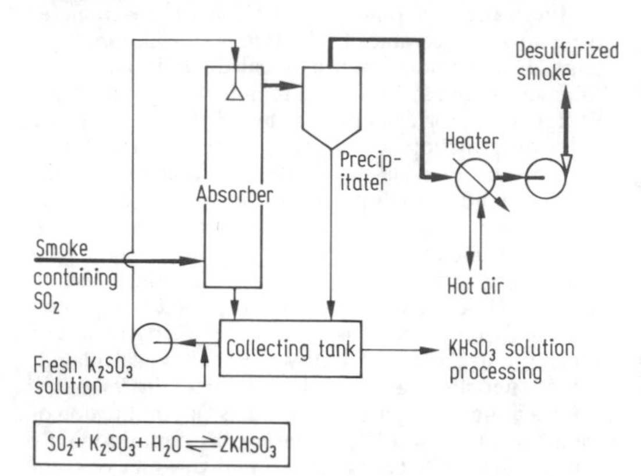

Flue gas desulfurization by scrubbing with potassium sulfite solution

Because of their high efficiency, absorption methods are most often used, for example, scrubbing with aqueous potassium sulfite solution (Fig.). Adsorption methods have the advantage that ash is simultaneously precipitated. The disadvantages are that the temperature of the smoke is reduced, and that the adsorber has to be regenerated. Some adsorption processes use activated charcoal. The limestone additive method, which can also be used with brown coal, consists of adding ground limestone to the coal. The desulfurizing reaction occurs between 900 and 1200 °C, forming $CaSO_4$.

Flue gas tester: a device used mainly to determine the contents of carbon monoxide and hydrogen in flue gases. F. can operate, for example, on the principle of relative heat conductivity (see Gas chromatography).

Fluid: a general term for all states intermediate between the ideal gas and the solid. It includes dense gases, liquids, melts and supercritical phases. Finely divided solid particles suspended in gases or liquids (e.g. fluidized beds, magnetic fluids) can be considered F. systems.

Fluid chromatography: chromatographic methods in which a highly condensed gas in its critical temperature range is used as the mobile phase.

Fluid-coking process: see Cracking.

Fluidification: a process for converting salts which tend to clump into easily flowing, non-clumping powders. F. is used, for example, with fertilizers (nitrates), cement, sodium chloride and ammonium chloride. Small amounts of ***fluidifiers*** are added to the slightly hygroscopic salt; the fluidifier is a scarcely soluble, water-absorbing substance such as kieselguhr, silica gel, zinc oxide (ZnO), magnesium oxide (MgO), aluminum stearate or another stearate. This makes the powder flow readily and thus easy to transport and to process.

Fluidization: a method of moving materials and heat for reactions of large amounts of solid. It was developed in 1921 by Winkler and in its original form was used for coal gasification (*Winkler process*).

If a stream of gas, such as air, oxygen or nitrogen flows up through a layer of fine-grained material, such as coal, coke or gypsum, the solid particles remain at rest if the gas flow rate is low; this is called a *fixed bed*. At medium gas flow rates above the fluidization limit, the material behaves like a boiling liquid, with a more or less defined surface (***fluidized bed***). As the gas velocity continues to increase, the fluidized bed begins to roil, until finally the gas flow rate exceeds the rate of fall of the solid, and it is carried up as a suspension. The chemical reaction occurs in the core of the fluidized bed.

F. has the following technological advantages: rapid heat exchange in the fluidized bed, a large heat transfer capacity, a large surface of the material, and the possibility of moving the solid material from one vessel to another in the same manner as a liquid. The resulting reaction products are continuously drawn off the top; ashes and other residues fall to the bottom and are removed by a screw. F. is used for reactions of gases or easily volatilized liquids with solids. It is used, for example, in the Winkler generator to make generator or water gas, in catalytic cracking, in the Kellogg synthesis (see Fisher-Tropsch synthesis), in the roasting of sulfide ores with air, for reduction of oxide ores, in lime burning and for reduction of gypsum to calcium sulfide.

Fluidized bed: a method in chemical technology in which a gas or liquid carrier stream holds a finely divided solid in motion. Depending on the size and shape of the solid particles, and on the technical processes, this motion can vary considerably. The fluidized state is characterized by an equilibrium between effects of gravity and friction between the flowing fluid and the solid aprticles. The F. of a finely divided solid behaves as a fluid, so that material can be continuously removed from it and added to it. The F. makes possible an excellent exchange of heat and oxygen; disadvantages can be the erosion of the solid particles and the reactor. F. are used in Cracking (see) and the gasification of solid fuels (see Generator).

Fluidized bed reactor: a device for carrying out a chemical reaction or physical operation (such as drying, heating or cooling) between a fluid phase (usually a gas) and a fluidized, fine-grained solid (see Fluidization, see Cracking). F. are used wherever high rates of material and heat transfer are needed.

Flumethasone pivalate: see Adrenal cortex hormones.

Fluocinolone acetonide: see Adrenal cortex hormones.

Fluohm process: see Petroleum.

Fluorene: a condensed aromatic hydrocarbon with a reactive CH_2 group in position 9. F. crystallizes in colorless leaflets; m.p. 116 °C, b.p. 294 °C. It is insoluble in water, but is soluble in ethanol, ether, benzene, chloroform and carbon disulfide. F. is found in coal tar, and is extracted from it. It can be synthesized by thermal dehydrogenation of diphenylmethane, by catalytic dehydrogenation of 2-methylbiphenyl or by the Pschorr reaction (see). At position 9, F. can be easily oxidized (***fluorenone***) or condensed with aldehydes (***benzalfluorene***). The products are systems with crossed conjugation of double bonds and, like fulvene, they are dark yellow. In electrophilic substitutions, reactions occur first at position 2 and then at 7. These positions are the sites of halogenation, nitration and sulfonation. F. and its derivatives are used to make pigments, drugs and pesticides.

Fluorescein, ***resorcinphthalein***: a Xanthene pigment (see). F. occurs in a labile yellow modification, which converts at 250 to 260 °C to a stable red modification. F. is synthesized by heating resorcinol with phthalic anhydride in the presence of a sulfuric acid catalyst.

Sodium fluorescein

The F. available commercially is the disodium salt of F., ***sodium fluorescein (uranin)***. It has an intense yellow-green fluorescence, which can be seen distinctly even at a dilution of 10^{-11} mol l^{-1}. F. is important as a starting material for the eosine dyes, as an indicator, and as a stain to color the water in emergencies at sea, to discover underground connections between aquifers, and for studies of metabolism in human beings, animals and plants. For example, a few ml of a 5% aqueous solution of F. injected into the human body is rapidly spread through all tissues and organs except those where the blood flow is impaired. These areas can be detected by fluorescence measurements with long-wave UV. F. was first made in 1871 by A. von Baeyer.

Fluorescence: a form of luminescence resulting from absorption of photons (photoluminescence). Formally, it is the reverse of light absorption, in that electrons in excited electronic states return to the ground state by re-emission of the energy of activation as radiation. F. can occur either in the ***optical*** or ***X-ray*** range (ultraviolet emission is considered optical F.). The light is emitted within 10^{-9} to 10^{-6} s after excitation, and has either the same energy as the excitation radiation (***resonance F.***) or lower energy (Stokes' rule). F. is distinguished from Phosphorescence (see), which occurs after a longer time (> 10^{-4} s). F. can occur in solid, liquid or gaseous, organic or inorganic substances. The number of fluorescing inorganic compounds is relatively limited (fluor spar, compounds of uranium and rare-earth metals, gaseous samples of alkali metals, among others); F. is much more common among organic compounds.

The following parameters can be used to characterize fluorescent emissions: 1) the *fluorescence spectrum* can be used to identify a sample and to elucidate its structure (*fluorescence spectroscopy*). 2) The intensity of the emission can be used to determine the concentration of the fluorescent substance (*fluorimetry* or *fluorescence photometry*). Because of concentration-dependent deactivation processes, calibration curves must be used for quantitative determinations. 3) The decay time is the time required, after excitation has been terminated, for the intensity to decline to 1/e. The decay of F. follows an exponential curve. 4) The degree of polarization of F. can be observed by irradiation of an oriented sample with polarized light.

The most important applications of F. are atomic F. (see Atomic fluorescence spectroscopy), X-ray fluorescence (see) and ***molecular F.*** The relationship between the absorption and F. spectra of molecules can be seen in Fig. 1.

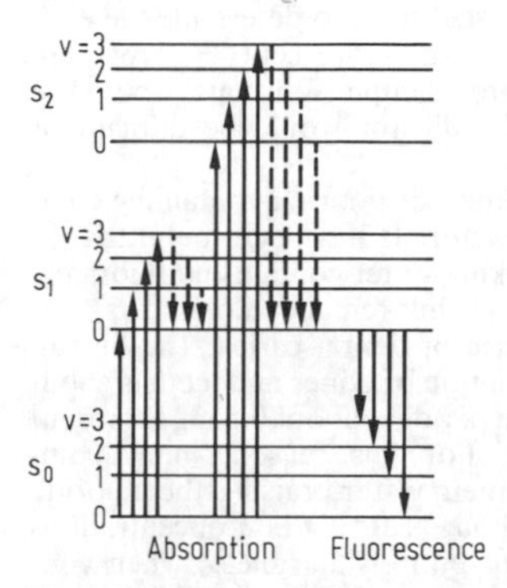

Fig. 1. Energy level diagram showing absorption and fluorescence transitions

The absorption of light by an electron in the lowest vibrational state of the electronic ground state (S_0 excites it into one of the vibrational levels of the first excited electronic state (S_1) or into a higher excited state (S_2, S_3, etc.). Non-radiative transitions occur among the vibrational states of a given electronic excitation level, so that the electron eventually occupies the lowest vibrational state of S_1. Light emission by F. occurs between this level and one of the vibrational states of the electronic ground state. This is the reason that the emitted F. light has a longer wavelength than the absorbed light. Otherwise, absorption and F. lines are nearly mirror images of one another (Fig. 2).

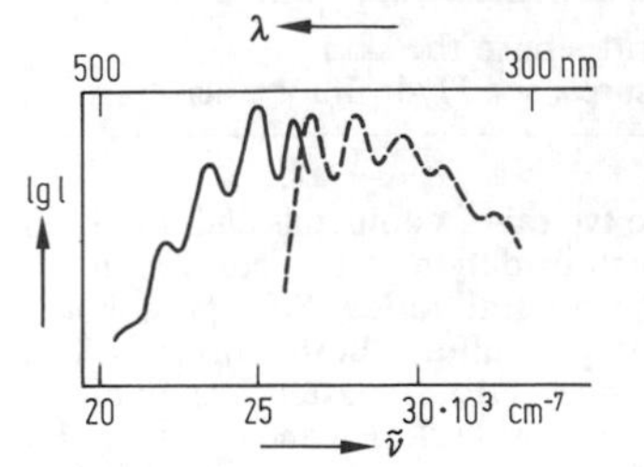

Fig. 2. Absorption spectrum (solid line) and fluorescence spectrum (dashed line) of anthracene. λ = wavelength, $\tilde{\nu}$ = wavenumber, I = intensity

Since most substances do not fluoresce, it must be concluded that there are other ways for electrons in excited states to return to the ground state, and that these may compete with F. Some important examples are loss of excitation energy by collision and induction of photochemical reactions, especially predissociation (cleavage of weak bonds by the activation energy). The conversion of electronic excitation energy into vibrational energy is also important.

Knowledge of F. spectra is very important in the theory of excited states (see Photochemistry). F. spectra permit identification of a substance and its quantitative determination. A number of applications of F. utilize its extremely low detection limits; for example, aqueous solutions of fluorescein, a particularly strongly fluorescing substance, are used to detect leaks or as markers to aid rescues on the high seas.

Fluorescence titration: volumetric methods in which the Indicators (see) used to determine the endpoint of the titration are fluorescent. F. are often used in strongly colored sample solutions, where the change in a colored indicator would be difficult or impossible to see.

Fluoridation: addition of fluorine-containing compounds to drinking water. It has been found that in areas where the drinking water contains no fluorides, the population (mainly children and adolescents) suffers a higher incidence of dental caries. The human requirement for formation of bones and teeth is about 1.5 to 1.8 mg fluoride per day; about 0.5 mg is present in the normal diet. For this reason, many communities fluoridate their water, raising the fluoride concentration to about 1.0 mg/l. Concentrations above 2 mg/l are not entirely harmless; when consumed regularly, such water produces bone damage and dental fluorosis (dark coloration of the dental enamel). The following compounds are used for F., either in dissolved or solid form: sodium hexafluorosilicate, Na_2SiF_6, sodium fluoride, NaF, or hydrogen fluoride, HF.

Fluorides: ionic, covalent or coordination polymeric compounds containing fluorine as the electronegative bond partner. Electropositive alkali and alkaline earth metals form ionic F. of the types MF and MF_2. There are also some hydrogen fluorides, e.g. potassium hydrogen fluoride, KHF_2; the linear hydrogen difluoride anion $[FHF]^-$ can be converted to angular anions, $H_2F_3^-$ ($[FHFHF]^-$) or $H_3F_4^-$, which are present in the acidic fluorides $KF \cdot 2HF$ and $KF \cdot 3HF$.

> Hydrogen fluorides burn the skin.
> Countermeasures: see Hydrofluoric acid.

Covalent F. are typical of nonmetals and transition metals in their high oxidation states, e.g. boron trifluoride, BF_3, silicon tetrafluoride, SiF_4, phosphorus pentafluoride, PF_5, sulfur hexafluoride, SF_6, uranium(VI) fluoride, UF_6, tungsten(VI) fluoride, WF_6, rhenium(VII) fluoride, ReF_7, and osmium(VII) fluoride, OsF_7. The transition metals in intermediate oxidation states form coordination polymer F., e.g. iron(II) fluoride, FeF_2 or cobalt(II) fluoride, CoF_2. The F. of most metals are soluble in water. However, CuF_2, PbF_2, CaF_2, SrF_2, BaF_2, RaF_2, and lanthanoid and actinoid fluorides of the types LnF_3, AnF_3 and ThF_4 are either insoluble or only slightly soluble in water. Many metal and non-metal F. form ***fluoro complexes*** with alkali fluorides, e.g. M_3AlF_6, MBF_4, M_2SiF_6; M_3FeF_6 and M_2ReF_6.

Fluorination: see Halogenation.

Fluorine, symbol ***F***: an element of group VIIa of the periodic system, the Halogens (see); a nonmetal, with one isotope; *Z* 9, atomic mass 18.99840, valence I, density (at b.p.) 1.108, m.p. - 219.61 °C, b.p. - 187.52 °C, standard electrode potential (F^-/F_2) + 2.87 V.

Properties. F. is a slightly green gas with a penetrating odor. It is highly caustic. Liquid F. is pale yellow.

> Fluorine is extremely poisonous and, among other things, burns the skin.
> Countermeasures: see Hydrofluoric acid.

With regard to its reactivity, F. is unique among the elements; it is the most electronegative element, has a high electron affinity and is unusually reactive. It reacts directly and very easily with nearly all elements, except for nitrogen, oxygen and the lighter noble gas elements. It combines with hydrogen even at room temperature and in the dark, exploding or self-igniting. It reacts vigorously with phosphorus or sulfur even at - 180 °C, and it reacts with finely divided carbon at room temperature, developing a flame. Alkali and alkaline earth metals ignite in a stream of F., forming the fluorides MF or MF_2. Other metals, such as copper, magnesium or aluminum, and alloys, such as steel or monel metal, are attacked by F. only on the surface; the formation of metal fluoride protective layers prevents further reaction. Nickel does not react with F. below 600 °C, and platinum metals and gold must be heated to red heat to react with F. Compact carbon reacts with F. at 250 to 300 °C. F. is capable of converting elements into their fluorides in the maximum oxidation states, for example, PF_5, SF_6, IF_7, VF_5, UF_6 or ReF_7. F. reacts with chemical compounds at elevated temperatures. It always combines with the more electropositive component of the compound. Organic compounds react with F. extremely vigorously. F. reacts with water to form hydrogen fluoride, oxygen, hydrogen peroxide and ozone; in the presence of alkalies, oxygen difluoride is formed. Quartz is not attacked by carefully dried F. at room temperature, but above 150 °C, amorphous silicon dioxide and quartz glass react to form silicon tetrafluoride and oxygen.

Analytical. Fluoride can be determined by electroanalysis using electrodes sensitive to fluoride ions. It can also be determined gravimetrically through the formation of calcium fluoride, CaF_2. In the presence of other anions which can precipitate with Ca^{2+}, fluoride can be precipitated as lead chloride fluoride and determined titimetrically after redissolving this salt in nitric acid.

F. makes up $6.6 \cdot 10^{-2}$% of the earth's crust, and is thus one of the more common elements. It is in 14th place in the scale of relative abundances. The most important F. mineral is fluorite (fluorspar), CaF_2; it is still the most important source of industrial F. and its compounds. Recently, natural phosphates containing fluoroapatite, $Ca_5(PO_4)_3F$ (3 to 4% F. content) have received attention, because this mineral represents the largest world reserves of F. F. is also found in cryolite, $Na_3[AlF_6]$, topaz, $Al_2[SiO_4](F,OH)_2$ and lepidolith, $(K,Li)Al_2[AlSi_3O_{10}](F,OH)_2$. Fluorite deposits which have been subject to radioactivity can also contain small amounts of elemental F. as crystal inclusions. F. is one of the trace elements, lack of which leads to deficiency symptoms in plants or animals. Soils contain about 110 ppm, and plants about 0.5 to 2 ppm F. (dry weight). For animals and human beings, the F. content of bones ($\approx$ 0.9 to 2.7 g kg^{-1}) and tooth enamel ($\approx$0.2 to 1.2 g kg^{-1}) is of particular importance. The fluoride requirement of human beings is provided mainly by the drinking water. Because a deficiency of F^- promotes caries, fluoride (0.6 to 1.2 ppm) is added to the water supplies in many areas where the natural content is low.

> Extreme caution is appropriate in all work with fluorine. The choice of materials for reaction vessels and armatures must be made very carefully. Care must be taken that all parts of the apparatus which come into contact with fluorine are clean and dry; in particular, traces of oils, fats and other organic substances must be scrupulously avoided. Small amounts of impurity can lead to explosive oxidation and fluorination reactions which, at sufficiently high temperatures, can lead to reaction of the whole apparatus (including steel and ceramics), sometimes as a fire.

Production. F. can be obtained only by electrolysis, due to its extreme reactivity. Anhydrous KF/HF melts (KF · HF, m.p. 217 °C; KF · 2HF, m.p. 72 °C; KF · 3HF, m.p. 66 °C) are used with copper, magnesium, monel metal, nickel, steel or carbon vessels and electrodes. The following variants are used: *low-temperature process*: electrolyte, KF · 2.9-6.7HF, working temperature 15 to 50 °C, nickel anodes; *medium-temperature process*: KF · 1.8-2.5HF, 70 to 130 °C, carbon anodes; *high-temperature process*: KHF_2, 245 to 310 °C, graphite anodes. Today the most frequently used method is the medium temperature process; the anodes are made of graphite-free carbon. The F. produced in this way is sold commercially in steel bottles.

Applications. The main industrial use of F. today is in the production of uranium hexafluoride, which is used to separate the uranium isotopes ^{235}U and ^{238}U. F. is used in smaller amounts to produce sulfur(VI) fluoride as a dielectric. F is also of some interest as a possible rocket fuel. In the laboratory, it is used as the starting material for synthesis of many new fluoride derivatives.

Historical. F. was first isolated in 1886 by H. Moissan. The chemistry of F. was later greatly promoted by the work of O. Ruff and H.J. Emeléus. The increase in industrial F. chemistry in the recent past is associated with the broad application of fluorocarbons, fluoropolymers such as polytetrafluoroethene, and the technological significance of uranium hexafluoride. The name "fluorine" is derived from the mineral fluorite, which was used very early as a flux in metallurgy.

Fluorite type, ***fluorite structure*** : a common Structure type (see) for compounds with the general composition AB_2. The mineral fluorite (calcium fluoride, CaF_2), forms a crystal lattice in which the Ca^{2+} ions form a cubic face-centered lattice, and the F^- ions occupy the centers of the eight cubes into which the unit cell can be mentally divided (Fig.).

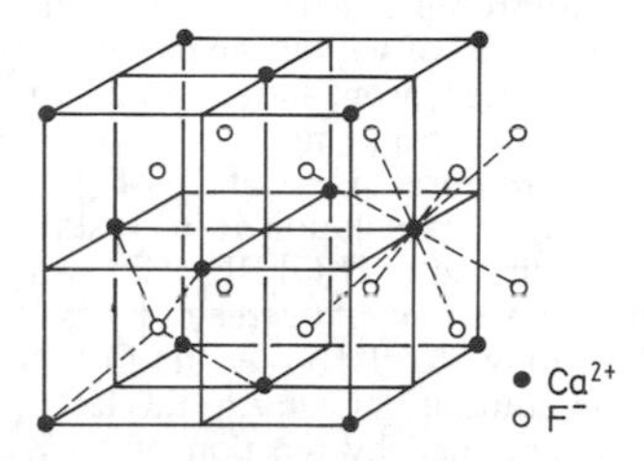

Fluorite structure (unit cell and coordination relation)

Each Ca^{2+} ion is surrounded cubically by 8 F^- ions, and each F^- ion is surrounded tetrahedrally by 4 Ca^{2+} ions. The occurrence of the F. depends critically on the ratio of ionic radii: r_A/r_B must be greater than or equal to 0.732. In addition to CaF_2, many oxides and sulfides of elements with large ionic radii, e.g. SrF_2, BaF_2, CeO_2, ThO_2, UO_2, crystallize in the F. In the ***antifluorite type***, the positions of metal and nonmetal are reversed (see Isotypism). Some compounds which crystallize in this structure are the alkali metal oxides $Li_2O...Rb_2O$ and sulfides $Li_2S...Rb_2S$, Mg_2Si, etc.

Fluoroacetic acid: F-CH_2-COOH, colorless and nearly odorless needles; m.p. 35 °C, b.p. 165 °C. F. is soluble in water and alcohol. It is a toxic component of certain poisonous plants. F. can be synthesized by reaction of iodoacetic acid with mercury fluoride, followed by hydrolysis of the fluoroacetate ester with calcium hydroxide. Because of its high toxicity, F. and some of its derivatives, such as its sodium salt and its methyl and ethyl esters, are used as pesticides. In addition, they are of some significance as chemical weapons.

> Fluoroacetic acid is extremely poisonous. It can be absorbed through the skin and leads to heart damage as well as violent convulsions. The toxicity of fluoroacetic acid is due to the fact that it is incorporated into a coenzyme-A compound, which is added to oxaloacetate by citrate synthase, forming fluorocitric acid. This is a strong inhibitor of aconitase, and blocks the citric acid cycle.

Fluoroazide: see Halogen azides.

Fluorobenzene: C_6H_5-F, a colorless liquid with an aromatic odor; m.p. -41.2 °C, b.p. 85.8 °C, n_D^{20} 1.4657. F. is nearly insoluble in water, but dissolves readily in most organic solvents. It is synthesized by the Schiemann reaction (see), or by pyrolysis of cyclopentadiene in the presence of chlorine- and fluorine-containing alkanes or alkenes. F. is used to prepare pesticides and drugs.

Fluoroborates, ***tetrafluoroborates***: the salts of tetrafluoroboric acid, HBF_4, which can be obtained in aqueous solution. The general formula of the salts is $M[BF_4]$, and their crystal structures and solubilities are largely analogous to the corresponding isoelectronic perchlorates. F. are obtained by reaction of the corresponding metal oxides, hydroxides or carbonates with aqueous fluoroboric acid solutions. Ammonium fluoroborate and F. of various metals, such as potassium, lead, cadmium, iron, nickel, zinc, tin and sodium, are used in electroplating.

Fluoroboric acid, ***tetrafluoroboric acid***: $H[BF_4]$, a compound obtained in aqueous solution or in crystalline form as the oxonium salt, $[H_3O][BF_4]$. F. is a strong acid which undergoes a slight degree of hydrolysis in water, according to $HBF_4 + H_2O \rightleftharpoons H[BF_3OH] + HF$ (the equilibrium is far to the left). F. is poisonous. It is obtained by reaction of 50% hydrofluoric acid with boric acid: $4\ HF + B(OH)_3 \rightarrow HBF_4 + 3\ H_2O$. It is used in the preparation of its salts, the Fluoroborates (see), as a soldering flux and preservative for glues, and for electrolysis baths.

Fluorocuprates: see Copper fluorides.

Fluorocarbons: compounds of fluorine and carbon used in macromolecular form as plastics and synthetic fibers. Their significance is due to their excellent resistance to chemicals and long-term exposure to high temperatures (up to about 260 °C). The best known F. are Polytetrafluoroethylene (see) and Polytrifluorochloroethylene (see).

1-Fluoro-2,4-dinitrobenzene: one of six possible fluorodinitrobenzene isomers. It forms bright yellow crystals which are insoluble in water; m.p. 25.8 °C, b.p. 296 °C. 1-F. is soluble in alcohol. It is synthesized by nitration of fluorobenzene. 1-F. is used in peptide chemistry to protect amino groups.

Fluorohydrocarbons: aliphatic and aromatic compounds in which hydrogen atoms are replaced by fluorine atoms. Perfluorinated derivatives are also called ***fluorocarbons***. F. are stable compounds, and their stability increases with the degree of fluorination. Multiply fluorinated F. are resistant to high temperatures and most of them do not burn, which greatly expands their range of uses. Mono- or difluoro compounds, on the other hand, have chemical properties similar to other, analogous halogen derivatives.

The ***chlorofluorocarbons*** are another important group of F.; in addition to fluorine, they have other halogen atoms, chiefly chlorine. These compounds are designated by the letter R (for refrigerant) and a number. The first digit of the number indicates the number of C atoms minus 1, the second digit indicates the number of H atoms plus 1, and the third, the number of F atoms. The number of Cl atoms can be calculated from the number of remaining bonds of which the carbon skeleton is capable. For example, 1,1,-dichloro-2,2,2-trifluoroethane, $CHCl_2$-CF_3, is called R 123. This system of nomenclature is not applicable to higher homologs.

F. are synthesized mainly by stepwise exchange of the halogen atoms in the reaction of chlorohydrocarbons with hydrogen fluoride in the presence of antimony(V) chloride, or by the Swarts reaction.

Applications. Because of their low toxicity and non-inflammablilty, F. are used in large amounts as propellants for sprays, as refrigerants and as solvents. They are also used as fire extinguishing compounds, in the production of foam plastics (e.g. styrofoam) and fluorocarbon resins; they are also used as lubricants and for impregnation.

The large-scale release of F. into the atmosphere is believed to be the cause of increasing depletion of the stratospheric ozone layer. Under the influence of UV light, the compounds release chlorine radicals which act as catalysts in the destruction of the ozone.

Fluorosilicates, ***hexafluorosilicates***: the salts of hexafluorosilicic acid, H_2SiF_6; the general formula is $M_2[SiF_6]$. The SiF_6 anion is octahedral; the Si-F distances are 170 pm. Alkali fluorosilicates (with the exception of $Li_2[SiF_6]$) are insoluble in water, while the other metal fluorosilicates (except $Ba[SiF_6]$) are readily soluble. The F. are poisonous and etch metal. They are obtained by reaction of metal hydroxides or carbonates with hexafluorosilicic acid. The most important F. in industry is that of sodium. Copper hexafluorosilicate, $CuSiF_6 \cdot 6H_2O$, and magnesium hexafluorosilicate, $MgSiF_6 \cdot 6H_2O$, are used to protect wood; the latter is also used to harden wood and waterproof it.

Fluorosilicic acid, ***hexafluorosilicic acid***: H_2SiF_6, a strong, dibasic acid which can be precipitated out of concentrated solution in the form of the oxonium salt, $[H_3O]_2[SiF_6]$ as hard, colorless crystals, m.p. 19 °C. Aqueous solutions of F. do not contain significant amounts of free hydrofluoric acid, and do not etch glass. When aqueous H_2SiF_6 solutions are evaporated, SiF_4 and HF are released as vapors; the vapor above a 13.3% solution contains a molar ratio of 1 SiF_4 to 2 HF, while the vapors above more concentrated solutions contain more SiF_4, and those above less concentrated solutions contain relatively more HF. F. is made by partial hydrolysis of silicon tetrafluoride: $3\ SiF_4 + 2\ H_2O \rightarrow 2\ H_2SiF_6 + SiO_2$. It is now an important byproduct of superphosphate or phosphoric acid production from fluoroapatite. F. is used to preserve wood, to clean, impregnate and harden masonry, to clean the surfaces of metals, to improve the surfaces of optical glass and to make its salts, the Fluorosilicates (see).

Fluostigmine, abb. ***DFP***: the fluoride of phosphoric acid diisopropyl ester. F. was originally developed as a Chemical weapon (see). It inhibits acetylcholinesterase (see Parasympathicomimetics) and is used to treat glaucoma.

Fluorosulfonic acid: same as Fluorosulfuric acid (see).

Fluorosulfuric acid, ***fluorosulfonic acid***: FSO_3H, colorless liquid; M_r 100.07, density 1.743, m.p. -87.3 °C, b.p. 165.5 °C. F. is a very strong acid made by combining sulfur trioxide and hydrogen fluoride; it is only slowly hydrolysed by water. It is used in organic synthesis as an acid catalyst and a convenient fluorination reagent. An equimolar mixture of F. with antimony pentafluoride, SbF_5, is called magic acid (see Superacids) and is used to protonate extremely weak bases.

Fluorouracil, ***5-fluorouracin***: an antimetabolite of the pyrimidine nucleotide bases uracil and thymine; it is used as a cytostatic. F. is made by radical fluorination of uracil with elemental fluorine.

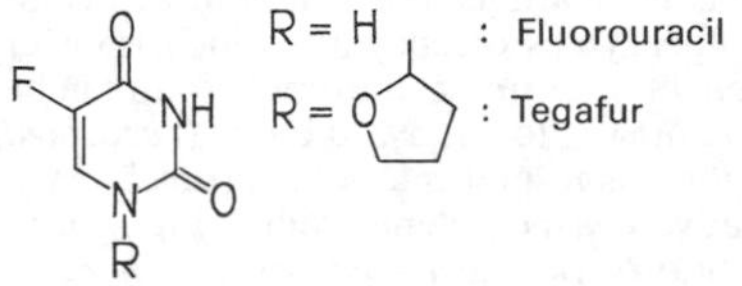

R = H : Fluorouracil
R = tetrahydrofuranyl : Tegafur

The active form is 5-fluoro-2'-deoxyuridine 5'-monophosphate, which inhibits thymidylate synthetase. This enzyme converts 2'-deoxyuridine 5'-monophosphate to thymidine 5'-monophosphate. The 1-tetrahydro-2-furanyl derivative *Ftorafur®* is used as a Prodrug (see).

Flurenol: see Morphactins.

Flushing process: a process for production of pigments. Pigments produced by wet chemistry, e.g. by precipitation, are not dried, but instead are kneaded together with organic paint solvents or binders, such as linseed oil, to which wetting agents have been added. The process displaces the water from the pigment mass, and the organic solvent replaces it. The result is flushed pigments in the form of a paste. The particle size of the pigments is nearly unchanged; during the process of drying the particles agglomerate to form larger particles.

Flush process: a process for production of dispersions in nonpolar media by direct transfer of colloidal particles which have been produced in the aqueous phase, for example, pigments. The particles are made hydrophobic by addition of surfactants, such as sodium salts of fatty acids. The fatty acids can be fixed to the particle surfaces by changing the pH and formation of the salts; this process is called *oleophilization*. It is more convenient to choose an oil-soluble, surface active compound, which, when the oil phase is dispersed in the aqueous phase, forms droplets with a surface charge opposite that of the solid particles. F. is important for adding pigments to paints and in the production of iron oxide or chromium oxide dispersions for coating magnetic tapes.

Fluvalinate: see Pyrethroids (table).

Fly ash: solid particles suspended in the flue gases resulting from combustion of solid fuels. The composition and size of F. particles varies. Most F. contains large amounts of silicon dioxide, along with magnesium, calcium, iron and aluminum oxides. F. settles onto buildings and vegetation in the environs of the emitter or, if fine enough to remain suspended in the upper atmosphere, it can provide nucleation centers for droplets in clouds.

Fm: symbol for fermium.

FMN: see Flavin nucleotides.

Foam: a fine or colloidal distribution of a gas (disperse phase) in a liquid, especially a surfactant solution, or a solid. Some examples are pumice, foam glass and polymer foams. The foam is formed in the presence of a surface active compound by forcing air into its solution through a suitable valve, by release of gases dissolved under pressure, or by generation of gas through a chemical reaction (carbon dioxide from acidified carbonates, for example). The foam must be stabilized by a Surfactant (see) or Foam stabilizer (see). F. with limited lifespans are used in Flotation (see). The ***foam stability*** of a surfactant solution is a measure of the dependence of foam height on time.

There are two types of F., spherical and polyhedral. ***Spherical F.*** consist of individual, independent bubbles which do not interact with each other. Their production is not dependent on the presence of a surface-active substance. The temporary stability of such foams depends on the viscosity of the dispersion phase. In solvents with low viscosity, they degrade within seconds (for example, carbonated water when the bottle is opened). When the viscosity is higher, the lifespan increases (e.g. saliva). When the liquid phase solidifies, e.g. when latex is foamed (*foam rubber*), the stability is practically unlimited. In ***polyhedral F.***, the individual layers of F. are separated by thin films. When they fill a volume completely, three polyhedral surfaces meet at each edge, and the edges are tetrahedrally arranged.

F. are formed in many technological processes, and are generally not desired; examples are in distillation, evaporation and mixing processes, gas scrubbing, and transport of liquids by pumps. They can be broken by Defoaming agents (see), by heating or cooling, or by shear.

Foam capacity: the height of foam formed by a surfactant solution at the time the foam is formed.

Foam concrete: see Plastic concrete.

Foam glass: a glass in which a high volume fraction of closed pores are created by gas-generating substances during the melting or sintering process. It can be made from a mixture of raw materials which contains, for example, calcium carbonate or aluminum as a gas generator; this is rapidly melted and cooled without clarification. A high porosity is also achieved by bubbling gas into the melt. Another method of making F. starts with ground glass, with the foaming provided by wet grinding or by addition of carbon to dry-ground glass. At high temperatures, the carbon reacts to form carbon dioxide in the melt. If no gas generator is added, the product is a porous ***sintered glass***.

F. with a very low crude density (up to 95% hollow volume), high thermal insulating capacity and relatively high strength is used in many ways in vehicles, industry and home construction.

Foam inhibitors: same as Defoaming agents (see).

Foam plastics: porous, very light substances made of various plastics, such as epoxide resins, urea resins, polyurethane, polystyrene, polyethylene or polyvinyl chloride. They are synthesized either by incorporation of air into an aqueous solution or dispersion of a resin, or by propellants. In this case a propellant is worked into the plastic which either evaporates when heated (e.g. pentane) or decomposes, forming a gas (e.g. azoisobutyrodinitrile or dinitrosopentamethylene tetramine) which expands the plastic. Premolded plastic foam parts can be made by placing the plastic (for example, a pentane-containing polystyrene granulate) in a steam-heated mold before it is foamed. A distinction is made between soft and hard F., which can be further subdivided into open-pored and closed-pored F. Soft F. are used for all types of padding and as sponges, while hard F. are used as low-density (0.005 to 0.1 g cm^{-3}) insulators in refrigerators and buildings. F., especially polystyrene, are also used to make toys, packaging, disposable cups, floats (for boats, buoys, fishnets, etc.). Because they burn easily, F. are often provided with fire-retarding compounds.

Foam rubber: a highly elastic, porous and lightweight rubber with a density of 0.10 to 0.25 g cm^{-3}; there are several methods of making it directly from latex. 1) The latex mixture containing gas-producing foaming agents, surfactants and heat sensitizers is converted to an aqueous foam by vigorous stirring, then coagulated in a metal mold at a temperature of 60 to 70 °C. 2) The latex mixture is foamed by oxygen released enzymatically from hydrogen peroxide and coagulated by passing carbon dioxide through it, or by addition of sodium fluorosilicate. 3) The latex mixture is prevulcanized and saturated with nitrogen under high pressure in an autoclave. When the pressure is released, the mixture is blown up by the escaping nitrogen and is fixed by vulcanization. Synthetic fibers, fiberglass or similar materials can be added to give the F. more strength. F. is used mainly as a cushioning material for furniture and vehicle interiors.

Foam stabilizers, ***boosters***: compounds which increase the lifespan of foams by increasing the surface viscosity and elasticity of the foam lamellae. F. may be, for example, long-chain amine *N*-oxides or alkanoic acid-*N,N*-bis(hydroxyalkyl)amides.

Folate synthetase: an enzyme present in microorganisms, especially bacteria, which catalyses the synthesis of folic acid from 4-aminobenzoic acid, methylpterin and glutamic acid. Sulfonamides, as antimetabolites of 4-aminobenzoic acid, act as competitive inhibitors; this is the basis of the antibacterial effect of the sulfonamides.

Folic acid, ***pteroylglutamic acid***: a member of the B_2 group of vitamins. It consists of pteroic acid linked to L-glutamic acid by an amide bond. *Pteroic acid (vitamin H')* consists of 6-methylpterin and 4-aminobenzoic acid. F. is orange-yellow and sensitive to light. It has no odor or taste and is nearly insoluble in water and ethanol, but it is soluble in acids and bases; $[\alpha]_D^{20}$ + 18-23° (in 0.1 M NaOH), λ_{max} 256, 283 and 365 nm (in 0.1 M NaOH). F. occurs widely in nature, and is most abundant in liver, yeast and vegetables. Plants and microorganisms contain the compound in conjugates, in which up to 7 L-glutamate groups are linked by peptide bonds involving the γ-carboxyl groups. F. can be synthesized by condensation of 6-formylpterin with *N*-(4-aminobenzoyl)-L-glutamic acid and hydrogenation of the resulting azomethine structure. The biochemically active form of F. is ***tetrahydrofolic acid***, in which the double bonds between N-5 and C-6, and between C-7 and N-8 are hydrogenated. Various derivatives of this compound act as coenzymes of enzymes which transfer one-carbon units. For example, a formyl group can be activated by binding it to the N-10 atom ("*activated formate*"), and in this form it is used in purine biosynthesis. Alternatively, a methylene group is introduced between N-5 and N-10 ("*activated formaldehyde*"), and transferred as a hydroxymethyl group, for example in the biosynthesis of serine from glycine.

F. hypovitaminoses are extremely rare. F. is used together with vitamin B_{12} in the treatment of various forms of anemia caused by lack of the enzyme which releases F. from its conjugates. *Leucovorin*, the calcium salt of 5-formyltetrahydrofolic acid, is used as an antidote for methotrexate when this compound is used in cancer therapy.

Historical. F. was first isolated from spinach leaves in 1947 by Mitchell, Snell and Williams.

Folin's reagent: 1) reagent of sodium 1,2-naphthoquinone 4-sulfonate used to detect amino acids, with which it forms colored condensation products. It is used for paper chromatographic determination of amino acids. 2) Reagent of sodium tungstate or sodium molybdate, phosphoric acid and lithium sulfate, which reacts with phenols to give a blue color.

Follicle hormones: same as Estrogens (see).

Follicle-stimulating hormone: same as Follitropin (see).

Follicle-stimulating hormone releasing hormone: see Liberins.

Folliliberin: see Liberins.

Follitropin, ***follicle-stimulating hormone***, abb. ***FSH***: a gonadotropin which stimulates the initial growth of the follicles in the ovary in females, and spermogenesis in males. It is released in response to gonadoliberin. F. is a glycoprotein composed of two polypeptide chains, α- and β. The separated chains have no biological activity, but when recombined under suitable conditions, they regain about 70% of the original activity.

Folpet: see Fungicides.

FO method: see Woodward-Hoffmann rules.

Food chemistry: a discipline concerned with the structure, properties and reaction kinetics of foods and the reactivities of the components of foods. F. is concerned with the composition of foods and changes

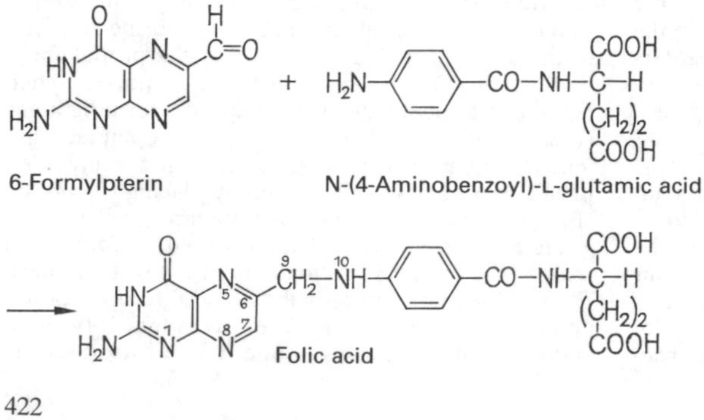

6-Formylpterin + N-(4-Aminobenzoyl)-L-glutamic acid → Folic acid

in their nutritional value due to internal and external factors. The field relates to nutritional physiology, hygiene, toxicology, food technology and law.

Food colorings: substances used to color foods or preparations of F. in carriers or solvents. ***Natural organic F.*** include riboflavin, curcumin, various carotinoids, anthraquinone, anthocyanin and porphyrins. ***Artificial organic F.*** include tartrazine, quinoline yellow, azorubin, naphthol red S., Ponceau 4R, erythrosin, indigotin, patent blue and brilliant black BN. ***Inorganic pigments***, such as chalk, titanium dioxide, iron oxide yellow, red and black are used for special purposes, such as painting pressed sugar decorations.

The laws of different countries vary considerably as to the amounts of F. permitted and the foods to which they may be added. Those F. which are legally permitted are thought to be harmless after extensive animal tests; however, the lists of permitted compounds are subject to change as new data is accumulated.

Foreign materials: in food chemistry, a term for substances which are not naturally present in foods, but are introduced intentionally or by accident; they are ingested along with the food or beverage. Vitamins, spices, flavorings and aromas of natural origin, air and alcohol added to a food product are not considered F. F. are classified as Additives (see) or Contaminants (see). They are added during the processing of foods or are picked up from the environment, as residues of agricultural chemicals or drugs used on animals, or as residues from machinery. There are strict regulations concerning F. which may impair health.

Forge coal: see Coal.

Formaldehyde, ***methanal***: H-CHO, the simplest aliphatic Aldehyde (see), and the first member of this homologous series; m.p. - 92 °C, b.p. - 21 °C. F. is a colorless gas with a pungent odor; it is readily soluble in water and many organic solvents, such as alcohol and ether.

> Formaldehyde is a strong irritant for the mucous membranes of the nose and throat and for the external skin. When concentrated vapors are inhaled, it can lead to eczema, vomiting and lung edema. Aqueous solutions of formaldehyde are also very poisonous. They can lead to necroses in the upper digestive organs and to kidney damage, respiratory distress and heart irregularity. Death often occurs due to perforation of the stomach, but only after many hours. Antidotes are urea or ammonium carbonate solutions and activated charcoal.

F. undergoes most of the typical aldehyde reactions, and differs from other aldehydes only in a few special reactions. The reason for this is the special structure and thus the size of the formaldehyde molecule; the carbonyl group is bound only to two hydrogen atoms.

Like most aliphatic aldehydes, F. tends to polymerize in the presence of acids or bases, forming long-chain or cyclic products, the polyoxymethylenes. In aqueous solutions, F. is present mainly as the hydrate, $HO\text{-}CH_2\text{-}OH$. Evaporation of such solutions leads to formation of a white, amorphous, linear polymer mixture called *paraformaldehyde*: $HO\text{-}CH_2\text{-}O\text{-}(CH_2\text{-}O)_n\text{-}CH_2\text{-}OH$. For more about higher polymers of F., see Polyoxymethylene. In the presence of acids, distillation of a F. solution leads to formation of a cyclic trimer, *trioxymethylene (1,3,5-trioxane)*;

1,3,5-Trioxane

formation of cyclic tetraoxymethylene has also been observed. Heating or hydrolysis of these polymers releases F. An irreversible polymerization which involves formation of C-C bonds is observed in the reaction of calcium hydroxide with a very dilute formaldehyde solution. This reaction leads to formation of a mixture of sugars, formose, which is difficult to separate. It consists mainly of hexoses. Since F. does not contain an α-H atom, it undergoes a Cannizzaro reaction in highly concentrated solutions in the presence of alkali or alkaline earth hydroxides. This reaction is a disproportionation to methanol and formic acid. F. reacts with ammonia to form hexamethylenetetraamine (urotropine). Compounds with reactive H atoms react with F. to form hydroxymethyl compounds. For example, F. and acetaldehyde form tris-(hydroxymethyl)acetaldehyde, which reacts with excess F. to form pentaerythritol:

$$3\,CH_2{=}O + CH_3{-}CH{=}O \rightarrow (HO{-}CH_2)_3C{-}CHO \xrightarrow{CH_2O} (HO{-}CH_2)_4C + HCOOH.$$

In the synthesis of phenol plastics, aromatic H atoms are replaced by hydroxymethyl groups in the same way. Similarly, the H atoms in nitromethane or acetylene can be replaced by hydroxymethyl groups: $H\text{-}C{\equiv}C\text{-}H + H_2C{=}O \rightarrow H\text{-}C{\equiv}C\text{-}CH_2\text{-}OH$ (propargyl alcohol).

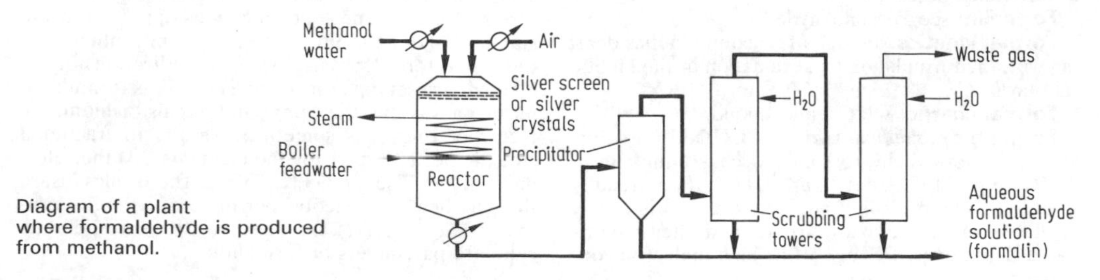

Diagram of a plant where formaldehyde is produced from methanol.

In the Mannich reaction (see) and in the synthesis of aminoplastics, reactive H atoms in amines or urea are replaced in the primary positions by hydroxymethyl groups. Some other important preparative reactions of F. are the Blanc reaction (see) and the Grignard reactions (see) to form primary alcohols.

Production. F. is most often synthesized from methanol by two different methods: 1) dehydration of methanol in the presence of a silver or copper catalyst (Fig.). The silver catalyst is used either in the form of coarse silver crystals or silver screens. The primary step in the reaction is the dehydrogenation of the methanol: $CH_3OH \rightarrow CH_2O + H_2$, $\Delta_R H = +84$ kJ/mol. Air is added to burn off the hydrogen: $H_2 + 1/2\ O_2 \rightarrow H_2O$, $\Delta_R H = -243$ kJ/mol. The overall equation is thus: $CH_3OH + 1/2O_2 \rightarrow CH_2O + H_2O$, $\Delta_r H = -159$ kJ/mol. Because the reaction is highly exothermic, the amount of air is regulated so a reaction temperature of 600 to 650 °C can be maintained. The methanol turnover is then only about 85%, so that the unreacted alcohol is returned to the reactor. Complete reaction of the methanol in one step is possible if a temperature around 700 °C is used and water is added. The hot reaction gases are immediately cooled to about 100 °C, and the resulting F. is washed out with water in several absorption steps. A concentrated *formalin solution* (about 40 M% F. in a water-methanol solution) is produced in a subsequent distillation step.

2) In the purely oxidative method, the methanol is oxidized in the presence of a Fe_2O_3/MoO_3 catalyst to F. Methanol vapor and a large excess of air are passed over a catalyst in a pipe furnace at 350 to 450 °C. The methanol reacts completely in this method. After the reaction gases are cooled to about 100 °C, the F. is washed out of the gas stream with water. Dimethyl ether can be used instead of methanol in this process. Direct oxidation of methane to F. has so far not been successful, due to enormous technical difficulties. In some plants in the USA, C_3/C_4 hydrocarbon mixtures are oxidized to F.

Applications. F. is used mainly to synthesize polymeric materials, particularly phenol plastics, aminoplastics and polyformaldehyde. Some other important products of F. are the polyols, for example pentaerythritol and trimethylolpropane, which are very important for the production of polyesters. Ethynylation of F. followed by hydrogenation produces butane-1,4-diol. F. is also used as a disinfectant and preservative, as a hardener for glue, casein products and gelatins, as a reducing agent in vat dyeing, as a deodorant, as a starting material for synthetic tannins, to fix anatomical preparations and to combat molds. In the laboratory, F. is used for a number of different syntheses.

Formalin: see Formaldehyde.

Formal kinetics: an area of reaction kinetics dealing with the formulation and evaluation of the kinetic rate laws.

Formal charge: see Chemical bond.

Formamide, ***methanamide***: $O{=}CH\text{-}NH_2$, a colorless and odorless, hygroscopic, viscous liquid; m.p. 2.5 °C, b.p. 210 °C (dec.), n_D^{20} 1.4472. F. is readily soluble in water, alcohol and acetone, and is nearly insoluble in ether, hydrocarbons and their chlorinated derivatives. Because of its high dielectric constant, F. is an excellent ionizing solvent for many inorganic salts. At room temperature it is only slowly hydrolysed to formic acid and ammonia. When heated, it is cleaved chiefly into carbon monoxide and ammonia. In the presence of suitable catalysts, however, its main decomposition products are hydrocyanic acid and water: $O{=}CH\text{-}NH_2 \rightarrow HCN + H_2O$. F. is synthesized industrially by reaction of methyl formate with ammonia, or directly from carbon monoxide and ammonia in the presence of sodium methanolate. It is used to make formic acid, formates and hydrocyanic acid. Because it is bifunctional, F. is particularly suitable for synthesis of N-containing heterocycles, such as imidazoles or pyrimidines. It is also used as a solvent and catalyst.

Formate: a salt or ester of formic acid; the general formula is $H\text{-}COOM^I$ or H-COOR, where M^I is an ammonium ion or a monovalent metal cation, and R is an aliphatic, aromatic or heterocyclic group.

The salts are formed by reaction of ammonia, ammonium hydroxides, metal hydroxides, or ammonium or metal carbonates with formic acid. One method used industrially is the reaction of carbon monoxide with the hydroxides of alkali or alkaline earth metals; the first step in the reaction is the addition of carbon monoxide to water, forming free formic acid: $CO + H_2O \rightleftharpoons H\text{-}COOH$. In the second step, the formic acid is neutralized by the corresponding hydroxide. The most important F. are the sodium, potassium, calcium, nickel and aluminum salts. These are used for etching, impregnating and preserving.

The esters of formic acid are made by direct esterification with alcohols, or by the reaction of carbon monoxide with alcohols. Some important *formic acid esters* are the methyl and ethyl esters, methyl formate and ethyl formate. These are used as solvents and as intermediates in the synthesis of other formic acid derivatives and other organic compounds.

Formate-potash process: see Potassium carbonate.

Formation, energy of: see Formation reaction.

Formation, enthalpy of: see Formation reaction.

Formation, entropy of: see Formation reaction.

Formation, heats of: see Formation reaction.

Formation reaction: in thermodynamics, a reaction leading to the formation of a compound from the stable forms of its elements. It is of no consequence whether the reaction can actually be carried out in this fashion. In chemical thermodynamics the molar heats of formation (molar energy of formation $\Delta_F E^\circ$ and molar enthalpy of formation, $\Delta_F H^\circ$) and the molar second law functions for formation (entropy of formation, $\Delta_F S$, free energy of formation $\Delta_F A^\circ$, and Gibbs free energy of formation, $\Delta_F G^\circ$) are defined in terms of the F. The following rules apply: 1) in the standard state, the internal energy and enthalpy of the stable forms (physical state, modification) of the element are set equal to zero; 2) the F. is formulated in such a way that the compound has a stoichiometric factor of one (this sometimes results in fractional stoichiometric factors for the elements); 3) the reference system is set at 298 K (25 °C). These rules insure that the heats of reaction per mole of formula turnover ($\Delta_R E$ or $\Delta_R H$) are equal to the corresponding standard parameters of formation.

Examples:

a) $6\,C_{graphite} + 3\,H_{2\,gas} \rightarrow C_6H_{6\,liq}$

$\Delta_R H^\circ = \Sigma(i)\nu_i\rho H_i^0$

$= H_{C_6H_6}^{\ 0} - 6\,H_C^{\ 6} - 3H_{H_2}^{\ 0} = \Delta_F H_{C_6H_6}^{\ 0}$.

b) $S_{solid,\,\alpha\,modification} + 1/2O_{2\,gas} + Cl_{2\,gas} \rightarrow SOCl_{2\,liq}$

$\Delta_R E^0 = H_{SOCl_2}^{\ 0} - H_S\alpha^0 - 1/2H_{O_2}^{\ 0} - H_{Cl_2}^{\ 0} = \Delta_F E_{SOCl_2}^{\ 0}$.

Since the internal energies of the elements are set equal to zero by definition, it follows that the molar internal energy E_i^0 or the molar enthalpy H_i^0 of each compound i is equal to its standard energy of formation $\Delta_F E_i^0$ or its enthalpy of formation $\Delta_F H^i{}_0$.

Therefore, the standard parameters of formation can be used instead of the molar parameters E_i^0 and H_i^0.

If the F. do not actually occur, the reaction data are calculated with the aid of Hess' Law (see) from the heats of combustion, hydrogenation or fluorination.

Formazane pigments: red pigments containing the formazane chromophoric group.

$$-C\begin{matrix}\nearrow N \searrow NHR \\ \searrow N \nearrow\!\!\!= NR\end{matrix}$$

Formazane group

They are made by coupling diazonium salts with the phenylhydrazones of aldehydes. A particularly important F. is ***triphenylformazane***, which is formed from benzaldehyde phenylhydrazone and benzene diazonium chloride, and is used for the synthesis of 2,3,5-triphenyl-2H-tetrazolium chloride. Triphenylformazane forms dark red, light-sensitive crystals with a metallic sheen. It is soluble in chloroform, ether and ethyl acetate, giving red solutions, and is used in histochemistry for measuring the activity of dehydrogenases, the germination capacity of seeds and as a disinfectant.

F. forms valuable dyes by chelate formation with metal atoms, such as cobalt, copper, nickel, cadmium, magnesium and zinc; for example, the 1:2 cobalt complex is used on wool and synthetic fibers to give a blue-gray color. When irradiated with light, some F. are reversibly rearranged out of the red *syn* form into the two unstable, yellow *anti* forms (photochromic dyes), so that the F. are used as camoflage colors in the military.

Formic acid, ***methanoic acid***: H-COOH, the simplest monocarboxylic acid. F. is a colorless, mobile, volatile liquid with a pungent odor; m.p. 8.4 °C, b.p. 100.7 °C, n_D^{20} 1.3714. It is caustic and irritates the skin and mucous membranes; contact with F. can cause very painful, slowly healing lesions. F. is very soluble in water, alcohol, ether, acetone and benzene. It is the strongest aliphatic monocarboxylic acid, with a pK_a (≈ 3.8) considerably lower than that of acetic acid (≈ 4.8). F. can be distilled without decomposition at normal pressures; at high temperatures it decomposes in the presence of catalysts into carbon monoxide and water, or into carbon dioxide and hydrogen: H-COOH → CO_2 + H_2. The vapor of F. can be ignited in air and burns with a blue flame; mixtures of this vapor with air are explosive in the range from about 14 to 33 vol.%. With water it forms an azeotropic mixture which boils at 107.3 °C and contains 77.5% F. In addition to the typical carboxylic acid reactions, F. displays a number of aldehyde reactions because it contains a formyl group; these can dominate its behavior. For example, in contrast to other carboxylic acids, F. is a strong reducing agent. It causes metallic silver or mercury to precipitate from ammoniacal silver or mercury(II) nitrate solutions. These reactions are often used to identify F. The salts and esters of F. are called Formates (see). F. is found both free and bound, e.g. as esters, in plants and animals. It is present in the poison glands of ants and bees, in caterpillar hairs, fir needles, nettles and many fruits. It is also present as a salt in human urine.

F. is produced industrially from sodium hydroxide and carbon monoxide at 210 °C and pressures of 0.6 to 0.8 MPa:

$$NaOH + CO \rightarrow HCOONa \xrightarrow[-Na^+]{+H^+} H\text{–}COOH$$

F. is also produced in large quantities from methyl formate and ammonia via formamide; the latter is converted by sulfuric acid into F. and ammonium sulfate. F. is a byproduct of the production of acetic acid by hydrocarbon oxidation. Small amounts of F. can be produced by heating oxalic acid with glycerol, or directly from carbon monoxide and water, at high pressures and in the presence of catalysts. Dilute F. is concentrated in one of several ways, depending on its water content: treatment with salts which bind water of crystallization, azeotropic distillation with toluene or butyl or amyl formates, or fractional distillation. F. is used in large amounts to coagulate latex, as an impregnating agent and mordant in the textile and leather industries, as a solvent, decalcifying agent, preservative for fruits and fruit juices, to disinfect beer and wine barrels, and to make formates.

Historical. A. was discovered around 1670 by Fisher in the red wood ant, *Formica rufa*. It was later purified from the ants by distillation and named for them.

L-Forming process: see Reforming.

Formose: a mixture of sugars formed by reaction of alkalies with formaldehyde solutions. It contains straight and branched monosaccharides of varying chain lengths and configurations, including DL-glucose and arabinose.

Formula: a symbolic representation of the composition and, in some cases, the structure of a molecule of a chemical compound. There are several kinds of F., and those for inorganic and organic compounds are constructed according to different principles. The simplest type of F. is the ***empirical F.***, which represents only the stoichiometry, e.g. H_2O. The order of the element symbols in the empirical F. for inorganic compounds is not the same as the order for organic compounds. The more electropositive component of a salt, the cation, is listed first. If an inorganic compound contains several electropositive or electronegative components, the symbols are arranged within each group in alphabetical order (*Hill's system*). For binary compounds of non-metals, the components are listed in the following order: Rn, Xe, Kr, B, Si, C, Sb, As, P, N, H, Te, Si, S, At, I, Br, Cl, O, F. For example, silicon tetrafluoride is SiF_4. Elements which are not contained in this list are indicated in the order shown by the arrows in the figure.

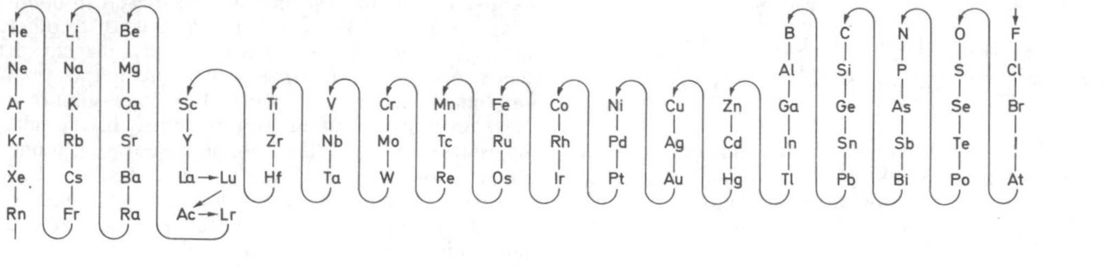

Order of the elements in formulas.

If one of the elements is found only in the figure, and the other in the above list, the one in the list is considered the more electronegative. If three or more elements are joined in a row, the order of their symbols in the F. is the same as the order of the atoms in the molecule, e.g. HOCN for hydrocyanic acid and HONC for fulminic acid. If two or more different atoms or groups are linked to a single central atom, the symbol for the central atom is placed first, followed by the symbols for the others in alphabetical order, as in $PBrCl_2$. In the F. of inorganic acids, the hydrogen is always listed first. In the F. of intermetallic compounds, the symbols of the metallic components (Sb counts as a metal) are arranged in alphabetical order, unless the ionic character is to be stressed. In the F. of inclusion compounds, for example, the symbols of the nonmetals (in the order given above) follow those of the metals. The number of identical atoms is given in the F. as a subscript to the element symbol; the symbols for a group of atoms are enclosed in parentheses, and where more than one of this group is present, the number is indicated as a subscript after the parenthesis. Examples: $Ca_3(PO_4)_2$ for calcium phosphate, $C_6H_7NO_2S$ for benzene sulfamide. The number of molecules of water of crystallization or other loosely bound molecules is indicated by an arabic numeral in front of the molecule symbol, e.g. $Na_2SO_4 \cdot 10H_2O$. If the compound consists of discrete molecules ("defined molecules" as opposed to an ionic crystal like NaCl), the ***molecular F.*** is given which corresponds to the actual molecular mass, e.g. S_2Cl_2 for disulfur dichloride, not SCl. If the relative molecular mass changes with the temperature, the aggregate state, etc., the simplest possible F. is used, e.g. S, P or NO_2 instead of S_8, P_4 or N_2O_4, unless the size of the molecule is to be expressly indicated. The bonding of atoms in a molecule is given by the ***structural formula***. The atoms which share electron pairs are joined by dashes; for example, the structural F. for water is H-O-H, for sulfuric acid,

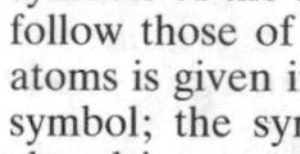

$$\begin{matrix} O \\ O \end{matrix} \!\gg\! S \!<\! \begin{matrix} O\text{–}H \\ O\text{–}H \end{matrix}$$

In ring structures, such as Benzene (see), each C is indicated by a vertex of the polygon.

A still better picture of the bonding within a molecule is given by an ***electron F.***, in which the outer electrons of an atom are indicated by dots.

For example, atomic chlorine is $:\ddot{\underset{\cdot\cdot}{Cl}}\cdot$, and water is H–$\overline{O}$–H. In the structural formula, the free electron pairs not involved in bonding can be indicated by dashes, e.g. H:$\underset{\cdot\cdot}{\ddot{O}}$:H. Electron formulas of ionized compounds, such as $[Na]^+\left[:\ddot{\underset{\cdot\cdot}{Cl}}:\right]^-$, are simplified to ***ionic F.***, in which the ionic charge, rather than the electrons, is indicated by a plus or minus as a superscript, e.g. Na^+Cl^-. The electron F. can be further simplified by leaving out all electrons and charges, and depicting compounds of higher order by a ***complex F.***, such as $H_2\left[\begin{matrix} O \\ OSO \\ O \end{matrix}\right]$ for sulfuric acid; here the F. of the complex ion is summarized, e.g. as $[SO_4]$. The spatial arrangement of the atoms in the molecule can be shown by a ***stereo F.*** in which heavy lines symbolize bonds which project above the plane of the paper, and dashed lines symbolize bonds which project below the plane of the paper. Solid lines of normal thickness indicate bonds in the plane. Wavy lines indicate bonds of indeterminate direction. The structure can also be indicated by a prefix, such as *a-symm., catena, cis, fac, mer, nido*, η or μ, as in *cis*-$[PtCl_2(NH_3)_2]$. Different forms of *polymorphic compounds* are identified by the crystal system (cub. = cubic, c. = body centered, f. = face-centered, tetr. = tetragonal, o-rh = orthorhombic, hex. = hexagonal, trig. = trigonal, mon. = monoclinic, tric. = triclinic), as in AuCd(cub.).

A general denotation of *crystalline phases* of variable composition (see Nomenclature, sect. II I) can be indicated by the symbol ≈ (read "circa") placed in front of or over the formula, e.g. ≈FeS, C uZn. If the variable composition of a phase is due to partial or complete substitution, the atoms or groups which can substitute for each other are indicated in parentheses, separated by a comma. Thus (Cu,Ni) indicates the entire range from pure copper to pure nickel. This system is more complete if the composition variables are indicated in the formula. For example, the homogeneous phase between copper and nickel can be indicated by Cu_xNi_{1-x}, and the phase between LiCl and $MgCl_2$ by $Li_{2x}Mg_{1-x}Cl_2$. The F. show that the total number of atoms in the unit cell is constant. A certain composition of the phase can be indicated by giving the actual value of x in parentheses directly after the F., e.g. $Li_{4-x}Fe_{3x}Ti_{2(1-x)})O_6$ ($x = 0.35$). An empty space in the structure (symbolized by $\square$) and interlattice spaces (symbolized by $\triangle$) can be described by placing these symbols after the element symbols, as in the γ-modification of Fe_2O_3 $(Fe_{8/3}{}^{III}\square_3)O_4$. Spaces in lithium magnesium chloride are indicated by $Li_{2x}Mg_{1-x}\triangle_{1-x}Cl_3)$; and a silver bromide crystal with Frenkel defects by $(Ag_{1-\delta}\square)$. $(Ag_\delta.\triangle)Br$. Electrons and positive holes associated with an excess of positive or negative charge in a field are indicated as e^- and v^+, respectively. To indicate that two types of defect occur simultaneously, the symbol $\between$ is used, as in iron(II) oxide with an iron deficiency:

$$(Fe^{II}_{1-3x}Fe^{III}_{2x} | \square \between \triangle \text{ cat }_{3x})O.$$

Formula unit: see Formula mass.

Formula weight: see Formula mass.

Formula mass: the relative formula mass, formerly called the ***formula weight***, is symbolized F_r. It is the sum of the relative atomic masses of all atoms which constitute the substance formula of a compound. The term F. is applied to substances which do not consist of discrete molecules, such as ionic, metallic and macromolecular substances, or those with unknown structural units. The "unit" of such a non-molecular substance is the *formula unit*, that is, the sum of the atoms indicated by the chemical formula; this formula usually indicates the stoichiometry in terms of the smallest possible integers. The relative F. is thus the mass of a formula unit relative to the atomic mass unit (see Atomic mass).

Formula reaction: the amount of substance reacting according to the stoichiometric reaction equation. For example, the elemental F. for the reaction $N_2 + 3H_2 \rightleftharpoons 2NH_3$, $\Delta_R H = -92.4$ kJ mol^{-1} is 1 molecule of nitrogen reacting with 3 molecules of hydrogen to form two molecules of ammonia. The molar F. indicates the reaction of the corresponding molar amounts, and is measured in moles. Heats of reaction are given per mole F.; for example, $\Delta_R H = -92.4$ kJ mol^{-1} is the reaction enthalpy for formation of two moles NH_3. Reaction rates can also be defined as the rate of the F. (see Kinetics of reactions).

Formyl-: a term for the atomic group -CHO in a molecule, for the radical CHO· and for the cation CHO^+.

Formylation: preparative methods for introducing a formyl group, -CHO, into an organic compound. The reaction is usually an electrophilic substitution, in which the reactive reagents are essentially derivatives of formic acid. Some typical examples are the Vilsmeier-Haack reaction (see), the Gattermann reaction (see) and the Gattermann-Koch synthesis (see).

Formylbenzene: same as Benzaldehyde (see).

Formyl group: the functional group -CHO, which is present in aldehydes, formic acid and some derivatives of formic acid. In an aldehyde, the F. may also be termed the ***aldehyde group***.

Formylsulfisomidin: see Sulfonamides.

Formyl violet: same as Acid violet (see).

Förster cycle: an indirect method for determining the equilibria of excited species, for example, for determining pK_a values for excited molecules. It is based on the thermodynamic data of the ground state (e.g. the pK_a value) and spectroscopic data on the undissociated compound and the ion of the dissociated compound.

Fortrel: see Synthetic fibers.

Fosetyl-Al: see Fungicides.

Fosfestrol: the tetrasodium salt of diethylstilbestrol *O,O'*-diphosphoric acid ester developed as a Prodrug (see) of diethylstilbestrol. It is used as a cytostatic with prostrate carcinoma. The active form is diethylstilbestrol, which has estrogen activity.

Fosfomycin: an antibiotic; chemically, a propylene oxide phosphonic acid. F. is obtained from cultures of *Streptomyces fradiae, viridochromogenes* and *wedmorensis*; it inhibits biosynthesis of the bacterial cell wall.

H_3C — O — PO_3H_2

Fourier synthesis: see X-ray structure analysis.

Fourier spectrometers, ***Fourier transform technique***: A technique for taking spectra introduced around 1960 which is fundamentally different from conventional spectroscopy (see Spectral instruments). The spectra are not measured one point at a time at individual wavelengths, wavenumbers or frequencies, but instead the sample is irradiated with polychromatic radiation and the entire range of frequencies is recorded at once. The result is not a series of signals at different wavelengths, but an interferogram made by superposition of many frequencies. The Fourier transform is a mathematical method with which it is possible to sort out the interferogram into its components. These can then be represented in the form of a traditional spectrum, in which the signal intensity is plotted against position.

The method requires a computer and a spectroscopic device which is very different from conventional spectrometers; for example, no dispersive component (grating, prism) is needed.

The main advantages of F. over conventional spectrometers are that the time required to record a spectrum is much shorter, and that the signal-noise ratio is improved. It is convenient to use F. in spectral regions where intensity is limited, such as the far infrared (see Infrared spectroscopy), and in ^{13}C-NMR (see NMR spectroscopy). However, there are also applications in other types of spectroscopy such as Ion-cyclotron resonance (see) or Microwave spectroscopy (see). When F. is used in NMR spectroscopy, the sample is irradiated at short intervals with a radiofrequency impulse, which contains all excitation frequencies (***Fourier pulse-transform technique***).

Four-membered heterocycles: cyclic carbon compounds with four atoms in the ring, one or two of which are heteroatoms. Some well known examples are oxetane, 1,2-dioxetane and azetidine.

FP: see Ignition.

Fr: symbol for Francium.

Fractional precipitation: see Precipitation; see Fractionation.

Fractionation: 1) in general, stepwise separation of the components of a mixture under certain conditions of temperature, pressure or concentration. The separated portions are called ***fractions***. F. is applied with all separation processes, especially distillation, crystallization, condensation, sublimation and sorption processes.

2) In macromolecular chemistry, the separation of a polydisperse substance into various fractions of more or less uniform molecular masses. The uniformity of a fraction depends on the resolution of the separation process.

Equilibrium processes:

a) *Fractional precipitation*. A precipitating solvent (which must be miscible with the solution solvent) is added to a solution of polymer until cloudiness is observed. After a time, this settles as a second liquid phase which contains the higher-molecular-mass polymer. It is removed, and more precipitating solvent is added to the supernatant. This process is repeated until a series of fractions is obtained; these can be used to determine the distribution function. The mass of the precipitated polymer and the average molecular mass of the polymer are determined, the molecular mass by one of the standard methods (viscosimetry, osmometry).

b) *Methods in which fractions are not separated*. These methods are rapid and provide a qualitative impression of the distribution function. In *turbidimetric F.*, the turbidity is measured as a function of the amount of precipitating agent added. In *volumetric F.*, the volume of the resulting second phase is determined as a function of the added precipitant. In *gravimetric F.*, a balance pan is suspended just over the bottom of the precipitation vessel. The precipitate settles onto it, and the amount is recorded as a function of the added solvent.

c) *Solvent fractionation* is based on the fact that in a homologous series of polymers, under otherwise equal conditions, the solubility of the polymer decreases with increasing degree of polymerization. If a polymolecular substance is first exposed to a poor solvent, only the low-mass fractions will dissolve. If it is then exposed to a better solvent, higher molecular-mass fractions will dissolve. The more finely the solvent properties are graduated, the better the separation can be. In practice, the polymeric substance is applied to an inert carrier and this is added to a graduated series of solvent mixtures. After a time, the solvent is removed (e.g. by evaporation) and the residue is weighed. In the Fuchs method, the substance is placed on a metal foil, usually aluminum.

d) *Determination of the sedimentation equilibrium in the Ultracentrifuge* (see).

e) The *distribution process* is based on the fact that any substance will be distributed between two mutually immiscible solvents in proportion to its relative solubilities in them. A similar distribution occurs between liquid and solid solvents (see Zonal melt process).

f) *Gradient elution* These methods can depend on temperature (thermoextraction) or pressure gradients (see Distribution chromatography).

Adsorption methods are based on the fact that the adsorption to suitable adsorbants depends on the molecular mass. This makes possible the separation, for example, of polyesters in benzene solution on urea columns (see Chromatography).

F. of very high polymers can be done by Permeation chromatography (see). Field-flux fractionation is suitable for some mixtures and for high molecular masses (up to 10^{12} g mol^{-1}).

Kinetic methods:

a) *Measurement of diffusion rates* depends on the fact that these are inversely proportional to the molecular mass; see Diffusion.

b) *Use of membranes with graduated porosities*. Certain membrane materials can be used as ultrafilters, or the time course of osmotic pressure can be measured.

c) *Utilization of thermodiffusion*. There are separation tube methods analogous to isotope separation in a Clusius spearation tube. Statistical methods are based on the Soret effect, which is a partial separation of molecules of different size within a solution.

d) *Rates of solubilization* in various solvents are also dependent on molecular mass and can be measured.

e) *Measurements in the ultracentrifuge* are based on the sedimentation of particles suspended in a liquid; the centrifugal force is proportional to the mass of the particle. There are various methods: determination of the sedimentation rate, of the sedimentation coefficients, or the Archibald method, which considerably shortens the otherwise time-consuming ultracentrifuge studies.

Fragmentation: ionic 1,2-elimination in which the β-substituent is a group rather than a proton or other single atom: a-b-c-d-X → a-b + c=d + X. The electrofugic group must be stabilized by inductive, resonance and/or hybridization effects, while the nucleofugic groups are the same as in normal 1,2-elimination; however, they become more highly unsaturated in F. (formation of alkenes, alkynes, azomethines, nitriles, carbonyl compounds, carbon dioxide, carbon monoxide or nitrogen). F. is often initiated by the action of a base, and can occur by either a synchronous or an asynchronous mechanism. The stereochemical requirements for a synchronous F. are the same as for E2 elimination.

Francium, symbol ***Fr***: a radioactive element from Group Ia of the periodic system (see Alkali metals); *Z* 87, 20 known isotopes with mass numbers from 204 to 224 (except 216) with half-lives from a few miliseconds to a few minutes. The longest-lived isotope is ^{223}Fr, with a half-life of 22 minutes; only this isotope has been detected in nature. The valence of Fr is I, m.p. 27 °C, b.p. 677 °C. The total amount of Fr on earth is estimated to be 30 g. F. is formed in a minor pathway of natural actinium decay; the main pathway is β^- decay to thorium, but about 1% of actinium nuclei undergo α-decay to F., which in turn forms radium by β emission: ^{227}Ac(-, $_2^4$He)^{223}Fr(-, β^-)^{223}Ra. The F. isotope 221 is a member of the radioactive decay series of neptunium, and decays with a half-life of 4.8 minutes to astatine: ^{221}Fr(-, $_2^4$He)^{217}At. Very small amounts of F. can be synthesized by bombarding thorium with protons, or radium with neutrons. The chemical behavior of F. has been little studied. Before its discovery, it was known as eka-cesium, and as expected, it resembles cesium. For example, it forms an insoluble perchlorate $FrClO_4$ and a hexachloroplatinate, Fr_2PtCl_6.

F. was discovered in 1939 by M. Curie as a minor product of natural actinium decay, and was named in honor of her native land.

Franck-Condon principle: a statement that during the time of an electron transition, the relative positions of the atomic nuclei in the molecule are unchanged. Since transitions from one electronic state to another are very much more rapid ($\approx 10^{-15}$ s) than nuclear vibration (period about 10^{-13} s), the atomic distance remains essentially constant during the electronic transition. The F. explains the possibilities for changes in the vibrational quantum number (see Infrared spectroscopy) during an electronic transition.

Frangula emodin: see Anthraglycosides.

Frasch process: see Sulfur.

Fraunhofer lines: see Spectral analysis.

Free energy: a thermodynamic parameter of state symbolized by a (Helmholtz free energy); it characterizes the ability of a system to perform work and is defined as $a = e - Ts$, where e is the internal energy, T is the temperature in K and s is the entropy. If the system undergoes an isothermal process associated with a change in energy Δe, Δa is the energy which can be converted to work if the process is reversible. The remainder (the "bound" energy) is converted to heat: $q_{rev} = T\Delta s$; $\Delta e = \Delta a + T\Delta s$. Here Δs is the change in entropy within the closed system.

Another parameter of state related to enthalpy is the Gibbs free energy (Gibbs free enthalpy) $g = h - Ts$. a is the more natural parameter to use for isochoric processes; g is more suitable for isobaric processes.

a and g depend on the external conditions, that is, on the state variables. For pure substances, the only possible kind of work is change of volume against a pressure, so the the differential equations $da = de - Tds - sdT$ and $dg = dh - Tds - sdT$ apply. In reversible processes, $de = dq_{rev} - pdv$; $dh = dq_{rev} + vdp$; and $dq_{rev} = Tds$. Combining the above equations for da and dg, we obtain: $da = -sdT - pdv$ and $dg = -sdT + vdp$.

When substances are mixed, the F. of the mixture is not equal to the sum of the F. for the pure components. Per mole of substance i in the mixture, there is a partial molar Helmholtz or Gibbs F. of mixing $\bar{A}_i = \mu_i = RT \ln x_i$, or $\bar{G}_i = \mu_i = RT \ln x_i$ (see Partial molar parameters). The two parameters are equal and are also called the Chemical potential (see) μ_i. x_i is the molar fraction of substance i in the mixture, R is the general gas constant, and T is the temperature of the mixture. For the total mixture, $da = -sdT - pdv + \sum_i \mu_i dn_i$ and $dg = -sdT + vdp + \sum_i \mu_i dn_i$ (fundamental Gibbs equations).

A process will occur spontaneously so long as the system is capable of doing work, that is, if $\Delta a < 0$ or $\Delta g < 0$ (see Thermodynamics, second law; Equilibrium). Since the two types of F.e. are parameters of the system itself, they are used for the thermodynamic derivation of equilibrium relationships.

Free enthalpy: abb. g, see Free energy.

Free enthalpy function: the relatively seldom used thermodynamic function $(g - h)/T = -(\Phi + h/T)$, where g is Gibbs free energy, h is enthalpy, Φ is Planck's function, h/T is the enthalpy function and T is the absolute temperature.

Free path: see Kinetic gas theory.

Free valence: see Hückel method.

Freeze drying: see Drying.

Freezing: the conversion of a vapor or liquid to the solid state by cooling it below its sublimation point (desublimation) or freezing point. When a solution is cooled below the freezing point of the pure solvent, this can begin to freeze out in pure form, which increases the concentration of the solute in the remaining liquid phase. The temperature drops during this process until the eutectic point is reached. The process of F. is then completed at the constant eutectic temperature, which is the minimum for this phase change.

Gases are dried in cold traps or adsorption traps where water vapor and/or other condensable components are removed by cooling to - 20 to - 70 °C.

Freezing curve: see Melting diagram.

Freezing-out technique: see Desalinization.

Freezing point: see Melting point.

Freezing point lowering, ***melting point lowering***: the lowering of the freezing point of a solution compared to that of the pure solvent. For ideal dilute solutions, Raoult's second law (see Raoult's law) applies: $\Delta T = T_{0,1} - T = E_F \cdot \bar{m}_2$, where $E_F = RT_{0,1}^2 M_1 / \Delta_p H$, T and $T_{0,1}$ are the freezing points of the solution and pure solvent, $\bar{m}_2$ is the molality of the solute and E_F is the ***cryoscopic constant***, which has a definite value for each solvent. It depends on the freezing point $T_{0,1}$, the molar mass M_1 and the molar enthalpy of melting, $\Delta_p H$ of the solvent. This equation applies only when pure solvent crystals are formed on freezing, that is, when mixed crystals do not form.

Cryoscopy is the determination of molar masses of solutes on the basis of the F. Boiling point elevation (see) is closely related to F.

Freezing range: see Glass state.

Freezing temperature: same as Glass transition temperature (see).

Freiberg solubilization: see Tin oxides.

Frenkel defects: see Crystal lattice defects.

Frequency factor: see Arrhenius equation.

Frequency sweep: see NMR spectroscopy.

Freundlich isotherms: see Adsorption.

Frictional corrosion: see Wear corrosion.

Friedelane: see Triterpenes.

Friedel-Crafts reaction: a reaction, first carried out by C. Friedel and J.M. Crafts in 1877, for the alkylation and acylation of aromatic hydrocarbons in the presence of anhydrous aluminum chloride.

1) ***Friedel-Crafts alkylation*** is a method for introducing alkyl groups into aromatic hydrocarbons or their derivatives, or into heteroaromatics, using alkyl halides, alkenes, alcohols or esters with inorganic or organic acids in the presence of Lewis acids (such as aluminum chloride, iron(III) chloride, tin(IV) chloride, boron trifluoride or zinc chloride) or proton acids (e.g. hydrogen fluoride, sulfuric acid or phosphoric acid):

$+C_2H_5X\ (AlCl_3)$, $-HX$; $+CH_2{=}CH_2\ (AlCl_3)$ → C_6H_5–CH_2–CH_3

Friedel-Crafts alkylation is an electrophilic substitution. The electrophilic reagent in the reaction of alkyl halides is either a complex of the Lewis acid and the alkyl halide or a free carbenium ion:

$$R{-}Cl + AlCl_3 \rightleftharpoons \overset{\delta+}{R}{-}\overset{\delta+}{Cl}{-}\overset{-}{Al}Cl_3 \rightleftharpoons R^+ + AlCl_4^-$$

$$\xrightarrow[-AlCl_4^-]{+C_6H_6} [C_6H_6(H)(R)]^+ \xrightarrow{-H^+} C_6H_5R$$

Since alkyl groups are first-order substitutents – they increase the activity of the aromatic system and thus promote the electrophilic second substitution – the monoalkyl aromatic compounds formed as primary products react further, sometimes forming highly alkylated products which are often difficult to remove. Another disadvantage of this F. is that when intermediate carbenium ions appear, rearrangement products can be formed. For example, benzene and 1-bromopropane form mainly cumene, and not n-propylbenzene as expected. Friedel-Crafts alkylation is very important in industry, especially with alkenes as alkylating agents, in the synthesis of ethylbenzene, cumene, alkylphenols and butylnaphthalene.

2) ***Friedel-Crafts acylation***: process for producing aromatic or aliphatic-aromatic ketones from aromatic hydrocarbons and their substituted derivatives, heteroaromatic compounds, carbonyl halides and anhydrides and carboxylic acids in the presence of Lewis acids:

$$C_6H_6 + R{-}CO{-}Cl \xrightarrow[-HCl]{(AlCl_3)} C_6H_5{-}CO{-}R$$

Friedel-Crafts acylation is also an electrophilic substitution reaction on the aromatic system. The electrophilic acylation reagent, formed, e.g. from the acyl halide and aluminum chloride, is a 1:1 complex of the reactants, or an acyl cation:

$$R{-}C(=O){-}Cl + AlCl_3 \rightleftharpoons R{-}C(Cl){=}\overset{+}{O}{-}\overset{-}{Al}{-}Cl_3 \rightleftharpoons R{-}\overset{+}{C}{=}O\overset{-}{AlCl_4}$$

The formation of the σ-complex and the subsequent cleavage of the proton is the same as in the alkylation mechanism. Since acyl groups are second-order substituents, a second substitution is not possible.

The following are closely related to Friedel-Crafts acylation: the Gattermann-Koch synthesis (see), the Gattermann reaction (see), the Vilsmeier-Haack reaction (see), the Houben-Hoesch reaction (see) and the Fries rearrangement (see).

Friedländer synthesis: a synthesis of quinoline by condensation of 2-aminobenzaldehyde or 2-aminoacetophenone with ketones which have at least one methyl or methylene group in the α-position:

$$2\text{-}(R^1CO)C_6H_4NH_2 + H_2C(R^2){-}C(=O){-}R^3 \xrightarrow{-2H_2O} \text{quinoline (4-}R^1\text{, 3-}R^2\text{, 2-}R^3)$$

The F. can be carried out in the presence of acid or base. When β-ketocarboxylic acids are used, the cyclization occurs under very mild reaction conditions, with simultaneous decarboxylation.

Fries rearrangement: rearrangement of phenol esters to acylphenols in the presence of aluminum chloride, boron trifluoride, zinc chloride, etc. At reaction temperatures below 100 °C, the products are mainly 4-acylated compounds; above 150 °C mainly 2-acylphenols are formed:

$$C_6H_5{-}O{-}CO{-}R \xrightarrow{(AlCl_3)} \begin{cases} <100\,^\circ C: & R{-}CO{-}C_6H_4{-}OH\ (4\text{-}) \\ >150\,^\circ C: & 2\text{-}(R{-}CO)C_6H_4{-}OH \end{cases}$$

The phenol esters of aliphatic carboxylic acids can generally be rearranged more easily than the corresponding aromatic esters. Solvents are not necessary for F., although in many cases the rearrangement is carried out in nitrobenzene, carbon disulfide or chlorobenzene under milder conditions.

Frit: a silicate, glassy or glassy-crystalline sintering or fusion product made of a mixture of glass-forming and glass affecting materials (e.g. quartz, feldspar, clays, borax, alkali metal and alkaline earth metal carbonates). A granulate is made by quenching the product in water or on a metal substrate. F. are important in the production of enamels, fusion colors and stoneware glazes, and as filter materials in the form of compressed fine granulates.

Fröhde's reagent: see Alkaloids.

Frontal technique: in chromatography, a technique in which the sample is continuously applied.

Frontier orbital: see Woodward-Hoffmann rules.

Front octane rating: see Octane rating.

Frozen state: see Equilibrium.

Fru: abb. for D-fructose.

Fructose, ***fruit sugar, levulose***, abb. ***Fru***: the most important ketohexose (see Monosaccharides). Commercial β-D-fructopyranose is a white, crystalline powder which is readily soluble in water and moderately soluble in ethanol; $[\alpha]_D^{20}$ in mutarotational equilibrium, - 91 to - 93°. D-F. is found in free form in honey and, in association with glucose, in most fruits. In the form of furanosides, it is a component of sucrose, raffinose and various polysaccharides (fructans), and can be obtained by hydrolysis of sucrose or inulin. F. tastes sweeter than sucrose. In vivo, it is converted by hexokinase to fructose 1-phosphate, which can be converted to glucose 1-phosphate by an isomerase. F. is used as a test for liver function, in protective therapy for the liver, and as a sugar substitute (see Sweeteners).

β-D-Fructopyranose

β-h-Fructosidase: same as Invertase (see).

Fruit sugar: same as D-Fructose (see).

FSH: see Follitropin.

FSP: abb. for flame spectrophotometry.

FT technique: abb for Fourier transform technique.

Fuc: abb. for fucose.

Fuchsin, ***rosaniline, magenta***: a triphenylmethane pigment. F. forms greenish-yellow crystals with a metallic sheen; these dissolve slowly in water and alcohol to give an intense red color. F. is used mainly to color paper and toys.

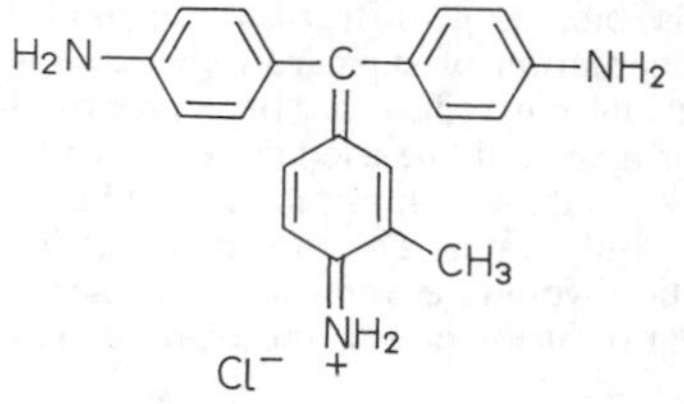

Fuchsin sulfurous acid was formerly used as a reagent to detect aldehydes. This color reaction is based on the formation of a colorless addition compound between F. and sulfurous acid; in the presence of an aldehyde, the addition compound is destroyed and the red color of F. reappears.

F. was first prepared in 1858 by A.W. von Hofmann; after mauve, it was the second most important pigment.

Fuchsone: the parent compound of a series of acid triphenylmethane pigments, the fuchsone pigments.

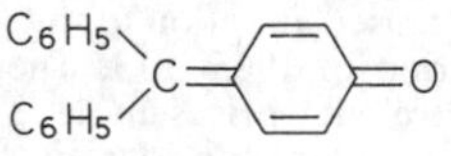

Fuchsonimine pigments: synthetic triphenylmethane pigments in which fuchsonimine acts as chromophore. F. include fuchsin.

Fucolipids: glycosphingolipids (see Sphingolipids) which contain fucose. This class of compounds includes, for example, the Blood group substances (see).

Fucose: abb. ***Fuc, 6-deoxygalactose***: a 6-deoxysugar. L-F. is a component of vegetable gums and mucins, hemicelluloses, blood group substances, fucosans (polysaccharides found in algae) and some glycolipids, the fucolipids. D-F. is found in some plant glycosides.

Fuel: a substance which, on reaction with oxygen, releases heat. The chemical energy of some F. can be directly converted to mechanical work; these are used in engines of various kinds. Most F. are organic, consisting mainly of carbon and hydrogen. The carbon accounts for most of the heat of combustion, and the hydrogen determines the ease of ignition and flammability. F. may also contain oxygen, nitrogen and sulfur, as well as smaller amounts of water and minerals which are converted to ash by combustion.

The chemical energy of some F. can be converted directly to mechanical energy in an engine or turbine. In general, the oxygen is obtained from the air, although pure oxygen is used in rocket engines. These F. include the gases propane, butane and, more recently, hydrogen, the liquids gasoline, petroleum, alcohols, and hydrazine, and solid coal dust. The quality of a F. for a combustion engine is determined essentially by its rate of vaporization and combustion, its ignition properties, heating value and anti-knock characteristics.

Heating values and tons coal equivalent of natural fuels

Fuel	Minimum heating value [kJ/kg]	TCE
Wood	$14.70 \cdot 10^3$	0.500
Peat	$12.60 \cdot 10^3$	0.429
Crude brown coal	$7.96 \cdot 10^3$	0.271
Brown coal briquets	$20.10 \cdot 10^3$	0.686
Brown coal coke	27.70	0.951
Hard coal	29.30	1.000
Gasoline	$43.60 \cdot 10^3$	1.486
Diesel fuel	$42.70 \cdot 10^3$	1.457
Heating oil (light)	$42.70 \cdot 10^3$	1.457
Petroleum	$41.87 \cdot 10^3$	1.443
Natural gas	$32.30 \cdot 10^3$	1.090
City gas	$16.10 \cdot 10^3$	0.550
Liquefied gas	$45.90 \cdot 10^3$	1.564

Carburetor F. are those used in engines, especially Otto cycle engines, with carburetors. The most important of these is Gasoline (see); it should burn rapidly and completely, without any residues and without causing knocking. Its resistance to knocking is given by its Octane rating (see); Antiknock agents (see) or components with high octane ratings, e.g. aromatics, *tert*-butyl methyl ether (MTBE), diisopropyl ether and alcohols, are added to gasoline to prevent knocking. Addition of 10 to 20% MTBE to gasoline increases its octane rating by 2 to 5 units. In the early days, petroleum fractions in the boiling range of 45 to 200 °C (straight-run gasoline) with octane ratings of 55 to 65 were adequate for nearly all engines. Modern four-cylinder engines, however, require octane ratings of 90 to 95. This level can be achieved only by modern processes of petroleum refining (see Reforming, Isomerization, Hydrocracking, and Cracking). High-quality F. can also be made synthetically, by alkylation and oligomerization of lower hydrocarbons, or by conversion of synthesis gas (see Fischer Tropsch synthesis) or methanol (see Mobil Oil process) to gasolines.

Most commercial gasolines are mixtures of straight-run gasoline with more knock-resistant branched or aromatic hydrocarbons and antiknock agents.

The emission of CO, NO_x, lead and hydrocarbons in vehicle exhaust has become a significant environmental problem. The amount of lead in gasoline is therefore limited by law in many countries.

Amount of lead permitted in gasoline

Country	g lead/l
Belgium	0.84
West Germany	0.15
Czechoslovakia	0.70
Denmark	0.70
East Germany	0.40
France	0.64
United Kingdom	0.55
Austria	0.40
Soviet Union	0.40

Catalytic afterburners have been developed to remove most of the pollutants. Since the catalysts are poisoned by lead, vehicles equipped with these must use unleaded F. Since 1975, new cars sold in the USA have been required (in many states) to have catalytic converters. In *single-bed converters*, carbon monoxide and unburned hydrocarbons are oxidized to carbon dioxide and water. In *double-bed converters*, NO is also removed according to the equation NO + CO $\rightarrow CO_2 + 1/2\ N_2$. The catalysts are generally 0.1 to 0.2% platinum or palladium on aluminum oxide.

Aviation gasoline differs from automobile gasoline in having a lower boiling limit (about 145 °C) and a higher octane rating ($\geqq 100$). ***Racing gasolines*** are special mixtures with low-boiling, very knock-resistant components with high heats of vaporization. They usually contain more than 50% ethanol (octane rating 99) or methanol (octane rating 98). Other additives may include acetone, glycols and ether.

Gaseous F. (liquefied petroleum gas, abb. LPG) is carried as Liquefied gas (see) in steel bottles or special tanks in the vehicle. Because of its high energy content and its lack of polluting emissions, hydrogen is being considered as a possible F. for motor vehicles.

Diesel F. (diesel oils) are also mixtures of hydrocarbons, but they are more dense than gasoline and boil at higher temperatures (about 180 to 370 °C). The starting materials for diesel F. are petroleum fractions and residues from refining higher-boiling fractions (see Hydrocracking, Cracking, Visbreaking). An important criterion for quality of diesel F. is its ignitability, which is given by the Cetane rating (see). The cetane ratings of typical diesel F. are between 30 and 60. Ignition accelerators can be added to improve the ignitability of diesel F.; these are high-energy, oxygen-releasing compounds such as alkyl nitrites, nitrates or peroxides.

To prevent blockage of the fuel lines and sieves of diesel vehicles at low temperatures, diesel F. are partially deparaffinized (see Paraffin), or medium petroleum fractions (see Kerosene) which boil between 180 and 240 °C are added. The content of aromatics must not exceed 25%, because high amounts would lead to formation of residues upon combustion.

Rocket F. consist of a F. component and an oxygen carrier; they can be either solid or liquid. *Double-base F.* are solid F. mixtures of glycerol trinitrate and cellulose nitrate; the molecules of these compounds contain both oxygen and a combustible portion. *Composite F.* are solid mixtures of two substances, one oxygen carrier (e.g. ammonium nitrate or perchlorate) and one combustible (e.g. high-polymer plastics such as polyurethane). Liquid rocket fuels are used in systems in which the F. and the oxygen carrier are stored separately. The F. components may be hydrogen, hydrocarbons, alcohols, hydrazine, amines or boranes. The oxygen carrier can be oxygen, ozone, hydrogen peroxide, nitric acid, nitrogen peroxide, tetranitromethane or oxygen-containing fluorine compounds. Some liquid rocket F. ignite spontaneously on contact with the oxygen component (*hypergols*); examples are hydrazine and nitric acid. The two components must therefore be stored separately. 80% hydrogen peroxide can be used as an independent F. (cold F.) if it is split by a catalyst (heavy metal salt) into oxygen and superheated steam (*catergols*). It is also possible to use powdered aluminum, beryllium or lithium with liquid oxygen carriers (*lithergols*).

Alternative F.. Because of the increase in price of petroleum and the limited supplies, many countries have established programs for production of non-petroleum F. At present, the most promising alternatives are synthetic hydrocarbons, alcohols and hydrogen. The most interesting of the alcohols is ethanol, since it can be produced from carbohydrate-containing plant materials (biomass). In Brazil, for example, ethanol from fermentation of sugar cane bagasse is already used as an alternative F. It has been proposed for years that hydrogen could be used directly as a F. It has a high energy content, ignites readily and burns cleanly, producing only water. The raw materials for hydrogen production would be water and the energy supplied by nuclear or fusion power; these are seen as inexhaustible reservoirs.

Historical. F. were developed concurrently with the engines which burn them. At the beginning of this development, volatility was the only criterion for the quality of a F., but now there are many criteria of quality, applicable not only to the performance of the engine, but also to economy, availability and environmental impact of the F. From 1885 to 1915, the interaction between the engine and the F. was the main consideration. Starting around 1915, and until about 1965, the performance of the engine was the center of attention. As a result of the great increase in production of motor vehicles starting in the 1960's, environmental problems caused by the large numbers of engines in centers of population caused attention to shift to these problems in the period of 1965-1973. The sudden increase in world petroleum prices in 1973 made economy a much more important consideration from that time on.

Until 1910, carburetor technology required use of light gasoline with components boiling below 85 °C. In order to cover the rising demand for F., the top of the boiling range was raised in less than 30 years from 85 °C to 230 °C. This development led to many difficulties, such as poor starting, knocking, deposits and dilution of the engine oil by the F., which had to be resolved. In 1921, an "enriched F." was introduced in Germany. This was the first F. not based on petroleum; it consisted of 50% benzene, 25% tetralene and 25% ethanol.

Fuel cell: an electrochemical current source in which the reactants are separately and continuously conducted to the two electrodes. The electrochemical reaction and the removal of products also occur continuously. So long as the reactants are supplied, F. can theoretically function indefinitely, providing electricity until the components wear out. They do not need to be regenerated, as storage batteries do.

The name F. was coined because these systems are in principle suitable for deriving electricity directly from the combustion of the traditional fuels, coal, petroleum or natural gas, with added oxygen.

A requirement for the functioning of a F. is that the oxidation of the fuel and the reduction of the oxidizing agent must occur rapidly. This requirement limits the number of fuels and oxidants which are suitable for use in F. Some fuels which may be used are hydrogen, formaldehyde, ammonia, carbon monoxide, natural gas and methane, or liquids such as hydrazine, methanol or ethanol. Possible oxidants include oxygen or air, fluorine, chlorine, nitric acid, hydrogen peroxide and bromine. The most promising of this selection are hydrogen as fuel and oxygen as oxidant. The ***hydrogen-oxygen F.*** has been technically perfected and is used, e.g. in space vehicles, as a current source. The cell reaction in the hydrogen-oxygen F. is the highly exothermic combination of the hydrogen with the oxygen to form water, which is the sum of the electrode reactions for the anodic oxidation of the hydrogen and the cathodic reduction of the oxygen. In acidic medium, the equations are:

$H_2 \rightarrow 2\,H^+ + 2e^-$ (anode)
$1/2\,O_2 + 2\,H^+ + 2\,e^- \rightarrow H_2O$ (cathode)

$H_2 + 1/2\,O_2 \rightarrow H_2O$ (cell)

These equations indicate only the overall reactions; the actual process is much more complicated. The presence of a suitable catalyst, e.g. platinum, palladium nickel or a metal complex, is essential.

Fuel gases: combustible industrial gases and mixtures of gases which are either synthesized by conversion of solid or liquid fuels (see City gas, Generator gas, Water gas, Liquefied gases) or are of natural origin (see Natural gas).

Fugacity, symbol p_i^*: in thermodynamics, an hypothetical pressure of real gases introduced in analogy to the activity by G.N. Lewis in 1901 to make the ideal gas laws apply to real gases: $p_i^* = f_i^*\,p_i$. Here p_i is the partial pressure and f_i^* is the fugacity coefficient. If the real gas approaches ideal behavior, f_i^* goes to 1.

Fugu poison: see Tetrodotoxin.

Fullerenes, ***buckminsterfullerenes, buckyballs***: large molecules of carbon in the form of hollow spheres. F. are formed when graphite is sublimed in a helium atmosphere, and have the formulas C_{60}, C_{70}, C_{76}, C_{84}, C_{90} and C_{94}. The F. are named after Buckminster Fuller, an architect who designed polyhedral domes based on hexagonal and pentagonal faces; the structure of the F. is reminiscent of these domes. The most thoroughly studied F. is C_{60}, which has a diameter of 700 pm and icosahedral symmetry. Its appearance is similar to that of a soccer ball, as it is made up of 20 hexagons and 12 pentagons. All the C atoms in it are equivalent; the mean C-C distance is 141 pm, which is almost the same as in graphite. As in graphite, each C atom is sp^2 hybridized, and forms a σ-bond with each of its 3 neighbors. Since the atoms lie on the surface of a sphere, the mean sum of the angles is reduced to 348°, and the C atom forms a flattened pyramid with its three neighbors. Both surfaces of the sphere are covered by π-electron clouds. Fullerite crystals can be extracted with benzene from the soot formed by sublimation of graphite and subsequent condensation. The crystals are platelets with a metallic sheen and low density (1.65 g cm^{-3}), in which the C_{60} units are in a close-packed arrangement. The F. are a new, third modification of carbon.

The C_{60} molecule is capable of many reactions, and these are currently the object of intense research. It reacts with the alkali metals K and Rb, the halogen fluorine, free radicals and Grignard reagents. C_{60} molecules can incorporate other atoms into the interior of the sphere, such as lanthanoids. The reactions lead to amazing properties. Although C_{60} is an electrical insulator, K_3C_{60} is a superconductor, while K_6C_{60} is again an insulator. When F. react with organic reducing agents, ferromagnetism is generated at low temperatures.

Fulminates: derivatives of fulminic acid, HCNO. Ionic F. contain the linear, resonance-stabilized anion $[CNO]^-$, which is one of the Pseudohalides (see). ***Sodium fulminate***, NaCNO, crystallizes in colorless needles which are isomorphic to sodium cyanate and sodium azide. Friction or heating causes them to explode violently. Some other important metal fulminates are Mercury fulminate (see) and Silver fulminate (see). With transition metals, the fulminate anion forms complexes in which the ligand forms coordinate bonds via the carbon, M-CNO.

Fulminating gold: a dirty yellow or brown, amorphous, explosive gold compound of unknown structure. It is formed by the reaction of ammonia with tetrachloroauric(III) acid or gold(III) hydroxide.

Fulminic acid: HCNO, a colorless, unstable compound, m.p. - 10 °C, which forms a weak acid in water ($pK \approx 5$). F. is unstable even at low temperature. It is made by treating an aqueous, ice-cooled solution of sodium fulminate with dilute sulfuric acid. The Fulminates (see) are of particular importance. F. was discovered in 1823 by J. von Liebig.

Fulvenes: yellow to red hydrocarbons with cross-conjugated double bonds. They are derived from cyclopentadiene by condensation with aliphatic or aromatic aldehydes or ketones. Like the azulenes, F. are hydrocarbons with non-alternating conjugation systems, and they have significant dipole moments. For example, 6,6-dimethylfulvene ($R = R = CH_3$) has a moment of 1.44 D. F. have low resonance energies (about 50 kJ mol^{-1}) and are therefore rather reactive. They act as intermediates between aromatic and unsaturated aliphatic compounds. They are readily hydrogenated and undergo autooxidation and polymerization. On the other hand, they undergo "aromatic" substitution reactions on the five-membered ring.

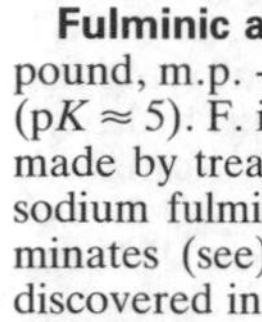

Fumarase: an enzyme which catalyses the formation of L-maleic acid from fumaric acid by addition of water. F. (M_r 194,000) consists of 4 identical subunits, contains 1784 amino acid residues, and does not require a cofactor for its activity.

Fumaric acid, ***trans-butenedioic acid, (E)-butenedioic acid***: an unsaturated aliphatic dicarboxylic acid which is the π-diastereoisomer of Maleic acid (see). F. crystallizes in colorless, monoclinic prisms; m.p. 300-302 °C (in a closed tube), sublimation at 200 °C. It is soluble in boiling water and alcohol, and practically insoluble in most organic solvents. F. is much less acidic than maleic acid. Its salts and esters are called *fumarates*. In free form, F. is present in many plants, e.g. fumitory (*Fumaria*), Iceland moss, mushrooms and lichens. Fumarate is also an important intermediate in the citric acid cycle.

F. is synthesized mainly by isomerization of maleic acid in the presence of hydrochloric acid, but it can also be made by industrial fermentation from glucose or starch. It is used in the synthesis of polyesters and alkyde resins, and as an additive for baked goods and drinks.

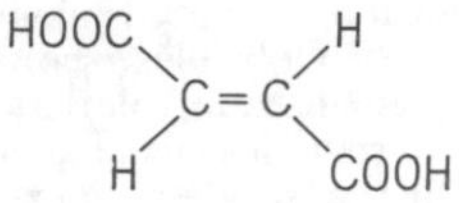

HOOC H
C=C
H COOH

Fumigant: a gaseous or readily vaporized preparation used as an Insecticide (see), Nematocide (see) and/or Rodenticide (see) to kill pests in soils or enclosed spaces; they are also used to control vermin in dwellings and to protect stored goods.

Functional group, ***characteristic group***: in organic compounds, an atom or group of atoms which is not linked by a direct carbon-carbon bond (exceptions are -C≡N, >C=X, where X = O, S, Se, Te or NH, and -COOH). They undergo characteristic reactions. Important F. include the following groups: hydroxyl, -OH; mercapto, -SH; amino, $-NH_2$; -carbonyl, >C=O; carboxyl, -COOH; and sulfonyl, $-SO_3H$. For further discussion, see Nomenclature, sect. III B and III G.

Fungicides: compounds used to combat fungi; fungal diseases in human beings and animals are treated with Antimycotics (see).

F. are used mainly to protect plants, in which the large majority of diseases are due to pathogenic fungi (84% of the agriculturally significant plant diseases in Europe are caused by fungi, and the rest by bacteria and viruses.) Worldwide, the harvest losses due to fungi are estimated at 12% of the potential harvest.

F. may be applied to the leaves of the plant, the soil or the seeds (tubers, etc.) before planting. Leaf F. are applied as dusts or sprays to the green parts of the plant; soil F. are applied to the soil in liquid, dust or granulate form. The fungicidal effect in the soil occurs mainly via the vapor phase of the chemical compound, which diffuses through the soil air. Treatment of seeds, tubers or bulbs with F. prevents diseases of the sprouting plant due to organisms in the soil or in the seed (tuber, bulb). The seeds may be dipped in or wetted by an aqueous F. preparation or coated with a dust. The classic agents used in this way are based on organomercury compounds, but these are very toxic, and have been forbidden in some countries. Their use is subject to legal limitations in others. Combination preparations of less toxic organic F. with specific effects have been substituted.

F. may be classified as ***protective, curative*** or ***eradicative***, depending on the phase of development in which they act: before infection, during the incubation time or after the disease has become visible.

A new and very important line of development is the introduction of ***systemic F.*** (see Systemic pesticides). These are true chemotherapeutics, and protect new growth from infection. Some natural products with fungicidal effects (e.g. griseofulvin, cycloheximide, blasticidine, polyoxines) have systemic properties and have been used for a long time. The Phytoalexins (see) are natural systemic F.

Many chemical classes are represented among the F. Of the ***inorganic F.***, only copper salts and elemental sulfur are still used, and these in minor amounts. The historically important *Bordeaux mixture* (copper-potassium salts) was the first (1885) practical F. in vineyards. Preparations based on copper(II) oxide chloride, $3Cu(OH)_2 \cdot CuCl_2$ are used today on grapevines and potatoes. Polydisperse *sulfur* is used mainly against true mildew.

Organometallic F. are usually volatile and act in the gas phase. *Organomercury compounds* have a very broad action spectrum and are highly effective; they are mostly used to treat seeds. Arylmercury compounds, such as *PMA* (abb. for phenylmercury acetate, C_6H_5-O-CO-CH_3) are somewhat less toxic to human beings and animals than the highly toxic alkylmercury compounds such as *Panogen*® (methylmercury cyanoguanidine, CH_3-Hg-NH-C(:NH)-NH(CN). The most important *organotin F.* are *fentin hydroxide*, $(H_5C_6)_3$Sn-OH (m.p. 118-120 °C) and *fentin acetate*, $(H_5C_6)_3$Sn-O-CO-CH_3 (m.p. 122-123 °C). These are used as protective and curative leaf F.; they also act as insecticides through their anti-feeding activity (see Insect repellents). *TBTO* (abb. for bis-(tributyltin) oxide, $[(H_9C_4)_3Sn]_2O$) is an industrial F. and bacteriocide used to protect wood and other organic materials.

The ***dithiocarbamate F.*** (see) are broad-spectrum leaf F.

The ***haloalkylmercaptoimide*** F. release thiophosgene, which inactivates the thiol groups of the fungal cell. These F. are used, for example, in orchards and vineyards. Most of these compounds contain an *N*-trichloromethanesulfonyl group, $-SCCl_3$. An example is *captan* (m.p. 178 °C) which is synthesized by reaction of tetrahydrophthalimide with trichloromethanesulfenyl chloride (perchloromethyl mercaptan). *Folpet* (m.p. 177-180 °C) contains the analogous phthalimide derivative, *dichlorofluanide* (N,N'-dimethyl-N'-phenyl-N'-(fluorodichloromethylmercapto)sulfamide, $(H_3C)_2$N-SO_2-N(C_6H_5)-S-$CFCl_2$, m.p. 106 °C). It is used against botrytis (e.g. in strawberries).

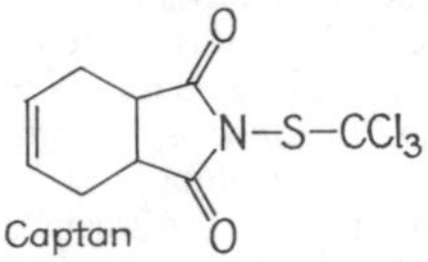

N–S–CCl_3

Captan

The ***quinone F.*** react with the SH groups of fungal proteins, sometimes also with amino groups; some of this group are used to treat seeds: *chloranil* (2,3,5,6-tetrachloro-1,4-benzoquinone, m.p. 292 °C), *dichlone* (2,3-dichloro-1,4-naphthoquinone, m.p. 193 °C) and *benquinox* (benzoylhydrazone or quinoximide, C_6H_5-CO-NH-N=C_6H_4=NOH, m.p. 195 °C.

Aromatic hydrocarbon F. with chloro, nitro, amino, hydroxy and/or alkoxy substituents can have very high activities, and can sometimes act selectively against fungal pathogens; some have a curative effect. They are used mainly as soil F. and to treat seeds and stored grain. Some examples are *PCNB*, pentachloronitrobenzene, m.p. 144 °C, *TCNB*, 2,3,5,6-tetrachloronitrobenzene, m.p. 99 °C, and *HCB*, hexachlorobenzene, m.p. 231 °C; the last of these is highly persistent, and its use has come into question. *PCP*, pentachlorophenol, m.p. 191 °C, *dichlorane*, 2,6-dichloro-4-nitraniline, m.p. 192-194 °C, *etridiazol*, 5-ethoxy-3-trichloromethyl-1,2,4-thiadizol, m.p. 19.9 °C, and diphenyl, C_6H_5-C_6H_5, m.p. 69-71 °C are other examples. Diphenyl is used to treat the skins of citrus fruits, to preserve them; it may be combined with 2-phenylphenol.

Dinitrophenyl ester F. are used to protect against and eradicate true mildew; they are also used as acaricides in orchards and vineyards, and on ornamental plants. Some examples are *dinocap* (see Acaricides, table) and *binapacryl* (2-*sec.*-butyl-4,6-dinitrophenyl β,β-dimethylacrylate, m.p. 69 °C). Most of the quinoline F. are derivatives of 8-hydroxyquinoline. A salt of sulfuric acid is formed; it is used in grape culture and soil fumigation; the copper salt is also used.

Some important ***phosphoorganic F.*** are: ***triamphos*** [(5-amino-3-phenyl-1,2,4-triazol-1-yl)-bis(dimethylamino)-phosphine oxide], a systemic F. which is also an insecticide and acaricide; *ditalimfos* (*N*-diethoxythiophosphorylphthalimide), used against true mildew in fruit, vegetable and ornamental culture; *carbophosphonate* (*O*-2,4,5-trichlorophenyl-*N*-α-carbethoxyethylamido-methylthiophosphonate), used for seed treatment, *pyrazophos* [(2-*O,O*-diethylthionophosphoryl)-5-methyl-6-carbethoxypyrazolo (1,5a)-pyrimidine], a systemic F., *fosetyl-Al* [aluminum-tris-*O*-ethylphosphonate] with preventive and curative effects against *Phytophthora* and *Peronospora* species.

A representative of the ***guanidine F.*** is ***dodine*** [$C_{12}H_{25}$-NH-C(:NH)-$NH_2 \cdot CH_3COOH$, m.p. 136 °C]; it is a curative agent against scab diseases in fruits. The ***carboxamide F.*** (see) and ***carboximide F.*** (see) are derivatives of carboxylic acids.

F. with systemic properties are to be found in the following groups (see separate entries on each): ***Benzimidazole F., Pyrimidine F., Azole F., Morpholine and Piperazine F.***, and ***acylalanine F.***.

Fungistatic: see Antimycotics.

Furalaxyl: see Acylalanine fungicides.

2-Furaldehyde: same as Furfural (see).

Furan: a heteroaromatic compound rich in π-electrons. F. is a colorless, readily ignited liquid with a chloroform-like odor; m.p. - 85.6 °C, b.p. 31.4 °C, n_D^{20} 1.4214. It is readily soluble in alcohol, ether, acetone and benzene, but is insoluble in water. It is stable in bases, but is decomposed by concentrated acids, which form esters with it. Dilute acids hydrolyse the ring, yielding 1,4-dicarbonyl compounds.

The oxygen atom is sp^2 hybridized, so that F. is a cyclic conjugated system; with six delocalized electrons, it is aromatic. Its chemical reactions are those of a 1,3-diene. It tends to undergo addition reactions, and can be made through the Diels-Alder reaction of maleic anhydride. Substitution reactions also occur under mild conditions; for example, the reaction with bromine produces 2,5-dibromofuran, and nitration with acetyl nitrate gives 2-nitrofuran. Sulfonation with sulfur trioxide/pyridine yields furan-2-sulfonic acid.

F. can be determined qualitatively by the spruce shaving test; a shaving of spruce wood moistened with hydrochloric acid turns emerald green in the presence of F. vapor. F. is present in the first runnings from distillates of spruce and beech tar. It is synthesized industrially by decarboxylation of furfural in the presence of zinc manganese chromate(III) at 400 °C, or by decarboxylation of furan-2-carboxylic acid. F. is the parent compound of many oxygen-containing natural products, such as the furocoumarins and griseofulvin; furan derivatives are components of aromas, pheromones and drugs.

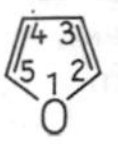

Furan 2-carbaldehyde: same as Furfural (see).

Furan-2-carboxylic acid, ***pyromucic acid***: a heterocyclic carboxylic acid; colorless crystals, m.p. 133-134 °C, b.p. 230-232 °C. It is soluble in water, alcohol and ether. F. is made by dry distillation of mucic acid. Oxidation of furfural also yields F. It is found in small amounts in the body.

Furanose: see Monosaccharides.

Furan resins: synthetic resins formed by linear polymerization of furfuryl alcohol. The dark brown to black resins are soluble in acetone, acetates and aromatics.

They are used as binders to harden the sand forms and cores used for casting metals. F. reinforced with glass fibers are very resistant to heat and flames.

Furazanes: see Oxadiazoles.

Furazolidone: a nitrofuran derivative used to treat bacterial infections of the intestines.

Furfural, ***2-furaldehyde, furan 2-carbaldehyde***, formerly ***furfurol***: a heterocyclic aldehyde. F. is a colorless, steam-volatile oil; m.p. - 36.5 °C, b.p. 162 °C, n_D^{20} 1.5261. It is soluble in water, alcohol, acetone and benzene, and is insoluble in alkanes and glycerol. It forms an azeotrope with water which boils at 97.9 °C and contains 9.2% F. In the air, F. slowly turns brown due to resin formation; aqueous acids accelerate this process. Basic reagents, such as amines, soda or hydroquinone, stabilize F. F. is very irritating to the mucous membranes, and exposure leads to inflammation of the eyes and loss of vision. Inhalation of large amounts can cause lung edema. Like all Aldehydes (see), F. is very reactive and undergoes the typical addition and condensation reactions. Because it does not have an α-hydrogen atom, F. also reacts similarly to the aromatic aldehydes (see Cannizzaro reaction, Perkin reaction, Benzoin condensation). The chief test for F. is the furfural reaction, in which a red-violet compound is formed with aniline salts. This reaction is simultaneously used to detect pentosans, because F. can be obtained from these compounds by their reaction with mineral acids. In addition, F. is present in small amounts in wood vinegar, fusel oil, coffee oil and beer wort. It is obtained industrially from plant materials containing pentosans, such as oat husks or corn cobs. F. is used as a solvent in petroleum refining, for cellulose esters and leather dyes, and is the starting material for synthesis of phenol-aldehyde synthetic resins, pigments, bacteriocides and polyamide fibers.

Furfurol: obsolete term for Furfural (see).

Furfuryl alcohol, *2-hydroxymethylfuran*: a heteroaromatic substituted primary alcohol. It is a colorless, toxic liquid with a bitter taste; m.p. - 29 °C, b.p. 170-171 °C, n_D^{20} 1.4845. It is miscible with water in any proportions and is readily soluble in ethanol and ether. It gives an intense, blue-green spruce shaving reaction. F. is sensitive to air oxygen and must be stabilized by addition of butylamine, piperidine or urea. Traces of concentrated mineral acids lead to complete polymerization, which occurs explosively. When boiled with dilute acids, F. forms levulinic acid; catalytic hydrogenation leads to tetrahydrofurfuryl alcohol. F. is present in clove oil. It is synthesized from furfural, either by catalytic hydrogenation on an industrial scale, or by the reaction of alkali hydroxide with furan-2-carboxylic acid (see Cannizzaro reaction). F. is used by itself and in combination with other monomers, especially phenols, for synthesis of amber-colored, tough casting and pressing resins; it is also used as a solvent for phenol polymers and as a dispersing agent for insoluble pigments in dyeing. In the Second World War, F. was used as a rocket fuel.

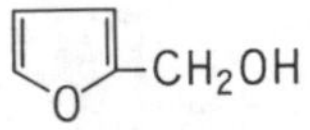

$-CH_2OH$

Furfuryl mercaptan, ***2-mercaptomethylfuran***: an oil with an unpleasant odor; b.p. 160 °C, n_D^{20} 1.5329. It is soluble in organic solvents but insoluble in water. F. dissolves in dilute alkalies, and on heating with dilute mineral acids, it polymerizes. It is synthesized by heating a solution of S-furfurylisothiouronium chloride which has been neutralized with a dilute aqueous solution of sodium hydroxide. With heavy metals, such as nickel, F. forms relatively insoluble *mercaptides*.

Furil: di-α-furyldiketone, a 1,2-diketone corresponding to benzil. F. forms yellow needles; m.p. 165-166 °C. It is readily soluble in chloroform, and barely soluble in water, alcohol or ether. F. is converted to ***furilic acid*** (di-(α-furyl)glycolic acid) on boiling with dilute potassium hydroxide. It is obtained by oxidation of furoin with nitrobenzene in sodium methylate solution.

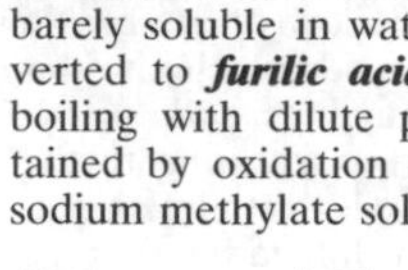

Furmethamide: see Carboxamide fungicides.

Furocoumarins: compounds found in plants of the *Apiceae* and *Rutaceae* families, where they are especially abundant in the fuits. There are two forms, furo-[2',3',7,6]- and furo-[2',3',7,8]-coumarin. F. can be intercalated between the nucleotide bases of nucleic acids (see Intercalation), and when these complexes are irradiated, they react irreversibly to form cyclobutane products with thymine residues. Therefore, F. are phototoxic and mutagenic. Some of them, such as methoxysalene (xanthotoxin), are used therapeutically.

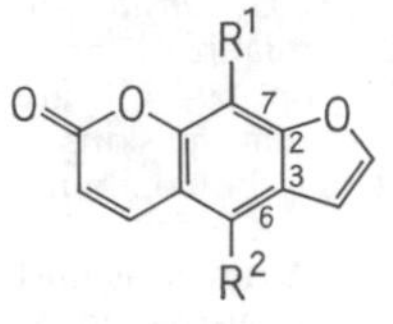

Furo-[2',3':7,6]-cumarins

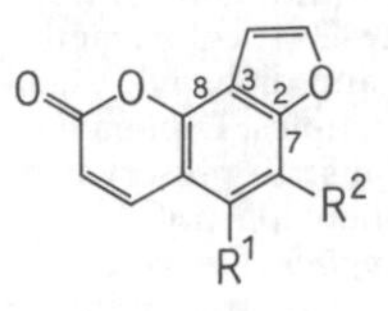

Furo-[2',3':7,8]-cumarins

Furo-[2',3',7,6]-coumarins	R^1	R^2	Furo-[2',3',7,8]-coumarins	R^1	R^2
Psoralen	H	H	Isopsoralen	H	H
Xanthotoxin[a]	OCH_3	H	Isobergapten	OCH_3	H
Xanthotoxol	OH	H	Sphondin	H	OCH_3
Bergapten	H	OCH_3	Pimpinellin	OCH_3	OCH_3
Bergaptol	H	OH			

a also methoxalen

Furoin, **α,α'-*furoin***: $C_{10}H_8O_4$, a furan derivative which forms colorless crystals. In the absence of air, it can be distilled without decomposition; m.p. 138-139 °C. F. is relatively insoluble in water, but is soluble in boiling alcohol and toluene. In the presence of bases, it is relatively easily oxidized to furil. It can be synthesized by Acyloin condensation (see) of furfural.

Furosemide: see Diuretics.

Furostan: see Saponins.

Furoxans: see Oxadiazoles.

Fursultiamin: see Vitamin B_1.

Fusaric acid, ***5-butylpyridine-2-carboxylic acid***: an acid formerly isolated from the culture media of various fungi of the genus *Fusaria* (wilt fungi). F. causes plants to wilt.

Fused ring systems: a term for bi-, tri- or polycyclic hydrocarbon compounds in which two rings share two carbon atoms (ortho-condensed systems), as in naphthalene, anthracene or phenanthrene.

Fusel oil: a mixture of alcohols produced as a byproduct of alcoholic fermentation. The main components are propan-1-ol, isobutanol and isomeric amyl alcohols. These are formed by enzymatic degradation of the yeast proteins and the plant products being fermented (potatoes, grain, molasses, etc.). F. is present in cheap spirits, and causes headaches and nausea, since the above alcohols are more toxic than ethanol. F. is used as a source of certain alcohols, especially propan-1-ol, and is also used to produce fruit esters.

Fusinite: see Macerals.

Fustic, ***yellowwood***: the stem wood of *Chlorophora tinctoria* from tropical America. F. contains the natural dyes morin and maclurin, which can be extracted from the chipped wood with water. The extract was formerly used to dye wool with alumina, chromium or iron mordants; it was also used to dye silk and leather.

G

g: symbol for Gibbs free energy.

Ga: symbol for gallium.

Gabriel synthesis: a method for production of primary amines by hydrolysis of *N*-alkylphthalimides in the presence of acids or alkalies:

$$C_6H_4(CO)_2NK \xrightarrow[-KCl]{+R-Cl} C_6H_4(CO)_2N-R \xrightarrow{+2H_2O} C_6H_4(COOH)_2 + RNH_2$$

The *N*-alkylphthalimides are readily available by reaction of potassium phthalimide with alkyl halides. In many cases, the yield can be significiantly increased by cleavage with hydrazine hydrate. In this hydrazinolysis, the cyclic phthalic hydrazide is formed in addition to the primary amine.

Gadolinium: symbol ***Gd***: chemical element in the Lanthanoid group (see) of the periodic system, a heavy, rare-earth metal; *Z* 64, with natural isotopes with mass numbers 158 (24.8%), 160 (21.90%), 156 (20.47%), 157 (15.68%), 155 (14.73%), 154 (2.15%), 152 (0.20%), atomic mass 157.25, valency III, m.p. 1312 °C, b.p. 3000 °C, standard electrode potential (Gd/Gd^{3+}) - 2.397 V.

G. is a silvery white, ductile metal which exists in two modifications. Below 1262 °C, the stable form is hexagonal (density 7.886), and above the conversion temperature, the lattice is cubic body-centered. Gd has unusual superconducting properties, is feromagnetic, and has a very large absorption cross section for thermal neutrons. It makes up about $5.9 \cdot 10^{-4}$% of the earth's crust. Its oxide is an ytter earth, and therefore Gd is found mainly associated with the lighter lanthanoids, as the silicate or phosphate. It is used to a certain extent in nuclear technology for control rods, and in metallurgy to improve the ductility and stability of stainless chrome steels. Further information on its properties, analysis, production and history are given in the entry on Lanthanoids (see).

Gadolinium compounds: compounds in which gadolinium is always found in the +3 oxidation state. The general properties and production of G. are discussed under Lanthanoid compounds (see). Some important G. are: ***gadolinium(III) chloride***, $GdCl_3$, colorless, monoclinic prisms, M_r 263.61, density 4.52, m.p. 609 °C; ***gadolinium(III) nitrate***, $Gd(NO_3)_3 \cdot 6H_2O$, colorless, triclinic crystals, M_r 451.36, density 2.332, m.p. 91 °C; ***gadolinium(III) oxide***, Gd_2O_3, white, M_r 362.50, density 7.407; ***gadolinium(III) oxalate***, $Gd_2(C_2O_4)_3 \cdot 10H_2O$, monoclinic crystals. Terbium-activated gadolinium-yttrium phosphate is used as a luminophore.

Gal: abb. for Galactose.

Galactans: homopolysaccharides made up of galactose or anhydrogalactose. The G. include the carrageenans, agar and various vegetable gums and slimes.

galacto-: a prefix for indicating a certain configuration (see Monosaccharides).

Galactomannans: Heteropolysaccharides consisting of D-galactose and D-mannose. G. are found mainly in the seeds of legumes. They are used in the food industry as thickeners and gelling agents.

Galactosamine, ***chondrosamine*** abb. ***GalN***: 2-amino-2-deoxygalactopyranose, a 2-aminosugar with the *galacto* configuration (formula, see Galactose). In the form of N-acetylgalactosamine (GalNAc), G. is a component of mucopolysaccharides and other proteoglycans.

Galactose, abb. ***Gal***: a monosaccharide, an aldohexose which is a stereoisomer of glucose. G. is commercially available as α-D-galactopyranose. It forms colorless, slightly sweet crystals which are readily soluble in water and slightly soluble in ethanol; $[\alpha]_D$ +79 to 82°. G. occurs naturally in both D- and L-forms. D-G. is a component of glycolipids, oligosaccharides such as lactose and raffinose, and polysaccharides (e.g. galactins, galactomannans). L-G. is present in some polysaccharides, e.g. agar and carrageenans. G. is obtained by acid-catalysed hydrolysis of lactose. It is used in medicine to test liver function by measurement of its galactose turnover.

CH₂OH, HO, O, OH, H,OH, R

Galactose (R = OH)

Galactosamine (R = NH_2)

β-Galactosidase: see Disaccharidases.

Galalith: see Synthetic horn.

Galbanum: see Gum resins.

Gallamine triethiodide: see Muscle relaxants.

Gallates: see Gallic acid.

Gallein, ***alizarin violet, pyrogallophthalein***: 4,5-dihydroxyfluorescein, a quinone pigment. G. forms red-brown cyrstals which contain 1 molecule of crystal water; when heated to 180 °C, these are converted to the anhydrous form. G. is made by heating pyrogallol or gallic acid and phthalic anhydride to 190-200 °C. G. is used as a dye for treated wool, cotton and silk. It is also used in the synthesis of the anthraquinone pigment ***cerulein***, which is formed from G. in the presence of conc. sulfuric acid. G. is also used as an acid-base indicator in the pH range 3.8 to 6.6 (color change from brownish yellow to pink) and for phosphate determination in urine.

Gallic acid, ***3,4,5-trihydroxybenzoic acid***: a phenolcarboxylic acid which occurs widely in plants, especially as a component of the Gallotannins (see), from which G. can be obtained by acid-catalysed hydrolysis. G. forms colorless crystals; m.p. 225 to 250 °C (dec.). It is soluble in hot water, slightly soluble in cold water and practically insoluble in chloroform or benzene. G. is a strong reducing agent; solutions of its alkali salts are used to absorb oxygen. Esters of G. (*gallates*) are used as antioxidants. G. forms a blue-black color with iron(III) ions; this is used to make inks (*iron-gallate inks*). G. forms intermolecular esters (see Depsides).

Gallium, symbol ***Ga***: element in group IIIa of the periodic system, the Boron-aluminum group (see), a metal; *Z* 31, with natural isotopes with mass numbers 69 (60.4%) and 71 (39.6%), atomic mass 69.72, valency III, rarely I, Mohs hardness 1.5, density 5.904, m.p. 29.78 °C, b.p. 2403 °C, electrical conductivity 2.5 Sm mm^{-2} (at 0 °C), standard electrode potential (Ga/Ga^{3+}) - 0.560 V.

Properties. G. is a soft metal with a silvery sheen. It crystallizes in a rhombic lattice, but because of a tendency to supercool, it is usually a liquid at room temperature. The wide range at which it can be liquid makes it suitable for high-temperature thermometers. In the transition to the solid state, it expands by about 3.1% of its volume.

The chemistry of G. is similar to that of both aluminum and zinc. It is stable to air and water at room temperature because, like aluminum, it is passivated by a thin oxide layer. At higher temperatures, it is converted in air to gallium(III) oxide, Ga_2O_3. In strong acids, it dissolves with generation of hydrogen, and formation of gallium(III) salts. Alkali hydroxides dissolve G. to form gallate solutions and hydrogen, e.g. $Ga + NaOH + 3\,H_2O \rightarrow Na[Ga(OH)_4] + 3/2\,H_2$. G. reacts with the halogens to form gallium(III) halides GaX_3, and with sulfur, selenium and tellurium to form gallium(III) chalcogenides. It forms 1:1 compounds with phosphorus, arsenic and bismuth, e.g. gallium arsenide, GaAs; these have semiconductor properties similar to those of silicon or germanium.

Analysis. The most reliable qualitative indicator of G. is its spectrum (violet lines at 417.1 and 403.1 nm). It can be determined quantitatively by complexometric titration; small concentrations are most conveniently determined by atomic absorption spectroscopy.

Occurrence. G. makes up 10^{-4}% of the earth's crust. It is found as a trace element in bauxite and sphalerite (zinc blende). The mineral with the highest G. content (0.7%) is germanite, $3\,Cu_2S \cdot FeS \cdot 2\,GeS_2$.

Production. G. is produced by electrolysis of aqueous, alkaline gallate solutions. Very pure G. for semiconductors is made by electrolysis of gallium chloride which has been purified by zone melting.

Applications. G. is used as a doping element, and especially in the form of gallium arsenide (see Gallium compounds), it plays an important role in the semiconductor industry. It is used as a thermometer liquid and to make low-melting alloys used to transport heat.

Historical. The existence and properties of the element eka-silicon were predicted by Mendeleyev in 1871. In 1875, Lecoq de Boisbaudran discovered G. spectroscopically in a zinc blende from Pierrefitte in France (Latin, Gallia) and named it for his native country.

Gallium compounds: Gallium is usually trivalent in its compounds. Some typical examples are ***gallium(III) chloride***, $GaCl_3$, deliquescent crystals, M_r 176.03, density 2.47, m.p. 77.9 °C, b.p. 201 °C, produced by heating elemental gallium in a stream of chlorine; and ***gallium(III) hydroxide***, $Ga(OH)_3$, M_r 120.74, obtained as an amorphous precipitate by adding bases to gallium(III) salt solutions. This is an amphoteric compound, dissolving in acids to yield Ga^{3+} salts, and in strong bases to give gallates. At 440 °C, it is dehydrated to ***gallium(III) oxide***, Ga_2O_3. Five modifications of this compound are known. The rhombic α-Ga_2O_3, M_r 187.44, density 6.44, is converted at 600 °C to a monoclinic β-form, which melts around 1900 °C. ***Gallium(III) arsenide***, GaAs, forms dark gray, cubic crystals, M_r 144.64, m.p. 1238 °C, which is produced by melting the elements together. It is a compound semiconductor, used in the construction of very fast computers which use very low amounts of energy.

Gallium(I) compounds can be synthesized at high temperatures in the gas phase; examples are ***gallium(I) oxide***, Ga_2O, and ***gallium(I) chloride***, GaCl. These compounds tend to disproportionate into gallium and gallium(III) compounds. Derivatives which are often named as gallium(II) compounds, such as gallium sulfide, GaS, or gallium halides, GaX_2, actually do not contain Ga^{2+} ions; they are $Ga^{I}Ga^{III}$ derivatives. For example, "gallium dichloride" has the ionic structure $Ga^{I}[Ga^{III}Cl_4]$.

Gallotannins, ***hydrolysable tannins***: a group of tannins based on Gallic acid (see). Gallic acid is present mainly as an intermolecular ester (depside) or as hexahydrodiphenic acid, a C-C dimer formed by oxidative coupling of two molecules of gallic acid. Gallic acid or polymeric phenol carboxylic acids derived from it are bound to monosaccharide or alditols by ester bonds, or less frequently, by glycosidic bonds. The G. include the tannin of rhubarb, and the galls of Chinese and Turkish sumach species.

GalN: abb. for galactosamine.

Galvanic cell: same as Electrochemical current source (see).

Galvanic potential: see Galvanic voltage.

Galvanic voltage, ***g***: The difference between the internal electric potential at an initial point in one phase and the internal electric potential at an end point in a second phase with a different composition when the two phases are in contact: $g^{I,II} = \Phi^{I} = \Phi^{II}$. The internal electric potential Φ is called the *galvanic potential*.

Gamboge: see Gum resin.

Gamma ray resonance spectroscopy: see Mössbauer spectroscopy.

Gamma ray spectroscopy: the study of γ-radiation (see Electromagnetic spectrum) emitted by excited atomic nuclei. Like the electrons of an atom, the components of the nucleus can exist only in quantized energy states. By absorbing energy – for example during a nuclear reaction – nuclei can jump from their ground state to excited states. The lifetime of excited nuclear states is very short; the nuclei return to the ground state within 10^{-12} to 10^{-20} s, emitting the energy of excitation as a γ-ray. G. is not the same as gamma ray resonance spectroscopy (see Mössbauer spectroscopy).

Ganglion blockers, ***ganglioplegics***: compounds which inhibit nerve impulse conduction in the sympathetic and parasympathetic ganglia. They were formerly used to treat severe hypertonia, but because of their severe side effects, they are no longer used. Bisquarternary ammonium compounds in which the two nitrogen atoms are linked by a saturated aliphatic carbon chain of 5 to 6 atoms block ganglia; the carbon chain may also contain a heteroatom. Some examples are *hexamethonium bromide*,

$$[(CH_3)_3\overset{+}{N}-(CH_2)_6-\overset{+}{N}(CH_3)_3]\ 2\ Br^-$$

and *azamethonium bromide*

$$[(CH_3)_2\overset{+}{N}(C_2H_5)-CH_2-CH_2-N(CH_3)-CH_2-CH_2-\overset{+}{N}(CH_3)_2(C_2H_5)]\ 2\ Br^-$$

Certain tertiary or secondary amines with large numbers of methyl groups also act as G. An example is *pempidine*.

CH3 CH3 CH3 N CH3 CH3

Pempidine

Gangliosides, ***sialoglycosylsphingolipids***: glycosphingolipids (see Sphingolipids) with one to four terminal sialic acid residues. In G. with several sialic acid residues, the oligosaccharide is branched. G. isolated from bovine brain or rat liver are obtained as amorphous powders, which are soluble in methanol/chloroform (1:1) or water and other polar solvents. The fatty acid component of G. is usually stearic acid. The G. belong to the ganglio- or neolacto-series of glycosphingolipids. Abbreviations for the G. are usually taken from the IUPAC-IUB recommendations of 1977, and the symbolism of Svennerholm is used (Table). In the former system, a Roman numeral indicates the number of monosaccharide units after the ceramide to which the sialic acid group(s) (e.g. NeuAc) is(are) bound; a superscript shows the bonding site on the sugar. In the Svennerholm system, G = ganglioside, M = monosialo, D = disialo, and so on. An arabic number indicates the position taken by the G. on a thin-layer plate.

The G., as acidic glycolipids, are located in the outer half of the cell membrane. Their concentration is especially high in the brain. The ganglioside pattern of the outer cell membrane is organ- and species-specific. GM1, GD1a, GD1b and GT1b are most abundant at nerve ends. G. are receptors for biologically active compounds, such as serotonin, tetanus toxin, cholera toxin, thyreotropic hormone and other hormones. The receptor G. for serotonin has the structure NeuAc(2→4)Gal(β1→4)GlcCer.

Nomenclature of some gangliosides

IUPAC-IUB	Svennerholm
II^3NeuAc-LacCer	GM3
$II^3(NeuAc)_2$-LacCer	GD3
II^3NeuAc-$GgOse_3$Cer	GM2
II^3NeuAc-$GgOse_4$Cer	GM1
$II^3(NeuAc)_2$-$GgOse_4$Cer	GD2
IV^3NeuAc, II^3NeuAc-$GgOse_4$Cer	GD1a
$II^3(NeuAc)_2$-$GgOse_4$Cer	GD1b
$IV^3(NeuAc)_2$, II^3NeuAc-$GgOse_4$Cer	GT1a
IV^3NeuAc, $II^3(NeuAc)_2$-$GgOse_4$Cer	GT1b

Gas: a state of matter in which the average thermal energy of the atoms or molecules is greater than the interaction energy between them. The distances between particles can change at random, so G. do not have a fixed shape. They completely fill any available space, and can be mixed with one another in any proportions. A G. is homogeneous and isotropic. Only in extremely large volumes (e.g. the earth's atmosphere) or in very strong gravitational fields (which can be generated in an ultracentrifuge, for example), will a density distribution arise due to sedimentation equilibria.

G. can be converted to the liquid or solid states by cooling and/or compression (*condensation*). A G. above the boiling point of the corresponding liquid is also called a Vapor (see). In the supercritical range (see Critical point), a G. cannot be liquefied, because there is only a single, uniform supercritical phase. The range of existence of the various states of matter and phases is shown in a Phase diagram (see).

The thermal and caloric properties of G. are described by the equation of state (see State, equation of). An ***ideal G.*** is an idealized, limiting state of matter characterized by the absence of attractive or repulsive forces between the particles of the G.; these particles act as mass points, that is, they have no volume and no internal degrees of freedom (see Kinetic gas theory). For such G., the thermal equation of state of an ideal G. applies strictly. It is approximately applicable to noble gases, hydrogen, nitrogen, oxygen and other nonpolar substances at high temperatures and low pressures.

Real G. deviate from the ideal gas laws, because there are forces of interaction between their particles. Their behavior can be approximately described by adding correction terms to the thermal equation of state for an ideal gas (see van der Waals equation). The internal energy and enthalpy of real G. depend on the volume and pressure of the G. (see Joule-Thomson effect).

Because of the high ionization energy and relatively low electron affinities of isolated atoms and molecules, charge transfers, that is, formation of ions, are highly endothermal processes. G. therefore consist of neutral molecules or atoms under normal conditions, and ions are observed in a G. only at very high temperatures of several thousand K, or after transfer of energy (e.g. photochemically or by an electric discharge; see Plasma). Reactions in G. thus tend to occur primarily homolytically, via radicals, or as multi-center reactions. Since the density of a G. is low, reactions in it tend to have low yields per unit of volume and time. In industry they are therefore generally carried out at high pressure.

Gas adsorption chromatography: see Gas chromatography.

Gas analysis: analysis of mixtures of gases.

Qualitative G. is used to determine the components of a mixture of gases. Reactive gases, such as ammonia, chlorine, carbon dioxide and sulfur dioxide, can be absorbed in water, and then analysed by the usual methods of qualitative inorganic chemistry. Nonreactive gases, such as hydrocarbons, nitrogen and hydrogen, are usually determined by gas chromatography. Some gases can be recognized very simply using Gas test tubes (see).

Quantitative G. is used to determine the amount of each gas in the mixture, and the methods can be either chemical or physical. The chemical methods are based on specific absorption of the components by various solutions; the volume decrease in the sample after each absorption gives a direct measure of the volume percent of that component. This ***volumetric G.*** is often carried out in an Orsat apparatus (see). Reactive gases can also be absorbed in suitable solutions and then determined by ordinary chemical methods, for example alkalimetry (see Neutralization analysis) for hydrogen chloride and other acidic gases. Some physical methods of G., for example, fractional condensation or distillation, have been made obsolete by the introduction of gas chromatography, which is now the most important method of G. The oxygen content in smoke and combustion gases can be monitored continuously by potentiometry; a galvanic chain with oxygen-conduction solid electrolytes is used. Infrared absorption can also be used to determine the concentrations of a few gases, e.g. in mines, for continuous monitoring of methane, carbon monoxide and carbon dioxide concentrations in the air.

Gas bottle: see Bottled gas.

Gas chromatography, abb. ***GC***: a chromatographic method with a gaseous mobile phase used for analysis of gaseous mixtures of substances. In ***gas-solid chromatography*** (abb. GSC), the stationary phase is an active solid (adsorbant), while in ***gas-liquid chromatography*** (abb. GSL), the stationary phase is a liquid fixed to an inert carrier material. G. depends almost entirely on Elution methods (see), either isothermal or temperature-programmed.

Methods of G. The apparatus consists of a pressure bottle with a pressure-reducing valve, a unit to measure and regulate the flow of gas, the sample inlet, a separation column with a thermostat, and a detector with an amplifier, recorder and integrater, and often with a computer. The most commonly used *carrier gases* are hydrogen, helium, nitrogen or argon; the flow rate is regulated with a fine needle valve and is measured by a rotation meter or soap film flow device. The sample is injected with a microsyringe through a membrane, or with an automatic pipet which can deliver 1 to 10 μl of liquid or 1 to 10 ml of a gas.

The stationary phases used in GSC are adsorbants such as silica gel, aluminum oxide, activated charcoal, molecular sieves (4A, 5A and 13X) or porous polymers. The grain sizes of these materials are 0.1-0.2 or 0.2-0.3 mm. In GLC, a granular carrier material based on kieselguhr (Chromosorb W., Celite, Chromaton N, Chezasorb, Sterchamol) or alumina (Porolith) is used. A number of liquids are available for use as the stationary phase; they differ with respect to polarity, selectivity and thermal stability (Table 1).

Table 1. Liquid phases and their uses

Liquid	Polarity	Temp. range [°C]	Uses
Alkanes			
Squalane	np	20–150	Aliphatics, aromatics,
Paraffin oil	np	50–200	separation by
Apiezon L	np	50–300	boiling points
Silicones			
Methyl			
DC 200	np	20–250	Alkanols, alkanals,
SE 30	np	50–300	alkanones, aromatics,
OV-1	np	100–350	separation by
NM 1–200	np	20–250	boiling points
Phenylalkyl			
DC 550	mp	20–225	Amino acid derivatives,
SE 52	mp	100–300	aromatics, heterocycles
OV-17	mp	20–350	
NM 5–400	mp	20–300	
Cyanoalkyl-			
XE 60	sp	20–275	Halogenated hydrocar-
OV 225	sp	20–275	bons, pesticides
Carborane methyl			
Dexsil 300	np	50–400	
Polyglycols			
Carbowax 400	mp	20–120	Alkanols, alkanals,
Carbowax 20 M	mp	50–250	alkanones, aromatics,
Ucon 50 HB	mp	20–200	alkenes
Esters			
Dinonyl-phthalate	sp	20–130	Fatty acyl esters,
Reoplex 400	sp	50–200	halogen compounds,
Ethylene glycol adipate	sp	100–200	amino acid derivatives
DEGS	sp	20–200	
N compounds			
β, β′-Oxi-dipropiono-nitrile	sp	20– 80 50–150 150–300	Aromatics, esters amines
7,8-Benzo-quinoline	mp		
Versamide 900	mp		
Liquid crystals			
4,4-Azoxy-phenetol	mp	140–165	Structural isomers

np = nonpolar or weakly polar
mp = medium polarity
sp = strong polarity

The capacities of liquid phases are indicated by their retention indices (see Retention) for selected test substances (Rohrschneider indices). The *detector* registers the separated components by one of several physical methods (Table 2); these differ in sensitivity and selectivity. The quantity recorded is the time differential (Fig. 2).

Table 2. Detectors for gas chromatography

Detector	Limit of detection in g s^{-1}	Applications
Heat conduction detector	10^{-6}	Usable for any substance
Flame ionization detector (FID)	10^{-12}	Selective for compounds with CH_2groups
Thermoionization detector (TID)	10^{-10}	Selective for halogen, N and P compounds
Flame photometer detector (FDP)	10^{-8}	Selective for S and P compounds
Electron addition detector (EAD)	10^{-14}	Sensitive for substances with high electron affinity

The modern high-performance version of G. is ***capillary G.***, in which an open tubular capillary column of 0.2 to 0.5 mm diameter and 50 to 200 m length gives a capacity of 10^5 to 10^7 theoretical plates. The stationary phase is a film on the interior wall of the capillary. Because their capacities are very small (10^{-2} to 10^{-3} µl), capillary G. requires special sample injection techniques and sensitive detectors (Fig. 1). The method is used, for example, in the analysis of natural products, petroleum products and isomer mixtures (Fig. 2).

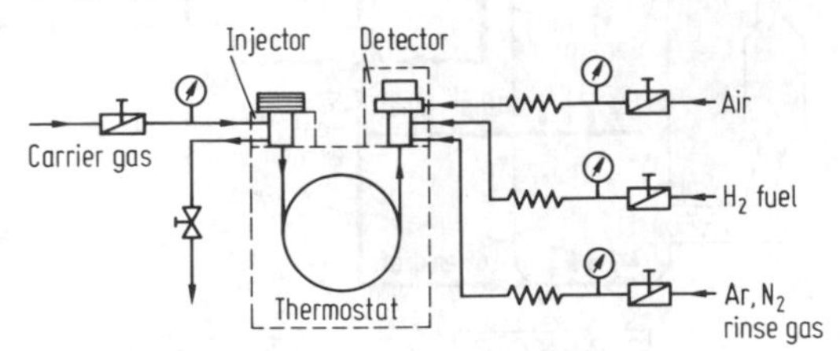

Fig. 1. Gas chromatograph with capillary column and flame ionization detector.

In addition to analytical G., there is ***preparative G.***, which is used to separate substances in gram amounts. For this, columns with inner diameters of 6 to 10 mm are used, and the separated substances are caught in cold traps. To obtain larger amounts of substance, the separation cycle is automatically controlled and is repeated several to many times.

For qualitative analysis, the chromatograms are evaluated by comparison of substances, by mixing with known substances or by combination with other methds (e.g. IR spectroscopy). For quantitative analysis, the *peak area* is measured. The peak area is proportional to the amount of substance. It is determined by an approximate method (height times width at half-height), or it is integrated mechanically or electronically. By multiplication by the appropriate factors, the peak area can be converted to mass or volume percent. The limit of detection of G. is about 10^{-15} g substance.

Historical. In 1952, James and Martin first separated a mixture of fatty acids on a silicone phase using the principle of gas-distribution chromatography. In 1958, Golay introduced the use of open capillary columns.

Gas coal: see Coal.

Gas constant: symbol R, a natural constant defined by the molar volumes V_0 of ideal gases under standard conditions (see Standard state), that is, at atmospheric pressure p_0 = 101.325 kPa and 0 °C or T_0 = 273.15 K:

$$R = \frac{p_0 V_0}{T_0} = \frac{101325\ \mathrm{Pa} \cdot 22.414 \cdot 10^{-3}\ \mathrm{m^3\ mol^{-1}}}{273.15\ \mathrm{K}} = 8.3143\ \mathrm{JK^{-1}\ mol^{-1}}.$$

In the now obsolete units, R = 0.0820561 atm K^{-1} mol^{-1} = 1.9858 cal K^{-1} mol^{-1}. The quotient $R/N_a = k_B$ is the Boltzmann constant, where N_A is Avogadro's number.

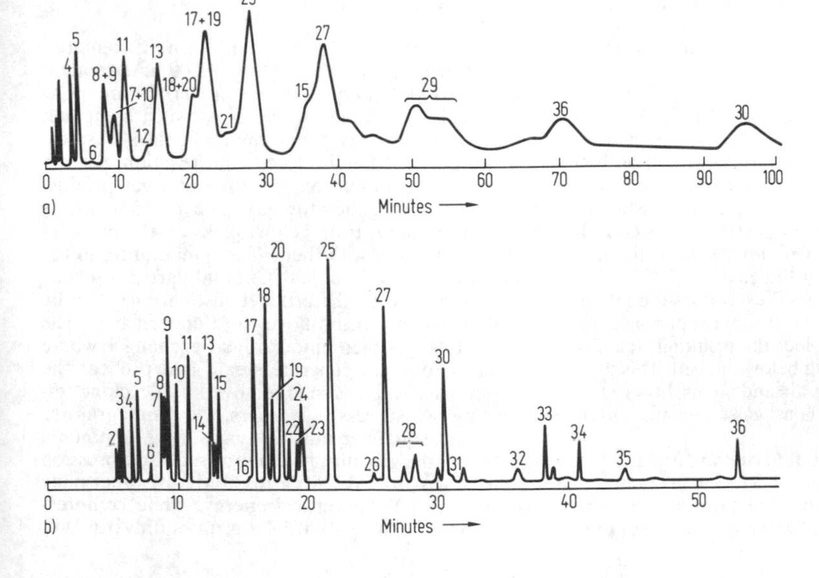

Fig. 2. Gas chromatogram of a light petroleum. *a* packed column, column length 3 m, inner diameter 6 mm, liquid phase dioctylphthalate on sterchamol; *b* capillary column, column length 50 m, inner diameter 0.3 mm, liquid phase, squalane.

Gas degeneration: see Fermi-Dirac statistics.

Gas detection device: an apparatus used with Gas test tubes (see) to detect gases, especially harmful gases, in the air. A hand bellows is used to pull a fixed volume of air through a gas test tube at the site being monitored. A scale on the gas test tube indicates the amount of gas directly. G. are used mainly to monitor the workplace concentrations of harmful substances. G. can be used for other purposes, such as detecting leaks in gas lines and containers.

Gas diffusion process: see Isotope separation.

Gas distribution chromatography: see Gas chromatography.

Gas electrode: an electrochemical electrode in which the potential-determining step involves a gaseous element. The equilibrium between the gaseous element and its ions in solution occurs on the metallic phase of the electrode (usually platinum). The metal dips into the ion solution and a stream of the gas is bubbled into the solution and around the electrode. An electrode voltage is established which depends on the pressure of the gas and the activity of the ions. The Hydrogen electrode (see) and the Oxygen electrode (see) are typical gas electrodes; the halogen electrodes are analogous.

Gas exchange: in water treatment, the driving out or enrichment of gases in water by a concentration gradient; for example, enrichment with oxygen and driving out of carbon dioxide.

Gas flame coal: see Coal.

Gas fractionation: the separation of a mixture of gases into its components by liquefaction (see Gas liquefaction) and subsequent distillation. An example is the separation of air into oxygen and nitrogen (see Air fractionation). There are other methods which can be used instead of or in combination with distillation, such as partial condensation, absorption, adsorption, permeation and thermodiffusion.

Gas generator: a plant for gasification of solid fuels such as coal or coke. A G. can produce Generator gas (see) by forcing air through the fuel, Water gas (see) by forcing air and steam in alternation through it (only the gas produced from steam is captured), and Synthesis gas (see) and City gas (see) by forcing a certain mixture of oxygen and steam through the fuel.

a) In a ***countercurrent G.***, air and fuel flow past each other in a countercurrent. The ash drops out as a liquid slag. To obtain a fluid slag, limestone or dolomite is added, as in iron smelting. Low-quality, low ash coal can be used as fuel. This type of G. has a very high capacity, and the coal is completely consumed. An air-fed G. can generate up to 6000 m^3 gas (standard conditions) per hour. A disadvantage is the high exit temperature (900 °C) of the gas.

b) In a ***fixed bed G.***, the fuel rests on a solid lattice made up of pipes through which water is pumped. Its walls can also be water-cooled; the resulting steam is then blown into the G. from below the bed. This type of G. is run by alternating air and steam flows, and can produce 8,000 to 10,000 m^3 gas (standard conditions) per hour.

c) The continuously operating ***rotating bed G.*** consists of a vertical shaft furnace, 1.5 to 3 m in diameter. In the older G. of this type, the furnace walls are lined with brick, but in the newer ones, the lower part has a cooling layer around the lower, hotter part, which prevents the ash from baking onto the walls. The fuel is loaded through the top, and at the bottom is an ash pan with a grid in the middle. This whole structure is rotated by a motor. The rotation stirs the fuel and removes the ashes. The ash pan is filled with water to seal the furnace against the atmosphere. The reaction gas enters through slits in the rotating grid. It can be either pure air or air and steam simultaneously. High-capacity G. have an additional, fixed grid on the sides to increase the rate at which reaction gas can be pumped in. The gas exit is above the fuel level. A rotating bed G. can produce up to 14,000 m^3 (standard conditions) gas per hour.

In a ***carbonization G.***, a carbonization shaft is set above the rotating bed generator, so that liquid products can be obtained in addition to the gas.

d) The ***pressure gas G.*** works in the same fashion, but here the G. is fed with pressurized (3 MPa) oxygen instead of air. The advantages of pressure gasification are increased gas production capacity, which thus permits a smaller size G. to be used.

e) The ***Winkler G.*** (Fig. 1), named for its inventer, can process unsorted coal.

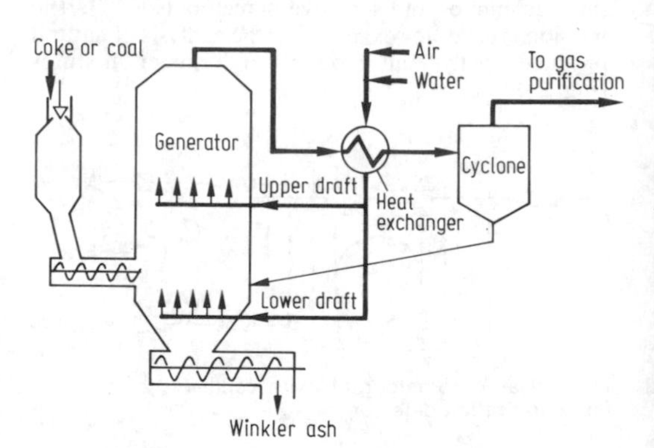

Fig. 1. Winkler generator.

This type of G. is 15 m high and 5.5 m in diameter. The production rate is 60,000 to 70,000 m^3 (standard conditions) per hour. The raw materials are powdered fuels with large surface areas, such as lignite, lignite coke and very young mineral coals. Gasification occurs in a fluidized bed; the reaction gas is a mixture of steam and oxygen for water-gas production, or air for generator gas. The reaction gas is blown into the G. from below, picks up the particles of fuel and reacts with them. The temperature in the fluidized layer is about 950 °C. Solid particles picked up from the bed by the product gases are gasified by reaction gas blown in above the fluidized bed. The hot gas (900 °C) then enters a heat exchanger where its heat is used to generate steam and preheat the reaction gas. The fly ash is removed in a cyclone followed by two successive washers, the disintegrators.

For subsequent synthesis stages (e.g. ammonia synthesis), the gas must be compressed. Compression of 1 mol gas from 0.1 MPa to 20 MPa, isothermally consumes 3.7 Wh electrical energy, while compression from 2 MPa to 20 MPa consumes only 1.6 Wh.

For this reason, all modern G. operate at pressures of 2 to 3 MPa (e.g. pressurized Winkler G.).

The carbon is much better utilized if the process heat for the G. is generated by a high-temperature nuclear reactor instead of by partial combustion of the coal (Fig. 2).

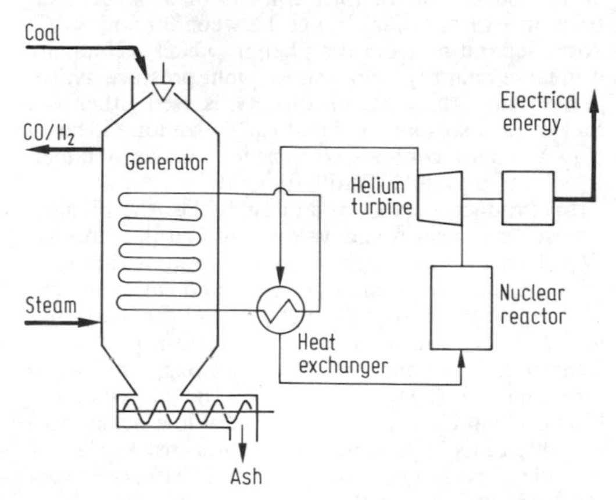

Fig. 2. Coupling of a generator to a high-temperature nuclear reactor.

Coupling of a plasma generator to the gasification generator is also advantageous (Fig. 3).

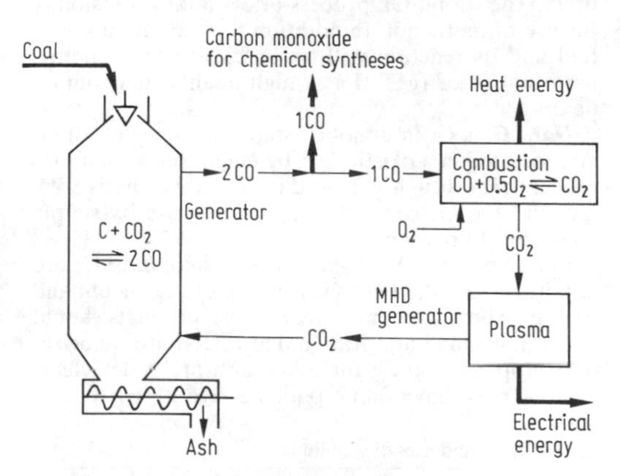

Fig. 3. Plasma gasification plant.

Gas law: in the broad sense, the thermal and caloric equations of state (see State, equation of) of gases; in the narrow sense, the thermal equation of state of an ideal gas, $pv = nRT$.

Gas liquefaction: the conversion of a gas or mixture of gases to a liquid. G. is achieved 1) through cooling to the pressure-dependent saturation temperature or dewpoint and further removal of heat; or 2) by processes of compression, cooling and decompression. The gas is converted to a liquid or a two-phase liquid-gas system from which the liquid is separated. To achieve the necessary temperature, the gas may have to be cooled. This is always the case when the critical temperature of the substance is lower than the ambient temperature, for example, in the case of air (-140.7 °C).

There are two possible methods of achieving G. at low temperatures: 1) use of a separate refrigerator or 2) integration of a heat-removing cycle in the process itself. The second type of liquefaction apparatus can be considered a refrigerator with an open coolant circulation, in which the liquid product is removed rather than recirculated.

G. is used industrially with chlorine, sulfur dioxide, halogenated hydrocarbons, ammonia, butane, propane, carbon dioxide and other gases which boil at even lower temperatures. The requirements for capital and energy are particularly high for natural gas or methane (saturation temperature of -161.5 °C at normal pressure), air or oxygen/nitrogen (-183 °C/ -196 °C), hydrogen (-253 °C) and helium (-269 °C).

G. is used to prepare gases for transport and storage, for use as coolants, and for gas fractionation by means of low-temperature distillation. The most important of these processes is Air fractionation (see) for production of oxygen, nitrogen and noble gases. Other important processes are the separation of hydrocarbon mixtures to obtain ethylene and propylene, the fractionation of residual synthesis gases and the separation of helium from natural gas.

Gas-liquid chromatography: see Gas chromatography.

Gas odorizing: the addition of small amounts of a substance with a strong odor to fuel gases which have no smell of their own. G. is intended to make leaks and unintentional gas flow (such as an unlit burner on a stove) immediately noticeable. Mercaptans, tetrahydrothiophene and other odorous compounds are used for G. In countries where natural gas is widely used, odorizing of public supplies is often required by law.

Gas oil: a medium petroleum fraction boiling between 240 and 360 °C (see Petroleum processing). The density at 20 °C is 0.85 to 0.88, and the heat content is 40,000 kJ kg^{-1}; the flame point is 80 °C. G. is used as a diesel fuel and light heating oil; it was formerly used to make oil gas.

Gasol: a mixture of hydrocarbons with chain lengths C_3 and C_4, formed by the Fischer-Tropsch synthesis.

Gasoline: a mixture of low-boiling hydrocarbons consisting chiefly of n-, iso- and cycloalkanes and aromatic compounds with 5 to 10 C atoms. G. is a colorless, flammable liquid; its density at 15 °C is between 0.72 and 0.80 g cm^{-3}, depending on its composition. G. evaporates readily and forms explosive gas mixtures with air. This is the basis of its use as a Fuel (see) in Otto-cycle engines. The heating value of G. is between 41,000 and 43,000 kJ kg^{-1}. G. is not miscible with water, but is readily soluble in alcohol, ether and halogenated hydrocarbons. ***Light G.***, which boils between 20 and 80 °C, and ***heavy G.***, which boils between 100 and 180 °C, are the most important starting materials for production of high-quality fuels.

Production. 1) G. is obtained by Distillation (see) of petroleum (the fraction boiling between 50 and 200 °C, straight-run G.). The G. content of petroleum varies from 5 to 85%; on the average, it is between 10 and 20%. The largest part of this *straight-run G.* is refined to high-quality G. fuels. It cannot be used directly for automobile engines because its octane rating is too low.

2) ***Natural gas G.*** is obtained from "wet" natural gases by adsorption on activated charcoal or silica gel. This consists of low-boiling hydrocarbons, about 90% of which boil below 135 °C. It has a high front octane rating, and is therefore mixed with G. from other sources which do not have the necessary concentration of low-boiling hydrocarbons.

3) The Cracking (see) of higher petroleum fractions yields ***crack G.***, which has a very good octane rating, because in the cracking process, especially catalytic cracking, hydrocarbons are dehydrogenated to alkenes, isomerized to isoalkanes and converted to aromatics. The largest portion of G. produced today is obtained by cracking.

4) Reforming (see) of straight-run G. produces a ***reformed G.*** which has an octane rating of 85 to 95. While cracking converts mainly the higher-boiling petroleum fractions to G., reforming involves major structural changes in the molecules of heavy G. fractions, which are thus converted to high-quality fuel.

5) A ***pyrolysis G.*** with an octane rating about 100 can be isolated from the liquid products of Pyrolysis (see); these are byproducts of ethene production. This G. is rich in aromatics, but it also contains thermally unstable compounds, such as diolefins and vinyl aromatics, which must be removed by selective Hydrogenation (see).

6) A major byproduct (up to 25%) of petroleum processing is refinery gas. These gaseous hydrocarbons were formerly used only as heating gas, but they are now used as as starting materials for chemical syntheses, or the olefins in them are converted to ***oligomer G. (polymer G.)***. For this purpose, the propene-propane (P-P) and butene-butane (B-B) fractions obtained by decomposition of refinery gas are used. The propene and C_4 olefins are polymerized on acid catalysts, e.g. phosphoric acid on kieselguhr, acidic ion exchangers and zeolites, at a reaction temperature around 220 °C and a pressure of 2 to 6 MPa. From propene, the main products are tri- and tetrapropene; butenes yield di-, tri- and tetraisobutenes. Mixed oligomers of propenes and butenes are heptenes of various structures. Tri- and tetrapropene and di- and tri-isobutene are made into oxoalcohols (see Oxosynthesis) as well as fuel components; the oxoalcohols are precursors of softeners and tensides.

7) ***Alkylated G.*** is a high-performance fuel with an octane rating > 95; it is made by catalytic alkylation of isobutane with butenes, propene or ethene. Hydrogen fluoride or sulfuric acid is used as catalyst. Isobutene is alkylated by an ionic chain reaction. Isobutane, which is needed for the alkylation, can be obtained by isomerization of butane: $H_3C\text{-}CH_2\text{-}CH_2\text{-}CH_3 \rightarrow H_3C\text{-}CH(CH_3)\text{-}CH_3$. Anhydrous aluminum chloride, $AlCl_3$, in the presence of hydrogen chloride, is used as catalyst at 80 to 90 °C and 1.5 to 2 MPa pressure.

8) G. can be obtained by hydrogenation of hard and soft coal, peat, pitch, high- and low-temperature tar. The methods were introduced in Germany on a large scale during the 1930s. Because of the high consumption of hydrogen and the expense of building and operating high-pressure plants, coal hydrogenation was not economical even then, so that most hydrogenation plants were converted to work with carbonization tars or residual oils from petroleum refining. However, Coal hydrogenation (see) is again becoming significant.

9) Carbonization of soft coal produces about 20% of a light oil (***carbonization G.***) with a boiling range of 30 to 180 °C; this in turn consists of 30% of a G. fraction with a boiling range between 30 and 80 °C. After separation from the phenols which accompany it in large quantity, and a subsequent pressure hydrogen refining, this carbonization G. is used either as a fuel or as a solvent. Refined carbonization G. has a high aromatic content, so benzene can be obtained from the 70 to 80 °C distillation cut.

10) Production of G. and diesel fuel by the Fischer-Tropsch synthesis was developed in Germany before World War II. Iron and cobalt are used as catalysts, with thorium, magnesium, calcium and aluminum oxides as additives. At 160 to 200 °C and pressures of 2 to 3 MPa, synthesis gas reacts to form a mixture of saturated, unbranched hydrocarbons: $n\text{CO} + (2n+1)H_2 \rightarrow C_nH_{2n+2} + nH_2O$. The G. from the Fischer-Tropsch synthesis has a very low octane rating, due to its high content of n-paraffins. In view of the price development for petroleum, efforts are being made to improve this process.

11) ***G. from methanol.*** The Mobil Oil process (MTG process) has recently been developed to produce G. hydrocarbons from methanol with high yields and very selectively. Because of its high aromatic content, the G. produced in this way has an octane rating of 95. The Mobil Oil process offers a third possibility for use of methanol, in addition to its direct use as a fuel and its reaction with isobutene to form methyl *tert*-butyl ether (MTBE), a high-quality fuel component.

Hard G. is G. in a "solid" state: tiny droplets of G. are enclosed in a plastic film by means of an emulsion process. The tough gel produced in this way is 95% G., and the G. can be separated for use by simple mechanical pressing.

In addition to fuels, ***special*** or ***boiling limit*** G. are obtained from the light G. fraction of petroleum and similar hydrocarbon mixtures. These products should not contain any impurities and have definite vaporization properties, pleasant odors and are water clear. Test G. must have flame points above 21 °C.

Boiling limits and uses of gasolines

Terms	Boiling limits °C	Applications
Light petroleum (Petroleum ether)	40– 70	Solvent, medicament (see Petroleum ether)
	65– 95	Extraction of edible oils
Extract gasoline	80– 95	Solvent for rubber
	80–125	Extraction solvent
Lighting gasoline	30– 85	For lighting
Cleaning gasoline	100–140	In dry cleaning and textile industries
Solvent gasoline	130–180	Paint and oilcloth industries
Test gasoline	140–200	Solvent and thinner for paint and varnish
Terpentine substitute	150–195	For shoe and floor polishes
Mineral terpentine oil	140–225	For paints and varnish

Gas-phase permeation: see Permeation.

Gas pipets: glass apparatus filled with a solid or liquid absorptive material. In chemical gas analysis, the gas to be determined is absorbed in the G., then the decrease in volume is determined in a gas buret. The old-fashioned G. consists of two spherical vessels connected by a tube, for example, ***Hempel pipets***. ***Orsat pipets***, as used in the Orsat apparatus (see), are more modern. In a few cases, glowing wire or explosion pipets are still used to determine hydrogen or hydrocarbons. In the ***glowing wire pipet***, the gas mixture is ignited in the presence of oxygen or air and burned. In the ***explosion pipet***, the gas mixture is ignited by an electric spark. The carbon dioxide produced by the combustion can be determined volumetrically.

Gas processing, also called ***gas purification***: processes for removal of moisture, contaminating gases (see Deoxo process), vapors and dust from technical gases. The first stage of G. is *cooling* and *condensation*, in which undesired components, such as steam and hydrocarbons, are mostly precipitated out of the gas, and the gas is cooled to the temperature needed for further G. In indirect cooling (surface condensation), the gas is passed over pipes in which a coolant (e.g. water) is flowing, and gives up its heat on the walls of the pipes. This type of cooling is used for gas which contains valuable condensates. In direct cooling (mixed condensation), the heat is removed by spraying the gas with water in a cooling tower; the gas and coolant mix. Direct cooling is used mainly with tar-free gases which need to be freed only of steam and fly ash.

After cooling and condensation, impurities and undesired components remaining in the gas are removed by various methods. There are various kinds of installation for Dust removal (see); tar is removed by precipitators, electrofilters, centrifugal scrubbers with tar as the wash liquid, or Venturi scrubbers with tar oil as wash liquid. Naphthalene is removed by static or dynamic scrubbers used with coal tar oil (anthracene oil) or by shock scrubbers (which spray cold water into the hot gas). Sulfur oxides are removed in scrubbers with fresh or circulating water (sometimes with addition of alkali), and soot is removed with water or oil scrubbers. Hydrogen sulfide is removed in dry scrubbers with iron or carbon-containing Gas purification masses (see). For high capacities, wet scrubbing processes, based on either extraction or physical methods, are used. An example is the pressurized water scrubber, which is sometimes also used to remove carbon dioxide (see Synthesis gas). Hydrogen cyanide is removed by the same methods of dry or wet scrubbing used for hydrogen sulfide, or with water; hydrocarbons are removed by scrubbers with wax oil or activated charcoal absorbants. They can be separated from fuel gases by pressurizing them at low temperature. Gas is dried by cooling it to 2 to 3 °C, then compressing it and absorbing or adsorbing the water. Organic sulfur compounds can be partially removed from fuel gases by oil scrubbing, activated charcoal or chemical reaction, for example, with sodium hydroxide. The carbon monoxide can be partially removed by Conversion (see). Nitrogen oxide can be partially removed as a side reaction of conversion, by gasoline production on activated charcoal, dry gas desulfurizing or in the course of tar removal by electrofiltration.

Gas purification: see Gas processing.

Gas purification mass: a product used to remove hydrogen sulfide from fuel and synthesis gases at temperatures between 20 and 60 °C, and to remove organic sulfur compounds from synthesis gas at temperatures around 300 °C.

Bog iron ore, iron hydroxide sludge from installations which remove iron from water, acidic sludges, substances containing activated carbon (e.g. activated charcoal) and activated roasted iron pyrites are used as G. to remove hydrogen sulfide. Iron-oxide containing masses are used to remove organic sulfur compounds.

Spent, sulfur-charged G. is roasted in the same manner as pyrites, in stage furnaces, to produce sulfuric acid or alkaline sulfite solutions. Sulfur can be obtained by extraction of spent G. with carbon disulfide.

Gas scrubbing: a process of thermal separation used to purify gases or isolate desired components from mixtures of gases. All processes of G. are governed by the physical laws of Absorption (see). G. is used to purify natural and synthesis gases, and also to recover valuable compounds, such as acetylene synthesized by the Sachse process.

Industrial gas scrubbing processes

Scrubbed component	Scrubbing liquid	Process
Carbon dioxide, carbon oxygen sulfide, hydrogen sulfide, hydrocyanic acid	Water	
	Potassium carbonate	
	Methanol	Rectisol
	Sulfosolvane solution	Sulfosolvane
	Sulfolane/ diisopropanolamine	Sulfinol
Steam from natural gas	Ethylene glycol	
Benzene from coking gas	Anthracene oil	
Acetylene from the gas stream of high-temperature pyrolysis	Acetone	
	N-Methylpyrrolidone	
	N,N-Dimethylformamide	
Residual carbon monoxide from ammonia synthesis	Ammonia in copper carbonate solution	

Gas silicate concrete: see Binders (construction materials).

Gas-solid chromatography: see Chromatography.

Gas test tubes: glass tubes, about 12 cm long and with an inner diameter of 5 to 6 mm, which are graduated and contain fine-grain absorbents. They are used with Gas detection devices (see) for detection and quantitative determination of a gas; the gas for which the G. is designed is printed on it. A certain volume of air or other gas is pulled through a G.; if the gas for which it tests is present, a colored zone is formed. The length and color of the zone depend on the type of gas and its concentration. As a rule, the confidence limits for quantitative determination with G. are ± 25%, and the standard deviations, 10 to 15%.

Gas theory, kinetic: a theory which predicts the macroscopic properties of gases, e.g. pressure, temperature or transport coefficients, from the properties of the molecules and the forces of their interactions. Conversely, information on molecular interactions can be inferred from the macroscopic effects.

The starting point of the G. is the model of the ideal gas, in which the molecules move completely randomly and independently of each other. Intermolecular forces are ignored. Interactions occur only when the particles collide with each other or with the limiting walls. The free flight time is very long in comparison to the collision time, so that collisions of more than two particles are extremely rare. This allows for important simplifications with respect to the kinetic thoery of dense gases or liquids. The collisions are elastic, and the conservation laws of classical mechanics apply. They change the velocities and kinetic energies of the colliding particles, so that a distribution pattern of velocities and energies of the molecules arises. In simple G., this is an equilibrium distribution, the Maxwell-Boltzmann distribution. The pressure of the gas is the result of the collisions of the gas particles with the vessel walls. It is the sum of the momenta which are transferred to the wall per unit of time and area: $p = {}^1N m\overline{v^2}/3$, where ${}^1N$ is the particle density, i.e. the particle number per unit of volume, m is the particle mass and $\overline{v^2}$ is the mean square velocity (see Maxwell-Boltzmann velocity distribution). Since the product of the particle density and the molar volume V is Avogadro's number: $N_A = {}^1NV$, and the product of the molecular mass m and Avogadro's number is the molar mass $M = mN_A$, it follows from the above equation that: $pV = N_A m\overline{v^2}/3 = M\overline{v^2}/3$.

If this result of the G. combined with the macroscopic equation of state for one mole of an ideal gas, $pV = RT$ (where R is the gas constant, T the absolute temperature), $M\overline{v^2}/3 = RT = 2/3E_{kin}$ or $E_{kin} = (3/2)RT$, where $E_{kin} = M\overline{v^2}/2 = N_A m\overline{v^2}/2$ is the mean kinetic energy of 1 mol of particles. It is directly proportional to the temperature of the gas. In simple G., the molecules are assumed to be rigid spheres, and therefore to have only three degrees of translational freedom. According to the *principle of equipartition of energy*, each degree of freedom has the same energy, i.e. 1/2 RT per degree of freedom and mole. This result is used in the theoretical interpretation of the molar heat of gases.

The molecules move through space undisturbed by any forces and thus in straight lines, with constant velocity. Their direction and speed change only through collision, which means that their paths are random zigzags. The distance between two collisions is called the *free path*, and the average for a very large number of identical particles (aggregate mean) or over a long period for the same molecule (time mean) is the *mean free path* $\bar{l}$. The average number of collisions per time unit is the *mean collision frequency (collision number)* $\overline{Z}$. The product of the mean free path and collision frequency is the *mean velocity* $\overline{v} = \overline{l}\overline{Z}$.

From the Maxwell-Boltzmann velocity distribution we have $\overline{v} = (8kT/\pi m)^{1/2}$, where $k = R/N_A$ is the Boltzmann constant. For the average collision frequency, $\overline{Z} = \sqrt{2}\,{}^1N\pi d^2\overline{v}$ can be derived, and for the mean free path, $l = 1/(\sqrt{2}\,{}^1N\pi d^2)$, where d is the molecular diameter. From the above equations, average velocities can be calculated: for the oxygen molecule at 273 K, $\overline{v} = 425\ \mathrm{m\ s^{-1}}$, and for the hydrogen molecule, $\overline{v} = 1700\ \mathrm{m\ s^{-1}}$. The mean free paths are $6.5 \cdot 10^{-8}$ m for O_2 and $11 \cdot 10^{-8}$ m for H_2 at standard pressure, while at a pressure of 0.1 Pa this increases to $11 \cdot 10^{-2}$m for H_2. The mean free path is closely related to the ***transport coefficients*** of ideal gases: for the coefficient of viscosity η, G. provides the equation $\eta = m{}^1N\overline{v}l/3$, for the self diffusion coefficient D, the value $D = \overline{v}l/3$, and for the heat conductivity coefficient λ, the equation $\lambda = m^{2/}{}^1Nc_v\overline{v}l/3$, where c_v is the specific heat at constant volume (see Molar heat). According to these equations, the viscosity and heat conductivity of ideal gases are independent of pressure and increase with $\sqrt{T}$, while the diffusion coefficient is proportional to $\sqrt{T}$, but also decreases with increasing pressure.

Real gases deviate to a greater or lesser extent from this behavior. This can be taken into account by introducing attractive and repulsive interaction potentials $\phi\,(r)$. In the ***Chapman-Enskog model***, a repulsive potential $\phi\,(r) = \mathrm{const.}/r^n$ is introduced, where r is the distance between particles and n is an empirical parameter which is always > 2. In the ***Sutherland model***, a repulsive potential $\phi\,(r) = \infty$ is introduced for $r < d$, and $\phi\,(r) = \mathrm{const.}/r^n$ for $r > d$. The ***Lennard-Jones potential*** $\phi\,(r) = \mathrm{const.}[(d/r)^{n_1}\text{-}(d/r)^{n_2}]$ (where n_1 and n_2 are empirically determined integral exponents) takes into account both attractive and repulsive forces. The ***Enskog formula*** $\eta = 0.499m{}^1N\bar{v}l$ is in much better agreement with experiment. From the Sutherland model, the mean free path $l = l_\infty(1+CT)^{-1}$, where l_∞ is the mean free path of the ideal gas, and l is the mean free path when interactive forces are taken into account, and C is the Sutherland constant. In this model, for example, the viscosity is $\eta = A\sqrt{T}\,(1+CT)^{-1}$, where A is a substance constant.

Historical. After its origin in the work of Bernoulli (1738), G. was developed in the middle of the 19th century by König and Clausius. They developed the concept of the free path and interpreted heat as mechanical motion of molecules. In the second half of the 19th century, a strictly statistical interpretation was developed by Boltzmann and Maxwell, which was later extended by Einstein, von Smoluchowski and Langevin. The important developments of the 20th century have been expansion to include non-equilibrium distributions and denser media, and the introduction of quantum statistical methods.

Gastrin: a peptide hormone secreted by the mucous membrane of the pyloric region of the stomach; it stimulates hydrochloric acid secretion in the stomach, enzyme secretion by the pancreas and peristalsis. ***Gastrin I*** (human) is Pyr-Gly-Pro-Trp-Leu-(Glu)$_5$-Ala-Tyr-Gly-Trp-Met-Asp-Phe-NH_2; ***gastrin II*** (human) has an *O*-sulfate on the tyrosine in position 12. All species of vertebrates secrete gastrins I and II, in various ratios. Surprisingly, the *C*-terminal, *N*-protected peptide has the same physiological effect as the native hormone, and about 1/10 the activity (per mole). The pentagastrin Boc-β-Ala-Trp-Met-Asp-Phe-NH_2 (Peptavlon®) is used clinically in diagnosis of gastric secretion. The precursor of G. does not have an *N*-terminal pyroglutamate group (Pyr); instead there is a glutamine group here which is converted to pyroglutamate in the course of releasing G. from the precursor.

The precursor of G. is thought to be progastrin, which has a relative molecular mass around 10,000. The shorter-chain ***human big gastrin*** consists of 34 amino acid residues, which in addition to the G. sequence contains an *N*-terminal 17 amino acids: Pyr-Leu-Gly-Pro-Gln-Gly-Pro-Pro-His-Leu-Val-Ala-Asp-Pro-Ser-Lys-Lys. Since the discovery of other G., human gastrin I has been named ***human little gastrin I*** (Hg-17 I), and the 14-amino-acid G. isolated from certain G.-producing tumors is called ***human minigastrin I*** (Hg-14 I).

The name G. was coined in 1905 by Edkins. He hypothesized that his extracts of pyloric mucous membrane stimulated release of hydrochloric acid, and thus initiated digestion by pepsin.

Gastrin inhibitory peptide, abb. ***GIP***: a linear polypeptide which inhibits all gastric secretion. Its *N*-terminal region has certain sequence homologies with glucagon, secretin and vasoactive intestinal peptide.

Gastrointestinal hormones: peptide hormones secreted by the gastrointestinal tract; they are involved in regulation of the digestive process. They are not secreted by glands, but by specialized cells in certain tissue regions. Some important representatives are gastrin, secretin, motilin, cholecystokinin, gastrin inhibitory peptide and vasoactive intestinal peptide. The G. have also been detected in the brain, nerve fibers and nerve endings.

Gattermann-Hopff reaction: a method for synthesis of aromatic carboxylic acid amides by reaction of arenes with carbamoyl chloride in the presence of aluminum chloride:

$$R\text{–}C_6H_5 + Cl\text{–}CO\text{–}NH_2 \xrightarrow{(AlCl_3)} R\text{–}C_6H_4\text{–}CO\text{–}NH.$$

Gattermann-Koch synthesis: synthesis of aromatic aldehydes by reaction of carbon monoxide and hydrogen chloride with aromatic hydrocarbons in the presence of aluminum chloride and copper(I) chloride:

$$R\text{–}C_6H_5 + CO + HCl \xrightarrow{(AlCl_3, CuCl)} R\text{–}C_6H_4\text{–}CHO$$

It was originally thought that the unstable formyl chloride was formed first from the carbon monoxide and hydrogen chloride, and this then served as the formylating reagent with the aluminum chloride. More recent studies have shown that the formylating agent is formed directly from the components:

$$CO + HCl + AlCl_3 \rightarrow CH{=}OAlCl_4.$$

Substituted aromatic hydrocarbons are formylated preferentially at the 4-position. Phenols and phenol ethers do not undergo G.

Gattermann reaction: method for substitution of the diazonium group in diazonium salts with halides, cyanides, rhodanides, etc. in the presence of freshly precipitated metallic copper:

$$C_6H_5\text{–}\overset{+}{N}{\equiv}N|\, X^- \xrightarrow{(Cu)} C_6H_5X + N_2.$$

The G. is a variant of the Sandmeyer reaction which is easier to carry out on preparative scale and occurs under milder conditions.

Gattermann synthesis: method for producing aromatic aldehydes by reaction of aromatic hydrocarbons, phenols or phenol ethers with hydrogen chloride and hydrogen cyanide in the presence of aluminum chloride or zinc chloride:

$$R\text{–}C_6H_5 + HCN + HCl \xrightarrow{(AlCl_3)} R\text{–}C_6H_4\text{–}CH{=}\overset{+}{N}H_2Cl^-$$

$$\xrightarrow[-\,NH_4Cl]{+\,H_2O} R\text{–}C_6H_4\text{–}CHO.$$

In this reaction, the formylation reagent is thought to be a complex of the primary formimidochloride and the catalyst:

$$H\text{–}C{\equiv}N + H\text{–}Cl \rightleftharpoons Cl\text{–}CH{=}NH \underset{-\,AlCl_3}{\overset{+\,AlCl_3}{\rightleftharpoons}} [\overset{+}{H}C{=}NH \longleftrightarrow$$

$$HC{\equiv}\overset{+}{N}H]\, AlCl_4^-.$$

The aldehyde group is highly selectively introduced into the 4-position relative to the activating substituent already present in the ring. If this position is blocked, the substitution occurs at the 2-position. By using zinc cyanide instead of hydrogen cyanide, R. Adams improved the reaction (1923) by avoiding the use of the very poisonous anhydrous hydrogen cyanide. The G. is a variant of the Gattermann-Koch synthesis (see).

Gauche conformation: see Stereoisomerism.

Gaussian error distribution: see Distribution function.

Gay Lussac laws: First G., see State, equations of, 1.1. Second G., see Thermodynamics, first law. Third G., see Chemical volume law of Gay-Lussac.

GC: abb. for gas chromatography.

GC/MS coupling: see Mass spectroscopy.

Gd: symbol for gadolinium.

Ge: symbol for germanium.

Gel: a colloidal system in which the components form a network or honeycomb-like structure in the dispersion medium. There are two types of G., macromolecular G. and coagulation structures.

1) ***Macromolecular G.*** are formed when their solutions cool, or due to limited swelling on addition of other solvents. They can be formed either from organic or from inorganic molecules. Macromolecular G. formed through limited swelling usually consist of cross-linked polymers. In macromolecular G. formed by addition of other solvents, the molecules are held together by non-covalent bonding. Organic macromolecular G. differ from inorganic G. in obeying Hook's law for deformation under mechanical stress over wide ranges of stress.

2) ***Coagulation structures*** are formed by slow coagulation of colloidal particles (oxides, latexes, etc.). A film of liquid remains between the particles. Coagulation structures can be transformed back to the sol state by mechanical stress such as shaking or stirring (see Redispersion). When the stress is removed, the G. forms again. This reversible sol-gel-sol transformation is called Thixotropy (see). The structures formed by drying of G. of both types are called ***xerogels*** (e.g. gelatins, glue, kieselguhr). ***Lyogels*** shrink on storage, even in moist air (see Syneresis).

Gelatin: a mixture of proteins obtained by partial hydrolysis of the collagen of bones, connective tissues and skin (M_r 60,000-90,000). G. is insoluble in alcohol and ether. It swells in aqueous solution, and when cooled, it sets as a gel. The most abundant amino acids are glycine (over 30%), proline, alanine and hydroxyproline. To produce G., chopped and cleaned raw materials, usually the skins and bones of cattle, are demineralized by treatment with dilute hydrochloric acid, followed by calcium carbonate suspension. After neutralization, they are extracted with water at 50 to 80 °C. The extract is dried as a powder, tablets, leaves or cubes, and is used in the food industry to prepare jellies, mayonnaise, dessert powders, puddings, etc. It is used in the photographic industry as the carrier for light-sensitive emulsion (photogelatin), in bacteriology for growth media and in pharmacy for production of G. capsules. Because it does not contain certain essential amino acids, the nutritional value of G. is limited.

Gelatin dynamite: see Dynamite.

Gel electrophoresis: see Electrophoresis.

Gel filtration: see Permeation chromatography.

Gelinite: see Macerals.

Gelite: see Coal.

Gel permeation chromatography: same as Permeation chromatography (see).

Gelseminic acid: same as Scopoletin (see).

Generator gas: an important gas for chemical syntheses and fuel, made by gasifying solid fuels in a generator. It is a lean gas, with a heating value of 3200 to 6800 kJ m^{-3}. In the oxidation zone of the generator, the solid fuel is burned in the presence of air or air and steam: $C + O_2 \rightleftharpoons CO_2$. The glowing fuel decomposes the water to hydrogen and oxygen; the oxygen reacts further with the carbon: $C + H_2O \rightleftharpoons CO + H_2$. In the layers above the oxidation layer, there is a lack of free oxygen, so a reduction occurs: $CO_2 + C \rightleftharpoons 2CO$. Depending on the fuel, temperature and gasification mixture, the product is (in mass %): 5 to 11% carbon dioxide, 11 to 18% hydrogen, 21 to 30% carbon monoxide, 47 to 56% nitrogen, and small amounts of methane and higher hydrocarbons. G. is used as a synthesis gas, or as a fuel, e.g. to heat metallurgical and glass furnaces. Wood is sometimes gasified, in specially constructed small generators (wood gas).

Gene synthesis: the chemical and enzymatic synthesis of genes. The current purely chemical methods of Oligonucleotide synthesis (see) are not adequate for the synthesis of DNA segments with known base sequences carring the information for other nucleic acids or proteins. Only by using enzymes to lengthen the chains of chemically synthesized oligo- and polynucleotides is it possible to synthesize structural genes. There are two basic strategies for synthesis.

1) ***DNA ligase method***. The basis of this method is enzymatic linkage of relatively short, chemically prepared oligonucleotides (10 to 15 bases) into double-stranded DNA segments. The sequences of the synthetic oligonucleotides must be chosen so that they include the entire sequence of each strand of the double-stranded product. In addition, the oligonucleotides of the two strands must overlap, by 4 to 5 bases, in such a way that their arrangement along the double strand is dictated by complementarity. The 5'-hydroxyl group of each oligonucleotide segment is phosphorylated by an oligonucleotide kinase, and then the 5'-phosphate group of each segment is linked to the 3'-hydroxyl group of the next segment by DNA T4-ligase. This is the strategy developed in 1970 by Khorana and coworkers for the synthesis of the gene for alanine tRNA from yeast. The DNA sequence they synthesized was implied by the primary structure of yeast alanine tRNA. This G. stimulated other research groups to take up the problem of G., in order to make possible the production of eukaryotic proteins. The genes for somatostatin, angiotensin II, both chains of insulin and interferon were landmark syntheses. For example, the gene for human leukocyte interferon was constructed from 67 chemically synthesized oligonucleotides. The 517-base-pair product was of very low purity, but it was possible by cloning (see Genetic engineering) to enrich and isolate it. This is a great advantage of oligonucleotide synthesis over peptide synthesis: even a mixture of sequentially correct gene fragments which would be absolutely useless to a chemist can be selected by genetic techniques and reproduced by cloning.

2) ***DNA polymerase methods***. As synthetic techniques for polynucleotides on polymeric carriers were improved (see Oligonucleotide synthesis), it became possible to construct oligonucleotides of 30 to 40 bases. The sequences of these are chosen so that their ends can form overlapping duplexes about 10 base pairs in length. This produces a molecule of alternating single- and double-stranded sections; it is accepted by DNA polymerase I (Klenow enzyme) as a substrate. In the presence of the four nucleoside 5'-triphosphates, dATP, dGTP, dCTP and dTTP, this enzyme fills in the single-stranded sections with the correct complementary sequences. It is obvious that the enzyme-catalysed fill-in stragegy is far more economical than the DNA ligase strategy, and that with further improvements in the chemical synthesis of polynucleotides on carriers, it will be possible to synthesize longer genes for proteins, or specifically designed operators, promotors and other regulatory sequences.

Genetic code: see Code.

Genetic engineering: methods of genetic manipulations of cells. G. is used to introduce foreign sequences of DNA (genes) into cells so that they synthesize proteins of which they were not previously capable. G. is of great interest for production of biological substances (see Biotechnology).

There are two basic processes in G.: recombination in vitro and transformation in vivo. The goal of recombination is to couple a DNA fragment encoding a protein of interest with a suitable carrier, creating a *recombinant DNA* (abb. *rDNA*). The necessary changes are achieved by *cutting* and *splicing*. The first step is to obtain the desired DNA fragment, either by automated total synthesis (see Gene synthesis), by synthesis from isolated messenger RNA using reverse transcriptase, or by isolation from chromosomal DNA. For the third method, the chromosomal DNA must first be fragmented, either by shear forces or, more commonly, by *restriction endonucleases*. These are bacterial enzymes which cleave the DNA at recognition sequences specific to the enzyme in use. The resulting mixture of DNA fragments must then be separated, and the desired sequence isolated from the rest. The isolated gene must then be incorporated into a vector which will carry it into the cell. The vectors are *plasmids* (extrachromosomal DNA from bacteria) or viruses which have been constructed for the purpose. To make selection of transformed cells possible, the plasmid usually carries genes for resistance to a certain antibiotic. When the mixture of transformed and untransformed cells is plated onto a medium containing the antibiotic, only the transformed cells grow.

Transformation begins with the transport of the recombinant DNA into the cell, for which special methods are often required to bring it past the cell membrane. Within the cell, the recombinant DNA may either be incorporated into the genome, or it can exist independently, as a plasmid or extrachromosomal DNA. The foreign gene must be transcribed and translated in the cell (gene expression) to produce the desired protein. Therefore, the recombinant DNA must also carry regulatory elements (operator and promoter) which insure mRNA synthesis (see Nucleic acids).

For genetically engineered production strains, host cells are required in which the genetics and metabolism are well known, and which thrive in fermenters. It is also important that the transformed host cell should be stable, that is, retain its new genetic material for as long as possible. At first, bacteria were usually chosen as hosts, especially *Escherichia coli* and *Bacillus subtilis*. However, *E. coli* does not secrete proteins. Yeasts are of considerable biotechnological interest. The first to be used was *Saccharomyces cerevisiae*. In 1982, a membrane protein from coliform bacteria was the first marketed product of genetic engineering; it is used as a vaccine against diarrheal diseases in pigs and cattle. In 1983, genetically engineered human insulin was introduced. It is likely that by the turn of the century, vaccines, hormones (especially insulin, somatostatin and growth hormone), interferons, some enzymes and blood components (human serum albumin, immunoglobulins, factor VIII) and antibiotics will be produced commercially from engineered organisms. G. can also be used for biotechnological production of xenobiotics and plastics, for introducing genes into crop plants (one goal is to introduce nitrogen fixation genes into grains) and in the energy sector (production of ethanol, methane and hydrogen). Another goal is introduction of genes for missing enzymes into the cells of patients with hereditary metabolic diseases, and reinjection of the engineered cells into their donor. To reduce the risk of inadvertent release of dangerous engineered organisms into the environment, strict safety regulations have been established for all such experimental work.

Genin: the non-sugar component of a Glycoside (see).

Genistein: see Flavones.

Gentamicins: a group of aminoglycoside antibiotics formed by *Micromonospora purpurea*. The aglycon of G. is 2-deoxystreptamine; two aminomonosaccharides are bound to it glycosidically. The G. applied therapeutically consists of components C_1, C_{1a} and C_2. The G. have a broad action spectrum and are used mainly in cases of severe general infection with gram-negative bacteria. Sisomicin has a similar structure.

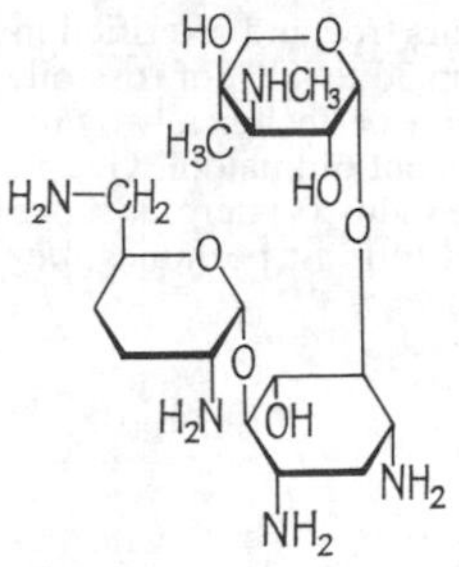

Gentamicin C_{1a}

Gentian violet: a mixture of tetra-, penta- and hexamethylparafuchsin. G. is used in Gram staining (see) and is often equated with methyl violet.

Gentianose: O-β-D-fructofuranosyl-(2→1)-α-D-glucopyranosyl-(6→1)-β-D-glucopyranoside, a nonreducing trisaccharide found in plants of the *Gentianaceae* family.

Gentiopicroside: see Iridoids.

Gentisic acid, ***2,5-dihydroxybenzoic acid, hydroquinone carboxylic acid***: a structural isomer of dihydroxybenzoic acid. G. forms colorless crystalline needles; m.p. 205 °C. It dissolves readily in hot water, alcohol and ether, but only slightly in benzene. When heated above its melting point, it decomposes into hydroquinone and carbon dioxide. G. is a metabolic product of various *Penicillium* species, and has an antibiotic effect. It can be synthesized by oxidation of salicylic acid with potassium persulfate, or by a modified Kolbe synthesis (see) from hydroquinone. G. and its amide are used in treatment of rheumatism.

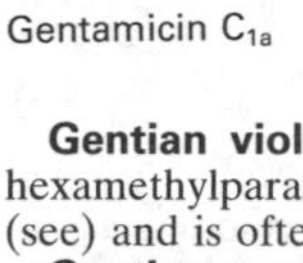

Gentisyl alcohol, ***2,5-dihydroxybenzyl alcohol***: a colorless, crystalline compound, m.p. 100-101 °C. G. is soluble in water, alcohol and ether, slightly soluble in chlorinated hydrocarbons and insoluble in petroleum ether. It is a metabolic product of the mold *Penicillium patulum* and is effective against gram-positive bacteria.

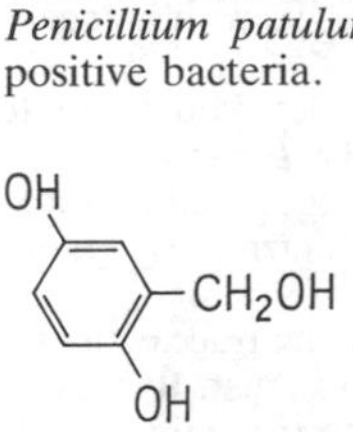

Geranial: see Citral.

Geraniol: 6-*cis*-2,6-dimethylocta-2,6-dien-8-ol, an acyclic monoterpene alcohol. The 6-*trans*-form is called ***nerol***. G. and nerol are colorless, oily liquids which smell like roses; they are optically inactive, and the b.p. of G. is 230 °C at 130 kPa. G. tends to oxidize, which leads to a less pleasant odor. G. is usually accompanied by nerol. It occurs free and esterified in many essential oils, making up 30 to 40% of rose oil, for example. G. can be synthesized technically by reduction of citral or rearrangement of linalool. G. and nerol are used in the perfume industry; nerol is more expensive. G. is a component of most perfumes, because of its rose fragrance.

Geranium oil, ***pelargonium oil***: a colorless essential oil; it can also be greenish to brownish. G. has a pleasant, rose-like odor. It consists chiefly of geraniol (up to 40%), citronellol, linalool, phenylethyl alcohol, terpineol and menthol. It is obtained from the leaves of geraniums by steam distillation. The yields of distillation are about 15 to 18 kg oil per hectare of crop land, or 0.05 to 0.15% of the leaves. The leaves are harvested several times a year. G. is used in perfumes and soaps.

Germanium, symbol ***Ge***: a chemical element in group IVa of the periodic system, the Carbon-silicon group (see); a semimetal; *Z* 32, with natural isotopes with mass numbers 70 (20.52%), 72 (27.43%), 73 (7.76%), 74 (36.54%) and 76 (7.76%). Its atomic mass is 72.59, valence IV, rarely II, density 5.35, m.p. 837.4 °C, b.p. 2830 °C.

Properties. Pure G. is a gray, very brittle semimetal which crystallizes in a diamond lattice and has a metallic sheen. Its electrical resistance increases as the temperature falls and as its purity increases. It is similar in its chemistry to silicon. Like silicon, it is generally tetravalent, and its compounds are tetrahedral. They are largely comparable to those of Si in structure and behavior. Germanium(II) compounds are not very stable; they are reducing agents or tend to disproportionate. G. is stable in air at room temperature, and is oxidized to germanium dioxide, GeO_2, only at red heat. Ge is not attacked by dilute acids and bases. With oxidizing acids, such as concentrated nitric or sulfuric acid or aqua regia, it forms the insoluble germanium dioxide. It reacts with halogens to form germanium tetrahalides.

Analytical. Qualitatively, G. is detected as white germanium sulfide, GeS_2, in the H_2S separation step. It is determined gravimetrically as germanium dioxide, GeO_2, or, in very small concentrations, photometrically as phenylfluorene.

Occurrence. G. is a widely occurring trace element, making up $6.7 \cdot 10^{-4}$% of the earth's crust. It is found in higher concentrations only in the rare minerals argyrodite, $4\,Ag_2S \cdot (Ge, Sn)S_2$ and germanite, $3\,Cu_2S \cdot FeS \cdot 2GeS_2$. Certain zinc ores, and especially anthracite coal, contain recoverable amounts of G.

Production. The starting materials for production of G. are germanite, zinc processing residues and fly ash from burning some types of anthracite coal. After a pretreatment, G. is separated from these materials in the form of the volatile germanium tetrachloride, $GeCl_4$. The chloride is hydrolysed with water to yield germanium dioxide, and this is reduced with hydrogen to elemental G. The G. used in semiconductors is purified by zone melting to a degree of purity so high that there is only one foreign atom for every 10^{12} Ge atoms.

Applications. G. is transparent to infrared radiation, and is therefore used for windows in IR spectrometers, etc. It is an important semiconductor material, which is usually doped with arsenic or gallium, and is used in the production of transistors and other electronic components. It is also a component of certain special alloys.

Historical. G. was discovered in 1886 by the chemist Winkler, who noticed a constant discrepancy of 7% in his analysis of argyrite. He determined that this was due to an element which had previously been unrecognized. Thorough studies of the physical and chemical behavior of the new element established its identity with eka-silicon, which had been predicted in 1872 by Mendeleyev.

Germanium compounds: a key role in the chemistry of G. is played by the germanium(IV) halides, which are prepared by reaction of halogens with germanium. ***Germanium tetrafluoride***, GeF_4, M_r 148.58, subl.p. -37 °C, is a colorless, poisonous gas, which reacts with metal fluorides to form hexafluorogermanates, $M^I_2[GeF_6]$. ***Germanium tetrachloride***, $GeCl_4$, M_r 214.41, density 1.8443, m.p. -49.5 °C, b.p. 84 °C, is a colorless liquid which fumes in air; it can also be prepared by reaction of germanium dioxide with conc. hydrochloric acid. The reaction of germanium with $GeCl_4$ yields the colorless powder ***germanium dichloride***, $GeCl_2$, which is a strong reducing agent and, at higher temperatures, disproportionates into $GeCl_4$ and Ge. The reduction of $GeCl_4$ with lithium alanate yields the gaseous ***monogermane***, GeH_4, M_r 76.62, m.p. -165 °C, b.p. -88.5 °C, which decomposes to its components above 350 °C. GeH_4 is the first of the ***germane*** or ***germanium hydride*** series, Ge_nH_{2n+2}, which are somewhat more stable in the presence of oxygen and to hydrolysis than are the corresponding silanes. Hydrolysis of sodium or calcium germanide, NaGe or CaGe, yields ***polygermanes***, high-molecular weight compounds with the composition $[GeH]_n$ and $[GeH_2]_n$. Hydrolysis of $GeCl_4$ or reaction of conc. nitric acid with germanium yields ***germanium dioxide***, GeO_2, M_r 104.59, which has two modifications. One is hexagonal and comparable to quartz (m.p. 1115±4 °C) and the other is tetragonal and isomorphic with tinstone (m.p. 1086±5 °C). Like silicon dioxide, GeO_2 forms a glass, which has a very high index of refraction. Glasses containing GeO_2 are therefore used in optics. GeO_2 dissolves in aqueous alkalies, forming the corresponding ***germanates***, e.g. $GeO_2 + 4\,NaOH \rightarrow Na_4GeO_4 + 2\,H_2O$. There are many germanates comparable to the Silicates (see); these are based on condensed polyanions. $GeCl_4$ is a starting material for a wide variety of ***organogermanium compounds***, which can be synthesized by reaction with Grignard reagents. They are of interest because of their pharmacological effects.

Germ removal: in sewage treatment, the removal of germs (microorganisms), especially the pathogens. a) *Sterilization* is the destruction of all microorganisms, e.g. by boiling. b) *Pasteurization* is the destruction only of pathogenic germs by heating, e.g. in the sludge from community sewage treatment plants. c) *Disinfection* is the reduction of the number of germs to a very low level. The process is essentially the same as disinfection in water treatment.

Gestagens, ***corpus-luteum hormones***: a group of female sex hormones derived from pregnane; they contain 21 C atoms and oxygen functions on C atoms 3 and 20. They differ from the adrenal cortex hormones, which are also pregnane derivatives, in having fewer oxygen functions. The most important natural G. is ***progesterone*** (4-pregnene-3,20-dione), which is formed by the corpus luteum in the second half of the menstrual cycle and by the placenta during pregnancy. Progesterone is biosynthesized from cholesterol, via pregnenolone. It forms colorless crystals, m.p. 128 °C (from ethanol/water), 121 °C (from pentane/hexane). $[\alpha]_D^{20}$ +192° (ethanol).

The G. are responsible for preparation of the proliferated uterine mucosal lining for the implantation of the fertilized egg, and for maintenance of pregnancy. They prevent further ovulations by inhibition of gonadotropin release.

Progesterone is used to treat menstrual irregularities, usually in combination with estrogens. It has practically no effect when taken orally; it is applied parenterally as a solution in oil. Its effects are enhanced by introduction of an α-OH group on C-17; acylation of this OH group delays degradation of the hormone, and thus prolongs its effect. *Hydroxyprogesterone acetate* has a gestagenic effect when applied orally; *hydroxyprogesterone caproate* is applied parenterally as a depot preparation. The 17α-hydroxyprogesterone derivative ***chlormadinone*** is used mainly as a gestagen component in contraceptives. 19-Nortestosterone derivatives which have a 17α-ethynyl or other alkyl or alkenyl residue and analogs are increasingly being used for contraceptives. Some examples are ***norethisterone*** and ***levonorgestrel***; the latter is the D(-)-form of the racemic synthetic norgestrel, and has the same configuration as the natural steroids.

Getter: a substance used to capture gases, usually by a chemical reaction. G. can be pure metals, such as cesium, barium, calcium, magnesium, zirconium or hafnium, alloys, such as titanium-zirconium alloys, and compounds, such as zirconium hydride and titanium hydride. Activated charcoal is also used. G. are used to remove the last traces of undesired gases, such as oxygen or nitrogen, from an evacuated object such as a light bulb, television tube or x-ray tube.

g-Factor: see Electron spin resonance spectroscopy.

GH: see Somatotropin.

Gibberellins: a group of phytohormones. Chemically they are cyclic diterpene acids with a hydrocarbon skeleton called *gibbane*. The G. may contain 19 or 20 C-atoms. Individual compounds are indicated by a capital letter with a numerical subscript; the most readily available is the C_{19} ***G. A₃***, called ***gibberellic acid***. This forms colorless crystals (m.p. 235 °C) which are soluble in ethanol, acetone and alkalies, but practically insoluble in water. Gibberellic acid is obtained from the fermentation liquor of the fungus *Fusarium moniliforme* (formerly *Gibberella fujikuroi*); it is not very stable. The G. promote germination, flowering and internodal lengthening in growing plants. Chlorocholine chloride (see) is a G. antagonist used in agriculture.

gibberellic acid

Gibbs adsorption equation: see Adsorption.

Gibbs-Duhem equation: in thermodynamics, an equation showing the change in the Partial molar parameters (see) $\overline{Y}_i$ of all components of a mixture when its composition changes by a differential amount, under otherwise constant external conditions (*p*, *T*

R = H : Progesterone
R = OH : Hydroxyprogesterone
R = $OCOCH_3$: Hydroxyprogesterone acetate
R = $OCO(CH_2)_4CH_3$: Hydroxyprogesterone caproate

Chlormadinone acetate

R = CH_3 Norethisterone
R = C_2H_5 Levonorgestrel

constant): $\sum_i x_i d\bar{Y}_i = 0$, where x is the molar fraction of component i. A change in the partial molar parameter of a single component thus causes those of all the other components to change as well. For example, for the chemical potential μ of a binary mixture of substances 1 and 2, $x_i d\mu_1 + x_2 d\mu_2 = 0$, or, (because $x_2 = 1 - x_1$), $x_1\, d\mu_1 = (1 - x_1)d\mu_2 = 0$. If one divides by dx_1, it follows that $x_1 d\mu_1/dx_1 + (1 - x_1)d\mu_2/dx_1 = 0$. Thus if the dependence of the chemical potential of one component on the composition is known, e.g. $d\mu_1/dx_1$, that of the second component can be calculated using the G. Since $\mu_i = \mu_i + RT\ln f_i x_1$, the same calculations are possible for activity coefficients f_i; on this basis the activity coefficients may be determined experimentally (see Activity).

Gibbs-Duhem-Margules equation: see Duhem-Margules equation.

Gibbs' fundamental equation: see Thermodynamic potentials.

Gibbs heat function: an older term for Enthalpy (see).

Gibbs-Helmholtz equations: equations relating the concepts of free energy and free enthalpy in the Thermodynamics (see) of chemical reactions. For isobaric reactions, $\Delta_R H = \Delta_R G + T\Delta_R S$, and for isochoric reactions, $\Delta_R E = \Delta_R F + T\Delta_R S$. Here $\Delta_R H$ and $\Delta_R E$ are the reaction enthalpy and energy, respectively, $\Delta_R G$ and $\Delta_R F$ are the free reaction enthalpy (Gibbs free energy) and isochoric free energy (Helmholtz free energy), and $\Delta_R S$ is the reaction entropy, all for 1 formula mole turnover at constant temperature.

If the reaction is carried out irreversibly and isobarically, the entire energy difference $\Delta_R H$ between the products and reactants is obtained in the form of heat. If the reaction is carried out reversibly, the energy difference is split into two portions. One of them, $\Delta_R G$, can be released as work (*reaction* or *useful* work), and the other, $Q_{rev} = T\Delta_R S$ is the reversible heat. The situation is analogous, in isochoric reactions, for $\Delta_R E$ and $\Delta_R F$. Since $(\delta\Delta_R G/\delta T)_p = -\Delta_R S$ and $(\delta\Delta_R F/\delta T)_v = -\Delta_R S$, the G. can also be written in the form $\Delta_R H = \Delta_R G - T(\delta\Delta_R G/\delta T)_p$ or $\Delta_R E = \Delta_R F - T(\delta\Delta_R F/\delta T)_v$. The G. also apply to the standard reaction parameters (see Standard state), e.g. $\Delta_R G^o = \Delta_R H^o + T(\delta\Delta_R G^o)/\delta T)_p$. Since $\Delta_R G^o$ is linked to the equilibrium constant (see Mass action law) by the equation $\Delta_R G^o = -RT\ln K$, this equation can be used as a starting point in the thermodynamic calculation of K (see Ulich approximations).

Gibbs phase law: a law derived in 1878 by J.W. Gibbs from thermodynamics, using the equilibrium conditions: the number of *coexistent phases* P in a system is given by: P + F = C + 2, where F is the number of "degrees of freedom", that is, the number of state variables which can be changed in a certain interval without causing one of the coexisting phases to disappear or a new one to appear. C is the number of independent chemical substances of which the system consists. If reactions can occur in a multi-component system, the number of reaction equations must be subtracted from the number of chemical components to give the number of independent components relevant to P.

Examples: 1) For single-component systems, e.g. water, C = 1. If there is only one phase (P = 1), e.g. water vapor, the system has two degrees of freedom, according to the G., i.e. the temperature and pressure can be changed independently without changing the phase state. If two phases coexist (P = 2), e.g. vapor and liquid, the system has only one degree of freedom. If the temperature is chosen arbitrarily, the vapor pressure is established according to the evaporation equation (see Phase diagram). For three phases in equilibrium, F = 0, i.e. there is no variable of state which can be arbitrarily chosen. The system is stable only at a single point, the *triple point*.

2) Binary systems, e.g. salt/water, lead/silver: C = 2. These systems always have one degree of freedom more than the single-component systems above: the composition of the mixed phases. At points with no degrees of freedom, the *quadruple points*, four phases are in equilibrium. At eutectic and peritectic points, the gas, liquid and two solid phases coexist; at manotectic points, the gas phase coexists with two liquid and one solid phase (see Phase diagram).

3) Non-stoichiometric system of $CaCO_3$, CaO and CO_2: The three chemical components can react according to $CaO + CO_2 \rightleftarrows CaCO_3$. The number of independent components is 2. The conditions listed under 2) apply.

It must be remembered that application of the G. requires that thermodynamic equilibrium has been reached. Especially when several solid phases are involved, a long time may be required for the system to reach equilibrium.

Gilding: the creation of thin layers of gold on metallic or nonmetallic objects, for decoration or as a corrosion protection. Contact metals in electronic devices are protected against corrosion by ***galvanic G.***. The electrolysis bath contains gold as a cyanide complex and excess potassium cyanide. Some metals can only be galvanically gilded after application of an intermediate layer of copper or brass. Objects may also be gilded chemically, by dipping in a gold solution; in ***contact G.***, the solution also contains metallic zinc as a sacrifice metal. In ***rubbed G.***, which is used to repair defective spots, the gold compound is applied to the object as a paste which also contains a reducing agent. Gold can also be applied by plating: rolling a thin gold layer on a suitable base, or a very thin layer of gold leaf can be glued to a smooth base.

GIP: see Gastrin-inhibiting peptide.

Girard's reagents: ***Girard's reagent T*** is trimethylammonium acetic hydrazide chloride, $(CH_3)_3N\text{-}CH_2\text{-}CO\text{-}NH\text{-}NH_2\text{-}Cl^-$, and ***Girard's reagent P*** is $C_5H_5N\text{-}CH_2\text{-}CO\text{-}NH\text{-}NH_2Cl^-$, pyridinium acetic hydrazide chloride. They are water-soluble reagents for detecting aldehydes and ketones, with which they react to form acylhydrazones. Several hormones were isolated using G.

Girbotol process: a process for removing hydrogen sulfide from gases (coking, natural, city or synthesis gases); it is similar to the Sulfosolvane process (see).

Girdler process: a process for removing hydrogen sulfide from gases (coking, natural, city or synthesis gases).

Gitaloxigenin: see Purpurea glycosides.

Gitaloxin: see Purpurea glycosides.

Gitoxigenin: see Purpurea glycosides; see Lanata glycosides.

Gitoxin: see Purpurea glycosides; see Lanata glycosides.

Glacial acetic acid: see Acetic acid.

Glaser reaction: a method for producing diacetylenes by oxidative coupling of acetylenes with copper(II) salts in the presence of ammonia or pyridine:

$$2R{-}C{\equiv}CH \xrightarrow[-2H^+,\,-2Cu^+]{+2Cu^{2+}} 2{-}C{\equiv}C{-}C{\equiv}C{-}R.$$

Application of the G. to longer-chain diynes produces cycloalkapolynes, which can be used as intermediates in the production of annulenes.

Glass: an amorphous solid formed from a melt by freezing in of the liquid structure during cooling or quenching. G. differs from crystalline materials in having no defined melting or freezing temperature, but instead a transformation or solidification range in which it does not undergo a phase change (change in state of matter). The structure of G. is disorderly, which leads to isotropism of the properties. A G. is thermodynamically unstable compared to a crystalline form of the same substance.

There are various kinds of G.: oxide, chalcogenide, metal, element and salt G. By far the most important are the oxidic silicate G., while element and salt G. are so far only of scientific interest.

1) Most ***oxide G.*** are silicate G., in which the most important oxide component is silicon dioxide, SiO_2.

Properties and structure of oxide G.: tensile strength, 70 to 200 $N \cdot mm^{-2}$, compression strength 800 to 2000 $N \cdot mm^{-2}$, bending breaking point, 50 to 80 $N \cdot mm^{-2}$, elasticity modulus 60,000 to 90,000 $N \cdot mm^{-2}$, density 2.2 to 6 $g \cdot cm^{-3}$, transformation temperature 450 to 750 °C (silica glass, 1100 °C), specific heat 0.35 to 1.00 $J \cdot g^{-1}K^{-1}$, heat expansion coefficient, 30 to 50 $\cdot 10^{-}K^{-1}$ (silica glass, 5 to 6 $\cdot 10^{-7}$ K^{-1}), heat conductivity 0.009 to 0.013 $J \cdot cm^{-1} s^{-1} K^{-1}$, index of refraction, 1.33 to 2.1, specific electrical resistance, 10^{10} to 10^{19} ohm · cm. The viscosity is the most important physical parameter for the process of melting and forming glass; it decreases rapidly with increasing temperature, due to breaking up of the silicate network structure. At room temperature, the viscosity is 10^{17} to 10^{19} Pa · s, and in the transformation range it is 10^{12} to $5 \cdot 10^{13}$ Pa · s. At viscosities of $5 \cdot 10^4$ to 10^8 Pa · s, the G. can be shaped by blowing, and between 10^3 to $5 \cdot 10^6$ Pa · s, by machine molding. In its melting range, G. has viscosities between 1 and 15 Pa · s. Its microstructure is characterized by two theories: the network theory of Zachariasen and Warren, which assumes an irregular cross-linking of polyhedra (e.g. SiO_4 tetrahedra) as structural elements, and the crystallite theory of Lebedjev, which postulates a certain fraction of ordered regions (tiny crystallites) in the G. surrounded by an irregular glass matrix. The state of a real G. depends on its thermal history and lies somewhere between these two models; the more slowly the melt is cooled, the greater the degree of order. Most industrial G. have a heterogeneous structure, due to Subliquidus separation (see). However, since these regions of separation are usually very small (< 0.5 μm), these G. are transparent.

Composition. Oxide G. consist of three groups of components, some of which are first synthesized from various compounds (hydrates, nitrates, carbonates, sulfates) by thermal decomposition during the melting process. a) Glass formers generate the network; examples are silicon dioxide, SiO_2, boron trioxide, B_2O_3, phosphorus pentoxide, P_4O_{10}, and for special optical G., beryllium fluoride, BeF_2; b) Glass converters, basic components which break up the network and thus make the melting temperature fairly low; these include lithium oxide, Li_2O, sodium oxide, Na_2O, potassium oxide, K_2O, calcium oxide, CaO and barium oxide, BaO; c) Glass formers and converters, which can either create or interrupt the network, depending on the type and concentration: e.g. aluminum oxide, Al_2O_3, magnesium oxide, MgO, lead oxide, PbO and oxides of elements in the b groups of the periodic system.

Synthesis. The production of silicate G. involves the following steps: mixing, melting, clarification and homogenization, molding, cooling and post-solidification processing. The starting materials for G. are pulverized, dried and stored in silos until use, at which time they are weighed out and mixed in ball mills or other mixers. Optimum melting depends on the presence of a certain degree of moisture in the quartz sand (3 to 4%), which improves mixing and prevents corrosive and erosive reduction of grain size during heating. Other essentials are preservation of a certain spectrum of grain size, the presence of a minimum amount of gas-generating raw materials to give a good clarification, and the use of 20 to 30% recycled glass to accelerate the melting process. As the raw materials are heated, the begin to undergo sintering and solid-solid reactions; after the melt phase has formed, the reactions continue as solid-liquid reactions. Before silicate formation occurs, there are changes in modification in some components (e.g. quartz) at relatively low temperatures; these increase their reactivity (*Hedvall effect*). A requirement for the generation and conversion of silicates is the thermal decomposition of gas-generating raw materials, which occurs with increasing temperature in this order: hydrates, hydroxides, nitrates, carbonates, sulfates. The condensed oxide decomposition products are then available for the formation of silicates. At first simple silicates are formed, which then fuse to form polymeric, irregular networks. During clarification, the residual gases are removed from the melt. This is accelerated by temporary reduction of the viscosity by raising the temperature, application of oxygen-releasing compounds (arsenic or antimony oxides, alkali metal nitrates) and bubbling air through the melt.

The oldest method of melting G. still in use is the discontinuous *pot furnace* method. The raw materials are placed in round, open or covered vessels (pot furnaces) which are 0.5 to 1.5 m in diameter and 0.6 to 0.8 m tall. They are melted at 600 to 1600 °C. The pots are indirectly heated in ovens, in which the atmosphere is usually oxidizing (especially for lead G.) but may be mildly reducing. Pot furnaces are still important, even in modern G. works, for the production of special G. (e.g. optical, tempered, colored or translucent G.). Smaller amounts of special G. can be melted in inductively heated crucibles which hold up to 50 l melt and permit homogenization by stirring. Mass-produced G. is made in a continuous *vat melt*. The shapes and sizes of the vats, which are lined with firebricks, vary widely and depend on the capacities, type of G. and methods for further processing. Large vats can accomodate up to 5000 t and their output can be more than 300 t/d. The vat has several distinct areas, the melting area, the clarification area and the working area. The powdered, briquetted or granulated mixture of raw materials is usually applied automatically through openings in the top of the vat to form a very thin layer on top of the melt. Care is taken to keep the unmelted mixture in the melting area, where it is melted by U-flame or cross heating (the flames are led from the walls across the melt. Both regenerative firing and gas burners are used; the latter are placed directly below the glass melt (unit melters). The degassing of the mixture, which occurs in the clarification area, is often promoted by a supplementary electrical heater. The G. flows into the working area through an opening below the melting level; here it is kept at the working temperature by separate heating. All-electrical vats are being used increasingly for hard (high-melting) G. In them, carbon or molybdenum electrodes are introduced into the melt from the floor of the vat, and the melt acts as the electrical resistor. This method gives high melting capacities, and a further advantage is that the relatively cooler surface prevents evaporative losses. Electrical melting is also the most energy efficient method, taking only about half the energy of conventional methods.

Both manual and mechanical methods of working the melt are used. Only special G. are worked manually, usually by means of a *glass blower's pipe* (an iron tube with a wooden handle); the end of the pipe is dipped into the melt and the glass which adheres to it is blown into a sphere. This is then shaped by turning, drawing, cutting and blowing into iron forms to make objects of various shapes. Mass-produced G. are worked mechanically. ***Hollow G.*** is made by pressure blowing, suction blowing or air-stream blowing; the following steps are involved: removal of the melt from the vat and transport to the blowing machine, formation of the preliminary mass, formation of the hollow mass, molding, and transport to the cooling aggregate. In series blowers, the objects are blown out of a continuous, molten G. ribbon.

Plate glass can be made from the vat melt by pouring and rolling (***profile glass***), or by mechanical drawing. In the mechanical process, a ribbon of glass is drawn vertically out of a special vat and pulled upwards over rollers and coolers (*Fourcault* or *Fourcault-Asahi method*). An extremely productive method is the *float glass process* in which a G. ribbon floats on a metal melt which is heated from below. Simultaneous heating from above produces a plate glass of high surface quality which does not need to be polished (G. for mirrors, furniture and motor vehicles).

G. rods are made by drawing the G. out of the melt, either vertically or horizontally, as a rod of the desired diameter. Special cooling apparatus is not needed (Koroljov process). In the *Danner rod drawing method*, a strand of G. is wrapped in a spiral on a rotating, inclined firebrick cone. It is drawn horizontally off the end of the cone as a rod, or pressurized air blown through a channel in the cone is used to shape it into a tube.

In gas or electrically heated tunnel furnaces, the cooling process must be conducted rapidly enough so that the G. does not crystallize, but slowly enough so that so that stresses which could cause the G. to break are avoided.

G. can be worked after cooling by secondary molding or by treating its surface. *Blowing* is a secondary shaping of simple, prefabricated glass objects using a flame; it is used to make complicated G. apparatus. In *molding*, the cooled G. is reheated and pressed into molds. G. can be *hardened* by blowing cold pressurized air across it while it is still warm. The resulting surface *quenching*, which can also be achieved by dipping the G. into water or oil, leads to formation of tension-pressure zones. These prevent splintering when the G. is broken. The surface can also be hardened by treatment in a salt melt (e.g. potassium nitrate, KNO_3); the exchange of the smaller cations of the G. (Na^+) for the larger cations of the melt (K^+) generates pressure stress in the G. surface. This pressure causes micro cracks to heal. Plate and mirror G. may be *sanded* and *polished*. Loose quartz sand, corundum or silicon carbide grains are used for sanding, and various grain sizes of iron(III) oxide is used for polishing. *Engraving* for decoration of the G. is done by sanding with fixed grains. *Etching* to give a matte surface is done with hydrogen fluoride, HF, or hydrofluoric acid. On the other hand, a mixture of hydrofluoric and sulfuric acids gives a clear surface (*acid polish*). G. can be *colored* by melting a dye paste onto it, or by diffusion of silver or copper compounds into the surface. The dye pastes (color oxides) are often applied to the surface by a screening method. Thermoflex plates are made by vapor deposition of silicon(II) oxide, SiO. This G. absorbs or reflects a large fraction of heat radiation.

Types of G. In addition to the G. used in buildings, jars and bottles for foods and medicines, and in the laboratory, there is a wide variety of special G. for various applications. ***Optical G.*** are used as lenses, prisms, mirrors and plane parallel plates in optical instruments, such as eye glasses, microscopes, telescopes, searchlights, etc. These G. can be made by grinding and polishing solid blocks, or by drop or profile casting methods. Such instruments require good transparency from the ultraviolet through the visible and into the infrared, and their refractive and dispersive properties are highly significant. ***Crown G.*** has low refraction and dispersion, while ***flint G.*** have large indices of refraction and large dispersion. These G. have long been known; the work of O. Schott (Jena) with oxide compounds, and sometimes also the fluoride and phosphate compounds, of the elements beryllium, lanthanum, thorium, cerium, tantalum, niobium, vanadium, tungsten, cadmium and others has added a wide selection of optical G. with a large range of refractive and dispersive properties.

The ***apparatus G.*** used in the chemical industry, households and laboratories must have a high degree of chemical resistance. In acidic and neutral media, the alkaline glass components are leached out, but the remaining silica gel protects the glass layers below it from further corrosion. Alkaline reactants attack the silicate network, and the G. must contain components (e.g. zirconium dioxide, ZrO_2) which form passivating layers on the surface in the presence of alkaline chemicals. ***Thuringer G.*** is still often used as an apparatus G.; it is a soda-lime G. which can be easily worked but is very resistant to chemicals. ***Jena G.*** is a low-alkali borosilicate G. which has a small coefficient of thermal expansion, and thus can withstand large changes in temperature. *Normal G.* contains zinc oxide, and has proven itself as a thermometer G. ***Suprax G., Duran G.*** and ***Geräteglas 20*** are thermally and chemically resistant G. suitable for laboratory and pharmaceutical use. ***Supremax G.*** is a laboratory G. with very high thermal and chemical stability; it is an alkali-free alkaline-earth aluminoborosilicate G. with a transformation temperature almost 200 K higher than those of the other G. ***Rasotherm G.*** has a very low expansion coefficient ($\alpha = 31 \cdot 10^{-7}K^{-1}$) and a silicon dioxide content of 80%; it is used for chemical apparatus and pipes.

Colored G. are transparent G. which absorb part of the visible spectrum because of the presence of colored components. These G. may be either optical or technical; the optical G. have defined ranges of transparence and are used as ***filters***. The color is provided by heavy metal ions, e.g. copper(II) oxide, CuO, gives a blue to blue-green c!lor; cobalt(II) oxide, CoO, gives a dark blue to dark violet; iron(II) oxide, FeO, bluegreen; iron(III) oxide, Fe_2O_3, yellowgreen to brown; chromium(III) oxide, Cr_2O_3, green, uranium(IV) oxide, UO_2, orange. The color is often achieved by colloidal precipitation of noble metals after tempering, as in ***ruby G.*** (gold or copper). Colored G. are often made as ***coated G.***, that is, they consist of a thin layer of colored G. over a thicker, colorless G. base.

Translucent G. can also be made as coated G. The cloudiness is provided by components such as phosphates, calcium fluoride, CaF_2, or tin(IV) oxide, SnO_2, which form tiny regions of crystalline precipitation that scatter the light (***milk G., opaque G.***).

Radiation shielding G. are used most often for heat protection (***heat protection G.***). Those which contain iron(II) oxide, FeO, are light bluegreen, and have a good absorption capacity for infrared radiation. ***Thermoflex G.*** are coated with thin layers of brownish silicon(II) oxide, SiO and copper, which absorb and reflect the heat rays. G. with a high lead(II) oxide content (up to 90%) are used to provide shielding from radioactive rays.

Safety G. is used in industry, laboratories, households and motor vehicles. ***Wire G.*** is made by rolling a wire mesh into the viscous glass melt. Single-layer safetly G. is very hard and breaks without splintering; it is made by physical (quenching the hot G.surface) or chemical hardening (ion exchange) of plate G. Multi-layer safety G. consists of two or more thin layers of G. glued together by an organic polymeric material (polyacrylates, polyvinyl acetates, polybutyrals).

Foam glass (see) is used as a thermally resistant, noninflammable thermal insulation.

2) ***Chalcogenide G.*** are based on the sulfides, selenides and tellurides, for example of arsenic, germanium, lead and thallium; they are transparent to IR radiation. Because the chalcogenide ions are highly polarizable, most of these G. are deeply colored. Because their components are easily oxidized, the melting process is done under inert gas, and usually in a closed vessel. There is interest in the possibility of using the chalcogenide G. as semiconductors.

3) ***Metal G.*** are amorphous metal alloys, which often consist of a metalloid component (boron, carbon, silicon or phosphorus) and one or several metallic components. The glassy, amorphous state can be achieved only by rapid rates of cooling of the metal melts ($\approx 10^6$ K s^{-1}). These are attained by anvil stamping of metal droplets, or by melt spinning a stream of liquid onto a cooled roller. It is now possible to make metal foils as thin as 100 μm which are very ductile, have high tensile strength and magnetic permeability, low coercivity, exceptional resistance to corrosion and temperature-independent electrical conductivity. They can therefore be used as transformer and magnetic shielding materials, and in instruments for measurement and control.

Glass ceramics: a group of materials with excellent properties obtained by partial, controlled crystallization between the transformation and liquidus temperatures of a glass produced from a melt (see Vitrocerams). The main components of the first G., produced in 1957 by Stookey, were silicon dioxide, SiO_2, aluminum oxide, Al_2O_3, magnesium oxide, MgO, lithium oxide, Li_2O, and titanium dioxide, TiO_2. This composition was subsequently varied in many ways, including production of silicate-free phosphate G.

The basic systems for G. are glasses with separation structures (see Subliquidus separation) in which the size of the droplets or penetration regions can be varied by tempering and is large enough to permit formation of crystals of defined dimensions. Crystallization nuclei are formed by precipitation of substances which are only slightly soluble in the glass, such as oxides of titanium, zirconium, tin, molybdenum, tungsten, chromium, iron, cerium and zinc, or fluorides or phosphates; the silicate crystals grow epitaxially upon these nuclei. The main crystalline phase consists of very small particles (< 3 μm), leading to high mechanical strength of the G. Cracks which form under stress are interrupted at the phase boundaries, so that G. can be worked with shaving tools. The strength of G. can be even further increased by the presence of crystalline spinel phases, such as nickel spinel, $NiAl_2O_4$. Inverse, iron-containing spinels, e.g. Fe_3O_4, make the G. ferrimagnetic. G. with minimal thermal expansion coefficients can be made by precipitation of β-spodumen, $LiAlSi_2O_6$ and β-eucryptite, $LiAlSiO_4$. An important new application of phosphate-containing G. is in bioactive and biocompatible implantation materials (bone replacements); the main crystalline phase of these materials is fluoroapatite or hydroxyapatite, $Ca_5(PO_4)_3F/OH$, or a phosphate glass which does not contain silicon dioxide.

Glass electrode: the most important type of indicator electrode (see Electrode) for potentiometric determination of pH. The G. usually consists of a spherical glass membrane of a special type of glass, filled with a buffer solution of known pH (pH_i). An internal reference electrode, usually a calomel or silver/silver chloride electrode, dips into this buffer solution. When the G. is dipped in a solution with unknown pH

(pH_x), a potential difference V_g forms across the two phase boundaries at the surfaces of the glass membrane; from the Nernst equation (see), at 25 °C this is $V_g = 0.5916\ (pH_i - pH_x)$ (in volts). To measure V_g, another electrode is needed as a reference electrode dipped in the sample solution. If two identical reference electrodes are used, they form a symmetrical glass electrode measuring series, which has advantages over an asymmetrical measuring series.

In practice the pH is determined with a pH meter or potentiometer which permits the pH to be read directly. For this it is necessary to calibrate the glass electrode measuring series. First, using a buffer at the same pH as the internal buffer solution, the *asymmetry potential* is determined; with a second buffer, at a pH as far as possible from the first, the *slope*, i.e. the change in potential of the series caused by a pH change of one unit, is determined. The apparatus is adjusted for both values, after which the pH of an unknown can be read. This calibration must be repeated from time to time, depending on the quality of the G.

G. can also be made in other shapes, such as cylinders, points, needles, surfaces or microelectrodes for measurements under special conditions or in living organisms. The G. is commonly combined with the external reference electrode in a single probe. The pH-sensitive glass membrane surface of any G. must be soaked in water for a time before use; this allows a gel layer to form on the glass surface which is needed for proper functioning.

G. have considerable advantages over other electrodes for determination of pH. They are not impaired by electrode poisons, such as proteins, heavy metal ions, sulfide ions or redox agents. However, G. cannot be used in acidic solutions containing fluoride ions, because these destroy the glass membrane. In very acidic or alkaline solutions, the earlier G. deviated from the linear relationship of EMF to pH; these were called acid or alkali errors. With the introduction of other types of G., there are now commercially available G. which no longer display these errors, and function perfectly in the range of pH 1 to 13, sometimes even to 14.

Starting from the alkali errors of certain types of glass, G. have been developed in which the potential depends on the activity of sodium or potassium ions in a solution. These G. are not very significant, because their potential still depends on the pH as well, and they can be used to measure sodium or potassium ion activities only at constant pH. They have been made obsolete by development of Ion-selective electrodes.

Glass silk: see Fiberglass.

Glass solder: a glass used to solder glass to metal or ceramic materials. It is based on boron oxide (B_2O_3) and lead(II) oxide (PbO), which make up 40 to 90% and also contains small amounts of aluminum oxide, Al_2O_3, and silicon dioxide, SiO_2. Its softening temperature is between 350 and 450 °C.

Glass state: an amorphous state of solids which is obtained by supercooling a melt. The transition of a substance from the liquid to the solid-crystalline state requires the presence of crystal nuclei capable of growth and a sufficiently rapid rate of crystal growth. Crystallization (see) may still not occur, even below the freezing temperature, if the rate of nucleus formation or of crystal growth is too slow. In this case the particles (atoms, ions or molecules) do not assume crystal lattice positions, but freeze on further cooling into a distribution typical of a liquid (see States of matter) (Fig.). A glass can therefore be defined as a *frozen, supercooled melt*. Like all amorphous substances, glass lacks a defined melting point; the transition between the G. and the liquid state is continuous. The physical properties also change continuously rather than abruptly. The position and extent of the *freezing* and *softening range* depend on the rate of cooling or warming. The region of especially rapid – but not discontinuous – change in properties is called the *transformation range*.

Whether the G. is reached as a melt is cooled or crystallization occurs depends both on the system and on the experimental conditions. A number of oxides, such as SiO_2, BeO, As_2O_3 and P_2O_5 are especially prone to formation of glasses. Glasses can also be formed by many other compounds, elements (sulfur, selenium, phosphorus and carbon), organic polymers and aqueous solutions (e.g. hydrogen peroxide, perchloric acid, potassium hydroxide, etc.). However, even the highest rates of cooling so far achieved, 10^6 K s^{-1}, are not sufficient to cause glassy solidification of sodium chloride or aluminum oxide.

Substances in the G. are extremely important materials (see Glass).

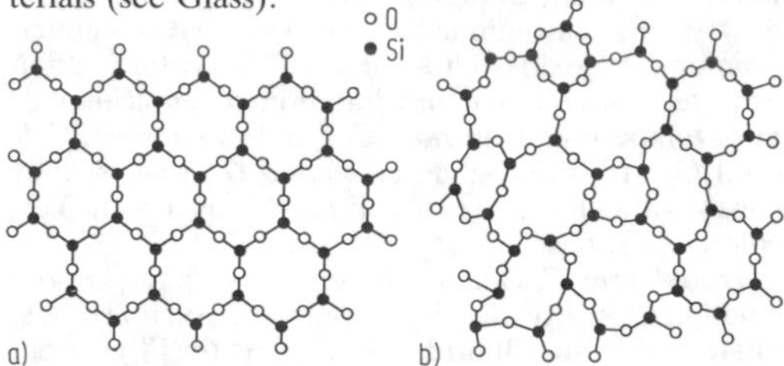

Two-dimensional model of the arrangement of atoms in (a) crystalline SiO_2 and (b) glassy SiO_2.

Glass temperature: same as Glass transition temperature (see).

Glass transition temperature, ***glass temperature, freezing temperature*** T_g: the temperature at which a polymeric, amorphous material is converted from the elastic-plastic to the glassy state, and conversely. Although crystalline materials have an exact melting point, amorphous substances change when heated from a hard, glassy state to a soft, plastic one. In this transition interval (transformation interval), many of the physical properties of the material change, for example its specific volume (Fig. 1).

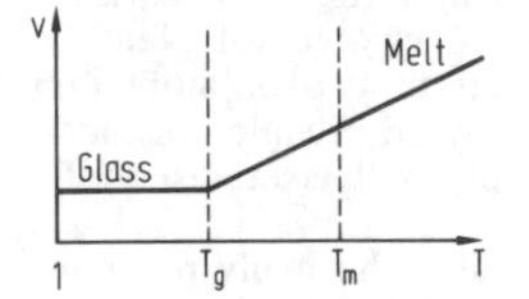

Fig. 1. Change of the specific volume of amorphus polymers with temperature. T_g, glass transition temperature; T_m, melting temperature.

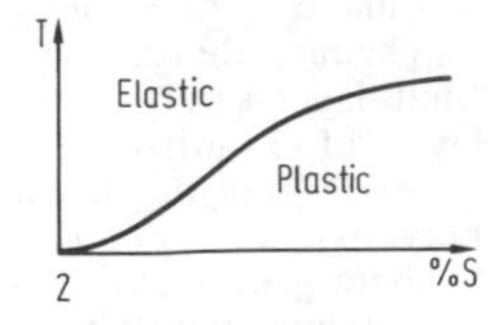

Fig. 2. Dependence of glass transition temperature on sulfur content.

The number of weak bonds, such as hydrogen bonds, has a large effect on the G. Many polymers with polar or polarizable groups, e.g. polyacrylonitrile, polyvinylalcohol and polyacrylates, have high G. These G. are reduced by softeners, and increased by addition of cross-linking substances, for example, during vulcanization of natural rubber with sulfur. However, vulcanization not only cross-links the material, but also introduces polar groups, which makes the effect even stronger (Fig. 2).

The behavior of polymers at different temperatures can be described in terms of molecular mobility. Below the G., the molecular mobility is limited. As heat is added, the chain segments begin to move in the range of the G. (micro-Brownian motion), although the macromolecules are still linked to each other at isolated points. Therefore, a polymer above the G. has rubber elasticity in response to short-term applied forces. Further heating of a polymer above its G. reduces the number of contact points between the chain segments, until all are gone. The macromolecules are now freely mobile (macro-Brownian motion). In amorphous polymers, the temperature at which this motion begins is called the flow temperature; in crystalline polymers, it is called the melting temperature. The G., like the melting temperature T_m, depends on the structure of the polymer; for this reason, the G. and the melting temperature are related. In symmetric polymers, the ratio of the two is about 0.5 (table). The internal mobility of a polymer chain can be increased by incorporation of heteroatoms. Polysilicon rubber, for example, has a G. of 123 °C, which is due to unhindered rotation around the Si-O bonds in the chain. Conversely, the mobility of the chains is greatly limited by the presence of bulky groups or cyclic structural units. For example, polymethacrylates have higher G. than polyacrylates, and polyimides have higher G. than polyamides.

Glass transition and melting temperature of various polymers

	T_g in K	T_m in K	T_g/T_m
Polyethylene	203	408	0.49
Polyoxymethylene	223	454	0.49
Polyvinylidene chloride	256	471	0.54

Glass wool: see Fiberglass.

Glauber's salt: see Sodium sulfate.

Glaze: a thin glassy layer on a ceramic product. It can be transparent or opaque, colorless or colored, shiny or matte. For a G., glass raw materials (see Glass) are melted, then granulated by quenching with water or by rolling between metal rollers, ground, and mixed with clay and sometimes colored oxides (see Ceramic pigments) to form a slip. This is applied to the ceramic object by dipping or spraying, then baked on. To avoid cracking of the glaze, its thermal expansion coefficient must be adapted to that of the ceramic. There are two types of G., ***crude G.***, made from raw materials which are insoluble in water, and ***frit G.***, in which some water-soluble starting materials are used. G. are also classified according to their melting ranges. ***High-temperature*** (> 1200 °C) G. (feldspar G.) are usually crude G., while ***muffle G.*** (< 1200 °C) are often frit G. containing boric acid or lead oxide components.

Glaze pigments: see Ceramic pigments.

GLC: see Chromatography; see Gas chromatography

GlcA: see Aldonic acids.

GlcN: abb. for D-glucosamine.

GlcNAc: abb. for *N*-acetylglucosamine (see D-Glucosamine).

GlcUA: abb. for D-glucuronic acid.

Gliadin: see Prolamins.

Glibenclamid: see Antidiabetics.

Gln: abb. for glutamine.

Globulins: a group of globular proteins found in all animal and plant cells, and also in serum and milk. The G. include many enzymes and most glycoproteins. The G. have higher molecular masses than the albumins, and are relatively insoluble in water. However, they are readily soluble in dilute salt solutions. They are classified, according to their electrophoretic mobility, as alpha, beta and gamma G. They can be fractionally precipitated by addition of ammonium sulfate. Albumins present in the same materials remain in solution.

Glospan: see Synthetic fibers.

Glucagon: a single-chain peptide consisting of 29 amino acids: His-Ser-Gln-Gly-Thr-Phe-Thr-Ser-Asp-Tyr-Ser-Lys-Tyr-Leu-Asp-Ser-Arg-Arg-Ala-Gln-Asp-Phe-Val-Gln-Trp-Leu-Met-Asn-Thr. The structure was elucidated in 1956 by Bromer, and the first total synthesis was achieved in 1968 by Wünsch. G. has a high degree of sequence identity with secretin and vasoactive intestinal peptide (VIP), and a lower degree of identity with gastrin-inhibiting peptide (GIP), especially in the *N*-terminal region. According to Wünsch, the four hormones should be grouped in a glucagon family.

G. is formed in the A cells of the islets of Langerhans in the pancreas, and increases the blood sugar level. A drop in blood sugar is the physiological stimulus for the release of G. It stimulates both glycogenolysis and gluconeogenesis, and also has a lipolytic effect. Like adrenalin, G. is an antagonist of insulin. It is used to treat hypoglycemic states (insulin overdose or hyperinsulinism) or glycogen storage diseases. G. also increases the contractibility of the heart muscle and the heart rate.

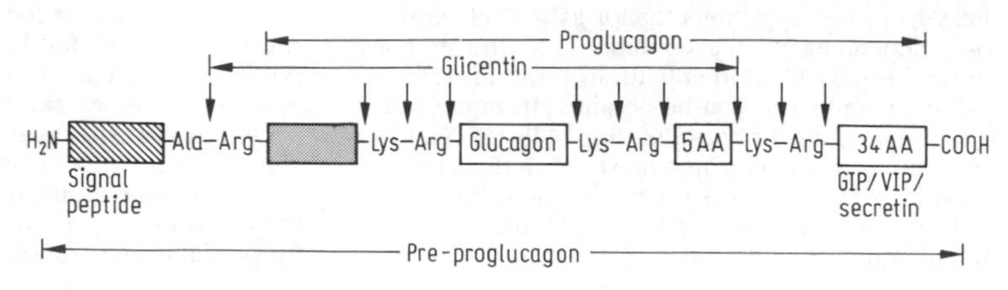

The biological precursor of G. is ***pre-proglucagon***, the amino acid sequence of which was deduced from the nucleotide sequence of the cDNA in 1980/81 by Lund et al. (Fig.). The *N*-terminal methionine is followed by a signal sequence (see Signal hypothesis) of 24 amino acids. The sequence of G. is located in the middle of ***proglucagon***, the *C* terminus of which is a 34-amino-acid chain which is homologous to GIP, vasoactive intestinal peptide and secretin. The arrows in the figure indicate proteolytic cleavage sites.

Glucaric acid: see Aldaric acids.

Glucinium: obsolete term for Beryllium.

Glucitol, ***sorbitol***: an alditol with the *gluco*-configuration. D-G. is a white, crystalline powder, very soluble in water, but scarcely soluble in ethanol. D-G. is made by catalytic hydrogenation of D-glucose. It occurs naturally in many fruits, especially those of the *Rosaceae*. It is present in 5 to 10% concentration in the fruit of the mountain ash (*Sorbus aucuparia*). D-G. is converted to fructose in the body. For this reason, and because of its sweet taste, it is used as a sugar substitute for diabetics (see Sweeteners). D-G. is also used in pharmaceuticals as a moisture-retainer, and is also widely used in the food industry, as a moistener and preservative. It is the starting material for synthesis of ascorbic acid and is used as a diuretic.

gluco-: prefix indicating a certain configuration (see Monosaccharides).

Glucocorticoids: see Adrenal cortex hormones.

Glucogitaloxin: see Purpurea glycosides.

Glucokinse: see Kinases.

Gluconolactone: D-glucono-1,5-lactone, gluconic acid δ-lactone; see Aldonic acids.

Glucosamine, ***chitosamine***, abb. ***GlcN***: 2-amino-2-deoxyglucopyranose, a 2-aminosugar with the *gluco* configuratiôn. D-G. has the chemical properties of a monosaccharide, e.g. it reduces Fehling's solution. It forms a crystalline, water-soluble hydrochloride with hydrochloric acid. D-G. is obtained by acid-catalysed hydrolosis of chitin. It occurs in animals and microorganisms only as ***N-acetylglucosamine***, abb. ***GlcNAc***, which is not basic. GlcNAc is a component of chitin, mucopolysaccharides and murein.

D-Glucosamine (R^1, R^2 = H)

N-Acetyl-D-glucosamine (R^1 = H, R^2 = $COCH_3$)

Muramic acid (R^1 = CH(COOH)–CH_3; R^2 = H)

Glucoscillaren A: see Scilla glycosides.

D-Glucose, abb. ***Glc, grape sugar, dextrose***: the most widely occurring monosaccharide, an aldohexose. It is a stereoisomer of galactose. D-G. crystallizes from water or hot ethanol as α-D-glucopyranose, depending on the conditions, as the monohydrate, m.p. 83 °C, or in anhydrous form, m.p. 146 °C. β-D-Glucopyranose can be obtained from hot water-ethanol mixtures or pyridine; m.p. 148-155 °C. Commercial G. is α-D-glucopyranose, a white, crystalline powder, which dissolves readily in water, but is insoluble in ethanol; $[\alpha]_D^{20}$ +52 to +53° after mutarotational equilibrium has been established.

Haworth projection

α-D-Glucopyranose in the preferred C_1 conformation

At equilibrium, the sugar is 64% β- and 36% α-pyranose. The initial value in water is +112.2° for the α-anomer and +18.7° for the β-anomer. D-G. tastes less sweet than sucrose (see Sweeteners).

D-G. is a central component of carbohydrate metabolism, and as such is ubiquitous in organisms. Free D-G. is found in fruits, honey and blood (about 0.1% concentration; blood sugar). This blood sugar concentration is regulated by the hormones insulin and glucagon. An elevated blood sugar level, as occurs in untreated diabetes mellitus, leads to excretion of the sugar in the urine. A very low blood sugar level (hypoglycemia) leads to shock. There are standardized tests for detection of D-G. in urine (Biophen-G test strips, Benedict method), blood (glucose oxidase, *o*-toluidine, hexacyanoferrate(III) methods) and in the cerebrospinal fluid (*o*-toluidine method); see Monosaccharides.

D-G. is a component of many glycosides and of the disaccharides sucrose, lactose and maltose. Polysaccharides consisting of D-G. are called glycans; they include starch, amylose, amylopectin, glycogen, cellulose and dextrans.

D-G. is obtained by hydrolysis of starch with about 0.3% acid at 120 °C in an autoclave. After cooling, it is decolored with activated charcoal and concentrated. When the syrup is cooled and inocculated with seed crystals, impure G. crystallizes out; it must be refined. D-G. can also be obtained from cellulose from wood.

Biochemistry. The phosphate esters of D-G. (glucose 6-phosphate, glucose 1-phosphate) play a central role in carbohydrate metabolism. In heterotrophic organisms, glucose is synthesized from products of the citrate cycle. Glyceraldehyde 3-phosphate and dihydroxyacetone phosphate are first condensed in an aldol reaction to form fructose 1,6-bisphosphate, which is then dephosphorylated and isomerized to glucose 6-phosphate. In photosynthetic plants, D-G. is synthesized by reduction of CO_2 (see Photosynthesis). The reversal of gluconeogenesis is glycolysis, in which D-G. is broken down to glyceraldehyde phosphate and dihydroxyacetone phosphate and then to lactic acid. D-G. is stored in the form of starch, amylose and amylopectin in plants, and as glycogen in animals.

Applications. D-G. is used in medicine in intravenous feeding. It is used on a large scale, usually in the form of hydrolysis products of polysaccharides, as a starting material for biotechnological and chemical syntheses, such as alcoholic fermentation and synthesis of ascorbic acid or glucitol (sorbitol).

Glucose isomerase: an enzyme which isomerizes glucose to fructose. It is found in many microorganisms, but commercial application is largely limited to the G. from *Streptomyces sp.* Immobilized G. is used for partial isomerization of the glucose syrup resulting from hydrolysis of starch by α-amylase and glucoamylase. The product mixture contains 42% fructose, 50% glucose and 8% other sugars and, because of its high fructose content, it is sweeter than sucrose. The world production of this new "liquid sugar" is now about 4 million tons per year, and is increasing rapidly.

Glucosidases: see Disaccharidases.

Glucosinolates, ***mustard oil glycosides***: glycosides of 1-thio-D-glucose found in plants of the families *Brassicaceae, Capparidaceae* and *Resedaceae*. Enzymatic cleavage of the G. yields glucose, sulfate and alkylisothiocyanates (mustard oils). Some important examples of G. are *sinigrin* (R = CH_2=CH-), found in black mustard (*Brassica nigra*), radishes and horse radishes, and *sinalbin* (R = HO-C_6H_5-), found in white mustard (*Sinapis alba*). Sinigrin is present at about 1% concentration in the seeds of black mustard.

Glucoverodoxin: see Purpurea glycosides.

Glucurone: the 3,6-lactone of D-glucuronic acid. G. forms colorless crystals which are readily soluble in hot water, but are nearly insoluble in ethanol; m.p. 175-178 °C (dec.), $[\alpha]_D$ +19° (in water). G. is a byproduct of isolation of D-glucuronic acid.

D-Glucuronic acid, abb. ***GlcUA***: a uronic acid. D-G. is usually syrupy, but it will crystallize out of ethanol (colorless crystals); m.p. 165 °C. D.-G. is readily soluble in hot water and ethanol, $[\alpha]_D$ +36.3° (in water, at mutarotational equilibrium). It is a component of mucopolysaccharides and plant slimes. It is biosynthesized by the oxidation of UDP-glucose to UDP-glucuronic acid. This "active" D-G. is used in the animal and human body to bind many endogenous and exogenous substances (e.g. drugs), especially alcohols, phenols, amines and carboxylic acids. The glycosides or esters formed are called ***glucuronides***. The conjugation reaction takes place in the liver and makes possible the excretion of the compound through the kidneys.

COR^1, OH, HO, R^2, OH

R^1 = OH
R^2 = OH

Glue: a nonmetallic substance which can bind solids together by adhesion and cohesion without significantly changing the structure and other properties of the bodies so joined. The main component of a G., its base, is generally a polymer or monomers which polymerize as the G. sets; it is the primary agent of adhesion. G. are used in the liquid state, so the base must be liquid, dissolved or emulsified in a solvent, or molten. In addition to solvent, a G. can contain fillers, stretchers, hardeners, accelerators, softeners, wetting agents, stabilizers and thinners.

The strength of a G. joint depends on the adhesion of the molecules of the G. to the surface of the material to be glued and the internal cohesion of the G. When the glued joint is subjected to mechanical stress, force is transmitted from one side of the joint, through the layer of G. and to the other side. Adhesion represents an energy per surface area. It can be thought of as work which overcomes a resistance. Because it is caused by various forces, there are various types of adhesion. ***Mechanical adhesion*** is caused by penetration of cracks and pores of the material by the G. This model adequately explains adhesion to porous and fibrous surfaces, such as paper, wood, brick, concrete and similar materials, but not to glass or polished metals. *Specific adhesion* describes adhesion due to physical and thermodynamic processes at the surface and to formation of chemical bonds. The molecular-physical interpretation of adhesion is based on the dipole character of the molecules (polarization theory). It predicts that the G. and the glued materials will stick only if their polarities are approximately the same. The distance between the glue and glued surface molecules must be less than $5 \cdot 10^{-10}$ m. Dipolar molecules develop an adhesion which is inversely proportional to the cube of the distance between them. Studies of intermolecular forces at surfaces led to development of the electrostatic adhesion theory. This is based on the existence of an electrical double layer in the surface layers of the G. and glued parts. The electrostatic interactions correlate with the adhesion. The formation of the electric layers is due to the diffusion of charge carriers, which are in thermodynamic equilibrium at the contact sites. Another model is based on the diffusion of molecules and molecular segments of the G. toward the glued part, or from the part toward the glue. The diffusion of molecules or segments across the phase limit leads to a "material bond". This model has been especially useful in the interpretation of plastic cements. The thermodynamic model is based on the observation that any solid surface exerts forces which bind foreign materials to this surface. In fact, there are forces acting between the components of the solid which hold it together. In the interior of the solid, these are saturated, but at the surface, the outward-directed forces attract the adjacent phase, leading to adhesion. Quantitative data for this model have been obtained by measurement of the surface tension by wetting experiments. In a few cases, adhesion is due to formation of chemical bonds between the G. and the surface of the glued material. This proposal is supported, for example, by the sulfur bridges between metals and vulcanized rubber. However, experimental evidence for chemical bonds in the surface layers has been found in only a few systems. The variety of interpretations of adhesion sketched here indicate the complexity of the phenomenon. None of the models is capable of completely explaining it.

Cohesion is a force which holds a substance together. It is generated in a G. primarily by the main chains of the molecules. Thus the internal strength of a G. film depends on the size of its molecules. On the other hand, adhesion decreases with increasing molecular mass, so that G. are characterized by a molecular size which maximizes the strength of the bond.

There are several kinds of G.: traditional G., emulsion G., cements, reaction G. and molten G.

Traditional ***glues*** are aqueous, usually colloidal solutions of natural or synthetic polymers. Starch, cellulose and occasionally plant gums are the most common plant raw materials. These glues are used for paper, cartons and textiles. Office G. and the G. used to apply labels to bottles are quite often made of starch or dextrin, a chemically or thermally degraded starch. Very viscous solutions of G. are called ***paste***. Wallpaper paste and the paste used for cigarette filters are made of methyl ethers of cellulose.

Collagen and casein G. are prepared from animal proteins. Collagen is obtained from hides, bones and leather scraps. Casein is obtained from sour skim milk. Casein G. consists of powdered casein, calcium hydroxide, inorganic salts and sometimes preservatives. They are prepared by mixing with cold water. Animal G. are still used to some extent to glue wood and in plywood. Collagen G. is specially prepared for matches, colored paper, emory boards and sandpaper.

Phenol-, resorcinol- and especially urea-formaldehyde resins are preferred for synthetic G. These are used as water-soluble products. Addition of a hardener causes the condensation to go to completion in the G. layer, producing quite stable joints. These G. are used to make plywood and to apply veneers.

In ***emulsion G.***, the G. base is the dispersed phase. The solvent is water. The most important G. bases are polyacrylate, polyvinylacetate and natural and synthetic rubbers. The solid content is 40 to 60%, which is significantly more than in cements. The emulsions are stabilized by additives; for example, polyvinylacetate emulsion by polyvinyl alcohol, and natural rubber emulsion by ammonia. The properties of these G. are usually improved by fillers. Softeners, resins and solvents are used to modify the emulsion of polyvinylacetate. Rubber and polyacrylate emulsions are used for adhesive tapes and joining compounds. Polyvinylacetate emulsions are used to glue furniture and in bookbinding.

Cements are solutions of polymers in organic solvents. Nearly all soluble plastics and elastics are suitable bases for cements. Natural and synthetic rubber, polyurethane, polyacrylate, polyvinyl acetate, polyvinyl ether, polystyrene, cellulose nitrate and chlorinated polyvinyl chloride are the most widely used polymers. To increase the adhesiveness of rubber solutions, zinc oxide, ZnO, or resin is often added. The resistance of rubber to heat and aging is greatly improved by vulcanization with sulfur or cross-linking with isocyanates. Cements are used mainly to glue plastics. For this reason, the choice of solvent is important. The solvent can dissolve the surface of the plastic and thus enhance diffusion into the surface, leading to better adhesion. Dimethylformamide, dichloroethene, dichloromethane and acetone are very effective solvents.

Reaction G. are G. which harden by means of a chemical reaction; they are used on metals, ceramics and other non-porous materials. A liquid, low-molecular-weight compound is converted by the reaction to a polymeric solid. The most important of these compounds are epoxide, phenol and unsaturated polyester resins, polyurethanes, silicon rubber and polyacrylates. These G. are prepared as two-component systems, or sometimes now as one-component systems. Silicon rubber is especially suitable for joining glass, ceramics and enamels. Phenol resins are used with polyvinylformaldehyde in a *redox process* for temperature-resistant joining of metals. *Superglues* are one-component acrylate compounds used in fine mechanics and optics. The glued joints achieve a tensile strength which can compete with that of soldered joints, although they are not as resistant to bending stress.

Molten G. are applied to the joint surfaces in the molten state, and bind as they freeze. Ethylene-vinyl acetate copolymers, fatty acyl polyamides, polyester and ethylene waxes are suitable raw materials. They form a homogeneous melt, are not sensitive to overheating and have a narrow softening or melting range. Molten G. have a very rapid rate of hardening, in the range of a few seconds. This property makes them useful for gluing cartons and bags on rapidly working packing machines. They can also be used for varnishing small areas on furniture, hemming leather and textile edges, and for gluing the backs onto books. A disadvantage of these G. is that it is complicated to prepare and apply them.

Glutaconic dialdehyde, ***pent-2-enedial***: OCH-CH_2-CH=CH-CHO, an unsaturated dialdehyde which exists almost completely in the enol form. G. can be synthesized by hydrolysis of N-substituted pyridinium salts. It is used for special organic syntheses.

Glutamine, abb. ***Gln***: the γ-semiamide of glutamic acid and a proteogenic amino acid (formula and physical properties, see Amino acids, Table 1). Sprouting seeds contain high concentrations of G.; it has an important role there in ammonia detoxication. G. is split into glutamic acid and ammonia by glutaminase; this reaction is the source of ammonia in urine. G. is also involved in the biosynthesis of the purine skeleton.

Glutamic acid, abb. ***Glu***: α-aminoglutaric acid, HOOC-$CH_2CH_2CH(NH_2)COOH$, a proteogenic Amino acid (see) found in nearly all proteins. It is particularly abundant in the proteins of grains, where, for example, it comprises 45% of the glutelins. It was discovered in wheat gluten in 1866 by Ritthausen. L-G. is now the most important amino acid in the world economy; the annual production is more than 340,000 t. The original method of production, acid hydrolysis of wheat gluten, is of no economic significance. DL-G. is produced on industrial scale by the Strecker synthesis with β-cyanopropionaldehyde. The α-aminoglutaric dinitrile formed as an intermediate is converted to sodium DL-glutamate, from which free DL-G. is obtained by acidification with sulfuric acid. The racemic mixture is separated by spontaneous crystallization: a supersaturated aqueous solution of DL-G. is seeded with L-G. crystals, which causes a large fraction of the L-G. to crystallize out. The D-G. remaining in the mother liquor is racemized with acid or base, and returned to the process. The biotechnological production of G. by fermentation, using *Corynebacterium glutamicus*, has become economically important. The bacteria are given glucose or molasses as a carbon source, and are grown in sterilized tanks at 35 °C with aeration and addition of ammonia. After 40 h, the G. can be isolated from the culture filtrate. The yield is 50 kg G. per 100 kg glucose. More than 90% of the L-G. produced is used in food as monosodium glutamate (MSG), which intensifies the specific flavors of foods.

In metabolism, G. is involved in transamination reactions with many α-ketoacids, which are reversibly converted to the corresponding amino acids. Decarboxylation of G. produces γ-aminobutyric acid. In medicine, G. is used to treat states of exhaustion and muscle diseases.

Glutaraldehyde, ***pentanedial, glutaric dialdehyde***: $OHC\text{-}CH_2\text{-}CH_2\text{-}CH_2\text{-}CHO$, a colorless, non-viscous oil, b.p. 187-189 °C (dec.). G. is readily soluble in water and can be steam distilled. It the presence of water, it forms a polymeric, glass-like structure, which is split into the monomeric G. by vacuum distillation. Like all Aldehydes (see), G. is very reactive and undergoes the typical addition and condensation reactions. It is synthesized by the diene synthesis from acrolein and vinyl methyl ether via a dihydropropane derivative; this can be isolated and cleaved to G. with boiling water. G. is the starting material for carbocyclic and heterocyclic compounds. It is used as a disinfectant and preservative, tanning agent and, in dilute solutions, as a corrosion preventive.

Glutaric dialdehyde: same as glutaraldehyde.

Glutaric acid, ***pentanedioic acid***: $HOOC\text{-}(CH_2)_3\text{-}COOH$, a saturated, aliphatic dicarboxylic acid. G. crystals are colorless needles or prisms, m.p. 99 °C, b.p. 302-304 °C (dec.). G. is soluble in water, alcohol, ether and chloroform. When heated, it forms the cyclic glutaric anhydride. G. is found in the juice of immature sugar beets and the wash water from raw sheep's wool. It can be synthesized by oxidation of cyclopentane, cyclopentanol or cyclopentanone, or by Malonic ester synthesis (see) from sodium malonic ester and methylene iodide. G. is used as an intermediate for organic syntheses.

Glutaric anhydride, ***tetrahydropyran-2,6-dione***: the compound forms colorless crystals, m.p. 56 °C, b.p. 286-288 °C. It is soluble in hot ether. G. is synthesized by heating glutaric acid with a water-extracting reagent such as acetic anhydride, acetyl chloride, thionyl chloride or phosphorus oxide chloride. It is used in the production of softeners and polymers, and in the synthesis of pyridine derivatives.

O O O

Glutathione: γ-L-glutamyl-L-cysteinylglycine, a naturally occurring tripeptide with a γ-peptide bond. It is found in all cells of higher animals. It is an important activator for various enzymes, protects lipids against autooxidation, and is a component of a transport system for amino acids in certain animal tissues (γ-glutamate cycle). It is biosynthesized by an enzyme-catalysed two-step reaction which requires ATP; ribosomes are not involved.

Glutelins: a group of globular proteins which occur together with prolamines in grains. They are soluble in water only at extreme pH. The G. contain a large amount of glutamic acid (up to 45%). The chief examples are ***glutenin*** from wheat and ***orycenin*** from rice.

Glutenin: see Glutelins.

Glutethimide: see Hypnotics.

Gly: abb. for glycine.

Glycan: same as Polysaccharide (see).

Glyceraldehyde, ***2,3-dihydroxypropanal***: $HOCH_2\text{-}CHOH\text{-}CHO$, the simplest aldose (see Monosaccharides). G. exists in the D-, L- and DL-forms. The DL-form (m.p. 145 °C) is formed by partial oxidation of glycerol. The dextro-rotatory D-form is the reference substance for the configuration of optically active compounds. G. is soluble in water, and barely soluble in alcohol and ether. It reduces Fehling's solution. In water, it forms an equilibrium with dihydroxyacetone. A mixture of G. with a small amount of dihydroxyacetone is called *glycerose*; in aldol reactions, it gives a mixture of hexoses, including fructose. D.-G.-3-phosphate is an important intermediate in glycolysis and other metabolic pathways.

Glyceroglycolipids: see Glycerolipids.

Glycerol, ***propane-1,2,3-triol***:

$CH_2\text{–}CH\text{–}CH_2$
OH OH OH

the simplest and most important triol; a viscous, sweet-tasting, hygroscopic liquid with no color or odor; m.p. 18.6 °C, b.p. 290 °C (dec.), n_D 1.4746. G. can be steam distilled; it is miscible with water and lower alcohols in all proportions, but is not miscible with ether. G. is very reactive; with dehydrating reagents, it forms acrolein.

G. forms ionic *glycerates* with alkali metals, and *metal complexes* with heavy metal ions. Oxidation yields various products (Fig.), depending on the conditions. Under mild conditions, such as the presence of bromine water, a mixture of glyceraldehyde and 1,3-dihydroxyacetone is formed. These two compounds can be interconverted via an enediol (Lobry-de-Bruyn-van-Ekenstein rearrangement).

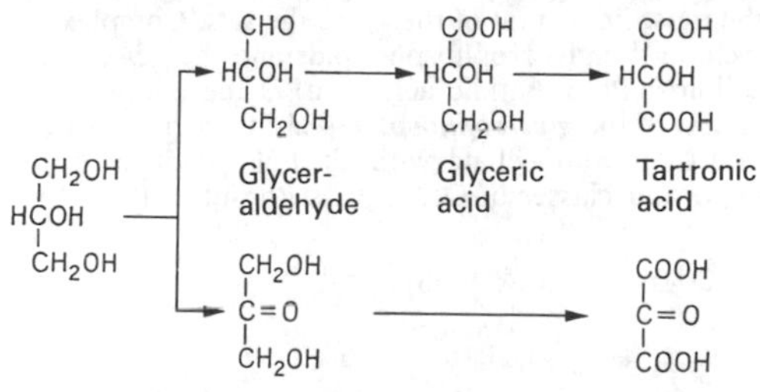

Oxidation products of glycerol.

G. can be esterified with many inorganic and organic acids. High temperatures and oxygen excess favor the triesters, some of which, the natural triacylglycerols (see Fats) are of great biochemical, industrial and nutritional importance. Esterification with phthalic acid and other dicarboxylic acids produces polyesters, which are widely used as alkyde resins, for example, in paints. In the presence of a mixture of concentrated nitric and sulfuric acids, G. is converted to glycerol trinitrate. With hydrogen chloride, G. reacts to form a mixture of isomeric chloropropane diols, and with boric acid, it forms a complex, relatively strong acid, which can be titrated.

Occurrence. G. occurs very widely as a component of fats and many phospholipids. In the digestive tract, enzymes called lipases hydrolyse the fats to G. and fatty acids; the free G. is absorbed and metabolized. In free form, G. has been found only in traces in the small intestine and blood plasma. In algae, it is present in the form of glycosides.

Production. G. is synthesized industrially from allyl chloride or allyl alcohol by addition of hypochlorous acid, HOCl, and subsequent alkaline hydrolysis. It is also produced by saponification of fats, which in industry is done with steam at about 200 °C and 2 MPa. Alcoholic fermentation can be diverted, by addition of sulfites, so that it produces a relatively high fraction of G., and this is also utilized for production.

Applications. G. is an important starting material for many technical syntheses. It is used in large amounts as an antifreeze, hydraulic fluid, salve base, textile additive, softener and working fluid in heating systems. Because of its hygroscopic properties, it is added to soaps, tobacco, skin-care products, toothpastes and printer's inks. It is used as a sweetener of liquors and essences. The distillation residue from processing of G. is used to make shoe polish and electrical insulation.

Glycerolipids: derivatives of glycerol containing at least one lipophilic *O*-acyl, *O*-(1-alkenyl)- or *O*-alkyl group. The most common are the *O*-acyl derivatives, the ***ester glycerolipids***. The most important group of G. with 1-alkenyl groups are the Plasmalogens (see). Alkoxyglycerolipids are found mainly as alkyldiacylglycerols, and they are abundant in fish liver oils, but small amounts are also found in mammals. G. with isoprenoid alkyl groups are found in halobacteria.

The G. are classified according to the substituent in position 3 of the glycerol (table) as ***simple (neutral) G.*** and ***complex G.***. The simple G. include the mono- di- and triacylglycerols (mono-, di- and triglycerides); the most important of these are the fats. Complex G. include the glycerophospholipids and the glyceroglycolipids. Phosphatidic acid (see) is the parent compound of the ***glycerophospholipids***; further esterification of the phosphate with alcohols produces other important classes of G., such as Phosphatidylcholine (see), Phosphatidylethanolamine (see), Phosphatidylserine (see), Phosphatidylglycerol (see) (see also Cardiolipin) and Phosphatidylinositol (see). ***Glyceroglycolipids*** are the glycosides of mono- or diacylglycerols.

CH_2OR^1 | R^2OCH | $CH_2—R^3$ $\quad R^1, R^2$ = Acyl

R^3	
OH	Diacylglycerol (diglyceride)
OAcyl	Triacylglycerol (triglyceride; fat)
OPO_3H_2	Phosphatidic acid
$OPO(O)(O^-)CH_2CH_2N^+(CH_3)_3$	Phosphatidylcholine (lecithin)
$OPO(O)(O^-)CH_2CH_2N^+H_3$	Phosphatidylethanolamine
$OPO(O)(O^-)CH_2CH(N^+H_3)COOH$	Phosphatidylserine
$OPO(O)(O^-)CH_2CHOHCH_2OH$	Phosphatidylglycerol
O—β—D—Gal	Monogalactosyldiacylglycerol
O—β—D—Gal—β—D—Gal	Digalactosyldiacylglycerol
SO_3H_2	Sulfolipids

(Gal = Galactose)

The natural G. are generally optically active, although triacylglycerols are only optically active if they carry different acyl groups at positions 1 and 3. The specific rotation of these compounds is 3 to 6°. Glyceroglycolipids contain additional chiral centers. The configuration of the G. is indicated by a stereospecific numbering (abb. sn) in which the C-1 atom is that atom which is on top in the vertical Fischer projection if the hydroxyl group in position 2 is on the left. Natural G. are derived from sn-glycerol 3-phosphate (formerly called L-glycerol 3-phosphate or D-glycerol 1-phosphate), which is also their biological precursor.

The chloroplast membranes of higher plants and algae contain, in addition to phosphatidylglycerol, almost entirely monogalactosyldiacylglycerols, digalactosyldiacylglycerols and ***sulfolipids***, which are highly surface active and acidic. Glyceroglycolipids with other sugar groups are found in bacteria.

Glycerol trinitrate, ***nitroglycerol, nitroglycerin, trinitroglycerol, TNT***: $CH_2O(NO_2)$-$CHO(NO_2)$-$CH_2O(NO_2)$. The terms "nitroglycerol" and "nitroglycerin" are confusing, because the compound is not a nitro-compound with a C-N bond; it is an ester of nitric acid. G. is an oily, colorless to pale yellow, odorless and highly explosive liquid; density 1.596, b.p. above 180 °C (dec.), n_D^{12} 1.4786. It is slightly soluble in water, cold alcohol, light petroleum and carbon disulfide; it is miscible in any proportions with ether. G. is very poisonous, and inhalation of larger amounts causes headaches, sensation of heat and acceleration of the pulse. When pure, G. can be stored indefinitely, but it explodes very violently on impact or overheating (rate of detonation, 7450 m s^{-1}). The decomposition approximately follows the equation 4 $C_3H_5(ONO_2)_3 \rightarrow 12\ CO_2 + 10\ H_2O + 6\ N_2 + O_2$. G. is produced by dropping glycerol into well-chilled nitrating acid (concentrated sulfuric and nitric acids).

G. is a very powerful explosive and is very sensitive to impact. It is so dangerous that it is used only in mixtures of explosives and in production of nitroglycerol powders. An important characteristic is its ability to form a gelatin in the presence of small amounts of collodium wool. Powdered explosives contain 4 to 6% G. to increase their strength and capacity to detonate. "Semiplastic" explosives contain 8 to 15% G. If higher concentrations of G. are used, it must be gelatinized to prevent sweating out. Explosives containing more than 20% G. are plastic.

Because of its high melting point, G. is now used almost exclusively in mixtures with ethylene glycol dinitrate, because frozen G. is dangerous.

G. has been used medicinally for its dilatory properties to treat angina pectoris and other blood vessel diseases. Because of the speed of its action, it is an excellent medicament for use in acute attacks of angina pectoris.

Glycerol trinitrate powder: see Smokeless powder.

Glycerophospholipids: see Glycerolipids.

Glycerose: see Glyceraldehyde.

Glycide esters, ***α,β-epoxycarboxylic acid esters***: esters of 3-hydroxypropylene oxide produced by the glycide ester synthesis. The ***glycide acids*** are obtained by hydrolysis of G.; when heated with dilute acids, they lose CO_2 and form homologous aldehydes or ketones with one CH_2 group more than the parent acid.

Glycide ester synthesis: same as Darzens-Erlenmeyer-Claisen condensation (see).

Glycidol, ***2,3-epoxypropan-1-ol***: an oxirane derivative and a monovalent alcohol; a colorless liquid with b.p. 166-167 °C, n_D^{20} 1.4338. G. is readily soluble in water, alcohol and ether; it is insoluble in benzene.

$$H_2C-CH-CH_2OH \text{ (with O bridging } C_1 \text{ and } C_2\text{)}$$

G. can be considered an especially reactive derivative of glycerol (anhydride), because the oxirane ring is readily cleaved. Reaction with water produces glycerol. Glycerol ethers are formed with alcohols, and aminoglycerols with amines. G. is made by reaction of epichlorohydrine with anhydrous potassium acetate. It is used mainly for synthesis of complicated glycerol derivatives.

Glycide acids. see Glycide esters.

Glycine, abb. ***Gly; glycocoll, aminoacetic acid***: the simplest proteogenic amino acid (formula and physical properties, see Amino acids, Table 1). G. is present in nearly all proteins, and together with alanine, makes up a large proportion of silk fibroin and gelatins, from which it was first isolated in 1820 by Braconnet. Most G. is synthesized industrially from aminoacetonitrile by acid or alkaline hydrolysis. Aminoacetonitrile is made by slowly adding ammonia to an equimolar mixture of formaldehyde and hydrogen cyanide at 40 °C: $HCNO + HCN + NH_3 \rightarrow H_2N\text{-}CH_2\text{-}CN + H_2O$. The synthesis of G. by ammonolysis of chloroacetic acid also produces secondary and tertiary amines as byproducts. These side reactions can be suppressed by use of a mixture of ammonia and hexamethylenetetraamine as the ammonolysis reagent.

G. is an important component of infusion solutions for intravenous feeding. It promotes healing of wounds and is used in the treatment of muscle diseases and liver damage. *N*-Alkyl derivatives of G., such as *N*-methylglycine (sarcosine) are intermediates in amino acid metabolism. G. itself is an important starting material for biosynthesis of porphyrins and many other natural products.

The derivative ***glyphosine***, $HOOC\text{-}CH_2N(CH_2PO_3H_2)_2$, is an important growth regulator in plants.

Glycocoll: same as Glycine.

Glycogen, ***animal starch***: animal reserve polysaccharide, a colorless powder which gives a brown to violet color with iodine solution. It is an α-D(1→4)-glucan with α(1→6) branching and a very high molecular mass. It is similar to the amylopectins of Starch (see), but more highly branched. G. is degraded enzymatically to maltose and glucose, or by acid catalysis to glucose. It comprises 0.5 to 2% of the muscle mass, and up to 20% of the liver mass.

Glycol, ***1,2-diol***: a dialcohol with properties similar to those of monoalcohols. However, G. are more soluble in water and are more viscous. Ethylene glycol (see) and Propylene glycol (see) are well known examples. The reactions typical of alcohols, such as the Criegee cleavage (see) and the Malaprade reaction (see) can, in some cases, be carried out on only one of the hydroxyl groups, but in other cases, both hydroxyls can react. Thus a large number of oxidation products, esters and ethers of G. are known; some are industrially important as organic intermediates, wetting agents, emulsifiers, solvents and starting materials for laundry products.

Glycol acetates: the mono- and diacetates of ethylene glycol. ***Glycol monoacetate, ethylene glycol monoacetate***, $HO\text{-}CH_2\text{-}CH_2\text{-}O\text{-}CO\text{-}CH_3$, is a colorless liquid; m.p. - 80 °C, b.p. 187-189 °C, n_D^{20} 1.4209. Glycol monoacetate is soluble in water, alcohol and ether. It is formed, along with the diester, by the reaction of glycol with acetic acid, or by addition of acetic acid to ethylene oxide.

Glycol diacetate, ethylene glycol diacetate, $CH_3\text{-}CO\text{-}O\text{-}CH_3$, a colorless, mobile liquid; m.p. - 31 °C, b.p. 190 °C, n_D^{20} 1.4159. Glycol diacetate is soluble in water, alcohol and ether. It is produced industrially from ethylene oxide and glacial acetic acid or acetic anhydride in the presence of sulfuric acid. Small amounts can also be made by the reaction of ethylene dibromide with potassium acetate in glacial acetic acid. G. are used mainly as solvents.

Glycolaldehyde, ***hydroxyacetaldehyde***: $HO\text{-}CH_2\text{-}CHO$, the simplest α-hydroxyaldehyde. G. forms colorless, sweet-tasting crystals; m.p. 96-97 °C. It is readily soluble in water and alcohol and tends to dimerize to a cyclic form. Like all Aldehydes (see), G. undergoes addition and condensation reactions. It is found naturally in elder leaves, and is synthesized by oxidation of vinyl acetate with hydrogen peroxide in the presence of osmium tetroxide.

Glycol diacetate: see Glycol acetates.

Glycol dinitrate: same as Ethyleneglycol dinitrate.

Glycolide: the Lactide (see) of glycolic acid.

Glycolipids, ***glycosyllipids***: complex lipids which contain mono- or oligosaccharide groups. In glyceroglycolipids (see Glycerolipids), the carbohydrate component is bound glycosidically to the primary hydroxyl group of mono- or diacylglycerols, and in sphingoglycolipids (see Sphingolipids), to the hydroxyl group of the sphingoid. The fatty acid esters of carbohydrates, such as those in Cord factor (see) are also considered to be G.

Glycol monoacetate: see Glycol acetates.

Glycol cleavage: see Criegee cleavage; see Malaprade reaction.

Glycolysis: in the narrow sense, the most important pathway for anaerobic degradation of glucose to lactic acid. Many organisms obtain chemical energy in the form of adenosine triphosphate (ATP) by means of G.: glucose + 2 ADP + 2 phosphate → 2 lactic acid + 2 ATP + 2 H_2O. This pathway for anaerobic utilization of glucose is believed to be the oldest biochemical mechanism for energy production; it apparently permitted development of living organisms in an oxygen-free atmosphere. The ability to degrade glucose anaerobically has been retained by most higher organisms, and most animals can use this form of energy production during short periods of oxygen deficiency.

In the broader sense, the term G. includes the catabolic pathway by which glucose is converted to pyruvic acid; under aerobic conditions, the pyruvic acid is converted to acetyl-coenzyme A and then completely degraded to carbon dioxide and water via the citric acid cycle. The complete pathway, G. plus the citric acid cycle, can be summarized by: glucose + 36 ADP + 36 phosphate + 36 H^+ + 6 $O_2 \rightarrow$ 36 ATP + 6 CO_2 + 42 H_2O.

In G., glucose is first phosphorylated to glucose 6-phosphate, which is isomerized to fructose 6-phosphate. This is phosphorylated to fructose 1,6-bisphosphate, then split by aldolase into glyceraldehyde 3-phosphate and dihydroxyacetone phosphate, which is isomerized to glyceraldehyde 3-phosphate. The glyceraldehyde 3-phosphate is dehydrogenated and phosphorylated in a single step by glyceraldehyde 3-phosphate dehydrogenase, yielding NADH and 1,3-diphosphoglyceric acid. This compound is able to transfer a phosphate group to ADP, yielding ATP and 3-phosphoglyceric acid, which is isomerized to phospho*enol*pyruvic acid. Phospho*enol*pyruvate transfers its phosphate group to another molecule of ADP. Thus for each molecule of glucose, two molecules of glyceraldehyde 3-phosphate are formed, and each molecule of glyceraldehyde 3-phosphate yields two molecules of ATP, or four in all. The two molecules of ATP required to phosphorylate the glucose and fructose 6-phosphate must be subtracted from the yield of ATP, leaving two ATP produced for each molecule of glucose degraded.

Up to pyruvic acid, the pathways of Alcoholic fermentation (see) and G. are the same. In G. in the narrow sense, the pyruvic acid is reduced to lactic acid by lactate dehydrogenase: CH_3-CO-COOH + NADH + $H^+ \rightarrow CH_3$-CHOH-COOH + NAD^+. The NAD^+ required by glyceraldehyde 3-phosphate dehydrogenase is regenerated in this step.

Glycoproteins: proteins with covalently bound oligo- or polysaccharide chains. The properties of G. are similar to those of proteins. The carbohydrate components are most commonly N-acetyl-D-glucosamine, D-mannose and D-galactose (Fig.). N-glycosidic bonds are relatively stable to acids and bases, but the O-glycosidic bonds are easily cleaved by acid catalysis. The O-glycosidic bonds to serine and threonine readily undergo β-elimination in alkaline media, forming α,β-unsaturated amino acid residues. A G. can contain more than one carbohydrate chain. Many enzymes, especially hydrolases, are G., as are most plasma proteins (albumins, immunoglobulins, proteohormones, interferons, etc.) and membrane proteins.

NeuAc-α(2→6)-*Gal*-β(1→4)-*GlcNAc*-β(1→2)-*Man*-α(1→3)
NeuAc-α(2→6)-*Gal*-β(1→4)-*GlcNAc*-β(1→2)-*Man*-α(1→6)
→ *Man*-β(1→4)-*GlcNAc*-β(1→4)-*GlcNAc*-β-*Asn*

Oligosaccharide residue of a glycoprotein, bound to an asparagine residue (Asn); NeuNAc = N-acetylneuraminic acid: Gal = galactose, GlcNAc = N-acetylglucosamine; Man = mannose; — possible termination points.

Binding of polysaccharide residues to glycoproteins

Bond type	Sugar-amino acid bond	Occurence
Nβ-glycoside	...Glc—β-NH—CO—CH_2—CH(NH...)(CO...)	Many glycoproteins
O-β-glycoside	...Gal—β-O—CH(CH_2NH_2)—CH_2CH_2CH(NH...)(CO...)	Collagen
	...Xyl—β-O—CH_2—CH(NH...)(CO...)	Heparin, chondroitin, other mucopolysaccharide
O-α-glycoside	...GalNAc—α-O—CH(R)—CH(NH...)(CO...)	R = H, CH_3 Immunoglobulins Membrane proteins

Glycosidases: group-specific hydrolases which cleave glycosidic bonds in carbohydrates, glycoproteins and glycolipids. Some examples are the Amylases (see), Disaccharidases (see), Cellulases (see), Invertase (see) and Lysozyme (see).

Glycosides: derivatives of monosaccharides formed by condensation of the glycosidic hydroxyl group with an alcoholic or phenolic hydroxyl group (***O-glycosides***), thiol group (***S-glycosides***) or NH group (***N-glycosides***, including the ***nucleosides***). They can be considered acetals, thiacetals or aminoacetals. The ***C-glycosides***, e.g. pseudouridine, showdomycin or barbaloin, however, are not analogs of acetals.

G. are classified as *holosides* or *heterosides*. In the holosides, the glycosyl group is attached to the OH group of a second monosaccharide; this group includes oligo- and polysaccharides. In heterosides, the glycosyl group is bound to an non-saccharide, called the *aglycon* (*genin*).

G. are split by acid-catalysed hydrolysis into their components; the stability of the glycosidic bond depends on the structure of the mono- or oligosaccharide (configuration, ring size, substituents) and aglycon. The G. of 2-deoxysugars are very labile; the N-G. of π-deficient heteroaromatics, such as pyrimidine nucleosides, are relatively stable. G. which tend to β-elimination are also alkali-labile. Enzymes which split G. are called glycosidases.

G. are optically active. The *Hudson rules* say that α-D-G. rotate polarized light to the right, while β-D-G. rotate it to the left. However, these rules fail for the pyrimidine nucleosides.

Occurrence. Nucleosides, as components of nucleic acids and some coenzymes, are ubiquitous. Other animal G. include glycolipids, glycoproteins and proteoglycans, and glucuronides. Many antibiotics are G., such as the aminoglycoside, macrolide and nucleoside antimetabolite antibiotics. The most diverse group of G., however, are the secondary metabolites of plants. Examples are the cardiac G. (G. of cardenolides and bufadienolides), the saponins and G. of steroid alkaloids, G. of flavonoids, phenol G. (e.g. arbutin or phlorizin), anthraglycosides, the widely occurring cyanogenic G. (e.g. the G. of mandelic acid – amygdalin, prunasin and sambunigrin – and linamarin, α-hydroxyisobutyronitrile-β-D-glucoside). The aglycons of cyanogenic G. are cyanohydrides which are formed in the plant from amino acids such as phenylalanine, tyrosine and valine. The cyanogenic G. are cleaved enzymatically to the corresponding sugars and cyanohydrins, which immediately break down to release HCN.

Synthesis. A direct reaction between the glycosidic OH group of a mono- or oligosaccharide and a nucleophilic group of the aglycon is not possible. According to E. Fischer, monosaccharides can react with alcohols to form G. in the presence of mineral acids, although the products are mixtures of anomers, and of pyranosides and furanosides. Therefore, G. are almost always synthesized from glycosyl halides (halogenoses) in which the OH groups are protected by acetyl, benzoyl, benzyl, toluoyl or other easily removed groups. The hydrogen halide formed by the reaction is removed by small amounts of alkalies or silver carbonate in the medium. In the synthesis of nucleosides, the silver salt of the aglycon is often used as a starting material. Some of the glycosyl halides used are 2,3,4,6-tetra-O-acetyl-α-D-glucopyranosyl bromide, 2,3,5-tri-O-benzoylribofuranosyl chloride, or 2,3,5-tri-O-benzylarabinofuranosyl chloride. Sugar anhydrides or acyl oxonium salts are also used; the latter are produced from acylglycosyl halides. Biosynthetic pathways start with sugar nucleotides, which are derivatives of aldose 1-phosphates, e.g. uridine diphosphoglucose (UDPGlc).

Glycosyllipids: same as Glycolipids (see).

Glycyrrhizic acid: an extremely sweet tasting triterpene saponin from the roots and underground stolons of liquorice (*Glycyrrhiza glabra*). G. comprises 3 to 9% of the root; a thickened extract of liquorice is used as an expectorant and taste-improving component of many cough remedies. The sapogenin, ***glycyrrhetic acid***, is derived from β-amyrin.

Glyoxal, ***ethanedial, oxaldehyde***: OHC-CHO, the simplest dialdehyde, and a yellow, crystalline compound which smells somewhat like formaldehyde; m.p. 15 °C, b.p. 51 °C, n_D^{20} 1.3826. G. is readily soluble in water and anhydrous organic solvents. When stored, it soon forms the colorless *polyglyoxal*, which does not melt, but carbonizes at 200 °C. When it is heated, it is reconverted to the monomeric form. As a dialdehyde, G. is very reactive, undergoing the reactions typical of Aldehydes (see) twice per molecule. G. is produced by oxidation of ethanol or acetaldehyde with nitric acid, or by oxidation of ethylene glycol on a copper oxide contact in the vapor phase. G. is used mainly in the form of a 30% aqueous solution, or as a bisulfite adduct. Its reactions are similar to those of formaldehyde with hydroxyl and amino groups of cellulose and protein, and it thus reduces their sensitivity to water. G. is therefore used in the textile industry to make cloth which resists wrinkling and shrinkage during washing, and to improve the tensile strength of threads. In the paper industry, it is used to increase the resistance of paper to water; and in leather processing, G. is often used together with formaldehyde as a tanning agent. It is also the starting material for pharmaceutical products with imidazole or hydantion structures.

Glyoxaline: see Imidazole.

Glyoxalic acid: same as Glyoxylic acid.

Glyoxylic acid, ***oxoethanoic acid, glyoxalic acid, formylformic acid***: OHC-COOH, the simplest oxocarboxylic acid. G. forms unstable, highly hygroscopic crystals; m.p. 98 °C (52-53 °C has also been reported). It is readily soluble in water and barely soluble in alcohol and ether. With water it forms a stable, crystalline hydrate, $(HO)_2CH-COOH$; m.p. 70-75 °C. As a bifunctional compound, G. undergoes the reactions typical of aldehydes and carboxylic acids. It is found in free form in green fruits, in rhubarb and gooseberries, and in the green leaves of many plants. G. has a central position in the glyoxylic acid cycle of microorganisms and plants. It can be synhesized by oxidation of glyoxal, hydrolysis of dichloroacetic acid or by cathodic reduction of oxalic acid. G. is used to make vanillin, pharmaceutical products and pesticides.

Glyphosphates: see Herbicides.

Glyphosphines: see Glycine.

Glyptal resins: see Alkyde resins.

GMP: see Guanosine.

Gold, ***aurum***, symbol ***Au***: a chemical element from Group 1b of the periodic system, the Copper group (see). G. is a noble metal and has only one natural isotope; *Z* 79, atomic mass 196.9665, valence III, I, Mohs hardness, 2.5, density 19.32, m.p. 1063.0 °C, b.p. 2660 °C, electrical conductivity, 48.1 Sm mm^{-2}, tensile strength 127.5 N mm^{-2}, standard electrode potential (Au/Au^+) + 1.691 V.

Properties. G. is a reddish-yellow ("golden yellow") metal which crystallizes in cubic close-packing. The crystal habit is usually dodecahedral or octahedral, or less often, cubic. Its electrical and thermal conductivities are about 70% of those of silver, and it is very ductile. For example, G. can be hammered into blue-green, translucent leaflets which are 0.0001 to 0.0002 mm thick. Wires less than 0.01 mm in diameter can be drawn cold. Because it is very soft, G. can be polished to a high sheen. As a typical noble metal, it has a high oxidation potential, and does not react with water, acids or bases, dry or moist air. Oxygen does not react with G. even at high temperatures, and caustic alkali melts do not solubilize it. Only very strong oxidizing agents, such as aqua regia or other combinations of hydrogen halide acids with oxidizing agents (free halogens, hydrogen peroxide, chromic acid) are able to dissolve G. Strong complexing agents, such as the cyanide ion, are able to dissolve G. in combination with atmospheric oxygen. This solubilizing process is due to the shift in the Au/Au^+ oxidation potential caused by the high thermodynamic stability of the resulting dicyanoaurate(I) ion. Chlorine and bromine react with G. to form the corresponding trihalides; chlorination occurs at high temperatures, but bromination takes place even at room temperature. Reducing agents precipitate elemental G. from solutions of its salts, often with the formation of intensely colored red-purple, blue or violet, or brown-black gold colloids as intermediates.

The gold(I) ion does not exist in aqueous solutions, because of its strong tendency to disproportionate: $3\,Au^{+}_{aq} \rightarrow 2\,Au + Au_{aq}^{3+}$. Gold(I) compounds can be synthesized only if they have very low solubilities, as the gold(I) halides do, or if their complexes are thermodynamically very stable. Gold(III) compounds are complexes with coordination number 4, in which the Au^{3+} ion (electron configuration $5d^8$) has a square-planar ligand field.

Analytical. Qualitative and quantitative analyses of G. are usually based on reduction of its compounds to the metal. The formation of Purple of Cassius (see) is a very suitable method of detection.

Occurrence. G. is a rare element, making up only about $5 \cdot 10^{-7}\%$ of the earth's crust. Its concentration in the oceans is from 0.008 to 4 ppb. The G. minerals calaverite (cremerite), $(Au,Ag)Te_2$, sylvanite $(Au,Ag)Te_4$, and foliated tellurium, $(Pb,Au)(S,Te,Sb)_{1-2}$ are very rare; most G. is found in the metallic state. This "elemental" G. is not chemically pure, but is alloyed with silver, and often contains copper and iron impurities. Silver-containing G. is found in veins in quartz; in the course of weathering processes, this is washed away by water and is found in the form of dust, grains or nuggets in the beds of streams.

Mining. The oldest method of mining is *panning*, which utilizes the difference in density between G. and the minerals around it. The G.-containing sands are washed over a rough surface, which catches the G. grains. They are then picked out by hand. G. is still separated by gravity, mainly in combination with cyanide extraction. Another method developed in the 19th century is *amalgamation*, in which the ground G.-containing material flows over an amalgam plate (slanted copper sheets covered with a surface layer of mercury); the G. is bound as an amalgam and later separated from the mercury by distillation.

Because it offers the highest yields, *cyanide extraction* is extensively used today, often combined with older methods. If sulfide minerals are present in the G. ore, the ore is enriched by flotation before cyanide extraction. Then 1 kg lime/t ore and 0.15 kg sodium cyanide/t ore are added to the crushed and wet-ground ore; with good aeration of the suspension, the G. dissolves within 12 h: $4\,Au + 8\,NaCN + O_2 + 2\,H_2O \rightarrow 4\,Na[Au(CN)_2] + 4\,NaOH$. The selectively dissolved G. is then precipitated by addition of zinc shavings: $2\,Na[Au(CN)_2] + Zn \rightarrow Na_2[Zn(CN)_4] + 2\,Au$. Impurities such as lead, zinc and iron are converted via their oxides into silicates and borates, by roasting and addition of fluxes. The residue is silver-containing crude G. This crude G. is purified to 99.7-99.8% by treatment with hot concentrated sulfuric acid, which dissolves the silver as silver sulfate. Further purification to more than 99.9% purity can be achieved by electrolysis; the G. to be purified is used as the anode, and tetrachloroauric acid is the electrolyte. The G. deposited at the cathode is of highest purity. Such electrochemical refining is especially advantageous when the crude G. contains platinum metals, which are enriched in the anode sludge (ruthenium, osmium, iridium) or in the electrolyte (platinum, palladium).

Applications. A large portion of the world's supply of G. serves as monetary reserves and is thus not available for further processing. Most of the G. which is processed is used for jewelry. G. is also used for dental prostheses, coins and medals. The electronics industry is using an increasing amount. Colloidal G. is used to make G. ruby glass. Most G. is used in the form of Gold alloys (see).

Historical. G. was the first metal known to human beings. It was worked into jewelry very early; the first archaeological finds have been dated at the end of the Neolithic. The oldest extensive finds which can be reliably dated are from predynastic Egypt (about 4000 BC). The Sumerian city of Ur (in the southern part of modern Irak) yielded rich, technologically and artistically advanced G. objects from the 3rd millenium BC. From the earliest times, the Egyptians obtained G. from the sands of the Nile; their Nubian G. mines were worked starting aruond 1950 BC. The native civilizations of Mexico and South America had an enormous wealth of G. from their highly developed mining industry. The use of G. as money began very early, about 2000 BC; the first G. coins were minted around 700 BC in the Greek areas of western Asia Minor.

Gold alloys: metallic systems based on gold, which may also contain silver, copper, nickel, palladium, platinum, cadmium, indium, tin and/or zinc. In the solid alloy, gold forms mixed crystals with most of these elements. The industrial uses of pure gold are limited, because it is very soft; the alloys possess good mechanical properties and retain the malleability and high resistance to corrosion of gold. G. are used for jewelry, watch housings, coins, medals, religious objects, art, dental prostheses, electrical conductors and in the chemical industry.

1) Gold for jewelry and coins (see Coinage alloys) are distinguished on the basis of color (red, yellow, green, white) and carat number. Alloys containing large amounts of copper are red; alloys containing about equal amounts of copper and silver are white; white gold contains nickel or palladium. Yellow gold, 14 carat, contains 59% gold, 4 to 34% silver and 7.5 to 37.5% copper. White gold, 14 carat, contains 59% gold, 10 to 16% nickel, 15 to 25% copper and 5 to 8% zinc or 59% gold, 10 to 20% palladium, 10 to 30% silver and 0 to 10% copper.

2) Dental alloys (see) contain at least 50 mol-% gold plus platinum, which gives high corrosion resistance.

3) Chemical industry: spinnerets are made of hardened gold-platinum alloys with 50 to 70% gold. Gold-silver-palladium alloys with 30% each gold and palladium are used as seals. Crucibles for analytical laboratories are made of gold-platinum alloys with 10% platinum.

4) Electronics and electrical industry. G. are used as contact materials in switches, roller plates, or galvanic coatings on contacts subject to corrosion. Gold-manganese alloys are used in resistance thermometers. Thermoelements consist of gold-palladium alloys with 35 to 45% palladium (negative side) and platinum-iridium or platinum-palladium alloys (posi-

tive side). The G. used as contacts to semiconductors contain small amounts of phosphorus, arsenic, antimony, bismuth, barium, aluminum, gallium or indium. Gold-tin, gold-germanium, gold-copper, gold-silver-copper, gold-nickel, gold-copper-nickel and gold-palladium alloys are used as solders.

Some alloys containing little or no gold are called "gold" because of their color. Nürnberger gold, for example, is a copper alloy which contains only 5% gold.

Gold chlorides. ***Gold(I) chloride***: AuCl, a pale yellow powder which is insoluble in water; M_r 232.42, density 7.4. AuCl is obtained in pure form by heating $AuCl_3$ to 190-200 °C in an HCl atmosphere. In alkali chloride solutions, it reacts to form alkali chloroaurates(I); the anion $[AuCl_2]^-$ has been shown to have a linear structure.

Gold(III) chloride: $AuCl_3$, forms red-brown crystalline needles; M_r 303.33, density 3.9, m.p. 287 °C (under Cl_2 pressure). $AuCl_3$ is formed by the reaction of chlorine with gold at 225 to 250 °C; it is obtained as the sublimate. It is a dimer, in which the $AuCl_3$ units are linked to form planar Au_2Cl_6 via chlorine bridges. $AuCl_3$ is readily soluble in water, forming a red-brown solution of $H[AuCl_3OH]$. With hydrochloric acid, it forms ***tetrachloroauric(III) acid***, $H[AuCl_4] \cdot 4H_2O$, a light yellow crystalline powder which tends to deliquesce in air; M_r 411.85. The acid is readily soluble in water, and soluble in alcohol and ether. It is very conveniently produced by dissolving gold in aqua regia followed by crystallization. Tetrachloroauric(III) acid is used as a caustic in medicine, in galvanic gold plating, for gold toner baths in photography, to make gold ruby glass, for purple glaze colors for decoration of porcelain and for production of Cassius' gold purple.

Gold complexes: coordination compounds of gold (see Coordination chemistry). Characteristic coordination numbers are 2 for gold(I) and 4 for gold(III). Formation of linear gold(I) complexes, AuL_2, stabilizes the oxidation state (+1), which is otherwise unstable; the pronounced tendency of gold(III) to form square planar complexes AuL_4 dominates its chemistry. G. are applied, for example, in gold production by cyanide leaching, or in galvanic gold plating (see Sodium disulfitoaurate(III)).

Gold cyanides: ***Gold(I) cyanide***: AuCN, yellow, six-sided crystalline platelets; M_r 222.98, density 7.12. AuCN has a linear chain structure (...-Au-C≡N-Au-C≡N-...). It is obtained by decomposition of ***potassium dicyanoaurate(I)***, $K[Cu(CN)_2]$, with acids. This complex forms colorless crystals, M_r 288.10, density 3.45. It is readily soluble in water, soluble in alcohol and insoluble in acetone. It is obtained by reaction of fulminating gold or gold(III) hydroxide with sodium cyanide, or by anodic dissolution of gold in a potassium cyanide solution using an ion-exchange membrane as a diaphragm. Alkali dicyanoaurates(I) contain the linear anion $[Au(CN)_2]$, which is exceptionally stable (stability constant $K = 10^{38}$). It plays an important part in gold extraction by cyanide leaching. Potassium dicyanoaurate(I) is used in large amounts for galvanic and chemical gold plating. This is traditional in jewelry manufacture (galvanic hard gold plating), and is now also used widely for electronic components. ***Potassium tetracyanoaurate(III)***, $[Au(CN)_4] \cdot 1.5H_2O$, is obtained by reaction of a neutral gold(III) chloride solution with hot conc. potassium cyanide solution in the form of colorless, crystalline platelets.

Gold doublé.: same as Talmi gold (see).

Golden sulfide of antimony: see Antimony sulfides.

Gold(III) oxide: Au_2O_3, a brown compound, M_r 411.93. G. loses oxygen above 150 °C. It is amphoteric, and can be converted by alkali hydroxide solutions to aurates $[AuO_3]^{3-}$. It is obtained by precipitation of the yellow gold hydroxide out of tetrachloroaurate solutions with alkali hydroxide solutions; this hydroxide is converted upon drying to G.

Goldschmidt process: same as Aluminothermal process.

Goltix®: see Herbicides.

Gomberg-Bachmann reaction: synthesis of symmetrical and unsymmetrical diaryl compounds by reaction of aryl diazonium salts with aromatic hydrocarbons in the presence of dilute sodium hydroxide solution:

$$C_6H_5\text{-}\overset{+}{N}{\equiv}N\text{-}Cl^- + C_6H_5\text{-}R \xrightarrow[-\,NaCl]{+\,NaOH} C_6H_5\text{-}C_6H_4\text{-}R + H_2O + N_2.$$

This reaction goes via the corresponding aryl diazohydroxide, which decomposes in nonpolar solvents, forming aryl radicals, nitrogen and hydroxyl radicals: $C_6H_5\text{-}N{=}N\text{-}OH \rightarrow C_6H_5\cdot + N_2 + OH\cdot$.

The aryl radical then reacts with the added aromatic compound to form the diaryl compound.

Gomberg reaction: reaction of triphenylchloromethane with finely divided silver in benzene, in the absence of air. Relatively stable and long-lived triphenylmethyl radicals are formed in the solution, and give it a yellow color:

$$2(C_6H_5)_3C\text{-}Cl \xrightarrow[-2AgCl]{+2Ag} 2(C_6H_5)_3C\cdot \rightleftharpoons (C_6H_5)_2C{=}C_6H_4(H)\text{-}C(C_6H_5)_3$$

Triphenylmethyl radical — Dimer

Addition of acetone or evaporation of the benzene leads to formation of a colorless dimer, which Gomberg thought was hexaphenylethane. It was later discovered, however, that the triphenylmethyl radicals stabilize by forming 3-diphenylmethylene-6-triphenylmethylcyclohexa-1,4-diene. The equilibrium between the dimeric molecules and the triphenyl radicals depends on the concentration, the solvent and the temperature. A one percent solution in benzene, for example, contains 25 to 30% of the triphenylmethyl radical form at 80 °C.

Gonadoliberin: see Liberins.

Gonadotropic hormones: same as Gonadotropins (see).

Gonadotropins, ***gonadotropic hormones***: protein hormones which regulate the production of sex hormones in the gonads. G. are glycoproteins produced either in the anterior pituitary (see Follitropin, Lutropin, Prolactin) or, in the placenta (see Placental hormones). The G. are produced by both sexes and stimulate either testes or ovaries.

Gonadotropin releasing hormone: see Liberins.

Gonane: see Steroids.

Goplat: a platinum-containing gold-silver alloy with graduated hardness. The concentration of gold + platinum is greater than 50 mol %, to achieve the necessary chemical stability. Palladium, copper and zinc are also present.

Gore-Tex: see Synthetic fibers.

Gossypol: a substance found in cottonseed oil. G. inhibits the lactodehydrogenase X in sperm, and thus leads to infertility in men. G. is a dimeric sesquiterpene of the cadinane type.

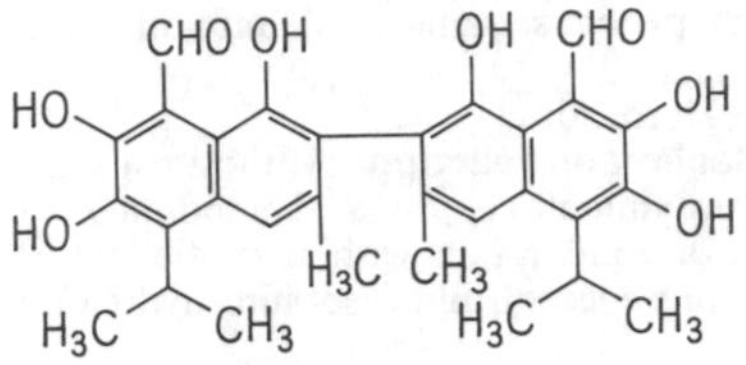

GOT: see Aminotransferases.

GPT: see Aminotransferases.

Gradient elution: see High-performance liquid chromatography.

Grading: a mechanical process to separate mixtures of solids according to their grain size (sieving) or mass (see Sorting). Sieving machines contain grates or screens with square or round holes or slits which are moved vertically or horizontally to sort the material according to size. Sorting by mass is based on the principle of equal rates of fall; the grains are separated by dropping them through a liquid or gas medium.

In ***wet grading***, the mixture to be separated is applied as a suspension, usually in water, to a free-fall sorter (e.g. the sorting cones for ore G.), mechanical sorters (e.g. scratch band or rake sorter in water treatment), counter-current sorters or centrifugal force sorters (hydrocyclones). If dry material is sorted in a stream of gas, the process is called Sifting (see).

Graham-Bunsen effusion law: see Effusion.

Graham's salt: see Sodium phosphates.

Gram atom: see Mole.

Gram equivalent: see Mole.

Gram formula unit: see Mole.

Gramicidins: a group of peptide antibiotics produced by *Bacillus brevis*.

Gramicidin S, cyclo-(-Val-Orn-Leu-D-Phe-Pro)$_2$, was isolated in 1944 by the Soviet workers Gause and Brazhnikova, and two years later, its structure was elucidated by Synge. In 1956, Schwyzer and Sieber achieved its total synthesis, which was the first chemical synthesis of a peptide antibiotic. Gramicidin S acts against gram-positive, but not gram-negative bacteria. Conformation studies led to a three-dimensional model with an antiparallel pleated sheet structure (Fig.).

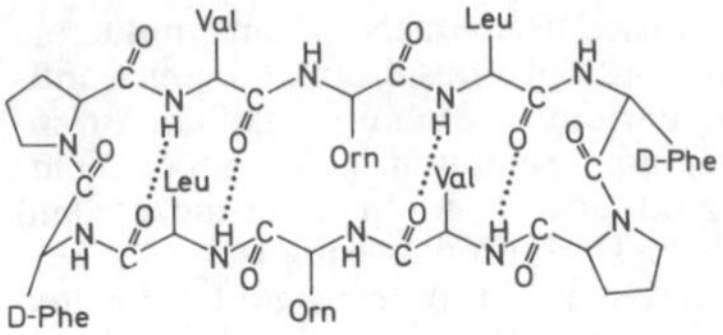

Three-dimensional model of gramicidin S

The biosynthesis of G. S was investigated by Lipmann, who showed that it is based on the principle of *S*-aminoacyl activation with a preliminary ordering on an enzyme matrix, a pattern which is also followed in the biosynthesis of other peptide antibiotics. Gramicidin S synthetase consists of enzyme I (M_r 280,000), which catalyses the formation of enzyme-bound acyl-AMP derivatives of the amino acids Pro, Val, Orn and Leu and their transfer to the SH-group of a covalently bound 4'-phosphopantetheic acid group, and enzyme II, which activates L-Phe in a similar manner, but also simultaneously catalyses its epimerization to the D-isomer. Now the proline group esterified to enzyme I is transferred to the free amino group of the D-Phe bound to enzyme II, forming Pro-D-Phe-S-EII. The analogous coupling of the amino acids Val, Orn and Leu creates the enzyme II-bound pentapeptide, which is then doubled to form the decapeptide and simultaneously cyclized.

Gramicidins A-C are linear pentadecapeptides with *N*-terminal formyl groups and *C*-terminal β-ethanolamide groups. Depending on the *N*-terminal amino acid (Val or Ile), the antibiotics are classified either as valine-G. A-C or isoleucine-G. A-C. The following structure is that of valine-G. A; G. B has a Phe replacing Trp^{11}, while G. C has a Tyr in that position: HCO-Val-Gly-Ala-D-Leu-Ala-D-Val-Val-D-Val-Trp-D-Leu-Trp-D-Leu-Trp-D-Leu-Trp-NH-CH$_2$-CH$_2$-OH. The alternating arrangement of L- and D-amino acids is worthy of note. It is believed that the linear G. aggregate around K^+-, Na^+ and other monovalent cations and form a pore in the membrane through which the cations are transported. As ionophores, the G. induce transport of monovalent cations through biological membranes (mitochondria and erythrocyte membranes), as well as through synthetic phospholipid bilayers.

G. are used as local antibiotics to combat infection by gram-positive bacteria.

Gramine, ***3-(dimethylaminomethyl)indole***: an intermediate in the synthesis of Tryptophan (see). G. forms colorless crystals; m.p. 139 °C. It is soluble in alcohol, ether and chloroform. G. is readily synthesized by aminoalkylation of indole.

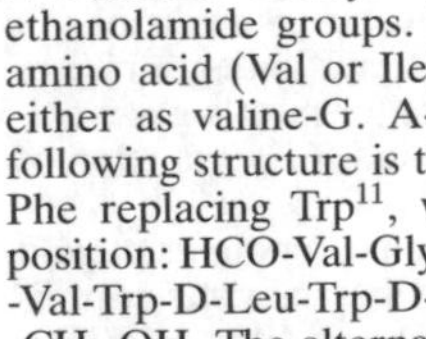

Graminicide: see Herbicides.

Gram molecule: see Mole.

Gram staining: a method for staining isolated bacteria developed by the Danish pharmacologist and pathologist H.C.J. Gram (1853-1923). In G. with gentian violet, the blue-black stain is fixed to the surfaces of those bacteria which contain certain nucleoproteids (gram-positive bacteria, such as staphylococci, streptococci, pneumococci, anthrax, diphtheria, tuberculosis and erysipeloid bacteria). All other bacteria are stained red by a subsequent treatment with fuchsin (gram-negative bacteria, such as gonococci, meningococci, coli, typhus, dysentery, bazillus and bubonic plague bacteria). Some bacteria do not fit into this scheme; the tetanus bacterium, for example, is gram-positive in young cultures and gram-negative in older cultures. The difference in staining of the two types of bacteria is due to structural differences in their cell walls.

Granulate: a grainy substance in which the grains are of uniform size, but have no uniform geometric form and have uneven surfaces.

Granulation, ***pelletizing***: production of a grainy material (granulate, pellets) with as uniform grain size as possible. There are several processes: 1) conversion of fine-grained material into coarser grains (G. of cement and fertilizer, pelletizing of ores); 2) breaking coarser pieces into a granulate (G. of plastics or pharmaceuticals); and 3) production of a granulate from a molten material (e.g. with blast furnace slag).

G. is used to avoid problems with fine grain size in further processing and transport of the material, or when a uniform grain size is required for further processing.

Graphite: the modification of Carbon (see) which is stable at room temperature. It is dark gray, d. 2.25, Mohs hardness 1, sublimation at 3652 to 3697 °C, b.p. 4827 °C. The hexagonal lattice of G. consists of planar layers (see Graphite structure). The structure accounts for its high electrical conductivity, color and the ease with which the crystal can be cleaved and the layers shifted relative to each other, i.e. the lubricating effect of G.

Although it is more reactive than diamond, G. is still fairly unreactive. It can be converted to diamond by subjecting it to very high pressure and temperature. Incorporation of ions or molecules into the space between the layers produces ***graphite compounds*** in which the layered structure of the G. is largely retained. True covalent bonds can be formed between these groups and the carbon atoms, in which case the C atoms are sp^3 hydridized. The layers are puckered, their spacing is increased, and the compound is non-conductive. These are the conditions in graphite fluoride $(CF)_n$ or graphite oxide $(C\text{-}C\text{-}O)_n$, for example. In addition, there are inclusion compounds of G. with halogens, alkali metals, etc., in which the layer spacing is merely increased. The conductivity is often increased in this case.

G. is found in nature. It is produced industrially by the *Acheson process*, in which coke is heated with quartz (silicon dioxide) to about 2500 °C in an electric furnace. Silicon carbide is formed as an intermediate:

$$SiO_2 + 2C \longrightarrow Si + 2CO;$$

$$C_{coke} + Si \xrightarrow{2000°C} SiC \xrightarrow{2200°C} C_{graphite} + Si$$

G. is used as an electrode material, and, because of its chemical resistance and good heat conductivity, to make crucibles, heat exchangers, etc. G. suspended in oil is used as a lubricant and corrosion protection. Solid G. is used as the "lead" in pencils and as a moderator in nuclear reactors.

Graphite structure: the crystal structure of the thermodynamically stable carbon modification graphite. It consists of plane layers in which each C atom has 3 nearest neighbors at a distance of 142 pm; the atoms are arranged in regular hexagonal rings. The layers are stacked at a distance of 335 pm, in such a way that half the C atoms are directly over or under another C atom, while the other half are either over or under the middle of a six-membered ring in the next layer up or down (stacking sequence ABAB of hexagonal graphite; Fig.).

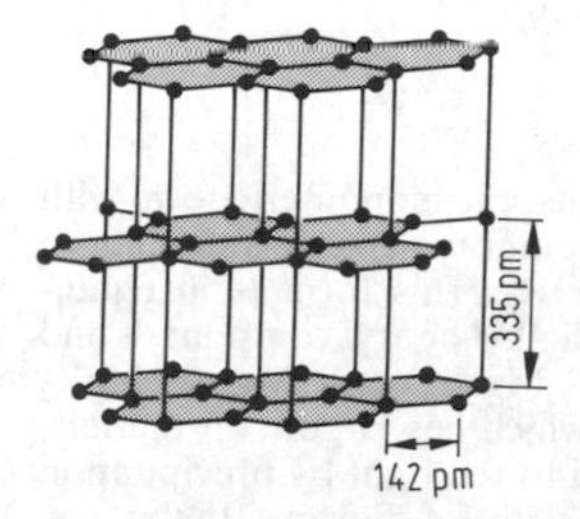

Crystal structure of hexagonal graphite

An ABCABC stacking sequence is also possible, but occurs more rarely (rhombohedral graphite). The layered lattice of G. with strong covalent bonds within the layers and weak van der Waals forces between the layers is the basis for the marked anisotropy of its physical properties.

Graphitization of gray iron: selective corrosion of gray iron in aqueous solutions in which the ferrite in the mixture goes into solution, leaving a structure of graphite and parts of the phosphide eutectic mixture. The corrosion product, which consists of the graphite and phosphide, has the same shape as the original solid. The microcorrosion element consists of the ferrite as the anode, the corrosion product as the cathode, and the aqueous solution.

Gravimetric analysis: a type of quantitative analysis in which addition of a suitable reagent to the sample solution causes the substance of interest to form a very insoluble precipitate. This precipitate is separated from the sample by filtration, washed and dried or roasted to convert it into the stoichiometrically defined form in which it is weighed. Its mass is then determined by weighing on an analytical balance. The mass and known composition of the compound then permit calculation of the absolute mass of the substance to be determined. G. is a very precise absolute method of quantitative analysis which does not require any standardization. Its disadvantages are that it is very time consuming and is not suitable for serial analyses.

Gravimetric factor: see Stoichiometry.

Green PLX, ***naphthol green B***: a water-soluble, acidic nitroso pigment with a grass-green color; chemically, it is the iron-sodium salt of 1-nitrosonaphth-2-

ol-6-sulfonic acid. It is used to dye wool and to make aquarell paints.

GRH: see Liberins.

Grignard compounds: magnesium organic compounds formed from halogenated hydrocarbons and magnesium in the presence of ethers. The first reactions of this type were carried out in 1898 by P. Barbier, and were later thoroughly studied by V. Grignard. The formation of G. is interpreted as a radical reaction of the halogenated hydrocarbon on the surface of the magnesium. The structure of the G. was for a long period the object of extensive studies. Chemical and spectroscopic observations indicate that the G. in concentrated solutions very probably have the following dimeric composition:

$$R_2Mg(\mu\text{-}Cl)_2Mg(OR_2)_4$$

At lower concentrations, the monomeric form with two coordinately bound ether molecules is thought to be present: $R\text{-}Mg\text{-}X \cdot 2OR_2$. This form is in equilibrium with magnesium dialkyl or aryl compounds and magnesium halide: $2\ R\text{-}Mg\text{-}X \rightleftharpoons R_2Mg + MgX_2$ (Schlenk equilibrium), which lies largely on the left, but which can be shifted to the right by precipitation of the magnesium halide with 1,4-dioxane. In this way dialkyl or diaryl compounds of magnesium can be obtained. Regardless of the actual structure, G. can be represented in chemical equations describing their reactions, the Grignard reactions (see), as R-Mg-X.

Grignard reactions, ***Grignard syntheses***: addition reactions of Grignard compounds (see) to the carbonyl double bond and other polar multiple bonds, such as

$$>C{=}O,\quad >C{=}S,\quad >C{=}N{-},\quad >S{=}O,$$

-N=O or -C≡N, which were investigated chiefly by V. Grignard (1871-1935). C=C double and C≡C triple bonds react with Grignard compounds only when they are components of a conjugated system with one of the above functional groups and are activated in this way. In the G., the magnesium halogen group is formally a cation which adds to the more electronegative element, and the hydrocarbon group is formally an anion which adds to the more positive element of the polar multiple bond:

$$>C{=}O + R{-}MgX \longrightarrow R{-}\overset{|}{\underset{|}{C}}{-}O{-}MgX.$$

Subsequent hydrolysis of the addition product forms the corresponding alcohol by splitting out Mg(OH)X:

$$R{-}\overset{|}{\underset{|}{C}}{-}O{-}MgX + H_2O \longrightarrow R{-}\overset{|}{\underset{|}{C}}{-}OH + Mg(OH)X.$$

In the sense of this generalized formulation, *formaldehyde* reacts with Grignard compounds to form primary alcohols:

$$CH_2{=}O + R{-}MgX \longrightarrow R{-}CH_2{-}O{-}MgX \xrightarrow[-\,Mg(OH)X]{+\,H_2O} R{-}CH_2{-}OH.$$

Higher *aldehydes* form secondary alcohols under the same reaction conditions. *Ketones* and *carboxylate esters* react with Grignard compounds to form tertiary alcohols. Carboxylate esters form ketones in this reaction, but they cannot be isolated, because they react further with the Grignard reagent, even more quickly than the original ester:

$$\underset{\displaystyle OR}{R{-}\overset{}{C}{=}O} \xrightarrow{R^1MgX} \underset{\displaystyle OR}{R{-}CR^1{-}OMgX} \xrightarrow{-\,Mg(OR)X}$$

$$RR^1C{=}O \xrightarrow{+\,R^1MgX} RR^1_2C{-}OMgX \xrightarrow[-\,Mg(OH)X]{+\,H_2O} RR^1_2C{-}OH.$$

In the reaction of formate esters, the aldehydes formed as intermediates can be isolated. This reaction course can be enhanced by excess ester. The reaction of Grignard reagents with carboxylic halides, carboxylic anhydrides, *N*-substituted carboxamides and salts of carboxylic acids is similar to the reaction with carboxylate esters, but in these cases the ketone intermediates can be isolated.

The reaction of *carbon dioxide* with Grignard compounds at low temperatures produces the corresponding carboxylic acids in good yields:

$$O{=}C{=}O \xrightarrow{+\,RMgX} R{-}CO{-}OMgX \xrightarrow[-\,Mg(OH)X]{+\,H_2O} R{-}COOH.$$

Similarly, *sulfur dioxide* can convert Grignard compounds into sulfinic acids:

$$O{=}S{=}O \xrightarrow{+\,RMgX} R{-}SO{-}OMgX \xrightarrow[-\,Mg(OH)X]{+\,H_2O} R{-}SO{-}OH.$$

In the reaction of *carbon disulfide* with with Grignard compounds, the primary addition products can be converted to dithiocarboxylate esters by alkylation:

$$S{=}C{=}S \xrightarrow{+\,RMgX} R{-}CS{-}SMgS \xrightarrow[-\,MgX_2]{R^1X} R{-}CS{-}SR^1$$

Nitriles react with Grignard compounds via ketimines, which can sometimes be isolated, to form ketones:

$$R{-}C{\equiv}N \xrightarrow{+\,R^1MgX} RR^1C{=}NMgX \xrightarrow[-\,Mg(OH)X]{+\,H_2O}$$

$$RR^1C{=}NH \xrightarrow[-\,NH_3]{+\,H_2O} RR^1C{=}O.$$

The reaction mechanism of the G. on polar multiple bonds can be described as follows, using the reaction of a ketone with the Grignard reagent as an example:

$$R_2C{=}O + R^1(X)Mg(OR^2_2)_2 \underset{+R^2_2O}{\overset{-R^2_2O}{\rightleftharpoons}} R_2C{=}O{\rightarrow}Mg(R^1)(X)(OR^2_2)$$

$$\xrightarrow[-R^2_2O]{+R^1MgX\cdot 2OR^2_2} [R_2C{=}O{\rightarrow}Mg(R^1)(X)(OR^2_2) \cdots R^1Mg(X)(R^2_2O)] \xrightarrow{-2R^2_2O} R^1R_2C{-}O{-}Mg{-}R^1 \cdots X{-}Mg{-}X$$

$$\underset{+R^1MgX}{\overset{-R^1MgX}{\rightleftharpoons}} R^1R_2C{-}O{-}MgX \xrightarrow[-Mg(OH)X]{+H_2O} R^1R_2C{-}OH$$

The synthesis of a series of hydrocarbons from Grignard compounds is achieved with the Wurtz-Grignard reaction (see) using reactive halogen hydrocarbons. Hydrocarbons are also formed by the reaction of Grignard compounds with reactants which contain an active hydrogen atom, e.g. water, alcohols, phenols, enols, carboxylic acids and primary and secondary amines: R-MgX + H-OH → R-H + Mg(OH)X. In the Cerevitinov method (see), this reaction is used for quantitative determination of active H atoms. Solvents which can be used for G. include diethyl ether, dibutyl ether, anisole, tetrahydrofuran and methylene chloride.

Grignard syntheses: see Grignard reactions.

Grimm-Sommerfeld phases: see Grimm-Sommerfeld rule.

Grimm-Sommerfeld rule: compounds of type AB with a sufficiently high covalent component in their bonding and 8 valence electrons per atom pair crystallize preferentially in Wurtzite type (see) or Zinc blende type (see) crystals. The ratio of bonding electrons to atoms, 4:1, can occur either when both elements A and B are in the 4th main group of the periodic system (e.g. carbon as diamond, SiC, GeSn) or when they are equal distances to the right and left of the 4th main group (e.g. AlP, BeO, CuBr). Intermetallic and semiconducting compounds which obey the G. are also called *Grimm-Sommerfeld phases*.

Griseofulvin: an antibiotic with a spirane structure, synthesized by *Penicillium griseofulvum* and *P. patulum*. G. can be used therapeutically as a systemic antimycotic. Only the natural (+)-G., with the configuration 2S,6'R, is biologically active. It is a yellow crystalline powder; m.p. 219 °C, $[\alpha]^{19}_{546}$ +417° (in acetone). It is slightly soluble in acetone, ethyl acetate and benzene, and practically insoluble in water. Because it is poorly absorbed after oral application, G. is administered as micronized preparations with small particle sizes.

Grisuten®: see Synthetic fibers.

Ground state: the lowest-energy state of an atomic nucleus, atom or molecule. Higher-energy states are called excited states (see Excitation). At a given temperature T, the majority of particles of a given species will be in the G. if their mean thermal energy $kT/2$ (k = Boltzmann constant) is much lower than the energy of the first excited state. In practice, this means that at room temperature, most molecules are in the *electronic* and *vibrational* G., but most are in excited *rotational* states.

Groundwater enrichment: methods to equalize the water table, which can rise or fall for various reasons. To increase the source of water for water treatment plants, the water table can be raised by increasing filtration through river banks. This is done by backing up watercourses in the source region of the groundwater, or by Infiltration (see) through percolation basins. However with this method, the risk of colmation (stoppage of a porous rock layer by sedimentation and/or filling of pores and cracks by solids suspended in the percolating water, or generated by precipitation of dissolved minerals) is very high.

Growth hormone herbicides, ***aryloxyalkanoic acid herbicides***: Herbicides (see) containing derivatives of phenoxyfatty acids. The search for synthetic compounds which would influence the growth of plants in similar fashion to the natural *auxin* (β-indolacetic acid) led, during the 1940's, to the discovery that halogenated phenoxyacetic acids are selective herbicides. In higher concentrations, these substances not only inhibit growth, but can kill plants. The monocotyledons have higher tolerance to the G., which can thus be used to control broad-leaved weeds in monocotyledenous crops, chiefly grains. The G. interfere with a large number of metabolic processes, but the key role seems to be their effect on nucleic acid metabolism. The G. are translocated in the plant.

Table 1. Growth hormone herbicides

$R{-}C_6H_4{-}O{-}(CH_2)_n{-}COOH$

Name	R	n	m.p. (°C)
2,4-D	2-Cl, 4-Cl	1	140.5
MCPA	2-CH_3, 4-Cl	1	118-119
2,4,5-T	2-Cl, 4-Cl, 5-Cl	1	155
MCPB	2-CH_3, 4-Cl	3	100

One advantage of the G. is their low toxicity for warm-blooded animals; the oral LD_{50} is 375-800 mg kg^{-1} rat. Another is that their synthesis is relatively inexpensive. The G. are made as free acids, salts of alkali metals or amines, esters, amides or other derivatives, each with different solubilities and volatility, and thus also with different herbicidal properties.

1) ***G. based on phenoxyacetic acid*** (Table 1, $n = 1$) are usually synthesized by reaction of the appropriate sodium phenolate with α-chloroacetic acid. Both 2,4-D and MCPA are used to control dicotyledenous weeds in grain and in pastures and lawns. 2,4,5-T also kills woody plants, and is therefore used as an arboricide on land not used for crops, to prepare land for planting and to thin forests. The reported long-term effects of 2,4,5-T, especially teratogenicity, are due to dioxin contaminants in one of the materials used for synthesis. When not properly prepared, 2,4,5-trichlorophenol can contain 2,3,7,8-tetrachloro-1,4-dioxin.

2) ***G. based on phenoxypropionic acid*** (Table 2).

Table 2. Growth hormone herbicides

Name	R (R–C₆H₄–O–CH(CH₃)–COOH)
Dichlorprop (2,4-DP)	2-Cl, 4-Cl
Mecoprop (MCPP)	2-CH_3, 4-Cl
Diclofop	(4) Cl–C₆H₃(Cl)–O–
Fluaziprop	(4) F_3C–C₅H₃N–O–

These G. are synthesized by reaction of the appropriate alkali metal phenol with α-chloropropionic acid. In addition to the classic examples, 2,4-DP and MCPP, which are used as post-emergence controls in grains, this group includes p-aryloxyphenoxypropionic acid derivatives with 2 to 8 times higher activity, and with a broader spectrum of activity.

3) ***G. based on phenoxybutyric acid*** (Table 1, $n = 3$) are usually synthesized by reaction of the corresponding substituted phenol with γ-butyrolactone. The resulting γ-phenoxybutyric acid derivative is converted to the herbicidally active phenoxyacetic acid derivative by β-oxidation in the plant. The selectivity of these herbicides is due to the fact that the crop legumes, unlike the weeds, do not have the enzymes needed to oxidize the phenoxybutyric acid. Thus only the weeds are damaged. The most important example is MCPB, which can be regarded as a precursor of MCPA. It is used to control weeds in alfalfa, clover and peas.

Growth hormone (human): same as Somatotropin.

Growth regulators: organic substances which, in very small amounts, affect the rates of growth, development and metabolism in crops. The G. are used to counteract the negative effects of many environmental factors, and they can eliminate or reduce certain difficulties arising from the development or morphologies of crop plants. They make it possible to utilize fully the yield potentials of cultivated plants and to increase the effectiveness of such intensive cultivation practices as irrigation and fertilization. G. also make possible increases in yield per hectare, simplification of harvesting and mechanization of various production processes.

According to this definition, the G. include not only those substances which serve the plant as nutrients or energy carriers, but pesticides as well. Transpiration inhibitors and substances which promote maturation, defoliation or drying are in an intermediate position. Some of these substances affect the opening of the stomata or the course of natural ageing, and thus are G.

Characteristics of the G. are 1) the specificity of their effects with respect to plant species, strain, organ, state of development and the process of development affected. The same active material at the same dose can have quantitatively and qualitatively different effects, depending on these conditions. 2) The effects are highly dependent on concentration; too high or too low a dose, accidental double treatment and other variations in application give undesirable results. 3) The effects depend greatly on the weather. In unfavorable weather, reduced uptake of the substance or poor growth of the plant can either prevent the desired effect, or can lead to damage and side effects; that is, the weather factors change the concentration of the substance which is actually present in the plant. 4) The toxicology and persistance time of residues have a major influence on the applicability.

The G. are classified in various ways. A distinction can be made between natural and synthetic Phytohormones (see), or between growth stimulators (auxins, gibberellins, cytokinins) and inhibitors (ethene, abcissic acid). However, a clear division into stimulators and inhibitors is not possible, because the same substance can either inhibit or promote growth, depending on its concentration, the species of plant, the treated organ and the stage of development. The natural growth hormones are Auxins, Gibberellins, Cytokinins, Abcissic acid and Ethene (see separate entry on each).

Most synthetic substances in practical use are derivatives of fluorene-9-carboxylic acid for inhibition of growth of grass and shrubs, and induction of parthenocarpy in purely female cucumbers, 2,3,5-triiodobenzoic acid, abb. TIBA, for growth inhibition, e.g. of soy and vegetables; 2-chloroethyltrimethylammonium chloride (chlormequat) for straw stabilization, especially of wheat; succinic mono-(*N*, *N*-dimethylhydrazide) (SADH, daminozide) for inhibition of shoot growth, induction of blossoming in fruit trees, promotion of self-thinning of fruit, increasing the winter resistance in members of the cabbage family, and increasing hop yields; *N*-(3-chlorophenyl)isopropylcarbamate (CIPC, chlorpropham) for inhibition of sprouting in stored potatoes; *N,N'*-bisphosphonomethylglycine (glyphosine, Polaris®) for increasing the sucrose content of sugar beets; *N*-methyl-1-naphthylcarbamate (carbaryl, Sevin®, bercema-NMC®) for thinning apples.

GSC: see Chromatography; see Gas chromatography.

Gua: abb. for guanine.

Guaiacol, ***2-methoxyphenol, pyrocatechol monomethyl ether***: colorless crystals with an aromatic odor; m.p. 28 °C, b.p. 205 °C. G. is only slightly soluble in water, but is soluble in organic solvents. It gives a green color reaction with iron(III) chloride. G. is present in large amounts in beech tar. It is synthesized by methylation of pyrocatechol, or by diazotizing *o*-anisidine and boiling the diazonium salt. G. is used as an intermediate in the synthesis of vanillin, and in medicine, for treatment of catarrh of the respiratory tract.

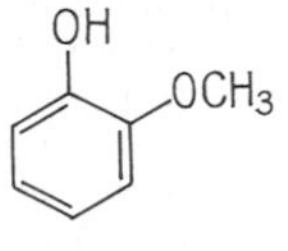

OH
OCH_3

Guaianolides: a group of sesquiterpene lactones; an example is Matricin (see).

Guaiazulene, ***1,4-dimethyl-7-isopropylazulene***: one of the azulenes; a blue-violet, crystalline hydrocarbon; m.p. 31.5 °C, b.p. 167-168 °C at 1.6 kPa. G. is found in geranium oil and can also be made from guaiol, a natural, bicyclic sesquiterpene alcohol, by dehydration with sulfur. G. is used as a mild antiseptic.

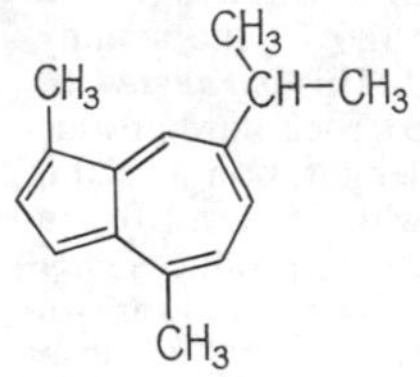

CH_3
CH_3 $CH-CH_3$
CH_3

Guaiazulen

Guanethidine: a guanidine derivative with an azocine ring which is used as an Antihypertensive (see). G. alters the membranes of postganglionic sympathetic nerves, reducing their storage capacity for noradrenalin. This is excreted from the vesicles and degraded, without acting as a transmitter.

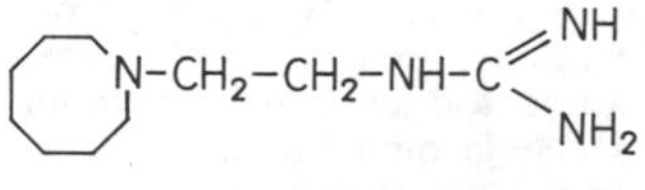

$N-CH_2-CH_2-NH-C(=NH)NH_2$

Guanidine, ***iminourea***: $H_2N\text{-}C(NH_2)\text{=}NH$, forms unstable, colorless, hygroscopic crystals; m.p. around 50 °C (not sharp). G. is soluble in water and alcohol, and insoluble in ether. Its aqueous solutions are strongly basic. G. is slowly hydrolysed in water, forming ammonia and urea. It forms stable ***guanidinium salts*** with acids. G. is a very reactive compound; for example, nitration produces Nitroguanidine (see). It reacts with dicarbonyl compounds and their derivatives to form *N*-containing heterocycles of various ring sizes; these are important in the synthesis of drugs. For example, β-chlorovinylaldehydes and G. react to form 2-aminopyrimidines.

G. compounds, such as arginine, creatine and creatinine, occur widely in nature. G. can be synthesized from cyanamide or dicyanamide and ammonium salts: $N{\equiv}C\text{-}NH_2 + NH_4Cl \rightarrow [H_2N{=}N(NH_2)_2]^+Cl^-$.

Guanine, abb. ***Gua***: 2-amino-4-hydroxypurine. G. forms colorless needles or is an amorphous powder; m.p. 365 °C (dec.). It dissolves in acids and bases to form salts; it is barely soluble in water, ethanol or ether. It can be detected by the murexide reaction. G. is found in large amounts in guano, the excrements of Peruvian seabirds. It can also be obtained from yeast nucleic acids or by the Traube purine synthesis. G. is a component of guanosine and its nucleotides, and of nucleic acids.

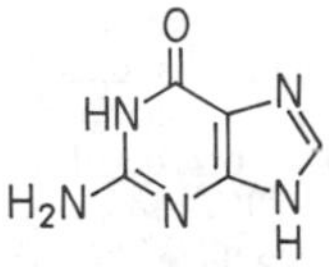

O
HN N
H_2N N N
H

Guanosine, abb. ***Guo***: 9-β-D-ribofuranosylguanine, a nucleoside. As the dihydrate, it forms colorless crystals which lose their water around 110 °C; m.p. about 240 °C (dec.). G. is only slightly soluble in cold water, and is insoluble in ethanol and ether. It dissolves in dilute acids and alkalies, forming salts. G. is a component of the ribonucleic acids. The nucleotides guanosine 5'-mono-, di- and triphosphate play an important role in the transduction of signals from cell membranes to the nuclei. In the biosynthetic pathway, the first guanine derivative is ***guanosine monophosphate (GMP)***, which is formed from inosine monophosphate. GMP is used as a flavor enhancer. G. is added to blood-stabilizing solutions because it increases the survival time of the erythrocytes.

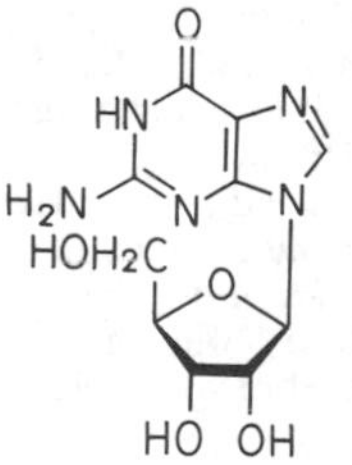

O
HN N
H_2N N N
HOH_2C O
HO OH

Guanosine 3',5'-phosphate: see Cyclic nucleotides.

Guignet green: see Chromium(III) oxide hydrate.

Guinier method: see X-ray structure analysis.

Guldberg rule: see Critical point.

Gum: the water-soluble fraction of a Gum resin (see).

Gum arabic: see Gum resin.

Gum resin: a hardened solution of a resin in a rubber-like substance, such as occurs in the latex of some plants. The resins are the alcohol-soluble components, and the gum is the water-soluble part. The following L. are well known: ***ammonia resin*** (***ammoniacum***) is obtained from various umbelliferae, for example *Dorema ammonicum*. The resin has an unpleasant smell and tastes bitter; it is used as a drug and in making putty. ***Gum arabic (acacia gum, mimosa gum)*** is collected from tropical acacias, e.g. *Acacia arabica*. It consists mainly of acid salts of arabinic acid and small amounts of tannins, sugars and enzymes. Its aqueous solutions are tough, sticky yellow to brown solutions with an insipid taste; they are used as glues for various purposes. ***Asafetida*** is obtained from various umbelliferae, such as *Ferula foetida* and *Ferula nartex*. It has a garlic-like odor and bitter taste, and was formerly used as a medicine to calm nerves and relieve cramps. ***Euphorbium*** is present in the spurge *Euphorbium resinifera*. The ether-soluble resin is used today mainly as a skin stimulant, mainly in veterinary medicine. ***Galbanum*** is collected from umbelleriferae of the genus *Ferula*. This resin has a spicy smell and taste, and consists of about 60% alcohol-soluble resins, 20% gum and essential oils. It is used as an additive to skin-stimulating bandages, and as a putty. *Gamboge* is obtained from a few species of gamboge. It is a greenish yellow, odorless and poisonous G. consisting of 60 to 80% garcinolic acids and 15 to 20% water-soluble gum. When alkalies are added, the aqueous solution becomes dark red. Gamboge is still used as a gold-yellow aquarell pigment. ***Mastix*** is obtained from the Mediterranean shrub *Pistacia lentiscus* in the form of yellowish grains with a weak, aromatic odor. Mastix consists mainly of free resins acids and resenes, along with essential oils and bitter substances. It is used to make varnish and various glues. A solution of mastix in volatile organic solvents (mastisol) is used in medicine to make adhesive bandages. ***Myrrh*** is a secretion of the balsam *Commiphora abyssinica* in the form of yellow to brown grains with an aromatic odor and bitter taste. In addition to 2.5 to 10% essential oils, it contains varying amounts of resin (including myrrholol) and gums. It is used as an aroma substance in perfumes. Tincture of myrrh, an alcoholic solution of myrrh, can be used as a mouthwash to prevent inflammation. ***Traganth*** (***tragacanth***) is secreted by various papillonidaceae in the form of hard, yellowish white lumps which have neither odor nor taste, and swell in water to a gelatinous, translucent mass. Traganth is used as a glue, as a component of varnishes, and as a binder in pharmaceuticals. ***Incense (olibanum)*** is a secretion of the balsam species *Boswellia carteri*; it is found as pea-sized, round or elongated, yellowish white grains which have a weak balsam odor and a bitter taste. It is used chiefly for ritual purposes.

Gun cotton, ***pyroxylin***: highly nitrated cellulose with a nitrogen content of 12.5 to 13.5%, used as an explosive. It is a white, cotton-like substance which burns extremely rapidly when ignited. It is soluble or at least swells in acetone; with 30% water content, it is not sensitive to impact or flame ignition, but can be made to explode with an ignition cap. Pressed G. has a detonation velocity of 6800 m s^{-1}, and is used to make smokeless powder.

Gunpowder: the propellent for conventional firearms; see Smokeless powder, Black powder. ***Monobasic powder*** is a single compound, e.g. cellulose nitrate, ***dibasic powder*** is a mixture of two, such as cellulose nitrate and glycerol nitrate powder, and ***tribasic powder*** is a mixture of three, such as cellulose nitrate, glycerol trinitrate and nitroguanidine.

Guo: abb. for Guanosine.

Gur dynamite: see Dynamite.

Gutta percha: a brown, spongy product related to natural rubber. It is made by drying the latex of the gutta percha tree, *Isonandra gutta*, which grows in the Malayan Archipelago. The basis of G. is a hydrocarbon known as ***gutta***, which consists of isoprene units $(C_5H_8)_n$. Unlike rubber, these molecules display the *trans* configuration. G. contains more resins than natural rubber, and is not elastic. It melts with decomposition at 150 °C, and is soluble in chloroform, carbon disulfide, petroleum ether, acetone, benzene, etc. G. can be vulcanized and is used for insulation of submarine cables and other electrical cables, and to make chemically resistant putty.

Gutzeit test: a preliminary test for the presence of arsenic. As in the Marsh test (see), arsenic compounds react with nascent hydrogen to form arsane. Moist silver nitrate turns yellow in the presence of arsane, through the formation of the addition compound $Ag_3As \cdot 3AgNO_3$. The presence of antimony or phosphane interferes with the G.

Gypsum: chemically, calcium sulfate dihydrate, $CaSO_4 \cdot 2H_2O$. Partial or complete dehydration leads to products which are used mainly as Binders (see) in construction materials. Around 120 °C, the semihydrate, $CaSO_4 \cdot 1/2H_2O$ is formed. This is ***roasted G.***, which sets with water in the air to a rock-hard product of fibrous G. crystals. Further dehydration at 120 to 180 °C yields ***plaster of Paris***, which is used for interior moldings. Heating to 180-200 °C produces ***plaster gypsum***, which is usually mixed with sand for construction purposes. When heated to 500 °C, G. loses the ability to bind into a solid mass. Above 1200 °C, ***estrich gypsum (flooring plaster)*** is formed; it has hydraulic properties, hardens slowly and forms very hard products. At temperatures above 1300 °C, anhydrous calcium sulfate begins to decompose to a solid solution of calcium oxide in calcium sulfate; such solutions are able to harden with water and are used as ***mortar gypsum*** or ***construction gypsym***. ***Marble gypsum*** is a double-burned product; between the two roasting steps, the G. is soaked with alum solutions. In addition to its use as a binder, G. or anhydrous calcium sulfate (anhydrite) is used as a raw material for production of sulfuric acid and cement (gypsum sulfuric acid process), ammonium sulfate, ceramic molds, modelling plaster, artificial marble, plaster casts and as a filler in paper. G. and its less hydrated forms (semihydrate, anhydrite) are obtained as by-products, often in very pure form, from various industrial processes, such as production of hydrogen fluoride from fluorspar and sulfuric acid extraction of phosphates.

Gypsum-sulfuric acid process: see Sulfuric acid.

H

h: unit symbol for Enthalpy (see).

H: 1) symbol for hydrogen. 2) °H, see Hardness, 2.

Haber-Born cycle: a thermodynamic cyclic process for compounds which have ionic crystal lattices. The H. links the molar energy of formation $\Delta_F E$ of the solid (see Formation reaction) with the sublimation and ionization energies, $\Delta_S E$ and $\Delta_I E$ of the metal, the dissociation energy $\Delta_D E$ and the electron affinity $\Delta_E E$ of the nonmetal, and the molar lattice energy $\Delta_L E$ released when the ionic lattice is formed from gaseous ions. For example, the H.-B. has the following form for potassium chloride:

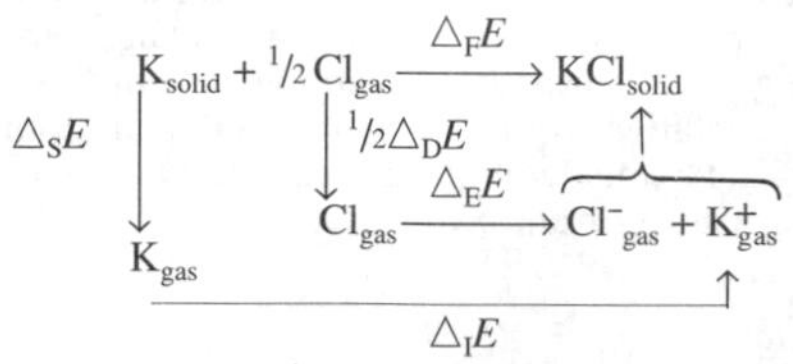

$$\begin{array}{lcccl} & K_{solid} + {}^1/_2\,Cl_{gas} & \xrightarrow{\Delta_F E} & KCl_{solid} & \\ \Delta_S E \downarrow & & \downarrow {}^1/_2\Delta_D E & \uparrow & \\ & & Cl_{gas} \xrightarrow{\Delta_E E} & Cl^-_{gas} + K^+_{gas} & \\ & K_{gas} & \xrightarrow{\Delta_I E} & \uparrow & \end{array}$$

Since the total energy change must be zero when a cyclic process operates, $\Delta_S E + 1/2\Delta_D E + \Delta_E E + \Delta_i E + \Delta_L E - \Delta_F E = 0$. Thus the H.-K. permits calculation of lattice energies if all other parameters are known.

Haber-Bosch process: see Ammonia.

Habit: see Crystal.

Hafnium, symbol ***Hf***: a chemical element in group IVb of the periodic system, the Titanium group. Hf is a heavy metal, *Z* 72, with natural isotopes with mass numbers 180 (35.44%), 178 (27.08%), 177 (18.39%), 179 (13.78%), 176 (5.15%), 174 (0.18%); atomic mass 178.49, valence usually IV, density 13.31, m.p. 2150 °C, b.p. 5400 °C, electrical conductivity 3.26 Sm mm^{-2}, standard electrode potential (Hf/HfO^{2+}) - 1.57 V.

Properties. The chemical and physical properties of H. are very similar to those of zirconium, its next lighter homolog in group IVb. This is due mainly to their nearly identical atomic and ionic radii: r_M = 145.4 pm (Zr); 144.2 pm (Hf); $r_{M^{4+}}$ = 74 pm (Zr), 75 pm (Hf). Pure Hf is a very shiny, ductile and rather soft metal, which can readily be drawn, rolled and worked. H. crystallizes in a hexagonal lattice, and is converted at 1760 °C into a cubic high-temperature modification. It develops a thin, hard and nearly impenetrable oxide layer when exposed to air; this inhibits its reactivity, and Hf is therefore not attacked by water or alkalies, or by cold hydrochloric, nitric or dilute sulfuric acids. Hot sulfuric acid, aqua regia and hydrofluoric acid dissolve it. Dry chlorine reacts with H. even at low temperatures, with flames, forming hafnium tetrachloride, $HfCl_4$. At higher temperatures, it is oxidized by oxygen to HfO_2, and reacts with other nonmetals, such as nitrogen, carbon, boron and silicon, to give stable inclusion compounds: the carbide, HfC, nitride, HfN, boride, HfB, and silicide, $HfSi_2$.

Analysis. Hf is determined in zirconium minerals and compounds by x-ray and mass spectroscopy, and by neutron activation analysis.

Occurrence. Hf makes up $2.8 \cdot 10^{-4}$% of the earth's crust. It always accompanies zirconium, the minerals of which contain 1 to 5% Hf. The mineral cyrtolite contains a relatively high content of 5.5% HfO_2, while the most favorable ratio of H. to zirconium is found in the scandium mineral thortveitite: 1 to 2% ZrO_2 is accompanied by an almost equal amount of HfO_2.

Production. H. production is a necessary byproduct of zirconium production, because the use of zirconium in nuclear reactors requires complete removal of the H. The two elements are separated by ion exchange and liquid-liquid extraction. The extraction utilizes the better solubility of hafnium isothiocyanate, $HfO(NCS)_2$ in methyl isobutyl ketone; it is run as a countercurrent process. The zirconium-free H. is made water-soluble by addition of sulfuric acid, then precipitated with ammonia as hafnium dioxide. It is chlorinated in the presence of coal at 950 °C, and the resulting hafnium tetrachloride is reduced with magnesium at 850-860 °C in an argon atmosphere. Magnesium and magnesium chloride are removed from the hafnium sponge by vacuum distillation, and the metal is finally remelted in an electron beam furnace.

Applications. Most of the H. produced at present is used for control elements in nuclear reactors, because of its high neutron absorption cross section. A major advantage of H. over other neutron absorbers is that the control rods are not "burned" during service; the hafnium isotopes formed by neutron capture also have high absorption cross sections. H. is also used as an alloy element in high-melting, heat-resistant alloys of niobium, tantalum, molybdenum or tungsten; it is added to about 2% concentration to improve the solidity of the alloy. H. foils are used in limited amounts to make flash cubes for photography.

Historical. H. was one of the last naturally occurring elements to be discovered. G. von Hevesey and D. Coster discovered the previously unrecognized element number 72 in 1922 in zirconium minerals by use of x-ray spectroscopy and application of Mosely's law. The discoverer named the new element in honor of the site of its discovery, Copenhagen (Latin Hafnia).

Hafnium compounds: The properties and syntheses of H. correspond to those of the homologous zirconium compounds, but they have been less thoroughly studied than the latter. The +4 oxidation state is by far the most common. Examples: ***haf-***

nium(IV) oxide, HfO_2, M_r 210.49, density 9.68, m.p. 2812 °C, b.p. about 5100 °C; ***hafnium(IV) chloride***, $HfCl_4$, M_r 320.20, subl.p. 319 °C; ***hafnium carbide***, HfC, M_r 190.54, density 12.00, m.p. 4160 °C, b.p. 5400 °C; ***hafnium nitride***, HfN, M_r 192.50, m.p. 3305 °C.

Hagedorn-Jensen test: see Monosaccharides.

Half-life: in kinetics, the time $t_{1/2}$ required for the concentration of a reactant to drop to half its initial value c_0. For a first-order reaction, $t_{1/2} = \ln 2/k$, and for an *n*th-order reaction, $t_{1/2} = (2^{n-1}-1)/(n-1)kc_0^{n-1}$, where k is the rate constant. The H. of first-order reactions is thus indirectly proportional to the rate constant, but in all other cases it also depends on the initial concentration. If the concentrations of several reactants appear in the rate equation, there is an H. for each of them. However, the composition of the reaction mixture must be such that the consumption of the individual component down to $c_0/2$ is stoichiometrically possible.

Radioactive decay always occurs by a first-order process, and it is customarily characterized by indication of H. rather than the rate constant.

The time for consumption of short-lived intermediates is often characterized by the mean Lifetime (see), which is closely related to the H.

Half cell: an electrochemical electrode consisting of at least two phases. One is an electron (or defect electron) conductor, and the other is an ion conductor. A galvanic cell is made by combination of two H.

Half-height width: the width of an analytical signal at the position half way between the maximum and base of the peak.

Haloalkanes, ***alkyl halides***: saturated, aliphatic, halogen-containing hydrocarbons. The systematic names of the H. consist of a prefix indicating the halogen and the name of the alkane, e.g. chloromethane, CH_3-Cl, 2-bromopropane, CH_3-CHBr-CH_3, 1-bromo-2-chloroethane, Br-CH_2-CH_2-Cl. Alternatively, a systematic radicofunctional name is constructed as the alkyl-substituted halide, e.g. methyl chloride, CH_3-Cl, isopropyl chloride, $(CH_3)_2$CH-Cl, methylene bromide, CH_2Br_2. The radicofunctional system cannot be used for compounds with different halogen atoms in the molecule. Many H. are colorless liquids with a characteristic sweetish odor. Some halomethane and haloethane derivatives are gases, and derivatives with larger molecules are solids. In the absence of hydrophilic groups, they are insoluble in water, but are soluble in most organic solvents. The halogen atoms in H. can normally be relatively easily exchanged for other nucleophilic reagents, so that other functional alkane derivatives, such as amines, thiols, alcohols, thiocyanates, etc., can readily be made from H. The exchange of one halogen atom for another can also occur under the conditions of the Finkelstein reaction. Alkenes and alkynes can be synthesized by dehydrohalogenation of suitable H. Reducing agents, the Wurtz reaction (see) and the Grignard reaction (see) can convert H. to alkanes.

The most important industrial methods for synthesis of H. are halogenation of alkanes and esterification of alcohols with hydrogen halides: R-OH + HX → R-X + H_2O. H. can also be made by addition of hydrogen halides to alkenes and alkynes, by reaction of alcohols with phosphorus halides or thionyl chloride, or by the Hunsdiecker reaction (see). H. are used as solvents, fire extinguishers, coolants, propellant gases for spray cans, anaesthetics and as alkylating agents in organic syntheses.

Haloalkanoic acids: see Halogen carboxylic acids.

Haloarenes, ***aryl halides***: aromatic halogenated hydrocarbons in which one or more halogen atoms are bound directly to the aromatic ring. The names of the H. are constructed 1) from the names of the halogen atom and the root for the aromatic hydrocarbon, e.g. bromobenzene, 1,4-dichlorobenzene, 1-bromo-6-chloronaphthalene, or 2) in radicofunctional nomenclature, from the radical name of the aromatic system and the name of the halide, e.g. phenyl iodide. The radicofunctional nomenclature is not commonly used, and it cannot be applied to compounds with different kinds of halogen atoms. H. are liquid or crystalline, depending on the size of the molecules. They are stable, and have pleasant, somewhat sweet odors. They are practically insoluble in water, but dissolve easily in most organic solvents. In contrast to the Alkyl halides (see), the halogen atoms in H. are relatively tightly bound to the aromatic system. In the presence of electron-withdrawing substituents, e.g. nitro, nitroso or cyano groups in the 2 or 4 position relative to the halogen, a halogen atom can be replaced by nucleophilic attack with an amino, hydroxy, mercapto or alkoxy group:

O_2N–$C_6H_3(NO_2)$–F + H_2N–NH_2 ⟶ O_2N–$C_6H_3(NO_2)$–NH–NH_2 + HF

H. are synthesized by halogenation of arenes in the presence of aluminum or iron(III) halides or by the Sandmeyer reaction (see) or Schiemann reaction (see) from arene diazonium salts. H. are important intermediates for production of dyes, pharmaceuticals, pesticides, textile conditioners, aromas and preservatives. They are also used as solvents and cleaners.

Halodecarboxylation: a synthetic method for converting monocarboxylic acids to halohydrocarbons. One possible sequence is the reaction of a carboxylic acid with lead(IV) acetate and calcium chloride in boiling benzene:

$$\text{R–COOH} \xrightarrow{[\text{Pb(OOCCH}_3)_4/\text{CaCl}_2]} \text{R–Cl.}$$

The Hunsdiecker reaction (see) is another variant. Some of the reaction steps in H. proceed by radical mechanisms.

Haloforms: trihalogen derivatives of methane, with the general formula CHX_3 (X = halogen atom). Fluoroform (CHF_3), chloroform ($CHCl_3$), bromoform ($CHBr_3$) and iodoform (CHI_3) are known.

Haloform reaction: the synthesis of a haloform by reaction of the hypohalogenite with a compound which contains a CH_3-CO- or $CH_3CH(OH)$-group. The Iodoform test (see) is an important variant of the H.

Halogen: The name "halogen" is derived from the Greek for "salt former". The H. are the elements of the 7th main group of the periodic system: fluorine (F), chlorine (Cl), bromine (Br), iodine (I) and astatine (At). H. react with electropositive metals, accepting a single electron per atom and thus entering the - 1 oxidation state. The electronegativity and reactivity of the H. decrease with increasing atomic mass, while the melting and boiling points increase, the color darkens and the strength of the X-X bond in the X_2 molecule decreases with increasing mass. The tendency to form halogen hydrides HX, and the stability of the hydrides, decrease with increasing atomic number of the H., while the acidity of the hydrohalogenic acids increases markedly from hydrofluoric to hydroiodic acid. Oxygen acids, in particular of the types HXO, HXO_3 and HXO_4, are formed by chlorine, bromine and iodine. The oxidizing strengths of the subhalogenous acids HOX and the halogenic acids HXO_3 decrease with increasing atomic number of the halogen; subhalogenous acids are generally stronger oxidation agents than halogen(V) acids. Halogen oxides, except for iodine(V) oxide, I_2O_5, are endothermic and most are relatively unstable. Fluorine is somewhat exceptional because of its high electronegativity. The association of hydrogen fluoride (hydrogen bonding) accounts for its high melting and boiling points. The solubilities of metal fluorides are often distinctly different from those of the other metal halides; for example, the alkaline earth metal fluorides are insoluble in water, while the other alkaline earth metal halides are extremely soluble. Another example is the water solubility of silver fluoride, compared to the extremely low solubilities of the other silver halides.

Properties of the elements of the 7th main group of the periodic system

	F	Cl	B	I	At
Nuclear charge	9	17	35	53	85
Electron configuration	[He] $2s^2 2p^5$	[Ne] $3s^2 3p^5$	[Ar] $3d^{10} 4s^2 4p^5$	[Kr] $4d^{10} 5s^2 5p^5$	[Xe] $4f^{14} 5d^{10}$m $6s^2 6p^5$
Atomic mass	18.998 40	35.453	79.904	126.904 5	210
Atomic radius [pm]	72	99	114	133	140
Ionic radius of X^- [pm]	136	181	195	216	227
Electronegativity	4.10	2.83	2.74	2.21	1.96
Electron affinity [eV]	3.448	3.613	3.363	3.063	≈3.1
Standard electrode potential (X^-/X_2) [V]	+2.87	+1.359 5	+1.065 2	+0.535 5	+0.25
Density [g cm^{-3}]	1.108	1.565	3.14	4.942	
m.p. [°C]	−219.61	−101.00	−7.25	113.6	≈ 300
b.p. in [°C]	−187.52	− 34.06	58.78	185.24	≈ 370

Halogenated refrigerants: see Refrigerants.

Halogenation: in the narrow sense, methods for synthesis of inorganic and organic halogen compounds; in the wider sense, any type of bond formation between an element or compound and halogen atoms. Chlorination and bromination can be done by similar methods, but fluorination and iodination require special techniques. The halogens react with most elements and compounds with regularly varying intensities. *Chlorination* of nitrogen, noble gases, oxygen, carbon or iridium must occur indirectly, via other compounds. Many hydrogen-containing compounds can react directly with chlorine to form chlorine-containing compounds; or these can be formed by addition reactions with chlorine or hydrogen chloride (e.g. ammonia, organic compounds). Certain oxides (carbon monoxide, nitrogen oxide, sulfur dioxide) can also be chlorinated by addition reactions. Some sulfides (e.g. carbon disulfide) and many metal oxides can be converted to the corresponding chlorine compounds by substitution of chlorine for the sulfur or oxygen atom. *Bromination* can, in principle, be carried out by the same methods as chlorination. Bromine reacts with hydrogen compounds, such as hydrocarbons, by addition and substitution mechanisms, usually ionic. Fluorination occurs by radical chain reactions, which are often so highly exothermal that they are explosive and uncontrollable (for example, the carbon skeletons of organic compounds are cleaved). Therefore, organic compounds are more practically fluorinated by substitution reactions using special fluorination reagents (e.g. bromine fluoride, antimony(V) fluoride, cobalt(III) fluoride, hydrogen fluoride, xenon difluoride, fluoroxytrifluoromethane). One method for complete replacement of the H atoms in hydrocarbons by fluorine is *electrofluorination*, in which the hydrocarbons dissolved in anhydrous hydrogen fluoride are electrolysed in special cells. *Iodination* is the most difficult type of H., due to the low reactivity of iodine. Replacement of hydrogen by iodine is possible only in special cases, e.g. in certain activated CH_2- or CH_3- groups. Iodine monochloride, phosphorus(III) iodide and acyl iodides are used as iodination reagents.

The following methods are suitable for the H. of organic compounds.

1) Direct H. with free halogens. These reactions with alkanes occur by an S_R mechanism. The regioselectivity increases in the order F < Cl < Br. Fluorination occurs explosively; primary, secondary and tertiary CH compounds are attacked with approximately equal probability. The products are mixtures of singly and multiply fluorinated alkanes. Chlorination is initiated thermally, photochemically or by initiators. Sulfuryl chloride can be used instead of chlorine. The products contain one or more chlorine atoms, e.g. monochloromethane, CH_3Cl,

dichloromethane, CH_2Cl_2, chloroform, $CHCl_3$ and carbon tetrachloride. If a limited amount of chlorine is present and the reaction time is also limited, monosubstitution products can be obtained selectively. The bromination of alkanes is analogous to chlorination; iodination is endothermal and therefore cannot take place in a chain reaction; instead, an equilibrium is reached. H. of arene rings takes place by an S_E mechanism, in which one or more H atoms are replaced by chlorine or bromine. First order substituents promote the reaction and direct the entering substituents to the 2- and 4-positions. Bromination of phenol is especially easy; 2,4,6-tribromophenol is formed with 3 moles bromine. Fluorine is too reactive for S_E substitutions of aromatic compounds, and destroys the hydrocarbon skeleton, while iodine is too unreactive and it is necessary to remove the hydrogen iodide product with an oxidizing agent such as mercury oxide or nitric acid. The site of substitution and mechanism of the H. of alkylarenes depend on the reaction conditions: *SSS conditions* (from the German "Sonnenlicht, Siedehitze, Seitenkette" = sunlight, boiling heat, side chain) promote an S_R mechanism and lead to halogenation of the side chain; for example, chlorine and toluene react to give benzyl chloride, benzal chloride and benzotrichloride. *KKK conditions* (from the German "Kälte, Katalysator, Kern" = cold, catalyst, nucleus) promote an S_E mechanism and yield ring-halogenated products; for example, bromination of toluene yields 2- and 4-bromotoluene. Carboxylic acids, aldehydes and ketones can be halogenated in the same way as alkanes and alkylarenes. In the presence of strong acid, the α-halogen compound is formed. Chlorination of alkenes and cycloalkenes at high temperatures leads, in a substitution reaction, to allyl chlorides; chloroaddition is not observed under these conditions.

2) Selective iodination of arenes by boiling arene diazonium salts in the presence of sodium or potassium iodide.

3) The Schiemann reaction (see) for synthesis of fluorobenzenes.

4) Chlorination or bromination of arenes by the Sandmeyer reaction (see).

5) Chlorination or bromination of carboxylic acids by the Hell-Volhard-Zelinsky reaction (see).

6) Bromination of alkenes and cycloalkenes by the Wohl-Ziegler reaction (see) with N-bromosuccinimide. A side-chain bromination of alkylarenes by a radical mechanism is also possible with this reagent.

7) The Finkelstein reaction (see) for halogen exchange in halogen alkanes, using metal halides.

8) The Hunsdiecker reaction (see) of silver salts of carboxylic acids with chlorine or bromine to prepare halogen alkanes.

9) Substitution of an OH group in alcohols with a halogen by an S_N mechanism, forming an alkyl halide. The replacement of the OH group in the carboxyl group requires energetic conditions; the products are acyl halides.

10) Addition reactions in which halogen, hydrogen halogen or hypohalous acid is added to an unsaturated compound. In principle, chlorine, bromine and iodine can react with any unsaturated compound to form a 1,2-dihalogen compound. Fluorine can only be used in combination with carbon tetrachloride or dichloromethane at low temperatures. In conjugated unsaturated compounds, either 1,2- or 1,4- addition can occur. The reactivity of the triple bond is lower than that of the double bond; addition occurs in steps, forming first the dihalogen and then the tetrahalogen adduct. Addition of chlorine to arenes occurs by an S_R mechanism; a mixture of isomeric hexachlorocyclohexanes is formed from benzene. Addition of hydrogen halides to non-activated aliphatic double bonds follows the Markovnikov rule (see). Terminal and exocyclic double bonds are more reactive than interior and endocyclic double bonds. Hydrogen halides add to α,β-unsaturated ketones and carboxylic acids to form β-haloketones or β-halocarboxylic acids (exception: α-bromopropionic acid from hydrogen bromide and acrylic acid). Hydrogen halides do not add to aromatic double bonds. The addition of hydrogen halides to alkynes is industrially important in the production of vinyl halides, e.g. vinyl chloride by addition of hydrogen chloride to ethyne. The addition of hydrogen fluoride to alkynes is also possible. The products of addition of hypohalous acids to alkenes and cycloalkenes are halogen hydrines.

Halogen azides: unstable compounds with the formula XN_3. They are made by reaction of halogens, X_2, with sodium azide: $NaN_3 + X_2 \rightarrow XN_3 + NaCl$. ***Fluoroazide***, FN_3, is a yellow-green gas which slowly decomposes at room temperature into $(NF)_n$ and N_2. ***Chloroazide***, ClF_3, is a colorless, explosive gas; ***bromoazide***, BrN_3, is a colorless, explosive liquid; and ***iodoazide***, IN_3, is a colorless, extremely explosive solid.

Halogen carbonyl compounds: a group of organic compounds which contain a readily substituted halogen atom in addition to the carbonyl group. The bifunctional α-halocarbonyl and β-halovinylcarbonyl compounds are important for many syntheses, especially of heterocycles. Because the H. can be substituted in many ways, there are numerous different synthetic methods. For example, α-haloketones can be made by halogenation of ketones which have α-H atoms:

$$R\text{–}CO\text{–}CH_3 \xrightarrow[-\,HBr]{+\,Br_2} R\text{–}CO\text{–}CH_2\text{–}Br.$$

The β-halovinylaldehydes can be made easily by the reaction of dimethylformamide and phosphorus oxygen chloride with aliphatic or aromatic-aliphatic ketones:

$$R\text{–}CO\text{–}CH_3 \xrightarrow{DMF/POCl_3} R\text{–}CCl{=}CH\text{–}CHO.$$

Many H. are very unstable, lacrimatory compounds which can be stored for longer periods only if they are very cold.

Halogen carboxylic acids: ***carboxylic acids*** in which one or more of the hydrogen atoms attached to the carbon atoms have been replaced by halogen atoms. The hydrocarbon part of the molecule can be aliphatic, aromatic or heterocyclic. The most important H. are those with a saturated aliphatic group and the halogen atom on the C atom to which the carboxyl group is linked (α-H.). These are named as ***haloalkanoic acids***, with the position of the halogen atom indicated. Because of the - I effect of the halogen atoms, the H. are more acidic than the corresponding unsubstituted carboxylic acid, and the C-X bonds are more strongly polarized, due to the - I effect of the carboxyl group, and thus more reactive than the C-X bonds in alkyl halides. The compounds can be synthesized by direct halogenation of the carboxylic acids with fluorine, chlorine or bromine, or by the Hell-Vollhard-Zelinsky reaction (see) of carboxylic acid halides with chlorine or bromine. Halogenation of the carboxylic acid in the presence of a strong Brœnsted or Lewis acid proceeds by an ionic mechanism and yields α-H. selectively, while under radical reaction conditions mixtures of α-, β-, γ- and other H. are obtained in statistical mixtures. β-H. can be made selectively by anti-Markovnikov hydrogen halide addition to α,β-unsaturated carboxylic acids. The halogen-substituted benzoic acids can be made by electrophilic halogen substitution of benzoic acid or its derivatives. H. are versatile starting materials for syntheses. Their hydrolysis in S_N reactions leads to hydroxycarboxylic acids or lactones. In α-H., S_N2 reactions with retention of the carboxyl group, due to neighboring group participation, lead to α-lactones. Hydrolysis of β-H. and their derivatives usually occurs by an elimination-addition mechanism. At high temperatures, this reaction stops at the level of the α,β-unsaturated carboxylic acid, or a competing reaction leads to a heterolytic fragmentation.

$$\overline{\underline{|X}}-\underset{\displaystyle R}{CH}-CH_2-C(=O)-\overline{\underline{O}}|^- \longrightarrow \overline{\underline{|X|}}^- + R-CH=CH_2 + CO_2$$

α-H. are the starting materials for the Reformatsky reaction (see) for synthesis of β-hydroxycarboxylic acids, and for the Darzen glycide ester synthesis (see Darzen-Erlenmeyer-Claisen condensation). Some important H. are chloroacetic, dichloroacetic, trichloroacetic and trifluoroacetic acids.

Halogen hydrines: reactive organic compounds with the general formula R-CHX-CH(OH)-R. H. are formed by addition of halogens to alkenes in aqueous solution, or directly, by addition of hypohalites: $R-CH=CH_2 + HO-X \rightarrow R-CH(OH)-CH_2-X$. In addition, H. can be produced by cleavage of epoxides with hydrogen halides. The chlorohydrines are an important group of H.

Halogen hydrocarbons: derivatives of hydrocarbons in which one or more H atoms are replaced by halogen atoms. The H. are subdivided into Haloalkanes (see) or alkyl halides, haloalkenes, haloalkynes and Haloarenes (see) or aryl halides. Most H. are colorless, water-insoluble liquid or crystalline compounds with faintly sweet odors. Only a few of the lower homologs are gases under normal conditions. The strength of the carbon-halogen bond varies, leading to large differences in reactivity in nucleophilic substitutions. The halogen atoms in the alkyl halides can be easily exchanged for nucleophilic reagents, forming the corresponding derivatives. In non-activated vinyl halides or aryl halides, however, the halogen atoms can undergo nucleophilic substitution only under drastic reaction conditions. The H. are relatively easy to synthesize and are an important group of starting materials for the production of further functional substance classes.

Halogen oxygen fluorides: very reactive, covalent, polar compounds of the types XFO_2 (X = Cl, Br, I), XF_3O (X = Cl, I), XFO_3 (X = Cl, Br, I), XF_3O_2 (X = Cl, I) and IF_5O. Their reactivity is comparable to that of the halogen fluorides. H. are used as oxidizing and fluorinating reagents; they react with fluoride donors and acceptors to form mixed oxofluorohalogen anions or cations. One important H. is ***perchlorylfluoride***, $ClFO_3$, a poisonous gas, m.p. -147.8 °C, b.p. - 46.7 °C, which is thermally stable up to 500 °C and is resistant to hydrolysis. $ClFO_3$ is obtained by solvolysis of potassium perchlorate with a mixture of hydrogen fluoride and antimony(V) fluoride: $KClO_4 + 2\,HF + SbF_5 \rightarrow FClO_3 + KSbF_6 + H_2O$. It is used as a selective fluorination reagent to introduce fluorine in $-CH_2$ groups and to introduce ClO_3 groups.

Haloperidol: see Butyrophenones.

Halothane, ***1-bromo-1-chloro-2,2,2-trifluoroethane***: $BrClCH-CF_3$, a colorless, non-combustible liquid with a sweetish, pleasant smell, which is usually stabilized with a small amount of thymol; b.p. 50.2 °C. H. is insoluble in water, and soluble in ether and petroleum ether. It is made by chlorination of 1-bromo-2,2,2-trifluoroethane, or by addition of hydrogen bromide to chlorotrifluoroethene, followed by catalytic isomerization by the Hoechst process. H. is used as an inhaled anaesthetic.

Hamaker constant: in colloid chemistry, a substance-specific parameter which depends on the polarizability and ionization energy of the interacting colloidal particles.

Hamilton operator: see Atom, models of.

Hammer scale: see Scale.

Hammett equation: an LFE equation (see) for aromatic compounds introduced in 1937 by L.P. Hammett. It relates the rate of a reaction or the position of its equilibrium to the nature of the substitutent. The H. is $\lg (k_i/k_0) = \varrho \cdot \sigma_1$. Here k_0 and k_i are the rate constants for the unsubstituted compound and the compound with substituent i. (In the H. for the position of the equilibrium, equilibrium constants K_i and K_0 are used instead of rate constants.) ϱ is the reaction constant and has a certain value for each type of reaction. The Substituent constant (see) σ_1 is determined from the electrolytic dissociation constants K_0 and K_i for the unsubstituted and substituted benzoic acid according to the equation: $\sigma_i = \lg (K_i/K_0)$. Two constants are obtained for each substituent, σ_p and σ_m, depending on the position (*meta* or *para*) of the substituent relative to the carboxyl group. The H. cannot be used for ortho compounds because of steric effects on the reaction center. The reaction constant ϱ is a measure of how much the reaction responds to changes in the electron density at the reaction center. It is positive if withdrawal of the electrons accelerates the reaction (nucleophilic reaction) and negative if increasing the electron density promotes the reaction (electrophilic reaction). Correlations between rate or equilibrium constants or parameters proportional to them on the basis of the H. have made essential contributions to systematization of organic chemistry.

Hammond principle: an empirical rule relating the enthalpies of reaction and activation of a chemical reaction. If the transition state of a chemical reaction has almost the same enthalpy as an adjacent state on the reaction coordinate (reactant, intermediate or product), then the interconversion of these two states requires only minor change in the molecular structure. In exothermal reactions, the reactants and transition state have similar structure, and the reaction enthalpy gives no indication of the activation enthalpy (and thus the reaction rate). On the other hand, in a highly endothermal reaction, the structure of the transition state is very similar to that of the product, and the reaction enthalpy is directly proportional to the activation enthalpy; that is, the more stable products are formed more rapidly.

Hantzsch pyridine synthesis: synthesis of a substituted pyridine by condensation of a β-ketocarboxylic acid ester with aldehyde and ammonia. In the first step of the synthesis, the β-ketoester and ammonia form a β-aminocrotonate ester, and the aldehyde forms an α-alkylidene or arylidene β-ketoester with the β-ketoester. The two intermediates, which can also be produced separately and added to the H., then react in the sense of a Michael addition (see),

$$ROOC{-}CH_2{-}CO{-}R + RCHO + NH_3 + H_2C(COOR){-}CO{-}R \xrightarrow{-3H_2O} \text{dihydropyridine} \xrightarrow{(HNO_2)} \text{pyridine}$$

forming a δ-aminoketone. This cyclizes to form the dihydropyridine derivative, which can be converted to the pyridine derivative by oxidation with nitrous acid.

Hantzsch thiazole synthesis: method for production of thiazoles by condensation of α-haloketones with thioamides:

$$R{-}CO{-}CHX{-}R + H_2N{-}C(=S){-}R \xrightarrow{-H_2O,\,-HX} \text{thiazole}$$

The thiazole ring system is important in biology as a structural element of thiamin (vitamin B_1).

Hantzsch-Widman names: see Nomenclature, sect. III A 8 and III F 2a.

Hapto system: see Organoelement compounds.

Hard alloys: alloys containing 1 to 3% carbon, 25 to 35% chromium, 32 to 67% cobalt, 4 to 20% tungsten and some iron; they can be worked mainly by casting and welding. They can only be forged if the carbon content is less than 1.5% and the tungsten content is less than 10%. H. are known as Stellite® (see), and are used as plating on machine tools for stamping metals, or as coatings to prevent wear on armature parts, scissor blades and machine parts.

Hard chrome plating: see Chrome plating.

Hard eloxation: see Anodic oxidation.

Hardening: 1) in plastics, especially duroplastics, a phase in processing. After the material has been poured into a mold, it is made nearly insoluble and unmeltable by a chemical process, which usually generates cross-linked structures. The process is initiated by addition of hardeners or by heating.

2) With alloys, heat treatment to increase their hardness and strength.

Hard metals: sintered alloys consisting of one or more hard phases (WC, TiC and TaC) embedded in a binder phase, usually cobalt, or more rarely, nickel (see Cermets). The carbides are responsible for the great hardness of these materials and their resistance to wear; they form mixed crystals among themselves. The binder provides a poreless joining of the carbide grains, and gives the H. the necessary ductility. H. are used mainly as *cutting metals* for shaving tools, as cutting blades on cold presses, stamping machines and deep drawing machines, and as drawing stones and rollers. H. are also used for drill bits in mining, turbine blades and valves in jet engines.

Hardness: 1) in materials testing, the resistance which the material offers to penetration by another, harder material. In technical measurements of hardness, a distinction is made between static and dynamic methods. In *static methods*, a test tool is slowly pressed into the sample under a constant load. The H. is determined from the magnitude of the test force and the size of the remaining impression (see Brinell hardness; Vickers hardness; Rockwell hardness; Small-load hardness; Micro hardness). The static methods include the scratch test (see Mohs hardness scale). *Dynamic methods* can be classified in two groups: a) dynamic plastic methods: the test tool is beaten into the material (impact hardness, e.g. with a Poldi hammer), and the analysis is similar to that in static methods. b) Dynamic elastic methods: the recoil of a test tool is measured. In the Shore skleroscope, the tool falls through a glass tube onto the sample; in the duroscope, it is part of a pendulum and strikes the material from the side. The measured values are called Shore H., skleroscope H. or duroscope H.

2) ***H. of water, total H.***, the sum of all alkaline earth ions in water. These are mainly calcium and magnesium; barium and strontium are less often present. The H. was formerly measured in *hardness degrees* (°H), which have different values in different countries.

In the effort to achieve a uniform international scale for water H., H. degrees have been replaced by the millival (mval). Due to the difference in molecu-

A selection of hardness degrees

Country	1°H =
France	5.6 mg CaO/l
Germany (FRG and DRG)	10 mg CaO/l
Great Britain	8.0 mg CoO/l
USA	9.6 mg CaO/l
USSR	1 mg Ca^{2+}/l
International	1 mval = 27.8 mg CaO/l

lar masses, an equivalent mass of MgO is 0.719 times the equivalent mass of CaO. Modern methods of analysis permit a precise determination of the Mg^{2+} and Ca^{2+} concentrations (see Water hardness, determination of). In natural waters, the concentration of calcium is higher than that of magnesium.

The H. of water is the sum of carbonate and noncarbonate H.

1) The ***carbonate H.*** is also called ***temporary H.*** because when the water is boiled, the associated free carbonic acid escapes (see Lime-carbonic acid equilibrium) and as a result the calcium and magnesium hydrogencarbonates precipitate. The carbonate H. is often defined (not completely correctly) as the fraction of the alkaline earth ions equivalent to the carbonate and hydrogencarbonate dissolved in the water. It is better to use the ***acid consumption*** (alkalinity) in mval/l instead (see m-Value). The carbonate H. can only be calculated from the acid consumption if m ⩽ total H. and there is no substance other than carbonate and hydrogencarbonate ions which neutralizes acid.

The ***noncarbonate hardness***, also called ***permanent H.***, is the fraction of alkaline earth ions in excess of the carbonate H. These cations are balanced by sulfates, chlorides, nitrates, phosphates, silicates, humates and other anions.

The H. of water is important in many industrial applications. The formation of boiler scale in hot-water and steam boilers and pipes causes the heat-transfer surfaces to overheat and wastes energy. The production of textiles, paper, preserved fruits and vegetables and beer requires soft water. The H. of water inhibits softening of dried legumes in cooking, impairs the taste of coffee and tea, and precipitates fatty acids from soaps as a grayish scum which interferes with washing. For methods of softening water, see Water softening.

Classification of water hardness according to degrees of hardness (total hardness)

	mval	°H (USA)
Very soft	0 – 1.5	0 – 4.0
Soft	1.5– 3.0	4.0– 8.0
Medium	3.0– 6.0	8.0–16.0
Hard	6.0–10.0	16.0–26.9
Very hard	>10.0	>27

Hard rubber: see Rubber.

Harkin's rule: see Elements 1.

Harmonic oscillator: see Infrared spectroscopy.

Harmonic vibrations: see Infrared spectroscopy.

Harries reaction: a method of alkene cleavage by reaction of alkenes with ozone in light petroleum, chloroform or carbon tetrachloride. The ozonides formed by H. can be converted to acids and/or ketones, or aldehydes and/or ketones by oxidative or reductive ozonolysis:

$$R^1R^2C{=}CHR^3 \xrightarrow{+O_3} R^1R^2C\overset{O-O}{\underset{O}{\ }}CHR^3 \begin{cases} \xrightarrow{Ox.} R^1R^2C{=}O + R^3{-}COOH \\ \xrightarrow{Red.} R^1R^2C{=}O + R^3CH{=}O \end{cases}$$

The reaction mechanism of H. has been shown by recent studies to involve a primary ozonide, which is formed by 1,3-dipolar cycloaddition of ozone to the C=C double bond. Decomposition of the unstable primary ozonide forms a carbonyl compound and a dipolar intermediate, which reacts to form the secondary ozonide. The cleavage of the secondary ozonide, the *ozonolysis* occurs in the same way as before. The H. is used to elucidate the structures of complicated unsaturated compounds.

Primary ozonide

Secondary ozonide

Mechanism of the Harries reaction

Hartshorn, salt of: a mixture consisting mainly of ammonium hydrogencarbonate, NH_4HCO_3, and ammonium carbamate, $H_2NCOONH_4$, used as a baking powder. When heated to about 60 °C, it decomposes into ammonium, carbon dioxide and water (the NH_4HCO_3). The CO_2 acts as a leavening.

Hashish: see Cannabinoids.

Hastelloy®: a group of nickel alloys (H, A, B, C, D, G) for the highest-stress applications in chemical technology. The most commonly used alloys are H. B and H. C. H. B consists of 26 to 30% molybdenum, 4 to 7% iron and the rest nickel, and is stable to attack of strongly reducing acids, such as hydrochloric acid; H. C consists of 15 to 17% molybdenum, 14 to 16% chromium, 4 to 7% iron, 3 to 4% tungsten and the rest nickel, and is stable to the effects of bleaching lyes, iron chloride solutions, chromic and hydrofluoric acids. It is used for valves and pumps.

Haworth synthesis: a phenanthrene synthesis. First naphthalene reacts with succinic anhydride to form an aryl-substituted γ-oxocarboxylic acid, which is reduced to γ-arylbutyric acid. This is cyclized to form a cyclic ketone, reduced to the hydroaromatic, and dehydrated with selenium to give the aromatic hydrocarbon, the phenanthrene.

This method can be applied to other aromatic systems. The aromatic anhydrides of various dibasic aliphatic carboxylic acids and their substitution products can be used to make other condensed systems and their derivatives.

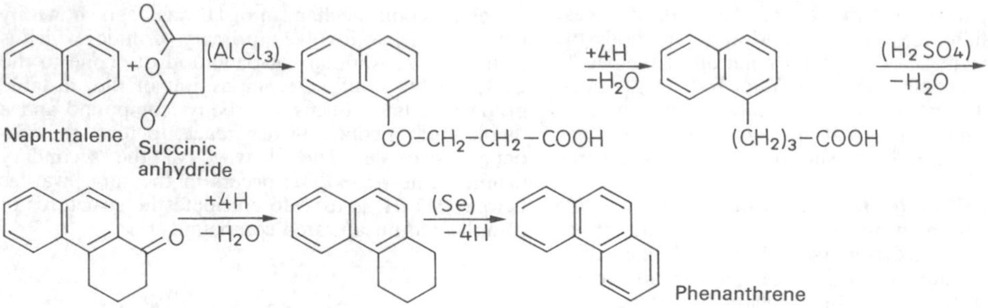

Naphthalene Succinic anhydride (AlCl3) CO–CH2–CH2–COOH +4H −H2O (CH2)3–COOH (H2SO4) −H2O +4H −H2O (Se) −4H Phenanthrene

Hb: see Hemoglobin.

HB: abb. for Brinel hardness.

HbO$_2$: see Hemoglobins.

HCB: see Fungicides.

HCCP: abb. for hexachlorocyclopentadiene.

HCCP dimer insecticides: a group of chlorinated hydrocarbon insecticides derived from dimerized hexachlorocyclopentadiene.

An example which is relatively harmless to the environment is *Kelevan* (Fig.), a colorless, crystalline substance; m.p. 91 °C (dec.), which has a vapor pressure at 20 °C < 1.3 Pa. It is readily soluble in organic solvents except for petroleum ether and hexane, is practically insoluble in water, and is hydrolysed in acidic and alkaline media.

Kelevan is an ingestion toxin, which is used mainly to combat potato beetles, and is effective even with resistant populations. The acute oral toxicity (LD_{50}) for warm-blooded animals is between 240 and 290 mg/kg rat.

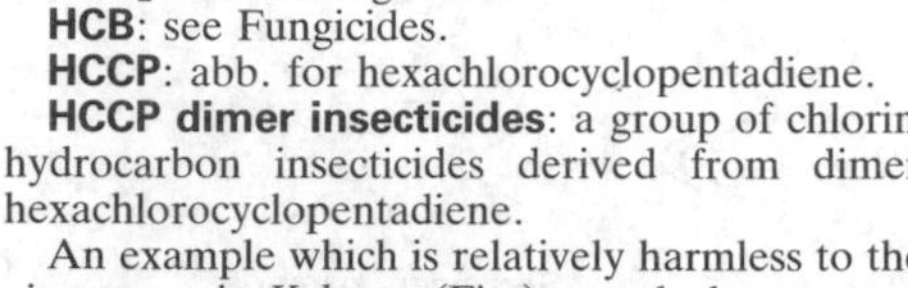

OH Cl CH2—CO—CH2—CH2—COOC2H5 Cl Cl Cl Cl Cl Cl Cl Cl Cl

HCG: see Choriogonadotropin.

HCH: abb. for 1,2,3,4,5,6-Hexachlorocyclohexane.

H-coal process: see Coal hydrogenation.

HCS: see Choriomammotropin.

Hcy: see Hemocyanin.

HDPE: see Polyethylene.

He: symbol for helium.

Heat: a form of energy, the energy of random motion of atoms and molecules. To heat a substance means to increase the kinetic energy of its molecules. This idea, which goes back to R. Mayer, Joule, Boltzmann, Clausius and Helmholtz, is called the *kinetic* or *mechanical* theory of H. R. Mayer first showed that H. is a form of energy and that different forms of energy can be interconverted (see Heat equivalent).

According to the first law of thermodynamics, $de = dq + dw$, where dq is the H. exchanged, dw is the work done, and de is the change in the internal energy of a system. Addition of H. thus leads to an increase in the internal energy of a material system. If $dw = 0$, the temperature of the system is increased, until finally, at a certain temperature, the substance can be converted to a different aggregate state. To go from a lower- to higher-energy aggregate state, a substance must absorb the Phase change heat (see). During the time that this H. is being transferred, the temperature of the substance remains constant.

The Reaction heat (see) is the H. released or absorbed by a chemical reaction.

The SI unit of H. is the joule (1 J = 1 Ws = 1 m^2 $kg\,s^{-2}$). The calorie was formerly the scientific unit of H. (1 cal = 4.1868 J).

Heat capacity: see Molar heat.

Heat content: see Enthalpy.

Heat equivalent: a conversion factor for converting calories into the energy units of mechanics or electricity. 1 cal = 4.1868 J = 4.1868 $m^2\,kg\,s^{-2}$. In the SI system, all energies are expressed in joules, so that conversion is no longer necessary. The H. was first measured in 1842 (J.R. Mayer) and 1843 (J. Joule). Recognition that heat is equivalent to other forms of energy was the basis for formulation of the law of conservation of energy.

Heat exchange: the transfer of heat from a higher-temperature medium to a lower-temperature one. In ***direct H.***, the two media mix, and the temperature equilibrates (e.g. when hot steam is passed through water). In ***indirect H.***, there is a wall separating the two media (e.g. in the form of a heating or cooling coil).

A ***heat exchanger*** is a device in which H. can occur. They are nearly always based on indirect H..

Heating baths: devices used for even, indirect and economical heat transfer from a heat source to a reaction vessel. An intermediate medium is used; it must have good self-convection and also a high thermal conductivity; it protects the reaction vessel from local overheating. H. are usually electrically heated, and are often kept at a constant temperature by a thermostat. They are used, for example, for distillation and gentle evaporation of heat-sensitive materials. In many chemical reactions, it is necessary to maintain the reaction temperature exactly.

There are various types of H. 1) In an ***air bath***, the reaction vessel is heated by hot gases from a gas flame or the heat of an infrared radiator.

2) ***Solid heating baths*** are usually filled with sand, iron filings or graphite. They can be used at temperatures up to about 350 °C. Temperature regulation in such baths is difficult.

3) ***Liquid baths*** are the most suitable because of their good self convection. The choice of liquid de-

pends on the desired highest temperature (table). The temperature of the liquid should be 10 to 20 °C higher than the desired internal temperature of the flask.

a) ***Waterbaths***. Water and saturated salt solutions are very good bath liquids, due to their high specific heats, relatively good heat conductivity, low viscosity and chemical and physiological inertness. A disadvantage is their relatively low boiling boints.

Steam baths without a water container are also used. The steam is passed in through openings in a funnel-like metal container in which the reaction vessel is placed.

b) ***Oil baths***. Because high-boiling oils (up to about 300 °C) release cracking products, oil baths should not be heated to more than 50 °C below their flame points. The large expansion of the oil, about 1/6 or its original volume, must also be considered.

Salt baths can be used in the range from 150 to 680 °C. Some disadvantages are the agressivity of the salt melt and the possibility of explosion in the presence of organic substances. A mixture of 53% potassium nitrate, 7% sodium nitrate and 40% lithium nitrate can be used between 150 and 450 °C. Copper(II) chloride and magnesium chloride are suitable for high temperatures; $CuCl_2$ can be used above 630 °C, and $MgCl_2$ above 712 °C.

d) ***Metal baths.*** Because of their low melting points (as low as 250 °C) and their good heat conductivity, some alloys are good bath liquids; for example, Lipowitz', Wood's and Rose's metals, and solder alloys. Lead is also used as a bath.

e) ***Other bath liquids***. Concentrated sulfuric acid is often used in melting point apparatuses for heat transfer. A mixture of diphenyl ether and diphenyl is used in industry because of its relative safety; it can be used up to its azeotrope point at 225 °C. In closed systems under a pressure of 0.6 MPa, it can be used up to 350 °C. Chlorinated diphenyl boils at 340 °C and is not combustible.

Application ranges of the most common heating bath liquids

Useful up to °C	Bath liquid
100	Water
105	Cold-saturated solution of sodium carbonate
108	Cold-saturated solution of sodium chloride
120	Cold-saturated solution of sodium nitrate
135	Cold-saturated solution of potassium carbonate
170	Glycerol
180	Cold-saturated solution of calcium chloride
180	High-pressure water, 1.2 MPa
200	Paraffin oil
200	Conc. sulfuric acid
220	Mepasin
240	Diglycol
250	Paraffin
255	Diphenyl
270	Triethylene glycol
300	Silicon oils
300	Steam cylinder oil
340	Chlorinated diphenyl
350	Diphenyl under 0.6 MPa
360	Silicic acid pentyl ester
360	Tetraaryl silicate (TAS)

Heating gases: fuel gases used for heating. Hydrogen, carbon monoxide, methane, ethane, benzene and other hydrocarbons and mixtures of hydrocarbons can be used as H.

Heating oil: a mixture of hydrocarbons obtained from Petroleum processing (see) as a pure distillate (***light H.***) or as a residual oil (***heavy H.***), or by mixing residual oil with a thin oil. H. can also be obtained by distillation of coal tar. H. has a heating value of 32,000 to 36,000 kJ kg^{-1}. Because of the high sulfur content of heavy H., removal of sulfur from it is very important for protection of the environment (see Hydrorefining).

Heating value: the heat of reaction released by complete combustion of 1 kg solid or liquid fuel, or 1 N m^3 gaseous fuel. It is given in kJ kg^{-1} or KJ m^{-3}. In the United States, the British Thermal Unit (Btu) = 1054 J is used to indicate the H. per ton of coal or barrel of oil. For those fuels which contain water (c.g. heating oil, gasoline, diesel oil, methanol), the ***lower*** and ***upper*** H. are differentiated. The lower H. is determined with the water present as steam, and the upper H. includes the heat of condensation of the water. Combustion is an exothermic reaction, but the H. are usually given with a positive sign, which is contrary to the thermodynamic convention.

Heat treatment: a process or a combination of several processes in which the temperature of a solid is changed in order to change certain of its properties. In the process, surrounding substances may lead to chemical changes in the material. The change in properties may be due to changes in its structure, concentrations of its components or arrangement of these components. For example, H. of silicate rock converts the quartz to its high-temperature α-cristoballite phase; normal roasting of supereutectic steels leads to α-γ-conversion of the iron-carbon mixed crystals; quenching hardens steel by diffusionless conversion of the γ-iron-carbon mixed crystals into martensite; recrystallization and new crystallization of plastomers and elastomers is due to changes in structure and arrangement, and to crystallization of amorphous regions; and heat hardening of steels occurs after the material has had carbon added.

Shaping of heated objects or processes for surface protection which involve heating are not considered H.

Heavy gasoline: see Gasoline.

Heavy hydrogen: same as Deuterium (see).

Heavy metal: a metal with a density greater than 5 g cm^{-3}. The H. include 70 metals, including the metallic modifications of arsenic, antimony and tellurium. The lightest of the H. is europium (density 5.245 g cm^{-3}), and the heaviest is osmium (density 22.61 g cm^{-3}). The H. include most of the commonly used metals, such as iron, copper, lead, zinc, tin, nickel, chromium, tungsten, molybdenum, cadmium, cobalt, niobium, tantalum, plutonium, uranium, vanadium, mercury, silver, gold, platinum and the lanthanoids.

The opposite of a H. is a Light metal (see).

Heavy oil: see Tar.

Heavy water: same as Deuterium oxide (see).

Hedvall effect: see Solid-state reactions.

Heisenberg uncertainty relationship: see Atom, models of.

Helianthin: same as Methyl orange (see).

Helicenes: see Stereoisomerism 1.1.

Helicity: see Stereoisomerism, 1.

Helindone dyes: a group of vat dyes used on wool and furs. They are very fast to light and washing.

Heliogen blue B: see Phthalocyanin pigments.

Heliogen pigments: same as Phthalocyanin pigments (see).

Heliotropin: same as Piperonal (see).

Helium, symbol ***He***: a chemical element from the 8th main group of the periodic system, the Noble gases (see). The atomic number of He is 2, its atomic mass is 4.0026, valency 0, density 0.1785 g l^{-1} at 0°C, m.p. -272.2°C at 2.6 MPa, b.p. -268.9°C, critical temp. -268.93°C, pressure for crystallization, 0.226 MPa, critical density, 0.069. The natural isotopes have mass numbers 4 (100%) and 3 (0.00013%), and there are synthetic isotopes with mass numbers 5, 6 and 8.

Properties. H. is colorless, odorless, tasteless and nontoxic. It has a very low boiling point, and is always monoatomic. Solid H. can be produced, also at extremely low temperature, at a minimum pressure of 0.226 MPa. There are two forms of liquid He: ***He I*** converts at 2.186 K and 0.383 MPa into the low-temperature modification, ***He II***, which has a number of remarkable physical properties designated by the term ***superfluidity***. Its heat conductivity is about 300 times that of silver, making it the best heat conductor known. The viscosity of He II is so low that it has practically no internal friction (*superfluid state*). It is therefore able to flow through narrow capillaries or cracks much faster than gaseous He.

H. is extremely unreactive; no compounds are known.

Analytical. Because of its unreactivity, only physical methods can be used to analyze He. It is separated from other gases by diffusion and adsorption techniques, especially those of gas chromatography. It is identified mainly by spectral analysis.

Occurrence. He makes up $4 \cdot 10^{-7}$% of the earth's crust, and 0.00046% of the atmosphere. The He content of higher atmospheric layers is higher. Natural gas often contains a significant volume percent of He, in extreme cases more than 10%. H. is contained in some minerals as the product of radioactive decay, and this trapped gas can be used to determine the age of the rocks in which it is found. H. has also been detected in the sun and other stars, where it is formed by fusion of hydrogen nuclei. This is the process which generates the energy of the sun and stars. H. is the second most abundant element in the universe, after hydrogen.

Production. Commercial helium is extracted from natural gas. After a chemical pre-purification (removal of carbon dioxide and hydrogen sulfide by an alkaline wash), the gas is cooled to - 200°C, at which temperature all other substances have liquefied and the H. remains as a gas. H. can also be obtained by liquefaction of air. After the nitrogen and oxygen have condensed, He and neon remain as gases, which can be separated by fractional adsorption and desorption on activated charcoal.

Applications. H. was formerly used as a carrier gas for dirigibles, since its lift is 92% of that of hydrogen, and it is not combustible. In addition, the loss of He by diffusion is lower than that of H_2. The cheaper, and also incombustible, mixture of 26% hydrogen and 74% He was also used. Now He is important as a protective gas for certain chemical reactions and in welding. It is also used as a carrier for doping of semiconductors and in gas chromatography, and liquid He is used as a coolant for superconducting magnets. A mixture of 21% oxygen and 79% He is used as compressed gas for divers to breathe, since He is less soluble in blood than nitrogen, and does not cause "bends" (air embolisms) when the diver approaches the surface quickly. Asthmatics can breathe more easily in helium/oxygen mixtures, since the pressure required to aerate the lungs is only half as great as with normal air.

Historical. H. was discovered in 1868 by P.J.C. Janssen through spectral analysis of sunlight. It was first thought that the absorption line of He was the yellow sodium D line, but Janssen disproved this and determined that the line was due to an element which was not known on earth. J.N. Lockyer and E. Frankland suggested the name "helium" for this element, from the Greek "helios" for "sun". The element was first discovered on earth in 1898, by W. Ramsay. W.F. Hillebrand found that cleveite releases an inert gas when heated, and Ramsay identified it as He.

Helix: a secondary structure in biopolymers and highly ordered synthetic polymers in which the polymeric chain winds around a common axis. A H. can be either right- or left-handed, and may consist of one or more strands. A single-stranded H. is characterized by the number p of monomeric units per q turns, after which the structure repeats itself; or by the number of monomeric units per turn. The α-H., which occurs frequently in proteins, is therefore also called an 18_5 H. (18 amino acids per 5 turns leads to the initial spatial arrangement), or as a 3.6_{13}-H. (3.6 amino acids per turn; the subscript gives the number of atoms, including the hydrogen bonds, which form the ring seen in cross section). A H. with a non-integral number of monomers per turn is called a ***non-integral H.***. The distance between turns (rise) is a function of the identity period, i.e. the distance between monomeric units in equivalent relative positions, or the height of the monomeric unit. Single-stranded H. are observed in proteins and polypeptides, polysaccharides and synthetic polymers such as polyvinyl compounds. ***Multiple-stranded H.***, in which more than one polymer strand wind around a common axis, are found in nucleic acids (***double helix***, see Nucleic acids) fibrillar proteins and various synthetic polynucleotides. In the ***triple helix*** of collagen, three polypeptide chains wind around a single axis to form the protofibril.

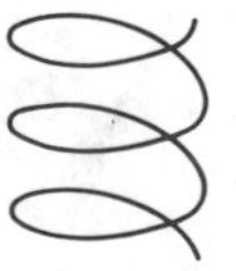

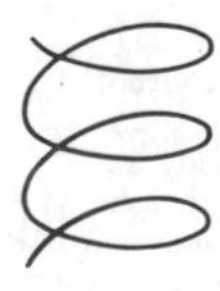

Right-handed helix   Left-handed helix

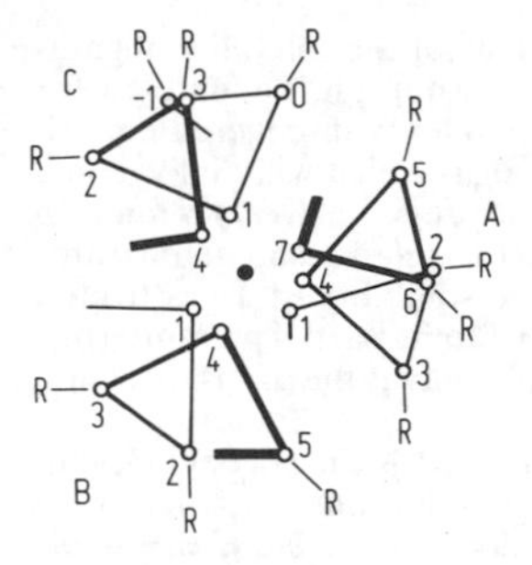

Fig. 2. Cross-sectional view of the triple helix of collagen (chains A, B and C)

Hell-Volhard-Zelinsky reaction: a method for production of α-halogen substituted aliphatic carboxylic acids from carboxylic acids and elemental halogens in the presence of phosphorus(III) halides. In this reaction, the acyl chloride is the primary product; because of its highly activated α-CH_2 group, it can be converted by halogenation to an α-halogencarboxylic acyl halide. This reacts with more carboxylic acid to form α-halogen carboxylic acids and acyl halides, which in turn are halogenated in the α-position:

$$R{-}CH_2{-}CO{-}OH \xrightarrow{(PX_3)} R{-}CH_2{-}CO{-}X \xrightarrow{(X_2)} R{-}CHX{-}CO{-}X; \; RCHX{-}CO{-}X + R{-}CH_2{-}CO{-}OH \longrightarrow RCHX{-}CO{-}OH + R{-}CH_2{-}CO{-}X$$

Hemateine: see Hematoxylin.

Hematin: see Hemins.

Hematin chloride: see Hemins.

Hematoxylin, ***hydroxybrasilin***: the pigment of logwood. It crystallizes as transparent, colorless to yellowish prisms of trihydrate; at 120 °C the crystals lose water of crystallization and then decompose. H. is nearly insoluble in cold water, but dissolves readily in hot water, alcohol or ether. In the light, H. turns red, even in the absence of air, and without any change in chemical composition. It dissolves in alkali metal hydroxide solutions with a purple-red color, but the solution rapidly turns blue-violet in the air. In the presence of oxygen, H. is oxidized to the actual pigment ***hemateine***. In the living wood, H. is probably present as the glycoside, which decomposes into the sugar and H. on storage. It has now been largely replaced as a textile dye by coal tar dyes, but it is still important for staining histological preparations, because it stains cell nuclei sharply. H. is also important in analytical chemistry, where it is used for colorimetric detection and quantitative analysis of various metals and ammonia.

Heme: an iron(II) complex of a porphyrin. H. are found in nature almost exclusively as prosthetic groups of hemoproteins, which are active in oxygen transport and storage (hemoglobins) or as oxidoreductases (cytochromes, catalases, peroxidases). The most widely occurring H. is ***H. b (protoheme)***, in which the porphyrin component is protoporphyrin IX. H. b is found in hemoglobins, cytochrome b, catalases and peroxidases. ***H. a*** and ***c*** are found in the cytochromes, and microorganisms contain still other H. Free H. are very quickly oxidized to Hemins (see). With nitrogen bases such as imidazole, pyridine or quinoline as additional ligands for the central iron atom, H. form ***hemochromogens***, which have characteristic absorption bands.

Hemellitene: see Trimethylbenzenes.

Hemerythrin, abb. ***Hery***: an iron(II) protein found in marine invertebrates; it has the same function as hemoglobins in higher animals. The iron is not bound to ligands. The oxygen-free form is colorless, and the oxygen-charged form, ***oxyhemerythrin***, is blue-violet.

Hemiacetal: an organic compound with the general formula $R^1R^2C(OH)(OR^3)$. H. are the primary addition products of aldehydes and alcohols; most of them are unstable and the reaction proceeds to yield an acetal. The H. formed by intramolecular addition of a hydroxyl group to a carbonyl group in a sugar are very important in the chemistry of carbohydrates.

Hemicelluloses: a group of polysaccharides which are found in wood, along with cellulose; their relative molecular masses are from 8,000 to 10,000. H. consist of hexoses (mannose, glucose, galactose) and pentoses (xylose, arabinose). The hexoses are more abundant in the H. of conifers, and the pentoses in the H. of broad-leaved trees. The main components of H. are β(1→4)-mannans and β(1→4)-xylans. H. are insoluble in water, but unlike cellulose, they dissolve in dilute acid and in bases.

The H. obtained from processing of wood are partly degraded; they are used to produce ethanol and yeast for animal feed.

Hemicyanins: see Cyanin pigments.

Hemins: Fe(II) complexes of porphyrins. H. are very readily formed from hemes by oxidation. Treatment of blood hemoglobin with glacial acetic acid and saturated sodium chloride solution produces the H. of protoporphyrin IX. As ***hematin chloride*** (***chlorohemin***), this substance precipitates as blue-black crystals (***Teichmann's crystals***) with a metallic sheen. In alkali, the corresponding hydroxide, ***hematin***, is formed. As the prosthetic groups of cytochromes, catalases and peroxidases, H. are in redox equilibrium with hemes.

Hemioxonols: see Polymethine pigments.

Hemochromogens: see Hemes.

Hemocyanin, abb. ***Hcy***: an oxygen-carrying metalloprotein found in the blood of molluscs and arthropods. It has the same function as hemoglobins in higher animals. H. contains copper. The oxygen-free form is colorless, and the oxygen-charged form (***oxyhemocyanin***) is dark blue.

Hemoglobins, abb. ***Hb***: a collective term for oxygen-carrying and storing hemoproteins. The H. contain heme as prosthetic group. The most common is heme a (protoheme). Some worms contain ***chlorohemoglobin***, in which the heme (chloroheme) has a vinyl residue instead of a formyl group. The intracellular H. of tissues and muscles in vertebrates is called ***myoglobin***. A related protein, ***legoglobin (leghemoglobin)***, is found in the root nodules of legumes. Lower animals contain chromoproteins in addition to H., such as Hemocyanin (see) and Hemerythrin (see).

In all H., the heme is bound to a peptide chain. The peptide chain of sperm whale myoglobin contains 153 amino acids. The H. in the blood of mammals, birds, amphibia and bony fishes consists of 4 hemes and 4 peptide chains; they are thus tetramers which form quaternary structure (Fig.). The peptide chains are identified by Greek letters (α, β, γ, δ and ε).

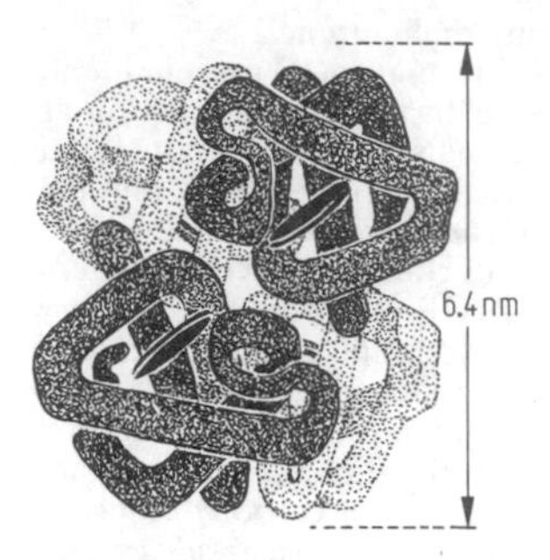

Quaternary structure of hemoglobin showing the spatial arrangement of the α and β chains. The black disk represents the heme group

There are two sets of identical chains; for example, the H. of normal adult human beings is 97.5% hemoglobin A_1 (HbA_1), with the composition $\alpha_2\beta_2$ and 2.5% HbA_2, which is $\alpha_2\delta_2$. The α-chains contain 141 amino acid residues, and the β-chains, 146. The tertiary and quaternary structures of all H. are very similar. The protein portion of the H. can be relatively easily separated from the heme by acids; the heme is simultaneously oxidized to a hemin. In the H., the heme is bound in a pocket formed by hydrophobic amino acids. The Fe(II) atom lies in the center of the porphyrin ring and makes four bonds with the ring. The 5th coordination site is occupied by a histidine residue in the H. The 6th coordination site can reversibly bind a molecule of O_2, without oxidizing the central iron atom. The binding of oxygen converts the orange-red H. to carmine-red ***oxyhemoglobin*** (abb. ***HbO_2***). The absorption bands are changed by O_2 binding (Hb: 414.556 to 565 nm; HbO_2: 425, 539 to 545, 576-578 nm). H. can bind other ligands than O_2; binding to CO and NO is toxic. Since the affinity of mammalian H. for CO is about 200 times its affinity for O_2, even a relatively low concentration of CO in the air can occupy enough H. to prevent adequate oxygen transport. Nitrite causes oxidation to ***methemoglobin***, which contains hemin and cannot bind O_2. Methemoglobin can be reduced to H. The H. are oxidized in the body to open-chain tetrapyrroles, the Bile pigments (see).

Hemoproteins: conjugated proteins containing Heme (see) as prosthetic group.

Hempel pipette: see Gas pipettes.

Hendecane: same as Undecane (see).

Henna: a red natural pigment. It is obtained from the powdered leaves and stems of the henna shrub *Lawsonia inermis*, which is a member of the *Lythraceae* family. H. has been used since ancient times to color fingernails and hair red. By combining H. with tannins, such as extract of nut galls, various colors can be obtained. The most abundant pigment in H. is ***lawsone***, 2-hydroxy-1,4-naphthoquinone; m.p. 192 °C.

Henry-Dalton law: a law for the solubility of gases in liquids at constant temperature; it is a limit law for the ideal case. In 1803, W. Henry discovered empirically that the amount of a gas which will dissolve in a liquid is proportional to its pressure (***Henry's law***). In 1807, J. Dalton extended this relationship to mixtures of gases, showing that the solubility of a gas from a gas mixture is proportional to its partial pressure (this is the H.). Both laws apply only if the gas is ideal and the solution dilute.

The H. is $\overline{m}_i = AP_i$, where P_i is the partial pressure of substance i over the solution and $\overline{m}_i$ is its molal concentration in the solution. A is called ***Henry's solubility*** or ***absorption coefficient***. It is in units of mol kg^{-1} Pa^{-1}, and depends on the nature of the gas and solvent, and on the temperature. As the temperature increases, A and thus the solubility decrease. For real gases and solutions, p_i must be replaced by the fugacity, and $\overline{m}_i$ by the activity.

For historical reasons, the H. is much more commonly written in the form $a_i/m_S = \alpha p_i$. Here a_i is the amount of gas dissolved (in cm^3 or g) in the mass m_S of solvent, and α is the ***technical solubility coefficient***. It is given in the units cm^3 g^{-1} atm^{-1}, which are not SI units.

Henry's law: see Absorption; see Henry-Dalton law.

Heparin: an acidic mucopolysaccharide with a less clearly defined structure than other mucopolysaccharides. It consists mainly of *N*-acetyl-D-glucosamine, glucuronic acid and iduronic acid, and is assigned the general structure GlcNAc-α(1→4)Glc UA-α(1→4)GlcNAc-α(1→4)-L-IdoUA-α$(1\rightarrow4)_n$. Heparin sulfate contains sulfate groups which are bound in ester linkage to the hydroxyl groups in position 6 or in amide linkage to the amino groups of glucosamine. β-Heparin is an obsolete term for Dermatan sulfate (see). The molecular mass of H. is 17,000 to 20,000. It can be isolated from animal organs such as heart, lung or liver. Because of its acidity, it forms ionic compounds with proteins. In this way, it can inhibit the conversion of prothrombin to thrombin, and thus prevents coagulation of blood. H. is formed in mast cells and basophilic granulocytes, and is used for prophylaxis and therapy of thromboembolitic diseases. It adsorbs to various surfaces, e.g. of plastic or glass, and thus prevents initiation of blood coagulation by these materials. Equipment for certain surgical procedures, such as heart operations, is heparinized.

Heparinoids: substances which inhibit blood coagulation in the same manner as heparin. Chemically, they are acidic polymers, especially sulfates of polysaccharides.

Hepar sulfuris: same as Liver of sulfur (see).

Hepar test: a preliminary test to detect the presence of sulfur in organic or inorganic samples. The sample is melted with sodium carbonate and reduced, converting sulfur compounds into sodium sulfide. When placed on an acidic silver plate, the mixture forms a black spot of silver sulfide.

Heptachlor: a Cyclodiene insecticide (see).

Heptadecanoic acid: same as Margarinic acid (see).

Heptanal: same as Enanthaldehyde (see).

Heptane: an alkane hydrocarbon with the general

formula C_7H_{16}; there are nine structural isomers (table). The H. are colorless, easily inflammable liquids which burn with sooty flames. They are insoluble in water, but are readily soluble in ethanol, ether and other organic solvents. The H. are found in petroleum, shale and tar oils, and in terpentine oils from American pines. They are also formed by cracking processes. H. are usually obtained by extraction or extractive distillation, and are purified by azeotropic distillation with ethanol. Branched H. with high octane ratings are made industrially by isomerization of H. or alkylation of propene with isobutane (see Butane). H. are used as solvents.

Properties of the heptanes

		m.p. [°C]	b.p. [°C]	n_D^{20}	Octane rating
Heptane	$CH_3(CH_2)_5CH_3$	− 91	98.4	1.3876	0
Isoheptane (2-Methylhexane)	$CH_3-CH(CH_3)-CH_2-CH_2-CH_2-CH_3$	−118	90.0	1.3848	46.4
3-Methylhexane	$CH_3-CH_2-CH(CH_3)-CH_2-CH_2-CH_3$	−119	92.0	1.3886	55.8
2,2-Dimethylpentane	$CH_3-C(CH_3)_2-CH_2-CH_2-CH_3$	−124	79.2	1.3826	95.6
3,3-Dimethylpentane	$CH_3-CH_2-C(CH_3)_2-CH_2-CH_3$	−135	86.1	1.3909	86.6
2,3-Dimethylpentane	$CH_3-CH(CH_3)-CH(CH_3)-CH_2-CH_3$	·	89.8	1.3919	88.5
2,4-Dimethylpentane	$CH_3-CH(CH_3)-CH_2-CH(CH_3)-CH_3$	−119	80.5	1.3815	83.8
3-Ethylpentane	$(CH_3-CH_2)_2CH-CH_2-CH_3$	−118	93.5	1.3934	69.3
2,2,3-Trimethylbutane	$CH_3-C(CH_3)_2-CH(CH_3)-CH_3$	− 25	80.9	1.3894	100

Heptane-1,7-dicarboxylic acid: same as Azelaic acid (see).

Heptanedioic acids: same as Pimelic acid (see).

Heptanoic acid, ***enanthic acid***: CH_3-$(CH_2)_5$-COOH, a saturated, aliphatic monocarboxylic acid. Its trivial name is no longer acceptable in IUPAC nomenclature. H. is an oily, colorless liquid with a rancid odor; m.p. -10.0 °C, b.p. 223 °C, n_D^{20} 1.4170. It is slightly soluble in water, and readily soluble in alcohol and ether. H. is a component of fusel oil and various essential oils, including the oils of violet blossoms and calamus (acorus or sweet flag) root.

Heptanol, ***heptan-1-ol***: $CH_3(CH_2)_5CH_2OH$, a monoalcohol, and the most important of the 36 possible isomeric C_7 alcohols. *Heptan-2-ol* and *heptan-3-ol* are also of some practical importance. H. is a colorless, oily liquid; m.p. -35 °C, b.p. 177 °C, n_D^{20} 1.4241. It is slightly soluble in water, but is miscible with alcohol and ether. Its esters are found in plant oils. H. is made by reduction of enanthaldehyde (heptanal), and therefore one of its names is *enanthic alcohol*. It is used for alkylation reactions.

Heptanones: aliphatic ketones derived from heptane. ***Heptan-2-one***, CH_3-CO-$(CH_2)_4$-CH_3, a colorless liquid, m.p. -35.5 °C, b.p. 151.4 °C, n_D^{20} 1.4069, found in essential oils. ***Heptan-3-one***, CH_3-CH_2-CO-$(CH_2)_3$-CH_3, is a colorless liquid; m.p. -39 °C, b.p. 147 °C, n_D^{20} 1.4057. ***Heptan-4-one***, CH_3-$(CH_2)_2$-CO-$(CH_2)_2$-CH_3, a colorless liquid; m.p. -33 °C, b.p. 144 °C, n_D^{20} 1.4069. H. are typical ketones with respect to their properties and reactions.

Heptenes: isomeric alkene hydrocarbons with the general formula C_7H_{14}. There are 27 structural isomers and additional *E,Z*-isomers, some of which are important in industry (table). The H. are colorless, readily inflammable liquids which burn with sooty flames. They are practically insoluble in water, but are readily soluble in ethanol and ether. They are found in the gasoline fractions of petroleum and can be isolated from this source. They are also obtained by catalytic hydrogenation of the heptynes or dehydration of heptanols. H. are used mainly for oxo synthesis; for example, isooctanol is made from isoheptene. Hept-1-ene can also be cyclized to toluene on a chromium oxide contact.

Heptoses: see Monosaccharides.

Heptyl aldehyde: same as Enanthaldehyde (see).

Heptynes: alkyne hydrocarbons with the general formula C_7H_{12}. The most important of fourteen structural isomers is ***hept-1-yne***, HC≡C-CH_2-$(CH_2)_3$-CH_3; m.p. - 81 °C, b.p. 99.7 °C, n_D^{20} 1.4087. The H. are colorless, readily inflammable liquids which burn with sooty flames. They are readily soluble in ethanol and ether, and are nearly insoluble in water (for their properties, synthesis and analysis, see Alkynes).

Herbicides: compounds used to kill plants or parts of plants, generally applied for weed control. Since the introduction of selective organic H. in the 1940's

Properties of selected heptenes

		m.p. [°C]	b.p. [°C]	n_D^{20}
Hept-1-ene	$CH_2=CH-CH_2-CH_2-CH_2-CH_2-CH_3$	−119	93.6	1.3998
Hept-2-ene Z-form	$CH_3-CH=CH-CH_2-CH_2-CH_2-CH_3$	–	98.5	1.4060
Hept-2-ene E-form		−109.5	98	1.4045
Hept-3-ene Z-form	$CH_3-CH_2-CH=CH-CH_2-CH_2-CH_3$	–	95.8	1.4059
Hept-3-ene E-form		−136.6	95.7	1.4043
2-Methyl-hex-1-ene	$CH_2=C(CH_3)-CH_2-CH_2-CH_2-CH_3$	–	91.7	1.4040
2,3-Dimethyl-pent-1-ene	$CH_2=C(CH_3)-CH(CH_3)-CH_2-CH_3$	−134.8	84.3	1.4033
2,3,3-Trimethyl-but-1-ene, Triptene	$CH_2=C(CH_3)-C(CH_3)_2-CH_3$	−109.9	77.9	1.4025

4) The ***halogen carboxylic acid H.*** are effective that of insecticides. The harvest losses in the world's crops due to weeds are about 9%.

Nonselective herbicides are unspecific, and are used to destroy all vegetation, e.g. on industrial land, railroad rights of way and along roads. ***Selective herbicides*** are used on crops where they are more toxic to the weeds than to the cultivated plants. Since one substance is usually not capable of controlling all the weed species which affect a crop, combinations are often used. ***Contact H.*** kill any part of the plant with which they come in contact, while ***systemic H.*** are taken up into the plant and transported (see Systemic pesticides). Systemic H. can be subdivided into ***foliage H.***, which are taken up by the leaves, and ***soil H.***, which enter the roots. H. may be applied before planting, before the plants emerge, or after they emerge from the soil.

The H. have a variety of mechanisms of action. *Photosynthesis inhibitors* and *respiratory inhibitors* are fairly common; these interfere with the primary metabolic processes of the plant. Other compounds interfere with nucleic acid or protein biosynthesis, inhibit key enzymes, mimic plant hormones and thus interfere with the regulation of growth and differentiation, affect membrane permeability, etc.

The *selectivity* may arise from application methods which prevent the H. from coming into contact with the crop plants, or it may be based on morphological differences between crop plants and weeds, on differences in metabolism which permit some species to detoxify the H., or on differences in the metabolic fate of the H.: some species convert it to a toxic substance while others do not.

Development of resistance to H. has so far not been serious. Trends in research are toward development of highly active, selective H. which require only a few grams per hectar, and are thus relatively harmless to the environment; toward application of phytotoxic natural products, e.g. microbial toxins, or synthetic imitations of such products. Another direction is increasing the resistance of crops to H. by simultaneous application of antidotes.

The H. belong to many different chemical classes. 1) Most nonselective H. are ***inorganic H.***, such as chlorates, which are effective against many types of plant, have long persistence, and have a long-term effect on roots. They are still occasionally used to kill potato plants, as is sulfuric acid. Ammonium sulfamidate is an arboricide used against weedy trees.

2) The classic ***nitrophenol H.*** include DNOC, i.e. the ammonium salt of 4,6-dinitro-o-cresol. It is used as a contact H. for post-emergence, selective control of annual weeds in grain. DNOC rapidly penetrates the leaf and causes cell damage by plasmolysis, thus causing the plant to dry out. Dinoseb (DNPB) is 2,4-dinitro-6-sec.-butylphenol. It is less toxic to mammals and is easier to apply. This is even more true of its acetate ester, dinoseb acetate, which is used as a post-emergence H. in grain and legumes.

3) The oldest ***diphenylether H.*** is nitrofen, 2,4-dichloro-4'-nitrodiphenyl ether. It is made by reaction of sodium 2,4-dichlorophenolate with 4-chloronitrobenzene, and is a contact and soil H. It acts as a photosynthesis inhibitor, and is used against weeds in grains, cotton, potatoes and vegetables.

and 1950's, the volume of H. applied has exceeded against weedy grasses. Trichloroacetic acid (TCA) is used as its sodium or amine salts as a pre-planting H. and causes leaf chloroses. TCA is obtained by oxidation of chloral, which can be used as an H. in almost the same fashion. 2,2-Dichloropropionic acid is also used for weed control in forests, pastures and irrigation ponds. It is produced by α-halogenation of propionic acid or propionitrile. Chlorfenprop consists of esters of 1-chloro-2-p-chlorophenylpropionic acid. For its technical synthesis, 4-chloroaniline is diazotized and the diazonium chloride is added to the acrylic ester with cleavage of N_2. This H. is used as post-emergence control of wild oats in various crops.

5) ***Thiocarbamate H.*** are used primarily as pre-emergence soil H. The most important of them are eptam $(n\text{-}C_3H_7)_2N\text{-}CO\text{-}S\text{-}C_2H_5$ and diallate $(i\text{-}C_3H_7)_2N\text{-}CO\text{-}S\text{-}CH_2\text{-}CCl{=}CHCl$.

An economically very important group of H. are the ***phenylamide H.***, which include the Acylaniline H. (see), Carbanilate H. (see) and Urea H. (see).

7) The breakthrough in the chemical control of weeds in grain crops was the development of the ***Growth hormone H.*** (see) based on aryloxyalkanoic acids (phenoxyfatty acids). This latter group includes 2,4-D (2,4-dichlorophenoxyacetic acid, or ester) and 2,4,5-T (2,4,5-trichlorophenoxyacetic acid, or ester). 2,4-D, 2,4,5-T, tordon, and Agent Orange (a mixture containing 2,4,5-T) were used as defoliants in the Vietnam War.

8) After the phenylamide H., the ***Triazine H.*** (see) (derivatives of the six-membered heterocycle 1,3,5-triazine) are economically the most important. The *triazinone H.* are derivatives of 1,2,4-triazine. These derivatives of 4-amino-1,2,4-triazin-5-one are highly effective and selective. Metribuzin (Sencor) is used as a pre- and post-emergence H., mainly on potatoes and soybeans. Metamitron (Goltix) is a highly selective soil and leaf H. used on sugar beets.

Metribuzin

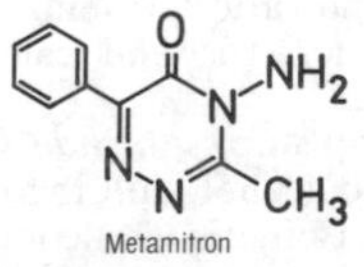

Metamitron

Some important ***aromatic carboxylic acid H.*** are *fenac* (2,3,6-trichlorobenzeneacetic acid), *dicamba* 3,6-dichloro-2-methoxybenzoic acid, *amiben* (2,5-dichloro-3-aminobenzoic acid), *tordon* (4-amino-3,5,6-trichloropicolinic acid), and *naptalam* (*N*-1-naphthylphthalamidic acid). These H. are used mainly for control of weeds in grains and other crops. A derivative of this group is *chlorothiamide* (2,6-dichlorothiobenzamide), which is used in orchards and vineyards as a pre-emergence control. The two substituted benzoic nitriles ioxynil and bromoxynil are used as contact H. to control weeds in grains.

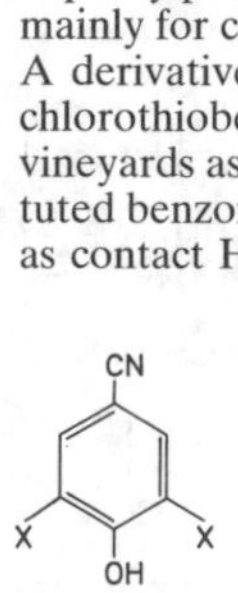

X = I Ioxynil
X = Br Bromoxynil

10) A more recently discovered group of H. are the ***nitroaniline H.***, which are derived from 2,6-dinitroaniline. They are used mainly as pre-planting H. and are selective for weedy grasses, e.g. in cotton and soybeans.

Table. Uracil herbicides

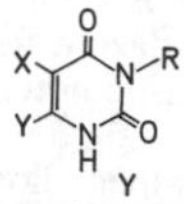

Name	X	Y	R	m.p. in °C
Lenacil	$-CH_2-CH_2-CH_2-$		(H)	315···317
Bromacil	Br	CH_3	$-CH(CH_3)-C_2H_5$	158···159
Terbacil	Cl	CH_3	$-C(CH_3)_3$	175···177

11) The ***organophosphate H.*** include the nonselective contact H. *Buminafos* used in Germany, the selective pre-emergence H. *isophos* used in the former Soviet Union, and the systemic, nonselective post-emergence foliage H. *glyphosate* used in the USA.

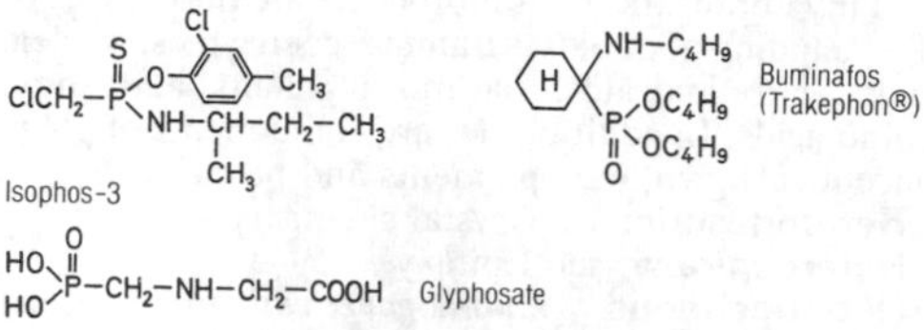

Isophos-3 Buminafos (Trakephon®) Glyphosate

Organophosphate herbicides

12) Aminotriazole (3-amino-1,2,4-triazole) is one of the oldest five-membered heterocycles used as a non-selective H.; it is especially effective against deep-rooted weeds and grasses. It is made industrially from cyanamide and hydrazine, via aminoguanidine, which is reacted with formic acid.

13) The ***pyridazine H.*** are six-membered heterocycles with two nitrogen atoms in the ring. The main example is *toxopyrimidine (pyramin)*, which is used as a pre-emergence H. in sugar beets. It is taken up through the roots, and is applied as a pre-emergence control. *Maleic acid hydrazide* (1,2-dihydro-3,6-pyridazinedione) is used for growth inhibition of grasses and inhibition of branching in tobacco.

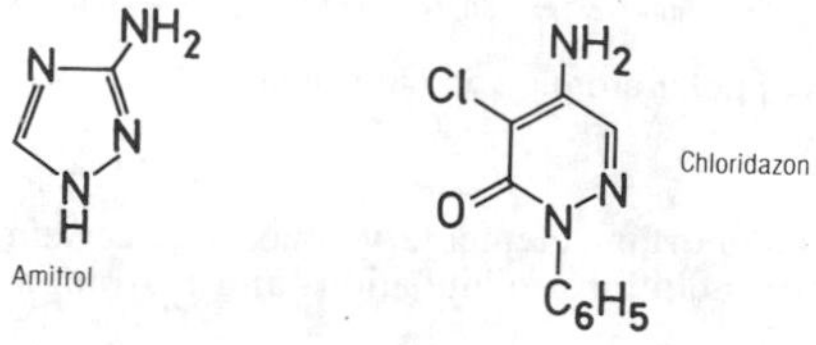

Amitrol Chloridazon

14) The ***uracil H.*** are also six-membered heterocycles, pyrimidine derivatives with internal urea structures. Their activity is based on inhibition of photosynthesis. The main representative is *lenacil*, which is used as a selective, pre-emergence soil H. for sugar beets, usually in combination with carbanilate H., but also by itself, e.g. for strawberries and spinach.

Table. Nitroaniline herbicides

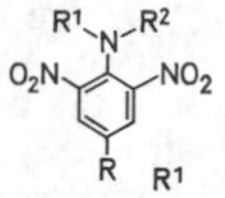

Name	R	R^1	R^2
Trifluralin	F_3C-	$n-C_3H_7$	$n-C_3H_7$
Nitralin	H_3C-SO_2-	$n-C_3H_7$	$n-C_3H_7$
Benfluralin	F_3C-	$n-C_4H_9$	C_2H_5
Oryzalin	H_2N-SO_2-	$n-C_3H_7$	$n-C_3H_7$

15) The ***bipyridylium H.*** are six-membered heterocycles containing one nitrogen atom; they are obtained by alkylation of 2,2'-bipyridyl or 4,4'-bipyridyl with 1,2-dibromoethane or 1,2-dichlormethane. The main examples are *diquat* and *paraquat*. These act rapidly in the presence of light; they attack only the above-ground parts of the plant. In the soil, the "quats" are adsorbed and rapidly degraded.

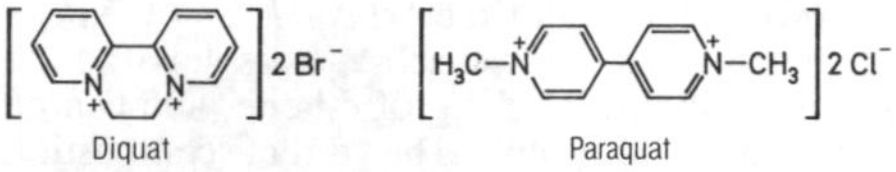

Diquat Paraquat

Bipyridylium herbicides

Diquat acts mainly against dicotyledenous plants, and is also used to desiccate potato foliage and plantings for seeds; paraquat is used for combatting both weeds and weedy grasses.

Heroin, ***diamorphine, diacetylmorphine***: the O,O-diacetyl derivative of morphine. It forms white, bitter-tasting platelets; m.p. 173 °C, $[\alpha]_D^{25}$ -166 °C (in methanol). H. is easily produced by acetylation of morphine; it is a strong analgesic, but is also one of most highly addictive drugs. Its medicinal use is proscribed.

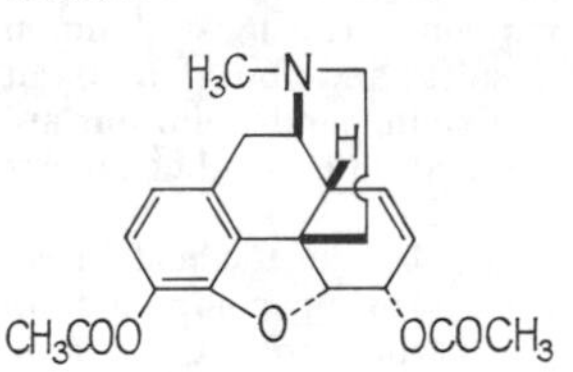

Hery: see Hemerythrin.

Herzig-Meyer method: a method for quantitative determination of *N*-alkyl groups. The amine is converted to alkylammonium iodide by reaction with hydroiodic acid; when heated, alkylammonium iodide cleaves off alkyl iodide: $R\text{-}NH\text{-}CH_3 + HI \rightarrow R\text{-}NH_2\text{-}CH_3I \rightarrow R\text{-}NH_2 + CH_3I$. The alkyl iodide can be converted to silver iodide by reaction with silver nitrate, and the amount is then determined gravimetrically.

Hesperitin: see Flavone.

Hess' law: formerly called ***law of constant heat sums***: a law established in 1840 by Hess, which can be formulated as follows: If a reaction can occur by more than one pathway from the same reactants to the same products, the sum of the heats of reaction for intermediates along each pathway is the same. H.

follows directly from the fact that the reaction energies $\Delta_R E$ and enthalpies $\Delta_R H$ are changes in thermodynamic parameters of state, and as such, they cannot depend on the path of the process. H. is the basic law of Thermochemistry (see), and makes possible the determination of enthalpies $\Delta_F H$ and energies $\Delta_F E$ of formation from experimentally measured heats of combustion, $\Delta_C H$ and $\Delta_C E$, and calculation of reaction enthalpies which cannot be measured.

Example:

$$2C + 2H_2 \xrightarrow{1} C_2H_4 \xrightarrow[+3O_2]{2} 2CO_2 + 2H_2O \quad (\text{path } 3: 2C + 2H_2 \rightarrow 2CO_2 + 2H_2O)$$

According to H., the sum of the reaction enthalpies of steps 1 and 2, $\Delta_R H_1$ and $\Delta_R H_2$ is equal to the reaction enthalpy $\Delta_R H_3$ along path 2: $\Delta_R H_1 + \Delta_R H_2 = \Delta_R H_3$. Step 1 is the formation of ethene, so the reaction enthalpy is the enthalpy of formation of ethene: $\Delta_R H_1 = \Delta_F H_{C_2H_4}$. Since reaction 1 cannot be made to proceed in a calorimeter, $\Delta_R H_1$ is not experimentally measurable. However, $\Delta_R H_2$ is equal to the combustion enthalpy $\Delta_C H_{C_2H_4}$ of ethene, and can be measured calorimetrically. The direct combustion reaction 3 can also be measured, and corresponds to formation of 2 mol each of CO_2 and H_2O from the elements: $\Delta_R H_3 = 2\ \Delta_F H_{CO_2} + \Delta_F H_{H_2O}$. Thus the enthalpy of formation of ethene is calculated from H. to be $\Delta_F H_{C_2}H_4 = 2\ \Delta_F H_{CO_2} + 2\ \Delta_F H_{H_2O} - \Delta_C H_{C_2}H_4$.

H. is also applicable to Phase-change heats (see). The transition of a solid into the gas state can occur directly by sublimation or indirectly by melting followed by evaporation. H. says that the sublimation enthalpy is equal to the sum of the enthalpies of melting and evaporation (at the same temperature).

Heteroaromatic: see Heterocycles.

Heteroatom: a term for any atom other than a carbon atom which is a member of the ring in a heterocyclic organic compound. The most common H. are nitrogen, oxygen, sulfur and phosphorus, but boron, silicon, arsenic, selenium and tellurium are not infrequently incorporated into synthetic compounds.

Heteroauxin: same as Indol-3-ylacetic acid (see).

Heterocycles: cyclic hydrocarbon compounds in which the ring consists of carbon and at least one other element (see Heteroatoms), usually N, O or S. H. vary with respect to ring size, the number and combination of the heteroatoms, the fusion of the rings and the bonding within the ring.

The various sizes of H. are discussed separately under the following entries: Three-membered H., Four-membered H., Five-membered H., Six-membered H. and Seven-membered H. In addition, there are larger rings and macrocyclic H., such as crown ethers, Phthalocyanins and Porphyrins. By far the most common and important H. in science, technology and nature are the five- and six-membered H.

The possibilities for synthesis are nearly unlimited with regard to the number and combinations of heteroatoms; elements of Groups Va and VIa, such as phosphorus, arsenic, selenium and tellurium are being incorporated with increasing frequency. Elements of other main groups, such as silicon and boron, are also being explored. There are limits to the stability of the H. prepared by the synthetic methods now available; these depend on the ring size and bonding. In addition to simple heterocyclic system (***monocyclic H.***), there are ***fused*** and ***heterocyclic fused*** or ***condensed H.*** For example, Pyrrole (see) is a component of the fused benzene systems Indole (see) with one benzene ring and Carbazole (two benzene rings). Combination of Imidazole (see) with Pyrimidine (see) produces the fused heterocyclic system Purine (see).

In addition to the trivial names allowed or recommended by IUPAC, there is a system for naming and numbering complicated heterocyclic systems consisting of two, three or more carbo- or heterocyclic rings (see Nomenclature, sect. III F). Like carbocyclic compounds, H. can be ***saturated H.***, ***unsaturated H.*** or ***aromatic H. (heteroaromatics)***. The last group are five- or six-membered rings stabilized by a π-electron sextet; they are readily formed because of their stability, and many syntheses can be developed for them. On the basis of their special reactivities, a distinction is made between π-Deficiency heterocycles (see) and π-Excess heterocycles.

Many important natural products are derivatives of H., including alkaloids, vitamins, coenzymes, nucleic acids, carbohydrates, natural pigments and some amino acids. In addition, H. are components of pharmaceuticals, synthetic pigments and pesticides.

Heterodesmic: see Crystal chemistry.

Heteroepitaxy: see Epitaxy.

Heterogeneous, ***inhomogeneous***: in physical chemistry, a material system consisting of several phases, for example a liquid and a solid. The antonym is Homogeneous (see).

Heterogeneous biocatalysis: see Biotechnology.

Heterolysis: cleavage of a bond in such a way that both bonding electrons remain on one of the products. The H. of polar covalent bonds yields two ions with opposite charges. H. of donor-acceptor bonds can yield neutral products. H. occur primarily in solution, via internal ion pairs (contact ion pairs) with a common solvation sheath, then external ion pairs (pairs separated by solvent) which share a solvation sheath, but are separated by a few solvent molecules between the two ions (Fig.).

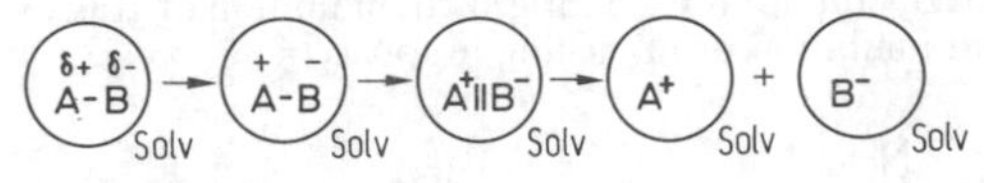

Heterolysis of polar atomic bond in solution. Solv = solvent sheath

H. is an important elementary process in certain substitutions, additions, eliminations and rearrangements.

Heteropolar bond: see Chemical bond.

Heteropolyanions: see Nomenclature, sect. II D.

Heteroside: see Glycosides.

Heterotropic groups: see Topic groups.

Heterotypic mixed crystal formation: same as Isodimorphism.

Heterotypism: see Isotypism.

HETP value: see Filler.

Heumann indigo syntheses: two methods for production of indigo from aniline or anthranilic acid and chloroacetic acid. The first, dating from 1890, goes via phenylglycine and indoxyl to indigo. In the more commonly used second method (1893), the starting material is anthranilic acid, which is converted to phenylglycine 2-carboxylic acid by reaction with chloroacetic acid. Cyclization of the phenylglycine carboxylic acid in an alkali melt around 200 °C produces indoxyl 2-carboxylic acid, which on heating loses CO_2 and forms indoxyl. An alkaline solution of indoxyl is oxidized to form indigo.

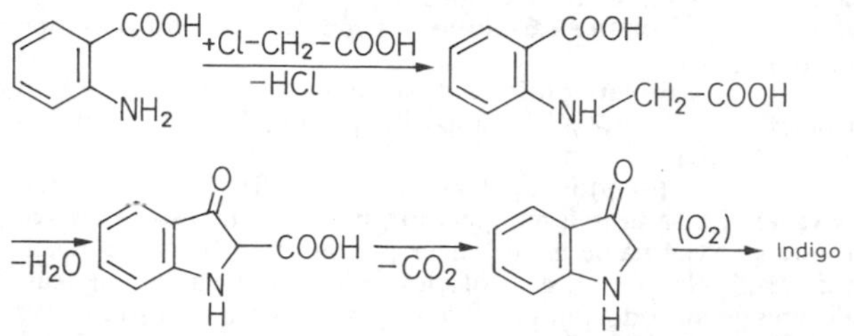

Second Heumann indigo synthesis

Heusler alloy: see Manganese bronze.
Hexacene: see Acenes.
Hexachlorobenzene, ***perchlorobenzene***: a colorless, crystalline compound; m.p. 230 °C, b.p. 322 °C. H. is insoluble in water, but is soluble in ether, benzene and chloroform. It is formed by complete chlorination of benzene or by chlorination of toluene at 400 to 450 °C (in this case carbon tetrachloride is also formed). H. is used as an intermediate in the synthesis of fungicides, as a softener and for organic syntheses.

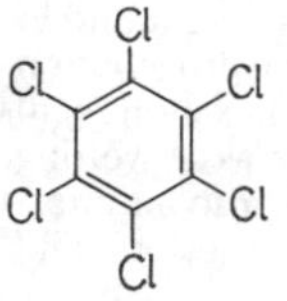

1,2,3,4,5,6-Hexachlorocyclohexanes, abb. ***HCH, benzene hexachlorides***, abb. ***BHC***: $C_6H_6Cl_6$, chlorine addition products of benzene. 1,2,3,4,5,6-H. occur in eight diastereoisomeric forms, five of which are formed by a radical chain reaction between chlorine and benzene promoted by UV light. The solid mixture smells intensely musty, and consists of 60 to 70% α-, 5 to 12% β-, 10 to 14% γ-, 6 to 10% δ- and 3 to 4% ε-HCH. The individual isomers can best be determined by gas chromatography. The γ-isomer is an important insecticide (see Lindane).

Hexachlorocyclopentadiene, abb. ***HCCP***: C_5Cl_6, a slightly yellowish liquid with a sharp odor; density 1.7119, m.p. 11.5-12 °C, b.p. 239 °C, n_D^{20} 1.5647. H. is an intermediate in the synthesis of HCCP dimer and diene insecticides. It is synthesized either by chlorination of monomeric cyclopentadiene with aqueous hypochlorite solution in a two-phase process, or by subjecting cyclic C_5 hydrocarbons to high-temperature chlorination (≈ 500 °C) on a suitable contact.

Hexachloroethane, ***perchloroethane***: CCl_3-CCl_3, colorless crystals with a camphor-like odor; m.p. 186-187 °C (in a fusion tube), subl.p. 185 °C. H. occurs in three modifications. The rhombic form which is stable at room temperature is converted at 46 °C to the triclinic form; above 71 °C this is transformed to the cubic form. H. is insoluble in water, but is soluble in alcohol, ether and benzene. It is synthesized by addition of chlorine to tetrachloroethane around 120 °C. H. is used as a chlorination reagent, as a component of moth repellents, as a substitute for camphor in the production of celluloid and to make smoke bombs.

Hexachlorophene: a dinuclear halogenated phenol which acts as a strong disinfectant and is used in cosmetics (e.g. deodorants). To prevent toxic effects from absorption through the skin, there are upper limits on allowable concentration.

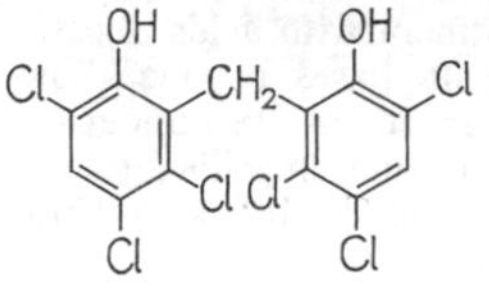

Hexachlorophene 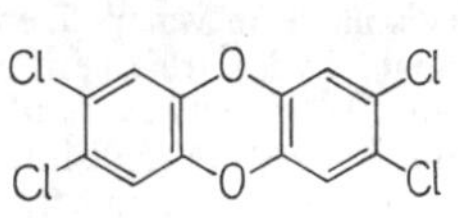 Dioxin

H. is synthesized from 2,4,5-trichlorophenol by condensation with formaldehyde. The synthesis of 2,4,5-trichlorophenol may lead to formation of dioxin as a byproduct; this is a highly toxic substance which is not subject to biotransformation or excretion from the body.

Hexachloroplatinic(IV) acid: $H_2PtCl_6 \cdot 6H_2O$, brown to red-orange, highly deliquescent prisms; M_r 517.92, density 2.43. When heated to about 300 °C, H. decomposes to platinum(IV) chloride and hydrochloric acid. It is made by dissolving platinum in aqua regia or chlorine-containing hydrochloric acid. Addition of potassium chloride to an aqueous solution of H. yields potassium hexachloroplatinate. H. is converted by alkali hydroxide solutions to yellowish-white hexahydroxoplatinic(IV) acid, $H_2Pt(OH)_6$. It is used to make other platinum salts and platinum asbestos.

Hexacosanoic acid, ***cerotinic acid***: CH_3-$(CH_2)_{24}$-COOH, a saturated fatty acid. It forms colorless crystals; m.p. 88 °C. H. is insoluble in water, but dissolves in ether and chloroform. It is found chiefly in esterified form in beeswax.

Hexacosan-1-ol: same as Ceryl alcohol (see).

Hexadecanoic acid: same as Palmitic acid (see).

Hexa-2,4-diendioic acid: same as Muconic acid (see).

(E),(E)-Hexa-2,4-dienoic acid: same as Sorbic acid (see).

Hexafluorosilicates: same as Fluorosilicates (see).

Hexafluorosilicic acid: same as Fluorosilicic acid (see).

Hexagonal: see Crystal.

Hexahydrobenzene: same as Cyclohexane (see).

Hexahydropyridine: same as Piperidine (see).

Hexahydropyrimidine-2,4,5,6-tetrone: same as Alloxan (see).

Hexahydropyrimidine-2,4,6-trione: same as Barbituric acid (see).

Hexamethonium bromide: see Ganglion blockers.

Hexamethylene: same as Cyclohexane (see).

Hexamethylenediamine, ***1,6-diaminohexane***: $H_2N\text{-}(CH_2)_6\text{-}NH_2$, an aliphatic diamine which forms colorless, fuming crystals with a silky sheen; m.p. 41-42 °C, b.p. 204-205 °C. H. is soluble in water, alcohols and hydrocarbons, and gives the reactions typical of a primary amine. It forms stable mono- and diacid salts with inorganic acids. H. is synthesized mainly by catalytic hydrogenation of adipic dinitrile; it is used in the form of the bisadipic acid salt (*AH salt*) for production of polyamides and polyurethanes.

Hexamethylenetetraamine, ***methenamine, Urotropin***®: a tetraazaadamantane formed by concentrating an ammoniacal formaldehyde solution. H. is a tetracyclic compound which forms colorless, shiny, rhombic crystals which sublime without melting. It is very soluble in water. Treatment with acids releases formaldehyde; treatment with bases produces ammonia. H. is used as a buffer. It was used under the name *Urotropin*® to disinfect urine; in acidic urine, it was supposed to form formaldehyde. It has no therapeutic applications at present.

$$6\,H\text{-}CHO \xrightarrow[-6H_2O]{+4NH_3} (CH_2)_6N_4$$

Hexamine: 1) short name for dipicrylamine. 2) short name for hexamethylenetetraamine.

Hexanal: same as Capric aldehyde.

Hexane: an alkane hydrocarbon with the general formula C_6H_{14}; there are five structural isomers (table). The H. are colorless, combustible and readily inflammable liquids. They are barely soluble in water, but readily soluble in ethanol and ether. Along with the pentanes, they are important components of petroleum ether. They are present in petroleum, and are obtained from it by distillation; they are also produced by cracking processes. Straight-chain H. is separated from the technical mixture of isomers as a urea inclusion compound or by use of molecular sieves (zeolites). It can also be synthesized by electrolysis of a concentrated aqueous solution of potassium butyrate. The branched H. are important as high-octane fuel components, and can be obtained by alkylation of ethylene with isobutane (see Butane), or by catalytic isomerization of *n*-H. H. are widely used as organic solvents, especially for dissolving and extracting fats and oils.

Hexane-2,5-dione: same as Acetonylacetone (see).

Hexanedioic acid: same as Adipic acid (see).

2,3,6,2′,4′,6′-Hexanitrodiphenylamine: same as Dipicrylamine (see).

Hexanoic acid, ***caproic acid***: $CH_3\text{-}(CH_2)_4\text{-}COOH$, a saturated monocarboxylic acid. The use of the trivial name is no longer permitted in IUPAC nomenclature. H. is a colorless, oily liquid with an unpleasant odor; m.p. - 9.5 °C, b.p. 205 °C, n_D^{20} 1.4163. It is slightly soluble in water, and readily soluble in alcohol and ether. It displays the reactions typical of monocarboxylic acids. It is found in nature mainly in the form of its esters, which are present in butter, essential and fatty oils, and the leaves and fruits of many plants. H. can be obtained in mixtures with other fatty acids by saponification of fats. It can also be made by oxidation of linoleic acid or hexanol, or by a malonic ester synthesis. H. is used in the synthesis of aromas and flavorings, and for organic syntheses.

Hexanol, ***hexan-1-ol***: $CH_3(CH_2)_4CH_2OH$, the most important of the 16 possible isomeric C_6 monoalcohols. H. is a colorless liquid; m.p. -47 °C, b.p. 157.5 °C, n_D^{20} 1.4178. It is not miscible with water, but is soluble in ethanol and ether. Its esters are found in plant oils. H. is obtained by catalytic hydrogenation of hexanoic acid or by reduction of hexanoic acid esters with sodium. It is used mainly as a wetting agent and for alkylation reactions. It is also an intermediate in organic syntheses, e.g. of esters and ethers.

Hexenal, ***trans-2-hexenal***: $CH_3\text{-}CH_2\text{-}CH_2\text{-}CH{=}CH\text{-}CHO$, a volatile defense secretion of certain plants which protects them against microorganisms. H. is also a growth regulator. In low concentrations, it stimulates callus formation, and in higher amounts, it stimulates shoot formation. Its biosynthesis is thought to occur by peroxidation of linolenic acid.

Hexenes: isomeric hydrocarbons with the general formula C_6H_{12}. There are 13 structural isomers and additional *E,Z*-isomers of hex-2-ene, hex-3-ene,

Properties of the hexanes

		m.p. [°C]	b.p. [°C]	n_D^{20}	Octane rating
Hexane	$CH_3(CH_2)_4CH_3$	− 95	68	1.3748	26.0
Isohexane, 2-Methylpentane	$CH_3{-}CH(CH_3){-}CH_2{-}CH_2{-}CH_3$	−154	60.3	1.3714	73.5
3-Methylpentane	$CH_3{-}CH_2{-}CH(CH_3){-}CH_2{-}CH_3$	−118	63.3	1.3765	74.3
Neohexane (2,2-Dimethylbutane)	$CH_3{-}C(CH_3)_2{-}CH_2{-}CH_3$	−100	49.7	1.3688	93.4
2,3-Dimethylbutane	$CH_3{-}CH(CH_3){-}CH(CH_3){-}CH_3$	−129	58	1.3750	94.3

Properties of selected hexenes

		m.p. [°C]	b.p. [°C]	n_D^{20}
Hex-1-ene	$CH_2=CH-CH_2-CH_2-CH-_2-CH_3$	−139.8	63.3	1.3837
Hex-2-ene Z-form	$CH_3-CH=CH-CH_2-CH_2-CH_3$	−141.3	68.8	1.3977
E-form		−133	67.9	1.3935
Hex-3-ene Z-form	$CH_3-CH_2-CH=CH-CH_2-CH_3$	−137.8	66.4	1.3947
E-form		−113.4	67.1	1.3943
2-Methylpent-1-ene	$CH_2=C(CH_3)-CH_2-CH_2-CH_3$	−135.7	60.7	1.3920
3-Methylpent-1-ene	$CH_2=CH-CH(CH_3)-CH_2-CH_3$	−153	51.1	1.3841
2-Methylpent-2-ene	$(CH_3)_2C=CH-CH_2-CH_3$	−135	67.3	1.4004
2-Ethylbut-1-ene	$CH_2=C(C_2H_5)-CH_2-CH_2-CH_3$	−131.5	64.7	1.3969
3,3-Dimethylbut-1-ene	$CH_2=CH-C(CH_3)_3$	−115.2	41.2	1.3763
2,3-Dimethylbut-2-ene, Tetramethylethene	$(CH_3)_2C=C(CH_3)_2$	− 74.3	73.2	1.4122

methylpent-2-ene and 4-methylpent-2-ene (table). The H. are colorless, flammable liquids which burn with very sooty flames. They are slightly soluble in water, and dissolve readily in many organic solvents. Some of the H. are present in the products of coal distillation and petroleum cracking. They are made by partial hydrogenation of the corresponding hexynes, or by dehydration of branched hexanols. They are used as intermediates in organic syntheses.

Hexite: short name for Dipicrylamine (see).

Hexobarbital: see Barbitals, table.

Hexogen: see 1,3,5-Trinitro-1,3,5-triazacyclohexane.

Hexokinase: see Kinases.

Hexoses: see Monosaccharides.

Hexyl: short name for Dipicrylamine (see).

Hexylaldehyde: same as Capric aldehyde (see).

Hexylamine, ***1-aminohexane***: CH_3-$(CH_2)_5$-NH_2, a primary aliphatic amine. H. is a colorless liquid with an amine-like odor; m.p. -19 °C, b.p. 130 °C, n_D^{20} 1.4180. It is slightly soluble in water, but readily soluble in alcohol and ether. H. can be synthesized by Hofmann degradation of enanthamide or by hydrogenation of caproic nitrile. It is used in the production of drugs and dyes.

Hexylresorcinol, ***4-hexylresorcinol***: a compound used as a disinfectant for wounds and as an antimycotic. It is synthesized by Friedel-Crafts acylation of resorcinol with caproyl chloride and subsequent reduction of the carbonyl group to a CH_2 group.

Hexynes: alkyne hydrocarbons with the general formula C_6H_{10}. There are seven structural isomers, of which the following are relatively important: ***hex-1-yne***, $HC\equiv C$-CH_2-CH_2-CH_2-CH_2-CH_3, m.p. -131.9 °C, b.p. 71.3 °C, n_D^{20} 1.3989; ***hex-2-yne***, CH_3-$C\equiv C$-CH_2-CH_2-CH_3, m.p. -89.6 °C, b.p. 84 °C, n_D^{20} 1.4138; ***hex-3-yne***, CH_3-CH_2-$C\equiv C$-CH_2-CH_3, m.p. -103 °C, b.p. 81.5 °C, n_D^{20} 1.4115. The H. are colorless, inflammable liquids which burn with sooty flames. They are nearly insoluble in water, but dissolve readily in ethanol and ether. Some are formed by pyrolysis of hard coal. In the laboratory, they can be synthesized by alkylation of mono- or disodium acetylides.

Hf: symbol for hafnium.

HF spectroscopy: abb. for high-frequency spectroscopy.

Hg: symbol for mercury.

High-frequency conductometry: a method of analysis based on measurement of the conductivity of an electrolyte and the dependence of its conductivity on concentration. In contrast to conductometry (see), H. uses high-frequency alternating voltages (>0.1 MHz) as excitation signal. The high frequencies make it possible to apply the electrodes to the outer wall of the sample cell, that is, the sample cell is integrated into the circuit as a capacitor. In this way, polarization phenomena and poisoning of electrodes by surface-active substances can be avoided. If H. is used to recognize the end point of a titration, the method is ***high-frequency conductometric titration***.

High-frequency drying: see Drying.

High-frequency spectroscopy, ***HF spectroscopy***: a collective term for study of the energy states of material systems and the transitions between them which occur by absorption of electromagnetic radiation with frequencies between a few kilohertz and a few gigahertz (see Electromagnetic spectrum). These include a variety of different methods such as NMR spectroscopy (see), ESR spectroscopy (see) nuclear quadrupole resonance spectroscopy (see) and microwave spectroscopy (see).

High-performance liquid chromatography, abb. ***HPLC*** (also called ***high-pressure liquid chromatography***): liquid chromatography under optimum conditions. The particle size of the stationary phase is 5 to 10 µm (Fig 1).

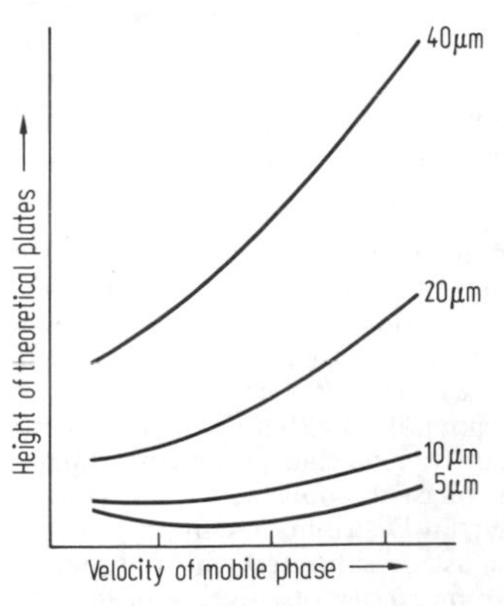

In an ideal column, the height of column required for each theoretical plate is equal to the particle diameter; in industrial columns, the height is three times that value. HPLC was developed from column chromatography, but it is more rapid, has a higher separation capacity and the limit of detection is lower.

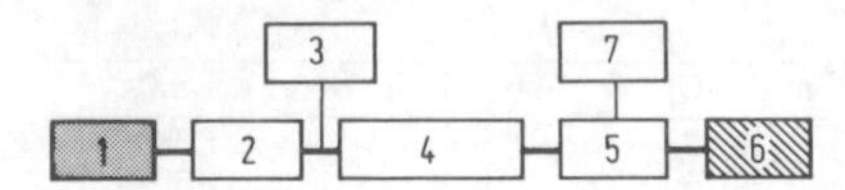

Fig. 2. Components of an HPLC apparatus: *1*, solvent reservoir; *2*, high-pressure pump; *3*, sample inlet; *4*, separation column; *5*, detector; *6*, sample collector; *7*, data evaluation system

The *pump* can be a short-stroke or long-stroke piston pump or a piston membrane pump with a capacity of 0.1 to 10 ml min^{-1} at pressures up to 60 MPa. The pumps should deliver the eluent with as little pulsation as possible and with a constant delivery volume. Gradient elution (binary or ternary gradients) is very important; the gradients can be created either on the low-pressure side, in a mixing chamber, or on the high-pressure side, by more pumps. The *sample* can be injected either in the low-pressure region, by septum injection, or at higher column pre-pressure by use of a sample loop. "Stop-flow" devices are used for sample injection without pressure and with stopped eluent flow. The *separation column* is made of stainless steel or, for lower pressures, glass; the inner diameter is between 2 and 4 mm, and the length is 10 to 50 cm. Packing the column requires great precision. Particles with diameters > 20 μm are packed dry, while those < 20 μm in diameter are packed wet. The *detector* continuously measures some physical property of the eluate, and gives a signal proportional to the concentration of the substance in the eluate. The eluate flows through a cell which has a volume between 5 and 20 μl. The most commonly used detectors are UV detectors and differential refractometers. Other types of detector and their sensitivities are listed in Table 1.

Table 1. Detectors for high-performance liquid chromatography

Detector	Sensitivity [g ml^{-1}]
UV detector	10^{-10}
Refractometer	10^{-7}
Adsorption detector	10^{-9}
Transport FID detector	10^{-7}
Conductivity detector	10^{-8}
Fluorescence detector	10^{-9}

The *solvents* or *eluents* used differ with respect to their dissolving capacities, dielectric constants, viscosities and refractive indices. The choice of eluent depends on the separation desired. For Adsorption chromatography (see) on polar adsorbents (silica gel, aluminum oxide), nonpolar or slightly polar solvents are used (hexane, methylene chloride); with nonpolar phases (reversed phases), polar solvents (water, methanol) are used. In Distribution chromatography (see), the eluent must be saturated with the separation liquid. *Eluotropic series* of solvents have been established to help with the selection of solvents; there are separate series for hydrophilic and hydrophobic adsorbents (Table 2). Mixtures of solvents are required in many cases. For optimum separation of very complex substrate mixtures, the elution strength of the eluent is varied continuously throughout the run (*gradient elution*).

Table 2. Eluotropic series of solvents for HPLC

Hydrophilic	Adsorbent	Hydrophobic
Water	↑	Hexane
Methanol		Benzene
Ethanol		Ethyl acetate
Propan-1-ol	Eluting strength	Diethyl ether
Ethyl acetat		Propan-1-ol
Acetone		Acetone
Diethyl ether		Ethanol
Benzene		Methanol
Hexane		Water

Porous materials (silica gel, aluminum oxide) with large specific surface areas are used as carriers. Porous layer beads (abb. PLB) with a porous layer (1 to 3 μm) on a hard, impenetrable core (glass) are sometimes used. Chemically modified carriers are made by reaction of the silanol groups of the silica gel with organic or organosilicon compounds. The organic group is covalently bound to the surface of the carrier. Alkylsilanes with 2 to 18 carbon atoms and no functional groups are used as nonpolar stationary phases for reversed-phase chromatography (abb. RPC). C_8 and C_{18} phases are most common (e.g. Lichrosorb RP 8 and RP 18, or PLB Perisorb RP 8 and RP 18). HPLC can be carried out as adsorption, distribution, ion-exchange, permeation or affinity chromatography. The method is widely applied in all fields of chemistry, biochemistry and pharmacy (Fig. 3).

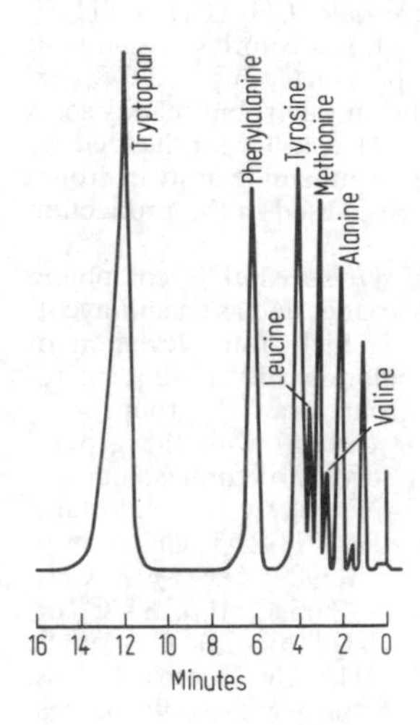

Fig. 3. Chromatogram of an amino acid mixture

High-performance thin-layer chromatography: see Thin-layer chromatography.

High pressure chemistry: that area of chemistry dealing with reactions at pressures above 20 MPa. Some well-known processes of H. are the synthesis of ammonia and the synthesis of methanol from carbon monoxide and hydrogen. Diamonds are made from graphite at pressures up to 40 GPa and temperatures up to 4000 °C.

High-pressure liquid chromatography: same as High-performance liquid chromatography (see).

High-severity cracking: see Pyrolysis.

High-spin complex: see Coordination chemistry, see Ligand field theory.

High-temperature chemistry: the chemistry of processes between about 800 °C and the highest temperatures at which atoms are still linked by chemical bonds. This definition of H. is somewhat arbitrary. For gaseous systems, high-temperature reactions can be defined on a thermodynamic basis as those processes in which the change in free enthalpy ΔG becomes negative only at high temperatures, that is, they occur spontaneously only at high temperatures. The ΔG values are defined by the equation of state $\Delta G = \Delta H - T\Delta S$; they depend on both the temperature and the change in entropy ΔS, which increases rapidly with increasing temperature. This definition of high-temperature chemical reactions also applies, with some limitations, to reactions in condensed (solid and liquid) phases. In such cases, however, it is often necessary to supply thermal energy (activation) for physical processes (e.g. diffusion) in order to make the chemical reaction possible. In other words, the high temperatures are needed to initiate solid- or liquid-phase reactions, or to allow them to occur at an acceptable rate, even though the reactions themselves are exothermal. For reactions in condensed phases, therefore, it is necessary arbitrarily to limit the range of high-temperature reactions.

H. is concerned, among other things, with energy structures and valence states at high temperatures, for example, with bonding and dissociation energies. The thermodynamic aspects of H. apply particularly to thermochemistry and the treatment of gas equilibria and thermodynamic diagrams of state. Some examples of typical high-temperature reactions are reactions in thermal plasma (plasma chemistry), chemical transport reactions and solid-state reactions in the narrow sense (solid-solid reactions). All High temperature materials (see) are made by chemical reactions at high temperatures. Many of the reactions in metallurgy (e.g. iron and aluminum production), silicate technology (e.g. production of cement, ceramics and glass) and the chemical industry (e.g. ammonia combustion, electrothermal methods of producing calcium carbide and phosphorus) occur at high temperatures. The generation of high temperatures is often dependent on high-temperature processes; for example, temperatures above 4500 °C can be achieved by combustion of dicyanogen, $(CN)_2$, or zirconium powder.

High-temperature corrosion: see Metal corrosion, chemical.

High-temperature materials: metals or metal borides, carbides, silicides, nitrides or oxides which are stable to thermal and mechanical stress at temperatures above 1500 °C under inert conditions.

The metallic H. are especially subject to corrosion by components of the atmosphere and other media. Their chemical resistance can be significantly increased by coating them (e.g. iron materials) with high-temperature borides, carbides or nitrides. H. are used in metallurgy, the chemical industry and silicate, electrical, laboratory, household and space technologies.

Industrially important high-temperature materials

Material	Composition	m.p., subl., temp. (S) and dec. temp. (D) in K
Carbon	C	3900 (S)
Titanium	Ti	1940
Tantalum	Ta	3270
Chromium	Cr	1825
Molybdenum	Mo	2895
Tungsten	W	3685
Iron	Fe	1813
Rhodium	Rh	2235
Platinum	Pt	2042
Zirconium boride	ZrB_2	3330
Chromium boride	CrB	3025
Boron carbide	B_4C	2720
Silicon carbide	SiC	2570 (D)
Titanium carbide	TiC	3520
Zirconium carbide	ZrC	3450
Tantalum carbide	TaC	4150 (D)
Chromium carbide	Cr_3C_2	2220
Molybdenum carbide	Mo_2C	2570 (D)
Tungsten carbide	WC	3170 (D)
Uranium carbide	UC	2860
Molybdenum silicide	$MoSi_2$	2120
Boron nitride	BN	3270 (S)
Aluminum nitride	AlN	2470
Silicon nitride	Si_3N_4	2200 (S)
Titanium nitride	TiN	3220
Beryllium oxide	BeO	2790
Magnesium oxide	MgO	3070
Aluminum oxide	Al_2O_3	2320
Silicon oxide	SiO_2	1978
Cerium oxide	CeO_2	2870
Thorium oxide	ThO_2	3320
Titanium oxide	TiO_2	2100
Zirconium oxide	ZrO_2	2970

High wet modulus fibers: see Modal fibers.

Hill reaction: see Photosynthesis.

Hill system: see Formula.

Hinsberg separation: separation of primary, secondary and tertiary amines by reaction with benzene sulfonylchloride. Primary and secondary amines form sulfonamides, but tertiary amines cannot react because they lack an H atom bound to the N atom:

$$R{-}NH_2 + Cl{-}CO_2{-}C_6H_5 \xrightarrow[-\,HCl]{} R{-}NH{-}SO_2{-}C_6H_5$$

$$R_2{-}NH + Cl{-}CO_2{-}C_6H_5 \xrightarrow[-\,HCl]{} R_2N{-}SO_2{-}C_6H_5$$

Further separation is possible because the presence of an acidic H atom in the sulfonamides of primary amines makes them soluble in basic solutions; the secondary sulfonamides are insoluble:

$$R{-}NH{-}SO_2{-}C_6H_5 + KOH \xrightarrow[-\,H_2O]{} R{-}N^-{-}SO_2{-}C_6H_5{-}K^+.$$

The amines are recovered from the sulfonamides by acid-catalysed hydrolysis. 4-Toluenesulfonamides are often made instead of the benzenesulfonamides because the former crystallize better.

Hipolan®: see Synthetic fibers.

Hippuric acid, ***N-benzoylglycine, N-benzoylaminoacetic acid***: C_6H_5-CO-NH-CH_2-COOH, a very acidic substitution product of glycine. H. crystallizes in colorless and odorless prisms, m.p. 190-193 °C. It is readily soluble in boiling water and alcohol, and is insoluble in petroleum ether and benzene. H. is formed in the kidneys of herbivores and excreted in the urine. This process binds the benzoic acid formed by degradation of aromatic amino acids and removes it from the organism. H. can be synthesized from benzoyl chloride and glycine.

Hirsutidine: see Anthocyans, table.

His: abb. for histidine.

Histamine, ***4-(β-aminoethyl)imidazole***: a biogenic amine which forms colorless, hygroscopic crystals with m.p. 86 °C. It is readily soluble in water and ethanol. H. is usually sold as the dihydrochloride, m.p. about 245 °C (dec.); it is formed in the body by decarboxylation of histidine. It is found in mast cells, skin and lung tissue, and reduces blood pressure, causes the smooth muscles of the bronchi, intestines and uterus to contract, and stimulates glandular secretion, especially of gastric juice. H. is a mediator in allergic reactions. Its effects are counteracted by antihistamines. In clinical chemistry, H. is used to study the gastric juice reaction.

$CH_2-CH_2-NH_2$ (imidazole ring, N–H)

Histidine, abb. ***His***: imidazolylalanine, a proteogenic amino acid which is particularly abundant in hemoglobin (structure and physical properties, see Amino acids, table 1). H. is part of the active centers of many enzymes, and is a significant buffer in the physiological pH range. H. is produced almost entirely by acid hydrolysis of hemoglobin (6 M HCl, 24 h). It is used in medicine to treat allergies and anemias.

Histones: a group of globular proteins associated with the DNA in eukaryotic chromosomes. They are highly basic, due to a large content of lysine (up to 27%) and arginine (up to 15%). The H. provide a skeleton on which DNA is tightly coiled; loosening of the DNA-H. association may be required for transcription.

Histrionicotoxins: piperidine alkaloids of the frog *Dendobates histrionicus*. The H. are very toxic, reacting specifically with acetylcholine receptors.

Hittorf's transfer number: see Transfer number.

HMO method: same as Hückel method (see).

HMPA: see Chemical sterilizers.

Ho: symbol for holmium.

Hock process, ***cumene-phenol process***: a process for production of phenol and acetone. It is now the preferred process for phenol production, in spite of the production of 0.62 t acetone for each ton of phenol, because it is very economical. In 1980, about 1.8 million tons phenol were produced, worldwide, by the H.

The H. is carried out in three steps. The first is alkylation of benzene with propene to form isopropylbenzene (cumene):

$$C_6H_6 + CH_2{=}CH{-}CH_3 \xrightarrow{AlCl_3} C_6H_5{-}CH(CH_3)_2.$$

The second step is oxidation of the cumene to cumene hydroperoxide: $C_6H_5\text{-}CH(CH_3)_2 + O_2 \rightarrow C_6H_5\text{-}C(CH_3)_2OOH$. The oxidation is carried out continuously in a bubble-plate reactor at approx. 130 °C and 0.2 to 0.5 MPa. Copper, cobalt or manganese salts are used as catalysts. Because there are undesirable secondary products, the reaction is allowed to go to only 15-30% of completion. In a subsequent distillation step, the cumene hydroperoxide is enriched to about 90% and the unreacted cumene is led back into the oxidation chamber. The hydroperoxide is stabilized with sodium carbonate, because otherwise it might decompose prematurely.

The third step is cleavage of the cumene hydroperoxide with acid. To facilitate removal of the high heat of reaction (ΔH = - 218 kJ/mol), the cleavage

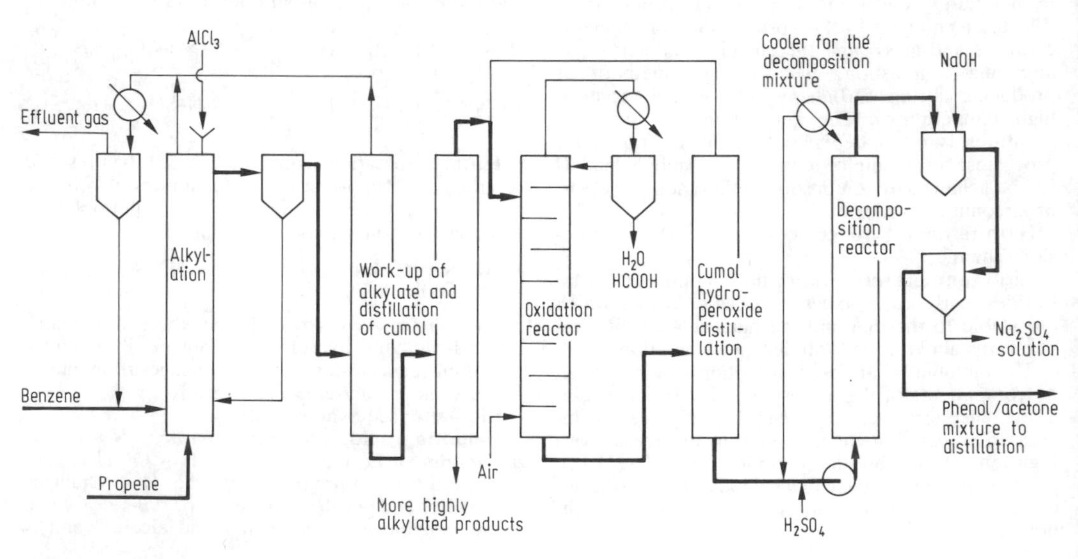

(Hock rearrangement) is carried out in a phenol-acetone mixture, such as is produced by the synthesis:

$$C_6H_5\text{–}C(CH_3)_2OOH \xrightarrow{H^+} C_6H_5\text{–}OH + CH_3\text{–}CO\text{–}CH_3.$$

After neutralization of the acid, the reaction mixture is distilled. The yield of phenol, in terms of the cumene reacted, is at least 90%. Byproducts of the process are acetophenone, α-methylstyrene, α-cumylphenol, and a *cumene-phenol tar*. The α-methylstyrene can be converted back to cumene by hydrogenation, or it can be purified and used in copolymerization with styrene. α-Cumylphenol can be thermally cleaved to α-methylstyrene and phenol. The tar is usually burned.

Toluene can also be converted to m- and p-cresol by the H. In another variant of the H., the benzene is alkylated with ethene rather than propene. In this case the end products are phenol and acetaldehyde.

Hofmann degradation of quaternary ammonium hydroxides: a method studied by A.W. Hofmann (1818-1892) for thermal cleavage of tetraalkylammonium hydroxides at 100 to 200 °C; in general, the products are tertiary amines, alkenes and water:

$$R_3\overset{+}{N}\text{–}CH_2\text{–}CH_3OH^- \longrightarrow R_3N + CH_2{=}CH_2 + H_2O$$

In the pyrolysis of tetramethylammonium hydroxide, the products are trimethylamine and methanol, because the structure does not permit formation of an alkene.

$$(CH_3)_4\overset{+}{N}OH^- \longrightarrow (CH_3)_3N + CH_3OH$$

The H. was very important in structure elucidation of natural products, in which linear or cyclic nitrogen bases are often obtained as structure fragments. The nitrogen-containing compound is first converted to the corresponding quaternary ammonium iodide by complete or exhaustive methylation. The ammonium hydroxide is then formed by reaction with silver hydroxide, and cleaved pyrolytically. Cyclic nitrogen compounds are usually converted to linear products; for example, penta-1,4-diene, trimethylamine and water are formed from *N*-methylpiperidine. The penta-1,4-diene primary product isomerizes under the reaction conditions to penta-1,3-diene or 1-methylbutadiene: $CH_2{=}CH\text{-}CH_2\text{-}CH{=}CH_2 \rightarrow CH_3\text{-}CH{=}CH\text{-}CH{=}CH_2$.

Hofmann degradation of acid amides: degradation of primary carboxylic acid amides to primary amines by reaction with alkali hypochlorites or bromides according to the following reaction mechanism:

$$R\text{–}CO\text{–}NH_2 \xrightarrow[-\,HBr]{+\,Br_2} R\text{–}CO\text{–}NH\text{–}Br \xrightarrow[-H_2O]{+\,KOH}$$

$$R\text{–}CO\text{–}N^-\text{–}Br\,K^+ \xrightarrow[-\,KBr]{} R\text{–}N{=}C{=}O \xrightarrow{+\,H_2O}$$

$$R\text{–}NH\text{–}CO\text{–}OH \rightarrow R\text{–}NH_2 + CO_2.$$

The primary product of reaction between bromine and the amide is an *N*-bromoamide, which is converted to the salt in the presence of base. It was originally thought that this salt decomposed, forming a nitrene, which rearranged to form an isocyanate by anionotropic migration of the alkyl or aryl group:

$$R\text{–}CO\text{–}\underline{\bar{N}}\text{–}BrK^+ \xrightarrow[-\,KBr]{} R\text{–}CO\text{–}\underline{\bar{N}} \xrightarrow{\approx R|} R\text{–}N{=}C{=}O.$$

More recent studies have indicated that the bromide cleavage is simultaneous with the nucleophilic R migration, so that no free nitrene is formed:

$$R\text{–}C(=O)\text{–}\bar{N}^-\text{–}Br \xrightarrow{-Br^-} R\text{–}\overset{-}{\underline{\bar{N}}}\text{–}\overset{+}{C}{=}O \leftrightarrow R\text{–}N{=}C{=}O$$

The resulting isocyanate is converted by water into the unstable carbamic acid, which rapidly decays into carbon dioxide and a primary amine. The H. thus permits conversion of carboxylic acid amides into amines with one C atom fewer.

Hofmann elimination product: see Elimination.

Hofmann-Martius rearrangement: rearrangement of *N*-alkylated aniline hydrochlorides or hydrobromides to form 4-alkylanilines by heating to 200-300 °C:

$$C_6H_5\text{–}\overset{+}{N}H_2R\ Cl^- \xrightarrow{250\,°C} 4\text{-}R\text{–}C_6H_4\text{–}\overset{+}{N}H_3\ Cl^-$$

If the 4-position of the starting compound is blocked by another substituent, the rearrangement

$$\text{N-methylpiperidine} \xrightarrow{+CH_3I} \text{N,N-dimethylpiperidinium } I^- \xrightarrow[-AgI]{+AgOH} \text{N,N-dimethylpiperidinium } OH^- \xrightarrow{-H_2O} CH_2{=}CH\text{–}CH_2\text{–}CH_2\text{–}CH_2\text{–}N(CH_3)_2 \xrightarrow{+CH_3I} CH_2{=}CH\text{–}(CH_2)_3\text{–}\overset{+}{N}(CH_3)_3\ I^-$$

$$\xrightarrow[-AgI]{+AgOH} CH_2{=}CH\text{–}(CH_2)_3\text{–}\overset{+}{N}(CH_3)_3\ OH^- \longrightarrow CH_2{=}CH\text{–}CH_2\text{–}CH{=}CH_2 + N(CH_3)_3 + H_2O$$

yields the 2-alkylaniline. The intermolecular mechanism was confirmed by trapping reactions.

Hofmann isonitrile reaction: same as Isonitrile reaction (see).

Hofmann mustard oil reaction: the production of isothiocyanates (mustard oils) by reaction of primary amines and carbon disulfide with mercury(II) chloride above room temperature:

$$R{-}NH_2 + CS_2 \rightarrow R{-}\overset{\oplus}{N}H_2{-}C(=S){-}\bar{\underline{S}}|^{\ominus} \xrightarrow{+RNH_2} R{-}NH{-}C(=S){-}\bar{\underline{S}}|^{\ominus}\ \overset{\oplus}{R}NH_3$$

$$\xrightarrow[-\ R\overset{\oplus}{N}H_3Cl^{\ominus}]{+\ HgCl_2} R{-}NH{-}C(=S){-}S{-}Hg{-}Cl \xrightarrow{-HgS,\ -HCl} R{-}N{=}C{=}S$$

Isothiocyanate
Mustard oil

Hollow cathode lamp: see Atomic absorption spectroscopy.

Hollow fiber: see Ion chromatography.

Hollow form: see Dendrite.

Holmium, symbol ***Ho***: a member of the Lanthanoid group (see) of the periodic system; a rare-earth metal with only one natural isotope; a heavy metal; *Z* 67, atomic mass 164.9304, valence III, density 8.78, m.p. 1470 °C, b.p. 2720 °C, standard electrode potential (Ho/Ho^{3+}) - 2.319 V.

H. is a silvery white, ductile metal with hexagonal close-packed crystals. It makes up $1.1 \cdot 10^{-4}$% of the earth's crust, and is always found in association with the other rare-earth metals. The main source of H. is gadolinite. It is used in the form of Cerium mixed metal (see). Its other properties, analysis, production and history are discussed in the entry on Lanthanoids (see).

Holmium compounds: most of these compounds are brownish yellow in color and contain the metal in the +3 oxidation state. Some important examples are: ***holmium(III) chloride***, $HoCl_3$, M_r 271.29, m.p. 718 °C; ***holmium(III) fluoride***, HoF_3, M_r 221.93, m.p. 1143 °C; ***holmium(III) oxide***, Ho_2O_3, M_r 377.86; ***holmium(III) oxalate***, $Ho_2(C_2O_4)_3 \cdot 10H_2O$, M_r 774.10.

Holocellulose: all the polysaccharide plant fiber which is insoluble in water (cellulose and insoluble hemicelluloses). Broad-leaf woods contain 72-79% H., and conifer woods contain 60 to 73%. H. can be determined by various methods of treating the chipped, lignified plant material. The H. content is an important parameter for paper and cardboard production, but it must be remembered that none of the methods of determining H. can entirely separate the lignin and polysaccharide components of the wood.

Holoenzyme: see Enzyme.

Holoside: see Glycoside.

Homatropin: see Atropin.

Homeopolar bond: see chemical bond.

Homeotypism: see Isotypism.

HOMO: see Woodward-Hoffmann rules.

Homoaromaticity: a concept developed by Winstein (1959) which explains the stability of cyclic polyenes in which the ring conjugation is interrupted at one or more sites by sp^3 centers. In such systems, the p_{π}-orbitals flanking the sp^3 center overlap to form a $(4n+2)$-π-electron system. The ion $C_8K_9^+$ can be taken as an example of a 6π-homoaromatic system; the ion is also called the homotropylium ion (Fig. 1).

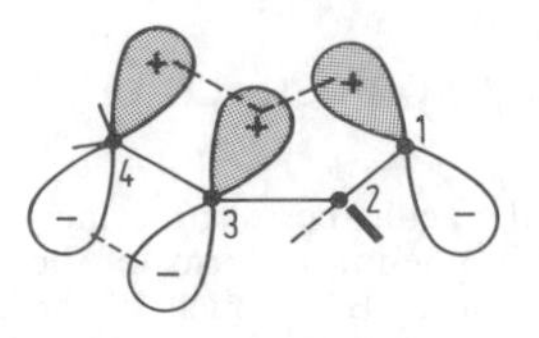

Fig. 1. Homotropylium ion

Overlapping of the p_{π} orbitals at centers 1 and 7 (Fig. 1) leads to a cyclic delocalized π system with six electrons. The formation of this type of structure agrees with the data from ^{1}H-NMR spectroscopy. The ring current shields the endo-proton (H_b) while the signal of the exo proton (H_a) appears at a lower field strength. The non-classical electron delocalization by overlapping of π-orbitals of C atoms which are not linked by a σ-bond is called *homoconjugation*. It makes possible the maintenance of conjugation past an sp^3 center. As the interaction of a carbenium ion with a β-olefinic group shows, overlapping between C atoms 1 and 3 (Fig. 2) is limited by geometric constraints to just one of the lobes of the p-orbital on each atom. The concept of H. can also be applied to anions and neutral systems.

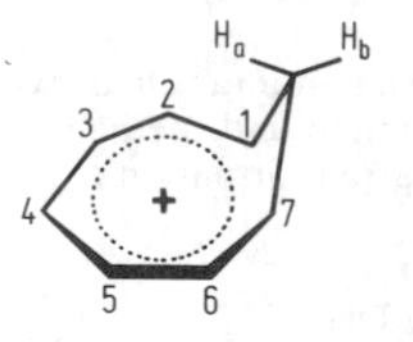

Fig. 2. Homoconjugation

Homocol: see Isocyanins.

Homoconjugation: see Homoaromaticity.

Homodesmic: see Crystal chemistry.

Homoepitaxy: see Epitaxy.

Homogeneous: in physical chemistry, the adjective is applied to a material system consisting of a single phase (solid, liquid or gas). The antonym is Heterogeneous (see); see also Isotropism.

Homogenization: the even distribution of the individual components of a material system over its entire volume, e.g. by mixing, stirring or kneading. Pulverization effects can also occur in the process. The desired result of H. is that all partial volumes have the same composition. H. is necessary for taking representative samples for purposes of analysis.

Homogeneity range: see Berthollides.

Homolysis: dissociation of a bond in such a manner that the bonding electron pair is dissociated, forming atoms and/or radicals. H. is an elementary process which passes through a transition state only when the radicals formed by it are stabilized by conjugation. It occurs mainly in the gas phase, and is initiated by thermolysis, photolysis, radiolysis, electrolysis or electron transfer in redox reactions. The energy supplied must be equal to or greater than the bond dissociation energy. Nonpolar or only slightly polar bonds undergo homolytic cleavage preferentially. H. is very important in certain technological processes, such as cracking, radical polymerizations and substitution reactions.

Homotopic groups: see Topic groups.

Homotropilidene, ***bicyclo[5,1,0]octa-2,5-diene***: a hydrocarbon with fluctuating bonds. H. was the first organic molecule discovered which can only be described by the average of two identical structures. It undergoes a constant, rapid, reversible valence isomerization (topomerization). From nuclear resonance studies, it was determined that at - 50 °C, H. undergoes a Cope rearrangement once a second; at 180 °C, it undergoes this valence isomerization about 1000 times per second. Bullvalene (see) is an extreme example of a molecule with fluctuating bonds.

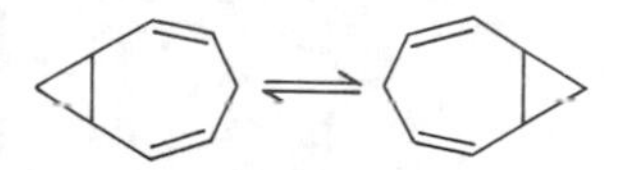

Hopane: see Triterpene.

Hordein: see Prolamines.

Hormones: substances which mediate communication between the cells of multicellular organisms. The H. of plants, called Phytohormones (see) are treated in a separate entry. H. may be classified according to their chemical structures as peptides or proteins (see Peptide hormones), amino acids or biogenic amines and their derivatives (see Neurotransmitters, Thyroid hormones), steroids (see Sex hormones, Adrenal corticosteroids, Ecdysteroids), isoprenoids (see Juvenile hormones) or derivatives of unsaturated fatty acids (see Eicosanoids).

H. are synthesized in nerve cells (neurohormones, neurotransmitters), endocrine glands (glandular H.) and in delocalized, non-specialized cells (tissue H.). The ***neurohormones*** include the hypothalamus hormones. In invertebrates, neurohormones probably play an even greater role than they do in vertebrates. It appears that in insects, neurosecretory peptides direct the secretion of H. by the prothoracic glands (see Ecdysteroids) and the corpora allata (see Juvenile hormones). The ***glandular H.*** of the vertebrates are summarized in the table. The known glandular H. of the invertebrates are juvenile H. and ecdysteroids. The ***aglandular H.*** of vertebrates include the gastrointestinal Peptides (see), Plasmakinins (see), Substance P (see), the Angiotensins (see), the Eicosanoids (see) and various differentiation, stimulating and growth factors, many of which have only been characterized functionally. Some of these peptide H. (e.g. gastrointestinal peptides and substance P) have also been detected in the central nervous system. The ***endogenous ligands*** of certain drug receptors, such as the opiate receptors (see Endorphins) and benzodiazepine receptors must also be classified as H. It is suspected that the endogenous ligands of the benzodiazepine receptors are β-carboline 3-carboxylic acid ethyl ester or the purines inosine and hypoxanthine.

Mechanisms. There is a hierarchy of H. The neurohormones produced by the hypothalamus inhibit (see Statins) or promote (see Liberins) the release of H. from the anterior hypophysis (pituitary). These H. regulate the release of H. from the peripheral glands. The peripheral H. regulate the production of the hypophysial H. in a feedback mechanism; nervous feedback also occurs. The H. are transported from the sites of their secretion to the sites of their action in the body fluids. Those H. which are released very close to their targets, so that diffusion provides adequate transport, are called ***diffusion activators*** or ***local H.***. These include neurotransmitters and tissue H. If a substance released by one cell acts on its immediate neighbor, one speaks of a *paracrine* effect; in *endocrine* mechanisms, the body fluids first transport the H. to the site of its action.

Biochemically, there are three stages in information transfer by H.: 1) biosynthesis or release (if the H. is stored); 2) specific interaction with receptors on or in the target cell, which leads to a specific biological response; and 3) removal of the H. from its site of action by enzymatic degradation (e.g. by acetylcholinesterase or monoamine oxidase) or back transport.

The ***H. receptors*** in the target cell can be localized on the cell membrane or in the interior of the cell. Most H. react with membrane-bound receptors, that is, most of them do not enter the cell. Only the steroid and thyroid H. have intracellular receptors. Some of these receptors have been isolated by affinity chromatography and characterized. Binding studies with agonists and antagonists have shown that most receptors have multiple forms which differ in their localization, structure and mechanisms of actions, even though their specificity for the H. is the same.

There are various mechanisms of H. action. After a steroid H. has bound to a receptor in the cytoplasm, the H.-receptor complex is translocalized to the nucleus, where a derepression leads to increased protein synthesis. Receptors bound to the cell membrane are coupled to an effector system. The best known of these is the *adenylate cyclase system*. Binding of the H. (first messenger) to the receptor activates an adenylate cyclase, which forms the Cyclic nucleotide (see) cAMP from AMP. cAMP then affects the cell as a second messenger, mainly by activating a variety of protein kinases. These kinases activate (or repress) specific target enzymes by phosphorylating them; the result is a change in the metabolic pattern of the cell. β-Adrenomimetics, glucagon and histamine (H_2 receptor), among others, have been shown to act via cAMP. Other receptors (including the muscarinic acetylcholine receptor and the α-adrenoreceptor) initiate cleavage of inositol phosphates and elevation of the intracellular concentrations of calcium ions and cGMP. Still other receptors, including the nicotinic acetylcholine receptor, are coupled to ion channels. The differences in mechanism are reflected in the differences in response times between binding of the H. to the target cell and appearance of the effect. The receptors coupled to ion channels react within milliseconds. The receptors which release a second messenger have reaction times of seconds to minutes, while the reaction with intracellular receptors may not be apparent for hours after arrival of the H., and may persist for days.

Production. Until recently, the peptide H. had to be isolated from the glands of animals by extraction, fractional precipitation and chromatography or electrophoresis. The most important of the human H.

used therapeutically (insulin, somatotropin) are now produced by genetically engineered microorganisms (see Gene technology). Oligopeptide H. and their analogs are synthesized. Steroid H. are made by partial or complete chemical synthesis.

Determination and applications. H. concentrations are now usually determined by immunological methods (radioimmunoassay, enzyme immunoassay). The H. are applied therapeutically in cases of deficiency and in veterinary medicine. Mixtures of estrogens and gestagens are used as contraceptives. Glucocorticoids are used to reduce inflammation.

Historical. The term H. was coined in 1905 by Starling, and applied to substances which could stimulate the activity of organs via the blood. The first H. isolated was adrenalin (Takamine, Aldrich, 1901); it was synthesized in 1903 (Stolz). In 1921, Banting and Best isolated a crude insulin, and insulin was the first protein to be crystallized (Apel, 1926). Its structure was elucidated in 1955 (Sanger). In 1926, the structure, and a little later, the synthesis of thyroxin were reported (Harrington). In 1929, Doisy and Butenandt independently isolated estrone from the urine of pregnant women; this was the first steroid H. isolated. In the 1930s and 1940s, the most important corticosteroids were isolated and their structures elucidated (Kendall, Reichstein and others). The first peptide H. synthesized was oxytocin (Du Vigneaud, 1954). In 1962, the first hypothalamus neurohormone was isolated (Guillemin). The first invertebrate H. was isolated in 1954: ecdysone from silkworms (Karlson, Butenandt).

Vertebrate glandular hormones

Endocrine gland	Hormone	Structure	Effects
Anterior hypophysis	Corticotropin	Polypeptide (39 AA)	Stimulates production and secretion of corticosteroids
	Follitropin	Glycoprotein (2 SU)	Stimulates growth of follicles or spermatogenesis
	Lutropin	Glycoprotein (2 SU)	Stimulates production of estrogens or androgens
	Prolactin	Protein	Stimulates production of gestagens and milk
	Thyrotropin	Glycoprotein (2 SU)	Stimulates formation and secretion of thyroid hormones
	Somatotropin	Protein (190 AA)	Growth hormone, promotes bone growth and protein synthesis
	Lipotropin	Polypeptide	Stimulates fatty acid synthesis
Mid-hypophysis	Melanotropin	Polypeptide	Regulates pigmentation in skin of animals
Posterior hypophysis	Oxytocin	Nonapeptide	Causes uterine smooth muscles to contract
	Vasopressin	Nonapeptide	Causes smooth muscles of blood vessels to contract; antidiuretic
Epiphysis (pineal body)	Melatonin	N-Acetylmethoxy-tryptamine	Regulates pigmentation in animals; role in photoperiodic processes
Pancreas	Insulin	Polypeptide of 2 chains (21 + 30 AA)	Reduces blood sugar level
	Glucagon	Polypeptide (29 AA)	Insulin antagonist, mobilizes glucose formation from glycogen
Thyroid gland	Thyroxin Triiodo-thyronin	Iodine-containing nonproteogenic AA	Accelerate metabolic processes
	Calcitonin	Polypeptide (32 AA)	Reduces blood plasma Ca^{2+} promotes Ca^{2+} incorporation in bones
Parathyroids (Epithelial bodies)	Parathyrin (Parathormone)	Polypeptide (84 AA)	Mobilizes Ca^{2+} and phosphate from bones
Placenta	Choriogonadotropin	Glycoprotein	Stimulates gonads; detection in urine diagnostic of pregnancy
	Choriomammotropin	Protein and prolactin	About the same as somatotropin
Corpus luteum	Relaxin	Polypeptide	Widens birth canal
	Gestagens, progesterone	C_{21} steroid	Shifts endometrium from proliferative to secretory phase
Follicles	Estrogens	C_{18} steroids	Cause development of female sex organs
Testes	Androgens	C_{19} steroids	Promote growth of male sex organs; stimulate protein synthesis (extragenital anabolic effect)
Adrenal cortex	Corticoids	C_{21} steroids	Regulate carbohydrate and protein metabolism (glucocorticoids) or electrolyte balance (mineralocorticoids)
Adrenal medulla	Adrenalin derivative	Phenylethylamine level	Increases blood sugar

Abb.: AA = amino acid(s); SU = subunits

Hornet toxin: see Bee toxin.

Hot gas recycle process: see Fischer-Tropsch synthesis.

Hot potash process: a process for removing carbon dioxide from Synthesis gas (see).

Houben-Hoesch reaction: synthesis for hydroxyaryl and alkoxyaryl ketones by reaction of phenols and phenol ethers with nitriles in the presence of hydrogen chloride and zinc chloride:

$$Ar{-}H + R{-}C{\equiv}N \xrightarrow{(HCl,\, ZnCl_2)} Ar{-}C({=}NH){-}R \xrightarrow{+ H_2O(H^+)} Ar{-}CO{-}R + NH_3$$

In the first reaction step, the nitrile and hydrogen chloride form an electrophilic imide chloride complex in the presence of a Lewis acid:

$$R{-}C{\equiv}N + HCl + AlCl_3 \rightarrow R{-}\overset{+}{C}{=}NH \cdot AlCl_4^-.$$

This complex adds electrophilically to the activated aromatic compound to form the imino compound which can be hydrolysed to the ketone in the presence of water. The H. is mechanistically closely related to the Gatterman reaction (see). When α,β-unsaturated or various β-substituted nitriles are used, the reaction does not proceed in the sense of the H.

Houdresid process: see Cracking.

Houdriforming process: see Reforming.

Houdryflow process: see Cracking.

Household cleansers: products used mainly for manual cleaning of hard surfaces of glass, organic polymers, enamel, stone, ceramic, metal or wood. The main components of typical H. are anionic or nonionic detergents, builders, either alkalies or mineral acids, abrasives such as quartz, alumina, marble or plaster powder, bleaches and disinfectants. Specialized cleansers are available for certain types of dirt and surfaces. 1) ***All-purpose cleansers*** are intended for all washable surfaces, often including floors. These are usually liquid, neutral detergent products containing disinfectants and bleaches. 2) ***Window cleaners*** are dilute aqueous solutions, usually of isopropanol, with <1% detergent and often a small amount of ammonia; they are used to clean windows and other glass surfaces. 3) ***Scouring products*** are solid or liquid surfactants containing abrasives with grain sizes between 20 and 50 μm. They often also contain disinfectants and bleaches, and are used to clean very dirty, mechanically durable surfaces. 4) ***Sanitary cleansers*** are acidic products containing sodium hydrogensulfate or phosphoric acid used to clean toilets, wash basins and tiles. They are not suitable for enamel. Disinfection is better done with alkaline solutions containing hypochlorite. 5) ***Drain cleansers*** are essentially sodium hydroxide, and usually do not contain surfactants. 6) ***Metal cleaners*** consist of detergents and abrasives with grain size < 20 μm, along with corrosion-inhibiting components. 7) ***Silver polish*** contains additional sulfur compounds, such as thiourea. 8) ***Grill*** and ***oven cleaners*** are based on strong alkalies, detergents and water-miscible solvents, especially glycol ethers. 9) ***Plastic cleansers*** are detergent-containing products adjusted for the low-energy surfaces of organic polymers; they may also contain bleaches.

HPLC: abb. for high performance liquid chromatography.

HPTLC: see Thin-layer chromatography.

HRA: see Rockwell hardness.

HRB: see Rockwell hardness.

HRC: see Rockwell hardness.

hR_F value: see Paper chromatography.

HSAB concept: see Acid-base concepts.

HTP process: abb. for high-temperature pyrolysis process, a process used to cleave light gasoline or crude oil into ethyne (acetylene) and ethene (ethylene) by rapidly heating it to 2700 °C, then rapidly cooling it to 1300 °C (see Pyrolysis).

Huang-Minion reduction: a variant of the Wolff-Kishner reduction of aldehydes and ketones in which the required hydrazone is not isolated, but is immediately cleaved into the corresponding hydrocarbon and nitrogen. This reaction is done in a high-boiling solvent around 200 °C using potassium hydroxide.

Hückel benzene: same as Benzvalene (see).

Hückel method, ***HMO method*** (for Hückel molecular orbital): an approximation method developed by E. Hückel on the basis of molecular orbital theory; it is used for treatment of π-electron systems. The H. starts from the σ-π separation of the molecular orbitals into the σ-molecular skeleton and the π-electron system for separate treatment. The π-system is described by multicenter molecular orbitals Φ, which are approximated according to the MO-LCAO method by a linear combination of the atomic orbitals χ with π symmetry: $\Phi = \sum_{i=1}^{n} c_i\chi_i$. The Hückel atomic orbitals are assumed to be normalized, $\int \chi_i\chi_i \, d\tau = 1$, and mutually orthogonal, $\int \chi_i\chi_j \, d\tau = 0$ $(i \neq j)$. The Hückel operator H_H is a one-electron Hamiltonion operator which is not further specified. The interaction integral $\int X_i \,|\, \hat{H} \,|\, X_i \, d\tau = H_i = \alpha_i$ (*Coulomb integral*) and $\int X_i \,|\, \hat{H} \,|\, X_j \, d\tau = H_{ij} = \beta_i$ (*resonance integral*) are the parameters of the process and are fitted to suitable experimental data. If i and j are directly linked centers, $H_j = \beta_{ij}$; otherwise β_{ij} is set equal to zero. The secular determinants (Hückel determinants) then simplify and can be given directly from the linkage scheme (topology) of the atoms. The Hückel eigenvalues ε_j and the n sets of Hückel coefficients (eigenvectors) $c_{ij} \ldots c_{nj}$ $(j = 1, \ldots n)$ which determine the n molecular orbitals are obtained from the solution of the secular equation system. The filling of the energy levels ε_j by successive electrons follows Hund's rule and the Pauli principle. The following important parameters are derived from the Hückel coefficients: The sum $\Sigma(j = 1, n) b_j c_{rj}^2 = q_r$ is called the *charge order*, and gives the π-electron charge at atom r. b_j is the occupation number of the jth molecular orbital. The parameter $p_{rs} = \Sigma(j = 1, n) b_j c_{rj} c_{sj}$ is called the *bond order* and is a relative measure of the strength of the π-bond between the atoms r and s. The residual bonding ability of an atom r in the π-system is defined by the *free valence* $F_r = \sqrt{3} - \Sigma(s) p_{rs}$. The summation is over the bond orders of all π-bonds ending at r. Centers with high F_r values are preferred sites of attack for radical reactions. Because of the simplifications introduced into it, the H. is more easily comprehended than other molecular orbital theories. Because of this, it has significantly influenced the thought and concept formation of modern chemistry.

Hückel rule: see Aromaticity.
Hückel topology: see Aromaticity.
Hudson's rules: see Glycosides.

Hughes-Ingold rule: a relationship between the polarity of reacting species, the polarity of the solvent and the free solvation enthalpy. The larger the charge and the less distributed it is over the molecule, the higher the free solvation enthalpy of ions and polar species will be. In reactions in which charges arise or are concentrated as the reactants enter the transition state, increasing the polarity of the solvent leads to greater solvation of the transition state relative to the reactants, thus reducing the activation enthalpy and increasing the rate of the reaction. For example, this effect occurs in S_N2 reactions between neutral molecules, between neutral molecules and ions, or between ions when both are positive or both are negative. Conversely, reactions in which charges are neutralized or more widely distributed over the reacting species as they enter the transition state are slowed down as the solvent polarity is increased.

Hume-Rothery phases: see Intermetallic compounds.
Huminite: see Macerals.
Humus coal: see Coal.
Hund's rule: see Multiplet structure.

Hunsdiecker reaction: method of converting the silver salts of aliphatic carboxylic acids into alkyl halides with one carbon atom fewer. Chlorine, bromine or iodine is used in carbon tetrachloride or trichloroethylene:

$$\text{R–COOAg} \xrightarrow[-\text{AgBr}]{+\text{Br}_2} \text{R–COOBr} \longrightarrow \text{R–Br} + \text{CO}_2$$

In principle, the H. can also be applied to aromatic compounds.

HV: abb for Vickers hardness.
HWM fibers: see Modal fibers.

Hyaluronic acid: an acidic mucopolysaccharide with the structure [GlcNAc-β(1→4)-GlcUA-β(1→3)]$_n$D; it is found in the form of proteoglycans. H. is found in cartilage, joint fluid, the umbilical cord and in the vitreous humor of the eye. It can be obtained in pure form from the vitreous humor or umbilical cord. H. is cleaved by hyaluronidases, enzymes found in bacteria, the saliva of leeches and spermatozoa.

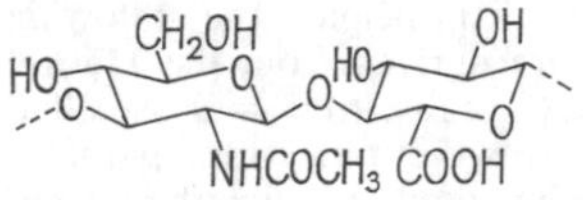

Hybridization: a linear combination (addition or subtraction) of atomic orbitals to form *hybrid orbitals* which have a high degree of directional orientation and therefore readily form localized bonds. The concept of H. was developed by L. Pauling and is especially useful in the description of bonding in carbon compounds and the complex compounds of the transition metals. The carbon atom, for example, is known to be tetravalent. In its ground state, it has the configuration $2s^2 2p^2$, in which there are only two unpaired electrons. In order to form four bonds, it must enter an excited state in which one of the 2s electrons is elevated into the empty 2p orbital; the energy required for the transition is called the *promotion energy*. The strength of a bond depends on the degree to which the orbitals of the bonding electrons from the two atoms overlap. Since s-orbitals are spherical, they are not able to overlap orbitals from other atoms to as great a degree as p-, d- and f-orbitals, which are more concentrated in a single direction. It might therefore be expected that the carbon atom would form three relatively strong bonds using the 3 2p electrons, and one weaker one with the 2s electron. However, it is found experimentally that all four bonds have the same strength. Four energetically equivalent sp^3 hybrid orbitals, each of which is occupied by one electron, can be obtained by linear combination of the 2s orbital with the three 2p orbitals. This state, the *valence state*, is denoted by $(sp^3)^4$. The H. itself is not a physical process, but is only a transformation of atomic orbitals which does not change the total energy (Fig. 1).

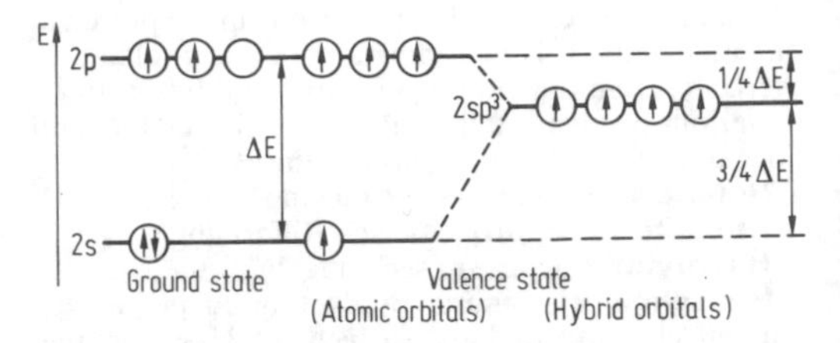

Fig. 1. sp^3-Hybridization of the carbon atom

The sp^3 hybrid orbitals point to the corners of a tetrahedron; the angle between any two of them is 109°28'. The promotion energy required for the C atom to enter the valence state is more than compensated by the reduction in total energy when it forms four bonds rather than the two which would be possible from the ground state.

In similar fashion, linear combination of an s- and a p-orbital on a single atom yields two sp-hybrid orbitals, which lie at an angle of 180° from each other. The linear structure of the BeH_2 molecule, for example, can be explained by assuming the occurrence of sp-hybridization. If an s-orbital combines with two p-orbitals of the same atom, three sp^2-orbitals are formed which point toward the vertices of an equilateral triangle; the angle between any two of them is 120°. sp^2 hybridization is therefore invoked to explain the bonding of trigonal-planar molecules such as BF_3. The spatial orientations of sp-, sp^2- and sp^3-hybrid orbitals are shown in Fig. 2.

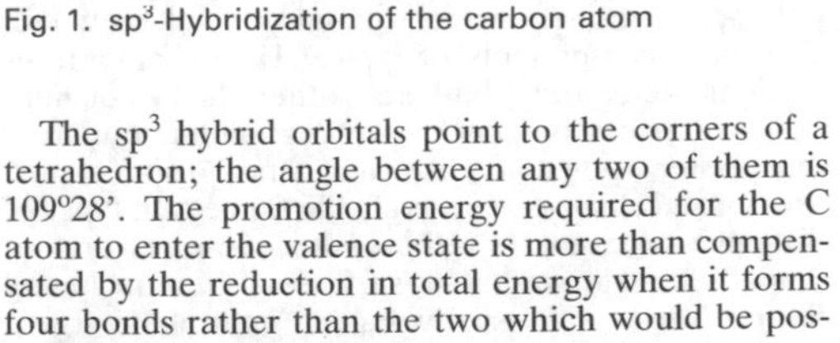

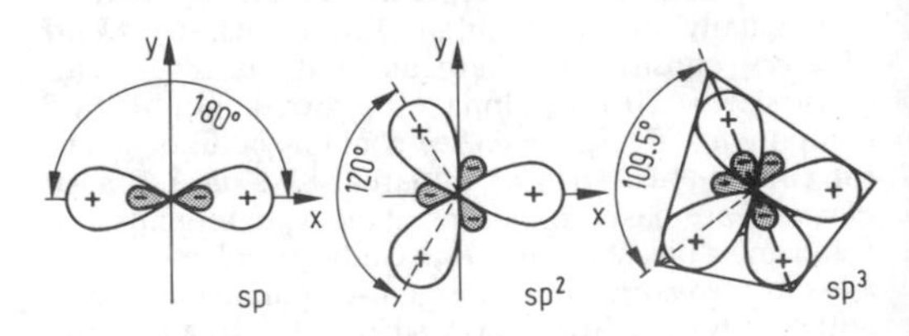

Fig. 2. sp-, sp^2- and sp^3-hybrid orbitals

By the principle of maximum overlapping (see Molecular orbital theory), the orientation of a hybrid orbital indicates its preferred bonding direction. The structures of molecules with multiple bonds, e.g. ethyne and ethene, can also be explained using sp- and sp^2 hybrid orbitals. H. is not limited to s- and p-orbitals; in elements in the third and higher periods, d-orbitals can be included. In the main group elements, s-, p- and d-orbitals from the same shell can combine, while in the transition metals, hybridization occurs between s- and p- orbitals and the d-orbitals of the next lower (inner) shell. The following geometric arrangements are obtained for hybrid orbitals involving d-orbitals: sp^2d, square planar; sp^3d, trigonal planar or square pyramidal; sp^3d^2, octahedral.

Hybrid orbitals: see Hybridization.

Hydantion, ***imidazolidine-2,4-dione, 2,4-dioxotetrahydroimidazole***: the cyclic ureide of glycine. H. forms colorless needles; m.p. 220 °C. It is readily soluble in hot water and alcohol, and very nearly insoluble in ether. It is found in the juice of sugar beet sprouts and in molasses. It was first obtained by A. von Baeyer from allantoin. Special syntheses for H. have been worked out, for example the Bucherer reaction. It can be made by addition of potassium cyanate to glycine followed by heating the product with hydrochloric acid. H. is the parent compound of various drugs used against epilepsy, such as ***phenytoin***.

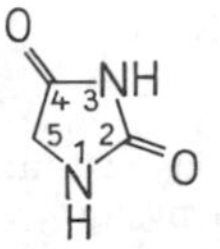

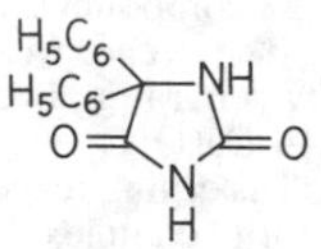

Phenytoin

Hydnocarpic acid: see Chaulmoogric acid.

Hydramine cleavage: see Ephedrin.

Hydrargyrum: see Mercury.

Hydrate: an addition or complex compound in which water molecules are bound through electrostatic dipole-dipole or dipole-ion interaction (see Hydration).

The salts of transition metals are very prone to formation of H. The water molecules are most often bound directly to the metal ions, but less frequently they are bound to anions, coordinated or occupy other positions in the crystal lattice.

Hydration: 1) interaction between water molecules and solutes such as ions, molecules and colloids. H. is a special case of solvation; its products are called Hydrates (see). It is important for the stability of hydrophilic colloids and for the water solubility of salts (see Solution). The H. of ions is due to an electrostatic ion-dipole interaction. The strength of this interaction is directly proportional to the charge on the ion, and inversely proportional to its radius. Cations are generally more highly hydrated than anions; the formation of hydrates in crystals is usually limited to cations. The number of water molecules bound to an ion is called the ***hydration number***. For cations, hydration numbers of 4, 6 and 8 are observed. The water solubility of compounds is generally determined by the relationship between lattice and hydration energies. Ionic compounds are water soluble if the sum of hydration energies of cations and anions is greater than the lattice energy. H. of colloidal particles stabilizes them; hydrophilic colloids stabilized by H. can in some cases be converted to gels containing large amounts of water.

2) addition of water to an unsaturated organic compound. There is no sharp line between definitions 1 and 2, nor between definition 2 and Hydrolysis (see). Acid-catalysed H. of alkenes and cycloalkanes produces saturated alcohols, for example, ethanol from ethene, *tert.*-butanol from isobutene. The acid-catalysed H. of alkynes, cycloalkynes and allenes yields as primary products unsaturated alcohols; these rearrange into the tautomeric carbonyl compounds. For example, $H\text{-}C{\equiv}C\text{-}H + H_2O \rightarrow CH_2{=}CH\text{-}OH \rightleftharpoons CH_3\text{-}CHO$; $H_2C{=}C{=}CH_2 + H_2O \rightarrow H_2C{=}C(CH_3)\text{-}OH \rightleftharpoons (CH_3)_2CO$. The H. of carbonyl compounds produces hydrates (1,1-diols) which are unstable. Various azanaphthalenes with several nitrogen atoms and their cations can be hydrated.

Hydraulic liquids: mineral oil fractions suitable for use in hydraulic systems. In systems where fire could occur, oil/water emulsions or water/oil emulsions or solutions of organic compounds (e.g. glycols) in water are used. Phosphoric acids, silicones, fluorocarbons and other inert liquids can be used as H. for special applications.

Hydrazides 1) inorganic compounds with the general formula $M^INH\text{-}NH_2$, in which an H atom of hydrazine is replaced by a monovalent metal atom. H. are made by reaction of metals (alkali metals) or their amides or hydrides with hydrazine. H. are sensitive to air and moisture, and tend to explode.

2) ***Acyl hydrazines (carboxylic acid hydrazides)*** are carboxylic acid derivatives with the general formula $R\text{-}CO\text{-}NH\text{-}NH_2$. The compounds have properties similar to those of the carboxamides. These basic, reducing agents (reduction of Tollen's reagent and Fehling's solution) crystallize readily. They form salts with strong acids, such as HCl. H. are obtained by reaction of hydrazine with carboxylic acid esters, anhydrides or halides, or by heating hydrazine salts of carboxylic acids. The reactions of H. with aldehydes and ketones yields insoluble *hydrazones*. *Girard reagents* are ionic H. which form water-soluble hydrazones, e.g. with hormones. H. form Acyl azides (see) with nitrous acid. Isonicotinic hydrazide (INH) is used as a tuberculostatic. Phthalic hydrazide is formed by hydrazinolysis of *N*-alkylphthalimides (see Gabriel synthesis). 3-Aminophthalic hydrazide is luminol.

Hydrazide imides: see Amidrazones.

Hydrazine: $H_2N\text{-}NH_2$, a colorless, poisonous liquid which fumes in moist air; M_r 32.05, density 1.011, m.p. 1.4 °C, b.p. 113.5 °C. The stable conformation of H. is the *gauch*-form, which minimizes the interaction between the free electron pairs (Fig.).

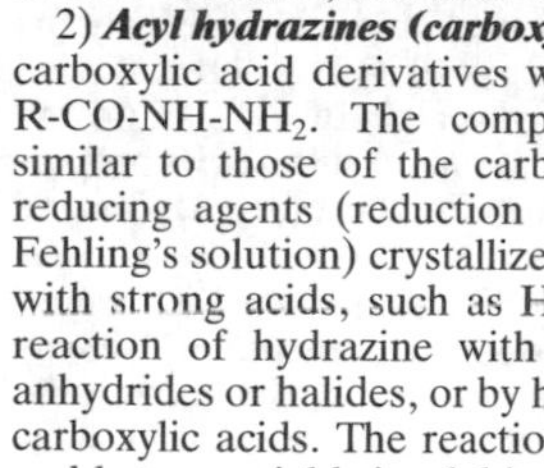

anti gauche

Newman projections of hydrazine

At high temperatures, H. decomposes, often explosively, into ammonia and nitrogen: $3 N_2H_4 \rightarrow 4 NH_3 + N_2$. It burns in air, releasing a large amount of heat: $N_2H_4 + O_2 \rightarrow N_2 + 2 H_2O$. It is miscible with water in any proportions, forming a Hydrazine hydrate (see) which can be distilled without decomposing: $N_2H_4 \cdot H_2O$. H. gives a basic reaction; its base constants are $K_{B1} = 8.5 \cdot 10^{-7}$ and $K_{B2} = 8.9 \cdot 10^{-16}$. It is thus not as strong a base as ammonia. With strong acids, it forms two series of ***hydrazinium salts*** with the general composition $[H_2N\text{-}NH_3]^+X^-$ and $[H_3N\text{-}NH_3]^{2+} \cdot 2X^-$. H. is a much stronger nucleophile than ammonia. Aqueous solutions of H. are strongly reducing; for example, they can precipitate silver from silver salt solutions. The reaction of H. with nitrous acid yields hydrazoic acid: $N_2H_4 + HNO_2 \rightarrow HN_3 + 2 H_2O$. H. is synthesized by the *Raschig process*, in which ammonia is oxidized with sodium hypochlorite in the presence of gelatin. Chloramine is formed as an intermediate: $NH_3 + NaOCl \rightarrow NaOH + Cl\text{-}NH_2$; $NH_3 + Cl\text{-}NH_2 + NaOH \rightarrow H_2N\text{-}NH_2 + NaCl + H_2O$. The gelatin has two functions: to catalyse the formation of H. by the equations shown here, and to bind metal ions which would otherwise catalyse the side reaction of chloramine with H. (This yields ammonium chloride and nitrogen: $2 NH_2Cl + H_2N\text{-}NH_2 \rightarrow 2 NH_4Cl + N_2$.) H. is used together with various oxidizing agents, such as nitric acid or dinitrogen tetroxide, as rocket fuel. It is a raw material for organic syntheses, especially of pigments, drugs and pesticides. It is also used for reduction of carbonyl compounds (see Wolff-Kishner reduction).

Hydrazine hydrate: $H_2N\text{-}NH_2 \cdot H_2O$, colorless, highly refractive hygroscopic liquid with a fishy odor; M_r 50.07, density 1.03, m.p. - 40 °C, b.p. 118.5 °C (9.8 kPa). H. is obtained by the reaction of sodium hydroxide solution with hydrazinium sulfate. Its properties and uses are discussed in the entry Hydrazine (see).

Hydrazine yellow: see Flavazine, see Tartrazine.

Hydrazinium chlorides. ***Hydrazinium monochloride***: $[H_2N\text{-}NH_3]Cl$, colorless, water-soluble needles; M_r 68.51, m.p. 89 °C. It is formed by neutralization of one mole hydrazine with one mole hydrochloric acid. ***Hydrazinium dichloride***, $[H_3N\text{-}NH_3]Cl_2$, colorless, water-soluble crystals; M_r 104.97, density 1.42, is obtained by the reaction of two moles hydrochloric acid with one mole hydrazine. It melts at 198 °C, giving off one equivalent of hydrochloric acid.

Hydrazinium salts: see Hydrazine.

Hydrazinium sulfates. ***Dihydrazinium sulfate***, $[H_2N\text{-}NH_3]_2SO_4$, colorless, hygroscopic crystals, M_r 162.18, m.p. 85 °C, is obtained by reaction of hydrazine with sulfuric acid at a molar ratio of 2:1. It is a common commercial form of hydrazine. ***Hydrazinium sulfate***, $[H_3N\text{-}NH_3]SO_4$, M_r 130.13, density 1.37, m.p. 254 °C, is formed as rhombic crystals which are relatively insoluble in cold water by the reaction of equimolar amounts of hydrazine and sulfuric acid. It is also produced directly by the Raschig synthesis of Hydrazine (see). It is used to make hydrazine hydrate and azides, as a reducing agent and in organic synthesis.

Hydrazinolysis: a reaction in which the substituent X is displaced from a compound R-X by hydrazine, $H_2N\text{-}NH_2$: $R\text{-}Cl + H_2N\text{-}NH_2 \rightarrow R\text{-}NH\text{-}NH_2 + HCl$. The course of H. is analogous to that of Hydrolysis (see) or Aminolysis (see).

Hydrazoic acid, ***azoimide***: HN_3, a colorless, poisonous, caustic liquid which tends to explode when heated or on impact; M_r 43.03, m.p. - 80 °C, b.p. 35.7 °C. H. is a weak acid in aqueous solution ($K_a = 1.2 \cdot 10^{-5}$). It forms ionic azides with electropositive metals, such as sodium or potassium, and reacts with silver nitrate to make explosive silver(I) azide, AgN_3. H. is synthesized by reaction of hydrazine and nitrous acid at 0 °C in ether: $N_2H_4 + HNO_2 \rightarrow N_3H + 2 H_2O$.

Hydrazone: a monocondensation product of an aldehyde or ketone with hydrazine or a substituted hydrazine: $R^1R^2C{=}O + H_2N\text{-}NH\text{-}R^3 \rightarrow R^1R^2\text{-}C{=}N\text{-}NH\text{-}R^3 + H_2O$.

H. prepared from mono- and dinitro-substituted phenylhydrazines and carbonyl compounds crystallize readily and have sharp melting points. They are very often used to isolate and identify ketones and aldehydes.

Hydride: in the widest sense, any element-hydrogen compound; in a narrow sense, however, a compound of hydrogen with a more electropositive element. There are ionic, covalent and metallic H.

Ionic H. are formed by alkali and alkaline earth metals (except beryllium and magnesium). These are colorless, crystalline substances in which metal cations and H. anions occupy the sites of the crystal lattice. This is confirmed by the fact that hydrogen is generated at the anode when molten ionic H. are electrolysed. These compounds are strongly basic and reducing agents. They react with proton donors, including water, by generating hydrogen; for example, $CaH_2 + 2 H_2O \rightarrow Ca(OH)_2 + 2 H_2$. For this reason, they are also used as drying reagents for many aprotic solvents. They form complex H. with covalent H. such as B_2H_6, AlH_3 or GaH_3, e.g. $LiH + AlH_3 \rightarrow LiAlH_4$ (see Lithium alanate, Sodium boranate). Both ionic and complex H. are widely used to convert element halides to the corresponding element H. or in organic synthesis for reduction of carbonyl compounds. The ionic H. are made by reaction of hydrogen with the molten alkali or alkaline earth metal at 500-700 °C.

Covalent H. are the H. of beryllium, magnesium and elements of groups IIIa and IVa (except for carbon and the electronegative elements). Their hydridic nature can be recognized by the generation of hydrogen with proton donors. The H. of group IVa elements have the expected tetrahedral structure (see Silanes), but in the H. of beryllium, magnesium, boron, aluminum and gallium, the electron deficiency situation leads to formation of two-electron, multicenter bonds and the formation of dimeric, oligomeric (see Boranes) or polymeric (see Aluminum hydride) structures. The latter are converted to complex H. by reaction with ionic H. Covalent H. are synthesized by reaction of the element halides with lithium hydride.

Most of the ***metallic H.*** are derivatives of the d- and f- elements with highly variable atomic ratios, which are often not integral (non-stoichiometric compounds); to a certain extent, they have metallic properties, such as electrical conductivity. The hydrogen occupies holes in the metal lattice, and causes it to expand slightly. The H. nature of the hydrogen is revealed by the fact that some of these substances, for example uranium hydride, react with acids to generate hydrogen:

$$UH_3 + 3\,HCl \xrightarrow{300°C} UCl_3 + 3\,H_2.$$

Metallic H. are formed directly from the elements; the stoichiometric ratios depend greatly on the H_2 pressure and the temperature.

Hydride ion donor: a compound which releases a hydride ion. H. include metal hydrides, e.g. lithium hydride, or mixed metal hydrides, such as lithium aluminum hydride or sodium borohydride. The strongest H. is lithium triethylborohydride ("super hydride"). In certain reactions (Meerwein-Pondorf-Verley reduction, Oppenauer oxidation) the alkoxide ions of secondary alcohols or aldehydes act as H. in the reaction complex.

H. which can release carbanions are called ***cryptobases***.

Hydridoborates: same as Boranates.

Hydride transfer mechanism: a mechanism of redox reactions of certain cryptobases (see Hydride ion donors). In H., the hydride ions do not exist in free state, because the substrate-H bond is broken simultaneously with the reaction of the carbonyl compound. A cyclic transition state has been formulated for the H. (Fig.). Some examples of reactions with H. are the Meerwein-Ponndorf-Verley reduction of ketones, the Oppenauer oxidation of secondary alcohols, the Cannizzaro reaction and the Claisen-Tiščenko reaction.

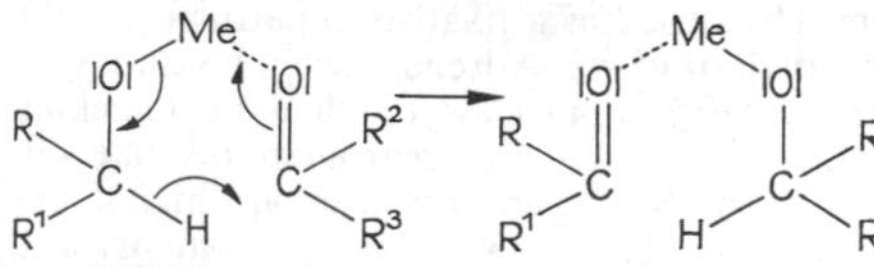

Me IOI IOI R R² R¹ C H C R³ → Me IOI IOI R R² R¹ C H C R³

Hydrindene: same as Indane (see).

Hydroaromatics: saturated mono- or bicyclic hydrocarbon compounds obtained by complete hydrogenation of aromatic hydrocarbons. An example is decalin.

Hydroboronation: addition of BH functional compounds to alkenes or alkynes. A typical example is the addition of diborane, B_2H_6, or organoboron hydrides to isobutene:

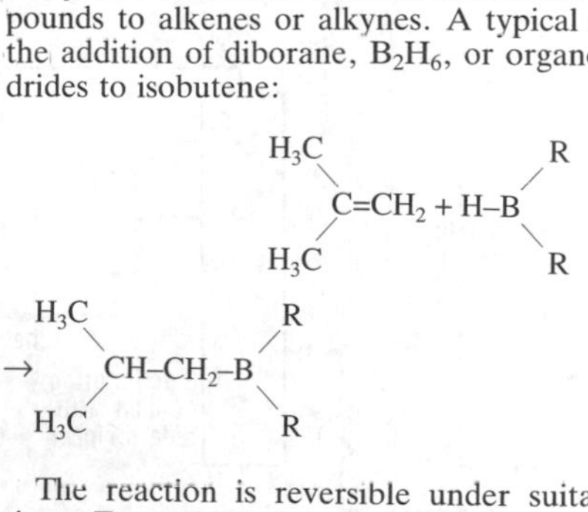

$$(H_3C)_2C{=}CH_2 + H{-}BR_2$$

$$\rightarrow (H_3C)_2CH{-}CH_2{-}BR_2$$

The reaction is reversible under suitable conditions. Formally, it is an anti-Markovnikov *cis*-addition. Its synthetic utility arises from the fact that the boron-carbon bond can be split again by thermal treatment or oxidation, and provides a pathway to interesting products. For example, the H. can be used to isomerize alkenes by migration of the double bond to the end of the chain, or to convert alkenes site-specifically into alcohols; the OH group will always bind to the C-atom with the larger number of H atoms.

Hydrocarbon: an organic compound consisting only of carbon and hydrogen. If all the bonds in the molecule are single bonds, it is said to be saturated; such compounds are Alkanes (see) or Cycloalkanes (see). Unsaturated H. contain double bonds (see Alkenes, Cycloalkenes, Dienes and Polyenes) and/or triple bonds (see Alkynes and Cycloalkynes); if the double bonds form an aromatic system, one has benzoid H. such as Arenes (see) or non-benzoid aromatic H. such as Azulenes (see) or Annulenes (see).

The different categories of H. differ considerably in their physical and chemical properties, the ways in which they can be synthesized, and in their chemical reactivity. For example, alkanes and cycloalkanes are relatively unreactive, compared to alkenes or other unsaturated H., which can undergo many addition and polymerization reactions. Aromatic H. tend not to undergo addition, but are susceptible to substitution reactions. There are preliminary analytical tests which can distinguish among the categories; for example, alkanes burn with bright flames while arenes burn with sooty flames, that is, incompletely. Double and triple bonds in unsaturated H. can add bromine, leading to bleaching of a dilute bromine solution. Arenes do not react under these conditions, but they are characterized by nitration with a mixture of nitric and sulfuric acid (nitrating acid).

Hydrocarbon resins: synthetic resins made by autoreaction of hydrocarbons (other than olefins) in the presence of aluminum chloride or sulfuric acid as catalyst. The relative molecular masses are below 2000 and the softening point is above 200 °C. The H. are subdivided into petroleum resins, terpene resins and cumarone-indene resins, depending on their structure. They also include the reaction products of xylene and formaldehyde, the ***xylene-formaldehyde resins***.

The H. are made by heating high-boiling fractions of gasoline pyrolysis (pyrolysis oil) or the isoprene-free C_5 fraction from gasoline pyrolysis in the presence of aluminum chloride. However, even in the C_5 fraction, the cyclopentadiene fraction should be as high as possible. The H. are soluble in most organic solvents, e.g. esters, ether, chlorohydrocarbons and aromatics. They are used chiefly in paints and printing inks.

Hydrochloric acid: HCl, an aqueous solution of hydrogen chloride and a strong mineral acid (pK -6.1). Because it releases hydrogen chloride, concentrated H. fumes in moist air, and for this reason, very concentrated H. is also called fuming H. Commercial grades of H. are as follows: ***fuming H.***, about 40% HCl, density 1.19 to 1.20; ***concentrated H.***, about 24 to 35% HCl, density 1.12 to 1.18; and ***dilute H.***, about 12.5% HCl, density 1.06 to 1.065. In laboratory practice, "dilute HCl" is about 7% HCl, density 1.035. A mixture of fuming H. and chloric acid is called Euchlorin (see). Pure H. is a clear, mobile liquid. Crude H. is usually more or less yellow, due to traces of iron which are present as hexachloroferric(III) acid,

$H_3[FeCl_6]$. The hydrogen chloride content of H. can easily be determined from its density, and there is coincidentally a simple rule for calculating percentage HCl from density: doubling the first two digits after the decimal point gives the approximate percentage.

Density:	1.03	1.06	1.12	1.16	1.21
Percent HCl:	6	12	24	32	42

H. forms an azeotropic mixture of 20.22% HCl, 79.78% H_2O which boils at 108.5 °C. If concentrated H. is cooled to - 24.9 °C, a trihydrate, $HCl \cdot 3H_2O$, crystallizes out. In contrast to hydrogen chloride, H. conducts electric current very well, because HCl reacts with water to form oxonium and chloride ions: $HCl + H_2O \rightarrow H_3O^+ + Cl^-$. The salts of H. are called Chlorides (see). H. reacts with reactive metals to generate hydrogen. A mixture of H. and nitric acid is Aqua regia (see), which dissolves nearly all metals, including most noble metals.

H. is present in human gastric juice at a concentration of 0.15 to 0.2%, where it promotes digestion and inhibits the growth of harmful bacteria. Traces of H. or hydrogen chloride are present in volcanic gases.

If hydrochloric acid vapors are inhaled over a period of time, they can lead to lung inflammation or bleeding. H. applied to the skin causes painful burns, which form blisters if the acid is allowed to remain for any length of time. It should be immediately washed off with a strong flow of water, and the burn should then be washed with soda solution. Oral consumption of concentrated hydrochloric acid causes painful burns in the mouth, throat, esophagus and stomach, leading to hoarseness, respiratory distress, weakening of the heart and, in severe cases, to death.

Antidotes for internal burns are consumption of milk or a magnesium oxide slurry, use of emetics or pumping the stomach.

H. is made by a counterflow of rising hydrogen chloride gas and water trickling down the walls of a tower. It is the cheapest industrial acid and is used widely in industry, for example to make metal chlorides, to process ores, in soldering, pickling and etching of metals, for removal of boiler scale, in production of wood sugar according to Bergius, for protein hydrolysis, in the synthesis of organic chlorine compounds, to precipitate fatty acids from soap solutions and in making glues. H. is an important laboratory reagent.

Hydrochloric acid group: see Analysis.

Hydrochlorothiazide: see Diuretics.

Hydrocinnamaldehyde, ***3-Phenylpropanal***: C_6H_5-CH_2-CH_2-CHO, a colorless oil with a jasmine-like odor; b.p. 223 °C. H. is insoluble in water and soluble in alcohol. It is the main component of Ceylon cinnamon oil, and is synthesized by reduction of cinnamic aldehyde. It is used in perfumes.

Hydrocinnamyl alcohol, ***3-phenylpropan-1-ol***: C_6H_5-CH_2-CH_2-CH_2-OH, an aromatic-substituted alcohol; a colorless, viscous liquid; b.p. 253-254 °C, n_D^{20} 1.5357. H. is slightly soluble in water and readily miscible with ethanol, ether and glacial acetic acid. In the form of esters, it is present in natural balsams and waxes. It is synthesized by hydrogenation of cinnamyl alcohol, e.g. with sodium amalgam. H. is used in perfumes.

Hydrocinnamic acid, ***3-phenylpropionic acid***: C_6H_5-CH_2-CH_2-COOH, a phenyl-substituted aliphatic monocarboxylic acid. H. forms colorless needles; m.p. 50-50.5 °C, b.p. 280 °C. It is steam volatile, and dissolves readily in hot water and most organic solvents. H. is formed by decomposition of proteins and is present in the rumen of ruminants. It is synthesized by reduction of Cinnamic acid (see) with sodium amalgam. H. is used as a fixative in perfumes.

Hydrocortisone: see Adrenal cortex hormones.

Hydrocracking: a process of catalytic Cracking (see) of medium and vacuum petroleum distillates in the presence of hydrogen. Because the higher petroleum fractions have low hydrogen contents, and therefore tend to precipitate a large amount of coke on the catalyst, it is advantageous to carry out catalytic cracking in an atmosphere of hydrogen. H. is a combination of catalytic cracking and catalytic hydrogenation, so the catalysts used must be at least bifunc-

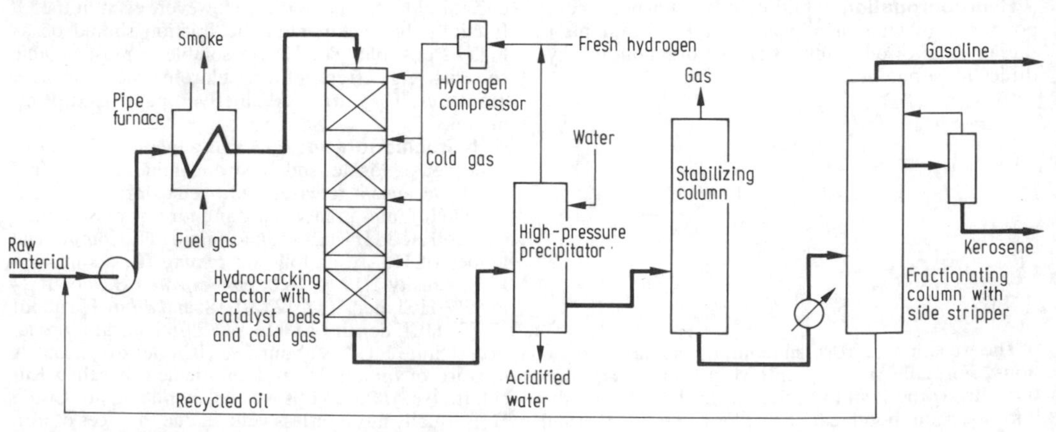

Cyclic transition state in the hydride transfer mechanism

tional. To achieve this, hydrogenation-active oxides and sulfides of molybdenum, palladium, tungsten or nickel are applied to acidic, cleavage-active carriers. H. requires a temperature between 300 and 420 °C, and pressures between 10 and 20 MPa. The desired products are gasoline or diesel fuel. A variant of H. is *selectoforming*, a process by which the less knock-resistant paraffins can be removed from the gasoline fractions. Here platinum-containing molecular sieves are used as catalysts.

H. was developed from cumulative experience with coal hydrogenation, and did not become important as a process for refining petroleum until about 1960. It is an alternative to catalytic cracking (FCC process). The advantage of H. is mainly that it can be used with higher-boiling distillates, such as gas oil and vacuum distillate, and that the products do not require further refining. Fused aromatics are degraded under the conditions of H.

H. can be run as a one- or two-step process. In the one-step process (Fig.), the catalysts must be resistant to sulfur, because hydrorefining occurs simultaneously with the H. In the two-step process, the starting material is hydrorefined at 350 to 420 °C in the first step, although some cleavage reactions also occur. The actual H. occurs in the second stage, at temperatures between 300 and 380 °C. H. reactions are highly exothermal, so cold gas must be fed into the reactor to control the temperature.

The H. stripper product is freed of the light gas components (methane, ethane, propane, butane) in a stabilization column, and then further processed in a fractionation column. The sump product of the fractionation column can be returned to the H. process.

Hydrocyanic acid: HCN, the nitrile of formic acid. B. is a colorless, non-viscous liquid with an odor of bitter almonds. It burns with a violet flame and is miscible with alcohol, ether or water in any proportions; m.p. - 13 to - 13.4 °C, b.p. 25.6 to 26.5 °C. Pure H. is stable in the absence of air, and its stability is increased by addition of oxalic acid, calcium chloride, etc. It forms formic acid and ammonia upon hydrolysis: $H\text{-}CN + 2\,H_2O \rightarrow H\text{-}COOH + NH_3$. H. is a very weak acid, which can be released from its salts, the Cyanides (see) even by carbonic acid. Gaseous H. (***hydrogen cyanide***) is a colorless gas which, in 6 to 40% mixtures with air, will explode upon ignition.

Hydrocyanic acid is one of the strongest and most rapidly acting poisons. The lethal dose is 60 mg. In either liquid or gaseous state, it penetrates the body through the stomach, skin or lungs and within a few seconds it paralyses cellular respiration by blocking cytochrome oxidase. At lower doses, death occurs by respiratory paralysis, and is preceeded by violent excitation, anxiety and convulsions. Even traces of hydrogen cyanide produce dizziness, headaches and scratchy throat; it must be remembered that not everyone is able to perceive the characteristic bitter almond odor. Cyanides cause the same symptoms of poisoning.

A patient can be saved from cyanide poisoning only if first aid (fresh air, artificial respiration, etc.) is followed with therapy. This is based on the fact that hydrogen cyanide is not bound by the divalent iron of hemoglobin, but it is readily bound by the trivalent iron of methemoglobin. The amount of hemoglobin in the body is greater than the amount of cytochrome oxidase. Some of it is converted to methemoglobin by injection of an oxidizing agent (e.g. nitrite); it can then bind chemically up to 20 times the lethal dose of CN^- and neutralize its effects. Because the hydrogen cyanide is gradually released from the methemoglobin, sodium thiosulfate is injected along with the nitrite. The enzyme rhodanase catalyses the conversion of cyanide and thiosulfate to rhodanide, which is harmless.

In H. the carbon atom is ambivalent; that is, in undissociated H. it is positively polarized and electrophilic, while the cyanide ion is nucleophilic. Many addition reactions are based on this property. For example, addition of H. to acetone, followed by saponification and dehydration of the acetocyanohydrins produces methacrylic acid, while addition of H. to ethyne or ethylene oxide produces acrylonitrile.

Occurrence. H. occurs widely in nature, for example, in amygdalin from bitter almonds and in other cyanogenic glycosides. It also occurs free, for example, in manioc roots, and small amounts are present in tobacco smoke and crude gas.

Production. In the laboratory, H. is obtained by the action of acids on cyanides. Scheele discovered H. in 1782 by the reaction of potassium hexacyanoferrate(II) and dilute sulfuric acid; it may also be produced by heating ammonium formate with phosphorus pentoxide. It is synthesized technically by water cleavage from formamide at 300 °C on an aluminum oxide contact, or by the Andrusov process, in which ammonia and methane are oxidized on a platinum/rhodium contact around 1000 °C: $2\,CH_4 + 3\,O_2 + 2\,NH_3 \rightarrow 2\,HCN + 6\,H_2O$.

Applications. H. is used in the chemical industry for the synthesis of many compounds, such as methacrylic acid and acrylonitrile, the starting materials for plastics and synthetic fibers. H. is used as a rodenticide for fumigating closed rooms. In the First World War, it was tested as a chemical weapon.

Hydrocyclone: a device for separation of solids and liquids, and for sorting solids by use of centrifugal

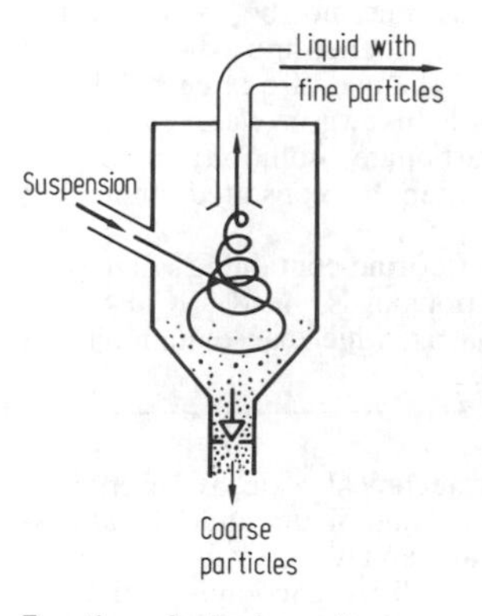

Function of a hydrocyclone

force. H. has the advantage over a centrifuge that it is simply built and has no moving parts. The suspension to be sorted is fed tangentially into the cylindrical part of the H. at a pressure of 20 to 200 kPa. The heavier particles are removed from the lower section, while the lighter particles are carried by the current upwards along the cyclone axis and are removed from the top.

The H. can also be used to concentrate a suspension, or for preliminary removal of water.

Hydrofining process: see Hydrorefining.

Hydrofluoric acid: the aqueous solution of hydrogen fluoride, HF. It is a weak acid (p*K* 3.17) which reacts with many metals, generating hydrogen and forming metal fluorides. It does not react with gold or platinum, and attacks lead only superficially. A monohydrate, $HF \cdot H_2O$ ($[H_3O]F$), can be isolated from concentrated solutions. H. containing 35.37% HF forms an azeotrope which boils at 112 °C. A characteristic of H. is its ability to etch glass, forming silicon tetrafluoride: $SiO_2 + 4\ HF \rightarrow SiF_4 + 2\ H_2O$. Consequently, it must be stored and used in vessels and apparatus made of platinum, lead, polyethylene or polytetrafluoroethylene.

> Hydrogen fluoride, hydrofluoric acid, hydrogen fluoride salts and elemental fluorine cause severe, deep and painful skin burns; tissues and bones are completely destroyed. Therefore extreme caution must be exercised in dealing with these substances.
>
> Burns (except those in the eyes) should be rinsed immediately with water, and it is then best to soak them with 2 to 3% ammonia solution for 1 to 2 hours with short interruptions, or to treat them with compresses soaked in this solution. Treatment must begin immediately after contact; later efforts to neutralize the acid or salts are not successful. Other useful first aid measures are treatment with 20% aqueous magnesium sulfate, 5% sodium hydrogencarbonate solution, slurries of magnesium oxide in water or a 10% calcium gluconate solution. After thorough first aid and external neutralization, the burn can be further treated with sulfonamide gel or alcoholic pyrolenin solution. Injection of calcium gluconate below the burn or cortisone treatment can speed healing. The burned areas must not be treated with fatty salves, and open wounds must not be exposed to ammonia! If the eyes are affected, they should be rinsed with lukewarm water and 1% sodium hydrogencarbonate solution; an opthamalogist should then be consulted immediately.
>
> Chronic uptake of fluorine-containing salt dusts and small concentrations of HF lead to CaF_2 deposits in the bone matrix, which cause immobility and pain (fluorosis).

F. is obtained on a technical scale by reaction of fluorite with sulfuric acid and oleum under an atmosphere of sulfur trioxide and water vapor: $CaF_2 + H_2SO_4 \rightarrow 2\ HF + CaSO_4$. This reaction is carried out in modern plants in rotating furnaces at approximately 300 °C; the crude product contains sulfuric acid, SiF_4, H_2O and CO_2 in addition to F. It is purified by precipitation and repeated distillation. Very pure F. is obtained by repeated distillation in apparatus made of polytetrafluoroethylene. Recently, phosphate minerals such as fluoroapatite, $Ca_5(PO_4)_3F$, have become more important as raw materials for production of F. and its products. Wet chemical processing of these minerals to phosphoric acid or superphosphate releases F. and SiF_4 as byproducts, which are further processed via hexafluorosilicic acid.

Applications. In 1725, Matthäus Pauli etched glass with F. produced from fluorite and sulfuric acid; the process was also carried out by Schwanhardt around 1760 in Nürnberg. Glass is still etched on a large scale to produce television screens, light bulbs and other glass products. For a matte etching of glass, mixtures of F. and fluorides, such as NaF, KHF_2 or NH_4HF_2, are also used. F. is also used to etch metals and to remove the residues of molding sand from cast pieces and to remove enamel scraps. F. is used on a large scale to produce metal fluorides, such as NaF, AlF_3 and Na_3AlF_6, and for production of fluorocarbons.

Hydroformylation: same as Oxosynthesis (see).

Hydrogen, symbol ***H***: a chemical element, *Z* 1, with the following natural isotopes: ***protium, light H.***, mass number 1 (99.985%); ***Deuterium*** (see), ***heavy H.***, mass number 2 (0.015%); and ***Tritium*** (see), mass number 3 (10^{-16}%); atomic mass 1.00797, valency I, density 0.0899 g l^{-1} at 0 °C and 0.1 MPa, density of the liquid at b.p. 0.070, m.p. -259.14 °C, b.p. -252.8 °C, crit. temp. -239.9 °C, crit. pressure 1.29 MPa, crit. density 0.031, standard electrode potential 0.00 V ($H_2 \rightleftharpoons 2\ H^+ + 2e$).

Properties: At room temperature, H. exists as molecular ***dihydrogen***, H_2. It is a combustible, colorless, odorless and tasteless gas. Its low molecular mass and its related high mean particle velocity make it diffuse very easily and give it a high heat conductivity; H_2 also behaves nearly as an ideal gas. H. is only slightly soluble in water and other solvents. However, some metals can dissolve large amounts of it; palladium, for example, can take up more than 800 times its own volume of H.

The nuclear spins of the two atoms in a molecule of H_2 can either be parallel (***ortho H.***; o-H_2) or antiparallel (***para H.***; p-H_2). These two forms are in a temperature-dependent equilibrium: o-$H_2 \rightleftharpoons$ p-H_2, ΔH = -1.66 kJ mol^{-1}. At room temperature, about 75% of H_2 is o-H_2 and 25% is p-H_2. As the temperature drops, the relative amount of p-H_2 increases. The two components can be separated by gas chromatography, and differ with respect to some physical properties, but their chemical behavior is identical.

Homolytic cleavage of molecular H. into atomic ***monohydrogen*** requires a large input of energy: $H_2 \rightleftharpoons 2\ H$, ΔH = - 432 kJ mol^{-1}. This is applied in the form of thermal energy in an electric arc. The resulting atomic H. rapidly recombines to H_2, releasing the energy. Temperatures over 4000 °C are reached in this way (*Langmuir torch*). Electric discharges into H_2 under reduced pressure can also produce monohydrogen (*Wood method*). In a Wood apparatus (Fig.), the dihydrogen is passed through the discharge region at 0.01 to 0.13 kPa.

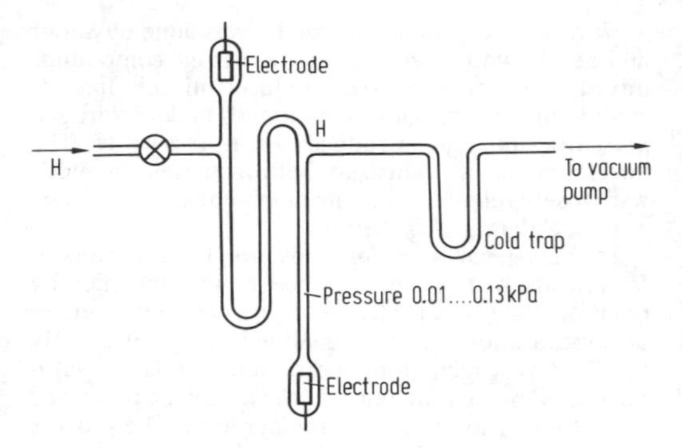

Wood apparatus for preparation of atomic hydrogen

A high flow rate makes it possible to bring the atomic H, which has a half-life on the order of a few tenths of a second, into a second region where it can react with another substance.

H. typically forms covalent bonds with the elements of medium and high electronegativity. The bond is often highly polarized with a positive partial charge on the H, but free H^+ ions do not exist in chemical systems. Because of the high polarizing effect of the proton, it is immediately bound by a suitable electron-pair donor (see Hydronium ion, Acid-base concepts, Hydrogen bonds). On the other hand, H is able to form a hydride ion H^- (see Hydrides).

Dihydrogen is relatively inert, due to its high bond energy. Its reactions usually require activation by heat or radiation energy, or the presence of a suitable catalyst. However, H_2 reacts very violently with elemental fluorine even at - 250 °C. After photochemical excitation, it reacts explosively with chlorine in a radical chain reaction, forming hydrogen chloride: $H_2 + Cl_2 \rightarrow 2\,HCl$, $\Delta H = -185$ kJ mol^{-1} (*chlorine detonating gas reaction*). The combination of H with bromine or iodine to form hydrogen bromide or iodide requires higher temperatures or catalysis. H. also reacts by a radical chain reaction when it combines with oxygen to form water: $2\,H_2 + O_2 \rightarrow 2\,H_2O$, $\Delta H = -286$ kJ mol^{-1}. After thermal initiation, the reaction occurs explosively with the realease of a great deal of heat (*detonating gas reaction*). This reaction is used in torches for welding and cutting metals. H reacts with other nonmetals when heated, forming the corresponding element-hydrogen compounds. The catalytic combination of H with nitrogen to form ammonia is especially important. When heated, H reacts with strongly electropositive metals, forming ionic hydrides. Many metal oxides, e.g. molybdenum(VI) oxide, are reduced to the metals by H at high temperature: $MoO_3 + 3\,H_2 \rightarrow Mo + 3\,H_2O$. Many organic compounds are hydrogenated, especially in the presence of metal catalysts (nickel, palladium, platinum).

Monohydrogen is much more reactive than dihydrogen, so that it reduces many metal oxides, sulfides or halides to the metals, even at room temperature. Arsenic, phosphorus and sulfur form the corresponding hydrogen compounds. At low temperature, monohydrogen combines with oxygen to form hydrogen peroxide; at higher temperatures, water is formed.

Analysis. H can be determined through the detonating gas reaction or by the water which is formed. Organically bound H is determined by Elemental analysis (see) or by the Cerevitinov method (see). Gas mixtures containing H are examined by Gas analysis (see).

Occurrence. H is the most abundant element in the universe. It makes up 0.81% of the earth's crust (by mass); this corresponds to 15.4 atom %. Considered in terms of numbers of atoms, H is the third most abundant element, after oxygen and silicon. H is found in elemental form in small amounts as a component of natural gas, volcanic gases, etc., but it forms the major portion of the highest atmospheric layers. It is present in enormous quantities as water. It is also a component of the hydrocarbons of natural gas and petroleum, and of countless organic compounds.

Production. At present, the most important method of H production is reaction of hydrocarbon mixtures from natural gas or cracked petroleum with water (*steam reforming*, see Synthesis gas). The reaction of steam with coal (water gas, see Synthesis gas) and separation of the H which is formed by cracking of petroleum are also important industrial methods. H is also obtained by electrolysis of aqueous salt solutions (e.g. as a byproduct of Chloralkali electrolysis). In the laboratory, H is most easily made by the reaction of acids with reactive metals, such as zinc: $Zn + 2\,H_3O^+ \rightarrow Zn^{2+} + H_2 + 2\,H_2O$, or by the reaction of sodium hydroxide and aluminum: $Al + NaOH + 3\,H_2O \rightarrow Na[Al(OH)_4] + 3/2\,H_2$. The reaction of ionic hydrides, such as calcium hydride, with water also produces H: $CaH_2 + 2\,H_2O \rightarrow Ca(OH)_2 + 2\,H_2$.

Applications. Large amounts of H are consumed in the synthesis of ammonia from atmospheric nitrogen. H is also the starting material for industrial syntheses of hydrogen chloride, methanol, higher alcohols and aldehydes (see Oxosynthesis), triethylaluminum, hydrocarbons, etc. It is used to hydrogenate numerous organic compounds, to harden fats and to reduce metal oxides in smelters. H is an important energy carrier, and is used as a fuel, for example in rocket engines and fuel cells. It is used as a protective gas for production of very pure components, e.g. in microelectronics, and as a carrier gas in gas chromatography.

Historical. After H had been produced in the 16th century by Paracelsus, by reaction of iron with mineral acid, the combustible gas was recognized as an element in 1766 by Cavendish. He also observed in 1781 that H combined with oxygen to form water. In 1783, Lavoisier decomposed water vapor with glowing iron. Electrolysis of water was first achieved in 1789. In 1929, Bonhoeffer and Hartek prepared pure parahydrogen.

Hydrogenation: in general, addition of hydrogen to a substance. The reaction usually occurs at elevated temperatures and pressures in the presence of catalysts (***catalytic pressure hydrogenation***). Addition of hydrogen to C-C multiple bonds and other unsaturated systems, such as nitroso-, nitro- and carbonyl compounds, azomethynes and nitriles is of great significance, both in the laboratory and in industry.

Catalytic H. are generally "*cis*-additions". A rough distinction is made between ***refining*** and ***destructive***

H., depending on the extent of the reaction. Destructive H. is always linked to hydrorefining. Most H. reactions are exothermic, that is, at low temperatures the equilibrium lies on the side of the products. The reverse reaction, Dehydrogenation (see) does not become significant until the temperature is raised above 400 °C. Several reactor types are used in which the heat of reaction is removed: those have coolant pipes in the reactor space, stage furnaces with cold gas intake, and cyclone reactors.

The product gas from gasoline cracking contains ethyne, propyne, propadiene and butynes in addition to the ethene and propene which are useful for further reaction. To remove the contaminating compounds, the gas fractions are subjected to ***selective H.***. In the presence of a Pd catalyst, only C-C triple bonds are hydrogenated to C-C double bonds; the isolated and conjugated C-C double bonds are retained. For example, $CH{\equiv}C\text{-}CH_3 + H_2 \rightarrow CH_2{=}CH\text{-}CH_3$; $CH_2{=}C{=}CH_2 + H_2 \rightarrow CH_2{=}CH\text{-}CH_3$.

Selective H. is carried out between 60 and 200 °C, and at a hydrogen pressure of 0.5 to 2.5 MPa. The production of cyclohexanone by selective H. of phenol is also possible.

Industrial hydrogenation reactions are always done in the presence of catalysts. These are divided into metallic (Co, Ni, Pt, Cu, Ag), oxide (Cr_2O_3, MoO_3, WO_3, Fe_2O_3, Cu_2O, ZnO) and sulfide (CoS, NiS, ZnS, MoS_2, WS_2) catalysts. Activated charcoal, aluminum oxide and barium sulfate are used as carriers for the catalysts. A property of all hydrogenation catalysts is that highly active and mobile hydrogen atoms are formed on the surface by chemisorption. Because of their incompletely occupied inner electron shells, the transition elements, especially those of Group VIIIb and copper, have the greatest hydrogenation activity. Some industrially important H. are: H. of carbon monoxide to Methanol (see), of fats (see below), Hydrorefining (see), Hydrocracking (see) and Coal hydrogenation (see). In organisms, H. occurs in the presence of enzymes (hydrogenases). Metal complexes of rhodium, ruthenium and iridium are used for *homogeneous phase H.*. *Asymmetric H.* (see Asymmetric synthesis) is possible with specially modified catalysts, such as palladium on polyamino acids.

Industrially important hydrogenation reactions

Class of compound	Hydrogenation product	Catalyst
Aromatic rings		
Benzene	Cyclohexane	Pd/Al_2O_3
Phenol	Cyclohexanol	Ni/Al_2O_3
Pyridine	Quinoline	
Carbonyl compounds		
Butyraldehyde	Butanol	Nickel
Acetone	*iso*-Propanol	Copper chromite
Nitriles		
Acetonitrile	Ethylamine	Cobalt/nickel
Adipic acid	Hexamethylene-diamine	
Nitro compounds		
Dinitrobenzene	Aniline	Copper
Dinitrotoluene	Diaminotoluene	Nickel/palladium
Carboxylic acids and their esters		
Benzoic acid	Hexahydrobenzoic acid	Platinum

Historical. The first catalytic H. was done by Satier and Sendi, who passed vapors of organic compounds mixed with hydrogen over reduced nickel. Ipatieff studied the H. of organic compounds under hydrogen pressure, and thus introduced high-pressure H. The development of industrial hydrogenation methods was closely related to the development of catalysts by Adkin, Adam and Raney.

2) ***Hydrogenation of fats***: oils are glycerol esters of fatty acids with several double bonds; the melting point of a fat (glycerol triester) depends on the degree of unsaturation of its constituent fatty acids. By catalytic hydrogenation, the double bonds are saturated and the melting point of the fat can be increased so that it is a solid at room temperature. The process is very important for the production of margarines and soaps. In the Normann process (1907), oils are first filtered and neutralized; they are then heated to 120 to 180 °C in an autoclave in the presence of pressurized hydrogen. Nickel-carrier catalysts are most commonly used. The degree of H. is determined from the iodine number. After the reaction, the mixture is cooled to about 80 °C, and the catalyst is removed by pressure filtration. Since the acid number of the product is increased by H., it is usually treated with an alkali hydroxide and fuller's earth. The process can be designed to run continuously.

Hydrogen bond: the interaction between a group X-H (proton donor) and the lone electron pair of an atom Y (proton acceptor) X-H...|Y, where X and Y must both be strongly electronegative. Since this bond involves four electrons and three atomic centers, it is an electron-excess H. (see Three-center bond). In *electron deficiency H.*, the bond is formed with a partner with an electronic deficiency rather than with the proton acceptor Y; examples are seen in diborane and the dimers of other hydrides (see Three-center bond). To a slight degree, the elements chlorine, sulfur and carbon can also form H. bonds. The energy of H. is usually 20 to 40 kJ mol^{-1}, and is thus intermediate between that of a true covalent bond and that of van der Waals interactions. In normal X-H...Y hydrogen bonds, the distance between the atoms X and Y is about 275-300 pm. Unusually strong H. with energies of about 200 kJ mol^{-1} occur in hydrated protons, $H_9O_4^+$ and in the salt KHF_2. If the atoms X and Y are part of the same molecule and in a sterically favorable arrangement, an *intramolecular H.* can be formed, e.g. in 2-hydroxybenzoic acid. The H. is essentially electrostatic, $\overset{\delta\ominus}{X} - \overset{\delta\oplus}{H} - \overset{\delta\ominus}{Y}$, as has been shown by infrared, ultraviolet and nuclear resonance spectroscopy, and can be explained by the coordinative divalency of the H^+ ion. The formation of strong H. explains the high melting and boiling points of the hydrides of fluorine, oxygen and nitrogen, and the density anomaly of water. H. are very important in the structures and interactions of biological molecules, such as proteins and nucleic acids.

Hydrogen brittleness: damage to unalloyed and alloyed steels at room temperature due to uptake of atomic hydrogen, which causes brittleness, formation of bubbles near the surface of the material, cracks and breaks. Bubbles and transcrystalline interior cracks near inclusions and parallel to the direction of milling are called *hydrogen-induced cracks*. Transcrystalline or intercrystalline cracks perpendicular to the direction of tension are called *hydrogen-induced tension-crack corrosion*. The hydrogen comes from electrochemical corrosion, electrolysis processes, pickling, cathodic protection, welding with moist protective gas or additives, and chemical reactions with compounds which release hydrogen. Significant damage arises during production, storage, transport and processing of petroleum and natural gas which contain hydrogen sulfide. In general, increased hardness of the steel, cold riveting, and the presence of indentations and other sources of tensile stress increase the susceptibility to H. Intermediate structures and martensite are more susceptible than pearlite. The probability of H. increases as the pH of the corrosion medium falls.

In the presence of pressurized hydrogen, at temperatures above 230 °C, hydrogen diffusing into unalloyed steels reacts with cementite to form methane. As carbon is lost from the steel and subcrystalline regions separate, the steel becomes brittle and cracks. This *pressurized hydrogen attack* occurs in hot, high-pressure synthesis reactors, in which the partial pressure of hydrogen is high, if resistant steels are not used.

Hydrogen bromide: HBr, a colorless gas with a pungent odor which fumes in moist air; M_r 80.90; one liter (at 0 °C and 100 kPa) weighs 3.6443 g. The density at the boiling point is 2.160; m.p. - 86.82 °C, b.p. - 66.73 °C. HBr is extremely soluble in water; at 0 °C, 100 g H_2O dissolves 221.2 g HBr. The aqueous solution is called ***hydrobromic acid***. It is a very strong acid (pK -8.9) and a stronger reducing agent than the otherwise very similar hydrochloric acid. In the air, HBr gradually turns yellow-brown. Concentrated hydrobromic acid contains 69% HBr. Dilute hydrobromic acid cannot be concentrated to this point by distillation, because it forms an azeotropic mixture of 47.63% HBr and 52.37% H_2O which boils at 124.3 °C. At low temperatures, hydrates can be obtained: $HBr \cdot 4H_2O$, m.p. -56 °C; $HBr \cdot 3H_2O$, m.p. -48 °C; $HBr \cdot 2H_2O$, m.p. -11.3 °C; $HBr \cdot H_2O$, m.p. -4 °C.

The salts of hydrobromic acid are called Bromides (see). The toxicity of H. is similar to that of hydrogen chloride, and the toxicity of hydrobromic acid is similar to that of hydrochloric acid. H. is made by hydrolysis of phosphorus trichloride: $PBr_3 + 3\ H_2O \rightarrow 3\,HBr + H_3PO_3$, or by a direct reaction of red phosphorus, bromine and a small amount of water. Very pure HBr can be made from the elements: $Br_2 + H_2 \rightleftharpoons 2\ HBr$; a relatively low temperature is required to shift the equilibrium to the right, and a platinum sponge or activated charcoal is used as catalyst. In industry, HBr is formed as a byproduct of bromination reactions of organic compounds, e.g. the synthesis of bromobenzene or bromoacetone.

Hydrogen chloride: HCl, a colorless gas with a pungent odor; it fumes strongly in moist air. M_r 36.461, density (at b.p.) 1.187, m.p. -114.22 °C, b.p. -85.05 °C, crit. temp. +51.3 °C, crit. pressure 8.301 Pa. In the solid state, H. forms a molecular lattice with hydrogen-bridged zig-zag chains. It is extremely readily dissolved by water, with the release of a considerable amount of heat: 1 vol. H_2O at ordinary temperatures dissolves about 450 vol. HCl. The aqueous solution is called Hydrochloric acid (see). H. is also readily soluble in ether, alcohol and other organic solvents. At red heat, H. is a very aggressive gas, which can convert to the chlorides even those metals which are not attacked by hydrochloric acid, e.g. silver: $2\ Ag + 2\ HCl \rightarrow 2\ AgCl + H_2$. Pure, liquefied H., in contrast to hydrochloric acid, does not attack most metals, oxides, sulfides or carbonates. H. forms a cheesy white precipitate of silver chloride with silver nitrate.

H. can be produced in the laboratory by the reaction of sulfuric acid with salt: $NaCl + H_2SO_4 \rightarrow HCl + NaHSO_4$. Pure H. is produced on an industrial scale by the reaction of chlorine and hydrogen from chloralkali electrolysis. In this way, formation of explosive H_2/Cl_2 mixtures is prevented. Large amounts of H. are produced as a byproduct of industrial chlorination of hydrocarbons. H. is used in the production of hydrochloric acid, metal chlorides, chlorosulfonic acid, vinyl chloride from ethyne, ethyl chloride from ethene and for isomerization of aliphatic hydrocarbons.

Hydrogen corrosion type: see Electrochemical metal corrosion.

Hydrogen economy: a proposal for an energy system in which hydrogen would serve as a secondary energy carrier (Fig.). This idea could become important when few fossil fuels are still available, or when the production of CO_2 must be curtailed to prevent a disastrous increase in the average surface temperature of Earth. The electrolysis of water plays a central role in this scheme; the required electrical energy would be supplied from some other source, such as the sun, wind, tides, or nuclear fission. A variation of the idea would be based on direct solar photolysis of the water. The hydrogen would be a form in which energy could be conveniently stored and transported.

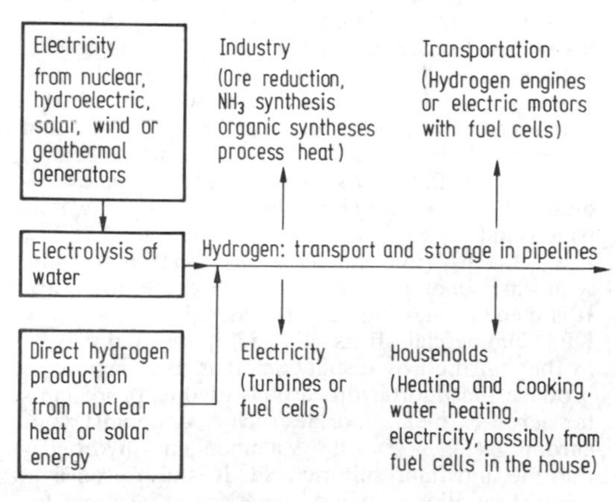

Hydrogen electrode

Hydrogen electrode: a gas electrode consisting of platinated platinum foil past which extremely pure, gaseous hydrogen flows; the foil dips into a solution containing protons (hydrogen ions). The potential is determined by the process 2 H^+ + 2 e^- → H_2. The Nernst equation for the electrode potential is

$$E = E^\circ + \frac{RT}{F} \ln \frac{a_{H^+}}{\sqrt{P_{H_2}}}$$

Here E^o is the standard electron potential, R the gas constant, T the temperature in Kelvin, and F the Faraday constant. The potential of the H. thus depends on the activity of the hydrogen ion a_{H^+} and the pressure of the hydrogen gas p_{H_2}. An H. in which the activity of the hydrogen ions is 1 mol l^{-1} and the pressure of the hydrogen gas is 101.3 kPa (1 atmosphere) is called a ***standard hydrogen electrode***. The electrode potential of this electrode is set equal to zero by convention, and serves as the reference point for establishing electrochemical potential series.

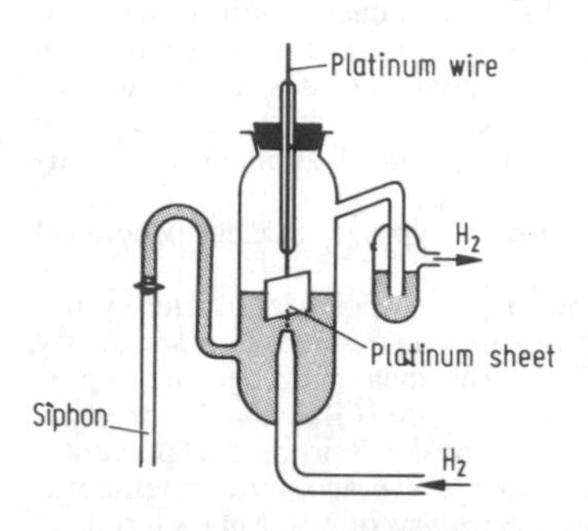

Hydrogen fluoride: HF, a colorless, non-viscous, poisonous, and caustic (see Hydrofluoric acid) liquid which fumes in air; M_r 20.01, density (at b.p.) 0.991, m.p. -83.55 °C, b.p. +19.51 °C, crit. temp. +230.2 °C. HF molecules have a distinct tendency to associate by forming hydrogen bonds. In the crystalline state, the compound is made up of long, zig-zag chains (Fig.), and association of molecules in the liquid state is responsible for its high melting and boiling points. In the vapor phase, near the boiling point, there is an equilibrium between hexameric $(HF)_6$ and monomeric HF molecules; the position of this equilibrium depends on the temperature and pressure. Only above 90 °C is HF monomolecular. HF is one of the most stable diatomic molecules. Liquid HF is an excellent solvent, similar to water, which undergoes dissociation according to 3 HF ⇌ H_2F^+ + HF_2^- (ion product = 10^{-11}). HF is miscible with water in all proportions (see Hydrofluoric acid). It is made by heating acid fluorides: KHF_2 → HF + KF. Commercial HF is ≈ 99.5% pure, and can be further purified by distillation. It is used mainly to produce the fluorohydrocarbons used as propellants for aerosols, metal fluorides such as uranium(IV) or chromium(III) fluoride, ammonium hydrogenfluoride and fluorosulfuric acid. It is also used as a catalyst for alkylation and isomerization reactions for production of anti-knock fuels, as a sulfur remover for gas oils and as a solvent in chemical laboratories.

Hydrogen iodide: HI, a colorless, poisonous gas with a pungent odor; M_r 127.9124, density (at b.p.) 2.799, m.p. - 50.80 °C, b.p. - 35.36 °C. I. is very soluble in water and forms a very strong acid in aqueous solution (pK - 9.3). With water, I. forms an azeotropic mixture which boils at 126.7 °C; it has an HI content of 56.7%. I. is relatively easily oxidized, and is therefore a strong reducing agent. It is slowly oxidized in the presence of air, forming iodine, so that hydrogen iodide solutions in water turn brown, and do so rapidly in the light. I. is obtained by the reaction of iodine with red phosphorus, followed by further reaction of the resulting phosphorus(III) iodide with water: 2 P + 3 I_2 → 2 PI_3; PI_3 + 6 H_2O → 6 HI + 2 H_3PO_3. A moderately concentrated aqueous solution of I. can be obtained by passing hydrogen sulfide through an aqueous iodine suspension: I_2 + H_2S → 2 HI + S. The salts of I. are called Iodides (see). Hydroiodic acid is used, for example, for quantitative alkoxy group determination according to Zeisel-Vieböck.

Hydrogen peroxide: H_2O_2, colorless liquid miscible with water in any proportions; M_r 34.01, density 1.4422, m.p. - 0.41 °C, b.p. 150.2 °C. H. slowly decomposes in an exothermal reaction, forming water and oxygen: H_2O_2 → H_2O + 1/2 O_2; ΔH = - 98.3 kg mol^{-1}. This decomposition is accelerated by dissolved heavy metal salts, finely divided noble metals, various metal oxides and alkalies. Therefore chelate formers, usually phosphoric acid, are added to commercial H. to bind the catalytically acting metal ions and thus to stabilize the H. The angular structure of the H_2O_2 molecule (Fig.) is the result of a minimization of the interaction of the free electron pairs on the oxygen atoms.

The O-H bond is highly polarized, and as a result, liquid H. is highly hydrogen bonded. It is a weak acid (pK_a = 11.8), but is considerably more acidic than water. The salts of H. are called Peroxides (see), and are based on the O_2^{2-} ion. Substitution of an alkyl or aryl group for one or both H atoms yields the Hydroperoxides (see) R-O-O-H, the dialkyl peroxides, R-O-O-R, the peroxocarboxylic acids, RC(O)OOH (see Peroxo compounds) and the diacyl peroxides, RC(O)-OO-C(O)R. H. can act either as an oxidizing or a reducing reagent, but the oxidizing effects are more prominent. For example, in acid solution it oxidizes hydrogen sulfide to sulfur, sulfides to sulfates, arsenious acid to arsenic acid, Fe^{2+} ions to Fe^{3+} ions or iodide to iodine. In the process, H. is reduced to water, as in 2 I^- + H_2O_2 + 2 H^+ → I_2 + 2 H_2O. The oxidizing capacity is also the basis of the bleaching effect of H. Strong oxidizing agents release oxygen from H., e.g. 5 H_2O_2 + 2 MnO_4^- + 6 H^+ → 2 Mn^{2+} + 5 O_2 + 8 H_2O. H. forms ***peroxide hydrates*** with a number of salts; these are compounds in which the water of crystallization is partly replaced by H_2O_2 (see, for example, Perborates).

H. can be detected by the conversion of the colorless titanium oxygen sulfate, $TiOSO_4$, into orange peroxotitanium complexes.

H. is now made on an industrial scale chiefly by the *anthraquinone process*, in which anthrahydroquinone or one of its 2-alkyl derivatives is oxidized by atmospheric oxygen. The products of this reaction are a 100% yield of H. and anthraquinone, which is returned to the process after hydrogenation to anthrahydroquinone. The sum of this reaction sequence corresponds to a synthesis of H. from hydrogen and oxygen. The oxidation of isopropanol with oxygen to form H. and acetone (*isopropanol process*) is also utilized industrially. H. was formerly synthesized by anodic oxidation of sulfuric acid or ammonium sulfate solution, followed by hydrolysis of the peroxodisulfuric acid or peroxodisulfate, via peroxomonosulfuric acid, to sulfuric acid or sulfate and H.: $H_2S_2O_4 + H_2O \rightarrow H_2SO_5 + H_2SO_4$; $H_2SO_5 + H_2O \rightarrow H_2SO_4 + H_2O_2$. However, this process has become much less common. The production of H. by reaction of ionic peroxides with acids, e.g. $BaO_2 + H_2SO_4 \rightarrow BaSO_4 + H_2O_2$, is now of historical interest only.

H. is used as a bleach in the production of textiles, cellulose and paper. Large amounts are used in the synthesis of peroxide hydrates, e.g. perborates and percarbonates, which are used in laundry products. H. is used as a powerful oxidizing agent in organic syntheses and as an initiator of polymerization reactions. It is also used in treatment of sewage, corrosion protection, hair cosmetics, etc.

Hydrogen selenide: H_2Se, colorless, poisonous gas which smells like rotting radishes; M_r 80.98, density of the liquid at -41.5 °C, 2.004, m.p. -60.4 °C, b.p. -41.5 °C. As an endothermal compound, S. tends to decompose into its elements ($H_2Se \rightleftharpoons Se + H_2$, ΔH = -30 kJ mol^{-1}). It is a strong reducing agent. In the presence of moisture, it is oxidized by atmospheric oxygen to red selenium. When heated, it burns to selenium dioxide. S. is a weak dibasic acid (pK_1 = 3.72, pK_2 = 12); its salts are the Selenides (see).

Hydrogen selenide is a strong poison, which enters the body chiefly through the respiratory tract. When small amounts are inhaled, it causes headaches and nausea. Higher concentrations irritate the mucous membranes of the eyes and respiratory passages ("selenium cold"). Subsequent results are bronchiopneumonia and damage to liver, spleen and kidneys. It is treated symptomatically.

S. is obtained by direct synthesis from the elements at temperatures above 400 °C, or by the reaction of acids with selenides, e.g. $MgSe + 2\,HCl \rightarrow H_2Se + MgCl_2$.

Hydrogen sulfide, *monosulfane*: H_2S, a colorless, combustible and very poisonous gas which even at very low concentrations has an unpleasant odor of rotten eggs; M_r 34.08, m.p. - 85.5 °C, b.p. - 60.7 °C, density of the liquid 0.9968 at - 194.6 °C, crit. temp. 100.4 °C, crit. pressure 8.89 MPa, crit. density 0.349.

Properties. The H_2S molecule is bent (bond angle H-S-H 92.25°). Because the electronegativities of the two elements are similar, the polarity of the bond is slight and hydrogen bonds are not significant in H. This is manifest in its low melting and boiling points (compared to those of water). H. is moderately soluble in water: 100 g water dissolves 0.539 g H_2S at 10 °C, and 0.398 g at 20 °C. H. is a weak dibasic acid, and undergoes partial dissociation in aqueous solution: $H_2S + H_2O \rightleftharpoons HS^- + H_3O^+$; $HS^- + H_2O \rightleftharpoons S^{2-} + H_3O^+$ (pK_1 6.92, pK_2 12.90). The degree of dissociation can be arbitrarily adjusted by changing the pH of the solution, and this is the basis for analytical separation of cations in the H_2S separation step in qualitative analysis schemes. The extremely low sulfide ion concentrations in acidic solution are sufficient to precipitate the very insoluble elements of the H_2S group, while the solubility products of the somewhat more soluble sulfides of elements in the NH_4HS groups are not reached until the solution is made alkaline with ammonia. H. forms two series of salts, the hydrogensulfides, M^IHS and the Sulfides (see) M^I_2S. In the air, H. burns to sulfur or sulfur dioxide, depending on the conditions: $2\,H_2S + O_2 \rightarrow 2\,S + 2\,H_2O$; $2\,H_2S + 3\,O_2 \rightarrow 2\,SO_2 + 2\,H_2O$. These reactions are industrially important for production of sulfur by the Claus process. H. reacts with fluorine to give hydrogen fluoride and sulfur hexafluoride; with the other halogens, e.g. chlorine, it reacts to form the hydrogen halide and sulfur: $H_2S + Cl_2 \rightarrow 2\,HCl + S$. Concentrated sulfuric acid is reduced to sulfur dioxide by H_2S, which in turn is oxidized to sulfur: $H_2SO_4 + H_2S \rightarrow S + 2\,H_2O + SO_2$. With sulfur trioxide, H. undergoes a coproportionation to water and sulfur: $3\,H_2S + SO_3 \rightarrow 4\,S + 3\,H_2O$. In the presence of moisture, especially when it is warm, H. combines with most metals to form sulfides. For example, the tarnishing of silver is due to surface formation of black silver sulfide, Ag_2S. H. also reacts with many metal oxides when hot, forming sulfides and water: $M^I_2O + H_2S \rightarrow M^I_2S + H_2O$.

Analysis. H. can be recognized from its odor; it colors lead acetate paper black, due to formation of lead sulfide, PbS. It precipitates yellow cadmium sulfide, CdS, from a cadmium acetate solution. For quantitative analysis, the stream of gas is passed through a solution of silver nitrate, cadmium acetate or copper acetate. H. is absorbed by precipitation of the metal sulfide followed by titration of the excess Ag^- ions with NH_4SCN solution or the Cd^{2+} or Cu^{2+} ions complexometrically. There are also many reagents for colorimetric determination of H. (*p*-aminodimethylaniline, heavy metal salts, etc.). Gas chromatography can be used to analyse very small amounts of H_2S.

Occurrence. H. is present in volcanic gases and dissolved in the water of sulfur springs. It is also formed by decay of protein-containing materials. H. is a component of eutrophic waters, as a result of anaerobic microbial reduction of sulfates.

Hydrogen sulfide is a strong poison. Its acute toxicity is equal to that of hydrogen cyanide. However, because of its intense odor, the gas can be detected at concentrations two orders of magnitude below the toxic concentration. On the other hand, at high concentrations of H_2S, the sense of smell is blocked.

Acute poisoning by hydrogen sulfide leads to inflammation of the eye mucosae, to coughing and lung edema. Higher concentrations cause convulsions, headaches, cyanosis and unconsciousness. Death occurs by respiratory paralysis. As countermeasures, artificial respiration, administration of analeptics and removal of the lung edema by diuretics are suggested.

Production. In the laboratory, H. is produced by the reaction of acids with sulfides, usually hydrochloric acid with iron sulfide in a Kipp apparatus: $FeS + 2\,HCl \rightarrow FeCl_2 + H_2S$. Another method is to heat equal parts of sulfur and paraffin to 100 °C. Very pure H. is obtained by passing a mixture of sulfur vapor and hydrogen through a tube heated to 600 °C. Industrial H. is produced by separation from synthesis gases, e.g. by the Sulfosolvane process (see).

Applications. H. is used in qualitative and quantitative analysis as a precipitation reagent for metal ions. The H. produced in industry is used to obtain sulfur or sulfur dioxide.

Hydrogen sulfide group: see Analysis.

Hydrogen telluride: H_2Te, a colorless gas with an unpleasant odor; M_r 129.62, density of the liquid, 4.49, m.p. - 48.9 °C, b.p. - 2.2 °C. H. is poisonous (see Tellurium). The compound is highly endothermal and unstable; it tends to decompose into the elements: $H_2Te \rightarrow H_2 + Te$, $\Delta H = -100$ kJ mol^{-1}. Even atmospheric oxygen is able to oxidize H. to tellurium. H. dissolves in water, and in solution is a weak acid ($K_1 = 2.64$, p$K_2 = 8.80$). The salts, the ***tellurides***, M^I_2 are more stable than T. and are used in its synthesis, e.g. $Al_2Te_3 + 6\,HCl \rightarrow 3\,H_2Te + 2\,AlCl_3$. Some tellurides are found in nature; they are similar in their behavior to the selenides.

Hydrolases: one of the main groups of Enzymes (see). Some well-known examples (see the separate entry for each) are: Arginase, L-Asparaginase, ATPases, Esterases, Glycosidases, Proteases and Urease.

Hydrolysis: 1) the reaction of a compound with water, now also called ***protolysis***. This can be considered a reversal of neutralization:

$$\text{acid} + \text{base} \underset{\text{hydrolysis}}{\overset{\text{neutralization}}{\rightleftharpoons}} \text{salt} + \text{water}$$

H. explains the alkaline or acidic reactions of aqueous solutions of salts of strong acids with weak bases, or of weak acids with strong bases. Application of the mass action law to the H. reaction yields an H. constant characteristic of the salt. The magnitude of this K_h is proportional to the degree of H.

2) the cleavage of covalent bonds by water, as a special case of solvolysis. Some important H. reactions are cleavage of esters and other derivatives of carboxylic acids, and of di- and polysaccharides into monosaccharides.

Hydrolyzable tannins: same as Gallotannins (see).

Hydrometallurgy, ***wet metallurgy***: a collective term for all metallurgical processes used to obtain metals (especially nonferrous metals) from their ores by means of aqueous metal salt solutions. The pure metals are recovered from the solutions by electrolysis, or metal compounds are obtained by precipitation.

Hydronalium: an Aluminum alloy (see).

Hydrone pigments: a subgroup of sulfur pigments. There are H. of every color. They are made by reaction of a mixture of sodium polysulfide and sulfur with indophenols. The most important of the group is ***hydrone blue*** (Fig.), a blackish powder with a coppery sheen; it dissolves in conc. sulfuric acid to give a dark blue color. Hydrone blue can be reduced to a yellow vat dye with sodium dithionite, $Na_2S_2O_4$; this dye colors cotton bluegreen to violet. The fastness of hydrone blue to light, washing, acid and chlorine have made it a popular dye for working clothes; it has to some extent also replaced indigo and indanthrene dark blue.

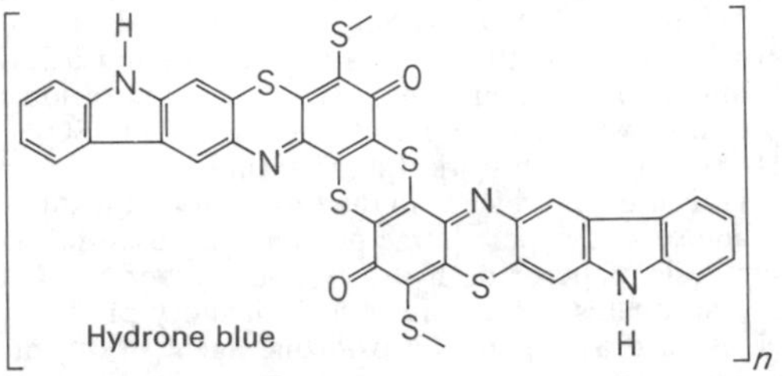

Hydrone blue

Hydronium ion, ***oxonium ion***: H_3O^+, the hydrated proton. In aqueous solution, it is further hydrated, approximately to $H_9O_4^+$. The H. concentration is of great importance in characterizing an aqueous solution (see Ion product, pH).

Hydronium ion conductivity: Hydronium and hydroxide ions have extremely high limiting conductivities, due to a special mode of conduction. Whereas metal ions must move through the solution in order to carry a current, a positive charge can be transferred from a hydronium ion to a water molecule by transformation of a hydrogen bond into a covalent bond and a covalent bond into a hydrogen bond. This is achieved by a shift of bonding electrons through a quantum mechanical tunneling effect, and can occur much more rapidly than diffusion of a charged particle. A similar mechanism is postulated for hydroxide ions: transfer of a proton from a neutral water molecule to one of its neighbors creates an hydroxide ion; the positive charge moves through the water by a shifting of bonds until it meets an hydroxide ion which absorbs it.

Hydroperoxides: organic compounds with the general formula R-O-OH. H. are often the first products of autooxidation of organic compounds, for example, in peroxide formation from ethers. They are important in industry as intermediates in paraffin oxidation. Some H. can be isolated in substance. The cumene hydroperoxide of the Hock synthesis, which is cleaved into phenol and acetone on a large scale, is especially important.

Hydrophilic: an adjective meaning "attracted to water".

Hydrophobic: an adjective meaning "repelled by water".

Hydrophobic interaction: an interaction between bipolar molecules, such as surfactants, in water. It is caused by the contrast between the strong attraction of water molecules for each other, and the relatively slight attraction between water molecules and organic groups such as $-CH_3$, $-CH_2$ or $-C_6H_5$. The result is that these groups are displaced out of the water phase, either by adsorption to the phase boundary, or by formation of micelles.

Hydroquinone, ***1,4-dihydroxybenzene***: a diphenol which forms colorless needles; m.p. 173.3 °C, b.p. 287 °C. It is readily soluble in water, ethanol and ether, and is a strong reducing agent. It reacts at room temperature with ammoniacal silver nitrate solution, precipitating silver. In alkaline solution, it absorbs atmospheric oxygen. In moist air, crystals of H. turn reddish brown due to autooxidation. A solution of H. mixed with an iron(III) chloride solution turns blue for a short period. H. is readily oxidized to 1,4-benzoquinone by a number of mild oxidizing agents. H. and 1,4-benzoquinone form a reversible redox system via a Semiquinone (see):

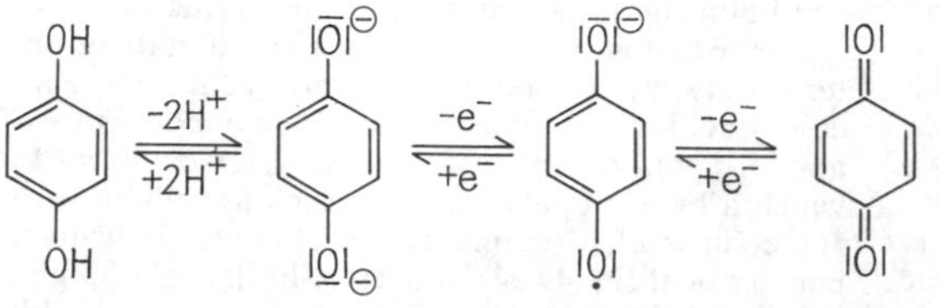

Hydroquinone $\underset{+2H^+}{\overset{-2H^+}{\rightleftharpoons}}$ $\underset{+e^-}{\overset{-e^-}{\rightleftharpoons}}$ Semiquinone $\underset{+e^-}{\overset{-e^-}{\rightleftharpoons}}$ 1,4-Benzoquinone

In the reduction of quinones, deeply colored 1:1 charge-transfer complexes are formed as intermediates. These can also be produced directly from their component quinones and dihydroxyarenes. They are called *quinhydrones*, and are stabilized by hydrogen bonds. The quinhydrone of H. and 1,4-benzoquinone is an emerald-green, crystalline compound (m.p. 171 °C).

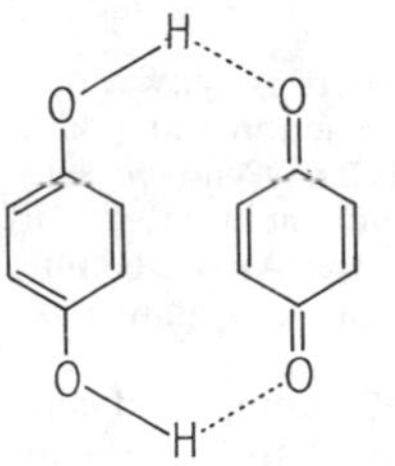

Quinhydrone

Quinhydrone H. exists free and as a component of the glycoside arbutin in various species of plant. It is made industrially by reduction of 1,4-benzoquinone with sulfurous acid, or electrochemically by anodic oxidation of benzene to 1,4-benzoquinone, followed by cathodic reduction to H.

H. is used mainly as a photographic developer, as a disinfectant and as an oxidation inhibitor, e.g. for fats and plant oils, or to make wood stains.

Hydroquinonecarboxylic acid: same as Gentisic acid.

Hydroquinone-β-D-glucopyranoside: same as Arbutin.

Hydrorefining: a process for removing olefins, sulfur, oxygen and nitrogen compounds from petroleum fractions by catalytic hydrogenation. H. can be carried out in the gas, liquid or droplet phase in the presence of a catalyst at temperatures up to 450 °C and hydrogen pressures between 2 and 20 MPa. It is used for the following purposes:

1) Pre-treatment of raw materials for catalytic processes, such as reforming. Before heavy gasoline enters the reformer, the nitrogen, sulfur and oxygen compounds must be removed from it.

2) Improvement of quality of products, such as removal of sulfur compounds from heating, diesel and lubricating oils.

3) Conversion of intermediate products to end-products, for example in production of carburetor fuel components by hydrogenation of cracked gasoline. Reactions:

$$R^1\text{–}S\text{–}R^2 + 2\,H_2 \longrightarrow R^1H + R^2H + H_2S$$

$$R^1\text{–}O\text{–}R^2 + 2\,H_2 \longrightarrow R^1H + R^2H + H_2O$$

$$(R^1R^2)N\text{–}R^3 + 3\,H_2 \longrightarrow R^1H + R^2H + R^3H + NH_3.$$

H. is especially important for removal of sulfur from heating oils. When sulfur-containing oils are burned, especially residual oils from refineries, sulfur dioxide is released. Sulfur dioxide is harmful to human lungs, coniferous forests and (as acid precipitation) fresh-water biota. Modern methods of H. are able to reduce the sulfur contents of residual oils to 0.5%. Sulfides and oxides of nickel, cobalt, tungsten or molybdenum on γ-aluminum oxide are used as catalysts for H.

The hydrogenation reactions produce hydrogen sulfide, water and ammonia. They are exothermic, but because of the low concentration of hetero compounds in the hydrocarbon fractions, no special measures are needed to remove the reaction heat from the reactor.

There are two major methods of H:

a) High pressure refining. In *low-temperature hydrogenation*, raw materials with low asphalt contents, such as tars, shale oil and petroleum are treated with hydrogen in the presence of a solid contact; the reaction occurs in a mixed liquid and gas phase. The asphalts and resins are converted to hydrocarbons, while oxygen, sulfur and nitrogen compounds are removed. The high-molecular-weight components are

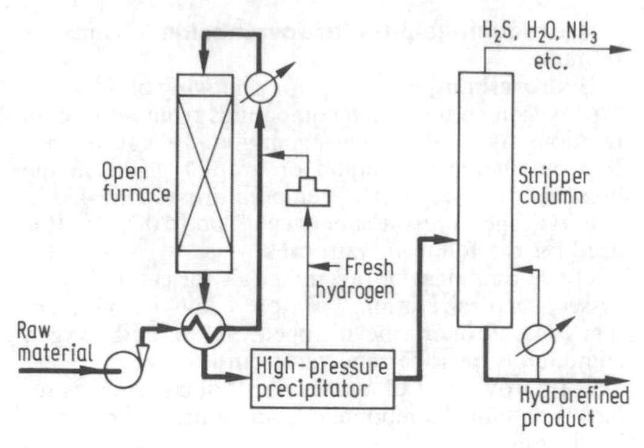

Diagram of a hydro refinery

not degraded, because the temperatures are low, so that the products are chiefly lubricating oils and paraffins, along with small amounts of diesel oil. The temperatures are between 280 and 340 °C, and the hydrogen pressure is about 30 MPa. Tungsten sulfide or tungsten-nickel sulfide catalysts are used on an aluminum oxide carrier.

The *medium-temperature hydrogenation* process is applied to the same raw materials, but at somewhat higher temperatures (340 to 370 °C). The end products in this case are mainly gas oil and gasoline. Pure tungsten sulfide is used as the catalyst.

b) Medium pressure refining. This process was developed in 1927 in Germany, but it was not applied on a large scale in the USA until after the Second World War. Distillation and cracking products in all boiling ranges from gasoline to distillates of petroleum, lubricating oils and coal are used in medium-pressure refineries. The deleterious impurities, such as alkenes, oxygen, sulfur and nitrogen compounds, are almost completely removed from the raw materials. This method has the advantage over sulfuric acid refining that the products are obtained in higher yields and are of better quality. Temperatures between 200 and 430 °C are used; the pressure is between 0.3 and 7 MPa. As the temperature increases, the refining effect is more intense, but at the same time, hydro cracking reactions begin to take place. The oxides of molybdenum and cobalt on clay carriers are the best contacts, but nickel and tungsten sulfides are also used. The contact is regenerated by coke precipitation and burning in air. The medium pressure refining processes began to be developed when large amounts of cheap hydrogen became available from the reforming processes. Some typical processes of this type are the *unifining* and *hydrofining* processes, of which there are many variations. There is also the *trickle process*, in which the raw material is a liquid which flows over the catalyst. In the *autofining process*, hydrogen does not need to be added, because under the reaction conditions, naphthenes are dehydrogenated to aromatics, and provide sufficient hydrogen for the hydrogenation. The product is simultaneously refined and reformed.

Medium-pressure refining also includes pressure refining of crude gasoline; in this step the gasoline is freed of impurities, especially thiophene, without the cumbersome sulfuric acid treatment. The process runs at 400 °C with cobalt oxide/molybdenum oxide catalysts.

Hydrothermal synthesis: the production of crystalline compounds from aqueous solutions in the supercritical state. H. is based on the fact that water in the supercritical temperature and pressure range (T_{crit} = 647 K, p_{crit} = 22.060 MPa) is a much better solvent than under normal conditions. This effect can be used for growing crystals of substances with low solubility in water. For example, under suitable conditions in an autoclave (temperature around 675 K, pressures up to 250 MPa), it is possible to grow single quartz crystals of considerable size at a rate of about 1 mm per day. These crystals are superior to natural quartz crystals with respect to purity and perfection of the crystal lattice. Other substances which can be produced in pure crystalline form through H. are gemstones such as corundum, beryll and emerald, and asbestos. H. corresponds in many cases to the natural conditions under which minerals are formed.

Hydrotropism: an increase of the solubility in water (or glycol) of insoluble or slightly soluble substances by addition of polar organic compounds. For example, lignin can be dissolved in water or glycol to which a hydrotropic compound such as sodium benzoate (in water), sodium xylenesulfonate (in water), benzoic acid (in glycol) or acetylsalicylic acid (in glycol) has been added. The dissolving of wood with hydrotropic salts is called *hydrotropic solubilization*, and the lignin which can be precipitated from such a solution by simple dilution with water is called *hydrotropic lignin*.

Hydroxamic acids: carboxylic acid derivatives with the general formula R-CO-NH-OH. H. exist in a tautomeric equilibrium:

$$R-C(=O)-NH-OH \rightleftharpoons R-C(-OH)=N-OH$$

The separation of the two isomers is only rarely possible. H. have a weakly acidic reaction and a dark violet iron(III) chloride reaction. They can be synthesized by reaction of carboxylic acid esters or halides with hydroxylamine or by the Angeli-Rimini reaction. H. can be converted to primary amines by the Lossen degradation (see).

Hydroxide: in the broad sense, a compound containing an OH group which can be derived formally from the combination of an oxide with water, e.g. $Ca(OH)_2$, $OP(OH)_3$, $Si(OH)_4$, $Fe(OH)_3$. In the narrow sense, the usually basic or amphoteric ***metal hydroxides***. The water-soluble H. of alkali metals, calcium, barium, strontium and thallium(I) dissociate in water, releasing *hydroxide ions* (formerly *hydroxyl ions*) OH^-, which are responsible for the basicity of the solution. Thus sodium hydroxide, for example, would be considered a salt rather than a base according to the Brœnsted definition (see Acid-base concepts). Most of the H. of metals further to the right in the periodic system are amphoteric, in the sense that they form metal salt solutions with acids, and hydroxo complexes or oxoanions with bases. A typical example of an amphoteric H. is aluminum hydroxide, which reacts with hydrochloric acid to form an aluminum chloride solution: $Al(OH)_3$ + 3 HCl → Al^{3+} + 3 Cl^- + 3 H_2O; and with sodium hydroxide to form a tetrahydroxoaluminate solution: $Al(OH)_3$ + NaOH → Na^+ + $[Al(OH)_4]^-$. With the exception of the alkali metal hydroxides, H. are dehydrated at elevated temperatures to form the metal oxides, e.g. 2 $Fe(OH)_3$ → Fe_2O_3 + 3 H_2O.

Hydroxide hydrates: see Oxide hydrates.

Hydroxyacetaldehyde: same as Glycolaldehyde (see).

Hydroxyaldehydes: organic compounds which contain one or more alcohol groups, -OH, in addition to the aldehyde group -CHO. Examples are acetaldol, CH_3-CH(OH)-CH_2-CHO or glycolaldehyde, HO-CH_2-CH(OH)-CHO.

Hydroxyanthraquinones: derivatives of anthraquinone in which one or more of the H atoms bound to the aromatic ring are replaced by hydroxyl groups, -OH. H. are stable, crystalline compounds which are yellow, red or brown in color. Most of them are barely soluble in water, and somewhat more soluble in alcohol and ether. They are phenolic in character; the strength of this effect depends on the position of the OH group. The H atoms of the OH groups which are adjacent to a carbonyl group, that is, in position 1, 4, 5 or 8, form hydrogen bonds to the carbonyl oxygen atoms. Compounds of this type readily form complexes with heavy metal ions. On the other hand, the reactivity of the OH groups is reduced markedly by the H-bond; the acidity of the compound drops, and alkylation and acylation reactions are much more difficult. OH groups in positions 2, 3, 6 and 7, on the other hand, display typical phenolic behavior. H. are synthesized either by ring-closing reactions between phthalic anhydrides and phenols in the presence of acid, or by substitution of OH groups for suitable substituents in an anthraquinone system. In the Bohn-Schmidt reaction (see), even hydrogen atoms are oxidatively replaced by OH groups to form H. H. are used as dyes themselves, and also as intermediates in the synthesis of a broad palette of related dyes.

1-Hydroxyanthraquinone

2-Hydroxybenzaldehyde: same as Salicylaldehyde (see).

Hydroxybenzene: same as Phenol (see).

Hydroxybenzenesulfonic acids: same as Phenolsulfonic acids (see).

Hydroxybenzoic acids, ***phenolcarboxylic acids***: three isomeric compounds, of which Salicylic acid (see) (2-hydroxybenzoic acid) is the most important.

3-Hydroxybenzoic acid: colorless crystals; m.p. 202-203 °C. It is soluble in water, ethanol and ether. It is made by sulfonation of benzoic acid in position 3 with sulfuric acid, followed by alkali fusion. It gives a brown color reaction with iron(III) chloride.

4-Hydroxybenzoic acid: colorless crystals, m.p. 214.5-215.5 °C. It is readily soluble in hot water and ethanol. It crystallizes out of aqueous solutions as the monohydrate. 4-H. is found in plants, and can be extracted from various natural resins. The acid gives a dark red-brown color reaction with iron(III) chloride. It is synthesized industrially by the Kolbe-Schmitt synthesis (see) from sodium phenolate and carbon dioxide under pressure; this is similar to the synthesis of salicylic acid. 4-H. is an intermediate for organic syntheses of pigments, drugs, etc. Its esters are used as preservatives for foods and to make cosmetics.

2-Hydroxybenzyl alcohol: same as Salicyl alcohol (see).

Hydroxybiphenyls: three isomeric phenyl-substituted phenols: ***2-hydroxybiphenyl***, m.p. 58-60 °C, b.p. 286 °C; ***3-hydroxybiphenyl***, m.p. 78 °C, b.p. 300 °C; ***4-hydroxybiphenyl***, m.p. 165-167 °C, b.p. 305-308 °C (subl.). H. form colorless crystals which are barely soluble in water, but dissolve easily in alkali hydroxide solution and most organic solvents. They are found in crude phenol, from which they can be isolated. They are synthesized by the same methods as phenols. They are bacteriocidal and fungicidal, and are used to make disinfectants and preservatives. The sodium salts of H. are sometimes used as color accelerators in the dyeing of PVC and polyester fibers.

2-Hydroxybiphenyl 3-Hydroxybiphenyl 4-Hydroxybiphenyl

Hydroxybrasilin: same as Hematoxylin (see).

3-Hydroxybutanal: same as Acetaldol (see).

Hydroxybutanedioic acid: same as Malic acid (see).

3-Hydroxybutan-2-one: same as Acetoin (see).

Hydroxybutanoic acids, ***hydroxybutyric acids***: the three isomeric hydroxy-*n*-butanoic acids. They are steam volatile and dissolve readily in water, alcohol and ether. ***2-H., α-hydroxybutyric acid***: CH_3-CH_2-CH(OH)-COOH, a colorless, crystalline compound; m.p. 44-44.5 °C (RS form), b.p. 266 °C (dec.). It is formed by hydrolysis of 2-chlorobutanoic acid. ***3-H., β-hydroxybutyric acid***: CH_3-CH(OH)-CH_2-COOH, a colorless, crystalline compound; m.p. 48-50 °C (RS form), b.p. 94-96 °C at 13.3 Pa. The com pound is synthesized by hydrolysis of β-butyrolactone. 3.-H. can easily be converted to crotonic acid by intramolecular dehydration. ***4-H., γ-hydroxybutyric acid***:, HO-CH_2-CH_2-CH_2-COOH, a colorless, syrupy liquid; m.p. 15-18 °C, b.p. 178-180 °C (dec.). It can be synthesized by hydrolysis of γ-butyrolactone. 4-H. stabilizes the blood pressure and relieves pain.

Hydroxybutyric acids: same as Hydroxybutanoic acids (see).

Hydroxycarboxylic acids: carboxylic acids in which one or more of the hydrogen atoms in the hydrocarbon part has been replaced by a hydroxyl group. The hydroxyl group can be bound to an aliphatic, aromatic or heterocyclic group. The names for H. with alcoholic hydroxyl groups is based on the name of the unsubstituted carboxylic acid; the position(s) and number of hydroxyl group(s) are given by prefixes. This system is also used in conjunction with the trivial names of the carboxylic acids, e.g. lactic acid, 2-hydroxypropionic acid. Most H. are liquid (except for glycolic acid, a solid), and are very soluble in water. They can usually be synthesized by catalytic hydrogenation of the oxocarboxylic acid esters, followed by hydrolysis of the ester, by acid hydrolysis of cyanohydrins, which are easily made from aldehydes or ketones by addition of HCN, by hydrolysis of halogen carboxylic acids, by the Reformatsky reaction (see), oxidation of hydroxyaldehydes, or the benzilic acid rearrangement. The properties of H. are typical of compounds containing their functional groups. Thermal dehydration may involve cooperative reactions of the hydroxyl and carboxyl groups, with the products dependent on the relative positions of the two functional groups. Intermolecular dehydration of α-H. forms 3,6-dialkyl-1,4-dioxane-2,5-diones (see Lactides); at a higher temperature, these are decarbonylated to form aldehydes:

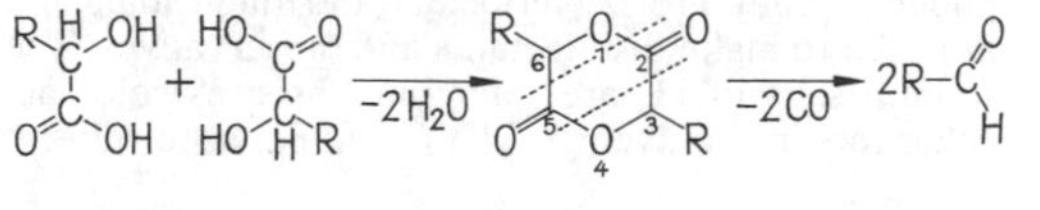

In contrast, the dehydration of the β-H. is intramolecular, and yields α,β-unsaturated carboxylic acids. Intramolecular dehydration of γ- or δ-H., which in some cases requires only acidification of their salts, forms γ- and δ-lactones, respectively. These can be considered internal esters of these carboxylic acids. Longer-chain H., in which the distance between the hydroxyl and carboxyl groups is even greater, form polyesters by intermolecular loss of water. α-H. are split by thermal treatment in the presence of dilute mineral acids (e.g. HCl, H_2SO_4) into formic acid and an aldehyde or ketone. H. can be reduced to carboxylic acids with hydrogen iodide. A few H. are central to metabolic processes, e.g. lactic, malic, tartaric and citric acids. Salicylic acid and gallic acid are important phenol carboxylic acids. The Kolbe-Schmitt synthesis (see) is a general method for preparation of these compounds; there are also special methods for preparation of individual phenol carboxylic acids.

Hydroxycinnamic acids: colorless to slightly yellow crystalline compounds (table) which are barely soluble in cold water, but dissolve readily in ethanol. H. are found in free form esterified or as glycosides in many fruits and vegetables.

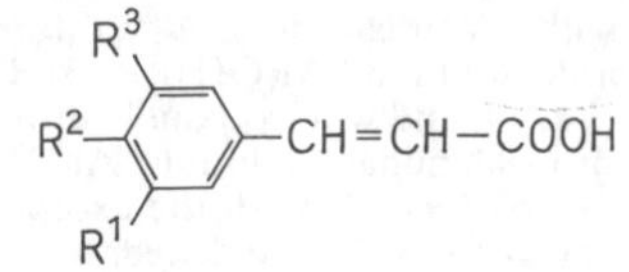

Trivial name	R^1	R^2	R^3	m.p. in °C
4-Coumaric acid	H	OH	H	210–213
Ferulic acid	OCH_3	OH	H	168–169
Caffeic acid	OH	OH	H	195–198
Sinapic acid	OCH_3	OH	OCH_3	191–192

2-Hydroxycinnamic lactone: same as Coumarin (see).

2-Hydroxycyclohepta-2,4,6-trienone: same as α-Tropolone (see).

2-Hydroxy-2,2-diphenylethanoic acid: same as Benzilic acid (see).

2-Hydroxyethylamine: see Ethanolamines.

Hydroxyfumaric acid: see Oxaloacetic acid.

Hydroxyhydroquinone, ***benzene-1,2,4-triol***: an unsymmetrical triphenol. H. forms colorless crystals; m.p. 140-141 °C. Its chemistry is similar to that of its isomers pyrogallol and phloroglucinol, but it is not industrially important.

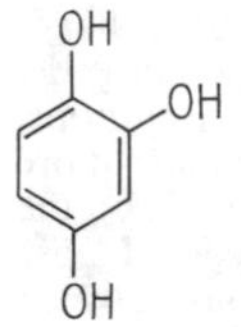

5-Hydroxyiminobarbituric acid: same as Violuric acid (see).

α-Hydroxyisobutyronitrile: same as Acetone cyanohydrin (see).

Hydroxyketone: a hydroxycarbonyl compound which has, in addition to the ketogroup >C=O, one or more alcoholic groups -OH in the molecule. Depending on the relative positions of the CO- and OH-groups, the H. can be a 1,2- or α-H, a 1,3- or β-H., 1,4- or γ-H., etc. 1,2-H. with the general formula R-CO-CH_2-OH are called *ketols*. H. are present in nature in great abundance and variety, especially among the carbohydrates (the ketoses). Other H. are the acyloins, e.g. acetoin, benzoin and furoin.

Hydroxykynurenine: see Kynurenine.

Hydroxylamine: H_2N-OH, colorless, hygroscopic crystals which are readily soluble in water and ethanol; M_r 33.03, density 1.204, m.p. 33.05 °C, b.p. 56.5 °C. The isolated compound tends to decompose explosively. Therefore the use of ***hydroxylammonium salts***, $[H_3N\text{-}OH]^+X^-$, is preferable. These salts can be obtained by the reaction of H. with strong acids. The salts, like the free base ($K_B = 6.6 \cdot 10^{-9}$) are strong reducing agents and are poisonous. They form Oximes (see) with aldehydes and ketones, and Hydroxamic acids (see) with carboxylic acid halides or esters. H. can be synthesized electrolytically by reduction of nitrates or nitrites. H. or its salts are used in organic synthesis, e.g. in the industrial production of caprolactam.

Hydroxylases: see Oxygenases.

Hydroxylation: introduction of hydroxyl groups into alkenes to form 1,2-diols (glycols). Treatment of alkenes or cycloalkenes with dilute alkaline solutions of potassium permanganate, $KMnO_4$, produces glycols via cyclic intermediates (Baeyer's test); at the same time, the violet potassium permanganate is converted to a brown manganese(IV) oxide hydrate precipitate. Osmium(VIII) oxide can be used instead of potassium permanganate; more recently the less toxic ruthenium(VIII) oxide has been used.

For C hydroxylation, see Biotransformation.

Hydroxyl group: the atomic grouping -OH, which can occur as a substituent in inorganic or organic compounds, such as hydroxylamine, NH_2-OH, methanol, CH_3-OH, phenol, C_6H_5-OH or acetic acid, CH_3CO-OH. The presence of the H. in a hydrocarbon compound increases its solublity in water (hydrophilic effect); in some types of compounds, it generates acidity.

Hydroxyl ion: see Hydroxides.

Hydroxymaleic acids: see Oxaloacetic acid.

Hydroxymalonic acids: same as Tartronic acids (see).

4-Hydroxy-2-mercaptopyrimidine: same as 2-Thiouracil (see).

Hydroxymethylene ketone: a β-ketoaldehyde which exists in the enol form. Like all aliphatic 1,3-dicarbonyl compounds, β-ketoaldehydes undergo keto-enol tautomerism:

$$\underset{A}{\text{R–C(OH)=CH–CHO}} \rightleftharpoons \underset{B}{\text{R–CO–CH}_2\text{–CHO}} \rightleftharpoons \underset{C}{\text{R–CO–CH=CH–OH.}}$$

In β-ketoaldehydes, the equilibrium is almost entirely on the side of form C.

2-Hydroxymethylfuran: same as Furfuryl alcohol (see).

3-Hydroxy-2-methyl-4H-pyran-4-one: same as Maltol (see).

3-Hydroxymethylpyridine, ***Radecol***®: a compound obtained by lithium alanate reduction of nicotinic acid esters; it has a dilating effect on blood vessels.

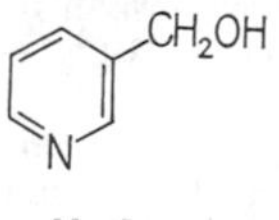

CH_2OH

N

Hydroxynaphthalenes: same as Naphthols (see).

α-Hydroxynitriles: same as Cyanohydrins (see).

α-Hydroxyphenylacetic acid: same as Mandelic acid (see).

2-Hydroxy-2-phenylethanoic acid: same as Mandelic acid (see).

4-Hydroxyphenylethylamine: same as Tyramine (see).

Hydroxyprogesterone: see Gestagens.

Hydroxyprogesterone acetate: see Gestagens.

Hydroxyprogesterone capronate: see Gestagens.

Hydroxypropanedioic acid: same as Tartronic acid (see).

2-Hydroxypropanoic acid: same as Lactic acid (see).

2-Hydroxypropane-1,2,3-tricarboxylic acid: same as Citric acid (see).

α-Hydroxypropionic acid: same as Lactic acid (see).

β-Hydroxypropionic lactone: same as propiolactone (see).

8-Hydroxyquinoline, ***oxine***: a complexing agent for multivalent metal ions (see Oxinates); its activity as a disinfectant and antimycotic is due to its ability to bind metals essential to the microorganisms. 8-H. forms yellow crystals with a phenol-like odor; m.p. 75-76 °C, b.p. 267 °C. It can be synthesized by sulfonation of quinoline to quinoline-8-sulfonic acid, followed by alkali fusion, or by reaction of 2-aminophenol with acrolein according to Skraup. Halogen-substituted and sulfonated derivatives may have a more specific complexing behavior and are used as antiseptics, for example 8-hydroxy-7-iodoquinoline-5-sulfonic acid.

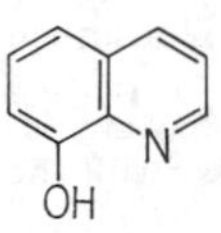

N

OH

Hydroxysuccinic acid: same as Malic acid (see).

Hydroxytoluene: same as Cresol (see).

β-Hydroxytricarboxylic acid: same as Citric acid (see).

Hygrin: see Alkaloids.

Hygroscopic: attracting water. A H. substance absorbs water from the air, which always contains some water vapor. If the substance is a solid, it then usually deliquesces. This property is especially strong in compounds which dissolve very readily in water, such as calcium chloride and phosphorus(V) oxide.

α-Hyodeoxycholic acid: 3α,6α-dihydroxycholanoic acid, a bile acid which is found as a glycoconjugate in porcine bile. When crystallized out of glacial acetic acid, α-H. forms colorless crystals which are slightly soluble in water and readily soluble in organic solvents; m.p. 197 °C, $[\alpha]_D$ +8° (ethanol). 3β,6α-dihydroxycholanoic acid, which is also present in porcine bile, is called β-H.

Hyoscine: same as Scopolamine (see).

Hyoscyamine: see Atropine.

Hyp: 1) abb. for 4-hydroxyproline (see Proline). 2) abb. for hypoxanthine.

Hyperchromic shift: see UV-VIS spectroscopy.

Hyperconjugation: interaction between the π electrons of multiple bonds and the σ electrons of neighboring C-R bonds (where R = H or an alkyl group). These can be interpreted as delocalization. H. was interpreted quantum mechanically by Mulliken (1939) in terms of molecular orbital theory. According to this interpretation, linear combinations of the equivalent molecular orbitals of the three C-H bonds of a methylene group can be constructed so that one of them has π symmetry. This can participate formally in the π system, in that an additional π orbital with two electrons is made available. Today, H. is used mainly to explain the stabilization of carbenium ions by adjacent alkyl groups. For example, the ethyl cation, CH_3-CH_2^+, is stabilized by 140 kJ mol^{-1} with respect to the methyl cation CH_3^+ by H. of the unoccupied p_π. orbital on the carbenium carbon atom. The introduction of more methylene groups on the carbenium carbon atom increases the stability; the *tert.*-butyl cation is the most stable of this series.

Hyperfine structure: a splitting of spectral lines beyond the normal fine structure (multiplet structure). In contrast to the fine structure of the spectral lines, which is due to spin-orbital coupling in the electronic shell, H. is due to interactions between the electrons and the nucleus. There are two causes of H., which may appear singly or together: 1) an interaction of the electrons with the magnetic dipole moments or quadrupole moments of the nucleus; and 2) an isotope effect due to the presence of an isotope mixture in the corresponding element. The hyperfine splittings are very small, lying in the range from 10 to 1 pm.

Hypergols: see Fuels.

Hyperm®: same as Permalloy® (see).

Hypersorption process: an industrial process for separation of hydrocarbons. The mixture is adsorbed on activated charcoal, then desorbed by passing steam across the charcoal. The hydrocarbons are desorbed according to their molecular mass, with the smallest molecules leaving first.

Hypertensin: same as Angiotensin (see).

Hypertonic solution: see Osmosis.

Hypnotics, ***sleep inducers***: compounds which suppress the central nervous system more strongly than Sedatives (see); they promote falling asleep and lead to a sleep-like state. In low doses, H. act as sedatives. Some have only short-term effects, so that they induce sleep but do not prolong it, while others cause sleep to last all night. The length of the action depends on the rates at which they are metabolized and eliminated and on the way they are distributed in the body. Barbitals, piperidine diones, such as pyrithyldione (didropyridine) and glutethimide, and quinazolin-4-one are important H.

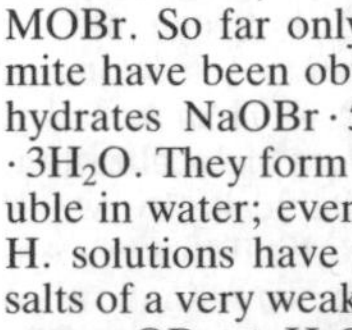

Glutethimide

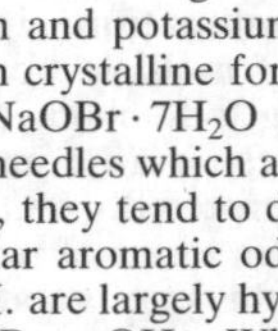

Didropyridine

Hypobromites, ***bromates(I)***: the salts of hypobromous acid, HOBr; they have the general formula MOBr. So far only sodium and potassium hypobromite have been obtained in crystalline form, as their hydrates $NaOBr \cdot 5H_2O$, $NaOBr \cdot 7H_2O$ and $KOBr \cdot 3H_2O$. They form yellow needles which are very soluble in water; even at 0 °C, they tend to decompose. H. solutions have a peculiar aromatic odor. As the salts of a very weak acid, H. are largely hydrolysed in water: $OBr^- + H_2O \rightleftharpoons HOBr + OH^-$. When heated, dissolved H. disproportionate to bromide and bromate: $3\ MoBr \rightarrow 2\ MBr + MBrO_3$. H. is obtained, along with bromides, by passing bromine through alkali hydroxide solutions: $Br_2 + 2\ MOH \rightarrow MOBr + MBr + H_2O$. Aqueous solutions of H. are sometimes used as oxidizing agents.

Hypobromous acid, ***bromic(I) acid***: HOBr, a weak acid known only in the form of its aqueous solutions (p*K* 8.66). Even at 30 °C, solutions of H. decompose, more rapidly in the light, by releasing oxygen. They are therefore strongly oxidizing. The anhydride of H. is dibromine oxide (see Bromine oxides), which is also very unstable. The salts of H. are called Hypobromites (see). H. is made by shaking bromine water with mercury oxide: $2\ Br_2 + H_2O + 2\ HgO \rightarrow HgO \cdot HgBr_2 + 2\ HOBr$.

Hypochlorites: the salts of hypochlorous acid, HOCl; they have the general formula MOCl. As salts of a very weak acid, H. are hydrolysed in solution: $OCl^- + H_2O \rightarrow 2\ HOCl + OH^-$. The hypochlorous acid formed in this way is responsible for the strong oxidizing effects of aqueous solutions of H., and therefore their bleaching effects, especially those of the alkali hypochlorites. H. is obtained along with chlorides by the reaction of chlorine with alkali hydroxide solution in the cold: $Cl_2 + 2\ MOH \rightarrow MClO + MCl + H_2O$. Alkali hypochlorites are also formed by chloralkali electrolysis performed in the cold, if a diaphragm is not present. In this case, the chlorine formed at the anode reacts with the alkali hydroxide formed at the cathode.

Hypochlorous acid, ***chloric(I) acid***: HOCl, a very weak acid (p*K* 7.537) known only in aqueous solution. When concentrated, such solutions release dichlorine oxide, Cl_2O, the anhydride of H. HOCl solutions have a faintly yellow color and a peculiar odor. They decompose slowly in the dark and very rapidly in sunlight: $HOCl \rightarrow HCl + 1/2O_2$. Further reaction of the nascent oxygen with H. also yields chloric(V) acid: $HClO + 2\ O \rightarrow HClO_3$. The release of oxygen is responsible for the very strong oxidizing effect of H. The salts of H. are called Hypochlorites (see). H. is prepared by passing chlorine through an aqueous suspension of mercury(II) oxide: $2\ Cl_2 + 2\ HCO + H_2O \rightarrow 2\ HOCl + HgO \cdot HgCl_2$.

Hypochromic shift: see UV-VIS spectroscopy.

Hypodisulfates: see Dithionates.

Hypodisulfites: see Dithionites.

Hypodisulfurous acid: see Dithionites.

Hypofluorous acid: HOF, a white compound which is stable only at low temperatures; M_r 36.01, m.p. - 117 °C. H. decays at 25 °C with a half-life of less than 1 h according to the equation: $2\ HOF \rightarrow 2\ HF + O_2$, or $2\ HOF \rightarrow F_2O + H_2O$. It reacts with water to yield hydrogen peroxide: $HOF + H_2O \rightarrow H_2O_2 + HF$.

Hypoiodous acid, ***iodic(I) acid***: HOI, a very unstable and very weak acid (p*K* 10.64). HOI is formed by the reaction of iodine with a suspension of mercury(II) oxide in water. Its salts are more stable than HOI. The ***hypoiodites*** or ***iodates(I)***, MIO, are formed by the reaction of iodine with cold alkali hydroxide solutions; however, these tend to disproportionate to iodide and iodate(V) even at room temperature.

Hyponitrites: the salts of Hyponitrous acid (see).

Hyponitrous acid: $H_2N_2O_2$, colorless, extremely explosive leaflets which are readily soluble in water and alcohol; M_r 62.03, molecular structure HO-N=N-OH. In aqueous solution, H. decomposes to dinitrogen oxide and water: $H_2N_2O_2 \rightarrow N_2O + H_2O$. As a weak dibasic acid, it forms two series of *hyponitrites*: the very unstable acidic hyponitrites $M^IHN_2O_2$ and the somewhat more stable neutral hyponitrites $M^I_2N_2O_2$. These salts are obtained from nitrites by reduction with sodium amalgam. H. can be released from its salts by the action of hydrochloric acid.

Hypophosphates: the salts of Hypophosphoric acid (see).

Hypophosphites: the salts of Hypophosphorous acid (see).

Hypophosphoric acid, ***hypodiphosphoric acid***: $H_4P_2O_6 \cdot 2H_2O$, deliquescent, water-soluble rhombic crystals; M_r 198.01, m.p. 70 °C, dec. 100 °C. As a tetrabasic acid, H. forms four series of hypophosphates. A characteristic of both the acid and its salts is the direct P-P bond (Fig.) in the molecule. H. and the hypophosphates are not as strong reducing agents as phosphorous acid and its salts. In aqueous solution, H. disproportionates to form phosphorous acid and orthophosphoric acid: $H_2P_2O_6 + H_2O \rightarrow H_3PO_3 + H_3PO_4$. H. is obtained via hypophosphate by oxidation of finely divided red phosphorus with sodium chlorite or hydrogen peroxide, e.g. $2\,P + NaClO + 8\,H_2O \rightarrow Na_2H_2P_2O_6 + 2\,HCl$. The acid is released as the dihydrate by ion exchange; the dihydrate can be dehydrated to yield the free acid, $H_4P_2O_6$, m.p 54.8 °C, using phosphorus(V) oxide.

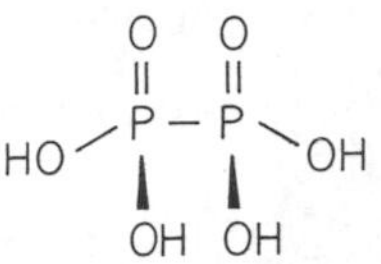

Hypophosphorous acid: H_3PO_2, colorless crystals, soluble in water, alcohol and ether; M_r 66.00, density 1.493, m.p. 26.5 °C, dec. 130 °C. H. has a tetrahedral structure with two hydrogen atoms bonded directly to the phosphorus. As a result, it is monobasic and forms only a single series of salts, the *hypophosphites* $M^IH_2PO_2$. The acid and its salts are strong reducing agents and are able, for example, to reduce Cu^{2+} ions all the way to copper hydride, CuH. H. is obtained via hypophosphite by heating white phosphorus in a solution of barium or calcium hydroxide, e.g. $2\,P_4 + 3\,Ba(OH)_2 + 6\,H_2O \rightarrow 2\,PH_3 + 3\,Ba(H_2PO_2)_2$; the free acid is formed by further reaction with sulfuric acid. Hypophosphites are used as reducing agents in nonelectrolytic nickel plating.

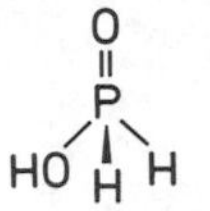

H_2PO_2 tetrahedron

Hyposulfurous acid: see Sulfoxylic acid.

Hypothalamus hormones: 1) neurohormones formed in the hypothalamus which promote or inhibit the release of anterior pituitary hormones. The Liberins (see) promote the formation and secretion of the anterior pituitary hormones, while the Statins (see) inhibit the release of somatotropin, prolactin or melanotropin.

2) In a broader sense, all the neuropeptides formed in the hypothalamus. In addition to the liberins and statins, there are the neurohypophyseal hormones oxytocin and vasopressin, substance P., neurotensin, proctolin, etc.

Hypotonic solution: see Osmosis.

Hypoxanthine, abb. ***Hyp***: 6-hydroxypurine. H. forms colorless needles; m.p. 150 °C (dec.). It is barely soluble in cold water, but it dissolves in acids and bases by forming salts. H. is an intermediate in biological degradation of purines to uric acid. The H. derivative inosine 5'-monophosphate is the starting material for biosynthesis of the guanine and adenine derivatives.

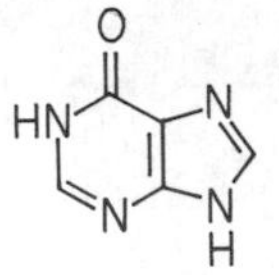

Hypsochromicity: a change in color due to a shift of the absorption bands toward shorter wavelengths when certain atomic groups, e.g. methyl, ethyl and some alkyl residues, are introduced into a molecule. Opposite: Bathochromicity (see).

I

I: 1) Symbol for iodine; 2) abb. for inosine; 3) symbol for retention index.

Ibuprofen: an aralkylcarboxylic acid antiphlogistic.

IC: see Photophysical processes.

Ice: the solid form of water. The freezing point is 0 °C at a pressure of 101.3 kPa (760 Torr). Depending on its air content, I. has a density between 0.86 and 0.92 kg dm^{-3}, which is lower than that of liquid water at 0 °C. The heat of melting is 334 kJ kg^{-1}, which is quite high. I. thus has a high cooling capacity, and is often used to keep perishable items cold during transport. The melting or freezing point increases with pressure. I. should not be confused with Dry ice (see).

Ice dyes: a group of development dyes for which the coupling component is soaked into the fibers (e.g. cotton), and then developed with an ice-cold diazonium solution. The reverse procedure (diazotization of the fiber followed by coupling) is also possible.

Ichthammol: see Shale oil sulfonates.

ICP: see Inductive coupled plasma.

ICR: abb. for ion-cyclotron resonance.

ICSH: see Lutropine.

Idaein: see Cyanidin.

Ideal crystal: see Crystal.

Ideal flow pipe: same as Pipe reactor (see).

Ideal gas: see Gas.

Ideal gas law: see State, equation of 1.1.

Ideal mixture: see Mixture.

Ideal structure: see Crystal.

Identification: establishment of the identity of a substance with one which has already been described in the literature. When two substances are identical, all their chemical and physical properties will be the same. In I., therefore, the procedure is to gather as much information as possible on the unknown substance. This can include properties such as melting and boiling points, density, refractive index, optical rotation, spectroscopic data and other physical data. Comparison of the data with published data reveals whether substance has already been described. If not, it is a new compound and its structure must be determined (see Structure elucidation). A necessary prerequisite for an I. is that the sample be pure.

Identity operation: see Symmetry.

Idose: see Monosaccharides.

Idoxuridine: 5-iododeoxyuridine, a pyrimidine nucleoside antimetabolite used locally as a virostatic for *Herpes simplex* infections of the cornea.

IEE: see Photoelectron spectroscopy.

I-effect: abb. for inductive effect.

IEP: abb. for isoelectric point.

IFN: abb. for interferon.

Ig: abb. for immunoglobulin.

Ignition: the beginning of a combustion, which begins as soon as one area of a combustible material in contact with air or oxygen reaches a certain temperature, the ***ignition temperature***.

With liquid combustibles, a distinction is made between the flame point and the fire point. These temperatures are determined in an apparatus consisting of a heatable cylindrical vessel and an ignition flame which is mounted a certain distance above the top of the cylinder and above the surface of the liquid. This flame can be rotated. The *flame point* of a material is the temperature at which it first flames briefly when the ignition flame is directed over it; the *fire point* is the temperature at which the flame burns indefinitely after ignition. The higher the boiling point of the liquid, the greater the difference between flame point and fire point. The flame point is used as a means of comparing the relative flammabilities of combustible liquids and is the basis for establishing safety regulations concerning the storage and transport of such liquids.

Ile: abb. for isoleucine.

Imazalil: see Azole fungicides.

Imidazole, ***1,3-diazole***: a heterocyclic nitrogen compound. I. forms colorless crystals; m.p. 90-91 °C, b.p. 257 °C. It is readily soluble in water, alcohol, ether, chloroform and pyridine, and only slightly soluble in benzene. I. forms metal salts by replacement of the H atom of the NH group, but it is also a stronger base than pyrazole and yields stable salts with acids (protonation of the tertiary N atom). It is very resistant to oxidizing and reducing agents. The completely hydrated form of I. is called ***imidazolidine***. Bromination of I. yields 2,4,5-tribromoimidazole, while nitration occurs in positions 4 and 5. I. couples with diazonium salts in the 2 position, if it is unsubstituted. In the reaction with acyl chlorides in alkaline medium, ring cleavage occurs and derivatives of diaminoethylene are formed. I. can be synthesized from glyoxal, ammonia and formaldehyde, and this is the source of the obsolete name ***glyoxaline***; it can also be synthesized by heating bromoacetaldehyde diglycol acetal with formamide to 180 °C and passing ammonia through the mixture. Today, it is obtained primarily by decarboxylation of the readily availabe imidazole-4,5-dicarboxylic acid with copper(II) oxide. I. is an important parent compound of natural products, such as the alkaloid pilocarpine, histidine and histamine. It is also a component of the purines. I. and its derivatives are used as intermediates in the synthesis of plastics, pigments, pharmaceutical products, pesticides and textile conditioners.

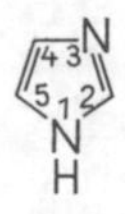

Imidazolidine: see Imidazole.

Imidazolidine-2,4-dione: same as Hydantion (see).

Imidazolidine-2,4,5-trione: same as Paranabic acid (see).

Imidazoline: the dihydro compound of imidazole. The position of the double bond is variable, giving 2-I., 3-I. and 4-I. Derivatives of I. are used as hardeners for epoxide resins and as emulsifiers.

Imide halides: Derivatives of the iminol form of carboxamides with the general formula R-C(X)=NH, X = halogen. The I. are formed by addition of hydrogen halides to nitriles:

$$R{-}C{\equiv}N + HX \longrightarrow [R{-}C{\equiv}\overset{+}{N}H \longleftrightarrow R{-}\overset{+}{C}{=}NH]X^- \rightleftharpoons R{-}C(X){=}NH$$

Where X = Cl, the equilibrium is on the side of the nitrilium chloride. Imide bromides (X = Br) and iodides (X = I) are more stable. Imide chlorides can also be synthesized from *N*-aryl- or *N*-alkylamides and phosphorus(V) chloride. The reduction of *N*-aryl-imide chlorides with $SnCl_2$/HCl yields Schiff's bases.

Imines, ***imino compounds***: 1) compounds containing the structural element C=NH, as in aldimines, R-CH=NH and ketimines, R-C(=NH)-R; 2) cyclic secondary amines, e.g. ethylenimine,

$CH_2{-}CH_2$ bridged by N–H

3) term for acyclic secondary amines which contain functional groups with higher priority, e.g. 4,4'-iminodibenzoic acid:

HOOC–C_6H_4–NH–C_6H_4–COOH

Iminium compounds: same as Immonium compounds (see).

Imino compound: same as Imine (see).

Imino group: the functional group -NH-, in which the two bonds on the N atom can be shared with one C atom >C=NH or two different C atoms >C–NH–C<

Iminourea: same as Guanidine (see).

Imipramine: see Dibenzodihydroazepine.

Immedial pigments: a large group of Sulfur pigments (see) made by heating sulfur with sodium polysulfide.

Immobilization: see Biotechnology.

Immonium compounds, ***iminium compounds***: ionic organic compounds containing the general structural element $>C{=}\overset{+}{N}<\ X^-$. They are formed, for example, as intermediates in the Vilsmeier-Haack reaction (see) and by the synthesis of ketones from nitriles and Grignard reagents. I. are generally very prone to hydrolysis; in the presence of water the amino group is converted to an aldehyde or ketone. In a few cases, I., especially in the form of the perchlorates, are stable and can only be hydrolysed at higher temperatures.

Immunoassay: a sensitive method of analysis based on the competitive binding of an antibody to an antigen. The antibody is labelled by covalent attachment to a radioactive isotope (usually ^{125}I), an enzyme or a fluorescent dye molecule. In ***radioimmunoassay***, the antigen is bound to a plastic microculture well, and the amount of labelled antibody which binds to the antigen is measured. (In a competitive assay, the unlabelled unknown antibody displaces the labelled antibody.) In an ***ELISA***, or enzyme-linked immunosorbent assay, the antibodies are labelled by covalent bonding to an enzyme. The amount of bound antibody is determined from the catalytic activity of the enzyme.

Immunoglobulins, abb. ***Ig***: specific vertebrate defense proteins which combine with the antigen which elicited their synthesis. I. are present in the blood, lymph and body secretions.

Immunosuppresives: compounds which suppress an immune reaction. They are used in conjunction with organ transplants to prevent rejection of the organ, and in autoimmune diseases in which there is an immune response to the body's own substances. Cytostatics, such as cyclophosphamide and azathioprin, can be used as I., as can the glucocorticoids and antilymphocyte sera.

IMP: see Inosine.

Impinger: a wash bottle filled with a special absorption solution. It is used to enrich gaseous trace substances from the atmosphere (a specific volume of air is pulled through the solution). It is used in Air analysis (see).

Impregnating agents: solutions or emulsions used to treat solids to make them water-tight, fire-resistant (see Fire retardants) or resistant to rot and pests. I. are used on textiles to make them water resistant. Washable I. used with cellulose fibers are pyridinium compounds; with wool, stearic acid-chromium complexes; and with synthetic fibers, silicones. Paraffin or wax emulsions, metal soaps and aluminum salts are used where washing is not required. Dry I. used in dry cleaning are mainly metal soaps.

In: symbol for indium.

Incendiaries: chemical substances which, on combustion, develop high temperatures, burn for a long time, cling and are difficult to quench. They are used as chemical weapons. ***Firebombs*** are an important class of I.; they may weigh 0.5 to 500 kg. I. may consist of Napalm (see), metallic mixtures (pyrogel), thermite and thermite mixtures, white phosphorus or magnesium alloys.

Inconel®: nickel-chromium alloys (see Nickel alloys) with high strength at high temperatures which are resistant to oxidizing and reducing agents. The alloy of 80% nickel and 20% chromium is used as a thermal conductor and electrical resistance. Because of its resistance to sulfur-containing flue gases, the alloy of 80% nickel, 14% chromium and 7% iron is used as steel pipe in the construction of chemical furnaces. I. is also resistant to nitrating and carbonizing furnace atmospheres.

Indamines: amino derivatives of *N*-phenylquinonediimine. The parent compound of the I. is ***phenylene blue***.

$H_2N-C_6H_4-N=C_6H_4=NH$

Phenylene blue

The I. are synthesized by oxidative coupling of a *p*-phenylenediamine containing at least one unsubstituted amino group and an aromatic amine, or by condensation of 4-nitrosodimethylaniline with an aromatic amine in acidic solution:

$$R_2N-C_6H_4-N=O + C_6H_5-NR_2 \xrightarrow[-H_2O]{+HX} R_2\overset{+}{N}=C_6H_4=N-C_6H_4-NR_2\,X^-$$

I. are not suitable as textile dyes, because they are sensitive to acids. However, they are important intermediates for the synthesis of azine dyes. An example of the I. is Bindschedler's green.

Indane, ***hydrindene***: a fused hydrocarbon with one aromatic and one alicyclic C_5 ring. I. is a liquid; m.p. -51.4 °C, b.p. 178 °C, n_D^{20} 1.5378. It is insoluble in water, but is soluble in ethanol and ether. I. is found in coal tar. It can be synthesized by hydrogenation of indene.

Indanone: a ketone derived from indane. ***Indan-l-one***, colorless crystalline leaflets; m.p. 42 °C, b.p. 129 °C at 1.6 kPa. It is synthesized by an intramolecular Friedel-Crafts reaction of 3-phenylpropionyl chloride in the presence of aluminum chloride. ***Indan-2-one***, colorless crystalline needles; m.p. 59 °C, b.p. 110-112 °C at 1.73 kPa. It can be made by oxidation of indene with hydrogen peroxide in formic acid. The I. are readily soluble in alcohol and ether. They are used mainly for organic syntheses.

Indanthrene dye: a term indicating quality for a selection of dyes from all chemical classes, especially the anthraquinone vat dyes, which are very fast to light and washing, and are very brilliant. The name "indanthrene" is a composite of "indigo" and "anthracene"; the first I., ***indanthrene (indanthrene blue RS®)*** gives an indigo-like color and is a derivative of anthracene.

Indanthrene bordeaux RR® and ***indanthrene brilliant orange GR®***, for example, are carbonyl dyes; ***indanthrene brilliant pink R®*** is an Indigo dye (see) and ***Indanthrene brilliant blue®*** is a phthalocyanine dye.

Indazole, ***lH-indazole, l,2-benzodiazole, benzo-[c]pyrazole***: a heterocyclic compound which forms colorless cyrstals; m.p. 147-149 °C, b.p. 267-270 °C at 99 kPa.

I. is steam volatile and dissolves in alcohol, ether and hot water. It couples in the 3-position with diazonium salts, and bromination also occurs at this position. Alkylation produces mixtures of isomeric 1- and 2-alkylindazoles. I. is formed by heating 2-hydrazinocinnamic acid to the melting point or by nitrosylation of *o*-toluidine in the presence of acetic anhydride/glacial acetic acid and subsequent heating of the reaction product in benzene. I. is the parent compound of numerous condensed systems, especially natural products, and is used in the synthesis of pigments for polyacrylonitrile fibers.

Indene: a condensed hydrocarbon with one aromatic and one unsaturated C_5 ring. I. is an oily liquid, m.p. - 1.8 °C, b.p. 182.6 °C, n_D^{20} 1.5768. It is insoluble in water, but is soluble in ethanol and ether. It is found in coal tar and is extracted from it industrially. The appropriate fraction is heated, and the resulting sodium indene (a salt) is isolated and decomposed by steam distillation. I. is not very stable and readily polymerizes to indene resin, which is used industrially as cumarone resin. The double bond of the five-membered ring can be saturated with nascent hydrogen (sodium and alcohol) to form indane. Oxidation with nitric acid yields phthalic acid. Condensation with aldehydes or ketones leads to fulvenes.

Independent ion migration, law of: a law discovered in 1873 by F. Kohlrausch: in an ideal dilute solution, where there is no interaction between ions, the motion of individual ions in an electric field is independent of the type of counterion present; each type of ion has an individual Migration rate (see). The limiting conductivity Λ_∞ is the sum of the limiting ionic conductivities l: $\Lambda_\infty = l_+ + l_-$. Since the limiting ion conductivity is the product of ion mobility and the Faraday constant, it also holds that: $\Lambda_\infty = F(u_+ + u_\infty)$ $cm^2 \cdot \Omega^{-1} \cdot eq^{-1}$, where F is the Faraday constant and u_+ and u_- are the migration rates. Given the equations $\Lambda_{eq} = \alpha \cdot \Lambda_\infty$ (see Ostwald's dilution law) and

$$\Lambda_{eq} = \frac{\kappa\,1000}{n_e \cdot c}\ cm^2\,\Omega^{-1}\,eq^{-1},$$

it follows that: $\Lambda_{eq} = \alpha F(u_+ + j_-)$ and

$$\Lambda_{eq} = \frac{a n_e c F}{1000}(u_+ + u_-)\,\Omega^{-1}.$$

Thus it is possible to determine ion mobilities from conductivity values, and also to determine the value of Λ_{inf}.

Indicator: in chemistry, a substance which shows, usually by a change in color, the presence of certain concentrations of ions. I. are added to solutions, or strips of filter paper soaked in the I. (***indicator papers***) are dipped into the solution.

1) ***pH indicators (acid-base indicators, neutralization indicators)*** show the H_3O^+ concentration or pH of a solution. They are weak acids or bases in a protolysis equilibrium determined by the pH of the aqueous solution: $InH + H_2O \rightleftharpoons In^- + H_3O^+$, where InH is the indicator acid, and In^- is the indicator base. Application of the mass action law to this equilibrium, after rearrangement and taking of logarithms, yields $pH = pK_i + \log(c_{In}/c_{InH})$. Indicator acid and base are different colors. Their relative concentrations are determined by the pH of the solution, as shown in the equation. If $pH > pK_i$, the equilibrium is largely on the side of the indicator acid and one sees the color of InH. If the pH of the solution is equal to the negative logarithm of the dissociation constant of the indicator acid, $pH = pK_i$, the concentration ratio c_{In}/c_{InH} is exactly 1, and a mixed color is seen; this is the changing point of the I. In order for the color of either the indicator acid or base to be clearly visible, that component must be present in about 10-fold excess over the other, i.e. c_{In}/c_{InH} must be 10 or 0.1. Thus the I. does not indicate a sharp pH value, but changes over an interval of about 2 pH units. This interval is determined by the pK_i (Table 1), and should be close to the pH at the equivalence point of an acid-base titration (see Neutralization analysis). There are suitable indicator systems for every pH range, and mixtures of different I. (***mixed indicators***) are also used to make the change easier to recognize and thus to narrow the change interval (Table 2).

2) ***Metal I. (metallochrome I.)*** are substances which indicate the amount of certain metal ions in a solution. Most are chelate-forming organic dyes which form complexes with the metal ions; the color of the complex differs clearly from that of the metal-free I. For use in a complexometric titration, the I. must form a complex with the metal which is less stable than that of the EDTA complex. If an I. is added to a metal salt solution, the color of the I. complex will be seen (Table 3). Addition of the EDTA standard solution leads to formation of the EDTA-metal complex, which is more stable than the I.-metal complex, at the expense of the I.-metal complex. At the equivalence point, the color of the free I. is seen.

3) ***Redox I.*** are used to indicate a certain redox potential in a solution. Most are organic substances which have different colors in different oxidation states. The I. system is present in the solution as a corresponding redox pair which is characterized by a certain standard electrode potential (Table 4). Be-

Table 1. The most common pK_i indicators (arranged in order of increasing change intervals)

Indicator	Change range pK_i	Color: Acid side	Color: Base side
Thymol blue	1.2 ... 2.8	Red	Yellow
Dimethyl yellow	2.9 ... 4.8	Red	Yellow-orange
Bromphenol blue	3.0 ... 4.6	Yellow	Purple
Methyl orange	3.1 ... 4.4	Red	Yellow-orange
Bromcresyl green	3.8 ... 5.4	Yellow	Blue
Methyl red	4.4 ... 6.2	Red	Yellow-orange
Bromcresol purple	5.2 ... 6.8	Yellow	Purple
Bromthymol blue	6.0 ... 7.6	Yellow	Blue
Neutral red	6.8 ... 8.0	Bluish red	Yellow-orange
Cresol red	7.0 ... 8.8	Orange	Purple
Thymol blue	8.0 ... 9.6	Yellow	Blue
Phenolphthalein	8.2 ... 9.8	Colorless	Red-violet
Thymolphthalein	9.3 ... 10.5	Colorless	Blue
Alizarin yellow GG	10.0 ... 12.1	Light yellow	Brownish yellow
Tropaeolin 0	11.0 ... 13.0	Yellow	Brownish yellow
2,4,6-Trinitrobenzoic acid	12.0 ... 13.4	Colorless	Red-orange

Table 2. Some common mixed indicators for pH

Mixture	Composition	Range	Color change
Unitest I. Soln. I	?	1 ... 11	Red to blue
Unitest I. Soln. II	?	1 ... 5	Red to yellow-orange
Unitest I. Soln. III	?	5 ... 9	Yellow to blue
Mortimer's I.	3 parts bromcresol blue (0.1 g in 100 ml ethanol), 1 part methyl red (0.2 g in 100 ml ethanol), by volume	5.1	Wine red to green
Tashiro's I.	2 parts methyl red (0.03 g in 130 ml ethanol), 3 parts methylene blue (0.1 g in 100 ml ethanol), by vol.	5.2	Red-violet to green; gray at pH 5.2
	1 part cresol red sodium salt (0.1 g in 100 ml water), 3 parts thymol blue (sodium salt, 0.1 g in 100 ml water), by volume.	8.3	Yellow to violet pink at pH 8.3
	2 parts phenolphthalein (0.1 g in 100 ml 50% ethanol), 1 part α-naphtholphthalein (0.1 g in 50% ethanol), by volume.	9.6	pale pink to violet green at pH 9.6

Table 3. Important metal indicators

Indicator	Metal ions	pH range	Color change
Eriochrome black T	Mg^{2+}, Ca^{2+}, Zn^{2+}, Cd^{2+}, Mn^{2+}	10	Red to blue
Murexid	Cu^{2+}, Ni^{2+}, Co^{2+}, Ca^{2+}	9...11	Yellow to violet
Xylenol orange	Bi^{3+}, Hg^{2+}, Pb^{2+}, Cd^{2+}, Zn^{2+}, Co^{2+}	1...6	Red to yellow
Pyrocatechol violet	Bi^{3+}, Th^{4+}	2...6	Yellow to blue
Chromazurol S	Al^{3+}, Fe^{3+}, Cu^{2+}, Zn^{4+}	2...6	Blue-violet to orange

cause the concentrations of the two members of the redox pair change continuously during the titration, their electrode potential also changes, and as a result the equilibrium of the I. pair also changes. At the equivalence point of the titration, the electrode potential of the system being titrated has either risen or dropped so far that the I. equilibrium lies either on the side of the oxidized or the reduced form, and its color is clearly visible. It must be remembered that the I. change is also influenced by a number of other factors, e.g. the pH of the solution.

Table 4. Important redox indicators

Indicator	Color change		$V°$ in volts at pH 7
Neutral red	Red-violet	to colorless	– 0.29
Safranin	Blue-violet	to colorless	– 0.29
Indigodisulfonic acid	Blue	to yellow	– 0.11
Methylene blue	Blue	to colorless	0.01
Biphenylamine	Blue-violet	to colorless	0.76
Ferroin	Blue	to red-orange	1.06

4) ***Adsorption indicators*** are dyes used in precipitation analysis which are adsorbed by the precipitate, due to the change in its electric charge, and this is associated with a color effect. For example, eosine is added in the argentometric iodide determination to aid recognition of the endpoint. It has a red-orange color when free in solution, and is red-violet when adsorbed.

Indicator electrode: see Electrode.

Indigo: formerly, the most important organic pigment. Pure I. is a dark blue powder with a coppery sheen; m.p. about 392 °C (dec.). At higher temperatures, I. sublimes into copper-colored prisms. It is insoluble in alcohol, ether, benzene, dilute acids and bases, and is slightly soluble in water, amyl alcohol, paraffin, and light petroleum. It is somewhat more readily soluble in chloroform, aniline or nitrobenzene. It dissolves in cold concentrated sulfuric acid to give a green solution which turns blue when heated; see Indigo dyes.

I. is found in some plants (*Indigofera* species) in the form of the glucoside, indican. It can be synthesized by the Heumann indigo synthesis (see). A new, continuous process starts from aniline and ethylene oxide, which are converted via β-hydroxyethyl aniline and its disodium salt to indoxyl and then oxidized to I.

Synthetic I. has distinct advantages over the natural product: it has a more intense color, and this does not depend on the time of harvest or location of the indigo plantation. The product always consists 100% of dye substance. Synthetic I. has essentially replaced natural I., and even the use of synthetic I. has declined in favor of other vat dyes.

Indigo dyes, ***indigoids***: a group of dyes which are structurally similar to Indigo (see). The I. are carbonyl pigments with the chromophore shown in the figure.

General structural formula of the indigo dyes.

Indigo dye	X	R^4	R^5	R^6	R^7	$R^{4'}$	$R^{5'}$	$R^{6'}$	$R^{7'}$
Indigo	NH	H	H	H	H	H	H	H	H
Antique purple	NH	H	H	Br	H	H	H	Br	H
Brilliant indigo B	NH	H	Cl	H	Cl	H	Cl	H	Cl
Thioindigo	S	H	H	H	H	H	H	H	H
Indanthrene brilliant pink R	S	CH_3	H	Cl	H	CH_3	H	Cl	H
Ciba violet 3B	NH/S	H	Br	H	H	H	Br	H	H

The most important industrial compounds are the halogenated derivatives and some methyl, methoxy and benzo compounds of I.

The I. are used mainly for printing cloth; they are considered to be vat dyes. Up to 1871, they were developed by microbial processes. In the vat process, which is now achieved with sodium dithionite, indigo white (leukoindigo) is formed; this dissolves in dilute bases to form salts (see Indigosols, Fig.).

Indigoids: same as indigo dyes.

Indigosols: disulfuric acid esters of indigo dyes formed by reaction of the leuko compounds of the indigoids with chlorosulfuric acid or sulfur trioxide (dissolved in pyridine or dimethylaniline).

1) $+ 2 ClSO_3H$ 2) $+ 2 NaOH$ / 1) $- 2 HCl$ 2) $- 2 H_2O$

Indigo white, leukoindigo — Indigosol

The I. are water-soluble and stable to alkalies and air oxygen. In acid solution, they slowly decompose, reforming the leukoindigo compounds. The I. have made a new method of dyeing wool possible:

The fibers are impregnated with the I. solution and then oxidized with nitrous or chromic acid; saponification occurs simultaneously. The I. are commercially available under various trade names.

Indigotin: see Indoxyl.

Indigo white: see Indigo dyes.

Indium, symbol ***In***: a chemical element in Group IIIa of the periodic system, the Boron-aluminum group (see). It is a metal, *Z* 49, with natural isotopes with mass numbers 113 (4.28%) and 115 (95.72%). The atomic mass is 114.82, valence III, rarely I, Mohs hardness 1.2, density 7.30, m.p. 156.61 °C, b.p. 2080 °C, electrical conductivity 11.2 Sm mm^{-2} (at 0 °C), standard electrode potential (In/In^{3+}) - 0.338 V.

Properties. I. is a silvery white, very soft metal which can be cut with a knife. It crystallizes in a tetragonal lattice, and its chemical behavior is very similar to that of gallium. Although I. occurs most often in the +3 oxidation state, the metal is stable to air and water at room temperature, and is oxidized by oxygen to indium oxide, In_2O_3, only when heated. I. forms indium salt solutions and hydrogen in the presence of strong acids, but does not react with alkalies. Sulfur, selenium, phosphorus and nitrogen oxidize I. at high temperatures to binary compounds of In^{3+}.

Analytical. Qualitatively, I. is most conveniently detected by spectroscopy (it has an indigo-blue line at 451.1 nm and a violet line at 410.1 nm). It can be determined quantitatively by complexometric titration with EDTA; trace analysis is done by atomic absorption spectroscopy.

Occurrence. I. is a very rare element, making up only 10^{-5}% of the earth's crust. It is found in low concentrations as the sulfide in sphalerite (zinc blende) and also in copper shale.

Production. I. is obtained as a byproduct of zinc production. After removal of impurities, indium hydroxide is precipitated and converted to indium oxide by roasting. The oxide is reduced to the metal with hydrogen. The metal can also be precipitated electrolytically from aqueous salt solutions.

Applications. I. is used as an alloy component, because it is able to improve the strength and corrosion resistance of many metals. I. and indium arsenide are used in the semiconductor industry to make transistors, rectifiers, etc.

Historical. I. was discovered in 1863 by C.F. Reich and H.T. Richter in Frieberg zinc blende. It was given the name I. because of the indigo-blue spectral line (the Greek word "indikon" means "blue color").

Indium compounds: As would be expected from its position in the periodic system, indium is usually trivalent in its compounds; sometimes, however, it is monovalent. When the pH of an aqueous In^{3+} salt solution is increased above 3.5, a colorless, amorphous precipitate of ***indium(III) hydroxide***, $In(OH)_3$, is formed. $In(OH)_3$ (M_r 165.84) is dissolved by strong acids to form indium salt solutions; with aqueous alkalies it forms ***hydroxoindates***, for example, $Na_3[In(OH)_6]$. Above 150 °C, $In(OH)_3$ is dehydrated to ***indium(III) oxide***, In_2O_3, M_r 277.64, density 7.18. The anhydrous ***indium(III) chloride***, $InCl_3$, M_r 221.18, density 3.46, m.p. 586 °C (sublimation begins at 300 °C) crystallizes in colorless, deliquescent leaflets. It is made by combustion of elemental indium in a stream of chlorine.

Indium(I) compounds are somewhat more stable than the comparable gallium(I) compounds. ***Indium(I) chloride***, InCl, can be formed from the elements. It forms reddish, deliquescent crystals, M_r 150.27, m.p. 225 °C, b.p. 608 °C. It disproportionates in water to elemental indium and $InCl_3$. ***Indium(I) oxide***, In_2O, can be prepared by thermal decomposition of In_2O_3.

Indium(I,III) compounds are often (incorrectly) called indium(II) compounds. ***Indium(I,III) chloride***, $InCl_2$, M_r 185.73, m.p. 235 °C, b.p. 550-570 °C, forms rhombic crystals. It is obtained by the reaction of hydrogen chloride with indium.

Indoanilines: amino derivatives of *N*-phenylquinonemonoimine. The parent compound of the I. is ***indoaniline***.

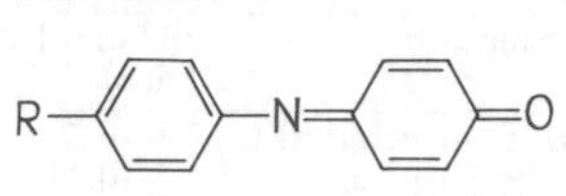

Indoaniline: R = NH_2
Indophenol: R = OH

I. can be made by oxidative coupling of phenols with *p*-phenylenediamines containing at least one unsubstituted amino group. I. are quinoid pigments, which are used, for example, in color photography in the form of the blue and blue-green indoaniline pigments. I. are also important intermediates for the synthesis of sulfur pigments.

Indole, ***benzo[b]pyrrole***: a heterocyclic compound consisting of ortho-fused benzene and pyrrole rings.

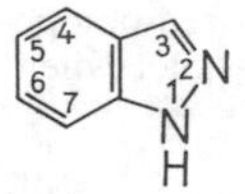

I. crystallizes as colorless leaflets and has an odor of feces; very dilute, carefully purified I. has a fresh, flower-like odor similar to jasmine; m.p. 52.5 °C, b.p. 254 °C. It is soluble in alcohol, ether, chloroform and benzene, and is relatively soluble in water. I. is steam volatile and a weak acid. Electrophilic substitution reactions occur in the 3-position if this is available, and in the 2-position only if the 3-position is occupied. If both the 2- and the 3-positions are occupied, the hydrogen atom in position 6 is replaced. I. is converted by mild oxidants to indigo. The most important method for synthesis of I. is the Fischer indole synthesis (see); however, I. can also be made by the Madelung synthesis (see) or from 2-nitrotoluene and diethyl oxalate (Reissert, 1897).

5-Hydroxyindoles are made by condensation of β-aminocrotonate esters with 1,4-benzoquinone (Nenitzescu, 1929). I. is obtained industrially from the heavy oil fraction of coal tar, or by reduction of indoxyl with zinc powder.

I. is found in the blossom oils of jasmine and wallflowers (*Cheiranthus*), and in locust tree blossoms. It is also a component of decaying proteins. Many derivatives of I., such as scatole, tryptophan, indican, serotonin, the indole alkaloids, toad poisons, isatin, indoxyl, indigo, indolacetic acid, etc., are important in industry or physiology. I. is used in perfumery to synthesize artificial jasmine and neroli oils, and it is the starting material for synthesis of drugs and pigments.

Indole alkaloids: see Alkaloids, table.

Indoline, ***2,3-dihydroindole, 1H,2H,3H-benzo[b]pyrrole***: the dihydro derivative of indole. It is a colorless liquid; m.p. 228-230 °C, n_D^{20} 1.5923. I. is soluble in ether, acetone and benzene, and insoluble in water, although it is steam volatile. It is made by hydrogenation of indole with Raney nickel at 100 °C and 9.81 MPa, or by treatment of 2-(2-chloroethyl)aniline hydrochloride or 2-(2-aminophenyl)ethanol with aqueous sodium hydroxide. Some important derivatives of I. are oxindole (indolin-2-one) and indoxyl (indolin-3-one).

Indoline-2,3-dione: same as Isatine (see).

Indolin-2-one: same as Oxindole (see).

Indolin-3-one: same as Indoxyl (see).

Indolizine, ***pyrrocoline, pyrrolo[1,2]pyridine***, a benzene fused ring system of the pyrrole type. I. forms colorless crystal leaflets which smell like naphthalene; m.p. 75 °C, b.p. 205 °C. I. is readily soluble in organic solvents; in warm, dilute acids, it dissolves with gradual decomposition. It is volatile in steam. The hydrochloric acid solution colors a spruce shaving red, and the solutions in benzene have a strong violet fluorescence. I. is formed by heating 2-methylpyridine with α-bromoacetaldehyde and boiling the reaction product with sodium hydrogencarbonate solution. Another method is decarboxylation of indolizine-2-carboxylic acid, which is made from 2-methylpyridine by reaction with α-bromopyruvic acid.

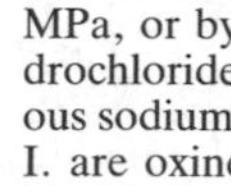

Indol-3-ylacetic acid, ***β-indolylacetic acid, heteroauxin***: a phytohormone. I. forms crystalline leaflets, m.p. 165-166 °C. It is readily soluble in alcohol, ether and acetone, but only slightly soluble in water and chloroform. I. occurs widely in higher plants; it promotes extension growth and many developmental processes. In high concentrations, however, it is poisonous to plants. I. is readily synthesized from indole and chloroacetic acid; from gramin via the corresponding indolylacetonitrile; or from indole and sodium glycolate. I. is determined colorimetrically after reaction with fructose and cysteine in sulfuric acid.

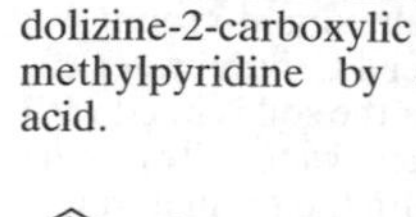

Indometacin: a derivative of indole with a very strong antipyretic effect. It is used to treat chronic polyarthritis.

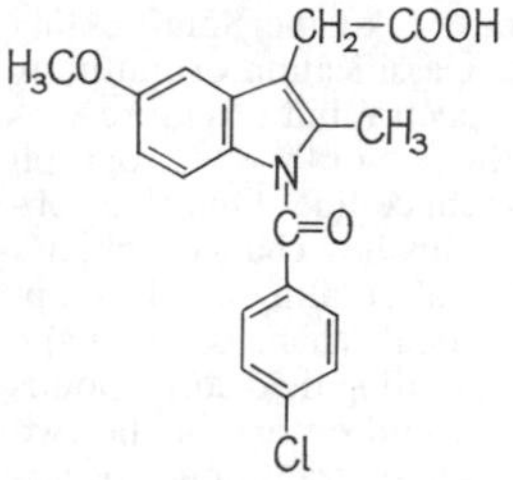

Indoor air quality: a term related to the concentration of various pollutants indoors. Government regulations generally have been established only for places of work, and for specific pollutants. Both semiquantitative and quantitative methods are available; in the latter, a defined amount of air is passed over an absorption or collection device. The sample is usually subjected to a microanalytical process (see Air analysis).

The maximum acceptable concentration (or maximum allowable concentration), abb. MAC is a legally established limit for each pollutant. For ready comparison, the MAC should always be given in mg m^{-3}, although in the United States, parts per million (ppm) are often used. The concentration of a gas or vapor can also be given in vol. %. For example, if 100 cm^3 carbon monoxide is present in 1 m^3 air, its final concentration is 100 ppm CO (or 0.01 vol.%). The final concentration is the amount of substance present after mixing, dissolving or distribution in the medium. The following formula can be used to calculate the content of a substance of molecular mass M_r and molar volume V at 25 °C and 101.3 kPa (= 760 Torr) in mg m^{-3}:

$$\frac{\text{ppm} \times M_r}{\text{constant for 1 mol gas at 25°C and 101.3 kPa}} = \text{mg m}^{-3}$$

Applied to the above example (100 ppm or 0.01 vol. % CO), the formula gives the following:

$$\frac{100 \cdot 28}{24.45} = 114.5 \text{ mg m}^{-3}$$

MAC values are based on experiments with animals and experience with human beings.

Indophenin reaction: see Thiophene.

Indophenols: hydroxyl derivatives of *N*-phenylquinonemonoimine. The parent compound of the I. is ***indophenol*** (for the formula, see Indoaniline). I. can be made, for example, in the Liebermann reaction (see) by the reaction of nitrous acid and phenols. They are structurally closely related to the indamines and indoanilines. Indophenol has many colored derivatives, the ***indophenol pigments***, a group of blue to blue-green pigments which are similar in their properties to the indamines. Like the latter, indophenol pigments have both a quinoid and a benzoid ring in the molecule. The indophenol pigments are important starting materials for the synthesis of sulfur pigments, and they are also used in color photography.

INDOR technique: (abb. for internuclear double resonance) a special double resonance technique used in NMR spectroscopy (see). Like the Spin-tickling technique (see), it permits classification of mutually coupled nuclei in complex spectra, but it is more sensitive and selective. In I., the pen of the spectrograph is set at the peak of a resonance line. One then observes how the intensity of this line changes when a second, variable frequency alternating field is applied. The intensity of the peak changes when the frequency of the second alternating field corresponds to transitions which begin or end on one of the two levels between which the observed transition is occurring.

Indoxyl, ***indolin-3-one***: an important indole derivative. I. forms bright yellow crystals; m.p. 85 °C. It is soluble in water, alkalies, alcohol and ether; the aqueous solutions have a yellow-green fluorescence. I. undergoes keto-enol tautomerism; the equilibrium is almost completely on the side of the keto form, however. In alkaline solution, it is oxidized by atmospheric oxygen to ***indigo (indigotin)***. I. is obtained from aniline and chloroacetic acid (see Heumann's indigo syntheses). I. is contained as a glycoside in indigo-yielding plants. In the urine, it is found as ***indoxylsulfuric acid*** or bound glycosidically to glucuronic acid.

Induced dipole moment: see Polarization.

Induction period: an initial phase in certain reactions during which the rate of the reaction is very low or nearly zero. Although reactions usually begin with a maximum rate of consumption of the reactants, and gradually slow down as the reactants are consumed, reactions with an I. achieve the normal rate of reaction only after the I. I. occur, for example, when a quasistationary concentration of short-lived intermediates must be built up (see Stationary state principle, see Complex reaction), or in heterogeneous reactions during the time required for seed formation of the new phase, and in autocatalytic and oscillating reactions.

Induction process: see Steel.

Inductive coupled plasma, ***ICP***: an atomization and excitation source used in atomic spectroscopy. High-frequency energy is transfered by inductive coupling to a flow of inert gas (such as argon) which contains the sample as an aerosol. The frequencies are between 27 and 90 MHz, and the generators have power outputs between 1 and 10 kW. The energy heats the argon to 10,000 K, causing vaporization of the aerosol and excitation of the resulting free atoms and ions so that they emit light. The intensity of this light is measured and related to the concentration of the emitting atoms.

I. generates higher temperatures than arcs and spark discharges. The plasma parameter (see Plasma) is considerably steadier than with those excitation sources. I. has been developed since 1960, and has made a significant contribution to the progress of spectroscopic multielemental analysis.

Inductive effect, ***induction effect***, abb. ***I effect***: the electrostatic effect of a covalently bound substituent X on the change in charge distribution in a molecule when a hydrogen atom is substituted for X. The I. is due to the polarity of the substituent bond, and to the formation of a bond dipole, which can have various effects on the rest of the molecule. If the polarization due to X is oriented along the bond, one speaks of the *I. in the narrow sense*, but if there is a direct electrostatic interaction between the potential field of the bond dipole and the other charged parts of the molecule, it is a *field effect*. The two types of effect are difficult to distinguish experimentally.

Substituents with higher electronegativity than hydrogen withdraw electrons, and therefore have a ***-I-effect***, e.g. $-N(CH_3)_3$, $-NO_2$, halogens, -OH, -SH and $-NH_2$. A ***+I-effect*** is displayed by substituents which are less electronegative than hydrogen, and therefore repel electrons, e.g. $-SiH_3$, $-SiR_3$, $-BR_2$. The I. in aliphatic compounds can be determined quantitatively by measurement of the σ^* substituent constant (see Taft equation).

Inductive forces: see Van der Waals's forces.

Induline: an Indulin pigment (see).

Induline black: see Nigrosine.

Induline pigments, ***indulines***: a group of blue-black azine pigments. They are used to make black spirit varnishes, stains and polishes. Water-soluble I. can be used to dye wool, to produce water-soluble wood stains and leather polishes and dressings.

Inert: not undergoing reactions with other substances.

Inertinite: see Macerals.

Inertite: see Coal.

Inert pair effect: see Boron-aluminum group.

Infiltration: a process of artificial Groundwater enrichment (see) in which a basin in the soil is lined with fine sand and filled with surface water. This percolates into the soil, and mixes with the ground water. The undissolved components of the surface water are retained by the upper layers of the basin; at the same time, a biological lawn (of microorganisms) forms and removes most of the dissolved organic substances. There is a temperature equilibration between the ground and surface water, so that especially during the warm seasons, the infiltrated surface water is cooled. The capacity of the basin diminishes with time, but it can be renewed by removal and replacement of the upper sand layer. If infiltration is not necessary on a year-round basis, grass fields can be used for I.; these do not need to be regenerated.

If the geohydrological conditions of the aquifer are suitable, I. can also be used to store groundwater.

Inflammation: the beginning of a combustion of a mixture of gases or vapors upon ignition. In general, also the beginning of a chemical reaction which occurs spontaneously with production of heat and light at the ***inflammation temperature***.

Infrared spectroscopy, ***IR spectroscopy***: If a suitable sample of a substance is irradiated with infrared radiation, some of the frequencies will be absorbed, while others will be transmitted. If the percentual absorption A or the transmission T of the sample is plotted against the wavenumber $\tilde{\nu}$ or wavelength λ, the result is an IR spectrum (Fig. 1). The absorption of IR radiation by a molecule changes its rotational and vibrational energy in accordance with the equation $\Delta E = h\nu$ (E = energy, h = Planck's constant).

Thus the frequencies at which the sample absorbs give an indication of its molecular structure and also makes possible qualitative and quantitative determinations. I. has been widely used in chemical analysis since about 1950.

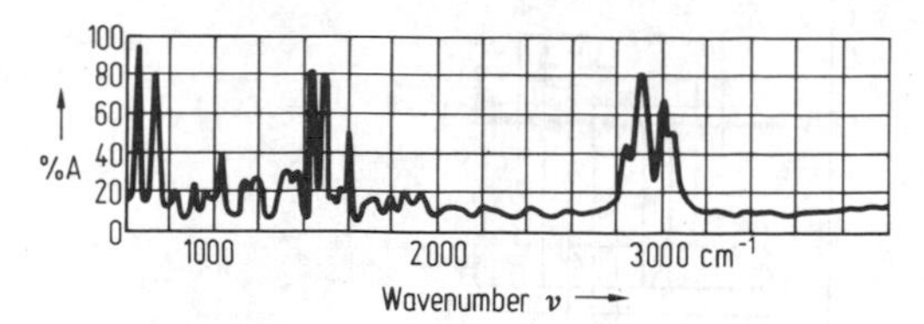

Fig. 1. IR spectrum of polystyrol

Theoretical basis. The atoms in a molecule are not rigidly bound to each other, but can oscillate around their equilibrium positions. The oscillations can be approximated by the laws of classical physics if the atoms are represented by spheres and the bonding forces by springs (Fig. 2).

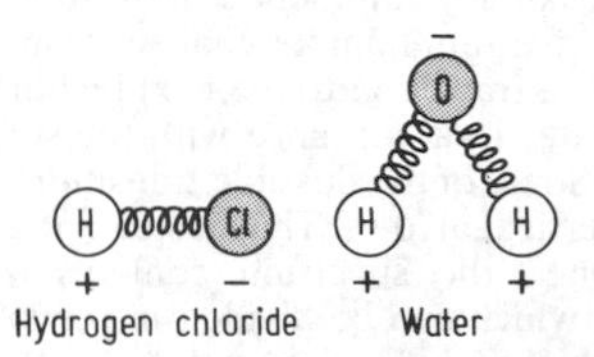

Fig. 2. Mechanical models of molecules.

A molecule of N atoms has $3N$ degrees of freedom of motion. 3 degrees of freedom are accounted for by translation and 3 by rotation in non-linear molecules (2 by rotation in linear molecules), so that for non-linear molecules, there remain $3N - 6$ degrees of freedom for vibration ($3N - 5$ for linear molecules).

The theoretically possible number of vibrational modes ($3N - 6$ or $3N - 5$) are not usually observed in IR spectra. There are certain selection rules: infrared light is only absorbed if a dipole moment which changes during the vibration interacts with the oscillating electric vector of the electromagnetic radiation. This occurs if the dipole moment of the molecule at one extreme of the vibrational motion is different from the dipole moment at the other extreme. In this case, there is an *IR-active vibration* which is observed in the IR spectrum. Vibrations which do not fulfill this condition are IR-inactive, and can only be observed in the Raman spectrum (see Raman spectroscopy). Further simplification of the IR spectrum can occur if certain vibrations are degenerate, i.e. they have the same frequency, but occur in different spatial directions, as in the CO_2 molecule (Fig. 3). To a first approximation, such a system can be treated as a *harmonic oscillator*. In a diatomic molecule, for example, the vibrational frequency is given by

$$\nu = \frac{1}{2\pi} \sqrt{\frac{k}{\mu}}$$

which can be derived from Hooke's law; ν is the vibrational frequency in Hz, k is the force constant of the bond in N m^{-1}, and

$$\mu = \frac{m_1 \cdot m_2}{m_1 + m_2}$$

is the reduced mass.

Vibration form	Number of waves	Vibration stimulus IR	Raman
	$\tilde{\nu}_1$	−	+
	$\tilde{\nu}_2$	+	−
	$\tilde{\nu}_3 = \tilde{\nu}_4$	+	−
	$\tilde{\nu}_4 = \tilde{\nu}_3$	+	−

+ − Vibrations perpendicular to the paper plane

Fig. 3. Vibrations of the CO_2 molecule.

If this equation is solved by substitution of the corresponding values, the frequencies fall in the infrared range.

The potential energy of a harmonic oscillator (also called the *oscillator potential*), plotted as a function of the distance between the vibrating atoms, gives a parabolic potential curve (Fig. 4). For a classic harmonic oscillator, any vibrational amplitude is allowed.

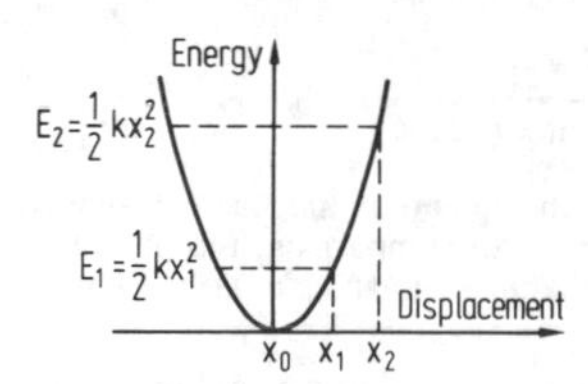

Fig. 4. Potential curve of the classic harmonic oscillator (x_1 and x_2 are arbitrary displacements).

However, this does not agree with the experimental observation of discrete vibrational frequencies in the IR spectrum. A solution was found in quantum mechanics: solution of the Schroedinger equation for a one-dimensional harmonic oscillator indicates that the total energy E can take only the values $E = h\nu(v + 1/2)$, where $v = 0, 1, 2, 3,...$ is the vibrational quantum number. Thus according to quantum mechanics, a harmonic oscillator can have only certain energies which differ by $h\nu$ (Fig. 5). Transitions from a lower to a higher vibrational level involve absorption of electromagnetic radiation of the same frequency as the frequency of the mechanical oscillation. Transitions between energy states which are not adjacent to each other are forbidden by the selection rules, as is coupling between different vibrations.

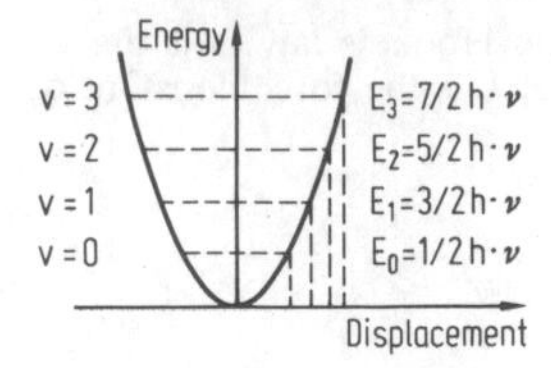

Fig. 5. Quantum mechanical harmonic oscillator.

The quantum mechanical calculation yields the important result that the lowest possible energy (ground state energy) of the harmonic oscillator is not zero, but $E_0 = h\nu/2$. Since the harmonic oscillator will certainly be in the ground state at absolute zero, and will still posess its ground-state energy, this is also called the *zero-point energy*. The finite zero-point energy of the harmonic oscillator is a requirement of the Heisenberg uncertainty principle, because if $E_0 = 0$, both the position and the energy would simultaneously have precise values, namely zero. A classical particle, however, would rest at its equilibrium position $x = 0$ at $T = 0K$.

The harmonic oscillator is an idealized model of real oscillators, which are more accurately represented by the *anharmonic oscillator*, for which the potential curve is shown in Fig. 6. The various vibrational states are not equidistant, transitions can occur between states which are not adjacent, and interactions between different vibrations are possible.

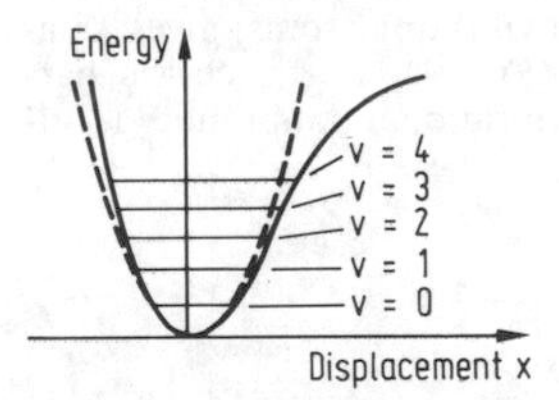

Fig. 6. Potential curve of the quantum mechanical anharmonic oscillator (solid curve); for comparison, the potential curve of the harmonic oscillator is shown as a dashed line.

The energy difference between two neighboring vibrational levels is $h\Delta\nu(1 - 2\nu x)$, where x is the anharmonicity constant, which usually has a value $<$ 0.05. Transitions between the ground state $v = 0$ and the second or higher excited states give rise to harmonic vibrations which have two or more times the frequency of the ground state vibration. Harmonics have much lower probabilities than the fundamental frequencies, so their absorption occurs at low intensity.

Interactions between various molecular vibrations lead to combination bands. These result from absorption of infrared radiation with a frequency equal to the sum of or difference between two fundamental frequencies; the probability of excitation of a combination vibration is comparable to that of harmonic vibrations. The theory discussed so far applies to pure vibrations. Actually, however, there is no such thing as a pure vibrational spectrum, because every vibrational transition is accompanied by rotational transitions. Since the energy required for a vibrational transition is always sufficient to cause rotational transitions as well, the two occur simultaneously, and it is more accurate to speak of *rotational-vibrational spectra* (Fig. 7).

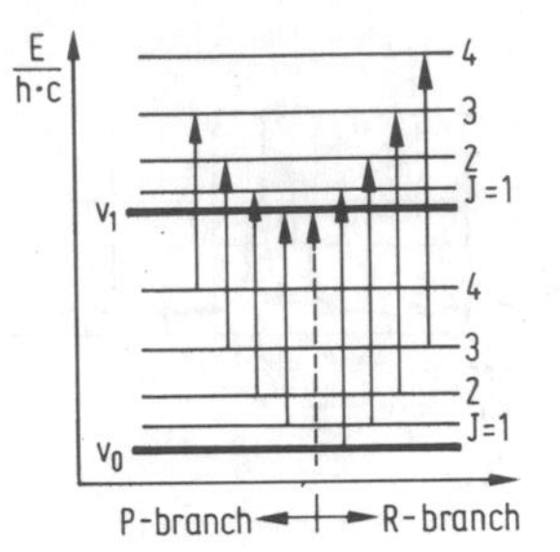

Fig. 7. Term scheme for a rotational-vibrational spectrum.

The two bold-face lines in an energy diagram represent the two lowest vibrational states of the molecule. Several rotational states are associated with each vibrational state; these are represented by lighter lines with the rotational quantum numbers J indicated. When the vibrational quantum number changes from $v = 0$ to $v = 1$, there is a transition to the next higher or lower rotational state, in accordance with the selection rule $\Delta J = \pm 1$. Some of the possible transitions are indicated by the vertical arrows. Thus instead of a single "vibrational line" the spectrum contains a whole system of lines which can be divided into two parts. The part for which $\Delta J = +1$ is called the *R-branch*, while the part for which $\Delta J = -1$ is called the *P-branch*. In the R-branch, the energy of the rotational transition is added to that of the vibrational transition, while in the P-branch, the rotational energy is subtracted from the vibrational. The 0-0 line between the two branches is absent, because it is forbidden by the selection rules for changes in the rotational quantum number. The intensity distribution between the two branches corresponds to the thermal distribution of the molecules among the rotational states. The position of the absorption in the spectrum is essentially determined by the vibrational energy difference, while the fine structure is due to the rotational states. A collection of absorption lines which arise from one vibrational transition, but different rotational transitions, is called an *absorption band* (Fig. 8).

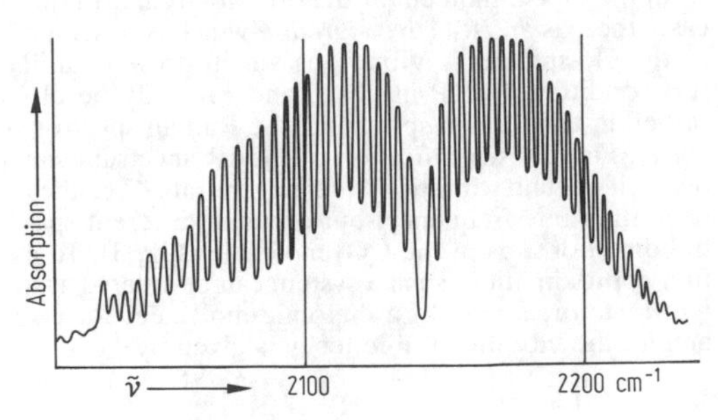

Fig. 8. Absorption band in the rotational-vibrational spectrum of carbon monoxide.

Apparatus. Modern spectrophotometers for the infrared range are dual-beam instruments (see Spectral instruments) with which it is possible to take an IR spectrum in a few minutes. With special-purpose instruments, recording times of fractions of a second can be achieved. Since the entire IR range cannot be measured with a single instrument, it has been divided into ***near, medium and far IR*** (NIR, intermediate IR, FIR); instruments for these areas differ in their components (Table 1).

Table 1. Components of IR spectrometers

	NIR	medium IR	FIR
Wavenumber $\bar{\nu}$ [cm^{-1}]	12,500	4,000	200
Wavelength λ [μm]	0.8	2.5	50
Light source	Tungsten ribbon lamp	Nernst rod, SiC or ceramic rod	Hg high-pressure burner
Optical material	Quartz	NaCl, KBr, LiF, CsBr	
Dispersive system	Prism, Grating	Prism, Grating	Grating
Detector	PbS photocell	Thermoelement Golay cell Bolometer	Golay cell

Lately Fourier spectrometers (see) have been increasingly used in the IR range. They are significantly more sensitive than traditional dispersion spectrometers and permit a more rapid recording of spectra. They are most suitable for situations in which intensity is limiting, e.g. in the FIR.

Properly prepared samples can be analysed in solid, liquid and gaseous forms. The amount of substance needed to take an IR spectrum is small, a few milligrams, and with special apparatus this amount can be considerably reduced.

Applications of IR spectroscopy. 1) *Structure elucidation* is the most common application (Fig. 9). Of all the possible vibrations in a molecule, a few are, to a first approximation, localized in individual bonds or groups of atoms, while others include the whole molecule (skeletal vibrations).

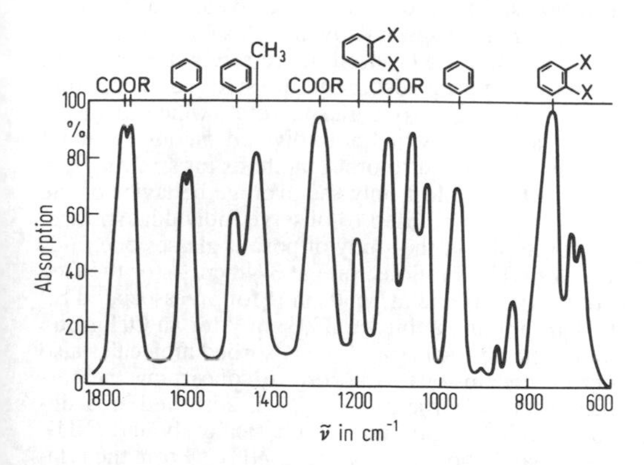

Fig. 9. IR spectrum of phthalic dimethyl ester.

Localized vibrations include Valence vibrations (see) and Deformational vibrations (see). Many of them, especially the valence vibrations, can be used to detect functional groups. Atoms of different elements have different masses and different types of bond have different force constants, which lead to different values of ν in the equation

$$\nu = \frac{1}{2\pi}\sqrt{\frac{k}{\mu}}$$

For example, the C≡C bond has a higher force constant (is less easily stretched or compressed) than either the C=C or C-C bond. The force constants of triple, double and single bonds stand in the ratio of approximately 3:2:1, and thus their vibrational frequencies stand in the ratio $\sqrt{3}:\sqrt{2}:\sqrt{1}$. The vibrational frequency is also determined by the reduced mass μ of the atoms; if *k* is constant, ν is proportional to $1/\sqrt{\mu}$. Thus as the mass of the atoms increases, the vibrational frequency decreases. Due to the low mass of hydrogen, for example, all vibrations in which H is involved have very high frequencies. Such considerations indicate the range of wavenumbers greater than 1500 cm^{-1} in which functional groups absorb (Table 2). Such localized vibrations for many functional groups have been published in tables. The expected ranges have been derived, for the greatest part, purely empirically. Certain groups of atoms, e.g. an -OH or

>C=O

group, always absorb at certain wavenumbers (and sometimes at several wavenumbers) which are largely independent of the group's position in the molecule. If such "characteristic frequencies" for a particular atomic group has been observed in a sufficient number of known compounds, and the spectrum of a substance with an unknown structure also contains bands at those frequencies one can safely conclude that it contains the atomic group in question.

Table 2. Expectation ranges for characteristic frequencies

3700	2800	2400	1900	1500 cm^{-1}
X–H valence vibrations		X≡Y valence vibrations	X=Y valence vibrations	Deformational, skeletal vibrations; Valence vib. of heavy atoms

2) *Identification of substances.* From Table 2 it can be seen that an IR spectrum below 1500 cm^{-1} has a large number of bands including the valence vibrations of heavy atoms, e.g. those of carbon-halogen bonds. Although it is difficult to assign bands to particular vibrations in this region, the absorption here is characteristic of the molecule as a whole, and it is therefore called the "fingerprint region". It is eminently suitable for establishing the identity of a substance with an authentic sample. Two pure substances are the same if their IR spectra match band for band. The IR spectrum has the advantage over other physical characteristics, such as boiling and melting points,

density, and refractive index, that it offers a much larger number of possible points of comparison. Substances with similar structures have similar IR spectra, but they are never identical.

3) *Quantitative IR analysis* has been replaced in the past few years by a number of other methods, including gas chromatography and UV spectroscopy. However, it is still important for certain problems. Here the large number of bands in the IR spectrum is advantageous, because it makes it easier to find bands for analysis which are not affected by the absorption bands of the other components of the mixture being analysed. Multi-component analyses are also possible. Quantitative analysis can be based either on standard curves or on the Lambert-Beer law.

The precision of the method is about 1%; this depends on the type of spectrometer, the instrumental parameters chosen and the properties of the substances being analysed. The precision can be improved by digital registration and computer analysis of the data.

When the Lambert-Beer law is applied, it must not be forgotten that in the IR, the band intensity depends on the spectral slit width of the spectrometer. True values are obtained only if the slit width is less than 1/5 the half-height width of the band being measured. This is usually the case in grating instruments, but is usually not true in prism instruments.

4) The *study of hydrogen bonds* is one of the oldest applications of I. and is still one of the most important. The effects, which are discussed here using the OH group as an example, are also observed with other proton donors. The OH valence vibration of a free, non-associated OH group is found between 3590 and 3650 cm^{-1}. Formation of hydrogen bonds causes a shift towards smaller wavenumbers, while at the same time the OH band becomes broader and increases in intensity. The stronger the H-bond, the larger the shift. Since association is an equilibrium reaction, the proportion of free and hydrogen-bound OH groups depends on the concentration and temperature. For example, in the spectrum of cyclohexanol in carbon tetrachloride, bands are observed at 3615 to 3620 cm^{-1} (non-associated OH groups), 3485 cm^{-1} (dimer) and 3320 cm^{-1} (larger aggregates); their relative intensities depend on the concentration (Fig. 10). Inter- and intramolecular H-bonds can easily be distinguished: intermolecular bonds are broken as the solution becomes more dilute, while intramolecular bonds are not.

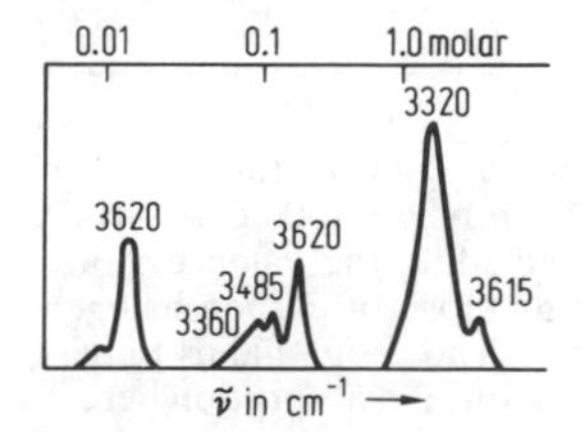

Fig. 10. Concentration dependence of the IR spectrum of cyclohexanol in carbon tetrachloride.

Infrared dichroism. IR spectra which are made with polarized radiation yield valuable information on the nature of vibrations and the orientation of groups within a sample. In conventional I., both the molecules of the sample and the electric vector of the radiation are randomly oriented, so no polarization effects are observed. However, studies on single crystals or aligned polymers show that the band intensities are highly dependent on the polarization direction of the radiation. The intensity A of a given band is measured with polarized light, once parallel to the orientation of the sample, and once perpendicular. The dichroic ratio $R = A_{\perp}/A_{\parallel}$ is obtained; it depends on the degree of orientation of the sample and the angle between the transition moment of the vibration and the orientation direction.

6) Using the *matrix isolation technique*, IR spectra of reactive intermediates can be obtained. The products to be analysed are imbedded in a solid matrix of nitrogen, carbon dioxide, argon or xenon by cooling a gaseous mixture of a small amount of the sample in the gas to the temperature of liquid hydrogen (20 K) or helium (4K). For example, the radical ClCO was identified on the basis of its absorption bands at 1880, 570 and 281 cm^{-1}. It was formed by photolysis of HCl in a CO_2 matrix at 14K; the Cl atoms released by the light reacted with the CO_2 of the matrix to form the ClCO radical.

7) *Spectroscopy in the near IR*. Numerous harmonic and combination vibrations of groups containing hydrogen, such as N-H, C-H, O-H, are observed in the NIR, and these may be used for quantitative determination of such groups (e.g. of alkenes with terminal double bonds or *cis*-disubstituted double bonds) or for studies of hydrogen bonds.

8) *Spectroscopy in the far IR*. The applications of the medium IR are extended in the FIR to the valence and deformation vibrations of molecules which have larger atomic masses and therefore absorb in the lower frequency range. This is important, for example, in the study of metal complexes, because vibrations between the metal and complexing reagent can be detected. The weak intermolecular interactions in many systems lead to vibrations in the FIR: for example, the intermolecular vibrations of H-bonds are found at frequencies from 100 to 200 cm^{-1}; N-J valence vibrations in charge-transfer complexes of iodine with substituted pyridines are between 65 and 95 cm^{-1}.

9) Study of adsorbents and adsorbed molecules. Adsorption studies are done with conventional IR spectrometers using special cuvettes. Usually the absorption spectrum is measured, more rarely the reflection spectrum. Typical adsorbents are silica gel, aluminum oxide, zeolite, porous glasses, etc., which are used either alone or with finely divided metals or metal oxides. In contrast to other methods for studying surfaces, which reflect only the average behavior of the system, I. can be used to observe individual species. For example, in the study of porous glasses one finds various OH vibrations: one at 3748 cm^{-1} for the free Si-OH group; one at 3650 cm^{-1} for an associated Si-OH group, and a third at 3703 cm^{-1} for an OH group on boron. The spectra of the adsorbed molecules also permit conclusions to be drawn concerning the surface of the adsorbent, e.g. NH_3 is adsorbed on a silicon-aluminum catalyst as physically bound NH_3, coordinately bound NH_3 and as NH_4^+. From the relative intensity of the corresponding bands, the ratio of

Lewis to Brønsted centers on the catalyst surface was found to be 4:1.

IR studies have made major contributions on the mechanism of reduction of olefins on nickel catalysts. Ethene adsorbed on the catalyst displays bands at 2800 and 2890 cm^{-1}, which are assigned to an Ni-CH_2-CH_2-Ni group.

10) The *attenuated total reflectance (ATR) technique* is used to study samples which have such limited transparancy to IR that they cannot be studied by conventional techniques, for example coatings on a wide variety of materials and solids such as foils, fibers, plastics, leather and rubber. This method is based on the fact that light impinging on the hypotenuse surface of a prism is totally reflected. In the process it penetrates the adjacent sample to a small degree. If this sample contains a substance which absorbs IR radiation, the total reflected radiation will be attenuated at the wavelengths at which absorption occurs. A plot of the intensity of the reflected radiation as a function of the wavelength is a spectrum which largely corresponds to the absorption spectrum of the substance. The effect can be intensified by multiple reflection (Fig. 11a and b).

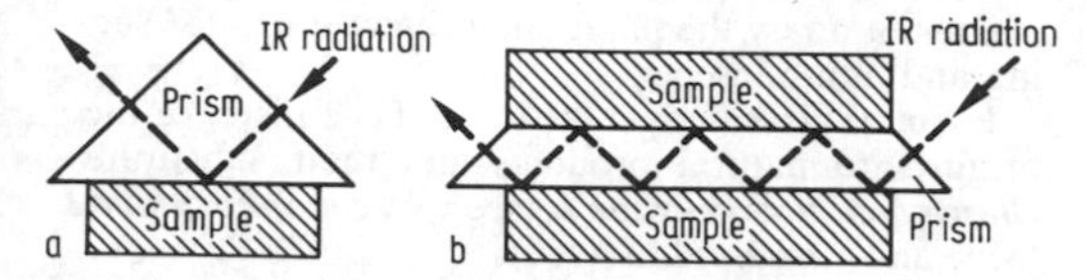

Fig. 11. Methods of attenuated total reflection: a) single, b) multiple

Infusorial earth: same as Kieselguhr.

Ingold rule: a rule discovered by Ingold and Hughes concerning the effect of solvent type on the rates of chemical reactions (***solvent effect***): if the polarity in the transition state of a reaction is greater than in the reactants, the rate of the reaction is increased in a more strongly solvating (polar) solvent. If charges in the transition state are compensated or delocalized, increasing the solvation ability of the solvent inhibits the reaction.

The theory of transition states (see Kinetics of reactions, theory) provides a basis for the I.

Ingold-Taft method: a method for separating inductive and steric effects of a substituent. It is based on the assumption that in the acid hydrolysis of substituted ethyl acetates, the effects of substituents are entirely steric, while in saponification, both an inductive and a steric effect are present. The inductive substituent constants are proportional to the difference between the logarithms of the rate constants for saponification and acid hydrolysis of substituted ethyl acetates k_x (unsubstituted ethyl acetate has rate constant k_{CH_3}): $\lg (k_x/k_{CH_3})_{saponification} - \log(k_x/k_{CH_3})_{acid\ hydrolysis} = 2.48\ \sigma^*$. The proportionality factor 2.48 (reaction constant) makes the σ^* values comparable to the Hammett σ constants.

Steric substituent constants E_s are a measure of the steric hindrance of the reaction center. They are obtained from the ratio of the rate constants for acid hydrolysis of substituted ethyl acetates k_x and unsubstituted ethyl acetate: $\lg(k_x/k_{CH_3})_{acid\ hydrolysis} = E_s = \lg(k_x/k_{CH_3})_{saponification} - 2.48\ \sigma^*$. The more voluminous the substituent, the more negative is E_s [E_s(-CH_3) = 0], and the more slowly the acid hydrolysis occurs. The I. can be applied to reactions of *ortho*-substituted benzene derivatives.

INH: abb. for isoniazid.

Inhabitant equivalent: see Sewage.

Inhibition: 1) the delay or slowing of a chemical reaction caused by added substances, the Inhibitors (see). I. is sometimes called *negative catalysis*.

2) see Corrosion protection.

Inhibitor, a substance which slows ar stops a chemical reaction. The nature and mechanism of an I. depend on the mechanism of the reaction inhibited. I. of radical chain reactions (see Complex reactions) are substances which form relatively stable intermediates with free radicals which are not capable of chain propagation. These include, e.g. nitrogen monoxide, alkenes, phenols, aromatic amines and nitroso compounds. Antioxidizing agents extend the life of oxygen-sensitive compounds, such as oils, fats, fuels or plastics, and stabilizers protect unstable compounds by acting as I. ***Corrosion I.*** are substances which form thin protective layers on the surfaces of metals by adsorption or reaction, and which strongly inhibit corrosion.

Inhomogeneous: same as Heterogeneous (see).

Initial explosive: see Explosives.

Initial ignition: see Explosion.

Initiation: the induction of a kinetically inhibited chemical reaction, causing it to occur at a measurable rate. I. can occur through temperature increase (thermal I.), light (photochemical I.), sound waves (hypersonic I.), input of mechanical energy (e.g. impact, friction – see Tribochemistry), addition of a catalyst (see Catalysis), radical formation (see Initiator) or a detonation shock (see Detonation).

Initiator: a substance which causes a chemical reaction to start by providing the necessary high-energy particles in a preliminary reaction. In contrast to a catalyst (see Catalysis), the I. is consumed and does not open a new reaction pathway. The I. is not involved in the reaction in stoichiometric amounts. I. are involved in the beginning of radical chain reactions (see Reactions, complex), e.g. radical polymerizations and halogenations. Substances with low dissociation energies, which easily form radicals, serve as I. in these reactions. Typical I. are peroxides (e.g. dibenzoylperoxide) and azo compounds (e.g. azo-bis-isobutyronitrile).

Inium compounds: see Onium compounds.

Ino: abb. for inosine.

Inoculation process: see Water softening.

Inorganic benzene: same as Borazine (see).

Inorganic rubber: see Phosphorus nitride dichloride.

Inosamines, ***aminodeoxyinositols***: inositols in which one or more hydroxyl groups have been replaced by amino groups. I. are components of aminoglycoside antibiotics.

Inosine, abb. ***Ino*** or ***I***: 9-β-D-ribofuranosylhypoxanthine, a nucleoside. The dihydrate forms colorless crystals; m.p. 90 °C. I. is found in meat and meat extracts, yeast and other biological materials. ***Inosine 5'-monophosphate (IMP, inosinic acid)*** is the first purine derivative in the biosynthetic pathway leading to adenosine and guanosine nucleotides. Inosinic acid was first isolated from meat extracts by Liebig.

Inositols: hexahydroxycyclohexanes. In the older nomenclature, various prefixes such as *epi-*, *chiro-* or *myo-* indicate the individual stereoisomers. In the modern nomenclature, the positions of the hydroxyl groups which lie above the plane of the cyclohexane ring (Fig.) are indicated first, followed by a slash, and then the positions of those below the plane. Although the I. contain 6 asymmetrical C atoms, all except the pair 1D-*chiro-* and 1L-*chiro-* are *meso*-forms.

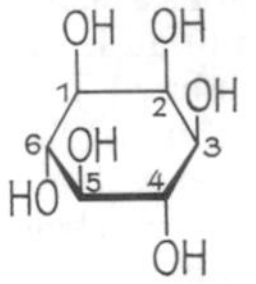

OH OH 1 2 OH 6 OH 3 HO 5 4 OH

(1,2,3, 5/4,6)-Inositol

The most widely occuring I. is ***(1,2,3,5/4,6)-I.***, also called ***myo-*** or ***meso-I.***. Myo-I. is a growth factor for some microorganisms, and is also called ***bios I***. It is considered part of the vitamin B complex, although it is not a vitamin for vertebrates. In eukaryotic cells it is present mainly as phosphatidyl inositol. Higher plants contain I. in sphingolipids as well. Inositol hexaphosphoric acid (*phytic acid*) occurs widely in fungi, bacteria and the fruits, seeds and leaves of higher plants. It is present as a calcium-magnesium salt, which is also called *phytin*. Phytic acid is a chelating agent and is used in the food industry.

In-out isomerism: isomerism at the bridgehead atoms of bicyclic compounds with large rings. Where there are more than six ring atoms in each bridge, and nitrogen is the bridgehead atom, or there are more than seven ring atoms and carbon is the bridgehead, the free electron pairs or protons can stand either inside or outside the ring system. The isomers can be interconverted by rotation around a single bond (see Atropic isomerism).

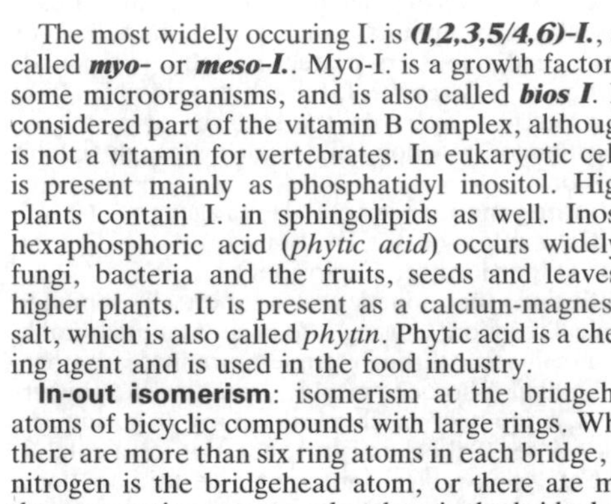

$(CH_2)_n$ H-C C-H $(CH_2)_n$ $(CH_2)_n$ *out-out* $(CH_2)_n$ C-H H-C $(CH_2)_n$ $(CH_2)_n$ *in-in* $(CH_2)_n$ C-H C-H $(CH_2)_n$ $(CH_2)_n$ *in-out*

Insecticides: biologically active compounds used to kill insects. I. are the largest and most important group of pesticides used against animal pests, most of which are insects. I. have vast agricultural and economic importance; until the early 1970's they were used in larger amounts than any other type of pesticide; however, they have now been surpassed by Herbicides (see). The significance of the I. lies in their contribution to agricultural productivity, and thus to the world food supply, and in their contribution to protection of human and animal health. They are also used to protect stored foods and other organic products, such as textiles.

Although effective preparations are now available, there is a constant research effort to prepare new ones, partly because of the development of resistance by the target insects (see Resistance breakers) and partly because of ecological and toxicological concerns (see Ecological chemistry). The ideal goals for a modern I. include broad activity against pests, but lack of toxicity to useful insects and vertebrates, lasting effects in the field but insignificant residues, no development of resistance in the target species and low cost. Some of these ideals are mutually exclusive, so the trend in research at present is to search for new basic principles of crop management. Some promising lines of research are based on use of insect hormones and hormone-mimicking compounds, Pheromones (see), Chemical sterilizers (see), bacterial, fungal or viral toxins, insecticidal Natural product substitutes (see) and biological control mechanisms.

On the basis of their biological mechanisms, the I. can be classified as Contact poisons (see), Ingestion poisons (see) and Respiratory poisons (see); or they can be classified on the basis of their range of activity as *universal*, specific and selective I. The *specific I.* which act on one to a few species of pest, and the *selective I.*, which do not kill the useful animals, are the agents of choice in programs of Integrated pest management (see). Although most I. are effective as topical agents, *systematic I.* (see Systemic pesticide) are taken up by the plant and protect it against sucking and biting insects.

I. can be classified on the basis of chemistry or their origin into natural products and their substitutes, ***chlorinated hydrocarbon I.*** (see), ***Organophosphate I.*** (see) and ***carbamate I.*** (see).

In addition to these broad and important classes of I., there are numerous examples which are classified in other groups of compounds: the most important of the ***thiocyanate I.*** are the *Lethanes®*, e.g. C_4H_9-O-$(CH_2)_2$-O-SCN. Because of their rapid contact effect, they are used against household pests as well as in agriculture.

The first synthetic organic larvicide, Phenothiazine (see), is one of the ***heterocyclic I.***.

An example of a ***xanthogenate I.*** is *dixanthogen*, C_2H_5O-C(S)-S-S-C(S)-OC_2H_5, which is effective against body lice, fleas, sarcoptic mites and similar parasites.

The ***indanedione I.*** include *pindone*, 2-isovaleryl-1,3-indanedione. It is used as an insecticidal component in fly sprays, but a more important application is based on its anticoagulating properties, which make it useful as a rodenticide.

Nitrophenol I. have long been used as contact and ingestion poisons, especially DNOC (see). Because they are also toxic to plants, they are used chiefly as winter sprays, e.g. in mixtures with mineral oils.

The ***acylurea I.*** have a new type of mechanism; the best known of this group is *diflubenzuron (Dimilin®)*, an *N*-(4-chlorophenyl)-*N'*-(2,6-difluorobenzoyl)urea. These I. act exclusively as ingestion poisons which interfere with the chitin metabolism of caterpillars and other larvae, preventing formation of cuticle. As a result, the larvae do not survive molting.

Historically, the oldest I. are the ***inorganic I.*** (first generation I.), which are now rarely sold. The most important of this type of I. is calcium arsenate, $Ca_3(AsO_4)_2$, which was used against potato beetles and other insects. However, it is now used almost exclusively on non-food crops such as cotton.

Insect repellents: substances used to drive off annoying or harmful insects. I. are used mainly on human beings and animals to repel annoying insects, which may carry infectious diseases. The I. act in the vapor phase which surrounds the treated object, body part or textiles. Synthetic I. are more effective than the essential oils and fumigants they have replaced for this purpose, and their cosmetic properties are also better. Combinations of the compounds shown in the table are usually applied.

Important insect repellents

Name	Formula	Melting or boiling point [°C]
Dimethyl-phthalate	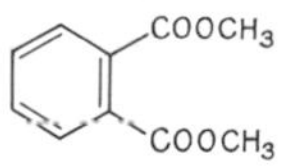 $COOCH_3$, $COOCH_3$	b.p. 282–285
Dimethyl-carbate	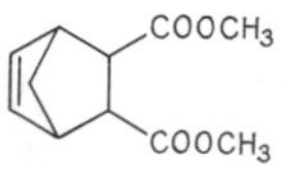 $COOCH_3$, $COOCH_3$	m.p. 38
DEET	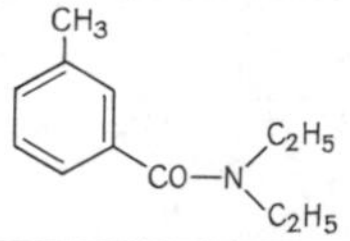 CH_3, $CO{-}N(C_2H_5)_2$	b.p. 111 (at 133 PA) m.p. 21 b.p. 150
Pyridine 2,5-dicarboxylic acid di-n-propyl ester	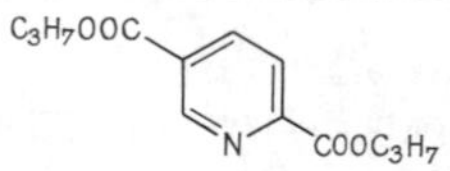 C_3H_7OOC, N, $COOC_3H_7$	(at 133 Pa)
2,3; 4,5-Bis-(butylene)-tetrahydrofurfural	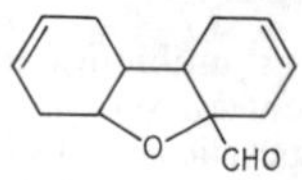 O, CHO	b.p. 307
2-Ethylhexane-1,3-diol	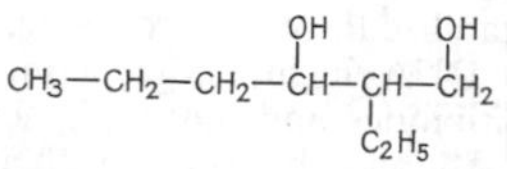 $CH_3{-}CH_2{-}CH_2{-}CH(OH){-}CH(C_2H_5){-}CH_2OH$	b.p. 244

The use of I. is also a goal in agriculture, to prevent feeding of insects. It is advantageous to use pesticides which are also antifeedants, such as organotin fungicides (e.g. fentin).

Insect toxins: toxins produced by insects and used for attack or defense. The number of insect species which produce I. is estimated to be 50,000. Some well known examples are the toxins of bees, wasps and hornets (see Bee toxins) and Ant toxins (see). The active components of I. are aldehydes, aliphatic acids (e.g. formic acid), terpene derivatives, quinones, steroid derivatives and proteins, polypeptides or free amino acids. These components usually occur in mixtures with synergistic effects.

Insertion: the insertion of a coordinatively unsaturated compound (e.g. carbene, alkene) into a metal-carbon, metal-hydrogen or metal-halogen bond. I. is typical of carbenes. The mechanism of coordinative polymerization (***polyinsertion***) can be represented as an I. in which a monomer previously bound coordinatively to the metal inserts into the metal-carbon bond of the catalyst.

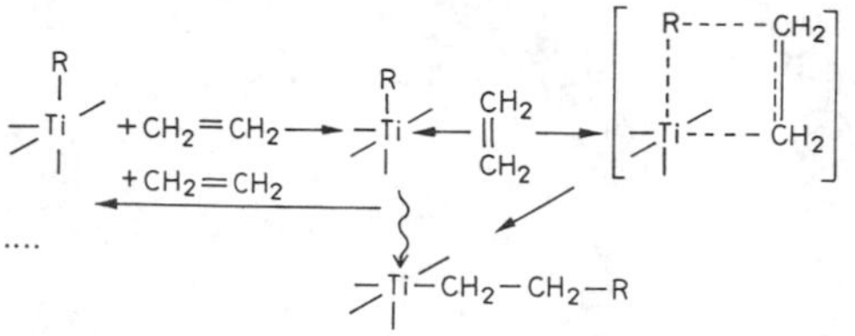

$$R{-}Ti \;+\; CH_2{=}CH_2 \longrightarrow R{-}Ti{\leftarrow}\,\begin{matrix}CH_2\\ \|\\ CH_2\end{matrix} \longrightarrow \left[\begin{matrix}R\text{- - - -}CH_2\\ \vdots \qquad \|\\ Ti\text{- - - -}CH_2\end{matrix}\right] \longrightarrow Ti{-}CH_2{-}CH_2{-}R \;\xrightarrow{+CH_2=CH_2}\; \ldots$$

Insulin: a polypeptide hormone formed in the B-cells of the islets of Langerhans in the pancreas; it causes the blood glucose level to decrease. I. consists of two peptide chains, the A chain with 21 amino acids and an intrachain disulfide bond, and the B chain with 30 amino acids. The two chains are held together by two interchain disulfide bonds. In spite of differences in amino acid sequence, the I. of different species have approximately equal biological activities in typical test systems, such as the convulsion test in mice, glucose oxidation in epididymal fatty tissue or isolated fat cells. Monomeric I. has a relative molecular mass around 6000; depending on the conditions it may form dimers or hexamers. A hexameric arrangement is present in the insulin crystal, with two coordinatively bound zinc atoms. A complex of I., zinc and protamine is used therapeutically.

Until recently, all commercial I. was extracted from porcine and bovine pancreas by extraction with 70% ethanol acidified to pH 1 to 2 with hydrochloric acid. However, human insulin genes have now successfully been incorporated into microorganisms, so that fermentive production is possible and can be expected completely to replace extraction methods, since there is always a danger of antigenicity of animal material.

The biosynthesis of I. proceeds via ***pre-proinsulin***, which contains an *N*-terminal signal sequence. This is removed to yield ***proinsulin***, which passes through the endoplasmic reticulum to the Golgi apparatus. Here it is split into I. and C-peptide, and is then stored in vesicles in the presence of zinc ions. Normally, the stored I. is released into the blood when the blood glucose level rises.

The first successful total synthesis of I. was reported in 1963 by Zahn, and a little later by Chinese and American groups.

The spatial structure of I. was elucidated in 1969 by D. Hodgkin, and the sequence first proposed in 1955 by Sanger was confirmed in 1974 by Rittel et al., who synthesized human I. in such a manner that disulfide exchange was not possible.

I. is the only hormone which specifically reduces the blood glucose level. It increases the permeability of cells to glucose, and also to amino acids, lipids and K^+ ions. I. inhibits both lipolysis and the synthesis of certain enzymes involved in gluconeogenesis. A lack of insulin causes diabetes mellitus; the causes of this disease are not completely understood.

The role of I. in the etiology of diabetes was discovered by Banting and Best in 1921; the hormone was first applied in treatment of diabetes by Banting (Nobel Prize, 1923).

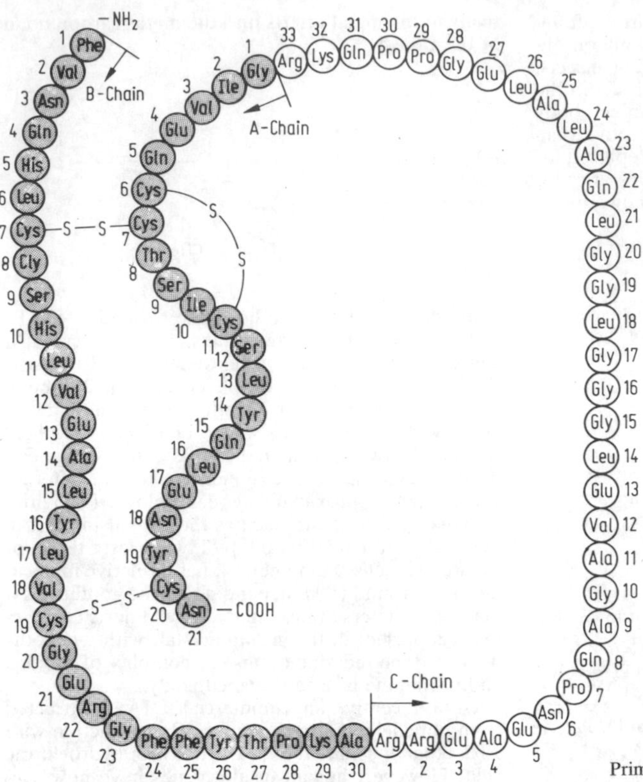

Primary structure of porcine proinsulin.

Integral reactor: primarily a research reactor, in which the turnover of the reactants in a direct run can be set as high as desired. The structure of such a reactor is often very simple, which is an advantage. A disadvantage, however, is that it is often difficult to maintain a constant temperature. The I. is used mainly for studies of reactions with heterogeneous catalysis.

Integrated pest management: a combination of chemical and nonchemical strategies for controlling pests of crop plants. I. uses only the amount of chemical pesticides actually required, in contrast to the original practice of prophylactic applications whether or not the pests were present. I. makes use of such devices as pheromone traps to determine the time of emergence of adult insects, selective insecticides and biological control.

Intercalation: insertion of foreign molecules between lamellar structures. Some examples from mineralogy and inorganic chemistry are the uptake of ammonia or water by clays, of chalcogenides such as niobium diselenide or molybdenum disulfide by alkali metals, or of various electron donors or acceptors by graphite. In biochemistry, insertion of lipophilic substances into the phospholipid bilayer of the membrane or of large planar ring systems between the stacked bases of nucleic acids is also called I. The DNA molecule is stretched by I., leading to changes in its hydrodynamic and optical properties, and interfering with its biological activity. Compounds capable of I. into DNA act as antibiotics and mutagens. The structural requirement for such a reagent is a planar carbocyclic or heterocyclic ring system with an area of at least 2.8 nm^2 and a height less than 0.34 nm. Basic substituents stabilize the complex. Compounds which intercalate in DNA include the actinomycins, anthracycline antibiotics and many pigments, such as acridines (e.g. ethidium bromide) or their derivatives (e.g. acriflavin, proflavin).

Intercrystalline corrosion: see Metal corrosion, electrochemical.

Interferons, abb. ***IFN***: species-specific proteins consisting of a protein portion of about 160 amino acid residues and often a specific, glycosidically bound carbohydrate component. The I. are synthesized in animal and human cells after induction by exogenous factors. At present 18 different human I. are known; they are classified as IFNα, IFNβ and IFNγ types. I. have antiviral, antitumor, immune and cell regulatory effects. The antiviral effect, in combination with the antigen-antibody reaction, is a primary defense mechanism against many viruses. After the virus enters the cell, its interferon genes are induced, leading to synthesis of specific IFN messenger RNA and subsequent protein synthesis. The I. are bound to specific receptors on the surface of other cells, inducing synthesis of intracellular proteins which block synthesis of virus proteins. Since the I. can inhibit cell replication, there is hope that they can be used to treat cancer. Partial results, however, have so far been obtained only in the treatment of osteosarcoma, a disease which most often affects young people. I. also affect the immune reaction and are applied, e.g. in infections and organ transplants. Human I. was first obtained from tissue cultures of human leucocytes, fibroblasts and lymphoblasts which had been induced by infection with viruses. However, they are now available from engineered strains of *Escherichia coli* into which the human I. genes have been inserted.

Interhalogen compounds: covalent, polar compounds of the types XY, XY_3, XY_5 and XY_7 (X = the more electropositive, heavier halogen; Y = the lighter halogen), which can be made by reactions between different halogens. The reactivity and tendency of XY compounds to decompose increases with increasing distance between the halogens X and Y in the periodic system. Iodine monofluoride, IF, can thus be obtained only at low temperatures. On the other hand, compounds of the type XY_n (n = 3, 5, 7) are the more stable, the more electropositive X is and the more electronegative Y is. The I. containing the highest proportion of Cl is the dimeric iodine trichloride $(ICl_3)_2$, and a heptafluoride is known only for iodine (IF_7). The stability and volatility of the higher fluorides increase with increasing fluorine content. I. of the XY_3 type have T-shaped molecules. IF_5 is square pyramidal, and IF_7 is pentagonal bipyramidal.

Molecular interhalogen compounds

Type XY:			Type XY_n:		
ClF			ClF_3	ClF_5	
BrF	BrCl		BrF_3	BrF_5	
IF	ICl	IBr	IF_3	IF_5	IF_7
				$(ICl_3)_2$	

Interlattice occupation: see Crystal lattice defects.

Interleukins: carbohydrate-free proteins produced by antigen-activated thymus cells (T-cells); as lymphokines, they regulate cellular immunity. Interleukin 2 (IL-2), for example, consists of 153 amino acid residues (M_r 17632), and acts as a T-cell growth factor. It increases the activity of natural killer cells, which recognize and kill cells infected with virus and certain tumor cells. I. are important in tissue culture of T-cells.

Intermolecular bond: see Chemical bond.

Intermolecular forces: see Van der Waals forces.

Intermolecular interaction: see Chemical bond.

Internal complexes: neutral metal complexes with anionic chelating ligands. In an I., the chelate ligand both neutralizes the charge of the central ion and saturates its coordination capacity. Some typical I. are formed from ligands such as diacetyldioxime 8-hydroxyquinoline, α-aminocarboxylic acids and acetylacetone. Some biochemically important I. are the porphyrins heme and chlorophyll.

Intermediate, ***transient***: a short-lived species formed in a complex reaction. The most important I. are radicals and radical ions; in organic chemistry, there are in addition carbenium ions, carbanions and carbenes, while in photochemistry electronically excited states are important.

Whether an I. is short-lived depends greatly on the reaction conditions. Certain species which are quite short-lived in the medium temperature range may be present in considerable equilibrium concentrations at extremely high temperatures (e.g. OH radicals, hydrogen atoms and C_2 in flames) or they may have a significant lifetime at very low temperatures (see Matrix isolation technique).

I. can be detected by separation from the reaction mixture after the reaction has been interrupted, by independent synthesis, capture reactions and physical methods, mainly spectroscopic methods.

Intermediate phases: same as Intermetallic compounds (see).

Intermetallic compounds, ***intermetallic phases, (metallic intermediary phases)***: alloys of metallic elements consisting of a single type of crystal with a lattice different from those of either of the elements. If the lattice can be continuously changed, by changing the chemical composition, into that of one of the elements, the crystal is a (metallic) Mixed crystal (see). The boundary between I. and mixed crystals is often difficult to delineate, and, for example, mixed crystals with a Superstructure (see) are often counted as I.

I. differ in several essential points from normal chemical compounds. Their stoichiometric composition usually does not correspond to normal bonding theory, and cannot be derived from it (e.g. KPb_2, $NaAu_2$, Cu_5Zn_8). However, there are also cases in which the stoichiometry does correspond to the usual valence rules (e.g. Ag_3Al, AlFe, NaTl), but thir behavior is not fundamentally different. Many I. have no clearly defined stoichiometry, and are stable over a wide range of variable compositions (see Berthollides). For these reasons, the terms intermetallic or (metallic) intermediary phases are often preferred to I.

The peculiarities of the I., especially their compositions, can largely be explained from the geometric principles on which their lattices are built; these are the same as apply to the metallic elements: the tendency towards high coordination numbers, higher packing density and three-dimensional atom linkage. While 12 is the largest possible coordination number for pure metals (cubic and hexagonal close packing), I. can have higher coordination numbers, up to 16, because of the different atomic radii of the components.

The above principles are most clearly expressed in *Laves phases* which, with over 300 binary representatives, form the largest group of I. They have the general composition AB_2, and crystallize in the three closely related structure types $MgCu_2$, $MgZn_2$ and $MgNi_2$. The $MgCu_2$ type is cubic; the Mg atoms form a Diamond structure (see) among themselves, and the empty 1/8-cubes are occupied by Cu_4 tetrahedra. The $MnZn_2$ type is hexagonal with a Wurtzite structure (see) of the Mg atoms and Zn_4 tetrahedra linked to each other. The $MgNi_2$ structure is hexagonal, and has a structure intermediate between the other two. The stability of the lattice depends entirely on the ratio of atomic radii. In the sphere-packing model of Laves phases, the A spheres touch only A, and the B spheres touch only B spheres. Here the ratio of radii is $\sqrt{3}$:$\sqrt{2}$. The coordination numbers are 16 for the A atoms and 12 for the B atoms, which gives an average total coordination number of 13 1/3.

Another large group of I. are the *Hume-Rothery phases*. Here the structures are determined by the *valence electron concentration*, that is, the ratio of the number of valence electrons to the number of atoms. If the ratio is 3/2, a cubic body-centered lattice forms (e.g. the β-phase of the Cu-Zn (brass) system with the composition CuZn); if it is 21/13, a complicated cubic lattice forms with 52 atoms in the unit cell (γ-brass phase, Cu_5Zn_8); if 7/4, a hexagonal close packing (ε-brass phase, $CuZn_3$). All three phases have large ranges of homogeneity. The I. which are determined less by geometry than by electronic considerations also include the transition metals in the α-manganese lattice, e.g. $Mg_{34}Al_{24}$ and the σ-phases.

In *zintl phases*, the metallic character is less strongly evident. Zintl phases (in the narrower sense) occur when one component is an alkali or alkaline earth metal, and the other is no more than 4 periodic table positions removed from a noble gas. Zintl phases have a large ionic component of bonding, sometimes also covalent components, and frequently their compositions obey valence rules, with narrow phase widths, e.g. Mg_2Si, Mg_3As_2, Li_4Sn, Li_3Bi. They crystallize in structure types which are typical of ionic lattices, and display correspondingly lower coordination numbers (8, 6 or even 4). Zintl phases have both metallic and semiconductor properties.

Grimm-Sommerfeld phases are not I. in the narrow sense, although they are often included among them, but are definitely semiconductors (see Grimm-Sommerfeld rules).

For an overview of the tremendous variety of the I., it is helpful to arrange them by lattice type, and these in turn by their coordination numbers. Such an arrangement also approximately follows the change in bond character; the following summary thus goes from the pure metallic Laves phases to the semiconducting Grimm-Sommerfeld phases.

Examples of intermetallic compounds

1) Compounds with coordination numbers greater than 12

Laves phases ($MgZn_2$, $MgCu_2$, $MgNi_2$)
$CaZn_5$ type ($BaPt_5$, $SrAg_5$)
μ-phases (Co_7Mo_6, Fe_7W_6)
$CuAl_2$ type ($AuNa_2$, $FeSn_2$, $RhPb_2$, $ZnTh_2$)
β-W type ($AuTi_3$, CoV_3, $IrMo_3$, $SnNb_3$)
σ-phases (CoCr, FeV, $MoRu_3$, Rh_2Ta_3)
α-Mn type ($Al_{12}Mg_{17}$, $MoRe_4$)

2) Hume-Rothery phases

β-phase (β-CuZn, Ag_3Al, Cu_4In)
β'-phase (β'-CuZn)
γ-phase (Cu_5Zn_8)
ε-phase (CuZn)

3) Inclusion compounds

AlB_2 type ($ThAl_2$, $CaGa_2$, UHg_2)
$MoSi_2$ type (WSi_3, $AlCr_2$, $MgHg_2$)
FeSi type (AuBe, GaPt, ε-CrSi)

4) Zintl phases

CsCl type (LaHg, AlFe, LiTi)
CaF_2 type ($AuAl_2$, $PbMg_2$, $SnMg_2$)
NaCl type (AsCl, BiPu)
NiAs type (AuSn, SoSb, IrPb)
NaTl type (AlLi, InNa)

5) Grimm-Sommerfeld phases

Zinc blende (wurtzite type) (AlAs, GaAs)

I. have so far not attained any great industrial significance, but are of considerable scientific interest. Their most interesting physical properties are those which differ significantly from those of the pure components. Among the most striking of these is their brittleness, which is the main hindrance to their industrial use. Their melting points can be much higher than those of the components; for example, $CeAl_2$ melts at 1738 K, compared to cerium at 1078 K and aluminum at 933 K. The electric resistance of I. depends on the metallicity of the bonding. It can be on the same order of magnitude as those of metallic elements and have a positive temperature coefficient, e.g. $CuMg_2$; on the other hand, other I. have typical semiconductor properties, e.g. Mg_2Sn. Superconductivity has been demonstrated in a number of I., sometimes even when the metallic components do not display this property, e.g. $CaAu_5$. It is of interest that I. composed of nonferromagnetic elements can be ferromagnetic, e.g. $ZrZn_2$ and $SmAl_2$.

Intermetallic phases: same as Intermetallic compounds (see).

Intermolecular reactions: reactions which occur between two or more molecules of the same type or different types.

Internal energy, symbol *e*: a state parameter in thermodynamics which indicates the total energy within a system. This includes the kinetic energy of thermal motion, bond energy, vibrational, rotational and electronic energies of excited states and energies of intermolecular interactions. Energies which apply to the entire system in the environment, e.g. its potential energy in a gravitational field or the kinetic energy from a translational motion of the entire system, are not included in the I. On the other hand, the nuclear energy and the relativistic energy corresponding to the Einstein equation $E = mc^2$ is part of I. Since nuclear reactions are not part of chemistry, and the mass defects in chemical reactions are too small to measure, these contributions to the I. can be summarized in thermodynamics as an additive constant. Phenomenological thermodynamics, as a continuum theory, is not capable of producing absolute values of the I. of systems. In principle, absolute values can be obtained using statistical thermodynamics, but in practice they are approximate at best. However, for thermodynamic changes of state, only changes in the I. are significant.

Internal pressure: see Van der Waals equation.

International Union of Pure and Applied Chemistry, abb. ***IUPAC***: the international scientific union of national organizations of chemists. Its goals are: working out of recommendations for international regulation, standardization or codification of areas of chemistry; cooperation with other international organizations whose work involves chemistry; promotion of contact between chemists of member countries; and promotion of progress in the field of chemistry. At present, the members are the National Adhering Organizations (NAO) of 44 countries, which are represented by chemists named by them (Titular or Associate Members). The representatives of the NAO form the IUPAC Council. The scientific work is carried out in seven divisions (physical, inorganic, organic, macromolecular, analytical, applied and clinical chemistry), which are subdivided into several commissions, and further into sections. The divisions and commissions meet at least every two years during the general assembly of the IUPAC, at the end of which the Council meets. Congresses are organized by the NAO, and take place at the same time as the general assembly. In addition, there are 20 to 30 IUPAC-sponsored conferences each year. The results of scientific efforts are published in the IUPAC journal *Pure and Applied Chemistry*. The magazine *Chemistry International* reports on the various events in the life of the IUPAC.

The IUPAC was founded on 22 July 1919 as the successor of the International Association of Chemical Societies (Association internationale des Sociétés chimiques). From 1930 to 1947 it was called the International Union of Chemistry (IUC).

The Secretariat of the IUPAC is located at Bank Court Chambers, 2-4 Pound Way, Cowley Centre, Oxford OX4 3YF, UK.

Interstitial cell stimulating hormone: see Lutropin.

Intramolecular reactions: reactions which occur within a single molecule, such as isomerization, ring closure, etc.

Intrinsic factor: see Vitamin B_{12}.

Introns: see Nucleic acids.

Inulin: a reserve polysaccharide in which about 30 fructofuranose molecules are joined in $\beta(1\rightarrow2)$ linkage. I. is a white powder which is nearly insoluble in cold water, but it dissolves readily in hot water. It is degraded in vivo to fructose. I. is found in many plants, especially in the *Asteraceae*, instead of starch. It is obtained from dahlia tubers or chicory root. Highly purified I. is used to test liver function. I. is used to produce fructose.

Invar®: a nickel-iron alloy (see Nickel alloys) with 35 to 37% nickel, and the rest iron. I. has linear thermal expansion coefficient of $\alpha_{0,100°C} = 1.5 \cdot 10^{-6}$ m/m · K, and is used for watches, balances and other high-precision instruments.

Inverse voltammetry, ***inverse polarography***: the most sensitive of the electrochemical analysis techniques. The apparatus is similar to that for Polarography (see). However, a stationary mercury electrode (hanging droplet or film) is used instead of a dropping mercury electrode. The working electrode is first polarized for a defined length of time, so that the depolarizer can precipitate, forming an amalgam. To maximize the matter transport to the electrode, the electrolyte is stirred. After this enrichment period (1 to 4 min) and a short stirring of the electrolyte, the polarization voltage is continuously changed to more positive values. Eventually a voltage is reached at which the previously precipitated depolarizer is redissolved. The resulting current peak is proportional to the concentration of the depolarizer in the solution. The sensitivity can be increased further if the redissolving process (stripping) is done using the difference-pulse polarographic method instead of a linear voltage function. This method, DPASV (for Differential Pulse Anodic Stripping Voltammetry), is especially suitable for ultra-trace analysis of heavy metals.

Inversion: 1) a change in the sign of an optically active compound in the course of a reaction, e.g. sucrose I. Here D(+)-sucrose is hydrolysed to D(+)-glucose and D(−)-fructose. The resulting mixture (***invert sugar***) rotates polarized light to the left, because D(−)-fructose rotates polarized light more strongly to the left than D(+)-glucose does to the right. The original (+) rotation is thus converted to a (−) rotation.

2) ***I. of configuration*** occurs in stereospecific reactions at a chirality center (see Stereoisomerism 1.1). If the configuration is retained, one speaks of *retention*. I. of configuration occurs, for example, in bimolecular nucleophilic substitution reactions (S_N2). Retention occurs, for example, in internal nucleophilic substitution reactions (S_N1) or in S_N reactions in which neighboring groups are involved. Non-stereospecific reactions at a chirality center (via a symmetric intermediate, such as a radical or carbocation) lead to Racemization (see). I. of configuration of only one chirality center among several is called *epimerization* (see Stereoisomerism 1.2.1).

3) I. at a center of symmetry, see Symmetry.

Inversion spectra: absorptions in the microwave spectrum (see Microwave spectroscopy) due to inversions of pyramidal molecules. The best known I. is that of ammonia, which was observed as early as 1934. It arises because there are two atomic arrangements of the molecule, which can be interconverted by an "inversion" of the nitrogen through the plane of the hydrogen atoms. Because the molecule at the same time can be in any of numerous rotational states, the spectrum consists of many lines; in the case of ammonia they lie in the range from 0.55 to 1.33 cm^{-1}. It is possible to obtain information on the height of the potential barrier to the inversion from the splitting of the lines.

Invertase: β-fructosidase, a sucrose-cleaving enzyme found in microorganisms and higher plants. The I. from yeast is a dimeric glycoprotein, M_r 270,000.

Invert soaps: surfactant compounds which are quaternary ammonium salts. In soaps, the anion is the surface-active portion of the molecule; in I., it is the cation which is surface active. Because of their quaternary structure, these compounds are also called *"quats"*. The N atom is usually bound to a long-chain alkyl group and three methyl groups or two methyl groups and a benzyl group. The positively charged N atom with the methyl groups is the hydrophilic part of the molecule, while the long-chain alkyl group is the hydrophobic part. The N atom can also be part of a ring, or it can be replaced by a phosphorus atom. Often the I. are not pure compounds, but mixtures in which the alkyl chains are of various lengths. Other structural elements, such as carbamides or esters can also be present in the hydrophobic part. Unlike soaps, I. remain effective in the presence of polychenar cations. I. and soaps form insoluble salts. The I. are used widely as disinfectants and preservatives, e.g. in eye medications. They have antibiotic activity against a large number of bacteria; gram-positive bacteria are particularly susceptible. In addition, I. act as antimycotics and can inactivate viruses. Some examples are benzododecinium bromide, alkonium bromide and cetylpyridinium chloride.

Invert sugar: an equimolar mixture of D-glucose and D-fructose. I. is obtained by acid-catalysed hydrolysis of sucrose. Because fructose is highly levorotatory, the sign of the optical rotation of the solution changes during the course of the hydrolysis, i.e. an inversion occurs. I. is present in honey at about 70% concentration. It is used to make artificial honey and to keep foods moist.

In vitro: (Latin, meaning "in glass") an adjective applied to experiments done under artificial conditions, e.g. in a test tube.

In vivo: (Latin, meaning "in life") an adjective applied to experiments in a living cell or organism.

Iodate: a salt of iodic acid, HIO_3, with the general formula MIO_3. I. are more stable than chlorates or bromates, but like these compounds, they are strong oxidizing agents. When mixed with combustible substances, I. explode easily on impact. Alkali iodates are obtained by dissolving iodine in hot alkali hydroxide solutions, or by anodic oxidation of alkaline iodide solutions. The slight amounts of sodium and calcium iodates found in Chile saltpeter are an important starting material for production of iodine.

Iodic acid: HIO_3, transparent, colorless, rhombic crystals; M_r 175.93, density 4.650, m.p. 110°C. I. is very soluble in water and is a medium strong acid (pK 0.804); it is a strong oxidizing agent and is the only halogen(V) acid of the type HXO_3 which can be isolated in anhydrous form. I. is obtained by oxidation of iodine with strong oxidizing agents such as conc. nitric acid, hydrogen peroxide, ozone or chlorine. If the oxidation is done with chlorine, hydrochloric acid is formed simultaneously: $I_2 + 5\ Cl_2 + 6\ H_2O \rightarrow 2\,HIO_3 + 10\ HCl$; the HCl must be removed by addition of silver oxide to pull the equilibrium toward the products. I. can be released from iodates by reaction with sulfuric acid: $MIO_3 + H_2SO_4 \rightarrow HIO_3 + MHSO_4$.

Iodides: metal salts of hydroiodic acid, and also covalent compounds of iodine with nonmetals, including organic compounds such as alkyl or aryl iodides. Alkali and alkaline earth metals form ionic, water-soluble I., MI or MI_2, while a few heavy-metal iodides are insoluble in water, e.g. silver(I) iodide, AgI (yellow), copper(I) iodide, CuI (colorless), mercury(II) iodide, HgI_2 (red) and thallium(I) iodide, TlI (yellow). There are also covalent, hydrolysable I., including phosphorus(III) iodide, PI_3, and silicon(IV) iodide, SiI_4.

Iodination: see Halogenation.

Iodine, symbol ***I***: an element in group VIIa of the periodic system, the Halogens (see); a nonmetal, with only one natural isotope, Z 53, atomic mass 126.9045, valence - I, +I, +III, +IV, +V, +VII, density 4.942, m.p. 113.6°C, b.p. 185.24°C, standard electrode potential (I^-/I_2) +0.5255 V.

Properties. I forms gray-black, semiconducting, rhombic crystals with a metallic sheen; even at room temperature they are somewhat volatile, forming a vapor of violet I_2 molecules. If I is heated fairly slowly, it will completely sublime below the melting point. It has a characteristic pungent odor, and the vapors are poisonous. Solid I forms a layered lattice, in which the intermolecular distance between the atoms of neighboring I_2 molecules is 349.6. This is a very short distance, and delocalization of electrons within the layers leads to two-dimensional semiconductor properties and the metal-like sheen of I. In nonpolar solvents, such as carbon disulfide, chloroform or tetrachloromethane, I. dissolves as molecules, giving a violet solution. On the other hand, the red solutions of I. in aromatic hydrocarbons and the brown solutions in donor solvents such as diethyl ether, acetone, dioxane and pyridine, contain charge-transfer complexes of the I. with the solvent molecules, $I_2 \cdot D$. I. is only very slightly soluble in water (0.022 g in 100 g H_2O), and gives a weak, brownish yellow color. However, it dissolves very readily in potassium iodide solution, forming dark brown potassium triiodide, KI_3. I. is chemically very similar to the other halogens, but its reactions are less vigorous. It reacts vigorously with a number of elements, including iron, mercury, sulfur, phosphorus, antimony, silicon and nickel, forming iodides. A characteristic of I. is its ability to form cationic compounds in the +1 and +3 oxidation states. For example, iodine(I) perchlorate is can be made by reaction of I. with silver perchlorate in a benzene solution: $I_2 + AgClO_4 \rightarrow IClO_4 + AgI$; iodine(III) perchlorate is obtained by reaction of the same substances in ether at - 85°C: $2\ I_2 + 3\ AgClO_4 \rightarrow I(ClO_4)_3 + 3\ AgI$. Iodine(I) compounds can be stabilized by Lewis bases, e.g. $[IPy_2][ClO_4]$ and $[Ipy_2][NO_3]$.

Analysis. I is characterized by formation of an intense blue inclusion compound with starch (see Iodometry). The iodide ion, I^-, can be detected by reaction with silver nitrate to form yellow silver(I) iodide, AgI. Iodine also forms a dark red mercury(II) iodide, HgI_2, and yellow lead(II) iodide, PbI_2. Elemental I. can be determined quantitatively by titration with sodium thiosulfate, while iodide is determined by argentometry or gravimetrically as AgI.

Occurrence. I makes up $6.1 \cdot 10^{-5}$% of the earth's crust, and is thus one of the least abundant elements. It is found only as its compounds in nature; the most important iodine deposits are the saltpeter deposits in Chile, and natural waters (from deep wells and brines from petroleum and natural gas wells). Chile saltpeter can contain up to 0.3% I in the form of sodium iodate, $NaIO_3$, or calcium iodate (lauterite), $Ca(IO_3)_2$. Water from deep layers used for I. production contain up to 50 ppm I.; the brines from petroleum deposits can contain up to 100 ppm. I. also occurs widely in rocks and soils (about 5 ppm). Seawater contains about 0.002% I., mainly in organic form. Various marine organisms, such as kelp and algae, corals and sponges can enrich I. up to 0.45% of their dry matter.

I. is an important bioelement; plants contain about 0.1 ppm. It is essential for the human body, as it is a component of the thyroid hormones thyroxin and triiodothyronin. The daily human requirement is about 2 mg. I deficiency leads to goiter, and in severe cases, to cretinism.

Production. I. is enriched in the mother liquors

from processing Chile saltpeter. The iodate present in the liquor is reduced with sulfurous acid to I: 2 HIO_3 + 5 $H_2SO_3 \rightarrow I_2$ + 5 H_2SO_4 + H_2O. The precipitated I. is filtered out and purified by multiple sublimation steps. To obtain I from iodide-containing waters, it is first oxidized with chlorine to I_2, then isolated and purified by repeated absorption and desorption, reduction and oxidation steps. A small amount of I. is still isolated from seaweed.

Application. I. is used as an antiseptic and to stop bleeding (see Iodine, tincture of). Considerable amounts of I. are used in the synthesis of drugs used to treat abnormal thyroid function. Iodides are added to animal feeds as trace element sources. I. and its compounds are used in photochemistry, preparative and analytical chemistry, and organoiodine compounds are used as x-ray contrast media. The nuclide ^{131}I is obtained from nuclear reactors; it is a β-emitter with a half-life of 8.04 d and is used in medicine.

Historical. I. was first isolated from the ashes of seaweed in 1811 by Coutois. In 1815, Gay-Lussac demonstrated that it is an element, and it was named after the color of its vapor (the Greek word "ioedides" means "violet").

Iodine azide: see Halogen azides.

Iodine bromide: see Iodine halides, see Interhalogen compounds.

Iodine chlorides: see Iodine halides, see Interhalogen compounds.

Iodine charcoal: granulated activated charcoal containing iodine. It is used to pick up spilled mercury.

Iodine cinnabar: see Mercury iodides.

Iodine fluorides: see Iodine halides, see Interhalogen compounds.

Iodine halides: very reactive Interhalogen compounds (see) obtained by reaction of iodine with the lighter halogens. The I. have general formulas IX (X = F, Cl, Br), IF_n (n = 3, 5, 7) and $(ICl_3)_2$. ***Iodine monofluoride***, IF, is a chocolate-brown solid, dec. above 0°C; ***iodine monochloride***, ICl, is a dimorphous compound. α-ICl forms ruby-red needles, m.p. +27.38°C, and β-ICl (metastable) forms red-brown, rhombic platelets, m.p. 13.9°C, b.p. 97.4°C. ***Iodine monobromide***, IBr, forms red-brown crystals, m.p. +41°C, b.p. +116°C, and ***iodine trifluoride***, IF_3, is a yellow powder (at - 78°C); m.p. - 28°C (dec.). ***Iodine trichloride*** $(ICl_3)_2$ forms yellow crystalline needles, m.p. 101°C (at 1.6 MPa), b.p. 77°C (dec.). ***Iodine pentafluoride***, IF_5, is a colorless liquid, m.p. +9.42°C, b.p. 104.48°C; ***iodine heptafluoride***, IF_7, colorless gas, m.p. +6.45°C, b.p. 4.77°C.

Iodine number: see Fats and fatty oils.

Iodine oxides: ***Diiodine tetroxide***: I_2O_4, yellow, grainy compound, M_r 317.81, density 4.2, m.p. 130°C. When heated to about 135°C, it undergoes a disproportionation reaction: 5 $I_2O_4 \rightarrow$ 4 I_2O_5 + I_2. I_2O_4 reacts with alkali hydroxide solutions to form iodide and iodate. It is synthesized by a slow reaction of hot, concentrated sulfuric acid with iodic acid. Structurally, I_2O_4 is probably iodosyl(III) iodate(V).

Diiodine pentoxide, iodine(V) oxide: I_2O_5, white, crystalline powder, M_r 333.80, density 4.799, m.p. ≈ 300°C (dec.). When heated to about 300°C it decomposes into the elements. It can be considered the anhydride of iodic acid, with the structure O_2IOIO_2 (∢IOI 139.2°, terminal I-O distances, 180 pm, bridge IO distance, 194 pm): $I_2O_5 + H_2O \rightleftharpoons 2\,HIO_3$. I_2O_5 is obtained by heating iodic acid to about 250°C; it is the only exothermal halogen oxide.

Diiodine heptoxide, iodine(VII) oxide, I_2O_7, an orange, polymeric solid, M_r 365.81, formed by dehydration of periodic acid with concentrated sulfuric acid. When heated to 100°C, it is converted to I_2O_5 according to the equation $I_2O_7 \rightarrow I_2O_5 + O_2$.

Iodine red: see Mercury iodides.

Iodine spirits: a dark, red-brown liquid which smells like iodine; it is made by dissolving certain amounts of iodine and potassium iodide in 80% ethanol. I. is used to disinfect wounds, but because of possible allergic reactions, it is now rarely used.

Iodine, tincture of: a dark brown liquid which smells like iodine; density 0.898 to 0.902. I. is an alcohol solution of iodine which contains 7% iodine and 3% potassium iodide; it is used in medicine to disinfect wounds.

Iodoacetic acid: I-CH_2-COOH, a colorless, crystalline compound; m.p. 83°C. I. is soluble in water and alcohol. It can cause severe burns on the skin. It is synthesized by reaction of chloroacetic acid with potassium iodide in aqueous solution. It is used in organic syntheses and in biochemistry to inhibit certain enzymes.

Iodobenzene, ***phenyl iodide***: C_6H_5-I, a colorless liquid which turns brown in the air, due to precipitation of iodine; m.p. - 31.3°C, b.p. 188.3°C, n_D^{20} 1.6200. I. is barely soluble in water, but dissolves readily in alcohol, ether, acetone and benzene. It can be made by iodination of benzene in the presence of nitric acid or by the Sandmeyer reaction (see) from benzene diazonium salts. I. is used in the synthesis of iodine-containing x-ray contrast materials.

Iodoform, ***triiodomethane***: CHI_3, forms yellow, hexagonal platelets with a penetrating, sweetish odor; m.p. 123°C, b.p. about 218°C. I. is practically insoluble in water, but is soluble in ether, acetone, carbon disulfide and chloroform. It is steam volatile. It decomposes readily in the presence of light. I. can be synthesized by the Iodoform test (see); industrially, it is made by electrolysis of alkali iodides in alcohol-water or acetone-water mixtures. I. is sometimes still used as an antiseptic for treating cuts, and as a non-sulfur vulcanizing material for rubber.

Iodoform test: a reaction used to detect the presence of an acetyl group, CH_3-CO- (e.g. in acetone) or a 1-hydroxyethyl group, CH_3-CH(OH)- (e.g. in ethanol). Iodine and potassium hydroxide react with these functional groups to form iodoform, which is a yellow, water-insoluble compound. For example: CH_3-CH(OH)-R + I_2 + 2 KOH → CH_3-CO-R + 2 KI + 2 H_2O; CH_3-CO-R + 3 I_2 + 3 KOH → CI_3-CO-R + 3 KI + 3 H_2O; CI_3-CO-R + KOH → HCI_3 + R-COOK. The I. can be used with compounds which are fairly insoluble in water in the presence of a solubilizer such as dioxane. This reaction is a variant of the Haloform reaction (see).

Iodomethane: same as Methyl iodide (see).

Iodometry: method of redox analysis based on the corresponding redox pair iodine/iodide. With a standard potential E_0 = +0.536, iodide is a weak oxidizing agent. Strong reducing agents, e.g. tin(II), arsenicIII), thiosulfate, sulfide and sulfite, can be titrated directly with a standard iodine solution. Strong oxidizing agents, e.g. bromate, chlorate, chlorine, chromate, iodate or permanganate react with iodide to form iodine, which can be determined by titration with thiosulfate standard solution. Both iodine and thiosulfate standard solutions must be calibrated, which is done by titration against the titrimetric standard substance arsenic(III) oxide or potassium dichromate. The potassium dichromate is converted to an equivalent amount of iodine by an excess of iodide. A starch solution is almost always used as an indicator in I. Starch forms a dark blue addition compound with triiodide ions, I_3^-. Depending on the method, the endpoint is recognized by the appearance or disappearance of this blue color. I. can also be used in organic chemistry to determine aldehydes, mercaptans and acetone.

Iodonium compounds: iodine-containing organic compounds with the general formula $R^1\text{-I-}R^2X$, where R^1 and R^2 are most often aromatic systems and X is an OH group or a monovalent acid group. Aliphatic I. have not been thoroughly studied. Diaryl iodonium hydroxides are strong bases which form rather stable salts with acids. I. can be synthesized by reaction of iodobenzene with iodyl benzene in the presence of silver oxide and water:

$$C_6H_5\text{–}IO_2 + C_6H_5\text{–}IO. \xrightarrow[-\,AgIO_3]{+\,AgOH} C_6H_5\text{–}I\text{–}C_6H_5OH.$$

The iodine released from this reaction (in the form of iodate) comes only from the iodyl benzene. I. can also be made by reaction of aromatic compounds with iodic acid; it is thought that in this case the iodylating reagent is iodous acid: $2\ CH_3\text{-}C_6H_5 + HIO_2 \rightarrow CH_3\text{-}C_6H_4\text{-}I^+\text{-}C_6H_4\text{-}CH_3\text{-}OH^- + H_2O$. I. can be used as arylating reagents for compounds with active methylene groups.

Iodosobenzene, ***iodosylbenzene***: $C_6H_5\text{-IO}$, a yellow, amorphous powder which decomposes explosively when heated to 210 °C. This spontaneous decomposition is due to disproportionation of iodyl benzene formed from I.; it is very unstable at this temperature. I. is somewhat soluble in alcohol and water, and insoluble in ether, acetone and benzene. The bonding in I. is best described by a zwitterionic structure: $C_6H_5\text{–}\overset{+}{I}\text{–}\overset{-}{\underline{O}}|$. The semipolar compound can also be represented by this formula: $C_6H_5\text{-I}\rightarrow O$. I. dissolves in acids to form salts. It can be made by oxidation of iodobenzene with ozone or another strong oxidizing agent; a more effective synthesis is the reaction of iodobenzene and chlorine, followed by hydrolysis of the primary benzene iodosochloride:

$$C_6H_5\text{–}I + Cl_2. \rightarrow C_6H_6\text{–}I^+\text{–}Cl\ Cl^-. \xrightarrow{(OH)^-}$$
$$C_6H_5\text{–}I^+\text{–}OH\ Cl^- \rightarrow C_6H_5\text{–}IO + HCl.$$

Iodosyl benzene: same as Iodosobenzene (see).

Iodoxybenzene: same as Iodyl benzene (see).

Iodyl benzene, ***iodoxybenzene***: $C_6H_5\text{-}IO_2$, crystalline needles; m.p. 236-237 °C (explosive decomposition). I. is soluble in water and glacial acetic acid, but is insoluble in alcohol, chloroform and benzene. It is formed by disproportionation of iodosylbenzene, which also yields iodobenzene: $2\ C_6H_5\text{-IO} \rightarrow C_6H_5\text{-}IO_2 + C_6H_5\text{-I}$. I. can also be formed by oxidation of iodobenzene or iodosobenzene with hypochlorites or Caro's acid.

Iomeglamic acid: see X-ray contrast materials.

Ion: an electrically charged atom, group of atoms or molecule. Like electrons, I. can carry electric current (see Ionic conductor). Positively charged I., which are formed from neutral particles by release of electrons, are called *cations*. Negatively charged I., *anions*, are formed by uptake of electrons by neutral particles. A relative measure for the size of an I. is the Ionic radius (see). Many elements can form I. with different charges (*ionic valences*). The tendency to go from one valence to another causes a potential to develop between electrodes which are dipped in aqueous solutions of differently charged I. (see Nernst equation); this phenomenon was used to establish a potential series of ionic reactions.

I. are symbolized by plus or minus signs written as superscripts, e.g. Na^+, Br^-. For I. with more than one unit of charge, the symbol can be written with either as many plusses or minuses as there are units of charge (e.g. Fe^{+++}), or with an Arabic numeral indicating the number and a plus or minus sign (e.g. Fe^{3+}).

The formation of I. occurs either by Dissociation (see) or Ionization (see) with addition of an amount of energy called the ionization energy. The numerical value of this energy in electron volts (eV) is the same as the numerical value of the ionization voltage. This voltage is the potential drop through which an electron must be accelerated in order to ionize an atom; on the average, it is about 10 eV. The exact value depends on the type of atom or molecule and the number of electrons to be removed or provided. In dilute aqueous solutions, many substances are completely ionized. Salts in the solid state usually form ***ionic lattices*** (see Lattice type). In the solid, the electrostatic attraction between the ions produces very strong binding forces (***ionic bonds***; see Chemical bonding). The ionization can be demonstrated with an electroscope. In the Wilson cloud chamber, I. can be visualized through the trail of condensation nuclei which they generate.

Ion chromatography: a variant of Ion-exchange chromatography (see) in which the electrolyte background of the mobile phase is suppressed by post-column derivatization. Specific reagents (e.g. alizarin red S) are used which react only with the sample ions to form complexes which can be detected with high sensitivity, or the separation column is connected to a second ion exchange column (*suppressor*) in which the eluant is neutralized or captured. For example, hydrochloric acid, which is very commonly used to separate cations, can be suppressed by an ion exchanger which is converted from its OH form to its Cl form and releases water. More recent methods make use of semipermeable *ion exchange membranes (hollow fibers)* as suppressors. Even shorter reaction columns and higher exchange capacity is obtained when an electric field is applied to accelerate ion transport in an electrochemical suppression chamber.

A distinction is made between ***cation*** and ***anion chromatography*** (Fig. 1a and 1b). In addition, there is a simple form of anion chromatography in which a

10^{-3} to 10^{-4} molar solution of phthalic acid is used as eluant in the pH range from 4 to 6. Because of its low electrical conductivity, no suppressor column is needed.

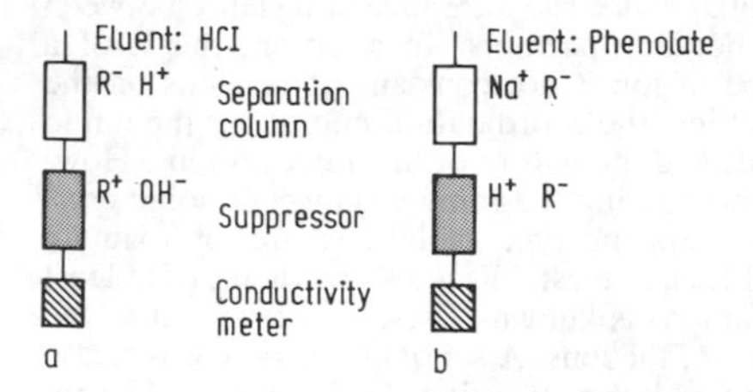

Fig. 1. *a*, Cation chromatography; *b*, anion chromatography.

The separated ions are detected by conductivity measurements or electrochemical detectors. The apparatus is essentially the same as in High-performance liquid chromatography (see).

I. is used in routine wet chemical analyses for ion determinations in drinking water, sewage and biological fluids (Fig. 2). Ion concentrations can be measured in the range from parts per hundred (sea water) to parts per billion (boiler feedwater).

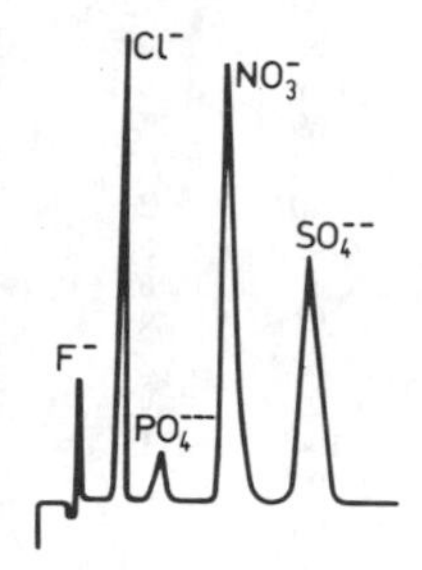

Fig. 2. Anion chromatogram of a water sample.

Ion cyclotron resonance, abb. ***ICR***: a spectroscopic technique for studying ion-molecule reactions at low kinetic energy and low pressures. The pathway of a charged particle in a magnetic field B is a circle in a plane which is perpendicular to the plane of the magnetic field. The circular frequency of this motion is called the natural cyclotron frequency ω_c: $\omega_c = v/r = eB/m$, where e is the charge on the ion, m is its mass, v is its velocity and r the radius of the circular pathway. If an electromagnetic alternating field E_1 with frequency ω_1 is applied perpendicular to the magnetic field, resonance occurs if $\omega_1 = \omega_c$, i.e. the ions absorb energy from E_1 and are accelerated to a greater velocity, and thus the radius of their path increases. By changing the magnetic field at constant radiofrequency and measuring the absorption, a plot of the absorption versus the magnetic field strength B (and thus against m/e) can be made; this is the mass spectrum of the ions present in the sample cell. At a frequency $o_1/2\pi = 153.57$ kHz and a variation of the magnetic field between 0.01 and 1.5 T, ions with masses from 1 to 150 can be observed.

Double resonance methods are especially important. Here the sample is irradiated with two frequencies which simultaneously bring two ions with different m/e values (e.g. reactant and product) into resonance. The method can be used to study ionic structure, unimolecular decomposition reactions, autoionization, thermodynamic stability, electron and proton affinities, collision-induced fragmentation, and so on.

Ion-exchange chromatography: a form of liquid-solid Chromatography (see) based on the reversible formation of heteropolar bonds between the fixed ions (F) bound to the matrix (M) of an ion-exchanger and mobile counter ions (C). If an ionic mixture passes through an *ion-exchange column*, neutral molecules or ions with the same charges as the fixed ions are eluted, while the species with the charge opposite to that of the fixed ions compete with the counterions for binding sites. The ions with higher charges than the counterions are bound and retained. Along the entire length of the column, specific concentration ratios between the old and new counterions in the ion-exchanger and eluant are established; this is the *ion-exchange equilibrium* $(MF\text{-}C_1) + C_2 \rightleftharpoons (MF\text{-}C_2) + C_1$, which is associated with an *exchange constant* $K_E = [MF\text{-}C_2][C_1]/[MF\text{-}C_1][C_2]$.

Corresponding to the positions of their exchange equilibria, the ions bound to the fixed ions are eluted differentially with suitable bases or acids, that is, the substances have different migration rates. The retention time of a component is affected by a number of factors, including the size and charge of the ion, the pH of the mobile phase, the absolute concentrations and types of ionic species in the mobile phase and the column temperature. Thus it is possible to separate even ions with the same charge and with very similar chemical properties.

The separation is also affected by adsorption and distribution, and, when exchangers based on dextran gels are used, by differential permeation (molecular sieve effects).

In ***ion-exclusion chromatography***, ionized fractions of the sample mixture are differentially ionized by suitable choice of the pH, the ionic strength, organic solvents, etc. The fractions are separated by repulsion from fixed ions with the same charge.

Chelate resins carry fixed ions in pairs which are able to form complexes with transition metal ions. In ***ligand-exchange chromatography***, such resins are used as ion exchangers; at the end of the run, the resin must be regenerated by introduction of protons.

Redox chromatography or ***electron-exchange chromatography*** is done on exchangers which carry reversibly oxidizible or reducible fixed ions.

I. is used for separation of proteins, nucleic acids, peptides, amino acids, carbohydrates and lipids.

Ion exchangers: substances, usually solid, which can reversibly exchange ions for other ions. ***Cation exchangers*** consist of high-molecular-weight, polyvalent anions with mobile cations. Their "exchange active groups" are hydroxyl (-OH), sulfonate ($-SO_3H$) or carboxyl (-COOH) groups. ***Anion exchangers*** are macromolecular polyvalent cations containing mobile anions. Here the exchange-active groups are usually singly or multiply substituted amino groups, $-NH_2R$, $-NH(R)_2$ and $N(R)_3$. I. are synthesized by introduction of functional groups into polycondensation resins (phenoplastics, aminoplastics) and especially polymers (polystyrol). Macroporous resins, with pore diameters up to 10 nm, are of especial interest. For example, if a cation I. is brought into contact with a sodium chloride solution, the following reaction occurs: $R\text{-}SO_3^-H^+ + Na^+Cl^- \rightleftharpoons R\text{-}SO_3Na + HCl$ (R is the exchanger skeleton). In an anion exchanger, the following reaction occurs: $[R\text{-}N(CH_3)_3]^+OH^- + Na^+Cl^- \rightleftharpoons R\text{-}N(CH_3)_3Cl + NaOH$.

If the capacity of the I. is exhausted, it can be treated with hydrochloric acid or sodium hydroxide to regenerate the original H^+ or OH^- form. The ease of regenerating I. accounts for their great importance in nature and industry.

Inorganic I. include natural silicates with a certain structure, the Zeolites (see), and synthetic compounds based on silica, the permutites; however, the latter are now of historical interest only. Because of their structure, these inorganic I. can only be used as cation exchangers. *Coal exchangers* are organic I. which are made by sulfonation and partial oxidation of coal. They are similar in structure to the natural humus substances in soil, and are also cation exchangers only.

Ion-exchange membranes are a special type of I. which is very important for electrodialysis. The cation and anion exchange membranes alternate in an electrodialysis cell used to deionize water. There are homogeneous membranes, which consist of the I. resin in a thin sheet, and heterogeneous membranes, in which finely ground I. resin is embedded in a plastic or elastic binder.

Applications. The oldest and still most common use of I. is in Water treatment (see). Water can be either softened or completely deionized with I. In analytical chemistry, I. are used: 1) to replace one ion by another; 2) to remove ions (contaminating ions) which interfere with analysis; 3) to determine trace elements; and 4) to separate ions with similar properties, e.g. uranium and plutonium ions, or ions of the rare-earth metals. In preparative chemistry, I. are used: 1) as catalysts for esterification; 2) to treat seawater by electrodialysis; 3) to enrich trace elements and purify radioactive waste water; 4) to purify industrial solutions, e.g. in sugar refineries; 5) to produce free acids and bases; 6) to isolate and separate organic compounds, especially natural products such as amino acids, antibiotics, alkaloids, pigments, etc; and 7) to improve drinking water.

Ion exclusion chromatography: see Ion exchange chromatography.

Ionic charge: see Valence.

Ionic conductor: a substance which conducts electric current by means of mobile ions. The type includes electrolyte solutions, molten salts and solid electrolytes. The charge transport is always coupled to matter transport. See Electrical conductivity.

Ionic equivalent conductivity: see Migration rate.

Ionic flotation: see Flotation.

Ionic formula: see Formula.

Ionic lattice: see Lattice type.

Ionic limit conductivity: see Migration rate.

Ionic radius: a relative measure of the size of an ion, when it is considered a rigid sphere. As in atoms (see Atomic radius), the electrons in ions cannot be assigned sharply delimited locations in space, so that a rigid sphere is only a rough approximation for an ion. Only the internuclear distances in molecules or crystals can be measured. Because the ions present in ionic crystals are of different types, it is not possible arbitrarily to divide the internuclear distance between the two bonding partners. In addition, the I. of a given type of ion is not constant; it depends on the type of lattice, the coordination number of the ion in that lattice and the nature of the adjacent ions. However, by comparing the ionic distances in ionic crystals with the same type of lattice, an approximate system of I. can be established if the radius of at least one of the ions is known. These I. are known as the *lattice radii* of the ions. A set of lattice radii was established by Goldschmidt, using the value $r_F = 133$ pm for the F^- which was determined by Wasastjerna from molar refraction studies. The values obtained by Pauling by quantum mechanical calculations, using the NaCl lattice with octahedral coordination as a reference, agree well with the Goldschmidt values (Table).

Ionic radii according to Goldschmidt (G) and Pauling (P) in pm

Ion	G	P	Ion	G	P
H^-	154	208	Fe^{2+}	76	75
F^-	133	136	Co^{2+}	70	72
Cl^-	181	181	Ni^{2+}	68	69
Br^-	196	195	Cu^{2+}	92	·
I^-	219	216	Zn^{2+}	69	74
O^{2-}	132	140	Cd^{2+}	92	97
S^{2-}	174	184	B^{3+}	20	20
Li^+	68	60	Al^{3+}	45	50
Na^+	98	95	Ga^{3+}	60	62
K^+	133	133	In^{3+}	81	81
Rb^+	148	148	Se	68	81
Cs^+	167	169	Y^{3+}	90	93
Ti^{2+}	80	·	La^{3+}	104	115
Ti^{3+}	69	·	Ce^{3+}	103[1])	·
Ti^{4+}	64	68	Gd^{3+}	94[1])	·
Cu^+	95	96	Ta^{5+}	68	·
Ag^+	113	126	Lu^{3+}	85[1])	·
Be^{2+}	30	31	Zr^{4+}	87	80
Mg^{2+}	65	65	Hf^{4+}	82	81
Ca^{2+}	94	99	Nb^{5+}	69	·
Sr^{2+}	110	113	Ho^{3+}	89[1])	·
Ba^{2+}	129	135	Mo^{6+}	62	·
Mn^{2+}	80	80	W^{6+}	62	·

[1] Radii according to Templeton and Dauben.

Using the lattice radii, the distance between ions in the crystal lattice can be calculated approximately, as the sum of the radii of adjacent ions. From the ratio of the radii of the cations and anions of an ionic crystal, the coordination numbers and lattice type can be predicted. Within a group of the periodic system, the I. increase on going from the top to the bottom, and within a period, with increasing atomic number. As a result, the radius of each main group element ion in the first period is approximately the same as that of the second-period element in the next higher group (diagonal relationship). For example, Li^+ and Mg^{2+} have approximately the same I. Such pairs tend to form isotypic mixed crystals.

Ionic valence: see Valence.

Ion implantation: see Doping.

Ionium: see Thorium.

Ionization: the complete removal of an electron from an atom or molecule. The energy required to remove the electron is the ***ionization energy***, and it can be added in thermal, electromagnetic or kinetic form. If sufficient energy is added, more electrons can be removed from a cation, generating highly charged positive ions A^{n+}. $A \rightleftharpoons A^+ + E^-$; $A^+ \rightleftharpoons A^{2+} + e^-$; $A^{2+} \rightleftharpoons A^{3+} + e^-$, etc. The energies of these sequential I. are called the first, second, third, etc. ionization energies. These energies depend on the electron configurations of the elements. The figure shows the first ionization energies of the elements as a function of their atomic numbers.

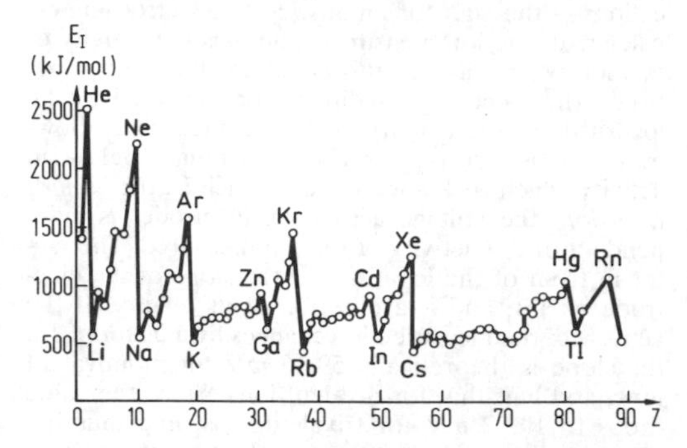

Dependence of the first ionization energy, E_1, on the atomic number Z.

The alkali metals have the lowest ionization energies, because they attain a stable octet shell by losing one electron. They therefore tend to form cations and to bond ionically. The noble gases have the highest ionization energies. The relative minima in the elements of the 3rd and 6th main groups are due to the fact that I. produces relatively stable electron configurations with closed or half-occupied shells.

Ion migration: the motion of the ions of an electrolyte solution under the influence of an electric field (see Migration rate).

Ionomers: polymers in which the molecular chains are cross-linked by ionic bonds. The I. obtained from copolymers of ethene with unsaturated carboxylic acids or of butadiene with methacrylic acid are commercially important. The carboxyl groups of the side chains are neutralized by addition of bases. The cations from the base (e.g. sodium, potassium, magnesium or zinc) form ionic bonds with the carboxyl groups, although the cations are not linked to the anions one by one. Instead, island-like clusters of ionic groups are formed (Fig.). At room temperature, the I. are like duroplastics.

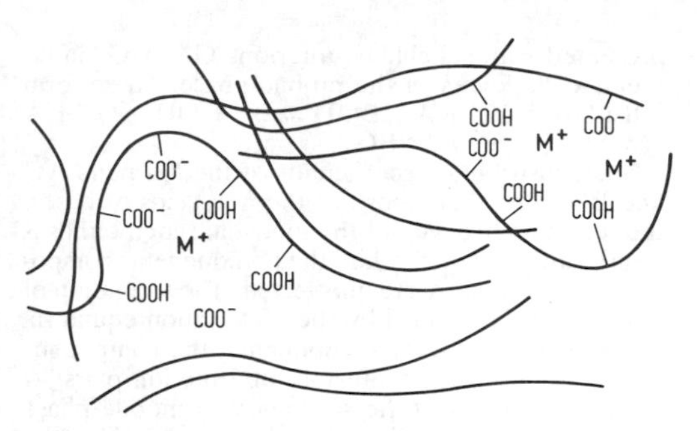

Structure of an ionomer

When the temperature is increased, the counterions are redistributed, so that the molecular chains are no longer fixed, but can shift relative to one another. Therefore at high temperatures, the I. act as thermoplastics.

Ionones: a group of multiply unsaturated ketones derived from sesquiterpenes. I. are pale yellow oils with varying, violet-like odors. ***Pseudoionone*** is acyclic. I. are found in some essential oils, but are relatively rare. Pseudoionone can be synthesized from citral or dihydrolinalool. Acid-catalysed cyclization of pseudoionone leads to α- and β-I. The I. are used in the production of perfumes; β-I. is also used as a starting material for vitamin A synthesis.

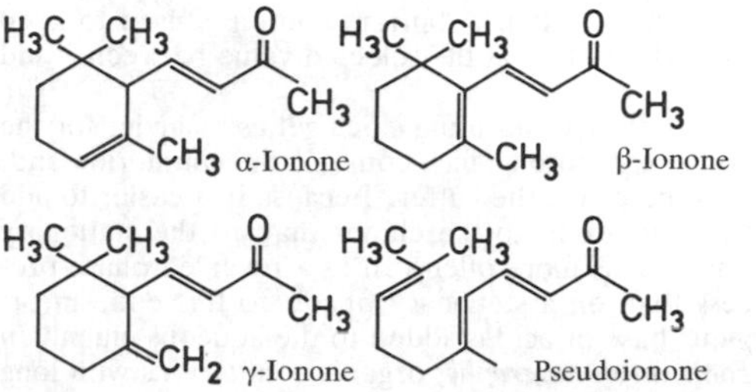

α-Ionone β-Ionone

γ-Ionone Pseudoionone

Ionophore: a compound with a lipophilic exterior and hydrophilic interior large enough to accomodate an ion. When present in a lipophilic membrane (e.g. a biological membrane), I. form cage-like complexes (cryptates, inclusion compounds) with ions which act as mobile carriers and transport the ions through the membrane. ***Pseudo-I.*** are thought to form stationary pores in the membrane. This group includes gramicidin and alamethicin. The I. which are formed by microorganisms are called *ionophoric antibiotics*. These I. can be either cyclic or acyclic. The cyclic ionophoric antibiotics include depsipeptides (valinomycin, enniatins, monamycin), peptides (adamanide) and tetraesters (antibiotics of the actin series, such as nonactin). Acyclic ionophoric antibiotics, which include various peptides (gramicidin, alamethicin), nigericin and antibiotics A-23187 and X-537A, wrap around the ion as pseudocycles. For example, two molecules of A-23187 form inter- and intramolecular hydrogen bonds and make a pseudo-cycle which can enclose divalent cations. I. can bind cations and anions. The compounds which preferentially transport K^+ (valinomycin: $K^+ > Na^+$) or calcium (calcium ionophores, such as the antibiotics A-23187 or X-573A: $Mn^{2+} > Ca^{2+} > Sr^{2+} > Mg^{2+}$) are very important in cell biochemical studies. The ionophoric antibiotics are not used in medicine. Of the synthetic I. the most interesting are Crown ethers (see) and kraken compounds.

Ionophoresis: see Electrophoresis.

Ioxynil: see Herbicides.

Ion-pair chromatography: a method for separating substances which are dissociated in aqueous solution. The acids HA or bases BOH to be separated are presented with suitable counterions G^+ or G^- in either the stationary or the mobile phase. These form salts: $HA \rightleftharpoons H^+ + A^-$, $BOH \rightleftharpoons B^+ + OH^-$, $G^+ + A^- \rightleftharpoons AG$, $B^+ + G^- \rightleftharpoons BG$.

Since distribution coefficients of the ion pairs AG and BG differ from those of the free acids or bases, the retention behavior of the ionogenic compounds is significantly changed, while the nonionogenic components in the mixture are unaffected. The formation of ion pairs is determined by the dissociation equilibria of the acidic and basic components, the compounds which produce the counterions and the ion pairs. A change in the pH of the stationary or mobile phase can affect the dissociation of the ion pair, and thus the system can be selectively optimized for isolation of the substance of interest. In order to adjust the pH for optimum separation, the compounds which produce the counterions should be dissociated over a broad pH range, i.e. strong acids (perchloric or alkylsulfonic acids) or the salts of strong bases (quaternary ammonium salts). A buffer solution is used to keep the pH constant at the selected value between 2 and 8.

I. is carried out using silica gel as a carrier for the aqueous solution which contains the counterion and, in some cases, the buffer. Because it is easier to add the counterion to the eluant than to the stationary phase, I. is more often run as a reversed phase process than on a stationary phase; to this end, an organic base or acid is added to the aqueous eluant. In ***soap chromatography***, organic counterions with long carbon chains are used.

I. is used to separate biogenic amines and their metabolic products, pharmaceuticals, carboxylic acids and pigment intermediates. In contrast to Ion-exchange chromatography (see), I. is capable of separating mixtures of ionic and nonionic components.

Ion-selective electrode, ***ion-sensitive electrode***: an indicator electrode (see Electrode) which contains an active material. At the phase boundary between this material and the sample solution, a potential difference arises which depends on the activity of a certain ion in this solution. The potential difference can only be measured if the I. is combined with a reference electrode to form a circuit. Then, in the ideal case, the voltage V (emf) between the two electrodes depends on the activity of the ion to be determined as described by the Nernst equation. At 25 °C, $V = V^o + (0.059/z_i)\log a_i$, where V^o is the standard electrode potential in a solution with $a_i = 1$, + is the sign for cations, - the sign for anions, z_i the electrochemical valence of the ion measured, and a_i is the activity of this ion. According to this equation, the voltage between the electrodes is directly proportional to the logarithm of the activity of the ion measured. However, this is true only in a limited range. Below an activity which is characteristic for each I., the *limit of detection*, the voltage across the electrodes is independent of the activity of the ion measured. The factor in front of the logarithm is the *slope* of the electrode circuit, and is a criterion of its quality. If the activity of the measured ion changes by a factor of 10, the slope is theoretically 59.16 mV for monovalent ions, and half that for divalent ions. Since the value drops to 19.72 mV for trivalent ions, and thus becomes rather small compared to the typical precision of +0.1 mV, I. are made mainly for mono- and divalent ions. In practice, the slopes are smaller than the theoretical values, and reach at best 90 to 98% of the latter.

The relative electrode voltage of a circuit with I. depends not only on the activity of a certain type of ion, but is also influenced by the presence of other ions.

I. are similar in construction to Gas electrodes

Table. The most important ion-selective electrodes and their properties

Ion-selective electrode	Type (1)	Active material	Range	Interfering Ions
Ammonium	PVC	Nonactin/monactin	$10^{-1} \dots 10^{-5}$	K^+
Barium	PVC	Ba^{2+} carrier	$1 \dots 10^{-6}$	
Bromide	PC	$AgBr/Ag_2S$	$1 \dots 5 \cdot 10^{-6}$	I^-, CN^-, S
	SC	AgBr	$1 \dots 10^{-6}$	I^-, CN^-, S^{2-}
Cadmium	PC	CdS/Ag_2S	$10^{-1} \dots 10^{-6}$	Ag^+, Hg^{2+}, Cu^{2+}, Fe^{3+}, Tl^+, Pb^{2+}
Calcium	PVC	Ca^{2+} carrier	$1 \dots 10^{-6}$	
	PVC	Ca salt of diorganophosphoric acids	$1 \dots 10^{-5}$	Zn^{2+}
Chloride	PC	$AgCl/Ag_2S$	$1 \dots 10^{-5}$	Br^-, I^-, CN
Copper	SC	Cu_2Se	$1 \dots 10^{-6}$	Ag^+, Hg^{2+}, Fe^{3+}
	PC	Cu_2S/Ag_2S	$1 \dots 10^{-7}$	Ag^+, Hg^{2+}, Fe^{3+}
	PC	$Cu_2Se/Ag_2/Se$	$1 \dots 10^{-7}$	Ag^+, Hg^{2+}
Cyanide	PC	AgI/Ag_2S	$10^{-2} \dots 10^{-6}$	I^-, S^{2-}
Fluoroborate	PVC	$[Ni(o\text{-}phen)_3]^{2+}$ (2)	$10^{-1} \dots 10^{-5}$	I^-, ClO_4^-
Fluoride	SC	LaF_3	$1 \dots 10^{-6}$	OH^-
Iodide	PC	AgI/Ag_2S	$1 \dots 5 \cdot {}^{-8}$	S^{2-}
Lead	PC	PbS/Ag_2S	$10^{-1} \dots 10^{-7}$	
Lithium	PVC	Li^+ carrier	$10^{-1} \dots 10^{-5}$	
Nitrate	PVC	Tetraalkylphosphonium nitrate	$1 \dots 10^{-5}$	ClO_4^-, I^{-1}, BF_4^-, tenside anions
Perchlorate	PVC	$[Ni(o\text{-}phen)_3]^{2+}$ (2)	$10^{-1} \dots 10^{-5}$	Tenside anions
Silver	PC	Ag_2S	$1 \dots 10^{-7}$	Hg^{2+}
Sodium	PVC	Na^+ carrier	$10^{-1} \dots 10^{-6}$	
Sulfide	PC	Ag_2S	$1 \dots 10^{-7}$	
Thiocyanate	PC	$AgSCN/Ag_2S$	$1 \dots 5 \cdot {}^{-6}$	I^-, Br^-, $S_2O_3^{2-}$, S^{2-}

(1) PC = polycrystalline membrane; PVC = PVC matrix membrane; SC = single-crystal membrane.
(2) $[Ni(o\text{-}phen)_3]^{2+}$ = tris-(o-phenanthroline)nickel (II) ion.

(see), but the active material, usually a disk, is glued or cemented to the end of a glass or plastic tube. (In spite of its thickness, this disk is called a membrane.) The tube contains an internal solution and an internal reference electrode. The active material may be in a solid (homogeneous) or liquid membrane. The former consists either of part of a single crystal or of polycrystalline material. I. with this type of membrane often do not contain an internal solution or reference electrode, but instead have a solid contact to the voltage detector; in this case they are called ***"all-solid-state I."*** Liquid membranes contain the active material in solution. They are either immiscible with water, or they are mixed with PVC as a type of softener, and are thus converted to gummy, solid membranes. Immiscible liquid membranes must be supported by a porous material. Because the PVC-matrix membranes are easy to attach to ordinary electrodes, they are becoming more and more important. The active components of such membranes are either ion-exchangers, e.g. quaternary, long-chain ammonium salts, or neutral organic molecules (e.g. valinomycin) which are able to form complexes with metal ions; these are dissolved in an organic solvent with a high dielectric constant.

Measurement using an I. is similar to the technique used with pH meters, and sometimes the same instruments can be used with either type of electrode. Most often, the activity of the free ion is measured in solution, but it is possible to determine concentrations, if the instrument is suitably calibrated.

The importance of measurements with I. is increasing constantly. They can be used in nearly all types of analysis, and have advantages over other methods.

Ion-sensitive electrodes: same as Ion selective electrodes (see).

IPC: see Carbanilate herbicides.

Ipecacuanha alkaloids: see Alkaloids.

Iprodione: see Carboximide fungicides.

Ipsenol: see Pheromones.

Ipso-substitution: (Latin "ipso" = "self") Electrophilic substitution on aromatic compounds in which the leaving group is not a proton, but another atom or group. I. occurs only when the leaving substituent can be stabilized, e.g. as an alkene:

$$CH_3-C_6H_4-CH(CH_3)_2 \xrightarrow{+NO_2^+AcO^-} CH_3-C_6H_4-NO_2 + CH_2{=}CHCH_3$$

10 % ipso-Substitution

Another example is substitution of SO_3H groups by nitro groups (e.g. production of picric acid by nitration of phenol-2-sulfonic acid or phenol-4-sulfonic acid).

Ipso-addition of an electrophilic reagent is more common than the products indicate. If the leaving substituent has no tendency to leave, the electrophilic reagent changes places with the H atom in a neighboring position. The proton is then the leaving group, and the product appears to be that of a normal electrophilic substitution.

Ir: symbol for iridium.

Irgalan dyes: a group of synthetic dyes which do not contain the typical anionic hydrophilic groups, such as sulfo- or carboxyl groups, but instead are chromium complexes of azo dyes. The I. are used as dyes for wool from a neutral or weakly acidic bath.

Iridium, symbol ***Ir***: an element from group VIIIb of the periodic system, a noble metal from the subgroup of heavy Platinum metals (see); *Z* 77, with natural isotopes with mass numbers 193 (62.7%) and 191 (37.3%); atomic mass 192.2, valency usually III, IV or I, more rarely II, V, VI, 0 or - I, density 22.65, m.p. 2454 °C, b.p. 4530 °C, electrical conductivity 21.2 Sm mm^{-2}, standard electrode potential (Ir/Ir^{2+}) +1.1 V.

Properties. I. is a very hard, brittle, silvery white metal which can be distilled in an electric arc. Like its lighter homolog, rhodium, I. is insoluble in acids, even aqua regia. However, it dissolves in alkali fusions and in a hot mixture of hydrochloric acid and sodium chlorate. When heated red hot in the air, I. powder is oxidized to iridium(IV) oxide, IrO_2. I. reacts with chlorine to the iridium chlorides $IrCl_3$ and $IrCl_4$; the relative amounts of the two products depends on the reaction temperature and the degree of dividedness of the metal. The most common oxidation states of I. are +III, +IV and +I. Compounds in which I. is in the +V or +VI oxidation state are limited to fluorine derivatives such as $[IrF_6]^-$ and IrF_6; iridium carbonyls arc iridium(0) complexes.

Occurrence. I. makes up about 10^{-7}% of the earth's crust. It is found in association with other platinum metals. The minerals sysserskite (iridiosmium) and newkanskite (osmiridium) are isomorphic mixtures of I. and osmium with admixtures of platinum, ruthenium and rhenium.

Production. I. is obtained in the course of enrichment and isolation of the Platinum metals (see). As a result of the processes used, I. is obtained by crystallization of ammonium hexachloroiridiate(IV); this is reduced to I. with hydrogen at high temperatures.

Applications. I. is so brittle that it can be used only in the form of alloys. Platinum-iridium alloys containing up to 30% Ir are very hard and chemically resistant, and are used in many ways (see Platinum alloys, see Dental alloys).

Historical. I. was discovered along with osmium in 1804 by Tennant, in the course of analyzing platinum. Its name derives from the many colors of its salts: the Greek "iridios" means "rainbow-colored".

Iridium compounds: iridium is most commonly found in the oxidation states +III, +IV, and +I, while the other possible oxidation states 0, +II, +V and +VI are less important. ***Iridium(IV) oxide, iridium dioxide***, IrO_2, M_r224.20, density 11.665, is a black compound which crystallizes in a rutile lattice. It is made by heating iridium in a stream of oxygen. The most important iridium halide, ***iridium(III) chloride***, $IrCl_3$, forms olive green or red-brown crystals, M_r 298.56, density 5.30, m.p. 763 °C (dec.); it is made by chlorination of the metal at about 660 °C. ***Iridium(IV) chloride***, $IrCl_4$, is a dark brown to black, hygroscopic compound, M_r 334.01. The +VI oxidation state is represented, for example, by the octahedral ***iridium(VI) fluoride***, IrF_6, a very reactive, yellow compound, M_r 306.19, m.p. 44 °C, b.p. 53 °C. The tetrameric ***iridium(V) fluoride***, $(IrF_5)_4$, M_r 1148.84, m.p. 104 °C, can be made by fluorination of iridium at about 360 °C; the black ***iridium(III) fluoride***, IrF_3, M_r 249.21, is obtained by reduction of iridium(VI) fluoride with iridium. In addition to the above binary I., there are many ***iridium complexes***: octahedral ***fluoroiridiates(V)***, $M[IrF_6]$, represent the +5 oxidation state. The hexachloro and hexafluoroiridiates(IV) are very important; one of them is the dark red ***ammonium hexachloroiridiate(IV)***, $(NH_4)_2[IrCl_6]$ which is isomorphic with $(NH_4)_2PtCl_6$. This complex is only barely soluble in water and plays a role in the production of iridium. Many octahedral, diamagnetic iridium(III) complexes are known, e.g. ammine complexes of the types $[Ir(NH_3)_6]X_3$, $[Ir(NH_3)_5X]X_2$ and $Ir[(NH_3)_4X_2]X_2$, and chloro complexes such as $M_3[IrCl_6]$, $IrCl_n(NH_3)_{6-n}Cl_{3-n}$ and $[IrCl_3(C_5H_5N)_3]$. Reduction of iridium(III) chloride in alcohol in the presence of π-acceptor ligands produces square planar iridium(I) complexes, of which the Vaska complex (see) has been most thoroughly studied. The ***iridium carbonyls*** are obtained by reduction of iridium(III) chloride with carbon monoxide: $[Ir_4(CO)_{12}]$, canary-yellow, trigonal crystals, M_r 1104.94, m.p. 210 °C (dec.) and $[Ir_6(CO)_{16}]$, red crystals, M_r 1601.40. Iridium complexes can be used in homogeneous catalysis, especially for hydrogenation reactions.

Iridoids: monoterpenes, usually with a cyclopentane ring. The parent compound is iridodial, which is formed from geranyl pyrophosphate. The derivative I. may have 10, 9 or only 8 C atoms and an open cyclopentane ring (seco-iridoids). I. are most often found in the form of glucosides, mainly in plants of the class *Magnoliatae*. They include the bitter substances ***loganin*** from the buck bean, ***gentiopicroside*** from gentian root and the Valepotriates (see) from valerian root.

Iris oil, ***violet root oil***: a solid, essential oil (m.p. 35-50 °C) which smells like violets or dried violet roots; it is obtained by petroleum ether or benzene extraction and/or steam distillation of the chopped roots of certain types of iris. I. contains mainly pelargonaldehyde, caprylaldehyde and benzaldehyde, various irones (these are ketones similar to the ionones), acetophenone, 3,4-dimethoxyacetophenone, furfural, naphthalene and higher fatty acid esters. I. is used in perfumery and to make soaps.

Iron symbol ***Fe***: a chemical element from group VIIIb of the periodic system, the Iron group (see). It is the most abundant and technologically most important heavy metal; *Z* 26, with natural isotopes with mass numbers 56 (91.66%), 54 (5.82%), 57 (2.19%), 58 (0.33%), atomic mass 55.847, valency usually II or III, less commonly IV, V, VI, 0, -II, Mohs hardness 4.5, density 7.873, m.p. 1539 °C, b.p. 3070 °C, electrical conductivity, 10 Sm mm $^{-2}$, standard electrode potential (Fe/Fe^{2+}) - 0.4402 V.

Properties. Pure I. is a silvery white, relatively soft and rather reactive metal with a high magnetic susceptibility. It occurs in three enantiotropic modifications. ***α-I.*** (cubic body-centered, lattice constant 286.6 pm) loses its ferromagnetic properties at 770 °C and becomes paramagnetic; this state is also called ***b-I.***. Further conversion into ***g-I.*** (lattice constant 364.7 pm) and cubic body-centered ***d-I.*** occurs at 928 or 1398 °C. The physical and chemical properties of I. are strongly affected by addition of other elements. For example, carbon-containing I., unlike pure I., has permanent magnetism. The carbon content of I. determines the ductility, hardness and brittleness of the metal produced on a technical scale (see Steel). The properties of I. are also very significantly modified by addition of alloying metals. In dry air, and in water which is free of oxygen and carbon dioxide, I. is unreactive, due to the presence of a continuous oxide layer. It is passivated by concentrated sulfuric or nitric acid, but dissolves in dilute acids with the generation of hydrogen. Moist air and water containing carbon dioxide attack I. to form rust, that is, iron(III) oxide hydrate, $Fe_2O_3 \cdot H_2O$, or $FeO(OH)$. Because the resulting oxide layer is soft and porous, the rusting process can progress indefinitely (see Corrosion). Concentrated sodium hydroxide attacks I. even in the absence of air. The I. goes into solution with the formation of hydroxoferrate(II). Finely divided (pyrophoric) I. readily ignites in air. When heated, I. can reduce hot steam to hydrogen: $3\ Fe + 4\ H_2O \rightarrow Fe_3O_4 + 4\ H_2O$. At room temperature, dry chlorine does not attack I., although moist chlorine does; at higher temperatures, dry chlorine reacts to form iron(III) chloride, $FeCl_3$. I. also reacts with other nonmetals, such as phosphorus, silicon, sulfur and carbon, at elevated temperature. Because of the methods of production, household objects of I. or steel are always alloys of I. with these nonmetals and manganese. In its compounds, I. is usually in the +2 or +3 oxidation states. It is a good complex former, and octahedral iron(II) complexes are especially important.

I. is an extremely important bioelement, and is present in all living cells. It takes part in electron-transfer reactions (see Ferredoxins) in processes such as photosynthesis, nitrogen fixation and respiration (see Hemoglobins).

Analytical. In the classical sulfide separation step for qualitative analysis, I. precipitates out of the ammonia solution as black iron(II) sulfide, FeS. The presence of iron(II) or iron(III) ions is indicated by formation of the dark red dipyridine complex $[Fe\ dipy_3]^{3+}$ or the dark red iron(III) thiocyanate complex $[Fe(NCS)_n(H_2O)_{6-n}]^{3-n}$, or precipitation of Prussian blue (see) from Fe^{3+} and $[Fe(CN)_6]^{4-}$ or Fe^{2+} and $[Fe(CN)_6]^{3-}$. I. can be determined gravimetrically by precipitation as iron(III) hydroxide, $Fe(OH)_3$, followed by roasting to iron(III) oxide, Fe_3O_3. It can also be determined by iodimetry or magnanometry.

Occurrence. I. is the most abundant heavy metal in the earth's crust, at a concentration of 4.7%. Since elemental I. also makes up about 90% of the core of the earth, it is the most abundant element in the planet as a whole. In addition, the existence of numerous I. meteorites indicates that I. is abundant in the entire solar system. Most of the I. ores on earth are oxides or sulfides. The most important are magnetite, Fe_3O_4, hematite, Fe_2O_3, limonite, $Fe_2O_3 \cdot H_2O$ ($FeO(OH)$), siderite, $FeCO_3$ and pyrite FeS_2. I. occurs widely in soils, with iron(III) oxide and iron (III) oxide hydrate giving soils their characteristic red, brown or yellow coloration.

I. is present in small amounts in the biosphere, where it plays important catalytic roles. The human body contains 4 to 5 g I., 75% in the form of hemoglobin. Plants contain up to 300 ppm I. I. deficiency can cause chloroses in plants; this is often due to an excess of lime in the soil, which binds the I. so that it cannot be absorbed by the roots.

Production. The large-scale production of I. by reduction of oxide iron ores with coke in a blast furnace (see Iron, crude) produces crude iron, which contains about 4% carbon. Most of this is further processed to Steel (see), which contains less than 1.7% C. Chemically pure I. is also produced on an industrial, but much smaller scale, by vacuum treatment of steel, by thermal reduction of iron pentacarbonyl, reduction of iron(III) chloride with hydrogen, or cathodic precipitation from aqueous solution using a soft iron anode and ammonium iron oxalate or iron(II) chloride electrolytes. Very pure I. is made from electrolyte or carbonyl I. by zone melting.

Applications. I. is the most important metal used in industry, and is usually used in the form of Steel (see). Pure I. is used in powder metallurgy and magnet technology.

Historical. Archeological discoveries of I. objects with high nickel contents indicate a very early use of meteorite I. The technique of I. production in a bloomery hearth (direct process) was probably discovered in the Caucasus. I. appeared in significant amounts about 1500 B.C., and the Iron Age began in central and northern Europe about 500 B.C. The first small blast furnaces were built in the Middle Ages. In the 18th century, charcoal was replaced by anthracite or coke as the heating and reducing material. The large demand for I. and its alloys led in the 2nd half of the 19th century to new methods of steel production by the Bessemer and Thomas converters and regenerative firing (Siemens-Martin smelting furnace). More recently, electric furnaces with electric arc or induction heating have been adopted.

Iron(II) acetate: $Fe(CH_3COO)_2 \cdot 4H_2O$, light green, monoclinic, water-soluble needles; M_r 246.00. I. is synthesized by reaction of finely divided iron with wood vinegar and is used as a dye mordant, in the synthesis of dyes and in calico printing.

Iron alums: double sulfates of iron(III) sulfate with alkali sulfates with the general composition $M^IFe \cdot (SO_4)_2 \cdot 12H_2O$. The most important iron alum is ***potassium iron alum***, $KFe(SO_4)_2 \cdot 12H_2O$, which crystallizes in colorless to pale violet octahedra; M_r 599.32, density 1.83, m.p. 22 °C. It is made by oxidation of a sulfuric acid solution of iron(II) sulfate with nitric acid in the presence of potassium sulfate, and is used as a tanning agent and mordant in dyeing, and in photographic printing.

Iron(III) ammonium oxalate: same as ammonium oxalatoferrate(III).

Iron carbide: see Cementite.

Iron-carbon alloys: the most important industrial iron alloys (see Iron, crude; Iron, cast and Steel). In the liquid state, iron can dissolve up to 6.67% carbon, but in the solid state it can dissolve much less. The solubilities can be read off the *iron-carbon diagram* (Fig.). The excess dissolved carbon is precipitated as iron carbide, Fe_3C (cementite), or, if the rate of cooling is extremely slow, as graphite.

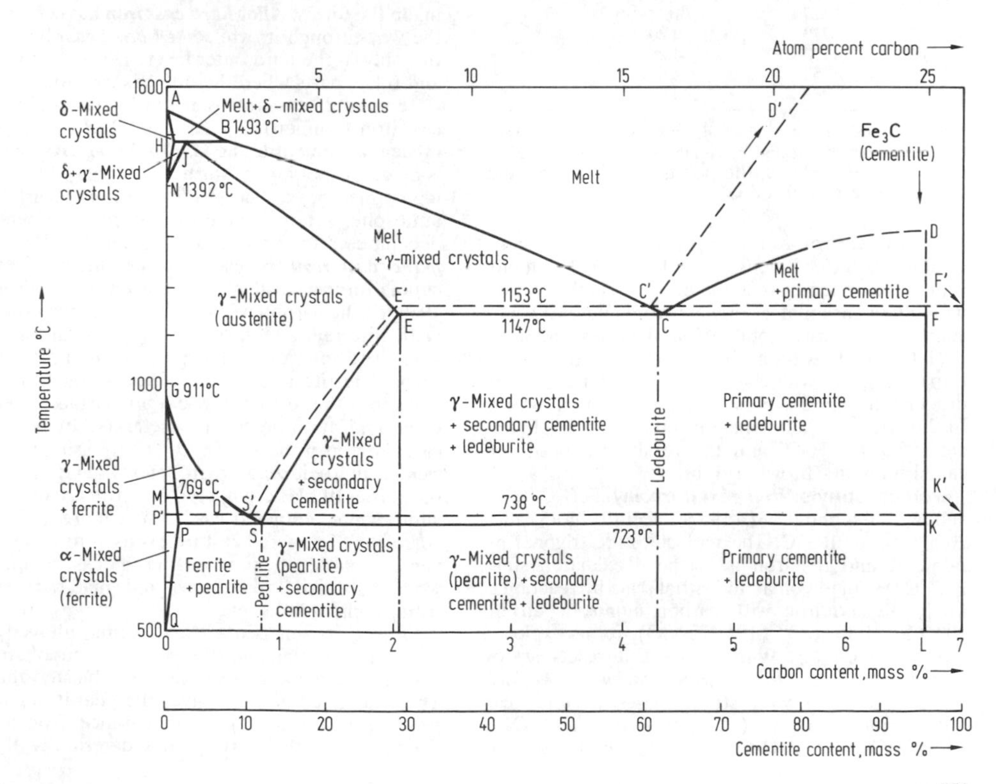

The I. have different structures, depending on the amount of carbon. 1) The basic structures are Ferrite (see), Austenite (see) and Cementite (see). 2) The following are mixtures of the basic structures: Pearlite (see), the eutectic mixture formed at 0.80% carbon content, consists of ferrite and cementite. Ledeburite (see) is formed as the eutectic mixture from a melt of cementite and austenite or pearlite; it contains 4.3% C. The possible combinations of the structure types at different temperatures and compositions can also be read off the iron-carbon diagram. 3) Hard structures. If I. are quenched from temperatures above the austenite region in water, oil or air, the metastable hard structure of Martensite (see) is formed; this can be converted by suitable annealing, via the intermediates Troostite (see) and Sorbite (see), into ferrite and cementite.

Point	Temp. in °C	Carbon conc. in mass %
A	1539	0
B	1493	0.51
C	1147	4.30
D	?	6.67
E	1147	2.06
F	1147	6.67
G	909	0
H	1493	0.10
J	1493	0.16
K	723	6.67
L	20	6.67
M	760	0
N	1388	0
O	760	0.65
P	723	0.02
Q	20	10^{-4}–10^{-5}
S	723	0.80
C'	1153	4.25
E'	1153	2.03
S'	738	0.69

Selected points in metastable and stable systems of iron-carbon. The points indicated by capital letters refer to the metastable system Fe-Fe_3C; the points C', E', and S' refer to the stable system Fe-C.

Iron(II) carbonate: $FeCO_3$, gray white, trigonal crystals; M_r 115.85, density 3.8. I. is insoluble in pure water. It is made by reaction of iron(II) salt solutions with alkali carbonates. In water containing CO_2, it is converted to water-soluble ***iron hydrogencarbonate***, $Fe(HCO_3)_2$. I. is found in nature as siderite (sparry iron), or mixed with clay, as clay ironstone. Iron hydrogen carbonate is present in some mineral waters. In the air, these solutions precipitate iron(III) hydroxide, $Fe_2O_3 \cdot xH_2O$, and the resulting deposits are called limonite (brown iron ore).

Iron carbonyls: ***Iron pentacarbonyl***, $Fe(CO)_5$, is a yellow, oily liquid, M_r 195.90, density 1.45, m.p. 20.5 °C, b.p. 103 °C. The molecules are trigonal pyramidal, and the iron is in the 0 oxidation state. $Fe(CO)_5$ is made on an industrial scale by reaction of finely divided iron with carbon monoxide around 200 °C and 15 to 30 MPa. $Fe(CO)_5$ forms explosive mixtures with air. With halogens it reacts at low temperature to form ***iron carbonyl halides***, $Fe(CO_4)X_2$, and with strong bases to form carbonylferrate(−II), $[Fe(CO)_4]^{2-}$: $Fe(CO_5) + 4\,OH^- \rightarrow Fe(CO_4^{2-} + 2\,H_2O + CO_3^{2-}$. Protonation of $[Fe(CO)_4]^{2-}$ yields ***iron carbonyl hydride***, $H_2Fe(CO)_4$, a colorless to faintly yellow liquid with a repulsive odor; M_r 169.85, m.p. - 68 to - 70 °C. Above -10 °C, it decomposes, developing a faint red color. In the presence of light, $Fe(CO)_5$ is converted to the dinuclear carbonyl $Fe_2(CO)_9$, which forms golden yellow, pseudohexagonal leaflets; M_r 363.79, density 2.08, m.p. 100 °C (dec.). When heated above 150 °C, I. lose CO and form elemental iron (see Carbonyl iron). The trinuclear carbonyl, $Fe_3(CO)_{12}$, forms dark green, monclinic prisms, M_r 503.64, density 1.996, m.p. 140 °C (dec.); it is obtained from $[Fe(CO)_4]^{2-}$ by oxidation.

Iron, cast: iron-carbon alloys with 2 to 4% carbon, silicon and similar components which cannot be worked hot or cold; they can be shaped only by casting and strain-relieving working. Cast iron has high compressional strength, but low tensile strength or elasticity. ***White, mottled*** and ***gray*** cast iron are distinguished on the basis of the appearance of a fresh break. Rapid cooling leads to formation of metastable cementite, Fe_3C (tempered cast iron, hard cast iron), while slow cooling leads to a more stable structure with precipitation of graphite (*second melt cast iron*; gray cast iron). *First-melt cast iron* is obtained directly from the blast furnace. The casting skin gives it a relatively high chemical resistance, which can be significantly improved by alloy additives, such as silicon and nickel.

In hard cast iron, the carbon is completely or partially bound in metastablc form as cementite, the formation of which is promoted by a low silicon and high manganese content in the iron melt, or by rapid cooling in the forms. ***Alloy hard cast iron*** has cementite all the way through it, while ***shell hard cast*** has cementite only in the hard outer layer. I. is used for rollers and other parts where hardness is essential.

Tempered cast iron is made by tempering white cast iron (temper crude cast) in which the carbon is present as cementite, Fe_3C. ***Black tempered cast iron.*** is made by melting in a furnace (cupola, flame, rotating or electric) at 950 °C under a non-decarbonizing atmosphere (largely free of oxygen); the cementite decomposes to form tempering carbon. ***White tempered cast iron.*** is usually made directly from the cupola furnace casting; it is roasted in the air at 980-1060 °C; the carbon released by decomposition of the cementite reacts with oxygen to form carbon monoxide. The interior of the piece contains tempering carbon, ferrite and pearlite; the carbon from it diffuses into the ferritic, low-oxygen surface layer. The decarbonization process is accelerated by placing the cast parts in iron ore (Fe_2O_3) or by using steam-enriched air (oxidizing environment). The mechanical properties of black tempered cast iron are more even, while white tempered cast iron is more resistant to corrosion. Tempered cast iron is used to make motor vehicle parts, such as engine blocks, crankcases, wheel hubs and brake drums, and small parts such as fittings, chain links, etc.

Gray cast iron is made by melting alloyed crude iron, casting scrap and steel scrap in a cupola furnace with 10% coke and 2% lime; air is blasted through. The embedded graphite gives the pearlite or ferrite-pearlite iron alloy a gray appearance. The form in which the graphite precipitates determines the mechanical properties of the iron. Lamellar graphite is formed in the presence of silicon; spherical graphite in the presence of added magnesium (up to 0.5%). Spherical graphite greatly increases the strength of the alloy. Because of its very low shrinkage (1%), gray cast iron is used to make machine parts.

Iron chlorides. ***Iron(II) chloride***: $FeCl_2$, white, sublimable, very hygroscopic, monoclinic prisms; M_r 126.75, density 3.16, m.p. 674 °C. $FeCl_2$ is made by heating iron in a stream of hydrogen chloride, or by reduction of iron(III) chloride. The pale green, monoclinic hexahydrate, $FeCl_2 \cdot 6H_2O$, M_r 234.84, crystallizes out of aqueous solution; it can also be made by directly dissolving iron in hydrochloric acid. This compound is actually dichlorotetraaquairon(II) hydrate, $[FeCl_2(H_2O)_4] \cdot 2H_2O$. $FeCl_2$ is used as a reducing agent, as a mordant in dyeing, and for pharmaceuticals.

Iron(III) chloride, $FeCl_3$, forms dark green, hexagonal crystals which have a metallic sheen and are reddish brown by transmitted light. Its M_r is 162.61, density 2.898, m.p. 306 °C. $FeCl_3$ can sublime. At 400 °C, the vapor density corresponds to the formula Fe_2Cl_6, and at 800 °C, $FeCl_3$ molecules are present. $FeCl_3$ is soluble in water and organic solvents, such as ethanol, acetone or diethyl ether. It crystallizes out of water as hydrates; the iron(III) chloride available commercially is the yellow hexahydrate, which exists in the solid state as the chloro complex $[FeCl_2(H_2O)_4]Cl \cdot 2H_2O$, M_r 270.30, m.p. 37 °C. $FeCl_3$ reacts with chloride ions to form chloro complexes, $[FeCl_4]^-$, $[FeCl_6]^{3-}$ and $[Fe_2Cl_9]^{3-}$. $FeCl_3$ is very hygroscopic and deliquesces in air to form an oily liquid (iron oil). In aqueous solution, $FeCl_3$ is hydrolysed, forming colloidal, red-brown iron(III) hydroxide, $Fe_2O_3 \cdot xH_2O$ via a yellow-brown hydroxo complex. $FeCl_3$ is made by passing chlorine over scrap iron at approximately 650 °C; it is also a byproduct of chlorinating solubilization of iron-containing oxide ores. Its hydrates are obtained by the reaction of iron(III) oxide and hydrochloric acid. $FeCl_3$ is used as a mordant in dyeing and textile printing, as an oxidation agent and chlorine transfer reagent, in the work-up of copper and silver ores, to make printed switches, to etch copper plates for printing, in photoengraving and to make inks. It is also important in water treatment. Because $FeCl_3$ coagulates protein, cotton soaked in a solution of it can be used to stop bleeding.

Iron(III) chloride reaction: see Phenols.

Iron(III) chromate: $Fe_2(CrO_4)_3$, yellow microcrystalline compound which is slightly soluble in water; M_r 459.61. I. is used as a pigment (siderin yellow) for water, lacquer and porcellain paints, and in metallurgy to make stainless steels.

Iron complexes: coordination compounds of iron. The most common types are octahedral iron(II) and iron(III) complexes. The spin type of these complexes depends on the field strength of the ligands; most are of the high-spin type, e.g. $[Fe(H_2O)_6]^{n+}$ (n = 2,3); the less common low-spin type is represented by $[Fe(CN)_6]^{n-}$ (n = 4,3). Tetrahedral iron(III) complexes such as $[FeCl_4]^-$ are also relatively stable, while tetrahedral iron(II) complexes, such as $[FeCl_4]^{2-}$ can be made only in nonaqueous media. π-acceptor ligands such as CO stabilize the lower valence states of iron (see Iron carbonyls); iron pentacarbonyl represents the rare trigonal-bipyramidal coordination type. Oxo complexes (see Ferrates) are examples of I. in which the central atom is in the +V or +VI state. Iron complexes such as Hemoglobins (see) and Ferredoxin (see) play important roles in basic biological processes.

Iron, crude, ***pig iron***: an iron with a high content of carbon which was melted into it in a blast furnace. It is used as a raw material for making cast iron or steel. Because it contains more than 1.7% carbon, it cannot be plastically deformed, either hot or cold. Depending on its composition, it is classified as foundry, steel, Thomas, Bessemer or specular crude iron; by far the largest part of the crude iron made is steel crude iron. Crude iron is also classified according to the appearance of a broken surface as gray, white or mottled. In *gray crude iron* (foundry iron), the carbon is present as graphite; in *white crude iron*, it is bound as iron carbide, Fe_3C. In *mottled crude iron*, there are gray spots on white surfaces (this is common in Thomas crude iron); the carbon has precipitated as both graphite and iron carbide.

Crude iron is made by smelting iron ores or, in smaller amounts, from scrap, slag or ash. The blast furnace is loaded with ore, additives and coke, and preheated air is blown into it. The carbon in the coke first burns to carbon dioxide: $C + O_2 \rightarrow CO_2$, which is reduced to carbon monoxide by the glowing coke: $CO_2 + C \rightleftharpoons 2\,CO$. The reactions in the blast furnace are arranged as a countercurrent, that is, the rising gases react with the sinking solid load. The ore and coke added at the top is first dried by the hot gases, then enters the zone in which water of hydration and CO_2 from the carbonates are driven off. In the temperature range from 400 to 1000 °C, the carbon monoxide in the furnace gas reduces the material (zone of indirect reduction). The most important reactions here are: $3\,Fe_2O_3 + CO \rightarrow 2\,Fe_3O_4 + CO_2$; $2\,Fe_3O_4 + 2\,CO \rightarrow 6\,FeO + 2\,CO_2$; $FeO + CO \rightarrow Fe + CO_2$. The iron(II) oxide, FeO, can be reduced indirectly only to a slight degree, while the manganese(II) oxide, MnO, which is also present in the ore, cannot be indirectly reduced to Mn at all. Therefore, only a small amount of metallic iron (sponge iron) forms in the zone of indirect reduction.

In the zone of direct reduction, the temperature is above 1000 °C. Here the reaction equations are $FeO + C \rightarrow Fe + CO$; $P_2O_5 + 5\,C \rightarrow 2\,P + 5\,CO$; $SiO_2 + 2\,C \rightarrow Si + 2\,CO$; $MnO + C \rightarrow Mn + CO$. There is no sharp boundary between the zones of indirect and direct reduction. The iron absorbs carbon, which reduces its melting point. The crude iron collects at the bottom of the furnace, and separates from the slag, which is less dense. The crude iron and slag are tapped at regular intervals from openings which are at different levels. The crude iron is poured out at 1250 to 1500 °C into ingots in which it is either transported to a steel plant or to a casting machine. The slag is granulated either by pouring directly from the furnace into a water trough, or it is caught in pans and transported to a granulator, a pumice plant or a casting bed.

Iron ores are increasingly being processed before they are put into the furnace; they are broken, sieved, ground, roasted, enriched in iron and granulated.

Composition of crude iron types in %.

Type of crude iron	Carbon	Silicon	Manganese	Phosphorus	Sulfur
Foundry crude					
a) Hematite (low phosphorus)	3.5–4.2	2.0–3.0	0.7–1.5	0.08–0.12	up to 0.04
b) medium phosphorus	3.5–4.2	2.0–3.0	up to 1.0	0.5 –1.0	up to 0.07
c) high phosphorus	3.5–4.2	1.8–2.5	up to 0.8	1.4 –2.0	up to 0.06
Steel crude					
a) low manganese	up to 0.4	up to 1.0	up to 1.0	up to 0.2	up to 0.05
b) high manganese	3.5–4.5	up to 1.0	2.0– 3.0	up to 2.0	up to 0.04
Thomas crude	3.2–3.6	0.2–0.5	0.5– 1.5	1.8–2.2	up to 0.06
Besemer crude	3.5–4.0	1.0–2.5	0.5– 2.0	up to 0.1	up to 0.03
Specular iron	4.0–5.0	up to 1.0	5.0–30.0	up to 0.15	up to 0.04

They are enriched by sedimentation of a suspension, magnetic separation (if necessary, this is preceded by magnetizing roasting), refining or flotation. The fine-grained concentrates and finely ground ores are sintered or pelletized, or sometimes made into briquets, before they are put in the furnace.

Since coke is a very expensive raw material, various measures are taken to reduce the consumption of it: a) improvement of indirect reduction. While direct reduction requires a high heat consumption and consumes carbon directly, indirect reduction consumes carbon monoxide and releases heat. To conserve metallurgical coke, the furnace is run in such a way that the proportion of indirect reduction is as high as possible (50 to 60% of the total reduction is normal; as high as 70% has been achieved). b) Elevation of the blast air temperature to 1350 °C. c) Elevation of the iron content in the furnace load; today values as high as 60% may be achieved. d) Addition of oxygen to the blast air. e) Operation of the furnace with elevated gas pressure at the top. f) Injection of additional fuel (oil, natural gas, coal dust) through the blast vents or through special vents in the lower part of the furnace. g) Use of 100% self-slagging sinter or pellets. These materials alread contain the additive (lime) required for formation of a normal, basic slag h) Good sorting of the raw materials, i.e. uniform grain size of the ore and additive.

Worldwide, specific coke consumption is usually 450 to 650 kg/t crude iron. The above measures have been able, in isolated cases, to reduce the consumption to about 350 kg/t iron. At the same time, any conservation of coke increases the capacity of the furnace.

Iron(III) formate: red crystals, $Fe(HCOO)_3 \cdot H_2O$; M_r 208.92. I. is hydrolysed in water to iron(III) hydroxide and formic acid, and is added to silage to establish acidity.

Iron group: the subgroup of group VIIIb which contains the 3d transition metals iron, cobalt and nickel. The stability of the +2 oxidation state of these elements increases from iron to nickel; the +3 oxidation state is also important for iron and cobalt. The maximum valence decreases from +6 for iron to +4 for nickel. These metals are excellent complex formers; iron(II), iron (III), cobalt(III) and nickel(II) complexes in particular are extremely numerous. The I. metals form typical metal carbonyls with carbon monoxide.

Properties of the iron group elements

	Fe	Co	Ni
Atomic number	27	28	26
Electron configuration	$[Ar]3d^64s^2$	$[Ar]3d^74s^2$	$[Ar]3d^84s^2$
Atomic mass	55.847	58.9332	58.70
Atomic radius [pm]	116.5	116.2	115.4
Ionic radius (M^{2+}) [pm]	76	74	69
Electronegativity	1.64	1.70	1.75
Density in g cm^{-3}	7.873	8.89	8.908
m.p. [°C]	1539	1492	1452
b.p. [°C]	3070	3100	2730
Standard electrode potential (M/M^{2+}) [V]	−0.4402	−0.277	−0.250

Iron hydroxides. ***Iron(II) hydroxide***: $Fe(OH)_2$, a white compound which can be made by precipitation of iron(II) salt solutions with alkali hydroxides in the absence of air. In the presence of air it is rapidly converted via dark gray-green and black intermediates to iron(III) hydroxide. $Fe(OH)_2$ is soluble in acids, but it also reacts with concentrated boiling alkali hydroxide solutions to a slight extent, giving hydroxoferrates(II), $M_4[Fe(OH)_6]$.

Iron(III) hydroxide, $Fe_2O_3 \cdot xH_2O$ iron(III) oxide (hydrate) is a hydrogel which is made by precipitation from iron(III) salt solutions with alkali hydroxide or ammonia; when it is dried, it undergoes condensation reactions to form crystalline α-Fe_2O_3. This compound occurs in nature as hematite.

Iron(III) hydroxide is used as a pigment called *iron oxide yellow* or *iron yellow* (see Iron oxides). When freshly precipitated iron(III) oxide hydrate is heated in 2 M sodium hydroxide with superheated steam, it forms ***iron(III) oxide hydroxide***, α-FeO(OH), M_r 88.85, density 4.28, which is found in nature as goethite. γ-FeO(OH) is an unstable modification of FeO(OH); it appears as rust in the course of corrosive processes. This modification is found in nature as lepidocrolite. The high adsorption capacity of freshly precipitated iron(III) oxide hydrate is utilized, for example, in water treatment.

Iron(II) iodide: FeI_2, gray or dark red, hexagonal crystals which are readily soluble in water; M_r 309.69, density 5.315, m.p. 177 °C. The tetrahydrate, $FeI_2 \cdot 4H_2O$, is obtained in the form of green crystals by dissolving iron in iodine water; the anhydrous form is made by roasting iron powder with iodine. I. mixed with sugar syrup is used in medicine as iodine iron syrup to prevent goiter.

Iron(III) nitrate: $Fe(NO_3)_3$ crystallizes out of aqueous solution as the enneahydrate, $Fe(NO_3)_3 \cdot 9H_2O$, in the form of monoclinic, slightly violet crystals; M_r 404.02, density 1.684, m.p. 42 °C, or as the cubic hexahydrate $Fe(NO_3)_3 \cdot 6H_2O$, M_r 349.95, m.p. 35 °C. The compound is made by dissolving iron in 20 to 30% nitric acid. I. is used as a mordant for dyes, to weight silk and color it black, to tan hides and in the synthesis of Prussian blue. Because I. precipitates protein, it is used in medicine as an astringent in cases of gastric and intestinal bleeding.

Iron nitrides: Fe_4N and Fe_2N are formed during nitrogen hardening on the surface of steel. Chemical and thermal treatment leads to diffusion saturation of the surface of the steel with nitrogen, forming a hard layer which consists primarily of Fe_2N.

Iron(II) oxalate: $FeC_2O_4 \cdot 2H_2O$, pale yellow, rhombic crystals; M_r 179.90, density 2.28. I. is obtained by reaction of aqueous iron(II) salt solutions with oxalic acid or alkali oxalates. It is used to make iron(II) oxide and as a developer in photography.

Iron oxides. ***Iron(II) oxide***: a black compound ("Wustite phase") with the composition $Fe_{0.9}O$ to $Fe_{0.95}O$. It is obtained by oxidation of iron with oxygen or steam above 566 °C. This oxide is stable above 566 °C, and metastable at room temperature. It can be made in the laboratory by thermal decomposition of iron(II) oxalate in vacuum; the product is a pyrophoric powder: $FeC_2O_4 \rightarrow FeO + CO + CO_2$.

Iron(II,III) oxide, Fe_3O_4, is the most stable iron oxide, a black, ferromagnetic compound with an inverse spinel structure, $Fe^{II}Fe^{III}{}_2O_4$; M_r 231.54, density 5.18, m.p. 1538 °C (dec.). Fe_3O_4 is made by passing steam over iron below 566 °C: $3\ Fe + 4\ H_2O \rightarrow Fe_3O_4 + 4\ H_2$, or by strongly heating iron(III) oxide. It is formed as "hammer scale" as the iron splinters flying off hammered iron burn. Fe_3O_4 is found in nature in large deposits as magnetite, and is a very valuable ore. It is used as electrode material for chloralkali electrolysis, as a glass pigment and polish, as a filler and as a pigment known as *iron oxide black* or *iron black*.

Iron(III) oxide, Fe_2O_3, is a dimorphic compound; M_r 159.69, density 5.12 to 5.24, m.p. 1565 °C. It occurs in nature as various types of hematite. The rhombohedral, paramagnetic α-Fe_2O_3 is the stable form. The cubic γ-Fe_2O_3 is formed by cautious oxidation of Fe_3O_4 according to $2\ Fe_3O_4 + O_2 \rightarrow 3\ Fe_2O_3$; when heated above 300 °C, it converts to α-Fe_2O_3. α-Fe_2O_3 is also obtained by heating iron(III) oxide hydrate or iron(III) salts of volatile acids. When heated above 1200 °C in air, Fe_2O_3 splits off oxygen to become Fe_3O_4. Fe_2O_3 is used in large amounts as a pigment known as *iron oxide red* or *iron red*. Roasted Fe_2O_3 is very hard, and can be used as a polish for glass, metal and gemstones (polishing red, rouge).

The *iron oxide pigments* discussed above, as well as iron oxide yellow (see Iron hydroxides) are very resistant and color fast. They are suitable for use with any type of binder, and as coloring additives for cement, building materials, plastics and ceramics. Iron oxide red is also used extensively in rust-protection paints for steel constructions and ships.

Iron phosphates. ***Iron(II) phosphate***: $Fe_3(PO_4)_2 \cdot 8H_2O$, is a colorless compound when pure. It occurs in nature as vivianite, and can be made by precipitation of iron(II) sulfate with sodium phosphate. If iron is dissolved in phosphoric acid, it forms colorless needles of $Fe(H_2PO_4)_2 \cdot 2H_2O$.

Iron(III) phosphate, $FePO_4$, forms a number of hydrates, such as $FePO_4 \cdot 4H_2O$, a yellowish powder which is insoluble in water, and $FePO_2 \cdot 2H_2O$, which forms yellow, rhombic crystals. $FePO_4$ is formed by the reaction of iron(III) salt solutions with alkali monophosphate.

Iron phosphides: binary iron-phosphorus compounds of the type FeP, M_r 86.82, density 6.07; Fe_2P, M_r 142.67, density 6.56, Fe_3P, M_r 198.51, density 6.74. I. are present in phosphorus-containing crude iron and are responsible for its brittleness.

Iron black: see Iron oxides.

Iron red: see Iron oxides.

Iron(III) rhodanide: same as Iron(III) thiocyanate (see).

Iron scale: see Scale.

Iron silicates. ***Iron(II) silicates*** are compounds formed by fusion of iron(II) oxide with quartz. They occur widely in nature; in addition to pure iron(II) silicates, such as fayalite (Fe_2SiO_4) and grunerite ($FeSiO_3$), there are many isomorphic mixtures with silicates of other metals, such as olivine, $(Mg,Fe)_2SiO_4$ and hedenbergite $(CaFe)(Si_2O_6)$.

Iron(III) silicates are found in nature only in the form of double silicates, e.g. as akmite, $NaFe[Si_2O_6]$ and andradite, $Ca_3Fe_2[SiO_4]_3$.

Iron(III) sulfate: $Fe_2(SO_4)_3$, a white compound; M_r 399.87, density 3.09, m.p. 480 °C (dec.). $Fe_2(SO_4)_3$ crystallizes out of aqueous solution in the form of hydrates with 3, 6, 7, 9, 10 or 12 molecules of water. It is hydrolysed to basic sulfates in water, especially warm water. It forms colorless to pale violet Iron alums (see) with alkali sulfates. $Fe_2(SO_4)_3$ occurs in nature as rhombic coquimbite, $Fe_2(SO_4)_3 \cdot 9H_2O$, and as monoclinic quenstedtite, $Fe_2(SO_4)_3 \cdot 10_2O$. It can be synthesized by dissolving iron(III) oxide in sulfuric acid, or by oxidation of $FeSO_4$ with nitric acid. It is used to solubilize copper ores, to make alums, as a flocculating and etching agent for aluminum, as a mordant in textile dyeing, in printing calico and in the production of iron pigments.

Iron sulfides. ***Iron(II) sulfide***: FeS, forms brass-yellow to black-brown crystals; M_r 87.91, density 4.74, m.p. 1193 °C. FeS is trimorphic, forming hexagonal crystals at room temperature. It is only slightly soluble in water, and reacts with acids to form hydrogen sulfide. FeS is made by fusing sulfur or FeS_2 with iron, or by precipitation of aqueous iron(II) salt solutions with ammonium sulfide. FeS is found in nature as pyrrhotine and magnetopyrite. Meteorites contain FeS as troilite.

Iron(IV) sulfide, FeS_2, occurs in nature in two modifications, as pyrite and markasite. Iron pyrite is brass-yellow and has a metallic sheen. The rhombic, bipyramidal markasite is more reactive, and is converted to pyrite when heated; M_r 119.97, density 5.0. Pyrite contains both Fe^{2+} and disulfide anions, $Fe_2{}^{2-}$. Roasting of pyrite is an industrially important method of producing sulfur dioxide: $2\ FeS_2 + 11/2\ O_2 \rightarrow Fe_2O_3 + 4\ SO_2$. FeS_2 can be synthesized by reaction of FeS with sulfur, or by reaction of iron oxide or chloride with hydrogen sulfide at red heat.

Iron(III) sulfide, Fe_2S_3, is a black compound which can be obtained by precipitation of aqueous iron(III) salt solutions with sulfide ions; M_r 297.87. Fe_2S_3 is found in nature in the form of double sulfides, e.g. chalcopyrite, $Cu_2S \cdot Fe_2S_3$, bornite, $3\,Cu_2S \cdot Fe_2S_3$ and smythite, $GeS \cdot Fe_2S_3$.

Iron(III) thiocyanates, ***iron(III) rhodanides***: thiocyanate complexes of iron(III) formed by reaction of iron(III) salt solutions with alkali thiocyanates in aqueous solution. The water-soluble complexes in the series $[Fe(NCS)_n(H_2O)_{6-n}]^{3-n}$ are dark red. Formation of these complexes is a specific and very sensitive test for the presence of iron(III).

Iron triad: the elements iron, ruthenium and osmium, which make up one of the three short columns of group VIIIb in the periodic system. The elements in the I. are able to enter the highest oxidation states available to the metals of the 3d, 4d and 5d series (+6 for Fe, +8 for Ru, Os). They are also characterized by the existence of stable, octahedral metal(II) complexes, $M^{II}L_6$, tetrahedral oxometallates(VI) $[MO_4]^{2-}$ and homologous metal carbonyls $M(CO)_5$ and $M_3(CO)_{12}$.

Iron vitriol: see Iron sulfates.

Iron yellow: see Iron hydroxides.

Irreversible: 1) in thermodynamics, an adjective describing processes which proceed via non-equilibrium states, and which generate entropy and lose the capacity to perform work. They cannot be undone without simultaneous changes in the state of the environment. All spontaneously occurring processes in nature (e.g. mixing of two substances, heat conduction, energy conversion by friction) are I. Antonym: Reversible (see).

Irreversible thermodynamics: see Thermodynamics.

IR spectroscopy: abb. for Infrared spectroscopy (see).

Irving-Williams series: see Coordination chemistry.

Isatin, ***indoline-2,3-dione***: the lactam of isatic acid. I. forms red-orange prisms; m.p. 203-205 °C (subl.). It is readily soluble in hot alcohol, and soluble in hot water and benzene, but only slightly soluble in cold water. Two tautomeric forms of I. are known (desmotropism). I. can be synthesized from ***isatic acid***, which is made in two steps from 2-nitrobenzoyl chloride. When heated, it is dehydrated to form I. I. is synthesized industrially by oxidation of synthetic indigo with nitric acid. It is a starting material for various pharmaceutical products, such as laxatives. I. derivatives are used to synthesize indigoid pigments. I. is used in analytical chemistry to detect copper ions, mercaptans and thiophene.

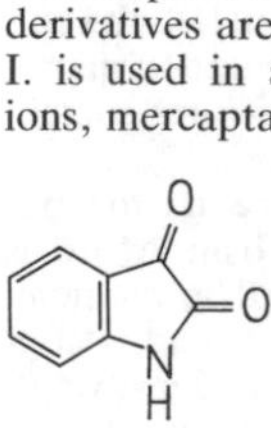

O
O
N
H

ISC: see Photophysical processes.

Isenthalp: a curve in a diagram connecting those points through which a system passes in the course of a thermodynamic process when the enthalpy is held constant (see Joule-Thomson effect).

Isenthalpic valve effect: same as Joule-Thomson effect (see).

Isentrop: see Adiabat.

Isoalloxazine, ***10-H-benzo[g]pteridine-2,4-dione***: an isomer of alloxazine, the parent compound of the flavins. It is a light brown to egg-yolk yellow powder which has a yellow-green fluorescence in solution. Unlike the tautomeric alloxazine, I. is stable only if it has an alkyl or aryl group on position 10. The derivatives of I. are readily converted to colorless products by hydrogenation, and are equally readily converted back to I. under conditions of dehydrogenation. This property of the I. is the basis for the activity of the flavin enzymes. When the hydrogen in postion 10 is replaced by a ribose group, the compound is riboflavin, one of the vitamin B_2 group.

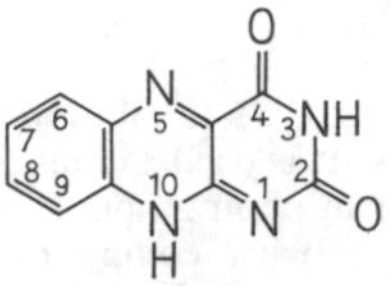

O
N 5 4 3 NH
7 6
8 9 10 1 2
N N O
H

Isoamylamine: see Biogenic amines.

Isoamyl ether: same as Diisoamyl ether (see).

Isoamyl mandelate: see Mandelates.

Isoamyl nitrite, ***isopentyl nitrite***: $(CH_3)_2CH\text{-}CH_2\text{-}CH_2\text{-}O\text{-}NO$, an unstable, colorless, oily liquid with a fruity smell; b.p. 99.2 °C, n_D^{20} 1.3918. I. is slightly soluble in water, but readily soluble in organic solvents. It is obtained by reaction of isoamyl alcohol with sodium nitrite in aqueous solution, in the presence of dilute mineral acid and under cooling with ice. It is hydrolysed by mineral acids, so it is used instead of sodium nitrite for reactions in organic solvents. I. was formerly used as a drug for coronary illness. It has more significance at present as an antidote to cyanide and hydrogen sulfide poisoning; the I. promotes formation of methemoglobin, which has a stronger affinity to these poisons than do the respiratory chain proteins they inactivate.

Isobaldric acid: see Valeric acids.

Isobar: a curve in a diagram which joins all points through which a system passes in the course of an ***isobaric process***, that is, at constant pressure.

Isobenzofuran: see Benzofurans.

Isobergaptene: see Furocumarins.

Isobutane: same as 2-methylpropane (see Butanes).

Isobutene: same as 2-methylpropene (see Butenes).

Isobutyraldehyde, ***2-methoxypropanal, dimethylacetaldehyde***: $(CH_3)_2CH\text{-}CHO$, a colorless liquid with a pungent, penetrating odor; m.p. - 65.9 °C, b.p. 64.5 °C, n_D^{20} 1.3730. I. forms an azeotropic mixture with water which boils at 60.5 °C. It is soluble in water and most organic solvents. Like all Aldehydes (see), it undergoes typical addition and condensation reactions. I. is subject to autooxidation, which forms isobutyric acid. It is found in tea leaves and the oil of the Jeffrey pine. I. is synthesized industrially by hydroformylation of propene, oxidation of isobutene with hydrogen peroxide or oxygen, rearrangement of methylallyl alcohol with sulfuric acid or aluminum silicate, or by dehydrogenation of isobutanol. It is used in the synthesis of isobutyric acid, vulcanization accelerators, synthetic resins, softeners, aromas, amino acids, pharmaceuticals, etc.

Isobutyrates: see Butyric acids.

Isochore: a curve in a diagram connecting all points through which a system passes during an ***isochoric process***, that is, at constant volume.

Isocol: see Isocyanins.

Isocrotonic acid, ***cis-but-2-enoic acid, (Z)-but-2-enoic acid***: an unsaturated, aliphatic monocarboxylic acid, the Z-diastereomer of Crotonic acid (see). I. is a colorless liquid; m.p. 15.5°C, b.p. 169°C. It can be made by isomerization of the stable crotonic acid.

```
H         H
 \       /
  C = C
 /       \
H3C        COOH
```

Isocyanates: derivatives of isocyanic acid, H-NCO. 1) the inorganic I. are discussed with the Cyanates (see).

2) The organic I. have the general formula R-NCO; R can be aliphatic, aromatic or heterocyclic. The electronic distribution can be described by the following resonance structures:

$$R–\underline{N}=\overset{+}{C}–\underline{O}|^{-} \leftrightarrow R–N=C=\overline{\underline{O}} \leftrightarrow R–\underline{N}–\overset{+}{C}=\overline{\underline{O}}.$$

The *aliphatic I.* are unstable, usually colorless liquids, and their vapors are irritating to the eyes and respiratory organs. In the absence of moisture, the I. can be stored for long periods. Methyl isocyanate was responsible for the disaster in Bhopal, India, in 1984.

Aromatic I. are more stable. They react very easily with compounds which contain acidic H atoms, forming the corresponding addition products: $R^1\text{-N=C=O} + \text{H-}R^2 \rightarrow R^1\text{-NH-CO-}R^2$. Some of the primary addition products are stable, and these are industrially important compounds, e.g. Urethanes (see). Others are unstable, and react further to form stable end products. For example, I. react with water to give *N,N'*-disubstituted ureas:

$$\text{R NCO} + H_2O \rightarrow \text{R–NH–COOH} \xrightarrow{+\text{R–NCO}}$$
$$\text{R–NH–CO–NH–R} + CO_2.$$

Alcohols and phenols form urethanes with I.: R-N=C=O + HO-R → R-NH-CO-OR. Primary and secondary amines, carboxylic acids and hydrogen halides add similarly to I. I. trimerize at elevated temperatures to yield substituted 1,3,5-triazine-2,4,6-triones, the esters of isocyanuric acid. In boiling aqueous lyes, I. are first converted to carbamic acids, which are then decarboxylated to yield primary amines.

I. are synthesized by reaction of cyanate salts with alkyl halides, or by the Hofmann degradation of carbamides, or by Lossen degradation of hydroxamic acids. The most common industrial synthesis of I. is from primary amines and phosgene: $\text{R-N}H_2$ + Cl-CO-Cl → R-NCO + 2 HCl.

I. are used to make *N*-substituted urethanes and ureas. The polyurethane plastics made from di- and polyisocyanates and diols are especially important.

Isocyanides: same as Isonitriles (see).

Isocyanins: synthetic pigments which form a subgroup of the Quinoline pigments (see). The I. are synthesized by heating a mixture of α-methylquinoline alkyl halides with α-methylquinaldine alkyl halides in alcoholic solution. The best known of the I. is ***Pinaverdol*** (see). Some others are ***pinachrome***, synthesized from 6-methoxyquinoline ethyl iodide and 6-ethoxyquinaldine ethyl iodide; ***ethyl red***, synthesized from quinoline ethyl iodide and quinaldine ethyl iodide; ***orthochrome***, synthesized from 6-toluquinoline ethyl iodide and 6-toluquinaldine iodide; ***pinachrome violet***, synthesized from 6-dimethylamino quinoline iodide with 6-dimethylaminoquinaldine ethyl iodide; and ***homocol, isocol*** and ***pericol***, made from bromoquinolines, bromoquinaldines and arylsulfonate esters. All I. are used as sensitizers in photography.

Isocyanic acid: see Cyanates.

Isocyclic: a term applied to cyclic carbon rings which consist entirely of carbon atoms; if another type of atom is present in the ring, it is Heterocyclic (see).

Isodimorphism, ***heterotypic mixed crystal formation***: the formation of mixed crystals of substances with heterotypic crystal types (see Isotypism). For example, magnesium crystallizes in hexagonal close packing, while lithium crystals are cubic close packed. However, the two metals are largely miscible (heterotypically miscible). I. is always limited to a certain range of concentrations, and the lattice type is always that of the component present in excess.

Isoelectric focusing: the separation of substances, especially proteins, on the basis of their isoelectric points (IP) by Electrophoresis (see) on a gel in which a stable pH gradient is created by a mixture of carrier ampholytes (buffers). The gradient goes from a low pH at the anode to a higher pH at the cathode. If the compounds are applied to the gel at pH 7, those with IP > 7 have a positive net charge, and those with IP < 7 have a negative net charge.

In the electric field, the positively charged molecules migrate toward the cathode and into the region of higher pH, while the negatively charged molecules move toward the anode and the lower pH region. Eventually each species of molecule reaches the pH corresponding to its IP and loses its net electric charge. It therefore stops moving and accumulates as a narrow band in the gel.

I. is used primarily as a flat bed method (see Thin-layer chromatography). It is a powerful method which permits separation of components differing by only 0.01 pH unit in their IP. I. is used analytically to determine isoelectric points, to check purity or to study the composition of samples; it can also be used preparatively to obtain very pure fractions for research purposes.

Isoelectric point, abb. ***IEP***: the hydrogen ion concentration (pH) at which an amphoteric electrolyte (see Ampholyte) has an equal number of positive and negative charges. The IEP is specific for each amphoteric electrolyte, which has characteristic properties at this point. Often its solubility and viscosity are at a minimum at the IEP. The electrophoretic mobility of the ampholyte drops to zero. If the ampholyte is a colloid, its rate of swelling also reaches a minimum, and it often precipitates. The IEP can be calculated from the dissociation constants of the acidic (K_a) and basic (K_b) groups and of water (K_w):

$$I = \sqrt{\frac{K_a \cdot K_w}{K_b}}$$

Isoenzymes: multiple forms of an enzyme which have the same substrate specificity, but genetically encoded differences in amino acid sequence of the protein. Because of differences in isoelectric points (IEP), the I. can be separated by electrophoresis or ion exchange chromatography.

Isoeugenol, ***2-methoxy-4-propenylphenol***: a colorless liquid, m.p. = - 10 °C, b.p. 266 °C, n_D^{20} 1.5726 (Z form), n_D^{20} 1.5784 (E form). I. smells like eugenol, but its aroma is weaker and more pleasant. It is slightly soluble in water, and readily soluble in ethanol and ether. I. is found in various essential oils, e.g. ylang-ylang and nutmeg oils. It is less widely distributed than eugenol, but can be produced from the latter by heating with potassium hydroxide to 230 °C (double bond shift). I. is used in perfumes and as a preservative for foods. It is also an intermediate in the synthesis of vanillin by oxidative degradation.

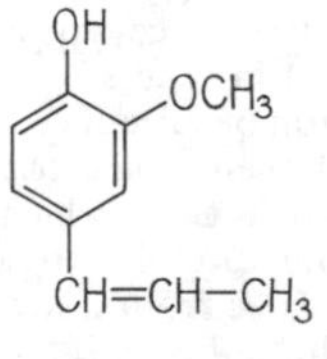

OH
OCH_3
$CH{=}CH{-}CH_3$

Isoflavone: see Flavone.

Isoglycans: branched Polysaccharides (see).

Isoindoline-1,3-dione: same as Phthalimide (see).

Isolated double bonds: carbon-carbon double bonds separated by at least two intervening single bonds; they thus do not interact (see Dienes).

Isoleucine, abb. ***Ile***: L-α-amino-β-methylvaleric acid, an essential amino acid. It is found in many proteins, especially hemoglobin, casein and serum proteins (formula and physical properties, see Amino acids). I. has two asymmetric C atoms, and therefore exists in a *threo* and an *allo* form, each of which can form a racemate. Only L-(*threo*)-I. is biologically active. The Strecker synthesis from 2-methylbutyraldehyde and hydrocyanic acid yields all 4 stereoisomers in equal amounts. After removal of *allo*-I., L-I. can be obtained by enzymatic hydrolysis of *N*-acetyl-DL-I. The amino acid can also be produced by fermentation using *Brevibacterium flavum* and carbohydrates as the carbon source. After 72 h, the culture filtrate contains 14 to 17 g I. per liter. In alcoholic fermentation, I. is converted to 2-methylbutanol(1), which is a component of the optically active amyl alcohol produced by fermentation.

Isomax process: a hydrocracking process for obtaining valuable, low-boiling C_2 to C_4 hydrocarbons from de-asphaltized residual oils without producing a large amount of gasoline fractions. The I. consists of two steps: in the first, the sulfur and nitrogen compounds are removed, while the hydrocracking, isomerization and dealkylation reactions occur in the second.

Isomerases: a major group of Enzymes (see), e.g. Glucose isomerase (see).

Isomeric shift: see Mœssbauer spectroscopy.

Isomerism: the phenomenon that two or more chemical compounds with different physical and chemical properties have the same net formula and the same molecular mass, but different structural formulas. In ***structural isomerism***, the atoms are arranged differently within the molecules; the isomers have different structural formulas. For example, the net formula $C_4H_{10}O$ describes both butanol, $CH_3(CH_2)_3OH$ and diethyl ether, CH_3-CH_2-O-CH_2-CH_3. The larger the number of atoms in the molecule, the more possibilities there are for I. In ***stereoisomerism***, the isomers have the same structural formulas but do not have the same three-dimensional structure.

Isomerization: an intramolecular transposition of atoms or groups of atoms; it is often reversible. I. can be expected whenever atoms develop unsaturated coordination sites in the course of a reaction. An electron sextet with or without a positive charge on an atom is the cause for *sextet I.* If the atom from which the mobile group departs is adjacent to the atom to which it becomes attached, the reaction is called a ***nucleophilic 1,2-rearrangement*** (internal nucleophilic substitution). The direction of the reaction is affected by the relative stability of the species with the electron sextet, steric factors, neighboring group effects and solvent effects.

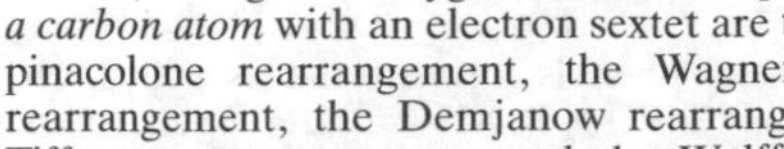

$$|\underline{A}(Y)-\overline{\underline{B}}-X \xrightleftharpoons{-X^{\ominus}} |\underline{A}(Y)-\overline{B} \rightleftharpoons \overline{\underline{A}}-\overline{\underline{B}}-Y \longrightarrow \text{Products}$$

Nucleophilic 1,2-rearrangement. Addition of a nucleophile to A or elimination of a substituent from B.

The key atom with the electron sextet can be a carbon, nitrogen or oxygen. Some examples for *I. on a carbon atom* with an electron sextet are the pinacol-pinacolone rearrangement, the Wagner-Meerwein rearrangement, the Demjanow rearrangement, the Tiffeneau rearrangement and the Wolff rearrangement. Examples of *I. on a nitrogen atom* with an electron deficiency are Hoffmann degradation, Lossen degradation, the Schmidt reaction, Curtius degradation and the Beckmann rearrangement. Examples of *I. on an oxygen atom* with an electron deficiency are the Hock phenol synthesis and the Baeyer-Villiger oxidation. The tendency of the mobile group to move depends on its nucleophilic strength. Sextet rearrangements proceed toward the lower-energy species with the electron sextet by a reaction pathway which requires the least changes in the substituents involved in the reaction.

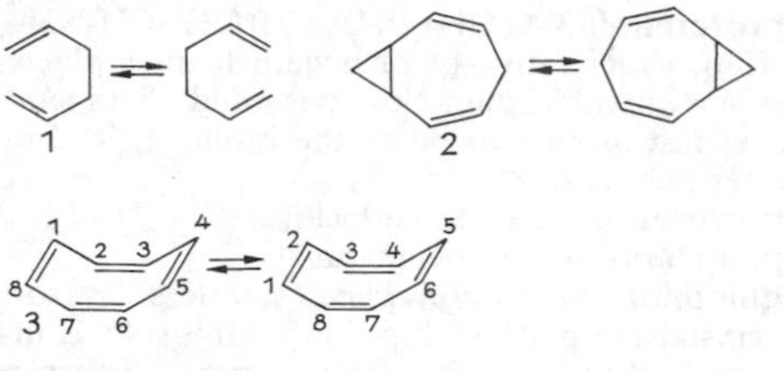

Degenerate rearrangements of *1* hexa-1,5-diene; *2* 3,4-homotropilidene; *3* cycloocta-1,3,5,7-tetraene.

Degenerate I. is a term applied when the reactant and product are identical (same energy content and structure). Some examples of this are the Cope rearrangements of hexa-1,5-diene, 3,4-homotropilidene or bullvalene. Cycloocta-1,3,5,7-tetraene undergoes a degenerate valence I. (valence tautomerism). Such degenerate I. are detectable by NMR spectroscopy.

Isomorphism, ***isotypic mixed crystal formation***: the phenomenon that different substances which crystallize in the same type of crystalline lattice (see Isotypism) can form mixed crystals. I. requires that the size and bonding characters of the components are not significantly different. For example, sodium chloride, NaCl, and silver chloride, AgCl are isomorphic, as are magnesium carbonate, $MgCO_3$, and calcium carbonate, $CaCO_3$. By contrast, while sodium chloride and lead sulfide, PbS, are isotypic, the bonds in their lattices are not isomorphic. I. was discovered in 1819 by Eilhardt Mitscherlich.

Isoniazide, abb. ***INH, isonicotinic hydrazide***: a compound used widely as a tuberculostatic. It forms white crystals which are soluble in water and hot ethanol, but are only slightly soluble in chloroform; m.p. 170-174 °C. I. is a basic compound which is protonated on the terminal amino group of the hydrazide structure. It is made by hydrazinolysis of isonicotinate esters.

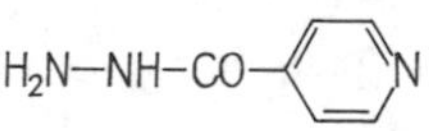

H_2N–NH–CO–(pyridin-4-yl)

Isonicotinic acid, ***pyridine 4-carboxylic acid***: colorless, crystalline needles; m.p. 319 °C; it can sublime at 2 kPa and 260 °C. I. is soluble in boiling water, slightly soluble in cold water, and insoluble in alcohol, ether and benzene. It is formed by oxidation of γ-picoline with potassium permanganate. I. is used mainly to synthesize isonicotinic hydrazide.

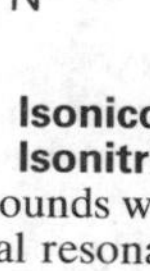

COOH

Isonicotinic hydrazide: same as Isoniazide (see).

Isonitrile, ***isocyanide***: one of a group of compounds with the general formula R-NC. The cannonical resonance structures are

$$R{-}\overset{+}{N}{\equiv}\overset{-}{C}l \leftrightarrow R{-}\overline{N}{=}Cl$$

I. are derived from hydroisocyanic acid, H-NC. They are very poisonous, colorless compounds with an unpleasant, sharp odor; some are crystalline. Their boiling points are lower than those of the isomeric nitriles. In alkaline solution, I. are rather resistant to hydrolysis. However, in dilute mineral acids, they are easily split into formic acid and primary amines:

$$R{-}\overset{+}{N}{\equiv}C + H_2O \longrightarrow R{-}NH_2 + HCOOH$$

When hydrogenated, I. are converted to secondary amines. The reaction of sulfur or mercury(II) oxide with I. yields isothiocyanates or isocyanates. Addition of chlorine to I. yields isonitrile dichlorides:

$$R{-}N{\equiv}C + Cl_2 \longrightarrow R{-}N{=}CCl_2$$

When the I. are heated to temperatures above 200 °C, they are isomerized to the more stable nitriles. I. are synthesized by cleavage of water from *N*-monosubstituted formamides, from primary amines and chloroform (see Isonitrile reaction) or by reaction of haloalkanes with silver cyanide. I. are used mainly for organic syntheses, e.g. in the Passerini reaction for synthesis of substituted α-hydroxycarboxylic acid amides or in the Ugi reaction for synthesis of *N*-substituted α-aminocarboxylic acid amides.

Isonitrile reaction, ***Hofmann's isonitrile reaction***: A reaction for detecting primary amines which, when mixed with chloroform and potassium hydroxide in alcoholic solution, form bad-smelling isonitriles. The primary reaction produces dichlorocarbene, which then reacts with the amino compound:

$$CCl_2{=}R{-}NH_2 + CCl_2 \longrightarrow R{-}NC + 2HCl$$

Hydrazines which have an aliphatic substituent on one side react in the same way, and interfere with the amine test.

Isonitrosoketone, ***α-oximinoketone***: a monooxime of an α-diketone or an α-ketoaldehyde with the general formula R^1-CO-CR^2=NOH. α-I. are the stable tautomeric forms of α-nitrosoketones with α-H atoms. α-I. can be made by the reaction of nitrous acid or one of its esters, e.g. isoamylnitrite, with a ketone with an α-methyl or α-methylene group:

$$R^1{-}CO{-}CH_2{-}R^2 \xrightarrow[-ROH]{+RONO} R^1{-}CO{-}CR^2{=}N{-}OH.$$

α-I. are used, for example, in the synthesis of α-dicarbonyl compounds, diazoketones and pyrazines.

Isooctane: see Octanes.

Isopentane: see Pentanes.

Isopentylnitrite: same as Isoamylnitrite (see).

Isopeptide bond: see Peptide.

Isophos: see Herbicides.

Isophthalic acid, ***benzene-1,3-dicarboxylic acid***: colorless needles, m.p. 348 °C (in a closed tube). It can be sublimed without decomposition. I. is practically insoluble in water, but is soluble in alcohol and glacial acetic acid. It undergoes the reactions typical of the dicarboxylic acids. I. can be synthesized by oxidation of *m*-xylene with potassium permanganate, and it is used in the synthesis of alkyde resins. Fig. p. 560.

Isophthalic acid

Iso-Plus process: see Reforming.

Isopolyanions: see Nomenclature, section II D.

Isoprenaline: 3,4-$(OH)_2C_6H_3$-CH(OH)-CH_2NH-$CH(CH_3)_2$, a phenylethylamine derivative with sympathicomimetic effects. It is used to treat asthma and also has a heart-stimulating effect.

Isoprene, ***2-methylbuta-1,3-diene***: CH_2=C(CH_3)-CH=CH_2, an unsaturated hydrocarbon with two conjugated double bonds. It is a colorless liquid; m.p. -146 °C, b.p. 34 °C, n_D^{20} 1.4219. I. is insoluble in water, but is soluble in ethanol and ether. It is the basic structural component of the terpenes and their many derivatives. It can be obtained from natural rubber by pyrolysis. Its industrial importance comes from its ability to polymerize to rubber. There are various methods of synthesis: 1) dehydrogenation of 2-methylbutane and 2-methylbutene on chromium oxide/aluminum oxide catalysts; 2) dehydrogenation of 3-methylbut-1-en-3-ol, which can be synthesized from acetone and ethyne. An intermediate is 3-methylbut-1-yn-3-ol; its triple bond is partially hydrogenated. 3) Pyrolysis of 2-methylpent-1-ene with loss of methane. This synthesis starts with propene, which can be dimerized to 2-methylpent-1-ene with trialkyl aluminum compounds.

I. can be dimerized in a [4+2]cycloaddition (Diels-Alder reaction). The product is the racemic dipentene 1-isopropenyl-4-methylcyclohex-3-ene; the levorotatory optically active form of this compound is (-)limonene, which can be converted to monocyclic terpenes. Otherwise, I. is only of systematic interest as the basic unit of the terpenes and steroids; that is, the majority of known terpenes consist of I. units linked head-to-tail (*isoprene rule*, Ruzicka, 1921), but they cannot be synthesized directly from I.

Isoprene rule: see Terpenes.

Isopropanolamine: see Aminopropanols.

Isopropenyl...: a term for the atomic group CH_2=C(CH_3)- in a molecule, the radical CH_2=C(CH_3) · or the cation CH_2=CH$(CH_3)^+$.

Isopropylacetaldehyde: same as Isovaleraldehyde (see).

Isopropyl acetate: see Propyl acetate.

Isopropyl alcohol: see Propanols.

Isopropylamine, ***2-aminopropane*** $(CH_3)_2$CH-NH_2, a primary aliphatic amine. I. is a colorless, combustible liquid with an ammonia-like odor; m.p. -101.2 °C, b.p. 33-34 °C, n_D^{20} 1.3742. It is very soluble in water, alcohol and ether, and displays the typical amine reactions. It is made by aminating reduction of acetone in the presence of catalysts. I. is used to make drugs, pesticides and special soaps. It is also used as a solvent.

4-Isopropylbenzaldehyde: same as Cuminaldehyde (see).

Isopropylbenzene: same as Cumene (see).

Isopropyl chloride: see Chloropropanes.

1-Isopropyl-4-methylbenzene: same as p-Cymene (see).

Isoproterenol: 3,4-$(OH)_2C_6H_3$-CH(OH)-CH_2NH-$CH(CH_3)_2$, a derivative of phenylethylamine which acts as a sympathicomimetic. It is used therapeutically against asthma to dilate the bronchii; it also stimulates the heart.

Isoproturone: see Urea herbicides.

Isopsoralene: see Furocumarins.

Isoquinoline, ***benzo[c]pyridine***: colorless, hygroscopic crystals; m.p. 26 °C, b.p. 243 °C. It is soluble in alcohol and ether, and insoluble in water. I. is basic and forms salts with acids. Its chemical properties are very similar to those of the quinolines. *N*-Benzoylisoquinolinium salts yield Reissert compounds and peracids oxidize I. to *N*-oxides. Nitration and sulfonation occur mainly in the 5-position, while bromination and mercurization occur in the 4-position.

The Čičibabin reaction yields 1-aminoisoquinoline. Oxidation with potassium permanganate produces mixtures of phthalic acid and cinchomeronic acid (pyridine-3,4-dicarboxylic acid). There are many methods of synthesizing I. and its derivatives. Some of the most useful are the Bischler-Napieralski synthesis (see), the Pictet-Spengler synthesis (see), the Pictet-Gams synthesis (see) and the Pommeranz-Fritsch synthesis (see). I. occurs in coal and bone tars. The opium alkaloids are derived from it. It is used in the synthesis of pigments, insecticides, malaria medications and vulcanization accelerators.

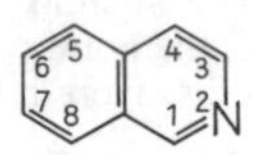

Isosafrol, ***1,2-(methylenedioxy)-4-propenylbenzene***: a colorless liquid with a typical anis odor; m.p. -22 °C, b.p. 77-79 °C at 500 Pa (*Z*-form), m.p. 6.8 °C, b.p. 253 °C (*E*-form). I. is insoluble in water, but soluble in ethanol and ether. It is made by heating safrol with alcoholic potassium hydroxide solution under pressure. It is used to perfume soaps and to make piperonal (oxidation with cleavage of the C-C double bond).

Isosbestic point: in an absorption spectrum of two substances in equilibrium with each other, the wavelength at which the two substances have the same molar extinction coefficients. At this point, the extinction remains constant, even if the position of the equilibrium shifts, because the total concentration is constant.

Isoselenocyanates: see Selenocyanates.

Isosorbide dinitrate: the dinitrate ester of sorbide. Sorbide (1,4:3,6-dianhydrosorbitol) is obtained by acid-catalysed dehydration of D-sorbitol and is esterified to I. with nitric acid. It is a long-acting coronary drug and is used as prophylaxis for angina pectoris. Fig. p. 561.

O_2N–O ... O–NO_2 Isosorbide dinitrate

Isoster: a curve in a suitable diagram which connects all points through which a mixed-phase process can proceed without changing the composition of the mixture.

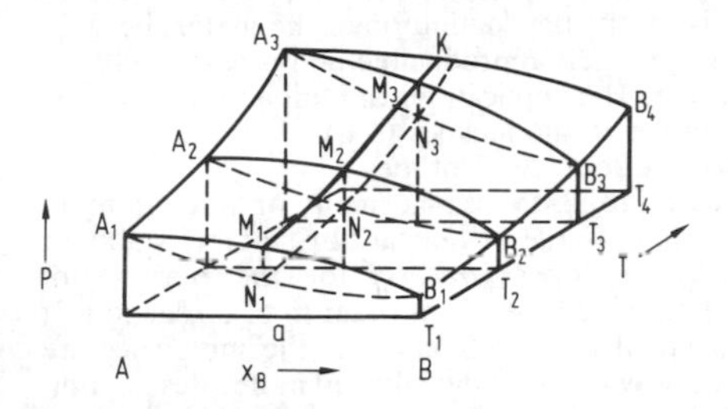

Isosters: schematic pTx diagram for a binary mixture with no mixing gaps in the liquid state and without extreme values in the vapor pressure curves of the mixture.

In liquid-vapor equilibria, the I. give the vapor pressure as a function of the temperature at a constant mole ratio in the liquid phase. They are obtained, for example, by drawing sections through a pTx diagram at constant molar fractions x (Fig.). For an adsorption process, the ***adsorption isoster*** shows the temperature dependence of that concentration of adsorbed substance which leads to a constant degree of coverage of the adsorbent.

Isotachophoresis: a method of Electrophoresis (see) in which the sample is injected between two electrolyte solutions. This avoids broadening of the zones of the separated substances by the carrier electrolytes. One of the electrolytes contains the "lead ion", which has a larger mobility than any of the sample ions, while the other electrolyte contains the "trailing ion", which has the lowest mobility. The polarity of the electric field is arranged so that the lead ion will migrate ahead of the sample ions. At a certain migration time, an equilibrium is reached, and all the ions move with the same rate in the electric field, separated into a number of sharp zones arranged in the order of their mobilities. The components are detected by UV detectors or thermoelements (Fig.). The resolution of the method is very high; for analytical purposes, capillary or thin-layer isotachophoresis is most often used.

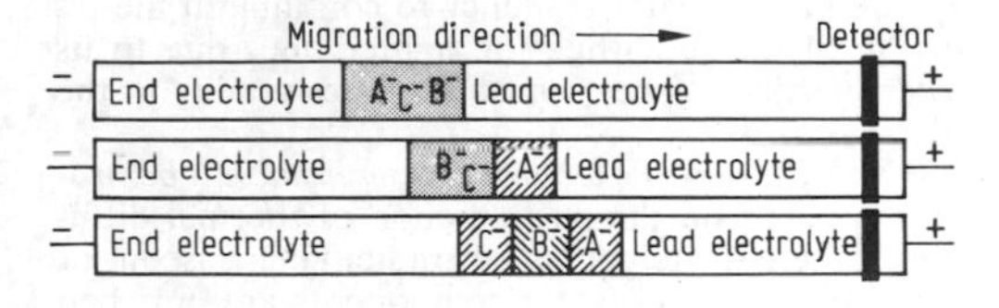

Isotachophoresis. A, B, C: sample ions; D, detector.

I. is used to separate ionic compounds such as proteins, peptides, phosphates, strong and weak acids, metal ions, etc., on either a preparative or an analytical scale. I. can also be used to determine ionic mobilities.

Isotactic: see Polymers.

Isotherm: a curve in a diagram of an ***isothermal process*** which shows the dependence of one physical parameter on another at constant temperature. Often the laws governing such behavior are called I., for example, Boyle's law, the van't-Hoff reaction isotherm or the various adosrption isotherms (see Adsorption). In the sense of this definition, the rate laws in kinetics are also I.

Isothermal process: see Adiabatic process.

Isothiazole: a heteroaromatic compound, and a colorless liquid which smells like pyridine; b.p. 115 °C, n_D^{20} 1.5324. It is soluble in organic solvents, but is only slightly soluble in water. Although the isomeric thiazole and derivatives of I. have been known for a long time, I. itself was first synthesized in 1956. The synthesis starts with 2-chlorobenzaldehyde and requires 13 steps. 5-Aminobenzoisothiazole and its oxidation product, isothiazole-4,5-dicarboxylic acid, are intermediates. Nitration or sulfonation of I. in the presence of sulfur trioxide yields derivatives substituted in position 4. Chlorination or bromination also occurs most readily at the 4 position. I. forms defined 1:1 adducts with Lewis acids. It can undergo quaternization reactions, but these go more slowly than with pyridine. Because of the similarity of derivatives of pyridine and I., the latter is used in the synthesis of analogs of natural products and pharmaceuticals. I. and its methyl derivatives are excellent solvents.

Isothiocyanates: derivatives of isothiocyanic acid, H-NCS. 1) the inorganic I. are considered to be Thiocyanates (see).

2) The organic I. with the general formula R-NCS are also called ***isothiocyanic acid esters (mustard oils)***. Depending on R (aliphatic, aromatic or heterocyclic group), these are oily or crystalline compounds which are lacrimators and have pungent odors. They are found naturally in both free and glycosidic form (see Glucosinolates) in the seeds of the mustard family. They include, for example, Allyl isothiocyanate (see), phenylethyl and benzyl isothiocyanates. I. can be converted to amines by hydrolysis in the presence of acids or bases, or by reducing agents: $R\text{-}N{=}C{=}S + 2\,H_2O \rightarrow R\text{-}NH_2 + CO_2 + H_2S$. They are synthesized by isomerization of thiocyanates, by the Hofmann mustard oil reaction (see) or by reaction of primary amines with thiophosgene: $R\text{-}NH_2 + Cl\text{-}CS\text{-}Cl \rightarrow R\text{-}NCS + 2\,HCl$. I. have bacteriocidal and fungicidal effects. They are used in a limited way as disinfectants for the skin and wounds. They are also used as intermediates for organic syntheses.

Isothiocyanic acid: see Thiocyanates.

Isothiocyanic acid allyl ester: same as Allyl isothiocyanate (see).

Isothiocyanic acid phenyl ester: same as Phenyl isothiocyanate (see).

Isothionaphthene: see Benzothiophenes.

Isotonic: a term for a solution which has the same osmotic pressure (see Osmosis) as a comparison solution. For example, physiological saline is I. with human blood.

Isotons: atomic nuclei with the same number of neutrons (N) but different nuclear charge (atomic number) and mass numbers. Three I. with N = 32 are the nuclides ^{59}Co, ^{60}Ni and ^{61}Cu (atomic numbers 27, 28 and 29, respectively).

Isotopes: atomic nuclei with the same nuclear charge number (atomic number) but different mass numbers due to different numbers of neutrons in the nuclei. I. of the same element, e.g. neon 20 and neon 22 (^{20}Ne and ^{22}Ne) differ not only in mass, but also in their nuclear volume, spin quantum numbers and magnetic quantum numbers (see Atom model). The I. of light elements also display slight differences in chemical behavior.

I. are classified as ***stable*** or ***radioactive***. Most elements found in nature are mixtures of I. which are present in different abundances. At present 267 stable and about 1500 radioactive I. are known.

According to ***Aston's isotope rule***, elements with even atomic numbers almost always have several stable I., e.g. selenium, with atomic number 34, has the I. ^{74}Se, ^{76}Se, ^{78}Se, ^{80}Se and ^{82}Se, while elements with odd atomic numbers have only 1 or at most 2 stable I. Fluorine, for example, with atomic number 9, has only ^{19}F, while chlorine, with atomic number 17, has the I. ^{35}Cl and ^{37}Cl. Elements with atomic numbers greater than 83 have only radioactive I.

Studies have shown that elements consisting of several I. almost always display the same relative abundances of the I. both in terrestrial sources and in meteorites. Exceptions to this rule are observed only when I. of an element are the products of natural radioactive decay, as in the case of lead. This effect is used for determination of age of minerals. Variations in the I. ratios of oxygen and hydrogen are also observed, due to geological processes which have differentiated between I. The result has been a (slight) uncertainty in the exact atomic masses of these elements.

Isotope analysis is carried out primarily by mass spectroscopy. The term ***isotope effect*** is used inclusively for all differences in physical or chemical properties of elements which are due to the relative mass differences of I. These include differences in heat conductivity, melting and boiling points, viscosity, capillarity, chemical reactivity, reaction rates, absorption or emission spectra, etc. Some of these effects are used for Isotope separation (see). The optical I. effects are shifts in the spectral lines of different I. of an element, and in the changes in absorption spectra of molecules when atoms of one I. are replaced by those of another I. of the same element. I. effects can be calculated using statistical thermodynamics and quantum mechanics. The most pronounced are those observed in the deflection of ions in electric and magnetic fields, because here the mass differences of the I. have a direct effect. This effect is utilized in mass separators which can completely separate I.

Applications of stable I. are found in chemistry, physics, biology, medicine and materials research. They are used as Tracers (see) to label various compounds. Optical I. shifts are used in elucidation of complicated molecular structures. Deuterium oxide is used in large amounts in nuclear reactors, because it is an effective moderator for fast neutrons. Gaseous boron trifluoride $^{10}BF_3$ (with the less abundant boron isotope ^{10}B) is used to fill neutron counting tubes and to coat plates for tracking nuclear reactions. An emission line of the heaviest natural isotope of krypton, ^{86}Kr (previously enriched to 99.5%) was chosen as the basis for the new definition of the meter, because it is very precisely reproducible and is essentially unchangeable. The applications of radioactive I. are discussed under Radionuclides (see).

Isotope effect: see Isotopes.

Isotope labelling: replacement of one or more atoms in a molecule or reactant by suitable stable or radioactive isotopes (*tracers*) for the purpose of elucidating the course or mechanism of a reaction. It is sufficient if a small fraction of the molecules are labelled, provided that the labelled molecules are uniformly dispersed among the otherwise identical unlabelled molecules. The presence of the labelled atoms in the products often indicates what the mechanism must have been. Often the participation of the solvent in the reaction can be demonstrated or excluded by I.

Isotope peak: see Mass spectroscopy.

Isotope separation: separation (or more accurately, enrichment) of individual isotopes present in polynuclidic chemical elements by physical or chemical methods. I. is essential for the preparation of nuclear fuels. It is accomplished by utilization of isotope effects (see Isotopes). The *enrichment factor q* is obtained by dividing the ratio of isotope abundances (a'/b') after I. by the corresponding ratio before I.: $q = (a'/b')/(a/b)$ In order to reduce the costs of I., the following methods are often combined or used sequentially; the enrichment factor of the total process is the product of the individual enrichment factors.

1) In the *gas diffusion process* (*diffusion process*), differences in the rates of diffusion of gaseous compounds of the isotopes to be separated are exploited. The gas mixture diffuses through porous walls or into a vapor stream; the compound containing the lighter isotope moves faster than the heavier one if the pressure is low and the pore diameter is small enough so that the mean free path of the gas molecules is on the same order of magnitude as the pore diameter. The separation effect must be multiplied by using numerous separation steps in sequence.

2) The *separation jet* developed by E.W. Becker exploits the differences in thermal velocities in gaseous mixtures of isotopes. A stream of gas is expelled through a jet at supersonic speed. The heavier isotope has a greater tendency to continue in the direction of the jet, while the lighter one, due to its lower inertia, escapes somewhat more readily to the side.

3) The *Klusius-Dickel process* (*separation tube process*) is based on the phenomenon of thermal diffusion. If there is a temperature gradient in a gaseous or liquid mixture, the lighter components are enriched in the hotter region, and the heavier ones in the

cooler end. The separation effect is extremely small, but it is magnified by combination of thermal diffusion with thermal convection.

4) The *centrifugal process* makes use of a gas centrifuge in which the heavier isotope is enriched at the outer wall of the centrifuge, because the centrifugal force is proportional to the mass.

5) *Countercurrent distillation* is used for separation of the isotopes of some lighter elements. This makes use of slight differences in vapor pressures of the isotopes or their compounds. For example, deuterium (the heavier isotope of hydrogen) can be obtained on an industrial scale by low-temperature distillation of liquified hydrogen. When water is distilled (especially under reduced pressure), the oxygen isotopes ^{17}O and ^{18}O and deuterium are enriched in the residual water. Distillation has become important, in the form of molecular distillation, even for isotopes of heavier elements, such as mercury.

6) In the *electrolysis* of water, light hydrogen is more readily generated at the cathode, so that deuterium is enriched in the non-electrolysed residual water. ^{6}Li can be enriched by electrolysis of molten lithium chloride. The effect is due to the difference in mobility of the isotopic ions in an electric field.

7) The *chemical exchange process* is also industrially important. It is based on the fact that the equilibrium constants K (see Mass action law) of isotope exchange reactions are often significantly different from 1. For example, for the exchange of nitrogen isotopes between gaseous nitrogen monoxide NO and aqueous nitric acid HNO_3 ($^{15}NO + H^{14}NO_3 \rightleftharpoons {}^{14}NO + H^{15}NO_3$),

$$K = \frac{[^{14}NO][H^{15}NO_3]}{[^{15}NO][H^{14}NO_3]} = 1.06$$

The heavier nitrogen isotope ^{15}N is thus enriched in the liquid nitric acid phase. This separation effect is magnified by using a counter-current process. In the large-scale processes for enrichment of deuterium by chemical exchange in systems such as hydrogen/water, hydrogen sulfide/water or hydrogen/ammonia, the phase change for hydrogen is uneconomical, but it can be avoided by using the *hot-cold process* which makes use of the temperature dependence of such isotope exchange equilibria. The hydrogen/water exchange must be accelerated by a catalyst suspended in water, e.g. platinum on active charcoal.

8) Nearly complete separation of isotopes occurs in the mass spectrometer (see Mass spectroscopy), but in ordinary equipment, the amounts of the individual isotopes obtained are too small to be weighed. Therefore especially high-intensity devices (mass separators) have been developed to separate isotopes in weighable amounts.

9) The development of high-power and tunable lasers has made another method of I. possible. Because there is an isotope-dependent shift in atomic spectra, monochromatic laser light can be used to excite only a single isotope in a mixture; the excited atoms are then separated from the others.

Methods 1) to 7) are statistical processes, in which an average effect is obtained; methods 8) and 9) are non-statistical, and in principle allow almost complete separation in a single step.

Isotropy: the independence of physical properties and behavior of a substance on direction. Amorphous substances such as gases, liquids and glasses are isotropic, so long as no external forces act on them. Crystals, however, are generally anisotropic (see Anisotropy). Some solids are isotropic with respect to some properties and anisotropic with respect to others. For example, cubic crystals are isotropic in their optical behavior, i.e. the velocity of light through the crystal and its absorption of light do not depend on the orientation of the crystal with respect to the direction of the incident light. On the other hand, the rate of crystal growth is not the same in all directions, even for cubic crystals. Polycrystalline substances in which the crystallites are completely randomly oriented are isotropic, in contrast to single crystals of the same material, because the anisotropy is averaged out. This phenomenon is more accurately termed *quasi-isotropy* (see Polycrystalline material, Texture).

Isotypic mixed crystal formation: same as Isomorphism (see).

Isotypism: Different substances crystallize in the same lattice type; this phenomenon is called I. In the strict sense, two sybstances are isotypic only if they have analogous summary formulas, the same symmetry (space group) and considerable similarity in atomic arrangement. This last requirement means occupation of the same positions with respect to similar atomic coordinates and similar lattice parameters ($a_0{:}b_0{:}c_0$, α, β, γ). Comparison of the isotypic substances sodium chloride, NaCl, magnesium oxide, MgO, and lead sulfide, PbS, indicates that neither the absolute size of the lattice components nor the nature of the bonding is important; I. is primarily a geometric concept.

The term *homeotypism* has been coined for crystal structures which do not completely fulfill the above conditions, but are still very similar in structure. For example, C (diamond) and ZnS (zinc blende), $CaCO_3$ (calcite) and $(Ca,Mg)(CO_3)_2$ (dolomite), SiO_2 (quartz) and $AlPO_4$ (berlinite) are homeotypic. Often the term I. is used in a broader sense than defined above, and substances which are actually only homeotypic are called I., if the structural analogy is complete.

If the places of the anions and cations are exchanged in isotypic crystal types, it is called *antiisotypism*. For example, thorium dioxide, ThO_2, crystallizes in the Fluorite type (see), while lithium oxide, LiO_2, forms an antifluorite lattice. Structure types which have little or no similarity are called heterotypic (*heterotypism*).

Isovaleraldehyde, ***3-methylbutanal, isopropylacetaldehyde***: $(CH_3)_2CH\text{-}CH_2\text{-}CHO$, a colorless liquid with an intense, unpleasant odor which stimulates coughing; m.p. - 51 °C, b.p. 92.5 °C, n_D^{20} 1.3902. I. is slightly soluble in water, but dissolves readily in alcohol and ether. In the presence of hydrogen chloride at - 20 °C, I. readily trimerizes. It undergoes the addition and condensation reactions typical of Aldehydes (see). I. is present in many essential oils, and is synthesized by dehydrogenation of isoamyl alcohol. It is used to synthesize isovaleric acid, drugs, vulcanization accelerators and especially in the perfume industry.

Isovaleric acid: see Valeric acids.

Isovisco curves: lines of equal viscosity during a change in the ratio of surfactant to electrolyte in aqueous solution.

Isoxazol: a heteroaromatic compound; a colorless liquid with an odor similar to that of pyridine; b.p. 95-96 °C, n_D^{20} 1.4928. It is readily soluble in organic solvents, but barely soluble in water. I. is a weak base. When it is treated with alcoholate solution, the ring is cleaved, and the alkali salt of cyanoacetaldehyde is obtained. Halogenation, nitration and sulfonation occur in the 4 position. I. is obtained by the reaction of hydroxylamine with propargylaldehyde or its diethyl acetal.

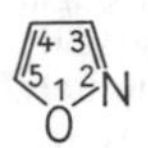

Istrona®: see Synthetic fibers.

Itaconic acid, ***methylenesuccinic acid, cis-methylenebutenedioic acid, prop-2-ene-1,2-dicarboxylic acid***: $CH_2=C(COOH)-CH_2-COOH$, an unsaturated dicarboxylic acid. I. forms colorless, hygroscopic crystals; m.p. 175 °C (161-162 °C also reported). It cannot be distilled without decomposition. I. is readily soluble in alcohol, is less soluble in water, and is practically insoluble in ether, benzene and chloroform. It undergoes the reactions typical of dicarboxylic acids and activated C=C double bonds. I. is a metabolic product of certain molds, e.g. *Aspergillus itaconicus*, and can be produced by industrial fermentation from sugar, molasses, etc. I. is also synthesized industrially from citric acid by pyrolysis, or by heating a concentrated aqueous solution. I. is used as an intermediate in the production of plastics, softeners and additives.

IUPAC: abb. for International Union of Pure and Applied Chemistry.

J

Jablonski diagram: see photophysical processes.

Jahn-Teller theorem, ***Jahn-Teller effect***: a prediction, based on group theoretical considerations, that a nonlinear molecule is a stable geometric structure only if its ground state is not degenerate. The degeneracy of the ground state of a molecular system can be relieved by a symmetry reduction associated with a static deformation, or by an oscillation among several structures with nondegenerate ground states (*dynamic Jahn-Teller effect*). It follows from the J., especially for complexes, that the regular coordination polyhedra (see Ligand field theory) with degenerate ground states will be distorted in order to relieve the degeneracy. The J. does not predict the size and direction of the distortion.

Jambolen®: see Synthetic fibers.

Janus green: see Janus pigments.

Janus pigments: a group of azo pigments, mostly di- or triazo pigments, made by diazotization of basic pigments (e.g. safranin) and coupling with β-naphthol. ***Janus green*** is an important representative of the group.

Japan lacquer: see Resins.

Japan wax, ***Japanese beeswax***: a plant wax obtained from various sumac fruits which grow in Japan and California. The pale yellow, hard mass (m.p., 52 to 54 °C, density about 1) has a tallow-like, slightly rancid odor. It dissolves easily in alcohol, ether and light petroleum. The main components of J. are glycerol palmitate and free palmitic acid. It is used to make candles, varnish and shoe polish.

Jasmine oil: an essential oil prepared by enfleurage or cold solvent extraction of jasmine blossoms. The main components are benzyl acetate (up to 65%), linalyl acetate, linalool, indole and benzyl alcohol. It is used in perfumes.

Jervanine: see Steroid alkaloids.

Jervaic acid: same as Chelidonic acid (see).

Jestryl®: see Carbachol.

JH: abb. for Juvenile hormone (see).

jj coupling: see Multiplet structure.

Jellyfisch toxins:toxins released from the sting capsules of jellyfish. The toxins of two species are dangerous to human beings, those of the sea wasp *Chinorex fleckeri* and of the Portugese man o'war, *Physalia physalis*. The toxic species live in a strip between 45° north latitude and 30° south latitude. Their toxins are a complex mixture of poisonous substances, including serotonin, which causes intense pain and damages the heart, tetramethylammonium hydroxide, which blocks nerve impulses (paralysis) and poisonous peptides and proteins. Contact with the above species is usually fatal.

Jet fuel: see Fuel.

Joule: the SI unit of energy: 1 kg m^2 s^{-2}. 1 J = 0.2389 cal = 10^7 erg = $6.242 \cdot 10^{18}$ eV = 1 newton meter = 1 watt second.

Joule-Thomson effect: when a real gas expands without doing external work, its temperature changes. This phenomenon was discovered by J. P. Joule (1818-1889) and W. Thomson, later Lord Kelvin (1824-1907). The cause of the J. is the presence of attractive and repulsive forces between the particles of a real gas, which make the internal energy e and the enthalpy h of real gases dependent on the volume v and the pressure p: $(\delta e/\delta v)_T \neq 0$ and $(\delta h/\delta p)_T \neq 0$ (see Thermodynamics, first law and Equations of state, real gases).

The change in enthalpy dh of a homogeneous substance is given by the equation: $dh = \delta q + \delta w = (\delta h/\delta T)_p dT + (\delta h/\delta p)_T dp$ (see Thermodynamics, first law). If a gas expands through a frictionless valve without doing external work, in a well-insulated apparatus, $\delta w = 0$ because the process is carried out adiabatically, and $\delta q = 0$, i.e. $dh = 0$ (isoenthalpic process). From the above equation it follows that $(\delta T/\delta h)_h \equiv \delta = (\delta h/\delta p)_T/(\delta h/\delta T)_p = [T(\delta v/\delta T)_p - v]n/C_p$. The differential $(\delta T/\delta p)_h$ is the ***differential Joule-Thomson coefficient*** δ. It determines the nature and magnitude of the temperature change δT when a gas expands ($\delta p < 0$). If it is positive, the gas cools ($\delta T < 0$), otherwise, the gas is warmed. Since $(\delta h/\delta T)_p = nC_v D$, where n is the molar amount and C_v is the molar heat capacity, the sign of δ depends only on the partial differential $(\delta h/\delta p)_T$. If the attractive forces between the gas molecules outweigh the repulsive ones, the enthalpy increases as the molecules separate; $(\delta h/\delta p) < 0$ and δ is positive. If the repulsive forces are stronger, the signs are reversed.

For ideal gases, $(\delta h/\delta p) = 0$ (see Gay-Lussac laws), and thus $\delta = 0$. The isoenthalpic expansion occurs with no change in temperature. For most real gases, δ is positive under normal conditions (the attractive forces are stronger). It decreases as the temperature and pressure increase, and reaches zero at the inversion temperature T_i; above this point it is negative. The van der Waals equation of state yields $T_i = 2\,T_B = 2a/Rb$, where T_B is the Boyle temperature, R is the gas constant, and α and b are van der Waals constants.

The J. is the basis of liquefaction of gases. Gases with inversion temperatures below room temperature were formerly called "permanent gases"; examples are helium with $T_i \approx 35$ K and hydrogen with $T_i \approx 193$ K at average pressures. These gases must be cooled to temperatures below their Boyle temperatures before expansion leads to further cooling.

Juglone: 5-Hydroxynaphtho-1,4-quinone; m.p. 155 °C. J. forms reddish yellow crystals and is practically insoluble in water. It is a component of staining oils prepared from the leaves and husks of walnuts.

Juniper oil: a colorless, sometimes pale green essential oil. It is mobile when fresh, but upon storage becomes a viscous gel. J. consists mainly of camphene, α-pinene, cadinene, *p*-cymol and terpineol, and various other alcohols. It is obtained by steam distillation of crushed juniper berries, in a 1 to 2% yield. The remaining parts of the shrub yield juniper needle oil, juniper wood oil and juniper bark oil. J. is used mainly in the production of liquors (e.g. gin), and in various areas of human and animal medicine.

Juvabione: see Juvenile hormones.

Juvenile hormones, abb. ***JH***: a group of insect endocrine hormones formed in the corpora allata; they promote differentiation of the larva but inhibit imagos. J. are analogs of vertebrate growth hormone (somatotropin). Natural J. are sesquiterpenes. ***Farnesol***, isolated from the larvae of the meal beetle *Tenebrio molitor* has some J. activity. Derivatives of farnesol carboxylic acid methyl ester with J. activity were isolated from larvae of *Hyalophoro cecropia* (the main components are C_{17}JH or JH-I, and C_{18} JH or JH-II) and *Manduca sexta* (C_{16}-JH).

Many compounds isolated from plants or synthesized chemically have J. activity, such as the sesquiterpene ***juvabione***. The structures of J. are being studied with the goal of developing insecticides with higher selectivities. However, insects react to exogenous J. only during a relatively short phase of development. Larvae and imagos are not affected in their development or activity.

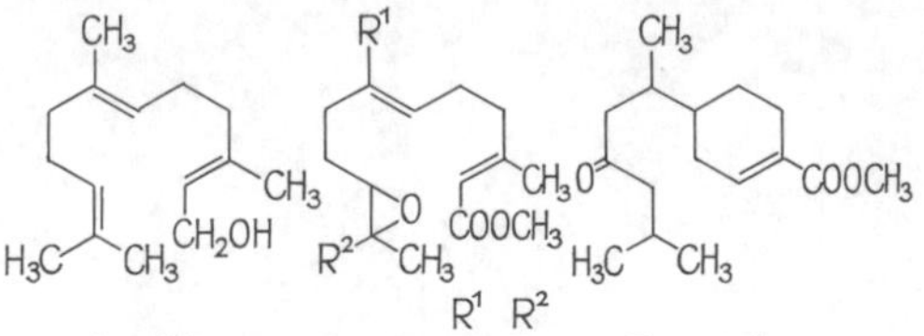

Juvabione

	R^1	R^2
C_{18} - JH:	Et	Et
C_{17} - JH:	Me	Et
C_{16} - JH:	Me	Me

Farnesol

Juvenile water: a mineral-containing water which comes from great depths (from magma) and enters the water cycle. It is formed by chemical processes in the magma, not by seepage of surface waters into the depths.

K

k: 1) symbol for a rate constant. 2) k or k_B, symbol for the Boltzmann constant.

K: symbol for potassium.

Kachkaroff process: see Sulfuric acid.

Kaempferol: see Flavones.

Kairomones: see Semiochemicals.

Kaldo and rotor process: see Steel.

Kallidin: see Plasma kinins.

Kanamycins: a group of aminoglycoside antibiotics obtained from culture filtrates of *Streptomyces kanamyceticus*. There are various components (K. A, B, C), all of which contain 2-deoxystreptamine as the aglycon; two amino sugars are linked glycosidically to the aglycon. The K. are effective against gram-negative bacteria and are used in treatment of tuberculosis. Like streptomycin, they have neurotoxic side effects. ***Amikacin*** is a derivative of K. A in which an amino group on the 2-deoxystreptamine is acylated.

Kanekalon®: see Synthetic fibers.

Kaolin: an important raw material for ceramics. It is a finely divided sedimentary rock formed by weathering of aluminosilicates, especially feldspars. Pure K. is white, but it is usually discolored by impurities (iron and manganese compounds). K. used for fine ceramics (porcelain) must fire to a white body. K. consists mainly of the minerals kaolinite, dickite and nakrite (all $Al_4(OH)_8Si_4O_{10}$) and halloysite, $Al_4(OH)_8Si_4O_{10} \cdot 4H_2O$. It also contains small amounts of quartz, mica-like minerals (e.g. muscovite), rutile, zircon, magnetite and residues of feldspar. Crude kaolin is treated by elutriation, flotation, electrophoresis and electroosmosis to obtain technical K.; depending on its quality, this material is used to make porcelain and other silicate ceramics. The non-plastic K. is used as a filler for paper to give it a smooth, absorbent surface.

Karl-Fischer solution: a solution for volumetric water determination in inorganic and organic substances, originally consisting of iodine, sulfur dioxide and pyridine in a molar ratio of 1:1.5:7 to 1:2.5:5 in anhydrous methanol. Although the K. contains both the oxidizing agent iodine and the reducing agent sulfur dioxide, they do not react so long as the necessary water is absent. In the presence of water, the reaction occurs immediately: $SO_2 + I_2 + 2\,H_2O \rightarrow SO_4^{2-} + 2\,I^- + 4\,H^+$. The actual course of the reaction in the K., which also involves the pyridine and methanol, is much more complicated and so far has not been completely elucidated. The titer of the K. decreases by about 0.4% per day, and must therefore be recalibrated frequently. This can be done either with methanol with a precisely known water content, or with substances with defined contents of water of crystallization, for example, oxalic acid dihydrate. In commercial products, the two redox components are separated to achieve an adequate storage life. K.I, a solution of sulfur dioxide and pyridine in anhydrous methanol, is then used as the solvent for the sample, while K.II, a solution of iodine in methanol, is used as the titrant. Other solvents, e.g. 2-methoxyethanol, and other amines, such as diethanolamine, can greatly improve the constancy of the titer of the K. Such one-component K. display a loss of titer of only 7.1% after 6 months.

The determination of water with the K. is now carried out by dissolving the liquid or solid sample in anhydrous methanol, or the water content of a solid sample is extracted with anhydrous methanol, and this sample is then titrated with the one-component K. until the color of the solution turns from pale yellow to brown. Great care must be taken to avoid moisture. The endpoint can be more precisely determined by the dead-stop method (see Amperometry).

Karplus equation: this equation gives the dependence of the vicinal coupling constant J (see NMR spectroscopy) on the dihedral angle Φ. From the Fig. it can be seen that J is at a maximum at $\Phi = 180°$. It is somewhat smaller at 0° and disappears at $\Phi = 90°$.

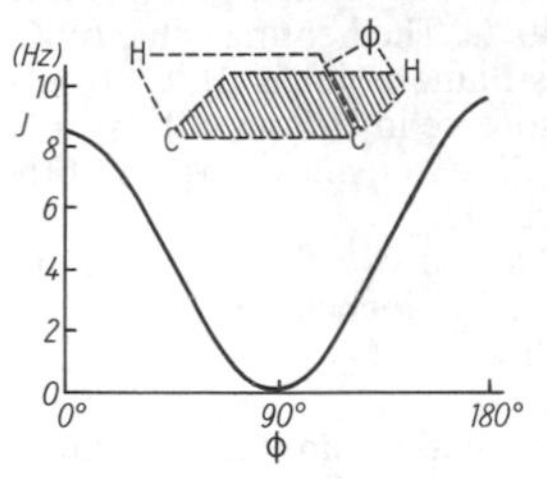

Graph of the Karplus equation. J is the coupling constant, and Φ is the dihedral angle.

Certain information on the stereochemistry of the molecule can be obtained from the relationship between the dihedral angle and the vicinal coupling constants. For the chair form of cyclohexane, the coupling constants depend on the equatorial and axial arrangement of the coupling protons; the following vicinal coupling constants are observed: $J_{ax\text{-}ax}$ = 8 to 14 Hz, $J_{ax\text{-}eq}$ = 0 to 6 Hz, $J_{eq\text{-}eq}$ = 0 to 6 Hz.

Kassinin: see Tachykinins.

kat: see Enzymes.

Katal: see Enzymes.

K-capture: see Radioactivity.

Kebuzone: see Phenylbutazone.

Keesom forces: see van der Waals forces.

Kelevan: see HCCP dimer insecticides.

Kellogg synthesis: see Fischer-Tropsch synthesis.

Keratins: a group of structural proteins found in wool, hair, claws, hoofs, horns and feathers, but also in skin and connective tissues. ***α-K.*** have a high cysteine content (11%) and an α-helix structure; **β-*K.*** do not contain cysteine and form antiparallel pleated sheet structures.

In the α-K., three α-helical polypeptide chains spiral around each other to form a *protofibril* with a diameter of 2 nm. Electron-microscopic photographs indicate that 11 of these protofibrils are twisted into a rope-like *microfibril* (8 nm diameter); and the microfibrils in turn are twisted around each other to form *macrofibrils* (diameter 200 nm). In wool, the macrofibrils lie parallel to the fiber axis in the dead cells of the hair; these cells have a final diameter of 20,000 nm. The α-K. fibers can stretch; when moist, they can be extended to twice their original length. This stretching is associated with a change in conformation into a β-chain structure with parallel polypeptide chains. The elasticity of the helix is due to its cross-linking by disulfide bridges.

A very important β-K. is ***silk fibroin***, in which the sequence -Gly-Ser-Gly-Ala-Gly-Ala- is repeated many times. Neighboring chains of β-K. are stabilized in the β-pleated sheet structure by hydrogen bonds. Association of pairs of polypeptide chains leads to an extensive protein complex, which is held together by ***sericin***, a second silk protein which is water-soluble. The relative inability of silk to stretch and its great flexibility are due to the strong covalent bonds between the extended peptide chain and the weak van der Waals bonds between the pleated sheets.

Keracyanin: see Cyanidine.

Kerogen: see Oil shale.

Kerosene: a medium petroleum fraction produced by top distillation (distillation at normal pressure); it boils between 180 and 240 °C. The heating value of K. is 41,000 kJ kg^{-1}, and its flame point is 21 °C. In ultraviolet light, K. can fluoresce in various colors. It is used as a Fuel (see) for diesel engines, airplane turbines and small space heaters.

Ketal: a term for the acetal of a ketone; it is no longer permitted in IUPAC nomenclature.

Ketamine: see Narcotics.

Ketazine: see Azine 2).

Ketene: a compound with cumulative double bonds in the arrangement $R^1R^2C=C=O$. These compounds can be considered internal anhydrides of monocarboxylic acids. Depending on the nature of R^1 and R^2, they are classified as ***aldoketenes*** (R^1 = alkyl, aryl and R^2 = H) and ***ketoketenes*** (R^1,R^2 = alkyl, aryl). The simplest compound of this type is called ketene; it is a colorless, very poisonous gas, which can be made by pyrolysis of acetone. The names of K. are composed of the names of the groups R^1 and R^2 and "ketene", for example, diphenyl ketene. K. are formed by dehalogenation of α-haloacyl halides with zinc, or by dehydrohalogenation of acyl halides with tertiary amines. In the Wolff rearrangement (see), a K. is formed as an intermediate. K. are very reactive; as acylation reagents they react with water to give carboxylic acids, with alcohols to give carboxylic acid esters, with carboxylic acids to give carboxylic acid anhydrides, and with ammonia to give amides. As unsaturated compounds, they undergo addition reactions; in a reversal of their formation, they add, for example, to bromine or hydrogen bromide. They undergo [2+2]-cycloaddition with compounds with double or triple bonds, forming carbocyclic and heterocyclic compounds. An example is the dimerization of K. to diketenes. Photolysis of K. yields *carbenes*.

$$2\,CH_2{=}C{=}O \longrightarrow \begin{array}{l} CH_2{=}C{-}O \\ \;\;|\;\;\;\;\;\;\;| \\ H_2C{-}C{=}O \end{array}$$

Diketene

Ketimide-enamine tautomerism: a tautomeric equilibrium between a stabilized ketimide (e.g. β-iminonitrile) and the corresponding enamine:

$$R{-}CH_2{-}\underset{NH}{\overset{\|}{C}}{-}\underset{R'}{CH}{-}C{\equiv}N \rightleftharpoons R{-}CH_2{-}\underset{NH_2}{C}{=}\underset{R}{C}{-}C{\equiv}N$$

Ketimide Enamine

Ketimines, ***ketimides***: compounds with the general formula $R^1R^2C=NH$. Unlike the aldimines, K. cannot be synthesized from ketones and ammonia. Acetone, for example, reacts with ammonia to form di- and triacetonamine. K. can be made by addition of Grignard compounds to nitriles:

$$R^1{-}C{\equiv}N + R^2{-}MgX \rightarrow R^1R^2C{=}N{-}MgX \xrightarrow{(H_2O)} R^1R^2C{=}NH.$$

However, this method produces only stable K., in some cases as the hydrochlorides. In most cases, hydrolysis of the Grignard addition product leads immediately to the corresponding ketone.

Ketoacids: see Oxocarboxylic acids.

Ketoaldehydes: organic compounds containing both aldehyde -CHO and keto groups $-C=O$. An example is methylglyoxal, CH_3CO-CHO.

β-Ketobutyric acid: same as Acetoacetic acid (see).

Ketocarboxylic acids: see Oxocarboxylic acids.

Keto-enol tautomerism: A compound containing a carbonyl group, with the general formula R'-CO-CH_2-R, can exist in two forms, which are in equilibrium:

$$\underset{\text{Keto-Form}}{CH_3{-}CO{-}CH_2{-}CO{-}CH_3} \rightleftharpoons \underset{\text{Enol-Form}}{CH_3{-}\underset{OH}{C}{=}CH{-}CO{-}CH_3}.$$

The conversion from one tautomeric form to the other occurs by intramolecular migration of a proton between a C-atom and the O-atom, and simultaneous shifting of the double bond.

The position of the equilibrium is determined mainly by the stability of the tautomers. The enol form is stabilized by acidity of the methylene (methyl) group and by conjugation. In compounds of the type R'-CO-CH_2R, the equilibrium is almost completely on the side of the keto form, but in β-dicarbonyl compounds, it is shifted more or less to the enol form. External factors which affect the position of the equilibrium are the nature of the solvent and the temperature. Attainment of equilibrium can be catalysed by acids or bases. In most cases, the enol form can be detected qualitatively by a reaction with iron(III) chloride (red color), and quantitatively by bromine titration (bromine addition to the C=C double bond). In special cases, e.g. acetoacetate and acetylacetone, the keto and enol forms differ sufficiently in solubility, melting and boiling points to be isolated in pure form.

β-Ketoglutaric acid: same as α,α-Acetone dicarboxylic acid (see).

Keto group: see Carbonyl group.

Ketol: see Hydroxyketone.

Ketomalonic acid: same as Mesoxalic acid (see).

Ketone: an organic compound with the keto-group

$$>C=O$$

in the molecule. Depending on the substituents, the K. are classified as aliphatic, aromatic and heterocyclic. In addition, there are mixed aliphatic-aromatic, aliphatic-heterocyclic and aromatic-heterocyclic K. Pure aliphatic saturated K. (***alkanones***) include, e.g. propanone (acetone) CH_3-CO-CH_3 and butanone (methyl ethyl ketone) CH_3-CH_2-CO-CH_3. Examples of aromatic K. are diphenyl ketone (benzophenone) C_6H_5-CO-C_6H_5. Pure heterocyclic K. are relatively rare. Some examples of mixed K. are methyl phenyl ketone (acetophenone) C_6H_5-CO-CH_3 and 2-thienyl methyl ketone (2-acetylthiophene) CH_3-CO-$C_4.H_3S$. In addition to the above open-chain K., there are compounds of the type in which the carbonyl group is a component of isocyclic or heterocyclic rings, e.g. cyclohexanone, β-tetralone and chromanone-4.

The systematic names of the K. are formed either from the name of the hydrocarbon skeleton plus the suffix "-one", or from the names of the two hydrocarbon residues on either side plus the term "ketone". In the first case, the position of the C-atom of the carbonyl group is indicated, where isomers are possible, by the smallest posible number inserted before the "one" ending:

$$CH_3-CO-CH_3$$
Propanone
(Dimethyl ketone)

$$CH_3-CO-CH_2-CH_3$$
Butanone
(Methyl ethyl ketone)

$$CH_3-CH_2-CH_2-CO-CH_3$$
Pentan-2-one
(Methyl propyl ketone)

$$CH_3-CH_2-\underset{\underset{O}{\|}}{C}-CH_2-CH_3$$
Pentan-3-one
(Diethyl ketone)

In addition, trivial names are still used for many K.

The lower-molecular-mass K. are non-viscous liquids; they become less volatile as the molecular mass increases. The higher K. are crystalline. In contrast to the aldehydes, K. are not reducing agents, and, due to the low reactivity of the carbonyl group, they do not tend to polymerize.

In general, K. display the same addition and condensation reactions as the Aldehydes (see). However, they do not form ketals with alcohols, nor do they usually react with ammonia; an exception is the formation of di- or triacetonamine by acetone and ammonia. Some reactions which are important in synthesis of K. are the Baeyer-Villinger oxidation (see), the Willgerodt-Kindler reaction (see) and the Clemmensen reduction (see). In addition to chemical reactions, spectroscopic methods can be used for characterization of the K. (see Aldehydes, Analytical section).

Most K. found in nature belong to the terpene series, e.g. menthone, pulegone and camphor, the steroids (e.g. testosterone and progesterone) and the alkaloids, e.g. pseudopelletierin.

K. can be synthesized by oxidation or dehydrogenation of secondary alcohols. Chromium(VI) oxide and acetone are used as oxidizing or dehydrogenation reagents in the Oppenauer oxidation (see), or *N*-bromosuccinimide may be used:

$$R_2CH-OH \xrightarrow{Ox.} R_2C=O.$$

Another synthetic possibility is the Grignard reaction (see), in which K. are synthesized from carboxylate esters or nitriles and the Grignard reagent. In analogy to the synthesis of aldehydes, K. are formed by pyrolysis of calcium, barium or thorium salts of carboxylic acids, with the exception of formic acid:

$$(R-CO-O)_2Ca \rightarrow R_2C=O + CaCO_3$$

Industrially, K. are formed from carboxylic acids by decarboxylation and water removal at 300 °C on a manganese(II) oxide contact, or by Friedel-Crafts acylation.

K. are used as solvents, components of artificial resins, and as starting materials for drugs, dyes, synthetic fibers, perfumes, etc.

Ketones

Name	Formula
Acetone	$CH_3-CO-CH_3$
Methyl ethyl ketone	$CH_3-CO-CH_2-CH_3$
Diethyl ketone	$CH_3-CH_2-CO-CH_2-CH_3$
Methyl vinyl ketone	$CH_3-CO-CH=CH_2$
Acetophenone	$C_6H_5-CO-CH_3$
Deoxybenzoin	$C_6H_5-CO-CH_2-C_6H_5$
Benzophenone	$C_6H_5-CO-C_6H_5$
Chalcone	$C_6H_5-CO-CH=CH-C_6H_5$

Ketone-aldehyde resins: synthetic resins formed by condensation of ketones with aldehydes. The most commonly used aldehyde is formaldehyde, but acetaldehyde and furfural are also used. Aside from aliphatic ketones such as acetone, the most commonly used ketones are cyclic compounds such as cyclohexanone, methylcyclohexanone and cyclopentanone. Cyclohexanone reacts with aldehydes, especially formaldehyde, to give resin-like compounds. For resin synthesis, 1 mol cyclohexanone is heated to 180 °C in an autoclave with 1.3 to 1.6 mol formaldehyde in the presence of a basic catalyst, such as methanolic potassium hydroxide. The ketone alcohols which form under these conditions readily condense with water loss. The properties of the resulting resins depend on the starting materials and reaction conditions; they are used for paints, as binders for composite materials, as insulators and for certain special purposes, such as synthesis of substitute amber products and organic glasses. The paint resins are very resistant to light and have no odor. They are also miscible with cellulose nitrate, fatty oils and, in some cases, alkyde resins and other paint components.

Ketone cleavage: a reaction of β-ketoacid esters with dilute acids or bases, forming β-ketoacids, which are converted to ketones by elimination of carbon dioxide:

$$CH_3-CO-CH_2-COOC_2H_5 \xrightarrow[-\ C_2H_5OH]{+H_2O}$$

$$CH_3-CO-CH_2-COOH \rightarrow CH_3-CO-CH_3 + CO_2$$

Ketone hydrates: addition products of ketones and water with the general formula $R\text{-}C(OH)_2\text{-}R$. Most K. are very unstable compounds, with the addition equilibrium largely on the side of the reactants (see Erlenmeyer rule). Like the aldehyde hydrates, K. are stabilized by strong electron-withdrawing substituents, as in ninhydrin, for example.

Ketone resins: synthetic resins made by autocondensation of ketones, usually cyclohexanone, methylcyclohexanone or cyclopentanone. The condensation takes place at 200-220 °C in the presence of alkaline catalysts, such as methanolic potassium hydroxide, or in acidic media, usually under pressure. K. are used in paints in combination with fatty oils, cellulose nitrate, synthetic rubber, etc.

Ketoses: see Monosaccharides.

γ-Ketovaleric acid: same as Levulinic acid (see).

Ketoximes: see Oximes.

Key fragments: see Mass spectroscopy.

kg molarity: see Composition parameters.

Kieselguhr, ***diatomaceous earth, infusorial earth***: natural silica gel. K. is a sedimentary rock which formed in prehistoric inland seas or marshes, mainly from the shells of diatoms, organic components and sand. The product consists of 70 to 90% silicon dioxide, 3 to 12% water and various organic and inorganic substances. It is heated in rotating furnaces to obtain a highly porous K. which is used as an insulating material because of its low heat conductivity. It has a large surface area and is therefore used as an adsorbent, filter aid and carrier material for catalysts. It is also used as a filler in paper and rubber.

Kinases: enzymes which transfer phosphate groups, usually from ATP, to hydroxyl groups of carbohydrates or proteins (serine or tyrosine residues). Some important examples are ***hexokinase***, which phosphorylates glucose, fructose and mannose in the 6 position; ***glucokinase***, which catalyses formation of glucose 6-phosphate in the liver, and ***phosphofructokinase*** (K_m 130,000), which is a key enzyme in glycolysis; it converts fructose 6-phosphate to fructose 1,6-bisphosphate. The activities of many cellular enzymes are regulated by their degree of phosphorylation, so the many specific ***protein kinases*** are major regulators of cell metabolism.

Kinetic control: a term for the experimental observation that the result of a process is controlled by kinetic rather than thermodynamic laws. The reason is inhibition of the progress toward thermodynamic equilibrium; the inhibition may be so great that it actually prevents the system from reaching equilibrium. In synthetic organic chemistry, especially, reaction rates are relatively low, so that if parallel reactions occur, the product is favored which is formed most rapidly (see Selectivity). A typical example is the selective catalysis of hydrogenation of carbon monoxide. Depending on the type of catalyst, the preferred product can be low molecular weight hydrocarbons, methanol, solid paraffins, higher alcohols or acetic acid. The opposite of K. is Thermodynamic control (see).

Kinetic modelling: the quantitative description of the time courses of complex reactions, usually with the aid of a computer. K. is different from the formal kinetics of complex reactions (see Kinetics of reactions). The starting point is construction of a kinetic model which consists of a series of steps (elementary reactions) and the formulation of the rates of formation (balance equations) for each reactant, intermediate and product. The next step varies: 1) If the rate constants of all the reaction steps are known, the concentration-vs-time curves for all components are calculated by numerical integration of the differential equations, and compared with the experiment. The degree to which the calculations agree with the experiment is a criterion for the adequacy of the model, and indicates whether it can be extrapolated to conditions which have not been measured. If this is not the case, the model must be changed. 2) If not all kinetic data are known, a well-documented kinetic model, and thus the system of differential equations, is chosen. Suitable methods of iteration are used to fit the numerically calculated concentration curves to the experimental curves, and in this way the kinetic constants, or some of them, are determined (parameter estimation methods).

Kinetic rate law: see Kinetics of reactions.

Kinetics of reactions, ***chemical kinetics***: The area of physical chemistry concerned with the rates of chemical reactions, which are studied experimentally (kinetic methods), described mathematically (formal kinetics) and explained theoretically (theoretical K.). A distinction is made between elementary and complex reactions (see Reactions, elementary, and Reactions, complex). Together with thermodynamic data and transport parameters (diffusion, heat conduction, hydrodynamics), K. are the basis of calculations for optimizing industrial reactions.

A central concept of K. is ***reaction rate r***, which indicates how rapidly the components A, B, C, D... of a system not in chemical equilibrium approach Equilibrium (see). If the reaction $|\nu_A|A + |\nu_B|B + \ldots \rightarrow \nu_C C + \nu_D D + \ldots$, the reaction rate is defined as:

$$r = \frac{1}{\nu_i}\frac{dc_i}{dt}$$

Here ν_i are stoichiometric factors, which have a negative sign for reactants and a positive sign for products, c_i are the concentrations of the components i, and t is the time. Sometimes the differentials dn_i/dt or dN_i/dt are used as measures of the reaction rate. (n_i is the number of moles, while N_i is the number of particles of component i in the system). In contrast to the defined reaction rate, they are extensive parameters, i.e. they are proportional to the size of the system.

The reaction rate depends on the concentrations of the reactants, catalysts, the solvent, temperature, pressure, the area of the limiting surface in heterogeneous reactions, and so on. The function $r = f(c_i)$ is called the *kinetic rate equation*. Reactions which go to completion are often described well by simple exponential equations of the form

$$r = \frac{1}{\nu_i}\frac{dc_i}{dt} = kc_A^{n_A} c_B^{n_B}$$

The sum of the exponents $\Sigma n_i = n$ is called the ***reaction order***, and the proportionality factor k is the Rate constant (see). From theoretical models, it follows that the reaction orders n_i of elementary reactions are small natural numbers which agree with the stoichiometric factors $|\nu_i|$ of the reaction equation (see below). However, the reaction order of a complex reaction does not necessarily agree with the stoichiometric factors. Non-integral orders can occur, as can rate equations which are not exponential functions and for which therefore no reaction order can be defined.

Integration of the rate equations yields functions $c_i = f(t)$ which explicitly describe the experimentally measurable concentration-vs.-time curve. For example, if the reaction follows a first order rate equation with respect to substance A:

$$r = \frac{1}{\nu_A}\frac{dc_A}{dt} = kc_A,$$

integration leads to $c_A = c_{A0} \exp -|\nu_A|kt$, where c_{A0} is the initial concentration of A at time $t = 0$. On the other hand, for an nth order rate equation

$$r = \frac{1}{\nu_A}\frac{dc_A}{dt} = kc_A^n,$$

integration yields

$$\frac{1}{c_A^{n-1}} = \frac{1}{c_{A0}^{n-1}} + (n-1)\,|\,\nu_A\,|\,kt$$

or

$$c_A = c_{A0}\,[(n-1)c_{A0}^{n-1}\,|\,\nu_A\,|\,kt + 1]^{1/(1-n)}$$

The changes in the other components follow from the stoichiometry (Fig.).

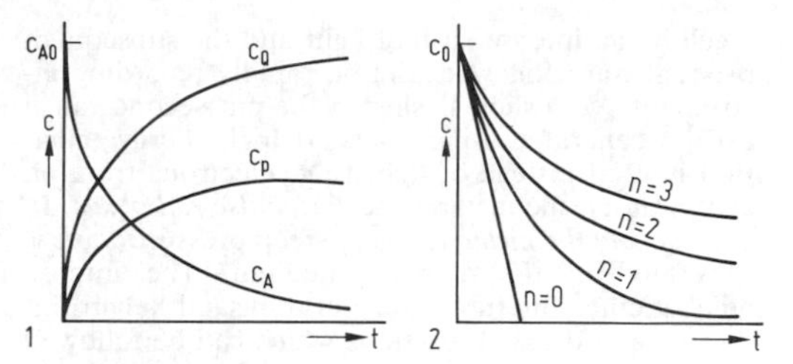

Fig. 1. Course of a second-order reaction 2A → P + 2Q; Fig. 2. Reaction courses for different rate equations under the same initial conditions.

The *half-life* $t_{1/2}$ is the reaction time after which half the initial amount of reactant has been consumed, i.e. when $c_i = 1/2\, c_{i0}$. For a first order rate equation, $t_{1/2} = \ln 2/(|\nu_i|k)$, and for an nth order rate quation, $t_{1/2} = (2^{n-1}-1)/[(n-1)|\nu_A|kc_{A0}^{n-1}]$. A number of simple rate equations and their half-lives are shown in the table.

The rate of a chemical reaction usually increases rapidly with temperature. This is described by the Arrhenius equation (see), which was established empirically by S. Arrhenius for the temperature dependence of rate constants.

Experimental methods. Measurement of reaction kinetics involves determination of the changes in concentration of reactants and products, and also of intermediates if possible. All time-resolved data can thus be evaluated kinetically. For relatively slow reactions (> 10^2 s), *static methods* are suitable; in them the reaction mixture is kept at constant temperature in a reactor and the changes are determined by sampling or analytically within the mixture. For faster processes (10^2 to 10^{-3} s), streams of the components are mixed immediately before they enter the reaction volume, flow through it, and are analysed after they leave the reactor (*flow methods*). Very rapid processes can no longer be studied by mixing the reactants, because the mixing times are at least 10^{-3} s. Instead, a method is used in which the equilibrium is disturbed by a sudden pulsed or periodic change in an external parameter (e.g. pressure or temperature). A rapid analysis method is used to follow the establishment of the new equilibrium in the system. If very small changes are used, the methods are *relaxation methods*. Large perturbations are used, e.g. in flash photolysis, pulse radiolysis and the impact method.

In *flash photolysis*, high concentrations of photochemically excited or cleaved species are generated in

Simple kinetic rate laws and half-lives (for $|\nu_i| = \pm 1$)

Order	Rate equation	Integrated form	Half-life
0	$-dc/dt = k$	$c_0 - c = kt$	$c_0/(2k)$
1	$-dc/dt = kc$	$c = c_0 e^{-kt}$	$\ln 2/k$
2	$-dc/dt = kc^2$	$1/c = 1/c_0 + kt$	$1/(kc_0)$
	$-dc_1/dt = kc_1c_2$	$\frac{c_1c_{02}}{c_2c_{01}} = \exp[(c_{01} - c_{02})kt]$	$\frac{\ln(2 - c_{01}/c_{02})^{a)}}{(c_{02} - c_{01})k}$ $\frac{\ln(2 - c_{02}/c_{01})^{b)}}{(c_{01} - c_{02})k}$
n	$-dc/dt = kc^n$	$\frac{1}{c^{n-1}} = \frac{1}{c_0^{n-1}} + (n-1)kt$	$\frac{2^{n-1} - 1}{(n-1)\,kc_0^{n-1}}$

[a] Half-life for reactant 1

[b] Half-life for reactant 2

a cell by an intense flash of light and the subsequent processes are followed using a rapidly recording instrument. Very short flashes in the picosecond range can be generated using lasers. If high-energy radiation is used instead of light (e.g. electrons from an accelerator), the method is called *pulse radiolysis*. In the *impact tube method*, a very steep pressure wave is generated in the reaction medium. The intense adiabatic heating (up to several thousand kelvin) induces rapid thermal reactions which can be followed by measurement techniques applied perpendicular to the direction of propagation of the pressure wave.

Theory. Theoretical K. has the goals of interpreting the detailed course of elementary reactions on a molecular basis and calculating the rate constants. Models are based on the following ideas: a) A macroscopic elementary chemical reaction is the result of many microscopic physical processes. The particles must collide; both reactive and nonreactive collisions are possible. In nonreactive collisions, energy is exchanged and the particles change direction, but nothing more. If the energy transfer involves only translational degrees of freedom, the collision is elastic, but if internal degrees of freedom are involved, it is inelastic. b) Reactive collisions usually require a *potential barrier* to be crossed. When the particles of the reactants and products are pictured as lying on a potential energy surface, they occupy minima of energy which are separated by an area of higher potential energy. This is due to the generation of repulsive forces when two particles approach each other. The height of the potential barrier is called the *threshhold energy* E_0; it usually lies between 20 and 400 kJ mol˙. There are also reactions without potential threshholds ($E_0 \approx 0$), e.g. the combination of two radicals or oppositely charged ions, or some ion-molecule reactions. If there is a potential barrier, a pair of colliding particles must have an amount of energy (including the energy of the internal degrees of freedom) which is greater than or equal to the threshhold energy in order for the collision to be reactive. In addition, their spatial orientation must be such that bond rearrangement can occur. c) For most reactions, the number of nonreactive collisions is very much larger than the number of reactive collisions. This leads to a random, homogeneous distribution of the particles in the reaction volume, and to an equilibrium distribution of their energies. Concepts of theoretical K. based on these ideas are called *equilibrium theories*. In very rapid processes, e.g. reactions without threshhold energies or in systems with high energies (e.g. reactions in high-temperature and radiation chemistry), these conditions are not always met, so that non-equilibrium methods must be applied to their theoretical treatment.

Starting from these basic ideas, a number of theoretical models have been developed with varying degrees of simplicity and generality.

1) ***Simple collision theory*** is based on the concepts of kinetic gas theory (see Gas theory, kinetic). It assumes rigid, spherical particles with no degrees of internal freedom which collide according to the laws of classical mechanics. The number of collisions between the reacting particles is calculated, and the fraction of them with a relative energy $\geqslant E_0$ is determined by the Boltzmann factor $\exp(-E_0/RT)$. The most important results are: a) for mono-, bi- and trimolecular elementary reactions, the rate equations are first, second and third order, respectively. Monomolecular reactions (see) must be considered separately. b) Collision theory yields the rate constant $k = PBT^{1/2}\exp(-E_0/RT)$, where for a bimolecular reaction A + B → products, the constant B is given by B = $\sigma(8\,RT/\pi\mu)^{1/2}N_A$, where R is the gas constant, T the temperature in kelvin, N_A Avogadro's constant and $\mu = M_AM_B/(M_A+M_B)$ is the reduced molar mass of A and B. $\sigma = (r_A+r_B)^2\pi$ is the *collision cross section*, i.e. the surface area within which two spherical particles A and B with radii r_A and r_B will collide or at least touch in passing. P is called the steric factor; it is a correction factor indicating the fraction of the collisions in which real, non-spherical particles are oriented spatially in such a way that bond rearrangement is not possible. It can therefore have only positive values $\leqq 1$. For many gas reactions, the above equation yields rate constants of the correct order of magnitude if P is set equal to 1. However, there are also often large deviations. This equation is very similar to the Arrhenius equation, where $PBT^{1/2}$ corresponds to A and E_0 to E_A (A is the frequency factor, and E_A is the Arrhenius activation energy).

2) A more general treatment of bimolecular reactions and nonreactive collision processes in dilute gases is possible using the ***Boltzmann collision equation***. After a series of simplifications, it yields the expression $k = \iint \mathbf{v}\sigma_R f_A(\mathbf{v}_A) f_B(\mathbf{v}_B) d\mathbf{v}_A d\mathbf{v}_B$. Here $\mathbf{v} = \mathbf{v}_A - \mathbf{v}_B$ is the difference between the vectorial velocities of the reactants A and B, and $f_A(\mathbf{v}_A)$ and $f_B(\mathbf{v}_B)$ are its distribution functions. σ_R is the reactive *scattering* or *reaction cross section* of the two reactants. Like the collision cross section, it can be understood as an effective area which particle A presents to the colliding particle B. If the particles meet within this area, the collision is reactive. Since the reaction cross section includes the energetic conditions of the collision, especially the existence of a potential barrier, it depends on the energy of the colliding particles and is generally much smaller than the geometrically defined collision cross section. Scattering cross sections can be determined experimentally with crossed Molecular beams (see), or by trajectory calculations (see below). In addition, there are physically based models for estimating reaction cross sections of certain types of reaction. The above equation can also be used to calculate k values for non-equilibrium reactions if non-equilibrium distribution functions are used.

3) ***Trajectory calculations*** give more precise information on the motion of a colliding pair of particles. The trajectory is the change in the distance between the pair under the influence of the interaction potential. In the model of a potential surface, the trajectory corresponds to the path of a point on the potential surface, which can be obtained from quantum chemical calculations. Classical mechanics are assumed to apply to the motion. The exact initial conditions, i.e. the relative energy of the translation, internal excitation states, geometric configuration and direction of motion, of the colliding pair are given, and from these the equations of motion of classical mechanics are solved stepwise. The results indicate the changes of the bonding distances on the potential path, whether

the collision is reactive or not, and the state of excitation of the particles after the interaction. By averaging a large number (usually several hundred to a thousand) of such calculations, one obtains the reaction probability $P = \Delta N_R/N$ (N is the number of pairs calculated and ΔN_R is the number of reactive collisions) and the reaction cross section $\sigma_R = P\sigma = P(r_A+r_B)^2\pi$.

4) At present, the most dependable theory of reactions is the ***Eyring theory (theory of the transition state or activated complex***), which was developed from 1931 to 1935 by Pelzer, Wigner, Evans, Polanyi and Eyring. It can be applied to reactions with a potential barrier if the relative energies are not too high, both in the gas phase and in solution. This theory considers the collision complex at the maximum of the potential barrier as a special configuration and calls this the *transition state* (*activated complex*). This complex has a lifespan of a few oscillation periods (about 10^{-12} s), which is enough for an equilibrium distribution of energy to be established between the internal degrees of freedom. It proceeds on the following assumptions: a) Thermodynamic functions can be defined for the transition state. b) The transition state is in equilibrium with the reactants, i.e. the following reaction scheme applies: $AB + C \rightleftharpoons [A...B...C]^{+} \rightarrow A + BC$, where $[A...B...C]^{+}$ is the transition state. c) The reaction rate is proportional to the concentration of the active complex. The proportionality factor $\nu^{+} = k_BT/h$ follows from the postulate that a vibrational degree of freedom, namely the one along the reaction coordinate, assumes the character of a translation which leads to a cleavage of the old bond and formation of the new one. k_B is the Boltzmann constant, T the absolute temperature, and h is Planck's constant. It follows from c) that $r = k_BT/hc^{+}$, where c^{+} is the concentration of the transition complex; it can be obtained by application of the mass action law to b): $K^{+} = c^{+}/c_{AB}c_C$). From this one obtains $r = (k_BT/h)^{+}c_{AB}c_C$, i.e. a second order rate equation for the bimolecular reaction. Application to mono- and trimolecular reactions $[A] \rightleftharpoons [A]^{+} \rightarrow$ products or $A + B + C \rightleftharpoons [ABC]^{+} \rightarrow$ products leads by analogy to first and third order rate equations, since $c^{+} = K^{+}c_A$ or $c^{+} = K^{+}c_Ac_Bc_C$. Thus Eyring's theory also leads to first, second and third order rate equations for mono-, bi- and trimolecular reactions, just as simple collision theory does.

The factor $(k_BT/h)K^{+} = k$ corresponds to the macroscopic rate constant k. The equilibrium constant K^{+} is accessible via two paths. The equations of phenomenological thermodynamics $\Delta G = -RT \ln K^{+}$ and $\Delta G^{+} = \Delta H^{+} - T\Delta S^{+}$ yield the rate constant $k = (k_BT/h) \exp(\Delta S^{+}/R) \exp(-\Delta H^{+}/RT)$ (***Eyring equation***). Here ΔS^{+} is the *activation entropy* and ΔH is the *activation enthalpy*. In addition, K^{+} can be formulated using the State sums (see) of statistical thermodynamics. For a bimolecular reaction,

$$K^{+} = \frac{Z^{+}}{Z_{AB}Z_C} N_A \nu \exp(-E_0/RT)$$

so that

$$k = \frac{k_BT}{h} = \frac{Z^{+}}{Z_{AB}Z_C} N_A \nu \exp(-E_0/RT)$$

Here Z^{+}, A_{AB} and Z_C are the state sums of the activated complex and the reactants AB and C. N_A is Avogadro's number and ν is the volume of the reaction mixture. E_0 is the molar difference between the zero point energies of the transition state and the reactants, and corresponds to the height of the potential threshhold. If the geometry of the transition state is known (e.g. from quantum chemical calculations), the state sums can be calculated from molecular data of the three components. It must be taken into account that one degree of vibrational freedom along the reaction coordinate must be separated from Z^{+}, in order to establish the decay frequency $\nu^{+} = k_BT/h$.

5) In principle, the same considerations apply to ***reactions in solution***. An important difference is that the particle density is much higher, which has two main consequences: a) the transport of the reacting particles to each other is more difficult. In many reactions, it controls the reaction rate (see Diffusion control). Simple diffusion models yield $k = 4\pi(r_A+r_B)(D_A+D_B)N$ for the rate constants of diffusion-controlled bimolecular reactions; r_A and r_B are the radii, and D_A and D_B are the diffusion coefficients of the reacting particles. b) There are strong interactions between the molecules of the solvent and the reactants, especially solvations, which change the energy relationships. In transition state theory, these can be taken into account by using Activities (see) instead of concentrations in formulating the equilibrium between the activated complex and the reactants. In this case the Eyring equation becomes

$$k = \frac{k_BT}{h} e^{\Delta S^{+}/R} e^{-\Delta H^{+}/RT} \frac{f_Af_B}{f^{+}},$$

where f_A, f_B and f^{+} are the activity coefficients of the reactants A and B and the transition state. If the interactions of the reactants with the solvent are stronger than that of the transition state, $(f_Af_B/f^{+}) < 1$, the reaction is delayed; in the opposite case, it is accelerated (see Ingold's rule).

6) Macroscopic results of theoretical K. are predicted by the nonlinear theory of irreversible processes (see Thermodynamics).

Kinetin: see Cytokinins.

Kininogens: see Kinins.

Kinins: a group of peptides which reduce blood pressure. K. also increase capillary permeability and, in vitro, cause contraction of the smooth muscles of the intestine, uterus and bronchi. The K. are formed from larger precursors, the ***kininogens***. The most important of the K. are the plasma kinins. K. are also found in lower animals; for example, the skins of various amphibia contain bradykinin, [Thr[6]]bradykinin and C-terminal shortened or lengthened derivatives of amphibian bradykinin. Some examples are the bradykinyl pentapeptide from *Rana nigromaculata* (-Val-Ala-Pro-Ala-Ser), the bradykinyl tetrapeptide from *Bombina orientalis* (-Gly-Lys-Phe-His) and ***Phyllokinin*** from the frog *Phyllomedusa rohdei*:

Arg-Pro-Pro-Gly-Phe-Ser-Pro-Phe-Arg-Ile-Tyr-SO_3H

Polisteskinin, a component of wasp venom, contains an *N*-terminal lengthened bradykinyl sequence.

Kirchhoff's law: a relationship established in 1858 by R. Kirchhoff which describes the temperature dependence of reaction energies and enthalpies:

$(\delta\Delta_R H/\delta T)_p = \Delta_R C_p$ or $(\delta\Delta_R E/\delta T)_v = \Delta_R C_v$. $\Delta_R H$ and $\Delta_R E$ are the molar reaction enthalpy and energy, respectively, $\Delta_R C_p$ and $\Delta_R C_v$ are the change in heat capacities of the reaction mixture at constant pressure or volume, when one formula mole turnover occurs.

For ideal mixtures, the additivity relationship (see Thermodynamics, first law) applies to calculation of $\Delta_R C_p$ and $\Delta_R C_v$.

With K., it is possible to calculate the reaction enthalpy at one temperature if it is known at another:

$$\Delta_R H_{T2} = \Delta_R H_{T1} + \int_{T_1}^{T_2} \Delta_R C_p \mathrm{d}T.$$

To a first approximation, $\Delta_R C_p$ can be set equal to a constant for the temperature range under consideration: $\Delta_R H_{T2} = \Delta_R H_{T1} + \Delta_R C_p(T_2 - T_1)$.

If more precision is required, $\Delta_R C_p$ must be calculated from the temperature functions of the molar heats of the substances involved, using the additivity relationship.

Kjeldahl nitrogen determination: a method introduced in 1883 by J. Kjeldahl for quantitative determination of the nitrogen content of organic substances. The sample is heated with conc. sulfuric acid and catalysts, such as mercury and copper salts or selenium, in a long-stemmed glass flask. The organic substances are decomposed, and the nitrogen from amines, amides and heterocyclic compounds is converted to ammonium ions. However, if the nitrogen is present in the form of azo-, nitro- or nitroso-groups, these must first be reduced, because otherwise the nitrogen content determined by the K. will be too low. After the organic substance has been decomposed and diluted with water, an excess of sodium hydroxide is added, and the ammonia released by it is distilled into a flask containing a known amount of standard sulfuric acid solution. The excess acid remaining after the reaction with the ammonia can then be determined volumetrically, and from this, the nitrogen content can be calculated. The K. is still very important, because it is a simple method of determining protein content of feeds.

Kjeldahl process: see Proteins.

Knapsack process: see Acetic anhydride.

Knocking: a sound made by internal combustion engines (gasoline and diesel) as a result of abnormal combustion; it is accompanied by a reduction in performance, increased mechanical stress on the engine and overheating of pistons and cylinders. The most common form of K. in gasoline engines is ***ignition K.*** (***acceleration K.***) caused by pre-reaction due to higher temperature and pressure in the fuel-air mixture. ***Pinging*** is caused by surface ignition, which is usually due to deposits in the combustion chamber. A special form of surface ignition is the continued running of the engine (***dieseling***) after the ignition has been turned off. When the compression ratio is greater than 9:1, ***residue ignition*** may occur. This is also due to deposits in the combustion chamber, and occurs most readily at high rpm. ***High-speed K.*** occurs when the engine is run for a long period at high rpm with a low octane fuel. The results are usually much worse than ignition K., because at high speed the sound of the air blowing past the car may drown out the noise in the engine, and damage to the engine is often the first sign of low fuel quality.

Theoretical. During combustion of the fuel-air mixture in the engine, a flame front moves out from the ignition spark at a rate of 15 to 30 m/s. This raises the pressure in the combustion chamber and causes the piston to do work. K. occurs when reactions occur in the gas ahead of the flame front. These reduce the self-ignition temperature of the gas mixture, so that it detonates. The result is an intense local elevation in the pressure and a significantly lower pressure in the normally burning gas-air mixture. There are thus two regions of different pressure in the cylinder, and their equalization causes the knocking sound. A more recent hypothesis is that during combustion, a counter current causes microwave turbulences in the flame front. These accelerate the motion of the front to 50 to 100 m/s. As a result, the mass flux on the hot side of the front exceeds the speed of sound; thus according to this hypothesis, the K. is a type of sonic boom. Additives and suitable design of the combustion chamber are used to prevent K. The resistance of a fuel to K. is characterized by its octane rating.

Knoevenagel condensation: a condensation of aldehydes and ketones with compounds which have high CH acidity, e.g. malonic acid, malonate esters, cyanoacetate esters, malonic acid dinitrile and β-diketones, in the presence of basic catalysts, e.g. piperidine, pyridine or ammonium acetate: R-CHO + $CH_2(CN)_2 \rightarrow$ R-CH-C$(CN)_2$ + H_2O.

In the condensation with malonic acid in pyridine with piperidine as catalyst, the primary condensation step is usually followed by decarboxylation to form the corresponding α,β-unsaturated carboxylic acid. K. can be considered a special variant of the aldol reaction. It is very important for the synthesis of numerous organic compounds.

Knoop hardness: see Micro hardness.

Knoop-Oesterlin reaction: a method for producing α-amino acids by catalytic hydrogenation of α-ketoacids in the presence of ammonia:

$$\mathrm{R{-}CO{-}COO^- + NH_3 \rightarrow R{-}C({=}NH){-}COO^-}$$
$$\xrightarrow{H_2/Pd} \mathrm{R{-}CH(NH_2){-}COOH}.$$

Knopf cell: an Electrochemical current source (see) and a secondary cell. The K. is a modified Nickel-cadmium battery (see). It is gas-tight, can be operated in any position, and requires no maintenance. In order to produce a gas-tight cell, gas development during the discharge must be avoided. This is achieved by using a cadmium electrode which has a higher capacitance than the nickel electrode. In this way only oxygen is formed at the anode, without cathodic hydrogen generation. The cell is designed so that the oxygen can reach the cathode.

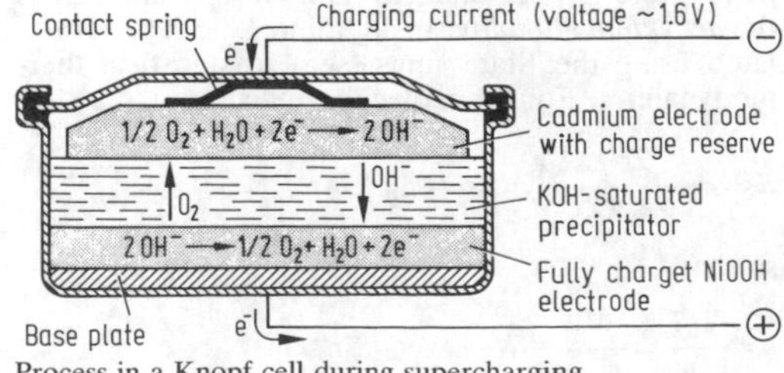

Process in a Knopf cell during supercharging.

Now oxygen reduction takes place at the cathode, and the gas pressure cannot increase.

Knorr pyrrole synthesis: a method for making substituted pyrroles from ketones with α-methylene groups or, more effectively, from β-ketoacid esters, and isonitroso compounds under reducing conditions. The isonitrosocarbonyl compounds are first converted to the corresponding amino compounds, which then react with the ketone or β-ketoacid ester to form the cyclic pyrrole derivative:

$$R-CO-C(R)=N-OH \xrightarrow[-H_2O]{+4H} R-CO-CH(R)-NH_2$$

$$R-CO-CH(R)-NH_2 + H_2C(COOR)-CO-R \xrightarrow{-2H_2O} \text{pyrrole (R, COOR, R, R; N–H)}$$

Since α-aminoketones are susceptible to autocondensation, forming pyrazines, byproducts must be expected from the K.

Knudsen diffusion, ***Knudsen flow***: see Diffusion.

Koch reaction: method for stepwise formation of monocarboxylic acids by carbonylation of alkenes followed by decomposition of the intermediate product with water:

$$CH_3-CH=CH_2 \xrightarrow{+H^+} CH_3-C^+H-CH_3 \xrightarrow{+CO}$$

$$(CH_3)_2CHC^+=O \xrightarrow[-H^+]{+H_2O}$$

$$(CH_3)_2CH-COOH.$$

In contrast to carbonylation of alkenes in the presence of water, which requires extreme reaction conditions, the K. can be carried out at 0 to 50 °C and pressures of 5 to 10 MPa.

Kohlrauch's square root law: see Electrical conductivity.

Kölbel-Engelhardt process: see Fischer-Tropsch synthesis.

Kolbe nitrile synthesis: process for production of nitriles by reaction of alkyl halides with alkali cyanides: R-Br + KCN → R-CN + KBr.

Small amounts of isonitrile formed as a byproduct of K. can be removed by shaking with dilute hydrochloric acid.

Kolbe-Schmitt synthesis: a method for producing salicylic acid from sodium phenolate and carbon dioxide at 125 °C and $5 \cdot 10^5$ to $6 \cdot 10^5$ Pa.

$$C_6H_5ONa \xrightarrow{+CO_2} o\text{-}HO-C_6H_4-COONa \xrightarrow[-Na^+]{+H^+} o\text{-}HO-C_6H_4-COOH$$

The reaction can be interpreted as an electrophilic aromatic substitution. The primary attack of the carbon monoxide on the sodium phenolate (chelate mechanism) is formulated as follows:

[Scheme: sodium phenolate chelated with CO_2 (O–Na···O=C=O) → cyclohexadienone intermediate with Na chelate, C=O and H]

Through the action of other sodium phenolates, the proton is then split off to form the disodium salt of salicylic acid. With potassium phenolate, which forms chelates less readily, the main product is 4-hydroxybenzoic acid; this supports the idea that the K. occurs by a chelate mechanism.

Kolbe synthesis: a reaction for producing alkanes by electrolysis of carboxylic acid salts in concentrated aqueous solutions. The reaction mechanism of the K. can be interpreted as the formation of an unstable radical through anodic oxidation of one electron from the carboxyl group. This carboxyl radical loses CO_2 and is thus converted to a short-lived alkyl radical, which very rapidly dimerizes to form the corresponding alkane:

$$R-CO-O^- \xrightarrow[-e^-]{} R-CO-O^\cdot \xrightarrow[-CO_2]{} R^\cdot\ ;\ 2R^\cdot \rightarrow R-R$$

At the cathode, the metal cation is discharged, and is immediately converted to hydrogen and metal hydroxide. Alkanes with medium to long chains can be obtained in preparative quantity by means of the K. Alkanes with odd numbers of carbon atoms can be made by using mixtures of different carboxylic acid salts.

Konovalov's rules: regularities in boiling and vapor-pressure diagrams for binary mixtures. 1) The vapor pressure and the boiling temperature of a binary mixture reach their maximum and minimum when the compositions of the liquid and vapor are the same, that is, when there is an azeotropic point. 2) If the fraction of the more volatile component of a binary mixture is increased, the vapor pressure of the mixture rises and its boiling point drops. 3) The composition of coexisting phases always changes in the same direction.

Konstantan®: a resistance alloy of 45% nickel, 1% manganese and the rest copper; it is used for thermoelements and precision electrical resistances.

Koppers process: see Sewage treatment.

Koppers-Totzek process: see Synthesis gas.

Kornberg enzyme: a DNA polymeraze isolated from *Escherichia coli* which, in cell-free systems, catalyses the synthesis of DNA from nucleotide triphosphates in the presence of a template DNA. The first synthesis in vitro of a natural deoxyribonucleic acid was achieved in 1967 using the K.

Kornblum rule: a rule applicable to nucleophilic substitution reactions between ambident reactants. In a reaction occurring by an S_N1 mechanism, the attack is made on the reactant with the greatest electron density, while in an S_N2 reaction, the reactant with the greatest nucleophilicity is attacked. This rule was proposed by N. Kornblum on the basis of his studies on the reactions of alkyl halides with alkyl nitrites:

S_N1– Reaction: $R-X + |\bar{O}-\bar{N}=\bar{O} \longrightarrow R-O-N=O + X^-$

S_N2– Reaction: $R-X + [\bar{O}\cdots\bar{N}\cdots\bar{O}]^- \longrightarrow R-NO_2 + X^-$

The validity of the K. was later confirmed in reactions with other ambident nucleophiles, e.g. cyanide, rhodanide and 1,3-dicarbonyl anions.

Kováats index: same as Retention index; see Retention.

Kovar: a proprietary name for an iron-nickel-cobalt alloy consisting of 20% nickel, 17% cobalt and the rest iron. Its thermal expansion coefficient is α = 4.3 to $5.3 \cdot 10^{-6}$ K^{-1}, which is close to that of hard glass, so the alloy is used for melt-in electrical connections in apparatus.

Kr: symbol for krypton.

Krafft point: a triple point at which the solid form (in some cases hydrated) of an ionogenic surfactant is in equilibrium with dissolved (hydrated) individual molecules and micelles.

Krebs cycle: same as Citric acid cycle (see).

Kröhnke reaction: a method for synthesis of aldehydes based on the fact that organic halogen compounds with a certain structure react with pyridine to form pyridinium salts, and these react with 4-nitrosodimethylaniline to form nitrones. These are then hydrolysed in acid to the aldehyde and an hydroxylamine derivative. The method is especially suitable for production of aldehydes which are sensitive to oxidation, starting from compounds with the general structure R-CH_2-X (R = aryl or hetaryl; X = Cl, Br, I), R-CO-CH_2-X and R-CH=CH-CH_2-X (R = alkyl or aryl). Saturated aliphatic aldehydes cannot be made via the nitrones.

R–CH_2–X + pyridine → R–CH_2–N^+(pyridinium) Cl^-

+ ON–C_6H_4–$N(CH_3)_2$ / NaOH, – pyridine / NaCl / H_2O → R–CH=N(→O)–C_6H_4–$N(CH_3)_2$ (Nitrone)

+ H_2O/H^+ → R–CHO (Aldehyde) + $(CH_3)_2N$–C_6H_4–NHOH

Krypton, symbol ***Kr***: a Noble gas (see), a member of group 0 or VIII of the periodic system; *Z* 36, with natural isotopes with mass numbers 78 (0.35%), 80 (2.27%), 82 (11.56%), 83 (11.55%), 84 (56.90%) and 86 (17.37%). Atomic mass 83.80, valency 0 (II with respect to fluorine), density 3.736 g/l at 0°C, m.p. -156.6°C, b.p. - 152.3°C, crit. temp. - 63.8°C, crit. pressure 5.43 MPa.

Properties. K. is a colorless, odorless and tasteless, monoatomic, nontoxic gas. It is extremely unreactive. K. forms a crystalline hydrate and a hydroquinone clathrate. There are a few true valence compounds: a relatively unstable krypton(II) fluoride, KrF_2, and a few complex compounds derived from it (see Krypton compounds).

Analytical. K. is detected by spectral analysis after gas chromatographic separation.

Occurrence. K. makes up 0.0001% of the atmosphere, and $1.9 \cdot 10^{-8}$% of the earth's crust. Volcanic vapors contain some K.

Production. The starting material for industrial production of K. is air, from which the relatively high-boiling elements K. and xenon are washed out with liquid air or nitrogen. The resulting mixture of liquid products is separated by distillation and, if necessary, by fractional adsorption and desorption from activated charcoal.

Applications. The most important application for K. is in light bulbs. Because its thermal conductivity is lower than that of nitrogen, K. permits a higher internal temperature or a smaller glass bulb. K. is also used as a filler for Geiger counter tubes. It is used in medicine for x-ray diagnostics.

Historical. K. (from the Greek "krypton" for "hidden") was discovered in 1898 by Ramsay and Travers in the course of fractional distillation of crude argon.

Krypton compounds. ***Krypton(II) fluoride***, KrF_2, is a colorless, tetragonal solid which is obtained by high-energy irradiation of a Kr-F_2 mixture. The KrF_2 molecule is linear. It rapidly decomposes into its components at room temperature. KrF_2 is a strongly oxidizing fluorination reagent, and is capable, for example, of converting chlorine to chlorine trifluoride, ClF_3, or chlorine pentafluoride, ClF_5. It forms *complex salts* with a few Lewis acids, e.g. $KrF^+[SbF_6]^-$, $KrF^+[Sb_2F_{11}]^-$ or $KrF^+[AuF_6]^-$, which are often more stable than KrF_2.

Ku: symbol for Kurčatovium, the name proposed for element 104 by its discoverers in Russia.

Kuhn-Roth method: a method of quantitative analysis of C-bound methyl groups by oxidative degradation to acetic acid. A mixture of chromic and sulfuric acids is used as the oxidizing agent, and excess chromic acid is reduced after the reaction. The acetic acid is separated and determined by titration. K. can be used to determine acetyl and ethoxy groups. It is not suitable for compounds in which the methyl groups are bound to aromatic systems, or for higher fatty acids.

Kuprodur®: copper-nickel-silicon alloy containing 1% silicon, 2% nickel and the rest copper. K. is used as a conductor because of its strength and good electrical conductivity.

Kurcatovium, symbol ***Ku***: a chemical element which can only be produced artificially; it is a member of group IVb of the periodic system, the Titanium group (see), *Z* 104. K. was the first transactinide element to be synthesized; it was obtained in 1964 in the Dubna (Russia) nuclear research center by irradiation of plutonium with neon nuclei accelerated to 115 MeV: $^{242}Pu + ^{22}Ne \rightarrow ^{260}Ku + 4\,^{1}n$. Kurcatovium 260 has a half-life of 0.1 s. Other isotopes with mass numbers 257 (α-emitter, 4.5 s half-life), 259 (α-emitter, half-life 3 s), 261 (α-emitter, half-life 65 s) were obtained in 1968/1969 in Berkeley (USA); the Berkeley research group suggested the name ***Rutherfordium*** and symbol Rf for the new element.

Kureha process: see Pyrolysis.

Kurrol's salt: see Sodium phosphates.

Kuss process: see Zinc.

Kynurenine, ***3-anthraniloylalanine***: in vivo, a degradation product of tryptophan. In the human body, it is an intermediate in the biosynthesis nicotinic acid. In plants, quinoline derivatives are synthesized as ***hydroxykynurenine***. Fig. p. 577.

Kyotorphin: a neuroactive peptide isolated from

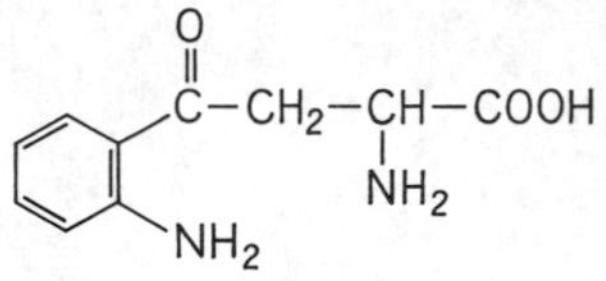

Kynurenine

the rat hypothalamus: L-tyrosyl-L-arginine. K. was named for the site of its discovery, Kyoto, Japan. K. has an analgesic effect, but it does not interact with the opiate receptors. Instead, it is thought to act by releasing an enkephalin and stabilizing it from degradation.

L

L: configuration symbol; see Stereoisomerism, 1.1.

La: symbol for lanthanum.

Laccases: copper-containing oxidases which convert 2- and 4-diphenols to phenoxy radicals by a radical mechanism. The phenoxy radicals then react further nonenzymatically. L. are essential for the biosynthesis of lignin. The L. from the Japanese lacquer tree *Rhus vernicifera* contain 2 Cu^+ and 2 Cu^{2+} ions, which give the enzyme its blue color.

Lactalbumin: see Albumins.

Lactam: a cyclic carboxamide containing the atomic grouping -CO-NH. It is formed by intramolecular condensation of a γ-, δ-, ε-, etc. aminocarboxylic acid with loss of water, or by a Beckmann rearrangement of a ketoxime. β-L. (azetidin-2-ones) are also formed by cycloaddition of ketones and azomethines. Special methods have been developed for large-scale synthesis of the most important of these compounds (see ε-Caprolactam). In solution, L. are in tautomeric equilibrium with ***lactims*** (see Lactam-lactim tautomerism). In the solid state, the lactam forms predominate (see Barbituric acid, see Uric acid). The well known simple L. γ-butyrolactam and δ-valerolactam are called pyrrolidin-2-one and piperidin-2-one, respectively, in IUPAC nomenclature.

β-Lactam antibiotics: a group of antibiotics which contain a β-lactam ring (2-azetidinone). Monocyclic β-L., i.e. compounds containing only the β-lactam ring are called ***monobactams*** (from *mono*cyclic *bac*terially produced β-lac*tam* antibiotics). The monobactams include nocardicin A and aztreonom. Most L. have a second ring condensed to the β-lactam ring.

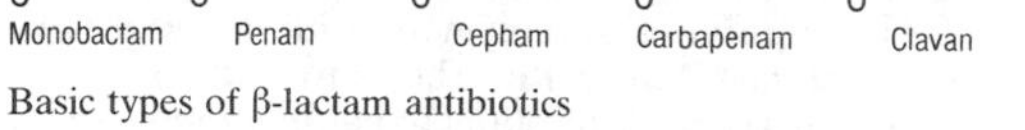

Monobactam Penam Cepham Carbapenam Clavan

Basic types of β-lactam antibiotics

The most important groups of L. are the Penicillins (see), which are representatives of the penam type, and Cephalosporins (see), which represent the cepham type. An example of the carbapenam type is *thienamycin*, which has a relatively broad action spectrum. *Clavulanic acid* is a clavan derivative. The L. are synthesized mainly by molds (*Penicillium, Aspergillus, Cephalosporium*), but also by bacteria (*Streptomyces clavuligerus*, clavulanic acid; *Streptomyces cattlaya*, thienamycin; *Nocardia uniformis*, nocardicin A). The biosynthesis of L. starts from L-amino acids. During the biosynthesis, the configuration is inverted, so that the centers of chirality of the incorporated amino acids have the D-configuration. The biological activity of these compounds is due to inhibition of biosynthesis of bacterial cell walls. The β-lactam ring is very easily cleaved by nucleophilic reagents, and enzymatic hydrolysis by β-lactamases (β-lactamase, penicillinase) gives the bacteria resistance to the antibiotic. Clavulanic acid and the carbapenam derivative olivanic acid are inhibiters of β-lactamase.

Lactam-lactim tautomerism: a special case of amide-iminol tautomerism in lactams. The equilibrium is usually on the side of the lactam form (Fig.). L. is observed in some heterocycles, especially in hydroxy-substituted nitrogen heterocycles (barbituric acid, uric acid, etc.).

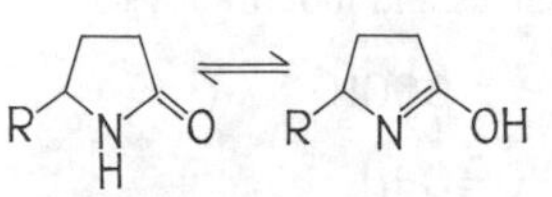

Lactam form Lactim form

Lactase: see Disaccharidases.

Lactate: a salt or ester of lactic acid. ***Antimony L.*** can be used as an emetic or as a mordant in dyeing of fabric; ***silver L.*** is used as an antiseptic in bandages; ***iron L.*** is used as a nutritional supplement to supply iron. ***Sodium***, ***potassium*** and ***calcium*** L. are used in small amounts in the preparation of baked goods and jellied meats. ***Ethyl L.*** is used as a special-purpose solvent.

Lactate dehydrogenase, abb. ***LDH***: an enzyme which catalyses the hydrogenation of pyruvic to lactic acid; one of the oxidoreductases. LDH is found in all animal cells and microorganisms. The heart and liver contain high concentrations of LDH, which consists of 4 subunits (M_r 140,000). There are two types of subunit, H and M (for "heart" and "muscle") which can combine with each other in any proportions, so that there are five possible isoenzymes: H_4, H_3M, H_2M_2, HM_3 and M_4. These can be separated because their isoelectric points are different. Elevated LDH levels in serum are indicators of heart infarct or hepatitis.

Lactic acid, ***α-hydroxypropionic acid, 2-hydroxypropanoic acid***: CH_3-CH(OH)-COOH, an α-hydroxycarboxylic acid. L. contains an asymmetric C atom and is thus chiral. It exists in an (RS) form (formerly DL form), and the two optically active R and S forms. The various forms of L. occur as metabolic products in plants and animals. L. is readily soluble in water and alcohol, and slightly soluble in chloroform and benzene. ***(RS)-L.*** is a colorless to yellowish, hygroscopic, syrupy liquid; m.p. 18 °C, b.p. 122 °C at 2 kPa. ***(S)(+)-L.*** and ***(R)(-)-L.*** form colorless crystals; m.p. 26 °C (53 °C also reported). The salts and esters of L. are called *lactates*.

The reaction of Fenton's reagent with L. yields pyruvic acid. Stronger oxidizing agents convert L. to oxalic acid and carbon dioxide, or to acetic acid and carbon dioxide, depending on the conditions. The reaction of L. with conc. sulfuric acid involves dehydration and leads to acetaldehyde and carbon monoxide. When L. stands for a long period or is heated, it loses water to form lactoyllactic acid, which is converted by further dehydration to polylactic acid. The presence of L. in wine, sauerkraut, sour milk, pickles, cheese, silage, etc. is due to fermentation of sugars by lactic acid bacteria. (+)-M. is formed as a product of glycolysis in muscle tissue during anaerobic conditions. Increased muscle activity leads to an elevation of the L. concentration in the blood. L. is produced mainly by ***lactic acid fermentation*** of lactose, maltose or glucose: $C_{12}H_{22}O_{11} + H_2O \rightarrow 4\ CH_3\text{-}CH(OH)\text{-}COOH$. In addition to this classic process, L. is also synthesized chemically, e.g. by hydrolysis of lactic nitrile. L. and some of its derivatives, especially the lactates, are versatile substances used as mordant and decalcifying reagents for wool and leather dyeing, as an additive to textile conditioners and galvanic baths, in non-alcoholic beverages, as preservatives, disinfectants and additives to salves and mouthwashes.

(R)-(-)-Lactic acid (S)-(+)-Lactic acid

Lactides, ***3,6-dialkyl-1,4-dioxane-2,5-dione***: heterocyclic compounds formed by condensation of two molecules of an α-hydroxycarboxylic acid, with intermolecular water loss. For example, the 3,6-dimethyl derivative (Fig.) is formed by heating lactic acid for a long period.

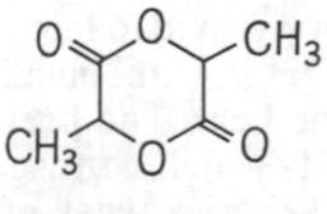

Lactim: see Lactam.

Lactoflavin: same as Vitamin B_2 (see).

Lactogenic hormone: see Prolactin.

Lactone: a cyclic carboxylic acid ester (internal ester) containing the atomic group -CO-O-. The size of the ring is indicated by the Greek letter prefix. α-L. (oxiran-2-ones) are stable only at low temperatures. Two well known β-L. are propiolactone (oxetan-2-one), which can be made from ketene and formaldehyde, and diketene. γ-L. have five-membered rings and occur widely; they are formed from γ-hydroxycarboxylic acids. δ-L., formed from δ-hydroxycarboxylic acids, have six-membered rings. ε-Hydroxycarboxylic acids generally do not form L.; instead they form linear polyesters. The systematic names for L. are obtained from the systematic names for the alkanoic acids by replacing the ending -ic acid with -olide. The numbers of the C atoms joined by the O atom are indicated as a prefix. Trivial names are also still used, e.g. β-propiolactone (1,3-propanolide), γ-butyrolactone (1,4-butanolide) or δ-valerolactone (1,5-pentanolide). The L. are liquids or solids with low melting points, and most of them are soluble in water, ethanol and ether. Their properties are similar to those of ordinary esters. Treatment of the L. with aqueous bases opens the ring, forming the salt of the parent hydroxycarboxylic acid. In acidic solutions, there is often an equilibrium between the L. and the hydroxycarboxylic acid. Lactams (see) are formed by treatment of L. with ammonia. L. are found as structural elements in natural products, for example, in aroma substances such as coumarin in woodruff, exaltolide (1,15-pentadecanolide) in angelica root or ambrettolide (1,16-hexadec-7-enolide) in musk. Natural L. are also used as drugs; the digitalis glycosides are an important example. Another important L. is mevalonic lactone, which is a central intermediate in the biosynthesis of isoprenoid natural products.

Lactose, ***milk sugar***: 4-(β-D-galactopyranosyl)-D-glucopyranose, a disaccharide. Commercial α-L. is a white, crystalline powder with a somewhat sweet taste. It is soluble in water and nearly insoluble in ethanol; $[\alpha]_D^{20}$ in mutarotational equilibrium +52-53°. L. is present at 4 to 5% concentration in cow's milk and 5.5 to 7.5% in human milk. It is obtained from the whey, after the proteins have been removed by heat denaturation and flocculation. L. is not fermented by baker's yeast. In the small intestine it is split enzymatically into glucose and galactose and resorbed, in infants and those adults who have not ceased production of β-galactosidase. The presence of this enzyme in the guts of adult mammals is rare. Among human beings, it is typical only in Europeans and some Africans who have traditionally kept cattle; adults of other ethnic groups are generally intolerant of L. L. is used in the pharmaceutical industry as a filler and binder in tablets and capsules, and in the food industry in the production of infant and diet foods.

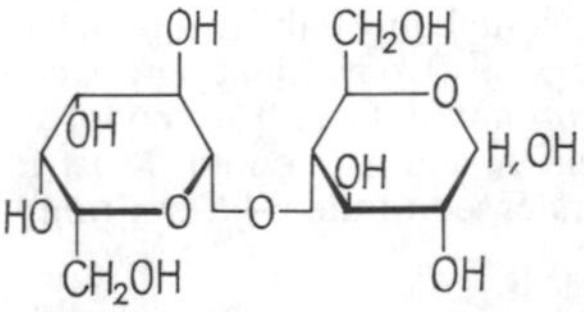

Lactotropin: same as Prolactin (see).

Ladder polymers: polymers with a ladder-like double-stranded structure. They are synthesized by polycondensation of tetrafunctional monomers or by dehydrogenating cyclization of polymers such as polyacrylonitrile. The pyrrones, made by polycondensation of aromatic tetracarboxylic anhydrides with aromatic tetraamines, are especially interesting. These include polyimidazonebenzophenanthroline and polytetraazapyrene.

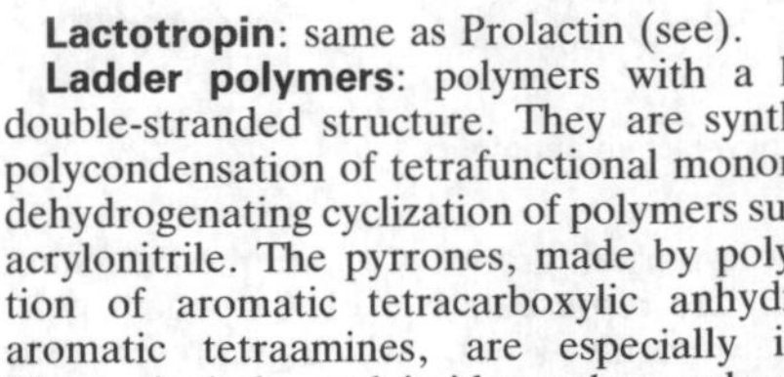

Polyimidazobenzophenanthroline

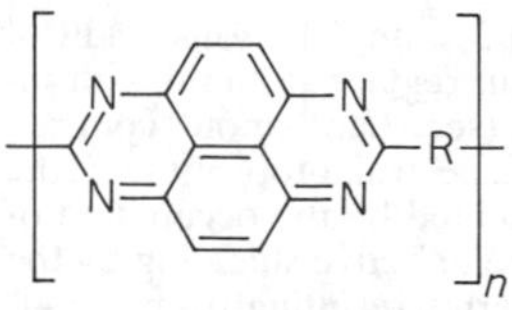

Polytetraazapyrene

Because of their double-stranded structure, L. are very stable at temperatures up to 500 or 600 °C. Because they are resistant to radiation damage, they are used in nuclear technology and space vehicles. Some L. are semiconductors.

Ladenburg benzene: same as Prismane (see).

Lambda convention: see Nomenclature, III F 4.

Lambert-Beer law, ***Bouguer-Lambert-Beer law***: The basic law of absorption spectroscopy obtained by combining the Bouguer-Lambert law (see Absorption) and Beer's law (see): $E = \log 1/T = \log I_0/I_T = \varepsilon c d$, where E is the extinction at a wavelength λ, d is the thickness of the sample, c is the molar concentration of the sample, and ε is a proportionality constant known as the molar absorption or molar extinction coefficient. In SI units, ε has the dimensions $m^2 mol^{-1}$, where c is in mol dm^{-3} and d is in mm. In the literature, the dimensions liter mol^{-1} cm^{-1} are used; they result from measuring the concentration in mol/liter and the thickness of the layer in cm. The ε values obtained using SI units are a factor of 10 smaller than the values obtained in traditional units. According to the L., the molar absorption coefficient ε at constant external conditions and at a given wavelength is independent of the concentration. It is a parameter which is used to characterize absorption. If the relative molecular mass M_r of the sample is not known, the expression $E_{1\%}^{10mm}$ is often used; it refers to a layer 10 mm thick and a 1% solution. The equation

$$E_{1\%}^{10\,mm} = \frac{10\,\epsilon}{M_r}$$

relates this expression to the molar absorption coefficient.

The L. is precisely true only for monochromatic light and ideal dilute solutions. In non-ideal solutions, ε is not independent of the concentration, because the state of the absorbing substance is affected by intermolecular interactions such as association, dissociation, complex formation, etc. If L. applies, the extinction E is directly proportional to the concentration c.

The L. is applied using the equations $\varepsilon = E/cd$ (1) and $c = E/\varepsilon \cdot d$ (2). To evaluate eq. (1), the extinctions E at various wavelengths λ are measured for a known concentration c and layer thickness d, and the calculated values of ε are plotted against the wavelength λ. This produces the absorption spectrum of the substance, which can be used in characterizing or identifying it. Since the ε values at different wavelengths often differ by orders of magnitude, $\log \varepsilon$ is often plotted instead of ε.

Eq. (2) permits the concentration c of a substance to be calculated from the measured extinction, if ε is known. Mixtures of several absorbing substances can also be analysed quantitatively, if the individual components do not affect each other. For a mixture of n components, one obtains an equation with n unknowns (c_1, c_2, c_3...): $E = (\varepsilon_1 c_1 + \varepsilon_2 c_2 + \varepsilon_3 c_3)d$. These can be solved if the extinction is measured at n different wavelengths, so that one has n different equations with n unknowns. The results of such multicomponent analyses are more accurate if the molar absorption coefficients of the individual compounds at the chosen wavelengths are distinctly different from each other.

Laminar cells: see Colloids.

Laminates: materials consisting of layers of cloth, paper, cardboard, wood, fiberglass or asbestos impregnated with condensed resins (laminating resins). They are produced by simultaneously heating and pressing the layers together. The laminating material is usually a phenol, melamine, polyester or epoxide resin, or an aminoplastic. In some cases, solutions of silicone resins are used instead of these plastic resins.

Lamoxactam: see Cephalosporins.

Lanata glycosides: cardiac glycosides from the foxglove *Digitalis lanata*. Because its dried leaves contain a relatively large amount of the glycosides, 0.4 to 1.0%, the plant is used as a source. To date, more than 60 L. have been isolated from it. The most important of them are shown in the table. In contrast to the Purpurea glycosides (see), the primary L. (***lanatosides***) have an acetyl group on the terminal digitoxose group and crystallize more readily. The terminal glucose is subject to enzymatic removal to form the secondary glycosides. Usually the glycoside mixture is worked up by additional mild alkaline saponification (to remove the acetyl group) to give the tridigitoxosides. Of these, ***digoxin*** is therapeutically most significant. Digoxin is about 65% absorbed after oral administration; the elimination rate is about 20% per day. ***Lanatoside C*** is also used; however, it is less readily absorbed than digoxin. The analysis and effects are discussed under Purpurea glycosides (see).

Acetyl
Digitoxose – Digitoxose – Digitoxose – Glucose

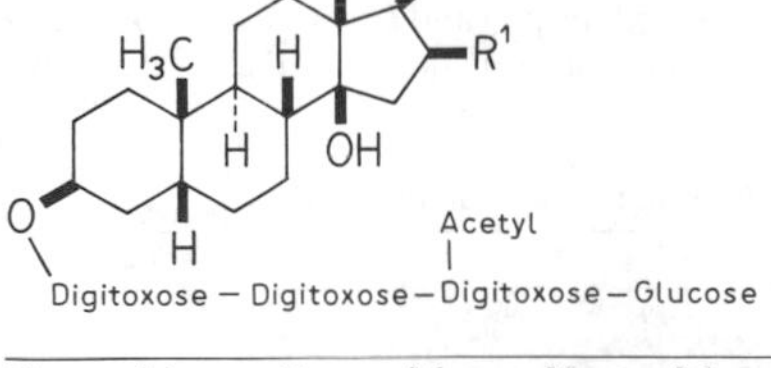

Name of the glycoside	Name of the aglycon	Name of the tridigitoxoside	R^1	R^2
Lanatoside A	Digitoxigenin	Digitoxin	H	H
Lanatoside B	Gitoxigenin	Gitoxin	OH	H
Lanatoside C	Digoxigenin	Digoxin	H	OH

Lanatosides: see Lanata glycosides.

Landanosine: see Alkaloids.

Langmuir adsorption isotherm: see adsorption.

Langmuir-Hinschelwood mechanism: see Catalysis.

Langmuir torch: see Hydrogen.

Lanolin: a yellow, salve-like mass which does not

become rancid and is very difficult to saponify; m.p. 36-41 °C. L. is a complicated mixture of various fatty acids, cholesterol, lanolin alcohol and other hydroxy compounds. It is obtained from sheep's wool, which contains up to about 10% L. L. is used mainly as a base for salves and other pharmaceutical products, an additive to soaps, a treatment for textiles and leather products and a rust preventative.

Lanosterol: 8,24-lanostadien-3β-ol, a tetracyclic triterpene. L. forms colorless crystals which are soluble in chloroform and ethanol, but practically insoluble in water; m.p. 138 to 140 °C, $[\alpha]_D$ +60°. L. is formed in animals and fungi as an intermediate of steroid biosynthesis. It is one of the main sterols in wool grease; the others are cholesterol and dihydrolanosterol.

Lanosterol

Lanthanide: a term for the Lanthanoids (see), use of which is now discouraged because the ending -ide signifies the anion component of a compound.

Lanthanoid: one of the fourteen metallic elements following lanthanum in the periodic system: cerium (Ce), praseodymium (Pr), neodymium (Nd), promethium (Pm), samarium (Sm), europium (Eu), gadolinium (Gd), terbium (Tb), dysprosium (Dy), holmium (Ho), erbium (Er), thulium (Tm), ytterbium (Yb) and lutetium (Lu). The L. are often represented by the common symbol ***Ln***. Along with the elements scandium, yttrium and lanthanum, they comprise the Rare earth metals (see).

Properties. L. are characterized by occupied $5s^2$, $5p^6$ and $6s^2$ orbitals, (sometimes also $5d^1$) and stepwise completion of the 4f shell. Their chemical and physical properties are therefore very similar.

The most common oxidation state for the L. is +3, but cerium, praseodymium and terbium also occur in the +4 state, and europium, samarium and ytterbium have been found in the +2 state. The ionic radii of Ln^{3+} decrease markedly, in regular fashion, with increasing atomic number (see Lanthanoid contraction). The magnetic and spectral properties of the lanthanoid ions are determined by the occupation of the 4f shell. Because of the effective shielding by the $5s^2$ and $5p^6$ orbitals, the terms resulting from the $4f^n$ configuration are little affected by the environment; instead, the magnetic and optical behavior of the L. ions are determined by the terms resulting from spin-orbital coupling to give defined values of the total angular moment J. The absorption lines of the Ln^{3+} ions are very sharp; these f-f transitions are quite different from the broad d-d bands of the d-transition metal complexes, which are affected by ligand field effects. Table 2 shows the relations between ground state and color of the Ln^{3+} ions.

Table 2. Ground states and colors of the Ln^{3+} ions

Ion	Ground state	Color	Ion	Ground state
Gd^{3+}	$^8S_{7/2}$	Colorless	Lu^{3+}	$^1S_0$
Ce^{3+}	$^2F_{5/2}$	Colorless	Yb^{3+}	$^2F_{7/2}$
Pr^{3+}	$^3H_4$	Green	Tm^{3+}	$^3H_6$
Nd^{3+}	$^4I_{9/2}$	Reddish	Er^{3+}	$^4I_{15/2}$
Pm^{3+}	$^5I_4$	Pink; yellow	Ho^{3+}	$^5I_8$
Sm^{3+}	$^6H_{5/2}$	Yellow	Dy^{3+}	$^6H_{15/2}$
Eu^{3+}	$^7F_0$	Pale pink	Tb^{3+}	$^7F_6$

All the L. are silvery-white, reactive metals which are strong reducing agents (standard electrode potentials ε_0 (Ln/Ln^{3+}): - 2.833 V for Ce to - 2.25 V for Lu). They react with water and dilute acids with evolution of hydrogen. L. react with nonmetals such as oxygen, chlorine, nitrogen, hydrogen and carbon at elevated temperatures to form oxides Ln_2O_3, chlorides $LnCl_3$, nitrides LnN, non-stoichiometric hydrides with compositions LnH_2 to LnH_3 and ionic carbides such as $Ln_2(C_2)_3$, LnC_2 and Ln_2C_3. The fluorides LnF_3 are not obtained by reaction of the elements; because of their insolubility they are conveniently made by precipitation from weakly acidic aqueous solutions. Precipitation of the oxalates $Ln_2(C_2O_4)_3 \cdot nH_2O$ from dilute nitric acid solutions of the L. is used as a quantitative and rather specific method for separation of these elements. The separation of the L. from each

Table 1. Properties of the lanthanoid elements

	Ce	Pr	Nd	Pm	Sm	Eu	Gd	Tb	Dy	Ho	Er	Tm	Yb	Lu
Nuclear charge	58	59	60	61	62	63	64	65	66	67	68	69	70	71
Electronic configuration*	$4f^1 5d^1 6s^2$	$4f^3 6s^2$	$4f^4 6s^2$	$4f^5 6s^2$	$4f^6 6s^2$	$4f^7 6s^2$	$4f^7 5d^1 6s^2$	$4f^9 6s^2$	$4f^{10} 6s^2$	$4f^{11} 6s^2$	$4f^{12} 6s^2$	$4f^{13} 6s^2$	$4f^{14} 6s^2$	$4f^{14} 5d^1 6s^2$
Atomic mass	140.12	140.9077	144.24	145	150.36	151.96	157.25	158.9254	162.50	164.9304	164.26	168.9342	173.04	174.97
Atomic radius [pm]	182.5	182.0	181.4	181.0	180.2	199.5	178.7	176.3	175.2	174.3	173.4	172.4	194.0	171.8
Electronegativity	1.06	1.07	1.07	1.07	1.07	1.01	1.11	1.10	1.10	1.10	1.11	1.11	1.06	1.14
Density [g cm^{-3}]	6.773	6.475	7.003	7.22	7.536	5.245	7.886	8.253	8.559	8.78	9.045	9.318	6.972	9.843
m.p. [°C]	798	935	1 016	1 168	1 072	826	1 312	1 356	1 407	1 470	1 522	1 545	816	1 675
B.p. [°C]	3 257	3 017	3 127	2 730	≈1 900	1 439	3 000	2 480	2 335	2 720	2 510	1 725	1 193	3 315

*) Electrons outside the Xe shell.

other was formerly an extremely laborious process of repeated crystallization, precipitation or decomposition of suitable salts; it has become much easier with modern extraction processes and is achieved by combinations of ion exchange and complex formation (see Lanthanoid separation).

Analytical. The L. are separated from other elements by utilizing the insolubility of their oxalates. Modern physical methods such as atomic adsorption and x-ray spectroscopy, spectral and flame photometry are used for quantitative determination of the individual L. in mixtures.

Occurrence. The L. as a group make up between about 0.01 and 0.02% of the earth's crust. L. with even atomic numbers are more abundant, accounting for 10^{-4} to 10^{-3}% of the crust, compared to 10^{-4} to 10^{-5}% for the individual odd-numbered elements. The most abundant of the L. is cerium, at $6.8 \cdot 10^{-3}$%. ^{147}Promethium is found only as a decay product of uranium fission; its half-life is 2.62 years; other Pm isotopes exist only in the laboratory. The longest-lived of these is ^{145}Pm, with a 17.7 year half-life.

The L. are always found together, and are accompanied by the other rare-earth elements lanthanum and yttrium (ionic radii 115 and 93 pm), which have similar atomic radii. The interchangeability of the L. in crystals depends to some degree on their ionic radii: silicate or phosphate minerals such as cerite, $(Ce_3)M^{II}H_3Si_3O_{13}$ (M^{II} = Ca, Fe), or monacite, $CePO_4$, contain, in addition to lanthanum, mainly the lighter L. (cerite earths, cerium to samarium). The heavier L. (Eu to Lu) and yttrium (ytter earths) are more abundant in minerals such as xenotim, YPO_4, thorveitite $(Y,Sc)_2Si_2O_7$, thalenite, $Y_2Si_2O_7$, and gadolinite (ytterbite), $Y_2M^{II}{}_3Si_2O_{10}$ (M^{II} = Fe, Be). Another lanthanoid mineral is bastnesite (Ce, La, Dy)$\cdot CO_3F$. L. are extracted from monacite sands which arise through weathering of primary monacite deposits.

Extraction. The most important ore for L. is prepared monacite sand, which contains about 60% cerite earths and 5% ytter earths, in addition to about 5% thorium oxide, ThO_2. The concentrated ore is usually extracted with sulfuric acid; the L. along with the thorium are precipitated as oxalates, and the thorium is removed by treatment of the mixed oxalates with ammonium oxalate, which forms water-soluble ammonium oxalothorate with the thorium oxalate. Next the L. oxalates are heated to glowing, which converts them to the oxides, and are dissolved in acid. Treatment of this solution with sodium sulfate gives a preliminary separation. Further separation is achieved by extraction of nitric acid solutions with tributyl phosphate or dimethylsulfoxide, and the metals are then purified by cathodic deposition on mercury followed by fractional dissolving. They can also be purified by a combination of ion exchange and complex formation. For preparation of pure L. metals, the salts are first converted to the chlorides $LnCl_3$ or fluorides LnF_3, which are then reduced by heating with a metal such as calcium.

Applications. Some L. are used in nuclear technology. Because of their high electron-capture cross sections, gadolinium, europium, samarium and dysprosium are used as components of the alloys for control rods in nuclear reactors. In metallurgy, L. are used as micro-additions to alloys, either in pure form or as Cerium alloys (see); they affect the shape and distribution of non-metal inclusions in the alloy, and thus its ductility, hardness, malleability and corrosion properties. L. are used to make flints. Alloys with cobalt, of the $LnCo_5$ and Ln_2Co_{17} types (Ln = Sm, Pr, Ce), are important magnetic materials. Glass is given a green color by Pr, a bright red color by Nd, and a blue color by a mixture of the two; these glasses are prized for artistic purposes. Neodymium is an important component of ultraviolet-screening glasses. Some L. are fluorescent (e.g. Eu, Ce, Tb and Er) and are used as activators for high-capacity luminophores based on rare-earth oxides, oxide sulfides, vanadates and silicates.

Historical. In 1794, Gadolin discovered a new oxide in the mineral ytterbite (gadolinite) which had been found in 1788 in Sweden; this oxide was named ytter earth by Ekeberg. In 1803, Berzelius and Klaproth independently discovered another new oxide, which was named cerite earth. For a long time, the two earths were thought to be pure substances, until Mosander succeeded, between 1839 and 1843, in separating them into three components each. For one component of each mixture he retained the old name, and named the elements on which the other oxides were based erbium and terbium, "didyme" and lanthanum. However, only four of the rare earths he had purified were pure metal oxides. From the erbium oxide described by Mosander, Marginac isolated ytterbium oxide in 1878, Nilson isolated scandium oxide in 1879, and Cleve isolated holmium and thulium oxides in the same year. This holmium oxide was also found to be a mixture; Lecoq de Boisbaudran demonstrated the presence of dysprosium in it in 1886. He also discovered samarium oxide, in 1879, in the didyme oxide described by Mosander. In 1896, Demarçay discovered europium oxide in samarium oxide; in 1880, Margnac isolated gadolinium oxide from the same material. The isolation of didyme into neodymium and praseodymium was achieved by Auer von Welsbach (1885). The last of the naturally occuring L., lutetium, was discovered in ytterbium oxide in the years 1905 to 1907 by Auer von Welsbach and Urbain. Promethium, which occurs naturally only in the tiniest traces, was discovered in 1945 by Marinsky, Glendenin and Coryell among the cleavage products of uranium.

Lanthanoid compounds: generally have analogous compositions, similar physical and chemical properties and can be synthesized by the same methods. Most are in the +3 oxidation state, but the +4 and +2 states are sometimes seen. The hydroxides $Ln(OH)_3$ can be precipitated from aqueous solutions of the hydrated lanthanoid(III) ions $[Ln(H_2O)_n]3+$ (n = 6-9) by addition of alkalies. The basicity of the lanthanoid hydroxides decreases from the beginning of the series (with basicities similar to that of $Ca(OH)_2$) to lutetium (basicity of $Lu(OH)_3$ is similar to that of $Al(OH)_3$). Roasting the hydroxides produces the lanthanoid(III) oxides, Ln_2O_3, which can also be made by a direct reaction of most lanthanoid metals with oxygen. The exceptions are cerium, which is oxidized by oxygen to CeO_2, and praseodymium and terbium, which form the Ln(III,IV) mixed oxides with the compositions Pr_6O_{11} and Tb_4O_7, for example. The lanthanoids react with hydrogen at high temperatures to form non-stoichiometric hydrides, LnH_2-LnH_3. These are black, pyrophoric compounds which decompose in water. Most are ionic. The reactions of lanthanoids with nitrogen yield nitrides with NaCl structure, and their reactions with carbon yield ionic carbides of the types $Ln_2(C_2)_3$, LnC_2 and Ln_2C_3. When the oxides are heated with suitable chlorination reagents, such as phosgene or ammonium chloride, the lanthanoid chlorides, $LnCl_3$, are formed. These have hexagonal UCl_3 layered structures (for Ln = Ce to Gd), or cubic $AlCl_3$ layered structures (Ln = Tb to Lu). In these structures, the central atoms have coordination numbers of 9 and 6, respectively. Reaction of fluorine with the lanthanoids yields the trifluorides, LnF_3; CeF_3 can react further with F_2 at room temperature to form CF_4. TbF_3 reacts at 300 to 400 °C with F_2, forming TbF_4. The trifluorides LnF_3 are precipitaed from acidic aqueous solutions. The precipitation of lanthanoid(III) oxalates, $Ln_2(C_2O_4) \cdot nH_2O$ from dilute nitric acid is used to separate the lanthanoids, and also for their gravimetric determination. Precipitation reactions with hydrogen carbonate or phosphate yield lanthanoid(III) carbonates or phosphates. The hydrates $LnCl_3 \cdot 7H_2O$ (Ln = Ce to Nd) and $LnCl_3 \cdot 6H_2O$ (Ln = Nd to Lu) can be obtained from aqueous solutions of the lanthanoid(III) chlorides. The very water-soluble nitrates also crystallize in the form of their hydrates, $Ln(NO_3)_3 \cdot 6H_2O$. Double nitrates such as $2\,Ln(NO_3)_3 \cdot 3Mg(NO_3)_2 \cdot 24H_2O$ or $Ln(NO_3)_2 \cdot NH_4NO_3 \cdot 4H_2O$ are important for separation of the lanthanoids by fractional crystallization, while double sulfates such as $Ln_2(SO_4)_3 \cdot 3Na_2SO_4 \cdot 12H_2O$ are used to separate the cerium group (Ln = Ce to Eu, and La) from the yttrium group (Ln = Gd to Lu, also Y).

Binary Ln(IV) compounds exist in the solid state for cerium (CeO_2, CeF_4), praseodymium ($Na[PrF_5]$, $Na_2[PrF_6]$ and PrF_4) and terbium (TbF_4, $M_n[TbF4+n]$). Only the orange Ce^{4+} ion is stable in aqueous solution; Ce(IV) salt solutions are used as oxidizing agents in volumetric analysis (see Cerimetry). Reduction of europium(III), samarium(III) and ytterbium(III) with zinc, magnesium or sodium amalgam leads to colorless, blood red or yellow Eu^{2+}, Sm^{2+} and Yb^{2+} ions; the crystalline dichlorides, $LnCl_2$ (Ln = Eu, Sm, Yb) are obtained by reduction of the trichlorides with hydrogen or the corresponding metallic lanthanoid.

The numerous ***lanthanoid complexes*** very frequently display coordination numbers > 6, especially 7, 8 and 9. Some examples are: [Ln(β-diketonate)$_3$L], $[Ln(EDTAH)(H_2O)_4] \cdot 3H_2O$, [Ln(pyridine oxide)$_8$]$(ClO_4)_3$, Ln en_3Cl_3, $[Ln(NCS)_3(OP(C_6H_5)_3)_4]$.

L. have many applications. Europium, cerium, terbium and erbium compounds are used as activator for phosphors. Derivatives of cerium, praseodymium and neodymium are used to color or bleach glass. L. improve the optical properties of glasses, and are used in the production of solid and liquid lasers.

Lanthanoid contraction: a marked and continuous decrease in ionic radius with increasing nuclear charge in the lanthanoid series. The radii are as follows: Ce^{3+}, 103.4 pm; Pr^{3+}, 101.3 pm; Nd^{3+}, 99.5 pm; Pm^{3+}, 98.1 pm; Sm^{3+}, 96.4 pm; Eu^{3+}, 95.0 pm; Gd^{3+}, 93.8 pm; Tb^{3+}, 92.3 pm; Dy^{3+}, 90.8 pm; Ho^{3+}, 89.4 pm; Er^{3+}, 88.1 pm; Tm^{3+}, 86.9 pm; Yb^{3+}, 85.8 pm; Lu^{3+}, 84.8 pm. The L. is caused by the increasing nuclear attraction for the 4f electrons, which do not completely screen the nucleus from one another. It is responsible for the progression in basicity of the lanthanoid hydroxides $Ln(OH)_3$, the stability of lanthanoid complexes, and the positions of the ion exchange equilibria, all of which are utilized in separating these elements.

The L. is also responsible for the great similarity between the ionic radii of the 5d metals hafnium, tantalum and tungsten and those of the homologous 4d metals zirconium, niobium and molybdenum: Zr^{4+}: 74 pm Nb^{5+}: 69 pm Mo^{6+}: 62 pm Hf^{4+}: 75 pm Ta^{5+}: 68 pm W^{6+}: 62 pm.

This effect is the basis for the great similarity in the chemistry of these pairs of elements.

Lanthanoid separation: because of their close physical and chemical similarities, the lanthanoid elements are difficult to separate. The insolubility of their oxalates is used to separate lanthanoids and other rare earths from other elements. Further separation of lanthanoid mixtures was formerly achieved by fractional crystallization, precipitation and decomposition of suitable compounds. Fractional crystallization depends mainly on solubility differences between double nitrates, sulfates, carbonates and oxalates of the lanthanoids; the process is laborious. Fractional precipitation of lanthanoid(III) hydroxides, $Ln(OH)_3$, depends on the decreasing basicities and solubility products of the hydroxides due to the decreasing ionic radii of progressively heavier elements. When base is added to Ln^{3+} solutions, the more weakly basic ytter earths precipitate before the cerite earths. Fractional decomposition of the lanthanoid(III) nitrates also makes use of the progression in basicity of the lanthanoid oxides. Nitrates of weaker bases decompose more readily on heating than those of stronger bases, so that when lanthanoid(III) nitrates are heated, the nitrates of heavier lanthanoids and yttrium are decomposed before those of the lighter lanthanoids.

The progression of ionic radii (see Lanthanoid contraction) is the basis for modern separation methods, which are carried out on industrial scale. In solvent extraction with tributylphosphate from strong nitric acid solution, the decrease in ionic radius is correlated with an increasing tendency to form complexes, and this is the basis for fractional extraction. The radius of the hydrated ions $[Ln(H_2O)_a]^{3+}$ increases from cerium to lutetium; correspondingly, the tendency of these ions to exchange on a cation exchanger increases from lutetium to cerium. Since the tendency to form complexes increases in the opposite direction, from cerium to lutetium, a combination of ion exchange and complex formation amplifies the separation. In practice, a lanthanoid-loaded exchange column is eluted with a suitable complexing agent such as α-hydroxyisobutyric acid or citric acid/ammonium citrate; the lanthanoids are eluted in the order lutetium, ytterbium ... praseodymium, cerium. A change in oxidation state can also be used to separate some lanthanoids. Europium and ytterbium can be obtained by reduction to the divalent state with zinc powder: $2\,Ln^{3+} + Zn \rightarrow 2\,Ln^{2+} + Zn^{2+}$ (Ln = Eu, Yb), followed by precipitation of these elements as the sulfates from sulfuric acid solution. Cerium and terbium can be separated from lanthanoid mixtures by oxidation with peroxodisulfate to the tetravalent state: $2\,Ln^{3+} + S_2O_8^{2-} \rightarrow 2\,Ln^{4+} + 2\,SO_4^{2-}$ (Ln = Ce, Tb). Ce^{4+} is precipitated from nitric acid solution as $(NH_4)_2[Ce(NO_3)_6]$.

Lanthanum, symbol ***La***: an element in group IIIb of the periodic system, the Scandium group (see); a rare-earth, heavy metal, *Z* 57, with natural isotopes with mass numbers 139 (99.911%) and 138 (0.089%, β-emitter, $t_{1/2}$ $1.3 \cdot 10^{11}$ a); atomic mass 138.9055, valence III, density 6.162, m.p. 920 °C, b.p. 2454 °C, electrophilic conductivity 17.6 Sm mm^{-2}, standard electrode potential (La/La^{3+}) -2.522 V.

Properties. L. is a silvery-white, ductile metal which occurs in three modifications; below 310 °C it exists as hexagonal crystals, between 310 and 868 °C as cubic face-centered crystals, and above 868 °C as cubic body-centered crystals. Below 6K, L. is superconducting. It is a rather reactive metal, which tarnishes in air even at room temperature, ignites at 350 to 450 °C, is attacked by water slowly when cold and more rapidly when heated. At elevated temperatures, it reacts with nonmetals, such as halogens, carbons, nitrogen, hydrogen and sulfur, to form the corresponding binary compounds. In its compounds, L. is generally in the +3 oxidation state.

Analytical. L. in rare-earth mixtures is generally determined by x-ray spectroscopy. L. can be precipitated, along with the other rare-earth metals, as its oxalate.

Occurrence and production. L. makes up $1.7 \cdot 10^{-3}$% of the earth's crust, and is always found together with other rare-earth metals, especially cerium and the other lanthanoids, in minerals such as cerite, monacite (turnerite) and orthite. The main ore for production of L. is monacite sand. The rare-earths are dissolved in sulfuric acid, then precipitated as the oxalates. L. is isolated from the mixed oxalate by a combination of precipitation, ion-exchange and extraction processes (see Lanthanoid separation). It is precipitated from pure L. solution as lanthanum(III) oxalate, which is roasted to lanthanum(III) oxide, and this is converted either to lanthanum(III) chloride by a mixture of carbon in a stream of chlorine at high temperature, or to lanthanum(III) fluoride by reaction with hydrogen fluoride in a rotating furnace. The metal is finally obtained by melt electrolysis of lanthanum(III) chloride or by reduction of lanthanum(III) fluoride with calcium/magnesium.

Application. L. is used as a microcomponent in alloys, often together with other lanthanoids in the form of Cerium mixed metal (see). The lanthanum-cobalt alloy $LaCo_5$ is used as a ferromagnet, and the homologous lanthanum-nickel alloy, $LaNi_5$, as a hydrogen storage alloy.

Historical. L. was discovered in 1839 by Mosander in impure cerium nitrate. The name is derived from the Greek "lanthanein", which means "to be hidden".

Lanthanum compounds: compounds of lanthanum, often with the general formula LnX_3. They are characteristically colorless and diamagnetic. ***Lanthanum(III) chloride***, $LaCl_3$, forms colorless, deliquescent, hygroscopic crystals, M_r 245.27, density 3.842, m.p. 860 °C, b.p. 1750 °C. It is made by reaction of a lanthanum(III) oxide carbon mixture with chlorine at high temperature. $LaCl_3$ tends to form complexes. With chloride ions it forms the complex anion $[LaCl_6]^{3-}$. $LaCl_3$ crystallizes out of water as the colorless, triclinic heptahydrate, $LaCl_3 \cdot 7H_2O$. ***Lanthanum(III) fluoride***, LaF_3, m.p. 1493 °C, b.p. 2330 °C, is formed from lanthanum(III) oxide and hydrogen fluoride. It reacts with alkali fluorides to form fluoro complexes of the types $M[LaF_4]$ and $M_3[LaF_6]$. Precipitation of lanthanum(III) salt solutions with alkali hydroxides, carbonates and oxalates produces the following water-insoluble compounds: ***lanthanum(III) hydroxide***, $La(OH)_3$; ***lanthanum(III) carbonate***, $La_2(CO_3)_3 \cdot 8H_2O$, colorless crystals, M_r 601.97, density 2.6 to 2.7; ***lanthanum(III) oxalate***, $La_2(C_2O_4)_3 \cdot 9H_2O$, colorless tetragonal crystals, M_r 704.02. If these compounds are heated, they produce ***lanthanum(III) oxide***, a white powder, M_r 352.82, density 6.51, m.p. 2750 °C, b.p. 4200 °C. It is slightly soluble in water, and in aqueous solution is slowly converted to lanthanum(III) hydroxide. La_2O_3 is the strongest base among the rare earths. Reaction of La_2O_3 with nitric acid yields ***lanthanum nitrate***, $La(NO_3)_3 \cdot 6H_2O$, colorless, deliquescent crystals, M_r 433.01. Binary lanthanum-nonmetal compounds are made by reaction of lanthanum with the corresponding nonmetal around 1000 °C; examples are ***lanthanum carbide***, LaC_2, yellow crystals, M_r 162.93, density 5.02, which is hydrolysable into C_2H_2, C_2H_6, C_2H_4 and C_4H_{10}; ***lanthanum boride***, LaB_6, metallic, purple compound, M_r 203.78, density 2.61, m.p. 2210 °C; ***lanthanum sulfide***, La_2S_3, orange, hexagonal crystals, M_r 374.01, density 4.911, m.p. 2100 to 2150 °C. ***Lanthanum hydride***, with the approximate composition LaH_3, is formed from lanthanum and hydrogen at about 240 °C.

Very pure lanthanum(III) oxide is used in making special optical glass, for example, for high-quality photographic lenses. Other L., when activated with europium or samarium, serves as lumiphores.

Laporte rule: selection rule according to which electron transitions between states of different multiplicity are forbidden (see UV-VIS spectroscopy).

Large-angle grain boundary: see Polycrystal.

Larixic acid: same as Maltol (see).

Lasalocide: same as Antibiotic X-537A.

Lassaigne test: a test for qualitative detection of nitrogen in organic compounds. The sample is heated to red heat with metallic sodium, which converts organic nitrogen into cyanide. This can be detected after conversion to hexacyanoferrate(II) with iron(III) ions as Prussian blue (see).

Latent heat: see Phase-change heat.

Latent image: an invisible image formed by the primary photochemical reaction in photography; it is made visible by development. In the silver-halide-based Photographic process (see), the L. consists of latent image nuclei. These are silver clusters of 4 to 10 silver atoms within the silver halide grains. The L. is stable in this form; the silver clusters act as catalysts in the development process and greatly accelerate the reduction of the silver halide crystal to elemental silver. This is the source of the large amplification effect (10^9) and the high sensitivity of the silver halide materials.

Latex: originally, a term for the milky sap of the rubber plant (see Rubber). Today it is also applied to aqueous suspensions of synthetic rubber and other polymers, such as polyvinyl chloride and polyvinyl acetate.

Latex paints: Paints based on Latex (see); they are resistant to weathering and wear, and are used in industry and construction.

Lattice: see Crystal.

Lattice defect: same as Crystal lattice defect (see).

Lattice energy: the energy released when the isolated components of a crystal come together to form the crystal lattice. It is usually referred to a mole of the resulting crystal substance, and then called the ***molar L.***, E_L. Since the L. by the above definition is associated with an exothermal process, it has a negative value, in accordance with thermodynamic convention. However, the L. is also often related to the reverse process, the isothermal conversion of a crystal into the ideal gas of its components, in which case it is seen as an endothermal parameter with a positive sign.

The L. is a measure of the stability of a crystal structure. It is therefore closely related to other crystal properties, such as melting temperature, compressibility, thermal expansion, hardness and other mechanical properties. If Polymorphism (see) is present, the modification with the largest value of L. is usually, but not always, thermodynamically the most stable. It must be observed that the free energy is the determining factor in the stability of a state, and it includes the entropy as well as the internal energy. However, there are at present no reliable rules for calculating the entropy of lattices with different symmetries.

The L. can be calculated only in those cases in which the transition between the two reference states, the crystal and the gas of the free components, is unequivocal and its associated energy can be measured. The possibilities for such measurements are quite different for differnt types of lattice, and are limited by the formation of ion pairs, atom clusters, molecular associations and other more highly aggregated gas particles. However, the L. can often be determined indirectly from thermochemical and spectroscopic data (see Habor-Born cycle). For theoretical calculations of the T., suitable values for the interaction potentials between the lattice components are used. For *ionic lattices*, the calculation must include both the Coulomb interaction and the repulsive forces which occur when the ions approach each other closely (see Chemical bond). According to Born, for a closest approach R and charges z_+ and z_- for ions with unequal charges, the L. is given by

$$E_L = -\frac{N_A z_+ z_- e^2 A}{4\pi\epsilon_0 R} + \frac{N_A B}{R^n}$$

Here N_A is Avogadro's constant, e is the unit charge on the electron, A is the Madelung constant, a characteristic of each type of lattice, ε_0 is the electric field constant, n is the Born repulsion exponent ($n > 1$; depending on the ion pair, it is between 6 and 12) and B is a substance constant. For the equilibrium distance R_0 in the crystal, there is a minimum in the potential energy; from the corresponding condition $dE_L/dR = 0$, it follows that:

$$E_L = -\frac{N_A z_+ z_- e^2 A}{4\pi\epsilon_0 R}\left(1 - \frac{1}{n}\right)$$

The values of the L. calculated using these equations agree well with the experimental results for typical ionic substances, such as the alkali halides. Differences greater than about 5% indicate considerable covalent contributions to the bonding and greater deviations from the spherically symmetrical charge distribution of the ions, due to polarization. These results are of interest for discussion of the binding forces in the crystal. As can be seen from the equation, the decisive factors for the L. of ionic crystals are the lattice symmetry, the distances, and above all, the charges on the ions. For example, the L. for sodium chloride, NaCl, is -765 kJ mol^{-1}; for sodium bromide, NaBr, -731 kJ mol^{-1}; for magnesium oxide, MgO, -3925 kJ mol^{-1}, and for thorium oxide, ThO_2, -20,100 kJ mol^{-1}.

For *metallic lattices* the L. can be defined on the basis of the electron gas model as that energy released when the gaseous cations and electrons come together to form the metal crystal. It can be determined experimentally as the sum of the sublimation and ionization energies, or it can be calculated from an equation which is analogous to the one used to calculate the L. of an ionic crystal. In spite of the great simplifications which these models represent, there is surprisingly good agreement between the theoretical and experimental results, especially for the alkali metals. For example, the L. for sodium is -604 kJ mol^{-1}, and for copper, -1087 kJ mol^{-1}.

In the calculation of the L. for *van der Waals lattices*, which are the basis of molecular crystals, all possiblities for various types of intermolecular forces must be considered. There are frequently large differences between the theoretical values obtained in this way and the experimental values derived from the sublimation energies; these differences are due to the difficulty of determining the repulsive forces. The L. of van der Waals lattices are between 3 and 15 kJ mol^{-1} for noble gases, and between 40 and 80 kJ mol^{-1} for organic molecular crystals.

Lattice forces: see Crystal chemistry.

Lattice holes: see Crystal lattice defects.

Lattice radius: see Ionic radius.

Lattice type: 1) in the narrow sense, same as Structure type (see). 2) In the broader sense, a term for characterization of a crystal lattice in terms of its components and the bonding between them. There are ***ionic*** (e.g. sodium chloride), ***atomic*** (e.g. diamond), ***molecular*** (e.g. naphthalene) and ***metal lattices*** (e.g. copper). A different classification of L. depends on the presence of firm bonding among a limited number of atoms to form molecules (***molecular lattice***, as in naphthalene) or among an unlimited number of atoms to form a unidimensional or ***chain lattice*** (as in gray selenium or silicon disulfide), a two-dimensional ***layer lattice*** (as in graphite or molybdenum sulfide) or a three-dimensional ***coordination lattice*** (e.g. diamond, sodium chloride or copper).

Laughing gas: see Nitrogen oxides.

Laundry products: products used to wash textiles in washing machines. The main components of L. are anionic and nonionic detergents (8-15% by mass), especially alkylbenzene sulfonates and fatty alcohol ethoxylates, long-chain soaps (up to 5%) as foam inhibitors, condensed inorganic phosphates, such as pentasodium triphosphate, as builders (these are now usually mixed with zeolites to reduce the phosphate load in waters; together they make up 30-40%), bleaches (8 to 20%), compounds to inhibit graying, such as carboxymethylcellulose (< 2%), corrosion inhibitors, such as sodium silicate (2 to 7%), compounds to prevent clumping of the product, such as sodium sulfate (up to 20%), white toners, and in some cases enzymes, bleach activators and perfumes.

Powdered L. are most commonly made by pressure valve pulverization at a pressure up to 8 MPa from liquid and solid components, which are prepared in the highest concentration which can still be moved (*slurry*). The pulverizing is usually done in a countercurrent of hot air (entrance temperature up to 300 °C) in a tower up to 30 m high. The temperature-sensitive components, such as perborate, enzymes and perfumes, are added afterwards to the "tower powder". In the hot spray method, evaporation of water from the interior of the externally solidified slurry droplets (*beads*) produces hollow spheres, which make the product pour well and prevents dust. Spray mixing processes consume less energy. They utilize the ability of some of the components to form hydrates; they include mixing in of solid components, spraying of liquid components onto the mix, and agglomeration. In either method, energy can be saved by utilization of neutralization reactions, such as neutralization of sulfonic acids with sodium lye.

Liquid L. are gaining in popularity. They have been developed from special products for fine laundry by use of more nonionic surfactants, more soluble builders, solubilizers and specialized additives.

Laundry enzymes: enzymes used in laundry products to remove stains, generally protein stains from milk, egg yolk, chocolate, blood or grass. These are specially developed enzymes which are stable in alkaline solution; most are proteases. Amylases are of minor importance.

Laundry presoak products: products which loosen stubborn soil by wetting and swelling processes. L. usually consist of a mixture of surfactants made strongly alkaline by soda or sodium silicate.

Lauraldehyde, ***dodecanal, n-dodecylaldehyde***: CH_3-$(CH_2)_{10}$-CHO, colorless, crystalline leaflets, b.p. 184-185 °C at 13.3 kPa. L. is insoluble in water, but is readily soluble in alcohol and ether. Its reactions are the same as those of other Aldehydes (see), forming typical addition and condensation products. In the presence of air, or when hydrogen chloride is passed through it, L. is converted to a cyclic, trimeric form. L. is present in the oils of bitter oranges, rue and Okinawa pines. It is made industrially by dehydrogenation of lauryl alcohol in the presence of copper or silver catalysts. L. is used mainly in perfumes and in the production of detergents and emulsifiers.

Laurates: see Lauric acid.

Laurel leaf oil: a light yellow, essential oil with a spicy, somewhat sweet odor. The main component is 50% cineol; (-)-α-pinene, α-phellandren, (-)-linalool, geraniol, eugenol and (-)-α-terpineol are also present. L. is obtained in 1 to 3% yield by steam distillation of laurel leaves. A less valuable oil, ***laurel berry oil***, can be isolated in 1% yield from the berries. L. is used as a flavoring for foods.

Lauric acid, ***dodecanoic acid***: CH_3-(CH_2)-COOH, a saturated fatty acid. L. crystallizes in colorless needles with a silky sheen; m.p. 44 °C, b.p. 298 °C. It is steam volatile, and dissolves readily in alcohol and ether, but is only very slightly soluble in water. The salts and esters of L. are called *laurates*. The glycerol ester of L. is present in laurel oil, coconut oil, palm heart oil and milk fat, and can be extracted from these esters by hydrolysis. L. is used as a stabilizer for PVC and as an intermediate for production of softeners and emulsifiers.

Lauryl alcohol, ***dodecan-1-ol***: CH_3-$(CH_2)_{10}$-CH_2OH, crystallizes in colorless leaflets; m.p. 24-26 °C, b.p. 255-259 °C. L. is practically insoluble in water, but is soluble in alcohol and ether. It is a component of many plant oils and can be obtained on an industrial scale by catalytic hydrogenation of the glycerol esters of lauric acid, which are abundant in coconut and palm oils. L. is used in the production of surfactants, for example in the form of sodium lauryl sulfate.

Lavender oil: a colorless, often yellowish to yellow-greenish, essential oil with a pleasant, sweet, flowery odor and a somewhat bitter taste. The main component of L. is (-)-linalyl acetate (at most 60%); other components are linalool, geraniol, borneol, cineol, thymol, caryophyllene, cumarin and pinene. L. is made by steam distillation of the herb, which is harvested when in bloom. A number of subspecies of true lavender are used. The largest production is in Provence (France). L. is used in perfumes, and in human and veterinary medicine as an external medicament against rheumatic diseases.

Laves phases: see Intermetallic compounds.

Lavsan®: see Synthetic fibers.

Law of conservation of elements: see Stoichiometry.

Law of conservation of mass: see Stoichiometry.

Law of constant heat sums: see Hess' law.

Law of constant proportions: see Stoichiometric proportions, law of.

Law of equivalent proportions: see Stoichiometric proportions, law of.

Law of independent ion migration: see Independent ion migration, law of.

Law of multiple proportions: see Stoichiometric proportions, law of.

Lawrencium, symbol ***Lr***: a short-lived, radioactive element which can be obtained only by nuclear reactions; a transuranium member of the Actinoids (see), *Z* 103. The known isotopes have the following mass numbers (decay types; half-life in parentheses): 255 (α; 22s); 256 (α; 31 s); 257 (α; 0.6 s), 258 (α; 4.2 s); 259 (α; 5.4 s); 260 (α; 3.0 min). The atomic mass (the most stable isotope) is 260, valency III.

Little is known about the chemistry of L., but the existence of a stable Lr^{3+} ion has been confirmed. L. was first obtained in 1961 by Ghiorso, Sikkeland, Larsh and Latimer by irradiating californium (isotope mixture of the mass numbers 249 to 252) with accelerated boron ions (^{10}B, ^{11}B) at the Lawrence Radiation

Laboratory in Berkeley, California. One of the possible reactions is $^{252}Cf + ^{11}B \rightarrow ^{257}Lr + 6\ ^{1}n$.

Lawsone: see Henna.

Laxatives: substances or preparations which initiate or promote emptying of the intenstines. L. stimulate intestinal peristalsis and thus accelerate the passage of the contents. The effect is achieved by several mechanisms, such as irritation of the intestinal mucosa, increasing the internal pressure in the intestine and lubricating the stool. Chemically, the L. can be classified as follows: 1) hydroxyanthracene derivatives, such as the anthraglycosides present in various plants; 2) synthetic phenolic compounds, sometimes with esterified OH groups, such as phenolphthalein and bisacodyl; 3) saline L. such as sodium sulfate; 4) plant oils, such as castor oil; 5) lubricating paraffin hydrocarbons; 6) osmotically active sugars or sugar alcohols, such as lactose and sorbitol.

Layer chromatography: see Chromatography.

Layer lattice: see Lattice type.

LCD: see Liquid crystals.

LC/MS coupling: see Mass spectroscopy.

LDAC process: see Steel.

LDPE: see Polyethylene.

LD process: see Steel.

LDP process: see Steel.

Leaching: see Extraction.

Leaching test: see Penetration.

Lead, ***plumbum***, symbol ***Pb***: a chemical element from the 4th main group of the periodic system, the Carbon-silicon group (see); a heavy metal, *Z* 82, natural isotopes with mass numbers 204 (1.48%), 206 (23.6%), 207 (22.6%) and 208 (52.3%). Atomic mass 207.19, valences II and IV, Mohs hardness 1.5, density 11.3437, m.p. 327.50 °C, b.p. 1740 °C, electrical conductivity 5.2 Sm mm^{-2} (at 0 °C), standard electrode potential (Pb/Pb^{2+}) -0.1263 V.

Properties. L. is a soft, ductile metal with a white shine on a freshly cut surface. It crystallizes in a single modification with a cubic face-centered lattice. In its compounds, L. is di- or tetravalent; the +2 oxidation state is more stable, corresponding to its position in the carbon-silicon group. Derivatives of Pb^{4+} are thus strong oxidizing agents. This makes the non-existence of Pb(IV) compounds with readily oxidized ligands understandable. For example, in the Pb(IV) halide series, the iodide and bromide are unknown. In the group, the tendency to form element-element bonds decreases on going from carbon to lead, so that in contrast to the lighter elements, L. is not known to form hydride chains. The tendency of lead(IV) compounds to hexacoordination is very pronounced, as can be seen from the structure of the solid lead (IV) fluoride and in the formation of hexahydroxo- and hexachloroplumbates(IV), $M^{I}_2[Pb(OH)_6]$ and $M^{I}_2[PbCl_6]$, respectively. In the air, the surface of the metal is attacked and forms a layer of basic lead carbonate, $Pb(OH)_2 \cdot 2PbCO_3$. This gives it the familiar matte gray color and protects it from further corrosion.

Finely divided L. is pyrophoric and burns to lead(II) oxide, PbO, which reacts further with excess oxygen at high temperatures, forming lead(II,IV) oxide, Pb_3O_4 (minium, red lead). L. is rapidly dissolved in concentrated nitric acid: $3\ Pb + 8\ HNO_3 \rightarrow 3\ Pb(NO_3)_2 + 2\ NO + 4\ H_2O$. Hydrochloric, sulfuric, phosphoric and hydrofluoric acids do not attack L. in spite of its negative standard electrode potential, due to the high hydrogen overvoltage at L. and the insolubility of the corresponding salts ($PbCl_2$, $PbSO_4$, $Pb_3(PO_4)_2$, PbF_2). L. reacts with sulfur when heated to form lead sulfide, PbS, and is oxidized by fluorine and chlorine to form the tetrahalides PbF_4 and $PbCl_4$.

Lead and its compounds are strong poisons. As dust or vapor they enter the body through the respiratory tract, and soluble lead compounds enter through the gastrointestinal tract. Tetraethyl lead is lipid-soluble and is very rapidly absorbed through the skin. The symptoms of acute lead poisoning are nausea, stomach and intestinal cramps, cardiac arrest, and under some circumstances, circulatory collapse. First aid is the administration of 2 to 3% sodium sulfate solution and activated charcoal. Later the excretion of the L. can be accelerated by infusion of $Na_2CaEDTA$. Tetraethyl lead acts mainly on the central nervous system, causing headaches, excitability, visual impairment and epileptic-like convulsions. Clothing which has been wetted with tetraethyl lead should be removed at once and the affected parts of the body washed thoroughly. Here again, treatment with $Na_2CaEDTA$ is helpful.

Indications of chronic lead poisoning are a shale-gray line on the gums (lead line), digestive upsets, anemia, pain in the joints, painful colic, etc. Even relatively low concentrations of lead cause severe damage to the central and peripheral nervous system over long periods. For this reason, the lead content of auto exhaust from leaded gasolines is a serious public health hazard in large cities.

Analytical. L. is separated as black lead sulfide, PbS, in the qualitative analytical separation series. The PbS is dissolved in dilute nitric acid, and the Pb^{2+} is identified as lead sulfate, $PbSO_4$, or lead chromate, $PbCrO_4$. The classic method of quantitative analysis is gravimetric determination as $PbSO_4$ or $PbCrO_4$. However, complexometric titration with EDTA is less time-consuming. Instrumental methods, such as atomic absorption spectroscopy and x-ray fluorescence analysis, are available for low Pb concentrations.

Occurrence. L. makes up $2 \cdot 10^{-4}$% of the earth's crust. It is occasionally found in the elemental state in nature. The most important mineral is galenite (lead glance), PbS. There are also minerals in which the L. is present as the carbonate-phosphate, chromate, molybdate or tungstate.

Production. Lead glance is almost the only ore processed. There are two methods of converting it to elemental L. In the *roasting reduction process*, the lead sulfide is roasted to lead(II) oxide, and this is then reduced in a blast furnace by carbon monoxide formed by burning coke: $PbO + CO \rightarrow Pb + CO_2$. In the *roasting reaction process*, the sulfide is roasted only until the lead oxide is formed in the correct stoichiometric proportion to react with the remaining sulfide to L. and sulfur dioxide: $2\ PbO + PbS \rightarrow 3\ Pb + SO_2$.

The crude lead formed by either method usually contains significant amounts of other elements as impurities. Arsenic, antimony and tin are removed as lead arsenate, antimonate and stannate through oxidative melting. Copper can be separated as a high-melting alloy, which also absorbs the sulfur (liquation). The removal of Silver (see) from crude lead is especially important. Electrolytic refining is also used.

Applications. L. is an important metal used mainly in the form of alloys (see Lead alloys). Pure L. is used in lead storage batteries, for radiation protection (absorption of x-rays and γ-particles), for making tetraethyllead (see Organolead compounds) and lead white, as a sheathing for cables, cladding for containers or pipes which come in contact with hot, concentrated sulfuric acid, and so on.

Historical. L. was known in antiquity. The Greeks mined it on Cyprus and Rhodes, and the Romans had mines in Spain, Gaul, Britannia and Germania. The Romans used L. mainly for water pipes and cooking vessels. Some lead compounds, such as minion, were also known in antiquity.

Lead acetates. ***Lead(II) acetate, lead diacetate***, $Pb(CH_3COO)_2$, is a colorless powder which dissolves readily in water; M_r 325.28, density 3.25, m.p. 280 °C. The monoclinic trihydrate which crystallizes out of water, $Pb(CH_3COO)_2 \cdot 3H_2O$, tastes sweet, and is therefore also called ***lead sugar***. $Pb(CH_3COO)_2$ is obtained by the reaction of acetic acid and lead(II) oxide, PbO. It is used in dyeing and cloth printing, in the production of other lead compounds, and of lead acetate papers.

Lead(IV) acetate, lead tetraacetate, $Pb(CH_3COO)_4$, forms colorless, monoclinic crystals; M_r 443.37, density 2.778, m.p. 175 °C. The salt is hydrolysed to lead(IV) oxide and acetic acid by water. It is produced by reaction of lead(II,IV) oxide with glacial acetic acid in the presence of acetic anhydride. The acetate is used in organic syntheses as a selective oxidizing agent. A special application is the Criegee cleavage. The L. are poisonous (see Lead).

Lead acetate paper: a reagent paper for detection of hydrogen sulfide or sulfides by formation of brownish black lead sulfide. It is made by soaking filter paper with 1% lead acetate solution, then drying it.

Lead-acid battery: an Electrochemical current source (see). In the charged condition, the positive pole of the L. consists of lead dioxide, and the negative pole of lead. Aqueous sulfuric acid is used as the electrolyte. The electrode reactions can be described as follows:

$$\text{Anode: } PbO_2 + 4\,H^+ + SO_4^{2-} + 2\,e^- \underset{\text{Charge}}{\overset{\text{Discharge}}{\rightleftharpoons}} PbSO_4 + 2\,H_2O$$

$$\text{Cathode: } Pb + 2\,H^+ + SO_4^{2-} \underset{\text{Charge}}{\overset{\text{Discharge}}{\rightleftharpoons}} PbSO_4 + 2\,H$$

Thus the net reaction of the cell is:

$$PbO_2 + Pb + 2\,H_2SO_4 \underset{\text{Charge}}{\overset{\text{Discharge}}{\rightleftharpoons}} 2\,PbSO_4 + 2\,H_2O$$

It can be seen from the net reaction that sulfuric acid is consumed in the discharge process. The charge of the battery can therefore be determined from the density of the electrolyte. The terminal voltage of the cell is approximately 2 V. If three of these cells are connected in series, a 6-V battery is made, and a 12-V battery consists of six cells. The L. is used mainly as a starter battery in automobiles.

Lead alloys: alloys of lead, primarily with antimony, tin, arsenic and copper. Addition of up to 11% antimony (***hard lead***) significantly improves the mechanical properties of the lead. These L. are used in storage batteries, cable sheaths, and acid-resistant claddings in the chemical industry. Lead-antimony-tin alloys (up to 81% Sn) are used mainly as sliding bearings. They are also used as type metal in printing and as solders.

Lead azide: $Pb(N_3)_2$, a colorless crystalline powder which is barely soluble in water; M_r 291.23, detonation point 350 °C. L. is poisonous (see Lead). It explodes on impact or heating, decomposing into nitrogen or lead. It is made by mixing aqueous solutions of lead(II) nitrate and sodium azide, conventionally in the presence of dextrin. The dextrin is added to produce small, round crystals; large crystals can explode upon breaking, even under water. The precipitated L. is used after drying, and without further purification, as an ignition explosive which is extremely sensitive to friction and impact. It is usually used in a mixture with lead trinitroresorcinate, to increase the ignitability of the latter.

Lead bronze: copper lead alloy with 10 to 25% lead and 5 to 10% tin. Since lead is not soluble in copper, the alloys are produced by casting or sintering. L. are excellent materials for bearings.

Lead(II) carbonate: $PbCO_3$, colorless powder or rhombic crystals, M_r 267.20, density 6.6, decomposition at 315 °C to lead(II) oxide and carbon dioxide. L. is barely soluble in water (solubility product $K_s = 3.3 \cdot 10^{-14}$ at 18 °C). L. is poisonous (see Lead). It is made by addition of Na_2CO_3 or K_2CO_3 solution to Pb(III) salt solutions. If the aqueous suspension is boiled, or if the precipitation occurs in a hot solution, basic L. (lead white) is formed: $Pb(OH)_2 \cdot 2\,PbCO_3$.

Lead chlorides. ***Lead(II) chloride, lead dichloride***, $PbCl_2$, colorless, rhombic crystals; M_r 278.10, density 5.85, m.p. 501 °C, b.p. 950 °C. $PbCl_2$ is barely soluble in cold water (solubility product $K_s = 1.86 \cdot 10^{-5}$ at 18 °C) and it is therefore precipitated, for example, by addition of hydrochloric acid to a solution of lead(II) nitrate or acetate. In the presence of excess chloride ions, chloroplumbates(II) form, with the composition $M^I[PbCl_3]$ and $M^I_2[PbCl_4]$. The molten chloride hardens upon cooling to a horn-like mass (phosgenite). It is used to make lead chromate.

Lead(IV) chloride, lead tetrachloride, $PbCl_4$, is a yellow, oily liquid; M_r 349.00, density, 3.18, m.p. -15 °C. $PbCl_4$ decomposes upon heating, sometimes explosively, to $PbCl_2$ and chlorine. It is made by the reaction of conc. hydrochloric acid and lead(IV) oxide, or of chlorine with $PbCl_2$. It is sometimes used for the Criegee glycol cleavage reaction. The L. are poisonous (see Lead).

Lead chromate: $PbCrO_4$, an orange-yellow, monoclinic crystalline powder; M_r 323.18, density 6.12, m.p. 844 °C. L. is poisonous (see Lead). It is nearly insoluble in water (solubility product K_s 1.8 · 10^{-14} at 18 °C), but it dissolves in acids or bases. In the presence of hydrogen sulfide, L. gradually turns black as lead sulfide forms. It is made by addition of a dichromate solution to a Pb(II) salt solution. L. is contained in pigments (see Chrome yellow, Chrome orange).

Lead cyanamide: see Cyanamide.

Lead fluorides. ***Lead(II) fluoride, lead difluoride***, PbF_2, is a colorless, rhombic crystalline powder; M_r 245.19 °C, density 8.24, m.p. 855 °C, b.p. 1290 °C. PbF_2 is nearly insoluble in water (solubility product K_s = 3.2 · 10^{-8} at 18 °C), but it dissolves in nitric acid. It is formed in the reaction of hydrofluoric acid and lead carbonate or hydroxide.

Lead(IV) fluoride, lead tetrafluoride, PbF_4, colorless crystals; M_r 283.18, density 6.7, m. about 600 °C. The lattice of solid PbF_4 contains $[PbF_6]$ octahedra which are linked in infinite chains via two shared edges. The compound is produced by oxidation of lead(II) fluoride with fluorine.

The L. are poisonous (see Lead).

Lead(II) hydroxide: $Pb(OH)_2$, colorless, amorphous powder, M_r 241.20, converts at 145 °C to lead (II) oxide, PbO, with loss of water. L. is poisonous (see Lead). It is nearly insoluble in water, but dissolves in acids to form the corresponding Pb(II) salts, and in alkali hydroxide solutions to form hydroxoplumbates(II), $M^I[Pb(OH)_3]$. It is formed by addition of hydroxides to Pb(II) salt solutions.

Lead(II) nitrate: $Pb(NO_3)_2$, colorless cubic or monoclinic crystals which dissolve readily in water; M_r 331.20, density 4.53. L. decomposes at 470 °C to lead(II) oxide, nitrogen dioxide and oxygen: $Pb(NO_3)_2 \rightarrow PbO + 2\,NO_2 + 1/2\,O_2$. L. is poisonous (see Lead). It is made by dissolving lead, lead(II) oxide or basic lead carbonate in nitric acid. It is the starting material for synthesis of other lead compounds, is used as an oxidizing agent in the textile industry, and to make textile treatments, matches, etc.

Lead oxides. ***Lead(II) oxide, lead monoxide***, PbO, M_r 223.19, occurs in two modifications, the red, tetragonal PbO, density 9.53, m.p. 886 °C, and the yellow, rhombic PbO, density 8.00, which is stable above 488 °C. PbO is insoluble in water. It is amphoteric, but its basic character predominates. It therefore dissolves in acids to form Pb(II) salt solutions, and in strong bases to form hydroxoplumbates(II), e.g. $Na[Pb(OH)_3]$. With carbon dioxide and water, it forms basic lead(II) carbonate, $Pb(OH)_2 \cdot 2PbCO_3$. PbO is made industrially by oxidation of molten lead with atmospheric oxygen (***litharge***) or by thermal decomposition of lead(II) nitrate, carbonate or hydroxide. If this is done below the melting temperature of PbO, it is obtained as a loose, yellow powder (***massicot***). PbO is used to make lead white, red lead (minium) and lead glass.

Lead(IV) oxide, lead dioxide, PbO_2, is a dark brown, tetragonal, crystalline powder which is nearly insoluble in water; M_r 239.19, density 9.375. It decomposes at 290 °C into lead(II) oxide and oxygen. It is amphoteric, but its acidic character predominates. It dissolves in alkali hydroxide solutions to form hydroxoplumbates(IV), for example, $PbO_2 + 2\,NaOH + 2\,H_2O \rightarrow Na_2[Pb(OH)_6]$. It combines with strongly basic oxides in the melt to form orthoplumbates(IV): $M^I_4[PbO_4]$. PbO_2 is a strong oxidizing agent, which is capable of oxidizing the chloride ion to chlorine or Mn^{2+} to permanganate. It is made by reaction of lead(II) acetate with calcium chloride, by reaction of minium with dilute nitric acid ($Pb_3O_4 + 4\,HNO_3 \rightarrow PbO_2 + 2\,Pb(NO_3)_2 + 2\,H_2O$), or electrolytically. PbO_2 is used as a versatile oxidizing reagent in the laboratory and in industry. It is used in combination with red phosphorus to make matches, and in fireworks.

Lead(II,IV) oxide, red lead, minium, Pb_3O_4, is actually lead(II) orthoplumbate, $Pb_2[PbO_4]$. It is an orange-red, tetragonal crystalline powder, M_r 685.57, density 9.1, which is nearly insoluble in water. Under O_2 pressure, the m.p. is 830 °C; under normal pressure, it decomposes at 500 °C into PbO and oxygen. The industrial synthesis consists of heating PbO in a stream of air to 450-500 °C. Pb_3O_4 is used mainly as a pigment for rust preventing paints, and in glass, ceramics and batteries.

The L. are poisonous (see Lead).

Lead pigments: lead compounds which can be used as pigments. They include lead white (both carbonate and sulfate lead whites), red lead (see Lead oxides), various lead chromates (which can also be considered chromium pigments) and lead cyanamide (see Cyanamide). Lead silicochromate, calcium plumbate and calcium barium plumbate are also L. The L. are poisonous (see Lead).

Lead plating: the creation of lead layers on steel objects. Because pure lead and lead-antimony alloys are resistant to corrosion, chemical apparatus can be protected from corrosion by hot leading: a layer of lead is melted onto a tin intermediate layer which provides the necessary bonding to the steel.

Lead styphnate: same as Trinitroresorcinyllead (see).

Lead sugar: see Lead acetate.

Lead(II) sulfate: $PbSO_4$, colorless, monoclinic or rhombic crystals; M_r 303.25, density 6.2, m.p. 1170 °C. L. is barely soluble in water (solubility product K_s = 1.06 · 10^{-8} at 18 °C) and it therefore precipitates when a sulfate-containing solution is added to a lead(II) salt solution. It is found in nature as anglesite. L. is used as a paint pigment under the name sulfate lead white, or also in the form of mixed crystals with other lead compounds (see, e.g. Chrome yellow).

Lead(II) sulfide: PbS, a black, amorphous powder, or shiny black, cubic crystals; M_r 239.25, density 7.5, m.p. 1114 °C. L. is insoluble in water (solubility product K_s = 3.4 · 10^{-28} at 18 °C) and forms when hydrogen sulfide is passed into a lead(II) salt solution. This reaction is used for the qualitative identification of Pb^{2+} or S^{2-} ions, and to separate lead in the H_2S step of qualitative analysis schemes. L. is found in nature as galenite (lead glance).

Lead white, ***carbonate lead white, basic lead(II) carbonate***: $Pb(OH)_2 \cdot 2\,PbCO_3$, colorless amorphous powder or hexagonal crystals, insoluble in water; M_r 775.60, density 6.14. At 400 °C, it decomposes to lead(II) oxide, carbon dioxide and water. L. has a good covering capacity, a pretty sheen and binds well. For these reasons, and in spite of its toxicity (see Lead) and sensitivity to H_2S (formation of black lead(II) sulfide), it is used as a valuable white pigment. There are several methods of producing it, all of which consist essentially of allowing lead or lead(II) oxide to react with acetic acid in the presence of air; the resulting basic lead(II) acetate is converted to L. by passing carbon dioxide through it. In white paints, the L. is often mixed with other pigments (titanium(IV) oxide, lead(II) sulfate, zinc oxide, barium sulfate), which improve the quality of the paint.

Leaning material: a non-plastic raw material for ceramics, usually added in finely divided form. L. reduce the plasticity and shrinkage due to the clay minerals, and thus prevent cracking during drying and firing. Quartz, feldspar, ceramic scraps or separately fired crude chamotte can be used as L.

Leather care products: products which leave a protective waxy film on the leather surface, or which penetrate to a greater or lesser degree. Like Floor care products (see), they are classified as pure oil-base products and emulsion products, and they are made up as the same basic materials as floor products.

Shoe polishes generally consist of a solid oil base, in which the proportion of hard waxes is higher than in floor care products, so that the wax film does not soften in bright sunlight and bind dust. ***Leather soaps*** are colorless, paste-like emulsions consisting of potassium or sodium soaps, which can be complemented with synthetic cleansers. To improve the cleansing effect, solvents such as tetraline, gasoline or glycerol are included. Leather soaps also contain mild abrasives, such as kaolin, alumina, talc or ground chalk. Beeswax, montane wax or a similar synthetic product is added to give shine.

Specialty products. 1) *L. for fine shoes*. The leather used for fashionable shoes is usually covered with an emulsion based on polyurethane, polyester and mixed polymers, which form a thin film on the surface of the leather, and this layer must not be damaged by solvents in the L. Therefore, emulsion L. with a large water content are most suitable. An oil-in-water emulsion is best, because it permits a cleaning action and improves the hydrophobic character of the wax film. The pH of these L. should be neutral, in order not to destroy the surface of the leather through repeated applications. This type of shoe polish might consist of hard wax, hard paraffin, emulsifier, water, gasoline, a small amount of perfume, preservatives, wetting agents, oil and water-soluble pigments and silicon oil, which improves the hydrophobic effect.

2) *L. for leather garments and luggage* are also based on emulsions with a low percentage of waxes. The proportion of organic solvents, such as gasoline, is only 10 to 20%. These L. contain, in addition to water and emulsifiers, various softeners and silicon oil.

3) *L. for heavy shoes*, such as boots. Here leather impregnating compounds, leather oils and leather fats are used. *Leather impregnation agents* are vegetable and animal fats dissolved in alcohol, gasoline or chlorinated hydrocarbons, or emulsified in water. ***Leather oils*** and ***leather fats*** consist mainly of saponifiable oils and fats of animal or vegetable origin. Leather fat is homogeneous and has a salve-like consistency. It contains, for example, fish liver oil, wool grease, vaseline, tallow and linseed oil. Leather oil is a clear liquid consisting, for example, of fish liver oil and vaseline oil. These L. reduce the permeability of the leather to water, but do not make it shiny. A certain amount of shine is obtained with oil-based L.; their wax components are saturated with plant or animal oils so that the leather can absorb fat from the wax film.

4) *L. for rough leather* are alcoholic solutions of pigments. Rough leather can also be cared for with powders containing kaolin as a base, together with pigments, waxes and oils.

Leather dyes: To dye with an aqueous solution, anionic dyes such as the sodium salts of sulfonic acid derivatives of various synthetic pigments are frequently used, but cationic dyes are also used, e.g. chlorides of dye bases, sulfur dyes solubilized with sodium sulfide, Na_2S, metal complex dyes for spray dyeing and reactive dyes. Depending on the degree of dispersion and their affinity to the fibers, these dyes will color either the upper layer or penetrate the entire cross section of the leather (pressure dyes). In some cases, true chemical bonds are formed with the proteins of the leather, and in others, the color is due to a color lake, for example with chromium salts or with vegetable tannins in the leather.

Surface paints contain inorganic or organic pigments which are anchored by means of proteins (e.g. casein), cellulose nitrate or other polymers to the surface of the leather. These paints must be provided with softeners to prevent cracking and flaking when the leather is bent. The most rapid and economical method for coloring leather is a printing process; rollers are used to apply the color very evenly.

Leather substitutes: artificial leathers can be single-layered or made up of two or more different layers. Most consist of a layer of felt or knit fabric to which a layer of polyvinyl chloride or other plastic (e.g. polyacrylate, polyisobutylene, polyester or polyurethane) has been fused. They are used to some extent to make shoes, but mainly to make luggage, clothing and upholstery.

Synthetic leathers are intended to imitate natural leather as far as possible. They are made of structured fiber collagen obtained from leather scraps. Synthetic leather is not yet used on a large scale.

Leaving group: in substitution reactions, the substituent which is replaced. The lower the energy of the L., the more readily its bond to the reaction center is cleaved, and thus, the more readily the reaction can occur. L. which can exist as molecules (such as the diazonium group, which forms nitrogen) or anions with low polarizability (basicity) have a particularly strong tendency to leave.

Leblanc process: see Soda.

Le Chatelier's principle: a formula developed by H. Le Chatelier and K.F. Braun describing the effects of external change on thermodynamic equilibria: if one of the state variables, e.g. temperature or pressure, is changed, the equilibrium of a chemical system changes in such a way as to reduce or counteract the effect of the change. Mathematically, L. is expressed by the dependence of the equilibrium constant K on temperature and pressure (see Mass action law): $(\delta \ln K/\delta T)_p = \Delta_R H/RT^2$ (van't Hoff reaction isobar) and $(\delta \ln K/\delta P)_T = -\Delta_R V/RT^2$, where T is the temperature, p is the pressure, R is the general gas constant, $\Delta_R H$ is the molar enthalpy of the reaction and $\Delta_R V$ is the molar volume change.

From these equations, it follows that for exothermal reactions ($\Delta H < 0$), an increase in temperature leads to a decrease in K, that is, the equilibrium is shifted in the direction of the starting materials. For endothermal reactions ($\Delta_R H > 0$) leads to an increase in K and thus the equilibrium shifts in the direction of the products. Decreases in temperature cause shifts in the opposite directions. Similarly, an increase in pressure shifts the equilibrium to the right if $\Delta_R H < 0$, and to the left if $\Delta_R H > 0$; that is, always in the direction of the smaller reaction volume.

L. is used both in the laboratory and in the chemical industry to determine the reaction conditions under which the highest equilibrium concentration of products can be achieved, for example, in ammonia synthesis.

Lecithin: see Phosphatidylcholine.

Leclanché cell, ***dry cell***: an Electrochemical current source (see); at present it is the most common type of primary cell.

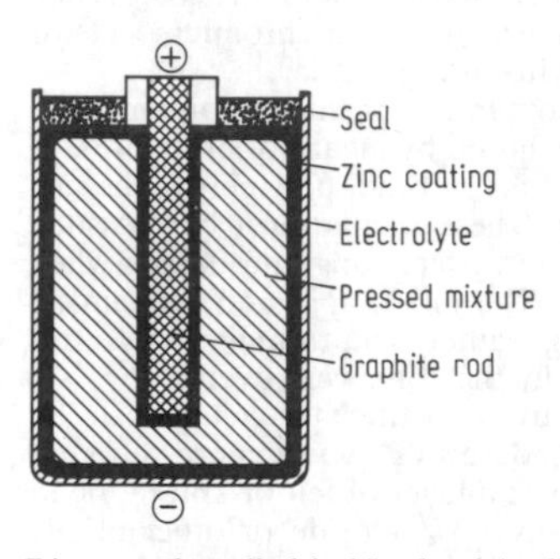

Diagram of a cylindrical Leclanché cell.

A zinc mantle, which serves as the anode, encloses a moist mass of ammonium chloride and zinc chloride. The cathode is a compressed mixture of manganese dioxide (MnO_2), graphite and/or acetylene black and electrolyte; in the center of this mixture is a graphite rod which serves as a conductor. The electrode reactions are extremely complex, but can be approximately summarized by the following partial equations:

Anode: $Zn \rightarrow Zn^{2+} + 2e^-$
Cathode: $2\,MnO_2 + 2\,NH_4^+ + 2\,e^- \rightarrow Mn_2O_3 + H_2O + NH_3$.

An L. has a potential of 1.25 to 1.50 V.

Lectins: glycoproteins, usually from plants, which bind specifically to carbohydrates and cell-surface carbohydrate groups. Because certain L. are capable of agglutinizing erythrocytes and other cell structures, they are also called ***phytohemagglutinins***. The molecular interactions between L. and the carbohydrates are comparable to those between antigens and antibodies. However, an important difference is that the L. are not produced by the plant in response to the presence of the sugar residues.

L. which bind *N*-acetylglucosamine interfere with chitin formation in the synthesis of fungal cell walls, and thus protect the plant from infection. In the nightshade family, L. fix bacteria to the cell walls of the infected plants. In legumes, L. recognize and bind the nitrogen-fixing, symbiotic bacteria. The agglutination of malignant cells (but not of their normal precursors) is used in tissue culture to detect transformation.

Some well-known L. are ***concanavalin A*** from jack beans and ***vicilin*** from the bean *Phaseolus aureus Roxb*. Concanavalin A consists of four subunits with 238 amino acids each; each subunit contains a Ca^{2+} and a Mn^{2+} ion. Vicilin is a glycoprotein, containing 0.2% glucosamine and 1% mannose.

Ledeburite: a metallographic term for the eutectic Iron-carbon alloy (see) containing 4.3% C and solidifying out of the melt at 1147 °C. Above 723 °C it consists of austenite and cementite, and below 723 °C, of perlite and cementite. L. is very hard and brittle.

Lederer-Manasse reaction: a reaction of phenols with formaldehyde which yields various addition and condensation products, depending on the conditions and relative amounts of reactants. Under neutral conditions, phenol and formaldehyde yield a mixture of 2- and 4-(hydroxymethyl)phenol. In the presence of acid catalysts, the hydroxymethylphenols formed as primary products can react with excess phenol to yield dihydroxydiarylmethanes. Under basic conditions and in the presence of excess formaldehyde, mainly bis- and tris(hydroxymethyl)phenols are formed. L. is the basis for synthesis of phenol resins.

Legal acetone tests: method for determination of acetone, e.g. in the urine, by reaction of sodium nitroprusside in alkaline. A cherry-red color appears; addition of acetic acid causes it to turn violet.

Leghemoglobin: see Hemoglobins.

Legumelin: see Albumins.

Leithner's blue: see Cobalt blue 1.

Lemon grass oil: a reddish yellow to brownish red, mobile, essential oil with a strong odor and a lemon-like taste. Its main components are citral, dipentene, limonene, myrcene, geraniol (both free and esterified), linalool, methylheptenol, nerol, farnesol, capric aldehyde and methylheptenone. L. is obtained by steam distillation or distillation of freshly harvested tips of lemon grass; the yield is 0.5 to 0.8%. L. is used mainly for isolation of pure citral. It is also used in perfumery, to make soaps and cosmetics and, in small amounts, as a flavoring for liquors and candies.

Lemon oil, ***citrus oil***: a faintly yellow essential oil which smells and tastes like lemons. It consists mainly of camphene, citral, citronellal, (+)-limonene, α- and β-pinene, terpines, terpineol and linalyl and geranyl acetate. C. also contains octyl, nonyl, decyl and lauric aldehydes. It is obtained from lemons by pressing the peels, or by steam distillation of the fruit. C. is used mainly as a flavoring for foods, but also in perfumes and in pharmaceutical and cosmetic preparations. In microscopy, C. is used to lighten the preparation.

Lenacil: see Herbicides.

Lennard-Jones potential: see Kinetic gas theory.

Lepidin, ***4-methylquinoline***: a colorless liquid; b.p. 264 °C, n_D^{20} 1.6206. It is soluble in alcohol, ether, acetone and benzene, slightly soluble in water and steam volatile. Oxidation with potassium permanganate yields 4-methylpyridine-2,3-dicarboxylic acid, while oxidation with chromic/sulfuric acid yields quinoline 4-carboxylic acid. When L. is heated with hydrogen peroxide in glacial acetic acid, 4-methylquinoline-1-oxide is formed. L. is synthesized by the Skraup synthesis (see); it is also a decomposition product of the cinchona alkaloids. L. is used to synthesize cyanine pigments.

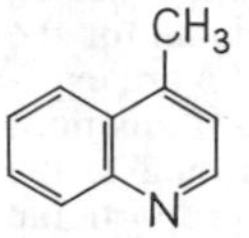

CH_3

N

Lethane®: see Insecticides.

Letsinger synthesis: see Solid-phase peptide synthesis.

Letter metal: see Type metal.

Leu: abb. for leucine.

Leucine, abb. ***Leu***: α-aminoisocaproic acid, $(CH_3)_2CH\text{-}CH_2\text{-}CH(NH_2)COOH$, an essential Amino acid (see), found in relatively large amounts in serum albumin and serum globulins. L. is isolated from protein hydrolysates, especially hemoglobin hydrolysates. After the hydrolysate has been neutralized, L. is obtained by salt formation with 2-bromotoluene 5-sulfonic acid. L. activates the internal secretory glands. It is converted by bacterial degradation to isoamylamine; alcoholic fermentation produces 3-methylbutanol(1), a component of fermentation amyl alcohol, from L.

Leucine aminopeptidase: see Aminopeptidases.

Leuckart reaction: cleavage of aromatic xanthogenic acid esters with alcoholic potassium hydroxide solution to form thiophenols:

$$C_6H_5\text{–}S\text{–}CS\text{–}OC_2H_5 \xrightarrow{+H_2O} C_6H_5\text{–}SH + COS + C_2H_5\text{–}OH$$

Leuckart-Wallach reaction: a method for reductive alkylation of amines with aldehydes or ketones and formic acid as reducing agents: $R\text{-}NH_2 + O = CH\text{-}R + HCOOH \rightarrow R\text{-}NH\text{-}CH_2\text{-}R + CO_2 + H_2O$. L. is best suited for production of tertiary amines, since in the synthesis of primary and secondary amines, the corresponding more highly alkylated compounds are always present in the reaction mixture as byproducts.

Leucosin: see Albumins.

Leucovorin: see Folic acid.

Leukobases: compounds formed by reduction of certain pigments (especially the triarylmethane series), or also the synthetic precursors of the pigments. In contrast to the Leuko compounds (see) of vat dyes, the L. are not directly absorbed by fibers and also cannot be oxidized by hanging the fibers in the air. The salts of L., however, are directly absorbed to wool and silk, and subsequent oxidation produces the dye.

Leukochrome: see Catecholamine.

Leuko compounds: literally "white compounds", the reduction products of the Vat dyes (see), which, however, are not necessarily white. The dye is usually reduced to an L. to make it water-soluble. The most commonly used reducing agent is sodium dithionite, $Na_2S_2O_4$.

Leuko indigo: see Indigo pigments.

Leukopterin: see Pterin.

Leukotrienes, abb. ***LT***: hormones formed from multiply unsaturated fatty acids, especially arachidonic acid, by the enzyme lipoxygenase. There are four types, A, B, C and D. A subscript indicates the number of double bonds; for example, LTB_4 is a 5,12-dihydroxyeicosa-6,8,10,14-tetraenoic acid (abb. 5,12-DHETE). L.C and D are thioethers. L. are secreted by certain cells, such as leukocytes and mast cells, and cause contraction of the smooth muscles, especially those of the bronchia. They are members of the group of *slow-reacting substances*) (abb. SRS). LTA is believed to be the biosynthetic precursor of the other L.

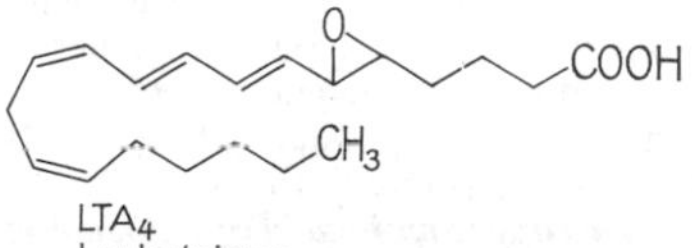

O COOH

CH_3

LTA_4
Leukotriene

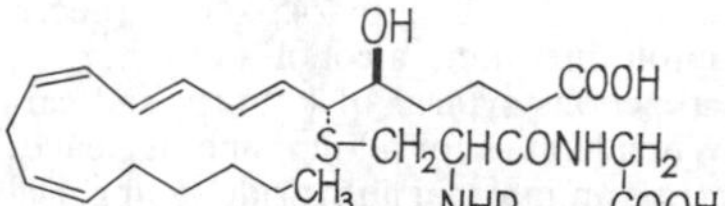

OH COOH

$S\text{–}CH_2CHCONHCH_2$

CH_3 NHR COOH

LTD_4: R = H

LTC_4: $R = CO(CH_2)_2CHCOOH$ with NH_2

Levan: a polysaccharide made up of fructofuranose; it is produced by sweet grasses and various bacteria. Bacterial L. have very high molecular masses, greater than 10^6.

Levarterenol: same as Noradrenalin (see).

Leveler: in dyeing, a substance which slows the penetration of the dye into the fibers, and in this way is supposed to give an even ("level") color. There are fiber-affine, dye-affine and indifferent E. ***Fiber-affine L.*** are adsorbed to the fibers more rapidly than the dye. They occupy the dye-binding sites and only slowly release them for the dye, so that it can be evenly taken up by the fibers. Sulfonates, sulfonic acids and fatty acid condensates are most often used for this. ***Dye-affine E.*** form compounds with the dye which are not adsorbed by the fibers, and these only gradually decompose to allow the dye to be slowly and evenly adsorbed. Ethoxylated fatty alcohols and alkylphenols (polyglycol ethers) are used here.

Levodopa: see Antiparkinson's drugs.

Levoglucosan: a Sugar anhydride (see).

Levonorgestrel: see Gestagens.

Levorphanol: a synthetic morphinane derivative which lacks the ether bridge of Morphine (see) and is thus a tetracyclic compound. L. is the (-)-enantiomer and has the same configuration as morphine at atoms 9, 13 and 14. Due to its phenolic OH group and the tertiary amine structure, L. is amphoteric. The O-methyl ether of the (+)-enantiomer is *dextromethorphan*. The (-)-enantiomer of L. is a highly effective analgesic. The (+)-enantiomer of L. has a pronounced cough-suppressing action, and dextromethorphan is still more effective. L. is addictive. Fig. p. 594.

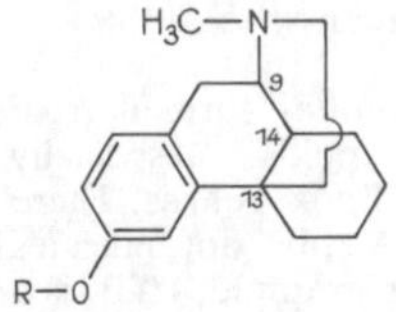

R = H; (−)-form: Levorphanol
R = CH_3; (+)-form: Dextromethorphan

Levothyroxin: see Thyroid hormones.

Levulinaldehyde, ***pentan-4-on-1-al***: CH_3-CO-CH_2-CH_2-CHO, a γ- or 1,4-ketoaldehyde. L. is a mobile, colorless liquid with a pungent odor; m.p. < -21 °C, b.p. 186-187 °C (dec.), n_D^{20} 1.4257. It is steam volatile and miscible with water, alcohol and ether in any proportions. L. undergoes the addition and condensation reactions typical of the Aldehydes (see). It is formed by ozonolysis of isoprene rubber.

Levulinic acid, ***4-oxopentanoic acid, γ-ketovaleric acid***: CH_3-CO-CH_2-CH_2-COOH, a colorless, crystalline compound; m.p. 37.2 °C, b.p. 245-246 °C (dec.). It is readily soluble in water, alcohol and ether. L. undergoes the reactions typical of ketones and carboxylic acids. When heated for a long time, it cleaves off water and forms an internal anhydride, α-angelica lactone:

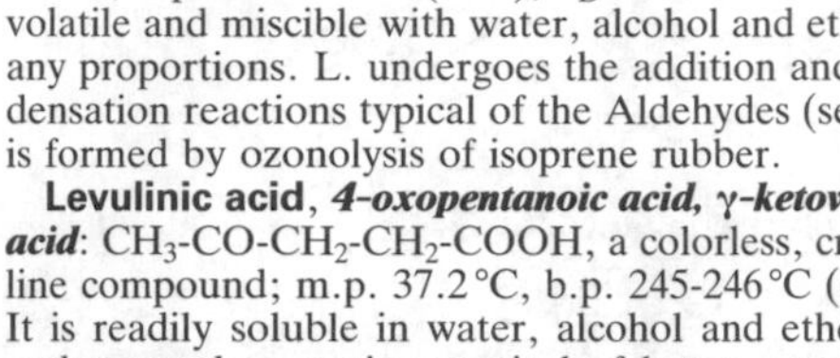

L. can be made by heating fructose with concentrated hydrochloric acid. Industrially, it is prepared mainly from starch. L. is used to prepare pharmaceuticals, softeners and plastics.

Lewis acid-base definition: see Acid-base concepts.

Lewisite: see Organoarsenic compounds, see Chemical weapons.

LFE relation: abb. for ***linear free enthalpy*** (or ***free energy***) relation: one of a group of empirical equations relating the structure of molecules to certain thermodynamic or kinetic properties. Such correlations between structure and reactivity have been observed from experiment. In general, they are applicable only within a single group of substances, or among closely related substances, but they are very useful in systematizing the multitude of reactions, especially in organic chemistry. There are LFE R. for equilibrium and rate constants, polarographic midstep potentials, infrared absorbance frequencies, chemical shifts in NMR and solvent effects on reaction rates.

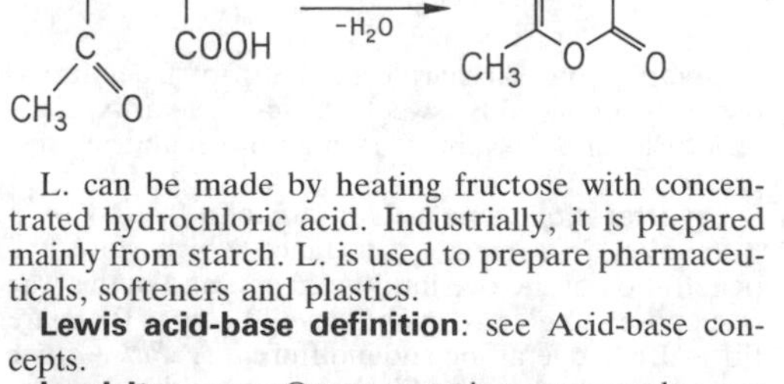

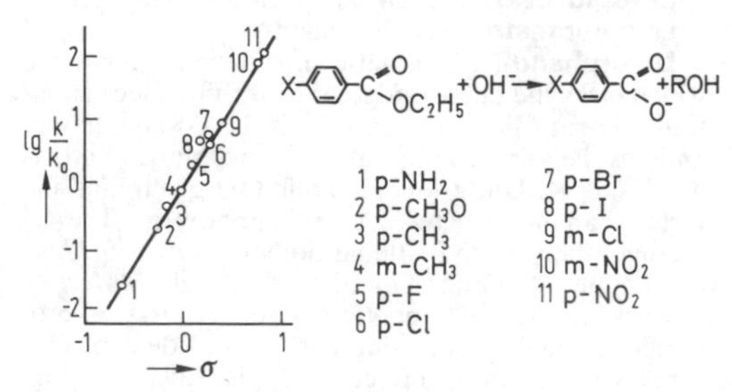

Hammett correlations for the reactions.

Most LFE R. have the form lg $k_1 = a$ lg $k_2 + b$. Here k_1 and k_2 are, for example, rate constants for the equivalent reactions in which one component has either substituent (1) or (2), or rate constants for the same reaction in different solvents. a and b are experimentally determined constants. Similar equations exist for equilibrium constants K. According to the second law of thermodynamics, lg K is proportional to the free energy of reaction, and according to transition state theory, lg k is proportional to the free energy of activation. Therefore these LFE correlations establish linear relationships between changes in free energy.

Some important LFE R. are the Hammett equation, the Taft equation, the Edwards equation and the Brœnsted catalysis law.

LFSE: abb. for ligand field stabilization energy; see Ligand field theory.

LH: see Lutropin.

Li: symbol for lithium.

Liberins, ***releasing hormones, releasing factors***: peptide neurohormones synthesized in the small-cell nuclear regions of the hypothalamus. They reach the anterior pituitary in the blood, and there stimulate secretion of the anterior pituitary hormones (e.g. thyrotropin, gonadotropins, somatotropin). The L. are synthesized in response to excitation of the nerve cells. In the pituitary, they act via the adenylate cyclase system. For the most part, each pituitary hormone has a corresponding L., but follitropin and lutropin are released by a single L. The following L. are known:

Thyroliberin, thyrotropin releasing hormone, abb. ***TRH***, pyroglutamyl-L-histidyl-L-prolinamide, Pyr-His-Pro-NH_2, stimulates the anterior pituitary to synthesize and secrete the thyroid-stimulating hormone Thyrotropin (see). It is not toxic and can be applied intravenously, intramuscularly and even per os. TRH is used in the diagnosis and therapy of thyroid disorders. Its reported antidepressive effect on human beings is disputed, as is its stimulating effect on prolactin secretion. An analog, [3-Me-His^2]TRH, has 8 times the biological activity of TRH.

TRH was isolated simultaneously by the groups of Guillemin and Schally (1969/70), from sheep and pig hypothalamus, respectively. Each group used about 300,000 hypothalami; both determined the structure of the hormone.

Gonadoliberin, gonadotropin releasing hormone, abb. ***GRH***, Pyr-His-Trp-Ser-Tyr-Gly-Leu-Arg-Pro-Gly-NH_2, promotes synthesis and secretion of lutropin and follitropin. Antibodies against synthetic GRH inhibit secretion of both LH and FSH. ***Luliberin, luteinizing hormone releasing hormone***, abb. ***LRH***, and ***folliberin, follicle-stimulating hormone releasing hormone*** are chemically identical. However, the existence of two L. cannot be excluded. GRH and synthetic analogs are used for diagnosis and therapy of male and female infertility. The hormone has broad application for control of fertility in animal breeding. GRH antagonists are of interest as a possible starting point for the development of long-term contraceptives. GRH was also discovered, and its structure elucidated, by Schally and Guillemin.

Corticoliberin, corticotropin releasing hormone, abb. ***CRH***, a peptide of 41 amino acid residues; it stimulates synthesis and secretion of corticotropin.

Somatoliberin, somatotropin releasing hormone, abb. ***SRH***, Val-His-Leu-Ser-Ala-Glu-Glu-Lys-Glu-Ala, a decapeptide which stimulates the pituitary to synthesize and secret somatotropin. The identity of the synthetic peptide with the natural hormone has not yet been confirmed.

Prolactoliberin, prolactin releasing hormone, abb. ***PRH***, stimulates formation and secretion of prolactin. Although TRH stimulates an increase in prolactin secretion in some animals and in human beings, TRH and PRH are not identical.

Melanoliberin, melanotropin releasing hormone, abb. ***MRH***, stimulates the secretion of melanotropin. It is postulated that the hexapeptide Cys-Tyr-Ile Gln-Asn-Cys derived from oxytocin is MRH.

Librium®: see Benzodiazepines.

Lidocain: see Local anesthetics.

Lieben's iodoform test: reaction for the detection of acetone in urine. The acidified, acetone-containing urine is distilled; the distillate is brought to alkaline pH and mixed with iodine solution. Yellow iodoform, which has a characteristic odor, precipitates from the solution:

$$R\text{-}CO\text{-}CH_3 + 3\,I_2 \rightarrow R\text{-}CO\text{-}CI_3 + 3\,HI;$$
$$R\text{-}CO\text{-}CI_3 + NaOH \rightarrow CHI_3 + R\text{-}CO\text{-}ONa$$

The L. is not specific for acetone, since ethanol and compounds which contain an acetyl group CH_3-CO also react in this way. The reaction can also be carried out with chlorine or bromine, and is known in general as the haloform reaction.

Liebermann reaction: a detection reaction for phenols using nitrous acid. It can also be used to detect nitrites and nitroso groups. Phenol in conc. sulfuric acid reacts with a sodium nitrite solution, forming a blue-green indophenol sulfate. When the solution is diluted with water, this is converted to a red indophenol. Addition of alkali hydroxide solution generates a dark blue indophenolate salt.

Liebig cooler: see Coolers.

Lifetime, ***average L.***: the time τ after which the initial concentration c_0 of a substance has been reduced to c_0/e (where e is the base of the natural logarithms) by a chemical or physical process. The L. is most often indicated for short-lived intermediates and excited states. If the decay of such a species occurs by a first-order process (see Kinetics of reactions), τ is equal to the reciprocal of the rate constant k: $\tau = 1/k$, and is related to the half-life $t_{1/2}$ by the equation: $\tau = t_{1/2}/\ln 2$. If the decay follows an nth-order rate equation, $\tau = t_{1/2}(e^{n-1}-1)/(2^{n-1}-1)$.

Ligand: a molecule or ion which is attached to a central atom by coordinate bonding (see Coordination chemistry, see Nomenclature, sect. II G 1).

Ligand exchange chromatography: see Ion-exchange chromatography.

Ligand field strength parameter: see Coordination chemistry.

Ligand field theory: a treatment of the changes in the electronic structure of the central atom of a complex due to the electric field generated by the ligands. The ligands are represented by point charges or point dipoles. This approach is sometimes called *crystal field theory* in the literature, with the term L. also including the electronic structure of the ligands as described by Molecular orbital theory (see). However, the first definition of L. seems more appropriate, especially for interpretation of properties of isolated complex ions in solution. Its expansion on the basis of the MO theory is called the *MO theory of complexes*.

The optical and magnetic properties of transition metal complexes can be explained by L. The addition of ligands leads to splitting of the energies of the degenerate d-orbitals of the transition metal ion, because the symmetry of the ligand arrangement is different from the spherical symmetry of the free ion. The splitting of the d-energy levels for a given ligand field can be qualitatively explained, in the one-electron approximation, by the directionality of the d-orbitals of the central ion (see Atom, models of) and the resulting differential electrostatic repulsions between the d-electrons and the ligands.

The splitting of the d-levels in octahedral, tetrahedral and square planar ligand arrangements are shown in the figure. The energy difference $E(e_g)$-$E(t_{2g}) = \Delta_{oct}$ or $E(t_2)$-$E(e) = \Delta_{tetr}$ is called the (ligand-)*field strength parameter* in the octahedral or tetrahedral field.

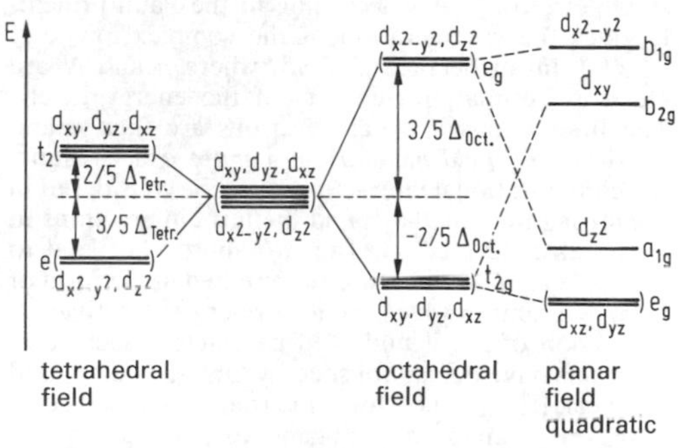

Splitting of a D-electronic state in different ligand fields.

The parameter Dq is often used instead of Δ; $\Delta = 10\,Dq$. The occupation of the split d-orbitals by electrons leads to a gain in energy, which is called the *ligand field stabilization energy (LFSE)*. The field strength parameter Δ depends on the distance between the central ion and the ligand, and on the charges or dipole moments of the ligands. It is usually determined by spectroscopy, and the wavenumber unit in cm^{-1} is given.

If the ligands are arranged according to increasing Δ values, i.e. according to increasing strength of the ligand field, the order is essentially the same for any central ion (*spectrochemical series*): $I^- < Br^- < Cl^- < SCN^- < F^- < OH^- < H_2O < NH_3 < NO_2 < CN^-$.

For a transition metal ion with several d-electrons, it is possible as a first approximation to generate configurations by occupation of one-electron states of the central ions, ignoring the inter-electronic interactions. This can be done in several ways, depending on the size of the field strength parameter Δ. For example, a strong ligand field causes a large splitting of the d levels of an octahedral Fe^{3+} complex, and the five d-electrons of the Fe^{3+} ion therefore occupy the lower-energy t_{2g} levels with some spin pairing. The total spin of the complex is $S = 1/2$. (*low spin complex*). The complex $[Fe(CN)_6]^{3-}$ is an example of this case. With a weak ligand field, e.g. in $[FeF_6]^{3-}$, the energy difference between the e_g and t_{2g} levels is smaller than in $[Fe(CN)_6]^{3-}$. The energy required for occupation of the higher e_g levels is less than the energy gained by unpairing of the electron spins (Hund's rule), so that in the $[FeF_6]^{3-}$ complex, the t_{2g} and e_g orbitals are first occuplied singly and with electrons of parallel spin. The total spin of the $[FeF_6]^{3-}$ complex is ($S = 5/2$) (*high spin complex*). The magnetic properties of the complexes can be interpreted by such approximations. The electron interactions and thus the term system of a transition metal complex can be accomodated in two ways: 1) if the interaction between individual d-electrons is large compared to the effect of the ligand field (*weak field method*), the configuration d^N of the central ion gives the terms ^{2S+1}L of the free ion (see Multiplet structure). The action of the ligand field on the terms of the free ion yields the term system of the complex ion. 2) If the effect of the ligand field is large compared to the electron interaction, the influence of the ligand field on the d^N configuration is considered first. The occupation of the single-electron states according to the Pauli principle gives the configuration of the complex ion, e.g. $t_{2g}{}^n e_g{}^{N-n}$ in an octahedral field, where n and N are electron occupation numbers of the energy levels. The interactions between electrons are then calculated (*strong field method*). For more exact results, the configurational interactions must be considered in the strong field method, and the term interactions in the weak field method. The two approaches lead to the same results, which are represented in the form of term diagrams in which the term energies are given as a function of the ligand field parameter. Such term diagrams have been published by Orgel, who applied the weak field method for some transition metal complex ions. Tanabe and Sugano carried out similar studies using the strong field method. In the *Tanabe-Sugano diagrams*, the term energy in the form E/B is given as a function of Δ/B. The parameter B is a measure of the electron interaction, and is called the *Racah parameter*.

As early as the 1930's, Pauling had developed a covalent model for treatment of complex compounds based on valence structure theory (see Coordination chemistry).

Ligases: one of the main groups of Enzymes (see). Some important L. are the DNA ligases (see) and Pyruvate carboxylase (see).

Ligator atom: see Coordination chemistry.

Light conversion: see Photosynthesis.

Light gasoline: see Gasoline.

Lighting gas: see City gas.

Lighting oil: same as Kerosene (see).

Light metal: a metal with a density lower than 5 g cm^{-3}. There are 15 L.: the alkali metals, the alkaline earth metals (except for radium), aluminum, yttrium, scandium and titanium. The lightest L. is lithium (density 0.534 g cm^{-3}) and the heaviest is titanium (density 4.506 g cm^{-3}). The most important L. in industry are aluminum and magnesium, but titanium and beryllium, and the two rare-earth metals yttrium and scandium are becoming increasingly important. In industry, the term L. is often applied to alloys of the L. as well. The opposite of L. is a Heavy metal (see).

Light quanta: same as Photons (see).

Lignans: phenolic substances from plant or wood extracts. They consist of two phenylpropane units linked in the β-position, and thus are structurally similar to lignin. L. can be extracted from the plant material with hot water, ether, benzene, toluene, polar solvants or alkali hydroxide solutions. They are formed mainly at the boundary between the heartwood and outer wood, and are deposited in the wood. The L. content of wood can vary between 1% and a maximum of 30%. L. have fungicidal, insecticidal and antioxidizing effects.

Lignin: a biopolymer found in higher plants. L. acts as a matrix for cellulose and hemicelluloses in wood, which contains 20 to 35% L. L. is an amorphous, optically inactive, three-dimensional heteropolymer. It is not soluble in any ordinary solvent unless it is degraded. It is biosynthesized by oxidative polymerization of hydroxylated and partially methylated cinnamyl alcohols, of which the most abundant is coniferyl alcohol.

CH_2O-
|
$CH-$
|
$CH-$
R, O

Bonding sites on coniferyl alcohol ($R = OCH_3$) which may form cross links in lignin.

The proportions of the individual monomers vary from one source to the next. The L. of pteridophytes and gymnosperms generally contain a majority of guaiacyl groups (4-hydroxy-3-methoxylphenyl or guaiacyl L.) The L. of conifers, for example, contains an average of 14% 4-hydroxyphenyl, 80% guaiacyl and 6% syringyl (4-hydroxy-3,5-dimethoxyphenyl) groups. The L. of deciduous trees contains about equal proportions of guaiacyl and syringyl groups (guaiacyl-syringyl L.), but contains very few 4-hydroxyphenyl groups. Among the dicotyledonous plants, the herbs contain fewer syringyl groups than the woody species. The wood of monocotyledonous plants generally contains about 10% 4-hydroxyphenyl groups.

The monomers are biosynthesized by the shikimic acid pathway via phenylalanine, cinnamic acid and the corresponding substituted cinnamic acids (see Hydroxycinnamic acids). Polymerization of the monomers is initiated by peroxidases, which catalyse formation of resonance-stabilized aroxyl radicals. These polymerize non-enzymatically by pairing. The process links the monomers by C-C and C-O-C bonds, leading to a network structure. Hemicelluloses are bound to the structure via ether bridges. The figure indicates the preferred bond sites.

The biosynthesis of L. can be considered a dehydrogenation polymerization, in which enzymatic phenol dehydrogenation and non-enzymatic polymerization are linked. The responsible enzymes are called *laccases* (phenol dehydrogenases), and they are also used for "artificial" production of L. from coniferyl alcohol.

The production of cellulose from wood requires removal of L. In the current industrial processes, the wood is treated with sodium hydroxide, sodium sulfide and sodium sulfate and boiled (*alkaline extraction* or *sulfate process*), or it is treated with hydrogen sulfite solution (*acid extraction* or *sulfite process*). The sulfite process generates low-molecular-weight, water-soluble lignin sulfonic acids and lignin sulfonates, which are not readily biodegradable. These products are used as additives to road surfacing materials or for making briquets.

Lignin sulfonic acid: the main component of Sulfite liquor (see).

Ligroin: a term for a gasoline fraction in the middle boiling range of about 120 to 135 °C. The term L. does not guarantee a particular composition. Sometimes light gasoline is also called L. It is used as a solvent for fats, oils, resins and gums.

Lime: a term applied to several calcium compounds. L. itself is Calcium carbonate (see), $CaCO_3$. By heating (burning), L. is converted to ***burnt L.***, CaO (see Calcium oxide). Treatment of burnt L. with water (slaking the L.) forms ***slaked L.***, $Ca(OH)_2$ (see Calcium hydroxide). ***Water lime*** is calcium oxide which contains more than 10% silicon dioxide, aluminum oxide and iron impurities, but is slaked by water. ***Hydraulic L.*** contains more than 15% of these impurities and is not slaked; it hardens in the air or under water.

Lime-carbonic acid equilibrium: an important parameter in water preparation. The equilibrium between free, dissolved carbonic acid and the carbonate hardness of the water (see Hardness) depends on pH and temperature. The equilibrium is important for de-acidifying processes in water purification. A distinction is made between free and bound carbonic acid in water. Tillmans was able to show that a certain amount of free carbonic acid (associated carbonic acid) is required to maintain the carbonate hardness of water in solution. If there is too little carbonic acid in solution, the calcium carbonate precipitates. If there is an excess of free carbonic acid, the water is aggressive and can cause considerable damage to building materials such as concrete and iron.

Lime excess process: see Water softening.

Lime milk: see Calcium hydroxide.

Lime paints: paints containing slaked lime, with or without addition of water-soluble glues as binders, and lime-resistant pigments.

Lime saltpeter: see Calcium nitrate.

Lime sandstone: see Binders (building materials).

Lime-soda process: see Water softening.

Lime water: see Calcium hydroxide.

Limit concentration: a parameter for the sensitivity of qualitative detection reactions. The L. is the smallest concentration of substance, in g ml^{-1}, which can be detected with certainty. The negative decadic logarithm of the G. is sometimes indicated as the sensitivity exponent pD. A L. can also be defined statistically for quantitative methods of analysis.

Limit conductivity: see Electrical conductivity.

Limit current: the maximum current flow during an electrolysis. It is reached when reactions occurring before or after the current-carrying reaction become rate-limiting, or when the transport processes determine the rates of the electrode reactions. At this point, an increase in the electrolysis voltage does not increase the electrolysis current, unless the increase is so great that it forces further electrode reactions to occur. The L. can be controlled, for example, by diffusion. Diffusion-controlled L. are used as analytical signals in polarography.

Limit hydrocarbon: see Alkanes.

Limit of detection: a parameter for the sensitivity of qualitative detection reactions. It is the smallest amount of substance in µg which can be reliably detected under exactly defined conditions. The L. for a given volume can be calculated from the Limit concentration (see). A L. can also be defined statistically for quantitative analysis methods.

Limit structures: see Resonance.

Limonene: a monocyclic monoterpene hydrocarbon, a colorless oil with an odor like lemon; b.p. 175-176 °C at 102 kPa. L. is practically insoluble in water, but is soluble in alcohol and ether. It is readily oxidized in the presence of air and light. L. exists as a d- and an l- form, and occurs very widely in essential oils. d-L. can be obtained by fractional distillation from orange peels or carroway oil, and l-L. from noble fir cone oil. The racemate is known as *dipentene*. L. is used in large amounts in the perfume industry, and is often used to fake essential oils.

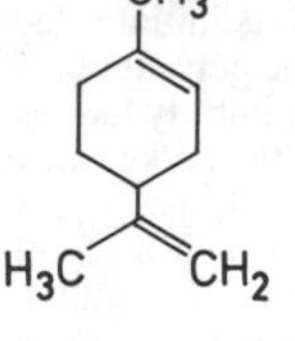

CH_3

H_3C CH_2

Linalool: 3,7-dimethylocta-1,6-dien-3-ol, an acyclic monoterpene alcohol. L. is a colorless, optically active oil with a flowery odor. It is slightly soluble in water, and soluble in alcohol; b.p. 198-199 °C at 130 kPa. L. is found in free form or as an ester in many essential oils. l-L. is the main component of spearmint oil, and d-L. of coriander oil. ***Linalyl acetate*** is present in relatively larger amounts in lavender, lemon and bergamot oils. L. is readily isomerized in the presence of acid to geraniol and nerol. It is used in the perfume industry and is the starting material for the synthesis of citral and ionone.

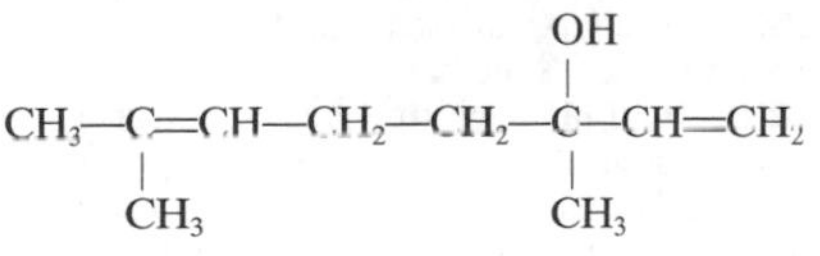

$$CH_3{-}\underset{\displaystyle CH_3}{\underset{|}{C}}{=}CH{-}CH_2{-}CH_2{-}\overset{\displaystyle OH}{\overset{|}{\underset{\displaystyle CH_3}{\underset{|}{C}}}}{-}CH{=}CH_2$$

Lincomycin: an antibiotic formed by *Streptomyces lincolnensis*; it has an S-α-glycoside structure. L. is most effective against gram-positive bacteria, and is used when the pathogens are resistant to penicillin and erythromycin. A partially synthetic derivative of L. is ***clindamycin***, which has the other configuration at C-7.

Lindane, ***γ-HCH***: the γ-isomer of 1,2,3,4,5,6-Hexachlorocyclohexane (see), and a chlorinated hydrocarbon insecticide. L. is a colorless crystalline powder, which has practically no odor; m.p. 112.9 °C, b.p. 176 °C at 1.33 kPa, vapor pressure $1.2 \cdot 10^{-4}$ Pa at 20 °C. It is soluble in acetone, chloroform, benzene, xylene, dichloroethane and ether, slightly soluble in alcohol, and dissolves to 10 ppm in water. L. is stable in neutral media, but in alkaline media, HCl is split off. By definition, L. contains at least 99% γ-HCH. It is synthesized by additive chlorination of benzene to technical hexachlorocyclohexane, from which L. is isolated by extraction and recrystallization, for example with methanol.

$3Cl_2$, hν; 15...20%; γ-HCH

The positions of the Cl atoms are aaa eee, where a = axial and e = equatorial.

The isolation of γ-HCH concentrates leaves ≈ 90% waste isomers, which are either transported to a toxic waste dump or are further processed, usually by way of the alkaline or pyrolytic dehydrochlorination to 1,2,4-trichlorobenzene (TCB).

γ-HCH is the only HCH isomer with excellent insecticidal properties. It is similar in its mechanism and action spectrum to DDT (see). Because it is more volatile, however, it acts as a respiratory poison as well as a contact and feeding poison. Its initial effect is greater than that of DDT, but its persistence is lower, as is its tendency to accumulate in fatty tissues. The acute oral toxicity is LD_{50} = 170-300 mg/kg rat. It is metabolized in warm-blooded animals via pentachlorocyclohexene to chlorophenols, which are coupled to glucuronic acid and then excreted. For human beings, oral ingestion of 14 g in an oily solvent can be fatal. L. is used to protect crops and to fumigate empty storage spaces; it is also used to combat soil and hygienic pests and ectoparasites in human and veterinary medicine.

Lindemann glass: glass with relatively light oxide components, such as lithium borate or lithium beryllium borate glass, used for x-ray windows in x-ray tubes.

Lindemann-Hinshelwood mechanism: see Monomolecular reactions.

Linde process: see Gas liquefaction.

Linear free enthalpy (or energy) relationships: see LFE relationships.

Line defects: see Crystal lattice defects.

Line spectrum: see Spectrum.

Linewidth: the finite width of a spectral line, which is independent of the resolution of the spectral device used to measure it. It can be described mathematically, e.g. by a Cauchy or Gauss function. The observed L. depends on the natural L. and external factors.

The *natural L.* is determined by the lifetime of the corresponding energy state. The Heisenberg Uncertainty Principle can be expressed as $\Delta E \Delta t \geqq h/2\pi$ (1). Substitution of the equation $E = h\nu$ gives the expression $\Delta\nu = 1/2\pi\Delta t$ (2). Δt is the uncertainty in the time of the emission process, ΔE is the uncertainty in the amount of energy emitted, $\Delta\nu$ is the natural linewidth and h is Planck's constant. For atoms, Δt is on the order of 10^{-8} s, so the natural width of a spectral line is about 10^{-5}. If long-lived metastable states are involved, the natural L. can be still narrower. The lifetime of an excited state can be decreased by external factors; according to Eq. (2), this causes a *broadening* of the lines. The lifetime is shortened by collisions of the particles with one another (collisional broadening). Lorentz broadening, which is caused by collisions with foreign particles, can be distinguished from resonance broadening, which is caused by collisions between identical particles. Collisional broadening is a function of the pressure and temperature. Line broadening is also observed as a result of the thermal motion of the emitting particles (Doppler effect). The Doppler broadening $\Delta\lambda_D$ depends on the temperature T, the wavelength λ and the mass m of the emitting particles: $\Delta\lambda_D \propto \lambda \cdot \sqrt{T/m}$ (3). To obtain sharp lines, one must work at low pressures and as low temperatures as possible.

α-Linolenic acid: all-*cis*-9,12,15-octadecatrienoic acid, abb. α-Lnn or 18:3(9,12,15), $H_3C(CH_2CH = CH)_3(CH_2)_7COOH$, a multiply unsaturated essential fatty acid. α-L. is a colorless liquid; m.p. -11 °C, b.p. 225 °C at 1.3 kPa. It is practically insoluble in water, and is very readily oxidized by oxygen. α-L. is the main component of drying oils. Linseed oil contains 40 to 62% α-L.; hemp, nut and soy oils are also rich in it.

Linoleic acid: all-*cis*-9,12-octadecadienoic acid, abb. Lin or 18:2(9,12), $H_3C(CH_2)_3(CH_2CH = CH)_2(CH_2)_7COOH$, a multiply unsaturated essential fatty acid. L. is a colorless liquid, m.p. -5 °C, b.p. 224 °C at 1.3 kPa. It is practically insoluble in water, readily soluble in organic solvents, and easily oxidized. L. is present in drying and semidrying plant oils. Linseed oil contains 16 to 25% L., poppyseed oil about 60% and sunflower seed oil about 50%. L. is used to make quickly drying paints and emulsifiers, and is used therapeutically for dermatoses.

Lino metal: see Type metal.

Linurone: see Urea herbicides.

Lipases: enzymes of the carboxylase group (see Esterases) which catalyse the cleavage of neutral fats (triacylglycerols) into fatty acids and glycerol or di- or monoacylglycerols. L. are abundant in fatty tissues, stomach and pancreas, and in ricinus seeds. They are activated by bile acids and calcium ions. The stomach L. initiates the process of liquefying fats by partial degradation. Degradation is completed in the small intestine by the pancreatic L. (M_r 35,000) with taurocholate as activator.

Lipids: lipophilic derivatives of long-chain monocarboxylic acids, the fatty acids. In ***simple L.***, the fatty acids are esterified to alcohols. This group of highly lipophilic compounds includes the Waxes (see), which are esters of lipophilic alcohols, and the Fats (see) or fatty oils, which are esters of glycerol. ***Complex L.***, also called lipoids, contain other components in addition to the fatty acids and alcohol, such as phosphoric acid or its esters (see Phospholipids), or mono- or oligosaccharide groups (see Glycolipids). Complex L. are subdivided into Glycerolipids (see) and Sphingolipids (see) on the basis of the alcohol component (Fig. 1).

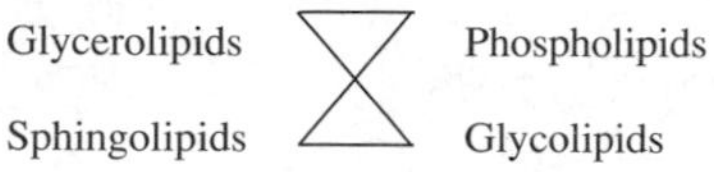

Glycerolipids — Phospholipids
Sphingolipids — Glycolipids

Complex L. are amphiphilic substances, consisting of a lipophilic component, namely the alkyl chains of the fatty acids and the sphingoid, and a hydrophilic component ("head group"), the phosphoric acid or carbohydrate group (Fig. 2).

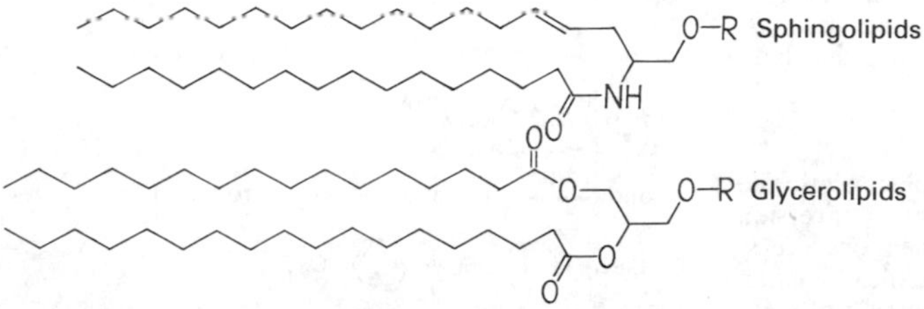

O–R Sphingolipids
NH
O–R Glycerolipids

Fig. 2. General structures of complex lipids. In phospholipids, R = phosphoric acid or phosphate group; in glycolipids, R = mono- or oligosaccharide group.

The hydrophilic component can be neutral (as in phosphatidylcholine, sphingomyelin and neutral glycolipids) or acid (as in phosphatidylserine, phosphatidylglycerol and acid glycolipids). Complex L. can exist in various mesomorphic states.

L. with low polarity, such as the triacylglycerols, are readily soluble in aliphatic or aromatic hydrocarbons, but only slightly soluble in methanol. On the other hand, polar L., such as phospholipids or glycolipids, are only slightly soluble in hydrocarbons. Most phospholipids are only slightly soluble in cold acetone, while the di- and triacylglycerols dissolve readily. These differences in solubility make rough separation possible. The melting points of the L. are not highly characteristic of individual compounds, due to the occurrence of polymorphic forms and mixed crystal formation. Specific optical rotations are very small and not well suited to characterization.

α-Lipoic acid: the disulfide of 6,8-dithiooctanoic acid; a coenzyme. Native α-L. is bound to the protein by an amide linkage of the carboxyl group. After hydrolysis of the amide bond, e.g. with 6 N mineral acid, the α-L. can be extracted with organic solvents. Only the (+) form is biologically active.

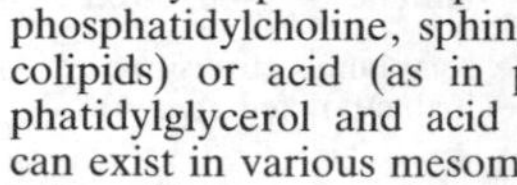

S—S
COOH

In vivo, α-L. acts as a redox system in oxidative decarboxylations. Reduction of the disulfide bridge oxidizes an "active acetaldehyde" bound to thiamin pyrophosphate to an acetyl group which is linked as a thioester to one of the mercapto groups of the reduced α-L. It is subsequently transferred to coenzyme A. α-L. is synthesized by most organisms, but it is a growth factor (vitamin) for a few bacteria and protozoa.

Lipoids: see Lipids.

Lipophilic: readily soluble in fats, fatty oils and lipid-like substances.

Lipopolysaccharides: components of the cell walls of gram-negative bacteria. L. can be extracted from the bacteria with 45% aqueous phenol at 65 °C. According to O. Westphal, the L. consist of three structurally and functionally different components, covalently linked to each other: the O-specific chain, a linking core oligosaccharide and the lipid A. The O-specific chain consists of repeating oligosaccharide units; the nature of this chain varies from one strain to the next, and gives the bacterial strain its antigenic specificity ("serotype"). The core oligosaccharide (R-core) contains a few characteristic carbohydrates, such as L-glycero-D-mannoheptose and 3-deoxy-2-oxo-D-mannooctanoic acid in *Salmonella* species. Lipid A contains disaccharide blocks; in *Salmonella, Shigella, Serratia, Pseudomonas* and *Selenomonas*, these consist of two β(1→6)-linked glucosamines; each disaccharide unit is linked to the next by a diphosphate bridge. Fatty acids are linked to the hydroxyl and amino groups of the disaccharides; some of these fatty acids have acyloxy groups in the 3-position. The L. are the most important surface antigens (O-antigens) in the bacteria, and act as endotoxins. The active component of the L. is lipid A, which also anchors the L. in the membrane.

Lipoproteins: see Proteins.

Liposomes: closed lamellar vesicles which form in aqueous suspensions of various lipids or lipid mixtures. L. formed by mixtures of phosphatidylcholine (egg lecithin), cholesterol and an acidic component, such as phosphatidic acid, are of particular importance. The L. which are formed by shaking the suspension are multilamellar, and have diameters of 100 to 3000 nm; when treated by ultrasonic vibration, they are converted to unilamellar L. (diameter about 50 nm). Unilamellar L. of larger diameter (up to 500 nm) can be created by addition of detergents to the lipid mixture followed by dialysis (cholate dialysis method). Active substances can be incorporated into both the lipophilic and the enclosed hydrophilic phase. L. serve as model membranes for cell biological research and may be developed as systems for application of drugs.

Lipothyronin: see Thyroid hormones.

Lipotropic hormone: same as Lipotropin (see).

Lipotropin, ***lipotropic hormone***, abb. ***LPH***: a polypeptide hormone with a lipolytic effect. L. is an anterior pituitary hormone discovered in 1964; it consists of β- and γ-lipotropin. Porcine β-L. consists of 91 amino acids, and the *N*-terminal sequence 1-58 corresponds to the primary structure of γ-L. Bovine β-L. consists of 93 amino acids and differs significantly in the *N*-terminal region from porcine β-L. β-L. itself has no known hormonal effects in human beings, and is thought to be a prohormone of β-endorphin and β-melanotropin. β-L. is biosynthesized from the common precursor pro-opiomelanocortin.

Lipowitz metal: an alloy named for its inventer; it consists of 50% bismuth, 25% lead, 13% tin and 12% cadmium. L. melts at 70 °C and is used as a solder for materials which are damaged by high temperatures. It is also used in melting fuses and as a heating bath liquid.

Liptinite: see Macerals.

Liptodetrinite: see Macerals.

Liquefaction: the conversion of a substance to the liquid state. This can be 1) the isothermal conversion of a solid to the liquid state by adding heat or increasing the pressure (see Melting) or by chemical reaction (see Coal hydrogenation); 2) the conversion of a gas to the liquid state by cooling and Condensation (see). The L. of gases and air is applied in gas liquefaction and in refrigerators.

Liquid air: see Gas liquefaction, see Air fractionation.

Liquid crystals, ***crystalline liquids, mesophases***: organic compounds in a state of order intermediate between that of a three-dimensionally ordered crystal and that of an ordinary, unstructured liquid. L. have properties both of the crystalline state (e.g. anisotropy of many properties, elasticity) and of the liquid (e.g. low viscosity in some cases, lack of independent form). L. were discovered in 1888 by F. Reinitzer and O. Lehmann, and were first investigated by them.

There are two types of L.: thermotropic and lyotropic.

1) ***Thermotropic L.*** are a state entered by pure substances or mixtures in a certain temperature range bounded by the melting point of the crystalline phase and the *clarification point*, at which the L. is transformed into an isotropic liquid.

a) *L. with rod-shaped molecules*. About 13,000 organic compounds with liquid crystalline properties are known; they are characterized by long, rod-shaped molecules with the general formula R^1-C_6H_4-M-C_6H_4-R^2, i.e. two benzene rings, each with a para-substituent, are connected by a middle group M (or directly). The middle group can be, e.g. -CH=N-, -CH=CH-, -C≡C-, -N=N-, -N=N(O)-, -COO-, -CH=CH-COO-, -C_6H_4- or another ring system. The p-substituents on the two benzene rings may be the same or different, e.g. -F, -Cl, -CN, -NO_2, -C_nH_{2n+1}, -CO--C_nH_{2n+1}, -OOC-C_nH_{2n+1}. Recently, compounds have been discovered in which one or more benzene rings are replaced by cyclohexane or five- or six-membered heterocyclic rings. Furthermore, elements such as Hg, Si, Sn, S and Se can be incorporated into the molecule. For practical applications, low-melting L. are needed; the existence range of the liquid-crystalline state can be extended by means of melting-point lowering in mixtures. A summary of representative examples of various classes of compounds with rod-shaped molecules is shown below.

1) CH_3-C_6H_4-N=N(O)-C_6H_4-OCH_3

solid ⟷(391) nematic ⟷(408) isotropic-liquid

2) C_6H_5-C_6H_4-C_6H_4-C_6H_4-C_6H_5

solid ⟷(674) nematic ⟷(718) isotropic-liquid

3) C_5H_{11}-C_6H_4-C_6H_4-CN

solid ⟷(297) nematic ⟷(308.5) isotropic-liquid

4) $C_6H_{13}O$-C_6H_4-(pyrimidine)-C_6H_{13}

solid ⟷(304) nematic ⟷(333.6) isotropic-liquid

5) C_5H_{11}-(H)-COO-C_6H_4-CN

solid ⟷(319) nematic ⟷(352) isotropic-liquid

6) C_4H_9O-C_6H_4-CH=N-C_6H_4-C_2H_5

solid ⟷(313) smectic ⟷(324) nematic ⟷(338) isotropic-liquid

Thermotropic liquid-crystalline compounds (transition temperatures are shown above the double-pointed arrows).

There are several types of L., all of which may occur in a single substance; in this case they are separated by phase transitions (polymorphism of L.). Each type corresponds to a certain structure (Fig. 1). In *nematic L.*, the long axes of the molecules, which in the isotropic liquid are randomly distributed, are mostly parallel to one another; their direction then determines the preferred direction of the system.

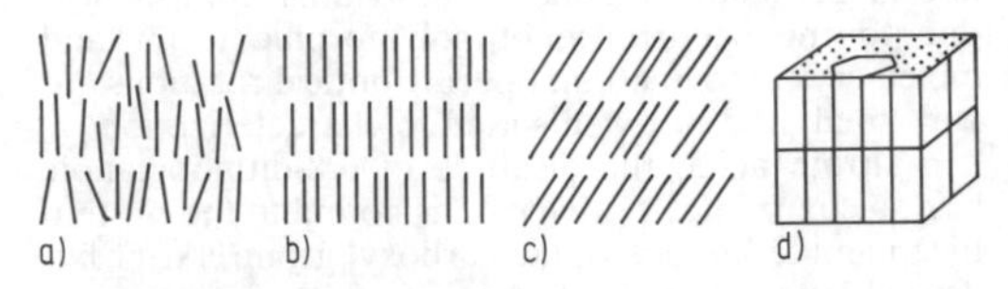

Fig. 1. Schematic representation of the states of order in liquid crystals: a) nematic phase, b) smectic A-phase, c) smectic C-phase, d) smectic B-phase.

Cholesterolic L., a sub-type of nematic L., consist of quasi-nematic layers arranged at a certain angle to each other, which results in a spiral structure. In *smectic L.*, the molecules are parallel to each other and are arranged in layers. There are at least 10 different types of smectic L., which differ in the arrangement of the molecules within the layers. For example, in smectic A the molecules are perpendicular to the plane of the layer, but there is no fixed distance between them; in smectic C the molecules are inclined with respect to the plane of the layer; in smectic B, the molecules are arranged within the layers in a two-dimensional, hexagonal lattice. Some types of smectic L. have recently been found to have three-dimensional order, and are thus very close to crystals. The various structures can be demonstrated by X-ray studies. L. with the same structure are generally completely miscible, and have characteristic optical properties, which are observed in the polarization microscope as textures. These are used for recognition and classification of the L.

It was shown by molecular statistical calculations that in the L., the parallel arrangement of the molecules arises mainly from the repulsive forces between the rod-shaped molecules. The attractive forces (dispersion force and orientation force) amplify this effect and are also responsible for the arrangement of the molecules in layers. L. have elastic properties which can be described using continuum theory.

The anisotropic molecular shape and arrangement causes anisotropy of many physical properties of L. Some examples of this are anisotropy of optical properties, which lead to optical double refraction, of electrical n conductivity, dielectricity and diamagnetism.

Applications of the L. Until about 1960, L. were the province of a small number of specialists, but since then a growing number of applications have been found for them. Because of their diamagnetic anisotropy, the preferred direction of nematic L. orients itself parallel to magnetic fields; this is used in NMR, ESR and UV/VIS spectroscopy to study the anisotropy of the properties of dissolved substances. The preferred direction orients either parallel or perpendicular to the direction of an electric field, depending on the sign of the dielectric anisotropy. L. which are homogeneously oriented by epitactic interaction (see Epitaxis) with specially treated electrodes can be reversibly reoriented, and can thus be used in optoelectronic components (displays) for electronic watches, pocket calculaters, digital displays and small display screens. Such liquid crystal displays (LCDs) are shown diagrammatically in Fig. 2; they use very little energy.

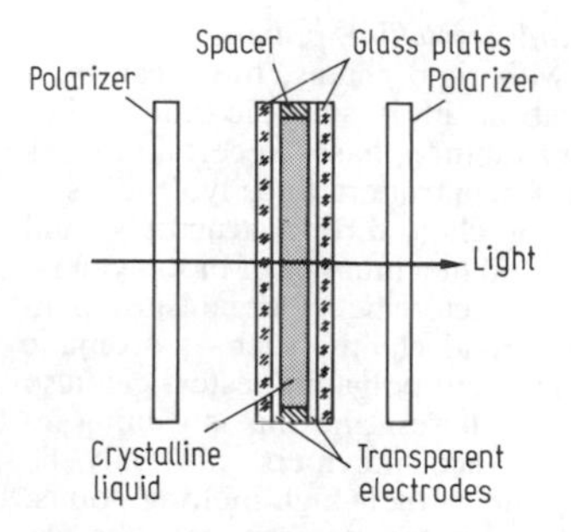

Fig. 2. Diagram showing the general structure of liquid crystal displays.

In gas chromatography, use of L. permits maximum separations. Cholesterolic L. reflect light selectively; the color depends on the temperature, so they can be used for measurement of surface temperatures (thermographics) for medical and industrial purposes.

In some cases, rod-shaped molecules do not display nematic or smectic layered structures, but instead form micellar aggregates of parallel molecules. The micellar aggregates form a three-dimensional structure with cubic symmetry. Such *cubic mesophases* are optically isotropic and highly viscous.

b) *L. with disk-shaped molecules*. As has been known since about 1980, disk-shaped (discotic) molecules (Fig. 3) can stack in columns, which in turn produce various liquid crystalline structures through their mutual orientation (*discotic L.*). Simple parallel orientation of columns which can still be shifted results in discotic nematic phases; however, discotic phases with hexagonal or tetragonal arrangements of the columns have also been found (Fig. 4). Discotic L. also have anisotropy of properties.

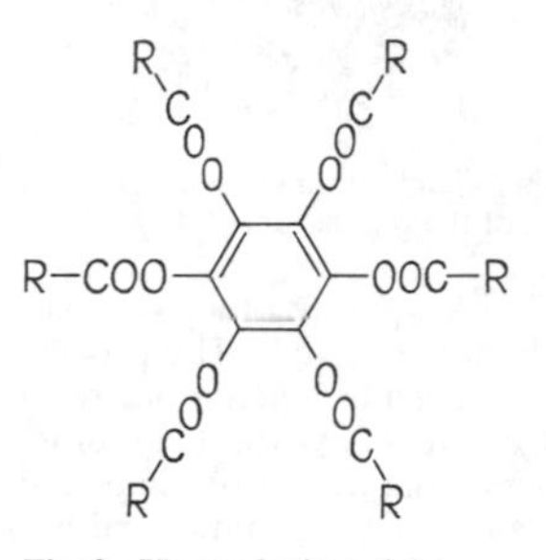

Fig. 3. Hexasubstituted benzene as an example of a disk-shaped (discotic) molecule.

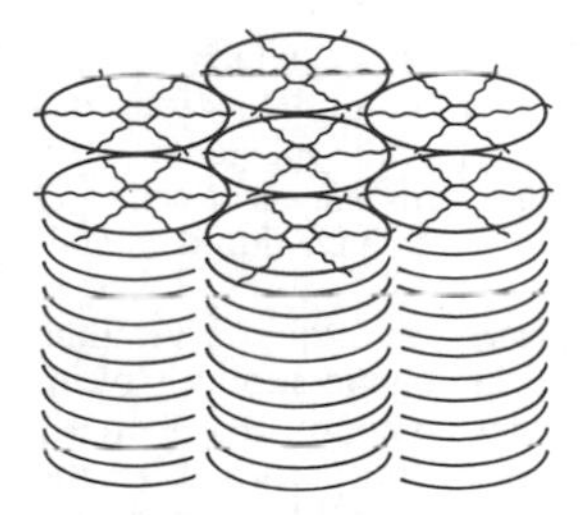

Fig. 4. Hexagonal arrangement of the columns in a discotic phase.

Lyotropic L cannot be formed from pure substances, but arise through interaction of polar substances with certain solvents in certain concentration and temperature ranges. Lyotropic liquid-crystalline phases can be formed by amphiphilic molecules consisting of a polar head group and a non-polar, long-chain paraffin residue. Some typical examples are potassium palmitate, lecithin, octylamine or dodecylsulfonic acid. The structures of lyotropic L. are characterized by the arrangement of amphiphilic molecules in layers, cylinders or spheres with their head groups interacting with the polar solvent (water, alcohols, chloroform). If the solvent is outside the layers, cylinders or spheres, the structures are "normal"; if the water is in the interior, they are called "inverse". An example of a layered structure is the neat phase (Fig. 5a), in which molecular double layers are formed with the paraffinic chains pointing towards each other and partly intertwined. The layers are separated by water. The middle phase (Fig. 5b) consists of cylinders with head-groups outward and solvent between them. Lyotropic K. can also occur in systems of three or more substances, and in mixtures of amphiphilic polymers (e.g. polyhexadecylacrylate) with solvents.

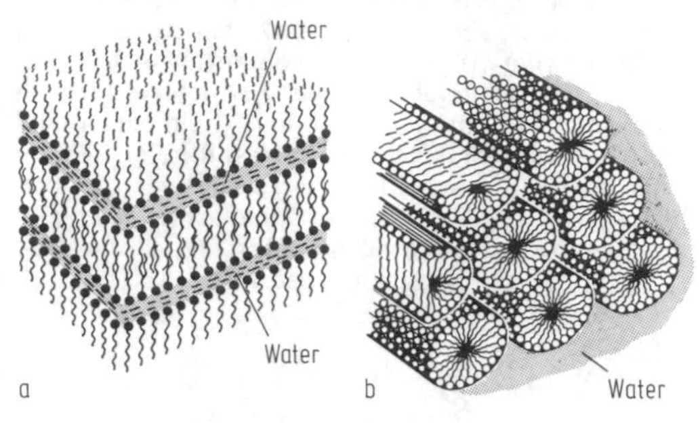

Fig. 5. Structures of lyotropic liquid crystals: a) structure of the neat phase, b) structure of the middle phase.

Lyotropic L. can easily incorporate other substances which are not soluble in the solvent. The process of washing using soap or detergents depends partly on this ability. Lyotropic L. are very important in biological systems. Muscle fibers, mixtures of proteins and polypeptides and some body fluids in animal and human bodies are liquid crystalline. Lecithin is an important component of all plant and animal cells, especially of nerve tissue, and is liquid crystalline in mixtures with water and cholesterol esters.

Cell membranes have a structure which is very similar to that of the lysotropic L.; model membranes are therefore made from amphiphilic substances with appropriate solvents. Certain viruses, cancer cells and the contents of red blood cells also have liquid crystalline properties.

Liquid crystalline polymers. The liquid crystalline state can also be observed in polymers, and is of increasing interest for the development of polymeric industrial materials with special properties. Like the lower-molecular-weight L., polymeric L. can be classified as thermotropic or lyotropic systems.

1) *Thermotropic liquid crystalline polymers*. The liquid crystalline ordering of the macromolecules found in molten linear polymers (e.g. polyethylene, polydiethylsiloxane) leads to favorable conditions for subsequent crystallization of the polymer in the production of very strong fibers and foils (*orientation crystallization*). The main route to thermotropic L. is the chemical bonding of polymeric chains with molecules of low-molecular-weight liquid crystals. The mesogenic groups (those responsible for the formation of mesophases) can be incorporated into the molecule either as part of the main chain or in the form of side chains (Fig. 6). Polymers with mesogenic groups in the main chain are obtained by polycondensation or co-polycondensation of different bifunctional compounds which consist of both flexible and rigid (mesogenic) groups. Such "semirigid" polymers, when they enter a thermotropic liquid-crystalline phase, create favorable conditions for production of fibers with very high tensile strength. Polymers with mesogenic side chains (*comb polymers*) can be produced either by polymerization of corresponding liquid crystalline monomers or by addition of molecules of low-molecular-weight L. to polymer chains under conditions similar to polymerization conditions. The immediate linkage of the mesogenic group to the main chain (Fig. 6b) often does not lead to formation of thermodynamically stable liquid-crystalline phases. Instead, the linkage of rigid mesogenic groups via a flexible *spacer*, which permits both a certain distance from the main chain and a certain degree of dependence of motion on the main chain (Fig. 6c), is decisive for thermodynamic stability of the L. The main chain consists, e.g. of polyethylene, polysiloxanes or polyacrylates; the spacers are usually alkylene groups, and the common mesogenic groups are Schiff's bases, cyanobiphenyl groups or cholesterol esters. The comb polymers form homogeneous nematic, cholesterolic and smectic phases of high viscosity.

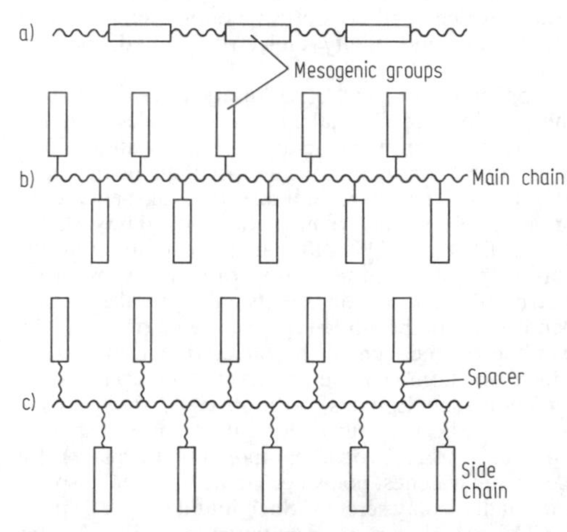

Fig. 6. Diagram of liquid-crystalline polymers with mesogenic groups: a) in the main chain, b) in side chains (without spacers) and c) in the side chains (with spacers).

2) *Lyotropic liquid crystalline polymers*. In solutions of polymers with rigid chains, there can be a spontaneous separation into isotropic and anisotropic, i.e. liquid crystalline, phases in certain ranges of temperature and concentration. Such lyotropic systems have nematic or cholesterolic structures, and have been demonstrated in a number of biopolymers; they are probably characteristic of organisms. Synthetic polymers with rigid chains, such as aromatic polyamides, polyesters and polyamidoesters can also form liquid-crystalline phases, and this is gaining industrial importance. Such polymers can only be worked in solution, due to their high melting points. If liquid-crystalline phases can be produced, the fibers spun from such solutions have much higher tensile

strength than those spun from isotropic solutions. For example, the Kevlar fibers developed in the USA, which are based on polyphenyleneterephthalamides, have higher specific strength than steel or glass fibers.

Liquid gases: hydrocarbons, especially propane, C_3H_8, and butane, C_4H_{10}, which are gases at normal temperature and pressure, but are liquefied even at moderate pressures (up to 2 MPa). L. are produced in all petroleum refining and in processing of "wet" natural gas. They are transported in pipelines or tankers, and are stored in pressure tanks or steel bottles (see Bottled gas).

L. are used as heating gases in industry and for household gas in areas where there are no gas pipelines. They are also used for welding and cutting metals and for street lighting. There is renewed interest in the use of L. as substitutes for liquid fuels in motor vehicles.

Liquid-liquid chromatography: see Chromatography.

Liquid phase peptide synthesis: a method of peptide synthesis using soluble polymeric carrier. The basic principle is the same as that of Solid-phase peptide synthesis (see) (Merrifield synthesis), but a soluble polymer is used to avoid carrying out the reactions in a heterogeneous system. The polymers used are soluble polystyrene, polyoxyethylene (POE), or polyethylenimine. The L. was considerably improved by use of POE as the C-terminal protective group for the growing polypeptide chain, and separation of low-molecular-weight starting materials and reagents. A further improvement is the crystallization method, in which a suitable organic solvent (diethyl ester, for example) is used to precipitate the peptidyl-POE rapidly and quantitatively in the form of a helical structure which avoids inclusions. In spite of the homogeneous reaction phase, it is necessary, to achieve optimum rates with L., to use excess acylation reagents and also secondary coupling. L. has been automated. A decided disadvantage of the method is the decreasing solubility of the peptidyl-POE with increasing chain length, and the handling of the resulting viscous solutions.

So far, L. has not become so widely used as the Merrifield synthesis.

Liquid-phase permeation: see Permeation.

Liquid-solid chromatography: see Chromatography.

Liquidus curve: see Melting diagram.

Litharge: see Lead oxides.

Lithergols: see Fuels.

Lithium: symbol ***Li***: a chemical element from group Ia of the periodic system (see Alkali metals); light metal, *Z* 3, with natural isotopes with mass numbers 6 (7.42%) and 7 (92.58%), atomic mass 6.941, valence I, Mohs hardness 0.6, density 0.534, m.p. 180.54 °C, b.p. 1317 °C, electrical conductivity 11.8 Sm mm^{-2} (at 0 °C), standard electrode potential (Li/Li^+) -3.045 V.

Properties. L. is a soft metal with a silvery sheen on a freshly cut surface. However, in moist air the surface is rapidly covered with a matte gray layer of lithium hydroxide. L. must therefore always be stored under an inert liquid, usually kerosene. In completely dry air, it is stable at room temperature. L. crystallizes in a cubic body-centered lattice. It is harder than sodium, but can be pressed into wire and hammered into thin leaf. Lithium vapor is largely monoatomic, but immediately above the boiling point there is a small fraction of Li_2 molecules. L. dissolves in liquid ammonia to give a dark blue solution.

The behavior of L. is determined by the ease with which it loses its valence electron (ionization potential 5.392 eV) and the resulting tendency to form colorless Li^+ cations. L. is a strong reducing agent, and is one of the most electropositive elements, as would be expected from its position in the periodic system. In its chemical behavior, L. is less similar to its heavier homologs in the Alkali metal group (see) than it is to magnesium in the alkaline earth metal group (diagonal relationship in the Periodic system (see)). This is due mainly to the very small ionic radius of the Li^+ cation (60 pm), which is close to that of the Mg^{2+} cation, and which is responsible for the lattice energy of Li salts, the solvation properties of the Li^+ cation and its ability to form bonds with covalent character. The similarity of its behavior to that of magnesium can be seen clearly in the thermal instability of the carbonates and nitrates, the high stability of "normal" lithium oxide, Li_2O, the ability of L. to react at high temperatures with nitrogen to form lithium nitride, Li_3N, with carbon to form lithium carbide, Li_2C_2, and with silicon to form lithium silicide, Li_6Si_2. The similarity extends to the fact that lithium fluoride, LiF, lithium carbonate, Li_2CO_3, and lithium phosphate, Li_3PO_4, are relatively insoluble in water, in contrast to the corresponding salts of the heavier alkali metals. The ability of Li to form covalent bonds is expressed in an extensive chemistry of Organolithium compounds (see), which is extremely important not only in organoelement chemistry, but also in organic synthesis in general.

If L. is heated in dry air, it burns with a bright red flame to lithium oxide, Li_2O. Its reaction with water is comparable to those of its heavier homologs, forming lithium hydroxide, LiOH, but is distinctly less vigorous. Li reacts vigorously with halogens to form the corresponding lithium halides. At high temperatures, L. combines with hydrogen to form lithium hydride, LiH. Molten L. attacks silicon oxide and silicates, and is therefore capable of dissolving glass.

Analysis. L. and its compounds give a flame an intense red color. The lines at 670.8 nm and 610.4 nm are used for qualitative spectroscopic detection. For quantitative analysis, L. can be precipitated as lithium phosphate, $Li_3PO_4 \cdot 1/2H_2O$ or as lithium aluminate, $2\,Li_2O \cdot 5Al_2O_3$. Flame photometric determination, atomic absorption spectroscopy and ion-sensitive electrodes are especially useful for analysis. Before the actual analysis, L. may be enriched by extraction of lithium chloride, LiCl, with an organic solvent (e.g. pentanol).

Occurrence. Although L. makes up 0.006% of the earth's crust, which is a significant amount (27th most abundant element on earth), it is considered one of the less abundant elements. It occurs very widely, but rarely in high concentrations. The most important lithium minerals are spodumen, $LiAl[Si_2O_6]$, amblygonite, $LiAlPO_4(F,OH)$, lepidolite, $KLi_2Al(F,OH)_2[Si_4O_{10}]$ and petalite, $(Li,Na)AlSi_4O_{10}$. In addition, L. is present in various mineral springs, brines and seawater. As a soil trace element, L. is important for the growth of plants and can be detected in the ashes of plants.

Production. L. is prepared by melt electrolysis of a eutectic mixture of lithium chloride, LiCl, and potassium chloride, KCl, at 400 to 500 °C in a Downs cell (see Sodium). The LiCl is prepared by solubilizing lithium minerals, usually silicates, with soda. The L. is separated as lithium carbonate, Li_2CO_3, purified and converted to the chloride.

Applications. Elemental L. is used as an alloy component. It significantly improves the mechanical properties of aluminum, lead and zinc, and increases the resistance of magnesium to corrosion. It is used as a deoxidant in copper casting. The high specific heat, 3.28 J g^{-1} K^{-1}, of L. is utilized in nuclear reactors, where it serves as a heat transfer liquid and coolant. For this purpose, pure ^7Li is used, because it has a low capture cross section for neutrons. In contrast, the nuclei of ^6Li react with thermal neutrons. This reaction has become very important in nuclear technology for the production of tritium: $^6Li(n,\alpha)^3H$. L. is used as an electrode material for dry cells and as a catalyst for *cis*-polymerization of isoprene. It is a starting material for the synthesis of organolithium compounds, which in turn are important in organic synthesis. In medicine, lithium salts are used in the therapy of nervous disorders.

Historical. L. was discovered in 1817 by Arfvedson in the mineral petalite. The name lithium (from the Greek "lithos" = "stone") was derived from the fact that in contrast to the heavier alkali metals, which were detected in plant ashes, it was discovered in a mineral. Metallic L. was first prepared by Bunsen and Matthiesen in 1855, by melt electrolysis.

Lithium alanate, ***lithium aluminum hydride***: $LiAlH_4$, a colorless powder when pure; but usually a gray powder. It is soluble in ether, benzene, chloroform and light petroleum, M_r 37.95, density 0.917; decomposes above 125 °C. As a typical metal hydride, L. reacts vigorously with protic solvents such as water, alcohols, and carboxylic acids. For example, $LiAlH_4 + 4\,H_2O \rightarrow LiOH + Al(OH)_3 + 2\,H_2$. It is made by reaction of lithium hydride with aluminum chloride: $4\,LiH + AlCl_3 \rightarrow LiAlH_4 + 3\,LiCl$. L. is an excellent reducing agent and is widely used in organic synthesis. For example, aldehydes, ketones, carboxylic acyl chlorides and esters are specifically reduced to the corresponding alcohols, while amides and nitriles are reduced to the amines. C=C double bonds are not attacked. In organoelement chemistry, halides are often converted to the hydrogen compounds, e.g. $2\,RPCl_2 + LiAlH_4 \rightarrow 2\,RPH_2 + LiCl + AlCl_3$ (R = aliphatic or aromatic group).

> Lithium alanate reacts violently with water, and the hydrogen released by the reaction forms an explosive mixture with air. Therefore, when the compound is used, water must be strictly excluded. Excess lithium alanate can be removed from reaction mixtures by dropwise addition of ethyl acetate.

Lithium aluminum hydride: same as Lithium alanate (see).

Lithium boranate, ***lithium borohydride***, formerly ***lithium hydrodiborate***: colorless orthorhombic crystals, soluble in ether; M_r 21.78, density 0.666, dec. 275 °C. It decomposes in water. L. is synthesized by the reaction of lithium hydride with a borate ester (or boron trifluoride): $4\,LiH + B(OCH_3)_3 \rightarrow LiBH_4 + 3\,LiOCH_3$. L. is used in organic synthesis for selective conversion of aldehydes, ketones, carboxylic acyl chlorides and esters to the corresponding alcohols. Carboxamides and nitriles are not attacked.

Lithium borohydride: same as Lithium boranate (see).

Lithium bromide: LiBr, colorless, cubic crystals which are readily soluble in water; M_r 86.85, density 3.464, m.p. 550 °C, b.p. 1265 °C. L. crystallizes out of water as one of several hydrates, depending on the temperature. It dissolves to a considerable extent in ethanol and acetone. It is synthesized by reaction of lithium carbonate and calcium bromide. L. is used as a drying agent in air conditioners and medically as a tranquilizer.

Lithium carbonate: Li_2CO_3, colorless, monoclinic crystalline powder; M_r 73.89, density 2.11, m.p. 723 °C. L. is only slightly soluble in water, and its solubility decreases somewhat with rising temperature: at 0 °C, 100 g water dissolves 1.54 g, and at 100 °C, 0.73 g. The solution is weakly basic. In water containing CO_2, L. dissolves to form ***lithium hydrogencarbonate***, which cannot be isolated in solid form. L. is produced by adding a soluble carbonate to a solution of a lithium salt. It is synthesized industrially, for example, by solubilizing spodumen in a soda solution at 200 °C. L. is the starting material for production of many other lithium salts. It is added to ceramic clays as a mineralizer; it reduces the firing temperature and increases the solidity of the product. It is used as an electrode coating in electric welding. In medicine, it is used in the treatment of psychoses.

Lithium chloride: LiCl, colorless, deliquescent, very hygroscopic, cubic crystals; M_r 42.39, density 2.068, m.p. 605 °C, b.p. 1325 °C to 1360 °C. L. is readily soluble in water and crystallizes out of aqueous solution as a mono- di- or trihydrate, depending on the temperature. It is also soluble in ethanol, glycerol, acetone and many other donor solvents. L. is poisonous. It is produced by reaction of hydrochloric acid with lithium carbonate and is used in heating baths, as an additive to solders (especially for aluminum), in air conditioners as a drying agent and in fireworks to achieve dark red flame colors.

Lithium citrate: $C_3H_4(OH)(COOLi)_3 \cdot 4H_2O$, colorless, water-soluble powder; M_r 281.98. L. is used to treat gout and in neurology.

Lithium deuteride: see Lithium hydride.

Lithium fluoride: LiF, colorless, cubic crystals; M_r 25.94, density 2.365, m.p. 845 °C, b.p. 1676 °C. L. is slightly soluble in water, but is dissolved in excess hydrofluoric acid, forming ***lithium hydrogenfluoride***, $LiHF_2$. It is synthesized by evaporation of a solution of lithium carbonate in hydrofluoric acid. L. is a component of welding fluxes for light metals and is used to make glazes and enamels. LiF single crystals are transparent to infrared and ultraviolet radiation, and are therefore used to make prisms, cuvette windows, etc. for UV and IR spectroscopes.

Lithium hydride: LiH, colorless, hard, cubic crystalline mass; M_r 7.95, density 0.82, m.p. 680 °C. L. is a typical ionic metal hydride (see Hydrides). When melted, it conducts electricity, with evolution of hydrogen at the anode. L. decomposes in the presence of protic media, such as water, with a vigorous evolution of hydrogen: $LiH + H_2O \rightarrow LiOH + H_2$. L. is synthesized by passing hydrogen over heated lithium. It is a strong reducing agent. Many element halides are converted to the hydrides by L. It is also used as a reducing agent in organic synthesis. It is a starting material for production of lithium alanate and lithium boranate. ***Lithium deuteride***, ^6LiD (made up of lithium-6 and hydrogen-2, or deuterium) is an essential component of hydrogen bombs.

Lithium hydridoborate: see Lithium boranate.

Lithium hydroxide: LiOH, colorless, tetragonal crystals, moderately soluble in water and alcohol; M_r 23.95, density 1.46, m.p. 450 °C, dec. 924 °C; it crystallizes out of the very basic aqueous solution as the monohydrate, $LiOH \cdot H_2O$. L. is obtained by electrolysis of aqueous lithium chloride solutions. It is used to absorb carbon dioxide in respiratory apparatus, submarines, etc., as an additive to batteries and photographic developers, and to make special lubricants.

Lithium iodide: LiI, colorless, cubic crystalline mass, usually brown because of iodine released by reaction with air. It is readily soluble in water; M_r 133.84, density 3.49, m.p. 449 °C, b.p. 1180 °C. L. crystallizes out of water as the trihydrate, $LiI3H_2O$ (m.p. 73 °C). L. is used in photography and medicine.

Lithium nitrate: $LiNO_3$, colorless, deliquescent crystals which dissolve very readily in water; M_r 68.94, density 2.38, m.p. 264 °C. It crystallizes out of water as the trihydrate $LiNO_33H_2O$. L. is obtained by decomposition of lithium carbonate with nitric acid: $Li_2CO_3 + 2\,HNO_3$ -> $2\,LiNO_3 + CO_2 + H_2O$. It used in a mixture with strontium nitrate to produce a red color in fireworks.

Lithium nitride: Li_3N, gray, amorphous powder; M_r 34.82, dec. 840 to 850 °C. It is decomposed by water to form NH_3 and lithium hydroxide, LiOH. It is formed when nitrogen reacts with hot lithium.

Lithium oxide: Li_2O, colorless, cubic crystalline mass; M_r 29.88, density 2.013, m.p. above 1770 °C. L. dissolves in water to form lithium hydroxide. It is the main product of combustion of lithium; ***lithium peroxide***, Li_2O_2 is formed in smaller amounts. L. is also formed by thermal decomposition of lithium carbonate, nitrate or hydroxide. It is used mainly in silicate technology to make glazes.

Lithium perchlorate: $LiClO_4$, colorless crystals soluble in water and alcohol; M_r 103.69, density 2.428, m.p. 236 °C, dec. 430 °C. L. crystallizes out of aqueous solution as the trihydrate, $LiClO_4 \cdot 3H_2O$. It is obtained from lithium chloride and perchloric acid. As a solid, oxygen-rich oxidation agent, it is a component of rocket fuels.

Lithium peroxide: see Lithium oxide.

Lithium phosphate: Li_3PO_4, colorless, rhombic crystal powder; M_r 115.79, density 2.537, m.p. 837 °C. Unlike other alkali metal phosphates, L. is only slightly soluble in water, and when phosphate is added to a lithium salt solution, it precipitates out as the semihydrate, $Li_3PO_4 \cdot 1/2H_2O$.

Lithium stearate: $C_{17}H_{35}COOLi$, a colorless powder soluble in water and alcohol; M_r 290.41, m.p. 221 °C. L. is used in the production of special lubricants and in waxes and pencils.

Lithium sulfate: $Li_2SO_4 \cdot H_2O$, colorless, monoclinic crystals; M_r 127.96, density 2.06, m.p. 845 °C. L. is one of those salts which become less soluble in water as the temperature increases. At 20 °C, 100 g water dissolves 34.8 g, and at 100 °C, 29.5 g. L. is obtained by dissolving lithium carbonate in sulfuric acid.

Lithopone: a nontoxic white pigment with a high opacity. L. is a mixture of zinc sulfide and barium sulfate, made by combining aqueous zinc sulfate and barium sulfide solutions: $ZnSO_4 + BaS \rightarrow ZnS + BaSO_4$. The amount of zinc sulfide can be varied by addition of zinc chloride. The technical qualities of the pigment are improved by filtration, drying, roasting and quenching the precipitate. L. is the most important white pigment for interior paints. Its stability to light can be improved by addition of cobalt salts. L. is also used to color or whiten plastics, elastics, linoleum, oilcloth, putty, cement and artificial stone.

Lithotype: see Coal.

Litmus: a natural pigment obtained from archil (dyer's moss; *Rocella* sp.) by addition of ammonia. The chromophore consists of phenoxazone and phenoxazine derivatives. L. is used to make litmus paper, which is used as an acid-base indicator. In acidic media, L. is red, and basic media, dark blue. The color change occurs between pH 4.5 and 8.3.

Litorin: see Bombesin.

Liver of sulfur, ***hepar sulfuris***: a mixture of potasium polysulfides (mainly K_2S_5), potassium sulfate, K_2SO_4, and potassium thiosulfate, $K_2S_2O_3$, formed by melting potassium carbonate and sulfur together. S. is used to prepare medicinal sulfur baths.

LLC: see Chromatography.

LLDPE: see Polyethylene.

Ln: a symbol often used to indicate a lanthanoid.

Lobelia alkaloids: see Alkaloids.

Lobelin: see Alkaloids.

Lobry de Bruyn-Van Ekenstein rearrangement: see Monosaccharides.

Local analysis, ***point analysis*** the determination of the qualitative or quantitative composition of a small, defined region of the surface of a solid sample. L. is achieved by special physical methods, such as laser micro-spectral analysis, secondary ion mass spectroscopy, electron-beam microprobe methods and Auger microprobes. Electron-beam microprobes can be used to determine the composition of an area 1 µm in diameter. L. is becoming every more important in materials testing and microelectronics.

Local anesthetics: see Anesthetics, local.

Löffler's reagent: a solution made from 30 ml saturated alcoholic methylene blue solution, 1 ml 1% potasium hydroxide solution and 100 ml distilled water. L. is used to stain blood cells and parasites in blood smears.

Loganin: see Iridoids.

Logwood, ***campeche wood***, the reddish brown heartwood of *Haematoxylon campechianum*, which is native to Central America and the West Indies. The extract of L. contains the natural dye Hematoxylin (see) and forms a blue-violet dye lake with alumina. With chromium salts it forms a blue-black to black lake, with iron salts, black, and with tin salts, a red-violet lake.

L. is used to dye textiles, most often silk, and furs; L. sawdust or an extract (noir réaduit) is used.

London forces: see Van der Waals bonding forces.

Long-range order: see States of matter.

Lorenz-Lorentz equation: see Polarization 2.

Loschmidt number, N_L: the number of particles (molecules or atoms) in 1 g of a substance. $N_L = N_A/M_r$, where N_A is Avogadro's constant and M_r is the relative molecular mass of the substance in question.

Lossen degradation: methods for converting hydroxamic acids or their *O*-acyl derivatives to primary amines with one carbon atom fewer. Hydroxamic acids can be obtained from carboxylic acid esters and hydroxylamine; the L. reaction is responsible for their thermal decay in inert solvents, or in the presence of strong mineral acids, acetic anhydride or thionyl chloride. The reaction involves simultaneous loss of water and migration of the alkyl group as an anion to the N atom, forming an isocyanate; this reacts with water to form a carbamic acid which loses CO_2 to form the amine:

$$\text{R–CO–NH–OH} \xrightarrow[-H_2O]{} \text{R–N=O} \xrightarrow[-CO_2]{+H_2O} \text{R–NH}_2$$

Lotka model: see Oscillating reaction.

Low-calorie foods: diet foods which have much lower energy contents than comparable ordinary foods. E. are prescribed for coronary-circulatory and metabolic disorders (obesity, diabetes mellitus), and are also suitable for prevention and cure of overweight. The reduction in energy content is achieved by reducing the contents of sugar and/or fats, for example, by using lower-energy substitutes, such as alginates, starch hydrolysis products, fillers, water and artificial sweeteners. The concentrations of vitamins and essential minerals should not be decreased, and the content of protein should be no more than proportionally reduced. The reduction in energy content should be at least 50% for mayonnaise, 40% for soft drinks, 40% for margarines, and 20% for baked products.

Lowry method: see Proteins.

Low severity cracking: see Pyrolysis.

Low-spin complex: see Coordination chemistry, see Ligand field theory.

Low-temperature chemistry: see Cryochemistry.

Low-temperature, high-pressure hydrogenation: see Hydrorefining.

Low-voltage electrophoresis: see Electrophoresis.

LPG: see Fuels.

LPG operation: see Reforming.

LPH: see Lipotropin.

Lr: symbol for lawrencium.

LRH: see Liberins.

LSC: see Chromatography.

LS-coupling: see Multiplet structure.

LSD: abb. for Lysergic acid diethylamide (see).

LT: abb. for Leukotrienes (see).

LTH: see Prolactin.

Lu: symbol for lutetium.

Lubricant: 1) (textiles) a material used to make fibers (especially wool) softer for better spinning. Emulsified mineral oils are often used, but other products, e.g. phosphates, are used because they can be easily removed without any special aids or detergents. In the past, oleic acid was used, because it can easily be washed out again with a soda solution. Removal of the L. is important, because otherwise the yarn or cloth cannot be evenly dyed.

2) in general, organic or inorganic substances which diminish direct rubbing of rotating or sliding machine parts. Good L. must have the following properties: a) they must adhere well to the sliding surfaces, and the film must not rub off at high pressure. b) The higher the pressure, the more viscous the L. must be. The viscosity, which changes with the speed of the sliding parts and the temperature (summer and winter oils) must not drop off too rapidly as the temperature rises (good viscosity-temperature behavior). c) They must have very low volatility and high flame points which exclude the possibility of burning at the temperatures encountered. d) For machines which work at low temperatures, they must have a sufficiently low Stock point. e) They must be very pure. f) The stability of an L. must be adequate at the temperatures it encounters.

Organic L. are subdivided into oils and greases. ***Lubricating oils*** are usually Mineral oils (see) with varying viscosities. They are obtained from certain petroleum fractions, after these have been purified by refining and their Stock points sufficiently reduced by removal of paraffin with urea (see Urea separation). These mineral oils are used with additives, for example as compressor, steam-cylinder, engine, gear and hydraulic oils. ***Synthetic lubricating oils*** have been developed for very demanding applications. Special oils, such as refrigerator oils, are made by cationic polymerization of ethene (see Polyethylene). Polymers of higher alkenes are sometimes used as lubricating oils, especially polymers of cogasine fractions with high alkene contents. However, these products must be stabilized by antioxidants, e.g. 2,6-di-*tert*.-butyl-*p*-cresol, because they still contain double bonds and are not stable in the presence of oxygen. These special L. have excellent viscosity-temperature behavior, and are used in aircraft.

Polyether oils are used as special lubricating oils, but particularly as hydraulic and brake fluids.

Ester oils are used mainly as engine oils for aircraft; the basic types in use are: 1) esters of primary alcohols with dicarboxylic acids; 2) esters of di- and polyols with monocarboxylic acids; and 3) esters of neopentyl polyols with monocarboxylic acids.

The lubricating oils with the best viscosity-temperature behavior are the silicon oils, which are also the basis for silicon greases. Silicone based L. can be used at very high temperatures because of their stability, and, because of their low Stock points, at very low temperatures as well. Substituted ferrocene compounds are thermally stable up to about 460 °C.

Polyethylene glycol can be used as a water-soluble lubricating oil.

Greases are colloidal suspensions of soaps in lubricating oils. The liquid components consist of mineral or synthetic lubricating oils of various viscosities. Naphthene-based oils are more suitable than paraffin-based, because the latter tend to undergo phase separation. The solid phase consists of soaps, fatty-acid, resin-acid, sulfonic-acid or naphthenic-acid salts of lithium, sodium, potassium, calcium, magnesium, aluminum or lead. Greases also contain stabilizers, such as casein, collagen, rubber, polyethylene oxide and substances to protect them from aging. Molybdenum(IV) sulfide is added to high-pressure greases. Water is also present in the greases. Stauffer grease (see) is the best known of the greases. Greases based on silicone can be very useful, as they can be used in the temperature range between -60 and +200 °C, are mechanically very stable and are very resistant to aging.

Greases resist permanent deformation, and are used wherever an oil lubrication is not possible.

Inorganic L. are less significant than organic L., because their range of application is much narrower. They are used mainly for special purposes. Some examples are molybdenum(IV) sulfide, titanium(IV) sulfide or graphite, which are used alone or mixed with mineral oils or paraffin. Syrupy phosphoric acid is useful because it is resistant to aggressive gases. A useful product can be made by dissolving 10 g metaphosphoric acid and 2 g boric acid in 100 ml water, then concentrating the solution to 25 ml and adding 1 ml 85% phosphoric acid. Concentrated sulfuric acid is sometimes used for lubricating glass apparatus, but its hygroscopic properties are a disadvantage. Under some circumstances, molten potassium rhodanide, potassium nitrate or kaolin are suitable inorganic L. in the temperature range from -75 to +360 °C.

Luciferase: an enzyme which, in the presence of oxygen, ATP and magnesium ions, catalyses the dehydrogenation of luciferin to oxiluciferin. Over 90% the energy released by the reaction is radiated in the form of visible (blue) light (bioluminescence).

Luciferin: see Bioluminescence.

Ludwig-Soret effect: see Thermodiffusion.

Luliberin: see Liberins.

Lumiflavin: see Flavins.

Luminescence: emission of light as a result of a previous non-thermal energy transfer. L. is a "cold light" and may occur in the wavelength range from a few nm to 200 nm, which includes the visible, near ultraviolet and infrared. The selective character of the emission is a result of the underlying discrete energy transitions in the atoms or molecules of gases, liquids and solids following absorption of energy. According to ***Stokes' rule***, the wavelength of the emission maximum is equal to or (usually) longer than that of the exciting radiation. However, if the emitting substance absorbs an additional excitation quantum from the vibrational (heat) energy of the lattice, it may break this rule (***anti-Stokes L.***).

If the light emission continues after the excitation energy has been turned off (post-emission), the phenomenon is called Phosphorescence (see). If the temperature of the phosphor is decreased, its storage capacity for light is considerably increased. Fluorescence (see) is an emission which does not have a post-emission component capable of being frozen in. Light emitted during the excitation cannot be classified as fluorescence or phosphorescence, and it is more correct to speak of L. in this case.

There are several types of L.: ***Bioluminescence*** is responsible for light emission by biological systems (deep sea animals, bacteria, fireflies, fungi, decaying wood or foliage); it is produced by enzyme-catalyzed oxidation of certain substances (luciferins). The enzymes are called luciferases. The required energy is supplied by the oxidation process, so that bioluminescence is a special case of chemiluminescence. ***Chemiluminescence*** occurs in exothermic reactions in which part of the reaction energy is converted to light (for example, the glowing of white phosphorus caused by slow oxidation of its vapor). ***Electroluminescence*** is a special case of chemiluminescence which occurs in some electrode reactions. The electrode reactions produce high-energy intermediate products (usually radical ions) which are stabilized in a light-emitting process. L. produced by excitation with UV, X-ray or visible radiation is called ***photoluminescence***. When cathode or electron beams striking a substance cause L., it is called ***cathodic L.*** In the boundary layer between p- and n-type semiconductors, electrons from the p-phase can migrate into the boundary layer. If these electrons are sufficiently accelerated by an external electric field, they can excite electrons from the valence band into the conduction band upon entering the n-phase (***electroluminescence***, or in particular, ***collision ionization L.***). Many substances, during the process of pulverization, give rise to a weak L. called ***triboluminescence***; compression of excited sodium chloride or fluorite crystals also gives rise to triboluminescence.

L. plays an important role in technology, which makes use of the emissions of synthetic compounds, the Phosphors (see).

Luminol, ***3-aminophthalic hydrazide, 5-amino-2,3-dihydrophthalazine-1,4-dione, 5-amino-1,4-dihydroxyphthalazine***: a compound which gives an intense blue chemiluminescence under alkaline oxidation. L. forms yellow crystals; m.p. 329-332 °C. It is soluble in ethanol, glacial acetic acid and alkalies, but insoluble in water. It is synthesized in the form of its hydrazine salt by boiling 3-nitrophthalic acid diethyl ester with excess hydrazine hydrate, or by reduction of 5-nitrotetrahydrophthalazine-1,4-dione with aqueous ammonium sulfide solution, hydrogen in the presence of palladium and animal charcoal, or tin(II) chloride and hydrochloric acid.

NH_2 O NH NH O

Lumisterol$_2$: see Vitamin D.

LUMO: see Woodward-Hoffmann rules.

Lupane: see Triterpenes.

Lupinin: see Alkaloids.

Lurgi pressure gasification: see Synthesis gas.

Luteinizing hormone: same as Lutropin (see).

Luteinizing hormone releasing hormone: see Liberins.

Luteolin: see Flavones.

Luteotropic hormone: same as Prolactin (see).

Lutetium, symbol ***Lu***: an element of the Lanthanoid (see) series of the periodic system, a heavy metal, Z 71, with natural isotopes with mass numbers 175 (97.39%), 176 (2.61%, β-emission, $t_{1/2}$ $3.3 \cdot 10^{10}$ a); atomic mass 174.97, valency III, density 9.843, m.p. 1675 °C, b.p. 3315 °C, standard electrode potential (Lu/Lu^{3+}) -2.255 V. The name ***cassiopeium***, symbol ***Cp*** was formerly used in some countries.

Lu is a silvery-white metal which is relatively stable in air. It crystallizes in a hexagonal lattice. Lu makes up $7 \cdot 10^{-5}$% of the earth's crust, and occurs in nature only in the company of other rare earth metals, especially yttrium. It is relatively commonly present in such minerals as samarskite and yttrotantalite; it is present in monacite at about 0.003%. L. is used only in the form of Cerium mixed metal (see). For other properties, analytical, production and historical, see Lanthanoids.

Lutidines: a term for the six possible isomers of dimethylpyridine. ***2,3-L., 2,3-dimethylpyridine***, is a colorless liquid, b.p. 163-164 °C, n_D^{20} 1.5057. It is soluble in water, alcohol and ether. ***2,4-L., 2,4-dimethylpyridine***, is a slightly yellowish liquid; m.p. 159 °C, n_D^{20} 1.5010. It is readily soluble in water, alcohol, ether and acetone. ***2,5-L., 2,5-dimethylpyridine***, is a colorless liquid; m.p. - 16 °C, b.p. 157-159 °C, n_D^{20} 1.4953. It is soluble in water, ether and acetone. ***3,4-L., 3,4-dimethylpyridine*** is a colorless liquid; b.p. 163-164 °C, n_D^{20} 1.5096. It is soluble in alcohol, ether, acetone and chloroform. ***3,5-L., 3,5-dimethylpyridine*** is a colorless liquid; b.p. 171.5 °C, n_D^{20} 1.5061. It is soluble in ether and acetone.

The L. are found in the basic fraction of coal tars and in Scottish oil shale. The individual isomers can be separated by fractional distillation. When the L. are oxidized, they form the corresponding dicarboxylic acids: quinolinic, lutidinic, isocinchomeronic, dipicolinic, cinchomeronic and dinicotinic acids.

Lutropin, ***luteinizing hormone***, abb. ***LH***; identical to ***interstitial cell stimulating hormone***, abb. ***ICSH***: a gonadotropin which promotes production of estrogen in the female and thus affects the ripening of the follicles in the ovary. Ovulation is initiated by L. acting with follitropin, and L. also promotes testosterone production by the interstitial cells of the testes in the male. L. is released in response to gonadoliberin; it is a glycoprotein consisting of two polypeptide chains. The α-chain is homologous to those of several other hormones. The β-chain is specific to the hormone; it can also combine with the α-chains of follitropin, thyrotropin or choriogonadotropin to form an active L.

Lyases: a main group of Enzymes (see). Some examples are Aldolases (see), Aspartase (see), Decarboxylases (see), Fumarase (see) and Enolases (see).

Lye: a solution of Sodium hydroxide (see) or Potassium hydroxide (see).

Lyophilic: solvent attracting.

Lyophobic: solvent repelling.

Lyogel: see Gel.

Lyotropic: see Liquid crystals.

Lys: abb. for lysine.

Lysergic acid: see Ergot alkaloids.

Lysergic acid diethylamide, abb. ***LSD***: a partially synthetic derivative of the Ergot alkaloids; m.p. 83 °C, $[\alpha]_D^{20}$ +30 °C (in pyridine). L. forms colorless crystals which dissolve readily in ethyl acetate, chloroform and benzene. L. is one of the strongest known hallucinogens. As little as 0.03 to 0.05 mg can generate a schizophrenia-like state in a human being.

Lysine, abb. ***Lys***: α-,ε-diaminocaproic acid, H_2N-$(CH_2)_4$-$CH(NH_2)$-COOH, a limiting essential Amino acid (see). L. occurs free and also in proteins; it is especially abundant in protamines, serum albumins and globulins, and in fibrin. The amounts of L. in grain proteins (wheat, barley, rice, maize) and other vegetable proteins are small.

L. is synthesized by the lactam method, in which caprolactam is converted via several intermediates into DL-α-aminocaprolactam. The latter is the hydrolysed to DL-L. The direct and complete conversion of DL-α-aminocaprolactam into L-L. by microbial L-aminocaprolactam hydrolase, coupled with enzymatic racemization of the remaining D-L., is economically important. Fermentation methods using *Glutamicus* or *Brevibacterium* mutants are also important; in 24 to 72 h reaction time, 40 to 50 g L-L./l culture filtrate is produced. This corresponds to a yield of 25 to 30%, relative to the glucose in the medium as carbon source. L-L. is used as an additive to feeds and to improve the nutritional value of proteins. Addition of 0.1 to 0.3% L-L. to their feed causes chicken and pigs to grow considerably faster.

Lysol: see Cresols.

Lysolecithin: see Phosphatidylcholine.

Lysozyme, ***muramidase***: an enzyme which catalyses hydrolysis of the β-1,4-bond between *N*-acetylglucosamine and *N*-acetylmuramic acid in the bacterial cell wall. High concentrations of L. are found in vertebrate tears, saliva and mucous membranes; they protect the organism against entry of bacteria. Animal L. consist of 129 amino acids. The three-dimensional structure is characterized by four disulfide bonds and 42% α-helix.

M

m: 1) abb. for molal. 2) m, symbol for mass. 3) m-, see Arenes. 4) *m*-, abb. for *meta*, see Nomenclature, sect. III D. 5) $\tilde{m}$, symbol for molality.

M: 1) abb. for mass number. 2) abb. for molar. 3) M, symbol for molar mass. 4) M_r, symbol for relative molecular mass. 5) In chemical formulas, symbol for metal.

MAC: see Indoor air quality.

Mace oil: see Nutmeg oil.

Macerals: the microscopic components of coals, comparable to the minerals comprising rocks, except that they are not single substances. Their physical properties can also be nonuniform. The M. of coals are formed by the various organs and tissues of the plants from which the coals developed in the course of carbonization (see Coal). In soft brown coals, where the process of carbonization has not gone too far, the M. often still show recognizable structures of the original plants. As carbonization proceeds, these structures are increasingly degraded.

The chemical, physical and technological properties of a coal of a given degree of carbonization are determined by the amounts of the M. comprising it and their associations. M. from the same type of plant and with similar compositions are grouped in *maceral groups* (Table).

The most important macerals of brown coal

Maceral group	Maceral
Huminite	Textinite
	Ulminite
	Attrinite
	Densinite
	Gelinite
	Corpohuminite
Liptinite	Sporinite
	Cutinite
	Resinite
	Suberinite
	Alginite
	Liptodetrinite
	Chlorophyllinite
Inertinite	Fusinite
	Semifusinite
	Sclerotinite
	Macrinite

The most important macerals of hard coals.

Maceral group	Maceral
Vitrinite	Telinite
	Collinite
Exinite	Sporinite
	Cutinite
	Resinite
	Alginite
Inertinite	Micrinite
	Macrinite
	Semifusinite
	Fusinite
	Sclerotinite

Typical associations of M. which have similar properties with respect to refining technology are defined as microlithotypes, or are classified in microlithotype groups.

Maceration: a method of solid-liquid extraction used to extract a substance from a solid. M. is used, for example, to extract substances from coal (see Coal extraction) or the active substances from medicinal plants.

Macrinite: see Macerals.

Macrolana®: see Synthetic fibers.

Macrolide antibiotics: a group of antibiotics produced by actinomycetes. They contain a macrocyclic lactone ring formed from a branched, polyhydroxy-fatty acid. Most M. give basic reactions, due to the presence of glycosidically bound amino sugars. The M. are classified as small (12, 14 or 16-membered rings) or large (usually 38 ring atoms) depending on the size of the lactone ring. The erythromycin (e.g. erythromycin, oleandomycin) and carbomycin antibiotics (e.g. carbomycin) are among the smallest M. The large M. are represented by the Polyene antibiotics (see). The small M. are active primarily against bacteria, and the larger ones against fungi. Ansamycins (see) and some Ionophores (see) are included among the M.

Macromolecules: molecules consisting of many constituent (monomeric) subunits and having a relative molecular mass of more than 10,000. M. made by Polymerization (see), Polycondensation (see) and Polyaddition (see) of monomeric starting materials are also called Polymers (see); natural M., such as polysaccharides, proteins and nucleic acids, are called biopolymers. M. can also consist of inorganic units, for example silicates.

Maddrell's salt: see Sodium phosphate.

Madelung constant: a numerical factor which expresses the energetic stabilization experienced by a given number of isolated ion pairs when they join together to form an ionic lattice with a certain structure. The M. for each Structure type (see) has a value which depends only on the lattice symmetry and not on the chemical composition of the ionic crystal. When the M. is calculated using a mathematical series, all the geometric relationships which a given ion has with other ions in the lattice are taken into account. The M. is used in the calculation of the molar Lattice energy (see) E_0 according to Born:

$$E_O = -\frac{-N_A z_+ z_- e^2 A}{4\pi\epsilon_0 R_0}\left(1-\frac{1}{n}\right)$$

Here N_A is Avogadro's constant, z_+ and z_- are the charges on the cation and anion, e is the unit electric charge, ε_0 is the electric field constant, R_0 is the equilibrium distance between neighboring ions with different charges and n is the Born repulsion exponent. The M. for the most important structure types of compounds with the compositions AB and AB_2 are given in reduced form in the table.

Madelung constants (reduced form) for various structure types

Structure type	A_{red}
Isolated ion pair AB	1.000
Cesium chloride CsCl	1.763
Sodium chloride NaCl	1.748
Wurtzite ZnS	1.641
Zinc blende ZnS	1.638
Fluorite CaF_2	1.679
Rutile TiO_2	1.605

In contrast to the other numerical values often given in the literature for M., which include the charges and stoichiometric coefficients of the ions, the A_{red} values depend uniquely and solely on the geometry. They can therefore be compared to predict the energetic effects of forming various crystal structures. However, the structure type into which a given ionic substance actually crystallizes depends on the ratio of the radii of its components (see Radius ratio). After multiplication by $m/2$ (m is the number of ions per formula unit, e.g. $m = 2$ for NaCl, $m = 3$ for CaF_2), the A_{red} values can be inserted into the above equation to calculate the molar lattice energies.

Madelung synthesis: synthesis of indole by cyclization of *N*-acylated *o*-toluidines in the presence of basic condensing agents, such as sodium alcoholate or potassium amide.

CH_3 ... N(H)–CO–R $\xrightarrow{-H_2O}$ indole (N–H, 2-R)

Mafenide: see Sulfonamides.

Magenta: same as Fuchsine (see).

Magic acid: see Superacids.

Magnesia: ***M. usta, burnt, caustic*** or ***calcined M.*** is the same as Magnesium oxide (see); ***M. alba*** is the same as basic Magnesium carbonate (see).

Magnesia cement: see Magnesium chloride.

Magnesia process: see Potassium carbonate.

Magnesium, symbol ***Mg***: a chemical element from the second main group of the periodic system, the Alkaline earth metals (see); a light metal, *Z* 12. The natural isotopes have mass numbers 24 (78.70%), 25 (10.13%) and 26 (11.17%); atomic mass 24.312, valency II, Mohs hardness 2.0, density 1.74, m.p. 648.8 °C, b.p. 1107 °C, electrical conductivity, 23.2 Am mm^{-2} (at 18 °C), standard electrode potential (Mg/Mg^{2+}) -2.375 V.

Properties. M. is a light metal with a silvery shine; due to formation of a thin oxide skin, the surface gradually becomes matte. It crystallizes in hexagonal close-packed array. M. is very ductile and can be rolled into thin foils and drawn into wire. Its electron configuration is $[Ne]3s^2$, and corresponding to its position in the periodic system, M. is electropositive. It tends to give up both its valence electrons, forming colorless Mg^{2+} cations. This is indicated by the high negative standard electrode potential; Mg is a strong reducing agent. It is able to reduce aqueous acids; the H_3O^+ ions are reduced to hydrogen and magnesium salt solutions are formed, e.g. $Mg + 2\,HCl \rightarrow Mg^{2+} + 2\,Cl^- + H_2$. M. reacts analogously with alcohols, forming magnesium alkoxide solutions and hydrogen, e.g. $Mg + 2\,HOCH_3 \rightarrow Mg(OCH_3)_2 + H_2$. From the standard electrode potential, M. should also be able to react with neutral water at low temperatures, but it does not, because it is protected by formation of a thin, insoluble layer of magnesium hydroxide. This protective layer can be destroyed by addition of a mercury salt, which forms a magnesium amalgam and changes the M. lattice. This increases the reactivity of the metal. M. also dissolves in boiling water, generating hydrogen and magnesium hydroxide, $Mg(OH)_2$. In aqueous solution, the Mg^{2+} cations are highly hydrated. Magnesium salts also tend to crystallize out of aqueous solution in the form of stable hydrates.

If M. is heated in the air, it burns with a bright, white flame to magnesium oxide, MgO, and magnesium nitride, Mg_3N_2. At room temperature, M. is stable to atmospheric oxygen due to formation of a thin oxide skin. M. is oxidized by halogens to the corresponding magnesium halides.

> Magnesium and many of its alloys can burn, a fact which must be kept in mind when working with it. Magnesium fires must be extinguished with dry sand, dry carnallite or special hand extinguishers. Under no circumstances should water, carbon dioxide or carbon tetrachloride be used, because they react explosively with the burning metal.

The strong reducing effect of M. is seen in its reactions with many element oxides, from which it releases the elements at high temperature, e.g. $2\,Mg + SiO_2 \rightarrow 2\,MgO + Si$, or $2\,Mg + CO_2 \rightarrow 2\,MgO + C$. M. also reacts with alkali metal hydroxides and oxides, often explosively. This behavior, which is comparable to aluminothermism, is due to its strong tendency to form magnesium oxide; this in turn is due to a high lattice energy. Its tendency to oxidize makes the use of M. as a building material difficult, and necessitates protection from corrosion. For this purpose, the passifying oxide layer is augmented, e.g. by treating the magnesium object with dichromate solution or nitric acid, or by anodic oxidation. Alkaline solutions or fluoride baths also cause formation of a stable protective layer of magnesium oxide or magnesium fluoride.

Analysis. In qualitative analysis, M. is found in the filtrate of the ammonium carbonate group, from which it is precipitated as magnesium ammonium phosphate, $MgNH_4PO_4 \cdot 6H_2O$, or as magnesium hydroxide, $Mg(OH)_2$. M. can be determined quantitatively by classical gravimetry, by precipitation of magnesium ammonium phosphate, which is weighed as such or is converted to magnesium diphosphate, $Mg_2P_2O_7$. Complexometric determination with EDTA is less time-consuming. Atomic absorption spectroscopy is also used, especially for low Mg^{2+} concentrations.

Occurrence. M. makes up 1.93% of the earth's crust, and is thus the 8th most abundant element. It is always found in the bound state in nature, usually in the form of silicates, carbonates, chlorides or sulfates. Some common minerals are serpentine, $[Mg_6(OH)_6][Si_4O_{10}]$, serpentine asbestos, $[Mg_6(OH)_6][Si_4O_{11}] \cdot H_2O$, olivine, $(Mg,Fe)_2[SiO_4]$, talcum, $[Mg_3(OH)_2][Si_4O_{10}]$, magnesite, $MgCO_3$, dolomite, $MgCO_3 \cdot CaCO_3$, carnallite, $MgCl_2 \cdot KCl \cdot 6H_2O$, bischofite, $MgCl_2 \cdot 6H_2O$, kieserite, $MgSO_4 \cdot H_2O$, kainite, $KCl \cdot MgSO_4$, and schoenite, $K_2SO_4 \cdot MgSO_4 \cdot 6H_2O$. Magnesium salts are found in all natural bodies of water, as weathering products of these minerals. Together with calcium salts, they account for the hardness of water. Large amounts of magnesium salts are dissolved in seawater, making up 15% of the total salt content (this corresponds to an average content of 1.3 Mg^{2+} per kg water). Some mineral springs (bitter waters) contain high concentrations of magnesium sulfate.

M. plays an important role in biological processes. It is a component of chlorophyll and is involved in phosphorylation processes, photosynthesis, the citric acid cycle and other enzymatic processes in plants. In the human body, M. acts as an activator in the degradation of sugar and as a antagonist of calcium. An elevated Mg^{2+} concentration in the blood reduces the excitability of the nerves and muscles, an effect which can extend to paralysis and anesthesia. Injections of calcium salts are used as a countermeasure.

Production. Most M. is made by melt electrolysis of anhydrous magnesium chloride; a smaller amount is made by electrothermal processes. The raw materials are magnesite, dolomite, carnallite, end liquors from potassium salt production or sea water.

Magnesium carbonate is burned to magnesium oxide, which is reduced in the presence of chlorine and carbon to anhydrous magnesium chloride: $MgO + C + Cl_2 \rightarrow MgCl_2 + CO$. The chlorine is recovered in the electrolysis step. If dolomite is used as the raw material, calcium oxide as well as magnesium oxide is obtained from the burning process. The calcium oxide is removed by treatment with magnesium chloride solution, which converts it to calcium chloride: $MgO + CaO + MgCl_2 + 2\,H_2O \rightarrow 2\,Mg(OH)_2 + CaCl_2$. The hydroxide is dehydrated to the oxide by heating. In the production of $MgCl_2$ from seawater, the magnesium hydroxide is first precipitated with lime milk (calcium hydroxide), then processed as above. The magnesium chloride hexahydrate, $MgCl_2 \cdot 6H_2O$ obtained by evaporation of the final liquor from potassium salt, can also be used as a source of magnesium. However, to prevent partial elimination of hydrogen chloride and formation of the oxygen chloride, $2\,MgCl_2 \cdot 6H_2O \rightarrow MgO \cdot MgCl_2 + 2\,HCl + 11\,H_2O$, it must be heated in a dry stream of hydrogen chloride. To reduce its melting point, calcium fluoride or other alkali metal or alkaline earth metal fluorides are added to the $MgCl_2$; the liquid mixture is electrolysed at 650-720 °C under a protective atmosphere. Chlorine develops at the graphite anode, and M. is formed at the iron cathode. It floats on the salt melt and is removed from time to time. The energy expenditure is about 17.5 kWh kg^{-1} Mg.

The electrothermal method is based on reduction of the magnesium oxide with carbon at about 2000 °C in an electric furnace: $MgO + C \rightarrow Mg + C$. Only at this temperature or above is the equilibrium of the reaction far enough to the right. To prevent the back reaction, the mixture is quenched, usually in a hydrogen atmosphere.

In addition to these methods, other reducing processes are used, in which MgO is reduced with calcium carbide, ferrosilicon, aluminum, etc.

Applications. Most of the M. produced is used in Magnesium alloys (see), which have very useful ratios of densities and mechanical stability. Because it does not react readily with fluorine, M. is used in the construction of electrolysis apparatus. It is used in metallurgy as an antioxidant and desulfurizing agent. It is also used as a reducing agent in the production of metals such as uranium, titanium, zirconium and beryllium from their halides. Very pure M. is increasingly used in cathodic corrosion prevention (magnesium sacrifice anodes) for pipes, tanks, steel construction, etc. Magnesium powder is used in mixtures for flash bulbs and magnesium flares. In organic synthesis, M. is used to produce organomagnesium compounds (Grignard compounds). It is also used to dry alcohols.

Historical. Although magnesium compounds, such as magnesium sulfate, have been known and used in medicine for centuries, M. was first recognized as an element in 1755 by Black. It was first produced as the metallic element in 1808 by Davy, by melt electrolysis. In 1857, St. Claire, Deville and Caron made large amounts of M. for fireworks in France, and thus gave the metal its first economic significance.

Magnesium acetate: $Mg(CH_3COO)_2 \cdot 4H_2O$, colorless, deliquescent, monoclinic crystals which are readily soluble in water and ethanol; M_r 214.46, density 1.454, m.p. 80 °C (in its own water of crystallization). The compound is obtained by dissolving magnesium carbonate in acetic acid and is used to make glazes and in dyeing.

Magnesium alloys: alloys which in addition to large proportions of magnesium contain chiefly aluminum, manganese, zinc, copper, cerium, zirconium and silicon. The alloys are more resistant to corrosion than magnesium, and have better mechanical properties. M. which contain 90 to 95% magnesium are known as electron metals. Their low densities (about 1.8) and their good working properties make M. important materials in all applications where strength and low mass are needed (aircraft and rockets, vehicles, machinery, etc.). Magnesium is also used as a component of Aluminum alloys (see), where it both improves the strength and reduces the sensitivity of the metal to alkalies. Nickel and iron are improved by alloying with magnesium. Many M. can burn (see Magnesium).

Magnesium ammonium phosphate: $MgNH_4PO_4 \cdot 6H_2O$, colorless, acid-soluble, rhombic crystals; M_r 245.41, density 1.711. M. is nearly insoluble in water (solubility product $K_s = 2.5 \cdot 10^{-13}$ at 25 °C). When roasted, M. is converted to magnesium diphosphate, $Mg_2P_2O_7$. M. is formed by adding a phosphate solution to an ammoniacal solution of a magnesium salt; its precipitation is used for qualitative and quantitative analysis of magnesium.

Magnesium bromide: $MgBr_2 \cdot 6H_2O$, colorless, water-soluble, hexagonal crystals; M_r 292.22, density 2.00. M. is found in small amounts in the minerals bischofite and carnallite, and in seawater. It is formed by evaporation of a solution of magnesium carbonate and hydrobromic acid. The anhydrous salt (m.p. 700 °C) is obtained by heating the hydrate in a stream of dry hydrogen bromide. M. is used as a tranquilizer.

Magnesium carbonate: $MgCO_3$, colorless, trigonal crystals, M_r 84.32, density 2.958. M. crystallizes out of water as the penta-, tri- or monohydrate, depending on the temperature. In nature it is found as magnesite and as a component of dolomite. Around 900 °C, M. is converted to magnesium oxide: $MgCO_3 \rightarrow MgO + CO_2$. M. is formed as a precipitate (solubility product $K_s = 2.6 \cdot 10^{-5}$ at 12 °C) when alkali carbonates are added to an aqueous solution of a magnesium salt in the presence of excess carbon dioxide. It is often used as a raw material for the production of other magnesium compounds (see Magnesium).

When neutral M. is boiled in water for a long time, it is converted into ***basic M. (magnesia alba, magnesia white)***, $4\,MgCO_3 \cdot Mg(OH)_2 \cdot 4H_2O$, a colorless, loose, water-insoluble powder. This is found in nature as hydromagnesite, and is also synthesized by precipitation of magnesium sulfate solutions with soda. It is used medically as a mild laxative, to neutralize stomach acid and to make powders for treating cuts and cleaning teeth. It is also used as an antidote to poisoning by arsenic trioxide, acids and metal salts, as a white pigment, a filler for rubber, paper and paints, a fire-proof material and to clarify liquids.

Magnesium chloride: $MgCl_2 \cdot 6H_2O$, colorless, hygroscopic, very water-soluble, monoclinic crystals; M_r 203.31, density 1.569. M. is found in nature as bischofite, and as a component of carnallite. Seawater also contains significant amounts of M. In addition to the hexahydrate, there are hydrates which contain 2,4,8 and 12 mol water of crystallization. If M. is heated, it is both dehydrated and loses some hydrogen chloride: $2\,MgCl_2 \cdot 6H_2O \rightarrow MgO \cdot MgCl_2 + 2\,HCl + 11\,H_2O$. The elimination of hydrogen chloride is avoided if the compound is dehydrated in a stream of dry hydrogen chloride. The hexagonal, anhydrous $MgCl_2$ melts at 714 °C and boils at 1412 °C. Its production is described in the entry on Magnesium (see). Anhydrous $MgCl_2$ is the starting material for production of magnesium by electrolysis. Mixtures of conc. magnesium chloride solution and heat-treated magnesium oxide rapidly solidify to a very hard mass of magnesium oxygen chlorides, which is called *magnesia cement*. The hygroscopic property of M. is utilized in its industrial application as a dust binder. It is also used as a frost-protecting material, a fire extinguisher, melting salt, impregnating agent for wood protection and a finish component for textiles. The large amounts of magnesium chloride end liquors produced in the potassium industry are not widely utilized and, as waste waters, they create a significant environmental burden on the surface waters into which they are dumped.

Magnesium fluate: same as Magnesium hexafluorosilicate (see).

Magnesium fluoride: MgF_2, colorless, tetragonal crystals; M_r 62.31, m.p. 1261 °C, b.p. 2239 °C. M. is rather insoluble in water (solubility product $K_s = 7.1 \cdot 10^{-9}$ at 18 °C). It occurs in nature in the form of the rare mineral sellaite. It is formed when a fluoride solution is added to a solution of a magnesium salt. When a thin layer of M. is vaporized onto a lens, it prevents reflections. Single crystals are used for optical purposes. M. is also used as an additive to welding fluxes and in the electrolysis and refining of aluminum.

Magnesium hexafluorosilicate, formerly ***magnesium silicofluoride, magnesium fluate***: $MgSiF_6 \cdot 6H_2O$, colorless, water-soluble, poisonous, trigonal crystals; M_r 274.48, density 1.788. M. is obtained by dissolving magnesium oxide or magnesium carbonate in an aqueous hexafluorosilicic acid solution: $MgCO_3 + H_2SiF_6 \rightarrow MgSiF_6 + CO_2 + H_2O$. Aqueous solutions of M. are used as thickeners for concrete: the calcium oxide in the concrete reacts with M. to form small crystals of calcium fluoride and silicic acid, which fill in the pores in the concrete. M. is also used to seal masonry, to provide a neutral base for paint and to preserve wood.

Magnesium hydride: MgH_2, colorless, tetragonal, crystalline powder; M_r 26.33. Above 280 °C, M. decomposes into its components. It reacts vigorously with water, forming magnesium hydroxide and elemental hydrogen: $MgH_2 + 2\,H_2O \rightarrow Mg(OH)_2 + 2\,H_2$. M. can be conveniently made by thermal decomposition of diethylmagnesium: $Mg(C_2H_5)_2 \rightarrow MgH_2 + 2\,H_2C{=}CH_2$. M. is used in organic synthesis as a reducing agent and basic catalyst. It is also used as a drying agent.

Magnesium hydrogencarbonate, formerly ***magnesium bicarbonate***: $Mg(HCO_3)_2$, a water-soluble compound which cannot be isolated in pure form; when its aqueous solutions are evaporated, it decomposes into magnesium carbonate, carbon dioxide and water: $Mg(HCO_3)_2 \rightarrow MgCO_3 + CO_2 + H_2O$. It is formed by the reverse of the above equation, when water containing CO_2 reacts with magnesite or dolomite. Together with dissolved calcium hydrogencarbonate, it is responsible for temporary water hardness.

Magnesium hydroxide: $Mg(OH)_2$, colorless powder or hexagonal crystals; M_r 58.33, density 2.36. M. is nearly insoluble in water (solubility product $K_s = 1.2 \cdot 10^{-11}$ at 18 °C). It is dissolved by acids, even by such a weak acid as an ammonium chloride solution. Above 350 °C it is converted to magnesium oxide: $Mg(OH)_2 \rightarrow MgO + H_2O$. M. occurs naturally as brucite. It can be synthesized by adding strong alkalies to magnesium salt solutions. M. is used as a starting material for making other magnesium compounds and as a weakly basic antidote to oral poisoning by mineral acids.

Magnesium nitrate: $Mg(NO_3)_2 \cdot 6H_2O$, colorless, deliquescent, monoclinic crystals which are readily soluble in water and alcohol; M_r 256.41, density 1.636, m.p. 89 °C. M. is obtained by dissolving magnesium carbonate or oxide in nitric acid.

Magnesium nitride: Mg_3N_2, a yellow-green, amorphous powder; M_r 100.95, density 2.712, dec. around 800 °C. M. is converted to magnesium hydroxide and ammonia by atmospheric moisture: Mg_3N_2 +

$6\ H_2O \rightarrow 3\ Mg(OH)_2 + 2\ NH_3$. It is formed by the reaction of nitrogen with magnesium at red heat.

Magnesium oxide, ***magnesia, magnesia usta, burned magnesia, caustic magnesia, calcined magnesia***: MgO, colorless, cubic crystalline powder; M_r 40.31; density 3.58; m.p. 2852 °C, b.p. about 3600 °C. Depending on the conditions of production, M. is either unreactive with water or slowly reacts to form magnesium hydroxide, $Mg(OH)_2$. M. occurs in nature as periclase. It is synthesized by heating magnesium carbonate (magnesite), magnesium hydroxide or magnesium nitrate. M. is an intermediate in the production of magnesium chloride for the electrolytic production of Magnesium (see). M. heated to very high temperatures is used as sintered magnesia for heat-resistant laboratory apparatus or bricks (magnesia bricks). M. is also used to make magnesia cement (see Magnesium chloride). In medicine, it is used to treat excess stomach acidity and poisoning by acids.

Magnesium perchlorate: $Mg(ClO_4)_2$, colorless, hygroscopic powder, M_r 223.21, density 2.12, dec. above 250 °C. M. is obtained by adding magnesium oxide or carbonate to dilute perchloric acid. It crystallizes out of the aqueous solution as the rhombic hexahydrate, $Mg(ClO_4)_2 \cdot 6H_2O$, which can be dehydrated by heating it in vacuum in the presence of phosphorus pentoxide or by passing dry air over it. M. is an excellent drying agent for many gases.

Magnesium phosphate: 1) in the narrow sense, a magnesium salt of orthophosphoric acid, H_3PO_4. a) ***Magnesium dihydrogenphosphate, primary magnesium phosphate, magnesium tetrahydrogenphosphate***, $Mg(H_2PO_4)_2 \cdot 3H_2O$, colorless, water-soluble, hygroscopic powder; M_r 272.36, used as a component of flame protectants. b) ***Magnesium hydrogenphosphate, secondary magnesium phosphate***, $MgHPO_4 \cdot 3H_2O$, colorless rhombic crystals which are slightly soluble in water and readily soluble in acids; M_r 174.34, density 2.123; it also forms a monoclinic heptahydrate. The secondary phosphate is converted to magnesium diphosphate at 550 to 650 °C: $2\ MgHPO_4 \rightarrow Mg_2P_2O_7 + H_2O$. It is used in ceramics and as a laxative. c) ***Magnesium phosphate, trimagnesium phosphate, tertiary magnesium phosphate***, $Mg_3(PO_4)_2$, colorless, rhombic crystalline powder; M_r 262.88, m.p. 1184 °C; also crystallizes with 5, 8 and 22 moles water. The phosphate is insoluble in water and is found in nature as bobierrite. It is used as a fertilizer and to treat excess stomach acid.

2) In the wider sense, magnesium salts of condensed phosphoric acids. The simplest example is ***magnesium diphosphate, magnesium pyrophosphate***, $Mg_2P_2O_7$, a colorless monoclinic crystalline powder which is relatively insoluble in water, but dissolves in acids; M_r 222.57, density 2.559, m.p. 1383 °C. The diphosphate is formed by heating magnesium hydrogenphosphate to 550-650 °C; it is used in the production of ceramics, glass and enamel.

Magnesium silicates: magnesium salts of silicic acids of varying structures. M. occur widely and are components of many minerals (see Magnesium). Some important representatives of this group are ***magnesium orthosilicate***, Mg_2SiO_4, colorless, orthorhombic crystals which are nearly insoluble in water; M_r 140.71, density 3.21, m.p. 1910 °C, and ***magnesium metasilicate***, $[MgSiO_3]^n$, colorless monoclinic crystals which are nearly insoluble in water; formula mass 100.4, density 3.192, dec. 1557 °C. The latter compound is made up of long-chain linear and cyclic polysilicate anions. Both compounds are found in nature. They are used as sorption materials (e.g. in thin-layer chromatography), as filter aids, fillers in plastics and rubber, and so on.

Magnesium silicofluoride: see Magnesium hexafluorosilicate.

Magnesium sulfate: $MgSO_4$, a colorless, rhombic, crystalline powder; M_r 120.37, density 2.66, dec. 1124 °C. M. is readily soluble in water, and its solubility increases rapidly with increasing temperature: 100 g water dissolves 30.9 g $MgSO_4$ at 10 °C, and 68.0 g at 100 °C. M. crystallizes out of water in a number of different hydrates, depending on the temperature. Between -3.9 and 1.8 °C, the dodecahydrate, $MgSO_4 \cdot 12H_2O$ crystallizes out; between 1.8 and 48.3 °C, the rhombic heptahydrate, $MgSO_4 \cdot 7H_2O$ (***Epsom salt***), between 48.3 and 68 °C, the monoclinic hexahydrate $MgSO_4 \cdot 6H_2O$, and above 68 °C, the monohydrate, $MgSO_4 \cdot H_2O$, which also is found in nature as kieserite. There are also a number of unstable hydrates and many double salts (e.g. kainite, schoenite). Epsom salt is made from kieserite , a byproduct of the potasium industry. It is dissolved in warm water, and when the solution cools, the heptahydrate crystallizes out. It is used to weight textiles, as a flame retardant and as a mordant for dyeing. It is used in medicine as a mild laxative.

Magnesium thiosulfate: $MgS_3O_2 \cdot 6H_2O$, colorless, water-soluble, rhombic crystals; M_r 244.53, density 1.818, at 170 °C, loss of 3 moles hydrate water. M. is obtained by adding sulfur to a solution of magnesium sulfite prepared by saturating a magnesium hydroxide suspension with sulfur dioxide. M. is used as a muscle relaxant, to reduce blood pressure and as a laxative.

Magnesium type, ***magnesium structure***: a Structure type (see) in which the atoms are in a hexagonal close-packed array (see Packing of spheres). In the ideal case, each atom has 12 nearest neighbors at exactly the same distance, and the axis ratio of the hexagonal unit cell is $c_0/a_0 = 1.633$ (c_0 and a_0 are lattice constants). The packing density is 0.74. Most representatives of the M. have axis ratios which are very close to the ideal. When the ratio is smaller, the bonding forces between the layers of spheres are stronger than within the layers, and when it is larger, the converse is true. In both cases, the space filling decreases. More than 25% of all metallic element modifications crystallize in the M. In addition to the prototype magnesium, there are, for example, beryllium, most lanthanoids, titanium, zirconium, zinc, cadmium and the noble gas helium.

Magnetic balance: see Magnetochemistry.

Magnetic circular dichroism: see Chiroptic methods.

Magnetochemistry: the study of the relationship between the magnetic properties of a substance and its structure. The most important of these properties are diamagnetism and paramagnetism.

If a body is placed in a homogeneous magnetic field, it may have one of two effects (Fig. 1): it either spreads apart the magnetic field lines which pass through it (diamagnetism) or it brings them closer together (paramagnetism). These phenomena can be explained as follows: If a closed electric circuit is moved toward a pole of a magnet, a current is induced in the circuit, and the direction of the current is such that the magnetic field it generates is opposed to the first magnetic field (Fig. 2).

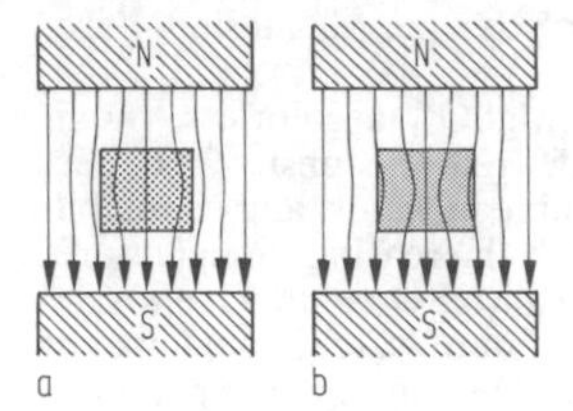

Fig. 1. Diamagnetic (a) and paramagnetic (b) substances in a homogeneous magnetic field.

This is known in the theory of electricity as Lenz's law, and it applies to the orbits of electrons in atoms as well. As a result, the field lines of the external and the induced internal magnetic field partially cancel each other, so that there is a dilution of the field lines within the body; it is diamagnetic. The *diamagnetic effect* occurs in all substances and is a universal property of matter. It is independent of temperature, because the motion of the electrons is not affected by the temperature. Diamagnetic substances placed in inhomogeneous magnetic fields are pushed into the regions of lower field strength, i.e. away from the poles. A second effect, the *paramagnetic effect*, is superimposed on the diamagnetic effect if the atoms or molecules of the substance have a permanent external magnetic moment. Every electron in an atom or molecule gives rise to a circular current, and thus has a magnetic moment, but in general the fields of the individual electrons cancel each other. This is always the case for the electrons in closed shells. The magnetic moments of electrons in partially occupied shells do not completely cancel, and the atoms, ions or molecules containing them have permanent magnetic moments; they are tiny unit magnets. In the absence of an external magnetic field, the unit magnets are randomly oriented because of thermal motions, but if an external field is applied, they orient themselves so that the north pole of the unit magnet faces the south pole of the external field, and vice versa. This orientation creates an additional magnetic field which is aligned in the same direction as the external field and the number of field lines in the interior of the body increases: it is paramagnetic. Unlike diamagnetism, paramagnetism is dependent on temperature, because the thermal motion counteracts the alignment of the unit magnets in the north-south direction. The higher the temperature, the lower the orientation effect. In an inhomogeneous magnetic field, a paramagnetic substance is pulled in the direction of increasing field strength.

Determination of magnetic properties. If a substance is placed in a magnetic field of field strength H, and the magnetization in the direction of the field is called M, $M = X_{vol}H$, where X_{vol} is the *susceptibility* per unit volume. Since the magnetization M has the same dimensions as the field strength H, the proportionality factor X_{vol} is dimensionless. However, the numerical value of X_{vol} depends on the system of measures used, and to avoid confusion, the system is indicated after the susceptibility value, e.g. X_{H_2O} $+0.72 \cdot 10^{-6}$ (CGS) $= 9.05 \cdot 10^{-6}$ (SI). Most of the X_{vol} values in the literature are given in the CGS system; these values can be converted to the SI system by multiplication by 4π. For practical reasons, the magnetization is often related to 1 g rather than a unit volume and defines the gram susceptibility $X_g = X_{vol}/\varrho$ where ϱ is the density. The atomic or molar susceptibility is obtained from the gram susceptibility by multiplication by the atomic or molar mass ($X_{atom} = X_g \cdot$ atomic mass; $X_{mol} = X_g \cdot$ molar mass). The susceptibility is a measure of the ability of the substance to absorb magnetic field lines. Diamagnetic substances, which repel the field lines, have negative susceptibilities; paramagnetic substances, which concentrate the field lines, have positive susceptibilities. Sometimes the permeability μ is used instead of the susceptibility X; the two are related by the equation $\mu = 1 + X$. For diamagnetic substances, $\mu < 1$, and for paramagnetic, $\mu > 1$.

Susceptibility can be measured using a magnetic balance. An inhomogeneous magnetic field is generated by an electromagnet with cone-shaped pole pieces. The substance to be examined is placed in a cylindrical glass tube and suspended, from a sensitive balance, between the pole pieces of the magnet. When the electromagnet is turned on, the apparent weight of the sample changes as it is drawn into or repelled from the magnetic field. The apparent change in weight is proportional to the susceptibility. The magnetic balance is usually used for relative measurements in which the susceptibility of the unknown is related to that of a known substance, e.g. water. In this way both the susceptibility of the sample holder and other systematic errors are compensated.

The diamagnetism of a substance indicates that it contains closed electron shells. Diamagnetism is observed in all noble gases and ions with closed shells, such as Na^+ and Cl^-; it is also observed in all atoms and ions of the transition metals with 18 electrons, such as Pd, Cu^+ and Zn^{2+}, which have fully occupied *d* levels in addition to full *s* and *p* orbitals. Finally, all atoms and ions which have only 2*s* electrons outside their completely filled shells (e.g. B, Be, Zn, Ca), are diamagnetic. Nearly all organic molecules are diamagnetic. The significance of diamagnetic susceptibility for chemists is that the value for a molecule is, to a first approximation, the sum of empirically determined individual values (*increments*) for the atoms and certain structural elements. The table shows a selection of these increments, which were first published by the French chemist Pascal. When there are several possible structural formulas for a substance (i.e. isomerism is possible) calculation of the susceptibility using these increments and comparison with the measured value makes it possible to choose the correct formula. This method has now largely been replaced by more powerful spectroscopic methods, however.

Although all substances have some diamagnetic behavior, paramagnetic behavior occurs only when the magnetic moments of the individual electrons do not

Atomic and structural elements, after Pascal, in 10^{-6} CGS units

Atomic	increments	Structure	increments
H	-2.93	>C=C<	+5.45
C	-6.0	-C≡C-	+0.8
N	-5.6	Benzene	-1.44
O	-4.6	Naphthalene	-8.02

compensate for each other. The molar susceptibility X_{mol} of a paramagnetic substance thus consists of 2 parts, a diamagnetic and a paramagnetic: $X_{mol} = X_{dia}+X_{para}$. The two parts not only have different signs, but differ in their absolute magnitudes; paramagnetic susceptibilities are 10 to 1000 times larger than diamagnetic. This means that if paramagnetism is present, it always more than compensates for the diamagnetism of the substance. The two parameters can be separated by calculating the diamagnetic molar susceptibility by means of an increment system, and subtracting it from the measured total susceptibility; the remainder is the paramagnetic molar susceptibility. The X_{para} values determined in this way are especially significant in the chemistry of complex compounds and free radicals.

In *complex chemistry*, the paramagnetic susceptibility can often be used to determine the electronic arrangement in the central atom, and thus its valency. This is important in the chemistry of transition metals, the cations of which usually have more *d* orbitals than are needed to accomodate all their *d* electrons. Thus there are several possible arrangements of the *d* electrons which differ with respect to their magnetic moments.

The magnetic moment μ in units of the Bohr magneton (see) μ_B is calculated from the measured paramagnetic susceptibility X_{para} using the equation:

$$\mu = \sqrt{\frac{X_{para}\,3kT}{N_A}}$$

where k is the Boltzmann constant, T the absolute temperature and N_A is Avogadro's number. In the simplest case, this magnetic moment can be predicted by ignoring the orbital moments of the electrons and substituting the number of unpaired electrons in the complex, n, in the equation $\mu = \sqrt{n(n+2)}\,\mu_B$; n can be derived from ligand field theory. The numerical values of the magnetic moments as a function of n are:

n =	1	2	3	4	5	
μ =	1.73	2,83	3.88	4.90	5.92	μ_B

Using paramagnetic measurements it is possible to detect free radicals, since these have unpaired electrons. For example, the existence of triphenylmethyl radicals in solution was demonstrated by their paramagnetism. The electronic structure of the oxygen molecule, in which the ground state is a triplet state with two parallel electron spins, was also recognized on the basis of its paramagnetism. However, the magnetic balance can be used to detect only stable radicals in reasonable concentrations. Lower concentrations of radicals can be detected by Electron spin resonance spectroscopy (see), which is also suitable for detection of paramagnetic compounds.

Ferromagnetism is not a property of atoms or molecules, but of crystals, which arises by coupling of atomic spin moments throughout a large crystalline region. Because many electrons are involved, the magnetic moments are large, and they can be aligned with an external magnetic field in the same way as the paramagnetic moments of atoms or molecules. Like the paramagnetic susceptibility, ferromagnetic susceptibility has a positive sign, but its magnitude is much greater.

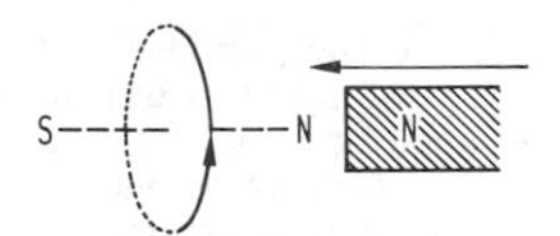

Fig. 2. Induced current and its associated magnetic field.

Magnetogyric ratio: see NMR spectroscopy.

Magnetooptical rotatory dispersion: see Chiroptic methods.

Magnus' salt: see Platinum complexes.

Maillard reaction: a reaction of reducing oligosaccharides with amino acids or amines. It starts with formation of glycosylamines, which then undergo Amadori rearrangement. An M. occurs when glycoproteins or mixtures of carbohydrates and proteins are heated. The brown color developed in many foods during storage is due to products of the M.

Majolica: silicate ceramic products with porous, non-white body, which is covered by a transparent lead glaze or a white, opaque glaze containing SnO_2; this is decorated with colored paints. The raw materials are limestone and iron-containing clay. The dried pieces are fired at 900 to 1000 °C. After application of the glaze and decoration, a second firing at 900 °C is required. The name is derived from the Mediterranean island Mallorca.

Malachite green: 4-dimethylaminofuchsone dimethylimonium salt, a basic, blue-green triphenylmethane pigment (fuchsonimine pigment). M. is readily soluble in water and alcohol. It is sold as water-soluble, green dye salts, e.g. as the oxalate (metallic, shiny green leaflets) or as the chlorozinc double salt (brass-colored, prismatic crystals). M. is made industrially from dimethylaniline, benzaldehyde and hydrochloric acid or sulfuric acid at 100 °C. First the colorless leukobase is formed, and then oxidized to M. in a hydrochloric acid solution with lead dioxide or manganese dioxide.

Because of its clear and very dark bluish green color, M. is still widely used as a dye, in spite of its lack of fastness to light, acids and alkalies, especially for printing silk and treated cotton, and to dye leather and paper. It is also used in paints. M. is used as an acid-base indicator, with a color change at pH 0.1 to 2.0 (from yellow to bluegreen). It is also used to stain histological sections and bacterial spores.

M. was first synthesized in 1877 by O. Fischer; almost at the same time (1878), Döbner obtained M. by condensation of dimethylaniline with benzotrichloride.

$(CH_3)_2\overset{+}{N}$=C_6H_4=C(C_6H_4—$N(CH_3)_2$)(C_6H_5) Cl^-

Malaprade reaction: a method of oxidative cleavage of glycols and α-aminoalcohols with periodic acid to form aldehydes and ketones:

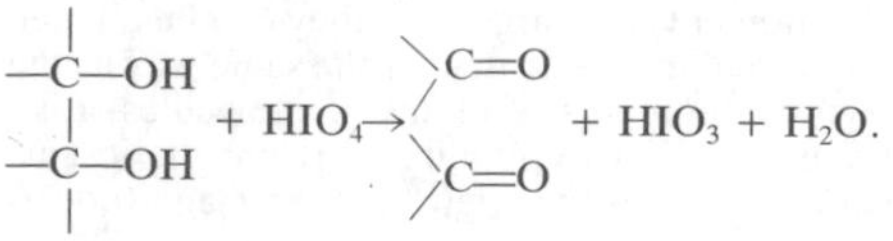

$$\begin{matrix} | \\ -C-OH \\ | \\ -C-OH \\ | \end{matrix} + HIO_4 \rightarrow \begin{matrix} \rangle C=O \\ \\ \rangle C=O \end{matrix} + HIO_3 + H_2O.$$

The M. probably goes via a cyclic periodic acid ester, in analogy to the cleavage of glycols with lead tetraacetate. Because of the simple and mild reaction conditions, M. is very important in carbohydrate chemistry, especially for elucidation of structures.

Malates: see Malic acid.

Malathion: *S*-[1,2-bis-(ethoxycarbonyl)ethyl]-*O*,*O*-dimethyldithiophosphate, an Organophosphate insecticide (see), and an oily liquid which is a strong contact poison. D_4^{25} 1.23, m.p. 3 °C, b.p. 156-157 °C at 93 Pa, vapor pressure at 20 °C, 16.3 mPa. M. is readily soluble in organic solvents, less soluble in mineral oils, and in water to a concentration of 145 ppm at 25 °C. It is hydrolysed to various degrees by acids and bases. M. is converted by oxidizing agents and under physiological conditions to the PO compound malaoxon, which has a higher cholinesterase-inhibiting effect. M. is synthesized industrially by addition of dimethyldithiophosphoric acid to diethyl maleic acid. It is applied mainly against sucking insects and spider mites, scale insects, fruit maggots and disease vectors. Its properties make it suitable for producing ultra-low-volume (ULV) concentrates for aerial applications.

Maleates: see Maleic acid.

Maleate resins: see Alkyde resins.

Maleic acid, ***cis-butenedioic acid, (Z)-butenedioic acid***: an unsaturated, aliphatic dicarboxylic acid which is the *p*-diastereoisomer of Fumaric acid (see). M. forms colorless prisms, m.p. 130.5 °C (139-140 °C and 143.5 °C also reported). It cannot be distilled without decomposition, and when it is heated it is very easily dehydrated to form ***maleic anhydride***. M. is soluble in water, alcohol, acetone, ether and glacial acetic acid. It is very irritating to the skin and mucous membranes. Irradiation with UV or heating to about 60 °C in the presence of catalysts causes M. to rearrange to fumaric acid. M. displays the reactions typical of dicarboxylic acids. Its salts and esters are called *maleates*. In contrast to fumaric acid, it does not form an acyl chloride. Due to the presence of the C=C double bond, M. can also undergo the usual addition reactions, forming saturated derivatives. For example, addition of chlorine leads to dichlorosuccinic acid, and catalytic hydrogenation yields succinic acid. Unlike fumaric acid, M. is not found in nature. It is made industrially by hydrolysis of maleic anhydride, which is obtained by catalytic oxidation of benzene or but-2-ene. M. is used in the synthesis of plastics, mixed polymers, pharmaceuticals and preservatives.

Historical. In 1838, Liebig demonstrated that M. and fumaric acid have the same chemical composition. He first thought that fumaric acid was a polymer of M. In 1887, Wislicenus discovered that the two acids are structurally isomeric ethene dicarboxylic acids. This was the first experimental evidence of the *cis-trans*-isomerism predicted by van't Hoff.

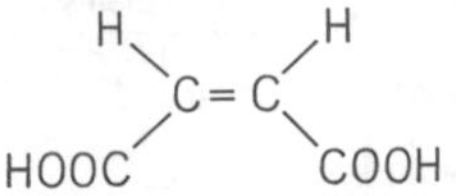

$$\begin{matrix} H & & & & H \\ & \diagdown & & \diagup & \\ & & C=C & & \\ & \diagup & & \diagdown & \\ HOOC & & & & COOH \end{matrix}$$

Maleic anhydride, ***2,5-dihydrofuran-2,5-dione***: colorless needles; m.p. 60 °C, b.p. 197-199 °C. It is soluble in ether and chloroform, in water with formation of malic acid and in alcohol with formation of maleic acid esters. With conjugated, unsaturated compounds, such as butadiene or cyclopenta-1,3-diene, M. acts as a dienophile in the Diels-Alder reaction. It is very irritating to the skin, and especially to the mucous membrane of the eye. M. is synthesized by catalytic vapor phase oxidation of benzene with atomospheric oxygen in the presence of vanadium pentoxide. Recently, C_4 hydrocarbons from petroleum fractions, such as butane and butene, have become the preferred starting materials. M. is used as an intermediate in the synthesis of unsaturated polyester resins, maleic and fumaric acids, agricultural chemicals (see Maleic hydrazide) and tetrahydrofuran. It is also used as an additive to drying oils and softeners, lacquers and paints.

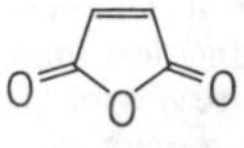

Maleic hydrazide, ***1,2,3,6-tetrahydropyridazine-3,6-dione***: colorless crystals; m.p. 296-298 °C (dec.). It is slightly soluble in alcohol and hot water. M. is synthesized from maleic anhydride and hydrazine hydrate. It is a selective herbicide for control of grasses; it acts by inhibiting mitosis in the vegetative points of the shoot tips. The result is dwarfed plants with short stems and chlorophyll-rich leaves.

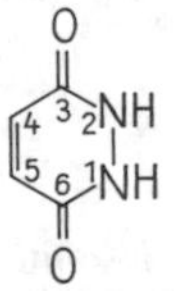

Malformin: a heterodetic, cyclic pentapeptide produced by *Aspergillus niger*. It has both antibacterial and cytotoxic activity, and causes malformation in higher plants.

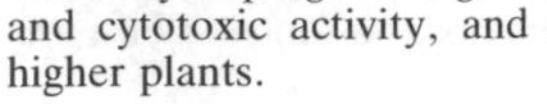

→D-Cys–D-Cys–Val–D-Leu–Ile→ (cyclic; Cys–Cys disulfide bridge)

Malic acid, ***hydroxybutanedioic acid, hydroxysuccinic acid***: $HOOC-CH_2-CH(OH)-COOH$, a hydroxydicarboxylic acid. A. can exist in two optically active, enantiomeric forms and as the racemate:

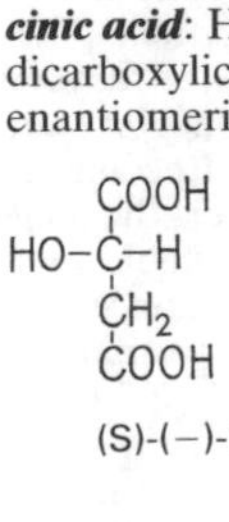

(S)-(−)-Malic acid	(R)-(+)-Malic acid
COOH; HO–C–H; CH₂; COOH	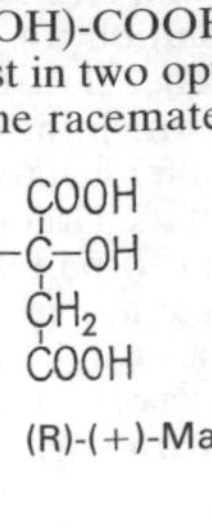 COOH; H–C–OH; CH₂; COOH

The (R)-(+) and the (S)-(-) forms form crystalline needles; m.p. 100.5 °C. The (RS) form is a colorless, crystalline compound, m.p. 128.5 °C. M. is readily soluble in water and alcohol, but only slightly soluble in ether. The salts and esters are called *malates*. (S)-(-)-M. is found free in sour apples, quinces, gooseberries, mountain ash berries and barberries. It is also part of the citric acid cycle. (S)-(-)-M. can be synthesized in pure form by racemate separation with optically active compounds. The (RS)-M. is synthesized by reaction of silver oxide with (RS)-bromosuccinic acid or by addition of water to maleic acid. In the reaction of (S)-(-)-M. with phosphorus(V) chloride, a Walden inversion (see) produces (R)-(+)-chlorosuccinic acid; this inversion of configuration was observed for the first time in this reaction. M. is used to impregnate packing materials for foods and as a flavoring.

Malodinitrile: same as Malonic dinitrile (see).

Malonaldehyde: same as Malonic dialdehyde.

Malonates: see Malonic acid.

Malonic dialdehyde, ***propanedial, malonaldehyde***: $OHC\text{-}CH_2\text{-}CHO$, colorless, hygroscopic needles; m.p. 72-74 °C. M. is very unstable and readily polymerizes to a glassy form. Like all typical 1,3-dicarbonyl compounds, M. undergoes keto-enol tautomerization, and exists practically completely in the enol form: $OCH\text{-}CH_2\text{-}CHO \rightleftharpoons OHC\text{-}CH{=}CH\text{-}OH$.

M. is fairly acidic, with a pK_a of 5.0 at 25 °C in water. It dissolves readily in aqueous bases, forming salts. The typical addition and condensation reactions of the Aldehydes (see) can occur twice per molecule of M. It can be synthesized by hydrolysis of 3-dialkylaminoprop-2-enals:

$$R_2N\text{–}CH{=}CH\text{–}CHO \xrightarrow[-R_2NH]{+H_2O} OHC\text{–}CH_2\text{–}CHO.$$

Because of its bifunctionality, M. is used chiefly as a C_3 unit for synthesis of heterocycles.

Malonic acid, ***propanedioic acid***: $HOOC\text{-}CH_2\text{-}COOH$, an aliphatic dicarboxylic acid. M. crystals are colorless and odorless, and melt at 135.6 °C with decomposition. They can be sublimed in vacuum without decomposition. M. is readily soluble in water and pyridine, less soluble in alcohol and ether, and insoluble in benzene. When the solid is heated above its melting point, or the solution above 70 °C, or when exposed to UV light, M. decomposes to acetic acid and carbon dioxide. This reaction is not typical of aliphatic dicarboxylic acids, but resembles the behavior of β-ketocarboxylic acids. Because it is activated by the neighboring carboxyl groups, the CH_2 group is highly reactive, and this is utilized in numerous syntheses, e.g. in the Knoevenagel condensation (see) and in Malonic ester syntheses (see). The reaction of phosphorus(V) oxide with M. produces its internal anhydride, carbon suboxide. Halogen reagents can react either with the CH_2 group or the carboxyl groups of M.; mono- or dichloromalonic acid is formed with sulfuryl chloride, and mono- or dichlorides are formed with thionyl chloride. The salts and esters of M. are called *malonates*. M. does not occur widely in nature; its potassium salt is present in the juice of sugar beets. It is formed by oxidation of malic acid, or synthesized by reaction of monochloroacetic acid with potassium cyanide, followed by hydrolysis of the primary cyanoacetic acid:

$$Cl\text{–}CH_2\text{–}COOH \xrightarrow[-Cl^-]{+CN^-} NC\text{–}CH_2\text{–}COOH \xrightarrow[-NH_3]{2\,H_2O} HOOC\text{–}CH_2\text{–}COOH$$

M. is used to produce carboxylic acids, ketones, aromas and malonic esters.

Malonic dinitrile, ***malodinitrile, propanedinitrile***: $N{\equiv}C\text{-}CH_2\text{-}C{\equiv}N$, colorless, poisonous crystals; m.p. 31 °C, b.p. 218-219 °C. M. is soluble in water, alcohol, ether and acetone, but only slightly soluble in chloroform or benzene. It is a very reactive compound, which can react with either the CH_2 group or the CN groups. Under the conditions of the Knoevenagle condensation (see), M. forms condensation products with aldehydes and ketones: $R\text{-}CH{=}O + CH_2(CN)_2 \rightarrow R\text{-}CH{=}C(CN)_2 + H_2O$. M. can be synthesized by dehydration of cyanoacetamide with phosphorus pentachloride or phosphorus oxygen chloride, or from acetonitrile and cyanogen chloride: $N{\equiv}C\text{-}CH_3 + Cl\text{-}C{\equiv}N \rightarrow N{\equiv}C\text{-}CH_2\text{-}C{\equiv}N$. It is used as an intermediate in the synthesis of pigments, pesticides and other organic compounds.

Malonic ester: same as Diethylmalonate.

Malonic ester syntheses: syntheses of saturated and unsaturated mono- and dicarboxylic acids, α-aminocarboxylic acids, barbituric acid and alicyclic compounds from malonic diethyl ester. Because of the electron withdrawing effect of the two ethoxylcarbonyl groups, malonic ester is a CH-acidic compound. With sodium alkoxylate it forms ***sodium malonic ester***, which reacts with alkyl halides to form ***alkylmalonic esters***. Hydrolysis and decarboxylation of these yield monocarboxylic acids:

$$H_2C(COOC_2H_5)_2 \xrightarrow[-C_2H_5OH]{C_2H_5\bar{O}^-Na^+} Na^+\,{}^-CH(COOC_2H_5)_2 \xrightarrow[-NaX]{+RX} RHC(COOC_2H_5)_2 \xrightarrow{\text{Hydrolysis}}$$

$$RHC(COOH)_2 \xrightarrow{\text{Decarboxylation}} R\text{–}CH_2C(=O)OH$$

Replacement of the two H atoms of malonic ester yields disubstituted acetic acids. Reaction of sodium malonic ester with a halocarboxylic acid yields a dicarboxylic acid in similar fashion. Dicarboxylic acids are also produced by reaction of 2 mol malonic ester with α,ω-dihaloalkanes, followed by hydrolysis and decarboxylation of the tetracarboxylic ester. The reaction of 1 mol α,ω-dihaloalkane, 1 mol malonic ester and 2 mol sodium methanolate produces cyclic carboxylic acids with three- to seven-membered rings.

In the Knoevenagel condensation (see), unsaturated carboxylic acids are produced from malonic ester and aldehydes. The base-catalysed reaction of malonic ester with vinylogenic carbonyl compounds (e.g. mesityl oxide) is called a Michael addition; the product is a dihydroresorcinol (e.g. dimedone). The synthesis of α-amino acids starts from oximinomalonic ester (see Amino acids). The reaction of malonic ester or its mono- or dialkyl derivatives with urea yields the ureides of malonic acid, the barbiturates.

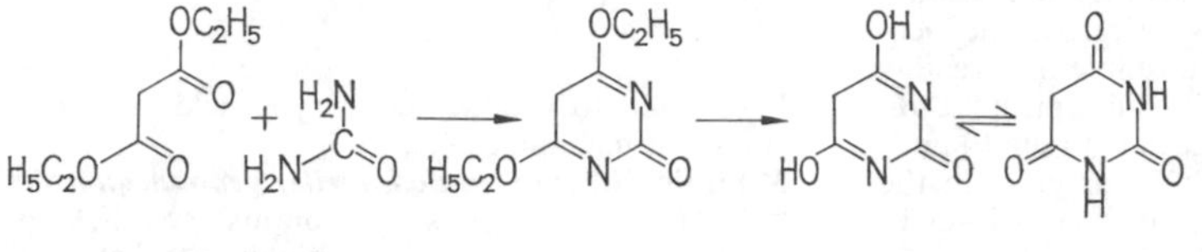

The skeleton is barbituric acid.

Malonic mononitrile: same as Cyanoacetic acid (see).

Maltase: see Disaccharidases.

Maltol, ***larixinic acid, 3-hydroxy-2-methyl-4H-pyran-4-one***: red crystals which smell like caramel; m.p. 163-164 °C (subl.). It is readily soluble in chloroform and hot water, but poorly soluble in ether and benzene. M. is found in larch bark and in wood pitch. It is formed by pyrolysis of starch and cellulose, during baking and frying processes, and during the roasting of malt. M. intensifies the sensation of sweetness, and when added to sweet foods reduces the amount of sugar needed.

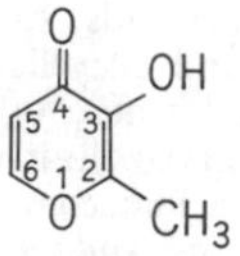

Maltose, ***malt sugar***: 4-(α-D-glucopyranosyl)-D-glucopyranose, a reducing disaccharide. M. is a degradation product of starch which is present in the malt (sprouted and roasted grain) used to make beer.

Malt sugar: same as maltose.

Malvidin: 3,5,7,4'-Tetrahydroxy-3,5-dimethoxyflavylium cation, an aglycon of various Anthocyanins (see). Some well known glycosides of M. are ***malvin***, (3,5-di-β-glucoside) from the wild mallow (*Malva sylvestris*), and ***Oenin*** (3-β-glucoside), the pigment of purple grapes and the blossoms of various species of cyclamen and primrose.

Malvin: see Malvidin.

Man: abb. for D-mannose.

Mandarine: see Naphthol orange.

Mandelates: the salts and esters of mandelic acid with the general formula C_6H_5-CH(OH)-COOR, where R is either a monovalent metal ion (or corresponds to 1/2 of a M^{II} ion), an ammonium ion or an organic group.

Some important salts are ***ammonium mandelate***, C_6H_5-CH(OH)-$COONH_4$, light-sensitive, hygroscopic crystals which are readily soluble in water and alcohol, but do not dissolve in ether; and ***calcium mandelate***, (C_6H_5-CH(OH)-COO$)_2$Ca, a white crystalline powder which is slightly soluble in water, alcohol and ether.

The esters are formed by reaction of mandelic acid with alcohols in the presence of strong mineral acids. Some important examples are ***benzyl mandelate***, C_6H_5-CH(OH)-$COOCH_2$-C_6H_5, colorless crystals which are slightly soluble in water and readily soluble in alcohol and ether; m.p. 95 °C; and ***isoamyl mandelate***, C_6H_5-CH(OH)-$COOC_5H_{11}$, an oily liquid which is insoluble in water; b.p. 155 °C at 1.6 kPa.

Mandelic acid, ***2-hydroxy-2-phenylethanoic acid, α-hydroxyphenylacetic acid, phenylglycolic acid***: C_6H_5-CH(OH)-COOH, a phenyl-substituted α-hydroxycarboxylic acid. The optically inactive (RS) form forms colorless, light-sensitive crystals; m.p. 121.3 °C. M. is readily soluble in water, alcohol and ether, slightly soluble in chloroform and insoluble in petroleum ether. It dissolves in base solutions by forming salts. The salts and esters of M. are called Mandelates (see). (RS)-M. is formed by the reaction of dilute alkali hydroxide solutions with α,α-dichloroacetophenone, followed by acidification, or by hydrolysis of mandelic nitrile. The (RS) acid can be separated into its optical antipodes via formation of alkaloid salts. It is easier to obtain the optically active forms by hydrolysis of naturally occurring hydrogen cyanide glycosides. ***(R)(-)-M.*** (m.p. 133 °C) is formed by acid hydrolysis of amygdalin under mild conditions; ***(S)(+)-M.*** (m.p. 133 °C) is formed by hydrolysis of the glycoside sambunigrin, which is found in elder leaves.

Mandelic nitrile, ***2-hydroxy-2-phenylethanenitrile, benzaldehyde cyanohydrin***: C_6H_5-CH(OH)-CN, a yellowish, poisonous, oily liquid; m.p. 20 °C (DL form). M. decomposes on distillation. It is insoluble in water, but is soluble in ethanol, ether and chloroform. It can be converted to mandelic acid by acid-catalysed hydrolysis. When heated, it is split into benzaldehyde and hydrogen cyanide. Optically inactive M. is formed by addition of hydrogen cyanide to benzaldehyde in the presence of small amounts of potassium carbonate or potassium hydroxide. The optically active L(+)-M. is present in sambunigrin, while D(-)-M. is found in amygdalin and prinazin. A 5% aqueous-alcoholic solution of M. was formerly used as a remedy for colds; it was called bitter-almond water.

Maneb: see Dithiocarbamate fungicides.

Manganates: in the narrow sense, ***oxomanganates*** in the oxidation states +4 to +7. ***Manganates(IV)***, formerly also called ***manganites***, are obtained by fusion of manganese(IV) oxide with basic oxides. They can also be thought of as double oxides, when alkaline earth oxides are used, for example 2 MO · MnO_2 ($M_2[MnO_4]$), MO · MnO_2 ($M[MnO_3]$) or MO · $2MnO_2$ ($M[Mn_2O_5]$). ***Manganates(V)***, $M^I_3[MnO_4]$, are blue-green to turquoise-colored compounds. Potassium manganate(V) is an industrially important intermediate in the production of potassium permanganate, and barium manganate(V) is used as a pigment, barium blue, in the form of mixed crystals with barium sulfate. ***Manganates(VI)***, $K^I_2[MnO_4]$, are bright green compounds. The most important of them in industry, ***potassium manganate(VI)***, $K_2[MnO_4]$, is an intermediate in the production of potassium permanganate. ***Manganates(VII), permanganates***, $M^I[MnO_4]$, are violet compounds. Potassium permanganate is produced on a large scale and is used in a wide variety of applications.

M. in the broad sense include anionic manganese complexes, including, for example, chloromanganates(IV), $M_2[MnCl_6]$, oxalatomanganates(II), $M_3[Mn(C_2O_4)_3]$, cyanomanganates(II), $M_4[Mn(CN)_6]$, and carbonylmanganates(-), $M[Mn(CO)_5]$.

Manganese: symbol ***Mn***: a chemical element of group VIIb in the periodic system, the Manganese group (see); a heavy metal, with one natural isotope, *Z* 25, atomic mass 54.9381, valence -I to +VII, with II, IV and VII especially common, Mohs hardness 6, density 7.44, m.p. 1244 °C, b.p. 2032 °C, electrical conductivity 2.6 Sm mm^{-2}, standard electrode potential (Mn/Mn^{2+}) -1.18 V.

Properties. M. is a silvery gray, hard and very brittle metal. There are four known modifications; the cubic body-centered α-M. is stable at room temperature. β-M. is cubic primitive and stable above 727 °C; γ-M. is cubic face-centered and stable above 1095 °C; and δ-M. is cubic body-centered and stable above 1104 °C. M. is dissolved by acids, and slowly even by water, with evolution of hydrogen. It burns when heated in the air to manganese(II,III) oxide, Mn_3O_4. With other nonmetals, M. reacts only at elevated temperatures. It burns in a stream of chlorine to manganese(II) chloride, $MnCl_2$, and reacts with fluorine to yield manganese(II) fluoride, MnF_2, and manganese(III) fluoride, MnF_3. Above 1200 °C, it reacts with nitrogen to form Mn_3N_2, and also forms compounds with boron, carbon, silicon, phosphorus and sulfur. M. itself is not ferromagnetic, but it forms ferromagnetic Heusler alloys (see Manganese bronze). Like the elements which preceed it in the 3d row, vanadium and chromium, M. is able to exist in many oxidation states. The -1 to +7 oxidation states are represented, for example, by the compounds $M[Mn(CO)_5]$, $Mn_2(CO)_{10}$, $Mn(CO)_5Cl$, $MnCl_2$, Mn_2O_3, MnO_2, $M_3[MnO_4]$, $M_2[MnO_4]$ and $M[MnO]$. Octahedral and tetrahedral structures are most common among the manganese complexes.

M. is a biologically important trace element which has a significant role in metabolic regulation, activation of various enzymes and in respiratory chain phosphorylation. It also stimulates biosynthesis of cholesterol and has an important role in photosynthesis.

Chronic uptake of manganese dust in manganese processing plants leads to poisoning; the symptoms are kidney, liver and metabolic diseases and severe nerve damage (manganism).

Analysis. Preliminary tests for M. are oxidative melts and phosphorus salt and borax beads. An oxidative melt turns green through formation of manganate(VI), and the beads give an oxidizing flame a violet color from manganese(III) phosphate or borate. Addition of hydrogen sulfide to an ammoniacal manganese(II) salt solution leads to precipitation of flesh-colored manganese(II) sulfide. In nitric acid solution, manganese(II) is oxidized by lead dioxide to manganate (VII): $2\,Mn^{2+} + 5\,PbO_2 + 4\,H^+ \rightarrow 2\,MnO_4^- + 5\,Pb^{2+} + 2\,H_2O$. In alkaline solution, the same oxidation state can be reached by oxidation with bromine in the presence of copper sulfate: $2\,Mn^{2+} + 5\,Br_2 + 16\,OH^- \rightarrow 2\,MnO_4^- + 10\,Br^- + 8\,H_2O$. For gravimetric analysis, Mn^{2+} is precipitated as ammonium manganese(II) phosphate, NH_4MnPO_4, and roasted to manganese diphosphate, $Mn_2P_2O_7$. Manganese(II) is determined volumetrically by titration with permanganate according to Volhard-Wolff.

Occurrence. M. makes up 0.065% of the earth's crust. The most important M. ore is pyrolusite, MnO_2. Some others are braunite (psilomelane), $3\,Mn_2O_3 \cdot MnSiO_3$, hausmannite, Mn_3O_4, manganite, γ-MnO(OH), rhodochrosite, $MnCO_3$, rhodonite, $MnSiO_3$, manganosite, MnO, alabandite (manganese blende), MnS and hauerite (manganese pyrite), MnS_2. Manganese ores are often associated with iron, for example, in bixbyite and siderite. Large amounts of M. are present in the manganese nodules in the deep sea. M. is often present in iron-containing waters, in amounts from 0.5 to 5 mg/l. As a trace element, M. is present in soils up to about 850 ppm. As a bioelement, it is present in all living cells. The dry material from plants contains about 50 ppm M. Manganese deficiency in plants can cause chloroses and dry spot disease.

Production. Very pure M. is obtained electrolytically from manganese sulfate solution which contains ammonium sulfate; it is deposited onto a steel cathode. The classic process is aluminothermal production, in which manganese(IV) oxide, MnO_2, is first converted to manganese(II,III) oxide, Mn_3O_4, and manganese(III) oxide, Mn_2O_3, by roasting, because MnO_2 reacts too violently with aluminum. The mixture of oxides is then reduced with aluminum. Recently, more use has been made of silicothermal processes, in which low-iron manganese ores or manganese slag concentrates are reduced with silicon to silicomanganese in an electric arc refining furnace. The liquid metal is then degassed in vacuum. Ferromanganese, which is industrially more important than pure M., is made by reduction of iron-manganese ores with coke.

Application. Only a relatively small fraction of the world production of M. is used as the pure metal, for making austenitic steels and nonferrous alloys. 90% of the M. ores are processed to ferromanganese, which is used in the iron and steel industry as a prealloy (see Manganese alloys).

Historical. M. was first prepared in 1774 by Bergmann, Scheele and Gahn.

Manganese acetates. ***Manganese(II) acetate***: $Mn(CH_3COO)_2 \cdot 4H_2O$, pink, monoclinic crystalline platelets which are very soluble in water; M_r 245.08, density 1.589. The compound is made by reaction of manganese(II) carbonate with acetic acid; the anhydrous form can be made by reaction of manganese(II) nitrate with acetic anhydride. It is used as a drying agent, in tanning and the textile industry as a catalyst, and as a fertilizer.

Manganese(III) acetate: $Mn(CH_3COO)_3 \cdot 2H_2O$, cinnamon brown crystals with a silky sheen; they are soluble in glacial acetic acid and are decomposed by water. This compound is made by oxidation of manganese(II) acetate with potassium permanganate, $KMnO_4$, or chlorine in hot glacial acetic acid. It is used in the production of other manganese(III) compounds.

Manganese alloys: alloys in which manganese is the primary component. M. are not widely used materials. ***Ferromanganese***, a carbon-containing manganese-iron alloy with 70 to 90% manganese, is used to remove sulfur from steel melts and deoxidize them. Manganese contents up to 16% increase the strength of the steels, and 12 to 20% manganese can replace the nickel in austenitic steels. Copper-manganese alloys (see Manganese bronze) and manganese-copper-nickel alloys have unusual physical properties. An alloy of 72% manganese, 10% nickel and 18% copper is used in bimetallic thermostats because of its high thermal expansion coefficient of $27 \cdot 10^{-6}$ m K^{-1}.

Manganese black: see Manganese oxides.

Manganese bronze: Copper-manganese alloy containing a maximum of 15% manganese. M. are heat resistant and have high electrical resistance. M. which also contain nickel or aluminum (see Manganin®, see Nickelin®), display only very small changes in electrical resistance with temperature, and are therefore useful as resistor materials. The strongly ferromagnetic *Heusler alloy* is an alloy of 25 to 30% manganese, 25% tin and the rest copper.

Manganese brown: see Manganese oxide hydrates.

Manganese carbide: Mn_3C, shiny, crystalline needles; M_r 176.83, density 6.89. M. decomposes in water: $Mn_3C + 6\,H_2O \rightarrow 3\,Mn(OH)_2 + CH_4 + H_2$. It is obtained at 1600 °C from the elements. Further M. are Mn_4C, Mn_7C_3 and $Mn_{23}C_6$.

Manganese(II) carbonate: $MnCO_3$, raspberry-red, rhombohedral crystals; M_r 114.95, density 3.125. M. also exists as a colorless monohydrate, $MnCO_3 \cdot H_2O$, and in the form of basic carbonates of varying composition. It is split into manganese(II) oxide, MnO, and carbon dioxide by heating, and when roasted in the air it is converted to manganese(II,III) oxide, Mn_3O_4. M. occurs in nature as rhodochrosite. It is synthesized via the monohydrate by addition of sodium hydrogencarbonate, $NaHCO_3$, to CO_2-saturated manganese(II) salt solutions. The monohydrate is converted to the non-hydrated form by heating it under pressure and under the solution, in the absence of air. M. is used as a fertilizer additive, and as a paint pigment, under the name manganese white.

Manganese carbonyl, ***dimanganese decacarbonyl***: $Mn_2(CO)_{10}$, golden yellow, monoclinic crystals, which can sublime in vacuum; M_r 389.97, density 1.81, m.p. 155 °C. The dinuclear carbonyl (Mn-Mn distance 297.7 pm) is obtained by the reaction of manganese(II) iodide with carbon monoxide in the presence of dialkylzinc. Reaction of M. with halogens produces manganese carbonyl halides, $Mn(CO)_5X$, while alkali metal M. in liquid ammonia are reduced to carbonylmanganates, $M[Mn(CO)_5]$. These are the salts of ***manganese carbonyl hydride***, $HMn(CO)_5$, a colorless liquid; M_r 196.02, m.p. -24.6 °C.

Manganese(II) chloride: $MnCl_2$, bright pink, very hygroscopic crystals; M_r 125.84, density 2.977, m.p. 652 °C, b.p. 1190 °C. M. is very soluble in alcohol and water, and the most important of the hydrates is the tetrahydrate, $MnCl_2 \cdot 4H_2O$, M_r 197.91, density 2.01, m.p. 58 °C. M. is obtained by the reaction of manganese, ferromanganese or manganese(II) carbonate with hydrogen chloride at high temperature, or by heating the hydrates in a stream of hydrogen chloride. The tetrahydrate is used in the production of corrosion-resistant manganese alloys and as a starting material for production of manganese brown.

The chlorides of tri- and tetravalent manganese are not stable, but ***chloromanganate(III)***, $M_2[MnCl_5]$, and ***chloromanganate(IV)***, $M_2[MnCl_6]$, are known.

Manganese complexes: coordination compounds of manganese. The magnetically normal manganese(II) complexes are of particular importance. The pale-pink aqueous solutions of manganese(II) salts contain the octahedral manganesehexaaqua ion $[Mn(H_2O)_6]^{2+}$, the absorption spectrum of which is characterized by weak intercombination bands. Both octahedral high-spin complexes, including ammine complexes, $[Mn(NH_3)_6]X_2$, and, in the presence of ligands with high ligand field strengths, octahedral low-spin manganese(II) complexes, such as $M_4[Mn(CN)_6]$ are known. In nonaqueous solutions with halides, one can obtain yellow-green tetrahedral manganese(II) complexes, $M_2[MnX_4]$ or red or red-brown, magnetically normal, octahedral manganese(III) complexes such as $[Mn(H_2O)_6]X_3$ or $M_3[Mn(C_2O_4)_2]$. Carbonyls, such as $[Mn_2(CO)_{10}]$ and $M[Mn(CO)_5]$, contain the Mn in the 0 or -1 oxidation state (see Manganese carbonyls), while the manganese carbonyl halides, $Mn(CO)_5X$, and the industrially important methylcyclopentadienyl manganese tricarbonyl are examples of manganese(I) complexes.

Manganese group: Group VIIb of the periodic system, including manganese (Mn), technetium (Tc) and rhenium (Re). Because of the similarity in their atomic and ionic radii, technetium and rhenium are very similar to each other, and somewhat different from manganese. The heavy metals of the M. are distinctly similar to their neighbors in the periodic system. The tendency to form the maximum oxidation state, +7, increases with increasing atomic mass; for technetium and rhenium, this oxidation state is the most stable, while the +2 state is typical for manganese. Manganate(VII), MnO_4^-, is a strong oxidizing agent. The M. metals form dinuclear carbonyls of the $M_2(CO)_{10}$ type.

Manganese(II) hydroxide: $Mn(OH)_2$, colorless, hexagonal leaflets; as a precipitate from manganese(II) salt solutions with alkali hydroxide solutions, an ivory-colored compound; M_r 88.95, density 3.258. When heated, M. is converted to manganese(II) ox-

Properties of the Group VIIb elements

	Mn	Tc	Re
Atomic number	25	43	75
Electron configuration	[Ar] $3d^5 4s^2$	[Kr] $4d^6 5s^1$	[Xe] $4f^{14} 5d^5 6s^2$
Atomic mass	54.9381	97	186.207
Atomic radius [pm]	117.1	127.1	128.3
Electro-negativity	1.60	1.36	1.46
Standard electrode potential (M/MO_4^-) [V]	+0.741	+0.472	+0.368
Density [g cm]$^{-3}$]	7.44	11.49	21.03
m. p. [°C]	1244	2250	3180
b. p. [°C]	2032	4877	5870

ide, MnO, by loss of water; in the air, it is oxidized to manganese(IV) oxide hydrate. It occurs occasionally in nature as pyrochroite.

Manganese(II) nitrate: $Mn(NO_3)_2$, pale pink compound, M_r 178.95, which crystallizes out of water as a mono-, tri-, tetra- or hexahydrate. M. can be obtained by reaction of manganese(II) carbonate with nitric acid, and the anhydrous form can be prepared by reaction of the monohydrate with dinitrogen pentoxide, N_2O_5.

Manganese oxides. ***Manganese(II) oxide***: MnO, gray-green, regular crystals which are insoluble in water; M_r 70.94, density 5.43, m.p. 1650 °C. MnO is found in nature as manganosite. As a basic oxide, it is soluble in acids, with formation of manganese(II) salts. MnO is obtained by reduction of manganese dioxide or by heating manganese carbonate in a stream of hydrogen or nitrogen; it is used as a fertilizer and to make oxide ceramic magnetic materials.

Manganese(II,III) oxide, red manganese oxide: Mn_3O_4 ($Mn^{II}Mn^{III}{}_2O_4$), a crystalline compound which occurs in nature as hausmannite. It has a spinel structure (Mn^{2+} in the tetrahedral and Mn^{3+} in the octahedral lattice sites); M_r 228.81, density 4.856, m.p. 1705 °C. Mn_3O_4 is formed when Mn_2O_3 is heated above 950 °C, and is used in very pure form as a magnetic material and semiconductor.

Manganese(III) oxide: Mn_2O_3, is a brown compound which occurs in nature as braunite and as a component of the pigment Umber (see). It is formed by heating MnO_2 in air above 550 °C; M_r 157.87, density 4.50. Mn_2O_3 is basic, and is soluble in acids with formation of manganese(III) salts; disproportionation reactions to Mn^{2+} and MnO_2 are also observed. Mn_2O_3 is used as a magnetic material and semiconductor.

Manganese(IV) oxide, manganese dioxide: MnO_2, is a dark brown to black compound; M_r 86.94, density 5.026. Nearly pure MnO_2 is found in nature as pyrolusite (β-MnO_2); there are other forms of natural and synthetic MnO_2 with compositions between $MnO_{1.7}$ and $MnO_{<2.0}$. These contain varying amounts of manganese in lower oxidation states, foreign ions and water, and varying degrees of crystallinity. While β-MnO_2 is highly crystalline and combines nearly ideal stoichiometry with low reactivity, γ-MnO_2 is nearly amorphous and very reactive. MnO_2 is insoluble in water, but forms a red-brown, unstable solution with conc. hydrochloric acid which contains $MnCl_4$. Sulfurous acid is completely oxidized to dithionate by MnO_2 even in the cold: $MnO_2 + 2\,H_2SO_3 \rightarrow Mn^{2+} + S_2O_6^{2-} + 2\,H_2$; this reaction is used to remove MnO_2 residues, e.g. from glass walls of apparatus. MnO_2 catalyses the cleavage of oxygen from hydrogen peroxide, and promotes evolution of oxygen when potassium chlorate is heated. Above 535 °C, MnO_2 splits off oxygen; Mn_3O_4 is obtained via Mn_2O_3 as the product of roasting MnO_2. In industry, MnO_2 is produced in large quantities by anodic oxidation of manganese(II) sulfate solutions, by reduction of potassium permanganate, by oxidation of manganese(II) salts or by thermal decomposition of manganese(II) nitrate: $Mn(NO_3)_2 \rightarrow MnO_2 + 2\,NO_2$. Natural and synthetic MnO_2 is used mainly for production of dry cells, and also to make ferrites, as a raw material for production of manganese and ferromanganese, to make pigments and drying agents for paints, as a laboratory oxidizing agent and in pyrotechnics, as a catalyst for oxygen transfer. MnO_2 is used as "glass-maker's soap" to make colorless glass melts. As manganese black, it is used as a lime paint pigment and to color cement.

Manganese(VII) oxide: Mn_2O_7, is a dark red oil, with a greenish, metallic sheen when viewed from the side and a peculiar odor; M_r 221.87, density 2.396, m.p. 5.9 °C. Mn_2O_7 can be distilled below 0 °C, but when heated it decomposes explosively. It dissolves in carbon tetrachloride as single molecules, and reacts with cold water to give manganese(VII) acid. It is obtained by the reaction of concentrated sulfuric acid with dry, powdered potassium permanganate.

Manganese oxide hydrates. ***Manganese(III) oxide hydrate***, $Mn_2O_3 \cdot H_2O$, can also be considered ***manganese(III) oxide hydroxide***, MnOOH. It exists as black-brown crystals also found in nature and known as manganite. The compound is made by reaction of manganese(III) salt solutions with hydroxides, followed by drying the resulting precipitate at 100 °C. Products obtained from the reaction of manganese(II) chloride solutions with calcium chloride and calcium hydroxide are called bister and used as brown paint pigment. ***Manganese(IV) oxide hydrate, manganese dioxide hydrate*** is a brown to brown-black powder with variable water content. It is more reactive than MnO_2. The hydrate is obtained by oxidation of manganese(II) salts or by reduction of permanganates; it is used as an oxidizing agent in making varnish and paints.

Manganese sulfates. ***Manganese(II) sulfate***: $MnSO_4$, forms very stable, nearly white crystals which are extremely soluble in water; M_r 151.00, density 3.25, m.p. 700 °C. It crystallizes out of aqueous solution as pink hydrates. Below 9 °C, one obtains the monoclinic heptahydrate, $MnSO_4 \cdot 7H_2O$, between 9 and 26 °C, the triclinic pentahydrate (***manganese vitriol***), $MnSO_4 \cdot 5H_2O$, between 26 and 27 °C, the rhombic or monoclinic tetrahydrate, $MnSO_4 \cdot 4H_2O$ (the monoclinic form is metastable, but crystallizes over a wider temperature range, 35-40 °C, and is the common commercial form of manganese sulfate), and, above 27 °C, the monoclinic monohydrate, $MnSO_4 \cdot H_2O$. The heptahydrate and monohydrate are occasionally found in nature as mallardite and szmikite, respectively. $MnSO_4$ forms a hexaammine complex, $[Mn(NH_3)_6]SO_4$, and forms double salts with alkali sulfates of the type $M_2[Mn(SO_4)_2] \cdot H_2O$. Anhydrous $MnSO_4$ can be obtained by evaporating manganese oxide or manganese salts of volatile acids in the presence of concentrated sulfuric acid. It is one of the most important manganese compounds; it is the starting material for electrolytic manganese and manganese dioxide, fungicides such as Maneb®, manganese soaps and pigments. It is used as a fertilizer for manganese-poor soils and in printing and dyeing textiles.

Manganese(III) sulfate: $Mn_2(SO_4)_3$, dark green, extremely hygroscopic, crystalline needles which are hydrolysed by water; M_r 398.06, density 3.24. $Mn_2(SO_4)_3$ is obtained by the reaction of manganese(III) oxide with concentrated sulfuric acid; this produces red crystals of $Mn_2(SO_4)_3 \cdot H_2SO_4 \cdot 4H_2O$, which are heated to produce $Mn_2(SO_4)_3$.

Manganese(IV) sulfate: $Mn(SO_4)_2$, black crystals, M_r 247.0, which are soluble in sulfuric acid (H_2SO_4 concentration greater than 50%), giving a brown solution. It is hydrolysed by dilute sulfuric acid and water to manganese dioxide hydrate. $Mn(SO_4)_2$ is obtained by oxidation of $MnSO_4$ with potassium permanganate in a sulfuric acid solution at a temperature between 50 and 60 °C; it is used in industry as an oxidizing agent.

Manganese sulfides. ***Manganese(II) sulfide***: MnS, occurs in three modifications pink green and orange); M_r 87.00, density 3.99. Pink MnS is obtained by precipitation of an ammoniacal manganese(II) salt solution with hydrogen sulfide; when this stands under the solution, it is converted to the more stable green modification, which also occurs in nature as alabandite (manganese blende). Another naturally occurring manganese sulfide, hauerite, is ***manganese(II) disulfide***, MnS_2, which crystallizes in a pyrite lattice; M_r 119.07.

Manganese vitriol: see Manganese sulfates.

Manganese white: see Manganese(II) carbonate.

Manganic(VII) acid, ***permanganic acid***: $HMnO_4$, a very unstable compound which can be obtained only at low temperatures; M_r 119.94. With a pK_a of -2.5, it is a strong acid. It forms a dihydrate $[H_5O_2] \cdot MnO_4$. M. is the only manganic acid which is known in the free state; the acids of manganese in the +4 to +6 oxidation states are hypothetical, and are known only in the form of their salts, the Manganates (see).

Manganin®: an alloy of 12% manganese, 4% nickel and the rest copper used to make precision electrical resistors.

Manganites: see Manganates.

Mannans: polysaccharides containing mannose. M. are generally not homopolymers, but gluco-, xylo-, galacto- or arabinomannans, which serve as reserve polysaccharides in many land plants.

Mannich bases: products of the Mannich reaction (see). Their common structural element is the aminomethyl group $R_2N\text{-}CH_2\text{-}$.

Mannich reaction: synthesis of β-aminoketones from CH-acidic compounds (e.g. ketones with α H atoms, aldehydes, aliphatic nitro compounds, HCN, ethyne, phenols, thiophene, pyrrole, indole), carbonyl compounds (e.g. formaldehyde) and primary or secondary amines:

$$R\text{–}CO\text{–}CH_3 + CH_2\text{=}O + HNR_2 \xrightarrow[-H_2O]{} R\text{–}CO\text{–}CH_2\text{–}CH_2\text{–}NR_2.$$

The M. is also called ***aminomethylation*** because an aminomethyl residue, $R_2N\text{-}CH_2\text{-}$ is introduced into the CH-acidic component. The products of M. are called ***Mannich bases***. The decisive step in the M. is the primary reaction between the amine and formaldehyde to form an aminomethyl cation. This can only occur if the amine is more nucleophilic than the CH-acidic component, since otherwise there will be an aldol reaction between the formaldehyde and the methylene compound. In the second step of the M., the aminomethyl cation reacts with the carbanion of the CH-acidic compound to form the Mannich base:

$$R_2NH + CH_2O \xrightleftharpoons{+\overset{+}{H},\,-H_2O} \left[CH_2\text{=}\overset{+}{N}R_2 \leftrightarrow \overset{+}{C}H_2\text{–}\bar{N}R_2\right]$$

$$R\text{–}CO\text{–}\overset{-}{\underline{C}}H_2 + \overset{+}{C}H_2\text{–}NR_2 \rightleftharpoons R\text{–}CO\text{–}CH_2\text{–}CH_2\text{–}NR_2$$

The M. can be catalysed either by acids or bases. It is very dependent on the pH of the reaction medium. The most favorable conditions are obtained by using a salt of the amine. Since in principle all reactive H-atoms in either the methylene component or the amine can react, one obtains a single product only when the reactants have only one reactive H-atom apiece. Because of the diversity of possible starting compounds, the M. provides access to various classes of compounds, e.g. the alkaloids gramine and tropinone and their derivatives, the Mannich bases of phenols and naphthalene, which are used industrially in vulcanization of polychlorobutadiene, and the falicain type drugs.

The Pictet-Spengler synthesis (see) is a modified M. in which intramolecular aminomethylation makes it possible to synthesize isoquinolines from β-arylethylamines and formaldehyde. The M. can be carried out under physiological conditions, and plays an essential role in the biosynthesis of some alkaloids.

D-Mannitol: an alditol. D-M. is a white, crystalline powder, m.p. 156-168 °C. It is readily soluble in water, and very slightly soluble in ethanol. D-M. occurs widely in plants, e.g. in manna, olives, mushrooms and celery. It may be excreted in the urine. M. is obtained by catalytic hydrogenation of glucose or invert sugar; the D-glucitol which is formed as a by-product is removed by recrystallization. D-M. is used as a softener, sweetener and as a nutritional additive to bacterial growth media. D-M. is used in medicine as a diuretic and in diagnosis.

D-Mannose, abb. ***Man***: an aldohexose epimeric to glucose. M. forms colorless crystals, which taste sweet but leave a bitter aftertaste. It is readily soluble in water, but only sparingly in ethanol. m.p. 131 to 133 °C, $[\alpha]_D$ +14.2° (in water, at mutarotational equilibrium). M. is a natural component of polysaccharides, the mannans, from which it can be obtained by acid-catalysed hydrolysis. It is used in preparing nutrient media for bacteria.

Margaric acid, ***heptadecanoic acid***: $CH_3\text{-}(CH_2)_{15}\text{-}COOH$, a long-chain, saturated fatty acid forming colorless leaflets, m.p. 62.0 °C, b.p. 277 °C at $1.33 \cdot 10^4$ Pa. M. is insoluble in water, but soluble in hot alcohol, ether and acetone. It is very rare in nature. It is prepared synthetically by oxidation of paraffin or by a Grignard reaction (see) from cetylmagnesium bromide and carbon dioxide.

Marignac process: see Niobium.

Marijuana: see Cannabinoids.

Markovnikov rule: a rule formulated in 1869 by Markovnikov describing the course of ionic addition of hydrogen halides to C=C double bonds in asymmetric alkenes. The halogen atom, as the more negative part of the reagent, adds to the C atom which has the lower number of H atoms. For example, when hydrogen halide (H-X) adds to propene, 2-halopropane is formed in accordance with this rule: CH_3-$CH=CH_2$ + H-X → CH_3-CHX-CH_3. The course of the reaction can be described by the following steps:

$$CH_3{-}CH{=}CH_2 \longleftrightarrow CH_3{-}\overset{+}{C}H{-}\overset{-}{C}H_2 \xrightarrow{H^+}$$

$$CH_3{-}\overset{+}{C}H{-}CH_3 \xrightarrow{+X^-} CH_3{-}CHX{-}CH_3$$

The addition of water, alcohols and carboxylic acids also occurs with formation of the Markovnikov products. In the presence of peroxides or other radical formers, the addition occurs contrary to the M. (***peroxide effect, anti-Markovnikov orientation***), that is, hydrogen halide and propene form 1-halopropane, the anti-Markovnikov product. In this case, the reaction occurs by a radical mechanism in which the orientation of the addition is reversed.

Marquis reagent: see Alkaloids.

Marsh test: a preliminary test for the presence of arsenic. Arsenic compounds are reduced to volatile arsanes by nascent hydrogen, which is generated by the reaction of dilute sulfuric acid and zinc. When heated in the absence of oxygen, the arsane decomposes to arsenic and hydrogen; the arsenic forms a black mirror on the glass wall of the Marsh apparatus. The M. is very sensitive, with a limit of detection of 1 μg arsenic. However, antimony and germanium compounds interfere with it, because they also form metal mirrors.

Martensite: a metallographic term for the metastable conversion product of Austenite (see) formed by rapid cooling (see Iron-carbon alloys). It greatly increases the hardness of steel. M. has a tetragonal, distorted cubic body-centered unit cell. The amount of distortion depends on the carbon content.

Martius yellow, ***naphthalene yellow***: 2,4-dinitronaphth-1-ol, a nitropigment. M. forms lemon-yellow needles, m.p. 130 to 133 °C. The 7-sulfonic acid of M. is known as Naphthol yellow S (see). M. is used to dye wool, silk, gelatins, cotton, synthetic fibers and paper. It was first prepared in 1864 by Martius.

Masking reagent: a reagent used in analysis to convert interfering ions to stable, colorless and soluble complexes so that they do not interfere with qualitative or quantitative analysis of a given ion. Both inorganic ions and organic compounds can be used as M., for example cyanide or fluoride ions, triethanolamine, tartaric acid and other hydroxycarboxylic acids, sulfosalicylic acid or complexones. The use of M. can greatly improve the selectivity of analytical methods; simultaneous application of several M. can make the analysis specific for a single ion. M. are used particularly in spot analysis and complexometry.

Mass action constants: see Mass action law.

Mass action law: a law formulated in 1867 by C.M. Guldberg and P. Waage which describes the equilibrium concentration of a chemical reaction. For ammonia synthesis, for example, $N_2 + 3H_2 \rightleftharpoons 2\,NH_3$, the M. states: $c_{N_2}c_{H_2}c_{NH_3\text{-}2} = K_c$; in words, at equilibrium, the product of the concentrations of the products, divided by the product of the concentrations of the reactants, is equal to a constant, the ***equilibrium constant*** or ***mass action constant*** K_c. The stoichiometric coefficients appear as exponents of the corresponding concentrations.

The M. applies in this form only to ideal solutions. If there are interactions between the molecules, one observes deviations from M.; K_c is no longer constant, but varies, depending on the starting concentrations. Activities (see) a_i are used in place of concentrations to correct for these effects.

For the general chemical reaction $|\nu_A|A + |\nu_B|B \underset{-1}{\overset{1}{\rightleftharpoons}} \nu_P$ P + ν_QQ the M. in strict from is

$$\frac{\alpha_P^{\nu_P}\,\alpha_Q^{\nu_Q}}{\alpha_A^{|\nu_A|}\,\alpha_B^{|\nu_B|}} = K_\alpha.$$

Here ν_i are stoichiometric factors which appear as exponents of the corresponding activities. The ν_i of the reactants are negative, according to the thermodynamic sign convention, but in the M. their absolute values are used.

The equilibrium constant is characteristic of the chemical reaction, and does not depend on the amounts and proportions of the reactants. It does depend on the temperature and pressure, however. If the above reaction equation is written in the opposite direction: $|\nu_P| + |\nu_Q Q| \rightleftharpoons \nu_A A + \nu_B B$, the numerator and denominator in the M. are exchanged, because it is a convention always to represent the right side of the equation in the numerator. The new equilibrium constant K_a is then equal to the reciprocal $1/K_a$ of the previous constant.

By definition, the activities of pure substances are equal to 1. This means that if a mixed-phase reaction occurs, the activity of the condensed phase does not enter the M. For example in the Boudouard equilibrium $CO_2 + C \rightleftharpoons 2\,CO$; $K_a = a^2_{CO}/a_{CO^2}$, because $a_C = 1$.

If the M. is applied to reactions in ideal solutions, the activities are $a_i = f_i x_i$, but since the activity coefficients f_i in an ideal solution are 1, the M. can be expressed in terms of the mole fractions x_i. To indicate this, the equilibrium constant is then written K_x. It is possible to convert the mole fractions into other measures of concentrations, e.g. concentrations c, molarities m_i, partial pressures p_i in gas reactions or mole numbers n_i, so the M. can be formulated with any of these measures of concentration. The corresponding equilibrium constants have different numerical values; to indicate which is meant, a subscript (c, m, p, n) is used. The following relationships apply: a) for ideal systems, $K_a = K_x$; b) for gas reactions in ideal mixtures, $K_x = K_{pp}\text{-}\Sigma\nu_i = K_c(RT/p)\Sigma\nu_i$; c) for ideal dilute solutions, $K_x = K_c(M_1/\varrho_1)\Sigma\nu_i = K_m(M_1)\Sigma\nu_i$. In these expressions, $\Sigma\nu_1 = \nu_P + \nu_Q - |\nu_A| - |\nu_B|$, p is the total pressure, R is the gas constant, T the absolute temperature, M_1 the molar mass and ϱ_1 the density of the solvent. For reactions in which the number of moles does not change, i.e. $\Sigma\nu_i = 0$, all K values are the same and unitless.

The M. can be derived thermodynamically from the *van't Hoff reaction isotherm* $\Delta_R G = \Delta_R G^0 + RT\Sigma\nu_i \ln a_i$ (see Thermodynamics, application d of the second law). At equilibrium, the Gibbs free energy for the reaction $\Delta_R G = 0$, and it follows that $K_a = e^{-\Delta_R G^0/RT}$, where $\Delta_R G^0$ is the free energy under standard conditions. This equation is the starting point for thermodynamic calculation of equilibrium constants (see Ulich's approximation). It is also used to derive the *van't Hoff reaction isobar* for the temperature dependence of the equilibrium constant at constant pressure:

$$\left(\frac{\delta \ln K_x}{\delta T}\right)_P = \frac{\Delta_R H^0}{RT^2}$$

the *van't Hoff reaction isochore* for the temperature dependence at constant volume:

$$\left(\frac{\delta \ln K_x}{\delta T}\right)_V = \frac{\Delta_R E^0}{RT^2}$$

and the pressure dependence of K_x at constant temperature:

$$\left(\frac{\delta \ln K_x}{\delta P}\right)_T = \frac{\Delta_R V^0}{RT}$$

where $\Delta_R H^0$ is the molar reaction enthalpy, $\Delta_R E^0$ is the molar reaction energy, and $\Delta_R V^0$ is the change in volume of the mixture when 1 mole reacts (see Le Chatelier's principle), all under standard conditions.

The M. was derived kinetically by Guldberg and Waage. In the equilibrium reaction formulated above, the rate equation $r_1 = k_1 c_A^{|\nu_A|} c_B^{|\nu_B|}$ applies to the forward reaction, and the equation $r_{-1} = k_{-1} c_P^{|\nu_P|} c_Q^{|\nu_Q|}$ where r is the rate of the reaction and k is the rate constant (see Reaction kinetics). A prerequisite for the formulation of the rate laws on the basis of the stoichiometric equation is that both partial reactions are Elementary reactions (see). Using the principle of microscopic reversibility (see Equilibrium), however, it can be shown that this type of derivation can also be applied to complex reactions. During the time that equilibrium is being reached, the effective, measurable rate $r = r_1 - r_{-1}$. At equilibrium, $r = 0$, i.e. $r_1 = r_{-1}$. The forward and back reactions occur at equal rates, so that there is no macroscopic concentration change. Equilibrium is thus a dynamic condition. It follows that

$$k_1 c_A^{|\nu_A|} c_B^{|\nu_B|} = k_{-1} c_P^{|\nu_P|} c_Q^{|\nu_Q|}$$

or

$$\frac{c_P^{\nu_P} c_Q^{\nu_Q}}{c_A^{|\nu_A|} c_B^{|\nu_B|}} = \frac{k_1}{k_{-1}} = K_c$$

i.e. the M. with the equilibrium constant K_c. If one observes that the Eyring equation holds for both rate constants, one obtains

$$K_c = k_1/k_{-1} = e^{-(\Delta G_1^{\neq} - \Delta G_{-1}^{\neq})/RT} = e^{(\Delta S_1^{\neq} - \Delta S_{-1}^{\neq})/R} = e^{-(\Delta H_1^{\neq} - \Delta H_{-1}^{\neq})/RT},$$

where $\Delta G^{\neq}$, $\Delta S^{\neq}$ and $\Delta H^{\neq}$ are the Gibbs free energy, the entropy and the enthalpy of activation, respectively, of the forward (1) and back (-1) reactions. Comparison with the equation $K_x = e^{-\Delta_R G^0/RT}$ reveals the following relationships between the thermodynamic reaction parameters and the kinetic activation parameters:

$$\Delta_R G^0 = \Delta G_1^{\neq} - \Delta G_{-1}^{\neq},\ \Delta_R H^0 = \Delta H_1^{\neq} - \Delta H_{-1}^{\neq} \text{ and } \Delta_R S^0 = \Delta S_1^{\neq} - \Delta S_1^{\neq} - \Delta S_{-1}^{\neq}.$$

The M. can be used to improve the yield of a reaction, either by a) changing the concentrations (activities) of the reactants (K constant) or b) changing the temperature and/or pressure and thus changing K. a) If the concentration of one reactant is increased, the numerator in the M. must also increase. Since K is constant, the reaction is driven in the direction of the products until the fraction reaches the value of K; that is, the yield, as a percentage of the reactant not present in excess, increases. This is economically important, since it allows better utilization of an expensive reactant; the less expensive reactant is used in excess. The same effect can be achieved by removal of products from the reaction mixture. b) K can be increased or decreased by changing the temperature or pressure of the reaction, and thus the equilibrium can be shifted in the direction of the products or the reactants (see Le Chatelier's principle).

For applications of the M., see pH, Dissociation, Ostwald's dilution rule, Solubility product, Hydrolysis, Buffers. The M. plays a central role in all areas of chemistry and chemical industry. It is the basis for analytical methods, makes possible the calculation of expected yields, and permits yields of syntheses to be favorably manipulated.

Mass concentration: see Composition parameters.

Mass crystallization: see Crystallization.

Mass diffusion: a kinetic process used to separate gases with different diffusion rates in a gaseous separation medium. The separation effect and the changes in concentration of the components are achieved in a continuous flow of vapor from an inlet to a condenser. The lighter gas components diffuse more quickly against the flow of vapor and become enriched in the center of the column.

M. is used to enrich stable isotopes of noble gases and to separate mixtures of other gases.

Mass fraction: see Composition parameters.

Massicot: see Lead oxides.

Massieu's function: symbol Ψ, the relatively seldom used thermodynamic function $\Psi = -f/T = s - s/T$, where f is the Helmholtz free energy, s is the entropy, s is the internal energy and T is the absolute temperature. It is analogous to Planck's function (see).

Mass number, abb. ***M***, the total number of nucleons (protons and neutrons) in an atomic nucleus.

Mass percent: see Composition parameters.

Mass spectroscopy: a method for determining the masses and abundances of charged particles. The principle is to generate ions from atoms or molecules, to separate the ions on the basis of their mass-to-charge ratios (*m*/*e*) and to generate a plot of the ion abundances (the ***mass spectrum***) using a suitable recording device.

The origins of M. go back to Thomson (1910) and Aston (1919), who used the method to detect isotopes (^{20}Ne, ^{22}Ne). The use of M. did not become widespread until the 1950s, when its applicability to structural elucidation of organic and inorganic compounds was recognized. Today M. has become a routine method of analytical chemistry, and mass spectrometers are commercially available. In addition to structure elucidation, M. is used for trace analysis, exact determination of atomic masses and isotope abundances, determination of energies of ionization, formation and dissociation, determination of elemental composition of organic compounds, study of ion reactions in the gas phase, etc.

Apparatus. A mass spectrometer consists of four functional units (Fig. 1): the sample inlet system, the ion source, the separation chamber and the detector.

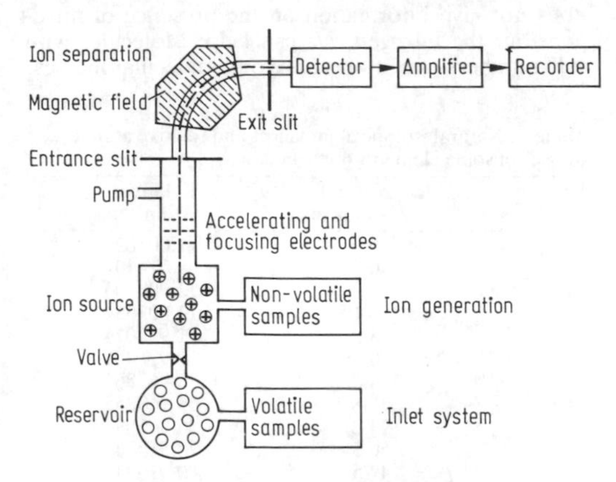

Fig. 1. Schematic diagram of a mass spectrometer.

1) *Sample inlet* The sample must be a gas at a pressure of about 10^{-4} Pa. Under these conditions, the mean free path of the ions is between 10 and 100 m, while the path through the mass spectrometer is about 1 m. In this way, collisions between the ions or neutral particles, which would interfere with exact mass separation, are avoided. The inlet system is used to convert a sufficient amount of the sample substance into the vapor phase so that the desired pressure in the ionization chamber is attained. Volatile substances are first vaporized in a storage container, from which they are released through a valve into the ionization chamber. Less volatile substances which are still able to produce a vapor pressure of 10^{-4} Pa are introduced directly into the ion source. Non-volatile substances can either be converted into volatile derivatives or examined using special ionization techniques (solid state mass spectroscopy; see below).

2) In the *ion source*, the sample is ionized, the ions are accelerated and formed into a beam. Ionization is usually achieved by electron collision ionization (see below), Chemical ionization (see), Field ionization (see) or Photoionization (see). In principle, either positive or negative ions can be formed, and these can be either singly or multiply charged:

$$M \xrightarrow{-e} M^{+};\ M \xrightarrow{-2e} M^{++};\ M \xrightarrow{+e} M^{-}$$

However, singly charged positive ions are created most often by all ionization techniques, and are the most commonly examined. The mass spectrometric study of negative ions (*electron addition mass spectroscopy*) is important only for special purposes (e.g. determination of molar masses of natural products). Of the ionization techniques listed above, *electron collision ionization* is most commonly used. When an energetic electron from a cathode ray source collides with an atom of the gaseous sample, it knocks an electron out of an atomic orbital, leaving a positive ion. The energy of the ionizing electrons can be varied, but is usually set at 70 eV, because this gives a favorable yield of singly charged positive ions. The ions are accelerated by a few kilovolts potential drop, and are focused on the entry slit of the separation chamber by an electromagnetic lens.

3) In the *separation chamber*, the ions are separated on the basis of their mass-to-charge ratio *m*/*e* by statistical or dynamic methods. The most common method at present makes use of a magnetic sector field apparatus, in which the separation is achieved by a strong magnetic field (statistical method). The dynamic methods are used in time-of-flight mass spectrometers, in which the separation is based on the mass-dependent velocity of the ions, and in quadrupole mass spectrometers, in which the ion beam is guided between four rod-shaped poles of an alternating electric field. Only those ions which change their direction in synchrony with the alternating field are able to pass the mass filter.

4) At present there are three main types of *ion detectors*: a) in the *Faraday trap*, the charge which is given up by the ions is conducted to ground via a high-ohmic resistor. The voltage drop across the resistor is proportional to the number of ions, and can be recorded after suitable amplification. With a light pen, spectra can be recorded in a few seconds. b) In *secondary electron amplifiers*, the ions impinge on a dynode and cause the emission of several secondary electrons, which are then further amplified. c) The use of a *photographic plate* as recorder requires that ions of different masses be focused in a single plane. The machines used for this purpose are the mass spectrographs.

Because much of the information obtained from a mass spectrometer can easily be digitalized, the direct coupling of mass spectrometers to computers for analysis and feedback control is becoming more and more common.

Representation of the mass spectra. Because a mass spectrum can contain peaks with extremely large differences in intensity, it is often necessary to use recorders with several sensitivity ranges (e.g. 1:10:100). Because spectra recorded in this way are somewhat confusing, it is customary to reproduce mass spectra for analysis and data storage as tables or line spectra. In a line spectrum, the abundances of the ions are plotted against their *m/e* values, with the value for the most intense peak, the basis peak, being set arbitrarily at 100% (Fig. 2).

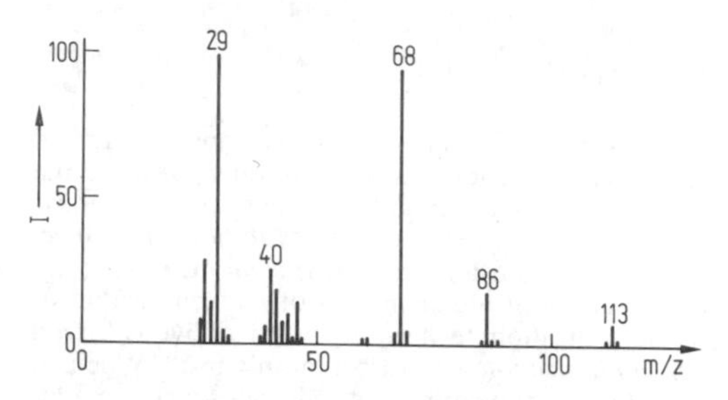

Fig. 2. Mass spectrum of ethyl cyanoacetate.

Applications of M. 1) *Structure elucidation*. M. can provide information on the relative molecular mass, empirical and structural formulae of a compound. Since almost all molecules contain an even number of electrons, they form radical cations with an unpaired electron, $M^{+\cdot}$, when ionized by loss of an electron. The process of molecular ion formation begins as soon as the electron collision energy is equal to the ionization energy of the molecule, which can be measured in this way. The ionization energies of molecules generally lie between 6 and 15 eV. If more energy is transferred by a collision process than is required for ionization, the molecule falls into fragments. *Fragmentation* of the molecular ion (and of ions derived from it) occurs in such a way that ions with odd numbers of electrons break into radicals and particles with even numbers of electrons (these can be ions or neutral particles); ions with even numbers of electrons form particles which also have even numbers of electrons: $M^{+\cdot} \rightarrow A^+ + R\cdot$, $M^{+\cdot} \rightarrow B^+ + N$, $A^+ \rightarrow C^+ + N$.

Formation of fragment ions requires a minimum energy, the *appearance energy*. This exceeds the ionization energy of the compound by an amount equal to the dissociation energy of the broken bond. The uncharged fragments (radicals and neutral particles) produced in the process are removed from the mass spectrometer by the diffusion pump which maintains the vacuum.

Molecular mass determination. The peak of the singly ionized molecule (molecule peak) is usually the last large peak in the mass spectrum, and is accompanied only by isotope peaks at M+1, M+2 or higher masses. If it can be clearly recognized, it yields directly the relative molecular mass of the compound. In singly focusing mass spectrometers, this determination gives an integral mass (nominal mass). At low resolution, the ions $CO^{+\cdot}$, $N_2^{+\cdot}$ and $C_2H_4^{+\cdot}$, all of which have $m/e \approx 28$, cannot be distinguished. At higher resolution, such as is attained in a double-focusing apparatus which determines mass to 4 decimal places, they can be distinguished ($CO^{+\cdot}$, 27.9949; $N_2^{+\cdot}$, 28.0062; $C_2H_4^{+\cdot}$, 28.0312). On the basis of these more precise values, the formula of the compound can be determined. There can be difficulty in recognizing the molecule peak if the molecule ion is very unstable and rapidly breaks down, so that it appears in the mass spectrum only at very low intensity, or not at all.

Isotope peaks and empirical formula. Molecule peaks (and also fragment peaks) are often accompanied by peaks of lower intensity, the isotope peaks. These arise because most of the elements in molecules consist of more than one isotope (Table 1). The isotope ratios of the elements determine the intensity ratios of these peaks. All singly charged ions which contain carbon display an isotope peak one mass unit higher, due to the natural content of ^{13}C (1.1%) in carbon. The intensity of this peak, for an ion with *n* C atoms, is $n \cdot 1.1\%$ of the ^{12}C peak. For example, the mass spectrum of benzene contains, in addition to the molecule peak at $m/e = 78$, an isotope peak at $m/e = 79$ with an intensity of 6.6% of the ^{12} peak. Chlorine, bromine and sulfur also give typical isotope peaks, although these are two mass units higher than the molecule peak (Table 1). Thus isotope peaks indicate the number of atoms of a given element present in the compound, and this is important data for construction of the empirical formula if it cannot be determined from high-resolution M. The isotope distribution does not give information on the presence of nitrogen, but the *nitrogen rule* can help: Molecules with odd numbers of N atoms have odd molecular masses.

Table 1. Natural isotope abundances and relative atomic masses of some elements important in M

Isotope	Natural Abundance	Relative Atomic mass
^{1}H	99.985	1.007825
^{2}H	0.015	2.014102
^{12}C	98.9	12.000000
^{13}C	1.1	13.003354
^{32}S	95.0	31.972074
^{33}S	0.76	32.971461
^{34}S	4.2	33.967865
^{35}Cl	75.8	34.968855
^{37}Cl	24.2	36.965896
^{79}Br	50.5	78.918348
^{81}Br	49.5	80.916344

Structural formula. The fragmentation pattern of a compound gives clues to its structure because there are certain known patterns of fragmentation. Molecules tend to fragment in energetically favorable ways. For example, fragment ions such as the tropylium ion or the tert-butyl cation, in which the positive charge is stabilized by resonance or inductive effects, are favored, as are those which form by elimination of a low-energy neutral particle (e.g. CO or HCN). In addition, the strength of the bond which has to be broken plays an important role. C-C bonds are more easily split than C-H bonds, for example. These and other regularities are expressed in a series of more or less empirical *fragmentation rules*, which must be used in structure elucidation. In simple fragmentation the bonds present in the original molecule are broken, but in addition, fragmentation often involves rearrangements which produce fragments which were not present in the original molecule. For example, the mass spectrum of isobutane, $(CH_3)_2CH\text{-}CH_3$, contains a peak for the ethyl cation, $C_2H_5^+$, which arises through rearrangement. In analysis of a mass spectrum for structural elucidation, one first considers fragments in the low mass range (*key fragments*), which often indicate the presence of certain structural elements (Table 2). For example, the $CH_2{=}OH^+$ ion, with mass 31, is typical of alcohols and ethers. In the higher mass range, the masses of cleaved-off radicals or neutral particles can be inferred from the mass differences between two peaks. The information obtained from the key fragments and mass differences is usually sufficient to suggest a formula, which must then be examined to see whether the fragmentation to be expected from the postulated structure can lead to the experimental mass spectrum (Fig. 2 and 3).

Table 2. Important key fragments

m/e	Fragment	Indicates
29	$(C_2H_5)^+$	Alkyl groups
30	$(NO)^+$	Nitro compounds
30	$(CH_2{=}NH_2)^+$	Amines
31	$(CH_2{=}OH)^+$	Alcohols, ethers
39	$(C_3H_3)^+$	Aromatics
43	$(C_3H_7)^+$	Alkyl groups
43	$(CH_3CO)^+$	Acetyl groups
45	$(COOH)^+$	Carboxylic acids
51	$(C_4H_3)^+$	Aromatics
57	$(C_4H_9)^+$	Alkyl groups
59	$(COOCH_3)^+$	Methyl esters
77	$(C_6H_5)^+$	Aromatics

Construction of the structural formula is often aided by peaks from metastable ions which disintegrate during their flight from the ion generation to the detection area. These are not registered as sharp lines with the mass of the original ion, but instead form broad peaks of low intensity at the apparent mass $m^* = m_2^2/m_1$, where m_1 is the mass of the original ion, and m_2 is the mass of the product ion. The appearance of metastable ions can be taken as evidence for a one-step fragmentation from m_1 to m_2.

Table 3. Characteristic mass differences M–X between molecules and fragment ions

M–X	Neutral unit lost	Indicates
M–16	O	Nitro compounds Sulfoxides
M–16	NH_2	Amides
M–17	OH	O-containing compounds
M–17	NH_3	Amino compounds
M–18	H_2O	Alcohols, ketones
M–26	C_2H_2	Aromatics
M–27	HCN	Nitriles, N-heterocycles
M–28	CO	Aryl ketones, quinones
M–29	C_2H_5	Alkyl compounds
M–30	NO	Nitrogen-containing aromatics
M–36	HCl	Chlorine compounds
M–44	CO_2	Esters, anhydrides

	Reaction	*m/z*
	$(N{\equiv}C \overset{a}{\vert} CH_2 \overset{b}{\vert} C({=}O) \overset{c/d}{\vert} O \overset{e}{\vert} CH_2-CH_3)^{+\cdot}$	*113*
a)	$M^{+\cdot} - HCN \longrightarrow (CH_2-C({=}O)-O-CH_2{=}CH_2)^{+\cdot}$	*86*
b)	$M^{+\cdot} - (\cdot C({=}O)-O-C_2H_5) \longrightarrow (CH_2-C{\equiv}N)^+$	*40*
c)	$M^{+\cdot} - (\cdot O-C_2H_5) \longrightarrow (N{\equiv}C-CH_2-C({=}O))^+$	*68*
d)	$M^{+\cdot} - (N{\equiv}C-CH_2-\dot{C}({=}O)) \longrightarrow (O-C_2H_5)^+$	*45*
e)	$M^{+\cdot} - (N{\equiv}C-CH_2-C({=}O)-\dot{O}) \longrightarrow (CH_2-CH_3)^+$	*29*

Fig. 3. Fragmentation scheme for ethyl cyanoacetate.

Two methods which permit the further disintegration of a metastable ion to be followed are gaining importance: *DADI*, for *d*irect *a*nalysis of *d*aughter *i*ons, and *MIKES*, for *m*ass analyzed *i*on *k*inetic *e*nergy *s*pectrometry. These methods make use of modified high-resolution mass spectrometers, in which a magnetic analyser is used to filter out the desired metastable ion; its disintegration is then followed in an electrostatic analyser. This type of M. is being further refined by addition of a third or even a fourth analyser (MS/MS).

2) M. is extremely important in qualitative and quantitative analysis, as a micro-method capable of detecting trace substances in the nanogram range and concentration differences as slight as 1 part per million. In qualitative analysis, the mass spectrum of the sample is compared with that of a standard; here it is essential to maintain constant operating conditions in the apparatus. Quantitative mass spectroscopic analysis depends on the proportionality of the ion beam, measured, e.g. as peak height, to the partial pressure of the sample in the ion source. The intensity of a peak from the sample is compared with that of a standard (external or internal).

In mixtures, the peak height I at mass number k is the sum of the contributions from the individual components, so that mixtures can also be analysed: $K_k = \Sigma E_{ik} p_i$. Here E_{ik} is the calibrated value for component i at the mass numbers k and p_i is the partial pressure of component i. With n equations of this form for n different masses, n unknown p_k can be calculated.

M. is used in biochemistry, clinical chemistry, toxicology, environmental studies and other areas, in addition to pure chemistry. In these applications, problems of trace detection are usually most prominent. M. is also widely used to analyse complex mixtures of substances, e.g. in hydrocarbon chemistry. Here, in many cases, the emphasis is less on the detection of individual compounds than on identifying classes of compound.

3) Solid state mass spectroscopy. In the techniques so far described, gaseous or volatile samples are ionized, but this branch of M. is concerned with the formation and analysis of gaseous ions from non-volatile solids, primarily inorganic solids such as metals, semiconductors, and minerals. Special ion sources are required to release and ionize the atoms or molecules from these substances; some common techniques are *thermal ionization*, in which high temperatures are used to release ions from inorganic solids, and *dis-*

charge ionization, in which the sample is incorporated into one of the electrodes of a spark chamber. In *secondary ion mass spectroscopy* (*SIMS*), ions are generated from a pure gas (noble gas or oxygen) in a primary ion source; these are accelerated and focused on a sample. This bombardment with ions releases secondary ions which are analysed in the mass spectrometer. In *laser mass spectroscopy*, laser light is used to ionize and volatilize solid samples. In *field desorption mass spectroscopy*, the sample is applied directly to the anode. Application of a strong electric field causes positive ions to be ejected from the anode.

4) Special applications. The use of M. has been greatly expanded by coupling with gas chromatography (GC-MS coupling). Here the components emerging from the column of a gas chromatograph are fed directly into the mass spectrometer for analysis. The most important use of this method is the identification of very small amounts of substances present in mixtures; the sensitivity extends into the parts per billion range. M. is also coupled to liquid chromatography (LC-MS coupling). For the study of highly polymeric substances, *pyrolysis mass spectrometry* is especially useful. The polymers are first pyrolysed, and their fragments are then examined by mass spectroscopy.

Mast-cell granulating peptide, abb. ***MCD peptide***: a heterodetic, cyclic, branched 22-peptidamide with two intrachain disulfide bonds; it is a component of the peptide toxin of bee venom which inhibits inflammation. The positive effects of bee venom on rheumatism are due to MCD peptide.

Ile–Lys–Cys–Asn–Cys–Lys–Arg–His–Val–Ile–Lys–Pro–His–Ile–Cys–Arg–Lys–Ile–Cys–Gly–Lys–Asn–NH_2

Mastication: see Rubber.

Mastix: see Gum resins.

Matches: small sticks of wood, cardboard, rolled or paraffinized paper or other material with an ignition mass at one end (the head). This mass is rubbed on a surface, generally on the box or packet, to ignite it. The flame is then transferred to the matchstick. Wooden M. are generally treated with monoammonium phosphate, $NH_4H_2PO_4$, or diammonium phosphate, $(NH_4)_2HPO_4$, to keep the coal from continuing to glow after the flame is blown out, and are dipped in paraffin. Most matches now in use are *safety matches*, which carry an ignition mixture invented in 1848 by R. C. Bœttger. It consists of a mixture of potassium chlorate, $KClO_3$, as oxidizing agent, sulfur as the flame carrier, catalysts which prevent explosive burning, fillers, glues and colorings. Safety M. must not contain white phosphorus. They are ignited by rubbing on an abrasive surface (on the box) which consists of red phosphorus, antimony trisulfide, Sb_2S_3, manganese(IV) oxide, MnO_2, or pyrite (iron(II) sulfide), FeS_2 and synthetic glues. When the match is rubbed on this surface, a small amount of the red phosphorus is rubbed off and converted to white phosphorus, which reacts with the potassium chlorate. The reaction is a limited chain reaction, which must be carefully adjusted.

The first M. had heads made of white phosphorus, potassium chlorate, gum arabic and some other ingredients. Because the combustion temperature of this mixture was not sufficient to ignite the wood, it was treated with sulfur (***sulfur M.***). These M. could be ignited by rubbing on any rough surface. However, because white phosphorus is poisonous, they were later forbidden.

Strike-anywhere M. contain tetraphosphorus trisulfide, P_4S_3, in addition to the ingredients in the heads of safety M., and thus combine igniter, oxygen carrier, flame carrier and catalysts. These matches are very sensitive to friction, and are not widely used.

Multi-strike M. ("permanent" M.) can be used repeatedly. They consist of a rod of potassium chlorate, $KClO_3$, metaldehyde, $(CH_3CHO)_x$, manganese(IV) oxide, MnO_2, monomethylmethacrylate, CH_2-C(CH_3)-$COOCH_3$, cellulose nitrate, pentaerythritol, $C(CH_2OH)_4$ and iron(III) sulfate, $Fe_2(SO_4)_3$. The rod is in a tube similar to a lipstick tube, and can be used a number of times.

Materials testing: determination of the mechanical, electrical, thermal and chemical properties of materials. The goals of M. are to determine the probability of occurrence of catastrophic breaks, to aid in the correct selection of materials, to monitor the quality of the materials produced, to analyse plant failures, and to determine guidelines for construction of parts and apparatus. M. is often divided into destructive and nondestructive methods.

1) ***Destructive M.***. a) In *mechanical technological M.*, the methods can be classified as static or dynamic. The resistance of a material to changes in form and breakage, and its ability to change form, are determined. b) In *metallographic M.* (see Metallography), finely polished and etched surfaces and their impressions, break surfaces and thin sections are examimed by light and electron microscopy to determine their structures.

2) ***Nondestructive M.***. The tests are made on unfinished pieces or finished parts without affecting their use. Such methods are therefore designed to work without interrupting production or without dismounting the parts.

If a test is designed to provide early recognition of damage, it is called *technical diagnosis*.

a) *Radiographic M.* permits recognition of internal defects such as separation, cavities or inclusions, by means of x-rays, gamma rays or thermal neutrons. b) *Acoustic M.* depend on the differences in the speed of travel of ultrasonic impulses through the material. The sound emitted by a part when it is struck can also give an indication if cracks are present. Leaks can be located by means of the ultrasonic noises made by the leaking material. The formation and propagation of cracks can be detected by sound emission analysis (SEA). c) In *magnetic* and *electrical* M., defects near the surface, such as cavities and inclusions, can be recognized by polymagnetization or current flooding, with the resulting inhomogeneities in the magnetic field being recorded with ferromagnetic materials. In magnetic induction tests, alternating magnetic fields are used to generate eddy currents in the object, and these interact with a test coil.

There are physical, chemical and technological methods of M.

1) ***Physical M.***: a) Determination of *electrical properties*: electrical conductivity, break-through potential, dielectric constant, dielectric losses.

b) *Magnetic properties*: coercive force, induction, remanence, permeability, magnetization curve.

c) *Thermal properties*: specific heat, heat conductivity and heat transfer. The energy loss due to mechanical vibration is caused by material-specific effects, and is also determined.

d) *Optical properties*: the interaction of the materials with light and high-energy radiation are determined.

2) ***Chemical M.***: a) *Chemical behavior*: see Corrosion testing. b) *Chemical composition*: both chemical and physical methods are used to determine the chemical composition. Rapid qualitative and semiquantitative results are provided by the spark test, drop analysis and spectroscopy. Quantitative determination of the composition requires an accurate sampling technique, which must be adapted to the purpose and the nature of the test.

3) ***Technological M.***: These are production floor tests which permit immediate decisions on the use of a material or machine part for a given purpose, without exact measurements. Both cold and warm tests can be applied.

Matricin: a prochamazulene of the guaianolide type, a sesquiterpene lactone present in camomile blossoms and yarrow. In weakly acidic solution, M. gives rise to the corresponding acid, ***guaiazulene carboxylic acid***, which is converted at about 50 °C, that is, during steam distillation of the camomile blossoms, to Chamazulene (see).

Matrix effects: factors which depend on the composition of the matrix, that is, materials surrounding and accompanying a sample, and which affect analysis of sample components. In spectroscopic methods, the wavelength of a line is not significantly affected by the composition of the sample, but its intensity is often dependent on both the concentration of the element of interest and on the composition of the sample. Ignorance of M. in quantitative analyses can thus lead to false results. M. can be largely eliminated by the choice of conditions (added substances, temperature changes), and this is part of optimizing the analytical method.

M. play a particularly important role in atomic spectroscopy, because the sample undergoes a number of processes which can be described as temperature-dependent equilibria. They can be formulated, schematically, as follows:

Vaporization (MeX) $\rightleftharpoons$ MeX
Dissociation MeX $\rightleftharpoons$ Me + X
Excitation Me $\rightleftharpoons$ Me*
Ionization Me $\rightleftharpoons$ Me^+ + E^-.

Vaporization and dissociation are considered chemical M., while thermal excitation and ionization are physical M., although it is difficult to draw a line between the various types of M. At plasma temperatures below 4000 K, it is mainly chemical M. which are important, while physical M. are more significant above 4000 K.

Vaporization. The matrix can either promote or hinder vaporization. For example, when the matrix contains anions which form relatively nonvolatile compounds with the metal to be determined, vaporization will be inhibited; this occurs with calcium if sulfate or phosphate is present in the matrix. This effect can be eliminated by increasing the flame temperature, adding other cations, such as strontium or lanthanum, which bind the sulfate or phosphate more strongly than calcium, or by adding complexing agents such as EDTA or 8-hydroquinoline, which form stable and volatile complexes with calcium.

Dissociation occurs in flame plasmas simultaneously with vaporization. The dissociation constant can be affected, for example, by the presence of certain anions. The equilibrium NaCl $\rightleftharpoons$ Na + Cl is shifted to the left in the presence of chlorine, and the concentration of atomic sodium decreases. In addition, chemical reactions such as Me + O $\rightleftharpoons$ MeO or Me + 2 OH $\rightleftharpoons$ $Me(OH)_2$ can take place in the flame; these also are subject to equilibria. Oxidation occurs mainly in "lean" flames, where the oxygen is present in excess of the fuel. The oxide molecules can give unspecific molecular emissions or absorptions, thus increasing the background. Such effects can sometimes be eliminated by changing the ratio of fuel to oxygen.

Excitation and *ionization*. These are affected mainly by cations. The mass action law gives the following expression for the ionization of a metal Me: $[Me^+][e^-]/[Me] = K$. Here $[e^-]$ is the electron pressure. If another metal in the sample is also ionized, this increases the electron pressure and the above ionization equilibrium is shifted towards to neutral metal atom. This effect can be removed by addition of ionization suppressers (spectrochemical buffers). These are readily ionized substances which lead to a relatively high concentration of electrons in the flame, and thus suppress ionization of the sample. In electrically generated plasmas, the change in electron pressure can lead to a change in the plasma temperature by changing its conductivity.

Matrix isolation technique: a method for study of very unstable, short-lived chemical compounds and intermediates. The compounds of interest are generated in a gas stream and, together with an inert substance such as a noble gas, hydrocarbon or ketone, frozen out on a surface cooled with liquid helium, hydrogen or nitrogen. The intermediates can also be generated photochemically by irradiation of the compounds embedded in the solid matrix. Since the subsequent processes occur very slowly at the low temperatures, the mean lifetimes of the unstable compounds are much longer, and their concentrations are thus much higher than under normal conditions. This makes it possible to record their spectra and to study their low-temperature reactions.

Matthias strips: see Paper chromatography.

Mauveine, ***mauve, N-phenylphenosafranine, aniline purple, Perkin violet***: a basic azine pigment. M. forms shiny black crystals which are soluble in alcohol, making a violet solution. Acids turn M. red-purple. M. was first synthesized in 1856 by W.H. Perkin, and was the first synthetic pigment. It is no longer used industrially, but some of its sulfonates are used to dye wool. Fig. p. 630.

$[C_6H_5-NH-\ldots-NH_2]Cl^-$ (with N^+–C_6H_5)

Mauveine

Maxwell-Boltzmann distribution: the statistical law describing the particle velocities in a gas at thermal equilibrium. It was first calculated by J.C. Maxwell, and given a theoretical basis in 1872 by L. Boltzmann.

A gas contains an immense number of particles (under standard conditions, $2.69 \cdot 10^{22}$ dm^{-3}) which move randomly. The molecules constantly collide with each other, causing their directions and velocities to change. A *velocity distribution* arises, that is, small numbers of molecules have very low or very high velocities, while most of them have intermediate velocities. This velocity distribution can be described by a Distribution function (see); because the number of particles is very great, it can be formulated as mathematically continuous.

If the molecules have no internal degrees of freedom, and if classical mechanics is valid (rather than quantized exchange of energy or quantized conservation of energy and momentum), the M. can be derived in several ways:

$$dN_v/N = (2/\pi)^{1/2}\,(m/kT)^{3/2} e^{-mv^2 r/kT} v^2\,dv = f(v)\,dv.$$

Here k is Boltzmann's constant, T is the absolute temperature, m is the mass of the particles, v is the absolute value of the particle velocity, and N is the total number of particles in the gas. dN_v is the number of particles with velocities in the differential velocity interval between v and $v + dv$. Therefore dN_v/N is the fraction of the particles which lie within this interval, or the probability for the occurrence of this velocity.

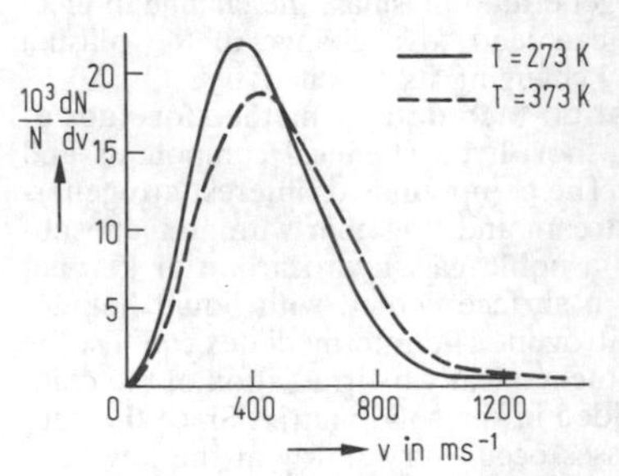

Velocity distribution of oxygen.

The figure shows that the largest fraction of the molecules have velocities in the neighborhood of the *most probable velocity* v_p, i.e. of the maximum of the curve. From $dF(v)/dv = 0$, it follows that $v_p = \sqrt{2kT/m}$. Since the curve is not symmetrical, v_p is not the *mean velocity* $\bar{v}$; the latter is somewhat larger. By taking the average $\bar{v} = \int_0^\infty .vf(v)dv$, we obtain $\bar{v} = \sqrt{8kT/\pi m} = \sqrt{4/\pi}v_p$. A third important parameter is the *mean square velocity* $\overline{v^2}$, which is obtained in the same way as $\bar{v}$, by averaging: $\overline{v^2} = \int_{-\infty}^{\infty} .v^2f(v)dv$. The result of the calculation is $\overline{v^2} = 3kT/m = 3v_p{}^2/2$. Both v_p and $\bar{v}$ increase with the temperature by the factor $\sqrt{T}$, that is, the distribution is shifted to higher velocities. At the same time, the distribution becomes broader and flatter (Fig.). This is understandable, since M. is normalized, that is, the sum of all fractions is equal to 1: $\int_{-\infty}^{\infty} .F(v)dv = 1$. Therefore the area under the curve must be equal to 1 for every temperature, and a broadening of the curve must lead to a flattening.

A basic assumption of statistics (see Thermodynamics, III) states that for a physical parameter, the mean for a very large number of particles at a given time is equal to the mean for a single particle over a very long time period. Thus the M. can also be understood as a probability distribution of the velocities that an individual molecule has over a very long time period in the gas. Because its velocity is constantly being changed by collisions, it is sometimes very high, sometimes very low, and is on the average equal to $\bar{v}$; mean square velocity is $\overline{v^2}$. The mean kinetic energy $E = m\overline{v^2}/2$ is therefore $E = 3/2kT$. Multiplication by Avogadro's number N_A gives the kinetic energy E per mol: $E = 3/2RT$, where $R = k \cdot N_A$ is the general gas constant. From $\delta E/\delta T = C_v$, it follows that the Molar heat capacity (see) of the gas $C_v = 3/2R$ or $1/2R$ for each degree of translational freedom.

If the rate v in the M. is replaced by the kinetic energy $\varepsilon = mv^2/2$ or $v = \sqrt{2e/m}$, the result is the energy distribution in the equilibrium state: $dN_\varepsilon/N = 2\pi^{-1/2}(kT)^{-3/2}\varepsilon^{1/2}e^{-\varepsilon/kT}d\varepsilon$.

The M. can be demonstrated directly with atomic or molecular beams; the particles are sorted according to their flight times using a rotating cylinder (Stern, 1920) or according to the Fizeau principle (Lammert, 1929).

Mayer's reagent: see Alkaloids.

Mazut, *masut*: 1) the high-boiling residue from petroleum distillation. It is subjected to hydrocatalytic cracking to make gasoline. Some M. is used for lubrication and some as a fuel (e.g. for steam boilers).

2) Pyrolysis tars and similar, high-boiling fractions formed by secondary processes.

MCD: see Chiroptic methods.

MCD peptide: abb. for mast-cell degranulating peptide.

McLafferty rearrangement: a rearrangement which often occurs as a result of fragmentation in a mass spectrometer (see Mass spectroscopy); it leads to formation of intense peaks. It can occur only when there is an H atom in a position γ- to a double bond (Fig.). In the spectrum of n-pentylbenzene, for example, the M. leads to an intense peak at mass number 92.

91

MCPA: see Growth substance herbicides.

MCPB: see Growth substance herbicides.

MCPP: see Growth substance herbicides.

Md: symbol for mendelevium.

Mebendazol: an Anthelminthic (see).

Mebenil: see Carboxamide fungicides.

MEC: see Emission.

Mechanochemistry: see Tribochemistry.

Mecocyanine: see Cyanidine.

Meclofenoxate: $4\text{-Cl-}C_6H_4\text{-O-}CH_2\text{-CO-O-}CH_2\text{-}CH_2\text{-N}(CH_3)_2$, a compound used as a nootropicum, that is, as a means of increasing intellectual capacity in cases of cerebral insufficiency and in states of tiredness or exhaustion.

Meconic acid: see Opium.

Mecoprop: see Growth substance herbicides.

Medazepam: see Benzodiazepines.

Medium-temperature hydrogenation: see Hydrorefining.

Meerwein-Ponndorf-Verley reduction: a method for reduction of aldehydes to primary alcohols or of ketones to secondary alcohols using aluminum isopropylate or ethylate in benzene or toluene:

$$R^1\text{-C=O} + R^2\text{-CH}_2\text{-OH} \xrightleftharpoons{(R^2\text{-CH}_2\text{-O})_3\text{Al}} R^1\text{-CH}_2\text{-OH} + R^2\text{-CHO}$$

The carbonyl compound resulting from the reduction, e.g. acetone or acetaldehyde, can be removed from the reaction mixture, thus pushing the equilibrium in the direction of the products.

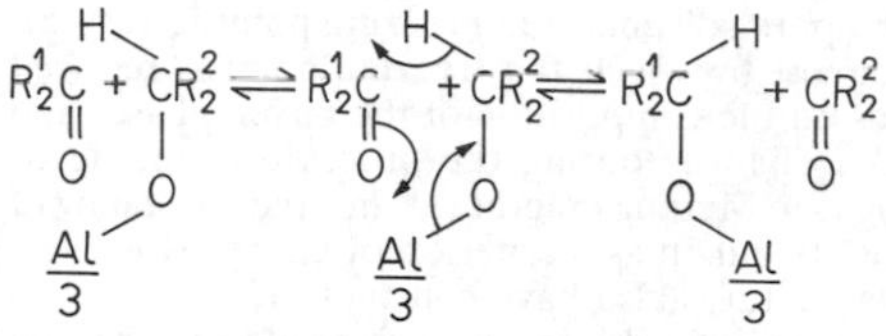

Reaction mechanism of the Meerwein-Ponndorf-Verley reduction.

The significance of the M. is that it is specific for carbonyl groups and does not affect C=C double bonds, nitro groups, halogen atoms, etc. The reverse of this reaction is the Oppenauer oxidation (see).

Meerwein salt: see Oxonium salts.

M-effect: abb. for mesomeric effect.

Mefrusid: see Diuretics.

Meisenheimer complex: a term for a σ-complex which is formed in a nucleophilic substitution reaction on an arene by an addition-elimination mechanism. In a few cases, M. have been isolated, for example in the reaction of 2,4,6-trinitroanisol with potassium ethanolate:

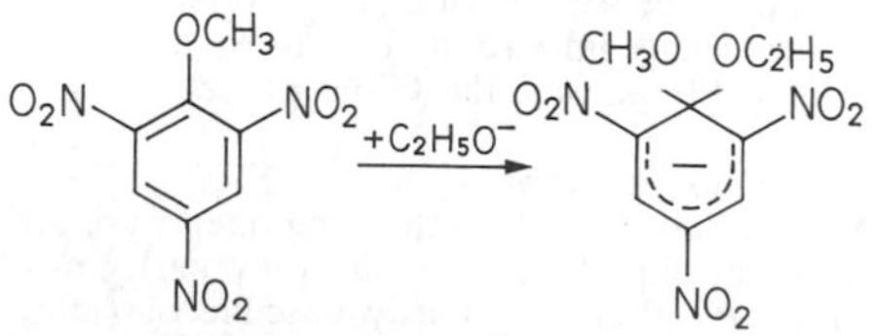

Melamine, ***cyanuric acid triamide, 2,4,6-triamine-1,3,5-triazine***: monoclinic prisms, m.p. 345 °C (dec.). It is readily soluble in hot water, but only very slightly soluble in alcohol, ether or cold water. M. can be sublimed. Like cyanuric acid, it can exist in two tautomeric forms. It was discovered in 1843 by J. von Liebig.

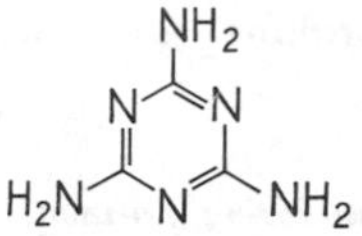

For a long time, M. was synthesized only by trimerization of dicyanodiamide. However, a simpler method is the reaction of urea in the presence of aluminosilicates: $6\ OC(NH_2)_2 \rightarrow C_3N_3 \cdot (NH_2)_3 + 6\ NH_3 + 3\ CO_2$. M. is an important intermediate in the production of melamine resins.

Melamine resins: hard plastics of the amide resin type, made by polycondensation of melamine with formaldehyde. Melamine can add up to six formaldehyde molecules to its three free amino groups. The resulting hexaaminomethanol compound then condenses, with loss of water and formation of ether bridges, into a cross-linked synthetic resin. The M. are superior to both dicyanodiamide and urea resins with respect to resistance to water and heat. In the unhardened state, they are fine, water-soluble powders, which are converted by heating to 120 to 165 °C to insoluble and non-melting products with a light color and good light stability. The M. are mostly made into molded plastics incorporating fillers such as sawdust, short-fiber cellulose, textile fibers and shreds, asbestos fibers and mineral powders. They are used for temperature-resistant household articles, such as dishes and containers of all kinds, handles, buckles, buttons, light switches, etc. Paper and cloth impregnated with M. solutions are pressed together into sheets used as mechanically resistant panelling for interior decoration. M. are also used to impregnate washable and non-wrinkling cloth and to protect paper from water. A melamine glue is superior to the urea-resin product, and is used for wood. For paint bases compatible with alkyde resins and cellulose nitrate, M. is condensed in the presence of an alcohol, such as butanol.

Melanins: high-molecular-weight pigments formed by oxidation of phenols. Natural M. are insoluble and amorphous, so that a detailed elucidation of their structure is extremely difficult. Structures have been suggested on the basis of products formed in vitro, through autooxidation or enzymatic oxidation of the corresponding phenols. The natural M. are classified as eumelanins, pheomelanins and allomelanins. ***Eumelanins*** contain nitrogen, and when fused with alkali, they yield 5,6-dihydroxyindole; oxidation with permanganate yields substituted pyrrol-2,3-dicarboxylic acids. They are thought to be polymers of 5,6-dihydroindole. Eumelanins are found as pigments of skin, hair, feathers and malignant melanomas. They are usually present as protein conjugates; often they are associated in mammalian hairs and bird feathers with ***pheomelanins***. These are red pigments which contain sulfur in addition to nitrogen. They are polymers of benzo-1,4-thiazine, which is formed by attack of cysteine on 2-benzoquinones. ***Allomelanins*** are compounds which do not contain nitrogen and are found in plant seeds or spores. Upon alkali fusion, these black pigments yield diphenols such as pyrocatechol or 1,8-dihydroxynaphthalene, which are linked together in the polymer by C-C and C-O-C bonds.

Melanocyte stimulating hormone: same as Melanotropin.

Melanoliberin: see Liberins.

Melanostatin: see Statins.

Melanotropin, ***melanocyte-stimulating hormone***, abb. ***MSH***: a peptide hormone formed under the control of the hypothalamus hormones melanoliberin and melanostatin. In cold-blooded vertebrates, M. stimulates the spreading of pigments in the melanocytes, which causes the skin to darken and adapt to the environment. Although the biological significance of M. in birds and mammals is not well understood, α-M. has strong effects in a number of mammalian tissues, and to some extent, also in human beings. M. is biosynthesized from the common precursor pro-opiomelanocortin. In addition to α- and β-M., a third MSH sequence was discovered in the precursor, γ-M. The following sequences were determined for the three bovine M.: α-MSH, Ser-Tyr-Ser-Met-Glu-His-Phe-Arg-Trp-Gly-Lys-Pro-Val-Gly; β-MSH, Tyr-Lys-Met-Glu-His-Phe-Arg-Trp-Gly-Ser-Pro-Pro-Lys-Asp; and γ-MSH, Tyr-Val-Met-Gly-His-Phe-Arg-Trp-Asp-Arg-Phe-Gly.

Some structural similarities with a fourth potential M. sequence in pro-opiomelanocortin (Ser-Met-Gly-Val-Arg-Gly-Trp) support the assumption that a series of gene duplications has occurred. α-MSH is an N^{α}-acetyltridecapeptidamide, which has the same sequence as ACTH(1-13). The length of β-M. depends on the species. M. sequences have been observed in various regions of the central nervous system, where they have neurological functions.

Melanotropin release inhibiting hormone: see Statins.

Melanotropin releasing hormone: see Liberins.

Melatonin, ***N-acetyl-5-methoxytryptamine***: an epiphysis hormone which is biosynthesized from the amino acid L-tryptophan. In amphibians, M. causes the skin to lighten. Its function in mammals is still unclear. M. has a role in photoperiodic processes and has an anti-gonadotropin activity.

$CH_2-CH_2-NHCOCH_3$
H_3CO
N
H

Melipax®: see Camphechlor.

Mellithic acid, ***benzenehexacarboxylic acid***: $C_6(COOH)_6$, colorless crystalline needles with a silky sheen; m.p. 286-288 °C (in a closed tube). M. is readily soluble in water and alcohol. It is very stable, and is not attacked by concentrated nitric acid, bromine or chlorine. It dissolves without decomposition in boiling concentrated sulfuric acid. The reaction of thionyl chloride with M. produces a dianhydride, and with acetyl chloride, a stable trianhydride is formed. The aluminum salt of M., $Al_2C_{12}O_{12} \cdot 18H_2O$, is found in brown coal as the mineral mellite. M. is formed by the oxidation of hexamethylbenzene or graphite with potassium permanganate, or, in low yields, from oxidation of charcoal with concentrated nitric acid.

Mellitin: Gly-Ile-Gly-Ala-Val-Leu-Lys-Val-Leu-Thr-Thr-Gly-Leu-Pro-Ala-Leu-Ile-Ser-Trp-Ile-Lys-Arg-Lys-Arg-Gln-Gln-NH_2, a linear 26-peptidamide which is the main component of the peptide toxin in bee venom. Its hemolytic effect and surface-tension-reducing activity are due to the distribution of hydrophobic amino acids in the *N*-terminal region, and the hydrophilic amino acids in the *C* terminal part. M. is thus an "inverted soap" on a peptide basis. The biosynthetic precursor is ***pre-pro-mellitin***, a peptide of 70 amino acids. M. is identical to the sequence 44-69 of this precursor, from which it is released by an endopeptidase. At the same time, the *C*-terminal glycine group is converted to an amide group.

Melt-in alloys: alloys with the same thermal expansion coefficients as glasses and ceramics. They can be melted into a glass or ceramic apparatus and give a sealed electrical connection. M. are based on iron-nickel, iron-chromium or iron-nickel-cobalt; examples are Invar® (see), Platinite (see), Kovar (see) and Fernichrome (see).

Melting point, ***melting temperature***, abb. ***m.p.***: the temperature at which a substance is converted from the solid to the liquid state. The M. is the same as the ***freezing point*** of the liquid phase. Pure substances have a constant M., while solutions have lower M. than the pure solvents (see Freezing point lowering). The amount by which the freezing point is reduced depends on the composition of the mixed phase, and therefore changes during the process of melting or freezing (see Melting diagram). Thus the constancy of a M. can be taken as a criterion of purity. However, eutectic mixtures also have constant M.

Noncrystalline substances, such as glasses or polymers, do not melt at a distinct temperature, but rather over a range of temperatures. At the *softening point*, they lose their rigidity, collapse into themselves and then gradually enter the liquid state. Many compounds are thermally decomposed before they reach their M. (***decomposition point***, abb. ***dec.***).

In contrast to the Boiling point (see), the M. is only slightly dependent on pressure (for an explanation, see Clausius-Clapeyron equation). For most substances, it increases with increasing pressure; only a few substances, notably ice, have the reverse tendency.

Melting point diagram: a phase diagram showing the relationship between the solid and liquid states, the temperature, the number of coexisting states and the concentrations of the components; usually the pressure is held constant. Thus ***melting curves*** are isobars. In practice, the M. of binary and ternary systems are especially important. For all M. which include only stable phases, the Gibbs phase law (see) applies.

1. ***M. of binary systems***

1.1 M. of systems in which the components are not miscible in the liquid state, or are only partly miscible, and are either only partly miscible or (more often) immiscible in the solid state. Fig. 1 is an example of a system in which the components are completely immiscible in the liquid and solid states under normal conditions. In phase space *A*, the two liquids exist as a heterogeneous system. At the melting temperature M_1 of the first component, the entire amount of substance 1 freezes. In phase space B, crystals of substance 1 coexist with liquid substance 2. At the melting temperature M_2 of the second substance, the entire amount of substance 2 freezes, so that in phase space C, the two types of crystals are present in a heterogeneous mixture.

Most liquids which are immiscible or only partly miscible just above the melting point are completely miscible above the critical separation temperature. However, in many cases, the critical separation temperature cannot be reached at normal pressures, because the components evaporate instead. In this case high pressure is needed to prevent evaporation.

1.2 Systems in which the components are completely miscible in the liquid state. There are many such systems; they differ in the behavior of the solids.

1.2.1. *Complete miscibility of the components in the solid state* (Fig. 2). The components form mixed crystals in any ratio; an example is the gold/silver system. There is no eutectic point (see below). The mixed crystals lie in phase space C. In phase space B, which is characterized by the melting points M_1 and M_2 of the two components, the mixed crystals are in equilibrium with the homogeneous melt of the two components; phase space A contains the components, still completely miscible, in the liquid state. Phase space B is separated from C by the ***freezing*** or ***solidus*** (Latin "solidus" = "solid") curve, and from A by the ***solubility*** or ***liquidus*** (Latin "liquidus" = "liquid") curve.

Fig. 2 shows a formal analogy to the Boiling point diagram (see). In the M., too, the solidus and liquidus curves can have maxima (e.g. the lead/thallium system) or minima (e.g. the gold/copper system).

1.2.2. *Complete immiscibility of the components in the solid state; formation of a eutectic mixture* (Fig. 3): Phase space A contains liquid components in a homogeneous mixture. At M_1 and M_2, the pure components begin to crystallize, so that the melt is depleted in the crystallizing component. The compositions of the solutions which are in equilibrium with substance 2 or substance 1 at various temperatures

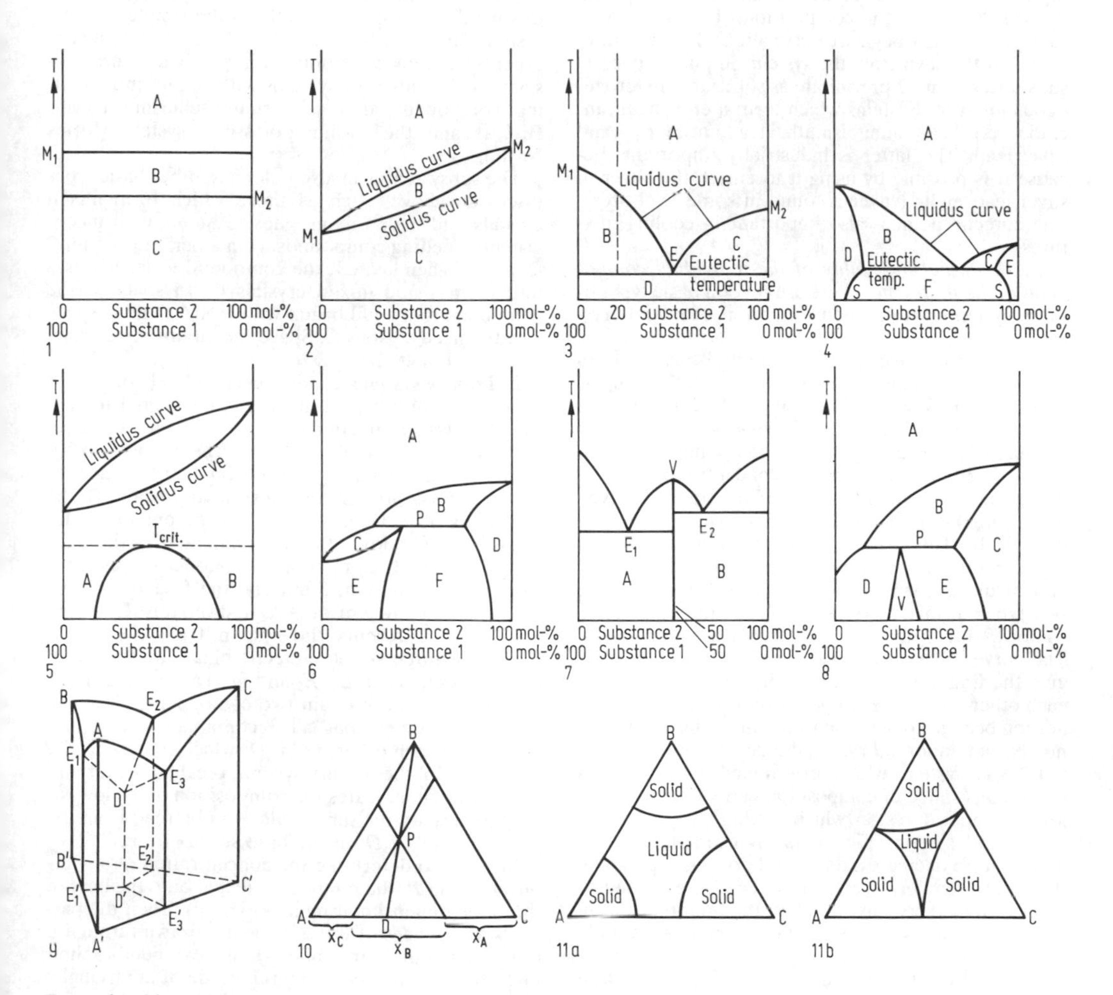

Types of melting point diagrams.

can be read along the liquidus curves M_1 - E or M_2 - E. From either side, the curve approaches the *eutectic temperature*, the temperature at which the remaining liquid has the concentration of the *eutectic point E*; this solution is called the *eutecticum*. (In the case of aqueous salt solutions, the terms *cryohydrate* and *cryohydratic point* are used.) At the eutectic point, there are 4 phases in equilibrium: the two pure components, the melt and the gaseous phase of the components, which appears because all substances always have a finite vapor pressure. At the eutectic temperature, the eutecticum solidifies, as a pure substance does, at a single temperature. In phase spaces *B* and *C*, there is one pure solid component in equilibrium with the mixed melt.

If, for example, a solution which contains 20 mol% of substance 2 and 80 mol% of substance 1 is cooled, substance 1 begins to crystallize out when the temperature reaches point *a* along the liquidus curve M_1-E. If the temperature drops further, substance 1 continues to crystallize out, and the composition of the liquid phase changes along the curve M_1-E, going towards *E*. When the composition of *E* is reached, substance 2 also begins to crystallize. If even more heat is withdrawn from the system, in phase space *D* substances 1 and 2 precipitate as solids in a heterogeneous mixture. Systems which form a eutecticum include, e.g. cadmium/bismuth, lead/antimony, and silver/lead. The latter is industrially important, because it is possible by using it to enrich the silver in silver-lead melts (silver is found in some lead ores). The eutectic point is also important in cooling mixtures.

1.2.3. *Partial miscibility of the individual components in the solid state*. There are several basic types of these systems with limited formation of mixed crystals.

1.2.3.1. An eutecticum with limited miscibility in the boundary regions often forms, e.g. in the copper/silver system (Fig. 4). Phase space *A* again contains the fully miscible melts, while spaces *D* and *E* are occupied by mixed crystals of components with limited miscibility. The limiting concentrations, up to which the two components are still miscible, are given by the intersections of the liquidus curves with the eutecticals. The regions *B* and *C* contain melts in equilibrium with one form of mixed crystal each; the eutecticum is characterized by being formed by the two types of mixed crystals in their limiting concentrations, rather than by the two pure components. The curves *S* are called solution lines, because they give the limits of solubility of the two substances in each other. When the temperature drops, further separation occurs along the solution lines; this phenomenon has an important role in the duraluminum effect.

1.2.3.2. Systems which form mixed crystals over a continuous range of composition can still separate at lower temperatures, at which mixing gaps exist; the gold/nickel system (Fig. 5) is an example. As the temperature decreases, the mixed crystals separate at the critical temperature T_{crit} into a mixture of the two types of mixed crystals *A* and *B*; the limiting concentrations of the two types of mixed crystals are indicated by the mixed crystal ranges.

1.2.3.3. In systems in which the mixing gap extends into the temperature range at which the mixed crystals melt (Fig. 6), a *peritecticum P* forms, in which the mixed crystal type *E* separates into a melt and mixed crystal type *D*. In space *F*, the mixed crystals *E* and *D* coexist, while *C* and *J* contain the mixed crystal types in equilibrium with the melt. Phase space *A* contains the completely miscible melts. The diagram has a formal similarity to systems which contain incongruently melting compounds (Fig. 8).

1.2.4. *Systems which form compounds* (see Intermetallic compounds) are subdivided into congruently melting compounds, in which the compound is converted directly into the melt or solution, and incongruently melting compounds, which decompose into their components before the melting or dissolving process begins. The M. of a congruently melting compound always displays a maximum. Fig. 7 shows the M. of a compound of 50 mol% of substance 1 and 50 mol% of substance 2 with a maximum *V*. In this case there are 2 eutectic points E_1 and E_2; substance 2 and the compound of both substances belong to E_1, while substance 1 and the compound belong to E_2. If one thinks of the M. as divided into two parts *A* and *B*, this would correspond exactly to the simple eutectic system. In this fashion, any M. of a system with congruently melting compounds can be divided into subsystems. Examples of systems with congruently melting compounds are the bismuth/tellurium (forms Bi_2Te_3) and the sodium/potassium system (forms Na_2K).

The subsystems can also belong to other basic types discussed above, such as those which form mixed crystals and have mixing gaps. Systems with incongruently melting compounds form a perieutecticum *P* (Fig. 8). When heated, the compound *V* decomposes into melt *A* and mixed crystals *C*. In space *D* the compound is in equilibrium with melt *A*, and in space *E* with mixed crystals *C*. Space *B* contains mixed crystals *C* and melt *A*.

2. ***Ternary systems*** cannot be completely displayed in a plane, but require a three-dimensional model. Fig. 9 shows a projection of such a M., with an eutecticum, projected onto a plane. The model has an equilateral triangle *A'B'C'* as its base; the vertices of the triangle represent the pure components. If the composition of the mixture E_1' (components *A* and *B*) is plotted along the segment *A'B'*), the face $AA'B'BE_1$ of three-dimensional model corresponds to this binary system. Similarly, the face $AA'C'CE_3$ represents the system of A,C and mixture E_3, while $CC'B'BE_2$ represents the system B,C, E_2. Fig. 9 shows the three simple eutectic binary systems with the binary eutectica E_1, E_2 and E_3. The outer surfaces of the figure thus contain two components each; the interior, however, has all 3 components in equilibrium with each other. Point *D*, which represents the eutecticum of the ternary system, is called the ternary eutecticum; it indicates the composition of the lowest-melting ternary mixture (which can be read from the projection *D*' of *D* on the base surface).

In order to determine the concentration of any arbitrary point *P*, the point is projected onto the base of the three-dimensional figure (Fig. 10). Then the parallels to the sides of the triangle are drawn through *P* (any two parallels are enough), and extended to their intersections with the nonparallel side of the triangle. If the molar fractions 0 to 1 are plotted along this nonparallel side, the segments between the intersections of the parallels show the molar fractions of the components *A*, *B* and *C*. The segment corresponding to each component is opposite the vertex representing that component (see Fig.). Along line *BD*, the ratio of component *A* to *C* is constant; only the ratios of *A* to *B* and *B* to *C* change. For a partial representation of a ternary system in a plane, isotherms (planes in which the temperature is constant) are cut through the three-dimensional figure (Fig. 11). These isotherms, if stacked sequentially, would give the complete space-temperature diagram. The lines in the interior thus give the parts of the limiting surfaces of the individual phase-spaces at the temperature of the isotherm; they are analogous to the elevation lines on a map. The temperature in Fig. 11a is higher than that in Fig. 11b ($T_2 < T_1$); T_1 already exceeds the binary eutectic temperature. The temperature of a ternary eutecticum is generally lower than that of any of the corresponding binary eutectica.

In addition to this simple type of M. of a ternary system, any imaginable combination of binary eutectica, extended to ternary systems, can occur. This usually makes a graphical representation so complicated that the relationships can only be visualized with the aid of a spatial model.

Some examples of ternary systems with eutecticum formation are bismuth/lead/tin, potassium nitrate/sodium nitrate/lithium nitrate and potassium carbonate/potassium chloride/sodium fluoride. Some ternary systems with complete miscibility of the components are copper/manganese/nickel and cesium chloride/rubidium chloride/potassium chloride.

The representation of ***quaternary systems*** is even more difficult, although only a few are nearly completely understood.

Melting salts: compounds used to thaw snow, ice or frost from streets and sidewalks. The salts are scattered on the surface to be thawed, or an aqueous solution (e.g. mother liquors from the potash industry) is sprayed on. $MgCl_2$, $MgSO_4$, NaCl, $CaCl_2$, etc. are used. Direct contact with the salts or their effects on the soils near roads can kill or damage plants.

Melting tube: same as Bomb (see).

Membrane equilibrium: the isothermal equilibrium between two solutions which are separated from each other by a semipermeable membrane. At osmotic equilibrium (see Osmosis), the membrane is permeable only to the solvent; in the Donnan equilibrium, it is permeable to the solvent and smaller ions, but not to ions of colloidal dimensions.

Membrane process: see Chloralkali electrolysis.

Menaquinone: see Vitamin K.

Menaquinones: see Ubiquinones.

Menadiol: see Vitamin K.

Menadione: see Vitamin K.

Mendelevium, symbol ***Md***: a short-lived, radioactive element which can be obtained only through nuclear reactions. It is a member of the Actinoids (see), *Z* 101, with isotopes with the following mass numbers (in parentheses the decay type, half-life and nuclear isomers): 248 (α, K-capture, 7 s); 249 (α, K, 24 s); 250 (α, K, 52 s); 251 (α, K, 4.0 min); 252 (K, 2.3 min); 254 (K, 10 min, 2); 255 (K, α, 27 min); 256 (K, α, 1.25 h), 257 (K, α, 5.0 h), 258 (α, 55 d), atomic mass of the most stable isotope, 258, valency III, II, standard electrode potential (Md^{2+}/Md^{3+}) -0.15 V.

The Md^{3+} ion, which can be demonstrated in solution, can be reduced by strong reducing agents, such as zinc powder, Cr^{2+} or V^{2+}, to Md^{2+}. M. was first synthesized in 1955 by Ghiorso, Harvey, Choppin, Thompson and Seaborg by irradiation of einsteinium 253 with energetic α-particles; about 10^{-12} g was made: $^{253}Es + ^{4}He \rightarrow ^{256}Md + ^{1}n$. The most stable nuclide, ^{258}Md, is obtained by reaction of einsteinium 255 with α-particles: $^{255}Es + ^{4}He \rightarrow ^{258}Md + ^{1}n$.

Mensvutkin reaction: a reaction of tertiary amines with alkyl halides to form quaternary ammonium halides: $R_3N + R\text{-}X \rightarrow R_4NX^-$.

p-Menthane: see Monoterpenes.

Menthol: 1-methyl-4-isopropylcyclohexan-3-ol, M_r 156.3, monocyclic monoterpene alcohol. Four stereoisomeric pairs of M. (a D- and L-form of each) are known; each pair differs from the others with respect to the relative arrangement of the methyl, hydroxyl and isopropyl groups on the cyclohexane ring. They are called M., *iso*-M., *eno*-M. and *neoiso*-M. D-(-)-M. is the main component of peppermint oil. It forms colorless crystals with a characteristic odor; m.p. dependent on the modification, $[\alpha]_D^{20}$ -47.0 to -51.0° (ethanol). M. is only slightly soluble in water, but is very soluble in ethanol. In addition to M., peppermint oil contains at least 3% of its acetic and valeric acid esters, which make an essential contribution to its aroma. The stereoisomers of M. smell musty. Industrial syntheses of DL-M. start from thymol, carvacrol or ketones of the *p*-menthane series, such as pulegone, piperitone or menthone. M. has an antiseptic effect and low concentrations stimulate the cold receptors of the skin and mucous membranes. It is used as a flavoring for foods, toothpastes and mouthwashes. D-, L- and DL-M. are used.

CH_3 OH H_3C CH_3

Menthone: *p*-menthan-3-one, 1-methyl-4-isopropylcyclohexan-3-one, a monocyclic monoterpene ketone; a colorless liquid which smells like peppermint. Four optically active forms of M. are found in essential oils: the d- and l-forms of M. and *iso*-M. M. is usually a mixture of different isomers. It can be isolated from the oils as the oxime or semicarbazone. M. is used to make menthol and artificial essential oils.

Mepacrine: a basic substituted aminoacridine derivative, which was the first Antimalaria agent (see) which was effective against schizonts in the blood. Because it led to a yellow coloration of the skin, it was later replaced by chloroquine.

Mepasine: a Cogasine (see) freed of alkenes by hydrogenation at 320 °C and 20 MPa pressure on a nickel-tungsten-sulfide contact. ***Mepasine I*** has a boiling point of 180 to 230 °C, while ***mepasine II*** boils between 230 and 320 °C. M. is further processed by chlorination and sulfochlorination to make detergents, softeners, etc.; it is also used as a heating bath liquid.

Mephenesin: see Muscle relaxants.

Mephytal: see Barbitals.

Meprobamate: a biscarbamic acid ester of a divalent alcohol. M. was the first tranquillizer (see Psychopharmaceuticals) to be widely used.

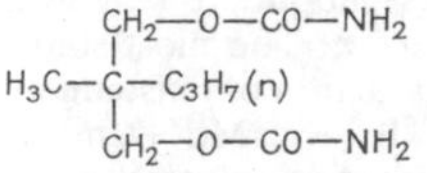

```
       CH2—O—CO—NH2
       |
H3C—C—C3H7(n)
       |
       CH2—O—CO—NH2
```

Mercaptals: same as Thioacetals (see). The term M. is no longer permitted in IUPAC nomenclature.

Mercaptans: same as Thioalcohols (see).

Mercaptides: same as Thiolates (see).

2-Mercaptobenzothiazole: a derivative of Benzothiazole (see), which forms colorless to light yellow crystals; m.p. 180-182 °C. It is readily soluble in acetone, slightly soluble in alcohol, ether and benzene and soluble in alkali and alkali carbonate solutions. In the laboratory, 2-M. can be made from 2-aminothiophenol and carbon disulfide. On an industrial scale, it is synthesized from aniline, carbon disulfide and sulfur under pressure at 250 °C. It is used as a vulcanization accelerator, corrosion protectant, fungicide and as a reagent for cadmium and other heavy metals.

Mercaptodimethur, ***methiocarb***: a Carbamate insecticide (see), acaricide and molluscicide.

Mercaptoacetic acid: same as Thioglycolic acid (see).

2-Mercaptomethylfuran: same as Furfuryl mercaptan.

6-Mercaptopurine: an antimetabolite of the purine nucleoside bases used as a cytostatic and immunosuppressant. 6-M. is formed in the organism from its derivative *azathioprine*, which also acts as an immunosuppressant. 6-M. inhibits various enzymes of nucleic acid metabolism.

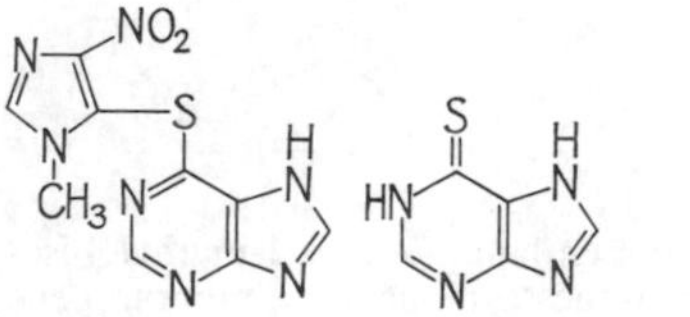

Azathioprine 6-Mercaptopurine

Mercerization: alkaline treatment of cotton yarns or cloth under tension. The treatment increases the tearing strength by about half and the dye affinity about one third; the shine is also increased. The cloth is given a more pleasant and fuller feel. Usually the yarn or cloth is treated for about 30 s in a 20 to 25% sodium hydroxide bath at a temperature of about 10 °C. The process was developed in 1844 by J. Mercer.

Mercurization: see Organomercury compounds.

Mercury, symbol ***Hg*** (for hydrargyrum): an element of Group IIb of the periodic system, the Zinc group (see); a heavy metal, *Z* 80, with natural isotopes with mass numbers 202 (29.80%), 200 (23.13%), 199 (16.84%), 201 (13.22%), 198 (10.02%), 204 (6.85%) and 196 (0.146%). The atomic mass is 200.59, valencies I and II, density 13.534, m.p. -38.84 °C, b.p. 356.59 °C, electrical conductivity 1.04 Sm mm^{-2}, standard electrode potential (Hg/Hg^{2+}) +0.8500V.

Properties. M. is the only metal which is liquid at room temperature. In the solid phase, it is soft and ductile, forming hexagonal close-packed crystals which are even more distorted from the ideal than those of zinc and cadmium. In the gas phase, M. is atomic. Some important and characteristic physical properties are its silvery white shine, its high surface tension, its large thermal expansion coefficient, its high density and high vapor pressure (saturation pressure at room temperature is 0.0013 mm) and its emission of UV light when it is excited by an electric discharge.

In the pure state, M. is stable at normal temperatures. However, impure M. is rapidly covered with a gray oxide skin in the presence of moist air. Above 300 °C, M. reacts with oxygen to form mercury(II) oxide, HgO; it also reacts readily with sulfur and halogens. Its position in the electrochemical potential series is such that M. is not attacked by nonoxidizing acids. Nitric and concentrated hot sulfuric acids dissolve it. Many metals react with M. to form alloys (amalgams; see Mercury alloys). M. is found in the monovalent and divalent states in its compounds. The compounds of monovalent M. have the general formula Hg_2X_2; they are linear compounds with direct Hg-Hg bonds. Mercury(II) compounds of the type HgX_2 are often monomolecular, and also have linear structures. A characteristic of mercury(II) compounds is stability and ease of formation of mercury-nitrogen and mercury-carbon bonds, which are found in amidomercury derivatives and Organomercury compounds (see). In mercury(II) complexes, the most common coordination numbers are 4 and 6, with tetrahedral and octahedral ligand arrangements, respectively.

> Mercury vapor and soluble mercury compounds are very poisonous. Air saturated with mercury vapor contains 15 mg M. per m^3. To avoid chronic mercury poisoning, it is essential to store mercury in closed vessels and to handle it only in well ventilated rooms. Spilled mercury should be cleaned up at once, for example, by picking it up with mercury tongs, sprinkling it with iodinated charcoal, or by amalgamation with zinc or copper. It must be remembered that organomercury compounds, many of which are volatile, are highly poisonous. The symptoms of chronic mercury poisoning are mild bleeding of the gums, later excitability and loss of memory, headaches and digestive upsets, and finally severe damage to the nervous system. Because mercury is only very slowly excreted from the body, treatment of mercury poisoning lasts for a long time. If vomiting cannot be induced after an acute oral poisoning, the stomach must be rinsed with a suspension of animal charcoal, and then magnesium sulfate is administered as a laxative. 1,2-Dithioglycerol can often effect a complete restoration. Sodium thiosulfate can also be used.

Analytical. Monovalent M. in the form of mercury(I) chloride, Hg_2Cl_2, is precipitated by addition of hydrochloric acid, and turns black when ammonia is poured over it. Hg^{2+} ions can be precipitated from a hydrochloric acid solution with hydrogen sulfide, in the form of mercury sulfide. If mercury compounds are heated with soda in a closed tube, a gray mercury mirror is precipitated in the cooler parts of the tube; when rubbed with a crystal of iodine, this is converted to yellow or red mercury(II) iodide. μg amounts of M. can be detected by the color reaction of diphenylcarbazide (blue to violet color). A rapid complexometric method of analysis is to use excess EDTA at pH 10 and to back titrate with Mg^{2+}.

Occurrence. M. makes up about 10^{-5}% of the earth's crust, and is thus one of the less common elements. The most important mercury ore is cinnabar, HgS, in which M. is sometimes also found as small droplets of the metal. M. is also present in a number of fahlerz minerals and zinc blende. Some M. ores of mineralogical interest are levinstonite ($Hg[Sb_4S_7]$), mercury horn silver (Hg_2Cl_2), tiemannite (HgSe), coloradoite (HgTe), and coccinite (Hg_2I_2).

Production. The industrial production of M. usually starts from cinnabar. The ores are heated in blast furnaces with air blasting at 350 to 400 °C. The M. released according to the equation $HgS + O_2 \rightarrow Hg + SO_2$ is trapped in water-cooled condensors made of glazed stoneware and is collected in cement-lined iron tanks. Some of the M. vapor condenses, together with mercury salts, fly ash, soot and tar, to form mercurial soot ("stupp"), which is mixed with lime and pressed out. Other methods of M. production start from the reaction of mercury sulfide with iron: $HgS + Fe \rightarrow FeS + Hg$, or allow mercury sulfide to react with calcium oxide: $4\,HgS + 4\,CaO \rightarrow 4\,Hg + 3\,CaS + CaSO_4$. M. is further purified by filtration through cloth, washing with dilute nitric acid and distillation. Repeated vacuum distillation produces M. with a purity of 99.999%.

Applications. M. is used to fill thermometers, barometers and manometers. It is also used in diffusion pumps, current rectifiers, standard cells and electrodes. The standard electrode for polarography is the dropping M. electrode. M. lamps are used as sources of UV light, and high-pressure M. lamps are used as street lights. M. is very important for making Mercury alloys (see), e.g. in chloralkali electrolysis. It was formerly important for extraction of gold and silver by the amalgamation method, but this use has declined. M. is used in dentistry to make Dental alloys (see). Mercury compounds are frequently used as catalysts in organic synthesis. However, because of the toxicity of M. compounds, their use in medicine and for pest control in agriculture has declined markedly.

Historical. M. was known in Europe in antiquity, and it was also known how to obtain it from cinnabar. Because of its red color, cinnabar has been prized in all times; it has been shown to have been used in ancient China. The Romans obtained it from the mines of Almaden in Spain, which still exist. M. had an extremely prominent place in alchemy. The element was considered to be the principle of the volatile and motile (quicksilver) and was named for the messenger god Mercury. It was thought to be a component of all metals, and was thus held to be the key for transmutation of elements. The production of M. received a strong impetus from the introduction of the amalgamation methods for production of noble metals; this occured in the 6th century for gold and in the 15th century for silver. Even in classical antiquity, M. was known to be a poison, but the curative effects of cinnabar were also known. M. compounds were introduced into medicine in the period of iatrochemistry, especially by Paracelcus. For example, finely divided M. was contained in the "gray salve" used to treat skin diseases and syphilis. "Yellow salve" containing mercury oxide as the active ingredient was used for inflamation of the edges of the eyelids, and "mercury precipitate salve", contained $HgNH_2Cl$ as the active ingredient. Hg_2Cl_2 was used as a laxative.

Mercury alloys, ***amalgams***: Alloys of mercury with many metals. M. have a special place among alloys, because some of them are plastic or liquid at room temperature. Liquid M. are true solutions of the alloy elements in mercury. Plastic M. are suspensions of solid particles in mercury or saturated mercury solutions. The reason for their plasticity can be an excess of mercury with respect to the composition of the solid phase, incomplete formation of compounds, or the melting of a low-melting intermetallic phase. Solid M. are intermetallic phases, which are often mixtures with the alloy components or their mixed crystals. Some transition metals, such as iron, are not capable of forming amalgams. M. are made industrially by powder metallurgical techniques in which the powdered alloy components are mixed with the mercury. They may also be produced galvanically, by electrolytic precipitation of the metals on the mercury or amalgam cathode, or by reaction of sodium amalgam with the salt solution of a less reactive metal. Some industrially important M. are amalgams of tin, copper and noble metals (see Dental alloys), gold, silver or tin amalgams for gilding or silvering and production of mirrors. Thallium amalgams are used to extend the range of measurement of mercury thermometers down to -58 °C. Sodium amalgam is an intermediate product in the amalgam method of chloralkali electrolysis. Alkali and zinc amalgams are used in the laboratory as reducing agents.

Historical. Vitruvius (1st century BC) was aware of the dissolving capacity of liquid mercury for gold and silver. In the Middle Ages, the majority of the European production of mercury was used for producing noble metals by the amalgam process. In the 16th century, the Venetians (Murano glass works) began producing mirrors with tin amalgam. In the 18th and 19th centuries, gold amalgam was used for gilding. M. have been used in dentistry for about 100 years.

Mercury(II) acetate: $Hg(OOCCH_3)_2$, white crystals; M_r 318.76, density 3.27. M. is poisonous (see Mercury). It is soluble in water and ethanol. M. is made by reaction of mercury(II) iodide with hot, 50% acetic acid. It is used as a catalyst in organic syntheses.

Mercury bromides. ***Mercury(II) bromide***: Hg_2Br_2, colorless, shiny, tetragonal leaflets; M_r 561.00, density 7.307, subl.p. 345 °C. The Hg-Hg distance in the linear molecule is 258 pm. Hg_2Br_2 is obtained by precipitation of mercury(I) nitrate out of aqueous solution with potassium bromide.

Mercury(II) bromide: $HgBr_2$, silvery-shining leaflets (from water) or colorless, orthorhombic prisms (from alcohol); M_r 360.41, density 6.109, m.p. 236 °C, b.p. 322 °C. Both in the solid and in the gas, the compound exists as linear $HgBr_2$ molecules. $HgBr_2$ is less soluble in water and less dissociated than mercury(II) chloride. It is obtained by the reaction of bromine with mercury, from mercury(II) sulfate and sodium bromide, or by treatment of mercury oxide with hydrobromic acid.

Mercury chlorides. ***Mercury(I) chloride***, formerly ***calomel***: Hg_2Cl_2, colorless crystals which are fibrous after sublimation; M_r 472.09, density 7.15, subl.p. 400 °C. Hg_2Cl_2 is barely soluble in water, but dissolves easily in benzene or pyridine. When ammonia solution is poured over it, it turns black due to a disproportionation reaction: $Hg_2Cl_2 + 2\,NH_3 \rightarrow H_2N\text{-}Hg\text{-}Cl + Hg + NH_4Cl$. The Hg-Hg distance in the linear Hg_2Cl_2 molecule is 253 pm. Hg_2Cl_2 turns dark when exposed to light. It is obtained by combining a solution of a mercury(I) salt with hydrochloric acid or a soluble chloride, by sublimation of an equimolar mixture of mercury and mercury(II) chloride, or by reduction of mercury(II) chloride. Hg_2Cl_2 is used in medicine as a mild laxative, to make calomel electrodes, in printing cloth, in galvanoplastics, in porcelain painting to apply gold, and to make green flares.

Mercury(II) chloride, sublimate: $HgCl_2$, white crystals soluble in water, alcohol and ether; M_r 271.50, density 5.44, m.p. 276 °C, b.p. 302 °C. $HgCl_2$ is a largely covalent, linear molecule with an Hg-Cl distance of 228 pm. Its solutions in water are only slightly dissociated, and it forms chloro complexes with chloride ions: $[HgCl_3]^-$ and $[HgCl_4]^{2-}$. Reducing agents reduce it to Hg_2Cl_2 or mercury. Its behavior with ammonia is very characteristic: with gaseous NH_3, it forms rhombohedral ***diamminemercury chloride ("meltable precipitate")***, $[Hg(NH_3)_2]Cl_2$, M_r 305.56, m.p. 300 °C; while with aqueous ammonia it forms a white ***"nonmelting precipitate"***, $[HgNH_2]Cl$ (M_r 252.07, density 5.70). This compound is built up of zig-zag $[HgNH_2]$ cation chains.

Cation structure of $[HgNH_2]^+$.

The chloride of Millon's base (see), $[Hg_2N]Cl$, is another product of the reaction of M. with ammonia.

> Mercury(II) chloride is a strong oral poison; as little as 0.2 to 0.4 g is fatal for an adult.
> First aid: see Mercury.

$HgCl_2$ is made on an industrial scale by heating mercury sulfate with sodium chloride. It is used in the chemical industry as a catalyst for the synthesis of vinyl chloride, in the synthesis of other mercury compounds, and as a depolarizer in dry cells. Although it is still sometimes used as a fungicide, its use in human and veterinary medicine, as an insecticide or as a seed protectant has greatly declined due to its toxicity.

Mercury(II) cyanide: $Hg(CN)_2$, colorless, tetragonal crystals which are readily soluble in water; M_r 252.63, density 3.996. M. is very poisonous (see Mercury). The crystal lattice contains nearly linear NCHgCN molecules, and the compound is also a nonelectrolyte in aqueous solution. With alkali cyanides, it forms tetrahedral tetracyanomercurate(II), $[Hg(CN)_4]^{2-}$. M. is made by dissolving mercury(II) oxide, HgO, in hydrocyanic acid, or by heating HgO with sodium cyanide in aqueous solution. Like mercury(II) chloride, M. was formerly used in medicine as an antiseptic, but it is now used only to disinfect metal instruments.

Mercury diuretics: see Diuretics.

Mercury fluorides: ***Mercury(I) fluoride***, Hg_2F_2, yellowish, light-sensitive crystals; M_r 439.18, density 8.73, m.p. 570 °C (dec.). Hg_2F_2 forms a linear molecule with a Hg-Hg distance of 243 pm. It is obtained by dissolving freshly precipitated mercury(I) carbonate in hydrofluoric acid.

Mercury(II) fluoride, HgF_2, is a white, ionic compound crystallizing in a calcium fluoride lattice; it dissociates in water and undergoes hydrolysis; M_r 238.59, density 8.95, m.p. 645 °C (dec.). HgF_2 is obtained by reaction of Hg_2F_2 with chlorine: $Hg_2F_2 + Cl_2 \rightarrow HgF_2 + HgCl_2$, by the reaction of fluorine with mercury or its salts, or by disproportionation of Hg_2F_2 at 450 °C. The dihydrate $HgF_2 \cdot 2H_2O$ is formed when mercury oxide is dissolved in a moderately concentrated hydrofluoric acid solution. HgF_2 is used as a mild fluorinating reagent, for example in the synthesis of organic fluorides.

Mercury fulminate: $Hg(CNO)_2$, a colorless, poisonous (see Mercury) explosive compound which crystallizes out of water as the hemihydrate $Hg(CNO)_2 \cdot 1/2H_2O$, M_r 284.62, density 4.42. M. is a coordination polymer in which the fulminate C atoms serve as bridges. It is not soluble in cold water, but is slightly soluble in hot water; when heated, or upon impact or friction, or when conc. sulfuric acid is poured onto it, M. explodes. It is made by dissolving mercury in concentrated nitric acid and pouring this solution into ethanol.

M. is used as an ignition compound for caps, but has been largely replaced by lead azide, which has a stronger initial impact.

Mercury iodides: ***Mercury(I) iodide***: Hg_2I_2, yellow, tetragonal crystals which are insoluble in water; M_r 654.99, density 7.70, subl.p. 140 °C. Hg_2I_2 precipitates when potassium iodide is added to a solution of mercury(I) nitrate, and is purified by sublimation under reduced pressure. An Hg-Hg distance of 269 pm is observed in the linear Hg_2I_2 molecule.

Mercury(II) iodide, HgI_2, forms red, tetragonal crystals which are barely soluble in water; M_r 454.90, density 6.36. Red HgI_2 is used as a pigment under the name ***iodine red*** or ***iodine cinnabar***. At 127 °C, HgI_2 is reversibly converted to a rhombic yellow modification (see Thermochromism); density 6.094, m.p. 259 °C, b.p. 354 °C. HgI_2 forms tetrahedral tetraiodomercurate(II), $K_2[HgI_4]$ with aqueous potassium iodide solution (M_r 876.41). Potassium hydroxide added to an aqueous solution of $K_2[HgI_4]$ is a very sensitive reagent for ammonia (see Nessler's reagent). Heavy metal tetraiodomercurates(II), such as $Ag_2[HgI_4]$ and $Cu[HgI_4]$, display thermochromism: the bright yellow silver compound is reversibly converted to an orange modification at 35 °C, and the red copper compound is converted at 71 °C to a black form. Like HgI_2, these complexes can be used to measure temperatures. HgI_2 is made by rubbing mercury with iodine, or by precipitation of the iodide out of a mercury(II) nitrate solution.

Mercury nitrates: ***Mercury(I) nitrate***, $Hg_2(NO_3)_2 \cdot 2H_2O$, colorless, monoclinic crystals which are very soluble in water; in the air they weather with loss of water; M_r 561.22, density 4.79, m.p. 70 °C. The crystal lattice consists of linear $[H_2O\text{-}Hg\text{-}Hg\text{-}OH_2]^+$ cations (Hg-Hg distance 254 pm) and more distant nitrate anions.

Mercury(II) nitrate, $Hg(NO_3)_2 \cdot 1/2H_2O$, colorless, water-soluble crystals, M_r 333.61, density 4.39, m.p. 79 °C. The compound is highly dissociated in aqueous solution; it is obtained by dissolving mercury in nitric acid. It can be made in anhydrous form by reaction of mercury(II) oxide with nitrogen tetroxide. The compound reacts with ammonia to form the nitrate of Millon's base, $[Hg_2N]NO_3$. It is used to make other mercury(II) compounds and to nitrate organic molecules.

Mercury(II) oxide: HgO, rhombic crystals; there are two forms, due to differences in grain size and lattice defects, a yellow one and a red one; M_r 216.59, density 11.14. The lattice contains zig-zag chains with a Hg-O distance of 203 pm; the O-Hg-O portions of the chains are linear.

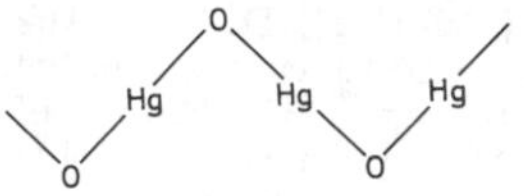

Structure of HgO.

If M. is heated above 400 °C, it is split into its components. It is only very slightly soluble in water, and gives a very weakly basic reaction. M. (red form) is found in nature as the very rare mineral montroydite. M. precipitates in the yellow form when alkali hydroxide solution is added to a mercury(II) salt solution; the red form is obtained by thermal decomposition of mercury(I) nitrate:

$$Hg_2(NO_3)_2 \xrightarrow{335°C} 2\,HgO + 2\,NO_2.$$

M. is used as a starting material for other mercury(II) compounds, and as a component of paints for ship hulls and porcelain. Recently red M. has become important in the production of small batteries with HgO anodes and zinc or cadmium cathodes. These are used for digital watches, hearing aids, pocket calculators, etc.

Mercury(II) rhodanide: same as mercury(II) thiocyanate.

Mercury sulfate: ***Mercury(I) sulfate***, Hg_2SO_4, forms colorless, monoclinic prisms which are only slightly soluble in water and dilute sulfuric acid; M_r 497.24, density 7.56. Hg_2SO_4 is formed by reaction of mercury(I) nitrate with dilute sulfuric acid or by reduction of mercury(II) sulfate with sulfur dioxide. It is used as a catalyst in the oxidation of organic compounds with fuming sulfuric acid, in Kjeldahl nitrogen determinations and in standard electrochemical cells.

Mercury(II) sulfate, $HgSO_4$, forms colorless crystalline leaflets; M_r 296.65, density 6.47. $HgSO_4$ is formed by reaction of mercury or mercury oxide with sulfuric acid. It is used in the chemical industry as a catalyst, for example, for acetaldehyde synthesis from ethyne.

Mercury sulfate electrode: a type II electrode. It is built similarly to the Calomel electrode (see). Calomel and potassium chloride are replaced by the insoluble mercury(I) sulfate and sulfuric acid, respectively. The M. is especially suitable for measurements in sulfuric acid solutions. If 0.5 molar sulfuric acid is used as the electrolyte, the electrode potential is +0.697 V.

Mercury(II) thiocyanate, ***mercury(II) rhodanide***: $Hg(SCN)_2$, colorless, crystalline needles; M_r 316.75. M. is only slightly soluble in water, but dissolves readily in alcohol. With alkali thiocyanate, it forms thiocyanatomercuriates(II), $M[Hg(SCN)_3]$ and $M_2[Hg(SCN)_4]$. When heated, M. puffs up dramatically, forming "pharaoh's snakes". It is obtained from aqueous solution by combining mercury(II) nitrate with alkali rhodanide. M. is used, for example, as a photographic intensifier, and in the determination of creatine.

Mercury process: see Chloralkali electrolysis.

Merinova®: see Synthetic fibers.

Merocyanins: see Polymethine pigments.

Merrifield synthesis: same as Solid phase peptide synthesis (see).

Mersalyl: see Diuretics.

Mescaline: 3,4,5-trimethoxyphenethylamine, a colorless oil; M_r 211.3, m.p. 35-36 °C, b.p.$_{12}$ 180 °C. It is soluble in water and ethanol, and practically insoluble in ether. In the air, M. forms a crystalline carbonate. It is a component of the peyote cactus *Lophophora (Anhalonium) williamsii*. The chopped, dried cactus is used as an intoxicant. M. is a hallucinogen and produces illusions, especially of color.

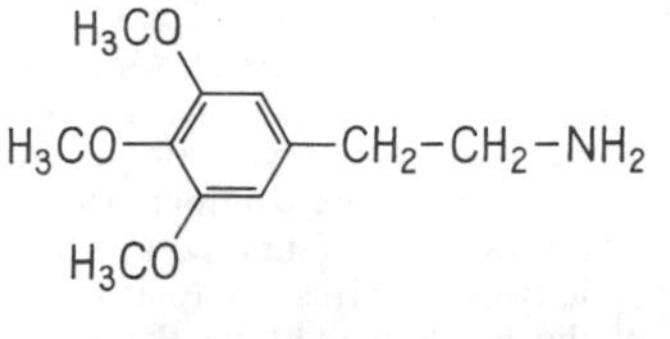

Mesitylaldehyde, ***2,4,6-trimethylbenzaldehyde***: a colorless, crystalline compound; m.p. 14 °C, b.p. 237 to 240 °C. Like all Aldehydes (see), M. undergoes addition and condensation reactions. It is synthesized by the Gattermann-Koch synthesis (see) or the Gattermann synthesis (see) from mesitylene. M. is used mainly for organic syntheses.

Mesitylene: see Trimethylbenzenes.

Mesityl oxide, ***4-methylpent-3-en-2-one***: $CH_3\text{-}CO\text{-}CH{=}C(CH_3)_2$, an aliphatic, unsaturated ketone. M. is a colorless liquid which smells like peppermint;

m.p. -52.9 °C, b.p. 129.7 °C, n_D^{20} 1.4440. M. is slightly soluble in water, and readily soluble in alcohol and ether. It is formed by dehydration during distillation of diacetone alcohol in the presence of iodine, or directly from acetone in the presence of dry hydrogen chloride. M. is used mainly as a solvent for paints, but also for organic syntheses.

meso-: see Stereoisomerism 1.2.1.

Mesobilin: see Bile pigments.

Mesobilinogen: see Bile pigments.

Mesobilirubin: see Bile pigments.

Mesophases: same as Liquid crystals (see).

Mesoporphyrin: see Porphyrins.

Mesoxalic acid, ***ketomalonic acid***: HOOC-CO-COOH, the simplest ketodicarboxylic acid. M. exists only in hydrated form as dihydroxymalonic acid, HOOC-C$(OH)_2$-COOH (exception to the Erlenmeyer rule (see)). Mesoxalic acid hydrate forms colorless, deliquescent crystals; m.p. 121 °C. It is soluble in water, alcohol and ether. When heated above its melting point, or when the aqueous solution is boiled, M. decomposes into glyoxalic acid and carbon dioxide. M. is present in sugar beets and alfalfa leaves, and is a cleavage product of uric acid.

Mesterolone: see Androgens.

Mestranol: see Estrogens.

Met: abb. for methionine.

meta-: see Arenes, see Nomenclature, sect. III D.

Metabisulfites: see Disulfites.

Metabolites: chemical compounds which take part in metabolism or are produced by it.

Metal: a chemical element with certain properties in the solid and liquid states. M. have shiny surfaces, are opaque, are good conductors of heat and electricity (electrical conductivity increases with decreasing temperature), have magnetic behavior, form crystal lattices with high coordination numbers, are ductile and are able to form Alloys (see). Aside from mercury, all M. are solids at room temperature, and they can be plastically deformed by rolling, hammering, pressing, drawing, etc. Their immense technological importance is based on their ductility, conductivity and capacity to form alloys. In the gas state, M. are monoatomic. Their properties can be explained by the fact that in the metallic crystalline lattice, the outer electrons are only loosely bound, and therefore are readily displaced. More about metallic bonding is given in the entry on Chemical bonds (see). One result of the ease with which outer electrons are released is the good electrical conductivity of metals; at temperatures near absolute zero, the conductivity of some M. suddenly becomes infinite (superconductivity). The shine and opacity of the M. is due to absorption or scattering of the incident light by the electrons. The plastic deformability is due to the fact that the positive ions in the lattice are densely packed, and are held together by the mobile electrons ("electron gas"). The densely occupied lattice planes are able to slide past each other fairly easily. In addition, one ion can be replaced by a different species of ion (see Mixed crystal). The M. have the tendency to give up their outer electrons when they form chemical bonds with nonmetals. This tendency is especially strong in the alkali metals with 1 valence electron, and in the alkaline earth group, with 2 valence electrons. The loss of electrons leads to formation of positive metal ions (cations) which arrange themselves in an ionic lattice with the negative nonmetal ions (anions); this is the pattern in metal halides, oxides and sulfides. The oxides of the M. are generally basic compounds, and form hydroxides with water; a few are amphoteric, such as lead(II) oxide, PbO. Only a few are acid anhydrides, such as manganese(VII) oxide, Mn_2O_7.

Of the currently known 109 chemical elements, three fourths are M.: the elements of groups Ia and IIa, all of the transition elements, the lanthanoids and actinoids. With the exception of hydrogen, M. can be defined as those elements in which the number of electrons in the outer shell is not greater than the principal quantum number of that shell. (Even hydrogen can be a metal at extremely high pressure.) There is a diagonal in the periodic system, running from beryllium to polonium, which forms the boundary between M. and Nonmetals (see). However, a sharp separation between M. and nonmetals cannot be made, because there are some elements, such as germanium and antimony, which have some metallic and some nonmetallic properties, and are therefore called ***semimetals***.

The M. can be classified on various bases. From their positions in the periodic system, they can be divided into main-group and transition group metals. The main-group metals can be subdivided into A and B M.; the A M. are "true M.", while the B. group elements may have complicated structures and differ in some of their properties from the "true M.". In the transition metals, the outer d bands are not occupied by the maximum of 10 electrons per atom. M. can also be classified on the basis of density into Light M. (see) (D. < 5) and Heavy M. (see) (D. > 5). The lightest M. is lithium (D. 0.534 g cm^{-3}), and the heaviest is osmium (D. 22.61 g cm^{-3}). On the basis of their tendency to oxidize, M. are divided into Noble M. (see) and reactive M. There are several groups of M. with very similar properties, including the Alkali M. (see), the Alkaline earth M. (see), the Platinum M. (see) and the Rare-earth M. (see). In industry, a distinction is made between iron and its alloys and nonferrous M., which include all other M. The most important M. are aluminum, beryllium, cadmium, chromium, cobalt, copper, gold, iron, lead, lithium, magnesium, molybdenum, mercury, nickel, platinum, plutonium, potassium, scandium, silver, sodium, tantalum, tin, titanium, tungsten, uranium, vanadium, yttrium, zinc and the lanthanoids.

Only the noble metals, copper and meteoritic iron are found in the elemental state in nature. Most of the M. are found in ores of widely varying compositions, in the form of their sulfides, oxides, carbonates, sulfates, chlorides, bromides, fluorides and silicates. Of these, the compounds with oxygen and sulfur are the most important.

Metal analysis: analysis of metals and alloys; all methods of analysis can be used in M.

Metal-aromatic complexes: see Organoelement compounds.

Metalaxyl: see Acylalanine fungicides.

Metal baths: see Heating baths.

Metal carbonyls: carbon monoxide complexes with transition metals from groups VIb to VIIIb of the periodic table or with vanadium; the metals are in the zero oxidation state. There are mononuclear and polynuclear M. Mononuclear M. formed according to the 18-electron rule include octahedral chromium, molybdenum and tungsten compounds of the type $M(CO)_6$; iron, trigonal bipyramidal ruthenium and osmium complexes of the type $M(CO)_5$, and tetrahedral $Ni(CO)_4$. Unlike the above diamagnetic M., the octahedral $V(CO)_6$ is a paramagnetic complex with an unpaired electron. The simplest polynuclear M. (Fig.) are dicarbonyls formed by cobalt, manganese, technetium and rhenium: $Co_2(CO)_4$ and $M_2(CO)_{10}$, where M is Mn, Tc or Re.

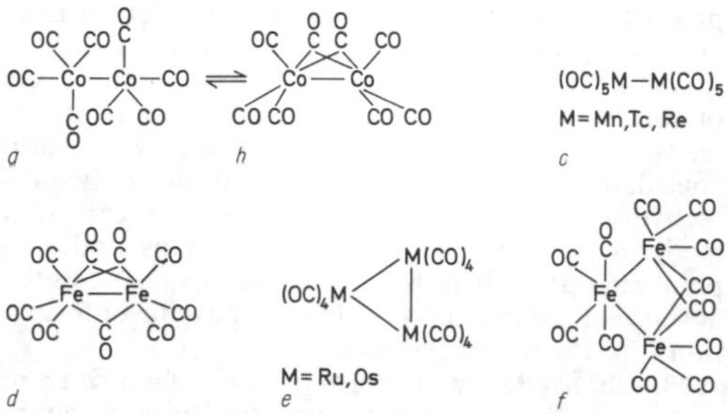

Structures of some important metal carbonyls.

As shown in the figure (a, b), bonding isomers have been demonstrated for dicobalt octacarbonyl in solution, but for the above dimetal carbonyls, $M_2(CO)_{10}$, (c), CO bridges can be excluded. Iron forms a dinuclear carbonyl, $Fe_2(CO)_9$, in which there are three bridge CO groups (d). The ruthenium and osmium trinuclear carbonyls of the $M_3(CO)_{12}$ type have trigonal structures with all the CO ligands in terminal positions (e), while $Fe_3(CO)_{12}$ has less symmetric structure (f). Tetranuclear carbonyls, $M_4(CO)_{12}$, are known for M = Co, Rh and Ir, and are the most stable binary M. for the latter two metals.

M. can undergo many reactions. Reductive base reactions yield ***metal carbonyl hydrides***, for example, $Fe(CO)_5 + 2\,NaOH \rightarrow Na[HFe(CO)_4] + H^+ \rightarrow H_2Fe(CO)_4 + Na^+$.

Reduction of M. with alkali metals yields alkali carbonylmetallates, for example, $Mn_2(CO)_{10} + 2\,Li \rightarrow 2\,Li[Mn(CO)_5]$. Metal carbonyl hydrides are weak acids. The pK of manganese carbonyl hydride, $HMn(CO)_5$, for example, is about 7; and iron carbonyl hydride has $pK_1 \approx 4$, $pK_2 \approx 13$. Polynuclear M. can be oxidized by halogens to ***metal carbonyl halides***, for example, $Mn_2(CO)_{10} + Br \rightarrow 2\,Mn(CO)_5Br$. Metal carbonyl halides can also be synthesized by direct reaction of metal halides with carbon monoxide, e.g. $2\,PtCl_2 + 2\,CO \rightarrow [Pt(CO)Cl_2]_2$. The bonding in M. is discussed under Coordination chemistry (see).

M. are made in some cases by direct reaction of the metal with carbon monoxide, e.g. $Ni + 4\,CO \rightarrow Ni(CO)_4$, at high temperatures. They are more commonly synthesized by reductive carbonylation, e.g. $OsO_4 + 9\,CO \rightarrow Os(CO)_5 + 4\,CO_2$; $WCl_6 + 3\,Fe(CO)_5 \rightarrow W(CO)_6 + 3\,FeCl_2 + 9\,CO$; $2\,CoCO_3 + 2\,H_2 + 8\,CO \rightarrow Co_2(CO)_8 + 2\,CO_2 + 2\,H_2O$. Polynuclear carbonyls are formed by photolysis or thermolysis reactions from simpler binary M., e.g.

$$2\,Fe(CO)_5 \xrightarrow[70°C]{h\nu} Fe_2(CO)_9 + CO;$$

$$2\,Co_2(CO)_8 \longrightarrow Co_4(CO)_{12} + 4\,CO$$

Historical. The first M. discovered was $Ni(CO)_4$, reported in 1888 by Mond and Langer. W. Hieber made outstanding contributions to the development of M. chemistry.

Metal ceramics: see Powder metallurgy.

Metal cleansers: see Household cleansers.

Metal corrosion, chemical: the corrosion of metals in a medium which conducts electricity poorly or not at all (e.g. hot, dry gases, non-conducting liquids), so that oxidation and reduction occur in one process. However, corrosion caused by liquid metals is also considered a form of M.

High temperature corrosion: If oxidizing gases, e.g. oxygen, sulfur, halogens, hydrogen sulfide, ammonia and/or water vapor, are present, Scale (see) forms on the metal surface. Scale is solid, and often non-porous, so that gas does not penetrate it. Depending on the medium, the scale can consist of oxides, sulfides, nitrides and/or chlorides. If the reaction continues in spite of the scale layer, it is because at least one of the reactants is diffusing through the crystal lattice of the scale. The scale produced by oxidation of pure metals which change valence by only one unit is uniform. If an element can exist in several oxidation states, the scale consists of several layers, e.g. $FeO/Fe_3O_4/Fe_2O_3$. In reactions with H_2O, NH_3 and H_2S, hydrogen is formed in addition to the oxides, sulfides and nitrides, and it causes side effects. Steam corrosion of steels occurs in boilers, while the reaction with hydrogen sulfide is a problem in hydrogenation of sulfur-containing petroleum. In high temperature corrosion, gaseous products can be formed, e.g. $Fe(CO)_5$ in the reaction of unalloyed steels with carbon monoxide under high pressure. In high-pressure reactors for synthesis of ammonia and methanol, and for hydrogenation of petroleum, the hydrogen reacts with the cementite in the steel (see Hydrogen brittleness).

When liquid metals react with other metals, alloy formation leads to corrosion. For example, the construction materials in sodium-cooled breeder reactors are uniformly corroded in this way. Austenitic Cr-Ni steels suffer an intercrystalline attack by liquid copper, zinc, lead or aluminum (*liquid metal brittleness, solder brittleness*). Electrically non-conducting liquids can also produce corrosion, as demonstrated by the selective corrosion of unalloyed steels by iron pentacarbonyl and the explosive reaction of aluminum with phenol above 60 °C.

Metal corrosion, electrochemical: the Corrosion (see) of a metal in the presence of an electrolyte, when electric charge carriers are involved. The total reaction consists of two simultaneous partial reactions, one anodic, and producing electrons, and one cathodic, which consumes electrons. If hydrogen is generated in the cathodic reaction, the E. is of the *hydrogen corrosion type*. The hydrogen, which is formed as atoms, e.g. during pickling in acids, can diffuse into the crystal lattice of the unalloyed steel and form H_2 at lattice defects. The bubbles of H_2 cause the steel to become brittle (see Pickling brittleness, Hydrogen brittleness). In the *oxygen corrosion type* of reaction, the cathodic reaction consists of reduction of oxygen to hydroxyl ions or water.

An electrolytic *corrosion element* is a heterogeneous mixed electrode at which electrochemical corrosion occurs. There are two electrodes which are connected by metal and electrically. The anode of a corrosion element is the surface region of a metal from which metal ions are released into the electrolyte by the corrosion. The electron-consuming reaction occurs at the cathodic surface area. The anode and cathode of a corrosion element can be formed in several ways:

1) By different metals (contact corrosion elements, e.g. Cu/Fe, with Fe as the anode);

2) By heterogeneities in a metal or on its surface; different parts of the structure (e.g. ferrite and cementite in unalloyed steel; ferrite becomes the anode); deformations and mechanical tensions (grain boundaries, displacements and stress centers act as anodes); coating layers and deposits (a layer of scale on unalloyed steel acts as cathode, while the steel below it becomes the anode; local corrosion can also occur when rust from elsewhere is deposited on a surface);

3) Heterogeneities in electrolytes. Concentration elements are formed in electrolyte-filled cracks in the metal itself or between parts of a structure. When the concentration of oxygen in the electrolyte varies, Aeration elements (see) form in the crack; the metal surface regions closer to lower oxygen concentrations act as anodes (crack corrosion).

4) Differences in the physical conditions. Differences in temperature on a metal surface generate thermogalvanic elements, in which the warmer region acts as anode. Differences in rates of transport of electrolyte cause concentration elements to form.

In *macrocorrosion elements*, the size of the anode and cathode surfaces are visible to the unaided eye (e.g. in contact corrosion elements, aeration elements, and hole-producing corrosion). *Microcorrosion elements* are generated by inhomogeneities in the structure of the material (alloy structure). Examples are selective corrosion of brass by Zinc loss (see), Graphitization of gray iron (see), selective solution of ferrites and *intercrystalline corrosion* of rust- and acid-resistant chromium-nickel steels in which the corrosion occurs primarily on the grain boundaries of the material.

When a corrosion element forms, there is no external current flow; the exchange current densities of the anodic and cathodic reactions are equal. This leads to a *mixed potential*, the *rest* or *corrosion potential*. The exchange current density of the anodic reaction is called the *dissolution* or *corrosion current density*, and, according to Faraday's law, it is directly proportional to the mass of metal which has gone into solution. If an equal external current is sent through the corrosion element, it becomes an *electrolytic corrosion element*. If the current is a wandering current, *wandering current corrosion* occurs at the sites where it leaves the metal.

Corrosion tendency and electromotive force (EMF) series. If the thermodynamic potential of the metal undergoes a negative change when the metal is ionized, the corrosion reaction occurs spontaneously. The tendency of metals to be corroded corresponds approximately to their positions in the electromotive force series. In practice, the thermodynamic potential and the position in the emf series is not a measure of the corrosion reaction which actually occurs. The reactions at the electrodes of the corrosion elements are subject to inhibition, which can be overcome only when the resting potential is overcome (see Polarization). In de-aerated water and non-oxidizing acids, the penetration reaction at the cathode, that is, discharge of the hydrogen ion, is inhibited. For this reason, lead does not dissolve in dilute sulfuric acid to produce hydrogen in the absence of depolarizing agents (O_2, H_2O_2, MnO_2, HNO_3), in spite of its position in the emf series. The formation of coating layers of corrosion products (e.g. AlO(OH) on aluminum) leads to an additional ohmic resistance, which also considerably reduces the rate of corrosion. If the protective layer is gradually removed by the corrosion medium, the course of the corrosion corresponds to a parabola rather than the straight line expected in the absence of induction time and protective layer formation (see Corrosion testing).

Metaldehyde: in the broad sense, a polymer of acetaldehyde; in the strict sense, the cyclic tetramer of acetaldehyde. m.p. 246 °C. M. forms white, prismatic crystals and is insoluble in water, but soluble in benzene and hot chloroform. It is formed from acetaldehyde in the presence of sulfuric acid at low temperatures. At temperatures above 100 °C it depolymerizes. M. is used as a compressed fuel in place of alcohol and as a slug and snail poison.

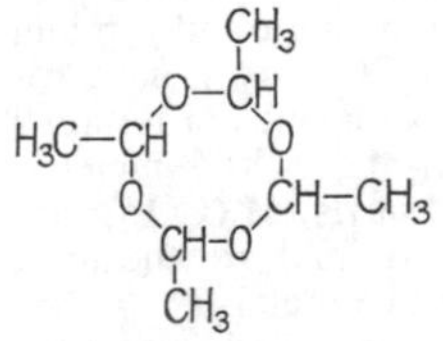

Metal glue: see Glue.

Metal-halogen exchange: see Organolithium compounds.

Metal indicators: see Indicators.

Metal lattice: see Lattice type.

Metallic bond: see Chemical bond.

Metallization: the application of a metallic coating to a metallic or nonmetallic object. M. is done for decoration, to create an electrically conducting surface, to make mirrors, and, in the case of metal objects, to create a metallic Protective layer (see), to prevent corrosion, to restore the dimensions of worn parts and to make a mechanically more resistant surface. Cadmium is used to protect unalloyed steels from corrosion; chromium as a corrosion-preventing layer, to prevent wear and for decoration; gold for decoration and corrosion protection; copper and nickel to make intermediate layers on metallic objects so that a chromium surface layer can be applied; silver to make glass mirrors and for decoration; and zinc and tin for corrosion-protection layers on unalloyed steels. See Lead plating, Cadmium plating, Chrome plating, Gilding, Copper plating, Nickel plating, Silver plating, Zinc plating and Tin plating. The metal can be applied by Plating (see), spraying (see Metal spraying), dipping, vapor deposition, chemical plating and Electroplating (see).

Metallization of plastics: the application of a metallic coating to plastic parts, to achieve a decorative effect or special physical properties, such as electrical conductivity. Metal layers can be applied to plastics by vapor deposition of the metal, especially aluminum, in high vacuum, by chemical silver plating with reducing silver salt solutions, by decomposition of metal carbonyls (especially for nickel layers), by electroplating (after application of a conductive layer), or by a metal spraying process. Vacuum metallization gives very thin layers (about 1 μm thick), while galvanizing produces a multilayered structure up to 30 μm thick. Galvanizing is used mainly with alkylbenzene sulfonate polymers and polypropylene.

In ***vapor deposition*** methods, the surface to be metallized is first coated with a base lacquer. The metal is in an electrically heated dish and is evaporated into the vacuum chamber at $1.33 \cdot 10^{-2}$ Pa; the process takes 10 to 20 minutes.

Continuous vapor deposition of metal onto films which are rolled into and out of the vacuum chamber at rates up to 500 m/min is used to make metallized foils. These are used, for example, to make decorative threads (Lurex threads) for use in textiles and infrared-reflecting sheets for window coverings, swimming pool covers and emergency rescue blankets.

Metallocene: see Nomenclature, sect. II.G 2; see Organoelement compounds.

Metalloids: see Nonmetals.

Metalloporphyrins: see Porphyrins, see Coenzymes.

Metalloproteins: see Proteins.

Metallurgy: the science and technology of producing metals from ores and secondary raw materials, and refining and processing them. Various techniques of producing metals are Pyrometallurgy (see), Hydrometallurgy (see), Electrometallurgy (see), Powder metallurgy (see), Vacuum metallurgy (see) and Oxygen metallurgy (see).

Metal soaps: the salts of all metals, except sodium and potassium, with higher fatty acids, resin and naphthenic acids. The corresponding sodium and potassium compounds are called Soaps (see). Depending on their composition, M. are amorphous, salve-like or liquid substances; in very pure state, they may be crystalline. Depending on the metal and the organic acid, they are colorless to dark red-brown. M. are much less soluble in water and organic solvents than soaps are. Water is partially bound by many M., so that they can swell and form gels, and then finally dissolve. The solubility of M. in mineral, animal and vegetable fats and oils is important.

M. are made industrially from alkali or ammonium soaps by precipitation with the corresponding metal salts. M. are often made by direct salt formation between the acid and the metal hydroxide, at a warm temperature, or by saponification of fats with the corresponding metal oxide or hydroxide. The palmitates, stearates, oleates, linoleates, ricinoleates, resinates, and naphthenates are of industrial importance.

Metal spraying, ***spray metallization***: a method for production of metallic Protective layers (see) on metallic and nonmetallic materials. The metal to be sprayed, e.g. zinc, aluminum, titanium, tantalum or nickel and chromium alloys, is put into a spray gun in wire or powder form. Inside the gun, it is melted by a gas-oxygen flame, an electric arc or some other heat source (e.g. a plasma beam) and sprayed onto the clean, roughened surface of the material, usually by means of pressurized air. M. is used to provide corrosion protection for metals, to restore the dimensions of parts which are too worn, to create an electrically conducting surface on non-conducting materials, to prepare molds for casting, and to preserve non-metallic materials.

Metamitron: see Herbicides.

Metanil yellow, ***victoria yellow O***: C_6H_5-NH-C_6H_4-N=N-$C_6H_5(SO_3Na)$, the sodium salt of 4'-anilidoazobenzene 3-sulfonic acid. M. is obtained from diazotized metanilic acid and diphenylamine. It is a yellow dye which is fast to light, but sensitive to acids.

Metaperiodate: see Periodic acid.

Metaphosphoric acids: cyclopolyphosphoric acids with the general formula $[HPO_3]_n$. The $[PO_4]$ tetrahedra are linked by oxygen bridges into cyclic molecules. A typical example of the M. is ***trimetaphosphoric acid, cyclotriphosphoric acid***, $[HPO_3]_3$ (Fig.).

Trimetaphosphoric acid.

Like other M., it cannot be purified, because it undergoes rapid isomerization. On the other hand, the salts of the M., the *metaphosphates* (see Phosphates) are often well characterized.

Metastable ions: see Mass spectroscopy.

Metastable state: see Equilibrium, see State.

Metformin: see Antidiabetics.

Methabenzthiazuron: see Urea herbicides.

Methacol: a slurry consisting of two-thirds pulverized hard coal and one third methanol. M. can be used instead of fuel oil in industrial plants.

Methacrylate: see Methacrylic acid.

Methacrylic acid, ***2-methylpropenoic acid***: CH_2 = $C(CH_3)$-COOH, a colorless liquid with an unpleasant odor; m.p. 16 °C, b.p. 161 °C, n_D^{20} 1.4314. M. is soluble in water, alcohol and ether. It polymerizes when heated or exposed to light, hydrochloric acid or perioxides. M. can be stored for a long time if stabilizers, such as hydroquinone, are added. Its salts and esters are called *methacrylates*. M. is present in camomile oil as an ester. It is synthesized by hydrolysis of acetone cyanohydrin and subsequent splitting out of water. M. is used to make polymethacrylates and various copolymers.

Methacrylic acid methyl ester: see Methyl methacrylate.

Methadone: a basic, substituted diphenylmethane derivative. M. is a strong analgesic; (-)-M. (*R*-configuration) has a considerably stronger effect than (+)-M. Commercially available M. is either the racemate or the (-)-enantiomer. M. satisfies the physical dependence of heroin addicts without causing intoxication and is used in maintenance programs for addicts.

β-Methallyl chloride, ***3-chloro-2-methylprop-1-ene***: $Cl\text{-}CH_2\text{-}C(CH_3){=}CH_2$, a colorless, poisonous and lacrimatory liquid; m.p. < -80 °C, b.p. 72.2 °C, n_D^{20} 1.4280. M. is barely soluble in water, but dissolves readily in most organic solvents. Its reactions are the same as those of Allyl chloride (see), that is, the chlorine atom can be easily replaced by other nucleophilic reagents. M. is made by chlorination of isobutene; the reaction mechanism is an addition-elimination:

$$CH_3\text{–}C(CH_3){=}CH_2 \xrightarrow[-Cl^-]{+Cl_2} [\overset{+}{C}H_3\text{–}C(CH_3)\text{–}CH_2\text{–}Cl]$$

$$\xrightarrow[-H^+]{} CH_2{=}C\text{–}(CH_3)\text{–}CH_2\text{–}Cl$$

M. is used in the synthesis of mixed polymers, pesticides and for organic syntheses.

Metham: sodium *N*-methyldithiocarbamate, a nematocide with fungicidal and herbicidal side effects. It is applied as an aqueous solution synthesized from methylamine, carbon disulfide and sodium hydroxide: $H_3CNH_2 + CS_2 + NaOH \rightarrow H_3C\text{-}NH\text{-}CS\text{-}SNa + H_2O$.

Methamidophos: see Organophosphate insecticide.

Methamphetamine: $C_6H_5\text{-}CH_2\text{-}CH(CH_3)\text{-}NH\text{-}CH_3$, the best known of the phenylethylamine stimulants (see Analeptics). It is addictive.

Methane: CH_4, the first of the alkane series, a colorless, odorless and non-toxic gas; m.p. -182.6 °C, b.p. -161.7 °C, D_{crit} 0.162, T_{crit} 190.7 K, p_{crit} 4.63 Mpa. M. is very slightly soluble in water, and is more soluble in organic solvents. It burns in air or oxygen with a pale flame. Mixtures with air or oxygen are easily ignited and explosive. Mixtures of M. with a two-fold volume of oxygen or a ten-fold volume of air are especially dangerous ("fire damp"). When M. is incompletely burned, it forms gas soot: $CH_4 + O_2 \rightarrow C + 2\,H_2O$. In direct sunlight, M. undergoes an explosive reaction with chlorine which precipitates carbon: $CH_4 + 2\,Cl_2 \rightarrow C + 4\,HCl$. In diffuse daylight, the H atoms of M. are replaced by Cl atoms in a stepwise reaction. This is very important for the industrial production of methylchloride, dichloromethane, chloroform and carbon tetrachloride. Bromine reacts as chlorine does, but much more slowly; iodine does not react. M. reacts with steam at 800 to 900 °C in the presence of nickel catalysts, forming synthesis gas: $CH_4 + H_2O \rightarrow CO + 3\,H_3$ (steam reforming). Comparable reactions with oxygen or steam occur at 1000 to 1200 °C with nickel catalysts or at 1400 to 1600 °C without catalysis: $CH_4 + 1/2O_2 \rightarrow CO + 2\,H_2$.

Hydrogen cyanide is obtained by partial oxidation of M. in the presence of ammonia (Andrussov process):

$$2\,CH_4 + 2\,NH_3 + 3\,O_2 \xrightarrow[1000°C]{(Pt/Rh)} 2\,HCN + 6\,H_2O.$$

Acetylene is produced industrially by high-temperature pyrolysis of M.; another industrially significant reaction is nitration to nitromethane.

Occurrence and production. M. is the main component of natural gas, and it is also found as *mine gas* in inclusions in anthracite seams. It is present with carbon dioxide in *marsh gas*, which is formed in the mud of lakes and swamps by anaerobic decomposition of cellulose. It is formed by similar processes in the rumens of ruminants or in the decomposition of feces (see Biogas). M. is also formed during coking or carbonization of coal, peat or wood, and in cracking processes in petroleum processing. For industrial use, it is separated in large amounts from natural gas. It is also produced by cracking or carbonization processes, or synthesized from carbon monoxide or carbon dioxide and hydrogen:

$$CO + 3\,H_2 \xrightarrow[200\text{ to }300°C]{(Ni)} CH_4 + H_2O;$$

$$CO_2 + 4\,H_2 \xrightarrow[400°C]{(Ni)} CH_4 + 2\,H_2O.$$

M. is also obtained as a fuel gas by decomposition of sludge from water-treatment plants, manure, straw or crop residues.

In the laboratory, M. is made by decomposition of aluminum carbide with water, by heating sodium acetate with sodium hydroxide, or by hydrolysis of methylmagnesium halides.

Uses of methane.

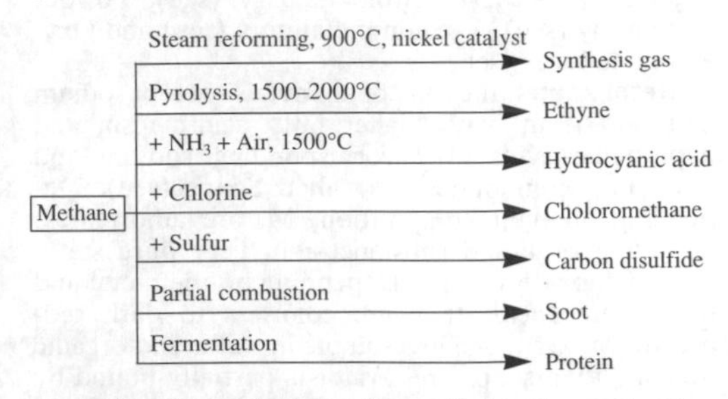

Applications. M. is used as a fuel and as a starting material for a number of important syntheses (Fig.).

Methanal: same as Formaldehyde (see).

Methanamide: same as Formamide (see).

Methanesulfonic acid: $CH_3\text{-}CO_3H$, a colorless liquid; m.p. 20 °C, b.p. 167 °C at 1.33 kPa. It dissolves in water with the release of heat. Boiling M. with thionylchloride produces either methanesulfonyl chloride (mesylchloride) or methanesulfonic anhydride, depending on the reaction conditions. Either derivative can be used for preparative syntheses. M.

is prepared from methyl iodide and potassium sulfite, or by reaction of dimethylsulfate with sodium sulfite.

Methanethiol, ***methylmercaptan***: CH_3-SH, a bad-smelling, poisonous, combustible gas; m.p. - 123 °C, b.p. 6.2 °C. M. is insoluble in water, but dissolves easily in ethanol and ether. It is formed by bacterial decomposition of proteins. When very dilute, M. smells like cooked cabbage, and has also been detected in the crusts of baked goods. It is a component of radish oil and onions. It is available commercially as a liquefied gas, and is an intermediate for organic syntheses, e.g. for methionine or in general for pesticides.

1,6-Methano[10]annulene, ***bicyclo[4,4,1]undeca-1,3,5,7,9-pentaene***: an aromatic hydrocarbon which, in contrast to [10]annulene, has no steric strain in a planar arrangement of the π-system. 1,6-M. is diatropic and shows a diamagnetic ring current in the magnetic field of a nuclear resonance spectroscope. It can be electrophilically substituted in position 2.

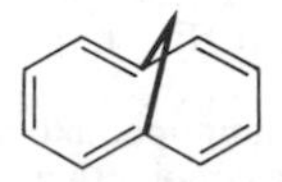

1,6-Methano [10] annulene.

Methanol, ***methyl alcohol***, obsolete names, ***wood alcohol, carbinol***: CH_3-OH, the simplest alcohol, a colorless fluid with an alcoholic smell and a burning taste; m.p. -97.8 °C, b.p. 64.7 °C, n_D^{20} 1.3287. M. is very reactive; it ignites easily and burns with a pale blue, transparent flame to carbon dioxide and water. M. is easily oxidized and esterified (see Alcohols). It is miscible with water in any proportions, and also dissolves in most organic solvents, such as ether, benzene and light petroleum. Magnesium is used to remove small amounts of water; it reacts with M. to give magnesium methanolate which consumes water as it is hydrolysed.

Methanol is highly toxic. Some is excreted unchanged by the human organism, and some is oxidized to formaldehyde and formic acid. The toxic effect is due to the formaldehyde, which precipitates or coagulates proteins, and thus causes severe damage. The retina is the first tissue to be affected, so that visual damage results and can lead to complete blindness. The first symptoms of a M. poisoning are headache, dizziness, eye pains and blurred vision. The lethal dose is about 25 g. Sodium lactate is given as an antidote.

A reaction for the qualitative analysis of M. is the formation of trimethylborate, which burns with a green flame. In the test, borax, concentrated sulfuric acid and M. are heated and the mixture is ignited.

Occurrence. M. is found widely in the plant kingdom, but usually only in small amounts, e.g. in the form of pure esters in essential oils or as methoxy groups in many substances, such as lignin and various alkaloids.

M. is produced commercially by the reaction of carbon monoxide and hydrogen (synthesis gas): CO + 2 $H_2 \rightleftharpoons CH_3OH$, ΔH = -92 kJ mol^{-1}. The pressure varies; high-pressure processes use 20 to 30 MPa, medium-pressure processes, 10 to 20 MPa, and low-pressure processes, 5 to 10 MPa. In older installations, pressures of 25 to 30 MPa and temperatures of 320° to 400 °C are used with a ZnO/Cr_2O_3 catalyst. Since 1970, lower pressures and temperatures have been adopted (5 to 10 MPa, 230 to 280 °C); this has been made possible by highly active copper catalysts (CuO/Cr_2O_3), which require a synthesis gas containing essentially no sulfur.

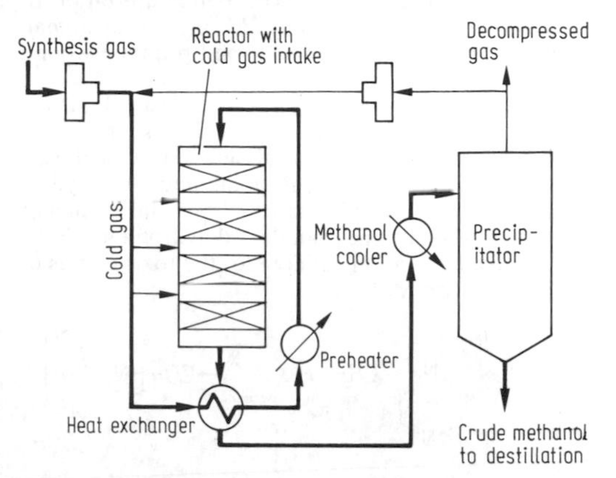

Diagram of a methanol plant.

The catalyst must contain no iron, cobalt or nickel, because these cause formation of methane. The formation of higher alcohols can be avoided by using a catalyst which contains no alkali metal or alkaline earth oxides. At present, the trend is towards medium-pressure processes, because these can use synthesis gas which contains slightly more sulfur.

Applications. M. is used for a wide variety of industrial syntheses: acetic acid, dimethyl esters, mono-, di- and trimethylamines, formaldehyde, methyl formate, monomethyl and dimethyl formamide. It is also used as a solvent, especially for resins, polyvinyl acetate, pigments, colophony, cellulose esters and many inorganic salts. M. is added to azeotropic distillations and is used as an extraction agent in petroleum chemistry, to purify natural gas, as a diluent in perfume and paints, as a methylating or denaturing agent, as an antifreeze, as a coolent or as a fuel. Experiments on the conversion of M. into gasoline hydrocarbons (see Mobil-Oil process) are in progress. M. is used as a carbon source for fermentative protein production (single-cell protein).

Methanoic acid: same as Formic acid (see).

Methaqualone: see Hypnotics.

Methemoglobin: see Hemoglobin.

Methenamine: same as Hexamethylenetetramine.

Methfuroxam: see Carboxamide fungicides.

Methimazole: see Thyreostatics.

Methionine, abb. ***Met***: α-amino-γ-methylmercaptobutyric acid, CH_3-S-$CH_2CH_2CH(NH_2)$-COOH, an essential Amino acid (see). Its concentration is low in many plant proteins, which limits their biological value. A defiency of M. leads to metabolic disorders which are expressed by slow growth, liver disorders and damage to the growth of skin and hair. M. is one of the few amino acids of which the D- and L-forms can both be utilized. For this reason, large-scale production of DL-M. is important. The most important synthetic process is the Strecker synthesis from β-methylmercaptopropionaldehyde, which is formed from acrolein and methylmercaptan. The final yields of the Strecker synthesis are 80 to 90% (based on acrolein). DL-M. is used in large amounts to increase the nutritional value of plant proteins in foods and feeds. Addition of as little as 0.1 to 1% to animal feeds gives more rapid growth, better utilization of feed and better meat quality. M. is used in human medicine to treat liver diseases and promote wound healing.

Methionyl-lysyl-bradykinin: see Plasma kinins.

Methomyl: a Carbamate insecticide (see).

Methotrexate: an antimetabolite of folic acid used chiefly as a cytostatic. M. is an inhibitor of dihydrofolic acid reductase, because it has a much higher affinity to the enzyme than does dihydrofolate. Folic acid is available as an antidote for the toxic effects of treatment with M.

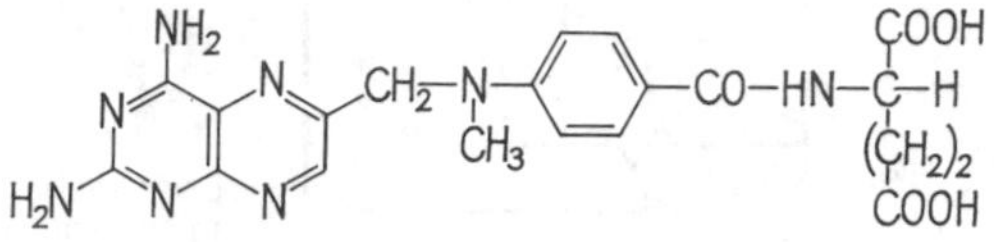

Methoxsalene, ***xanthotoxin, 8-methoxypsoralene***: a linear, fused furocoumarin derivative; the parent compound is *psoralene* (see Furocoumarins). M. is used for radiation protection and in photochemotherapy of psoriasis. When activated by long-wave UV light, M. reacts with the pyrimidine bases of DNA and in this way inhibits the excessive epidermal cell division which characterizes psoriasis.

Methoxyaniline: same as Anisidine (see).

4-Methoxybenzaldehyde: same as Anisaldehyde.

Methoxybenzene: same as Anisol.

4-Methoxybenzoic acid: same as Anisic acid (see).

Methoxychlor: 1,1,1-trichloro-2,2-bis(4-methoxyphenyl)ethane, an insecticidal structural analog of DDT; colorless crystalline powder, D. 1.41, m.p. 89 °C. M. is easily soluble in most organic solvents and practically insoluble in water. It is stable to light and air. In the presence of base it splits off HCl, but more slowly than DDT. The synthesis is analogous to that of DDT: condensation of methoxybenzene with chloral in the presence of sulfuric acid or boron trifluoride. M. is a significantly weaker insecticide than DDT. Due to its lower toxicity (oral LD_{50} = 5000 to 5800 mg/kg rat) and its failure to accumulate in fatty tissue, it is used as a contact and oral poison against biting insects, e.g. cherry fruit flies and rape pests; it is also used in veterinary practice against ectoparasites. The estimated fatal oral dose for an adult human being is 450 g.

Methoxymethane: same as Dimethyl ether (see).

2-Methoxynaphthalene, ***naphth-2-ol methyl ether***: white crystals with a floral, nerolin-accacia note; m.p. 73-74 °C, b.p. 274 °C. The substance is nearly insoluble in water and is also only slightly soluble in ether, benzene and carbon disulfide. It is made by the reaction of naphth-2-ol with methylating reagents, such as dimethylsulfate. 2-M. is stable in soaps, and is used mainly as a perfume in soaps.

2-Methoxyphenol: same as Guaiacol (see).

Methyl-: a term for the atomic group CH_3- in a molecule, the radical $CH_3\cdot$ or the cation CH_3^+.

Methyl acetate, ***acetic acid methyl ester***: CH_3-CO-OCH_3, a colorless liquid with a pleasant, ether-like odor; m.p. -98.1 °C, b.p. 57 °C, n_D^{20} 1.3595. M. is readily soluble in water and most organic solvents. It is synthesized by the reaction of acetic acid with methanol in the presence of a small amount of a strong mineral acid. It is used as a solvent for cellulose esters and ethers, resins, fats, oils and formaldehyde resins.

Methyl alcohol: same as Methanol (see).

Methylamines: the primary, secondary and tertiary methyl derivatives of ammonia. They are the simplest of the aliphatic amines. ***Monomethylamine (methylamine, aminomethane)***, CH_3-NH_3: m.p. -93.5 °C, b.p. -6.3 °C. ***Dimethylamine*** $(CH_3)_2NH$: m.p. -93 °C, b.p. 7.4 °C. ***Trimethylamine***, $(CH_3)_3N$: m.p. -117.2 °C, b.p. 2.9 °C, n_D 1.3631.

M. are combustible gases with an ammonia-like odor; they are readily soluble in water, alcohol, ether and glycols. They display the usual reactions of amines, and form water-soluble methyl ammonium salts with mineral acids. The reaction of nitrous acid with dimethylamine produces a yellow, poisonous nitrosoamine. Trimethylalkylammonium salts are formed from trimethylamine and alkyl halides.

> Methylamines burn the skin and mucous membranes. At high concentrations, contact with the mucous membranes of the eyes can produce temporary or permanent blindness. The inhalation of large amounts can lead to lung edema.

M. occur in plants and animals as metabolic products. Mono- and trimethylamine are present in bone oil and the distillation products of wood. Di- and trimethylamine are formed by decomposition of proteins in yeasts, cheese and fish. They are also present in sugar beet molasses.

M. are synthesized by reaction of methanol and ammonia in the presence of aluminum oxide catalysts at pressures of 1.5 to 3 MPa and 350 to 450 °C. The product mixture contains about 1 part monomethylamine, 2 parts dimethylamine and 7 parts trimethylamine, which are separated by fractional distillation. M. are used mainly in the chemical and pharmaceutical industries. They are used as intermediates for synthesis of dyes, drugs and pesticides. Monomethylamine is also used in the production of textile conditioners, sarcosine and dimethylurea. Dimethylamine is a starting material for dimethylformamide, dimethylacetamide, *N,N*-dimethylethanolamine, vulcanization accelerators and quaternary ammonium salts. It is also used as a softener in preparing leathers and as a solvent for synthetic fibers. Trimethylamine is used in the production of choline and choline salts, emulsifiers, disinfectants and quaternary ammonium salts.

N-Methylaminobenzene: same as *N*-Methylaniline (see).

N-Methylaniline, ***N-methylaminobenzene***: the *N*-monomethyl and *N,N*-dimethyl substituted anilines. ***N-Methylaniline (N-monomethylaniline, N-Methylaminobenzene)***, C_6H_5-NH-CH_3, is a secondary amine, and a colorless, poisonous liquid which is steam volatile; m.p. -57 °C, b.p. 196 °C, n_D^{20} 1.7714. It is slightly soluble in water, and dissolves readily in most organic solvents and in dlute mineral acids. *N*-M. is used as an anti-knock compound and as an intermediate in the synthesis of pigments. ***N,N-Dimethylaniline (N,N-dimethylaminobenzene)***, C_6H_5-N$(CH_3)_2$, is a yellow, poisonous liquid; m.p. 2.5 °C, b.p. 194 °C, n_D^{20} 1.5582. It is insoluble in water, but dissolves well in most organic solvents and in dilute mineral acids. *N,N*-Dimethylaniline readily undergoes electrophilic substitution in position 4 of the ring, and this is the basis for its use as a coupling component in the synthesis of pigments. *N*-M. is synthesized by reaction of aniline and dimethyl ether in the vapor phase on an aluminum oxide contact, or from aniline and methanol under pressure at 230 °C in the presence of hydrochloric or sulfuric acid. A crude separation of the *N*-M. is possible by fractional distillation, and further purification can be achieved by acylation of the *N*-monomethylaniline.

Methylation: a special case of alkylation, the introduction of methyl groups, CH_3-, into organic compounds. Some typical methylation reagents are dimethylsulfate, methanol, methyl halides and diazomethane. Diazomethane is especially suitable for M. of acidic compounds (phenols, carboxylic acids, enols); the methyl compound is formed under mild conditions with evolution of nitrogen.

Methylbenzene: same as Toluene.

Methylbenzyl bromide: same as Xylyl bromide (see).

Methylbenzoate: see Benzoates.

Methyl bromide, ***bromomethane***: CH_3-Br, a colorless, very poisonous gas which smells like ether; m.p. -93.6 °C, b.p. 3.6 °C. M. is slightly soluble in water, but readily soluble in alcohol and ether. It is made by reaction of hydrogen bromide with methanol, or of bromine and sulfur with methanol. It is used as a methylating reagent and as a pesticide.

3-Methylbutanal: same as Isovaleraldehyde (see).

2-Methylbutanoic acid: see Valeric acids.

3-Methylbutanoic acid: see Valeric acids.

2-Methylbutyric acid: see Valeric acids.

Methyl cellulose: a Cellulose ester (see).

Methyl chloride, ***chloromethane***: CH_3-Cl, the simplest aliphatic chlorinated hydrocarbon; it can be considered the methyl ester of hydrochloric acid. M. is a colorless, gas with an odor like ether; m.p. -97.1 °C, b.p. -24.2 °C. It burns with a green-edged flame. M. is slightly soluble in water, and more soluble in alcohol, ether, carbon tetrachloride and glacial acetic acid.

> Methyl chloride is a weak narcotic, and inhalation of larger amounts leads to vomiting, headache and paralysis, as well as causing damage to liver and kidneys. The danger of poisoning is greatly increased by the fact that methyl chloride causes little immediate irritation.

As in all Haloalkanes (see), the halogen atom can be readily replaced by other nucleophilic groups. The chlorine atom can be hydrolysed in boiling water, or more rapidly in alkalies, leading to formation of methanol from M. The most important industrial syntheses of M. are chlorination of methane at temperatures of 400 to 500 °C, which yields a mixture of chlorinated methane derivatives, and esterification of methanol with hydrogen chloride.

M. is used as a solvent, a coolant in small units, and a spray for pest management. It is also an important methylating reagent used in the synthesis of silicones, methyl cellulose, methyl amines, methyl thiols, methyl esters, dimethyl sulfoxide and methylated aromatics.

Methylcholanthrene: a fused aromatic hydrocarbon which forms bright yellow crystals; m.p. 188 °C, b.p. 280 °C at 10.7 kPa. It is insoluble in water, but is soluble in benzene and toluene. M. is found in coal tar, and is one of the most powerful carcinogens. It was first synthesized in 1933 by Wieland, by degradation of deoxycholanoic acid.

CH_3

Methylcinnamate: see Cinnamates.

Methylcyanide: same as Acetonitrile (see).

Methylcyclohexanones: the three isomeric ketones derived from cyclohexane. M. are colorless liquids which are slightly soluble in water, and readily soluble in most organic solvents. ***2-Methylcyclohexanone***: m.p. -13.9 °C, b.p. 165 °C, n_D^{25} 1.4483. ***3-Methylcyclohexanone***: m.p. -73.5 °C, b.p. 65 °C at 2 kPa, n_D^{25} 1.4456. ***4-Methylcyclohexanone***: m.p. -40.6 °C, b.p. 170 °C, n_D^{20} 1.4451. The M. are usually obtained in industry as a mixture of isomers synthesized by oxidation of technical methylcyclohexanol, and are used as a solvent in this form.

Methylcyclopentadienyl manganese tricarbonyl: $(CH_3C_5H_4)Mn(CO)_3$, a yellow liquid; M_r 218.08, density 1.39, b.p. 233 °C. M. is produced by reaction of biscyclopentadienylmanganese(II) (manganocene) with carbon monoxide, which forms $(C_5H_5)Mn(CO)_3$; this compound is then subjected to Friedel-Crafts alkylation and is used as an antiknocking material and as a combustion catalyst for heating oils.

5-Methylcytosine: colorless crystals; m.p. 270 °C (dec.). It is slightly soluble in water. 5-M. is present in nucleic acids at a concentration around 0.1%. The amount of it depends on the species and tissue; methylation of cytosine bases is believed to play a role in the regulation of transcription of DNA.

Methyldopa: S(-)-3-(3,4-dihydroxyphenyl)-2-methylalanine, a highly active Antihypertensive (see). Its effect is thought to be due mainly to its uptake into the central nervous system, where it is metabolized to α-methylnoradrenalin, and activates central α-receptors. In addition, methyldopa competes with dopa for peripheral α-receptors, and thus inhibits their transmission. This also reduces the blood pressure.

Methylene: 1) see Carbenes. 2) ***Methylene-*** is a term for the atomic group $-CH_2-$ in a molecule.

Methylene blue: the most important basic thiazine dye, chemically 2,7-bis(dimethylamino)phenthiazonium hydroxide, m.p. 190 °C. The chloride forms dark blue leaflets with a metallic sheen; they are soluble in water and alcohol, and emit a weak red-violet fluorescence in sunlight. The zinc chloride double salt forms brown crystalline needles or prisms with a coppery sheen. M. is synthesized industrially from 4-aminodimethylaniline, dimethylaniline and sodium thiosulfate by oxidation with chromic acid in the presence of zinc chloride. Because of its true color and its excellent fastness to washing and boiling, M. is used, in spite of its tendency to fade in the light, to dye cotton with a tannic acid mordant and silk. It is also used to enhance dull colors, to stain tuberculosis and leprosy bacteria and as a nuclear stain for microscopy. It is used as a vital stain which selectively stains the gray matter in the peripheral nervous system. M. can serve as a hydrogen acceptor for enzymatic dehydrogenations, forming colorless leukomethylene blue. This compound forms a reversible redox system with M. ***Methylene green (nitromethylene blue)*** is made by nitrating M.; it is used as a dye for cotton and also as a nuclear stain.

M. was first produced in 1876 by Caro (1834-1910) by oxidation of 4-aminodimethylaniline with iron chloride in the presence of hydrogen sulfide.

Oxidation + ½O_2 − H_2O
Reduction + 2H
Leuko compound
Methylene blue

Methylene bromide, ***dibromomethane***: CH_2Br_2, a colorless liquid; m.p. -52.5 °C, b.p. 97 °C, n_D^{20} 1.5420. M. is scarcely soluble in water, but dissolves readily in alcohol and ether. It is formed by the reaction of bromine with methylene chloride in the presence of aluminum. It is used as a solvent, fire extinguishing liquid and intermediate for organic syntheses.

cis-Methylenebutenedioic acid: see Itaconic acid.

Methylene chloride, ***dichloromethane***: CH_2Cl_2, a colorless, volatile liquid with a sweetish odor; m.p. -95.1 °C, b.p. 40 °C, n_D^{20} 1.4242. M. is scarcely soluble in water, but dissolves readily in most organic solvents. Under normal conditions, M. is very stable. It is obtained, in a mixture of chlorinated methane derivatives, by chlorination of methane or methyl chloride, and can be purified by fractional distillation. Because of its relatively low toxicity and noncombustibility, M. is a preferred solvent for fats, oils, alkaloids, waxes, chlorinated rubber and plastics. It is also used as a propellant gas for sprays, as a coolant and as a local anaesthetic.

3,4-Methylenedioxybenzaldehdye: see Piperonal.

Methylene green: see Methylene blue.

Methylene iodide, ***diiodomethane***: CH_2I_2, a colorless liquid which turns dark when it is stored in the presence of light and air, due to loss of iodine. Its m.p. is 6.1 °C, b.p. 182 °C, n_D^{20} 1.7425. M. is slightly soluble in water, but readily soluble in most organic solvents. It is formed by reduction of iodoform with sodium arsenite or from methylene chloride and sodium iodide. Because of its high density, 3.325 g cm^{-3}, M. is used in minerology for separating mineral mixtures. It is also used to determine the refractivity of gemstones.

Methylergometrin: see Ergot alkaloids.

β-Methylesculetin: same as Scopoletin (see).

Methyl ether: same as Dimethyl ether (see).

Methyl ethyl ketone, ***butan-2-one***: CH_3-CO-CH_2-CH_3, an aliphatic ketone. M. is a colorless liquid with an acetone-like odor; m.p. -86.3 °C, b.p. 79.6 °C, n_D^{20} 1.3788. M. is partially soluble in water, and is readily soluble in alcohol, acetone, ether and benzene. It forms an azeotropic mixture with water which boils at 73.4 °C and contains 88.6% M. M. is synthesized industrially by oxidation or dehydrogenation of *sec.*-butanol. After acetone, it is the most important industrial ketone. It is used mainly as a solvent for plastics, paints, resins and rubbers.

N-Methylformamide: O=CH-NH-CH_3, a colorless liquid; m.p. - 3.8 °C, b.p. 180-185 °C or 102-103 °C at 2.66 kPa, n_D^{20} 1.4319. N-M. is soluble in water, alcohol and acetone. It is formed by formylation of methylamine with formic acid esters. N-M. is used to make insecticides and in other organic syntheses.

N-Methylformanilide, ***N-methyl-N-phenylformamide***: O=CH-N(CH_3)C_6H_5, a colorless, oily liquid; b.p. 253 °C. N-M. is slightly soluble in water, but dissolves readily in numerous organic solvents. It is formed by the reaction of *N*-methylaniline with manganese(IV) oxide in chloroform. N-M. is used as a formylation reagent in the Vilsmeier-Haack reaction (see).

N-Methylglycine: same as Sarcosine (see).

Methylglyoxal, ***2-oxopropanal, pyruvic aldehyde***: CH_3-CO-CHO, the simplest α-ketoaldehyde. M. is a yellow liquid with a pungent odor and a tendency to polymerize; b.p. 72 °C. It is soluble in water and alcohol and undergoes the reactions typical of Aldehydes (see). M. is synthesized by oxidation of acetone with selenium dioxide or by acid cleavage of isonitrosoacetone. It is used in the production of drugs, pigments, insecticides, etc., and as a reducing agent.

Methyl green, ***Paris green***: a pigment, chemically heptamethyl-*p*-rosaniline chloride, $C_{26}H_{33}Cl_2N_3$. M. is a green, water-soluble powder. It is used in microscopy as a stain for nuclei and mitochondria, and to differentiate between diphtheria and other bacteria.

Methyl hydroxybenzoate: see Nipa ester.

3-Methylindole: same as Skatole (see).

Methyl iodide, ***iodomethane***: CH_3-I, a colorless liquid which turns brown if stored for a long period in the presence of air; m.p. -66.4°C, b.p. 42.4°C, n_D^{20} 1.5380. M. is barely soluble in water, but is readily soluble in most organic solvents. It is synthesized by reaction of dimethylsulfate with a concentrated potassium iodide solution, or by the reaction of iodine and phosphorus with methanol: $6\ CH_3OH + 2P + 3I_2 \rightarrow 6\ CH_3I + 2\ H_3PO_3$.

Methyl mercaptan: same as Methanethiol (see).

Methyl methacrylate, ***methacrylic acid methyl ester***: CH_2=C(CH_3)-CO-OCH_3, a colorless liquid with a pungent odor; m.p. -48°C, b.p. 100-101°C, n_D^{20} 1.4142. M. is soluble in alcohol, ether and acetone. Like most α,β-unsaturated carbonyl compounds, M. is very reactive and readily polymerizes when heated or in the presence of light or other radical formers. Water, alcohol, amines, etc. can be added to the C=C double bonds, producing β-substituted methyl isobutyrates. M. can be synthesized by esterification of methacrylic acid with methanol in the presence of sulfuric acid. It is used mainly to make high-quality polymethylmethacrylates (including Plexiglas) and various copolymers.

N-Methyl-N-nitrosourea, ***1-methyl-1-nitrosourea***: H_2N-CO-N(NO)-CH_3, an *N*-nitroso compound. N-M. forms colorless to bright yellow, unstable crystalline platelets; m.p. 123°C. It is practically insoluble in water, but dissolves readily in most organic solvents. It is made by reaction of urea with methylamine hydrochloride followed by nitrosylation. N-M. is used mainly to synthesize diazomethane. Like all *N*-nitroso compounds, N-M. is a very potent carcinogen.

N-Methyl-N-nitrosourethane: CH_3-N(NO)-CO-OC_2H_5, an unstable, yellowish red liquid; m.p. < -20°C, b.p. 65-66°C at 2.13 kPa. It is slightly soluble in water and readily soluble in alcohol and ether. N-M. can be synthesized by reaction of ethyl chloroformate with methylamine, followed by nitrosylation of the *N*-methylurethane product. It is used mainly to synthesize diazomethane. N-M. is a carcinogen.

Methyl orange, ***helianthin, orange III***: the sodium salt of 4'-dimethylaminoazobenzene-4-sulfonic acid. It forms yellow-orange leaflets which are very slightly soluble in water, and somewhat more soluble in pyridine. Aqueous solutions are yellow. When rubbed, the crystals turn dark red, but in contact with moist air, they again become yellow-orange. M. is synthesized by coupling diazotized sulfanilic acid with dimethylaniline and conversion of the acid into the sodium salt. The free acid forms violet prisms or leaflets, which are barely soluble in water, giving a yellow solution. Because of its sensitivity to acids and light, M. is no longer used as a dye, but it is used in titration as a pH indicator in the range from 3.1 to 4.4 (color change from red to yellow).

Methyloses: see Deoxysugars.

Methyl oxirane: same as Propylene oxide (see).

Methylparaben: see Nipa ester.

Methyl parathion: an Organophosphate insecticide (see) which is related to Parathion (see) in its chemistry and biological activity.

4-Methylpent-3-en-2-one: same as Mesityl oxide.

Methylpentoses: see Deoxysugars.

Methylpentynol: see Sedatives.

Methylphenobarbital: see Barbitals.

Methylphenols: same as Cresols (see).

Methyl phenyl ether: same as Anisol (see).

N-Methyl-N-phenylformamide: same as *N*-Methylformanilide (see).

Methyl phenyl ketone: same as Acetophenone (see).

2-Methylpropanal: same as Isobutyraldehyde (see).

2-Methylpropanoic acid: see Butyric acids.

2-Methylpropanols: see Butanols.

2-Methylpropenic acid: same as Methacrylic acid (see).

Methylpyridine: see Picoline.

N-Methylpyrrolidone, ***1-methylpyrrolidin-2-one***: a colorless, hygroscopic liquid; m.p. -23°C, b.p. 206°C, n_D^{20} 1.4684. It is soluble in water, alcohol, acetone, ether and benzene. N-M. is a weak base which can be synthesized from γ-butyrolactone and methylamine at 200°C. N-M. is stable in dilute acids and bases. It is used as a solvent for many substances which are difficult to dissolve, such as pigments, plastics, polymers, etc., and for technical gases, such as acetylene, butadiene and isoprene. It is a selective extraction solvent for aromatic hydrocarbons (see Arosolvane process), and is used to desulfurize gases.

2-Methylquinoline: same as Quinaldine (see).

4-Methylquinoline: same as Lepidine (see).

Methyl sterols: tetracyclic secondary alcohols which are the first stable intermediates in sterol biosynthesis. M. have three more methyl groups than the sterols, usually two in the 4-position and one in the 14-position. The most common M. is Lanosterol (see).

Methyl tertiary butyl ether: same as *tert.*-Butyl methyl ether.

Methyltestosterone: see Androgens.

Methylthiouracil: see Thyreostatics.

Methylthymol blue: see Bromphenol blue.

Methyl vinyl ketone, ***but-3-en-2-one***: CH_3-CO-CH=CH_2, the simplest α,β-unsaturated aliphatic ketone. M. is a colorless,, unstable, toxic liquid with a pungent and lacrimatory odor; b.p. 81.4°C, n_D^{20} 1.4081. It is soluble in water and most organic solvents. M. can be synthesized by addition of water to vinylacetylene, or better, from acetone and formaldehyde by a Mannich reaction (see) followed by thermolysis. It can also be synthesized from 4-hydroxybutan-2-one. M. is used for the preparation of pharmaceuticals, fungicides and other organic syntheses.

Methyl violet: a mixture of various violet triphenylmethane pigments (fuchsonimine pigments). M. consists of the hydrochlorides of *p*-rosanilines with varying degrees of methylation of the nitrogen, mainly tetra-, penta- and hexamethyl-*p*-rosaniline. M. forms chunks or powder with a metallic green shine; it dissolves in water to give a violet solution, but is also readily soluble in organic solvents. M. is made by oxidation of dimethylaniline in the presence of phenol and copper chloride. It is used to dye and print cotton (with a tannic acid mordant), synthetics, natural silk, wool, leather and paper, as a component of black dyes used on polyacrylonitrile, and to color fats. M. is also used as a pigment for inks, and as a stain for microscopy; it is part of Gram, vital and nuclear stains. It is used medicinally to eliminate worms.

Metixen: see Antiparkinson's drugs.

Metobromurone see Urea herbicides.

Metofenazate: see Phenothiazines.

Metoxurane: see Urea herbicides.

Metribuzine: see Herbicides.

Metronidazol: a nitroimidazole derivative used to treat trichomonad infections. It also has an antibacterial effect.

O_2N – N – CH_3 – N – CH_2-CH_2OH

Mevalonic acid: 3,5-dihydroxy-3-methylvaleric acid, $HOH_2C-CH_2-C(OH,CH_3)-CH_2-COOH$, the starting material for biosynthesis of Terpenes (see). M. has been isolated from microorganisms and higher plants. It tends easily to form a γ-lactone. Mevalonoactone has been isolated from plants as a glycoside.

Mevinphos: see Organophosphate insecticides.

Mexacarbate: a Carbamate insecticide (see).

Meziocillin: see Penicillins.

Mg: symbol for magnesium.

Micellar catalysis: catalysis of a reaction by micelles can occur by two mechanisms: decreasing the free enthalpy of the transition state relative to the ground states of the reactants, and by increasing their effective concentrations in the micelles due to solubilization or the binding of ions to the Stern layer (see Electrochemical double layer). Thermal reactions can be accelerated by factors of 10^2 to 10^3 by M.; the selectivity effects of micelles on partial reactions are more important, however.

Micellar colloids: see Colloids.

Micelle: an aggregate of tenside molecules which forms when the concentration of the tenside in water exceeds a characteristic concentration, the *critical micellar concentration* (CMC). The number of tenside molecules per M. is called the *aggregation number*; for ionic tensides it can be up to 100, and for nonionic tensides, up to 1000. The hydrophobic parts of the tenside molecules form the nucleus of the M., and the hydrophilic head groups form the hydrated outer layer (Fig.).

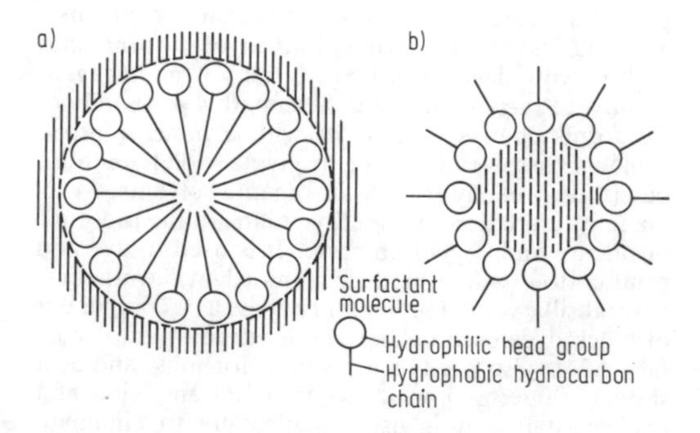

Micelles: a Micelle in water; *b* inverse micelle.

With ionogenic tensides, this outer layer is an electrical double layer consisting of tensides and counter ions. As the concentration of tenside increases, the shape of the M. changes from approximately spherical to cylindrical, and then the cylindrical M. fuse to form a lamellar (liquid crystalline) phase.

Inverse M. are M. with reversed orientation found in nonpolar solvents, such as hydrocarbons.

Michael addition: a reaction of CH-acidic compounds with activated C=C double bonds in the presence of a basic catalyst:

$$(ROOC)_2CH_2 + CH_2=CH-CO-R \xrightarrow{(OH^-)} (ROOC)_2CH-CH_2-CH_2-CO-R$$

Malonic acid esters, malonic dinitrile, cyanoacetic acid esters, acetoacetic acid esters, benzyl cyanides, etc. can be used as CH-acidic compounds. The reaction mechanism can be interpreted as an aldol addition. Withdrawal of a proton from the methylene component by the base produces a carbanion, which is a nucleophile and attacks the polarized C=C double bond:

$$(ROOC)_2CH_2 \xrightarrow[-H_2O]{+OH^\ominus} (ROOC)_2H\overset{\ominus}{C}|$$

$$(ROOC)_2H\overset{\ominus}{C}| + \overset{\oplus}{C}H_2-CH=C(-\overline{\underline{O}}|^\ominus)-R \longrightarrow$$

$$(ROOC)_2HC-CH_2-CH=C(-\overline{\underline{O}}|^\ominus)-R \xrightarrow[-OH^\ominus]{+H_2O}$$

$$(ROOC)_2HC-CH_2-CH_2-C(=\overline{O}|)-R$$

The M. is very versatile for organic syntheses, due to the wide range of reagents which may be used in it.

Michaelis-Arbuzov reaction: a method for synthesizing a phosphorus-carbon bond by reaction of an alkyl ester of phosphorous, phosphonous or phosphinous acid with an alkyl halide, e.g.

$$(EtO)_3P + BrCH_2C_6H_5 \longrightarrow \left[(EtO)_3\overset{+}{P}CH_2C_6H_5 \; Br^-\right]$$

$$\longrightarrow C_6H_5CH_2\overset{O}{\overset{\|}{P}}(OEt)_2 + Et-Br$$

This step produces an alkoxyphosphonium salt as an intermediate which, in a second step known as the *Michaelis-Arbuzov rearrangement*, is converted into the phosphoryl compound and the alkyl halide. The M. is used mainly to produce esters of phosphonic acid from trialkyl phosphites and alkyl halides.

Michaelis-Menten constant: see Enzymes.

Michaelis-Menten mechanism: in the kinetics of enzyme reactions, a simple reaction scheme which provides a good description of the experimentally observed rates of many enzyme reactions. The M. assumes the formation of an intermediate complex ES from the enzyme E and the substrate S; the intermediate then dissociates to the product P and the regenerated enzyme:

$$E + S \underset{-1}{\overset{1}{\rightleftharpoons}} ES \xrightarrow{2} P + E.$$

If one assumes that at the beginning of the reaction, the substrate concentration c_{oS} is much larger than that of the enzyme, c_{oE}, i.e. $c_{oS} \gg c_{oE}$, and that ES is in equilibrium with the reactants, $K = c_{ES}/(c_{oE} - c_{ES})c_{oS}$ the rate of the reaction $v = V_{max}c_{oS}(c_{oS}+K^{-1})$ with $V_{max} = k_2c_{oE}$. It follows for small substrate concentrations, $c_{oS} \ll K^{-1}$, that $v \approx V_{max}Kc_{oS}$. In words, the reaction rate increases linearly with the substrate concentration. At high substrate concentrations, $c_{oS} \gg K^{-1}$, on the other hand, $v = V_{max}$, that is, the reaction rate approaches an upper limit, which depends on c_{oE}, but not on c_{oS}.

Michler's ketone, ***4,4-bis(dimethylamino)benzophenone***: an aromatic ketone which crystallizes in colorless leaflets; m.p. 179 °C, b.p. > 360 °C (dec.). M. is only slightly soluble in water or ether, but dissolves readily in benzene, alcohol and pyridine. It is synthesized by reaction of dimethylaniline and phosgene in the presence of zinc chloride, or from dimethylaniline and carbon dioxide in the presence of aluminum chloride. M. is used for the synthesis of di- and triphenylmethane pigments and as an initiator for photopolymerization reactions.

$(CH_3)_2N-C_6H_4-C(=O)-C_6H_4-N(CH_3)_2$

Micrinite: see Macerals.

Microanalysis: analysis with very small samples; in ***semimicroanalysis***, sample sizes are 0.05 to 0.2 g, in ***ultramicroanalysis***, 1 to 10 micrograms, and in ***submicroanalysis***, 1 to 10 nanograms. When chemical methods are used, the small sample sizes make very precise laboratory technique necessary, and require the use of special apparatus, such as microbalances, microliter burets and microvessels.

In ***qualitative M.***, the methods of Drop analysis (see) are used.

Quantitative M. can be done with all the methods of ordinary analysis, only they must be modified for the purpose. Physical methods, such as electron-beam microprobes, are becoming more widespread.

M. is used when the amounts of sample available are very small, such as in forensic chemistry or when works of art are being tested for their authenticity. They are also used when the substances are very expensive or difficult to isolate, such as hormones, alkaloids or rare natural products.

Microbial toxins: see Toxins.

Microbiological corrosion: destruction of a material by bacteria and fungi. M. affects both metals and non-metallic, organic materials, and can occur in the atmosphere, soil or water. The microorganisms secrete enzymes which catalyse hydrolysis and oxidation of large molecules; the smaller products of these reactions diffuse into the cells and are metabolized.

The most common type of M. which occurs in the atmosphere is caused by molds. Especially in moist, warm climates, these microorganisms cause corrosion of wood, organic polymers, paints, leather and textiles.

In the soil and water, sulfate-reducing bacteria (*Sproribrio desulfuricans*) thrive under anaerobic conditions in the pH range of 5.5 to 8.5. These obtain their energy from reduction of sulfates to hydrogen sulfide; the atomic hydrogen needed for this reduction is formed by the cathodic reaction on the surface of iron: $H_2O + e^- \rightarrow H_{ad} + OH^-$. The reaction products are FeS and $Fe(OH)_2$ in mass ratios 1:2.4 to 1:3.4.

Protection against M. is provided by surface treatment of the materials with bacteriocides and fungicides.

Microcosmic salt: same as Ammonium sodium hydrogenphosphate (see).

Microcosmic salt beads: a preliminary test in qualitative inorganic analysis. If a small amount of ammonium sodium hydrogenphosphate is heated on a platinum or magnesium oxide rod in the flame of a bunsen burner, it forms a melted bead of sodium metaphosphate. Metal oxides or salts in the sample can be dissolved in the bead, forming the corresponding orthphosphates, which give the bead a characteristic color. This test can be used to indicate the presence of several elements. It is especially sensitive for cobalt, which gives a blue color.

Borax beads are used for an analogous test; borax is used to form the beads, and gives the same colors.

Microelements: see Trace elements 1).

Microhardness: the hardness of metals, measured with very small test pressures (as a rule, ≦ 2 N). The test body is often a diamond pyramid with a square base (see Vickers hardness), or with a rhombic base (*Knoop hardness*). Because the impressions are very small, they must be measured under high magnification (metal microscope). It is possible to determine the M. of individual parts of a structure or of very thin sheets.

Micronazole: an Antimycotic (see).

Microwave spectroscopy: portion of high-frequency spectroscopy which borders on the far infrared (see Electromagnetic spectrum) and includes the range of mm, cm and dm wavelengths. The spectra observed in this range include Rotational (see) and Inversion spectra (see) of gaseous molecules, and the ESR spectra (see Electron spin resonance spectroscopy) of paramagnetic substances. In addition, the resonance absorption of ferromagnetic and antiferromagnetic substances and cyclotron resonance (absorption by electrons and holes in semiconductors and metals in an external magnetic field) can be measured in this range. M. in the narrower sense is usually understood to include only the study of rotational spectra.

Microwax, ***hard petrolatum***: a waxy product which precipitates in oil pipes and tanks, and a mineral wax present in the lubricating oil distillates and distillation residues of heavy petroleum. It has a fine crystalline structure. It is separated from the oil and bleached with sulfuric acid, aluminum chloride or fuller's earth. M. can be white or yellow to dark brown. Its melting point is between 50 and 60 °C. It is similar to Mineral wax (see).

Middle oil: see Tar.

Middle phase: see Liquid crystals.

Migration rate: the rate at which the ions of an electrolyte move in an electric field. The conductivity of an electrolyte solution is due to the migration of charge carriers in the form of solvated ions (see Electrical conductivity). Cations are accelerated in the direction of the cathode, and anions in the direction of the anode. The Stoke's frictional force opposes the effects of the field, so that after a short initial period, during which the ions are accelerated, there is a constant M. of the ions v:

$$v = \frac{z \cdot e_0 \cdot E}{6\pi \cdot \eta \cdot r} \text{cm} \cdot \text{s}^{-1}.$$

Here z is the charge number, e_0 is the unit of charge, E is the field strength, η is the viscosity of the surrounding medium, and r is the radius of the solvated ion.

If the M. is related to a uniform field strength E of 1 V cm^{-1}, the parameter *ion mobility u* is obtained: $u = v/E$ cm^2 s^{-1} V^{-1}. The product of the ion mobility and the Faraday constant F is called the *ionic limiting conductivity (ionic equivalent conductivity) l*: $l = u F$ cm^2 Ω^{-1} eq^{-1}.

Ionic limit conductivities l in aqueous solutions at 25°C

Ion	l^+, l^- (Ω^{-1} eq^{-1} cm^2)
H^+	349.8
OH^-	197.0
K^+	73.5
NH_4^+	73.7
Mg^{2+}	53.0
Na^+	50.0
SO_4^{2-}	80.8
Cl^-	76.4
NO_3^-	71.5

The addition of the ionic limit conductivities of the anions and cations of an electrolyte yields the limit conductivity of the electrolyte (see Independent ion migration, law of). The extremely high limit conductivities of H^+ and OH^- ions are due to a special mechanism of conduction (see Extra conductivity).

MIH: see Statins.

Mikado dyes: a group of direct dyes based on diaminostilbene.

MIKES: see Mass spectroscopy.

Milk sugar: same as Lactose (see).

Miller indices: see Crystal.

Millon's base: $[Hg_2N]OH \cdot 2H_2O$, bright yellow, cubic crystals; M_r 167.22. The $[Hg_2N]^+$ cation is isosteric with silicon dioxide, SiO_2, and forms a three-dimensional network linked by covalent bonds; it has a crystoballite structure in which the void volumes are filled by hydroxide ions and water molecules. M. is formed from mercury(II) oxide and aqueous ammonia. It can be reversibly converted to the brown hydrates $[Hg_2N]OH \cdot H_2O$ and $[Hg_2N]OH \cdot 1/2H_2O$. The salts of M. are obtained by reaction with the corresponding acids. The orange-brown iodide of M. is formed during the very sensitive test for ammonia using Nessler's reagent.

Mine gas: see Methane.

Mineral: a chemically and physically homogeneous, natural part of the earth's crust. Nearly all M. are solid; only natural mercury, which occurs in tiny drops in rocks, is liquid. By far the largest number of M. are crystalline; only a very few are amorphous. Most M. are also inorganic substances. Objects with organic form and structure are not considered M. Amber and the fossil fuels (brown and hard coal, petroleum) are mixtures, and thus not M.

Volcanic glasses are considered rocks, because if they had cooled slowly, they would have crystallized into a mixture of M.

An effort is made to express the chemical composition of crystalline M. through chemical formulas. This is easy for minerals formed at low temperatures, because here the formation of mixed crystals is rare and easily understood. However, among M. which form above about 400 °C, mixed crystal formation is common and often very complicated, which is reflected by chemical formulas which may also be very complicated. The M. contain either chemical elements (sulfides, halides, oxides) or oxygen salts (nitrates, carbonates, sulfates, chromates, phosphates, borates, silicates). The most important physical properties of the M. are their hardness (usually reported in the Mohs scale), density, cleavability, shine, color, elasticity, simple and double refraction of light, and conductivity of heat, magnetism and electricity.

M. are classified according to their origins as magmatic, sedimentary or metamorphic. Well over 2000 M. are known. Some M. form large parts of the earth's crust and are considered to be rocks. Most rocks are mixtures of different M.

Some M. are important raw materials for production of metals, glass and ceramics. Many are used as starting materials for the chemical industry.

Mineral acids: in the original sense, a term for acids the salts of which are found in minerals. Today the term is applied mainly for nitric, hydrochloric and sulfuric acids, that is, the strong M.

Mineral fibers: a collective term for natural and synthetic inorganic fibrous materials. Asbestos (see) is a natural, crystalline M.; its use is rapidly declining for toxicological reasons. The importance of synthetic M. is increasing rapidly. These materials are made by melting minerals and spinning the melts into glass-like fibers (see the following separate entries: Rock wool, Fiberglass, Basalt fibers, Ceramic fibers, Silicagel fibers, Slag fibers).

Mineralizers: same as Crystalizers 2).

Mineralocorticoids: see Adrenal cortex hormones.

Mineral oils: a term for all liquid products which can be obtained from petroleum or coal tar by distillation, processing with selective solvents or hydrogenations, such as gasoline, kerosene, gas oils, fuel oils and lubricating oils. M. are similar in their physical properties to the oils from living organisms, but they consist of hydrocarbons rather than acylglycerols.

Mineral pigments: see Pigments.

Mineral spirits: (paints) a fraction of gasoline which boils between 150 and 180 °C, used to thin oil-base paints.

Mineral wax, ***earth wax, ozocerite***: a naturally occurring, very hard waxy substance; the solid component of many petroleum deposits. It consists mainly of alkanes, starting from $C_{18}H_{38}$. Crude M. is dark brown to black. It is freed of gangue by melting it over hot water. The various types of M. differ with respect to hardness, which is indictated by the liquefaction points. Good quality M. is found in the USA and Iran.

M. is purified by treating it with sulfuric acid, then with fuller's earth. The crude wax is heated with 20 to 25% sulfuric acid to 140-180 °C, until sulfur dioxide escapes. The pure wax separates from the black, acidic resin in cast iron basins. To obtain completely colorless M., the process must be repeated. The wax is then filtered through bone charcoal or a similar material. After as much has been pressed out of the charcoal as possible, the remainder is extracted with light petroleum.

Purified M. is called ***ceresine***; it has a density between 0.91 and 0.97, and melts between 60 and 80 °C. Ceresine is soluble in ether, chloroform and gasolines, and its physical properties are comparable to those of beeswax. Ceresine is plastic, kneadable and can be stuck onto a surface, although it is not as sticky as tallow. Mixed with terpentine oil, ceresine is used to make floor waxes, because it binds very well to wood. It is also used to make shoe polish, furniture polish, synthetic vaselines, modelling wax, etc. Very high quality (and very expensive) candles are made from it. Ceresine is commercially available as "ozerocite ceresine"; this product is usually extended with lower quality paraffin (sometimes up to 90%!).

Mineral wool: 1) same as Rock wool (see). 2) Same as Slag fibers (see).

Minium: see Lead oxides.

Miotisal®: see Paraoxon.

Mirror plane: see Symmetry.

Miscibility: same as Solubility (see).

Mitomycins: a group of antibiotics formed by *Streptomyces caespitosus*. They contain structural elements which are also present in synthetic alkylating agents, such as the aziridine ring and the carbamate group. They act as bifunctional nucleophilic alkylating agents. Mitomycin C is used as an antineoplastic.

Mitscherlich test: see Phosphorus.

Mixed acid: same as Nitrating acid.

Mixed crystal: a homogeneous crystalline mixture formed by incorporation of a certain amount of a substance (guest component) in the lattice of another substance (host lattice) without changing the structure type of the host lattice. M. formation is very common among metals and ionic substances.

There are two limiting cases for the incorporation of the guest components. 1) In ***substitution mixed crystals***, components of the host crystal are replaced by components of the guest, which thus occupy regular lattice positions. Substitution M. can form only when the size and shape of the components are nearly the same; the tolerance limit is between about 10 and 15%. In addition, the pure components must have the same or similar crystal lattice types. The substitution can occur in any arbitrary ratio (continuous mixed crystal formation) or be limited to certain concentration ranges (appearance of a mixing gap in the system). Substitution mixed crystals are formed, e.g., by copper and gold, potassium chloride and rubidium bromide, and barium sulfate and potassium permanganate; however, sodium chloride and potassium chloride are miscible only to a slight extent, because of the difference in cation sizes. The exchangeable components usually occupy symmetrically equivalent lattice positions, over which they are randomly distributed. In this case, the lattice dimensions obey Vegard's rule (see). Under suitable crystallization conditions, however, a Superstructure (see) can also be formed.

2) In ***inclusion mixed crystals***, the atoms of the guest component are incorporated into lattice holes in the host crystal and occupy interlattice positions. These M. form only when the difference in radii of the lattice components is extreme. Many hydrides, borides, carbides and nitrides of transition metals can be considered inclusion M. in which the small nonmetal atoms occupy the interlattice positions in the metallic lattices. Such structures are often characterized by high thermal resistance and extreme hardness (especially carbides and nitrides). Inclusion M. of iron with carbon have an important role in the hardening of steel.

Mixed phase: same as Mixture (see).

Mixing: a process used to achieve a uniform distribution of individual components throughout a mixture. M. is used to prepare mixtures, suspensions, emulsions and solutions. It increases the rates of reactions and improves the exchange of material and heat. The choice of method for M. and the apparatus used depend on the properties of the materials to be mixed (state of matter, density, viscosity, particle size, etc.). M. of gases usually depends on turbulent flow processes, e.g. in a Daniel valve, mixing jets or the wake of a propeller. Mechanical stirring is most often used to distribute gaseous, liquid or solid materials in liquids. However, pneumatic stirrers, mixing jets, colloid mills and homogenizers are also used. Plastic, viscous masses are mixed by kneading. Finely divided solids are mixed in drums, screw mixers, and cyclones. If M. is intended to occur at the same time as a chemical reaction, the reactor must be provided with M. devices.

In gases or liquids, M. can also occur without application of mechanical forces, that is, by diffusion. In this case, additional M. devices serve only to accelerate the process.

Mixing enthalpy: same as Mixing heat (see).

Mixing equations: see Stoichiometry.

Mixing gap: in a multi-component system of substances with limited miscibility, that region of composition in which a homogeneous mixture is not possible. When the components are mixed, two liquid, solid or supercritical phases with different compositions arise which are in thermodynamic equilibrium with each other. The miscibility is highly dependent on temperature, and weakly pressure-dependent. As a rule, the miscibility increases with increasing temperature, i.e. the M. becomes smaller. If the compositions of the two coexisting phases are plotted against the temperature in a phase diagram, the points are joined by a ***binodal curve***, which surrounds the two-phase region. If the binodal curve has a maximum, this is called the ***upper critical mixing point*** (Fig. 1). Above this point, the two components are infinitely miscible. If the binodal curve has a minimum, this is called the ***lower critical mixing point*** (Fig. 2). The components are infinitely miscible below this temperature. There are also mixtures which are incompletely miscible only in a limited temperature range (Fig. 3), and thus have both upper and lower critical mixing points. If two components are only slightly soluble in each other, two branches of the binodal curve run close to the edge of the phase diagram, and the M. is not limited above or below (Fig. 4).

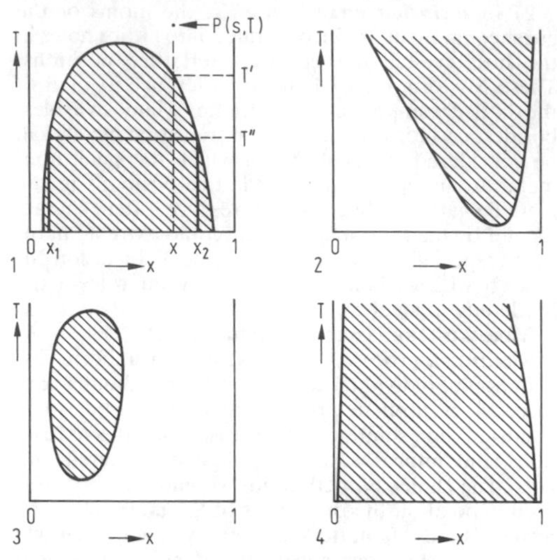

Types of mixing gaps (two-phase area is hatched).

If a homogeneous mixture with the composition $x_A = x$ (point P(s, T) in Fig. 1) is generated at a temperature above the critical mixing point and cooled, when it reaches the binodal curve (at T'), Separation (see) begins. If the system is cooled further to T'', two phases form with compositions $x_A = x_1$ and $x_A = x_2$. The amounts m_1 and m_2 of the two phases are given by the equation: $x(m_1 + m_2) = x_1m_1 + x_2m_2$, or $m_1/m_2 = (x_2 - x)/(x - x_1)$.

Mixing heats, ***Solution heats, mixing enthalpy, solution enthalpy***: those heats which are released or consumed when a real mixture is made without a chemical reaction. Since mixtures are nearly always made at constant pressures, the heats are changes in the Enthalpy (see).

The source of M. are differences between the interaction energies of molecules in the pure components and in the mixture. If the interaction energies in the pure components are greater than in the mixture, the mixture cools when it is made, and the M. is positive (e.g. when ethanol and heptane are mixed); in the opposite case, the mixture is warmed, and the M. is negative (e.g. when ethanol is mixed with water).

The size of the M. depends on the mixing ratio. The ***integral*** or ***mean molar M.*** is the M. per mol of mixture when a mixture of arbitrary composition is made from the pure components (see Mixture). The ***first M.*** is the heat which is exchanged with the environment if 1 mol of a component is mixed with a sufficient amount of another substance to produce an infinitely dilute mixture. The molar M. very close to the saturation limit is called the ***last M.***.

If a solid is dissolved in a liquid, the M. is usually called the heat of solution. When salts dissolve, their crystal lattices are destroyed. The energy required for this is the lattice energy. In the solution, the ions are solvated; in this case, the heat of solution is the difference between the solvation and lattice energies.

The distinction between M. and reaction heat cannot always be made. For example, solution of gaseous hydrogen chloride in water is not simply a mixing process, because an electrolytic dissociation of the HCl also occurs. The same is true for the mixing of concentrated sulfuric acid with water.

M. are measured calorimetrically. The two pure components are allowed to equilibrate in a heat-insulated vessel, then mixed and the resulting temperature change is measured.

Mixture, ***mixed phase***: a homogeneous, multicomponent system. The components in a M. are molecular disperse; they can be either atomic (certain chemical elements) or molecular. Depending on the number of components, the M. is binary, tertiary or quaternary. If the number of components is larger than 4, the number is given. If one component is dissociated, the ions are not considered separate components, i.e. a M. of water and potassium chloride is a binary M. If one component is present in great excess, the M. is also called a Solution (see). The composition of M. is given by the composition variables, e.g. molar amount, molar fraction or concentration.

Gases are miscible with one another in all proportions. Liquids and solids often demonstrate limited miscibility, which depends on the intermolecular interactions and the temperature. In solid M., which are also called Mixed crystals (see), the lattice type, lattice distances and the ionic or molecular radii are important determinants of miscibility. When M. are cooled, and sometimes when they are warmed, Separation (see) occurs. The temperature dependence of miscibility is shown in a Phase diagram (see).

In mixed-phase thermodynamics, an ***ideal M.*** is a mixed phase with physical properties y^M which are the sum of those of the pure components i (additivity): $y^M = \sum_i y_i = \sum_i n_iY_i$. Here n_i is the molar amount of the pure substance i. If the equation is divided by the sum of the molar amounts $\sum_i n_i$, $Y^M = \sum_i x_iY_i$, because $n_i/\sum_i n_i = x_i$, the molar fraction of component i in the M. $Y^M = y^M/\sum_i n_i$ is the *mean molar mixing parameter*. For example, the volume of an ideal binary M. $v^M = n_1V_1 + n_2V_2$ or $V^M = x_1V_1 + x_2V_2$. Here V^M is the mean molar volume of the mixed phase. In the production of an ideal M., there is therefore no volume change or caloric effect (see Mixing heats). An ideal M. can be made when the forces between the molecules are equal, or at least of the same magnitude, as those between the molecules in the pure components. Thus the ideal case is a limit which is approached by ideal gases or mixtures of substances which are chemically very similar, e.g. M. of a compound containing different isotopes ($^{14}NH_3$ and $^{15}NH_3$), M. of enantiomers, structural isomers (o- and p-xylene), compounds of homologous series (heptane and octane) or nonpolar substances.

The above additivity cannot, in principle, apply to parameters of state which are defined by the second law of thermodynamics, i.e. Entropy (see) and, Free energy (see), since mixing is an irreversible process. However, the mixing parameters can be calculated exactly for ideal M. (see Thermodynamics, second law). ***Real M.*** are all M. for which additivity does not hold for volume, pressure, heat capacity, internal energy or internal enthalpy. In such cases there are mixing effects such as volume changes or heats of mixing. These are caused by the differences in intermolecular force between different types of molecules. If the interaction energies between the molecules of the different components are greater than those between molecules of the pure components, volume contraction occurs and mixing heats are negative. The vapor pressure of the M. is smaller than predicted by Raoult's law. In the opposite case, the signs of the mixing effects are reversed. Thermodynamic treatment of real mixing effects is based on Partial molar parameters (see) $\bar{Y}_i$ instead of the molar parameters Y_i; these partial molar parameters are additive: $Y_{real}^M = \sum_i x_i \bar{Y}_i$. For example, the change in enthalpy on mixing two components in the ideal case is $H_{ideal}^M = x_1 H_1 + x_2 H_2$, and in the real case is $H_{real}^M = x_1 \bar{H}_1 + x_2 \bar{H}_2$. The difference $H_{real}^M - H_{ideal}^M = \Delta H^M$ is called the *mean molar mixing enthalpy*.

For second-law state parameters, it is useful to split the ***mixing functions*** $\Delta Y^M = Y_{real}^M - Y_{ideal}^M$ into two parts. One part, Y_0^M, appears when an ideal mixture is made, and the other part, the ***excess function*** Y^E, is a result of interaction: $Y^M = Y_0^M + Y^E$. For example, for the mixing function ΔG^M of the free enthalpy $\Delta G^M = RT \sum_i x_i \ln a_i = RT \sum_i x_i \ln x_i + RT \sum_i x_i \ln f_i$, i.e. $\Delta G_0^M = RT \sum_i x_i \ln x_i$ corresponds to the ideal part Y_0^M, and $G^E = RT \sum_i x_i \ln f_i$ corresponds to the excess part Y^E. For first-law parameters, that is e and h, there are no ideal mixing parts: in these cases the excess function is the same as the mean molar mixing heats ΔE^M and ΔH^M.

Mn: symbol for manganese.

Mo: symbol for molybdenum.

MO: abb. for molecular orbital.

Mobil-Oil process, ***MTG process*** (for methanol to gasoline): a process for converting methanol to hydrocarbons. The desired products are gasoline fuels and lower olefins such as ethene and propene. Methanol reacts on a molecular-sieve catalyst in a temperature range from 200 to 400 °C. The reaction is highly exothermic, so that removal of heat is a major problem.

The advantage of M. is mainly that cheap methanol from large plants can be converted to valuable fuels, thus avoiding the necessity of changing motors and the gas station system, as would be necessary if methanol were used directly as fuel.

Mobile phase: see Chromatography.

Möbius system: see Aromaticity.

Möbius topology: see Aromaticity.

Möbius process: see Silver.

Modelling: see Kinetic modelling.

Modification: see Polymorphism.

Modifier: a term for an additive to viscose or a viscose spinning bath. Together with a special spinning process (e.g. the high zinc sulfate two-bath process), M. permit production of very strong viscose silks and modal fibers. Polyamines, polyethylene glycols, ethylene oxide adducts to fatty acids, aliphatic or aromatic alcohols, amines, etc. are used as M.

Modulon®: see Synthetic fibers.

Mofex process: a process for isolating aromatics from reformed gasoline fractions by extraction with monomethyl formamide.

Mohr's salt: same as Ammonium iron(II) sulfate (see).

Mohs hardness scale: a scale for determining the hardness of materials introduced by the mineralogist C.F.E. Mohs (1773-1839). The M.H. consists of 10 standard minerals, each one representing a certain degree of hardness (1 is the softest, 10 is the hardest).

Mohs hardness scale

Talc	1	Apatite	5	Topaz	8
Gypsum	2	Feldspar		Corundum	9
Calcite	3	(orthoclase)	6	Diamond	10
Fluorspar	4	Quartz	7		

In the M., the scratch hardness, or resistance to scratching, is measured. Minerals with the same hardness do not scratch each other; each mineral scratches all minerals with lower hardness numbers. If a material can be scratched by a mineral with hardness number $n+1$, but not by a mineral with hardness n, it has a Mohs hardness of $n+1/2$. Minerals up to a hardness of 2 can be scratched with a fingernail; with hardness up to 5, they can be scratched with a knife. Materials with hardness of 6 or more can scratch window glass.

This type of hardness determination requires a great deal of experience, is not very exact, and is not compatible with the SI system. It is now used practically only for approximate indications in minerology.

Molality: see Composition parameters.

Molar absorption coefficient: see Lambert-Beer law.

Molar conductivity: see Electrical conductivity.

Molar dispersion: the difference between the molar refractions (see Refractometry) at two different wavelengths. M., like molar refraction, is the sum of atomic and bond increments.

Molar formula turnover: see Mole.

Molar extinction coefficient: see Lambert-Beer law.

Molar heat, ***molar heat capacity***: the heat which is required to increase the temperature of 1 mol of a substance 1 kelvin. The unit of M. is J mol^{-1} K^{-1} (formerly cal mol^{-1} K^{-1}). The M. of elements is also called ***atomic heat***. There are two M., one measured at constant external pressure p (C_p) and one at constant volume v (C_v). Heating at constant pressure is nearly always associated with an expansion of the substance (water between 0 and 4° C is an important exception), and thus with volume work $p\Delta V$, for which additional heat energy must be supplied. Therefore, in general, $C_p > C_v$, and the difference between the two parameters depends on the amount of the thermal expansion. In general, $C_p - C_v = \alpha^2 TV/\gamma$ Here α is the isobaric expansion coefficient, and γ is the isothermal compressibility coefficient. V is the molar volume, and T the absolute temperature. For ideal gases, the thermal equation of state yields $\alpha = 1/T$ and $\gamma = 1/p$, so that $C_p - C_v = R$, where R is the general gas constant. For liquids and especially solids, the difference between C_p and C_v is very much smaller, due to the small expansion coefficients.

The M. is related to the ***specific heats*** $c_{p,sp}$ or $c_{v,sp}$ and the ***heat capacities*** c_p or c_v of pure substances by the equations $C_p/M = c_{p,sp}$, $C_v/M - c_{v,sp}$, $c_p = nC_p = mc_{p,sp}$ and $c_v = nC_v = mc_{v,sp}$. Here M is the molar mass, while n and m are the number of moles or the mass of an arbitrary amount of substance.

The heat capacities are related to the molar internal energy E and the molar enthalpy H (see Thermodynamics, First law) by the equations: $(\delta E/\delta T_v = C_v$, $(\delta H/\delta T)_p = C_p$.

The M. of ideal gases can be estimated from the molecular structure using the kinetic gas theory (see Gas theory, kinetic). According to the principle of equipartition of energy, each translational and rotational Degree of freedom (see) contributes 1/2 R to C_v, and each vibrational degree of freedom contributes 1 R. It follows that $C_v = 1/2\ R(f_{trans} + f_{rot} + 2f_{vib})$, where f_{trans}, f_{rot} and f_{vib} are the number of translational, rotational and vibrational degrees of freedom, respectively.

Monoatomic gases (e.g. noble gases, mercury vapor) have only three degrees of translational freedom. It follows that $C_v = 3/2\ R = 12.5$ J mol^{-1} K^{-1}. Diatomic gases (e.g. H_2, N_2, O_2) have three degrees of translational freedom, two rotational degrees and one vibrational degree, so that $C_v = 7/2\ R = 29.1$ mol^{-1} K^{-1}. Triatomic bent molecules have three degrees each of translational, rotational and vibrational freedom, so $C_v = 6\ R \approx 50$ J mol^{-1} K^{-1}. However, these are limiting values for high temperatures. Rotational and vibrational energies are quantized, and can only be absorbed by molecules in integral energy packets $h\nu$. Therefore the rotational and vibrational states do not absorb energy or contribute to C_v until the mean thermal energy kT ($k = R/N_A$ = Boltzmann constant) reaches the order of magnitude of $h\nu$. Since the rotational frequencies ν_{rot} are generally smaller than the vibrational frequencies, the rotational excitation occurs at lower temperatures than the vibrational. For most di- and triatomic gases, the rotations are excited at room temperatures, but the vibrations are not, so that C_v lies at 5/2 R or 6/2 R, respectively, and increases as the temperature increases.

Since $C_p = C_v + R$, the ratio $\varkappa = C_p/C_v$ is $5/3 = 1.66$ for monoatomic, $7/5 = 1.40$ for diatomic and linear triatomic, and $8/6 = 1.33$ for bent triatomic molecules.

The M. of real gases are usually only a little greater than those predicted for ideal gases. Liquids have higher M. than gases and solids. For solids, the M. at constant pressure (C_p) are the only ones of interest, because measurement at constant volume to obtain C_v is nearly impossible. According to the *rule of Dulong and Petit* (1819), the C_p values of most solid elements in the medium temperature range are about 25 J mol^{-1} k^{-1}. However, this value is not reached by many light elements (carbon, boron, beryllium, silicon) until they are at much higher temperatures, while it is exceeded by others. The M. of solid compounds can be approximated by addition of those of the elements (*Neumann-Kopp rule*, 1831). For some elements (e.g. hydrogen, boron, carbon, nitrogen, oxygen), however, empirically determined values must be used.

At low temperatures, the M. are greatly reduced and, according to the third law of thermodynamics, they go to zero at absolute zero. Between 0 and 50 K, the *Debye T^3 rule* applies: $C_v = aT^3$, where a is a substance constant.

Theoretically, the M. of ideal solids can be understood in the same way as those of ideal gases. The crystal components have only three degrees of vibrational freedom in the three space directions. If they are completely excited, $C_v = 3\ R = 25$ J mol^{-1} K^{-1}, i.e. the value of the Dulong-Petit rule. If the quantized uptake of energy by vibrating systems at low temperature is taken into account, the Planck-Einstein equation results: $C_v = 3R(\Theta/T)^2 e^{\Theta/T}/(e^{\Theta/T}-1)^2$, where $\Theta = h\nu/k$. Here ν is a uniform vibrational frequency for the solid and Θ is called the characteristic temperature. If a spectrum of lattice frequencies is introduced, as suggested by Debye, the result is the Debye T^3 rule.

At phase boundaries, M. are not continuous. Because of their great practical significance, M. are measured in calorimeters and tabulated. Their temperature dependences are usually given in the form of exponential series $C_p = a + bT + cT^2 + \ldots$ C_v can be calculated using State sums (see).

Molar heat capacity: same as Molar heat (see).

Molarity: see Composition parameters.

Molar mass: see Mole.

Molar mass determination, formerly ***molecular weight determination***: the determination of the molar mass of a chemical compound which exists as defined molecules. In the past, a large number of methods for M. were developed which are now primarily of historical interest. M. was used together with Elemental analysis (see), which provides only the ratio of the elements in a compound, to determine the empirical formulas of molecular compounds.

At present, M. is achieved by the following methods:

1) The most rapid and most reliable method is Mass spectroscopy (see). It yields the molar mass with a precision of 0.01%, or in high-resolution devices, with a precision of 10^{-3} to 10^{-5}%. Because atomic masses are not integers, it is possible to determine the elemental composition from the molar mass alone; in other words, the empirical formula can be established by mass spectroscopy.

2) One group of methods for M. is based on application of the thermal equation of state (see State, equations of) for an ideal gas, $pv = nRT$. Since the molar amount n is the ratio of the mass m of the sample to its molar mass M, $n = m/M$, it follows from the gas law that $M = mRT/pv$. All such methods of M. have in common that the substance is evaporated at a constant temperature T. Of the three other parameters, p, v and m, one or two are fixed by the experimental arrangement and the other(s) is(are) measured.

Method of	Fixed parameters	Measured parameter
Dumas, 1827; Regnault, 1847	T, p, v	m
Meyer, 1878	T, p, m	v
Blackmann, 1908; Bodenstein, 1910; Menzies, 1911	T, v, m	p
Gay-Lussac, Hofmann, 1868	T, m	p, v

Treatment of vapors as if they were ideal gases leads to considerable systematic errors.

3) Another group of methods for M. of dissolved substances is based on the reduction in the vapor pressures of solutions according to Raoult's law

(freezing point lowering, boiling point elevation). In the Rast micromethod (1922), camphor is usually used as a solvent, due to its high cryoscopic constant $E_g = 40$ K g mol^{-1}. Even with small amounts of sample, a large reduction of freezing temperature can be observed.

4) Osmosis (see) is also used for M. The calculation of molar mass from osmotic pressure is based on the van't Hoff equation: $p = cRT = mRT/vM$. Because the osmotic pressure is very high even at low concentrations, this method is suitable for determination of large molar masses (natural products, natural and synthetic polymers).

5) Special methods are required for M. of high polymers, partly because these substances do not have uniform molar masses (except for proteins), and partly because the molar effects listed under 2) and 3) are too small to be measured readily.

Molar parameter: see Mole.

Molar polarization: see Polarization.

Molar refraction: see Polarization.

Molar rotation: see Polarimetry.

Molar volume: see Mole.

MO-LCAO rule: see Molecular orbital theory.

Mold releasers: usually liquid, readily melted or powdery substances applied to the surfaces of molds to make the molded object separate more easily from the mold, and to prevent sticking of the mold (e.g. sand) to the casting. Silicon oils, mineral oils, waxes (also suspensions), soap solutions, talcum, etc. are commonly used. For plaster molds, a coating of thermoplastic suspension is usually applied.

Mole, abb. ***mol***: adopted in 1971 as the 7th basic unit in the international system of units (SI) with the following definition: that quantity of a system containing as many elementary objects as there are atoms in 0.012 kg of the carbon nuclide ^{12}C.

The M. is a quantitative unit which is used instead of specifying the enormous numbers of elementary objects (molecules, atoms, ions, electrons, etc.) involved in chemistry. 1 mol = N_0 elementary objects; according to the best current measurements, N_A (Avogadro's number) is $6.022045 \cdot 10^{23}$. Decimal multiples and fractions of the M. are frequently used; for example, 1 kmol = 1000 mol and 1 mmol = 0.001 mol. The M. can be used with any type of elementary object, but it must be clearly indicated which type is meant. Amounts of substance can also be indicated in equivalents; the special unit *val* is sometimes used: 1 mol equivalents = 1 val. For example, if H_2SO_4 is treated as a diprotic acid, $n_{ev}(H_2SO_4) = 0.2$ val, $n(H_2SO_4) = 0.1$ mol and $n(1/2H_2SO_4)$ refer to the same amount of sulfuric acid. However, because equivalents can be misleading, their use is discouraged.

The ratio of an extensive parameter to the associated amount of substance n is an intrinsic or *molar parameter*; examples are the *molar mass* $M = m/n$ and the *molar volume* V (v/n). The molar mass $M(X)$ is related to the mass $m(X)$ of a particle of substance X by Avogadro's number (see), N_A: $M(X) = m(X)N_A$. The molar volume of an ideal gas under standard conditions is $V = 22.4136$ l mol^{-1}. Other molar parameters are the Faraday constant (see Faraday's laws) $F = 96485.44$ C mol^{-1}, the mol of electric charge, and the general gas constant $R = 8.31441$ J mol^{-1} K^{-1}. The term M. was formerly used as an abbreviation for "***gram molecule***", which was the molecular weight of that substance in grams. (The term "molecular weight" has been replaced by the term "molecular mass"). After the formulation of the modern definition of the M., this unit, along with the similarly defined gram atom, gram formula weight and gram equivalent, became obsolete.

Molecular beam: a beam of atoms, molecules or ions in a vacuum. The distances between the molecules within the beam are so large that collisions or interactions between them are negligible.

M. are generated by particles leaving a heated space through a circular opening or valve. There are special techniques which insure that nearly all the particles have the same kinetic energy, rotational or vibrational excitations, or the same orientation in space.

M. can be used, for example, to make high resolution spectra, to measure magnetic moments and to determine the average lifetime of excited molecules, or to study their interactions with surfaces. The ***crossed beam method*** makes use of two M. which cross each other at a right angle or some other angle. The particles in the two beams collide, and the scattered molecules are registered by a suitable detector, such as a mass spectroscope, as a function of the scattering angle. This technique gives information on the scattering cross section, that is, the probability of energy exchange or chemical reaction between the colliding particles (see Kinetics of reactions, theory).

Molecular colloid: see Colloids.

Molecular distillation: see Distillation.

Molecular formula: see Formula.

Molecularity: in kinetics of reactions, a term for the number of particles which interact by collision in an Elementary reaction (see).

Molecular lattice: see Lattice type.

Molecular mass, relative, symbol $\boldsymbol{M_r}$: formerly ***molecular weight***: the ratio of the absolute mass of a molecule to the atomic mass unit u (see Atomic mass). If the chemical composition of a substance is known, the M_r can be calculated as the sum of the relative atomic masses, A_r, of the atoms in the molecule: $M_r = \sum_i \nu_i A_{r,i}$, where ν_i is the stoichiometric coefficient and $A_{r,i}$ is the relative atomic mass of element i. The term M_r can only be applied to those substances which consist of molecules, such as water, H_2O, or carbon dioxide, CO_2. In all other cases, such as ionic substances, it must be replaced by the concept of relative Formula mass (see). The M_r (a dimensionless number, since it is a ratio) has the same numerical value as the molar mass (with the unit g mol^{-1}; see Mole) of the same molecular substance. Because the M_r is often needed for stoichiometric and other chemical calculations, tables of M_r values for many compounds are available.

Molecular model: see Stereochemistry.

Molecular orbital: see Molecular orbital theory.

Molecular orbital theory, *MO method*: a quantum mechanical approximation used in calculating the electronic structure and energy of molecules; the essentials were developed by Hund, Mulliken, Lennard-Jones and Hückel (1927-1929). As in the Central field model (see) of the atom, one considers only a single electron, which moves in an effective potential generated by the nucleus and the other electrons. The resulting one-electron state of the molecule is called a ***molecular orbital*** (abb. ***MO***), φ It extends over the entire molecule, and is characterized by an orbital energy. The occupation of the energetically lowest molecular orbitals in accordance with the Pauli principle yields the ground state electronic configuration of the molecule. Near a nucleus, the average potential is essentially determined by the potential of that atom. Therefore, molecular orbitals φ can be approximated by linear combinations of atomic orbitals X_i (*MO-LCAO rule*): $\varphi = \Sigma c_i X_i$. In the variation technique, the coefficients c_i are chosen so that the energy of the total system is a minimum. The principle of the M. is most easily demonstrated for a homonuclear, diatomic molecule, e.g. the H_2 molecule. Combination of the 1s orbitals of the two hydrogen atoms gives two molecular orbitals $\varphi_\pm = N_\pm(X_a \pm X_b)$, where $n_\pm$ is the normalization factor for the plus or the minus combination. The corresponding orbital energies ε_+ and ε_- are lower and higher, respectively, than the energy of the atomic orbital. Therefore φ_+ is called the *bonding*, and φ_- the *antibonding* molecular orbital. In the ground state of the hydrogen molecule, both electrons occupy the bonding molecular orbital (Fig. 1), and this explains the formation of a bond in the hydrogen molecule. The bonding energy is equal to twice Δε. The amount of splitting of the orbital energies is proportional to the degree of *overlapping* of the contributing atomic orbitals.

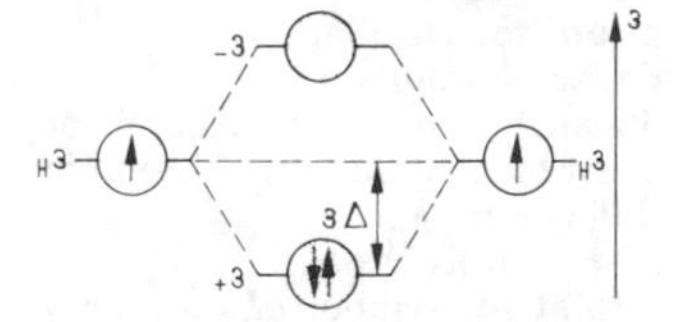

Fig. 1. Energy of the molecular orbitals for H_2.

Mathematically, the overlapping between the atomic orbitals X_a and X_b is expressed by the *overlapping integral* $S_{ab} = \int X_a X_b \, dv$. The degree of overlapping is large if the atomic orbitals have significant densities in the same volume of space. Other atomic orbitals can be combined to molecular orbitals in analogy with the 1s orbitals. There are various types of overlapping, depending on the symmetry and directional dependence of the angular portions of the wavefunctions (see Atom, models of). The combination of orbitals which are symmetric with respect to rotation around the bond axis yields σ-*molecular orbitals*. These are characterized by overlap in the region of the bond axis. If the bond axis is embedded in the plane of antisymmetry of the atomic orbitals, their interaction gives a π-*molecular orbital*. These have two regions of overlap away from the bond axis which are separated by a nodal plane. If there are four regions of overlap away from the bond axis, separated by two mutually perpendicular nodal planes, the result is a δ-*molecular orbital*. The bonds formed by these orbitals are called σ-, π- and δ-bonds. The corresponding antibonding orbitals are indicated by an asterisk, σ*, π* and δ*. In addition to the nodal planes of their constituent atomic orbitals, antibonding orbitals have a nodal plane between the two atoms and perpendicular to the axis joining them. Several combinations of atomic orbitals to molecular orbitals are given in Fig. 2. In the combination of two atomic orbitals with different symmetry about the bond axis, the overlap integral is equal to zero, so these cannot be used to form a molecular orbital (Fig. 3.).

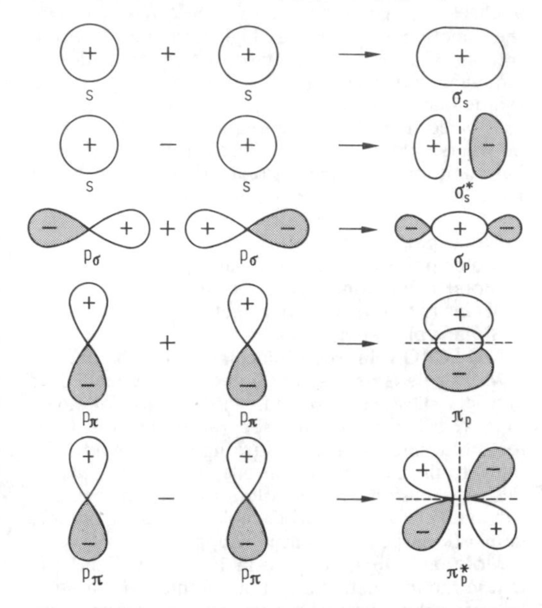

Fig. 2. Combinations of atomic orbitals to make molecular orbitals.

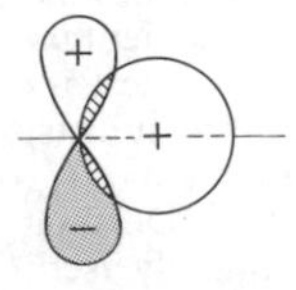

Fig. 3. Overlap zero for symmetry reasons (s = O).

The molecular orbital treatment based on the possibilities for overlapping of atomic orbitals can be extended to homonuclear diatomic molecules of elements of the second period (Li_2, B_2, C_2, N_2, O_2, F_2. If the axis joining the nuclei is taken as the z direction, the molecular orbital scheme shown in Fig. 4 is obtained. (For qualitative purposes, it is sufficient to consider only the overlap of orbitals with the same or similar energy.) The molecular orbitals are occupied in the order of increasing energy, following Hund's rule and the Pauli principle. According to this method, the total spin of the oxygen molecule should be $S = 1$ (Fig. 4), which is consistent with the experimentally observed paramagnetism of oxygen. This could not be explained by classical bonding theory.

The *bond order*, which is equal to the number of electron pairs in bonding orbitals minus the number in anti-bonding orbitals, serves as a qualitative measure of the bond strength in diatomic molecules. For oxygen, the bond order is 2, which corresponds in strength to a double bond.

MO-LCAO can also be applied to more complicated molecules, although the required calculations increase rapidly as the number of atomic orbitals increases. Molecular orbitals are delocalized over the entire molecule and reflect the symmetry of the molecule, making possible the interpretation of excitation and ionization processes. However, it is also possible to give an equivalent description of the electronic structure of the ground state using localized molecular orbitals (*bonding orbitals*); this is consistent with the classical concept of localized bonds, and is justified by numerous experimental results (see Additivity principle). Hybrid orbitals (see Hybridization) can also be used to construct molecular orbitals; in this way the structures of molecules with multiple bonds, such as ethene and ethyne, can also be easily explained. For molecules with conjugated double bonds, such as benzene, a localized description of the π-electron system in terms of two-center bonds is incomplete (see Resonance). The overlap of the p_π-orbitals in such molecules leads to a single π-bonding system. The simplest approximation is the Hückel method (see). The Roothaan-Hall method has adapted M. to computers, so that with a modern computer, MO calculations can be done even for large molecules.

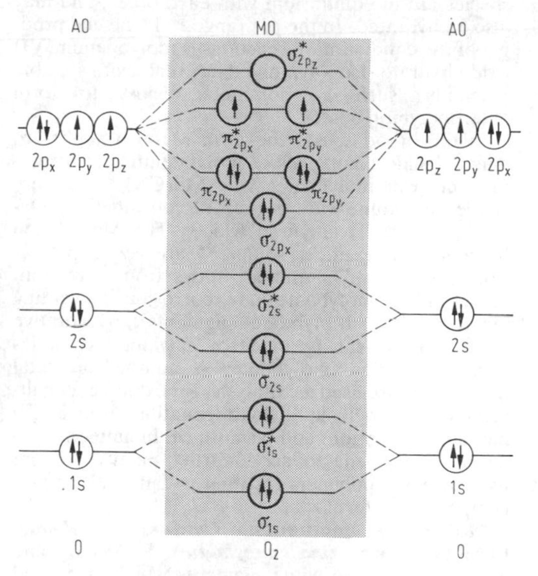

Fig. 4. Qualitative MO scheme for a diatomic molecule O_2 is used as an example.

Molecular peak: see Mass spectroscopy.

Molecular sieve: synthetic zeolites and porous glasses which have defined pore sizes and are able to separate mixtures of substances with different molecular diameters. They are used, for example, for drying gases and liquids, or selective separation of organic mixtures. Porous glasses can be made from sodium borosilicate glasses by Subliquidus separation (see); this process creates a penetration structure, and the sodium borate phase is then removed by leaching.

Molecular spectroscopy: studies of the spectra of molecules. Usually the absorption spectra are studied because excitation of molecules to the point of light emission frequently destroys them.

The total energy of a molecule is the sum of its translational, rotational, nuclear vibrational and electronic energies. For the last three forms of energy listed, there are discrete quantized energy states; transitions between them are mediated by interactions with electromagnetic radiation. The probabilities of transitions between these states are given by Selection rules (see). For allowed transitions, the intensity of the corresponding lines is high; for forbidden transitions, it is very low.

The energy ranges of the three types of molecular energy are quite different. Changes in the electronic state usually involve large amounts of energy compared to those required for vibrational energy, and these in turn are large compared to the energies of rotational transitions (Fig.). The corresponding molecular spectra lie in different spectral ranges (see Electromagnetic spectrum). If only the rotational state of the molecule changes, one observes a pure Rotational spectrum (see), which consists of lines in the far infrared and microwave range. Changes in the vibrational energy are associated with simultaneous changes in the rotational energy, leading to band spectra (see Spectrum) called rotational-vibrational spectra; these lie in the infrared range (see Infrared spectroscopy). A change in the electronic energy is linked to changes in both the vibrational and rotational states, so that again band spectra arise, in this case in the visible and ultraviolet (see UV-VIS spectroscopy).

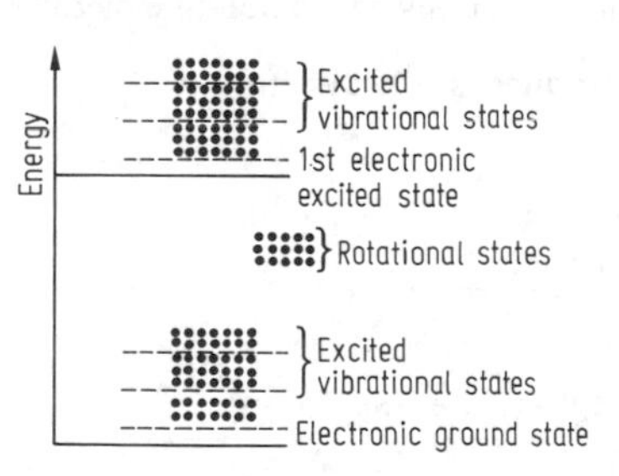

Term scheme of a molecule.

Since the average energy of thermal motion at room temperature is not sufficient to excite the electronic and vibrational states, molecules at room temperature are in the electronic and vibrational ground state. However, this thermal energy is high enough so that even without external excitation, molecules at room temperature occupy all possible rotational states. Some other important spectroscopic methods used to examine molecules are Raman spectroscopy (see), in which changes in vibrational energy are observed (as is also true in IR spectroscopy), NMR spectroscopy (see) and Electron spin resonance spectroscopy (see), in which realignment of nuclear and electron spins in external magnetic fields are measured, and Mass spectroscopy (see).

All molecular spectra are closely related to the molecular structure, and they are therefore used mainly for elucidation of structure. They are also suitable for qualitative and quantitative determinations of compounds.

Molecular weight: see Molecular mass, relative.

Molecule: a particle consisting of a limited number of atomic nuclei and electrons in a certain spatial arrangement. *Homonuclear M.*, e.g. H_2, O_2, P_4 or S_8, contain atoms of a single element, while *heteronuclear M.* contain more than one element (e.g. CO_2, NH_3, C_nH_{2n+2}). Most M. are electrically neutral, but there are various interactions between individual M., producing van der Waals forces. If the intermolecular interactions are weak, the substance exists as a gas. Strong interactions are present in liquids and solids. In the latter case, the solid consists of *molecular crystals*. M. consisting of thousands of atoms held together by covalent bonds are called *macromolecules*. In the liquid and solid phases, in the presence of counter ions, *molecular ions* are stable, but in the gas phase, they have only a limited lifetime. Molecular ions with opposite charges are attrated by electrostatic forces.

Complex M. are complicated structures built up of simple M. or ions; as a rule they have a positive or negative charge. The M. also include *free radicals*, which contain one or more unpaired electrons, and therefore have a total spin other than zero. As a rule, they are highly reactive. The relative Molecular mass (see) is the sum of the relative atomic masses of the atoms which make up the M.

Mole fraction: see Composition parameters.

Mole percent, abb. ***Mol%***: see Composition parameters.

Molex process: a process for isolating *n*-alkanes from petroleum fractions by adsorption to molecular sieves.

Molting hormones: see Ecdysteroids.

Molluscicides

Name	Formula
Clonitranilide	O_2N–(Cl-substituted ring)–NH–CO–(ring with Cl, OH)
Rafoxanide	Cl–(ring)–O–(ring with Cl)–NH–CO–(ring with I, I, OH)
Trifenmorph	$(C_6H_5)_3$C–N(morpholine, O)

Molluscicide: a biologically active compound used to combat molluscan pests, especially snails and slugs.

Drying agents such as burned lime (calcium oxide) or calcium cyanamide can be used. Metaldehyde, which acts as an oral and contact poison, is more widely used; it is applied in the form of baits. The Carbamate insecticide (see) mercaptodimethur is also effective. Clonitranilide and rafoxanide (salicylic acid anilides) and trifenmorph (Table) are used to combat the water snails which transmit bilharziosis (schistosomiasis), and also as veterinary remedies for liver flukes in cattle and sheep.

Molybdate: a term applied in a narrow sense to ***molybdates(VI), oxomolybdates(VI),*** salts of molybdenum acids, including derivatives of monomolybdic acid (see Molybdenum(VI) oxide hydrate), which does not exist in free form. In addition to these, M. of the types $M^I_2MoO_4$ and $M^{II}MoO_4$ are especially important. These M. contain discrete, tetrahedral MoO_4^{2-} ions which may be formed by dissolving molybdenum(VI) oxide, MoO_3, in alkali hydroxide solutions; they are stable in alkaline and neutral solutions. MoO_4^{2-} reacts in alkaline media with hydrogen peroxide, forming the red peroxomolybdate(V), $[Mo(O_2)_4]^{2-}$. When molybdate solutions are acidified, polycondensation processes lead to polymolybdates; of these the para- and metamolybdate ions $[Mo_7O_{24}]^{6-}$ and $[Mo_8O_{26}]^{4-}$ are especially important. These two species are in equilibrium with each other, and may also be hydrated. In the pH range < 1, the end product of the condensation reactions is molybdenum(VI) oxide hydrate. Isopolymolybdates react with suitable reactants, such as non-metal acids, to form heteropolymolybdates.

In the broad sense, the term M. also applies to anionic acido complexes of molybdenum, for example octacyanomolybdate(IV), $[Mo(CN)_8]^{4-}$, a very stable compound with eight-fold coordination. Oxomolybdates(VI) are used widely. The MoO_4^{2-} ion has a passivating effect, so that M. are used to protect iron, copper and aluminum alloys from corrosion. Some metallomolybdates are corrosion preventing pigments (see Molybdate pigments). Decorative black coatings can be made by galvanic techniques using ammonium molybdate. A number of metal molybdates are used as catalysts, for example, cobalt-nickel M. in petroleum hydrogenation, iron M. in methanol oxidation, and bismuth M. in ammono oxidation of propene to acrylonitrile. Finally, M. are used to make polymers resistant to burning and as a component of fertilizers.

Molybdate pigments: ***Calcium molybdate***, $CaMoO_4$, and ***zinc molybdate***, $ZnMoO_4$ and $5ZnO \cdot 7MoO_3$ are white pigments which give good protection against corrosion. Unlike anticorrosive lead pigments, these M. are nontoxic. ***Molybdate red*** is another M.; it consists of mixed crystals of lead molybdate and lead chromate. It is very stable to light, but is not very stable to acids and bases.

Molybdate red: see Molybdate pigments.

Molybdenum:, symbol ***Mo***: an element of group VIb of the periodic system, the Chromium group (see); a heavy metal, *Z* 42, with natural isotopes with mass numbers 98 (23.75%), 96 (16.50%) 92 (15.86%), 95 (15.70%), 100 (9.62%), 97 (9.45%) and 94 (9.12%); atomic mass 95.94, valence usually VI, also V, IV, III, II, 0, density 10.28, m.p. 2620 °C, b.p. 4825 °C, electrical conductivity 18.7 Sm mm^{-2}, standard electrode potential (Mo/Mo^{3+}) -0.20 V.

Properties. M. is a silvery white metal which is steel gray in powdered form. The strength of compact M. depends greatly on the purity and degree of cold working. It crystallizes in a cubic body-centered lattice and is very resistant to air and non-oxidizing acids, due to passivation. Oxidizing acids, such as hot, conc. sulfuric, conc. nitric or mixtures of hydrofluoric and conc. nitric acids, will dissolve M., while fused alkalies react only slowly with it. M. is attacked by atmospheric oxygen at higher temperatures. In the range below 550 °C, it first forms nonvolatile oxides as a blue, solidly attached surface layer. At temperatures above 600 °C, molybdenum(VI) oxide, MoO_3, is rapidly formed. While fluorine reacts with M. even on slight heating, forming molybdenum(VI) fluoride, MoF_6, molybdenum(V) chloride, $MoCl_5$, is formed only when the elements are heated under pressure. Nonmetals, such as carbon, boron or silicon, form hard and high-melting inclusion compounds with M. Similarly, molybdenum nitrides are formed from M. and ammonia. The reaction of carbon monoxide with M. under pressure and at elevated temperature gives molybdenum carbonyl, $Mo(CO)_6$.

The most important oxidation state of M., the +6 state, is present in the molybdates. In contrast to the chromates(VI), these are only weak oxidizing agents. The +5 and +4 oxidation states are also important. The +3 and +2 oxidation states are present in the cluster compounds molybdenum(III) chloride and molybdenum(II) chloride, and in a few anionic acido complexes, while π-acceptor ligands, such as carbon monoxide, are also able to stabilize the 0 oxidation state. M. is a good complexing agent; most of its complexes are anionic, and have coordination numbers from 6 to 8. Examples: octahedral fluorometallates, $[MoF_6]^{n-}$, (n = 1-3), pentagonal bipyramidal heptacyanomolybdate(III), $[Mo(CN)_7]^{4-}$, and the very stable octacyanomolybdate(IV), $[Mo(CN)_8]^{4-}$.

M. is an essential bioelement, which is a component of certain metalloflavin enzymes, the molybdenum enzymes. Examples are nitrogenase and nitrate reductase, which play essential roles in plant and microbial nitrogen metabolism. In animals, molybdenum compounds such as the flavin enzymes xanthine oxygenase and aldehyde oxidase act as respiration catalysts.

Analytical. The most common reaction for qualitative determination of M. is the Molybdenum blue (see) reaction. M. can be determined spectrophotometrically with the aid of tin(II) chloride thiocyanate, with which it forms blood-red isothiocyanatomolybdate(V); this is useful over a wide concentration range down to 1 ppm. Serial analyses can be done by atomic absorption spectrometry or X-ray fluorescence spectroscopy. For gravimetric determination of M., it is precipitated as lead molybdate and molybdenum(VI) oxide hydrate; the latter is roasted to molybdenum(VI) oxide, MoO_3, before weighing.

Occurrence. M. makes up $1.1 \cdot 10^{-4}$% of the earth's crust. The most important molybdenum ore is molybdenite, MoS_2. M. is also found in wulfenite, $PbWO_4$ and powellite, $CaMoO_4$. M. is present in copper shale at concentrations of 0.01 to 0.02%. It is an important trace element, and is present in soils at about 3 ppm; it makes up about 0.2 ppm of the dry weight of plants. As an enzyme component, M. is active in symbiotic nitrogen fixation by legumes and root-nodule bacteria, and in nitrate assimilation. The lack of M. in soils causes typical deficiency symptoms in higher plants.

Production. The molybdenum ore (molybdenite) is first concentrated by flotation, then roasted at 400 to 650 °C to molybdenum(VI) oxide. The resulting MoO_3 is then further purified by wet chemistry or sublimation. For chemical purification, the MoO_3 is dissolved in aqueous ammonia, forming ammonium molybdate: $MoO_3 + 2 NH_3 + H_2O \rightarrow (NH_4)_2MoO_4$. This complex is crystallized out, then heated in the air to yield MoO_3. For sublimation, MoO_3 is heated to 1200 to 1250 °C in an electrically heated rotating pipe furnace. The metal is produced by reducing MoO_3 or ammonium molybdate in a two-step process with hydrogen; MoO_2 is first formed around 500 °C, and then further reduced to M. at 1100 °C in the second step. The reduction process yields M. powder, which can be converted to compact M. by powder metallurgical processes or by vacuum electric arc melting. There are other methods of production. M. can be produced from the ore by an oxidizing fusion reaction with soda, forming sodium molybdate, which is converted to molybdenum oxide hydrate by addition of acid. Electrochemical processes exist, and low-grade ores can be solubilized by a chlorinating method; the molybdenum(V) chloride, $MoCl_5$, which is distilled off is reduced to M. with hydrogen. Very pure M. is made by thermal decomposition of molybdenum hexacarbonyl, $Mo(CO)_6$, at 350 to 400 °C. M. used to make molybdenum steel is not purified as such, but is used in the form of ***ferromolybdenum***, with 60 to 75% M. Ferromolybdenum is now produced primarily by metallothermal processes using reduction of a molybdenum-iron oxide mixture with ferrosilicon/aluminum.

Applications. M. is resistant to corrosion, is very strong even at high temperatures and has a high melting point; these qualities make it a suitable material for extreme conditions. It is used to make resistance wire for high-temperature electric furnaces, furnace reflector materials, anodes in electronic tubes, as a carrier for semiconductor elements and as an additive in making hard metals. M. is used in the glass industry as an electrode material for glass melting furnaces, and in the chemical industry for valves and other parts which must be highly resistant to corrosion. M. is an important alloy element (see Molybdenum alloys).

Historical. In 1778, Scheele obtained molybdic acid from molybdenite, and from the acid, Hjelm prepared the metal in powder form in 1781. Molybdenum steel was first made in 1913.

Molybdenum alloys: alloys in which molybdenum is the main element present. ***Ferromolybdenum*** is a molybdenum-iron alloy with 60 to 75% molybdenum used in the steel industry as a preliminary alloy. Molybdenum is added to steels to improve their tempering qualities, and resistance to heat and corrosion. The most important nonferrous alloys of molybdenum are nickel-molybdenum alloys (see Hastelloy®). A M. containing 0.5% titanium and 0.1% zirconium (TZM) is very resistant to heat and corrosion. It is used to make molds for pressure and injection molding, and valves for rockets and space vehicles. Because of its excellent corrosion resistance in the presence of molten zinc, an alloy of 70% molybdenum and 30% tungsten is used for zinc injection molds.

Molybdenum blue: a comprehensive name for bright blue molybdenum(VI)-molybdenum((V) mixed oxides. They can be prepared in amorphous form, with varying water contents, by cautious reduction of acidified molybdate solutions with tin(II) chloride, sulfurous acid or zinc; their general formula is $MoO_{3-x}(OH)_x$. Crystalline M. of the type Mo_3O_{3n-1} (e.g. Mo_9O_{26} or Mo_8O_{23}) can be prepared by reduction of molybdenum(VI) oxide with hydrogen at a temperature below 470 °C. Amorphous M. is used as a pigment, and the formation of M. from molybdate solutions is used as a qualitative test for molybdenum.

Molybdenum carbides. ***Dimolybdenum carbide***: Mo_2C, white, hexagonal crystals; M_r 203.89, m.p. 2687 °C, Mohs hardness 7. Mo_2C is chemically inert, but is attacked by nitric acid and aqua regia. It is formed by reaction of the elements at 1200 °C. ***Molybdenum carbide***: MoC, gray, shiny crystals, M_r 107.95, density 8.20, m.p. 2692 °C, Mohs hardness 7 to 8. MoC is soluble in hydrofluoric and concentrated nitric acids. It is made from molybdenum(VI) oxide and carbon at high temperatures. The M. are used to make hard metals.

Molybdenum carbonyl, *molybdenum hexacarbonyl*: $Mo(CO)_6$, highly refractive, volatile, rhombic crystals; M_r 264.01, density 1.96, m.p. about 180 °C (dec.). M. is formed by the reaction of carbon monoxide with molybdenum powder under high pressure, or by reaction of carbon monoxide with molybdenum(V) chloride in the presence of zinc as a reducing agent. It is used for vapor deposition of molybdenum on metals, ceramics or graphite.

Molybdenum chlorides: The most important M. is ***molybdenum(V) chloride, molybdenum pentachloride***, $MoCl_5$, which forms green-black, paramagnetic crystals; M_r 273.21, density 2.928, m.p. 194 °C, b.p. 268 °C. $MoCl_5$ is a dimer in the solid state, as $NbCl_5$ is (see Niobium chlorides); however, the dark red vapor is monomolecular. It is readily soluble in organic solvents such as carbon tetrachloride (in which it forms a dark red solution) or alcohol (dark green solution). $MoCl_5$ is very easily hydrolysed, and the primary product of hydrolysis is molybdenum oxygen trichloride, $MoOCl_3$. $MoCl_5$ is obtained from its elements under pressure, by heating molybdenum in a stream of chlorine, or by reaction of molybdenum(VI) oxide, MoO_3, with carbon tetrachloride under pressure. Chlorination of MoO_3 with thionyl chloride produces the black, diamagnetic ***molybdenum(VI) chloride, molybdenum hexachloride***, $MoCl_6$; M_r 308.66. This compound is extremely sensitive to water. Reduction of $MoCl_5$ with molybdenum or hydrogen at elevated temperature yields ***molybdenum(III) chloride***, $MoCl_3$, density 3.578, a compound which crystallizes in dark, red-brown needles and is insoluble in water. When heated to red heat, $MoCl_3$ disproportionates to yellow molybdenum(II) chloride, $MoCl_2$, density 3.718, and ***molybdenum(IV) chloride***, $MoCl_4$, a brown compound which is especially susceptible to oxidation and hydrolysis; M_r 237.75. $MoCl_2$ consists of hexameric clusters of the type $[MoCl]_8Cl_4$; the structure of the cations is a cubic structure with chlorine atoms at the vertices of the cube and Mo atoms in the centers of the faces.

Molybdenum(VI) fluoride, *molybdenum hexafluoride*: MoF_6, white, diamagnetic, very hygroscopic compound; M_r 209.93, density 2.55, m.p. 17.5 °C, b.p. 35 °C. M. reacts with fluorides to form fluoro complexes, $[MoF_7]^-$ and $[MoF_8]^{2-}$. M. is obtained by gently warming molybdenum in a stream of fluorine.

Molybdenum oxide: a compound of molybdenum with oxygen. The most important M. is ***molybdenum(VI) oxide, molybdenum trioxide***, MoO_3, a white powder which turns yellow when heated; M_r 143.95, density 4.69, m.p. 795 °C (beginning sublimation), b.p. 1155 °C. MoO_3 forms a layered lattice structure of MoO_6 octahedra which share edges and vertices. It is barely soluble in water, does not react with acids other than hydrofluoric and concentrated sulfuric acids, but does react with alkali hydroxide solutions to form alkali molybdates. MoO_3 reacts with dry hydrogen chloride at 150 to 200 °C, forming pale yellow, very volatile $MoO_2Cl_2 \cdot H_2O$; with fluorides, it forms fluoro complexes, $[MoO_3F_2]^{2-}$ and $[MoO_3F_3]^{3-}$. MoO_3 is an intermediate in industrial production of Molybdenum (see). In the laboratory, it can be made by roasting molybdenum disulfide or ammonium molybdate in air, or by heating $MoO_3 \cdot xH_2O$. MoO_3 is used on aluminum oxide carriers as a catalyst in coal hydrogenation, and for hydroforming, reforming and cracking petroleum. Reduction of MoO_3 with hydrogen at temperatures below 470 °C produces violet to blue-black intermediates Mo_nO_{3n-1}, such as Mo_9O_{26}, Mo_4O_{11}. ***Molybdenum(IV) oxide, molybdenum dioxide***, MoO_2, is a brown violet compound with a coppery sheen; M_r 127.95, density 6.47. The structure of MoO_2 is a highly distorted rutile lattice with pronounced Mo-Mo interactions. MoO_2 can also be made by cautiously heating molybdenum in the air. It is insoluble in acids and bases, is oxidized by nitric acid to MoO_3, reacts with chlorine to give the oxygen chloride MoO_2Cl_2, and forms oxo compounds with metal oxides, trigonal clusters of the type $[Mo_3O_8]^{4-}$.

Molybdenum(VI) oxide hydrate: $MoO_3 \cdot xH_2O$, often inaccurately called ***molybdic acid***: a yellow or white precipitate which forms when a molybdate solution is made very acidic. A defined yellow dihydrate, $MoO_3 \cdot 2H_2O$, is obtained by pouring a molybdate solution into 5 M hydrochloric acid; at 100 °C, it is converted to the yellow monohydrate, $MoO_3 \cdot H_2O$. There is also a white form of the monohydrate. M. does not contain discrete H_2MnO_4 molecules.

Molybdenum silicide, *molybdenum disilicide*: $MoSi_2$, gray crystals with a metallic sheen; M_r 152.11, density 6.31, m.p. 2050 °C. M. is chemically very inert and is soluble only in hydrofluoric acid. It is made from the elements at high temperatures. It is used to make wires for resistance elements for electric furnaces, combustion chambers, gas turbines, burners and high-temperature molds.

Molybdenum(IV) sulfide, *molybdenum disulfide*: MoS_2, soft, lead-gray, hexagonal leaflets with a violet sheen; M_r 160.07, density 4.80, m.p. 1185 °C. M. also exists in a second, rhombic modification. It is diamagnetic and a semiconductor. In the absence of air, M. is stable up to about 1300 °C. It is insoluble in water or dilute mineral acids. Above 315 °C, it is oxidized in air to molybdenum(VI) oxide, MoO_3. It reacts with hydrogen around 1100 °C, forming Mo_2S_3 and molybdenum; complete reduction to the metal occurs above 1500 °C. M. is found in nature as molybdite (molybdenum glance). It is isolated from natural sources by several flotation steps, with intermediate roasting steps at relatively low temperatures, followed by treatment with hydrofluoric acid around 600 °C. Because M. is very stable, it is obtained in fairly pure state. Hexagonal M. can be synthesized from the elements at 1100 °C; a long period of tempering is necessary to obtain crystalline preparations. Rhombic M. is formed by the reaction of MoO_3 with sulfur in a sodium or potassium carbonate melt at 900 °C; when heated for a long period, it converts to the hexagonal modification. Because M. is easily cleaved, is chemically and thermally stable and protects other materials from corrosion, it is an excellent lubricant in the temperature range between -185 °C and +450 °C. Unlike graphite and other lubricants, M. has better gliding properties at higher loads. It is also used as a catalyst.

Molybdic acid: see Molybdenum(VI) oxide hydrate.

MO methods: same as Molecular orbital theory (see).

Monalide: see Acylaniline herbicides.

Monastral pigments: same as Phthalocyanine pigments (see).

Monellin: see Sweeteners.

Monel metal®, ***Monel***®: a nickel alloy containing 63 to 70% nickel, 27 to 33% copper, 2% iron and 2 to 4% aluminum. M. is named for its discoverer, who obtained it for the first time as a natural alloy by melting suitable ores. Its resistance to corrosion is not due to the formation of a protective layer, which accounts for its good behavior even at high flow velocities of the corrosive medium. M. is used to build equipment in chemical plants and oil refineries.

Monensin: an ionophoric antibiotic. As the sodium salt, M. forms colorless crystals which are readily soluble in water and most organic solvents. It forms complexes with monovalent cations in the following order: $Ag^+ > Na^+ > K^+ > Rb^+ > Li^+ > NH_4^+$; these complexes are soluble in organic solvents. M. is weakly active in vitro against gram-positive bacteria.

Monoazo pigments: see Azo pigments.

Monobactams: see β-Lactam antibiotics.

Monocarboxylic acids: organic compounds with a carboxyl group -COOH in the molecule. Aliphatic M. are also called Fatty acids (see), because the higher homologs of this series are present in fats. The homologous series of saturated aliphatic M. can be symbolized by the formula C_nH_{2n+1}-COOH or $C_nH_{2n}O_2$. Many M. have trivial names related to their occurrence in free or bound form in animals and plants. The systematic names follow the same rules as those of other Carboxylic acids (see) (Table).

The lower homologs, from C_1 to C_3, are colorless liquids with pungent odors. The C_4 to C_9 compounds have a rancid smell. The higher homologs are solids with no odors. Like the saturated, aliphatic *n*-hydrocarbons, the M. display alternation of melting points. The M. with odd numbers of C atoms melt at lower temperatures than the next lower, even-numbered acids.

Unsaturated M. contain one or more C=C double bonds and/or C≡C triple bonds. The multiple bonds can be isolated or conjugated. Acrylic acid (see) is the simplest M., and in it the C=C and C=O group are conjugated. Vinylacetic acid is an M. with an isolated C=C double bond, which is readily converted to the isomeric and conjugated crotonic acid by bases or mineral acids:

$$\underset{\text{Vinylacetic acid}}{CH_2{=}CH{-}CH_2{-}COOH} \xrightleftharpoons[(OH^-)]{} \underset{\text{Crotonic acid}}{CH_3{-}CH{=}CH{-}COOH}$$

Saturated, unbranched, aliphatic monocarboxylic acids

Trivial name	Systematic name	Formula	m. p. in °C
Formic acid	Methanoic acid	H-COOH	8.4
Acetic acid	Ethanoic acid	CH_3-COOH	16.6
Propionic acid	Propanoic acid	CH_3-CH_2-COOH	–20.8
Butyric acid	Butanoic acid	CH_3-$(CH_2)_2$-COOH	–6.5
Valeric acid	Pentanoic acid	CH_3-$(CH_2)_3$-COOH	–34.5
Caproic acid*	Hexanoic acid	CH_3-$(CH_2)_4$-COOH	–9.5
Enanthic acid*	Heptanoic acid	CH_3—$(CH_2)_5$-COOH	–10.0
Caprylic acid*	Octanoic acid	CH_3-$(CH_2)_6$-COOH	16.0
Pelargonic acid*	Nonanoic acid	CH_3-$(CH_2)_7$-COOH	12.0
Capric acid*	Decanoic acid	CH_3-$(CH_2)_8$-COOH	31.5
Undecylic acid*	Undecanoic acid	CH_3-$(CH_2)_9$-COOH	28.0
Lauric acid	Dodecanoic acid	CH_3-$(CH_2)_{10}$-COOH	44.0
–	Tridecanoic acid	CH_3-$(CH_2)_{11}$-COOH	40.5
Myristic acid	Tetradecanoic acid	CH_3-$(CH_2)_{12}$-COOH	58.0
–	Pentadecanoic acid	CH_3-$(CH_2)_{13}$-COOH	52.1
Palmitic acid	Hexadecanoic acid	CH_3-$(CH_2)_{14}$-COOH	64.0
Margarinic acid	Heptadecanoic acid	CH_3-$(CH_2)_{15}$-COOH	62.0
Stearic acid	Octadecanoic acid	CH_3-$(CH_2)_{16}$-COOH	69.4
–	Nonadecanoic acid	CH_3-$(CH_2)_{17}$-COOH	69.4
Arachic acid*	Eicosoanic acid	CH_3-$(CH_2)_{18}$-COOH	75.3

* Trivial name not approved by IUPAC.

This type of isomerization is called three-carbon tautomerism.

The most important unsaturated fatty acid is Oleic acid (see). Some other well-known M. with multiple bonds are Cinnamic acid (see) and Propiolic acid (see), the simplest M. with a C≡C triple bond. The melting points of the M. are significantly reduced by branching or introduction of double bonds. Unsaturated M. with the Z configuration usually have lower melting points than the isomeric E compounds.

In aromatic M., the carboxyl group is bound directly to a C atom of the aromatic system. The simplest of these M. is Benzoic acid (see). Some other well-known examples of this type are α- and β-naphthoic acids. The many substituted aromatic M. are more important than the simple M. Examples are anthranilic acid, 4-aminobenzoic acid and salicylic acid.

Like the aromatic compounds, heterocyclic M. have a COOH group bound to a C atom of the ring system. Some examples are pyromucic (furan-2-carboxylic) and nicotinic (pyridine-3-carboxylic) acids. Like all other Carboxylic acids (see), M. are relatively easily deprotonated, and therefore give an acidic reaction. However, the dissociation equilibrium lies on the side of the undissociated form. In other words, M. are weak acids. Their acidity is greatly increased by introduction of electronegative substituents, especially at the α-carbon atom. For example the pK of acetic acid is 4.76, while that of trichloroacetic acid is 0.66. M. have lower boiling points than the corresponding alcohols and phenols. The reason is that the M. are able to form dimers, which are also present in the gas phase of many of them.

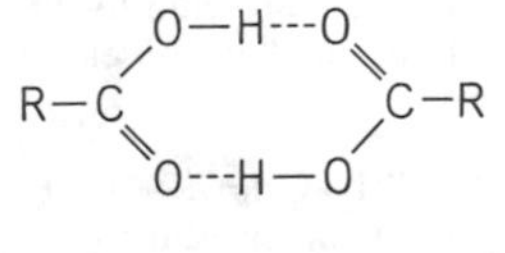

```
     O—H---O
    /       \\
R—C          C—R
   \\       /
     O---H—O
```

There are many derivatives of M. created by substitution of one or more hydrogen atoms in the organic residue or by conversion of the carboxyl group to an Ester (see), Amide (see), Acyl halide (see), etc.

Aliphatic M. are synthesized chiefly by oxidation of alkanes, primary alcohols or aldehydes. Higher M. can also be obtained by hydrolysis of lipids. The aromatic M. are synthesized mainly by hydrolysis of the corresponding nitriles and trihalomethyl compounds. In addition, there are many general and special methods for their synthesis.

Monochloroacetic acid: same as Chloroacetic acid (see).

Monochloropentanes: same Amyl chlorides (see).

Monoclinic: see Crystal.

Monocrotophos: see Organophosphate insecticides.

Monodentate: see Coordination chemistry.

Monolinurone: see Urea herbicides.

Monomer: a reactive chemical compound which has the capacity to form polymers. M. contain either olefinic double bonds (e.g. ethene, propene, butadiene, acrylonitrile and the ***vinyl monomers*** vinyl acetate, vinyl ether and vinyl chloride), or other reactive groups (e.g. ethene oxide, ethenimine and formaldehyde). M. are linked into Polymers (see) by the process of Polymerization (see), Polycondensation (see) or Polyaddition (see). The ability of M. to form polymers is utilized mainly to make plastics.

Monomolecular reaction: a chemical elemental reaction in which only one type of molecule reacts and generally follows a first order rate law. A schematic reaction equation is A → Products. Thermal and photochemical dissociations or isomerizations are typical M.

Thermal monomolecular gas-phase reactions follow first order kinetics at high pressure. At low pressures, the first order rate constant decreases as the pressure drops. The activation energies for these reactions are usually high. This activation energy is provided by absorbed light in photochemical processes. In thermal reactions, however, it can be obtained only by collisions between the molecules. Collision theory (see Kinetics of reactions, theory), however, requires second order rate laws for interactions between two particles. The *Lindemann-Hinshelwood mechanism* (1922) solves the problem with the following postulates: 1) If a molecule A acquires the sufficient energy through collisions to cross the potential barrier, this activated particle A* does not react immediately. Its energy is redistributed among its internal degrees of freedom, and there is a certain probability that the energy will be concentrated on the critical vibrational mode which can lead to a cleavage or isomerization. The activated molecule A* thus has a certain lifetime. 2) If this lifetime is long compared to the rate at which the energy distribution equilibrium in the system is established, A* will be in equilibrium and will decay by a first order process. 3) If A* reacts rapidly in comparison to the time required for equilibration, its formation becomes rate-determining, and this follows a second-order rate law. This is the case in gas reactions at low pressure and high decay constants for A* (high temperature, low activation energy). These ideas can be expressed formally as follows:

$$A^* \xrightarrow{1} \text{Products} \qquad \text{Reaction}$$

$$A + M \xrightarrow{2} A^* + M \qquad \text{Activation}$$

$$A^* + M \xrightarrow{-2} A + M \qquad \text{Deactivation}$$

Here M acts as the energy-transferring collision partner. It can be another particle of type A, or an inert gas molecule.

If a quasistationary concentration of A* (see Complex reaction) is assumed, the rate law is

$$-\frac{dc_A}{dt} = \frac{k_1 k_2 c_A c_M}{k_1 + k_{-2} c_M}$$

At high concentrations c_M, $k_1 \ll k_{-2} c_M$, and one obtains the first-order rate equation

$$-\frac{dc_A}{dt} = \frac{k_1 k_2}{k_1} c_M = k_\infty c_M.$$

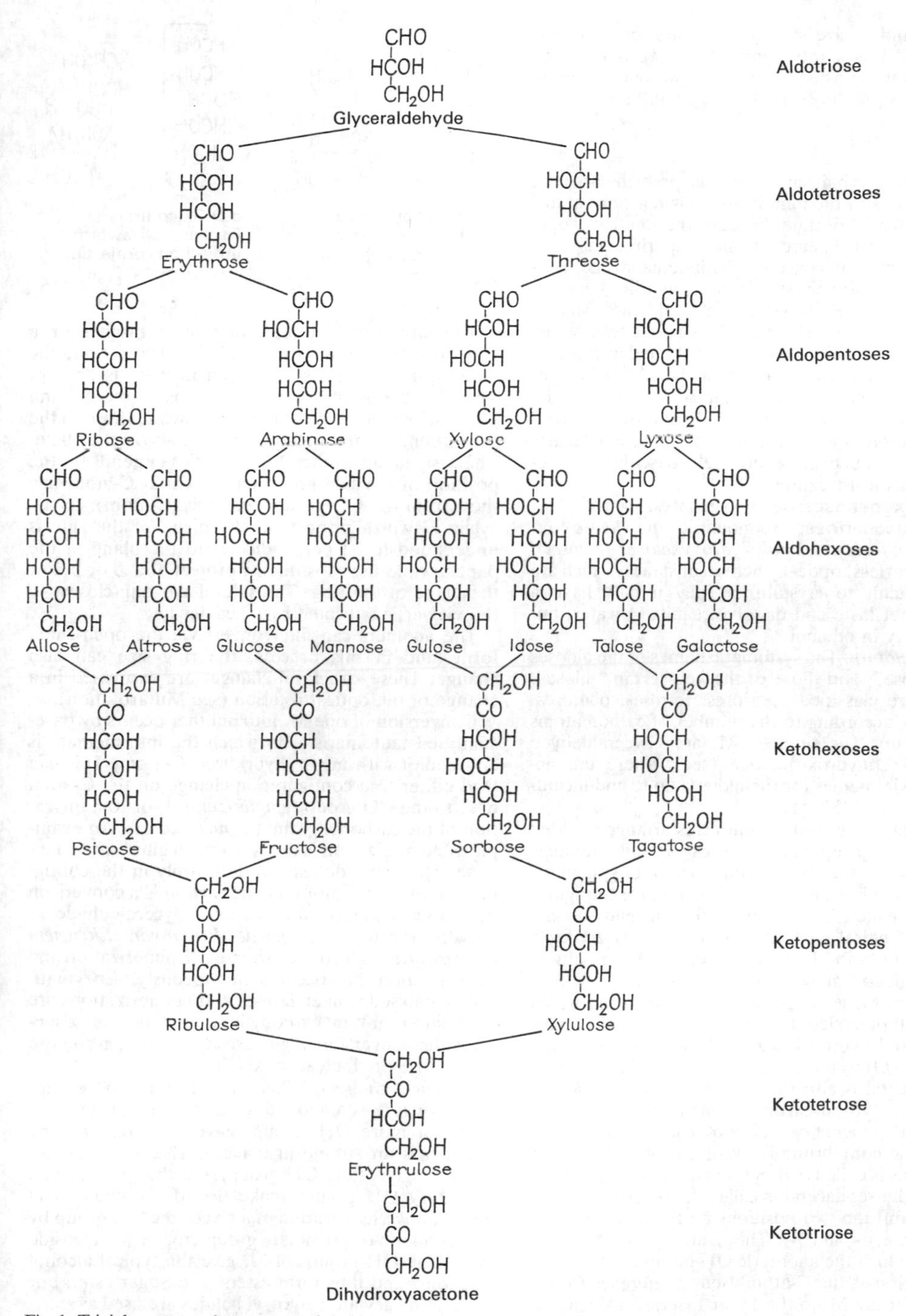

Fig. 1. Trivial names and structures of the monosaccharides.

Here c_A and c_M are the concentrations of substances A and M, k is a rate constant and k_∞ is the high-pressure rate constant. At very low concentrations, i.e. low gas pressures, $k_1 \gg k_{-2}c_M$, and thus

$$-\frac{dc_A}{dt} = k_2 c_A c_M$$

From this mechanism, it could be predicted that the rate constants of monomolecular gas reactions at low pressures would decline, because the rate law would change. This has been confirmed experimentally.

The theoretical concept of Lindemann and Hinshelwood has proven very fruitful, and has been extended. Rice, Ramsperger, Kassel and Marcus (*RRKM theory*) refined the ideas about energy distribution in the activated molecule A* and its reaction probability. It must be assumed that A* no longer reacts with a fixed rate constant k_1, but that this depends on the excess energy in A* (specific rate constant). The energy distribution among the molecules can also be much more precisely described by detailled statistical treatment.

Monooxygenases: see Oxygenases.

Monosaccharides: monomeric *polyhydroxyaldehydes* (*aldoses*) and *polyhydroxyketones* (*ketoses*). M. are colorless, optically active substances which are often difficult to crystallize. They usually have a weak, sweet taste and dissolve readily in water, but only slightly in ethanol.

Nomenclature. The systematic names of the aldoses end in "-ose", and those of the ketoses, in "-ulose". The M. are classified as trioses, tetroses, pentoses, and so on, according to the number of carbon atoms they contain. The simplest M. are Glyceraldehyde (see) and Dihydroxyacetone (see). There are accepted trivial names for the aldoses up to and including the hexoses (Fig. 1).

The numbering of the C-atoms is arranged so that the carbonyl group has the lowest possible number, C1 in the aldoses. For the construction of systematic names for the longer-chain M., the relative configuration is indicated by a prefix which depends on the number of chiral C-atoms: 1) glycero-, 2) erythro-, threo-, 3) arabino-, lyxo-, ribo-, xylo-, 4) allo-, altro-, galacto-, gluco-, gulo-, ido-, manno-, talo-.

Structure. M. generally adopt a cyclic form, which is the result of nucleophilic attack of a sterically favorably situated hydroxyl group (-OH) on the carbonyl group (-C=O) to form a semiacetal. The formation of the semiacetal is also the reason that the M. do not give certain typical carbonyl reactions, such as that with Schiff's reagent or addition of sodium hydrogensulfate. The equilibrium between the open-chain and cyclic forms is called *oxo-cyclo tautomerism*. The OH-group of the semiacetal is called "anomeric" or "glycosidic", and the two positions on the new center of chirality are α- and β-. The α-anomer is the compound in which the anomeric OH group has the same configuration as the C-atom used as reference for assignment of the M. to the D- or L-series. Depending on the ring size, the semiacetal is called a *pyranose* (prefix, pyranosyl-) or *furanose* (prefix, furanosyl-) (Fig. 2). In compounds with more than 4 centers of chirality, two or more of these prefixes are used. The presence of chiral centers makes the M. and their derivatives optically active.

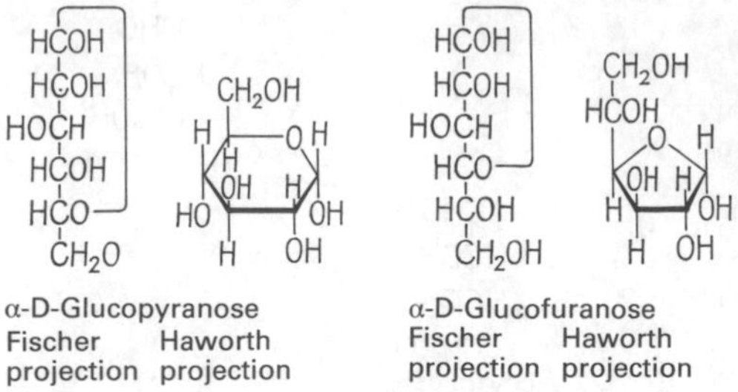

α-D-Glucopyranose
Fischer projection Haworth projection

α-D-Glucofuranose
Fischer projection Haworth projection

Fig. 2.

The structure is represented by a Fischer or a Haworth projection. In the Fischer projection, the atom with the highest oxidation number is on top. The C-C bonds of the carbon atoms in the column point below the plane of the paper, while those to the substituents on the right and left are above the plane. The assignment to the D- or L-series depends on the position of the OH group on the chiral C-atom farthest from the CO group (reference C-atom). In the cyclic Haworth projection, the plane of the ring is understood to be perpendicular to the plane of the paper, while the substituents project above or below the plane of the ring. The side of the ring closest to the viewer is indicated by a heavier line.

The anomers can interconvert via the open-chain form, and in this process, the ring size can also change. These structural changes are expressed in a change of the optical rotation (see Mutarotation).

Conversion of one M. into another occurs by base-catalysed tautomerism, in which the intermediate is an enediol with an sp^2-hybridized C-atom. This can lead either to a configuration change on the C-atom next to the CO group (*epimerization*), or to a migration of the carbonyl group (*isomerization*). An example of epimerization is the conversion glucose ⇌ mannose. The two aldohexoses differ only in the configuration of the C2 atom. Isomerization is a conversion of aldose ⇌ ketose, for example, glyceraldehyde ⇌ dihydroxyacetone (*Lobry-de Bruyn-van Ekenstein conversion*) or glucose ⇌ fructose. Epimerization and isomerization can occur simultaneously. Enzymatically catalysed isomerizations and epimerizations are very important in carbohydrate metabolism, especially the conversion of glucose ⇌ fructose, arabinose ⇌ ribulose or ribulose ⇌ xylulose.

Chemical changes. When one or more OH groups are lacking, the compound is a Deoxysugar (see), and if one or more OH- groups have been replaced by $-NH_2$, it is an Aminosugar (see). The potential CO group or glycosidic OH group, and the primary and secondary OH groups make the M. polyfunctional compounds. Substitution of a glycosidic OH group by an alcoholate or phenolate group creates a Glycoside (see). The OH groups of M. give the typical alcohol reactions, and thus form esters (see Sugar esters) or ethers. The acylated glycosyl halides are used as starting materials for glycoside syntheses (see Glycosides). M. form acetals (e.g. benzylidene compounds with benzaldehyde) or ketals (e.g. isopropylidene compounds with acetone) with aldehydes or ketones. Sugar anhydrides (see) are formed by intramolecular water loss between an alcoholic and a glycosidic OH group; loss of water from two alcoholic OH groups produces an Anhydrosugar (see). Condensation reactions of the CO group produce oximes (e.g. with hydroxylamine) or hydrazones (e.g. with phenylhydrazine). At a pH > 4-5, aldoses and ketoses react with phenylhydrazine to form Osazones (see). The Aldonic acids (see) are derived from the M. by oxidation of the terminal CO group; the Uronic acids (see) by oxidation of the primary OH group, and the Aldaric acids by oxidation of the CO group and the primary OH group. Periodate oxidation leads to a C-C cleavage. Reduction of the CO group produces an Alditol (see).

Analytical. Qualitative and quantitative methods depend on reduction or color reactions. 1) In *reduction tests*, the M. is oxidized, e.g. to an onic acid, and an oxidizing agent is reduced (Table 1). Some of these methods are also suitable for quantitative analysis of the M. 2) The *color reactions* are based on the formation of furfural derivatives. Removal of water by mineral acids produces furfural from pentoses, or 5-hydroxyfurfural from hexoses. In the presence of acids, furfural derivatives form colored condensation products with various reagents (Table 2). Some of these reactions can be made relatively specific for certain M.

Table 1. Sugar determination by reduction of metal ions

Metal	Reagent	Name of test
Cu(II)→Cu(I)	$CuSO_4$/tartrate/OH^-	Fehling
	$CuSO_4$/OH^-	Trommer
	$CuSO_4$/citrate/OH^-	Benedict
Ag(I)→Ag	$[Ag(NH_3)_2]^+$	Tollens
Bi(III)→Bi	Bi(III)/tartrate/OH^-	Nylander
Fe(III)→Fe(II)	$K_3[Fe(CN)_6]$	Hagedorn-Jensen

Table 2. Color reactions of sugars via furfural derivatives

Reactive partner of furfural	Reagent	Test
Active methylene groups	Anthrone/H_2SO_4	Anthrone reaction
Phenols	Orcinol/HCl/Fe^{3+}	Bial reaction
	γ-Naphthol/H_2SO_4	Molisch reaction
SH compounds	Cysteine/H_2SO_4	Dische reaction
Aromatic amines	Aniline phthalate	
	o-Toluidine	

Occurrence. The most widely naturally occurring M. are the aldohexoses D-glucose, D- and L-galactose and D-mannose, the aldopentoses D-ribose, D-xylose and L-arabinose, and the ketose D-fructose. The phosphate esters of D-xylulose, D-ribulose and D-sedoheptulose are important intermediates in carbohydrate metabolism. A few branched M. are found in microorganisms and plants.

Monosulfane: same as Hydrogen sulfide (see).

Monosulfane monosulfonic acid: same as Thiosulfuric acid (see).

Monoterpenes: terpenes consisting of two isoprene units, that is, of 10 C atoms. They can be acyclic, monocyclic or bicyclic. The hydrocarbons ocimen and myrcene, the alcohols geraniol, its *trans*-isomer nerol, linalool and citronellol, and the aldehydes citral and citronellal are examples of acyclic M. The cyclohexane derivatives of the p-menthane type make up the largest group of monocyclic M., which includes hydrocarbons, such as the terpinenes, dipentene, the phellandrenes and p-cymene, alcohols, such as menthol and terpineol, and ketones, such as menthone, pulegone, carbone and piperitone. Some compounds, such as p-cymene and the phenols carvacrol and thymols, are benzene derivatives. Cantharidine is an example of a cyclohexane derivative with a different type of structure. The iridoids and pyrethrins are cyclopentane derivatives. The most important carbon skeletons of the bicyclic M. are thujane, carane, pinane, bornane (formerly called camphane) and fenchane. M. are the main components of volatile oils and are important aroma substances. The strongest known aroma substance is the M. p-menth-1-ene-8-thiol, which is present in grapefuit juice. Those M. which are not esterified with fatty acids (e.g. geraniol) or glycosides (iridoids) can be steam distilled.

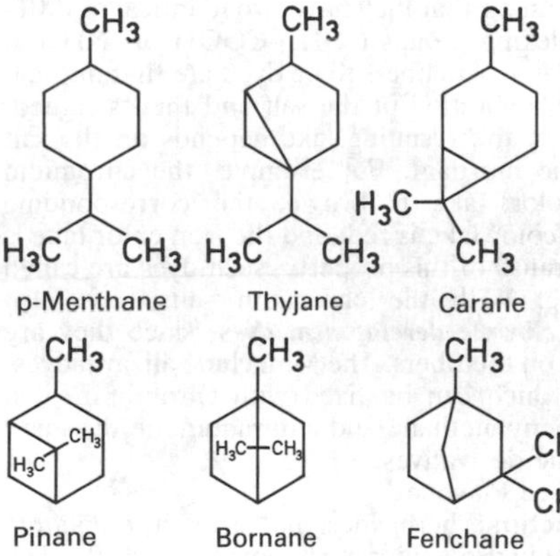

p-Menthane Thyjane Carane

Pinane Bornane Fenchane

Basic structures of mono- and bicyclie monoterpenes.

Monotropism: see Polymorphism.

Monovinyl acetylene: same as Vinyl acetylene (see).

Monsanto: sec Synthetic fibers.

Monsanto process: an Organic electrosynthesis (see).

Montage diagram: see Technological diagrams.

Montane resin: a dark colored resin which can be extracted from crude montane wax with organic solvents.

Montane wax: a fossil plant wax which is obtained from bitumen-rich crude brown coal by extraction with benzene or a mixture of benzene and alcohol. It is a brown to black product which melts at 80 to 90 °C, and is considered a true wax because of its chemical composition. It consists of esters of monobasic acids, such as montanoic acid, $C_{29}H_{56}O_2$, and also contains resin and asphalt. Crude M. can be purified by distillation and bleaching with nitric or chromic acid, and modified by partial esterification and saponification. Because of its hardness, high melting point and ease of polishing, M. is used in many ways, e.g. for shoe polish, floor wax and candles.

Monurone: see Urea herbicides.

MORD: see Chiroptic methods.

Mordant: a substance which fixes a dye to the textile; otherwise, the pigment would not be washfast. The M. is applied before, or sometimes after the dye, in a colloidal suspension which is adsorbed to the fibers.

Metal M. include aluminum salts (e.g. aluminum sulfate, basic aluminum acetate), chromium salts (e.g. chrome alum, chromium acetate, potassium dichromate), iron salts (e.g. iron acetate, basic iron(III) sulfate), tin salts (e.g. tin oxalate, tin acetate) and copper salts (e.g. copper sulfate). ***Tannin M.*** are fixed to the fibers with a subsequent treatment with potassium antimonyl tartrate; catechu, gambir and sumach are similar M. There are also ***synthetic organic M.***, such as catanols, and ***oil M.*** such as Turkish red oil. The last group are sulfonated hydroxyfatty acids or sulfonated unsaturated fatty acids.

Mordant dyes: a group of synthetic pigments which form colored complex compounds, or Lakes (see) with metal salts (usually chrome alum, tin, iron, aluminum or copper salts). All of these pigments have in common that they have two identical or different complexing groups (-OH, -COOH or $-NH_2$) at position 2 or 4. The fibers to be dyed are first impregnated with a solution of the salt and then steamed. The color of the resulting lake depends on the salt used as the mordant. For example, the chromium alizarin color lake is brown, the corresponding aluminum color lake is red, and the iron color lake is violet. Because of this property, such dyes are called polygenetic. With the chromium salts, one also speaks of ***chrome development dyes***, since they are developed on the fibers. The M. include alizarin dyes, azo dyes which can be fixed with chromium salts, some triphenylmethane and anthraquinone dyes and gallocyanine derivatives.

Morin: see Flavones.

Morphactins: herbicides and growth regulators based on 9-hydroxyfluorenecarboxylic acid-(9). The main representative of the group is the *n*-butyl ester (*flurenol*). The plant takes up the M. through its leaves and roots, and the compound inhibits growth. M. are usually applied as post-emergent herbicides, together with aryloxyalkanoic acids.

Morphine: an opium alkaloid which contains the morphine skeleton. The morphine base is not very water-soluble, has a bitter taste, and loses its water of crystallization at 120 °C. m.p. 254 °C (dec.), $[\alpha]_D^{20}$ -130.9° (in methanol). The hydrochloride is readily soluble in water and crystallizes with 3 molecules water of crystallization; m.p. 200 °C, $[\alpha]_D^{20}$ -97.9 °C (in water). M. has a pentacyclic ring system (Fig.); its tertiary N atom is part of a piperidine ring which is in the chair conformation.

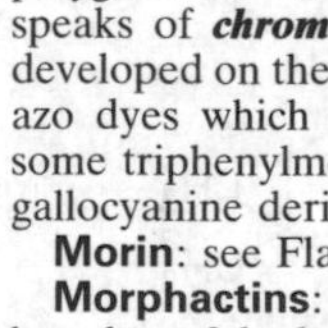

Morphinane

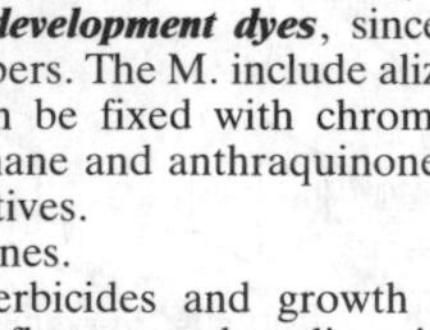

R = H: Morphine
R = CH_3: Codeine
R = C_2H_5: Ethylmorphine

M. contains 5 chiral C atoms, and its absolute configuration is 5R, 6S, 9R, 13S, 14R. M. contains a phenolic and an alcoholic hydroxyl group. It is an amphoteric compound, with the basic properties due to the tertiary N atom (pK_a = 7.89) stronger than the acidic properties due to the phenolic hydroxyl group (pK_a = 9.85). The isoelectric point, at which the base is most readily extracted into an organic solvent, is at pH 9.1. M. can be converted into other products, such as morphine N-oxide and dehydrodimorphine (pseudomorphine), by mild oxidizing agents. In the latter, a C-C bond forms between the C2 atoms of two M. molecules.

As a phenolic compound, M. gives a blue color with iron(III) chloride. With Marquis' reagent (a little formaldehyde in conc. sulfuric acid), it gives a violet color. This reaction and many others which take place in conc. sulfuric acid are due to the conversion of M. into Apomorphine (see). A suitable reaction for characterizing and quantitative analysis of M., for example in opium, is the formation of the insoluble 2,4-dinitrophenol ether with 2,4-dinitrochlorobenzene or 2,4-dinitrofluorobenzene. M. is present in opium at a concentration of 6 to 21%, and it is extracted from this source. M. is also extracted from poppy straw. It is a strong analgesic which acts on the central nervous system. It also suppresses coughing. An undesirable side effect is suppression of respiration. An overdose causes death by respiratory paralysis. M. leads to euphoria and is addictive.

Codeine is the 3-methyl ester of M.; it is a base and crystallizes with one molecule of crystal water; m.p. 155 °C, $[\alpha]_D^{20}$ -138° (in ethanol). The dihydrogen phosphate is used therapeutically; its m.p. is 235 °C, $[\alpha]_D^{20}$ -102° (in water. Codeine is more stable than M. It does not react with $FeCl_3$, but does give the same reactions in the presence of conc. sulfuric acid which lead to apomorphine. C. is present at a concentration of 0.3 to 6.5% in opium. It is now obtained mainly by selective methylation of M., e.g. with benzenesulfonate methyl ester, or phenyltrimethylammonium hydroxide. Under these conditions, the N atom is not quaternized. Codeine can also be partially synthesized from Thebaine (see). Codeine is a weak analgesic, but has a strong cough-suppressing activity. It is much less likely to cause addiction than M.

Ethyl morphine, Dionin®, is the ethyl homolog of codeine. It is used in optometry, and occasionally as a cough suppressant.

Historical. M. was the first alkaloid to be isolated; this was achieved in 1906 by Sertürner, who obtained it from opium. The structural formula was elucidated in 1927 by Robinson and Schœpf. The first total synthesis, which is of no practical significance, was achieved in 1952 by Gates and Tschudi.

Morpholine, ***tetrahydro-1,4-oxazine***: the anhydride of diethanolamine. M. is a colorless, hygroscopic liquid with a strong amine odor; m.p. -4.9 °C, b.p. 129 °C, n_D^{20} 1.4548. It is soluble in water and most organic solvents, and can be steam distilled. M. is a strong base which irritates the eyes, skin and mucous membranes. It is synthesized either by heating diethanolamine with 70% sulfuric acid, or dichlorodiethyl ether is reacted with ammonia. M. is a versatile intermediate in industry; it is used to make vulcanization accelerators, optical brighteners, corrosion protection agents and as a solvent for dyes, resins and waxes. M. is also used for de-gassing water. Derivatives with substituents on the N atom are used as pharmaceuticals, herbicides and pesticides.

Morpholine and piperazine fungicides: Systemic Fungicides (see) with piperazine or morpholine skeletal components. *Triforin* and *trimorphamide* are used mainly against true mildew and scab. *Tridemorph, aldimorph* and *dodemorph* are also used against mildew. *Fenpropimorph* is effective against rusts as well.

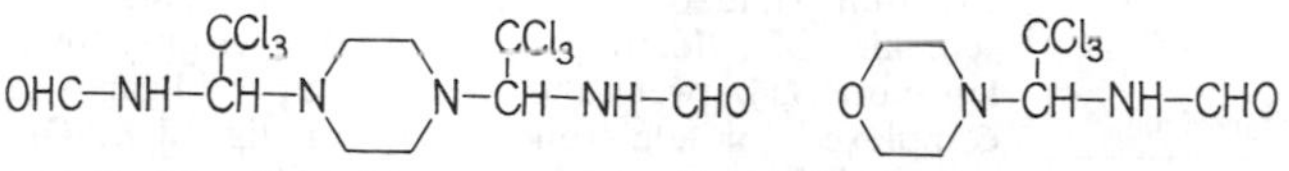

Triforin Trimorphamide

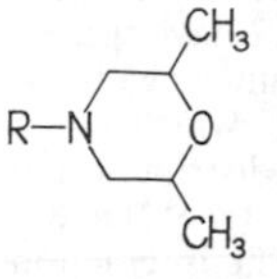

Tridemorph (R = branched $C_{13}H_{27}$ chain)
Aldimorph (R = n-$C_{12}H_{25}$)
Dodemorph (R = cyclic $C_{12}H_{23}$)
Fenpropimorph
(R= $(H_3C)_3C-C_6H_4-CH_2-CH(CH_3)-CH_2$)

Morphylan process: see Benzene.

Mortar: see Binders (building materials).

Mosaic gold: gold-colored, shiny, hexagonal leaflets consisting essentially of crystalline tin(IV) sulfide, SnS_2. It is used as a pigment for fake gilding.

Mosaic texture: see Crystal lattice defects.

Moseley's law: A regularity in x-ray spectra discovered in 1913 by H. Moseley. He found that the frequencies of comparable x-ray lines, for example the K_α-lines (see X-ray spectroscopy) increase monotonically with the nuclear charge number of the emitting chemical element. Mathematically, this is expressed by $\sqrt{\nu} = A(Z-S)$, where Z is the nuclear charge number of the atom and A and S are constants characteristic of the two lines. For example, if the square roots of the frequencies of the K_α-lines of different elements are plotted against Z, a straight line is obtained. The slope of this line and its intersection with the ordinate yield values for A and S. For the K series, $S = 1$. This value corresponds to the nuclear screening predicted by a simple shell model for x-ray spectra of the K series. Thus S is also called the *screening constant*.

M. was very important in the discovery of atomic structure; it was used to show that the basis of order in the periodic system is not atomic mass, as was first assumed, but nuclear charge (atomic numer). On this basis, the contradictions between chemical behavior and arrangement by atomic mass could be resolved for the elements iodine/tellurium, potassium/argon and nickel/cobalt. M. was also the key to discovery of still unknown elements and was an important support for the recognition of Bohr's atomic model.

Mössbauer spectroscopy, ***gamma ray resonance spectroscopy:*** the study of resonance absorption of monochromatic gamma rays by nuclei in solids, based on the Mössbauer effect discovered in 1957 by R. Mössbauer.

Theoretical. A resonance absorption occurs when a system absorbs energy quanta which are equal to the difference between two of its possible energy states. Resonance absorptions have long been known in various areas of Spectroscopy (see); an example is shown in Fig. 1.

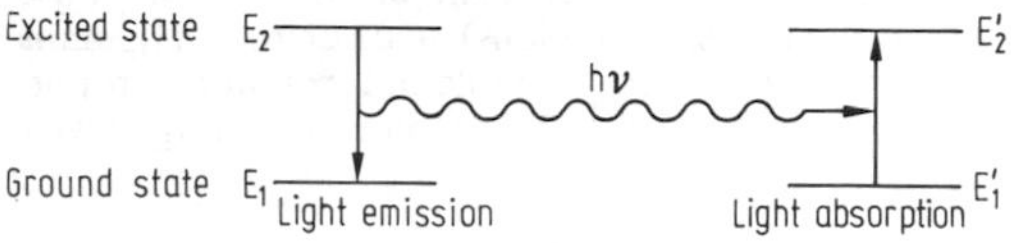

Excited state E_2 E_2'
$h\nu$
Ground state E_1 E_1'
Light emission Light absorption

The nuclear resonance absorption of gamma rays is thought to be analogous to the absorption of visible light by atoms or molecules in which electronic transitions occur. One would expect that a gamma ray emitted by a nucleus in an excited state with the energy E_2 as it returns to the ground state E_1 would be capable of absorption by another nucleus of the same isotope; the second nucleus would simultaneously be excited to the higher excited state. However, the experimental realization of this idea ran into difficulty. Excited atomic nuclei in atoms or molecules which are freely mobile, i.e. in a gas or liquid, experience a recoil when they emit a γ-quantum (Fig. 2), due to conservation of momentum.

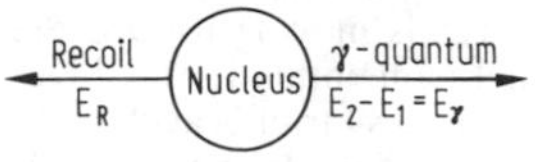

Recoil E_R Nucleus γ-quantum $E_2 - E_1 = E_\gamma$

Fig. 2. Recoil in the emission of a γ-quantum.

Thus the energy of the γ-quantum is not equal to the energy difference $E_2 - E_1$ between the two nuclear states, but is less by an amount equal to the recoil energy E_R transferred to the nucleus. Absorption of a γ-quantum is similarly affected. If the absorbing nucleus is in a mobile atom or molecule, it receives additional momentum from the absorption process, so that the energy of the γ-quantum must be larger than $E_2 - E_1$ by an amount equal to the recoil energy. As a result, absorption and emission do not occur at the energy $E_0 = E_2 - E_1$. Absorption occurs at shorter wavelengths, corresponding to $E_0 + E_R$, and emission at longer wavelengths, corresponding to $E_0 - E_R$. In other words, the absorption and emission lines are separated by the amount $2E_R$. Thus observation of a nuclear resonance absorption of gamma rays in gases and liquids, in which recoil effects appear, is nearly impossible. Mössbauer recognized that, for gamma radiation of moderate energy, there is a certain probability of recoil-less absorption or emission only if the nucleus is bound in a solid. Because the mass of the entire crystal lattice is many orders of magnitude larger than that of the atomic mass, the recoil energy is negligibly small. The relative frequency range $\Delta\nu/\nu$ of the γ-quantum emitted under recoil-less conditions is about 10^{-12} to 10^{-14}, and is thus very much smaller than the natural linewidth.

Instrumentation. The simplest Mössbauer spectrometer consists of a radiation source which emits γ-radiation under recoil-less conditions, an absorber or a scattering unit, which contains the sample (also under recoil-less conditions) a detector for gamma rays (counting tube, scintillation counter, proportional counter) and an electronic recording device (Fig. 3).

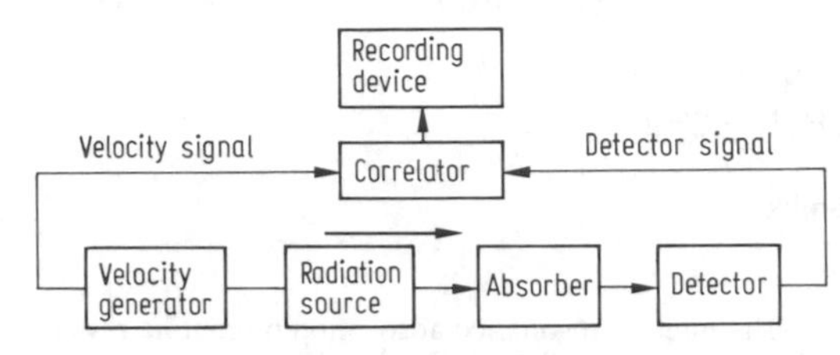

Fig. 3. Diagram of a Mössbauer spectrometer.

The detector can measure the radiation transmitted or scattered by the sample. Because the linewidth of the emitted gamma rays is extremely narrow, even very small energy differences between the radiation source and the absorber, such as could be caused by differences in the chemical or physical states of the nuclei in the two units, can interfere with the resonance. In order to achieve a resonance absorption, it is necessary to vary the frequency of the radiation source, and this is done by moving the source relative to the absorber. Due to the Doppler effect, the radiation emitted while the source is moving toward the absorber receives additional momentum in the direction of motion, and thus has a shorter wavelength; conversely, the radiation emitted while the source is moving away from the absorber has a longer wavelength. The relative velocity is changed continuously and periodically in a Mössbauer spectrometer, for example by placing the radiation source on an electromechanical vibrator which oscillates with a certain frequency.The number of γ-quanta reaching the detector is recorded as a function of the relative velocity.

The radiation source and absorber contain the same type of nuclei. Radioactive isotopes are used as radiation sources; they must be relatively stable and have narrow gamma lines with energies of about 100 KeV. Of the numerous isotopes now available, about 80 are suitable; ^{57}Fe and ^{119}Sn are especially good for M. Other conditions which favor observation of the Mössbauer effect are low temperatures, low energies of the γ-quanta and a certain degree of rigidity of the nuclei in the crystal lattice.

Spectral parameters and their information content. The resonance absorption of a nucleus can be shifted or split into several lines by the interactions of the nucleus with electric and magnetic fields generated mainly by the electron shells. The splitting is called hyperfine splitting. The spectral parameters most important to chemists are isomer shifts, quadrupole splitting and magnetic splitting.

Isomer shifts. If the nuclei in the radiation source and absorber are in the same chemical and physical state, the resonance absorption will be maximum when the relative velocity of the two is zero. If they are not equivalent, the energy of the emitted γ-quanta is varied by the Doppler shift until the absorption is maximum, and the required relative velocity v is called the "isomer shift", δ, or sometimes, the "chemical shift". It is caused by electric interactions between the nucleus and electrons, and is very sensitive to changes in the electron shell. Measurement of the isomer shift provides information on the electron density at the nucleus, and this in turn depends on the oxidation state of the atom and its bonding (e.g. screening of s-electrons by p- and d-electrons, covalence effects between the central atom and ligands in complexes, or electronegativity of ligands). The energy differences can be calculated from the isomer shift in mm/s; their magnitude is about 10^{+8} eV.

Quadrupole splitting. Although nuclei with nuclear spin quantum numbers $I = 0$ and $I = 1/2$ have spherically symmetrical charge distributions, nuclei with $I > 1/2$ have electric quadrupole moments which are a measure of the deviation of the charge distribution in the nucleus from spherical symmetry (see Nuclear quadrupole resonance spectroscopy). The interaction of the electric quadrupole moment of the nucleus in the ground or excited state with an inhomogeneous electric field causes the nuclear levels to split into $I+1/2$ substates (Fig. 4), and leads to line splitting in the spectrum.

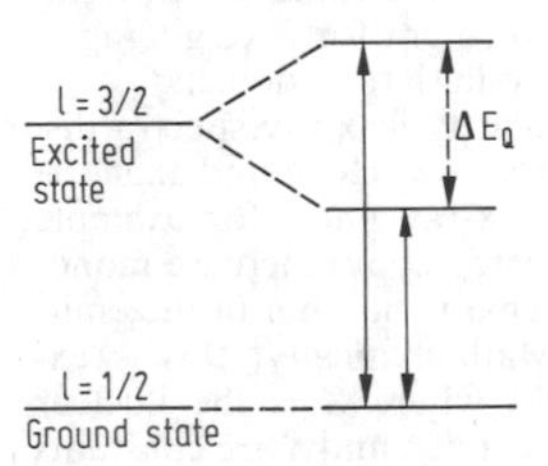

Fig. 4. Diagram of quadrupole splitting in the ^{57}Fe nucleus.

The energy difference between these lines is called the quadrupole splitting ΔE_Q; in general, like isomer shifts, it is on the order of 10^{+8} eV. The inhomogeneous electric field at the nucleus is caused by unsymmetrical arrangement of the electrons around the nucleus, which may arise, for example, because the orbitals are not completely filled or because the ligands on a complex ion are not identical. Quadrupole splitting therefore gives information on the symmetry of the electronic shell, including the coordination sphere in complexes.

Magnetic splitting. Nuclei with a spin quantum number $I > 0$ have a magnetic dipole moment (see NMR spectroscopy) which can interact with a magnetic field at the nucleus. This leads to splitting of the nuclear states into $2I+1$ substates (Fig. 5), and corresponding splitting of the spectral lines, which is called magnetic splitting, ΔE_M.

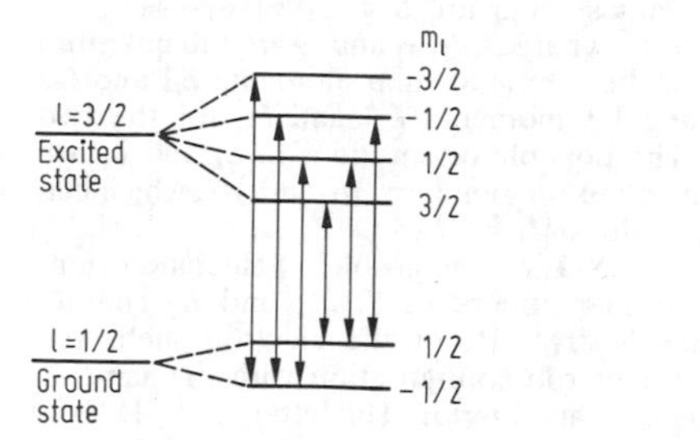

Fig. 5. Diagram of the magnetic splitting of the ^{57}Fe nucleus. The indicated transitions correspond to the selection rule $\Delta I = \pm 1$ and $\Delta m_I = 0, \pm 1$ m_I is the magnetic quantum number.

Magnetic fields at the nucleus can be generated by the atom itself (internal fields), but can also be created by external fields. Internal fields are generated, e.g. by the electrons of the atom, by conduction electrons of the lattice, or magnetic moments of neighboring atoms, and they depend on the chemical and physical properties of the substance. The magnitude of the magnetic splitting depends both on the magnetic dipole moment of the nucleus and on the strength of the magnetic field at the nucleus, so that it provides information on local magnetic fields within the crystal and thus on the magnetic properties of solids.

Isomer shift, quadrupole splitting and magnetic splitting can all occur simultaneously. For example, the Mössbauer spectrum of metallic iron consists of 6 resonance lines (Fig. 6) caused by magnetic splitting. The position of the lowest level in the excited state is also affected by the nuclear quadrupole interaction, as the lack of symmetry of the splitting shows.

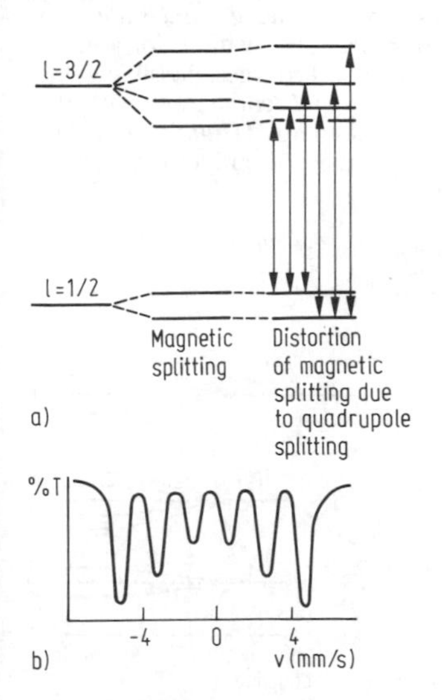

Fig. 6. Mössbauer spectrum of metallic iron: a) term scheme; b) spectrum.

Application. In chemistry, M. is used mainly as a non-destructive aid in elucidation of structure, and for studies of bonding.

For example, it was shown using M. that Prussian blue (which can be made from Fe^{3+} and $[Fe^{II}(CN)_6]^{4-}$, and Turnbull's blue (from Fe^{2+} and $[Fe^{III}(CN)_6]^{3-}$) give the same Mössbauer spectrum, and are thus identical, although they were long held to be different compounds.

The lack of quadrupole splitting indicates a symmetric electron distribution around the nucleus in question, while the appearance of such splitting indicates significant deviation from symmetry. $Na_2[Fe(CN)_6]$, for example, has an octhedral arrangement of its ligands around the iron atom, and therefore the iron ion gives a single Mössbauer line. On the other hand, the symmetry is disturbed in $Na_2[Fe(CN)_5NO]$, in which quadrupole splitting (a doublet) is seen.

M. is also used in nuclear physics, minerology, biology and medicine. From the Mössbauer spectra of biological systems containing ^{57}Fe (e.g. hemoglobin, enzymes), the oxidation state of the iron and changes in it can be determined; in addition, information is obtained on the structure of the coordination compounds.

M. is being used increasingly to study surface phenomena (adsorption, catalysis).

Mother liquor: the liquid remaining after a substance has been crystallized out of a solution. In addition to the residual substance, it generally contains the impurities which were associated with the substance.

Motilin: a gastrointestinal peptide which stimulates stomach motility and secretion of pepsin. The 22-peptide does not affect the secretion of acid. Its sequence, Phe-Val-Pro-Ile-Phe-Thr-Tyr-Gly-Glu-Leu-Gln-Arg-Met-Gln-Glu-Lys-Glu-Arg-Asn-Lys-Gly-Gln, is not related to that of any other gastrointestinal peptide hormone. M. has been synthesized chemically.

Motor oil: a lubricant for motors which may be adjusted for a single viscosity range or several. M. based on mineral oils consist of 3 to 5% aromatics, 20 to 45% cycloaliphatics and 50 to 70% alkanes and additives.

Motor octane rating: see Octane rating.

Mova: see Vinylacetylene.

Moving-bed: a term for methods of catalysis in which the catalyst is movable.

Moving-bed process: see Cracking.

Moving bed reactor: a reactor for solid-gas reactions in which the bed of solid is moved axially to permit continuous addition of unreacted solid and removal of product. The M. is used mainly for non-catalysed reactions, such as ore processing and combustion of coal.

m.p.: abb. for melting point.

MRH: see Liberins.

mRNA: see Nucleic acids.

MSH: see Melanotropin.

MTBE: see *tert.*-Butyl methyl ether.

MTG process: same as Mobil Oil process (see).

MTH process: see Hydrorefining.

MTU: see Thyreostatics.

Muconic acid, ***hexa-2,4-diendioic acid, buta-***

diene-1,4-dicarboxylic acid: HOOC-CH=CH-CH=CH-COOH, an unsaturated dicarboxylic acid with a conjugated π-electron system. M. can exist in three π-diastereoisomeric forms: (E),(E)-M., m.p. 305 °C (dec); (E),(Z)-M, m.p. 190-191 °C; and (Z),(Z)-M, m.p. 194-195 °C. The isomeric M. are readily soluble in glacial acetic acid, but are nearly insoluble in water. In alkaline solution, M. can be reduced to hex-3-enedioic acid with sodium amalgam. It is synthesized by dehydrohalogenation of 3,4-dihalogenadipic acid with alkali hydroxides, or by oxidation of hexa-2,4-dienedial with peroxy acids. M. is of physiological interest, because it is excreted by animals after they have consumed foods containing benzene.

Mucopolysaccharides: a group of proteoglycans (see Polysaccharides) of slimy consistencey which occur in animals. M. can give either acid or neutral reactions. The ***neutral M.*** include the Blood group substances (see). Most M. give an acid reaction, due to the carboxyl groups of uronic acids and sulfuric acid monoesters. The ***acidic M.*** are found in the ground substance of connective tissue. They are bound covalently to collagen, and thus form Glycoproteins (see). The basic structure of these M. is a sequential polysaccharide built up of disaccharide units, $[A(1\rightarrow4)-B(1\rightarrow3)]_n$, where A is an acetylated aminosugar (D-glucosamine, D-galactosamine) and B is usually an uronic acid (D-glucuronic, L-iduronic acid). Various other hexoses are also found in M. Some acidic M. have been shown by x-ray diffraction to contain helical segments. The most important acidic M. are chondroitin, chondroitin sulfate, dermatan sulfate, heparin, hyaluronic acid and keratan sulfate.

Müller-Kühne process: see Sulfuric acid.

Müller-Rochow process: see Polysiloxanes.

Multibody evaporation: same as Cascade evaporation.

Multidentate: see Coordination chemistry.

Multienzyme complex: see Enzymes.

Multilayer adsorption: see Adsorption.

Multiple-layer films: films which are formed on a carefully cleaned metal or glass surface by passing it repeatedly through a film of water-insoluble, surface-active substance on the surface of water. For example, alkaline earth salts of heptadecanoic acid form M. By repeating the dipping process, films with up to 3000 layers can be built up. M. are used as diffraction lattices for x-rays, or as calibration standards for measurements of thickness.

Multiplet: see Multiplet structure.

Multiplet structure: a more refined model of the energetic states in a multi-electron atom than that given by the Central field model (see). The M. takes into account the electrostatic interaction between electrons and the spin-orbital interaction. The central field model makes use of single-electron states characterized by the quantum numbers n, l, m_l and m_s. In this approximation, the total orbital angular momentum $\boldsymbol{L}_L$ for a multiple-electron atom is calculated as the sum of the orbital angular momenta L_l, and the total spin $\boldsymbol{L}_S$ as the sum of the spins L_s of the individual electrons. Only the valence electrons n have to be considered, because filled electron shells have zero angular momentum. The quantum numbers L and S determine the absolute values of the vectors L_L (total orbital angular momentum) and L_S (total spin). For a system with two electrons, the possible L values are $L = |l_1-l_2|, |l_1-l_2|+1, ..., l_1+l_2-1, l_1+l_2$; the values for S are $S = |s_1-s_2|, |s_1-s_2|+1, ..., s_1+s_2-1, s_1+s_2$, where l_1, l_2, s_1 and s_2 are the quantum numbers for the orbital angular momenta $\boldsymbol{L}_l1$ and $\boldsymbol{L}_l2$ and spin angular momenta $\boldsymbol{L}_s1$ and $\boldsymbol{L}_{s2}$ for the two electrons. The possible orientations of $\boldsymbol{L}_L$ and $\boldsymbol{L}_S$ are given by the quantum numbers M_L and M_S, which can take on the values $M_L = -L, -L+1, ..., L-1, L$ or $M_S = -S, -S+1,...,S-1, S$. The absolute value and orientation quantum numbers L, S; M_L and K_S characterize states $\Psi_{L,S,M}L,M_S$ of the electron shell. All states of an electron configuration with the same L and S value are called ***terms***. The letters S, P, D, F... are used to indicate the terms with $L = 0,1,2,3....$ To indicate the S value of a term, it is customary to indicate the ***multiplicity*** $2S+1$ as a left-hand superscript to the term symbol. Terms with multiplicities $2S+1 = 1$, 2, 3, etc. are called *singlet* ($2S+1 = 1$), *doublet* ($2S+1 = 2$), *triplet* ($2S+1 = 3$), and so on. A term includes $(2L+1)(2S+1)$ states. Consider a system of two non-equivalent electrons (n_1+n_2), e.g. the configuration $2p^13p^1$. Here L and S can have the values $L = 0, 1, 2$, and $S = 0, 1$. The resulting terms are ^1S, ^1P, ^1D; ^3S, ^3P, ^3D. For equivalent electrons ($n_1 = n_2$), the impossibility of distinguishing electrons and the Pauli principle limit the number of terms (Table), because the quantum numbers m_l and m_s cannot be arbitrarily combined. An electron or a missing electron ("hole") in a subshell makes the same contribution to the total orbital momentum and total spin. The explicit consideration of the electron interaction lifts the degeneracy of the terms of an electron configuration. Each term ^{2S+1}L remains $(2L+1)(2S+1)$-fold degenerate, because the electron interaction does not change the spherical symmetry of the atom, and all states with the same L and S, but different M_L and M_S have the same energy. The energies of the terms can only be obtained by quantum mechanical calculations or experimental measurements. A rule for determining the term with the lowest energy, the *ground state*, has been derived from empirical observation: for a given configuration, the term with the largest possible value of S and the highest value of L compatible with the maximum S has the lowest energy (*Hund's rule*). For example, the ^3P term of the p^2 configuration (Fig.) has the lowest energy.

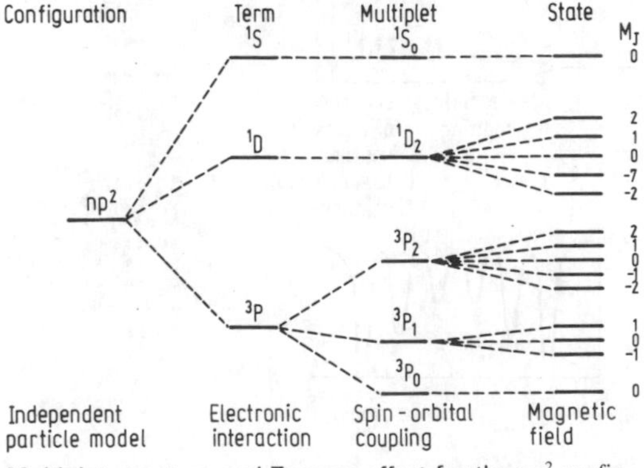

Multiplet structure and Zeeman effect for the np^2 configuration.

In the independent-particle model, Hund's rule is often stated in a simplified form: in the ground state, all energetically degenerate orbitals are first occupied by single electrons with parallel spins (same spin quantum number m_s) before any are occupied by two electrons.

In addition to the electrostatic interactions between electrons in an atom, there are magnetic interactions (*spin-orbital coupling*). These produce small energy differences between states with different orientations of orbital and spin angular momenta. All states with particular values of L and S, which differ only in the relative orientations of $\boldsymbol{L}_L$ and $\boldsymbol{L}_S$, constitute a ***multiplet***. The splitting of the terms in a magnetic field is called the *fine structure*. The fine-structure levels are determined by the quantum numbers $J = L+S, L+S-1,... |L-S|$ which determine the absolute value of the total angular momentum $\boldsymbol{L}_J$. These are written as a subscript to the term symbol: $2S+1L_J$. For $S \leq L$, there are $(2S+1)$ fine-structure levels, and for $L \leq S$, $(2L+1)$ levels. The directional quantization of the total angular momentum $\boldsymbol{L}_J$ is determined by the quantum number $M_J = -J, -J+1,...,J-1, J$. Thus spin-orbital coupling leads to a splitting of terms into components $^{2S+1}L_J$ with different energies, each one of which is $(2J+1)$-fold degenerate. For example, for the ^3P term resulting from the p^2 configuration ($L = S = 1$) has three components $^3P_2$, $^3P_1$ and $^3P_0$, because J can have the values 2, 1 and 0. The coupling scheme, in which first the individual orbital angular momenta $\boldsymbol{L}_l$ and spins $\boldsymbol{L}_s$ are combined to the total orbital angular momentum $\boldsymbol{L}_L$ and total spin $\boldsymbol{L}_S$ of the electron cloud, and then these are combined to give the total angular momentum $\boldsymbol{L}_J$, is called the *LS* or *Russell-Saunders coupling*. It is a good approximation for light and medium-heavy atoms. For heavy atoms, the opposite form of coupling (*jj coupling*) is significant; here the orbital and spin angular momenta of each electron are combined to a total angular momentum $\boldsymbol{L}_j$, and then these are combined to give the total angular momentum of the atom $\boldsymbol{L}_J$. If an atom is brought into an homogeneous external magnetic field, each fine-structure level $^{2S+1}L_J$ is split again into $2J+1$ equidistant levels (Zeeman effect). This completely lifts the degeneracy. The figure shows the various splittings of the p^2 configuration.

Table. Possible terms from configurations with equivalent electrons

Configuration	s	p, p^5	p^2, p^4	p^3
Term	^2S	^2P	^1S, ^1D, ^3P	^2P, ^2D, ^4S

Multiplicity: see Multiplet structure.

Multi-step distillation: same as Cascade evaporation.

Mur: abb. for muramic acid.

Muramidase: same as Lysozyme (see).

Muramic acid, abb. ***Mur***: 2-amino-3-O-1-(S)-carboxyethyl-2-deoxyaldehydo-D-glucose, a lactate ester of D-Glucosamine (see). In the form of N-acetylmuramic acid, abb. MurNAc, M. is a component of murein, the peptidoglycan of bacterial cell walls.

Murein: a peptidoglycan which forms the matrix of the bacterial cell wall. M. makes up about 50% of the dry mass of the cell wall in gram-positive bacteria, and about 10% in gram-negative bacteria.

--GlcNAc–MurAc–GlcNAc–MurAc–GlcNAc–MurAc-- Glycan chain
Tetra- or pentapeptide
Branch point
Direct linkage of the peptide chain
--MurAc–GlcNAc–MurAc–GlcNAc–MurAc–GlcNAc--

Scheme of the structure of murein.

The network of M. is composed of linear glycan chains, cross-linked by oligopeptide chains. The glycan is a block polymer of β(1→4)-linked N-acetylglucosamine and N-acetylmuramic acid: $(GlcNAc\text{-}MurAc)_n$. The carboxyl group of the N-acetylmuramic acid forms an amide bond to a tetrapeptide, or, more rarely, a pentapeptide, which contains both L- and D-amino acids. The peptide chains are in turn cross-linked by another peptide chain. The branch points are provided by diamino acids, such as L-lysine, 2,2'-diaminopimelic acid or D-glutamine. The glycan chains are cleaved in animals and plants by the enzyme lysozyme. The β-lactam antibiotics interfere with the biosynthesis of M.

Murexid: the ammonium salt of purpuric acid. The monohydrate is a red-brown powder. M. is nearly insoluble in cold water, but somewhat more soluble in hot water. It is insoluble in alcohol and ether. The reddish purple aqueous solution turns blue on addition of alkali. M. is formed in the test for uric acid (see Murexid reaction). M. was first isolated as the red-purple pigment from the snail *Murex purpurea*. It is used as a metal indicator in complexometry.

Murexid reaction: a reaction used to detect uric acid and other purine derivatives by formation of murexid. Uric acid is mixed with concentrated sulfuric acid, then the reaction mixture is evaporated, producing alloxanthin, a compound of alloxan and dialuric acid which, on reaction with ammonia, is converted to murexid and makes the solution purple.

Muscarin: the most important active substance in the fly agaric mushroom, *Amanita muscaria*, in which it accounts for 0.0003% of the fresh weight. M. is also found in mushrooms of the *Inocybe* genus. Its crystals are highly hygroscopic; m.p. 181-182 °C, $[\alpha]_D^{20}$ +8° (in ethanol). M. is readily soluble in water and ethanol, and slightly soluble in ether and chloroform. The compound is a quaternary ammonium base which resembles acetylcholine both structurally and in its activity. M. affects only the smooth muscle and the excretory glands; only the (+)-(2S,3R,5S) form is active. Because of its strong affinity for acetylcholine receptors, M. is highly toxic. Fig. p. 674.

$$H_3C \quad CH_2-\overset{+}{N}(CH_3)_3$$

Muscarin

Muscle adenylic acid: see Adenosine phosphates.

Muscle relaxants: compounds which reduce the tone of striated muscles or paralyse them, either by a central or a peripheral site of action. ***Central M.*** reduce a pathologically elevated skeletal muscle tone and can be used to relieve muscle spasms. In higher concentrations, they also act as tranquilizers. The first central M. to become therapeutically significant was *mephenesin (Myocuran®)*, CH_2-OH-CHOH-CH_2-O-C_6H_4-2CH_3. ***Peripheral M.*** inhibit impulse transmission from the nerve endings to the muscle fibers, and thus paralyse the muscle. They are used, among other things, to produce complete immobility of the muscles in large operations in the chest and abdomen. There are 1st and 2nd order peripheral M. *First order M.* competitively displace acetylcholine from the receptor. Their effects can be reversed by an acetylcholinesterase inhibitor, which increases the concentration of acetylcholine. Some examples are tubocurarine chloride, gallamine triethiodide and pancuronium bromide. *Second order M.* are bound to the receptor, depolarize the cell membrane and maintain it in this state for a relatively long time. Some examples are decamethonium and suxethonium. The latter has a relatively short activity span, because it can be hydrolysed by esterases. No quaternary M. are effective after oral administration; they must be injected. *Tubocurarine chloride* is a component of curare (see Curare alkaloids); it is a bisbenzylisoquinoline alkaloid and has a quaternary N atom. The absolute configuration at its two centers of chirality is 1R,1'S; $[\alpha]_D^{20}$ +215° (in water), m.p. 275 °C (dec.). It was originally thought that tubocurarine chloride had two quaternary N atoms, and this suggested the synthesis of bis-quaternary M. with approximately the same distance (1.4 nm) between the two N atoms as in tubocurarine chloride. The synthetic compound *gallamine triethiodide* (*Tricuran®*) is a pyrogallol ether with three quaternary N atoms. *Pancuronium chloride* (*Pavulon®*) has a steroid skeleton and two quaternary N atoms incorporated into a piperidine ring. *Decamethonium*,

$$[(CH_3)_3\overset{+}{N}-(CH_2)_{10}-\overset{+}{N}(CH_3)_3]2X^-,$$

is no longer used because it is difficult to manage. Bischoline succinate, surethonium,

$$[(CH_3)_3\overset{+}{N}-CH_2-CH_2-O-CO-CH_2-CH_2-$$
$$CO-O-CH_2-CH_2-\overset{+}{N}(CH_3)_3]2X^-,$$

is used as the dichloride (Succicuran®) or dibromide (Myo-Relaxin®). (X= Cl or Br.)

M-value: in water chemistry, the amount of 0.1 M HCl (in ml) which must be added to 100 ml water to change the color of the methyl orange or Tashiro indicator. The total alkalinity or methyl orange alkalinity (MA) of the water is obtained by multiplication of the m-value by 2.8. The m-value always includes the P-value (see). At a p-value of 0, the m-value times 28 is equal to the carbonate hardness (in mg/l) of the water (see Hardness for the units used in different countries). Knowledge of the p-value and m-value permits calculation of the caustic alkalies (e.g. NaOH), carbonates (e.g. Na_2CO_3) and hydrogencarbonates (e.g. $NaHCO_3$) in boiler feedwater.

Mycolic acids: long-chain, α-branched fatty acids from the mycobacteria. Most are β-hydroxyfatty acids with the general structure R^1CHOH-CH(R^2)-COOH. In the genus *Mycobacterium* (e.g. *M. tuberculosis*), R^1 can be, for example, an alkyl group with two cyclopropane rings, and R^2 an alkyl group with 22 or 24 C atoms. The corresponding β-methoxy- and β-ketoacids are also found. Free and esterified M. are components of the waxes of mycobacteria.

Mycotoxins: toxic metabolic products of lower fungi. They are important as food contaminants (see Toxins). The Aflatoxins (see) produced by the mold *Aspergillus flavus* are a particular problem, due to their acute nephrotoxic and cytotoxic effects, and also because of their carcinogenicity. See Mushroom poisons.

Mydriatics: see Parasympathicolytics.

Myocuran®: see Muscle relaxants.

Myoglobin: see Hemoglobin.

Myo-relaxin®: see Myscle relaxants.

Myosin: a large protein (M_r 480,000) which is an essential component of contractile muscle fibers; it is normally associated with actin. M. consists of two polypeptide chains with 2000 amino acid groups each; there is a globular head piece at one end. The thick filaments in muscle myofibrils consist of 200 to 400 molecules of M. in a parallel array. The role of M. in muscle contraction is discussed in the entry on Actin (see).

Myricetin: see Flavones.

Myricyl alcohol, ***triacontan-1-ol***: $CH_3(CH_2)_{28}$ CH_2OH, a higher monoalcohol; it forms colorless crystals which are insoluble in water; m.p. 88 °C. M. is only slightly soluble in ethanol at room temperature, but is readily soluble in ether and benzene. It is found in esterified form in various natural waxes, for example, as the palmitic acid ester in beeswax. It has also been found in spruce needles, rice bran and rice sprouts.

Myristic acid, ***tetradecanoic acid***: CH_3-$(CH_2)_{12}$-COOH, a higher, saturated fatty acid. M. crystallizes in colorless leaflets; m.p. 58 °C, b.p. 250.5 °C at 13.3 kPa. It is insoluble in water, but is soluble in alcohol, ether and acetone. M. is one of the most widely occurring saturated fatty acids; its esters are found in nearly all animal and plant fats, and in many essential oils. M. can be obtained by saponifcation of fats, and purified by fractional distillation or crystallization. It is used to make soaps and fatty alcohols.

Myristyl alcohol, ***tetradecan-1-ol***: $CH_3(CH_2)_{12}$ CH_2OH, a monoalcohol, colorless crystals; m.p. 38 °C, b.p. 160 °C at 2.13 kPa. It is insoluble in water, but soluble in ethane and ether. Small amounts of M. are present in whale oil. M. is synthesized by high-pressure hydrogenation of myristic acid, and is used as a starting material for making surfactants.

Myrrh: see Gum resins.

Myrtle wax: a plant wax; m.p. 48 °C. It consists largely of glycerol palmitate. M. is secreted on the surface of the fruits of *Myrica* species. Chemically, it is not a wax but a fat.

N

N: 1) symbol for nitrogen. 2) abb. for normal (concentration). 3) N_A, see Avogadro's number.

Na: symbol for sodium.

Nacken-Kryopoulos process: see Crystal growing.

NAD: see Nicotinamide nucleotides.

NADH: see Nicotinamide nucleotides.

NADP: see Nicotinamide nucleotides.

NADPH: see Nicotinamide nucleotides.

Naled: see Organophosphate insecticides.

Nalidixic acid: a 1,8-naphthyridine derivative used to treat urinary tract infections, especially those involving gram-negative bacteria.

Name reaction: a chemical reaction named for its discoverer, for example, the Beckmann rearrangement, the Cannizzaro reaction or the Oppenauer oxidation.

Nandrolone: see Anabolics.

Napalm: (derived from naphthenic and palmitic acids) a jellied incendiary material which consists of a colloidal solution of small amounts of aluminum salts of naphthenic and palmitic acids or other higher fatty acids as thickeners in hydrocarbons (such as gasoline or benzene). N. is used in bombs, grenades, rockets and flamethrowers. It clings, burns for a long time and develops high combustion temperatures (800-1200 °C). It is difficult to extinguish and causes extensive fires. N. with phosphorus or sodium additives ignites spontaneously.

N. with added magnesium, aluminum or asphalt is called *pyrogel*. Pyrogels burn at temperatures as high as 2000 °C.

Naphtha: an obsolete name for petroleum.

Naphthacene: see Acenes.

Naphthalene: a condensed aromatic hydrocarbon of the acene series consisting of two fused benzene rings. N. crystallizes in colorless, rhombic leaflets which have a characteristic odor (moth balls); m.p. 80.6 °C, b.p. 218 °C, n_D^{85} 1.5898.

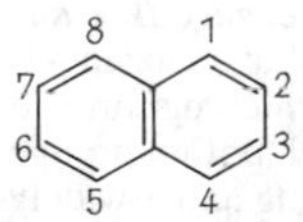

N. is insoluble in water but can be steam distilled. It is very soluble in boiling ethanol, and in ether, benzene and other organic solvents. It sublimes and burns with a bright flame.

A π-electron sextet can be formally assigned to only one of the two rings of N., so that its aromatic character is less pronounced than that of benzene. One of the two rings can easily be hydrogenated with nascent hydrogen to 1,2- or 3,4-dihydronaphthalene or to 1,2,3,4-tetrahydronaphthalene (tetralene). Oxidation with air in the presence of vanadium(V) oxide yields phthalic anhydride; reaction with chromium(VI) oxide in glacial acetic acid yields 1,4-naphthoquinone. Like benzene, N. can be electrophilically substituted. The first substitution is usually kinetically controlled and regionally selective for the 1-position (α-position). Thus, e.g. in chlorination, the products are 95% 1-chloro- and only 5% 2-chloronaphthalene. Nitration produces almost exclusively 1-nitronaphthalene, and the main product of sulfonation below 80 °C is naphthalene-1-sulfonic acid; above 120 °C, the main product is naphthalene-2-sulfonic acid. Energetic reaction of the two acids with sulfuric acid produces naphthalene-1,5- and -1,6-disulfonic acid, or naphthalene-2,6- and -2,7-disulfonic acid, respectively. These are are very important intermediates in the industrial production of azo dyes. Naphthalene sulfonic acids are often converted to naphthols by "alkali melting". Overall, the situation in the second substitution of N. is much more complicated than with benzene, as it may be either homonuclear (in the previously substituted ring) or heteronuclear (in the other ring).

N. is found in small amounts in some types of petroleum. N. is formed during dry distillation of anthracite coal, due to thermally promoted cyclization and dehydrogenation reactions. It comprises about 6% of anthracite coal tar, from which it is still extracted industrially. It may be synthesized according to Haworth from benzene and succinic anhydride, via numerous intermediates. This synthesis is not industrially significant.

N. is used on a large scale for production of phthalic anhydride, as starting material for dyes, pharmaceuticals and tanning agents, and as an insecticide.

Naphthalene yellow: same as Martius yellow (see).

Naphthalene oil: see Tar.

Naphthalenesulfonic acids: derivatives of naphthalene formed by sulfonation of the naphthalene ring in various positions. When the sulfonation is done around 60 °C, the sole product is ***naphthalene-1-sulfonic acid*** (m.p. 139 °C, anhydrous), while at 160 °C, the thermodynamically more stable ***naphthalene-2-sulfonic acid*** (m.p. 124 °C) is formed. Further sulfonation under severe conditions produces a mixture of naphthalene-1,5- and -1,6-disulfonic acids (from the 1-isomer) or naphthalene-2,6- and -2,7-disulfonic acid (from the 2-isomer). It is possible to introduce as many as three or four sulfo groups. The N. are used as dye components, because the hydrophilic sulfo groups make the dye water-soluble.

Naphthazarin: see 5,8-Dihydroxy-1,4-naphthoquinone.

Naphthenes: a term commonly used in petroleum chemistry for the cycloalkanes found in petroleum, such as cyclopentane and cyclohexane, and their alkyl derivatives.

Naphthenic acids: hydroaromatic carboxylic acids derived from cycloalkanes (naphthenes), particularly from cyclopentane and cyclohexane. They are found in naphthene-base petroleums, and are distributed over the entire boiling range of the oil. The highest concentrations are present in the kerosene and gas oil fractions, from which they are obtained industrially by extraction with sodium hydroxide. The salts of N., the ***naphthenates***, are important in industry.

Naphthionic acid, ***4-aminonaphthalene-1-sulfonic acid***: shiny, white needles which are readily soluble in water, slightly soluble in ethanol and insoluble in ether. N. are formed by the reaction of α-naphthylamine with conc. sulfuric acid at 130 °C. α-Naphthylamine sulfate can also be rearranged to form N. at 180 °C in vacuum. N. are used as coupling components in the synthesis of azo dyes (see Alphabet acids) and can also be used as photographic developers.

Naphthol, ***hydroxynaphthalene***: a derivative of naphthalene in which one, two or more H atoms are replaced by OH groups. The chemical properties of N. are similar to those of the phenols, although the N. are generally more reactive. They are only slightly soluble in cold water, and somewhat more soluble in ethanol, ether, benzene and aqueous sodium hydroxide solution. The most important compounds are naphth-1-ol and naphth-2-ol, which are found in small amounts in coal tar.

Naphth-1-ol (α-N., 1-hydroxynaphthalene): colorless, readily subliming crystalline needles which smell like phenol; m.p. 96 °C, b.p. 288 °C. Naphth-1-ol gives a violet color reaction with iron(III) chloride. It is also oxidized to 4,4'-bis-1-naphthol.

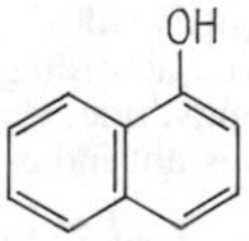

OH

Naphth-1-ol

This reaction involves radicals, which dimerize to give the stable product. Naphth-1-ol readily undergoes electrophilic substitution in position 4, or it can be oxidized to 1,4-naphthoquinone by chromium(VI) oxide. It is synthesized either by sulfonation of naphthalene in position 1, followed by alkali fusion of the naphthalene-1-sulfonic acid, or by treatment of 1-naphthylamine with dilute sulfuric acid at 180 °C. Various derivatives with substituents in positions 2 and 4 are important as color couplers for the blue-green pigment in color photography. Naphth-1-ol is also used to make azo pigments.

Naphth-2-ol (β-naphthol, 2-hydroxynaphthalene): colorless, rod-shaped crystals which produce leaflets after sublimation; m.p. 123 °C, b.p. 295 °C. In dilute solution, it gives a green color reaction with iron(III) chloride. It is also oxidized to 1,1'-bis-2-naphthol. 2-Napthol is synthesized by sulfonation of naphthalene in position 2 (at 165 °C) followed by alkali fusion. Naphth-2-ol is more reactive than the isomeric 1-compound and is a very important color coupler in the synthesis of azo dyes. It is also used as a starting material for synthesis of insecticides and fungicides, and as a preservative. Some of its ethers, such as nerolin, are valuable aroma substances.

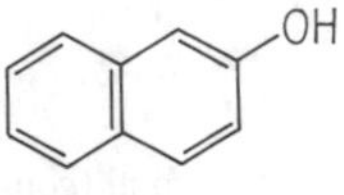

OH

Naphth-2-ol

Naphthol AS dyes: a group of azo development dyes. They are characterized by the use of ***naphthol AS*** (Fig.) as the coupling component. This adsorbs directly to cotton from an alkaline solution. Coupling with a diazonium compound then takes place at low temperature directly on the fiber; the N. are thus Ice dyes (see). The resulting colors are very fast to light, washing and acids, and are very intense.

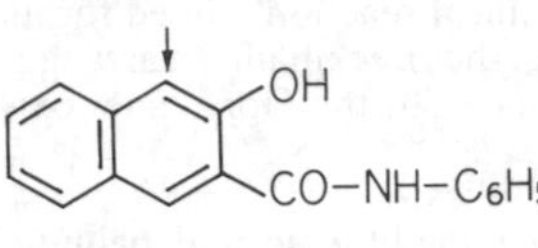

Coupling site

OH

CO–NH–C_6H_5

Naphthol green: a term for two nitroso pigments (formula, see Nitroso pigments).

1) Complex iron-sodium salt of 2-nitroso-1-naphthol-4-sulfonic acid. N. forms dark green leaflets which are readily soluble in water, giving a grass-green solution; they are only slightly soluble in alcohol. N. is used as a green dye for wool and pigment in aquarell paints, in complexometry, and as a green plasma stain in microscopy. Its unusually high capacity for absorption of infrared radiation is noteworthy. If small amounts of N. are added to an aqueous solution, the solution can be evaporated by the energy from sunlight falling on it.

2) ***Naphthol green B***: same as Green PLX.

Naphthol orange: a term for azo pigments made of diazotized sulfanilic acid and α- or β-naphthol. α-Naphthol orange (orange 1), the sodium salt of α-naphthol-azobenzene-4-sulfonic acid, is a red-brown, crystalline powder which is soluble in water and alcohol. The compound is used as an indicator; between pH 7.6 and 8.9 its color changes from brownish yellow to purple. ***β-naphthol orange (orange II, mandarine)***, the sodium salt of β-naphthol-azobenzene-4-sulfonic acid, forms yellow leaflets which dissolve in water to give an orange-red color. β-Naphthol orange is used as a barium lake in color printing and is widely used to make colored paper. β-N. is a commonly used dye.

Naphthol red: see Amaranth.

Naphthoquinones: Quinones (see) derived from naphthalene. ***1,4-(α-)naphthoquinone*** forms yellow crystalline needles with a typical, quinone-like, pungent odor; m.p. 128.5 °C. It is only slightly soluble in water, but is soluble in alcohol, ether, benzene and glacial acetic acid. 1,4-N. is steam volatile and sublimes. It is synthesized by oxidation of naphthalene, α-naphthol or 4-aminonaphth-1-ol by atmospheric oxygen in the presence of a catalyst or by chromic acid in glacial acetic acid. 1,4-N. is used to make anthraquinone and anthraquinone derivatives, as a regulator of polymerization in the production of synthetic rubber and acrylate resins, and as a corrosion inhibitor. Some important, naturally occurring 1,4-naphthoquinone derivatives are vitamins K_1 and K_2 and a number of natural products, including lawsone (2-hydroxy-1,4-naphthoquinone), juglone (5-hydroxy-1,4-naphthoquinone) and alkanin. There are many synthetic 1,4-N., such as 2-chloromethyl-, 2,3-dichloro-, 5-nitro- and various hydroxy-1,4-naphthoquinones. These are used as dye intermediates or pharmaceuticals.

1,2-(β-)naphthoquinone forms odorless, reddish yellow crystalline needles; m.p. 146 °C (dec.). This N. decomposes in the presence of water, and is soluble in alcohol, ether and benzene. It is synthesized by oxidation of 1-amino-2-hydroxynaphthalene, and is used as a reagent for resorcinol, as an intermediate in the synthesis of dyes, and as a stabilizer for transformer oils. The 2-monooxime is used as a reagent for zirconium and cobalt.

2,6-(amphi-)Naphthoquinone is a red, crystalline compound; m.p. 135 °C (dec.). It is soluble in alcohol and methanol. 2,6-N. is synthesized by oxidation of 2,6-dihydroxynaphthalene with lead dioxide in benzene, and it has a higher oxidizing capacity than the α- and β-isomers. Other N., such as 1,5-naphthoquinone, are known only in the form of derivatives or are of theoretical interest only.

Synthetic naphthoquinone dyes are used as pigment, suspension or vat dyes.

Naphthol yellow S, ***citronin A***: the disodium or dipotassium salt of flavianic acid (2,4-dinitronaphth-1-ol 7-sulfonic acid), a nitro pigment. N. is available commercially as a slightly yellow, crystalline powder; it dissolves readily in water, and slightly in alcohol. It was formerly used as a yellow dye for wool and silk, but because it is not very fast to washing and light, it is no longer important.

Naphthylamines, ***aminonaphthalenes***: the two aromatic primary amine derivatives of naphthalene in which the amino group is in the α or β position (position 1 or 2) on the aromatic skeleton. ***α-Naphthylamine (1-aminonaphthalene)*** forms colorless, unpleasant smelling crystals which turn dark red in the air; m.p. 50 °C, b.p. 300.8 °C. α-N. is slightly soluble in water, but dissolves readily in alcohol and ether. It is steam volatile and sublimes. It dissolves in mineral acids, forming salts. α-N. is formed by reduction of α-nitronaphthalene with iron and hydrochloric acid according to Bechamp. It is used as a starting material for numerous azo dyes and for production of aminonaphthalenesulfonic acids. ***β-Naphthylamine (2-aminonaphthalene)*** forms slightly reddish, odorless leaflets which are steam volatile; m.p. 113 °C, b.p. 306.1 °C. It is slightly soluble in cold water, and readily soluble in hot water, ethanol and ether.

NH_2

β-Naphthylamine

It dissolves in mineral acids, forming salts. β-N. is usually synthesized from β-naphthol by the Bucherer reaction. Large-scale industrial synthesis has been severely curtailed because of its carcinogenicity. β-N. is converted mainly to naphthylaminesulfonic acids and in this form, used for pigments.

α-Naphthylthiocarbamide: same as α-naphthylthiourea.

α-Naphthylthiourea, ***N-(1-naphthyl)thiourea, α-naphthylthiocarbamide***: C_{10}-NH-CS-NH_2, a poisonous, colorless, crystalline compound; m.p. 198 °C. α-N. is slightly soluble in water and cold alcohol, and slightly more soluble in hot alcohol. It is synthesized by reaction of α-naphthylamine and alkali thiocyanates, or of α-naphthylisothiocyanate and ammonia: $C_{10}H_7$-N=C=S + NH_3 → $C_{10}H_7$-NH-CS-NH_2. N. is a very effective rodenticide and is used under the name ***Antu*** against rats.

Naptalam: see Herbicides.

Narcotine, ***noscapine***: an isoquinoline alkaloid of the phthalide isoquinoline type; it is present at a concentration of 1.5 to 12.5% in opium. The natural N. is the α- or (-)-form; m.p. 176 °C, $[\alpha]_D^{20}$ - 198° (in chloroform). The absolute configuration is 1R,1'S. The diastereomeric β-narcotine has the opposite configuration at the C1' atom (1'R). N. suppresses coughing, but is no longer used in medicine.

O N–CH_3 O H_3CO H H O H_3CO O OCH_3

Narcotics: compounds which, in suitable doses, cause temporary inhibition of the central nervous system function; the results are loss of consciousness, insensitivity to pain, relaxation of muscle tension and defense reflexes, without loss of the vital functions of the medulla oblongata. N. are either inhaled or injected; rectal N. are not currently used.

The most important N. are ***inhalation N.***, that is, gaseous or volatile compounds which enter the lungs with the breath and are absorbed there. This group includes dinitrogen monoxide, N_2O, ethers such as diethyl and divinyl ethers, CH_2=CH-O-CH=CH_2, hydrocarbons such as cyclopropane and halogenated hydrocarbons such as halothane and chloroform. Of these compounds, halothane and dinitrogen monoxide are the most widely used. The others are no longer used for various reasons, such as the hepatotoxicity of chloroform and the danger of explosion, as well as the difficulty of dosing of diethyl ether.

Injection N. include thiobarbitals and *N*-methylbarbitals (see Barbitals) and some special compounds such as propanidide, ketamine (Velonarcon®) and etomidate (Radenarcon®). Injection N. are used at the beginning of anaesthesia and as short-term anaesthetics.

Historical. The first successful experiments in producing narcosis were carried out in the middle of the

10th century. After Davy had discovered in 1799 that inhaled N_2O alleviated pain, the dentist Wells began to use N_2O during operations in 1844. The first experiments with ether as a N. were carried out in 1846 by Long. Chloroform was first tested in animal experiments in 1847 by Fluorens, and was used a little later by Simpson in human beings.

Natural gas: a high-quality, combustible gas found in large deposits in the earth's crust. It is formed under conditions similar to those which produce Petroleum (see), and the two are therefore often found together. However, N. is also found alone, especially in porous rocks. The N. which is associated with petroleum is partially dissolved in the oil, to an extent which depends on the pressure in the deposit. (This can be as high as 100 MPa). It is released when the pressure on the oil is relieved.

Composition. Some N. consists almost exclusively of methane or other light hydrocarbon gases (***dry gas***). Other N. contain significant amounts of readily liquefied hydrocarbons, such as propane, butane, pentane and even higher saturated hydrocarbons (***wet gas***). In addition to hydrocarbons, N. can contain widely varying amounts of nitrogen, carbon dioxide and hydrogen sulfide. Oxygen and carbon monoxide are rarely found in N., and then only in traces. Some N. contain significant amounts of helium (up to 7%), and they are the most important source of Helium (see). The water vapor content of N. is not insignificant, and must be removed before transport.

Production N. passes through the pores of the rock to the borehole and rises to the surface under its own pressure. In most cases, the pressure in the deposit is sufficient to transport the N. through the pipelines to the consumers, that is, over a distance of several hundred kilometers. The deposit energy consists of 1) the pressure of the compressed gas (gas expansion can account for 100% of the energy), and 2) the pressure of water in adjacent rocks which presses into the deposit (water pressure, up to 10% of the energy). The yield of N. from a deposit is as high as 90% of the reserve. In deposits with extremely wet N. (gas condensate deposits), the higher-boiling fractions may condense, leaving a portion of the reserve as liquids in the pores of the rocks. In such cases, the yields are only 30 to 60%, unless the dry N. is pumped back into the deposit to prevent the drop in pressure and thus the condensation (recycling).

Storage. Limited amounts of N. can be stored above ground in gas tanks, but storage below the earth is more economical. In this case, an exhausted oil or gas deposit is refilled with gas; up to several billion m^3 can be stored in such reserves. Another very convenient form of storage is provided by caverns in salt deposits which are made by drilling and flushing with water. A single cavern can have a volume of 500,000 m^3 and store up to 50 million m^3 gas.

N. is transported mainly in Pipelines (see), in which the working pressure is 2 to 5 MPa. However, there are also high-pressure pipelines with up to 40 MPa pressure. Since 1965, large amounts of N. have been transported in liquid form in special tank ships. Intermediate storage for is provided by underground reservoirs.

Processing. Before transport and further processing, the N. must be dried and purified, because otherwise water can condense in the pipelines as gas hydrates or ice crystals. There are various methods of drying the gas. 1) In expansion drying, the cooling of the gas causes part of the water vapor to precipitate as liquid water. 2) Drying can be achieved by washing the gas with a suitable liquid, usually diethylene glycol and triethylene glycol. The wet glycol is led into a fractionation tower where the water is driven off. 3) Solid adsorbants, especially molecular sieves, silicic acid gels or activated aluminum oxide are used; they are reactivated by heating. Wet N. can be freed of the C_5 and C_6 hydrocarbons by adsorption with activated charcoal, or by a gas scrubbing process. Wet N. can also be treated by low-temperature condensation. The resulting methane is then further processed as dry N., and the C_2 to C_4 hydrocarbons are used for

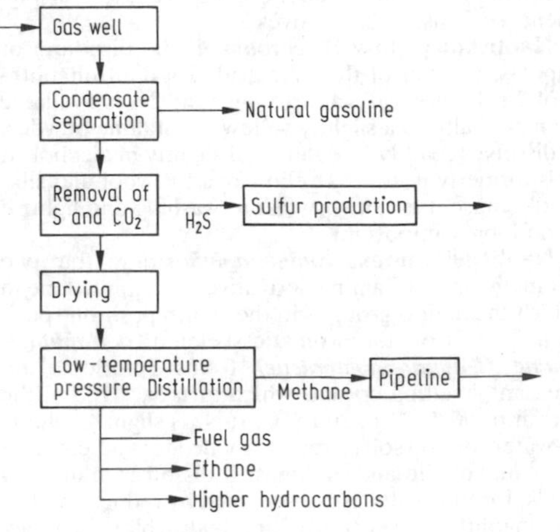

Diagram of natural gas processing.

Compositions of natural gases

Country	Deposit	Methane	Ethane	Propane	Butane	Pentane	CO_2	N_2	H_2S
Algeria	Hassi-R'Mel	79.6	7.4	1.4	3.6	0.2	5.1	–	
France	Lacq	69.3	3.0	0.9	0.6	0.8	9.3	0.2	15.8
Iran	Sarajeh	81.4	11.9	3.4	0.5	0.2	2.1	–	0.2
Canada	Medicine Hat	96.0	0.5	–	–	–	3.5	–	–
Netherlands	Gronigen	81.3	2.8	0.4	0.2	0.1	0.9	14.3	–
Germany	Salzwedel	34.6	0.4	–	–	–	0.3	64.7	–
Russia	Orenburg	83.8	5.2	4.3	1.05	0.8	1.0	5.55	1.3
USA	Laketon, Tex.	85.9	5.8	3.3	1.6	1.0	0.2	2.1	–

ethene production. The higher-boiling hydrocarbons are used in gasolines. H_2S is removed from the N. by a selective solvent (sulfolane/diisopropanolamine, sulfosolvane solution, diethylene glycol/triethanolamine, etc.). Sulfur can be obtained from the H_2S by the Claus process (see diagram).

Properties. The densities of N. vary from 0.7 g dm^{-3} for dry N. to 1.3 g dm^{-3} for a wet N. under standard conditions. Dry N. has a heating value of 16,000 to 32,000 kJ m^{-3} (under standard conditions), while wet N. has a value up to 50,000 kJ m^{-3}. Generally only dry N. is used in burners. N.-air mixtures can explode at a gas content of 5 to 14%.

Occurrence. The most important producing countries at present are the USSR, the USA, Canada, the Netherlands and Algeria. The world production rose from about 0.24 billion tce in 1950 to 1.83 billion tce in 1980. The estimated world reserves of N. are smaller than those of petroleum, about 500 billion tce. Of these, about 105 billion tce are considered proven reserves, and about 90 billion tce of these are recoverable (1 tce = 29310 MJ = 0.7 t petroleum = 380 m^3 N.) The largest reserves of N. are found in the USSR, the Near East, North America and North Africa.

Applications. N. was formerly burned (flared) at the tops of oil wells. Small amounts were used as heating gas in the oil fields. Now large amounts are used as fuel. Because N. is completely odorless, small amounts of an odorant are added so that leaks in pipes (especially in houses) will be noticed immediately. N. is also used as a propellant gas in gas turbines. Large amounts are used in the production of chemicals; N. is an important raw material for petroleum chemistry. It is converted by Pyrolysis (see) into ethyne, which is extremely important for many syntheses. Another important product of N. is soot, which is obtained by combustion of the N.

The most important chemical use of N. is in production of Synthesis gas (see). The following processes are important uses for methane, the main component of dry N.: steam reforming to carbon monoxide,

$$CH_4 + H_2O \xrightarrow[800°C]{Ni} CO + 3\,H_2;$$

electric arc pyrolysis to ethyne,

$$2\,CH_4 \xrightarrow[2000°C]{} CH{\equiv}CH + 3\,H_2;$$

Sachse-process for ethyne,

$$2\,CH_4 + 1.5\,O_2 \xrightarrow[1800°C]{} CH{\equiv}CH + 3\,H_2O;$$

production of carbon disulfide,

$$CH_4 + 4\,S \xrightarrow[650°C]{SiO_2} CS_2 + 2\,H_2S;$$

Fluohm process for hydrogen cyanide,

$$CH_4 + NH_3 \xrightarrow[1500°C]{} HCN + 3\,H_2;$$

Andrusov process for hydrogen cyanide,

$$CH_4 + NH_3 + 1.5\,O_2 \xrightarrow[800°C]{Pt} HCN + H_2O;$$

methane chlorination, e.g. to methyl chloride and carbon tetrachloride.

Historical. N. has been known for a very long time (the Eternal Fire of Baku), but it was used only in very limited amounts for heating. The first city lighting to use N. was installed in 1821 in Fredonia, New York. The modern era of N. economy began in 1891 with the first long-distance pipeline, 190 km long. The first transport of methane across the Atlantic took place in 1959 from Port Charles (USA) to Canvey in England on the "Methane Pioneer". At present, N. provides about 20% of the world's energy, and has had the fastest growth rate of any primary energy source in the time since the Second World War.

Natural products: a term for biogenic organic compounds. 1) ***Primary N.*** are found in all types of organisms and are the products of primary metabolism. This group includes various short-chain carboxylic acids, proteins, nucleic acids, carbohydrates and lipids. It should be noted that the structural details of these primary N. can differ from one type of organism to the next. 2) ***Secondary N.*** are found only in certain groups of organisms, some in only one or a few related species. These products are formed by secondary metabolism, as a rule, from primary N. precursors. Secondary N. are found mainly in plants and microorganisms. About 4/5 of the known N. have been isolated from plants. These include terpenes, steroids, alkaloids, numerous aromatic compounds (polyketides and N. derived from shikimic acid) and many chemically heterogeneous metabolic products from microorganisms which act as antibiotics against other microorganisms.

Natural product substitutes: synthetic substances designed to mimic natural products. In addition to such large-scale products as synthetic rubber and fibers, there are many drugs and pesticides designed to replace or improve on natural substances which may be expensive to produce, unstable when stored and have undesirable side effects. In addition to numerous pharmaceutical products (including antibiotics), N. with pesticidal activity have been developed, e.g. Pyrethroids (see) and Cartap (see).

Nb: symbol for niobium.

NBS: abb. for *N*-bromosuccinimide.

Nc powder: see Smokeless powder.

NCS: abb. for *N*-chlorosuccinimide.

Nd: symbol for neodymium.

NDGA: abb. for nor-dihydroguaiaretic acid.

Ne: symbol for neon.

Near infrared, ***NIR***: see Infrared spectroscopy.

"neat" Phase: see Liquid crystals.

Neber reaction: rearrangement of sulfonic acid esters of ketoximes with potassium ethylate and subsequent hydrolysis to α-aminoketones:

$$R{-}CH_2{-}\underset{\underset{NOH}{\|}}{C}{-}R' + C_6H_5{-}SO_2Cl \xrightarrow[-HCl]{} R{-}CH_2{-}\underset{\underset{NOSO_2C_6H_5}{\|}}{C}{-}R'$$

$$\xrightarrow[b)\ H_2O]{a)\ KOC_2H_5} R{-}\underset{\underset{NH_2}{|}}{CH}{-}CO{-}R' + C_6H_5SO_3H$$

α-Aminoketone

This rearrangement can be done a second time on the same compound, if the amino group is first protected by benzoylation, then the ketoxime and its sulfonic acid ester are prepared. Rearrangement and hydrolysis yields diaminoketone. The mechanism of the reaction is thought to involve an azirine as an intermediate.

Neckar process: see Water softening.

Nef reaction: cleavage of nitroalkanes with strong acids, forming aldehydes or ketones and dinitrogen oxide: 2 R-CH_2-NO_2 $\rightarrow$ 2 R-CHO + N_2O + H_2O. Primary nitroalkanes yield aldehydes, and secondary nitroalkanes, ketones, with yields between 80 and 85%.

Cyclic ketones can also be produced by the N. Pentoses can be converted to hexoses by addition of nitromethane and subsequent N.

Negative catalysis: see Inhibition.

Negative osmosis: same as Reverse osmosis.

Nematic: see Liquid crystals.

Nematocide: a biologically active compound used to combat nematodes. The greatest economic damage is caused by the root parasites (free-living root nematodes, root gall nematodes and cyst nematodes). These animals find ideal propagation conditions in monocultures of field crops, and they can cause so much damage that the harvest losses, both quantitative and qualitative, are very significant. In extreme cases, the plants die. N. must mix thoroughly with the soil and have a suitable persistence. Therefore gaseous fumigants and water-soluble or suspendable preparations are used; they are often general soil disinfectants. Many of the soil fumigants are phytotoxic, so that after fumigation of a field, it cannot be replanted for several weeks. Because large amounts of material are required, the treatments are expensive and are thus used only for truck and nursery crops. 1,2-Dibromomethane, mixtures of 1,3-dichloropropene and 1,2-dichloropropane (*DD®*), 1,2-dibromo-3-chloropropane (*Nemagon®*), Metham (see) and Dazomet (see) are the most commonly used soil fumigants. The last two compounds mentioned act by releasing methylisocyanate, H_3C-N=C=S ("methyl mustard oil"), in the soil; methylisocyanate itself is also a component of some N.

The carbamate insecticide aldicarb and various thiophosphoric acid esters (see Phospho-organic insecticides) act as N. and are distributed through the soil via the water capillary system.

Neocarcinostatin: a linear polypeptide of 109 amino acids which has cancer-inhibiting activities. N. is a component of ***carcinostatin***, a mixture of antibiotics produced by *Streptomyces carcinostaticus* which, in addition to N., contains three antibacterial substances A-C. ***Substance A*** is a polypeptide of 87 amino acids and contains a disulfide bridge. N. has been shown to have two disulfide bridges. Doses of 0.1 to 1.6 mg/kg are required for the cancer-inhibiting effect, which is due to the inhibition of DNA synthesis.

Neodymium, symbol ***Nd***: an element of the Lanthanide (see) group of the periodic table, a rare-earth, heavy metal; *Z* 60. The natural isotopes have the following mass numbers: 142 (27.16%), 144 (23.80%, α-emitter, $t_{1/2}$ $2.1 \cdot 10^{15}$ a), 146 (17.19%), 143 (12.18%), 145 (8.29%), 148 (5.75%) and 150 (5.63%). The atomic mass is 144.24, valence III, m.p. 1016 °C, b.p. 3127 °C, standard electrode potential (Nd/Nd^{3+}) -2.431 V.

N. is a silvery white metal which crystallizes in two modifications: hexagonal (density 7.004) at room temperature, and cubic body-centered above 868 °C. N. makes up $2.2 \cdot 10^{-3}$% of the earth's crust, and is one of the most abundant lanthanides. It is found mainly in cerium minerals, accompanied by other rare-earth metals, especially praseodymium. It is relatively highly enriched in cerite, monacite and orthite. Other properties, methods of analysis, production and history are discussed in the entry on Lanthanides (see). N. is most often used in a mixture with 15 to 25% praseodymium, under the name ***didymium metal***, to improve the mechanical properties of magnesium alloys. It is also used in the form of Cerium mixed metal (see).

Neodymium compounds: compounds in which neodymium is usually in the +3 oxidation state; in the case of the complex $Cs_3[NdF_7]$, obtained by fluorinating a mixture of neodymium(III) chloride and cesium chloride, Nd is in the +4 state. The general properties and production of N. are discussed under Lanthanoid compounds (see). Neodymium(III) compounds are reddish to violet in color. Some important examples are: ***neodymium(III) chloride***, $NdCl_3$, pinkish violet, M_r 250.60, density 4.134, m.p. 784 °C; ***neodymium(III) fluoride***, NdF_3, pale lavender, M_r 201.24, m.p. 1410 °C; ***neodymium(III) oxide***, Nd_2O_3, bright blue, M_r 336.48, density 7.24; and ***neodymium(III) oxalate***, $Nd_2(C_2O_4)_3 \cdot 10H_2O$, pink, M_r 732.69. Neodymium(III) oxide is used to dope glass ceramics, make solid lasers, color glass and enamel and make sunglasses.

Neoendorphin: Tyr-Gly-Gly-Phe-Leu-Arg-Lys-Tyr-Pro-Lys, an opiate peptide first isolated in 1979 from porcine hypothalamus extracts; it was called α-N. in analogy to the endorphin nomenclature. The N-terminal sequence of N. is the same as that of Leu-enkephalin, and before the discovery of pre-proenkephalin, it was thought to be a potential precursor of Leu-enkephalin. N. is biosynthesized together with dynorphin in the form of pre-prodynorphin. β-N. differs from α-N. in lacking the C-terminal lysine group.

Neoeserin®: see Neostigmine.

Neomycins: a group of aminoglycoside antibiotics made by *Streptomyces fradiae*. Mixtures of N. consisting chiefly of ***neomycin B*** are used therapeutically. The components of N. B are the aglycon 2-deoxystreptamine, D-ribose and two diaminomonosaccharide units. N. are very toxic, and are therefore used only for local application or, because they are scarcely absorbed, to treat certain intestinal infections.

Neon, symbol ***Ne***: an element of group 0 or VIIIa of the periodic system, the Noble gases (see); *Z* 10, mass numbers of the natural isotopes 20 (90.92%) and 22 (8.82%), atomic mass 20.183, valence 0, density 0.9002 g/l at 0 °C, m.p. -248.67 °C, b.p. -245.9 °C, crit. temp. -228.7 °C, crit. pressure 2.69 MPa.

Properties. N. is a colorless, odorless, tasteless, nontoxic monoatomic gas. It is extremely unreactive; no compounds of N. are known.

Analytical. N. is usually isolated by gas chromatography, and detected by spectral analysis.

Occurrence. N. makes up 0.0016 vol% of the air; it is estimated to make up $5 \cdot 10^{-7}$% of the earth's crust. A few natural gas deposits and volcanic vapors contain N. In the universe as a whole, N. is the third most common element, after hydrogen and helium.

Production. The industrial production of N. begins from air. When air is liquefied, N. and helium are left behind in the residual gas mixture. After any remaining nitrogen has been adsorbed with activated charcoal, the N. is separated from helium by cooling with liquid hydrogen.

Applications. N. is used almost exclusively as a filler gas for gas-discharge and glow-discharge lamps. Neon-filled high-voltage discharge tubes (neon lights) give a scarlet light and are used for advertizing. A slight amount of mercury vapor added to the neon almost completely suppresses its spectrum, and a warm, cornflower-blue light is emitted instead. Green lights are made by using a yellow glass tube with the mercury-neon mixture.

Historical. N. was discovered in 1898 by Ramsay in the course of fractional distillation of crude liquid argon. As the liquid evaporated, the first gas bubbles had a red spectrum, indicating the presence of a new element (Greek "neos" = "new").

Neopentane: see Pentane.

Neopentyl chloride: see Amyl chlorides.

Neostigmine, ***Neoeserni***®: $[3(CH_3)_3N\text{-}C_6H_4\text{-}O\text{-}CO\text{-}N(CH_3)_2]^+X^-$, the dimethylcarbamate of 3-hydroxyphenyltrimethylammonium bromide or methylsulfate. As a carbamate, it is cleaved in alkaline media to form dimethylamine, carbon dioxide and a 3-hydroxyphenyltrimethylammonium salt. N. is an inhibitor of acetylcholinesterase (see Parasympathicomimetics) and is used to treat atonia of the intestine and bladder, and glaucoma. It has the advantage over non-quaternary compounds, such as physostigmine, that it does not cross the blood/brain barrier in significant amounts.

Nephelometry: optical method for quantitative determination of the solid fraction in suspensions or aerosols from the intensity of light scattering. The Rayleigh equation relates the degree of scattering to the amount of scattering substance:

$$I \approx I_0 \frac{nd^6}{r^2\lambda^4},$$

where the scattered light intensity I is proportional to the incident intensity I_0, the particle number n and the sixth power of the diameter d of the particles; and inversely proportional to the square of the distance r from the lighted volume element and the fourth power of the wavelength λ of the incident light. Under constant conditions, if the size of the scattering particles is constant, the intensity of the scattered light is thus proportional to the number of scattering particles, and this relationship is the basis of quantitative analysis. The most important requirement for such analyses is the preparation of suspensions with reproducible particle size. The intensity of the scattered light can be measured either directly (scattered light measurement) or indirectly, from the decrease in intensity of the transmitted light (Fig.). The scattered light is usually measured (Tyndallometry) at a right angle to the incident light beam. In turbidity measurement, the fraction of the incident light which is scattered is determined from the difference between incident and transmitted light.

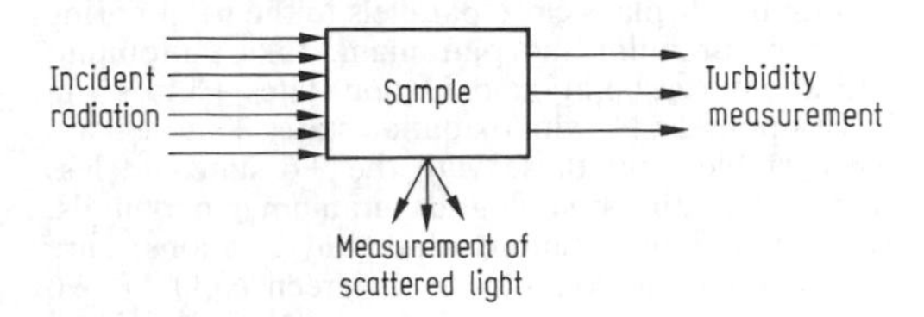

Nephelometry (schematic).

The instruments used in N. are called ***nephelometers***; however, most commercial photometers can be equipped to carry out N. N. is used mainly to determine very low concentrations; for example, chloride or sulfate concentrations in water are determined from the turbidity generated by addition of silver nitrate or barium chloride. In general, N. can be used to follow precipitation reactions and to determine the concentrations of colloidal solutions or suspensions, such as butterfat or proteins.

Neptunium, symbol ***Np***: a radioactive element from group IIIb of the periodic system, the Actinoid group (see); a heavy, transuranium metal; Z 93. The known isotopes have the following mass numbers (decay type, half-life, nuclear isomers in parentheses): 228 (spontaneous decay, 60 s); 229 (α, 4.0 min); 230 (K-capture, α, 4.6 min); 231 (K, α, 48.8 min); 232 (K, 14.7 min); 233 (K, α, 36.2 min); 235 (K, β^+, 4.40 d); 235 (K, α, 396.1 d); 236 (β^-, $1.29 \cdot 10^6$ a, 2); 237 (α, $2.14 \cdot 10^6$ a, 2); 238 (β^-, 50.8 h); 239 (β^-, 2.355 d); 240 (β^-, 65 min, 2); 241 (β^-, 16.0 min); The most stable isotope has atomic mass 237.0482, valence III, IV, V, also VI, VII, density 20.48, m.p. 639 °C, b.p. 3900 °C, standard electrode potential ($Np.Np^{3+}$) -1.856 V.

N. is a silvery white, ductile, rather reactive metal which exists in three modifications: orthorhombic, tetragonal and cubic; the conversion temperatures are 280 and 577 °C. N. is obtained by reduction of neptunium(III) fluoride with barium or lithium vapor above 1200 °C. The chemical behavior of N. is very similar to that of uranium and plutonium; those Np compounds in which the metal is in the +5 or +4 oxidation state are more stable than the corresponding uranium compounds.

N. was discovered in 1940 by McMillan and Abelson as a decay product of uranium 239, which was formed by absorption of slow neutrons:

$$^{238}U + {}^1n \rightarrow {}^{239}U \xrightarrow[23.54\ \text{min}]{} {}^{239}Np.$$

^{239}Np is important as an intermediate in the synthesis of plutonium 239. The long-lived isotope ^{237}Np is formed in uranium reactors by the action of fast neutrons on ^{238}U, so that it is now available on a kg scale:

$$^{238}U \longrightarrow {}^{237}U \longrightarrow {}^{237}Np.$$

^{237}Np is the first member of the neptunium decay series, decaying via a series of radioactive intermediates to the stable bismuth isotope ^{209}Bi. ^{237}Np is used to make ^{238}Pu.

N. was named for the planet Neptune.

Neptunium compounds: compounds in which neptunium displays close parallels to the neighboring actinoids uranium and plutonium. Like plutonium, neptunium can be in the oxidation states +3 to +7 in its compounds; N. with oxidation states +3 to +5 are more stable, and those with the +6 state are less stable than the homologous uranium compounds. Dioxoneptunium cations (neptunyl cations) are known for oxidation states +5 (green NpO_2^+), +6 (wine red NpO_2^{2+}) and +7 (green NpO_2^{3+}). Neptunium(III) compounds (purple) are stable in water, but are readily oxidized in air to neptunium(V) compounds. Neptunyl(VI) salts are stronger oxidizing agents than the isomorphic uranyl(VI) compounds. ***Neptunium(III) fluoride***, NpF_3, forms purple, hexagonal crystals, M_r 294.00, density 9.12, and is made by heating neptunium(IV) oxide to 500 °C with a mixture of hydrogen and hydrogenfluoride. ***Neptunium(III) chloride***, $NpCl_3$, white hexagonal crystals, M_r 343.36, density 5.38, m.p. 802 °C, is made by reduction of $NpCl_4$ with hydrogen at 450 °C. Addition of ammonia to neptunium(III) salt solutions leads to formation of dirty blue ***neptunium(III) hydroxide***, $Np(OH)_3$. ***Neptunium(IV) oxide***, NpO_2, apple-green, cubic crystals, M_r 269.00, density 11.11, is formed by heating neptunium(IV) hydroxide, nitrate or oxalate to 800 to 1000 °C. If NpO_2 is dissolved in hot concentrated sulfuric acid, and the solution is then concentrated, ***neptunium(IV) sulfate is obtained in the form of bright green crystals. Addition of oxalic acid or alkali oxalate to neptunium(IV) salt solutions causes neptunium(IV) oxalate*** to precipitate as the green hexahydrate, $Np(C_2O_4)_2 \cdot 6H_2O$. ***Neptunium (IV) chloride***, $NpCl_4$, reddish-brown, tetragonal crystals, M_r 378.81, density 4.92, m.p. 538 °C, is made by chlorination of neptunium(IV) oxide with tetrachloromethane around 530 °C. ***Neptunium(IV) fluoride***, light green, monoclinic crystals, M_r 312.99, density 6.8, are obtained by reaction of neptunium(IV) oxide with hydrogen fluoride, or by oxidation of neptunium(III) fluoride with a mixture of oxygen and hydrogen fluoride. Fluoroxidation of neptunium(IV) compounds in the presence of an alkali fluoride, MF, produces lavender ***fluoroneptunates(V)*** of the types $M[NpF_6]$, $M_2[NpF_7]$ and $M_3[NpF_8]$. Monoclinic, black-brown ***neptunium(V) oxide***, Np_2O_5, is obtained by heating neptunium(VI) oxide hydrate in vacuum to about 350 °C. ***Neptunium(VI) fluoride***, NpF_6, orange-brown, orthorhombic crystals, M_r 350.99, m.p. 54.4 °C, b.p. 55.76 °C, is obtained by reaction of neptunium(IV) fluoride with fluorine. The +7 oxidation state is represented by ***orthoperneptunates, orthoneptunates(VII)***, $M_5[NpO_6]$, which can be made by solid-state reactions. The ***metaperneptunates, metaneptunates(VII)***, $M_3[NpO_5]$, and the ***dioxoneptunium(VII) ion***, NpO_2^{3+}, are also +7 state compounds; they can be obtained by from neptunium(VI) compounds in aqueous solution by use of strong oxidizing agents. ***Lithium perneptunate***, $Li[NpO_6]$, is obtained by reaction of neptunium(IV) oxide with lithium oxide in an oxygen atmosphere at 430 °C.

Neradols: see Tannins.

Neral: see Citral.

Nernst approximation: see Ulich approximation.

Nernst distribution law: a relationship between the concentrations of a substance in two different phases (such as two immiscible liquids) in contact with each other, published by Nernst in 1891. At constant temperature, the ratio of the concentrations c_1 and c_2 of the substance in the two phases is a constant, which is called the ***Nernst distribution coefficient K_c***: $c_1/c_2 = K_c$. It follows that K_c does not depend on the absolute values of the concentrations or on the relative amounts of the two phases. The N. in the above form is completely valid only when the two mixed phases display ideal behavior. In addition, the substance cannot undergo dissociation or association in one of the phases. For real substances, the Activities (see) must be substituted for the concentrations.

The N. follows thermodynamically from the equilibrium condition $\mu_1 = \mu_2$, with $\mu_i = \mu^0{}_i + RT \ln a_i$, where μ_i is the chemical potential of the substance being distributed in phase i (1 or 2), $\mu^0{}_i$ is the standard potential, and a_i is the corresponding activity. The resulting equation is $\mu^0{}_1 + RT \ln a_1 = \mu^0{}_2 + RT \ln a_2$, or $a_1/a_2 = \exp[(\mu^0{}_2 - \mu^0{}_1)/RT] =$ constant. For ideal dilute solutions, this leads directly to the above form of the N.

The distribution coefficients must be determined experimentally. For example, the values obtained for distributions between an organic solvent and water at 25 °C were 581 for carbon disulfide, 366 for benzene and 75.6 for carbon tetrachloride.

The N. is the basis for analytical, preparative and industrial separation of substances by distribution methods such as extraction, phase-transfer reactions and Chromatography (see).

Nernst equation: an equation developed by W. Nernst in 1889; it describes the concentration dependence of relative electrode potentials (see Electrochemical equilibrium): $V = V^o + (RT/zF) \ln(a_{ox}/a_{red})$. V is the relative electrode potential, V^o is the relative electrode potential under standard conditions, R is the gas constant, T is the temperature in Kelvin, z the electrochemical valence, F the Faraday constant, a_{ox} and a_{red} the activities of the oxidized and reduced forms, respectively.

Nernst heat theorem: the third law of Thermodynamics (see).

Nerol: see Geraniol.

Nerolin: a collective term for various 2-alkoxynaphthalenes (naphth-2-ol alkyl ethers), e.g. 2-Methoxynaphthalene (see) or 2-Ethoxynaphthalene (see).

Neroli oil: same as Orange blossom oil (see).

Nerve gas: a nerve-damaging Chemical weapon (see).

Nervon: see Cerebrosides.

Nesmeanov reaction: a reaction of diazonium salts with mercury(II) chloride, followed by heating the resulting double salt with copper or iron powder to produce an arylmercury chloride: $[C_6H_5\text{-}N{\equiv}N]^+ Cl^- + HgCl_2 \rightarrow [C_6H_5\text{-}N]^+HgCl_3^- + 2\ Cu \rightarrow C_6H_5HgCl + 2\ CuCl + N_2$. Various organometal compounds of tin, lead, arsenic, antimony, bismuth and thallium can be made by varying the diazonium double salt. The reaction is apparently heterolytic; this is supported by the following reactions and the observed substituent effects:

$$\left[X-\langle\bigcirc\rangle-Sb(Cl)_5\right]^- Ar-\overset{+}{N}\equiv NI \xrightarrow[-FeCl_2]{+Fe}$$

$$\left[X-\langle\bigcirc\rangle-Sb(Cl)_3\right]^- Ar-\overset{+}{N}\equiv N \xrightarrow{-N_2} X-\langle\bigcirc\rangle-SbCl_3-Ar$$

The substituents X accelerate the reaction, and the stronger their electron donor effects are, the stronger the effect: $NO_2 < Cl < H < CH_3 < OC_2H_5$.

Nessler's reagent: a reagent for detection of ammonia. It consists of an alkaline, aqueous solution of potassium tetraiodomercurate, $K_2(HgI_4)$. If a few drops of N. are added to a sample solution, traces of ammonia cause a brown precipitate to form. The compound formed is $(Hg_2N)I$, which can be considered a mercury-substituted ammonium salt. The reaction is extremely sensitive, with a limit of detection of $2 \cdot 10^{-8}$ g NH_3 l^{-1}, and is therefore especially useful for detection of ammonia in drinking water.

N. can be used for quantitative determination of ammonia by spectrophotometry.

Neu: abb. for neuraminic acid.

Neumann-Kopp rule: see Molar heats.

Neuraminic acid, abb. ***Neu***: 5-Amino-3,5-dideoxy-D-galactononulopyrosoic acid, an amino sugar. Free N. is not stable, but cyclizes immediately to 4-hydroxy-5(1,2,3,4-tetrahydroxybutyl)-Δ^1-pyrrolidine-2-carboxylic acid. N. occurs in nature as glycosidically bound N-acyl-N. (see Sialic acid).

Neuroactive peptides: see Neuropeptides.

Neurohormones: see Neuropeptides.

Neurohypophysial hormones: a group of peptide hormones which are synthesized in the large-cell nuclei of the hypothalamus and transported to the neurohypophysis (posterior pituitary) by axonal transport. Vasopressin and oxytocin are important representatives.

Neuropeptides: peptides formed in peptidergic neurons which have signal functions for other cells. The category includes the ***neurohormones***, formed in neurosecretory nerve cells and secreted into the blood (and possibly liquor spinalis). Certain neurotransmitters also belong in this category. The term ***neuroactive peptides*** was coined from a pharmacological point of view; it includes all native peptides, their active partial sequences and synthetic analogs which induce neuronal responses. There are N. which prolong the extinction phase of various conditioned responses (for example, avoidance behavior which is either passively or actively learned by the experimental animal); others are involved in sleep regulation, pain perception (see Endorphins), have "memory transfer" activity (see Scotophobin) or are reported to improve intellectual productivity (corticotropin-(4-9) or (4-10) and Met(O)-Glu-His-Phe-D-Lys-Phe (ORG 2766).

Neurotensin: Pyr-Leu-Try-Glu-Asn-Lys-Pro-Arg-Arg-Pro-Tyr-Ile-Leu, a vasoactive peptide first isolated from bovine hypothalamus. In addition to biological effects typical of the plasma kinins, such as reduction of blood pressure and elevation of capillary permeability, contraction of intestine and uterus, N. increases the secretion of follitropin and lutropin, without affecting the release of somatotropin and thyreotropin; in addition, it has a hyperglycemic effect. The *C*-terminal sequence 9 to 13 contains the full intrinsic activity. The *C*-terminal segment of N. is quite similar to the sequence of the peptide ***xenopsin*** (Pyr-Gly-Lys-Arg-Pro-Trp-Ile-Leu) isolated from the skin of the frog *Xenopus laevis*; this peptide has about 20% of the effects of N. on blood pressure and blood sugar in rats.

Neurotransmitters: compounds released from the presynaptic nerve endings into the synaptic cleft; after diffusion across the synapse they transmit information to the postsynaptic cell. The postsynaptic nerve ending has an outer membrane provided with receptors; the postsynaptic cell can be a nerve, muscle or gland cell. Synapses are classified according to the type of N. active in them as ***cholinergic*** (acetylcholine), ***adrenergic*** (noradrenalin, adrenalin) or ***peptidergic*** (there are numerous peptide N.).

Neutral acid: see Acid-base concepts, section on Brønsted definition.

Neutral base: see Acid-base concepts, section on the Brønsted definition.

Neutralization: the reaction of an acid with a base which leads to a mutual cancellation of acidity and basicity. In the Arrhenian sense (see Acid-base concepts), acids and bases react to form salts and water, e.g. $NaOH + HCl \rightleftharpoons NaCl + H_2O$. The essential process is the combination of the H^+ and OH^- ions formed by dissociation of the acid and base into undissociated H_2O. The same process occurs in the N. of aqueous solutions of strong acids and bases, corresponding to the Brønsted definition, because the strong acid after protolysis in water is present in the form of H_3O^+ ions and its corresponding base, and the strong base is present as OH^- ions and the corresponding acid. Thus for all these reactions, the neutralization enthalpy ("heat of neutralization") is the same, $\Delta_N H = -57.5$ kJ mol^{-1}, regardless of which acid and base react. In reactions of stoichiometric amounts of strong acids and bases, after N. the H_3O^+ and OH^- concentrations are equal, the solution is neutral and has pH 7. Fundamentally, however, according to the Brønsted definition of acids and bases, the N. process represents a protolysis equilibrium in which the acid and base are converted into their corresponding partners: $HA + B \rightleftharpoons A^- + HB^+$. If a weak acid HA or a weak base B is involved in this reaction, the strong base A^- or the strong acid HB^+ will undergo a protolysis reaction with the water, and even when stoichiometric amounts of acid and base react, the protolysis will shift the pH away from the neutral.

Neutralization analysis, ***acid-base titration***: volumetric method of determining the acid or base concentration of a solution by titration with standard solutions of suitable bases or acids. The standard base solution is usually 0.1 N NaOH; the standard acid solution is 0.1 N HCl or 0.1 M H_2SO_4. In the neutralization reaction, H_3O^+ and OH^- ions combine to form water, which changes the pH value of the solution in the course of the titration. At the equivalence point, i.e. the endpoint of the titration, the amounts of acid and base are exactly equal, and the unknown amount of acid or base can be calculated from the volume of the standard solution required to reach this point. The equivalence point is recognized by use of an Indicator (see) or a potentiometric control (see Glass electrode). In many cases, protolysis of the neutralization products causes the pH of the equivalence point to be considerably different from 7, and this must be taken into account in the choice of indicator. For example, in the titration of acetic acid with NaOH standard solution, the high basicity of the acetate ion causes the equivalence point to lie in the basic range.

Neutralization coagulation: see Colloids.

Neutral point: see Ion product.

Neutral red, ***toluylene red***: a phenazine pigment which is commercially available as the hydrochloride. It forms a dark green powder which is readily soluble in water and alcohol, giving a red solution. N. is made by the reaction of 4-nitrosodimethylaniline hydrochloride with 3-toluylenediamine, followed by heating. It is used as a dye, but not to any great extent, because it is not very fast to light. However, it is used as an indicator; it changes color from red to yellow in the pH range 6.8 to 8.0. It is especially useful for examination of gastric juice and urine. It is also used as a desensitizer in photography, and as a vital and nuclear stain for microscopy.

Neutrocyanine pigments: see Polymethine pigments.

Neutron: see Elementary particles.

Neutron diffraction: a diffraction method based on the scattering of thermal neutrons. It is analogous to X-ray structure analysis (see), and is used to determine crystal structures. While x-rays are scattered off the electrons of the atoms, neutrons are scattered off the nuclei, except in the case of magnetic atoms, where the unpaired electrons also contribute to the diffraction of the neutrons. The ability of an element to scatter x-rays increases sharply and evenly with the atomic number. When the scattering of neutrons is plotted against atomic numbers, however, an irregular plot is obtained (Fig.). Different isotopes of the same element have different scattering properties.

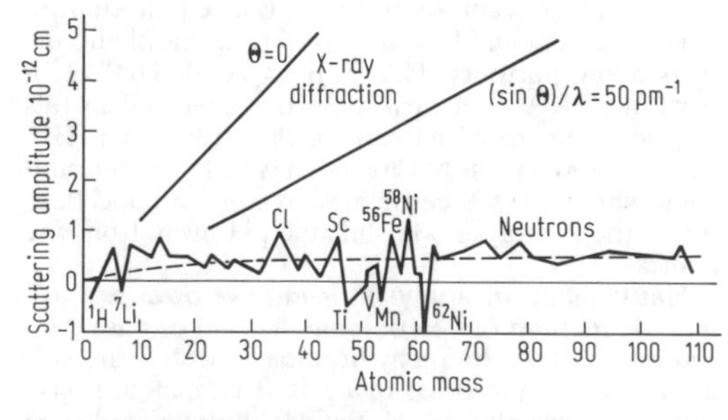

Neutron scattering amplitudes as a function of the relative atomic masses.

The neutrons are obtained from a nuclear reactor. The relationship between the mean square velocity v^2 of thermal neutrons and the temperature is given by $1/2mv^2 = 3/2\ kT$, where m is the mass, $1.674 \cdot 10^{-27}$ kg, k is Boltzmann's constant and T is the temperature in K. For 373 K, this equation yields $v = 3$ km s^{-1}, and the de Broglie relationship gives the wavelength 133 pm. This is only an average. Monochromatic neutron beams are obtained by reflection from a crystal surface.

The reflection intensities are measured with a boron trifluoride counting tube on a four-circle diffractometer (see X-ray structure analysis). The most important applications of N. are in the determination of magnetic structures, because the diffraction of magnetic atoms depends on the orientation of their magnetic moments with respect to the crystal plane. Therefore, N. gives information on the orientation of the atomic magnets. N. can also be used to distinguish heavy elements with closely similar atomic numbers. For example, it is not possible with x-ray analysis to determine the positions of iron and cobalt atoms in a unit cell of the alloy, because the atomic form factors of the two elements are too similar. However, the atomic form factors for neutron beams are in a ratio of 4:1 in this case. If both heavy and very light elements are present in a structure, the positions of the light elements (e.g. hydrogen) are difficult to determine by x-ray diffraction, but the scattering of neutrons off these elements is not so extremely different. Therefore both types of atoms give good maxima in a neutron Fourier synthesis.

Nevile-Winther acid: an Alphabet acid (see).

New blue: same as Bluing (see).

New fuchsin: a bluish red, basic triphenylmethane pigment. N. is soluble in water and methanol. It is made by heating diamino-2-ditolylmethane with 2-toluidine in the presence of hydrochloric acid and an oxidizing agent. N. is used to dye paper, leather and polyacrylonitrile, and to make paints.

New green: 1) same as Malachite green (see). 2) same as Schweinfurt green (see Copper(II) acetate arsenite).

Newman projection: see Stereochemistry.

New silver: an alloy of copper, nickel and zinc, with 10 to 25% nickel and 15 to 25% zinc; the remainder is copper. Because of the nickel, these alloys have a color similar to that of silver. They can be cast and polished to a high gloss. They are used to make tableware, sanitary devices, musical instruments and medical instruments; they are also used as hard solder (see Solder).

Newton's cooling law: an approximate relationship discovered by I. Newton describing the rate of cooling of a hot body by loss of heat to the environment. If the difference in temperature between the body (T) and the environment (T_e) is not too large, the rate at which heat is lost (dQ/dt) is directly proportional to the difference in temperature: $dQ/dt = \alpha A(T-T_e)$. If dQ is set equal to cdT, it follows that $dT/dt = -\alpha A(T-T_e)/c$. Here α is the heat transfer number, c is the heat capacity and A is the surface area of the body. αAc is called the ***cooling constant***. By integration, it follows that

$$T = T_0\, e^{-(\alpha A/c)(T-Te)t}$$

that is, the temperature decreases exponentially with the time *t*.

N. plays an important role in the cooling and tempering of materials, calorimetry and thermal analysis.

Newton's metal: an alloy of 53% bismuth, 21% cadmium and 26% tin with a melting point of 103 °C, named for the English physicist. N. is used for melt fuses, as a heating bath liquid, as soft solder and in dentistry.

New yellow: same as Fast yellow.

Ni: symbol for nickel.

Nickel, symbol ***Ni***: a chemical element from Group VIIIb of the periodic system, the Iron group (see). It is a heavy metal, *Z* 28, and its natural isotopes have the following mass numbers: 58 (68.27%), 60 (26.10%), 62 (3.59%), 61 (1.13%), 64 (0.91%). The atomic mass is 58.70, the valence is usually II, more rarely, III, IV, I or 0, Mohs hardness, 3.8, density 8.908, m.p. 1452 °C, b.p. 2730 °C, electric conductivity 14.5 Sm mm^{-2}, standard electrode potential (Ni/Ni^{2+}) -0.250 V.

Properties. Ni is a silvery white, shiny, polishable, ductile metal. It is highly resistant to corrosion, and can be welded, hammered, rolled into sheets and drawn into wires. Below 367 °C (Curie temperature), it is weakly ferromagnetic. N. occurs in two modifications, the hexagonal α-N. and cubic β-N. It is very resistant to water and air at room temperature, and it is therefore often used to protect iron parts by plating or electroplating. At room temperature, non-oxidizing acids dissolve N. very slowly; in dilute nitric acid it dissolves readily, but in concentrated nitric acid it is insoluble, due to passivation. Ni is resistant to alkali hydroxides even at 300 to 400 °C. Nickel wire burns in pure oxygen with a spray of sparks. It reacts at high temperatures with other nonmetals such as halogens, sulfur, phosphorus, arsenic, silicon and boron. Ni forms alloys with metals (see Nickel alloys). Hydrogen is absorbed by Ni, especially at higher temperatures, by lattice expansion. Nickel-aluminum alloys treated with sodium hydroxide yields a very active metal (***Raney nickel***) which is used as a hydrogenation catalyst.

Analytical. In the classical separation scheme, Ni precipitates with the $(NH_4)_2S$ group, as gray-black nickel(II) sulfide, NiS. This is oxidized via Ni(OH)S to Ni_2S_3, which is insoluble in dilute cold hydrochloric acid. It is specifically determined with an alcoholic dimethylglyoxime solution (Čugaev's reagent) by the formation of a scarlet nickel dimethylglyoxime complex, which is also suitable for gravimetric determination of nickel. Quantitative analysis is also carried out by complexometry.

Occurrence. Ni makes up 0.015% of the earth's crust. Elemental Ni is found in meteorites, alloyed to iron, and it is thought that the earth has an iron-nickel core. The most important minerals for extraction of Ni are nickel-containing magnetic pyrites, which contain Ni as pentlandite (iron nickel pyrite, $(Fe,Ni)_9S_8$) and silicate or oxide ores, which contain Ni as garnierite, $(Mg,Ni)_3Si_2O_5(OH)_4$. Some other important nickel ores are millerite, NiS, nickelin, NiAs, chloanthite, $NiAs_{2-3}$, gersdorffite, NiAsS and breithauptite, NiSb. As a trace element, Ni is found in soils at about 40 ppm, and in plants at about 0.5 ppm.

Production. The methods are complicated, because the nickel contents in the ores are low and the presence of related elements (iron, copper and cobalt) makes chemical-metallurgical concentration and refining difficult. About 70% of the ores are sulfidic (1 to 5% Ni, 1.5 to 4.5% Cu), and these are usually physically prepared to concentrate the nickel to 5 to 15%. The oxide ores containing 1 to 2% Ni contain no Cu (10% Fe), and can only be prepared chemically.

The following chemical-metallurgical concentration methods are used: nickel concentration in a sulfide melt (Ni-Cu-Fe-rock), nickel leaching with sulfuric acid or ammonia solution and nickel concentration in ferronickel or in an arsenide melt. The intermediates of nickel production (nickel stone, nickel liquor, nickel melt) are processed to pure nickel; ferronickel is used directly for steel production.

The sulfidic nickel concentrates are partially roasted, then melted to yield a crude ore. This is smelted and separated into a copper ore (≈ 30% Cu) and an enriched nickel ore (40 to 70% Ni, ≈ 10% Cu, 1% Fe, 20% S). This is purified by electrolysis, or by roasting, reductive smelting (crude nickel containing 15% Cu) and refining electrolysis (the roasting-reducing method). The enriched nickel ore can also be smelted to yield a Ni-Cu-Fe alloy, which is carbonylated at a pressure of 20 MPa at 200-250 °C (pressure carbonyl process). The Fe-Ni carbonyl is separated by distillation, and the tetracarbonylnickel is thermally decomposed to nickel powder and carbon monoxide.

The oxide ores can be reduced in rotating furnaces. Magnetic separation of the sintered product yields Ni-Fe lumps containing 6 to 8% Ni. After blasting, this is converted to a ferronickel with 40 to 60% Ni (Krupp-Renn process). Pressure leaching of the ores with sulfuric acid yields a $NiSO_4$ solution, from which NiS is precipitated, then dissolved under pressure in sulfuric acid. This is electrolysed to produce pure electrolyte nickel. For ammonia leaching, the finely powdered oxide ores must be reduced. The leaching process with ammonia or ammonium carbonate solution yields a solution of $[Ni(NH_3)_6]^{2+}$, which is thermally decomposed to nickel carbonate. This in turn is calcined to nickel(II) oxide, which can be purified by solution and electrolysis processes or by reduction.

Applications. Most of the Ni produced is used in stainless steels and nonferrous Nickel alloys (see). N. is used as a corrosion protection for more reactive metals (see Nickel plating) and as an intermediate layer in direct enamelling (see Enamel). Laboratory apparatus is made from pure Ni, and it is a alloy component in bimetallic strips for temperature sensing. On oxide carrier materials (SiO_2, Al_2O_3), Ni is an important component of catalysts used for many hydrogenation processes in industrial organic chemistry. In hard metals, Ni serves as a ductile binder.

Historical. Around 2000 BC, a nickel alloy similar to the modern new silver was used in China. The Greeks used nickel alloys for coins and other objects around 200 BC. In 1751, Cronstedt (re)discovered Ni as a new metal. Bergmann (1775) and Richter (1804) purified the metal and studied its compounds. Since 1824, Ni has been used in copper alloys with appearances similar to silver. In 1832, Faraday discovered electroplating with nickel. In 1870, Parkes discovered stainless steel alloys containing Ni.

Nickel acetate: $Ni(CH_3COO)_2 \cdot 4H_2O$, green prisms soluble in water and alcohol; M_r 248.86, density 1.774. Aqueous solutions of N. have a sweet taste. N. is formed by dissolving nickel carbonate or hydroxide in dilute acetic acid. It is used to a limited extent in galvanic solutions and as a treatment for cotton in preparation for dyeing.

Nickel alloys: alloys of nickel with manganese, chromium, copper, molybdenum and iron. Nickel forms mixed crystals with these metals which are highly resistant to change in shape, but highly formable, and highly resistant to corrosion. The largest fraction of the nickel produced today is used by the steel industry to produce non-rusting, acid-resistant and cold-drawable steels. The remaining production goes for coins (see Coinage alloys) and production of the following N. (nickel content above 50%). 1) ***Nickel-manganese alloys*** with 1.5 to 2.5% manganese are used for parts of electronic tubes and light bulbs; alloys with 3% manganese and a little aluminum are used as the "nickel" side of a NiCr-Ni bimetallic strip. 2. ***Nickel-chromium alloys*** retain their strength when heated and have high electrical resistance. They are unusually inert to corrosive aqueous media and hot gases in reducing or oxidizing atmospheres. Other components of these alloys are iron, molybdenum, cobalt, tungsten, aluminum and titanium. Some industrial alloys are the bimetallic strip alloy (10% chromium, the rest nickel), Inconel® (see), Corronel® (see) and the Superalloys (see), e.g. Nimonic® (see) and Ni-O-Nel 825® (see). 3) ***Nickel-copper alloys*** contain nickel as the main component (for alloys with less than 50% nickel, see Copper alloys). An important member of this group is Monel metal®, which has excellent resistance to corrosion. 4) ***Nickel-molybdenum alloys***. Alloys of the Hastelloy® (see) type were developed for the highest corrosion stresses. 5) ***Nickel-iron alloys*** include Invar® (see) and Permalloy® (see) and Ferronickel (see). 6) ***Composite materials*** include dispersion-reinforced sintered nickel (see TD nickel), heat resistant Hard metals (see), Cermets (see) and Superalloys (see).

Nickel arsenide type, ***Nickel arsenide structure***: a Structure type (see) occurring mainly in semiconducting compounds with the (approximate) composition AB. In nickel arsenide with the strictly stoichiometric composition NiAs, the As atoms occupy the lattice positions corresponding to a hexagonal close packing, and the Ni atoms occupy the octahedral holes. Each atom thus has 6 atoms of the other type as nearest neighbors.

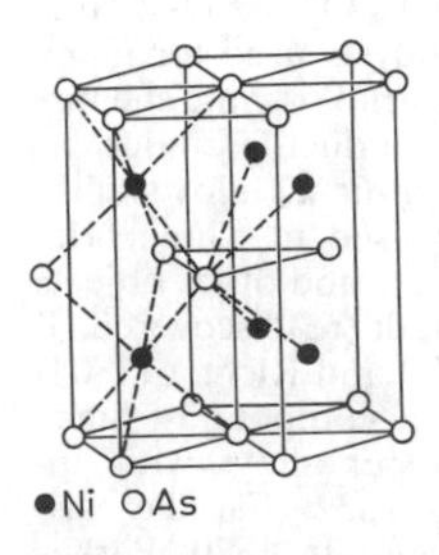

Nickel arsenide structure.

The Ni atoms have octahedral coordination; the As atoms have trigonal prismatic coordination (Fig.).

Many comopunds which crystallize in the N. are Berthollides (see), which deviate from the ideal composition in the range $A_{0.5}B_1$ to A_2B_1 because of vacancies or occupations of interlattice positions in the metal part of the lattice. There are many N. compounds, in which the metal component can be a transition metal from the titanium to the nickel group, and the nonmetal can be gallium, silicon, phosphorus or sulfur, or one of their higher homologs.

Nickel bromide: $NiBr_2$, yellow-brown, rhombohedral, hygroscopic crystals; M_r 218.53, density 5.098, m.p. 963 °C. When heated, N. first turns gold-yellow and then sublimes into yellow-gold leaflets. In the presence of water, it is converted to the trihydrate, $NiBr_2 \cdot 3H_2O$, M_r 272.57. N. is formed by the reaction of bromine with heated nickel powder.

Nickel bronze: a term for copper-nickel alloys. Industrial alloys contain 10 to 30% nickel, 0.4 to 1% iron, 0.5 to 1.5% manganese and the rest copper. N. are resistant to tension crack corrosion, and are used in electrical power plants, ships, chemical apparatus, the petroleum industry and desalinization plants for sea water.

Nickel-cadmium battery: an Electrochemical current source (see) and a secondary cell.

The anode in the charged state consists of nickel(III) oxide hydroxide (active component) and graphite powder or nickel flakes as conductor. The anode is encased in a perforated metal jacket. The cathode is formed as cadmium powder or sponge, and is similarly encased. The discharge and recharge processes are:

$$\text{Anode: } NiOOH + H_2O + e^- \underset{\text{recharge}}{\overset{\text{discharge}}{\rightleftharpoons}} Ni(OH)_2 + OH^-$$

$$\text{Cathode: } Cd + 2\,OH^- \underset{\text{recharge}}{\overset{\text{discharge}}{\rightleftharpoons}} Cd(OH)_2 + 2e^-$$

Potassium hydroxide is used as electrolyte. During discharge, insoluble metal(II) hydroxides form on both electrodes. The working potential of the N. is 1.2 V.

Nickel carbonate: $NiCO_3$, pale green, rhombic crystals which are insoluble in water; M_r 118.72. N. is formed by the reaction of sodium hydrogencarbonate with nickel(II) salt solutions; when sodium carbonate is added, basic N. is obtained. N. is used as a pigment for painting glass and ceramics and to make catalysts for hardening of fats. It is also a starting material for producing other nickel compounds.

Nickel carbonyl: same as Nickel tetracarbonyl (see).

Nickel chloride: $NiCl_2$, gold-yellow, shiny, easily sublimed, regular, scale-shaped crystals which are easily soluble in water and alcohol; M_r 192.62, density 3.55. At room temperature, the hexahydrate crystallizes out of aqueous solution: $NiCl_2 \cdot 6H_2O$, M_r 237.70. Other hydrates, such as the mono-, di- and tetrahydrate, are known. In non-aqueous solutions, N. forms blue, tetrahedral chloro complexes $[NiCl_4]^{2-}$ with chloride ions. It is easily made by heating nickel in a stream of chlorine, or by dehydration of the hydrates in a stream of hydrogen chloride. N. is used, for example, to color ceramics, to make nickel catalysts and for galvanic nickel plating.

Nickel complexes: coordination compounds of nickel. There are numerous nickel(II) complexes, of which the most abundant sort are the paramagnetic, octahedral compounds such as $[Ni(H_2O)_6]X_2$, $[Ni(NH_3)_6]X_2$ and $[Ni(diamine)_3]X_2$. Diamagnetic, square planar complexes such as tetracyanonickelates(II), $M_2[Ni(CN)_4]$ (see Nickel cyanides) and Nickeldiacetyldioxime (see) are also known. Tetrahedral, paramagnetic nickel(II) complexes such as tetrahalonickelates(II), $M_2[NiX_4]$, can be made from nonaqueous media. π-Acceptor ligands such as CO and CN^- stabilize complexes of nickel in lower oxidation states (see Nickel tetracarbonyl, Nickel cyanides), while octahedral fluoro complexes such as $K_3[NiF_6]$ and $K_2[NiF_6]$ are representatives of the rare +3 and +4 oxidation states.

Nickel cyanides. ***Nickel(II) cyanide***: $Ni(CN)_2 \cdot 4H_2O$, apple green, very poisonous compound, M_r 182.79. It precipitates upon reaction of nickel(II) salt solutions with alkali cyanides in a stoichiometric ratio; when heated to 180-200 °C, it is converted to the yellow-brown anhydrous salt $Ni(CN)_2$. Nickel(II) cyanide reacts with more alkali cyanide to give the gold-yellow ***alkali tetracyanonickelates(II)***, $M_2[Ni(CN)_4]$. The tetracyanonickelate(II) anions are square planar. In the presence of concentrated cyanide solution, alkali tetracyanonickelates(II) are converted to the trigonal bipyramidal pentacyanonickelates(II), $M_3[Ni(CN)_5]$, while reduction with potassium in liquid ammonia gives red cyanonickelate(I), $K_4[Ni_2(CN)_6]$ or yellow tetracyanonickelate(0), $K_4[Ni(CN)_4]$. Nickel(II) cyanide or tetracyanonickelates(II) are used in galvanic nickel plating.

Nickel diacetyldioxime, $Ni(C_4H_7N_2O_2)_2$, scarlet crystalline needles; M_r 288.94. N. is a water-soluble internal complex (Fig.). It is made by adding an alcoholic diacetyldioxime solution to a solution of nickel(II) salt in which either ammonia or dilute acetic acid is also present. N. is used for qualitative and quantitative (gravimetric) analysis of nickel.

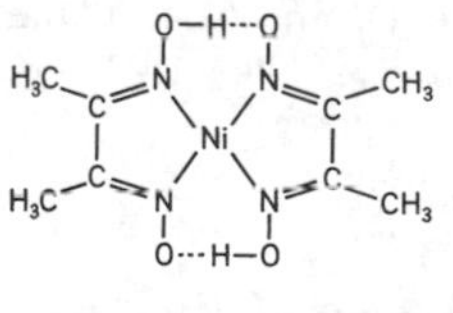

Nickel(II) dicyanamide: see Dicyanamides.

Nickel formate: $Ni(HCOO)_2 \cdot 2H_2O$, green crystals which are readily soluble in water; M_r 184.78, density 2.154. N. is formed by reaction of nickel carbonate with formic acid, and is used as a catalyst in fat hardening.

Nickel group: same as Nickel triad (see).

Nickel hydroxide: $Ni(OH)_2$, green, water-soluble compound; M_r 92.72, density 4.1. N. is readily soluble in acids, ammonia and ammonium salt solutions. When roasted, it is converted to nickel oxide, NiO; oxidation with potassium hypobromite converts it to nickel(III) metahydroxide, NiO(OH). N. is obtained as a voluminous precipitate when alkali hydroxide solution is added to a nickel salt solution.

Nickelin®: a resistance alloy of 30.5% nickel, 2.5% manganese and the rest copper.

Nickel-iron battery: an Electrochemical current source (see) and a secondary cell. The anode is the same as in the Nickel-cadmium battery (see). The cathode consists of iron powder or sponge and is enclosed in the same way as the anode is. The discharge and recharge processes are:

$$\text{Anode: } NiOOH + H_2O + e^- \underset{\text{recharge}}{\overset{\text{discharge}}{\rightleftharpoons}} Ni(OH)_2 + OH^-$$

$$\text{Cathode: } Fe + 2\,OH^- \underset{\text{recharge}}{\overset{\text{discharge}}{\rightleftharpoons}} Fe(OH)_2 + 2e^-$$

Potassium hydroxide is used as electrolyte. During discharge, insoluble metal(II) hydroxides form on both electrodes. The working potential of the N. is 1.3 V.

Nickel(II) nitrate: $Ni(NO_3)_2 \cdot 6H_2O$, emerald green, monoclinic crystals which are very soluble in water and alcohol; M_r 290.81, density 2.05, m.p. 56.7 °C. Some other hydrates are the enneahydrate $Ni(NO_3)_2 \cdot 9H_2O$, which is stable below -9 °C, the tetrahydrate, $Ni(NO_3)_2 \cdot 4H_2O$, which is stable above 65 °C, and the dihydrate, $Ni(NO_3)_2 \cdot 2H_2O$, which is stable above 85 °C. At higher temperatures, N. decomposes to form nickel oxide, NiO. N. is obtained by reaction of nickel with dilute nitric acid and is used to make brown colors for ceramics, and in nickel plating.

Nickel(II) oxide: NiO, gray-green, cubic crystals; M_r 74.71, density 6.67, m.p. 1990 °C. N. is not soluble in water, but dissolves readily in acids. When hot, it is easily reduced by hydrogen to the metal. It is obtained by roasting nickel hydroxide, nitrate or carbonate. N. is found in nature in the form of the green mineral bunsenite. It is used to make glass and enamel paints and nickel catalysts for hydrogenations.

Nickel plating: the application of a layer of nickel to a metal object to protect it from corrosion, or as an intermediate layer for application of chromium. It is sometimes also used for decoration. N. is carried out directly, or after an intermediate layer of copper has been applied. ***Galvanic N.*** makes use of nickel anodes and a bath electrolyte of nickel(II) salts and boric acid. In ***chemical N.***, the reducing agent is sodium hypophosphite, and the bath contains other compounds in addition to the nickel(II) salts. The applied layer contains phosphates as well as metallic nickel, so that it has almost no pores and binds very tightly. The first nickel deposited catalyses the further reaction. In ***spray plating***, alloys of nickel are powdered and sprayed onto the piece, which is then heated to the melting point of the alloy. This then diffuses into the surface of the coated object to a certain extent, creating a very tightly bonded protective layer.

Nickel(II) sulfate: $NiSO_4$, yellow, cubic crystals; M_r 154.51, density 3.68, m.p. 848 °C (dec.). N. is soluble in water, and crystallizes out of aqueous solution as the heptahydrate (***nickel vitriol***), $NiSO_4 \cdot 7H_2O$, in emerald green, rhombic crystals; M_r 280.88, m.p. 99 °C, density 1.98. There is also a hexahydrate which occurs in two modifications, tetragonal and monoclinic. Nickel vitriol is found in nature in the form of the very rare mineral morenosite. N. is obtained, for example, by reaction of nickel oxide or carbonate with sulfuric acid; it is used in galvanic technology. N. forms double salts with alkali and ammonium sulfate. The blue-green ammonium-nickel sulfate $(NH_4)_2SO_4 \cdot NiSO_4 \cdot 6H_2O$ is used for electroplating of nickel, and to blacken zinc and copper.

Nickel(II) sulfide: NiS, black compound; M_r 90.77, density 5.3 to 5.4, m.p. 797 °C. When precipitated from nickel(II) salt solutions with ammonium sulfide, N. is obtained in amorphous form (α-NiS). One of the crystalline modifications, γ-NiS, is found in nature as millerite. Upon standing, α-NiS loses its solubility in cold dilute hydrochloric acid, as it is converted to Ni_2S_3. N. is used to make catalysts, e.g. for reforming and hydrogenation processes.

Nickel tetracarbonyl: $Ni(CO)_4$, colorless, very poisonous liquid; M_r 170.25, density 1.32, m.p. -19.3 °C, b.p. 42.1 °C. N. is readily soluble in organic solvents, but not in water. When ignited, it burns with a bright flame, and mixtures of it with air are explosive. It also detonates upon contact with concentrated sulfuric acid. At 180 to 200 °C, N. decomposes into its components. It can be reduced with sodium in liquid ammonia or ether to give carbonyl nickelates such as $M_2[Ni_2(CO)_6]$, $M_2[Ni_3(CO)_8]$ or $M_2[Ni_4(CO)_9]$. N. is formed by the reaction of carbon monoxide with finely divided nickel at 50 to 100 °C. It is an important intermediate in the technical production of pure nickel by the Langer-Mond process, and is used as a CO-transferring catalyst.

Nickel titanium yellow: a bright yellow pigment, titanium(IV) oxide with a rutile structure into which nickel and antimony oxide are incorporated as the color-producing components. N. is synthesized in a thermal process above 800 °C from titanium(IV) oxide, nickel and antimony compounds. It meets the highest standards of light fastness, is stable in the presence of cement, lime, acids, alkalies, sulfites and sulfides, and is therefore suitable for light-colored wall paints and colored cement products.

Nickel triad, ***nickel group***: the elements nickel, Ni, palladium, Pd, and platinum, Pt, which form a short column in group VIIIb of the periodic system. The unreactive character of the metals increases from nickel to platinum. All the metals form numerous metal(II) complexes; the tetracyanometallates(II), $[M(CN)_4]^{2-}$, for example, are characterized by planar structures. Indeed, planarity is the rule among the palladium(II) and platinum(II) complexes. Within the N., the maximum oxidation states increase from nickel to platinum, and the stability of compounds in the higher oxidation states also increases in this direction.

Nickel vitriol: see Nickel sulfate.

Nicosamide: see Anthelminthics.

Nicotinamide: pyridine 3-carboxamide, a form of niacin. N. forms colorless crystals, m.p. 129-131 °C. It is readily soluble in water and ethanol, but barely soluble in ether. N. can be produced from nicotinic acid, its nitrile or esters by reaction with ammonia. Low concentrations are present, in the form of the N. nucleotides, in all organisms. A dietary deficiency of N. or nicotinic acid leads to pellagra, a condition leading to changes in the skin, digestive and nervous functions. N. was therefore also called ***PP factor*** (pellagra preventive factor). It can be synthesized in the human body from tryptophan, via kynurenine, but the amount of tryptophan in the diet is usually inadequate to meet the N. requirement. In microorganisms and plants, the synthesis goes by another pathway.

Nicotinamide adenine dinucleotide: see Nicotinamide nucleotides.

Nicotinamide adenine dinucleotide phosphate: see Nicotinamide nucleotides.

Nicotinamide mononucleotide: see Nicotinamide nucleotides.

Nicotinamide nucleotides: a group of coenzymes in hydrogen-transferring enzymes. The reactive group of N. is nicotinamide. In the oxidized forms, the nicotinamide is a pyridinium ion, and in the reduced forms, it is a dihydropyridine compound. The absorption spectra of the two forms are distinctly different. The N. include the following compounds: ***nicotinamide mononucleotide*** (abb. ***NMN***), ***nicotinamide adenine dinucleotide*** (abb. ***NAD***; obsolete term, ***diphosphopyridine nucleotide***, abb. ***DPN***) and ***nicotinamide adenine dinucleotide phosphate*** (abb. ***NADP***), obsolete term, ***triphosphopyridine nucleotide***, abb. ***TPN***).

The N. are coenzymes of carbohydrate and amino acid metabolism, and of the citrate cycle. The hydrogen transfer by these enzymes is stereospecific; the enzymes are classified as A or B depending on the side from which the substrate approaches the planar pyridine ring.

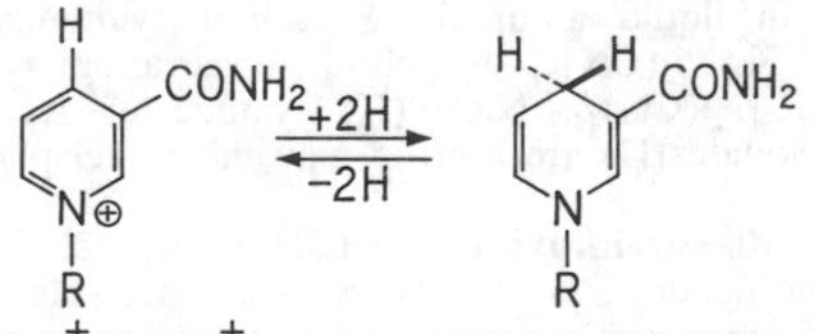

NAD^+, $NADP^+$ ⇌ (+2H / −2H) NADH, NADPH

NAD$^+$, oxidized form of NAD, M_r 663.4, λ_{max} = 260 nm ($\varepsilon = 18.0 \cdot 10^3$ mol^{-1} cm^{-1}), E_0 = 0.316 V (pH = 7.0).

NADH, reduced form of NAD, M_r 665.4, as the disodium salt, M_r = 709.4; λ_{max} = 229 and 340 nm ($\varepsilon = 6.22 \cdot 10^3$ mol^{-1} cm^{-1}).

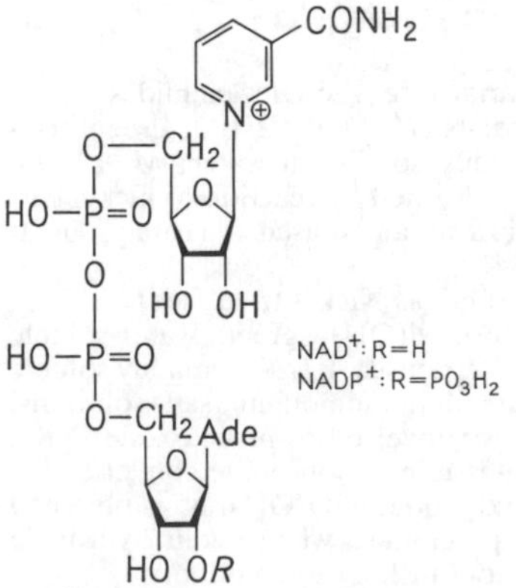

NAD^+: R=H
$NADP^+$: R=PO_3H_2

NADP$^+$, oxidized form of NADp, M_r 743.4, as the disodium salt, M_r = 787.4; absorption same as NAD^+, E_0 = 0.317 V (pH 7.0).

NADH, reduced form of NADP, M_r 745.4, as the tetrasodium salt, M_r = 833.4; absorption same as NADH.

Nicotine: 3-(N-methyl-α-pyrrolidyl)pyridine, the main alkaloid in tobacco and some other *Nicotiniana* species. N. is a colorless, oily liquid, m.p. -79 °C, b.p. 247 °C, $[\alpha]_D$ -169° (in water), miscible with water and volatile with steam. N. has a sharp, burning taste. The solution quickly turns brown in the air. N. is a relatively strong base, pK_1 = 10.96, pK_2 = 6.16. Natural N. has the (S) configuration. It is present in the leaves of *Nicotiniana tabacum* at 0.05 to 10% concentration. The plants contain other alkaloids in addition to N., such as nornicotine, anabasine, N-methylanabasine or nicotyrine. The alkaloids are synthesized mainly in the roots, from which they are transported into the above-ground organs. N.-poorer strains of *N. tabacum* are raised for tobacco, while N.-richer strains are raised for production of N. N. is highly poisonous; 50 mg can be fatal for a human being due to respiratory paralysis. When cigarette smoke is inhaled, the absorption of N. is practically quantitative. N. or N.-containing extracts were formerly important as insecticides, especially for aphids. Their use in open fields has dwindled due to the high toxicity, lack of stability and excessive water solubility.

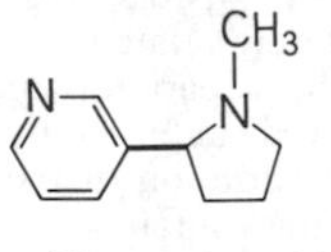

CH_3
N N

Nicotinic acid, ***pyridine 3-carboxylic acid, 3-carboxypyridine***: a heteroaromatic carboxylic acid. N. forms colorless crystalline needles; m.p. 236-237 °C (subl.). It is soluble in hot water and alcohol, and insoluble in ether. N. was first obtained by oxidation of the alkaloid nicotine with potassium permanganate. It is found in nearly all cells, where it is a component of coenzymes. It is not synthesized by mammals, which therefore require it in the diet; an adult human being must consume about 15 mg per day. Compounds with the biological activity of N. are known as niacin, which is one of the B_2 vitamins. Niacin is especially abundant in bread, kidneys, liver, muscle meat, milk, fish, yeast and fruits. A deficiency leads to a complex disorder (pellagra) which can be healed by high doses of the vitamin. N. is synthesized by oxidation of β-picoline with potassium permanganate or sulfuric acid in the presence of selenium. N. is used industrially to protect metals from corrosion and as a vitamin additive to flour and animal feeds. The sodium salts and esters are used medicinally to promote circulation.

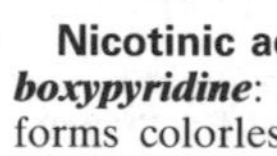

COOR
N
R = H: Nicotinic acid
R = $CH_3-CH_2-CH_3$: n-Propyl nicotinate
R = $CH_2-C_6H_5$: Benzyl nicotinate

Nifedipin: a 1,4-dihydropyridine derivative. It is synthesized from 2-nitrobenzaldehyde, acetoacetate and ammonia. N. is sensitive to light, which initiates an intramolecular redox reaction to form a compound with a nitroso group and an aromatic pyridine ring. N. is used to treat coronary disease.

Nigericin: an antibiotic from *Streptomyces* species. N. is a monoprotic acid which is readily soluble in organic solvents, but only slightly soluble in water. It acts as an Ionophore (see), but is rather toxic.

Nigrosines: a group of azine pigments made by heating a mixture of nitrobenzene, aniline and aniline hydrochloride with metallic iron or copper at 200 °C. ***Spirit black*** (spirit-soluble N.), ***indulin black*** (water-soluble N.) and fat-soluble N. are important types. The N. are used to dye and print cotton and silk, to blacken leather, in shoe polish and as specific stains for bacteria.

Nile blue: a term for various basic oxazine dyes which are soluble in water and alcohol. N. are used to dye treated cotton, to make paints and as stains for microscopic detection of cancer cells.

Nimonic®: a heat and scale resistant nickel-chromium alloy (see Nickel alloys) containing 60 to 80% nickel, 20% chromium, 0 to 20% cobalt; a Superalloy (see).

Ninhydrin reaction: see Amino acids.

Niobates: ionic reaction products of niobium(V) oxide with basic metal oxides. Metaniobate, $[NbO_3]^-$, diniobate, $[Nb_2O_7]^{4-}$, and the hexaniobates formed in aqueous solution, $[Nb_6O_{19}]^{8-}$, $[HNb_6O_{19}]^{7-}$ and $[H_2Nb_6O_{19}]^{6-}$, are especially important. The mineral niobite is an isomorphic mixture of iron(II) and manganese(II) metaniobate.

Niobe oil: see Benzoates.

Niobic(V) acid: see Niobium(V) oxide hydrate.

Niobium, symbol ***Nb***: an element of group Vb of the periodic system, the Vanadium group (see); a heavy metal with a single isotope, *Z* 41, atomic mass 92.9064, valency primarily V, but also IV, III and II; density 8.81, m.p. 2468 °C, b.p. 4742 °C, electrical conductivity 6.5 Sm mm^{-2}, standard electrode potential (Nb/Nb^{3+}) -1.099 V.

Properties. N. is a silvery white, ductile metal which crystallizes in a cubic body-centered lattice. Pure N. can be hammered, pressed and drawn, but even small amounts of carbon, nitrogen or oxygen impurities make it hard and brittle, so that it can then be shaped only at red heat. Chemically, N. is very inert, due to formation of a thin, tight oxide layer on the surface. It is not attacked by most acids, including aqua regia. Hydrofluoric acid and hot, concentrated sulfuric acid will dissolve it, and molten alkalies attack N., forming alkali niobates. At high temperatures, N. reacts with most nonmetals. Oxygen oxidizes it at red heat to niobium(V) oxide, while chlorine reacts with it even below 300 °C to form niobium(V) chloride. Nitrogen and carbon monoxide react with N. above 400 °C to form niobium nitrides, carbides and oxides. Hydrogen is also absorbed by N. in the temperature range between 350 and 500 °C, forming niobium hydride.

Analysis. Niobium, like tantalum, is now determined mainly by x-ray fluorescence spectroscopy or emission spectroscopy. Chemical separation of the two closely related elements can be achieved by paper chromatography or ion-exchange methods. The classical method for separation and detection of N. depends on precipitation of niobium(V) oxide hydrate; the precipitate is then subjected to alkali fusion with potassium carbonate or hydroxide to form potassium niobate. This is reduced with zinc or tin to form blue, low-valency niobium derivatives, which react with thiocyanate to form golden-yellow complexes. For quantitative determination, N. is precipitated as the oxide hydrate and roasted to niobium(V) oxide.

Occurrence. N. makes up $1.8 \cdot 10^{-3}\%$ of the earth's crust. Because of its close chemical similarity to tantalum and the nearly identical ionic radii of these elements ($r_{Nb^{5+}}$ 134.2 pm; $r_{Ta^{5+}}$ 134.3 pm), N. is always found in association with tantalum. Columbite $(Fe,Mn)(Nb,TaO_3)_2$ is an important source; depending on the predominant metal, it is also called niobite or tantalite. Another important N. mineral is pyrochlor, with the general formula $(Ca,Na)_2[Nb, Ta, TiO_3]_2(OH,F)$; it occurs in carbonatites, often together with apatite, and its composition depends on the deposit. The calcium and sodium in it can be replaced by other alkaline earth or rare earth metals, or by uranium and thorium.

Production. The classical method for production of N., and thus also for separation of N. and tantalum, is the *Marignac process*, which starts from columbite. The ore is solubilized with hydrofluoric acid. Addition of potassium salts to the filtrate leads to crystallization of the insoluble potassium heptafluorotantalate(V), $K_2[TaF_7]$, which is recrystallized for further purification. The more soluble potassium oxofluoroniobate(V), $K_2[NbOF_5]$, which remains in the filtrate, is then precipitated as the oxide hydrate, sometimes along with other metals. This oxide hydrate can be directly processed to ferroniobium.

For production of N. from pyrochlor-containing carbonatites, the ore is first enriched in pyrochlor by a flotation process; the enriched ore can contain up to 50% niobium(V) oxide. This can be used directly for production of ferroniobium, or it can be separated chemically. For this, the pyrochlor concentrate is solubilized with hydrofluoric acid and the filtered solution is extracted with methyl ethyl ketone. Niobium and tantalum fluoro complexes are extracted into the organic phase, from which they are recovered sequentially by re-extraction into water. N. is then crystallized out of the aqueous solution as niobium(V) oxide hydrate, and tantalum as potassium heptafluorotantalate.

For reductive preparation of N., niobium(V) oxide is prepared by dehydration of the oxide hydrate and is heated above 1500 °C in a vacuum in the presence of coal. Niobium carbide is formed as an intermediate, but decomposes around 1700 °C to N.: $Nb_2O_5 + 7\,C \rightarrow 2\,NbC + 5\,CO$; $5\,NbC + Nb_2O_5 \rightarrow 7\,Nb + 5\,CO$.

Another method starts from niobium(V) chloride; it is reduced with hydrogen in the gas phase, sometimes under plasma conditions. Very pure N. is also made by thermal decomposition of niobium(V) chloride. Finally, N. can be obtained electrochemically; a chloride fluoride electrolyte containing $NbCl_5$ is electrolysed at 850 to 900 °C under a protective atmosphere.

N. is purified by a method which utilizes its exceptionally high boiling point. It is heated in stages under high vacuum to its sintering temperature of 2500 °C, causing the impurities to evaporate and producing 99.9% pure N. 99.99% pure N. is obtained by zonal melting.

Applications. N. is used mainly in the form of ***ferroniobium*** and ***ferroniobium tantalum*** as an additive for stainless steels and special steels. N. is also an alloy component of nonferrous metal alloys, such as those containing chromium, nickel and cobalt. Niobium-containing, high-temperature alloys with extreme resistance to wear and corrosion are used in the construction of aircraft, rockets and gas turbines. N. is used as the casing for uranium fuel rods, because it has a low neutron capture cross section and is very resistant to the liquid metals used as coolants (potassium, sodium, lithium, lead, mercury, bismuth or tin). Because of its pronounced affinity for gas, N. is used in high vacuum apparatus as a getter. It is used in the construction of rectifiers, in tungsten-niobium thermoelements for use up to 2000 °C, and as a component of alloys for permanent magnets. N.-tin alloys are interesting superconductors; Nb_3Sn_2 has a lambda point of 22 K, while N. itself becomes superconducting below 9.13 K.

Historical. The development history of N. is closely entwined with that of tantalum. N. was discovered in 1801 in the form of its oxide by Hatchett; this oxide was mixed with tantalum oxide, however. It was first named columbium (until 1949 in the English literature) for the mineral of origin. In 1802, Ekeberg discovered minerals containing an oxide of a previously unknown element, for which he suggested the name tantalum. In 1844, Rose suggested the name niobium for the closely related element associated with tantalum. The separation of N. and tantalum was first achieved in 1866 by Marignac, who used fractional crystallization of the alkali fluorometallates. N. was first obtained in pure, ductile form in 1905.

Niobium borides: metal-like compounds of the composition Nb_3B_2, NbB, Nb_2B_4 and NbB_2; they are very hard and have high melting temperatures. The most important of this group is ***niobium diboride***, NbB_2, hexagonal, gray crystals; M_r 114.53, density 6.6, m.p. about 3000 °C. NbB_2 is supercconducting below 1.27 K, is resistant to nitric acid and aqua regia, but reacts with hydrofluoric acid. It is obtained, for example, by heating niobium(V) oxide with boron carbide. N. can be used as a component of electronic tubes or as a heat-resistant ceramic material.

Niobium carbide: NbC, gray-brown powder with a violet sheen; M_r 104.917, density 7.78, m.p. 3490 °C. N. exists as a stable phase in the niobium/carbon system, along with Nb_2C (superconducting below 9.18 K); its transition temperature to superconductivity is around 6 K. It is used as a partial substitute for tantalum in hard materials based on tungsten carbide/tantalum carbide and tungsten carbide/titanium carbide/tantalum carbide/cobalt. It has also been proposed that N. could be used as a ferro- or piezoelectric material or a ceramic; it can be used as a starting material or intermediate in the production of niobium.

Niobium chlorides: ***Niobium(V) chloride, niobium pentachloride***, $NbCl_5$, yellow crystalline needles; M_r 280.17, density 2.75, m.p. 204.7 °C, b.p. 254 °C. $NbCl_5$, like tantalum(V) chloride, $TaCl_5$, is a monomeric trigonal bipyramidal molecule in the gas phase;

in the solid and dissolved states, it forms dimeric molecules held together by chlorine bridges (Fig.).

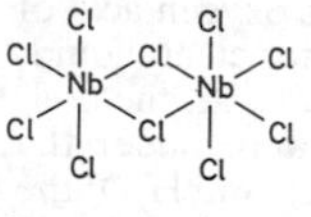

Structure of $NbCl_5$.

$NbCl_5$ is a Lewis acid and an effective Friedel-Crafts catalyst; it adds donor molecules D to form octahedral complexes $NbCl_5D$, and is hydrolysed by water, with loss of HCl, to niobium(V) oxide hydrate. It is made by heating niobium in a stream of chlorine, or by reaction of niobium(V) oxide with carbon and chlorine at about 400 °C: $Nb_2O_5 + 5\,C + 5\,Cl_2 \rightarrow 2\,NbCl_5 + 5\,CO$. The treatment of $NbCl_5$ with reducing agents, such as niobium, aluminum, cadmium, iron or hydrogen, produces chlorides of niobium in lower oxidation states. ***Niobium(IV) chloride, niobium tetrachloride***, $NbCl_4$, forms violet-black, orthorhombic crystals (M_r 234. 72). It is a polymeric compound in which $NbCl_6$ octahedra share edges. The other N. are $NbCl_3$, $NbCl_{2.67}$ and $NbCl_{2.33}$. The last of these is an octahedral metal cluster with the composition $[Nb_6Cl_{12}]Cl_2$. $NbCl_{2.67}$ is a cluster compound with the composition Nb_3Cl_8; it has a trigonal basic structure. $NbCl_3$ consists of mixed crystals Nb_3Cl_8/Nb_2Cl_8, with a homogeneity range of $NbCl_{2.76}$ to $NbCl_{3.13}$. Green Nb_3Cl_8 is obtained by reduction of $NbCl_5$ with aluminum at 350 to 500 °C, while $[Nb_6Cl_{12}]Cl_2$ is obtained from Nb_3Cl_8 at temperatures above 800 °C. If this reaction is carried out in the presence of potassium chloride, a complex salt is obtained: $K_4[Nb_6Cl_{12}]Cl_6$. The anion of this salt has an octahedral arrangement of the metal atoms, in which the octahedral edges are bridged by chlorine atoms, and terminal positions contain one chlorine atom bound to the niobium atom.

Niobium(V) fluoride, ***niobium pentafluoride***: NbF_5, colorless, very refractive crystals in which tetrameric $(MF_5)_4$ units are present (Fig.). In the gas phase its structure is monomeric trigonal bipyramids; M_r 187.90, density 3.293, m.p. 73 °C, b.p. 236 °C. N. is formed by treating niobium(V) oxide hydrate with anhydrous hydrogen fluoride acid: $NbCl_5 + 5\,HF \rightarrow NbF_5 + 5\,HCl$.

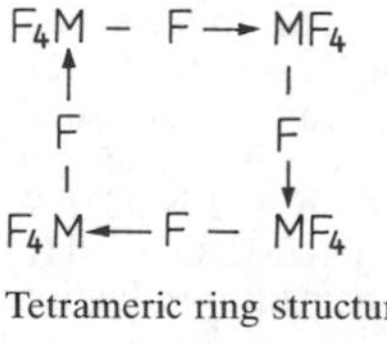

Tetrameric ring structure of solid NbF_5.

In aqueous solution, N. forms fluoroniobates(V) of varying composition, e.g. $[NbF_6]^-$, $[NbF_7]^{2-}$, $[NbOF_5]^{3-}$; these play an important role in the hydrofluoric acid solubilization of niobium-tantalum ore concentrates and in the separation of niobium and tantalum.

Niobium hydride: $NbH_{0.9}$, dark gray, acid-resistant compound which is superconducting below 14 K; density 6.6. N. is formed by uptake of hydrogen by niobium. It is used as a neutron moderator; like titanium and zirconium hydrides, it can be used as a soldering aid to join metallic or non-metallic materials such as ceramics or diamond.

Niobium oxides: ***Niobium(V) oxide, niobium pentoxide***, Nb_2O_5, a white compound; M_r 265.81, density 4.47, m.p. 1460 °C. Nb_2O_5 is an acid oxide and is thus not soluble in acids, except for hydrofluoric acid. If fused with alkali hydroxides or carbonates, it reacts to form Niobates (see). Nb_2O_5 is synthesized by dehydration of niobium hydrate; it is an important industrial intermediate in the production of niobium or niobium carbide. If Nb_2O_5 is heated in a stream of hydrogen to about 1000 °C, it is converted to dark gray ***niobium(IV) oxide***, NbO_2, M_r 124.90, density 5.90, which can be further reduced to ***niobium(III) oxide***, Nb_2O_3, M_r 233.81, m.p. 1780 °C, and ***niobium(II) oxide***, NbO, M_r 108.91. NbO is characterized by a strong metal-metal interaction; its conductivity approaches that of niobium.

Niobium(V) oxide hydrate: $Nb_2O_5 \cdot xH_2O$, colorless, gelatinous substance soluble in strong acids and bases; its water content varies. It is often also called ***niobic(V) acid***. When heated strongly, N. is converted to niobium(V) oxide, and like the latter, it is acidic. N. is obtained by adding acid to niobate solutions, or by hydrolysis of niobium(V) chloride. N. is an important industrial intermediate in the production of niobium.

Ni-O-Nel 825®: a nickel-chromium alloy (see Nickel alloys) consisting of 36 to 48% nickel, 1.5 to 3.0% copper, 15.9 to 23.5% chromium, 2.5 to 3.5% molybdenum, 0.6 to 1.2% titanium, and the rest iron. N. is a casting metal for pumps and armatures used in the fertilizer industry to transport sulfuric and phosphoric acid even in the presence of fluorine compounds.

Niotensides: non-ionic tensides; see Surfactants.

Nipa esters: alkyl esters of 4-hydroxybenzoic acid. These compounds are used as preservatives, especially in pharmaceuticals, but also in foods. The most commonly used is the methyl ester (methyl hydroxybenzoate, methylparaben, nipagin). The ethyl and propyl esters (propyl hydroxybenzoate, propylparaben, nipasol) are also used to some extent.

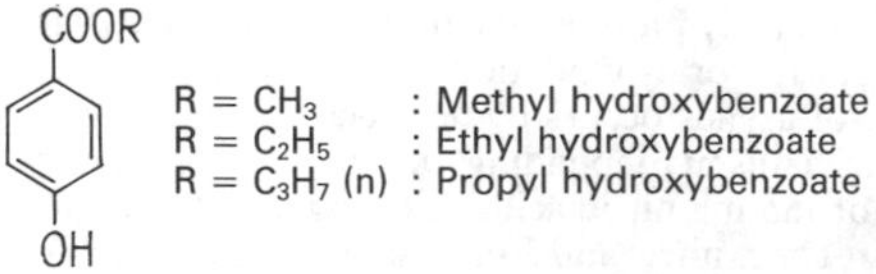

R = CH_3 : Methyl hydroxybenzoate
R = C_2H_5 : Ethyl hydroxybenzoate
R = C_3H_7 (n) : Propyl hydroxybenzoate

These compounds are synthesized by the reaction of potassium phenolate with CO_2 according to Kolbe-Schmitt, followed by esterification of the resulting 4-hydroxybenzoic acid.

Nipagin: see Nipa esters.

Nipasol: see Nipa esters.

NIR: abb. for near infrared (see Infrared spectroscopy).

Niton: see Radon.

Nitralin: see Herbicides.

Nitraniline red: same as Para red (see).

Nitrate: 1) a *salt* of nitric acid, with the general formula M^INO_3. All N. are based on the planar, resonance-stabilized nitrate ion.

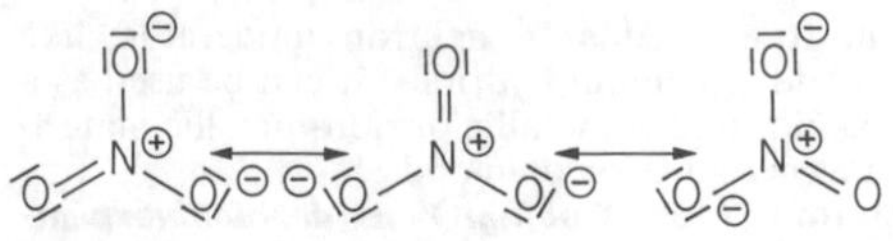

All simple inorganic N. are readily soluble in water. When heated, they decompose, releasing oxygen. Therefore, nitrate melts are strongly oxidizing. The alkali N. are converted to nitrites, e.g. $NaNO_3 \rightarrow NaNO_2 + 1/2\ O_2$. Heavy metal N. decompose to metal oxides, nitrogen dioxide and oxygen, e.g. $Pb(NO_3)_2 \rightarrow PbO + 2\ NO_2 + 1/2\ O_2$. In aqueous solution, N. are attacked only by very strong reducing agents, e.g. nascent hydrogen reduces N. to ammonia, NH_3. The N. are obtained by the reaction of nitric acid with metals or their oxides, hydroxides or carbonates.

2) an *ester* of nitric acid, with the general formula R-O-NO_2 (R = alkyl or aryl group).

Nitrate cellulose: same as Cellulose nitrate (see).

Nitrate reductase: a flavin enzyme which contains iron and molybdenum and catalyses the first step of nitrate reduction. N. consists of two subunits. One contains molybdenum and non-heme iron, and the other, which is induced by the substrate (nitrate), contains only iron. N. first reduces nitrate to nitrite, which is then further reduced to ammonia by ***nitrite reductase***, an enzyme which contains iron and SH groups.

Nitrating acid: a mixture of conc. nitric and conc. sulfuric acids; the relative concentrations vary. The amount of water in the N. depends on the reactivity of the organic compound to be nitrated, and can be reduced nearly to zero by use of oleum and fuming nitric acid. N. is used to make mono-, di- and polynitro compounds which are used in the production of dyes, drugs, explosives and pesticides.

Nitration: introduction of a nitro group, -NO_2, into an organic compound. N. of alkanes occurs by radical substitution (chain reaction) in the gas phase with nitric acid at temperatures around 400 °C. The products are isomeric nitroalkanes; oxidative cleavage of C-C double bonds also produces nitro compounds of the lower alkanes. Nitroalkanes can be obtained relatively pure by reaction of haloalkanes with sodium nitrite or by oxidation of tertiary alkylamines. The N. of arenes occurs by an electrophilic substitution mechanism; depending on the arene, concentrated or fuming nitric acid, nitrating acid (a mixture of 1 part conc. nitric and 2 parts conc. sulfuric acid) or nitric acid in glacial acetic acid is used as the nitrating reagent.

Nitrazepam: see Benzodiazepines.

Nitrenes: R-N, a highly reactive, uncharged intermediate with an electron sextet on the nitrogen. N. are the nitrogen analogs of the carbenes. They exist in a lower-energy triplet state and a singlet state. They are formed by photolysis or thermolysis of azides or hydrazoic acid: $R\text{-}N\text{-}N{\equiv}N \rightarrow R\text{-}N + N_2$.

N. stabilize themselves by various reactions, such as isomerization to imines by intramolecular H transfer, by intermolecular H abstraction (e.g. from the solvent) to form primary amines, intramolecular C-H insertion reactions to form five or six-membered rings, dimerization to form azo compounds (***aryl nitrenes***) and addition to alkenes to form aziridines, or addition to benzoid compounds with ring expansion to form 1-substituted azepines.

Nitric acid: HNO_3, the most stable oxygen acid of nitrogen. In pure form, N. is a water-clear liquid, density 1.522, m.p. -41.65 °C, b.p. 84 °C. At the boiling point, or even at room temperature under the influence of light, it decomposes to NO_2 and H_2O; the nitrogen dioxide dissolves in the N., colors it yellow to red, and gives it a pungent odor. 100% N. is not stable at room temperature. The most important forms available commercially are conc. HNO_3, approximately a 69.2% solution, and dilute N., which is about 25%. N. is miscible with water in all proportions. Depending on the concentration and temperature, it forms a monohydrate, $HNO_2 \cdot H_2O$ or a trihydrate, $HNO_3 \cdot 3H_2O$. The solution of 69.2% HNO_3 and 30.8% water is an azeotropic mixture, density 1.410, which boils at 121.8 °C. Conc. N. is a strongly oxidizing acid, which is capable of oxidizing arsenic to arsenic acid, H_3AsO_4, phosphorus to phosphoric acid, H_3PO_4, and sulfur to sulfuric acid, H_2SO_4. Metals, except for gold, platinum, iridium, compact rhodium and ruthenium, are dissolved by N. The metal is first oxidized, forming nitrogen monoxide. With silver, for example, the reaction is $6\ Ag + 2\ HNO_3 \rightarrow 3\ Ag_2O + 2\ NO + H_2O$; the metal oxide then immediately reacts with excess N. to form the salt and water: $Ag_2O + 2\ HNO_3 \rightarrow 2\ AgNO_3 + H_2O$. Since silver can be separated from gold in this way, N. was once called "separating water". A mixture of N. and hydrochloric acid, called Aqua regia (see), dissolves most noble metals. A few non-noble metals, such as aluminum, calcium, chromium and iron, do not dissolve in very concentrated N., due to passivation. However, they do dissolve in dilute N. with a vigorous reaction. Dilute N. does not oxidize the metals, so that hydrogen is evolved as the salt is formed. N. either oxidizes or nitrates organic compounds, the latter reaction occurring especially in mixtures with sulfuric acid. This mixture is therefore called Nitrating acid (see).

Analysis. N. is usually detected qualitatively by the ring test with iron(II) sulfate and concentrated sulfuric acid. If the N. is present in the form of a salt, it is released by the concentrated sulfuric acid, for example: $2\ NaNO_3 + H_2SO_4 \rightarrow 2\ HNO_3 + Na_2SO_4$. The N. is reduced by iron(II) sulfate to nitrogen monoxide: $2\ HNO_3 + 6\ FeSO_4 + 3\ H_2SO_4 \rightarrow 3\ Fe_2(SO_4)_3 + 4\ H_2O + 2\ NO$. Nitrogen monoxide adds to the excess iron(II) sulfate, forming nitroiron(II) sulfate: $FeSO_4 + NO \rightarrow Fe(NO)SO_4$. N. is determined quantitatively by reduction to ammonia with zinc (or aluminum) an alkaline solution: $4\ Zn + 3\ NaOH + 8\ H_2O + NaNO_3 \rightarrow 4\ NaZn(OH)_3 + 2\ H_2O + NH_3$. N. can also be detected as nitrous acid after reduction with zinc in acidic solution.

Production. N. is made on a large scale by catalytic ammonia combustion, the *Ostwald process* (Fig.). Ammonia is mixed with air an a 1:10 ratio, then preheated in a heat exchanger to about 200 °C. The gas mixture is very rapidly passed over a catalyst (screens of platinum or a platinum-rhodium alloy), in a contact furnace, generating nitrogen monoxide and water vapor: $4\ NH_3 + 5\ O_2 \rightarrow 4\ NO + 6\ H_2O$; $\Delta H \approx -900$ kJ mol^{-1}. The reaction temperature is about 800 °C.

Because the reaction is so exothermal, a large excess of air is needed to carry off the heat of reaction. The mixture passes through a heat exchanger where it generates high-pressure steam and is cooled to about 250 °C. It then passes through a water cooler where it is cooled to 30 to 40 °C. The time in the contact furnace must be very brief, and the heat must be removed rapidly, since nitrogen monoxide otherwise decomposes to nitrogen and oxygen. In practice, flow rates of 0.3 m/s are achieved; this corresponds to a time of about 0.2 ms on the catalyst screen. Nitrogen monoxide is subsequently oxidized to nitrogen dioxide by the excess oxygen in the reaction gas and by added fresh air: $2\,NO + O_2 \rightarrow 2\,NO_2$, $\Delta_R H = -113$ kJ mol^{-1}. The activation energy of this reaction appears to be negative, and it therefore occurs at reasonable rates only at temperatures below 50 °C. Nitrogen dioxide reacts with water in a series of absorption towers to form N.: $3\,NO_2 + H_2O \rightarrow 2\,HNO_3 + NO$; $\Delta_R H = -72$ kJ mol^{-1}. Simultaneously, the remaining nitrogen monoxide is oxidized with atmospheric oxygen to form nitrogen dioxide. In each tower, the dilute N. trickles down past the rising gas which was formed in a second tower from unreacted nitrogen dioxide and water. In practice, large numbers of such paired towers are connected in series. The residual gases are absorbed with lye, e.g. sodium hydroxide, to form a nitrate-nitrite solution, which is then treated with oxygen. This oxidizes the nitrite to nitrate. The N. formed from the acidic absorption is at a concentration of about 50%. It can be concentrated to about 68% by distillation. Further concentration is possible only by distillation in the presence of added sulfuric acid or phosphorus pentoxide. Nitric acid plants may be run at a pressure up to 9 MPa in the combustion and absorbtion stages. In modern plants, the combustion is carried out at atmospheric pressure and the absorption at approximately 3 MPa. The oxidation of 1 t nitrogen in the form of ammonia generates about 5 t high-pressure steam.

In the laboratory, N. is usually made by heating potassium nitrate with conc. sulfuric acid: $KNO_3 + H_2SO_4 \rightarrow HNO_3 + KHSO_4$. It was formerly made in similar fashion from Chile saltpeter, $NaNO_3$.

Applications. N. is extremely important in many branches of the chemical industry. It is used as the starting material for production of nitrates, to nitrosylate organic compounds, e.g. cellulose or glycerol, and as an oxidant for production of sulfuric acid by the nitrogen oxide method. In metal working, N. is used to pickle and etch. Most N. is processed to form calcium ammonium nitrate and to extract phosphates.

Historical. N. was used in the Middle Ages. It was first synthesized from saltpeter mixed with alum and copper sulfate, and later, from saltpeter and sulfuric acid. The structure of N. was first determined toward the end of the 18th Century by Lavoisier and Priestley.

Nitrides: binary compounds of nitrogen with electropositive elements, usually metals. N. are usually formed by direct combination of the elements at high temperatures, or by the reaction of ammonia with a metal oxide or halide. The N. of alkali and alkaline earth metals can also be made by thermolysis of the amides. There are three groups of N.: 1) ***Ionic N.*** are formed by highly electropositive elements such as lithium, the alkaline earth metals, zinc and cadmium. They form ionic lattices consisting of metal cations and nitride anions N^{3-}, and thus can be considered salts of ammonia. They react with water to form the metal hydroxide and ammonia, e.g. $Mg_3N_2 + 6\,H_2O \rightarrow 3\,Mg(OH)_2 + 2\,NH_3$. Nitrogen combines with less electropositive elements to form 2) ***covalent, "nonmetallic" N.***. Some examples are Boron nitride (see), BN, and Silicon nitride (see), Si_3N_4. The elements of groups IVb to VIb tend to form 3) ***nonstoichiometric N. (inclusion nitrides)***. In these compounds, most of which are very hard and have high melting points, the nitrogen atoms occupy interlattice positions in the metallic lattice.

Many covalent and nonstoichiometric N. are valuable materials, due to their inertness and resistance to high temperatures.

Nitrification, ***nitrate formation***: the oxidation of ammonia to nitrite to nitrate by bacteria living in the soil and water. N. is an important part of the nitrogen cycle in nature. Two groups of bacteria are involved in the process, the nitrite and the nitrate bacteria.

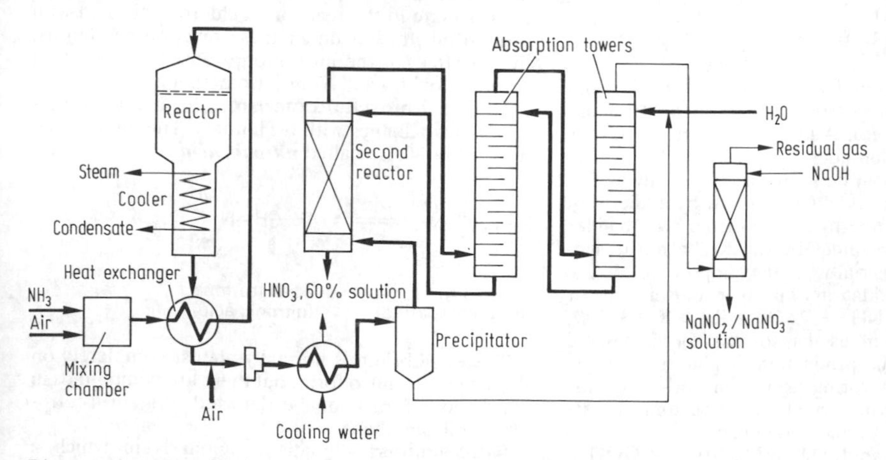

Diagram of a nitric acid plant.

The first group oxidizes ammonia to nitrite, and the second, nitrite to nitrate. In humus-containing soils, N. proceeds spontaneously so long as other effects, such as lack of oxygen or acidity, do not interfere.

In contrast to N., ***denitrification*** converts nitrate to gaseous molecular nitrogen which escapes from the soil to the atmosphere. This process is carried out by anaerobic dinitrification bacteria.

Nitriles: organic compounds with the general formula R-C≡N. They may be classified as aliphatic, aromatic or heterocyclic, depending on R. Compounds with the C≡N group can be named as follows: 1) for compounds in which the -C≡N group is the principal group, the name is composed of the root name of the hydrocarbon $R\text{-}CH_3$ and the suffix -nitrile, e.g. butanenitrile $CH_3\text{-}CH_2\text{-}CH_2\text{-}C{\equiv}N$. If the N. is considered as derived from the acid R-COOH, its name is the root for the hydrocarbon R + carbonitrile, e.g. thiophene-2-carbonitrile; or the syllables "-oic acid" or "-ic acid" are dropped from the trivial name of R-COOH and replaced by "-onitrile", e.g. $C_6H_5\text{-}C{\equiv}N$ is benzonitrile. 2) In compounds which contain a different principle group, the -C≡N group is indicated by the prefix "cyano-". 3) In radicofunctional nomenclature, the suffix "-cyanide" is added to the name for R-, e.g. propyl cyanide $CH_3\text{-}CH_2\text{-}CH_2\text{-}C{\equiv}N$.

Properties and reactions. Most N. are colorless crystalline or liquid compounds. Some low-molecular-weight saturated and unsaturated aliphatic N. are very poisonous, with the exception of acetonitrile, which is distinctly less poisonous. The reactivity of the N. is due chiefly to the reactive C≡N group. Aliphatic N. with α H-atoms also have the reactivity typical of CH-acidic compounds. Acid- or base-catalysed hydrolysis of N. yields carboxylic acids or carboxamides. N. are converted to primary amines by suitable reducing agents. Grignard compounds (see) react with N. to form ketones:

$$R^1\text{–}C{\equiv}N + R^2Mg\,X \rightarrow$$

$$R^1R^2C = N\text{–}Mg\,X \xrightarrow[-\,Mg\,(OH)\,X]{+\,H_2O}$$

$$R^1R^2C = NH \xrightarrow[-\,NH_3]{+\,H_2O} R^1R^2C = O$$

Aromatic substituted ketones can be made from phenols or phenol ethers and N. by the Houben-Hoesch reaction (see). Addition of hydrogen halides to N. produces haloimides.

Production. N. can be synthesized by dehydration of carboxamides or aldehyde oximes, by addition of hydrogen cyanide to ethynes, and from alkyl halides and alkali metal cyanides by the Kolbe nitrile synthesis (see). Industrially, a few important N. are made by ammonoxidation of propylene, isobutene or toluene: $R\text{-}CH_3 + NH_3 + 3/2\,O_2 \rightarrow R\text{-}C{\equiv}N + 3\,H_2O$.

Application. N. are used as solvents, pesticides and intermediates in the production of pharmaceuticals, dyes, perfumes, flavoring agents and other organic compounds. Some important N. are acetonitrile, acrylonitrile, benzylcyanide and benzonitrile.

Nitrilotriacetic acid, abb. ***NTE***: $N(CH_2\text{-}COOH)_3$, a Complexone (see). N. forms colorless, prism-shaped crystals, m.p. 250-259 °C (dec.). It is barely soluble in water, but readily soluble in alcohol; it dissolves in alkali hydroxide solutions, forming salts. N. is very important in complexometry, and is used, among other things, to separate lanthanum from the other rare-earth metals by chromatography. The sodium bismuth salt is used in medicine.

Nitrile oxides: compounds with the general formula R-CNO or R-C≡N→O, where R is an aliphatic, aromatic or heterocyclic group. The electron distribution in N. can be described with the aid of the following resonance structures:

$$R\text{–}C\text{–}\overset{\oplus}{N}{\equiv}\underline{O}\,|^- \longleftrightarrow R\text{–}\bar{C}\text{–}N{=}\underline{O} \longleftrightarrow R\text{–}C{=}\overset{\oplus}{\bar{N}}\text{–}\underline{O}\,|^-$$

Most N. are very unstable under normal conditions, and can be converted to furoxanes by spontaneous dimerization. Many aromatic N. readily rearrange when heated, forming the isomeric isocyanates: R-CNO → R-NCO. They are synthesized by dehydration of aldoximes, by removal of hydrogen chloride from hydroxamic acyl chlorides or by elimination of nitrous acid from nitrolic acids: $R\text{-}C(NO_2){=}NOH \rightarrow R\text{-}CNO + HNO_2$. N. are suitable reactants for 1,3-dipolar cycloadditions, especially for synthesis of heterocyclic compounds. They are also used to make pesticides and pharmaceuticals.

Nitrites: 1) the *salts* of nitrous acid, with the general formula M^INO_2. The resonance-stabilized anion is angular (O-N-O bond angle 115 °C; Fig.). The alkali and alkaline earth nitrites are made by passing an equimolar mixture of nitrogen monoxide, NO, and dioxide, NO_2, through the corresponding metal hydroxide solution or by reduction of the metal nitrates.

$$^-|\bar{O}\text{–}\underline{N}{=}O \longleftrightarrow O{=}\underline{N}\text{–}\bar{O}|^-$$

The *esters* of nitrous acid with the general formula RO-N=O (R = alkyl or aryl group).

Nitroacetic acid: $O_2N\text{-}CH_2\text{-}COOH$, forms crystalline needles; m.p. 93-93.5 °C. N. is soluble in alcohol, ether, benzene and chloroform. It is synthesized by heating nitromethane with magnesium methylcarbonate in dimethylformamide. N. is also formed as an intermediate in the reaction of chloroacetic acid with sodium nitrite, but under these reaction conditions, the - I-effect of the nitro group causes it to split off carbon dioxide, and form nitromethane.

Nitro-aci-nitro tautomerism: a form of tautomerism in nitroalkanes with α-H atoms. The enol or aci-nitro form is also called ***nitronic acid***.

$$R\text{–}CH_2\text{–}\overset{+}{N}(=O)\text{–}O^- \rightleftharpoons R\text{–}CH{=}\overset{+}{N}(OH)\text{–}O^-$$

Nitro form (nitro compound) — aci-Nitro form (nitronic acid)

The equilibrium is normally almost completely on the side of the nitro form, but there are compounds in which both forms can be detected, e.g. in 4-nitrophenylnitromethane.

Nitroalkanes: aliphatic compounds in which a nitro group, $-NO_2$, is directly bound to a C atom of an aliphatic hydrocarbon group. In the presence of more than one nitro group, the compound is a ***dinitroalkane***, ***trinitroalkane***, etc. The degree of substitution of the C atom to which the nitro group is bound is indicated by the designation ***primary N.*** ($R-CH_2-NO_2$), ***secondary N.*** (R_2CH-NO_2) or ***tertiary N.*** (R_3C-NO_2). The electronic effect of the nitro group is primarily an I-effect, with the result that the α-H atoms in primary and secondary N. are highly activated. N. of this type are in a tautomeric equilibrium between the nitro and enol or aci-nitro form (see Nitro-aci-nitro tautomerism).

Primary and secondary N. dissolve in bases, forming salts. In the presence of nitrous acid, primary N. are converted to colorless Nitrolic acids (see); secondary N. are converted in such solutions to the colored Pseudonitrols (see). This reaction is used to characterize aliphatic nitro compounds. Primary and secondary N. can be synthesized by reaction of alkyl halides with sodium or silver nitrate (the latter is better). Tertiary N. are made by oxidation of tertiary alkylamines with potassium permanganate. Industrially, N. are made by liquid or vapor phase nitration of alkanes. N. are of great industrial importance, being used as solvents, fuels, insecticides and explosives. They are also important intermediates in the synthesis of drugs, dyes, emulsifiers, plastics, etc.

Nitroanilines, ***aminonitrobenzenes***: the three isomeric nitroderivatives of aniline. ***o-Nitroaniline (2-nitroaniline)***: orange-yellow, steam-volatile leaflets or needles; m.p. 71.5 °C, b.p. 284 °C (dec.). It is only slightly soluble in water, but is soluble in chloroform, alcohol, acetone and benzene. ***m-Nitroaniline (3-nitroaniline)***: yellow needles, m.p. 114 °C, b.p. 305-307 °C (dec.). It is only slightly soluble in water, but is soluble in alcohol, ether and acetone. It is steam-volatile. ***p-Nitroaniline (4-nitroaniline)***, yellow needles, m.p. 148-149 °C, b.p. 331.7 °C. It is nearly insoluble in water, but dissolves in alcohol and ether.

NH_2

NO_2 (2-, 3- or 4-position)

N. are solid, yellow, crystalline and poisonous. Their color is due to the π-electron interaction between the amino and nitro groups. Interruption of this resonance, for example by blocking the free electron pair of the amino group by salt formation, leads to a hypsochromic shift. As a result, the salts of N. are colorless. N. are weak bases. Their salts hydrolyse in aqueous solution, regenerating the initial components. The methods for synthesis of N. depend on the position of the nitro group. o-N. and p-N. can be obtained through the reaction of ammonia with the corresponding chloronitrobenzenes. Another method is nitration of acylated aniline, which produces both o- and p-isomers. These can be separated after hydrolysis, using aqueous alkali hydroxide solutions. m-N. is synthesized by partial reduction of m-dinitrobenzene. N. are used mainly for the synthesis of azo dyes. Reduction of N. produces phenylenediamines. o-N. is used for photometric determination of vitamin C in fruits.

Nitrobarite: see Barium nitrate.

2-Nitrobenzaldehyde: bright yellow crystalline needles; m.p. 43-44 °C, b.p. 156 °C at 2 kPa. 2-N. is slightly soluble in water, and readily soluble in most organic solvents. It can be synthesized by oxidation of 2-nitrocinnamic acid with potassium permanganate. 2-N. undergoes the addition and condensation reactions typical of the Aldehydes (see). It is used in the determination of isopropanol, acetone and other compounds with acetyl groups. 2-N. is also used in the synthesis of triphenylmethane pigments.

NO_2

CHO

Nitrobenzene: $C_6H_5-NO_2$, the simplest aromatic nitro compound. N. is a colorless, highly refractive liquid with an odor like bitter almonds; m.p. 5.76 °C, b.p. 210.8 °C, n_D^{20} 1.5562. N. is steam volatile. It is slightly soluble in water, but dissolves very readily in alcohol, ether, benzene and acetone. It can be reduced to nitrosobenzene, phenylhydroxylamine, aniline, azoxybenzene, azobenzene or hydrazobenzene, depending on the conditions. In addition to these reductive conversions of the nitro group, it can also undergo electrophilic substitutions on the aromatic ring. These occur mainly in the 3- (*m*-) position, and are more difficult than in unsubstituted benzene. Further nitration yields 1,3-dinitrobenzene, bromination, 3-bromonitrobenzene and chlorosulfonation, 3-nitrobenzenesulfonyl chloride. Friedel-Crafts and Vilsmeier-Haack reactions fail because the N. is not reactive enough.

> Nitrobenzene is very toxic, leading to formation of methemoglobin and thus inhibiting oxygen transport. It can enter the body through the skin, the respiratory tract or the gastrointestinal tract. Chronic intoxication with lower concentrations can lead to anemia, liver and skin diseases. Higher concentrations produce headaches, stomach aches, vomiting and unconsciousness, often with cyanosis. More serious poisonings produce convulsions and respiratory paralysis, and can be fatal.
>
> The countermeasures in cases of poisoning are artificial respiration, pumping and rinsing the stomach, and blood transfusions. Do not use ethanol to clean skin which has been wetted with nitrobenzene!

N. can be obtained in good yields by reaction of nitrating acid with benzene. Other possible syntheses, such as oxidation of aniline or phenylhydroxylamine, are of no industrial importance. N. is used mainly to produce aniline and its derivatives. It is also an intermediate in the production of explosives, chloronitrobenzene, nitrobenzenesulfonic acid and dinitrobenzene, a solvent, and a mild oxidizing agent.

Nitrobenzoic acids: the three structurally isomeric mononitroderivatives of benzoic acid. The N. are colorless to yellowish, crystalline, steam-volatile compounds which can sublime in vacuum. They are relatively insoluble in water and benzene, but are somewhat soluble in methanol and ether. In alkali hydroxide solutions, they dissolve with salt formation. They undergo the reactions typical of monocarboxylic acids and of aromatic nitro compounds. N. are stronger acids than benzoic acid. They are obtained in a mixture by direct nitration of benzoic acid; the proportions are about 80% 3-N., 18% 2-N. and 2% 4-N. Synthesis of individual N. requires special methods.

COOH

NO_2 (2-, 3- or 4-position)

2-N. forms yellowish, triclinic needles; m.p. 148 °C. It is synthesized by oxidation of 2-nitrotoluene with potassium permanganate or nitric acid. It is toxic to lower organisms, and is therefore used to make pesticides. It can be used in analytical chemistry to detect mercury and tetravalent metal ions. Reduction of 2-N. produces anthranilic acid.

3-N., colorless, monoclinic prisms; m.p. 142 °C. It is made by direct nitration of benzoic acid or by nitration of benzaldehyde and oxidation of the resulting 3-nitrobenzaldehyde. It is used in the synthesis of azo pigments and in analytical chemistry to detect alkaloids and thorium.

4-N. forms colorless, monoclinic prisms; m.p. 242 °C. It is synthesized by oxidation of 4-nitrotoluene with potassium permanganate or nitric acid, and is used mainly to make 4-aminobenzoic acid. It is also an intermediate in the synthesis of pesticides and pharmaceutical preparations.

Nitrocellulose: an incorrect name for Cellulose nitrate (see).

Nitrocellulose powder: see Smokeless powder.

Nitrochloroform: same as Trichloronitromethane (see).

Nitro compounds: organic compounds containing one or more nitro groups, $-NO_2$, bound to the C atoms of aliphatic, aromatic or heterocyclic groups, or to the N atoms of nitramines. Some examples of N. are nitromethane, nitrobenzene, 4-nitropyridine and hexogen. Aliphatic and aromatic N. are important intermediates for the production of special fuels, pigments, explosives, pharmaceuticals, pesticides and plastics. A few N. are used as solvents. Many N., especially those with two or more nitro groups, can be very explosive. A few of these are used as explosives.

Nitro dyes: a group of yellow or orange acidic dyes in which a nitro group, $-NO_2$, is the chromophore. The N. are suitable for dyeing wool or silk, but they are not very fast to washing or light. Some examples are picric acid, naphthol yellow S, martius yellow and amido yellow E.

Nitroethane: CH_3-CH_2-NO_2, a primary nitroalkane. N. is a colorless liquid with a pleasant odor; m.p. -50 °C (-90° has also been reported); b.p. 115 °C, n_D^{20} 1.3917. It is only slightly soluble in water, but is readily soluble in alcohol, ether and acetone. It dissolves in alkali hydroxide solutions with salt formation. N. is formed by gas-phase nitration of propane in about 10% yield; the other products are nitromethane, 1- and 2-nitropropane. N. is used as a solvent and as an intermediate in the synthesis of pharmaceuticals, insecticides and surfactants.

Nitrofen: see Herbicides.

Nitroform, ***trinitromethane***: $CH(NO_2)_3$, a colorless liquid with a pungent odor; m.p. 26.4 °C, b.p. 50 °C at 6.65 kPa, n_D^{24} 1.4451. N. is soluble in water, acetone and alcohol. It dissolves in alkalies with salt formation. N. explodes very easily when heated above its melting point. It is formed by the reaction of nitric acid with ethene or ethyne, and is used to make explosives.

Nitrofural: a nitrofuran derivative which is applied externally to control or prevent bacterial infections. It is synthesized from 5-nitrofurfural and semicarbazide.

Nitrofurantion: a nitrofuran derivative with antibacterial effects; it is used mainly for urinary tract infections.

5-Nitrofurfural: a heterocyclic, bifunctional compound. 5-N. is formed by nitration of furfural. Some derivatives of 5-N., e.g. the semicarbazone, are bacteriostatic and bacteriocidal, and are used as chemotherapeutics.

Nitrogen, symbol ***N***: a chemical element in group Va of the periodic system, the Nitrogen-phosphorus group (see); a nonmetal, Z 7, natural isotopes with mass numbers 14 (99.63%) and 15 (0.37%), atomic mass 14.0067, valency usually III or V, but sometimes I, II or IV, density of gaseous N_2 1.2506 g/l, m.p. -209.86 °C, b.p. -195.8 °C, crit. temp. -147 °C, crit. pressure 3.35 MPa.

Properties. Under normal conditions, N. is a colorless, odorless and tasteless gas; it can be condensed to a colorless liquid at very low temperatures. It is slightly soluble in water (at 0 °C, 100 ml water dissolves 2.33 ml N.).

In accordance with its position in the periodic system, N. tends to form covalent bonds. With an electron configuration of $2s^2p^3$, it can complete an octet by forming three covalent bonds. Some typical derivatives, in which this bonding state is achieved, are ammonia, amines, hydrazine, hydroxylamine, etc. Each of these compounds has a trigonal pyramidal structure, a free electron pair, and can react as a base and as a nucleophile. It is able to accept a proton, which converts it to a four-fold coordinated, tetrahedral salt, the structure of which can best be described by an sp^3 hybridization on the central N atom. The nucleophilic potential of these compounds is expressed in their behavior towards alkyl halides or carboxylic acyl halides and esters, with which they form N-C bonds. As an element of the first period, N. is able to form $p_\pi p_\pi$ bonds and to act as a component of a conjugated system. In elemental dinitrogen, the nitrogen oxides, azomethines and azo compounds, for example, the N forms one or two π-bonds in addition to a σ-bond. In ionic nitrides such as Li_3N and Mg_3N_2, however, N displays ionic bonding.

Molecular ***dinitrogen***, N_2, has a bond dissociation energy of 942 kJ mol^{-1}, which makes it extremely unreactive. This value is an expression of the stability of the triple bond in N_2, which can be described either as the result of forming one σ- and two π-bonds between two sp-hybridized N atoms, or in terms of an

MO scheme. Reactions of elemental N. require considerable activation, which can be provided by addition of energy or suitable catalysts.

At high temperatures, electropositive elements, such as alkali and alkaline earth metals, reduce N_2 when hot to the corresponding nitrides. Above 1000 °C, N. reacts with calcium carbide to form calcium cyanamide; above 3000 °C, it reacts with oxygen to form nitrogen oxide, NO. The reaction with hydrogen to form ammonia, $N_2 + 3\ H_2 \rightleftharpoons 2\ NH_3$, $\Delta H = -92.5\ kJ\ mol^{-1}$, being exothermal, cannot efficiently be activated by heat, since this shifts the equilibrium to the left. It therefore requires catalysis for industrial utilization. Recently, molecular N. has been successfully bound to transition metals in special coordination compounds (nitrogen fixation).

Analytical. Organically bound N. is detected qualitatively by the Lassaigne test (see), and quantitatively by Elemental analysis (see) or the Kjeldahl nitrogen determination (see). Determination of elemental N. in gas mixtures is discussed under Gas analysis (see). Nitrate can be reduced to ammonia, which is distilled off and titrated with acid. In addition, both NH_3- and NO_3-sensitive electrodes are available.

Occurrence. 75.1 mass % (78.70 vol. %) of the earth's atmosphere is N_2. N. also makes up about 0.03% of the earth's crust. N.-containing minerals are relatively rare, although there are large deposits of Chile saltpepter, $NaNO_3$. In the living world, many classes of N. compounds are essential (proteins, nucleic acids, porphyrins, etc.). These are synthesized by plants and bacteria from inorganic compounds, NH_4^+ ions and NO_3^- ions, which they absorb from the soil or water. For optimum plant growth, the concentration of these ions in the soil must be maintained, with fertilizers if necessary. The ability to fix molecular nitrogen into organic compounds is limited to a few groups of microorganisms. The decay of animal and plant materials returns the organic N. derivatives to ammonia or ammonium salts, which are oxidized by certain soil bacteria to nitrites, and these to nitrates (see Nitrification), which are then available for reincorporation into plants.

Production. The most important industrial method of obtaining N. is fractional distillation of liquefied air. Another way to remove N. from the air is to bind the oxygen with coal, and to remove the resulting carbon dioxide with water. Pure N. can be obtained in the laboratory by heating aqueous ammonium nitrite solution: $NH_4NO_2 \rightarrow N_2 + 2\ H_2O$, or by thermal decomposition of sodium or barium azide.

N. is available commercially as the pressurized gas and in liquid form.

Applications. N. is the starting material for industrial synthesis of ammonia and calcium cyanamide. Because of its inertness, it is used as a protective gas for chemical reactions and welding, to fill light bulbs, etc. Liquid N. is used as a coolant.

Historical. Ammonium salts and nitrates were used by the Arabian alchemists. In 1771, Scheele discovered atmospheric N. Ammonia was first synthesized in 1774 by Priestley. The Frank-Caro process for calcium cyanamide production made atmospheric N. available for utilization at the beginning of the 20th century. Such important large-scale processes as the nitric acid synthesis of Birkeland-Eyde, Ostwald catalytic ammonia combustion and the Haber-Bosch process of ammonia synthesis were all developed within a few years, between 1905 and 1909.

Nitrogenase: an enzyme system which catalyses reduction of atmospheric nitrogen to ammonia; it is present in many bacteria and cyanobacteria. N. is a complex of two proteins, the *Fe protein* and the *Fe-Mo protein*. The Fe-protein contains 4 iron atoms in a 4 Fe-4 S cluster of the type $Fe_4S_4(SR)_4^{n-}$, which links the peptide chains of two identical subunits. The Fe-Mo protein contains 2 molybdenum, 33 iron and 27 sulfur atoms, and the binding site for nitrogen. A low-molecular-weight Mo-polypeptide complex acts as cofactor. In the active center of the N., an electron transfer to the nitrogen molecule occurs via a polynuclear iron and molybdenum complex. The electrons come from biological single-electron donors (ferredoxins), and are passed from one to another of several iron complexes before being transferred to the two molybdenum atoms, and then to the nitrogen. The catalytic process thus corresponds to a switch from single-electron to multi-electron processes. The development of a nitrogenase-like catalyst for large-scale ammonia synthesis under mild conditions is of great interest.

Nitrogen group: same as Nitrogen-phosphorus group (see).

Nitrogen halides: ***Nitrogen trifluoride***: NF_3, colorless, unreactive gas, M_r 71.00, m.p. -206.60 °C, b.p. -128.8 °C; it is formed by reaction of excess fluorine with ammonia according to $NH_3 + 3\ F_2 \rightarrow NF_3 + 3\ HF$, or by electrolysis of molten ammonium hydrogenfluoride.

Nitrogen trichloride, NCl_3, is a yellow, oily liquid, M_r 120.37, which is formed by chlorination of a saturated ammonium chloride solution. The compound is endothermal and explodes when the temperature is increased. It is saponified to ammonia, NH_3, and hypochlorous acid, HOCl, by water.

The reaction of concentrated NH_3 solution with elemental iodine forms a ***nitrogen triiodide-ammonia adduct***, $NI_3 \cdot NH_3$, which is usually known as ***nitrogen triiodide***. It forms black-brown, rhombic crystals; M_r 411.75, which explode on the slightest impact.

Nitrogen oxides. ***Dinitrogen oxide***, N_2O, is a colorless gas; M_r 44.01, m.p. -90.8 °C, b.p. -88.5 °C, crit. temp. 36.5 °C, crit. pressure 7.17 MPa. N_2O is isosteric with CO_2 and the azide ion, N_3^-, and is linear:

$$|N\equiv \overset{\oplus}{N}-\overset{\ominus}{\underline{\overline{O}}}| \longleftrightarrow \overset{\ominus}{N}=\overset{\oplus}{N}=\overline{O}$$

When cold, N_2O is relatively unreactive, but when hot it decomposes to nitrogen and oxygen. It therefore supports combustion. N_2O is obtained by heating ammonium nitrate: $NH_4NO_3 \rightarrow N_2O + 2\ H_2O$. When inhaled, it is intoxicating and disposes one to laugh (*laughing gas*). It is used as an anesthetic in medicine.

Nitrogen monoxide, nitrogen oxide, NO, is a colorless gas which is slightly soluble in water; M_r 30.01, m.p. -163.6 °C, b.p. -151.8 °C, crit. temp. -93 °C, crit. pressure 6.4 MPa. The NO molecule has an unpaired electron, but has little tendency to dimerize. Its bonding state is most readily described by the MO scheme for a diatomic molecule, from which it can be seen that the unpaired electron occupies an antibonding molecular orbital, so that the bond order is 2.5. As a result, this electron is readily removed from the molecule, forming the stable nitrosyl cation NO^+ (e.g. $NO^+HSO_4^-$). NO is immediately converted to brown nitrogen dioxide by oxygen: $2\ NO + O_2 \rightarrow 2\ NO_2$. With the halogens (except iodine), NO reacts to form nitrosyl halides NOX (X = F, Cl, Br; see Nitrosyl chloride). Strong oxidizing agents, such as chromic or hypochlorous acid, oxidize NO to nitric acid. NO is reduced to NO_2 by sulfur dioxide. Carbon and magnesium will burn in an NO atmosphere. In the laboratory, NO is made by reduction of dilute nitric acid with copper: $8\ HNO_3 + 3\ Cu \rightarrow 3\ Cu(NO_3)_2 + 2\ NO + 4\ H_2O$. Industrially, the gas is made by catalytic combustion of ammonia (Ostwald process; see Nitric acid). At very high temperatures, NO can also be formed by direct combination of the elements. The *Birkeland-Eyde process* in which air is passed through a zone of very high temperature generated by an electric arc, was used industrially for NO synthesis before the introduction of the Ostwald process (air combustion). NO is an intermediate in the industrial synthesis of nitric acid, and, in mixtures with NO_2, is used to synthesize nitrites.

> Nitrogen monoxide, NO, and nitrogen dioxide, NO_2, and mixtures of them (nitrous gas) are strong lung poisons. Inhalation causes irritation of the eyes, nose and throat mucous membranes, and also dizziness and headaches. Higher concentrations cause lung edema.
> Countermeasures: absolute rest and oxygen inhalation if needed. Medical supervision is required for at least 1 to 2 days.

Dinitrogen trioxide, N_2O_3, exists only in the solid state. It is a blue compound, M_r 76.01, m.p. -102 °C. With increasing temperature, the dissociation equilibrium $N_2O_3 \rightleftharpoons NO + NO_2$ shifts increasingly to the right. N_2O_3 forms when equimolar amounts of NO and NO_2 are mixed; N_2O_3 is the formal anhydride of nitrous acid. Mixtures of equal parts NO and NO_2 react with alkali hydroxide solutions to give the corresponding nitrites.

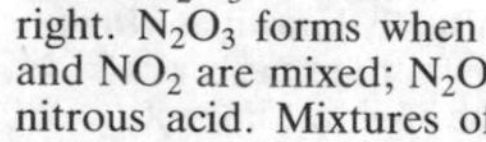

Nitrogen dioxide, NO_2, is a readily liquefied gas; M_r 46.01, density, 1.4494, m.p. -11.20 °C, b.p. 21.2 °C, crit. temp. 157.8 °C, crit. pressure 10 MPa. The paramagnetic, red-brown NO_2 is in equilibrium with the diamagnetic, colorless ***dinitrogen tetroxide***, N_2O_4: $2\ NO_2 \rightleftharpoons N_2O_4$, ΔH = -57 kJ mol^{-1}. The equilibrium position shifts to the right with falling temperature:

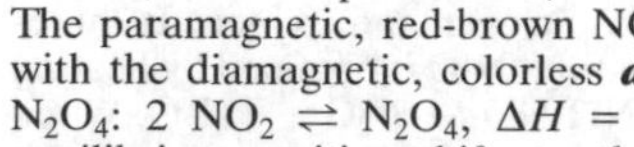

Above 200 °C, NO_2 dissociates into NO and O_2. It is a strong oxidizing agent. NO_2 or N_2O_4 acts as the mixed anhydride of nitrous and nitric acids. It forms nitrites and nitrates in alkali hydroxide solutions, e.g. $2\ NO_2 + 2\ NaOH \rightarrow NaNO_2 + NaNO_3 + H_2O$. If NO_2 is passed through water, it disproportionates to nitric and nitrous acids; in the acidic solution, the latter dissociates into NO, NO_2 and water. In the presence of air, NO is oxidized to NO_2, so that eventually all the NO_2 is converted to nitric acid: $2\ NO_2 + 1/2\ O_2 + H_2O \rightarrow 2\ HNO_3$. In the laboratory, NO_2 is made by heating heavy metal nitrates, e.g. $Pb(NO_3)_2 \rightarrow 2\ NO_2 + PbO + 1/2\ O_2$. Industrially, it is made as an intermediate in the synthesis of nitric acid by air oxidation of NO.

Dinitrogen pentoxide, N_2O_5, colorless, rhombic or hexagonal crystals; M_r 108.01, m.p. 30 °C, dec. 47 °C. N_2O_5 is the anhydride of nitric acid: $N_2O_5 + H_2O \rightarrow 2\ HNO_3$. It is formed in the reversal of the above reaction when HNO_3 is dehydrated with phosphorus(V) oxide. Solid N_2O_5 has the structure of ionic nitryl nitrate, $NO_2^+NO_3^-$, and is converted to the molecular form, O_2N-O-NO_2 at higher temperatures. It tends to explode.

Nitrogen-phosphorus group, ***nitrogen group, pnictogens***: the fifth main group of the periodic system includes the elements nitrogen (N), phosphorus (P), arsenic (As), antimony (Sb) and bismuth (Bi). The chemical and physical properties within the group vary widely (see). Nitrogen and phosphorus are typical nonmetals, and their oxides are acid anhydrides. Bismuth is a metal, and its oxide, Bi_2O_3, is basic. Arsenic and antimony are semimetals, with amphoteric oxides As_2O_3 and Sb_2O_3. The chemistry of the elements of the N. is determined by their common valence electron configuration ns^2p^3. A stable noble gas configuration is thus possible by acquisition of three electrons or loss of five. E^{3-} ions (E = element) exist, for example in Li_3N, Mg_3N_2, Ca_3P_2, etc. The tendency to form these ions decreases as the atomic number increases. Free +5 cations do not exist, because of the high ionization potentials. However, covalent compounds with electronegative partners exist in which the N. element has a +5 oxidation state (nitric acid and nitrates, P(V) halides, phosphoric acid and phosphates, etc.). Going from nitrogen to bismuth, the stability of these derivatives decreases and the +3 oxidation state becomes more important (*inert pair effect*, see Boron aluminum group). As a result, nitrites and phosphites, for example, are reducing agents, while bismuthate(V) is a strong oxidizing agent. As is generally observed in the main groups of the periodic system, the tendency to form ionic bonds increases with increasing atomic number; ionic compounds are known in which antimony and bismuth are in the +3 oxidation state (e.g. BiF_3, $Sb_2(SO_4)_3$, SbO^+ and BiO^+ salts). The three singly-occupied p-orbitals make the formation of three covalent bonds favorable; such bonding is observed, e.g. in ammonia, the amines, PH_3, $PHAl_3$, AsH_3, $AsCl_3$ and $BiCl_3$. These compounds have pyramidal structures. 4-Fold coordinated derivatives such as $[NR_4]^+$, $[AsR_4]^+$ and $[PR_4]^+$ (R = H or an organic group) are isosteric with the comparable neutral compounds of the elements of the 4th main group, and are tetrahedrally coordinated, as would be expected.

Nitrogen, as a first-period element, can form at most four bonds. The heavier elements can expand

Properties of the elements of the 5th main group of the periodic system

	N	P	As	Sb	Bi
Nuclear charge	7	15	33	51	83
Electron configuration	[He] $2s^2p^3$	[Ne] $3s^2p^3$	[Ar] $3d^{10}4s^2p^3$	[Kr] $4d^{10}5s^2p^3$	[Xe] $4f^{14}5d^{10}6s^2p^3$
Atomic mass	14.006 7	30.973 76	74.921 6	121.75	208.980 4
Atomic radius [pm]	75	110	122	143	146
Electronegativity	3.07	2.06	2.20	1.82	1.67
Density [g/cm³]	1.2506[a])	1.82[b])	5.727	6.684	9.80
m. p. [°C]	–209.86	44.1[b])	817[c])	630.5	271.3
b. p. [°C]	–195.8	280[b])	613 (Sbp.)	1750	1560 ± 5

a) Density of gaseous N_2 in g/l
b) White phosphorus, P_4
c) Under 2.8 MPa pressure

their valences to form octets, as they have energetically favorable d-levels which can be used in the formation of trigonal bipyramidal derivatives with coordination number 5 (for example, PF_5, AsF_5, $P(C_6H_5)_5$) or of octahedral derivatives with coordination number 6 ($[PCl_6]^-$, $[AsF_6]^-$, $[Sb(OH)_6]^-$. While nitrogen can form stable $p_\pi p_\pi$ bonds with itself, oxygen, carbon or boron (axo compounds, nitrogen oxides, HNO_3 and NO_3^-, Schiff's bases, borazenes), compounds of heavier N. elements with $p_\pi p_\pi$-double bonds are either not formed, or form only under very special structural conditions. However, phosphorus forms $d_\pi p_\pi$ bonds through interactions of its empty d-orbitals with filled p orbitals of suitable symmetry on the bond partner. Antimony and arsenic also form such bonds, but they are weak. This type of bond is responsible for the unusual stability of the —P=O bond, formation of which is the driving force in many reactions (see Phosphorous acid, Michaelis-Arbuzov reaction, Wittig reaction). Such bonds are also the basis of the π-acceptor properties of the P and As ligands, and thus for the high stability of the corresponding complex compounds.

Nitrogen rule: see Mass spectroscopy.

Nitrogen triiodide: see Nitrogen halides.

Nitroglycerol: same as Glycerol trinitrate (see).

Nitroglycerin powder: see Smokeless powder.

Nitroglycol: same as Ethyleneglycol dinitrate.

Nitro group: the functional group $-NO_2$. In nitro compounds, the N atom is are linked to a C atom of the hydrocarbon or heterocycle. NO_2-groups linked by an O atom in compounds of the type $R-O-NO_2$ are nitrates, or esters of nitric acid. The structure of N. can be described by the following canonical structures:

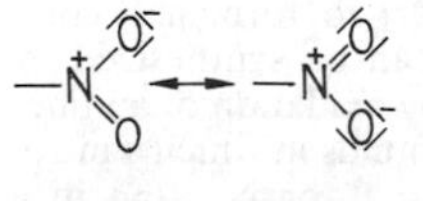

$$-\overset{+}{N}(=O)-O^- \longleftrightarrow -\overset{+}{N}(-O^-)=O$$

N. is thus a substituent with internal resonance. It exerts strong - I and - M effects on the hydrocarbon to which it is bound.

Nitroguanidine: $(NH_2)_2C=N-NO_2$, a colorless, crystalline compound; m.p. 239 °C (dec.). N. can exist in two forms:

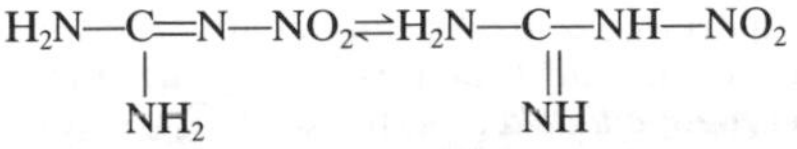

$$H_2N-C(NH_2)=N-NO_2 \rightleftharpoons H_2N-C(=NH)-NH-NO_2$$

Nitramine form   Nitrimine form

N. is barely soluble in ether and alcohol, and is soluble in hot water. It is formed by the reaction of cold, concentrated sulfuric acid with guanidine nitrate. N. is a powerful explosive, but the pure compound is rarely used as such. Its detonation velocity at density 1.2 is 6775 m s^{-1}. Mixed with cellulose nitrate and glycerol trinitrate, N. is used in large amounts as gunpowder.

Nitroguanidine powder: see Smokeless powder.

Nitrolic acids: reaction products of primary nitroalkanes with nitrous acid: $R-CH_2-NO + O=N-OH \rightarrow R-C(NO_2)=N-OH + H_2O$. N. are colorless, crystalline compounds which are soluble in ether and form red alkali salts in alkali hydroxide solutions:

$$R-C(NO_2)=N-OH \xrightarrow[-H_2O]{+NaOH} [R-C(NO_2)=N-\overline{O}|^-]^- Na^+$$

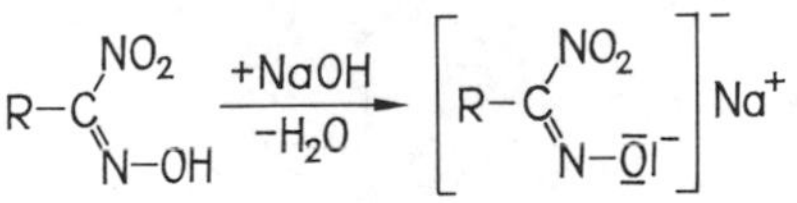

Nitromethane: CH_3-NO_2, the simplest aliphatic nitro compound. N. is a colorless, combustible, oily liquid which slowly turns brown in storage; m.p. -28.5 °C, b.p. 100.8 °C, n_D^{20} 1.3817. It is slightly soluble in water, and readily soluble in alcohol, ether and acetone. A 0.01 M aqueous solution of N. at 25 °C has a pH of 6.4. Like all nitroalkanes with α-H atoms, N. can exist in equilibrium with its *aci*-form (see Nitro-*aci*-nitro tautomerism), but the equilibrium for N. is almost completely on the side of the nitro form. N. dissolves in alkali hydroxide solutions with formation of the alkali salts, the anion of which is resonance stabilized:

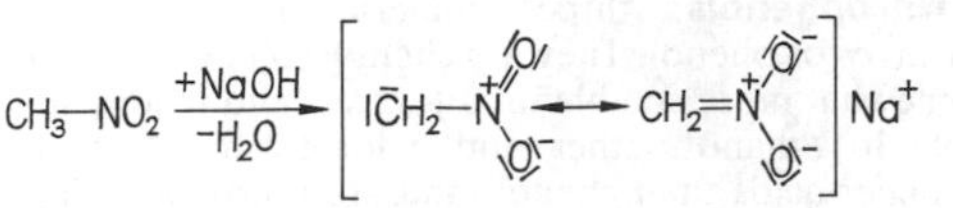

$$CH_3-NO_2 \xrightarrow[-H_2O]{+NaOH} [|\overline{C}H_2-\overset{+}{N}(=O)-O^- \longleftrightarrow CH_2=\overset{+}{N}(-O^-)-O^-] Na^+$$

Because of its CH acidity, N. can undergo reactions similar to the aldol reactions of aldehydes and ketones. For example, it reacts with benzaldehyde to form 2-nitro-1-phenylethanol:
$C_6H_5-CHO + CH_3-NO_2 \rightarrow C_6H_5-CH(OH)-CH_2-NO_2$.

Loss of water from N. forms o-nitrostyrol. N. can also undergo aminomethylation in the sense of the Mannich reaction (see). It is synthesized industrially from propane in the gas phase. The reaction mixture is then separated by fractional distillation. N. can also be synthesized from chloroacetic acid and sodium nitrite, or from methyl iodide and silver nitrite. It is used as a solvent for inorganic compounds, cellulose nitrate and cellulose acetate, fats and oils, acylnitrile and vinyl resins; it is also a reactant in the synthesis of fuels, explosives, pesticides, *N*-methylhydroxylamine and monomethylamine.

Nitromethylene blue: see Methylene blue.

Nitron®: see Synthetic fibers.

Nitronaphthalene: a nitro derivative of naphthalene. Only α- or 1-N., β- or 2-N., 1,5-dinitronaphthalene and 1,8-dinitronaphthalene are of moderate industrial importance. Nitration of naphthalene with concentrated nitric acid produces mainly the bright yellow ***1-nitronaphthalene***, m.p. 61.5 °C, subl.p. 304 °C. It is insoluble in water and soluble in alcohol, ether, benzene and pyridine. A byproduct is the colorless, steam volatile ***2-nitronaphthalene***; m.p. 79 °C, b.p. 165 °C at 2 kPa. It is insoluble in water, but is readily soluble in alcohol, chloroform and ether. When nitrating acid reacts with 1-N., ***1,5-dinitronaphthalene*** (colorless crystals, subl.p. 219 °C) and ***1,8-dinitronaphthalene*** (bright yellow crystals, m.p. 173 °C) are formed; these can be separated by recrystallization from pyridine. The N. are used mainly to synthesize aminonaphthalenes and thus for the synthesis of pigments.

Nitrone: the *N*-oxide of a Schiff's base. ***Aldonitrones*** (R^1 or R^2 = H) and ***ketonitrones*** (R^1 and R^2 alkyl or aryl groups) can occur in the *Z*- or *E*-form.

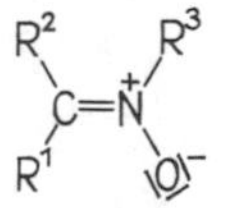

Aldonitrones exist primarily in the *Z*-form. N. are readily crystallized and form adducts with strong or Lewis acids. Under suitable conditions, they can be rearranged into carboxamides or oxime ethers. Oxaziridines can be made by photochemical isomerization of N. N. are synthesized by dehydration of hydroxylamines, by condensation of ketones or aldehydes with *N*-monosubstituted hydroxylamines, or by reaction of CH-acidic methylene compounds with aromatic nitroso compounds. They are important reactants in syntheses by 1,3-dipolar cycloaddition.

Nitronic acids: see Nitro-*aci*-nitro tautomerism.

Nitropenta: see Pentaerythritol tetranitrate.

Nitrophenols: three isomeric mononitro derivatives of phenol. They are slightly soluble in water, somewhat more soluble in hot water, and readily soluble in ethanol, ether and chloroform. They are stronger acids than phenol, and are poisonous. Contact with the skin or inhalation of vapors or dust should be avoided.

2-Nitrophenol (o-nitrophenol), yellow, steam-volatile needles; m.p. 45-46 °C, b.p. 216 °C. It is produced by reaction of 2-chloronitrobenzene with alkali hydroxide, or by nitration of phenol. The latter reaction also produces 4-N., from which 2-N. is separated by steam distillation. 2-N. is used as an indicator (colorless at pH 5, yellow at pH 7), and to make drugs, perfumes and photochemicals.

3-Nitrophenol (m-nitrophenol), nearly colorless crystals; m.p. 97 °C, b.p. 194 °C at 9.3 kPa. It is synthesized by diazotization of 3-nitroaniline and boiling the diazonium salt. It is used as an indicator (colorless at pH 6.6, yellow at pH 8.6) and to make 3-aminophenol.

4-Nitrophenol (p-nitrophenol), colorless crystals which sublime when heated strongly; m.p. 115 °C, b.p. 279 °C (dec.). It is made by hydrolysis of 4-chloronitrobenzene or by nitration of phenol and removal of the 2-N byproduct. 4-N. is used as an indicator (colorless pH 5.6, yellow pH 7.6) and as an intermediate in the production of pigments, pesticides, photochemicals, etc.

4-Nitrophenylhydrazine: red-orange leaflets or needles; m.p. 158 °C (dec.). 4-N. is soluble in hot water, ether, alcohol and chloroform. It is made by diazotization of 4-nitroaniline and reduction of the diazonium salt with sodium hydrogensulfite. 4-N. is used to characterize aldehydes and ketones, with which it forms nicely crystallizing 4-nitrophenylhydrazones. It is also used in the dye industry to make pyrazolone dyes.

Nitrosamines: organic compounds with the general formula $R^1R^2N\text{-}N{=}O$. N. are yellow to orange, oily or solid compounds which are only slightly soluble in water and are cleaved by the action of dilute acids into the starting materials. N. are converted into *N,N*-disubstituted hydrazines of the type $R^1R^2N\text{-}NH_2$ by reducing agents such as sodium amalgam. Aromatic N. form 2- and 4-nitroso compounds in the presence of mineral acids (Fischer-Hepp rearrangement). They are formed by the action of nitrous acid on aliphatic or aromatic amines: $R^1R^2NH + O{=}N\text{-}OH \rightarrow R^1R^2N\text{-}N{=}O + H_2O$.

They can be used to purify and identify secondary amines. Secondary amides react similarly with nitrous acid, forming the corresponding N., for example *N*-methyl-*N*-nitrosourea and *N*-methyl-*N*-nitrosourethane.

Nitrose process: see Sulfuric acid.

Nitrosobenzene: $C_6H_5\text{-}N{=}O$, the simplest aromatic nitroso compound; colorless crystals with a pungent odor; m.p. 68-69 °C, b.p. 57-59 °C at 2.39 kPa. N. is insoluble in water but is soluble in alcohol, ether and benzene. In solution and in its melts, N. looks green, because it is present in monomeric form. It undergoes the reactions typical of the Nitroso compounds (see). The nitroso group reacts easily, for example, with primary amines and reactive methylene compounds, with loss of water, to form azo compounds and azomethines. N. can be synthesized by reduction of nitrobenzene, or by oxidation of aniline.

Nitroso compounds: compounds in which a nitroso group $\text{-}N{=}O$ is bound to a C or N atom in a molecule. ***C-Nitroso compounds*** include aromatic compounds such as nitrosobenzene, 4-nitroso-*N,N*-dimethylaniline and 1-nitrosonaphthalene as well as aliphatic compounds such as nitrosomethane and 2-methyl-2-nitrosopropane. Crystalline *C*-N. are usually in a colorless, dimeric form. Monomers are intensely colored; aromatic N. tend to be green, and aliphatic N., blue. N. with α-H atoms are easily converted to oximes: R-CH_2-N=O → R-CH=N-OH. *C*-N. are synthesized mainly by Nitrosylation (see), but they can also be made by oxidation of substituted *N*-phenylhydroxylamines or primary amines with potassium dichromate and sulfuric acid. N. react, for example, with primary amines to form azo compounds: Ar-N=O + H_2N-R → Ar-N=N-R + H_2O. They condense with CH acidic compounds in an Ehrlich-Sachs reaction (see) to form azomethines. N. are used to make pigments, as vulcanization accelerators and as polymerization inhibitors.

Some important ***N-nitroso compounds*** are N-nitrosoamine-nitrosoureas (see, e.g., *N*-Methyl-*N*-nitrosourea) and nitrosourethanes (see, e.g., *N*-methyl-*N*-nitrosourethane). These are usually bright yellow to greenish oils or crystalline compounds. They can be synthesized by nitrosylation of secondary amines or amides. *N*-N. are used as stabilizers, polymerization inhibitors, drivers for polymerization processes, pesticides and drugs.

> *N*-Nitroso compounds of the nitrosamide type are highly irritating to the skin, and inhalation of the vapors can cause severe bronchitis. Nitroso compounds are potent carcinogens; the aliphatic *N*-nitroso compounds are the most dangerous in this respect. Therefore any work with nitroso compounds must be done extremely carefully, and all safety regulations must be observed.

Nitrosodicyanomethanides: derivatives of oximinomalonic dinitrile, HON=C$(CN)_2$. Ionic N. contain the planar, resonance-stabilized anion $[ON=C(CN)_2]^-$, which is a Pseudohalide (see). Because of the C$(CN)_2$-O analogy (see Pseudochalcogenides), it displays pronounced parallels to nitrite. The ***alkali nitrosodicyanomethanides*** MONC$(CN)_2$ (M = Na, K, Rb, Cs) crystallize in yellow needles. With silver nitrate, the anion forms ***silver(I) nitrosodicyanomethanide***, AgNOC$(CN)_2$. The anion is an ambivalent ligand which preferentially forms coordinate bonds via the nitroso nitrogen; however, with f elements and main group metals in organometal derivatives of the type R_3MONC$(CN)_2$ (M = Sn, Pb), coordinate or covalent bonds with the oxygen are preferred. N. can be synthesized by reaction of malonic dinitrile with nitrous acid via silver nitrosodicyanomethanide:

$$(NC)_2CH_2 + ONOH \xrightarrow[-H_2O]{} (NC)_2C{=}NOH \xrightarrow[-H^+]{+Ag^+} AgNOC(CN)_2.$$

p-Nitrosodimethylaniline, ***N,N-dimethyl-4-nitrosoaniline***: (*p*)-NO-C_6H_4-N$(CH_3)_2$, green crystalline leaflets; m.p. 86 °C. *p*-N. is insoluble in water, but is soluble in alcohol and ether. It can be synthesized by nitrosylation of *N,N*-dimethylaniline with nitrous acid. *p*-N. is used in the synthesis of pigments and vulcanization accelerators.

Nitroso group: the functional group -N=O of the Nitroso compounds (see). The N. is one of the strongest chromophores in chemistry. It gives monomeric nitroso compounds, in which the N. is bound to a C atom, an intense color. The reactivity of the N. bound to a C atom is comparable to that of an aldehyde group, and for this reason the N. is considered a heteroanalogous carbonyl group.

α-Nitroso-β-naphthol, *1-nitrosonaphth-2-ol*: an orange, crystalline compound; m.p. is not sharp, up to 112 °C (dec.). α-N. is only slightly soluble in water, but dissolves readily in benzene and glacial acetic acid. It is used as a reagent for determination of cobalt, palladium, molybdenum, copper, iron, nickel, etc.

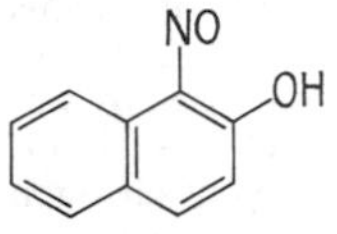

α-Nitroso-β-naphthol

β-Nitroso-α-naphthol is a yellow-green, crystalline powder; m.p. 162-164 °C (dec.). It is insoluble in water, readily soluble in benzene and glacial acetic acid, and is used for colorimetric determination of copper, cobalt and zirconium.

4-Nitrosophenol: ON-C_6H_4-OH, a tautomer of 1,4-benzoquinone monooxime; it crystallizes in bright yellow needles; m.p. 140 °C. When it is heated above 124 °C, it begins to decompose gradually, and turns brown. 4-N. is moderately soluble in water, and readily soluble in ethanol, ether and acetone; it also dissolves in alkalies to give a green-brown solution. It tends to decompose explosively, a process facilitated by alkalies. 4-N. irritates the skin and contact should be avoided. It is synthesized from phenol and nitrous acid, or by reaction of 4-nitrosodimethylaniline with sodium hydroxide solution. It is an intermediate in the synthesis of pigments, and is also used as a vulcanization accelerator.

Nitroso pigments: a group of pigments in which the nitroso group, -NO, acts as a chromophore (Fig.). The most important are pigment green B ($R^4 = R^6$ = H; *n* = 1) and naphthol green ($R^4 = SO_3H$, R^6 = H; *n* = 4). The hydroxynitroso compounds, which can also be considered quinone monooximes due to their tautomerism, are used mainly as metal complex pigments.

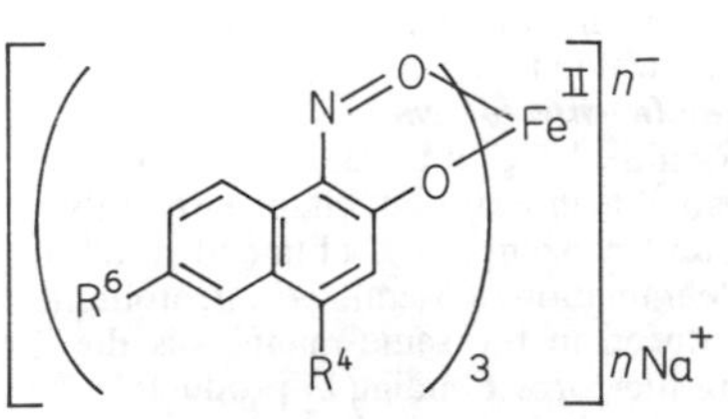

Nitrosylation: reaction introducing a nitroso group, -N=O, into an organic compound; in the process, a reactive H atom is forced out of the molecule as a proton: R-H + HO-N=O → R-N=O + H_2O. This is the principle of N. of reactive aromatics, such as phenols and tertiary aromatic amines, CH-acidic compounds, such as aliphatic ketones and β-dicarbonyl compounds, and primary and secondary amines and amides. The formation of nitrites can be considered the N. of aliphatic alcohols. When nitrosyl chloride is used, and under radical conditions, H atoms of even unreactive alkanes can be replaced. Nitrous acid and its esters are most commonly used as N. reagents. When nitrous acid is used, it is believed that the actual nitrosylating agent is dinitrogen trioxide which is in equilibrium with the acid: 2 HO-N=O $\rightleftharpoons$ O=N-O-N=O + H_2O.

The products of N. are Nitroso compounds (see), some of which are stable enough to be isolated.

Nitrosyl chloride: NOCl, the chloride of nitrous acid. A yellow-red, easily condensed gas; M_r 65.46, m.p. -64.5 °C, b.p. -5.5 °C. N. is decomposed by water according to the equation NOCl + $H_2O \rightarrow HNO_2$ + HCl. It is made by direct combination of nitrogen monoxide with chlorine: 2 NO + $Cl_2 \rightarrow$ 2 NOCl. N. is also formed by the reaction of conc. nitric with conc. hydrochloric acid, and is therefore a component of Aqua regia (see).

Nitrosyl complex: a transition metal complex with nitrogen monoxide, NO, as ligands. In many cases, the complex chemistry of NO is similar to that of carbon monoxide, CO, but in metal carbonyl nitrosyls, NO is always a three-electron donor. Thus complexes such as $Ni(CO)_4$, $Co(CO)_3NO$ and $Ni(CO)_2(NO)_2$ are isoelectronic and usually the MNO group is linear. NO acts as a one-electron donor in N. in which the MNO bond angle is close to 120°, for example in $[Co(NH_3)_5NO]^{2+}$ and $[IrCl_2(NO)(P(C_6H_5)_3)_2]$. Some other important N. are the nitrosyl pentacyanometallates, $[M(CN)_5NO]^{n-}$, e.g. sodium nitroprusside and $[Fe(H_2O)_5NO]^{2+}$, which is responsible for the brown ring reaction used to detect nitrate. Like CO, NO can act as a double or triple bridging ligand.

Nitrotoluenes: the three structurally isomeric mononitro derivatives of toluene. N. can be made by direct nitration of toluene, which produces a mixture of about 50% 2-N., 4% 3-N. and 37% 4-N. ***2-Nitrotoluene (o-nitrotoluene)*** is a steam-volatile, yellow liquid which exists in an unstable α-form and a stable β-form: m.p. -9.5 °C (α-form) and -2.9 °C (β-form), b.p. 221.7 °C, n_D^{20} 1.5450. 2-N. is only slightly soluble in water, but dissolves readily in most organic solvents. It can be reduced by various reagents to *o*-toluidine, *o*-tolylhydroxylamine, *o*-azoxytoluene, *o*-azotoluene and *o*-hydrazotoluene. Oxidation of 2-N. leads to 2-nitrobenzaldehyde and 2-nitrobenzoic acid.

3-Nitrotoluene (m-nitrotoluene) is a steam-volatile, yellowish liquid; m.p. 16 °C, b.p. 232.6 °C, n_D^{20} 1.5466. 3-N. is slightly soluble in water but dissolves readily in alcohol, ether and benzene. It is obtained in about 65% yield by deamination of 4-amino-3-nitrotoluene. 3-N. can be reduced in the same manner as the 2-isomer, forming the corresponding *m*-products.

4-Nitrotoluene (p-nitrotoluene) forms bright yellow crystals; m.p. 54.5 °C, b.p. 238.3 °C. It is nearly insoluble in water, but is readily soluble in most organic solvents. 4-N. is used as an intermediate for the production of *p*-toluidine, 4-nitrobenzoic acid, 2-chloro-4-nitrotoluene, 4-nitrotoluene-2-sulfonic acid, etc. N. are thus important raw materials for the synthesis of pigments, explosives, perfumes and polyurethane foams.

Nitrous acid: HNO_2, a weak acid which is stable only in cold, dilute aqueous solution; M_r 47.01. When the temperature is raised, S. dissociates into nitrous oxide and nitric acid: 3 $HNO_2 \rightarrow HNO_3$ + 2 NO + H_2O. In alkaline solution, it forms more stable salts, the Nitrites (see). As derivatives of nitrogen in the +III oxidation state, S. are distinctly amphoteric in their redox behavior. They are oxidized to nitric acid by strong oxidizing agents, e.g. 2 MnO_4^- + 5 NO_2^- + 6 $H^+ \rightarrow$ 2 Mn^{2+} + 5 NO_3^- + 3 H_2O; and strong reducing agents can reduce N. to nitrous oxide, NO, dinitrogen oxide, N_2O, hydroxylamine, H_2NOH, ammonia, NH_3, or even elemental nitrogen, N_2. Secondary amines react with N. to form nitrosoamines, while primary amines form diazonium salts with it. S. is obtained by acidifying aqueous nitrite solutions. It is used mainly in organic syntheses to make diazonium salts.

Nitrous gases: mixtures of Nitrogen oxides (see), primarily NO and NO_2.

Nixan process: see ε-Caprolactam.

NMN: see Nicotinamide nucleotides.

NMR spectroscopy, ***nuclear magnetic resonance spectroscopy***: Because of their intrinsic spin, some atomic nuclei have magnetic moments and are capable of alignment with respect to an external magnetic field. Different possible alignments are associated with different energy levels; N. is based on radiofrequency-induced transitions between these energy levels.

N. is a relatively new method, having been discovered in 1946 by Purcell and Bloch, but because of its great power it is applied in many areas of research, and the amount of structural information which can be obtained from it, especially for organic compounds, is greater than from any other single method now available.

Theoretical. The *spin p* of a nucleus is the vectorial sum of the spins of the nucleons (protons and neutrons) which comprise it. Like the other atomic parameters, *p* is quantized: $p = \sqrt{I(I+1)}\, h/2\pi$ (1). Here I is the nuclear spin quantum number, which can have half- or whole-integral values between 0 and 6, and h is Planck's constant. Since any rotating electric charge has an associated magnetic moment, the spin *p* of the nucleus is accompanied by a magnetic moment μ: $\mu = \gamma \cdot p$ (2). The proportionality constant γ is called the *gyromagnetic ratio* and is a characteristic parameter for each kind of atom. It can be seen from the nuclear properties of some important isotopes (Table 1) that all nuclei with even numbers of protons and neutrons (*g,g nuclei*) have the nuclear spin quantum number 0 and therefore have no magnetic moment; this group includes such important isotopes as ^{12}C, ^{16}O and ^{32}S. It includes about 60% of all stable atomic nuclei, which therefore cannot be examined by N. Only nuclei in which either the proton or the neutron number (or both) is odd have magnetic moments; these are integral for the *u,u nuclei* (odd proton number, odd neutron number, e.g. ^{14}N) and half-integral for the *g,u* and *u,g nuclei* (e.g. ^{13}C and ^{11}B). If a nucleus with a magnetic moment, which one can visualize as a tiny magnet, is brought into an external magnetic field of field strength H_0, it becomes aligned with the direction of the external field. Because of the atomic dimensions of this system, it is quantized, and only a limited number of orientations or spin states are possible for it, namely $2I+1$. For each of these, the projection of the spin angular moment on the field direction (z axis of the coordinate system), p_z, must be a half- or whole-integral of the unit of spin $h/2\pi$. The $2I+1$ possible projections are characterized by the magnetic quantum number m, which can take on any value from $+I, I-1, \ldots -I$: $p_z = m(h/2\pi)$ (3) (Fig. 1).

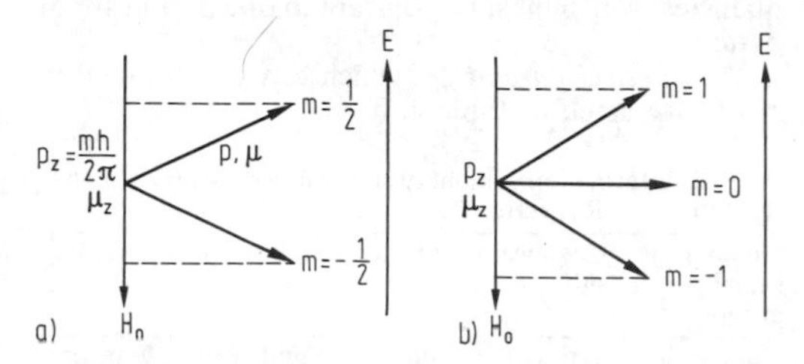

Fig. 1. Allowed orientations of the magnetic moments of nuclei with $I = ½$ (a) and $I = 1$ (b).

For the component of the magnetic moment μ in the direction of the field, z, this yields $\mu_z = \gamma p_z = \gamma m(h/2\pi)$ (4).

For nuclei with $I = 1/2$, which are most frequently used in N., there are two possible orientations with respect to the external field: parallel ($m = +1/2$) and antiparallel ($m = -1/2$); these differ in their energies and can be represented by a term scheme. The energy E of a magnetic dipole in an external magnetic field H_0 is

$$E = \mu_Z H_0 = \frac{\gamma m h H_0}{2\pi}$$

Thus the nuclear orientations shown correspond to potential energies of $E_+ = +1/2\ \gamma H_0(h/2\pi)$ and $E_- = -1/2\ \gamma H_0(h/2\pi)$. In contrast to the situation in other types of spectroscopy, these energy levels are not present in the substance, but are created by application of the external magnetic field. The distribution of the nuclei among the energy levels in thermal equilibrium is given by the Boltzmann equation. It can be shown that for ^1H nuclei at room temperature and in a magnetic field of 1.4 T (tesla), for every 1,000,000 nuclei in the higher-energy antiparallel arrangement, there are 1,000,003 in the lower-energy, parallel orientation; in other words, the difference in occupation is extremely small. Using electromagnetic radiation of suitable energy, transitions can be induced between the energy states shown in Fig. 2. The transitions consist of changes in the orientation of the nuclear magnetic moment in the external magnetic field. Since the energy difference between the two states $\Delta E = \gamma H_0(h/2\pi)$ (6), the relation between frequency and energy $\Delta E = h\nu$ gives the resonance condition $\nu = \gamma H_0/2\pi$ (7).

Fig. 2. Energy levels of a nucleus with $I = ½$ in an external magnetic field.

The resonance condition indicates the frequency of electromagnetic radiation at which a given nucleus (characterized by its gyromagnetic ratio) in an external field H_0 is in resonance (exchanges energy). For transitions between the spin states, there is a selection rule, $\Delta m = \pm 1$. It follows from eq. (6) that the spin states outside the magnetic field (H_0) have the same energy, and that the stronger the external magnetic field is, the larger their energy is (as well as the difference between their energies) (Fig. 3). The commonly used field strengths in N. lie between 1.4 and 9.4 T, which correspond in the case of the ^1H nucleus to resonance frequencies of 60 to 400 MHz, and these are in the range of radio waves.

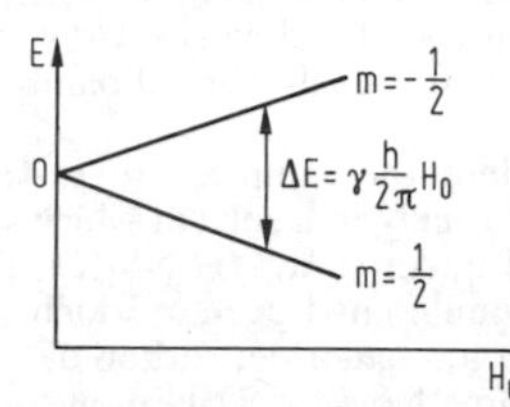

Fig. 3. Energy of the nuclear orientations as a function of magnetic field strength.

As can be seen from Table 1, the nuclei ^1H, ^{11}B, ^{19}F and ^{31}P have both high natural abundances and large gyromagnetic ratios. For reasons of sensitivity, these elements are most often studied with N., but more recently ^{13}C nuclei (see below) are also used. The most important of these are the ^1H nuclei, and the following refers mostly to these (^1H-NMR or PMR, for proton magnetic resonance).

Table 1. Properties of some nuclei

Nucleus	Natural Abundance [%]	Nuclear spin quantum number I [$\gamma \cdot 10^{-8}$]	Gyromagnetic ratio	Relative sensitivity*
^1H	99.98	1/2	2.673 8	1.00
^2D	0.016	1	0.410 4	0.01
^{11}B	80.42	3/2	0.858 3	0.165
^{12}C	98.9	0	0	.
^{13}C	1.1	1/2	0.672 4	0.016
^{14}N	99.63	1	0.192 6	0.001
^{16}O	99.76	0	0	.
^{19}F	100	1/2	2.516 1	0.834
^{31}P	100	1/2	1.082 8	0.066
^{35}Cl	75.4	3/2	0.262 8	0.005
^{55}Mn	100	5/2	0.662 5	0.175
^{209}Bi	100	9/2	0.429 8	0.137

* For the same number of nuclei and the same value of H_0, relative to $^1H = 1.00$ (H_0 is the external magnetic field strength).

Because there is only a slight difference in the numbers of nuclei occupying the different energy levels, the intensity of the observed signal is weak. In the course of the measurement the situation becomes even more unfavorable, because the upper level is gradually filled. If this has occurred, absorption can no longer be detected externally; the system is said to be "saturated". If not countered by relaxation processes, the process of saturation would limit the time during which NMR measurements could be made. Relaxation processes transfer energy non-radiatively to the neighboring atoms or molecules, which are collectively termed the "lattice". Thus one speaks of *spin-lattice relaxation*. The time course of this process is characterized by the *spin-lattice relaxation time* T_1, which is the time required for $1/e$ of the excess energy to be transferred. T_1 varies considerably between samples, from 100 to 10^{-4} s; in general short relaxation times are observed in liquids and gases. The spin-lattice relaxation time determines how high the intensity of the radiofrequency (RF) radiation can be without causing saturation. The form of the signals detected by an NMR spectrometer depends on the state (solid, liquid or gas) of the sample. Sharp signals are obtained from liquid samples (high-resolution N.), and this method is most significant for chemists. Crystalline compounds give broad signals (broad-band N.).

I) High-resolution N. Here the samples are examined in the liquid state, usually in a solvent which does not contain the nuclei under study. ^{1}H-NMR is done in solvents which contain no hydrogen or which are deuterated, e.g. carbon tetrachloride, carbon disulfide, chloroform, [$^2H_4$]methanol, [$^2H_5$]acetone, [$^2H_6$] dimethylsulfoxide, etc. The sample tubes are 5 mm in diameter, and for ^{1}H-NMR should contain about 10^{17} protons.

The main components of an NMR spectrometer (Fig. 4) are: 1) a magnet with a high field which is very homogeneous over the sample volume, 2) an RF generator with a stability equivalent to the high field homogeneity and 3) a very sensitive RF receiver.

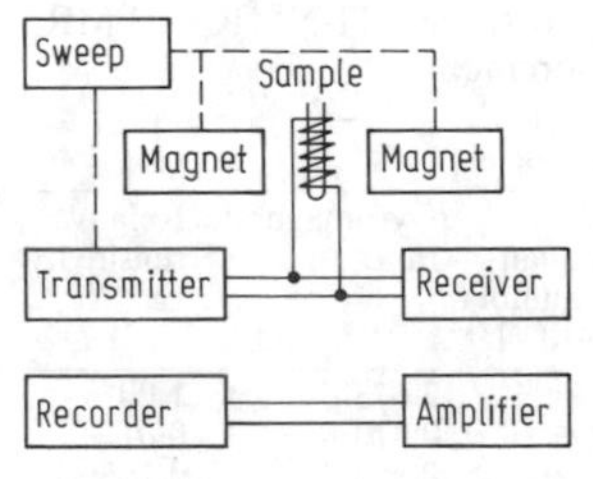

Fig. 4. Block diagram of an NMR spectrometer with continuous wave (CW) technology.

The sample is placed in the sample tube between the poles of the magnet. After the equilibrium Boltzmann distribution among energy states has been reached, the electromagnetic alternating field produced by the RF generator is transmitted to the sample via a coil around the sample holder. If the frequency and the external magnetic field H_0 meet the resonance condition (Eq. 7), energy is absorbed; this is registered by the RF receiver. The result is a graph, an ***NMR spectrogram***, in which the energy absorption is plotted against the frequency of the radiowave (Fig. 6).

Resonance can be achieved either by variation of the frequency of the alternating electromagnetic field (*frequency sweep*) or by variation of the magnetic field (*field sweep*). The two methods are both included in the concept "*continuous wave (CW) technique*", which expresses that in both, a physical parameter is continuously changed; they are fundamentally different from the pulse Fourier transform method (see below).

The main experimental problem with NMR instruments is the requirement for extreme constancy and homogeneity of the magnetic field or the radiofrequencies, which must be constant to one part in 10^8 or better.

The *spectral parameters* which can be observed in NMR are listed in Table 2.

Table 2. Information content of spectral parameters in high-resolution NMR spectroscopy

Number of signal groups	Chemical shift	Intensity ratio	Multiplicity	Coupling constant
Number of different atomic groups	Types of groups	Number of atoms per group	Number of neighboring nuclei	Type of neighboring nuclei and distance between them

Chemical shift δ. The gyromagnetic ratio is a characteristic parameter for each isotope (Table 3), so that for different types of nucleus in a constant external field, there will be different resonant frequencies which often differ by several MHz (Fig. 5).

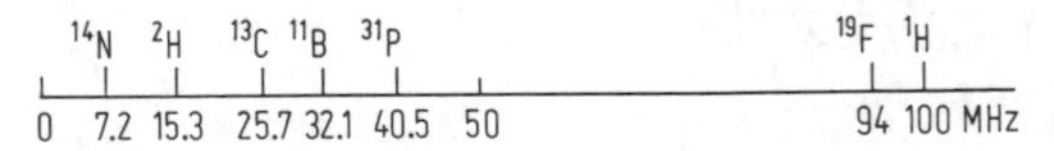

Fig. 5. Absorption frequencies of some nuclei at 2.35 T magnetic field strength.

It is much more important for the chemical applications of N., however, that there are slight differences in the resonant frequencies for nuclei of the same type in different chemical environments. For example, the ^{1}H-NMR spectrum of benzyl alcohol (Fig. 6) contains 3 signals which correspond to hydrogen nuclei bound in 3 different ways. The reason is that at the position of a nucleus, there is an effective mag-

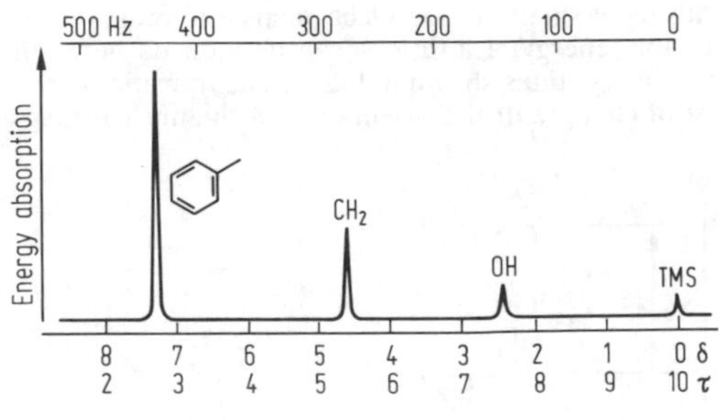

Fig. 6. ^{1}H-NMR spectrum of benzyl alcohol $C_6H_5-CH_2-OH$ (working frequency 60 MHz).

netic field H_{eff} which differs slightly from the external field H_0 because the nucleus is somewhat shielded by its electron shell: $H_{eff} = H_0(1-\sigma)$. The differential field $H_0\sigma$ is proportional to H_0.

The factor σ is the nuclear screening constant, which is very small, on the order of 10^{-5} for protons. For a magnetic field of 2.3 T, such as is used in many NMR spectrometers, the resonance frequencies lie between 100,000,000 and 100,001,500 Hz. In order to avoid using such large numbers to indicate positions of signals, they are given in reference to the absorption by a standard substance rather than in absolute frequencies. The reference substance which is now almost universally used is tetramethylsilane (TMS), whose absorption thus serves as the zero point of the frequency scale. It is usually added to solutions as an internal standard. Since different models of spectrometer work with different field strengths, and the resonance frequencies are dependent on H_0, the positions of signals must be indicated on a scale which is independent of the external field. This unit, called "chemical shift", δ, is obtained by dividing the difference between the sample signal ν_σ and the standard signal ν_s by the frequency of the standard ν_0 in the external magnetic field (working frequency):

Since ν_σ-ν_s is expressed in Hz, and ν_0 is in MHz, the dimensionless parameter δ is in units of 10^{-6} or parts per million (ppm). The chemical shift of TMS is thus by definition 0, and the chemical shifts of most nuclei in organic compounds are greater than 0, i.e. their signals appear in conventional NMR spectra to the left of the TMS signal. Chemical shifts are sometimes expressed in τ units, which are related to δ values by the equation:

$\tau = 10 - \delta$. (9)

Empirically established ranges of chemical shifts can be used to identify certain proton groupings, and are thus very helpful in the elucidation of unknown structures (Fig. 7). There are a number of terms used to express the position of resonance signals with reference to one another (Fig. 8).

The signal intensity in N. is considered to be the area under the resonance signal. Since signal intensities depend on the instrumental parameters of the spectrometer, their absolute values are not reported. Instead, the ratios of intensities between individual signals are reported. In modern NMR instruments, the signal surfaces are integrated and the integrated

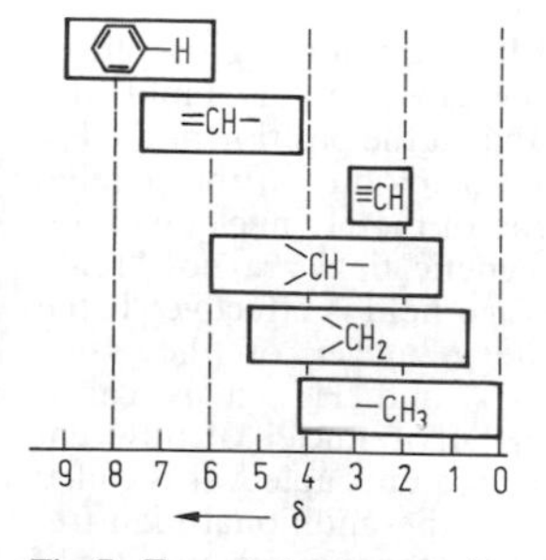

Fig. 7. Expectation ranges for chemical shifts of variously bound protons.

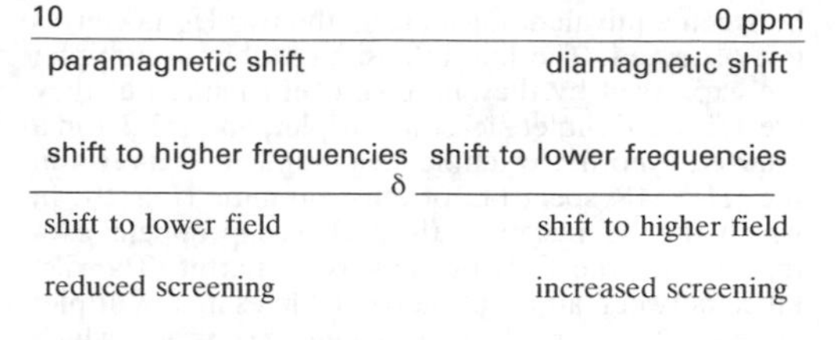

Fig. 8: Terms used to indicate the shift of an NMR signal.

steps are drawn onto the spectrum. The intensity ratio of the signals is then identical to the ratio of the step heights of the integrated curve. In ^{1}H-NMR, the intensity of a signal is normally directly proportional to the number of ^{1}H nuclei which give this signal. If the sample consists of a pure compound, the relative numbers of differently bound nuclei are given directly by the intensity ratios of the peaks (Fig. 9).

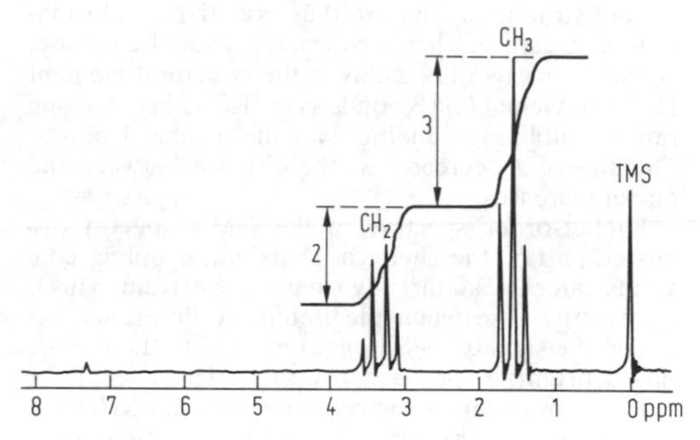

Fig. 9. ^{1}H-NMR spectrum of ethyl bromide, CH_3-CH_2Br. Working frequency, 60 MHz.

Multiplicity. In N., signals are often split into multiplets which cannot be explained by differences in chemical shifts. They are caused by coupling between the magnetic moments of neighboring nuclei. An example is shown in the ^{1}H-NMR spectrum of 1,1,2-trichloroethane shown schematically in Fig. 10. Because the nucleus H_A has spin, magnetic field arises around it. This field is transferred by interaction with the bonding electrons to the nuclei H_X, and appears as a small differential field acting on those nuclei. Since nucleus H_A is aligned either parallel or antiparallel to the external magnetic field, the differential field acting on the nuclei H_X is either positive or negative. Therefore, a signal arising from many H_X nuclei appears as a doublet. The two equivalent nuclei H_X create three additional fields at H_A, depending on which of three possibilities is realized: a) both spins add and are parallel to the external magnetic field; b) the two have opposite spins and compensate each other; or c) the two are antiparallel to the external field. The signal of H_A therefore is split into a triplet. In general, the multiplicity M of a signal is $M = 2nI+1$, where n is the number of neighboring magnetic nuclei and I is their nuclear spin quantum number. For nuclei like ^{1}H, ^{13}C, ^{19}F and ^{31}P, with $I = 1/2$, this relationship simplifies to $M = n+1$. Coupling between equivalent nuclei (e.g. the two H_X nuclei) is not observed. The line intensities within a multiplet are expressed by the binomial coefficients, i.e. they are 1:1 in a doublet, 1:2:1 in a triplet, and 1:3:3:1 in a quartet. Another example of multiplicity is shown in the ^{1}H-NMR spectrum of ethyl bromide (Fig. 9), in which the ^{1}H nuclei of the CH_3 group appear as a triplet, and the CH_2 protons as a quartet. The distance between adjacent individual lines in a multiplet are given by the nuclear spin coupling constant, which is called the *coupling constant J* for short. Since the frequency differences between the lines of a multiplet are caused only by the different orientations of the neighboring spins, they are independent of the working frequency of the apparatus and are measured in Hz. Their magnitude is affected by factors like the nature of and distance between the coupled nuclei, the nature of the bond between these nuclei and their geometrical relationship to each other (configuration, conformation). Just as for the chemical shifts, there are tables of empirically established coupling constants which give their magnitudes as a function of certain structures, and are thus essential for elucidation of structure. There are several types of coupling: geminal (via two bonds, as in the structural element H-C-H), vicinal (via 3 bonds as in H-C-C-H) and long range coupling extending over more than 3 bonds. The size of J decreases as the distance between the nuclei increases.

Higher-order spectra. In the NMR spectra discussed so far, the chemical shifts and coupling constants can be read directly from the spectrum. This is always the case when the frequency difference between the signals $\Delta\nu$ is large compared to their coupling constants J (first-order spectra). However, if $\Delta\nu$ is less than six times the coupling constant $\Delta\nu/J \leqq 6$, the specrum is higher-order, and the simple rules about multiplicity and line spacing and intensities in multiplets no longer apply. Exact spectral analysis of such spectra is only possible by laborious calculations. Higher-order spectra occur often in practical N., and make the analysis of spectra considerably more difficult.

Nomenclature of spin systems. In order to show the relations between the various nuclei or groups of equivalent nuclei in a spin system, all the coupled nuclei are assigned capital letters. Equivalent nuclei are given the same letter, and the number of them is shown as a subscript. If the chemical shifts of the coupled nuclei differ greatly, the nuclei are assigned letters from the beginning and end of the alphabet. If other nuclei are present with intermediate chemical shifts, they are assigned the letters M, N, O. If the chemical shifts of the nuclei are similar, which is the case in higher-order spectra, they are assigned letters in sequence.

Examples: in 1,1,2-trichloroethane (Fig. 10), there is an AX_2 or A_2X coupling, in ethyl bromide (Fig. 9) there is an A_3X_2 coupling, in 1-chloropropene, CH_3-CH=CHCl, an A_3XY system, and in ethyl alcohol, CH_3-CH_2-OH, an A_3M_2X system. In dichloroacetaldehyde, CCl_2H-CHO, there is an AX system, but in a disubstituted alkene RHC=CHR', in which the chemical shifts of the two protons are only slightly different, an AB system. The AB system produces a higher-order spectrum.

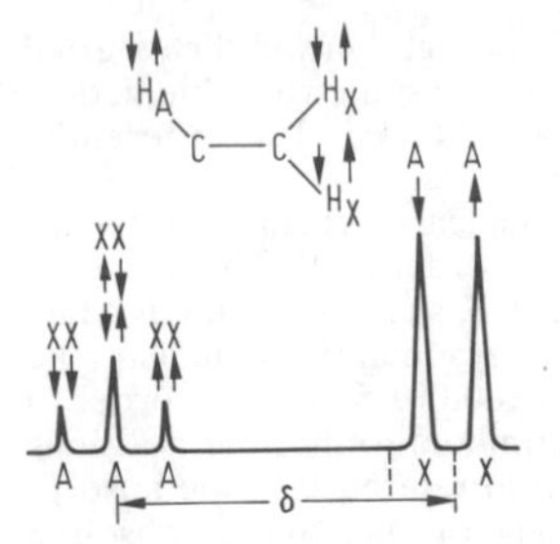

Fig. 10. Spin-spin coupling in 1,1,2-trichloroethane.

To simplify complicated spectra a series of methods have been developed. 1) *Increasing the magnetic field strength*. The differences between chemical shifts of two signals increase with increasing field strength, but the coupling contants are independent of the field strength, so that the conditions for first-order spectra are more likely to be met if the field strength of the spectrometer has a higher field strength. If a sample gives a complex spectrum at 60 MHz, it is usually significantly simpler at 100 MHz or 220MHz; however, the amount of improvement depends on the parameters of the spectrum in question.

2) *Spin uncoupling* (also called *double resonance*, because two frequencies are beamed onto the sample simultaneously). If two nuclei with spin 1/2 are coupled with each other in an AX system, the spectrum has 4 signals (Fig. 11). However, if each of the nuclei H_X is additionally irradiated intensely at its resonant frequency ν_X while the signal is being recorded, the original doublet for H_A is replaced by a singlet; the signal of H_A is spin-uncoupled.

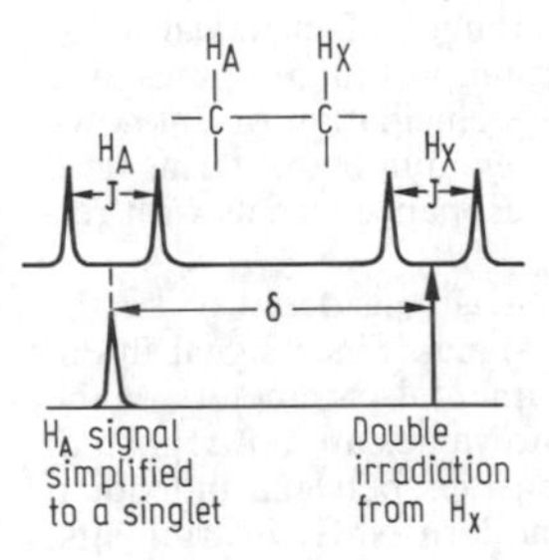

Fig. 11. Spin uncoupling in an AX system.

This can be explained by the following simplified model: the radiofrequency causes a very rapid flipping of the H_X nuclei, so that at the position of the H_A nuclei the two differential fields due to the parallel and antiparallel orientation of the H_X nuclei with respect to the external magnetic field are not "felt"; instead only a single average field is effective. If the uncoupling occurs between nuclei of the same isotope, e.g. between H_A and H_X, it is called *homonuclear uncoupling*, but if nuclei of different elements (e.g. ^{1}H and ^{19}F) are uncoupled, it is called *heteronuclear uncoupling*. If the additional high-frequency field contains all the resonance frequencies of the ^{1}H nuclei (*broad-band* or *noise uncoupling*), it can uncouple, e.g. all ^{1}H-^{13}C coupling, which is essential for the simplification of ^{13}C-NMR spectra. In addition to simplifying the spectrum, spin uncoupling gives information on the relationships between certain signals.

3) *Shift reagents*. It is possible to spread apart overlapping signals by using paramagnetic complexes of the rare-earth metals as shift reagents. Some commonly used shift reagents are $Eu(dpm)_3$, $Eu(fod)_3$, $Pr(dpm)_3$ and $Pr(fod)_3$, where dpm is the dipivaloyl methanato group, fod is the 1,1,1,2,2,3,3-heptafluoro-7,7-dimethyl-4,6-octanedionato group (Fig. 12).

dpm: $R = C(CH_3)_3$
fod: $R = CF_3-CF_2-CF_2-$

Fig. 12. Shift reagents. Me is a rare-earth metal.

The coordination number of the metal can be increased by interaction with free electron pairs from polar groups in the sample, such as -OH, $-NH_2$, >C=O, -O- and -COOR, and the sample can be reversibly complexed with the lanthanide complex ligand. In the complex, the local magnetic field of the paramagnetic metal ion acts on the nuclei of the sample, and shift the signals in its NMR spectrum. The closer the sample nucleus is to the paramagnetic metal ion, the greater the effect. For example, in the normal ^{1}H-NMR spectrum of hexanol, $HO-CH_2-CH_2-CH_2-CH_2-CH_2-CH_3$, the middle CH_2 groups give a non-resolved broad signal. After addition of $Eu(dpm)_3$, a clearly separated signal appears for each CH_2 group, and the distance of the signal from the position of the original signal depends on the distance of the CH_2 group from the coordination center (the OH group).

^{13}C-NMR spectroscopy The carbon skeleton of a molecule can be directly observed by studying the ^{13}C nuclei. The main difficulty with ^{13}C-NMR is its low sensitivity, compared to that of ^{1}H-NMR. This is because 1) the magnetic moment of the ^{13}C nucleus is smaller than that of the ^{1}H nucleus (its gyromagnetic ratio, as shown in Table 1, is a factor of 4 smaller) and 2) the abundance of ^{13}C in natural carbon is only 1.1%. The combined effect of these two factors is a decrease in sensitivity by a factor of 5000 to 6000 relative to ^{1}H-NMR. Routine measurements of ^{13}C-NMR spectra in non-enriched samples thus did not become possible until the the sensitivity of NMR spectrometers was significantly increased. The *sensitivity* is the ratio of signal strength *S* to the noise level *N*. The most important of the methods used to increase sensitivity are 1) increasing the sample volume relative to that for ^{1}H-NMR; 2) use of superconducting magnets to generate high field strengths; 3) spectral accumulation, i.e. a computer coupled to the spectrometer adds spectra which are obtained by repeated sweeps. This causes noise to average out, while signals are amplified. The signal/noise ratio improves with the square root of the number of recordings. 4) *Pulse Fourier transform (PFT) technique* (Fig. 13).

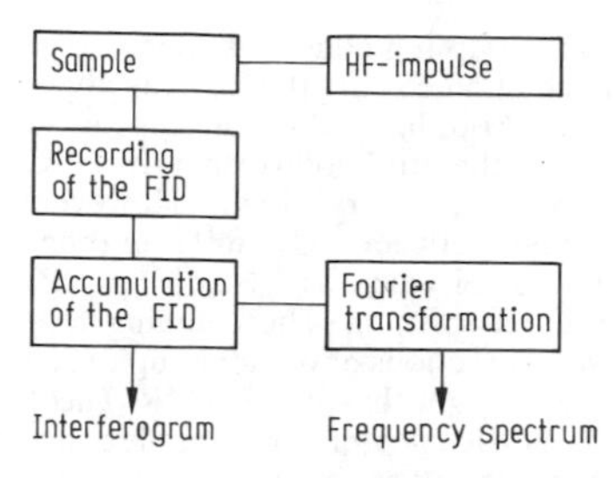

Fig. 13. Block scheme of a pulse-Fourier transform spectrometer.

All the ^{13}C nuclei are simultaneously excited with a short, high-frequency impulse ($\approx 10^{-6}$ s) of very high intensity; this causes the sample to become magnetized. After the HF impulse has ended, the nuclei return to the equilibrium of the Boltzmann distribution; one measures the free induction decay (FID) as a function of time. All the information present in a conventional spectrum is contained in the FID, although in a different form. Each part of the FID contains information about the entire spectrum. A mathematical operation, the Fourier transformation, is used to convert the interferogram obtained by plotting the decay of the induced magnetism against time into a conventional frequency spectrum. The transformation is done by a computer linked directly to the spectrometer. Since the time required for an impulse and subsequent recording of the FID is very short, the computer can be used to accumulate a number of interferograms before carrying out the Fourier transformation. The short time requirement thus permits more spectral accumulation and contributes significantly to increasing the sensitivity.

Characteristics of ^{13}C-NMR. The spectral parameters shown in Table 2 can be obtained from a ^{13}C-NMR spectrum in the same way as for an ^{1}H-NMR spectrum, but there are some differences. 1) The number of signal groups and chemical shift: the ^{13}C resonance range (about 0 to 300 ppm) is considerably larger than that of ^{1}H (0 to 10 ppm). As a result, even small differences in the bonding states of individual C atoms lead to clearly separated signals, and overlapping of signals is rare. *Example*: For vitamin B_{12}, 59 different resonance frequencies, corresponding to 59 non-equivalent C-atoms, are observed. The expectation range of the ^{13}C chemical shifts of variously bound C-atoms are summarized in Table 3; they permit prediction of the structure of the sample.

Table 3. Typical expectation ranges of the ^{13}C-chemical shifts (in ppm vs TMS)

Alkanes and alkyl compounds	0 ... 70
Alkynes	70 ... 90
Alkenes	80 ... 150
Aromatics	110 ... 160
Carbonyl compounds	160 ... 230
Carbonium ions	150 ... 300

2) Signal intensities: these are not always proportional to the number of nuclei in routine ^{13}C-NMR measurements. Quaternary C-atoms are especially prone to giving less than proportional intensities.

3) Multiplicity and coupling constants: because of the low natural abundance of the ^{13}C isotope, homonuclear ^{13}C-^{13}C couplings do not appear in routine spectra, since the probability that two ^{13}C nuclei are neighbors is very low. However, heteronuclear couplings between ^{13}C and ^{1}H (or other magnetically active nuclei, such as ^{15}N, ^{19}F or ^{31}P) lead to strong multiplet splitting. The coupling constants $J_{C,H}$ for these interactions have values up to 300 Hz, and are thus much larger than in ^{1}H-NMR. These couplings make ^{13}C-NMR spectra very complex, but they can be eliminated by the proton noise uncoupling mentioned above. Most ^{13}C-NMR spectra are recorded as noise-uncoupled, and in them there is a singlet for each ^{13}C nucleus (Fig. 14a).

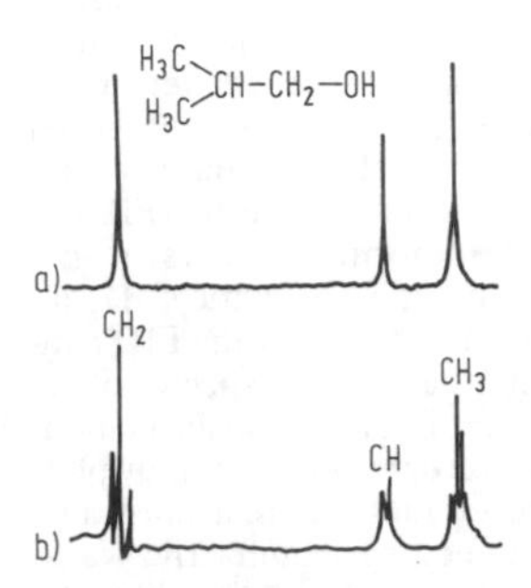

Fig. 14. ^{13}C-NMR spectrum of isobutanol: a) noise-uncoupled; b) off-resonance sprectrum.

However, suppression of the ^{13}C-^{1}H coupling eliminates valuable spectral information. For example, in the coupled spectrum, the ^{13}C signal for a CH_3 group is a quartet, due to coupling of the 3 bound H nuclei; a CH_2 group signal is a triplet, the CH group signal a doublet, and a quaternary carbon signal, a singlet. This information, which is very important for interpretation, can be partially recovered by the "off-resonance" method, in which a fixed frequency other than the proton resonance frequency is beamed in at high amplitude. Under these conditions, the ^{13}C-^{1}H couplings are visible, but they are greatly reduced in comparison to the non-uncoupled spectrum (by about 10 to 50 Hz). In this way the overlap of neighboring multiplets is largely eliminated, and an interpretable, first-order spectrum is obtained (Fig. 14b).

Overview of applications of N. The most important application is *elucidation of structure*, especially of organic compounds. The spectral parameters frequently give detailed information on the presence and number of certain atomic groups, as well as on their relative positions. In the case of chemical shifts and coupling constants, the information is obtained by comparison of the experimental results with empirically established tables. The number, intensity and multiplicity of the signals permit direct structural inferences to be drawn. However, complete structural analysis is usually not possible on the basis of NMR alone; results of elemental analysis, molecular mass determination, chemical reactivity, IR, UV and mass spectrum must also be considered. NMR spectra are of special importance for studies of the structures of high-molecular-weight polymers, because they give information on the tacticity, sequence, *cis/trans* arrangements and so on. Recently, N. has been used to solve numerous biochemical questions. With a special technique (*topical magnetic resonance*), studies can be made on the living organism, and individual organs, such as the liver, can be selectively measured. Due to the central role and relatively high concentrations of adenosine triphosphate in living tissues, ^{31}P-NMR measurements are highly significant.

Quantitative analysis: the basis for quantitative determinations is the proportionality between signal intensity and number of nuclei which give the signal. If certain signals coming from a mixture can be unequivocally assigned to individual components, their molar ratios are given directly by the spectrum, although it must be taken into account how many nuclei give the signal in question. Care must also be taken that the relative intensities of the signals are not affected by differences in relaxation times or by the nuclear Overhäuser effect. However, these effects can be largely eliminated by addition of *relaxation reagents*, e.g. paramagnetic complexes such as iron trisacetoacetate. Quantitative NMR studies are very important for isotope analysis, especially for determination of deuterium content. NMR measurements are very successful in *studies of dynamic processes* such as inter- or intramolecular migrations or exchange processes, which require activation energies between about 20 and 100 kJ mol^{-1}. Some examples are rearrangements in tautomers, conformational rearrangements, valence isomerization, inhibited internal molecular rotations and nitrogen inversions. In a dynamic process $A \rightleftharpoons B$, suppose that the chemical environment of a nucleus in component A is different than in B. If the mean time spent by the nucleus in each environment is long in comparison to the time required for a reorientation of the nuclear spin ($\approx 10^{-3}$ s), there will be two sharply separated signals (Fig. 15a). If the mean residence time is small in comparison to the transition time, however, the chemical environment of the nucleus changes during the reorientation process. The two signals then merge into a single sharp signal (Fig. 15d). The chemical shift of this signal is determined by the average chemical environment of the nucleus. For intermediate re-

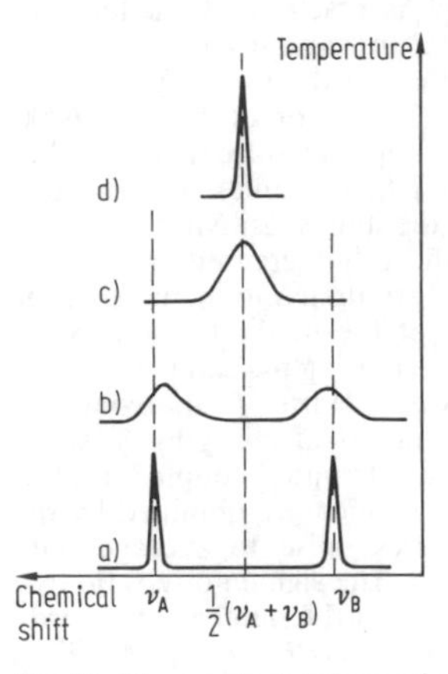

Fig. 15. Changes in signals as a function of rate of exchange.

sidence times, either one or two broad signals (Fig. 15b and c) will appear. A dynamic process is considered slow, on the NMR time scale, if one can see two separate signals in the spectrum, and fast if there is only an average signal. Since the rates of dynamic processes can be affected by the temperature, NMR measurements are carried out at different temperatures. N. can be used to determine the equilibrium concentrations of the components of a dynamic process, and kinetic parameters of the process in the form of rate constants and activation energy or free energy of activation.

II) ***Broad-band N.*** NMR spectra of solids differ from those of liquids in two major respects: 1) in many cases, dipole, quadrupole and other interactions, which are averaged out in the liquid by molecular motions, determine the line shapes of the spectrum. 2) The anisotropy of the interactions make the spectrum dependent on the orientation of the sample in the external magnetic field. As a result, the lines in spectra of solids are much broader than in those of dissolved substances, and may also be dependent on the angle. Therefore it is usually not possible to observe chemical shifts and multiplicities. However, internuclear distances and angles in the crystal can be measured. This is most important for determination of proton-nuclear distances in hydrogen-containing crystals, because these light nuclei cannot be detected by X-ray diffraction. For some solids, the linewidth suddenly decreases if the temperature increases above a certain threshhold. This indicates the beginning of certain molecular motions which quench the interactions with neighboring atoms.

With special techniques, it is possible to suppress the otherwise dominant dipole interactions, so that high-resolution spectra are obtained for solids.

NMR tomography, *Computer assisted tomography, CAT scan*: a technique for representing cross-sectional planes of proton distribution in intact biological systems. NMR-T. provides the same type of information as x-ray computer tomography, but has the advantage that the body is not exposed to damaging radiation. In NMR-T., inhomogeneous magnetic fields (see NMR spectroscopy) are used so that nuclei in different places have different NMR resonance frequencies. The NMR data are subjected to complicated mathematical treatment to obtain images of the object under examination. Such objects can be, for example, limbs or the entire human body. The intensity of an image point is, to a first approximation, related to the proton density at the site of measurement; this varies from one tissue to the next. If the PFT technique (see NMR spectroscopy) is used, and the time between two impulses is shorter than the relaxation time of the nuclei, the intensity also depends on the relaxation times of the nuclei. It can be shown that protons in different types of tissues have different relaxation times. For example, the relaxation times in tumors are longer than in normal tissues, so the method can be used to localize tumors.

No: symbol for nobelium.

Nobelium, *No*: a short-lived, radioactive chemical element which can be obtained only through nuclear reactions; it belongs to group IIIb of the periodic system, the Actinoids (see). Z 102, isotopes (decay type, half-life in parentheses): 251 (α, 0.8 s); 252 (α, 2.3 s); 253 (α, 1.6 min); 254 (α, 55 s); 255 (α, 3.3 min); 256 (α, 3.5 s); 257 (α, 26 s); 258 (spontaneous decay, $1.2 \cdot 10^{-3}$ s); 259 (α, 58 min); mass of the most stable isotope, 259, valencies II, II, standard electrode potential (No^{2+}/No^{3+}) +1.45 V.

Little is known about further physical and chemical properties of N. It forms a No^{3+} ion which is relatively easily reduced to No^{2+}. N. was discovered in 1957 in the Nobel Institute for Physics in Stockholm by bombardment of curium-244 with accelerated carbon nuclei, but this could not immediately be confirmed. In 1958, Ghiorso, Seaborg, Sikkeland and Walton repeated and confirmed the experiment in Berkeley, California, USA: $^{244}Cm + ^{12}C \rightarrow ^{252}No + ^{1}n$.

Nobel prize: a prize awarded annually for outstanding achievement in the fields of physics, chemistry, medicine, literature and peace. The prizes were established by the will of the Swedish chemist and industrialist Alfred Nobel (1833-1896); the funds are provided by interest on the endowment created from his estate. The prizes (accompanied by gold medals) are to be awarded, according to Nobel's will, to those persons "who in the past years have done the greatest good for mankind".

Nobel prizes in chemistry and other fields related to chemistry.

Year	Prize winner	Prize awarded for
1901	J. H. van't Hoff (1852–1911)	Discovery of the laws of chemical dynamics and of osmotioc pressure in dilute solution
1902	H. E. Fischer (1852–1919)	Synthesis of glucose and work on purine bodies
1903	S. A. Arrhenius (1859–1927)	Discovery of the theory of electrolytic dissociation
	H. A. Becquerel[1] (1852–1908)	Discovery of spontaneous radioactivity
	Marie Curie[1] (1867–1934) Pierre Curie[1] (1859–1906)	Research on the radiation phenomena discovered by Becquerel
1904	Lord J. W. s. Rayleigh[1] (1842–1919)	Studies on the densities of the most important gases and the discovery of argon
	Sir W. Ramsay (1852–1916)	Discovery of the noble gases and determination of their place in the periodic system

Nobel prize

Nobel prizes in chemistry and other fields related to chemistry.

Year	Prize winner	Prize awarded for
1905	A. von Baeyer (1835–1917)	Work on organic pigments and hydroaromatic compounds
1906	H. Moissan (1852–1907)	Discovery and purification of fluorine
1907	E. Buchner (1870–1917)	Biological-chemical studies and discovery of cell-free fermentation
1908	E. Rutherford (1871–1937)	Theory of radioactive decay of the elements
1909	W. Ostwald (1853–1932)	Work on catalysis, chemical equilibria and reaction kinetics
1910	A. Kossel[2] (1853–1927)	Research on the chemistry of cells and cell nuclei
	J. D. van der Waals[1] (1837–1923)	Work on the equations of state of gases and liquids
	O. Wallach (1847–1931)	Research on aromas and essential oils
1911	Marie Curie (1867–1934)	Discovery of radium and polonium, isolation and discovery of the properties of radium
1912	V. Grignard (1871–1935)	Discovery of the reactions named for him, which are of great importance in organic synthesis
	P. Sabatier (1854–1941)	Discovery of the catalytic action of finely divided metal in hydrogenation of organic compounds
1913	H. Kamerlingh-Onnes[1] (1853–1926)	Studies on the properties of bodies at low temperatures
	A. Werner (1866–1919)	Studies on the bonding of atoms in molecules
1914	M. von Laue[1] (1879–1960)	Discovery of the diffraction of x-rays in crystals
	T. W. Richards (1868–1928)	Accurate determination of the atomic masses of many chemical elements
1915	Sir W. H. Bragg[1] (1862–1942) W. L. Bragg[1] (1880–1971)	Research on crystal structures by x-ray diffraction
	R. Willstätter (1872–1942)	Work on plant pigments, determinations of chlorophylls
1917	C. C. Barkla[1] (1877–1944)	Discovery of the characteristic x-ray emissions of the elements
1918	F. Haber (1868–1934)	Synthesis of ammonia from the elements
1920	W. Nernst (1864–1941)	Thermochemical research
1921	F. Soddy (1877–1956)	Studies on the regularities of decay of radioactive elements and on the nature of isotopes
1922	N. Bohr[1] (1885–1962)	Research on the structure of atoms and their emissions
	F. W. Aston (1877–1945)	Discovery of many isotopes in several nonradioactive elements and of the law of integral masses
1923	F. Pregl (1869–1930)	Discovery of the quantitative microanalysis of organic compounds
1925	R. Zxigmondy (1865–1929)	Elucidation of the heterogeneous nature of colloidal solutions
1926	T. Svedberg (1884–1971)	Basic research on colloids
1927	H. O. Wieland (1877–1957)	Studies on the structure of bile acids and related substances
1928	A. Windaus (1876–1959)	Research on the structure of sterols and their relationship to the vitamins
1929	C. Eijkman[2] (1858–1930)	Discovery of the antineuritic vitamins
	L. V. de Broglie[1] (1892–1987)	Discovery of the wave nature of the electron

Nobel prizes in chemistry and other fields related to chemistry.

Year	Prize winner	Prize awarded for
	H. K. von Euler-Chelpin (1873–1964); A. Harden (1865–1940)	Studies on sugar fermentation and the enzymes involved
1930	Sir C. V. Raman[1] (1888–1970)	Work on the diffusion of light, discovery of the effect named for him
	H. Fischer (1881–1945)	Studies on hemes and chlorophyll, syntheses of heme
1931	O. H. Warburg[2] (1883–1970)	Discovery of the nature and functions of respiratory enzymes
	F. Bergius (1884–1949); C. Bosch (1874–1940)	Development of chemical high-pressure techniques
1932	W. Heisenberg[1] (1901–1976)	Founding of quantum mechanics
	I. Langmuir (1881–1957)	Work on adsorption and absorption at surfaces
1933	P. A. M. Dirac[1] (1902–1984); E. Schrödinger[1] (1887–1961)	Development of new forms of the atomic theory
1934	H. C. Urey (1893–1981)	Discovery of heavy hydrogen
1935	J. Chadwick[1] (1891–1974)	Discovery of the neutron
	Frédéric Joliot (1900–1958); Irène Joliot-Curie (1897–1956)	Synthesis of new radioactive elements
1936	Sir H. H. Dale[2] (1875–1968); O. Loewi[2] (1873–1981)	Proof of chemical transmission of nerve impulses to the target organs
	C. D. Anderson[1] (*1905)	Discovery of the positron
	P. J. W. Debye (1884–1966)	Studies on dipole moments and on the diffraction of x-rays and electrons in gases
1937	A. Szent-Györgyi[2] (*1893)	Discoveries in the field of biological oxidations, especially with regard to vitamin C and fumaric acid catalysis
	C. J. Davisson[1] (1881–1958); G. P. Thomson[1] (1892–1972)	Experimental discovery of electron diffraction in crystals
	W. N. Haworth (1883–1950)	Studies on carbohydrates and vitamin C
	P. Karrer (1889–1971)	Studies on the carotinoids and vitamnins A und B_2
1938	E. Fermi[1] (1901–1954)	Determination of new radioactive elements synthesized by neutron bombardment; discovery of the nucelar reactions initiated by slow neutrons
	R. Kuhn (1900–1967)	Work on carotinoids and vitamins
1939	G. Domagk[2] (1895–1964)	Discovery of the antibacterial effect of protosil
	E. O. Lawrence[1] (1901–1958)	Invention and development of the cyclotron and the results obtained with it
	A. F. J. Butenandt (*1903)	Isolation and structural elucidation of the sex hormones
	L. Ružička (1887–1976)	Work on polymethylenes and polyterpenes
1943	H. Dam[2] (1895–1976); E. Doisy[2] (*1893)	Discovery of the chemical structure of vitamin K

Nobel prizes in chemistry and other fields related to chemistry.

Year	Prize winner	Prize awarded for
	O. Stern[1] (1888–1969)	Development of the molecular beam method and discovery of the magnetic moment of the proton
	G. de Hevesy (1885–1966)	Work on the application of isotopes as markers in the study of chemical processes
1944	I. L. Rabi[1] (1898–1988)	Discovery of the resonance method of determining the magnetic properties of atomic nuclei
	O. Hahn (1879–1968)	Discovery of nuclear fission of uranium and thorium
1945	E. B. Chain[2] (1906–1979) Sir A. Fleming[2] (1881–1955) Sir H. W. Florey[2] (1898–1968)	Discovery of penicillin and its therapeutic use
	W. Pauli[1] (1900–1958)	Discovery of the Pauli principle of electron exclusion
	A. I. Virtanen (1895–1973)	Studies and discoveries in the field of agricultural and nutritional chemistry
1946	J. B. Sumner (1887–1955)	Discovery of the ability of enzymes to crystallize
1947	Carl F. Cori[2] (1896–1957) Gerty T. Cori[2] (1896–1957)	Research on intermediary metabolism, especially the degradation of carbohydrates in muscle
	Sir R. Robinson (1886–1975)	Work on plant alkaloids and blossom pigments
1948	P. M. S. Blackett[1] (1897–1974)	Further development of the applications of the Wilson cloud chamber, discoveries in nuclear physics and cosmic ray research
	A. W. K. Tiselius (1902–1971)	Studies on protein molecules
1949	H. Yukawa[1] (1907–1981)	Prediction of the existence of mesons, on which the theory of nuclear forces is based
	W. F. Giauque (1895–1982)	Application of adiabetic demagnitization of paramagnetic substances to reach extremely low temperatures
1950	P. S. Hench[2] (1896–1965) E. C. Kendall[2] 1886–1972 T. Reichstein[2] (*1897)	Work on adrenal hormones
	C. F. Powell[1] (1903–1969)	Development of the photographic methods for the study of nuclear processes, discovery of mesons
	K. Alder (1902–1958) O. Diels (1876–1954)	Development of diene synthesis
1951	Sir J. D. Cockcroft[1] (1897–1967) E. T. S. Walton[1] (*1903)	Studies in the area of transmutation of atomic nuclei by artificially accelerated atomic fragments
	E. M. McMillan (*1907) G. T. Seaborg (*1912)	Discovery of the transuranium elements neptunium and plutonium
1952	F. Bloch[1] (1905–1983) E. M. Purcell[1] (*1912)	Development of new methods for exact measurement of magnetic fields in atomic nuclei and related discoveries
	R. L. M. Synge (*1914) A. J. P. Martin (*1910)	Development of distribution and paper chromatography
1953	H. A. Krebs[2] (1900–1981)	Discovery of coenzyme A and its significance in metabolism

Nobel prizes in chemistry and other fields related to chemistry.

Year	Prize winner	Prize awarded for
	F. A. Lipmann[2] (*1899)	Discovery of the citric acid cycle
	F. Zernicke[1] (1888–1966)	Invention of the phase contrast microscope
	H. Staudinger (1881–1965)	Work on the chemistry of macromolecular substances
1954	M. Born[1] (1882–1970)	Statistical interpretation of quantum mechanics and the lattice theory of crystals
	L. Pauling (*1901)	Quantum mechanical studies in the field of chemical bonding
1955	A. H. Theorell[2] (1903–1982)	Discoveries on the nature and mechanisms of oxidation enzymes
	P. Kusch[1] (*1911)	Precise determination of the magnetic moment of the electron
	W. E. Lamb[1] (*1913)	Discovery of the fine structure of the hydrogen spectrum
	V. du Vigneaud (1901–1978)	Syntheses in the field of proteins, studies on biochemically important sulfur compounds
1956	Sir C. N. Hinshelwood (1897–1967) N. N. Semenov (1896–1986)	Research on the mechanisms of chemical reactions
1957	D. Bovet[2] (*1907)	Work on substances which inhibit the action of autogenous substances (especially on the striated musculature and blood vessel systems)
	Sir A. R. Todd (*1907)	Structural elucidation of the nucleic acids, chemical elucidation of vitamin B_{12} structure, and synthesis of a dinucleotide
1958	G. W. Beadle[2] (1903–1989) E. L. Tatum[2] (1909–1975)	Studies on the control of chemical structure in organisms by genes
1959	A. Kornberg[2] (*1918) S. Ochoa[2] (*1905)	Discovery of the mechanism of biological synthesis of ribonucleic and deoxyribonucleic acids
	O. Chamberlain[1] (*1920) E. Segrè[1] (1905–1989)	Discovery of the antiproton
	J. Heyrovsky (1890–1967)	Research in the area of polarography
1960	D. Glaser[1] (*1926)	Invention of the bubble chamber
	W. F. Libby (1908–1980)	Development of a method for use of ^{14}C to determine ages of objects
1961	R. Hofstadter[1] (*1915)	Studies on electron scattering in atomic nuclei research on the structure of nucleons
	R. L. Mößbauer[1] (*1929)	Discovery of the effect later named for him
	M. Calvin (*1911)	Studies on carbon dioxide assimilation in plants
1962	F. H. C. Crick[2] (*1916) J. D. Watson[2] (*1928) M. H. F. Wilkins[2] (*1916)	Discovery of the molecular structures of nucleic acids and their significance for information transfer in organisms
	L. D. Landau[1] (1907–1968)	Pioneering work on theories of liquids, especially liquid helium
	J. C. Kendrew (*1917) M. F. Perutz (*1914)	Studies of the structure of myoglobin and hemoglobin

Nobel prizes in chemistry and other fields related to chemistry.

Year	Prize winner	Prize awarded for
1963	Maria Göppert-Mayer[1] (1906–1972) H. Jensen[1] (1907–1973)	Development of theories of the shell structure of the atomic nucleus
	E. P. Wigner[1] (*1902)	Studies on the process of transmutation and fission of atomic nuclei which formed the basis for development of modern models of the nucleus
	G. Natta (1903–1979)	Regularities in polymerizations
	K. Ziegler (1898–1973)	Use of mixed catalysts for polymerization of ether
1964	Dorothy Crowfoot-Hodgkin (*1910)	Research on the chemical structures of complicated organic compounds, e.g. penicillin and vitamin B_{12}
1965	R. B. Woodward (1917–1979)	Synthesis and structure elucidation of natural products
1966	R. S. Mulliken (1896–1986)	Quantum chemical studies
1967	M. Eigen (*1927) R. G. W. Norrish (1897–1978) G. Porter (*1920)	Studies of extremely rapid reactions
1968	L. Onsager (1903–1976)	Thermodynamics of irreversible processes
1969	O. Hassel (1897–1981) D. H. R. Barton (*1918)	Studies on conformation
1970	L. F. Leloir (1906–1987)	Studies on carbohydrate-uridine diphosphate glucose
1971	W. Herzberg (*1904)	Elemental structure and geometry of free radicals
1972	C. B. Anfinsen (*1916) S. Moore (1913–1982) W. H. Stein (1911–1980)	Structure and mechanism of action of ribonuclease
1973	E. O. Fischer (*1918) G. Wilkinson (*1921)	Sandwich structure of π-complexes
1975	V. Prelog (*1906) J. W. Cornforth (*1917)	Stereochemistry of organic compounds
1976	W. N. Lipscomb (*1917)	Studies on the boranes
1977	I. Prigogine (*1917)	Thermodynamics of irreversible processes
1978	H. H. Mitchell (1886–1966)	Chemiosmotic hypothesis of electron-transport phosphorylation
1979	G. Wittig (1897–1987) H. C. Brown (*1912)	Contributions to preparative chemistry
1980	P. Berg (*1926)	Work on gene technology
	W. Gilbert (*1932) F. Sanger (*1918)	Methods for rapid and precise analysis of nucleotides
1981	K. Fukui (*1918)	Work on limit orbital model
	R. Hoffmann (*1937)	Work in theoretical chemistry

Nobel prizes in chemistry and other fields related to chemistry.

Year	Prize winner	Prize awarded for
1982	A. Klug (*1926)	Contributions to elucidation of complex biological systems
1983	H. Taube (*1915)	Work on reaction mechanisms of electron transfer in metal compounds
1984	R. B. Merrifield (*1921)	Development of methods which could be automated for the synthesis of polypeptides and polynucleotides
1985	H. A. Hauptman (*1917) J. Karle (*1918)	Development of direct methods for the determination of crystal structures
1986	D. R. Herschbach (*1932)	Detailed understanding of chemical reaction mechanisms
	Y. T. Lee (*1936) J. C. Polanyi (*1929)	Contributions concerning the dynamics of elementary chemical processes
1987	D. J. Cram (*1919) J.-M. P. Lehn (*1939) C. J. Pedersen (*1904)	Elucidating mechanisms of molecular recognition, which are fundamental to enzymic catalysis, regulation, and transport
1988	J. Deisenhofer (*1943) R. Huber (*1937) H. Michel (*1948)	Revealing the three-dimensional structure of closely-linked proteins that are essential to photosynthesis
1989	S. Altmann (*1939) T. R. Cech (*1947)	Discoveries about the active role of RNA in chemical cell reactions
1990	E. J. Corey (*1928)	Development of the theory and methodology of organic synthesis
1991	R. R. Ernst (*1933)	Development of high-resolution nuclear magnetic resonance (NMR) spectroscopy
1992	R. A. Marcus (*1923)	Development of the Marcus theory of electron-transfer reactions

[1]) Nobel Prize for physics. [2]) Nobel Prize for medicine and physiology.
In the years 1916, 1917, 1919, 1924, 1933 and 1940–1943, no Nobel Prizes were awarded in chemistry.

Noble gas: one of the elements of the 8th main group of the periodic system, including helium (He), neon (Ne), argon (Ar), krypton (Kr), xenon (Xe) and radon (Rn). They are monoatomic, colorless, noncombustible, nontoxic gases with no odor or taste, and low boiling points. Their physical properties change regularly with increasing atomic number (table). They are reasonably soluble in water and some organic solvents. Their electrical conductivity is considerably higher than that of other gases.

The term "noble" refers to the extreme chemical inertness of these elements, which is due to the high stability of their electron configurations: $1s^2$ for helium and ns^2p^6 for the others. The failure of these elements to form molecules is explained by the observation that an interaction between two noble gas atoms would require occupation of both a bonding and an antibonding molecular orbital by electron pairs.

Until about 20 years ago, it was believed that the N. were completely inert and unable to form stable compounds either among themselves or with other elements. Crystalline hydrates, e.g. of krypton and xenon, had been known since the turn of the century, and clathrates (inclusion compounds) of the N. had been isolated, for example with the N. atoms in the vacant spaces in a hydroquinone host lattice. Although true "valence compounds" of the N. had been predicted in 1927 by L. Pauling, they were first prepared in 1962. Working independently, N. Bartlett and R. Hoppe obtained xenon fluorides by reaction of elemental fluorine and xenon. During the next several years, a series of fluorides, oxide fluorides and oxides were prepared. Most were xenon compounds, but some compounds of krypton and radon were also prepared. A xenon dichloride has also been synthesized.

The thermodynamic stability of the N. compounds depends both on the relatively low ionization energies of the heavier N. and on the high electron affinity of fluorine and the low dissociation enthalpy of the F_2 molecule. Thus the krypton derivatives are much less stable than their xenon homologs, while radon has been reported to form the most stable compounds.

Furthermore, xenon dichloride, $XeCl_2$, is much more labile than XeF_2. The bonding in N. compounds depends on octet expansion and occupation of d-orbitals, or on formation of three-center, four-electron bonds. Their structures are adequately described by the VSEPR model (see Xenon compounds). More on the N. is to be found in the entries on the individual elements.

Noble gas configuration: see Electron configuration.

Nocardicin A: see β-Lactam antibiotics.

NOE: see Nuclear Overhauser effect.

Nomenclature, ***Nomenclature of chemical elements and compounds***: the system by which chemical elements and compounds are named; more narrowly, the rules published by the nomenclature commission of the International Union for Pure and Applied Chemistry (see) (***IUPAC***).

I) ***General***. The oldest names are trivial names which give no indication of systematic relationship. The number of these still allowed by the IUPAC rules is limited; they may also be used as roots for the derivation of names of chemical derivatives.

The structure of a chemical compound can be unequivocally indicated by the Formula (see) or by the systematic or semi-systematic name. The systematic name is the alphanumeric description of the three-dimensional structure of the molecule, which can also be represented by the formula.

In the process of naming, the structural formula is divided into certain structural units, which are indicated by name segments. These are then joined on a modular principle, using symbols to denote positions and configurations (Fig. 1). In order to achieve precision and the greatest simplicity possible, there are conventions for names, just as there are for formulae; these apply to the use, spelling and arrangement of parts of names and other symbols. There are several nomenclature systems for the systematic naming of a chemical compound; some of these still allow alternatives. Therefore the same compound can have various IUPAC names, the number of which increases with its complexity. In this book, in addition to trivial names, the recommendations of the IUPAC commission for binary or coordinative names of inorganic compounds are used; for organic compounds, the substitutional system is used.

II) ***N. of inorganic chemistry***. At present, the rules published by the IUPAC commission in 1970 for nomenclature of elements and inorganic chemistry apply. A revision of these rules is being prepared. The most important rules are summarized below.

A) ***Elements***. It is desirable that the names of the elements should be as similar as possible in the various languages, but this goal has not yet been reached. The symbols for the Elements (see) are the same in all languages, however, and all newly discovered elements (see Periodic system) are to be given 3-letter

Properties of the noble gases	He	Ne	Ar	Kr	Xe	Rn
Nuclear charge	2	10	18	36	54	86
Electron configuration	$1s^2$	[He] $2s^2p^6$	[Ne] $3s^2p^6$	[Ar] $3d^{10}$ $4s^2p^6$	[Kr] $4d^{10}$ $5s^2p^6$	[Xe] $4f^{14}$ $5d^{10}$ $6s^2p^6$
Atomic mass	4.002 6	20.183	39.948	83.80	131.30	(222)
Atomic radius [pm]	93	131	174	189	209	214
lst ionization potential [eV]	24.587	21.564	15.759	13.999	12.130	10.748
Density [g l^{-3}]	0.178 5	0.900 2	1.784	3.736	5.887	9.73
m. p. [°C]	–272.2*	–248.67	–189.2	–152.6	–111.9	–71
b. p. [°C]	–268.93	–245.9	–185.7	–152.3	–107.1	–61.8

* at 2.6 MPa

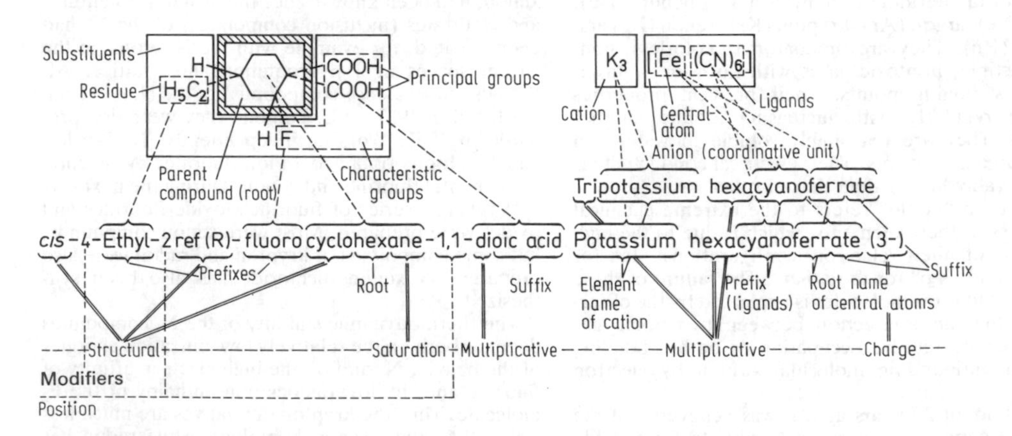

1. Construction of the name from the structural formula.

symbols. All isotopes of an element should have the same name; an exception is that the names protium, deuterium and tritium may continue to be used for the isotopes of hydrogen. For other elements, the isotopes are distinguished by the mass numbers, e.g. oxygen-18 (symbol: ^{18}O).

For the elements copper, gold, iron, lead, silver, tin and tungsten, the names of derivatives based on the element name are based on the Latin names of these elements: cuprum, aurum, ferrum, plumbum, argentum, stannum and wolframum, e.g. aurate rather than goldate. For nickel, niccolate is preferred to nickelate (but nickelate is commonly used).

The names of some compounds of sulfur, nitrogen and antimony are derived from the Greek "Theion", the French "azote" and "stibium" (another Latin name for antimony).

2) Collective names for groups and subgroups. The names halogens (F, Cl, Br, I and At), chalcogens (O, S, Se, Te and Po), alkali metals (Li to Fr), alkaline earth metals (Ca to Ra) and noble gases continue to be used; halides (or halogenides) and chaldogenides are compounds of halogens and chalcogens, respectively. The use of "pnicogen" (N, P, As, Sb and Bi) and "pnictides" is discontinued. The elements Sc, Y and La to Lu are rare-earth metals; the name lanthanoids is recommended for La to Lu, and the names actinoids, uranoids and curoids are to be used analogously. The term "metalloid" should no longer be used; elements are classified as metals, semi-metals and non-metals.

B) ***Names of compounds***. 1) Trivial names which are not confusing, such as soda and caustic lime, may continue to be used in industrial and popular literature. Terms like water, ammonia, hydrazine, ozone, white phosphorus and the names of many oxoacids are allowed trivial names.

2) The systematic name of a compound is formed by indicating the components and their proportions. Many compounds are binary, i.e. consist of two components, or they can be treated as if they were binary. a) Indication of the *components of a compound*. The name of the electropositive component, that is, the atom or group listed first in the formula, is not changed. If there are two or more electropositive components, these are listed in alphabetical order, except for hydrogen, which is always listed last, e.g. ammonium sodium hydrogenphosphate tetrahydrate. If the electronegative component consists of one or more atoms of the same element, the suffix "-ide" is added to the root of its Latin name and this word is added to that of the electropositive component, e.g. NaCl is sodium chloride rather than chlorosodium; CCl_4 is carbon tetrachloride rather than tetrachlorocarbon. If the electronegative component is heteropolyatomic, its name ends in "-ate". The name of the complex is derived from those of the constituent atoms, e.g. disodium tetraoxosulfate (Na_2SO_4), disodium trioxosulfate (Na_2SO_3). In a polyatomic group there is generally a characteristic or central atom (Cl in ClO^-, I in ICl_4^-). In this case the name of the complex is formed from the name of the central or characteristic element, modified to end in "-ate". In exceptional cases, the endings "-ide" and "-ite" are used (cf. II C 3). If there are two or more electronegative components, they are listed in alphabetical order.

b) The *stoichiometric proportions of components* may be indicated by the Greek numerical prefixes (mono, di, tri, tetra, penta, hexa, hepta, octa, ennea, or nona, deca, hendeca and dodeca) preceeding the names of the elements without hyphens, e.g. PCl_5 is phosphorus pentachloride. The end vowels of prefixes should not be elided before initial vowels of element names, except for compelling linguistic reasons. The prefix "mono-" can be omitted except where confusion would arise. Beyond 10, the prefixes can be replaced by Arabic numerals. When the number of complete groups of atoms must be designated, especially when the name includes a numerical prefix with another significance, the multiplicative numerals (Latin bis, Greek tris, tetrakis, etc.) are used. The group to which the multiplicative numeral refers is enclosed in parentheses, e.g. $Ca[PCl_6]_2$ is calcium bis(hexachlorophosphate). The stoichiometric proportions of the components can also be given by indicating the oxidation level of an element with Roman numerals in parentheses immediately after the name of the element (Stock notation); the Arabic 0 is used for zero. This system can be used with both cations and anions. Another method is the Ewens-Bassett system, in which the charge of an ion is indicated by an Arabic numeral, followed by the sign of the charge, in parentheses immediately following the name of the ion, e.g. iron(2+) chloride for $FeCl_2$.

3) Hydrides. Binary hydrogen compounds may be named systematically, e.g. hydrogen sulfide. Volatile hydrides, except those of Group VII and of oxygen and nitrogen, may be named by citing the root name of the element followed by the suffix "-ane". If the molecule contains more than one atom of that element, the number is indicated by a Greek prefix, e.g. diborane, polysulfane. Trivial names like water, ammonia, hydrazine, phosphine, arsine, stibine and bismutine are retained. Caution must be exercised to avoid conflict with names of saturated six-membered heterocyclic rings, e.g. trioxane and diselenane or allowed trivial names for other structures (indane).

C) ***Names for ions and radicals*** 1) Cations. a) *Monoatomic cations* have the same name as the element, e.g. Cu^+ is the copper(I) ion. The same principle holds for polyatomic cations corresponding to radicals for which special names exist (cf. II C 3), e.g. NO^+ is the nitrosyl cation. b) *Polyatomic cations* (except for those treated under c to e) which are formed by addition of other ions, neutral atoms or molecules (ligands) are considered complexes (cf. II G). If a polyatomic cation is derived from a monoatomic anion by addition of protons, the name is formed by adding the ending "-onium" to the root of the name of the anion element, e.g. PH_4^+ is the phosphonium ion. The H_3O^+ ion, that is the simplest hydrated proton, is called the oxonium ion. If the degree of hydration is not significant, it may be called the hydrogen ion. c) *Nitrogen-containing cations*. The name ammonium (not nitronium) is used for the NH_4^+ ion. Substituted ammonium ions derived from nitrogen bases with the ending "-amine" are given the ending "-ammonium", e.g. $HONH_3^+$ is the hydroxylammonium ion. d) For *nitrogen bases whose names do not end in "-amine"* and for non-nitrogen bases, the cation name is formed from the name of the base and the ending "-ium", e.g. $N_2H_5^+$ is the hydrazinium ion. e) If the cations are formed by addition of protons to acids with polyatomic anions, the word "acidium" is added to the name of the anion, e.g. $H_2NO_3^+$ the nitrate acidium ion.

2) Anions. a) The names of *monoatomic anions* are formed by the name of the element (sometimes abbreviated) and the termination "-ide", e.g. Cl^- is the chloride ion; S^{2-}, the sulfide ion; P^{3-}, the phosphide ion; and C^{4-}, the carbide ion. b) The names of *polyatomic anions* are generally formed from the root name of the central atom and the ending "-ate"; atoms and groups bound to the central atom are treated as ligands (II G 2), e.g. $[Sb(OH)_6]^-$ is the hexahydroxoantimonate(V) ion. Oxygen as a ligand is often ignored, and the oxidation state can be omitted if this does not lead to confusion. Some polyatomic anions of pseudobinary compounds end in "-ide", e.g. HO^- is the hydroxide ion; O_2^{2-}, the peroxide ion; S_2^{2-}, the disulfide ion; and N_3^-, the azide ion; and H_2N^- is the amide ion. Some polyatomic anions containing oxygen which are derived from acids with lower oxidation states (cf. E 2a) are given the ending "-ite", e.g. NO_2^- is the nitrite ion; SO_3^{2-}, the sulfite ion; and ClO_2^-, the chlorite ion. The use of the prefixes "hypo-" and "per-" is limited to a few anions.

3) Radicals. A radical is a group of atoms which occurs repeatedly in a number of different compounds. If the radical contains oxygen or other chalcogens, they have special names ending in "-yl", e.g. HO, hydroxyl; CO, carbonyl; NO, nitrosyl; NO_2, nitryl; PO, phosphoryl; SO, sulfinyl or thionyl; SO_2, sulfonyl or sulfuryl; CrO_2, chromyl; ClO, chlorosyl; ClO_2, chloryl. Names such as bismuthyl and antimonyl are not approved because the compounds do not contain BiO and SbO groups. The substitution of other chalcogens for oxygen is indicated by the prefixes "thio-", "seleno-", etc., e.g. CSe, selenocarbonyl. In compounds such radicals are always treated as positive components, regardless of unknown or controversial polarity, e.g. NOCl is nitrosyl chloride.

D) ***Iso- and hetero-polyanions***. 1) Isopolyanions. The stoichiometric names for salts with polyanions formed according to B 2a contain no structural information. If the anions are formed by condensation of molecules of a monoacid containing the characteristic element in the oxidation state corresponding to its Group number, they are named by indicating with Greek numerical prefixes the number of atoms of the characteristic element. The number of oxygen atoms does not have to be indicated if the charge of the anion or the number of cations is given, e.g. $S_2O_7^{2-}$ is the disulfate(2) ion. If the oxidation number differs from the Group number, it is indicated by the Stock number (cf. II B 2b), e.g. $[O_2HAs\text{-}O\text{-}AsO_3H]^{2-}$ is the dihydrogendiarsene(III,V)(2-) ion. Cyclic and straight chain structures may be distinguished by means of the prefixes *cyclo* and *catena*, although the latter may usually be omitted. If the bridge oxygen is replaced by another group, the location of this group is indicated by μ and named by the corresponding prefix, e.g. $[O_3P\text{-}S\text{-}PO_2\text{-}O\text{-}PO_3]^{5-}$ is the 1,2-μ-thiotriphosphate(5-) ion.

2) Heteropolyanions with a chain or ring structure. Dinuclear anions are named by treating the anion which comes first in alphabetical order as the ligand on the characteristic atom of the second. Longer chains are named similarly (unless the chain contains an obvious central atom) beginning with the end group which comes first in alphabetical order and treating the chain with (n-1) units as the ligand on the other end group, e.g. $[O_3Cr\text{-}O\text{-}AsO_2\text{-}O\text{-}PO_3]^{4-}$ is the (chromatoarsenato)phosphate(4-)ion. 3) Condensed heteropolyanions in which a network of coupled octahedra surrounds the central atom are named by indicating the number and names of the octahedra before the name of the central ion, e.g. $(NH_4)_3PW_{12}O_{40}$ is triammonium dodecawolframophosphate.

E) ***Acids***. The accepted N. of acids is the result of a long development, and it would be difficult to systematize acid names without drastic alteration. According to the present rules, useful names are retained, but an effort should be made to give new compounds systematic names.

1) Binary and pseudobinary acids. Acids giving rise to the "-ide" anions (cf II C 2) are named as binary and pseudobinary hydrogen compounds, e.g. hydrogen chloride, hydrogen sulfide, hydrogen azide.

2) Acids derived from polyatomic anions may be named analogously to the binary and pseudobinary acids, with the anions treated as complexes. If is customary, however, to use the terms "-ic acid" and "-ous acid" corresponding to the terminations "-ate" and "-ite", respectively, for those acids giving rise to anions with names ending in those suffixes. a) *Oxoacids*. For the oxoacids, the "ous-ic" notation is used to distinguish between different oxidation states in many cases. The "-ous acid" names are restricted to the following "-ites":

HNO_2	nitrous acid
$HNOO_2$	peroxonitrous acid
H_3AsO_3	arsenious acid
H_2SO_3	sulfurous acid
$H_2S_2O_5$	disulfurous acid
$H_2S_2O_4$	dithionous acid
$H_2S_2O_2$	thiosulfurous acid
H_2SeO_3	selenious acid
$HClO_2$	chlorous acid (and similarly for the other halogens)

Oxoacids with still lower oxidation states can be indicated by the prefix "-hypo-" in the following cases:

HOCl	hypochlorous acid (and similarly for the other halogens)
$H_2N_2O_2$	hyponitrous acid
$H_4P_2O_6$	hypophosphorous acid

The prefix "per-" is used to designate a higher oxidation state, but is retained only for $HClO_4$, perchloric acid, and corresponding acids of other halogens.

The prefixes "ortho-" and "meta-" are used to distinguish acids differing in the "content of water". The prefix "meta-" should only be used when the structure is not known; the acids are named systematically according to the rules for isopolyacids:

H_3BO_3	orthoboric acid
H_4SiO_4	orthosilicic acid
H_3PO_4	orthophosphoric acid
H_5IO_6	orthoperiodic acid
H_6TeO_6	orthotelluric acid
$(HBO_2)_n$	metaboric acid
$(H_2SiO_3)_n$	metasilicic acid
$(HPO_3)_n$	metaphosphoric acid

The following are also allowed:

H_2CO_3	carbonic acid
HOCN	cyanic acid
HNCO	isocyanic acid
HONC	fulminic acid
HNO_3	nitric acid
H_2NO_2	nitroxylic acid
$H_4P_2O_7$	diphosphoric acid or pyrophosphoric acid
H_2PHO_3	phosphonic acid
$H_2P_2H_2O_5$	diphosphonic acid
HPH_2O_2	phosphinic acid
H_3AsO_4	arsenic acid
H_3AsO_3	arsenious acid
$HSb(OH)_6$	hexahydroxoantimonic acid
H_2SO_4	sulfuric acid
$H_2S_2O_7$	disulfuric acid
$H_2S_2O_6$	dithionic acid
H_2SO_2	sulfoxylic acid
H_2SeO_4	selenic acid
H_2SeO_3	selenious acid
H_2CrO_4	chromic acid
$H_2Cr_2O_7$	dichromic acid
$HClO_3$	chloric acid
$HBrO_3$	bromic acid
HIO_3	iodic acid
H_2MnO_4	manganic acid
H_2TcO_4	technetic acid
H_2ReO_4	rhenic acid

b) *Peroxoacids* The prefix "peroxo-" in conjunction with the trivial name of an acid indicates replacement of -O- by -O-O-, e.g. H_2SO_5, peroxomonosulfuric acid, $H_2S_2O_8$, peroxodisulfuric acid. c) *Thioacids*. The prefix "thio-" indicates that an oxygen atom of an oxoacid has been replaced by sulfur, e.g. $H_2S_2O_3$, thiosulfuric acid. If more than one oxygen atom can be replaced by sulfur, the number of sulfur atoms must be indicated, e.g. H_2CS_3, trithiocarbonic acid. The prefixes "seleno-" and "telluro-" are used in like manner. d) *Acids with ligands other than oxygen or sulfur* are named as coordination compounds (cf. II G), e.g. H_2SiF_6, hexafluorosilicic acid. For all other acids not listed above, the names are formed systematically; for many acids with trivial names, systematic names are preferable, e.g. H_2MnO_4, tetraoxomanganic(VI) acid, to distinguish it from H_3MnO_4, tetraoxomanganic(V) acid.

3) Functional derivatives of acids are compounds which are obtained by replacing OH or sometimes O by other atoms or groups. Functional names for these compounds are still used, although not recommended by the IUPAC. The names of the *acid halides* are derived from the corresponding acid, if it has a name, e.g. SO_2Cl_2, sulfuryl chloride. In other cases, the compounds are named systematically as halide oxides, e.g. $MoCl_2O_2$, molybdenum dichloride dioxide. *Anhydrides* of inorganic acids are named as oxides, e.g. N_2O_5 as dinitrogen pentoxide rather than nitric anhydride. *Esters* of inorganic acids can be named as the salts, e.g. $(H_3C)_2SO_4$, dimethylsulfate. Names of *amides* can be derived from the names of acids by replacing "acid" by "amide", or from the names of acid radicals, e.g. $PO(NH_2)_3$ is phosphoric triamide or phosphoryl triamide. If not all the hydroxyl groups of the acid are replaced by H_2N groups, the names end in "-amidic acid", e.g. NH_2SO_3H, sulfamidic acid. Compounds with triply bound nitrogen are named as *nitrides*, e.g. $(PCl_2N)_3$ is trimeric phosphorus dichloride nitride, not trimeric phosphonitrile chloride. The use of the terms "nitrile" and "nitrilo" in purely inorganic names is no longer allowed.

F) ***Salts and salt-like compounds***. 1) Simple salts are binary compounds and are named according to the rules for this category (cf. II B 2a, II C 1, II C 2), e.g. KCl, potassium chloride, $NaClO_2$, sodium chlorite, $(NH_4)_2SO_4$, ammonium sulfate, $KClO_4$, potassium perchlorate, $Na_2S_2O_8$, sodium peroxodisulfate.

2) Salts which contain acidic hydrogen are named by adding the word "hydrogen" immediately before the name of the anion (cf. II B 2a), e.g. $NaHCO_3$, sodium hydrogencarbonate, LiH_2PO_4, lithium dihydrogenphosphate.

3) Double salts, triple salts, etc. In these names, all cations should be listed before any anions; the names are formed according to the rules for simple salts. With the exception of hydrogen, the cations and the anions are listed in alphabetical order. The stoichiometric proportions of the components are indicated by numerical words. The number of complex anions is indicated by the prefixes "bis", "tris", etc., because "di", "tri", etc. are used in the names for polyatomic anions, e.g. $Na_6ClF(SO_4)_2$, hexasodium chloride phosphide bis(sulfate).

4) The names of the oxide and hydroxide salts should be formed like the names of double salts in which O^{2-} and HO^- anions are present, e.g. BiClO, bismuth chloride oxide, $CuCl_2 \cdot 3\ Cu(OH)_2$ or $Cu_2Cl(OH)_3$, dicopper chloride trihydroxide.

5) Double oxides and hydroxides. Expressions such as "mixed oxide" or "mixed hydroxide" are no longer recommended; instead the compounds should be named as double, triple, etc. oxides, e.g. Cr_2CuO_4, chromium(III) copper(II) oxide. If needed, the structure is indicated in parentheses and in italics after the name, e.g. $NaNbO_3$, sodium niobium trioxide (*perovskite type*). If the type name is identical to the mineral name, the word "type" is omitted, e.g. $FeTiO_3$, iron(II) titanium trioxide (illmenite).

G) ***Coordination compounds and complexes***. Any compound which is formed by addition of one or more molecules or ions to one or more central atoms can be treated as a complex compound. Thus many simple and well-known compounds can be named by the same rules as the complex compounds.

1) Formulae and names for coordination compounds in general. The characteristic atom of the complex is called the *central atom*, and atoms, radicals or molecules bound to it are called *ligands*. In the names of complex compounds, the central atom is named after the ligands (in the formula, the symbol of the central atom preceeds those for the anionic, neutral or cation ligands). The oxidation state of the central atom can be given by a Stock number, the ionic charge according to Ewens-Basset, or it is to be inferred from the stoichiometric proportions of the components (cf. II B 2b). Complex anions are given the ending "-ate"; the names of complex cations and neutral components have no special endings. The ligands are listed in alphabetical order.

2) Names for ligands. The names of anionic ligands are derived from the names of the corresponding anions to which the suffix "-o" is added. Exceptions are the monoatomic halide ions, oxide, hydroxide, peroxide, hydroperoxide, sulfide, hydrosulfide, cyanide,

methanolate and methanethiolate ions, e.g. Cl^- is chloro, not chlorido, O^{2-} is oxo, not oxido, HS is mercapto, not hydrogensulfido. Examples: $Na[Au(OH)_4]$ is sodium tetrahydroxoaurate(III), $K[CrF_4O]$ is potassium tetrafluorooxochromate(1-). The names of coordinately bound neutral molecules or cations are not changed, except for water and ammonia; these are called aqua and ammine in complexes, e.g. $[Cr(H_2O)_6)]Cl_3$, hexaaquachromium(3+) chloride, $[Cu(NH_3)_4]SO_4$, tetraamminecopper(II) sulfate. The radicals NO and CO are treated as neutral ligands and indicated by nitrosyl and carbonyl, respectively, if they are bound directly to a metal atom, e.g. $Na[Co(CO)_4]$, sodium tetracarbonylcobaltate(-I). For ligands with several possible ways to form coordinate bonds, the italicized symbols of the coordinated atoms are added to the names of the ligands, e.g. $-NH_2-CH_2-COO^-$-*O,N*-; $(H_3C-CO)_2CH$-pentane-2,4-dionato-C^{3-}. In names of complexes with unsaturated molecules or groups, the binding of all atoms to the central atom is indicated by η before the name of the ligand, e.g.$[Ni(C_5H_5)_2]$, bis(η-cyclopentadienyl)nickel. Binding via a single atom is indicated by σ, e.g. di(η-cyclopentadienyl)(σ-cyclopentadienyl)nitrosylmolybdenum. Cyclopentadienyl complexes in general are called *metallocenes*; the trivial name for bis(η-cyclopentadienyl)iron, $Fe(C_5H_5)_2$, is ferrocene. Introduction of other trivial names like cymantrene and cytizel should be avoided. Di- and poly-nuclear compounds with bridge-forming groups are indicated by placing μ before the bridge-forming group, e.g. $[(CO)_3Fe(CO)_3Fe(CO)_3]$, tri-μ-carbonyl-bis(tricarbonyliron). Extended structures are named analogously, e.g. the ion $[...Cl-CuCl_2-Cl-CuCl_2...]^{n-}$, catena-μ-chloro-dichlorocuprate(II). Di- and poly-nuclear compounds with metal-metal bonds are indicated by the symbols of the directly bound metal ions appended to the names, e.g. $(CO)_3Co(CO)_2Co(CO)_3$, di-μ-carbonyl bis(tricarbonylcobalt)(Co-Co). The geometric arrangement of aggregates of identical atoms (clusters) is indicated by the prefixes "triangulo-", "quadro-", "tetrahedro-", "octahedro-" etc.

H) ***Addition compounds***. Since according to the IUPAC rules, the names of anions end in "-ate", this ending should not be used for addition compounds, with the exception of the term "hydrate" for a compound which contains water of crystallization. The names of addition compounds are formed by connecting the names of individual compounds by spaced hyphens and indicating the number of molecules after the name by Arabic numerals separated by the solidus, e.g. $CaCl_2 \cdot 6\ H_2O$ is calcium chloride hexahydrate or calcium chloride-water (1/6); $AlCl_3 \cdot 4\ C_2H_5OH$ is aluminum chloride-ethanol (1/4).

I) ***Crystalline phases with variable composition***. Intermediate crystalline phases of a two-component (or multi-component) system which can vary markedly in composition are called berthollides. It is preferable to use Formulae (see) for berthollides and solid solutions. If it is necessary to use words, the direction of the deviation is given, e.g. iron(II) sulfide (iron deficient) or molybdanum dicarbide (excess of carbon).

J) ***Polymorphism***. The usage of Greek letters or Roman numerals to indicate polymorphic modifications, such as α-iron or ice-I, continues for trivial names. For chemical purposes, the crystal system is indicated after the name or formula, e.g. zinc sulfide(cub.).

III) ***N. of organic chemistry***. Due to the enormous variety of possible structures, the systematic N. of organic chemistry is much more complex than that of inorganic chemistry. The oldest names, which have in many cases been retained to the present, are purely *trivial names*. They are often indicative of the source (lactic acid, citric acid), characteristics (methyl orange, picric acid), use (tannic acid) or the name of the discoverer (Michler's ketone), or they may be purely arbitrary (barbituric acid). Trivial names are still being invented for compounds of unknown structure, or for those whose systematic names are cumbersome (see I, General).

Systematic names are made up of special syllables, and often include alphabetical or numerical symbols, and alphanumeric prefixes, e.g. pentane, oxazole. In ***semisystematic names***, only part of the name has systematic significance, e.g. benzene, cholesterol. Most names in organic chemistry are semisystematic.

A) There are several ***methods of forming names***. 1) *Substitutive names*: those in which hydrogen is replaced by a group or by another element. For some substituents there are no suffixes, only prefixes (e.g. chloro, nitro; Table 2). Examples: ethanal, nitrobenzene. 2) *Replacement names*: for compounds in which C, CH or CH_2 is replaced by a hetero atom, a prefix ending in "a" is used ("a" names; Table 8a), e.g. 2,7,9-triazaphenanthrene; for compounds in which an oxygen is replaced by sulfur, selenium or tellurium, the prefix "thio-", "seleno-" or "telluro-" is used, e.g. dithioacetic acid. 3) *Subtractive names* are used for compounds in which specified atoms have been removed; these are indicated by the suffixes "-ene" and "-yne" and by prefixes such as "anhydro-", "dehydro-", "deoxy-" or "nor-". 4) *Additive names* signify addition between molecules or atoms, e.g. styrene oxide. 5) *Conjunctive names* are formed by placing together the names of two molecules, which are understood to be linked by loss of one hydrogen atom from each; e.g. napthaleneacetic acid. 6) *Radicofunctional names* are composed of the name of a radical and the name of a functional class, e.g. acetyl chloride, ethyl alcohol. 7) *Fusion names* for cyclic systems are formed by use of a linking "o" between the names of two ring systems, denoting that the two systems are fused by two or more common atoms, e.g. benzofuran. 8) *Hantzsch-Widman names* are formed from a prefix or prefixes to denote one or more hetero atoms and a suffix to denote ring size and degree of unsaturation (Table 9) e.g. oxirane, 1,3-thiazole.

Special N. have been developed for some classes of structures or compounds (structures built up of identical units, free radicals, ions, isotope-modified compounds, polymers, stereoisomers, etc.). The principle known as nodal nomenclature treats the structural formula as a graph. Special N. have arisen historically for many groups of natural products (e.g. alkaloids, amino acids, cyclitols, carbohydrates, nucleic acids, nucleotides, peptides, steroids, terpenes, vitamins, the use of abbreviations and symbols), and these are retained.

The following sections are based on the definitive

and preliminary rules published by the IUPAC Commission on the N. of Organic Chemistry in 1979. For the most part, substitutive N. is described.

B) ***Principle of name construction***. Substitutive names are constructed in the following steps: 1) determination of the principle group. Characteristic groups can be indicated either by prefixes or by suffixes (Table 1), although there are some which can be indicated only by prefixes (Table 2). If there is more than one characteristic group, the one which is listed first in Table 1 is chosen as the principal group. 2) Determination of the parent compound. The parent compound is that part of the compound (hydrocarbon chain or ring system) which carries the principal group. There is a series of selection rules which are followed if several parts of the compound could be chosen, or there is no principal group. The most important are that the parent compound is a) the system with the largest number of principal groups; b) a ring system; b_1) a heterocycle; b_2) a carbon cycle; c) the longest chain. 3) Naming the parent compound and the principal group. The parent compound is named according to the rules listed in sections C to F. The principal group, of which several may be present, is indicated by a suffix (Table 1) after the name of the parent compound. 4) Naming of substituents. Characteristic groups other than the principle group, and acyclic, carbocyclic and heterocyclic radicals can only be indicated by substitutive prefixes before the name of the parent compound.

Table 1. Prefixes and suffixes used for characteristic groups, in the order of decreasing priority (substitutive nomenclature)

Functional class	Formula of the characteristic group[1]	Prefix	Suffix
Anions		-ato-[2]	-ate[2]
Cations		-ido-[3]	-ide[3]
		-(on)io- -onia- (replacement nomenclature)	-(on)ium
Carboxylic acids	$-COOH$	carboxy-	-carboxylic acid
	$-(C)OOH$	–	-oic acid
Peroxycarboxylic acids	$-C(=O)-OOH$	peroxy-	-peroxycarboxylic acid
	$(R'-C)(=O)-OOH$ [4]	peroxy-	(root for $R-CH_3$[4]) + -peroxycarboxylic acid
Thiocarboxylic acids	$-C(=O)-SH$	thiocarboxy-	-thiolcarboxylic acid
	$-C(=S)-OH$	–	-thiocarboxylic acid
	$-C(=S)-SH$	cithiocarboxy-	-dithiocarboxylic acid
	$-(C)(=O)-SH$	–	thiolic acid
	$-(C)(=S)-OH$	–	thionic acid
	$-(C)(=S)-SH$	–	dithioic acid
Sulfonic acids	$-SO_3H$	sulfo-	-sulfonic acid
Sulfinic acids	$-SO_2H$	sulfino-	sulfinic acid
Sulfenic acids	$-SOH$	sulfeno-	-sulfenic acid
Phosphoric acids	$-PO(-OH)_2$	phosphono-[5]	-phosphonic acid
Phosphinic acids	$=PO(-OH)$	phosphinico-	-phosphinic acid
Phosphonous acids	$-P(-OH)_2$		-phosphonous acid
Esters of carboxylic acids		$-(R')-COOR$[4] -oxycarboxy-	name for R-[4] + name for R′- (+ root for R′-H[4]) + -carboxylate[6] or (root for R′-H[4] +) -carboxylic acid + radical name for R-[4] + ester
	$-(R'-C)OOR$[4]	–	radical name for R-[4] (+ trivial root for $R'-CH_3$[4] +) -ate or radical name for R-[4] (+ root for R′-H[4] +) -oate or (root for $R'-CH_3$[4] +) oic acid + radical name for R-[4]) + ester

Functional class	Formula of the characteristic group[1]	Prefix	Suffix
Carbonyl halides	$-C(=O)Halogen$	haloformyl- –	-carbonyl halide oyl halide oxyhalide
Amides	$-(C)(=O)Halogen$		
	$-C(=O)NH_2$	carbamoyl- –	-carboxamide (trivial root of R′–COOH[4] +) -amide
	$(R'-C)(=O)NH_2$ 4)		
Hydrazides	$-C(=O)NH-NH_2$	hydrazinocarbonyl-	-carbohydrazide
	$-(C)(=O)NH-NH_2$	–	-ohydrazide
Amidines	$-C(=NH)NH_2$	amidino- or carbamimidoyl-	-carbamidine
	$-(C)(=NH)NH_2$	–	-amidine
Sulfonamides	$-SO_2NH_2$	sulfamoyl-	-sulfonamide
Nitriles	$-C\equiv N$ $-(C)\equiv N$	cyano- –	-carbonitrile -nitrile
Aldehydes	$-CHO$ $-(C)HO$	formyl- oxo-	carbaldehyde -al
Thioaldehydes	$-CHS$ $-(C)HS$	thiformyl- -thioxo	-carbothialdehyde -thial
Ketones	$>(C)=O$	oxo-	-ol
Alcohols	$-OH$	hydroxy-	-ol
Phenols	$-OH$	hydroxy-	-thiol
Thiols	$-SH$	mercapto-	-thiol
Hydroperoxides	$-O-OH$	hydroperoxy-	–
Amines	$-NH_2$	amino-	-amine
Imines	$=NH$	imino-	-imine
Phosphines	$-PH_2$	phosphino-	-phosphine
Ethers	$-OR^{4)}$	radical name for R-[4] + -oxy-	–
Sulfides (thioethers)	$-SR^{4)}$	radical name for R-[4] + -thio-	–
Peroxides	$-O-OR^{4)}$	radical name for R-[4] + -dioxy-	–

[1] The radicals R′ and C-atoms enclosed in parentheses are not denoted by the prefix or suffix.
[2] For anions of acids, alcohols or phenols.
[3] For anions which arise by loss of protons from carbon atoms.
[4] R- and R′-: acyclic, carboxyclic or heterocyclic radicals.
[5] This prefix may only be used for the unsubstituted radical. Names for complex radicals are derived from the radical name phosphoryl for P(O)≡. The general prefix for use in biochemical nomenclature is phospho-, if the bond is to a heteroatom and not to carbon.
[6] When used in names for salts of carboxylic acids, the characteristic group denoted by -carboxylate has priority over the cation.

Names for complex substituents are formed according to the same rules which apply to compounds, except that characteristic groups may only be indicated by prefixes (Table 2). If identical substituents occur in more than one place, this is indicated by numerical prefixes. 5) Establishment of numbering. Numbering is assigned on the principle of lowest possible locants. In series of locants, the first difference is decisive, i.e. 2,3,8 is lower than 2,4,4. *Acyclic hydrocarbons* are numbered from one end to the other. The numbering of *cyclic compounds* is indicated for the individual systems. If there are several possibilities for numbering the parent compound, the atoms are numbered so that the lowest locant is assigned to the characteristic groups in the following order of priority: a) indicated hydrogen; b) the principal group; c) multiple bonds (double bonds have priority over triple bonds); c) substituents named as prefixes, including the hydro prefix; e) the first prefix listed in alphabetical order. In *substituents*, the site of bonding to the parent compound is given the lowest possible locant, after which the rules for numbering parent compounds are applied. In acyclic compounds the site of attachment always has the number 1. In cyclic radicals, the indi-

Table 2. Characteristic groups cited only as prefixes in substitutive nomenclature.

Characteristic group	Prefix
$-Br$	bromo-
$-Cl$	chloro-
$-ClO$	chlorosyl-
$-ClO_2$	chloryl-
$-ClO_3$	perchloryl-
$-F$	fluoro-
$-I$	iodo-
$-IO$	iodosyl-
$-IO_2$	iodyl-
$-I(OH)_2$	dihydroxyiodo-
$-IX_2$	e.g. $X = -COCH_3$, diacetoxyiodo-
$=N_2$	diazo-
$-N_3$	azido-
$-NO$	nitroso-
$-NO_2$	nitro-
$=N(O)OH$	*aci*-nitro-
$-OR$	R-oxy-
$-SR$	R-thio- (similarly R-seleno and R-telluro)

cated hydrogen has priority over the bond site. 6) Formation of the entire name. The suffix for the principal group comes after the name of the parent compound; prefixes come before it. Locants are given before the associated name segment. The order of prefixes is alphabetical, except for those prefixes which cannot be separated from the segment (indicating ring formation, ring cleavage, changes in size, fused rings, "a" terms, isomers, indicated hydrogen, bridge groups and hydro-). Prefixes for simple substituents are ordered according to the letters in the names; multiplicative affixes are not counted (e.g. 2,3,6,7-tetra*b*romo-1,4-di*m*ethylnaphthalene). Prefixes for complex substituents are ordered according to the letters in their complete names. Within the complex prefix, the names of the secondary substituents are also arranged alphabetically (e.g. 1-*c*hloro-3-(*e*thylmethylamino)-2,7-di*m*ethoxyfluorene-4-carboxylic acid, 1-(2-hydroxy*e*thyl)-8-(2-hydroxy*p*ropyl)-1-naphthoic acid). If several prefixes consist of the same root words, then the first one to be listed is the one which has the lowest locant at the first site at which there is a difference (e.g. 2-(2-nitrophenyl)-7-(3-nitrophenyl)naphthalene).

C) ***Aliphatic (acyclic) hydrocarbons***. The saturated hydrocarbons with the general formula C_nH_{2n+2} are called *alkanes* (*paraffins, limit hydrocarbons*). The first 4 unbranched members of the series ($n = 1$ to 4) are called methane, ethane, propane and butane. All other names are formed from the root of a Greek or Latin numerical word for n and the ending "-ane", e.g. for $n = 12$, dodecane. The names of unsaturated aliphatic hydrocarbons, *alkenes* (*olefins, ethylene hydrocarbons*) and *alkynes* (*acetylenes, acetylene hydrocarbons*) are formed from the root of the corresponding saturated compound, the locants for the multiple bonds (cf. III B 5) and the endings "-ene", "-adiene" (1 or 2 double bonds, respectively), "-yne", "-adiyne" (1 or 2 triple bonds), "-enyne" (1 double and 1 triple bond), etc.. Numbers as low as possible are given to double and triple bonds, even though this may at times give "-yne" a lower number than "-ene", e.g. $H_3C\text{-}CH{=}CH\text{-}CH_3$, but-2-ene; $H_3C\text{-}CH{=}CH\text{-}C{\equiv}CH$, pent-3-ene-1-yne. *Saturated hydrocarbon radicals* are given the root of the corresponding alkane and the endings "-yl-" (monovalent radical), "-ylidene-" (bivalent terminal carbon) or "-ylidyne-" (trivalent terminal carbon). The names of bivalent radicals derived from normal alkanes by removal of a hydrogen atom from each of the two terminal carbon atoms of the chain are given the ending "-ylene-". For *saturated* and *unsaturated radicals* the endings "-diylidene-" (terminal bivalent carbon atom), "-diylidyne-" (terminal trivalent carbon), "triyl-", "tetrayl-", etc. (three or more disjunct monovalent carbon atoms, two of them terminal) are applied to the names of the corresponding hydrocarbons. Examples: $\text{-}CH_2\text{-}CH_2\text{-}CH_2\text{-}$, propane-1,3-diyl-; $=C{=}C{=}C{=}$, propadiendylidene-. Names for branched-chain hydrocarbons are constructed by listing the name of the radicals before the names of the main chains. Numbering is described in sect. III B 5. The following *trivial names* are retained for hydrocarbons: $H_3C\text{-}CH\text{-}(CH_3)_2$, isobutane; $H_3C\text{-}CH_2\text{-}CH(CH_3)_2$, isopentane; $C(CH_3)_4$, neopentane; $H_3C\text{-}CH_2\text{-}CH_2\text{-}CH(CH_3)_2$, isohexane $H_2C{=}CH_2$, ethylene $H_2C{=}C{=}CH_2$, allene; $HC{\equiv}CH$, acetylene; $H_2C{=}CH\text{-}C(CH_3){=}CH_2$, isoprene. For radicals, $\text{-}CH(CH_3)_2$, isopropyl; $\text{-}H_2C\text{-}CH\text{-}(CH_3)_2$, isobutyl; $\text{-}CH(CH_3)\text{-}CH_2\text{-}CH_3$, *sec*-butyl; $\text{-}C(CH_3)_3$, *tert*-butyl; $\text{-}H_2C\text{-}CH_2\text{-}CH(CH_3)_2$, isopentyl; $\text{-}CH_2\text{-}C(CH_3)_3$, neopentyl; $\text{-}C(CH_3)_2\text{-}CH_2\text{-}CH_3$, *tert*-pentyl; $\text{-}H_2C\text{-}CH_2\text{-}CH_2\text{-}CH(CH_3)_2$, isohexyl; $\text{-}CH{=}CH_2$, vinyl; $\text{-}CH_2\text{-}CH{=}CH_2$, allyl; $\text{-}C(CH_3){=}CH_2$, isopropenyl; $=CH_2$, methylene; $\text{-}C{=}$, methyne-; $=C(CH_3)_2$, isopropylidene; $=C{=}CH_2$, vinylidene; $\text{-}CH{=}CH\text{-}$, vinylene are retained.

D) ***Monocyclic hydrocarbons*** are compounds in which a saturated or unsaturated chain of carbon atoms is closed into a ring. The class names are *cycloalkanes, cycloalkenes*, etc. The names, including those for the radicals, are derived from those of the corresponding acyclic compounds by adding the prefix "cyclo-". Examples:

CH_3 C_2H_5 (4) (3) (5) 1 (2) (6)CH_2

Cyclopentene | 3-Ethyl-2-methyl-cyclohexene | Cyclohexa-2,4-dien-1-yl | Fulvene

The trivial name fulvene is retained. The compounds o-, m- and p-menthanes are Terpenes (see). The names for *aromatic, monocyclic hydrocarbons* are derived from the semitrivial name benzene; the class name (including polycyclic compounds) is *arenes*. Locants for di- tri- and tetrasubstituted products with the same substituents are "o-" or "1,2-" (for ortho-), "m-" or "1,3-" (for meta-); "p-" or "1,4-" (for para-); "vic-" = "1,2,3-" or "1,2,3,4-" (for vicinal-); "asymm-" = "1,2,4-" or "1,2,3,5-" (for assymmetric-); and "symm-" = "1,3,5-" or "1,2,4,5-" (for symmetric-). Example:

CH_3 CH_3 CH_3

vic- or 1,2,3-Trimethylbenzene

The monovalent or bivalent radical of benzene is called "phenyl" or "o-", "m-" or "p-phenylene"; the class names are *aryls* or *arylenes*. The following *trivial names* are retained:

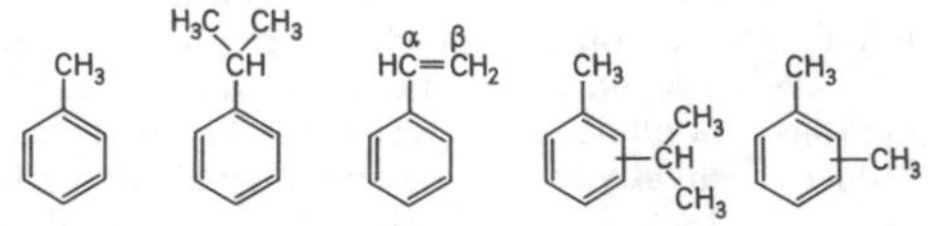

IUPAC: Toluene, Cumene, Styrene, o-, m- or p-Cymene, o-, m- or p-Xylene

Toluol, Cumol, Styrol, o-, m-, p-Cymol, o-, m-, p-Xylol

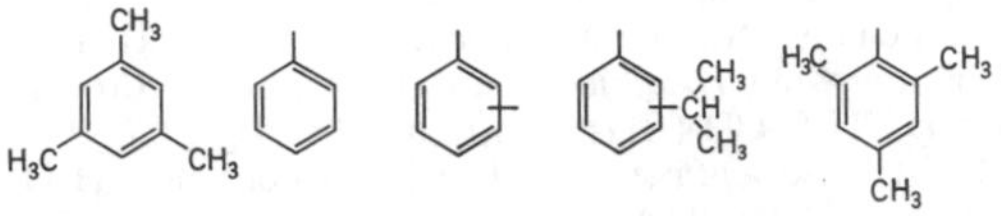

Mesitylene-, Phenyl-, o-, m- or p-Phenylene, o-, m- or p-Cumenyl-, Mesityl-

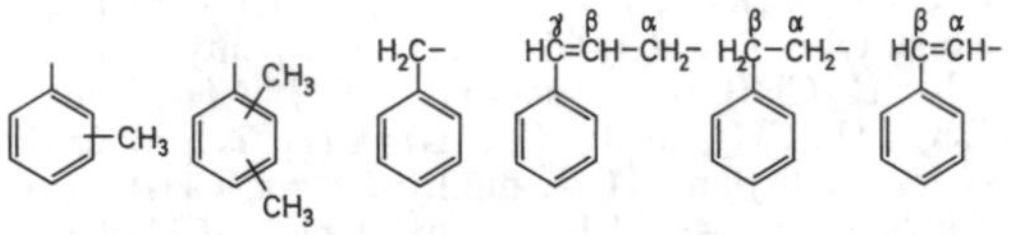

o-, m- or p-Tolyl-, 2,3–,2,4–, 2,5–,2,6–, 3,4– or 3,5–Xylyl-, Benzyl-, Cinnamyl-, Phenethyl-, Styryl-

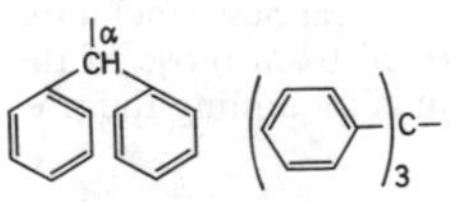

Benzhydryl-, Trityl-

E) ***Polycyclic hydrocarbons***. If two or more rings are directly bonded, they can be classified according to the number of ring atoms common to two rings as *ring sequences* (*diphenyloids*) with no common atoms, *spiro-systems* (*spiranes*) with 1, *ortho-* or *ortho-* and *peri-fused systems* with two neighboring common atoms, and *bridge systems* with two or more common atoms.

1) Ring sequences (diphenyloids) can be named as cycloalkyl- or aryl-substituted cyclic aliphatic or aromatic systems. If two or more identical rings are linked, the prefixes "bi-", "ter-", "quater-", etc. are affixed to the parent or radical name of the corresponding ring. Examples:

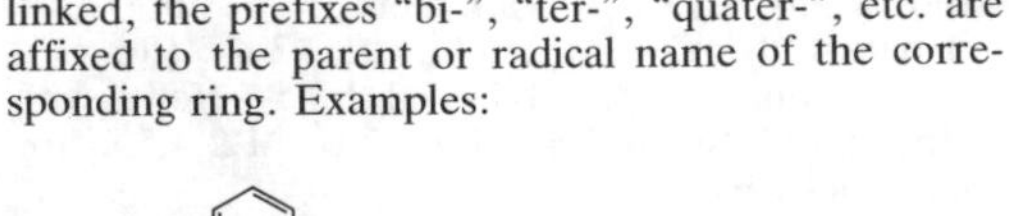

2-Phenylnaphthalene, Tercyclopropane, p-Terphenyl

Ring sequences of benzene rings can only be named with the root "-phenyl".

2) Spiro-systems (spiranes). Spirocyclic systems can be named in two ways: a) the prefix "spiro-", "dispiro-", etc. corresponding to the number of spiro atoms is appended to the name of the hydrocarbon with the same number of C-atoms, and the numbers of carbon atoms linked to the spiro atom in each ring, beginning with the smaller terminal ring, are indicated in brackets placed between the spiro prefix and the hydrocarbon name. b) the names of the individual ring systems are linked by the term "spiro". The numbering according to a) begins in the smaller terminal ring adjacent to the spiro atom and proceeds in the direction which gives the spiro atom the smallest possible locant. When b) is used, the numbering of the individual systems is retained (locants indicated in parentheses, below).

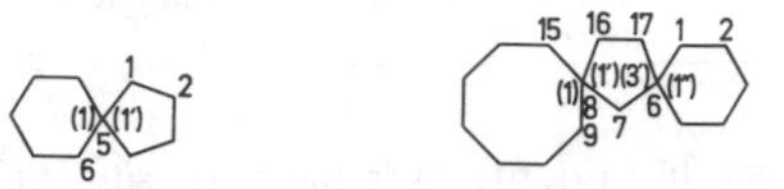

a) Spiro [4.5] decane
b) Cyclohexane-spiro-cyclopentane

a) Dispiro [5.1.7.2] heptadecane
b) Cyclooctane-spiro-cyclopentane-3′-spiro-cyclohexane

3) Ortho- or ortho- and peri-fused systems. Polycyclic compounds with the maximum number of non-cumulative double bonds in which two rings have only two atoms in common are called *ortho-fused*; those in which a ring shares only two atoms with each of two or more rings in a sequence of fused rings are called *ortho- and peri-fused*. Trivial names with the numbering indicated are allowed for 35 polycycles (Table 3). The names for polycycles without allowed trivial names are constructed by taking the name of the largest possible component ring system as the base component and prefixing designations of the other components. The attached components should be as simple as possible. The type of fusion is indicated in square brackets by indicating the locants of the atoms of the fused component and the side of the base component (1,2 = a). The prefixes designating attached components are formed by changing the ending "-ene" of the parent hydrocarbon into "-eno". Exceptions are the following abbreviated prefixes: acenaphtho-, anthra-, benzo-, naphtho-, perylo- and phenanthro-. For monocyclic systems with the maximum number of non-cumulative double bonds, the prefix is formed by adding "a" to the root, e.g. cyclopenta-. The numbering of the fused system is independent of that of the individual systems; the following conventions are observed: the individual rings are drawn as follows;

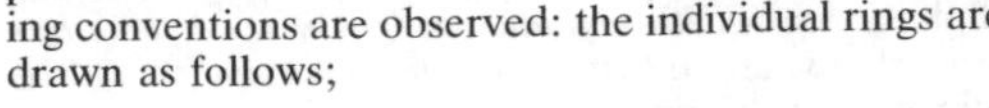

or , , or , , or , .

and the polycyclic system is oriented so that the greatest number of rings are in a horizontal row and a maximum number of rings are above and to the right of the horizontal row. If two or more orientations meet these requirements, the one is chosen which has as few rings as possible in the lower left quadrant.

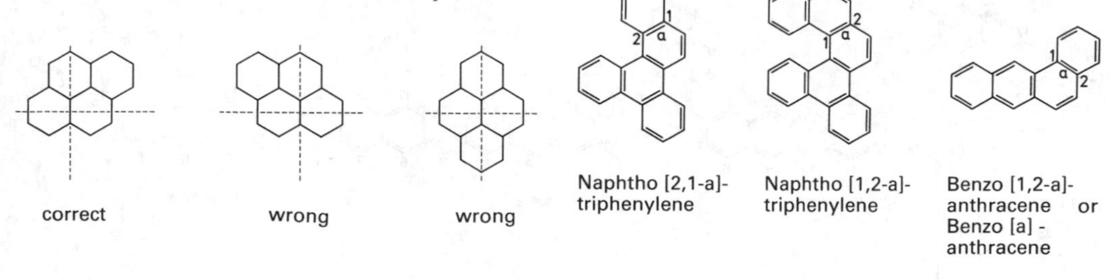

The oriented system is numbered clockwise, beginning with the carbon atom which is the farthest to the left in the upper right ring and omitting atoms common to two or more rings. Atoms common to two or more rings are designated by adding the letters "a", "b", "c", etc. to the number of the position immediately preceeding.

Compounds of more than 4 linear fused benzene rings are named by a numerical prefix and the suffix "-acene", e.g. pentacene.

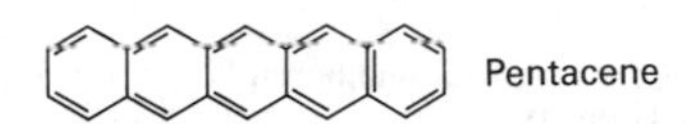

Table 3. Allowed trivial names for *ortho*- and *ortho*- and *peri*-fused hydrocarbons

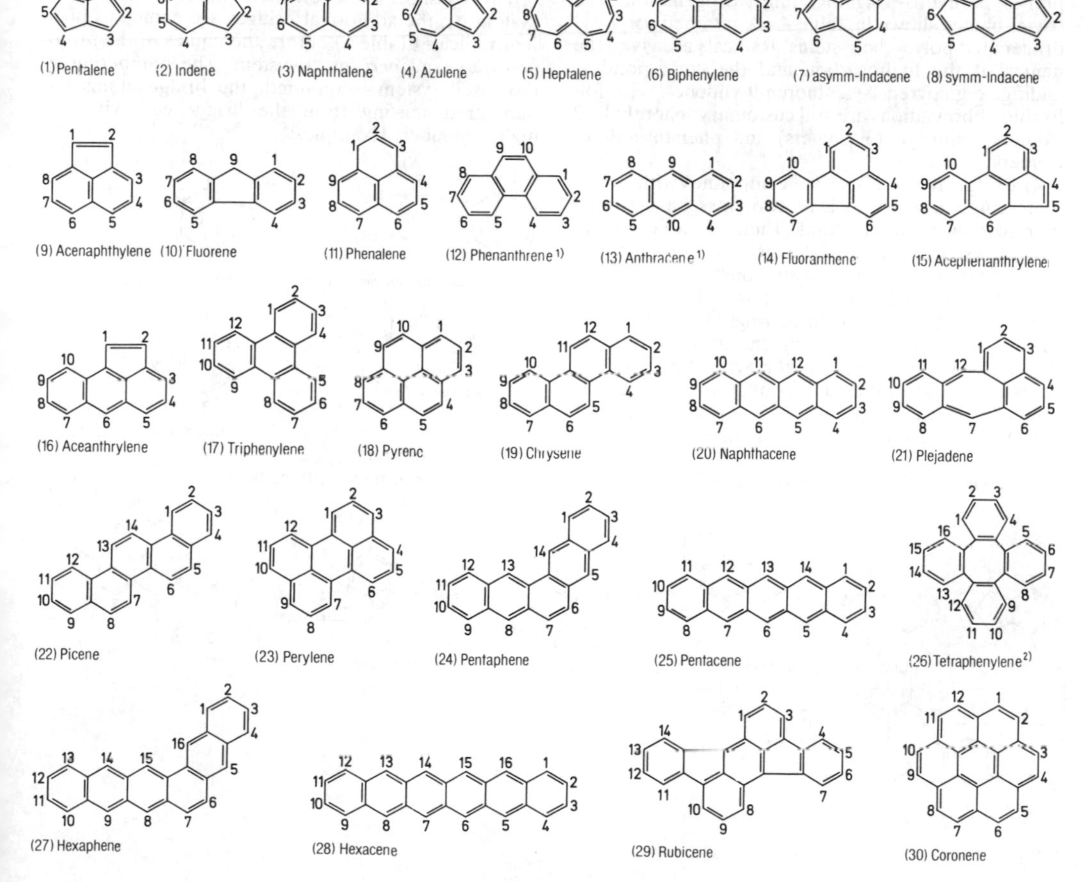

(31) Trinaphthylene[2]

(32) Heptaphene

(33) Heptacene

(34) Pyranthrene

(35) Ovalene

[1] Exception to the rules for systematic numbering
[2] Only for the indicated isomers

The names of *ortho-* and *ortho-* and *peri-*fused systems with fewer than the maximum number of noncumulative double bonds are formed from a prefix "dihydro-", "tetrahydro-" etc. followed by the name of the corresponding fully reduced hydrocarbon. The prefix "perhydro-" signifies full hydrogenation. The trivial names shown in Table 4 are retained for 7 hydrogenated polycyclic systems. Radicals are given the names of the hydrocarbon and the corresponding ending, e.g. pyren-1-yl, fluoren-9-ylidene-. The following abbreviations are still customary: naphthyl- (2 isomers), anthryl- (3 isomers) and phenanthryl- (5 isomers).

4) Bridge hydrocarbons are compounds with two or more rings in which at least two rings have two or more common carbon atoms. There are a number of bridge systems with trivial names among the terpenes. Systematic names may be constructed either a) using the extended von Baeyer nomenclature or b) as *ortho-* or *ortho-* and *peri-*fused rings. In a) the name consists of a prefix indicating the number of rings (bicyclo-, tricyclo-, etc.), a characteristic in square brackets indicating the number of atoms between the bridgeheads (in descending order) and the name for the hydrocarbon with the same number of C-atoms. The numbering begins at a bridgehead and proceeds via the longest possible path back to this atom, then follows the main (longest) bridge. Secondary bridges are then numbered in decreasing order (the positions of the secondary bridges are indicated in the characteristic by superscripts on their carbon numbers). In system b), the additional bridges are named as bivalent radicals (Table 5) before the names of the *ortho-* or *ortho-* and *peri-*fused system. The numbering of the fused system is retained; the bridge atoms are numbered starting from the bridgehead with the higher number. Examples:

Bicyclo[2.2.1]heptane (Trivialname: Norbornane)

Tricyclo[$2.2.1.0^{2,6}$]heptane

4a,8a-Ethanoanthracene

1,4-Dihydro-1,4-propanonaphthalene

Table 4. Allowed trivial names for partially hydrogenated *ortho-* and *ortho-* and *peri-*fused hydrocarbons

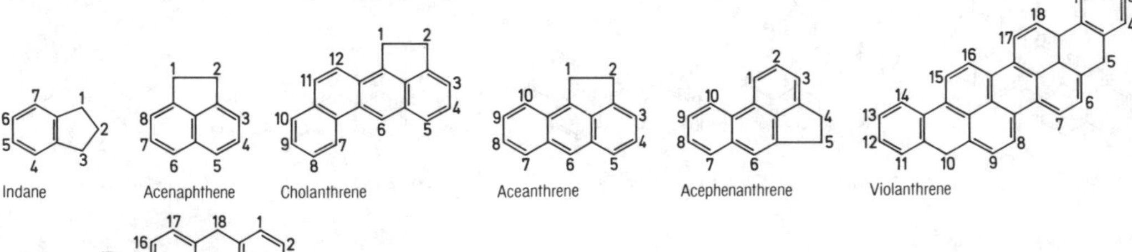

Isovialanthrene

Table 5. Examples for bridge prefixes (including prefixes for hetero bridges)

Azimino	$-N{=}N-NH-$	Ethino-	$-C{=}C-$
Azo-	$-N{=}N-$		
Benzeno- (o-, m-, p-)	$-C_6H_4-$		
Biimino-	$-NH-NH-$	Furano- (in examples 3 and 4)	
Butano-	$-CH_2-CH_2-CH_2-CH_2-$		
Epidioxy-	$-O-O-$		
Epidithio-	$-S-S-$	Imino- (or Epimino-)	$-NH-$
Epimino- (or Imino-)	$-NH-$	Iminoethano-	$-NH-CH_2-CH_2-$
Epithio-	$-S-$	Iminomethano-	$-NH-CH_2-$
Epoxy-	$-O-$	Methano-	$-CH_2-$
Epoxyethano-	$-O-CH_2-CH_2-$	Methaniminomethano-	$-CH_2-NH-CH_2-$
Epoxyethenothio-	$-O-CH{=}CH-S-$	Methanothiomethano-	$-CH_2-S-CH_2-$
Epoxyimino-	$-O-NH-$	Methanoxymethano-	$-CH_2-O-CH_2-$
Epoxymethano-	$-O-CH_2-$	Metheno-	$-CH{=}$
Epoxymethanoxy-	$-O-CH_2-O-$	1-Methyl-ethano-	$-\underset{CH_3}{CH}-CH_2-$
Epoxymetheno-	$-O-CH{=}$		
Epoxynitrilo-	$-O-N{=}$		
Epoxythio-	$-O-S-$	Nitrilo-	$-N{=}$
Epoxythioxy-	$-O-S-O-$	Propano-	$-CH_2-CH_2-CH_2-$
Ethaniminomethano-	$-CH_2-CH_2-NH-CH_2-$	Propeno-	$-CH{=}CH-CH_2-$
Ethano-	$-CH_2-CH_2-$	Thiomcthanothio-	$-S-CH_2-S-$
Etheno-	$-CH{=}CH-$	Thioximino-	$-S-O-NH-$

Table 6. Allowed trivial and semitrivial names for heterocycles with the maximum number of non-cumulative double bonds.

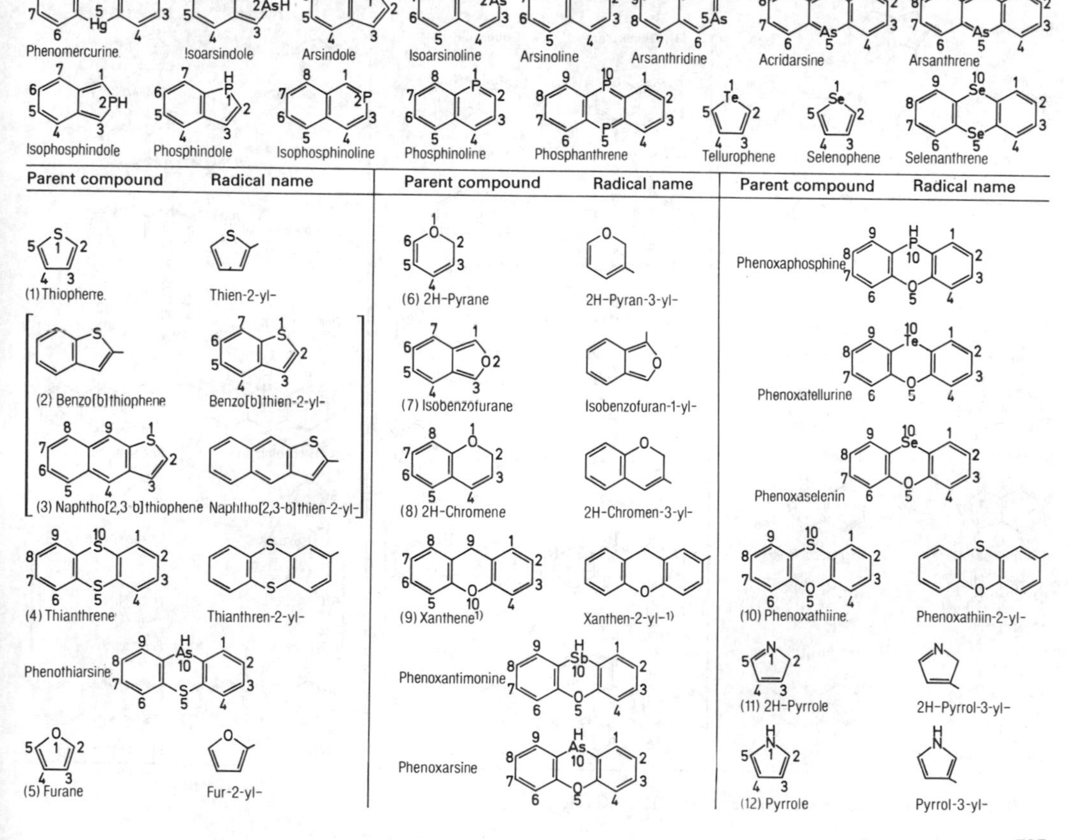

Nomenclature

table 6 cont.

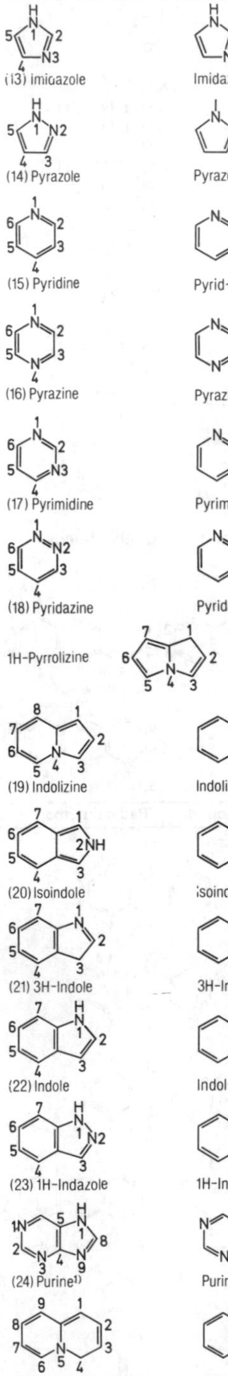

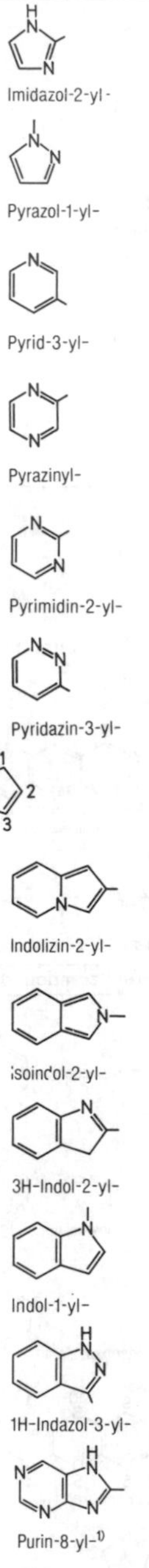

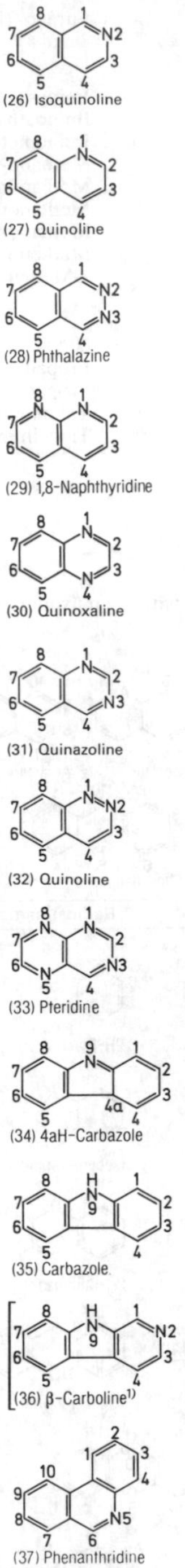

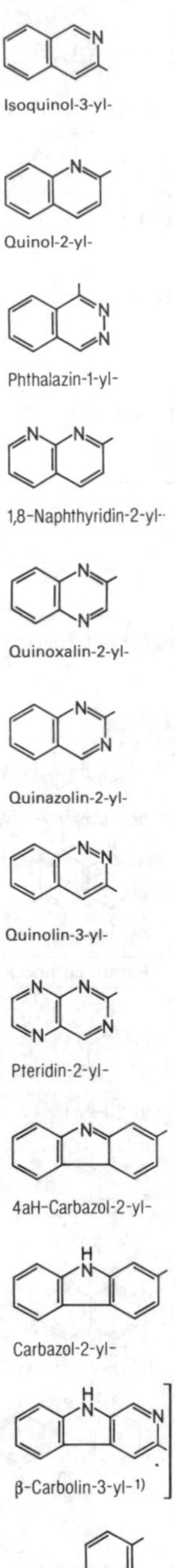

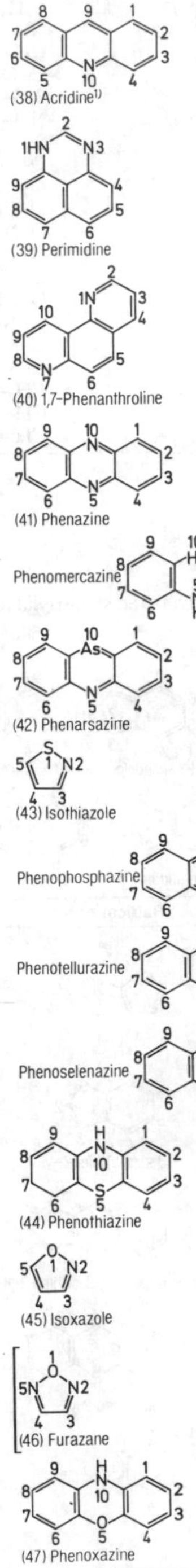

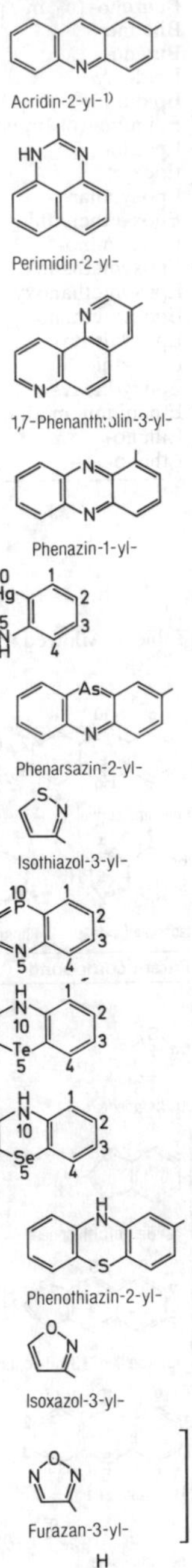

Parent compound	Radical name
(13) Imidazole	Imidazol-2-yl-
(14) Pyrazole	Pyrazol-1-yl-
(15) Pyridine	Pyrid-3-yl-
(16) Pyrazine	Pyrazinyl-
(17) Pyrimidine	Pyrimidin-2-yl-
(18) Pyridazine	Pyridazin-3-yl-
1H-Pyrrolizine	
(19) Indolizine	Indolizin-2-yl-
(20) Isoindole	Isoindol-2-yl-
(21) 3H-Indole	3H-Indol-2-yl-
(22) Indole	Indol-1-yl-
(23) 1H-Indazole	1H-Indazol-3-yl-
(24) Purine[1]	Purin-8-yl-[1]
(25) 4H-Quinolizine	4H-Quinolizin-2-yl-
(26) Isoquinoline	Isoquinol-3-yl-
(27) Quinoline	Quinol-2-yl-
(28) Phthalazine	Phthalazin-1-yl-
(29) 1,8-Naphthyridine	1,8-Naphthyridin-2-yl-
(30) Quinoxaline	Quinoxalin-2-yl-
(31) Quinazoline	Quinazolin-2-yl-
(32) Quinoline	Quinolin-3-yl-
(33) Pteridine	Pteridin-2-yl-
(34) 4aH-Carbazole	4aH-Carbazol-2-yl-
(35) Carbazole	Carbazol-2-yl-
(36) β-Carboline[1]	β-Carbolin-3-yl-[1]
(37) Phenanthridine	Phenanthridin-3-yl-
(38) Acridine[1]	Acridin-2-yl-[1]
(39) Perimidine	Perimidin-2-yl-
(40) 1,7-Phenanthroline	1,7-Phenanthrolin-3-yl-
(41) Phenazine	Phenazin-1-yl-
Phenomercazine	
(42) Phenarsazine	Phenarsazin-2-yl-
(43) Isothiazole	Isothiazol-3-yl-
Phenophosphazine	
Phenotellurazine	
Phenoselenazine	
(44) Phenothiazine	Phenothiazin-2-yl-
(45) Isoxazole	Isoxazol-3-yl-
(46) Furazane	Furazan-3-yl-
(47) Phenoxazine	Phenoxazin-2-yl-

[1] Exception! Numbering does not agree with the IUPAC rules.

Table 7. Allowed trivial and semitrivial names for partially hydrogenated heterocycles (use in fusion names is not allowed)

Parent compound	Radical name	Parent compound	Radical name	Parent compound	Radical name
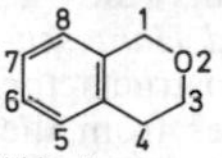 (1) Isochromane	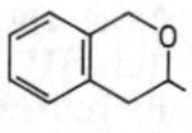 Isochroman-3-yl-	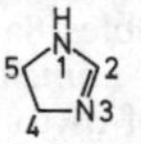 (6) Δ^2-Imidazoline[1]	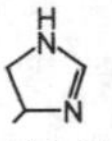 Δ^2-Imidazolin-4-yl-[1]	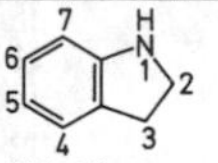 (11) Indoline	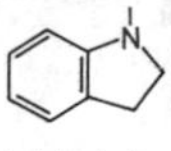 Indolin-1-yl-
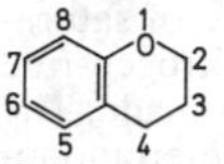 (2) Chromane	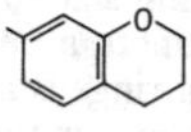 Chroman-7-yl-	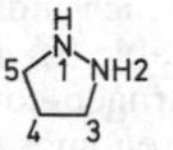 (7) Pyrazolidine	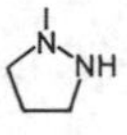 Pyrazolidin-1-yl-	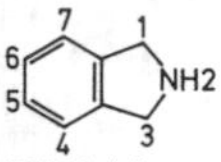 (12) Isoindoline	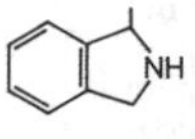 Isoindolin-1-yl-
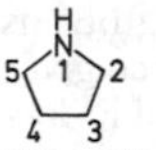 (3) Pyrrolidine	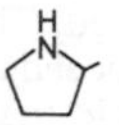 Pyrrolidin-2-yl-	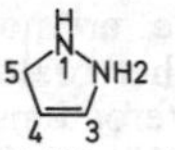 (8) Δ^3-Pyrazoline	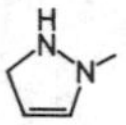 Δ^3-Pyrazolin-2-yl-	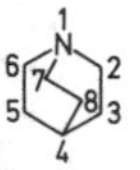 (13) Quinuclidine	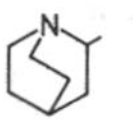 Quinuclidin-2-yl-
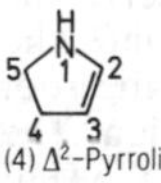 (4) Δ^2-Pyrroline[1]	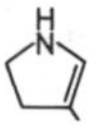 Δ^2-Pyrrolin-3-yl-[1]	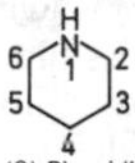 (9) Piperidine	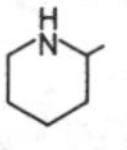 Piperid-2-yl-[2]	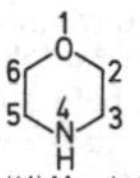	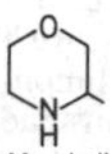
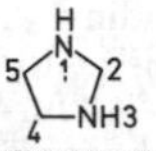 (5) Imidazolidine	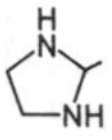 Imidazolidin-2-yl-	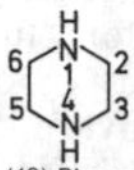 (10) Piperazine	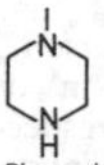 Piperazin-1-yl-	(14) Morpholine	Morpholin-3-yl-[3]

[1] Δ indicates the position of the double bond if locants for heteroatoms are required. The starting point is indicated by a superscript.
[2] When the radical is bound via an N-atom, the name "piperidino-" is used.
[3] When the radical is bound via an N-atom, the name "morpholino-" is used.

Heterocycles. In the broad sense, heterocyclic compounds are ring systems which contain at least two different elements; in the narrower sense, they are structures which contain atoms of elements other than carbon (hetero atoms) as members of a ring containing carbon atoms.

1) Trivial names. The ring systems listed in Table 6 are arranged in order of increasing priority for choice as parent compound in names of fused systems. The partial list with 47 trivial and semitrivial names previously included in the IUPAC rules was extended in the 1979 edition by reference to the "more complete list in Section D" to include another 27 rings. 4 compounds (numbers 2, 3, 36, 46; listed in Table 6 in brackets) were not included because they are not used in fusion names. In the order of priority, compounds 36 and 46 follow compound 14 as 14a and 14b. The trivial and semitrivial names listed in Table 7 are still allowed, but may not be used as base components in naming fused systems.

Table 8. Elements and the "a" terms used in replacement nomenclature in order of decreasing priority

Element	"a"-term	Element	"a"-term	Element	"a"-term	Element	"a"-term	Element	"a"-term
F	Fluora	Sn	Stanna	Ir	Irida	La	Lanthana	U	Urana
Cl	Chlora	Pb	Plumba	Fe	Ferra	Ce	Cera	Np	Neptuna
Br	Broma	B	Bora	Ru	Ruthena	Pr	Praseodyma	Pu	Plutona
I	Ioda	Al	Alumina	Os	Osma	Nd	Neodyma	Am	America
At	Astata	Ga	Galla	Mn	Mangana	Pm	Prometha	Cm	Cura
O	Oxa	In	Inda	Tc	Techneta	Sm	Samara	Bk	Berkela
S	Thia	Tl	Thalla	Re	Rhena	Eu	Europa	Cf	Californa
Se	Selena	Zn	Zinca	Cr	Chroma	Gd	Gadolina	Es	Einsteina
TE	Tellura	Cd	Cadma	Mo	Molybda	Tb	Terba	Fm	Ferma
Po	Polona	Hg	Mercura	W	Wolframa	Dy	Dysprosa	Md	Mendeleva
N	Aza	Cu	Cupra	V	Vanada	Ho	Holma	No	Nobela
P	Phospha	Ag	Argenta	Nb	Nioba	Er	Erba	Lr	Lawrenca
As	Arsa	Au	Aura	Ta	Tantala	Tm	Thula	Be	Berylla
Sb	Stiba	Ni	Nickela	Ti	Titana	Yb	Ytterba	Mg	Magnesa
Bi	Bisma[1]	Pd	Pallada	Zr	Zircona	Lu	Luteta	Ca	Calca
C	Carba	Pt	Platina	Hf	Hafna	Ac	Actina	Sr	Stronta
Si	Sila	Co	Cobalta	Sc	Scanda	Th	Thora	Ba	Bara
Ge	Germa	Rh	Rhoda	Y	Yttra	Pa	Protactina	Ra	Rada

[1] Sometimes written "bismuta".

Table 8b. "a" Terms for heteroatoms in heterocycles

Element	Valence[1]	"a" Term
fluorine	I	fluora
chlorine	I	chlora
bromine	I	broma
iodine	I	ioda
oxygen	II	oxa
sulfur	II	thia
selenium	II	selena
tellurium	II	tellura
nitrogen	III	aza
phosphorus	III	phospha[2]
arsenic	III	arsa[2]
antimony	III	stiba[2]
bismuth	III	bisma[2]
silicon	IV	sila[2]
germanium	IV	germa[2]
tin	IV	stanna[2]
lead	IV	plumba[2]
boron	III	bora[2]
mercury	II	mercura

[1] Other valencies can be indicated by the bond number; λ^n (n = number of bonds) (lambda convention).
[2] Until the 1979 revision, the terms "phosphor-", "arsen-" and "antimon-" had to be used before the ending "-in" or "-ine", instead of "phospha-", "arsa-" and "stiba", respectively. The saturated compounds corresponding to phosphorine and arsenine were called "phosphorinane" and „arsenane". When lead, germanium, silicon or zinc was present as a heteroatom, the ending "-ane" could not be used. In these cases, the prefix "perhydro-" was used with the name of the corresponding unsaturated compound.
[3] Sometimes "bismuta" was used.

Table 9. The expanded Hantzsch-Widman system (1982 revision) for indicating ring size and degree of saturation

Atoms in ring	Unsaturated heterocycles[1]	saturated heterocycles[2]
3	-irene[3]	-irane[3,4]
4	-ete	etane[4]
5	-ole[6]	-olane[4]
6	-in, -inin	ane, -inane
7	-epin	-epane
8	-ocin	-ocane
9	-onon	onane
10	-ecin	-ecane

[1] The term is used if the ring contains the maximum number of non-cumulative double bonds possible (and at least one is possible) with the heteroatom having the valence indicated in Table 8b.
[2] The term is used if the ring system contains no double bonds or none is possible. Until the 1982 revision, saturation for N-containing rings with 6 or more members was indicated by "perhydro-" prefixed to the name of the corresponding unsaturated compound.
[3] Until the revision of 1982, the term for unsaturated, 3-membered, N-containing heterocycles was "-irine"; it may continue to be used for rings which contain only nitrogen.
[4] The traditional terms "-iridine", "-etidine" and "-olidine" are preferred for naming N-containing rings.
[5] Until the 1982 revision, the ending "-etine" was used for N-containing heterocycles with a double bond; the term "-etene" was used for those not containing N.
[6] Until the 1982 revision, N-containing heterocycles with one double bond were indicated by "-oline", and those not containing N, by "-olene".

2) Heteromonocycles. The systematic names can be constructed by one of two principles. Both use names ending in "-a" for the hetero atoms (the "a" term; Tables 8a and 8b). a) The *extended Hantzsch-Widman nomenclature* (cf. III A 8) constructs the names of 3- to 10-membered heterocycles from the "a" term for the heteroatom (Table 8b) and a root word which indicates ring size and degree of saturation (Table 9). The distinction between nitrogen-free and nitrogen-containing rings is no longer made. The special endings for 4- and 5-membered heteromonocycles with one double bond, when more than one are possible, are no longer used. The numbering begins with a heteroatom and proceeds in a way which gives the heteroatoms as a whole the lowest possible locants. When there are various possibilities, the heteroatoms with the highest priorities receive the lowest possible locants. For 6-membered rings the root words "-inine" or "-inane" are used if the root words "-ine" or "-ane" could lead to confusion. The following short terms for radicals are still allowed: "furyl-" for "furanyl-", "pyridyl-" for "pyridinyl-", "piperidyl-" for "piperidinyl-", "thienyl-" for "thiophenyl-", "furfuryl-" for "fur-2-methyl-", "fur-furyliden-" for "fur-2-methylene-", "thenylidene-" for "thienylmethylidine-". The names "piperidino-" and "morpholino-" are preferred to the radical names "piperid-1-yl-" and "morpholin-4-yl-".

b) *Replacement nomenclature*. The names are constructed from the "a" term for the heteroatom (Table 8a) and the name of the corresponding hydrocarbon (obtained by the formal replacement of each heteroatom by

$$>CH_2,\ \geqslant CH \text{ or } -\overset{|}{\underset{|}{C}}-$$

depending on its valency – 2, 3 or 4. Examples:

H_2Si ring: a) Δ^2-Silolene b) Silacyclopent-2-ene

H Si ring: a) Siline b) Silabenzene

3) Heteropolycycles. For some heteropolycycles there are trivial names (Tables 6 and 7). The systematic names for a) *ring sequences* (cf. III E 1), b) *spiro systems* (cf. III E 2), c) *ortho-* and *ortho-* and *peri-*fused heterocycles (cf. III E 3) and d) *bridge systems* (cf. III E 4b) can include allowed trivial names for heterocycles. Short names for fused rings are "furo-" for "furano-", "thieno-" for "thiopheno-", "imidazo-" for "imidazolo-", "pyrido-" for "pyridino-", "pyrimido-" for "pyrimidino-", "quino-" for "quinolino-", "isoquino-" for "isoquinolino-". Alternatively, heteropolycycles can be named by replacement, starting from the names of the corresponding hydrocarbons (cf. III E 1, 2 and 4). If the corresponding hydrocarbon does not contain the maximum number of non-cumulative double bonds but does have a trivial name (Table 4), the replacement name can be based on this trivial name. If this is not possible, the compound produced by introducing the heteroatom is considered the heterocyclic parent compound. The hydrocarbon from which it is for-

mally derived is named as the form with the maximum number of non-cumulative double bonds.

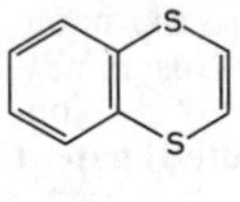

derived from

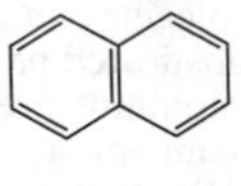

In system c), the heteroatoms are given the lowest possible locants. A heteroatom at a site of fusion is counted in the numbering around the ring. In systems a), b) and d), the numbering of the basic system is retained. Examples:

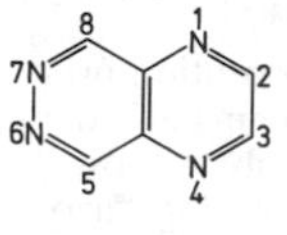

Pyrazino [2,3-d] pyridazine

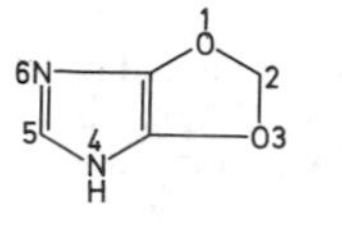

4H-1,3-Dioxolo [4,5-d] imidazole

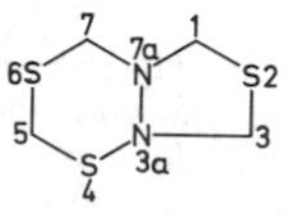

2,4,6-Trithia-3a,7a-diazanidene

4) Treatment of variable valency in organic chemistry (lambda convention). The distinction between heteroatoms which can occur in two or more valencies is unified by the *lambda convention*, according to which the non-standard valency is indicated by λ^n (*n* is the bond number of a skeleton atom) in front of the name of the parent hydride. The symbol δ^c (*c* is the number of double bonds) has been suggested for indication of the number of cumulative double bonds. Examples:

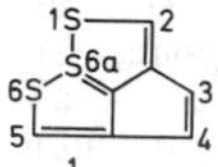

1,6,6aλ^4-Trithiacyclopenta [cd] pentalene

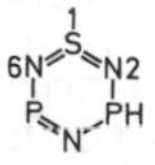

1$\lambda^4\delta^2$,2,4,6,3,5-Thiatriazadiphosphorine

G) ***Functional derivatives***. The various classes of compounds in organic chemistry are indicated by prefixes and suffixes for the characteristic groups (Table 1) appended to the name of the basic compound. Some classes of compound can only be named by functional class names in the radicofunctional system of nomenclature. The nomenclature of *organoelement compounds* (III G 14) is a border area between inorganic and organic chemistry. Compounds may be named either by the substitutive or the additive principle; rules for naming of individual compounds within a class can only be sketchily indicated in the following.

1) Halogen derivatives can only be indicated by prefixes in substitutive nomenclature, e.g. H_3C-Br, bromomethane (radicofunctional name: methyl bromide). The allowed trivial names are: HCF_3, fluoroform; $HCCl_3$, chloroform; HCI_3, iodoform; $COCl_2$, phosgene; $CSCl_2$, thiophosgene.

2) Alcohols and phenols are indicated by the substitutive suffix "-ol". In radicofunctional nomenclature, the class name "alcohol" is used; the term "carbinol" is no longer customary. Examples of trivial names are: glycerol (glycerin), geraniol, phytol, pinacol, glucitol, menthol, borneol, phenol, cresol (3 isomers), xylenol (6 isomers), cavarcrol, thymol, pyrocatechol, resorcinol, hydroquinol, pyrogallol, picric acid, styphnic acid. Names for groups bound via oxygen are formed from the root name and the suffix "-oxy-". The following short names are retained for radicals: "methoxy-" for "methyloxy-", "ethoxy-" for "ethyloxy-", "propoxy-" for "propyloxy-", "isopropoxy-" for "isopropyloxy-", "butoxy-" for "butyloxy-", "isobutoxy-" for "isobutyloxy-", "*sec*-butoxy-" for "*sec*-butyloxy-", "*tert*-butoxy-" for "*tert*-butyloxy-" and "phenoxy-" for "phenyloxy-".

3) Ethers are named as alkoxy/aryloxy-substituted hydrocarbons or heterocycles in substitutive N.; the radicofunctional class name is "ether", e.g. H_5C_2-O-CH_3, methoxyethane (radicofunctional name, ethyl methyl ether). Examples of trivial names are: anethol, anisol, eugenol, guaiacol, phenetol, veratrol.

4) Aldehydes. The substitutive name is formed from the corresponding root name and the suffix "-al". The -CHO group directly bound to a cyclic compound is called "carbaldehyde". The trivial names are derived from those of the corresponding acids by substitution of "-aldehyde" for "-ic acid" or "-oic acid". Examples are acrylaldehyde, cinnamaldehyde. Other examples of trivial names are citral, vanillin, piperonal.

5) Ketones are indicated by the suffix "-one" on the root name. In radicofunctional nomenclature, the functional class name "ketone" is used for the -CO-group, together with the names of the radicals bound to it. Aromatic diketones are called *quinones*. Examples: H_5C_2-CO-CH_3, butan-2-one (radicofunctional name, ethyl methyl ketone). Some examples of trivial names are chalcone, deoxybenzoin, benzil. For some N-containing cyclic ketones, the following short names are allowed in addition to the systematic names: pyrid-2-one, pyrid-4-one, quinol-2-one, quinol-4-one, isoquinol-1-one, acrid-9-one, oxazol-4-one, pyrazol-4-one, pyrazol-5-one, isoxazol-4-one, thiazol-4-one. If there is no chance of confusion, the ending "-idinone" in the names of completely saturated, N-containing ketones may be contracted to "-idone", e.g. piperid-4-one, thiazolid-4-one. A number of classes derived from carbonyl compounds are summarized in Table 10.

6) Carboxylic acids. The conversion of a -CH_3 group into a -COOH group is expressed by placing the suffix "-oic acid" on the root of the original compound. A -COOH group attached to a parent compound is indicated by the suffix "-carboxylic acid". The following are the only permitted trivial names: a) *for saturated aliphatic monocarboxylic acids*: formic, acetic, propionic, butyric, isobutyric, valeric, isovaleric, pivalic, lauric, myristic, palmitic, and stearic acids; for *saturated aliphatic dicarboxylic acids*, oxalic, malonic, succinic, glutaric, adipic, pimelic, suberic, azelaic and sebacic acids. The following are examples of permitted trivial names for c) *unsaturated aliphatic carboxylic acids*: acrylic, propiolic, methacrylic, crotonic, isocrotonic, angelic, tetrinic, tiglinic, sorbic, geranic, oleic, elaidic, linoleic, elaostearic, linolenic, stearic, erucic, brassic, maleic, fumaric, citraconic, mesaconic, itaconic, glutaconic,

Table 10. Characteristic groups derived from carbonyl groups

Group	Formula	Example
imine	⟩C=NH	acetonimine
oxime	⟩C=N–OH	benzal oxime
hydrazone	$\rangle C=N–NH_2$	acetone hydrazone
semicarbazone	$\rangle C=N–NH–CO–NH_2$	acetone semicarbazone
azine	⟩C=N–N=C⟨	acetone azine
acetal	$\rangle C\langle$ $O–C_2H_5$, $O–C_2H_5$	acetaldehyde diethylacetal
acylal	$\rangle C\langle$ $O–CO–CH_3$, $O–CO–CH_3$	ethylidine diacetate

muconic, aticonic, aconitic acids; for *isocyclic carboxylic acids*, camphoric, benzoic, phthalic, isophthalic, terephthalic, naphthoic (2 isomers), toluic (3 isomers), hydratropic, cinnamic acids; e) for *heterocyclic carboxylic acids*: furoic, thenoic, nicotinic, isonicotinic acids. The following trivial names should no longer be used: caproic, enanthic, caprylic, pelargonic, capric, undecylic, arachinoic, behenic, lignoceric, cerotic, mellisic acids.

7) Nitriles. The names are formed from the names of the corresponding carboxylic acids by changing the syllables "-oic acid" or "-ic acid" to "-onitrile". They can also be formed from the name of the corresponding hydrocarbon with the suffix "-nitrile". The -C≡N group bound directly to a cyclic compound is called "-carbonitrile", e.g. valeronitrile or pentanenitrile, thiophene-2-carbonitrile.

8) Amines. Names for *primary amines* are formed from the root or radical name of the corresponding hydrocarbon and the word "-amine". *Secondary* and *tertiary amines* are named as substitution products of primary amines, e.g. hexan-2-amine or 1-methyl-pentylamine. Some allowed trivial names are: aniline, anisidine (3 isomers), phenetidine (3 isomers), toluidine (3 isomers), xylidine (6 isomers), adenine, benzidine.

9) Azo compounds. a) *Symmetric compounds* are named by placing the prefix "azo-" before the name of the unsubstituted parent molecules. b) Unsymmetric azo compounds are named by placing "azo" between the complete names of the (substituted) parent molecules. Examples:

a) 1,2′-Azonaphthalene b) Naphthalen-2-azobenzene

10) Hydrazines are named a) as substitution products of hydrazine. b) When the substituents are identical, the name can be formed analogously to those of azo compounds (cf. III G 9a) by replacing "azo-" with "hydrazo-". Examples:

a) *N*-Methyl-*N*-phenylhydrazine b) 1,2′-Hydrazonaphthalene

11) Diazonium and related compounds. a) Names of compounds which contain the group $RN_2^+X^-$ are formed from the name for RH, the suffix "-diazonium" and the name of the ion X^-, e.g. C_6H_5-$N_2^+Cl^-$ is benzenediazonium chloride. b) Names for compounds with the group RN=NX are formed from the name for RH, the affix "-diazo-" and the name for the atom or group X, e.g. C_6H_5-N=N-OH is benzenediazohydroxide. c) RN=N-OM compounds are named as metal diazotates, e.g. C_6H_5-N=N-O-Na is sodium benzenediazotate. d) The N_2 group bound to a carbon is indicated by the prefix "diazo-", e.g. H_2CN_2 is diazomethane. e) If a spiro system is formed by the group -N=N-, the prefix "azi-" is used, e.g. 3-azi-5β-androstan-17-one.

12) Organosulfur compounds. 1) Compounds with bivalent sulfur. a) *S-analogs* of compounds with O functions are indicated by use of "thio-" with the name. Multiple substitution of S for O is indicated by a numerical prefix, e.g. HS-CH_2-CH_2-SH is dithioethylene glycol, $H_3C[CH_2]_4CSSH$ is hexanedithioic acid. There are two types of thiocarboxylic acids, -thio-S-acids (-COSH) and -thio-O-acids (CSOH). b)

N-Containing group[1]	Modifying syllables	N-containing carboxyl group	Modifying syllables
–(C)(=NH)OH	-imid-	–C(=NH)OH	-carbimid-
$–(C)(=N–NH_2)OH$	-hydrazon-	$–C(=N–NH_2)OH$	-carbohydrazon-
–(C)(=N–OH)OH	-hydroxim-	–C(=N–OH)OH	-carbohydroxim-
–(C)(=O)NH–OH	-hydroxam-	–C(=O)NH–OH	-carbohydroxam-

[1] The C-atom in parentheses is included in the root name of the acid.

Examples

$H_3C–CH_2–CH_2–C(=NH)OH$ Butyrimidic acid

$HN=C(HO)–(CH_2)_5–C(=NH)OH$ Heptanediimidic acid

Pyrrole-2-carbimidic acid

$H_3C–(CH_2)_5–C(=N–NH_2)OH$ Cyclohexanecarbohydrazonic acid

$H_3C–(CH_2)_3–C(=N–OH)OH$ Pentanehydroximic acid

2H-pyrane-3-carbohydroxamic acid

Heptanehydrazonic acid

Table 11. Suffixes and prefixes for characteristic groups with bivalent sulfur

Char. group	Formula	Name as Main group	Name as Substituent	Radicofunctional	Other
Thiols	–SH	-thiol	mercapto-	hydrosulfide[1]	
	$-S^-$	-thiolate	sulfido-		
	$-S_n-H$[2)]			hydropolysulfide[3]	
	$-(S)_n-H$[4)]				polysulfane[3]
Sulfides	–S–		-thio-	sulfide[5]	
	$-S_n-$		-polythio-[3]	polysulfide[3]	
	$-(S)_n-$		-polysulfanyl[3]		polysulfane[3]
Thioaldehydes	–C(=S)H	-thial			
	–(C)(=S)H	carbothialdehyde		thioformyl-	
Thioketones	>(C)=S	-thione	thioxo-[6]		

[1] The name "mercaptan" is no longer allowed.
[2] The structure of S_n is unknown.
[3] If n is known, the corresponding numerical prefix replaces "poly-".
[4] The chain of S-atoms is unbranched.
[5] The name "thioether" is no longer allowed.
[6] The prefix "thiono-" is no longer allowed.

Table 12. Suffixes and prefixes for the characteristic groups of oxygen acids of sulfur with organic radicals bound directly to the sulfur

Characteristic	Suffix	Prefix
	-sulfonic acid	sulfo-
	-sulfinic acid	sulfino-
–S–OH	-sulfenic acid	sulfeno-

Thiols and related compounds are named using the suffixes listed in Table 11, or (if there is another principle group) by prefixes. Examples: $H_3C\text{-}CH_2SH$ is ethanethiol (substitutive N.) or ethyl hydrosulfide (radicofunctional N.); the name ethylmercaptan is no longer permitted. $HS\text{-}C_6H_4\text{-}COOH$ is p-mercaptobenzoic acid. 2) Compounds in which the sulfur has a valency of 4 or 6. a) *Acids of sulfur with organic radicals bound directly to the S atom* are named using the affixes listed in Table 12, e.g. $C_6H_5\text{-}SOH$. Benzenesulfenic acid; $HOSO_2\text{-}C_6H_4\text{-}COOH$, 4-sulfobenzoic acid. Exceptions: naphthionic acid, sulfanilic acid, taurine. The names of the radicals are formed using "-sulfonyl-" ($-SO_2-$) or "-sulfinyl-" (-SO-): $H_3C\text{-}CO$ is methylsulfinyl-. Exceptions: tosyl- (only for the p-isomer), mesyl, tauryl-. b) *Acids of sulfur and its derivatives with radicals bound to the S-atom via an oxygen* are named as substitutions of the corresponding acids (cf. II E 2). In English, the derivation of anion names from binary nomenclature is preferred: $H_3C\text{-}O\text{-}SO_2\text{-}O\text{-}CH_3$: sulfuric acid dimethyl ester or dimethyl sulfate. The names for radicals bound via O or N are composed from the prefix names of their structural elements, or by adding "o-" to the anion name: $HO\text{-}SO_2\text{-}O\text{-}$ is hydroxysulfonyloxy- or sulfato-. The naming of *sultones* and *sultames* as heterocycles is becoming the preferred form:

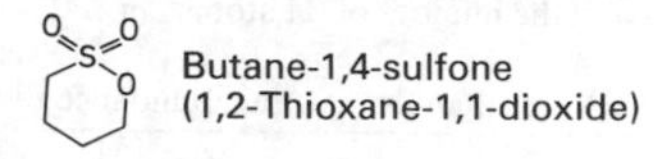

13) Organoselenium and organotellurium compounds are named analogously to the corresponding sulfur compounds, by replacing the syllables "sulf-" and "thi-" by "selen-" or "tellur-", e.g. $H_5C_2\text{-}Se\text{-}CH_2\text{-}COOH$, ethylselenoacetic acid; $H_5C_2\text{-}Te\text{-}C_2H_5$, diethyltelluride.

14) The nomenclature of other organoelement compounds has not yet been unified, either internationally or within countries. In 1973 the following preliminary rules were published by the IUPAC for nomenclature of organic compounds which do not consist entirely of carbon, hydrogen, oxygen, nitrogen, halogen, sulfur, selenium and tellurium. The compounds are named as derivatives of the hydrides or acids, or by the nomenclature for coordination compounds. a) *Organometallics* in which the metal atom is bound only to an organic radical and hydrogen are denoted by indicating the radicals in alphabetical order before the name of the metal. Hydrogen must always be indicated by the prefix "hydrido-": $(C_{10}H_9)ReH_2$ is (dihydronaphthyl)dihydridorhenium. As an alternative, the compound can be named as a substitution product of the corresponding hydride: $(C_2H_5)_4Pb$ is tetraethyl lead. b) *Organophosphorus compounds and their arsene, antimony and bismuth analogs* are named as derivatives of the hydrides (phosphine, arsine, stibine, bismuthine) or the acids (phosphoric, arsenic; phosphinic, arsinic; phosphonic, arsonic; phosphinous, arsinous; phosphonous, arsonous; phosphorous, arsenious), or as coordination compounds, e.g. $(C_6H_5)_2PCl$ as chlorodiphenylphosphine, diphenylphosphinous acid chloride or chlorodiphenylphosphorus; $C_6H_5BiCl_2$ as dichloro(phenyl)bismuthine or dichloro(phenyl)bismuth. c) *Organoboron compounds*. The N. of this group is built on that of the borohydrides. The name of BH_3 is borane, and the higher boron hydrides are also called boranes. The number of boron atoms is indicated by a Greek numerical prefix (except for nona, undeca and icosa). The number of hydrogen atoms is indicated by an Arabic numeral

placed after the name in parentheses: B_4H_{10} is tetraborane(10), $B_{20}H_{16}$ is icosaborane(16). Polyboranes in which skeleton atoms have been replaced by other elements are called *heteroboranes*: $B_{10}C_2H_{12}$ is dicarbadodecaborane(12). Substitution products, radicals and so on are named as usual: $(H_3C)_3B$ is trimethylborane, -HB-BH- is diborane(4)-1,2-diyl-.

H) ***Special topics***. 1) *Chains and rings with regular series of hetero atoms* are named using the corresponding "a" terms (except for boranes, sect. III G 14c) in analogy to the hydrocarbons, e.g. HS-S-S-SH is tetrasulfane, $H_2N(NH)_7NH_2$ is nonaazane. 2) Names for free radicals of hydrocarbons are the same as those for the corresponding bound radicals. For oxygen radicals, the ending "-oxy" is changed to "-oxyl". Nitrogen radicals are named by adding "-yl" or "-ylene" to the name of the base. Examples: $H_3C\cdot$, methyl; $H_3C\text{-}CH_2\text{-}O\cdot$, ethoxyl; $(H_3C)_2N\cdot$, dimethylaminyl; $H_3C\text{-}N$:, methylaminylene. Hydrazine radicals are called "hydrazyls". 3) Cations (cf. II C 1) which arise formally by addition of protons to a hetero atom are named using "-ammonium", "-oxonium", "-sulfonium", "-selenonium", "-telluronium", "-chloronium", "-bromonium", "iodonium". (The prefixes are formed by changing the "-um" to "-o", e.g. "ammonio-"). Examples: $[(H_3C)_4N]^+$ is tetramethylammonium-, $[(H_3C)_2S(CH_2)_6N(CH_3)_3]^{2+}Cl_2^-$ is trimethyl[6-(dimethylsulfonio)hexyl]ammonium dichloride. If the heteroatom is indicated in the name of the base compound, or if the cation is formed by addition of protons to an unsaturated hydrocarbon, the suffix "ium" is added, e.g. $[(H_2N)_3C]^+$ is guanidium; $[C_2H_5]^+$ is ethenium. Names for cations which arise formally by removal of electrons from a radical are formed from the radical names and the ending "-ium", e.g. $^+CH_2\text{-}CH_2^+$ is ethane-1,2-diylium ("carbenium" can be used for H_3C^+.) *Cation radicals (radical cations)* are given the ending "-iumyl", e.g. $[C_2H_6\cdot]^+$ is ethaniumyl.

4) Anions (cf. II C 2) which are formed by loss of a proton from the characteristic group of acids or alcohols are named by changing the "-ic acid" to "-ate", the "-ol" to "-olate" (or "-oxide"), "-thiol" to "-thiolate" (or "-sulfide"). The ending "-ide" is added to the root name of *hydrocarbon anions* or to the radical name of *anion radicals*. Prefix names are formed from the name of the anion and "-o". Examples: $H_3C\text{-}CH_2\text{-}O^-$ is ethanolate, $Na_2^+[O_3S\text{-}C_6H_4\text{-}COO]^{2-}$ is disodium 4-sulfonatobenzoate. Exception: the *salts of acetylene* continue to be called acetylides.

5) In compounds with two or more ionic centers with the same charge, one structural element must be named as a prefix. For cations the order of priority is carbenium ions, hetero cations in the order of listing in Table 8 ("a" terms), cyclic structures before acyclic structures. For anions the priority established for characteristic groups (Table 1) applies, e.g.

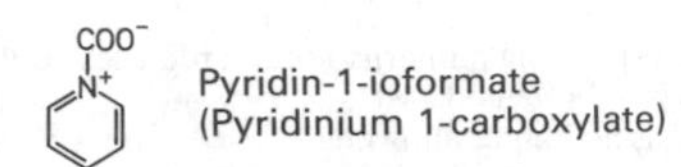

(1-Methylpyridin-3-io)-tropylium dibromide

6) The names of compounds with two or more ionic centers with different charges are formed by taking the name of the cationic structure as a prefix before the anion, or the anionic structure is named as a suffix on the name of the cation, e.g.

COO⁻ / N⁺ — Pyridin-1-ioformate (Pyridinium 1-carboxylate)

7) Isotope-modified compounds. In 1981, the IUPAC published a general system of nomenclature for compounds, ions, radicals and other species in which the nuclide composition differs from the natural relative abundances. (The "natural composition" is defined in the list of relative atomic masses published every two years by the IUPAC.) The definitive name includes the type and if needed the positions and numbers of labelled nuclides, either as the first element of the name or, preferably, before the term indicating the labelled nuclide. A distinction is made between *isotope-unmodified*, *isotope-modified* and *isotope-depleted* compounds. The isotope-modified compounds are subdivided into isotope-substituted and isotope-labelled; the latter category is further divided into specific, selectively and non-selectively labelled. To indicate *general* or *uniform* labelling, the syllables "gen" or "unf" are placed before the names of inorganic compounds; for organic compounds, the capital letters "G" or "U" are used. For metastable nuclides, the letter "m" after the mass number is used: ^{133m}Xe. The various types of formulae and names are summarized in Tables 13 and 14.

Table 13. Types of isotope-modified inorganic compounds

Type of modification	Example: formula and name	Comments
Substitution	$SiH_3{}^2H$ $(^2H_1)$silane	Each molecule contains only one ^2H atom.
Specific labelling	$SiH_3[^2H]$ $[^2H_1]$silane	The total ^2H content is greater than the natural abundance; the excess ^2H is present in mono-substituted molecules.
Selective labelling	$[1\text{-}^2H]Si_2H_6$ $[1\text{-}^2H]$disilane	The total ^2H content is greater than the natural abundance; the excess ^2H is present in two or more substituted molecules; there can be more than one ^2H in a given molecule, but all are at the indicated position.
	$[^2H_{1;3}]SiH_4$ $[^2H_{1;3}]$silane	The ^2H content is greater than the natural abundance; the excess ^2H is found in molecules which have either 1 or 3 ^2H atoms.
Non-selective labelling	$[^2H]Si_3H_8$ $[^2H]$trisilane	The ^2H content is greater than the natural abundance; the excess ^2H is distributed over all positions and the number of ^2H atoms per molecule is random.
Isotope deficient	$[def^{29}Si]$trifluorosilane	$[def^{29}Si]SiHF_3$ The ^{29}Si content is lower than the natural abundance.

Table 14. Types of isotope-modified organic compounds

Type of compound	Formula	Name
Unmodified	H_3C-CH_2-OH	Ethanol
Isotope substituted	$^2H_3C-CH_2-O^2H$	(2,2,2−$^2H_3$)Ethan[^2H]ol or [O,2,2,2−$^2H_4$]Ethanol
Selectively labelled	a) [O,2−^2H]H_3C-CH_2-OH b) [2,2−$^2H_{2;2}$,$^{18}O_{0;1}$]H_3C-CH_2-OH	a) [O,2−^2H]Ethanol b) [2,2−$^2H_{2;2}$,$^{18}O_{0;1}$]Ethanol
Nonselectively labelled	[^2H]H_3C-CH_2-OH	[^2H]Ethanol
Isotope deficient	[def^{13}C]H_3C-CH_2-OH	[def^{13}C]Ethanol

Historical. The historical development of the N. cannot be considered separately from that of the symbols and formulae. In the earliest times, substances were named arbitrarily. Their compositions and structures were completely unknown, although ideas about the structure of matter developed early (atomism, the four-element idea). In the 16th century, Paracelsus attempted to replace "alchimie speculativa" by interpretation of scientific phenomena. The early N. was cumbersome, and there were multiple trivial names for the same substances. In the 17th and 18th century, the constantly growing fund of knowledge made some ordering of nomenclature necessary. There were several attempts to develop a systematic N. (Bergman, Macquer, Guyton de Morveau and others). On 18 April 1787, Lavoisier presented a scheme for systematically naming inorganic substances to the Academie in Paris. To accompany this new N., Hassenfratz and Adet suggested a new system of symbols which was the direct precursor of the element symbols of Berzelius (1814); the earlier systems (Boyle, 1661; Stahl, 1697; Geoffroy, 1718; Bergman, 1775; Dalton, 1808) had proven unsatisfactory.

The developing chemical theories (dualistic electrochemical theory, substitution theory, radical theory, type theory, etc.) were the basis for the systematics of classes of substances and the naming of individual compounds (Berzelius, 1820; Liebig and Wöhler, 1832; Dumas, 1834; Gerhardt, 1839). These have been retained in the various principles of nomenclature, which have been extended by addition of special rules to include nomenclature of poly- and heterocyclic compounds (Hantzsch, 1886; Widman, 1888; extended by Bredt, 1896; Baeyer, 1900; Stelzner, 1921; Patterson, 1925). With the establishment of the periodic system of the elements (Mendeleyev, Meyer) and the development of structural chemistry and structural formulae (Couper, 1858, Butlerow, 1861, Kekulé, 1865), a need was seen for the reform of the N. of organic compounds. This was undertaken at the International Congress of Chemists in Paris in 1889 and in the Geneva Nomenclature, 1892. Due to the incompleteness of the rules (cyclic compounds received only brief treatment, and heterocycles were not treated at all) and to objections (some of them unfounded), the IUPAC established a commission for the N. of organic chemistry in 1922. At the N. conference in 1930 in Lüttich, the commission made the "functional groups", i.e. the most reactive part of the molecule, rather than the parent compound (Geneva N.) the basis of the name. It also questioned the necessity of an official or register number. The work of the following N. conferences (Lucerne, 1936; Rome, 1938) were again concerned with development of official names. The Commission for inorganic N., which was established in 1921, proposed rules in 1939 which were published in 1940. The present rules are based on the 1947 (London) extension and revision of these rules. The systems for naming chemical compounds are continuously improved by the individual nomenclature commissions, which, after public discussion of suggested changes, publish new and revised rules.

Nonactin: an ionophoric antibiotic produced by species of *Actinomyces*; chemically, it is a 32-membered cyclic ester consisting of two molecules each of D- and L-nonactinic acid. N. can be thought of as a naturally occuring crown ether. It is accompanied by the antibiotics monactin, dinactin and trinactin.

Nonadiabatic reaction: see Potential surface.

Nonanal: same as Pelargonaldehyde (see).

Nonanedioic acid: same as Azelaic acid (see).

Nonanes: alkane hydrocarbons with the general formula C_9H_{20}. The most important of 35 structural isomers is ***nonane***, $CH_3(CH_2)_7CH_3$, m.p. -53.5, b.p. 151 °C, n_D^{20} 1.4054. It is a colorless, combustible liquid which is miscible with ethanol and ether, but insoluble in water. N. are present in petroleum and are formed by coal hydrogenation. Branched N., especially 2,2,4- and 2,2,5-trimethylhexane, are byproducts of isooctane production, and are components of gasoline.

Nonanol, ***nonan-1-ol***: $CH_3(CH_2)_7CH_2OH$, a monovalent alcohol, the most important of the isomeric C_9 alcohols. N. is a pale yellow liquid which smells like lemon oil; m.p. -5 °C, b.p. 212 °C, n_D^{20} 1.4333. It is insoluble in water, but is miscible with ethanol and ether in any proportions. It is made by reduction of pelargonaldehyde (nonanal) or ethyl pelargonate, and is used in perfumes.

Nonanoic acid, ***pelargonic acid***: CH_3-$(CH_2)_7$-COOH, a higher saturated monocarboxylic acid. The trivial name is no longer permitted in IUPAC nomenclature. N. is a colorless, oily liquid; m.p. 12.0, b.p. 255 °C, n_D^{20} 1.4343. It is slightly soluble in water and somewhat more soluble in alcohol and ether. N. is irritating to the skin and mucous membranes. It undergoes the reactions typical of the monocarboxylic acids. It is present in free or bound form in North American petroleum, lavender oil, the volatile oil from *Pelargonium roseum* and in rancid fats, where it arises by oxidation of oleic acid. N. can be synthesized by the malonic ester synthesis from heptyl bromide. It is used to make alkyde resins, softeners and perfumes.

Noncarbonate hardness: see Hardness, 2.

Nonferrous metals: a collective term for all metals other than iron. See Metals.

Nonhistone proteins: see Nucleic acids.

Nonionics: see Surfactants.

Nonmetals: chemical elements which do not have the properties characteristic of metals. All N. are found in the representative groups of the periodic table. They include hydrogen, helium, neon, argon, krypton, xenon, radon, boron, carbon, silicon, nitrogen, phosphorus, arsenic, oxygen, sulfur, selenium, tellurium, fluorine, chlorine, bromine, iodine and astatine. The N. are characterized chemically by high ionization energies for their valence electrons, and a tendency to achieve the noble gas configuration by adding electrons, that is, by forming anions. Except for most noble gases, fluorine and astatine, all N. form oxides, and these are generally acidic. The N. are poor thermal and electrical conductors. The semimetals (see Metals) have properties intermediate between those of the metals and N.

Nonpolar bond: see Chemical bond.

Non-silver-based imaging systems: light-sensitive systems for imaging based on photochemical reactions other than those of the silver halides. A number of possibilities are known, but only a few are actually applied, including photopolymerization, electrophotography, diazotypography, photochromism, etc. There are both amplified and non-amplified processes which may be used. Some primary photochemical processes which seem suitable for N. include photocatalysis, oxidative and reductive pigment bleaching, radical-generating processes initiated by photolysis of perhaloalkanes, formation of fluorescent species from non-fluorescent compounds and photochemical redox reactions and disproportionations of inorganic salts (lead(II) salts, Hg_2I_2, copper(I) halides, thallium halides, titanium dioxide) and coordination compounds.

The major applications of the N., because of their low sensitivity and inability to reproduce color, are monochromatic reproduction and copying techniques.

Nonylaldehyde: same as Pelargonaldehyde (see).

Noradrenalin, ***norepinephrin, levarterenol***: a phenylethylamine derivative which acts as a transmitter in sympathetic nerves (see Sympathicomimetica). Its *N*-methyl derivative adrenalin (epinephrin) is the hormone of the adrenal medulla. Both compounds are synthesized from L-tyrosine. N. and adrenalin contain a chiral C atom, and only the R(-)-enantiomers have biological activity. N. forms white crystals; m.p. 217-218 °C, $[\alpha]_D^{25}$ -37.3 °C (in dilute hydrochloric acid); it is slightly soluble in water, alcohol and ether, but readily soluble in dilute acids and bases. The racemate melts at 191 °C.

Adrenalin has similar properties: m.p. 215 °C (dec.), $[\alpha]_D$ -50.7° (in dilute hydrochloric acid); the hydrochloride melts at 206 °C and is very water soluble.

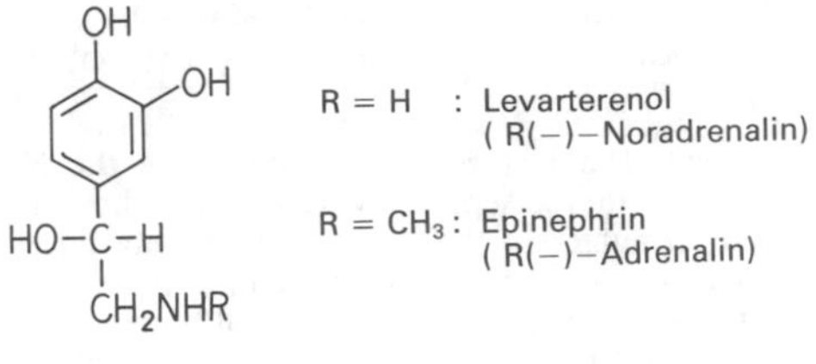

R = H : Levarterenol (R(−)−Noradrenalin)

R = CH_3 : Epinephrin (R(−)−Adrenalin)

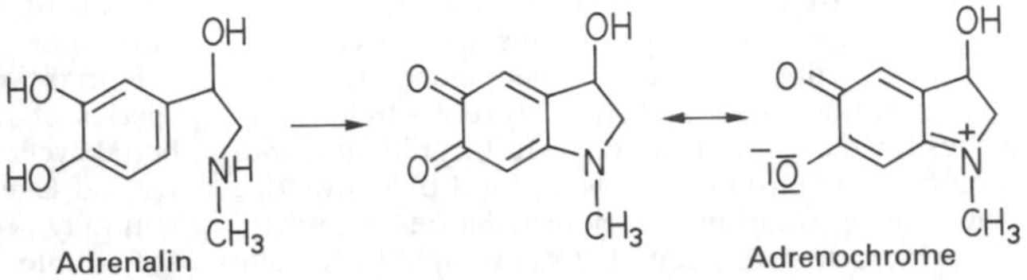

Adrenalin → Adrenochrome

The chemical synthesis of N. begins with pyrocatechol, which is reacted with chloroacetyl chloride to give the α-chloroketone. This reaction occurs partly via the phenol ester and a subsequent Fries shift. The amino ketone is made by reaction with NH_3 or methylamine, and the aminoalcohol is obtained by catalytic reduction of the aminoketone. The product is the racemate which, when separated with (+)-tartaric acid, yields N. or adrenalin.

Because of their pyrocatechol structures, N. and adrenalin are easily colored by reactions in air and light. For example, adrenalin forms the dark red ***adrenochrome***, a bicyclic quinone compound which can also be formulated as *quinonimine*, with a betaine structure. As phenolic compounds, N. and adrenalin give colored compounds with iron(III) ions, and can be coupled to azo dyes with diazonium salts.

N. and adrenalin are not effective upon oral administration. They are added to injection solutions of local anaesthetics because of their vasoconstrictive effects. Adrenalin is occasionally injected intracardially in cases of cardiac arrest.

Historical: adrenalin was isolated from the adrenal medulla in 1900 by von Fürth and in 1901 by Takamine and Aldrich. It was synthesized in 1904 by Stolz. N. was first recognized as a neurotransmitter by von Euler in 1948.

Norbornadiene, ***bicyclo[2,2,1]hepta-2,5-diene***: an unsaturated derivative of the hydrocarbon norbornane (see Bicycloalkanes). N. is readily synthesized by a Diels-Alder reaction between cyclopentadiene and ethyne and is used in the synthesis of insecticides. Its reaction with hexachlorocyclopentadiene, for example, leads to aldrin. N. is formed from the isomeric quadricyclane by UV irradiation.

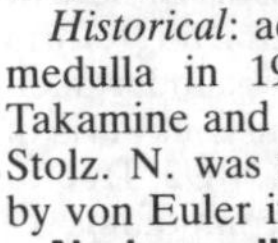

Norbornane: see Bicycloalkanes.

Norcarane: see Bicycloalkanes.

Nordihydroguaiaretic acid, abb. ***NDGA***: a natural antioxidant; m.p. 184-185 °C. N. is a grayish powder which is readily soluble in ethanol and ether, and dissolves in alkalies with the development of red color.

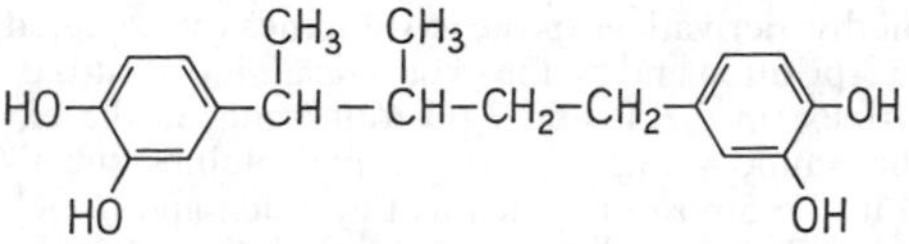

It is isolated from the Mexican plant *Larree divaritica*, or it can be made by partial or total synthesis. The starting material for partial synthesis is guaiac resin. N. is used as an antioxidant in pharmaceuticals.

Norethisterone: see Gestagens.

Norharmane: see Rauwolfia alkaloids.

Normality: see Composition parameters.

Normal potential: the relative electrode voltage under the standard conditions of 25 °C, 101.3 kPa and a 1 normal solution. The concept of N. has been replaced by the standard electrode potential, in which the definition is based on the activity rather than the normality of the solution. The terms N. and standard potential are sometimes used interchangeably in the literature. See Electrochemical equilibrium.

Normal solution: see Composition parameters.

Normalization condition: in quantum mechanics, the requirement placed on the wavefunction $\Psi(x,y,z)$ that the integral of the square of its absolute value $|\Psi|^2 = \Psi^*(x,y,z)\Psi(x,y,z)$ over all of space must be equal to 1 (see Atom, models of). The N. is a result of the statistical interpretation of the wavefunction; $|\Psi|^2$ is the probability density. The N. is usually given in the form:

$$\int_{-\infty}^{\infty}\int_{-\infty}^{\infty}\int_{-\infty}^{\infty} \psi^*(x, y, z)\, \psi(x, y, z)\, dx\, dy\, dz = \int_{\infty}^{\infty} |\psi|^2\, d\tau = 1$$

If a wavefunction $\phi(x,y,z)$ does not fulfill the N., i.e.

$$\int_{-\infty}^{\infty} |\varphi|^2\, d\tau = A,\ A \neq O,$$

the normalized wavefunction is given by the equation $\Psi(x,y,z) = A^{-1/2}\phi(x,y,z)$, where $A^{-1/2}$ is the *normalization factor*. The requirement that a wavefunction be normalizable means that it must be single-valued, continuous and finite.

Norpseudoephedrin: see Appetite suppressants.

Norrish type I reaction: a photochemical decarbonylation reaction in an aldehyde or ketone; a C-C bond is broken next to the carbonyl group. The reaction occurs via an n → π* excited S_1 or T_1 state. Cleavage of the C-CO bond produces an acyl and an alkyl radical. The acyl radical splits off carbon monoxide, forming another alkyl radical, and the alkyl radicals recombine:

$$R^1\text{–CO–}R^2 \xrightarrow{h\nu} R^1\text{–CO} + {}^{\cdot}R^2\ (\text{or } {}^{\cdot}R^1 + {}^{\cdot}\text{CO–}R^2)$$

$$\xrightarrow{-CO} R^1\text{–}R^2.$$

N. are particularly effective in the gas phase, and can take place in solution only if the radicals are sufficiently stable (due to resonance or steric stabilization). Lower aliphatic aldehydes and ketones without γ-H atoms undergo N., as do cyclic and *tert.*-butyl ketones.

Norrish type II reaction: a photochemical fragmentation of an aldehyde or ketone with an H atom in the γ-position to form an alkene and a shorter aldehyde or ketone. The mechanism of the reaction involves a cyclic transition state. Intramolecular abstraction of the γ-H atom creates a hydroxybiradical, which decays by splitting the C_α-C_β bond and thus forming an alkene and the enol of the smaller carbonyl compound.

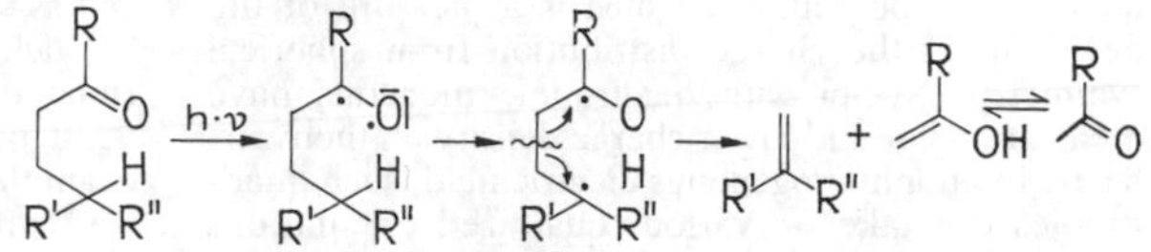

The N. is initiated by n → π* excitation and occurs in either the gas phase or in solution, via an S_1 or T_1 state. Aliphatic aldehydes and ketones with γ-H atoms and higher alkyl aryl ketones with γ-H atoms can undergo the N.

Norethylene®: see Synthetic fibers.

Nortropane: see Tropane.

Noscapin: see Narcotin.

Novobiocin: an antibiotic formed by *Streptomyces* species; it has a glycosidic coumarin structure. Its monosodium salt is used to treat infections by gram-positive staphylococci.

Novolakes: semiliquid Phenol resins (see).

Noxiptilin: see Dibenzodihydrocycloheptadienes.

Np: symbol for neptunium.

NQR spectroscopy: same as Nuclear quadrupole resonance spectroscopy (see).

nRNA: see Nucleic acids.

NTE: abb. for nitrilotriacetic acid.

Nuarimol: see Pyrimidine fungicides.

Nuclear charge number: see Atomic number.

Nuclear induction: same as NMR spectroscopy (see).

Nuclear magnetic resonance spectroscopy: same as NMR spectroscopy (see).

Nuclear magneton: the unit of nuclear magnetic moments. $\mu_N = eh/4\pi m_p = 5.050824 \cdot 10^{-27}$ J T^{-1}, where e is the unit of electric charge, h is Planck's constant and m_p is the mass of the proton. The Bohr magneton is related to the N. by the equation $\mu_N = 1/1836\ \mu_B$, because the mass of the proton is 1836 times larger than that of the electron.

Nuclear Overhauser enhancement, abb. ***NOE***: an enhancement of the intensity of certain signals in an NMR spectrum by use of the double resonance technique (see NMR spectroscopy). If two protons H_A and H_B are close enough so that a dipole interaction between them is possible, application of an additional alternating magnetic field with a frequency equal to the resonant frequency of H_B will increase the area under the signal of H_A. The cause of this is a decrease in the spin-lattice relaxation time for H_A. The NOE can always be detected when magnetic nuclei are within 0.3 nm of each other. It is therefore of great value for determining molecular geometries. A practical consequence of the NOE is that in double resonance experiments, the intensities are no longer necessarily proportional to the number of nuclei.

Nuclear quadrupole resonance spectroscopy, ***NQR spectroscopy, Quadrupole resonance spectroscopy***: a method of high-frequency spectroscopy in which a high-frequency alternating electromagnetic field is used to induce transitions between energy levels which arise from the interaction of the nuclear electric quadrupole moment with an inhomogeneous electric field. Although nuclei with nuclear spin quantum numbers of $I = 0$ or 1/2 have a spherically symmetrical distribution of positive charge, nuclei with $I > 1/2$, e.g. B, Cl, Br, I, N, Sb and Sn, have an electric quadrupole moment eq, which is a measure of the deviation of the charge distribution from spherical symmetry. Nuclei with quadrupole moments have areas of higher and lower charge density on their surfaces. In an inhomogeneous electric field, such quadrupoles can take on various quantized orientations which differ in their energies. By irradiating them with suitable energy quanta, it is possible to induce transitions between the quadrupole energy levelş and to detect them; the resonance frequencies are in the meter and decimeter range.

N. is related to NMR spectroscopy (see). An essential difference between the two methods is that in NMR an external magnetic field is required to split the various energy levels, while no external field is required in N., because the splitting occurs in the inhomogeneous field generated by the atoms and molecules in the crystal. In order to measure NQR spectra, a spectrometer is used which is very similar to an NMR spectrometer, except that there is no external magnet. The position of resonance frequencies depends on the electric field gradient q at the nucleus. The parameter eQq is called the *nuclear quadrupole coupling constant*.

N. is used mainly to examine bonding and charge distribution in ionic crystals and to study imperfections in crystal structure.

Nuclear resonance spectroscopy: same as NMR spectroscopy (see).

Nuclear spin: see NMR spectroscopy.

Nuclear spin coupling constant: see NMR spectroscopy.

Nuclear spin resonance spectroscopy: same as NMR spectroscopy (see).

Nucleic acids: biopolymers in which nucleosides are linearly joined by 3',5'-phosphodiester bonds. In ***ribonucleic acids*** (abb. ***RNA***), the sugar component of the nucleosides is ribose, while in ***deoxyribonucleic acids*** (abb. ***DNA***), the sugar is deoxyribose.

Each monomeric unit consists of a sugar phosphate linked N-glycosidically to a ***nucleoside base***; in the polymer, the bases protrude approximately at right angles to the sugar-phosphate chain which makes up the backbone of the molecule. The most common nucleoside bases are the purines adenine and guanine and the pyrimidines cytosine and uracil (in RNA) or thymine (in DNA). The less common "rare bases" are methyl derivatives of the pyrimidines (methylated at the 3-position and on the exocyclic amino group) or of purines (in the 1- and 7-positions and on the exocyclic amino group), or other pyrimidines substituted in the 5-position, such as thiouracil and the C-glycoside Pseudouridine (see). Methylation of DNA seems to play a role in the regulation of transcription. The rare bases (which actually are not rare) are most abundant in transfer RNA, where they may make up 10% of the molecule.

In the nomenclature of the N., a "nucleoside" is a base linked to a sugar, while a "nucleotide" is a base linked to a sugar phosphate.

N. can be hydrolysed by acid catalysis. The phosphate group on the 5'-OH is preferentially cleaved, so the primary products are the 3'-phosphates. Further hydrolysis releases the nucleosides and then the free bases. DNA is more easily hydrolysed than RNA. N. give an acidic reaction, and are usually present in the cell as salts of basic proteins.

Nucleic acid bases, nucleosides and nucleotides are isolated from hydrolysates of nucleic acids, usually of yeast nucleic acids, which are relatively cheap. For example, treatment of N. with trichloroacetic acid at 0 °C produces an acid-soluble nucleotide fraction from which adenosine phosphates can be isolated. The nucleotides are isolated from the hydrolysates by precipitation and ion-exchange chromatography. Hydrolysis with aqueous ammonia or pyridine at elevated temperature and pressure yields nucleosides.

Deoxyribonucleic acid. Most DNA is found in the chromosomes, which in eukaryotic cells are found in the nuclei. Small amounts of DNA are found in the chloroplasts and mitochondria of eukaryotic cells,

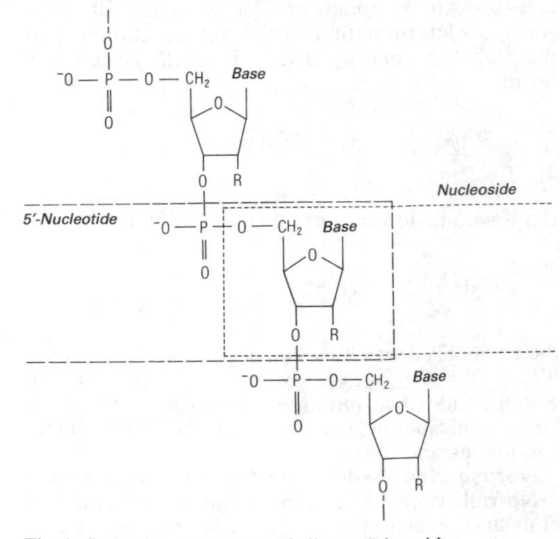

Fig. 1. Secondary structures of ribonucleic acids.

Table 1. Components of nucleic acids

Bases		Ribosides		
Name	Abb.	Name	Abb.	One-letter abb.
Adenine	Ade	Adenosine	Ado	A
Guanine	Gua	Guanosine	Guo	G
Cytosine	Cyt	Cytidine	Cyd	C
Thymine	Thy	Ribosylthymine	Thd	T
Uracil	Ura	Uridine	Urd	U

and in plasmids in prokaryotic cells. The raw material isolated from the nucleus is called ***chromatin***; it contains DNA, small amounts of RNA and proteins - the very basic histones and a wide variety of non-histone proteins. These components are tightly packed in the chromatin. The DNA is an extremely long molecule which, if extended, might be as much as 2 m long. It is wound into a series of bead-like structures on matrices provided by the histones; the DNA in the beads (nucleosomes, ν-bodies) is more resistant to nucleases than the DNA between them, so that nuclease treatment releases relatively intact nucleosomes of about 200 base pairs.

The ***Watson-Crick model*** of DNA is a *double helix* of antiparallel strands held together by hydrogen bonds between complementary base pairs (2 between A and T, 3 between C and G). This model, proposed in 1953, was based on three lines of evidence. 1) The chemical studies of Todd had shown that DNA is a chain of nucleosides linked by 3',5'-phosphodiester bridges. 2) The analyses of Chargraff had shown that the ratios A/T and G/C are always 1 (constant base ratios), while the other combinations can vary (*Chargraff's rules*). 3) The fiber x-ray diagrams of Astbury and Wilkins could be interpreted as indicating a helical arrangement. It was later discovered that the two strands are held together not only by base pairing, but by base stacking as well. A major objection to this model was that, in order to replicate, the two strands must separate, either by breaking and rejoining the sugar-phosphate backbone, or by untwisting. The latter would require a very high rate of rotation of a long, extensively folded molecule within the viscous environment of the nucleoplasm or bacterial cytoplasm, and seemed improbable to many. (Actually, bacterial DNA is usually ring-shaped and one of the strands must be broken to allow the daughter strands to separate.)

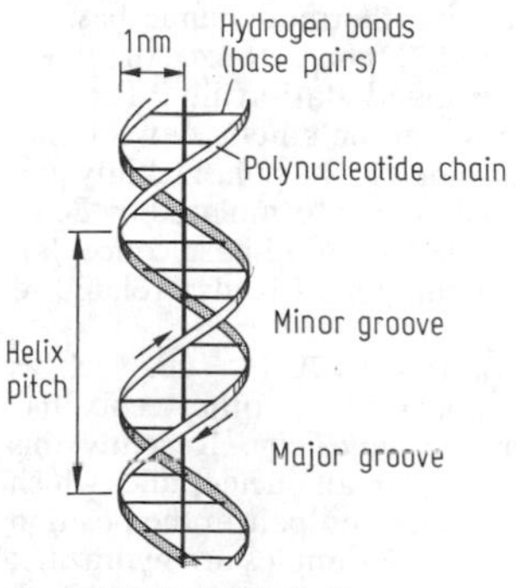

Fig. 2. DNA double helix.

It now appears that helicases, enzymes which untwist the molecule by breaking and reannealing one strand of the duplex, as well as limited rotation, are involved in replication.

The DNA double helix can exist in various forms (A, B, C, D and Z); their structural parameters are given in Table 2. The B-structure is the major one present in native DNA, which is highly hydrated (the sodium or lithium salt has a 90% water content). During preparative procedures, the B-form is converted to the A or C form, depending on the water content and electrolyte concentration. Hybrids of complementary DNA and RNA are in the A-form, regardless of the ionic strength. Unlike the A, B, C and D forms, the Z form ("Z" for "zig-zag") is a left-handed helix. It was discovered in polynucleotides such as poly(dG-dC) · poly(dC-dG) or poly (dG-dT) · poly(dC-dA). In the Z form, the purine nucleosides are in the *syn*-conformation, which results from rotation around the glycosidic bond of the *anti*-conformation in which purines usually are found. The pyrimidines, however, remain in the *anti*-conformation, and this causes the zig-zags in the backbone. The transition from the B to the Z form appears to be promoted by methylation of bases. Short stretches of Z-DNA have been observed in natural DNA and they may be involved in regulation of transcription.

Table 2. Structural parameters of the various forms of the DNA double helix

	A	B	C	D	Z
Rise of the helix [nm]	1.1	1.0	0.933	0.8	1.2
Numer of bases per turn	28.2	34	30.7	24	44.5

Ribonucleic acids. RNA is found mainly in the cytoplasm, primarily in the ribosomes, the site of protein synthesis. The most abundant types in cells are ***messenger RNA*** (abb. ***mRNA***), ***ribosomal RNA*** (abb. ***rRNA***) and ***transfer RNA*** (abb. ***tRNA***); there are also ***small nuclear RNA*** molecules which, in association with proteins, are responsible for the processing of newly transcribed RNA, and nuclear RNA, which is a precursor of mRNA. mRNA is a copy of the genes which are translated into proteins. Its precursor (in eukaryotes) is a direct transcript of the DNA and contains untranslated segments (introns), which must be removed by splicing before the mRNA associates with the ribosomes. When attached to a ribosome, mature mRNA acts as a template for the alignment of amino acyl-tRNA molecules (see Proteins, translation). tRNA molecules act as adaptors which convert the sequence of nucleotide bases in the mRNA into the sequence of amino acids in the protein. The cell has enzymes which specifically bind each amino acid to the corresponding tRNA (e.g. $tRNA^{phe}$ + Phe + ATP → Phe-$tRNA^{Phe}$ + ADP + P_i), at the same time activating it (supplying the energy which will later be required to form the peptide bond with the next amino acid in the polypeptide). The tRNA molecule has a specific triplet of bases, the anticodon, which recognizes and binds to the complementary codon on the mRNA molecule. The correct alignment of the mRNA and tRNA is facilitated by the rRNA in the ribosome. The RNA of RNA viruses serves both as a messenger in the infected cell and as a template for synthesis of complementary DNA (by reverse transcriptases).

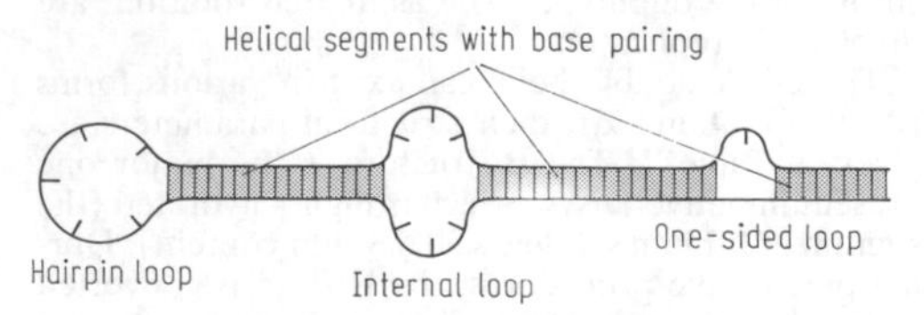

Fig. 3. Structure of a ribonucleic acid (R = OH) or deoxyribonucleic acid (R = H) (schematic).

RNA, unlike DNA, is generally single-stranded. Only a few RNA viruses of plants and animals contain double-stranded RNA, the structure of which is similar to the A form of DNA. Single strands of RNA typically fold back upon themselves wherever complementary sequences are available (Fig. 2). tRNA molecules, consisting of about 80 nucleotides, are relatively small, with M_r about 25,000. They fold into a "cloverleaf structure" in which complementary regions form helical, double-stranded sections. The single-stranded loops at the ends of the helical sections make up the "leaves" of the "cloverleaf". The rare bases are found in these single-stranded segments. mRNA is much larger, usually having a M_r around 10^6. Its secondary structure is much more complicated than that of DNA. rRNA exists largely as double-helical structures.

Sequence analysis. Long segments of DNA must be broken into smaller fragments before their sequence can be analysed. This is done with enzymes called restriction endonucleases, which recognize specific sequences of nucleosides and cleave the molecule only in the context of those sequences. The restriction fragments are then separated, usually by chromatography, and are sequenced by the end-labelling techniques of Maxam and Gilbert or Sanger, Nicklen and Coulson.

Biosyntheses. DNA is the repository of genetic information (genome) in all cells. Its synthesis is called *replication*; the synthesis of RNA on a DNA template is called *transcription*. Both processes require the presence of a template (DNA), the four nucleoside triphosphates and polymerases. All of the DNA polymerases so far isolated work only in the 5' to 3' direction, but it is clear from autoradiography that the synthesis must occur simultaneously from the 3' end of one strand and the 5' end of the other. The 5'-3' strand, which must be replicated in the 3' to 5' direction, is replicated in *Okazaki fragments* about 2000 base pairs in length. These are later linked by a DNA ligase.

The synthesis of RNA is regulated by a number of DNA-binding proteins, including the RNA polymerase, which must become attached to the DNA at a specific site called the promoter. This region may be blocked by a regulatory protein which binds specifically to it; such a protein is called a repressor. In many bacteria, one promoter can control transcription of a group of genes called an operon; usually the enzymes encoded by the operon all function in a specific biosynthetic or catabolic pathway. The regulatory mechanisms in eukaryotic cells are more complex, and in addition to simple blocking of promoters by repressors, there are sequences of DNA which enhance the transcription of genes "downstream" (in the 3' direction), or "upstream", and others which enhance transcription on the complementary strand. It is thought that the latter two types must act by changing the coiling of the molecule or its interaction with histones.

Nucleophiles: electron-pair donors which can react with electrophilic substrates. N. are bases or Lewis bases. They can be negative ions, such as halides, HO^-, RO^-, $RCOO^-$, HS^-, CN^-, NO_2^-, N_3^-, SCN^-, $S_2O_3^{2-}$, uncharged compounds with free electron pairs, such as H_2O, H_2S, NR_3, ROH, R_2O, compounds with olefinic double bonds, such as $R_2C=CR_2$, and aromatic compounds, such as benzene and substituted benzenes. The strength of a N. is related to that of R_3C^+. Its reactivity is called Nucleophilicity (see).

Nucleophilicity: the attraction of a nucleophile for a nucleus or an electron deficiency; its ability to provide an electron pair for bonding to an electrophile. N. is determined by electronic and steric factors, and depends on the polarizability and basicity of the key atoms of the nucleophile. The degree to which N. depends on polarizability or basicity is determined essentially by the solvent: in dipolar, aprotic solvents, the basicity is the determining factor for N., while in protic solvents, the N. is determined by the polarizability. Therefore, the strength of N. varies considerably in solvents of different types. For example, $CN^- > CH_3COO^- > F^- > N_3^- > Cl^- > Br^- > I^- > SCN^-$ in dimethylformamide (dipolar, aprotic), but $SCN^- > CN^- \approx I^- > Br^- \approx HO^- > Cl^- > I^- \approx CH_3COO^-$ in protic solvents. For this reason, it is not possible to establish a scale of N. which applies to all substrates and solvents. For some applications, quantitative equations for N. have been established using the LFE equations. An equation relating the kinetic N. n of a nucleophile X to that of water in S_N2 reactions with methylbromide at 25 °C in water/acetone mixtures is the *Swain-Scott equation*: $\lg(k_X/k_{H_2O}) = n \cdot S$, where S is the susceptibility factor. For typical substrates reacting by an S_N2 mechanism, S is about 1. This equation applies only to S_N2 reactions on saturated C atoms. Using a four-parameter equation (Edwards equation), the effect of structure of both the nucleophile and the electrophile on the rate of S_N2 reactions can be described: $\lg(k_X/k_{H_2O}) = \alpha P_i + \beta H_i$, where P_i is the relative polarizability of the nucleophile, and H_i is its relative thermodynamic basicity (proton basicity). α and β characterize the electrophile; they are determined statistically. The Edwards equation indicates that the kinetic Lewis basicity of a reagent is determined by its polarizability and proton basicity; this concept was formulated by Pearson as the HSAB principle (see Acid-base concepts). The ease of oxidizing a compound is also related to its N.

Nucleophiles with values of $n_{CH_3I} > 10$, such as vitamin B_{12}, cobaloxime(I) and other cobalt(I) chelates, display ***supernucleophilicity***. Recently this property has been attributed to all nucleophiles which have an atom with a free electron pair in the position α to the nucleophilic center. Examples are hydrazine, the peroxy anion, and phenylhydroxylamine. Such species usually react more rapidly than normal nucleophiles.

Nucleophilic reactions: reaction between nucleophilic reagents and electrophilic substrates. N. are substitution reactions on saturated hydrocarbon atoms and aromatic compounds; they proceed by an S_N mechanism (see Substitution 1.). They proceed as addition reactions on alkynes and polar double bonds (carbonyl compounds, thiocarbonyl compounds, nitriles, azomethyne, nitroso compounds, sulfonyl compounds). The products are often unstable, so that an elimination follows to stabilize the product. The mechanism of N. with carbonyl groups is similar to the S_N mechanism: the point of attack for the nucleophilic reagent is the atom with the lowest electron density; at the same time, a bonding electron pair is displaced and localized on the carbonyl oxygen atom. The negative charge on the oxygen atom is stabilized by an electrophilic species. The rates of N. depend on the nucleophilicity of the nucleophile, the polarization of the carbonyl group (positive partial charge on the C atom of the carbonyl group) and the electrophilic species which stabilizes the negative charge on the carbonyl oxygen atom. Nucleophilic reagents are bases or Lewis bases, CH-acidic compounds or cryptobases. N. on vinylogenic carbonyl compounds (see Vinylogic principle) are similar to those on carbonyl compounds (see Michael addition).

Nucleophilic substitution: see Substitution, 1.

Nucleoproteins: see Proteins.

Nucleoside antibiotics: a group of antibiotics derived structurally from pyrimidine and purine nucleosides, which are components of nucleic acids. Both the sugar and the base component can be chemically modified. The ribose moiety can lack the hydroxyl group in position 3 (e.g. in 3'-deoxyadenosine) or it can be replaced by an amino or acylamino group (e.g. in Puromycins (see)). The configuration of the hydroxyl group in position 2 can be altered, as in the arabinosyl nucleosides (spongonucleosides), for example, spongoadenosinc, abb. Ara-A. The most common N. are analogs of the purine nucleosides. The normal purine bases adenine and guanine are replaced by 6-dimethylaminopurine (e.g. puromycin), pyrrolopyrimidine (e.g. tubercidin) or pyrazolopyrimidine (e.g. formycin and coformycin). The most important of the pyrimidine nucleoside analogs is 5-azacytidine. The C-glycosides are a special group (see Showdomycin). The antibiotic effect of the N. is due to inhibition of nucleic acid or protein synthesis.

Nucleosides: in the narrow sense, the N-ribosides and N-2'-deoxyribosides of purine and pyrimidine bases found in nucleic acids. In the broader sense, other *N*-ribosides, including the components of nicotinamide nucleotides and vitamin B_{12} are considered N.; the term may even be extended to any *N*-glycosides (see Glycosides) of heterocycles. Some derivatives of the nucleic acid N. act as antimetabolites (see Nucleoside antibiotics).

Nucleotidases: see Phosphatases.

Nucleotides: phosphate esters of nucleosides. Ribonucleotides can have phosphate groups in the 2', 3' or 5' positions, while deoxyribonucleotides can be esterified only in the 3' or 5' positions. In cyclic phosphate diesters, two hydroxl groups of a nucleoside are linked to a single phosphate group (see Cyclic nucleotides). Nucleoside bisphosphates have two phosphate groups on two different hydroxyl groups, for example, adenosine 3',5'-bisphosphate. The 5'-phosphates can form anhydrides with acids; the anhydrides with phosphoric acid are nucleoside diphosphates, and those with pyrophosphoric acid are nucleoside triphosphates (see Adenosine phosphates). ***Adenosine-3'-phosphate-5'-phosphosulfate*** ("active sulfate", abb. ***PAdoPS***) and the ***nucleoside diphosphate sugars*** are also anhydrides. The anhydride compounds have high energies of formation, and the energy released by their hydrolysis is used to drive energy-requiring biological reactions. The adenosine phosphates are the "energy currency" of the cell, used in the majority of reactions. PAdoPS is used in syntheses which require incorporation of a sulfate group, and the nucleoside diphosphate sugars, especially uridine diphosphate sugars, are the donors of monosaccharide units in the synthesis of polysaccharides. Mixed anhydrides of adenosine monophosphate with carboxylic acids are the donors of acyl residues to coenzyme A. in vivo. The N. also include a few coenzymes, such as the Flavin nucleotides (see) and Nicotinamide nucleotides (see). These coenzymes contain pyrophosphate bridges.

In the formulas of ***oligonucleotides***, the phosphate diester group (3'-5' bond) is shown by a hyphen or a p between the symbols for the nucleosides (see Nucleic acids); a terminal phosphate group is indicated by a p before (5') or after (3') the symbol, as in ApGpUp or A-G-Up. The nomenclature of synthetic ***polynucleotides*** can be inferred from the following examples (IUPAC-IUB recommendations): Poly(A) is a polymer of adenosine monophosphate; poly(A-C) is a polymer with alternating A and C nucleosides; poly-$(A\text{-}C)_n$ is a polymer with *n* AC blocks; poly(A,C) is a polymer with statistically distributed A and C; poly-(A) · poly(U): is a duplex consisting of one strand of poly(A) and one of poly(U) held together by hydrogen bonds. If the two polynucleotide chains are not held together by hydrogen bonds, the mixture is denoted by a plus sign, and when there is no information about the interaction between the chains, a comma is used; for example, poly(dC)+poly(dT) or poly(A),poly(A,U).

N. are components of nucleic acids. Free N. act as energy storage molecules, to activate and transport substances, as second messengers (see Cyclic nucleotides) and as coenzymes. Synthetic oligo- and polynucleotides are used as model systems for nucleic acids and as genetic material for genetic engineering.

Nuclide: a form of an element with a fixed mass number; an isotope. The nuclear charge (number of protons) is constant for all atoms of a given element, but the number of neutrons, and hence the mass, can vary. Unstable N. decay with the emission of radiation (see Radionuclides).

Numa: see Synthetic fibers.

Nurex process: a process for isolating straight-chain alkanes from petroleum fractions using urea inclusion compounds.

Nürnberger gold: a copper alloy containing 6% silver, 5% gold and the rest copper. It is used to make jewelry.

Nutmeg oil: a colorless, nonviscous essential oil which eventually turns to a resin due to the action of atmospheric oxygen. It has a spicy odor like that of nutmeg. Large amounts of N. are poisonous, due to its content of myristicine. It consists mainly of (+)-pinene, (+)-camphene (together about 80%), dipentene, (+)-linalool, (+)-borneol, terpineol, geraniol, eugenol, isoeugenol, myristicine, safrol and free and esterified myristic acid. N. is obtained from the fruits of the nutmeg tree by steam distillation. An essential oil obtained from the seed coat of the nutmeg (4 to 15% oil content) is available under the name ***mace oil***, but it is only slightly different from N. Both oils are used to make cosmetic products (e.g. soaps), drugs, liquors (Chartreuse) and flavorings.

Nylander test: see Monosaccharides.

Nylon®: see Synthetic fibers.

Nystatin: a polyene antibiotic used as an antimycotic.

O

o-: 1) see Arenes. 2) abb. for *ortho-*; see Nomenclature, sect. III D.

O: symbol for oxygen.

Obidoxime chloride: an antidote to poisoning by acetylcholinesterase inhibitors (see Parasympathicomimetics) of the phosphate and phosphonate type. O. is similar in structure to Pralidoxime (see), and its effect is also similar, but stronger.

$$\left[HO-N{=}CH-C_5H_4\overset{+}{N}-CH_2-O-CH_2-\overset{+}{N}C_5H_4-CH{=}N-OH\right]\ 2Cl^-$$

Ochre: yellow to reddish, loose, earthy mixture of minerals, mostly iron oxide hydrate or iron oxide and clay. The O. are pigments and can be used in many ways, as they are very fast to light and water. They can be combined with any other pigments and all binders to make floorings, oilcloth, linoleum, paper, fake stones, etc.

Octadecanoic acid: same as Stearic acid (see).

Octahydroquinolizine: same as Quinolizidine (see).

Octamethylene: same as Cyclooctane (see).

Octanal: same as *n*-Caprylaldehyde (see).

Octane: one of the alkane hydrocarbons with the general formula C_8H_{18}. Of the 18 structural isomers (table), 3-methylheptane and 2,3- and 2,4-dimethylhexane have one asymmetric C atom each, and therefore exist as optically active forms or racemates. The O. are colorless, combustible liquids which are practically insoluble in water, but dissolve readily in ethanol and ether. O. are found in petroleum and the products of coal hydrogenation, and can be isolated by alternating fractional distillation and crystallization. Mixtures of isooctane and 2,3,-, 2,3,3- and 2,3,4-trimethylpentane are obtained by catalytic alkylation of butene with isobutane; the butane-butene fraction of cracking gases and cracking gasoline are used as starting materials. Isooctane can also be made by catalytic hydrogenation of isooctene, an industrial byproduct. The highly branched O. have high octane ratings, and are therefore used as components of special fuels, such as aviation gasoline. ***Isooctane*** is the standard on which the octane rating is based. Catalytic dehydrogenation of the O. introduces double bonds and simultaneously cyclizes the compounds to aromatic hydrocarbons, such as toluene, ethylbenzene and xylenes.

Octane-1,8-dicarboxylic acid: same as Sebacic acid (see).

Octanedioic acid: same as Sebacic acid (see).

Octane rating: an indicator of the quality of fuels which expresses the knock resistance. Knocking is caused by uneven and premature combustion proces-

Physical properties of the octanes

Name	Formula	m.p. [°C]	b.p. [°C]	n_D^{20}	Octane rating
n-Octane	$CH_3(CH_2)_6CH_3$	− 57	125.7	1.3975	− 17
2-Methylheptane	$CH_2-CH(CH_3)-CH_2-CH_2-CH_2-CH_2-CH_3$	−109	117.7	1.3949	23.8
3-Methylheptane	$CH_3-CH_2-CH(CH_3)-CH_2-CH_2-CH_2-CH_3$	−121	118.9	1.3985	35.0
4-Methylheptane	$CH_3-CH_2-CH_2-CH(CH_3)-CH_2-CH_2-CH_3$	−121	117.7	1.3979	39.0
3-Ethylhexane	$CH_3-CH_2-CH(C_2H_5)-CH_2-CH_2-CH_3$		118.5	1.4016	52.4
2,2-Dimethylhexane	$CH_3-C(CH_3)_2-CH_2-CH_2-CH_2-CH_3$	−121	106.8	1.3935	77.4
2,3-Dimethylhexane	$CH_3-CH(CH_3)-CH(CH_3)-CH_2-CH_2-CH_3$		115.1	1.4011	78.9
2,4-Dimethylhexane	$CH_3-CH(CH_3)-CH_2-CH(CH_3)-CH_2-CH_3$		109.4	1.3953	69.9
2,5-Dimethylhexane	$CH_3-CH(CH_3)-CH_2-CH_2-CH(CH_3)-CH_3$	− 91	109.4	1.3925	55.7
3,3-Dimethylhexane	$CH_3-CH_2-C(CH_3)_2-CH_2-CH_2-CH_3$	−126	111.6	1.4001	83.4
3,4-Dimethylhexane	$CH_3-CH_2-CH(CH_3)-CH(CH_3)-CH_2-CH_3$		117.7	1.4041	81.7
3-Ethyl-2-methylpentane	$CH_3-CH(CH_3)-CH(C_2H_5)-CH_2-CH_3$	−115	115.7	1.4040	88.1
3-Ethyl-3-methylpentane	$CH_3-CH_2-C(CH_3)(C_2H_5)-CH_2-CH_3$	− 91	118.4	1.4081	88.7
2,2,3-Trimethylpentane	$CH_2-C(CH_3)_2-CH(CH_3)-CH_2-CH_3$	−112	109.8	1.4028	99.9
2,2,4-Trimethylpentane (isooctane)	$CH_3-C(CH_3)_2-CH_2-CH(CH_3)-CH_3$	−107.4	99.2	1.3915	100
2,3,3-Trimethylpentane	$CH_3-CH(CH_3)-C(CH_3)_2-CH_2-CH_3$	−100.7	114.8	1.4075	99.48
2,3,4-Trimethylpentane	$CH_2-CH(CH_3)-CH(CH_3)-CH(CH_3)-CH_3$	−109.2	113.4	1.4042	95.9
2,2,3,3-Tetramethylbutane	$CH_3-C(CH_3)_2-C(CH_3)_2-CH_3$	−100.7	106.5	1.4695	

ses (see Antiknock compounds). By international convention, pure heptane has an O. of zero, and pure isooctane (2,2,4-trimethylpentane) has an O. of 100. The O. gives the percentage of isooctane in a mixture of *n*-heptane and isooctane which has the same knock resistance as the fuel in question. The O. of a fuel is measured in a standardized test motor. There are three common methods, yielding the ***motor O.***, ***research O.*** and ***street O.***, respectively. The ***front O.*** is the O. of the fraction of the fuel which boils below 100 °C. The O. given at a service station is the research O.

Which O. gives the best knock value for a given engine depends on the properties of the engine, the load on it and the way it is driven. For moderate speeds, the research O. is usually applicable, while the motor O. applies better at high speed, on hills and during acceleration. Since the knocking conditions in a test engine cannot be directly applied to the behavior of an engine on the road, the street O. was developed to approximate the behavior of an engine under practical conditions. In general, the street O. is between the motor and research O. The difference between research and motor O. should be as small as possible in good quality gasolines; this difference is called the *fuel sensitivity*.

Octanoic acid, ***caprylic acid***: CH_3-$(CH_2)_6$-COOH, a saturated monocarboxylic acid. The use of the trivial name is no longer permitted in IUPAC nomenclature. O. is a colorless, oily liquid with a rancid odor; m.p. 16.5 °C, b.p. 239.3 °C, n_D^{20} 1.4285. O. is only slightly soluble in water, but is soluble in alcohol, chloroform and acetone. It displays the reactions typical of monocarboxylic acids. O. is found free or esterified in butter, essential and fatty oils, and in the leaves, fruits and seeds of many plants. It can be obtained by saponification of coconut oil or by paraffin oxidation. The esters of O. are used as solvents, softeners, aromas and flavorings. A number of its salts are used in the production of antimycotic salves and powders, and others are used as flotation agents.

Octanols, ***octyl alcohols***: monoalcohols with the general formula $C_8H_{17}OH$, of which 89 structural isomers are possible. Most of these are known. The O. are either insoluble in water or only slightly soluble, but are soluble in alcohol and ether.

n-Octanol (octan-1-ol), $CH_2(CH_2)_6CH_2OH$, a colorless liquid with an aromatic odor; m.p. -16.7 °C, b.p. 194-195 °C, n_D^{20} 1.4293. *n*-O. is found free and as the butyrate and acetate in essential oils. It is produced industrially by catalytic hydrogenation of capric esters or reduction of these esters with sodium and ethanol, and is used as a solvent for paints.

Octan-2-ol, CH_3-$(CH_2)_5$-CH(OH)-CH_3, a colorless liquid; DL-form, m.p. -38.6 °C, b.p. 179-180 °C, n_D^{20} 1.4203. The optically active isomers have $[\alpha]_D^{20}$ 9.8° (ethanol). This O. is obtained industrially from ricinus oil; in the laboratory it can be made by a Grignard reaction of acetaldehyde with *n*-hexylmagnesium bromide. It is used as a solvent for resins and lacquers, and to make softeners.

Isooctanol (2-ethylhexan-1-ol), a colorless liquid, CH_3-$(CH_2)_3$-$CH(C_2H_5)$-CH_2OH, b.p. 180-182 °C, n_D^{20} 1.4318. Isooctanol is one of the most important higher alcohols in industry. It is readily synthesized by aldol condensation of *n*-butyraldehyde with subsequent catalytic hydrogenation of the resulting unsaturated C_8 aldehyde. It can also be made by oxosynthesis from isoheptene. The phthalate, acrylate and adipate of isooctanol are synthesized on a large scale and are used as softeners, especially for PVC products.

Octant rule: an empirically derived rule that predicts the sign of the Cotton effect (see Chiroptic methods) for a considerable number of compounds from their structures. The structure of the molecule is projected onto a system of three mutually perpendicular planes which divide the space into 8 octants. The substituents contribute an amount to the sign of the Cotton effect which depends on their positions in the octants. The O. is used for steric assignment of substituents, especially in optically active ketones, which are commonly encountered in natural products.

Octenes: isomeric hydrocarbons from the alkene series with the general formula C_8H_{16}. There are more than 60 structural isomers, some of which are industrially important (Table). The O. are colorless liquids which are readily ignited; they burn with sooty flames. They are practically insoluble in water, but dissolve readily in ethanol and ether. An especially important product is technical ***isooctene***, essentially a mixture of 2,4,4-trimethylpent-1-ene and 2,4,4-trimethylpent-2-ene, obtained from isobutene by reaction with 70% sulfuric acid. It is catalytically hydrogenated to isooctane, or is converted to *p*-xylene by catalytic dehydrocyclization. It is also used to make isononanol and higher alkylation products of benzene and phenol.

Physical properties selected octenes

Name	Formula	m.p. [°C]	b.p. [°C]	n_D^{20}
Oct-l-ene	$CH_2{=}CH{-}CH_2{-}CH_2{-}CH_2{-}CH_2{-}CH_2{-}CH_3$	−101.7	121.3	1.4087
Oct-2-ene (*Z*)-form	$CH_3{-}CH{=}CH{-}CH_2{-}CH_2{-}CH_2{-}CH_2{-}CH_3$	−100.2	125.6	1.4150
Oct-2-ene (*E*)-form		− 87.7	125.2	1.4132
Oct-3-ene (*Z*)-form	$CH_3{-}CH_2{-}CH{=}CH{-}CH_2{-}CH_2{-}CH_2{-}CH_3$	−126	122.9	1.4135
Oct-3-ene (*E*)-form		−110	123.3	1.4126
Oct-4-ene (*Z*)-form	$CH_3{-}CH_2{-}CH_2{-}CH{=}CH{-}CH_2{-}CH_2{-}CH_3$	−118.7	122.5	1.4148
Oct-4-ene (*E*)-form		− 93.8	122.3	1.4114
2-Methylhept-1-ene	$CH_2{=}C(CH_3){-}CH_2{-}CH_2{-}CH_2{-}CH_2{-}CH_3$	− 90.1	118.2	1.4120
2,4,4-Trimethylpent-1-ene, (diisobutylene)	$CH_2{=}C(CH_3){-}CH_2{-}C(CH_3)_2{-}CH_3$	− 93.5	101.4	1.4086
2,4,4-Trimethylpent-2-ene	$(CH_3)_2C{=}CH{-}C(CH_3)_2{-}CH_3$	−106.3	104.9	1.4160

Octet principle: see Chemical bond.

Octogen: see 1,3,5,7-Tetranitro-1,3,5,7-tetraaza-octane.

Octyl alcohols: same as Octanols (see).

Octyl aldehyde: same as *n*-Capric aldehyde (see).

Odoroside H: see Purpurea glycosides.

O-esters: see Thioesters.

Off-resonance technique: see NMR spectroscopy.

Oil: a general term for liquid organic compounds which are combustible and insoluble in water, but are soluble in organic solvents. See Fats and fatty oils; Essential oils; Mineral oils.

Oil barrier: a special device to contain oil spills on the ocean, in harbors, lakes or rivers. They can be either flexible or rigid. ***Flexible O.*** are usually made of coated cloth. Individual sections can be connected to make a barrier of any desired length. Simple flexible O. are provided with floats and weights so that they stand upright in the water. ***Double hose O.***, which are used if there are waves, consist of an upper hose filled with air and a lower one filled with water. ***Rigid O.*** can also be used when the water is flowing rapidly. They consist of floats and braces which secure the walls. When the flow rate is greater than 1.0 m/s, barges are also used to make a barrier. To keep a spill within a harbor without preventing the exit of ships, perforated pipes are laid on the bottom of the harbor and pressurized air is forced through them. The rising curtain of bubbles restrains the oil.

Oil bath: see Heating bath.

Oil binder: a natural or synthetic material which is capable of absorbing and immobilizing mineral oils. O. must be both oleophilic and hydrophobic. They are spread by ships or aircraft in the form of powders, granulates, flakes or blocks to pick up oil spills. See Oil contamination.

Oil contamination: the contamination of surface or ground waters, sewage systems or soil by mineral oils such as petroleum, gasoline, machine oil, etc. O. can occur through leakage or wreckage of oil wells, tankers, pipelines, storage tanks, and so on. Because of their chemistry and physical properties, mineral oils can cause serious damage under certain conditions. Even very small amounts of petroleum products can make large amounts of water unusable for drinking or industry. The average threshhold for odor perception is a concentration of 1 ppm. Oil which leaks out of damaged tankers on the high seas, or is pumped into the ocean by ships, can cause heavy pollution of large areas of ocean or beaches. On an open water surface, the oil spreads very rapidly and forms a film which prevents gas exchange between the surface of the water and the air. Oil which penetrates the soil can cause severe pollution of groundwater.

The chemical and bacterial degradation of mineral oils in soil, ground water and open waters (see Self purification of waters) is extremely slight, so that damage can be corrected only with very great effort and expense, if at all. Plants are damaged by toxic contact effects. Fish often die of insufficient oxygen, and water birds because their feathers are fouled and stick together. Because most aromatic hydrocarbons are carcinogenic, human consumption of oil-polluted water can lead to serious illness.

Strict safety precautions have been established to prevent O. in oceans and other waters. In case of accidents, the oil is picked up by special pumps or harbor-cleaning boats, or Oil binders (see) are spread out to absorb the oil. Oil barriers (see) are used to localize oil spills, for example in harbors.

Oil green: same as Chrome green (see).

Oil paints: paints in which the binder consists mainly of a plant oil (linseed, wood, ricinus, soy, oiticica, sunflower, poppy, etc.) which has been purified and processed (boiled, dehydrated, polymerized). The formation of the paint film is due to uptake of oxygen and cross linking the oil molecules. The speed of the process depends on the degree of unsaturation of the fatty acids in the oil, and can be accelerated by adding siccatives. O. are prepared by working pigments and additives into the oil.

The compositions of O. depend on their intended use; for interior, exterior or rust protection. Interior O. usually contain a high proportion of oil and little resin; they dry quickly, give a shiny surface and are not weather-resistant. Exterior O. contain a lower proportion of oil and more resin; they dry slowly, are moderately shiny and are weather resistant. O. are usually thinned with terpentine oil and petroleum distillates. They are generally applied by brushing, less frequently by spraying or dipping.

Oil removal: in water treatment, the removal of oil from the waste water from petroleum refineries, coal briquette factories, etc. or from the cooling water from various machines, which contains small amounts of lubricants. Oily deposits in boilers or their plumbing cause machine failures, because they conduct heat about a thousand times less efficiently than steel, and in combination with phosphate sludge, they can lead to stoppage of the pipes. The oil content of boiler feedwater for modern high-pressure steam generaters must not exceed 1 to 3 mg/l. Oil is removed mechanically (this always requires a second stage of O.), or chemically, by activated charcoal. When the oil content is greater than 10 mg/l, a mechanical preliminary O. is always necessary; this is achieved in settling tanks, impact-surface or centrifugal deoilers. In chemical O. (which is only rarely used), aluminum sulfate or iron chloride solution is added to the condensate. At an optimum pH of 6 to 7, the aluminum or iron hydroxide precipitate which forms adsorbs the oil, which is then retained by filtration.

The most common method is absorption by activated charcoal; its effectiveness depends on the water temperature. A good activated charcoal yields a condensate with an oil content of < 1 mg/l and at a water temperature of 100 °C, it can absorb up to 35% of its mass of oil. At 50 to 60 °C, it will absorb only 20 to 25% of its mass. The oil-saturated charcoal cannot be regenerated.

Internationally, the use of *flotation filters* for treatment of condensates is becoming more common. This process has the additional advantage that other impurities, such as rust particles, are also trapped and filtered off. In addition, some of these installations are now fully automated, and due to steady progress in the development of filter aids, they are superior to activated charcoal.

Oil sand, ***tar sand***: bituminous, black mixtures of sand and oil, containing 5 to 18% petroleum. They were formed during the Cenozoic period and are found all over the earth. The oil can be separated from the sand by hot water flotation, centrifugation and hot steam injected directly into the deposit. The oil is not yet economically attractive, but its large-scale utilization in future can be expected.

Oil shale, ***bituminous shale***: a dark colored rock formed by solidification of sapropel. It is rich in bitumen and fossils, clay and lime minerals. It burns with a sooty flame and releases a sulfurous, resinous odor. O. are found in various sedimentary formations, and can eventually produce petroleum. The portion of the bitumen which is not soluble in organic solvents (stable bitumen) is called ***kerogen***. Because its composition is relatively constant, the type of processing used for an O. depends on its inorganic components, primarily its content of limestone. In the thermal process, the stable bitumen is first deoxidized and desulfured, and then depolymerized by increasing the temperature, thus forming a soluble bitumen. This is finally cracked to tar, oil and gas.

Oil tar: see Tar.

OKC resin: a stabilizer (antioxidant) for synthetic rubber and Buna latex; it is based on terpene resin.

Oleanane: see Triterpenes.

Oleandomycin: a macrolide antibiotic with an oxirane structural element in its 14-membered aglycon. O. is labile to acids; its triacetate, in which all the free OH groups are esterified (*Troleandomycin*) is used against infections by gram-positive bacteria.

Oleanolic acid: a pentacyclic triterpene sapogenin derived from β-amyrin. O. is a component of saponins, resins and some vegetable waxes.

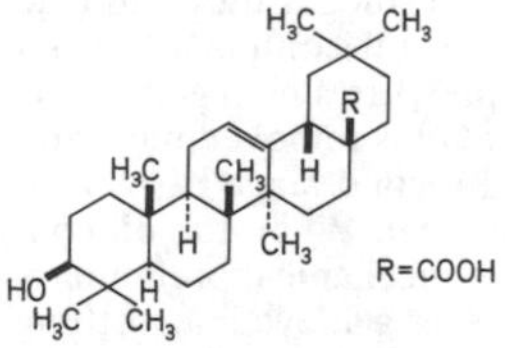

R=COOH

Olefins: same as Alkenes (see).

Olefin complex: see Organoelement compounds.

Oleic acid: *cis*-octadec-2-enoic acid, abb. Ole or 18:1(1), $H_3C\text{-}(CH_2)_7\text{-}CH{=}CH\text{-}(CH_2)_7\text{-}COOH$, a monounsaturated fatty acid. Under certain conditions, it is converted to the thermodynamically more stable *trans*-form, Elaidic acid (see). O. is a colorless or slightly yellow oil; m.p. 13-16 °C. It is nearly insoluble in water, but is miscible with ethanol. O. is the most abundant fatty acid in fats and fatty oils. Olive oil contains 80 to 85% O., and peanut oil contains 55 to 63%. Animal fats contain 35 to 50% O. Technical O. is the residue remaining after saponification of fats and removal of most of the saturated fatty acids by distillation and crystallization. This oil contains 70 to 75% O. as well as linoleic acid and saturated fatty acids. It is used as a textile conditioner and in the production of soaps, bandages and liniments.

Oleophilization: see Flush process.

Oleum, ***fuming sulfuric acid***: an oily liquid, usually brown, with a high density. It is a mixture of sulfuric acid, H_2SO_4, disulfuric acid, $H_2S_2O_7$ and sulfur trioxide, SO_3. O. has a strong dehydrating and oxidizing effect. It often carbonizes organic substances, and this is the source of the color of slightly impure O. When it is heated, O. releases sulfur trioxide, due to thermal cleavage of the disulfuric acid: $H_2S_2O_7 \rightarrow H_2SO_4 + SO_3$. O. is a byproduct of industrial sulfuric acid synthesis by the contact method. It is used as a source of SO_3, and as a sulfonization or oxidizing agent.

Olex process: a process for separation of linear alkenes ($< C_{18}$) from petroleum fractions by adsorption to molecular sieves (see Parex process).

Oleyl alcohol, ***octadec-9-en-1-ol***: $CH_3(CH_2)_7CH{=}CH(CH_2)_7CH_2OH$, an unsaturated, monovalent alcohol; a bright yellow liquid; m.p. 15 to 16 °C, b.p. 177-183 °C at 0.4 kPa. O. is insoluble in water, but is soluble in ethanol and ether. It is found in whale oil and wool grease, and can be synthesized from oleic acid by a special process which preserves the double bond. O. is an important raw material for surface-active substances, such as alkyl polyglycol ethers.

Oligodynamism: the germ-killing effect of extremely small amounts of a heavy metal or heavy metal salt in water containing bacteria. Copper and silver (see Silvering) are especially effective. The cumasina and catadyne processes are based on the oligodynamic effects of heavy metal salts (see Disinfection).

Oligomer: see Polymer.

Oligomer gasoline: see Gasoline.

Oligonucleotide: see Nucleotide.

Oligonucleotide synthesis: chemical preparation of oligonucleotides by successive coupling of internucleotide bonds (phosphodiester bonds) between the 5'-hydroxyl group of one nucleoside and the 3'-hydroxyl group of a second nucleoside. The basic principle of O. is the linkage of the hydroxyl group of one nucleoside (hydroxy or nucleoside component) with the phosphate group of a nucleotide (phosphate or nucleotide component), using a suitable activating reagent to esterify them. The two monomers must be linked in the desired sequence. For construction of a defined oligonucleotide, it is necessary to block reversibly all functional groups which are capable of undergoing undesired side reactions during the formation of the internucleotide bond. The presence of the 2'-hydroxyl group in ribose makes the synthesis of oligoribonucleotides more complicated, since the extra hydroxyl group must be selectively blocked. However, it is the synthesis of oligodeoxyribonucleotides which is of greatest interest in connection with Gene synthesis (see).

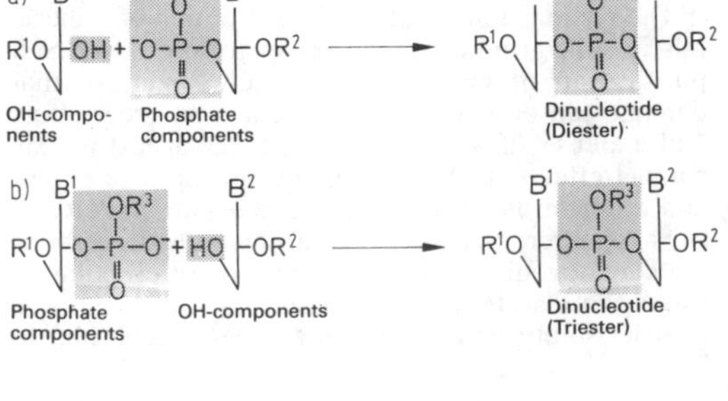

a) OH-components + Phosphate components → Dinucleotide (Diester)

b) Phosphate components + OH-components → Dinucleotide (Triester)

In principle, there are two methods of O., the ***phosphodiester*** (Fig. 1 a) and the *phosphotriester* methods (Fig. 1 b). The absence of the negative charge on the triesters makes them nonpolar, so that they dissolve readily in nonpolar solvents, and also prevents undesired reactions with certain activating reagents. Although oligonucleotides with chain lengths of 10 to 20 bases could be synthesized by the diester method, it is less well suited to the construction of gene segments. The yields are low, and the reaction times are long; in addition, the purification of the intermediates by column chromatography is laborious and causes losses. Even though the solvent extraction and high-pressure liquid chromatography techniques alleviate these problems somewhat, the triester method is still more effective. Variants of the triester method have been developed to work on solid carriers, such as are used in Solid-phase peptide synthesis (see).

Protective groups are required for the primary 5'-hydroxyl group and the secondary hydroxyl group in the 3'-position (and, in the case of oligoribonucleotide syntheses, for the 2'-hydroxyl), for primary amino groups in various nucleotide bases, and for the negative charge of the phosphate ester. The dimethoxytrityl (DMTr) group is most commonly used as the protective group (R^1) for the primary hydroxyl function; it can be removed with aromatic sulfonic acids, trifluoroacetic acid or $ZnBr_2$. Base-labile acetyl or benzoyl groups (R^2) are usually applied to the secondary 3'-hydroxyl group. The amino groups on the bases (B) are usually protected by the anisoyl group on cytosine, benzoyl on adenosine and isobutyryl on guanosine; these are normally removed by ammonolysis. Some commonly used phosphate protective groups (R^3) are the β-cyanoethyl group, which can be removed by β-elimination, and chlorosubstituted phenyl esters with various deblocking methods.

The choice of *activating reagent* for internucleotide bond formation depends on the synthesis strategy. Dicyclohexylcarbodiimide (DCC), mesitylene sulfonylchloride (MS) and triisopropylbenzene sulfonylchloride (TPS) are good for the diester method; for the triester method, 1-(mesitylenesulfonyl)-3-nitrotriazole (MSNT) is preferred.

Although there are many variants of O., the two described below are most widely used.

1) ***Phosphotriester polymer synthesis*** (Fig. 2). The phosphate component 1 is used as the triethylammonium salt of the phosphodiester, which is soluble in pyridine.

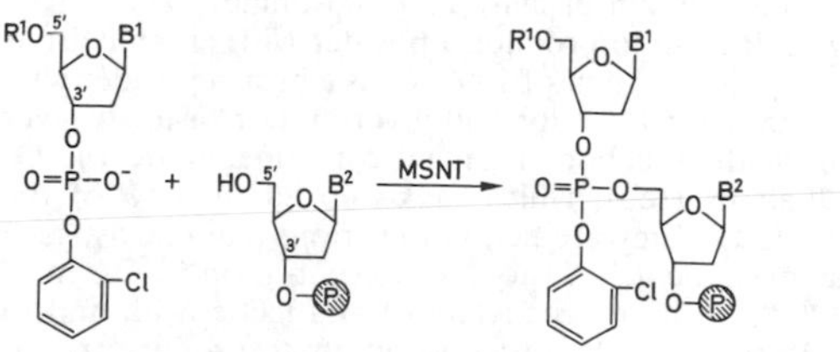

It reacts with the carrier-bound nucleoside component in the presence of the coupling reagent MSNT, to form a polymer-bound dinucleotide, 3. If hydroxyl component 2 is not completely reacted, the 5'-hydroxyl function is acetylated ("capping" step) to prevent formation of wrong sequences. Then the protective group R^1 is removed from compound 3, and the next partially protected nucleotide is linked. With MSNT as activating reagent, the coupling time for each step is about 45 min at 45 °C. About 60 minutes are required for a cycle. For the construction of longer oligonucleotides, it is more convenient to add protected dimers or trimers, which are prepared in solution by the triester method; this considerably reduces the synthesis time. The polymeric carrier is a copolymer of polystyrene and 1% divinylbenzene, or a polyamide or silica gel. As for Solid-phase peptide synthesis (see), there are various commercial semiautomatic and programmed "DNA synthesizers" available. The reaction sequences for coupling and selective cleavage of the 5'-hydroxyl protective group is repeated until the desired sequence has been made on the carrier; then all protective groups and the carrier are split off, and the product is purified by HPLC or polyacrylamide gel electrophoresis. The identity of the product is confirmed by sequence analysis or by a two-dimensional "fingerprint", that is, a two-dimensional separation of the hydrolysis product after incubation with snake venom diesterase.

2) ***Phosphite triester polymer synthesis*** (Fig. 3). In this method, a nucleotide phosphite ester, 5, is formed, then oxidized to the phosphotriester with iodine. The partially protected deoxynucleoside phosphoramidites, 4, prepared from the corresponding nucleoside and chloro(*N,N*-dimethylamino)-methoxyphosphane, $CH_3O\text{-}P(Cl)\text{-}N(CH_3)_2$, are extremely reactive and give nearly quantitative reactions in coupling times of only 5 minutes. A reaction cycle takes about 30 minutes. Silica gels can be used as polymeric carriers. After the synthesis, the protective groups are removed, and the oligonucleotide is split off the carrier and purified as described under 1). Two additional methods of purification are "reversed phase" and gel chromatography.

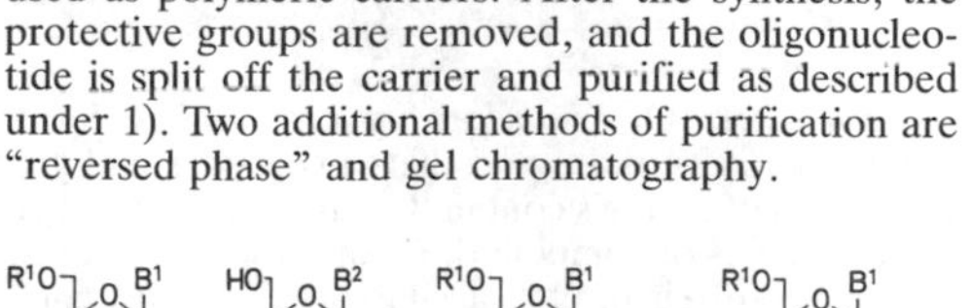

The two methods of synthesis on polymeric carriers permit rapid construction of oligonucleotides, which are the starting materials for gene synthesis and are also used for other molecular biological purposes. For example, they are used as probes for the isolation of mRNA, as starters for sequencing DNA and as linker sequences for genetic engineering operations.

Oligopeptide: see Peptide.

Oligosaccharide: a carbohydrate with 2 to 10 glycosidically linked monosaccharide units. The O. are colorless, crystalline, optically active substances which taste sweet. They are soluble in water and insoluble in alcohols and organic solvents. They can be degraded to monosaccharides by acid- or enzyme-catalysed hydrolysis. The number of monosaccharide units is indicated by a Greek prefix, as in *disaccharide* or *tetrasaccharide*.

There are two types of bond in the O. 1) The glycosidic hydroxyl group -OH of one monosaccharide reacts with the glycosidic hydroxyl group of another monosaccharide. O. of this type (dicarbonyl bond or trehalose type) have no free glycosidic OH group and are therefore nonreducing and unable to form further glycosidic bonds. This type includes sucrose, trehalose and raffinose. 2) The glycosidic OH group of one monosaccharide reacts with an alcoholic OH group of another (monocarbonyl bond). This type of O. have another free glycosidic OH group and react as monosaccharides do; they are reducing, display mutarotation and form glycosides. This group includes, for example, maltose, cellobiose, gentiobiose and lactose. The various structures are formed by the structures of the monosaccharide components, the configuration of the glycosidic bond (α or β) and the position of the alcoholic OH group. α and β (1→4) and (1→6) are the most common types. Natural O. contain mainly hexoses.

The most important oligosaccharides

Trivial name	Structure
Sucrose	β-D-Fruf (1→1)-α-D-Glc
Trehalose	α-D-Clc (1→1)-α-D-Glc
Maltose	α-D-Glc (1→4)-D-Glc
Cellobiose	β-D-Glc (1→4)-D-Glc
Gentiobiose	β-D-Glc (1→6)-D-Glc
Lactose	β-D-Gal (1→4)-D-Glc
Raffinose	α-D-Gal (1→6)-α-D-Glc (1–>2)-β-D-Fruf

Abb. Glc, glucose; Gal, galactose; Fru, fructose; f, furanosyl.

Occurrence. O. are relatively rare in free form. Many plant glycosides and the gangliosides and glycoproteins of animals contain O. glycosidically linked to other types of compound. O. are also obtained by partial hydrolysis of polysaccharides. The most common plant O. are sucrose, raffinose and gentianose. Animal sources of free O. are milk (lactose) and honey.

Oligosaccharide antibiotics: same as Aminoglycoside antibiotics (see).

Ommatins: see Ommochromes.

Ommins: see Ommochromes.

Ommochromes: a group of acidic natural oxazine pigments; they may be red, brown, yellow or violet. O. are found primarily in arthropods (crustaceans, insects, arachnids, cephalopods). Many O. have a characteristic redox behavior, which can be used to identify them: in the reduced state they are red to violet, and in the oxidized state, yellow-brown.

Natural O. can be divided into two groups: the high-molecular-weight, alkali-stable ***ommines*** and the low-molecular-weight, alkali-labile ***ommatines***. At present, the most thoroughly studied ommatin is the yellow-brown ***xanthommatin*** (Fig.), which was isolated in 1954 from the emergence secretions of butterflies and flies.

Xanthommatin

Omnidel®: see Herbicides.

OMPA: see Organophosphate insecticides.

Onic acids: see Aldonic acids.

Onium compounds: ionic compounds containing a coordinatively saturated cation formed by addition of protons or other positive groups to a neutral central atom; examples are ammonium nitrate, NH_4NO_3, triethylphosphonium iodide, $(C_3H_5)_3 \cdot PH^+I^-$ (see Phosphonium salts), triethyloxonium tetrafluoroborate, $(C_2H_5)_3O^+[BF_4]^-$ (see Oxonium salts), trimethylsulfonium tetrafluoroborate, $(CH_3)_3S^+[BF_4]^-$. The corresponding derivatives of nitrogen bases whose names do not end in -amine are called ***inium compounds***; some examples are hydrazinium sulfate, $[N_2H_6]SO_4$, pyridinium bromide, $[C_5H_6N]^+Br^-$, and anilinium chloride, $[C_6H_3NH_3]^+Cl^-$.

Although the cations of triply coordinated carbon, R_3C^+, are called Carbenium ions (see), and its salts are called carbenium salts, the term "carbonium" should be reserved for the derivatives of the $[CH_5]^+$ and $[R_5C]^+$ cations, in accordance with the above definition.

Onsager's reciprocity equation: see Thermodynamics II.

Opacity: a measure of the degree to which an optical system is impenetrable to light. The O. is the reciprocal of the Transparency (see).

Operator: 1) see Atom, models of. 2) see Nucleic acids.

Operon: see Nucleic acids.

Ophiobolins: see Sesterterpenes.

Opiate peptides: same as Endorphins (see).

Opium: the dried latex of unripe seed capsules of the opium poppy, *Papaver somniferum*. There are various cultivated races of the opium poppy, which vary considerably with respect to the amount and kinds of Opium alkaloids (see) they contain. The green, developed fruit capsules of the poppy are scratched to obtain the latex. After a number of hours, the latex flows out, dries in the air and turns brown. It is scratched off and kneaded together. One capsule yields about 0.02 g O. By mixing products with various alkaloid contents, O. is produced with a certain content of Morphine (see); medicinal O. should contain at least 12% morphine. O. is mixed with lactose to produce a powder with an anhydrous morphine content of 10%. O. is a brown powder with a characteristic odor and a very bitter taste. It gives an acidic reaction. Its main components are the O. alkaloids (see), which make up 20 to 30% of the crude O. They are bound to various plant acids, such as meconic acid (3 to 6% concentration in O., m.p. 120 °C with decarboxylation) and lactic acid, and to sulfuric acid. O. is used primarily to produce the alkaloids morphine and codeine. It is an analgesic, but is no longer used for this purpose. Because of its antidiarrhoic effect, it was formerly prescribed as a tincture for diarrhea. A large fraction of illegally produced O. is processed for smoking, or is used as a source of morphine, from which heroin is made. O., morphine and heroin are addictive.

Opium alkaloids: alkaloids present in opium and various other poppy species. So far more than 40 different O. have been isolated; all belong to the isoquinoline group and are biosynthesized from two molecules of tyrosine. The most important types are 1) *morphinane type* (morphine, codeine and thebaine), 2) *benzylisoquinoline type* (papaverine) and 3) *Phthalide isoquinoline type* (narcotine). A variety of partially synthetic derivatives are also considered to be O., such as apomorphine, heroin and ethyl morphine, as well as a variety of other strongly analgesic substances which are no longer in use. The O., especially morphine, are responsible for the activity of opium.

Oppenauer oxidation: conversion of alcohols, especially secondary alcohols, into aldehydes or ketones using acetone or cyclohexanone and aluminum *tert.*-butylate or isopropylate:

$$R_2CH(OH) + O{=}C(CH_3)_2 \underset{}{\overset{\text{(Al-ter.butylate)}}{\rightleftharpoons}} R_2C{=}O + HOHC(CH_3)_2$$

It is assumed that the reaction passes through a cyclic intermediate. The alcohol to be oxidized first forms the aluminum alkoxide, and this in turn forms a complex in which a hydride is transferred from the alkoxide to the carbonyl group.

O. is used mainly to convert secondary alcohol groups into keto groups in delicate natural products of the steroid or alkaloid classes under mild reaction conditions.

Opsin: see Vitamin A.

Optical-acoustic spectroscopy: same as Photoacoustic spectroscopy (see).

Optical activity: the ability of a medium to cause the plane of transmitted linearly polarized light to rotate by a certain amount. O. can be a property of cyrstals or molecules. If it is a property of a crystal, then the crystal lattice is chiral (as in α- and β-quartz), but its components are achiral (SiO_2), and the O. disappears when the crystal melts or dissolves. When the O. is a property of the molecules, it remains when the substance is melted, vaporized or dissolved (see Stereoisomerism). O. is measured in a polarimeter (see Polarimetry).

Optical antipodes: see Stereoisomerism 1.1.

Optical bleach: same as Optical brightener (see).

Optical brightener, ***optical bleach, whitener***: chemical compounds which increase the whiteness of bleached materials, or which counteract yellowing of textiles, paper, paints, waxes, plastics, etc. by optical means. The effect is achieved by substances which absorb in the UV and fluoresce in the blue to blue-violet range. Since this range is complementary to yellow, the material appears whiter and brighter. O. must themselves be colorless in the visible, and, like dyes, they must readily be absorbed to the fibers.

The oldest O. is bluing (e.g. ultramarine); the types now in use are mainly derivatives of stilbene and benzidine or heterocyclic compounds (coumarins, benzimidazoles, cyanins, pyrazolines, etc.). The O. are usually added to about 0.1% concentration to laundry products; less commonly, they are applied to the textiles in the course of the manufacturing process.

Optical density: same as Extinction (see).

Optical purity: see Racemate.

Optical rotatory dispersion: see Chiroptic methods.

OP wax: a shine-producing wax obtained by bleaching and chemical modification of montane wax. O. is used to make various wax products and as a lubricant for polyvinyl chloride.

Orange: a name for several synthetic orange dyes, such as ***orange I*** and ***orange II*** (see Naphthol orange), ***orange III*** (see Methyl orange) and ***orange IV*** (see Tropeolin).

Orange blossom oil, ***neroli oil***: a yellowish, weakly fluorescent essential oil which turns reddish brown in sunlight; it smells strongly of orange blossoms and has a bitter taste. O. is composed of β-ocimene, (-)-α-pinene, (-)-camphene, dipentene, phenylethyl alcohol, (-)-linalool, α-terpineol, geraniol, nerol, capric aldehyde, methyl anthranilate and various other esters. O. is obtained by steam distillation, maceration or enfleurage of freshly picked blossoms of the bitter orange, *Citrus amara*. It is used in the synthesis of expensive perfumes, and is an important component of cologne.

Orbital: a single-electron function which describes the state of an electron in an electronic system. Single-electron functions of an atom are called ***atomic orbitals*** (see Atom, models of), and those of a molecule are ***molecular orbitals*** (see Molecular orbital theory).

Orbital radius: the maximum in the radial distribution of an orbital (see Atom, models of). The O. of the outermost occupied orbital agrees with the theoretical Atomic radius (see).

Orcein: a mixture of red-brown plant pigments obtained from various species of the lichen genus (*Rocella*) (dyer's lichen). It is the colored component of orseille, and consists mainly of α-, β- and γ-aminoorceines and α-hydroxyorcein. O. is used in histology to stain fundamental structures in animal tissue.

ORD: see Chiroptic methods.

Ord: abb. for orotidine.

Ore: 1) ***Ore minerals*** are those minerals which contain one or more metals in chemically bonded state. The nature of the chemical bonds, the metal content and the occurrence of the mineral are such that metal extraction is feasible.

2) Mineral raw materials which can be used for production of metals. They can be mineral aggregates or rocks; in addition to the ore minerals, they contain non-ore components (***gangue, tailings***). By special processing, the content of ore minerals can be increased by separating the materials into a concentrate and non-workable tailings.

O. are found in ore deposits. The value of the metal in an ore is the factor which determines whether the O. can be economically extracted.

Organic electrosynthesis: the production or reaction of organic compounds using electrolysis; either cathodic reduction or anodic oxidation can be used. In *direct* O., the organic substrate reacts directly at the electrode; if an inorganic mediator system is used, the process is *indirect* O. Some examples for cathodic reduction are: 1) hydrogenation of C=C double bonds; 2) reduction of nitro groups ($-NO_2 + 6\,e + 6\,H^+ \rightarrow -NH_2 + 2\,H_2O$); 3) cleavage of carbon-halogen bonds; 4) cathodic dimerization.

Anodic oxidation is used, e.g. 1) for oxidation of alkenes to epoxides; 2) for oxidation of aromatics to quinones; and 3) for dimerization of carbonic acids (Kolbe synthesis).

O. is of industrial significance for production of adipodinitrile by hydrodimerization of acrylonitrile (*Monsanto process*), and for production of tetraethyllead and various other products.

Organic glass: see Polymethacrylates.

Organic coatings: protective layers of polymeric organic substances which protect the underlying material from corrosion and prevent adhesion of solid and semisolid dirt. Most O. form thicker layers than corrosion paints (see Rust preventing paints), and are generated from reactive resin mixtures, thermoplastic sheets, natural or synthetic rubber or powders or emulsions. The protective layer must prevent access of corrosive media to the protected material, and must adhere well to it under chemical, thermal and mechanical loading. This is achieved by pretreating the surface (cleaning, roughening), and if necessary, by use of glues for sheets of plastic or rubber. Thermoplastic sheets of polyvinyl chloride, polyethylene or polytetrafluoroethylene, polybutadiene, polychlorobutadiene or vinylidene fluoride hexafluoropropylene copolymers can protect steel containers against acids, alkalies and corrosive salts. A 3 mm thickness of such a layer forms an effective barrier to diffusion. Polyvinyl and polyethylene coatings are also made as extrusion layers or adhesive ribbons for protection of underground pipelines. Tar and bitumen paints are also still used for this purpose.

Layers of molten, powdered epoxide resins (see Plastic sprays) reduce the frictional resistance in petroleum pipelines, which both increases the lifetime of the pipe and reduces the energy required for transport.

Organoaluminum compounds: Alkyl and aryl derivatives of aluminum, especially those of the types R_3Al, R_2AlX and $RAlX_2$ (X = H, halogen, NR_2, OR, etc.) (see Organoelement compounds). The most important type are the ***alkylaluminums***, and in particular, triethylaluminum (Fig.), which like most O. is dimeric. The Al-C-Cl 3-center, 2-electron bonds provide electron octets for both aluminum atoms. Triethylaluminum is synthesized by reaction of aluminum with ethyl chloride, producing aluminum sesquichloride, which is reduced with sodium: $Et_2AlCl_2AlEtCl + 3\,Na \rightarrow Et_3Al + Al + 3\,NaCl$. On an industrial scale, triethylaluminum is now produced by direct synthesis from aluminum, hydrogen and ethene: $2\,Al + 3\,H_2 + 6\,H_2C{=}CH_2 \rightarrow Al_2(CH_2CH_3)_6$.

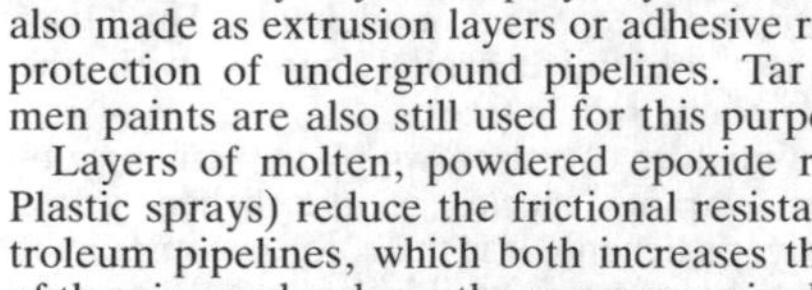

Triethylaluminium

Alkylaluminums are extremely sensitive to oxidation, igniting immediately on exposure to air. They react explosively with water, forming alkanes and aluminum hydroxide. The significance of these compounds is their ability to insert into alkenes. The insertion competes with elimination of the alkyl residue from the alkene, and in this way alkenes can be dimerized or oligomerized. The insertion of ethene in triethylaluminum is industrially important; an average alkyl chain length of about 14 C atoms is desirable. Oxidation and hydrolysis according to

$$Al(C_2H_5)_3 \xrightarrow{(\text{Ethene})} Al(C_{14}H_{29})_3 \xrightarrow{(O_2)} Al(OC_{14}H_{29}) \xrightarrow{(H_2O)} 3\;C_{14}H_{29}OH + Al(OH)_3$$

produces biologically degradable alcohols, which are suitable for production of detergents. Heat treatment of the long-chain alkylaluminums (Ni^{2+} catalysis) can also produce 1-alkenes; direct hydroylsis produces alkanes. Triethylaluminum is also a component of catalyst systems for alkene polymerization.

Organoarsenic compounds: alkyl or aryl derivatives of arsenic (see Organoelement compounds). Some important types are the ***arsines*** R_nAsH_{3-n}, ***arsonium salts*** $R_4As^+X^-$, and ***arsinic acids*** $R_2As(O)OH$. In both structure and reactivity, the O. show considerable analogy to the corresponding Organophosphorus compounds (see). Most O. are poisonous.

Most arsines are synthesized by reaction of organolithium or -magnesium compounds with arsenic(III) chloride, sometimes with subsequent reduction with $LiAlH_4$ or $NaBH_4$. Primary arsines can be made by reduction of the corresponding arsonic acids with zinc. Tertiary arsines react with alkyl halides to form arsonium salts; they form stable complexes with many transition metals. Aromatic arsonic acids are produced by reaction of sodium arsenite with arenediazonium salts (Bart reaction), e.g. $[C_6H_5\text{-}N_2]^+Cl^- + NaAsO_3 \rightarrow C_6H_5\text{-}AsO_3Na_2 + N_2 + NaCl$.

Some O. were important in the early years of chemotherapy (see Arsphenamine). Halogenarsines, e.g. 2-chlorovinylarsinedichloride $ClCH{=}CH\text{-}AsCl_2$ (Lewisite) or diphenylarsine chloride $(C_6H_5)_2Cl$ (Clark I; abb. for *Ch*lor*ar*sen*k*ampfstoffe) are highly toxic substances, as is diphenylarsinecyanide $(C_6H_5)_2As\text{-}CN$ (Clark II). They were used in the First World War as Chemical weapons (see).

Organoboron compounds: alkyl and aryl derivatives of boron (see Organoelement compounds) of the type R_3B, R_2BX and RBX_2, where X = H, halogen, OR, NR_2, etc. The Carboranes (see) are a special class of O. The lower trialkylborons are colorless, poisonous and sensitive to oxidation; they are gases or liquids which are pyrophoric in air. O. are stable to water. They do not tend to associate and are monomeric. Their molecular structure is trigonal planar, based on sp^2 hybridization of the boron. The boron-carbon bond is oxidatively cleaved by hydrogen peroxide or halogens, e.g. $R_3B + 3\,H_2O_2 \rightarrow 3$ R-OH + $B(OH)_3$, $R_3B + 3\,Br_2 \rightarrow 3$ R-Br + BBR_3. O. are obtained from the reaction of organolithium or organomagnesium compounds with boron halides or esters, for example, $BF_3 \cdot OEt_2 + 3$ R-MgBr $\rightarrow BR_3 + 3$ MgBrF + OEt_2, or by Hydroboronation (see). They are used in organic syntheses.

Organochromium compounds: alkyl- and arylchromium compounds and chromium complexes with π-binding ligands; see Organoelement compounds. An example of the alkylchromium compounds is the monomolecular $Cr(CH_2C(Si(CH_3)_3)_3)_4$. Methylchromium derivatives can be synthesized in complex-stabilized form, such as $Li_3[Cr(CH_3)_6] \cdot 3$ dioxane. ***Chromocene***, $Cr^{II}(h^5\text{-}C_5H_5)_2$, which can be made by reaction of $Cr(CO)_6$ with NaC_5H_5, is paramagnetic. It is isomorphic to ferrocene, but is much more reactive. A rather stable and important O. is Bisbenzene-chromium (see), $Cr(C_6H_6)_2$.

Organoelement compounds: compounds in which there are one or more direct carbon-element bonds with elements other than halogens, oxygen, sulfur or nitrogen; compounds with these elements are traditionally included in organic chemistry. The type of bond is unimportant (σ-, π- or aromatic π-bonds all occur). If the element is a metal, it is an ***organometallic compound***. O. do not include derivatives in which the carbon is bound to the element via nitrogen, oxygen, etc., i.e. salts of carboxylic acids, alkoxides, amides or esters of phosphoric acids.

Among the large number of O., there are naturally a variety of different types of bonds, which permit the following classification scheme:

1) ***Ionic O.***, the derivatives of the most electropositive elements, especially of the heavier alkali metals sodium, potassium, rubidium and cesium, are mostly nonvolatile compounds which are insoluble in hydrocarbons. Some examples are sodium acetylide, $Na^+\text{-}C{\equiv}CH$ and magnesium dicyclopentadienylide, $Mg^{2+}(C_5H_5)^-_2$.

2) ***Covalent O.*** are characterized by localized 2-center, 2-electron bonds. They are formed by the less electropositive elements gallium, indium, thallium, the elements of the IVth, Vth and VIth main groups, zinc, cadmium and mercury. Most are monomeric, volatile and soluble in hydrocarbons. Examples: boron trialkyls, R_3B, dibutyltin chloride, $(C_4H_9)_2SnCl_2$, ethanephosphonic acid, $C_2H_5P(O)(OH)_2$, diphenylmercury, $(C_6H_5)_2Hg$.

3) ***O. with electron deficiency structure***, are represented chiefly by the organic derivatives of beryllium, magnesium and aluminum, and to some extent, also by those of lithium and iron. Most are dimeric, oligomeric or polymeric compounds characterized by 3-center, 2-electron bonds. In such bonds, the center atom achieves or at least approaches the noble gas configuration. Examples: methyllithium, $(CH_3Li)_4$, dimethylberyllium, $[(CH_3)_2Be]_n$, and the trialkylaluminums, $(R_3Al)_2$.

There is overlapping between these three categories, as with lithium (ionic/electron deficiency) or boron (electron deficiency/covalent). While the above categorization of the organic derivatives of the main group elements is clearly related to their positions in the periodic system, an understanding of the 4) ***organotransition-metal compounds*** must be based on coordination chemistry, and it is convenient to subdivide these elements according to ligand type (e.g. alkyl, arene complexes, and so on). This justifies special treatment of this class of compound, but before that, a summary of the behavior of the first three types of compound will be presented.

O. of the representative elements (including the derivatives of zinc, cadmium and mercury). Stability and reactivity. Among series of compounds of the main group elements with comparable structures, the thermal stability generally decreases with increasing atomic number of the element. The chemical behavior is determined by the polarization of the element-carbon bond, or, in the case of ionic compounds, by the more or less freely existing carbanion. Most O. are strongly nucleophilic and strong bases. The reactivity is determined primarily by the polarity of the metal-carbon bond, i.e. the electronegativity of the element. Organolithium compounds are therefore generally more reactive than Grignard reagents. In addition, however, the nature of the organic residue is important, which becomes clear when reactivities of different organic derivatives of the same element are compared. For example, the stability of ionic O. increases with increasing stability of the carbanion or, in other words, with the acidity of the hydrocarbon part of the molecule. In covalent O., the reactivity is strongly affected by the solvent, which, if it is a good electron donor, coordinates the positive metal and decreases its electronegativity. In this way the bond polarity is decreased, and the nucleophilicity of the carbon is increased. This is the reason, for example, why organolithium compounds are more reactive in tetrahydrofuran or diethyl ether than in benzene or aliphatic hydrocarbons. As the electronegativity difference between carbon and the element decreases, the stability of the compound increases. One expression of this trend is that the bonds of elements of the IVth and Vth main groups with carbon are not split by air or water.

O. are typically highly susceptible to oxidation and the metal-carbon bond is easily hydrolysed. Many alkyl derivatives of boron, aluminum, beryllium, magnesium, zinc and the alkali metals ignite spontaneously in air, and must therefore be handled in an inert atmosphere. The reactive O. compounds are often explosively hydrolysed by water to the corresponding hydrocarbons and the metal hydroxide, e.g. $C_4H_9\text{-}Li + H_2O \rightarrow C_4H_{10} + LiOH$. For ionic O., this can be understood as the reaction of the strongly basic carbanion with the water; in covalent O., it involves nucleophilic attack of the water on the metal, followed by a proton transfer. Compounds with strongly polarized metal-carbon bonds, e.g. organyls of lithium, magnesium and sometimes aluminum, can react with suitable C=C, C=O and C=N double bonds by insertion, e.g. $C_6H_5\text{-}MgBr + O{=}C{=}O \rightarrow C_6H_5\text{-}COO^- MgBr^+$, and they are therefore useful reagents for organic syntheses (see, e.g. Grignard reactions, Reformatskij reaction).

Synthesis. Some ionic O. can be produced by a direct reaction of the metal with very acidic hydrocarbons, e.g. $Na + C_6H_5\text{-}C{\equiv}CH \rightarrow C_6H_5\text{-}C{\equiv}C^-Na^+ + 1/2\,H_2$. O. can also be made by reaction of more electropositive metals with the relatively easily synthesized organomercury compounds, e.g. C_6H_5-Hg-

$C_6H_5 + 2\ Li \rightarrow 2\ C_6H_5\text{-}Li + Hg$. The most important method is the synthesis from the alkyl or aryl halide and the metal, which is used for production of lithium and magnesium compounds, e.g. $CH_3\text{-}I + 2\ Li \rightarrow CH_3Li + LiI$; $C_6H_5\text{-}Br + Mg \rightarrow C_6H_5\text{-}MgBr$. The latter are very frequently used as starting materials for synthesis of other O., especially the organic derivatives of elements of the IIIrd to VIth main groups and of transition metals; the lithium or magnesium compound simply reacts with the corresponding element halide, simultaneously forming the metal halide and the alkylated or arylated element, e.g. $C_6H_5\text{-}MgBr + (CH_3)_2SiCl_2 \rightarrow (CH_3)_2C_6H_5SiCl + MgBrCl$; $C_4H_9\text{-}Li + (C_6H_5)_2 \cdot BCl \rightarrow (C_6H_5)_2B\text{-}C_4H_9 + LiCl$; $CrCl_3 + 3\ C_6H_5MgBr$ (in THF) $\rightarrow (C_6H_5)_3Cr(THF)_3 + 3\ MgBrCl$. Many O. can be made by addition of organoelement hydrides to C=C double bond systems. This method has proven especially useful in the synthesis of organoboron compounds (see Hydroboration) and organosilicon compounds (hydrosilylation), and in the chemistry of organophosphorus compounds, e.g.:

$(CH_3)_2C{=}CH_2 + H\text{-}BR_2 \rightarrow (CH_3)_2CH\text{-}CH_2\text{-}BR_2$; $(C_2H_5O)_2P(O)H + CH_2{=}CHCN \rightarrow (C_2H_5O)_2P(O)CH_2CH_2CN$.

Application. Organolithium and organomagnesium compounds are essential reagents in organic synthesis (see Grignard reagents), as are organozinc and organomercury compounds (see Reformatskij reaction). Organophosphorus compounds in the form of P-ylides are used in olefination of carbonyls (see Wittig reaction). The organophosphorus compounds are of great industrial interest as biocides, flame retardants, detergents, lubrication additives, etc. The organosilicon compounds are used mainly as Polysiloxanes (see) (silicones); organotin compounds as biocides and PVC stabilizers, tetraethyllead as a gasoline additive and aluminum alkyls for oligomerization of alkenes and production of 1-alkenes and alkanols, as well as components of catalyst systems for alkene polymerization. For detailed information on individual classes of compounds, see Organoaluminum compounds, Organoarsene compounds, etc.

Organotransition-metal compounds In the organic derivatives of the transition elements, the carbon can be bound by both σ- and π-bonds to the metal atom. The stability of the compound is strongly dependent on the electron configuration of the central metal atom, and can be influenced by coordination of other suitable ligands. An empirical rule is that complexes will be particularly stable if the sum of the electrons in s-, p- and d-orbitals of the metal is 18, i.e. the next noble gas configuration has been reached. The number of electrons donated by various organic ligands toward completion of the electron shell is shown in the table.

In the transition metal complexes of polyenes, aromatics, etc., indication of the number of organic ligands is not sufficient for a complete description of the compound, because not all π-bonding electron pairs are necessarily interacting with the metal atom. Therefore, in 1968, F.A. Cotton introduced the *hapto system* ("hapto" is the Greek for "to fasten" or "to bind"), in which the number of atoms of the ligand which are directly bonded to the central ion is indicated by the prefixes "monohapto-", "dihapto-", "trihapto-", etc. (h^1, h^2, h^3, etc.). For example, the allyl group can act as a σ-bonded monohapto-, a σ- and π-bonded mono- and dihapto- or as a π-bonded trihapto ligand (Fig. 1).

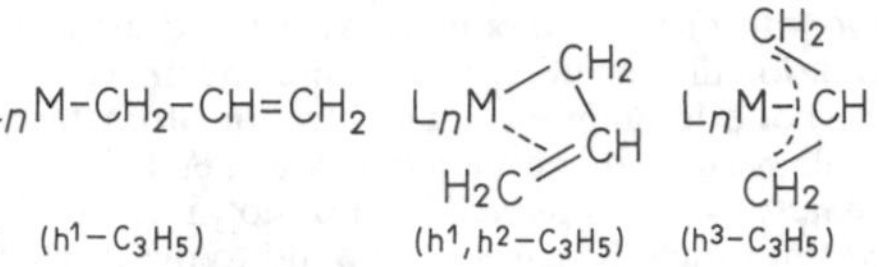

Classification of organic groups as ligands in organotransition-metal compounds

No. of electrons	Term	Example
1	-yl	CH_3, CF_3, $CH_2\text{-}CH{=}CH_2$, C_6H_5
2	-ene	$H_2C{=}CH_2$, $F_2C{=}CF_2$
3	-enyl	*n*-Allyl
4	-diene	Butadiene, cyclobutadiene
5	-dienyl	π-Cyclopentadienyl
6	-triene	Benzene
7	-trienyl	Cycloheptatrienyl
8	-tetraene	Cyclooctatetraene

1 ($h^1-C_3H_5$) ($h^1,h^2-C_3H_5$) ($h^3-C_3H_5$)

A complex in which the three different bonding possibilities of the cyclopentadienyl ligand are realized is (monohaptocyclopentadienyl)(trihaptocyclopentadienyl)(pentahaptocyclopentadienyl) nitrosylmolybdenum (Fig. 2). The Greek letter η is often used instead of the letter "h".

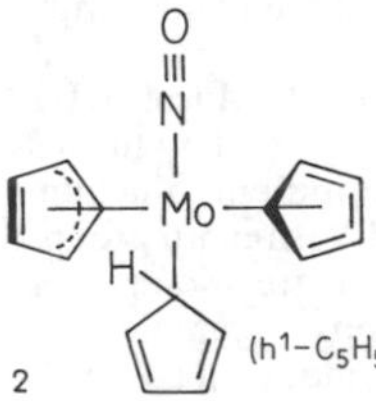

2 ($h^1-C_5H_5$)($h^3-C_5H_5$)($h^5-C_5H_5$)Mo(NO)

Structure and synthesis. ***One-electron ligands***. The thermal stability of transition metal derivatives with C-metal σ-bonds increases with increasing electronegativity of the carbon. For example, $(CO)_4Co\text{-}CH_3$ is stable only to -30 °C, but $(CO)_4\text{-}Co\text{-}CF_3$ can be distilled at 91 °C. Stability is also increased by the opportunity to form additional π-relationships. Therefore, transition metal aryls are generally more stable than alkyls; alkynyl complexes, in which the bonding is similar to that in metal carbonyls or cyanocomplexes (see Coordination chemistry) are not very sensitive to heat. For example, $[(Et_3P)_2Ni(C{\equiv}C\text{-}Ph)_2]$ melts without decomposition at 149-151 °C. The thermal lability of transition metal complexes containing longer alkyl groups is due primarily to the rapid olefin elimination by β-hydride shifting (reductive elimination) (Fig. 3).

$$\underset{L_nM-CH_2}{\overset{H}{\searrow}}CH-R \rightleftharpoons \left[\begin{array}{c}H---CHR\\ \vdots\quad\vdots\\ L_nM---CH_2\end{array}\right] \rightleftharpoons L_n\overset{H}{M}-\overset{CHR}{\underset{CH_2}{\|}} \rightleftharpoons L_nMH + CH_2{=}CHR$$

3

σ-Organotransition-metal compounds are obtained by alkylation or arylation of suitable coordination compounds with reactive organometallics: $[(Et_3P)_2PtCl_2] + CH_3MgI \rightarrow [(Et_3P)_2Pt(CH_3)I] + MgCl$; $[(Et_3P)_2PtCl_2] + 2\,CH_3Li \rightarrow [(Et_3P)_2Pt(CH_3)_2] + 2\,LiCl$. A second method is the reaction of complex transition metal anions with alkyl halides, e.g. $Na^+[Mn(CO)_5]^- + CH_3I \rightarrow [(CO)_5MnCH_3] + NaI$.

The Carbene complexes (see) and Carbyne complexes (see) of transition metals are of great current interest.

Olefin complexes The simplest synthesis is by substitution of the olefin for suitable ligands; in this way the oldest metal-olefin compound, Zeise's salt $K[PtCl_3(C_2H_4)]$ (see Organoplatinum compounds) was made in 1827 by reaction of ethene and potassium tetrachloroplatinate. Metal carbonyls are especially suitable for such exchange reactions, e.g. $Fe(CO)_5$ + butadiene $\rightarrow$ $(CO)_3Fe(CH_2{=}CHCH{=}CH_2) + 2\,CO$. The bonding of the olefin to the metal is a σ interaction of the bonding π-electron pair with a metal orbital of suitable symmetry, and a simultaneous back donation from filled metal d orbitals into antibonding molecular orbitals of the olefin (Fig. 4). Olefins with two or more isolated double bonds are bonded to the metal in analogous fashion. Delocalized π-molecular orbitals are involved in the bonding of conjugated dienes, e.g. that of the 4-electron ligand butadiene to iron (see above).

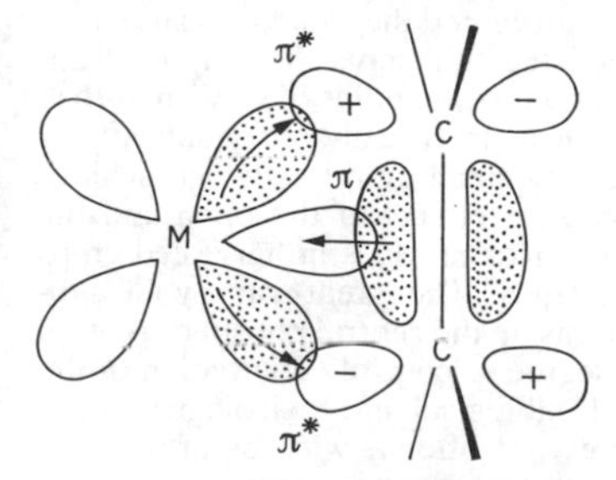

Fig. 4. Diagram of the bonding between ethene and a transition metal ion.

Alkyl complexes. As an example, let us take π-allylcobalt tricarbonyl (Fig. 5a), which is formed by the reaction of allylbromide and $Na^+[Co(CO)_4]^-$ via the σ-allyl complex $[(CO)_4Co\text{-}CH_2CH{=}CH_2]$. The latter eliminates CO to form the π-allyl compound. Another example is bis-π-allylnickel (Fig. 5b), which is made by reaction of nickel chloride with allylmagnesium bromide. The C-C distances in the allyl groups are identical, and the H atoms lie in the plane of the allyl carbon atom. The ability of the allyl residue to act either as a σ-bonded one-electron ligand or as a π-bonded three-electron ligand makes possible rapid σ,π-conversions (σ,π-equilibration), which is the basis of some technical processes.

CH
H2C CH2
Co
OC CO
CO
5 a

CH2
HC
CH2
Ni
H2C CH
b H2C

Cyclopentadienyl systems. Nearly all transition elements are known to form sandwich-like, pure bis-π-C_6H_5 complexes (***metallocenes***) as well as mixed compounds containing other ligands. Some examples are ferrocenes (Fig. 6a) and the π-cyclopentadienylmanganese tricarbonyl (Fig. 6b). The bonding of the C_5H_5 ligand to the transition metal atom is interpreted as an interaction of the π-electron sextet of the cyclopentadienyl anion with the empty d-orbitals of the metal. These compounds are often quite stable to heat, and permit reactions on the C_5H_5 ring which indicate its aromatic character. Metallocenes are usually obtained by reaction of metal halides with sodium cyclopentadienide, e.g. $FeCl_2 + 2\,NaC_5H_5 \rightarrow (h^5\text{-}C_5H_5)_2Fe + 2\,NaCl$.

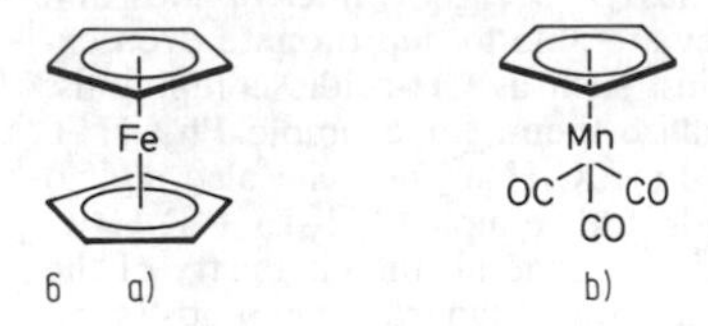

Arene complexes π-Electron systems like benzene, the tropylium cation or the cyclopentadienyl anion form aromatic complexes with many transition metals. A typical example is bis-π-benzenechromium (Fig. 7), in which the two arene rings are planar and parallel to one another. All C-C and C-Cr distances are the same.

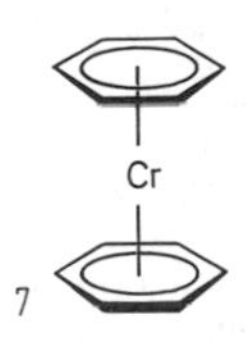

Applications. Organotransition-metal compounds have become extremely important for the activation of olefins in numerous industrially important syntheses; they are components of catalyst systems for hydrogenation, polymerization, oxidation and oxosyntheses of these compounds.

Organogermanium compounds: see Germanium compounds.

Organogold compounds: gold(I) and gold(III) alkyls and aryls. (see Organoelement compounds). O. were among the first transition metal organometal compounds to be discovered. They include both alkylgold(I) complexes, such as those of the type $CH_3AuP(C_6H_5)_3$, and numerous gold(III) derivatives, such as $LiAu(CH_3)_4$ and R_2AuX. The especially stable dialkylgold halides can be made by reaction of gold(III) halides with Grignard compounds; they are halogen-bridged dinuclear complexes. Dialkylgold cyanides, on the other hand, have tetrameric structures.

Organolead compounds: alkyl or aryl derivatives of lead, R_nPbX_{4-n}, where X = halogen, OR, Sr, etc. (see Organoelement compounds). ***Tetraalkylleads*** are of special interest. They are synthesized in a complicated reaction of organolithium or Grignard compounds with lead(II) chloride, e.g. $4\ CH_3Li + 2\ PbCl_2 \rightarrow (CH_3)_4Pb + 4\ LiCl + Pb$. They may also be synthesized directly from the alkyl halide and a lead-sodium alloy. The latter method is used to synthesize ***tetraethyllead***, $(C_2H_5)_4Pb$, a colorless, water-insoluble, oil which is extremely poisonous. M_r 323.44, density, 1.659, m.p. -136.80 °C, b.p. 91 °C at 2.5 kPa. Like tetramethyllead, tetraethyllead is used as an Antiknock agent (see) in gasolines.

Organolithium compounds: colorless, covalent Organoelement compounds (see) which are soluble in hydrocarbons. They are electron-deficiency structures which are very reactive, due to the high polarity of the carbon-lithium bond $\rangle C^{\delta -}\text{—}Li^{\delta +}$, and this can be increased by a suitable donor solvent, such as ether or THF. Thus O. are good nucleophiles and strong bases. They are able to deprotonate even extremely weak acids, such as CH-acidic compounds, and thus to metallize them, for example $Ph_3CH + C_4H_9\text{-}Li \rightarrow Ph_3C\text{-}Li + C_4H_{10}$. They can also add to polar double bonds, for example $C_6H_5\text{-}Li + R_2C{=}O \rightarrow C_6H_5R_2C\text{-}O^-Li^+$. In general, the chemistry of the O. is similar to that of the Grignard compounds (see), but they are distinctly more reactive than the organomagnesium compounds. They are extremely susceptible to oxidation, and are often pyrophoric in air. They react explosively with water.

Methyllithium, $CH_3\text{-}Li$, is a solid. It forms a tetrameric aggregate in which the methyl groups are arranged on the triangular surfaces of a tetrahedron consisting of 4 Li atoms. The aggregate is held together by four 4-center, 2-electron bonds, each involving 3 Li and 1 C atom. ***Butyllithium***, $C_4H_9\text{-}Li$, is a liquid at room temperature. In alkane solution it is a hexamer, and in ether, a tetramer.

O. are formed by direct synthesis from an alkyl or aryl halide and lithium, e.g. $C_4H_9\text{-}Cl + 2\ Li \rightarrow C_4H_9\text{-}Li + LiCl$. They can be made by the reaction of lithium with organomercury compounds, e.g. $(C_6H_5)_2Hg + 2\ Li \rightarrow 2\ C_6H_5\text{-}Li + Hg$. Aryllithium compounds can be made conveniently from alkyllithium and aryl halides by *metal-halogen exchange*, e.g.

$$\text{(2-Br-}C_6H_4\text{-}OCH_3) + C_4H_9\text{-}Li \rightarrow \text{(2-Li-}C_6H_4\text{-}OCH_3) + C_4H_9\text{-}Br$$

O. are extremely important as reagents for organic and organoelement synthesis. They serve as catalysts for alkene polymerization and as bases in the Wittig reaction.

Organomagnesium compounds: Organoelement compounds (see) derived from magnesium. There are two types, the dialkyl or diaryl derivatives R_2Mg, and the organomagnesium halides, R-Mg-X, which are known as Grignard compounds (see).

Organomercury compounds, ***mercury organic compounds***: alkyl- and arylmercury(II) Organoelement compounds (see) of the types HgR_2 and RHgX. They are made by reaction of sodium amalgam with alkyl or aryl halides, or by reaction of mercury(II) chloride with Grignard compounds: $RMgX + HgCl_2 \rightarrow RHgCl + MgXCl$; $RHgCl + R\ MgX \rightarrow HgR_2 + MgXCl$. Compounds with the general formula RHgX are crystalline solids, while alkyl- or arylmercury(II) compounds are extremely toxic liquids or low-melting solids with linear moleccules: $Hg(CH_3)_2$, b.p. 92.5 °C; $Hg(C_2H_5)_2$, b.p. 159 °C; $Hg(C_6H_5)_2$, m.p. (sublimation) 121.8 °C, b.p. 204 °C (at 1.3 kPa). O. are used as fungicides and in the production of other organometal compounds. Arylmercury derivatives of the type RHgX (X = e.g. CH_3CO_2, NO_3) can be made by an interesting direct reaction, called *mercurization*, of mercury(II) salts with aromatic compounds.

Organometal compounds: see Organoelement compounds.

Organophosphate insecticides: a group of insecticides chemically derived from phosphoric acid, $(HO)_3P{=}O$. The applicability of esters and other derivatives of phosphoric acid was discovered in the 1930s, as the industrialization of agriculture led to a demand for insecticides which outstripped the production of natural insecticides (see Nicotine, Pyrethrum). Since the introduction of Parathion (see) in 1947, the development, study and production of this group has been intense. The decline in the use of persistent Chlorinated hydrocarbon insecticides (see) since the 1960s has promoted the trend. Because most of them are esters, the O.i. have the major advantage, from the environmental point of view, that they are easily degraded, hydrolytically, enzymatically or biologically. The very small amounts which achieve the desired insecticidal effects in the open and the much reduced danger of residues in harvested crops are further advantages. The great variety of substituent combinations on the central phosphorus atom make possible a degree of controlled variation of the chemical, physical, biological and toxicological properties, so that the O.i. offer a wide assortment of products for pest control in agriculture and silviculture, and also in hygiene and health protection. They act with varying degrees of intensity as contact, ingestion or respiratory insecticides, and often as systemic agents as well (see Systemic pesticides).

The action of O.i. is similar in insects and warm-blooded animals. The main effect is usually phosphorylation and consequent inhibition of acetylcholinesterase; in addition, there are alkylation reactions, especially methylations, which offer a possibility for differentiation between insecticidal effect and toxicity to warm-blooded animals. Such toxicity ranges from high to low to almost absent.

The symptoms of poisoning with O.i. may include nausea, vomiting, stomach cramps, profuse sweating, fainting, headaches, partial loss of consciousness, tachycardia, hypertonia, mydriasis and respiratory arrest.

Therapy: fresh air, washing of affected skin with flowing water and soap. When ingested orally, induce vomiting (with Glauber's salt); do not administer castor oil, milk or alcohol! Specific counter measures: for treating shock, high doses of atropine; afterwards cholinesterase activators (e.g. obidoxime chloride).

Classification: O.i. can be chemically classified as phosphate esters and thioesters, dithiophosphates, phosphonates and thiophosphonates. The ***phosphate esters*** used as O.i. are made mainly by reaction of trialkylphosphites with α-halogenated carbonyl compounds, which leads to O,O-dialkyl-O-vinylphosphates. The main representative of this group is Dichlorvos (see). Other representatives are shown in Table 1.

$$(R^1O)_2P{-}OR^1 + X{-}C(R^2)(R^3){-}CO{-}R^4 \xrightarrow{-R^1X} (R^1O)_2P(=O){-}O{-}C(R^4){=}C(R^3)(R^2)$$

Table 1. Important phosphate ester insecticides

Name	Formula	peroral LD_{50} [mg/kg rat]
Dichlorvos (DDVP)	$(CH_3O)_2P(O){-}O{-}CH{=}CCl_2$	50 ... 80
Noled	$(CH_3O)_2P(O){-}O{-}CHBr{-}CBrCl_2$	450
Mevinphos	$(CH_3O)_2P(O){-}O{-}C(CH_3){=}CH{-}COOCH_3$	3,7
Monocrotophos	$(CH_3O)_2P(O){-}O{-}C(CH_3){=}CH{-}CO{-}NHCH_3$	21
Phosphamidon	$(CH_3O)_2P(O){-}O{-}C(CH_3){=}C(Cl){-}CO{-}N(C_2H_5)_2$	10
Chorfenvinphos (X = Cl)	$(C_2H_5O)_2P(O){-}O{-}C(C_6H_3Cl_2){=}CHX$ (2,4-dichlorophenyl)	24 ... 39
Bromfenvinphos (X = Br)		52 ... 64

The systemic insecticide and acaricide dimefox (oral LD_{50} = 5 mg/kg rat) can also be considered a derivative of phosphoric acid; chemically it is bis-(dimethylamino)phosphoryl fluoride. It is used mainly for aphids on hops. It is sometimes combined with the systemic insecticide OMPA (octamethyl pyrophosphoramide), a derivative of diphosphoric acid with LD_{50} (oral) = 10 mg/kg rat.

The most important of the ***thiophosphate esters*** have two alkyl ester residues and one aryl or heteroaryl ester component in the molecules. These compounds exist primarily in the thiono form, and are synthesized by successive reactions of phosphosulfochloride with alcohols and the alkali salt of a phenol or hydroxyheterocycle ($NaO{-}R^2$):

$$PSCl_3 + 2R^1OH \xrightarrow{-2HCl} (R^1O)_2P(S)Cl \xrightarrow[-NaCl]{+NaO-R^2} (R^1O)_2P(S)OR^2$$

The main representative is Parathion (see), which is shown with other representatives in Table 2.

Table 2. Important insecticidal thiophosphate esters $(R^1O)_2P(S)OR^2$

Name	R^1	R^2	peroral LD_{50} [mg/kg rat]
Parathion	$-C_2H_5$	4-nitrophenyl ($-C_6H_4{-}NO_2$)	6 ... 15
Methyl-parathion	$-CH_3$	4-nitrophenyl ($-C_6H_4{-}NO_2$)	12 ... 22
Fenitrothion	$-CH_3$	3-methyl-4-nitrophenyl (NO_2, CH_3)	242 ... 433
Bromophos	$-CH_3$	4-bromo-2,5-dichlorophenyl (Cl, Br, Cl)	3700 ... 6100
Diazinon	$-C_2H_5$	6-methyl-2-isopropylpyrimidin-4-yl (CH_3, $CH(CH_3)_2$)	100 ... 220
Demeton	$-C_2H_5$	$-(CH_2)_2{-}S{-}C_2H_5$	6 ... 12
Methyl-demeton	$-CH_3$	$-(CH_2)_2{-}S{-}C_2H_5$	40 ... 60

The industrial forms of thiophosphate esters with three (substituted) alkyl ester residues in the molecule usually contain, in addition to the thiono form, a substantial amount of the thiol form. This is generally more water-soluble, more resistant to hydrolysis, more active biologically, but also more toxic than the thiono form. Heat, water and other polar solvents promote the conversion to the thiol form, which also arises under plant-physiological conditions and has pronounced systemic properties. To synthesize these compounds, O,O-dialkylthiophosphoryl chloride is reacted with a substituted alkanol $HO{-}R^2$, usually an alkylthioalkanol:

$$(R^1O)_2P(S)Cl + HO{-}R^2 \xrightarrow[-HCl]{(OH^-)} (R^1O)_2P(S)OR^2 \text{ (Thionophosphate)} \rightarrow (R^1O)_2P(O)SR^2 \text{ (Thiolphosphate)}$$

The main representative of this group is methyl demeton (Table 2).

Methamidophos contains an O,S-dimethylthiophosphoramide (oral LD_{50} = 30 mg/kg rat); it is used as a systemic insecticide and acaricide. The N-acetyl derivative is acephate, which has a much lower toxicity (oral LD_{50} = 945 mg/kg rat).

The starting material for industrial synthesis of dithiophosphorate esters is dialkyldithiophosphoric acid, which is obtained by alkoholysis of P_4S_{10} and alkylated (e.g. Dimethoate – see –, disulfoton). Alternatively, its anion is added to a conjugated system (e.g. see Malathion):

$$P_4S_{10} \xrightarrow[-2H_2S]{8ROH} 4\ (RO)_2P(=S)SH$$

$$(R^1O)_2P(=S)SH \xrightarrow[-H_2O \quad -NaX]{1)\ +NaOH \quad 2)\ +XR^2} (R^1O)_2P(=S)-S-R^2$$

$$(R^1O)_2P(=S)SH \xrightarrow{R^3-CH=CH-R^4} (R^1O)_2P(=S)-S-CH(R^3)-CH_2(R^4)$$

The insecticidal dithiophosphate esters are characterized by high systemic activity, good depth activity, and are used to combat pests, especially in field and fruit crops (Table 3).

Table 3. Important insecticidal dithiophosphate esters $(R^1O)_2P(S)SR^2$

Name	R^1	R^2	peroral LD_{50} [mg/kg rat]
Malathion	$-CH_3$	$-CH(COOC_2H_5)-CH_2-COOC_2H_5$	1375 . . . 2800
Dimethoate	$-CH_3$	$-CH_2-CO-NHCH_3$	250 . . . 310
Methyl-azinphos	$-CH_3$	$-CH_2-N$ (4-oxo-benzo-1,2,3-triazine ring)	11 . . . 20
Phosmet (Imidan®)	$-CH_3$	$-CH_2-N$ (phthalimide ring)	230 . . . 299
Disulfoton (Disyston®)	$-C_2H_5$	$-(CH_2)_2-S-C_2H_5$	3 . . . 12
Thiometan	$-CH_3$	$-(CH_2)_2-S-C_2H_5$	85
Phorate	$-C_2H_5$	$-CH_2-S-C_2H_5$	2 . . . 4
Phencapton	$-C_2H_5$	$-CH_2-S-C_6H_3Cl_2$ (2,5-dichlorophenyl)	65 . . . 182

The most important insecticidal phosphonic and thiophosphonic esters are Trichlorfon (see) and EPN (see). They are characterized by a carbon-phosphorus bond in the molecule.

Organophosphorus compounds, ***phosphoorganic compounds***: compounds which contain one or more P-C bonds (see Organoelement compounds). However, the esters of phosphoric and phosphorous acids are often also called O., incorrectly, especially in industry. Because of the great variability of phosphorus with respect to its oxidation and coordination numbers, there are O. with widely varying structures. Some important classes are primary, secondary and tertiary phosphines, R_nPH_{3}-n, tertiary phosphine oxides, R_3PO, Phosphonium salts (see) $R_4P^+S^-$, phosphorylides, $R_3PCR_2 \leftrightarrow R_3P{=}CR_2$ (see Ylides), phosphonous RP(O)HOH and phosphonic acids $RP(O)(OH)_2$, phosphinous (sec. phosphine oxides) $R_2P(O)H$ and phosphinic acids, $R_2P(O)OH$. There are many derivatives of the organic phosphoric acids, including esters, halides, amides, their thio analogs, etc. The phosphoranes R_5P, monomeric and polymeric phosphazenes, $R_3P{=}NR$ or $[-R_2P{=}N-]_nD$ (for example, dichlorophosphazene; see Phosphorus nitride dichloride), and many heterocyclic compounds which contain phosphorus are of special interest. Quite recently, phosphaalkenes, $RP{=}CR_2$, and phosphaalkynes, $RC{\equiv}P$, have been synthesized. These compounds break the double bond rule and contain the phosphorus as part of the p_π-double or triple bond systems.

The P-C bond is formed by addition of PH functional phosphines or phosphites to C=C, C=O or C=N double bonds, as in PH_3P + Me-I $\rightarrow PH_3P^+$ MeI^-, $(EtO)_3P$ + Ph-CH_2Br $\rightarrow$ Ph-CH_2-P(O)$(OEt)_2$ + Et-Br (see Michaelis-Arbuzov reaction). Other possibilities are the reaction of phosphorus trichloride with aromatics under the conditions of the Friedel-Crafts reaction, e.g.

$$C_6H_6 + PCl_3 \xrightarrow{(AlCl_3)} C_6H_5\text{–}PCl_2 + HCl;$$

the reaction of Grignard reagents or lithium or aluminum organyls with phosphorus halides, and other methods.

O. differ in a few significant ways from their nitrogen analogs. The bond angles in phosphines are generally smaller than in amines. The inversion barrier for trivalent P compounds is about 140 kJ mol^{-1}, so that unsymmetrical *tert.* phosphines can be separated into optical isomers. Phosphines form stable complexes with many transition metal ions. They are stronger nucleophiles, but weaker bases than amines; the PH function is much more acidic than the NH group.

The behavior of O. is greatly affected, on the one hand, by the inability of phosphorus in the +3 valence state to take part in π-coordination systems, and on the other hand, by the tendency of tetracoordinated phosphorus to form stable $d_\pi p_\pi$ bonds. This determines the stereochemistry and reactivity of many P. heterocycles, is responsible for the high stability of the P ylides and phosphazenes, and explains why the formation of a phosphoryl function, -PO, contributes to the driving force of many reactions of the O. Penta-coordinated states of phosphorus are relatively stable and often undergo rapid isomerization (see Berry pseudorotation).

O. have great economic importance because of their versatility. Phosphates and thiophosphates (which are not O. in the strict sense!) are used as pesticides in agriculture and human hygiene. 2-Chloroethane phosphonic acid, $Cl\text{-}CH_2CH_2P(O)(OH)_2$, is used as a ripening accelerator and straw stabilizer. Other phosphonic acid derivatives are used as flame retardants for textiles and polyurethane foams, while *tert.*-phosphine oxides, such as Bu_3PO and Oct_3PO, are used as extraction solvents for various metals. Other O. are used as additives to lubricating oils, flotation agents, softeners and stabilizers. O. are also significant as Chemical weapons (see).

Organoplatinum compounds: alkyl and aryl platinum derivatives and platinum complexes with π-bonding ligands (see Organoelement compounds). The trimethylplatinum(IV) halides are exceptionally stable, tetrameric molecules: $((CH_3)_3PtX)_4$. $((CH_3)_3PtI)_4$ is orange, and is obtained by reaction of $PtCl_4$ with methylmagnesium iodide. Platinum(II) alkyls and aryls are colorless solids, stable in air, which are obtained by reaction of platinum(II) halides with organolithium compounds or Grignard reagents. ***Zeise's salt***, $K[PtCl_3C_2H_4]$, discovered by the Danish pharmacist C.W. Zeise, is obtained by reaction of $K_2[PtCl_4]$ with ethanol. It was was the first olefin complex to be discovered. It is characterized by a Pt-ethene π-interaction in which the bond from the Pt atom is perpendicular to the ethene C=C axis.

Organosilicon compounds: Alkyl or aryl derivatives of silicon, in particular R_4Si, R_3SiX, R_2SiX_2 and $RSiX_3$ (X = H, halogen, NR_2 OR, etc.). The silicon-carbon bond is stable to air and water. The industrially important alkyl chlorosilanes are obtained by direct synthesis from silicon and alkyl chlorides in the Müller-Rochow process. On a laboratory scale, the O. can be synthesized by reaction of halosilanes with organolithium or Grignard reagents, e.g. $HSiCl_3 + 3\ C_6H_5MgBr \rightarrow (C_6H_5)_3SiH + 3\ MgBrCl$. Polysiloxanes (see), commonly known as silicones, are economically important. Tetramethylsilane (TMS), $(CH_3)_4Si$, and hexamethyl siloxane (HMDS), $(CH_3)_3SiOSi(CH_3)_3$, are used as standards in NMR spectroscopy.

Organosodium compounds: see Organoelement compounds.

Organotin compounds: alkyl and aryl derivatives of tin, R_nSnX_{4-n} (X = H, halogen, OR, NR_2, etc.) (see Organoelement compounds). Stannylenes analogous to carbenes, R_2Sn:, appear as highly reactive intermediates in reactions of certain O. O. which carry small alkyl groups are often toxic. The carbon-tin bond is fairly stable to hydrolysis and oxidation. Unlike comparable carbon or silicon derivatives, organotin halides have a marked tendency to interact with donor molecules and to form pentacoordinated states, e.g.

R_3SnHal + Pyridine → $C_5H_5N\text{-}SnR_3\text{-}Hal$

Tetraalkyl- or tetraaryltin compounds are synthesized by reaction of organolithium compounds or Grignard reagents with tin tetrachloride, e.g. $4\ C_6H_5MgBr + SnCl_4 \rightarrow (C_6H_5)_4Sn + 4\ MgBrCl$. They can also be made by addition of organotin hydrides to C=C double bond systems (hydrostannylation). Organotin halides are of particular importance, because the halogen can be readily exchanged for other groups. They are synthesized in two ways: one is by redistribution reactions, in which tetraalkyltin compounds are heated with tin tetrahalides, with the stoichiometric ratio of the reactants determining the composition of the products, e.g. $3\ (CH_3)_4Sn + SnCl_4 \rightarrow 4\ (CH_3)_3SnCl$. The other, which is used especially on an industrial scale, is direct synthesis from tin or an alloy and the respective alkyl halide.

Organotitanium compounds: a group of Organoelement compounds (see) characterized by the presence of Ti-C bonds. Both alkyl- and aryltitanium compounds and titanium(III) and titanium(IV) complexes with cyclopentadienide ligands are known. Alkyltitanium(IV) compounds of the type TiR_4 are only stable when the R groups are stable to elimination reactions, for example, $\text{-}CH_2C_6H_5$ or $\text{-}CH_2Si(CH_3)_3$. Tetramethyltitanium, $Ti(CH_3)_4$, is a yellow compound which is stable only below - 40 °C. The red, crystalline bis(cyclopentadienyl)titanium dichloride, $(h^5\text{-}C_5H_5)_2TiCl_2$, m.p. 230 °C, is particularly significant. It is made by reaction of $TiCl_4$ with NaC_5H_5, and is used in combination with alkylaluminum compounds as a homogeneous catalyst in alkene polymerizations. The green-black tetrakis(cyclopentadienyl)titanium(IV), $(h^5\text{-}C_5H_5)_2(h^1n\text{-}C_5H_5)_2Ti$ is of theoretical interest, because it is characterized by a fluctuating structure.

Organozinc compounds: alkyl or aryl zinc compounds, ZnR_2 (see Organoelement compounds). O. are obtained, for example, by reaction of zinc with alkyl halides ($2\ Zn + 2\ RX \rightarrow ZnR_2 + ZnX_2$) or by reaction of zinc with mercury(II) arylenes ($Zn + HgR_2 \rightarrow ZnR_2 + Hg$). O. are covalent compounds with linear molecular structures; they are low-boiling liquids or low-melting solids, e.g. $Zn(CH_3)_2$, b.p. +46 °C, $Zn(C_2H_5)_2$, b.p. 117 °C. They react violently with water, and the lower zinc alkyls ignite spontaneously in air. Zinc dialkyls are of historical interest because they were the first organometallic compounds to be synthesized (E. Frankland, 1849). They are now used for selective alkylations.

Orientation forces: see Van der Waals forces.

Orientation polarization: see Polarization.

Orlean: a yellow to orange plant pigment extracted from the red seeds of the tropical bush *Bixa orellana*. O. contains three different pigments, of which the most important is Bixin (see). It is used to color foods such as butter and cheese; the red wax used on Edam cheese, for example, is dyed with O. It was formerly used to dye cotton, wool and silk.

Orlon®: see Synthetic fibers.

Ornamental silver: an alloy of silver and copper, usually containing 80% silver. Sterling silver contains 92.5% silver. O. can contain up to 40% nickel instead of copper.

Oro: abb. for orotic acid.

Orotic acid, abb. ***Oro, uracil 6-carboxylic acid***: colorless needles; m.p. 345-347 °C (dec.), barely soluble in cold water, slightly soluble in hot water, and soluble in aqueous alkali hydroxide solutions, where it forms salts. It is practically insoluble in alcohol and ether. O. was isolated from milk. It is used to treat liver diseases and is supposed to improve memory and learning.

Orotidine, abb. ***Ord***: orotic acid 1-β-D-ribofuranoside, a nucleoside. Orotidine 5'-monophosphate is the first nucleotide synthesized along the biosynthetic pathway of pyrimidines. Its immediate precursors are orotic acid and 5'-phosphoribosyl pyrophosphate; the orotic acid is formed from aspartic acid.

Orsat apparatus: a device for volumetric gas analysis by chemical means. The O. consists of a gas buret combined with several gas pipets in a frame. Sometimes it is equipped with a combustion furnace for nonabsorbable gases, such as hydrogen or methane. The samples are measured with the gas buret and forced through the pipets for absorption. The resulting decrease in volume is measured with the buret. The pipets are most often filled as follows: Pipet 1 (counting from the gas buret) contains 25% potassium hydroxide to absorb CO_2. Pipet 2 is filled with bromine water to absorb alkenes. Pipet 3 contains alkaline pyrogallol solution to absorb oxygen. Pipet 4 is filled with copper(I) chloride in an ammonia or hydrochloric acid solution. Further pipets can contain special absorption solutions. Gases which cannot be absorbed can be burned in the furnace after addition of oxygen or air. The combustion products, mainly carbon dioxide, can then be absorbed and determined as usual.

The O. is used mainly for rapid and simple analysis of generator and other fuel gases, and has not lost much of its significance even with the introduction of Gas chromatography (see).

However, the O. can be used only for discontinuous determination, so that its use in process control is limited.

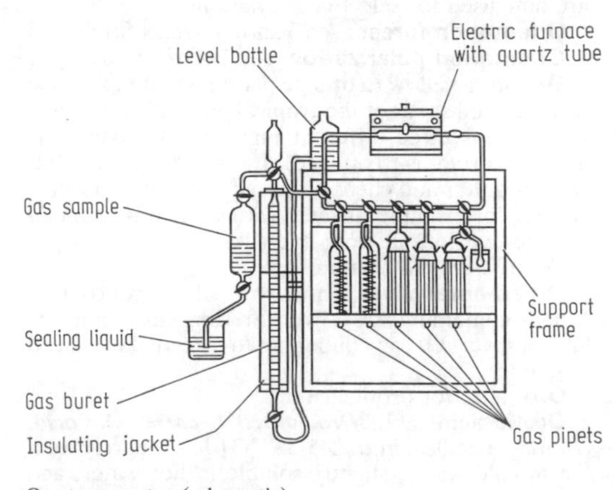

Orsat apparatus (schematic).

ortho-: see Arenes. See Nomenclature, sect. III D.

Orthocarbonates: salts M_4CO_4 or esters $C(OR)_4$ of orthocarbonic acid, $C(OH)_4$, which does not exist as such. O. are stable, colorless liquids with pleasant odors. In water or acids, O. are decomposed, forming alcohols and carbon dioxide. They can be synthesized by reaction of chloropicrin with alkali alcoholates: $Cl_3CNO_3 + 4\,NaOR \rightarrow C(OR)_4 + 3\,NaCl + NaNO_2$.

Orthochrome: see Isocyanins.

Orthochromatic: see Spectral sensitization.

Ortho esters: esters of the hypothetical ortho acids $R^1C(OH)_3$. The O. are named by using the prefix "ortho-" with the name of the related carboxylic acid, e.g. ethyl orthoformate. The O. are pleasant-smelling liquids which can be distilled without decomposition; they are stable to bases but are hydrolysed by acids to normal esters or to carboxylic acids and alcohol. O. can be made by special syntheses, for example by alcoholysis of imidoester hydrochlorides.

Ortho-fused systems: see Nomenclature, sect. III E 3.

Orthogonality: in quantum mechanics, the property of a wavefunction that the integral over all space of the product $\Psi^*_i(x,y,z)\Psi_j(x,y,z)$ (see Atom, models of) is equal to zero; that is, the overlap integral S_j of the two functions goes to zero. Mathematically, O. is given in this form: $\int_{-\infty}^{\infty}\int_{-\infty}^{\infty}\int_{-\infty}^{\infty} \Psi^*_i(x,y,z)\Psi_j(x,y,z)\,dx\,dy\,dz = 0$ for $i \neq j$. Certain quantum mechanical operators, including the hamiltonian operator, have the important property that their eigenfunctions to various eigenvalues are mutually orthogonal.

Orthoperiodates: see Periodic acid.

Orthorhombic: see Crystal.

Orton rearrangement: a thermally induced rearrangement of *N*-haloarylamines to 1,2- and 1,4-haloarylamines:

R = H, Allyl
X = Halogen

Orycenin: see Glutelins.

Oryzalin: see Herbicides.

Os: symbol for osmium.

Osazones: products of the reaction of monosaccharides with 3 molecules of phenylhydrazine. By-products are 1 molecule aniline and 1 molecule ammonia. Since the asymmetry of the C-atom next to the carbonyl group is lost in the formation of the O., epimeric monosaccharides form the same O. The O. are bright yellow compounds which crystallize readily; they played an important role in the elucidation of monosaccharide structure by E. Fischer.

HC=N–NHPh
|
C=N–NHPh
|
$(CHOH)_n$
|
CH_2OH

Osazones (Ph = phenyl) of hexoses (n = 3) and pentoses (n = 2)

Oscillating reaction: a reaction course of very complex reactions in which the reaction rate does not change monotonically, but shows periodic oscillations. It is a special case of a Dissipative structure (see). Requirements for the occurrence of oscillations are that the system is very far from equilibrium, and that the kinetics are described by highly nonlinear differential equations. The nonlinearity can be caused, e.g. by autocatalytic steps (see Catalysis), temperature changes in nonisothermal reactions, or rhythmic passivization of the electrodes in electrochemical processes. In each case, the effect is one which is called feedback coupling in cybernetics. The results of certain steps (e.g. changes of intermediate concentrations, temperature, electrode state) affect the rate constants or concentrations of the kinetic rate laws which apply to those steps, and thus accelerate or delay the reaction process.

O. are known in homogeneous solutions, in the oxidation of hydrocarbons, carbon monoxide or hydrogen sulfide in the gas phase, in the electrochemical dissolution of metals in acids, electrochemical membrane processes, certain heterogeneous catalytic reactions and many enzymatic reactions. The most intensively investigated example of an O. in homogeneous solution is the ***Belousov-Zabotinski reaction*** (1958), i.e. the oxidation of malonic or citric acid or a similar organic compound by bromate in sulfuric acid solution in the presence of cerium(IV) ions:

$$2\,BrO_3^- + 3\,CH_2(COOH)_2 + 2\,H^+ \xrightarrow{Ce^{4+}/Ce^{3+}} 2\,BrCH(COOH)_2 + 3\,CO_2 + 4\,H_2O.$$

After a certain induction period, a freshly prepared reaction mixture undergoes regular oscillations which can be made visible with ferroin as an indicator (red/blue). Measurements of the redox potential are also possible. In open reactors, the oscillations are not damped and have a very nearly constant frequency, but in a closed reactor they stop abruptly, long before equilibrium has been reached. The oscillation times are in the second to minute range, depending on the temperature and composition of the mixture. Spatial structures can also be observed.

The mechanisms of O. are still not exactly known, because many coupled steps are involved in them. However, there are various models, which with a suitable choice of rate constants and concentrations show oscillations. The simplest model for a reaction $A \rightarrow B$ is the ***Lotka model***:

$A + X \rightarrow 2\,X$ (1)
$X + Y \rightarrow 2\,Y$ (2)
$Y \rightarrow B$ (3)

X and Y are intermediates, partial steps (1) and (2) are autocatalytic reactions which cause the feedback.

In addition to oscillations, such systems also often demonstrate *bistability*. This is a condition in which there are two stable reaction states (one with a higher, the other with a lower reaction rate or intermediate concentration) available to the system. The transition from one to the other can be induced by an external disturbance (a stimulus). Bistabilities are also known in industrial reactors. They are clearly analogous to the excitation of nerve and muscle cells.

Oscillator potential: see Infrared spectroscopy.

Osmates: see Osmium compounds.

Osmium, symbol ***Os***: an element of group VIIIb of the periodic system, the Platinum metals (see); *Z* 76, with natural isotopes with mass numbers 192 (41.0%), 190 (26.4%), 189 (16.1%), 188 (13.3%), 187 (1.6%), 186 (1.6%), 184 (0.020%); atomic mass 190.2, valence II-VIII, 0, -II, density 22.61, m.p. 3050 °C, b.p. 5020 °C, electrical conductivity 11.7 Sm mm^{-2}, standard electrode potential (Os/Os^{2+}) +0.85 V.

Properties. O. is a very hard, brittle, blue-gray metal which crystallizes in cubic close-packed array; it is characterized by a high density and high melting and boiling points. A notable property is the ease with which it can be oxidized to osmium(VIII) oxide, OsO_4. Even at room temperature, osmium powder reacts slowly with air to form OsO_4, and at higher temperatures, the reaction becomes very rapid. At high temperature and low oxygen pressure, osmium(VI) oxide, OsO_3, is formed in addition to OsO_4. Oxidizing alkali melts convert O. to osmates (VI). Fluorine and chlorine react with O. above 100 °C to form various osmium halides. At high temperatures, O. combines rapidly with sulfur or phosphorus. O. forms alloys easily with the other platinum metals. O. and its lighter homolog ruthenium differ from the other elements of group VIIIb in being able to form stable tetroxides, MO_4, in which the +VIII oxidation state is achieved. Osmium(VI) compounds, such as osmium(VI) fluoride, OsF_6 and the osmates(VI), are also quite stable. Osmium(IV) compounds are also important, while osmium(III) and osmium(II) complexes are much less common than the corresponding ruthenium or iron compounds. The oxidation states 0 and - II are realized in osmium carbonyls and osmium hydrogencarbonyls, respectively.

Analysis. The volatile OsO_4 is used for detection and separation of O.

Occurrence. O. makes up about 10^{-6}% of the earth's crust. It is found together with platinum, and in isomorphic mixtures with iridium in the minerals sysserskite (iridosmium) and newjanskite (osmiridium).

Production. O. is produced along with the other Platinum metals (see). The solubilized ore solutions generally contain osmates(VI), which are converted by stronger oxidizing agents to OsO_4. This is distilled off and condensed directly, or it is absorbed by an alkali hydroxide solution, forming osmate. The salts are reduced to the metal with sodium boranate or another suitable reagent.

Applications. O. is used widely in the form of alloys with other platinum metals (see Platinum alloys). Filaments for metal filament lamps are also made of O.

Historical. Foureroy and Vauquelin discovered an insoluble residue when they dissolved platinum minerals in aqua regia. After solubilization with potassium carbonate and treatment with nitric acid, the residue produced vapors. These pungent smelling vapors, which attack the eyes and mucous membranes, were later recognized as osmium(VIII) oxide (the name "osmium" is from the Greek "osme" =

"odor"). In 1804, Tennant isolated the metal from the volatile oxide.

Osmium compounds: the most important binary osmium compound is ***osmium(VIII) oxide, osmium tetroxide***, OsO_4, which crystallizes in two modifications: colorless, monoclinic needles, density 4.906, m.p. 39.5 °C, and yellow needles, m.p. 41 °C, b.p. (of both forms) 130 °C, M_r 254.20. OsO_4 is volatile. It is moderately soluble in water, and readily soluble in organic solvents.

> Osmium(VIII) is very poisonous. The vapors have a characteristic, chloride oxide-like odor, and cause dangerous inflammation of connective tissue; they cause headaches and visual impairment.

OsO_4 is formed in small amounts when osmium powder is simply exposed to air; it is synthesized by heating osmium in a stream of oxygen between 500 and 800 °C. It is used as a catalyst for oxidation and hydroxylation reactions, and to burnish copper. The reaction of osmium with halogens produces a variety of osmium halides, depending on the conditions. Examples: ***osmium(IV) chloride, osmium tetrachloride***, $OsCl_4$, red-brown needles, M_r 332.01, b.p. 450 °C; ***osmium(III) chloride, osmium trichloride***, $OsCl_3$, brown, cubic crystals which dissolve readily in water and alcohol, M_r 296.56; ***osmium(II) chloride, osmium dichloride***, $OsCl_2$, dark brown, water-insoluble compound, M_r 261.11; ***osmium(VII) fluoride, osmium heptafluoride***, OsF_7, yellow, pentagonal bipyramidal molecules; the compound is very unstable and is formed from the elements at 600 °C and 40 MPa, M_r 323.2; ***osmium(VI) fluoride, osmium hexafluoride***, OsF_6, forms volatile, lemon-yellow crystals; M_r 304.2, m.p. 33.2 °C, b.p. 46 °C; ***osmium(V) fluoride, osmium pentafluoride*** $(OsF_5)_4$, blue-green, tetrameric molecules, M_r 1140.8, m.p. 70 °C, b.p. 225.9 °C; ***osmium(IV) fluoride, osmium tetrafluoride***, OsF_4, yellow, nonvolatile, and probably polymeric compound, M_r 266.2, m.p. 230 °C.

There are also numerous osmium complexes, which are formed through the following reactions: reactions of osmium with oxidizing alkali melts – such as alkali hydroxide/alkali nitrate – yield ***alkali osmates(VI)***, such as the dark violet, diamagnetic $K_2[OsO_4] \cdot 2H_2O$, which should be called ***dioxotetrahydroxoosmate(VI)***, $K_2[OsO_2(OH)_4]$. Osmium(VIII) oxide dissolves in strong alkali hydroxide solutions to form red ***osmates(VIII)***, $M_2[OsO_4(OH)_2]$. If concentrated ammonia is also present, the interesting ***nitridoosmates(VIII)*** complexes $M[OsO_3N]$, are formed; these have a covalent Os≡N triple bond. Other osmium complexes are formed by reaction of the osmium halides with alkali halides; for example the ***hexafluoroosmates(V)***, $M[OsF_6]$, and the ***hexachloroosmates(IV)***, $M_2[OsF_6]$, which are isomorphic with the hexachloroplatinates (IV) and are rather stable. Octahedral osmium(II) and osmium(III) complexes are much less common than the numerous, homologous iron(II) complexes. Some examples, however, are $M_4[Os(CN)_6]$, $[Osdipy_3]X_2$, $[Os(NH_3)_6]X_3$ and $M_3[OsCl_6]$. ***Osmium carbonyls*** are also known; these are analogous to the iron carbonyls: $Os(CO)_5$, colorless liquid, M_r 330.22, m.p. -15 °C; $Os_2(CO)_9$, yellow-orange crystals, M_r 632.44, m.p. 64-67 °C (dec.); $Os_3(CO)_{12}$; bright blue crystals, M_r 906.66, m.p. 224 °C (structures, see Metacarbonyls); $H_2Os(CO)_4$, colorless liquid, m.p. -15 °C. Osmium reacts with cyclopentadiene to form ***osmocene***, bis-(h^5-cyclopentadienyl)osmium(II), $(C_5H_5)_2Os$, colorless crystals, M_r 320.39, m.p. 230 °C.

Osmocene: see Osmium compounds.

Osmodiuretics: see Diuretics.

Osmosis: the passage of a fluid through a semipermeable membrane when it separates two solutions with different concentrations or a solution and the pure solvent. Films of high-polymer substances, animal skins, precipitates of copper hexacyanoferrate(II) or cell membranes can serve as semipermeable membranes. These are often supported on a carrier, such as a clay wall. The solvent moves towards the more concentrated solution, and if the vessel is closed, causes an increase in pressure there. The equilibrium pressure is called the ***osmotic pressure***.

Equilibrium is established when the chemical potentials of the solvent on the two sides are equal. If we consider the case of a solution in equilibrium with the pure solvent, the chemical potential of the pure solvent is μ_1^0 and that of the solvent in the solution is $\mu_1 = \mu_1^0 + RT\ln a_1 + V_1\Delta p$. Here a_1 is the activity of the solvent, V_1 is its molar volume, and Δp is the pressure difference between the two sides. The equilibrium condition $\mu_1^0 = \mu_1$ requires $RT\ln a_1 + V_1\Delta p = 0$, or $\Delta p = \pi = -(RT/V_1)\ln a_1$. In ideal dilute solutions, a_1 is equal to the molar fraction x_1. In addition, as an approximation, $\ln x_1 = \ln(1-x_2) \approx -x_2 \approx -c_2V_1$, where x_2 and c_2 are the molar fraction and the concentration of the dissolved substance (see Composition parameters). From this follows the law for the osmotic pressure of ideal, dilute solutions of nonelectrolytes, $\pi = c_2RT$ (*van't Hoff equation*) discovered in 1887 by J.H. van't Hoff. The osmotic pressure of a solution in equilibrium with the pure solvent is proportional to the molar concentration and the temperature, but does not depend on the type of solute or solvent. For a 1 molar solution, the ideal osmotic pressure would be π = 2.48 MPa (= 24.5 atm.). Since $c_2 = m_2/M_2v$, the measurement of osmotic pressure can be used to determine the molar mass. Because the measurement effects are very large, the method is used mainly for macromolecular compounds. An ***osmometer*** is an apparatus for measurement of osmotic pressure.

The osmotic pressure is one of the colligative properties, that is, the magnitude of the effect depends on the effective number of dissolved particles rather than the concentration. Association of the solute leads to a decrease in the number of particles, and dissociation to a larger number. For weak electrolytes, the increase in the number of particles can be taken into account by introduction of the degree of dissociation α via the van't Hoff factor $i = 1 + (\nu-1)\alpha$ (see Raoult's law), where ν is the number of ions into which the electrolyte dissociates. It follows that $\pi = ic_2RT$. However, especially with strong electrolytes ($\alpha = 1$), there are deviations, because the solution is no longer ideal. Therefore, on the suggestion of Bjerrum, an ***osmotic coefficient*** $f_0 = \pi_{real}/\pi_{ideal}$ is introduced; this is related to the activity coefficient f_1 of the solvent by the equation $f_0 = 1 + \ln f_1/\ln x_1$.

The phenomenon of O. was discovered in 1748 by the French physicist Nollet. In 1877, the botanist Pfeffer made quantitative measurements with the ***Pfeffer cell*** (Fig.).

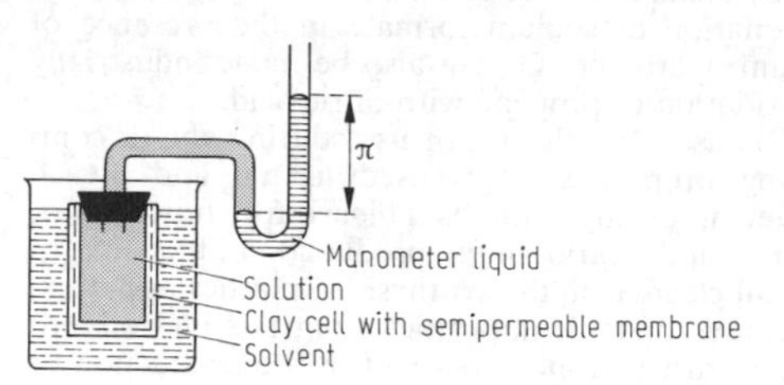

Fig. 1. Pfeffer cell.

O. plays an important role in organisms, making possible the internal pressure, or turgor, of plant cells. Solutions are called *hypertonic, isotonic* or *hypotonic*, depending on the magnitude of their osmotic pressure relative to that of cells bathed in them. Hypertonic solutions (with higher osmotic pressure than the cell) cause water to leave the cell, so that it shrivels; hypotonic solutions cause water to enter the cell, which therefore swells. Marine organisms live in a hypertonic solution (seawater) and must actively excrete salt to compensate for the osmotic loss of water; fresh-water organisms live in hypotonic solutions and must actively excrete water to maintain their physiological osmotic pressures. Osmotic pressures of 3 to 4 MPa are measured in leaf cells, and in mold cells, up to 20 MPa. Human blood has an osmotic pressure of 0.78 MPa; a physiological saline solution isotonic to blood contains 0.9 to 1.0 mass % sodium chloride.

Reverse O. is a separation process in which the pure solvent is transported out of a solution, so that the solution is enriched.

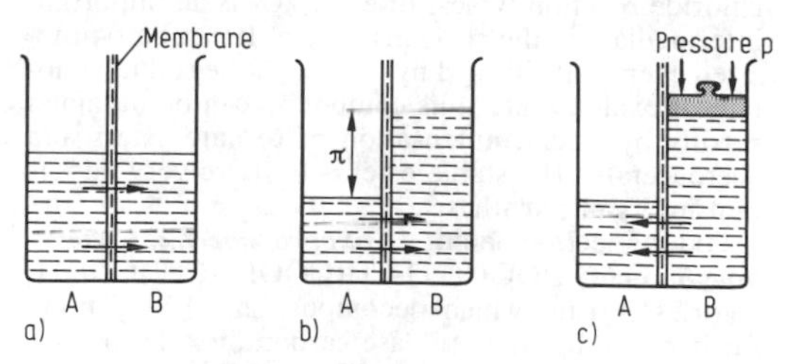

Fig. 2. Reverse osmosis.

If a solution B and pure solvent A are separated by a semipermeable membrane in a vessel (Fig. 2a), the pressure in B increases until the osmotic pressure π is too great to permit further accumulation of solvent (Fig. 2b). If pressure p is applied to B in excess of π, the solvent flows out of B and into A, causing B to become more concentrated (Fig. 2c).

Depending on the applied pressure and the size of the molecules retained, the principle is called Ultrafiltration (see) (0.3 to 2 MPa) or reverse O. (2 to 12 MPa). The essential component of the apparatus is the membrane, which is synthetic (e.g. cellulose acetate). These membranes are very thin (e.g. 0.02 μm) and are supported by a solid structure. These are built into modules (tube-shaped packets) in which the separation occurs. The flowthrough is proportional to the strength of the membrane; equipment with a flowthrough of ≈ 100 m^3 h^{-1} has been built.

The applications of reverse O. are constantly expanding: desalination of water, removal of detergents and recovery of metals from waste water, recycling water in textile, paper and photochemical industries, and processing used oil emulsions in the metal and chemical industries. In spacecraft, reverse O. is used to recover drinking water from urine.

Osmotic pressure: see Osmosis.

Osmotic coefficient: see Osmosis.

Ostwald dilution law: an equation discovered in 1888 by W. Ostwald for determination of dissociation constants of weak electrolytes from their equivalent conductivities at finite (Λ_{eq}) and infinite dilution (Λ_∞):

$$K_D = \frac{\Lambda_{eq}^2 \cdot c_0}{\Lambda_\infty(\Lambda_\infty - \Lambda_{eq})}$$

(see Dissociation 1, Electrical conductivity).

Ostwald process: see Nitric acid.

Ostwald's step rule: a rule formulated by W. Ostwald, which says that systems generally do not pass in a single step from a very high-energy, unstable state into the most stable state, but go via intermediate stages. Some examples are the transitions from gases to solids via liquids, precipitation of solid modifications out of a melt, or the stepwise transition from electronically highly excited atoms into the ground state. From a modern perspective, the O. is a principle which is often observed, but not always.

Ouabain: same as g-strophanthin. See Strophanthus glycosides.

Ovalbumin: see Albumins.

Overlap: see Molecular orbital theory.

Overlap integral: see Molecular orbital theory.

Overvoltage: see Polarization.

Oxadiazoles: a group of heterocyclic compounds containing one oxygen and two nitrogen atoms in a five-membered ring. There are four isomers which differ in the arrangement of the two N atoms relative to the O atom (always in position 1). ***1,2,3-O.*** can be obtained by reaction of α-amino-1,3-dicarbonyl compounds with nitrous acid.

1,2,3-Oxodiazole

These are also known as ***diazo oxides***. The Sydnones (see) are derivatives of the 1,2,3-O. ***1,2,4-O.*** are formed by reaction of amide oximes with carboxylic acids, or by Beckmann rearrangement (see) of glyoximes in the presence of phosphorus pentachloride. ***1,2,5-O.***, ***furazans***, are made by dehydration of the dioximes of 1,2-dicarbonyl compounds (glyoximes). The 1,2,5-oxadiazole N-oxides are also called ***furoxans***. ***1,3,4-O.*** are formed by dehydration of diacylhydrazines in the presence of water-withdrawing reagents, or by treating acylamidrazones with nitrous acid.

Substitution products of the O. are used as drugs and pigments.

Oxadi-π-methane rearrangement: a photochemical reaction of β,γ-unsaturated ketones to form saturated α-cyclopropyl ketones. The O. is a sigmatropic 1,2-acyl shift and a special case of the Di-π-methane rearrangement (see). The O. tends to occur via an excited triplet state.

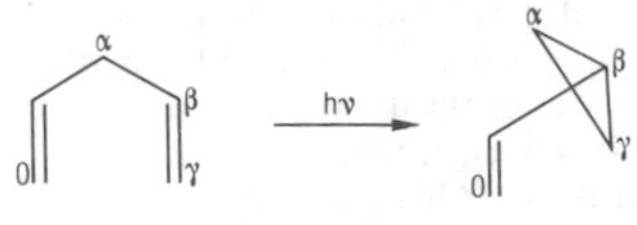

Oxaldehyde: same as Glyoxal (see).

Oxalene: see Pseudoazulenes.

Oxalates: the salts and esters of oxalic acid, with the general formula $R^1O\text{-}CO\text{-}CO\text{-}OR^2$, where R^1 and R^2 are monovalent metal ions (or a single divalent metal ion), ammonium ions or alkyl or aryl groups. R^1 and R^2 can be identical or different. As a dibasic acid, oxalic acid can also form acidic O., that is, one group R can be a hydrogen.

With the exception of the alkali O., the salts are only barely soluble in water, although most are readily soluble in strong acids. They are obtained by dissolving metal salts, such as carbonates, and oxalic acid in hot water. Some important O. are ammonium, potassium and calcium oxalates. These are used in the textile and leather industries, in analytical chemistry and in the production of electrodes.

The esters can be easily synthesized from oxalic acid and alcohols. The acidic esters are thermally unstable, strong acids, which form stable salts. Most of the neutral esters are oily, stable liquids which are readily soluble in alcohol, ether and benzene. They are used mainly as solvents for cellulose esters, cellulose ethers and resins, and for organic syntheses. Some important examples are dimethyl, diethyl and di-*n*-butyl oxalates.

Oxalic acid, ***ethanedioic acid***: HOOC-COOH, the simplest aliphatic dicarboxylic acid. O. crystallizes as the dihydrate in colorless, monoclinic prisms; m.p. 101.5 °C. When heated to about 100 °C, or on sublimation in vacuum, the dihydrate can be converted to the anhydrous form. This exists in both an α- and a β-form; m.p. 189.5 °C (α-O.) or m.p. 182 °C (β-O.). O. is soluble in water, alcohol, acetone and dioxane, and is insoluble in benzene and chloroform.

When heated in conc. sulfuric acid, it decomposes into carbon dioxide, carbon monoxide and water. It is quantitatively oxidized to carbon dioxide and water by potassium permanganate (manganometry). This reaction distinguishes O. from homologous dicarboxylic acids. O. is the strongest of the aliphatic dicarboxylic acids, with a pK value (≈1.2) which is significantly lower than that of malonic acid (≈2.8). Its salts and esters are called Oxalates (see). The insoluble calcium salt is often used for quantitative determination of either O. or calcium.

O. occurs widely in plants; its salts are found in sorrel, beet and spinach leaves, rhubarb and the bark and roots of many other plants. It is also found, as calcium oxalate, in the blood and urine of many mammals, including human beings.

O. can be synthesized in the laboratory, and also industrially, by oxidation of sucrose with concentrated nitric acid in the presence of vanadium(V) oxide, by alkali fusion of cellulose, or by thermal dehydrogenation of sodium formate in the presence of sodium hydroxide. O. can also be made industrially by oxidation of propene with nitric acid.

O. is used in industry for its reducing and/or complexing properties. It is used to tan and bleach leather, in dyeing cloth, as a bleach for straw, wood, feathers and stearin, in the production of blue ink, as a metal cleaner, in the synthesis of plastics and dyes, to remove metallic impurities from fats and oils, to remove rust and ink spots and in the production of syrup from starch. It is used in analytical chemistry as the primary standard for alkalimetry and manganometry, and to precipitate rare earths.

Oxalic acid is poisonous in high doses, because it interferes with calcium metabolism. Ingestion of 3 to 10 g oxalic acid can be fatal. First aid for poisoning is to drink a large amount of water, to which calcium hydroxide can be added.

Oxalic acid diamide: same as Oxamide (see).

Oxalic acid dichloride: same as Oxalyl chloride (see).

Oxaloacetic acid, ***oxosuccinic acid***: HOOC-CO-CH_2-COOH, a ketodicarboxylic acid which is not known in free form. There are two isomeric enol forms, which can be isolated separately under appropriate conditions of synthesis: ***hydroxymaleic acid***, the Z or *cis*-form; m.p. 152 °C, and ***hydroxyfumaric acid***, the E or *trans* form; m.p. 184 °C. The Z form is barely soluble in ether, while the E form is readily soluble; neither enol form is soluble in benzene. Their aqueous solutions give the iron(III) chloride reaction typical of enols. O. is an important intermediate in the citric acid cycle. It can be synthesized chemically by acid hydrolysis of the sodium enolate of oxaloacetate; this compound can be obtained readily by ester condensation of oxalate esters and acetoacetate. The stable esters of O. are very important in organic synthesis.

Oxalosuccinic acid, ***1-oxopropane-1,2,3-tricarboxylic acid***: HOOC-CO-CH(COOH)-CH_2-COOH, colorless prisms which decompose at 80.5 °C; upon further heating, they release carbon dioxide to form ketoglutaric acid. O. is readily soluble in water, alcohol and ether, and is only slightly soluble in chloroform and petroleum ether. It is formed as an intermediate in the citric acid cycle from isocitric acid; the reaction is catalysed by isocitrate dehydrogenase. Chemically, O. can be synthesized by condensation of succinic acid diesters with oxalic acid diesters, followed by hydrolysis of the resulting triester.

Oxalyl chloride, ***oxalic acid dichloride***: Cl-CO-CO-Cl, a colorless, very caustic liquid; m.p. -16 °C, b.p. 63-64 °C, n_D^{20} 1.4316. O. is soluble in ether, chloroform and benzene. In the presence of water, it is spontaneously hydrolysed to oxalic acid and hydrogen chloride. O. can be synthesized by the reaction of oxalic acid with phosphorus(V) chloride. It is used mainly for organic syntheses.

Oxalylurea: same as Parabanic acid (see).

Oxamide, ***oxalic acid diamide***: H_2N-CO-CO-NH_2, colorless needles; m.p. about 350 °C (dec.) (419 °C in a closed tube). O. is slightly soluble in water, and insoluble in ether. It is formed by the reaction of diethyl oxalate with ammonia, or by the hydrolysis of dicyanogen under very acidic conditions. O. is used as a depot fertilizer.

Oxamyl: a Carbamate insecticide (see).

Oxathiine fungicides: see Carbamide fungicides.

Oxazepines: triply unsaturated, seven-membered heterocycles with an oxygen and a nitrogen atom in the ring. The relative positions of the two heteroatoms vary, giving 1,2-O., 1,3-O. and 1,4-O.

Oxazine: a doubly unsaturated, six-membered heterocyclic compound with an oxygen and a nitrogen in the ring. There are eight possible isomers, of which the 1,3- and 1,4-O. are important. Substituted 4H-1,3-O. can be synthesized by cyclization of β-aminoketones with phosphorus pentachloride or polyphosphoric acid.

4H-1,4-Oxazine

4H-1,4-O. provides the skeleton of the phenoxazines from which the Oxazine pigments (see) are derived. The tetrahydro compound of 4H-1,4-O. is known as Morpholine (see).

Oxazine pigments, ***phenoxazine pigments***: a group of pigments derived structurally from the oxazine ring. They are very fast to light (formula, see Triarylmethane pigments). Some O. are *o*- and some are *p*-quinoid. They include various chrome and mordant dyes, such as gallocyanine and gallamine blue, the natural pigments in the actinomycin group and the ommochromes. Derivatives of dioxazine are used as violet pigments, such as sirius light blue.

Oxaziridine: a saturated, three-membered heterocyclic compound with an oxygen and a nitrogen atom in the ring. O. are structural isomers of the oximes, and can be made by epoxidation of azomethines with per acids. When heated, the O. rearrange to the isomeric nitrones.

Oxazole: a five-membered, heterocyclic compound with an oxygen and a nitrogen atom in the ring, an isomer of isoxazole. O. is a basic liquid which smells like pyridine; b.p. 60-70 °C, $n_D^{17.5}$ 1.4285. It is formed from hydroxyketosuccinates by heating them with formamide to 100-120 °C, followed by saponification of the oxazole-4,5-dicarboxylate and decarboxylation of the resulting oxazole-4,5-dicarboxylic acid.

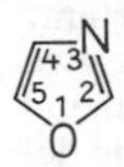

Oxazolidines: the completely saturated derivatives of the Oxazoles (see). These are stable compounds which can be distilled and recrystalized. They are hydrolysed by acids to ethanolamines and carbonyl compounds; in some cases water is able to hydrolyse them. O. are synthesized by condensation of aminoalcohols with aldehydes or ketones, and the resulting water is removed by azeotropic distillation. They can be used to carry out defined *N*-alkylation of ethanolamines.

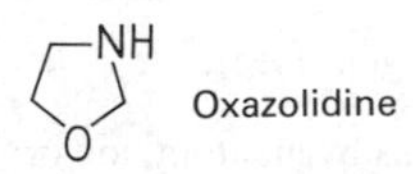
Oxazolidine

Oxazolines: the dihydro derivatives of Oxazole (see). Three tautomeric forms are known, of which the most important by far is 2-O, a colorless liquid which smells like pyridine; b.p. 98 °C. The other tautomers differ from 2-O. in the position of the double bond. 2-O. is synthesized by heating β-chloroethylformamide with alkalies in aqueous or alcoholic solution.

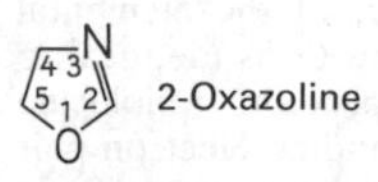
2-Oxazoline

Substituted 2-O. are valuable as starting materials for the synthesis of esters, amino acids and drugs. Derivatives with long-chain aliphatic groups in the 2 position are used as wetting agents, emulsifiers and dispersants. They are also used to make solvents, softeners and surfactants.

Oxazolinones, ***oxazolones***: oxo derivatives of the oxazolines. Of the possible isomeric derivatives, those of 2-oxazolin-5-one are important; they are also called ***azlactones***. 2-Oxazolin-5-ones are synthesized by dehydration of (β-hydroxyalkyl)amides by sulfuric acid. Azlactones or their 4-alkylidene and 4-benzylidene derivatives are intermediates in the Dakin-West reaction and in the Erlenmeyer amino acid synthesis.

2-Oxazolin-5-one

Oxazolones: same as Oxazolinones (see).

Oxepin: a triply unsaturated, seven-membered heterocyclic compound with an oxygen in the ring. At room temperature, O. exists only in a mixture with its valence tautomer, Benzoxide (see).

Oxetane, ***trimethyleneoxide, 1,3-epoxypropane***: a saturated, four-membered heterocycle with an oxygen in the ring. O. is a colorless oil with a pleasant odor; b.p. 48 °C, n_D^{20} 1.3961. It is soluble in water and organic solvents. O. are chemically very similar to the oxiranes, but they are less reactive, because the bond angle tension is lower. O. and its simple derivatives can be made by splitting hydrogen halides out of γ-haloalcohols. The *Paterno-Büchi reaction*, a photochemical cycloaddition of olefins to carbonyl compounds, is a general method for synthesis of O. Nucleophiles open the ring; for example, acid-catalysed hydrolysis produces 1,3-diols. Aromatic carbonyl compounds generally react with O. in the triplet state ($^3n, \pi$), while aliphatic carbonyl compounds can react either through a triplet state or an excited singlet state.

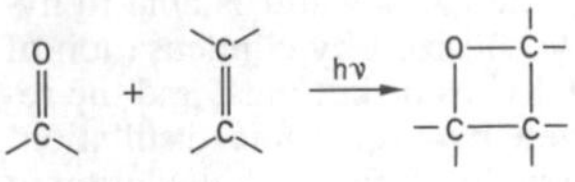

Paterno-Büchi reaction.

Oxidant: same as Oxidizing agent (see).

Oxidases: see Flavin enzymes.

Oxidation: the loss of electrons by an atom, ion or molecule, with a resultant increase in its Oxidation state (see); a part of a Redox reaction (see).

Oxidation process: in water and sewage treatment, processes used to remove odors, tastes and micro-impurities, using chlorine, chlorine dioxide, potassium permanganate, ozone, etc. (see Sewage treatment, Disinfection, Iron removal).

Oxidation state: the charge which would have to be assigned to an atom in a molecule if the molecule consisted of atomic ions. The O. of the elemental state is 0, and for simple ions, the O. is identical to the ionic charge. The O. of a complex ion or molecule is established by assigning the bonding electron pair of each bond to the more electronegative atom; if two identical atoms are bonded, each receives one electron. The sum of the ionic charges determined in this way must be equal to the total charge of the complex ion or molecule. The O. is written directly over the symbol for the element, with the sign of the charge in front of the numeral, e.g.

$$\overset{-1}{Cl}{}^{-}, \overset{+1}{H_2}\overset{-2}{O}, \overset{+1}{H_2}\overset{-1}{O_2}, H\overset{+3}{N}O_2, H\overset{+5}{N}O_3, \overset{+4}{S}O_3^{2-}, \overset{+6}{S}O_4^{2-}, \overset{+6}{Cr_2}O_7^{2-},$$

$$[\overset{+2}{Fe}(CN)_6]^{-4}$$

The O. is a formal parameter which does not indicate the actual charge on an atom or its bonding. However, it is essential for determining the stoichiometry of redox reactions.

Oxide: a binary compound of oxygen in its - 2 oxidation state, for example, aluminum oxide, Al_2O_3, sulfur dioxide, SO_2, copper oxide, CuO. The oxygen compounds of fluorine are thus not fluorine oxides, but oxygen fluorides, because fluorine is the more electronegative component. O. can be named as ***mon(o)oxides, dioxides, trioxides, tetr(a)oxides, pent(a)oxides***, etc., for example, carbon monoxide, CO, nitrogen dioxide, NO_2, sulfur trioxide, SO_3, osmium tetroxide, OsO_4, phosphorus pentoxide, P_2O_5; however, it is clearer to indicate the oxidation state of the oxide-forming element, as in iron(III) oxide, Fe_2O_3, chlorine(IV) oxide, ClO_2, or chromium(VI) oxide, CrO_3. The term ***sesquioxide***, for an O. in which the element and oxygen are in the stoichiometric ratio 2:3, e.g. Fe_2O_3, is obsolete. ***Double oxides*** are the O. of two elements in a single compound, e.g. iron(II) chromium(III) oxide, $FeO \cdot Cr_2O_3$; ***mixed O.*** contain the oxide-forming element in more than one oxidation state, e.g. lead(II,IV) oxide (minium), Pb_3O_4. ***Suboxides*** are compounds in which the content of oxygen is lower than would be expected from the typical valence of the oxide-forming element; the best known example is carbon suboxide, C_3O_2. The Peroxides (see) and Superoxides (see) do not fit the above definition, and are treated separately.

The O. can be classified by structure or reactivity. The most electropositive elements form ***ionic O.***, compounds in which the metal cations and the oxide anions O^{2-} occupy the positions in an ionic lattice; examples are lithium oxide, Li_2O, magnesium oxide, MgO, aluminum oxide, Al_2O_3 and manganese(IV) oxide, MnO_2. The O. of strongly electronegative elements are covalent. In these ***covalent O.***, the structures are formed either by element-oxygen $p_\pi p_\pi$ multiple bonds (as in CO, CO_2, NO_2, SO_2), or the oxygen acts as a bridge atom and creates chains, rings and cages (e.g. SeO_2, solid SO_3 and P_4O_{10}). The unusual composition of ***nonstoichiometric O.***, especially of derivatives of the transition elements, e.g. $TiO_{0.64\text{-}1.27}$, is due to the fact that the metal in the compound is present in more than one oxidation state, e.g. $Fe_{0.93}O_{1.09} = Fe^{II}{}_{79}Fe^{III}{}_{14}O_{100}$.

Ionic O. generally have high lattice energies and are therefore often thermally and chemically very unreactive compounds (e.g. MgO, Al_2O_3). Soluble ionic O. react with water to release OH^- ions: $O^{2-} + H_2O \rightarrow 2\,OH^-$; they are therefore called ***basic O.***. This group includes the O. of the alkali and alkaline earth metals (except BeO), and many transition metal oxides. They react with acids to form the corresponding salts, e.g. $CaO + 2\,HNO_3 \rightarrow Cs(NO_3)_2 + H_2O$. Many covalent O. react with water to form acids; these are the ***acid O.***. The formation of an acid can be understood as addition of water to the element-oxygen double bond, e.g. $CO_2 + H_2O \rightleftharpoons H_2CO_3$, or as hydrolysis of the element-oxygen single bond, e.g. $P_4O_{10} + 6\,H_2O \rightarrow 4\,H_3PO_4$. A few elements form ***amphoteric O.***, e.g. BeO, ZnO, Al_2O_3. These act as bases in their reactions with strong acids, e.g. $ZnO + H_2SO_4 \rightarrow Zn^{2+} + SO_4^{2-} + H_2O$, but with strong bases, they react as acids, e.g. $ZnO + 2\,OH^- + H_2O \rightarrow [Zn(OH)_4]^{2-}$. Many O. are found in nature as minerals, e.g. hematite, Fe_2O_3, magnetite, Fe_3O_4, kassiterite, SnO_2, quartz, SiO_2, etc.

Oxide aquate: see Oxide hydrate.

Oxide ceramics: oxidic, non-silicate materials which are produced by ceramic technologies. Organic plastifiers (e.g. paraffins) are often used in forming them; these evaporate or are thermally decomposed and oxidized during the firing process. Oxide ceramics are usually finely divided metal oxides with high melting temperatures (see High-temperature materials), such as aluminum oxide, Al_2O_3, beryllium oxide, BeO, magnesium oxide, MgO, titanium oxide, TiO_2, zirconium oxide, ZrO_2 and thorium oxide, ThO_2, which are solidified by Sintering (see) during firing (e.g. sintered corundum, sintered magnesia). Mixed oxides, such as ferrites and spinell, $MgAl_2O_4$, are of considerable interest as magnetic and semiconductor materials. In addition to the sintering processes, mixed oxide systems undergo solid state reactions during firing. In technical diagnostics, semiconducting solid electrolytes (see Semiconductors) made from mixed oxide materials have become important.

Oxide electrode: an electrochemical, type-two Electrode (see). As a rule, its solid phase is a metal and its insoluble oxide. The potential of an O. is determined by the electrochemical equilibrium between the metal and the metal oxide; hydroxide ions are involved. The electrode potential thus depends on the activity of the hydronium ion via the ion product of water. O. are therefore used as indicator electrodes in measurement of pH. The most commonly used of them are the Antimony electrode (see) and the bismuth electrode.

Oxide halides, formerly ***oxyhalides***: compounds in which a metal or nonmetal is bound simultaneously to oxygen and a halogen, e.g. bismuth oxide iodide, BiOI, molybdenum dioxide dichloride, MoO_2Cl_2. Most O. are halides of oxygen acids; for example, phosphorus oxide chloride, $POCl_3$, is the trichloride of phosphoric acid.

Oxide hydrates, ***aqua oxides***: compounds of metal oxides with water; they were classified in four groups by O. Glemser (1961): 1) ***Hydroxides*** (see); 2) ***hydroxide hydrates***, which are hydroxides with a certain number of water molecules in the lattice, e.g. $LiOH \cdot H_2O$; 3) ***oxide hydrates*** in the narrow sense are compounds in which H_2O is present in the lattice as water of crystallization, e.g. $WO_3 \cdot H_2O$, $V_2O_5 \cdot H_2O$; 4) ***oxide aquates*** are finely divided oxides which have adsorbed varying amounts of water, e.g. titanium dioxide aquate, $TiO_2 \cdot H_2O$. This compound is precipitated when titanium tetrachloride is hydrolysed. The brown precipitates which form when iron(III) salt solutions react with ammonia or sodium hydroxide also fall into this category.

Oxidimetry: same as Redox analysis (see).

Oxidizing agent, ***oxidant***: a substance which can oxidize other substances; it takes electrons from the other reactant (acts as an electron acceptor) and is itself reduced (see Redox reactions).

Oxidizing melt: a melt of potassium nitrate and sodium carbonate. O. are used, for example, in qualitative analysis as a preliminary test for chromium(III) compounds, which it converts to yellow chromate(VI), e.g. $Cr_2O_3 + 3\,KNO_3 + 2\,Na_2CO_3 \rightarrow 2\,Na_2CrO_4 + 3\,KNO_2 + 2\,CO_2$, or for manganese(II) or manganese(IV) compounds, which it converts to green manganate(VI), e.g. $MnSO_4 + 2\,KNO_3 + 2\,Na_2CO_3 \rightarrow Na_2MnO_4 + 2\,KNO_2 + Na_2SO_4 + 2\,CO_2$. The sample to be examimed is melted together with a mixture of KNO_3 and Na_2CO_3 in the flame of a bunsen burner.

Oxidoreductases: a main group of Enzymes (see) which includes Alcohol dehydrogenase (see), Cytochrome oxidase (see), Dehydrogenases (see), Flavin enzymes (see), Lactate dehydrogenase (see) and Luciferase (see).

Oxidotropism: see Acid-base concepts.

Oximes: isonitroso compounds formed by condensation of aldehydes (***aldoximes***) or ketones (***ketoximes***) with hydroxylamine:

$$R^1R^2C{=}O + H_2N{-}OH \xrightleftharpoons{(H^+)} R^1R^2C{=}N{-}OH + H_2O.$$

O. are readily hydrolysed back to the starting compounds. They crystallize well, and for this reason are often used to characterize aldehydes and ketones. They are weak acids and form salts with aqueous bases. Formation of the O. from a carbonyl compound and hydroxylamine hydrochloride is often used for quantitative analysis of aldehydes and ketones; the hydrochloric acid released by the reaction is titrated with sodium hydroxide. This reaction is known as the *oxime titration*, and it can also be used to determine the equivalent mass of the carbonyl compound. O. of ketones rearrange to form amides in the presence of acid catalysts, such as sulfuric acid or phosphorus(V) chloride (see Beckmann rearrangement). This reaction is very important for the conversion of cyclohexanone oxime into ε-caprolactam in the industrial synthesis of polyamide fibers. The dehydration of Z-aldoximes by acetic anhydride produces nitriles which are used as intermediates in organic syntheses.

α-Oximino ketones: same as α-isonitroso ketone.

Oxinates: complex metal salts of 8-hydroxyquinoline, which forms nearly insoluble chelates with a series of metals such as magnesium, zinc, chromium, manganese, iron, cobalt, nickel and silver. These can be used for the quantitative determination of the metals by one of the following methods: 1) precipitation of the metal with oxine, followed by roasting to form the metal oxide and weighing the oxide; 2) drying and weighing the oxinate; 3) titration of the oxinate bound to the metal with bromine; or 4) colorimetric determination of the complex.

Because the O. have different solubilities at different hydrogen ion concentrations, oxine can also be used to separate cations, for example, copper and cadmium.

Oxindole, ***indolin-2-one***: an isomer of Indoxyl (see) and the parent compound of the oxindole alkaloids. It forms colorless needles; m.p. 127 °C at 3.06 kPa. It is soluble in alcohol, ether, hot water, alkalies and sulfuric acid. When heated with zinc powder, it is converted to indole. O. is formed by the reduction of 2-nitrophenylacetate with tin and hydrochloric acid, or by the reaction of aniline with chloroacetyl chloride.

Oxine: same as 8-Hydroxyquinoline (see).

Oxinol: a mixture of *tert.*-butanol and methanol (1:1) which is used as a fuel or substitute fuel.

Oxirane: same as Ethylene oxide (see).

Oxiranes: same as Epoxides (see).

Oxo acids: same as Oxocarboxylic acids (see).

3-Oxobutanoic acid: same as Acetoacetic acid (see).

Oxocarboxylic acids, ***oxo acids***: carboxylic acids with an additional carbonyl group. O. can be either ***aldehyde*** or ***keto carboxylic acids*** (***keto acids***). Many have trivial names. The systematic IUPAC name is based on that of the unsubstituted carboxylic acid, and the carboxylic group is indicated by the prefix "oxo-", as in 3-oxobutanoic acid (see Acetoacetic acid). The simplest method of synthesis of α-O. is the reaction of the acyl chloride with cyanide to form the α-nitrile, which is then hydrolysed. Another method is by hydrolysis of α,α'-dichlorocarboxylic acids. β-O. are obtained via their esters by the Claisen condensation (see). The reaction of succinic anhydride with the corresponding Grignard compounds gives γ-O., and δ-O. can be obtained by Acid cleavage (see) of 2-alkylcyclohexane-1,3-diones. A general synthetic method for O. is oxidation of the corresponding hydroxycarboxylic acids. In addition to these general methods, special syntheses have been developed for the most important O. Glyoxylic, pyruvic, acetoacetic and levulinic acids are some of the most important O. Some oxodi- and oxotricarboxylic acids are intermediates in the tricarboxylic acid (citric acid) cycle in cells (oxaloacetic and α-ketoglutaric acids). Some important naturally occurring O. are the unsaturated (E)-9-oxo-dec-2-enoic acid and the Prostaglandins (see). Most O. are unstable and susceptible to decarboxylation. Their esters are used extensively in syntheses (see Acetoacetates). In general, O. give the typical carbonyl reactions.

Oxo compounds: compounds in which oxygen is linked by a double bond to an atom of another element, or those in which the O^{2-} ion is present as a ligand, e.g. aldehydes, RCHO, ketones, $R_2C=O$, amine oxides, R_3N^+-O^-, phosphine oxides, $R_3P=O$, sulfoxides, $R_2S=O$, sulfones, R_2SO_2. According to Werner, most oxygen-containing inorganic acids are also O., e.g. H_2SO_4, H_3PO_4, $HClO_3$, as are their salts.

Oxo-cyclo tautomerism, ***ring-chain tautomerism***: a tautomeric equilibrium between 1,4- and 1,5-hydroxycarbonyl compounds and their cyclic semiacetal forms. O. is found in monosaccharides and is responsible for some of their special properties.

CH_3–CH(OH)–CH_2–CH_2–CHO ⇌ (cyclic five-membered hemiacetal, CH_3, OH); HO–CH_2–CH_2–CH_2–CH_2–CHO ⇌ (cyclic six-membered hemiacetal, OH)

Oxoethanoic acid: same as Glyoxylic acid (see).

Oxo group: same as Carbonyl group (see).

Oxolane: same as Tetrahydrofuran (see).

Oxonium ion: same as Hydronium ion (see).

Oxonium salts: 1) onium compounds derived from the hydronium (oxonium) ion H_3O^+. Some or all of the hydrogen atoms are replaced by alkyl or aryl groups. ***Monoalkyl oxonium salts*** or ***primary O.***, $ROH_2^+X^-$, are formed by protonation of alcohols by strong acids; ***dialkyl oxonium salts*** or ***secondary O.***, $R_2OH^+X^-$ are formed in the same way from ethers. ***Trialkyl oxonium salts*** or ***tertiary O.***, $R_3O^+X^-$, are formed in various ways. An example is triethyloxonium tetrafluoroborate, $(C_2H_5)_3O^+[BF_4]^-$, which can be synthesized from boron trifluoride, diethyl ether and epichlorohydrin. The tertiary O. are excellent alkylation reagents and are widely used in preparative organic chemistry.

2) Same as flavylium salts; see Flavanoids.

Oxonol pigments: see Polymethine pigments.

3-Oxopentanedioic acid: same as α,α'-Acetone dicarboxylic acid (see).

4-Oxopentanoic acid: same as Levulinic acid (see).

Oxophenarsine: see Arsphenamine.

2-Oxopropanoic acid: same as Pyruvic acid (see).

2-Oxopropanol: same as Methylglyoxal (see).

1-Oxopropane-1,2,3-tricarboxylic acid: same as Oxalosuccinic acid (see).

Oxosuccinic acid: same as Oxaloacetic acid (see).

Oxo synthesis, ***hydroformylation***: a method for producing aldehydes (alkanals) from olefins, carbon monoxide and hydrogen by lengthening the chain by one C atom (Fig. 1). Except for ethene, which can only form propanal, the alkene is always converted to both the straight-chain (n-) and iso-aldehyde; the n-aldehyde is favored.

$$R-CH=CH_2 + CO + H_2 \xrightarrow{\text{catalyst}} R-CH_2-CH_2-CHO \;+\; R-CH_2-CH(CHO)-CH_3$$

Cobalt salts are almost the only catalysts used. In the SHELL low-pressure method, trialkylphosphines are also present. Rhodium complex catalysts are used only in special cases, due to their high prices. Under the conditions of the O., metallic cobalt or cobalt salts, e.g. cobalt naphthenate, acetate or hydroxide, form octacarbonyldicobalt, which reacts with hydrogen to form hydrogen tricarbonylcobalt, the actual catalyst. The structurally isomeric aldehydes form via alkyl tri- and tetracarbonylcobalt (Fig. 2).

$$Co(CO)_8 + H_2 \rightleftharpoons 2HCo(CO)_4$$
$$2HCo(CO)_4 \rightleftharpoons HCo(CO)_3 + CO$$

+CH_3–CH=CH_2 (Propene) → CH_3–CH=CH_2 · $HCo(CO)_3$

Left branch: CH_3–CH($Co(CO)_3$)–CH_3 → (+CO) → CH_3–CH(CO–$Co(CO)_3$)–CH_3 → (+H_2, –$HCo(CO)_3$) → CH_3–CH(CHO)–CH_3

Isobutyraldehyde

Right branch: CH_3–CH_2–CH_2–$Co(CO)_3$ → (+CO) → CH_3–CH_2–CH_2–CO–$Co(CO)_3$ → (+H_2, –$HCo(CO)_3$) → CH_3–CH_2–CH_2–CHO

Butyraldehyde

The ratio of n- to iso-aldehyde can be influenced by the process conditions and modification of the catalyst. The O. consists of several steps: hydroformylation, catalyst separation, catalyst processing and isolation of the products (Fig.).

CO and H_2 in a ratio of 1:1 to 1:1.2 and the olefin enter the reactor at 110 to 180 °C and about 20 MPA; the catalyst is added separately. The heat of reaction is removed by bundles of pipes in the reactor, or by external cooling. The products leaving the reactor include alcohols and condensation products of the aldehyde and formate esters, but the main products are the aldehydes. These are separated by distillation. The catalyst is separated from the mixture and fed back into the process.

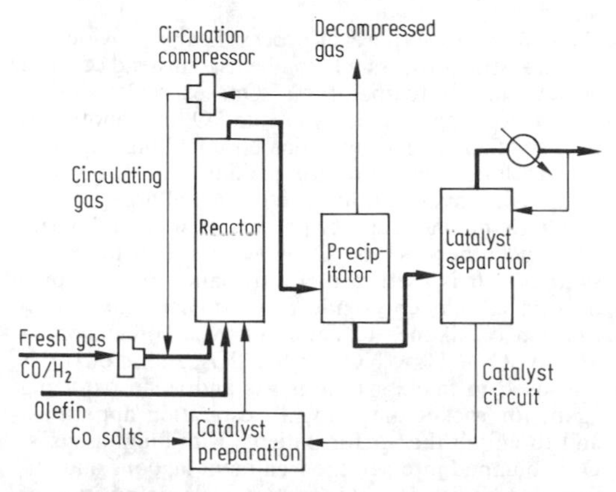

Diagram of an oxo installation.

The aldehydes are often not purified, but are led immediately into another reactor, where they are hydrogenated to the corresponding alcohols. The most important oxo products are butanal and the C_{12} to C_{18} alcohols, which are formed by hydrogenation of the primary products.

Historical The reaction principle of the O. was discovered in 1938 by Roelen, when he observed the formation of propanal by the reaction of ethene with CO and H_2 on a cobalt-thorium catalyst. The first oxo installation (10 kt year^{-1}) was built in 1945 to produce surface-active alcohols. At present, the world capacity for oxo production is about 6 million t year^{-1}.

Oxo aldehydes and subsequent products

Olefin	Aldehyde	Products
Ethene	Propanol	Propanol, propylamine, propionic acid
Propene	Butanal	Butanol, butylamine, 2-ethylhexanal, 2-ethylhexanol, trimethylpropane, butyric acid
	iso-Butanal	iso-Butanol, iso-butylamine iso-butyric acid neopentylglycol
C_{12} to C_{14} olefins	C_{12} to C_{14} aldehydes	Surfactant alcohols, softener alcohols

Oxycarboxin: see Carbamide fungicides.

2,2-Oxydiacetic acid: same as Diglycolic acid (see).

Oxygen, symbol ***O***: an element of group VIa of the periodic system, the chalcogen group (see Oxygen-sulfur group); a nonmetal, *Z* 8, with natural isotopes with mass numbers 16 (99.759%), 17 (0.037% and 18 (0.204%); atomic mass, 15.9994, valency II, density of the solid at -252.5 °C, 1.426; density of the liquid at -183 °C, 1.149; m.p. -218 °C, b.p. -182.962 °C, crit. temp. -118.4 °C, crit. pressure 5.06 MPa, standard electrode potential, 1.229 V ($2\,H_2O \rightleftharpoons O_2 + 4\,H^+ + 4\,e$).

Properties. Under normal conditions, O. is a colorless, odorless and tasteless gas. It condenses at very low temperatures to a light blue liquid or to light blue cubic crystals. O. is slightly soluble in water (at 0 °C and 0.1 MPa, 4.91 ml O_2 dissolves in 100 ml water).

O. is the most electronegative element of group VIa. With a valence electron configuration $2s^2p^4$, it is able to take up two electrons, forming oxide anions O^{2-}; these occur widely as components of ionic oxides (e.g. Al_2O_3, MgO, CaO, Fe_2O_3, MnO_2). A noble gas configuration can also be achieved by formation of two covalent bonds. All compounds in which this bonding state is present (e.g. water, alcohols, ethers, siloxanes) are bent at the O. atom and it has two free electron pairs. The bonding situation is usually described in a simplified sense by an sp^3 hybridization at the O atom. However, the actual molecular geometries are better predicted by the assumption of non-equivalent hybrid orbitals, in which the two electrons used for bonding are in somewhat higher orbitals, and the free electron pairs are in somewhat lower orbitals than the sp^3 orbital. The free electron pairs are responsible both for the weakly basic properties of O. compounds and for their donor behavior with respect to Lewis acids. Such interactions lead to derivatives of triply coordinated O. (e.g H_3O^+, $(C_2H_5)_2O\text{-}BF_3$, trialkyl oxonium salts, transition metal aquo complexes) or four-fold corrdinated O. (e.g. $OZn_4(OOCCH_3)_6$). With the formation of four covalent bonds, the possibilities are exhausted for this type of O. compound.

As an element of the first period, O. is able to form $p_\pi p_\pi$ multiple bonds. Examples of this type of bond are found in dioxygen, O_2, CO, CO_2, NO, NO_2 and carbonyl compounds. Double bonds of O. to elements outside the first period such as to P in phosphine oxides and phosphates, to S in sulfoxides and sulfates and to Cl in chlorate and perchlorate, are due to $d_\pi d_\pi$ interactions.

In all covalent compounds (except for the fluorine derivatives, such as oxygen fluoride OF_2), the high electronegativity of O. leads to bond polarity, which is often pronounced. This is responsible, especially in organic chemistry, for the ease with which reaction centers carrying an O function act as nucleophiles.

Under normal conditions, the stable form of O. is dioxygen, O_2. There is in addition a trioxygen, O_3, called Ozone (see). The bonding situation and chemical properties of ***dioxygen*** can be readily understood from the MO scheme for the O_2 molecule (see Molecular orbital theory). In the ground state, there are two unpaired electrons in antibonding π^* orbitals (*triplet oxygen*, $^3O_2$, term symbol $^3\Sigma_g{}^+$. This explains the radical character and paramagnetism of the O. molecule. O. can be converted to an excited state by photochemical activation of energy transfer from suitable sensitizers; or it can be achieved by the reaction of sodium hypochlorite with hydrogen peroxide or by decomposition of potassium peroxochromate or various peroxides. In this excited state, *singlet O.* $^1O_2$, term symbol $^1\Delta_g$, the two highest-energy electrons are paired in a π^* state, which is diamagnetic and is more reactive than ordinary triplet oxygen. It is used in preparatory chemistry as a stereospecific oxidizing agent (for more information, see Singlet oxygen). The bond order in O_2 is 2. The uptake of electrons leads, depending on the nature of the reducing agent, to O_2^- ions with a bond order of 1.5 (see Superoxide), O_2^{2-} ions with a bond order of 1 (see Peroxides) or to breakage of the O-O bond, forming O^{2-} ions (see Oxides).

The high dissociation enthalpy of the O_2 molecule, 498 kJ/mol, is the reason that molecular O. is relatively unreactive and its reactions usually require thermal or photochemical activation. O. combines directly or indirectly with all elements except helium, neon and argon. It forms oxides with many elements, e.g. $2\ Mg + O_2 \rightarrow 2\ MgO$, $P_4 + 5\ O_2 \rightarrow P_4O_{10}$, $2\ H_2 + O_2 \rightarrow 2\ H_2O$. Many compounds react, usually exothermally, with O. to form the oxides of their component elements, e.g. $CH_4 + 2\ O_2 \rightarrow CO_2 + 2\ H_2O$, $H_2S + 3/2\ O_2 \rightarrow SO_2 + H_2O$. Other compounds are oxidized by molecular O. in melts (e.g. $Cr_2O_3 + 2\ Na_2CO_3 + 3/2\ O_2 \rightarrow 2\ Na_2CrO_4 + 2\ CO_2$) or in solution (e.g. $R_3P + 1/2\ O_2 \rightarrow R_3P{=}O$). Atmospheric O. plays a part in the corrosion of metals.

O. is necessary for respiration in organisms. An adult human being consumes about 20 l O. per hour when at rest. This amount increases several-fold when strenuous work is being done. Air with an O. content less than 7% leads to loss of consciousness, and less than 3% O. causes death by suffocation. On the other hand, pure O. can be inhaled for a short period with no ill effects, so long as the pressure does not exceed 0.1 MPa.

Analytical. High concentrations of O. can be easily recognized, because a glowing sliver of wood dipped into the gas mixture flares up and burns with a bright flame. O. can be determined qualitatively and quantitatively by gas chromatography. It can also be determined by absorption into alkaline pyrogallol solution (see Gas analysis). There are a number of other instrumental methods for determining the O_2 contents of gas mixtures or solutions by physical measurements. Polarographic and galvanic measuring cells and methods based on the paramagnetism of molecular O. are of particular importance. In Elemental analysis (see), determination of O. is one of the less well studied methods.

Occurrence. O. makes up 50.5% of the earth's crust, and is thus the most abundant element on earth. It makes up 23.2% by mass (20.9% by volume) of the air. Most of the O. on earth is bound to hydrogen in the form of water, and more is present in the ore- or rock forming oxides and salts of O. acids. It is present, for example, in magnetite, Fe_3O_4, limestone, $CaCO_3$, dolomite, $CaCO_3 \cdot MgCO_3$, anhydrite, $CaSO_4$, gypsum, $CaSO_4 \cdot 2H_2O$ and in many silicates. In spite of the high oxygen consumption by many combustion processes in industry, transportation and households, respiration, decay and weathering, etc., the continuous formation of O. by photosynthesis keeps the O. content of the atmosphere nearly constant. O. can also be detected in space.

Production. In the laboratory, O. can be made by electrolysis of dilute sulfuric acid or thermal decomposition of potassium chlorate, $KClO_3$, potassium nitrate, KNO_3, or barium peroxide, BaO_2. For example, $KNO_3 \rightarrow KNO_2 + 1/2\ O_2$. The gas is obtained industrially by fractional distillation of liquefied air; adsorption processes based on the differential adsorption of N_2, O_2 and Ar in the cavities of zeolites or carbon molecular sieves are also used. In addition, O. is obtained in smaller amounts by electrolysis of water. It is commercially available as compressed gas or liquid.

Applications. Wherever process heat is provided by a combustion process, O. can be used instead of air to achieve higher temperatures. One example is autogenous welding with acetylene and O. and another is the use of an oxyhydrogen flame for melting quartz or high-melting metals, or for making synthetic gems. Large amounts of O. are used in metallurgy, for example in the blast furnace process, to melt and purify iron and other metals. O. is also used in industrial synthesis. It is used in olefin oxidation (e.g. in production of ethylene oxide from ethylene), oxychlorination of alkenes to chloroalkanes, conversion of SO_2 to SO_3, NH_3 to NO_2, H_2S in SO_2, and so on. O. is also used to make synthesis gas and as an oxidizing agent for rocket fuels, to fill respiration apparatus, and to enrich the air for patients with lung diseases. O. is pumped into sewage treatment lagoons and eutrophied bodies of water to hasten decomposition of the organic material in the water.

Historical. O. was discovered in 1772 by Scheele, and, independently in 1774 by Priestley. Scheele obtained O. by heating saltpeter, and Priestley by heating mercury oxide and red lead (minium). In 1774, Lavoisier recognized the participation of O. in oxide formation, combustion and respiration.

Oxygenases: enzymes which catalyse the incorporation of oxygen into organic compounds. ***Dioxygenases (oxygen transferases)*** are those enzymes which introduce both atoms of molecular oxygen, O_2, into the substrate, forming two adjacent hydroxy groups, while ***monooxygenases (hydroxylases, mixed-function O.)*** incorporate only one oxygen atom. The other is used to reduce a cosubstrate, forming water. An example of a monooxygenase is Phenylalanine hydroxylase (see). By hydroxylation, the O. increase the solubility and mobility of compounds, especially aromatic compounds, steroids, drugs and other xenobiotics.

Oxygenation: special case of oxidation in which molecular oxygen is introduced into an organic compound, e.g. $R_3CH + O_2 \rightarrow R_3C\text{-}O\text{-}O\text{-}H$ (see Autoxidation).

Oxygen blasting process: see Steel.

Oxygen complexes: see Dioxygen complexes.

Oxygen corrosion: see Electrochemical metal corrosion.

Oxygen electrode: a gas electrode consisting of oxygen flowing over a platinized platinum sheet which dips into a solution containing hydroxide ions. The potential-determining process is given by the following equation: $O_2 + 4e + 2\ H_2O \rightarrow 4\ OH^-$. The electrode potential of the O. is poorly reproducible, due to polarization phenomena, so the O. is currently used only for research purposes.

Oxygen fluorides: ***Oxygen difluoride***, OF_2, is a colorless, very poisonous gas, which attacks the respiratory organs. It can be condensed at - 145.3 °C to a yellow liquid; M_r 53.996, density $_{-183}$ 1.719, m.p. -223.8 °C. The OF_2 molecule has an angular structure (∢FOF 103.7°). The F-O distance is 140.5 pm. OF_2 is somewhat soluble in water, and the solutions are not acidic, but are oxidizing, e.g. of hydrogen halides: $OF_2 + 4\,HX \rightarrow 2\,X_2 + H_2O + 2\,HF$. OF_2 is obtained by passing fluorine through an alkali solution (about 0.5 M): $2\,F_2 + 2\,OH^- \rightarrow OF_2 + 2\,F^- + H_2O$.

Dioxygen difluoride, O_2F_2, is a yellow compound stable only at low temperatures; M_r 70.0, density $_{-157}$ 1.736, m.p. -163.5 °C, b.p. (extrapolated) -57 °C (dec.). The O_2F_2 molecule has a chair structure, and is a strong fluorinating and oxidizing reagent. It is made by the action of a high-voltage glow discharge on an equimolar mixture of F_2 and O_2 under very low pressure. Mixtures with alcohols or hydrocarbons explode even at very low temperatures. O_2F_2 reacts with fluoride acceptors such as boron trifluoride, forming dioxygenyl salts: $O_2F_2 + BF_3 \rightarrow [O_2][BF_4] + 1/2\,F_2$.

Oxygen halide, formerly ***oxide halide*** or ***oxyhalide***: a compound in which a metal or nonmetal is bound to both oxygen and halogen atoms, for example bismuth oxygen iodide, BiOI, molybdenum dioxygen dichloride, MoO_2Cl_2. The O. are usually the halides of oxygen acids; phosphorus oxygen chloride, $POCl_3$, for example, is the trichloride of phosphoric acid.

Oxygen metallurgy: a collective term for all metallurgical reactions in which oxygen-enriched air or pure oxygen is used. Oxygen-enriched air is used, for example, in blast furnaces, converters and cupellating furnaces for iron and steel production. Pure oxygen is used in Refining (see), that is, to remove undesired elements from crude iron (oxygen refining); in this case a stream of oxygen is blown across the surface of the melt or into the bath (oxygen blasting, Siemens-Martin furnaces, electric furnaces). For the Siemens-Martin process and electric furnace process, an oxygen purity greater than 99.5% is required, and for the oxygen blasting process, better than 99.9% oxygen is needed for quality (see Steel).

O. makes it possible to shorten the time required for steel refining and improve the quality of the product. It has not been widely applied in non-ferrous metallurgy.

Oxygen pressure gasification: see Synthesis gas.

Oxygen-sulfur group: the sixth main group of the periodic system includes the elements oxygen (O), sulfur (S), seleneium (Se) and tellurium (Te), the *chalcogens*, and polonium (Po). There are definite trends in the physical and chemical properties of these elements (table). The non-metals oxygen and sulfur exist under normal conditions as O_2 molecules and S_8 rings, respectively. Selenium and tellurium are semimetals. Selenium forms Se_8 rings or spiral chains, while there is only one known modification of tellurium, consisting of Te chains. Polonium is a radioactive metal; the half-life of the longest-lived natural isotope, ^{210}Po, is 138.4 days. The oxides of sulfur, selenium and tellurium are acid anhydrides. Polonium(IV) oxide is amphoteric.

The chemical behavior of the elements of the O. is determined by their common ns^2p^4 valence electron configuration. Although a noble gas configuration can be achieved by acquisition of two electrons, and polonium can also form a cation (Po^{4+}), most of the compounds of these elements are covalent. Oxygen is limited to the use of its 2s and 2p levels, and achieves the neon configuration by forming two covalent bonds (as in H_2O, H_2O_2, alcohols, ethers, SiO_2, etc.) or one double bond (as in CO_2, NO_2, SO_3 and carbonyl compounds). The oxidation state in these compounds is usually -2, but in exceptional cases (H_2O_2 and peroxides), it is -1. Its preferred coordination number is also 2, although oxygen can form derivatives with coordination number 1 (e.g. CO, CO_2), 3 (e.g. H_3O^+, $(C_2H_5)_3O^+[BF_4]^-$) and more rarely, 4 (e.g. $OBe_4(OOCCH_3)_6$). Sulfur, selenium and tellurium can use their d-orbitals to form bonds, especially with electronegative elements; by octet expansion they can form, for example, a 12-electron configuration, as in SF_6, SeF_6 and H_2SO_4. Some important oxidation states for these three elements are - 2 (e.g. sulfides, selenides, tellurides), +4 (e.g. SO_2, SeO_2, sulfites, selenites, $SeCl_4$, $TeCl_4$) and +6 (e.g. SF_6, SeF_6, H_2SO_4, sulfates, selenates). The stability of the highest oxidation number decreases with increasing atomic number. Thus sulfites are reducing agents, sulfates are largely unreactive in redox reactions, and both selenates and tellurates are strong oxidizing agents. The coordination numbers of the heavier elements of this group are quite variable; compounds with coordination numbers from 1 to 6 are known, and for tellurium, coordination numbers of 7 (e.g. $[TeF_7]^-$) and 8 (e.g. $[TeF_8]^{2-}$) are known.

Sulfur, as an element of the 2nd period, is unusual with regard to its ability to form $p_\pi p_\pi$ double bond systems (see Sulfur), but double bonds in selenium and tellurium compounds should always be interpreted as $d_\pi p_\pi$ interactions.

The electronegativity of the elements decreases markedly on going from oxygen to tellurium. The high electronegativity of oxygen is the cause of strong hydrogen bond formation in compounds which contain OH groups (e.g. water, carboxylic acids).

Properties of the elements of the 6th representative group of the periodic system

	O	S	Se	Te	Po
Nuclear charge	8	16	34	52	84
Electron configuration	[He] $2s^2p^4$	[Ne] $3s^2p^4$	[Ar] $3d^{10}\,4s^2p^4$	[Kr] $4d^{10}\,5s^2p^4$	[Xe] $4f^{14}\,5d^{10}\,6s^1p^4$
Atomic mass	15.999 4	32.064	78.96	127.60	(209)
Atomic radius [pm]	73	102	117	135	
Electronegativity	3.50	2.44	2.48	2.01	1.76
Density [g cm^{-3}]	1.149[a])	2.07[b])	4.81	6.25	9.196[c])
m. p. [°C]	−218.4	119.0	217[d])	452	252
b. p. [°C]	−182.962	444.674	684.9	1390	962

a) liquid O_2 at −183 °C; b) α-sulfur; c) α-polonium; d) β-selenium

Oxyhalide: see Oxygen halide.

Oxyhemerythrin: see Hemerythrin.

Oxyhemocyanin: see Hemocyanin.

Oxyhemoglobin: see Hemoglobins.

Oxyliquite: same as Oxyliquite explosives (see).

Oxyliquite explosives, ***oxyliquites, liquid air explosives***: explosives which are made using liquid air or liquid oxygen and carbon carriers. Cartriges made of absorbant paper and filled with absorbant, carbon-rich material (such as acetylene soot or cork powder) are dipped in oxygen-rich liquid air near the site where they are to be used. They are ignited in the conventional manner with ignition caps (with lead azide charge) or electrically. The strength of the O. is comparable to that of powerful ammonium nitrate explosives; their advantage is the safety of the undipped cartriges and the fact that duds which do not explode become harmless after the oxygen diffuses away. However, use of liquid oxygen is not without risk, and after dipping, the cartriges are sensitive to impact, friction and sparks. The explosiveness of the cartriges decreases continuously as the oxygen evaporates (in a borehole, they are almost non-explosive after about 25 minutes. At the same time, the carbon monoxide content of the smoke increases, so the cartriges must be used very soon after they are dipped. For these reasons, O. are used only in very limited applications.

Oxymercurization: production of organomercury compounds by electrophilic *trans* addition of mercury(II) acetate to alkenes: $(CH_3)_2C{=}CH_2 + Hg(OCOCH_3)_2 + H_2O \rightarrow (CH_3)_2(OH)C\text{-}CH_2\text{-}HgOCOCH_3 + CH_3COOH$.

Oxynervon: see Cerebrosides.

Oxytetracycline: see Tetracyclines.

Oxytocin: one of the neurohypophysial peptide hormones. O. is a heterodetic cyclic peptide, the structure of which was elucidated in 1953 by du Vigneaud, and was synthesized one year later. It was the first peptide hormone to be chemically synthesized. O. promotes contraction of the smooth muscles of the uterus, and stimulates contraction of the breast glands (milk ejection). It has a small amount of vasopressin activity.

Cys–Tyr–Ile–Gln–Asn–Cys–Pro–Leu–Gly–NH_2

O. is synthesized in the nucleus paraventricularis of the hypothalamus, in the form of an oxytocin-neurophysin precursor. It is transported, inside neurosecretory vesicles, via the tractus paraventriculo-hypophyseus to the posterior pituitary, where it is stored, and, after proteolytic release of the prohormone form, it is secreted into the blood. Secretion can be induced by psychological or tactile stimulation of the genitals or sucking at the breast. O. is partially responsible for initiation of labor pains. A large number of analogs have been prepared in the course of extensive studies of structure and activity. Those with prolonged or dissociated effects are of interest, as are analogs which are more active than the native hormone (e.g. [Thr^4]oxytocin) or those which can be used as inhibitors (antagonists). O. is prepared industrially by chemical synthesis.

Ozocerite: same as Mineral wax (see).

Ozone, ***trioxygen***: a high-energy form of elemental oxygen; M_r 47.9982, density of the liquid at -183 °C, 1.571, m.p. -182.7 °C, b.p. -111.9 °C, crit. temp. -5.16 °C, crit. pressure 6.7 MPa, crit. density 0.54, standard electrode potential 2.07 V ($O_3 + 2\,H^+ + 2\,e \rightleftharpoons O_2 + H_2O$).

Properties. O. is a blue, poisonous gas with a characteristic odor; upon cooling, it is condensed to a dark blue liquid which finally crystallizes to blue-black crystals. The O. molecule is bent (bond angle 116.5°) and can be described by resonance structures.

$$^{\ominus}O{-}\overset{+}{O}{=}O \longleftrightarrow O{=}\overset{+}{O}{-}O^{\ominus}$$

Its solublity in water is slight (100 ml water dissolves 49.4 ml O. at 0 °C). O. is more soluble in halogenated hydrocarbons. As an endothermal compound, O. slowly decomposes to dioxygen: $O_3 \rightarrow 3/2\ O_2$, $\Delta H = -145$ kJ mol^{-1}. This decay is accelerated by heating, irradiation and various catalysts, e.g. MnO_2 or PbO_2. Pure O. or concentrated solutions can explode, even at low temperatures. O. is one of the strongest oxidizing agents known. For example, it is capable of converting sulfides to sulfates, ammonia to nitric acid and carbon to carbon dioxide. With the exception of gold, platinum and iridium, metals are converted by O. to their highest oxides. Compact silver turns black in the presence of O., due to formation of silver(II) oxide, AgO. O. reacts with alkali metal hydroxides to form Ozonides (see). O. destroys many organic materials, so it cannot be passed through a rubber hose, for example. Combustible substances, such as ether, alcohol and lower hydrocarbons, ignite in O. or O.-rich oxygen. Cautious treatment of unsaturated organic compounds with O. is used to ozonize them (see Harries reaction).

> At high concentrations, ozone burns the eyes and mucous membranes of the respiratory tract. Lower organisms are killed by ozone.

Analytical. The simplest method of qualitative and quantitative analysis of O. makes use of the effect of the gas on a potassium iodide solution: $O_3 + 2\,KI + H_2O \rightarrow I_2 + 2\,KOH + O_2$; the iodine can be determined volumetrically or calorimetrically. Physical chemical methods are based on the characteristic absorption bands of O_3 in the IR and UV ranges.

Occurrence. O. is formed wherever atomic oxygen is formed and can react with molecular oxygen: $O + O_2 \rightarrow O_3$. As a result of homolysis of O_2 through UV radiation from the sun, and recombination of the resulting atomic oxygen with O_2, O. is always present in the earth's atmosphere. At the surface of the earth, the concentration is $3 \cdot 10^{-6}$ vol. %, and reaches a maximum of about $2 \cdot 10^{-5}$ vol. % at 20 to 25 km altitude. This O. layer absorbs the hard ultraviolet radiation, and is therefore of greatest importance for life on the earth. In the same reaction, O. is formed near quartz lamps or radioactive preparations, or as a result of electric discharges, as in thunderstorms. Decomposition of peroxides or electrolysis of sulfuric or perchloric acid also produces O.

To produce O., silent electric discharges in an oxygen stream are used; this is the principle of the *Siemens ozonator*. This consists essentially of two concentric pipes with electrically conducting coatings (Fig.). A high-voltage alternating current generates silent electric discharges in the intermediate space, through which the oxygen is flowing. Part of the O_2 is thus converted to O. With good cooling to avoid undesired decomposition, an oxygen mixture with up to 15% ozone can be obtained.

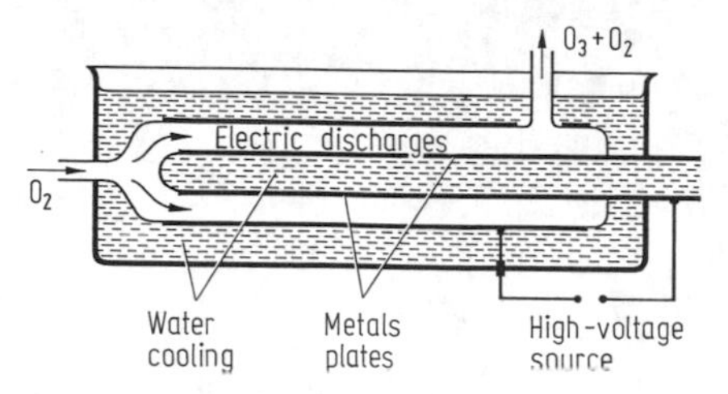

Diagram of the Siemens ozonator.

Applications. The most important application for O. is treatment of drinking water and sewage. Its ability to kill and inhibit the growth of microorganisms is utilized in preservation of foods, removal of odors (e.g. in rooms where foods are processed or refrigerated storage rooms), and for disinfection. In synthetic chemistry, O. is used principally in the Harries reaction.

Ozonides: 1) ionic compounds with the general formula MO_3 (M = K, Rb, Cs) formed by the reaction of ozone with solid alkali metal hydroxide, e.g. $3\,KOH + 2\,O_3 \rightarrow 2\,KO_3 + KOH \cdot H_2O + 1/2\,O_2$. KO_3 is a red-orange solid which slowly decomposes to KO_2 and oxygen. An ammonium ozonide is also known. 2) The 1,2,3-trioxolanes formed by ozonization of alkenes (primary ozónides) and their rearrangement products, the 1,2,4-trioxolanes (secondary ozonides); see Harries reaction.

Ozonization: 1) disinfection of drinking water with ozone. 2) ***Ozone cleavage***, reaction of unsaturated compounds with ozone (see Harries reaction).

P

p: 1) see Arenes. 2) abb. for *para-*; see Nomenclature sect. III D.

P: 1) symbol for phosphorus. 2) symbol for parachore.

Pa: symbol for protactinium.

PA: abb. for polyamide.

PAB: abb. for *p*-Aminobenzoic acid (see).

PABA: abb. for *p*-aminobenzoic acid. See Aminobenzoic acids.

Packings: in chemical industry, materials which serve to create a larger active surface, thus accelerating reactions and promoting the exchange of heat and matter between gases and liquids. P. come in a wide variety of shapes and materials. Short, hollow cylindrical pieces of stoneware, porcelain, glass, carbon, plastics, steel, copper, etc. are used; examples are ***Raschig*** and ***Pall*** rings (Fig.). Saddle-shaped bodies of ceramics, spheres of lead, ceramics or copper rings of wire mesh, helices of glass or wire, and so on, are also used. Perform grid and Pyrapack packings are very effective. These are vertical contact packings consisting of mats made up of triangular pyramids. P. are used, for example, in reactions between gases and liquids, for mixing gases and liquids, adsorption and desorption processes, distillation of liquid mixtures, extraction, heat exchange processes, precipitation of dust and dirt, degassing, defogging and deoiling.

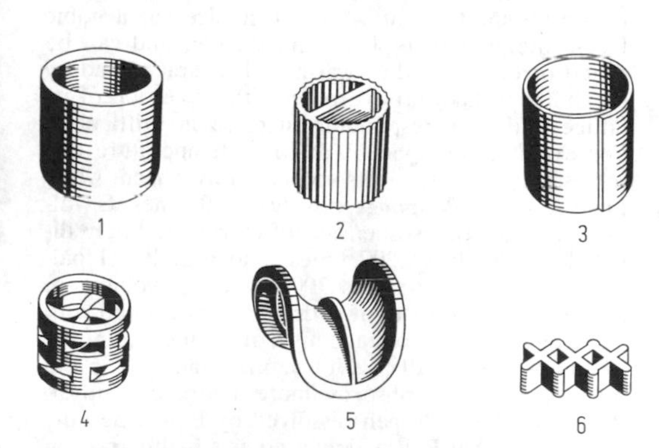

Packing shapes: *1-4*, packing rings (*1,2* and *3*, Raschig rings; *4*, Pall ring), *5*, saddle body; *6*, radiation body.

Depending on the application, P. are placed in tall towers (e.g. in the lead chamber process) or in packed columns (e.g. in distillation and absorption columns).

The geometric height of a P. layer corresponding to a single concentration step (see Theoretical plate number) is called the ***height equivalent to a theoretical plate***, or ***HETP*** value.

Packing coefficient: same as Packing density.

Packing density, ***packing coefficient***: the ratio of the volume occupied by atoms to the total volume of a structure. The P. is a measure for the amount of volume occupied by matter in a solid. The regular packing of identical spheres is taken as a model for the atomic arrangement in metals (see Packing of spheres); for this model, P. is calculated by $P = V_S/V_C$ (V_S is the volume of the spheres in a unit cell, and V_C is the volume of the cell). The maximum value of P. in this case is 0.74. In an ionic crystal, the structure is that of a close packing of the larger ion, with the hollows between occupied by the smaller type of ion. In this case, higher values for P. are possible. The larger the number of nearest neighbors for the crystal components, the larger P. will be. The principle of close packing also applies to molecular crystals. Here the effective molecular volumes can be calculated using an incremental system for the volumes of atoms and atomic groups; in most cases, the values for the P. are between 0.68 and 0.74.

Packing of spheres: spatial arrangement of rigid spheres of the same size in which neighboring spheres touch. The packing model makes possible a purely geometric description of the crystal structures of elements which consist of atoms (metals, noble gases), and of simple ionic substances. The P. adopted usually has a high density and a large coordination number. The primitive cubic lattice with 8 spheres at the vertices of the unit cell has a low packing density, 0.52, and a low coordination number, 6, for each sphere. The cubic body-centered lattice has a density of 0.68 and coordination number 8 (Fig. 1; see Tungsten type). The densest P. is obtained by starting from close-packed plane layers of spheres. For this there is only one possible geometric arrangement with hexagonal symmetry, in which each sphere is surrounded by 6 others, which touch it. If two such layers A and B are stacked so that each sphere of A sits in the depression between three spheres of B, and conversely, two types of hollow spaces with different sizes and geometries are formed between the two layers (Fig. 2): tetrahedral holes T are surrounded by 3 spheres of

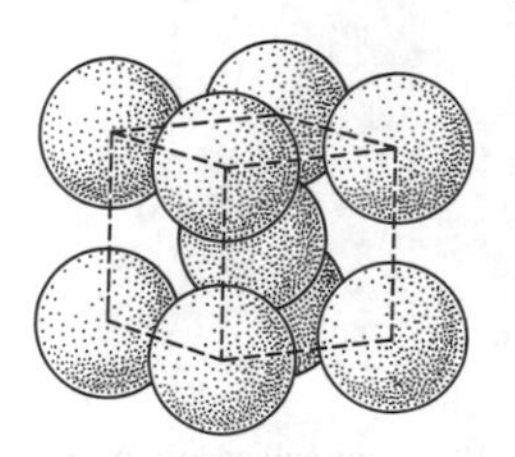

Fig. 1. Cubic body-centered packing

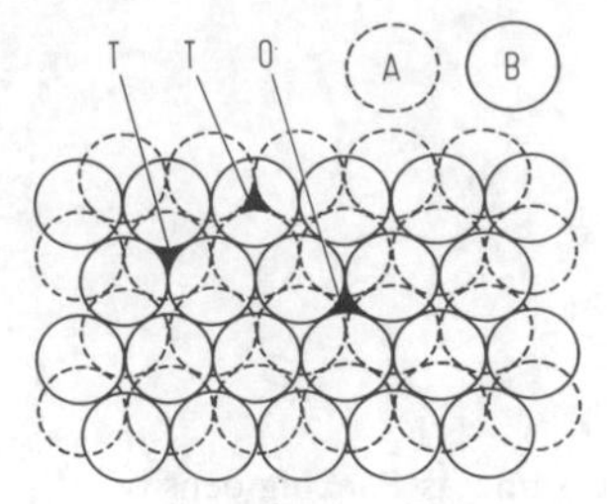

Fig. 2. Close packing of two layers with tetrahedral (T) and ocathedral (O) holes.

one layer and 1 sphere from the other layer, and octahedral holes O are surrounded by 3 spheres from each layer. If the radius of the spheres is *r*, a smaller sphere with a radius no larger than 0.255*r* fits in the tetrahedral hole, and a sphere with a radius no larger than 0.414*r* fits the octahedral hole.

There are two possibilities for the placement of the next layer: 1) the spheres of the third layer are directly over those of the first layer, so this layer is also denoted by A. 2) The spheres of the third layer lie over the octahedral holes; this new position is denoted C. Both result in *close packing* with a density of 0.74 and coordination number 12 for each sphere. There are infinitely many possible orders for further close packing; the simplest are most common. *Hexagonal close packing* has the periodic stacking sequence ABAB... (Fig. 3; see Magnesium type), while *cubic close packing* has the sequence ABCABC... (Fig. 4, see Copper type). However, there are also complicated close packings (see Polytypism), irregularities in the normal stacking order (stacking defects) and structures with statistically random stacking sequences.

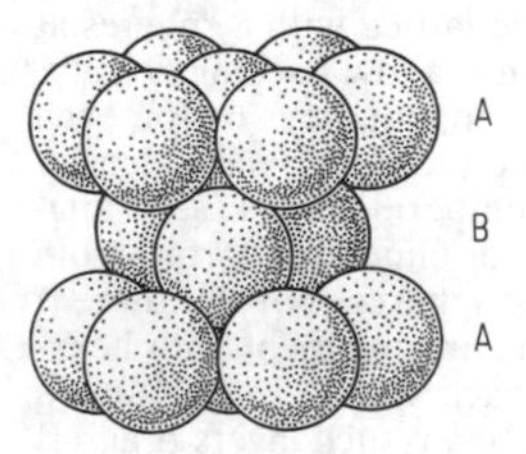

Fig. 3. Stacking sequence ABAB . . . of hexagonal close packing.

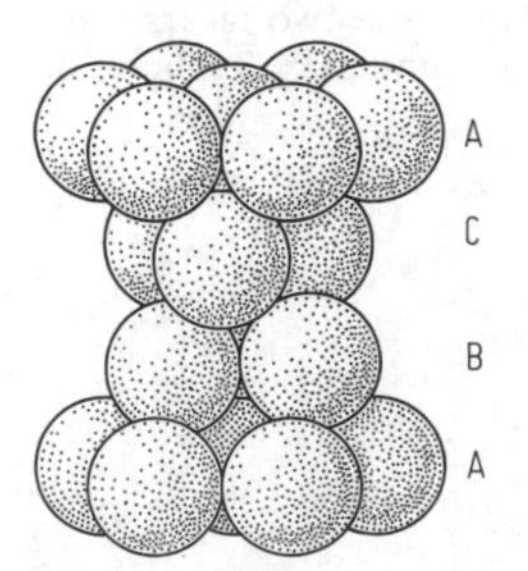

Fig. 4. Stacking sequence ABCABC . . . of cubic close packing.

If the tetrahedral and octahedral holes are considered as lattice sites which can be occupied by smaller spheres, the model of P. can also be used successfully to interpret the structures of ionic crystals.

PAF, ***platelet activating factor***: a lipid which at a concentration of 10^{-11} to 10^{-10} M activates blood platelets (thrombocytes). This lipid has been identified as 1-*O*-alkyl-2-*O*-acetyl-*sn*-glycero-3-phosphocholine, where the alkyl is hexadecyl or octadecyl. Only the optically active form indicated here is active.

Paint: a liquid, paste or powder applied to a surface by brushing, spraying, dipping or other method. It consists of pigments, solvents, binders, drying agents, fillers and additives (such as skin inhibitors). P. are classified on the basis of the binder as oil, synthetic resin, silicate and waterglass P. P. based on synthetic resins and dispersions are also called latex P.

Paint removers: compounds which can be used to remove old layers of paint, varnish, etc. Caustic P. contain alkaline compounds such as sodium hydroxide, trisodium phosphate, ammonia and sodium carbonate. Dissolving P. are mixtures of solvents: acetone, halogenated hydrocarbons and aromatics. Combined P. contain both caustic and dissolving components.

Palladiates: see Palladium compounds.

Palladium, symbol ***Pd***: an element from group VIIIb of the periodic system, and one of the light Platinum metals (see); *Z* 46, mass numbers of the natural isotopes 108 (26.46%), 106 (27.33%), 105 (22.33%), 110 (11.72%), 104 (11.14%) and 102 (1.020%); atomic mass 106.42, valency usually II, rarely IV, 0; density 12.02, m.p. 1552 °C, b.p. 2930 °C, electrical conductivity 10.07 Sm mm^{-2}, standard electrode potential (Pd/Pd^{2+} +0.987 V.

Properties P. is a silvery gray-white, shiny and corrosion-resistant metal which crystallizes in a cubic face-centered lattice. It is very ductile, and can be rolled or drawn cold or warm. It is characterized by its ability to take up hydrogen with expansion of its lattice and a corresponding increase in brittleness. For example, compact P. at room temperature dissolves about 600 times its volume of hydrogen, while finely divided ***P. sponge*** dissolves 850 times its volume, an aqueous suspension of extremely finely divided P. dissolves 1200 times, and a colloidal palladium solution dissolves 3000 times its volume of hydrogen. The uptake of hydrogen is associated with a decrease in the paramagnetism of the metal, until at a composition of $PdH_{0.66}$, it becomes diamagnetic and remains so as it absorbs more hydrogen (up to $PdH_{0.85}$). The hydrogen dissolved by P. is especially reactive, so that P. is widely used as a hydrogenation catalyst. A corollary of the ready uptake of hydrogen is the extreme permeability of hot sheet P. for hydrogen; it can be used as a method of purifying hydrogen. As a noble metal, P. is not dissolved by hydrofluoric or phosphoric acid, but it is the only platinum metal which dissolves in nitric acid. The best solvent for P. is aqua regia. Oxygen and chlorine oxidize P. at a dark red heat to palladium(II) oxide, PdO, or chloride, $PdCl_2$. P. is divalent in most of its compounds, but is sometimes tetravalent; square planar palladium(II) complexes are most common, and octahedral palladium(IV) complexes are known. Palladium(0) complexes occur in compounds with π-acceptor ligands such as $K_4[Pd(CN)_4]$ and $Pd(PF_3)_4$.

Occurrence and production. P. makes up about $5 \cdot 10^{-7}$% of the earth's crust. It is found together with other platinum metals. Palladium minerals, such as stibiopalladinite, Pd_3Sb, and braggite, (Pt,Pd,Ni)S, are rare. P. is present at concentrations up to 8% in certain gold-bearing sands in Brazil. The important starting materials for production of P. are the copper and nickel ores of Sudbury (Canada) and South Africa. In the course of processing and separating these Platinum metal (see) ores, P. is precipitated as ammonium hexachloropalladate, $(NH_4)_2[PbCl_6]$, and is then reduced to the metal. Considerable amounts of P. are obtained as a byproduct of electrolytic refining of ores, e.g. in the anode sludge from copper refining.

Applications. P. is used as a catalyst for hydrogenation and in the watch industry. Its alloys with gold are prized as jewelry (white gold) and are used as materials for chemical apparatus, electrical and electronic equipment (see Gold alloys). P.-silver alloys (see Silver alloys) and P.-copper alloys are used for electrical contacts. P. is also used in Titanium alloys (see).

Historical. P. was discovered in 1803 by Wollaston; it was named for the planetoid Pallas.

Palladium compounds: The most important binary palladium compound is the red-violet ***palladium(II) chloride***, $PdCl_2$, M_r 177.31, density 4.0, m.p. about 500 °C (dec.). This compound exists in two forms: α-$PdCl_2$ has a chain structure with a planar arrangement of the Cl ligands (Fig.),

Structure of α-$PdCl_2$

and is obtained by chlorination of palladium at 500 °C, while β-$PdCl_2$ has hexameric molecules $(PdCl_2)_6$. $PdCl_2$ crystallizes out of aqueous solution as the dihydrate, $PdCl_2 \cdot 2H_2O$, M_r 213.34, in the form of red-brown needles. It forms red, planar complexes of the type $M_2[PdCl_2]$ with alkali chlorides, e.g. ***potassium tetrachlorpalladate***, $K_2[PdCl_4]$, M_r 326.42, density 2.67, m.p. 105 °C (dec.). $PdCl_2$ in aqueous solution is reduced to palladium by carbon monoxide, a process which is used to detect CO: $PdCl_2 + H_2O + CO \rightarrow Pd + 2\,HCl + CO_2$. In the Wacker process, $PdCl_2$ is used as a catalyst for oxidation of alkenes.

Palladium(II) oxide, PdO, forms tetragonal, black crystals; M_r 122.40, density 8.70, m.p. 870 °C. It can be obtained by fusing $PdCl_2$ with sodium nitrate at 600 °C, or by reaction of palladium with oxygen at red heat. ***Palladium(II) oxide hydrate*** is obtained by hydrolysis of palladium(II) salts; it forms a brown precipitate. This precipitate reacts with sulfuric or nitric acid to give palladium(II) sulfate, $PdSO_4 \cdot 2H_2O$, or palladium(II) nitrate, $Pd(NO_3)_2 \cdot 2H_2O$; finely divided palladium also reacts with these acids to give the same products. ***Palladium(IV) oxide hydrate***, $PdO_2 \cdot nH_2O$, is a strong oxidizing agent which is precipitated out of hexachloropalladiate solutions by alkali hydroxide solutions. It is a dark red compound, which is converted to PdO when heated to 200 °C.

Palladium(II) iodide, PdI_2, is obtained as a black precipitate, M_r 360.21, density 6.003, m.p. 350 °C (dec.) when potassium iodide is added to palladium(II) salt solutions. ***Hexachloropalladic(IV) acid***, $H_2[PdCl_6]$, is obtained by dissolving palladium in hot aqua regia; the acid can be characterized by formation of insoluble red salts, e.g. potassium hexachloropalladiate(IV), $K_2[PdCl_6]$, M_r 397.32, density 2.738, or ammonium hexachloropalladiate(IV), $(NH_4)_2[PdCl_6]$, M_r 355.20, density 2.418. Brick-red ***palladium(IV) fluoride***, PdF_4, M_r 182.39, is made by reaction of palladium(II) bromide, $PdBr_2$, with bromine(III) fluoride, followed by reaction of the resulting $Pd^{II}Pd^{IV}F_6$ with fluorine. Brown, tetragonal ***palladium(II) fluoride***, PdF_2, M_r 144.40, density 5.80, is obtained by reaction of palladium with fluorine at dark red heat. $Pd^{II}[Pd^{IV}F_6]$, a black, hygroscopic compound, is also obtained by treating palladium(II) chloride, $PdCl_2$, with fluorine at 200-250 °C.

Palladium(0) complexes, such as $K_4[Pd(CN)_4]$ and $Pd[P(C_6H_5)_3]_4$, are tetrahedral compounds which can be obtained by reaction of $K_2[Pd(CN)_4]$ with hydrazine in the presence of triphenylphosphine.

Palliag®: same as Sipal® (see).

Palmitic acid, ***hexadecanoic acid, cetylic acid***: CH_3-$(CH_2)_{14}$-COOH, a saturated monocarboxylic acid; it forms colorless needles or platelets, m.p. 64.0 °C, b.p. 267 °C at 13.3kPa. P. is insoluble in water, but dissolves readily in ether, chloroform or hot alcohol. It has the typical reactions of a carboxylic acid. Its salts and esters are called *palmitates*. Many of the salts are of industrial importance as Metal soaps (see). P. is the most widely occurring fatty acid in nature. In the form of its glycerol esters, it is present in nearly all fats and fatty oils, and it is the chief component of palm fat. P. is present in beeswax as the myricyl ester, and in spermaceti as the cetyl ester. It is obtained mainly by saponification of fats, but it is also a product of soft coal distillation. It is used to produce soaps and lubricants. Its aluminum salt is used in the synthesis of napalm.

Palmityl alcohol, ***cetyl alcohol, hexadecan-1-ol***: $CH_2(CH_2)_{14}CH_2OH$, forms colorless crystals which are insoluble in water; m.p. 49 °C. P. is soluble in alcohol and ether. It is found in nature in spermaceti, as its palmitate. It is synthesized in industry by catalytic hydrogenation of palmitic acid esters, and is used as a textile conditioner, an emulsifier and a component of cosmetics.

PAM: see Pralidoxime.

Pamachine: see Primaquine.

PAMBA: see Antifibrinolytics.

PAN: 1) abb. for polyacrylonitrile; 2) abb. for Peroxiacetyl nitrate.

Panchromatic: see Spectral sensitization.

Pancuronium chloride: see Muscle relaxants.

Panogen®: see Fungicides.

PAN fibers: see Polyacrylonitrile fibers.

Panthenol: see Pantothenic acid.

Pantolactone: see Pantothenic acid.

Pantothenic acid: a B_2 vitamin, an amide of R(+)-2,4-dihydroxy-3,3-dimethylbutanoic acid (R(+)-pantoic acid) and 3-aminopropanoic acid (β-alanine) found very widely in plants and animals. P. is an unstable, light yellow oil; it is soluble in water and ethyl acetate and only slightly soluble in benzene and chloroform. The more stable calcium calt, $[\alpha]_D^{20}$ +27° (in water), is used therapeutically.

I $HOH_2C-C(CH_3)_2-CH(OH)-CN$ II

R = COOH : Pantothenic acid
R = CH_2OH: Dexpanthenol

In the synthesis of P., 2,4-dihydroxy-3,3-dimethylbutane nitrile (I) is converted to *pantolactone* (II) by acid hydrolysis. This yields racemic P. upon aminolysis with β-alanine. The racemic mixture is separated to give biologically active P. P. is used mainly in the form of salves or sprays to treat non-healing wounds, burns and mucous membrane inflammation. It is also present in a number of cosmetics. The alcohols *panthenol* (racemate) and *dexpanthenol* (R-form) are used for the same purpose.

P. is a component of coenzyme A, which plays an important part in fat, carbohydrate and protein metabolism. The human daily requirement for P. is 8 to 10 mg. The vitamin is present in relatively high concentrations in yeast and egg yolk. Deficiency symptoms are not known in human beings. In experimental animals, deficiency of P. can lead to changes in the skin and depigmentation of the hair (it was therefore named "anti-gray-hair factor").

Historical. In 1933, P. was isolated from mammalian liver by Williams. The structure elucidation and synthesis were achieved in 1940 by Williams, Kuhn, Wieland and Reichstein. P. was recognized as a component of coenzyme A in 1946 by Lipmann and Kaplan.

Papain: an endopeptidase (see Proteases) obtained from the latex of the papaya (*Carica papaya* L.) which catalyses hydrolysis of peptide bonds of basic amino acids, leucine or glycine. P. consists of 212 amino acid residues (M_r 23,350, IEP 8.75), with the catalytically active cysteine residue in position 25. As a thiol enzyme, P. is activated by mercaptoethanol, and inhibited by reagents which block SH groups, such as monoiodoacetic acid. P. is used in preparations to promote digestion and as a meat tenderizer.

Papaverin: an opium alkaloid which is present at 0.1 to 4.5% concentration in opium; m.p. 147 °C. Its hydrochloride, m.p. 220 °C (dec.) is used. As an isoquinoline derivative, P. is a weak base, pK_a 5.93, so its hydrochloride gives a very acidic reaction. Commercial P. is now entirely synthetic; it is a spasmolytic which acts peripherally, and is used to treat spasms in the stomach, intestine, bile and urinary systems. P. is not addictive. It was first isolated from opium in 1848 by Merck. The structure was elucidated in 1900 by Goldschmidt, and confirmed by synthesis in 1911 by Pictet and Gams.

Paper chromatography: a method for separation of small amounts of substances. P. is a form of Distribution chromatography (see) in which the stationary phase is the film of water adhering to the fibers of filter paper. The mobile phase is an organic solvent or a mixture of solvents. The chromatograms are developed in closed chambers which are saturated with the solvent vapor and guarantee a constant water content of the paper. The amounts of substance separated are between 5 and 50 µg, and are applied to a starting line in 0.1 to 1% aqueous solutions. Interfering proteins or salts must first be removed from samples of biological origin. The diameter of the starting spot should not be larger than 0.5 mm. After a run of several hours, one obtains substance spots at various distances from the starting line; these are visible or can be made visible.

Technique. P. is called "ascending", "descending" or "horizontal", depending on the direction in which the solvent moves (Fig.). In ***ascending P.***, the orgin is above the surface of the solvent, which is drawn into the paper by capillary action and can rise to about 30 cm. In ***descending P.***, the paper strip hangs from a trough containing the solvent, and the origin is below the trough. The solvent descends under the influence of gravity, and the paper strip can be as long as desired.

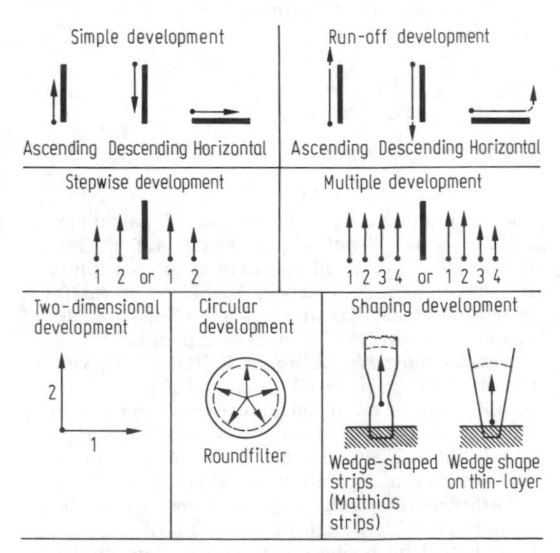

Paper chromatography: types of development

Using descending P., *run-off chromatograms* can be developed; these have longer running times and make difficult separations possible. In ***horizontal P.*** the solvent flows in the plane of the paper due to capillarity. This type of P. is generally done in circular format on a round filter paper. The sample is applied in a circle around a center, and the solvent is dropped onto this center; the solvent front and the substances spread out in nearly perfect circles around the center. The advantages of this method are short running times and high separation resolution. A modification of this method is shaping development, in which wedge-shaped strips (*Matthias strips*) are developed by the ascending method. These are segments of a round filter, and instead of spots, the substances form segments of circles. By combination of suitable solvents with step-wise, multiple or two-dimensional development, the separations can be significantly improved.

Chromatography papers are made from cotton cellulose. They have a fiber direction, and differ with respect to absorbancy, thickness and mass. For some purposes, special papers may be used, such as chemically modified papers (acetylated papers), impregnated papers, papers with ion-exchange properties and glass-fiber papers. Chromatography papers must be carefully stored and protected from impurities. The solvents are specially purified organic solvents which are either completely or partially miscible with water, and special buffer solutions. Some standard solvent mixtures are butanol/glacial acetic acid/water, phenol/water and pyridine/butanol/water. For a given mixture of substances, the best solvent mixture is the one which yields the greatest separation of the individual spots.

Evaluation. The separated compounds can be detected through their own colors, or colorless substances are sprayed with suitable reagents with which they react to form colored products. For example, amino acids, peptides or amines are developed with ninhydrin, reducing sugars with silver nitrate/ammonia, organic acids with acid-base indicators, and phenols with diazotized sulfanilic acid. The separated substances can also be localized by observing them in UV light (fluorescence or fluorescence quenching), and radioactively labelled compounds can be found by autoradiography.

Qualitative evaluation is based on the different migration rates of the compounds in certain solvents. The position of a substance on the chromatogram is indicated by the $\boldsymbol{R}_F$ (retention factor), which is the ratio of the distance moved by the spot to the distance moved by the solvent front. For simplicity, the hR_F ($= R_F \cdot 100$) is now usually indicated. If there is no solvent front (*flow-off chromatogram*), a comparison substance X is applied to the paper. The ratio of the distance moved by the unknown to that moved by X is then given as the R_X (or R_{st}, from ratio standard) value.

Quantitative analysis is based on the relationship between spot size and concentration. This can be done by comparison on the paper, photometrically or, after extraction of the spot, by measurement of the absorption.

Preparative P. can be done with thicker papers (cardboard).

Applications. P. is used in analysis of amino acids, peptides, proteins and carbohydrates, especially in biological fluids, in food analysis and pharmaceutical quality control, and in the analysis of natural products.

Historical. The beginning of P. lies in the capillary analyses of F. Runge (1822), and experiments by Groppelsröder (1861) with paper strips. The modern method was developed in 1944 by R. Consden, A.H. Gordon and A.J.P. Martin.

Paper electrophoresis: see Electrophoresis.

para-: see Arenes; see Nomenclature, sect. III D.

Parabanic acid, ***oxalylurea, imidazolidine-2,4,5-trione***: the cyclic ureide of oxalic acid. P. forms colorless crystals; m.p. 243-245 °C (dec.), and sublimes at 100 °C. It is soluble in water and alcohol and can be synthesized from oxalyl chloride and urea.

O
HN NH
O O

Paracetamol: see Phenacetin.

Parachore: a measure of the molar volumes of liquids at a given surface tension:

$$P = \frac{M}{d_1 - d_v} \gamma^{1/4}$$

where γ is the surface tension, M is the molar mass, d_1 the density of the liquid and d_v is the density of the vapor. P is nearly independent of temperature, and can be calculated by addition of atomic and bond increments. In the past, the P. was used for elucidation of structures of compounds. For example, the P. measured for the trimer of acetaldehyde, paraldehyde, corresponds to that calculated for a six-membered ring consisting of 3 carbon and 3 oxygen atoms, with no double bonds. This was later determined to be the structure of paraldehyde. The method has now been largely replaced by more exact methods, especially spectroscopic methods.

Paraffin: 1) a mixture of saturated hydrocarbons consisting mainly of alkanes with chain lengths C_{14} to C_{30}. Depending on the molecular mass, P. is a wax-like mass or a viscous or mobile liquid. It has no odor or taste, is nontoxic, does not stick and is an electrical insulator. P. is insoluble in water and slightly soluble in ethanol; it is readily soluble in gasoline, ether, chloroform and carbon disulfide. It is very resistant to attack by sulfuric acid, nitric acid and bromine. P. is obtained mainly from petroleum, or from the fractions from Fischer-Tropsch synthesis which boil above 400 °C. It can also be obtained from soft coal carbonization tar, either by direct distillation or by catalytic hydrogenation as ***TTH paraffin***. Both TTH and ***Fischer-Tropsch*** paraffins are extremely pure (low sulfur) and have few branched chains, so they are very suitable as raw materials for the synthesis of fatty acids (see Paraffin oxidation).

Petroleum P. is a byproduct of deparaffinization of high-boiling petroleum fractions or lubricating oils; these must be freed of P. to reduce their stock points. There are several methods of *deparaffinization*: 1) Treatment of lubricating oils or high-boiling petroleum fractions with a mixture of methyl ethyl ketone or acetone and technical benzene or toluene; the P. is dissolved at room temperature and crystallizes out at -30 °C. 2) Extraction with propane, which acts as a solvent and, due to its low boiling point and heat of vaporization, it cools the solution to the temperature required for recrystallization of the P. 3) Reaction with urea (see Urea separation); the P. is purified by hydrorefining, washing with sulfuric acid and sodium hydroxide, decolored with activated charcoal or fuller's earth, and often washed again with alcohol. 4) Adsorptive separation using molecular sieves, for example, by the Parex process (see Adsorption).

Hard P. is a solid, colorless and nearly odorless mass. It consists of saturated hydrocarbons with 20 to 30 carbon atoms; m.p. 50 to 62 °C. Hard P. is insoluble in water and alcohol, soluble in ether, chloroform and benzene. Hard P. can be purified by vacuum distillation, extraction and refining. It is used mainly to make candles, floor and shoe polishes, lubricants and wax paper; it is also used as an electrical insulator. ***Soft P.*** has a melting point of 42 to 44 °C; it is used to make matches, impregnate paper and wax strings. It is also used in the leather industry and to make medicinal salves and packings.

The P. produced by Fischer-Tropsch synthesis include ***slack wax***, with a melting point between 35 and 40 °C, ***medium P.***, m.p. 44-46 °C, and ***macroparaffin***, m.p. 90-95 °C (used as an impregnating agent, polish and electrical insulator). ***Microparaffin*** is a P. with a particle size of 1 to 5 μm; due to numerous branchings and rings in its molecules, it contains a high percentage of carbon and is especially tough, plastic and cold-resistant. Microparaffins are obtained from crude oils or by enrichment from macroparaffins; they are used chiefly in the form of aqueous emulsions as impregnating agents. ***Paraffin oil (vaseline oil)*** is a mixture of liquid hydrocarbons (isoalkanes, naphthenes) obtained by hydrorefining and sulfuric acid refining of petroleum or lignite coal tar P. P. oils are classified according to viscosity: paraffinum subliquidum has a viscosity greater than 120 cSt at 20 °C, and paraffinum perliquidum has a viscosity lower than 60 cSt at 20 °C. The latter is used as a fine lubricant for clocks, sewing machines, etc., as a heating bath liquid, in the concoction of salves and as a stool softener to prevent constipation.

2) same as Alkane (see).

Paraffin oil: see Paraffin.

Paraffin oxidation: a process for producing fatty acids by oxidation of alkanes (paraffins) with chain lengths between 20 and 30 carbons. The oxidation is carried out in the presence of 0.3 mole % manganese salts as catalysts at a temperature of 100 to 120 °C in acid-resistant reactors. After about 20 hours reaction time and a paraffin turnover of 30 to 50%, the oxidation is stopped and the reaction products are worked up.

The product mixture includes both unreacted alkanes and fatty acids of various chain lengths, lactones, keto and hydroxy acids, esters, etc. The reaction mixture is first neutralized with sodium hydroxide, and the unreacted alkanes are returned to the oxidation reactor. The byproducts and residual paraffins are removed by placing the crude soap solution under 2.5 MPa pressure and heating it to 180 °C, then 330 to 370 °C. The crude fatty acid solution is next acidified with mineral acid (hydrochloric or sulfuric); the crude fatty acids precipitate and are washed with water to remove the short-chain acids. The mixture is then distilled in vacuum.

The first runnings are hydrogenated to alcohols which, when esterified to phthalic or adipic acid, are used as softeners. The next distillation fraction, the soap fatty acids, are converted to soft soap, soap powder and laundry soaps; with the addition of natural fats, they are also used in toilet soaps. A large fraction of the production is also hydrogenated to fatty alcohols, which are then processed, e.g. to alkyl sulfates. The last distillation fraction of fatty acids is partially returned to the process, and partly combined with the first runnings to make lubricants. The distillation residue, a solid, black product, can be used as an insulator or protective paint.

To oxidize alkanes in the C_{10} to C_{20} range to secondary alcohols without chain cleavage, higher temperatures are used than with normal P and boric acid is added. The boric acid causes the *sec*-alkylhydroperoxide intermediates to decompose to *sec*-alcohols. In addition, the *sec*-alcohols are esterified by the boric acid, and thus protected from further oxidation. The resulting mixture of *sec*-alkyl boric acid esters is readily saponified to boric acid and *sec*-alcohols in the presence of water. The alcohols are converted to *sec*-alkyl sulfates or are reacted with ethylene oxide or propylene oxide to form non-ionic surfactants.

Paraffin slack wax: see Paraffin.

Paraflow: an alkylated naphthalene used to improve the Stock points of paraffin-based lubricating oils.

Paraformaldehyde: a polymeric mixture derived from Formaldehyde (see).

Parafuchsin, ***pararosaniline, pararosaniline hydrochloride, para magenta***: crystals with a green sheen (with 4 mol crystal water) which are slightly soluble in cold water, more soluble in hot water and soluble in alcohol, where they give a red solution. The compound is synthesized by reaction of aniline with *p*-toluidine in the presence of nitrobenzene, which in the above acidic melt acts as an oxidizing agent.

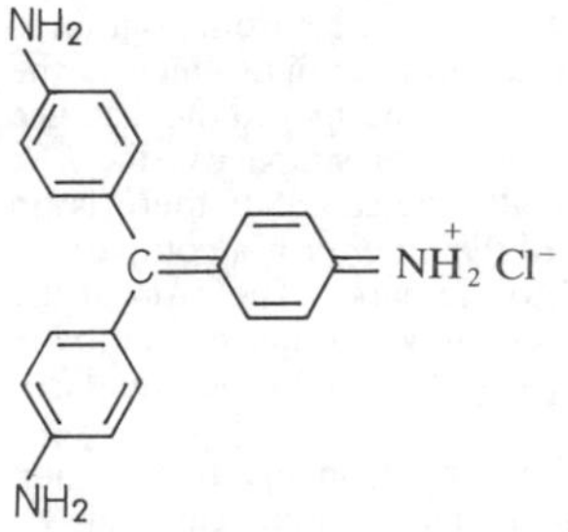

P. was the first synthetic aniline dye, and is the parent compound of the triphenylmethane pigments. It is now used mainly as a component of marine blue pigments and as a chemotherapeutic against protozoan infections.

Paraldehyde, ***2,4,6-trimethyl-1,3,5-trioxane***: a cyclic trimer of acetaldehyde. P. is a colorless liquid, m.p. 12.6 °C, b.p. 128 °C. It is only slightly soluble in water, but dissolves readily in alcohol and ether. P. is formed from acetaldehyde in the presence of small amounts of sulfuric acid at 20 °C, and it can be converted back to acetaldehyde by heating with acids.

Because of its stability, P. is a suitable storage form of acetaldehyde. It is used in medicine as a sedative and to induce sleep. It is also used as a solvent and to promote vulcanization.

Parallel reaction: a Complex reaction (see).

Paramagenta: same as Parafuchsin (see).

Paramagnetic electron resonance spectroscopy: same as Electron spin resonance spectroscopy.

Paramagnetism: see Magnetochemistry.

Paraoxon, ***Diethyl-p-nitrophenyl phosphate, E 600***: 4-(NO_2)-C_6H_4-O-P(O)$(OC_2H_5)_2$, a reactive phosphate ester which inhibits acetylcholinesterase by irreversible phosphorylation (see Parasympathicomimetics). P. is an oily liquid and is used to treat glaucoma.

Paraquat: see Herbicides.

Parared, ***nitroaniline red***: 1-(*p*-nitrophenylazo)-2-naphthol, the first synthetic dye comparable to alizarin; m.p. 248-252 °C. P. is generated in the fibers, which are pretreated by impregnating them with alkaline β-naphthol solution, by coupling with cold, diazotized *p*-nitroaniline solution. P. serves as a substitute for alizarin and is used for dyeing and printing cotton.

Pararosaniline: same as Parafuchsin (see).

Parasympathicolytics: substances which inhibit the post-ganglionic effects of the neurotransmitter acetylcholine. P. are used mainly as spasmolytics. The prototype natural product is Atropine (see). Other P. include atropine methobromide, scopolamine butyl bromide and homoatropine. P. are used in opthalmology as pupil dilators (*mydriatics*).

Parasympatheticomimetica: substances which excite the parasympathetic nerves. Acetylcholine (see) acts as neurotransmitter in the postganglionic parasympathetic nerve synapses; for this reason the P. are also called *cholinergics*. P. can act either as analogs of of acetylcholine (***direct P.***) or as inhibitors of the enzyme which degrades acetylcholine (***indirect P.***). The latter are also called *acetylcholinesterase inhibitors*. P. stimulate the smooth muscles and various glands, e.g. the salivary glands. They inhibit the generation and transmission of impulses to the heart. They cause the pupil of the eye to contract and decrease the interior pressure of the eye; for this reason they are used to treat glaucoma. Various compounds are used for their stimulatory effects of the smooth muscles of the intestines and bladder. Some direct P. are acetylcholine, carbachol and pilocarpine; acetylcholinesterase inhibitors include physostigmine, neostigmine, pyridostigmine, diethyl *p*-nitrophenyl phosphate (phosphacol), and fluostigmine.

Parathion: O,O-Diethyl-O-4-(nitrophenyl)thiophosphate, an organophosphate contact, oral and respiratory insecticide (thiophosphorate esters). P. is a liquid, D_4^{25}. 1.265, m.p. 6 °C, m.p. 150 °C at 80 Pa, vapor pressure at 20 °C $0.75 \cdot 10^{-3}$ Pa. Solubility in water, 24 ppm at 25 °C, rapid hydrolysis in alkaline medium. Oxidizing agents or enzymatic reactions in plants, insects and warm-blooded animals produce *paraoxon* [O,O-diethyl-O-4-(nitrophenyl)phosphate], which is more toxic than P.

P. is synthesized by reaction of diethylthiophosphoryl chloride with sodium *p*-nitrophenolate. P. was developed by G. Schrader and named by him *E 605*; it was the first organophosphate brought into large-scale production (1947/48). It is used widely to combat biting and sucking insects in field crops, forestry, fruit and truck plantations.

Parathormone: same as Parathyrin.

Parathyrin, ***parathormone, parathyroid hormone***, abb. ***PTH***: a polypeptide hormone from the parathyroid which regulates the calcium and phosphate metabolism. P. is a linear polypeptide consisting of 84 amino acid residues. Human P. differs from that of bovine or porcine P. in 11 positions. Not all of the 84 amino acids are necessary for biological activity. Chemical synthesis of human P. was first reported in 1982 by Sakakibara. Biosynthesis occurs via the precursor protein preproparathyrin.

When the normal calcium level of the blood drops, P. is secreted by the parathyroid gland. It elevates the Ca^{2+} concentration of the blood and stimulates phosphate excretion through the urine, so that the phosphate level in the blood sinks. Calcitonin has the opposite effects.

Parex process: a method for extracting paraffins with chain lengths C_{10} to C_{20} from petroleum fractions by selective adsorption of the paraffins on molecular sieves. The P. is a cyclic adsorption-desorption process carried out in the gas phase at temperatures of 300-400 °C and pressures of 0.5 to 1.0 MPa. The starting material is a mixture of n- and iso-paraffins which is preheated with a carrier gas (hydrogen-containing refinery gas), then passed over adsorber A (Fig.) containing the molecular sieve. The paraffins are adsorbed to the molecular sieve, and the product leaving the adsorber is distilled to separate it from the carrier gas and desorption reagent. It is then used for production of high-grade diesel fuel.

When the capacity of the molecular sieve in adsorber A is exceeded, the n-paraffin concentration in the exit gas rises; this is a signal for an automatic shift

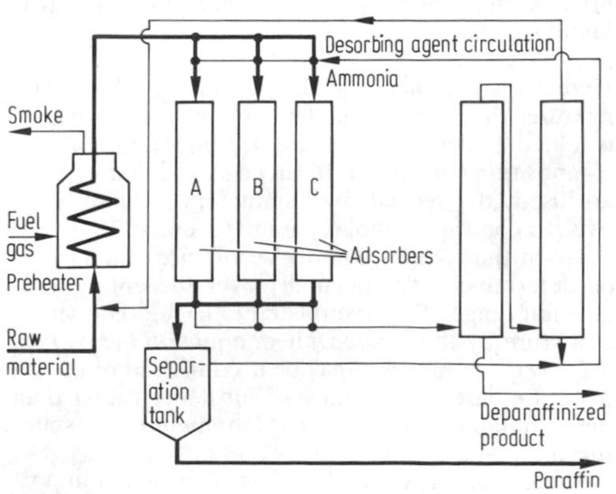

The Parex process (schematic).

to adsorber B or C. Adsorption in one adsorber and desorption (with ammonia) in the other two occurs simultaneously. The gas mixture leaving the desorption process consists of n-paraffins and desorption reagent; the desorption reagent is removed and returned to the desorption circulation. Three adsorbers are needed to maintain continuous operation, because the desorption time is approximately twice the time required for adsorption. To achieve a desirable product purity, the adsorbers are flushed between working phases.

Pargylin: a monoamine oxidase inhibitor which causes blood pressure to drop and acts as a thymeretic.

$$C_6H_5-CH_2-N(CH_3)-CH_2-C\equiv CH$$

Paris green: same as Methyl green (see).

Parkerization: see Phosphatization.

Parkes process: see Silver.

Paromomycins: a group of aminoglycoside antibiotics which are structurally closely related to Neomycins (see). The P. are synthesized by *Streptomyces rimosus*, and are poorly absorbed. They are used chiefly for intestinal infections.

Parr bomb: see Solubilization.

Partial molar parameters: calculated parameters used in thermodynamic descriptions of real mixtures. The properties y of ideal mixtures are equal to the sums of the molar properties Y_i of the components: $y = \sum_i n_i Y_i$, where n_i is the molar amount of substance i in the mixture (see Thermodynamics, first law). The parameter Y_i can be the molar volume, heat, enthalpy, etc. These parameters are not strictly additive in real mixtures, because interactions occur between the components. For the parameters defined by the second law of thermodynamics, entropy and free energy (Helmholtz or Gibbs), the parameters are not additive even in ideal solutions, because the spontaneous mixing process contributes to them. There are two possible ways to treat the properties of such mixtures: 1) by introducing mixing and excess functions (see Mixing) as corrections for the total mixture, or 2) by applying these corrections to the molar contributions of individual components in such a way that they become additive. The P. arise from the second approach. They are defined by the equation: $(\bar{Y})_i = (\delta y/\delta n_i)p, T, n_j$.

Here δy is the change in the property y of the total mixture when a differential amount δn_i of substance i is added at constant pressure p, temperature T and composition (sum of n_j). It can be shown that when P. are used, the second-law parameters are also additive. In contrast to molar parameters of state, P. depend on the composition of the mixture and have to be determined experimentally over the entire composition range. They are related to each other within a mixture by the Gibbs-Duhem equation (see). *Partial specific parameters* may be used instead of the P.; these are based on the mass of substance rather than the number of moles (see Parameters of state, specific).

The Chemical potential (see) is of particular importance.

Partial pressure: the contribution of one component of a gaseous mixture to the total pressure. For ideal gases, the P. of substance i is equal to the pressure p_i which it would display if present by itself in the available volume. It can be calculated from the equation of state (see State, equation of) for ideal gases: $p_i = n_i RT/v$, where n_i is the number of moles of substance i in the volume v of the total gas mixture. In special cases, the P. of individual components can be measured directly by manometry, if a semipermeable membrane is available which is permeable only to this type of particle. For example, hydrogen can easily pass through hot walls made of platinum or palladium. Dalton's law (see) applies to mixtures of ideal gases.

Based on Dalton's law, the P. of a real gas may be defined by the equation $p_i = x_i p$, where p is the total pressure and x_i is the mole fraction of this gas in the mixture. The P. p_i defined in this way is not identical with the pressure which the real gas would display if it were the only gas present in the available volume. It is generally lower than this ideal value, because there are usually attractive forces in the mixed gas which reduce the pressure.

Another way to treat real effects is to use the Fugacity (see).

Partial syntheses: see Biotechnology.

Parting: the separation of more noble metals from less noble ones. It can be achieved with acids, as in the separation of gold and silver by means of hot concentrated sulfuric acid (*affination*) or hot nitric acid (*quartation*); in each case, the silver forms a soluble salt and the gold remains solid. Gold and silver can also be separated by electrolytic P. (Möbius process; see Silver).

Parting compound: a solid or liquid film which prevents two adjacent surfaces from sticking together. Silicones, waxes, metal soaps, fats, polymers (polyethylene, polyamides, polyvinyl alcohol), fluorinated hydrocarbons, or inorganic substances (talcum, mica) are used as P.

PAS: 1) abb. for *p*-aminosalicylic acid; 2) abb. for photoacoustic spectroscopy.

Paschen-Back effect: see Zeeman effect.

Passerini reaction: production of an α-acyloxyamide by reaction of an isonitrile with a carboxylic acid and an aldehyde or ketone. In reactions with ammonia or a primary amine, the product is the corresponding bisamide.

$$R-\overset{+}{N}\equiv\overset{-}{C}I + R^1R^2C{=}O + R^3-COOH \longrightarrow R-NH-C(=O)-CR^1R^2-O-C(=O)-R^3$$

Passivator: 1) same as Inhibitor (see); 2) see Passivity.

Passivization agent: a non-pigmented priming paint based on a low-viscosity solution of air-drying oils and resins. It soaks into the rust remaining on a steel surface after cleaning and passivizes the steel surface by means of the lead soaps it contains. P. can be used when a complete removal of rust (see Surface pretreatment, Rust) is not possible for technical or economic reasons. Its protective action lasts for a time comparable to that of a Penetrating agent (see).

Passivity: the phenomenon that certain reactive metals, e.g. iron, chromium, nickel, lead, aluminum, and cobalt, or alloys of reactive metals, are much less reactive under certain conditions than would be expected from their position in the electromotive force series. The transition from the active to the passive state (passivation) can occur *electrochemically* or *chemically*; it occurs because a passive layer forms of the surface of the metal. Redox systems which cause a chemical passivation are called *passivators*. Iron is passivated by concentrated nitric acid; its aqueous solutions are passivated by chromates, nitrites, molybdates, tungstates and pertechnetates.

Paste: see Glue.

Pasteurization: a process, named for L. Pasteur, for improving the short-term stability of foods by heat treatment; because the temperatures are relatively mild and the treatment short, only the vegetative stages of microorganisms, including pathogens, are killed. Heat resistant spores survive P., in contrast to Sterilization (see). Enzymes which reduce the quality and storage life of the food are also inactivated. P. is ordinarily carried out between 62 and 85 °C, depending on the food being treated. For example, it may be heated to 71-74 °C for 15 to 40 s, 85 °C for 10 to 14 s, or to 130-150 °C for up to 2 s. P. is generally applied to milk and milk products, fruit juices, beer, pickles, etc.

Patchouli oil: a highly viscous, yellow to dark brown essential oil with an intense, musty odor. It consists mainly of patchouli alcohol, $C_{15}H_{26}O$ (up to 50%), and also contains large amounts of sesquiterpenes, eugenol, cinnamaldehyde and benzaldehyde. P. is obtained from the leaves of *Pogostemon patchouli*, and is used for perfumes; in Southeast Asia where the plant is native, the oil is used to protect textiles from insects.

Patent blue: various types of water-soluble and usually also alcohol-soluble, acidic triphenylmethane pigments with bright colors but low fastness to light. They are used to make colored paper and paints.

Paterno-Büchi reaction: see Oxetane.

Patina: a basic copper carbonate of varying composition. The main component is $CuCO_3 \cdot Cu(OH)_2$. P. is formed in moist air as a green layer on objects made of copper or copper alloys and is responsible for the color of copper roofs on historical buildings. It protects the copper under it from further reaction.

Pattinson process: see Silver.

Pauli principle: see Electron configuration.

Pauli reaction: see Proteins.

Pb: symbol for lead.

PB: abb. for polybut-1-ene.

PBI: abb. for polybenzimidazene.

PCNB: see Fungicides.

p-Conductor: see Semiconductors.

PCP: see Fungicides.

PCTFE: abb. for Polytrifluorochloroethylene.

Pd: symbol for palladium.

PE: abb. for polyethylene.

Peak: a maximum in a diagram.

Peak surface: see Gas chromatography.

PeCeU®: see Synthetic fibers.

Pearlite: the structure of technical iron in a eutectic mixture (see Iron-carbon alloys) with 0.80% carbon; a mixture of ferrite and cementite. The cementite can be finely dispersed in strips (lamellar cementite) or grains (grainy cementite). P. is not stable above 723 °C and is converted to Austenite (see).

Peat: a black to light brown, combustible organic material. P. consists of 55-60% carbon, 6% hydrogen and 33% oxygen. Fresh P. may contain 90% water, and air-dried P. still contains 17 to 35% moisture. The dry matter of P. is about 50% huminic acid and a small amount of waxes and resins, which can be isolated by extraction. P. is a precursor of coal, and is the most recent combustible humus deposit. The oldest layers of the oldest known P. deposits are about 25,000 years old. The heating value of P. is 9400 to 16300 kJ kg^{-1}, and very dry P. with a low ash content can have a value as high as 17600 to 22600 kJ kg^{-1}. P. is used as a fuel in some countries, particularly the former Soviet Union, Ireland, Finland and Switzerland. When it is heated in the absence of air, it produces a low-value ***peat gas***, a watery distillate which contains mainly ammonia, acetic acid and methanol. P. can also be carbonized to yield a tar and coke (see Carbonization), or gasified to yield a relatively high quality generator gas.

Pectin: see Pectin substances.

Pectinic acid: see Pectin substances.

Pectin substances: vegetable heteropolysaccharides found in association with other polysaccharides, such as cellulose and hemicelluloses, in the primary cell walls and in the cytoplasm.

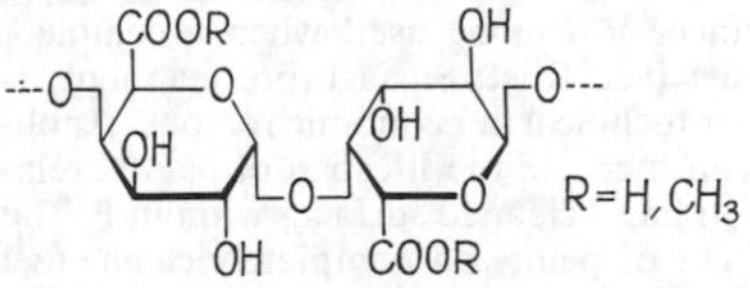

$R = H, CH_3$

They are found in the plant as high-molecular-weight, insoluble ***protopectin***, which contains calcium and magnesium ions. The skeleton of the P. is a linear α(1→4)-D-polygalacturonic acid, to which galactan, araban or galactoxylan residues can be linked. These ***pectinic acids*** are polyelectrolytes, which form gels with divalent cations. The salts are called ***pectinates***. The pectinic acids of fruits are esterified with methanol, to a greater or lesser degree, and these partial esters are called ***pectin***. Pectin is obtained from solid residues remaining after the juice has been pressed from fruit; it is sold commercially as a white or brownish powder, or as a liquid concentrate. Pectin forms stable gels in weakly acidic media in the presence of sucrose or glycerol. It is used as a gelling agent in jams and jellies.

PEEK: abb. for poly ether ether ketone.

Pelargonaldehyde, ***nonanal, nonyl aldehyde***: CH_3-$(CH_2)_7$-CHO, a colorless liquid which smells strongly of oranges; b.p. 78 °C around 400 Pa. P. is slightly soluble in water, and readily soluble in alcohol. It undergoes the reactions typical of the Aldehydes (see). P. is present in lemon, tangerine and cinnamon oils, and can be synthesized by reduction of pelargonic acid in the vapor phase or by catalytic hydrogenation of oleic acid ozonide. It is used as a perfume.

Pelargonidine, ***pelargonidium chloride***: 3,5,7,4'-tetrahydroxyflavylium chloride, the aglycon of many

Anthocyans (see). P. hydrates form red-brown platelets or prisms which are readily soluble in hot water and alcohol; m.p. > 350 °C. P. can be obtained by hydrolysis of ***pelargonin***, its 3,5-β-diglucoside. It is present in orange dahlias, red geraniums and purple asters.

Pelargonin: see Pelargonidine.

Pelargonic acid: same as Nonanoic acid (see).

Pempidine: see Ganglion blockers.

Penam: see Penicillins.

Pelletizing: a process for increasing the grain size of solids by mechanical compaction of fine-grained material. The resulting ***pellets*** have a standard size, in contrast to the products of other methods of increasing grain size (gruanulates, agglomerates). The solidity of the pellets is due to cohesion and adhesion forces between the molecules, and can be increased by addition of binders. P. without binders is used chiefly for soft coal, but is also used to improve the properties of sawdust, peat, feeds and other materials. P. with binders is used mainly for coal dust, ores and plastics. Pitch and sulfite waste liquors and various inorganic compounds are often used as binders. Compression of the pellets is done at various pressures, up to 300 MPa.

Penetrating agent: a pigmented or non-pigmented priming paint based on a low-viscosity solution of air-drying oils and/or resins. It penetrates deep into dry, porous rust, isolates the corrosion product and, if it contains zinc chromate pigment, passivates the steel surface. P. can be used when a complete removal of rust (see Rust; Surface pretreatment) is not possible for technical or economic reasons. Application of P. can increase the lifespan of paints, relative to those on hand-cleaned surfaces without P. The protective effect of paints on completely clean steel surfaces is not matched, however.

Penetration: in Ecological chemistry (see), a measure of the degree to which a chemical has been washed into the soil. Studies of P. apply mainly to estimates of contamination of agricultural soils by xenobiotics, such as pesticides, and of the danger of contamination of ground and surface waters. In model experiments with soil columns, the depth of P. is measured as a function of a simulated amount of precipitation, and the degree of contamination of the water leaching out of the soil is recorded (leaching test). In the field, lysimeters are used in the study of P. A field leaching test is also used to determine the rate at which water penetrates soil, in order to determine whether a septic system can be built in that area.

Penetration complex: see Coordination chemistry.

Penicillins: a group of β-lactam antibiotics which contain the bicyclic skeleton *penam*. The P. are derivatives of 6-aminopenicillanic acid in which the amino group is acylated. 6-Aminopenicillanic acid contains 3 chiral C-atoms (C2, C5 and C6). The biosynthesis of the P., like that of the cephalosporins, begins with L-2-aminoadipic acid, L-cysteine and D-valine; after formation of an acylated 6-aminopenicillanic acid, the L-2-aminoadipic acid group is replaced by another carboxylic acid, such as phenylacetic acid (benzylpenicillin). The P. are formed by molds of the genuses *Penicillium*, *Aspergillus* and *Trichophyton*.

P. are widely used antibiotics; the most important microbially synthesized P. is ***benzylpenicillin*** (***penicillin G***). Byproducts such as P. F and P. K are of no practical importance. Benzylpenicillin is an amorphous, white, unstable powder; $[\alpha]_D^{20}$ +282° (ethanol). The more stable alkali salts are used, especially the sodium salt, m.p. 215 °C (dec.), $[\alpha]_D^{25}$ +301° (water). Benzylpenicillin is produced industrially by high-yield strains, especially of *penicillium chrysogenum*, in submersion culture under aerobic conditions. To repress formation of other P., phenylacetic acid is added as a precursor; the water in which corn has soaked contains phenylacetic acid and may be used. The mycelium is removed from the solution, which is then acidified and extracted, e.g. with butyl acetate; the free acid P. enters the organic phase. The sodium salt of the P. is then made by neutralization of the acid.

Benzylpenicillin kills gram-positive bacteria (streptococci, pneumococci and those staphylococci which do not form β-lactamase) and gram-negative cocci, e.g. gonococci. It is relatively nontoxic, and can be used in high doses. Its lack of chemical stability is a disadvantage; in particular, its instability in acid precludes oral application. In addition, it is rapidly eliminated, and is effective only against a limited spectrum of bacteria. Resistant strains are not uncommon, and some patients are allergic to benzylpenicillins. In order to produce a sufficiently high blood level over a longer period, aqueous solutions or oil suspensions of insoluble salts of benzylpenicillin, e.g. the procaine or benzathine (N,N'-bisbenzylethylenediamine) salts, are injected. P. stable to acid and/or β-lactamase and P. with broader action spectra have been synthesized; these have different acyl groups than benzylpenicillin. This is done by addition of other monosubstituted acetic acids as precursors, but mainly by acylation of 6-aminopenicillinanic acid which is obtained by enzymatic deacylation of benzylpenicillin. Recently, 6-aminopenicillanic acid has been prepared chemically from benzylpenicillin by the Delfter process.

The most important partially synthetic P. include the phenoxypenicillins, which resist acid, but not β-lactamase; an example is ***phenoxymethylpenicillin*** (***penicillin V***). The isoxazolylpenicillins, e.g. ***cloxacillin*** and ***dicloxacillin***, are resistant to both acid and β-lactamase. The broad-spectrum aminobenzylpenicillins such as ***ampicillin*** and ***amoxicillin***, resist acid but not β-lactamase. The α-carboxybenzylpenicillins, e.g. ***carbenicillin*** and acylureidopenicillins, e.g. ***azlocillin*** and ***mezlocillin***, are resistant to neither acid nor β-lactamase, but they are used for infections with difficult bacteria, such as *Pseudomonas aeruginosa* and some *Proteus* species. For better absorption from the gut, esters of P. are sometimes used; these are enzymatically cleaved into the active free acids in the blood.

Historical. In 1928, A. Fleming noticed that growth of staphylococci on an agar plate which had been contaminated by a mold was inhibited. The antibiotic agent was named penicillin after the mold, which was identified as *Penicillium notatum*. In cooperation with the "Oxford group" including Chain and Florey, American scientists isolated and began to produce the P. at the beginning of the 1940's. The structure was elucidated by the work of Chain, Robinson, du Vigneaud and others. A total synthesis has been achieved, but it is of no practical significance.

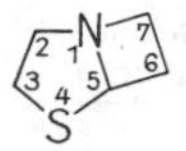

Penam

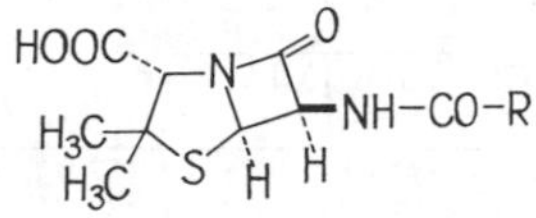

Derivative of 6-aminopenicillanic acid

Name	R
Benzylpenicillin	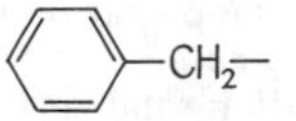 $C_6H_5-CH_2-$
Penicillin K	$H_3C-(CH_2)_5-CH_2-$
Penicillin F	$H_3C-CH_2-CH=CH-CH_2-$
Phenoxymethyl-penicillin	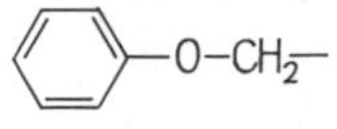 $C_6H_5-O-CH_2-$
Cloxacillin (R = H) Dicloxacillin (R = Cl)	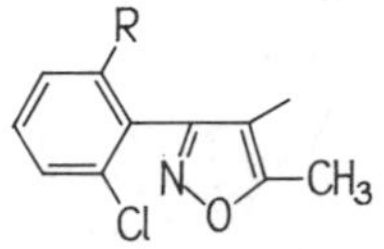 R, Cl, N, O, CH_3
Ampicillin (R = H) Amoxicillin (R = OH)	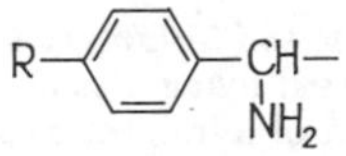 $R-C_6H_4-CH(NH_2)-$
Carbenicillin	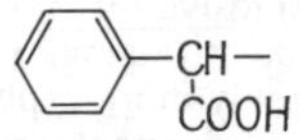 $C_6H_5-CH(COOH)-$
Azlocillin (R = H) Mezlocillin ($R = SO_2CH_3$)	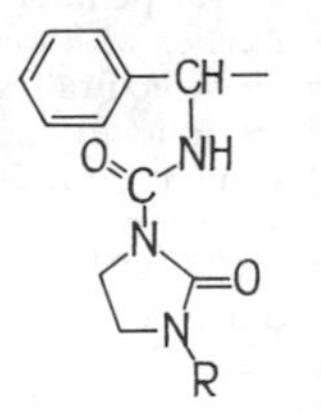 CH–, NH, O=C, N, =O, N, R

Pentacene: see Acenes.

Pentachlorophenol: 2,3,4,5,6-pentachlorophenol, colorless, crystalline needles; m.p. 191 °C, b.p. 310 °C (dec.). P. is practically insoluble in water, but is soluble in ethanol and ether. It gives a weakly acidic reaction, and is dissolved in alkali hydroxide solutions, forming phenolate solutions. It is synthesized by the reaction of sodium hydroxide and methanol with hexachlorobenzene. P. is very toxic to fungi, algae, sponges and bacteria, and is therefore used to impregnate wood. This can be done either by pressure impregnation with a solution of P. in kerosene, or by use of the water-soluble sodium pentachlorophenolate, which penetrates the wood well. The active P. is then released by treatment with carbon dioxide. The compound can also be used to impregnate leather, straw, dry glue and textiles to prevent mildew.

1,15-Pentadecanolide: same as Exaltolide (see).

Pentaerythritol: see Pentaerythrol.

Pentaerythritol tetranitrate, ***pentrit***, often incorrectly called ***tetranitropentaerythritol***, abb. ***nitropenta***: the tetranitrate of pentaerythritol, which is made by a crossed Cannizzaro reaction from acetaldehyde and formaldehyde, then esterified with nitric acid to P. P. is a powerful explosive, with a detonation velocity of 8300 m s^{-1}. To make it easier to compress and reduce its sensitivity to impact, P. is usually phlegmatized by addition of 5 to 10% montane wax. It is used mainly to make compressed charges and high-power gun shells; it is also used to make secondary charges for blasting caps and to fill fuses. In medicine, P. is used as a long-acting coronary medication; it is suitable for prevention of attacks of angina pectoris.

Pentaerythrol, ***pentaerythritol, 2,2-bis(hydroxymethyl)propane-1,3-diol***: water-soluble, colorless crystals; m.p. 261-262 °C. It is obtained by condensation of acetaldehyde with formaldehyde in aqueous alkaline solution. The higher polymers formed by this reaction can easily be separated because they are insoluble in water. P. is used as a softener, and its esters have many applications. The *phthalates* are used to make alkyde resins, *various monocarboxylates* are used as raw materials for laundry products, wetting agents and emulsifiers. The *nitrate* (pentaerythritol tetranitrate) is a powerful explosive, and also an effective relaxant for use in cases of angina pectoris and asthma.

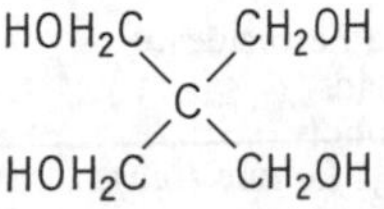

$$C(CH_2OH)_4$$

3,5,7,2',4'-Pentahydroxyflavone: same as Morin (see).

Pentamethylenediamine: see Cadaverine.

Pentamidine: an example of the bisamidines used to combat trypanosome infections.

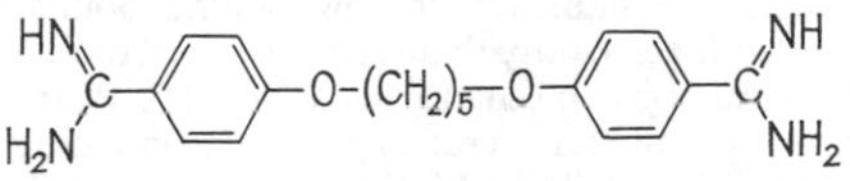

$$(H_2N)(HN=)C-C_6H_4-O-(CH_2)_5-O-C_6H_4-C(=NH)(NH_2)$$

Pentanal: same as Valeraldehyde (see).

Pentane: a hydrocarbon with the general formula C_5H_{12}. There are three structural isomers, ***pentane, isopentane*** and ***neopentane*** (Table). The P. are colorless liquids which ignite readily and burn with sooty flames. They are practically insoluble in water, but dissolve in ethanol and ether. They are found in natural gas, refinery gas and cracking gas, and can be separated by fractional distillation. They are main components of petroleum ether, along with the hexanes. Isopentane is produced on a large scale from P. by catalytic isomerization. Low-temperature isomerization can be catalysed by aluminum chloride, and high-temperature isomerization by metals, especially platinum. Some other industrially important reactions of the P. are alkylation of isopentane with alkenes to produce branched, high-octane hydrocarbon mixtures (see Butane), catalytic dehydrogenation to pentenes, pyrolysis to low-molecular weight alkenes, chlorination to amyl chlorides or more highly chlorinated P., and nitration to C_1 to C_5 nitroalkanes. The P. are not only important components of gasoline, but are excellent solvents for fats, oils and some resins.

Physical properties of the pentanes

Name	Formula	m.p. [°C]	b.p. [°C]	n_D^{20}	Octane rating
Pentane	$CH_3(CH_2)_3CH_3$	–129.7	36.1	1.357 9	62
Isopentane (2-methylbutane)	CH_3–$CH(CH_3)$–CH_2–CH_3	–160	27.9	1.353 7	90.3
Neopentane	CH_3–$CH(CH_3)_2$–CH_3	–20	9.5	1.347 6	80.2

Pentanedial: same as Glutaraldehyde (see).

Pentanedioic acid: same as Glutaric acid (see).

Pentanediols: Dialcohols of the pentane series. There are 6 structural isomers derived from *n*-pentane, 7 from 2-methylbutane, and so on; the total number of isomers is very large. The best known and most important is *pentane-1,5-diol*, $HOCH_2(CH_2)_3CH_2OH$, a colorless, very viscous liquid; m.p. -18 °C, b.p. 260 °C, n_D^{20} 1.4498. It is readily soluble in water, low-molecular-weight alcohols and acetones. It is made from tetrahydrofurfuryl alcohol by hydrogenating cleavage on a copper-chromium catalyst. It is used to make polyurethanes, protective films for glass and as a brake fluid and lubricant. The diesters are used as softeners and glues.

Pentane-2,4-dione: same as Acetylacetone (see).

Pentan-4-one-1-al: same as Levulinaldehyde (see).

Pentanochlor: see Acylaniline herbicides.

Pentanoic acid: see Valeric acids.

Pentanols: same as Amyl alcohols (see).

Pentan-3-one: same as Diethyl ketone (see).

Pentapeptides: see Peptides.

Pentele: see Nomenclature, sect. II.A.2.

Pentene, ***amylene***: one of the isomeric alkene hydrocarbons with the general formula C_5H_{10}. There are five structural isomers, of which one, pent-2-ene, exists in an E and a Z form (Table). The P. are colorless, inflammable liquids with low boiling points, which burn with very sooty flames. They are insoluble in water, but dissolve in ethanol and ether. The P. are contained in coal tar, cracking gases, cracking gasoline and shale oils. 2-Methylbut-1-ene and 2-methylbut-2-ene are formed by pyrolysis of rubber. P. can be obtained from cracking gasoline by fractional distillation, but they can also be made by catalytic hydrogenation of the pentynes or dehydrogenation of the pentanes. Dehydration of the amyl alcohols or dehydrochlorination of the monochloropentanes are also industrially important. The pentene mixture made from fermentation amyl alcohols is called ***commercial amylene*** or ***fusel oil amylene*** (b.p. 35-39 °C). P. are used in the synthesis of amylphenols, thiophene derivatives, isoprene, pentanols and polymers.

Pent-2-enedial: same as Glutaconic dialdehyde.

Pentetrazole, ***pentamethylenetetrazole***: a heterocyclic compound used as an analeptic. P. forms colorless crystals which dissolve very easily in water and ethanol; m.p. 58 °C. It can be made by reaction of cyclohexanone with hydrazoic acid, HN_3, but is most often made from ε-caprolactam; first the cyclic imide acid methyl ester is formed, this is reacted with hydrazine, and the product is reacted with nitrous acid.

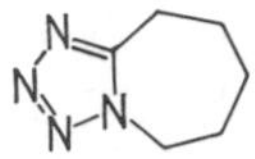

N N N N

Pentoses: see Monosaccharides.

Pentrinites: dynamites containing more than 10% pentaerythritol tetranitrate (pentrite), which replaces part of the explosive oil and makes these dynamites especially stable to aging. However, it also makes them considerably more expensive.

Pentrite: abb. for pentaerythritol tetranitrate.

Pentynes: alkynes with the general formula C_5H_8. There are three structural isomers, which differ in their reactivities, due to the position of the triple bond (see Alkynes): ***pent-1-yne***, $HC{\equiv}C$-CH_2-CH_2-CH_3, m.p. -90 °C, b.p. 40.2 °C, n_D^{20} 1.3852; ***pent-2-yne***, CH_3-$C{\equiv}C$-CH_2-CH_3, m.p. -101 °C, b.p. 56 °C, n_D^{20} 1.4039; and ***3-methylbut-1-yne***, $H{\equiv}C$-$CH(CH_3)_2$, m.p. -89.7 °C, b.p. 29.4 °C, n_D^{20} 1.3723. The P. are readily ignited, colorless liquids which burn with sooty flames. They are scarcely soluble in water, but dissolve readily in ethanol and ether. They are made in the laboratory by alkylation of the mono- or disodium salt of ethyne. They have no particular industrial importance.

Peonidin: see Anthocyanins.

Peonin: a plant pigment; one of the Anthocyans (see).

Peppermint oil: a colorless, essential oil which smells like peppermint. The refreshing effect of P. is

Physical properties of the pentenes

Name	Formula	m.p. [°C]	b.p. [°C]	n_D^{20}
Pent-1-en	CH_2=CH–CH_2–CH_2–CH_3	–138	30	1.371 5
Pent-2-en Z-Form	$(CH_3)(H)C$=$C(CH_2$–$CH_3)(H)$	–151.4	36.9	1.383 0
Pent-2-en E-Form	$(CH_3)(H)C$=$C(H)(CH_2$–$CH_3)$	–136	36.,3	1.379 3
2-Methyl-but-1-en	CH_2=$C(CH_3)$–CH_2–CH_3	–137.5	31.2	1.337 8
2-Methyl-but-2-en	$(CH_3)_2C$=CH–CH_3	–133.8	38.6	1.387 4
2-Methyl-but-3-en	$(CH_3)_2$CH–CH=CH_2	–168.5	20.1	.

due to a very high content of (-)-menthol (50 to 66%); it also contains menthone (9 to 12%) and menthyl esters (3 to 21%). P. is obtained by steam distillation of dried peppermint plants in yields up to 0.5%. It is the starting material for production of menthol. P. is used as a flavoring and to treat stomach upsets and respiratory diseases.

Pepsin: an acid endopeptidase (see Proteases) found in the stomach. It catalyses the cleavage of peptide bonds, preferentially those with aromatic amino acids. Proteins are broken down into mixtures of polypeptides, called ***peptones***. The relative molecular masses (M_r) of these products are between 300 and 3000. P. is a phosphoprotein consisting of 327 amino acid residues (M_r 34,500); it arises from the inactive precursor, ***pepsinogen*** (M_r 42,500) by autolysis in the presence of hydrochloric acid. The cleavage products released by autolysis include an inhibitor peptide which must be degraded before the enzyme achieves full activity.

Peptide antibiotics: a group of antibiotics consisting of oligopeptides, usually cyclic. They contain both proteogenic and nonproteogenic amino acids, including D-amino acids, and sometimes ester as well as peptide bonds. The resistance of P. to proteases is due both to their cyclic structures and the high content of nonproteogenic amino acids and bond structures. Their structures are often so complicated that production by chemical synthesis is impractical. Most P. are synthesized by bacteria, although some are produced by streptomycetes or lower fungi. Their antibiotic effects are due to interference with various metabolic processes, such as biosynthesis of proteins, nucleic acids or cell wall components, or with membrane activities. P. are not synthesized on ribosomes, but by *S*-aminoacyl activation and transfer on specific enzymes, as demonstrated by the synthesis of gramicidins in bacteria.

P. are active against both gram-positive and gram-negative bacteria. Because many are also toxic to vertebrates, only a few types are used systemically, including penicillins, gramicidins, tyrocidins, polymixins, etc.

Peptide bond: the most important type of covalent bond between the amino acyl residues in peptides and proteins. Formally, a P. is an *N*-substituted amide group formed by reaction of the activated carboxyl group of one amino acid with the nucleophilic amino group of a second amino acid. Because the electron pair on the N atom is conjugated with the carbonyl group, the P. has a resonance-induced partial double-bond character.

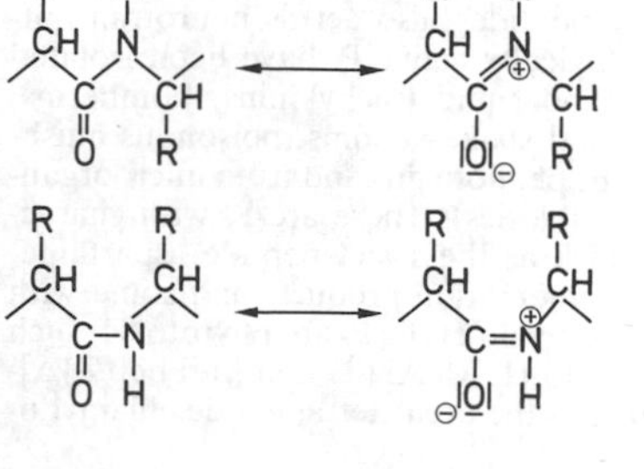

I

II

Rotation around the C-N axis is therefore hindered, giving rise to two different planar arrangements (Fig.), *trans*-P. (I) and *cis*-P. (II). The *trans*-form is the more stable; the energy difference is 33.5 kJ mol^{-1}. In the cyclic dipeptides (2,5-dioxopiperazines), however, only *cis*-P. are possible. In cyclic tripeptides, too, all three P. must be in the *cis*-form, but cyclotriprolyl is the only cyclic tripeptide with three *cis*-P. which it has been possible to synthesize. Although *trans*-P. predominate in natural peptides and proteins, *cis*-P. do occur; in each case, a proline is involved in the bond.

Peptides: Organic compounds consisting of two to approximately one hundred amino acyl residues covalently linked by Peptide bonds (see). The number of amino acyl residues in short P. is indicated by a numerical prefix (di-, tri-, etc.); for longer P., the number is indicated by an Arabic numeral instead of a prefix: 11-peptide, for example, instead of undecapeptide. ***Oligopeptides*** contain fewer than 10 aminoacyl residues, and the boundary between ***polypeptides*** and proteins, which cannot pass through natural membranes, lies at a relative molar mass of about 10,000 (about 100 amino acyl residues).

For the purposes of systematic nomenclature, the P. are regarded as acylamino acids; the amino acid whose carboxyl group is part of the P. bond, is given the suffix -yl. Only the terminal amino acid with a free carboxyl group retains its original trivial name, as in

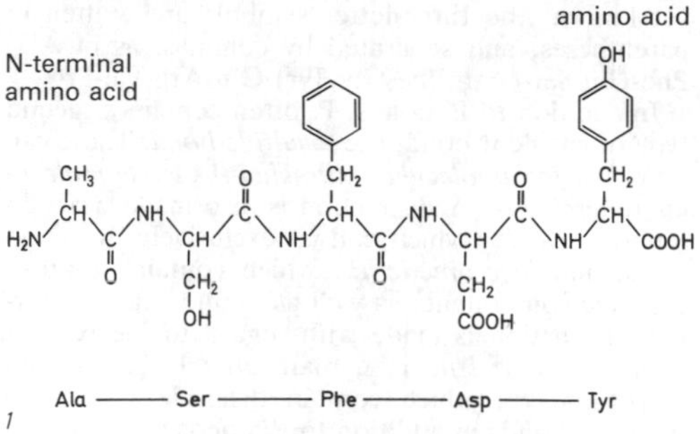

C-terminal amino acid

N-terminal amino acid

Ala — Ser — Phe — Asp — Tyr

1

alanyl-seryl-phenylalanyl-asparagyl-tyrosine. With the three-letter abbreviations for the amino acids, this pentapeptide is designated by Ala-Ser-Phe-Asp-Tyr. When the name is written horizontally, the amino acid with the free α-amino group is always on the left; it is called the *N-terminal amino acid*; the terminal amino acid with the free carboxyl group, on the right end of the name, is the *C-terminal amino acid* (Fig. 1) The formula Ala-Ser-Phe-Asp-Tyr symbolizes the pentapeptide regardless of its ionization state. The terminal amino and carboxyl groups can be indicated by addition of an H or OH, respectively (H-Ala-Ser-Phe-Asp-Tyr-OH), which allows a simple representation of the ionized state. The abbreviated notation presumes that trifunctional amino acids with additional amino or carboxyl functions (Lys, Orn, Glu, Asp) are linked by α-peptide bonds. A special designation is required for ω-peptide bonds, as shown in Fig. 2 for the γ-peptide bond in the naturally occurring tripeptide glutathione.

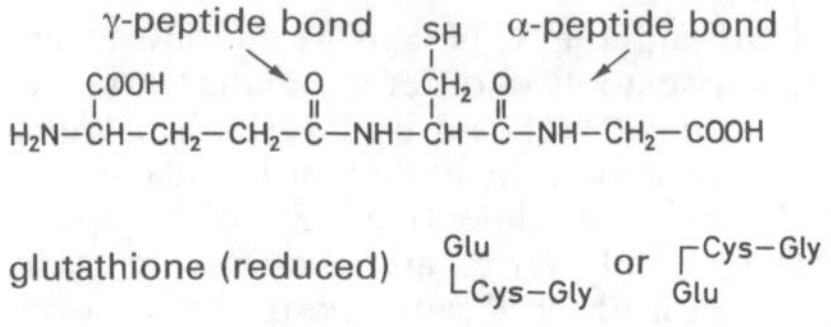

```
γ-peptide bond       SH      α-peptide bond
              ↘       |      ↙
   COOH          O    CH2  O
   |             ||   |    ||
H2N–CH–CH2–CH2–C–NH–CH–C–NH–CH2–COOH
```

glutathione (reduced) Glu ┗Cys–Gly or ┏Cys–Gly Glu

A P. bond between the α-amino group of lysine and the side-chain carboxyl group of glutamic or aspartic acid is called an *isopeptide bond*, e.g. in N^{ε}-γ-glutamyllysine:

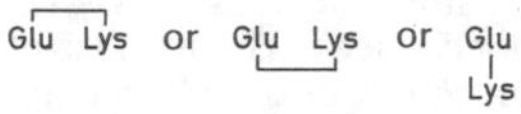

Glu Lys or Glu Lys or Glu | Lys

In the abbreviated notation, side chain substitutions are indicated by an abbreviation for the substituent placed above or below the three-letter symbol for the substituted amino acid, or it is placed in parentheses immediately after the symbol. The fully protected pentapeptide N^{α}-benzyloxycarbonyl- L-alanyl-O-*tert.*-butyl-L-seryl-L-phenylalanyl-L-asparagyl (β-benzyl ester)-O-benzyl-L-tyrosine methyl ester, for example, is denoted by Z-Ala-Ser(But)-Phe-Asp(OBzl)-Tyr(Bzl)-OMe.

The number and sequence of the amino acids in a P. make up its *primary structure*. If the sequence is known, the three-letter symbols are written in order and linked by hyphens. If part of the sequence is not yet known, the three-letter symbols are written in parentheses, and separated by commas, as in Ala-Phe-Glu-Ser-(Asn,Phe,Gly,Tyr)-Glu-Arg-Val-Pro.

In addition to P. bonds, P. often contain a second type of covalent bond, the *disulfide bond*. These can be either *intramolecular (intrachain)* or *intermolecular (interchain)*. A distinction is also made between *homeomeric P.*, which consist exclusively of amino acids, and *heteromeric P.*, which contain non-proteogenic components as well as amino acids. A further distinction is made with regard to the type of bonding: *homodetic P.* contain only P. bonds, and *heterodetic P.*, which contain disulfide, ester and thioester bonds in addition to P. bonds (Fig. 4).

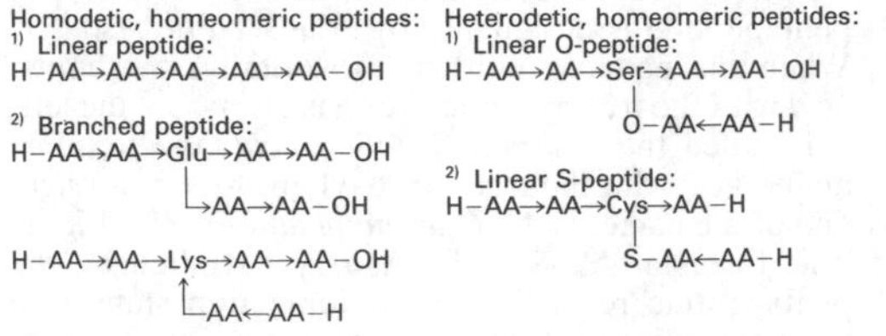

```
Homodetic, homeomeric peptides:        Heterodetic, homeomeric peptides:
1) Linear peptide:                      1) Linear O-peptide:
H–AA→AA→AA→AA→AA–OH                     H–AA→AA→Ser→AA→AA–OH
                                                  |
2) Branched peptide:                              O–AA←AA–H
H–AA→AA→Glu→AA→AA–OH
             └→AA→AA–OH                 2) Linear S-peptide:
                                        H–AA→AA→Cys→AA–H
H–AA→AA→Lys→AA→AA–OH                              |
         ↑                                        S–AA←AA–H
         └AA←AA–H

3) Cyclic peptide:                      3) Cyclic peptide (disulfide):
AA→AA→AA→AA→AA                          AA→AA→AA→AA→AA→AA
↑              ↓                        ↑                ↓
AA←AA←AA←AA←AA                          AA←Cys—S—S—Cys←AA

4) Cyclic branched peptide:             4) Cyclic branched peptide (peptide lactone):
AA←AA←Lys←AA←AA–H                       H–AA→AA→Asp→AA→AA
↓      ↑                                         ↓      ↓
AA→AA→AA                                         AA–CO–O–Thr–OH
```

Fig. 4. Schematic structures of homeomeric peptides. AA = amino acid.

The bond direction of the P. bond is indicated by an arrow with its point directed toward the nitrogen of the peptide bond; this is generally needed only when formulas for cyclic P. are written on two lines. The Depsipeptides (see) are heterodetic P.

Composite P. are P. with heterocomponents linked covalently to the chain by amino, carboxyl or other side-chain functions. Examples are glyco-, lipo-, phospho- and chromopeptides.

Synthetic analogs of natural P. are named according to the following IUPAC-IUB guidelines. When an amino acid is replaced by a new amino acid, the full name of the new amino acid and its position are written in square brackets before the trivial name of the P.; for example, [4-threonine]oxytocin, abb. [Thr^4]oxytocin. Multiple substitutions are treated similarly. When a P. is extended, the additional amino acyl residue is indicated in the terminal position in the usual fashion, e.g. glycyl-vasopressin, abb. Gly-vasopressin for an *N*-terminal addition, or vasopressyl-alanine, abb. vasopressyl-Ala, for the *C*-terminal addition. When the new amino acid is inserted into the chain, this is indicated by the prefix "endo" with the corresponding position indicator; for example, if valine is inserted between the 6th and 7th amino acids of bradykinin, the new P. is denoted by endo-6a-valine-bradykinin, abb. endo-Val^{6a}-bradykinin. Elimination of an amino acid is denoted by indication of the position and the prefix "des", e.g. des-4-glycine-bradykinin or des-Gly^4-bradykinin. Side chain substitutions on the amino group or carboxyl group are indicated in the usual fashion, e.g. N^{ε}11-alanyl-corticotropin, abb. N^{ε}11-Ala-corticotropin, or C^{γ}3-valyl-corticotropin, abb. C^{γ}-3-Val-corticotropin. Partial sequences of P. with established trivial names are denoted by indicating the positions of the first and last amino acids and the Greek numerical indicator for the number of amino acids in the partial sequence, as in bradykinin-(5-9)-pentapeptide.

Occurrence. P. are ubiquitous, and the range of their physiological functions is unusually broad. Many P. act as hormones; they are synthesized in the hypothalamus (oxytocin, vasopressin, releasing hormones, release-inhibiting hormones), pituitary (corticotropin, melanotropin, thyreotropin, gonadotropins, prolactin, somatotropin), pancreas (insulin, glucagon), thyroid (calcitonin), parathyroids (parathromone), gastrointestinal tract (gastrin, secretin, motilin, cholecystokinin), neurons (neurohormones), leukocytes (interferons, interleukins, etc.), bone marrow, epithelia (various growth factors) and quite probably in other tissues as well. In the past few years, dozens of P. have been discovered in the cells of the immune and nervous systems (see Neuropeptides), and it has been found that some of the immunoregulatory peptides also act as neurotransmitters. Pharmacologically active P. have been isolated from amphibia and squid (tachykinins, bombesins, caerulein), bee and snake venoms, poisonous mushrooms (amatoxin, phallotoxin) and from microorganisms (peptide antibiotics). There are P. with characteristic tastes, such as the sweet peptide aspartame, bitter P. from fermentation products and some with definite flavors. Certain cyclic P. are phytotoxic, such as tentoxin, cyclo-(L-MeAla-L-Leu-MePhe[(Z)Δ]-Gly; others, such as the cyclic tetrapeptide chlamydocin, have cytostatic effects. Muscle tissue contains very simple P., such as carnosine (β-Ala-His) or anserine (β-Ala-MeHis). The tripeptide glutathione is found in all cells of higher animals.

Biosynthesis. Most P. are synthesized in the same way as proteins, often in the form of precursor proteins which are subjected to specific proteolytic cleavage to release the bioactive P. However, there are many short-chain P. which differ structurally from the P. released from protein precursors. These often contain nonproteogenic amino acids (β-alanine, γ-aminobutyric acid, D-amino acids, N^{α}-alkylated amino acids, etc.), and non-peptide bonds, and have cyclic or branched-cyclic structures. Such P. are synthesized by *S*-aminoacyl activation on multienzyme systems. Their structural variations protect them from rapid proteolytic degradation.

General properties. P. are intermediate forms between amino acids and proteins. They have high melting or decomposition points, because they crystallize out of neutral solutions in the form of dipolar ions and are incorporated into an ionic lattice. The acid-base properties and solubilities of linear P. depend on their amino acid sequences, because their ampholyte nature depends on the number and distribution of the available basic and acidic groups, and the solubility is also affected by the hydrophobic side chains. The free *N*-terminal amino group and the *C*-terminal carboxyl group have the same chemical reactivities as the corresponding functional groups on free amino acids. They develop blue or blue-violet color with ninhydrin, and this is used for chromatographic or electrophoretic detection of free P. Unlike amino acids, however, the P. give a Biuret reaction (see). P. are hydrolysed to the amino acids by acids, alkalies or proteolytic enzymes. Amino acid analysis gives a quantitative determination of their components; the primary sequence can be determined by the standard methods of sequence analysis (see Proteins). The spatial arrangement of a P. molecule, its *conformation*, is determined by physical chemical methods. Spectroscopic methods are suitable for conformational analysis of P. in solution, while x-ray crystal structure analysis has been applied in several cases (gramicidin S, insulin). In general, conformation studies on cyclic P. are more successful than those on linear P., which can have extremely flexible structures. The study of the topochemical properties of active P. is important for an understanding of P.-receptor interactions.

Isolation of P. which normally occur in very low concentrations, and usually together with proteins, carbohydrates, lipids, nucleic acids, etc., is extremely difficult. After the cells are broken, a variety of enrichment and separation techniques can be applied, such as ultrafiltration, column chromatography, ion exchange chromatography, electrophoresis, preparative high-pressure liquid chromatography, and countercurrent distribution. Although the main tools for isolation, purification and structure elucidation of P. were developed between 1944 and 1954, the techniques for assaying them were lacking; P. often occur in nanogram quantities or less. The development of radioimmunoassays (RIA) in the 1970s made it possible, for example, to detect picogram quantities of a P. hormone in a milliliter of blood. Analogs of the native active P. were systematically constructed to determine the sequence segments responsible for the biological effect (active center), receptor binding, immunological behavior and transport of the P. Modification of native P. is also used to obtain longer-acting or more easily administered analogs. The large number of P. analogs used in such work can only be produced by Peptide synthesis (see).

Peptide hormones: a group of hormones which, chemically, are oligopeptides, polypeptides or proteins. Although a distinction is sometimes made between P. and protein hormones, it is not a chemically necessary one. The smallest P., thyroliberin, consists of only three amino acid residues, while others, such as thyrotropin, follitropin, lutropin and choriogonadotropin, are high-molecular-weight glycoproteins. P. are secreted by the hypothalamus, hypophysis, pancreas, thyroid and parathyroid glands and, during pregnancy, the placenta. In addition to the glandular P., there are numerous tissue hormones. The P. secreted by nerve cells, the neurohormones, are closely related to peptide neurotransmitters. Most P. are synthesized from precursor proteins or polypeptides by cleavage at specific sites. The primary structures of a number of ***preprohormones*** have been deduced from the nucleotide sequences of their genes. These contain an N-terminal sequence rich in hydrophobic amino acids, the signal sequence (see Proteins), which guides the nascent protein into and through the membrane of the endoplasmic reticulum. Once the preprotein is in the lumen of the endoplasmic reticulum, the signal sequence is removed by a specific enzyme, leaving the ***prohormone***. At this time the carbohydrate portion of proteoglycan hormones is added, and then the biologically inactive prohormone is stored until an external stimulus causes it to be released. It attains full biological activity only through further proteolytic modification. This pattern is followed, for example, by proinsulin, kininogens and angiotensinogens, which are considered first order prohormones. There is a second order of prohormones which includes bioactive P. containing within their sequences P. with other biological information. For example, corticotropin contains the sequence of α-melanotropin, β-lipotropin contains both β-melanotropin and β-endorphin, and oxytocin contains melanostatin, a tripeptide amide.

The action of P. on their target cells is mediated by specific receptors in the cell membranes, which often have adenylate cyclase activity. A second group of P. receptors activate the inositol phosphate system as "second messengers". After a certain biological effect has been initiated, the P. are quickly inactivated by proteolysis.

Some P. are commercially available through chemical synthesis: oxytocin, vasopressin, corticotropin, etc., but for others, it can be expected that genetically engineered microorganisms will become the major source.

Peptide synthesis: a several-step chemical process for controlled linking of amino acids to form peptides. The P. serve 1) to confirm primary structures of peptides and proteins determined by sequence analysis; this is often the most reliable method for definitive proof of structure; 2) for determination of structures responsible for the biological activity of a natural peptide by comparison of synthetic analogs with native peptides; 3) for chemical alteration of a natural peptide in order to change its pharmacological effect; 4) for industrial production of biologically active peptides and their analogs; and 5) for production of model peptides for physical-chemical studies, determination of antigenicity and as artificial substrates in enzymology.

P. can be achieved under mild reaction conditions only if the carboxyl function of one reactant is activated, and it yields a defined product only if all functional groups not involved in the desired peptide bond are temporarily blocked by suitable protective groups. P., that is the joining of a peptide bond, is a several-step process (Fig. 1).

selective blocking

carboxy components

activation

amino components

coupling

selective deblocking

1

The first step is selective blocking of the functional groups of the amino acids, which also releases them from the unreactive zwitterion structure. The amino acid which reacts with its carboxyl function, the *carboxy component*, is reversibly blocked at its amino function, while the second amino acid, the *amino component*, is protected at its carboxyl group. The second step is activation of the carboxy component, followed by formation of the peptide bond. These two reactions may be done in a single step or in succession. The next stage is selective removal of the protective groups, unless the dipeptide is the end-product and is completely deblocked at this point. The partially blocked dipeptide derivatives are then used as either carboxy or amino components in subsequent coupling reactions. P. is complicated by the fact that nine of the twenty proteogenic amino acids have, in addition to the carboxy and amino groups, reactive side chain functions which must also be selectively protected (Ser, Thr, Asp, Glu, Lys, Arg, His, Tyr and Cys). *Intermediate protective groups* are those groups which are used to block terminal amino and carboxy groups, and they must be removed under conditions which do not affect the *constant protective groups* on the side chains. Only at the end of the synthesis are the latter protective groups removed. Some protective groups are listed in the table.

Selected protective groups

	Abb.	Ways of removing
Amino-group protectors		
Benzyloxycarbonyl	Z-	HBr/AcOH; H_2/Pd; Na/liq. NH_3
tert.-Butyloxycarbonyl-	Boc-	HCl/AcOH; CF_3COOH
Fluorenyl-9-methoxy-carbonyl-	Fmoc-	Morpholine, 2-aminoethanol, liq. NH_3
2-[Biphenylyl-(4)]-propyl-2-oxycarbonyl-	Bpoc-	80% AcOH
Carboxyl-group protectors		
Methyl esters	-OMe	alk. hydrolysis
Ethyl esters	-OEt	alk. hydrolysis
Benzyl esters	-OBzl	NaOH; H_2/Pd
tert.-Butyl esters	-OBu^t-	CF_3COOH; HCl/AcOH
Side-group protectors		
S-Acetamidomethyl-	Acm-	Hg^{2+} (pH 4)
O-tert.-Butyl	Bu^t	CF_3COOH
N^{Im}-Dinitrophenyl-	Dnp-	2-Mercaptoethanol (pH 8)
N^G-Tosyl-	Tos-	HF
O-Benzyl-	Bzl-	H_2/Pd; HF

The carboxyl group is activated by introduction of electronegative substituents (X) which reduce the electron density on both the carbonyl O- and C-atoms, so that nucleophilic attack of the amino component is promoted. The *coupling reaction* should, under ideal conditions, occur very rapidly without racemization or side reactions, and in high yields, when equimolar amounts of the carboxy and amino components are added. The fact that more than 130 coupling methods have been reported shows that there is at present no ideal method. In spite of this, only a few methods are of practical significance, including the azide, mixed anhydride, active ester, dicyclohexylcarbodiimide (Fig. 2) and dicyclohexylcarbodiimide/additive methods.

Although there is no great difficulty in the choice of protective groups and coupling methods for di- and tripeptides, careful planning is essential for construction of long peptide chains with defined amino acid sequences. The *strategy of P.* is the series of couplings of the amino acids to form the peptide; the goal may be achieved either by step-wise chain lengthening or by segment condensation. Such P. can occur either in homogeneous solution (conventional P.) or in a second phase, as in Solid-phase peptide synthesis (see) (Merrifield synthesis), Liquid-phase peptide synthesis (see) and Alternating solid-liquid phase peptide synthesis (see). The *tactics* of P. is the optimum combination of protective groups and coupling methods for each peptide bond. In conventional P., the maximum and minimum protection tactics are the extremes of a range of possibilities which have both advantages and disadvantages. Maximum blocking of side chain functions guarantees a minimum of side reactions, and permits a greater variety of protective groups and coupling methods. However, the associated solubility problems limit this tactic. The syn-

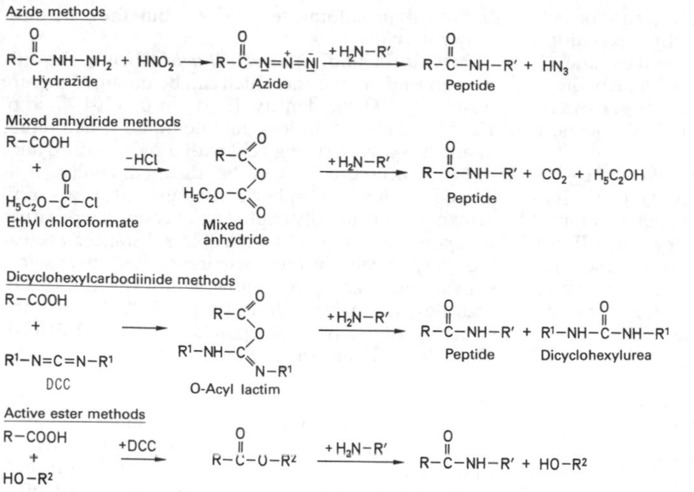

Azide methods

$$R-\overset{O}{\overset{\|}{C}}-NH-NH_2 + HNO_2 \longrightarrow R-\overset{O}{\overset{\|}{C}}-\underline{N}=\overset{+}{N}=\overset{-}{\underline{N}}| \xrightarrow{+H_2N-R'} R-\overset{O}{\overset{\|}{C}}-NH-R' + HN_3$$

Hydrazide — Azide — Peptide

Mixed anhydride methods

$$R-COOH + H_5C_2O-\overset{O}{\overset{\|}{C}}-Cl \xrightarrow{-HCl} R-C(=O)-O-C(=O)-OC_2H_5 \xrightarrow{+H_2N-R'} R-\overset{O}{\overset{\|}{C}}-NH-R' + CO_2 + H_5C_2OH$$

Ethyl chloroformate — Mixed anhydride — Peptide

Dicyclohexylcarbodiinide methods

$$R-COOH + R^1-N=C=N-R^1 \longrightarrow R-C(=O)-O-C(=N-R^1)-NH-R^1 \xrightarrow{+H_2N-R'} R-\overset{O}{\overset{\|}{C}}-NH-R' + R^1-NH-\overset{O}{\overset{\|}{C}}-NH-R^1$$

DCC — O-Acyl lactim — Peptide — Dicyclohexylurea

Active ester methods

$$R-COOH + HO-R^2 \xrightarrow{+DCC} R-\overset{O}{\overset{\|}{C}}-O-R^2 \xrightarrow{+H_2N-R'} R-\overset{O}{\overset{\|}{C}}-NH-R' + HO-R^2$$

Fig. 2. Selected methods of peptide synthesis. R = side chain of carboxy component, R' = side chain of amino component, R^1 = cyclohexyl, R^2 = side chain of the activated alkyl or aryl group.

theses of α-bungarotoxin, with 74 amino acids, and ribonuclease A, with 124, were achieved in this way, and demonstrate the advantages. The minimum protection tactic was used sucessfully in the synthesis of the S-protein of ribonuclease A (103 amino acids); only the ε-amino and thiol functions were masked, but only *N*-hydroxysuccinimide esters, the *N*-carboxyanhydride and the azide methods could be used for coupling. In general, neither extreme tactic is utilized, and the choice of degree of protection depends on the sequence being synthesized.

Although some small proteins have been constructed by P., the limit for economically useful syntheses is around 100 amino acids. For some purposes, ***semisynthesis*** is an alternative. Here fragments of native proteins are used as intermediates for construction of new proteins with modified sequences. For some coupling reactions, proteolytic enzymes are also used. ***Enzymatic P.*** makes use of the reversibility of protease-catalysed reactions; toward the end of the 1970's it became interesting for synthesis of biologically active peptides. Protease-catalysed coupling has several advantages over chemical methods, such as the lack of racemization, simple procedures at room temperature, and no need for protection of side chains. The disadvantages are the lack of universal applicability, due to the primary and secondary specificities of known proteases, and the difficulty of predicting the results. However, enzymes have also been used for selective cleavage of protective groups. P. is useful for modifying peptides for molecular biological studies and medicinal use. Combination of chemical and enzymatic methods is likely to be utilized in future, along with the genetic engineering of important biologically active peptides and proteins.

Historical. The first P. was carried out in 1881 by Theodor Curtius in Leipzig; he later introduced the azide method, which gave further impetus to the development of P. The pioneer work for the development of P. was done around the turn of the century by Emil Fischer and his school in Berlin. After these first studies, about 30 years passed before Max Bergmann and Leonidas Zervas in Dresden introduced the modern phase of P. by developing the first reversibly cleavable protective group, the benzyloxycarbonyl group. After the mixed anhydride method was introduced in 1955 by Theodor Wieland, the total synthesis of oxytocin was achieved by Vincent du Vigneaud; this was the first biologically active polypeptide to be synthesized. The introduction of new protective groups and methods of coupling, and new methodological principles, such as the Merrifield synthesis (1962) led the way to synthesis of small proteins (ribonuclease, proinsulin, etc.).

Peptidoglycans: see Polysaccharides.

Peptization: same as Redispersion (see).

Peptoids: see Peptides.

Peptolides: see Depsipeptides.

Peptone: see Pepsin.

Peracetic acid: CH_3-CO-OOH, a combustible, colorless liquid with a characteristic, pungent odor. In pure form, P. is explosive at higher temperatures and in the presence of organic compounds. It is made from acetic acid or anhydride and concentrated H_2O_2 solution. Commercial solutions, contain 40% P. For use as a disinfectant for skin and apparatus, a 0.5 to 1.0% solution is used. The effect is due to the release of atomic oxygen.

Per acids: 1) ***Inorganic P.*** are oxygen acids of elements in the +7 oxidation state, such as perchloric acid, $HClO_4$, periodic acid, H_5IO_6 or perrhenic acid, $HReO_4$. The term P. is also often applied to the ***peroxo acids***, in which an oxygen atom is replaced by the group -O-O-, e.g. peroxodisulfuric acid, HO_3S-O-O-SO_3H. 2) ***Organic P.*** are named as ***peroxy acids*** according to IUPAC nomenclature; they are carboxylic acid derivatives with the general formula R-CO-OOH. Their names are formed by adding the prefix per- or peroxy- to the name of the carboxylic acid, e.g. peroxyacetic or perbenzoic acid. P. are strong oxidizing agents and tend to explode. When heated in acidic solution, they form radicals, which decompose into alcohols with one C atom fewer. P. are synthesized from the corresponding carboxylic acids by reaction with H_2O_2 in strongly acidic solution. They are used as initiators for radical polymerizations, as oxidizing agents in epoxidation and hydroxylations, in the Baeyer-Villiger oxidation (see) and as catalysts of sulfoxidation. Esterification of P. with alcohols or of carboxylic acids with peroxides yields ***per esters***, which are also used as initiators for radical polymerizations.

Perbenzoic acid, ***peroxybenzoic acid***: C_6H_5-CO-OOH, a crystalline compound, m.p. 41-42 °C. It is slightly soluble in water, but dissolves readily in most organic solvents. Like all other peroxy acids, P. is a strong oxidizing agent. When heated, or upon impact, it can explode. P. can be synthesized from benzoic acid and hydrogen peroxide in methane sulfonic acid, or from dibenzoyl peroxide and sodium methanolate:

$$C_6H_5\text{–CO–O–O–CO–}C_6H_5 \xrightarrow[-\,C_6H_5\text{–COOR}]{+RO^-}$$

$$C_6H_5\text{–CO–O–O}^- \xrightarrow{+H^+} C_6H_5\text{–CO–O–OH}.$$

P. is used chiefly as an oxidizing agent for synthesizing epoxides, *N*- and *S*-oxides and as a disinfectant.

Perborates: a common name for borates in which the water of crystallization is partially replaced by hydrogen peroxide; examples are sodium metaborate-hydrogen peroxide trihydrate and sodium tetraborate-hydrogen peroxide nonahydrate.

Perborax: see Sodium perborate.

Perbromates, ***bromates(VII)***: the salts of perbromic acid, $HBrO_4$, with the general formula $MBrO_4$. The Br-O distance in the tetrahedral BrO_4^- ion is 161 pm. The P. were first described rather recently (1968). They can be obtained by oxidation of bromates, $MBrO_3$, with very strong oxidizing reagents, such as xenon difluoride, XeF_2, or fluorine, in 5 M sodium hydroxide solution: $BrO_3^- + F_2 + H_2O \rightarrow BrO_4^- + 2\,HF$. Dilute sulfuric acid releases ***perbromic acid*** from aqueous solutions of P.; a hydrate, $HBrO_4 \cdot 2H_2O$, of this acid can be crystallized. Perbromic acid is less volatile than $HClO_4$, but like perchloric acid, it is very strong (p$K \approx -10$). In dilute solution, P. are oxidizing agents, but act very slowly. Pure $KBrO_4$ is stable to 275 °C, NH_4BrO_4 to about 170 °C.

Perchlorates, ***chlorates(VII)***: the salts of perchloric acid, $HClO_4$, with the general formula $MClO_4$. The perchlorate ion, ClO_4^-, is isosteric with SO_4^{2-}, PO_4^{3-}, SiO_4^{4-} and AlF_4^-; it has a tetrahedral structure with Cl-O distances of 150 pm. As the most stable oxygen compounds of chlorine, most P. can be stored almost indefinitely in solid or dissolved form. When heated, they lose O_2 and are converted to chlorides: $MClO_4 \rightarrow MCl + 2\,O_2$. Most P. are water-soluble; potassium, rubidium and cesium perchlorate are only slightly soluble in cold water, however, but are more soluble in warm water. P. are synthesized from chlorates by cautious heating, according to the equation: $4\,MClO_3 \rightarrow 3\,MClO_4 + MCl$, or by anodic oxidation of alkali chlorates. P. are used to make perchlorate explosives.

Perchlorate explosives: explosives consisting of potassium perchlorate or ammonium perchlorate and combustible substances. They are less sensitive to friction than chlorate explosives, but they are now only rarely used.

Perchloric acid, ***chloric(VII) acid***: $HClO_4$, the only oxygen acid of chlorine which can be obtained in pure form: M_r 100.46, density 1.761, m.p. -101 °C, b.p. 120.5 °C. P. is a colorless, mobile liquid which fumes in air; it is a very strong acid with a p$K \approx -10$. Under reduced pressure, it can be distilled. When it is heated under atmospheric pressure, it turns red-brown and eventually explodes; at room temperature it decomposes slowly. Oxidizable substances are explosively burned in the presence of P. P. is miscible with water in any proportions, and forms a series of hydrates; the monohydrate (m.p. 49.905 °C) has the structure of an oxonium perchlorate, $[H_3O][ClO_4]$. The salts of P. are called Perchlorates (see).

> Anhydrous perchloric acid often explodes without any recognizable external impetus. It causes poorly healing sores on the skin, and therefore, any which reaches the skin must be rinsed off under a rapid flow of water. Afterwards, 1% hydrogen carbonate solution should be dropped onto the sore.

P. is synthesized by cautiously heating a mixture of potassium perchlorate and concentrated sulfuric acid to 160 °C: $KClO_4 + H_2SO_4 \rightarrow HClO_4 + KHSO_4$. The resulting aqueous solution is distilled in vacuum. P. is used as an oxidizing agent and as a reagent for detecting potassium. It is usually sold as a 72% solution, which represents an azeotropic mixture boiling at 203 °C.

Perchlorobenzene: same as Hexachlorobenzene (see).

Perchloroethane: same as Hexachloroethane (see).

Perchloroethene: same as Tetrachloroethene (see).

Perchloroformic acid methyl ester, ***chloroformic acid trichloromethyl ester, perchloromethyl formate, diphosgene***: Cl-CO-$OCCl_3$, a colorless, liquid with a suffocating odor; b.p. 127 °C. P. is soluble in most organic solvents. It slowly decomposes in water, forming carbon dioxide and hydrogen chloride. It reacts with ammonia to form ammonium chloride and urea. P. readily decomposes into two molecules of phosgene. It was therefore used in the First World War as a chemical weapon; the symptoms of poisoning are the same as those of phosgene poisoning. P. can be rapidly detoxified by hexamethylenetetraamine. It is synthesized by radical chlorination of chloroformic acid methyl ester:

$$Cl\text{-CO-}OCH_3 + 3\,Cl_2 \rightarrow Cl\text{-CO-}OCCl_3 + 3\,HCl.$$

Perchloromethylformate: same as Trichloromethyl chloroformate (see).

Perchloromethylmercaptan, ***trichloromethanesulfenyl chloride***: CCl_3SCl, a bright yellow, poisonous, oily liquid which irritates the eyes; b.p. 147-148 °C. P. is insoluble in water, but is soluble in chloroform and carbon tetrachloride. It is made industrially from carbon disulfide, sulfur dichloride and water. It is used as an intermediate in the synthesis of insecticides, dyes, thiophosgene and fuel additives. It is also used as a vulcanization accelerator. In World War I, P. was used for a short time as a weapon.

Perchloryl fluoride: see Halogen oxygen fluorides.

Per esters: see Per acids.

Perfluoroethene: same as Tetrafluoroethene (see).

Perforator: see Extraction.

Performic acid, ***peroxyformic acid***: H-CO-O-OH, the simplest peroxy acid. P. is a colorless liquid with a pungent odor, which is explosive, and is therefore usually synthesized and used in dilute solution; m.p. -18 °C, b.p. 50 °C at 13.3 kPa (m.p. for 90% P.). P. is soluble in water, alcohol and ether. In aqueous solution, it is very rapidly hydrolysed, unlike the other peroxy acids, forming formic acid and hydrogen peroxide. P. is a strong oxidizing agent. It is synthesized by reaction of formic acid with hydrogen peroxide in the presence of concentrated sulfuric acid.

Pericol: see Isocyanins.

Pericyclic reaction: see Woodward-Hoffmann rules.

Periodic acid: ***Orthoperiodic acid***, H_5IO_6, crystallizes in colorless prisms; K_m 227.96, m.p. 128.5 °C. P. is a relatively weak acid (pK_1 1.64, pK_2 8.36, pK_3 14.98) and a very strong oxidizing agent. When heated in vacuum to about 100 °C, P. is converted to ***metaperiodic acid***, HIO_4. P. is made by anodic oxidation of iodic(V) acid. The *salts* of P. are derived from either ortho- or metaperiodic acid; in the absence of water, the *orthoperiodates* M_5IO_6 are fairly stable, while *metaperiodates*, MIO_4, tend to explode when heated. Acidic salts, $M_{5-n}H_nIO_6$ (n = 2,3), are obtained from aqueous solutions by oxidation of iodate with chlorine or hypochlorite in alkaline solution at about 100 °C: $IO_3^- + ClO^- + 2OH^- \rightarrow H_2IO_6^{3-} + Cl^-$.

Periodic system, ***periodic table***: a table which shows the periodicity in the chemical properties of elements when they are arranged according to increasing nuclear charge number (atomic number). The P. is a natural classification system; the arrangement of elements into groups and periods on the basis of their chemical properties reveals underlying principles of atomic structure.

Early attempts at systematizing the elements were made by Döbereiner, Beguyer de Chancourtois and Newlands. The modern system was developed independently in 1869 by D.I. Mendeleyev and L. Meyer, who listed the elements in a table vertically in the order of increasing atomic mass. Elements with similar properties were written on the same horizontal line; this arrangement was rotated by 90° to reach the modern convention. Where there was a conflict between atomic mass and chemical properties, the element was assigned its position on the basis of chemical properties; thus tellurium (Te) and iodine (I) were placed correctly (according to modern insight), although their masses would have indicated a different placement. Similar irregularities were later discovered for nickel (Ni)/cobalt (Co), potassium (K)/argon (Ar) and thorium (Th)/protactinium (Pa). In addition, there were gaps in the series, for which Mendeleyev predicted the existence of yet undiscovered elements. He assigned these the names Eka-boron, Eka-aluminum and Eka-silicon, and made detailed predictions of their properties. The astonishing accuracy of these predictions, which were confirmed a few years later by the discoveries of scandium, gallium and germanium, was taken as convincing evidence that the table was based on fundamental properties of nature.

According to modern atomic theory, the number of electrons in a neutral atom is equal to its nuclear charge, and it is the arrangement of the electrons in a definite, periodic pattern (see Electron configuration, table) which is responsible for the periodicity in chemical properties of elements.

Hydrogen and helium constitute the 1st period, in which the 1s electron shell is filled. The second and third horizontal rows (the 2nd and 3rd periods) represent the filling of the 2s and 2p states (lithium to neon) and the 3s and 3p states (sodium to argon). The spaces left between beryllium and boron, magnesium and aluminum have no physical significance; they merely allow the columns in the table to be properly aligned. In the 4th period, potassium ($4s^1$) and calcium ($4s^2$) are followed by the 3d transition elements scandium to zinc. From gallium to krypton, the 4p state is filled. This arrangement is repeated in the 5th period with the 5s, 4d and 5p elements. In the 6th period, lanthanum ($6s^25d^1$) is followed by the 4f elements cerium to lutetium; these are the lanthanoids, which, however, are listed separately for the sake of visual clarity. The 5f elements thorium to lawrencium, the actinoids, are treated similarly.

The vertical columns correspond to representative groups Ia to VIIIa and transition groups Ib to VIIIb. The representative groups include those elements in which last electrons are in an s or p state (s- and p-elements.). Transition elements are those in which the atoms are constructed, formally, by incorporation of the last electron into a d state. According to this definition, zinc, cadmium, mercury, manganese and technetium would actually be s-elements, but they are included with the transition elements on the basis of their other properties.

The groups have historical names: the 1st representative group is Alkali metals (see), the 2nd is Alkaline earth metals (see), the 3rd is the Boron group (see), the 4th is the Carbon group (see), the 5th is the Nitrogen group (see), the 6th is the chalcogens (see Oxygen-sulfur group), the 7th is the Halogens (see) and the 8th is the Noble gases (see). The 1st transition group (coinage metals) is the Copper group (see), the 2nd is the Zinc group (see), the 3rd includes the Rare earth metals (see), the Lanthanoides (see), actinium and the Actinoides (see); the rare-earth metals scandium, yttrium, lanthanum and actinium are also grouped as the Scandium group (see). The 4th transition group is the Titanium group (see), the 5th is the Vanadium group (see), the 6th is the Chromium group (see), the 7th is the Manganese group (see), and the 8th includes the metals of the Iron group (see) and the Platinum metals (see).

Many of the chemical and physical properties of the elements change in periodic fashion with increasing atomic number. In the P., this can be seen by comparison of the elements in a period or in a group; as the atomic number increases, there is a gradient of properties.

In the representative elements, there is a steady decrease in atomic radius from left to right along a period. This is due to the increasing nuclear charge and the fact that the additional electrons are introduced into the same shell. Within a group, the atomic radius increases with increasing atomic number. Even though the nuclear charge is increasing, in this case the added electrons go into higher shells with larger radii. The *ionic radii* follow the same pattern. As would be expected, the ionic radius decreases in the series Na^+, Mg^{2+}, Al^{3+} or Si^{4-}, P^{3-}, S^{2-}, Cl^-. The *ionization energy* is also a periodic function of the atomic number. The energy needed to release an electron from the neutral atom increases from left to right along a period, due to the increasing nuclear charge and decreasing radius, which together increase the coulomb attraction of the nucleus. However, there are minima in the trends for p^1 and p^4 elements, which indicate the unusual stability of the s^2 and p^3 states. Within the groups, the ionization energies decrease from top to bottom as the ionic radii increase. The *electron affinities* are controlled by the same effects; changes in these affinities going through the P. are the results of coulomb interactions and the stability of certain electron configurations. As would be expected, the electron affinities of elements on the right side of the table, the halogens and chalcogens, are high.

The atomic or ionic radii affect the interaction between nucleus and electron shells, which in turn affect a number of other properties. For example, the *electronegativity* is highest in elements with the smallest atomic radii and highest atomic numbers; electronegativity increases from left to right in the periods and from bottom to top in the groups. Thus the elements with the highest electronegativity are at the upper right of the P.: fluorine, oxygen and chlorine. The lowest electronegativities are observed in the heavy alkali and alkaline earth metals.

The chemical properties also show clear periodicities. Stable *oxidation states* are formed when the number of electrons accepted or donated produces the same configuration as that of the nearest noble gas. In view of the ionization energies and electron affinities discussed above, it can be seen that the elements on the left tend to donate electrons and form cations (*electropositive elements*), while the elements on the right form anions by accepting electrons (*electronegative elements*). This is expressed by the sign of the standard electron potentials of the elements. Thus the representative elements on the left side of the P. are reducing agents, while those on the right are oxidizing agents. The elements from the middle of the P. (3rd to 6th main groups) tend to form covalent rather than ionic bonds.

The number of electrons which can be donated is indicated by the group number, which represents the maximum positive oxidation state of the element. This was formerly expressed as the valence with respect to oxygen. The maximum negative oxidation number is the difference between 8 and the group number; it corresponds to the valence of the electronegative elements with respect to hydrogen.

The elements on the left side of the P. are metals. Moving toward the right through the main groups, the *non-metal character* increases; at the same time, the element oxides change from strong base-formers (Na_2O, CaO) or amphoteric compounds (Al_2O_3) to acid anhydrides (CO_2, NO_2, SO_3). The acidity of binary hydrides increases from left to right in the periods (NH_3, H_2O, HF) and from top to bottom in the groups (H_2O, H_2S, H_2Se; or HF, HCl, HBr, HI). Within a group, the acidity of homologous oxygen-containing acids generally decreases with increasing atomic number of the central element, as in H_2SO_3, H_2SeO_3 and H_2TeO_3.

The differences between first and second elements in a representative are greater than those between second and third, third and fourth, etc. elements. The properties of first elements are often more similar to those of the second element of the succeeding group than to those of the second element in their own group, that is, lithium is very similar to magnesium, beryllium to aluminum, boron to silicon, and so on. These are called *diagonal relationships* in the P.; they are due to comparable charge-to-radius ratios in the two members of each pair of elements or their ions.

The variation in properties of transition group elements is not so pronounced, due to the fact that they all have a d^ns^2 electron configuration. Predictable deviations occur only in chromium, molybdenum, tungsten, copper, silver and gold. Within the individual transition series, the electron occupation changes only in the second shell from the outside.

The atomic and ionic radii of transition group elements follow the same trends as those of the representative elements. The decrease in ionic radii observed within a transition series continues in the immediately following p-elements, and causes an unexpected change in the behavior of these elements. For example, the irregularities in the ionization potentials or electronegativities of the elements of the 3rd and 4th main groups are explained by this effect. The tendency of the heavier representative elements to form oxidation states with two units less than the maximum oxidation state is at least partly due to this same effect.

All transition group elements are metals; most are dense and have good electrical and thermal conductivity. Most are electropositive, dissolving in non-oxidizing acids with evolution of hydrogen. Only a few are "noble", that is, they have positive standard electrode potentials and are attacked only by strong oxidizing acids: copper, silver, gold, mercury and the platinum metals ruthenium, osmium, rhodium, iridium, palladium and platinum.

In agreement with their electron configuration, all transition group elements occur in the +2 oxidation state. Because they are able to activate d-electrons to form bonds, they are extremely flexible with respect to their oxidation states. The maximum oxidation number is given by the group number, and corresponds to the sum of the s- and d-electrons, except for the d^{10} (groups Ib and IIb) and d^8 (group VIIIb). The group Ib and IIb elements tend to use only s-electrons to form bonds, and the only group VIIIb elements which display a valence of 8 are ruthenium and osmium. Most transition metal ions are colored, and their compounds are often paramagnetic.

The f-elements differ from one another only in the

occupation of the third electron shell from the outside, and thus their physical and chemical properties are very similar. However, periodic and aperiodic properties can be recognized among the Lanthanoids (see) (see Rare earth metals). The Actinoids (see) also display regular changes in properties which are due to shell occupation.

Peric acid: an Alphabet acid (see).

Peritecticum: see Melting diagram.

Perkin reaction: synthesis of α,β-unsaturated carboxylic acids from aromatic aldehydes and anhydrides of aliphatic monocarboxylic acids in the presence of the sodium or potassium salts of the corresponding carboxylic acid (usually sodium acetate). In the simplest case, heating of benzaldehyde with acetic anhydride and sodium acetate to 170-180 °C produces cinnamic acid:

$$C_6H_5-CHO+(CH_3CO)_2O \xrightarrow{CH_3COONa} C_6H_5CH=CH-COOH+CH_3COOH$$

In addition to aromatic aldehydes, aromatic or aliphatic-aromatic ketones and heterocyclic aldehydes, e.g. furfural, can serve as carbonyl components; the CH-acidic carboxylic acid can be, e.g. malonic, cyanoacetic or phenylacetic acid. The reaction mechanism is the same as an aldol reaction, in which the acid anhydride or the CH-acidic acid is the methylene-active component:

$$(CH_3CO)_2O \xrightarrow[-CH_3COOH]{+CH_3COO^-} {}^{\ominus}CH_2-CO-O-CO-CH_3 \xrightarrow{C_6H_5CHO}$$

$$C_6H_5-CH(O^{\ominus})-CH_2-CO-O-CO-CH_3 \xrightarrow[-CH_3COO^-]{+CH_3COOH} C_6H_5-CH(OH)-CH_2-CO-O-CO-CH_3$$

$$\xrightarrow{-H_2O} C_6H_5-CH=CH-CO-O-CO-CH_3 \xrightarrow[-CH_3COOH]{+H_2O} C_6H_5-CH=CH-COOH$$

Perkin violet: same as Mauvein (see).

Perlon®: see Synthetic fibers.

Permalloy®, ***Supermalloy®***, ***Hyperm®***, ***Ultraperm®***: a nickel-iron alloy (see Nickel alloys) consisting of 75 to 89% nickel and the rest iron. P. is a soft magnetic material, and has a high initial permeability; it is magnetically saturated even in the earth's magnetic field. The magnetization curve climbs sharply with the field. P. is used in microphone transmitters in radio and in broad-band telephone cables.

Permanent blue: same as Bluing (see).

Permanent green: see Chromium(III) oxide hydrate.

Permanent white: see Barium sulfate.

Permanent polishes: see Self-polishing emulsions.

Permanganate: see Manganate.

Permanganic acid: same as Manganic(VII) acid (see).

Permanganometry: method of redox analysis in which permanganate ion is the titrator. This is a strong oxidizing agent in acidic solution, with a standard potential E_0 = +1.51 V; it is reduced to manganese(II). In neutral or weakly alkaline solution, the reduction goes only to the +4 state of manganese, forming manganese(IV) oxide. Under special conditions, the reduction of the permanganate ion can also lead to the +3 and +6 oxidation states of the manganese.

Potassium permanganate is usually used to prepare the standard solution. Because the solution is not stable, it must be calibrated and frequently checked against the titration standard sodium oxalate.

No indicator is needed to recognize the endpoint in P., because a slight excess of permanganate ion over the equivalence point gives the solution a pale violet color. P. is used mainly to determine inorganic reducing agents, because it can enter uncontrollable side reactions with organic substances. Although P. was formerly very popular, because it requires no indicator, it has become less significant compared to other redox analyses. This is also due to the instability of the standard solutions, the fact that the reaction is not unequivocal, and the interference by chloride ions.

Permeation: a membrane process in which the components of a mixture are separated by means of the differences in their rates of migration through a membrane (see Semipermeable membrane). ***Permeation methods*** include Dialysis (see), Electrodialysis (see), Osmosis (see), Electroosmosis (see), Ultrafiltration (see) and P. in the gas or liquid phase.

In ***gas-phase permeation***, the gas molecules are transported selectively through a membrane by solution diffusion driven by a pressure difference. It is used on an industrial scale for separation of helium from natural gas and for separating hydrogen from the product gas streams from various syntheses. Hollow polyester fibers can be used to separate hydrogen from methane by P.

Liquid-phase permeation is a separation of homogeneous liquid mixtures on the basis of the different solubilities of the components in the membrane, and on the basis of their different diffusion rates through the membrane. The permeation process can be subdivided into absorption, diffusion through the membrane, exsorption and evaporation of the permeant from the limiting surface on the low-pressure side.

Liquid-phase permeation is used, for example, to separate mixtures of liquids which boil at nearly the same temperature, or to remove water from organic liquids.

Permeation chromatography, ***gel permeation chromatography***, incorrectly called ***gel filtration***: method of separating molecules on the basis of their sizes and shapes. The method depends on the differential permeability of gels to molecules of different size; molecules which are larger than the pore size of the gel are not able to penetrate it. These are eluted first from a column, while smaller molecules permeate the stationary phase to a greater or lesser extent, depending on their size and shape. The elution of the separated substances occurs in the order of decreasing molecular size (Fig. 1).

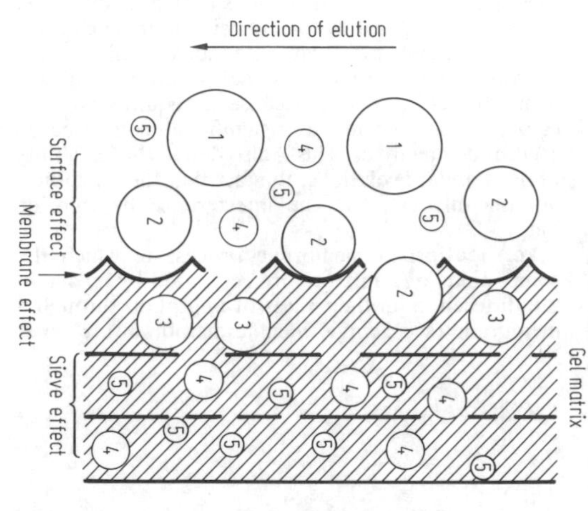

Fig. 1. Permeation chromatography.

P. can be carried out in a thin layer or a column. The choice of the mobile phase depends on the nature of the substances to be separated and the character of the gel matrix (hydrophilic or hydrophobic). *Gels* consist of dispersed substances and a dispersion medium, which in P. is also part of the mobile phase. The gel structure of such *xerogels* is a three-dimensional network of solvated macromolecules, which are linked by covalent and non-covalent bonds (H-bonds, dipole-dipole interactions, colloidal forces). The solid products formed by drying such gels can be resolvated and swell to reestablish their characteristic network. Xerogels include *polydextran gels*, which are made by cross-linking the natural polysaccharide dextran (Fig. 2), and cross-linked polyacrylamide or polystyrol gels.

Aerogels are rigid, preformed and non-swelling matrices which retain their porosity even when dried. Some examples are porous silica gel, porous glass, xerogel-aerogel hybrids and *agarose gel*. This last consists of agarose molecules from seaweed which are cross-linked by H-bonds.

Gels are characterized by their grain size, degree of cross-linking, and the nature of the matrix. An important parameter for elution behavior in P. is the *distribution coefficient* K_d, which is independent of the geometry and packing density of the bed. As a rule, the useful range for K_d is between 0 and 1; the larger K_d values are equivalent to a higher penetration of the gel matrix.

P. is used in biochemistry, clinical chemistry and industrial laboratories to remove low-molecular-weight substances from solutions of polymers such as proteins, polypeptides or carbohydrates, to determine molecular masses of high-molecular-weight compounds, to desalinate solutions and to isolate proteins on a preparative scale.

Permeation effect: see Electrophoresis.

Permethrin: see Pyrethroids.

Permutites: synthetic, inorganic cation exchangers (see Ion exchangers) based on silicates; they are not stable to acids or bases. Since only certain cations can exchange with P. in a neutral medium, such as Na^+, Ca^{2+}, Al^{3+} and Fe^{3+}, P. are no longer of great interest. They have been almost completely replaced by ion exchangers based on zeolites and synthetic resins, which are capable of completely deionizing water.

Perovskite type, ***Perovskite structure***: a common Structure type (see) for ternary compounds with the general composition ABX_3. In the crystal lattice of perovskite, calcium titanate ($CaTiO_3$), the Ca^{2+} ions occupy the vertices of cubic unit cells. The O^{2-} ions are in the centers of the faces, and the Ti^{4+} ion is in the center of the cube (Fig.). The Ti^{4+} ion is surrounded octahedrally by 6 O^{2-} ions, while the Ca^{2+} ion is surrounded by 12 O^{2-} ions. Thus the Ca^{2+} and O^{2-} ions together form a cubic close-packed structure (see Packing of spheres).

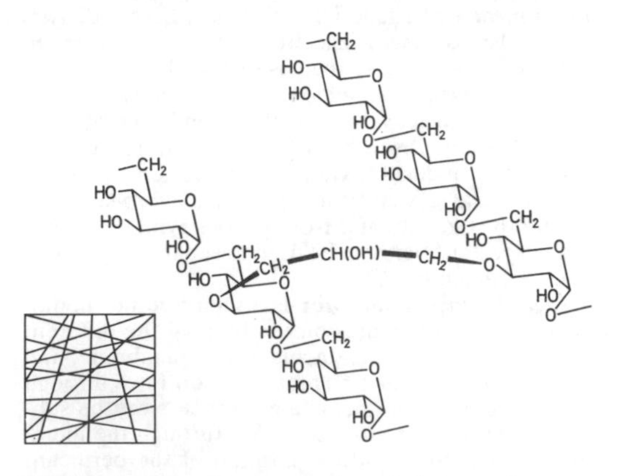

Fig. 2. Permeation chromatography: polydextran gel.

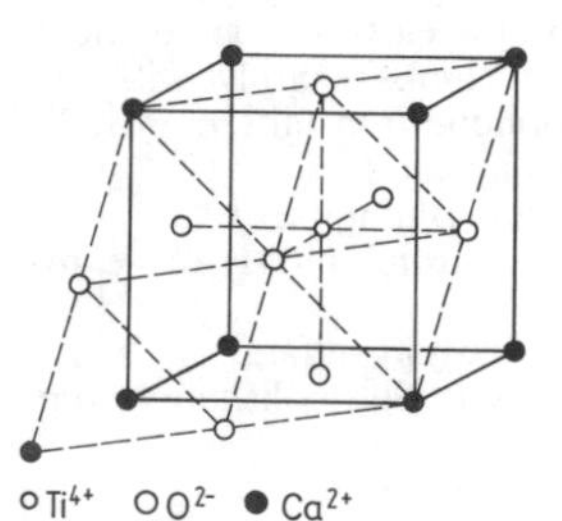

Unit cell of the perovskite lattice

For an ABX_3 compound to crystallize in the P., the A and X ions must be of similar size, and the B ion must be considerably smaller. In addition to $CaTiO_3$, this structure is found, e.g. in KIO_3, $SrZrO_3$, $BaSnO_3$, $KMnF_3$ and many other compounds of suitable composition.

Peroxidase: an enzyme found in higher plants and animal tissues; it catalyses the dehydrogenation of various substrates using hydrogen peroxide as oxidizing agent, and reducing it to water. P. is a glycoprotein M_r 40,000 and contains one mole protoheme as the prosthetic group. The best known P. is the easily crystallized horseradish P., which can be separated from its heme component by acetone-hydrochloric acid treatment.

Peroxide: a compound containing the O_2^{2-} ion or the covalent -O-O- group (see Peroxo compounds), in other words, derivatives of oxygen in the -1 oxidation state. The difference in bond type is also the basis for classification as covalent and ionic P. ***Ionic P.*** form an ionic lattice. They are formed mainly by the alkali and alkaline earth elements. The peroxide ion, O_2^{2-} is formed by occupation of the two singly occupied π^* orbitals of the O_2 molecule by two more electrons (see Molecular orbital theory). Ionic P. react with water, due to the very basic nature of the O_2^{2-} ion, by protolysis to hydrogen peroxide and the corresponding metal hydroxide, e.g. $Na_2O_2 + 2\,H_2O \rightarrow 2\,Na^+ + 2\,OH^- + 2\,H_2O_2$. The reaction of barium peroxide with sulfuric acid was formerly used to make hydrogen peroxide: $BaO_2 + H_2SO_4 \rightarrow BaSO_4 + H_2O_2$. When heated, most P. split off oxygen to form the oxides, e.g.

$$BaO_2 \xrightarrow{(700°C)} BaO + \frac{1}{2}\,O_2.$$

Only the alkali metal peroxides are relatively stable to heating, so they can be melted almost without decomposition. With easily oxidized substances (carbon, sulfur, aluminum powder, organic compounds, etc.), they react explosively, often even at room temperature. The oxidizing effect is the basis for use of ionic P., e.g. sodium or barium peroxide, to make bleaches or ignition caps, to regenerate air in respiration devices, to solubilize refractory materials in the laboratory, etc.

The most important ***covalent P.*** is hydrogen peroxide, H_2O_2. This group of P. also includes the *organic P.*, R-O-O-R, which are often formed by autooxidation of organic compounds, e.g. when air and light react with ether, or in the reaction of hydrocarbons, carbonyl compounds and fats with atmospheric oxygen or pure O_2. Certain organic O. can also be synthesized from haloalkanes and sodium peroxide. Organic P. are usually explosive. They are used as initiators for radical substitution reactions and polymerizations. The undesired formation of P. in solvents, such as diethyl ether, tetrahydrofuran, dioxane, tetralene, etc., often causes serious accidents.

Peroxide effect: see Markovnikov rule.

Peroxide hydrate: see Hydrogen peroxide.

Peroxo acids: see Per acids.

Peroxo compounds, formerly ***peroxy compounds***: compounds in which the O_2^{2-} ion is present as a ligand, or compounds in which an -O group is replaced by an -O-O group. P. are usually obtained from the corresponding oxo compounds by partial or complete exchange of peroxo groups for oxygen atoms. Some well known examples are Peroxomonosulfuric acid (see), $HO_3S\text{-}OOH$, and Peroxodisulfuric acid (see), $HO_3S\text{-}O\text{-}O\text{-}SO_3H$, their salts, and the Peroxochromates (see).

Peroxochromates: there are two main types, the red, paramagnetic ***peroxochromates(V)***, $M_3[CrO_8]$, and the blue, diamagnetic ***peroxochromates(VI)***, $MH[CrO_6]$. In P.(V), the central atom is surrounded dodecahedrally by 8 oxygen atoms; these compounds are derived from chromates(V) by replacement of the tetrahedrally arranged oxide ligands by peroxo groups. In P.(VI), two oxygen atoms of the chromate(VI) are replaced by O_2 groups (Fig.). Potassium peroxochromate(V) forms dark red-brown prisms which are relatively stable at room temperature. However, at 170°C, the compound decomposes, and at higher temperatures, the reaction is explosive. The blue-violet prisms of potassium peroxochromate(VI) explode when heated, upon impact or contact with conc. sulfuric acid. A slow decomposition is observed even at room temperature. P.(V) are synthesized by the reaction of 30% hydrogen peroxide with alkaline chromate solutions under ice chilling; P.(VI) are obtained by cautious addition of 30% hydrogen peroxide to acidic chromate solutions, also with ice chilling.

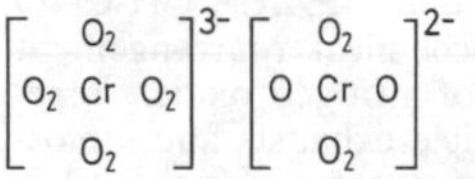

$$\left[\begin{matrix} & O_2 & \\ O_2 & Cr & O_2 \\ & O_2 & \end{matrix}\right]^{3-} \qquad \left[\begin{matrix} & O_2 & \\ O & Cr & O \\ & O_2 & \end{matrix}\right]^{2-}$$

Structures of the peroxochromates $[CrO_8]^{3-}$ and $[CrO_6]^{2-}$.

Peroxodisulfates, ***peroxosulfates***, formerly ***persulfates***: the salts of peroxodisulfuric acid, $H_2S_2O_8$, with the general formula $M^I_2S_2O_8$. Most P. are water-soluble. They are based on the peroxodisulfate ion $S_2O_8^{2-}$, in which the two tetrahedral sulfur atoms are linked via a peroxo group. P. are strong oxidizing agents. They are able to oxidize Mn^{2+} ions, in the presence of catalytic Ag^+ ions, to permanganate: $2\,Mn^{2+} + 5\,S_2O_8^{2-} + 8\,H_2O \rightarrow 2\,MnO_4^- + 10\,SO_4^{2-} + 16\,H^+$. When heated dry, the P. are converted to disulfates with loss of oxygen. They are obtained by electrolysis of concentrated aqueous solutions of the corresponding sulfates at high current densities. Ammonium peroxodisulfate (see) and Potassium peroxodisulfate (see) are industrially important.

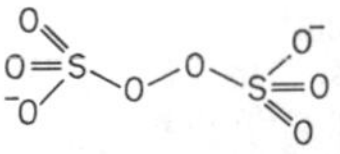

$$^-O\text{-}S(=O)_2\text{-}O\text{-}O\text{-}S(=O)_2\text{-}O^-$$

Peroxodisulfuric acid, ***peroxosulfuric acid***: $H_2S_2O_8$, colorless, hygroscopic crystals; K_m 194.14, m.p. 65 °C (dec.). P. is a strong oxidizing agent which reacts explosively with organic substances. It dissolves exothermally in water with rapid hydrolysis to sulfuric acid and peroxomonosulfuric acid, which in turn is decomposed to sulfuric acid and hydrogen peroxide: $H_2S_2O_8 + H_2O \rightarrow H_2SO_4 + H_2SO_5$; $H_2SO_5 + H_2O \rightarrow H_2SO_4 + H_2O_2$. This reaction was formerly important for production of hydrogen peroxide. P. is a strong, diprotic acid; its salts are the Peroxodisulfates (see). P. is made by anodic oxidation of sulfuric acid, or by reaction of Caro's acid with chlorosulfuric acid: $H_2SO_5 + ClSO_3H \rightarrow H_2S_2O_8 + HCl$.

Peroxomonosulfuric acid, ***Caro's acid***: H_2SO_5, colorless, hygroscopic crystals; M_r 114.08, m.p. 45 °C (dec.). P. is derived structurally from sulfuric acid by substitution of an OOH group for an OH group. It decomposes to sulfuric acid and oxygen. It is hydrolysed in water to sulfuric acid and hydrogen peroxide: $H_2SO_5 + H_2O \rightleftharpoons H_2SO_4 + H_2O_2$. P. is a strong oxidizing agent,

O=S(=O)(OH)OOH

which is able, for example, to oxidize aniline to nitrobenzene. It is a strong monoprotic acid. Its salts, the ***peroxomonosulfates***, M^IHSO_5, are not stable. P. is made by reaction of chlorosulfuric acid with hydrogen peroxide according to the equation $H_2O_2 + ClSO_3H \rightarrow H_2SO_5 + HCl$, or by hydrolysis of peroxodisulfuric acid: $H_2S_2O_8 + H_2O \rightarrow H_2SO_5 + H_2SO_4$.

Peroxonitrous acid: see Nitric acid.

Peroxosulfates: same as Peroxodisulfates (see).

Peroxosulfuric acid: same as Peroxodisulfuric acid (see).

Peroxyacetyl nitrate, abb. ***PAN***: $CH_3COO \cdot ONO_2$, one of the photochemical rearrangement products from the reaction of nitrogen oxides, reactive hydrocarbons from vehicle exhausts and atmospheric oxygen in the presence of solar radiation. P. is an aggressive substance, and is a component of Los Angeles type smog which causes a burning sensation in the eyes (see Smog).

Peroxy acids: see Per acids.

Peroxybenzoic acid: same as Perbenzoic acid (see).

Peroxy compounds: see Peroxo compounds.

Peroxyformic acid: same as Performic acid (see).

Perrhenates: see Rhenium compounds.

Persistance: the stability of organic chemicals in the environment. The term P. was first used for the chlorinated hydrocarbon insecticides, and is now used by the FAO/WHO to characterize the speed with which agrochemicals, especially pesticides, are completely degraded to inorganic compounds. The P. of an organic chemical depends on its ability to be degraded by biological and abiotic agents in the environment. It is an important criterion for the ecological-chemical behavior of the chemical.

Persulfates: see Peroxodisulfates.

Pertechnetiates: see Technetium compounds.

Peru balsam: a brownish yellow, viscous mass with a vanilla-like odor. It is soluble in alcohol, but only slightly soluble in ether or petroleum ether. P. contains up to 70% benzyl esters of cinnamic and benzoic acid; it also contains resins, free cinnamic acid, vanillin and coumarin. It is obtained from the trunk of the Peru balsam tree *Myroxylon balsamum*, var. *pereira*, which grows in Central America. P. is used as a component of hair tonics and sun oils, and as a fixative in perfumery and soaps. In the form of a salve, P. is used as an antiseptic and as a granulation promoter in medicine.

Perylene: a *peri*-condensed aromatic hydrocarbon. P. crystallizes out of toluene in the form of yellow crystals, and out of benzene as bronze-colored leaflets; m.p. 278 °C, b.p. 497 °C. P. sublimes at 350-400 °C.

It is insoluble in water, slightly soluble in ether and ethanol, readily soluble in chloroform and carbon disulfide. P. is synthesized by heating naphthalene with aluminum chloride. It is used to synthesize pigments (***perylene pigments***).

PES: abb. for Photoelectron spectroscopy.

Pesticides: chemicals used to protect crop and ornamental plants from animal pests, microbial diseases and weeds, and to protect human beings and animals from insects and other animal pests which may carry diseases, consume stored foods or other goods, or simply irritate their hosts. The importance of P. can be appreciated from the estimated loss, world-wide, of about 35% of the possible harvest to plant disease and animal pests. About 12% of these losses are due to microbial plant diseases, about 14% to insects, and about 9% to weeds. Thus measures to protect the harvest are of immense importance, in light of the rapidly growing world population. In countries with highly developed agrarian production, the yields per hectar have increased greatly in the past few decades, both through use of fertilizer and by reduction of losses due to various pests. Chemical P. have been and are supported by crop rotation, the development of varieties resistant to disease and sound cultvation practices.

In modern, intensive agriculture, ***chemical P.*** are of great economic importance. The cost-benefit ratio is generally very favorable; the cost of the chemicals is returned at least two-fold, and often more than tenfold. The new, energy-conserving cultivation technologies, such as low-tilling or no-tilling practices, depend on P. The world production of P. has at least doubled in the time between 1970 and 1980. At first, Insecticides (see) were produced in the largest amounts, but several years ago Herbicide (see) production surpassed insecticide production, and now accounts for about two fifths of the world P. production. Fungicides (see) account for about one fifth, insecticides for about one third, and the remainder is distributed among the other types (Acaricides, Nematicides, Molluscicides, Piscicides, Avicides, Rodenticides, Fungicides, Bacteriocides, Viricides; see separate entry for each.) The crop requiring the largest amounts of P. is cotton, followed closely by maize (together they account for about 40%); rice, wheat and potatoes use about 7 to 8% of the world production each, and more than 16% of the P. are used on fruits, including citrus fruits.

In addition to chemical P., there are physical practices (such as burning fields to destroy weed seeds), which, however, are now only of local importance. Biological methods have been strongly promoted in many countries in the past few years; these include such measures as massive release of sterilized males (see Autocide process, see Chemosterilants) and pheromone traps. The combination of biological and chemical methods for optimum results is the goal of Integrated crop management (see).

P. are by nature poisons. However, the goal is to minimize their toxicity to human beings and useful flora and fauna. To reduce the risk of acute toxicity or eventual chronic effects, most countries have established extensive regulations concerning their application, storage, disposal, etc. As a result, P. are now among the best studied of the xenobiotics (see Ecological chemistry).

Before a P. is certified for use, it must be subjected to an extensive *toxicological characterization*. In addition to acute and chronic toxicity, its accumulation in organisms and in food chains must be studied. The maximum acceptable residue concentration is established in each country; these are based on the acceptable daily intake values determined by a group of experts working for the WHO/FAO. This amount is the daily dose of a substance (in mg/kg body mass) which has no deleterious effects if consumed daily for a lifetime. The *ecological chemical characterization* of a substance is required to determine the dynamics of its residues: metabolism in plants, animals and microbes, persistance in and penetration of the soil, and hydrolysis, photolysis and microbial degradation in water. The *persistance* is one factor determining the time which must pass between the last application to a crop and its harvest. Regulations have also been developed concerning exposure of farm workers to various material. The toxicity of a new material to bees, fish, the prey organisms of fish and water animals, and to birds must also be determined. The officially approved P. are published in national registries, together with regulations concerning their application; the active material is usually listed along with the common name recommended by the International Standards Organization (ISO).

The production of highly effective, safe P. has led to a sharp rise in the costs for research and development. At the beginning of the 1980's, it was estimated that the average cost (world-wide) was $20 million for each product, and development times of 8 to 10 years are normal. The success rate of research to develop new P. has steadily declined as the requirements have become more stringent. New methods and strategies have therefore been developed to rationalize the search for new P., including the use of quantitative analyses of the relationship between structure and activity, and synthesis of biologically active substances suggested by knowledge of metabolic regulatory processes in the target organisms.

The search is not over when an effective substance has been discovered; in most cases, it must be mixed with extenders, surfactants and possibly other active materials. It must be formulated in such a way as to be applicable with existing equipment (sprayers, etc.) and the toxicology and environmental impact of the complete formula must then be determined.

P. are applied in a number of different forms: *dusts* with 0.1 to 10% active material and the remainder finely ground filler (talcum, kaolin, etc.); *granulates*, *wettable powders* (these contain 250 to 800 g active material per kg, along with surfactants to allow preparation of an aqueous suspension), *pastes* and *flowables*. In liquid form, there are *emulsifiable concentrates*, which contain 10 to 80% active material in a solvent which is generally not miscible with water; they also contain emulsifying agents to permit the preparation of an aqueous emulsion. A variety of other emulsions and solutions in organic solvents are also used.

Historical. Even in antiquity, readily obtained chemicals were used as P., e.g. sulfur and copper against fungal diseases, vinegar to treat grain, arsenic, sulfur and "pyrethrum powder" against insects, lime, wood ashes and chalk against storage pests. In the 18th century, tanbark, sulfur, tobacco extract, ashes and chimney soot were used to combat insects; later nicotine, pyrethrum extracts, extracts of the legume *Derris elliptica* (e.g. rotenone), cryolite, and compounds of mercury, thallium, antimony, selenium and boron were used.

The era of scientific research in plant diseases began in the 19th century with the studies of de Bary on the biology and physiology of fungi (1853), and by Kühn on the role of fungi in infection and disease (1858).

The fungicidal effects of lime-sulfur and Bordeaux solution (from copper sulfate and lime milk) were discovered. Wood poles for electrical and telephone lines and railroad ties were protected by creosote, tar oils and salts of mercury, copper, cadmium, arsenic and zinc. In 1892, the contact effect of dinitroorthocresol (DNOC) was recognized, and this "yellow spray" was the first synthetic organic compound used as a P.

The modern era of chemical P. began in the 1930s to 1950s. In 1939, Paul Müller discovered the insecticidal effects of DDT, which was used during and after the end of World War II to combat disease vectors, especially the *Anopheles* mosquitoes which carry malaria. Organophosphate compounds were recognized as insecticides in 1937 by G. Schrader. In 1941, the selective herbicide 2,4-D was introduced; it is toxic only to dicotyledons and is used to control broadleaf weeds in grain fields. The fungicidal effect of dithiocarbamates was discovered in the 1930s by Tisdale and Williams. Since the end of the 1970s, even such resistant pathogens as the potato rot organism have been controlled by fungicides (e.g. metaxyl).

Pethidine: a derivative of piperidine which is relatively easily synthesized. It is formed by the reaction of the CH-acidic benzyl cyanide with a nitrogen-mustard gas derivative in the presence of sodium amide, which leads to formation of a piperidine ring. The cyano group is saponified and then esterified.

$H_3C{-}N(CH_2{-}CH_2{-}Cl)_2 + H_2C(C_6H_5)(CN)$

$\longrightarrow$ 1-methyl-4-phenylpiperidine-4-CN $\longrightarrow$ 4-COOH $\longrightarrow$ 4-COCl $\longrightarrow$ 4-$C(=O){-}O{-}C_2H_5$

Synthesis of pethidine

P. was the first highly effective synthetic analgesic, and was introduced in 1939.

Petrolatum: a microcrystalline, soft, bright yellow to ochre colored wax; density 0.82 to 0.85, m.p. 45 to 48 °C. P. can be obtained by extraction or low-temperature crystallization of petroleum residues. It consists mainly of the hydrocarbons $C_{17}H_{36}$ to $C_{21}H_{44}$. Its molecular mass is between 450 and 1000. P. is used in cosmetics as a salve base.

Petroleum: an oily liquid found in the earth, consisting mainly of hydrocarbons. P. contains paraffins, naphthenes and aromatic hydrocarbons as well as compounds of sulfur, oxygen and nitrogen. It also contains small amounts of inorganic compounds and trace elements, such as iron, aluminum, manganese, nickel, vanadium, molybdenum, copper and other elements or their compounds. A number of classifications for P. have been developed. For example, they can be classified according to the predominant hydrocarbon components as methane oils, naphthene oils, aromatic oils, methane-naphthene oils, etc. It was formerly also very common to classify oils according to their "bases", that is, the residues remaining after distillation. There were paraffin-base, asphalt-base and mixed-base P. The following classification is based on the boiling point-density relationships of two key fractions of the P. (fraction I boils at 250 to 275 °C under normal pressure, and fraction II at 275-300 °C at 5.4 kPa):

	Density at 15°C (D_{15}) of fraction I g cm^{-3}	Density at 15°C (D_{15}) of fraction II g cm^{-3}
Paraffin-base P.	<0.825	<0.876
Mixed-base P.	0.825-0.860	0.876-0.934
Naphthene-base P.	>0.860	>0.954

Naphthene-base P. usually contain large amounts (up to 25%) of aromatics.

Formation. The formation of P. is still not completely understood. Although the organic origin of P. is generally recognized, an inorganic source is often proposed. The occasional occurrence of P. or natural gas deposits in magmatic or metamorphic rocks is taken as evidence for an inorganic origin. However, the economically important deposits of this type all contain organic hydrocarbons which have collected in porous and cracked parts of magmatic and metamorphic rocks.

The formation of P.-bearing rocks requires the conversion of plant and animal remains, especially plankton, into low-molecular-weight compounds in the absence of air. This formation of sapropel occurs in shallow seas or lakes. The presence porphyrins in P. shows that both plants and lower animals give rise to P., because metal complexes (V, Ni, Fe) of porphyrins are key metabolic pigments (chlorophyll, heme, etc.). At relatively low temperatures, anaerobic bacteria reduce complicated organic compounds to substances similar to those in P. A byproduct of this microbial action is methane. At the same time, the sapropel muck hardens to rock. The P. usually migrates, under the pressure gradient in the water phase, or of the overlying rock, through pores or cracks into porous or cracked storage formations. These can be sands, sandstones or other composites, or dolomites or limestones. This process is called *primary migration*. During *secondary migration*, the P. is finally collected in a "trap" in the storage formation, that is, in the region of lowest potential. This can be, for example, the dome of an anticlinal fold or a similar formation. Its further migration is prevented by impermeable rock layers, such as clays or salts or tectonic formations. Water which migrates with the P. collects below it in the deposit, due to its greater density, and seals the deposit on the bottom. Above the P. there is often a cap of Natural gas (see) under high pressure (Fig.). The oldest P. deposits are in Cambrium deposits, that is, in layers which were formed about 500 million years ago.

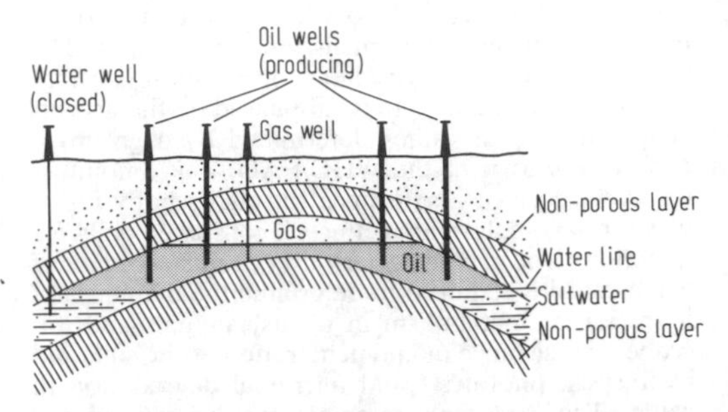

Petroleum deposits in an anticlinal fold. As the oil is depleted, the boundary water and gas cap move closer together.

The distribution of P. in geological formations is shown in the following table:

Stratigraphic unit	Petroleum %
Cenozoic	25.5
Neogenic, Paleogenic	
Mesozoic	67.6
Cretaceous	39.0
Jurassic	28.3
Triassic	0.1
Paleozoic	6.8
Permian, Carboniferous, Devonian	3.7
Silurian, Ordovician, Cambrium	3.1
Weathered crystalline and metamorphic formations	< 0.1

Prospecting. The search for P. and natural gas begins with a careful examination of surface features and underground layers for features which indicate the possibility of deposits. This is done by geological charting and field geophysical studies, including magnetic, gravimetric, electric, telluric, geothermal or seismic measurements. Some of these methods are carried out on the ground, some by soundings in the oceans or from aircraft or satellites. Seismograms of the ground waves induced by underground explosions give indications of the depth and extent of layers. In gravimetric measurements, very sensitive instruments (precision gravimeters) are used to detect anomalies in the gravitational field, which are clues to the structures below the ground, such as the presence of salt deposits; the flanks of salt deposits can act as traps for

P. deposits. Much progress has been made recently with the construction of charts based on photographs from the air or space, with cameras which record at a number of different wavelengths. Geochemical methods are very important; the presence of relatively large amounts of methane or other hydrocarbons in the soil samples can indicate the presence of natural gas or P. deposits farther down. Thus gas analysis of soil samples is useful. Similarly, the increased growth of hydrocarbon-utilizing microorganisms above P. (natural gas) deposits can also be an indicator. If all prospecting methods indicate a probability of a deposit, a test well is drilled. The search is ended when P. is discovered in a well, or when all test wells are "dry".

Exploration. In the exploration phase, further test wells are drilled in a newly discovered deposit to clarify its geological structure and the properties of the oil-bearing layer (porosity, penetrability). This phase is concluded with the calculation of the P. reserve.

Exploitation is the phase in which wells are drilled and oil is produced. For the last several decades, only *rotary drilling* has been used. The drill bit is fastened at the end of a pipe. This is driven either above the ground, by a revolving platform or a power flushing head, or at the bottom of the well itself, by a turbine or screw motor powered by the liquid used to flush out the well. Electric-powered motors for use in the well have not been successful.

The well is flushed with liquids to which clay, usually bentonite, and other materials are added to achieve certain properties such as gel strength, viscosity, thixotropy, thermostability and resistance to alkalies found in the well. Modern flushing liquids contain either no solids or very little, and are based on polymers. At the high pressures expected in the deposits, material must be added to increase the density of the liquid, usually finely powdered baryte. The flushing liquid has a number of functions: to clean the bottom of the borehole, to remove the rock from the bore, to cool and lubricate the drill, to support the walls of the hole and prevent collapse, and to prevent inflow of liquids from porous, permeable layers. In modern drilling, the flushing liquid, which is placed under high pressure, also takes part in the erosion of the rock (erosion rotary boring). It is pumped through the drill pipes, flushes the bottom of the of the well, and rises on the outside around the pipe, carrying with it the debris from drilling. The sides of the hole are stabilized by a cement slurry pumped into the well. To prevent blow-outs (uncontrolled outflow of P. and natural gas under high pressure), the borehole must be provided with a capping system (preventer). In exploratory drillings, test cores from layers suspected of carrying oil must be obtained and examined in the laboratory. Geophysical measurements on the borehole and other tests give quantitative information on the presence of deposits and their exploitability.

The present state of the technology permits holes to be drilled to about 15,000 m. The deepest hole on earth at present is on the Kola Peninsula, Russia; it had reached 12,000 m in 1984.

Drilling at sea or under extreme climatic conditions (permafrost and deserts) creates special conditions and additional problems. There are several approaches to drilling at sea (artificial islands, platforms, semisubmersibles, drilling ships), the use of which depends on the depth of the water. As a rule, oil wells at sea are five to ten times as expensive as those on land. The greatest water depths at which wells have been drilled are about 3000 m.

Since 1965, the techniques of drilling for oil have become more and more scientific. This is seen in increased drill power, better information obtained on the layers penetrated by the hole, and in increased security against accidents, especially blow-outs. Progress has been made in the measurement techniques for monitoring the process of drilling and the geophysics of the hole.

The energy required to produce P. (to overcome frictional losses in the pores of the oil-bearing rock) comes from the energy of the deposit itself. It consists of 1) energy of the gases dissolved in the oil (gas release pressure, up to 95% of the total), 2) energy of the expanding gas cap (gas pressure, up to 25% of the total), 3) energy of the groundwater (up to 10%) and 4) gravitational energy (up to 10%). Usually several types of energy are involved.

Production using the deposit energy alone (primary production) can recover only 20 to 30% of the oil in the deposit. With secondary production methods, at most about 40% of the oil can be extracted. Secondary production methods include pumping water (usually salt water) into the outer areas of the field, or pumping gas into the gas cap or into the oil itself. Injection wells and spent production wells are used for this purpose. The gas injected can be either the gases found with the P. in the field, natural gas or waste gases. Secondary methods should be applied from the beginning of production in the field, to prevent a loss of pressure in the deposit. Tertiary production methods are used in addition to reduce the resistive forces in the pores. Some examples are Surfactant-polymer flooding (see), which reduces the surface tension between water and oil, hot-water/steam flushing to reduce viscosity by heating the oil to 80-120 °C and mixed flushing by injection of lighter oil fractions (diesel oil). With tertiary production, the yields can be as high as 60% of the reserve. Experimentally, the heaviest fractions have been burned in situ, in efforts to increase yields. In this technique, the viscosity of the oil decreases ahead of the combustion front; near the front, the heavy hydrocarbons are cracked, permitting easier flow to the production wells. However, the technique is very difficult to control, and is still in the experimental stage.

The transport of P. from the botton of the well to the surface by means of the energy in the deposit is only possible in the first years of production (eruptive production). Later, pumps are sunk into the wells; they are driven from the surface by pistons. If compressed gas is available, the specific density of the oil can be decreased by injection of gas into it (gas lift production).

Shallow deposits can be mined, but this is limited to a few special situations. The oil flows out of oil sands or shales into the bottom of the mine, and is pumped from there, or the oil sand is removed by strip mining and extracted aboveground with 90 °C water (Canada).

Offshore production of oil is gaining in importance. In 1980, 680 million t P. were produced offshore by 37 countries on all continents. This production is undertaken mainly on the continental shelves, which generally have the same geological structures as the land. Prospecting and drilling are the same in principle as on land; but depending on the depth of the water, they are carried out from anchored or floating drilling platforms or drill ships. Drilling is possible in much deeper water than production; at present, the maximum depth for production is about 600 m under water.

Transport. Freshly produced P. can contain considerable amounts of dissolved gases, and often it contains water and solids as well. The gas and impurities must be removed before transport. The most important means of transportation are Pipelines (see) on land, and tankers at sea; these have been built with capacities greater than 500,000 t. Oil spills resulting from tanker accidents can cause great damage, and are cleaned up using surfactants, synthetic foams and iron-containing powders.

Storage. P. is stored above ground in large steel tanks, which may hold up to 150,000 m^3 (Cordoan Refinery, owned by Shell, in Venezuela). It is more economical to store large amounts of P. underground, for example, in salt caverns, which can have volumes up to 500,000 m^3.

Reserves. The assured reserves, those known with precision, are estimated at present at about 430 billion tce. The presently known, recoverable reserves are 130 billion tce (1 tce = 0.7 t P.).

Estimates of the total reserves vary widely, between 1850 and 2300 tce. These figures do not include the P. in oil shales and sands, which account for more than 1400 tce.

World production of petroleum in million tons

1860	0.07	1970	2275
1880	4.1	1975	2645
1900	19.8	1979	3127
1920	97.0	1980	2979
1940	294.2	1981	2866
1950	523.3	1982	2755
1960	1053	1983	2726

The most important oil producing countries are those on the Persian Gulf (Kuwait, Saudi Arabia, Iraq, Iran and some smaller states), which own more than two thirds of the presently known reserves of P. Other large deposits are located in the former USSR (Volga-Ural area, North Caucasus, Azerbaijan, Western Siberia, Western Kazachstan) in the USA (Texas, Alaska, Louisiana, Oklahoma, California), Venezuela, Mexico and Africa (Libya, Algeria, Nigeria). In the Far East, P. is produced mainly in China, Indonesia and Burma. Smaller deposits are located in many countries, and in the North Sea.

Chemical composition. In addition to hydrocarbons, P. contain varying amounts of sulfur, nitrogen and oxygen compounds, and small amounts of iron, aluminum, manganese, nickel, vanadium, molybdenum, copper and other compounds. The average composition (in % by mass) is

Carbon 85-90
Hydrogen 10-14
Nitrogen 0.1-0.5
Sulfur 0.2-0.3
Oxygen 0-1.5

The hydrocarbons are largely saturated. There are only traces of alkenes, and the amounts of aromatic hydrocarbons vary widely. Straight-chain, branched and cyclic alkanes (naphthenes) are also present in widely varying relative amounts. The molecular masses of the hydrocarbons range from that of methane, the simplest compound, to over 10,000 for some asphalt molecules. Most molecules have masses between 200 and 800. The cycloalkanes consist of single rings or are polycyclic; five- and six-membered rings are most common. The most common aromatic hydrocarbons are indanes, tetralines, diphenyls, acenaphthenes, fluorenes and isoprenes. The higher-boiling fractions contain large amounts of polycyclic aromatics. Sulfur in P. is mostly in the form of thiols, disulfides or cyclic sulfides; thiophene structures are present in the higher-boiling fractions. In general, the sulfur content increases as the boiling point increases. Elemental sulfur is rare, but free hydrogen sulfide is common. Oxygen is present in the carboxyl groups of isoprenoid naphthenic acids, which are derived from sterols. However, these compounds are found only in certain P. Oxygen is also present in phenols and ethers, especially in the larger molecules of the distillation residues. Nitrogen compounds are present mainly as substituted pyridines, quinolines and carbazoles in the higher boiling ranges.

Chemical and physical properties. Crude oil is a thin to viscous liquid, light to dark brown, often with a blue to green fluorescence. The odor can be pleasantly aromatic, but if sulfur compounds are present, it can be unpleasant and garlic-like. The density of P. is usually 0.8 to 0.9 g cm^{-3}; the naphthene-base oils are denser. The heat of combustion is 40,000 to 46,000 kJ kg^{-1}. The flame-point temperature depends on the proportion of volatile substances; it can be below 0 °C or as high as 70 °C. The economically most important components of crude oil boil between 50 and 350 °C. The viscosity is higher in oils with higher densities and higher boiling points. Crude oils are not soluble in water, but give stable emulsions. They are slightly soluble in ethanol, and dissolve readily in chloroform, carbon tetrachloride, ether and aromatic solvents.

Typical composition and properties of crude oils

Source	Boiling begins [°C]	Density at 20°C [g cm^{-1}]	Sulfur content [%]
Muchanov Former USSR	38	0.822	0.99
Romaschkino Former USSR	46	0.862	1.68
Matzen (Austria)	150	0.918	0.23
Mexico	125	0.970	
Iran	35	0.845	1.78

Applications. P. can be used in crude form only as a fuel for steam generators. Its enormous potential is only realized after refining (see Petroleum processing).

Historical. Because of its combustibility, low-viscosity P. was used as a fuel and lighting oil even in antiquity. Natural asphalt was also known in the second century BC and was used as mortar by the Assyrians and Babylonians. The Chinese are said to have produced by P. 1700 BC, and to have used it for lighting. Production of P. began in Baku around 1700. The first modern oil well was drilled in 1859 by E. Drake, in Titusville, Pennsylvania, USA. Since then, several million wells have been drilled, and of them, about 750,000 are producing at present. P. provides about 40% of the world's energy at present. Although this fraction will decline in the future, it is predicted that the absolute production will rise for a time, peaking in the 1990's. This, however, will depend on developments in the price of oil.

Petroleum chemistry: the chemistry of processes used to convert petroleum or natural gas to starting materials and intermediates for production of plastics, elastics, fibers, laundry products, pigments, textile conditioners and so on. The processes used to produce fuels and lubricants are not included in P.

Growth of demand for petroleum chemical products

Product	Production in billion tons		
	1970	1975	1980
Ethene	17	35	55
Propene	8	13	21
Butadiene	2.9	4	6
Benzene	7.5	13	31
Ethyne	3.0	3.0	3.0

Petroleum coke: a very low-ash coke produced as a residue of petroleum processing, especially cracking. P. is used to make electrodes (***electrode coke***, abb. ***E-coke***) because of its low ash content.

Petroleum distillation: see Petroleum processing.

Petroleum ether, ***light petroleum***: a very low-boiling gasoline obtained by distillation of crude petroleum. P. consists mainly of pentanes and hexanes, and boils between 40 and 70 °C; its density is 0.655 to 0.675. Mixtures with air explode when ignited! P. is used as a solvent for extraction of essential oils from flowers and leaves, in the textile industry to burn off the fine hairs on textiles, as a solvent for resins, fats and oils (e.g. in spot remover) and, after special purification, as a treatment for wounds.

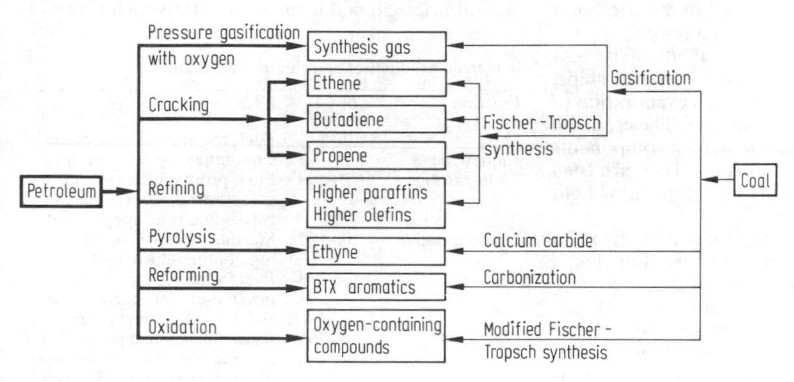

Production of basic chemicals from petroleum and coal

The most important starting materials for P. are methane, ethene, propene, butenes, butadiene, the BTX aromatics and sometimes also ethyne, since it is also used in large amounts to produce aliphatic products (see Acetylene chemistry), and some is produced from natural gas or petroleum fractions. Up to the end of the 1940s, Coal chemistry (see) played a major role in production of intermediates for the chemical industry. As petroleum refining became more powerful, coal chemistry became less significant. At the same time, as petroleum was used to produce primary reagents, especially the lower olefins and aromatics (table), the chemical industry underwent a structural change. The majority of the aliphatic compounds produced today are made from C_2 to C_3 olefins; ethene is the most important precursor and, along with sulfuric acid, chlorine and methanol, is one of the major products of the chemical industry.

As petroleum prices rose in the mid-1970s, coal chemistry became more significant. At present, the production of primary chemicals from coal is still too expensive, so that before 1990 there is not likely to be any significant decline in P.

Petroleum processing: in the narrow sense, the chemical-industrial processing of petroleum; in the wider sense, Cracking (see), Reforming (see), Hydrorefining (see) and Hydrocracking (see) are included. Crude petroleum contains gases (e.g. methane, ethane and hydrogen sulfide), solids and salts dissolved in water (e.g. alkali and alkaline earth chlorides). It is first degassed. The crude oil, which emerges from the production well at high pressure, is depressurized and led into separators. These are vertical or horizontal cylinders containing sieve-like impact screens. The dissolved gas escapes in these tanks; it can be used for fuel or as a raw material for chemistry. The gas, which often contains valuable gasoline and liquid gas components, is processed in special separation units.

After degassing, the oil emulsion is fed into measuring tanks, where the amount produced is determined and the more readily separated emulsions settle out solids and water. The actual dehydration takes place in a separate unit. It can be accelerated by heating the oil to 40-60 °C, which reduces its viscosity. If this step does not completely separate the emulsion of oil and water, de-emulsifiers are added; for example, sulfonates, esters of high-molecular-weight alcohols and polybasic acids, or condensation products of ethylene oxide. The emulsion can also be disrupted by the action of alternating current at 15,000 to 30,000 V; the water is collected into larger droplets which precipitate. Ultrasound can also be used to de-emulsify. Removal of water also removes the salts, since most of the salts are dissolved in the aqueous phase. Solid salts suspended in the petroleum are removed by addition of condensate. Removal of salts is an important step in P., because in the distillation steps, magnesium chloride decomposes at high temperatures, releasing highly corrosive hydrogen chloride.

Crude oils also contain small amounts of ash-forming compounds, such as iron, aluminum, magnesium, manganese, nickel and vanadium salts. These are enriched in the heavy distillation residues. The vanadium compounds, in particular, lead to the corrosion of heat-resistant steels at high temperatures.

The de-gassified, dried and de-salted oil is next stabilized, if it is to be transported in tanks or ships, to prevent losses of 0.5 to 2% through evaporation of low-boiling components in hot climates. The crude oil is heated to 100-120 °C, and the lighter components are separated by pressure distillation. They are then separated by another distillation step into light gasoline, butane and propane.

The actual refining of the preprocessed petroleum is done by Distillation (see) in a refinery. For this, a fractionation column with a side outlet is needed, or a distillation unit consisting of several columns (Fig.). The oil is heated to about 400 °C and separated in the fractionating column which it enters directly after leaving the heater. The first distillation step (*atmospheric distillation*) takes place at normal pressure. The hot, partially vaporized petroleum is injected into the bottom of the column, and the vapors rise. In this way, the raw material is separated into fractions with different boiling limits. The fractions are drawn off at different levels of the column, usually into smaller side columns (stripper columns) where small amounts of low-boiling components are removed by spraying steam directly into them. The low-boiling components are returned to the main column.

The distillation residue (top residue) remaining at the bottom of the column contains the valuable lubricating oil fractions. These can be separated only by vacuum distillation, because the high temperatures which would be required at normal pressure would lead to their decomposition (see Cracking). In *vacuum distillation*, lubricating oils of varying viscosity are obtained (light to heavy machine oils), and bitumen remains as a residue. The top residues from oils which are not suitable for making lubricants can be used as fuel oils or as starting materials for cracking (see Visbreaking).

Refining. Before they can be further processed, practically all petroleum products must be refined to remove disruptive components. The first generally used method was treatment with sulfuric acid, but this has now been almost completely replaced by selective methods. Sulfuric acid is still used to a certain extent only in the distillation of lubricating oils. Unsaturated compounds, aromatics, resins, etc. are absorbed by the sulfuric acid and form an acid tar which is insolu-

The main products of petroleum distillation

Fraction	*T*[°C]	Uses
Gas Liquefied gas	20	Fuel, raw material for chemical syntheses (see Natural gas)
Light gasoline	20-100	Fuel, production of synthesis gas, starting material for pyrolysis and reforming
Heavy gasoline	100-180	Starting material for reforming, motor vehicle fuel
Kerosene	180-240	Fuel for turbine aircraft engines, diesel fuel, starting material for making surfactants
Gas oil	240-360	Diesel fuel, starting material for cracking, formerly also for oil gas production
Vacuum distillate	360-500	Motor oils, machine oils, starting material for cracking and hydrocracking
Vacuum residue	500	Lubricants, industrial heating oil, raw material for cooking, pressure gasification of oil and visbreaking

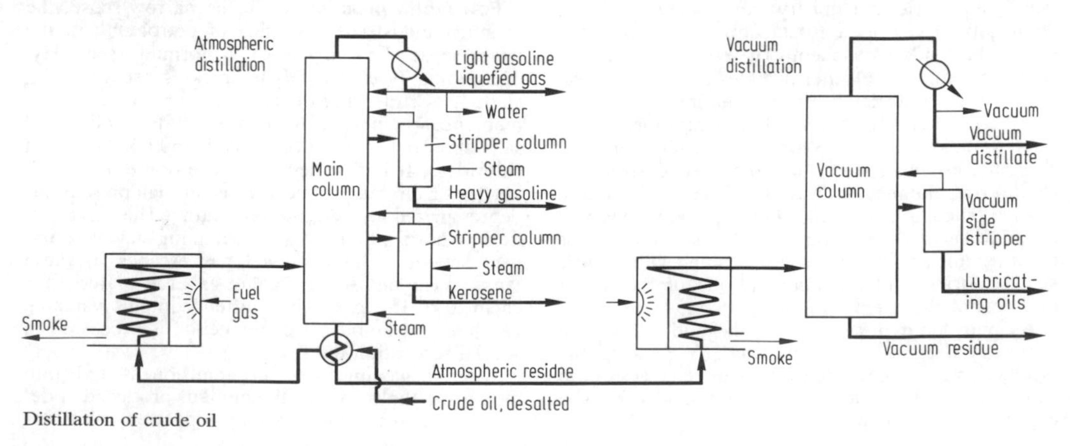

Distillation of crude oil

ble in oil. The sulfuric acid treatment must be followed by treatment with fuller's earth, which removes the remaining high-molecular-weight products, most of which are intensely colored. Acidic components (fatty acids, naphthenic acids, phenols and to a slight extent, mercaptans) are removed by treating the product with aqueous sodium hydroxide, which converts these compounds to water-soluble alkali salts. Today this process is usually run continuously in scrubbing towers, or in such a way that the product and lye are mixed in circulating pumps and then allowed to settle or are centrifuged.

Effective removal of all sulfur compounds can be achieved only by catalytic pressurized hydrogen refining (see hydrorefining). This also has the advantage that the sulfur is obtained as hydrogen sulfide.

Diesel fuels and lubricating oils must often have Paraffin (see) removed to lower their stock points; otherwise the use of these products in cold weather could cause considerable problems.

High-molecular-weight compounds (asphalt) which tend to precipitate can be removed from lubricating oils by treating them with aliphatic hydrocarbons, such as propane, in which the asphalt compounds are insoluble (in contrast to the lubricants).

The viscosity-temperature behavior of lubricants are improved by Extraction (see) with a suitable solvent, such as liquid sulfur dioxide, phenol or creosol. This removes aromatic compounds which make the viscosity especially sensitive to temperature.

Historical. Although petroleum and a few of its products (tar, vaseline) were known in antiquity, it did not become an important substance until the 19th century. In 1810, a "naphtha" was produced in Galicia by distillation in liquor stills; it was used to remove fat from wool and for lighting. After the first oil well was drilled in 1859 by Drake (Pennsylvania, USA), enough petroleum was available to supply primitive refineries modelled on tar stills. At first, distillation was discontinuous, and to increase the yield of lighting oil, it was combined with a thermal cracking process. In 1875, R. Nobel introduced continuous distillation of petroleum in Baku. Refining of petroleum fractions with sulfuric acid to improve the quality of the lighting oil was introduced around the turn of the century. With the introduction of electric lighting, the demand for petroleum dropped markedly. However, the increasing numbers of automobiles made the high-boiling fractions more and more important, so that thermal cracking processes were of greater interest. The first large-scale catalytic cracking process was introduced in 1936 by Houdry. Catalytic reforming using hydrogen was developed independently in 1939/40 in Germany and the USA.

Petrurgy: the processes of melting, phase formation (e.g. crystallization) and the formation of sedimentary and magmatic rocks.

Petunidin: see Petunin.

Petunin: an Anthocyanin (see) plant pigment, the 3,5-di-β-glucoside of petunidin. P. is the pigment of blue petunias. Its aglycon is ***petunidin*** (3,5,7,4',5'-pentahydroxy-3'-methoxyflavylium chloride).

Pevlen®: see Synthetic fibers.

Pfeffer cell: see Osmosis.

PFT technique: abb. for pulse Fourier Transform technique; see NMR spectroscopy.

PGE: see Prostaglandins.

PGF: see Prostaglandins.

pH, ***proton activity exponent***: an expression for the negative decadic logarithm of the proton activity, pH = $-\log a_{H^+}$, introduced in 1909 by Sörensen. Protons are always solvated in aqueous solution; therefore, pH = $-\log a_{H_3O^+}$. For dilute solutions, the activities can be approximated by molar concentrations: pH ≈ $-\log c_{H_3O^+}$. The product of the H_3O^+ and OH^{-1} ion concentrations is constant (see Dissociation constant), with a value of $c_{H_3O^+} \cdot c_{OH^-} = K_w = 10^{-14}$. Therefore, the pH of a neutral aqueous solution, where $c_{H_3O^+} = c_{OH^-}$, is 7. Acidic solutions have pH < 7, and basic solutions have pH > 7. Values of pH < 0 and pH > 14 are entirely possible. The pH of a solution is determined experimentally with a suitable pH color indicator (see Indicators) or potentiometrically, with a pH meter (see Glass electrode). The pH of aqueous acid and base solutions can be approximately calculated from their molar concentrations. For very strong acids, which are completely dissociated, $c_{H_3O^+} = C_{acid}$ (C = total concentration), i.e. pH = $-\log C_{acid}$. For a 0.001 M solution of HCl, the H_3O^+ concentration is 10^{-3} mol l^{-1}, pH = 3. For dilute solutions of weak acids, the pH is given by the equation pH = $(pK_a - \log C_{acid})/2$. Thus a 0.001 M solution of acetic acid has a pH of 3.88. Similarly, for very strong bases, protolysis in water is complete, and $C_{bas} = c_{OH^-}$; therefore, the pH = 14 - pOH = 14 + log C_{base}. A 0.001 M solution of NaOH thus has a pH of 11. The pH of aqueous solutions of weak bases is given by the equation pH = $14 - (pK_b - \log C_{base})/2$, so that 0.001 M NH_3, for example, has a pH of 10.10.

The pH is extremely important for the course of many chemical and biological processes. Buffer solutions (see) are used to establish and maintain suitable pH values.

Phallacidin: see Phallotoxins.

Phallicin: see Phallotoxins.

Phalloidin: see Phallotoxins.

Phalloin: see Phallotoxins.

Phallotoxins: bridged, heterodetic, cyclic heptapeptides which, together with the amatoxins, are the most important of the toxins in the death cap fungus *Amanita phalloides* (Fig.). The P. include ***phalloidin, phalloin, phallicin*** and ***phallacidin***.

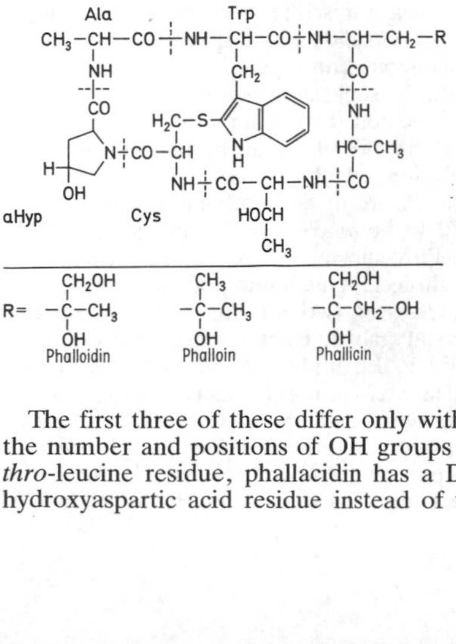

Ala, Trp, aHyp, Cys

$CH_3-CH-CO-NH-CH-CO-NH-CH-CH_2-R$

R = $-C(CH_2OH)(OH)-CH_3$ Phalloidin; $-C(CH_3)(OH)-CH_3$ Phalloin; $-C(CH_2OH)(OH)-CH_2-OH$ Phallicin

The first three of these differ only with respect to the number and positions of OH groups on the *erythro*-leucine residue, phallacidin has a D-*erythro*-β-hydroxyaspartic acid residue instead of the D-threonine, and in the neighboring position, a valine instead of an alanine. The toxicity of the P. depends on the cycloheptapeptide structure and on the thioether bridge of the trypthione middle part. In the mouse, the LD_{50} of the P. are around 2 mg/kg. The P. act after a relatively short time by destroying the endoplasmic reticulum of the hepatocytes. It is interesting that the effect of antamanide, which is also found in the death cap mushroom, antagonizes the toxic effect of phalloidin when it is applied soon enough. The structures and syntheses of the P. were reported by Th. Wieland et al.

Pharmaceutical chemistry: the study of the chemistry of drugs, pharmaceutical aids and diagnostic materials. This includes the synthesis of these materials, isolation of biologically active natural products which can be used as drugs, and development of methods for identification, quantitative determination and purity testing for these materials. P. also includes studies on the stability of drugs and the mechanisms of their decomposition. An essential concern of P. is the development of new biologically active compounds which can be used as drugs, and the elucidation of structure-activity relationships. In part, the study of Biotransformation (see) is also one of the tasks of P.

Pharmacognosy: the branch of pharmacology dealing with medicinal plant and animal products, and with the plants and animals from which these products are made. The science has recently been redefined as the study of *biogenic drugs*, which includes medicinal plants and isolated biogenic drugs, as well as the microbial, plant or animal sources of these compounds.

Pharmacokinetics: the study of the quantitative time course of absorption, distribution, metabolism and elimination of drugs.

Phase: a chemically pure substance (***pure P.***) or mixture (***mixed phase***) with constant or continuously varying macroscopic properties, such as density, concentration, index of refraction, extinction coefficient, heat capacity, etc. Different P. are separated by ***phase boundaries*** at which the properties change abruptly. Continuous variation of certain properties within a P. can be caused by force fields (e.g. gravity) or disequilibrium (e.g. concentration gradients in solid or liquid solutions).

Gases are always miscible, and thus form a single P., but liquids and solids may display several phases. In a heterogeneous (multiple-phase) system, there may be spatially separated regions with the same properties; these constitute a single P. An example is the individual crystals of salt at the bottom of a saturated salt solution, which together comprise the solid P. If different P. are in thermodynamic equilibrium, they are said to be ***coexistent P.*** and they are in ***P. equilibrium***. Pure substances are generally capable of existence in three P., gas, liquid and solid. However, if the substance forms two or more modifications with differing crystal habits, each of these constitutes a separate solid P. Examples are provided by sulfur, carbon, quartz, etc. Helium forms two liquid P., normal He^{I} and, below 2.186 K, superfluid He^{II}. Above the Critical point (see), there is only a single uniform supercritical phase instead of a gas and a liquid P. The ranges of existence of coexistent P. are shown in Phase diagrams (see). The number of such P. can be determined by the Gibbs phase rule (see). Conversion of a substance from one P. into another (***phase change***) is associated with an energy change (see Phase change heats).

In general, the effects of P. boundaries on the thermodynamic properties of the total system can be ignored, but this is not true for highly disperse systems, such as colloids, in which the energies are very dependent on the size of the P. boundaries. The enrichment of foreign substances at the P. boundaries is called Adsorption (see).

Phase change: the movement of a substance from one phase into another. The diagram shows various possible P.:

$$\text{Gas} \underset{\text{Evaporation}}{\overset{\text{Condensation}}{\rightleftharpoons}} \text{Liquid}$$

$$\text{Gas} \underset{\text{Sublimation}}{\overset{\text{Condensation}}{\rightleftharpoons}} \text{Solid}$$

$$\text{Liquid} \underset{\text{Melting}}{\overset{\text{Solidification (crystallization)}}{\rightleftharpoons}} \text{Solid}$$

$$\text{Solid modification I} \underset{\text{change}}{\overset{\text{Modification}}{\rightleftharpoons}} \text{Solid modification II}$$

P. of pure substances are isothermal processes. Because a change in the state of matter represents a sudden change in the interactions between the particles, P. are associated with an exchange of heat with the environment (see Phase change heats).

P. begin with the formation of seeds of the new phase. The formation of seeds is often inhibited, so that at the beginning the process is delayed, and supersaturation or superheating may occur.

P. are used in many ways to purify materials; mixtures of two coexisting phases of a pure substance are also used to maintain a constant temperature, because the P. is strictly isothermal. For example, a mixture of ice and water is used for 273.15 K, or $CO_{2(solid)}$ and $CO_{2(gas)}$ for 194.7 K.

Phase-change energy: see Phase-change heats.

Phase-change enthalpy: see Phase-change heats.

Phase-change heats (older term, ***latent heats***): Energy released or consumed when a substance passes from one phase to another. P. are related to molar amounts (***molar P.***) in chemistry. As a rule, phase changes are observed at constant pressure, usually atmospheric pressure. The P. in this case is an enthalpy (***molar phase-change enthalpy*** $\Delta_P H$), and is equal to the difference in molar enthalpies H of the pure substance in the final (") and initial (') phases: $\Delta H_p = H'' - H'$. If the phase change occurs at constant volume (which is experimentally very difficult to achieve with condensed substances), the P. is a change in the internal energy $\Delta_P E$. (***phase-change energy***. If the direction of the phase change is reversed, only the sign of the P. is changed. For example, the molar melting enthalpy is equal to the negative of the molar enthalpy of crystallization (Table 1). From the first law of Thermodynamics (see), it follows that $(\delta h/\delta n)_{p,T} = \Delta_P H$ or $(\delta e/\delta n)_{v,T} = \Delta_P E$, where $\Delta_P E$ is the molar P. at constant volume of the system. P. are dependent on temperature; this dependence is given by Kirchhoff's law (see).

Table 1. Types and terminology of phase-change heats

Process	Name Direction (1)	Direction (−1)
Solid $\underset{-1}{\overset{1}{\rightleftharpoons}}$ liquid	Heat of melting	Heat of crystallization
Liquid $\underset{-1}{\overset{1}{\rightleftharpoons}}$ gas	Heat of vaporization	Heat of condensation
Solid $\underset{-1}{\overset{1}{\rightleftharpoons}}$	Heat of sublimation	
Solid $\underset{-1}{\overset{1}{\rightleftharpoons}}$ solid	Heat of modification change	

If heat is added to a pure substance at constant pressure, its temperature remains constant during a phase change until the entire amount of the substance has been converted into the new phase. The added P. causes a discontinuous change in the internal energy and enthalpy (Fig., Table 2) because of the sudden changes in the interaction energies between the particles and the state of order in the two phases. The ratio $\Delta_P H/T_{pc} = \Delta_P S$ is the ***molar phase-change entropy***, a measure of the change in order. In mixed phases, changes in phase generally are not isothermal (see Phase diagrams).

P. can be measured calorimetrically. Heats of evaporation and sublimation are more readily available from the temperature dependence of the vapor pressure, if the Clausius-Clapeyron equation (see) is used. The ***Pictet-Trouton rule*** can be used for rough estimation: the molar ***vaporization entropy*** $\Delta_V S$, i.e. the ratio of the molar enthalpy of evaporation $\Delta_V H$ and the boiling temperature T_b at standard pressure (0.1 MPa) is approximately the same for all liquids: $\Delta_V S = \Delta_V H/T_b \approx 84$ to 92 J K^{-1} mol^{-1}. There are deviations from this rule among substances with very low boiling points ($T_b < 170$ K), molten metals and associating substances.

Dependence of the molar enthalpy H of a pure substance on the temperature T. T_{pc}, temperature of the modification change I $\rightleftharpoons$ II; T_m, melting point; T_b, boiling point; $\Delta_{pc}H$, ΔH_{melt}, $\Delta_V H$, molar phase-change energies for the change in modification, melting and evaporation.

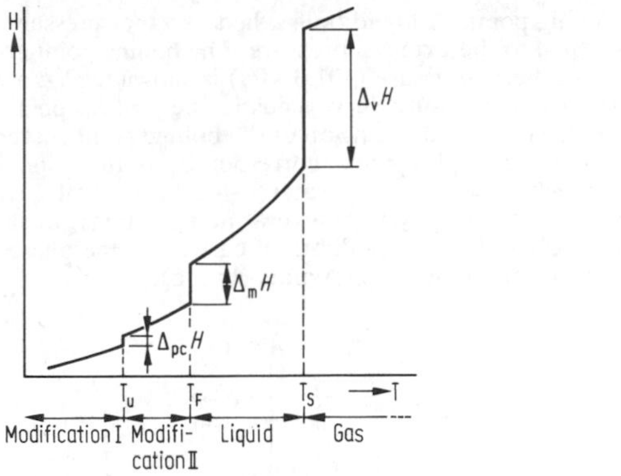

Richard's rule: The entropy of melting $\Delta_{melt}S = \Delta_{melt} H/T_m$ for monoatomic elements is between 7 and 14 J K^{-1} mol^{-1}. $\Delta_{melt}H$ is the molar melting enthalpy, and T_m is the melting temperature.

Phase diagram, ***state diagram***: term for diagrams which show the phase compositions of systems of one or more substances as functions of state parameters (pressure, volume, temperature, composition or entropy). The ranges in which the different phases can exist are bounded by the equilibrium curves of the phase changes. The number of coexisting phases is described by the Gibbs phase rule, and the boundary curve by the applicable equilibrium equation. P. are prepared by equilibrium measurements, e.g. of vapor pressure, boiling points or melting points, calorimetric determinations of phase-change heats, thermal analysis or thin-layer studies of solid mixtures. P. include p,T, p,v and T,s diagrams, vapor-pressure, boiling-point and melting-point diagrams.

1) ***P. of pure substances*** are usually presented as the dependence of equilibrium pressures p on temperature T. These consist of vapor-pressure (liquid-gas), sublimation pressure (solid-gas) and melting-point-pressure (solid-liquid) curves). All three curves meet at a point, the triple point (Fig. a). At this point, all three phases coexist in equilibrium, that is, it is the melting point of the pure substances under its own vapor pressure. For water, the triple point is 273.16 K (0.01000 °C), and 610 Pa (4.58 Torr). The Clausius-Clapeyron equation (see) applies to all three equilibrium curves. It follows from that equation that the melting points of most substances increase with pressure. Water is an exception, because the molar volume of ice is greater than that of water at the

Table 2. Molar phase-change enthalpies of pure substances

Substance	Modification	change	Melting	Evaporation		
	T in °C	$\Delta_P h$ in kJ/mol	T in °C	$\Delta_P H$ kJ/mol	T in °C	$\Delta_P H$ in kJ/mol
O_2	−250	0.0937				
	−229	0.7431	−219	0.445	−183	6.819
H_2O	—	—	0	6.007	100	40.66
C_6H_6	—	—	5.53	9.837	80.1	30.76
NaCl	—	—	800	28.80	1461	170
$Fe_{\alpha\text{-}\gamma}$	906	0.878				
$Fe_{\alpha\text{-}\delta}$	1401	0.46	1535	15.5	2735	354

melting point. A liquid boils when its vapor pressure is equal to the external pressure. The boiling point at atmospheric pressure (101.3 kPa) is shown in Fig. a. The vapor pressure curve ends at the Critical point (see). The dotted extension of the boiling point curve below the triple points corresponds to the vapor pressure of supercooled water, a metastable state. If pure substances exist in several modifications, their P. also contain the equilibrium curves for the phase changes and more triple points (Fig. b).

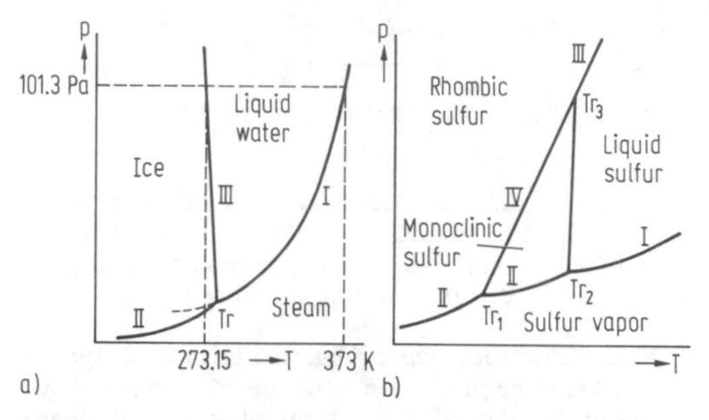

Phase diagrams: a) Diagram of state for water (schematic); *I*, vapor pressure curve; *II*, sublimation curve; *III*, melting pressure curve; Tr, triple point. b) Diagram of state of sulfur (schematic); *IV*, equilibrium curve S_{monocl}; S_{rhomb}.

P. of multicomponent systems. In binary mixtures of substances, according to the Gibbs phase rule the composition is needed as a further degree of freedom to describe the phase structure. Since the complete P. can then no longer be represented in a plane, one state parameter is kept constant and the interactions of the other two are plotted. The Melting-point diagram (see) and Boiling-point diagram (see) show the dependence of the melting or boiling temperatures on the composition, when the pressure is held constant. The Vapor pressure diagram (see) shows the vapor pressures of mixtures at constant temperature as a function of composition. The composition is usually given in molar fractions or percents, but mass fractions or percents are sometimes used (see Composition parameters).

P. of mixtures with more than two components require a special form of representation, e.g. triangular coordinates for ternary systems. If the components cannot be mixed in any arbitrary ratio, the P. contains Mixing gaps (see).

Phaseollin: see Phytoalexins.

Phase I reactions: see Biotransformation.

Phase II reactions: see Biotransformation.

Phase transfer catalysis, abb. ***PTC***: a method of preparing a reagent for nucleophilic substitution reactions which may be used, for example, in alkylation of -O-H, -S-H, -N-H and activated C-H compounds, carbene reactions, Wittig-Horner reactions or oxidation reactions. Nucleophilic substitutions by P. occur by the following mechanism:

Aqueous phase $Na^+Y^- + Q^+X^-$
Phase boundary
Organic phase $[Q^+Y^-] + RX \rightarrow [Q^+X^-] + RY$

The catalyst cation Q^+ passes into the organic phase with the anion Y^-, which is the nucleophilic reagent for the substitution reaction. Here the weakly solvated ion pair reacts rapidly. Afterwards, Q^+ returns to the aqueous phase with anion X^- (the leaving group of the nucleophilic substitution). The catalyst cations Q^+ are onium salt cations, usually ammonium or phosphonium salts with long-chain ogranic groups, or alkali metal cations complexed with crown ethers.

The advantages of the P. over conventional synthetic processes are, among others, that anhydrous aprotic solvents are not needed, the reaction times are shorter and the temperatures are lower, aqueous sodium hydroxide can be used instead of alkali metal alkoxides, the reaction products are more easily worked up, and the selectivity and product ratios are shifted by suppresion of side reactions.

Phase width: see Berthollides.

Phe: abb. for phenylalanine.

α-Phellandrene: *p*-mentha-1,5-diene, a monocyclic, monoterpene hydrocarbon. α-P. is a colorless oil with a pleasant odor; b.p. 173-176 °C. It is a very unstable terpene which is easily converted to a resin. P. is found as its d-form in dillweed, fennel and angelica root oils, and as its l-form in star anis and eucalyptus oils. It is used in perfumes.

CH_3

H_3C CH_3

Phenacetin: 4-ethoxyacetanilide, a derivative of aniline and phenetidin; a white, crystalline substance, m.p. 135 °C. P. can be formed by nitration of chlorobenzene to 4-nitrochlorobenzene, substitution of the activated Cl atom by an ethoxy group to form 4-nitroethoxybenzene, reduction of the nitro group and acetylation of the resulting amino group. It was introduced in 1887 as an analgesic.

R—O—C_6H_4—NH—$COCH_3$

R = C_2H_5 : Phenacetin
R = H : Paracetamol

However, if taken regularly in high doses, it can produce kidney damage. *Paracetamol* is 4-hydroxyacetanilide (a white, crystalline substance); m.p. 168 °C; it was introduced into therapy after it was discovered to be one of the metabolic products of P. and a pain-killing effect was ascribed to it.

Phenacyl chloride: same as ς-Chloroacetophenone (see).

Phenanthrene: a condensed aromatic hydrocarbon with angular fusion of the rings. P. forms colorless, shiny crystalline leaflets with a blue fluorescence; m.p. 101 °C, b.p. 340 °C, n_D^{20} 1.5943. It is insoluble in water, soluble in hot methanol and ethanol, and readily soluble in chloroform, ether, acetone and benzene.

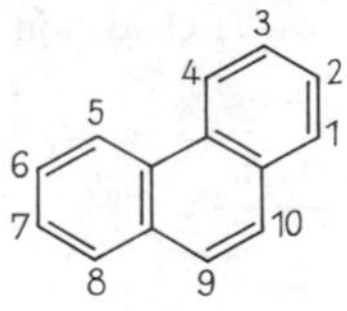

Formally, two π-electron sextets can be identified in the ring system of P. This means that P. is more strongly aromatic than anthracene. The bond between C atoms 9 and 10 is nearly a true double bond, so that P. can undergo addition reactions in addition to electrophilic substitutions. Catalytic hydrogenation in the presence of a copper-chromium oxide catalyst yields 9,10-dihydrophenanthrene, and oxidation forms 9,10-phenanthrenequinone. The reaction with bromine first produces the 9,10-dibromo adduct, which easily loses hydrogen bromide to yield 9-bromophenanthrene.

P. is found in coal tar, from which it is still obtained industrially. It is also formed by high-temperature treatment of a mixture of biphenyl vapor and ethene. It can be synthesized by the Pschorr synthesis from 2-nitrobenzaldehyde and phenylacetic acid, or, according to Haworth, from naphthalene and succinic anhydride.

P. and its derivatives are important in the production of certain pigments, drugs and synthetic resins. A few natural products, including vitamin D, cholesterol, steroid hormones, saponins, digitalis glycosides and morphine alkaloids contain the phenanthrene skeleton.

Phenanthroquinone, ***9,10-dihydrophenanthrene-9,10-dione***: a quinone derived from phenanthrene. P. forms odorless, orange needles with m.p. 208-210 °C, sublimation above 360 °C. P. gives the reactions typical of an α-diketone. For example, when heated in aqueous alkalies, it undergoes the benzil-acid rearrangement to form 9-hydroxyfluorene 9-carboxylic acid.

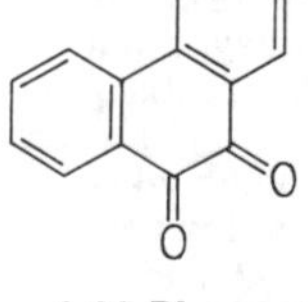

1,10-Phenanthroline, ***o-phenanthroline***: a heterocyclic compound derived from pyridine. It forms colorless needles; m.p. 117 °C, b.p. 300 °C. 1,10-P. is soluble in hot water, alcohol, acetone and benzene. It can be obtained by the Skraup synthesis, in which *o*-phenylenediamine is heated with glycerol, nitrobenzene and concentrated sulfuric acid. It forms nearly insoluble complexes with a number of heavy metal ions, which can be used for the quantitative determination of these metals. The iron(II) complex (see Ferroin) serves as a redox indicator.

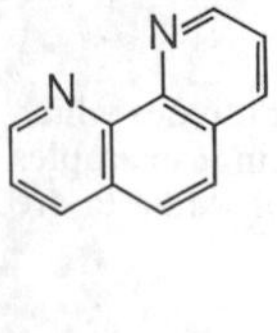

Phenazine, ***dibenzopyrazine***: a heterocyclic compound. P. forms yellow needles; m.p. 171 °C, b.p. 360 °C. It is slightly soluble in alcohol, ether and benzene, and insoluble in water. P. sublimes readily. It is a weak base, and dissolves in concentrated acids by forming salts. P. is synthesized by condensation of *o*-benzoquinone with *o*-phenylenediamine.

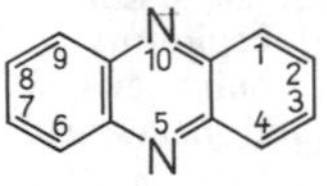

Phenazine pigments: same as Azine pigments (see).

Phenazone: see Pyrazolone.

Phendimetrazin: see Appetite suppressants.

Phenes: condensed aromatic hydrocarbons in which the benzene rings are fused at an angle. See, for example, Phenanthrene and Chrysene.

Phenetidine, ***aminophenetols, ethoxyanilines, aminophenyl ethyl ethers***: the three isomeric ethoxy derivatives of aniline. ***o-Phenetidine (2-ethoxyaniline)***, m.p. < -20 °C, b.p. 232.5 °C, n_D^{20} 1.5560. ***m-Phenetidine (3-ethoxyaniline)***, b.p. 248 °C. ***p-Phenetidine (4-ethoxyaniline)***, m.p. 2.4 °C, b.p. 254 °C, n_D^{20} 1.5528.

P. are oily liquids which rapidly turn brown in the presence of light and air. They are insoluble in water, but dissolve readily in most organic solvents. P. are blood poisons and can be absorbed through the skin. They can be synthesized by reduction of the corresponding nitrophenetols. They are used in the synthesis of pigments and pharmaceuticals. The 4-isomer is used, for example, for the synthesis of phenacetin and artificial sweeteners.

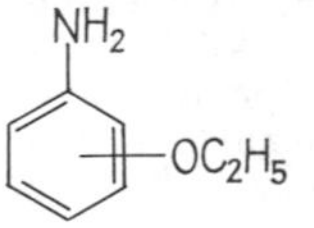

NH_2 — OC_2H_5 (Position 2, 3 or 4)

Phenetol, ***ethyl phenyl ether, ethoxybenzene***: C_6H_5-OC_2H_5, a colorless liquid with a pleasant odor; m.p. -29.5 °C, b.p. 170-172 °C, n_D^{20} 1.5076. P. is insoluble in water but readily soluble in ethanol and ether. It is synthesized by the reaction of diethyl sulfate and phenol in alkaline medium. P. can be used as an intermediate for organic syntheses and as a component in perfumes.

Phenkapton: see Organophosphate insecticides; see Acaricides.

Phenmedipham: see Carbanilate herbicides.

Phenobarbital: see Barbitals.

Phenol, ***hydroxybenzene***: the parent compound of the Phenols (see). P. crystallizes in colorless needles which turn reddish in air and tend to deliquesce; m.p. 43 °C, b.p. 181.7 °C, n_D^{20} 1.5408. P. is hygroscopic and dissolves in either a small amount of water or a very large amount. In between, there is a mixing gap. The aqueous solution is called *carbolic acid*. P. is readily soluble in ethanol and aqueous alkali hydroxide solutions, but is only slightly soluble in alkanes.

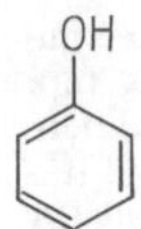

P. is a strong protoplasm toxin; the lethal dose for a human being is 10 to 30 g. It burns the skin, but there is no sensation of pain, because it is also an anaesthetic. Acute poisonings lead to delirium, respiratory paralysis and cardiac arrest. Chronic poisoning leads to kidney damage with albuminuria and hematuria. Other P. have similar effects.

In very dilute solution, P. acts as an antiseptic and disinfectant. It is a weak acid; its salts are the Phenolates (see). P. is very reactive, and can form esters and ethers. Some well-known ethers are, for example, anisol and phenetol, and some important esters are the phosphate ester, which is used as a softener, and the acetate, which can rearrange to 2- or 4-hydroxyacetophenone (see Fries rearrangement). A large number of derivatives can be made by electrophilic substitution (see Phenols). These include chloro- and bromophenols, nitrophenols, picric acid and its salts, phenoxyacetic acids, phenolsulfonic acids, salicylic acid and salicylaldehyde, and phenol-formaldehyde synthetic resins. Catalytic hydrogenation of the benzene ring produces cyclohexanol, and further reaction leads to ε-caprolactam.

Analytical. The classical analytical detection reactions are the iron(III) chloride color reaction (violet color) and the Liebermann reaction with sodium nitrite and sulfuric acid, which yields a red indophenol. This compound forms a blue alkali salt. Further reactions, see Phenols.

Occurrence and production. P. is found in tars and the waste water from coking and cracking plants. However, the amounts produced from these sources is insignificant compared to the amounts synthesized. The introduction of the hydroxyl group to the ring is indirect in all industrial P. syntheses. The first five of the following methods are the most common: 1) Acid cleavage from cumene hydroperoxide (see Hock process). 2) Alkali fusion of the sodium salts of arene sulfonic acids at about 300 °C: C_6H_5-SO_3Na + 2 NaOH → C_6H_5-ONa + Na_2SO_3 + H_2O. Free P. is obtained by addition of acid. (This was the first industrial synthesis, from 1889). 3) Hydrolysis of haloarenes, especially chlorobenzene, with sodium hydroxide or sodium carbonate at 200-250 °C and 2-5 MPa in a flow reactor with a length greater than 1000 m: C_6H_5-Cl + 2 NaOH → C_6H_5-ONa + NaCl + H_2O. The main disadvantage of this method is that for each mole of P., a mole of chlorine is required, which appears as NaCl and interferes with the process. A much less expensive method, in terms of chlorine consumption, is the Raschig process (see). 4) *Toluene-benzoic acid process.* Toluene is oxidized to benzoic acid in a first step, and then oxidatively decarboxylated to carbolic acid in a second step. 5) *Scientific Design process.* A mixture of cyclohexanone and cyclohexanol, such as is produced by cyclohexane oxidation, is dehydrogenated to P. at 400 °C. 6) Hydrolysis of arene diazonium salts with boiling water ("cooking" of diazonium salts): C_6H_5-N≡N^+Cl^- + H_2O -→ C_6H_5-OH + N_2 + HCl. The diazonium salts are easily made from aromatic amines. 7) Oxidation of aryl magnesium halides:

$$C_6H_5\text{–}MgX \xrightarrow{O_2} C_6H_5\text{–}OOMgX \xrightarrow{C_6H_5\text{–}MgX}$$

$$2\,C_6H_5\text{–}OMgX \xrightarrow{H_2O} 2\,C_6H_5OH.$$

There are several newer methods which may be of interest in the future: 1) acetoxylation of benzene as an alternative to oxychlorination: C_6H_6 + CH_3 COOH + $1/2O_2$ → C_6H_5-O-CO-CH_3 + H_2O; C_6H_5-O-CO-CH_3 + H_2O → C_6H_5-OH + CH_3-COOH. 2) In analogy with the Hock process (see), ethene rather than propene is used to alkylate the benzene. Cleavage of the corresponding hydroperoxide on a nickel complex catalyst yields P. and acetaldehyde.

Applications. P. is used as the starting material for synthesis of many organic intermediates (see above), pigments, insecticides, herbicides, drugs, wood protecting materials, tannins, lubricants and explosives.

Historical. P. was discovered in 1834 in coal tar by Runge. In 1889, the first industrial synthesis on a large scale was begun.

Phenolaldehydes: aromatic aldehydes which contain hydroxyl groups in addition to the aldehyde group on the aromatic ring, for example, salicylaldehyde, 3- and 4-hydroxybenzaldehyde and protocatechualdehyde. With the exception of salicylaldehyde, P. are crystalline compounds which are readily soluble in most organic solvents. Only those which contain several hydroxyl groups dissolve to a significant extent in water. In the 2-hydroxybenzaldehydes, there is a relatively strong intramolecular hydrogen bond, which is responsible for the lower melting and boiling points, higher vapor pressures and lower heats of combustion of isomeric P. relative to the 2-hydroxy compounds. These P. are volatile with steam and form chelates with heavy metal ions. The P. can be synthesized by the Gattermann synthesis (see) or the Reimer-Tiemann reaction (see).

Phenolates: salts of the phenols formed by replacement of the H atom of the hydroxyl group by metals. The alkali salts are formed by treatment of phenols with alkali hydroxide solutions. They are much more stable than the corresponding alkali alcoholates, but they are largely hydrolysed in aqueous solution: $C_6H_5OH + NaOH \rightleftharpoons C_3H_5 \cdot O^- Na^+$. As a result, they give a basic reaction. Phenolate ions are nucleophilic reagents and play a part in the synthesis of phenol ethers. The formation of P. depends on the solubility of various phenols in alkali hydroxide solutions. Mineral acids, carboxylic acids and carbon dioxide release the phenols from P. solution. The decomposition of P. solutions by carbon dioxide is utilized in industry and for analysis, for example, for separation of phenols and carboxylic acids. Solutions of P. are also used in industry for extractions.

Phenolcarboxylic acids: see Hydroxybenzoic acids.

Phenol esters: see Phenols.

Phenol ethers: see Phenols.

Phenol ether aldehydes: organic aldehydes which have an alkoxy group on an aromatic ring; examples are anisaldehyde, vanillin and piperonal. Most P. are pleasant smelling compounds which are often used as flavorings and perfumes. They can be obtained synthetically by the Gattermann synthesis (see) or the Vilsmeier-Haack reaction (see).

Phenolphthalein, ***3,3-bis(4-hydroxyphenyl) phthalide***: the lactone of 4',4",α-trihydroxytriphenylmethane-2-carboxylic acid, which cannot be isolated as such. P. forms colorless, odorless crystals; m.p. 262-263 °C. It is soluble in alcohol, ether, acetone, chloroform, pyridine and alkali hydroxide solutions, but insoluble in water. It is made by condensation of 2 mol phenol with phthalic anhydride in the presence of concentrated sulfuric acid. The resulting 4',4",α-trihydroxytriphenylmethane-2-carboxylic acid splits off water under the acidic reaction conditions and is converted to the lactone, P. In acidic and neutral media, phenolphthalein solution is colorless, but in dilute alkalies, there is a color change to carmine red in the pH range 8.2 to 10. The colorless lactone is first hydrolysed to the 4',4",α-trihydroxytriphenylmethane-2-carboxylic acid structure, then splits off water to form a quinoid structure. The red color is due to the anion, which contains resonance-stabilized benzoid and quinoid ring systems. The color change can be reversed by adding acid. In concentrated sodium hydroxide solutions, P. is decolored, because it forms the trisodium salt of the benzoid carbinol form. Because of its color change from colorless to red, P. is used as an indicator in alkalimetry, as a 1% alcoholic solution or as an indicator paper (not applicable for ammonia). In electronics, it is used in the form of a pole reagent paper. In medicine, P. is used as a laxative. It was first synthesized in 1871 by A. von Baeyer.

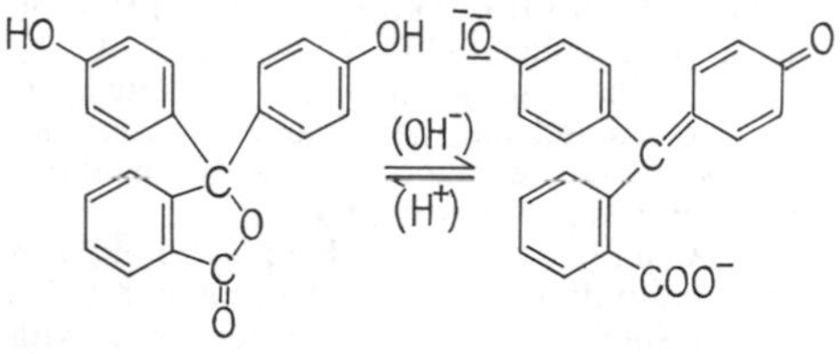

$$\xrightleftharpoons[(H^+)]{(OH^-)}$$

colorless red

Phenol resins, ***phenoplastics***: hard plastics made by polycondensation of phenol (for high-quality P.), cresols (for laminated materials), xylenols (for low-quality P.) or a mixture of these with aldehydes, especially formaldehyde. The condensation reaction goes via the 1,2- and 1,4-methylphenols as intermediates; more phenol units are then added with loss of water. This is followed by formation of more hydroxymethyl groups (CH_2OH) by addition of formaldehyde to the new phenols, and so on, to form linear or branched macromolecular products:

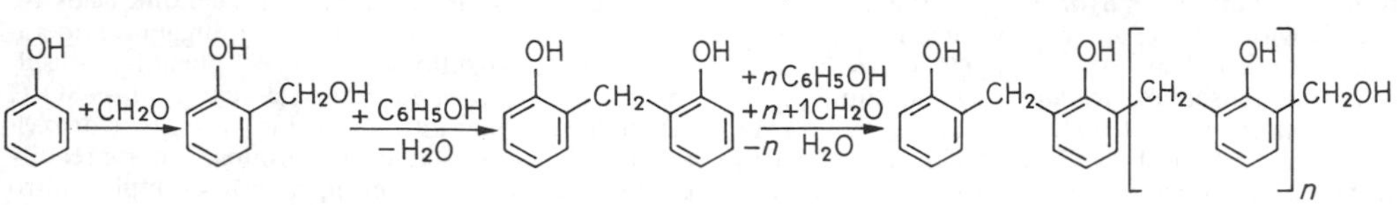

$$\xrightarrow{+CH_2O} \quad \xrightarrow[-H_2O]{+C_6H_5OH} \quad \xrightarrow[-n\ H_2O]{+nC_6H_5OH,\ +n+1CH_2O}$$

Since the formation of hydroxymethyl groups can occur only at the 2 or 4 position relative to the OH group of the phenol, 1,3-cresol, 1,3,5-xylenol or resorcinol (1,3-dihydroxybenzene) are the preferred starting materials. The type of P. produced depends on the starting materials and the reaction conditions, such as condensation time, relative amounts of reactants, reaction and type of catalyst, so the properties of the product can be tailored to the application.

1) In acidic condensation, the phenols and formaldehyde react in a 5:4 ratio in the presence of a small amount of mineral acid, usually sulfuric acid, and the products are linear structures which do not have any free hydroxymethyl groups. Such products are called ***novolakes***; they can be hardened by addition of hexamethylenetetraamine, which decomposes when heated to form formaldehyde (which permits renewed formation of hydroxymethyl groups and thus a cross-linking of the linear chains) and ammonia (which, as an alkaline compound, supports this further condensation).

2) In alkaline condensation, the phenols are mixed with formaldehyde in a 1:1 to 1:2.5 ratio in the presence of alkaline catalysts, such as ammonia, sodium hydroxide, alkaline earth oxides or amines. This reaction occurs in three steps. First the hydroxymethyl groups are gradually converted to methylene bridges, $-CH_2-$, by loss of water. These link the molecular components together and solidify them. At temperatures up to 100 °C, the product is the soluble, readily melted thermoplastic A resin (resol), which when heated to 150 °C undergoes further condensation to yield B resin (resitol). This is barely soluble, and is elastic-plastic only when heated. The last stage is reached by hardening at temperatures up to 200 °C; the C-resin (resite) consists of molecules cross-linked in three dimensions. It is therefore completely insoluble and cannot melt. Hexamethylenetetraamine is often added to the B resin, which gives added strength to the hardened product by increasing the amount of cross linking.

Properties. The P. have no odor or taste, but since they usually still contain small amounts of free phenol, they are not totally harmless. These unreacted phenolic components also cause the products to darken with time, and for this reason P. are usually dyed dark brown or black before further processing. They are not attacked by water, organic solvents, acids or dilute alkalies. Their density is 1.25 g/cm^3, and their hardness is the same as that of copper. They are poor conductors of heat and electricity.

Applications. The novolakes obtained by acid condensation are used mainly for paints and impregnating compounds, in production of floor coverings and, in combination with fillers and hexamethylenetetraamine, as hardeners for pressed boards. P. made by alkaline condensation are used to some extent as casting plastics; the resins are poured into forms in the resol stage, and then hardened by slow heating to translucent objects. These may also be colored. For most purposes, up to 50% filler is added to the P. to improve its mechanical properties; asbestos, cotton, cellulose, sawdust, magnesium oxide or magnesium stearate can be used as fillers. Resol or resitol and the fillers are made into parts such as knobs, containers, housings for radios and cameras, automobile bodies. Resitol is added as binder to make pressed wood, and other pressed materials. A hardened mixture of resol and asbestos is used in many applications in the chemical industry. A special paper impregnated with resitol can be used for water-resistant gluing of wood. A mixture of resol, benzyl alcohol, quartz sand and acid is used as an acid-resistant putty which is often used in the chemical industry. P. are used as glues; both cold-hardening and hot-hardening varieties are available.

In general, the cold-hardening P. glues are dissolved in solvents and harden when a highly acidic hardener is added; heating is not necessary. The hot-hardening varieties are dissolved in water without hardener and must be heated to 125-160 °C to harden them.

In the paint industry, ***plastified P.*** are used; these are made by partial ether formation on the hydroxymethyl groups and are used as bake-on paints. Modified P. combined with colophony, glycerol and abietic acid are used as paints which mix well with drying oils and can be used in nitro, spirit and oil paints. The ***alkylphenol resins***, obtained by condensation of 4-alkylphenols with formaldehyde, are also soluble in oil. Finally, P. are also used as ion-exchange resins.

Historical. A. v. Baeyer carried out the condensation of phenol and formaldehyde in acidic solution as early as 1885. In 1909, L.H. Baekeland made the first pressing powders by stepwise condensation. The production of P. was started in 1912 under the trade name Bakelite. The P. were the first synthetic plastics.

Phenols: organic hydroxy compounds in which one or more OH groups are bound directly to sp^2-hybridized C atoms of arene rings. The parent of this class of compounds is Phenol (see). Compounds such as phenol, *o*-, *m*- and *p*-cresol, thymol, carvacrol and α- and β-naphthol are monophenols; pyrocatechol, resorcinol and hydroquinone are diphenols, and phloroglucinol and pyrogallol are triphenols. The wide variety of P. arises from the variety of arenes, from which P. are derived formally by substitution of one or more H atoms; they can also actually be synthesized in this fashion. Most P. have trivial names. In the formation of systematic names, the hydroxyl group is considered a substituent of the arene ring. For example, hydroquinone is 1,4-dihydroxybenzene, and α-naphthol is 1-hydroxynaphthalene.

Properties. The P. are solid, crystalline compounds, many of which have a characteristic odor (as phenol itself does). Their solubilities in water vary, depending on the number of hydrophilic hydroxyl groups. P. with several hydroxyl groups are generally miscible with water in any proportions. P. are readily soluble in ethanol and ether. Many P. act as disinfectants and antiseptics, usually at high dilution. In higher concentration, they are toxic (see Phenol). In contrast to other alcohols, P. are weakly acidic, which can be easily shown for aqueous solutions with indicator paper. It can be shown that those P. which are not soluble in water are acidic by dissolving them in alkali hydroxide solutions and precipitation with mineral acids. The salts of P. are called Phenolates (see).

Phenol has a pK_a value of about 10. It is thus less acidic than carbonic acid and can therefore be precipitated from phenolate solutions with carbonic acid. This reaction generally distinguishes P. from more acidic carboxylic acids, which can be released from their salts only by acids with $pK_a < 3$. Many known P. give color reactions with aqueous iron(III) chloride solutions, due to the formation of iron complexes (phenol, violet; cresols, blue; pyrocatechol, green); these can be used to detect these compounds.

Reactions. These compounds can undergo reactions of the phenolic hydroxyl group (I) and of the arene ring system (II), which is destabilized by the OH group and readily undergoes all manner of electrophilic substitution reactions. The first group of reactions (I) includes the following, in addition to formation of salts:

1) *Acylation*. P. react readily with acyl chlorides to form *phenol esters*. Acid anhydrides can also be used, but not carboxylic acids:

$$C_6H_5–OH + R–COCl \xrightarrow[-HCl]{} C_6H_5–O–CO–R.$$

Phenol esters are especially interesting because they can undergo Fries rearrangements to form *o*- or *p*-hydroxyphenylketones.

2) *Alkylation*. In strongly alkaline media, P. react with various alkylation reagents such as iodomethane, or dimethyl or diethyl sulfate, to give *phenol ethers (alkyoxyarenes)* with the general formula R-O-R', where R is an aryl group and R' is an alkyl group, for example, $C_6H_5\text{-}O^- + CH_3I \rightarrow C_6H_5\text{-}OCH_3 + I^-$. P. can also be converted to methyl aryl ethers with diazomethane. These compounds are insoluble in water and resistant to bases. The methoxy and ethoxy arenes are important as aromas.

3) *Halogen substitution* of the hydroxyl group can be achieved with phosphorus(V) chloride, but it also produces phosphoric acid esters. For example, with phenol only a small amount of chlorobenzene is formed; the main product is triphenylphosphate: $C_6H_5\text{-}OH + PCl_5 \rightarrow C_6H_5\text{-}Cl + (C_6H_5O)_3PO$. On the other hand, with picric acid, the only product is picryl chloride (2,4,6-trinitrochlorobenzene).

The reactions of the arene group (II) include the following electrophilic substitutions in the *o*- and *p*-positions: 1) *Bromination*. The reaction of P. with bromine produces 2,4,6-tribromophenol (detection reaction for phenol). 2- or 4-bromophenol is obtained, for example, by working in carbon disulfide at 0 °C. 2) *Nitration*. Reactions with nitric acid yield 2- or 4-nitrophenol, 2,4-dinitrophenol or 2,4,6-trinitrophenol (picric acid), depending on the concentration of the nitric acid. 3) *Sulfonation*. In reactions with concentrated sulfuric acid, phenol-2-sulfonic acids are formed at room temperature. At higher temperatures, phenol-4-sulfonic acids or phenol-2,4-disulfonic acids are formed. Sulfonic acid groups $-SO_3H$ can thus be easily introduced into various hydroxylarenes. They are good leaving groups, and are readily exchanged for other groups, for example, nitro groups. In addition, the sulfonic acid groups increase the water-solubility, which is very important in the dye industry. 4) *Friedel-Crafts reactions* can be used to alkylate or acylate the arene ring. 5) P. also react with diazonium salts to form azo pigments (*azo coupling*), with formaldehyde to form hydroxymethylphenols (see Lederer-Manasse reaction) and then phenol-formaldehyde synthetic resins, with formaldehyde and secondary amines to form aminomethylphenols, or with chloroform and sodium hydroxide to give 2-formyl compounds (see Reimer-Tiemann reaction). The Gattermann reaction (see) can be used to introduce an aldehyde group, and the Kolbe-Schmitt synthesis (see) yields salicylic acid from phenol and carbon dioxide in the presence of sodium hydroxide.

In addition to the above reactions in groups I and II, the P. can be hydrogenated to alicyclic alcohols, e.g. cyclohexanol from phenol, and there are various possibilities for oxidation, e.g. by means of ammoniacal silver nitrate solution, potassium dichromate and sulfuric acid (Jollens' reagent) or peroxide sulfate in alcoholic solution (Elbs reaction). Thus diphenols with *o*-hydroxyl groups are oxidized to *o*-quinones, and those with hydroxyl groups in the *p*-position, to *p*-quinones.

Analytical. Preliminary tests are the determination of the acidity, the iron(III) chloride reaction and the reducing properties of polyphenols by Jollens' reagent. The compounds are identified as benzoic acid, 4-nitrobenzoic acid or 3,5-dinitrobenzoic acid esters after reaction with the corresponding acyl chlorides (see Alcohols). The IR spectra display bands in the range of 1050-1300 cm^{-1} (C-O valence vibrations), and from 3200 to 3700 cm^{-1} (O-H valence vibrations). The position of the O-H valence vibration depends on the association of the hydroxyl groups; if intramolecular hydrogen bonds are present, the bands are between 3420 and 3590 cm^{-1}, if there is intermolecular association, the bands are between 3200 and 3550 cm^{-1}, and if the hydroxyl groups are free, the bands are between 3590 and 3650 cm^{-1}. In the UV range, P. absorb between 210 and 230 nm and between 270 and 280 nm. The mass spectra of P. generally have a strong molar peak. A typical fragment is the hydroxytropylium ion (M_r = 107).

Occurrence and isolation. P. are present in coal tars, and are extracted from them on an industrial scale. Because of the great demand, however, there are also methods for synthesis (see Phenol).

Phenolic hydroxyl groups are found in many natural products, such as gallic acid, cannabidiol and cannabinol (the psychoactive components of the resin of the *Cannabis sativa* plant), in the yellow, red and blue flavone and anthocyanin flower pigments, and in tannins. These compounds are usually present in the plant cells as glycosides, that is, the phenolic hydroxyl groups are linked to sugar molecules by acetal (glycosidic) bonds.

Applications. P. are important starting materials and intermediates in organic synthesis chemistry and for large-scale industrial processes. For example, one pathway for synthesis of polyamide fibers starts with hydrogenation of phenol to cyclohexanol, and proceeds via cyclohexanone, cyclohexanone oxime and caprolactam. Various diphenols are used in the synthesis of phenol-formaldehyde synthetic resins. Phenol and naphthol are used as couplers for azo, azomethine and indophenol dyes. Phenol, cresols and thymol are used as disinfectants and antiseptics. Finally P. are used in many ways in the synthesis of drugs.

Phenolsulfonic acids, ***hydroxybenzenesulfonic acids***: three isomeric phenol monosulfonic aicds with the general formula $HO\text{-}C_6H_4\text{-}SO_3H$. ***Phenol-2-sulfonic acid***: colorless crystals which gradually turn dark in the light; m.p. about 50 °C (anhydrous compound). The acid is readily soluble in water, and is soluble in ethanol. When heated above its melting point, it decomposes. It gives a blue-violet color with iron(III) chloride. This P. can be made by sulfonation of phenol with concentrated sulfuric acid at room temperature. ***Phenol-3-sulfonic acid***: colorless needles, m.p. 140 °C. The acid is soluble in water and ethanol. It is made by reaction of benzene with conc. sulfuric acid to obtain benzene-1,3-disulfonic acid, followed by treatment with sodium hydroxide solution of an intermediate concentration. The acid is an intermediate for synthesis of tannins, pigments and disinfectants. ***Phenol-4-sulfonic acid***: colorless, hygroscopic needles without a defined melting point. It is synthesized by sulfonation of phenol with conc. sulfuric acid around 100 °C. This acid is used as an antiseptic and to combat parasites.

Phenoplastics: same as Phenol resins.

Phenosolvane process: see Sewage treatment.

Phenothiazines: compounds containing the tricyclic ring system ***phenothiazine*** (Fig.). Phenothiazine forms pale yellow leaflets (m.p. 185 °C, b.p. 370 °C) and is NH acid. It is synthesized by heating diphenylamine with sulfur to 170-190 °C in the presence of iodine or aluminum chloride as catalyst.

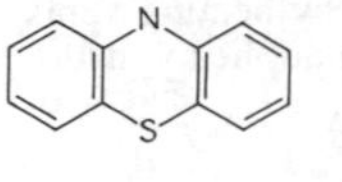

P. are sensitive to oxidation. In the presence of air and light, the solutions are readily discolored, especially if heavy metal ions are present. This property can be utilized for identification.

The phenothiazine ring system is common in pigments, drugs and insecticides. P. are used therapeutically as neuroleptics and sometimes as antiemetics and antihistamines. Low-activity P., such as *promazine* and *chlorophenethazine* are used as antiemetics. *Promethazine* and *dioxopromethazine* are used as antihistamines. All P. used as drugs contain an alkyl group as substituent R (see table) and some contain a substituent such as Cl, CN, CF_3 or $CO\text{-}(CH_2)_2\text{-}CH_3$ as X. The compounds are prepared by basic alkylation of the correspondingly substituted parent compound. For example, promazine can be made by reaction of phenothiazine with the basic alkyl halide $(CH_3)_2N\text{-}CH_2\text{-}CH_2CH_2Cl$ in the presence of sodium amide or sodium hydride.

P. are used as pesticides, for example to combat intestinal worms in sheep.

Table of phenothiazines, p. 812.

Phenothiazin
R = H
X = H

Name	X	R
Promazine (Sinophenin®)	H	$-(CH_2)_3-N(CH_3)_2$
Promethazine (Prothazin®)	H	$-CH_2-CH(CH_3)-N(CH_3)_2$
Chlorpromazine (Propaphenin®)	Cl	$-(CH_2)_3-N(CH_3)_2$
Chlorphenethazine (Marophen®, Elroquil®)	Cl	$-(CH_2)_2-N(CH_3)_2$
Butaperazine (Tyrylen®)	$-CO-(CH_2)_2-CH_3$	$-(CH_2)_3-N(C_4H_8)N-CH_3$
Metofenazate (Frenolon®)	Cl	$-(CH_2)_3-N(C_4H_8)N-CH_2-CH_2-O-C(=O)-C_6H_2(OCH_3)_3$

Phenoxazine, ***dibenzo-4H-1,4-oxazine***: the parent compound of the phenoxazine pigments (see Oxazine pigments). It forms colorless crystals, m.p. 156 °C, and decomposes upon boiling. It is soluble in alcohol, ether, benzene and glacial acetic acid, but insoluble in water. P. is produced by heating a mixture of 2-aminophenol and 2-aminophenol hydrochloride.

Phenoxazine pigments: same as Oxazine pigments (see).

Phenoxyacetic acid: C_6H_5-O-CH_2-COOH; colorless, crystalline needles which smell like honey; m.p. 98 to 99 °C, b.p. 285 °C (dec.). P. is slightly soluble and dissolves easily in glacial acetic acid, alcohol and ether. It is highly antiseptic. It is synthesized by reaction of chloroacetic acid with phenol under alkaline conditions, followed by acidification of the reaction mixture. P. is used as an intermediate in the synthesis of perfumes, pigments and pesticides, and is used itself as a fungicide.

Phenoxybenzene: same as Diphenyl ether.

Phenoxyethanol, ***1-Hydroxy-2-phenoxyethane***: C_6H_5O-CH_2-CH_2-OH, a monoalcohol, and a colorless, viscous liquid; m.p. 14 °C, b.p. 245 °C. P. is slightly soluble in water, and readily soluble in ethanol and ether. It is made industrially by reaction of phenol with ethylene oxide in alkaline solution, and is used in organic synthesis and in perfumes.

Phenoxyfatty acids: see Growth hormone herbicides.

Phenoxymethylpenicillin: see Penicillins.

Phenoxypolymers: polyadducts of diglycide ether and hydroquinone. The terminal epoxide groups are stabilized by addition of monofunctional phenols. P. are used mainly as coatings, but can also be molded.

$$\left[-C_6H_5-O-CH_2-\underset{OH}{CH}-CH_2-O-\right]_n$$

Phenprocoumon: a mono-4-hydroxycoumarin derivative used as an anticoagulant. Its maximum effect is reached after about 60 h, and the effect lasts for 120 to 140 h. *Warfarin*, which is used as a rodenticide, has a similar structure and effect.

Phenyl-: a term for the atomic group C_6H_5- in a molecule, the radical $C_6H_5\cdot$ and the cation $C_6H_5^+$.

Phenylacetaldehyde, ***phenylethanal***: C_6H_5-CH_2-CHO, an oily, colorless liquid; m.p. 33-34 °C, b.p. 193-194 °C, n_D^{20} 1.5293. P. is slightly soluble in water but dissolves readily in most organic solvents, such as ether, ethanol and chloroform. It is very reactive, and undergoes the addition and condensation reactions typical of Aldehydes (see). Under normal conditions, P. polymerizes to form high-molecular-weight compounds. It is synthesized by oxidation of 2-phenylethanol or β-phenyllactic acid, or by Hofmann degradation of cinnamide: C_6H_5-CH=CH-$CONH_2$ → [C_6H_5-CH=CH-NH_2] → C_6H_5-CH_2-CHO. Because its odor is similar to that of hyacinths, P. is used as a perfume.

Phenyl acetate, ***acetic acid phenyl ester, O-acetylphenol***: CH_3-CO-OC_6H_5, a colorless liquid with a phenol-like odor; b.p. 195.7 °C, n_D^{20} 1.5035. E. is slightly soluble in water, but readily soluble in alcohol, ether and chloroform. P. is produced by acetylation of phenol with acetic anhydride or acetyl chloride in the presence of an acidic or basic catalyst. It is used mainly as a solvent.

Phenylacetic acid, ***2-phenylethanoic acid***: C_6H_5-CH_2-COOH, colorless, crystalline leaflets; m.p. 77 °C, b.p. 265.5 °C. P. is slightly soluble in water, but readily soluble in alcohol, ether, acetone and hot water. Impure P. has an unpleasant, clinging odor. It is found in the form of esters in some essential oils. P. is formed in the body as a product of phenylalanine metabolism and can be excreted in the form of an amide with glycine. It can be synthesized by hydrolysis of benzyl cyanide or by a Grignard reaction (see) between benzylmagnesium bromide and carbon dioxide. It is used in the synthesis of penicillin, aromas and drugs.

Phenylacetone, ***1-phenylpropan-2-one***: C_6H_5-CH_2-CO-CH_3, an aliphatic ketone with an aromatic substituent. P. is a colorless liquid with a pleasant odor; m.p. -15 °C, b.p. 216.5 °C, n_D^{20} 1.5168. It is slightly soluble in water, but dissolves readily in most organic solvents. P. can be synthesized by condensation of benzyl cyanide with ethyl acetate, followed by ketone cleavage, or from phenylacetic acid and acetic acid in the presence of thorium oxide. It is used in organic syntheses.

Phenylacetonitrile: same as Benzyl cyanide (see).

β-Phenylacrylic acid: same as Cinnamic acid (see).

Phenylalanine, abb. ***Phe***: α-amino-β-phenylpropionic acid, an essential, proteogenic amino acid (formula and physical properties, see Amino acids). P. can be made from phenylacetaldehyde by the Strecker synthesis. The resulting DL compound is converted to L-P. by enzymatic hydrolysis of the *N*-acetyl derivative. There is an asymmetric synthesis of L-P. starting from benzaldehyde, which is condensed with hydantoin. Hydrogenation of the condensation product with an optically active catalyst yields L-P. in 90% optical yield. P. was first isolated, in 1879, from lupine sprouts by E. Schulze.

Phenylalanine hydroxylase: a monooxygenase which is found in the liver (see Oxygenases). It converts phenylalanine to tyrosine by incorporating an oxygen atom. A lack of P. leads to phenylketonuria, a hereditary disease which, if untreated, leads to mental retardation. A diet low in phenylalanine prevents the symptoms. The urine of newborns is now routinely tested to detect the disease from the excessive excretion of phenylpyruvic acid.

3-Phenylallyl alcohol: same as Cinnamyl alcohol (see).

Phenylamide herbicides: see Herbicides.

Phenylamine: same as Aniline.

N-Phenylaminobenzene: same as Diphenylamine.

N-Phenylalanine: same as Diphenylamine.

Phenyl bromide: same as Bromobenzene (see).

Phenylbutazone: a derivative of pyrazolidine-2,5-dione (Fig.) formed by condensation of hydrazobenzene, C_6H_5-NH-NH-C_6H_5 with n-butylmalonic acid diesters, ROOC-CH(C_4H_9)-COOR. P. forms a water-soluble sodium salt with an enolate structure. It was first developed as a solubilizing agent for aminophenazone (see Pyrazolone), but is itself a strong antipyretic and is used as such therapeutically.

Kebuzone is one of numerous derivatives of P., which is supposed to be better tolerated.

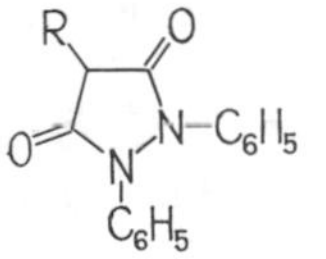

R = CH_3–CH_2–CH_2–CH_2 : Phenylbutazone
R = CH_3–CO–CH_2–CH_2 : Kebuzone

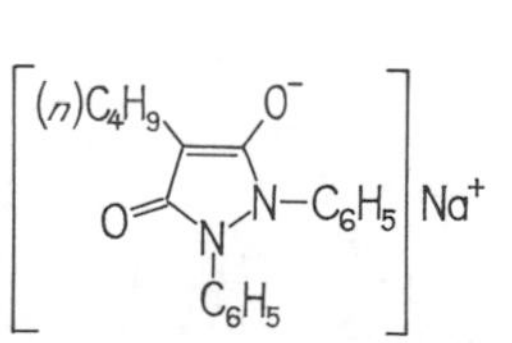

Sodium salt of phenylbutazone

4-Phenylbut-3-en-2-one: same as Benzylacetone (see).

N-Phenylcarbamic acid ethyl ester: same as *N*-phenylurethane.

2-Phenylchromane: same as Flavane (see).

2-Phenylchroman-4-one: see Flavones.

2-Phenylchromone: same as Flavone.

3-Phenylchromone: same as Isoflavone (see).

Phenylcyanide: same as Benzonitrile (see).

2-Phenyl-2,3-dihydro-4H-1-benzopyran-4-one: see Flavones.

Phenylene blue: see Indamines.

Phenylenediamine: see Diaminobenzene,

Phenylene thiourea fungicides: see Benzimidazole fungicides.

Phenylethanal: same as Phenylacetaldehyde (see).

2-Phenylethanoic acid: same as Phenylacetic acid (see).

Phenylethene: same as Styrene (see).

Phenylethylamine: see Biogenic amines.

Phenylethyl alcohol, ***2-phenyl-ethan-1-ol***: C_6H_5-CH_2-CH_2-OH, a liquid with a pleasant, rose-like odor; b.p. 220 °C, n_D^{20} 1.5315. P. is soluble in water and ethanol, is present in rose and orange blossom oils, and can be extracted from them. It can be synthesized by reduction of ethyl phenylacetate or from ethylene oxide and benzene or phenylmagnesium bromide. P. is used in the composition of perfumes.

N-Phenylglycine, ***N-phenylglycocoll, anilinoacetic acid***: C_6H_5-NH-CH_2-COOH, a colorless, crystalline compound, m.p. 127-128 °C. N-P. is soluble in water and alcohol. It can be synthesized from aniline and chloroacetic acid, or from aniline, formaldehyde and hydrogen cyanide via anilinoacetonitrile. N-P. is an important intermediate in the the Heumann indigo synthesis (see).

Phenylglycolic acid: same as Mandelic acid (see).

N-Phenylglycocoll: same as *N*-Phenylglycine (see).

Phenylhydrazine: C_6H_5-NH-NH_2, a colorless, oily liquid with a faintly aromatic odor; m.p. 19.8 °C, b.p. 243 °C, n_D^{20} 1.6084. P. is only slightly soluble in water, but dissolves readily in most organic solvents. It is a blood poison and has a strong sensitizing effect on the skin. In sensitive persons, it can cause eczema. So far, no carcinogenic effect in human beings has been observed. P. turns yellow to brown in the presence of air, due to oxidation. It is synthesized by reduction of benzene diazonium chloride with sodium hydrogensulfite or tin(II) chloride. P. is used for qualitative analysis of aldehydes, ketones and sugars, with which it forms crystalline Phenylhydrazones (see) and Osazones (see). It is also used in the synthesis of pigments, heterocyclic compounds and drugs.

Phenylhydrazones: Condensation products of phenylhydrazine with aldehydes and ketones; general formula R^1R^2C=N-NH-C_6H_5. P. generally crystallize well and have sharp melting points, and for this reason are used for isolation, purification and characterization of aldehydes and ketones. The P. of sugars can be obtained by addition of equal amounts of the reactants in the cold; with excess phenylhydrazine, Osazones (see) are formed. P. can be decomposed into the starting materials by boiling with dilute mineral acids. The 4-nitro- and 2,4-dinitrophenylhydrazones are often used instead of the P. for identification of aldehydes and ketones, due to the better crystallization of the nitro compounds.

N-Phenylhydroxylamine: C_6H_5-NH-OH, colorless needles; m.p. 83-84 °C. *N*-P. is slightly soluble in water, and readily soluble in alcohol, ether, hot benzene and chloroform. It is a strong skin poison and is synthesized by reducing nitrobenzene with zinc in acetic acid or ammonium chloride solution. *N*-P. is used for organic syntheses.

Phenyl iodide: same as Iodobenzene (see).

Phenylisocyanate: C_6H_5-N=C=O, the phenyl ester of isocyanic acid. P. is a colorless, lacrimatory liquid; m.p. -33 °C, b.p. 55 °C at 1.73 kPa, n_D^{20} 1.5368. It is soluble in ether. In the presence of moisture, it is decomposed; it reacts with ethanol to form phenylurethane. P. can be synthesized by Curtius degradation (see) of benzazide or by reaction of aniline with phosgene. P. is used as a reagent to detect alcohols and amines, to make pharmaceutical products and as a softener.

Phenylisothiocyanate: C_6H_5-N=C=S, a colorless, poisonous liquid with a mustard-like odor; m.p. 21 °C, b.p. 221 °C, n_D^{23} 1.6492. P. is steam volatile. It is insoluble in water, but is soluble in alcohol, ether and chloroform. P. undergoes the reactions typical of Isothiocyanates (see). It can be synthesized from aniline and thiophosgene or from *N*-phenyldithiocarbamates and esters of chloroformic acid.

$$C_6H_5\text{–NH–CS–S}^- + \text{Cl–CO–OR} \xrightarrow[-\text{Cl}^-]{} $$
$$C_6H_5\text{–NH–CS–S–CO–OR} \rightarrow C_6H_5\text{–NCS} + \text{ROH} + \text{COS}.$$

P. is used for organic syntheses and to identify compounds with acidic H atoms.

Phenylmercury borate: an organomercury compound used to disinfect the skin and mucous membranes, and as a preservative for eye medications. The compound is sometimes formulated as the formic acid monoester and used together with phenylmercury hydroxide, C_6H_5-Hg-OH.

3-Phenylpropanal: same as hydrocinnamaldehyde (see).

Phenylpropane derivatives: a large group of secondary metabolic products of plants. The P. include the phenylacrylic acids (derivatives of cinnamic acid), coumarin derivatives, phenylallyl derivatives, lignanes and lignin. Many of the phenylcarboxylic acids in plants are also derived from the P.

1-Phenylpropan-2-one: same as Phenylacetone (see).

3-Phenylpropenal: same as Cinnamaldehyde (see).

(E)-3-Phenylpropenic acid: same as Cinnamic acid (see).

3-Phenylpropionic acid: same as hydrocinnamic acid (see).

N-Phenylurethane, ***N-phenylcarbamic acid ethyl ester, ethyl-N-phenylcarbamate***: C_6H_5-NH-CO-OC_2H_5, colorless crystals with an aromatic odor; m.p. 52-53 °C, b.p. 238 °C (dec.). *N*-P. is insoluble in cold water, but dissolves easily in alcohol and ether. It can be made from ethyl chloroformate and aniline, or from phenylisocyanate and ethanol: C_6H_5-N=C=O + C_2H_4-OH → C_6H_5-NH-CO-OC_2H_5. It was formerly used as a medicine to reduce fever and an antirheumatic.

Phenytoin: see Hydantoin.

Pheophorbides: see Chlorophylls.

Pheophytins: see Chlorophylls.

Pherogram: see Electrophoresis.

Pheromones: substances which are used for communication between individuals of a species. P. affect sexual behavior, aggregation, alarm behavior and other behaviors of the animals. They are produced in extremely small quantities, often from secondary metabolic products of the host plants of the animals. They are usually complex mixtures, which can be separated by gas or high-pressure liquid chromatography. The best studied of the P. are from ***insects***. P. elicit either an immediate, but relatively short-lived reaction from the recipient, or a long-term physiological change. The first type of P. act via the odor receptors, as "*releasers*", and the second type are orally active "*primers*". It is thought that P. are phylogenetically very old, and that they are functional precursors of hormones. In recent years, they have been utilized as ***attractants*** to lure pest species of insects into traps.

H₃C ... CH₂OH (1)
R ... O–COCH₃ (2: R = CH₃; 3: R = (CH₂)₂CH₃)
H₃C(C=O) ... COOH (4)
H₃C ... (lactone) (5)
H₃C(CH₃) ... O–COCH₃ (6)
H₃C ... CH₂OH (7)
8, 9, 10, 11, 12

Selected pheromones

Chemically, most P. so far studied are either isoprenoid or non-isoprenoid, saturated or unsaturated alcohols, acids, esters, aldehydes or hydrocarbons. Most are acyclic. The non-isoprenoid compounds are thought in most cases to be derived from fatty acids; this group includes most sexual pheromones of female moths, such as *bombykol* (1), and bombykal from the silkworm moth, *Bombyx mori*; (Z)-9-dodecenyl acetate (2) from *Parolobesia viteana*; (Z)-9-tetradecenylacetate (3), a component of the sexual pheromone of *Spodoptera frugiperda* and *Prodenia eridanis*. In some compounds of this group, e.g. bombykol, a few molecules are often enough to elicit a reaction from the male. The active P. in *"queen substance"* from the honeybee, *Apis mellifera*, is (E)-oxo-2-decenoic acid (4). Queen substance, when fed to workers, prevents them from building new queen cells. The P. has also been isolated from other bees. The "queen substance" of the hornet *Vespa orientalis* was identified as γ-hexadecalactone (5). The odor from the broken stinger of a honeybee attracts other bees. The "sting P." was identified as isopentyl acetate (6). The sexual P. of the housefly, *Musca domestica* is (Z)-9-tricosene (7). Because of the great economic damage done by bark beetles, their P. have been rather intensively studied. A combination of monoterpene alcohols was identified as the attractant of the bark beetle *Ips paraconfusus*; the mixture consists of (+)-cis-verbenol (8), (8)-2-methyl-6-methylen-7-octen-4-ol (*ipsenol*, 9) and (+)-2-methyl-6-methylen-2,7-octadien-4-ol (*ipsdienol*, 10). The aggregation pheromone of the spruce bark beetle, *Ips typographus*, was identified as a mixture of stereoisomers of 2-ethyl-1,6-dioxaspiro[4.4]nonane (*chalcogran*, 11).

P. containing nitrogen are rare, but such compounds as *danaidon* (12) have been found in butterflies of the *Danaid* family. However, the precursors of these P. are alkaloids obtained from the plants on which the insects feed.

Phillips process: see Polyethylene.

Phlegmatization: the reduction of the sensitivity of explosives to impact or vibrations by addition of water, oil, paraffin, wax, etc. The addition of these substances inhibits transport of heat from one crystal of the explosive to the next. P. can also be used in highly exothermal reactions to keep the reaction temperature under control.

Phlobaphenes, ***tannin reds***: components of a few condensed tannins with high molecular weights. For example, catechols can condense with one another by opening the oxygen bridge of the heterocyclic ring and forming new phenolic hydroxyl groups. P. are soluble in hot water and precipitate on cooling of the aqueous solution as red-brown, amorphous solids. They can be made soluble in cold water by sulfite treatment, which is sometimes used in the production of tannin extracts.

Phloretin: see Phlorhizin.

Phlorhizin, ***phloridzin***: a phloretin-2'-β-D-glucoside which forms colorless crystals; m.p. 110 °C, $[\alpha]_D^{20}$ -53.8° (in 50% ethanol). P. is readily soluble in alcohol, acetone and dilute alkali hydroxide solutions. It dissolves in boiling water, but is practically insoluble in ether, benzene or chloroform.

HO OH OH Glc-O O

P. is found, for example, in the bark of apple and cherry trees. When it is boiled with acid, P. is split into phloretin and glucose. The aglycon *phloretin* is a dihydrochalcone (see Flavanoids). P. inhibits the transport of aldoses through biological membranes.

Phloridzin: same as Phlorhizin (see).

Phloroglucinol, ***benzene-1,3,5-triol***: a symmetrical phenol with three hydroxyl groups. It crystallizes with two moles of water of crystallization in the form of colorless platelets or leaflets; m.p. 113 to 116 °C (hydrated form), or 218-219 °C (anhydrous); it sublimes with decomposition. P. is scarcely soluble in water, and readily soluble in ethanol and ether. It gives a violet color reaction with iron(III) chloride in dilute aqueous solution. Since it gives an intense red color with lignin and hydrochloric acid, it is used to detect lignin, e.g. in paper. P. reacts both as an hydroxy compound (e.g. it is acetylated by acetic anhydride) and as a tautomeric tricarbonyl compound; it can form a trioxime with hydroxylamine. P. is found in the form of phloridin, a glucoside, in the bark of apple and plum trees. It can also be considered a component of flavone and anthocyanin pigments. P. can be obtained by alkali fusion of benzene-1,3,5-trisulfonic acid or by hydroxylation of resorcinol with alkali hydroxides. It is used in the synthesis of pigments, phenol plastics and pharmaceuticals.

OH HO OH

Phloxin: see Eosine pigments.

Pholedrin: $4(OH)C_6H_4$-CH_2-$CH(CH_3)$-NH-CH_3, a phenylethylamine derivative with sympathicomimetic effects. It is a vasoconstrictor used for hypotonia and to stimulate the circulation.

Phorates: see Organophosphate insecticides.

Phorbol: a diterpene alcohol; its esters are components of croton oil and act as cocarcinogens.

Phoron, ***2,6-dimethylhepta-2,5-dien-4-one, diisopropylideneacetone***: $(CH_3)_2C$=CH-CO-CH=C$(CH_3)_2$, a doubly unsaturated, aliphatic ketone. P. crystallizes in yellow-green crystals which smell somewhat like geraniums; m.p. 28 °C, b.p. 197.8 °C. It is nearly insoluble in water, but dissolves readily in alcohol and ether. It is easily cleaved by mineral acids. P. can be synthesized by the reaction of hydrogen chloride with acetone. P. is used as a solvent for cellulose nitrate and lacquer, and or organic syntheses.

Phosgene, ***carbon oxygen chloride, carbonyl dichloride***: $COCl_2$, a colorless gas with an odor like hay; M_r 98.92, density (liq.) 1.392, m.p. -104 °C, b.p. 8.3 °C, crit. temp. 181.9 °C, crit. pressure 5.79 MPa.

As the chloride of carbonic acid, P. reacts with water gradually, forming CO_2 and hydrochloric acid, HCl. It reacts instantaneously with ammonia to form urea, H_2N-CO-NH_2, and ammonium chloride, NH_4Cl.

> Phosgene is a nasty lung poison. After inhalation, an interval of time passes, often several hours, before hydrolysis of the $O{=}CCl_2$ to CO_2 and HCl leads to toxic lung edema. Symptoms: depending on the concentration, irritation of the mucous membranes of the eyes, nose and throat, increasing respiratory distress, tortured coughing, cyanosis. Therapy: complete rest; hexamethylenetetramine for prophylaxis.

The compound is synthesized industrially by reaction of carbon monoxide with chlorine in the presence of activated charcoal. P. is used mainly for the industrial synthesis of isocyanates and polycarbonates, for halogenation and as a versatile reagent for organic synthesis. During the First World War, it was used as a Chemical weapon (see).

Phosgenite: see Lead chlorides.

Phosmet: see Organophosphate insecticides.

Phosphaalkenes: see Organophosphorus compounds.

Phosphaalkynes: see Organophosphorus compounds.

Phosphamidon: see Organophosphate insecticides.

Phosphane, ***phosphine, monophosphane***: PH_3, a colorless and extraordinarily poisonous gas with an odor like garlic. It has a pyramidal molecule; M_r 34.00, m.p. -133.8 °C, b.p. -87.7 °C. PH_3 is slightly soluble in water. In the air, it burns to form phosphorus(V) oxide, but when pure it is not pyrophoric. It is a very weak base (pK_b = 28) and is therefore bound in phosphonium salts only by very strong acids. The salts are almost completely hydrolysed to PH_3 by water. On the other hand, PH_3 can act as an acid. For example, if it is passed through a solution of sodium in liquid ammonia, it is instantly converted to sodium phosphide, $NaPH_2$ and hydrogen (see Phosphides). PH_3 is a strong reducing agent, and is able to reduce many metal salts to the free metals. Its alkyl and aryl derivatives are called ***phosphines*** (see Organophosphorus compounds). PH_3 is made by the reaction of water or dilute acid with metal phosphides, e.g. $Ca_3P_2 + 6\,H_2O \rightarrow 2\,PH_3 + 3\,Ca(OH)_2$, or by disproportionation of white phosphorus in alkali hydroxide solutions, e.g. $P_4 + 3\,KOH + 3\,H_2O \rightarrow PH_3 + 3\,KH_2PO_2$. P. is used as a pesticide, and as a doping gas in semiconductor technology. ***Diphosphane, diphosphine***, P_2H_4, is a colorless liquid; M_r 65.98, m.p. - 99.8 °C, b.p. 52 °C. P_2H_4 is pyrophoric. It is usually formed as a byproduct of production of PH_3, and is responsible for the self-ignitability of technical phosphane. In addition to PH_3 and P_2H_4, there are many known P. with the composition P_nH_{n+2} and P_nH_n, which have chain or cyclic structures.

Phosphatases: a widely occuring group of esterases which catalyse the cleavage of phosphate monoesters. The P. are most abundant in the liver, spleen, pancreas and digestive tract. Most are dimeric esterases, and have a catalytic serine group in the active center. The ***acid P.*** (pH optima up to 5) include the P. from liver (M_r 16,000), erythrocytes (M_r 10,000) and prostate (M_r 102,000). The ***alkaline P.*** have pH optima of 7 to 8, and are most abundant in the mucous membranes of the small intestine (M_r 140,000) and placenta (M_r 120,000). The alkaline P. from *Escherichia coli* consists of two chains with an M_r of 43,000 each. The alkaline P. require two zinc atoms per subunit and Mg^{2+} ions for activity. Alkaline P. is inhibited by ethylenediamine acetic acid, inorganic phosphate and L-phenylalanine. The acid P. are inhibited by fluoride ions; D-tartaric acid is a very specific inhibitor of the acid P. from the prostate. The alkaline P. are important in nutritional physiology, because they are involved in numerous reactions of carbohydrate, protein and lipid metabolism. The P. of liver, for example, catalyses the hydrolysis of glucose 6-phosphate, and thus provides free glucose for energy. The determination of acid P. in serum is a diagnostic aid when prostate cancer is suspected, and the alkaline P. are used in diagnosis of bone, liver and gall bladder diseases, such as bone tumors, hepatitis and obstructive jaundice. In such diseases, the serum protease activities are increased several-fold.

The P. also include the highly specific ***nucleotidases*** which cleave 5'- or 3'-nucleotides to the corresponding nucleosides.

Phosphates: 1) the ***salts*** of phosphoric acids; in the narrow sense, ***orthophosphates (monophosphates)***, the salts of orthophosphoric acid, H_3PO_4. Depending on the number of substituted hydrogen atoms, P. are classified as ***primary P., dihydrogen phosphates***, $M^IH_2PO_4$, ***secondary P., hydrogenphosphates***, $M^I_2HPO_4$, and ***tertiary P., neutral P.***, $M^I_3PO_4$. All primary P. and secondary and tertiary alkali P. are soluble in water; however, many heavy metal phosphates are only slightly soluble. When roasted, primary and secondary P. lose water and form polyphosphates or neutral diphosphates. The orthophosphates are made by reaction of metal carbonates or hydroxides with phosphoric acid. Insoluble P. can be precipitated out of the corresponding metal salt solutions by addition of alkali phosphate. The most important use of orthophosphates is as fertilizers.

In the broad sense, P. are the salts of condensed phosphoric acids; this definition includes ***condensed P.*** based on polyanions in which $[PO_4]$ tetrahedra are linked via oxygen bridges. There are three types of condensed P.

a) ***Polyphosphates***, $M^I_{n+2}[P_nO_{3n+1}]$, are obtained by heating primary P., usually in the form of a glass. Polyphosphates are always a mixture of salts with anions of different chain lengths.

$$^{-}O-P(=O)(O^{-})-\left[O-P(=O)(O^{-})\right]_n-O-P(=O)(O^{-})-O^{-}$$

Polyphosphate anion

Polyphosphates up to a chain length of $n = 12$ can be separated by chromatography, and can often be isolated as uniform, crystalline compounds.

b) ***Metaphosphates***, $[M^IPO_3]_n$ (n = 2-8). These salts are based on cyclic polyanions, such as the trimetaphosphate anion. Metaphosphates are always components of the glassy phosphate mixtures obtained by heating primary P. and can be separated from these mixtures.

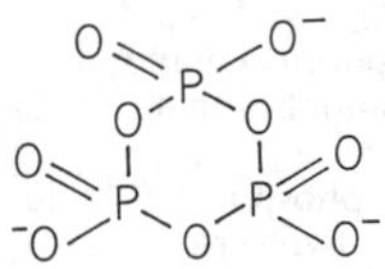

Trimetaphosphate anion

c) ***Cross-linked P., ultraphosphates***. These polyanions contain phosphorus atoms which are linked via three oxygen bridges to other $[PO_4]$ tetrahedra. They are obtained as a glassy mass by heating phosphate mixtures in which the ratio of metal to phosphorus is lower than 1. Such mixtures can be obtained, for example, by dehydration of mixtures of dihydrogen monophosphates and phosphoric acid. On the properties and applications of the various types of P., see Sodium phosphate, Calcium phosphate, etc.

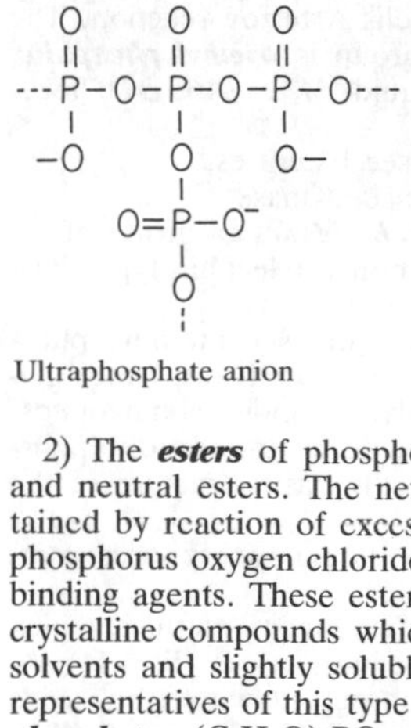

Ultraphosphate anion

2) The ***esters*** of phosphoric acid. There are acid and neutral esters. The neutral triesters can be obtained by reaction of excess alcohol or phenol with phosphorus oxygen chloride in the presence of acid-binding agents. These esters are colorless liquids or crystalline compounds which are soluble in organic solvents and slightly soluble in water. Some simple representatives of this type of compound are ***triethyl phosphate***, $(C_2H_5O)_3PO$, a colorless liquid, m.p. -56.4 °C, b.p. 215-216 °C, n_D^{20} 1.4053. It is used as an ethylating reagent and as a starting material for synthesis of insecticidal phosphate esters. ***Tributyl phosphate***, $(CH_3\text{-}(CH_2)_3\text{-}O)_3PO$, is a colorless liquid, b.p. 289 °C, n_D^{25} 1.4224. This P. is used as a softener for cellulose esters and as a solvent. ***Triphenyl phosphate***, $(C_6H_5O)_3PO$, forms colorless crystals; m.p. 50-51 °C, b.p. 245 °C at 1.465 kPa. It is used as a substitute for camphor in the production of cellulose, and as a softener in the film and lacquer industries. The tricresyl phosphates are especially important.

Of the many synthetic phosphate esters and thioesters, some are used as insecticides, e.g. parathion, and as chemical weapons. Their toxicity is due to phosphorylation (formation of phosphate esters in the organism) and inhibition of the enzyme acetylcholinesterase. On the other hand, certain phosphate esters are biochemically important as metabolic intermediates, e.g. in alcoholic fermentation and in gluconeogenesis, and as components of nucleotides (including AMP, ADP and ATP), the phosphatides (e.g. lecithins and cephalins) and the phosphoproteins in which phosphoric acid is a prosthetic group.

Phosphatidases: same as Phospholipases (see).

Phosphatidic acid: 1,2-diacylglycero-3-phosphoric acid, abb. Ptd, the parent compound of the most abundant glycerophospholipids (see Glycerolipids). P. can be partially or completely synthesized in the laboratory; it is obtained as colorless, waxy and hygroscopic masses. Free P. is not stable, but decomposes autocatalytically. Its salts are stable, however. P. makes up about 1 to 5% of the total phospholipids of cells. It is formed by the action of the enzyme phospholipase D on other glycerophospholipids. P. can be isolated from cabbage or spinach leaves.

Phosphatidylcholine, abb. ***PtdCho***; the natural form is also called ***lecithin***: 1,2-diacyl-*sn*-glycero-3-phosphocholine, the most common glycerophospholipid. The commercially available P. include totally and partly synthetic P. with defined fatty acid esters and melting points as well as lecithins isolated from biological materials, especially ***egg lecithin*** and ***soybean lecithin***. These isolated P. are colorless when freshly isolated, but turn yellow to brown in the air. They are waxy and hygroscopic. P. is soluble in chloroform or chloroform/methanol mixtures and benzene and is less soluble in acetone. It forms highly ordered supermolecular structures in aqueous suspension (model membranes).

Because of their heterogeneous fatty acid composition, P. of biogenic origin do not have defined melting points, and because of their content of unsaturated fatty acids, they tend to peroxidize in the air. In egg lecithin, the major fatty acid component is oleic acid.

Composition of egg and soybean lecithin (in %)

	Egg lecithin	Soybean lecithin
Palmitic acid	35-37	17-21
Stearic acid	9-15	4-6
Oleic acid	33-37	12-15
Linoleic acid	12-17	53-57
Linolenic acid	0.5	6-7
Arachidonic acid	3.7	—
Phosphatidylcholine	73	30
Phosphatidylethanolamine	15	22
Phosphatidylserine	—	3-4
Phosphatidylinositol	1	18
Sphingomyelin	2-3	—
Glycolipids	—	13

The phospholipid fraction of chicken eggs contains varying amounts of other lipids in addition to P., especially phosphatidylethanolamine and sphingomyelin. P., especially the cheaper soybean lecithin, are used to make chocolates and baked goods, as emulsifiers, in margarine, and so on.

The enzyme phospholipase A_2 cleaves the acyl group from the 2-position, forming ***lysolecithin***. Lysolecithin is highly hemolytic, that is, it destroys the membranes of erythrocytes.

Phosphatidylethanolamine: a glycerophospholipid (see Glycerolipids). P. isolated from biological material, such as chicken eggs, is colorless at first, but rapidly turns yellow. It is obtained as an oil or wax which is readily soluble in benzene or chlorform, and poorly soluble in ether, acetone or ethanol. P. is associated with phosphatidylcholine in nearly all organisms. P. isolated from resin contains a high proportion of unsaturated fatty acids.

Phosphatidylglycerol: a glycerophospholipid (see Glycerolipids). P. can be isolated from spinach leaves or microorganisms, or it can be obtained by partial synthesis from egg lecithin as a viscous oil. It is readily soluble in chloroform or benzene. Up to 70% of the total phospholipids of plants are P., but animal tissues contain only small amounts.

Phosphatidylinositols: a group of glycerophospholipids (see Glycerolipids) which contain *myo*-inositol in the head group. The P. include 1-phosphatidylinositol (abb. PtdIns), 1-phosphatidylinositol-4-phosphate and 1-phosphatidylinositol-4,5-bisphosphate. Bacteria, especially mycobacteria, also contain phosphatidylinositolmannoside. P. isolated from biological material have a specific rotation of +5.5°. Unlike most other glycerophospholipids, they are soluble in water. About 2 to 12% of the total phospholipids of eukaryotic cells are P. In general, 1-phosphatidylinositol is most abundant, while the 4-phosphate and 4,5-bisphosphate are found in only small amounts. Oligophosphates are found in erythrocytes.

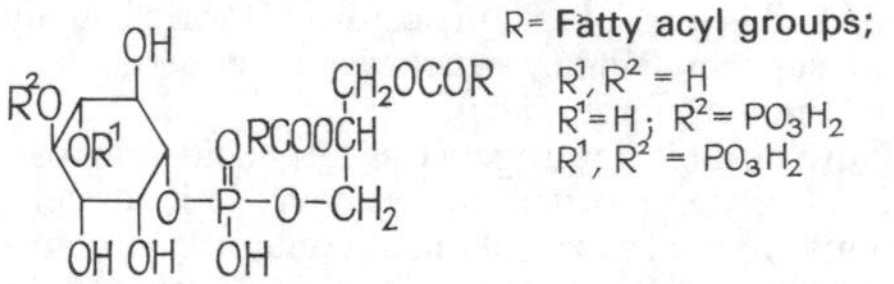

R= Fatty acyl groups;
$R^1, R^2 = H$
$R^1 = H;\ R^2 = PO_3H_2$
$R^1, R^2 = PO_3H_2$

Phosphatidylserine: an acidic glycerophospholipid (see Glycerolipids). P. can be isolated from the brain as a yellowish, easily oxidized oil which is soluble in chloroform and ether and insoluble in ethanol and acetone. P. can exist in several ionogenic forms, which depend on the pH.

Phosphatization: the creation of phosphate layers, from 5 to 20 μm thick, mainly on steel, but also on other metals such as zinc and aluminum. The process provides temporary protection against corrosion and a surface to which paint layers will bind (see Surface preparation). It can also create a lubricating layer which protects the metal during stamping, and increases the lifetime of the tools. The layers can be made by dipping, spraying or electrochemically. Heavy-metal phosphates in phosphoric acid are generally used.

Phosphazenes: see Organophosphorus compounds.

Phosphides: compounds of phosphorus with metals. There are ***ionic P.*** derived formally by substitution of metal cations for the protons of phosphane, PH_3, and ***alloy-like P.***. The alkali and alkaline earth P. and zinc and aluminum phosphide belong to the first category. These react wtih water or dilute acids to give phosphane, e.g. $Mg_3P_2 + 6\ H_2O \rightarrow 2\ PH_3 + 3\ Mg(OH)_2$. Most of the alloy-like P. are high-melting, heat-resistant compounds which are dense and hard, and often show good thermal and electrical conductivity. They include the transition metal P. Their composition varies between M_3P and MP_3. They are not attacked by water, dilute acids or bases. The P. are obtained by the reactions of metals, metal oxides or metal halides with phosphorus, or by reduction of the corresponding phosphates with carbon or hydrogen.

Phosphine: see Phosphane.

Phosphine oxides: see Organophosphate compounds.

Phosphines: see Organophosphorus compounds.

Phosphinic acids: see Organophosphate compounds.

Phosphites: 1) the *salts* of phosphorous acid, H_3PO_3. As a dibasic acid, it forms two series of salts, the ***primary P., hydrogenphosphites***, $M^IH_2PO_3$, and ***secondary P.***, $M^I_2HPO_3$. Except for the alkali phosphites and calcium phosphite, all P. are slightly soluble to insoluble in water; all are strong reducing agents.

2) The ***esters*** of phosphorous acid: ***trialkyl phosphites*** $(RO)_3P$ and ***dialkyl phosphites***, $(RO)_2P(O)H$. The triesters are obtained by reaction of excess alcohol with phosphorous trichloride in the presence of a base which binds the hydrochloric acid as it is produced, e.g. $PCl_3 + 3\ HOC_2H_5 + 3$ amine $\rightarrow$ $P(OC_2H_5)_3 + 3$ amine · HCl. The triesters are important synthesis reagents in organophosphorus chemistry, and are used in particular to synthesis phosphonate esters in the Michaelis-Arbuzov reaction. The most important of this group is ***triethyl phosphite***, $P(OC_2H_5)_3$, a colorless liquid, M_r 166.16, D. 0.9629, b.p. 157.9 °C.

Phosphodiesterases: see Esterases.

Phosphofructokinase: see Kinases.

Phospholipases, *phosphatidases*: a group of esterases which cleave ester bonds in lecithin-type phosphoglycerols.

Phospholipids: complex lipids containing phosphate ester groups. P. are subdivided into glycerophospholipids (see Glycerolipids), sphingophospholipids (see Sphingolipids) and inositolphospholipids on the basis of the basic structure of the lipid.

Phosphonic acids: see Organophosphorus compounds.

Phosphonium salts: compounds with the composition $[R_4P]^+X^-$. The simple inorganic P. (R = H) obtained by the reaction of phosphane with strong acids are unstable. For example, ***phosphonium iodide***, $[PH_4]I$, forms colorless, deliquescent, tetragonal crystals, M_r 161.91, density 2.86, subl.p. 61.8 °C. It is synthesized according to $PH_3 + HI \rightleftharpoons [PH_4]I$, and, due to the low basicity of phosphane, it is hydrolysed in water in a reversal of the equation. Organic P. (R = aliphatic or aromatic group), which are most easily obtained by quaternizing primary, secondary or tertiary phosphines with alkyl halides, are stable, crystalline compounds which are of great importance in organic syntheses, for example as starting materials for the Wittig reaction (see).

Phosphoproteins: see Proteins.

Phosphor, *luminophor*: a synthetic inorganic crystalline compound which can emit light after absorbing energy (UV, X-rays, daylight, corpuscular radiation);

see Luminescence. Natural P. also occur. Only a few synthetic compounds emit light without doping (e.g. magnesium tungstate, calcium tungstate, yttrium vanadate, uranium salts and barium tetracyanoplatinate). Therefore most are doped with foreign atoms which serve as activators and sensitizers. The most important host materials are salts of the alkaline earth metals and their periodic table neighbors, such as phosphates, silicates, aluminates, sulfides, and the halides of alkali and alkaline earth metals. The foreign ions used as *activators* (especially manganese, antimony, lead, tin and rare earth metals) occupy cation sites in the crystal lattice.

Mechanism. After absorption of a quantum of radiation, an electron moves from one of the surrounding anion sites in the lattice to the central cation, and generates a state with high potential energy. The activation energy required to establish equilibrium is taken from the potential energy of the excited state, or it may be supplied by heating the P. (thermoluminescence). As a result, the energy $h\nu$ emitted when the electron jumps back into the valence band from the conduction band is smaller than that absorbed $h\nu'$. If the spectral positions of the exciting radiation and the absorption band do not coincide, highly absorptive foreign atoms (e.g. lead, tin, antimony, cesium) can be introduced as sensitizers; these absorb the energy and pass it on. If the excitation energy is incompletely transferred, the sensitizers themselves emit light, and the luminescence spectrum contains more bands.

Production. The P. used in technology are produced synthetically. Since impurities and lattice defects act as quenching centers, the maximum concentration of impurities should be at least two orders of magnitude lower than the concentration of the activators; in manganese-activated phosphors the manganese concentration is 10^{-2} g-atom per mol compound. The symbol used to denote, for example, a manganese-activated zinc silicate phosphor is $Zn_2SO_4:Mn^{2+}$. Solid crystalline P. are used as thin layers on transparent substrates. They are applied as suspensions (in butyl acetate or water) which are sprayed or rinsed onto the substrate. Adhesion to the substrate is improved by addition of a binder such as nitrocellulose, carboxycellulose or water glass.

Applications. *Phosphor lamps* contain combinations of calcium halogenphosphate:Sb^{3+},Mn^{2+} and strontium magnesium phosphate:Sn^{2+}. More recently, combinations of yttrium oxide:Eu^{3+}, cerium magnesium aluminate:Tb^{3+} and barium magnesium aluminate:Eu^{2+} have been used. For black and white *television tubes*, a mixture of $ZnS:Ag^{+},Ni^{2+}$ and $Zn,CdS:Ag^{+}$ is used; in *color tubes*, yttrium oxide:Eu^{3+} is used for red, yttrium oxide sulfide:Tb^{3+} for green and $ZnS:Ag^{+}$ for blue. In *X-ray machines*, $Ba_3(PO_4)_2:Eu^{2+}$ is used in amplifier layers, and yttrium gadolinium phosphate:Tb^{3+} in the viewing screen. Some other applications are mercury vapor lamps, light-emitting diodes, radar technology, scintillation counters, night lights and radiation dosimeters.

Phosphoranes: see Organophosphorus compounds.

Phosphorescence: a form of photoluminescence (see Luminescence) in which the time between excitation and re-emission of the absorbed radiation is longer than in Fluorescence (see) (> 10^{-4} s). After the excitation has ended, there is a *post-emission* which may last for such a short time that it is difficult to distinguish the phenomenon of P. from fluorescence. On the other hand, in extreme cases, it can last for several hours or even days. Phosphorescent substances can therefore be used as light accumulators which store the energy of light and only gradually release it.

P. is observed most frequently in solid organic compounds (phosphors), although it also occurs in solid solutions of organic compounds. Some substances phosphoresce in liquid media or in the gas state. Like other forms of light emission, P. is the result of electron transitions from higher to lower energy states. It occurs (Fig.) when an electron goes by non-radiative processes from an excited state to a lower, metastable state. From there, the return to the ground state is accompanied by emission of P. radiation, which has a longer wavelength than the excitation radiation. If the metastable state has a long lifetime, as is the case for a few solids, the emission can occur a considerable length of time after the exciting radiation has been turned off.

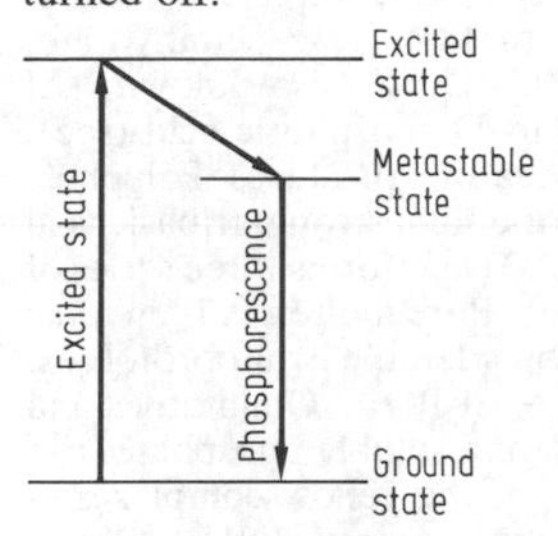

Energy scheme for phosphorescence phenomena.

There are two types of P.: 1) Many inorganic substances phosphoresce in the solid state. Strong P. is associated with the presence of certain impurity atoms in the crystal lattice, the *activators*. Some examples are alkaline earth sulfides, zinc sulfide and cadmium sulfide activated by atoms of heavy metals, such as bismuth, copper, silver, lead, iron or nickel. The P. is a function of the energy state of the crystal as a whole. Various electron transitions can contribute to the P. of these solids; an essential role is played by traps in the crystal lattice which capture and hold the electrons for a time. These traps are created by the activators and have energies in the forbidden zone between the valence band and the conduction band. In order for trapped electrons to move into the conduction band, from which they can return to the valence band by emitting light, energy must be supplied, e.g. in the form of heat. In substances which have long fluorescence times, the energetic distance between the traps and the conduction band is large compared to the mean thermal energy, so that this transition is greatly delayed. This interpretation is in agreement with the fact that P. can be accelerated by heating the sample or irradiating it with infrared light.

2) P. in organic compounds is a property of molecules, and its wavelength is largely independent of the

environment. During excitation, the electrons are raised from the ground state to the first excited state (singlet state). Reversal of this process causes fluorescence. P. occurs when there is a non-radiative transition from the singlet state into a metastable state (a triplet state with two unpaired electron spins). In such a transition, there is a combination of two electron states with different spin multiplicities, associated with spin flipping by an electron. The transition from this metastable state to the ground state requires another spin flip. Triplet-singlet and singlet-triplet transitions are in principle forbidden, but due to spin-orbital coupling, they do occur, with a very low probability. For this reason, the metastable state has a long lifetime, and the emission, when it occurs, is P. rather than fluorescence.

The spectral composition of P. is determined by *phosphorescence spectroscopy*, which is used to study atomic and molecular structures, especially in excited states (see Photochemistry), as well as the structures of solids. P. is exploited industrially in the use of phosphors.

Phosphoric acid, more precisely, ***orthophosphoric acid***: H_3PO_4, hygroscopic, water-soluble, rhombic crystals; M_r 98.00, density 1.834, m.p. 42.35 °C, tetrahedral molecular structure. A semihydrate, $H_3PO_4 \cdot 1/2H_2O$, m.p. 29.3 °C, is also known. At 213 °C, P. is dehydrated to Diphosphoric acid (see), $H_4P_2O_7$, and stronger heating produces Polyphosphoric acids (see). P. is a medium strong, tribasic acid (pK_1 -2.0, pK_2 6.8, pK_3 12.3) and forms three series of salts, the Phosphates (see). Pure molten P. is an ionic conductor, due to considerable autoprotolysis: $2H_3PO_4 \rightleftharpoons [P(OH)_4]^+ + H_2PO_4^-$. Qualitative and quantitative analysis of P. and soluble phosphates can be done by precipitation of the yellow complex ammonium dodecamolybdatophosphate, $(NH_4)_3[PMo_{12}O_{40}]$. Alkalimetric titration of P. is also possible. The acid is synthesized industrially by two methods. ***Wet P.*** is obtained by the reaction of strong acids, usually sulfuric, with crude phosphates. Starting with fluoroapatite, the reaction follows the equation: $Ca_5(PO_4)_3F + 5\,H_2SO_4 + 10\,H_2O \rightarrow 3H_3PO_4 + HF + 5\,CaSO_4 \cdot 2H_2O$. The dilute P. obtained in this way is relatively impure. It can be purified, especially by extraction, and this is done, but it is expensive. More than 80% of the wet P. produced is used in the fertilizer industry. ***Thermal P.*** is obtained by the "dry" process in which white phosphorus is first burned to phosphorus(V) oxide, and this is added to water: $P_4O_{10} + 6\,H_2O \rightarrow 4\,H_3PO_4$. The high purity of this acid permits its use in foods and beverages, for example, to acidify fruit juice drinks and cola, and to make phosphates which are used in foods or feeds. P. is also used to phosphatize metal surfaces, to make phosphates (e.g. for laundry products), etc.

Phosphorous acid, H_3PO_3, M_r 82.00, forms colorless, hydroscopic, water-soluble crystals with D. 1.651, m.p. 73.6 °C, dec. 200 °C. The derivatives of P. are based on either of two tautomeric forms:

$$P(OH)_3 \rightleftharpoons HP(=O)(OH)_2$$

The free acid and the simple salts contain tetrahedral anions with one hydrogen atom bound directly to the phosphorus atom. The preference for this structure is a result of the exceptional thermodynamic stability of the phosphoryl group —P=O. Therefore the acid is only dibasic; it forms two series of salts, the primary and secondary Phosphites (see). By contrast, esters derived from both tautomeric forms exist: there are pyramidal phosphite triesters and tetrahedral phosphite diesters. On heating, P. disproportionates to phosphorous acid and phosphane: $4H_3PO_3 \rightarrow 3\,H_3PO_4 + PH_3$. In aqueous solution, P. acts as a reducing agent. It is obtained by hydrolysis of phosphorus trichloride: $PCl_3 + 3\,H_2O \rightarrow H_3PO_3 + 3\,HCl$.

Phosphorous acid esters: see Phosphites, 2.

Phosphorus, symbol ***P***: a chemical element from group Va of the periodic system, the Nitrogen-phosphorus group (see). P. is a non-metal, Z 15, with one natural isotope but many synthetic ones, atomic mass 30.97376, valence III, IV and V.

Properties. P. exists in three allotropic modifications: white, red and black. All three basic types exhibit forms with differing crystal structure, and thus differing density and reactivity.

The longest known, industrially most important, most volatile and most reactive modification is the cubic, wax-like, translucent ***white P.***, or, in the impure state, ***yellow P.***; density 1.82, m.p. 44.1 °C, b.p. 280 °C. White P. is usually sold in rods, which are very readily soluble in carbon disulfide and phosphorus trichloride (100 g CS_2 dissolves more than 1000 g P.), and moderately soluble in benzene, ether and carbon tetrachloride. It has a significant vapor pressure at room temperature, and is steam volatile. At -77 °C, the cubic α-form is converted to a hexagonal β-form. Both forms of white P., P. solutions and P. vapor contain P_4 tetrahedra (Fig. 1). The bond angle of 60 °C resulting from this structure accounts for the high reactivity.

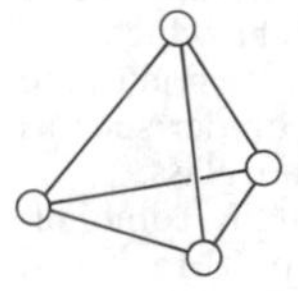

Fig. 1. Molecular structure of white phosphorus.

White phosphorus is a strong poison; the lethal dose for oral intake is 0.05 to 0.5 g. It is absorbed through the digestive tract, or through the lungs if the vapors are inhaled. The symptoms of poisoning are nausea, stomach pains, vomiting, diarrhea and, under some circumstances, circulatory collapse. First aid: Rinse the stomach with 0.05 to 0.1% potassium permanganate or 1% copper sulfate solution. Burning phosphorus causes very painful burns which do not heal easily. Such burns should be thoroughly rinsed with 2% sodium carbonate or 2% copper sulfate solution.

Black P., density 2.70, is thermodynamically the most stable modification at room temperature. It exists in three crystalline and one amorphous form. Orthorhombic black P. is obtained by subjecting white or red P. to high pressure, or by heating white P. at ordinary pressure in the presence of mercury and a crystallization nucleus of black P. Black P. is polymeric, insoluble, almost noncombustible, and in general, extremely unreactive. It is a semiconductor. The crystal lattice consists of infinite, puckered double layers (Fig. 2), in which each P atom is linked pyramidally to three neighbors (bond angle 100°).

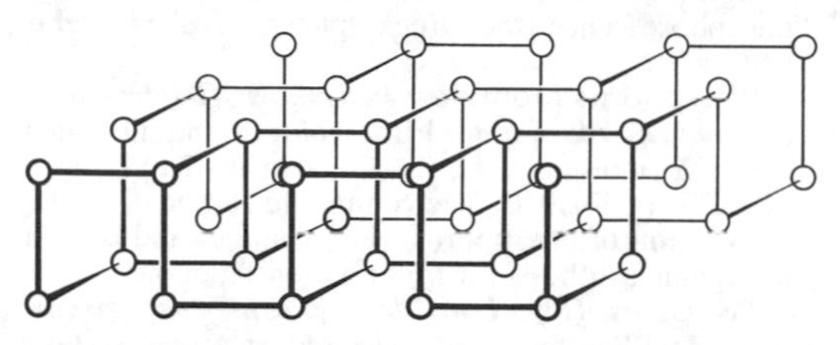

Fig. 2. Lattice of black phosphorus.

The term ***red P.*** is applied tp a number of different crystalline and amorphous forms of P. The densities vary between 2.0 and 2.4, and the melting points between 585 °C and 610 °C. Red P. is obtained by heating white P. to 260-400 °C for several hours. Commercial red P. is largely amorphous. Recrystallization out of molten lead produces monoclinic ***Hittorf's (violet) P.***, a polymeric form which is cross-linked in three dimensions (Fig. 3). If a solution of white P. is boiled in phosphorus tribromide, the scarlet ***Schenck's P.*** precipitates. The differences between the several red forms are due to differences in the grain sizes, the fraction which is crystalline, and the nature of the lattice, but also to impurities, especially from the varying degree of saturation of boundary groups by foreign atoms (Hal, O, OH).

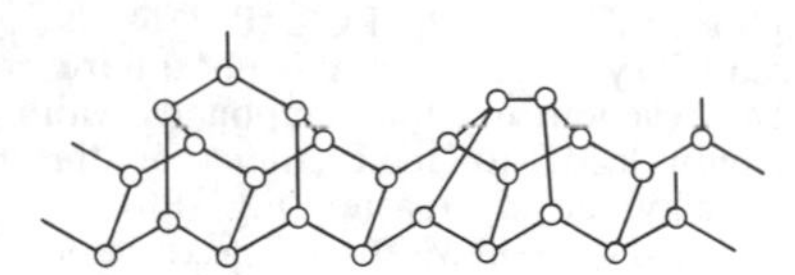

Fig. 3. Structure of violet phosphorus.

P. is able to form an unusually wide variety of bonds. Compounds of P. arc known with coordination numbers of 1 to 6. The electronic configuration of the element is $[Ne]3s^2p^3$. As would be expected, the three unpaired p-electrons are able to form three covalent bonds, and the corresponding compounds, such as PH_3, $PHal_3$, $P(C_4H_9)_3$ or $P(OC_2H_5)_3$ have pyramidal structures. The bond angle at the P atom is generally smaller than in comparable nitrogen derivatives (most are around 100°, in PH_3, 93°). This suggests that the orbital of the free electron pair is largely s in nature, and explains the fact that trivalent P compounds are much less basic than their N analogs. Nitrogen compounds are also more polarizable, which makes them better nucleophiles than the corresponding P. compounds. The energetically favorable position of the P. d level makes a 3s → 3d promotion possible, and thus the formation of pentacovalent compounds (PF_5, PCl_5, $P(C_6H_5)_5$, etc.). These derivatives are trigonal-bipyramidal in structure. The formation of octahedral P anions, such as $[PCl_6]^-$, can be interpreted in terms of contributions from the d orbital. The tetravalent state is realized in the phosphonium salts $[R_4P]^+X^-$. The phosphonium cation is isosteric with the ammonium ion, and, as would be expected, has a tetrahedral structure.

Although the double bond rule forbids formation of stable $p_\pi p_\pi$ double bonds in P., an element of the second period of 8, or the participation of such double bonds in π-conjugation systems, it has recently become possible to synthesize p_π double and p_π triple bond systems of P. under special steric and electronic conditions (examples: phosphorine, Ph-P=CH-NMe_2, *t.*-Bu-C≡P). On the other hand, derivatives of P. with $d_\pi p_\pi$ double bonds are common and very stable; examples are $(HO)_3P{=}O$, $Cl_3P{=}O$, $R_3P{=}O$, $R_3P{=}NR$, and P-ylids (R is an alkyl or aryl group). The phosphoryl group →P=O is preferred; its bond energy of 540 kJ mol^{-1}, provides the driving force in many reactions of phosphorus (see Phosphorous acid, Michaelis-Arbuzov reaction and Wittig reaction).

Finely divided ***white P.*** is pyrophoric in air. Compact pieces ignite at 50 °C and burn to phosphorus(V) oxide: $P_4 + 5\ O_2 \rightarrow P_4O_{10}$. For this reason, P_4 is always stored under water. Under certain conditions, white P. displays a bright green shine in the air; this is due to a gas-phase oxidation of phosphorus vapor, via P_4O_6, to P_4O_{10} (chemiluminescence). P_4 also combines with sulfur, the halogens and many metals, in exothermal and often vigorous reactions, to form the phosphorus sulfides, phosphorus(II) and phosphorus (V) halides and phosphides. In hot concentrated alkali hydroxide solutions, white P. disproportionates to form phosphane and hypophosphite, e.g. $P_4 + 3\,KOH + 3\ H_2O \rightarrow PH_3 + 3\ KH_2PO_2$. White P. is a strong reducing agent; for example, it precipitates silver, copper and lead out of aqueous solution. Disulfur dichloride is reduced to sulfur, and potassium iodate to iodide: $P_4 + 6\ S_2Cl_2 \rightarrow 4\ PCl_3 + 12\ S$; $P_4 + 12\ KIO_3 \rightarrow 4\ K_3PO_4 + 6\ I_2 + 10\ O_2$. While P_4 is inert to water at room temperature, it reacts with it above 200 °C to form phosphane, PH_3, and phosphorous acid, H_3PO_3. Above 1000 °C, P. reduces water in the vapor phase to hydrogen: $P_4 + 10\ H_2O \rightarrow P_4O_{10} + 10\,H_2$.

Red P. is fairly nontoxic, and is insoluble in the usual solvents. Its reactivity is intermediate between that of the white and black modifications. It is not pyrophoric, and reacts only very slowly with moist air, forming orthophosphoric acid. Concentrated nitric and sulfuric acid also oxidize red P. to orthophosphoric acid: $P + 5\ HNO_3 \rightarrow H_3PO_4 + H_2O$, or $4\ P + 8\,H_2SO_4 \rightarrow 4\ H_3PO_4 + S + 7\ SO_2 + H_2O$. When red P. is rubbed with a strong oxidizing agent such as potassium chlorate, it explodes violently. It combines directly with O_2, S_8, the halogens and many metals, but the reactions are much less vigorous than those of white P. Red P. is inert with respect to alkali hydroxide solutions.

When heated to a high temperature, all the modifications form a vapor consisting of P_4 molecules.

Above 800 °C, these begin to dissociate into P_2 molecules. At 1700 °C, the vapor consists of about equal parts of P_2 and P_4 molecules.

Analysis. For either qualitative or quantitative analysis, the P. in a compound is first oxidized to orthophosphate, which is detected by reaction with ammonium molybdate: $H_3PO_4 + 12\ (NH_4)_2MoO_4 + 21\ HNO_3 \rightarrow (NH_4)_3PMo_{12}O_{40} + 21\ NH_4NO + 12\,H_2O$. This precipitation reaction can also be used for quantitative gravimetric determination of P. Volumetric methods of determining phosphate are based on precipitation with Bi^{3+} or La^{3+} standard solutions and back titration with EDTA. The *Mitscherlich test* had a certain significance in forensic chemistry as a means of detecting white P. For this test, the sample (such as the stomach contents of a poisoned individual) is heated with water in a distillation apparatus. The white P. is volatile with steam, and when it first comes into contact with air in the condenser, it can be recognized by its luminescence. The structures of phosphorus compounds can be studied by ^{31}P nuclear resonance spectroscopy.

Occurrence. P. makes up about 0.11% of the earth's crust, and is found in nature in the form of various phosphates. Apatites, $Ca_5(PO_4)_3(F, Cl, OH)$, especially fluoroapatite, and phosphorite (which has an analogous structure but is usually combined with $CaCO_3$) are economically important. Some other phosphorus minerals are wavellite, $Al_3(PO_4)_2(F, OH) \cdot 5H_2O$, vivianite, $Fe_3(PO_4)_2 \cdot 8H_2O$ and turquoise, $CuAl_6[PO_4)(OH)_2] \cdot 4H_2O$. At present, the annual world production of crude phosphate is about 100 million tons, of which more than 90% is converted to phosphate fertilizers. It is estimated that if this level of production is maintained, the presently known deposits will be exhausted after about 1000 years.

Many phosphates are essential for organisms. Hydroxylapatite, $Ca_5(PO_4)_3OH$, is the main component of bone and tooth minerals. In addition, phosphates are components of nucleic acids, and nucleoside phosphates serve as energy carriers in metabolic processes.

Production. The starting material for industrial production of white P. is apatite or phosphorite, which is reduced in electric furnaces at 1400 to 1500 °C in the presence of silicon dioxide and coal: $2\,Ca_3(PO_4)_2 + 6\,SiO_2 + 10\,C \rightarrow 6\,CaSiO_3 + 10\,CO + P_4$. The P. escapes as a gas; it is condensed and collected under water and removed.

Applications. More than 80% of the white P. is oxidized to phosphorus(V) oxide, which is then converted to phosphoric acid and various phosphates. Much of the remainder is converted to phosphorus trichloride, PCl_3 and phosphorus(V) sulfide, P_4S_{10}; these in turn are the basis for many pesticides, softeners, extraction solvents, flame retardants, additives to lubricants, etc. Red P. is used in the production of matches.

Ecology. The use of large amounts of polyphosphates in laundry products, heavy fertilization of fields and the phosphates in human and animal excrements have led to an enrichment of PO_4^{3-} in many lakes and rivers. Since phosphate is normally the limiting nutrient for plant growth in these waters, its presence leads to increased algal growth (blooms). The algae blocks light penetration to greater depths and thus assimilation of CO_2 and release of O_2 in the deep water. As a result, decay processes predominate. Restoration of the waters requires a great decrease in the phosphate content, which can be achieved by appropriate treatment of sewage, for example, by precipitation of the phosphate with iron(III) or aluminum salts.

Historical. The discovery of P. is ascribed to the Hamburg apothecary and alchemist Hennig Brand, who, in 1669, obtained a substance which glowed in the dark by distillation of urine which had been evaporated to dryness. He then isolated the first samples of the element, which was named for its phosphorescence (the Greek "phosphoros" = "light-bearing").

Phosphorus bromides: ***Phosphorus(III) bromide, phosphorus tribromide***, PBr_3, colorless liquid which fumes in moist air; M_r 270.70, density 2.852, m.p. -40 °C, b.p. 172.9 °C. The compound can be obtained by reaction of phosphorus with bromine, and is used in organic synthesis for bromination reactions.

Phosphorus(V) bromide, phosphorus pentabromide, PBr_5, forms yellowish red, rhombic crystals; M_r 430.52, dec. > 100 °C. The lattice of PBr_5 consists of $[PBr_4]^+$ cations and Br^- anions. The P. react with water to form hydrogen bromide, HBr, and phosphorous acid, H_3PO_3 or phosphoric acid, H_3PO_4.

Phosphorus bronze: a copper-zinc alloy to which phosphorus has been added to remove oxides. P. is a high-quality cast alloy used mainly in construction of armatures.

Phosphorus chlorides: ***Phosphorus(III) chloride, phosphorus trichloride***, PCl_3, colorless liquid which fumes in moist air and has a pungent odor; M_r 137.33, density 1.574, m.p. -112 °C, b.p. 75.5 °C. PCl_3 reacts with oxygen to form phosphorus oxygen chloride, $POCl_3$, with sulfur to form phosphorus thiochloride, $PSCl_3$. The acyl chloride is split into phosphorous acid, H_3PO_3, and hydrochloride acid by water. PCl_3 is obtained by combusiton of white phosphorus in a chlorine atmosphere, or by chlorination of red phosphorus suspended in PCl_3. Excess chlorine converts PCl_3 to phosphorus(V) chloride, PCl_5. PCl_3 is the starting material for synthesis of organic phosphorus compounds. The reaction of these compounds with Grignard reagents leads to replacement of the chlorine by an alkyl or aryl group, e.g. $PCl_3 + 3\,C_6H_5MgBr \rightarrow P(C_6H_5)_3 + 3\ MgBrCl$. The reaction of excess alcohol in the presence of acid-binding media leads to triphosphites, e.g. $PCl_3 + 3\ C_2H_5OH + 3\,NR_3 \rightarrow P(OC_2H_5)_3 + 3\ [HNR_3]^+Cl^-$. These in turn are essential synthetic units in organophosphorus chemistry (see Michaelis-Arubzov reaction). PCl_3 is also used to chlorinate organic compounds.

Phosphorus(V) chloride, phosphorus pentachloride, PCl_5 is colorless when completely pure, but is usually obtained as light green tetragonal crystals which fume in moist air; M_r 208.24, m.p. 166.8 °C (under pressure), subl.p. 162 °C. Solid PCl_5 and its solutions in polar solvents consist of tetrahedral $[PCl_4]^+$ cations and octahedral $[PCl_6]^-$ anions. In the vapor phase and in benzene or carbon disulfide solution, PCl_5 molecules have a trigonal bipyramidal structure. Above 200 °C it is largely dissociated into chlorine and PCl_3. PCl_5 reacts with water to form phosphorus oxygen chloride, which reacts further to form orthophosphoric acid and hydrogen chloride: $PCl_5 + H_2O \rightarrow POCl_3 + 2\ HCl$; $POCl_3 + 3\ H_2O \rightarrow H_3PO_4 + 3\ HCl$. PCl_5 reacts with phosphorus(V) oxide to form phosphorus oxygen chloride: $6\ PCl_5 + P_4O_{10} \rightarrow 10\ POCl_3$. Similarly, P_4S_{10} forms phosphorus thiochloride, $PSCl_3$. Alcohols are converted to alkyl halides by PCl_5: R-OH + PCl_5 → R-Cl + $POCl_3$ + HCl; and carboxylic acids to acyl chlorides: R-COOH + PCl_5 → R-COCl + $POCl_3$ + HCl. PCl_5 reacts with ammonium chloride to form phosphorus nitride chlorides $[PNCl_2]_n$. PCl_5 is obtained by chlorination of PCl_3 and is used chiefly as a chlorination agent in organic syntheses.

> The phosphorus chlorides and the phosphorus oxygen chlorides are very irritating to the skin, and especially the eyes. Safety glasses should be worn when working with these chemicals. Inhaled vapors are gradually hydrolysed to phosphorous acid and hydrochloric acid, which cause severe damage to the respiratory passages and lungs. If large amounts of the halides are inhaled, the patient may feel well for several hours and then develop severe lung edema. First aid: inhalation of steam or alcohol vapors.

Phosphorus fluorides: ***Phosphorus(III) fluoride, phosphorus trifluoride***, PF_3, colorless gas; M_r 87.97, m.p. -151.5 °C, b.p. -101.5 °C. ***Phosphorus(V) fluoride, phosphorus pentafluoride***: PF_5, colorless gas; M_r 125.97, m.p. -83 °C, b.p. -75 °C. PF_3 has a trigonal bipyramidal configuration in the gas state. It is obtained by fluorination of the corresponding phosphorus chlorides.

Phosphorus(III) iodide, ***phosphorus triiodide***: PI_3, orange, hexagonal crystals, M_r 411.68, density 4.18, m.p. 61 °C. P. is synthesized by the reaction of elemental iodine with white phosphorus dissolved in carbon disulfide.

Phosphorus nitride dichloride, ***dichlorophosphazene***: oligomeric compounds with the composition $[PNCl_2]_n$, formed by the reaction of phosphorus(V) chloride and ammonium chloride in a molten mixture or in a heated solution of tetrachloroethane. Individual compounds (n=3-7) can be isolated from the reaction mixture as crystalline solids. ***Cyclic trimeric P.*** $[PNCl_2]_3$, colorless, rhombic crystals, M_r 347.66, density 1.98, m.p. 114 °C, b.p. 256.5 °C, has a planar, six-membered ring structure with delocalized P-N-P $d_\pi p_\pi$ bonds (Fig.). The halogen atoms can be replaced by alkoxy, amino, aliphatic or aromatic residues by suitable reagents. Tempering of cyclic P. at 300 °C leads to formation of high-molecular-weight chains with elastic properties. These derivatives are therefore sometimes called ***inorganic rubber***.

Phosphorus oxides: ***Phosphorus(III) oxide, phosphorus trioxide***, P_4O_6, colorless, poisonous, monoclinic crystals with an unpleasant odor; M_r 219.89, density 2.135, m.p. 23.8 °C, b.p. 175.4 °C. The P_4O_6 molecule is tetrahedral (Fig. 1). It is unstable when heated and disproportionates to P_4O_8 and red phosphorus. In the air it is rapidly oxidized to P_4O_{10}. P_4O_6 is the anhydride of phosphorous acid and reacts with water according to $P_4O_6 + 6\ H_2O \rightarrow 4\ H_3PO_3$. The compound can be made by burning white phosphorus in an oxygen-deficient atmosphere; it is used as a standard in ^{31}P NMR spectroscopy.

Phosphorus(V) oxide, phosphorus pentoxide, P_4O_{10}, a colorless, extremely hygroscopic powder; M_r 283.88, density 2.39, sublimes at 359 °C. When rapidly heated to 580-585 °C, it melts. P_4O_{10} consists of cage-like tetrahedral molecules (Fig. 2). The most unusual property of the compound is its extreme tendency to absorb water, with which it forms orthophosphoric acid: $P_4O_{10} + 6\ H_2O \rightarrow 4\ H_3PO_4$. It acts as a dehydrating agent for many other compounds, converting many acids to their anhydrides, e.g. $4\ HNO_3 + P_4O_{10} \rightarrow 2\ N_2O_5 + 4\ HPO_3$. It reacts with amides to form nitriles, with malonic acid to yield carbon suboxide. With phosphorus(V) halides, it forms phosphorus oxygen halides, e.g. $P_4O_{10} + 6\,PCl_5 \rightarrow 10\ POCl_3$. P_4O_{10} is formed by combustion of white phosphorus in the presence of excess oxygen. It is an intermediate in the production of thermal phosphoric acid. In the laboratory, P_4O_{10} is used as a drying and dehydrating agent.

In addition to the above two P., there are a number of compounds which are less thoroughly characterized. The structures of these are derived from P_4O_6 by successive conversion of trivalent P atoms into phosphoryl groups. For example, P_4O_8 contains two trivalent and two pentavalent P atoms.

Phosphorus oxygen chloride: $POCl_3$, a colorless liquid which fumes in moist air; M_r 153.33, density 1.675, m.p. 2 °C, b.p. 105.3 °C. On its toxicity, see Phosphorus chlorides. The halogen atoms of P. can be easily exchanged for nucleophilic agents. Water reacts with P. to form phosphoric acid and hydrochloric acid. The reaction of alcohols or phenols with P. leads to phosphates, and reaction with primary or secondary amines yields phosphamides. Grignard reagents replace one or more of the chlorine atoms with alkyl or aryl groups. On a laboratory scale, P. is obtained by the reaction of phosphorus(V) chloride with oxalic acid: $PCl_5 + (COOH)_2 \rightarrow POCl_3 + 2\ HCl + CO + CO_2$. In industry, the direct oxidation of phosphorus trichloride with oxygen or, more commonly, the reaction of PCl_5 with phosphorus(V) oxide or sulfur dioxide is used: $PCl_5 + SO_2 \rightarrow POCl_3 + SOCl_2$. P. is the starting material for synthesis of many organic phosphoric acid derivatives, and is widely used in organic synthesis for phosphorylation and chlorination reactions.

Phosphorus sulfides: ***Phosphorus(V) sulfide, phosphorus pentasulfide***, P_4S_{10}, yellow, combustible, triclinic crystals; M_r 444.54, density 2.09, m.p. 286-290 °C, b.p. 513-151 °C. The molecular structure is the same as that of phosphorus(V) oxide, P_4O_{10}. P_4S_{10} is converted to orthophosphoric acid, H_3PO_4 and hydrogen sulfide, H_2S, by water. In the air it burns to P_4O_{10} and sulfur dioxide, SO_2. P_4S_{10} is obtained by fusion of stoichiometric amounts of red phosphorus (in industry, white phosphorus is also used) and sulfur. It is the starting material for production of dithiophosphates, which in turn are used to make pesticides, lubricating oil additives, flotation agents, etc. In the laboratory, P_4S_{10} is used to convert carbonyl to thiocarbonyl compounds (thionization). In addition to P_4O_{10}, P. with the compositions P_4S_9, P_4S_7, P_4S_5, P_4S_4 and P_4S_3 are known.

Phosphorus thiochloride: $PSCl_3$, colorless, highly refractive liquid with an unpleasant, pungent odor; M_r 169.40, density 1.688, m.p. -35 °C, b.p. 125 °C. With respect to its reactivity, P. is directly comparable to Phosphorus oxygen chloride (see). It is made by oxidation of phosphorus(III) chloride with sulfur in the presence of aluminum chloride. P. is the starting material for synthesis of thiophosphoric acid derivatives.

Phosphorylases: enzymes which catalyse the cleavage of glucoside bonds in carbohydrates by incorporation of inorganic phosphate. The terminal glucoside group is released as glucose 1-phosphate.

Photoaddition: the addition of a reagent to a substrate by light absorption, forming either an acyclic or cyclic product (photocycloaddition). In the broadest sense, the electrocyclic reaction is a P. In the P. of halogens, hydrogen halides, polyhalogen alkanes, hydrogen sulfide and thiols, C-H acidic compounds (e.g. aldehydes), thiocarboxylic acids, sulfenyl chloride, alcohols, amines, etc. to alkenes and alkynes, the reagent is photochemically split into radicals, which add. Photochemically initiated cycloadditions governed by the Woodward-Hoffmann rules (see) are especially important.

Photoacoustic spectroscopy, abb. ***PAS, optical-acoustic spectroscopy***: spectroscopic methods which are based on a conversion of electromagnetic radiation into sound waves; it is used primarily to make absorption spectra in the UV, VIS and IR ranges.

P. depends on the photoacoustic effect, which was first observed in 1880 by Bell. When the gas in a closed absorption cell absorbs light from a pulsed beam, it becomes excited; the energy of excitation (electronic or vibrational excitation) is converted by collisions of the gas molecules into thermal energy. In a closed system, this leads to increases in pressure, which occur with a frequency corresponding to the modulation frequency of the incident light. The modulation frequency is chosen to correspond to frequencies in the acoustic range; these are picked up by a sensitive microphone or are converted by a suitable pressure-sensitive instrument into electrical signals and recorded. The intensity of the sound signal corresponds to the amount of light absorbed. The pressure changes in the sample gas are recorded as a function of the wavelength of the incident light, producing an optical absorption spectrum. Since scattered and reflected radiation does not interfere with the measurement, P. is particularly useful for studies of samples, such as solids, in which a large amount of scattering is often a problem. The non-radiative relaxation of the solid produces periodic heat waves, which are converted by a surrounding gas into pressure waves which are transmitted to the pressure recorder.

Applications. P. is used to study the structures of samples which cannot be measured by absorption. Since only the layers of sample close to its surface contribute to the signal, the method is especially suitable for studying surfaces; variation of the modulation frequency makes it possible to sample different layers of an object. P. is often used to study minerals, semiconductor materials, in clinical chemistry or for direct study of spots on thin-layer or paper chromatograms.

P. has been systematically developed since about 1970, and has become increasingly important during this time.

Photoanation: see Photosubstitution.

Photoaquation: see Photosolvation.

Photobromination: see Photohalogenation.

Photocatalysis: a collective term for several types of process. 1) Photocatalytic reactions or photo-initiated catalytic reactions in which a catalyst C is photochemically generated. The main reaction is thermal:

$$A \xrightarrow{h\nu} C,\ S \xrightarrow{C} P.$$

for example, photochemical generation of cyanide ions from cyano complexes of molybdenum, which act as catalysts in benzoin condensation. 2) Photoactivated catalytic reactions in which a pseudocatalyst C' is formed in the actual photochemical process and is required for each reaction cycle:

$$A \xrightarrow{h\nu} C',\ C' + S \rightarrow P + A$$

An example would be special carbonylation reactions of alkenes with mixed cobalt carbonyl complexes. 3) Catalysed photoreactions, in which the reaction of an excited molecule S is catalysed by a species C in the ground state:

$$S \xrightarrow{h\nu} S \xrightarrow{C} P.$$

An example is the reduction of benzophenone with isopropanol with the sodium salt of alcohol as the catalyst. 4) Sensitized photoreactions, see Photosensitization.

Photochemical information storage systems: light-sensitive systems which are suitable for recording and storing information. Some important examples are the Photographic process (see; see also Color photography), based on the light sensitivity of the silver halides, and the Non-silver based imaging systems (see).

Photochemical reactions: chemical reactions which are initiated by absorption of UV, visible or IR light. P. can occur via various types of electronic excited states, either singlet or triplet; σ,π* or π,π* states. Because the excited state has a different electronic structure (affecting dipole moment, acid-base properties, redox properties and molecular structure) from the ground state, the products of P. are usually more selective than those of thermal reactions. P. occur in three stages: a) absorption of light by interaction of the photons with the absorbing species and formation of an electronic excited state; b) photochemical primary process, starting from the excited state; and c) secondary processes, which usually occur as "dark reactions". Adiabatic P. may be represented (using the Born-Oppenheimer approximation) as taking place on a potential energy hypersurface of the excited state extending from the excited reactants to the excited products (Fig. 1).

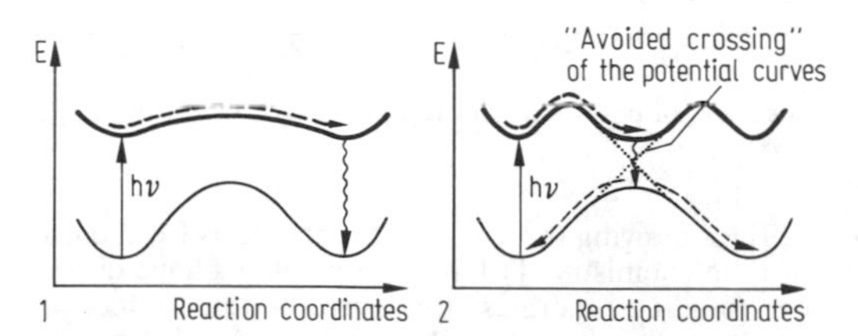

Photochemical reactions: 1) adiabatic reaction, 2) Diabatic reaction. → Radiative transition; ⇝ Nonradiative transition. Lower, ground state, upper, exicted state.

A non-adiabatic P., on the other hand, begins on the potential energy hypersurface of an excited state but ends on another hypersurface, usually that of the ground state (Fig. 2) as the result of a non-radiative transition between the two potential energy hypersurfaces in the region of an "avoided" crossing ("funnel"), which occurs when potential energy hypersurfaces with the same symmetry approach and give way to each other rather than crossing (e.g. in *cis-trans*-isomerization of olefins.) For concerted P., the Woodward-Hoffman rules (see) predict possible outcomes and their stereochemical courses (e.g. in electrocyclic reactions, cyclo additions and sigmatropic shifts). For some P., regularities in certain reaction courses have been recognized on the basis of differences in the geometries of biradical singlet and triplet states ("narrow" geometry of the biradical singlet state leads, e.g., to *cis*-arrangement of biradical sites, while "wide" geometry of the biradical triplet state leads, e.g., to *trans*-arrangement of the biradical sites.) This explains the tendency of singlet state reactions to proceed stereospecifically, while triplet state reactions tend not to be stereospecific. Salem and Turro have classified non-concerted P. on the basis of state correlation diagrams in which the excited states of the reactants and intermediates are plotted in relation to an idealized plane of symmetry; all the nuclei involved in the reaction process are contained in this plane and are designated in terms of their symmetry with relation to it. The intermediate states can have zwitterionic character (heterolytic uncoupling of the bonding electron pair) or biradical character (homolytic uncoupling of the bonding electron pair). From such correlation diagrams, inferences may be drawn concerning the reaction mechanism and the influence of reaction conditions on it (e.g. differential stabilization of zwitterionic and biradical structures by polar solvents). The number of different σ- and π-radical centers which appear in the course of a P. is termed the *reaction topicity*. It is used to classify P. as bitopic (e.g. coplanar H abstraction by excited ketones, carbenes or aza-aromatics), tritopic (heterolytic electrophilic or nucleophilic photosubstitution), tetratopic (e.g. photo-Fries reaction) and so on. P. are often classified in analogy with thermal reactions, with the syllable "photo-" preceeding the type of reaction: photoaddition, photocycloaddition, photosubstitution, photoisomerization, photodissociation, photooxidation, photoreduction.

Photochemistry: a field of chemistry dealing with processes which occur in electronically excited states. In most cases, the excited states are the result of absorption of light (photons), but in some cases they can also be reached without light. Photochemical reactions (see) and Photophysical processes (see) begin in these excited states. It is a basic rule of P., formulated by Grotthus and Draper, that only the light absorbed by a substance is able to induce photochemical reactions.

P. makes use of spectroscopy, photophysics, quantum chemistry and synthetic and analytical chemistry. Its practical significance is greatest in the fields of information gathering and storage (see Photographic process), storage of solar energy (e.g. photosynthesis) and manufacturing (e.g. photooxidation, photopolymerization). ***Photobiochemistry*** has developed as a branch of P.; it is concerned with photosynthesis, the visual process, the effects of light absorption by biochemical structures, and mutations.

Photochromism: a reversible photochemical reaction in which a change in structure leads to a change in absorption spectrum (color). The back reaction is induced photochemically or thermally. The mechanism can be a photophysical process (triplet-triplet absorption), but as a rule it is a photochemical reaction, such as *cis-trans*-isomerism (stilbene derivatives, merocyanins), valence isomerization (spiropyrans), dimerization (anthracene derivatives), tautomerism (chromones, quinolones, *o*-nitrobenzyl derivatives), heterolysis (triphenylmethane pigments) or redox reactions (AgBr decomposition into Ag + 1/2 Br_2). This last reaction is often used in photochromic glasses (automatically darkening glass for eyeglasses). Photochromic materials are used in self-regulating filter materials, light masks in photography, UV dosimeters, advertizing and decoration, for reversible information recording and storage, data indicators, holographic storage plates, etc. However, photochromic materials suffer fatigue through side reactions, and in many cases the cycle numbers (one cycle = one forward reaction and one back) needed for technical applications, 10^4 to 10^6, cannot be achieved.

Photo-Claisen rearrangement: see Photo-Fries rearrangement.

Photoconductivity effect: see Photo effect.

Photo effect, ***photoelectric effect***: all phenomena in which mobile charge carriers are generated by the interaction of matter with electromagnetic or particulate radiation. The various types of P. are:

1) *External P. (Photoemission effect)*. The energy of the charge carriers is so great (exceeds the emission

threshold) that they leave the solid (e.g. alkali metal) and pass through the adjacent gas-filled space or vacuum to an anode.

2) *Internal P. (photoconductivity effect, semiconductor photoeffect)*. The charge carriers do not leave the solid. The internal P. is characterized by an increase in the electrical conductivity of the irradiated material (e.g. Se, GaAs, InP, GaP, CdS, Ge, Si). A special case is the *boundary layer photo effect* (*photovoltaic effect*). Here absorption of light by a semiconductor-semiconductor, semiconductor-metal or metal-electrolyte boundary creates pairs of charge carriers which are separated in the field of the boundary layer. This creates a voltage drop or a change in current in an adjacent circuit. The internal P. is utilized in photoresistors, photoelements, photodiodes and phototransistors.

3) *Atomic P. (photoionization)*, the ejection of electrons from atoms in gases, observed in high layers of the atmosphere, gas discharges or Geiger counters. The prerequisite is that the energy of the ionizing radiation is greater than the ionization energy of the atom. Radical cations are formed from molecules by photoionization (see Photoelectron spectroscopy).

4) *Nuclear photoelectric effect*. Very energetic radiation forms charge carriers and initiates nuclear reactions.

Photoelectric effect: see Photo effect.

Photoelectron spectroscopy, ***PES***: a method of measuring the energy distribution of electrons which are ejected by light quanta from atoms, molecules or a solid according to the equation $M + h\nu = M^+ + e^-$. A sample is irradiated with X-rays or UV light of known energy $h\,\nu$; absorption of this radiation causes emission of electrons with varying kinetic energy E_{kin}. The binding energy E_B is given by $E_B = h\nu - E_{kin} - \pi\eta$, where $\pi\eta$ is the emission work. In studies of gaseous substances, $\pi\eta$ does not occur, and the binding energies obtained can be set equal to the ionization energies or the energy levels of various electron states e. The photoelectron spectrum can be thought of as an energy diagram of the atom or molecule which has been rotated through 90°. P. is therefore one of the most important methods of obtaining experimental results which can be compared with quantum chemical calculations.

Depending on the radiation used to excite the electrons, one has *X-ray PES* (X-PES or XPS), in which internal electrons are excited, or *UV PES* (UPS), which is used to study valence electrons. XPS is also called *electron spectroscopy for chemical analysis (ESCA)* or *induced electron emission (IEE)*. The term PES is often used in the narrow sense to mean UPS. X-ray tubes are used as radiation source in XPS photoelectron spectrometers. The K_α line of aluminum (1487 eV) or magnesium (1253 eV) is often used. In UPS, the resonance lines of helium (21.21 eV) and He^+ (40.82 eV) are used as radiation sources. The samples are usually in the gaseous or solid state. The analysis of the electrons' kinetic energy is achieved by application of electrostatic or magnetic fields, which deflect the electrons by different amounts, depending on their kinetic energy. By changing the field strength, electrons of different kinetic energies can be successively deflected onto the detector and registered. In many cases, a secondary electron amplifier is used as detector. The measurements are made in high vacuum (down to 10^{-8} Pa) and at temperatures of -190 to +1800 °C. The number of electrons reaching the detector is plotted against their energy in the photoelectron spectrum (Fig.).

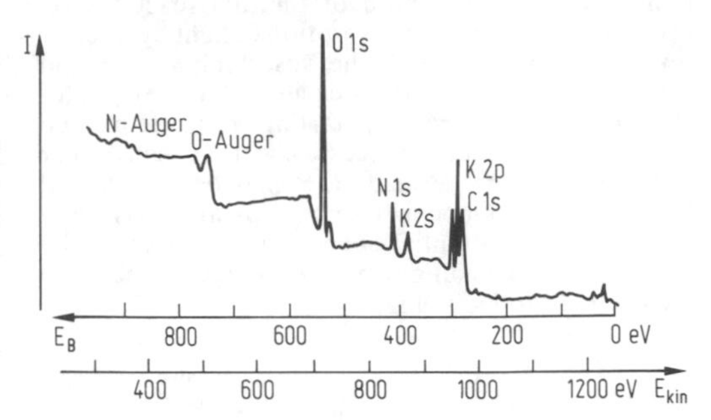

Spectrum of potassium nitrate KNO_3 (excitation with MgK_α rays).

The following three points are the basis for the use of P. in chemistry: 1) Different elements have different binding energies for corresponding electron levels, so that there are *characteristic signals* for each element; these can be used to identify the elements. This is shown in the figure, which is the photoelectron spectrum of KNO_3, excited by Mg K_α radiation. The various signals for O 1s, N 1s, K 2s, and K 2p excitation are clearly visible. The carbon which can be recognized is an impurity. 2) Atoms of the same element have slightly different binding energies as a result of their bonding in molecules; these are responsible for the *chemical shifts*. For example, in ethanal, CH_3-CHO, the 1s electrons of the two C atoms have different signals. In many cases, the chemical shift is also clearly affected by the oxidation state of the element (Table). Measurements of chemical shifts are done by preference on internal shells, because the interpretation is simpler here. 3) In solids, the information is obtained from *regions near the surface*, because the depth from which photoelectrons can escape is limited: it is no more than about 5 nm for X-ray excitation or about 1 nm for UV excitation. Electrons from greater depths lose their kinetic energy through inelastic scattering processes, and thus either do not reach the analyzer or have low energies; they appear as a scattering background if at all. P. is thus extremely sensitive to the nature of the surface, and can be used to study problems like adsorption or corrosion on surfaces, changes in catalyst surfaces, changes in polymeric fibers during chemical treatment, etc. A stepwise removal of the sample surface by a beam of argon ions (sputtering) opens possibilities for depth profile analyses.

Dependence of the chemical shift (in eV) on the oxidation state

Element	Oxidation level:								
	−2	−1	0	+1	+3	+4	+5	+6	+7
N (ls)		0		+4.5	+5.1		+8.0		
S (ls)	−2.0	0				+4.5		+5.8	
Cl (2p)		0			+3.8				+9.5

Additional signals (Fig.) in XPS are generated when the electron hole left by photoionization is filled by a non-radiative transfer of an electron (Auger electron) (see Auger electron spectroscopy).

Photoemission effect: see Photoeffect.

Photo-Fries rearrangement: opening of an α-bond in an excited aryl ether to form *ortho*- and *para*-hydroxyaryl ketones. In contrast to the Fries rearrangement (see), which is a thermal reaction, the P. is an intramolecular reaction which proceeds by radical pair formation. Allyl esters, anilides, *N*-arylcarbamates, arylsulfones, lactones and Phenol allyl ethers (see) react similarly to aryl ethers (***Photo-Claisen rearrangement***; see Claisen rearrangement).

Photographic process: the process of making permanent images of objects through the effects of radiation energy on light-sensitive layers. The most widely used systems are based on light-sensitive silver halides, but a number of other photochemical processes appear to be suitable for special applications (see last paragraph). The disadvantages of the silver halides are the relatively long interval between formation of the latent image (exposure) and the finished image, the complicated wet process for development, and the use of silver, which is relatively scarce, and therefore expensive. The advantages are: 1) light sensitivity over a wide spectral range which is unrivalled by other materials (amplification factors up to 10^9), 2) good picture quality, in terms of reproduction of color and detail; 3) versatility and adequate stability of the films and finished images; 4) applicability in both negative and positive processes.

The basis of the silver halide process is the photolysis of silver halides into silver and halogen:

$$AgX \xrightarrow{h\nu} Ag + 1/2\, X_2$$

X = Cl, Br, I). This reaction generates a latent image with a quantum yield < 1. The amplification process responsible for the extraordinary sensitivity of silver halide materials occurs during developing: the reduction of exposed silver halide grains is catalysed by the nuclei of the latent image. About 5 to 10 light quanta are required per silver halide microcrystal to form stable clusters of 4 to 10 silver atoms. These serve as development nuclei.

The light-sensitive layers used in the P. consist of an emulsion of silver halide microcrystals in a binder (gelatin) applied to a substrate of paper, glass, or film (polyester or acetylcellulose). The photographic emulsion is usually produced by simultaneous controlled addition of silver nitrate and alkali halide solutions to a gelatin solution. This process produces a very homogeneous (monodisperse) emulsion in which the gelatin acts as a protective colloid; it surrounds the colloidal silver halide microcrystals and prevents flocculation. For negative emulsions and x-ray films, silver bromide is used with an addition of 1 to 10% silver iodide to prevent cloudiness. Positive emulsions, microfilms and photographic papers contain mixtures of silver chloride and bromide with various compositions. The precipitation process is followed by a two-stage curing process. In the physical curing process (Ostwald curing), heat treatment is used to promote growth of larger grains at the expense of smaller ones, thus decreasing the number of grains. Afterwards, undesired salts (alkali nitrates) are washed out, and the emulsion is treated with sensitizing reagents. Compounds containing labile sulfur, gold salts and/or traces of suitable reducing agents are applied to create centers of sensitivity (lattice defects in the silver halide crystals, such as silver sulfide, vacancies, etc.). The silver halide microcrystals have a mean diameter of 1 μm, and the silver consumption is between 1 and 10 g of silver per m^2 of film. After the curing process, emulsion stabilizers (triazaindolizine derivatives) and clarifiers (heterocyclic mercapto compounds) are added to prevent cloudiness. Other additions to the emulsion are spectral sensitizers (polymethylene pigments), hardeners (polyfunctional organic compounds) and, for color materials, Color couplers (see). The final emulsion is poured onto the substrate in thin layers. Black and white films carry 1 to 3 layers of emulsion and support material. Color films have 4 to 14 layers, x-ray films 2 to 4 layers, and color films for immediate development, 15 to 18 layers.

Since silver halides are sensitive only to light with wavelengths shorter than 500 nm, the sensitivity range of photographic emulsions must be extended by Spectral sensitization (see) to longer wavelengths. The limit of spectral sensitization is around 1300 nm in the near infrared. The Latent image (see) is generated by a short exposure of the emulsion to light. Photoelectrons generated in the primary process are trapped in surface lattice defects until they react with silver ions, reducing them to silver atoms. These silver atoms themselves serve as electron traps from which latent image nuclei are generated by a series of electronic ($Ag_n^{\pm 0} + e^- \rightarrow Ag_n$-) and ionic steps ($Ag_n$- + $Ag^+ \rightarrow Ag_n$+1); when they reach a certain size ($Ag_n \geq 4$), these nuclei are stable and remain capable of development for years.

The process of generating the final, stable silver image begins with developing, in which the silver halide in exposed areas is selectively reduced to metallic silver through catalysis by the latent image nuclei (development nuclei) in the individual silver halide grains. Weak reducing agents with normal potentials between +0.6 and +0.8V, such as hydroquinone, aminophenols or aromatic diamines, in weak alkaline solution are used. This allows silver to precipitate at the silver nuclei, but silver in non-exposed grains is not reduced.

In chromogenic development (see Color photography), the oxidation product of the developer forms colored pigments with the color couplers in the emulsion. The reduced silver resulting from the developing step is oxidized to silver cations in a bleaching bath (potassium cyanoferrate(III), potassium dichromate), and is thereafter removed in the fixing step. In the fixing of either black and white or color film, the non-developed, insoluble silver halide is dissolved by a suitable complexing reagent (thiosulfate, rhodanide, cyanide). To obtain a positive image, a photographic material (paper or film) is exposed to the light passing through the negative, then developed and fixed in the same way as the negative.

With positive photomaterial, a positive image is made without the negative-image step. The unique characteristic of x-ray film is that the film is coated on both sides with an emulsion layer to insure that the energetic light quanta are absorbed.

Of the many photochemical reactions which might in principle be utilized in image formation, only a few are used widely, including photopolymerization, electrophotography and photochromism. Some of these processes are amplifying, and others are not. Some other photochemical primary processes which appear suitable for imaging are photocatalysis, oxidative and reductive pigment bleaching, photolysis of perhalogenated alkanes to produce radicals, formation of fluorescent species from non-fluorescent compounds, and photochemical redox reactions and disproportionation of inorganic salts, such as lead(II) salts, Hg_2I_2, copper(I) halides, thallium halides and titanium oxide, or coordination compounds.

The applications of non-silver-based P. are limited by their relatively low sensitivity and the inability of most of these systems to reproduce color. At present they are used mainly for copying of documents, microfilms and polygraphy.

Photohalogenization: photochemically initiated substitution of a halogen atom for a hydrogen atom in a molecule. ***Photochlorination*** and ***photobromination*** are of practical importance. P. occurs by a radical chain mechanism, in which the actual photochemical reaction is the formation of a primary radical by cleavage of a halogen molecule. The advantage of P. over the thermal reaction is the lower reaction temperature, which increases the selectivity. A high degree of selectivity is achieved by use of halogen carriers (e.g. *N*-bromosuccinimide, *N*-halogen-1,2,4-triazoles), and in the case of photochlorination, by use of suitable solvents (aromatic hydrocarbons, carbon disulfide). Photobromination is more selective than photochlorination.

Photoionization: see Photoelectric effect.

Photoisomerization: intramolecular rearrangement initiated by a photochemical excitation. All the important forms of thermal isomerization, e.g. *cis-trans-* and valence isomerization, tautomerism, etc., can also occur as P.; they are usually reversible. *cis-trans*-P. can be done with or without a catalyst (Br_2, I_2, carbonyls, Cu(I) salts act as catalysts), and with or without sensitization. In some cases, photochemical valence isomerization leads to relatively unstable compounds which cannot be obtained by thermal reactions (e.g. valence isomers of benzene). In some cases, the mechanism of photochromism is P.

In addition to P., there are photochemical rearrangements in which atoms can migrate. Such skeletal rearrangements are most often observed in aromatic and heterocyclic compounds (e.g. mesitylene $\rightarrow$ 1,2,4-trimethylbenzene; pyrazine $\rightarrow$ pyrimidine and five-membered heterocycles with two or three heterocycles undergo changes in their skeletons). Photochemically initiated sigmatropic rearrangements, such as the di-π-methane rearrangement, the Oxadi-π-methane rearrangement (see) and the Photo-Fries rearrangement (see) are also P. In metal coordination compounds, geometric isomerization, optical isomerization and bonding isomerizations can also be carried out photochemically.

Photokinetics: the study of the rate equations of photophysical processes and photochemical reactions. In P., the laws of thermal kinetics are generally followed. A few peculiarities of P., however, are: a) the application of the Bodenstein principle is justified in most cases, that is, the excited species are deactivated by various channels just as rapidly as they are formed. b) The lifetimes of excited states are often used in derivations of kinetics. c) The product yields are very sensitive to the quantum yields and the efficiency, which is a measure of the rate of a reaction step relative to the sum of all possible competing steps. d) In studies of processes in which energy transfer plays a role, consideration of quenching simplifies the kinetic problem. One obtains Stern-Volmer equations (see), which can be interpreted correspondingly. e) The temperature dependence of the reaction rate or the quantum yield is very low in most cases.

Photolysis: photochemically induced bond cleavage. The term P. is often incorrectly used to mean irradiation of a sample. In the P. of a molecule, homolysis, heterolysis, elimination or fragmentation can occur, with the formation of radicals, ions, atoms or smaller molecules. The advantage of photochemical bond cleavage is that the activation energy for the process is supplied entirely by the light absorbed, rather than by thermal excitation, so it can occur at a low temperature. Some examples of heterolytic P. are photodissociation of acids and bases, cleavage of the C-N bond in arene diazonium salts as competing reactions to the corresponding homolytic bond cleavage, cleavage of the C-halogen bond in *m*-substituted benzyl halides (where the *m*-group is an electron donor), and cleavage of the metal-ligand bond in coordination compounds. Examples of homolytic P. are the initiation reaction in Photohalogenation (see), or in Norrish type I reactions (see). Norrish type II reactions (see) and the reaction of azo compounds to form hydrocarbons and nitrogen are fragmentation P., while the photochemical reactions of diazonium salts and of azides are eliminations.

Photometry: light measurement. In the narrow sense, in chemistry, the determination of concentrations of solutes by measurement of their light absorption. The samples, which are usually colorless, are converted into colored compounds by reaction with suitable reagents. The Extinction (see) is determined, and the concentration is determined from the Lambert-Beer law (see). Since there are often deviations from the Lambert-Beer law, a calibration curve is usually required; the extinctions of several standard solutions of known concentration are plotted against concentration, and the concentrations of samples are read off the calibration curve.

The main components of a ***photometer*** are a light source, a device for selection of wavelength (filter, monochromator), cuvettes, a detector and an indicator system. The instrument may have one or two light beams (see Spectral instruments). The detector can be a photoelement, a photocell a Secondary electron multiplier (see), or the human eye. (The visual methods are now essentially obsolete.) Modern photometers use monochromatic radiation selected by a monochromator, because measurement at the wavelength of greatest light absorption gives the greatest sensitivity. Photometers can be built to work either at a single wavelength or at variable wavelengths.

The advantage of photometric measurements over

absorption spectroscopy (see UV-VIS spectroscopy) is that they are based on color reactions which are specific for the substance being analysed.

Photometric analysis is rapid, highly selective and gives standard deviations of 2 to 5%; hence it has a wide range of applications. It is especially important for determination of very small quantities, such as those found in trace metal analysis. The light absorption of solvated metal ions is generally too low for photometric analysis, so they are converted to colored complexes, usually with chelating reagents. These are quite selective, very stable, and most have rather high extinction coefficients, and thus permit high sensitivity (table). Many nonmetals can be determined photometrically in similar fashion. Color reactions are also known for numerous organic compounds; these are often not specific for individual compounds, but are typical of an entire class of compounds. Photometric measurements are also widely used in natural products chemistry, clinical chemistry, biochemistry, toxicology and environmental protection.

Special forms of P. are flame spectrophotometry, fluorophotometry (fluorometry) and nephelometry.

Table. Colored cation complexes for photometric analysis of metals

Ligand		Cations
Diphenylthiocarbazone	$C_6H_5-N=N-C(=S)-NH-NH-C_6H_5$	Ag, Bi, Co, Cu, Cd, Hg, Pb, Zn
o-Phenanthroline	[structure]	Co, Fe (II), Ce, Te
Dithiooxamide	$S=C(NH_2)-C(NH_2)=S$	Fe, Co, Cu, Ni
Benzoin oxime (Cupron)	$C_6H_5-CH(OH)-C(=NOH)-C_6H_5$	Cu, Mo
Dimethylglyoxime (Diacetyldioxime)	$CH_3-C(=NOH)-C(=NOH)-CH_3$	Fe (II), Ni
α-Nitroso-β-naphthol	[structure: NO, OH]	Co, Cu

Photomultiplier, ***secondary electron amplifier***: an apparatus which amplifies a stream of electrons by generating secondary electrons. In spectroscopy, P. in which a photocathode is linked to a secondary electron amplifier are used as very sensitive detectors (see Spectral instruments). The electrons released from the photocathode by photons of suitable energy are guided to a dynode where they release secondary electrons. These are guided to a series of other dynodes, resulting in cascade amplification of the secondary electrodes, until finally the electrons reach the anode (Fig.). With about 10 dynodes, the photoelectron stream is amplified about 10^8-fold.

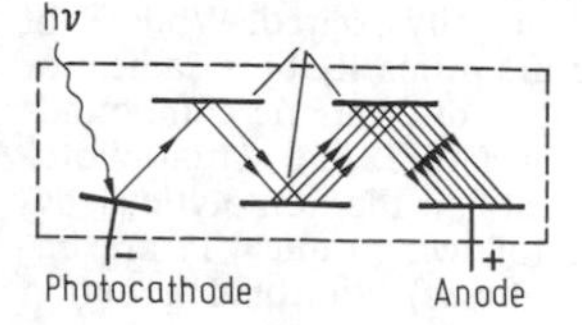

Photon, ***light quantum***: the smallest possible amount of electromagnetic energy which can be emitted or absorbed at a certain frequency or wavelength ($E = h\nu = hc/\lambda$) by a physical system. P. have wave-particle characteristics; they have zero rest mass, and momentum $p = h/\lambda$.

Photonitrosylation: see ε-Caprolactam.

Photooxidants: products of photochemical reactions among automobile exhaust and air components which irritate the eyes and are toxic to plants. P. are formed from nitrogen oxides, alkenes and atmospheric oxygen in the presence of sunlight. They are mixtures of ozone, peroxides, peroxiacetyl nitrate and free radicals.

Photooxidation: 1) the loss of one or more electrons from an atom or molecule as a result of photochemical excitation of this species. 2) same as Photooxygenation (see).

Photooximation: photochemically initiated reaction of nitrosyl chloride, NOCl (or a mixture of chlorine and nitrogen monoxide) with hydrocarbons to form oximes. P. is a radical substitution reaction which does not occur by a chain reaction. In the first step, nitrosoalkanes are formed, and in the HCl-saturated hydrocarbon solution, these rearrange to the oximes, if an H atom is present on the C atom to which the NO group is bound. P. is industrially significant in the case of cyclohexane and cyclododecane; the resulting oximes are converted to caprolactam and lauric lactam, respectively, by Beckmann rearrangement.

Photooxygenation, ***photooxidation***: the reaction of molecules with molecular oxygen upon absorption of light. There are two types.

1) *Type I reaction*: the substrate is photochemically activated, forming an initiator radical, which reacts in a radical chain reaction with molecular triplet oxygen, as in thermal autooxidation. These P. can often be sensitized, e.g. by carbonyl compounds. Some examples are the P. of primary and secondary alcohols (formation of carbonyl compounds) and of *tert.*-aliphatic amines (to imines and aldehydes).

2) *Type II reaction*: reaction of photochemically generated Singlet oxygen (see) with substrates in the ground state.

Photophosphorylation: see Photosynthesis.

Photophysical processes: transitions between the ground state and an excited state or between two excited states. P. may or may not be accompanied by emission of radiation.

1) The most important photophysical *radiative processes* are: a) "allowed" singlet-singlet excitation either from the ground state or from a lower to a higher singlet excited state; b) "forbidden" singlet-triplet excitation from the ground state to an excited triplet state; c) triplet-triplet absorption during a transition from a lower to a higher triplet state; d) Fluorescence (see) and e) Phosphorescence (see).

2) The most important non-radiative P. are: a) vibrational relaxation, i.e. the non-radiative deactivation of excited nuclear vibrations back to thermal equilibrium; b) internal conversion, abb. IC, the non-

radiative transition between states of the same multiplicity; and c) intersystem crossing, the transition between states of different multiplicity. These processes are represented in a *Jablonski diagram* (Fig.):

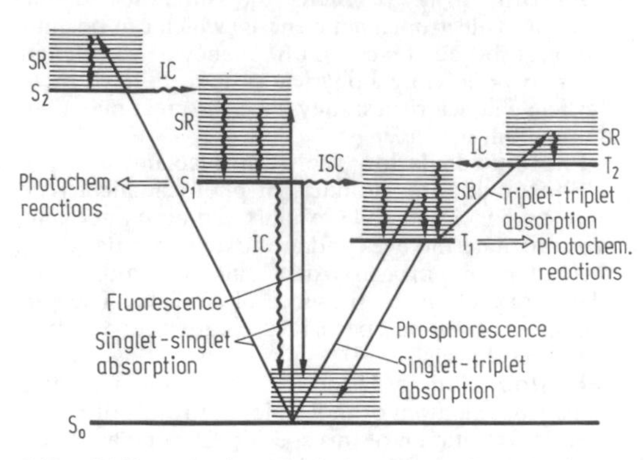

Jablonski diagram: photophysical processes of a molecule. → radiative process; ⇝ non-radiative process.

In non-radiative deactivation of excited states, the electronic energy is converted to vibrational energy of the nuclear skeleton, which is released to the environment as heat. The closer the electronic energy levels are to each other, the more rapidly these processes can occur. Although transitions between states with different multiplicity are forbidden, there are various mechanisms, such as spin-orbital coupling, which allow them to occur. In addition to intramolecular P., there are intermolecular P., such as the transfer of excitational energy between different species (photosensitization), which occur by various mechanisms of energy and electron transfer, and by Annihilation (see).

Photopolymerization: radical or ionic oligomerization or polymerization initiated by photochemical excitation. Photochemical dimerization of polymers with unsaturated groups is called *photo cross-linking*; it is used, for example, in light hardening of photo lacquers for making printing plates and microelectronic circuit boards. Polyvinylcinnamates, for example, can be used as *photo lacquers* for this purpose. P. can occur without sensitization (***type I P.***, with quantum yields usually < 1) or sensitized (***type II P.***, quantum yields often $> 1$7. Radical P. occur by a chain reaction mechanism, in which the actual photochemical reaction serves to initiate the chain. Polyesters of maleic, phthalic or acrylic acids with polyalcohols are often used as monomers. Ionic P. are initiated by photochemical cleavage of diazonium, sulfonium or iodonium salts of strong acids, which then catalyse the polymerization, e.g. of enol ethers or epoxides.

Photoreduction: 1) the uptake of one or more electrons by an electronically excited species; 2) the uptake of hydrogen or the substitution of hydrogen for functional groups in a photochemical reaction. P. can occur by either an electron transfer or an H transfer mechanism. The observed products are formed by disproportionation and/or dimerization. The P. of carbonyl compounds is usually carried out with aliphatic alcohols (e.g. isopropanol), yielding pinacols. With diketones and quinones, disproportionation leads to formation of enediols or hydroquinones. Pigments with acridine, phenazine, oxazine, thiazine or xanthene skeletons are reduced to the corresponding leukopigments. In this case, the reducing agents are *tert.*-amines, alcohols or their ethers, inorganic ions (e.g. Fe^{2+}, Sn^{2+}, etc.) or molecules with C-H bonds adjacent to carbonyl or carboxyl groups, double bonds or aromatic rings. Azomethines are photoreduced in the presence of carbonyl compounds to monomeric or dimeric reduction products (amines or diamines).

The P. of aromatic nitro compounds with primary or secondary alcohols or amines as reducing agents yields arylhydroxylamines in neutral and acid media, and azo- or azoxy compounds in basic media. Rearrangements and substitutions are possible side reactions. Aromatic hydrocarbons (e.g. naphthalenes, anthracenes, phenanthrenes) and heteroaromatics (e.g. acridine, phenazine) are photoreduced to the dihydro compounds with *tert.*-aliphatic amines, alcohols, tributylstannane or complex hydrides.

Photoresist: a lacquer used in the production of integrated circuits which permit high resolution and the use of very thin layers. P. are normally sensitive only to UV light, and are suitable either for negative or positive systems. They are used to transfer patterns to chips by a process similar to photography. In negative systems, Photopolymerization (see) in areas exposed to light reduces the solubility of the lacquer, and in positive systems, exposure to light makes the layer soluble to the etching compound. (Irradiation of quinone diazide derivatives linked to polymers with formation of carboxylic acids is discussed under Diazotypy (see)). After exposure to light, the substrate is etched. That part protected by the P. remains, raised above the etched surface.

Photosensitizer: see Photosensitization.

Photosensitization: a process in which a photochemical or photophysical change in a species occurs without its having absorbed light; instead, it is excited by light absorption of another species, the photosensitizer. In a sensitized photochemical reaction, the photosensitizer is not consumed. There are both singlet and triplet photosensitizers; the latter are more important, because the triplet states of many molecules cannot be directly excited (transitions between states of different multiplicity are forbidden). P. occurs by various mechanisms of energy or electron transfer. P. must be distinguished from Spectral sensitization (see) which is important in photography.

Photosolvation: Photosubstitution in which a ligand is replaced by a solvent molecule. The replacement of a ligand by a water molecule is called ***photoaquation***, and is of practical significance in the case of Reinecke's salt, which serves as an actinometer system in the long-wave spectral range (320-750 nm).

Photosubstitution: the substitution of an atom or group of atoms in a substrate when either the substrate or reagent is electronically excited. While P. of aliphatic compounds occurs primarily by a radical or radical chain mechanism, P. of aromatic compounds is either radical or ionic in nature. The actual photochemical reaction in radical P. is the homolytic cleavage of the reagent. The following radical P. are important (reagent in parentheses): Photohalogenation (see; Cl, Br_2); Photooximation (see; NOCl or NO + Cl_2); photosulfochlorination (SO_2 + Cl_2), the Barton reaction (see Alkyl nitrites) and photosulfoxidation (SO_2 + O_2).

P. on the rings of aromatic compounds occur by various mechanisms: a) *radical P.*, e.g. arylation of aromatics with aryl radicals from aryl halides, or amination of anthraquinone sulfonic acids with NH_2 radicals from hydrazine. b) In *nucleophilic P.*, electron donor substituents direct the reagent (any suitable nucleophile) into *orth/para* positions. Electron acceptor substituents direct the entering substituents to the *meta* positions. In a few cases, *ipso*-substitution occurs independently of the type of substituent present. Some examples of nucleophilic P. are nitration (NO_2^-), cyanation (CN^-), hydroxylation (OH^-) and amination (NH_3). c) *Electrophilic P.*, e.g. in H/D or H/T exchange in benzene derivatives and photodesulfonation of anthracene and anthraquinone sulfonic acids. In addition to these reactions, nucleophilic P. are possible in the side chains of benzene derivatives (photohydrolysis of benzyl chlorides or benzyl acetates). Photosolvation (see), photoanation (replacement of a water molecule by a nucleophile) or in general, ligand exchange can occur by P. on metal coordination compounds.

Photosynthesis: in the broad sense, any chemical reactions in which compounds are synthesized under the influence of light. In the narrow sense, the conversion of light energy into chemical energy in plants, mediated by Chlorophyll (see). The energy fixed annually by P. is $3 \cdot 10^{21}$J, leading to the production of $1.7 \cdot 10^{11}$ t dry biomass per year.

In higher plants and green algae, P. occurs in chloroplasts. These organelles are surrounded by an outer membrane, and have a complex system of internal membranes folded in stacks which are called thylakoids. The membranes are embedded in an aqueous matrix.

The process of P. in higher plants can be summarized by the equation:

$$6\,H_2O + 6\,CO_2 + 2824\text{ kJ (674 kcal)} \underset{\text{Respiration}}{\overset{\text{Photosynthesis}}{\rightleftharpoons}} C_6H_{12}O_6 + 6\,O_2$$

The reverse of P. is respiration.

The P. of plants occurs in two parts. 1) The first is a light-driven reduction of $NADP^+$ to NADPH, with conversion of H_2O to O_2: $NADP^+ + H_2O \rightarrow NADPH + H^+ + 1/2O_2$. This is also called the ***Hill reaction***. It is coupled to ATP synthesis and occurs in the membrane (***photophosphorylation***; Fig.). The oxygen in the atmosphere has been generated by the Hill reaction.

2) The second part of P. is enzymatic reduction of 6 CO_2 to 1 glucose (***Calvin cycle***) with simultaneous oxidation of 12 NADPH and consumption of 18 ATP. These reactions are independent of light (the dark reactions); they occur in the aqueous matrix. In CO_2 reduction, the energy originally stored as ATP is converted to stable carbohydrates, which can be used in metabolism.

Light conversion. Both chlorophylls and auxiliary pigments are used to absorb the light; the carotinoids are very important in this process. They permit utilization of a wider range of wavelengths than are absorbed by chlorophyll alone. There are two main populations of chlorophylls in the chlorplasts. About 99% of the chlorophyll is used as "antenna" pigment, which absorbs the light and transfers it to a special chlorophyll molecule in the reaction center. This molecule absorbs at long wavelengths; in ***photosystem I***, it is chlorophyll P 700, and in ***photosystem II***, chlorophyll P 682. It is excited to its first singlet state by absorption of a light quantum, and from this state it donates an electron to an acceptor A. A donor (D) is involved in the process of light-induced charge separation: D ChlA → D Chl^xA → D Chl^3A^- → D^+ChlA^-. Two quanta of light are required for the transport of a single electron. The electron acceptor is coupled via a series of redox pairs to ferredoxin, which can transfer the electron directly to the proton of the water, forming hydrogen. The electron transport occurs as a dark reaction; it does not require light once the chlorophyll has been activated. The electron transport chain consists of plastoquinone, various cytochromes, plastocyanin and ferredoxin, an iron-containing protein. In higher plants, two photosystems (II and I) are coupled in series; they have different chlorophylls in their reactive centers (P 682 and P 700), different electron donors (D_{II} and D_I) and different electron acceptors (A_{II} and A_I) (Fig.).

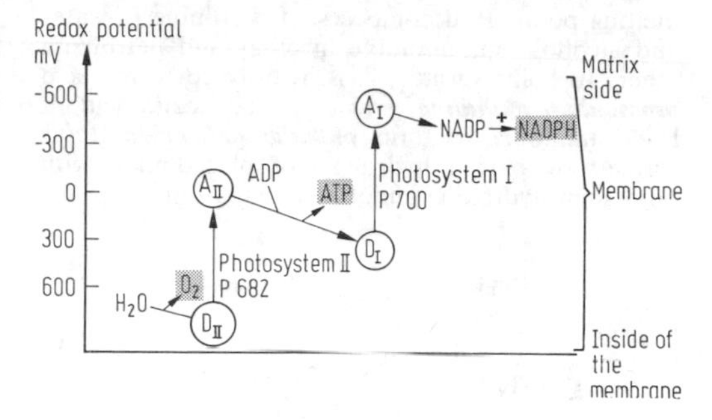

Compounds other than $NADP^+$ can accept the hydrogen. In cyanobacteria, nitrate can be reduced to NH_3 in a ferredoxin-dependent reaction. Potassium cyanoferrate(III) is an artificial electron acceptor which was extensively used to study the Hill reaction. Many inhibitors of the Hill reaction are known which attack the two photosystems at different sites. Some of these inhibitors are used as herbicides.

The second part of P. begins with CO_2 fixation in the Calvin cycle. The CO_2 is first bound to a C_5 body, ribulose bisphosphate, to generate a C_6 body. This is immediately split into two molecules of phosphoglyceric acid (C_3 bodies). These are reduced to trioses, from which hexoses (glucose) are synthesized. Phosphoglyceric acids are also converted, with consumption of ATP, back to the C_5 CO_2 acceptor.

In some plants, CO_2 is temporarily fixed by carboxylation of phosphoenolpyruvate to oxaloacetic acid and reduction to malate, reactions which do not require light. They can occur at night, when the stomata of the leaves are open and CO_2 is abundant in the mesophyll cells. During daylight hours when the plant's stomata are closed to prevent excessive water loss, the malate is decarboxylated to provide CO_2 for the Calvin cycle.

Photosynthetic bacteria (other than the cyanobacteria or "blue-green algae") have only a single photosystem, and do not produce O_2. The biliproteins phycoerythrin, phycocyanin and allophycocyanin (see Bile pigments) are involved in energy transfer.

Photosystems: see Photosynthesis.

Photo cross-linking: see Photopolymerization.

Photovitroceramics: see Vitroceramics.

Photovoltaic effect: see Photoeffect.

Phrenosine: see Cerebroside.

Phthalaldehydes: the isomeric diformyl derivatives of benzene. ***1,2-Phthalaldehyde*** is a bright yellow, crystalline compound; m.p. 56 °C. It can be made by reduction of the di-(*N*-methylanilide) of phthalic acid with lithium aluminum hydride. ***1,3-Phthalaldehyde (isophthalaldehyde)*** is a crystalline compound; m.p. 89 °C. It is usually synthesized by oxidation of *m*-xylene by chromic acid in acetic anhydride and sulfuric acid. ***1,4-Phthalaldehyde (terephthalaldehyde)*** is a crystalline compound; m.p. 117-119 °C. 1,4-P. can be synthesized by oxidation of *p*-xylylene glycol with nitric acid around 30 °C. All P. can be dissolved in most organic solvents. They are used mainly for special syntheses.

Phthalamic acid, ***phthalic acid monoamide***: colorless prisms, m.p. 148-149 °C When heated above the melting point, P. decomposes. It is soluble in water and alcohol, but insoluble in ether and petroleum ether. In boiling water, P. is hydrolysed to the acid ***ammonium phthalate***, and in glacial acetic acid, it loses ammonia to form ***phthalic anhydride***. P. is formed by partial hydrolysis of phthalimide with potassium hydroxide under mild conditions.

CO–NH2, CO–NH2

Phthalamide

CO–NH2, COOH

Phthalamic acid

Phthalamide, ***phthalic acid diamide***: a colorless, crystalline compound; m.p. 222 °C. P. cannot be distilled at normal pressure without decomposition, because when heated, it readily loses ammonia and is converted to phthalimide. P. is used in the synthesis of *O*-phthalodinitrile.

Phthalates: the salts and esters of phthalic acid, with the general formula 1,2-$C_6H_4(COOR)_2$, where R can be a monovalent metal or ammonium ion, or an aliphatic or aromatic substituent. The esters of phthalic acid are of the greatest industrial importance. As dibasic acids, phthalic acids can form neutral and acidic P. Most of the esters are colorless, water-insoluble liquids; they are used in large amounts as softeners, lubricants, solvents and intermediates in the production of resins. Some important ones are listed below: ***Phthalic acid dimethyl ester, dimethylphthalate***, 1,2-$C_6H_4(COOCH_3)_2$, a colorless, oily liquid which is readily soluble in most organic solvents; m.p. 0 °C (also reported as -46 °C), b.p. 282.4 °C, n_D^{20} 1.5138. It is used as a solvent for cellulose nitrate lacquers, as a substitute for camphor in the production of celluloid, and to denature alcohol. ***Phthalic acid dibutyl ester, dibutyl phthalate***, 1,2-$C_6H_4(COOC_4H_9)_2$; m.p. -62 °C, b.p. 340 °C, n_D^{20} 1.4911. Its solvent properties are similar to those of the dimethyl ester, and it is used as a gellation agent for explosives, as a solvent and as a softener. In addition to the P. with identical ester groups, compounds with different alcohol moieties are known, for example, ***phthalic acid benzyl butyl ester***, 1,2-C_6H_4-$(COOCH_2C_6H_5)(COOC_4H_9)$, a colorless liquid; m.p. -43 °C, b.p. 280-288 °C at 2.6 kPa. It is used as a cold-resistant, nonvolatile softener for lacquers and plastics.

Phthalate resins: see Alkyde resins.

Phthaleins: a group of synthetic pigments which belong to the diarylmethane and Triarylmethane pigment (see) classes. Some examples are phenolphthalein, fluorescein, eosine and erythrosine.

Phthalein synthesis: a method of synthesis of xanthene pigments from phthalic anhydride and 3-aminophenols or resorcine derivatives. The term P. is based on the synthesis of phenolphthalein by condensation of phthalic anhydride with phenol.

Phthalic acid, ***benzene-1,2-dicarboxylic acid***: colorless, monoclinic crystals, m.p. 191 °C (in a closed tube). When the compound is heated, it loses some water even below the melting point, forming the cyclic anhydride. P. is slightly soluble in cold water, chloroform and ether, but is readily soluble in hot water or alcohol. When it is heated to about 300 °C with sodium hydroxide, it decarboxylates to give benzoic acid. P. has properties typical of the dicarboxylic acids, and because of the proximity of the two carboxyl groups, it also has a few special reactions, such as the formation of a cyclic anhydride or imide. The salts and esters of P. are called Phthalates (see). P. is synthesized industrially by hydrolysis of phthalic anhydride. It is used as an intermediate in the production of pigments, polyesters and softeners.

COOH, COOH

Phthalic acid

O, O, O

Phthalic anhydride

Phthalic anhydride: colorless, sublimable needles with an aromatic odor; m.p. 131 °C, b.p. 295 °C. It is soluble in pyridine, ketones and halohydrocarbons, and only slightly soluble in cold water, alcohol and ether. When boiled in water for a long time, P. is quantitatively converted to phthalic acid. P. is made by melting phthalic acid. It can also be synthesized by oxidation of *o*-xylene or naphthalene with atmospheric oxygen in the presence of vanadium(V) oxide at 400-450 °C. P. is used in large amounts in the synthesis of pigments, such as phthaleine, rhodamine dyes and indigo, and also of softeners, polyester resins, lacquers and plastics.

Phthalic acid benzyl butyl ester: see Phthalates.

Phthalic acid dibutyl ester: see Phthalates.

Phthalic acid diethyl ester: see Phthalates.

Phthalic acid dimethyl ester: see Phthalates.

Phthalic diamide: see Phthalates.

Phthalide: the lactone of 2-hydroxymethylbenzoic acid. P. forms colorless needles; m.p. 75 °C, b.p. 290 °C, n_D^{99} 1.536. It is soluble in alcohol and ether, but insoluble in water. P. is formed by loss of water from hydroxymethylbenzoic acid. It is the parent compound of some phthalides found in nature; these compounds are derived from P. by substitutions for the hydrogen atoms of the methylene group, such as 3-butyl- and 3-butylenephthalide, the typical aroma compounds of the essential oils of celery and lovage.

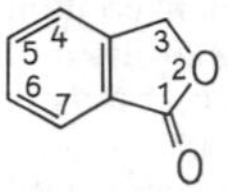

Phthalimide, ***phthalic acid imide, isoindoline-1,3-dione***: a white, crystalline powder; m.p. 238 °C. It is soluble in glacial acetic acid, pyridine and alkali hydroxide solutions, where it forms salts. It is insoluble in water and most organic solvents. P. is made industrially by heating phthalic anhydride with ammonia to 200 °C under pressure. The hydrogen atom of the NH group is activated by the neighboring carbonyl groups, so it can be replaced by alkali metals. ***Potassium phthalimide*** is important for the preparation of primary amines by the Gabriel synthesis. P. is also used to make anthranilic acid.

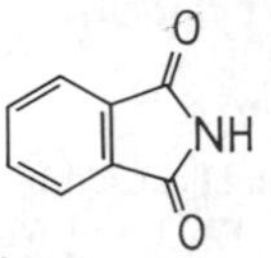

Phthalocyanins, ***tetrabenzotetraazaporphins***: a class of compounds with a heterocyclic skeleton which is structurally very similar to that of the porphins. The parent compound, ***phthalocyanin***, is a blue-green, very stable compound without a melting point. It is formed by heating phthalodinitrile in quinoline in the presence of ammonia or amines. The P. are the parent compounds of the Pthalocyanin pigments (see). Various metal-containing P., especially the iron(III) complexes, are important as oxidation catalysts. P. are of interest as semiconductors.

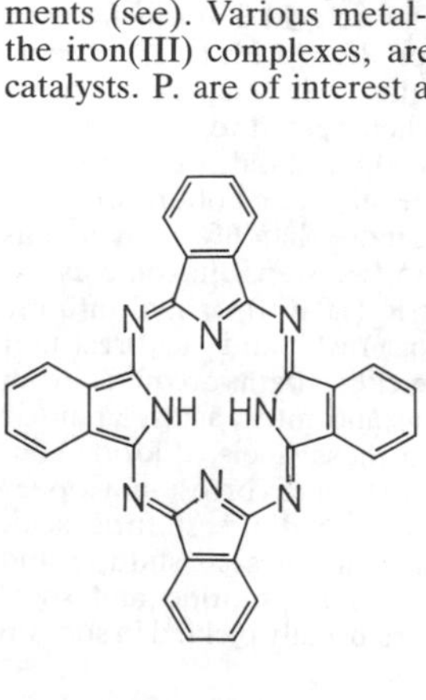

Phthalocyanin pigments, ***heliogen pigments, monastral pigments***: a group of synthetic aza[18]annulene pigments. They are derived structurally from phthalocyanin (***heliogen blue G***), which is no longer used as a dye. On the other hand, metal complexes of phthalocyanin (especially the complexes with copper, zinc, iron, cobalt and cadmium) are very important textile dyes and are also used in paints and plastics. The simplest of these compounds is ***copper phthalocyanin (heliogen blue B)***, which forms dark blue, very stable crystalline needles; these sublime at 500 °C without decomposition. They are attacked neither by hot hydroxide solutions nor by boiling hydrochloric acid, and are insoluble in ordinary solvents. Other colors can be obtained by sulfonation, chlorination or substitution of the aromatic rings by heterocycles. Partially sulfonated compounds of copper phthalocyanin are used as acid and direct dyes for textiles. The colored phthalocyanin complex can also be produced by a reaction on the fiber itself (ingrain color).

Phthalodinitrile, ***phthalodinitrile, 1,2-dicyanobenzene***: a colorless, crystalline compound, m.p. 141 °C, which cannot be distilled. P. is barely soluble in water, but dissolves readily in alcohol, ether and acetone. It is synthesized by dehydration of phthalamide with acetic anhydride, or by ammonooxidation of *o*-xylene. P. is an important intermediate in the production of phthalocyanin pigments.

Phthalylsulfathiazole: see Sulfonamides.

Phycocyanins: see Bile pigments.

Phycocyanobilin: see Bile pigments.

Phycoerythrins: see Bile pigments.

Phycoerythrobilin: see Bile pigments.

Phylloceruleine: see Ceruleine.

Phyllokinin: see Kinins.

Phyllomedusin: see Tachykinins.

Phylloquinone: see Vitamin K.

Physalemin: see Tachykinins.

Physisorption: see Adsorption.

Physostigmine, ***eserine***: an indole alkaloid. It is biosynthesized from tryptophan and is the primary alkaloid in calabar beans, the seeds of *Physostigma venenosum*, in which it is present at about 0.1% concentration. The base is dimorphic; m.p. 187 °C or 106 °C; $[\alpha]_D^{20}$ -76° (in chloroform). The relatively stable and non-hygroscopic salicylate, m.p. 186-189 °C, $[\alpha]_D^{20}$ -91 to 94° (in water) is used as a drug. As a carbamate, P. is labile to alkalies; it is hydrolysed to methylamine, carbon dioxide and physostigmol (eseroline), which contains a phenolic OH group. In the presence of oxidizing agents, it forms colored compounds with O-quinoidal structure. P. is an acetylcholinesterase inhibitor (see Parasympathicomimetics) and is used to treat glaucoma.

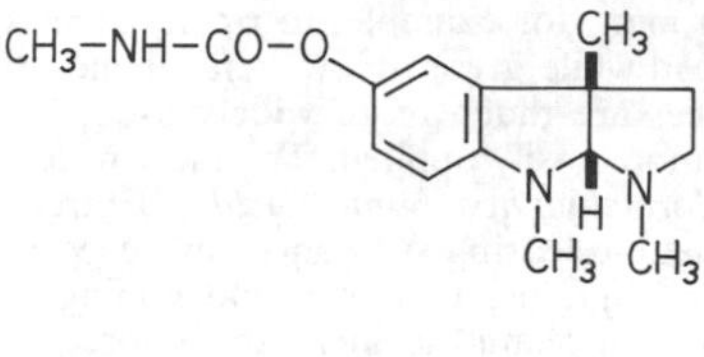

Phytin: see Inositol.

Phytic acid: see Inositol.

Phytoalexins: defense substances formed by plants after contact with pathogens; they inhibit the growth of the pathogenic microorganisms. The substances which induce formation of P. are called ***elicitors***; they include microbial polysaccharides, glycoproteins and peptides. Oligosaccharides released from the plant cell walls by the action of either the microorganisms or the plant's own enzymes also act as elicitors. The known P. include isoflavonoids (in legumes), sesquiterpenes (in nightshades), polyacetylenes (in asters) and dihydrophenanthrenes (in orchids). Some examples are *rishitin*, a sesquiterpene induced by potato rot, and also found in tomatoes, and *phaseollin*, an isoflavonoid produced by beans (*Phaseolus*).

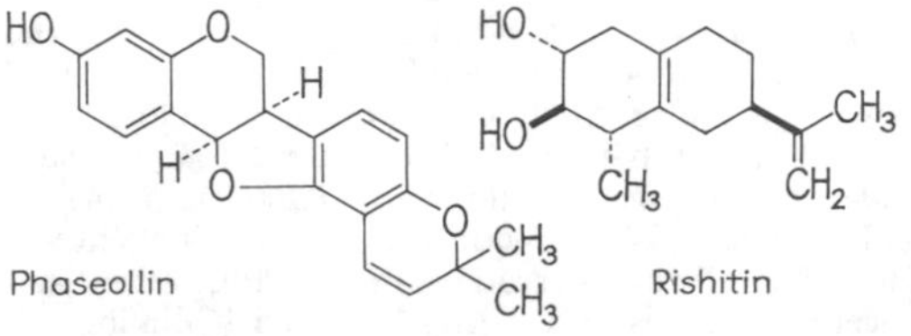

Phaseollin   Rishitin

Phytochrome: see Bile pigments.

Phytohemagglutinins: see Lectins.

Phytohormones, ***plant growth regulators***: intracellular regulatory substances of the higher plants (see Hormones). P. are not synthesized in special glands, but in various parts of the plant, and they generally have no specific target organs. They are thus more similar to animal tissue hormones than to the endocrine hormones. There are five groups of classic P.: Auxins (see), Gibberellins (see), Cytokinins (see), Abscissic acid (see) and ethene. In addition, responses to injury are regulated by specific oligosaccharins (oligosaccharides with specific sequences of sugar units), and it is possible that the mysterious "flowering hormone" is also an oligosaccharin. The classic P. are isoprenoids (gibberellins, abscissic acid and the isoprenoid C_5 unit of the cytokinins) or products of amino acid metabolism (auxins are products of tryptophan or phenylalanine metabolism, ethene is a product of methionine). Proteinaceous P. are not yet known. *trans*-2-Hexenal also has a P.-like effect. The effects of P. depend on many factors, such as the relative amounts of different P., the species of plant and its stage of development, and the environmental conditions. The cytokinins, auxins and gibberellins promote growth, while abscissic acid inhibits growth. The growth hormones probably influence ethene production. Ethene is the only known gaseous P.; it is formed mainly by ripening fruit and inhibits the growth of younger parts of the plant. It is used in closed rooms, for example, to ripen bananas which are shipped while green. However, ethene-releasing substances are much more widely used, because they are more easily applied. The most widely used is *2-chloroethylphosphonic acid* (Ethrel). Ethene and ethene-releasing substances are used as straw stabilizers in grains, to induce blossoming of fruit and ornamental plants, to increase the proportion of female blossoms (cucurbits), promote ripening (tomatoes, fruits), promote fruit drop (mechanical fruit harvesting), as defoliants, and to increase latex (rubber trees) and resin (pines) yields.

Phytol: 3,7,11,15-tetramethylhexadec-2-en-1-ol, an acyclic, unsaturated diterpene alcohol. P. is a colorless oil, b.p. 202-204 °C, which is practically insoluble in water and soluble in organic solvents. P. is a component of chlorophyll, in which it is bound by an ester linkage; it provides an hydrophobic anchor which holds the molecule in the thylakoid membrane. P. is also a component of vitamins K_1 and E.

Phytomenadione: see Vitamin K.

Phytosphingosine: see Sphingoids.

Phytosterols: neutral derivatives of cyclopentanoperhydrophenanthrene (gonane) which contain no nitrogen. They are considered to be sterols, and are found in nearly all plants. P. always accompany plant oils and fats, and are often present as alcoholic components of plant waxes. The most important sterols of conifers are β-sitosterol, stigmasterol, campesterol and β-sitostanol. Pine wood contains 0.04 to 0.08% P., and Douglas fir wood, 0.12% P. The ratio of β-sitosterol:β-sitostanol:campesterol varies in the individual species. Stigmasterol is present in the sterols of *Chamaecyparis* species.

The tall oil obtained from pine wood contains 2.5 to 4% P. (of which 1 to 3% is β-sitosterol), and tall pitch (the residue from tall oil distillation) contains 5 to 6%. Industrial methods for extracting crude sterol and β-sitosterol from tall oil have been developed. P., especially β-sitosterol, is used as a medication to prevent arteriosclerosis and as a starting material for synthesis of sex hormones, contraceptives and adrenal hormones.

Phytotoxins: see Plant toxins.

PIB: abb. for polyisobutylene.

Picloram: see Herbicides.

Pickling brittleness: brittleness of steel caused by diffusion of atomic hydrogen into the crystal during pickling (see Electrochemical metal corrosion). P. is suppressed by pickling inhibitors (see Pickling of metal).

Pickling of metal: treatment of metals with chemicals which remove nonmetallic substances or oxide layers to prepare a clean, polished metallic surface for application of paints, cold working or the decorative effect. Pickling is usually done in acids or lyes, or in special cases, alkali hydroxide melts with reducing agents as additives; an example is sodium hydride, NaH. Sulfuric or hydrochloric acid is normally used for non-alloy steels or for those with small proportions of non-ferrous metal. All temperable steels should be pickled only when heated to softness. The composition of the pickling bath should be adapted to the composition of the steel alloy, and often also to its intended use. Steels with more than 6% alloy metals must usually be pickled in two steps. In some cases, phosphoric (H_3PO_4), nitric (HNO_3) or hydrofluoric (HF) acid is used. For rust- and acid-resistant chromium-nickel steels, pickling baths are made from a mixture of hydrochloric and nitric acids, or nitric and hydrofluoric acids. In these steels, chloride-containing pickling baths easily lead to corrosion. Copper and its alloys are pickled with dilute sulfuric acid. Brass is often pickled in concentrated sulfuric and nitric acids with added sodium chloride and soot. Aluminum and its alloys are usually pickled in sodium hydroxide, then treated with dilute nitric acid. After pickling, all metal pieces are carefully rinsed with water.

To reduce the loss of metal in the process, due to dissolving and generation of hydrogen, pickling inhibitors are often added to strongly acidic pickling solutions. These keep the rate of dissolution of the metal as low as possible without slowing the dissolution of scale. Complete protection of the metal is not possible. Instead of sulfuric and hydrochloric acids, citric and formic acids are sometimes used in pickling. Aldehydes, amines, nitriles, nitrogen-containing heterocycles, mercaptans, thioethers and compounds in which nitrogen and sulfur are bound to a common carbon atom are used as pickling inhibitors. Urotropin is often used as an inhibitor when unalloyed or low-alloy steels are pickled in hydrochloric acid; dibenzylsulfoxide is used in sulfuric acid.

Organic wetting agents are added to the pickling solution to accelerate the removal of scale.

Picolines: three isomeric methyl derivatives of pyridine. The P. are colorless liquids soluble in water, alcohol and ether; they smell similar to pyridine. They are present in the light oil fractions of coal tar distillates, from which they can be isolated by fractional distillation. ***α-P., 2-methylpyridine***; m.p. -66.8 °C, b.p. 128.8 °C, n_D^{20} 1.4957. Oxidation of α-P. with potassium permanganate yields picolinic acid. It is an intermediate in the production of pigments and synthetic resins, and is used as a solvent for rubber. ***β-P., 3-methylpyridine***; m.p. -18.3 °C, b.p. 144.1 °C, n_D^{20} 1.5040. Oxidation with potassium permanganate gives nicotinic acid. It is used in the same ways as α-P., and also to make water repellent compounds and insecticides. ***γ-P., 4-methylpyridine***; m.p. 3.6 °C, b.p. 144.9 °C, n_D^{20} 1.5037. Oxidation of γ-P. with potassium permanganate produces isonicotinic acid. It is a good solvent for synthetic resins and is used as a starting material for isonicotinic hydrazide. The methylene groups of α- and γ-P. are about as reactive as the methyl group in 2,4-dinitrotoluene. When the P. are condensed with aldehydes, they form pyridine derivatives with unsaturated side chains.

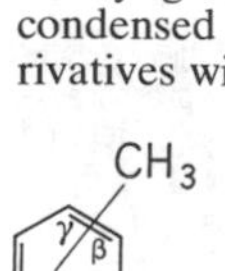

CH_3 γ β α N

Picolinic acid: ***pyridine-2-carboxylic acid***: a compound formed by oxidation of α-picoline with potassium permanganate. It forms colorless crystals; m.p. 136-137 °C (subl.). P. is soluble in hot water, alcohol and glacial acetic acid, and insoluble in ether and benzene. When heated above its melting point, it decomposes. It is an isomer of nicotinic and isonicotinic acids.

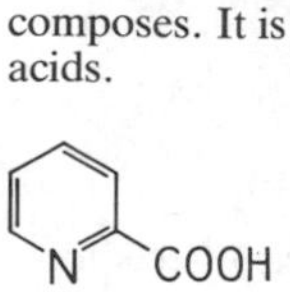

N COOH

Picramic acid, ***2-amino-4,6-dinitrophenol***: a red, crystalline compound; m.p. 168 °C. P. is slightly soluble in water, and is soluble in ethanol, benzene and glacial acetic acid. It is made by partial reduction of picric acid, and is used as a precipitating reagent for proteins, and in the synthesis of azo pigments.

Picramide, ***2,4,6-trinitroaniline***: a yellow-orange, crystalline compound; m.p. 192-195 °C. P. is nearly insoluble in water, but dissolves easily in benzene and acetone. It can be synthesized from picryl chloride and ammonia, and is used chiefly as an explosive.

Picrates: 1) salts of picric acid in which the H atom of the hydroxyl group is replaced by a metal. P. are very intensely colored, usually yellow. The ammonium and alkali salts are readily soluble in water, but the heavy metal P. are less soluble or insoluble. Some P. are highly explosive. Heavy metal P. are especially sensitive to impact and sparks. Lead picrate is used for ignition caps, and ammonium picrate as an explosive. The P. of organic bases have sharp melting points and are used to characterize the bases.

2) A term for molecular compounds of picric acid with arenes. They are formed when the components are heated in alcoholic solution; they crystallize very well and have sharp melting points, so they can be used for identification (picrate of phenanthrene melts at 143 °C). These P. are often charge-transfer complexes with intense colors.

Picric acid, ***2,4,6-trinitrophenol***: yellow, very bitter-tasting crystals; m.p. 122 °C. P. is barely soluble in cold water, but is soluble in boiling water and is readily soluble in ethanol and benzene. P. is a strong acid (pK_a 1.02), due to the cumulative effects of its electron-withdrawing substituents, which considerably increase the acidicty of the phenolic hydroxyl group (see Dinitrophenols). P. forms salts, the Picrates (see) with many organic bases. These crystallize well, and can thus be used to characterize the bases. P. also forms defined colored molecular compounds with condensed aromatic hydrocarbons; these are charge-transfer complexes, and are also called picrates. They are insoluble and have sharp melting points, and are thus useful for isolation and characterization of the corresponding arenes.

P. is poisonous. Contact with the skin and inhalation of vapors and dust must be avoided. Silk, wool, leather, human skin and other proteins are dyed yellow when treated with aqueous solutions of P. Thus P. is a dye (nitro dye), but it is not very fast and for this reason, as well as its toxicity, it is no longer used as such.

P. burns in the air when ignited, forming large amounts of smoke. When heated very rapidly it can explode violently, with a detonation rate of 7100 m/s. Thus the explosive force of P. is 10 to 15% greater than that of 2,4,6-trinitrotoluene (TNT), and it was used in World War I to fill grenades. However, because of its aggressivity towards metals and the formation of heavy metal picrates, which explode unpredictably, it has been replaced by TNT.

P. can be chemically converted to picramic acid and Picryl chloride (see), and from there to picramide. It is synthesized from phenol, by conversion to phenol-2,4-disulfonic acid and treatment with concentrated nitric acid, or from chlorobenzene via 2,4-dinitrochlorobenzene and 2,4-dinitrophenol, which is then nitrated again. P. is used in the dye industry to make picramic acid, and in organic analytical chemistry. Fig. p. 836.

Picryl chloride, ***1-chloro-2,4,6-trinitrobenzene***: colorless, crystalline needles; m.p. 83 °C. P. is barely soluble in water, but dissolves readily in alcohol, acetone and benzene. It is easily synthesized by reaction of picric acid with phosphorus(V) chloride.

Pictet-Gams synthesis: a ring-closing reaction of *N*-acetylated 2-hydroxy-2-phenylethylamines to form isoquinolines:

When the starting materials are heated with phosphorus pentoxide in boiling toluene, 2 moles water are split off. The classical example of this reaction was the synthesis of the alkaloid papaverine.

Pictet-Spengler synthesis: synthesis for the isoquinoline ring system starting from 2-arylethylamines and carbonyl compounds, usually aldehydes (R = alkyl or aryl):

Tetrahydroisoquinoline

It involves an intramolecular aminomethylation which leads to a tetrahydroisoquinoline. The intermediate imino compound can be isolated in some cases. This reaction is very suitable for synthesis of alkaloids, e.g. starting from tryptamine and forming carboline derivatives. As with the Mannich reaction, it is possible with the P. to maintain conditions which are physiological for plants.

Pictet-Trouton rule: see Phase change heats.

Piezoelectricity: the appearance of electrical charges on the surfaces of certain crystals when these undergo mechanical stress (pressure, tension or bending). The ***piezoelectric effect*** was discovered in 1880 by the brothers J. and P. Curie; it is observed in crystals which have an axis of polarity, but no center of symmetry. Quartz, tourmaline, zinc blend, d- and l-tartaric acid, sodium chlorate, sucrose and Seignett's salt are examples of compounds which display P. Of the 32 classes of crystals, 20 permit the appearance of P.

P. can be explained using quartz as a model (Fig. 2). A quartz crystal has a three-fold primary axis (crystallographic c-axis) and, perpendicular to it, three two-fold polar axes, the electrical axes (in Fig. 1, these are labelled E_1, E_2 and E_3). The quartz lattice

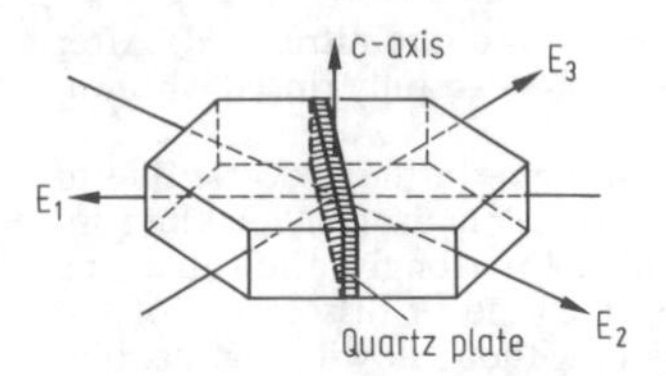

Fig. 1. Section of a piezoelectric quartz plate showing the crystallographic c axis and the electrical axes E_1, E_2 and E_3.

is considered for the sake of simplicity to consist of Si^{4+} and O^{2-} ions; the polarity of the two-fold axis arises because the two opposing ends of it are occupied by ions of opposite charge. One now cuts a section out of the crystal as shown in Fig. 1. If the quartz plate is compressed in the direction of one electrical axis (e.g. E_3), the two surfaces perpendicular to the axis become charged by an excess of one type of ion (positive or negative); this is the *longitudinal piezoelectric effect*, Fig. 2b. If the plate is stretched, the charges on the two surfaces are reversed. If the pressure or tension is applied perpendicular to the electrical axis (Fig. 2c), the charge shifts still cause the surfaces to become charged, but the signs of the charges are the opposite from those in the last case (*transverse piezoelectric effect*). The appearance of P. is reversible: if an electrical field is applied parallel to an electrical axis, the shift of the lattice components leads to a compression or dilation in the direction of the E axis (*inverse piezoelectric effect*).

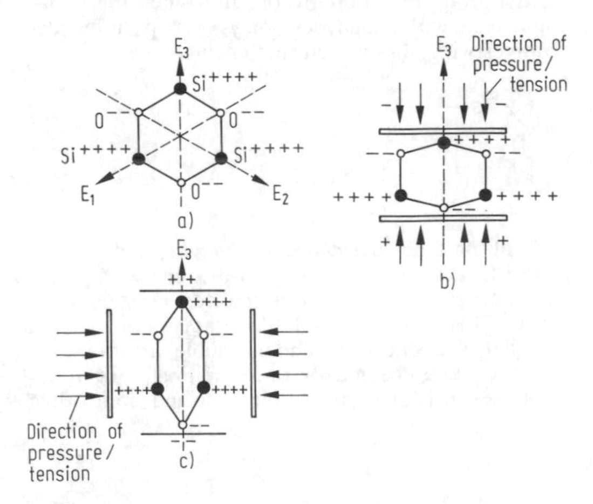

Both direct and inverse piezoelectric effects have important applications; quartz is the most commonly used piezoelectric material. Suitably cut quartz crystals are used as vibrational standards for quartz clocks and watches, transmitters and frequency meters.

They also serve as vibration generators for measurement of layer thicknesses, etc. Electromechanical converters used for piezoelectric measurements of pressure and tension changes are usually made of quartz or lithium niobate, Li_2NbO. *Piezoceramics* (especially titanates and niobates) have been developed for various applications; because of their polycrystalline structure, these are isotropic in their properties. Piezoceramics are used mainly as converters, filters and amplifiers in acoustical electronics.

Piezoceramics: see Piezoelectricity.

Pigment: a colored compound, often insoluble or only very slightly soluble in water, used as a coloring, especially in paints. The most important properties of white P. are their ability to lighten and to cover; of black, gray and other P., their ability to color and cover.

1) ***Inorganic P.*** a) Natural inorganic P. are obtained by washing, drying and grinding colored soils and minerals, e.g. ocher, umbra and shale black. b) Artificial inorganic or *mineral P.* are obtained by precipitation and roasting of inorganic starting materials, e.g. lead white, chromium yellow, red lead, iron oxide yellow, zinc white, titanium white, ultramarine and cobalt blue. c) Metallic P., also called *bronzes*, are obtained by powdering very thin metal foils, e.g. aluminum, brass and gold.

2) ***Organic P.*** a) Natural organic P. may be of animal or vegetable origin (e.g. madder, carmine). b) Synthetic organic P. are classified according to their chemical structure: azo P. (e.g. scarlet 3B), dioxazine P. (e.g. Sirius light blue), Phthalocyanin P., Triarylmethane P. etc. The Dye lakes (see) are a special form of P.

Processing and application. The P. are added to the material to be colored with or without a binder such as lime, unsaturated oils, lacquer or latex. P. are used to color linoleum and plastics and to make paints.

Pigment green B: see Nitroso pigments.

Pigment scarlet 3B: a water-soluble monoazo compound. It is made by diazotization of anthranilic acid and coupling with 2-naphthol-3,6-disulfonic acid. P. is used mainly to make colored lakes for colored paper, wallpaper and synthetic leather.

PIH: see Statins.

Pilocarpine: an imidazole alkaloid present, in 0.2 to 1% concentration, as the main alkaloid in the dried leaves of various species of *Pilocarpus*. P. is a colorless oil; its hydrochloride is used: m.p. 201 to 204 °C, $[\alpha]_D^{20}$ +89-93° (in water). Because of its lactone structure, it is destroyed in alkaline milieu.

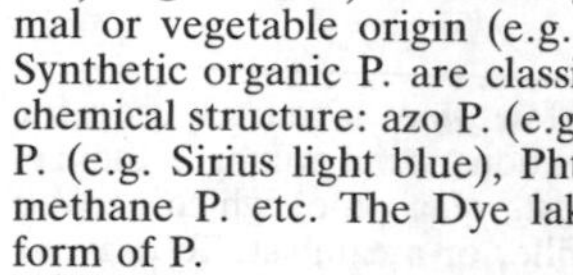

P. is a direct parasympathicomimetic, and is used to treat glaucoma in the form of eye drops and oils.

Pilot plant: a test installation intermediate in scale between the laboratory and a full-sized industrial plant.

Pimelic acid, ***heptanedioic acid***: HOOC-$(CH_2)_5$-COOH, a saturated dicarboxylic acid. P. forms colorless crystalline prisms; m.p. 106 °C, b.p. 212 °C at 1.33 kPa. It is slightly soluble in water, and dissolves readily in alcohol and ether. P. undergoes the reactions typical of saturated aliphatic dicarboxylic acids. At higher temperatures, it cleaves off water and carbon dioxide and cyclizes, forming cyclohexanone. H. can be synthesized by oxidation of cycloheptanone or heptane-1,7-diol, or by alkaline cleavage of hexahydrosalicylic acid, which can be made by catalytic hydrogenation of salicylic acid. P. can be used in the synthesis of polyesters and polyamides.

Pimocide: a diphenylbutylpiperidine derivative (Fig.). P. is used as a neuroleptic.

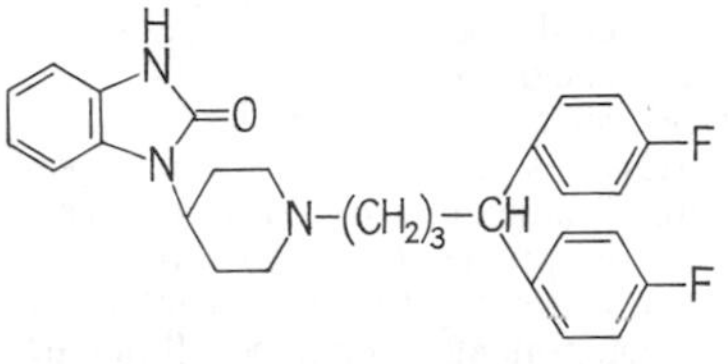

Pimpinellin: see Furocoumarins.

Pinachrome: see Isocyanins.

Pinacol, ***tetramethylethylene glycol***: a dialcohol, colorless needles; m.p. 41 °C, b.p. 174-175 °C.

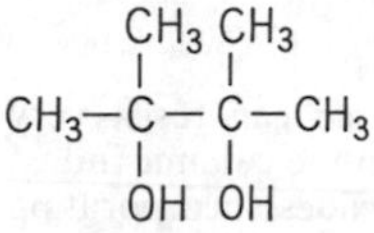

P. is slightly soluble in cold water, and readily soluble in ethanol, ether and hot water. It is obtained by electrolytic reduction of acetone, or by reduction of acetone with sodium amalgam. It can be readily dehydrogenated to 2,3-dimethylbutadiene, or rearranged to pinacolone.

Pinacoline: obsolete term for pinacolone.

Pinacolone, formerly ***pinacoline; 3,3-dimethylbutan-2-one***: CH_3-CO-C$(CH_3)_3$, an aliphatic ketone. P. is a colorless, steam-volatile liquid with an odor like peppermint; m.p. -49.8 °C, b.p. 106 °C, n_D^{20} 1.3952. P. is very slightly soluble in water, but dissolves readily in most organic solvents. It can be obtained through Pinacol-pinacolone rearrangement (see) from pinacol.

Pinacol-pinacolone rearrangement: a nucleophilic 1,2-rearrangement in which a hydroxl group is first protonated in an acidic medium, then water is cleaved off to form a carbenium ion. This is the rate-determining step. The carbenium ion is stabilized by migration of an alkyl or aryl anion. The electron hole which reappears in this step is compensated by loss of a proton. The rearrangement works especially well with ditertiary 1,2-diols, but it is also possible with secondary 1,2-diols, with the formation of the corresponding aldehydes. In cyclic 1,2-diols, which are easily obtained from cycloketones, the P. causes a ring expansion to form a spiro compound with a carbonyl group.

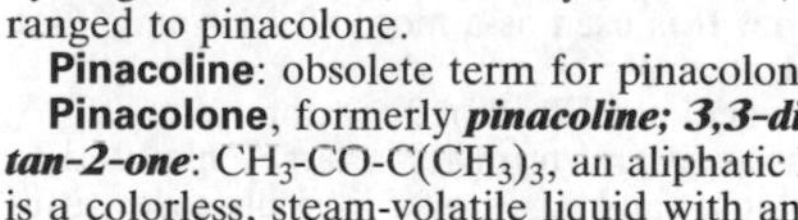

Pinacryptol yellow: a synthetic isocyanin pigment. Like ***pinacryptol green*** (an azine pigment), it is used as a desinsitizer for photographic film for bright-light development.

Pinane: see Monoterpenes.

Pinaverdol: 1,1',6-trimethylisocyanidine iodide, $C_{22}H_{21}N_2I$, a red-violet, water-soluble synthetic pigment, the best-known of the isocyanins. P. is used as a sensitizer in photography.

Pina white: a synthetic pigment made from water-soluble derivatives of anthraquinone or phenanthaquinone and sodium sulfite. It is used as a sensitizer for photographic film for bright-light development.

Pindon: a Rodenticide (see).

Pindone: an Insecticide (see).

Pine needle oil: see Spruce needle oil.

Pine oils: a series of essential oils similar to turpentine oils which are obtained by pyrolysis of the resinous roots of pines or spruces, along with wood vinegar, wood tar and charcoal. Purified P. contain the following main components: α-pinene, β-pinene, (+)-silvestrene (C-carene), limonene, cymol, cadinene, dipentene, toluol, saturated hydrocarbons (e.g. heptane) and camphene. P. are used for various industrial purposes, e.g. the production of lacquers and oil-base paints. Crude P. are used as disinfectants.

Pinging: see Knocking.

Pink pigments: pigments which can resist very high temperatures and are used in the ceramic industry. Chemically, they are tin oxides incorporating foreign metals in the crystal lattice, for example chromium in ***pink red*** or vanadium in ***pink yellow***.

Pink salt, ***ammonium hexachlorostannate***: $(NH_4)_2SnCl_6$, colorless, water-soluble cubic crystals, M_r 367.49, density 2.4. P. is obtained by reaction of a concentrated ammonium chloride solution with tin tetrachloride. It is used as a mordant in dyeing. (see Tin chlorides).

Pipe furnace: see Fixed-bed reactor.

Pipeline: a general term for a large pipe used to transport liquids and gases over long distances, usually from the site of production to a harbor or refinery. The most common use is for Petroleum (see), but fuels and gases (natural gas, city gas, coking gas, industrial gases) are also transported in P. Other substances transported in P. include oxygen, nitrogen, ethene, coal, ores, cement, limestone, sulfur, crude phosphates, carbon dioxide, ammonia, other chemicals, sand, sludge, gravel, ashes, water, etc.

The pipes for modern P. are steel and have diameters up to 2 m. They are laid above or below the ground, depending on the climate, or on the sea floor.

In P. for petroleum and gas, the pressure required to move the medium is generated at intervals (60 to 200 km, maximum 250 km) by pumping stations built into the P. As a rule, rotating pumps are used for liquids and piston or turbine compressors for gas. In P. for petroleum, screw-like channels can be built into the pipes to give the liquid a rotational motion which reduces friction. Some media can be transported in P. only by means of special measures. Coal, ores and other solids are ground to a certain grain size and made into slurries with water or oil; they are separated from the liquid by centrifuges at the destination. Cement is moved pneumatically (mixed with air). Sulfur is melted for transport and kept hot by hot water flowing in a jacket around the pipes.

The longest P. in the world, the "Friendship", leads from Russia (Kuibyschev) to Poland, with branches to Hungary, Czechoslovakia and East Germany (Schwedt). The total length of the P. is more than 5400 km; it has 20 pumping stations and can transport 45 million tons oil per year. Natural gas, ethene and liquid ammonia are also transported by very long P.; a 2500 km line from Algeria to Italy crosses the Mediterranian between Tunesia and Sicily. There is a 2450 km P. for liquid ammonia between Togliatti and Odessa (Ukraine).

Historical. The oldest use of P. is for water. The Chinese used bamboo pipes for water around 5000 BC. The first petroleum P. was built in 1865 in the USA; the pipes were made of wood.

Piperazine: a dibasic heteroaliphatic amine. The base crystallizes as the hexahydrate; m.p. 44 °C. It is very water-soluble. P. is used as a base or in the form of its salts (adipate, citrate, phosphate) as an anthelminthic for ascaride and oxyurene infections.

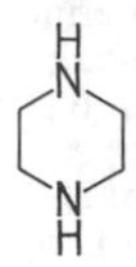

Pipe reactor, ***ideal flow pipe***: a reactor used for continuous chemical reactions; axial mixing in them is negligible. The P. consists of a pipe, which can be empty or filled with a filler or a catalyst. To prevent mixing of successive volume elements, high flow rates are necessary (for liquids, ω 0.5 m s^{-1} and for gases, ω 10 m s^{-1}). The P. is most suitable for rapid reactions and in cases where the desired product can undergo further reaction; the P. permits higher selectivity of product. P. can have lengths up to 3 km; the longer ones are constructed as a series of successive pipes (e.g. in the saponification of chlorobenzene with sodium hydroxide to make phenol).

Piperidine, ***hexahydropyridine***: a colorless liquid with an odor similar to that of ammonia; m.p. -9 °C, b.p. 106 °C, n_D^{20} 1.4530. P. is readily soluble in water and organic solvents. It is a strong base and forms complexes with heavy metal salts. P. is made by catalytic hydrogenation of pyridine. It is the parent compound of a series of alkaloids, the piperidine alkaloids, which includes piperine, coniine, cocaine and tropine. It is present in pepper, due to the hydrolysis of piperine. P. is used in industry as a solvent, catalyst for condensation reactions, antioxidant in lubricants, in the synthesis of pharmaceuticals and as a hardener for epoxide resins.

Piperidine alkaloids: see Alkaloids.

Piperine: piperic acid piperidide, the piperidine alkaloid responsible for the hot taste of black and white pepper (*Piperis nigri* and *Piperis alba*). P. is present in a concentration of 5 to 9% in pepper, and acts as a Synergist (see).

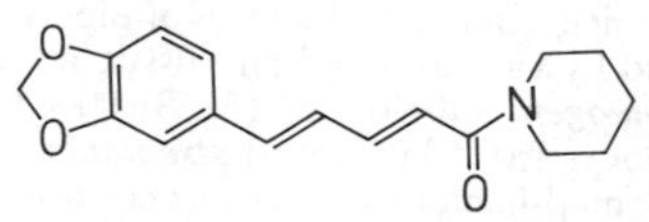

Piperonal, ***3,4-methylenedioxybenzaldehyde, heliotropin***: a colorless, crystalline substance with a pleasant odor; m.p. 37 °C, b.p. 263-264 °C (dec.). P. is steam volatile. In air and light it decomposes, turning brown. P. is present in essential oils, and is also a degradation product of the alkaloid piperine.

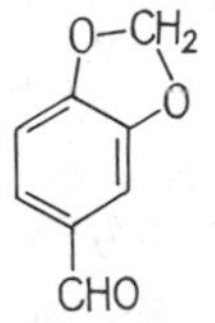

P. can be synthesized by oxidation of isoasafrol or by reaction of protocatechualdehyde with diiodomethane and sodium hydroxide. It is used mainly as an aroma substance in perfumes.

Piperonyl butoxide: see Synergist.

Piscicide: a biologically active substance used to kill fish, for example, Camphechlor (see) or Endosulfan (see). P. are sometimes used in fishery waters before valuable species are set out. 1,1'-Methylene-di-2-naphthol (Squoxin®) is a selective P. used in North American rivers to combat the Sacramento pike while protecting trout and salmon.

Pitch: a viscous to solid and brittle, sticky substance, usually brown to black, which is left as a residue from the distillation of tar and petroleum. P. is classified as ***soft P.*** (m.p. 35 to 50 °C), ***medium*** or ***briquet P.*** (m.p. 60 to 75 °C) and the brittle, easily pulverized ***hard P.*** (m.p. 75 to 90 °C). P. contains free carbon, high-molecular weight polycyclic compounds and soot-like components. It is used for waterproofing, electrical insulation and so on. Most of the P. from continuously working distillation processes is medium P., which is used as a binder for making coal briquets. Addition of heavy oil converts it to soft P., while air blasting converts it to hard P. Part of the P. is converted to liquid products by hydrogen cleavage. Prepared tars, used for streets, steel works and roofing paper, are made by dissolving P. in heavy tar oils. Paints and insulating compounds are made from P.-oil mixtures. Coking of P. produces a low-ash product which is used mainly for production of electrodes.

Pitting corrosion: A local corrosion leading to small pits and crater-like depressions in a metal, and eventually to holes all the way through the piece. It occurs when the surface layer has pores, or the passivity of the material surface is locally lost. The pores or active regions in a protective layer form very small anodes, while the large surrounding area of the protective layer acts as cathode in a macrocorrosion element. As a result, current densities at the anode are large and cause rapid corrosion.

If the redox potential of the corrosion medium is higher than the pitting potential of passivatable chromium steels and chromium-nickel steels, pinprick holes form in the presence of chloride, bromide or iodide ions. Because the chloride ion is the prototype of this type of corrosion, it is also called ***chloride ion corrosion***. P. can also occur in aluminum, copper, nickel and their alloys.

Pitzer tension: see Stereoisomerism 2.1.

Pivalic acid: see Valeric acid.

Piviacid®: see Synthetic fibers.

pK: see Acid-base concepts, section on Brønsted definition.

Placental hormones: Hormones formed by the placenta during pregnancy, including both peptide hormones (see Choriogonadotropin and Choriomammotropin) and a steroid hormone, progesterone (see Gestagens).

Planck's function: symbol φ, the relatively rarely used thermodynamic function $\varphi = -g/T - s - h/T$, where g is the free enthalpy, s is the entropy, h is the enthalpy and T is the absolute temperature. It is analogous to Massieu's function (see).

Plant hormones: see Phytohormones.

Plant toxins: toxins found in certain plants, e.g. Mushroom poisons (see). P. should not be confused with Herbicides (see), which are poisonous to the plant. For greater detail, see Poisons.

Plasma: a gas at a very high temperature (T 3000 K) containing neutral atoms and molecules, ions, electrons, radicals and tiny, incompletely vaporized particles of liquids or solids. The distribution of the particles among the possible forms depends on the temperature of the plasma and the properties of the substance, such as boiling point, dissociation energy and ionization energy. P. exist in flames, electric arcs, spark discharges and heated graphite cuvettes which are used in Atomic spectroscopy (see). The spectrum of light emitted or absorbed by a P. can contain, in addition to the line spectra of the atoms, band spectra and continua. The bands are due to the interactions of molecules and radicals with electromagnetic radiation (see Molecular spectroscopy), while the continua may be due to the solid particles or to collisions between ions and electrons. The band spectra and continua form the spectral background of the P. and often limit the sensitivity of spectroscopic methods. A P. is in thermal equilibrium if all the basic processes such as excitation, dissociation and ionization are in equilibrium with their corresponding reverse processes, so that there are no external losses of energy. The P. used for spectroscopic analysis are often in thermal equilibrium.

In the extreme case, the atomic nuclei in a P. are completely ionized, i.e. they have lost all their electrons. Occasionally the P. state is counted as the fourth state of matter, alongside the solid, liquid and gaseous states.

Plasma albumin: see Albumins.

Plasma expanders, ***plasma substitutes***: substances administered in solution to restore the blood volume in the body after extensive hemorrhage. They are intended to remain in circulation for a certain time, but then to be eliminated or degraded. Dextrans, modified gelatins, hydroxyethyl starches and polyvinylpyrrolidones are used as P.

Plasma gasification: see Synthesis gas.

Plasma kinins: the most important of the Kinins (see), which are tissue hormones. P. are released from the α-globulin fraction of the plasma (kininogens) by the plasma protease kallikrein. Kallikreins (also found in pancreas, kidneys and other organs) in turn are produced from inactive precursors and have various substrate specificities. Plasma kallikrein cleaves a Lys-Arg bond in the kininogen of ***bradykinin*** (kinin 9) to release Arg-Pro-Pro-Gly-Phe-Ser-Pro-Phe-Arg. ***Kallidin*** (kinin 10) is released by pancreatic kallikrein by cleavage of a Met-Lys bond: Lys-Arg-Pro-Pro-Gly-Phe-Ser-Pro-Phe-Arg. ***Methionyllysylbradykinin*** (kinin 11) is Met-Lys-Arg-Pro-Pro-Gly-Phe-Ser-Pro-Phe-Arg. All P. are very quickly degraded by kininases. The pharmacological effects of the three P. differ only quantitatively. The physiological function of the kinin system is still not completely understood. They regulate the blood flow through various vessel systems and capillary permeability, thus affecting the entire organism. They also play a role in pathophysiological processes. Since the first total syntheses (1960), more than 200 analogs of P. have been synthesized.

Plasmal: see Plasmalogens.

Plasmalogens: glycerophospholipids (see Glycerolipids) with a 1-alkenylether group on the glycerol group. The parent compound of the P. is ***plasmenic acid*** (2-acyl-1-alkenyl-sn-glycerol 3-phosphate); its derivatives plasmenylethanolamine, plasmenylserine, plasmenylcholine, etc., are analogous to the derivatives of phosphatidic acid.

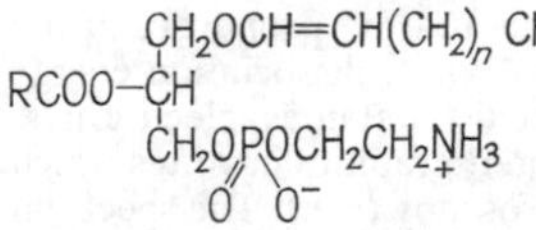

$$\begin{array}{l} \quad\quad\quad\;\; CH_2OCH{=}CH(CH_2)_n\,CH_3 \\ RCOO{-}CH \\ \quad\quad\quad\;\; CH_2OP(=O)(O^-)OCH_2CH_2\overset{+}{N}H_3 \end{array}$$

Plasmenylethanolamine
n = 13, 15; R = Fatty acid group

P. occur widely in plants, animals and microorganisms as the components of phospholipids. The highest concentrations are found in the heart, skeletal muscle and myelin (brain, nerves). The lecithin from bovine heart is very rich in P. P. was discovered by Feulgen, because of the violet color it develops when cell components are treated with fuchsin/sulfurous acid. Mild acid-catalysed hydrolysis of the enol ether group releases a mixture of the fatty aldehydes hexadecanal and octadecanal; this mixture was named *plasmal*, and the color reaction it causes is the *plasmal reaction*. Mild alkaline hydrolysis of plasmenylcholine yields 1-alkenyl-sn-glycero-3-phosphocholine (plasmalogen lysolecithin).

Plasma substitutes: same as Plasma expanders (see).

Plasmenic acid: see Plasmalogens.

Plasmids: extrachromosomal ring-shaped molecules of DNA found in bacteria; they are transmitted from one strain to the next, thus permitting exchange of genetic information. The ***R-plasmids*** or resistance factors are very significant, because they transfer resistance to certain antibiotics. Transfer between different species of bacteria has been observed.

Plasmin, ***fibrinolysin***: a serine protease which catalyses hydrolysis of peptide bonds and the esters of arginine and lysine. P. is responsible for the degradation of fibrin clots which can cause blockage of blood vessels. It is released by various blood and tissue activators from ***plasminogen***, a β-globulin (M_r 81,000). ***Urokinase*** from kidneys (M_r 53,000) and ***streptokinase*** (M_r 47,000) from β-hemolytic streptococci are well known activators. The latter is used medicinally.

Plastic: a synthetic, organic solid which can easily be shaped. Although they are closely related, Synthetic fibers (see) are generally not considered to be P. The silicones and fluorocarbons are also related to P. The key characteristic of P. is their Plasticity (see). There are two basic types. The ***thermoplastics*** can be repeatedly softened by heating, while the ***thermoset plastics*** are hardened after initial casting and cannot be remelted. Most thermoplastics are formed by polymerization (e.g. polyolefins and vinyl polymers), but some are polycondensates (e.g. polyamides). Most thermoset plastics are polycondensates or polyaddition products, such as phenol, urea, melamine, epoxide and polyester resins.

Properties. The P. have low densities, and most are good electrical and thermal insulators. They often have low frictional cofficients and high damping capacities; they are easily dyed, formed and worked, and can be welded and glued. Products made of P. do not need any special surface protection, they have no odor or taste, and some are stable to aggressive and corrosive chemicals, and other influences. Maintenance costs are low. Some disadvantages are their low modulus of elasticity, their low tensile strength, their large thermal expansion coefficients, aging, and the low temperatures at which they soften or decompose. Many P. begin to soften even at 80 to 100 °C, but this can be improved by increasing the degree of polymerization, mixed polymerization, incorporation of inorganic compounds or adding fillers.

Production. 1) P. can be made by conversion of natural substances, usually cellulose or proteins, into either thermoset plastics, such as artificial horn, or thermoplastics, such as cellulose esters.

2) In the synthesis of P., monomeric molecular precursors are linked into macromolecules by Polymerization (see), Polycondensation (see) or Polyaddition (see).

Processing of macromolecular raw materials to finished or semi-finished products almost always depends on heat and pressure, and is done in solution or suspension. Various substances, including softeners, fillers, stabilizers, lubricants, dyes, antistatics, flame retardants and microbe retardants (biocides) are added during *mixing* in rotating drums, stirrers or ball mills. The mixtures are pressed into forms, or crude films or impregnated ribbons are pressed into plates and blocks. During *extrusion*, heated and softened raw material is driven through an opening to give rods with uniform cross section, pipes, films and tubes. The extrusion process can be followed by *blowing* to make hollow objects, such as bottles or balloons. *Calendering* is the term for rolling the raw material between several rollers (calenders), forming films and ribbons. In *drawing, blowing* or *vacuum stamping*, P. plates or films are heated and shaped into nonplanar objects, such as bowls or bottles, by pressurized air or vacuum. In *spinning*, the dissolved or molten raw material is pressed through fine openings into a precipitation bath (wet spinning) or hot air (dry spinning) to form an infinite thread. Bristles and ribbons can also be produced in this way. In *casting*, the dissolved, molten or suspended raw material is poured into forms, onto a solid base or into a precipitation bath to form plates, blocks or films. Foams are formed in several ways: by using emulsifiers, by substances which generate gases, or by addition of a salt which is later dissolved out.

Overview of plastics

Chemical name	Ab.	Basic unit	Main applications
Modified natural products			
1) *Cellulose and its derivatives*			
Vulcan fiber	VF		Suitcases, stencils, fillers
Cellulose nitrate	CN	$-[cell-(ONO_2)_2]_n-$	Injection casting, paints
Celluloid			
Cellophane			Foils, plates
Cellulose acetate	CA	$-[cell-(O-COCH_3)_3]_n-$	Injection casting, safety films, foils, paints
Cellulose mixed esters			
Cellulose acetobutyrate	CAB		Window coatings, toilet articles, oil pipes, foils
Cellulose propionate	CAP		
Cellulose ethers			
Methyl cellulose	MC		Binders, dispersion agents
Ethyl cellulose	EC		Spinning spools, electrical materials, paints
2) *Casein-formaldehyde condensates*			
Artificial horn			Toys, jewelry
Synthetic plastics			
1) *Polymerization products*			
Pure polyolefins	PO		
Polyethylene	PE	$-[CH_2-CH_2]_n-$	Cable coatings, molded articles, pipes, foils, hoses
Polypropylene	PP	$-[CH_2-CH(CH_3)]_n-$	Electrical apliances, apparatus
Polyisobutylene	PIB	$-[CH_2-C(CH_3)_2]_n-$	Coating of textiles, oils, corrosion protection
Polybut-1-ene	PB	$-[CH_2-CH(CH_2-CH_3)]_n-$	Hot water pipes, chemical apparatus
Poly-4-methylpent-1-ene	PMB	$-[CH_2-CH(CH_2-CH(CH_3)_2)]_n-$	Special packaging foils
Ethylene copolymers			Special injection-molded objects, foils, plates
Ethylene-propylene copolymers			
Ethylene vinyl acetate copolymers	EVA		Rubber-like products
Polyvinyl chloride	PVC	$-[CH_2-CH(Cl)]_n-$	Floor coverings, molded objects, plates, foils, gutters, insulation
Polyvinylidene chloride	PVDC	$-[CH_2-CCl_2]_n-$	Threads, bristles, ropes, foils, filter and upholstery cloth

Plastic

Overview of plastics

Chemical name	Ab.	Basic unit	Main applications
Polybutadiene		$-[CH_2-CH=CH-CH_2]_n-$	Elastics
Polyisoprene		$-[CH_2-C(CH_3)=CH-CH_2]_n-$	Elastics
Polychloroprene	CR	$-[CH_2-CH=C(Cl)-CH_2]_n-$	Oil-resistant elastics
Polystyrene	PS	$-[CH_2-CH(C_6H_5)]_n-$	Cheap molded objects as foam, in buildings and packaging
Butadiene-styrene copolymers	BS		Office articles, coverings, housings for appliances
Acrylonitrile-butadiene-styrene copolymers	ABS		Construction materials, vehicles, household appliances, metalized plastics
Styrene-acrylonitrile copolymers	SAN		Motor vehicles, housings on electrical equipment
Polyacrylonitrile	PAN	$-[CH_2-CH(C{\equiv}N)]_n-$	Synthetic fibers
Polyacrylates $R = -H, -CH_3, -C_2H_5$, etc.		$-[CH_2-CH(C(=O)OR)]_n-$	Glues, paints, mortars, and concrete additives
Polymethacrylates $R = -CH_3$ $= -CH_2-CH_3$ $= -CH_2-CH_2-CH_3$	PMMA PMAA	$-[CH_2-C(CH_3)(C(=O)OR)]_n-$	Cast acrylic glasses, injected and extruded forms, lights for streets, offices, and dwellings, optical lenses, prostheses, paints, concrete additives, optical fibers
Polyvinylidene cyanide		$-[CH_2-C(C{\equiv}N)_2]_n-$	Special synthetic fibers
Polyoxymethylene Polyformaldehyde	POM	$-[CH_2-O]_n-$	Material for gears, pump housings, electrical housings
Polymeric fluorocarbons			
Polytetrafluoroethylene	PTFE	$-[CF_2-CFCl]_n-$	No-maintenance bearings, caulks, coatings
Polytrifluorochloroethylene	PCTFE	$-[CF_2-CF_2]_n-$	Pipes, profile rods, foils for high-frequency technology, chemical apparatus, laboratory devices
Polyvinylfluoride	PVF	$-[CH_2-CH(F)]_n-$	Non-stick coatings packaging, greenhouses
Polyvinylidene fluoride	VF_2 or PVDF	$-[CH_2-CF_2]_n-$	Injection castings, foils, electrical transformers
Polyvinylpyrolidone		$-[CH_2-CH(N\text{-pyrrolidone})]_n-$	Textile conditioners, binders, blood extenders
Polyethylene oxide		$-[CH_2-CH_2-O]_n-$	Nonionic surfactants, softeners, textile conditioners
Polyethylenimine		$-[CH_2-CH_2-NH]_n-$	Paper industry, textile conditioners
Polyvinylcarbazene	PVK	$-[CH_2-CH(N\text{-carbazole})]_n-$	Parts for electrical devices subject to high mechanical and thermal stress

Overview of plastics

Chemical name	Ab.	Basic unit	Main applications
Polyvinylacetate	PVAC	$-\!\!\{CH_2-CH(O-C(=O)-CH_3)\}_n$	Glues, paints, putty
Polyvinylalcohol	PVAL	$-\!\!\{CH_2-CH(OH)\}_n$	Glues, impregnation, special foils
Polyvinyl ethers		$-\!\!\{CH_2-CH(OR)\}_n$	
$R = -CH_3$ Polyvinyl methyl ether			
$R = -C_2H_5$ Polyvinyl ethyl ether			
$R = -C_3H_7$ Polyvinyl propyl ether			
Polyvinyl acetals		$-\!\!\{CH_2-CH-CH_2-CH\}_n$ (both CH linked via O to CH–R)	
$R = -H$ Polyformaldehyde acetal	PVFO		Insulation, coatings
$R = -CH_3$ Polyacetaldehyde acetal			
$R = -C_3H_3$ Polybutyraldehyde	PVB		Metal foils, bake-on paint, oil- and gasoline-resistant hoses
Polyamide 6	PA-6	$[-NH-(CH_2)_6-C(=O)-]_n$	Molded parts, synthetic fibers, foils
2) *Polycondensates*			
Phenol-formaldehyde condensates	PF		Paints, glues, pressboard
Urea-formaldehyde condensates	UF		Molded objects, plywood glue
Melamine-formaldehyde condensates	MF		Pressboard
Dicyanodiamide resins			Pressed objects, glues
Aniline resins			Paints, special pressed objects
Polyesters		$-\!\!\{CO-R-COO-R'-O\}_n$	
Linear, saturated polyesters			Synthetic fibers
Cross-linked, saturated polyesters			Alkyde resins
Unsaturated polyester resins	UP		Paints, casting resins, pressboards, glass-fiber, reinforced polyesters, allyl resin molded objects
Polyamide	PA		Injection casting, foils, synthetic fibers
Polyamide 6,6	PA-6,6	$[-NH-(CH_2)_6-N(H)-C(=O)-(CH_2)_4-C(=O)-]_n$	
Polycarbonates	PC	$[-O-C_6H_5-C(CH_3)_2-C_6H_5-O-CO-]_n$	Electrical housings, household objects, foils
Polysiloxane/silicones	SI	$[-Si(CH_3)_2-O-Si(CH_3)_2-O-]_n$	Silicon oils, pastes, lubricants, impregnating compounds
Polyphenylene oxide	PPO	$-\!\!\{C_6H_5-O\}_n$	Motor vehicles and electrical devices, medical apparatus
Polyalkene polysulfides		$-\!\!\{CH_2-CH_2-S_x\}_n$	Fire-resistant construction materials
Polyamides		$[-N(CO)_2C_6H_2(CO)_2N-C_6H_5-O-C_6H_5-]_n$ (pyromellitimide unit)	Space vehicles, nuclear energy plants, electronics

Overview of plastics

Chemical name	Ab.	Basic unit	Main applications
Polyhydantoins		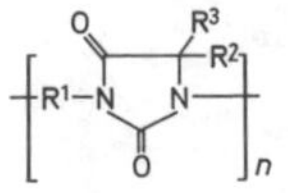	Special films for electrical insulation
Polyparabanic acids		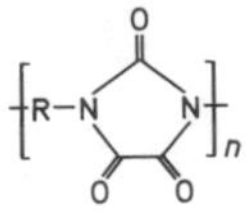	Coatings and glues
Polyisocyanurates		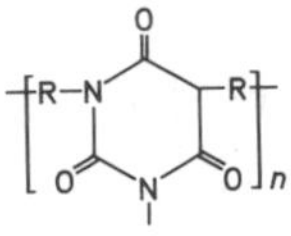	Light concrete for buildings and insulation in vehicles and ships
Polycarbodiimides		$-[R-N=C=N-R]_n-$	Insulation, coatings and molded objects
Triazine polymers Triazine A resins		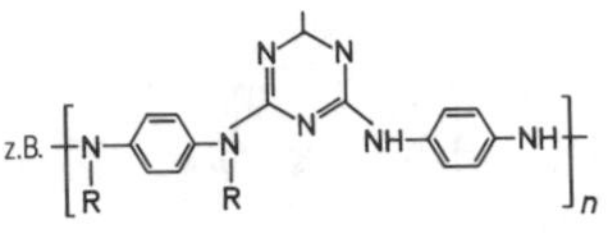 z.B.	Press-molded objects and laminates
Polyphosphazenes	PNF	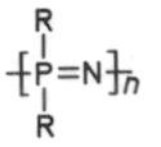 $-[P(R)_2=N]_n-$	Additives for elastics, caulks, damping bearings and arctic fuel lines
Polysulfones Polyphenylsulfone		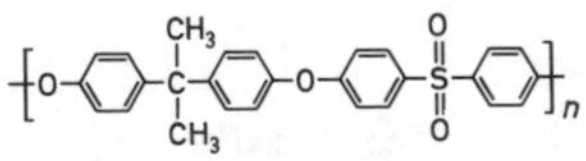	Coatings in the electrical industry and vehicles
Polybenzimidazene Poly-2,2′-(m-phenylene)-5,5′-benzimidazine	PBI	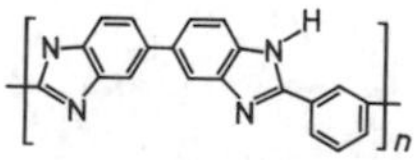	PBI fibers for protective clothing, joinings for heat protection in military and space applications
Polyphenylene quinoxaline	PPQ	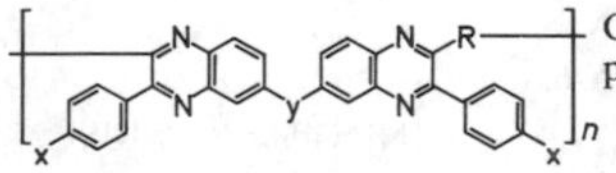	Coatings to withstand high temperatures, joinings
Poly-p-xylylene $R^1=R^2=R^3=R^4=X=H$ Poly-p-xylene		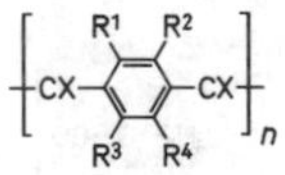	Extremely durable surface films
Polyphenylene sulfide	PPS	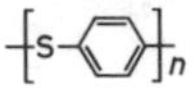 $-[S-C_6H_4]_n-$	Construction material, coatings on metals, household articles such as pots and pans
Polyphenylene		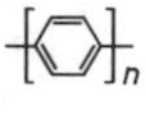 $-[C_6H_4]_n-$	Special material for electrical and space technology
Polyimidazopyrrolone (Pyrrone)		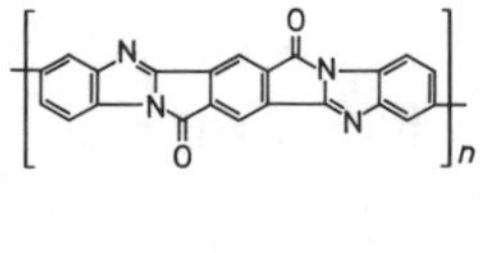	Special material used in electronics and nuclear and space technologies
3) *Polyadducts*			
Polyepoxides Epoxide resin precondensates	EP	$-[O-C_6H_5-C(CH_3)_2-C_6H_5-O-CH_2-CH(OH)-CH_2]_n-$	Casting resins, laminating resin, molded parts glue, paints, construction materials reinforced with fiberglass

Overview of plastics

Chemical name	Ab.	Basic unit	Main applications
Polyurethanes Linear polyurethanes	PUR	$-[CO-NH-R^1-NH-CO-O-R^2-O]_n-$	Injection casting, foils, synthetic fibers
Cross-linked polyurethanes			Glues, paint binders, coatings molded objects
Phenoxy polymers		$-[C_6H_4-O-CH_2-CH(OH)-CH_2-O]_n-$	Molded objects, coatings
Polyalkylene glycols e.g. Polyethylene glycol		$-[CH_2-CH_2-O]_n-$	Textile conditioners, surfactants, polyurethane precondensates
Chlorinated polyethers		$-[CH_2-C(CH_2Cl)_2-CH_2-O]_n-$	Injection casting, foils

Applications. Because they can be engineered to have almost any desired properties, P. are enormously useful, and world production of them has already exceeded that of such important materials as aluminum and copper. About 50% of P. are used in the metal processing and construction industries, about 40% in light industry and packaging, and about 10% in consumer goods.

Current developments are directed towards better methods of production, reduction of toxicity, lower residual monomer contents, production of P. which can withstand high temperatures, and of nonmetallic electrical conductors.

Historical. The oldest P. are cellulose derivatives, which came into production in the second half of the 19th century. Other converted natural products were developed around the turn of the century, such as the development of artificial horn from casein and formaldehyde. The first synthetic P. were developed about this time: Bakelite (from condensation of phenol with formaldehyde to make a synthetic resin) in 1909 by Baekeland, and the first experimental urea resins (urea and formaldehyde) and dicyanodiamide resins (dicyanodiamide and formaldehyde) in 1918 by John and Wallasch. Until about 1930, the chief developments concerned the chemistry and technology of cellulose and phenol resin plastics, and the first work toward polymerization was begun. E. Fischer and H. Staudinger coined the term "macromolecule". Around 1926, the first PVC was produced industrially, followed in 1928 by polymethacrylate, in 1930 by polystyrene, polyacrylonitrile and polyvinylacetate, in 1931 by polyvinylacetal, polyethylene oxide and polyisobutylene, in 1934 by polyvinylcarbazole and post-chlorinated PVC, in 1935 by high-pressure polyethylene, in 1936 by polyvinyl ethers, in 1937 by polyamide (after the preliminary work by Carothers in 1931) and polyurethane, in 1938 by polyethylenimine and in 1939 by polyvinylpyrrolidone and polytetrafluoroethylene. The following years brought a rapid increase in the production figures and further refinements in the types of P. (work of Ziegler, Natta, etc.). The P. developed since about 1960 include polypropylene, polycarbonates, low-pressure polyethylene, polyester and epoxide resins, polyformaldehyde and various sterically uniform polymers such as *cis*-1,4-polyisoprene, polybut-1-ene and poly-4-methylpent-1-ene, polyimides, polyimines, polyphenylene oxide, polysulfones, poly-*p*-xylene and ionomers.

Plastic concrete, *plastic mortar*: special concretes and mortar which have greater tensile strength and cohesiveness than pure cement concrete, while retaining the same compressional strength. P. contain synthetic resins which harden cold, mainly epoxide and polyester resins, which harden with fillers to nonabrading, chemically resistant surfaces. P. are used to make concrete pads in the open, in warehouses and factories, and to close cracks in pre-cast concrete. ***Plastic cement concrete*** contains 5 to 20% (by mass) of a plastic suspension in addition to cement and quartz sand. This plastic binder can be, for example, polyvinyl acetate. Such materials are used for seamless ribbons of new concrete, in highways, railroad platforms and stairs, for example. ***Foam concrete*** is a cement concrete with a plastic additive. Polystyrene spheres or polystyrene foam scraps are mixed with cement or cement mortar as a binder, and the product is used in the form of plates or blocks as a heat-insulating construction material.

Plastic explosives: explosives which are plastically deformable, kneadable and adhesive, which can be detonated by ignition. P. consist of one or more explosives, such as dynamite, 2,4,6-trinitrotoluene (TNT), pentaerythritol tetranitrate (PETN) and 1,3,5-trinitro-1,3,5-triazacyclohexane (hexogen), to which vaseline, rubber solution or a highly viscous oil is added. This makes the powdered explosive plastic. The power of the P. is somewhat less than that of the pure explosives. They are detonated by an ignition cap and fuse, which are pressed into the plastic mass at the site of use.

The peculiarity of the P. is that the pure explosives, which are very sensitive to temperature changes, are made insensitive to temperature changes, impact, friction and moisture by incorporation into P.; they can therefore be carried easily and inconspicuously, for example, under the clothing. At the site of application, they can be formed into grenades, pressed into hollows or stuck onto the object like chewing gum. For these reasons, they are favored by saboteurs and terrorists (***plastic bombs***), but are also used in quarries and by military forces.

Plasticity: the ability of solids to flow in response

to external forces, including the force of gravity on the mass of the solid. In plastic materials, flow occurs under lower stress than is required to break the material.

In plastic deformation of a metal, the atoms in the lattice are displaced, with rupture of their chemical bonds, but without complete loss of cohesion. The unit cell of the lattice is not changed, but larger portions of the lattice shift with respect to each other, until the atoms reach a new minimum in the lattice potential. This shifting, which is the cause of P. and associated properties (e.g. strength), depends on lattice defects and other irregularities in the crystal structure. Because of these lattice defects, atoms can leave their normal lattice sites and migrate to other lattice or interlattice sites. The result of such possibilities for rearrangement is sliding in crystallographically oriented P. (crystal P.); it occurs by means of displacement.

In addition to crystallographically oriented P., there is a non-oriented P. (banal P., amorphous P.), which does not involve sliding along crystal planes, but irreversible crystal distortion. At lower temperatures, amorphous P. is more common, and at higher temperatures, crystal P. In some alloys, P. is absent, which indicates that the bonding here is different from metallic; this leads to cleavage planes or breaks.

The technological utility of metals is based on the property of P., which makes it possible to shape them by hammering, rolling, drawing or pressing.

Plastic mortar: same as Plastic concrete (see).

Plastic spraying: a method of applying plastics by Flame spraying (see) onto metallic and nonmetallic materials. The material being applied, e.g. polyethylene or polyvinylchloride, is sprayed as a powder through a flame onto the pre-heated and roughened surface of the material (see Surface pretreatment). On the heated surface, the individual powder particles melt into a homogeneous layer, between 0.3 and 1.2 mm thick, depending on the type of plastic and the application.

Plastifiers: same as Softeners (see).

Plastilin: a modelling "clay" made of kaolin, zinc oxide, chalk, pigments, oils and waxes.

Plastoquinone: abb. ***PQ-9***: a benzo-1,4-quinone with an isoprenoid side chain which acts as a redox system in the chloroplasts of plants. It is related to ubiquinone, and plays a part in photosynthesis.

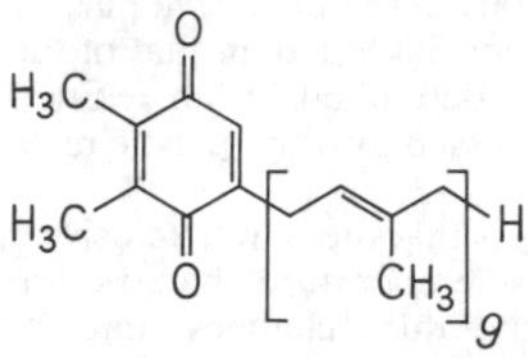

Platelet activating factor: see PAF.

Plate number. 1) ***Practical P.***, the number of exchange plates built into a distillation, extraction or sorption column; for example, bell, sieve, or lattice plates.

2) ***Theoretical P.*** The number of exchange plates (see Distillation) theoretically required for a separation. If it is assumed that the liquid and vapor phases of a distillation are in equilibrium at each practical plate, each plate provides one ***separation stage***. This means that at the first plate of a distillation column, the liquid has the molar fraction x_1 of the higher-boiling component, and is in equilibrium with a vapor with the molar fraction y_1 of this component.

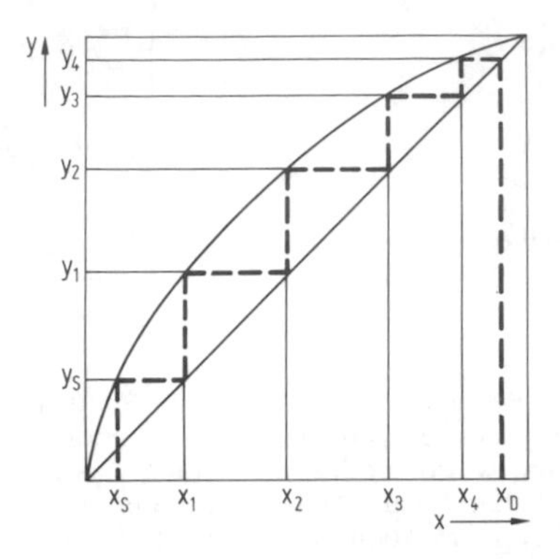

Determination of the theoretical plate number in an infinite reflux situation. x is the molar fraction of the more volatile component, x_s is the molar fraction of this component in the sump, x_r is its fraction in the reflux, and y is the molar fraction of the more volatile component in the vapor.

When this vapor condenses on the second plate, the resulting liquid contains the molar fraction x_2 of the more volatile component. It is in equilibrium with a vapor containing y_2 of the more volatile component, and this vapor condenses on the third plate. In the figure, the enrichment levels of the higher-boiling components are drawn in as separation stages. If the entire condensate (x_r) is returned as reflux (infinite reflux), the enrichment of x_s to x_r requires at least 5 separation stages, i.e. 5 theoretical plates. The number of separations determined in this fashion is called the ***minimum P.***. This is understood as the minimum P. required for the desired separation.

3) ***Equivalent P.*** In filled columns, in which there are no built-in plates, there are separation stages or transition units analogous to definition 2). These are equivalent to the theoretical P. of plate columns with approximately the same dimensions and operating conditions.

Platfining: a process of simultaneous hydrogenation of alkenes and removal of sulfur and nitrogen compounds from hydrocarbon mixtures which contain high proportions of aromatics.

Platforming process: see Reforming.

Plating: the combination of two or more metal layers (base and plate) to form a composite material. It serves to combine the desirable properties of the base metal (such as strength) with those of the plate (such as resistance to corrosion or wear, or electrical conductivity). P. is used, for example, to apply a layer of aluminum, copper, nickel, titanium or its alloys, or austenitic chrome-nickel steel to steel bands, sheets or pipes. It is also often used for decorative purposes: copper or silver ornaments are plated with gold (see

Talmi gold). P. can be achieved by casting or cast plating (e.g. in the production of special steel slabbing), solder plating (bonding of the metal layers by a readily melted and highly diffusible bonding metal), rolling (rolling two oxide-free ribbons or plates together at welding temperature), or by explosion plating (mechanical bonding of the base metal and plate by the pressure of an explosion, without formation of an alloy at the boundary between them). Large surface metallic protective layers can also be made by weld plating. Strictly speaking, metallic layers applied by spraying, dipping, coating or galvanically are not plate; they are generally much thinner. The protective layers on steel chemical apparatus should not be thicker than 1/10 the steel thickness; for corrosion or wear protection, the layer of plate should be 2 to 4 mm.

Platinit: a proprietary name for an iron-nickel alloy consisting of 40 to 46% nickel, the remainder iron. It has a thermal expansion coefficient of $\alpha = 5$ to $10 \cdot 10^{-6} K^{-1}$ and is used for melting into all types of glass. The name refers to the fact that this alloy can replace platinum as a melted in wire.

Platinum, symbol ***Pt***: chemical element from group VIIIb of the periodic system, and the primary example of a Platinum metal (see); a noble metal, *Z* 78, with natural isotopes with mass numbers 195 (33.8%), 194 (32.9%), 196 (25.3%), 198 (7.21%), 192 (0.78%) and 190 (0.0127%); the isotopes ^{190}Pt and ^{192}Pt are weakly radioactive, with half-lives of $6 \cdot 10^{11}$ and 10^{15} a, respectively; atomic mass 195.09, valencies usually II or IV, less commonly 0, I, III, V, VI; density 21.45, m.p. 1769.3 °C, b.p. 3830 °C, electrical conductivity 10.15 Sm mm^{-2}, standard electrode potential (Pt/Pt^{2+}) +1.5 V.

Properties. P. is a gray-white, shiny, very ductile metal which can be hammered and welded when hot. Its crystals are cubic face-centered and readily absorb hydrogen and oxygen in a manner which activates these elements, so P. is frequently used as a catalyst for hydrogenation and oxidation. The degree of uptake, especially of hydrogen, depends greatly on the degree of dividedness of the P. (see Platinum black, Platinum sponge). P. is more resistant to acids than palladium is, and is not attacked by pure mineral acids or potassium hydrogensulfate melts. On the other hand, P. is readily dissolved by aqua regia, and in the presence of oxygen, it will dissolve slowly even in hydrochloric acid; hydrogen chloride is also an oxidant for P. Fluorine and chlorine react with P. only above about 500 °C.

When P. laboratory equipment is used, it must be remembered that alkali peroxides, which are also easily formed from alkali hydroxides in the presence of P., attack P. rapidly. In addition, when it is hot, P. binds phosphorus, silicon, lead, arsenic, antimony, sulfur and selenium. Compounds of these elements must not be heated in platinum dishes under reducing conditions. Reducing flame gases also make P. brittle through inclusion of carbon or absorption of hydrogen.

Analysis. P. is detected by formation of yellow ammonium hexachloroplatinate, $(NH_4)_2[PtCl_6]$. It can be determined quantitatively by precipitating platinum sulfide or $(NH_4)_2[PtCl_6]$, then heating the precipitate to decompose it to the element; or it can be reductively precipitated directly from the solution. P. is also determined by atomic spectroscopy.

Occurrence. P. makes up about $5 \cdot 10^{-7}$% of the earth's crust. Like gold, it is found as the element, also in sands and alluvial deposits, usually associated with other platinum metals and iron, lead, copper, silver and gold. In such alluvial deposits, P. is present in the form of fine grains and leaflets. Pure P. minerals are very rare; examples are sperrylite, $PtAs_2$, and cooperite, PtS. The P.-containing ores of other metals are important sources of P., and some of the most important of these are the magnetic pyrites containing copper and nickel from the Sudbury district of Canada.

Production. Ore concentrates made by wet mechanical or flotation processes are further processed in a variety of ways. The P.-containing nickel and copper ores of Sudbury are are subjected to the acetone process: first the copper and nickel are removed, and then the residues are melted with lead and cupellated. The resulting alloy is treated with sulfuric acid, which dissolves the silver and part of the palladium. Most of the palladium, the remaining platinum metals and gold remain in the residue, which is then treated with aqua regia to dissolve the gold, platinum and palladium. Gold is precipitated from the solution by a reducing agent which does not reduce P., such as iron(II) sulfate. P. is then precipitated as $(NH_4)_2[PtCl_6]$, and is converted to the metal by heating.

The P. concentrates obtained from aluvial ores are processed by the wet method (aqua regia method): they are heated in aqua regia, which dissolves palladium, indium, rhodium and gold as well as P.; the residue contains osmiridium, small amounts of rhodium and P. and slag. Here, too, the P. is precipitated as $(NH_4)_2[PtCl_6]$. Repetition of the precipitation and dissolving steps is used to prepare a purified P. which still contains small amounts of iridium, rhodium, ruthenium and iron. For further purification, this metal is melted with lead, and the resulting alloy is treated with nitric acid. P. remains in the residue, from which it is extracted with aqua regia and again precipitated with ammonium chloride. Further refining eventually yields a P. with a purity of 99.999%.

Application. P. is used to make laboratory apparatus, anode materials for electrochemical processes, galvanic elements, electrical contacts, spinerets and jewelry. Since it has the same thermal expansion coefficient as glass, P. can be fused into glass and used as an electrode in glass apparatus. P. is used as a catalyst for hydrogenation, dehydrogenation, reforming, oxidation and cracking processes. It is also widely used in the form of its alloys (see Platinum alloys; Titanium alloys; Dental alloys).

Historical. P. was known to the Mayas of Central America before the 15th century, and was used with gold for jewelry. The Spanish gold-seekers regarded the heavy metal with a silvery sheen as an undesirable impurity in the gold, and named it "platina" (the diminutive form of the Spanish "plata" = "silver"). In 1735, P. was brought to Europe, and in 1759 was recognized by Watson as a new element.

Platinum alloys: alloys with the platinum metals as the main component. Platinum metals are miscible with one another in the liquid state. Platinum, iridium, palladium and rhodium form mixed crystals when they solidify. In the platinum-iridium and platinum-gold systems, there are mixing gaps at lower temperatures. Osmium and iridium also form alloys, as do platinum and ruthenium. Laboratory crucibles and electrodes are made of an alloy of 97% platinum and 3% iridium, or in some cases 99% platinum and 0.3% iridium or 95% platinum and 5% gold (***apparatus platinum***) because it holds its shape better than pure platinum. Injection needles are made of platinum-iridium alloys. The stability of platinum can also be increased by dispersion hardening through addition of zirconium dioxide, ZrO_2 (100 ppm). An alloy of 95% platinum and 5% rhodium is used as a catalytic screen in ammonia combustion. Melting dishes for special glasses and spinerets for synthetic fibers are made of platinum-rhodium or platinum-ruthenium alloys. P. are used for some applications in electronics, e.g. osmium and ruthenium alloys to make electrical contacts. The alloy consisting of 10 to 30% rhodium and the remainder platinum can be used at temperatures up to 1500 °C as a heat conductor. The most important thermocouple pair in industry is Pt/Rh-Pt, which permits measurements in the range of 800-1600 °C. The Pt/Rh arm consists of 90% platinum and 10% rhodium. Chlorine is produced at anodes consisting of 90% platinum and 10% iridium. ***Jeweler's platinum*** is an alloy of 96% platinum and 4% copper, 96% platinum and 4% lead, or, in a few cases, 90% platinum and 10% iridium. P. are also used as Dental alloys (see). Alloy combinations of ruthenium, iridium and osmium are used as the tips of pens. The standard meter bar kept in Paris consists of an alloy of 90% platinum and 10% iridium. Some of the platinum metals improve the resistance of Titanium alloys (see) to corrosion.

Platinum asbestos: a catalyst often used in laboratories; it is made by precipitation of platinum on asbestos. A solution of hexachloroplatinic acid in methanol is applied to asbestos which has been boiled with hydrochloric acid and roasted. The asbestos saturated with hexachloroplatinic acid is dried and roasted. P. was also used as an industrial catalyst in the past, but has now been replaced by cheaper catalysts for most applications.

Platinum black: very fine platinum powder which activates hydrogen and oxygen, and is therefore a good catalyst. P. is made by reduction of platinum chloride solution with formaldehyde and sodium hydroxide solution, followed by careful washing of the precipitate and vacuum drying over phosphorus(V) oxide.

Platinum brine: a colloidal platinum solution which can activate oxygen and hydrogen, and therefore act as a catalyst. P. is made by reduction of hexachloroplatinate in weakly alkaline solution with hydrazine hydrate in the presence of a protective colloid.

Platinum chlorides: ***Platinum(II) chloride***, $PtCl_2$, is known in two modifications: a reddish black, hexameric β-form and an olive-green α-form. The M_r is 266.00, density 6.05, m.p. 581 °C (dec.). $PtCl_2$ is obtained from the elements around 500 °C. It is insoluble in water, and forms complexes of the type $PtCl_2L_2$ with neutral ligands. With alkali chlorides, it reacts to form red, planar tetrachloroplatinates(II), $M_2[PtCl_4]$, in which the crystal lattice is formed by parallel stacking of the complex anions. An example is potassium tetrachloroplatinate(II), $K_2[PtCl_4]$, which forms tetragonal, red-brown crystals; M_r 415.11, density 3.38. ***Platinum(IV) chloride***, $PtCl_4$, forms red-brown crystals; M_r 336.90, density 4.303, m.p. 370 °C (dec.). $PtCl_4$ is readily soluble in water and acetone. It is obtained by reaction of platinum with chlorine at 250 to 300 °C, or by heating hexachloroplatinic(IV) acid to 300 °C. $PtCl_4$ reacts with alkali chlorides to form alkali hexachloroplatinates(IV), $M_2[PtCl_6]$; an example is potassium hexachloroplatinate(IV), $K_2[PtCl_6]$, yellow, cubic crystals; M_r 486.01, density 3.499, m.p. 250 °C. Reaction with acids yields Hexachloroplatinic(IV) acid (see).

Platinum complexes: coordination compounds of platinum. The most common are square planar platinum(II) complexes and octahedral platinum(IV) complexes, while tetrahedral complexes such as $Pt(P\{C_6H_5\}_3)_4$ and $Pt(PF_3)_4$ and the octahedral $[PtF_6]^-$ are examples of the less common oxidation states 0 and +5. Platinum(II) cationic, anionic and neutral complexes ($[PtL_4]^{2+}$, $[PtX_4]^{2-}$ and $[PtX_2L_2]$) are known in large numbers, especially the ammine complexes $[Pt(NH_3)_4]^{2+}$, $[Pt(NH_3)_3X]^+$, $[Pt(NH_3)_2X_2]$, $[Pt(NH_3)X_3]^-$ and $[Pt(NH_3)_4][PtCl_4]$ (*Magnus' salt*). Planar platinum(II) complexes played a significant role in the development of coordination chemistry; they were used as model compounds for studying the trans effect. *cis*-Platinum(II) ammine complexes, such as *cis*-dichlorodiammineplatinum(II), $[PtCl_2(NH_3)_2]$, inhibit cell division and are used in the chemotherapy of cancer. The extensive series of octahedral platinum(IV) complexes includes, for example, the range of hexaammine complexes $[Pt(NH_3)_6]X_4$ through all the intermediates such as $[Pt(NH_3)_5X]X_3$, $[Pt(NH_3)_4X_2]X_2$... to the hexachloroplatinates(IV), $M_2[PtCl_6]$.

Platinum fluorides: ***Platinum(VI) fluoride***: PtF_6, is a rather unstable, dark red compound; M_r 309.08, m.p. 61.3 °C, b.p. 69 °C. PtF_6 is one of the strongest oxidizing agents, and reacts, for example, with oxygen to form $O_2[PtF_6]$. Fluorination of platinum(II) chloride at 350 °C produces tetrameric, dark red ***platinum(V) fluoride***, $(PtF_5)_4$, M_r 1130.32, m.p. 80 °C, while fluorination of platinum(IV) bromide with bromine(III) fluoride yields brick-red ***platinum(IV) fluoride***, PtF_4, M_r 271.08.

Platinum metals: Noble metals of Group VIIIb of the periodic system, which usually occur together, often in the metallic state. The ***light P.*** include the 4d metals ruthenium (Ru), rhodium (Rh) and palladium (Pd), which have densities of approximately 12. The ***heavy P.*** include the 5d metals osmium (Os), iridium (Ir) and platinum (Pt); their densities are about 22. The world reserve of the P. is estimated to be about 13,000 tons. The content of primary deposits of P., mainly sulfide copper and nickel ores in which the P. are present as sulfides, is very low. Secondary deposits of P., which have arisen from the primary deposits through weathering and flotation processes, usually contain mixtures of metallic P. P. are obtained from their sulfides in nickel or copper ores by a complex series of steps which eventually produce hydrochloric acid solutions of the metals. The same product is obtained from treatment of ores which contain metallic P. Ruthenium and osmium are removed from such solutions in the form of tetroxides, MO_4, by oxidative distillation. The other P. are precipitated in the form of ammonium hexachlorometallates, $(NH_4)_2[MCl_6]$ (M = Pd, Pt) or $(NH_4)_3[MCl_6]$ (M = Ir, Rh). The tetroxides of ruthenium and osmium and the hexachlorometallates are finally converted to the metals by reduction.

Chemically, all of the P. are relatively inert noble metals. Iridium is the most inert and is not dissolved even by aqua regia; palladium, the most reactive, dissolves in nitric acid. All the P. are good complex formers. There are pronounced similarities between the homologous pairs of 4d and 5d elements. For example, ruthenium and osmium have a strong tendency to form tetroxides, MO_4, and their oxoanions $[MO_4]^{n-}$ are very stable. Rhodium and iridium are characterized by the formation of octahedral trivalent and planar monovalent complexes; and palladium and platinum form numerous square planar complexes of the divalent metals.

Platinum oxides: ***platinum(IV) oxide***, PtO_2, is a blackish-brown complex, M_r 227.03, density 10.2, for which the hydrates $PtO_2 \cdot nH_2O$ (n = 1-4) are known. PtO_2 is obtained by dehydration of hexachloroplatinic acid. When heated above 400 °C, it is converted by loss of oxygen to ***platinum(II) oxide***, PtO, which loses oxygen at 560 °C to form platinum.

Platinum sponge: porous, gray-white mass of very finely divided platinum, which activates hydrogen and oxygen, and is therefore an excellent catalyst for hydrogenation and oxygenation processes. P. is formed by roasting ammonium hexachloroplatinate(IV).

Platinum sulfides: ***Platinum(II) sulfide***, PtS, is a gray-black compound with a zinc-blende structure, M_r 227.15, density 10.04, obtained by precipitation of H_2S from platinum(II) salt solutions. ***Platinum(IV) sulfide***, PtS_2, is a black-brown compound with a cadmium iodide structure, M_r 259.22, density 7.66, which can be obtained by precipitation with H_2S from a solution of a platinum(IV) salt. PtS_2 is soluble in concentrated nitric acid and aqua regia, and forms a water-soluble thio salt in ammonium polysulfide solution.

PLB: see High-performance liquid chromatography.

Pleated sheet: see Proteins.

Plumbate: a compound containing lead in the form of a complex anion, for example, $M^I_4[PbO_4]$, $M^I_2[Pb(OH)_6]$, $M^I[Pb(OH)_3)]$ (see Lead oxides), or hexachloroplumbate(IV), $M^I_2[PbCl_6]$.

Plumbum: see Lead.

Plutonium: symbol ***Pu***: a radioactive transuranium element from the Actinoid (see) group of the periodic system; it is made in nuclear reactors. ^{239}Pu occurs naturally, but only in vanishingly small amounts in uranium minerals, such as pitchblende or carnotite, as a result of neutron absorption by uranium-238. P. is a heavy metal, Z 94, with known isotopes with the following mass numbers (the decay type, half-life and nuclear isomers in parentheses): 232 (K-capture, α, 34 min); 233 (K, α, 20.9 min); 234 (K, α, 8.8 h); 235 (K, α, 25 min, 2); 236 (α, 2.85 a, 2); 237 (K, α, 45.63 d, 2); 238 (α, 87.75 a, 2); 239 (α, $2.439 \cdot 10^4$ a, 2); 240 (α, $6.537 \cdot 10^3$ a, 2); 241 (β^-, α, 14.89 a, 2); 242 (α, $3.87 \cdot 10^5$a, 2); 243 (β^-, 4.955 h, 2); 244 (α, $8.26 \cdot 10^7$ a); 245 (β^-, 10.48 h); 246 (β^-, 10.85 d). The mass of the most stable isotope is 244, valency is usually IV, but III, V, VI and VII are known; density 19.737, m.p. 639.5 °C, b.p. 3200 °C, standard electrode potential Pu/Pu^{3+}) -2.031 V.

P. is the most important of the transuranium elements obtained by transmutation. It is a silvery-white, reactive metal which occurs in six allotropic modifications and can be obtained by reduction of plutonium(IV) fluoride with lithium, calcium or barium at 1200 °C. P. is not attacked by concentrated sulfuric or nitric acid or glacial acetic acid, probably due to passivation. However, it dissolves readily in hydrochloric or perchloric acid. P. reacts at high temperatures with nonmetals such as carbon, nitrogen, oxygen and hydrogen, forming binary compounds. These compounds and P. itself are extremely poisonous.

P. was the second transuranium element discovered; it was obtained in 1940 by Seaborg, McMillan, Kennedy and Wahl by bombarding uranium-238 with 16 MeV deuterons:

$$^{238}U \xrightarrow[-2n]{+d} {}^{238}Np \xrightarrow[50.8\,h]{-\beta^-} {}^{238}Pu.$$

Today the most important isotope is ^{239}Pu, which is formed by interaction of neutrons with uranium-238 and is made on an industrial scale:

$$^{238}U \xrightarrow{+n} {}^{239}U \xrightarrow[23.54\,min]{-\beta^-} {}^{239}Np \xrightarrow[2.355\,d]{-\beta^-} {}^{239}Pu.$$

The importance of ^{239}Pu is that it, like uranium-235, is fissionable by slow neutrons. Reactors designed so that they produce more fissionable material (^{239}U) from ^{238}U than was initially present (in the form of ^{235}U) are called breeder reactors. A breeder reactor of 10^6 kW power output produces about 1 kg P. daily. Conversion of ^{238}U to ^{239}Pu makes it possible to use all the uranium (of which only about 0.7% is fissionable ^{235}U) for power production. The long-lived isotopes ^{242}Pu and ^{244}Pu are produced by neutron irradiation from ^{239}Pu. The energy source in radionuclide batteries, e.g. in space vehicles or cardiac pacemakers, is the nuclide ^{238}Pu, which is obtained by neutron irradiation of neptunium-237:

$$^{237}Np \xrightarrow{+n} {}^{239}Np \xrightarrow[50.8\,h]{-\beta^-} {}^{238}Pu.$$

P. is obtained from spent fuel elements of nuclear reactors. It is separated from uranium, neptunium and various uranium cleavage products by utilizing the different stabilities of the oxidation states ($UO_2^{2+} > NpO_2^{2+} > PuO_2^{2+}$; $Pu^{3+} \gg Np^{3+}$, U^{4+}) together with extraction, precipitation and ion exchange methods. Methyl isobutyl ketone or tributyl phosphate is used as an extraction solvent. When P. is used as a nuclear fuel, it is in the form of its oxides PuO_2 or PuO_2/UO; the most promising nuclear fuels for future use are plutonium carbide, nitride and carbonitride. P. is used to build nuclear weapons. Like ^{235}U, ^{239}Pu explodes when its critical mass (5.6 kg) is exceeded.

The atom bomb dropped on Nagasaki in 1945 was a plutonium bomb.

P. was named for the planet Pluto.

Plutonium compounds: compounds in which plutonium displays close similarities to the actinoids neptunium and uranium. Of the possible oxidation states III to VIII, the IV state is distinctly preferred. The fire-red plutonyl salts, which contain the linear dioxoplutonium(V) cation PuO_2^{2+}, are stronger oxidizing agents than the isomorphic uranyl and neptunyl salts, while the yellow-green plutonium(IV) derivatives are weaker reducing agents than the corresponding neptunium(IV) and uranium(IV) compounds. In the oxidation states III and IV, plutonium shows distinct similarities to actinium (oxidation state III) and the actinoids thorium(IV), protactinium(V), uranium(VI) and neptunium(VII).

Plutonium(III) chloride, $PuCl_3$, forms emerald green, hexagonal crystals, M_r 348.36, density 5.70, m.p. 760 °C; it is obtained by reaction of chlorine with plutonium around 450 °C, or by chlorination of plutonium(IV) oxide with phosphorus pentachloride or a mixture of tetrachloromethane/chlorine around 280 °C. ***Plutonium(III) fluoride***, PuF_3, forms purple, hexagonal crystals isotypic with LaF_3; M_r 299.00, density 9.32, m.p. 1425 °C. It is precipitated from aqueous plutonium(III) salt solutions when alkali fluorides are added.

Plutonium(III) compounds are oxidized to plutonium(IV) compounds by atmospheric oxygen. The most stable plutonium oxide is ***plutonium(IV) oxide***, PuO_2, an olive-green compound isotypic with plutonium(IV) nitrate, oxalate or hydroxide, M_r 274.00, density 11.46, m.p. 1750 °C. It is used in nuclear reactors as a fuel. ***Plutonium(IV) fluoride***, PuF_4, forms pink, monoclinic crystals, M_r 317.99, density 7.0, m.p. 1037 °C; it precipitates as the hydrate, $PuF_4 \cdot 2.5H_2O$, when fluoride is added to a plutonium(IV) salt solution, and the anhydrous compound is formed by gentle heating. It forms fluoro complexes $M[PuF_5]$ with alkali fluorides. ***Plutonium(IV) sulfate*** crystallizes as a coral-red tetrahydrate, $Pu(SO_4)_2 \cdot 4H_2O$, M_r 506.18. Addition of ammonia to aqueous plutonium(IV) salt solutions yields pale green ***plutonium(IV) hydroxide***, $Pu(OH)_4$.

Some examples of other oxidation states of plutonium are listed below. Plutonium(V) is represented by fluoro complexes of the types $M[PuF_6]$ and $M_2[PuF_7]$. The pale violet ***plutonyl(V) salts, dioxoplutonium(V) salts***, PuO_2X, tend to disproportionate: $2\ PuO_2^+ + 4\ H^+ \rightleftharpoons PuO_2^{2+} + Pu^{4+} + 2\ H_2O$. ***Plutonyl(VI) salts, dioxoplutonium(VI) salts***, PuO_2X_2, are formed by the action of strong oxidizing agents on the corresponding plutonium(IV) compounds; examples are ***plutonyl acetate***, $PuO_2(OOCCH_3)_2 \cdot 2H_2O$ and ***plutonyl nitrate***, $PuO_2(NO_3)_3 \cdot 6H_2O$. These compounds correspond to the uranyl compounds UO_2X_2, and plutonates(VI) of the type $M_2[Pu_nO_{3n+1}]$ correspond to the uranates(VI). Reaction of ozone with a suspension of plutonium(IV) hydroxide produces red-gold ***plutonium(VI) oxide hydrate***, $PuO_3 \cdot 0.8H_2O$. ***Plutonium(VI) fluoride***, PuF_6, forms orange, orthorhombic crystals isotypic with UF_6 and NpF_6; M_r 355.99, m.p. 50.75 °C, b.p. 62.3 °C. These are obtained by the reaction of fluorine with PuF_4 around 500 °C. P. in the VII oxidation state are represented by ***plutonates(VII), perplutonates***, of the type M_5PuO_6 or M_3PuO_5, and by the bluegreen dioxoplutonium(VII) cation, PuO_2^{3+}. Li_5PuO_6, for example, is obtained by the reaction of Li_2O with PuO_2 in a stream of oxygen around 400 °C. Plutonium reacts with nitrogen and carbon to form plutonium nitride, PuN, and plutonium carbide, PuC, respectively; both have NaCl structure and are considered possible fuels for nuclear reactors.

Pm: symbol for promethium.

PMA: see Fungicides.

PMAA: see Polymethacrylates.

PMMA: see Polymethacrylates.

PMP: abb. for Poly-4-methylpent-1-ene.

PMR spectroscopy: see NMR spectroscopy.

PNC process: see ε-Caprolactam.

PNF: abb. for polyphosphazenes.

Pnictogens: elements of the Nitrogen-phosphorus group (see).

Po: symbol for polonium.

PO: abb. for Polyolefins.

Point analysis: same as Local analysis (see).

Point defect: see Crystal lattice defects.

Point group: see Symmetry.

Poison: a chemical substance which, in relatively modest quantity, damages the structure and function of an organism. The size of the damaging dose (see Toxicity) depends on the properties of the substance, the route of application, the application vehicle, the species, the characteristics of the individual organism and other factors. Because there is considerable diversity between individuals of a single species, and even more between species, the same dose of a compound may be poisonous to one organism and not to another.

P. are historically classified according to their sources as simple inorganic P. (e.g. arsenic trioxide, arsenic hydride, phosphorus hydride, ammonia, chlorine), Animal poisons (see), Plant poisons (see), and microbial poisons (see Toxins).

Another possible means of classification was based on the organ-specific effects of certain P., and thus included certain toxicological experiences or ideas about mechanisms of action. Liver P. included chlorinated hydrocarbons, alcohols, etc.; certain heavy metals were classed as kidney P.; derivatives of lysergic acid, tryptamine, cannabinol, psylocybin and biogenic amines as central-nervous-system P.; hydrogen cyanide, cyanides, arsenic and phosphorus hydrides and CO as blood P.; strong acids and bases and mustard gas as skin and cell P.; chlorine gas and phosgene as lung P., and halogenated aromatic compounds such as benzyl chloride or chloroacetophenone as irritants.

Modern classifications are based on the mechanism of action of the P. (see also the table):

1) *Oxygen-supply P.*. Interruption of the supply of oxygen to the organism, or its cellular energy metabolism, leads to a lack of energy in its tissues. Organs with high energy requirements, especially the central nervous system, are particularly sensitive to these P. The energy supply may be interrupted in several ways: a) denaturation of protein (irritant gases, lung gases), b) inhibition of hemoglobin (CO), c) formation of methemoglobin (nitrite, some nitro compounds, etc.).

Mechanism	Examples	Type of damage, symptoms	First aid
1) Poisons which interference with oxygen supply			
Protein denaturing	Lung and irritant gases such as – acids and anhydrides (HF, HCl, SO_2) – bases (NH_3, NaOH and CuO aerosols) – free halogens and halogen-releasing substances (e. g. $CoCl_2$)	Local or extensive damage to the lungs, excessive loss of capillary liquid to the tissues (edema) and lung alveoli (lung edema)	Rest, fresh air
Hemoglobin inhibition	CO	Reduction of blood transport capacity for O_2; headaches, ringing of ears, dizziness, confusion, unconsciousness	Fresh air
Methemoglobin formation	– Oxidizing agents (e. g. $NaClO_3$, $NaBrO_3$) – Nitrites and nitrates – some aromatic nitro compounds – some aromatic amino compounds	Oxidation of the Fe^{2+} in hemoglobin (direct Met-Hb formers) or formation of compounds which indirectly form Met-Hb	Therapy with reducing agents
2) Poisons which interfere with production of energy in tissues			
Blockage of cytochrome oxidase	Cyanides, sulfides	Inhibition of blood O_2 utilization and CO_2 removal conversion to rhodanide (thiosulfate therapy) complexation (cobalt salts)	Artificial respiration; with cyanides, bonding to Met-Hb formers (nitrites)
Uncoupling of oxidative phosphorylation	2,4-Dinitrophenol, 4,6-dinitro-o-cresol	Reduction of yield of high-energy phosphates and increased turnover of glucose and oxygen; complex symptoms	Symptomatic treatment
3) Blockage of sulfhydryl groups			
SH-groups blocked in enzyme systems (hexokinases, transaminases)	Some heavy metals and their compounds e. g. Pb, Cd, As, Hg	Ph, Hg and some of their cpds., inhibition of porphyrin synthesis; if chronic, inhibition of heme synthesis and spasms of smooth muscles. Generally complex presentation with damage to capillaries, kidneys, CNS gastroenteritis, etc.	Treatment whit complexing agents
	Some lacrimators	Complex action spectrum in acute intoxication; depends on site of application	Fresh air
4) Disturbances of the internal milieu			
Acid-base balance	Methanol	Metabolic conversion to formic acid via formaldehyde; acidosis, nausea, vomiting, danger of retinal nerve damage.	Inhibition of the metabolic conversion (ethanol therapy), elimination of P. (artificial kidney, exhalation) Alkali hydroxide therapy
	Oxalic acid	Binding with calcium ions, muscle and nerve impairment	Calcium thiosulfate therapy
5) Disturbances of the citric acid cycle			
Inhibition of aconitrate hydratase	Fluoroacetic acid and its derivatives, Substances which are metabolized to fluoroacetic acid	Muscle twitching, respiratory disturbances, pupil dilation, heart arrythmia, cramps	Ca ion therapy (only briefly effective)
6) Blockage of cholinesterase			
Inhibition of acetylcholinesterase with subsequent endogenous choline poisoning	Organic phosphate esters Carbamates	Nicotinic, muscarinic and central symptoms, such as narrowing of pupils, salivary flow, respiratory inhibition, heartbeat slowing, fibrillary muscle twitching, cramps, paralysis of central respiratory control	Direct cholinolytics (atropine), with some phosphate esters, reactivators of acetylcholinesterase (various pyridine salts and oximes)
7) Physical chemical damage to organs			
Protein denaturation	Strong acids and bases	Inhibition of enzymatic processes, hormonal regulation and damage to structural proteins	Rinsing with water
Damage to lipidcontaining structures	Organic solvents (alcohols, aromatic hydrocarbons, halogenated hydrocarbons)	Enrichment in nerves and fat tissues, paralysis or excitatory effect; sometimes further damage to organs (liver, kidneys, bone marrow, blood)	P. elimination (artificial kidneys, exhalation, forced diuresis, etc.)
8) Interference with transmission of genetic information			
Interference with nucleic acid metabolisms, control of protein synthesis	Many substances from very different classes e. g. some polycyclic hydrocarbons (3,4-benzpyrene), some heavy metals (arsenic, chromium), halogenated thioethers and their derivatives (bis-2-chloroethylsulfide)	Mutagenic, carcinogenic, teratogenic and/or embryotoxic effects	Prevention of exposure

2) *Interruption of tissue energy supply*. Some P. are able to reduce or completely suppress the ability of cells to utilize the oxygen provided by the blood. This can occur by a) blocking cytochrome oxidase (cyanides, sulfides), an enzyme which makes the oxygen available to the cells, or b) uncoupling oxidative phosphorylation (e.g. 2,4-dinitrophenol), that is, preventing formation of ATP from the energy released by oxidation of glucose or other energy metabolites.

3) *Blockage of sulfhydryl groups* (heavy metals). Many enzymes contain essential sulfhydryl groups. Blocking these groups inactivates the enzymes and

leads to a broad spectrum of toxic effects in many organs.

4) *Interference with the internal homeostasis*. The most significant are impairments of the acid-base balance and ion equilibria. One important example is the occurrence of intracellular acidosis in methanol poisoning. Disturbance of ionic equilibria (e.g. by oxalic acid) affects membrane transport processes and conduction of impulses by nerve cells. The symptoms are therefore concentrated in the nervous system and musculature (tetany, inhibition of heart contraction, drop in blood pressure).

5) *Interference with the citric acid cycle*. Some fluoro-organic compounds, including fluoroethanol, are converted by oxidation or hydrolysis to fluoroacetic acid, which enters the citric acid cycle. It inhibits aconitate hydratase, an enzyme which catalyses the conversion of citrate to isocitrate, and thus blocks the entire cycle.

6) *Blockage of cholinesterases*. Some organic phosphates and carbamates block cholinesterases. Toxicologically, the most important of these enzymes is acetylcholinesterase, which has a key function in the cholinergic synapses of the nervous system and at the motor endplates of the muscles. Death usually occurs because of paralysis of the respiratory center, accompanied by cramps. Some of the extremely toxic cholinesterase inhibitors are discussed under Chemical weapons (see).

7) *Physicochemical damage* of structures and organs. The damage may be due to a) protein denaturation (strong acids and bases) and/or b) damage to lipid-containing structures (many organic solvents). Protein denaturation changes the structure of the protein molecules and inactivates them; lipid-containing structures include cell membranes, the insulating sheaths around nerve cells, and intracellular structures and organelles. The destruction of the nerve-cell sheaths leads to paralysis or excitation of the central nervous system.

8) *Interference with the transmission of genetic information*. A large number of chemicals interfere with transmission of genetic information, due to interruption of the nucleic acid metabolism or to more subtle effects on regulation of the process. Extensive damage to the DNA leads to cell death, but less severe damage may be passed on to daughter cells (mutation). Under some circumstances, this leads to development of cancer. The effects of noxious chemicals during embryonic development, especially during the first trimester, can lead to characteristic malformations (teratogenesis). Various compounds are metabolized in the organism (forming epoxides or radicals) before the actual carcinogens are formed. Carcinogenesis may be due in some cases to alkylation or acylation reactions; some of the most highly teratogenic substances, the mustard derivatives, nitrosourea derivatives and methylhydrazine, are acylating agents. Agents which prevent mitosis, such as colchicine and thalidomide, are also carcinogens.

Treatment of acute intoxication begins with the removal of the P. from the organism and prevention of further absorption; next is preservation of the vital functions (symptomatic therapy), and administration of an Antidote (see) if indicated. To prevent further absorption of the P., residues of it must be removed from the body surface (skin, clothing, mouth, conjunctival sack) by thorough rinsing with water. P. is removed from the upper part of the digestive tract by inducing vomiting or pumping the stomach. Another possibility is oral administration of adsorbants, usually activated charcoal; for cationic P. such as paraquat or diquat, bentonite may be used. The chemical, physical and toxicological properties of the P. must of course be taken into account in these manoeuvers.

Elimination of the P. from the organism can be achieved by forced ventilation (alcohols, volatile solvents), forced diuresis (kidney P.; caution must be exercised not to damage the kidneys), laxatives, hemodialysis, peritoneal dialysis, hemoperfusion (adsorption to special activated charcoal preparations external to the circulation), and recently, hemo-ultrafiltration. The choice of elimination method depends on the nature of the P., the patient's condition and the local conditions. In some cases (burning of the gastrointestinal tract with strong acids or bases), surgical intervention is required. Symptomatic therapy is directed towards maintaining life functions (breathing, circulation) and countering specific symptoms (cramps, changes in blood pressure, hyperthermia, shock, heart fibrillation, arrhythmias of the heart, etc.) by pharmaceutical means. The possibilities for antidote therapy are often overestimated. Actually, the number of usable antidotes is limited.

Poisson equation: see Adiabatic process.

Polarimetry: optical methods based on detection of the rotation of the plane of linearly polarized light upon transmission through an optically active substance (see Optical activity). A clockwise rotation is considered positive (+; dextro-), while counterclockwise rotation is considered negative (-; levo-). An apparatus to measure this rotation is a ***polarimeter***.

The *specific rotation* $[\alpha] = 100\alpha/cd$, where α is the measured angle of rotation, d is the thickness of the medium (usually given in dm) and c is the concentration (in g per 100 ml) solvent). For many purposes, the *molar rotation* $[\alpha]_{mol} = M[\alpha]/100$ is used instead of the specific rotation. Here M is the molar mass, and $[\alpha]$ is a function of the temperature and wavelength, which are given as a superscript and subscript, respectively. $[\alpha]$ also depends on the solvent, and in some cases, the concentration. A specific rotation is therefore given in this form: $[\alpha]^{25}_{589} = 27.3^o$ in water (c = 0.130 g mol^{-1}). The D-line of sodium (589 nm) is most often used for the measurement, and is indicated thus: $[\alpha]_D$. The dependence of the rotation on the wavelength is called the optical rotatory dispersion (ORD; see Chiroptic methods).

There are no simple or general relationships between the specific or molar rotation observed at a wavelength and the structure of an optically active compound, although some empirical rules (superposition rule, shift rule) have been formulated.

Measurement of optical rotation is used mainly to characterize pure, optically active substances and for their quantitative analysis, which is based on the proportionality between α and c.

Polarization: ***1) Electrolytic P.*** the difference in the electrode potential measured when current is flowing and when it is not. The electrode potential when no current is flowing, i.e. in the unstressed state, is also called the *resting potential*. It is identical with the equilibrium galvanic potential only when an electrochemical equilibrium has been reached at the electrode, which is rarely the case. The difference between the equilibrium galvanic potential and the voltage when current is flowing is called the *overvoltage*. The distinction between P. and the overvoltage is not always made. The reason for both phenomena is inhibition of the transport processes or of the partial reactions which comprise the electrode reactions.

2) dielectric P. A shift in the charges on a molecule, atom or ion under the influence of an electric field. There are two types of P.: displacement and orientation. In ***displacement polarization P_D***, the electrons are shifted toward the positive pole of the field, and the nucleus toward the negative pole. These shifts create a dipole moment, the *induced dipole moment* μ_i, which is proportional to the field strength E: $\mu_i = \alpha E$. The proportionality constant α is the electric polarizability, which has units of C m^2 V^{-1} (coulomb meter^2 per volt). Polarizabilities were formerly given in the electrostatic system of units (α_μ, in which α_μ has the dimensions of a volume. The values are on the order of magnitude of molecular volumes (10^{-30} m^3). The equation $\alpha_\mu = \alpha/4\alpha\varepsilon_0$ is used to convert to SI units. ε_0 is the electric field constant. The polarizability of a molecule is anisotropic, i.e. it depends on the orientation of the molecule in the electric field. Usually only one value for α is given, reflecting the mean polarizability. The largest contribution to the displacement P. is the *electron shift P.*, P_E, while the *nuclear shift P.* (*atomic P.*) P_A accounts for only 10 to 15% of P_E. If a molecule with a permanent dipole moment μ_P is brought into an electric field, an ***orientation polarization*** P_O is observed in addition to the displacement P. This is due to the orientation of the molecule counter to the direction of the field. The displacement and orientation P. lead to a weakening of the field in the interior of the substance which is given macroscopically by the *dielectric constant* ε This is the ratio of the field strength in the vacuum to its strength in the substance. The total dielectric P. for a mole of substance is the ***molar P.***, measured in M^3 mol^{-1}, which is the sum of P_D and P_O. According to Debye, the following equation relates ε (a macroscopic parameter) to μ_p and α (microscopic parameters):

$$\frac{\varepsilon - 1}{\varepsilon + 2}\,\frac{M}{\text{P}} = \frac{1}{3\varepsilon_0} N_A \left(\alpha + \frac{\mu_p^2}{3kT}\right) = P_D + P_O + P,$$

where M is the molar mass, ϱ is the density of the substance, N_A is Avogadro's number, T is the absolute temperature, ε_0 is the electric field constant and k is Boltzmann's constant. For molecules without permanent dipole moments ($\mu_p = 0$), the contribution from the orientation P. is zero. Since the molar P. is proportional to $1/T$, the permanent dipole moment of a molecule can be determined from the temperature dependence of P. If molecules are placed in a high-frequency alternating field (e.g. visible or UV light), the inertia of the nuclei is too large to allow them to re-orient as rapidly as the field changes, so atomic and orientation P. do not occur. Only the electrons are shifted, which causes induction of a dipole moment which oscillates with the frequency of the light; this is characterized by the optical polarizability α_λ, which has a smaller value than the electric polarizability. The parameter relating electron-shift polarization which occurs in a high-frequency field to one mole of substance is called the *molar refraction* R_M. This is related to the refractive index n and the optical polarizability $\alpha\lambda_\lambda$ by the *Lorenz-Lorentz equation*:

$$\frac{n^2 - 1}{n^2 + 2}\,\frac{M}{\text{P}} = \frac{1}{3\varepsilon_0} N_A \alpha_\lambda = R_M$$

The molar refraction and the refractive index are dependent on the wavelength (see Refractometry). The molar refraction, extrapolated to infinite wavelength, is essentially the same as the displacement P. The molar refraction is an additive parameter (see Additivity principle) which can be calculated approximately, for compounds with covalent bonds, from the atomic and bond refractions (Table). It depends on the structure of the molecule, so for atomic refractions, the bonding state of the atom must be taken into account. For example, the atomic refraction for oxygen is different for hydroxyl, ether or carbonyl oxygens. Comparison of the calculated and experimentally determined molar refractions yields significant information about the molecular structure.

Atomic and bond refractions in M^3 mol^{-1} for the wavelength 589 nm

Atom or Bond	$R \cdot 10^6$	Atom or Bond	$R \cdot 10^6$
C	2.418	C–Cl	6.51
H	1.1	C–Br	9.39
O=	2.211	C–I	14.61
O⟨	1.643	C–O	1.54
F	0.997	C=O	3.32
Cl	5.967	C–S	4.61
Br	8.856	C=S	11.91
I	13.9	C–N	1.57
=	1.733	C=N	3.76
≡	2.398	C≡N	4.82
C–H	1.676	N–N	4.12
C–C	1.296	O–H	1.66
C≡C	5.87	N–H	1.76
C–F	1.44	C=C	4.17

Polarizability: a measure of the ease of relative displacement of the electrons and nuclei in a molecule under the influence of an electric field (see Polarization, 2).

Polarography: an electrochemical method in which current-voltage curves are analysed. A mercury dropping electrode is the working electrode. In ***direct current P.***, the apparatus consists of the measurement cell and an external, variable direct current cell. The measurement cell contains a conduction electrolyte, a depolarizer and two electrodes, a mercury dropping electrode and a reference electrode, usually calomel. The working electrode is first given an increasingly negative polarization. When the decomposition voltage of the depolarizers is reached, current begins to flow as a result of the electrode reaction. As the potential on the working electrode continues to become more negative, the current reaches a limit (*limit current*) determined by diffusion (*diffusion current*) or kinetic processes (*kinetic current*). The diffusion current is a measure of the concentration of the depolarizers. The potential at which the current has reached half its maximum value is called the *half-step potential* $E_{1/2}$; this is a qualitative indicator of the nature of the depolarizer.

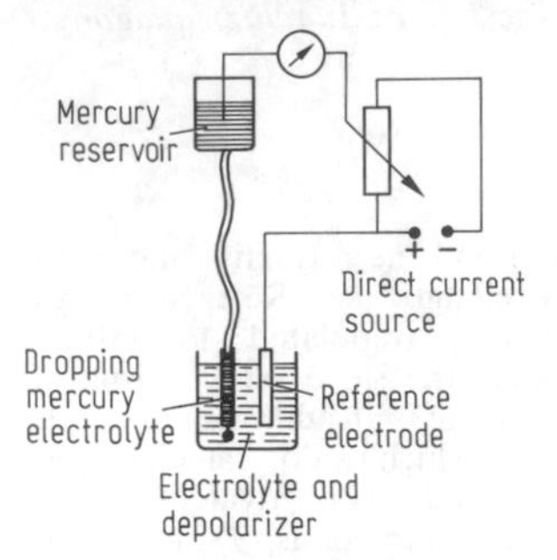

Fig. 1. Essentials of the circuit in a polarographic apparatus.

The measured current undergoes an oscillation due to the changes in the surface of the mercury drop. Polarographic analysis is suitable for inorganic ions or organic substances which can be reduced (or, more rarely, oxidized) at the surface of the mercury drops. Diffusion-controlled limit currents are most useful for quantitative analysis. Therefore a *conduction salt* is added to the analysis solution to take over charge transport. In this fashion, the diffusion of the depolarizers to the electrode becomes the rate-limiting step of the electrode reaction.

The "classic P." described so far can be used for depolarizer concentrations from 10^{-3} to 10^{-5} mol l^{-1}. Below this range, measurements are not possible because the capacitative (non-Faraday) current flowing through the cell is approximately equal in magnitude to the diffusion current. This interfering parameter is caused by the changes in the surface of the mercury drops and of the electrode potential. To circumvent the problem, direct current P. has been modified in various ways so that lower concentrations of depolarizer could be detected.

In ***alternating current P.***, the voltage slope applied in direct-current polarography is modulated by low-amplitude alternating current (frequency usually 50 to 60 Hz). Only the alternating portion of the current flowing through the cell is measured. Instead of the polarographic steps obtained with direct-current P., alternating-current P. produces current peaks.

Contact P. makes use of the different time courses of the diffusion and capacitative currents over the lifetime of a drop. The diffusion current increases with time, while the capacitative current declines. The current is measured only at the end of the drop's life, and in this way the contribution of the capacitative current is partly eliminated.

In ***pulse P.***, the working electrode is polarized only with single pulses of voltage. The measured current corresponds to the difference in the current flowing shortly before the drop falls and during the measurement interval. As in direct-current P., steps are observed.

In ***differential pulse P.***, the voltage slope of direct current P. is modulated by square, small-amplitude impulses. The measured signal is the difference between the direct current flowing during the measurement interval and the current flowing shortly before the pulse is applied. This is the most sensitive of the polarographic methods which do not make use of preenrichment. Modern polarographs are therefore designed especially for this technique.

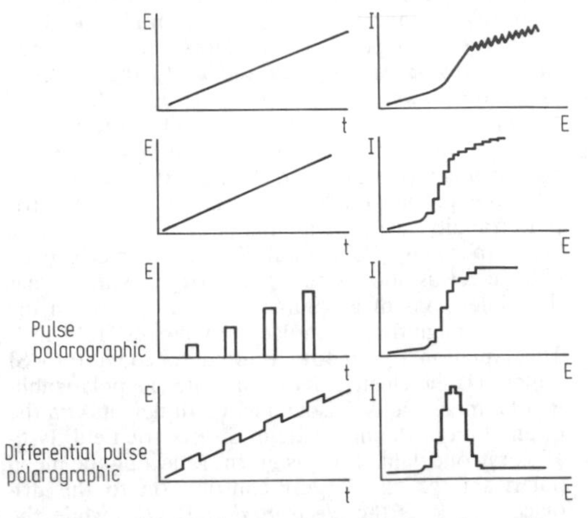

Fig. 2. Potential functions and signal curves for various polarographic methods. *E*, voltage; *I*, current strength; *t*, time.

Polfalan gray 3 BL: see Formazane pigments.

Poling: a metallurgical process used to refine molten metals. Fresh wood is introduced into the melt, leading to production of pole gases, that is, steam and decomposition products of the wood. These gases cause mixing of the metal, and after the process, the impurities can be skimmed off the surface of the melt. Both reduction (carbon, hydrocarbons) and oxidation (atmospheric oxygen) processes occur during poling. In copper refining, a distinction is made between tight P. (removal of SO_2) and raw P. (Cu_2O reduction). In modern plants, P. is achieved by passing oil or natural gas through the metal melt.

Pollutant: a substance generated by human activities which, above a certain concentration (dose) or exposure time, is harmful to human beings, animals, cultivated plants or natural ecosystems. The type of damage depends on the substance, damaged organism(s), dose, exposure time and other factors, and can include acute or chronic toxicity, delayed toxicity or impairment of normal productivity. P. can be present in the air, water, soil and foods. The sulfur oxides, especially SO_2 and SO_3, and nitrogen oxides, NO_x, are particularly significant air P. However, other atmospheric P. are becoming more important; these include organic compounds of sulfur (mercaptans), nitrogen (amines, pyridine derivatives), phosphorus (phosphates), oxygen (certain alcohols, ketones and aldehydes) and halogens (especially multiply chlorinated and fluorinated compounds). High-molecular-weight fused aromatic compounds are also significant. Air P. can arise from both direct emission (from power plants, the chemical industry and diffuse sources) and by migration from other regions of the environment.

Polonium, symbol ***Po***: a short-lived, radioactive element from group VIa of the periodic system, the oxygen-sulfur group (see); a metal, Z 84, with 34 known isotopes with mass numbers from 192 to 218 (half-lives from 10^{-7}s to 103 a). The most stable isotope has a mass of 209; valency II, IV, VI, density of cubic form, 9.196, m.p. 252 °C, b.p. 962 °C, standard electrode potential (Po/Po^{2+}) +0.9 V.

Properties. The P. isotopes 215 (actinium A) and 211 (actinium C) are part of the natural uranium-actinium decay series, while isotopes 218 (radium A), 214 (radium C') and 210 (radium F) are members of the natural uranium-radium decay series and P. 216 (thorium A) and 212 (thorium C') are members of the natural thorium decay series. The isotope 231 is part of the synthetic neptunium decay series. The longest-lived isotope is ^{209}Po, with a half-life of 103 years; the much more easily obtained isotope ^{210}Po decays by α-emission to lead, with a half-life of 138.4 days. At room temperature, P. crystallizes in a cubic modification (ideal cubic primitive lattice, α-P.) which is converted at 100 °C to a rhombohedral form (β-P.). The element is decidedly more metallic than tellurium. P. is less reactive than silver. The polonium(IV) oxide PoO_2 is amphoteric, but it is more basic than tellurium oxide. P. is chemically very similar to the lighter members of group VIa, and also displays some similarities to bismuth. Polonium compounds contain the element in the -II, II, IV and VI oxidation states. Halides with the compositions PoX_2, PoX_4 and PoX_6 are known, as are polonium dioxide, PoO_2, and trioxide, PoO_3. The polonites and polonates are derived from these oxides, and a few other compounds are also known.

Occurrence and production. P. is a component of uranium pitchblende (1000 t pitchblende contains 0.03 g P.) and when the ore is processed, the P. is enriched in the fractions containing bismuth. It can be separated from these fractions by fractional precipitation with hydrogen sulfide. The metal can be obtained from the resulting salt solutions by electrolysis. Today, ^{210}Po is obtained by irradiation of bismuth with neutrons in a nuclear reactor: $^{209}Bi(n; \gamma)\ ^{210}Bi(-; \beta^-)\ ^{210}Po$.

Applications. P. is used as a strong source of radiation, for example, in radiation chemistry, radiobiology and activation analysis. Mixed with beryllium, it serves as a neutron source.

Historical. P. was discovered in 1898 by Marie Curie-Sklodowska and Pierre Curie. M. Curie (Nobel Prize 1911) named the element in honor of her native land, Poland.

Polyacetal: same as Polyoxymethylene (see).

Polyacetylenes: see Polyethynes.

Polyacrylates, ***polyacrylic resins, polyacrylic acid esters***: thermoplastics made by polymerization of acrylic acids and their esters, which are colorless, viscous liquids or solids, depending on the degree of polymerization. The methyl, ethyl and propyl esters of acrylic acid are most often used. Dispersions of the lower aliphatic P. are used as softeners for other high polymers. P. together with fillers are used as insulation for cables, sizing, glues and in emulsion paints and paint bases. The mixed polymers with styrene, acrylonitrile and vinyl chloride are used as mortar and concrete additives. In the broader sense, Polymethacrylates (see) and mixed polymers based on acrylates are considered P.

$$\left[CH_2-\underset{\underset{R-O-C=O}{|}}{CH} \right]_n$$

Polyacrylic resins: same as Polyacrylates.

Polyacrylonitrile, ***PAN***: a macromolecular, non-thermoplastic compound which is used in the synthesis of synthetic fibers and rubbers (as a component of mixtures).

$$\left[CH_2-\underset{\underset{C\equiv N}{|}}{CH} \right]_n$$

P. is a white powder which decomposes at 350 °C without previously melting; density 1.14 to 1.16 g cm^{-3} at M_r 15,000 to 250,000. Due to intermolecular linkage of polymer chains via hydrogen bonding, P. is insoluble in most organic solvents, and has a high softening point. It will dissolve in dimethylformamide and dimethylsulfoxide, because these solvents weaken hydrogen bonds.

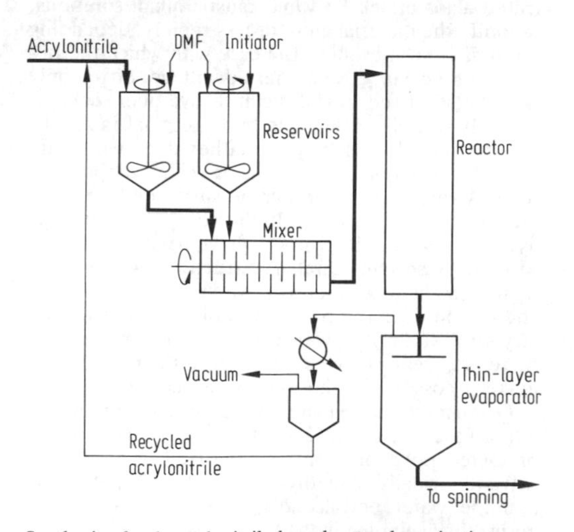

Synthesis of polyacrylonitrile by solvent polymerization.

Synthesis. P. is formed by radical polymerization of acrylonitrile, by either a precipitation or a solvent process. Solvent polymerization in dimethylformamide is of particular technical interest, because the polymer solutions can be spun directly and the solid does not have to be isolated (Fig.). At 20 to 40 °C and within 25 to 50 h, about 75% of the acrylonitrile is converted to P. Ammonium persulfate can be used as an initiator. The unreacted monomers are removed in vacuum, and returned to the process. The polymer solution can be immediately spun into polyacrylonitrile fibers, by either a wet or dry process. The process is also suitable for production of copolymers, e.g. with vinyl acetate, acrylic acid esters, vinylpyridine, etc. Acrylonitrile-styrene copolymers which still contain vinylcarbazene are very hard, and are used as a substitute for type metal. Copolymers of 40% acrylonitrile and 60% vinyl chloride are used as fibers for filters and sieves. When P. is cautiously heated to about 200 °C, neighboring nitrile groups cyclize and then are dehydrated, forming Conducting polymers (see). These are of great interest because they are highly resistant to temperature and conduct electricity. Glass-clear polymers resistant to tension and bending are obtained by copolymerization of acrylonitrile and methacrylates. The acrylonitrile-butadiene copolymers are very important as synthetic rubber.

Historical. Although P. were known for decades, they could not be spun into synthetic fibers until it was discovered, during World War II, that dimethylformamide is a suitable solvent. This was independently discovered in Germany by H. Rein and Houtz in the USA.

Polyacrylonitrile fibers: synthetic fiber materials consisting of polyacrylonitrile or mixed polymers containing at least 85% acrylonitrile.

Properties. P. have excellent elasticity, are very weather-resistant, stable to light, rotting and acid, are not attacked by insects and have better long-term heat resistance than polyamide fibers. Only concentrated acids attack P., while caustic alkali solutions, saponify the material more or less rapidly, depending on their concentrations. The ease with which the fibers can be cleaned is another advantage; dirt is only superficially fixed. It is difficult to dye fibers consisting only of polyacrylonitrile by traditional methods, but it is possible, if they are either dyed while still moist from spinning, or if basic or acidic compounds, such as vinylpyridine or styrene sulfonate, are polymerized into the material. In this case, acidic or basic dyes can be used. Recently, P. have been synthesized with an absorbant capillary structure, which gives them the capacity to absorb water rapidly (as wool and cotton do). The core of the fiber contains capillary hollows through which the water can move, and there are breaks in the outer layer of the fiber, which make it possible for the water to enter.

Production. P. are made by wet or dry spinning. In either case, a 20 to 30% solution of polyacrylonitrile or mixed polymer is first synthesized in a purified solvent, usually dimethylformamide. Dimethylformamide, water, polyalcohols, hydrocarbons, etc., can be used as spinning baths for the wet spinning process. Wet spinning is most suitable for production of staple fibers, because spinnerets with several thousand orifices can be used. The fibers are washed, stretched at high temperature, curled and cut. P. can also be spun into ***polyacrylonitrile silks (PAN silks)***.

Applications. Because of their favorable properties, the P. can be used for many types of clothing: sweaters, swim suits, outerwear and underwear, fake furs. An advantage is that the textiles dry almost wrinkle-free after washing. The weather resistance of the P. makes them useful for sails, tents, umbrellas, tropical clothing, curtains, fishing nets, filter cloth and artificial lawns. Textiles made of P. should be washed warm (up to 40 °C) with a fine-textile wash product. They can be dry cleaned. The important proprietary names are listed under Synthetic fibers (see).

Polyaddition: a type of chemical reaction in which different types of low-molecular-weight, polyfunctional compounds form macromolecules by molecular rearrangement. The linking of the chain is associated with migration of a hydrogen atom. The difference between P. and polymerization, which can formally be considered a type of P., is that in P. the intermediates are stable and can be isolated. Since the reaction occurs without splitting off of low-molecular weight compounds, the product of P. has the same stoichiometric formula as the reactants. Bifunctional starting materials produce linear macromolecules, while trifunctional monomers produce cross-linked products. Addition usually occurs via -OH, $-NH_2$, -COOH, -NCO or epoxide groups. The most important synthetic polyaddition products are epoxide resins and polyurethanes.

Polyalkene polysulfide: see Rubber.

Polyalkylene glycols, ***polyether polyols, polyalkylene oxides, polyethers***: Polyadducts of epoxides (e.g. ethylene oxide, propylene oxide) with mono-, di- and polyfunctional alcohols, carboxylic acids and other compounds with active hydrogen atoms. Opening of the epoxide ring is accelerated by addition of suitable catalysts (e.g. acids and bases). The chain length of the P. depends on the ratio of epoxide to alcohol. The lower-molecular-weight P. are also known as Cellosolve (see). P. are used as textile conditioners, components of surfactants (non-ionogenic surfactants) and as polyurethane precursors.

$$R{-}OH + n\,\underset{\diagdown O \diagup}{CH_2{-}CH_2} \rightarrow R{-}O{-}\left[CH_2{-}CH_2{-}O\right]_n{-}H$$

Polyalkylene oxides: same as Polyalkylene glycols.

Polyallomers: see Ethylene copolymers.

Polyamides, abb. ***PA***: thermoplastics which are characterized by periodic amide bonds, -CO-NH-. P. are formed by polycondensation of dicarboxylic acids with diamines, or of aminocarboxylic acids, or by ring-opening polymerization of lactams.

$$\left[-\overset{H}{\overset{|}{N}}-(CH_2)_x-\overset{O}{\overset{\|}{C}}-\right]_n$$

The P. are characterized by the number of C atoms in the starting monomers, with the number in the diamine being given first. For example, PA 6,6 = hexamethylenediamine + adipic acid, PA 6,10 = hexamethylenediamine + sebacic acid. The ratio of methylene groups to amide groups (CH_2:CONH) determines the physical properties of the P. (Table). The high mechanical strength of the P. is due to formation of intermolecular hydrogen bonds between the CO and NH groups:

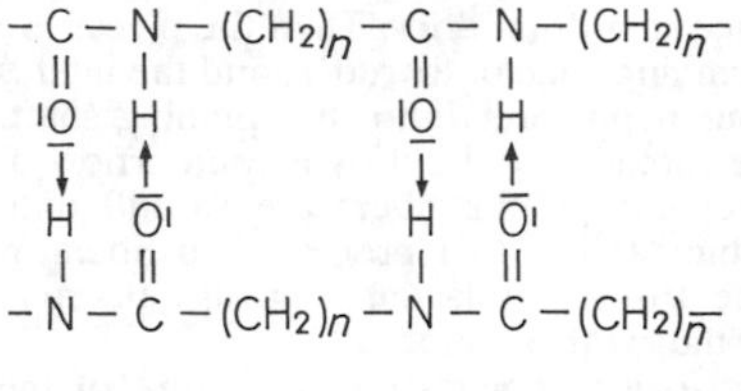

```
–C – N –(CH2)n – C – N –(CH2)n–
 ||   |           ||   |
 |O   H           |O   H
 ↓    ↑           ↓    ↑
 H    O|          H    O|
 |    ||          |    ||
–N – C –(CH2)n – N – C –(CH2)n–
```

Properties. P. are white to yellowish, and may be opaque or transparent. They have high tensile strength, and resist bending, impact and abrasion. They are elastic and resistant to many organic solvents, even at high temperatures. P. are good electrical insulators. They are somewhat affected by dilute acids and oxidizing agents. P. have very small softening ranges, remaining strong and solid until just below their melting points (215-250 °C, depending on the starting materials), and then become liquid. They can be spun, cast, pressed or worked with sharp tools. P. can be strengthened by stretching them in the temperature range just below their melting point. These properties are very important, especially in the production of P. fibers.

Synthesis. P. are formed 1) through polycondensation of higher ω-amino acids: n NH_2-$(CH_2)_x$-COOH → NH_2-$(CH_2)_x$-CO-[NH-$(CH_2)_x$-CO$]_{n-2}$-NH-$(CH_2)_x$-COOH + n-1 H_2O, where x is greater than 4. Typical ω-amino acids for starting materials are ω-aminoenanthic acid, NH_2-$(CH_2)_6$-COOH, ω-aminopelargonic acid, NH_2-$(CH_2)_8$-COOH and ω-aminoundecanoic acid, NH_2-$(CH_2)_{10}$-COOH.

2) A second method is ring opening polymerization of lactams with more than six atoms in the ring, especially ε-caprolactam, CO-$(CH_2)_5$-NH, in the presence of catalysts. Polyamide formation occurs by a stepwise or ionic chain growth reaction, depending on the catalyst. Water leads to stepwise growth (hydrolytic polymerization), while addition of alkalies leads to anionic chain growth. Anionic or "rapid" polymerization in the presence of alkaline catalysts, e.g. sodium or sodium carbonate, leads in a few minutes to high-molecular-weight products. The reaction mechanism is completely different than in hydrolytic polymerization. In addition, lactams can also be polymerized to P. by cationic mechanisms. In hydrolytic polymerization of caprolactam, there are three important reactions: a) hydrolysis of the caprolactam to 6-aminocaproic acid, b) addition of caprolactam to a polyamide molecule, and c) condensation of polyamides with different degrees of polymerization.

In industry, a distinction is made between discontinuous and continuous polymerization of ε-caprolactam. In the discontinuous process, the caprolactam is dissolved in 5 to 10% water, with addition of a stabilizer, such as acetic acid, and other ingredients, including a delustering agent such as titanium dioxide. This mixture is heated in an autoclave for several hours, in the absence of air, to 250-260 °C. After the reaction has ended, the viscous polyamide melt is pressed out through a valve in the bottom of the autoclave, and cooled as a ribbon, then cut into small pieces. In the continous process, the polymerization occurs in a pipe at atmospheric pressure. Molten lactam with a stabilizer and catalyst is continuously fed into the top of a vertical U- or N-shaped pipe heated to 240-280 °C; the pipe is several meters long. At the lower end, the polymer melt is continuously pumped out. The material passes through the pipe in 15 to 40 hours. The product can then be further processed in the same way as in the discontinuous process. Since the polymerization always stops at a temperature-dependent equilibrium, the lower-molecular-weight fractions (oligomers and lactam) must be extracted with water from the crude solid product, in which their concentration is about 10%.

3) The third method is polycondensation of dicarboxylic acids with diamines: n HOOC-$(CH_2)_x$-COOH + n H_2N-$(CH_2)_y$-NH2 ⇌ [OC-$(CH_2)_x$-CO-HN-$(CH_2)_y$-NH$]_n$ + 2n H_2O; here x and y must be greater than 3. The most frequently used dicarboxylic acids are adipic, HOOC-$(CH_2)_4$-COOH, and sebacic, HOOC-$(CH_2)_8$-COOH, while hexamethylenediamine is the most commonly used diamine. Since mixtures of various dicarboxylic acids and diamines can also be used as starting materials, the possible variations are very large. In the industrial synthesis of polycondensates from hexamethylenediamine and adipic acid, an AH salt is made first. This is heated in the absence of oxygen to about 220 °C, forming a condensation to a linear P. (***polyamide AH***). The water released by the condensation reaction is continuously distilled off, while the polymeric condensate is drawn off as a hot, viscous mass.

A number of valuable products can be made by variation in the structure of the dicarboxylic acids and diamines, especially by use of aromatic compounds such as pyromellitic anhydride and 1,4-phenylenediamine. By further condensation reactions, these can be converted to polymers which are resistant to high temperatures, such as Polyimide (see) and Polybenzimidazene (see).

Applications. P. are used mainly to make synthetic fibers (see Polyamide fibers). They can also be injection molded or melt cast into parts, such as bearings, gears, valve plates, pipes, housings, parts for electronic and optical equipment, films, nuts and bolts. P. reinforced with fiberglass is sometimes used for such parts. P. can be worked by extrusion, sintering, cutting tools and blowing. They are used to make pros-

Properties of the Polyamides

Polyamide	6	6,6	6,10	11
Water uptake in % at 100% rel. humidity	7	7	3	0.5
Melting temperature [°C]	220	260	225	190
Tensile strength [N/mm²]	60 ... 70	60 ... 80	60 ... 70	50 ... 60
Breaking elongation [%]	≤350	≤300	100 ... 200	90 ... 120
Elasticity modulus [N/mm²]	3000	3000	2200	1000
Ball pressure hardness [N/mm²]	100 ... 150	100	70	60
Density [g/cm³]	1.12 ... 1.15	1.13 ... 1.16	1.07 ... 1.09	1.03 ... 1.04

theses and as bone substitutes. Parts made of P. can be glued with special glues, and they can also be welded. Variants of P. are made by condensation of di- or triamines with higher-molecular-weight fatty acids; these are obtained by thermal treatment of unsaturated fatty acids, such as linoleic and linolenic, or their methyl esters, in the absence of oxygen. These P. are used in making paints, printing inks and glues.

Polyamide fibers: synthetic fibers made by spinning polyamides (PA). The most commonly used PA are PA 66 (from adipic acid and hexamethylenediamine), PA 6 (from ε-aminocaproic acid) and PA 11 (from ω-aminoundacanoic acid). The numbers indicate the number of C atoms in the molecules of the monomers. P. are made by spinning a melt; the hot PA coming out of the continuous synthesis pipe (see Polyamides) passes into a tub through which water is flowing, where it immediately solidifies to a ribbon. This is cut into fine pieces, which are washed with water and dried in a drum drier. Then the pieces are heated in a spinning head or extruder, and the melt is pumped through spinnerets into the spinning shaft. Here the fine threads are dried by a counterflow of air. After preparation, the fibers are wound onto spinning bobbins and further processed to fibers or silks. P. made from caprolactam can also be spun directly from the synthesis pipe.

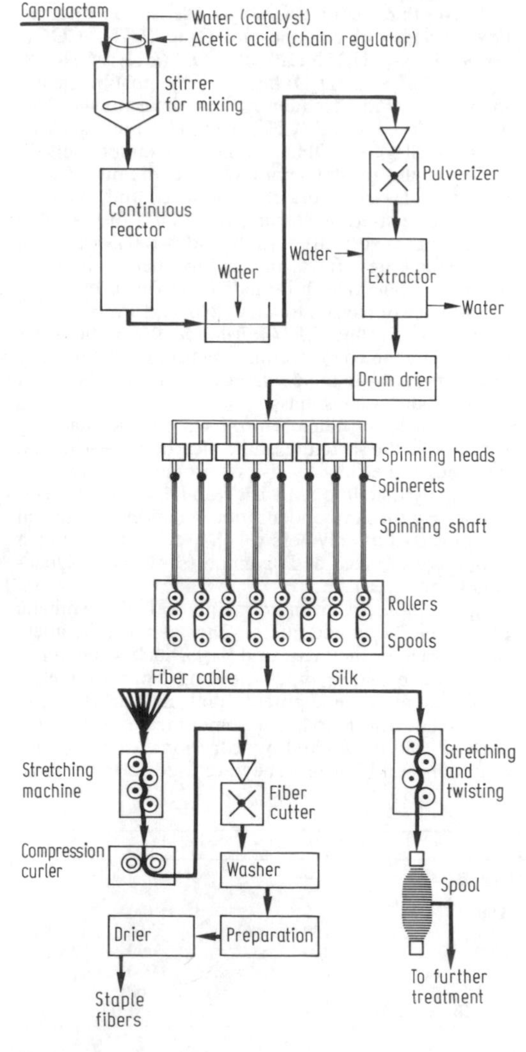

Further processing depends on the nature of the desired thread. The freshly spun threads can stretch too much and are too weak to be made into a silk. They are therefore stretched to three or four times their original length, then twisted together to make thicker threads. For production of fibers, the threads from 150 to 200 spinning bobbins are twisted into a cable, and after stretching, they are curled, then cut to the desired fiber length (staple). The staple fibers are washed, prepared and dried before being spun into yarns.

P. are used wherever long-wearing fibers are needed, especially for stockings, underwear, blouses and dresses, shirts and upholstery. Their enormous strength and elasticity makes them suitable for use in the industrial sector as well, e.g. for conveyor belts, filter cloth, ropes, nets, fishing line, electrical insulation, and tire cords. Like other synthetic fibers, P. can be used in mixtures with other types of fiber. Textiles made of P. can be washed in any kind of laundry product in warm (40 °C) water. They should be only lightly spun (not wrung out), or hung up dripping wet, if possible not in the bright sun or near a heat source. Polyamide textiles can also be dry cleaned. The most important trade names are listed under Synthetic fibers (see).

Historical. At the beginning of the thirties, W.H. Carothers worked out the basis for industrial production of the first P., which in the meantime has become known world-wide as nylon. A few years later, P. Schlack in Germany developed the second very important polyamide fiber, polycaprolactam.

Polybenzimidazene, ***polybenzimidazole***, abb. ***PBI***: a polycondensate of diphenyl isophthalate and 3,3'-diaminobenzidine. P. is similar to polyimide, but is not resistant to organic solvents. It is used as a metal glue, in composite materials for military and space technology, and for special fibers.

Polyblends: see Polymers.

Polybutadiene: see Rubber.

Polybut-1-ene, abb. ***PB***: a polyolefin made by low-pressure polymerization of but-1-ene, using organometallic mixed catalysts (Ziegler-Natta catalysts). Systems of titanium(III) chloride and organoaluminum compounds are used. The reaction is usually carried out as a suspension polymerization at a temperature of 20 to 100 °C and a pressure up to 1 MPa. Butene can also be polymerized neat and in the gas phase (see Polyethylene). P. can exist in syndiotactic, atactic and isotactic conformations, where the molecular chains have a helix structure (see Polymers).

$$\left[-\underset{H}{\overset{H}{C}}-\underset{\underset{CH_3}{CH_2}}{\overset{H}{C}}- \right]_n$$

With the known mixed catalysts, either isotactic or atactic P. can be produced. While atactic P. cannot be crystallized, there are a number of crystalline modifications of isotactic P., of which the tetragonal and hexagonal forms are of particular interest. The advantages of P. over other polyolefins are that it is more resistant to tension cracking, has a very good creeping behavior, and better temperature parameters, so that it is used by preference for hot-water pipes and armatures.

Polycarbodiimides: polycondensates obtained from polyfunctional isocyanates by polymerization with loss of carbon dioxide. They are used to make molded objects and films. Hard foam materials from P. are good heat and sound insulators.

$$\left[R-N=C=N-R \right]_n$$

Polycarbonates: abb. ***PC***: $[O\text{-}CO\text{-}O\text{-}R\text{-}O]_n$, polyesters which are made by polycondensation of carboxylic acid derivatives with diols. The most important are based on aromatic bishydroxy compounds, such as bisphenol A (2,2(4,4'-dihydroxydiphenyl) propane).

Properties. The densities of the P. are between 1.17 and 1.22. Their relative molecular masses range from 25,000 to more than 150,000. Nearly all P. are soluble in dichloromethane, and some are also soluble in aromatic hydrocarbons, esters and ketones. They display a slight uptake of water and water penetrability. They are resistant to salt solutions, dilute mineral acids and weak alkalies. In the absence of steam, acids or alkaline materials, they are stable up to temperatures above 300 °C in the molten state. They have excellent mechanical and electrical properties (Table), and are very stable to sunlight, weathering and radioactive radiation.

The P. are synthesized either by phosgenation or by transesterification processes. In the *phosgenation process*, bisphenol A is dissolved in methylene chloride at 20-40 °C and reacts with phosgene in the presence of pyridine, or it is dissolved in a concentrated aqueous alkali solution and reacts with a tertiary amine (e.g. triethylamine) in the presence of pyridine. The resulting hydrogen chloride is bound to the organic base. The P. dissolved in methylene chloride is precipitated, for example by adding aliphatic hydrocarbon, centrifuged off, dried and granulated.

In the *transesterification process*, the first step of synthesis produces diesters of phosgene and phenol in methylene chloride as solvent; these are readily purified. In the second step, the diester reacts with bisphenol A. in a stirred reactor to form a P. The phenol which is released is removed using a spiral evaporator and is returned to the process.

Applications. P. are extruded or injected into molds or are precipitated from solution to form foils, plates, tubes and molded parts. They can be glued, planed, pressed and metallized, and are used mainly for precision parts in fine machinery, in electronics and for high-quality household appliances. P. reinforced with glass fibers are used for housings for switches and measurement apparatus of various kinds.

Properties of polycarbonates

Density (g/cm^3)	1.17-1.22
Water uptake (%)	0.2-0.4
Melting range (°C)	≈230
Tensile strength (N/mm^2)	60-70
Notched bar impact strength (N/mm)	0.9-1
Elasticity modulus (N/mm^2)	2000-25 000

Polychlorobutadiene: see Rubber.

Polychlorocamphene: same as Camphechlor (see).

Polychloroprene: see Rubber.

Polycon®: a mixed textile woven from polyester, polyamide or polyacrylonitrile and cotton threads.

Polycondensation: a type of chemical reaction in which low-molecular weight starting compounds, which have at least two reactive groups in the molecule, combine to form macromolecules by splitting off a low-molecular-weight product, such as H_2O, HCl, NH_3, CH_3OH, etc. In contrast to polymerization, in which the macromolecules are formed by a mechanism of chain elongation, and in which the polymer has the same elemental composition as the monomers, P. occurs by a step-wise growth mechanism which changes the elemental composition of the polycondensate. The products are formed via stable intermediates which have the same reactivity as the monomers, so that during the P., monomers, oligomers and polymers are in equilibrium with each other and can be isolated. The polymers are gradually formed via dimers, trimers, etc. The average relative molar mass increases in the course of the reaction. The final value is determined by the equilibrium, which can be shifted in favor of products with high relative molar masses by removing the low-molecular weight reaction products. However, since this is usually difficult, the products of P. generally have average relative molar masses of 10,000 to 20,000. The monomers can be a single compound or two or more different ones .

If the monomers are linked into one-dimensional chains by P., the products are thermoplastics, such as polyamides. Most frequently, however, P. is used to make thermoset plastics, such as phenol resins and aminoplastics, in which the monomers are cross-linked in two or three dimensions.

P. can be interrupted at a given degree of condensation, and the resulting polycondensate processed into the desired shapes. Then the shaped mass is heated and the P. is continued, causing Hardening (see).

Polycrystalline material: a material which consists of many small crystals (*crystallites*, grains). Antonym: single crystal (see Crystal). The crystallites can be separate (crystal powder) or bound together (aggregate). Their diameter is usually between 10 and 500 µm. The boundary region between the crystallites of an aggregate is called the ***large-angle grain bound-***

ary. If the crystallites are completely randomly oriented, as is usually the case, the observed crystal properties are an average for all crystal directions, and the material is (quasi-)isotropic. However, if the crystallites have a preferred direction (see Texture), the material is more or less anisotropic. Polycrystalline aggregates are made, e.g. by pressing and sintering of crystal powders, by crystallization from a melt, solution or gas phase, or by plastic deformation of larger crystals. The majority of industrial materials are polycrystalline.

Polycyclic: a term for hydrocarbon compounds in which several carbon rings are fused.

Polydextran gel: see Permeation chromatography.

Polyelectrolytes: macromolecules containing groups which can dissociate; when dissolved in water, they separate into macroions and counter ions. Natural macromolecules can contain anionic or cationic groups or both. Proteins, ion exchangers, nucleic acids, polyacrylic acid, polyvinylpyridine and some polysaccharides are P.

Polyene antibiotics: a group of antibiotics produced by *Streptomyces sp.* which includes the macrolide antibiotics. A P. consists of a many-membered lactone ring formed from a polyunsaturated, polyhydroxy fatty acid with relatively few branches. The P. are classified according to the number of conjugated, *trans*-double bonds they contain: tetraenes (e.g. nystatin), pentaenes (e.g. filipin), hexaenes and heptaenes (e.g. amphothericin). The P. usually occur as mixtures of structurally similar compounds, which are difficult to separate and are also often amorphous. They are only slightly soluble in organic solvents. P. are effective against fungi and yeasts. They form a complex with cholesterol present in the plasma membranes of these organisms and disrupt them.

Polyene: an unsaturated aliphatic hydrocarbon with three or more conjugated double bonds in the molecule: $R\text{-}CH{=}CH\text{-}(CH{=}CH_2\text{-}CH)_n{=}CH\text{-}CH\text{-}R$, where $n = 1, 2, 3, \ldots$ The compunds are called hexatrienes, octatetraenes, decapentenes, etc. The P. state is characterized by alternating double and single bonds. As the number of double bonds increases, a partial equalization of the bonds occurs; however, this is only complete in an infinitely long chain. The conjugated system absorbs light, and as the number of double bonds increases, the longest-wave absorption maximum is shifted to longer wavelengths; long-chain P. absorb in the visible and hence are colored. The wavelength shift per double bond ($\approx$ 20 nm) is relatively small, however, compared to that in the polymethynes ($\approx$ 100 nm per triple bond).

Lycopin and β-carotene are examples of naturally occurring, colored P. The P. structures in zeaxanthin, bixin and crocetin are highly significant in determining the properties of these compounds.

P. are generally in the all-*trans*-configuration; they are very reactive and often very sensitive to oxygen. Some of the naturally occurring compounds are used as food colorings, while others have biological and medical significance, especially the carotinoids (including vitamin A).

Polyene pigments: natural and synthetic pigments in which the color is due to the conjugated double bond system of an unbranched methine chain, $\text{-}(CH{=}CH)_n\text{-}$. Some important examples of P. are the Carotenoids (see) and Polymethine pigments (see).

Polyester: a polymeric compound in which the monomers are linked via ester groups, -CO-O-. The P. are usually synthesized by polycondensation of bifunctional carboxylic acids or their derivatives with diols (Fig.).

$$n\,X\text{-}\underset{\underset{O}{\|}}{C}\text{-}R_1\text{-}\underset{\underset{O}{\|}}{C}\text{-}X + n\,HO\text{-}R_2\text{-}OH \rightleftharpoons n\,X\text{-}\underset{\underset{O}{\|}}{C}\left[R_1\text{-}\underset{\underset{O}{\|}}{C}\text{-}O\text{-}R_2\right]_n OH + (n-1)\,HX$$

$$X = \text{-}OH, \text{-}OR, \text{-}Cl, \text{-}O\text{-}\underset{\underset{O}{\|}}{C}\text{-}$$

$$HO\text{-}R_1\text{-}O\left[\underset{\underset{O}{\|}}{C}\text{-}R_2\text{-}\underset{\underset{O}{\|}}{C}\text{-}O\text{-}R_1\text{-}O\right]_n H$$

P. with widely varying properties can be made by varying the monomeric starting materials: 1) ***thermoplastic P.*** with relative molecular masses $M_r > 10{,}000$ are made from dicarboxylic acids, dicarboxylate esters and diols, or from polymerizable lactones or hydroxycarboxylic acids (Table). Polyethyleneglycol terephthalate (see) and polybutyleneglycol terephthalate are especially important as construction materials and for making Polyester fibers (see). In addition to these *homopolyesters*, thermoplastic *copolyesters*, e.g. *polyether ester block polymers* for elastics, copolyesters based on 1,4-bis(hydroxymethyl)cyclohexane for hard, clear injection molded articles, and copolyesters of varying composition for paint resins, melting glues and coating materials are becoming increasingly important. The Polycarbonates (see) are among the thermoplastic P.

2) ***Linear*** or ***slightly branched polyester condensates***, M_r 400 to 6000. Hydroxypolyesters (Fig.) can be made by suitable choice of reaction conditions (diol excess); these are important for the production of polyurethanes. The acid component in these compounds is most often adipic acid, but glutaric, succinic, sebacic and azelaic acids are also used. Any of the diols listed in the table may be used. To make branching modifications, glycerol, trimethylpropane, etc. are often used.

3) P. from di-, tri- and polyfunctional alcohols and polyfunctional aromatic carboxylic acids with $M_r <$ 10,000 are called Alkyde resins (see) (these are saturated polyester resins).

4) P. from polyfunctional alcohols and polyfunctional, unsaturated carboxylic acid derivatives, such as maleic anhydride, are classified as unsaturated polyester (UP) resins. They are used to make Polyester resins (see).

Synthesis. P. are most often made in a two-step process. In the first step, a polyester precondensate with M_r 100 to 2000 is obtained by esterification of the dicarboxylic acid with excess diol. In the second step, this is polycondensed to P. with $M_r > 10{,}000$ (see

Polyethyleneglycol terephthalate). Both these reactions are part of the stepwise growth reaction. Esterification, transesterification and polycondensation are equilibrium reactions which are accelerated by suitable catalysts. The equilibrium is shifted in favor of polyester formation by removal of the more volatile fractions (water, alcohol and diol) from the reaction mixture. Alternatives to the usual two-step process are reaction of ethylene oxide with dicarboxylic acids, polycondensation of dicarboxylic acid dichlorides with bisphenols with release of hydrogen chloride, and polycondensation of lactones in the presence of organometallic compounds.

Starting materials for synthesis of thermoplastic polyesters

Dicarboxylic acid or dicarboxylic acid derivative	Diol
Terephthalic acid	Ethyleneglycol
Isophthalic acid	Propane-1,2-diol
Adipic acid	Propane-1,3-diol
Azelaic acid	Butane-1,4-diol
Sebacic acid	1,4-bis (hydroxymethyl)-cyclohexane
Dodecanedioic acid	
Hexahydroterephthalic acid	Neopentylglycol
Isophthalic acid dimethyl ester	Hexane-1,6-diol
Hexahydroterephthalic acid dimethyl ester	
4-Hydroxybenzoic acid	
Piralolactone	
ε-Caprolactone	

Polyester-ether fibers: see Polyester fibers.

Polyester fibers: synthetic fibers consisting of at least 85% polyester. The preferred polyester for production of P. is Polyethyleneglycolterephthalate (see). The P. are resistant to light, weathering and wrinkling, and retain their form well. They take a good crease and can stretch. Textiles from P. are easy-care, but very difficult to dye. They are less resistant to abrasion than the polyamide fibers, but much more resistant than cellulose or polyacrylonitrile fibers. P. are spun from molten polyester cuttings or granulate as synthetic silk or stable fibers. They must be extruded at high temperatures (about 70 °C). P. are especially suitable for making textiles for suits and dresses, shirts, ties, curtains and sewing threads, either pure or mixed with wool, cotton or rayon. Ironed-in creases are retained when the garment is washed. Textiles from P. can be washed in warm water, but should not be spun dry or wrung out; they should be hung out to dry while dripping wet. They can be dry cleaned. P. are also used to make industrial fabrics, such as filters, conveyer belts, tire cords, drive belts, ropes and electrical insulation.

Polyester-ether fibers are fibers and silks made by polycondensation of hydroxycarboxylic acids with diols.

The most important trade names for P. are listed under Synthetic fibers (see).

Polyester resins, ***unsaturated P.***: hardenable synthetic resins which are Polyesters (see). They are made by polycondensation of polyfunctional unsaturated carboxylic acid derivatives (e.g. maleic anhydride) with polyfunctional alcohols (e.g. glycerol, propane-1,2-diol, etc.). To adjust the fraction of monomers containing reactive double bonds, mixtures of maleic anydride with dicarboxylic acid derivatives without reactive double bonds are used. The highly viscous, unsaturated P. formed in this way, $[CO\text{-}CH{=}CH\text{-}CO\text{-}O\text{-}R\text{-}O]_n$, has a molecular mass between 2000 and 5000, and is subsequently cross-linked. The cross-linking (hardening) occurs by radical copolymerization of the unsaturated P. with vinyl monomers, especially styrene and allyl esters. The unsaturated groups of the polyester molecule react with the styrene to yield a three-dimensional cross-linked product. Peroxides are used to initiate the radical polymerization when heated to 70-100 °C. If the hardening is to occur at a lower temperature, organic metal salts (e.g. cobalt salts) can be added as accelerators (cold hardening). It is also possible for P. to be cross-linked by UV irradiation (UV hardening) in the presence of sensitizers. The hardening releases heat, and does not split off smaller molecules. The hardened products are resistant to dilute acids and bases, and to most organic solvents, except for acetone and ethyl acetate, which cause them to swell.

Applications. The unsaturated P. are used as a casting plastic (e.g. to embed industrial parts and biological specimens), to make polyester concrete for construction, to make cable ends, buttons and ornaments, as metal glues and solvent-free paints (***polyester paints***). These are used mainly in the production of furniture with hard, shiny finishes. Unsaturated P. can also be hardened with fillers, such as chalk or kaolin, or reinforcers (textile or glass fibers). P. reinforced with glass fibers (fiberglass) have low densities and very high mechanical strength, and are used to make translucent sheets for walls and roofs, automobiles, boats and airplanes, crash helments, containers for chemicals, machine housings, sports articles, pipes and rods.

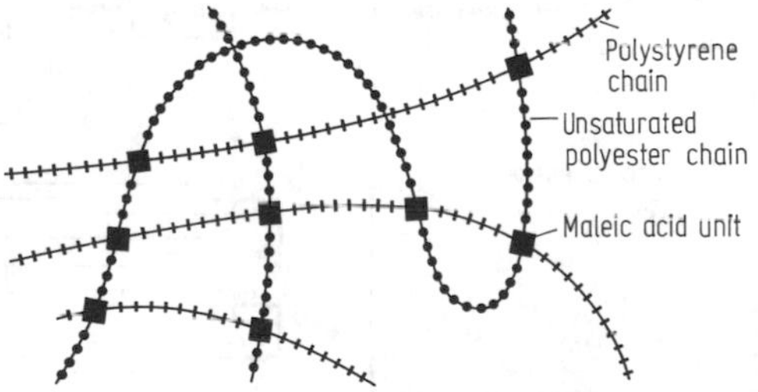

Polyether: same as Polyalkylene glycol (see).

Polyether ether ketone, abb. ***PEEK*** a linear, partly crystalline, aromatic thermoplastic which is resistant to high temperatures. Its melting temperature is 334 °C, and its glass temperature is 142 °C. P. is resistant to solvents, chemicals and radiation, and, up to 200 °C, can resist steam. It is synthesized at 320 °C from 4,4'-difluorobenzophenone and the dipotassium salt of hydroquinone; diphenylsulfone is used as solvent.

Polyether polyones: same as Polyalkylene glycols (see).

Polyethylene, abb. ***PE***: $[CH_2\text{-}CH_2]_n$, a polyolefin obtained by a chain growth reaction of ethene (ethylene). It is produced in very large amounts; along with polyvinylchloride and polystyrene, it is one of the most common plastics. P. can be obtained as a solid, a waxy product (***polyethylene wax***) or a very viscous oil (***polyethylene oils***), depending on the polymerization process. There are high, medium and low pressure methods; the properties of the products depend on the method (Table).

Properties of polyethylene produced by various methods

	High-pressure polyethylene	Medium-pressure polyethylene	Low-pressure polyethylene
Rel. molec. mass	$2 \cdot 10^4$-10^5	$2 \cdot 10^4$-$5 \cdot 10^5$	10^5-$3 \cdot 10^6$
Degree of branching; CH_3/1000		10- 35	1- 2
Crystallinity (%)	50- 70	80- 95	65- 80
Density (g/cm^3)	0.915-0.935	0.95-0.97	0.945-0.955
Melting range (°C)	105-115	134-136	125-130
Tensile strength (N/mm^2)	9- 13	24- 33	20- 26
Common names	LDPE (low-density polyethylene)	Medium-density polyethylene	HDPE (high-density polyethylene)
	Soft or branched P.		Hard or linear P.

1) *High pressure polymerization* of ethene occurs by a radical mechanism, and is initiated by very small amounts of oxygen or other radical-forming initiator. The industrial process is run at a temperature between 150 and 320 °C, and under a pressure of 100 to 300 MPa. Polymerization can occur in a pipe or stirred reactor. In the pipe-reactor process, carefully purified ethene (99.99%) is gradually brought to the desired high pressure, and passed at high velocity through a system of slightly inclined, high-pressure steel pipes which are several hundred meters long. The temperature is monitored by thermoelements, so that if a critical temperature is exceeded, the pressure of the reactor is reduced by opening a safety valve. In the first part of the pipe system, the polymerization is initiated by heating the pipes with hot, pressurized water. In the second part, the high heat of polymerization necessitates a cooling system. Since the reaction is not allowed to run to completion, the product contains monomers dissolved in it, and is a liquid. When the pressure is released, the polymer precipitates, and the unreacted ethene is fed back into the circulation. The yield is up to 23%. The relative molecular masses of the P. obtained in this way depend on the pressure and the amount of oxygen added to the mixture. The lower the oxygen content, the higher the degree of polymerization. The oxygen is incorporated into the polymer in the form of hydroxyl groups, -OH. This process yields branched polymers with relative molecular masses up to 50,000, and in a few cases, the molecules are even larger. A highly efficient process for producing P. in a pipe reactor is the Polymir 60 process (Fig. 1).

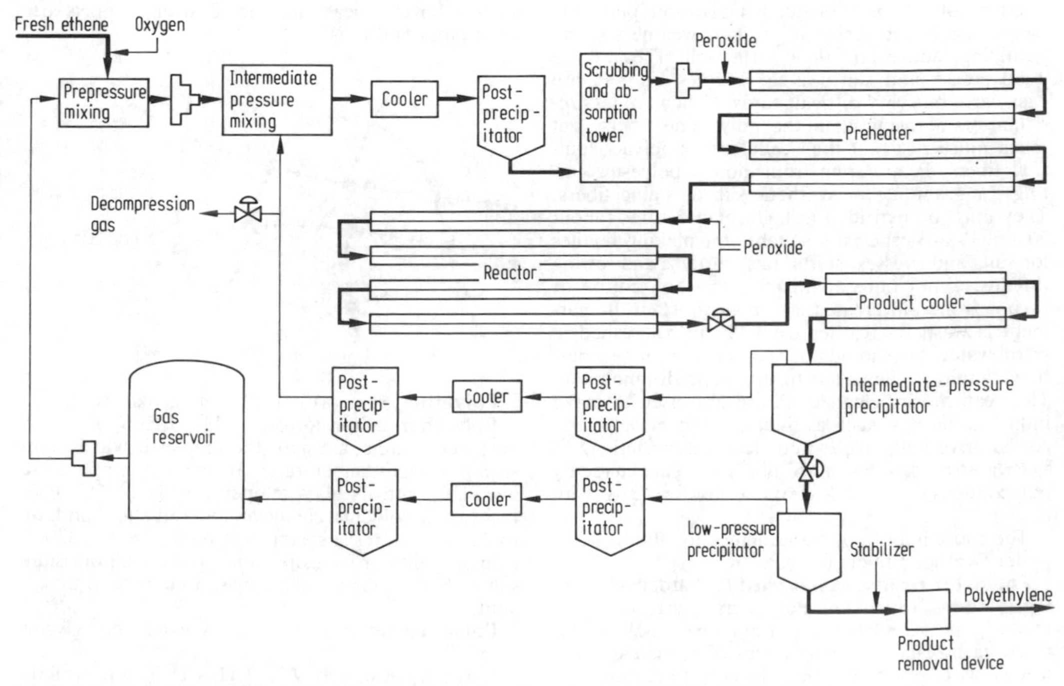

Fig. 1. Scheme for production of high-pressure polyethylene in a pipe reactor (Polymir 60 process).

In the stirred reactor method, the main problem is the tightness of the bearing boxes for the stirrer; for this reason, the motor is placed inside the high-pressure vessel (Fig. 2). Because of the problem of leakage, the temperature can be measured only at four sites on the inner reactor wall; therefore there is no temperature-controlled pressure-release valve; instead the reactor is provided with safety plates which break if the pressure becomes too high. Both types of reactor are used in industry; in the stirred reactor the ethene consumption is somewhat lower (18%).

$$\text{(O)(O)Cr–H} \leftarrow CH_2{=}CH_2 \xrightarrow{+CH_2=CH_2} \text{(O)(O)Cr–}CH_2\text{–}CH_3 \leftarrow CH_2{=}CH_2$$

$$\text{Me–R} \leftarrow CH_2{=}CH_2 \xrightarrow{+CH_2=CH_2} \text{Me–}CH_2\text{–}CH_2\text{–R} \leftarrow CH_2{=}CH_2 \xrightarrow{+CH_2=CH_2}$$

$$\text{Me–}CH_2\text{–}CH_2\text{–}CH_2\text{–}CH_2\text{–R} \leftarrow CH_2{=}CH_2 \xrightarrow{+CH_2=CH_2} \cdots$$

The resulting P. (***high-pressure P.***), like the other P., is tasteless and odorless, physiologically harmless and plastic even without addition of softener. It can be welded and has a paraffin-like surface. It is attacked by nitric acid, but by no other concentrated or dilute acid, base or salt solution. It is even resistant to fluorine and hydrogen fluoride. Its permeability to water vapor is very low. P. is insoluble in vegetable and animal oils and fats, but it is attacked by halogens and strong oxidizing agents at high temperatures. P. is dissolved or swollen by organic solvents, especially chlorinated hydrocarbons, at high temperatures. It is also dissolved by mineral oils. The crystalline fraction in high-pressure P. is about 75%. Complete crystallinity is prevented by branching in the molecules. The excellent electrically insulating properties of the material are due to its completely nonpolar structure, and to the its extremely low capacity for absorbing water.

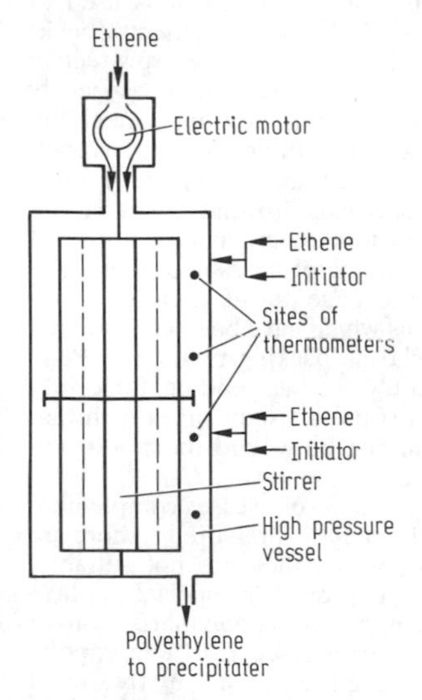

Fig. 2. Stirred reactor for production of high-pressure polyethylene.

2) *Medium-pressure processes* of ethene polymerization (Phillips process) are used mainly in the USA; they are run at temperatures between 50 and 180 °C and pressures of 2 to 3 MPa in the presence of chromium oxide catalysts with silicon dioxide or aluminum silicates as carriers. The process is a solution or precipitation method. Kinetic studies have shown that before polymerization occurs, the ethene is adsorbed by the catalyst and that the active centers contain chromium in a low valence state. The initiation reaction is an insertion into a chromium-hydrogen bond.

The solution process is a continuous reaction carried out at 130-180 °C and an ethene pressure of 2 to 3 MPa, in an organic solvent such as paraffin or cycloparaffin hydrocarbons. The P. which forms must be soluble in the solvent under the reaction conditions. In the precipitation method, a solvent is also used; the temperature is 60 to 110 °C, and the pressure is 2 to 3 MPa. The P. precipitates out under these conditions. Yields of 5 to 20 tons P. per kg catalyst are obtained, so that the catalyst does not have to be separated from the polymer. The P. obtained by this method is 85-95% crystalline. It is strong and elastic; its density is 0.96 kg cm^{-3}.

3) *Low-pressure polymerization* is a precipitation process in which organometal mixed catalysts (Ziegler-Natta catalysts) are used under atmospheric or slightly higher pressure. The temperatures are below 100 °C; and an inert solvent, usually a hydrocarbon (aromatic, alkane or cycloalkane), is used. The most useful industrial catalysts have proved to be combinations of trialkylaluminums (e.g. triethylaluminum) and alkylaluminum halides (e.g. Et_2AlCl, $EtAlCl_2$) and titanium(III) chloride, titanium(IV) chloride or titanium esters. Hydrogen is used as a chain-reaction regulator. With titanium(III) chloride catalysts, a P. with a relative molecular mass of a few million can be obtained. Lower molecular masses are obtained if titanium(IV) chloride is present in the reaction mixture. For this reason, the aluminum to titanium ratio in the catalyst system has a decisive influence on the relative molecular mass of the P. The following is a simplified version of the reaction used in low-pressure polymerization of ethene (Fig. 3): The catalyst components are mixed with ethene and solvent in a polymerization reactor (stirred or bubble column reactor). The resulting suspension of polymer is then treated with an alcohol (isopropanol, isobutanol) in order to deactivate the catalyst; the catalytic components are converted to soluble compounds and remain behind in the solvent when the polymer is removed by filtration. The P. is further processed by extraction, filtration and drying. The unreacted ethene is returned to the polymerization reactor.

The most recent generation of catalysts (3rd generation) are highly active Ziegler-Natta catalysts in which the transition metal compound (e.g. titanium(IV) chloride) is applied to a very pure,

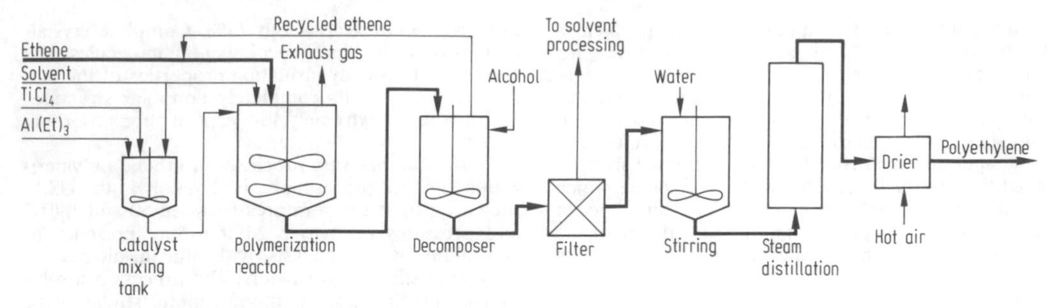

Fig. 3. Scheme for production of low-pressure polyethylene.

anhydrous magnesium compound. With these catalysts, *gas-phase polymerization* of ethene is possible. In this process, very pure ethene (in some cases together with a co-monomer) and hydrogen as a chain-reaction regulator are brought into a moving bed reactor. The catalyst is added continuously in the form of a dry powder carried on a stream of nitrogen. The polymerization occurs at a pressure around 3.5 MPa and a temperature of 85 to 100 °C. The unreacted ethene is circulated back to the reactor, and at the same time serves to carry off the heat of polymerization and to keep the catalyst bed in motion.

Low-pressure P. is about 85% cyrstalline. Its density is 0.95 g cm^{-3}; its electrical properties are very good, although the loss angle, which is important in applications in the high-frequency regions, is somewhat higher than that of high-pressure P. Tension cracks, which form in objects exposed to chemicals for long periods, are evidently less likely to form in low-pressure P. products than in high-pressure P. In general, low-pressure P. is superior to high-pressure P. because of its unbranched molecular structure, which makes it harder and gives it a higher melting range.

4) In *cationic polymerization*, the ethene is forced to dissolve in a solvent under a pressure of about 50 MPa, and in the presence of a catalyst, e.g. aluminum chloride, zinc chloride, boron trifluoride or tin chloride. In the process it polymerizes, with the release of a large amount of heat, into relatively low-molecular mass, highly branched products. When no more ethene can be absorbed, the solvent is driven off with steam and the residue is filtered. The method produces lubricating oils with a relative molecular mass around 400; because of their low freezing points and good viscosity-temperature behavior, their quality is excellent. These oils are used pure (as motor oils for aircraft and refrigerators) or in mixtures with other lubricants.

5) If ethene is polymerized *in solution* (methanol) with benzoyl peroxide as catalyst at 110 to 120 °C and at pressures up to 30 MPa, the products have relatively low molecular masses (2000 to 3000); they are a wax-like material with properties similar to those of hard carnauba wax.

Processing and applications. P. are worked around 200 °C. To reduce their tendency to oxidize at such temperatures, anti-oxidants are added. Large pieces of P. are made by extrusion. Continuous pipes can be made by pressing the plastic through circular openings, and they can then be blown to make bottles or balloons. Small objects can be made by injection casting. Films and sheets are not made on calander rollers, but by extrusion through slits or by cutting open a thin, blown-up tube. Cables, wires and easily corroded surfaces can be given a protective coating of P. The protective film can be applied to a large area by plastic sprayers. P. is usually used without a softener additive, because of its high degree of plasticity. To reduce the potential for formation of tension cracks, small amounts of polyisobutylene are added, or are copolymerized with the ethene to give internal softening. In some cases, addition of a certain amount of activated soot can be useful. P. can be made into a foam, ***P. foam***, by addition of a foaming agent, such as ammonium hydrogencarbonate. This material, with a density of only about 0.46 g cm^{-3}, is used for insulation.

Chlorosulfonation of P. produces an elastic material which, like rubber, can be vulcanized with alkaline earth oxides to products which resist ozone, chemicals and oils; these are used, for example, for floorings. P. is used to make insulators for electrical equipment and cables used in the high-frequency ranges. Because P. is readily colored, it is used to make various unbreakable household objects (buckets, cups, etc.). Bottles, tubes, valves, protective housings, armatures, joints, gears and especially pipes are made of P. Because of their impermeability to water vapor, ***polyethylene films*** (sometimes with added softeners, such as citrate esters) are used in packages for foods, medicines, frozen fruits, etc. The films can also be used to protect early plantings of vegetables from frost, as vapor barriers in buildings, and to prevent water seepage out of streams, ponds and water basins. Films which have been cross-linked by irradiation are used as packing materials. ***Polyethylene pipes*** are widely used as conduits for drinking water, because pressure-resistant pipes with diameters up to 150 mm can be wound on spools and then laid from the roll.

In spite of the good and in some cases complementary properties of high- and low-pressure P., there are many applications for which they are not suitable. Their qualities can be improved for special applications by mixing polymers, by copolymerization of ethene with vinyl monomers (see Ethylene copolymers) and by cross-linking the polymer by treatment with high-energy radiation or chemicals, such as dicoumyl peroxide. Cross-linked P. is much less subject

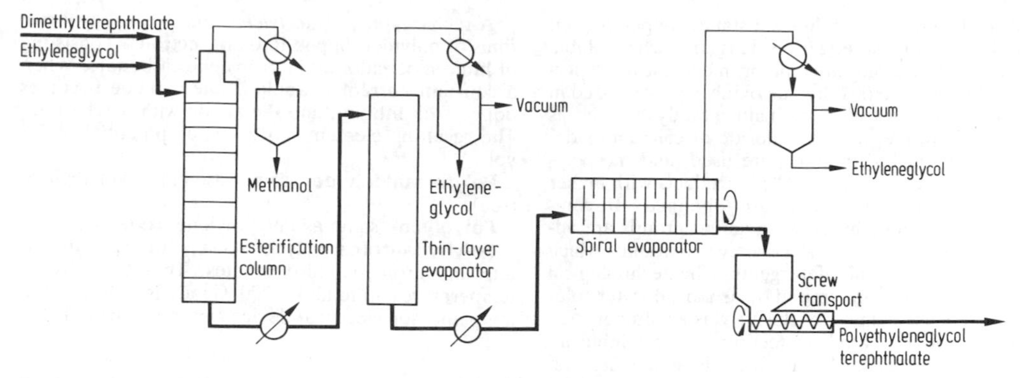

Continuous method for production of polyethyleneglycol terephthalate.

to cracking under tension, and is more resistant to oil and acid. Addition of about 10% alkenes, especially but-1-ene, produces P. with very low density: LLDPE (for low low density polyethylene). The coating of metal pipes, containers, etc. with P. to protect them from corrosion is important. Cables, ropes and cable hulls of P. which are resistant to seawater are made of P.

In view of the high per-capita consumption of P. for packaging, and the resulting quantities of P. trash, materials which can be degraded by light or organisms have been developed. However, the use of these materials could reduce the usefulness of recycled plastic.

Historical. The first industrial experiments on cationic polymerization of ethene were done in 1930. The high-pressure polymerization process was developed in the laboratories of ICI (Imperial Chemical Industries); it was patented in 1933 and was implemented in various countries starting in 1938. Low-pressure polymerization based on special catalysts developed by K. Ziegler was introduced in the years 1954/55.

Polyethylene fibers: see Polyolefin fibers.

Polyethylene glycol: same as Polyethylene oxide.

Polyethyleneglycol terephthalate, abb. ***PETP***: $[OOC\text{-}C_6H_4\text{-}COO\text{-}CH_2\text{-}CH_2]_n$, a linear polyester which is made by reaction of terephthalic acid with ethylene glycol, or by transesterification of dimethylterephthalate with ethylene glycol followed by polycondensation.

$$\text{HO-}\underset{\underset{\text{O}}{\|}}{\text{C}}\text{-C}_6\text{H}_4\text{-}\underset{\underset{\text{O}}{\|}}{\text{C}}\text{-OH} + 2\ \text{CH}_2\text{-CH}_2 \text{(-O-)} \quad (a),\ (b)$$

$$\downarrow$$

$$\text{HO-CH}_2\text{-CH}_2\text{-O-}\underset{\underset{\text{O}}{\|}}{\text{C}}\text{-C}_6\text{H}_4\text{-}\underset{\underset{\text{O}}{\|}}{\text{C}}\text{-O-CH}_2\text{-CH}_2\text{-OH} \quad (c)$$

$$\downarrow$$

$$[\text{OOC-C}_6\text{H}_4\text{-COO-CH}_2\text{-CH}_2]_n \quad (d)$$

Properties. P. is a solid with a melting point around 260 °C; it has a density of 1.38 to 1.41 g cm^{-3}., takes up 0.3 to 0.4% water (at a relative atmospheric humidity of 65%). Its tensile strength is 45 to 80 N mm^{-2}; it has a break expansion of 400 to 500%, and a notched-bar impact strength of 2.5 N mm^{-2}.

Synthesis. The transesterification of dimethylterephthalate is done in two steps. In the discontinuous process, the melted dimethylterephthalate reacts with ethyleneglycol in a stirred reaction at about 100 °C in the presence of catalysts (alcoholates or acetates of zinc, magnesium, lithium, etc.) to form a polyester precondensate. The methanol which is formed and the excess ethylene glycol are distilled off. In the second step, the polyester precondensate is heated to 250-280 °C under vacuum, and polycondenses. A monofunctional carboxylic acid, e.g. benzoic acid, is added as a chain terminator.

In the continuous process (Fig.), the dimethylterephthalate is transesterified to diglycolterephthalate in a column, and the methanol is distilled off over the head. The excess ethylene glycol is then driven off in a thin-layer evaporator. The precondensate is polycondensed in the spiral evaporator at about 280 °C under vacuum. A very pure P. can be obtained by reaction of terephthalic acid (1a) with ethylene oxide (1b) to form diethyleneglycolterephthalate (1c); this is polycondensed in a second step to P. (1d).

Applications. The main application for P. is in textiles (see Polyester fibers). P. foils can be used as film bases because they are very transparent. Highly viscous types of P. are used in the construction of machines and apparatus.

Polyethylenimine: $[CH_2\text{-}CH_2\text{-}NH]_n$, where n is 6 to 1000; a water-soluble polymer of ethylenimine. The polymerization of the monomer begins spontaneously in the presence of small amounts of acid or metallic catalysts. Because the monomer is very toxic, caution must be exercised in working with it. P. is not used in the plastic industry, but in the production of paper, an aqueous solution of the polymer is allowed to react with the paper slurry, which increases the wet strength of the paper considerably. Reaction products of ethylenimine with isocyanates and phosphorus oxygen chloride are used in the textile industry to make cloth water repellent or to soften it, and other products are used as fixatives in pigment printing. The reaction product of ethylenimine and acrylonitrile is used as a molding and gluing resin.

Polyethylene oxide, ***polyglycol, polyethylene glycol, carbowax***: a polymeric product which is either a viscous liquid or a wax-like substance, depending on the degree of polymerization. It is somewhat soluble in water. P. has no outstanding mechanical properties. The monomeric ethylene oxide is polymerized in the presence of acidic or alkaline catalysts, such as boron trifluoride, tin tetrachloride or calcium oxide. If alkaline earth carbonates are used, and the polymerization occurs above 100 °C, products with higher degrees of polymerization are obtained. If compounds containing hydroxy or amino groups are added, the products are surface-active comounds which are used as nonionic detergents. The technological applications of P. are limited by its partial water solubility. It is used almost exclusively as a softener, textile conditioner, wood protection, water-soluble lubricant, stabilizer for lubricating oils, emulsifier, additive to cellulose nitrate paints and in the production of plastic foils.

Polyethynes, ***polyacetylenes***: organic compounds with two or more conjugated carbon triple bonds -C≡C- in the molecule. P. are very unstable, and their instability increases with the number of conjugated triple bonds. While unsubstituted P. and those substituted on only one end of the molecule are so unstable that they explode even at low temperatures, those which are substituted on both ends of the chain are somewhat more stable. Especially in the presence of light, P. tend to polymerize. Diethynes (two conjugated triple bonds) form brown to red polymers, tetraethynes (four conjugated triple bonds) form dark blue polymers, and all P. with more than four conjugated triple bonds form black polymers. These have the same compositions as the starting materials, and are completely in soluble in any solvent. Polymerization usually occurs in substance. In contrast to polyenes, the P. are inert to oxygen. They readily take up hydrogen, and with partially poisoned catalysts, they can be selectively hydrogenated to form polyenes. Compounds which have double bonds in the molecule in addition to the conjugated triple bonds have somewhat different properties than pure P.; the ***diphenyl polyethynes***, for example. A few P. occur naturally; these contain double bonds in addition to the conjugated triple bonds. Matricaria esters, dehydromatricaria esters, lachnophyllum ester and erythrogenic acid are examples of P. which have been isolated from various Compositae. A few plant P. are antibiotics, for example mycomycin from *Nocardia acidophilus*, nemotin and nemotin A from basidiomycetes. Some toxins, such as enanthotoxin and cicutoxin from *Oenanthe crocata* and *Cicute virosa*, have been identified as P.

P. and polyethyne nitrile, $H[C{\equiv}C]_5$-C≡N, have been recognized as one of the main sources of carbon-containing compounds in interstellar space. As of 1982, the most complicated polyethyne molecule known in space is decapentyne nitrile, discovered by Canadian scientists by microwave spectroscopy.

The first P. produced industrially was ***cuprene***. It is synthesized by dissolving ethyne in hot oil in the presence of a copper powder. Dimerization of ethyne produces monovinylethyne (monovinylacetylene, abb. mova), trimerization yields divinylethyne (divinylacetylene, abb. diva). Polymers of ethyne have a high electrical conductivity, and are thus of growing significance in microelectronics.

A rechargable *polyacetylene battery* consists of two films of polymer dipped into an electrolyte consisting of lithium perchlorate in propylene carbonate. When a constant current is applied, the cathode becomes doped with lithium, and the anode with perchlorate. The resulting element has a resting potential of 3.8 volt.

Polyformaldehyde: same as Polyoxymethylene (see).

Polyglycol: same as Polyethylene oxide (see).

Polyhydantoins: polycondensates of bisglycine esters and aromatic diisocyanates. The P. can resist temperatures up to about 200 °C indefinitely. Insulating paints for electrical devices are made from P.

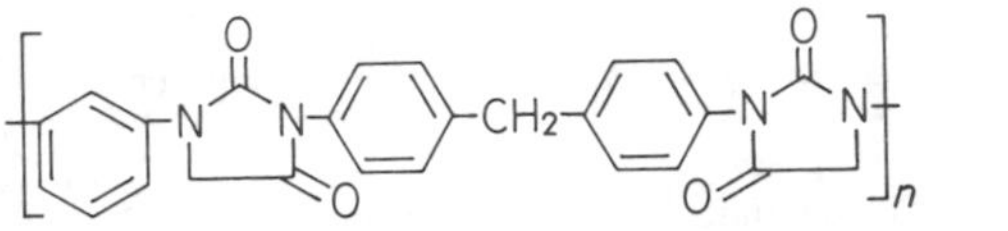

Polyimide: a polycondensate of polymellitic anhydride and 4,4'-diaminodiphenyl ether. P. can withstand temperatures in the range of -240 °C to 408 °C, so that it is very important as a special material for space and nuclear technologies. Addition of graphite to P. makes it suitable for production of packings and bearings which can withstand heavy loads. P. is resistant to organic solvents, but is attacked by strong bases. In electronics, it is used for printed circuits, capacitors and other components.

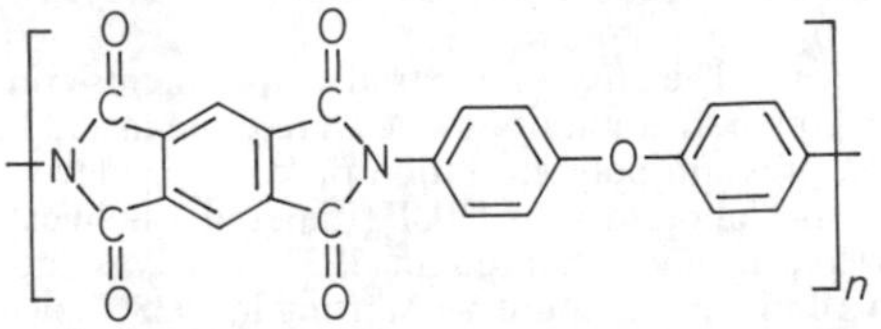

Polyimidazopyrrolone, ***pyrrone***: a polycondensate of aromatic tetracarboxylic acids and aromatic tetraamines. P. is a special material which is of interest in space technology because of its extreme stability to radiation.

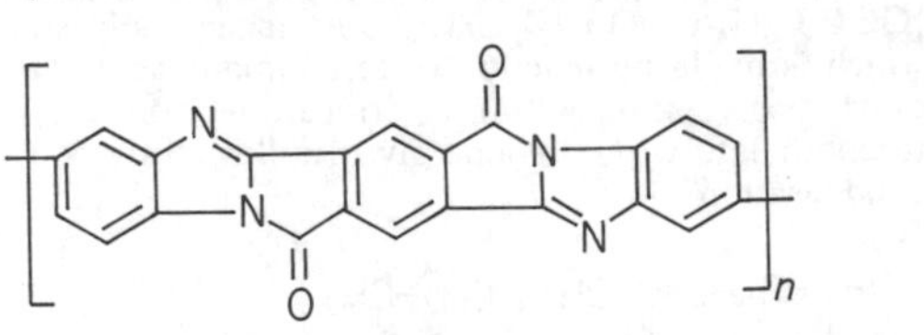

Polyiodides: salts with the general formula $M[I(I_2)_n]$ (n = 1-4). The most important P. is ***potassium triiodide***, KI_3. The triiodide ion is a linear, symmetrical structure. KI_3 is obtained by the reaction of aqueous potassium iodide solution with iodine; it is a dark red-brown compound. Polyiodide anions, such as I_5^-, I_7^- and I_9^-, are stable in the presence of cations with large volumes; these anions are loose aggregates of I_2 molecules with the I^- ion.

Polyisobutene: same as Polyisobutylene (see).

Polyisobutylene, ***polyisobutene***, abb. ***PIB***: a polyolefin obtained by cationic polymerization of isobutene at low temperatures, between about -70 and -50 °C. The polymerization is carried out in pressure vessels or in liquid hydrocarbons such as ethene, propane or butane in the presence of Friedel-Crafts catalysts, such as boron trifluoride. The isobutene in solution (e.g. in liquid propane) and the dissolved boron trifluoride are mixed on a steel conveyer belt, where polymerization begins instantaneously (Fig.).

$$n\,(CH_3)_2C{=}CH_2 \xrightarrow{H^+} \left[-C(CH_3)_2-CH_2-\right]_n$$

Method of producing polyisobutylene.

The resulting heat causes the propane to evaporate; it is collected, reliquefied and returned to solution. Depending on the degree of polymerization, P. are thin to thick oils or rubbery substances with relative molecular masses around 400,000. The oils are valued as lubricants, which can greatly improve the viscosity-temperature characteristics of other lubricants to which they are added. Because of their excellent resistance to oxidation and weathering, mixtures of the higher-molecular-weight P. with fillers (talcum or soot) are used to seal masonry against moisture and as a roofing compound. P. are also used to coat vats and pipes, to seal tunnels and domes, to impregnate textiles, insulate cables and make glues. The properties of P. can be significantly modified by copolymerization with butadiene, vinyl or acryl compounds. Copolymerization of isobutene and isoprene yields butyl rubber (see Rubber).

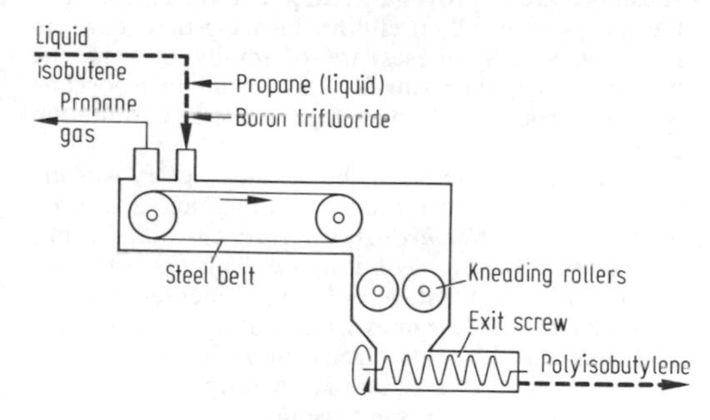

Process for polyisobutylene production.

Polyisocyanates: same as Polyurethanes (see).

Polyisoprene: see Rubber.

Polyketides: a large group of secondary natural products which are synthesized from acetate units via poly-β-carbonyl compounds, $[CH_2\text{-}CO]_n$, as unstable intermediates. Many P. become aromatic by loss of water. The P. include many fungal products, including the aflatoxins, tetracyclines, polyene antibiotics and griseofulvin.

Polymer: a term for a macromolecular substance composed of repeating atomic groups, called monomers. P. are made by polymerization, polycondensation or polyaddition, and can be broken down again by depolymerization. The number of monomers linked together in a polymeric molecule is called the degree of polymerization (see Polymerization, degree of). P. consisting of two monomers are called ***dimers***, those of three monomers are ***trimers***, and so on. P. with low degrees of polymerization are called ***oligopolymers (oligomers)***, and those with a degree of polymerization greater than 50 are called ***high polymers***. If the individual molecules of a P. have different sizes, the substance is called "polymolecular". The ***Biopolymers*** (see) include many natural products.

The shape of the P. can be ***linear, branched*** or ***cross-linked***. If two linear P. are linked by bridge bonds like the rungs of a ladder, it is called a ***Ladder polymer*** (see). ***Spiropolymers*** are linked via quaternary C atoms. P. which consist of a single type of monomer (not counting slight structural deviations at the ends or branchings) are called ***homopolymers***. If the P. are composed of more than one type of monomers, they are called ***copolymers (mixed P.)***. Purely physical mixtures of different P. are called *compounds* or *polyblends*. Depending on the arrangement of the monomers, copolymers are further subdivided into ***statistical*** and ***alternating copolymers***. In ***block P.***, a long block of monomer A is linked to a block of monomer B., while ***graft copolymers*** are branched P. in which the main chain consists of one type of monomer and the branch chains of another.

spiropolymers — isotactic polymers — syndiotactic polymers

P. which contain a pseudoasymmetric C atom in each monomer are classified according to the spatial arrangement (tacticity) of the atoms or substituents with respect to each other as ***atactic, isotactic*** and ***syndiotactic P.*** Isotactic and syndiotactic P. can be synthesized by stereospecific polymerization using certain catalysts, such as the Ziegler-Natta catalysts. They often have better properties than the ordinary atactic P., such as higher mechanical strength, crystallinity, density and thermostability. If the monomer has two centers of asymmetry, the P. is ***ditactic***, and its configuration is distinguished by the usual organic chemical criteria into *threo-* and *erythro*-forms. Since the properties of a P. depend very greatly on the configurational uniformity of the monomers, efforts are made to synthesize sterically uniform P. and to categorize the irregularities which appear. This is done using the tacticity in triads (three neighboring centers of isomerism) in the P. Because there are two different forms of each center of isomerism (called a and b here), the following triads are possible: aaa = bbb = isotactic triads with two isotactic bonds (ii); aba = bab = syndiotactic triads with two syndiotactic bonds (ss); and abb = baa = heterotactic triads with one isotactic and one syndiotactic bond (is). If long blocks of one type of triad alternate with long blocks of the other, the P. is a ***stereo block P.***. If there are double bonds in the chain (e.g. polybutadiene, polyisoprene), the monomers can be linked by *cis* or *trans* bonds, and therefore a distinction is made between ***cis*** and ***trans-tactic P.***

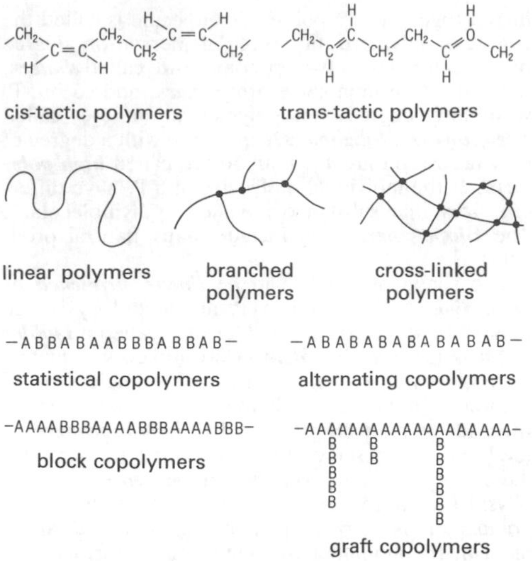

cis-tactic polymers trans-tactic polymers

linear polymers branched polymers cross-linked polymers

—ABBABAABBBABBAB— statistical copolymers

—ABABABABABABAB— alternating copolymers

—AAAABBBAAAAABBBAAAABBB— block copolymers

—AAAAAAAAAAAAAAAAAAA— graft copolymers

Development and application of P. By programmed temperature treatment of molded objects made of suitable P., one obtains ***graphitized P.***, which have long-term stability in air at temperatures up to 400 °C, and up to 4000 °C in nitrogen. These are used to produce carbon fibers, carbon glass and carbon foam. The course of the temperature treatment consists of a coking stage (up to 300 °C in the presence of oxygen), a carbonizing stage (up to 1000 °C in the presence of nitrogen) and a graphitization stage (up to 2000 °C under nitrogen).

To achieve desired properties, mixtures of P. are used. For example, soft, elastic polybutadiene and hard, brittle polystyrene are mixed in such proportions that a hard, impact-resistant material is obtained.

For future applications of P., continued improvements in their stiffness and elasticity, temperature stability and electrical and optical conductivity will be important.

P. in which the stiffness is emphasized are called ***high-modulus P.***. They are made from P. which form highly oriented, liquid-crystalline phases, or P. in which the chain systems undergo post-polymerization physical orientation in the direction of stress. Some examples are the carbon fibers and poly-*p*-phenyleneterephthalamide. Carbon fibers are also very stable to high temperatures, as are P. based on carbocyclic or heterocyclic ring systems, such as polyquinoxalines, polybenzimidazoles and polyimides. Long-term stability at temperatures up to 150 °C and short-term stability (for a few minutes) up to 600 °C have been achieved.

P. in the form of ion-exchange resins or membranes are used to separate mixtures of substances. In addition to the classical cellulose acetate membranes, membranes based on polyamides and polyimides are increasingly used; these are resistant to chlorine and hydrolysis.

The incorporation of chelate-forming groups, such as iminodiacetic acid, hydroxyquinoline or crown ethers, leads to P. which can be used for selective binding of metal ions in hydrometallurgy, or for reprocessing of nuclear fuels.

The use of the optical qualities of organic P. is still limited compared to that of glasses. The advantages are the lower density of the materials and the ease of producing them, e.g. lenses for simple cameras and sunglasses can be made by injection molding. The polytrifluoromethylstyrenes with low refractive indices and high dispersion are used for this purpose. The polyines have a special position among the P. with high electrical conductivity.

Photosensitive P. contain light-sensitive groups in the P. molecule. In copolymerization of *o*-nitrobenzyl acrylate and methylmethacrylate, for example, there is a light-induced rearrangement of the *o*-nitroaromatic ring, in which the free carboxyl function is introduced into the side chains. The P. becomes alkali-soluble in the sites exposed to light, and this is the basis for a non-silver photographic process. Copolymers of maleic anhydride and styrene with azo groups in the side chain undergo a reversible *cis-trans* photoisomerization, which is associated with a change in the index of refraction. Because of this property, the P. can be used as a high-resolution optical information carrier.

P. with chromophoric groups which can convert light to mechanical energy may eventually find applications. The conversion process is based on an isomerization, for example of azo groups or β-carotene groups of the chromophore, which leads to contraction of the P.

P. are gaining importance as carriers for fixation of active substances (pharmaceuticals, pesticides) and organic reagents (see Solid-phase peptide synthesis). The lifespan and transport properties of the active substance are improved by fixing it to the carrier; the method is especially useful for insuring that reagents for expensive syntheses are optimally utilized. In biotechnology, the fixing of enzymes and microorganisms to carriers is of great importance in continuous processes.

A new development in the synthesis of P. was introduced in 1982 with the first biotechnological production of a P. ***Polyhydroxybutyrate*** was made using the bacteria *Alcaligenes entrophus*. This new thermoplastic has the advantage of being biodegradable.

Polymerases: enzymes which catalyse the formation of macromolecules from simple units, such as DNA polymerase (see) and RNA polymerase (see).

Polymer gasoline: see Gasoline.

Polymeric electretes: polymeric materials with a permanent electric charge. They are made by introducing electron charges by electron bombardment of the polymeric film or, as in the case of polyvinylidene fluoride, by orienting polar groups in a preferred direction by application of an external electric field. P. are used as converters of mechanical and thermal signals to electrical signals (e.g. in microphones, loudspeakers and ultrasound detectors).

Polymerization: a type of chemical reaction in which many molecules of low-molecular-weight compounds (monomers) are combined into macromolecules (polymers) by a chain reaction mechanism of chain growth. A characteristic of a P. is that the percent elemental composition of the polymers is the same as that of the monomers. In other words, no products are split off during the P., as happens during Polycondensation (see), and the molecular groups are not rearranged as in Polyaddition (see). A P. is an exothermal reaction, because the polymer always has lower energy than the unsaturated monomers. To avoid explosions, the heat released by the reaction must be removed. The difference in density between monomers and polymer leads to a decrease in volume (e.g. of 14.7% in the case of styrene), and this is utilized to monitor the rate of the reaction. The most suitable for P. are those with multiple bonds or rings (Table). If multiple unsaturated starting molecules are used, the polymer is cross-linked in two or three dimensions and is a thermoset plastic. Polymers can be "custom made" with nearly any desired set of properties. For one thing, the characteristics of the products can be affected by the chemical structures of the monomers, or by subsequent chemical reactions of the polymers. On the other hand, it is also possible to create polymers with special properties by mixed P. of different monomers. A measure of the number of monomers combined in a macromolecule is the degree of polymerization (see Polymerization, degree of).

If a single type of monomer is polymerized, the process is known as ***homopolymerization***. Assuming that the monomer has an asymmetric structure, AB, it can be linked either ***"head-to-tail"***, or AB-AB-AB..., or "***tail-to-tail***" or "***head-to-head***", AB-BA-AB-BA.... In addition, a random order such as AB-AB-BA-BA-AB-BA... is possible.

P. is a ***chain growth reaction***, which can occur by a radical, coordinative or ionic mechanism, depending on the type of catalyst. P. is begun by an initiation reaction, and further growth of the chain occurs by a growth reaction. It is stopped by a termination reaction. If the reactive state is transferred to another molecule, the chain reaction can be restarted.

$$\left[CH_2{-}\underset{\displaystyle R{-}O{-}C{=}O}{\underset{|}{CH}} \right]_n$$

R	→	R*	Initiation reaction
R* + M	→	RM*	Growth reactions
RM* + M	→	RMM*, etc.	
RMM* + R	→	RMM + R*	Transfer reaction
RMM*	→	Inactive product	Termination reaction

R is the initiator and M are monomers.

In order for a polymer to form, the growth reaction must occur at a much higher rate than the transfer and termination reactions combined. ***Depolymerization*** is the reverse reaction of the growth reaction, that is, the cleavage of a monomer from the active polymer. P. and depolymerization are in equilibrium:

$$P_n^* + M \underset{\text{Depolymerization}}{\overset{\text{Polymerization}}{\rightleftharpoons}} P^*_{n+1}$$

$$n\,OCN{-}(CH_2)_6{-}NCO + n\,HO{-}(CH_2)_4{-}OH$$
$$[\,CONH - (CH_2)_6\,NHCO - O - (CH_2)_4 - O]_n$$

Here P^*_{n+1} is active polymer and M is the monomer.

For every equilibrium between polymer and monomers, there is a certain limiting temperature at which chain growth and loss of monomers are in equilbrium; above this temperature, no P. can occur (see Ceiling temperature).

Types of P. 1) In ***radical P.***, the reaction is initiated by a radical. The radicals can be generated photochemically, electrochemically, thermally or by a catalytic reaction involving radical-producing initiators, such as hydrogen peroxide, benzoyl peroxide, cumene hydroperoxide, perborates, persulfates, percarbonates, azodiisobutyric nitrile, etc. The chain growth reaction in radical P. depends on addition of a radical intermediate to the double bond of a monomeric molecule (Fig. 1a). In the transfer reaction, the radical state of the growing polymer is transferred to some other molecule in the system, such as unreacted monomers, solvent, initiator or other polymers, and the polymer loses its radical nature (Fig. 1b). The termination reaction usually occurs by disporportionation or combination of two radicals (Fig. 1c).

Radical P. always produces spatially random (*atactic*) polymers.

In ***ionic P.***, ionic catalysts are used which attract or repel the π-electron pair of the monomer double bond. In *anionic P.*, the chain growth reaction is initiated by the anion of the initiator molecule, or by transfer of an electron to the double bond of the monomer. Metal hydrides, sodium alcoholates, sodium amide, tertiary phosphines, Grignard compounds, etc. are used as initiators (Fig. 2a). During the growth reaction, a polymer carbanion reacts with monomers until they are consumed or the reaction is ended by a termination reaction (Fig. 2b). In nonpolar solvents, and in neat polymerization, the anionic P. are not free ions. In these cases, the growth reaction is carried by polarized metal-carbon bonds, which are partially associated (*pseudoanionic P.*). The formation of complexes causes a weakening of the metal-carbon bond, and an activation of the monomeric alkene. The subsequent reaction of the alkene is known as an insertion reaction, and this type of P. is *insertion P.* (Fig. 3). *Cationic P.* is initiated by addition of a cation to a C=C double bond. Proton acids and Lewis acids are used as initiators (Fig. 4a). The resulting carbenium ion can be stabilized by charge equilibration or it can react with further monomers (Fig. 4b). In cationic P., the charge equilibration between the active chain end and the counter ion plays an important role. The rates of ionic P. increase with increasing dielectric constants of the solvents.

3) In ***coordinative P.***, the carriers of chain growth are transition metal complexes; the incorporation of the monomer into the growing chain occurs in the coordination sphere of the metal complex. The first step in P. is the formation of a metal-carbon bond, into which further monomers are incorporated (Fig. 5). The transition metal complex compounds are synthesized from salts of transition metals in groups IIIb to VIIIb, especially the halides of titanium, vanadium, chromium, cobalt and nickel, by alkylation with metal alkyls of groups Ia to IIIa, chiefly aluminum alkyls. Coordinative P. is very important in

industry for synthesis of polypropylene and of polyethylene, by the low-pressure method.

$$\mathrm{R\cdot + \mathit{n}\,CH_2{=}CHR \rightarrow R{-}[CH_2{-}CH(R)]_{\mathit{n}-1}{-}CH_2{-}\dot{C}HR}$$

1a

$$\mathrm{R{-}[CH_2{-}CH(R)]_{\mathit{n}-1}{-}CH_2{-}\dot{C}HR + R^1{-}H \rightarrow R{-}[CH_2{-}CH(R)]_{\mathit{n}-1}{-}CH_2{-}CH_2R + R^1\cdot}$$

1b

$$\mathrm{R^1\cdot + R{-}[CH_2{-}CH(R)]_{\mathit{n}-1}{-}CH_2{-}\dot{C}HR \rightarrow R{-}[CH_2{-}CH(R)]_{\mathit{n}-1}{-}CH_2{-}CH(R){-}R^1}$$

1c

Buta-1,3-diene and isoprene are polymerized with the aid of these catalysts (Ziegler-Natta catalysts) to 1,4-*cis*-polybutadiene and 1,4-*cis*-polyisoprene (see Rubber).

Various cyclic monomers, such as cyclopentene, ethylene oxide, caprolactam and cyclic siloxanes, can be polymerized by ***ring-opening P.***. This can occur by an ionic or coordinative mechanism, as either a stepwise or chain growth reaction (Fig. 6).

$$\mathrm{C_2H_5O^-Na^+ + CH_2{=}CHR \rightarrow C_2H_5O{-}CH_2{-}C^-HR\ Na^+}$$

2a

$$\mathrm{C_2H_5O{-}[CH_2{-}CH(R)]_{\mathit{n}}{-}CH_2{-}C^-HR + CH_2{=}CHR \rightarrow C_2H_5O{-}[CH_2{-}CH(R)]_{\mathit{n}+1}{-}CH_2{-}C^-HR}$$

2b

$$\mathrm{[CH_2{-}CH(R)]_{\mathit{n}}{-}CH_2{-}CH(R){-}Me + CH_2{=}CH{-}R \rightleftharpoons}$$

$$\mathrm{[CH_2{-}CH(R)]_{\mathit{n}}{-}CH_2{-}CH(R){-}Me \leftarrow CH_2{=}CH{-}R \rightarrow [CH_2{-}CH(R)]_{\mathit{n}+1}{-}CH_2{-}CH(R){-}Me}$$

3

$$\mathrm{CH_2{=}CH(R) + HX \rightarrow CH_3{-}C^+HR + X^-}$$

4a

$$\mathrm{CH_3{-}C^+HR + X^- + CH_2{=}CH(R) \rightarrow CH_3{-}CH(R){-}CH_2{-}C^+HR\ X^- \xrightarrow{+\mathit{n}\,CH_2{=}CHR} Polymer}$$

$$\mathrm{CH_3{-}C^+HR + X^- \rightarrow CH_3{-}CH(R){-}X}$$

4b

$$\mathrm{(Cl)_4Ti(R)(CH_2{=}CH_2) + CH_2{=}CH_2 \rightarrow (Cl)_4Ti(CH_2{-}CH_2{-}R)(CH_2{=}CH_2)}$$

5

$$\mathrm{\underset{O}{CH_2{-}CH_2} + ROMe \rightarrow RO{-}CH_2{-}CH_2O^-\,Me^+ \xrightarrow{+\ \underset{O}{CH_2{-}CH_2}} RO{-}[CH_2{-}CH_2{-}O]_{\mathit{n}}{-}CH_2{-}CH_2O^-\,Me^+}$$

6

All P. which lead to spatially ordered polymers (*isotactic* and *syndiotactic* polymers) are known as ***stereospecific P.***. These are formed mainly by coordinative and anionic P.

Polymerization aids. In addition to the catalysts and initiators mentioned above, regulators, such as mercaptans, can also be used. These affect the degree of P. and with it, the solubility, viscosity, thermoplasticity and mechanical properties of the product. They can lead to chain termination at a certain chain length, or to branching and cross-linking.

Polymerization methods. In ***block P. (mass P.)***, liquid monomers are linked into solid polymers in the absence of solvents. Because of the difficulty of removing the heat of reaction, however, this method is rarely applied. In ***solvent P.***, the monomers are dissolved in a solvent in which the resulting polymer is also soluble. After the P. has ended, the polymers can be precipitated by addition of a solvent in which only the monomers are soluble. This form of P. has the advantage that the heat of reaction is more easily removed. The degrees of P. achieved are not very high, however, and depend on the nature and amount of solvent. If the polymer is insoluble in the solvent and precipitates during the P., the method is called ***precipitation P.***. In ***emulsion P.***, the monomers are emulsified in water using an emulsifying agent, and the polymer precipitates. Water-soluble substances are used as initiators, and the emulsifying agents are usually fatty alcohol sulfonates. If the monomers are distributed in the form of fine droplets in a non-miscible liquid, the method is called ***suspension, bead*** or ***grain P.***, because the polymers precipitate in the form of beads. To prevent adhesion between the monomer droplets, emulsion stabilizers or distribution substances, such as water-soluble, organic colloids (starch, gelatin or synthetic polymers) are added. In special cases, P. can also occur in the gas phase, as in the synthesis of polyethylene. In ***gas-phase P.***, a fluidized bed of polymer grains is suspended by the circulating monomers. The advantage of this method is that the heat of reaction can easily be removed, and the polymer is obtained in pure, dry form. A few crystalline monomers, such as trioxane, can also polymerize in crystalline state. Here the P. is generally initiated by high-energy radiation. ***Mixed*** or ***copolymerization*** is used to generate polymers with special properties; this starts from a mixture of different monomers. ***Heteropolymerization*** is the copolymerization of certain unsaturated compounds which are not capable of homopolymerization; an example is maleic acid. A special type of copolymerization is ***graft copolymerization***. This is the grafting of polymers onto pre-existing chains. The side chains thus consist of different monomers than the backbone chains. In contrast, ***block P.*** is the formation of polymeric chains in which blocks or sequences of identical monomers form the chain. For example, a block of monomer A is followed by a block of monomer B, and so on. The individual blocks are made by separate P. of the monomers, and are then copolymerized in a second step.

Forms of polymerization for selected monomers

Monomers	Polymerization type			
	Radical	Anionic	Cationic	Coordinative
Ethene	+			+
Propene				+
But-1-ene				+
Isobutene			+	+
Buta-1,3-diene	+	+		+
Isoprene	+	+		+
Styrene	+			
Acrylonitrile	+	+		
Vinyl chloride	+			
Acrylate	+			
Tetrafluoroethylene	+			

Polymerization, degree of: a measure of the number of monomeric units in the macromolecules of a polymer. Since the individual macromolecules of a polymeric substance are usually not of uniform length, only an average P. can be determined experimentally. This is done by determining the average relative molar mass M of a polymer, and dividing this number by the relative molar mass M_0 of the monomer. For example, polyethylene with a relative molar mass of 280,000 has a P. of 10,000 (the M_0 of ethene is 28).

Polymer isomeric substances: polymers consisting of the same kinds of monomeric units, but with different molecular structure; for example atactic polypropylene and isotactic polypropylene formed by stereospecific polymerization.

Polymethacrylates, ***polymethylacrylates, polymethacrylic acid esters***: thermal plastics produced by polymerization of esters of methylacrylic acid (methylpropionates). Some of the industrially more important members of this group are ***polymethacrylic acid methyl ester, polymethylmethacrylate, PMMA*** (R = $-CH_3$); ***polymethacrylic acid ethyl ester, PMAA*** (R = $-CH_2-CH_3$); ***polymethacrylic acid propyl ester, PMAP*** (R = $-CH_2-CH_2-CH_3$); and ***polymethacrylic acid isopropyl ester*** (R = $-CH-CH_3)_2$.

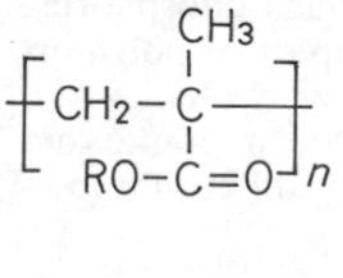

$$\left[-CH_2-\underset{RO-C=O}{\overset{CH_3}{\underset{|}{\overset{|}{C}}}}- \right]_n$$

Properties. P. are solid, hard and transparent, and are therefore sometimes referred to as ***organic glass (acrylate glass)***. Up to a point, they are also transparent to ultraviolet and x-rays, and are fully isotropic. They are unbreakable and very temperature-resistant (up to about 70 °C). They are resistant to water, acids, alkalies, gasoline, alcohol and mineral oils, but they dissolve in benzene and its homologs, esters and ketones. P. have lower specific weights than silicate glasses; they do not splinter and can be readily molded at higher temperatures. They can be dyed.

Production. The monomeric methacrylate is usually polymerized in the ingot, sometimes with an addition of a few percent dibutylphthalate as a softener. Peroxides, such as benzoyl, acetyl or hydrogen peroxide, are used as catalysts; however, the monomers can also be activated by heat or light. The resulting heat of polymerization must be carefully removed in order to obtain colorless and absolutely unstressed polymers. P. can also be polymerized in solution or emulsions.

Applications. P., especially polymethylmethacrylate, are used in many ways; they have become indispensible as materials for aircraft cockpits, safety glasses for vehicles, apparatus in chemical industry, building components and so on. They are used to produce household items, watch crystals, jewelry, sunglasses, equipment for the electrical industry, and so on. In medicine, polymethylmethacrylate is used as a material for false teeth and bone replacements. Lenses and light conductors are made of P. The solution and emulsion polymers are used as paints for automobile bodies and buildings. Aqueous emulsions of P., with added softeners, are used in the cosmetics industry and as sizing in textiles. P. is sold as semiprepared sheets and blocks which are used in many ways, in part due to the ease of working the material.

The industrial development of the P. began around 1912 with the work of O. Röhm.

Polymethinyl pigments: synthetic pigments which contain an odd-numbered chain of methinyl groups, -C=. Auxochromic groups are located on the ends of the methinyl chain; these are conjugated with each other via the methinyl groups (Fig. 1). P. are conjugated according to their charges. Some important ***cationic P.*** are the Cyanins (see), while the *oxonol pigments* are important ***anionic P.***. These can be considered vinyl homologs of the carboxylate anion in which the last two carbon atoms of the methinyl chain are components of a heterocycle (Fig. 2).

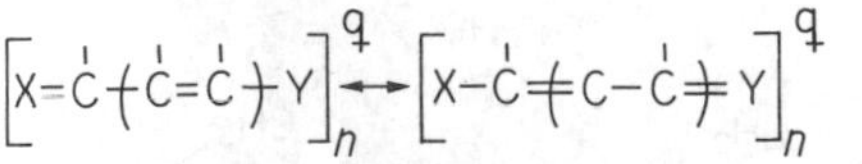

$$\left[X{=}\overset{|}{C}\left(\overset{|}{C}{=}\overset{|}{C}\right)_n Y\right]^q \longleftrightarrow \left[X{-}\overset{|}{C}\left({=}C{-}\overset{|}{C}\right)_n{=}Y\right]^q$$

Fig. 1. Odd-numbered polymethine chain.

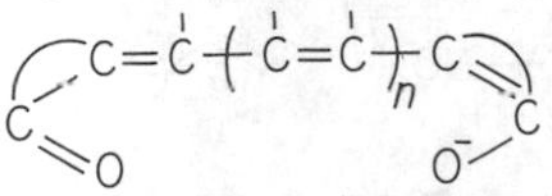

$$C{=}\overset{|}{C}\left(\overset{|}{C}{=}\overset{|}{C}\right)_n C \quad C{=}O \quad C{-}O^-$$

Fig. 2. Structure of oxonol pigments.

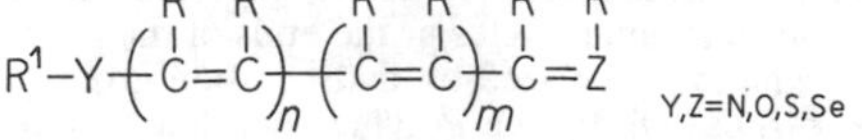

$$R^1{-}Y\left(\overset{R^3}{\overset{|}{C}}{=}\overset{R^4}{\overset{|}{C}}\right)_n\left(\overset{R^5}{\overset{|}{C}}{=}\overset{R^6}{\overset{|}{C}}\right)_m\overset{R^7}{\overset{|}{C}}{=}\overset{R^8}{\overset{|}{Z}} \qquad Y,Z = N,O,S,Se$$

Fig. 3. Structure of the neutrocyanine pigments.

In aza derivatives of anionic P., the methinyl groups are replaced by one or more azamethinyl groups -N=, for example in murexide. The ***neutral P.***, also called ***neutrocyanin pigments***, can be represented by the general formula shown in Fig. 3. These are subdivided into ***merocyanins, hemioxonoles*** and ***dequaternized cyanins***.

Ring closure between	Subgroup
R_2 and R_3; R_6 and R_7	Merocyanins
R_6 and R_7	Hemioxonoles
R_2 and R_3, R_7 and R_8	Dequaternized cyanins

The P. are usually made by separate synthesis of the polymethinyl chain and the heterocycles, which are then combined.

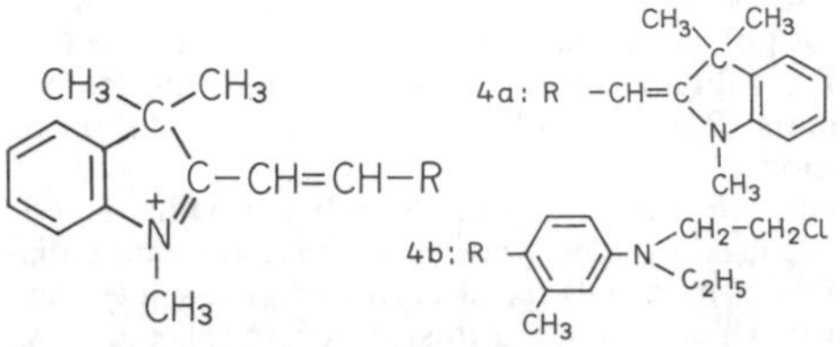

Fig. 4a. Astrophloxin FF; Fig. 4b. Astrazone red GB

The P. are very important spectral sensitizers in black-white and color photography. Astraphloxin FF (Fig. 4a) is used to dye wool a fluorescent, brilliant red, and astrazone red GB (Fig. 4b) is a dye for polyacrylonitriles. Celliton fast yellow 7G (Fig. 5), a merocyanin, is a suspension pigment. P. are important in laser technology as optical switches and as photochromes. For example, G-nitro-BiPS (G-nitro-1,3'3'-aminoethyl-*spiro*-2n-1-benzopyran-2,2'-indoline; Fig. 6) is used as a photochromic dye in data storage, as a temperature indicator in photochromic glasses and for decoration.

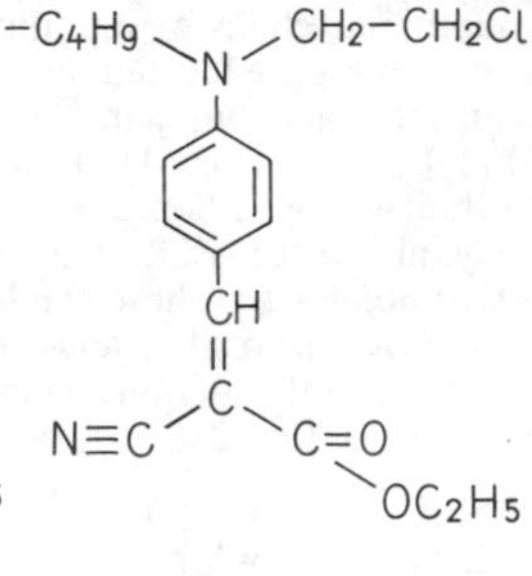

5

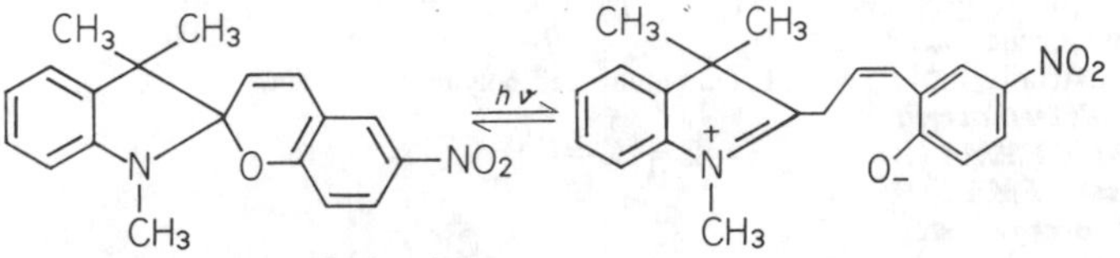

6 G-nitro-BiPS

The P. also include the Carotenoids (see), in which the auxochromic heterocycles at the ends of the polymethinyl chain are replaced by hydrocarbon groups.

Polymethylacrylate: see Polymethacrylate.

Polymethylene: $[CH_2]_n$, a linear alkane. P. can be synthesized conveniently by catalytic reduction of carbon monoxide with hydrogen (Fischer-Tropsch synthesis) or by catalytic decomposition of diazomethane. Without a catalyst, the Fischer-Tropsch method yields oils or waxes with relatively low molecular masses. On the other hand, ruthenium as catalyst yields products with relative molecular masses greater than 10,000, with melting points around 130 °C and densities around 0.98. The decomposition of diazomethane in the cold and in the presence of copper powder or boric acid esters produces polymers with high melting points, but because of their highly crystalline structures, they are more brittle than polyethylenes with similar relative molecular masses.

Poly-4-methylpent-1-ene: abb. ***PMP***: a polyolefin which is obtained by low-pressure polymerization of 4-methylpent-1-ene with organometallic mixed catalysts (Ziegler-Natta catalysts). P. is characterized by high transparency and a melting temperature over 200 °C, so that packaging films made from it can be filled with hot foods. In addition, P. is used as a building material for special purposes.

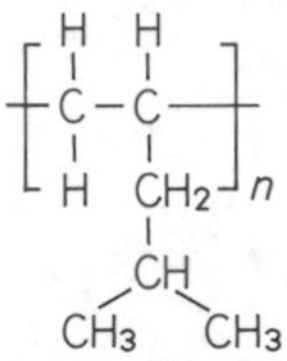

Polymir 60 process: see Polyethylene.

Polymolecularity: a property of substances which is due to nonuniformity in the masses of the molecules of a polymeric substance. With the exception of nucleic acids and proteins, both natural and synthetic polymers are formed by mechanisms which lead to a random distribution of molecular size, and these substances always exist in the form of polymolecular mixtures. In contrast to P., a colloid is said to have the property of polydispersity.

Polymorphism: the existence of different solid forms (*modifications*) of a compound. Although they have the same chemical composition, modifications have different structures and thus different physical and sometimes also chemical properties. If two, three or more modifications are known, the terms dimorphism, trimorphism and so on are used. The concept P. is often applied only to compounds, and the same phenomenon in elements is called *allotropism*. Some examples of allotropic modifications are rhombic and monoclinic sulfur, white, red and black phosphorus, graphite and diamond. Some examples of polymorphic compounds are zinc sulfide (zinc blende and wurtzite), calcium carbonate (calcite, aragonite and vaterite) and ammonium nitrate (five different modifications). The occurrence of the various modifications depends on external conditions (pressure, temperature, crystallization conditions).

The interconversion of modifications is called *transformation*. A distinction is made between enantiotropic pairs of modifications, which can interconvert (reversible conversion, *enantiotropism*), and monotropic pairs of modifications, in which the conversion can occur in only one direction (irreversible conversion, *monotropism*). The difference in behavior can be understood from the vapor pressure curves (Fig.).

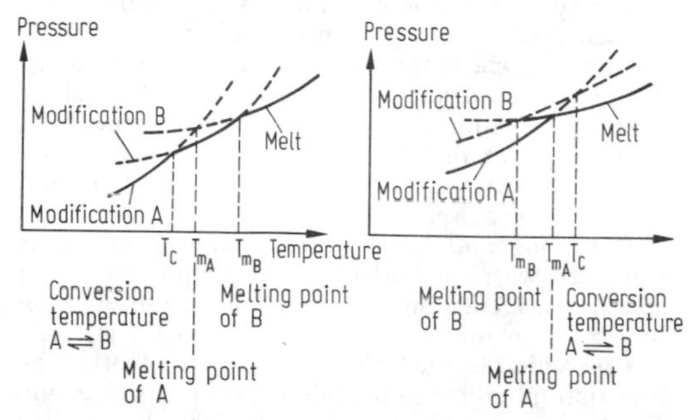

Polymorphism. Vapor pressure curves (schematic) *left* of an enantiotrophic pair and *right* of a monotropic pair of modifications.

Under any given pressure and temperature, other than the conversion points, only one modification is stable, the one with the lowest vapor pressure. The left figure shows the vapor pressure curves of two enantiotropic modifications A and B. As the melt cools, at the melting point T_{mB}, modification B crystallizes. When the conversion temperature T_c is reached, the vapor pressure curves intersect, and the two forms are in equilibrium. Below this temperature, modification A forms. When the substance is heated, the reverse sequence of phases leads to A-B melts. For example, the α-, γ- and δ-modifications of iron interconvert reversibly at 1183, 1661 and 1812 K. Often the conversion rate in the solid phases is so slow that even modifications which are unstable under the conditions can be kept for a long time (see Metastable state). For example, quartz glass, an amorphic modification of silicon dioxide, is unstable below 1978 K. However, quartz glass can exist indefinitely, even under moderate heating; it is thus in a metastable state. Only if the transformation rate is considerably accelerated by heating to glowing, especially in the presence of impurities, does the crystalline modification crystobalite appear ("deglassification").

If its temperature changes very slowly, the phase behavior of sulfur corresponds completely to the solid curve shown in the left figure. If rhombic α-sulfur (modification A) is very slowly heated, it is transformed at T_c = 368.6 K into monoclinic β-sulfur (modification B), which melts at 392.1 K (T_{mB}. However, if α-sulfur is rapidly heated, it melts, without previous conversion to β-sulfur, at T_{mA} = 387.1 K (dashed curve). If the melt is cooled, at T_{mB} β-sulfur first forms, and if the temperature then drops very slowly, α-sulfur is formed. However, β-sulfur can easily be supercooled, and remains as a metastable state for a short time even at room temperature before it is transformed into α-sulfur.

The figure on the right shows the corresponding behavior of a pair of monotropic modifications A and B. Here the transformation temperature T_c is above the melting point T_{mA}, and therefore the A modification melts before it reaches the conversion temperature. A. is stable over the entire temperature range of the solid state, and B is unstable. However, it is often possible to cool a melt to the point T_{mB}, at which point the metastable B modification is formed; as a result of its higher vapor pressure, it is converted to A sooner or later. A conversion in the opposite direction is in principle impossible. For example, with phosphorus only the direct transformations of the white into the red, and the red into the black modification can occur; white phosphorus cannot be obtained directly from the other modifications, but only by condensation of phosphorus vapor.

Since the melting point depends on the pressure, it can happen that a transformation which is monotropic under normal pressure becomes enantiotropic when the pressure is increased. This is the basis of the production of synthetic diamonds at a pressure of $p \geqq 5.5 \cdot 10^3$ MPa and a temperature $T \geqq 1730$ K.

Polymyxins: cyclic peptides containing fatty acids produced by *Bacillus polymyxa*; they have antibiotic effects against gram-negative bacteria. The P. are cyclic branched heptapeptides with an L-α,γ-diaminobutyric acid group (Dbu) in the branch position; it is linked by its γ-amino function to the carboxyl group of a threonine group to form the ring structure. Its α-amino function is bound to a tetrapeptide sequence. The terminal amino group is linked to a branched fatty acid group, either (+)-6-methyloctanoic acid (MOA; also called (+)-isopelargonic acid) or 6-methylheptanoic acid (also called isooctanoic acid, IOA) (Fig.).

Ⓡ – Dbu – Thr – Ⓧ – Dbn – Ⓨ – Ⓩ – Dbu – Dbu – Thr

Polymyxin	R	X	Y	Z
B_1	MOA	Dbu	D-Phe	Leu
B_2	IOA	Dbu	D-Phe	Leu
D_1	MOA	D-Ser	D-Leu	Thr
D_2	IOA	D-Ser	D-Leu	Thr
Colistin A = E_1	MOA	Dbu	D-Leu	Leu
Colistin B = E_2	IOA	Dbu	D-Leu	Leu
Circulin A	MOA	Dbu	D-Leu	Ile

Peptide antibiotics of the polymyxin family.

The P. are toxic to human beings. Because they are hydrolysed and not absorbed when administered orally, however, they can be used to treat infections of the gastrointestinal tract.

Polynose fibers: see Rayon.

Polynucleotide: see Nucleotides.

Polynucleotide ligases: same as DNA ligases (see).

Polyolefins: 1) collective term for organic compounds which contain several olefinic double bonds in the molecule, such as the polyenes.

2) abb. ***PO***, thermoplastic polymerization products of alkenes (olefins). Some industrially important P. are Polyethylene (see), Polypropylene (see), Polyisobutylene (see), Polybut-1-ene (see) and Poly-4-

methylpent-1-ene. The copolymers of ethylene with propylene and other alkenes and the copolymers of alkenes with vinyl monomers are also considered to be P.

Polyolefin fibers: synthetic fibers consisting of at least 85% polyolefin, especially polyethylene and polypropylene. ***Polyethylene*** and ***polypropylene fibers*** are obtained by melt spinning. The spun threads are cooled in air or water, then stretched to six times their original length. P. are elastic, insoluble in most organic solvents, resistant to acids and bases, and are very good electrical insulators. P. are used for rain and work clothing, fish nets, ropes, furniture upholstery, plush fabrics, electrical insulation, etc.

The most important trade names are listed under Synthetic fibers (see).

Polyoxymethylene, abb. ***POM, polyformaldehyde, polyacetal***: a thermoplastic formed by ionic polymerization of formaldehyde or its trimer, trioxane: $n\ H_2C{=}O \rightarrow [CH_2\text{-}O]_n$.

Properties. P. is very hard, stiff and tough down to -40 °C, retains its shape when hot, does not absorb significant amounts of water, and has good sliding, wear and electrical properties. It is not soluble in the usual organic solvents (except in perfluorinated alcohols or ketones); its resistance to gasoline (including methanol-containing gasoline) and weak bases is of technological interest. P. is attacked by strong acids and oxidants. Its density is 1.41 to 1.42 g cm^{-3}, its softening point is between 178 and 183 °C, its limit of bending stress is 105 to 120 N mm^{-2}, its impact cracking resistance is 6 to 8 N mm^{-1} and it can stretch by 20 to 40% under tension.

Production. The formaldehyde can polymerize by a cationic or anionic precipitation mechanism in aliphatic hydrocarbons, such as gasoline fractions, at 40 to 50 °C under normal pressure. Amines, phosphines, metal carbonyls, alcoholates, etc. are used as catalysts. The high heats of reaction (about 65 kJ mol^{-1}) are removed by wall and reflux cooling. The gaseous formaldehyde is purified in a prereactor by cooling (condensation of water) and formation of paraformaldehyde. Formaldehyde, solvent and catalyst are continuously fed into the polymerization reactor in proportions which lead to a polymer suspension containing up to 15% P. To prevent depolymerization, which occurs by cleavage of formaldehyde, the terminal hydroxyl groups must be acetylated, which is done at about 130 °C with acetic anhydride in the presence of sodium acetate and pyridine. The P. precipitates as a granulate (Fig.).

P. can also be produced by cationic polymerization of trioxane, which can polymerize in the absence of solvent and in the gas phase. Lewis acids and proton acids are suitable catalysts for this process. Recently, polymerization induced by irradiation of pure trioxane has become important. Here a very pure, molten trioxane is granulated and carried on a water-cooled belt through an electron beam. It is subsequently polymerized at 50 to 55 °C, the unreacted trioxane is removed with solvents and recovered, and the P. is stabilized by acetylation.

The semiacetal terminal groups can also be stabilized by carrying out the polymerization of the formaldehyde or trioxane in the presence of a small amount of cyclic ethers or acetals (copolymerization). The copolymerization is followed by thermal or chemical treatment which removes formaldehyde from the oxymethylene chain until a stable co-monomer unit is reached, and this becomes the stable end group.

Processing and applications. The P. granulate is stabilized with antioxidants, acid trappers and light stabilizers. Depending on the intended use, internal lubricants, pigments, fillers (e.g. glass fibers), etc. can be added to the polymer. Most P. is worked by injection molding. Rods, pipes, profiles, sheets and films are made by extrusion or blowing. P. are used for household appliances, in automobiles, machines, apparatus and plumbing.

Historical. The first high polymer of formaldehyde, eupolyoxymethylene, was produced in the 1920's by J. von Staudinger. It was of no industrial importance, however, because of its instability. Methods for industrial production of stable P. were first developed in 1950 in the USA.

Polyparabanic acids: polycondensates with structures similar to those of the Polyhydantoins (see). P. are obtained from oxamic acid esters and isocyanates, and are used as insulators, glues and for synthesis of special fibers.

Polypeptide antibiotics: a group of antibiotics produced by bacteria of the genus *Bacillus*; their structures are distinctly different from normal pep-

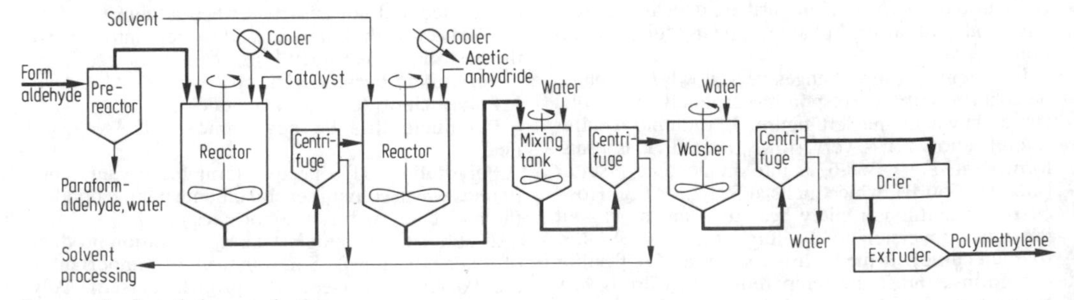

Process of polymethylene production.

tides. Most are cyclic, and their structures are not coded by DNA. They contain non-proteogenic amino acids, such as D-amino acids, *N*-methylated amino acids and ornithine. Their syntheses are catalysed by special enzymes or groups of enzymes, not by the regular protein-synthesis machinery of the cell. The molecular masses of the P. are between 350 and 3000. Most are too toxic for therapeutic applications. The group includes the gramicidins, tyrocidins, bacitracins and polymyxins.

Polypeptides: see Peptides.

Polyphenylethylene: same as Polystyrene.

Polyphosphazenes, abb. ***PNF***: inorganic polymers which contain alternating phosphorus and nitrogen atoms in the main chain. To synthesize P., phosphorus pentachloride is first made to react with ammonium chloride, forming hexachlorocyclotriphosphazene; when heated, this polymerizes to form polydichlorophosphazene. This polymer then reacts with fluorinated sodium alkoxides to yield a polyfluoralkylphosphazene. P. are used as sealers, damper bearings and arctic fuel lines.

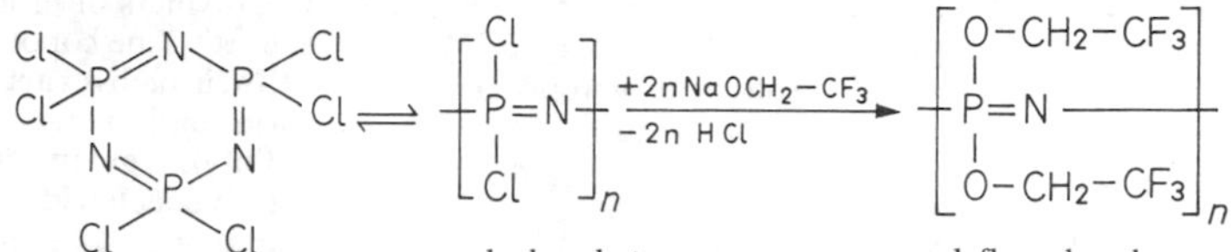

$+2n\,NaOCH_2{-}CF_3$ / $-2n\,HCl$

polyphosphates

polyfluorphosphazene

Polyphenylenequinoxalines, abb. ***PPQ***: polycondensates of aromatic tetramines and bis(1,2-dicarbonyl) compounds. P. are used for temperature-resistant coatings and as matrices for composite materials.

Polyphenylenes: polymers synthesized by dehydrogenation and polycondensation of benzene. P. can also be obtained by oxidative cationic polymerization of benzene with aluminum chloride and copper chloride as catalysts. P. are insoluble and do not melt; they can withstand temperatures up to 300 °C indefinitely. They are therefore used for special purposes in the construction of electronic equipment and space vehicles.

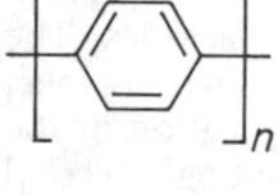

Polyphenylene oxide, abb. ***PPO***: a thermoplastic formed by polycondensation (oxidative dehydrogenation) of 2,6-disubstituted phenols. In order to prevent bonding at any but the 4-position, the 2-positions must be occupied. P. is stable at high temperatures, with a melting temperature of 230 °C; it is used to make medical instruments which require repeated sterilization, household equipment, vehicles and electrical equipment.

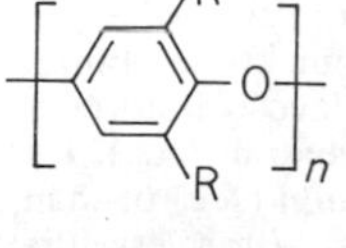

Polyphenylene sulfide, abb. ***PES***: Polythio-1,4-phenylene, a polycondensate of 1,4-dichlorobenzene and disodium sulfide: $n\ Cl{-}C_6H_4{-}Cl + nNa_2S \rightarrow [S{-}C_6H_4]_n + 2\,n\ NaCl$. P. is very stable to acids, bases and organic solvents. It is used to make parts, as a building material and to coat objects of glass and metal (e.g. pots and pans).

Polyphenylenesulfone: same as Polysulfone (see).

Polyphosphoric acids: linear polymers derived from orthophosphoric acid, H_3PO_4, by dehydration. In the chains, the $[PO_4]$ tetrahedra are linked by oxygen bridges. The simplest member of this group is Diphosphoric acid (see), $H_4P_2O_7$, and the next simplest is ***triphosphoric acid***, $H_5P_3O_{10}$ (Fig.).

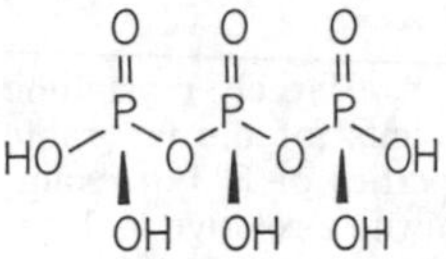

Triphosphoric acid

With the exception of diphosphoric acid, the individual P. are poorly characterized and cannot be isolated; this is in contrast to their salts, the polyphosphates (see Phosphates), many of which are well defined. Purification of the P. is prevented by a rapidly established equilibrium between orthophosphoric acid and P. of various degrees of condensation. In dilute aqueous solutions, P. are thus largely degraded to orthophosphoric acid. P. are stronger acids than orthophosphoric acid. They are made by dissolving phosphorus(V) oxide, P_4O_{10}, in phosphoric acid and the solutions are often characterized by their content of P_2O_5. P. are used in organic synthesis as catalysts for condensation and polymerization reactions.

Polypropylene, abb. ***PP***: a polyolefin formed by low-pressure polymerization of propene (propylene) using an organometallic mixed catalyst (Ziegler-Natta catalyst). P. can polymerize in three different structures.

Isotactic P. is industrially important, because the lower its atactic fraction, the better its properties are. For many applications, it is superior to polyethylene. Like most isotactic vinyl polymers, it has a helical structure (Fig. 1), in which the methyl groups point away from the axis and towards the outside. ***Atactic P.*** is a byproduct of P. production, and can be separated from the polymer by extraction with heptane; it has limited use as a glue, sealant, etc.

● CH_3 ● CH ○ CH_2

Fig. 1. Helical structure of polypropylene.

Isotactic polypropylene

Syndiotactic polypropylene

Atactic polypropylene

Properties. The formation of a largely crystalline structure in the polymer accounts for the favorable thermal and mechanical properties of P. For example, the softening point of a P. with a relative molecular mass greater than 100,000 is around 160 °C. The fibers made of P. are as strong as steel, although they are only 1/6 as dense as iron, 0.90 to 0.91 g cm^{-3}. The tensile strength of P. is 34 to 38 N mm^{-2} and its breaking strength is about 40 N mm^{-2}. The electrical properties of P. are comparable to those of polyethylene, as is its resistance to chemicals. At temperatures below 0 °C, P. becomes brittle, but this disadvantage can be removed by copolymerization with ethylene.

Production. P. is made by the precipitation technique at temperatures between 40 and 80 °C and a pressure of 0.5 MPa; the process is analogous to the low-pressure polymerization of ethylene (see Polyethylene). The relative molecular mass and stereospecificity of the polymer are affected by the temperature, monomer and catalyst concentrations and the nature of the catalyst. With the original titanium(III) chloride and diethylaluminum monochloride (first-generation) catalysts, the fraction of isotactic P. was between 85 and 90%. In the 1970's, catalyst systems were developed which contained, in addition to titanium(III) chloride and organoaluminum components, electron donors such as phosphorus compounds, α,β-unsaturated carbonyl compounds, etc. (second-generation catalysts). With these catalysts, 93 to 97% isotactic P. was obtained. The third generation catalysts are carrier catalysts consisting of titanium(IV) chloride on crystalline, anhydrous magnesium chloride, activated with triethylaluminum and electron donors. With these catalysts, the propene can be polymerized in the gas phase. 10 to 50 t P. are obtained per kg catalyst, and the isotactic fraction is between 93 and 97%, so that subsequent purification and separation of the catalyst is no longer necessary. In this technique, the heats of polymerization are carried off by expansion cooling.

Applications. P. is used for mechanical parts subjected to heavy use, such as parts for household devices, machines and motor vehicles. It can also be worked into high-quality films, pipes and fibers (see Polyolefin fibers). Because of its nonpolar structure, P. cannot be printed or dyed; it can be colored only by addition of pigments. Fiberglass-reinforced P. is a valuable construction material for automobiles. Copolymers of propene with ethylene (polyallomers, see Ethylene copolymers) are very important because of their high structural stability and low tendency to form tension cracks.

Polypropylene fibers: see Polyolefin fibers.

Polysaccharides, ***glycans***: carbohydrate polymers containing more than 10 monosaccharide molecules linked glycosidically. ***Homopolysaccharides*** consist of only one type of monosaccharide; ***heteropolysaccharides*** consist of serveral linked together. Homopolysaccharides are named by combining the root of the name of the monosaccharide with the suffix "-an", as in glucan, mannan, etc. Heteropolysaccharides often consist of a main chain of one type with side chains of another type. P. can be branched or unbranched. *Branched P.* can have either comb-like or tree-like branching (Fig.). The most common components of P. are the hexoses D-glucose, D-mannose, D- and L-galactose and D-fructose, the pentoses L-arabinose and D-xylose, the 6-deoxysugars L-fucose and L-rhamnose, the uronic acids D-glucuronic, D-galacturonic, D-mannuronic and D-iduronic acids and, in animals, the aminosugars D-glucosamine and D-galactosamine. Fructose and arabinose are present as furanosides, but all others form pyranosides. The linkage of the monomers in the P. differs with respect to the configuration (α or β) and the position of the hydroxyl group to which the preceeding glycosyl group is bound. Branching occurs when two or more glycosyl groups are bound to the OH groups of a single monosaccharide unit.

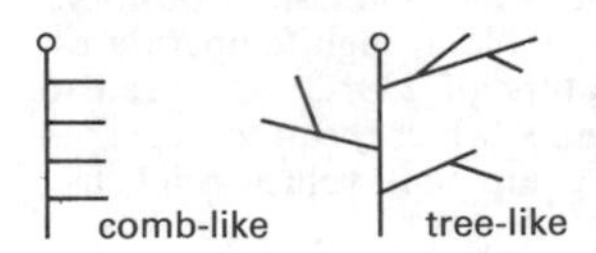

P. have two main functions, as reserve and skeleton substances. Most *reserve P.* are branched; they are the most important, relatively quickly mobilized energy reserves of the organism. Glucose is stored as Glycogen (see), Starch (see) or Dextran (see); less commonly, fructose is stored as Inulin (see) or mannose as Mannans (see). The *structural P.* are essential elements of cell walls and of the exoskeletons of arthropods. Plant cell walls contain lignin and the P. Cellulose (see), Hemicelluloses (see) and Pectin substances (see). The exoskeleton of the arthropods and the cell walls of fungi consist mainly of Chitin (see).

Complex P. contain, in addition to carbohydrates, components which are lipids (see Lipopolysaccharides), peptides (*peptidoglycans*, see Murein) or proteins *proteoglycans*, see Mucopolysaccharides).

Properties. As the molecular mass of the P. increases, its solubility in water, sweet taste and reducing power decrease. Unlike the oligosaccharides, the P. do not give the typical reactions of monosaccharides. The solubility of the P. in water depends both on the degree of polymerization and on their order. The typical skeletal P., cellulose and chitin, are highly ordered, unbranched β(1→4)-glucans, which are insoluble because of their strong intermolecular interactions. On the other hand, when these intermolecular interactions between the glucan chains are neutralized by hydration of the OH groups, the P. becomes soluble in water. This is especially true in P. in which irregular structure prevents a high degree of order, such as heteropolysaccharides or branched P. Chemical modification, such as esterification or ether formation, also changes the solubility. Thus the number of intermolecular hydrogen bonds between the glucan chains of cellulose is reduced, and the solubility of the polymer in water is increased, by methylation of 1 or 2 OH groups per glucose residue. By introduction of lipophilic substituents, P. derivatives which are soluble in organic solvents can be obtained. Native P., however, are insoluble in organic solvents. In relatively concentrated solutions, there are dynamic interactions between the chains of linear P. or P. which are not too highly branched. This leads to highly viscous solutions, in which the viscosity increases with increasing molecular mass. Some P. can form permanent noncovalent bonds between glycan chains, producing netlike structures and causing the P. solution to go from a sol to a gel state. Some widely used gellation agents are the carrageenans, alginic acids, pectin substances and agar. The cross-linked dextrans are synthetic gels based on P.

Polysiloxanes: see Silicones.

Polystyrene, ***polyvinylbenzene, polyphenylethylene***: a thermoplastic made by polymerization of styrene. The term ***styrene polymers*** includes ***polystyrenes*** (abb. PS), ***styrene-butadienes*** (abb. SB), ***acrylonitrile-butadiene-styrenes*** (abb. ABS) and ***styrene-acrylonitriles*** (abb. SAN).

$$\left[-CH_2-CH(C_6H_5)-\right]_n$$

PS are hard, shiny, glass-clear substances which are readily worked and are resistant to acids, bases, alcohols. They are dissolved by most organic solvents, and burn with a bright, sooty flame. SB are two-phase systems, in which the continuous phase consists of a PS plastic and the dispersed phase is a rubber (in general, polybutadiene). SB are shatter-resistant, but not as transparent as PS. ABS are also two-phase systems, in which the continuous phase is a copolymer of styrene and acrylonitrile. They crack and stretch less readily than PS. SAN copolymers contain 20 to 40% acrylonitrile; in some cases they contain up to 65%. Their mechanical proerties are better than those of PS (Table).

Properties of styrene polymers

	PS	SB	ABS	SAN
Density [g cm^{-3}]	1.05	1.05	1.05	1.08
Crack resistance [N mm^{-2}]	50- 65	20- 45		75- 85
Stretching before cracking [%]	2- 3	20- 50		5
Bending resistance [N mm^{-2}]	100-110	35- 80	30- 50	110-140
Impact resistance [kJ m^{-2}]	15- 20	40	10- 15	20- 25
Resistance to heat deformation [°C]	80- 90	70- 90	70-100	90-105
Electrical resistance [kV mm^{-1}]	100	50		50
Water uptake [%]	0.1	0.1	up to 1.5	0.2-0.6

Production. P. are synthesized mainly by radical polymerization of styrene, but anionic and coordinative polymerizations are also used to some extent. Coordinative polymerization using Ziegler-Natta catalysts yields isotactic P. with crystalline properties and a softening temperature above 200 °C. PS and SB plastics are obtained by radical chain reactions in the substance, suspension or solvent polymerization methods. In the *substance polymerization method*, the monomeric styrene itself acts as a solvent for the polymer, so that the reaction mixture remains homogeneous. For SB plastics, a rubber (usually polybutadiene) is dissolved in monomeric styrene. In the tower method (Fig. 1), styrene or styrene-polybutadiene solution is about 35% polymerized in two parallel stirred reactors at 80 °C, then transferred to the tower, where polymerization is completed (about 99%).

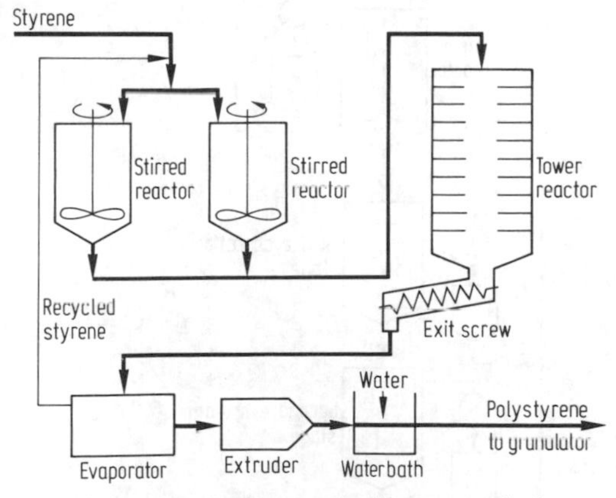

Fig. 1. Production of polystyrene by the tower method.

The P. is mixed with pigments and stabilizers as required, passed through an extruder, and cooled in a water bath. The endless ribbon of extrudate is then granulated. In the *suspension process*, the styrene is suspended in water using a suspending agent (polyvinyl alcohol, kaolin, etc.) and polymerized in the presence of initiators at 70 to 140 °C. In *solvent polymerization*, styrene is dissolved in ethyl benzene and polymerized in three successive stirred reactors (Fig. 2). The rate of polymerization is greatest in the first reactor and is progressively slower in reactors 2 and 3. Because the viscosity of the mixture is lowest in the first reactor, heat removal is also most efficient there. The solvent and residual monomers are removed in a vacuum apparatus.

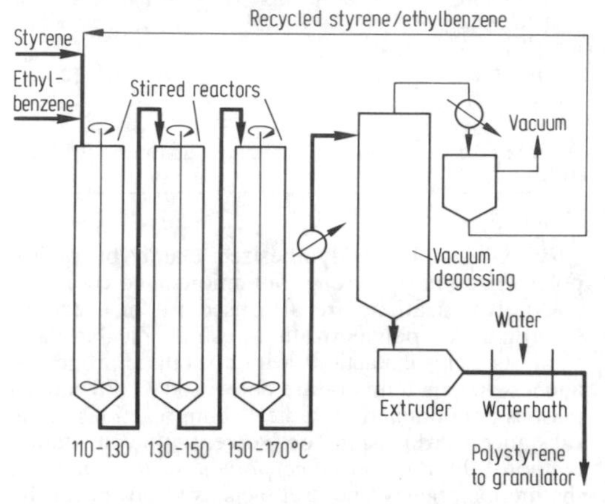

Fig. 2. Production of polystyrene by the solution polymerization technique.

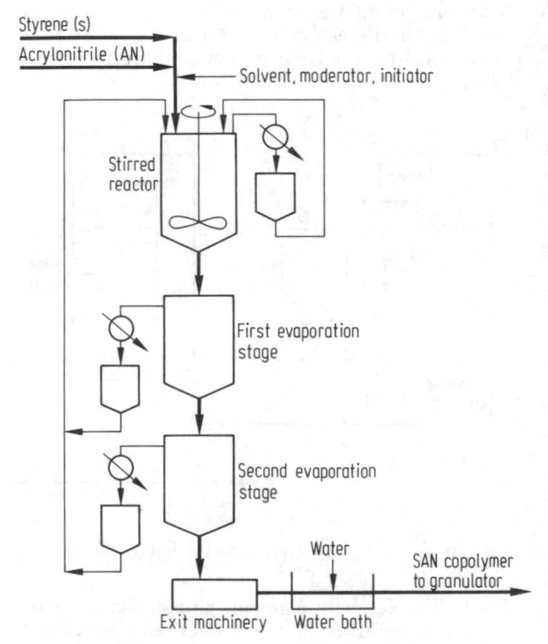

Fig. 3. Production of SAN copolymers using the solvent polymerization method.

ABS plastics are synthesized mainly by copolymerization of styrene with acrylonitrile in the presence of a polybutadiene, using the emulsion method. The product is a mixture of a styrene-acrylonitrile copolymer and a butadiene-styrene-acrylonitrile copolymer.

The most important of the styrene copolymers are the SAN copolymers. These are produced by continuous solvent polymerization (Fig. 3) of styrene and acrylonitrile at 80 to 160 °C in the presence of a solvent, chain regulators and starters in a stirred reactor. The heat of reaction is removed by evaporation of solvent and starting monomers. In two subsequent evaporation stages, the SAN copolymers are freed of solvent and unreacted monomers. These are returned to the polymerization reactor.

Applications and working. The most important applications of P. are in the electric industry, vehicle construction, furniture, building, packing and photography. Articles are formed by injection molding, and to some extent, by pressure molding. Injection molding using modern screw injection machines can be used to make parts weighing up to 30 kg. P. is used in the electrical industry to make insulation for wires and molded insulators. Emulsions of P. containing 20 to 30% softener, e.g. dibutylphthalate, are used for interior paints. The paint dries very quickly and is sufficiently wear-resistant for the purpose. If pigments and fillers are added to such an emulsion, the product is an emulsion paint of the type now widely used. Mixed polymers of styrene with fatty oils and alkyd resins yields products which are excellent paints. ***Porous P.*** or ***P. foam*** is obtained when a gas-producing substance, such as an azo compound, or a readily evaporating solvent, such as pentane, is added to the polymer. When the mixture is heated, the gas-producing substance decomposes, or the solvent evaporates, and the resulting bubbles and pores remain in the material when it is cooled and solidifies. Because of their very low densities (20 kg m^{-3}), P. foams are used for heat and sound insulation in construction of buildings and refrigeration. They are also used to make packaging, flower pots, buoys, life vests, life-saving rings, decorations and toys. They are used in agriculture to improve the texture of the soil and as surface covers for tennis courts or football fields. When cross-linked with divinylbenzene and sulfonated, P. yields a resin which can be used as an ion exchanger. Styrene can also be polymerized directly at the site of use. ABS plastics are used wherever high impact resistance and toughness of the product are important. Like PS plastics, ABS can be formed by pressing, sintering, gluing, welding and planing. In vehicles, vapor deposition and galvanic deposition of thin metal layers on ABS plastics has become very significant. ABS plates formed by extrusion are used in the construction of refrigerators. Extruded fibers are used in brushes, brooms and mats. SAN copolymers are used in housings for electric devices, kitchen appliances and so on.

Historical. P. were first produced in 1893 by a Berlin pharmacist, Simon, who incorrectly identified it as styrene oxide. P. has been produced industrially since 1930, and the copolymers of styrene with acrylonitrile have been known since 1936. In 1980, the world production of styrene polymers exceeded the 6 million ton level.

Polysulfanes: hydrogen compounds of sulfur with the general formula H_2S_n; the lower members of the group (up to about $n = 8$) have been isolated and characterized, while the higher compounds are known only in mixtures. The P. are yellow liquids, usually with a pungent odor. As the length of the chain increases, the color deepens and the viscosity increases. They slowly decompose into hydrogen sulfide, H_2S, and sulfur. This process is accelerated by traces of alkali, light and heat. The P. react with solid alkali or alkaline earth hydroxides to form Polysulfides (see). P. are made from alkali metal polysulfides by the reaction of excess concentrated hydrochloric acid or anhydrous formic acid.

Polysulfides: the salts of the polysulfanes H_2S_n, with the general formula $M^I_2S_n$, where $n = 2, 3, 4$ or 5, occasionally higher numbers. The polysulfide anions are linear in structure. They are decomposed to hydrogen sulfide and sulfur by acids, e.g. $Na_2S_4 + 2 H_2SO_4 \rightarrow 2 NaHSO_4 + H_2S + 3 S$. Under certain conditions, they can be converted to polysulfanes by treatment with excess acid. The alkali or alkaline earth metal polysulfides are obtained by melting the sulfides, hydroxides or carbonates with sulfur (e.g. see Liver of sulfur). Ammonium polysulfide solutions, prepared by digesting NH_4HS solution with sulfur, are used in qualitative analysis.

Polysulfone, ***polyphenylenesulfone***: a thermoplastic with good mechanical properties in the temperature range from - 70 to +150 °C. It is made by polycondensation of diane (bisphenol A) and 4,4'-dichlorodiphenylsulfone. The diphenylsulfone group in the polycondensate gives it high structural stability and resistance to oxidation. P. are stable to bases, acids and oils, and are difficult to ignite. They are used mainly for household machines and coverings, and in tools and motor vehicles.

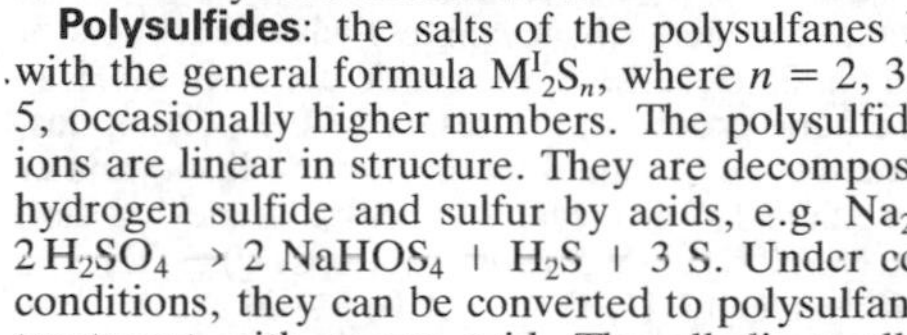

Polysulfur nitride, ***SN***: a linear polymeric compound consisting entirely of sulfur and nitrogen atoms (Fig.). P. is a good electrical conductor, with a conductivity comparable to those of metals. At a temperature of 0.3 K it becomes superconducting.

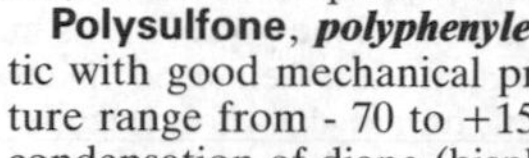

Polytetrafluoroethylene, ***teflon***®: $[C(F)_2\text{-}C(F)_2]_n$, a thermoplastic obtained by polymerization of tetrafluoroethylene; it is exceptionally resistant to chemicals and high temperatures.

Properties. P. consists of largely linear, polymeric chains and contains some crystalline regions. The unbranched structure and the exceptional strength of the carbon-fluorine bond are responsible for the favorable properties of the polymer. The material is as plastic as polyethylene, but does not become brittle until the temperature is lower than -270 °C. Its upper limit for use is about 260 °C. Above 300 °C, P. decomposes, and some of the degradation products are poisonous. When heated above 500 °C in a vacuum, P. is depolymerized and 90% returns to the monomeric state. Its density is 2.1 to 2.3 g cm^{-3}. P. is water repellent, noncombustible and has excellent electrical properties. It is not attacked by any organic solvent, hydrofluoric acid, nitric acid, hydrochloric acid, aqua regia or alkalies, even when heated. Only elemental fluorine and alkali metal fusions can degrade it.

Production. The starting material for production of P. is tetrafluoroethene, $CF_2{=}CF_2$, which polymerizes in the presence of peroxide catalysts in water and under pressure. Because a great deal of heat is released by the polymerization, the reaction must be thoroughly stirred and effectively cooled. The polymer is obtained as a white powder.

Processing and applications. Because of its high softening temperature, P. cannot be worked by the methods ordinarily used with thermoplastics. Instead, the crude powder is pressed into shape in cold molds, and then sintered in a furnace or by infrared irradiation. It can then be treated to remove tension in the piece and glued. For impregnating and production of corrosion-resistant coatings, P. is used in the form of an aqueous emulsion. Films of P. are made by shaving cylindrical blocks. Because P. is relatively expensive, products made from it are used mainly in special applications in the electrical, aircraft, spacecraft and chemical industries. It is made into membranes, insulation, pipes, pump cylinders, bearings, industrial textiles, etc. Because of its anti-sticking properties, P. is used to coat kitchen ware, making possible the preparation of food without fat for frying. The landing skids on aircraft for use in arctic regions are coated with P., to prevent freezing of the craft to the ground.

Polythionates, ***sulfane disulfonates***: the salts of polythionic acids, $H_2S_nO_6$ with the general formula $M^I_2[O_3S\text{-}S_n\text{-}SO_3]$ ($n = 1$ to 12). The alkali metal polythionates are stable, colorless, solid, water-soluble compounds. In acidic solution, the P. are rapidly degraded via polythionic acids to sulfur, sulfur dioxide and sulfates. The individual P. are synthesized by special processes which depend on the length of the sulfur chain. For example, sodium trithionate is obtained by oxidation of sodium thiosulfate with hydrogen peroxide: $2 Na_2S_2O_3 + 4 H_2O \rightarrow Na_2S_3O_6 + Na_2SO_4 + 4 H_2O$. Thiosulfate is oxidized by iodine to tetrathionate: $2 S_2O_3^{2-} + I_2 \rightarrow S_4O_6^{2-} + 2 I^-$. Sodium pentathionate can be made by reaction of sodium thiosulfate with sulfur dichloride, and sodium hexathionate can be made with disulfur dichloride: $2 Na_2S_2O_3 + S_2Cl_2 \rightarrow Na_2S_6O_6 + 2 NaCl$.

Polythionic acids, ***sulfane disulfonic acids***: oxygen acids of sulfur with the general formula $HO_3S\text{-}S_n\text{-}SO_3H$ ($n = 1$ to 12). The name of a P. is based on the total number of sulfur atoms in it, e.g. tetrathionic acid is $HO_3S\text{-}S_2\text{-}SO_3H$, or $H_2S_4O_6$. The free acids are unstable and decompose into sulfur, sulfur dioxide and sulfate. The aqueous solutions of P. are more stable, as are their salts, the Polythionates (see). ***Wackenroder's liquid*** is an aqueous solution containing mainly tetra- and pentathionic acids; it is obtained by passing hydrogen sulfide through an aqueous sulfur dioxide solution.

Polytriazines: same as Triazine polymers (see).

Polytrifluorochloroethylene, abb. ***PCTFE***: [C$(F)_2$-C(FCl)-C$(F)_2$-C(FCl)$]_n$, a high-quality thermoplastic made by radical polymerization of trifluoromonochloroethylene. The starting materials for the monomers are hexachloroethane and anhydrous hydrofluoric acid.

Properties. P. is very similar in its properties to teflon (see Polytetrafluoroethylene), but is somewhat less stable and harder. It swells or dissolves at high temperatures in chlorinated and fluorinated hydrocarbons, aromatics and esters; it is also attacked by oleum, chlorosulfuric acid, elemental fluorine, liquid chlorine and alkali metal fusions. It can be used in the temperature range between -100 °C and +160 °C. Its density is 2.1 g cm^{-3}. P. does not burn and is physiologically harmless.

Processing and applications. In contrast to teflon, P. can be worked on ordinary plastic processing machines, such as presses, injection casting machines and extrusion presses at 250 to 300 °C. Semifabricates of P. can readily be machined. P. is used mainly in high-frequency technology and the construction of chemical apparatus. It is used to make corrosion-resistant coatings for metal surfaces, chemical-resistant wire paints, chemical-resistant lubricants and cable sheathing. By mixed polymerization with softening monomeric components, rubber-elastic products can be made with the ability to stretch up to 500%. A mixed polymer of trifluoroethene and vinylidene fluoride is used for impregnation and as a corrosion resistant coating.

Polytropic process: a change in thermodynamic state in which the system is not completely isolated from the environment; there is some exchange of heat, and also a change in the temperature in the interior of the system. The limiting situations for P. are Adiabatic processes (see), in which no heat is exchanged with the environment ($\Delta q = 0$) and isothermal processes in which the heat exchange with the environment is complete, and the temperature in the interior of the system remains constant. For ideal gases, the thermal equation of state of a P. is the ***polytropic equation*** pV^k = const., or TV^{k-1} = const; p is the pressure, V is the molar volume, and T is the temperature. k is the index of polytropism. Its numerical value depends on the conditions, that is, on the degree of heat isolation. It is always between 1 and $\varkappa = C_p/C_v$ (Poisson equation). P. are very important in practice. They always occur when the adiabatic or isothermal limiting cases are not achieved.

Polytypism: a special case of Polymorphism (see) which describes the possibility of different stacking orders in layered structures. Polytypic modifications (*polytypes*) can occur in structures consisting of geometrically defined layers with the same atomic arrangement, or from a few different types of layers. The spatial structure is determined by the sequence of layers A, B, C, etc. If the same compound can display different layer sequences (e.g. ABAB... and ABCABC...), it is polytypic.

The phenomenon of P. was discovered in 1911 in silicon carbide, SiC, by H. Baumhauer. Here a layer of Si atoms and one of C atoms form a double layer, for which there are three possible positions A, B and C. The sequence of these three types of layers can vary greatly, and it determines the translation period of the crystal lattice in the stacking direction; this is also called the *stack height*. At present dozens of SiC polytypes are known, some of them with extremely large stack numbers (stack heights up to 150 nm). Another important example of P. is the compound zinc sulfide, for which there are two natural polytypes: the Zinc blende type (see) with three layers in the sequence ABC, and the Wurtzite type (see), with two layers in the sequence AB. P. have also been discovered in synthetic ZnS crystals.

Many structures display defects in the order of their lattices (stacking defects; see Crystal lattice defects).

Polyurethanes, ***polyisocyanates***: a group of thermoplastics which are formed by polyaddition of diisocyanates (also triisocyanates) with dialcohols or polyalcohols. Because of the amide bond -CO-NH- in the urethane group, the P. resemble the polyamides.

Linear P. are obtained by reaction of diisocyanates (e.g. hexane-1,6-diisocyanate) with butane-1,4-diol: n OCN$-(CH_2)_6-$NCO$+n$ HO$-(CH_2)_4-$OH $\rightarrow$ [CONH$-(CH_2)_6$NHCO$-$O$-(CH_2)_4-$O$]_n$.

The relative molecular masses of these linear P. are around 8000; the melting temperatures about 190 °C and the highest temperature for continued use is around 130 °C. Like the polyamides, they are used for molds, brushes, etc. P. with higher melting points are obtained from aromatic diisocyanates. The P. made from phenyldiisocyanate and ethylene glycol melts at 340 °C. 2,4-Toluenediisocyanate is often used in industry, with ethyleneglycol, butane-1,4-diol, hexanetriol, etc., or with polyesters like adipic acid diglycol ester, HO-CH_2-CH_2-OOC-$(CH_2)_4$-COO-CH_2-CH_2OH. Depending on the starting materials and the reaction conditions, the products are fibers, solids, rubber-like substances, foams, paints and glues for metals. Additional fillers such as carbon black, titanium dioxide and aluminum oxide improve the working qualities of the materials, from which one can make extruded products like threads, bands and hoses at temperatures of 40 to 100 °C. In the process of centrifugal molding, the liquid addition product is poured into a rotating form, where it hardens under constant rotation. This is used to make P. bicycle and auto tires with and without embedded fibers or metal reinforcers. P. are also used for components in the electrical industry and for scientific apparatus. P. with a small amount of methylhexane-1,6-diol added can be rolled into foils which are used as synthetic leathers.

Cross-linked P. are obtained by polyaddition of triisocyanate-diisocyanate mixtures with high-molecular-weight polyols. A cross-linking can also be achieved by addition of the di- and triisocyanate to branched polyesters. The properties of the cross-linked P. can be varied over a wide range, depending on the amount of cross-linking. Cross-linked P. are used as paint binders, glues, paint for PVC, molds, textile coatings, foils, etc. ***Polyurethane rubbers*** are made by addition of diisocyanates to linear polyesters (e.g. the polyester made from glycol and adipic acid). They are very resistant to abrasion, oil and gasoline, but they are not resistant to acids, bases and boiling water. A disadvantage is their incompatibility with other types of rubber. P. rubbers are used to make seals, drive belts, suspension components, membranes, shoe soles, heels, etc. These P. stick so well to textiles of natural or synthetic fibers that coatings made of them are very durable. ***Polyurethane foams*** are obtained if the polyaddition is done in aqueous emulsions; the carbon dioxide split off during the reaction causes bubbles in the material. These foams are used for heat and sound insulation, as padding, in upholstery, and as mattresses. A special P. foam can absorb up to 100 times its own mass of oil, and is used to clean up oil spills at sea and to remove oil in ground water. The foam is generated at the site of the spill, and after the oil has been pressed out of it, it can be reused.

Polyurethane paints are resistant to chemicals. Their hardness, elasticity, sheen, adhesiveness and electrical qualities are good, and they can be colored as desired. ***Polyurethane glues*** are obtained by reaction of triisocyanates, e.g. with polyesters. If linseed oil is converted to mono- and diesters with polyalcohols, then combined with diisocyanates, the product is a ***urethane linseed oil***, i.e. a rapidly drying oil which is exceptionally resistant to chemicals.

Polyuronides: polysaccharides which contain uronic acids and are therefore acidic. The pectin substances and alginic acids found in plants are P.

Polyvinylacetals: thermoplastics made by reaction of polyvinyl alcohol with aldehydes (or ketones). The most important P. are ***polyvinylformal, polyvinylacetal*** and ***polyvinylbutyral***, made with formaldehyde, acetaldehyde and butyraldehyde, respectively.

$$\left[CH_2{-}CH{-}CH_2{-}CH \right]_n$$

(the two CH groups are linked through O atoms to a CH–R group, forming a ring)

R = –H Polyvinylformal
R = $-CH_3$ Polyvinylacetal
R = $-C_2H_5$ Polyvinylbutyral

The properties of the P. depend on the degree of acetalization, i.e. on the number of aldehyde or ketone molecules which are added to the polyvinyl alcohol chain. Polyvinylbutyral is the most important of the P.; at an acetalization of 65%, its softening point is about 135 °C. It is not dissolved by benzene or its homologs, mineral oils, light petroleum, higher esters or fatty oils, but it is dissolved by mineral acids and alkalies.

Polyvinylacetal and polyvinylformal are used as paint bases, and polyvinylformal is used together with phenol resins for certain heat-setting lacquers and glue layers for light metal foils. Polyvinylbutyral is used as a priming coat for painting iron and aluminum, as a lacquer for metal foils and as the inner plastic film layer in safety glass.

Polyvinylacetate, abb. ***PVAC***: a clear, brittle plastic which is resistant to light and heat; density 1.16 to 1.18 g cm^{-3}. P. is readily soluble in lower alcohols, esters, ketones and chlorinated hydrocarbons. Its softening point depends greatly on its relative molecular mass, and can be as high as 180 °C. P. is obtained mainly by suspension polymerization of vinyl acetate in the presence of hydrogen peroxide or organic peroxides as catalysts, and with polyvinyl alcohol as a stabilizer for the suspension. Because of its low mechanical strength, P. cannot be used as a building material. Its uses are in lacquers and glues. Pure P. is used as a melt glue, and emulsions of P. with added polyvinyl alcohol are used as glues for plywood, etc. Solutions of P. with small amounts of added softeners are used together with cellulose nitrate as glues for paper, cardboard, leather, etc. P. is used as a lacquer base for spraying and dipping lacquers. It combines well with other lacquer components, such as cellulose nitrate, chlorinated rubber and phenol resins, and forms an elastic, adhesive, light-fast and clear film of lacquer. Aqueous emulsions of P. are used together with pigments and fillers to make emulsion paints for wood, masonry and stucco, as glue and as spackel. P. with low degrees of polymerization is used to make chewing gum.

Copolymers of vinyl acetate and vinyl chloride are in use as glues, lacquer bases and raw materials for plastic objects. Addition of vinyl acetate to polymerization mixtures of acrylates, maleic acid, fumaric acid, crotonic acid, vinyl laurate and vinyl stearate provides internal softening, and the products are useful for lacquers, paints and glues.

$$\left[CH_2{-}CH(O{-}C(=O){-}CH_3) \right]_n$$

Polyvinylbenzene: same as Polystyrene (see).

Polyvinylbutyral: see Polyvinylacetals.

Polyvinylcarbazene, ***polyvinylcarbazol***, abb. ***PVK***: a temperature-resistant thermoplastic; density 1.19. Its upper temperature limit for use is at 160 to 170 °C. P. is soluble in benzene and its homologs and various chlorinated hydrocarbons, but is resistant to acids, bases, polar solvents and mineral oils. P. is made by suspension polymerization of vinylcarbazene at temperatures of 180 °C with sodium chloride and some potassium chromate as catalysts. Because of its relatively high price and its special properties, P. is used only in special areas, particularly for insulators. Addition of P. to other plastic masses can increase their resistance to temperature; for example, copolymers with styrene can withstand boiling water. P. was developed in Germany in 1934.

$$\left[CH_2{-}CH(N{<}carbazole) \right]_n$$

Polyvinyl chloride, abb. ***PVC***: $[CH_2\text{-}C(H)Cl]_n$, an important thermoplastic, which is tasteless, odorless and difficult to burn; it absorbs little water and has good electrical properties. P. with low degrees of polymerization (M_r up to 30,000) are soluble in organic solvents. Products with higher degrees of polymerization ($M_r \geqq 100{,}000$) are much less soluble. They are resistant to concentrated and dilute alkalies, oils and aliphatic hydrocarbons, while oxidizing mineral acids such as concentrated sulfuric acid and nitric acid decompose them. A disadvantage of P. is its moderate stability to light, low temperature resistance (softening temperature 75 to 80 °C, deformation temperature 110-130 °C) and its poor heat conductivity. ***Hard PVC*** is produced without added softeners and fillers; when it contains no additives, it is physiologically safe. It can stretch 10 to 20%, and its density is 1.38 g cm^{-3}. Addition of suitable stabilizers gives P. greater resistance to aging and weathering. To increase the low impact resistance of hard PVC when it is cold, it is modified with chlorinated polyethylene or rubber. ***Soft PVC*** is obtained by addition of softeners, especially phthalic acid esters, and has different mechanical properties. Its density is about 1.30 g cm^{-3}, and its elasticity can be extended to 180%.

Production. Gaseous vinyl chloride is usually polymerized in suspension, under pressure, but emulsion techniques are also used. In the industrial process of *suspension polymerization* (Fig.), temperatures of 45 to 75 °C and pressures between 0.5 and 1.2 MPa are used; the process is usually discontinuous.

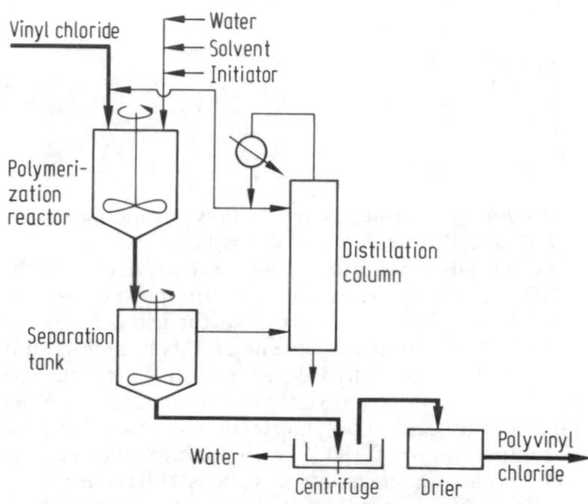

Process for production of polyvinyl chloride by suspension polymerization.

Water is used as a suspension medium, to which stabilizers such as talcum, calcium carbonate, polyvinyl alcohol and gelatin are added. Polymerization is initiated by peroxides, especially dilauryl peroxide and diisopropyl peroxide, or azobis compounds. *Emulsion polymerization* is also done in water, with added emulsifiers, peroxide accelerators (e.g. ammonium persulfate) and small amounts of electrolytes (e.g. phosphates). Very pure P., suitable for electrical applications, is obtained by use of pure water (no additives) with hydrogen peroxide as catalyst. Similar products are obtained by precipitation polymerization, in which pure, liquid monomer is polymerized in the presence of hydrogen peroxide.

Applications. To avoid loss of hydrochloric acid when the crude PVC powder is processed at 160 °C, stabilizers such as soda, alkali phosphates, lead silicate, calcium stearate or α-naphthylamine, are added. Transparent products are obtained by addition of organic compounds of tin, lead or other heavy metals, for example, dibutyltin laurate. Prerolled hard PVC is extruded to semi-finished products such as pipes and profile rods. Complicated finished parts of hard PVC, such as handles, covers, etc. are made directly from crude PVC powder by impact pressing and extrusion molding. The semi-finished products can be further processed by mechanical or thermal methods.

Since only low-molecular-weight P. are soluble in acetone-carbon disulfide mixtures, and can be spun, higher-molecular-weight P. is post-chlorinated to a chlorine content of 64% in a suspension of tetrachloroethane. The fibers made from unchlorinated P. are used in carpets or twines and ribbons. Post-chlorinated PVC is soluble in acetone and can be spun into water to make a fiber which is resistant to chemicals, light, weather, rot and insects and is non-inflammable; this can be further processed to make belts, ropes, fishing nets, etc. Solutions of post-chlorinated P. in acetone or other solvents are used as glues. P. with low degrees of polymerization are used as lacquer bases. PVC adhesive foils are also widely used in households and crafts.

In the chemical industry, hard PVC is used mainly as a raw material, but also to coat vessels and apparatus which are to be filled with aggressive liquids. It is used to make packaging, construction materials and consumer items.

Soft PVC is used to make conveyer belts, hoses, stoppers, soles, drive belts, insulation for wires, films and objects such as tubes and gloves. Synthetic leather for luggage and upholstery can be made by layering a PVC paste onto a textile base and gelling it at 170 to 190 °C. PVC floor coverings are made with fillers, such as chalk, ground shale, kaolin and barite, which increase their resistance to wear.

Hard PVC foam is a porous material; its structure is generated by gas released from a foaming agent mixed into the polymerization system; because of its low inflammability and water-repellent products, it is used in the construction of furniture and vehicles, as insulation and to make life preservers and life jackets. Soft PVC foam is made by addition of softeners; it is elastic and is used, for example, for cushions.

Copolymers of P. are made for special purposes. A copolymer of P. with about 10% vinyl acetate is suitable for spinning (from acetone) and to make plastic objects. It can also be used for glues and in lacquers; in this case maleic and fumaric acids are often copolymerized. Copolymers which contain up to 45% acrylonitrile in addition to P. are important for synthetic fibers. Transparent copolymers of vinyl chloride and acrylates are made into tubes and plates. Copolymers of vinyl chloride and vinylidene chloride are used for packing films, filters, wall coverings, etc.

Historical. Although the first solid polymers of vinyl chloride were known in 1835, PVC was first made industrially in 1916, after Klatte developed a process for it in 1912.

Polyvinyl esters: thermoplastics which are produced by radical polymerization of vinyl esters. Only polymers of vinyl acetate and vinyl propionate are of technological significance. The clear, colorless polymers are used as glues and in paints. For more detail, see Polyvinyl acetate.

Polyvinyl ether: $[CH_2\text{-}C(H)OR]_n$, thermoplastics which are obtained by polymerization of methyl, ethyl or isopropyl vinyl ether. Block, emulsion and solution polymerization techniques are used to make P. in industry. Depending on the molecular weight of the monomeric ether and the polymerization conditions, P. can be oily, sticky masses or viscous, tough, resinous products. Vinyl chloride, acrylonitrile and acrylic acid esters are copolymerized with vinyl ethers.

Polyvinyl methyl ether (R = $-CH_3$) is soluble in water, but it precipitates when heated. It is used as a settling or emulsifying agent and to heat-sensitize rubber emulsions. ***Polyvinyl ethyl ether*** (R = $-C_2H_5$) is insoluble in water and is used to prepare leather and to make glues and lacquers. ***Polyvinyl isobutyl ether*** (R = $-C_4H_9$) is also insoluble in water; it is used as a coating or impregnating adhesive, or as a film to seal and isolate cables. ***P. of larger monomers*** are waxy. In general, P. are used for textile finishes, as paint components and in adhesives, insulating and adhesive tapes, bandages, putty and chewing gum.

Polyvinyl ethyl ether: see Polyvinyl ether.

Polyvinyl fluoride, abb. ***PVF***: $[CH_2\text{-}C(H)F]_n$, a valuable thermoplastic obtained by polymerization of vinyl fluoride. P. is used chiefly to make highly transparent films. It is also used to make adhesive films to coat metals and PCF dispersions.

Polyvinylformal: see Polyvinylacetals.

Polyvinylidene chloride, abb. ***PVDV***: $[CH_2\text{-}CCl_2]_n$, a thermoplastic obtained by radical polymerization of vinylidene chloride (1,1-dichloroethene). The softening point of pure P. is near its decomposition point. Therefore, as a rule it is used in the form of a copolymer with addition of up to 20% vinyl chloride or 13% vinyl acetate and 2% acrylonitrile. Such copolymers have a density of 1.65 g cm^{-3} and a softening point of 100 to 120 °C. They are very resistant to wear, chemicals and water, cannot burn and are extremely impermeable to water vapor. They can be dissolved only by hot tetrahydrofuran, cyclohexanone, chlorinated hydrocarbons, hot mineral acids and hot ammonia, but by no other solvents or chemicals. The tendency of P. to crystallize can be countered by quenching the heated plastic material.

Processing and applications. After the addition of softeners, the crude thermoplastic can be extruded as threads, ribons or films, and strengthened by stretching the extruded material. P. films are used as packaging materials for foods. Threads made of P. or of mixed vinylidene chloride/vinyl chloride and/or acrylonitrile are made into work cloths, filter cloths, upholstery and drapery fabrics, conveyor belts and ropes; here the resistance to chemicals and to burning are advantages. Sieves, brushes, pipes, etc. are also made from the mixed polymers. Water-tight packing papers are obtained by coating the paper with an aqueous emulsion of vinylidene chloride-vinyl chloride copolymer.

Industrial synthesis of P. began in the USA and Germany in the years 1939-1940.

Polyvinylidene cyanide: $[CH_2\text{-}C(CN)_2]_n$, a hard, transparent thermoplastic which is exceptionally resistant to chemicals. Vinylidene cyanide is polymerized in the cold in the presence of water, alcohols, amines, etc. Usually copolymers of vinylidene cyanide and vinyl acetate or vinylidene chloride and methyl acrylate are used. P. and its copolymers are often used to make synthetic fibers which are resistant to water and heat. They are added to other chemical fibers during the process of spinning to increase their elasticity, softness and resistance.

Polyvinylidene fluoride: $[CH_2\text{-}CF_2]_n$, abb. ***PVF₂*** or ***PVDF***: a temperature-resistant thermoplastic which is obtained by polymerization of vinylidene fluoride (1,1-difluoroethene). P. is used mainly to make injection molded articles for use at temperatures between -60 °C and 170 °C. Pecause of their piezoelectric properties, films of P. have been used as polymeric converters and detectors in microphones, loudspeakers and ultrasound detectors (see Polymeric electretes).

Polyvinyl isobutyl ether: see Polyvinyl ethers.

Polyvinyl methyl ether: see Polyvinyl ethers.

Polyvinyl pyrrolidone: a thermoplastic which is obtained from *N*-vinylpyrrolid-2-one by block or solution polymerization. P. is readily soluble in water, alcohols, ethyl acetate, chloroform and methylene chloride, but will not dissolve in petroleum ether, mineral oils, aliphatic hydrocarbons and ethers. P. is used as an emulsifier and protective colloid; it forms a viscous aqueous solution. It is also important in the production of adhesives, textile treatments, etc. Aqueous solutions of P. which also contain small, precisely defined amounts of salts are used as blood plasma extenders which are well tolerated.

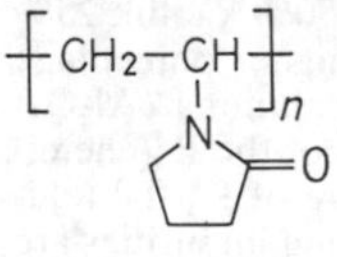

$$[-CH_2-CH(N\text{-pyrrolidone})-]_n$$

Poly-p-xylylene: a group of special plastics with the monomer:

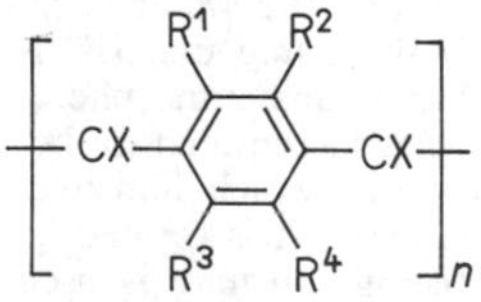

$$[-CX-C_6(R^1)(R^2)(R^3)(R^4)-CX-]_n$$

Poly-p-xylylene ($R^1 = R^2 = R^3 = R^4 = H = H$) is of great interest; it is synthesized from *p*-xylene at 550 °C under vacuum. Poly-*p*-xylylene is used as a paint-like coating for metal surfaces, and is stable up to 220 °C in this form. Layers of P. are impermeable to steam and gases, and are used in microelectronics because of their good dielectric properties.

POM: abb. for Polyoxymethylene.

Pomeranz-Fritsch synthesis: a synthesis for isoquinolines which starts from aromatic aldehydes and aminoacetals:

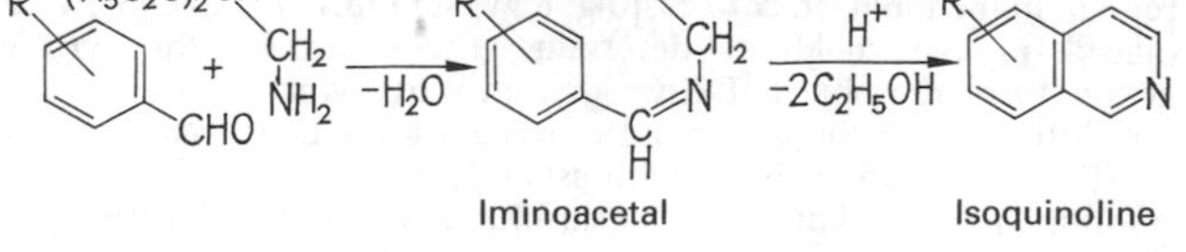

$$R\text{-}C_6H_4\text{-}CHO + (H_5C_2O)_2CH\text{-}CH_2\text{-}NH_2 \xrightarrow{-H_2O} R\text{-}C_6H_4\text{-}CH{=}N\text{-}CH_2\text{-}CH(OC_2H_5)_2 \xrightarrow[-2C_2H_5OH]{H^+} \text{R-isoquinoline}$$

Iminoacetal   Isoquinoline

The reaction mechanism is thought to be an intramolecular electrophilic aromatic substitution. Hydroxy and alkoxy groups in position 3, relative to the formyl group of the arene aldehyde, promote the reaction. With 3-nitrobenzaldehyde, the condensation to the iminoacetal occurs, but ring closure is not possible. A particularly effective reagent for ring closure is polyphosphoric acid. With the P. it is possible to make isoquinolines which are difficult to obtain by the Bischler-Napieralski or the Pictet-Spengler syntheses.

Ponceau pigments: an extensive group of pigments, most of them red. A well known representative of this group is ***Ponceau R*** (Fig.), which is synthesized from diazotized *m*-xylidine and 2-naphthol-3,6-disulfonic acid (R-acid). The P. are now used, among other things, to make colored paper and paints.

Ponceau R

Popril®: see Synthetic fibers.

Porcelain: a high-quality Ceramic (see) product.

1) ***Hard P.*** was invented in 1708/1709 by Böttger. It has a dense, white body with a relatively high proportion of glass. The raw materials are 50% kaolin, 25% quartz sand and 25% feldspar. A higher percentage of quartz or partial substitution of alumina (Al_2O_3) for the quartz increases the strength of the P. (chemical-industrial P.). In the manufacture of P., the feldspar is first ground and iron-containing impurities are removed. It is then wet ground together with the quartz sand; after an initial grinding the fine kaolin is added to the wet grinder. Most of the water is removed by pressing, and then the crude material is aged for up to several months in a moist atmosphere. In this process, the clay particles in the mixture become hydrated (swelling process), which improves the working properties of the mass. It is formed by turning, press molding or casting. *Turning* is used mainly for production of flat plates. A thin mound is placed on a rotating gypsum mold, and formed by a counter-rotating template. Such machines can produce up to 500 plates per hour. Simple *press molding* is used to make rotationally symmetric objects which can be further processed on a lathe, such as the insulators for electric power lines. For *casting*, alkaline materials are added to make a thin slurry, which is poured into a porous gypsum mold. The water is partially absorbed by the mold, so that the object stiffens. In the technique of hollow casting, the slurry is left in the mold for a time, then poured back out. The material next to the walls has stiffened enough to remain in the mold. In core casting, a gypsum form is placed in the mold to determine the internal geometry of the object. Before drying, parts which are difficult to shape in a mold (handles) are glued on. The objects are then dried (moist air, rocking and rotating table rapid driers). The shrinkage which occurs during drying must be taken into account when the molds are made. The objects are then fired (see C.); the chemistry of this process is complicated (sintering, kaolinite decomposition, mullite formation and crystallization, polymorphic coversions). The firing process also creates a glass phase, which gives P. a dense body. To make a porous body capable of absorbing a glaze, P. objects are first fired at relatively low temperatures (800-1000 °C), and then glazed. The Glaze (see) serves to improve the mechanical and electrical properties as well as the surface of the P. The pre-fired and glazed objects are then given a second firing at 1410 to 1420 °C. They can be decorated by painting either below or on top of the glaze (see Ceramic pigments), or thin layers of gold or silver may be applied.

Hard P. is used for tableware, kitchen ware and decorative objects, electrical insulators, laboratory apparatus and laboratory containers.

Soft P. includes Asian P. (china) and is now of little commercial importance. It has a high feldspar content (40% kaolin, 36% feldspar, 24% quartz), which makes it possible to fire it at 1320 °C. The lower temperature makes possible the use of many subglaze decorations, but its mechanical strength and resistance to both chemicals and temperature changes are much lower than those of hard P.

Bone P., which is still made in England, consists of 40 to 50% bone ashes and kaolin, flint and pegmatite. ***Seger P.***, containing 25% kaolin, 45% quartz and 30% feldspar was developed more than 100 years ago, and all modern soft P. are derived from it.

Pore gradient electrophoresis: see Electrophoresis.

Porphin: a cyclic tetrapyrrole in which 4 pyrrole rings are connected by =CH- groups. It forms dark red cyrstals, M_r 310.3. P. is a very stable heteroaromatic which does not decompose below 360 °C. Its derivatives are called Porphyrins (see). In ***porphinogens***, the four pyrrole rings of P. are linked by methylene groups.

Porphinogen: see Porphin.

Porphobilinogen: 2-aminomethyl-3-carboxymethyl-4-carboxyethylpyrrole. P. is formed from 2 mol δ-aminolevulinic acid, and serves as the starting material for the biosynthesis of natural porphyrins.

Porphyrin: a derivative of porphin; the nomenclature of P. is based on prefixes (table) indicating the number and types of substituents. The distribution of the substituents on the pyrrole rings is characterized by a roman numeral placed after the name. Most naturally occurring P. are derived from ***protoporphyrin IX***. ***Porphyrinogens*** are P. with completely hydrogenated bridges (-CH_2- instead of =CH-) between the pyrrole rings. These hexahydrohporphyrins are formed as intermediates in the enzymatic and chemical syntheses of P. A dihydroporphyrin partially hydrogenated in ring D is called ***chlorin***. A close relative of the P. is ***corrin***.

The biosynthesis of P. starts from δ-aminolevulinic acid and leads to the pyrrole derivative porphobilinogen, which is then cyclized either to the corrinoids or to ***uroporphyrinogen III***.

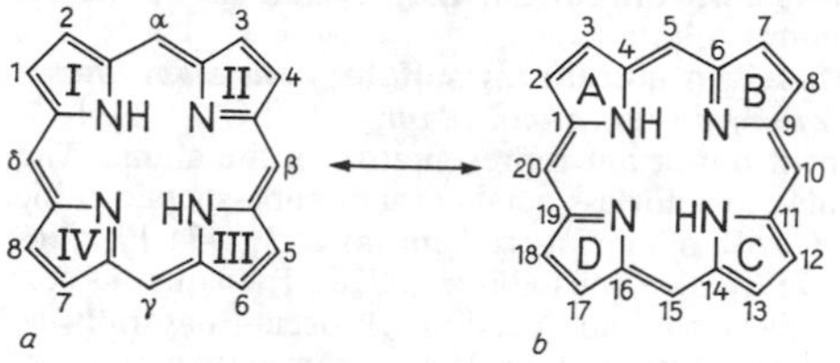

Porphin: *a* numbering according to H. Fischer; *b* numbering according to IUPAC-IUB.

Subsequent derivatives are produced by enzymatic conversions of the substituents in positions 2,3,7,8,12 and 18: carboxymethyl groups are converted to a methyl group and carboxyethyl groups to ethyl or vinyl. Subsequent oxidation of the ring system produces P.

The P. form chelate complexes with many metal ions. The stability of the complexes decreases in approximately the following order: Pt(II) > Ni(II) > Co(II) > Cu(II) > Fe(II) > Zn(II) > Mg(II). ***Metalloporphyrins*** are of enormous biological importance. The oxygen-carrying Hemoglobins (see) are Fe complexes of P., as are the electron-transferring Cytochromes (see). The Chlorophylls (see) are Mg complexes of chlorin. In bacterial chlorophylls, the tetrahydroporphyrins ***bacteriochlorin***, and isobacteriochlorin are also found. The vitamin B_{12} compounds are CO complexes of corrin. The colors of some tropical birds are due to Cu-porphyrin complexes.

Porphyrin	Substituents in position							
	2	3	7	8	12	13	17	18
Etioporphyrin I	M	E	M	E	M	E	M	E
III	M	E	M	E	M	E	E	M
Uroporphyrin III	Cm	Ce	Cm	Ce	Cm	Ce	Ce	Cm
Deuteroporphyrin IX	M		M		M		Ce	M
Coproporphyrin III	M	Ce	M	Ce	M	Ce	Ce	M
Protoporphyrin IX	M	V	M	V	M	Ce	Ce	M
Mesoporphyrin IX	M	E	M	E	M	Ce	Ce	M

Ce = carboxyethyl; Cm = carboxymethyl; E = ethyl; M = methyl; V = vinyl.

Porphyrinogens: see Porphyrin.

Portland cement: see Binders (building materials).

Potash: same as Potassium carbonate (see).

Potash alum: same as Potassium aluminum sulfate (see).

Potash lye: an aqueous solution of Potassium hydroxide (see).

Potash magnesia: same as Potassium magnesium sulfate (see).

Potash soaps: mixtures of potassium salts of long-chain fatty acids. The P. form a soft, translucent mass which is soluble in water and alcohol. They are used in cleansers and disinfectants.

Potassium, symbol ***K***: an element from group Ia of the periodic system (see Alkali metals), a light metal, *Z* 19, with natural isotopes with mass numbers 39 (93.23%), 40 (0.012%) and 41 (6.76%), atomic mass 39.0983, valency I, Mohs hardness 0.5, density 0.86, m.p. 63.65 °C, b.p. 774 °C, electrical conductivity 15.9 Sm mm^{-2} (at 0 °C), standard electrode potential (K/K^+) -2.924 V.

Properties. P. is a very soft metal with a silvery sheen on a fresh surface. It crystallizes in a cubic body-centered lattice. P. is weakly radioactive due to the presence of the isotope ^{40}K, which decays with a half-life of $1.28 \cdot 10^9$ years. It decays both by β^- emission to ^{40}Ca and by K-capture to ^{40}Ar. The latter process is used to determine the age of rocks. P. forms alloys with sodium and mercury; with suitable compositions, these are liquid at room temperature. P. dissolves in liquid ammonia to give a dark blue solution.

P. is one of the most reactive elements. Its large atomic radius, low ionization potential and high negative standard electrode potential characterize P. as a very electropositive element and a strong reducing agent (see Alkali metals). It tends to form K^+ cations; its tendency to form covalent bonds is slight. Freshly cut surfaces of P. exposed to moist air are rapidly covered with a layer of potassium oxide or hydroxide, K_2O or KOH. It must therefore be stored under petroleum. P. reacts so violently with water that the hydrogen evolved according to the equation $2\,K + 2\,H_2O \rightarrow 2\,KOH + H_2$ ignites spontaneously. P. also reacts with alcohols with evolution of hydrogen and formation of the potassium alkoxides. It reacts explosively with halogens (with iodine it must be heated) to form potassium halides. In a stream of oxygen, P. burns to potassium superoxide. The reaction of gaseous ammonia with molten P. yields potassium amide, KNH_2, and the reaction with hydrogen produces potassium hydride, KH.

The violent reaction of elemental potassium with water makes special precautions necessary in handling it. All apparatus and solvents must be carefully dried. Residues of metallic potassium should be disposed of by dissolving small portions in a large amount of ethanol.

Analysis. The simplest qualitative test for P. is the flame test (red-violet color) or spectroscopy. It has a red double line at 766.494 and 769.901 nm and a violet double line at 404.414 and 404.720 nm which are easily detected. P. forms a number of insoluble compounds, precipitation of which can serve as qualitative tests, and in some cases, for quantitative analysis as well. These include potassium hexachloroplatinate, K_2PtCl_6, potassium perchlorate, $KClO_4$, potassium hexanitrocobaltate, $K_3[Co(NO_2)_6]$ and potassium tetraphenyl boranate $K[B(C_6H_5)_4]$. P. can also be determined volumetrically, e.g. on the basis of ion exchange. Instrumental methods such as atomic absorption spectroscopy or ion-sensitive electrodes are to be recommended, especially for large numbers of analyses.

Occurrence. P. is one of the ten most abundant elements in the earth's crust, making up 2.59% of it. The most important minerals are orthoclase (kalifeldspar) $K[AlSi_3O_8]$, muscovite, $KAl_2(OH,F)_2[AlSi_3O_{10}]$ and sodium orthoclase, which consists of mixed crystals of potassium and sodium feldspar, $(K,Na)[AlSi_3O_8]$. The P. salts, which are composed of the minerals sylvinite, carnallite, cainite, etc., are of great economic importance. Weathering of K.-containing minerals releases K^+ ions into the water, but they are much more strongly adsorbed in the soil than Na^+ ions. Thus the concentration of K^+ in seawater is only 1/30 that of Na^+, although the two elements are of approximately equal abundance on earth.

K^+ plays an important role in plants. The roots are much more capable of taking up K^+ than other ions, and K^+ is essential for photosynthesis, respiration and other metabolic processes; they are also involved in maintenance of the osmotic pressure in the cells. Therefore a sufficient concentration of P. in the soil is necessary for normal growth of the plants, and also for their resistance to disease. Plant matter contains considerable amounts of P., which remains in the ash as potassium carbonate. K^+ is also essential for animals; it activates many processes such as glycolysis, lipolysis, protein synthesis or synthesis of acetylcholine. Since Na^+ ions inhibit these processes, the maintenance of a certain ratio of Na^+ to K^+ within the cell is very important. This also applies to maintenance of the right osmotic pressure and overall regulation of the water balance. Because the concentration of K^+ inside the cells is normally about 10 times greater than outside (and the reverse is true for Na), K^+ ions are constantly pumped into the cell and Na^+ ions are pumped out (sodium pump).

Production. P. is produced in similar fashion to Sodium (see); anhydrous K_2O is subjected to melt electrolysis. It may also be produced by adding sodium to a P. salt melt, usually potassium chloride, and this method is now very important. Calcium may also be used as a reducing agent. Other possible methods are reduction of potassium carbonate with carbon, $K_2CO_3 + 2\,C \rightarrow 2\,K + 3\,CO$, or the reaction of potassium fluoride with calcium carbide, $2\,KF + CaC_2 \rightarrow 2\,K + CaF_2$.

Applications. Because P. is difficult to handle, it is not widely used in industry, but only when its high reactivity (compared to sodium) is really necessary. Normally, the cheaper and less dangerous sodium is preferred. P. is used in organic syntheses and in the production of alkali metal photocells. Liquid K-Na alloys are used as coolants in some nuclear reactors.

Historical. Potassium carbonate (potash) obtained from plant ashes was known in antiquity. Elemental P. was first prepared in 1807 by H. Davy, who used melt electrolysis of potassium hydroxide. Further information is given under Sodium (see).

Potassium acetate:, CH_3COOK, colorless, hygroscopic, very water-soluble crystals with a silky sheen; M_r 98.14, density 1.57, m.p. 292 °C. P. is usually made by neutralizing acetic acid with potassium hydroxide or potassium carbonate. It is used in electroplating and photography. It is used in medicine as a diuretic.

Potassium aluminates: The colorless, anhydrous compounds $KAlO_2$ and K_3AlO_4 are obtained by fusing aluminum oxide and potassium carbonate or hydroxide. The hydroxyaluminates $K[Al(OH)_4]$ and $K_3[Al(OH)_6]$ can be made by dissolving aluminum hydroxide in potassium hydroxide solution. P. are used in the paper industry and as mordants for dyes.

Potassium aluminum alum: same as Potassium aluminum sulfate (see).

Potassium aluminum sulfate, ***potassium aluminum alum, potash alum, alum***: $KAl(SO_4)_2 \cdot 12H_2O$, the most important representative of the alums. The crystals are colorless octahedral or cube-shaped cubic lattices with a sour, astringent taste; M_r 474.19, density 1.757, anhydrous above 200 °C. The aqueous solution gives an acid reaction. P. occurs naturally as alunite and potash alum. It is produced by extraction of clay with hot sulfuric acid and addition of potassium sulfate. P. is used extensively in the paper industry, for tanning of hides and dyeing of textiles; in medicine it is used as a mild astringent.

Potassium amalgam: an alloy of potassium with mercury; with up to 1.5% K, such alloys are liquid (see Mercury alloys).

Potassium antimony tartrate, formerly ***tartar emetic***: $K[C_4H_2O_6Sb(OH)_2] \cdot 1/2H_2O$, colorless, water-soluble crystals; M_r 255.93. P. is formed by the reaction of potassium hydrogentartrate with antimony oxide. It was once used in medicine as an emetic.

Potassium bi-: see Potassium hydrogen-.

Potassium boron fluoride: same as Potassium tetrafluoroborate (see).

Potassium bromate: $KBrO_3$, colorless, water-soluble trigonal crystals; M_r 167.01, density 3.27, dec. above 430 °C with release of oxygen. P. can be made by disproportionation of bromine in hot potassium hydroxide solution; in industry it is made by anodic oxidation of potassium bromide. P. is a strong oxidizing agent. Mixtures of it with readily oxidized substances, such as carbon or sulfur, explode on impact or when heated. P. is used in volumetric analysis (bromatometry).

Potassium bromide: KBr, colorless, water-soluble, cubic crystals; M_r 119.01, density 2.75, m.p. 734 °C, b.p. 1435 °C. P. is made by a continuous electrolytic process. It is used in medicine as a component of sedatives and to make photographic plates and engravings.

Potassium carbonate, ***potash***: K_2CO_3, colorless, hygroscopic, monoclinic crystals which are very soluble in water; M_r 138.20, density 2.428, m.p. 891 °C. Various hydrates of P. can be isolated from the alkaline aqueous solution. If carbon dioxide is passed through a solution of P., the somewhat less soluble potassium hydrogencarbonate precipitates: $K_2CO_3 + CO_2 + H_2O \rightarrow 2\,KHCO_3$. This, however, is much more soluble in water than sodium hydrogencarbonate, which prevents application of the Solvay process to the production of P. Instead, the *magnesia process (Engel-Precht process)* is used: carbon dioxide is passed through a solution of potassium chloride in the presence of magnesium carbonate. The insoluble double salt $MgCO_3 \cdot KHCO_3$ precipitates and is separated at 60 °C under CO_2 pressure into $MgCO_3$ and soluble K_2CO_3. However, most P. is now obtained by carbonization of potassium hydroxide: $2\,KOH + CO_2 \rightarrow K_2CO_3 + H_2O$. P. is also produced by the *formate-potash process*, in which potassium sulfate and lime are treated under pressure with carbon monoxide: $K_2SO_4 + Ca(OH)_2 + 2\ CO \rightarrow 2\,HCOOK + CaSO_4$. The resulting potassium formate is oxidized to P. by heating it in air: $2\ HCOOK + O_2 \rightarrow K_2CO_3 + CO_2 + H_2O$. P. can also be obtained from ashes of wood, bagasse, wool grease, etc. In addition, P. is present in high concentrations in salt lakes (in the former USSR and USA) and in the Dead Sea. These natural sources are also exploited. P. is used in the synthesis of many other potassium compounds, to make soft soaps and to prepare cold asphalts for street repairs. It is also used in bleaches, dyeing, glass and ceramics.

Potassium chlorate: $KClO_3$, colorless, shiny, monoclinic platelets which are readily soluble in water; M_r 122.55, density 2.32, m.p. 356 °C. Above 400 °C, it gradually decomposes, first forming potassium chloride, KCl, and potassium perchlorate, $KClO_4$; the latter decomposes further to the chloride and oxygen. When heated rapidly above the decomposition temperature, P. can explode violently. It releases oxygen and is a strong oxidizing agent. Therefore, mixtures of P. with readily oxidized substances, such as sulfur, carbon, phosphorus, iodine or organic compounds, explode violently on impact or when heated.

> Potassium chlorate is poisonous. Amounts greater than 1 g attack the gastric mucous membranes. Countermeasures: pumping the stomach, administration of animal charcoal and emetics.

P. is formed, along with potassium chloride, when chlorine disproportionates in hot potassium hydroxide solution. A more economical process is one developed by Liebig, in which chlorine is passed through hot lime milk, and the P. is then precipitated by addition of potassium chloride. P. is produced industrially by anodic oxidation of a sodium chloride solution, followed by precipitation with potassium chloride. P. is used to make matches, fireworks, explosives, as a herbicide, disinfectant in toothpaste and, in very dilute solution, as a gargle solution.

Potassium chloride: KCl, colorless, cubic crystals; M_r 64.55, density 1.984, m.p. 770 °C, subl.p. 1500 °C. P. is readily soluble in water; in contrast to sodium chloride, it becomes more soluble as the temperature increases. It is found in nature as sylvinite. P. is obtained by neutralization of potassium hydroxide solution with hydrochloric acid, followed by evaporation of the solution. P. is obtained on a large industrial scale from crude potassium salts, and is used as a fertilizer. In addition, it is used to make potassium hydroxide, potassium carbonate and other potassium salts.

Potassium chromate: K_2CrO_4, M_r 194.20, yellow, very water-soluble, rhombic crystals which are isomorphic with potassium sulfate, K_2SO_4. K_2CrO_4 has a density of 2.732; at 667 °C the yellow form is converted to a red, hexagonal form, m.p. 968.3 °C. P. is poisonous (see Chromium). It is synthesized by reaction of potassium dichromate with potassium hydroxide (or carbonate): $K_2Cr_2O_7 + 2\ KOH \rightarrow 2\,K_2CrO_4 + H_2O$. P. is used in dyeing, printing, to make pigments, etc., but it has largely been replaced by the equivalent, but cheaper, sodium chromate. It is used as an indicator in volumetric analysis (Mohr chloride determination).

Potassium chrome alum: same as Potassium chromium(III) sulfate.

Potassium chromium(III) sulfate, ***potassium chrome alum, chrome alum***: $KCr(SO_4)_2 \cdot 12H_2O$, dark violet, octahedral-shaped cubic crystals; M_r 499.25, density 1.826, m.p. 89 °C (in their own water of crystallization). P. dissolves in water, forming a violet solution, which gradually turns green, as the violet chromium(III) hexaaqua ions are converted to green sulfatochromium(III) ions. P. is slightly poisonous (see Chromium). It is made by reaction of sulfur dioxide with a solution of potassium dichromate in sulfuric acid at 38 °C, or by addition of equivalent amounts of potassium sulfate to sulfuric acid solutions of chromium(III) sulfate. P. is used to tan hides, as a mordant in dyeing and textile printing, and to harden gelatins in photographic films.

Potassium cyanate: see Cyanates.

Potassium cyanide: KCN, colorless, hygroscopic, cubic crystals, M_r 65.119, density 1.553, m.p. 634.5 °C. P. is readily soluble in water, but only slightly soluble in alcohol. In the absence of air, moisture and carbon dioxide, it is stable. However, it decomposes in moist air even at room temperature, forming potassium carbonate and hydrogen cyanide. Aqueous solutions of P. are alkaline, due to hydrolysis, and smell like hydrogen cyanide. P. is presently produced in industry by neutralization of hydrogen cyanide with potassium hydroxide. However, P. has been largely replaced by the cheaper sodium cyanide, though it is still used, primarily in electroplating of various metals, especially silver and gold. P. is used in chemical analysis to mask interfering metal ions. The preferred method of disposing of potassium cyanide residues is the reaction with iron sulfate to form potassium hexacyanoferrate(II): $6\ KCN + FeSO_4 \rightarrow K_4[Fe(CN)_6] + K_2SO_4$.

> Potassium cyanide is extremely poisonous; as little as 50 mg, taken orally or absorbed through wounds in the skin, causes rapid death. See Hydrogen cyanide.
>
> Potassium cyanide must be sealed tightly for storage, because it reacts with carbon dioxide in the air, forming potassium carbonate and releasing hydrogen cyanide, which is poisonous.

Potassium dichromate: $K_2Cr_2O_7$, red-orange, water-soluble triclinic crystals; M_r 294.19, density 2.676, m.p. 398 °C. Molten P. is nearly black. Above 600 °C, P. decomposes to oxygen, potassium chromate and chromium(III) oxide. Especially in acidic solution, P. is a strong oxidizing agent, which is capable of oxidizing alcohols to aldehydes or carboxylic acids, chloride to chlorine, or sulfite to sulfate; at the same time, green Cr^{3+} ions are formed. P. is very poisonous (see Chromium). It is made by oxidative fusion of chromium(III) oxide with potassium carbonate and nitrate or oxygen, or by reaction of sodium dichromate with K^+ ions; the less soluble P. precipitates out of solution. P. is used in qualitative and quantitative analysis, to make pigments, as an oxidizing agent in photography, etc. However, in most technical applications it has been replaced by the cheaper sodium dichromate.

Potassium dicyanamide: see Dicyanamides.

Potassium disulfate, ***potassium pyrosulfate***: $K_2S_2O_7$, colorless crystals; M_r 254.33, density 2.512. P. is formed when potassium hydrogensulfate is melted. It is used in analysis to solubilize insoluble oxides, e.g. aluminum, iron(III) or chromium(III) oxide.

Potassium disulfite: $K_2S_2O_5$, colorless, water-soluble, monoclinic crystals; M_r 222.33, density 2.34, dec. 190 °C. P. is formed by heating potassium hydrogensulfite. It is used mainly to disinfect wine casks and fermentation tanks, but also in dyeing and printing.

Potassium ethylxanthate: see Potassium xanthate.

Potassium fluoride: KF, colorless, hygroscopic, cubic crystalline powder; M_r 58.10, density 2.48, m.p. 858 °C, b.p. 1505 °C. It also forms a dihydrate, $KF \cdot 2H_2O$. P. is poisonous. It is made by neutralization of aqueous hydrofluoric acid with potassium hydroxide or potassium carbonate. It is used in making enamel, as a wood protectant and in fermentation technology.

Potassium fluoroborate: same as Potassium tetrafluoroborate (see).

Potassium fluorosilicate: same as Potassium hexafluorosilicate (see).

Potassium formate: HCOOK, colorless, deliquescent, rhombic crystals; M_r 84.12, density 1.91, m.p. 167.5 °C. P. is formed by neutralization of aqueous formic acid with potassium hydroxide or potassium carbonate. It is an industrial intermediate in the formate-potash process (see Potassium carbonate).

Potassium hexafluorosilicate, ***potassium fluorosilicate***: K_2SiF_6, colorless, cubic or hexagonal crystals which are moderately soluble in water; M_r 220.25, density 2.665 (cubic). P. is made by neutralization of hexafluorosilicic acid with potassium hydroxide. It is used in the production of enamel and glass, as an opacifier.

Potassium hexanitrocobaltate(III), ***Fischer's salt***: $K_3[Co(NO_2)_6] \cdot H_2O$, a yellow, crystalline powder; M_r 470.29. The salt is practically insoluble in water; its formation is used for detection of potassium ions. It is used as the pigment ***cobalt yellow (aureolin)*** in paints.

Potassium hydrogencarbonate, formerly ***potassium bicarbonate***: $KHCO_3$, colorless monoclinic crystals, readily soluble in water; M_r 100.12, density 2.17. At 200 °C, $KHCO_3$ decomposes to form potassium carbonate, water and carbon dioxide: $2\ KHCO_3 \rightarrow K_2CO_3 + H_2O + CO_2$. It is formed by the reverse reaction when CO_2 is passed through an aqueous K_2CO_3 solution. P. is used as a starting material for various potassium compounds, including K_2CO_3, and in fire extinguishers.

Potassium hydrogensulfate, formerly ***potassium bisulfate***: $KHSO_4$, colorless, hygroscopic, rhombic crystals, very soluble in water; M_r 136.17, density 2.322, m.p. 214 °C. When heated strongly, the compound dehydrates to potassium disulfate. Its aqueous solutions are acid. P. is made by reaction of potassium sulfate with excess sulfuric acid: $K_2SO_4 + H_2SO_4 \rightarrow 2\ KHSO_4$. It is used as a solubilizer for insoluble oxides and as a water-binding agent in organic syntheses.

Potassium hydrogensulfide: KHS, colorless, hygroscopic, rhombic crystals, readily soluble in water; M_r 72.17, density 1.68-1.70, m.p. 455 °C. The crystals smell like hydrogen sulfide; aqueous solutions are basic. P. is made by passing hydrogen sulfide through potassium hydroxide solution. It is used to separate heavy metals.

Potassium hydrogensulfite, formerly ***potassium bisulfite***: $KHSO_3$, colorless, water-soluble crystalline powder; M_r 120.17. P. is made by passing sulfur dioxide through potassium hydroxide solution. It is used as a reducing agent.

Potassium hydrogentartrate: $KHC_4H_4O_6$, colorless, rhombic crystalline powder which is relatively insoluble in water; M_r 188.18, density 1.984. The aqueous solution is acid and tastes sour. P. is a component of the juices of grapes and other fruits, and precipitates on the walls of fermentation tanks as ***tartar***. It is used in a mixture with potassium hydrogencarbonate as baking powder, and is used as a mordant in dyeing and in electroplating.

Potassium hydroxide, ***caustic potash***: KOH, colorless, hard, translucent, hygroscopic mass, also rhombic crystals; M_r 56.11, density 2.044, m.p. 360.4 °C, b.p. 1320-1324 °C. K. is usually sold in the form of pellets. It dissolves exothermally in water; the solution is sometimes called potash lye. The solution is colorless, caustic and has a slippery feeling. Its concentration can be determined from its density:

% KOH	5	10	20	30	40	50
Density	1.041	1.082	1.176	1.287	1.411	1.538

> Potassium hydroxide is caustic. Protective glasses should always be worn when working with it. Body parts which come into contact with KOH should be washed thoroughly in water and 1% acetic acid solution. If the eyes are affected, a physician should be called immediately.

KOH is also soluble in alcohol (alcoholic potassium hydroxide solutions). Solid KOH or aqueous solutions react with the CO_2 from the air, forming potassium carbonate, K_2CO_3. For this reason, KOH must be protected from air, a precaution which is essential for standardized KOH solutions.

K. is obtained by evaporation of the solution obtained from Chloralkali electrolysis (see) of potassium chloride solution; it was formerly also made by caustification of potassium carbonate with calcium hydroxide: $K_2CO_3 + Ca(OH)_2 \rightarrow 2\ KOH + CaCO_3$. K. is used in alkali fusions in making dyes and pigments, as a drying agent and to absorb carbon dioxide. It is also a starting material for many other potassium compounds.

Potassium hypochlorite: KOCl, known only in solution; M_r 90.55. P. is formed by passing chlorine through a solution of potassium hydroxide. The aqueous solution is strongly oxidizing, and is used as a bleach. It is also known as ***eau de Javelle***.

Potassium iodate: KIO_3, colorless, water-soluble, monoclinic crystals; M_r 214.000, density 3.93, m.p. 560 °C. P. is usually synthesized by anodic oxidation of potassium iodide solution. It is a strong oxidizing agent, especially in acidic solution. Filter paper saturated with P. and soluble starch is used as ***potassium iodate starch paper***, for detection of reducing agents. Their presence is indicated by development of a blue color. P. is used as a primary standard in volumetric analysis.

Potassium iodide: KI, colorless, water-soluble cubic crystals; M_r 166.01, density 3.13, m.p. 681 °C, b.p. 1330 °C. P. is formed by neutralization of hydroiodic acid with potassium hydroxide or potassium carbonate, or by reduction of potassium iodate. It is used in the photographic industry, to iodize table salt, to make other iodine compounds and in volumetric analysis (iodometry). ***Potassium iodide starch paper*** is made by saturating filter paper with P. and starch; it is used to detect various oxidizing agents, which give it an intense blue color.

Potassium magnesium sulfate, ***potash magnesia***: $K_2SO_4 \cdot MgSO_4 \cdot 6H_2O$, colorless, monoclinic crystals; M_r 402.73, density 2.15. P. is found in nature as schoenite, and is made industrially by reaction of magnesium sulfate with potassium chloride. It is an intermediate in the production of Potassium sulfate (see) and is used as a special fertilizer.

Potassium manganate(VI): K_2MnO_4, dark green crystals; M_r 197.14. P. is stable in alkaline aqueous solution, but in neutral or acidic solution, it rapidly disproportionates to potassium permanganate and manganese(IV) oxide: $3\ K_2MnO_4 + 2\ H_2O \rightarrow 2\ KMnO_4 + MnO_2 + 4\ KOH$. This reaction can easily be recognized by the change in color of the solution from green to violet. P. is made by oxidative fusion of Mn^{2+} compounds or manganese(IV) oxide with potassium hydroxide and oxygen, potassium nitrate or potassium carbonate: $MnO_2 + KNO_3 + K_2CO_3 \rightarrow K_2MnO_4 + KNO_2 + CO_2$.

Potassium manganate(VII): same as Potassium permanganate (see).

Potassium nitrate: HNO_3, colorless, rhombic, very water-soluble crystals; M_r 101.11, density 2.109, transition at 120 °C from the rhombic to a trigonal form, m.p. 334 °C, dec. above the m.p. into potassium nitrite and oxygen. P. is an oxidizing agent, and mixtures with various oxidizable substances, such as coal, sulfur or organic compounds, explode when heated. This property has long been utilized in the production of gunpowder; here the potassium salt is preferred to the sodium because sodium nitrate is hygroscopic. P. is found in nature as ***saltpeter***. It is also formed by rotting of nitrogen-containing organic materials in the presence of potassium carbonate or potassium hydroxide. It was once made in this way in saltpeter plantations. Now P. is usually made by the reaction of potassium hydroxide or carbonate with nitric acid, or by reaction of sodium nitrate with potassium chloride. The resulting reciprocal salt pairs are separated, by allowing sodium chloride to crystallize out of hot solution and P. out of very cold solutions. P. is used as the oxidizing agent in black powder and in fireworks, and is also used to carry out oxidative fusions in the laboratory. It is used in the production of ceramics and glass, dyeing and printing, to make potassium nitrite, to treat tobacco and as a fertilizer.

Potassium nitrite: KNO_2, colorless to yellowish, deliquescent, very water-soluble crystals; M_r 85.11, density 1.915, m.p. 440 °C. P. is a strong oxidizing agent. It is made by heating potassium nitrate, preferably in the presence of a weak reducing agent such as lead: $KNO_3 + Pb \rightarrow KNO_2 + PbO$. P. is used for diazotization of dyes, and also in photography.

Potassium oleate: $C_{17}H_{35}COOK$, bright yellow, semisolid mass; M_r 320.56. P. is soluble in water and ethanol, and is used mainly to treat textiles.

Potassium oxalates: 1) ***Potassium oxalate, dipotassium oxalate, neutral potassium oxalate***: $K_2C_2O_4 \cdot H_2O$, colorless, water-soluble monoclinic crystals; M_r 184.24, density 2.127, transition above 100 °C to the anhydrous salt. P. is obtained by reaction of equivalent amounts of oxalic acid and potassium hydroxide or potassium carbonate. It is used in galvanizing; it was formerly used in photography to make iron oxalate developers. It is used in analytical chemistry, for example, for calcium determination.

2) ***Potassium hydrogenoxalate, monopotassium oxalate, acidic potassium oxalate***: KHC_2O_4, colorless, moderately water-soluble, monoclinic crystals; M_r 128.11, density 2.044. The salt is obtained by partial neutralization of oxalic acid with potassium hydroxide or carbonate.

3) ***Potassium tetraoxalate, superacid potassium oxalate***: $KHC_2O_4 \cdot H_2C_2O_4 \cdot 2H_2O$, the most important of the P. It forms colorless, water-soluble, triclinic crystals; M_r 254.20, density 1.836. Potassium tetraoxalate or its mixture with potassium hydrogenoxalate was called ***salt of sorrel***, as it was formerly obtained from the juice of the sorrel plant. The salt is now made by reaction of oxalic acid with potassium hydroxide or carbonate. It is used mainly in printing cloth, or to remove blood and rust stains from textiles; it forms colorless, soluble complexes with iron.

> Potassium oxalates are poisonous. As an antidote, use a suspension of chalk or lime water, which converts the oxalate into an insoluble calcium salt.

Potassium oxides: combustion of elemental potassium leads to one of three different oxides, depending on the partial pressure of oxygen. The nature of the oxygen anions in these oxides is discussed under Oxygen (see).

1) ***Potassium oxide***: K_2O, colorless, unstable, cubic crystals; M_r 94.20, density 2.32, dec. 350 °C. This oxide is formed when potassium is burned in insufficient oxygen, or by reaction of potassium superoxide with potassium.

2) ***Potassium peroxide***: K_2O_2, colorless, amorphous powder; M_r 110.20, dec. 490 °C. The peroxide dissolves in water with an exothermal reaction and evolution of oxygen: $K_2O_2 + 2\ H_2O \rightarrow 2\ KOH + H_2O_2$; $2\ H_2O_2 \rightarrow O_2 + 2\ H_2O$.

3) ***Potassium superoxide***: KO_2, yellow, cubic crys-

tals; M_r 71.10, density 2.14, m.p. 380 °C (dec.). The superoxide is the main product of combustion of potassium in a stream of oxygen. Its reaction with water is: $2\ KO_2 + 2\ H_2O \rightarrow 2\ KOH + O_2 + H_2O_2$.

Potassium perchlorate: $KClO_4$, colorless, rhombic crystals; M_r 138.55, density 2.52, decomposes above 400 °C to potassium chloride and oxygen. P. is slightly soluble in cold water or alcohol. At 20 °C, 100 g water dissolves 1.67 g P. Its precipitation serves as an indicator of the presence of K^+ ions. P. is found in small amounts in nature in caliche, the raw material of Chile saltpeter. P. is produced by heating potassium chlorate, which causes it to disproportionate according to $4\ KClO_3 \rightarrow 3\ KClO_4 + KCl$. Anodic oxidation of a potassium chlorate solution is a better method. In industry, P. can be made by reaction of sodium perchlorate with potassium chloride. It was formerly used as a component of safety explosives, and is still used in fireworks.

Potassium permanganate, ***potassium manganate(VII)***: $KMnO_4$, dark purple, rhombic crystals with a brown sheen; M_r 158.04, density 2.703. It dissolves in water to give a red-violet solution. P. is a strong oxidizing agent both in aqueous solution and in solid form. Its oxidation potential in solution depends strongly on the pH. Acidic $KMnO_4$ solutions can oxidize Fe^{2+} to Fe^{3+}, sulfite to sulfate, H_2O_2 to oxygen or oxalate to CO_2, while the Mn^{7+} is reduced to Mn^{2+}. For example, $2\ MnO_4 + 5\ C_2O_4^{2-} + 16\ H^+ \rightarrow 2\ Mn^{2+} + 10\ CO_2 + 8\ H_2O$. In alkaline solution, the P. is only reduced as far as manganese(IV) oxide: $MnO_4^- + 4\ H^+ + 3\ e \rightarrow MnO_2 + 2\ H_2O$. In very alkaline solution, P. loses oxygen and is converted via manganate(VI) to manganese(IV) oxide. Solid P. decomposes in the same way above 240 °C. P. reacts with organic compounds, such as ethylene glycol, benzaldehyde and mannitol, with the appearance of flames. The ignition of a mixture of P. with a few drops of glycerol is used as a safe way to ignite a thermite mixture. Many organic substances and biological materials reduce $KMnO_4$ solution to manganese(IV) oxide. P. is produced by anodic oxidation of potassium manganate(VI): $K_2MnO_4 + H_2O \rightarrow KMnO_4 + KOH + 1/2\ H_2$. P. is used as an oxidizing agent in organic synthesis; for example, methyl groups bound to aromatics are converted to carboxyl groups. P. is also used as a bleach, antiseptic, in making flash powders and in analytical chemistry.

Potassium peroxide: see Potassium oxides.

Potassium peroxodisulfate, ***potassium persulfate***: $K_2S_2O_8$, colorless, triclinic crystals which are only slightly soluble in water; M_r 270.33, density 2.477. Aqueous solutions are stable when cold, but when heated they evolve oxygen and convert to the sulfate. P. is a strong oxidizng agent; it is able, for example, to convert Mn^{2+} ions to permanganate. P. is made by anodic oxidation of potassium hydrogensulfate solutions. It is an intermediate in the production of hydrogen peroxide by the peroxodisulfate process. P. is used as an initiator in emulsion polymerization, and is a versatile oxidizing agent and bleach.

Potassium persulfate: same as Potassium peroxodisulfate (see).

Potassium phosphates. 1) in the narrow sense, potassium salts of orthophosphoric acid, H_3PO_4. a) ***Potassium dihydrogenphosphate, primary potassium phosphate***: KH_2PO_4, colorless, water-soluble, tetragonal crystals; M_r 136.09, density 2.338, m.p. 252.6 °C. The aqueous solution is acidic. When the compound is roasted, water is lost and the salt condenses to potassium meta- or polyphosphates. KH_2PO_4 is synthesized by mixing equimolar amounts of orthophosphoric acid and potassium hydroxide or carbonate. The reaction of hot phosphoric acid with potassium chloride also yields KH_2PO_4. It is an excellent fertilizer. Single crystals are used in electrooptics. b) ***Dipotassium hydrogenphosphate, secondary potassium phosphate***: K_2HPO_4, colorless, water-soluble, amorphous powder; M_r 172.18. The aqueous solution is weakly basic. When roasted, K_2HPO_4 is dehydrated to potassium diphosphate. It is formed by the reaction of orthophosphoric acid with potassium hydroxide in a molar ratio of 1:2. c) ***Tripotassium phosphate, tertiary potassium phosphate***: K_3PO_4, colorless, rhombic crystals which are readily soluble in water; M_r 212.38, D_4^{14} 2.564, m.p. 1340 °C. The aqueous solution is strongly basic. K_3PO_4 is formed by the reaction of phosphoric acid with potassium carbonate (or potassium hydroxide): $2\ H_3PO_4 + 3\ K_2CO_3 \rightarrow K_3PO_4 + 3\ CO_2 + 3\ H_2O$. It is used in laundry products.

2) In the broad sense, potassium salts of condensed phosphoric acids. a) ***Potassium diphosphate, potassium pyrophosphate***, $K_4P_2O_7 \cdot 3H_2O$, a colorless, water-soluble, hygroscopic powder; M_r 384.40, density 2.33. It is formed by heating dipotassium hydrogenphosphate and is used, for example, as a laundry product additive, textile conditioner and stabilizer for hydrogen peroxide. b) ***Potassium polyphosphates*** in the broad sense are products which are obtained by heating potassium dihydrogenphosphate. The ***potassium metaphosphates***, $(KPO_3)_n$, have cyclic polyanions, while the ***potassium polyphosphates*** have linear condensed polyanions. These structural conventions are not always observed in the labelling of commercial products. There are many defined compounds which correspond in their structures, properties and applications to the corresponding sodium metaphosphates and polyphosphates (see Sodium phosphates).

Potassium pyrosulfate: same as Potassium disulfate (see).

Potassium sodium tartrate, ***Seignette salt, Rochelle salt***: $KNaC_4H_4O_6 \cdot 4H_2O$, colorless, rhombic crystals which dissolve readily in water; M_r 282.23, density 1.790, m.p. 70-80 °C; above 200 °C, conversion to the anhydrous form. P. is produced by neutralization of potassium hydrogentartrate with sodium hydroxide. It is used to make Fehling's solution and in medicine, to treat digestive upsets. Single crystals of P. are piezoelectric, and are used in crystal microphones, recording microphones, etc. to convert mechanical to electrical oscillations.

Potassium sulfate: K_2SO_4, colorless, water-soluble, rhombic or hexagonal crystals; M_r 174.24, density 2.662, m.p. 1069 °C. P. forms double salts with the sulfates of trivalent metals; these are called Alums (see). They crystallize as the dodecahydrates, e.g. potassium aluminum alum, $KAl(SO_4)_2 \cdot 12H_2O$ (see Potassium aluminum sulfate). P. also forms double salts with calcium, magnesium and sodium sulfates, and is found in this form in potash deposits. P. is produced in industry by reaction of potassium chloride with magnesium sulfate; the first product is potassium magnesium sulfate, which must be removed and allowed to react with more potassium chloride: $K_2SO_4 \cdot MgSO_4 \cdot 6H_2O + 2\,KCl \rightarrow 2\,K_2SO_4 + MgCl_2 + 6H_2O$. P. is used to produce potassium aluminum sulfate, potassium peroxydisulfate and potassium water glass; it is used in the glass industry and as a potassium fertilizer for chloride-sensitive crops.

Potassium sulfide: $K_2S \cdot 5H_2O$, colorless, rhombic crystals soluble in water and alcohol; M_r 200.34, m.p. 60 °C (in its own water of crystallization). P. is obtained by mixing equimolar amounts of potassium hydrogensulfide and potassium hydroxide. The aqueous solution is basic and releases elemental sulfur, forming ***potassium polysulfides***, K_2S_n (n 3 to 5). The latter are used in dyeing, and are essential components of Liver of sulfur (see).

Potassium sulfite: $K_2SO_3 \cdot 2H_2O$, colorless, hexagonal crystals; M_r 194.30. The aqueous solution is weakly basic. P. is made by passing sulfur dioxide through a potassium hydroxide solution until added phenolphthalein indicator changes color. It is used as a reducing agent, e.g. in printing textiles, and in photo developers.

Potassium tartrate: $K_2C_4H_4O_6 \cdot 1/2H_2O$, colorless, monoclinic crystals which dissolve readily in water; M_r 235.28, density 1.984. P. is made by neutralizing potassium hydrogentartrate with potassium carbonate. It is used mainly to deacidify wine.

Potassium tetrafluoroborate, ***potassium fluoroborate, potassium boron fluoride***: KBF_4, colorless rhombic or cubic crystals which are only slightly soluble in water; M_r 125.91, density 2.498, decomposes above 350 °C into potassium fluoride and boron fluoride. P. is made by neutralization of fluoroboric acid with potassium hydroxide or carbonate, or by reaction of potassium fluoride with boric acid and hydrofluoric acid. It is used in soldering, in electrochemistry and as a component of sand molds for aluminum and magnesium melts.

Potassium tetrathionate: $K_2S_4O_6$, colorless, water-soluble, monoclinic crystals; M_r 302.46, density 2.296. P. is formed by the oxidation of potassium thiosulfate with iodine, e.g. in iodometry. It is used in bacterial cultures to prevent growth of foreign microorganisms.

Potassium thiocyanate: see Thiocyanates.

Potassium triiodide: see Polyiodides.

Potassium xanthate, ***potassium xanthogenate***: the usual term for ***potassium methylxanthate***, $CH_3OC(S)SK$, a colorless powder; M_r 160.30, density 1.558, dec. above 200 °C. P. is formed by the reaction of ethanol with carbon disulfide and potassium hydroxide. It is used mainly as a flotation agent.

Potassium xanthogenate: $C_2H_5O\text{-}CS\text{-}SK$, the potassium salt of the unstable ethylxanthogenic acid. It forms colorless to light yellow crystalline needles; m.p. 215.3 °C. P. is readily soluble in water, but is only slightly soluble in alcohol. It is insoluble in ether. P. is formed by the reaction of carbon disulfide with alcoholic potassium hydroxide or potassium alcoholate. It can be used to detect copper ions, with which it forms an intensely yellow precipitate. It is also used for qualitative determination of molybdenum, with which it develops a red-violet color. In industry, P. is used as a flotation agent in the processing of sulfide ores, and in grape culture, as an insecticide against aphids.

Potential: 1) see Chemical potential; 2) see Thermodynamic potential; 3) see Electrochemical equilibrium.

Potential surface: a surface which represents the interaction energy of a di- or polyatomic molecule or molecular aggregate as a function of the geometric arrangement of its nuclei. The definition of the P. depends on the uncoupling of nuclear and electronic motions (see Born-Oppenheimer approximation); due to the very different masses of nuclei and electrons, it is assumed that the electron distribution follows any change in the nuclear coordination, practically without inertia.

The P. represents the potential energy ε_{pot} for the motion of the atomic nuclei in the molecule. In principle, it can be calculated by quantum methods, by solving the Schrödinger equation for the entire system for each nuclear configuration. The lowest energy corresponds to a point on the P. of the ground state; higher eigenvalues are points on higher P. of excited states. P. can be calculated point by point by systematic variation of the nuclear coordinates.

The P. of a stable diatomic molecule AB depends only on the internuclear distance R_{AB}. It is thus a ***potential curve*** on the ε_{pot}-R_{AB} plane (Fig. 1). A triatomic molecule ABC requires three coordinates to establish the exact nuclear configuration, e.g. the bond distances R_{AB}, R_{BC} and R_{AC}, or R_{AB}, R_{BC} and the bond angle between these two bonds. If the latter is fixed, e.g. 180° in a linear molecule, the two distances R_{AB} and R_{BC} are sufficient, because R_{AC} is always given by the sum of the other two distances. The P. is then a surface in the ε_{pot}-R_{AB}-R_{BC} diagram (Fig. 2). Similarly, the P. of a reaction $AB + C \rightarrow A + BC$ can be represented if C approaches linearly in the direction of the AB bond. Here the undisturbed molecule AB lies in a potential energy minimum on the side of the reactants, where the distance R_{BC} is large; and on the product side, BC is in a minimum at large R_{AB}. The two minima are separated by a region of high potential energy (Fig. 3).

The curve from AB to BC along the path of minimum potential energy is called the *reaction coordinate*. In the theory of the transition state (see Kinetics of reactions, theory), the saddle point of the P., the highest energy value on the reaction coordinate, plays a critical role. If the reacting particles AB and C are in this saddle, they are called an activated complex or transition state. The difference in potential energies between the saddle point and the reactants is the *threshold energy*.

Systems of N atoms require 3N-6 nuclear coordinates to establish their configurations, and thus are represented by multidimensional surfaces in a 3N-5-dimensional space. Local minima correspond to stable or metastable compounds, and saddle points to the transition states. The calculation of sections of P. and their interpretation is a central problem in modern quantum chemistry, and an important basis for theoretical reaction kinetics.

Reactions which take place on a P. are *adiabatic*. If a collision causes the transition to the P. of an excited state, it is a *nonadiabatic* reaction.

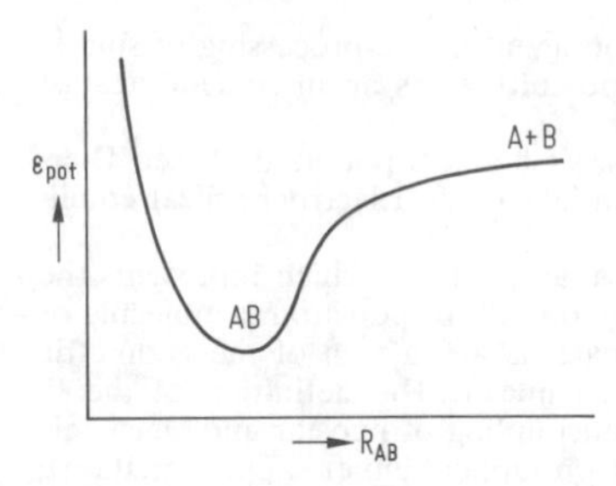

Fig. 1. Potential curve of a diatomic molecule AB.

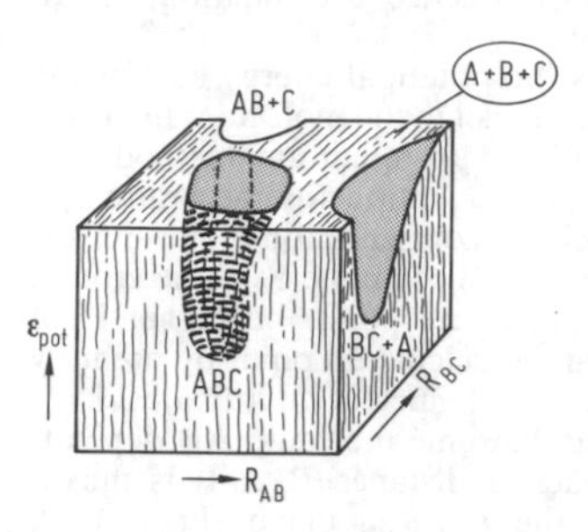

Fig. 2. Section through the potential surface of a linear triatomic molecule ABC.

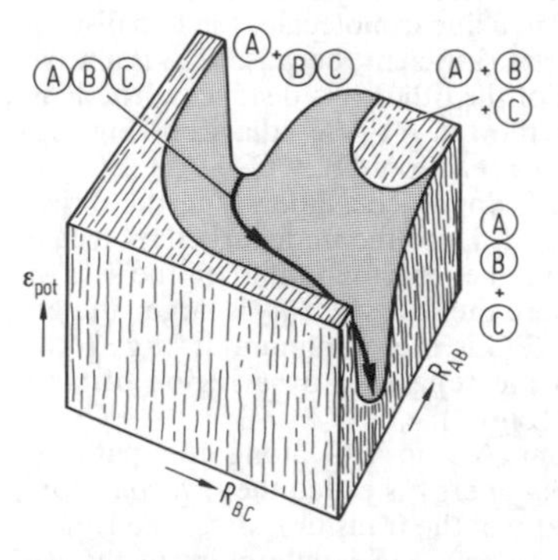

Fig. . Section through the potential surface for the reaction A + BC → AB + C.

Potentiometry: a method of electrochemical analysis based on the concentration dependence of the electrode voltage of an indicator electrode.

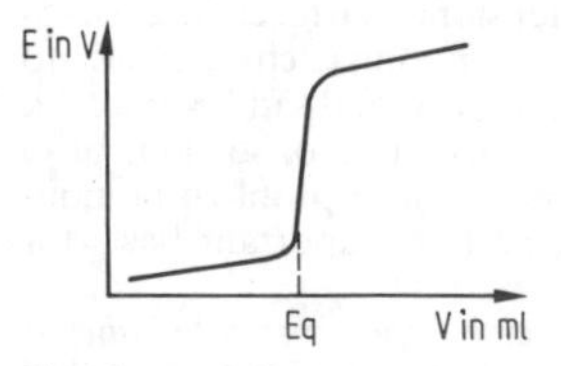

Signal curve of a potentiometric titration. E, potential; V, volume of standard solution; Eq, equivalence point.

Potentiometric measurements are usually made with an ion-sensitive electrode; the most common type are pH measurements, for which glass electrodes are usually used. There is a linear relationship between the potential of the glass electrode and the pH, or, more generally, between the potential of the indicator electrode and the decadic logarithm of the concentration of the species being measured (see Nernst equation).

The potentiometric apparatus consists of the indicator and reference electrodes, usually combined in one probe, and a high-resistance voltmeter.

If P. is used for endpoint recognition in volumetric analysis, the process is called a ***potentiometric titration***. The equivalence point is indicated by a sudden change in the electrode potential of the indicator electrode. Potentiometric titration is particularly well suited for acid-base and argentometric determinations.

Potentiostat: an electronic current source with which an electrolysis can be carried out with a constant potential on the working electrode. A three-electrode technique is used, that is, in the case of a cathodic reduction, a supplemental electrode is placed close to the cathode, in order to measure its potential. The measured value is transmitted to the P., causing it to regulate the clamped voltage in such a way that the measured value approaches a previously established control value. A constant potential at the working electrode during an electrolysis increases the selectivity of the electrode reaction.

Pott-Broche process: see Coal extraction.

Powder metallurgy, formerly ***metal ceramics***: a term for all metallurgical processes for production of parts from metals or metal compounds, with or without added nonmetallic components. The powder particles are usually bound together by pressure and subsequent sintering below the melting point of the main component.

P. is used, for example, for sintered hard metals, which consist of at least one carbide or mixed carbide (e.g. tungsten-titanium-tantalum carbide) and one binder metal (e.g. cobalt); they can be coated with titanium carbide or nitride to increase their resistance to wear. Other applications are contact alloys of tungsten and copper or silver and nickel for making switches; soft and hard magnetic materials; high-melting metals (e.g. tungsten, molybdenum, chromium and tantalum) used in the chemical industry, aircraft, reactors and space vehicles; sliding and friction tools of iron, bronze or copper-titanium with graphite, lead or magnesium oxide additives; oil- or polymer-impregnated self-lubricating sliding bearings; metal filters for purification of gases and liquids; and machine parts of alloyed or unalloyed steels.

Materials made simply by pressing a powder with a binder are also considered powder metallurgical products. These include soft magnetic cores made of carbonyliron powder pressed with an insulating organic binder (such as phenol resin); these are used in high-frequency cores. Permanent magnets are made with pre-alloyed iron-aluminum-nickel powder pressed into a synthetic resin.

Powder technique: see X-ray structure analysis.

Powerforming process: see Reforming.

PQ 9: see Plastoquinone.

PP: abb. for polypropylene.

PP factor: see Nicotinamide.

PPO: abb. for polyphenylene oxide.

PPS: abb. for polyphenylene sulfide.

Pralidoxime, ***PAM***: abb. for 2-pyridine aldoxime methyl iodide, bright yellow crystals; m.p. 224 °C. P. is used as an antidote for poisoning by acetylcholinesterase inhibitors (see Parasympathicomimetics) of the phosphate and phosphonate types (nerve poisons used as weapons, organophosphate insecticides), because it is able to release phosphorylated acetylcholinesterase by abstracting the phosphoryl or phosphonyl group.

Praseodymium, symbol ***Pr***: an element from group IIIb of the periodic system, the Lanthanoid (see) group; a rare-earth, heavy metal, with a single natural isotope, *Z* 59, atomic mass 140.9077, valency III, less often IV, m.p. 935 °C, b.p. 3017 °C, standard electrode potential (Pr/Pr^{3+}) -2.462 V.

P. is a silvery white, ductile metal which forms hexagonal crystals below 798 °C (density 6.475). Above the conversion temperature, it forms a cubic body-centered lattice (density 6.64). It makes up $5.2 \cdot 10^{-4}$% of the earth's crust, and is always found together with the other rare earth metals, especially cerium and neodymium. It is relatively enriched in minerals such as cerite, monacite and orthite. Other properties, analysis, production and history are discussed under Lanthanoids (see). P. is used to dope liquid and solid lasers and in the form of Cerium mixed metal (see).

Praseodymium compounds: compounds in which praseodymium is usually found in the III oxidation state, although some praseodymium(IV) derivatives are known. In addition to fluoro complexes of the types $Na[PrF_5]$ and $Na_2[PrF_6]$, which can be obtained by reaction of praseodymium(III) salts with sodium fluoride in a fluorine atmosphere at 300 to 500 °C, there are a blue-brown ***praseodymium(IV) oxide***, PrO_2, M_r 172.91, density 6.83, and a series of praseodymium(III)/(IV) mixed oxides, such as $Pr_{12}O_{22}$, $Pr_{11}O_{20}$ to Pr_6O_{10}, in which the fluorite lattice contains some unoccupied O lattice sites. Praseodymium(III) compounds are characteristically green, which was the source of the name of the element (the Greek "prasios" means "leek green"). They are discussed under Lanthanoid compounds (see). The P. are used to give a green color to glasses, or, together with neodymium compounds, a blue color; they are also used to make green and yellow sub-glaze paints for porcellain. Some important praseodymium(III) derivatives are: ***praseodymium(III) chloride***, green needles, M_r 247.27, density 4.02, m.p. 786 °C, b.p. 1700 °C; ***praseodymium(III) oxide***, yellow-green, M_r 329.81, density 7.07; ***praseodymium(III) oxalate***, $Pr_2(C_2O_4)_3 \cdot 10H_2O$, bright green crystals, M_r 726.03.

Prazosin: see Sympathicolytics.

Precipitation: The separation of a solid (rarely liquid) phase out of a solution after addition of gaseous, liquid or dissolved solid substances to the solution. The ***precipitation effect*** is due to a) reaction of the precipitating reagent with the substance to be precipitated, as in the addition of a silver nitrate solution to hydrochloric acid: $AgNO_3 + HCl \rightarrow AgCl\downarrow + HNO_3$; b) change of the pH by addition of acids, bases or water as in $[Ag(NH_3)_2]Cl + 2\ HNO_3 \rightarrow AgCl\downarrow + NH_4NO_3$; c) reduction of the solubility of the substance by the precipitating agent, as when ethanol is added to sodium chloride solution; d) the P. of metals from their salt solutions by more reactive metals is due to reduction; see Cementation. In ***fractional P.***, several substances are precipitated one at a time, through addition of the precipitating agent in portions. The most insoluble substance is precipitated first, and so on, in inverse order of solubility. The P. of macromolecules is discussed under Fractionation (see).

Reprecipitation is used to purify precipitates by dissolving them in fresh solvent and then causing them to precipitate again.

Precipitation analysis, ***precipitation titration***: a type of volumetric analysis in which the substance to be determined in the sample solution is precipitated with a suitable standard solution of a precipitation reagent. There are two conditions which must be met for a successful P.: 1) The precipitation must occur rapidly and lead to a strictly stoichiometric precipitation with low solubility and 2) the equivalence point must be readily determined. Since these conditions are seldom fulfilled by a single system, Argentometry (see) was the only method of P. for a long time. Using metal indicators (see Indicator), other P. can now also be carried out, e.g. the determination of sulfation by titration with barium standard solutions.

Precipitation titration: same as Precipitation analysis (see).

Prednisolone: see Adrenal cortex hormones.

Prednisone: see Adrenal cortex hormones.

Pregnane: the parent compound of steroids with 21 carbon atoms (formula, see Steroids). The natural gestagens and various adrenal cortex hormones are derived from P. P. forms colorless scales, m.p. 83.5 °C. It is soluble in methanol, chloroform, etc., and can be obtained by degradation of cholanoic acid.

Prenol: see Terpenes.

Prenylamine: see Coronary pharmaceuticals.

Prephenic acid: an unstable compound formed as an intermediate in the biosynthesis of many aromatic compounds by the shikimic acid pathway.

HOOC CH₂COCOOH

HO H

Preprocalcitonin: see Calcitonin.

Preprodynorphin: see Dynorphin.

Preproglucagon: see Glucagon.

Preprohormones: see Peptide hormones.

Preproinsulin: see Insulin.

Preprosomatostatin: see Statins.

Preservatives: 1) in food chemistry, compounds used to protect foods from degradation by microorganisms. As additives, P. must be harmless to health and must meet certain requirements for purity. The use of P. is legally regulated, with different requirements in different countries. The most common P. are formic acid and its sodium, potassium and calcium salts, 4-hydroxybenzoic acid ethyl, propyl and methyl esters and their sodium salts, propionic acid and its sodium, potassium and calcium salts, calcium acetate, sulfur dioxide and sulfurous acid, sodium and calcium sulfite, sodium and potassium hydrogensulfite, sodium and potassium disulfite, sorboyl palmitate, oligodynamic silver.

2) P. for textiles: see Impregnating agents.

3) P. for wood: see Wood preservatives.

Pressing: a mechanical procedure for separation of a liquid from a solid by application of pressure. The goal is to obtain a liquid and a solid with the lowest possible amount of residual liquid.

Pressure filtration: see Filtration.

Pressurized hydrogen attack: see Hydrogen brittleness.

PRH: see Liberins.

Priležhaev reaction: a method of producing oxiranes (epoxides) by reaction of peracetic acid, perbenzoic acid or monoperphthalic acid with an alkene. Chloroform, ether, acetone or dioxane can be used as solvent. The mechanism is an electrophilic synchronous *cis*-addition via a cyclic activated complex. Alkyl substituents promote the reaction in the order $CH_2{=}CH_2 < RCH{=}CH_2 < RCH{=}CHR < (R)_2C{=}C(R)_2$, but carboxyl groups inhibit the reaction. The acid-catalysed hydrolysis of oxiranes leads to *trans*-1,2-diols by ring opening. P. is valuable chiefly for nonvolatile, water-insoluble alkenes with which other epoxidation methods fail.

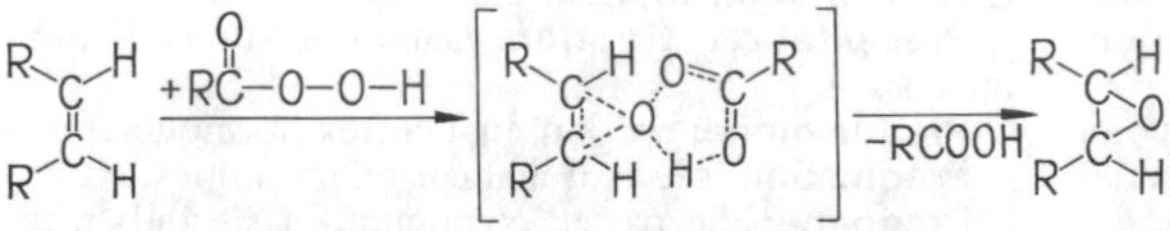

Primaquine: a basic substituted 8-aminoquinoline derivative used as an Antimalarial (see); it is effective against schizonts and gameotcytes.

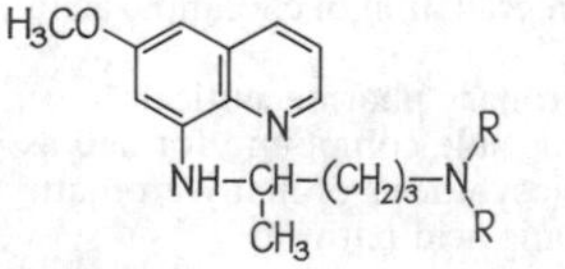

R = H : Primaquine
R = C_2H_5 : Pamaquine

It was developed as a better tolerated substance than *pamaquine* (*Plasmoquine®*), which was the first synthetic antimalarial.

Primary cell: an electrochemical current source which contains limited amounts of the reactants. When these have reacted, the cell can do no more electrical work and is discarded. There is no regeneration after it has been discharged. P. are widely used as batteries for flashlights, radios, calculators, and quartz watches (see Leclanché cell; Alkaline zinc-manganese dioxide cell; zinc-mercury oxide cell; zinc-silver oxide cell).

To achieve higher energy densities and specific energies in P., a number of systems have been proposed, in which anodes with very negative redox potentials (usually lithium) and cathodes with very positive redox potentials are used. In such cases, aqueous electrolytes cannot be used; aprotic electrolytes (e.g. propylene carbonate, ethylene carbonate, acetonitrile, combined with an appropriate conductive salt), salt melts or solid ionic conductors are used instead. Such P. have so far been developed only for very limited applications, such as microelectronic devices, heart pacemakers and weapons.

Primary ozonide: see Harries reaction.

Primary standard: a substance which is strictly stoichiometric, is not hygroscopic, and does not change its composition during prolonged storage. Standard solutions for use in volumetric analysis can be prepared by direct weighing of a certain amount of a P. into a volumetric flask and filling it to a certain volume with water. The titer of any other standard solution can be determined by titration against a precisely known amount of a suitable P. The following substances are suitable as P. for various methods of volumetric analysis: For neutralization, amidosulfonic acid, benzoic acid, potassium hydrogenphthalate, sodium carbonate. For precipitation, sodium chloride, silver nitrate. For redox analysis, arsenic(III) oxide, potassium dichromate, sodium oxalate. For complexometric analysis, lead(II) chloride, calcium carbonate.

Primary structure: see Biopolymers, see Peptides, see Proteins.

Primer: see Pheromones.

Primitive lattice: see Crystal.

Printers' inks: mixtures of pigment, binder and filler used to print paper, cardboard, metal, plastic, textiles, leather or wood. The naphthol AS pigments, metal complexes and basic dyes are used for color: varnishes of various viscosities, synthetic resins or cellulose nitrate are used as binders, and alumina hydrates are used as fillers. For ordinary black print, soot is added to the binder. Four-color printing requires yellow, red, blue and black P. Diarylide yellow, ruby-4B calcium (red), copper phthalocyanin pigments (blue) and soot (black) are commonly used for these colors. For ***silk screen printing***, liquid inks with 2 to 5% pigment are used. Fluorescent pigments can be used with silk screen techniques to give very bright colors. ***Transfer inks*** have dyes which sublime when heated instead of the ordinary pigments. These can be transferred from a printed paper to a textile using an iron. ***Magnetic P.*** are used to print checks, account cards and tickets. These are recognized and sorted by special machines. ***Safety inks*** prevent forgeries on documents and bonds, etc., by changing color.

Prismane, ***Ladenburg benzene, tetracyclo[2.2.0.0^{2,6}.0^{3,5}]hexane***: a valence isomer of benzene; a prism-like molecule, the structure of which was suggested as a structure of Benzene (see) by Ladenburg. P. is a colorless, unstable liquid which decomposes above 90 °C and tends to explode. It can be made by UV irradiation of Dewar benzene, with which it is in equilibrium (Dewar benzene is favored). P. can also be thermally converted to benzvalene.

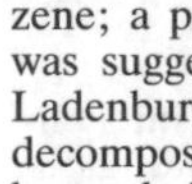

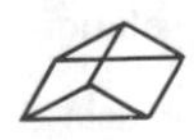

PRL: abb. for prolactin.

Pro: abb. for proline.

Probability density: same as Distribution function (see).

Procain: see Local anesthetics.

Procalcitonin: see Calcitonin.

Process technology: a branch of engineering which deals with the design, construction and operation of plants. The most important task of ***chemical P.*** is the scaling up of laboratory processes into economically optimized large-scale industrial plants. The step from the laboratory to the industrial scale may be made directly, or an experimental, intermediate-scale Pilot plant (see) may be built before the investment in a full-scale plant is made.

The central focus of chemical P. is the reaction (***reaction technology***), which is controlled by physical processes and technological measures. The individual working steps in the reaction are called *unit processes*. These are basic chemical processes, such as oxidation, neutralization, reduction, condensation, polymerization, etc. Chemical P. is concerned with the design of the apparatus, the flow of energy to or from it, the nature of the reaction, its equilibrium, pathway and rate.

An additional task is the design of processes for preparing raw materials for the reaction and working up its products, which generally involves physical processes. The individual working steps for these tasks are called *unit operations*. These are basic operations which do not change the chemical nature of the materials, including heat transfer, mixing, separation, shaping, transport, grinding, agglomerating, storing and packing.

The individual operations can be classified on the basis of the energy used in them. ***Mechanical P.*** involves calculation of the requirements for breaking, grinding, granulating, pressing and shaping the materials, emulsifying, kneading, mixing, stirring, suspending, foaming, and atomization for combining the materials, and decanting, filtering, flotation, sorting, clarifying, pressing, sifting and centrifuging the materials to separate them. ***Thermal P.*** calculates the requirements for freezing, condensing, melting, subliming, liquefaction and evaporation to form the substances, and absorption, adsorption, desorption, distillation, extraction, condensation, crystallization, sublimation, drying, evaporation and thermodiffusion to separate the substances. In addition to these general areas, the use of electromagnetic energy forms a third category of operations, such as electrophoresis, electroosmosis, electrodialysis, electrical gas purification and magnetic precipitation.

Recently another division of P. has become accepted in addition to the above; in it a distinction is made between ***process technology, systems technology*** and ***plant technology***.

For all possible production technologies of the chemical industry, P. must choose the most economical methods and design the production scheme, plant and apparatus.

Process water: see Water treatment.

Prochiral: see Topic groups.

Procion dyes: a group of light-fast and brilliant (intensely colored) synthetic reactive dyes. During the dyeing process, the dye molecule forms a covalent chemical bond to the cellulose fiber.

Proctolin: Arg-Tyr-Leu-Pro-Thr, an excitatory neurotransmitter from the intestinal musculature of insects. Extremely small amounts (10^{-9} mol/l) cause violent contractions of the end of the gut. P. was isolated by Brown from the cockroach *Periplaneta americana*; 125,000 cockroaches were required. The sequence was confirmed by the same laboratory in 1977 by total synthesis.

Procymidon: see Carboximide fungicide.

Prodrug: a substance which is converted into an active pharmaceutical agent in the organism. The conversion can be either enzymatic or non-enzymatic. The P. principle makes possible construction of pharmaceuticals with desired behavior. For example, the absorption of very water-soluble acids or alcohols can be improved by formation of lipophilic esters (see Penicillins). A depot effect is achieved by synthesis of less soluble derivatives (e.g. salts of penicillin which have low solubility); or insoluble drugs can be made water-soluble. In addition, P. may be less toxic than the drugs themselves, so that local damage can be reduced (e.g. cyclophosphamide). It was discovered that the active forms of some compounds used as drugs are actually formed only in the organism (see, e.g., Acetophenetidin, Phenylbutazone). The term P. applies to these compounds, too.

Progesterone: see Gestagens; see Adrenal cortex hormones.

Proglucagon: see Glucagon.

Proguanil: a biguanide derivative used as an Antimalarial (see). P. is effective against schizonts outside the erythrocytes, and damages the gametocytes. The active form is the dihydro-2-triazine derivative formed by dehydrogenating cyclization.

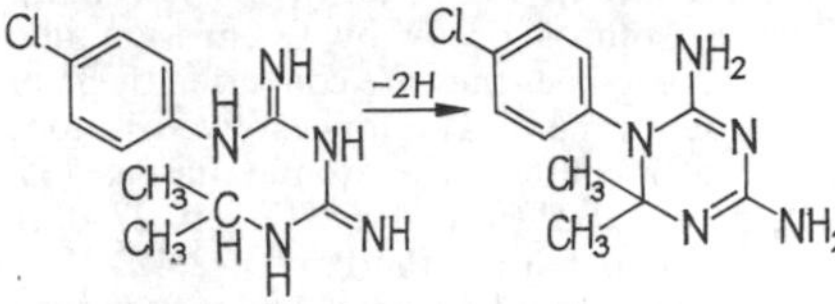

Prohormone: see Peptide hormones.

Proinsulin: see Insulin.

Projection formula: see Stereochemistry.

Prolactin, abb. ***PRL***: a hormone from the anterior pituitary consisting of a single polypeptide chain with 198 amino acid residues and three intrachain disulfide bonds. It is formed by selective cleavage of a longer polypeptide, pre-prolactin, which was discovered in 1977. Its biological effects on the target organs are mediated by adenylate cyclase. P. is relatively easily isolated from the pituitaries of cattle and sheep. Not until 1973 was human P. identified as a different hormone from somatotropin. Its formation is controlled by the interaction of the hypothalamus hormones prolactoliberin and prolactostatin. P. stimulates milk secretion from the mammary glands and stimulates production of the gestagens. It promotes growth, affects pigment metabolism and osmotic regulation and, in many animals, initiates nesting behavior. Its effect in males is still unclear. During pregnancy and lactation, P. is formed in larger amounts.

Prolactin release inhibiting hormone: see Statins.

Prolactin releasing hormone: see Liberins.

Prolactoliberin: see Liberins.

Prolactostatin: see Statins.

Prolamins: a group of globular proteins which occur together with the glutelins, mainly in grains. They are characterized by a high content of glutamic acid (30 to 45%) and proline (15%). They are not soluble in water, but will dissolve in 50 to 90% ethanol. The main P. are ***gliadin*** in wheat and rye, ***zein*** in maize and ***hordein*** in barley.

Proline, abb. ***Pro***: pyrrolidine-2-carboxylic acid, a proteogenic amino acid in which the amino group is part of a heterocyclic ring system (formula and physical properties, see Amino acids, Table 1). L-P. is found mainly in collagen, gliadin and zein. It is isolated from gelatins, acid hydrolysis of which yields about 15% P. L-P. is not essential for mammals, but it promotes growth in chicks. Because it does not fit into a helix, P. affects the tertiary structures of proteins which contain it. A proline antagonist is azetidine-2-carboxylic acid. Some important P. derivatives are ***4-hydroxyproline***, abb. ***Hyp***, found in animal connective and supportive tissues, and ***4-methylproline***, a component of antibiotics.

P. was first isolated in 1901, by E. Fischer, from a casein hydrolysate.

Promazine: see Phenothiazines.

Promethazine: see Phenothiazines.

Promethium, symbol ***Pm***: a radioactive, synthetic element found in only the smallest traces in nature. It belongs to the Lanthanoids (see) and is a rare-earth metal; *Z* 61, with the following isotopes (half-lives in parentheses): 141 (22 min), 142 (34 s), 143 (265 d), 144 (≈ 400 d), 145 (17.7 a), 146 (≈ 710 d), 146 (2.5 a), 148 (2 nuclear isomers, 42 d, 5.39 d), 149 (53.1 h), 150 (2.7 h), 151 (28 h), 152 (6 min), 154 (2.5 min). Most of these isotopes decay by β^--emission into samarium, but sometimes they are converted by β^+ or K-capture to neodymium. The longest-lived isotope of Pm is 145, a β^--emitter. The atomic mass is 145, valency III, density 7.22, m.p. 1168 °C, b.p. 2730 °C, standard electrode potential (Pm/Pm^{3+}) -2.423 V.

P. is a typical lanthanoid element, and is very similar to the neighboring elements neodymium and samarium. P. is always found in the III oxidation state in its compounds, which are intensely colored. ***Promethium(III) chloride***, $PmCl_3$ is blue-violet; ***promethium(III) fluoride***, PmF_3, is pink; ***promethium(III) hydroxide***, $Pm(OH)_3$, is rose-violet, and ***promethium(III) oxide***, Pm_2O_3, occurs in three modifications: blue-violet, rose-violet and coral red. P. is one of the indirect fission products of uranium, but can also be synthesized by neutron irradiation of neodymium or samarium. ^{147}Pm is of technological significance, as it is used for nuclear batteries, as an additive to lumiphores and is suitable because of its soft β emission for thickness measurements.

Historical. P. was discovered in 1945 by Marinsky, Glendenin and Coryell among the fission products of uranium, after earlier claims of the discovery of element 61 in natural sources by American and Italian groups had not been confirmed. These groups had suggested the names illinium and florentium, respectively. In 1965, extremely small traces of ^{147}Pm were discovered in lanthanoid concentrates of an apatite; its formation could be due to the effects of cosmic rays on ^{146}Nd.

Prometon: see Triazine herbicides.

Prometryn: see Triazine herbicides.

Promotion energy: see Hybridization.

Promoter: 1) see Catalysis, sect. III. 2) see Nucleic acids.

Pronase: a mixture of proteases obtained from *Streptomyces griseus*, which is used for complete hydrolysis of the peptide bonds in proteins and polypeptides. The yields of complete enzymatic hydrolysis are between 70 and 90%, but the acid-sensitive amino acids, notably tryptophan, are preserved.

Pro-opiomelanocortin: see Corticotropin.

Propachlor: see Acylaniline herbicides.

Propadiene: same as Allene (see Allenes).

Propane: CH_3-CH_2-CH_3, an alkane hydrocarbon, a colorless, odorless and combustible gas; m.p. ≈ -190 °C, b.p. -44.5 °C. It is readily liquefied (see Liquid gases), and is used for heating and sometimes also as a vehicle fuel. P. is slightly soluble in water, but dissolves readily in ethanol and ether. P. is obtained from natural gas, the exhaust gases of petroleum distillation and hydrogenation, and from cracking gases. Catalytic dehydrogenation of P. yields propene, and pyrolysis produces methane and ethane. Gas phase oxidation yields various oxygen-containing products such as propylene oxide or lower alcohols and aldehydes. In the presence of hydrogen bromide as catalyst, the main product under these conditions is acetone.

Propanal: same as Propionaldehyde (see).

Propane-1-carboxylic acid: see Butyric acids.

Propane deparaffinization: the removal of *n*-alkanes from lubricating oil fractions using propane as a solvent (see Paraffins).

Propanediol: same as Malonic dialdehyde (see).

Propane dinitrile: same as Malonic dinitrile (see).

Propanedioic acid: same as Malonic acid (see).

Propane-1,2,3-triol: same as Glycerol (see).

Propanil: see Acyl aniline herbicides.

Propanoic acid: same as Propionic acid (see).

Propanolamine: see Biogenic amines.

Propanolamines: same as Aminopropanols.

Propanols, ***propyl alcohols***: monoalcohols with the general formula C_3H_7OH; there are two structural isomers. The P. are miscible in any proportion with water, ethanol and ether. They are not very toxic and are good solvents.

Propanol (propan-1-ol), CH_3-CH_2-CH_2-OH, is a colorless, hygroscopic, combustible liquid with a pleasant odor; m.p. -126.1 °C, b.p. 97.2 °C, n_D^{20} 1.3850. It is obtained mainly by fractional distillation of fusel oil. It can be synthesized by catalytic hydrogenation of propargyl alcohol. It is made industrially by hydrogenation of propionaldehyde, which is made by the Oxo synthesis (see). It is used as a solvent, to make esters and as a disinfectant.

Isopropanol (propan-2-ol), CH_3-$CH_2(OH)$-CH_3, a colorless, combustible liquid which smells like ethanol; m.p. -89.5 °C, b.p. 82.3 °C, n_D^{20} 1.3776. It forms an azeotropic mixture with 12.1% water (b.p. 80.4 °C). It is made on a large scale from the propene from cracking gases, by addition of water in the presence of sulfuric acid. It is used as a solvent, antifreeze and antiseptic. By far the largest part is dehydrated to acetone.

1,3-Propanolide: same as Propiolactone (see).

Propanone: same as Acetone (see).

Propargite: an Acaricide (see).

Propargylaldehyde, ***prop-2-ynal***: $HC{\equiv}C\text{-}CHO$, the simplest aldehyde with a $C{\equiv}C$ triple bond. P. is a liquid which boils at 60 °C; its vapors irritate the mucous membranes. P. undergoes the reactions typical of Aldehydes (see). It can be synthesized by oxidation of propargyl alcohol. P. is used in organic syntheses.

Propargyl alcohol, ***prop-2-yn-1-ol***: $HC{\equiv}C\text{-}CH_2\text{-}OH$, an unsaturated monoalcohol, and a colorless liquid with a pleasant odor which irritates the skin; m.p. -48 °C, b.p. 114 °C, n_D^{20} 1.4322. P. is soluble in water, ethanol and ether. It forms an explosive silver compound. P. is a byproduct of large-scale industrial synthesis of butynediol from ethyne and formaldehyde according to Reppe, and is used as a starting material for syntheses of aliphatic C_3 compounds, such as glycerol. P. is also used as a corrosion inhibitor for chlorinated solvents and hydrochloric acid.

Propazine: see Triazine herbicides.

Propene, ***propylene***: $CH_3\text{-}CH{=}CH_2$, an alkene hydrocarbon. P. is a colorless, combustible gas; m.p. -185.3 °C, b.p. -47.7 °C. It can be liquefied at normal temperature if the pressure is raised to about 800 kPa. P. is slightly soluble in water, and readily soluble in ethanol and ether. It burns with a yellow, sooty flame. P. is very reactive and undergoes many addition reactions (chart).

Propene:
- $+ nC_3H_6$ → *Polypropylene*
- $+ nC_3H_6 + mC_2H_4$ → *Ethene-Propene copolymers*
- $+ nC_3H_6 + mCH_2{=}CH{-}CN$ → *Nitrile rubber*
- $+ NH_3 + 1{,}5\,O_2$, $-H_2O$ → $CH_2{=}CH{-}CN$ *(Acrylic nitrile)*
- $+ O_2$, $-H_2O$ → $CH_2{=}CH{-}CHO$ → ($+0{,}5\,O_2$) → $CH_2{=}CH{-}COOH$ *(Acrylic acid)*
- $+ Cl_2 + H_2O$, $-HCl$ → $CH_3{-}CH(OH){-}CH_2Cl$ → ($-HCl$) → $CH_3{-}CH{-}CH_2$ (epoxide, O) *(Propene oxide)*
- $+ C_6H_5{-}CH(OOH){-}CH_3$, $-C_6H_5{-}CH(OH){-}CH_3$ → $CH_3{-}CH{-}CH_2$ (epoxide, O) *(Propene oxide)*
- $+ Cl_2$, $-HCl$ → $CH_2{=}CH{-}CH_2Cl$ → ($+HOCl$, $-HCl$) → $CH_2{-}CH{-}CH_2Cl$ (epoxide, O) *(Epichlor hydrin)*
- → *Cumene-Phenol process* → *Phenol+Acetone*
- $+ H_2O$ → $CH_3{-}CH(OH){-}CH_3$ → ($-H_2O$) → $CH_3{-}CO{-}CH_3$ *(Acetone)*
- $+ CO/H_2$ → $CH_3{-}CH(CHO){-}CH_3$ $+ CH_3{-}CH_2{-}CH_2{-}CHO$ *(n-Butyraldehyde)*
- $+ C_3H_6$, *Al-Alkyles* → $CH_2{=}C(CH_3){-}CH_2{-}CH_2{-}CH_3$ → ($-CH_4$) → *Isoprene*
- $+ C_3H_6$, *Alkali metal catalyst* → $CH_2{=}CH{-}CH_2{-}CH(CH_3){-}CH_3$ → *Poly-4-methyl-pent-1-ene*
- $+ C_3H_6$, *Phosphoric acid catalyst* → *Di- und Tripropylene* → *Fuels*
- $+ C_3H_6$ → $CH_3{-}CH{=}CH_2$ / $CH_3{-}CH{=}CH_2$ → $CH_3{-}CH{=}CH{-}CH_3$ *(But-2-ene)* $+ CH_2{=}CH_2$ *(Ethene)*
- $+ 0{,}5\,O_2$, *$PdCl_2$ catalyst* → $CH_3{-}CO{-}CH_3$ *(Acetone)*

Chart. Industrially important reactions of propene.

Synthesis. P. is obtained exclusively as a byproduct of ethene synthesis (see Pyrolysis) and a few refinery processes, especially catalytic cracking. The work-up of pyrolysis gas yields a C_3 fraction which consists mainly of P., propane, propyne and allene. Propyne and allene are removed by selective hydrogenation to P. Then the mixture of propane and P. is distilled. Separation is best achieved by a double column process or by a heat pump process. In the double column method, the pressure in the first column is chosen so that the vapors coming off the top can be used to heat the second column (Fig.).

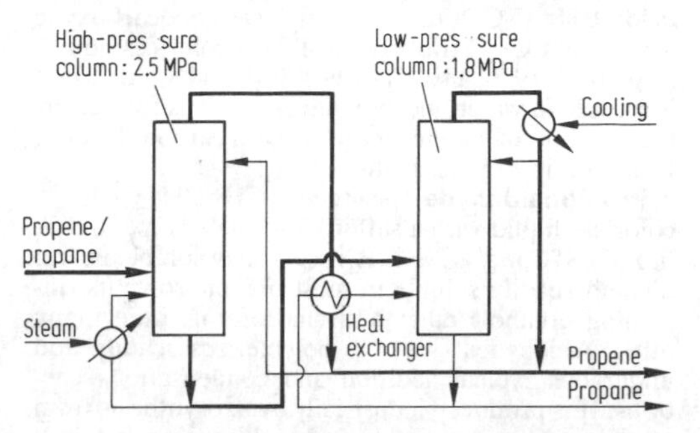

Isolation of propene by the double column method.

The distillation residue of the first column is led into the second column. Each stage of the distillation yields about half the total amount of P.

Applications. P. is used for the industrial syntheses of isopropanol, allyl chloride, propylene oxide, acrylonitrile, acrylic acid, cumene and other intermediates. It has also become very important as the starting material for synthesis of polypropylene.

Prop-2-enal: same as Acrolein (see).

Prop-2-ene-1,2-dicarboxylic acid: same as Itaconic acid (see).

Propene-1,2,3-tricarboxylic acid: same as Aconitic acid (see).

Propenic acid: same as Acrylic acid (see).

Prop-2-en-1-ol: same as Allyl alcohol (see).

Propen-2-yl acetate: $CH_2{=}C(CH_3)\text{-}O\text{-}CO\text{-}CH_3$, the enol acetate of acetone. P. is formed by the reaction of acetone with ketene: $CH_3\text{-}CO\text{-}CH_3 + CH_2{=}C{=}O \rightarrow CH_2{=}C(CH_3)\text{-}O\text{-}CO\text{-}CH_3$. In the presence of mineral acids, P. acts like a mixture of acetone and ketene. It is therefore often used as an acetylation reagent for enols.

Propham: see Carbanilate herbicides.

Prophyphenazone: see Pyrazolone.

Propiconazole: see Azole fungicides.

Propineb: see Dithiocarbamate fungicides.

Propiolactone, *β-hydroxypropionic acid lactone, 1,3-propanolide*: the internal ester of β-hydroxypropionic acid. P. is a colorless liquid with a strong, aromatic odor; m.p. - 33.4 °C, b.p. 162 °C (dec.), n_D^{20} 1.4105. It is soluble in water, alcohol, acetone and chloroform. P. is made by cycloaddition of formaldehyde and ketene in the presence of zinc chloride in an anhydrous medium. P. is very reactive. In the presence of sulfuric acid, it polymerizes to a high-molecular-weight polyester acid, and reaction with aqueous sodium chloride solution yields β-chloropropionic acid. The reaction with sodium hydrogensulfide leads to β-mercaptopropionic acid, and with ammonia, to β-alanine. P. is a versatile intermediate in organic synthesis. It is highly carcinogenic.

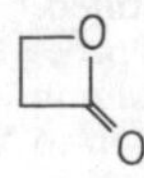

Propiolic acid, ***propinic acid, acetylenecarboxylic acid***: $CH\equiv C\text{-}COOH$, the simplest monocarboxylic acid with a C≡C triple bond. P. is a colorless liquid, m.p. 18 °C (9 °C also reported), b.p. 144 °C (dec.). It is soluble in water, alcohol and ether. P. is formed by the reaction of sodium acetylide with carbon dioxide. It is used in organic syntheses.

Propionaldehyde, ***propanal***: $CH_3\text{-}CH_2\text{-}CHO$, a colorless liquid with a suffocating smell; m.p. -81 °C, b.p. 48.8 °C, n_D^{20} 1.3636. P. is sparingly soluble in water and readily soluble in most organic solvents, including ethanol, ether and chloroform. Like many other Aldehydes (see), P. polymerizes readily and undergoes typical addition and condensation reactions. It is produced industrially by oxosynthesis from ethene, carbon monoxide and hydrogen on a cobalt contact, or by gas-phase oxidation of propane. In the laboratory, P. can be made by oxidation of n-propanol with chromic acid. P. is used as an intermediate in numerous organic syntheses, and is used in the plastics industry.

Propionates: the salts and esters of propionic acid with the general formula $CH_3\text{-}CH_2\text{-}COOM^I$ or $CH_3\text{-}CH_2\text{-}COOR$, where M^I is ammonium or monovalent metal, and R represents an aliphatic, aromatic or heterocyclic group.

The salts are formed by reaction of ammonium or metal hydroxides or carbonates with propionic acid, or by reaction of propionic acid with highly electropositive metals with generation of hydrogen. Ammonium, sodium and calcium propionate are commercially important; they are used as preservatives in foods and animal feeds, and as additives in vulcanization.

The esters of propionic acid are synthesized by the reaction of alcohols or olefins with the free acid. They are used as solvents and in the synthesis of perfumes.

Propionic acid, ***propanoic acid***: $CH_3\text{-}CH_2\text{-}COOH$, a saturated aliphatic monocarboxylic acid. P. is a colorless liquid with a pungent odor; m.p. -20.8 °C, b.p. 141 °C, n_D^{20} 1.3809. It is readily soluble in water and most organic solvents. It undergoes the reactions typical of aliphatic carboxylic acids. P. forms salts with alkali hydroxides, and esters with alcohols or alkenes. The salts and esters of P. are called Propionates (see). P. is found in coal tar and wood distillates, and in some essential oils. P. esters are found in some plants. P. is produced industrially by reaction of ethene, carbon monoxide and water, or by oxidation of higher alkanes or propionaldehyde. P. inhibits the growth of molds, and is therefore used as a preservative in foods and animal feeds. It is also used as an intermediate for production of cellulose esters, solvents, perfumes and pesticides.

Propiophenone, ***ethyl phenyl ketone***: $C_6H_5\text{-}CO\text{-}CH_2\text{-}CH_3$, a colorless liquid with a pleasant odor; m.p. 21 °C, b.p. 218 °C, n_D^{20} 1.5269. P. is slightly soluble in water, and readily soluble in most organic solvents. It can be made in a Friedel-Crafts reaction (see) from benzene and propionoyl chloride in the presence of aluminum chloride. P. is used as a synthetic intermediate in organic chemistry.

Propipocaine: see Local anesthetics.

Propiverin: see Spasmolytics.

Propoxur: a Carbamate insecticide (see).

Propranolol: see Sympathicolytics.

Propyl-: a term for the atomic grouping $CH_3\text{-}CH_2\text{-}CH_2\text{-}$ in a molecule, the radical $CH_3\text{-}CH_2\text{-}CH_2\cdot$ or the cation $CH_3\text{-}CH_2\text{-}CH_2^+$.

Propyl acetates: the two structurally isomeric propyl esters of acetic acid, ***n-propyl acetate***, $CH_3\text{-}CO\text{-}O\text{-}CH_2\text{-}CH_2\text{-}CH_3$, a colorless liquid with a fruity odor; m.p. -95 °C, b.p. 101.6 °C, n_D^{20} 1.3842, and ***isopropyl acetate***, $CH_3\text{-}CO\text{-}O\text{-}CH(CH_3)_2$, a colorless liquid with a fruity odor; m.p. -73.4 °C, b.p. 90 °C, n_D^{20} 1.3773. The E. are only slightly soluble in water, but dissolve readily in alcohol and ether. They are formed by esterification of acetic acid with *n*-propanol or *iso*-propanol. They are used as solvents and in the synthesis of aromas and flavorings.

Propyl alcohols: same as Propanols (see).

Propyl chloride: see Chloropropanes.

Propylene: same as Propene (see).

Propylene glycol, ***propane-1,2-diol***: $CH_3\text{-}CH(OH)\text{-}CH_2OH$, a divalent alcohol, a colorless, viscous, sweet-tasting liquid; b.p. 189 °C, n_D^{20} 1.4324. P. is miscible with water and ethanol in all proportions, but is relatively insoluble in ether. In industry, it is produced mainly from propene (propylene), via propylene oxide, by addition of water. P. is used in the synthesis of polyesters, epoxide resins, polyurethane foams, softeners and antifreezes. P. can also be used as a solvent, preservative and disinfectant. The isomeric *propane-1,3-diol* has similar properties; b.p. 213-214 °C, n_D^{20} 1.4389.

Propylene oxide, ***1,2-epoxypropane, methyloxirane***: a colorless, readily ignited liquid with an ether-like, sweet odor; b.p. 34.3 °C, n_D^{20} 1.3670.

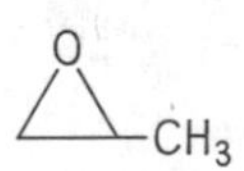

P. is soluble in water, alcohol and ether. It is made industrially by the Halcon process from propylene and hydroperoxides on titanium contacts. It is very irritating to the skin and mucous membranes; the vapor causes nausea and vomiting. P. is used as an intermediate in the synthesis of propylene glycol, glycerol and propanolamines. It is used as a solvent for cellulose acetates, cellulose nitrates, natural resins, etc. Hydration of mixtures of ethylene oxide and P. produces mixtures of glycols which are used as working fluids in freezers.

Propyl ether: same as Di-*n*-propyl ether (see).

Propylhydroxybenzoate: see Nipa ester.

2-Propylpiperidine: same as Coniine (see).

Propylthiouracil: see Thyreostatics.

Prop-2-ynal: same as Propargylaldehyde (see).

Propyne: $HC\equiv C\text{-}CH_3$, an alkyne hydrocarbon, and a colorless, easily inflammable gas; m.p. -101.5 °C, b.p. -23.2 °C, n_D^{-40} 1.3863. P. is slightly soluble in water, but miscible with ethanol and ether. It burns with a very sooty flame and has many properties comparable to those of ethyne (see Alkynes; Acetylene chemistry).

Propynic acid: same as Propiolic acid (see).

Prop-2-yn-1-ol: same as Proparyl alcohol.

Proscillaridin A: see Scilla glycosides.

Prostacyclin: see Prostaglandins.

Prostaglandins: a group of animal hormones derived from *prostanoic acid*. This compound and all natural P. have a cyclopentane ring bearing two adjacent aliphatic chains *trans* to each other. One of these chains ends with a carboxyl group. The P. of the E and F series are most important. P. of the E series (***PGE***) have an oxo function on C9 and an α-OH group on C11. P. of the F series (***PGF***) have a β-OH group on C11 and another OH group on C9. Compounds of both series have an OH group on the *S*-configured C15 atom. In compounds with the subscript 1, for example PGE_1, there is a double bond with *E* configuration at position 13; in compounds with the subscript 2, there an additional double bond with *Z* configuration at position 5. In compounds with the letter α appended to the subscript, the OH group on C9 is α; a β in the subscript means that this OH group is β. The PGA series is obtained from the PGE series by treatment with acids, and the PGB series, by treatment of PGE with base.

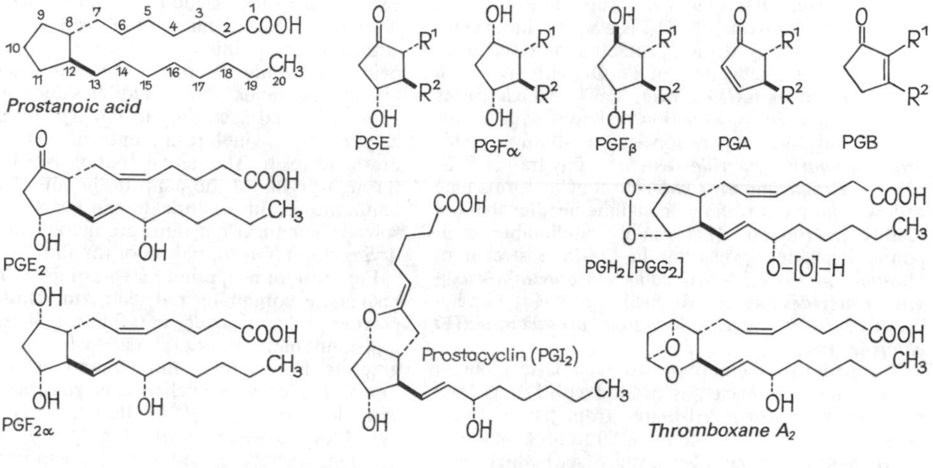

Prostanoic acid PGE PGFα PGFβ PGA PGB
PGE_2 PGH_2 [PGG_2]
$PGF_{2\alpha}$ Prostacyclin (PGI_2) *Thromboxane A_2*

The P. are found in all animal tissues, in low concentrations (< 10^{-7} ng g^{-1}). They are obtained by multi-step total syntheses. Many structural variants of the P. have been synthesized. Biosynthesis of P. begins from unsaturated fatty acids with 20 C atoms. For example, PGE_2 and $PGF_{2\alpha}$ are formed from arachidonic acid. Prostaglandin endoperoxides, PGH_2 and PGG_2, are intermediates formed by the action of molecular oxygen in the presence of prostaglandin cyclooxygenase. PGG_2 has an additional exocyclic hydroperoxide structure.

The P. have multiple physiological effects, but they are so rapidly metabolized that their use as drugs is difficult. $PGF_{2\alpha}$ (dinoprost) and PGE_2 (dinoprostone) are used locally to induce abortion or labor pains. In animal husbandry, P., especially derivatives, are used to synchronize heat in breeding animals. P. also have bronchospasmolytic effects and inhibit the secretion of gastric juice. They promote inflammation and intensify pain. The effects of antiphlogistics and many weak analgesics is due to inhibition of P. biosynthesis. *Prostacyclin*, PGI_2, is made from PGH_2 or PGP_2 by prostacyclin synthetase; it inhibits thrombocyte aggregation and dilates the coronary arteries. The very short-lived *thromboxane*, which is formed by thromboxane synthetase, on the other hand, promotes thrombocyte aggregation and constricts the coronary arteries. The first successful syntheses of stable prostacyclin analogs and specific thromboxane synthetase inhibitors have been reported.

P. were first detected in 1934 by von Euler and Goldblatt, in human seminal fluid. In the middle of the 1950s, Bergström et al. succeeded in isolating the first P. (PGE_1 and $PGF_{1\alpha}$) in crystalline form.

Prostanoic acid: see Prostaglandins.

Prosthetic group: see Enzymes.

Protactinium, symbol ***Pa***: a radioactive element in the Actinoid group (see) of the periodic system; a heavy metal, *Z* 91, with natural isotopes with the following mass numbers (decay type, half-life in parentheses): 231 (α, 3.25 · 10^4 a) and 234 (2 nuclear isomers, β^-, 1.18 min and 6.75 h). The following synthetic isotopes are also known: 216 (α, 0.20 s); 217 (α, short-lived); 222 (α, 5.7 ms); 223 (α, 6.5 ms); 224 (α, 0.95 s); 225 (α, 1.8 s); 226 (α, K-capture, 1.8 min); 227 (K, α, 38.3 min); 228 (α, K, 26 h); 229 (K, α, 1.4 d); 230 (K, β^-, α, 17.4 d); 232 (β^-, 14.7 d); 233 (β^-, 27.0 d); 235 (β^-, 24.2 min); 236 (β^-, 9.1 min), 237 (β^-, 8.7 min); and 238 (β^-, 2.3 min). The atomic mass is 231.0359, density 15.37, valence V, IV, less often III, II; m.p. 1568 °C, b.p. 4200 °C. P. is a malleable metal with a silvery sheen; it tarnishes in air and its chemical properties are more similar to those of the elements of the vanadium group than those of the other actinoids. As a member of the actinium decay series which begins with uranium-235, P. accompanies uranium in uranium minerals; because of the relatively short half-life of the longest-lived natural isotope, ^{231}Pa ($t_{1/2}$ $3.25 \cdot 10^4$ a), however, the P. content of uranium ores is very small. For example, Jáchymov pitchblende (from Czechoslovakia) contains a few hundred mg P. per ton uranium. The isotope ^{231}Pa is an α-emitter, which decays via ^{227}Ac and a series of other intermediates to the stable lead isotope ^{207}Pb (actinium lead, actinium D). Metallic P. is obtained by reduction of protactinium(IV) fluoride with lithium or barium at 1400 °C.

Historical. P. was discovered in 1918 by Hahn and Meitner, and independently from them, by Soddy and Cranston. The name P. was proposed by Hahn and Meitner. Before its discovery, P. was called eka-tantalum, and even later was assigned to group Vb of the periodic system.

Protactinium compounds: compounds in which protactinium is most often in the +V or +IV oxidation state. Protactinium(V) compounds have a strong tendency to hydrolyse, and are therefore difficult to maintain in solution. The reaction of hydrogen with protactinium at 250 to 300 °C yields black ***protactinium hydride***, PaH_3, which forms cubic crystals. ***Protactinium(V) oxide***, Pa_2O_5, is a white, basic compound produced by heating protactinium in the air to 500 to 1000 °C; reduction of Pa_2O_5 with hydrogen yields ***protactinium(IV) oxide***, PaO_2, which forms black, cubic crystals. Reaction of Pa_2O_5 with hydrofluoric acid and evaporation of the solution yields ***protactinium(V) fluoride*** as the dihydrate, $PaF_5 \cdot 2H_2O$. ***Protactinium(V) chloride***, $PaCl_5$, forms pale yellow, transparent, shiny crystalline needles that are readily hydrolysed. This readily sublimable compound is made by heating Pa_2O_5 in a stream of phosgene at 550 °C. A byproduct is ***protactinium oxygen trichloride***, $PaOCl_3$. Reduction of $PaCl_5$ with hydrogen at 600 °C yields yellow-green ***protactinium(IV) chloride***, $PaCl_4$.

Protamines: a group of strongly basic globular proteins found in nuclei, associated with DNA. They are gene repressors. P. isolated from fish and bird sperm have arginine contents of 80 to 85%.

Proteases: a group of Enzymes (see) which catalyse the hydrolysis of the peptide bond in peptides and proteins. ***Endopeptidases (proteinases)*** are P. which attack internal peptide bonds and yield products with varying chain lengths; examples are the Serine P. (see), Rennin (see), Pepsin (see), Papain (see), Bromelain (see) and Thermolysin (see). ***Exopeptidases*** split only terminal peptide bonds; examples are the Aminopeptidases (see), Carboxypeptidases (see) and Enterokinase (see). The P. are required for degradation of nutritional proteins in digestion. They are synthesized as inactive precursors (see Zymogens) and transported and stored in the presence of inhibitors. This prevents autodigestion of the P. and the cell containing it. The release of the P. at the site of action is achieved by specific cleavage of an activation peptide from the N-terminal end of the zymogen.

Protective colloid: a macromolecular solution used to stabilize lyophobic colloids. The colloidal particles become lyophilic by adsorption of the macromolecules, and their stability increases. For example, PVC dispersions are stabilized by methylcellulose, and photographic dispersions by gelatin.

Protective gas: a gas under which chemical processes which are sensitive to oxygen, water vapor or air can be carried out. For example, if the dust of a substance is capable of forming explosive mixtures with air (e.g. coal or sulfur), it can be ground under a P. to prevent explosions. For the same reason, pipelines and containers for combustible gases and liquids are rinsed with P. P. are also used to preserve chemical plants and parts of equipment. Nitrogen, carbon dioxide, exhaust or stack gases can be used as P. When metals are treated with heat, P. are used to prevent corrosion. Carbon dioxide, argon, helium and hydrogen are used in a number of welding processes to prevent oxidation of the metal.

Protective layers: layers of varying thickness which prevent rust or other forms of corrosion. There are several distinct types.

Rust-preventing paints act both by covering the surface of the metal, thus preventing water and oxygen from reaching it, and by inhibiting the oxidation reaction. As a rule, such paints are applied as a series of layers of different kinds of paint, which complement each other in their action. The total thickness of the layer is 0.1 to 0.3 mm; for objects exposed to the atmosphere, a minimum of 0.14 mm is needed. The paints are liquids or pastes consisting essentially of binders, pigments, fillers, additives and solvents. The paint is named according to the type of binder: oil, alkyde resin, vinyl resin, epoxide resin and polyurethane resin. The actual coating is a film created through drying of the paint in the air or at elevated temperature. In addition to the evaporation of the solvent, chemical reactions are involved (oxidatively drying paints) in formation of the film.

For outdoor use, primer layers of linseed oil-alkyde resin base containing red lead, zinc chromate, zinc powder and iron oxide, and finishing layers containing aluminum powder, lead white, iron oxide, graphite, iron glimmer and titanium dioxide are used. For protection against chemicals, chlorinated rubber, epoxide resins, chlorosulfonated polyethylene, polychlorobutadiene, vinyl polymers and polyurethane resins are used as bases, with iron glimmer, iron oxide, graphite, chromium oxide, silicon carbide or titanium dioxide as pigments. Cumaron resin, butyltitanate, phenol and silicon resins are used as bases in paints from protection during high-temperature use.

Nonmetallic inorganic protective layers are used either to protect objects from corrosion (enamel, cement, eloxal or bronze layers), or they provide a surface to which a corrosion-preventing paint can bind, and reduce the tendency of the metal to rust under its paint (phosphate layers). The effect of baked enamel layers is to shield the surface of the cast iron or steel below them, and depends on the chemical, thermal and mechanical stability of the coating. Cement layers are used to protect cast iron or steel water pipes inside and outside, both by physical coating and by passivation of the iron by the components of the cement. Bronzing and phosphate layers (see Phosphatizing) on steel, and chromate layers on zinc and cadmium are formed by chemical reaction with the metal surface. Anodic oxidation (see) or light metals (eloxal protective layers) increases the corrosion resistance of the treated metal.

Organic protective layers are made of organic polymers; the layers are usually thicker than those provided by rust paints (above). They are manufactured as spackel masses of reactive resin, thermoplastic foils, sheets of natural or synthetic rubber, powders or dispersions. The layer must prevent the corrosion medium from reaching the carrier, and be resistant to chemical, thermal and mechanical wear. This is achieved by good bonding to the surface (see Surface pretreatment), and if necessary, with glues for foils and sheets of rubber. If the coatings are bonded to the carrier at only a few points, diffusion is not reduced and the protective lifespan is less. A steel container can be protected against acids, bases and corrosive salts by layers of thermoplastic polyvinylchloride, polyethylene or polytetrafluoroethylene foils, or by sheets of polybutadiene, polychlorobutadiene or vinylidene fluoride/hexafluoropropylene copolymers. At a thickness of 3 mm, such layers effectively prevent diffusion. Polyvinyl and polyethylene layers are also used as extrusion layers or adhesive sheets for protection of underground pipes. Tar is also often used for this purpose.

Layers of melted epoxide powders (see Plastic sprays) reduce the frictional resistance in petroleum pipelines, which, in addition to lengthing the life of the pipe, reduces the energy required to transport the oil.

Metallic protective layers are applied by the following methods: *dipping in melt* to produce a layer of zinc or tin (see Zinc plating; Tin plating), *diffusion* to apply chromium or aluminum (see Chromization; Aluminization); *vapor deposition*, *rolled* or *explosion plating* for applying a plate, e.g. of chromium-nickel steel, nickel-molybdenum and nickel-molybdenum-chromium alloys, titanium, zirconium, silver; *spraying* (see Metal spraying, Flame spraying); *galvanization* or Electroplating (see) to produce P. of zinc, cadmium, nickel, chromium, copper, silver, gold, platinum or tantalum; *chemical metal precipitation* to produce nickel layers.

Calcium carbonate layers which form in water pipes protect them from corrosion. They form if the water contains a minimum amount of carbonate (see Lime-carbonic acid equilibrium; Hardness) and at least 5 mg/l oxygen. Carbonic acid, which dissolves calcium, has a decisive influence. If the water is neutralized with lime or sodium hydroxide or is filtered over half-burned dolomite, the carbonic acid is removed, and a protective layer can form in the pipes.

The equations for these stabilizing reactions are:

a) Hydrated lime: $Ca(OH)_2 + 2\ CO_2 \rightarrow Ca(HCO_3)_2$

b) Sodium hydroxide: $2\ NaOH + 2\ CO_2 \rightarrow 2\ NaHCO_3$

c) Half-burned dolomite:
$CaCO_3 + MgO + 3\ CO_2 + 2\ H_2O \rightarrow Ca(HCO_3)_2 + Mg(HCO_3)_2$.

If calcium or magnesium hydrogencarbonate compounds (equations a and c) are formed, the water is made harder, which is often undesirable if it is already hard. However, if it is soft, this can be an advantage.

The conditions necessary for producing a protective layer cannot always be achieved in practical situations, or there are other factors, such as the velocity of the water in the pipes, which lead to additional uncertainty. In such cases, phosphate or silicate additives are used to create a protective layer in spite of the aggressive qualities of the water.

Proteides: composite Proteins (see).

Proteinases: see Proteases.

Protein fibers: in the narrow sense, synthetic fibers made from proteins of animal or plant origin. In the broad sense, the term also includes natural fibers made of protein (wool and silk). E. can be made from regenerated plant proteins, such as the ***zein fibers*** from the maize protein zein, ***soy fibers*** from the soy bean protein glycinin, and ***ardein fibers*** from the peanut protein ardein. Animal proteins are also used, e.g. ***casein fibers*** from milk proteins and ***fibroin silks*** from regenerated fibroin from silkworm cocoon wastes.

The protein solutions are extracted with aqueous alkali, then processed to P. by wet or dry spinning. The polypeptide chains are oriented by stretching, and can be stabilized by treatment with formaldehyde.

The P. are usually used only in mixed textiles, because they are not very stable to water. They have excellent felting properties and are used to make hats, felt and upholstery.

Proteins: macromolecular Biopolymers (see) consisting largely or entirely of amino acids. P. have many different roles in organisms, and the number of P. made by an organism is related to its complexity. An *Escherichia coli* cell, for example, makes about 3000 different P., while a human being makes more than 100,000. Enzymes (see), Peptide hormones (see) and various DNA-binding P. are forms of P. responsible for the control of metabolism. ***Structural P.*** such as collagens, elastins and keratins are essential components of connective tissue, support tissues and intracellular structures, such as the cytoskeleton. ***Contractile P.***, especially actins and myosins, permit motion of cells, including muscle contraction. ***Immunoglobulins*** are ***defense P.*** ***Transport P.*** include hemoglobin and transferrin. There are many ***membrane P.***, which act as receptors for hormones or neurotransmitters, actively pump specific molecules or ions into or out of the cell, or serve as electron carriers in the processes of respiration and photosynthesis. Blood clotting depends on a series of enzyme P. which activate each other in a cascade, finally transforming soluble fibrinogen to fibrin, which forms the clot.

Classification. P. are classified in a number of different ways: according to origin (type of organism, tissue or cell organelle), function (enzyme, structural, transport, storage, receptor, etc.) or by solubility and shape, i.e. globular, fibrillar and membrane P. ***Globular P.*** are soluble in water and dilute salt solutions. They are approximately spheroidal; the exact folding of the polypeptide chain is not random, but is determined by hydrophobic interactions, and hydrogen, ionic and disulfide bonds between amino acid side chains. The solubility of these P. is due to hydrophilic amino acid residues located on the surface of the molecule. All soluble enzymes and many other biologically active P. are globular.

Fibrillar P. are practically insoluble in water or salt solutions. The polypeptide chains are parallel to each other and make up long fibers; this arrangement is seen in collagens, keratins and elastins. ***Membrane P.*** contain regions of hydrophobic amino acid residues, which often form helices within the membrane. Those which span the membrane or extend into the cytoplasm or extracellular space are very often glycoproteins; the carbohydrate residues are attached to those parts of the P. which extend into aqueous media.

Simple P. contain only proteogenic amino acids, while ***composite P. (conjugated P.)*** contain a covalently bound non-protein component. Some simple P. are Albumins (see), globins, Globulins (see), Glutelins (see), Histones (see), Prolamins (see) and Protamins (see) (all of these are globular P.) and the fibrillar skeleroproteins. The classes of composite P. are ***glycoproteins*** (containing carbohydrate), ***lipoproteins*** (containing triacylglycerols, phospholipids or cholesterol), ***metalloproteins*** (with metal ions in ionic or coordinate bonding), ***phosphoproteins*** (containing serine or threonine esterified to phosphate), ***nucleoproteins*** (containing nucleic acids) and ***chromoproteins*** (containing bound pigments; for example, chlorophyll or hemoglobin).

Structure. P. consist of 20 different kinds of Amino acids (see), also called proteogenic amino acids, linked together by α-peptide linkages. The sequence of the amino acids in the polypeptide is genetically determined, and it in turn determines the physical and catalytic properties of the P. The *primary structure* of a P. is the sequence of amino acids (Fig. 2a, see Insulin). As a rule, P. contain more than 100 amino acids in a polypeptide chain. This is too long for sequence determination by the Edman method (Fig. 2b), so the polypeptide must first be cleaved into fragments. The enzymes most commonly used for this are trypsin, chymotrypsin, pepsin, papain, subtilisin, elastase and thermolysin; cyanogen bromide is the most common chemical reagent. Cyanogen bromide specifically cleaves peptide bonds in which methionine provides the carboxyl group. The fragment peptides are then degraded, one amino acid at a time, and the resulting derivative of each cleaved-off amino acid is identified by chromatography. In the Edman method, the N-terminal amino acid reacts with phenylisothiocyanate, producing first a thiourea derivative (I), which reacts further to form a 2-anilinothiazolin-5-one (II), the phenylthiocarbamic acid (III) and finally, the 3-phenyl-2-thiohydantion

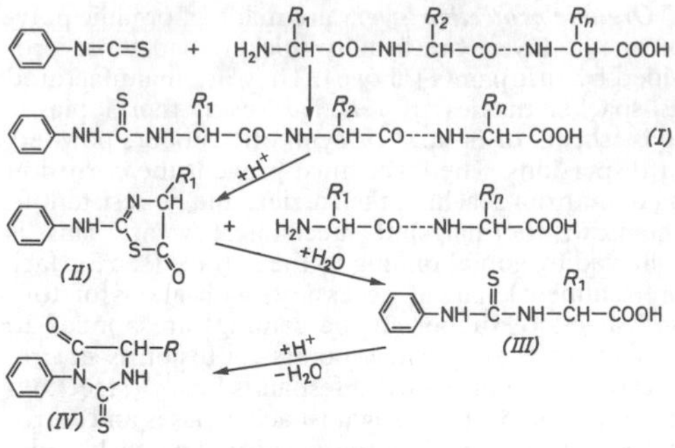

Fig. 1. Stepwise degradation of the polypeptide chain (Edman degradation).

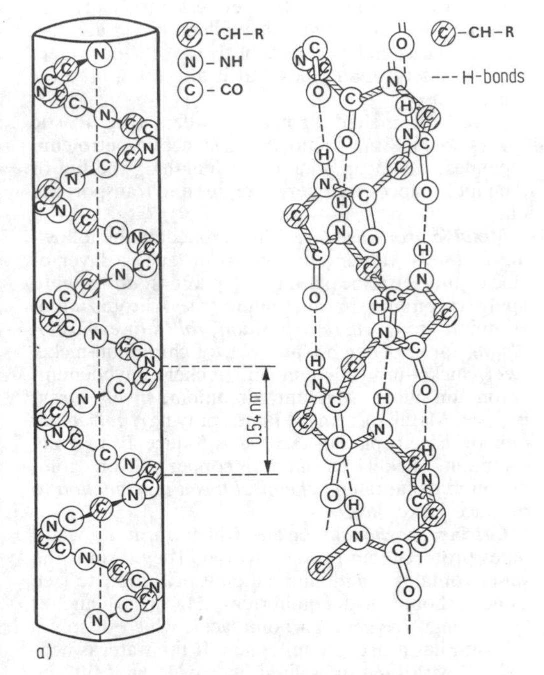

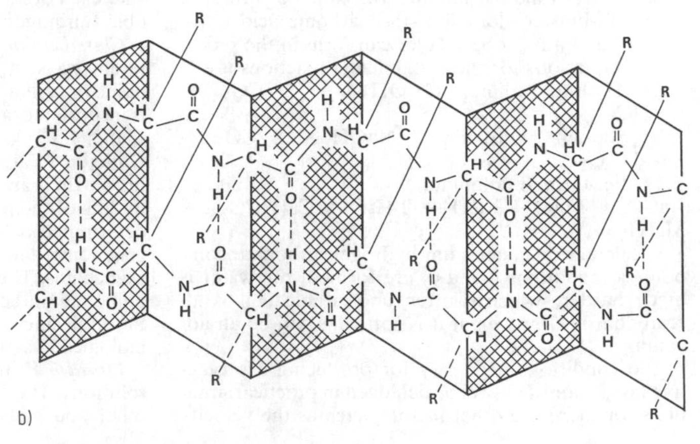

Fig. 2. Secondary structure of proteins. *a*, α-helix; *b*, antiparallel pleated sheet.

(PTH-amino acid, IV), which is is split off. The order of the peptides in the original polypeptide is determined by using two different reagents to generate two sets of peptides; overlapping sequences then indicate the original order.

The *secondary structure* is the specific folding of the polypeptide chain due to the formation of H-bonds between carbonyl oxygen and amide nitrogen atoms. If the H-bonds form within a chain, it forms a helical structure called an α-*helix* (Fig., see Helix). If the H-bonds are intermolecular, the resulting structure is called a *pleated sheet* (β-structure, Fig. 2). For example, myoglobin has a very large proportion of α-helix structure (> 75%), while silk fibroin is entirely β-structure.

The α-helix of hair keratin can be reversibly converted to a β-structure when stretched. The secondary structure of P. is examined by optical rotatory dispersion (ORD) and circular dichroism (CD).

Tertiary structure is the spatial folding of the polypeptide chain (Fig. 3) due to intramolecular interactions of side chain functions. This structure is often stabilized by disulfide bonds, and it fixes the positions of reactive amino acid residues, such as those in the active sites of enzymes or in the binding sites of receptor P. Tertiary structure is examined by x-ray studies of heavy-atom derivatives of the P. From the diffraction diagrams the electron density distribution functions are calculated, and these have now reached sufficient resolution (0.15 nm) to indicate the positions of individual atoms. It should be remembered that these rigid structures obtain only in crystals; in solution, P. are flexible and their shapes change continuously, though probably within fairly narrow limits.

Intermolecular interactions between two or more polypeptide chains, which can be identical or different, leads to stable oligomeric P. This association is known as *quaternary structure*. Most quaternary structure is based on non-covalent association; some examples are given in Table 1. Quaternary structures are studied by electron microscopy or by x-ray analysis (Fig., see Hemoglobin).

Table 1. Relative molecular masses M_r and subunits of some proteins with quaternary structure

Protein	M_r	Subunits Number	Subunits M_r
Hemoglobin	64,250	4 (α_2,β_2)	α: 15,130 β: 15,870
Tabacco mosaic virus	39,400,000	2130	17,530
Lactoglobulin	36,750	2	18,375
Enterotoxin	84,000	6	14,000
Ceruloplasmin	125,000	4 ($\alpha_2\beta_2$)	α: 16,000 β: 53,000
Nerve growth factor (Mouse)	26,520	2	13,260
Lactose repressor	150,000	4	39,000
Catalase	240,000	4	60,000
Alcohol dehydrogenase	141,000	4	35,000

Properties. The molecular mass M_r of a P. is determined by diffusion and sedimentation rates in the ultracentrifuge, measurement of light and x-ray small-angle scattering, osmotic pressure, electrophoresis and rate of migration through a dextran or polyacrylamide gel. The M_r of single polypeptide chains lie between about 10,000 and 50,000; polymeric P. have M_r in the range from about 50,000 to several million. The diameter of a P. molecule is between 2 and 100 nm, which makes them colloidal particles. They do not pass through dialysis membranes and do not form true solutions; they show Tyndall effects and have relatively high viscosities. Because of the large number of ionized groups in their molecules, the P. have high dipole moments. Because acidic and basic groups are both present, P. are ampholytes. The charge of the entire molecule depends on the pH of the solution: in very acidic media, the P. are polycations, and in very basic media, polyanions. The resulting positive or negative excess charge causes hydration and solubility to increase; hydration depends only on the absolute charge. The sign of the overall charge is responsible for the electrophoretic behavior of the P., that is, for the direction in which it moves in an electric field. At its Isoelectric point (see), a P. has

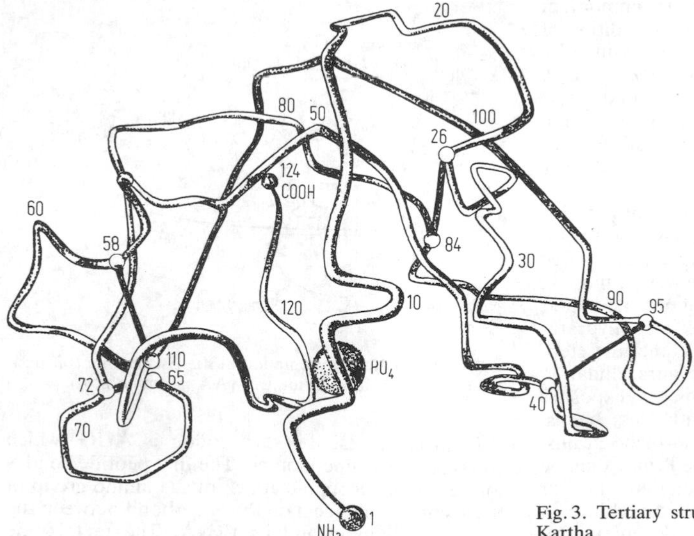

Fig. 3. Tertiary structure of ribonuclease A, according to Kartha.

no net charge, and in this zwitterionic form, its solubility and hydration are at a minimum. As a result of their ampholyte character, the P. act as buffers in biological systems. Because of their hydration, globular P. are able to enclose hydrophobic substances and prevent their precipitation. This protective colloid function is important for the stabilization of body fluids. Addition of weakly or non-polar solvents (e.g. alcohol or acetone) or of high concentrations of neutral salts to these fluids destroys the hydrate shells of P. and causes them to precipitate. As a rule, if P. are heated above 60 °C, they undergo large changes in structure which destroy their biological activity. This *denaturation* is due to destruction of their tertiary and quaternary structure. They can also be denatured by UV or x-rays, extreme pH, detergents such as 1% sodium dodecylsulfate, or reagents which destroy hydrogen bonds, usually 8 M urea or 6 M guanidine hydrochloride. If the denaturation is carried out in the presence of reducing agents, disulfide bonds are broken as well as non-covalent bonds; the polypeptide configuration is called a random coil. If the denaturation is reversible, the native configuration of the P. can be re-established (*renaturation*).

Analysis. P. can be qualitatively determined by precipitation with trichloroacetic, picric or perchloric acid, heavy metal ions (Cu, Fe, Zn or Pb salts) or by specific color reactions. In the *xanthoprotein reaction*, for example, a yellow color develops on reaction with concentrated nitric acid; in the *biuret reactions*, a purple color is generated by addition of copper sulfate to a highly alkaline protein solution; and in the *Pauly reaction*, a red color is formed by treatment of the alkaline solution with diazobenzene sulfonic acid. The *Lowry method* is most often used for quantitative determination of P. In it, a copper phosphomolybdanic acid complex is formed with the P.; this has an absorption maximum at 750 nm and is determined colorimetrically. Human or bovine serum albumin is used as a standard.

In the classic *Kjeldahl method*, the sample to be analysed is heated in concentrated sulfuric acid, which converts the nitrogen in the P. to ammonium sulfate. The ammonia is then released by addition of sodium or potassium hydroxide, and determined by back titration with acid. The *UV absorption* of a P. solution at 280 nm provides a direct and rapid measurement of concentration, but because the extinction is due to the presence of aromatic amino acid residues (Tyr, Trp), the molar extinction of each protein is different.

Isolation and purification. The isolation of proteins which are present in high concentrations, e.g. hemoglobin from erythrocytes or ovalbumin from egg white, is not particularly difficult. However, most P. are present in rather low concentrations, and their isolation from other proteins, lipids, carbohydrates and nucleic acids may be difficult. As the first step, cells must be broken to release the intracellular P. This is done by homogenization of tissues, exposure to supersonic vibration, shaking with glass beads, grinding frozen tissue with aluminum oxide grains, detergent treatment or proteases. The P. may then be extracted with a salt solution, glycerol, dilute acid or other solvent. It is often necessary to add inhibitors of proteases at this stage, to prevent endogenous proteases from destroying the desired P. The P. may then be coarsely fractionated by fractional ammonium sulfate precipitation or solvent extraction. Further purification is achieved by gel filtration, ion exchange or adsorption chromatography, preparative electrophoresis, electrofocusing or ion filtration chromatography, that is, by a combination of ion exchange and gel chromatography. The most elegant method is affinity chromatography, in which the specific binding of the desired P. to another biomolecule is utilized. A column is prepared in which the binding partner is covalently bound to a support such as dextran, and the cell homogenate is poured over it. For example, if a hormone receptor is to be isolated, the hormone is bound to the column, or an antibody against the desired P. is immobilized on the column. The bound P. is eluted from the column by a solution of its binding partner, or by a high-ionic strength medium which interrupts non-covalent bonds between the P. and its partner.

Biosynthesis. The sequence of amino acid residues for each P. is encoded in the DNA of the cell. The first step of P. biosynthesis is the *transcription*, or copying, of the coded sequence onto a molecule of *messenger RNA* (mRNA) with the complementary sequence of nucleotide bases. Each amino acid is activated by binding to a specific *transfer RNA* molecule (tRNA), in a step requiring hydrolysis of ATP (and thus, expenditure of energy). The tRNA carries a triplet of bases complementary to the codon (set of three bases) specific for its amino acid. The mRNA is *translated* (a polypeptide corresponding to it in amino acid sequence is synthesized) on the ribosome, a complex structure of proteins and ribosomal RNA. In a complex series of events, which require hydrolysis of GTP, the mRNA first binds to a dissociated ribosome, then the ribosomal subunits associate to form an active unit ribosome, which moves along the messenger until it locates the correct starting point for "reading" the "message", and loaded tRNA molecules are aligned with the first and second codons on the messenger (Fig. 4).

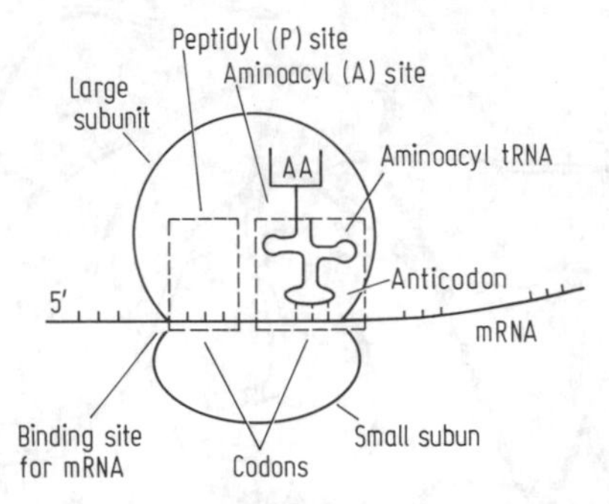

Fig. 4. Schematic representation of the *Escherichia coli* ribosome structure, from Lehninger (AA = amino acid).

In prokaryotes, the start codon is AUG, which codes for formylmethionine. The first peptide bond is formed by nucleophilic attack of the amino group of the second amino acid on the ester bond between the formyl-methionine and its tRNA. The methionine tRNA is released, and the dipeptide remains bound to the tRNA of the second amino acid. This is transferred to the site on the ribosome previously occupied by the methionine tRNA, and a new amino acyl tRNA enters the vacant site. At the same time, the ribosome moves along the messenger by three bases. The process is then repeated, with the growing polypeptide being transferred each time to the new tRNA and the previous tRNA being released (Fig. 5). Since the process adds amino acids sequentially to the carboxy terminus of the polypeptide, it is always the N-terminal section of the P. which is synthesized first. The end of the synthesis is signalled by a stop codon, that is, a triplet for which there is no corresponding tRNA.

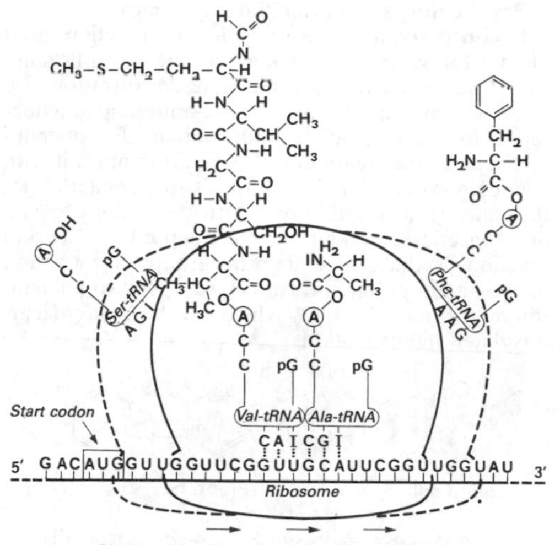

Fig. 5. Translation, schematically.

As the P. is synthesized, it adopts the most stable configuration; this is determined by the amino acid residues which are first incorporated. If the P. is to be inserted into or through a membrane, it is synthesized with a series of hydrophobic amino acids at its N-terminus. This *signal sequence* promotes insertion of the nascent polypeptide into an adjacent membrane of the endoplasmic reticulum. It is not known what causes nascent secretory proteins to pass clear through the membrane for processing and accumulation in the lumen of the endoplasmic reticulum; however, it is known that on the luminal side of the membrane, there is an enzyme which selectively removes the N-terminal signal sequence. Other enzymes may add methyl, hydroxyl, carbohydrate or phosphate groups in post-translational modification; many P. are cleaved at specific sites to produce the active products.

Biosynthesis of P. is both extremely precise and rapid; up to 100 peptide bonds per second are formed. It is regulated at the levels of transcription and translation. Individual steps are inhibited by certain antibiotics. For example, chloramphenicol inhibits translation by blocking the peptidyl transfer reaction, while rifamycin inhibits transcription by specific inhibition of RNA polymerases. The ribosomes and their attendant enzymes are similar in prokaryotes and eukaryotes, but there are enough differences that antibiotics which interrupt protein biosynthesis in prokaryotes may be harmless to eukaryotes, and conversely.

P. as nutrients. P. are an essential part of human and animal nutrition. Animal P., from eggs, milk and meat, have higher nutritional value than vegetable P. because they contain essential amino acids in sufficient amounts and in ratios which closely approximate the human requirements. Most plant P. contain too little lysine, methionine, tryptophan or threonine. This deficit can be corrected by addition of the limiting amino acid, or by combination of complementary P. (Table 2).

In the year 2000, it is estimated that the P. requirement of the earth's human population will have risen to 160 million tons, compared to 108 million tons in 1980. For animal nutrition, an increase in the same time period from 432 million tons to 640 million tons is expected. To meet the steadily increasing demand, not only will greater production of animal and vegetable foods be needed, but grains with better distributions of amino acids will be bred, and the production of single cell protein (SCP) is expected to increase. The biotechnology already exists for using microorganisms to convert raw materials such as hydrocarbons, alcohols, starch, molasses or cellulose into high-protein biomass. Some advantages of microbial production are the rapid increase in biomass and its high P. content, the amino acid spectrum of the P., and the independence of the process of climate or season. At present, large plants (100,000 t annual production) are operational in the former USSR,

Table 2. Essential amino acids present in proteins from various sources

Amino acid	% essential amino acids in dry matter						
	Wheat flour	Soy meal	Fish meal	Beef	Cows' milk	Nutritional yeast	Petroleum yeast
Leu	7.0	7.7	7.8	8.0	11.0	7.6	7.0
Ile	4.2	5.4	4.6	6.0	7.8	5.5	3.1
Val	4.1	5.0	5.2	5.5	7.1	6.0	8.4
Thr	2.7	4.0	4.2	5.0	4.7	5.4	9.1
Met	1.5	1.4	2.6	3.2	3.2	0.8	1.2
Lys	1.9	6.5	7.5	10.0	8.7	6.8	11.6
Arg	4.2	7.7	5.0	7.7	4.2	4.1	8.0
His	2.2	2.4	2.3	3.3	2.6	1.7	8.1
Phe	5.5	5.1	4.0	5.0	5.5	3.9	7.9
Trp	0.8	1.5	1.2	1.4	1.5	1.6	1.2

Italy, Great Britain and Japan; some of them utilize petroleum or methanol as sources of carbon and energy. Other large reserves of P. are found in the green parts of plants (leaf and grass P. make up 1/3 of the dry mass) and in insects, for example grasshoppers, which are 65% (dry mass) P.

Historical. In 1784, Fourcroy recognized proteins as an independent class of substances. In 1839, Mulder applied the name "protein", which had been coined by Berzelius. Around 1900, Hofmeister and Fischer elucidated the structural principle of the P., and Fischer confirmed this by trail-blazing synthetic studies. In the 1920's, the first P. were crystallized (ovalbumin, insulin). Between 1925 and 1930, Svedberg succeeded in determining the molar masses of a series of P. using the ultracentrifuge. In 1937, Tiselius introduced electrophoresis as an analytical method. In 1945, Brand reported the first complete amino acid analysis of a P. (lactoglobulin) by chemical and microbial methods. In 1950, Pauling and Corey proposed the α-helix structure for keratins. From 1951 to 1956, Sanger determined the primary structure of insulin. Between 1952 and 1960, Kendrow and Perutz developed methods of determining the three-dimensional structure of P. (x-ray crystal structures of myoglobin and hemoglobin). Insulin (Zahn, 1963) and ribonuclease (Merrifield, 1969) were the first P. to be synthesized by purely chemical methods. Since the beginning of the 1980's, the biotechnological production of P. such as insulin, interferons and growth hormone has become significant. These syntheses are achieved using genetically engineered microorganisms.

Proteoglycans: see Polysaccharides.

Proteohormones: see Peptide hormones.

Prothrin: see Pyrethroids.

Protium: the isotope of Hydrogen (see) with mass number 1.

Protocatecualdehyde, ***3,4-dihydroxybenzaldehyde***: a phenol aldehyde. P. is a bright yellow, crystalline compound; m.p. 153-154 °C. It is slightly soluble in water, but soluble in alcohol and ether. P. is the parent compound of several phenol ether aldehydes which are found in free or bound form in nature. It is an intermediate in the synthesis of piperonal and vanillin.

Protocyanin: see Cyanidin.

Protoheme: see Cytochromes.

Protolysis: see Hydrolysis, see Acid-base concepts, section on Brønsted definition.

Proton acceptor: see Acid-base concepts, section on Brønsted definition.

Proton donor: see Acid-base concepts, section on Brønsted definition.

Protoporphyrin: see Porphyrins.

Proximpham: see Carbanilate herbicides.

Process: see State.

Prussian blue, ***iron(III)hexacyanoferrate(II)***: $Fe^{III}_4[Fe^{II}(CN)_6]\cdot nH_2O$ (n = 14...16). A dark blue, water-insoluble compound made by combination of iron(III) salt solutions with potassium ferrocyanide (see Cyanoferrates). The crystals are cubic. The compound is a coordination polymer which contains octahedral coordinated iron(II) and iron(III) ions ($Fe^{II}C_6$ and $Fe^{III}N_6$). P. is used as a blue pigment in artist's paints, in ink, blueprint paper and as a blueing agent for laundry. "***Soluble prussian blue***", $KFe[Fe(CN)_6]$, is obtained as a colloidal suspension by reaction of iron(III) salt solutions with $K_4[Fe(CN)_6]$ in equimolar amounts.

Prussiates: see Cyanoferrates.

Prussides: see Cyanoferrates.

Prynachlor: see Acylaniline herbicides.

Pschorr reaction: a ·ring-closure reaction used mainly for synthesis of phenanthrene and fluorene. The decisive reaction steps are the diazotization of a suitable aromatic amine, e.g. 2-aminodiphenylmethane for fluorene or 2-aminostilbene for phenanthrene, and the treatment of the diazonium salt with copper powder. For the synthesis of phenanthrene, the starting materials are 2-nitrobenzaldehyde and phenylacetic acid, which are condensed in a Perkin reaction. Reduction of the nitro group, diazotization of the amino group and ring closure lead to phenanthrene 9-carboxylic acid, which can be easily decarboxylated by distillation.

Diazonium salt — Cu, $-N_2, -H^+$ → Fluorene

Pseudoasymmetry: see Stereoisomerism 1.1.

Pseudoazulenes: heteroaromatic compounds which are isoelectronic with azulene. Formally, they are derived from azulene by exchange of a C=C group in the cycloheptatriene ring of the azulene molecule for a heteroatom (O, S, N-R). They are correspondingly named ***oxalenes, thialenes*** and ***azalenes***.

Oxalene

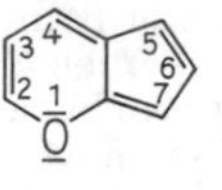

Thialene

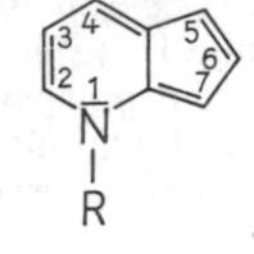

Azalene

The methods for synthesis of P. are quite different from those for azulenes. The molecular skeleton is obtained in a variety of ways, such as the reaction of quaternary salts with bases (azalenes), dehydrogenation of saturated compounds in which the ring structure is already present, or suitable condensation reactions. For example, thialenes can be synthesized in a single step from 2-mercaptobenzaldehyde and α,β-unsaturated ketones. In contrast to the azulenes, the P. are often not very stable, especially those with few substituents. In spite of this, many electrophilic substitution reactions of these compounds are known, such as bromination, nitration, Vilsmeier formylation, etc. The substitutions occur on the five-membered ring, and thus correspond to analogous reactions of the azulenes.

Some P. are of interest because they are present as structural elements in alkaloids such as semperverine, alstonine and cryptolenine, Some P. also have antitumor effects.

Pseudochalcogenides: a group of polyatomic, resonance-stabilized, divalent anions with symmetrical charge distributions which behave similarly to the chalcogen anions. Examples of the P. are the linear cyanamide ion, $[NCN]^{2-}$ and the nonlinear dicyanomethanide and tetracarbonylferrate ions, $[C(CN)_2]^{2-}$ and $[Fe(CO)_4]^{2-}$.

The concept of P. is based on close parallels in the chemical and physical properties of these ions: 1) Isosteric relationships between P. and pseudohalides corresponding to the isoelectric relationships between chalcogenides and halides: $[NCN]^{2-}$-$[NCO]^-$, $[C(CN)_2]^{2-}$-$[N(CN)_2]^-$, $[Fe(CO)_4]^{2-}$-$[Co(CO)_4]^-$; 2) existence of stable, basic, hydrogenpseudochalcogenides comparable to hydrogenchalcogenides, such as $NHCN^-$, $CH(CN)_2^-$, $HFe(CO)_4^-$; 3) acidity of the hydrogenpseudochalcogenides H_2Y: NH_2CN, $CH_2(CN)_2$, $H_2Fe(CO)_4$; 4) the high basicities of pseudochalcogenide anions Y^- which are comparable to those of the chalcogenides; 5) the existence of oxidizing molecular pseudochalcogens, such as azocarbodinitrile, NCN=NCN, tetracyanoethene, $(NC)_2C=C(CN)_2$, dodecacarbonyl triiron, $Fe_3(CO)_{12}$); 6) significant similarities in the properties of oxoanions and pseudochalcogen oxoanions, which are due in significant measure to the their conjugated bond systems; 7) the group electronegativities of the P., which are comparable to those of chalcogenides; and 8) the pronounced reactivity of the pseudochalcogenide anions Y^{2-} and YH^- with nonmetal oxides or chlorides to form pseudochalcogen oxoanions. The following equations show the parallels in behavior of oxide and cyanamide, or hydroxide and hydrogencyanamide, with respect to reaction with P_4O_{10} or $POCl_3$. $P_4O_{10} + 6\ Na_2O \rightarrow 4\ Na_3PO_4$; $P_4O_{10} + 6\ Na_2NCN \rightarrow 2\ Na_3PO_3NCN + 2\ Na_3PO_2(NCN)_2$. $POCl_3 + 6\ NaOH \rightarrow Na_3PO_4 + 3\ NaCl + 3\ H_2O$; $POCl_3 + 6\ NaNHCN \rightarrow Na_3PO(NCN)_3 + 3\ NaCl + 3\ NH_2CN$.

Pseudocumene: see Trimethylbenzene.

Pseudohalogens: see Pseudohalides.

Pseudohalides: a group of polyatomic, resonance-stabilized, monovalent anions in which the charge distribution is largely symmetrical; these anions are markedly similar to the halides in their behavior. The group includes linear anions X^- such as cyanide $[CN]^-$,fulminate $[CNO]^-$, cyanate $[NCO]^-$, thiocyanate $[NCS]^-$, selenocyanate$[NCSe]^-$, tellurocyanate $[NCTe]^-$ and azide $[NNN]^-$, nonlinear anions, such as dicyanamide $[N(CN)_2]^-$, dicyanophosphide $[P(CN)_2]^-$, tricyanomethanide $[C(CN)_3]^-$ and nitrosodicyanomethanide $[NOC(CN)_2]^-$, and anionic transition metal complexes, such as tetracarbonylcobaltate $[Co(CO)_4]^-$ and pentacarbonylmanganate $[Mn(CO)_5]^-$.

The concept of pseudohalogens is based on parallels in behavior and properties of pseudohalides and halides, of which the following are examples: 1) formation of water-insoluble silver(I), mercury(I) and lead(II) salts, AgX, Hg_2X_2 and and PhX_2; 2) the existence of ***hydrogen pseudohalide acids*** HX, such as hydrocyanic acid HCN, fulminic acid HNCO, hydrazoic acid HN_3, dicyanoketenimine (cyanoform), $HNCC(CN)_2$, hydrogen carbonylcobaltate $HCo(CO)_4$ and hydrogen carbonylmanganate $HMn(CO)_5$. These acids form ionic alkali salts, such as NaCN, KNCO, NaCNO, KNCY (Y = S, Se, Te), $NaN(CN)_2$, $KC(CN)_3$ and $NaCO(CO)_4$, covalent nonmetal derivatives, such as $P(CN)_3$, $Si(NCO)_4$ and $Si(N_3)_4$, and polymeric coordination complexes with transition metals, such as $Co(CN)_2$, $Fe(N(CN)_2)_2$ and $Ni(C(CN)_3)_2$. 3) Formation of many homogeneous and mixed pseudohalide complexes, such as $[MX_4]^{2-}$, $[MX_6]^{3-}$, $[MX_2L_2]$ and $[MX_3L_3]$. For cyanide and fulminate, coordination typically is based on the carbon, while for azide, cyanate, dicyanamide and tricyanomethanide, the ligands are typically bonded to the nitrogen. Cyanates NCY^- (Y = S, Se) and nitrosodicyanomethanide, which are ambivalent ligands, can form coordinate bonds via either nitrogen or chalcogen (S, Se or O). All P., like the halides, can act as polydentate ligands in bridge functions. 4) Reversible oxidation of some P. to the corresponding molecular ***pseudohalogens*** X_2 according to $2\ X^- \rightleftharpoons X_2 + 2e$ (X = CN, SCN, SeCN, $C(CN)_3$, $Co(CO)_4$ or $Mn(CO)_5$). 5) The existence of covalent ***interpseudohalogens*** such as cyanoazide, NC-N_3, cyanoisocyanate, NC-NCO, tetracyanomethane, NC-$C(CN)_3$ and phosphorus tricyanide, NC-$P(CN)_2$ and covalent ***halogen pseudohalides*** such as chlorocyanide, ClCN, chloroazide, ClN_3, chloroisocyanate, Cl-NCO, bromotricyanomethane, Br-$C(CN)_3$ and cobaltocarbonyl chloride, Cl-$Co(CO)_4$. 6) The group electronegativities of the P. are comparable to the halide electronegativities.

Historical. The concept of pseudohalogens goes back to L. Birckenbach (1925), and was based on thorough studies of individual P. Some examples are the synthesis of urea from ammonium cyanate by Wöhler (1828) and the introduction of the isomerism concept by Berzelius after Wöhler and Liebig had recognized in 1824 that silver fulminate and silver cyanate have the same composition.

Pseudonitrols: reaction products of secondary nitroalkanes with nitrous acid:

$$R_2CH\text{–}NO_2 + HO\text{–}N{=}O \rightarrow R_2C(NO_2)(N{=}O) + H_2O$$

In the crystalline state, the P. are in a colorless, dimeric form, but in the melt or in solution, they are intensely colored, blue or blue-green, which indicates the existence of a monomeric form.

Pseudoorder: apparent reaction order, lower than the actual order, which is observed when one reactant is present in very great stochiometric excess and its concentration is essentially constant during the period of measurement (see Kinetics of reactions). For example, for a second-order rate law $r = kc_A \cdot c_B$, if $c_{0A} \gg c_{0B}$, the initial concentration c_{0A} remains nearly unchanged during the entire reaction of component B. In this case, $k \cdot c_A$ can be set equal to k', and $r = k'c_B$. Here r is the reaction rate, k' is the pseudo-first-order rate constant and c_B is the concentration of B.

The second-order rate law has been converted to a pseudo-first-order rate law. P. is very common in solutions if one of the reactants is the solvent.

Pseudorotation: see Berry pseudorotation.

Pseudouridine, abb. ψ ***rd***, one-letter abb. ψ or Q: 5β-D-ribofuranosyluracil, a C-glycoside. P. is a component of tRNA (see Nucleic acids), where it is located in the T ψ-loop. A partial conversion of P. to 5-ribopyranosyluracil is catalysed by acid.

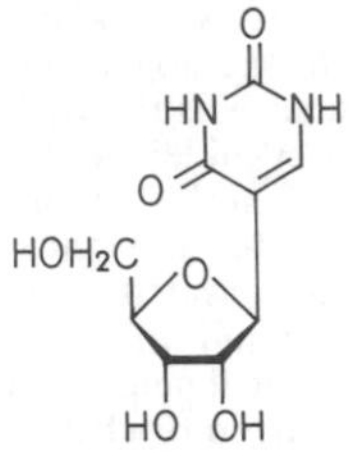

Psilocin: 2-dimethylaminoethyl-4-hydroxyindole; together with *psilocybin*, its phosphate ester, P. is the active compound in the Mexican drug teonanacatl, obtained from mushrooms of the genuses *Psilocybe, Stropharia* and *Conocybe*.

Psilocybin: see Psilocin.

Psoralene: see Methoxsalene.

Psychodysleptics: see Psychopharmaceuticals.

Psycholeptics: see Psychopharmaceuticals.

Psychopharmaceuticals: compounds which promote or inhibit higher nervous function, that is, which interfere with psychological and intellectual processes in human beings. P. affect the processing, association and storage of information, mood, affect and social behavior, etc. It is often not possible sharply to delineate the individual effects. A common classification divides these drugs into neuroleptics, tranquilizers and antidepressives. Neuroleptics and tranquilizers are sometimes lumped in the category *psycholeptics* as opposed to psychoanaleptics (psychostimulants). The psychoanaleptics include antidepressives and amphetamines.

1) *Neuroleptics (antipsychotics)* reduce the level of tension in the central nervous system, without significantly decreasing awareness and without hypnotic effect. They suppress aggressiveness, anxiety and excitation. Neuroleptics are used successfully in the treatment of true psychoses. Their side effects are sedation, blood-pressure reduction and antiemetic effects; the strength of these effects varies in different individuals. The latter are also used therapeutically. The most important classes of substances are basic, alkylated Phenothiazines (see), basic Butyrophenones (see), diphenylbutylpiperidines, such as pimozide, and Rauwolfia alkaloids (see).

2) *Tranquilizers (ataractics, psychosedatives)* have a weaker effect on emotional processes than do neuroleptics, and are used widely in non-psychotic states of excitation, tension and anxiety and related sleep disturbances. The most important classes are the Benzodiazepines (see), basic alkylated diphenylmethane derivatives, such as benactyzine, and carbamates of polyalcohols, such as meprobamate.

3) *Antidepressives* elevate the general mood and sometimes stimulate drives. *Thymoleptics* relieve or improve a depressive mood, but have little effect on normal moods. On the other hand, *thymeretics* lead to psychic stimulation and increase drive even when the mood is normal. The most commonly used thymoleptics are basic alkylated Dibenzodihydroazepines (see) and basic alkylated Dibenzodihydrocycloheptadienes (see). Thymeretics, e.g. pargyline, inhibit monoamine oxidase. This prevents degradation of the biogenic amines in the brain.

Psychodysleptics (hallucinogens) are not used in medicine. They cause temporary psychotic states in persons with normal moods. The best known of these substances are lysergic acid diethylamide, mescaline and psilocybin.

Psychosedatives: see Psychopharmaceuticals.

Pt: symbol for platinum.

PTC: abb. for Phase-transfer catalysis.

PtdCho: see Phosphatidylcholine.

Pteridine, ***pyrazino[2,3-d]pyrimidine***: a condensed heterocyclic compound. P. forms yellow leaflets; m.p. 139-140 °C, sublimes at 2.66 kPa and 125-130 °C. It is soluble in water and alcohol. P. is the parent compound of the ***pteridines***, a class of compounds found widely in nature, for example, as wing and eye pigments of insects. P. can be synthesized by condensation of glyoxal with 4,5-diaminopyrimidine.

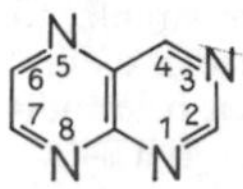

Pterin: 2-amino-4-hydroxypteridine (R^1, R^2 = H). Derivatives of P. are found in all organisms. They are biosynthesized from pyrimidine derivatives.

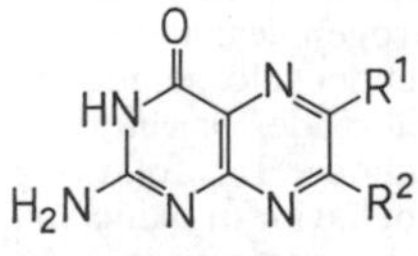

Pterin derivatives

Biopterin (R^1 = -CHOH-CHOH-CH_3; R^2 = H) is a growth factor for many microorganisms. ***Leukopterin*** (R^1 = OH; R^2 = H) is a wing pigment of the white cabbage butterfly, and ***xanthopterin*** (R^1, R^2 = OH) is found in the wings of the brimstone butterfly. P. are coenzymes of oxidoreductases. Folic acid (see) is an important derivative of P.

Pteroic acid: see Folic acid.

Pteroylglutamic acid: same as Folic acid (see).

PTFE: abb. for polytetrafluoroethylene.

PTH: abb. for parathyrin.

Pu: symbol for plutonium.

Puddled iron, ***ball iron***: iron with slag inclusions formed by the direct process and used to make steel in a Siemens-Martin or electric arc furnace.

Pulegone: *p*-menth-4(8)-en-3-one, a monocyclic, yellow liquid which smells like peppermint; b.p.$_{750}$ 224 °C. P. is contained in the oils of many labiates, usually as the d-form. It is generally obtained from pennyroyal oil, and is used to make menthol.

Pulse Fourier transform technique: see NMR spectroscopy.

Pulse polarography: see Polarography.

Pulse radiolysis: see Radiolysis, see Reaction kinetics.

Pupating hormone: see Ecdysteroids.

PUR: abb. for polyurethane.

Purification mass: see Gas-purification mass.

Purine: a bicyclic heterosystem; there are many natural P. occurring both in free form and as components of nucleosides and nucleic acids. The skeleton is ***purine (7H-imidazole[4,5-α]pyrimidine)***. This compound forms colorless needles, m.p. 217 °C. It is soluble in water and hot alcohol, and insoluble in ether and chloroform. It forms salts with acids or bases. The parent compound, which in unsubstituted form can exist in the tautomeric forms 9H-P. and 7H-P., has not yet been found free in nature.

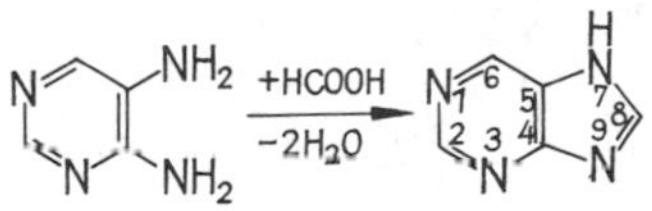

+HCOOH
−2H₂O

7H-Purine

The most important P. are adenine (6-aminopurine) and guanine (2-amino-6-hydroxypurine), which are components of nucleic acids. Adenine is also a component of adenosine diphosphate (ADP) and adenosine triphosphate (ATP), and of various coenzymes. Hypoxanthine, xanthine, uric acid and the alkaloids caffeine, theobromine and theophylline are also important. The last three compounds are present in coffee, cocoa (chocolate) and black tea.

P. can be synthesized by condensation of 4,5-diaminopyrmidine with anhydrous formic acid.

Purine alkaloids: a small group of alkaloids containing the Purine (see) skeleton. The most important of them are N-methyl derivatives of 2,6-dioxo-1,2,3,6-tetrahydropurine (xanthine). This group includes caffeine, theobromine and theophylline. Partially synthetic derivatives of xanthine, such as etofylline, and salts of natural xanthine derivatives, such as aminophylline, are used therapeutically.

Purisol process: a method for removing acidic components (carbon dioxide, hydrogen sulfide and carbon oxygen sulfide) from natural, city and synthesis gases. The P. is similar to the Sulfinol process (see); in P., the solvent for the acidic gases is N-methylpyrrolidone.

Puromycin: 6-dimethylamino-9-[3-(p-methoxy-L-β-phenylalanylamino)-3-deoxy-β-D-ribofuranosyl]-purine, a nucleoside antibiotic produced by *Streptomyces alboniger*. It forms colorless crystals; m.p. 175 to 177 °C, $[\alpha]_D^{20}$ -11° (ethanol). It is soluble in water at weakly acidic pH, and in ethanol. P. is highly effective against gram-positive bacteria, while gram-negative bacteria are only slightly inhibited. It is structurally similar to the terminus of phenylalanyl-tRNA, and is bound to the ribosomes, where it reacts to form peptidylpuromycin. Thus it blocks protein synthesis.

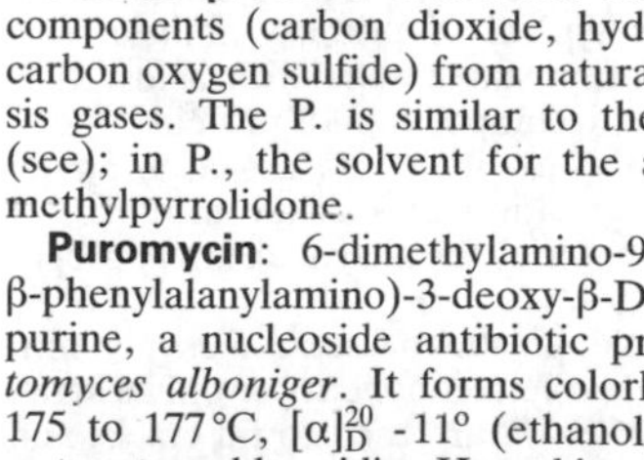

Purple of Cassius: a very stable, intensely purple colloidal suspension of gold. P. is made by reduction of very dilute gold(III) salt solutions with tin(II) chloride. The $SnO_2 \cdot xH_2O$ adsorbs the finely divided gold, forming a stable colloid. P. is used to detect gold and as a red pigment for painting glass and porcelain.

Purpurea glycosides: cardiac glycosides from the foxglove, *Digitalis purpurea*. The dried leaves contain 0.2 to 0.6% P.; about 30 different active compounds have been identified. The leaves also contain steroid compounds with no effect on the heart, the *digitanol glycosides*, and saponins such as digitonin.

The P. are extracted from cultivated varieties which yield relatively large amounts of digitoxin. The aglycons of the P. are digitoxigenin, gitoxigenin and gitaloxigenin, which are cardenolides. The most important native primary glycosides in the leaves of the plant are purpurea glycosides A and B; during drying, the terminal D-glucose groups are removed enzymatically to form the secondary glycosides digitoxin and gitoxin. In these compounds, a chain of three digitoxose groups is β-glycosidically linked to the aglycon. The digitoxose groups are linked to each other by β-(1→4) bonds. The table gives an overview of other P.

The P. are unstable compounds. As glycosides of deoxysugars, they are acid-labile, but a high percentage survives passage of the acid stomach. The compounds with 16β-O-formyl structures are especially readily cleaved. In acid, the 14β-OH group is also split off, together with a neighboring H atom, to form anhydro compounds. In the compounds derived from gitoxigenin, the 16β-OH group is also split off, forming a conjugated double bond system (dianhydro compounds). In a basic medium, the compounds can undergo isomerization in which the lactone ring is opened and a new ring is formed by incorporation of the 14β-O atom. The resulting iso compounds have no therapeutic effect. The P. and other cardenolides, such as the Lanata glycosides (see), are identified by their reactions in alkaline media with aromatic di- or trinitro compounds to form colored charge-transfer complexes. In this case, it is the activated methylene group of the lactone ring which reacts. The Baljet test uses picric acid, the Raymon test, 1,3-dinitrobenzene, and the Kedde test, 3,5-dinitrobenzoic acid. The compounds with 2-deoxysugars give red colors with xanthydrol or dixanthylurea. The compounds are identified and tested for purity by thin-layer chromatography on formamide-impregnated silica-gel plates.

Only *digitoxin* is therapeutically important; after oral administration, it is more than 80% absorbed. The daily elimination rate is less than 10%; therefore an overdose can easily lead to accumulation and intoxication. Digitoxin, like all cardiac glycosides, is used in cases of cardiac insufficiency. It increases the strength of the contractions and reduces their rate, and makes the heart less responsive to stimulus. This effect is attributed to a partial inhibition of the Mg^{2+}-dependent Na^+-K^+-ATPase system of the heart cell membranes.

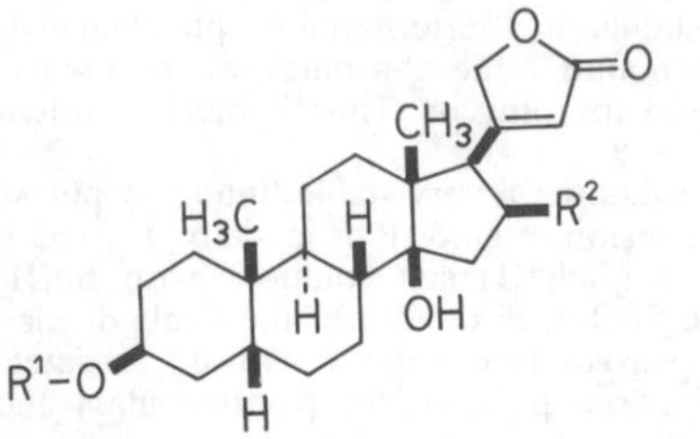

Name	R[1]	R[2]
Digitoxigenin	H	H
Gitoxigenin	H	OH
Gitaloxigenin	H	–O–C(=O)H
Digitoxin	–Digitoxose–Digitoxose–Digitoxose	H
Gitoxin	–Digitoxose–Digitoxose–Digitoxose	OH
Gitaloxin	–Digitoxose–Digitoxose–Digitoxose	–O–C(=O)H
Purpureaglycoside A	–Digitoxose–Digitoxose–Digitoxose–Glucose	H
Purpureaglycoside B	–Digitoxose–Digitoxose–Digitoxose–Glucose	OH
Glucogitaloxin	–Digitoxose–Digitoxose–Digitoxose–Glucose	–O–C(=O)H
Odoroside H	–Digitalose	H
Strospeside	–Digitalose	OH
Verodoxin	–Digitalose	–O–C(=O)H
Digitalinum verum	–Digitalose–Glucose	OH
Glucoverodoxin	–Digitalose–Glucose	–O–C(=O)H

Purpuric acid: see Murexide.

Purpurin: 1,2,4-trihydroxyanthraquinone, a synthetic pigment which dissolves readily in alkalies, giving a carmine red solution; m.p. 253-256 °C. P. is formed by oxidation of alizarin with manganese dioxide and sulfuric acid. Its properties as a dye are similar to those of alizarin, and it is used as a nuclear stain in histology. P. is found together with alizarin in madder roots.

Purpurogallin: trihydroxybenz-α-tropolone, a natural tropolone pigment. P. crystallizes in brick-red needles; m.p. 257 °C. It is relatively insoluble in most solvents. It can be obtained by oxidation of pyrogallol with potassium hexacyanoferrate(II), hydrogen peroxide and peroxidases.

Push-pull systems: molecules containing at least one atom or group with electron-pair-donor (EPD) properties and at least one with electron-pair-acceptor (EPA) properties. For example, the merocyanins can be considered push-pull polyenes. Donor groups can be, for example, substituents with +I and +M effects (e.g. $-NH_2$, -OH), while acceptors could be substituents with -I and -M effects (e.g. $-NO_2$, -CHO).

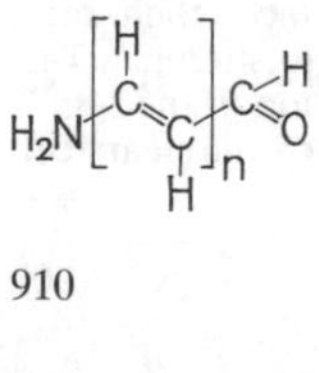

Putrescine, ***1,4-diaminobutane, tetramethylenediamine***: $H_2N\text{-}(CH_2)_4\text{-}NH_2$, an aliphatic, biogenic diamine (see Biogenic amines). P. is a colorless crystalline compound with an unpleasant odor; m.p. 27-28 °C, b.p. 158-159 °C, n_D^{20} 1.4969. It is soluble in water and alcohol, and barely soluble in ether. P. undergoes the reactions typical of primary amines, is a strong base and forms stable salts with acids. It is formed by bacterial decomposition of proteins. It can be synthesized from 1,4-dihalobutane via an amide stage, or by hydrogenation of succinic nitrile. It is used as a starting material for pharmaceutical preparations.

Putty: a kneadable, plastic to viscous mass containing fillers, but no volatile components or only a small percentage of volatiles. P. are used to fill cracks and holes, and less often, as glues. They harden mainly through polymerization or oxidation of the binder. Finely ground mineral sands (glass, quartz, shale, whiting, talc, feldspar or barite) or organic materials such as sawdust or ground rubber, or plastic powder, are used as fillers. P. are classified as inorganic or organic on the basis of the binder used in them.

1) Inorganic P.: ***Waterglass P.*** are resistant to all acids (except hydrofluoric), chlorine, chlorine bleaches, etc. up to a temperature of 1000 °C. ***Metal P.*** consist of zinc oxide, ZnO, antimony trioxide, Sb_2O_3, and manganese dioxide, MnO_2. ***Magnesia P.*** are made by mixing 2 parts magnesium oxide, MgO,

with one 1 part magnesium chloride, $MgCl_2$, and water; after warming, it hardens within a few hours. These P. are not resistant to water or heat. ***Zinc oxide P.*** consist of zinc oxide, ZnO, zinc chloride, $ZnCl_2$ and water or phosphric acid. They are stable in the presence of water and organic solvents, and are used in dental chemistry. ***Iron-rust cement*** consists of iron powder and weak acids or bases, which solidify by formation of the oxide (rusting). ***Borax P.*** are made from molten sodium borate, quartz powder, red lead and litharge.

2) Organic P.: ***Bitumen P.*** are resistant to water, bases and acids up to 100 °C. ***Resin P.*** consist of epoxide, phenol, formaldehyde, furan and silicon resins. They are often resistant to hydrofluoric acid. ***Oil putty*** is made of linseed oil and whiting, or litharge, etc. and used for windows. ***Glue P.*** consist of a bone glue or casein, chalk and sawdust. They are used mainly on untreated wood. ***Rubber P.*** consist of natural or synthetic rubber, organic solvents, lead compounds and fillers. They are used as rapidly hardening spackle which is highly elastic and resistant to water, acids and bases. ***Glyerol P.*** are mixtures of glycerol and metal oxides such as lead oxide, PbO. These harden in 10 to 30 minutes to hard, firmly adhering solids. They can bind different types of materials to each other, and are resistant to gaseous chlorine and nearly all acids and bases at moderate temperatures. ***Polychloronaphthalene P.*** are used in electrical devices because of their good insulating properties.

PVAC: abb. for polyvinyl acetate.

PVAL: abb. for polyvinyl alcohol.

PVC: abb. for polyvinyl chloride.

PVDF: abb. for polyvinylidene fluoride.

PVF: abb. for polyvinyl fluoride.

PVF$_2$: abb. for polyvinylidene fluoride.

PVK: abb. for polyvinylcarbazene.

P-value: in water chemistry, the amount of 0.1 M HCl required to reach the phenolphthalein end-point (in ml per 100 ml sample). Multiplication of the p-value by 2.8 gives the phenolphthalein alkalinity (PA); multiplication by 40 gives the alkalinity number (AN). In modern high-pressure and forced-circulation steam generators, the p-value required for protective alkalinity cannot exceed 0.1. See m-value.

Pyramine®: see Herbicides.

Pyran: a six-membered, heterocyclic compound with one oxygen atom and two C=C double bonds in the ring. Depending on the position of the double bonds, the compound is called ***2H-P.*** (***α-P.***) or ***4H-P.*** (***γ-P.***).

2H-Pyran

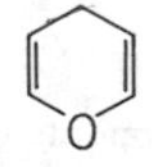

4H-Pyran

Only 4H-pyran is known to exist as such. It is a colorless oil, b.p. 80 °C, n_D^{20} 1.4550, which is soluble in alcohol, ether and bezene. Some important derivatives of the P. are the Pyrones (see), the chromenes (see Benzopyrans), Xanthene (see) and the Pyrylium salts (see).

2H-pyran-2-one: see Pyrone.

4H-pyran-4-one: see Pyrone.

4H-pyran-4-one-2,6-dicarboxylic acid: same as Chelidonic acid (see).

Pyranose: see Monosaccharides.

Pyrantel: an Anthelminthic (see).

Pyrazinamide: a compound used as a tuberculostatic.

Pyrazine, ***1,4-diazine***: crystals smell like pyridine; m.p. 54 °C, b.p. 115-116 °C, n_D^{61} 1.4953. P. is readily soluble in water, alcohol and ether, and is volatile with steam. It is obtained by catalytic dehydrogenation of piperazine. Derivatives of P. are synthesized by autocondensation of α-aminoketones or by condensation of 1,2-diamines with 1,2-dicarbonyl compounds and oxidation of the resulting dihydropyrazines. The pyrazine skeleton is present in many natural products, and pigments (azine pigments). Supstituted P. are found in dried mushrooms, the roasting aroma of coffee and cocoa, in peanuts, potato chips and rye crisp.

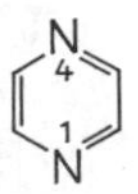

Pyrazino[2,3]pyrimidine: same as Pteridine (see).

Pyrazole, ***1,2-diazole***: colorless crystals with a pyridine-like odor; m.p. 70 °C, b.p. 186-188 °C. It is readily soluble in water, alcohol and ether. P. is weakly basic and is distinctly aromatic.

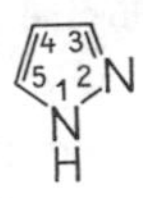

It is very stable in the presence or acids and oxidizing agents. When P. is reduced with sodium alcoholate, pyrazoline is formed; complete hydrogenation leads to pyrazolidine. Due to intermolecular hydrogen bonding, P. exists as dimers.

P. is made by condensation of hydrazine with propargylaldehyde acetal, or by addition of diazomethane to acetylene. The first natural pyrazole derivative to be isolated was the amino acid pyrazolylalanine from watermelon seeds. P. is also the parent compound of a series of antipyretic and antirheumatic drugs. Derivatives of P. are used as coupling components for the production of azo pigments.

Pyrazolines: the dihydro compounds of pyrazole. The position of the remaining double bond is indicated by a numerical prefix; the three possible tautomeric forms are 1-, 2- and 3-P. Derivatives of all three are known, and it has been shown by spectroscopic studies that 2-P. (4,5-dihydropyrazole) is the stable form, because of its monoimino character.

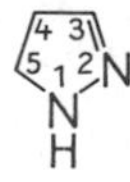

2-Pyrazoline

2-P. is a colorless liquid, unstable in air, with a cocoa-like smell; b.p. 144 °C, n_D^{17} 1.4796. It is readily soluble in water, alcohol and ether, and it is volatile in steam. 2-P. is made by reaction of diazomethane with ethylene, or of acrolein with hydrazine, or by reduction of pyrazole with sodium alcoholate. In general, the P. are stronger bases and more readily substituted and oxidized than pyrazole. The oxo derivative of 2-P., 2-pyrazolin-5-one, is used therapeutically.

2-Pyrazolin-5-one, ***pyrazolone***: the oxo derivative of 2-pyrazoline. 2-P. crystals are colorless needles, m.p. 165 °C, readily soluble in water and alcohol, and insoluble in ether. 2-P. is made by reaction of formylacetic acid ethyl ester with hydrazine sulfate. Some reactions of 2-P. can only be explained by tautomerism. For example, the coupling with diazonium salts occurs in the 4-position, while reaction with methyl iodide or dimethylsulfate leads to mixtures of O- and N-methylated products. 2-P. is the basic unit of a large group of antipyretics and antineuralgics, such as phenazone and aminophenazone, and of some industrially important Pyrazolone pigments (see).

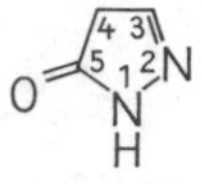

Pyrazolone: same as 2-pyrazolin-5-one.

Pyrazolone pigments: a group of azo pigments which contain phenylmethylpyrazolone or another pyrazolone derivative as a developing element. Some examples are eriochrome red (Fig.) and tartrazine. The P. are used in the paint industry and for chrome dying of wool.

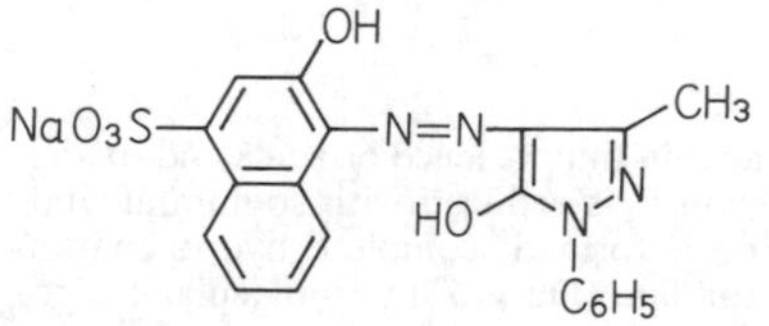

Pyrazolones: derivatives of pyrazol-5-one (see 2-Pyrazolin-5-one, Fig.) used as analgesics. The first pyrazolone derivative introduced as a weak analgesic was *phenazone*, m.p. 113 °C. It is obtained by condensation of ethyl acetoacetate, CH_3COCH_2-$COOC_2H_5$, with phenylhydrazine, $C_6H_5NHNH_2$, followed by methylation. Phenazone is quite soluble in water. However, because of its relatively weak analgesic effect and its side effects, it has been largely replaced by *aminophenazone*, m.p. 109 °C. The latter is produced from phenazone by nitrosylation, reduction of the nitroso group in the 4-position to an amino group, and dimethylation of this amino group. This compound is less water soluble than phenazone, and its use has been limited or forbidden in some countries because of possible side effects, such as a-granulocytosis and formation of alkylnitrosoamines.

A very water-soluble derivative is *analgin*, which is formed by condensation of 4-methylaminophenazone with formaldehyde and sodium hydrogensulfite. In acidic media, analgin is decomposed to 4-methylaminophenazone, formaldehyde and SO_2.

In *propylphenazone*, the dimethylamino group is replaced by an isopropyl group. This compound has a weaker effect than aminophenazone, but it is not likely to form alkylnitrosoamines.

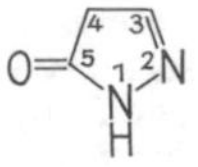

Pyrazol-5-one

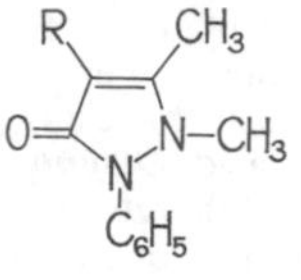

R = H : Phenazone
R = $N(CH_3)_2$: Aminophenazone
R = $N(CH_3)$–CH_2–SO_3^- Na^+ : Analgin
R = $CH(CH_3)_2$: Propylphenazone

P. may be identified by color reactions. Phenazone forms a green compound, 4-nitrosophenazone (4-nitrosoantipyrine), with nitrite in acidic solution. Aminophenazone gives a blue-violet color. Iron(III) chloride reacts with aqueous phenazone solution to form a dark red solution; aminophenazone and analgin give unstable blue colors. If an aqueous solution of aminophenazone reacts with silver nitrate, a violet color appears, and after a short time, metallic silver precipitates.

Pyrazophos: see Fungicides.

Pyrene: a condensed aromatic hydrocarbon. P. forms yellow, sublimable crystals with a blue fluorescence; m.p. 156 °C, b.p. 393 °C. It is insoluble in water, slightly soluble in boiling ethanol, and readily soluble in ether, benzene, toluene and carbon disulfide. P. is able to undergo electrophilic substitution reactions. The first reaction occurs at position 3, and the second substitution at position 8 or 10. P. is found in coal tar and is extracted from it with carbon disulfide. It is purified via the picrate, and is used in the synthesis of pigments.

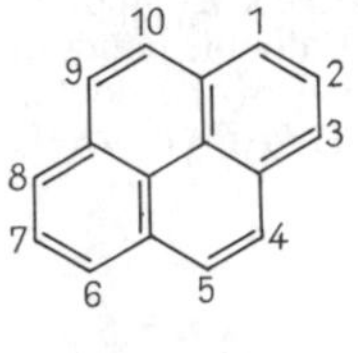

Pyrethrins: insecticidal components of Pyrethrum (see). The most important P. occurring as plant natural products are the optically active esters of (+)-*trans*-chrysanthemic acid or (+)-*trans*-pyrethrinic acid with (+)-pyrethrolone or (+)-cinerols (Table). The absolute configuration in all esters is 1*R*, 3*R*, 4*S*. The double bond in the side chain of the alcohol portion is *cis*; the one in the carboxylic acid part is *trans*. The P. are viscous oils which can be distilled in vacuum. Their contact with insects leads to a "knock-down" effect.

Table. The most important pyrethrins

Name	Formula
Pyrethrin I $R = CH_3$; Pyrethrin II $R = CH_3{-}O{-}CO{-}$	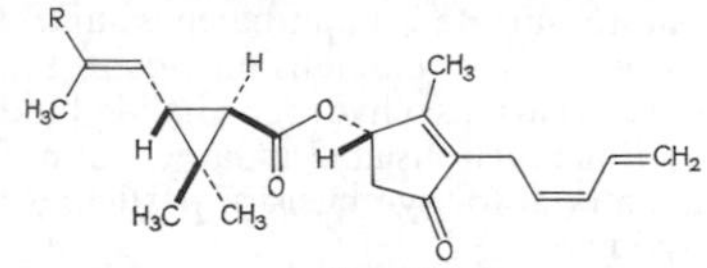 Pyrethrolone
Cinerin I $R = CH_3$; Cinerin II $R = CH_3{-}O{-}CO{-}$	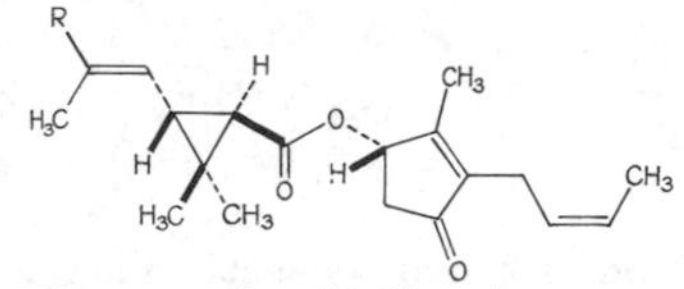 Chrysamemic acid Cinerolone

Pyrethroids: synthetic analogs of the pyrethrum insecticides. The high production costs and lack of stability of the Pyrethrins (see) stimulated efforts to discover molecular structures which are more readily synthesized and retain the low toxicity to warm-blooded animals, but have higher stability and greater insecticidal effects. The most important compounds of this type are listed in the table; most are produced industrially.

The first commercialized P. (1950) was *allethrin*, the allyl homolog of the natural cinerin I (see Pyrethrins, table). The more easily synthesized chrysanthemic esters *tetramethrin*, *prothrin* and *resmethrin* have very low toxicity for warm-blooded animals, and their insecticidal effects are equal or even considerably better than those of the natural products.

If the isobut-1-enyl group of chrysanthemic acid is replaced by a dihalogen vinyl group, P. are synthesized which are not only more stable to light but much more active (*permethrin, cypermethrin, decamethrin*). These P. are contact and oral poisons, and even in the open are effective for several weeks. They are used both in hygiene and in crop protection. *Decamethrin* has more than 1000 times the activity of natural pyrethrin and is produced in the optically active D,*cis* form ($Decis_8$). It is one of the most effective insecticides known, and is applied to fields at only 10 to 20g per hectar. In spite of its elevated acute toxicity (oral LD_{50} = 25 to 60 mg/kg rat), this product permits a radical reduction of environmental effects in comparison to traditional insecticides. In industrial synthesis, which is partly stereospecific, the corresponding cyclopropane carboxylic acids are esterified with m-phenoxybenzyl alcohols which have α-cyano groups as required. Industrially tested synthetic processes are available for both ester components.

In *fenvalerates* and *fluovalinates*, the cyclopropanecarboxylic acid group is replaced by structurally and sterically similar substituted carboxylic acids. This somewhat reduces the insecticidal effectivity, but the action spectrum and the use properties of these P. are several times better than those of the natural pyrethrins.

Table. The most important pyrethroids

Name	Formel
Allethrin	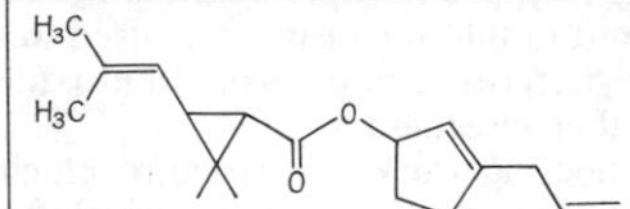
Tetramethrin	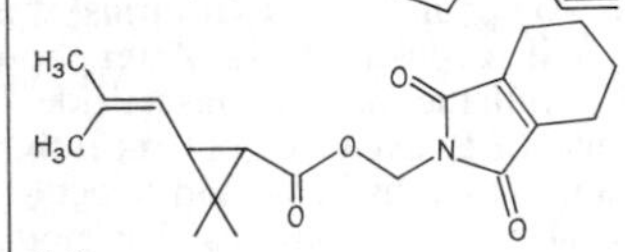
Prothrin	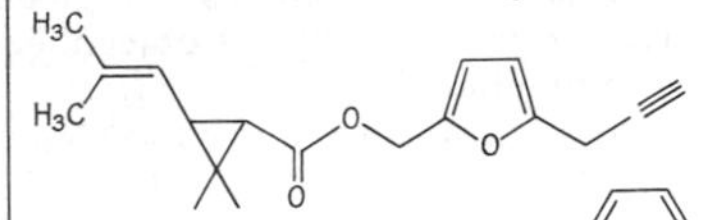
Resmethrin	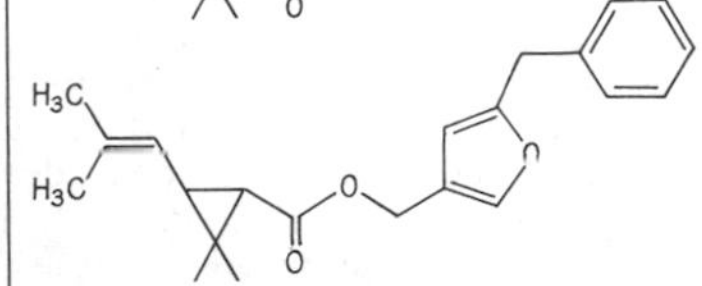
Permethrin	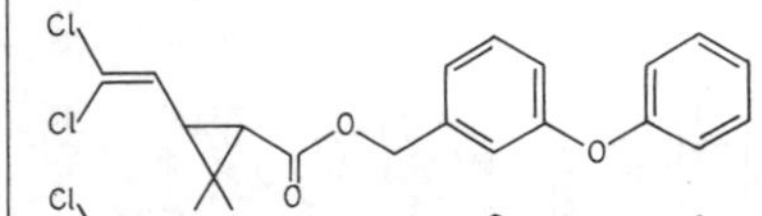
Cypermethrin	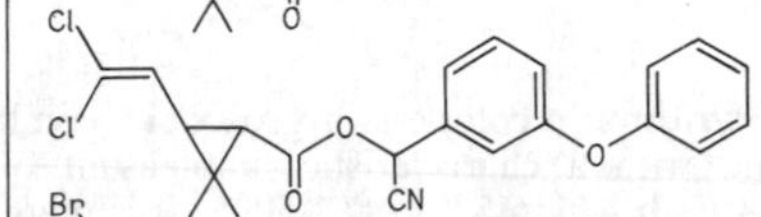
Decamethrin	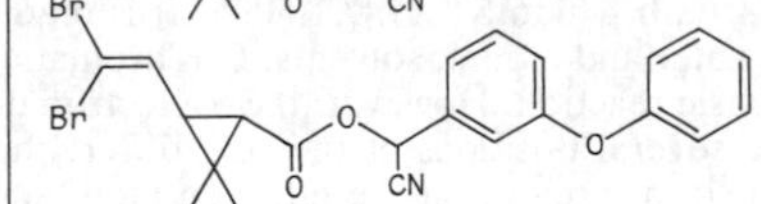
Fenvalerate	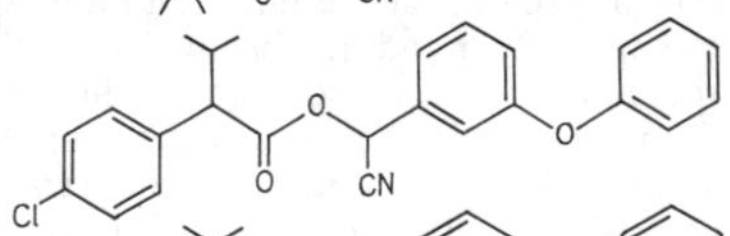
Fluvalinate	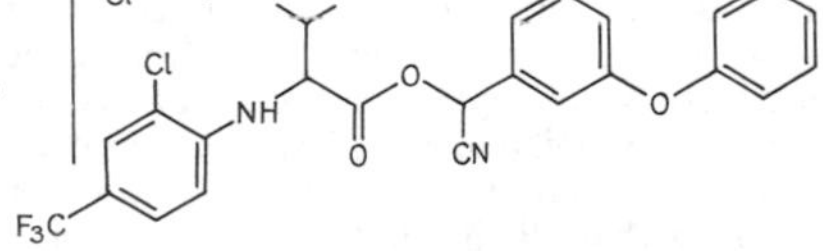

Pyrethrum: a mixture of insecticidal substances, the Pyrethrins (see) which are obtained by extraction from the blossoms of various chrysanthemums; it is sold as a concentrate with about 25% active substance. The crude extract is a dark brown, viscous oil which can be distilled under vacuum with decomposition. In the presence of light and air, or of alkalies, P. is oxidized and inactivated. The addition of antioxidants and synergists increases stability and storage life, as well as insecticidal activity and length of effectiveness. In addition to extraction, it is common to dry and grind the harvested blossoms, which contain 0.3 to 2% pyrethrins.

The extracts and powder are used in insecticidal preparations, mainly for use against flies, mosquitoes, fleas, ticks, lice, etc. Their use on field plantings is limited because of their short effective life under field conditions.

P. is a contact insecticide. Advantages are its low toxicity to warm-blooded animals (oral LD_{50} = 570 to 1500 mg/kg rat), and its rapid degradation in the environment and in the organism. P. is used in combination with synergists, e.g. piperonyl butoxide, and often with other insecticides.

The pioneering work on structure elucidation of the P. active compounds (see Pyrethrins) was done by Staudinger and Ruzicka (1924). After the discovery of modern synthetic organic insecticides, such as DDT, the use of P. declined, but recently it has increased again, due to hygienic and toxicological considerations. The pyrethrins are used as models for the synthesis of environmentally less objectionable insecticides (see Pyrethroids).

Pyridazine, ***1,2-diazine***: a colorless liquid with an odor similar to that of pyridine; m.p. ≈ 8 °C, b.p. 208 °C, n_D^{20} 1.5218. P. is readily soluble in water, alcohol and ether. It is a weak base and forms salts with acids. The ring is opened by oxidizing agents only under extreme conditions. P. is made by decarboxylation of pyridazine-4,5-dicarboxylic acid. Pyridazinones are of interest as herbicides.

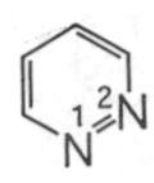

Pyridine: a colorless, hygroscopic, combustible liquid with a characteristic, unpleasant odor; m.p. -42 °C, b.p. 115.5 °C, n_D^{20} 1.5095. It is readily soluble in water and organic solvents. P. is aromatic and gives a basic reaction. However, the π-electron densities at the several positions of the ring differ; the electron density is greatest at the nitrogen atom and smallest at positions 2, 6 and 4. This is due to the fact that nitrogen is more electronegative than carbon. The pyridine molecule therefore has a dipole moment, and this affects its ability to undergo substitution reactions. Electrophilic substitutions occur very slowly and require extreme reaction conditions; halogenation, sulfonation and nitration occur in position 3. Nucleophilic substitution reactions occur more readily, especially in positions 2, 4 and 6. Thus the reaction with sodium amide yields 2-aminopyridine and 2,6-diaminopyridine (see Čičibabin reaction). Because of this property, P. is regarded as the parent compound of the π-deficiency heterocycles. Nitrobenzene displays similar reactivity; that is, the ring nitrogen affects the reactivity of P. in the same general way that the -M effect of the nitro group affects that of the benzene ring. P. is an electron-pair donor, and readily forms donor-acceptor complexes, e.g. with sulfur trioxide, zinc or copper chloride. It is a good solvent for organic compounds; with acids and alkylating reagents it forms pyridinium salts. Oxidation with hydrogen peroxide yields pyridine *N*-oxide.

P. is found in coal and bone tars, pyrogenous oils, shale oils, coffee oil, city gas and technical pentanol. It is obtained industrially from coal tar, by extraction with dilute sulfuric acid, precipitation with alkalies and distillation. It can also be synthesized from a mixture of formaldehyde/acetaldehyde and ammonia. Many natural products are derivatives of P., including coniine, piperine, nicotine, tropine and cocaine.

P. is used as a solvent, e.g. for anhydrous inorganic salts and many organic compounds. It is also used for determining the soluble components of hard coal, to separate nitrated naphthalenes and to purify anthracene, as a condensing agent for synthesis of phenol resins, as a hydrogen halide binding reagent in acylation reactions and, in large amounts, as the starting material for synthesis of pesticides based on 2,2'-bipyridine.

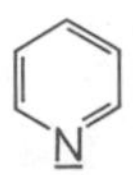

Pyridine-2-carboxylic acid: same as Picolinic acid (see).

Pyridine-3-carboxylic acid: same as Nicotinic acid (see).

Pyridine-4-carboxylic acid: same as Isonicotinic acid (see).

Pyridine-2,3-dicarboxylic acid: same as Quinolinic acid (see).

Pyridine-3,4-dicarboxylic acid: same as Cinchomeronic acid (see).

Pyridine-2,5-dicarboxylic acid di-n-propyl ester: an Insect repellent (see).

Pyridinium salts: quaternary salts of pyridine formed by addition of acids, alkyl halides, dialky sulfates, sulfonic acid esters and acyl chlorides to pyridine. They are a type of onium compound. Most crystallize well and are reactive in alkaline solution. Because of the strong (-)I effect of the quaternary nitrogen atom, methyl-substituted *N*-alkylpyridinium salts can undergo aldol reactions. *N*-substituted P. are split by sodium hydroxide solution into 5-hydroxypenta-2,4-dienal, which can condense with *N*-methylaniline to form 5-(*N*-methylanilino)penta-2,4-dienal (Zincke aldehyde); this can condense with cyclopenta-2,4-diene to form azulene.

Pyridine N-oxide, ***pyridine l-oxide***: colorless, deliquescent crystals; m.p. 65-66 °C, b.p. 146-147 °C at 1.73 kPa. It is formed by oxidation of pyridine with per acids or 30% hydrogen peroxide. P. is able to undergo nucleophilic substitutions in positions 2 and 4, and – in contrast to pyridine – electrophilic substitutions in position 4. The reaction of nitrating acid with P. gives an 85% yield of 4-nitropyridine *N*-oxide, which can be converted by treatment with phosphorus(III) chloride in chloroform to 4-nitropyridine. If the oxygen is removed by reductive cleavage, e.g. with Raney nickel in glacial acetic acid/acetic anhydride, the nitro group is simultaneously reduced, forming 4-aminopyridine. The reaction of 4-nitropyridine *N*-oxide with acyl halides yields 4-halopyridine *N*-oxides. These can be reduced, and in this way it is possible to make 4-substituted pyridine derivatives, which are otherwise difficult to synthesize.

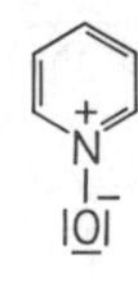

Pyridostigmine: see Parasympathicomimetics.
Pyridoxal: see Vitamin B_6.
Pyridoxal phosphate: see Coenzymes.
Pyridoxamine: see Vitamin B_6.
Pyridoxine: same as Vitamin B_6 (see).
Pyridoxol: same as Vitamin B_6 (see).

Pyrimethamine: a diaminopyrimidine derivative used as an Antimalarial agent (see). It kills schizonts and is used to prevent malaria; sometimes it is used with a sulfonamide. P. inhibits dihydrofolic acid reductase (see Trimethoprim).

Pyrimidine, ***1,3-diazine***: colorless crystals with a characteristic odor; m.p. 22 °C, b.p. 123-124 °C, n_D^{20} 1.4998. It is soluble in water, alcohol and ether, and forms salts with inorganic acids. The two possible Kekulé structures of P. have the same energy. Electrophilic and nucleophilic substitution reactions with P. are very slow. There are many syntheses of P., for example, reaction of barbituric acid with phosphorus oxygen chloride and reduction of the 2,4,6-trichloropyrimidine with zinc powder. P. is found as a component of many natural products, including the cleavage products of nucleic acids (uracil, thymine and cytosine), plant bases (pteridines, purines) and in vitamins B_1 and B_2. The biological degradation of P. leads to β-alanine. Some pharmacologically important derivatives are Barbituric acid (see) and alloxane. P. is used in the synthesis of pharmaceuticals, such as sulfonamides. It was first synthesized by the reaction of acetoacetate with amidines (Pinner, 1856).

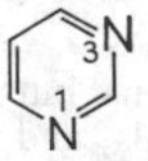

Pyrimidine fungicides: Systemic Fungicides (see) with pyrimidine skeletons. There are two types of P., the *ethirimol* and the *fenarimol* types.

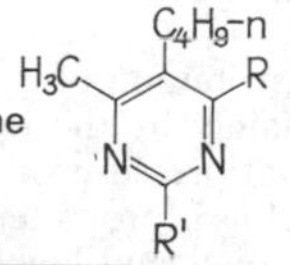

Table. Ethirimol-type pyrimidine fungicides

Name	R	R'
Ethirimol	OH	NHC_2H_5
Dimethirimol	OH	$N(CH_3)_2$
Bupirimate	$OSO_2N(CH_3)_2$	NHC_2H_5

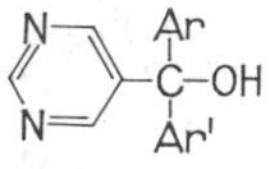

Table. Fenarimol-type pyrimidine fungicides

Name	Ar	Ar'
Fenarimol	2-Chlorophenyl	4-Chlorophenyl
Nuarimol	2-Chlorophenyl	4-Fluorophenyl
Triarimol	Phenyl	2,4-Dichlorophenyl

Pyrinuron: a Rodenticide (see).
Pyrithyldione: see Hypnotics.

Pyrocatechol, ***1,2-dihydroxybenzene***: a divalent phenol. P. crystallizes in colorless and odorless needles which sublime; m.p. 105 °C, b.p. 245 °C. It is readily soluble in water, ethanol and ether. In alkaline solution, it is strongly reducing; for example, it reduces an ammoniacal silver salt solution.

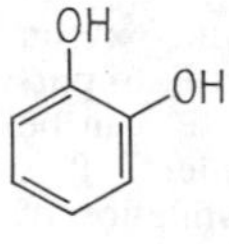

Pyrocatechol

P. gives a characteristic, emerald-green color reaction with iron(III) hydroxide in dilute solution. When alkali hydroxide solution is added, the color changes to dark red or violet. P. is poisonous and bacteriocidal. It is obtained in relatively large amounts from the products of brown coal carbonization, and can be separated from them. It is also found widely in the plant kingdom, and can be obtained by dry distillation (pyrolysis) of natural products such as wood, charcoal or lignin. It is synthesized by alkali fusion of phenol-2-sulfonic acid, hydrolysis of 2-chlorophenol with sodium hydroxide solution at a high temperature in the presence of copper(II) sulfate, or by oxidative degradation of salicylaldehyde with hydrogen peroxide and sodium hydroxide solution (see Dakin reaction). P. is very reactive (see Phenols). Methylation of one hydroxyl group produces the acidic ether guaiacol, and methylation of both, the neutral ether veratrol. It can be oxidized with mild oxidizing agents to 1,2-benzoquinone. P. is therefore used as a photographic developer. It is also used as a starting material for hair dyes, tannins and pharmaceutical preparations.

Pyrocatechol monomethyl ether: same as Guaiacol (see).

Pyroelectricity: the appearance of electric charges on opposing surfaces of certain crystals when they are evenly heated. The phenomenon was discovered on tourmaline, and can also be observed when the crystal is cooled (with a reversal of the signs of the charges.) P. can occur only in crystals which lack centers of symmetry, and can be observed in 10 of the 32 classes of crystals. The cause of P. is closely related to that of piezoelectricity.

The ***pyroelectric effect*** consists of two parts. The primary or actual pyroelectric effect is due to a change in dipole moments in response to the change in temperature (changes in the spontaneous polarization). The secondary pyroelectric effect is due to changes in charge densities caused by thermal expansion. Substances with large pyroelectric effects include triglycine sulfate, $(NH_2CH_2COOH)_3 \cdot H_2SO_4$, lithium germanate, Li_2GeO_3, barium titanate, $BaTiO_3$ and lead germanate, $Pb_5Ge_3O_{11}$.

Pyroelectric crystals are finding many applications, since they can be used to convert temperature changes into electric signals. They are used, for example, to make extremely sensitive radiation detectors (especially IR detectors).

Pyrogallol, ***1,2,3-trihydroxybenzene, benzene-1,2,3-triol***: a triphenol in which the hydroxyl groups are vicinal. P. forms colorless needles; m.p. 133-134 °C, b.p. 309 °C. It is soluble in water, ethanol and ether, gives a red reaction with iron(III) chloride, and is a strong reducing agent. In alkaline solution, it very rapidly absorbs oxygen, and is therefore used in gas analysis to absorb oxygen from mixtures of gases. P. is a blood poison, because it converts hemoglobin to methemoglobin and thus interferes with oxygen transport. Various oxidizing agents convert P. to purpurogallin (a benzotropolone derivative). P. can be considered a component of anthocyan pigments. P. is formed by decarboxylation during dry distillation of gallic acid. It is obtained by distilling a mixture of gallic acid and pumice. P. is used as a photographic developer, as an oxygen absorbant in gas analysis, in the synthesis of pigments and as a drug for skin diseases.

Pyrogallophthalein: same as Gallein (see).
Pyrogel: see Napalm.
Pyrogen indigo: see Sulfur pigments.
Pyroceramics: see Vitroceramics.
Pyrolysis: a method of petroleum processing yielding lower alkenes (olefins); it is one of the most important methods of producing primary petrochemicals.

Growth of demand for lower olefins in billion tons

Product	1970	1975	1980	1985
Ethene	17.0	35.0	55.0	62.0
Propene	8.0	13.0	21.0	27.0
Butadiene	2.9	4.0	6.0	9.3
Isoprene	0.25	0.4	1.0	2.1

1) Medium temperature P. In principle, all hydrocarbons with two or more carbon atoms are capable of producing lower olefins, especially ethene, propene, butene and butadiene. The higher the hydrogen content of the starting material, the greater the yield of olefins; therefore, ethene, propane and butane are the best starting materials, and these are used in the former USSR and USA, where natural gas is abundant. In most countries, however, light gasoline is used because natural gas is not available in sufficient quantities. The use of higher petroleum fractions (e.g. gas oil or heavy gasoline) is always associated with loss of lower olefins (Fig. 1).

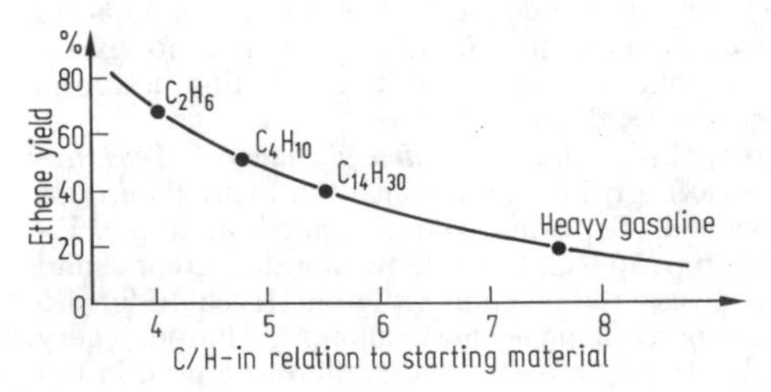

Fig. 1. Ethene yields as a function of C/H mass ratio of the starting material.

So far, there is no satisfactory industrial process for P. of crude oil.

Mechanism. Above 700 °C, the rate of decomposition of a paraffin hydrocarbon into two radicals, $R_1\text{-}CH_2\text{-}CH_2\text{-}CH_2\text{-}CH_2\text{-}R_2 \rightarrow R_1\text{-}CH_2\text{-}CH_2\text{-}CH_2\text{-}CH_2\cdot + R_2\cdot$, becomes high enough for industrial purposes. Fragmentation of the radicals leads to an olefin (e.g. ethene) and a correspondingly smaller radical (β-*cleavage*): $R_1\text{-}CH_2\text{-}CH_2\text{-}CH_2\text{-}CH_2\cdot \rightarrow R\text{-}CH_2\text{-}CH_2\cdot + CH_2{=}CH_2$. Reactions among the products can produce other olefins: $R_1\text{-}CH_2\text{-}CH_2\cdot + CH_2{=}CH_2 \rightarrow R_1\text{-}CH_2\text{-}CH_2\text{-}CH_2\text{-}CH_2\cdot \rightarrow R_1\text{-}CH_2\cdot + CH_2{=}CH\text{-}CH_3 \rightarrow R_1\cdot + CH_2{=}CH\text{-}CH_2\text{-}CH_3$. Unimolecular β-cleavage of a radical can occur several times, until a radical is formed which cannot undergo further β-cleavage, for example:

$$CH_3\text{-}CH(CH_3)\text{-}CH_2\text{-}CH_2\cdot \rightarrow CH_3\text{-}CH(CH_3)\cdot + CH_2{=}CH_2 \rightarrow CH_3\cdot + CH_2{=}CH_2$$

The very reactive methyl radical undergoes *bimolecular C-H cleavage*: $CH_3\cdot + R_1\text{-}CH_2\text{-}CH_2\text{-}CH_2\text{-}R_2 \rightarrow CH_4 + R_1\text{-}CH\cdot\text{-}CH_2\text{-}CH_2\text{-}CH_2\text{-}R_2$.

Unimolecular β-cleavage and bimoleclar C-H cleavage are the carriers of the chain reaction. It is terminated by disproportionation and radical recombinations, e.g. $R_1\cdot + R_2\text{-}CH_2\cdot \rightarrow R_1\text{-}CH_2\text{-}R_2$; $R_1\text{-}CH_2\text{-}CH_2\cdot + R_1\text{-}CH_2\text{-}CH_2\cdot \rightarrow R_1\text{-}CH{=}CH_2 + R_1\text{-}CH_2\text{-}CH_3$.

Benzene is not cleaved under these conditions, but benzene derivatives give rise to benzyl radicals, and these react to form undesirable condensation products, such as: $C_6H_5\text{-}CH_2R \rightarrow C_6H_5\text{-}CH_2\cdot + R\cdot$, $2\,C_6H_5\text{-}CH_2\cdot \rightarrow$ tetrahydroanthracene + H_2. To suppress these condensation reactions, water vapor is added to the hydrocarbons being pyrolysed (dilution effect).

The distribution of hydrocarbon cleavage is determined mainly by the temperature, time of reaction and partial pressure of hydrocarbons. Formation of lower olefins is promoted by high temperatures, short reaction times and low partial pressures. Two methods have become widely used, *low-severity cracking* below 800 °C with reaction times around 1 s, and *high-severity cracking* close to 900 °C with about 0.5 s reaction time.

Industrial P. methods. The main differences among methods for production of lower olefins are the means by which heat is applied and the ways of removing coke deposits: 1) In flow reactors, the reactants flow through pipes; 2) Granular heat conductors in moving or fluid bed reactors can be ceramics, sand or coke. 3) Methods with fixed heat conductors in regenerating reactors use porous and temperature-resistant ceramics 4) Gaseous heat conductors, such as superheated steam or exhaust gases, are used in open furnace-type reactors, while 5) liquid heat conductors, such as a salt melt or liquid lead, are used, in bubble reactors.

95% of the pyrolysis installations now in use are based on the principle of pipe reactor (Fig. 3) technology.

Table 2. Product distribution as a function of the hydrocarbon fraction used

Starting material	Furnace exit temp °C	H_2	CH_4	C_2H_4	C_2H_6	C_3H_6	C_3H_8	C_4	C_5
Ethane	825	3.0	7.4	42.9	37.3	2.3	0.9	1.1	4.5
Propane	800	0.9	26.4	32.2	6.8	11.9	10.8	3.4	7.4
Light gasoline	760	0.9	12.7	24.0	5.0	19.7	0.5	9.8	27.2
Heavy gasoline	745	0.8	11.7	22.0	2.3	11.8	0.6	10.2	40.4

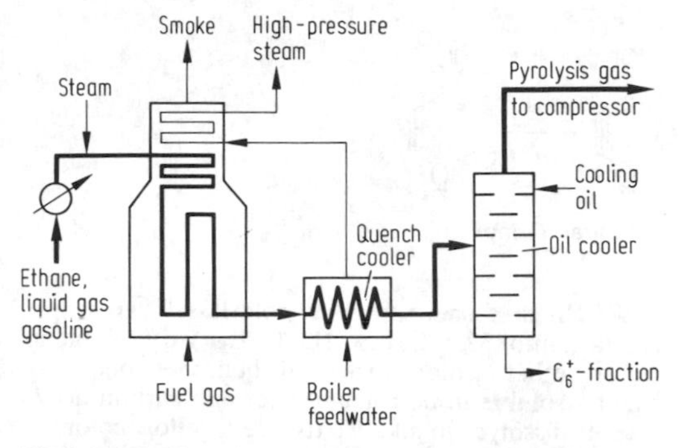

Fig. 3. Production of lower olefins by pyrolysis of hydrocarbons in a pipe furnace.

The chromium-nickel pipes in modern pipe furnaces are vertical and are directly heated to about 1050 °C at the hottest point by combustion of fuel gas or oil. The hot cracking gases are quickly cooled in quenchers to about 300 °C. In the cooling tank, the cleavage gas is cooled to a point just above its dew point; in this process, high-pressure steam is generated, and it makes a substantial contribution to the energy requirement of the plant. Further steps in the process are preliminary separation into gas and condensate, purification of the crude gas by removal of hydrogen sulfide, carbon dioxide and carbon oxygen sulfide by pressure scrubbing with 5 to 15% sodium hydroxide and drying the gas mixture using molecular sieves, aluminum oxide and diethylene glycol.

The drying must be thorough, to prevent precipitation of ice and gas hydrates in the following step, low-temperature, pressure distillation. This step is used to separate the cracking gas into its individual components (Fig. 4).

The low temperatures are generated using a coolant of ethene (≈ -100 °C) and propene (≈ -40 °C). The coolant vapors resulting from cooling the products (ethene and propene) are reliquefied and recycled. The products of distillation are usually not pure enough for their intended uses; ethyne, propyne and propadiene are generally present as contaminants. These are removed by selective hydrogenation or scrubbing processes.

To meet the demand for ethene and to extend the available raw material base, methods have been developed which use higher-boiling petroleum fractions (boiling above 250 °C). An interesting process is the *Kureha process* used in a technical installation in Japan since 1970. Crude oil from which asphalt has been removed is partially oxidized in a special reactor at temperatures as high as 2000 °C in the presence of steam. In addition to other products, this yields ethene and ethyne at a ratio of 1:1.

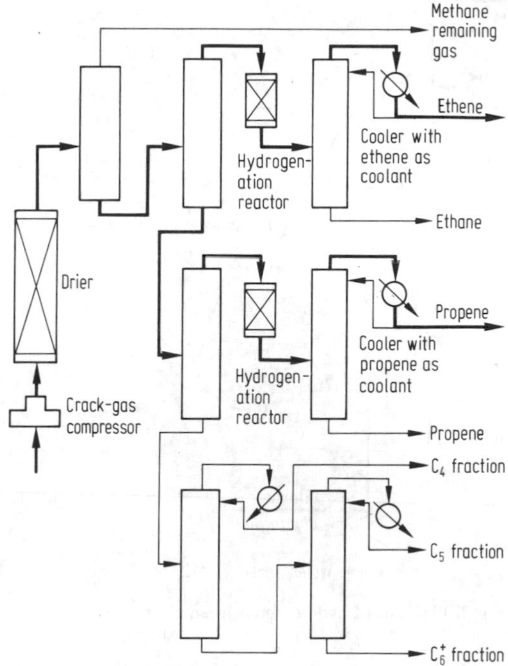

Fig. 4. Low-temperature pressure distillation of the pyrolysis gas. Hydrogenation reactors for selective hydrogenation of ethyne and propene.

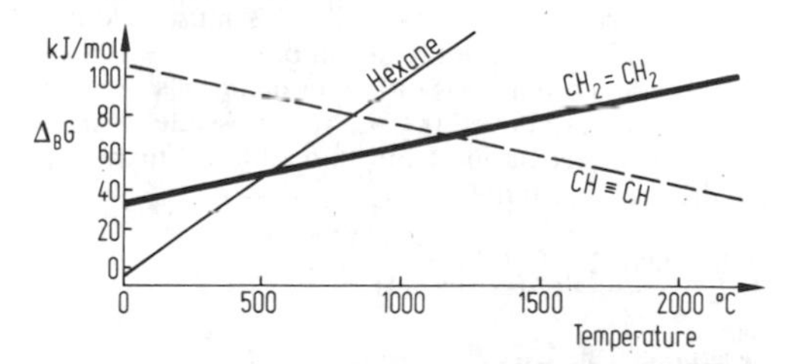

Fig. 5. Free energy of formation $\Delta_F G$ of ethyne, ethene and hexane.

2) High-temperature pyrolysis (HTP) processes are used to produce ethyne. Lower olefins are also produced, as in medium-temperature P., along with carbon monoxide and hydrogen, so that these processes also yield synthesis gas. Fig. 5 shows that as the temperature rises, the difference in the stabilities of ethene and ethyne becomes progressively smaller. Above 1300 °C, ethyne is more stable than ethene. For this reason, industrial methods for producing

ethyne from hydrocarbons require temperatures above 1500 °C. Heat is provided by *combustion of fuel gas with oxygen* (Fig. 6), *hydrogen-electric arc pyrolysis* and *electric arc pyrolysis* (Fig. 7).

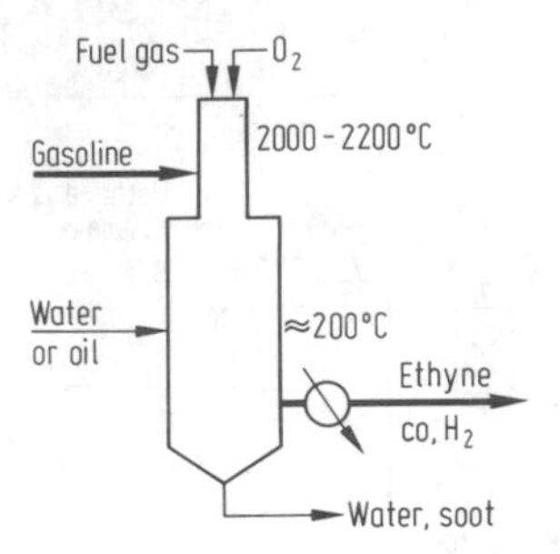

Fig. 6. High temperature pyrolysis to generate ethyne.

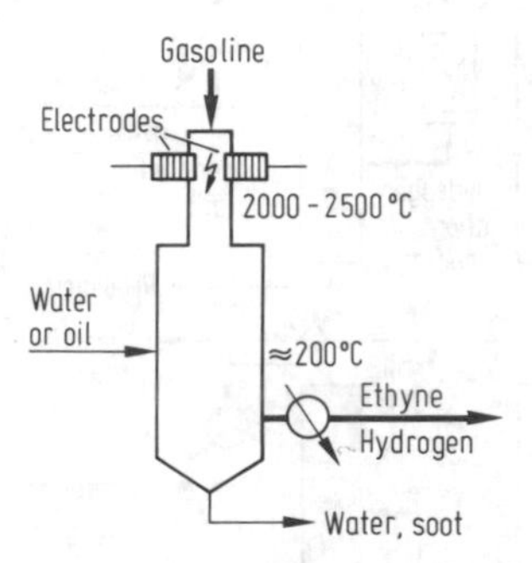

Fig. 7. Pyrolysis of hydrocarbons in an electric arc.

All these processes have in common the fact that the hydrocarbons are brought into the hot combustion gases or the electric arc. After a reaction time of a fraction of a second, the gas leaves the reactor and is immediately quenched to a temperature below 200 °C by a spray of water or oil. This must be done because ethyne very readily decomposes: $CH{\equiv}CH \rightarrow 2\,C + H_2$. Since synthesis of longer chains also occurs at temperatures above 2000 °C, it is possible to start with natural gas or methane and obtain ethyne by dehydrodimerization:

$$2\,CH_4 \xrightarrow{2000°C} CH{\equiv}CH + 3\,H_2.$$

Pyrolysis gasoline: see Gasoline.

Pyrometallurgy, ***dry metallurgy, melt metallurgy***: a general term for all metallurgical processes for producing and refining metals at high temperatures. Modern methods for production of iron and steel are exclusively methods of P. (see Iron, crude, and Steel). Some nonferrous metals are produced from aqueous solution (see Hydrometallurgy), in those cases where a combination of methods is not economically advantageous.

Pyrometer cones: same as Seger cones.

Pyromucic acid: same as Furan-2-carboxylic acid (see).

Pyrone: an oxoderivative of a pyran.

2H-Pyran-2-one, α-pyrone, coumaline is a colorless, water-soluble liquid which smells like coumarin; m.p. 5 °C, b.p. 208 °C, n_D^{22} 1.5277. α-Pyrone can be synthesized by decarboxylation of coumalic acid. The parent compound is not important industrially, but the α-pyrone ring is found in many natural products, for example Coumarin (see).

4H-Pyran-4-one

4H-Pyran-4-one, γ-pyrone: colorless, hygroscopic crystals; m.p. 32.5 °C, b.p. 215 °C. It is very soluble in water, ether, chloroform and benzene, but only slightly soluble in petroleum ether and carbon disulfide. It dissolves in alkalies to give a yellow color. γ-Pyrone is basic and cannot be converted to an oxime or the phenylhydrazone. This indicates that there are a number of resonance structures contributing to the ground state. γ-Pyrone is formed by pyrolysis of comanic acid (γ-pyrone-2-carboxylic acid) or chelidonic acid (γ-pyrone-2,6-dicarboxylic acid). Dehydration of 1,3,5-tricarbonyl compounds with phosphorus pentachloride also produces γ-pyrones. Some important natural derivatives of γ-pyrone are kojic acid, chelidonic acid and maltol; the chromones and xanthones are also important.

Pyrophosphates: same as Diphosphates (see).

Pyrophosphoric acid: same as Diphosphoric acid (see).

Pyrosulfuric acid: same as Disulfuric acid (see).

Pyrosulfurous acid: see Disulfites.

Pyrosulfates: same as Disulfates.

Pyrosulfites: same as Disulfites.

Pyroxylin: same as Gun cotton (see).

Pyrrocoline: same as Indolizine (see).

Pyrrol-2-aldehyde, ***pyrrole 2-carbaldehyde***: a heterocyclic aldehyde. P. is a colorless, crystalline, steam-volatile compound; m.p. 50 °C, b.p. 217-219 °C. P. can be synthesized from pyrrole by a Reimer-Tiemann synthesis (see). It is used as a starting material for pharmaceutical products, pigments and special organic syntheses.

Pyrrole: a five-membered heterocyclic ring compound containing one nitrogen atom. P. is a colorless, combustible, highly refractive liquid which smells like chloroform; m.p. -23.4 °C, b.p. 130-131 °C, n_D^{20} 1.5085. It is readily soluble in most organic solvents, but is only slightly soluble in water. It slowly turns brown in air, forming a resin. It is resistant to alkalies, but in the presence of dilute hydrochloric or sulfuric acid it forms ***pyrrole resins***.

Many syntheses for P. have been developed. It can be made 1) by the reaction of ammonia, primary amines or hydrazine with 1,4-dicarbonyl compounds (Paal, Knorr, 1885), 2) by reduction of a mixture of the monoximes of α-dicarbonyl compounds and β-ketocarboxylic acid esters or β-diketones with zinc in glacial acetic acid (Knorr, 1886), 3) by condensation of β-ketocarboxylic acid esters with α-haloketones in the presence of ammonia or primary amines (Hantzsch, 1890).

P. can also be synthesized in the laboratory by pyrolysis of the ammonium salt of mucic acid, or by the reaction of ammonia with furan in the presence of aluminum oxide. It is obtained industrially from bone tar, but the largest amount is made from but-2-yne-1,4-diol and ammonia under pressure. When P. reacts with potassium, or is boiled with potassium hydroxide solution, it forms ***potassium pyrrole***, which reacts with carbon dioxide to give the potassium salt of pyrrole 2-carboxylic acid. In the cold, potassium pyrrole reacts with chloroform and alkali hydroxide to form pyrrole 2-carbaldehyde (Reimer-Tiemann synthesis), but when warm this combination of reagents leads to ring expansion to 3-chloropyridine. Some other substitution reactions are chlorination to 2,5-dichloropyrrole, nitration to 2-nitropyrrole, sulfonation to pyrrole 2-sulfonic acid, formylation to pyrrole 2-carbaldehyde (see Vilsmeier-Haack reaction) and azo coupling. P. also reacts with maleic anhydride in a substitution (ene) reaction. Because of this property and its high reactivity with electrophilic reagents, P. is regarded as the parent compound of the π-excess heterocycles. Its reactions are similar to those of aniline or *N*-alkylated anilines, that is, the ring nitrogen of P. affects its reactivity in a manner similar to the (+)M effect of the amino or dialkylamino groups on a benzene ring. In the reduction with zinc powder and hydrochloric acid, or under catalytic hydrogenation, P. is converted first to pyrroline and then to pyrrolidine.

P. has a narcotic effect on the central nervous system, leads to severe, long-lasting hypothermia and, when injected, causes melanuria and local darkening of the skin. The vapor of P. colors a spruce shaving moistened with hydrochloric acid a characteristic red. Many natural products are derived from P. and its derivatives, such as the porphyrins, pyrrole pigments, bile pigments, alkaloids such as atropine, cocaine and nicotine, and the amino acids proline and hydroxyproline. P. was discovered in 1833 by Runge, in coal tar.

Pyrrole 2-carbaldehyde: same as pyrrol-2-aldehyde.

Pyrrole pigments: a group of pigments in which the molecule contains several (usually 4) pyrrole rings; an example is porphin. If the four pyrrole nuclei are linked by methine groups, the compound has a porphyrin skeleton, which is the basis of the Aza[18]annulene pigments (see).

Pyrrolidine, ***tetrahydropyrrole, tetramethyleneimine***: a colorless, hygroscopic liquid which smells like ammonia and fumes strongly in air; m.p. ≈ 58 °C, b.p. 88-89 °C, n_D^{20} 1.4431. It is readily soluble in water and organic solvents. P. is very basic, and forms crystalline salts with acids. It is made by catalytic hydrogenation of pyrrole, by reaction of tetrahydrofuran with ammonia in the presence of aluminum oxide, or by pyrolysis of 1,4-diamine hydrochlorides. The pyrrolidine ring is present in many natural products. P. is used as an intermediate in the synthesis of pharmaceuticals, fungicides, insecticides, plastics and vulcanization accelerators.

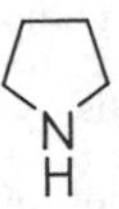

Pyrrolidine alkaloids: see Alkaloids.

Pyrrolidine-2,5-dione: same as Succinimide (see).

Pyrrolidin-2-one: same as Pyrrolidone.

Pyrrolidone, ***pyrrolidin-2-one, γ-butyrolactam, 4-aminobutyric acid lactam***: a pyrrole derivative; colorless crystals, m.p. 24.6 °C, b.p. 251 °C, n_D^{30} 1.4806. P. is readily soluble in water and organic solvents. It is obtained industrially from but-2-yne-1,4-diol, which can be made by the Reppe synthesis. But-2-yne-1,4-diol is hydrogenated to butane-1,4-diol, which is oxidized to butyrolactone and reacted with ammonia. If methylamine is used instead of ammonia, the product is *N*-methylpyrrolidone. P. is used as a high-boiling solvent and as a starting material for synthesis of polyvinylpyrrolidone.

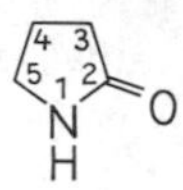

Pyrrolines: term for the three possible dihydropyrroles, which are named according to the position of the double bond: 1-, 2- and 3-P. 1- and 2-P. are unstable and tend to polymerize in the air or upon distillation. 3-P. is a colorless, hygroscopic liquid which fumes in the air and has an ammoniacal odor; b.p. 90-91 °C, n_D^{20} 1.4664. It is soluble in water and organic solvents. 3-P. is formed by reduction of pyrrole with zinc powder and glacial acetic acid, or with hydroiodic acid and red phosphorus.

Pyrrolizidine alkaloids: see Alkaloids.

Pyrrolo[1,2]pyridine: same as Indolizine (see).

Pyrrone: same as Polyimidazopyrrolone (see).

Pyruvate carboxylase: an enzyme which catalyses the addition of carbon dioxide to pyruvate, forming oxaloacetic acid. The reaction requires ATP and magnesium ions. The coenzyme is biotin.

Pyruvate decarboxylase, obsolete name, ***carboxylase***: an important enzyme in microorganisms and plants which catalyses the cleavage of pyruvate into acetaldehyde and carbon dioxide, using thiamin pyrophosphate as coenzyme. Mg^{2+} is required for activity. P. from yeast (M_r 190,000) consists of two identical subunits and is essential to the process of alcoholic fermentation.

Pyruvates: see Pyruvic acid.

Pyruvic acid, ***2-oxopropanoic acid***: CH_3-CO-

COOH, the simplest α-ketoacid. P. is a colorless liquid with a pungent odor, which gradually turns yellow on standing; m.p. 13.8 °C, b.p. 165 °C (dec.) n_D^{20} 1.4280. P. is readily soluble in water, alcohol, acetone and ether. The presence of two carbonyl groups and an active methyl group makes it very reactive. Even at room temperature, intermolecular condensation reactions lead to formation of various products. Bases, acids or metal ions catalyse these reactions, most of which are aldol condensations. P. is a relatively strong carboxylic acid, and its carboxyl group undergoes the reactions typical of this class. The salts and esters of P. are called *pyruvates*. The keto group can be converted to the oxime or hydrazone. P. reduces ammoniacal silver nitrate solution, and is oxidized to acetic acid and carbon dioxide in the reaction. Dilute sulfuric acid causes decarboxylation of P., leading to formation of acetaldehyde; in conc. sulfuric acid, carbon monoxide is split off to form acetic acid.

Due to the position of its carbonyl groups, P. is very suitable for synthesis of heterocyclic compounds. The salt (pyruvate) is an important intermediate in numerous metabolic processes, including glycolysis, the degradation of glucose. Under anaerobic conditions, P. is converted to lactate; under aerobic conditions, it is converted by oxidative decarboxylation to acetyl-coenzyme A, which enters the citric acid cycle and is subsequently converted to carbon dioxide and water by the respiratory chain, yielding a large amount of ATP. In alcoholic fermentation, P. is converted to acetaldehyde by pyruvate decarboxylase, and the acetaldehyde is converted by alcohol dehydrogenase to ethanol. Pyruvate is also involved in the metabolism of several amino acids; it is converted to alanine by transamination.

The most common technical synthesis of P. is dehydration of tartaric acid in the presence of potassium and sodium hydrogensulfate. It is also made by hydrolysis of acetylcyanide, or of 2,2-dichloropropionic acid. P. is used to make drugs and cosmetics.

Pyrvinium embonate: an Anthelminthic (see).

Pyrylium salts: cyclic oxonium salts derived from pyranes. They are obtained, for example, by the reaction of Grignard compounds with γ-pyrones and treatment of the resulting pyranols with strong acids. Similar treatment of chromones yields benzopyrylium salts, and xanthone yields dibenzopyrylium salts (xanthylium salts). The synthesis of pyrylium perchlorate was achieved by F. Klages in 1953. P. are used for organic syntheses, especially of pyridine and benzol derivatives.

Q

Q: 1) abb. for Coenzyme Q, see Ubiquinone. 2) abb. for pseudouridine.

Quadratic acid, ***3,4-dihydroxycyclobut-3-ene-1,2-dione***: a colorless and odorless, crystalline compound; m.p. 293 °C (dec.). Q. is slightly soluble in water, and is insoluble in most organic solvents. It undergoes the reactions typical of dicarboxylic acids. Its salts and esters are called *quadratates*. Q. and its derivatives can cause allergies. The compound is made in the reaction of perfluorocyclobutene with alcohol and subsequent hydrolysis of the 1,2-diethoxy-3,3,4,4-tetrafluorocyclobut-1-ene.

HO O
HO O

Quadricyclane, ***tetracyclo[3.2.0.0^{2,7}.1^{4,6}]heptane***: a tetracyclic hydrocarbon which is an isomer of norbornadiene. It is a colorless liquid consisting of molecules under high tension; b.p. 108 °C. Q. can be made by UV irradiation of norbornadiene; it returns thermally or photochemically to the starting material in a reversible reaction. Like cyclopropane, Q. undergoes various addition reactions, e.g. with bromine, hydrogen and acids; the products are derivatives of norbornene or norbornane.

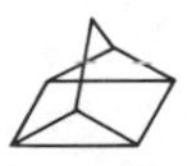

Quadruple point: a point in a two-substance system at which four different phases are in thermodynamic equilibrium; see Gibbs phase law.

Quadropole mass spectrometer: see Mass spectroscopy.

Quadropole resonance spectropy: see Nuclear quadrupole resonance spectroscopy.

Qualitative analysis: see Analysis.

Quantitative analysis: see Analysis.

Quantum chemistry: a field of theoretical chemistry dealing with the application of quantum mechanical approximation techniques to chemical problems. The main concern of Q. is the theoretical description of structural and energetic properties of atoms and molecules and their reactive behavior. The solution of the time-independent Schrödinger equation, $H\Psi = E\Psi$ (see Atom, models of) is a central theme of Q. H is the Hamiltonian operator of the atomic or molecular system, and the wavefunction Ψ describes a stationary state, that is, a state in which the system does not change. E is the energy of the system in this state. The mathematical character of the Schrödinger equation is such that it cannot be solved exactly for a multielectron system, and approximate solutions must be sought instead. By introducing various model Hamiltonian operators and approximate wavefunctions, many methods have been developed for approximate calculations, most of which depend on a variation technique or a perturbation calculation. The most successful of these approximations are the MO and VB methods.

Progress in the development of computers, especially in the last 15 years, have made possible ab-initio calculations, that is, calculations can be made within the chosen quantum mechanical approximation to high numerical precision, even in medium to large molecular systems with 40 to 80 electrons. Thus it is now possible to calculate properties of isolated molecular systems such as charge densities, dipole moments, spectroscopic, structural features and stabilities. At present, numerical Q. offers a true alternative to the precision experiment, especially for smaller molecular systems.

A major task of Q. is to explain phenomena, introduce concepts and generalize facts, thus contributing to theoretical understanding of chemistry. Model theories (see Hückel method, Ligand field theory and Woodward-Hoffmann rules) are outstanding examples of this work.

Quantum number: see Atom, models of.

Quantum yield, **Φ**: a parameter which expresses the effectiveness of photons in inducing a photophysical or photochemical process. The Q. is the number of results (e.g. reacted molecules) per photon absorbed by the system. A Q. can be calculated for photophysical processes (e.g. fluorescence or phosphorescence) or photochemical reactions. The ***differential Q.*** is $\Phi = dx/dt/I_a$, where dx/dt is the molar rate and I_a is the quantum flow of absorbed radiation. Integration of this parameter over time gives the ***integral Q.*** The Q. of a photochemical or photophysical process is at most 1, but is usually less. When a photoreaction initiates a thermal chain reaction, however, Q. is usually much greater than 1. The Q. is measured by an actinometer.

Quarks: see Elementary particles.

Quartation: see Separation.

Quartz fibers: see Flint glass fibers.

Quartz glass: see Flint glass.

Quasiisotropic: see Isotropism.

Quasiracemate: a crystallized, optically active 1:1 molecular compound of the D- and L-forms of two different, but structurally similar substances, such as D-(+)-methylsuccinic acid and L(-)-mercaptosuccinic acid. Q. are used to determine configurations.

Quaternary structure: see Biopolymers, see Proteins.

Quats: see Invert soaps.

Queen substance: see Pheromones.

Quencher: a species (e.g. a molecule) which inactivates an electronic excited state of another species.

Quenching: 1) the inactivation of an electronic excited species by species of the same or different types which are in the ground state. The process must be a nonradiative one.

2) see Pyrolysis.

Quercetin: see Flavones.

Quercetin-3-β-(6-O-α-L-rhamnosyl)-D-glucoside: same as Rutin (see).

Quiana: see Synthetic fibers.

Quinacridone: see Carbonyl fiber materials.

Quinaldic acid, ***quinoline 2-carboxylic acid***: a heterocyclic carboxylic acid, which forms a faintly yellowish, crystalline powder; m.p. 157 °C. It crystallizes from water in the form of the dihydrate, in crystalline needles; the anhydrous compound crystallizes from benzene. Q. decomposes at high temperatures to quinoline and carbon dioxide. It can be synthesized by the Reissert reaction (see Reissert compounds). Q. is used in the colorimetric determination of iron and in the gravimetric determination of various heavy metals, such as cadmium, copper, zinc and uranium.

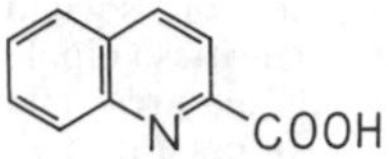

Quinaldine, ***2-methylquinoline***: a colorless liquid with a quinoline-like odor; m.p. -2 °C, b.p. 247.6 °C, n_D^{20} 1.8116. Q. is soluble in alcohol, ether and chloroform, and slightly soluble in water. The methyl group in position 2, relative to the ring nitrogen, is activated, so that condensation reactions with aldehydes and ketones are possible. Q. is obtained by a Friedländer synthesis (see) from 2-aminobenzaldehyde or by a Doebner-von Miller synthesis (see) by heating aniline and acetaldehyde in the presence of hydrochloric acid. The acetaldehyde is first converted to crotonaldehyde, which then reacts with the aniline to form Q. Q. is used in the production of dyes (quinoline yellow, quinoline red), as an additive to perfumes, nitro lacquers and lubricants. Halogen-substituted Q. are used in the synthesis of cyanin pigments.

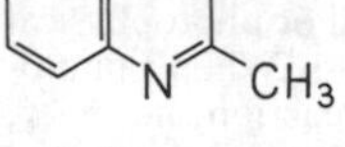

Quinazoline, ***benzo[d]pyrimidine***: an important benzodiazine. C. forms leaflets which smell like quinoline; they are soluble in water and many organic solvents; m.p. 48 °C, b.p. 241.5 °C. C. is chemically very similar to pyrimidine. Oxidation by potassium permanganate yields pyrimidine-4,5-dicarboxylic acid. C. is made by treating 2-aminobenzaldehyde with glyoxylic acid and ammonium acetate, followed by oxidation of the dihydro compound with potassium ferricyanide, or by reduction of bisformylamino-(2-nitrophenyl)methane with zinc or iron in hydrochloric acid solution. 4-Hydroxyquinazoline, or its keto form, ***quinazolone***, is the parent compound of various hypnotics, for example, methaqualone hydrochloride.

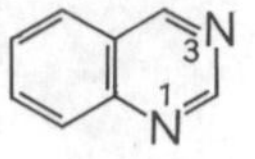

Quinazolone: see Quinazoline.

Quinhydrones: see Hydroquinones.

Quinhydrone electrodes: a redox electrode (see Electrodes) which is used as an indicator electrode in measurement of pH. It consists of a polished platinum wire dipped into a solution containing quinhydrone. The reaction which determines the potential is: $HO\text{-}C_6H_4\text{-}OH \rightleftharpoons O{=}C_6H_4{=}O + 2\,H^+ + 2e$. The relative electrode voltage of the Q. is

$$V = V^0 + \frac{RT}{2F} \ln \frac{a_Q a_{H^+}^2}{a_{hyd}},$$

or $V = 0.6996$ V $- 0.059$ pH. Here V^0 is the standard electrode voltage, R is the gas constant, T is the temperature in kelvin, F the Faraday constant, and a the activity. The Q. works in the pH range 1 to 7.9, and its potential is affected by other redox systems which are present.

Quinic acid: 1L-1(OH),3,4/5-tetrahydroxycyclohexane carboxylic acid. Q. forms colorless crystals; m.p. 162-163 °C, $[\alpha]_D^{20}$ -42 to -44° (in water). It is readily soluble in water, slightly soluble in ether and nearly insoluble in ether. When it is heated, it undergoes racemization. Q. is found widely in higher plants, either free or esterified with caffeic acid (see Chlorogenic acid). It was first isolated from cinchona bark, and is also found in citrus fruits, coffee beans and many temperate-zone fruits.

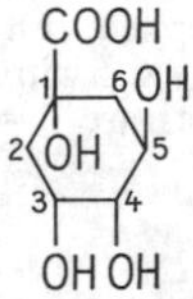

(−)-Quinic acid

Quinidine: see Cinchona alkaloids.

Quinine: see Cinchona alkaloids.

Quinol: a reaction product of a quinone with a Grignard reagent:

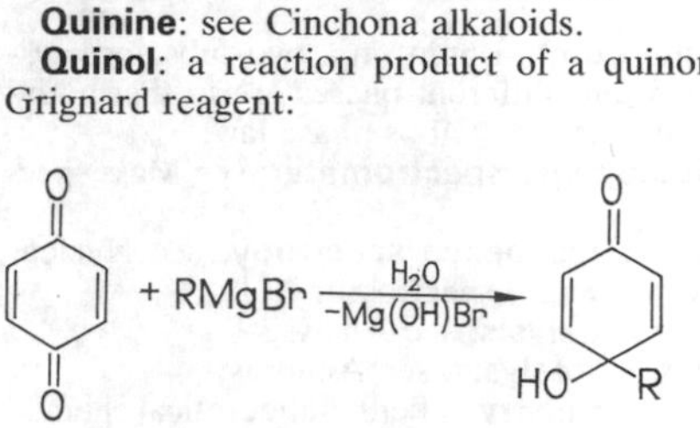

$$\text{Quinone} + RMgBr \xrightarrow[-Mg(OH)Br]{H_2O} \text{Quinol}$$

Q. readily undergo dienone-phenol rearrangement, forming substituted hydroquinones; which are thus easily synthesized in this fashion:

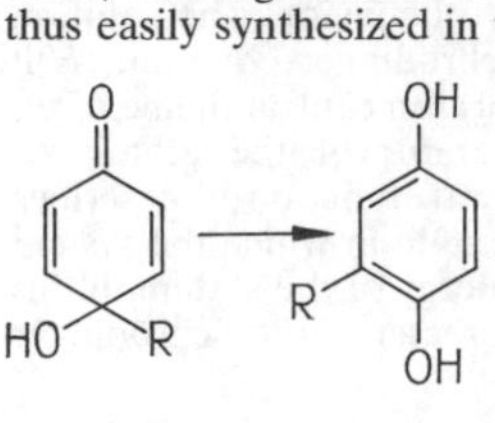

Quinoline, ***benzo[b]pyridine***: a heterocyclic compound consisting of fused benzene and pyridine rings. C. has a characteristic odor; it is a colorless liquid; m.p. -15.6 °C, b.p. 238 °C, n_D^{20} 1.6268. C. is readily soluble in alcohol and ether, and slightly soluble in water. It is basic and dissolves in acids, forming salts which crystallize readily. Alkyl halides quaternize the basic nitrogen atom, and metal salts form complexes. C. is the parent compound of the quinoline alkaloids; it is a strong protoplasm poison. It is synthesized by the Skraup synthesis (see). C. is a component of coal tar, and is obtained industrially by sulfuric acid extraction of the heavy oil fraction of coal tar distillation.

C. and its derivatives are important starting materials for drugs and dyes, the ***quinoline pigments***. These include, for example, quinoline yellow (from quinaldine and phthalic anhydride), which is used in color printing and making colored paper, and quinoline red, which is used as a sensitizer in photography. C. is added to cooling liquids and lubricants to prevent corrosion, and it is used in preparative organic chemistry as a catalyst of esterification. It is also used as a fungicide.

Quinoline alkaloids: see Alkaloids.

Quinoline 2-carboxylic acid: same as Quinaldic acid (see).

Quinoline pigments: see Quinoline.

Quinolinic acid, ***pyridine-2,3-dicarboxylic acid***: colorless, monoclinic prisms; m.p. 228-229 °C (reported also as 190-191 °C). In the presence of strong oxidizing agents, the benzene ring of quinoline is opened and Q. is formed.

It is partially soluble in water, but is nearly insoluble in organic solvents. When heated, it splits off carbon dioxide, making nicotinic acid. Q. is synthesized by oxidation of 2,3-dimethylpyridine by potassium permanganate, or by heating quinoline with sulfuric acid and hydrogen peroxide, or with sulfuric and perchloric acids.

Quinolizidine, ***octahydroquinolizine***: the parent compound of the quinolizidine alkaloids. The skeleton, Q., is a saturated, bicyclic heterocycle, which has not yet been synthesized in unsubstituted form.

Quinolizidine alkaloids: see Alkaloids.

Quinones: cyclic dicarbonyl compounds, in which the carbonyl groups, >CO, are conjugated with aromatic or olefinic systems. 1,2- or 1,4-*quinoid systems* are chromophores, and are responsible for the colors of the Q. Q. are oxidation products of diphenols. Their names are derived from the aromatic systems which are their parent compounds. There are benzoquinones, naphthoquinones, anthraquinones, phenanthrene quinones, etc. Depending on the relative positions of the carbonyl groups, they are 1,2- (*o*-), 1,4- (*p*-) or 2,6- (*amphi*-)quinones. Q. are crystalline, volatile compounds with colors from yellow to red. They have a somewhat pungent, characteristic odor. The reactivity of the Q. corresponds approximately to that of normal ketones. Although benzoquinones react similarly to α,β-unsaturated ketones, the aromatically conjugated compounds act more like diketones. This means that benzoquinones are less stable than other Q. Q. react with Hydroquinones (see) to form quinhydrones. Q. are strong oxidizing reagents which can be reversibly reduced to dihydroxyarenes.

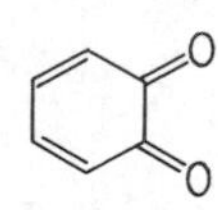

1,2-Benzoquinone 1,4-Naphthoquinone 2,6-(amphi-)Naphthoquinone

Q. are synthesized chiefly by oxidation of 2- or 4-disubstituted phenols, aminophenols or amines, or by direct oxidation of the aromatic hydrocarbons, except for benzene, by strong oxidizing agents such as chromic acid. Quinoid systems are the basic chromophores of many synthetic and naturally occurring pigments, the ***quinone pigments***, for example, the Anthraquinone pigments (see) and the Alizarin pigments (see). Quinonimine groups, such as $HN{=}C_6H_4{=}NH$, are found in various sulfur and indophenol pigments. The K vitamins contain substituted naphthoquinone systems. Some Q. are fungicides.

Quinone pigments: see Quinones.

Quinonimine pigments: a group of synthetic pigments derived structurally from ***quinonimine*** (quinone monoimine, $O{=}C_6H_4{=}NH$ or quinone diimine, $NH{=}C_6H_4{=}NH$). They include various sulfur pigments and indophenol pigments.

Quinoxaline, ***benzopyrazine***: a benzodiazine. Q. forms colorless crystals; m.p. 28 °C, b.p. 229.5 °C, n_D^{48} 1.6231. It is readily soluble in water, alcohol, ether and benzene. Q. is synthesized from *o*-phenylenediamine and glyoxal. It is the parent compound of a series of Azine pigments (see) and drugs.

Quinoxalines: see Azine pigments.

Quintozene: see Fungicides.

R

R: abb. for Alkyl- (see) or in general, for any organic group. 2) Configuration symbol (see Stereoisomerism 1.1). 3) *R*, symbol for the gas constant.

Ra: symbol for radium.

Racah parameter: see Ligand field theory.

Racemate, ***racemic mixture***: a mixture of equal amounts of two enantiomers (see Stereoisomerism) of a compound. R. do not rotate the plane of linearly polarized light; they are optically inactive. R. are indicated by placing "(D,L)-" or "(R,S)-" in front of the name of the compound. In the gas and liquid states, and in solutions, they have the same chemical and physical properties as the individual enantiomers, except for optical activity, physiological effects and reactions with chiral reagents. In the solid, they form either ***conglomerates*** of crystals (each enantiomer forming separate crystals), or ***racemic mixed crystals***, that is, solid solutions. ***Racemic compounds*** are equimolecular molecular compounds of the enantiomers, and they have properties different from those of the individual enantiomers. Naturally occurring chiral compounds are rarely R.; usually they consist of only one enantiomer. Chemical synthesis of chiral compounds from achiral starting materials or R. in the absence of optically active substances or other asymmetric influences yield R.

Separation of R.. There are several methods for separation: 1) Diastereomers. The R. reacts with one enantiomer of a chiral compound, yielding a mixture of two diastereomers. For example, if the R. of an acid, (R)-acid/(S)-acid, reacts with a pure enantiomeric base (R)-base, it forms the diastereomeric salts (R,R)-salt and (R,S)-salt, which can be separated. Hydrolysis yields the separated (R)-acid and (S)-acid. Tartaric, camphoric, camphorsulfonic and bromocamphorsulfonic acids are useful for separating bases and alcohols; suitable bases for separating acids are quinine, quinonine or brucine. This method is used, for example, to separate over 300 t (R,S)-menthol annually. 2) Molecular complexes or inclusion complexes, e.g. using urea. 3) Chromatography on optically active phases. 4) Physical separation of enantiomers from racemic mixtures of crystals. This method was used by Pasteur in 1848 to achieve the first separation of an R., sodium ammonium tartrate crystallized from aqueous solution at 27°C. 5) Enzymes, molds or bacteria can be used to metabolize one of the enantiomers, leaving the other.

The criterion for the effectiveness of the separation of an R. is *optical purity*, the ratio of the specific rotation of the separated compound to that of the pure enantiomer, expressed in percent.

Racemic acid: see Tartaric acid.

Racemization: the conversion of an optically active compound (see Stereoisomerism 1.1) into its racemate. R. occurs during all reactions at a chirality center which proceed via a symmetric intermediate such as a carbocation, radical, carbene or carbanion, i.e. non-stereospecifically. R. can also be catalysed by enzymes, called racemases.

Radecol®: see 3-Hydroxymethylpyridine.

Radedorm®: see Benzodiazepines.

Radenarcon®: see Etomidate.

Radepur®: see Benzodiazepines.

Radial density distribution function: see State of matter.

Radial part: see Atom, models of.

Radial distribution: see Atom, models of.

Radiation chemistry: an area of chemistry dealing with the chemical effects of high-energy (ionizing) radiation. As the radiation passes through a material, some of its energy is absorbed and leads to specific primary reactions, which are followed by secondary and further reactions. The result can be a complicated chemical change in the substance. The energy added to a substance of mass *m* by ionizing radiation is the *energy dose*. The SI unit is the Gray (Gy): 1 Gy = 1 J kg^{-1}. The original unit, the rad (rd), is no longer acceptable: 1 rd = 0.01 Gy = 10 mGy.

Reactions can be initiated by x-rays, gamma rays, fast neutrons, electrons and daughter nuclides from nuclear reactions. Gamma-ray sources for technical use may contain cobalt 60, which is produced in a reactor by neutron activation of cobalt, cesium 137, a fission product of uranium fuel rods, and fast electrons which are generated in special particle accelerators. The decisive factor is the effective range of the radiation; for ^{60}Co, for example, this is 1 m through organic liquids.

The practical use of R. has grown rapidly in the past few years. Its methods are now used for synthesis of a wide variety of low- and high-molecular-weight compounds; for example, for the technical production of ethyl bromide by addition of hydrogen bromide to ethylene, the synthesis of the insecticide hexachlorocyclohexane (gammexane), sulfation of hydrocarbons with chain lengths C_{10} to C_{20}, synthesis of C-P bonds in organophosphorus compounds, and for inducing polymerization reactions. Plastics are either degraded or cross-linked by radiation, depending on the dose. The cross-linking of polyethylene with electron accelerators at doses between 20 and 200 kGy yields products which resist melting up to 300°C. This process is used industrially to cross-link plastic films and cable insulation. *Electron-beam paint hardening* has become important; in it, the initiating molecules for the polymerization process which hardens the paint are created from the carbon double bonds in the binder. Solvents and photoindicators are not needed, in contrast to UV paint hardening. The resulting paint layers are very hard, dense and resistant to chemicals.

It cannot be overlooked that radiochemical reactions also cause damage to living or non-living matter (radiation damage).

Radical: 1) an electrically neutral species with magnetic moments due to the presence of an unpaired electron. R. are paramagnetic and in the ground state, they normally have a doublet multiplicity. They are formed by Homolysis (see) of a bond, absorbing the bond dissociation energy. Depending on how this energy is supplied, the process is classified in one of the following categories: 1) *Thermolysis* occurs at elevated temperatures and the conditions are chosen to correspond to the bond strength (between about 120 and 400 kJ mol^{-1}), as in cracking of hydrocarbons. 2) *Photolysis* occurs in the presence of light of the appropriate wavelength, e.g. in the cleavage of chlorine molecules. 3) *Radiolysis*, for example with γ-rays. 4) *Chemical or electrochemical cleavage* in redox processes by electron transfer. 5) *Cleavage by mechanical energy* in grinding processes.

R. have varying stabilities. Because the unpaired electron is included in the conjugation of the π-electron system, benzyl, trityl, allyl and acyl R. are more stable than alkyl R. Some R. obtain additional stabilization by steric hindrance of radical recombination exerted by solvent molecules.

R. can be demonstrated by the following methods: 1) ESR spectroscopy; 2) Paneth mirror method, in which the R. is passed over a lead or silver mirror and dissolves in it, forming volatile metal alkyls. These metal alkyls can be cleaved thermally at another site in the apparatus. 3) IR and UV/VIS spectroscopy at sufficiently high R. concentration. 4) Chemical reactions (e.g. polymerization) or their initiation. 5) Trapping reactions, e.g. trapping of carbon R. by atmospheric oxygen to form peroxides, or use of violet diphenyl picryl hydrazile (formation of colorless or yellow products which are suitable for quantitative colorimetric analysis).

The reactions of R. may retain their radical nature, as in decomposition, abstraction, addition, polymerization or rearrangement to lower-energy R., or the radical character may be lost, as in combination, recombination, disproportionation, reduction to anions or oxidation to cations (electron transfers).

Radical anions and cations are charged species with radical character.

2) The term R. is also applied to atoms or groups of atoms which remain together in the course of chemical reactions, such as the chromyl R. (group) CrO_2 or the uranyl R. UO_2 (see Nomenclature, sect. II C 3). In organic chemistry, aliphatic, carbo- or heterocyclic groups bound via a carbon group are also known as R. (see Nomenclature, sect. III C to III F) and are distinguished from Characteristic groups (see).

Radical anion: A ·, a negatively charged species with a magnetic moment; a molecular anion with an additional unpaired electron. R. are formed as intermediates in the reduction of compounds with high electron affinities. For example, in the reduction of naphthalene with sodium in ethanol, a green, ionic compound is formed. The ketyls formed as intermediates in the reduction of ketones with sodium or magnesium amalgam are also R. The ground state of these anions is usually a doublet. Many R. are colored. The unpaired electron occupies the lowest unoccupied molecular orbital of the neutral molecule. The long-wave absorption can be explained qualitatively as the small energy difference between this orbital and the next-higher unoccupied one.

Radical cation: A^{+}, a positively charged species with a magnetic moment; a molecular cation with an unpaired electron, but with a total of one electron fewer than the neutral species. R. are formed by ionization of molecules, e.g. by irradiation of molecules with electrons in a mass spectrometer or by polarographic oxidation (aromatic hydrocarbons). The ground state is usually a doublet. The unpaired electron occupies the highest occupied molecular orbital of the neutral molecule. The electron spectra of aromatic hydrocarbon R. are very similar to those of the radical anions.

Radical reactions: reactions in which radicals take part. The most important R. are radical substitution reactions (see Substitution 3.) and the addition of radicals to C-C double and triple bonds. Many radical addition reactions occur as radical chain polymerizations of unsaturated monomers. R. are usually initiated photochemically, or, more rarely, thermally. The compounds added include: 1) *Halogens*, usually excluding iodine, which is not very reactive. Chlorine can be added radically to benzene, forming stereoisomeric hexachlorocyclohexanes. 2) *Hydrogen bromine*, initiated by the presence of peroxides (peroxide effect, Markovnikov rule), adds contrary to the Markovnikov rule (see). 3) *Aldehydes* add to form long-chain ketones. 4) *Carbon tetrachloride* and mixed polyhaloalkanes. 5) *Alcohols* with the α-C atom (formation of primary alcohols by addition of methanol and secondary alcohols by addition of ethanol). 6) *Esters*. 7) *Hydrogen sulfide* and *thiols, thiolic acids* and *bisulfites*.

R. occur by the following mechanism:

x–y + Z· → x· + y–Z, where Z' is the initiator

x· + R(>C=C<) → R(>Ċ–C–x)

R(>Ċ–C–x) + x–y → y–C(R)–C–x + x·

Both radical steps are exothermal if X-Y = halogen-halogen, CCl_2-Cl, CCl_3-Br, CCl_3-H, Br-H, RS-H. This means that they can occur as chain reactions. Many photochemical cycloadditions which form three- and four-membered rings are R. The carbon radical postulated in this mechanism can react in similar fashion with other alkene molecules, leading to long-chain macromolecules.

Radical substitution: see Substitution, 3.

Radioactivity: the spontaneous conversion of radionuclides into stable nuclides with release of energy in the form of corpuscular or electromagnetic radiation. R. occurs both in natural (***natural R.***) and artificial (***artificial R.***) radionuclides. The R. of the radionuclides leads to a decrease in the number of atoms of the starting substance according to the equation $N = N_0 \cdot e^{\lambda t}$, where λ is the decay constant and N

is the number of atoms still present after the time t. The half-life $t_{1/2} = \ln 2/\lambda$ is often used to characterize radionuclides; this is the time after which the original number of atoms has been decreased by half.

1) ***Natural R.*** None of the elements with atomic number greater than 83 have stable nuclides. Polonium, astatine, radon, francium, radium, actinium, thorium, protactinium and uranium exist only as ***radionuclides***. The natural radionuclides with $Z > 81$ are members of three natural *decay series* which lead from uranium-235, uranium-238 and thorium-232 to various stable isotopes of lead (Fig. 1). Most other naturally occurring radionuclides have long half-lives; some examples are: ^{10}Be (β-emitter, $t_{1/2} = 1.6 \cdot 10^6$ years); ^{40}K (β$^-$, K capture, $1.28 \cdot 10^9$ years); ^{48}Ca (β$^-$, $> 1.1 \cdot 10^{18}$ years); ^{50}V (β$^-$ $> 1.2 \cdot 10^{16}$ years); ^{87}Rb (β$^-$, $4.7 \cdot 10^{10}$ years); ^{115}In (β$^-$, $6 \cdot 10^{14}$ years); ^{123}Te (β$^-$, $1.24 \cdot 10^{13}$ years); ^{138}La (β$^-$, $1.3 \cdot 10^{11}$ years); ^{142}Ce (α-emitter, $5 \cdot 10^{16}$ years); ^{144}Nd (α, $2.1 \cdot 10^{15}$ years); ^{147}Sm (α, $1.06 \cdot 10^{11}$ years); ^{148}Sm (α, $7 \cdot 10^{15}$ years); ^{149}Sm (α $\approx 4 \cdot 10^{14}$ years); ^{152}Gd (α, $1.1 \cdot 10^{14}$ years); ^{176}Lu (β$^-$, $3.3 \cdot 10^{10}$ years); ^{174}Hf (α, $2 \cdot 10^{15}$ years); ^{187}Re (β$^-$, $4.3 \cdot 10^{10}$ years); ^{190}Pt (α, $6.1 \cdot 10^{11}$ years); ^{192}Pt (α, $\approx 10^{15}$ years); ^{204}Pb (α, $1.4 \cdot 10^{19}$ years); ^{209}Bi (α, $2.5 \cdot 10^{17}$ years).

The naturally occurring radionuclides tritium (^{3}H) and carbon-14 (^{14}C) are formed in the high atmosphere by cosmic rays.

Naturally occurring radionuclides may emit α-particles (He nuclei), β$^-$ particles (electrons) or γ-rays (high-energy electromagnetic radiation). These types of radiation can be distinguished by their behavior in a magnetic field (Fig. 2). They are detected or counted using Geiger counters, ionization cloud chambers, scintillation or blackening of photographic plates. The changes in position in the periodic system resulting from radioactive emission is described by the Sody-Fajans law. When α-particles are emitted, the nuclear charge of the emitting atom is reduced by two units, while the mass number decreases by 4 units: the element moves two places to the left in the periodic system, e.g. ^{219}Po → ^{215}Pb + ^{4}He + energy. In β-emission, the atomic number is increased by one unit, because loss of a β-particle is associated with the conversion of a neutron into a proton, e.g. ^{211}Pb → ^{211}Bi + e$^-$ + energy. The element moves one place to the right in the periodic system. The emission of γ-radiation has no effect on the mass or atomic number; only the energy level of the nucleus is affected. The α-particles emitted from a radionuclide have defined

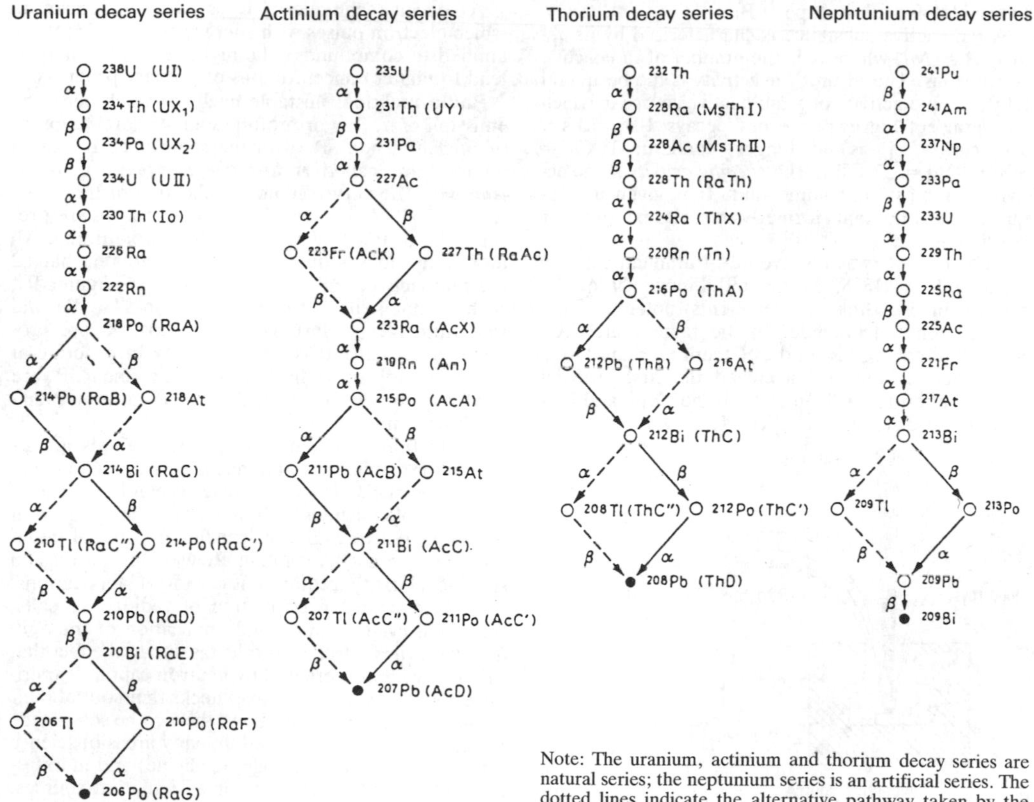

Note: The uranium, actinium and thorium decay series are natural series; the neptunium series is an artificial series. The dotted lines indicate the alternative pathway taken by the smaller number of nuclei. The historical terms are set in parentheses.

Fig. 1. The radioactive decay series.

kinetic energies. The particles ionize atoms close to their paths of flight, lose energy in this way, and after a certain distance come to rest. As a result, they give a characteristic, brush-like track in the cloud chamber. β-particles have continuous spectra. The emission of these particles is associated with emission of uncharged antineutrinos which are either massless or have an extremely low mass; the kinetic energy of the antineutrino added to that of the β-particle yields the maximum observed kinetic energy for β-particles in the decay in question.

2) ***Artificial R.*** is observed in nuclides produced by nuclear reactions. Synthetic radionuclides are generated in large amounts by neutron irradiation of materials in nuclear reactors. Some of them, e.g. ^{90}Sr or ^{137}Cs, are used as radioactive sources in industry, while numerous others are byproducts of nuclear reactors used for power and must be safely disposed of. Most artificial R. involves β- and γ-rays. α-emitters are relatively rare. Proton excesses in the nucleus are often relieved by positron emission (β^+) associated with emission of a neutrino; or by *K-capture*. K-capture is the absorption of an electron from the K shell by a proton, which reduces the nuclear charge by one unit and is accompanied by the emission of an X-ray quantum as an electron from a higher shell fills the vacancy in the K-shell (see Atom, models of). There is a fourth decay series, starting from the radioactive nuclide plutonium-241, and ending with the stable bismuth isotope ^{209}Bi.

A radioactive substance is characterized by its *activity* $A = N/t$, where N is the number of radioactive conversions. The SI unit for activity is the becquerel (Bq), i.e. the activity of a radioactive source in which an average of 1 atom per second decays: $1\ Bq = 1\ s^{-1}$. The curie (Ci) is no longer standard: $1\ Ci = 3.7 \cdot 10^{10} Bq = 37\ GBq$. The *specific acitvity* is the activity per mass or volume unit. It is given in becquerels per kilogram (formerly in curies per gram or per liter).

Historical. R. was discovered in uranium 1896 by Becquerel. In 1898, Marie and Pierre Curie succeeded in isolating the elements polonium and radium from pitchblende. In the same year, C.G. Schmidt discovered R. in the element thorium. Using α-particles, Rutherford achieved the first artificial transmutation in 1919. In 1934, Irène Curie and Frédéric Joliot discovered artificial R.

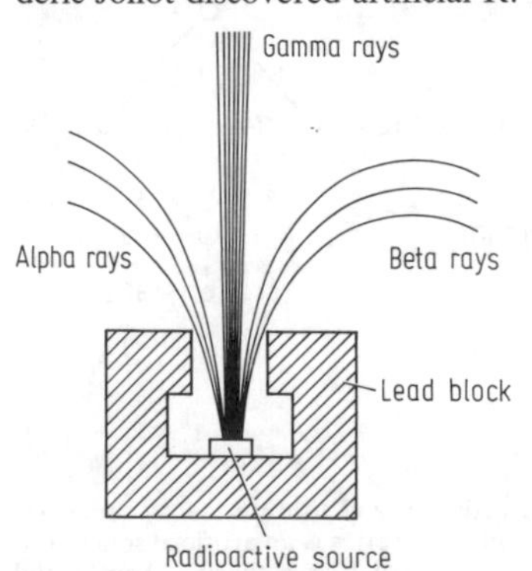

Fig. 2. Behavior of radiation in a magnetic field perpendicular to the plane of the paper.

Radiochemistry: the area of chemistry dealing with radionuclides. In ***general R.***, the chemistry of radionuclides in microconcentrations, the chemical properties of radioactive elements and nuclear reactions are studied. Many ultramicro methods have been developed in R., since radioactive isotopes are often available only in extremely small amounts. For example, reactions are carried out in capillary tubes, and quartz-fiber balances are used which are sensitive to one part in 10^8. ***Applied R.*** is concerned with the use of radionuclides in analysis, kinetics, biochemistry, industry, agriculture and medicine, with the production, enrichment and separation of radionuclides, with development of methods for synthesis of radiolabelled compounds, production of standardized preparations of radioactive substances, and with processing of spent fuel from reactors and other radioactive wastes. The methods of R. are based on the great sensitivity with which radiation can be detected. For example, with a simple counting tube it is possible to detect as little as $3 \cdot 10^{-16}$ g ^{32}P or $2 \cdot 10^{-9}$ g ^{14}C.

Radioelements: see Radioactivity.

Radioimmunoassay: see Immunoassay.

Radioindicator: see Tracer.

Radioisotope: see Radionuclides.

Radiolysis: chemical reactions initiated by ionizing radiation; the concept is limited to the electron shells. The primary products are radicals, radical ions, excited molecules and/or solvated electrons.

A special technique of R. is ***pulse radiolysis***, in which electron pulses with energies up to 40 MeV are applied to compounds. The higher electron energies lead to higher concentrations of primary products.

Radionuclides: unstable nuclides which decay by emission of α, β, γ or proton radiation, by K capture or nuclear fission. R. with the same nuclear charge number are called ***radioactive isotopes*** or ***radioisotopes***. The very common application of these two terms to all R. is incorrect. About 1500 R are presently known; of them, about 50 occur naturally. All natural nuclides with atomic numbers greater than 82 are radioactive, and in addition there are lighter R., such as potassium-40 (^{40}K), rubidium-87 (^{87}R) and samarium-147 (^{147}Sm). However, most R. are produced artificially. It is possible today to generate at least one radioactive nuclide of each element. R. are relatively simple to detect, due to the ionizing radiation they emit.

Production. The most important methods are irradiation of suitable stable nuclides with high-energy charged particles from a particle accelerator, irradiation of stable nuclides with slow or fast neutrons from a nuclear reactor, and nuclear fission. Special methods must be used to obtain R. with the purity and specific acitivity required for various applications, such as chemical precipitation of radioactive salts, fractional cyrstallization and application of the *Szilard-Chalmers effect*. This effect is due to the fact that radioactive atoms created by neutron capture experience a kickback effect which knocks them out of their chemical bonds and this allows them to be separated.

Applications. R. are used in many areas of science (chemistry, physics, biology, medicine) and in industry, either as Tracers (see) or as radiation sources. When they are used as *radiation sources*, the point of interest is the interaction of the ionizing radiation with the substance which is irradiated. R. are used as radiation sources in industry in many ways. For example, the hard gamma rays from ^{60}Co can be used to examine even thick objects for cracks and other defects. R. can also be used to measure the thickness of all types of rolled products. If a paste containing an R. is applied to the surface of a material, the R. remaining after removal of the paste indicates the sites of surface irregularities. The absorption or scattering of the radioactivity can also be used to monitor filling of closed containers, to check the heat-resistant cladding of furnace walls, to measure the density of geological layers or coal heaps, to determine the ash content of coal and the sulfur content of liquid petroleum products, to measure the thickness of pipe walls, to determine the water content of the soil, etc. In chemistry, R. are used in quantitative analysis to monitor the completion of precipitation reactions. An R. of the element to be determined is added to the solution before precipitation; precipitation is complete when no more radioactivity can be detected in the filtrate. Chemical exchange processes can also be studied using R.

In medicine, radiation sources are used in therapy of cancer, either externally or with needles (or the like) which contain an R. In some cases, an organ can be irradiated internally by oral application of a radioactive preparation. For example, Basedow's disease or thyroid cancer can be combatted by preparations containing ^{131}I, because the thyroid accumulates the iodine and is therefore irradiated internally.

Radium, symbol ***Ra***: a radioactive chemical element from group IIa of the periodic system, the Alkaline earth metals (see); a heavy metal, *Z* 88, with natural isotopes with the following mass numbers (the half-lives in parentheses): 213 (2.7 min), 214, (2.6 s), 215 (1.6 ms), 216 (< 1 ms), 217 (very short), 220 (23 ms), 221 (30 s), 222 (38 s), 223 (11.43 d), 224 (3.64 d), 225 (14.8 d), 226 (1600 a), 227 (41.2 m), 228 (5.77 a), 230 (1 h). The atomic mass is 226.0254, valence II, density about 5, m.p. about 700 °C, b.p. about 1140 °C.

Properties. R. is a metal with a silvery sheen which gradually turns black when exposed to air, due to formation of radium nitride. Its chemistry fits directly into the series of the alkaline earth metals. It is very similar to barium, but as might be expected, is even more reactive. The water solubility of R. salts is completely analogous to that of the comparable calcium or barium compounds. R. is a strong reducing agent, tending to form Ra^{2+} ions. It reacts very vigorously with water or acids, generating hydrogen. Volatile R. compounds give a carmine red flame color.

All isotopes of R. are radioactive. Isotopes 225, 227, 228 and 230 decay by β-emission, while all other isotopes are α-emitters. These processes are accompanied by γ-rays. The most stable isotope, ^{226}Ra, is a member of the natural uranium-radium decay series, being formed from uranium-238 and decaying to radon: ^{226}Ra (-, α) ^{222}Rn. The ^{222}Rn decays in several ways via nuclides which are often called R. A to R. G, finally forming a stable lead isotope. ^{224}Ra (ThX) is a member of the thorium decay series and decays to radon-220; ^{223}Ra (AcX) is a decay product of uranium-235 (actinium-uranium decay series); and ^{235}Ra is a member of the synthetic neptunium decay series (see Radioactivity). Because of their intense radiation, radium preparations glow in the dark.

The number of decays in 1 g R. per second, $3.7 \cdot 10^{10}$, was used as the unit for radioactivity and is equal to 1 curie: 1 Ci = $3.7 \cdot 10^{10}$ decays per s = $3.7 \cdot 10^{10}$ Bq.

Occurrence. R. is one of the rarest elements, making up about $7 \cdot 10^{-12}$% of the earth's crust. As a decay product of uranium, it is found in uranium minerals, e.g. pitchblende or carnotite. 1 t pitchblend with 60% uranium content contains about 0.15 g R. Small concentrations of R. are also found in some mineral springs and in seawater.

Production. To separate R. from uranium minerals, these are first solubilized with soda, and then barium salts are added. Next the R. is precipitated with the barium by addition of sulfuric acid. After conversion of the sulfates to the bromides, the less soluble radium bromide can be separated by fractional crystallization. The element is prepared by melt electrolysis of the bromide on a mercury cathode; this forms a R. amalgam, from which the pure metal can be recovered by distillation of the mercury at 500 to 700 °C in a hydrogen atmosphere. Spent fuel from uranium reactors is a rich source of R.

Applications. Because of its intense radiation, R. can be used in nuclear reactions, for example, in a mixture with beryllium, to generate neutrons; it is also used as a radiation source for medicine. However, it has largely been replaced by radionuclides of other elements which are more readily available from nuclear reactors.

Historical. R. (from the Latin "radius" = "ray") was detected in 1898 by M. Curie-Sklodowska, P. Curie, and G. Bémont in Jáchimov pitchblende. After laborious enrichment, they succeeded in preparing about 100 g pure radium bromide. The element was first prepared in 1910 by M. Curie-Sklodowska and A. Debierne by electrolysis of radium chloride.

Radium compounds: these have not been very intensively studied; they are largely similar to those of barium. ***Radium bromide***, $RaBr_2$, colorless to bright yellow, monoclinic crystals which are moderately soluble in water; M_r 385.82, density 5.79, m.p. 7.28 °C, subl.p. 900 °C. It forms a monoclinic dihydrate which becomes anhydrous at 100 °C. ***Radium carbonate***, $RaCO_3$, forms colorless, monoclinic crystals which are only slightly soluble in water, and ***radium chloride***, $RaCl_2$, forms colorless to bright yellow, water-soluble monoclinic crystals, M_r 296.91, density 4.91, m.p. 1000 °C. The monoclinic dihydrate is converted to the anhydrous form at 100 °C. ***Radium sulfate***, $RaSO_4$, forms colorless, rhombic crystals which are only slightly soluble in water.

Radium emanation: see Radon.

Radius ratio: the ratio r_A/r_B of the radii of spherical components (ions, atoms) A and B in a crystal. The letter A is usually applied to the smaller sphere, so that r_A/r_B is less than 1. The practical significance of the R. is that for primarily ionogenic crystals, the formation of certain structure types depends only on the geometry; it is determined by the relative rather than the absolute size of the ions. The crystal structures of ionic substances are characterized by the coordination polyhedra or coordination numbers.

According to the *radius ratio rule*, there is a critical value of r_A/r_B for each coordination number, and this can be calculated from the geometry. The critical value corresponds to the limiting case in which the coordinating B ions touch each other and the A ions; this is energetically the most favorable case. This ratio can be exceeded as the packing of the B ions is loosened, until the critical value for the next higher coordination number is reached. The ratio for a crystal cannot be lower than the limiting value, because in this case the A ion would no longer fill the space in the polyhedral holes, and the substance would go to a lower coordination number to stabilize the situation. The coordination number for the B ion follows from the stoichiometric composition of the compound, because the principle of electrical neutrality must be observed even in small regions of the crystal lattice. The predictions of the radius ratio rule for compounds with the general composition AB and AB_2 are summarized in the table.

The radius ratio rule is obeyed by many compounds, but it is also rather frequently broken, especially at low coordination numbers. The deviations are understandable, because the geometric derivation regards the ions as rigid spheres, and does not take into account the sometimes considerable deformation by the polarizing effect of the counterions.

Relationship between radius ratios and coordination

Composition	r_A/r_B	Coordination	Coordination polyhedra	Structure type
AB	0.732-1	[8] : [8]	Cube	CsCl
	0.414-0.732	[6] : [6]	Octahedron	NaCl
	0.225-0.414	[4] : [4]	Tetrahedron	ZnS
AB_2	0.732-1	[8] : [4]	Cube/tetrahedron	CaF_2
	0.414-0.732	[6] : [3]	Octahedron/triangle	TiO_2
	0.225-0.414	[4] : [2]	Tetrahedron/dumbbell	SiO_2

Radon, symbol ***Rn***, obsolete names, ***emanation, niton***: a radioactive element from Group VIII of the periodic system, the Noble gases (see). The atomic number is 86, and the natural isotopes have the following mass numbers (half-lives in parentheses): 204 (75s), 205 (1.8 min), 206 (6.5 min), 207 (11 min), 208 (23 min), 209 (30 min), 210 (2.42 h), 211 (15 h), 212 (25 min), 213 (19 ms), 215 (short), 216 (4.4 · 10-5 s), 217 ($5.4 \cdot 10^{-3}$s), 218 (formerly called astatine emanation, 0.035 s), 219 (formerly called actinium emanation or actinon, 4.0 s), 220 (formerly called thorium emanation or thoron, 55 s), 221 (25 min), 222 (formerly called radium emanation, 3.823 d), 223 (43 min) and 224 (1.9 h). Thus the most stable isotope has atomic mass 222. The valence of Rn is 0 (but probably II with respect to fluorine); density 9.73 g l^{-1} at 0 °C, m.p. -71 °C, b.p. -61.8 °C, crit. temp. 104.04 °C, crit. pressure, 6.2 MPa.

Properties. Rn is a colorless, odorless and tasteless noble gas, which is always monoatomic. It is chemically very inert. Following the trend of ionization energies within the noble gas group, and the increase in the stability of the element-fluorine compounds on going from krypton to xenon, it can be predicted that compounds of Rn with fluorine, oxygen and possibly other electronegative elements should be readily synthesized. However, so far only radon fluoride, probably RnF_2, has been synthesized by direct reaction of fluorine with Rn. A few complexes of the type $FRn[SbF_6]^-$ are derived from this fluoride. Other compounds of Rn are not yet known. The study of the chemistry of Rn and its compounds is difficult because of the short half-lives of all its isotopes. The longest is less than four days; it decays to polonium by emission of α-particles.

Analytical. R. is most conveniently separated from mixtures of gases by gas chromatography; it can be identified by spectral analysis or Geiger counter.

Occurrence. As a result of the short half-life of the Rn nucleus, the proportion of Rn in the earth's crust is extremely small; it is estimated at $6 \cdot 10^{-16}$%. The atmosphere contains $6 \cdot 10^{-18}$ vol. % Rn. Rn is an intermediate in the radioactive decay of uranium, thorium and actinium (see Radioactivity), and it is therefore found in small amounts in the neighborhood of uranium and thorium minerals. It has also been found in water from some springs.

Preparation Rn is obtained by pumping on the gas arising by radioactive decay of radium.

Applications. Rn is used in medicine, in radiation therapy.

Historical. ^{220}Rn was discovered in 1899 by Rutherford, and ^{222}Rn was found in 1901 by Dorn in the course of studies of the gases (emanations) produced by radioactive elements.

Raffinose: α-D-galactopyranosyl-(1→6)α-D-glucopyranosyl- (1→2)-α-D-fructofuranoside, a non-reducing trisaccharide found in many higher plants; it can be obtained from sugar beet syrup.

Rafoxanide: a Molluscicide (see) and veterinary medicine.

Railroad metal: see White metals.

Rainout: the inclusion of gaseous and particulate air pollutants in water droplets in clouds and their return to the earth with the resulting precipitation.

Raman lines: see Raman spectroscopy.

Raman spectroscopy: spectroscopy based on the *Raman effect* predicted in 1923 and independently discovered in 1928 by C.V. Raman and G.S. Landsberg and L.I. Mandelstam. In the scattering spectrum of molecules irradiated with monochromatic light, there are lines which are at different frequencies from those of the incident light, and which arise from molecular vibrations and rotations.

Generation of the Raman spectrum. If a sample is irradiated with intense, monochromatic radiation of frequency ν_o which is sufficiently distant from any absorption line of the substance, part of the incident light is scattered (Fig. 1).

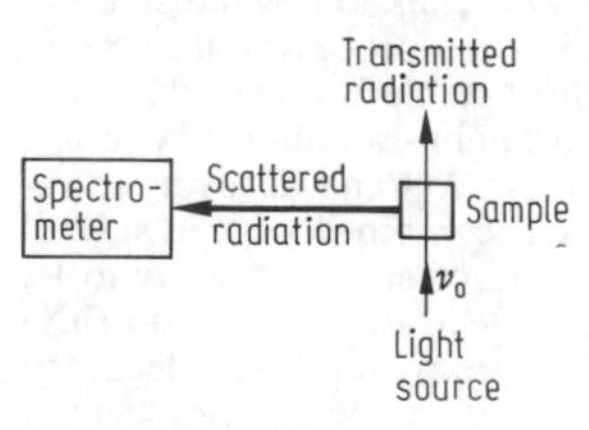

Fig. 1. Scattered radiation as the origin of the Raman spectrum.

The scattering intensity is proportional to ν^4, so that scattered light is observed only when visible or ultraviolet light is used. Spectral dispersion of the scattered light in a spectrometer reveals primarily the incident frequency ν_o (*Rayleigh scattering*, but in addition there are other frequencies (*Raman scattering*) located symmetrically above and below the *Rayleigh line*) (Fig. 2).

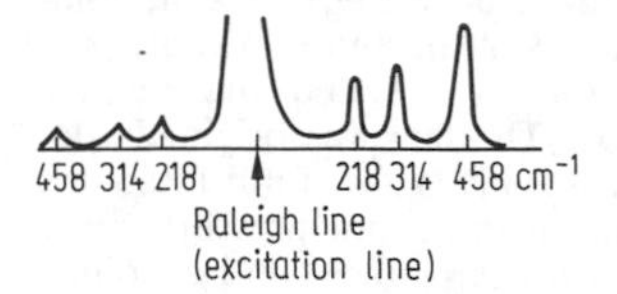

Fig. 2. Distance between the Stokes and anti-Stokes lines and the Rayleigh line in the scattering spectrum of carbon tetrachloride.

The difference in frequencies $\Delta\nu$ between the Raman and Rayleigh lines correspond to the vibrational frequencies ν_v of the molecule. The Raman lines which are at lower frequency or wavenumber than the Rayleigh line, and thus at lower energy, are called *Stokes lines*. They arise from transitions in which molecules which were in the vibrational ground state absorb energy from the excitation light and enter an excited vibrational state (Fig. 3a). The difference between the excitation (Rayleigh) line and the Raman line therefore corresponds to the vibrational frequency. Molecules which are already in an excited vibrational state can emit this energy through interaction with the excitatory radiation. The result is *anti-Stokes lines*, which have a higher frequency or wavenumber than the excitation line (Fig. 3b).

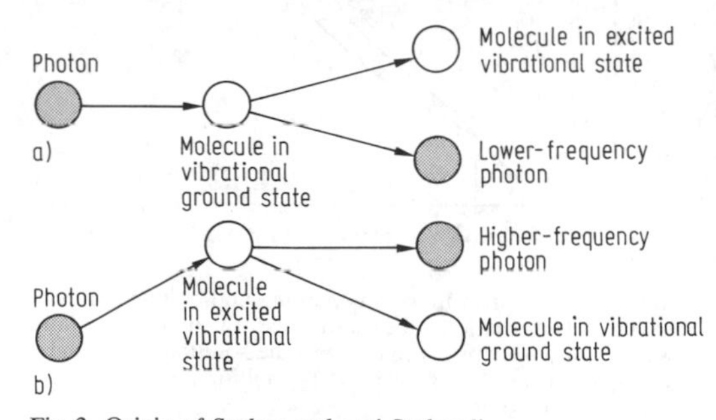

Fig. 3. Origin of Stokes and anti-Stokes lines.

Since most molecules are in the vibrational ground state, the intensity of the Stokes lines is greater than that of the anti-Stokes lines. The Raman spectrum of a substance is obtained when the intensities of the individual Raman lines are plotted against their distances from the excitation line, which are usually expressed in wavenumbers. The wavenumber differences observed with R. are independent of the excitation frequency used; they correspond to the positions of absorption bands observed in IR spectroscopy (see). Raman and IR spectra are therefore comparable, but are only partially identical, due to the difference in excitation conditions. Both IR absorption and the Raman effect are subject to certain selection rules. Only those vibrations which lead to a change in the polarizability α of the electron shell are Raman active and observed in the Raman spectrum. $\delta\alpha/\delta r \neq 0$: Raman active; $\delta\alpha/\delta r = 0$: Raman inactive, where r is the interatomic distance. Some vibrations can be seen in the Raman spectrum which are not seen in the IR spectrum, and conversely. Other vibrations may be both Raman and IR active or inactive. The distribution of the eigenvibrations of a molecule between the Raman and IR spectrum permit inferences to be drawn concerning its symmetry. In molecules with centers of symmetry, for example, a vibration can be either IR or Raman active, but not both. Therefore IR and Raman spectroscopy must be combined to give a complete picture of the vibrational behavior of a molecule. In R., not only frequency and intensity but also the depolarization degree p is important. When polarized laser light is used, a polarizing filter is placed after the sample in the light path, and for every line the intensity I is measured once with the filter parallel to the direction of polarization of the incident light, and once with it perpendicular. The depolarization degree is $p = I_\perp / I_\parallel$. It is important in the description of the symmetry of vibrations. For totally symmetric vibrations in isotropic molecules (e.g. carbon tetrachloride), $p = 0$; for less symmetric vibrations the value of p can be as high as 0.75.

Instrumentation. Raman spectra can be made from transparent, optically empty, non-fluorescent samples, especially liquids and solutions, and, with somewhat more difficulty, from gases and solids as well. In modern instruments, only a few milligrams of substance are needed for a spectrum.

The scattered light examined by R. is of relatively low intensity. The Rayleigh scattering is about 10^{-3} as intense as the excititation radiation, and the Raman scattering is only 10^{-6} to 10^{-7} as intense. The light sources must therefore be very bright, a maximum amount of the scattered light must be collected, and the detector must be as sensitive as possible. ***Raman spectrometers*** are one-beam instruments which measure the intensity of the scattered light directly.

There have been three stages in the development of Raman spectrometers. In the first, the light source was usually a mercury arc, which has strong emission lines at 253.7, 365.0, 405.7, 435.8, 546.1, 577.0 and 579.0 nm. All but the desired excitation line, usually the intense line at 435.8 nm, were filtered out of this emission spectrum. The detector was a photographic plate with which it was possible to collect the scattered radiation for a long period, sometimes up to several hours. In the second stage, the use of highly sensitive secondary electron amplifiers as detectors brought a great advance, because they significantly decreased the time required to record a spectrum, and made possible the construction of automatically recording Raman spectrometers.

The greatest improvement (3rd stage) was brought about by the use of suitable lasers as light sources; these are nearly ideal light sources for R. because of their high intensities and extreme monochromaticity. Lasers with various frequencies and also tunable lasers are used.

Applications. Since the introduction of lasers, R. has been more widely used in chemistry. Within the limits of the selection rules, it has essentially the same applications as infrared spectroscopy, so that here the two methods will only be compared. Glass or quartz cuvettes are used in R. which, unlike the alkali halide cuvettes used in IR spectroscopy, can be used with any solvent system (aqueous, acid, alkaline), which is particularly important for studies of biological systems (e.g. aqueous solutions of amino acids, body fluids) and of inorganic aqueous systems. Modern Raman spectrometers can be used to detect very low absorption frequencies (less than 250 cm^{-1}), which cannot be measured in conventional IR spectrometers and require special far-IR instruments. Because of the difference in selection rules, the Raman spectrum gives more information over the skeletal vibrations of a molecule, while the vibrations of polar groups appear very strongly in the IR spectrum. The Raman spectrum is often less confusing than the IR spectrum, because there are no combination and harmonic vibrations.

For structure elucidation, it is very useful that some important molecular vibrations which are inactive or have very low intensity in the IR are easily recognized in the Raman spectrum. This is true, for example, for symmetrically substituted C=C and C≡C bonds, about which the IR spectrum gives no information. There is a linear relationship between the number of scattering molecules and the intensity of light scattering, which can be used for quantitative measurements. However, since it requires a considerable effort to determine absolute scattering intensities, in most quantitative determinations relative intensities are determined instead and are then compared to a a standard curve to obtain quantitative results. Raman spectra and IR spectra can give the same results on a substance, and there is no general reason to prefer one over the other. On the contrary, the choice will depend on the problem under study. In many cases, especially in elucidation of structure, it will be an advantage to use both methods.

Ranatensin: see Bombesins.

Raney catalysts: see Catalysis, sect. III.1.

Raoult's law: a thermodynamic equation relating the partial vapor pressure p_i of substance i over a mixture to the composition of the liquid phase: $p_i = p_{0,i}x_i$. Here $p_{0,i}$ is the vapor pressure of substance i in the pure state, and x_i is its molar fraction in the liquid phase. If substance i forms a homogeneous mixture with the other substances, its vapor pressure is lower than $p_{0,i}$ by an amount which is proportional to its molar fraction x_i; at infinite dilution, $x_i = 0$.

In this form, R. is a limit law, which applies only to ideal mixtures, or approximately to very dilute solutions. If interactions between the components in the mixture are different from the interactions in the pure substances (real mixture), the Activity (see) $a_i = f_ix_i$ must be used instead of the mole fraction; here f_i is the activity coefficient. In this case, the R. is $p_i = p_{0,i}a_i = p_{0,i}x_if_i$. If the attractive forces predominate, the partial pressure will be lower than in the ideal case, and $f_i < 1$. If the repulsive forces dominate, the volatility is increased ($f_i > 1$).

Applications of R.: 1) If the mixture consists of two volatile components 1 and 2, the total vapor pressure p over the mixture is, according to Dalton's law (see), equal to the sum of the partial pressures: $p = p_1 + p_2$. In the ideal case, R. has the form $p_1 = p_{0,1}x_1$ and $p_2 = p_{0,2}x_2$. These relationships are the basis for vapor pressure diagrams and the separation of substances by distillation.

2) If the binary mixture contains a substance 2 which has no significant vapor pressure, the mixture is usually called a Solution (see), and substance 2 is the solute, while substance 1 is the solvent. In the ideal case, the total pressure $p = p_1 = p_{0,1}x_1 = p_{0,1}(1-x_2)$, because $p_2 = 0$. From this, it follows that $x_2 = (p_{0,1}-p)/p_{0,1})$, i.e. the vapor pressure p over the solution is lower than that over the pure solvent, $p_{0,1}$. The expression $(p_{0,1}-p)/p_{0,1}$ is called the relative *vapor pressure lowering*. The law was originally derived in 1882 by F.M. Raoult in this form (Raoult's first law). The lowering of the vapor pressure curve causes a Freezing point lowering (see) and a Boiling point elevation (see) (Fig.). If the shift in these two fixed points, Δp, is expressed in terms of the Clausius-Clapeyron equation (see), Raoult's second law is obtained: $\Delta T = Em_2$, where $E = RT^2M_1/\Delta_pH$, R is the general gas constant, T the freezing or boiling point, Δ_pH the associated molar phase change enthalpy and m the molar concentration of substance 2. E is called the molar freezing point lowering or boiling point elevation, or the cryoscopic or ebullioscopic constant. The two special forms of the R. are limit equations for the behavior of ideal dilute solutions. Deviations from ideal behavior can be corrected by introduction of activity coefficients. If the solute dissociates in solution, the number of dissolved particles, and thus the freezing point lowering or boiling point elevation, is increased (van't Hoff).

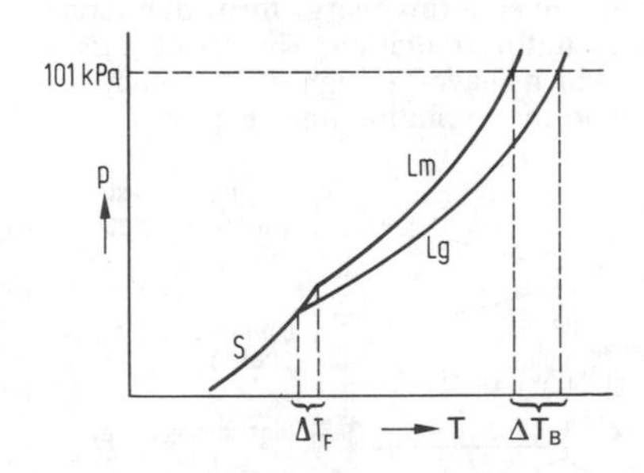

Effect of vapor pressure on the melting and boiling points of a solution (schematic). Lm, vapor pressure curve of the pure solvent, Lg, vapor pressure curve of the solution, S, sublimation curve of the solid solvent, ΔT_B, boiling point elevation, ΔT_F, freezing point.

Rare earth metals: the elements scandium (Sc), yttrium (Y) and lanthanum (La) (see Scandium group), of Group IIIb of the periodic system, and the fourteen Lanthanides (see) following lanthanum, in which the 4f shell is filled. Actinium and the following Actinides (see) also belong in Group IIIb, but because of their radioactivity they are generally not considered part of the rare earth group.

R. are soft, malleable and reactive metals which react at high temperatures with hydrogen, oxygen, nitrogen, carbon or halogens to form binary derivatives. They are usually trivalent in compounds, and mixtures of them are found in natural deposits. The R. are neither rare nor earth metals; however, relatively concentrated deposits of them are uncommon.

Rare earths: oxides of scandium, yttrium, lanthanum and the lanthanoids.

Raschig process: 1) a process for synthesis of phenol from benzene in two steps. In the first, benzene reacts at 250 °C with hydrochloric acid and air in the presence of activated aluminum hydroxide and copper salts as catalysts; the product is chlorobenzene: $C_6H_6 + HCl + 1/2O_2 \rightarrow C_6H_5Cl + H_2O$. This reaction is highly exothermic, so efficient means of removing the heat must be provided. In the second step, the chlorobenzene is saponified at 480 °C in the presence of copper compounds as catalysts, yielding phenol and hydrochloric acid: $C_6H_5Cl + H_2O \rightarrow C_6H_5OH + HCl$.

An advantage of the R. is that 97% of the HCl can be recycled. A disadvantage is that the moist hydrochloric acid is extremely agressive, and in the presence of chlorobenzene and phenol it attacks even special steels.

2) A process for producing Hydrazine (see).

Raschig rings: see Fillers.

Rate constant, ***reaction constant***: the proportionality factor in kinetic equations, denoted by k. It is the central parameter of reaction kinetics (see Kinetics of reactions), having a characteristic value for each reaction, at constant temperature and in a given solvent, which is independent of the concentrations of the reactants. The temperature dependence of the R. is given by the Arrhenius equation (see) on the Eyring equation (see Kinetics of reactions, theory.) The R. is closely related to the Half-life (see).

Rate-determining step: the step in a complex reaction which is significantly slower than the other elementary reactions, so that it determines the rate of the total reactions, and thus the rate equation (see Reaction, complex).

Rate equation: in kinetics, an equation which relates the rate of a reaction to the concentrations of the reactants; see Kinetics of reactions, see Reaction, complex).

Rauwolfia alkaloids: a group of alkaloids containing the β-carboline (norharmane) skeleton. β-Carboline forms colorless needles; m.p. 198 °C. There are many species of Rauwolfia, of which the most important is *Rauwolfia serpentina*.

There are three groups of R.: 1) quarternary anhydronium bases, such as *serpentine*, m.p. 205 °C, $[\alpha]_D^{20}$ +188° in water, which reduces blood pressure.

2) Tertiary indoline alkaloids, such as *ajmaline*, m.p. 160 °C, $[\alpha]_D^{23}$ +128 °C in $CHCl_3$. It is used therapeutcically as an antiarrhythmic. The derivative ***detajmium bitartrate***, obtained by quaternization, has a stronger effect.

3) Tertiary indole bases, such as *reserpine*, m.p. 263 °C, $[\alpha]_D^{23}$ -118 °C in chloroform. Of the approximately 40 R. so far isolated, reserpine has become the most important. It was isolated and its structure elucidated in 1952 by Schlittler. In 1956, its total synthesis was announced by Woodward. It is used to reduce blood pressure and as a neuroleptic. Its effect is based on a reduction of the storage capacity for catecholamines in the vesicles of the peripheral sympathetic nerve fibers and the corresponding structures of the central nervous system. Efforts have been made to separate the two effects by partial synthesis of derivatives, mainly with altered acid components.

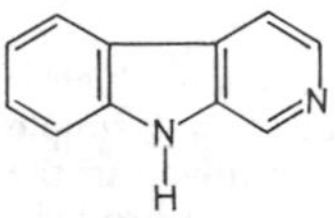

β-Carboline

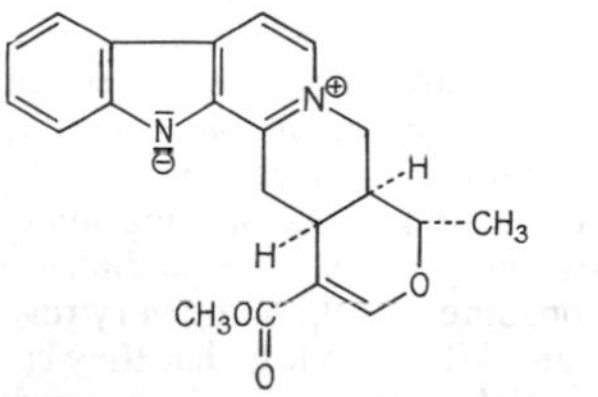

Serpentine

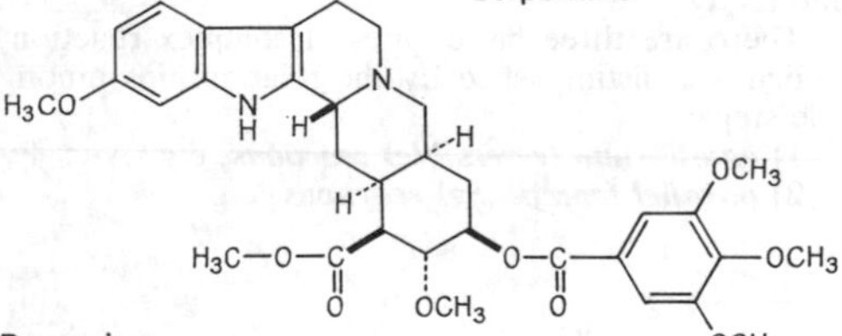

Reserpine

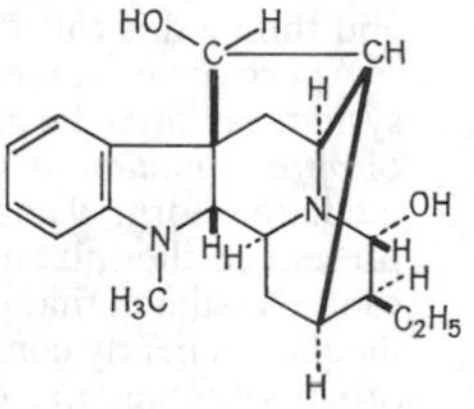

Ajmaline

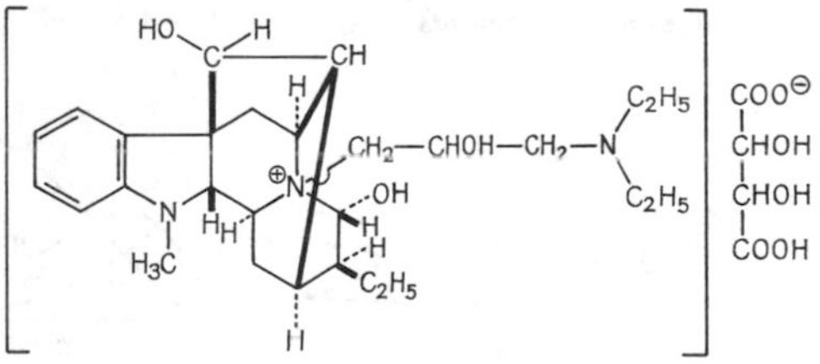

Detajmium bitartrate

Reactive components of reactive dyes.

Rayleigh lines: see Raman spectroscopy.

Rayleigh scattered radiation: see Raman spectroscopy.

Rayon: synthetic fibers made by regeneration of dissolved cellulose, for example, cuprammonia and viscose R.

Modified rayon materials: synthetic fibers made from regenerated cellulose by a modified viscose spinning process. They are subdivided into ***high wet modulus fibers (HWM)*** and ***polynose fibers*** (from the French "polymère non synthètique", non-synthetic polymer). M. are produced using certain spinning conditions with specific modifiers, and their molecular structure is very similar to that of cotton. The starting material is alkali cellulose with an average degree of polymerization of 450, or cellulose acetate fibers which have been converted to polynose fibers by partial hydrolysis under tension.

M. have a stability to moisture and curling which is similar to that of cotton. The HWM fibers can stretch more than the polynose fibers, but they cannot be mercerized with cotton so well. Special qualities can be obtained by spinning in additional substances, such as barium sulfate for x-ray contrast. Textiles made from M. are very resistant to wrinkling and shrinking. They dye readily, are easy to care for, non-irritating to the skin and comfortable. Mixed with cotton or other synthetic fibers, M. are used to make shirts, rainwear, sportswear, underwear and decoration fabrics.

The most important trade names are given under Synthetic fibers (see).

Rb: symbol for Rubidium.

Re: symbol for rhenium.

Reaction, complex: a chemical reaction which is not completed in one step; instead several simple reactions (see Reaction, simple) are involved in the conversion of the starting material into the product. In many cases, intermediates are formed and then consumed. If these are very reactive, their concentrations will be so low that they can be detected only by special methods, or their presence must be inferred indirectly.

There are three basic types of complex reaction, which are distinguished by the relationships among the steps:

1) ***equilibrium (reversible) reactions***, e.g. $A \rightleftharpoons P$;

2) ***parallel (competing) reactions***, e.g.

$$A \rightarrow P,\quad A \rightarrow Q,\quad A \rightarrow R$$

3) ***sequential reactions***, e.g. $A \rightarrow B \rightarrow C$.

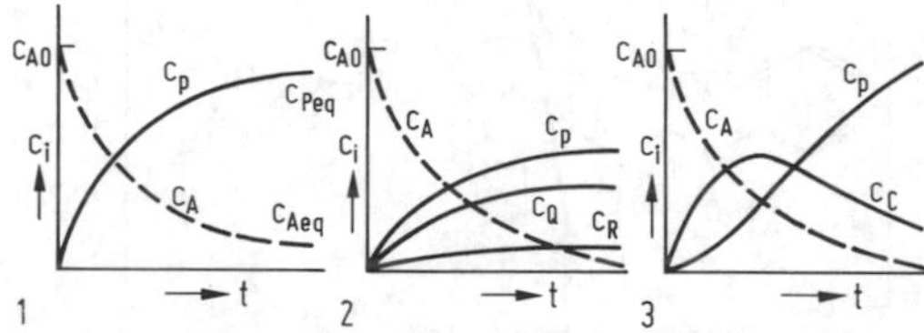

Plots of concentration vs. time for complex reactions: 1) Equilibrium, 2) parallel and 3) sequential reactions. $c_{P_{eq}}$ and $c_{R_{eq}}$ are equilibrium concentrations.

These basic types can also be coupled:

4) ***Sequential reactions with reversible steps***, e.g. $A + B \rightleftharpoons C \rightarrow P + Q$;

5) ***Sequential reactions with parallel steps***, e.g.

$$A{+}B \rightarrow C \rightarrow P,\quad C \rightarrow Q,\quad C \rightarrow R$$

6) ***Competing sequential reactions***, e.g. $A + B \rightarrow P + Q$ and $A + B \rightarrow R + S$;

7) ***Parallel reactions with reversible steps, e.g.***

$$A{+}B \rightleftharpoons P,\quad A{+}B \rightarrow Q$$

8) ***Chain reactions***, e.g. $A \rightarrow 2X$,

$$A + X \rightarrow Y + P \rightarrow X + Q.$$

(with X returned to the first step)

Catalytic reactions are always complex, because the catalyst forms an intermediate compound with the substrate (see Catalysis). If products of a reaction act as catalysts, which either accelerate or slow the reaction, it is called an ***autocatalytic reaction***. Such complex reactions can have anomalous behavior under certain conditions (see Oscillating reaction).

Formal kinetic treatment of complex reactions requires a mathematical description of the changes in the concentrations c_i of all reactants, intermediates and products i. It must be taken into account that the change in concentration dc_i/dt of each substance i is the sum of its rates of formation by all steps minus its rate of consumption by all steps j in which it is involved, i.e. $dc_i dt = \sum_i \nu_{ij} r_j$, where ν_{ij} is the stoichiometric factor for substance i in reaction j. The result for most complex reactions is complicated systems of differential equations, which are usually not linear and can be solved only in the simplest cases. They are always soluble if all the steps are first order. A few particularly simple cases are summarized in the table. If a closed solution is not possible, the system of equations must be numerically integrated in steps. There are various computer programs for these calculations. If the rate constants of all steps are known, the calculation gives the concentration progress curves for all the substances involved. If the rate constants have to be calculated, the calculated and experimentally measured progress curves have to be matched according to certain criteria for minimizing error. This approach is called parameter estimation, and the field is called Kinetic modelling (see).

Approximation methods. The differential equation system can often be greatly simplified by introduction of *approximation assumptions*: 1) If a reactive intermediate is formed and rapidly reacts further, it will be present in the mixture in very low concentration. It can be assumed that the concentration c_i of the intermediate is nearly constant over a large portion of the progress curve, i.e. $dc_i/dt = 0$ (***quasi-stationary*** or ***Bodenstein approximation***). 2) If a complex reaction includes reversible steps, in which the back reactions are very rapid in comparison to the other steps, it can be assumed that they are in chemical equilibrium. The concentration of the reactants can be approximately calculated from the mass action law (***quasi-equilibrium approximation***). 3) If one step in a sequential reaction is much slower than the others, the rate of product formation and the kinetic equation will be determined by this step (***rate-limiting step approximation***). Application of these approximations often yields simple kinetic equations for complex reactions, with integral or fractional orders. Therefore, if a reaction is second-order, for example, this fact is not a proof that it is bimolecular.

Kinetics of the simplest basic types of complex reactions (table).

1) ***Reversible reaction*** (Fig.): The Mass action law (see) can be derived kinetically from the progress curve.

2) ***Parallel reaction***: if all the parallel reactions have the same kinetics, the total rate constant k is the sum of the k values for the individual steps: $k = \Sigma k_i$. If no products are present at the beginning of the reaction, then at each point in time, the ratio of product concentrations is equal to the ratio of the rate constants (Fig. 2; see Wegscheider principle). The ratio of k values is also called Selectivity (see).

If several reactants B, C and D compete for a reactant A, the rate equation will contain the concentrations c_B, c_C and c_D in addition to c_A. The Wegscheider principle does not apply in this case, because the rates also depend on the arbitrarily variable concentrations of the various other components.

3) ***Sequential reactions*** of two first order steps: The concentration of intermediate B goes through a maximum (Fig. 3). At this point, B is formed and consumed at equal rates. If B is the desired product of a reaction, it must be interrupted at this point. From $dc_D/dt = 0$, it follows for the required reaction time

$$t_{max} = \frac{1}{k_2 - k_1} \ln \frac{k_2}{k_1}$$

$$c_{Bmax} = C_{OA}\,(k_2/k_1)^{1/(k_1/k_2-1)}$$

If a rate constant is known, e.g. k_1 from the initial phase of the reaction, k_2 can be calculated from t_{max}. Most chemical reactions are sequential; in addition, there are often irreversible steps or parallel reactions which lead to formation of byproducts.

4) ***Chain reactions***: Typical examples are polymerizations, radical halogenations of hydrocarbons, combustion, autooxidations and pyrolysis of many organic compounds. Chain reactions consist of three stages: chain initiation, propagation and termination. In the *chain initiation*, reactive intermediates, the chain carriers (e.g. radicals, atoms, ions or radical ions) are formed. In halogenations, the initiation is a thermal or photochemical dissociation of the halogen, e.g.

$$Cl_2 \xrightarrow{h\nu} 2\,Cl^{\cdot},$$

and in autooxidations, it is a redox process involving atmospheric oxygen, e.g. $RH + O_2 \rightarrow R\cdot + \cdot O\text{-}O\text{-}H$. The reaction can also be started by Initiators (see). In the phase of *chain propagation*, a chain carrier reacts with a substrate molecule, each time producing a new chain carrier which can enter the reaction again. The result is a reaction cycle in which the starting compounds are constantly consumed, but the primary chain carriers are regenerated. The chain ends with *chain termination*, in which chain carriers are destroyed. The chain termination can occur as a bimolecular combination of two radicals, or it can be a monomolecular adsorption of the chain carrier on the wall or a reaction with an Inhibitor (see) which converts the chain carrier into an unreactive intermediate. The nature of the termination reaction has a major effect on the form of the kinetic rate equation. The following simple chain scheme applies to monochlorination of hydrocarbons:

Initiation: $Cl_2 \rightarrow 2\,Cl\cdot$ (1)
Propagation: $Cl\cdot + RH \rightarrow HCl + R\cdot$ (2)
$R\cdot + Cl_2 \rightarrow RCl + Cl\cdot$ (3)
Termination: $2\,Cl\cdot \rightarrow Cl_2$ (4)

A chain cycle consists of steps (2) and (3), which regenerate $Cl\cdot$ so that it can react again according to reaction (2). Under conditions of quasi-stationary concentrations (Bodenstein approximation) of the chlorine atom $Cl\cdot$ and the alkyl radicals $R\cdot$, the rate law for this reaction is $dC_{RCl}/Dt = kc_{RH}c_{Cl_2}1/2$, where k is an effective rate constant composed of the rate constants of steps (1), (2) and (4) according to the equation $k = k_2(k_1/k_4)^{1/2}$. The *chain length* is the average number of chain cycles in which a chain carrier participates between its formation in the initiation

Reaction	Rate law	Integrated forms
1) $A \underset{2}{\overset{1}{\rightleftharpoons}} P$	$-\frac{dc_A}{dt} = \frac{dc_P}{dt} = k_1c_A - k_2c_P$	$c_P = c_{oA} - c_A = (c_{oA} - c_{eA})(1 - e^{-(k_1+k_2)t})$
2) $A \xrightarrow{1} P$, $A \xrightarrow{2} Q$, $A \xrightarrow{3} R$	$-\frac{dc_A}{dt} = (k_1 + k_2 + k_3)c_A$	$c_A = c_{oA}\,e^{-(k_1+k_2+k_3)t}$
	$-\frac{dc_Q}{dt} = k_2c_A$	$c_Q = \frac{k_2}{k_1 + k_2 + k_3}\,c_{oA}(1 - e^{-(k_1+k_2+k_3)t})$
3) $A \xrightarrow{1} B \xrightarrow{2} C$	$-\frac{dc_A}{dt} = k_1c_A$	$c_A = c_{oA}\,e^{-k_1t}$
	$-\frac{dc_B}{dt} = k_1c_A - k_2c_B$	$c_B = \frac{k_1}{k_2 - k_1}\,c_{oA}(e^{-k_1t} - e^{-k_2t})$
	$-\frac{dc_C}{dt} = k_2c_B$	$c_C = c_{oA}\left[1 + \frac{1}{k_1 - k_2}(k_2e^{-k_1t} - k_1e^{-k_2t})\right]$

Rate laws of the simplest complex reactions.

1) Equilibrium reaction, 2) parallel reaction, 3) sequential reaction; c_{oA}, original concentration of reactant A, c_{eA}, equilibrium concentration of A. The initial concentrations of all intermediates and end products are 0.

reaction and its destruction in the termination reaction. The average chain length can be calculated as the ratio of the chain propagation rate/initiation rate. Depending on the kind of chain reaction, chain lengths between about 1 and 10^6 may be observed. In polymerizations, the chain length determines the size of the resulting macromolecules, and thus the physical properties of the polymer. In photochemically initiated chain reactions, the quantum yields are considerably greater than one. For example, chain lengths of 10^4 to 10^5 have been observed for the chlorine detonating gas reaction, $H_2 + Cl_2 \rightarrow 2\ HCl$.

Branched chain reactions are a special type which are important in combustion reactions. The chain cycles contain additional branching steps in which two chain carriers are formed from one, and each can introduce a new cycle. In the detonating gas reaction $2\ H_2 + O_2 \rightarrow 2\ H_2O$, the reactions $H\cdot + O_2 \rightarrow \cdot OH + \cdot O\cdot$ and $\cdot O\cdot + H_2 \rightarrow \cdot OH + H\cdot$ are branching reactions. Another possibility is the decomposition of unstable intermediate products into two radicals (***degenerative branching***). For example, in the hydrocarbon oxidation, hydroperoxides ROOH are formed, which can easily be cleaved thermally according to: $ROOH \rightarrow RO^{\cdot} + \cdot OH$, thus forming new chain carriers. In such reactions, it often occurs that the rate of formation of chain carriers is greater than the rate of their destruction by termination reactions. The radical concentration, and thus the rate of the total reaction increases exponentially, causing a chain explosion (see Explosion). The situation is analogous to that of nuclear explosions, in which the neutrons play the role of chain carriers.

Reaction constant: same as Rate constant (see).

Reaction coordinate: see Potential surface.

Reaction cross section: see Kinetics of reactions (theory).

Reaction degree: see Turnover.

Reaction, elementary: a chemical reaction which occurs in one step, i.e. without formation of intermediates. The reaction is the statistical average of a large number of microscopic elementary processes in which the colliding particles have different energies. In reaction kinetics, elementary reactions are classified according to the number of particles which collide in them as mono-, bi- or trimolecular. ***Mono-*** or ***unimolecular reactions*** $A \rightarrow B + C$ include decomposition and isomerization reactions. In ***bimolecular reactions*** $A + B \rightarrow C + D$, two identical or different particles interact, e.g. in substitutions or additions. ***Trimolecular reactions*** $A + B + C \rightarrow D + \ldots$ require the collision of three identical or different particles to initiate the process. An example is a combination of atoms in which the bonding energy released by formation of a molecule must be transferred to a third particle M which carries it away: $H\cdot + H\cdot + M \rightarrow H_2 + M$. The collision of more than three particles is so improbable that reactions with higher molecularity than three practically do not occur. If the stoichiometry of the reaction requires interaction of more than three molecules, it will always occur in a series of several elementary reactions, and is thus a complex reaction (see Reaction, complex).

The theory of reaction kinetics can be used to demonstrate that monomolecular elementary reactions have first order rate equations; bimolecular reactions have second order, and trimolecular reactions have third order. This means that for elementary reactions, the rate equation can be deduced from the stoichiometry. The reaction orders relative to the individual reactants are the same as their stoichiometric coefficients. For complex reactions, this method does not apply.

Examples: The reaction $2\ NO + O_2 \rightarrow 2\ NO_2$ is now thought to be an elementary reaction. The rate equation is thus $r = kc_{NO}^2 c_{O_2}$. The reaction $N_2 + 3\ H_2 \rightarrow 2\ NH_3$ cannot be an elementary reaction, because four molecules would have to collide. Its rate equation therefore cannot be derived from the stoichiometry, but must be determined experimentally.

The application of the concepts of elementary reactions and molecularity is, strictly speaking, defined only for gas reactions, because the particles in condensed phases constantly interact with the environment. However, the concepts can be applied to reactions in solution, if the interactions with the solvent can be ignored.

Reaction energy: see Reaction heat.

Reaction enthalpy: see Reaction heat.

Reaction equation: see Chemical reaction equation.

Reaction heat: the heat ***q*** consumed or released by a chemical reaction. It is generally given for one formula mole turnover (***molar R.***), and uses the symbol *Q*. In industry, the heat is sometimes still given for a mass, such as kg or ton, of a solid or liquid, or for a volume of gas (see, e.g., Heating value). If heat is released by a reaction (*exothermic reaction*), *Q* is negative; if the reaction is *endothermic*, *Q* is positive. There is a difference between the molar R. Q_v of a reaction at constant volume, and Q_p of a reaction at constant pressure. Q_p includes the work done by the reacting system as its volume changes at constant pressure: $-p\Delta V$, so the two heats are related by $Q_v = Q_p - p\Delta V$.

In gas reactions where the number of moles changes, there are large changes in volume. If the ideal gas law (see State equation 1.1) is assumed to apply, it follows that $-p\Delta V = -\Delta nRT$, where Δn is the change in number of moles. For example, in the synthesis of ammonia, $N_2 + 2\ H_2 \rightarrow 2\ NH_3$, $\Delta n = -2$ and R (general gas constant) $= 8.314$ J mol^{-1} K^{-1}. Taking $T = 298$ K, we can calculate $-p\Delta V \approx 5$ kJ mol^{-1}. For reactions in the liquid and solid phases, the volume effects and thus the difference between Q_v and Q_p are about 10^3 times smaller.

R. are due to the differences in internal energies and enthalpies of the reactants and products. If the reaction is carried out irreversibly, i.e. without addition or release of usable work, the R. Q_v is equal to the molar ***reaction energy*** $\Delta_R U$, and Q_p is equal to the molar ***reaction enthalpy*** $\Delta_R H$ (see Thermodynamics, first law). These can be calculated as sums of standard enthalpies or energies of formation which are published in tables. R. are measured experimentally in calorimeters. They are temperature-dependent (see Kirchhoff's law). If there are several reaction paths, the various R. are related by Hess' law (see).

Reaction isobar: see Mass action law.

Reaction isochore: see Mass action law.

Reaction isotherm: see Mass action law.

Reaction kinetics: see Kinetics of reactions.

Reaction mechanism: term for the detailed course of a chemical reaction. A description of the R. includes the number and types of steps or elementary reactions involved, the particles which interact and the Intermediates (see) formed and consumed. The R. also includes a description of the electron redistributions occurring between the reactants and their spatial orientation.

Reactions which occur in a single step are called simple or elementary reactions, while those which involve more than one are called complex reactions (see Kinetics of reactions and Reactions, complex). R. can be classified according to the electron processes as electron-transfer (redox), synchronous, radical, ionic or polar. In addition, the nature of the overall process is of essence, e.g. R. of additions, substitutions, eliminations or rearrangements. The elucidation of R. is one of the chief goals of chemistry. Knowledge of R. makes possible control of reactions and suggests further synthesis possibilities. However, it must be remembered that when an R. is published, it is only a model which reflects the current state of knowledge.

Reaction order: see Kinetics of reactions.

Reaction rate: see Kinetics of reactions.

Reaction technology: see Process technology.

Reactive dyes: a group of dyes which contain reactive groups, usually heterocyclic ring structures with imide chloride structures, in addition to the actual chromophores. The reactive groups make possible the formation of covalent bonds between the dye and fiber molecules. The reactive groups are designated as follows (Fig.): *Type I* contains a chlorotriazinyl group, *type Ia* a chloropyrimidyl group, *type II* an activated vinyl group, *type III* an epoxide group and *type IIIa*, an ethylenimine group.

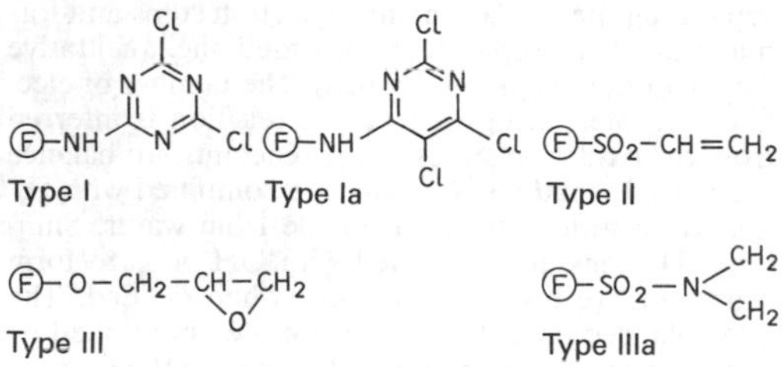

Ⓕ–NH–(dichlorotriazinyl) Type I
Ⓕ–NH–(trichloropyrimidyl) Type Ia
Ⓕ–SO_2–CH=CH_2 Type II
Ⓕ–O–CH_2–CH–CH_2 (epoxide, O) Type III
Ⓕ–SO_2–N(CH_2)(CH_2) Type IIIa

In principle, any organic pigment is a suitable chromophore for the R. The most commonly used are azo, pyrazolone, anthraquinone and phthalocyanin pigments which are linked by amino (-NH_2), hydroxy (-OH) or sulfo (-SO_3H) groups to the reactive group. In the dyeing process, the reactive component forms a true chemical bond with the free hydroxyl groups of cellulose, or with primary or secondary amino groups of wool, silk or polyamide fibers. The result is an especially stable color. The first R. for cotton were procoin dyes, introduced in 1956 by the ICI (England).

Reactivity: the ability of a substance to react with other substances. R. is not an intrinsic property of a substance, but always depends on its reaction partner, and also on the reaction conditions such as solvents, temperature or degree of distribution in heterogeneous reactions. Rate constants (see Kinetics of reactions) can be taken as quantitative measures of R.

An important problem in chemistry is the effect of electronic and spatial structure of a compound in its R. Such comparisons are especially useful within series of similar reaction types, homologous substances or analogous skeletons with different substituents. An example of such relationships between structure and R. are the LFE relationships (see). If a substrate has several reactive centers, the preferred position of attack by a reagent is significant (see Selectivity). It is often true, but not always, that the more reactive a reagent is, the less selective it is.

Reactor: 1) a plant for production of energy from nuclear reactions.

2) a device in which a chemical reaction is carried out. R. are classified as Stirred reactors (see) or Pipe reactors (see), depending on the principle of material flow in them. Heterogeneous reactions, e.g. gas-liquid or gas-solid reactions, are carried out in fluidized bed, moving bed, fixed bed, trickle-phase or thin-layer reactors. R. can also be classified on the basis of their thermal properties; in ***adiabatic R.***, there is no heat exchange through the walls, in ***isothermal R.***, the temperature is held constant, both spatially and in time, and in ***polytropic R.***, part of the reaction heat is transported through the walls to a coolant. Another system of classification distinguishes between differential, integral and cyclic R., depending on the turnover. Bioreactors (fermenters) and photoreactors are special types of R. Laboratory and pilot R. are often built to determine experimentally the optimum design and materials for R. construction.

Reagent: in general, any chemical substance used in a chemical reaction.

Reagent paper: a filter paper impregnated with indicators or reagents. R. used to show the pH are also called indicator papers (see Indicator). Recently such papers are available in which the indicator is chemically bound to the cellulose. These are called ***non-bleeding R.***, because the indicator does not wash out of them even when left in solution for a long time. They are not very susceptible to errors.

R. can be used as a simple means of detecting substances in solutions. A strip of the R. is either dipped into the solution, or a droplet of the solution is placed on the R., and the resulting color is observed. There are also R. which provide a semiquantitative indication of the amount of the substance by means of a color scale. In Droplet analysis (see), mainly home-made R. are used.

Recombination: see Gene technology.

Recrystallization: crystal growth within a Polycrystalline material (see) in which grain boundaries migrate and are reformed. R. is seen in metallic materials which have large lattice defects, due to previous plastic deformation, rapid quenching or other causes, if the materials are held for a long time above their R. temperature (tempered). The R. temperature depends mainly on the melting point, but also on the degree of deformation of the treated material, and is, for example, about 1500 K for tungsten, 700 K for iron and 300 K for tin. The starting point of R. is the formation of seeds of new, undistorted crystallites or grains, a process which tends to begin at highly deformed sites in the sample. Subsequent growth of the grain until the old lattice is completely consumed makes it possible for R. to remove the traces of a plastic deformation. R. can also occur as rapid, preferential growth of one crystallite at the expense of surrounding ones, so that R. can be used as a method for production of nearly perfect single crystals, e.g. of lead, iron or aluminum (see Crystal growing). Since the grain size is an important factor determining the hardness of a metal lattice, R. is very important for the mechanical properties of metallic materials. The connection between grain size and the degree of deformation and temperature is shown in R. diagrams.

Rectification, ***counter-durrent distillation***: the separation of mixtures of liquids with small differences between their boiling points. R. is a process of evaporation and recondensation; see Distillation.

Rectisol process: a process for removing acidic components (carbon dioxide, hydrogen sulfide and carbon oxygen sulfide) from natural, city and synthesis gases. The R. is similar to the Sulfinol process (see); methanol is used as a solvent for the acidic gases.

Recycling: 1) see Waste product; 2) see Natural gas.

Recycling reactor: a device for carrying out chemical reactions in which a fraction of the product flow leaving the reactor is cycled back into the reactor. R. are used where the reaction does not go to completion in one pass, or where the catalyst or solvent must be returned to the reactor.

Redispersion, ***peptization***: a spontaneous dispersion of particle aggregates in which the particles have not coalesced; instead, they are separated by a layer of dispersion medium. R. can be induced by reducing the electrolyte concentration in the medium (see Coagulation) so that the energy of Brownian motion (see) is sufficient to overcome the particle attraction.

Redox analysis, ***oxidimetry, redox titrations***: a type of volumetric analysis in which the unknown concentration of a reductant in solution is determined by addition of a standard solution of an oxidizing agent. The reverse process, using a standard solution of a reducing agent, is very rarely used. A R. can be carried out only if the standard electrode potentials of the redox system to be analysed and the titrator are markedly different. Because of the constantly changing concentration ratios in the solution, its electrode potential changes (see Redox reactions). This is often plotted as a titration curve. At the equivalence point, addition of a small amount of titrator standard solution causes a very large change in potential. The equivalence point can be indicated electrochemically (see Potentiometry) or a suitable redox indicator (see Indicators) can be used. In some cases, the colors of the reactants themselves can be used (permanganate or iodine).

Some frequently used oxidimetric methods are Permanganometry (see), titanometry (see), Iodometry (see), Cerimetry (see) and Bromatometry (see). Iodatometry, in which iodate ions are used as titrator, is less common.

Redox chromatography: see Ion exchange chromatography.

Redox electrode: see Electrodes.

Redox equilibrium: see Redox reaction.

Redox indicators: see Indicator.

Redox method: see Glue.

Redox reaction: a reaction in which electrons are exchanged. Since electrons do not exist free in aqueous solution, oxidation of one reactant (release of electrons) must always be accompanied by reduction (aquisition of electrons) of a second reactant:

Red	$\rightleftharpoons$ Ox	$+ e^-$
Fe^{2+}	$\rightleftharpoons Fe^{3+}$	$+ e^-$
Zn	$\rightleftharpoons Zn^{2+}$	$+ 2e^-$
$2\,Cl^-$	$\rightleftharpoons Cl_2$	$+ 2e^-$
H_2	$\rightleftharpoons 2\,H^+$	$+ 2e^-$

A substance which has been oxidized is potentially able to accept electrons, and thus to act as a reducing agent; similarly, a reduced substance is potentially an oxidizing agent. The two components which can be interconverted by transfer of electrons are called a *corresponding redox pair*:

Red_1	$+ Ox_2$	$\rightleftharpoons Ox_1$	$+ Red_2$	
Zn	$+ Cu^{2+}$	$\rightleftharpoons Zn^{2+}$	$+ Cu$	
$2\,I^-$	$+ Cl_2$	$\rightleftharpoons I_2$	$+ 2\,Cl^-$	
AsO_3^{3-}	$+ I_2 + H_2O$	$\rightleftharpoons AsO_4^{3-}$	$+ 2\,I^-$	$+ 2H^+$
$5\,H_2O_2$	$+ 2MnO_4^- + 6H^+$		$\rightleftharpoons 5\,O_2$	$+ 2Mn^{2+} + 8H_2O$

The analogy to proton transfer reactions and to the concept of corresponding acid-base pairs is apparent (see Acid-base concepts, Brönsted definition).

In the formulation of R., it must be remembered that the number of electrons released by one reactant is always equal to the number accepted by the other. When setting up complicated redox equations, especially when the R. is coupled to an acid-base or complex-formation reaction, it is wise to proceed by separating the oxidation and reduction steps and formulating them separately (provided the qualitative course of the reaction is known). The number of electrons transferred in each partial reaction is inferred from the oxidation states of the reactants. To balance the partial reactions, O^{2-} ions are combined with H^+ ions from water, or are available from water. Similarly, H^+ ions are combined with OH^- ions to form water, or are taken from water when needed. The two balanced partial reactions are then multiplied by whatever factors are required to equalize the number of electrons transferred, then added. For example, the reaction of oxalate with permanganate in acidic solution, yields carbon dioxide and manganese(II) ions; the above procedure leads to the following equations:

$$
\begin{array}{ll|l}
C_2O_4^{2-} & \rightleftharpoons 2\,CO_2 + 2e^- & \times 5 \\
MnO_4^- + 8\,H^+ + 5e^- & \rightleftharpoons Mn^{2+} + 4\,H_2O & \times 2 \\
\hline
5\,C_2O_4^{2-} + 2\,MnO_4^- + 16\,H^+ & \rightleftharpoons 10\,CO_2 + Mn^{2+} + 8\,H_2O &
\end{array}
$$

The position of the *redox equilibrium* and the question of which redox pair is able to oxidize or reduce the other are determined by the electrode potentials E_H of the individual redox systems. According to the Nernst equation, the latter depend on the standard electrode potentials $E_H°$ of the system and the concentration ratio in the solution:

$$E = E_H° + \frac{0.059}{z} \log \frac{c_{ox}}{c_{red}}.$$

Redox pairs with higher, positive electrode potentials act as oxidants of those with lower electrode optentials (see Standard electrode potentials). In the system $Zn/ZnSO_4$-$Cu/CuSO_4$, therefore, if the concentrations are comparable, the redox pair Zn/Zn^{2+} will be oxidized by the Cu/Cu^{2+} pair; the standard potentials of the two are $E_H°$(zinc) = -0.76 V, and $E_H°$(copper) = +0.34 V. This leads to a reduction in the concentration of Cu^{2+} ions and an increase in the Zn^{2+} concentration, so that the two electrode potentials approach each other. Equilibrium has been reached when the two potentials are equal. Then the reaction appears to stop. The concentration dependence of the electrode potentials also makes clear the relativity of the concepts of oxidizing and reducing agents. Under suitable conditions, the roles of the two can be reversed.

When determining the electrode potentials for reactions coupled with protolytic processes, which is usually the case if derivatives of oxoacids are involved in the redox process, one must remember that the Nernst equation includes not only the components immediately involved in electron transfer, but the entire mass action expression. For example, for the redox system $MnO_4^- + 8\,H^+ + 5\,e^- \rightleftharpoons Mn^{2+} + 4\,H_2O$, the Nernst equation is: $E_H(MnO_4^{-1}/Mn^{2+})$ = $E_H°(MnO_4^{-1}/Mn^{2+})$ +

$$\frac{0.059}{5} \log \frac{c_{Mn_4O} - c^8_{H_3O^+}}{c_{Mn^{2+}}}$$

From this, it can be seen that the electrode potential, and thus the oxidizing effect of the MnO_4^-/Mn^{2+} system is highly dependent on the pH. This is true for many similar cases, and a change in pH can in some circumstances lead to a complete reversal of the reaction direction. For example, in the acidic range, the equilibrium of the reaction

$$\overset{+3}{As}O_3^{3-} + I_2 + H_2O \rightleftharpoons \overset{5+}{As}O_4^{3-} + 2I^- + 2\,H^+$$

lies on the left side, but in basic solution, it lies on the right. This consideration makes it understandable that under strongly acidic conditions, chlorate reacts with chloride to form chlorine, while in basic solution, chlorine disproportionates to chloride and chlorate.

Disproportionation (see) and Coproportionation (see) are special forms of R.

Redox titration: same as Redox analysis (see).

Reduced parameter: a dimensionless state parameter established with reference to a standard state. The R. is the ratio of the value of the state parameter under arbitrary conditions to its value in the standard state. The formulation of laws with R. often makes comparison of the behavior of different substances or reactions easier. In thermal equations of state of gases and liquids, the critical point is chosen as the standard state. The R. are then $p_r = p/p_{crit}$, $v_r = v/v_{crit}$, and $T_r = T/T_{crit}$ (see Corresponding states).

In reaction kinetics, the initial concentration is used as the reference value: $c_r = c/c_0$, and the reduced time t_r is defined by the equation $t_r = c_0^{n-1}kt$, where n is the order of the reaction and k is the rate constant. In integrated form, the first-order rate equation is then $\ln c_r = -t_r$, and the nth-order rate equation is $c_r^{1-n} = 1 + (n-1)t_r$.

Reduced heat: see Entropy.

Reducing agent, ***reductor***: a substance which can reduce other substances, by donating electrons and itself being oxidized in the process (see Redox reactions).

Reductases: see Flavin enzymes.

Reduction: the uptake of electrons by atoms, ions or molecules, which leads to a lower Oxidation number (see); one part of a Redox reaction (see).

Reductor: same as Reducing agent (see).

Reevon®: see Synthetic fibers.

Reference electrode: see Electrode.

Reference EMF value: see Electrochemical equilibrium.

Reference (cell) voltage: see Electrochemical equilibrium.

Refining: 1) a term for any industrial process for purifying and improving natural or industrial products, e.g. petroleum, silver, gold, copper, oils, sucrose. In particular,

2) a basic process in steel production, in which crude iron is converted to steel by removal of the carbon and other undesirable elements, most of which are introduced with the raw materials (crude iron or scrap). Elements such as silicon, manganese and phosphorus, as well as carbon, are removed by oxidation. The products are released as gases (carbon monoxide) or as solids which are converted to slag (e.g. silicon dioxide, phosphorus pentoxide). Because of their high affinity for oxygen, however, valuable alloying elements such as chromium are also oxidized (burn off). The carbon monoxide produced by oxidation of the carbon is circulated through the melt and, especially in blast furnace processes, provides an equilibration of temperature and concentration, and also flushes out impurities (oxides and hydrogen).

The oxygen required for oxidation is provided by the iron monoxide content of the slag, the rust of the scrap iron, the iron oxide content (iron(III) oxide, Fe_2O_3, and iron(II,III) oxide, Fe_3O_4) from the fresh ore, or it is supplied by oxidizing flame gases. Oxygen can also be supplied directly in the form of air (*blasting*) or pure oxygen (*oxygen blasting*; see Oxygen metallurgy).

Refinery: a plant for processing petroleum (see Petroleum refining). By a combination of physical and chemical processes, the petroleum is converted to intermediates or endproducts. The most important process stages of a R. are a) atmospheric distillation of the crude oil; b) vacuum distillation of the residue of atmospheric distillation; c) catalytic reforming and distillation of the reforming products; d) hydrorefining of starting materials and endproducts; and e) catalytic cracking and distillation of the cracked products. These products are combined in a manner which adjusts the amounts of the various petroleum fractions to the market demand. A R. can be designed to produce mainly heating oil (*hydroskimming R.*) or to have a high output of gasoline.

Products of a heating oil or gasoline refinery

Products	Boiling point (°C)	Heating oil (%)	Gasoline (%7)
Gasoline	up to 180	21	48
Kerosine	180-250	7	7
Heating oil distillate	200-360	20	22
Residual heating oil	>360	43	11
Lubricating oils		1	2
Other products		8	10

A heating oil R. requires only atmospheric distillation, catalytic reforming and the associated supporting processes. However, if a high yield of gasoline is desired, further processes (e.g. catalytic and thermal cracking, vacuum distillation and hydrorefining) are needed.

In the past few years, *chemical R.* has been developed to yield such petrochemicals as olefins, diolefins, aromatics, methanol and ammonia.

Reflection spectroscopy, ***remission spectroscopy***: A form of spectroscopy in which the radiation reflected from a sample is measured. R. is used mainly with opaque and insoluble samples. The measured reflectivity of a sample consists of two parts: *regular reflection*, in which the radiation is reflected from the surface as described by the Fresnel equations, and *diffuse reflection*, in which the radiation is reflected uniformly in all directions. Diffuse reflection arises when the radiation penetrates the sample and returns to the surface after partial absorption and multiple scattering. Regular and diffuse reflection are idealized limiting cases; the total measured reflection from a sample consists of a mixture of both types. However, the regular reflection can be largely eliminated by special techniques such as measurement between crossed polarizing prisms or by dilution of the sample with a white standard (e.g. MgO).

The theory of diffuse reflection and its connection with absorption was developed mainly by Kubelka and Munk:

$$F(R_\infty) = \frac{(1 - R_\infty)^2}{2R_\infty} = \frac{\varepsilon \cdot c}{s}$$

where R_∞ is the diffuse reflectivity of an infinitely thick layer of the sample relative to a standard [$R_\infty = R$ (sample)/R (standard)], s is the scattering coefficient, ε is the absorption coefficient and c is the concentration. Since the scattering coefficient s is almost independent of wavelength, at least in the visible range, absorption and diffuse reflection spectra are very similar. In the reflection spectrum, $F(R_\infty)$ is plotted against the wavelength or wavenumber. Since $F(R_\infty)$ is proportional to c under constant external conditions (mainly, constant size of sample particles), the Kubelka-Munk equation can be used to carry out quantitative determinations.

Measurement of reflection spectra can be done in ordinary spectrophotometers (see Spectral instruments), since these are usually equipped with a reflection attachment. The reflection of the sample is compared to that of a standard such as MgO or milk glass. To avoid a dependence of the reflection on the angle of incidence of the radiation, the sample is illuminated with a photometer sphere (Ulbricht sphere) which is coated on the inside with a diffusely reflecting white layer (MgO) (Fig.). The sample is placed over the opening of the sphere and irradiated by the diffuse light from it. The reflected radiation is registered by the photocell, which is positioned to receive the radiation reflected perpendicularly from the surface of the sample. The spectrum is obtained by spectral resolution of the reflected light.

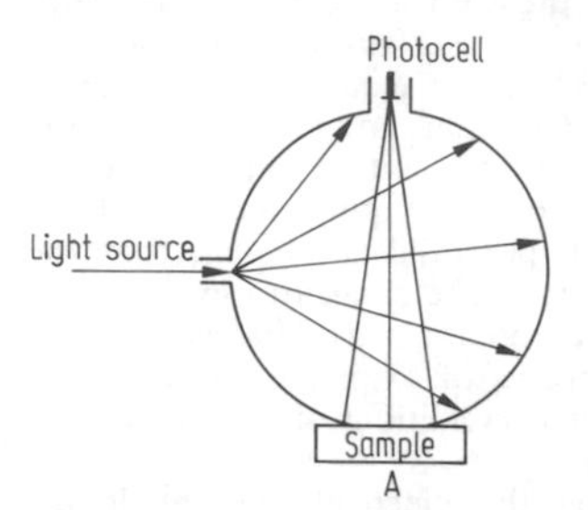

Diagram of a photometer sphere.

The applications of R. have increased steadily in the past few years, e.g. for study of pigments, plastics, textiles and paper. R. is used to study surfaces, catalysts, reactions of solids and also for quantitative analysis of paper and thin-layer chromatograms. The method of diminished total reflection has become very important in Infrared spectroscopy (see), because it permits examination of samples which are not transparent to IR.

Reflux: see Distillation.

Reflux cooler: see Coolers.

Reformatskij reaction: a reaction for synthesis of β-hydroxycarboxylate esters, starting from α-halogencarboxylate esters, carbonyl compounds and zinc in ether, benzene or toluene:

$$Br{-}CH_2{-}COOC_2H_5 \xrightarrow{+Zn} BrZn{-}CH_2{-}COOC_2H_5$$

$$\xrightarrow{+RCHO} R{-}CH(OZnBr){-}CH_2{-}COOC_2H_5 \xrightarrow[-Zn(OH)Br]{+H_2O}$$

$$R{-}CH(OH){-}CH_2{-}COOC_2H_5$$

Organozinc compounds are formed as intermediates; these add to the carbonyl groups. Unlike Grignard reagents, they do not react with the ester group, although otherwise the R. is comparable to the Grignard reaction. Aliphatic and aromatic aldehydes and ketones, cyclic and α,β-unsaturated aldehydes and ketones are suitable carbonyl reagents. α-Bromocarboxylic acid esters are used as the halogen carboxylate ester. Unsaturated carboxylate esters are easily obtained as products of the R., and these may then be hydrogenated to produce the saturated esters. The reaction is therefore important for syntheses in which the carbon chain is to be lengthened.

Reforming: a process in petroleum refining which changes the structures of hydrocarbons without significantly changing their molecular masses. A distinction is made between *thermal* and *catalytic R.*. Today only catalytic processes are used, in a hydrogen atmosphere at pressures of 1 to 5 MPa and temperatures around 500 °C. The purposes of R. are 1) to increase the octane rating of the starting material, and thus to produce a high-quality fuel, and 2) to form aromatics from 5- and 6-ring naphthenes.

Cracking processes serve mainly to produce smaller hydrocarbon molecules from high-boiling petroleum fractions, but in R. the main processes are isomerization, cyclization and dehydrogenation of hydrocarbons.

1) *Isomerization*: e.g.

dimethylcyclopentane $\underset{}{\overset{H^+}{\rightleftharpoons}}$ methylcyclohexane

2) *Cyclization*: e.g.

$CH_3-CH_2-CH_2-CH_2-CH_2-CH_3 \xrightarrow{H^+}$ methylcyclopentane $+ H_2$; $\xrightarrow{H^+}$ cyclohexane $+ H_2$

3) *Dehydrogenation*: e.g.

methylcyclohexane $\overset{Pt}{\rightleftharpoons}$ toluene $+ 3H_2$

In addition, heterocompounds present in the starting material are reduced to hydrogen sulfide, ammonia and water (see Hydrorefining). These compounds must be removed from the circulating hydrogen by gas scrubbing processes, since they would otherwise become enriched in the hydrogen. The octane rating of a straight-run gasoline is increased by R. from 30-50 to 85-95. The stripping products of gasoline reforming are important starting materials for production of benzene, toluene and the xylenes (BTX aromatics). Naphthalene, trimethylbenzenes and durene can also be isolated from the reformate.

Catalytic R., which was first used in 1939 in Germany, is characterized by several (3 to 6) successive reactors with fixed catalysts and circulating hydrogen (Fig.).

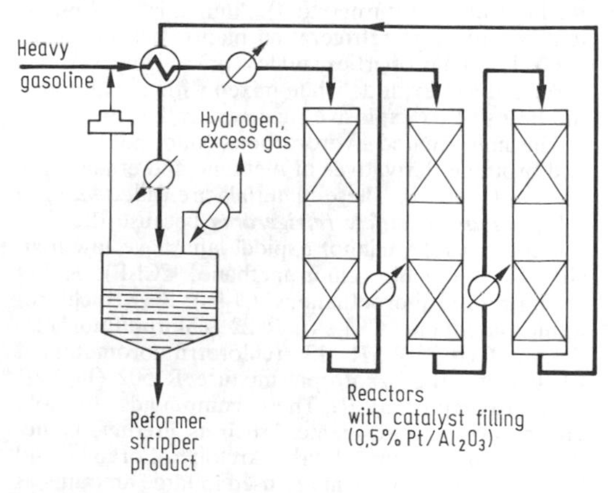

Catalytic reforming of heavy gasoline.

The prototype of solid-catalyst R. is the *DHD process* (Pressure-Hydrogen Dehydrogenation; the first "D" is for "Druck" = "pressure" in German) developed in 1939/40. The hydrogen delays precipitation of the high-molecular-weight byproducts on the catalyst; these would otherwise rapidly deactivate it. Small amounts of such compounds do precipitate on the catalyst, and are gradually converted to coke; they are removed by burning the catalyst with atmospheric oxygen at regular intervals (regeneration).

Since nearly all the reactions of R. are endothermic, the losses of heat in the reactors must be compensated by heaters which prewarm the gases entering them.

In the older R. process, 10 to 15% molybdenum(III) oxide on aluminum oxide $\varkappa$-Al_2O_3 was used as catalyst. Today platinum on aluminum oxide is used. Rhenium additives increase the lifetime of the catalyst.

The catalysts are sensitive to sulfur compounds, which are hydrogenated on them to form hydrogen sulfide. This in turn causes reversible damage to the active MoO_3 or Pt components; however, at higher sulfur concentrations, the damage is irreversible. Therefore, sulfur must be removed from the starting materials before R.; this is done today by catalytic hydrogenation using the hydrogen released by the R. itself. The sulfur content should not exceed 0.01% of the starting material.

The *L-forming process* also uses a Pt catalyst in three reactors (520 °C and 4 MPa). Similar processes have been developed, mainly in the USA, such as *platforming* (the first industrial use of platinum catalysts), *catforming*, *houdriforming*, *ultraforming* and *powerforming*. Some of these processes are provided with reserve reactors (swing reactors) which are switched into the process during regeneration of the catalyst in the other reactors; this permits continuous operation of the plant.

In the *iso-plus process*, catalytic and thermal R. are coupled; the gaseous alkenes produced thermally are catalytically polymerized in a further reactor, producing high-octane liquid hydrocarbons in the gasoline boiling range. The total product of the iso-plus process has an octane rating over 100.

A combination of catalytic R. and catalytic cracking can be achieved by use of acidic carriers (e.g. molecular sieves). Use of these catalysts produces large amounts of C_3 to C_5 paraffins (*LPG method*, abb. for liquid-petroleum gas).

Refractive index: see Refractometry.

Refractometry: optical measurements for determination of *refractive indices* and their applications.

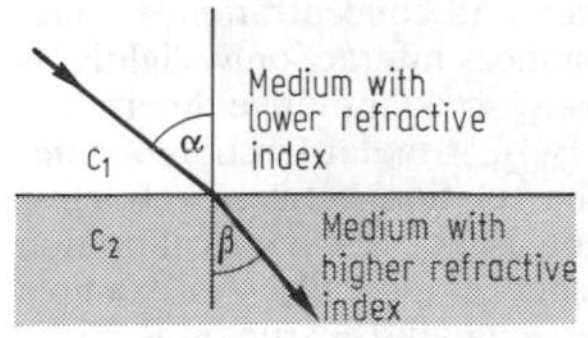

Light path in refraction

Principle. If monochromatic light (Fig.) passes from a medium in which the velocity of light is c_1 into another in which it is c_2, the beam of light is bent (refracted) according to Snell's law: $\sin\alpha/\sin\beta = c_1/c_2 = n_{21}$. Here n_{21} is the refractive index of substance 2 relative to substance 1. If the less dense medium is a vacuum or air, in which the velocities of light are practically the same, one obtains the absolute index of refraction of the denser substance. The refractive indices for most substances are between 1.3 and 1.7 (Table). Instruments for measuring the refractive index are ***refractometers*** The most common of them are the *Abbé* and the *Pulfrich refractometers*; these measure the limiting angle of total reflection. *Interference refractometers* depend on another principle, and are used for higher-precision measurements, especially on gases.

Refractive indices of some substances at 20°C

Substance	n_α (656.3 nm)	n_D (589 nm)	n_β (486.1 nm)	n_γ (434.1 nm)
Quartz	1.541 8	1.544 2	1.549 7	1.554 0
Diamond	2.410 8	2.419 0	2.440 8	2.449 9
Flint glass	1.620 2	1.625 2	1.637 8	1.648 1
Water	1.331 2	1.333 0	1.337 1	1.340 3
Cyclohexane	1.424 1	1.426 2	1.431 9	1.436 2
Benzene	1.496 5	1.501 2	1.513 2	1.522 9
Hexane	1.373 1	1.374 9	1.379 5	1.383 2
Carbon tetrachloride	1.457 6	1.460 3	1.467 1	1.471 9
Acetic acid	1.369 5	1.371 8	1.376 2	1.379 4
Ethyl acetate	1.370 5	1.373 0	1.376 8	1.379 9

The refractive index depends on the wavelength of the light and the temperature. Both parameters are therefore indicated, as a subscript and superscript, respectively, to the index *n*. n^{20}_{589} means that *n* was measured at 20 °C and a wavelength of 589 nm. In addition to the Na D line at 589 nm, various lines from the hydrogen spectrum are used for R., including the red H_α line at 656.3 nm, the green H_β line at 486.1 nm and the blue H_γ line at 434.1 nm.

For a large number of liquids, it has been found that increasing the temperature by 1 °C causes the refractive index to drop by $3.4 \cdot 10^{-4}$ to $5.5 \cdot 10^{-4}$. Therefore, in order to determine the refractive index to a precision of 10^{-4}, the temperature during the measurement must be held constant to ±0.2 °C.

Application. R. is used in analysis mainly as a means to identify substances and check their purity, e.g. the purity of fats and oils. Refractometers are used as detectors in separations of substances, e.g. by liquid chromatography. Flow-through refractometers, which give continuous measurement of the refractive index of a sample as it flows through the instrument, are used in process control. The relation between refractive index and concentration, in mixtures in which the substances interact only slightly, is given by $\nu_{tot} = \gamma_1 n_1 + \gamma_2 n_2$, so quantitative determinations are also possible [$\gamma_{1(2)}$ = molar fraction of component 1(2)]. When the relation between refractive index and concentration is not linear, a calibration curve must be established empirically. An example for a common quantitative analysis based on R. is the determination of the sugar contents in aqueous sugar solutions.

It has been discovered empirically that the molar refractions of organic compounds are equal to the sum of the individual atomic refractions (see Polarization).

Refrigeration technology: a branch of technology concerned with processes, methods and installations for achieving and maintaining temperatures below the ambient temperature. Many processes in chemical industry require low temperatures, including the liquefaction of gases, separation of mixtures of gases and liquids, freezing of liquids, regulation of reaction rates, and crystallization of salts out of solution. The power requirements for refrigeration in large chemical plants are often as high as several megawats. The cooling is needed over a wide range of temperature, from slightly below the ambient to just above absolute zero. The development of new areas of R. at very low temperatures is still very fluid (see Cryotechnology).

Refrigerant: the working substance of a Refrigerator (see). The R. goes through a cyclic process of evaporation at low temperature and low pressure, absorption of heat, and condensation at higher temperature and pressure, with loss of heat to the environment. A liquid R. must have a suitable vapor pressure curve, so that it undergoes the conversion from liquid to vapor in the desired temperature range. Its freezing point limits its working range on the low-temperature side; on the other side, its condensation temperatures must be far below its critical temperature. In most refrigerators, the low-pressure side is kept above 0.1 MPa, and the high-pressure side does not exceed about 2.5 MPa. Because of the varying requirements for temperature, the boiling temperatures of most R. are between -100 °C and +25 °C at atmospheric pressure. In addition to a suitable vapor pressure curve, a R. should not be combustible or explosive, poisonous, chemically unstable, corrosive or too expensive. In addition, it should have suitable thermodynamic and physical properties. There is no substance which fulfills all these requirements in ideal fashion.

Ammonia and halogenated hydrocarbons are used as R.; in industry, hydrocarbons and in a few cases even water are used as R. Water has a fairly high working range (it is much more commonly used as a non-circulating coolant, for example, in power plants where the cold water is taken from a river or lake and the hot water is returned to it). Ammonia is an important R. for large refrigeration plants, because of its good thermal properties and low price. However, it is toxic and can explode when mixed with air, although the limits of its explosive range are rather narrow.

The halogenated hydrocarbons (fluorine, chlorine and bromine derivatives of methane and ethane) are very widely used. These materials are called ***halogen refrigerants***, or ***safety refrigerants*** because they are not combustible, cannot explode and have low toxicity. R 11 (trichlorofluoromethane, CCl_3F), R 114 (dichlorotetrafluoroethane, $C_2Cl_2F_2$), R 12 (dichlorodifluoromethane, CCl_2F_2), R 22 (chlorodifluoromethane, $CHClF_2$), R 13 (chlorotrifluoromethane, $CClF_3$) and the azeotropic mixture R 502 (R 22/R 115) are important R. These compounds are sold under various trade names, such as Freon®, Genetron®, Frigen®, Fridohna®, Arcton®, Forane® and Kaltron®. Some of them are used in large amounts as solvents and carrier gases for sprays and foams; there is strong evidence that their presence in the upper atmosphere is responsible for the loss of ozone over the past few years. International efforts to curb their release into the atmosphere have begun.

Refrigerator: a machine for reducing the temperature of a system and keeping it low: a cold space, a coolant or materials. This is achieved by removal of heat at the low temperature, a process which requires the expenditure of energy (second law of thermodynamics). R. consist of condensers, expansion devices (valves or apparatus), a circulating medium and other apparatus (such as precipitators, collectors, driers, pumps, pipes, switches, thermostats, thermometers, etc.). The structure depends on the required temperature, the cooling capacity, the type of energy used and the technical and technological requirements. The most common type of R. is based on the circulation of a Refrigerant (see) which expands as it escapes into a low-pressure area through a valve. The cold vapor then absorbs heat from the space or material to be cooled as it flows through a system of cooling coils on its way to a compressor. Compression of the refrigant increases its temperature; this heat is removed to the exterior of the R. by means of fins or a fan. The compressed refrigerant is then returned to the expansion valve. The refrigerant can be a gas during the entire cycle, or it may be a liquid in the compressed stage and evaporate during the expansion stage. For special purposes, there are ***electrothermal and magnetic R.***. For the temperature range between +10 and about -100 °C, however, R. with mechanical compressors are by far the most common.

Regan®: see Synthetic fibers.

Regenerator: see Heat transport substances.

Regioselectivity: a preferential but not exclusive attack of a reagent at one of several non-equivalent sites. Some examples are electrophilic double substitution of benzene derivatives, Markovnikov or anti-Markovnikov additions, and eliminations (Saizev or Hofmann products) from unsymmetric compounds.

Reimer-Tiemann reaction: an aldehyde synthesis, first described in 1876, in which phenols react in regioselective fashion with chloroform and aqueous alkali hydroxide solution; the 2-position is reactive.

$+ ICCl_2^-$ ... CCl_2

Phenolate anion

$CHCl_2$ $\xrightarrow[-2Cl^-, -H_2O]{+2OH^-}$ CHO $\xrightarrow{+H^+}$ OH, CHO

Salicylaldehyde

The reaction yields salicylaldehyde in 60% yield from phenol itself. Dichlorocarbene is formed as an intermediate; it adds electrophilically to the ambident phenolate ion. If the 2-position of the phenol ring system is occupied, the formyl group is introduced in the 4-position, relative to the hydroxyl group. Phenol ethers do not undergo the R., but the corresponding aldehydes (with the formyl groups in position 2) can be made from pyrrole and indole. The reaction is more difficult when there is an electron-withdrawing group (-NO_2, -CN, -COOH) on the aromatic or heteroaromatic ring. Bromoform, iodoform or trichloroacetic acid can be used in the R. in place of chloroform.

Reinecke salt: see Chromiacs.

Reissert compounds: organic compounds obtained by addition of cyanide ions to *N*-benzoylquinolinium chloride. The R. decompose by hydrolysis into quinaldic acids (quinoline 2-carboxylic acids) and benzaldehyde. R. can readily be synthesized under mild reaction conditions. The hydrolysis occurs by a complicated mechanism with several intermediates.

N^+–COC_6H_5 + CN^- → (H, CN, N–COC_6H_5) $\xrightarrow[-NH_3, -C_6H_5CHO]{2H_2O/H^+}$ N, COOH

Reissert-compounds

Relaxation: the delayed reaction of a system to an external perturbation. If the perturbation is in the form of a square wave jump, the new equilibrium of the system is not reached immediately; the rate is described by an exponential law: $x_t = x_0 e^{-t/T}$. Here x_0 is the difference between the old and new equilibrium position, x_t is the difference between the present position and the new position, and T is the ***relaxation time***. T is related to the rate at which equilibration occurs. The phenomenon of R. is used to measure rate constants of very rapid chemical reactions (see Kinetics of reactions, experimental methods).

Relaxation effect: see Debye-Hückel theory.

Relaxation reagents: see NMR spectroscopy.

Relaxin: see Hormones.

Release inhibiting factors: same as Statins (see).

Release-inhibiting hormones: same as Statins (see).

Releaser: see Pheromone.

Releasing factors: same as Liberins (see).

Releasing hormones: same as Liberins (see).

Remission spectroscopy: same as Reflection spectroscopy.

Renaturation: see Protein.

Renin: an endopeptidase (see Proteases) formed in the kidney; it catalyses the specific cleavage of the Leu-Leu bond in angiotensinogen, thus releasing proangiotensin (angiotensin I), the precursor of angiotensin II. Angiotensin II is a powerful regulator of blood pressure. The relative molecular mass of R. is 43,000.

Rennin, ***Chymosin***: a pepsin-like endopeptidase (see Proteases) with high substrate specificity; M_r 30,700, pH optimum at 4.8. The enzyme is released from an inactive precursor (prorennin, M_r 36,200)

and it requires calcium ions as cofactors. R. is a milk-coagulating enzyme; its only substrate is the milk protein ϰ-casein, which it converts to insoluble para-ϰ-casein (M_r 22,000) and a *C*-terminal glycopeptide (M_r 8000). The action of R. destroys the function of ϰ-casein as a protective colloid.

Repair enzymes: enzymes which are able to repair damage to DNA caused by chemicals or radiation. The damaged sections are cut out and replaced by the correct sequences. R. systems include DNA polymerases (see) and DNA ligases (see).

Repellents: substances used to drive off annoying or deleterious animals. There are Bird repellents (see), Deer repellents (see) and above all, Insect repellents (see).

Replication: see Nucleic acids.

Reppe chemistry: see Acetylene chemistry.

Reppe syntheses: see Acetylene chemistry.

Repulsion energy: the energy resulting from a close approach of two atoms or molecules with closed shells, as a result of the Pauli principle. The R. at a distance R is usually given by the expression $\varepsilon_{rep} = A/R^n$ (n = 9-12) or $\varepsilon_{rep} = Be - CR$. Here A, B and C are experimentally determined constants. The R. is slight for large distances, but increases rapidly after R has decreased below a certain limit. It is the main factor determining the volume requirement of an atom or molecule.

Research octane rating: see Octane rating.

Resenes: highly unsaturated organic compounds found in natural resins.

Reserpine: see Rauwolfia alkaloids.

Reserves: in textile processing, a protective paste applied to the surface of cloth before it is dyed, to prevent coloration of that part of the surface which is covered. In ***over-printing reservation***, the cloth is impregnated with dye solution, and subsequent application of the reserver prevents fixation of the dye. In ***preprinting reservation***, the cloth is printed with a paste which prevents subsequent coloration of the printed areas, either chemically or physically. Resins, waxes and chemicals which convert the dye to an inactive form are used for preprinting reservation.

The best known R. is variamine blue reserve. After the cloth is treated with a naphthene compound, an acidic salt is printed onto it; when the cloth is later treated with variamine blue salt (a diazonium compound), coupling is prevented in the printed areas, because the coupling can occur only in an alkaline medium. Addition of vat dyes and sulfite to the printing paste is used to create colored reserves under the variamine blue; the diazonium compound of the variamine blue salt is reduced to a phenylhydrazine derivative which can be coupled. White and colored reserves are frequently used under aniline black. After the cloth is impregnated with aniline salt (aniline chlorohydrate) and dried, the white or colored reserve is printed on. Since aniline black forms by oxidation of aniline salt in a mineral acid medium, a white reserve can be achieved by printing on a reducing agent or an acid binder (sodium acetate, zinc oxide). Vat dyes are used for colored reserves.

Re/Si configuration: see Topic groups.

Resin: an organic solid or semisolid, usually amorphous and translucent and having a characteristic sheen. R. are characterized more by their similar physical properties than by their chemical similarities; they are supercooled melts, somewhat like glasses. They often consist of many similar substances with molecular sizes up to the macromolecular range. From this it follows that they do not have fixed melting points, but undergo a gradual transition from the liquid to the solid state. Pure R. have no odor or taste, are insoluble in water, but are soluble in ether, alcohols, various esters, essential and fatty oils and halogenated hydrocarbons such as chloroform and carbon tetrachloride. Most of them burn with bright, very sooty flames.

1) ***Natural R.*** can be classified on the basis of their chemistry, botanical sources or geographical distribution. Liquid R. or solutions of R. in essential oils are called *balsams*. Commercially, they are often named for their origin, e.g. Canada balsam (see), Peru balsam (see) or Japan lacquer. The natural R. are nearly all of plant origin (***tree resins, plant resins***) and are formed as such or mixed with terpentine oil or other oils, usually in the bark or trunks of certain trees, especially conifers (see Balata, Benzoin resin, Catechu, Dammar, Elemi, Sandarac, Tolubalsam, Gum resins). They may also be found in the fruits, e.g. bergamotte and Dragon's blood (see). The odor of these secretions is due to the oils mixed with the R.

Chemically, the natural R. are related to the terpenes and essential oils, and usually consist of complex mixtures of resinic acids, resin alcohols and phenols (resinols), phenols with tannin properties (resinotannols), highly unsaturated substances (resenes) and esters of resin acids. For example, the non-steam-volatile fraction of pine R. is a mixture of five isomeric diterpenes, of which abietic acid makes up the largest fraction.

The most important of the ***fossil R.*** is Amber (see); others are batu and Copals (see). The fossil R. are deposits which may have been created by the destruction of large forests.

Fresh R. are called recent R.; the balsam R. are the most abundant. These are obtained for the most part by artificial injury of conifers (secondary resin flow). Distillation of the crude balsam yields terpentine oil and, as a distillation residue after melting, colophony. The conversion products of natural rubber (see Caoutchouc) are also considered natural R.

The most important animal R. is shellack.

Many of the natural crude R. are first melted and filtered, then separated into various components by steam or vacuum distillation. They are used to make lacquers, varnishes, polishes, cosmetics, textile additives and drugs. As the production and variety of synthetic R. has increased, the natural R. have become less important.

2) ***Synthetic resins*** are organic products made from low-molecular-weight starting materials. These compounds are a type of plastic, and most are made by polycondensation. The most important are 1) Aldehyde resins (see), Ketone resins (see) and Ketone-aldehyde resins (see); 2) amidaldehyde resins, such as Sulfonamide resins (see), Dicyandiamide resins (see) and Urea resins (see); 3) Aminaldehyde resins, e.g. Aniline resins (see) and Melamine resins (see); 4) Epoxide resins (see); 5) Carbohydrate resins (see); 6) Hydrocarbon resins (see); 7) Phenol resins (see); 8)

Polyester resins (see); 9) Silicon resins (see Polysiloxanes); and 10) Furan resins (see).

Resin acids, ***resinoic acids***: acid components of resins; chemically they are di- or triterpene acids, phenylacrylic acids or phenylcarboxylic acids. The resins of gymnosperms contain mainly diterpene carboxylic acids such as Abietic acid (see) and pimaric acids (levo- and dextro-). Angiosperm resins contain mostly triterpene acids, e.g. siaresinolic acid in benzoin, masticadienonoic and Oleanolic (see) acids in mastic, or boswellic acid in frankincense (olibanum). Aromatic R., formed from intermediates of lignin biosynthesis, are found mainly in resins formed in response to injury or disease. The salts and esters of R. are called ***resinates***.

The R. are obtained from terpentine, a mixture of R. and terpenes. They can be hydrogenated to hydroresins or resin alcohols. R. are used in the production of resin soaps, sizing, paste for paper, stiffeners and so on.

Resinate: 1) ***Resin soap***: a salt of a resin acid. The applications of these soaps depend on the metal ion. Alkali resinates are made by saponification of colophony, which consists of about 80% resin acids, with alkali hydroxide or carbonate solutions. They are added in small amounts to fat soaps to improve their solubility and foaming ability. However, good soaps should contain little or no R., because they often make the soap sticky and also form insoluble compounds with the calcium and magnesium salts in hard water; these readily precipitate onto the fibers of laundry. The name resin soap often applies to a mixture of fat and resin soaps. R. with other metal ions than alkali are used in industry (see Metal soaps).

2) ***Resin ester***: usually a glycol, glycerol or pentaerythritol ester of a resin acid; they are used as components of paints. See Resinates.

Resinite: see Macerals.

Resinol: a basic component of natural resins consisting of resin alcohols and phenols.

Resinol acids: same as resin acids.

Resin soap: see Resinate.

Resistance breaker: a substance intended to prevent or reverse the development of resistance to pesticides in the target species. The regular use of Insecticides (see), Acaricides (see) and other pesticides acts as a selective factor favoring those individuals which are less susceptible to the pesticide, and the descendents of those individuals form a resistant population. An example of an R. is *N,N*-di-*n*-butyl-4-chlorobenzene, which is mixed with DDT to counteract DDT resistance.

Resitol: see Phenol resins.

Resmethrin: see Pyrethroid.

Resol: see Phenol resins.

Resonance: a theory developed independently by Robinson and Ingold (1926) and Arndt and Eistert (1924) to explain bonding in molecules in which the localization of double bonds between fixed atoms does not satisfactorily reflect the chemistry. The ground state of these molecules can be described by superposition of all the possible valence-dash formulas (*canonical or limiting structures*). The canonical structures do not represent any actual state of the molecule, but are only aids in conceptualizing the true electronic structure. The superposition of the canonical structures is indicated by a double-headed arrow (↔). Resonance theory is useful in explaining the bonding in molecules with conjugated double bonds, especially aromatic compounds such as benzene. For the benzene molecule, the essential canonical structures include not only the two Kekulé structures K_1 and K_2, but the three Dewar structures D_1, D_2 and D_3 (Fig.).

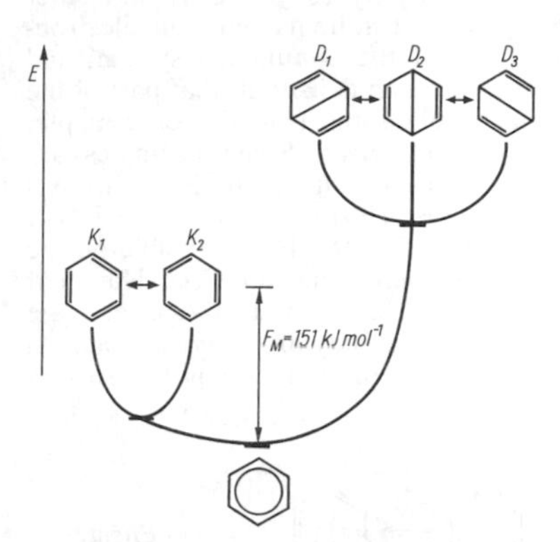

Resonance energy of benzene

The superposition of the canonical structures expresses the delocalization of the 6 π electrons in benzene over the entire molecule. The resulting equalization of the C-C bond lengths (140 pm) has been confirmed experimentally. This value is intermediate between the length of a single (154 pm) and an isolated double (133 pm) bond. The delocalization and the bond-length equalization are usually indicated in formulas by drawing a circle within the hexagon instead of double bonds (Fig.). The quantum mechanical treatment of the superposition of canonical structures (see Valence bond method) indicates that the energy of the resonant system is lower than that of the fictitious canonical structure (Fig.). It is understood that the high-energy Dewar structures contribute only a small amount to the reduction in energy. The difference in energy of the actual ground state of the molecule and the lowest-energy canonical structure is called the ***resonance energy***. It can be approximated by subtracting from the experimentally determined enthalpy of combustion or reduction the hypothetical energy content of the lowest-energy canonical structure, which is determined according to the additivity principle. R. is possible only when two or more canonical structures for the molecule exist in which the positions of the atoms are the same; only the electron distribution is different. The valence-dash formulas of the canonical structures can therefore be interconverted simply by shifting the valence electron pairs. In molecules with conjugated double bonds, maximum R. stabilization due to overlap of π-orbitals (see Molecular orbital theory) is possible only when there is a planar σ-bond structure. Since the canonical structures are only heuristic aids, R. must be strictly differentiated from Tautomerism (see), in which there is a true equilibrium between two structurally different molecular forms.

Resonance effect, ***mesomerism effect***, abb. ***M-effect***: the effect on the electronic structure of a conjugated molecular system of introducing a substituent which can participate in the conjugation with a lone electron pair or a double bond. If the substituent increases the charge in the resonant structure, as the amino group does in aniline (Fig. 1), this is called a ***+M effect***. Some other substituents with +M effects are the halogens (-X), hydroxyl (-OH) and ether groups ($-OCH_3$). If substituents have an electron-withdrawing effect on the conjugated system, and thus decrease the electron density in that part of the molecule, as in nitrobenzene (Fig. 2) for example, one speaks of a ***-M effect***. Some examples are the acetyl ($H_3C(O)C-$), nitrile (-CN), carboxyl (-COOH), aldehyde (-CHO) or nitro group ($-NO_2$). A quantitative estimate of the R. is possible on the basis of the σ-substituent constants (see Hammett equation). The resonance and inductive effects are called *substituent effects*. They are important aids in qualitative interpretation of the reactions and spectra of molecules.

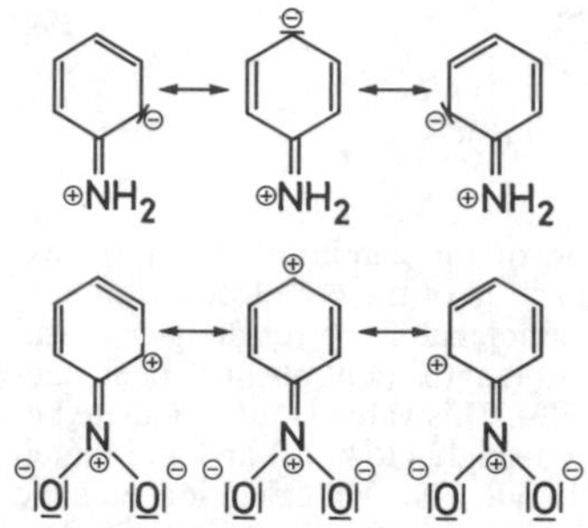

1) +M-Effect

2) −M-Effect

Resonance integral: see Hückel method.

Resonance-Raman effect: a special method used in Raman spectroscopy in which molecules are excited using the frequency of an electronic transition. The excited molecules can return to an excited vibrational state of the electronic ground state. The frequency of the emitted light is then $\nu_E - \nu_V$, where ν_E is the frequency of thc emitted radiation and ν_V is that of the pure vibrational excitation. The spectrum of emitted light is the same type as in the normal Raman spectrum. However, because the energy uptake and loss occur by different mechanisms, the selection rules are somewhat different. The most important difference from the normal Raman effect is that the lines are several orders of magnitude more intense. In addition, the vibrations which can be excited are intimately related to the electronic transition, so it is possible to obtain a spectrum of the chromophoric group alone.

Resorcinol, ***1,3-dihydroxybenzene***: a diphenol. R. forms colorless and odorless crystals; m.p. 110 °C, b.p. 281.4 °C. It is readily soluble in water, ethanol and ether, and it is a much weaker reducing agent with respect to an ammoniacal silver nitrate solution than are pyrocatechol and hydroquinone. R. gives a dark violet color with iron(III) chloride in dilute solution. It is poisonous in high concentrations; in dilute solutions it is antiseptic and is used in dermatology. R. is very easily hydrogenated, producing dihydroresorcinol (cyclohexane-1,3-dione). The addition of 2 H atoms occurs even with sodium amalgam in water. R. is usually synthesized by alkali fusion of benzene-1,3-disulfonic acid. It is the starting material for phenoplastics and pigments, especially triphenylmethane pigments. Nitration yields styphnic acid (2,4,6-trinitroresorcinol), m.p. 175 °C, which forms readily crystallizing styphnates with organic bases.

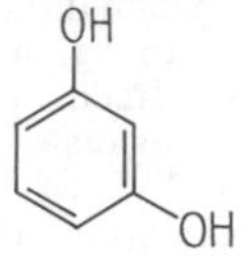

Respiration: the oxidation of organic nutrients by molecular oxygen to form low-energy end products and to generate ATP by respiratory chain phosphorylation of ADP. Acetyl-CoA, formed from pyruvic, fatty or amino acids, is oxidized in the citric acid cycle to carbon dioxide and hydrogen atoms (reducing equivalents in the form of NADH). NADH donates electron pairs to the respiratory chain, and they flow to the terminal electron acceptor, molecular oxygen. The *respiratory chain* is located in the mitochondria; its components are NADH dehydrogenase, iron-sulfur proteins, ubiquinone and the cytochromes B, c^+, c and a_3. Each component forms redox pairs with the adjacent components, being first reduced as it accepts electrons from a stronger reducing agent, then oxidized as it reduces the next component. Three of the redox steps are coupled to ATP-generating processes (respiratory chain phosphorylation).

Respiratory chain: see Respiration.

Respiratory poison: in the broad sense, a toxic substance which is taken into an organism via its respiratory system and causes temporary or permanent changes in one or more of the life functions of this organism. In the narrower sense, an insecticide which is inhaled by the insects.

Rest potential: see Polarization.

Restriction enzymes: see Genetic engineering.

Retention: 1) preservation of configuration in substitution reactions.

2) The "holding back" of a substance in chromatography; this is characterized by the ***R. index, I***. Some retention indices are, for example, the retention time, t_R, that is, the time between the appearance of the air peak and the substance peak in a gas chromatogram, and the retention volume, v_R, which is the volume of carrier gas which flows through the column between the air peak and the substance peak. In gas chromatography, the retention index difference, ΔI, of the same substance on different stationary phases, of which one is nonpolar, is a expression for the polarity of a dissolved compound, and also of the stationary phases.

Retinal: see Vitamin A.

Retinoic acid: see Vitamin A.

Retinol: see Vitamin A.

Retort graphite: a type of graphite which forms when a methane or city-gas flame meets a smooth surface at a temperature of about 1500 °C. In coking furnaces, it forms on the walls of retorts. R. is harder and denser than ordinary graphite, and is used mainly to make electrodes for arc lamps and galvanic cells.

Retro-Diels-Alder reaction: a reverse Diels-Alder reaction (see) in which the products of a Diels-Alder reaction form its starting materials, a diene and philodiene. For example, *endo*-dicyclopentadiene decomposes into the monomers when distilled.

180°C
≈20°C
2

Retropinacoline rearrangement: a nucleophilic 1,2-rearrangement in which the hydroxyl group of a secondary alcohol is first protonated, then cleaves off water to form a carbenium ion. This is stabilized by migration of an alkyl anion from an adjacent tertiary C atom, that is, with breaking and reforming of covalent bonds. The electron deficit, which also migrates, is satisfied by loss of a proton from a neighboring C atom and formation of an allene molecule.

$$R^2-C(R^1)(R^3)-CH(OH)-R^4 \xrightarrow{H^+} R^1R^2C=CR^3R^4 + H_2O$$

In this way, for example, 3,3-dimethylbutan-2-ol is converted to tetramethylethene (2,3-dimethylbut-2-ene). The Pinacol-pinacolone rearrangement (see) and the Wagner-Meerwein rearrangement (see) are similar reactions.

Reverse-phase chromatography: see High-performance liquid chromatography.

Reversible: an adjective used in thermodynamics to describe processes which can be made to go in the opposite direction at any time by application of differential changes in state variables (e.g. pressure or temperature), without causing a change of state in the environment. R. processes are idealized limiting cases, which occur so slowly that the systems in which they occur are always at thermodynamic equilibrium. The Entropy (see) s of the system therefore remains constant during a R. process; the equilibrium condition is $ds = 0$. R. processes make possible the extraction of a maximum of work from a process, or, in the other direction, require a minimum expenditure of work. All processes which do not fulfill this condition are Irreversible (see).

Examples: 1) From a heat reservoir at temperature T_2, an amount of heat q is transferred to a cooler reservoir at temperature T_1. If the heat transport is irreversible, through heat conduction, no work can be obtained from it. The change in entropy is $\Delta s = \Delta s_1 - \Delta s_2 = q/T_1 - q/T_2 > 0$, since $T_1 > T_2$. If this process is to be undone, work must be done on the system, e.g. by a heat pump. However, if an ideal heat machine (see Carnot cycle) is placed between the two heat reservoirs, the mechanical work $|a| = q\ (T_2 - T_1)T_1$ is done, and only the heat $q_1 = q - |a|$ is released to the reservoir at T_1. Since the Carnot machine works reversibly, the direction can be changed at any time. 2) If zinc powder is added to dilute copper sulfate solution, an irreversible redox reaction occurs: $Cu^{2+} + Zn \rightarrow Cu + Zn^{2+}$. The reaction enthalpy, $\Delta_R H = -217$ kJ mol^{-1}, is released entirely as heat. However, if a copper and a zinc strip are dipped into solutions of their sulfates, they form a galvanic element with electromotive force $E = 1.10$ V. If electric current is withdrawn from the element so slowly that the system is always in equilibrium (e.g. by using a counterpotential, or a very large resistance), the above reaction is R. The electric work $\Delta_R G = -ZFE = -2 \cdot 96487$ C mol$^{-1} \cdot 1.10$ V $= -212$ kJ mol^{-1} is done. Here F is the Faraday constant and $^{-1}$n is the charge number of the ions. The heat Q_{rev} in R. reactions corresponds to the difference $Q_{rev} = T\Delta_R S = \Delta_R H = \Delta_R G = -5$ kJ corresponding to the Gibbs-Helmholtz equation (see).

Reversibility, ***principle of microscopic reversibility***: the statistically based prediction that a macroscopic system is in equilibrium if and only if all microscopic elementary processes which can occur in it are in equilibrium. This implies that these processes are reversible and the rates of the forward and back reactions of any given process are equal to each other.

The principle of microscopic R. can be applied to coupled macroscopic equilibria (*simultaneous equilibria*), and can then be formulated as follows: if a macroscopic process consists of several coupled, reversible partial reactions, it is in equilibrium only if all the partial reactions are in equilibrium. This is also known as the ***principle of detailed equilibrium***. For chemical equilibria, it follows that the equilibrium constant K of the total reactions is equal to the products of the equilibrium constants K_i of the partial equilibria: $K = \Pi K_i$.

Rexforming process: see Reforming.

Rf: symbol for rutherfordium, see Kurčatovium.

R_f: see Paper chromatography.

Rh: symbol for rhodium.

Rhamnetin: see Flavanoids.

Rhamnose, ***l-deoxymannose***: a 6-deoxysugar. L-R. is found in glycosidic form in various plant slimes, hemicelluloses, cardenolides, bufadienolides and flavonoids.

Rhein: see Anthra glycosides.

Rheinpreussen-Koppers process: see Fischer-Tropsch synthesis.

Rhenium, symbol ***Re***: a chemical element in group VIIb of the periodic system, the Manganese group; a heavy, noble metal, Z 75, with natural isotopes with mass numbers 187 (62.93%, weakly radioactive, β-emitter, half-life $7 \cdot 10^{10}$ years) and 185 (37.07%), atomic mass 186.207, valence II to VII, less often -I, 0, +I, density 21.03, m.p. 3180°C, b.p. 5870°C, standard electrode potential (Re/ReO_4^-) + 0.368 V.

Properties. R. is a shiny white metal which crystallizes in a hexagonal close-packed array; it looks like platinum and is characterized by high density, high melting and boiling points and resistance to air oxidation. It is scarcely soluble in hydrochloric or hydrofluoric acid, but dissolves readily in nitric acid. Oxidizing alkali melts convert R. to green rhenates(VI), M_2ReO_4. When heated above 400°C with oxygen, R. forms rhenium(VII) oxide, Re_2O_7. When heated with fluorine or chlorine, it forms the hexahalides. Many rhenium compounds correspond formally to manganese compounds, but the higher oxidation states of R. are more stable than those of manganese, and the lower oxidation states are less stable. Re(II) compounds have been little studied.

Analytical. R. is precipitated out of strongly acidic solution with hydrogen sulfide as rhenium(VII) sulfide, which can be converted to the volatile rhenium(VII) oxide with oxygen. Reaction of Re_2O_7 with water yields perrhenic acid, which forms characteristic, crystalline salts. For example, tetraphenylarsonium perrhenate is suitable for gravimetric analysis of R. The hexachlororhenates(IV), M_2ReCl_6, are also suitable for microanalytical determination. Spectroscopic methods are also used.

Occurrence and production. R. is a rare metal, which makes up about 10^{-7}% of the earth's crust. Only a few minerals contain it in higher concentrations, for example, molybdenite (molybdenum glance) contains about 10^{-3} to 10^{-5}%. Some other minerals which contain R. are columbite, gadolinite, alvite and platinum ores. R. is present at about $5 \cdot 10^{-3}$% concentration in the smelting residues from Mansfield copper shale, and is extracted from them. The residues are subjected to oxidizing leaching, during which R. goes into solution as perrhenate. It is precipitated as potassium perrhenate and purified by recrystallization, and the metal is obtained by reduction with hydrogen. When molybdenite is processed, R. escapes as Re_2O_7 with the fly ash. This is trapped and converted to $NH_4[ReO_4]$, then reduced with hydrogen. R. can also be obtained by thermal decomposition of rhenium halides or electrolytically.

Applications. R. is used to make thermoelements which have high heating power, and is used in combination with platinum elements and tungsten. It is used as a material for pen points, electrodes, transistors and galvanic coatings, e.g. for jewelry. R. catalysts are used for hydrogenation, hydrocracking and reforming processes. Addition of R. to tungsten and molybdenum alloys improve their mechanical properties and resistance to high temperatures, so that such alloys can be used for furnaces, generators and in space vehicles. Re-Mo alloys are superconducting below 10 K.

Historical. Because of the scarcity of R. minerals, the element was discovered very late, in 1925, by W. Noddack and I. Tacke. Earlier efforts to find this heavy manganese homolog had failed. Noddack and Tacke discovered it by x-ray spectroscopy of enriched ores; they later isolated it as well.

Rhenium compounds: Rhenium exists in the +7 oxidation state in its compounds; the +3 to +6 states are also common, and the - 1 to +1 states are found in a few complexes.

The most important oxide of rhenium, ***rhenium(VII) oxide***, Re_2O_7, forms yellow crystals, M_r 484.40, density 6.103, m.p. 300.3 °C; it is formed when rhenium powder is heated in the air. Re_2O_7 is soluble in water, with formation of ***perrhenic acid***, $HReO_4$. The colorless salts of this acid, the ***perrhenates*** $M[ReO_4]$, are much weaker oxidizing agents than permanganates. ***Rhenium(VI) oxide***, ReO_3, forms red, cubic crystals, M_r 234.20, density 6.9 to 7.4, subl.p. 614 °C. It is made by reduction of Re_2O_7 with rhenium at 250 °C, while dark brown ***rhenium(IV) oxide***, ReO_2, M_r 218.20, density 11.4, is formed by the reaction of rhenium with Re_2O_7 at 600 °C, and ***rhenium(III) oxide*** is obtained in the form of its hydrate, $Re_2O_3 \cdot 3H_2O$ by hydrolysis of $ReCl_3$ with sodium hydroxide solution.

Black ***rhenium(VII) sulfide***, Re_2S_7, M_r 596.85, density 4.866, is formed when hydrogen sulfide is passed through a perrhenate solution. Its thermal decomposition yields black ***rhenium(IV) sulfide***, ReS_2, M_r 250.33, density 7.506, which can also be formed directly from the elements.

If rhenium is heated with fluorine under moderate pressure to 400 °C, the product is ***rhenium(VII) fluoride***, ReF_7, a bright yellow compound, M_r 319.21, m.p. 48.3 °C, b.p. 73.7 °C; with fluorides, it forms ***octafluororhenates(VII)***, $MReF_8$. At 125 °C, combination of the elements yields pale yellow, very volatile ***rhenium(VI) fluoride***, ReF_6, M_r 300.19, density 6.157, m.p. 18.7 °C, b.p. 33.7 °C, from which the ***octafluororhenates(VI)***, $M_2[ReF_8]$, are derived.

Rhenium reacts with chlorine at 600 °C to form green ***rhenium(VI) chloride, $ReCl_6$***, M_r 398.91, m.p. 29.5 °C. ***Rhenium(V) chloride***, $(ReCl_5)_2$, is a red-brown compound with a dimeric, octahedral structure. Chlorination of ReO_2 with thionyl chloride yields black ***rhenium(IV) chloride***, $ReCl_4$, a trimeric compound, M_r 984.03. The dark red, trimeric ***rhenium(III) chloride***, Re_3Cl_9, M_r 877.68, m.p. > 550 °C, is a triangular metal cluster in which the Re-Re distance is 248 pm, indicating a metal-metal double bond. In addition to chlorocomplexes derived from Re_3Cl_9, $[Re_3Cl_{9+n}]^{-n}$ (n = 1, 2 or 3), the chloro complexes of the $[Re_2Cl_8]^{2-}$ are worthy of note; the Re-Re distance observed here is only 224 pm, which indicates a bond order of 4 for the metal-metal bond.

Hydrogenation of perrhenates with strong reducing agents leads to ***nonahydrorhenate(VII)***, $[ReH_9]^{2-}$, a complex anion with a trigonal-prismatic basic structure with 3 H atoms in equatorial positions perpendicular to the surfaces of the prism. Reaction of Re_2S_7 or Re_2O_7 with carbon monoxide yields the colorless, monoclinic prisms of ***rhenium carbonyl, dirhenium decacarbonyl***, $Re_2(CO)_{10}$, M_r 652.51, m.p. 177 °C.

Rhenium trioxide type:, ***rhenium trioxide structure***: a Structure type (see) for compounds with the general composition AB_3. In the crystal structure of rhenium trioxide, ReO_3, the metal atoms occupy the vertices of a cubic unit cell, and the oxygen atoms occupy the centers of the edges. Each Re atom is surrounded octahedrally by 6 O atoms (Fig.).

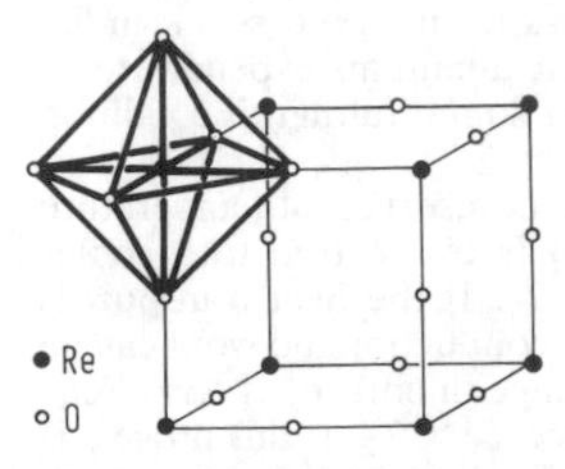

Unit cell of the rhenium trioxide lattice.

$TiOF_2$, $Sc(OH)_3$, a number of trifluorides (e.g. VF_3, FeF_3 and CoF_3), and other compounds crystallize in the R. or related lattices. The Perovskite type (see) is closely related to the R.

Rhodamine pigments: a group of xanthene pigments which can be made by condensation of substituted 3-aminophenols with phthalic anhydride. A typical example is ***rhodamine B*** (Fig.), which forms green crystals or a red-violet powder.

Rhodamine B

It is readily soluble in water and alcohol, and aqueous solutions are highly fluorescent. ***Rhodamine 6 G (brilliant pink, rosamine)*** differs from rhodamine B in that the carboxyl group is esterified and it has monoethylamino groups -NHC_2H_5 instead of the $(C_2H_5)_2N$-groups. Textiles dyed with R. have extraordinarily pure red nuances, but are very prone to fading in the light. R. are used in the paper industry, as a vital stain in microscopy, and to dye leather.

Rhodanides: same as Thiocyanates.

Rhodia®: see Synthetic fibers.

Rhodiacs: see Rhodium compounds.

Rhodinizing: see Rhodium.

Rhodium, symbol ***Rh***: a chemical element in group VIIIb of the periodic system, a member of the light Platinum metals (see), and a noble metal. R. is isotopically pure, *Z* 45, atomic mass 102.9055, valence usually III, less often I, IV, V, VI, 0, -I, density 12.41, Mohs hardness 6, m.p. 1960 °C, b.p. 3670 °C, electrical conductivity 23.1 Sm mm^{-2}, standard electrode potential (Rh/Rh^{2+}) +0.6 V.

Properties. R. is a silvery white metal which crystallizes in a regular, face-centered lattice; it can be distilled in an electric arc. R. is characterized by high chemical stability, and in compact form it is not dissolved by acids, even aqua regia. At red heat it is oxidized by atmospheric oxygen to rhodium(III) oxide, Rh_2O_3, and by chlorine to rhodium(III) chloride, $RhCl_3$. Molten potassium hydrogensulfate dissolves R., forming potassium rhodium alum, $KRh(SO_4)_2 \cdot 12H_2O$. Alkali hydroxide and alkali nitrate melts are suitable solubilization media for R., which is converted to Rh_2O_3 in them. Finely divided R. can be obtained by reduction of its salts (***rhodium black***), and is more reactive than the compact metal. It can be converted to rhodium(III) sulfate, $Rh_2(SO_4)_3$, by sulfuric acid, or by aqua regia to rhodium(III) chloride, $RhCl_3$. The chemistry of R. is determined largely by the stability and variety of octahedral, diamagnetic rhodium(III) complexes. In addition to the fluorides of R. in higher oxidation states, the square planar rhodium(I) complexes are of particular interest.

Analysis is based mainly on the insolubility of the rhodium(III) complexes $(NH_4)_3[RhCl_6]$ and $[RhCl(NH_3)_5]Cl_2$, and on characteristic reactions of R. with thionalide and sodium diethyl dithiocarbamate.

Occurrence and production. R. is a very rare element, making up only about 10^{-7}% of the earth's crust. It is always found in association with the other Platinum metals (see), and is one of the rarest of these. As a result of the laborious separation operations required for industrial extraction of platinum metal ores, R. is finally precipitated as ammonium hexachlororhodate, $(NH_4)_3RhCl_6$, which is reduced to the metal with hydrogen at high temperatures.

Applications. Because of its hardness, reflectivity and chemical resistance, R. is often used to make protective layers on the surfaces of metal parts, such as jewelry and decorative objects, contacts, silver mirrors, reflectors, and spinerets for fiberglass. The application of a thin R. layer by a galvanic process is called *rhodinizing*. Platinum-rhodium alloys (see Platinum alloys) are used widely, and R. is also added to Titanium alloys (see).

Historical. R. was discovered in 1803, along with palladium, by Wollaston; it was named for the red color (Greek "rhodeos") of many of its compounds.

Rhodium compounds: Rhodium is most often found in the +III oxidation state in its compounds. In addition to numerous octahedral rhodium(III) complexes, square planar rhodium(I) complexes are important. The +IV through +VI oxidation states are found mainly in fluorides and fluoro complexes, while polynuclear rhodium carbonyls represent the 0 oxidation state. Some important binary R. are: ***Rhodium(III) oxide***, Rh_2O_3, a gray compound crystallizing in a corundum lattice. It is made by heating rhodium powder in air; M_r 253.81, density 8.20. In hydrated form, Rh_2O_3 is obtained as a yellow pentahydrate, $Rh_2O_3 \cdot 5H_2O$, by addition of an alkali hydroxide solution to a rhodium(III) salt solution. ***Rhodium(III) chloride, rhodium trichloride***, $RhCl_3$, is a red-brown compound with the same structure as $AlCl_3$; M_r 209.28, m.p. 450 °C. It is formed by chlorination of rhodium at 700 °C. ***Rhodium(III) fluoride, rhodium trifluoride***, RhF_3, red, rhombic crystals; M_r 159.90, density 5.38, subl. p. > 600 °C. RhF_3 is obtained by fluorination of rhodium or rhodium(III) chloride at 500 to 600 °C; hydrates such as $RhF_3 \cdot 6_2O$ or $RhF_3 \cdot 9H_2O$ are obtained by addition of fluorides to rhodium(III) salt solutions. Blue ***rhodium(IV) fluoride, rhodium tetrafluoride***, RhF_4, M_r 178.90, is formed by reaction of $RhCl_3$ with bromine trifluoride, while dark red, tetrameric ***rhodium(V) fluoride, rhodium pentafluoride***, RhF_4, M_r 791.59, is formed by fluorination of rhodium at 400 °C. The very reactive, octahedral, black ***rhodium(VI) fluoride, rhodium hexafluoride***, RhF_6, M_r 216.90, m.p. 70 °C, is formed as a fluorination product of rhodium and can be trapped out of the gas phase. The ***rhodium(III) complexes*** are especially significant. All are octahedral, diamagnetic complexes which may be cationic, anionic or neutral. Some examples are: $[Rh(H_2O)_6]^{3+}$, formed by dissolving $Rh_2O_3 \cdot nH_2O$ in mineral acids, or by repeated evaporation of perchloric acid solutions of $RhCl_3$; rhodiumammine complexes (***rhodiacs***), such as $[Rh(NH_3)_6]X_3$; $[Rh(NH_3)_5Cl]Cl_2$, a yellow complex which, because of its insolubility, is suitable for separating rhodium from iridium; and red hexachlororhodates, $M_3[RhCl_6]$. The so-called soluble rhodium chloride is obtained by evaporation of a hydrochloric acid solution of hexachlororhodic(III) acid, $H_3[RhCl_6]$; it is an important commercial rhodium salt with an approximate composition $RhCl_3 \cdot 2.5H_2O$. When rhodium(III) chloride hydrate is heated in alcohol in the presence of π-acceptor ligands, the interesting square-planar ***rhodium(I) complexes*** are formed. An example is the dimeric, chlorine-bridged carbonyl chloride $[Rh(CO)_2Cl]_2$, which crystallizes in red needles. Another is the phosphine complex $[Rh(CO)Cl(P(C_6H_5)_3)_2]$. A similar type of complex is the trigonal bipyramidal rhodium(I) complex, for example $[RhH(CO)(P(C_6H_5)_3)_3]$. Examples of ***rhodium(O) complexes*** are the rhodium carbonyls $Rh_4(CO)_{12}$, dark red crystals, M_r 747.65, m.p. 150 °C (dec.), and $Rh_6(CO)_{16}$, black crystals, M_r 1065.51, m.p. 220 °C (dec.); see Metal carbonyls for structures.

Rhodium complexes can be used as catalysts of homogeneous hydrogenation and in oxosynthesis.

Rhodopsin: see Vitamin A.

Rhombic: see Crystal.

Rhombohedral: see Crystal.

Rhovyl®: see Synthetic fibers.

RIA: abb. for radioimmunoassay; see Immunoassay.

Ribo-: a prefix for a certain configuration of a sugar; see Monosaccharides.

Riboflavin: same as Vitamin B_2 (see).

Riboflavin 5'-phosphate: see Flavin nucleotides.

Ribonucleases: phosphodiesterases (see Esterases) which catalyse the cleavage of RNA molecules. R. which break the RNA molecule in the middle of the strand are endonucleases; those which remove only the terminal nucleotide are exonucleases. Some R. produce 5'-monoesters or oligonucleotides; others produce 3'-phosphorylated products.

Ribonucleic acids: see Nucleic acids.

D-Ribose, abb. ***Rib***: a pentose. R. forms colorless crystals which are readily soluble in water, but only slightly soluble in ethanol; $[\alpha]_D$ -23.7° (in water, at mutarotational equilibrium). D-R. is a component of the ribonucleic acids (see Nucleic acids), the ribonucleosides and ribonucleotides, some coenzymes (e.g. coenzyme A), the flavin nucleotides, nicotinamide nucleotides and vitamin B_{12}. D-R. can be obtained by hydrolysis of yeast nucleic acids or synthesized from arabinose.

Ribosomal RNA: see Nucleic acids.

Richard's rule: see Phase change heats.

Rich gas: a mixture of gases obtained by hydrogenation of carbon monoxide, by thermal cleavage of liquid or solid fuels in the presence of water pressure or by pressure gasification. R. is used as a fuel (city gas) or a synthesis gas. The main component is methane, and it also contains carbon monoxide, carbon dioxide and nitrogen.

Ricin: see Albumins.

Ricinolic acid: see Ricinus oil.

Ricinus oil, ***castor oil***: the cold pressed, fatty oil from the seeds of *Ricinus communis*. It consists of about 90% acylglycerols of *ricinolic acid* (12R-hydroxyoleic acid). In the small intestine, lipases catalyse its hydrolysis, forming ricinolic acid, which is irritating to the intestinal mucosa and therefore acts as a laxative. R. is added to skin medications and cosmetics due to its alcohol solubility and stability; it is also used as a special lubricant.

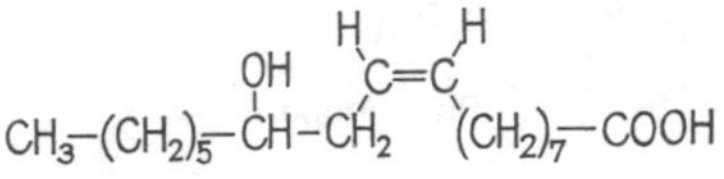

$CH_3-(CH_2)_5-CH(OH)-CH_2-CH=CH-(CH_2)_7-COOH$

Ricinolic acid

Rideal-Eley mechanisms: see Catalysis, sect. I.

Ridomil®: see Acylalanine fungicides.

Rifampicin: see Rifamycins.

Rifamycins: a group of antibiotics formed by *Streptomyces mediterranei* and their partially synthetic derivatives. They have a naphthohydroquinone structure. ***Rifamycin B***, for example, is a natural product from which ***rifamycin*** and ***rifampicin*** can be made; the latter is very important for treatment of tuberculosis and meningitis. R. inhibit RNA polymerases.

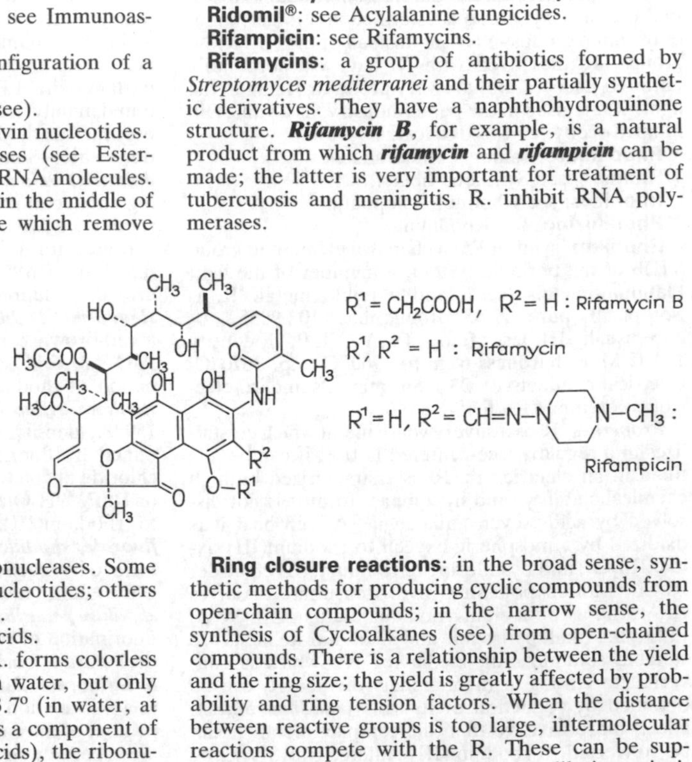

$R^1 = CH_2COOH$, $R^2 = H$: Rifamycin B

$R^1, R^2 = H$: Rifamycin

$R^1 = H$, $R^2 = CH{=}N{-}N\langle\ \rangle N{-}CH_3$: Rifampicin

Ring closure reactions: in the broad sense, synthetic methods for producing cyclic compounds from open-chain compounds; in the narrow sense, the synthesis of Cycloalkanes (see) from open-chained compounds. There is a relationship between the yield and the ring size; the yield is greatly affected by probability and ring tension factors. When the distance between reactive groups is too large, intermolecular reactions compete with the R. These can be suppressed by use of the Ruggli-Ziegler dilution principle. Often the reactive groups can be fixed close to each other, e.g. by formation of the salt of a dicarboxylic acid.

Ring contraction reactions: reactions which decrease the ring size of a cycloalkane due to a rearrangement:

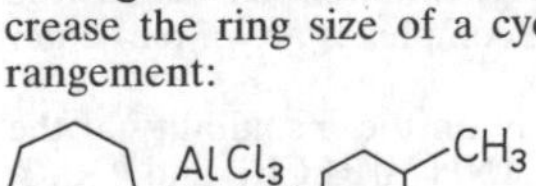

$$\text{cycloheptane} \xrightarrow[\Delta]{AlCl_3} \text{methylcyclohexane}$$

The Benzilic acid rearrangement (see) of cycloalkane-1,2-diones and the Favorskij rearrangement (see) are also R.

Ring expansion reactions: synthetic methods for increasing the number of ring atoms in a cyclic compound. Cycloalkanes and their derivatives rearrange when heated with aluminum chloride as shown:

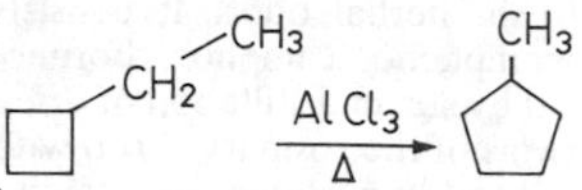

1

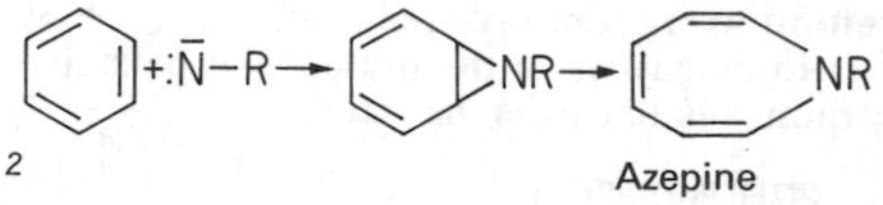

2 Azepine

An important R. is the reaction of cycloalkanes with diazomethane; for example, cycloheptanone can be relatively easily made from cyclohexanone in this way. Arene rings can be made by addition of carbenes or nitrenes and subsequent valence isomerization by a R. (Fig. 2).

Ring fusion rule: a rule in UV-VIS spectroscopy (see) which says that in fused aromatic rings, increasing the number of rings shifts the wavelength of absorbed light to longer wavelengths and increases the intensity of absorption. For example, anthracene absorbs in the UV, naphthacene is yellow (absorbs blue light) and pentacene is blue (absorbs red).

Ring inversion: the interconversion of stable conformations of cycloalkanes (see Stereoisomerism 2.2, Fig. 18). This process occurs via highly strained conformations, and therefore requires a relatively high activation energy.

Ring sequences: see Nomenclature, sect. III.E.1.

Ring tension: see Baeyer's tension theory.

Rinmann's green: same as Cobalt green (see).

Rishitin: see Phytoalexins.

Ritter reaction: the synthesis of primary amines from alkenes or alcohols and nitriles or hydrogen cyanide in the presence of concentrated sulfuric acid:

$$R{-}CN + CH_2{=}C(CH_3)_2 \xrightarrow{H_2O/H^+} R{-}CO{-}NH{-}C(CH_3)_3$$
$$\xrightarrow{H_2O/H^+} (CH_3)_3{-}C{-}NH_2 + R{-}COOH.$$

The R. makes possible the synthesis of amines with substituents which are difficult to obtain by other methods, such as *tert.*-alkyl groups. In addition to simple alkenes, unsaturated carboxylic acids and their esters can be used. The tertiary alcohols are very good in the R., but hydroxycarboxylic acid esters are also quite useful. The alkenes or alcohols can be reacted with aliphatic and aromatic nitriles and dinitriles, aldehyde cyanohydrins and potassium cyanide. The R. is acid-catalysed; a carbocation intermediate is formed which adds to the nitrile. Addition of water and cleavage of a proton leads to an amide, which is hydrolysed to the amine and carboxylic acid.

Ritz's combination principle: see Combination principle.

Rivanol®: same as Ethacridine (see).

RLCCC: see Countercurrent chromatography.

Rn: symbol for radon.

RNA: see Nucleic acids.

RNA polymerase, ***transcriptase***: an enzyme which catalyses the formation of ribonucleic acids (RNA) from ribonucleoside triphosphates on a DNA matrix. After the enzyme binds to the deoxyribonucleic acid (DNA), it unwinds a short section and initiates incorporation of the corresponding nucleotides. The R. moves down the DNA strand as it incorporates nucleotides into the growing RNA strand, which is released from the DNA-enzyme complex as it is synthesized. There is a specific R. for each type of RNA. The rifamycins inhibit bacterial R. (M_r about 500,000).

Robinson condensation: 1,4-addition of an enone to a carbonyl compound followed by aldol condensation to form a cyclic structure. The R. is important in the synthesis of natural products, especially steroids, for example:

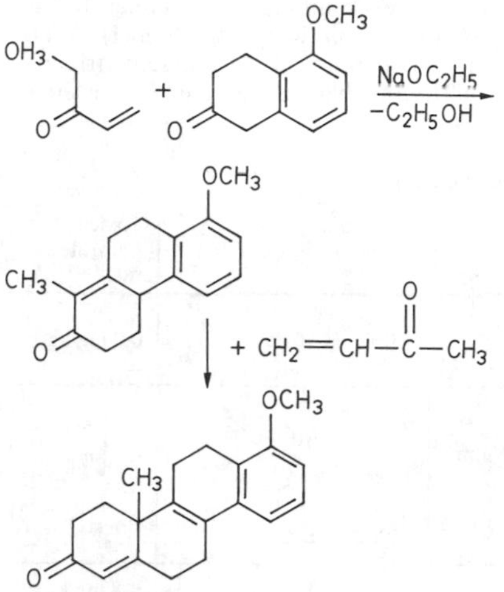

Rochelle salt: same as Potassium sodium tartrate.

Rockwell hardness, abb. ***HRB*** (for ball) or ***HRC*** (for cone): the hardness of a metal determined by the Rockwell method. The test body is a hardened steel sphere of about 1.59 mm diameter (HRB) for soft metals with 35 to 100 HRB, or a diamond cone with an apical angle of 120° (HRC) for hard metals with 20 to 68 HRC (such as hardened steels). The test body is pressed into the metal at 981 N (HRB) or 1471.5 N (HRC). Of this amount, 98.1 N is applied as prepressure. The penetration of the test body into the metal is recorded by a meter; after a short application of the total force (up to 30 s), the additional force is removed and only the pre-pressure is applied while the residual penetration depth *e* is read off. The residual penetration depth *e* in units of 0.002 mm is subtracted from 130 or 100 to give the R.: 130-*e* = HRB, 100-*e* = HRC. The ***Rockwell A hardness***, abb. ***HRA***, is usually measured for very hard and brittle substances, such as hard metals. The HRA is like the HRC, but a total force of only 588.6 N is applied.

Rock wool: a fibrous material obtained by melting sedimentary rocks, such as clay shale, sandstone, marl, limestone or dolomite. The synthesis is similar to that of fiberglass, with the fibers produced by blowing or spinning the melt off a rotating disk. The fibers are 8 to 15 μm in diameter, and are relatively inert. R. contains 50 to 75% silicon dioxide and aluminum oxide, and 25 to 50% calcium oxide and magnesium oxide. It can be used as a heat insulator at temperatures up to 700 °C.

Rocornal®: see Coronary drugs.

Rodenticide: a biologically active compound used to combat rodent pests, usually rats or mice. These animals consume and defile food and transmit diseases, including typhus, plague, spotted fever and amebic dysentery. The loss of food to rodents is estimated, for cereal grains and rice alone, to be 3 to 5% of the total production.

The R. is usually set out in a bait. *Acute R.* are completely effective when consumed once. The symptoms of poisoning, which usually occur very soon, can lead rats to avoid the bait. In addition, their high toxicity for warm-blooded animals makes them dangerous for domestic animals and human beings. Some long-used inorganic R. are thallium(I) sulfate, Tl_2SO_4, trizinc diphosphide, Zn_3P_2, arsenic trioxide, As_2O_3 and yellow phosphorus. Organic R. are shown in Table 1.

Table 1. Acute rodenticides

Name	Formula	Toxicity peroral (LD_{50}/rat)
Sodium fluoroacetate	$FCH_2—COONa$	0.2 mg/kg
ANTU (*a*-Naphthylthiourea)	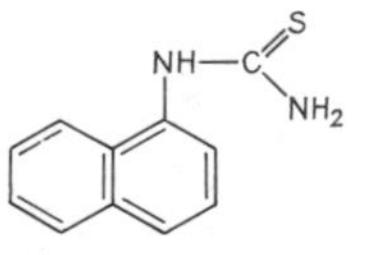 NH—C(=S)—NH₂ on naphthalene	7 mg/kg
Pyrinaron	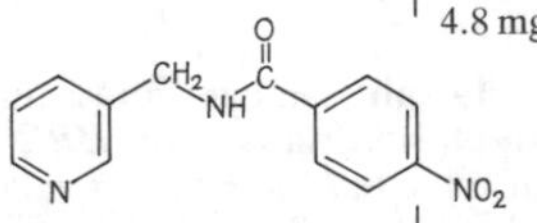	4.8 mg/kg
p-Chlorophenylsilatran	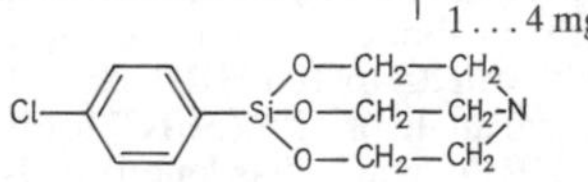	1 ... 4 mg/kg
Crimidine	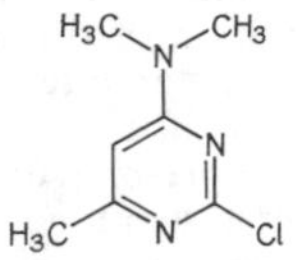	1.2 mg/kg

The *chronic R.* are Anticoagulants (see), which only kill the animals after repeated consumption of sublethal doses. They increase the permeability of the capillaries and cause death by internal hemorrhage. Development of resistance has been observed. These compounds are derivatives of coumarin and indanedione (Table 2), p. 953.

Rodents in closed spaces are also killed by gassing, e.g. with hydrogen cyanide, sulfur dioxide, phosphine, etc.

Rolitetracycline: see Tetracyclines.

Rosamine: see Rhodamine pigments.

Rosaniline: same as Fuchsin (see).

Rose bengal: see Eosine pigments.

Rosemary oil: a colorless, or sometimes yellowish essential oil with a fresh, herbal odor. It consists mainly of α-pinene, camphene, camphor, borneol and cineol. It is obtained by steam distillation of dried leaves, blossoms and twigs of the rosmary shrub with a yield up to 2%. R. is used in perfumes and soaps.

Rosenmund reduction: reduction of carbonyl chlorides to aldehydes in the presence of palladium on a barium sulfate carrier substance:

$$R–COCl \xrightarrow[-\,HCl]{+2H(Pd/BaSO_4)} R–COH$$

To prevent hydrogenation of the carbonyl group, the catalyst is deactivated with thiourea or 2-mercaptobenzthazole. As early as 1872, M. Sajcev prepared benzaldehyde in good yield from benzoyl chloride in a stream of hydrogen at 220 °C in the presence of a palladium catalyst. This reaction was taken up by K.W. Rosenmund and thoroughly studied. It can be applied to practically all available acyl chlorides.

Rose oil: a colorless essential oil which smells like roses. It is rather viscous, and solidifies below 25 °C to a translucent mass. It is obtained by various methods from the petals of Damascus roses and some of their crosses. The melting point varies between 16.5 and 23.5 °C, due to variation in the alkane content. R. contains 35 to 55% (-)-citronellol, 30 to 40% geraniol, 5 to 10% nerol, and small amounts of ethanol, phenylethyl alcohol, linalool, citral, carvone, eugenol and eugenol methyl ether, farnesol and higher aliphatic hydrocarbons and aldehydes. "Concrete" R. obtained by extraction with petroleum ether have high contents of phenyl ethyl alcohol (up to 45%). R. can also be obtained by maceration. The blossoms must be picked in the early morning hours, and the yield is only 0.02 to 0.03%, that is, from 1000 kg rose blossoms, 200 to 300 g R. can be isolated. For this reason, R. is by far the most expensive essential oil. It is used to make expensive perfumes, cosmetics and pharmaceuticals, and as a flavoring for liquors, candies and tobacco.

Rose oxide: ***tetrahydro-4-methyl-2-(2-methylpropenyl)2H-pyran***: a colorless oil which smells strongly like geraniums; b.p. 70 °C at 1.59 kPa. n_D^{20} (***cis-R.***) 1.4548, n_D^{20} (***trans-R.***) 1.4585. R. is an optically active compound. It is present in Bulgarian rose oil at a concentration of 0.1% and in Bourbon geranium oil. It is synthesized by photooxidation of citronellol in the presence of rose bengal as sensitizer.

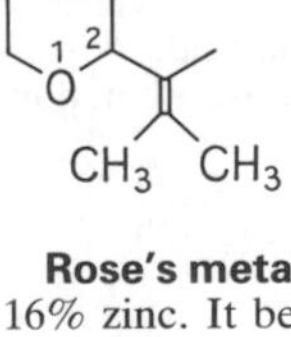

Rose's metal: alloy of 52% bismuth, 32% lead and 16% zinc. It begins to melt at 96 °C and is therefore used as the contact material for melt and delay fuses,

Table 2. Chronic rodenticides

Name	Formula	melting point
Cumarin type	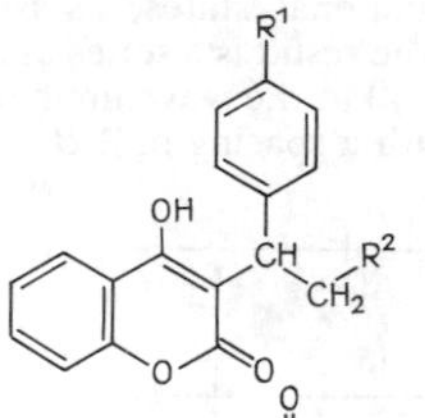 R^1; OH; CH; R^2; CH_2; O; O	
Warfarin	$R^1 = H$; $R^2 = -\overset{O}{\overset{\|}{C}}-CH_3$	161···162 °C
Coumachlor (Cumachlor)	$R^1 = Cl$; $R^2 = -\overset{O}{\overset{\|}{C}}-CH_3$	160···162 °C
Bromadiolon	$R^1 = H$; $R^2 = -\overset{OH}{\overset{\|}{CH}}-C_6H_4-C_6H_4-Br$	200···210 °C
Indanedione type	O; O; CO—R	melting point
Diphacinon	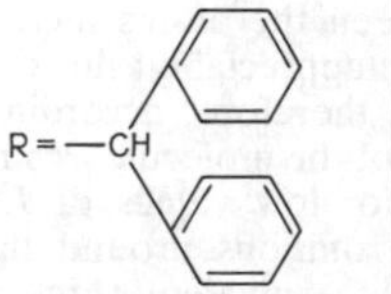 $R = -CH(C_6H_5)_2$	145···147 °C
Chlorphacinon	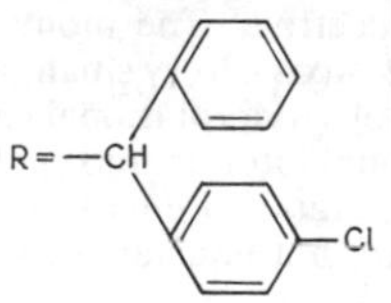 $R = -CH(C_6H_5)(C_6H_4Cl)$	140 °C
Pindon	$R = -C(CH_3)_3$	109···110.5 °C

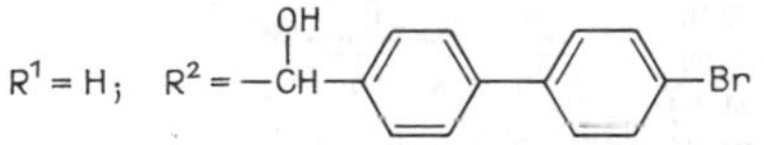

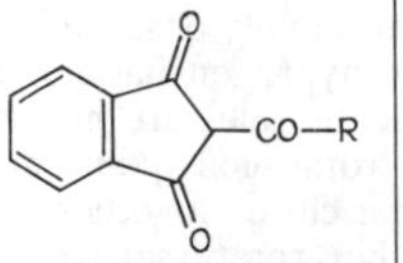

as heating bath liquid and as soft solder for materials which are sensitive to heat.

Rotamers: see Stereoisomerism 2.

Rotary axis: see Symmetry.

Rotary evaporator: a device for gentle evaporation of solutions. It consists of a heating bath in which a rotating flask is dipped. The liquid is distributed on the hot walls and can readily evaporate. Usually some degree of vacuum is applied to draw off the vapor.

Rotary process: see Petroleum.

Rotating band column: a distillation column with rotating internal structures.

Rotating crystal method: see X-ray structure analysis.

Rotation: see Symmetry.

Rotational inversion: see Symmetry.

Rotational reflection: see Symmetry.

Rotational-vibrational spectrum: see Infrared spectroscopy.

Rotation, axis of: see Symmetry.

Rotation barriers: see Stereoisomerism 2.

Rotation isomers: see Stereoisomerism 2.

Rotation-locular CCC: see Counter-current chromatography.

Rotation spectra: molecular spectra based on excitation of molecular rotations. They are observed in the far infrared (see Infrared spectroscopy) and the microwave range (see Microwave spectroscopy), and provide information on molecular structure.

Theoretical. Molecules with sufficient energy rotate around their main axes of rotational inertia. Because the rotational energy is quantized, only certain definite values are possible, which are expressed as discrete rotational energy states of the rotator. According to the quantum theory, only those rotational energies E_{rot} are possible which correspond to the equation

$$E_{rot} = \frac{h^2}{8\pi^2 I} J(J+1) \quad (1)$$

I is the moment of inertia of the molecule relative to the axis of rotation, J is the rotational quantum number, which can have the values 0, 1, 2, 3 ..., and h is Planck's constant. In a transition from a lower energy state to a higher one, the molecule absorbs the difference in energy in the form of electromagnetic radiation, provided it has a permanent Dipole moment (see), because it is otherwise unable to interact with electromagnetic radiation.

The geometric shape of the molecule, which is reflected by its moment of inertia, largely determines the appearance of the rotation spectrum. Linear molecules, which are characterized by a single moment of inertia, and radially symmetrical molecules (that is, molecules with an axis of three-fold or higher symmetry, e.g. CH_3Br, in which 2 of the main moments of inertia are the same) have relatively simple R. which consist of lines with nearly equidistant spacing. Radially asymmetric molecules (molecules with an axis of two-fold symmetry or lower symmetry) have complex R. which are difficult to interpret. Therefore the theoretical part of this discussion will be limited to the simplest case of a diatomic molecule in which the two atoms A and B are at a fixed distance r_1 apart; this molecule has the same rotational properties as a dumbbell (Fig. 1a). For further simplification, the dumbbell model can be replaced by the rigid rotator model, in which a mass

$M = \dfrac{m_A m_B}{m_A + m_B}$ rotates around a fixed axis at a distance r from it (Fig. 1b).

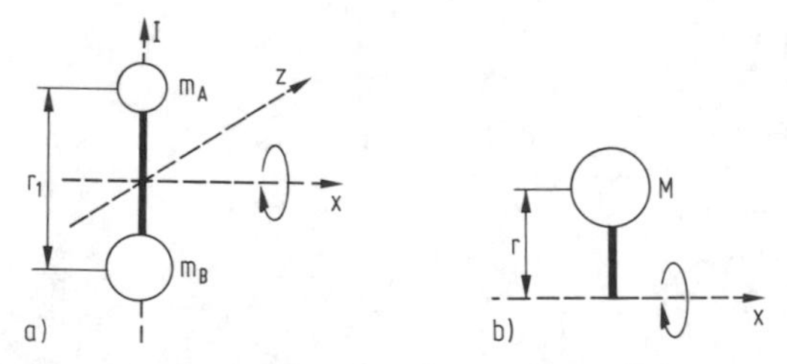

Fig. 1. Rotating dumbbell (a) and rigid rotator (b) models.

In molecular spectroscopy, eq. 1 is written

$$E_{rot} = h\,c\,B\,J(J+1) \qquad (2)$$

where the rotational constant

$$B = \frac{h}{8\pi^2 cI}$$

is a characteristic of each molecule (c = velocity of light). B can be determined directly from the spectrum, and has the dimensions of a wavenumber (cm^{-1}). If the term values obtained from eq. 2, that is, the energy values divided by $h \cdot c$, are represented graphically, the term scheme shown in Fig. 2 is obtained. Here the rotational quantum number J is indicated on the right, and the term values in units of B on the left. Each absorption line in the spectrum corresponds to a defined transition between terms; for absorption, the selection rule $\Delta J = +1$ applies. It must be remembered that at room temperature, the thermal energy is sufficient to excite the molecules to higher rotational states, and the absorptions therefore start not only from the ground state but from higher rotational states, as is shown in the term scheme. The result is a series of equidistant rotational lines (Fig. 2) at the wavenumbers 2 B, 4 B, 6 B, 8 B, ..., i.e. with a spacing of 2 B.

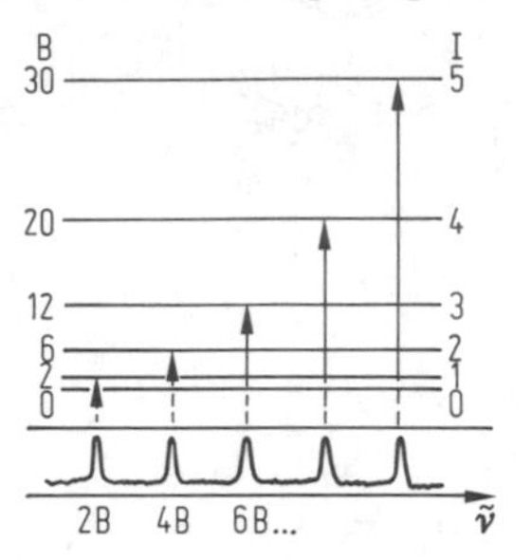

Fig. 2. Rotational term scheme.

Such R. are actually observed for diatomic molecules such as the hydrogen halides. Precise measurement reveals that the lines are not absolutely equidistant, but as the rotational quantum number J increases, they become closer together. This is due to the fact that at higher rotational excitation, the centrifugal force between the atoms increases the distance between them appreciably; this changes the moment of inertia and therefore, according to eq. 2, also E_{rot}. Treatment of the molecule as a rigid rotator is acceptable only for low values of J. For diatomic molecules, the rotations around the x and z axes are energetically equivalent (Fig. 1a), so they produce the same spectrum. The moment of inertia with respect to the y axis is very small, and as a result it is not possible to observe rotational excitation around the y axis. For multiatomic molecules with lower symmetry, the rotational energies around the main axes of inertia are not equivalent, so that the spectra are complex.

Sample preparation and measurement. R. can be made only in the gas state. To eliminate intermolecular interactions, the lowest possible pressure (between 1 and 100 Pa) is used. With high-temperature cells in which temperatures up to 1000 °C can be attained, nearly all substances which vaporize without decomposition can be analysed. R. are made using two different techniques. The short-wave portion of the spectrum is measured with optical spectrometers (see Spectral instruments), while the long-wave portion falls in the microwave range and is measured by the high-frequency technique. According to eq. 1 and 2, molecules with small moments of inertia, e.g. hydrogen fluoride, absorb in the far infrared, while molecules with large moments of inertia absorb in the microwave range.

Applications of R. *Analysis*. Because the spectra arise by rotation of the entire molecule around the axes of inertia, R. is characteristic of the molecule as a whole and can be used for identification, if a comparison spectrum from an authentic sample is available. Since there are no intermolecular interactions under the conditions of measurement, mixtures of substances can also be examined. The resolution capacity is very high; for example, two absorption series are obtained for methyl chloride, one for each of the chlorine isotopes ^{35}Cl and ^{37}Cl. Thus determination of isotope composition is possible.

Determination of interatomic distances and valence angles. Because the rotational constant B is directly related to the moment of inertia of the molecule (Eq. 1 and 2), for diatomic molecules with constant atomic masses, the interatomic distances can be calculated with high precision from the value of B. For more complicated molecules, determination of 3 moments of inertia is not sufficient for calculation of the interatomic distances and valence angle, because more equations are required to solve for more unknowns. These may be obtained by substituting different isotopes of the atoms in the molecule.

Determination of dipole moments. If a rotating dipole molecule is placed in an external electric field, the Stark effect (see) leads to splitting of the individual rotation lines into Stark components. The magnitude of the splitting is measured as the difference in the frequencies measured in the presence and absence of the external field, and depends both on the strength of the applied field and the size of the permanent molecular dipole moment. If the external field strength is known, the shift in frequency permits calculation of the dipole moment with a precision of about ± 1%.

Nuclear quadrupole coupling. Precise analysis of R. has revealed that some molecules have a hyperfine structure of their rotational lines which results from nuclear quadrupole coupling (see nuclear quadrupole resonance spectroscopy). This is the case if the molecule contains atoms with a nuclear spin quantum number $I > 1/2$, and thus have a nuclear quadrupole moment which interacts with an inhomogeneous electric field. The inhomogeneous electric field is generated by an asymmetrical charge distribution at the nucleus and is caused by the distribution of the electrons in the atom. Hyperfine structure measurements can thus give information about the electron distribution in the molecule.

Determination of rotational barriers. Interaction between the rotation of part of the molecule (e.g. methyl groups) and the rotation of the whole can lead to a splitting of the rotational lines into multiplets; the distance between multiplet lines depends on the height of the rotational barrier.

Rotaxans: see Topological isomerism.

Rotenoids: a group of isoflavone natural products. The best known example is *rotenone*, which is obtained from the roots of derris species grown in Southeast Asia and Africa, especially *Derris elliptica*. Rotenone is an excellent insecticide which is practically non-toxic for mammals.

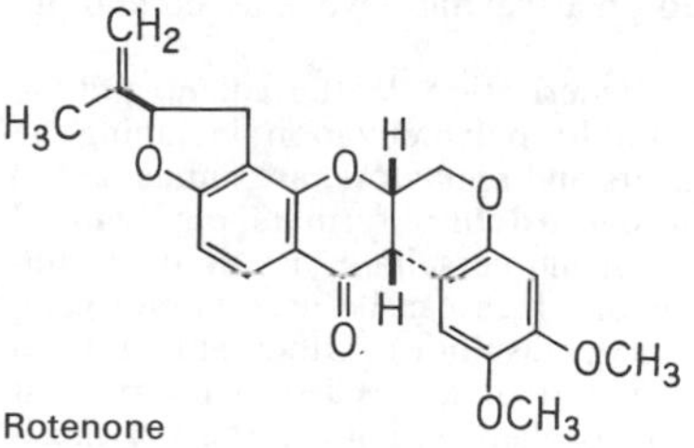

Rotenone

Rotor process: see Steel.

Roussin's salts: sulfur-bridged nitrosyl complexes of iron. Red R. have a dinuclear structure, $M^I_2[Fe_2S_2(NO)_4]$, and black R. have a tetranuclear structure, $M^I[Fe_4S_3(NO)_7]$.

RPC: see High-performance liquid chromatography.

RRKM theory: see Monomolecular reaction.

r,s-configuration: see Stereoisomerism 1.1.

R,S-configuration: see Stereoisomerism 1.1.

R_{St} value: see Paper chromatography.

Ru: symbol for ruthenium.

Rubber: the most important representatives of the Elastics (see). Originally only ***natural rubber*** was called R., but now all highly polymeric substances which are amorphous at room temperature, have a low glass temperature and a relatively low degree of cross-linking are called R. Thus polymers, polycondensates and polyadducts which are made highly elastic by cross-linking with isocyanates, peroxides or light are ***synthetic rubbers***

1) ***Natural R.*** is made from the latex of some tropical and non-tropical plants. The most important of these is the rubber tree (*Hevea brasiliensis*), which grows up to 30 feet tall. The average composition of the latex from this tree is 50 to 75% water, 25 to 40% R. dry matter, 1.5-2% protein, 1.5-2% resins, 1.5-2% carbohydrates and small amounts of fatty acids. The R. component is present in the form of extremely small droplets.

Structure and properties. Chemically, the natural R. are polyisopropenes of almost 100% *cis*-1,4-configuration (Fig. 1), while the closely related natural products gutta percha and balata have the *trans*-configuration.

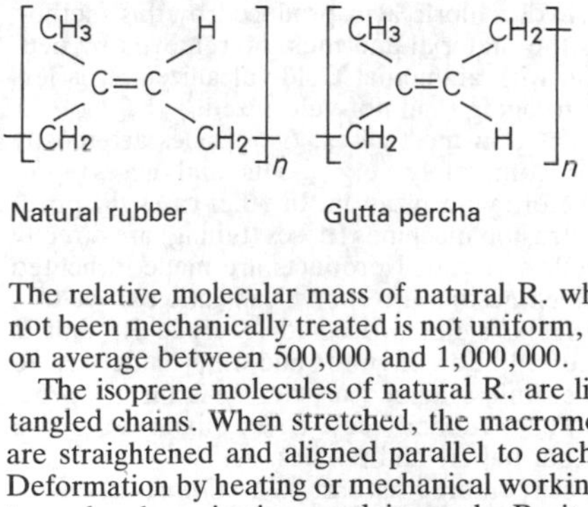

Natural rubber Gutta percha

The relative molecular mass of natural R. which has not been mechanically treated is not uniform, and lies on average between 500,000 and 1,000,000.

The isoprene molecules of natural R. are linked in tangled chains. When stretched, the macromolecules are straightened and aligned parallel to each other. Deformation by heating or mechanical working is due to a depolymerization, and in crude R. it is only partly reversible (plastic deformation). Vulcanization, the incorporation of sulfur into the double bonds, leads to sulfur bridges between the chains, and this reduces the mobility of the chain. Deformation now requires more force, is limited, and is reversible after the force is no longer applied (see Rubber elasticity).

R. is susceptible to oxidation. For example, complete ozonization degrades it to derivatives of levulinic acid. Even after long storage, atmospheric oxygen has little effect on R. If R. is cooled for a long period, it loses its elasticity through partial crystallization. Its density (at 20 °C) is 0.934 and it can be stretched 800 to 1000% before breaking. Non-vulcanized R. is soluble in benzene, gasoline, chlorinated hydrocarbons and low-molecular-weight fats. Vulcanized R. is not attacked by these solvents, which only cause it to swell. Aside from its high elasticity, R. has high mechanical resistance and a high resistance to tearing.

Production. The latex is acidified with dilute acetic or formic acid, causing the R. to coagulate. It is lifted out onto rollers, washed, and rolled out into brownish mats about 3 m long and 50 cm wide.

Vulcanization. In *hot vulcanization*, the crude R. is first mixed with finely powdered sulfur and fillers like soot, kaolin, calcium silicate or silica, and then mechanically worked. The actual vulcanization process begins when the mass is heated to temperatures between 100 and 180 °C. The time can be shortened to a few minutes by addition of vulcanization promoters, i.e. organic nitrogen or sulfur compounds such as secondary amines, xanthogenates, dithiocarbamates and thiazenes, or inoganic substances such as magnesium or zinc oxide, calcium hydroxide or antimony tri- or pentasulfide. ***Soft R.*** has a relatively low sulfur content of 5 to 10% (of the hydrocarbon in the crude R.), while ***hard R.*** contains 30 to 50% sulfur. The latter is an excellent electrical insulater and is chemically resistant to concentrated and dilute acids and bases; it can be dissolved in gasoline, benzene or plant or animal oils.

In *cold vulcanization*, disulfur dichloride is used instead of sulfur. The method is limited to thin-walled products, however, because the S_2Cl_2 can only penetrate a short distance into the R. In the industry, the product to be vulcanized is dipped for a few seconds into a solution of at most 6% S_2Cl_2 in gasoline, benzene, carbon disulfide or the like. The disadvantage is that the hydrochloric acid produced by this method damages the material and must be removed by neutralization with ammonia. Cold-vulcanized R. is less resistant to aging than hot-vulcanized.

Processing. In most cases, the articles are mostly (tires) or completely (e.g. boots and hoses) preformed before vulcanization. In other cases the products of extrusion machines (hoses, tubing) are directly vulcanized, or molded products are made in heated molds which accomplish the vulcanization as well.

Some latex is not processed to crude R., but is used, after addition of sulfur and other materials, to produce seamless dipped articles or is further processed to Foam rubber (see). Latex mixtures are also used to coat papers and textiles.

Chemically altered natural R. 1) ***Chlorinated R.*** is made by chlorination of crude R. at temperatures of 80 to 110 °C. It contains up to 65% chlorine and is readily soluble in polar solvents. Its density is 1.64. Due to its hardness, non-flammability and corrosion resistance, it is used to produce paints which are resistant to alkali hydroxides, acids and weathering. 2) ***R. hydrochloride*** is made by treating dissolved crude R. with dry hydrogen chloride. It is insoluble in polar solvents, and slightly soluble in aromatic and chlorohydrocarbons. The crude product, with a chlorine content of 30 to 34% has a density of 1.12 to 1.16. After drying with stabilizers, it is used to produce mechanically strong, chemical-resistant, water-vapor-tight, flexible, glass-clear foils which are used for packing. 3) ***Cyclized R.*** is obtained by heating crude R. with sulfonic acids or sulfo chlorides to about 145 °C. Depending on the degree of cyclization, its density is 0.97 to 1.02, and it is resistant to fats, dilute acids and alkali hydroxides, but it is dissolved by aliphatic and aromatic hydrocarbons. It is used to produce transparent, colorless films, as a water-vapor-proof coating, as an additive to printer's inks, and in larger amounts as a paint base which is compatible with phenol resins, drying oils and so on.

Synthetic R. is processed in a mixture with natural R. or by itself, so these products have other properties than natural R.

1) ***Polybutadiene R.*** The starting material is buta-1,3-diene, which is polymerized to polybutadiene with sodium; this was first achieved in the 1930's. The polymerization of buta-1,3-diene can occur by 1,2- or 1,4-linking; the 1,4-linkage can result in either a *cis*- or a *trans*-1,4-arrangement of the monomers (Fig. 2). The polybutadiene obtained by block polymerization consists of about 15% *cis*-1,4-, 25% *trans*-1,4- and 60% 1,2-polybutane.

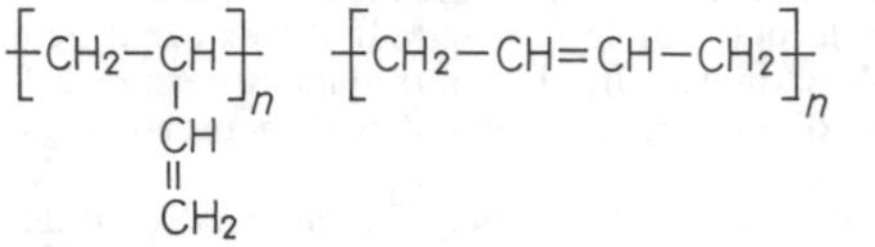

$[-CH_2-CH(-CH=CH_2)-]_n$ $\qquad$ $[-CH_2-CH=CH-CH_2-]_n$

1,2-linkage $\qquad$ 1,4-linkage

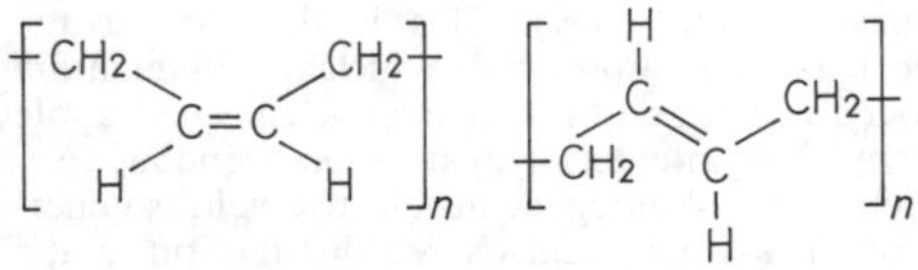

cis-1,4-linkage $\qquad$ *trans*-1,4-linkage

To produce more valuable types of R., the *cis*-1,4-fraction must be increased; it is possible by polymerization of butadiene in solution, using organometallic complex catalysts (Ziegler-Natta catalysts) based on nickel and cobalt, to produce butadiene with 98% *cis*-1,4-polybutadiene (stereorubber). This is superior to natural R., e.g. in its abrasion resistance. The solution polymerization of butadiene (Fig. 3) is done at 10 to 50 °C in stirred vats.

Afterwards, the polymer solution is fed into a mixer, in which the catalyst is deactivated and stabilizers are added to the polybutadiene. It is precipitated by addition of water in the precipitation vat and the solvent-water mixture is removed in the stripper. The polymer is separated on a shaking sieve, washed and put in a drier.

In emulsion polymerization, the butadiene and the substances required for polymerization, including the catalysts, stabilizers and regulators, are emulsified in water with resin soaps. Radical formers, e.g. benzoyl peroxide and potassium persulfate, are used as catalysts. One regulator which can be used is diproxide (diisopropylxanthogendisulfide). Other aids are sodium dimethyldithiocarbamate as a stopping reagent and sodium pyrophosphate as buffer. The polymeri-

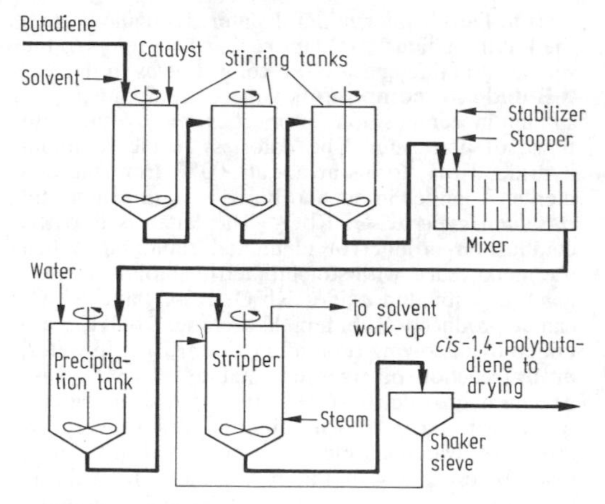

Production of *cis*-1,4-polybutadiene by solution polymerization.

zation is done at a low pressure around 0.6 MPa, and an exactly controlled temperature between 45 and 50 °C; the synthetic R. is obtained in latex form. The latex is worked up after addition of acetic acid and magnesium chloride as precipitants.

2) ***Styrene-butadiene R.*** is a copolymer of styrene and butadiene made by emulsion polymerization, especially at low temperatures (+5 to -10 °C). This ***cold R.*** has much better processing and wearing properties, and is used mainly in the tire industry for tire treads. The styrene content is between 25 and 75%; the product with 75% styrene has a high mechanical resistance. For better processing, the styrene-butadiene mixed polymers are stretched with oil. Polymerization is initiated by redox systems, e.g. cumene hydroperoxide mixed with iron(II) salts and amines.

3) ***Acrylonitrile butadiene R.*** is also produced by emulsion polymerization. The copolymer is highly resistant to chemicals, oil and aging. It is used mainly for fuel hoses, sleeves, seals, etc. The processing of polybutadiene and its mixed polymers to various R. articles is very similar to that of natural R.

Chlorination of polybutadiene in solvents like chloroform, ethylene chloride or tetrachloroethane yields a product which is used similarly to chlorinated R. for paints.

4) ***Polychloroprene, polychlorobutadiene*** is the polymer of chlorprene (2-chlorobuta-1,3-diene), which polymerizes very readily without addition of special catalysts. It is vulcanized without sulfur by heating with zinc oxide or magnesium oxide. The products are highly resistant to heat and oil, and are solid without addition of soot. Since they soften around 60 °C, they can easily be worked. They are used as protectors for cables, conveyor belts, coatings for vessels and pipes, oil hoses, glues, etc., but not for automobile tires.

5) ***Polyisoprene*** with a high *cis*-1,4-content has the same properties as natural R. It is made by solution polymerization in butane or pentane with organometallic complex catalysts, e.g. titanium trichloride/aluminum alkyls. Depending on the catalyst system, the *cis*-1,4- fraction is between 92 and 98%.

6) ***Butyl R., isobutene-isoprene R.*** is obtained by cationic mixed polymerization of isobutene with a small amount of isoprene or butadiene in a solvent such as methylene chloride at very low temperatures, around - 100 °C. Aluminum(III) chloride, boron trifluoride, etc. are used as catalysts. The products are resistant to weathering and oxidation, are not penetrated by gases, and are used mainly for inner tubes and tubeless tires, forms and linings of vessels.

7) ***Polyolefin R., ethylene-propylene R.*** is made by copolymerization of ethylene with 15 to 70% propylene. Copolymerization prevents crystallization, which is a disadvantage in the use of the homopolymers as R. Because this product is saturated, it cannot be vulcanized with sulfur, but must be cross-linked by radical formers, preferably peroxides. Copolymerization of the ethylene with propylene in the presence of a diolefin yields a terpolymer, ***ethylene-propylene-terpolymer R.***, containing some double bonds which permit vulcanization with sulfur. Ethylidene norbornene is an important third component. Terpolymer R. has a low density (0.87 g cm^{-3}), ages well and is highly resistant to abrasion.

8) ***Polyalkene polysulfide, thiokol*** is obtained by polycondensation of 1,2-dichloroethylene with sodium polysulfide. It is vulcanized with lead, zinc or magnesium oxide. Polysulfide R. has an unpleasant smell. It is resistant to organic solvents and lubricants, but only slightly resistant to acids and bases. Polysulfide R. with relative molecular masses of 10^2 to 10^3 are used in building as crack filler.

9) ***Polyurethane R.***: see Polyurethanes.

10) ***Fluorine R.*** is obtained by copolymerization of hexafluoropropylene with vinylidene fluoride. Hydrogen fluoride is cleaved off the copolymer using magnesium oxide. The double bonds left in the polymer are then cross-linked with diamines, e.g. hexamethylenediamine. This R. is resistant to chemicals and temperatures up to 200 °C.

11) ***Chlorosulfinated polyethylene***. Sulfochlorination of polyethylene introduces sulfonylchloride (SO_2Cl_2) groups into the polyethylene chains. The elastic, semi-hard R. is then vulcanized with epoxide resin or zinc oxide. This R. is very resistant to oxygen and chemicals.

12) ***Silicone R.*** is produced from dimethylpolysiloxane. The linkage of the polymer chains is initiated by peroxides, which form radicals from the methyl groups. Silicates and silica is used as filler. Silicone R. is used in the temperature range -60 to +250 °C, mainly as a temperature-resistant insulation and sealing material in machinery, ground vehicles and aircraft.

All products of natural and synthetic R. are included by the term Elastic (see).

Historical. R. was used by the Central Americans long before Columbus. In 1839, Goodyear discovered the process of hot vulcanization. In 1879, Bouchardat succeeded in making an elastic substance by treating isoprene with concentrated hydrochloric acid, and in 1907, Lebedejev synthesized R.-like products. Starting around 1926, usable synthetic R. were produced in larger amounts (at first, pure butadiene polymers, later the butadiene-styrene and butadiene-acrylonitrile mixed polymers, in 1931 polychlorobutadiene, 1937, butyl R.) In 1945, the cold polymerization which had previously been discovered in Germany was further developed in the USA ("cold rubber"). Silicone R. was first produced in the USA. In 1943, the world production of synthetic R. exceeded the production of natural R. for the first time.

Rubber elasticity ***entropy elasticity***: the elastic properties of a group of highly polymeric substances, the Elastics (see). The opposite of R. is Energy elasticity (see). R. was first observed in natural rubber, and can be characterized as follows: under relatively low external tension forces, the rubber-like substances stretch by 800 to 1000%, and simultaneously, their solidity increases 10 to 100-fold. The stretch can be maintained for a long period without a drop in the end tension. When released, the body returns within a fraction of a second to its original state.

Rubber hydrochloride: see Rubber.

Rubidium, symbol ***Rb***: a chemical element from group Ia of the periodic system (see Alkali metals); a light metal, *Z* 37, with natural isotopes with mass numbers 85 (72.15%) and 87 (27.85%), atomic mass 85.4678, valence I, Mohs hardness 0.3, density 1.532, m.p. 38.89 °C, b.p. 688 °C, electrical conductivity 8.8 Sm mm^{-2} (at 0 °C), standard electrode potential (Rb/Rb^+) -2.925 V.

Properties. R. is a soft metal with a silvery sheen on freshly cut surfaces; it crystallizes in a cubic body-centered lattice. ^{87}Rb is weakly radioactive, undergoing β^- decay to form ^{87}Sr. It is used for physical age determinations. The chemistry of R. is very similar to that of potassium, although it is more reactive (see Alkali metals). As a highly electropositive element and strong reducing agent, R. tends to form colorless Rb^+ cations. In the presence of oxygen, it ignites spontaneously and burns primarily to rubidium superoxide, RbO_2. It reacts explosively with water, forming rubidium hydroxide, RbOH, and hydrogen; and with alcohols it forms rubidium alkoxides, RbOR and H_2. Molten R. is exceptionally aggressive and attacks even glass and porcelain.

Analysis. Rb^+ cations display the same precipitation reactions as potassium cations. Precipitation with sodium tetraphenylboranate is used for quantitative determination. A valuable method for qualitative and quantitative analysis is spectral analysis. The element has a red line at 780.0 nm and two blue lines at 420.2 nm and 421.6 nm. Atomic absorption spectroscopy can also be used.

Occurrence. R. makes up 0.031% of the earth's crust, and is therefore the 16th most common element in it. It is always found with potassium in nature, in concentrations up to 1% in potassium minerals such as carnallite or lepidolite. Like potassium ions, Rb^+ ions are retained in the soil, so that the Rb concentration in seawater is very low.

Production. R. can be obtained by electrochemical methods similar to those used with the lighter alkali metals, but the metal is usually obtained by reduction of rubidium hydroxide with magnesium or calcium, or by reaction of rubidium dichromate with zirconium at 500 °C in a high vacuum: $Rb_2Cr_2O_7 + 2\ Zr \rightarrow 2\ Rb + 2\ ZrO_2 + Cr_2O_3$.

Applications. R. is used in photocells and as a getter metal in electronic tubes and high-voltage lamps.

Historical. R. was discovered in 1860 by Bunsen and Kirchhoff, who noticed its characteristic spectral lines in Dürkheimer mineral water. Its name is from the Latin "rubidus" = "dark red", which refers to the red spectral line and the red color it gives to flames.

Rubidium compounds: these compounds are similar in composition, properties and synthesis to those of potassium. The colorless, cubic rubidium hydride, RbH, forms at about 300 °C from the elements. It ignites in air and burns to form water and ***rubidium superoxide***, RbO_2. The latter is also the combustion product of elemental rubidium. When the metal reacts with stoichiometric amounts of oxygen, the rubidium oxides Rb_2O, Rb_2O_2 and Rb_2O_3 can be produced. Rubidium hydroxide, RbOH, is obtained by dissolving rubidium oxide, Rb_2O, in water, or by reaction of rubidium sulfate, Rb_2SO_4, with barium hydroxide, $Ba(OH)_2$. It is used as an electrolyte in galvanic cells. Rubidium halides are very hygroscopic, water-soluble, cubic crystals which can be made by dissolving rubidium carbonate, Rb_2CO_3, in the corresponding hydrogen halide acid. Rubidium salts are used as additives to the glass for television tubes, as single crystals in optoelectronics and as antiepileptics in medicine.

Rudotel®: see Benzodiazepines.

Ruggli-Ziegler dilution principle: method of carrying out ring closure reactions to form multiple-membered ring compounds A. To avoid the undesired intermolecular linking to form polymers B, the reaction is carried out at high dilution (Fig.).

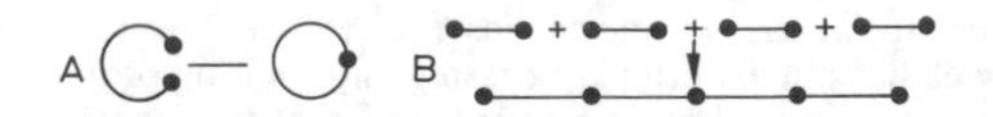

Although bifunctional compounds with terminal active groups preferentially form 5 and 6-membered rings by intramolecular reactions (see Dieckmann condensation), longer carbon chains will lead under the same conditions to a preponderance of intermolecular reactions (see Ring tension). The R. depends on the different reaction orders of the two types of reaction. Ring closure is monomolecular, and the probability that the two ends of a chain will encounter each other does not decrease as the dilution increases. In contrast, the number of collisions leading to reaction between different molecules decreases with increasing dilution, as this is a bimolecular reaction. Based on preliminary work by P. Ruggli in 1912, K. Ziegler was able in 1933 to obtain good results with the synthesis of cycloalkanones with 7, 8 and 14 to 33-membered rings by dinitrile cyclization (see Thorpe reaction). The cyclization takes place in the presence of the lithium salt of *N*-ethylaniline in a large volume of solvent, and for the above ring system, yields were between 85 and 90%. Experimentally, the required high dilution is more conveniently achieved by adding the components to be cyclized very slowly, over a period of hours or days, to a normal volume of solvent, than by using a large volume of solvent. The R. is applied in the synthesis of macrocyclic ethers and lactones.

Rung's chloralkali test: a test used for aniline, which gives a characteristic, red-violet color when it reacts with lime chloride.

Run-off chromatogram: see Paper chromatography.

Russel-Saunders coupling: see Multiplet structure.

Rust: a mixture of amorphous and crystalline oxides, hydroxides and oxide hydroxides of iron. The main components are γ-FeOOH (lepidocrocite), α-FeOOH (goethite) and Fe_3O_4 (magnetite). Iron sulfates, carbonates and chlorides are also included in the structure of the rust layers. R. is heterogeneous and contains many capillaries and pores. It forms from iron, unalloyed steels and steels with low nonferrous metal contents when moisture and oxygen are present (see Electrochemical metal corrosion). There are different types of R. with different compositions which form under different conditions. These differences can be seen in the color: yellow, brown or black. Sulfate and chloride ions and their iron salts have a large influence on the corrosion.

Because of the structure of R., the access of the corrosion medium to the surface of the iron or steel surface is not greatly inhibited, so that the underlying metal is not protecteed from further corrosion. Unreactive steels with low amounts of non-ferrous metals are an exception to this rule. Since the capillaries and pores or the rust layer are filled with corrosive substances, application of a protective layer (see Corrosion protection) to rusted iron or steel surfaces leads to rusting under the paint (see rust-preventing paints, under Protective coatings). For this reason, a thorough removal of the rust is necessary; (see Surface pretreatment).

Rust converter, ***rust stabilizer***: a liquid consisting mainly of aqueous phosphoric acid solution; it is used to convert tightly adhering rust into firmly anchored, insoluble tertiary iron(III) phosphate. To reduce the amount of unoxidized metal dissolved by the acid, inhibitors are added to R.; catalysts are added to reduce the time required, and wetting agents and solvents are added to clean and de-grease surfaces which are not sufficiently clean. An acid-hardening synthetic resin is often added to close the pores in the resulting phosphate layer. R. can be used to give rusted surfaces a temporary protection from further rusting (see Surface pretreatment).

Rust protection: see Corrosion protection.

Rust removal: see Surface pretreatment.

Rust stabilizer: same as Rust converter (see).

Ruthenates: see Ruthenium compounds.

Ruthenium, symbol ***Ru***: a chemical element of group VIIIb of the periodic system and one of the light Platinum metals (see); a noble metal, *Z* 44, with natural isotopes with mass numbers 102 (31.6%), 104 (18.7%), 101 (17.0%), 99 (12.7%), 100 (12.6%), 96 (5.52%), 98 (1.88%); atomic mass 101.07, valence maximum VIII, also II to VII, 0, -II, density 12.45, m.p. 2450 °C, b.p. 4150 °C, electrical conductivity 14.9 Sm mm^{-2}, standard electrode potential (Ru/Ru^{2+}) +0.45.

Properties. R. is a silvery white, brittle metal which crystallizes in a hexagonal lattice. It is very resistant to chemicals; acids do not attack it in the absence of atmospheric oxygen. When R. is heated in a stream of oxygen, it forms the ruthenium oxides RuO_4 and RuO_2. It reacts with chlorine at red heat to give ruthenium(III) chloride, $RuCl_3$. In oxidizing alkali melts – MOH/M_2O_2 or MOH/MNO_3 – R. dissolves to form ruthenates(VI), M_2RuO_2. R. is characterized by ready formation and high stability of the tetroxide, the existence of the hexa-, penta- and tetrafluoride, and the tendency to form octahedral complexes with the metal in the II or III oxidation state.

Analysis. R. is detected by formation of volatile ruthenium(VIII) oxide, RuO_4. Absorption of the volatile oxide by conc. hydrochloric acid gives a red-brown solution, which changes color to dark blue when zinc is added.

Occurrence. R. makes up about 10^{-7}% of the earth's crust, and is thus one of the rare metals. It is found with platinum in nature (see Platinum metals). The very rare mineral laurite, RuS_2, is a homolog of pyrite.

Production. R. is obtained as a byproduct of platinum production (see Platinum metals). A powerful oxidizing agent, such as chlorine or potassium permanganate, is added to solutions containing R., e.g. as potassium ruthenate(VI), $K_2[RuO_4] \cdot H_2O$, or potassium hexachlororuthenate, $K_3[RuCl_6]$. The solution is distilled, and RuO_4 which escapes is absorbed in hydrochloric acid. The ammonium hexachlororuthenate obtained by crystallization is reduced to R. with hydrogen at 800 °C.

Applications. R. is used to make precision layer resistors and thermometers. It is used as a hardening component in platinum and palladium alloys (see Platinum alloys) and is added to Titanium alloys (see).

Historical. R. was discovered in 1844 by the Russian chemist Claus, who named it in honor of Russia (Latin "Ruthenia").

Ruthenium compounds: The most important binary ruthenium compound is ***ruthenium(VIII) oxide, ruthenium tetroxide***, RuO_4, which forms gold-colored rhombic needles; M_r 165.07, density 3.29, m.p. 25.5 °C, b.p. 108 °C (dec.). It is slightly soluble in water, and readily dissolves in carbon tetrachloride. RuO_4 forms tetrahedral, covalent molecules, and is relatively volatile at room temperature. The compound is very poisonous; it smells like ozone and attacks the mucous membranes. RuO_4 reacts with alkali hydroxide solutions, generating O_2 and forming dark green ***perruthenates***, $M[RuO_4]$. These are converted by further loss of oxygen to ruthenates(VI), $M_2[RuO]_4$. RuO_4 is obtained as a sublimate by passing a stream of chlorine into a concentrated potassium ruthenate solution, or by heating ruthenium in a stream of oxygen to 800 °C. ***Ruthenium(IV) oxide, ruthenium dioxide***, RuO_2, is an indigo-blue, acid-resistant compound with a metallic sheen; it crystallizes in a rutile lattice. Its M_r is 133.07, density 6.97; it is obtained by heating ruthenium or $RuCl_3$ with oxygen to 1000 °C. RuO_2 is used in the production of switching circuits, and RuO_2-coated titanium electrodes are now used widely for chloralkali electrolysis. ***Alkali ruthenates(VI)***, $M_2[RuO_4]$, are formed by direct oxidation of ruthenium with oxidizing alkali melts, (MOH/MNO_3), e.g. ***potassium ruthenate(VI)***, $K_2[RuO_4] \cdot H_2O$, a green compound. Ruthenates(VI) dissolve in water to give red-orange solutions; when acidified, these disproportionate to a lower oxide and dark green, dissolved perruthenates, e.g. ***potassium perruthenate, potassium ruthenate(VII)***, $K[RuO_4]$, which forms lustrous black octahedra. ***Ruthenium(IV) sulfide, ruthenium disulfide***, RuS_2, is a gray-black compound which forms a cubic lattice; density 6.96, m.p. 1000 °C (dec.). It is formed by passing hydrogen sulfide into hot ruthenium salt solutions. RuS_2 is found in nature as laurite. ***Ruthenium(III) chloride, ruthenium trichloride***, $RuCl_3$, exists in two modifications: M_r 207.43, density 3.11. Black α-$RuCl_3$ is insoluble in water and alcohol; it forms a layered lattice isomorphic with chromium(III) chloride, $CrCl_3$. Brown, alcohol-soluble β-$RuCl_3$ has a polymeric octahedral structure. $RuCl_3$ is obtained by the reaction of chlorine with ruthenium sponge at 330 °C, and is converted to β-$RuCl_3$ at 700 °C. Ruthenium fluorides, RuF_n are known for n = 2-6. ***Ruthenium(VI) fluoride, ruthenium hexafluoride***, RuF_6, is a dark brown compound, M_r 215.06, m.p. 54 °C. ***Ruthenium(V) fluoride, ruthenium pentafluoride***, $(RuF_5)_4$, is a dark green compound forming tetrameric molecules; M_r 784.24, m.p. 86.5 °C, b.p. 227 °C. ***Ruthenium(IV) fluoride, ruthenium tetrafluoride***, RuF_4, is an ochre yellow compound, M_r 177.06. Ruthenium fluorides form octahedral ***fluoro complexes*** $M_n[RuF_6]$ (n = 1-4) with alkali fluorides. Ruthenium(IV) chloride is stable only in the form of its hexachloro complex $M_2[RuCl_6]$. Of the other ruthenium complexes, octahedral compounds such as $[Ru(NH_3)_6]^{2+}$, $[Ru(NH_3)_5OH_2]^{2+}$, $[Ru(H_2O)_6]^{2+}$, $[RuCl_{6-n}(NH_3)_n]^{n-3}$ and $[RuCl_6\text{-}n(H_2O)_n]^{n-3}$ are particularly important. The ruthenium(II) complexes are stronger reducing agents than the homologous iron(II) compounds. The aqua complex $[RuCl_3(H_2O)_3]$ is the starting material for synthesis of most other R.

The close relationship between the complex chemistry of ruthenium and that of iron and osmium when π-acceptors or π-ligands are present is demonstrated by the formation and properties of the following R.: ***ruthenium carbonyl***, $Ru(CO)_5$, is a colorless liquid, M_r 241.22, m.p. -22 °C; $Ru_3(CO)_{12}$, forms red-orange needles, M_r 639.27, m.p. 154 °C (dec.); $Ru_6(CO)_{18}$, red crystals, M_r 1110.51, m.p. 235 °C (dec.) (structures, see Metal carbonyls); ***ruthenium hydrogen carbonyl***, $H_2Ru(CO)_4$, a colorless liquid, M_r 243.11; ***ruthenocene***, $Ru(C_5H_5)_2$, M_r 232.05.

Of the ruthenium(II) complexes, the dinitrogen complex $[Ru(NH_3)_5N_2]X_2$ is worthy of mention, because its discovery (Allen, 1965) provided an entry into the complex chemistry of molecular nitrogen.

Rutherfordium: see Kurčatovium.

Rutherford's atom: see Atom, models of.

Rutile type, ***rutile structure***: a common Structure type (see) for compounds with the general composition AB_2. In the mineral rutile, a modification of titanium dioxide, TiO_2, the T^{4+} ions form a tetragonal body-centered lattice, and each is surrounded by a distorted octahedron of 6 O^{2-} ions; each O^{2-} ion has 3 Ti^{4+} ions in a planar arrangement as nearest neighbors. A ratio of radii r_A/r_B in the range 0.414 to 0.732 is necessary for such a [6]:[3] coordination to occur.

Compounds crystallizing in the R. are mainly a series of difluorides and dioxides of di- or tetravalent elements. Typical examples, in addition to TiO_2, are MgF_2, MnF_2, SnO_2, MnO_2 and MoO_2. Under high pressure, SiO_2 also forms a modification with R., *stishovite*.

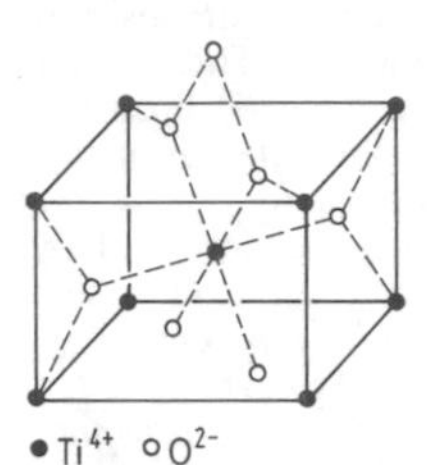

Rutin, ***quercetin-3***-β-(6-O-α-L-rhamnosyl)-D-glucoside, rutoside, antipermeability factor, vitamin P: a flavanoid glycoside (see Flavanoids). It forms yellow to greenish crystals which decompose on heating. R. is soluble in alkalies and pyridine, very slightly soluble in cold water and ethanol, and practically insoluble in ether or chloroform. It is present in buckwheat at a concentration of about 3% (dry weight), and eucalyptus trees. The flower buds of the East Asian tree *Sophora japonica* contain about 20% R. It is used to alter membrane permeability in hemorrhoids, varices and exanthemas.

Rutoside: same as Rutin.

R value: see X-ray structure analysis.

R_X value: see Paper chromatography.

Rydberg constant, ***R***: an atomic constant which appears in the series formula of atomic spectra (see Atomic spectroscopy):

$$R = \frac{2\pi^2 me^4}{h^3 c}$$

= 10,973,731.43±0.10 m^{-1}. Here m is the mass of the electron, h is Planck's constant, e the charge of the electron and c the velocity of light. R can be determined from atomic spectra. Physically, R is the ionization energy of the hydrogen atom divided by hc.

S

s: symbol for Entropy (see).

S: 1) symbol for sulfur; 2) configurational symbol, see Stereoisomerism 1.1; 3) S_E, see Substitution, 2; 4) S_N, see Substitution, 1.

Saarberg-Otto process: see Synthesis gas.

Sabotage poisons: chemicals intended for use against the unprotected hinterland to destroy human beings, domestic animals and plants. Poisons used for this purpose are found in many chemical classes, including the alkaloids, other toxins, fluoroacetic acid and its derivatives, defoliants (see Herbicides) and neurotoxins.

Saccharates: see Aldaric acids.

Saccharide: same as Carbohydrate (see).

Saccharin, ***o-sulfobenzoimide, 1,2-benzoisothiazol-3(2H)-one-1,1-dioxide***: a synthetic sweetener wich is 550 times sweeter than sucrose. S. forms colorless, very sweet crystals; m.p. 229 °C. It is soluble in alcohol, boiling water and alkali carbonate solutions, but sparingly soluble in cold water. S. is used in the form of its sodium salt. It has no nutritional value, and is excreted unchanged. S. is synthesized industrially, e.g. from toluene, which reacts with chlorosulfonic acid to form a mixture of 2- and 4-toluene sulfonyl chloride. The reaction of this mixture with ammonia gives a mixture of 2- and 4-toluenesulfamide. The 4-isomer can be removed by recrystallization from alcohol or by fractional precipitation of the alkaline solution with hydrochloric acid. The 2-toluenesulfamide is converted to the potassium salt of 2-sulfamidobenzoic acid by potassium permanganate; this salt is converted to S. by acidification.

S. is an important sugar substitute in the diets of diabetics and overweight persons. There is some evidence, however, that it is carcinogenic, and it has been replaced in many products by other artificial sweeteners.

$$\text{(structure: } O{=}S{=}O,\ N^-Na^+ \cdot 2H_2O,\ O)$$

Saccharin sodium salt

Sachse process: a process for making ethyne by autothermal cleavage of liquid gases, light gasoline and methane; the most common starting material is natural gas. The gas and oxygen are separately preheated to 500 to 600 °C, then mixed and brought to reaction in a special burner. Here the dehydrodimerization of methane: $2\ CH_4 \rightarrow CH{\equiv}CH + H_2$ ($\Delta_R H$ = +425 kJ mol^{-1}) is coupled to combustion of the resulting hydrogen, $2\ H_2 + O_2 \rightarrow 2\ H_2O$ ($\Delta_R H$ = -485 kJ mol^{-1}).

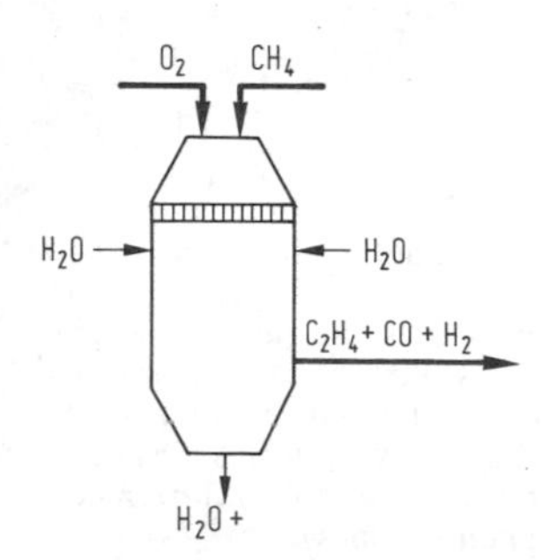

Production of ethyne by the Sachse process.

Safety explosives: a general term for industrial explosives which are relatively safe to handle; they do not ignite combustible mixtures of gas and air or dust and air in mines. This is achieved by reducing the energy of the explosive and the temperature of the flame by addition of inorganic salts, usually sodium chloride, which withdraw heat from the flame by melting.

Safflower: a natural dye made from the dried petals of the safflower, *Carthamus tinctorius*. S. consists of a red component, ***Carthamine***, and a yellow one, ***safflower yellow***; it is ued to color foods and cosmetics.

Safranine: see Azine pigments.

Safrol, ***1-allyl-3,4-(methylenedioxy)benzene***: a colorless to yellowish, mobile liquid with a fresh, spicy odor; m.p. 11.2 °C, b.p. 234.5 °C, n_D^{20} 1.5381. S. is insoluble in water, but is readily soluble in ethanol and ether. It is a component of various essential oils, mainly sassafras and camphor oil, and can be isolated in pure form from these sources. It is very reactive, undergoing change when heated or in the presence of atmospheric oxygen. In the presence of alkali hydroxides, S. isomerizes to Isosafrol (see) through a shift in the double bond. It is used in perfumes and as an intermediate in organic synthesis, e.g. in the production of insecticides and drugs.

$$\text{(structure: O–CH}_2\text{–O ring; } CH_2{-}CH{=}CH_2)$$

Safron: a natural product consisting of the dried stigmas of *Crocus sativus* flowers; this plant is native to Southern Europe. The active material is Crocin (see). S. is still used as a spice and to give foods a yellow color. It was formerly used as an analgesic and anticonvulsant, and as a yellow dye for textiles.

Salamander alkaloids: see Steroid alkaloids.

Salamander toxins: see Amphibian toxins.

Salamandrin: see Steroid alkaloids.

Salicyl alcohol, ***2-hydroxybenzyl alcohol, saligenin***: an aromatic, substituted alcohol; colorless, crystalline needles or leaflets which can sublime; m.p. 86-87 °C.

OH
CH₂OH

S. is readily soluble in boiling water, ethanol and ether. It gives a red-violet color reaction with iron(III) chloride. S. is present as the glucoside *salicin* in the bark and leaves of willow trees. It is synthesized by the reaction of formalin and sodium hydroxide solution with phenol. S. easily forms resin-like products with various reagents, and is an intermediate in the synthesis of phenoplastics.

Salicylaldehyde, ***2-hydroxybenzaldehyde***: an oily liquid with a spicy smell; m.p. -7 °C, b.p. 196.5 °C, n_D^{20} 1.5740. Because of the intramolecular hydrogen bonds in the molecule, its physical properties are markedly different from those of 3- and 4-hydroxybenzaldehyde. For example, unlike the two isomeric aldehydes, S. is steam volatile. It is slightly soluble in water, and very soluble in ethanol and ether.

S. gives a red-violet color with iron(III) chloride. It is much more resistant to autooxidation than benzaldehyde. Suitable oxidizing reagents convert S. to salicylic acid; reduction converts it to salicyl alcohol.

S. can be synthesized by oxidation of *o*-cresol with manganese(IV) oxide and sulfuric acid, by oxidation of salicyl alcohol with nitrosobenzene in the presence of copper, or by the Reimer-Tiemann reaction of phenol. S. is used in the synthesis of coumarin and pyrocatechol. It is also used as an intermediate in the production of drugs and dyes.

Salicylamide: see Salicylic acid.

Salicylic acid, ***2-hydroxybenzoic acid***: a phenol carboxylic acid which forms crystalline needles with a sweetish taste; m.p. 159 °C. It is slightly soluble in cold water, but dissolves readily in hot water, ethanol or ether; it is steam volatile. The S. molecule contains a relatively stable intramolecular hydrogen bond.

COR
OH

R = OH : Salicylic acid
R = OCH_3 : Methyl salicylate
R = NH_2 : Salicylamide

S. forms blue-violet chelate complexes with iron(III) ions. When heated, it is decarboxylated to phenol. S. can be synthesized by the Kolbe-Schmitt synthesis (see) (1885) from sodium phenolate and CO_2 at 120-140 °C and 500-600 kPa. S. can be acetylated to yield Acetylsalicylic acid (see). It has an antibacterial and keratolytic effect, and is therefore included in skin medications. It also has an antirheumatic effect and the sodium salt was formerly used therapeutically.

Esterification of S. with methanol yields methyl salicylate, a liquid with a characteristic odor, b.p. 224 °C. It is present in oils of wintergreen and cloves, and is a component of antirheumatic salves.

Salicylamide is made by ammonolysis of salicylates. It is a white, crystalline substance, m.p. 142 °C, and is used as an analgesic.

Salicylic acid methyl ester: see Salicylic acid.

Saligenin: same as Salicyl alcohol (see).

Salt: in the narrow sense, table salt (see Sodium chloride); in the wider sense, a group of heteropolar compounds (see Chemical bond) which form crystal lattices consisting of ions (ionic lattices). S. are characterized by hardness, high melting and boiling points, and low coefficients of thermal expansion or compression. If the H atoms of di- or polybasic acids are only partially substituted by salt-forming cations, the S. are acid (hydrogen salts), e.g. $NaHSO_4$. Mixed crystals of two or three S. are called double or triple S.; they are special cases of complex S. (see Coordination chemistry), See also Acid-base concepts and Nomenclature, sect. II F and II G.

Salt baths: see Heating baths.

Salt coal: see Coal.

Salt effect: collective term for the effects of foreign salts on the thermodynamics and kinetics of solution processes. Addition of the foreign salt changes the ionic strengths and thus the activity coefficients of all the components of the solution. In dilute solutions, these changes can be calculated using the Debye-Hückel theory. The results are changes in equilibrium concentrations, vapor pressures over solutions and solubilities (salting-out and salting-in effects). The rate constants of ion reactions $A^{Z_A} + B^{Z_B} \rightarrow$ products in the Debye-Hückel range are given by the equation $\log k = \log k_0 + 1.02\, Z_A Z_B I^{1/2}$, which was derived by Bjerrum and Brönsted from transition state theory (see Kinetics of reactions, theory). In this equation, k and k_0 are the rate constants in the presence and absence of the added salt, Z_A and Z_B are the charge numbers of A and B, and I is the ionic strength in units mol dm^{-3}. If the ions have the same charge, k increases linearly with $\sqrt{I}$, but if they have the opposite charge, k decreases.

Salting-in and ***salting-out effects***. In the saturated state, the saturation activity $a_{c,2}$ of solute (2) is constant: $a_{c,2} = f_{c,2}c_2$ = constant (if T, p are constant). Addition of a salt or other substance to the solution decreases or increases the activity coefficient $f_{c,2}$ of solute 2, and thereby increases or decreases the saturation concentration c_2. In the first case, the result is salting in; in the second, salting out. The solubilities of gases and solid non-electrolytes in salt solutions are usually decreased, and the effect is used in separations.

The terms salting in and salting out are also applied to mixtures of two liquids with limited miscibility. Here the third substance reduces or increases the Mixing gap (see).

Salting out: see Salting in.

Saltpeter: see Potassium nitrate.

Saltpeter blasting powder: an explosive similar to black powder, except that the potassium nitrate has been replaced by the cheaper sodium nitrate, and the charcoal by coal. S. is used in mining operations where tunnel collapse is possible.

Saluretics: see Diuretics.

Salvarsan®: see Arsphenamine.

Samandarin: see Amphibian toxins.

Samarium, symbol ***Sm***: a chemical element from group IIIb of the periodic system, the Lanthanoids (see); it is a heavy, rare-earth metal, *Z* 62, with natural isotopes with mass numbers 152 (26.6%), 154 (22.6%), 147 (15.1%, α-emitter, $t_{1/2}$ 1.06 · 10^{11} a), 149 (13.9%, α-emitter, $t_{1/2}$ 7 · 10^{15}a), 150 (7.4%), 144 (3.1%), atomic mass 150.36, valence +III, more rarely II, density 7.536, m.p. 1072 °C, b.p. 1900 °C, standard electrode potential (Sm/Sm^{2+}) -2.414 V.

S. is a lustrous, silvery metal which is relatively stable in air. It exists in two modifications: the lower-temperature rhombohedral form (density 7.520) is converted to the cubic, high-temperature modification (density 7.40) at 917 °C. It makes up about 6 · 10^{-4}% of the earth's crust, and is always found in nature associated with other rare-earth metals. It is most abundant as a companion of cerium in minerals such as bastnesite, monacite, cerite and complex rare-earth ores such as samarskite. Further properties, analysis, production and history are discussed under Lanthanoids (see). An important application of S. is the production of permanent magnet alloys of the types $SmCo_5$ and Sm_2Co_{17}. It is used to dope calcium fluoride for lasers and masers, and as a neutron absorber in nuclear reactors. S. is a component of Cerium mixed metal (see).

Samarium compounds: compounds in which samarium exists in the +III and +II oxidation states. The more stable samarium(III) compounds are typically yellow, while samarium(II) compounds tend to be blood-red to violet. The samarium(II) compounds are not stable in aqueous solution. Samarium(II) halides can be obtained by reduction of samarium(III) halides with hydrogen or samarium. The general properties and production of S. are discussed under Lanthanoid compounds (see). Some important S.: ***samarium(II) chloride***, $SmCl_2$, red-brown crystals, M_r 221.26, density 4.56, m.p. 740 °C; ***samarium(III) chloride***, $SmCl_3$, yellow to white, hygroscopic crystals, M_r 256.71, density 4.46, m.p. 678 °C; ***samarium(III) fluoride***, SmF_3, M_r 207.35, m.p. 1306 °C, b.p. 2323 °C; ***samarium(III) nitrate***, $Sm(NO_3)_3 \cdot 6H_2O$, pale yellow, triclinic crystals, M_r 444.46, density 2.375, m.p. 79 °C; ***samarium(III) oxalate***, $Sm_2(C_2O_4)_3 \cdot 10H_2O$, white crystals, M_r 744.91; samarium(III) oxide, bright yellow, density 7.43, m.p. 2300 °C, is used to improve the IR absorption of optical glasses and as a catalyst for dehydrogenation and dehydration of ethanol. Samarium salts are used as sensitizers for phosphors which can be excited by IR radiation.

Sampling: methods of taking samples for analysis from large amounts of material in such a way that the samples are representative of the composition of the total. Correct S. is absolutely essential for accurate analysis of large amounts of materials.

SAN copolymers: abb. for styrene-acrylonitrile copolymers; see Polystyrene.

Sandarac: a slightly yellowish vegetable resin which is readily soluble in alcohol and ether, but is insoluble in carbon disulfide and petroleum ether. It consists mainly of free resinic acids and essential oils. S. is obtained from the bark of the trunk of the sandarac tree, *Tetraclinis articulata*, which is cultivated in southern Spain and northwest Africa. S. is used to make paints and glues for porcelain, as a filler for teeth, and as a basis for bandaging plaster.

Sandlewood oil: a colorless, somewhat viscous essential oil with a weak but long-lasting balsam-like, sweet-wood odor and a slightly bitter, spicy taste. The most important component is santalol (about 90%); S. also contains santene, α- and β-santalenes, isovaleraldehyde and various other related compounds. S. is obtained from the wood of the sandlewood tree by steam distillation; the yield is 2.6 to 4.5%. It is used in making expensive cosmetics, such as perfumes and soaps.

Sandmeyer reaction: a reaction of arene diazonium salts with copper(I) salts of chloride, bromide, iodide, nitrite, cyanide and rhodanide, first described in 1844. It leads to loss of nitrogen and introduction of the anion atom(s) into the arene ring:

$$\text{Ar–N}\equiv\text{N} \xrightarrow[-N_2]{\pm X^- / Cu_2X_2} \text{ArX}$$

where X = Cl^-, Br^-, CN^- or SCN^-.

The significance of the S. is that it permits substitutions on the arene which are either not directly possible (e.g. iodine), or not in the desired position. A nitro group on the arene ring can easily be converted to an amino group, which determines the site of introduction of the substituent in the S. Byproducts of the S. are biphenyl derivatives, which indicate that the reaction proceeds by a radical mechanism:

$$C_6H_5\text{–N}\equiv\text{N}\,Cl^- \xrightarrow[-N_2,\,-Cu^{2+},\,-Cl^-]{+Cu^+;\ \text{heat}} C_6H_5\cdot \xrightarrow[-Cu^+]{+Cu^{2+},+Cl^-} C_6H_5\text{–Cl}$$

Diazonium salt — Phenyl radical — Cholorbenzene

The copper ion acts only as an electron donor or acceptor. Thus it is understandable why reversibly oxidizable substituents can be introduced in the absence of the copper ion; for example, iodide ions react directly with the diazonium salt to release nitrogen. In a sense, the iodide anion acts as its own catalyst. The arsenite group (see Bart reaction) acts similarly. The Gatterman reaction (see) is a variation of the S.

Sandwich complexes: see Organoelement compounds.

Sanger-reagent, ***2,4-dinitrofluorobenzene***: a very reactive substance in which the fluorine atom is readily replaced by nucleophilic amino groups. S. is therefore useful in sequence analysis of polypeptides.

Sanotensin®: see Guanethidine.

Sapogenin: see Saponins.

Saponification: in the broad sense, any hydrolytic

cleavage of an organic molecule, including, for example, acid-catalysed ester hydrolysis, or the hydrolysis of proteins into amino acids or of polysaccharides into simple sugars. In the narrow sense, however, it is the alkaline hydrolysis of carboxylic acid esters into the salt of the acid and the alcohol. The formation of the salt shifts the equilibrium of the reaction towards the products, so that a stoichiometric amount of the base is needed. The name of the process goes back to the splitting of fats into fatty acids and glycerol by boiling with alkali; the alkali salts of fatty acids are soaps.

Saponification number: see Fats and fatty oils.

Saponins: glycosides of tetra- or pentacyclic alcohols; they are among the most abundant secondary plant metabolites. S. are highly surface active and cause hemolysis, that is, they dissolve the membranes of erythrocytes. S. are very toxic to fish and amphibians. They form molecular complexes with sterols such as cholesterol.

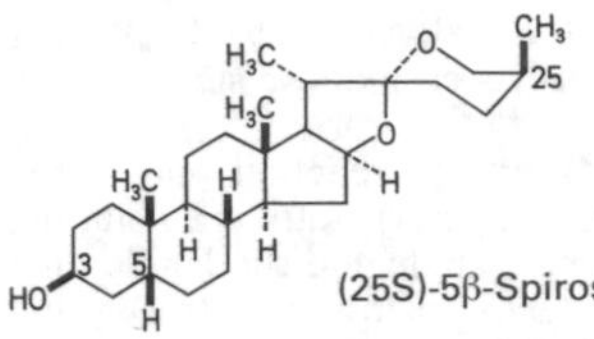

(25S)-5β-Spirostanol

The terms "monodesmoside" and "bisdesmoside" are applied to S. with one or two sugar groups, respectively, bound glycosidically to the aglycon. The sugar groups are often branched. Depending on the structure of the aglycon (***sapogenin***), the S. are classified as steroid or triterpene S. Most ***steroid saponins*** are the spirostane type; the configuration at the C5 atom (α- or β- series) and C25 atom (25S or 25R configuration) can vary. The steroid S. are monodesmosides with the sugar in the 3-position; they occur mainly in monocotyledons. An important dicotyledon steroid S. is Digitonin (see). Steroid S. of the furostane type are less common; when hydrolysed, they are readily converted by ring closure to the spirostane type. An example of this type is convallamarin, the S. of the lily-of-the-valley, *Convalleria majalis*. The glycosides of the Steroid alkaloids are also steroid S. ***Triterpene S.*** occur mainly in dicotyledons; most of them are derived from β-amyrin (see Triterpenes) and are acidic because of carboxyl groups on the triterpene genin or sugar (uronic acid). The sugar groups are bound glycosidically to the 3-position, but sometimes may also be esterified to the 17-position. Glycyrrhetic acid and oleanolic acids are triterpene sapogenins.

S. are soluble in water, slightly soluble in ethanol and practically insoluble in chloroform and ether. Some S. are used in industry as foaming agents and detergents. S. or S.-containing plant materials are used in medicine, in some cases because of their mucus-dissolving effects; liquorice is a good example (see Glycyrrhizic acid).

Sapropel: a fine-grained, gray to black sediment with a high content of fat and protein. It is formed in shallow seas or standing waters by accumulation of dead organic matter which decays in the absence of oxygen; most of the organisms are plankton, although higher animals are also occasionally present. S. hardens to ***sapropel rocks***, e.g. S. Coal (see). When calcium carbonate is present, it forms bituminous limestone; when clay is present, it forms oil shale, and when marls are present, copper shale. Combustible gases and tar can be obtained from S. by distillation.

Sapropel coal: see Coal.

Sapropel process: a biological method of conversion of organic substances in sewage sludge by anaerobic microorganisms. The sludges produced by Sewage treatment (see) are concentrated and subjected to an anaerobic decay process. In the absence of atmospheric oxygen, the organic substances are degraded to methane, ammonia, carbon dioxide and small amounts of hydrogen sulfide. The resulting gas can be used as a source of methane or as a fuel gas for heating or gas engines. The remaining sludge is dried and can then be used as fertilizer.

Sar: abb. for sarcosine.

Saran®: see Synthetic fibers.

Sarcosine, ***N-methylglycine***, abb. ***Sar***: CH_3-NH-CH_2-COOH. an intermediate in amino acid metabolism.

Sarin: see Chemical weapons.

SAS: see Alkane sulfonates.

Sassafras oil: a yellow to reddish-yellow essential oil which smells like safrol and has an aromatic taste. The main component of S. is safrol (about 80%), and phelandrene, α-pinene, (+)-camphor, eugenol and a mixture of various sesquiterpenes are also present. S. is made by steam distillation of roots, stumps and stem wood of a type of laurel; the yield is 1.5 to 2.0%. The bark of the roots has by far the largest oil content, 6 to 9%. S. is used mainly as a basis for perfumes, and for isolation of pure safrol.

Sassoline: see Boric acid.

Saturation slurry: same as Separation slurry (see).

Saxitonin: a toxin found in shellfish which occasionally causes poisonings along the North Atlantic coasts of Europe and North America, the Pacific coast from California to Alaska, and on the coasts of South Africa and Japan. The compound is produced by dinoflagellates of the genus *Gonyaulax* (*G. catenella, G. tamarensis*); increased consumption of these organisms by the shellfish leads to human poisonings. Like tetrodotototoxin, to which it is structurally related, S. is one of the strongest non-peptide toxins. It also resembles tetrodotoxin pharmacologically. The lethal dose for human beings is reported to be 1 mg.

Sb: symbol for antimony.

Sc: symbol for scandium.

Scale: oxide layer which forms on iron and its alloys (steel) at high temperatures during rolling, hammering and hardening processes (see Metal corrosion, chemical). S. consists mainly of iron(II,III) oxide, Fe_3O_4. The losses to scale during steel production may be as high as 4%. Stainless steels can be produced and worked without S. S. can be removed by various methods (see Surface pretreatment).

Scandium, symbol ***Sc***: an element from group IIIb of the periodic system, the Scandium group (see); a light, rare-earth metal, *Z* 21, with a single isotope, atomic mass 44.95592, density 2.985, valence III, m.p. 1539 °C, b.p. 2832 °C, electrical conductivity

15.0 Sm mm^{-2}, standard electrode potential (Sc/Sc^{3+}) -2.077 V.

Properties. S. is a silvery-white, relatively soft, ductile metal which exists in two modifications. The hexagonal modification is stable up to 1335 °C, and above this temperature converts to a cubic body-centered modification. S. is always found in the +III oxidation state in its compounds, and is more similar in its properties to yttrium and the lanthanoids than to aluminum or titanium. At room temperature, it is not attacked by atmospheric gases, but at higher temperatures it reacts with oxygen to form scandium(III) oxide, Sc_2O_3, or with chlorine to form scandium(III) chloride, $ScCl_3$. At high temperatures it also reacts with other nonmetals, such as hydrogen, carbon, nitrogen and boron, forming non-stoichiometric inclusion compounds such as ScH_2, ScC, ScN and ScB.

Occurrence. S. makes up about $5.1 \cdot 10^{-4}$% of the earth's crust, and is found in the mineral thortveitite, $(Y,Sc)_2Si_2O_7$. It accompanies other rare-earth metals in the corresponding phosphate or silicate ores (see Lanthanoids). It is present in small amounts in tungsten, tin and uranium ores, and in coal ashes.

Like other rare-earth metals, S. can be precipitated as the oxalate from weakly acidic solution. It can be separated from the accompanying group IIIb metals by precipitation with thiosulfate, by dissolving scandium(III) fluoride in a solution of ammonium fluoride, or by extraction with ether from an aqueous solution containing ammonium thiocyanate.

Metallic S. is prepared by electrolysis of $ScCl_3$/LiCl/KCl melts on a zinc cathode or by reduction of scandium(III) fluoride with calcium in the presence of zinc and lithium fluoride at 1000 °C. The Sc/Zn alloy obtained by either method is distilled to remove the zinc, and the remaining metallic Sc is further purified by high-vacuum distillation.

Historical. S. was discovered in 1879 by Nilson in Swedish gadolinite and euxenite, after Mendeleyev had predicted the existence of the element in 1871. The properties Mendeleyev had predicted for the new element (atomic mass, density, compositions and solubilities of important compounds) were very similar to those of S.

Scandium compounds: compounds of scandium, usually of the ScX_3 type. They are generally colorless and diamagnetic. ***Scandium(III) chloride***, $ScCl_3$, colorless crystals, M_r 151.32, density 2.39, m.p. 960 °C, sublimes at 800 to 850 °C and is prepared by heating a mixture of scandium(III) oxide and coal in a stream of chlorine, or by dehydration of the hexahydrate, $ScCl_3 \cdot 6H_2O$ in a stream of hydrogen chloride. ***Scandium(III) fluoride***, ScF_3, rhombohedral crystals, M_r 101.95, reacts with ammonium fluoride to form water-soluble, octahedral crystals of ammonium hexafluoroscandate(III), $(NH_4)_3[ScF_6]$. ***Scandium(III) hydroxide***, $Sc(OH)_3$, M_r 95.98, is obtained by precipitation from aqueous scandium(III) salt solutions with alkali hydroxide; precipitation of the same solutions with alkali oxalate yields ***scandium(III) oxalate***, $Sc_2(C_2O_4)_3 \cdot 5H_2O$, M_r 444.05. Scandium(III) hydroxide is soluble in concentrated alkali hydroxide solutions, with formation of hexahydroxoscandate(III), $M_3[Sc(OH)_6]$. ***Scandium(III) nitrate***, $Sc(NO_4)_3 \cdot 4H_2O$, M_r 303.03, forms colorless prisms. When it or scandium(III) oxalate is roasted, the product is ***scandium(III) oxide***, Sc_2O_3, a white, loose powder, M_r 137.91, density 3.864, m.p. 3100 °C.

Scandium group: the d-metals of group IIIb of the periodic system: scandium (Sc), yttrium (Y), lanthanum (La) and actinium (Ac). Scandium, yttrium and lanthanum are very similar to the Lanthanides (see), and are included with them in the Rare earth metals (see). Actinium, as a radioactive decay product of uranium, is somewhat different from the other metals of this group. All the S. elements are similar to, but are more reactive than aluminum; they become more electropositive with increasing atomic mass. Elements of the S. are found exclusively in the +3 oxidation state in their compounds. The hydroxides $M(OH)_3$ become increasingly basic with increasing atomic mass; and the solubilities of the sulfates $M_2(SO_4)_3$ decrease with increasing atomic mass, as is also observed with the sulfates of the alkaline earth metals.

Scattered current corrosion: see electrolytic metal corrosion.

Scattered light measurement: see Nephelometry.

Scattering cross section: see Kinetics of reactions (theory).

Scents: organic compounds or mixtures of such compounds with strong odors used to make perfumes, to perfume soaps, cosmetics and other products, and to add aroma to foods and beverages.

There are natural, semisynthetic and synthetic S. Natural S. are present in Essential oils (see), resins and balsams extracted from plants, and animal secretions, such as ambergris, musk, civet or castor. Semisynthetic S. are chemically altered natural S., such as linalyl acetate from linalool or ionone from citral. Synthetic S. are obtained from simple organic chemicals. They can be identical to natural S., e.g. geraniol or phenylethanol, or different, e.g. toluol musk. S. are characterized by the nature of the odor, its intensity and its durability.

Schäffer acid: an Alphabet acid (see).

Schardinger dextrins: see Dextrin.

Scheel's green: see Copper(II) arsenite.

Scheibler's reagent: see Alkaloids.

Schiemann reaction: introduction of fluorine atoms into arenes by cautious heating of arene diazonium fluoroborates:

$$C_6H_5\text{-}\overset{+}{N}{\equiv}N \; BF_4^- \xrightarrow[-N_2,\,-BF_3]{\text{Heat}} C_6H_5\text{-}F$$

Heat　　　　　　Fluorobenzene

This method can be used with single- and multiple-ring aromatic amines (e.g. naphthylamine, phenanthramine), heterocyclic amines (e.g. aminopyridine). Substituents on the arene ring influence the reaction by changing the solubility of the diazonium tetrafluoroborate and the decomposition temperature. Nitro, amino and alkoxy groups reduce the yield, as do high decomposition temperatures. In such cases, the use of arene diazonium hexafluorophosphates is advantageous.

Schiff's bases, ***azomethines***: organic compounds with the general formula $R^1R^2C{=}N\text{-}R^3$, where R^3 can be either aromatic or aliphatic. Compounds in which R^3 = phenyl are also called ***aniles***. S. are formed by the condensation of ketones or aldehydes with primary amines: $R^1R^2C{=}O + H_2N\text{-}R^3 \rightarrow R^1R^2C{=}N\text{-}R^3 + H_2O$. Most are stable compounds, especially when the C=N double bond is part of a conjugated system of π-bonds, as in benzaniline, $C_6H_5\text{-}CH{=}N\text{-}C_6H_5$, which is synthesized from benzaldehyde and aniline.

The hydrolysis of S. back to the starting compounds can be either acid or base catalysed. Because they crystallize readily, B. are often used to characterize carbonyl compounds. The aldimine group found in a number of S. is the characteristic chromophore of the azomethine pigments. The S. of aromatic 2-hydroxyaldehydes or ketones, which can be seen as the N-heteroanalogs of their parent compounds, form stable chelate complexes with transition metal cations. These can be used in gravimetric analysis. Many S. are important intermediate compounds in the synthesis of nitrogen-containing heterocycles, for example, in the isoquinoline syntheses of Pictet-Spengler and Pomeranz-Fritsch.

Schlipp's salt: same as Sodium thioantimonate (see).

Schmidt reaction: a method for conversion of carbonyl compounds to amines, amides or nitroles by reaction with sodium azide and sulfuric acid. A typical example is the reaction of carboxylic acids, via azides and isocyanates, to form amines:

$$R\text{-}C(=O)OH \xrightarrow{+H^+} R\text{-}C^+(OH)OH \xrightarrow{+HN_3} R\text{-}C(OH)(OH)\text{-}NH\text{-}\overset{+}{N}{\equiv}N \xrightarrow{-H_2O} R\text{-}C(=O)\text{-}NH\text{-}\overset{+}{N}{\equiv}N \xrightarrow{-N_2} R\text{-}C(=O)\text{-}\overset{+}{N}\text{-}H \xrightarrow{-H^+} R\text{-}N{=}C{=}O \xrightarrow[-CO_2]{+H_2O} R\text{-}NH_2$$

As in the Curtius degradation (see), Hofmann degradation (see) of amides and Lossen degradation (see), the carbon chain is shortened by one atom. In the S., dicarboxylic acids with more than one CH_2 group between the carboxyl groups yield the corresponding diamines, aldehydes yield nitriles and *N*-substituted formamides, and ketones form amides. Cyclic ketones are converted to lactams or, if a large excess of hydrazoic acid is used, substituted tetrazoles. Aliphatic and alicyclic carbonyl compounds react more readily than aromatic compounds. Thus if there are several different carbonyl compounds in a molecule, the S. can be carried out selectively.

Scholler-Tornesch process: see Wood sugar.

Schotten-Baumann reaction: acylation of alcohols and phenols with an acyl chloride in the presence of sodium hydroxide in aqueous solution, forming carboxylic acid esters: $R^1\text{-}OH + R^2\text{-}COOH + NaOH \rightarrow R^2\text{-}COOR^1 + NaCl + H_2O$. R^1 can be an alkyl or aryl group, while R^2 is usually an aryl group. To make the acylation as close to quantitative as possible, an excess of base and acyl chloride are used, for example, in the ratio 1:7:5. The method is applied to characterize hydroxyl compounds by their acyl derivatives.

Acylation in the presence of pyridine, which is a much gentler method, is more commonly used (Deninger, Einhorn, Ullmann, et al.). The acylating reagent in this method is the N-acylpyridine hydrochloride, which transfers the acyl group, leaving pyridine hydrochloride.

$$C_5H_5N \xrightarrow{C_6H_5COCl} [C_5H_5\overset{+}{N}\text{-}CO\text{-}C_6H_5]\,Cl^- \xrightarrow[-\text{Pyridine}\cdot HCl]{+R\text{-}OH} C_6H_5\text{-}COOR$$

Schottky defects: see Crystal lattice defects.

Schrödinger equation: see Atom, models of.

Schulze-Hardy rule: see Colloids.

Schweinfurt green: see Copper(II) acetate arsenite.

Schweitzer's reagent: see Copper(II) hydroxide.

Schwenk process: see X-ray structure analysis.

Scientific Design process: see Phenol.

Scilla glycosides: cardiac glycosides found in the squill *Urginea (Scilla) maritima*, of which there are two varieties, red and white. The main active compound of the white variety is ***scillaren A*** (0.06%), which is obtained by enzymatic removal of a glucose from ***glucoscillaren A***. Removal of a second glucose produces ***proscillaridin A***, scillarenin-3α-L-rhamnoside. Scillarenin, the aglycon of these compounds, is a bufadienolide. The absorption of proscillaridin A from the gut is about 30%, and the daily elimination is about 50%. The main active substance in the red variety of squill is ***scilliroside***, the 3-β-D-glucoside of the aglycone scillirosidin. Red squill can be used as a rat poison, because scilliroside has a very toxic effect on the central nervous system of the animals.

H3C, CH3, H, H, OH, O

Rhamnose – Glucose – Glucose

Proscillaridin A

Scillaren A

Glucoscillaren A

Scillaren A: see Scilla glycosides.

Scilliroside: see Scilla glycosides.

Sclerotinite: see Macerals.

Scopine: see Scopolamine.

Scopolamine, ***hyoscine***: an ester alkaloid in which the OH group of scopine is esterified to the carboxyl group of S-tropic acid. As an ester alkaloid, S. is readily hydrolysed, and the primary product, ***scopine***, undergoes further reaction between the OH group and the oxirane ring. S. is present in high concentration in the Australian *Duboisia* species. The hydrobromide is used therapeutically; it has m.p. 194 °C, $[\alpha]_D^{20}$ -15.7° (in ethanol) or -26° (in water).

S. has more central activity than atropine, inhibiting and paralysing the central nervous system even in low doses. It was discovered in 1888 by E. Schmidt.

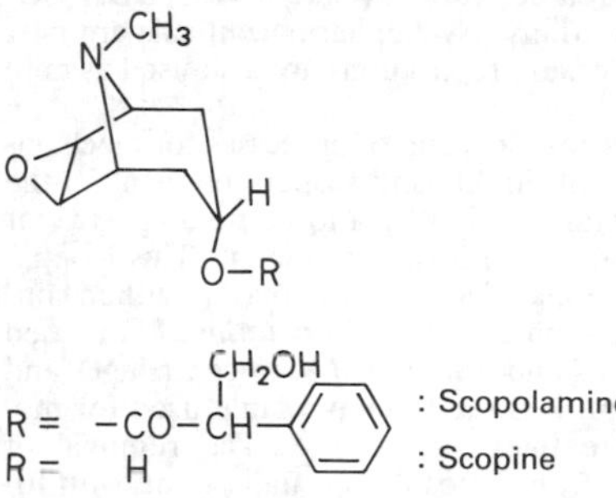

R = $-CO-CH(CH_2OH)-C_6H_5$: Scopolamine
R = H : Scopine

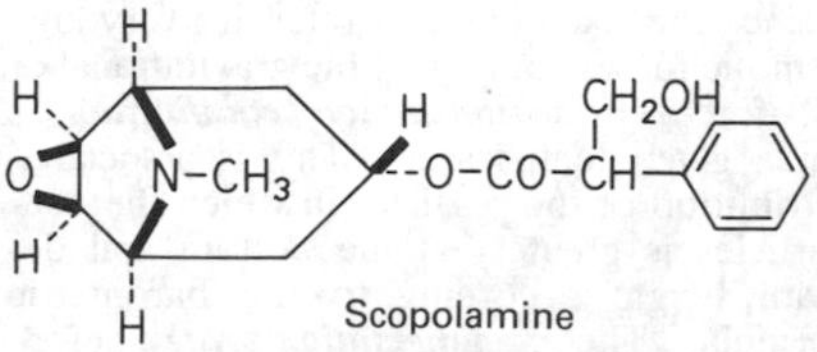

Scopolamine

Scopoletin, *gelseminic acid,* β-methylesculetin: a derivative of coumarin. S. forms colorless needles; m.p. 204-205 °C. It is readily soluble in chloroform and glacial acetic acid, less soluble in water and alcohol. S. occurs along with scopolamine in the roots of *Scopolia*, *Belladonna*, tobacco, oat seedlings, gelsemium roots and oleander.

Scorpion toxins: Scorpions have a fully developed venom apparatus, usually consisting of a sting equipped with two poison glands, which they use for offense and defense. The toxicity of S. for human beings varies widely, depending on the species and geographical origin of the scorpion. The sting of the dangerous species causes violent pains, which then yield to insensibility, excitation and seizures, visual impairment, anxiety, blood pressure oscillations and respiratory paralysis. The initial symptoms, in particular the pain followed by loss of feeling, are caused by serotonin. The neurotoxins in S. resemble those in the venom of adders (see snake toxins) with respect to molecular mass (6800 to 7200 g mol^{-1}, corresponding to 63-64 amino acid residues of known sequence, 4 disulfide bridges), amino acid composition (a large proportion of basic and aromatic groups) and effect.

In addition to the usual method of injection of the venom, certain scorpions have the capacity to spray an anal secretion which is a mixture of 84% acetic acid and 5% caprylic acid.

Scotophobin: Ser-Asp-Asn-Asn-Gln-Gln-Gly-Lys-Ser-Ala-Gln-Gln-Gly-Gly-Tyr-NH_2 a 15-peptide amide which is reported by Ungar to have "memory transferring" activity. S. was isolated from the brains of rats which had been trained to avoid darkness, a behavior which is anomalous for this species; rats are normally active in darkness and avoid light. When S. was injected into untrained rats, these animals also avoided darkness. According to Ungar, S. might be the first code word of a neural code system. Although his experiments have been confirmed in a few laboratories, there is a general reluctance to believe that the observations will be generalized.

Ungar's group also isolated ***ameletin***, a hexapeptide with the sequence Pyr-Ala-Gly-Tyr-Ser-Lys, which they report transfers a learned response to a sound signal. ***Chromodiopsin***, Pyr-Ile-Gly-Ala-Val-Phe-Pro-Leu-Lys-Tyr-Gly-Ser-Lys, is reported by the same group to transfer the ability of trained goldfish to distinguish colors.

Scrap metal: metal objects which have been worn or corroded, or metal wastes. ***New scrap (factory scrap)*** is generated by metal working (rolling, hammering, machining, stamping, etc.). Metallurgical S. is an important secondary raw material and should be recycled by remelting or chemical conversion.

Old scrap consists of metal objects which have become unusable (machine parts, used cars, etc.). It can be reworked by special techniques of refining, blasting, precipitation and electrolysis.

Scrap iron is used in large amounts in Steel (see) production.

Screening constants: see Screening, see Moseley's law.

Screening: reduction of the attractive interaction between an electron and the nucleus by the other electrons in the atom. In the central field model of the atom, the energy $\varepsilon_{n,l}$ of an electron depends on the quantum numbers n and l. Slater proposed analytical expressions for for the radial parts $R_{n,l}(r)$ of the single electron functions $\Phi_{n,l,m}l$:

$$\varepsilon_{n,l} = \frac{-e^2(Z-\sigma_{n,l})^2}{4\pi\varepsilon_0\, 2a_0 n'^2},$$

$$R_{n,l}(r) = N r^{n'-1}\, e^{-\frac{(Z-\sigma_{n,l})}{n'}\frac{r}{a_0}},$$

Here e is the charge of the electron, Z is the nuclear charge constant, ε_0 is the electric field constant, a_0 is the Bohr radius (see Atomic units) and N is the normalization factor. The ***screening constant*** $\sigma_{n,l}$ and the effective quantum number n' are determined by the following rules (*Slater rules*): 1) Electron shells with principle quantum numbers greater than n are ignored. 2) For every additional electron with the same n value, $\sigma_{n,l}$ is increased by 0.35 (but for n = 1, only 0.3 is added). 3) For l = 0 or 1 (s,p), each electron in the shell n-1 contributes 0.85 to $\sigma_{n,l}$, but for l = 2 or 3, the contribution is 1.0. 4) All electrons in still deeper shells contribute 1.0. The values of n' depend on n as shown in the table:

n =	1	2	3	4	5	6
n' =	1	2	3	3.7	4.0	4.2

The single electron functions $\Phi_{n,l,m}l = R_{n,l}(r)\, Y_{l,m}l(\Theta\varphi)$ with radial parts determined by the Slater rules are called *Slater orbitals*.

SDS electrophoresis: see Electrophoresis.

Se: symbol for selenium.

Sealing: see Anodic oxidation.

Sebacic acid, ***decanedioic acid, octane-1,8-dicarboxylic acid***: HOOC-$(CH_2)_8$-COOH, a saturated dicarboxylic acid. S. forms colorless platelets; m.p. 134.5 °C, b.p. 295 °C at 13.3 kPa. It is barely soluble in water, but dissolves easily in alcohol and ether. The salts and esters of S. are known as *sebacates*. It is synthesized by alkaline cleavage of ricinus oil or ricinoic acid, or by oxidation of stearic acid with dinitrogen tetroxide. S. is used in the production of synthetic resins, polyamides and fruit ethers. Its esters with straight-chain lower alcohols are used as softeners; its esters with branched or higher alcohols are used as lubricating oils.

Secale alkaloids: same as Ergot alkaloids (see).

Secondary cell, ***storage battery***: an Electrochemical current source (see) which, like a Primary cell (see) contains only limited amounts of the reactants. However, these can be recharged after use, that is, after the cell has discharged. The electrochemical reaction must be reversible, so that an electric current run through the cell reverses the electricity-producing reaction and regenerates the reactants. The best-known and most widely used S. is the Lead-acid battery (see) used in automobiles. Some other common S. are the Nickel-cadmium battery (see), the Knopf cell (see), the Nickel-iron battery (see), the Silver-cadmium battery (see) and the Silver-zinc battery (see). The Sodium-sulfur battery (see) is coming into use for special applications, such as electric cars and peak-load power plants.

Secondary electron amplifier: see Photomultiplier.

Secondary ozonide: see Harries reaction.

Secondary structure: see Biopolymers, see Proteins.

Second messenger: a chemical substance released or synthesized after a hormone molecule has bound to its membrane-bound receptor. The hormone or other chemical stimulus is the "first messenger", and a variety of first messengers stimulate their target cells by the same second messenger. At present, two main systems are known: the Cyclic nucleotides and inositol phosphates and calcium ions. In either case, the second messenger activates protein kinases which phosphorylate specific cellular proteins; beyond this, the mechanism of action has not been traced. It is thought that the phosphorylated proteins are responsible for the cellular responses to the hormones.

Secretin: a peptide hormone consisting of 26 amino acids, His-Ser-Asp-Gly-Thr-Phe-Thr-Ser-Glu-Leu-Ser-Arg-Leu-Arg-Asp-Ser-Ala-Arg-Leu-Gln-Arg-Leu-Leu-Gln-Gly-Leu-Val-NH_2. S. is formed in the mucous membrane of the duodenum, and stimulates the production and release of $NaHCO_3$-containing digestive juice by the pancreas. S. also increases the flow of bile, and is thought to limit the production of gastric juice. The *N*-terminal 12 residues are homologous to glucagon and vasoactive intestinal peptide (VIP).

S. was discovered in 1902 by Bayliss and Starling; it was the first peptide hormone to be discovered. It was isolated in pure form in 1961 by Jorpes and Mutt, and a little later, its structure was elucidated. Two total syntheses reported by Bodanszky confirmed the proposed primary structure. Shortened or partial sequences have no biological activity.

Sedatives: compounds which reduce the excitability of the central nervous system without significantly affecting its function. Many Hypnotics (see) can be used in low doses as S. Some compounds used primarily as S. include methylpentynol and bromoisovalum, an acylurea derivative, $(CH_3)_2$CH-CH(Br)-CO-NH-CO-NH_2. Many psychopharmaceuticals are now used as S. Valerian preparations are also used as mild S.

Sedimentation: a group of processes for mechanical separation of liquids and suspensions. In S., the solid particles are settled by a force field (gravity or centrifugal force) and form a sediment. This is separated from the clarified liquid above it, either continuously or discontinuously. ***Gravitational S.*** is used in settling tanks and ***centrifugal S.*** in centrifuges and hydrocyclones. S. is the basic principle used for mechanical pretreatment of sewage. The removal of solid particles from aerosols is usually done with filters, electrostatic precipitators or wet scrubbing, however, since the sedimentation rates are very low.

Since osmotic forces counteract the gravitational or centrifugal force, a ***sedimentation equilibrium*** is reached at a given temperature. This is associated with a distribution of the particles in which the density of particles is greatest at the bottom and decreases with height according to the barometric height formula. The ***sedimentation rate*** before equilibrium is reached increases with the diameter of the particles, and decreases with increasing viscosity of the dispersion medium. It is described by the *Stokes equation*:

$$v_0 = \frac{d\kappa^2(d_s - d_l)g}{18\eta}$$

Here d_κ is the diameter of the particles, d_s is the density of the particles, d_l is the density of the fluid, g is the gravitational constant and η is the dynamic tension.

The molecular mass of the particles can be determined from the density distribution at sedimentation equilibrium or the rate of sedimentation (***sedimentation analysis***). For determinations of molar mass of colloidal particles, where the ***sedimentation effect*** is very slight, stronger forces than the earth's gravitational field must be used; usually an ultracentrifuge is necessary.

Sedimentation analysis: a method of determining the size of solid particles with diameters between about 1 and 100 µm. The determination is based on the dependence of the sedimentation rate of the particles in a liquid medium (distilled water, cyclohexanol, ethylene glycol, isobutanol, etc.) on their diameters. If the sedimentation distance and time are measured, then in the range of laminar flow the particle size can be calculated by the Stokes friction law; in addition to the sedimentation parameters, the density difference between the particles and liquid medium and the viscosity of the medium must be known. The amounts of individual fractions can be determined by measuring the changes in concentration in the plane of measurement as a function of time. The analysis is usually carried out by the Andreasen pipet, Casagrande areometer, Sartorius sedimentation balance or Svedberg-Dumanski centrifuge method.

Sedimentation potential: see Zeta potential.

Sedoheptulose: D-*altro*-heptulose, a monosaccharide. S. is an intermediate in carbohydrate metabolism formed by the transketolase reaction from ribose 5-phosphate and xylulose 5-phosphate ($C_5 + C_5 = C_7 + C_3$). The other product is glyceraldehyde phosphate.

Seed crystal: see Crystal growing.

Seed crystal formation: see Crystallization.

Seger cones, ***pyrometer cones***: a series of indicators developed by H. Seger for control of firing of ceramics. They are blunt, triangular pyramids, about 6 cm tall, made of mixtures of silicon oxide, SiO_2, and aluminum oxide, Al_2O_3. They are placed behind a peephole in the kiln, and as they are heated, they eventually soften ("fall"). They are numbered to indicate increasing softening temperature, in the range of 600 to 2000 °C, which depends on the amount of Al_2O_3. (Higher concentrations of Al_2O_3 produce higher softening temperatures). With a series of S., the temperature can be determined accurately to ± 10 °C. They are used instead of conventional thermometers because the process of firing depends on both the temperature and the time the kiln has been heated. The process of softening the S., like that of firing the ceramics, takes time.

Seger cones

Seignett salt: same as Potassium sodium tartrate (see).

Selection rules: rules which determine which transitions between energy states are allowed and which are forbidden; an example would be certain changes in quantum numbers which are allowed or forbidden.

Selectivity: the capacity of a reaction, method or operation to choose among several possibilities. The concept of S. is defined and used differently in different areas of chemistry.

1) In chemical reactions which can lead to several different products, S. is the fractional yield of a given product. The larger this fraction is, the more selective the reaction. If the products are structural isomers, the term Regioselectivity (see) is used; if they are stereoisomers, the term Stereoselectivity (see) is applied. Quantitatively, the S. is defined either a) as the ratio of two product concentrations $S_{12} = c_1/c_2$ or b) as the ratio of the concentration c_1 of the desired product to the sum of all product concentrations, $S_1 = c_1/\Sigma c_i$. The value of S_{12} can lie between 0 and ∞, while that of S_1 can lie between 0 and 1. S_{12} is also called "relative reactivity". If the kinetics of formation of the different products are the same order, the rate constants k can be substituted for the concentrations (see Complex reaction): $S_{12} = k_1/k_2$ or $S_1 = k_1/\Sigma k_i$. Only in such cases is the S. of a reaction a constant independent of the initial concentrations and extent of the reaction; even then, it is affected by changes in the reaction conditions (temperature, catalyst, solvent).

2) In chemical analysis, a method is said to be selective if it can be used to determine one of several components of a sample independently of the others, i.e. if the analysis signal of the component i is not affected by the other substances k which are present. This is nearly always an approximation; the more selective a method is, the closer its partial sensitivities $\gamma_{ik} = \delta x_i/\delta c_k$ are to 0 for all i ≠ k. Here x_i is the calibration function, that is, the dependence of the analysis signal x_i for the component i to be determined on the concentrations of all components of the sample: $x_i = f(c_1, c_2,...c_n)$. Thus $\gamma_{ik} = 0$ means that component k has no effect on the analysis signal for component i. On the other hand, $\gamma_{ii} = \delta x_i/\delta c_i$ should be as large as possible. Thus, the S. of analytical methods can be quantitatively defined in terms of the sensitivities γ_{ii} and γ_{ik}.

3) In separation processes which depend on phase equilibria, especially extraction, extractive distillation and chromatography, S. is the effect of selective solvents on the separation of two components (1) and (2). It characterizes the effect of molecular interactions on the relative solubilities or volatilities of the two substances. S. is defined by the equation $S = f_1/f_2$, where f_1 and f_2 are the activity coefficients of the two components in the selective solvent. However, the suitability of a phase equilibrium for a separation depends not only on the S., but also on the solubility of the two substances in the solvent, i.e. on its capacity.

Selenates: the salts of selenic acid, H_2SeO_4. As a dibasic acid, it forms two series of salts, the ***acid S., hydrogenselenates***, M^IHSeO_4, and the ***normal S., neutral S.***, $M^I_2SeO_4$. The selenate anion, SeO_4^{2-}, has a tetrahedral configuration comparable to that of the sulfate ion. The S. are strong oxidizing reagents. They decompose when heated with release of oxygen to form selenites. The solubility of the S. is similar to that of the sulfates, e.g. lead selenate, $PbSeO_4$, and barium selenate, $BaSeO_4$, are essentially insoluble in water. S. are poisonous (see Selenium). They are obtained by anodic oxidation of selenites or by fusion of selenides with potassium nitrate.

Selenic acid: H_2SeO_4, forms colorless, hexagonal crystals, M_r 144.97, density 2.951, m.p. 58 °C. S. forms hydrates with water: $H_2SeO_4 \cdot H_2O$ (m.p. 26 °C) and $H_2SeO_4 \cdot 4H_2O$ (m.p. -51.7 °C). It is a strong dibasic acid ($pK_1 \approx -3$, $pK_2 \approx 2$), and two series of salts are derived from it, the acid and neutral Selenates (see). S. is a strong oxidizng agent. For example, a mixture of S. and hydrochloric acid releases active chlorine ($H_2SeO_4 + 2\,HCl \rightleftharpoons H_2SeO_4 + 2\,Cl + H_2O$), and, like aqua regia, it can dissolve gold or platinum as chlorocomplexes. S. is poisonous (see Selenium). It is obtained by anodic oxidation of selenous acid.

Selenides: derivatives of selenium in the - II oxidation state; the salts of hydrogen selenide, H_2Se. As a dibasic acid, hydrogen selenide forms two series of salts, the ***acid S., hydrogenselenides***, M^IHSe, and the ***neutral S.***, M^I_2Se. The water-soluble, colorless alkali metal selenides react with selenium to form red ***polyselenides***, $M^I_2Se_x$. Heavy metal selenides are usually colored, and, like sulfides, are scarcely soluble in water. A few S. are found in nature (see Selenium). They are obtained by combination of the elements or by passing hydrogen selenide into the corresponding metal salt solutions. Indium and gallium selenide are used in semiconductor technology, and zinc selenide is used as a phosphor for fluorescent screens.

Selenites: the salts of selenous acid, H_2SeO_3. As a dibasic acid, it forms two series of salts; the ***neutral S.***, $M^I_2SeO_x$, are stable compounds. The selenite ion SeO_3^{2-} found in them has the shape of a flattened pyramid. These salts are oxidizing, and are obtained by neutralization of selenous acid by the corresponding metal hydroxides or carbonates. The ***acidic S., hydrogenselenites***, M^IHSeO_3, exist only in dilute aqueous solution. When the solution is concentrated, they condense to ***diselenites***: $2\ HSeO_3^- \rightleftharpoons [O_2Se\text{-}O\text{-}SeO_2]^{2-} + H_2O$. In contrast to the nonsymmetrical disulfite ion $[O_2S\text{-}SO_3]^{2-}$, which has a direct S-S bond, the diselenite ion is symmetrical and the two Se atoms are linked by an oxygen bridge. S. are poisonous (see Selenium).

Selenium, symbol ***Se***: a chemical element from group VIa of the periodic system, the Oxygen-sulfur group (see), a semimetal, *Z* 34, with natural isotopes with mass numbers 74 (0.87%), 76 (9.02%), 77 (7.58%), 78 (23.52%), 80 (49.82%) and 82 (9.19%), atomic mass 78.96, valences IV, II, VI, Mohs hardness 2, density 4.81, m.p. 217 °C, b.p. 684 °C, standard electrode potential (Se^{2-}/Se) -0.78 V.

Properties. Like its homolog, sulfur, S. exists in several modifications. Rapid cooling of a S. melt yields ***black, glassy S.***, which consists of a mixture of Se_8 rings and long Se chains. Cyclooctaselenium can be extracted from this material with carbon disulfide. Two monoclinic, red forms crystallize out of the CS_2 solution: ***α- and β-monoclinic S.***; these consist of Se_8 molecules with the same structure as cyclooctasulfur molecules. When red S. is heated to about 150 °C, or when molten S. is slowly cooled, hexagonal ***gray, metallic S.*** is formed. This is insoluble in carbon disulfide, and the molecules are infinite spiral chains of Se. Gray S. is a typical semiconductor. It conducts very little current in the dark, but in the light its conductivity increases considerably.

The chemistry of S. is also closely similar to that of sulfur. Its preferred oxidation states are -II, +IV and +VI, and, in contrast to sulfur, the most stable state for S. is +IV. Se(VI) compounds, e.g. selenates, are strong oxidizing agents. S. combines with hydrogen at high temperatures to give hydrogen selenide, H_2Se. Many metals reduce S. to the corresponding metal selenides. When heated in air, S. burns to selenium dioxide, SeO_2. The halogens oxidize S. to various selenium halides, depending on the ratios of the reactants. For example, fluorine reacts with S. to form selenium tetrafluoride, SeF_4, or selenium hexafluoride, SeF_6; chlorine reacts to form diselenium dichloride, Se_2Cl_2 or selenium tetrachloride, $SeCl_4$. The intense green color which develops when S. is treated with concentrated sulfuric acid solutions is due to the formation of colored Se_8^{2+} cations.

> Selenium and its compounds are toxic. Orally ingested or inhaled elemental selenium or inorganic selenium compounds lead to gastrointestinal symptoms, drop in blood pressure and central nervous impairment. The sweat and breath of the patient develop a typical garlic-like odor. Countermeasures to oral intake are induced vomiting and pumping of the stomach; further treatment is symptomatic.

Analysis. S. is precipitated out of its compounds by reducing agents and dissolves in nitric acid. The resulting selenous acid is reduced with sulfur dioxide to red S. This dissolves in warm, concentrated sulfuric acid to give an intense green color: $8\ Se + 5\ H_2SO_4 \rightarrow Se_8^{2+} + 4\ HSO_4^- + 2\ H_3O^+ + SO_2$. For quantitative determination, selenites or selenates are reduced, e.g. with hydrazine hydrate, and the precipitated S. is determined gravimetrically. Small amounts of S. can be determined by calorimetric methods or atomic absorption spectroscopy.

Occurrence. S. makes up $9 \cdot 10^{-6}$% of the earth's crust. In nature, it is found in the form of selenides, usually in low concentrations as components of sulfide ores. Pure S. minerals, e.g. berzelianite, Cu_2Se, naumannite, Ag_2Se, and clausthalite, PbSe, are rare. S. is therefore obtained as a byproduct of processing sulfide ores. It is enriched, for example, in the fly ash from roasting of sulfidic materials and in the anode sludge from electrolytic copper refining. S. was formerly isolated from lead chamber sludge from sulfuric acid production.

S. is an essential trace element for plants and seems to have some significance in the metabolism of higher animals as well.

Production. The starting material is usually the anode sludge from copper electrolysis, which contains an average of 8 to 9% S. in the form of copper or silver selenide. This sludge or the fly ash from roasting operations is either melted with soda in the presence of air, leading to sodium selenite, Na_2SeO_3, and sodium selenate, Na_2SeO_4, or it is heated with concentrated sulfuric acid in the presence of sodium nitrate; the main oxidation products are then sodium selenite and selenous acid, H_2SeO_3. The cooled melts are dissolved in water, and red S. is precipitated from the solution by passing sulfur dioxide into it.

Applications. As a semiconductor, S. is used in rectifiers and photocells. It is an essential component of the photoconductive layers in electrophotographic copying (xerography). S. is also used to make pigments for both glass and ceramics, and as a decolorizer or colorant for glass and enamel. Glasses colored with S. are yellow to ruby red. S. is used in metallurgy as a component of alloys. Selenium sulfide, SeS_2, is added to some shampoos to prevent dandruff.

Historical. S. (from the Greek "selene" = "moon") was discovered in 1817 by Berzelius in the lead chamber sludge from a Swedish sulfuric acid plant.

Selenium oxides: ***Selenium(IV) oxide, selenium dioxide***, SeO_2, forms colorless, hygroscopic, monoclinic needles; M_r 110.96, density 3.95, subl.p. 340-350 °C. SeO_2 sublimes readily and is a strong poison (see Selenium). In contrast to sulfur dioxide, SO_2, which is monomeric, SeO_2 consists of infinite chains (Fig.).

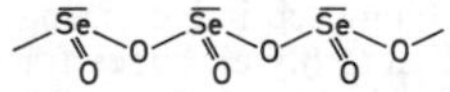

It dissolves in water to form selenous acid ($SeO_2 + H_2O \rightleftharpoons H_2SeO_3$), of which it is the anhydride. SeO_2 is an oxidizing agent, being reduced to red selenium. It is obtained by burning selenium in air ($Se + O_2 \rightarrow SeO_2$, ΔH = -235.6 kJ mol^{-1}), or by dehydration of selenous acid. SeO_2 is used as an oxidizing reagent for preparative organic chemistry to convert active methyl and methylene groups to carbonyl functions. It is used in industry as an additive to lubricants and for production of special glasses.

Selenium(VI) oxide, selenium trioxide, SeO_3, forms colorless, hygroscopic, cubic crystals which sublime in vacuum; M_r 126.96, density 3.6, m.p. 118 °C. SeO_3 is poisonous (see Selenium). It has a molecular structure comparable to ice-like sulfur trioxide (see Sulfur oxides), but the cyclic tetrameric forms seem to predominate. Above 180 °C, SeO_3 decomposes into SeO_2 and oxygen. It is a strong oxidizing agent. As the anhydride of selenic acid, it reacts violently with water: $SeO_3 + H_2O \rightarrow H_2SeO_4$. SeO_4 is obtained by boiling potassium selenate with sulfur trioxide ($K_2SeO_4 + SO_3 \rightarrow K_2SO_4 + SeO_3$), or by dehydration of selenic acid.

Selenocyanates: derivatives of selenocyanic acid (more correctly, isoselenocyanic acid), H-NCSe. Ionic S. contain the linear, resonance-stabilized, ambivalent anion $[NCSe]^-$, which is a Pseudohalide (see). ***Alkali selenocyanates*** are typical, colorless salts which can be obtained, for example, by oxidation of alkali cyanides with selenium. The selenocyanate anion can be reversibly oxidized to the pseudohalogen ***diselenocyanogen***, a yellow powder, m.p. 133 °C. Protonation of this compound yields ***isoselenocyanic acid***, HNCSe, which can be detected only in aqueous solution. The anion is an ambivalent complex ligand similar to the thiocyanate anion $[NCS]^-$. Depending on the reaction partner and conditions, it can form coordinate bonds via nitrogen (isoselenocyanate complexes, M-NCSe) or selenium (selenocyanate complexes, M-SeCN). In bridge positions, it can act as a bidentate ligand: M-NCSe-M'.

Selenous acid: H_2SeO_3, colorless, deliquescent, hexagonal crystals; M_r 128.97, density 3.004. When heated in dry air, S. releases water and is converted to selenium dioxide, SeO_2. In aqueous solution, it is a weak dibasic acid, $pK_1 = 2.62$, $pK_2 = 8.30$, and forms two series of salts, the acidic and neutral Selenites (see). S. is redox amphoteric, but its oxidizing effects are stronger. It is oxidized to selenic acid only by very strong oxidizing agents, such as chlorine: $H_2SeO_3 + Cl_2 + H_2O \rightleftharpoons H_2SeO_4 + 2\ HCl$. S. is poisonous (see Selenium). It is obtained by dissolving selenium dioxide in water, or by oxidizing selenium with dilute nitric acid.

Self-polishing emulsions: dispersions of waxes and plastics which contain no solvents. Dispersions containing mainly waxes leave soft films which can easily be polished, while a higher content of polymers gives a very resistant film which is difficult to polish. Dispersions with large amounts of polymer are preferred for floors in homes.

A S. with a high content of polymers would contain 15 to 50% polymer (40% typical), 1 to 10% wax, 0 to 3% resins, 2 to 7% softener/film binder, 1 to 3% emulsifier, perfumes, preservatives and water.

Combined cleansers and polishes are becoming ever more popular. Such products are added to the water used to wash a floor and contain both surfactants and dispersed wax or wax-like substances, or polymers.

Permanent polishes, which were formerly used as a substitute for S., were laquer-like products consisting of resin solutions or plastic dispersions. They cannot be used on modern plastic floor coverings, because the softeners in the floor migrate into the film of polish and make it sticky.

Self-purification of waters: the limited ability of bodies of water to remove pollutants by natural biological, chemical and physical processes. S. occurs primarily through the activity of microorganisms, plants and animals living in the water (this is the basis for biological Sewage treatment (see)). The organic pollutants entering the water are absorbed by microorganisms and metabolized; higher organisms live off the lower ones (food chains). If S. is to function, no toxins can be permitted to enter the water and sufficient oxygen must be available.

SEM: abb. for secondary electron multiplier.

Semibatch reactor: see Stirred reactor.

Semicarbazide, ***carbamic hydrazide, N-aminourea***: H_2N-CO-NH-NH_2, colorless crystals, m.p. 96 °C. S. is readily soluble in water and alcohol, but only very slightly soluble in ether and benzene. When heated, it decomposes, forming hydrazine and hydrazodicarbonamide. S. forms crystalline salts with mineral acids. With aldehydes and ketones, it reacts by elimination of water to form Semicarbazones (see). S. is synthesized by reaction of potassium cyanate with hydrazine hydrochloride, or reaction of hydrazine hydrate with urea: $H_2N\text{-}NH_2 + H_2N\text{-}CO\text{-}NH_2 \rightarrow H_2N\text{-}CO\text{-}NH\text{-}NH_2 + NH_3$. S. is used mainly in the form of the stable hydrochloride.

Semicarbazones: condensation products of aldehydes and ketones with semicarbazide: $R^1R^2C{=}O + H_2N\text{-}NH\text{-}CO\text{-}NH_2 \rightleftharpoons R^1R^2C{=}N\text{-}NH\text{-}CO\text{-}NH_2 + H_2O$. Because they crystallize readily and most of them have sharp melting points, the S. are used for detection and characterization of aldehydes and ketones. They are also often used for isolation and purification of these carbonyl compounds, because they are easily hydrolysed by dilute acid.

Semiconductor: an element or compound which has a specific electrical resistance under standard conditions which is intermediate between those of metals and insulators. The electrical conductivity increases rapidly with increasing temperature, concentration of foreign components (doping), radiation and mechanical stress, and depends on the composition of the main components and the nature of the surrounding atmosphere. There are two basic types of S., those with electron conduction and those with ionic conduction. The electron-conducting S. are of more industrial importance, but ion-conducting S. are gaining importance as solid electrolytes.

The mechanisms of the conduction process in electronic S. can be understood with the aid of the band model of electrical conductivity. In all substances, the electronic states are clustered in two bands, the lower-energy valence band and the higher-energy conduction band. In metals, the two levels overlap, that is, an electron can easily make the transition from one band to the other, and there is therefore a finite concentration of electrons in the conduction band. In insulators, the conduction band is at a very much higher energy level, and the electrons are essentially all in the valence band. The two bands are separated by a zone of forbidden transitions. A material can become a S. if it has additional energy states between the valence and conduction bands. If the energy difference between this level and the completely occupied valence band is slight, electrons from the valence band can jump to the new acceptor levels. This gives rise to electronic holes, or defect electrons, which contribute to the electrical conductivity. This type of S. is called a defect-electron, deficiency or p-conductor because its increase in conductivity is due to sites which are not occupied with electrons, and are thus in a sense positive (as in Cu_2O). If intermediate levels which are partially occupied with electrons are energetically close to the conduction band, the material can be conductive due to the ready electronic transition from the donor level to the conduction band. This is called excess-electron or n-type conduction (as in ZnO). In either case, the conduction is due to defect sites, and can be greatly increased by introduction of foreign atoms (dopants) into the basic system. By doping with 1 ppm foreign material (e.g. of germanium with antimony), the conductivity can be increased by several orders of magnitude (n-conducting germanium). In addition to the defect conduction which occurs when foreign atoms are incorporated (for example, by doping silicon or germanium with elements from group IIIa or Va), there is an intrinsic conduction which occurs when positive and negative charge carriers (holes and electrons) are present in equal concentration. This formation of pairs, which increases with increasing temperature, is in thermal equilibrium with recombination of the charge carriers.

S. are used as transistors, thermoelectric and photoelectric components, and in microelectronics.

Various oxides and oxide systems (see Oxide ceramics) display good conduction even at medium temperatures (e.g. ZrO_2). Both electron and ion conduction increase with temperature. These O^{2-}-conducting solid electrolytes are being used in increasing measure as batteries, as solid electrolyte probes for monitoring oxygen activity in steel melts and gas mixtures (combustion, flue and exhaust gases) and for measurement of equilibrium voltages in concentration chains of reactive metals. They are also used as oxide thermoelements and resistance elements.

Semiconductor photoeffect: see Photoeffect.

Semidine: a derivative of aminodiphenylamine. The o-S. are derived from 2-aminodiphenylamine, and the p-S., from 4-aminodiphenylamine. They are formed by Semidine rearrangement (see) of substituted hydrazobenzenes.

Semidine rearrangement: a one-sided Benzidine rearrangement (see) of 4-substituted and 4,4'-disubstituted hydrazobenzene in the presence of strong mineral acids. The products with 4-substituted hydrazobenzenes are o- and p-semidines, while only p-semidines are formed from 4,4'-disubstituted hydrazobenzenes:

p-Semidine

o-Semidine

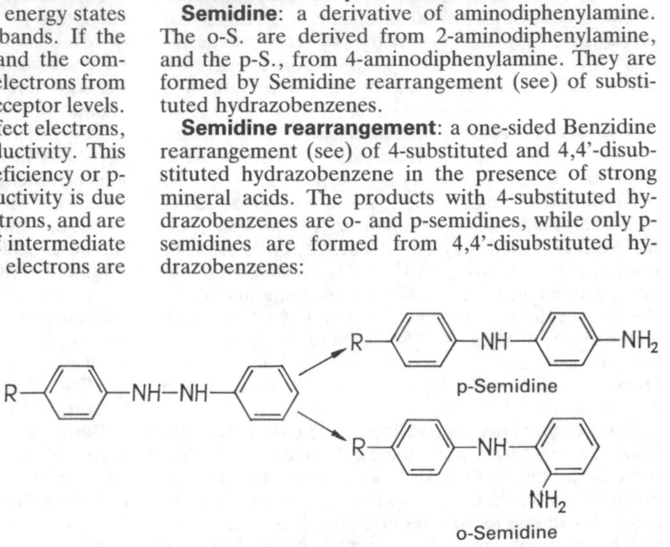

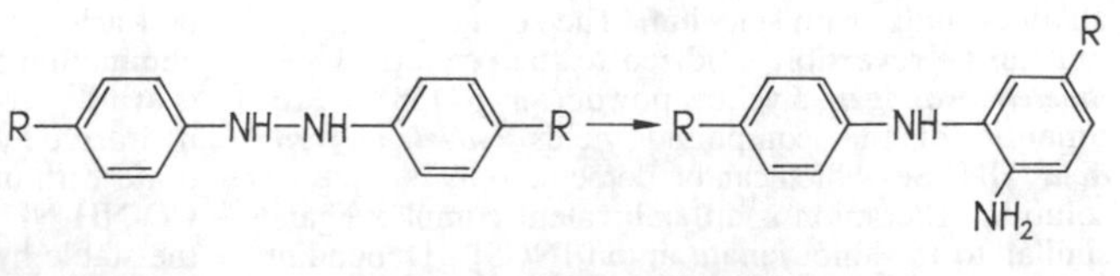

Semifusinite: see Macerals.

Semimetals: chemical elements which have an intermediate position between metals and nonmetals (see Metal).

Semiochemicals: substances which serve for transfer of information between organisms. Substances used for communication between individuals of the same species are called Pheromones (see), while those which communicate between different species are called ***allelochemicals***. This group is further subdivided into ***allomones***, which serve the interests of the sender, and ***kairomones***, give an advantage to the receiver.

Semipermeable membrane: a barrier which allows some components of an adjacent mixture to pass through it, while remaining impermeable to others. Substances pass through the membrane by dissolving in it, diffusing across it and passing out on the other side. Films of synthetic polymers, animal skins, parchment, ceramics with inclusions (e.g. copper hexacyanoferrate(II)) or cell membranes serve as S. They are essential to Osmosis (see), Dialysis (see) and the Donnan equilibrium (see). Heated palladium sheets can serve as S. for hydrogen, allowing its separation from other gases and measurement of its partial pressure. There are also membranes with pore structures which act as molecular sieves, allowing molecules below a certain size to pass while larger ones are retained. Examples of this type of membrane are made of cellulose acetate or films in which pores of defined size are created by bombarding them with heavy atomic nuclei followed by etching.

Semiquinones: relatively long-lived, resonance-stabilized radical ions. S. are formed in the course of redox processes by the uptake of a single electron by a quinone, or loss of an electron by a hydroquinone. They are stabilized by resonance in a conjugated π-electron system, but cannot always be isolated. The actual S. is the radical anion formed by electron uptake by the 1,4-benzoquinone:

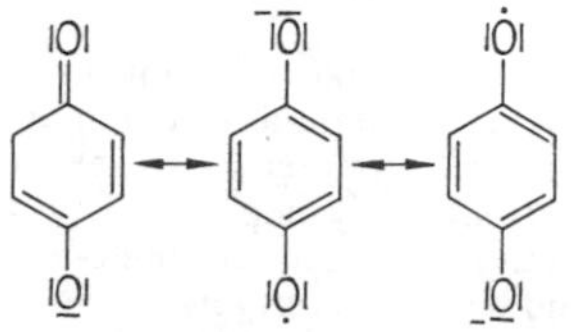

An example of an S. which can be isolated is the sodium salt of semiduroquinone, a derivative of durol. A well-known S.-like radical cation is *Wuster's red*, which is formed by oxidation of *N,N*-dimethyl-*p*-phenylenediamine; it is an intermediate in photographic development.

Sencor®: see Herbicides.

Senecio alkaloids: see Alkaloids.

Sensitizer: see Photosensitization, see Spectral sensitization, see Phytoalexins.

Separating water: see Nitric acid.

Separation: 1) The appearance of different phases in a previously homogeneous mixture of two or more components. S. occurs in liquid and solid mixtures if the components are not infinitely miscible, and a change in external conditions, e.g. temperature or pressure, brings the system into the range of the Mixing gap (see). In liquid mixtures, the beginning of S. can be recognized as a cloudiness which results because one phase is at first finely divided into extremely small droplets in the other. These later combine to form a macroscopically uniform phase region.

The different phases are in thermodynamic equilibrium; thus recrystallization of supercooled melts or release of supersaturation phenomena are not considered S. However, the S. of solid mixtures are often inhibited, so that a long tempering is needed before equilibrium is reached. The differential enrichment of heavier and lighter gas components in a strong gravitational field (e.g. in a gas centrifuge) is also not S., because no phase boundary forms. In supercritical areas (see Critical point), however, S. is possible.

2) the removal of single components from mixtures. The differences in physical or chemical properties of the substances are utilized, for example, solids can be separated from a suspension by centrifugation, filtration, sedimentation or decantation.

Separatory funnel, ***shaking funnel***: a glass device, usually pear-shaped, but sometimes cylindrical or spherical, provided with a ground-glass stopcock and used to separate mixtures of liquids.

Sephadex®: see Dextran.

Sepharose®: see Agar.

Sepia: a black to brown pigment secreted by the ink glands of cephalopods of the genus *Sepia* (squids). S. is fast to light and is nearly insoluble in water, somewhat more soluble in soda solution, and insoluble in alcohol. It is purified by dissolving it in potassium hydroxide solution, precipitation with hydrochloric acid, washing with water and drying. It was often used in aquarell painting, with gum arabic as a binder. Colored S. were obtained by mixing with madder lakes.

Sequential therapy: see Contraceptives.

Sequence analysis: see Nucleic acids.

Sequence rule: see Stereoisomerism 1.2.3.

Ser: abb. for serine.

Sericin: see Keratins.

Serine, abb. ***Ser***: α-amino-β-hydroxypropionic acid, a proteogenic amino acid which is especially abundant in silk fibroin (formula and physical properties, see Amino acids, Table 1). It can be synthesized chemically from ethyl acrylate. Bromine addition yields α,β-dibromopropionate. The β-bromine atom is then exchanged for a hydroxyethyl group by treatment with sodium ethylate. Subsequent aminolysis and acid hydrolysis produce DL-S. The racemate is separated by spontaneous crystallization of the supersaturated solution of the *p*-toluenesulfonate of DL-S. D-S. is an important starting material for synthesis of the antibiotic D-cycloserine. O-Diazoacetyl-L-serine (azaserine) is formed by *Streptomyces* strains and has activity against tumors. S. was discovered in 1865 by Cramer.

Serine proteases: an important group of endopeptidases (see Proteases) in which the active center contains a serine group with an unusually strongly nucleophilic hydroxyl group. In the course of the cleavage of the peptide bond, the cleavage peptide is temporarily bound to the enzyme through an ester linkage to the serine hydroxy group. Certain organophosphates (e.g. diisopropylfluorophosphate) or phenylmethanesulfonyl fluoride block the serine hydroxyl group and irreversibly inhibit the enzyme. Some important S. are Chymotrypsin (see), Trypsin (see), Elastase (see), Thrombin (see), Plasmin (see) and Subtilisin (see).

Serotonin: 3-(2-aminoethyl)-5-hydroxyindole, a vasoactive, biogenic amine which acts as a neurotransmitter. It forms colorless crystals and is soluble in water. S. is usually available as the hydrochloride.

Serpentine: see Rauwolfia alkaloids.

Serum albumin: see Albumins.

Sesamex: see Synergist.

Sesquiterpenes: terpenes consisting of 3 isoprene units, thus 15 C atoms. The S. display a wide variety of structures; they include linear and cyclic compounds with one, two or three rings. The basic structures are the acyclic alcohols Farnesol (see) and its isomer, nerolidol. The monocyclic hydrocarbons cadinene, caryophyllene and bisabolene are widely occurring components of essential oils. The plant family *Asteraceae* is particularly rich in bicyclic sesquiterpene lactones, such as santonin and the guaianolides. The latter group occurs very widely; an example is matricin. The Ionones (see) are derivatives of the S.

S esters: see Thioesters.

Sesterterpenes: terpenes consisting of 5 isoprene units, or 25 C atoms. The S. are a small group, first discovered in 1965. They are found in phytopathogenic fungi, insect cuticle waxes, sponges, and in scattered species of higher plants. The phytotoxic ***ophiobolins***, produced by the plant pathogens *Cochliobolus miyabeanus* and *Helminthosporium oryzae*, are of particular importance.

Settling: see Sedimentation.

Seven-membered heterocycles: cyclic carbon compounds with seven atoms in the ring, one or two of which (in general) are heteroatoms. Some well known examples are oxepine, thiepine, azepine, diazepine, benzazepine and benzodiazepine.

Severity: see Pyrolysis.

Sewage: collective term for water which has been used and thereby suffered deterioration in its quality. ***Household S.*** is waste water from homes, public buildings, etc.; it consists of toilet flushings, laundry, bath and dishwater, and it may also contain kitchen wastes from garbage disposal units. ***Community S.*** contains household S., waste water from various types of manufacturing and water from street drains. ***Industrial S.*** potentially carries the most toxic wastes from chemical or metallurgical processes. Cooling water and rainwater from street drains must also be considered S. if they cannot be reused without treatment or would cause damage in natural waterways.

About 80 to 90% of the water used in industry, agriculture and households is returned to the natural waterways; about 10 to 20% is lost by evaporation or percolation into the soil.

The amounts of city S. depend on the size of the community and the associated industrial and manufacturing capacity. The amount of industrial S. produced per ton of raw material or product varies widely. Modern methods of animal husbandry also produce large amounts of S.

Table 1. Amounts of sewage produced by industries

Industry	Reference unit	Average sewage production [m^3]
Coal briquet factory	1 t briquets	1
Slaughterhouse	1 animal	1.5
Dairy	1 m^3 milk	5
Potassium mine	1 t K_2O	15
Sugar factory	1 t beets	15
Laundry	1 t laundry	15
Petroleum refinery	1 t crude oil	17
Brewery	1 m^3 beer	20
Iron foundry	1 t crude iron	22
Sulfuric acid production	1 t SO_3	50
Synthetic fiber production	1 t viscose fibers	100
Textile dyeing plant	1 t cloth	100
Paper mill	1 t pulp	130
	1 t fine paper	400
Sulfite cellulose production	1 t cellulose	1000

In order to reduce the amount of water needed by an industrial installation and the resulting S. volume, efforts are made to recycle process water directly or after suitable treatment within the plant. This circulation often reduces the production costs and reduces the degree and types of S. damage to the environment.

Types of S. S. can contain inorganic or organic pollutants or both. The mining, metallurgical and and chemical industries produce mainly inorganic wastes, in some cases radioactive wastes. This type of S. contains heavy metal ions, halides, carbonates, sulfates and cyanides of the alkali and alkaline earth metals. Organic pollutants, on the other hand, are generated by production or processing of coal, petroleum, sugar beets, potatoes and grain, cellulose, paper, glues, gelatins and textiles. These organic S. contain an extremely wide variety of organic materials, so that in many cases qualitative analysis of the individual components is very difficult. Phenols, ketones, aldehydes, alcohols, acids, oils, fats, detergents, amines, carbohydrates and proteins are the most important types of pollutant.

Degree of pollution of S.. Inorganic pollutants in S. can be determined by the usual methods of quantitative analysis. For an organically polluted S., the exact determination of content is usually impractical (except for a few special methods, such as chromatography), so the chemical and biological oxygen demands are determined instead. The *Chemical oxygen demand* (abb. COD) (see) is usually determined by the consumption of permanganate or dichromate. The *biological oxygen demand* (abb. BOD) is determined by adding a certain amount of S. to a precisely measured sample of water containing oxygen and adapted bacteria. The sample is sealed against oxygen and incubated at a constant temperature of 20 °C. The bacteria in the sample consume the pollutants, using the dissolved oxygen in the sample. After a certain time, the BOD is determined, in mg/l, from the difference between the original and final oxygen contents. The time of incubation (in days) is indicated by a subscript, e.g. BOD_1, BOD_{24}. As a rule, the BOD_5 is reported (Table 2); the conversion factors are given in Table 3.

Table 2. BOD_5 values for some organic compounds

Substance [1% solution]	BOD_5 [mg oxygen/1]
Acetic acid	700
Methanol	960
Ethanol	1350
Starch	680
Phenol	1700

Another parameter derived from the BOD_5 value is the *inhabitant equivalent*. This is the amount of biologically degradable pollution – measured as BOD – generated daily per inhabitant. One inhabitant equivalent is the amount of pollution equivalent to a BOD_5 of 54 g (Table 4).

The inhabitant equivalent is also used in the evaluation of S. from industrial sources (Table 5).

The *sewage load* is a concept often used in water chemistry as a measure of the degree of pollution of water. This is the number of inhabitants per 1 l/s water flow in a river at average low-water flow rates. The inhabitant equivalents of the industrial S. must also be included. As a rule, the S. load refers to the organic, degradable pollutant content of the water.

Table 3. Conversion of BOD_n to BOD_5 at 20°C

n in days	1	2	3	4	5	10	15	20
BOD_n/BOD_5	0.30	0.54	0.73	0.88	1.00	1.32	1.42	1.46

Table 4. Composition of an inhabitant equivalent

Pollutants	BOD_5 in g
Sedimentable suspended substances	19
Non-sedimentable suspended substances	10
Dissolved substances	25

Table 5. Inhabitant equivalents of some industries

Industry	Reference amount	Inhab. equiv.
Sugar plant	1 t sugar beets	120- 400
Starch plant	1 t corn	800-11000
Cellulose plant	1 t cellulose	4000- 6000
Paper mill	1 t paper	100- 300

S. damage is direct or indirect damage caused by untreated or insufficiently treated S. It has been repeatedly observed that infiltration of S. has deleterious effects on the groundwater, endangering the drinking-water supplies of large cities. Direct S. damage can be, for example, corrosion, bad smells and foam on rivers or lakes. Water pollution endangers human health and causes silting and loss of fish. Uncontrolled dumping of S. can lead to loss of fish catches in several ways: 1) the S. may contain toxic substances which poison the flora and fauna of a body of water; 2) the decomposition of organic pollutants consumes the oxygen in the water and leads to anaerobic conditions which the fish cannot tolerate; 3) the fish absorb the pollutants and become inedible. Some common toxins found in S. are cyanides, hydrogen sulfide, phenols, ammonia, detergents, chlorine, radioactive wastes and salts of copper, chromium, cadmium and lead. Oxygen is depleted by proteins, alcohols, sugars, certain phenols, aldehydes, fatty acids, sulfides, sulfites and nitrites.

Water quality standards are used to evaluate the qualitative and quantitative properties of surface waters. Flowing and standing bodies of water are subject to different criteria. Flowing water is classified according to three criteria, with 6 quality classes in each (1 to 6):

1) Organic load and oxygen turnover. This is referred to the critical season for the water, and is affected by a variety of factors (for example, the increase in water temperature in summer leads to lower oxygen solubility; cf. Table 6). The saprobe system (classification of the microorganisms and small organisms) is used in determining the organic load (Table 7). Bacteriological studies can be used to complement this system.

Table 6. Solubility of oxygen in water as a function of temperature

Temperature	Solubility of O_2 [mg/1]
0	14.2
5	12.4
10	10.9
15	9.8
20	8.8
25	8.1
30	7.5

2) Salt load; classification by ecological criteria and economically sound uses (Table 8).

3) Other area-specific criteria; classification according to important contents, such as cyanide, phenols, nitrate.

This classification scheme gives a true picture of water quality. The assignment of the classes to applications (Tables 7 and 8, p. 976) can be done in bookkeeping form.

Standing water is classified according to three complexes of characteristics with 3 groups of characteristics each, and 5 quality classes:

1) Hydrographic and territorial characteristics, such as morphometry, hydrographic relationship between the drainage area and the water, and load.

2) Trophic characteristics, such as oxygen, nutrient and bioproduct relationships.

3) Salinity and other hygienically relevant characteristics, such as contents of ions, iron, manganese, phenols and detergents.

The assignment of waters to various applications is based on the summarized evaluation over the individual groups of characteristics.

Sewage damage: see Sewage.

Sewage load: see Sewage.

Sewage treatment: measures to improve and change the physical, chemical, and biological quality of sewage, to recover valuable components, and to make the purified water reusable. The recovery of valuable components is becoming more and more important as natural resources become scarcer (Table).

Methods. Sewage can be purified by physical, chemical or biochemical methods. Sieving, sedimentation, filtration and Flotation (see) are all physical methods. Others used for certain industrial sewage include evaporation, burning of the contents, freezing out, extraction, electrolysis, electroosmosis, electrodialysis and electrophoresis. However, chemical and biochemical processes are more commonly used, such as precipitation, flocculation, neutralization, purification by microorganisms, saponification and fermentation.

1) Purification of sewage with mainly inorganic components. This group includes the wastes from metal pickling plants, galvanization plants and radioactive wastes. They are generally treated by chemical processes.

a) Pickling plant sewage. Pickling is often used to improve the surfaces of metals. The oxygen-containing surface layers which would interfere with further processing are removed by treatment with strong acid, such as sulfuric, hydrochloric, nitric or hydrofluoric acid. The surface is then rinsed with water. When sulfuric acid is used on iron, for example, the wastewater contains about 100 g free sulfuric acid and 700 g iron(II) sulfate, $FeSO_4$, per liter. Many processes have been developed to treat such wastes by destroying the liquors or by recovering their contents,

Table 7. Classification of flowing water according to organic load and oxygen content

Class		1	2	3	4	5	6
Degree of saprobity		Oligosaprobic	β-mesosap.	α-mesosap.	polysaprobic	hypersap.	abiotic
Quality criteria							Poisoned waters with no biological capacity. Characterized by abiotic conditions, no capacity for self-purification, and toxicity.
Saprobe index S		$\leqq 1.75$	$\leqq 2.5$	$\leqq 3.25$	$\leqq 4.0$		
O_2 conc.	mg/l	$\geqq 7$	$\geqq 6$	$\geqq 4$	$\geqq 2$	< 2	
O_2 deficit	% air saturation	$\leqq 25$	$\leqq 40$	$\leqq 55$	$\leqq 75$	> 75	
BOD_2 (preferred for lower loads	% of O_2 concentration or mg O_2/l	$\leqq 40$ $\leqq 2$	$\leqq 65$ $\leqq 5$	$\leqq 90$ $\leqq 10$	$\geqq 90$ $\geqq 10$	> 90 > 10	
BOD_2 (preferred for heavy loads	mg O_2/l	$\leqq 4$	$\leqq 10$	$\leqq 20$	$\leqq 40$	> 40	
COD_{Mn} or	mg O_2/l	$\leqq 5$	$\leqq 10$	$\leqq 30$	$\leqq 50$	> 50	
COD_{Cr}	mg O_2/l	$\leqq 8$	$\leqq 25$	$\leqq 80$	$\leqq 120$	> 120	
NH_4^+	mg/l	$\leqq 0.5$	$\leqq 2$	$\leqq 4$	$\leqq 10$	> 10	
Dissolved sulfides and H_2S	mg/l				$\leqq 0.1$	> 0.1	
Supplemental criteria							
Total bacterial count	per ml	$\leqq 10^6$	$\leqq 3 \cdot 10^6$	$\leqq 10^7$	$\leqq 5 \cdot 10^7$	$> 5 \cdot 10^7$	
Colony number	per ml	$\leqq 10^3$	$\leqq 10^4$	$\leqq 10^5$	$\leqq 10^6$	$> 10^6$	
Coliform[1]	per ml	$\leqq 10^2$	$\leqq 5 \cdot 10^2$	$\leqq 5 \cdot 10^3$	$\leqq 5 \cdot 10^4$	$> 5 \cdot 10^4$	
Uses:		Treatment required					Unusable for any purpose except shipping; dangerous to the environment because of toxicity and corrosiveness.
Drinking water		Simple to normal[2]	Extensive[3]	Complicated[4]	Unusable	Unusable	
Swimming		No objections	Usable	Not advisable	Unusable	Unusable	
Fisheries		No limitation for salmonids	Conditional for salmonids, unlimited for cyprinids	Only conditional for cyprinids	Unusable	Unusable	
Industrial		No treatment or simple	Normal[2]	Extensive[3]	Complicated[4]	Unusable	
Cooling		No objections	Usable	Usable	Conditionally usable	Conditionally usable or unusable	
Irrigation Pollutants mainly organic		No objection	No objection	Usable	Conditionally usable	Conditionally usable or unusable	
Mainly hygienic load (fecal contaminants)		No objection	Usable	Conditionally usable	With limitations conditionally usable	Unusable	

[1] Colifirm bacteria (especially *Escherichia coli* are found in the intestines of human beings and many other vertebrates. They normally do not cause disease, but their presence indicates a high probability that pathogens of fecal origin are also present (typhus, paratyphus, dysentery). See Colititer.
[2] Simple to normal treatment required for surface water.
[3] Extensive technology is required for treatment.
[4] Complicated technology is required for treatment, and the treatment is at times very difficult.

the acids and metal salts. Neutralization of the liquor with ammonium hydroxide, lime, sodium hydroxide or sodium carbonate produces ammonium, calcium or sodium salts and iron(II) hydroxide; the latter is then oxidized to iron(III) oxide hydrate: $2\ FeSO_4 + 4\ NH_4OH \rightarrow 2\ Fe(OH)_2 + 2\ (NH_4)_2SO_4$; $2\ Fe(OH)_2 + 1/2O_2 + H_2O \rightarrow Fe_2O_3 \cdot 3H_2O$. Addition of sulfuric acid to iron(II) sulfate-containing liquors and crystallization at a temperature between 65 and 100 °C produces the monohydrate, $FeSO_4 \cdot H_2O$. If the liquor is cooled without addition of sulfuric acid, the heptahydrate, $FeSO_4 \cdot 7H_2O$, precipitates. After removal of most of the $FeSO_4$, the sulfuric acid solution is enriched with sulfuric acid (depending on the pickling method, 5 to 20% H_2SO_4 is used) and reused.

b) Wastewater from galvanic plants. Such wastes may contain cyanide, either as a simple salt or as part of a complex, and metals such as chromium in acidic solutions. The cyanide-containing wastes, such as potassium tetracyanozincate, $K_2[Zn(CN)_4]$, solutions, can contain as much as 100 mg CN per liter. They are detoxified by oxidation with potassium hypochlorite; the less poisonous cyanates are formed as intermediates: $K_2[Zn(CN)_4] + 4\ KOCl \rightarrow ZnCl_2 + 2\ KCl + 4\ KCNO$; $4\ KCNO + 2\ H_2SO_4 \rightarrow 2\ K_2SO_4 + 4\ HCNO$; $4\ HCNO + 4\ H_2O \rightarrow 4\ CO_2 + 4\ NH_3$. In

Table 8. Classification of flowing water according to salinity

Class			1	2	3	4	5	6
Criterion of water quality:								
Calcium	Ca^{2+}	mg/l	≦ 60	≦ 100	≦ 150	250	350	> 350
Magnesium	Mg^{2+}	mg/l	≦ 25	≦ 50	≦ 100	≦ 150	≦ 300	> 300
Sodium	Na^+	mg/l	≦ 30	≦ 70	≦ 150	≦ 500	≦ 500	> 500
Chloride	Cl^-	mg/l	≦ 50	≦ 100	≦ 250	≦ 500	≦ 1000	> 1000
Sulfate	SO_4^{2-}	mg/l	≦ 100	≦ 150	≦ 350	≦ 500	≦ 1000	> 1000
Total hardness as	CaO	mg/l	≦ 100	≦ 150	≦ 300	≦ 500	≦ 700	> 700
Carbonate hardness as	CaO	mg/l	≦ 70	≦ 120	≦ 250	not considered		
Total salt content		mg/l	≦ 350	≦ 750	≦ 1500	≦ 2500	≦ 4000	> 4000
Uses:								
Drinking water			As Table 7					
Swimming			As Table 7					
Fishing			As Table 7					
Process water			As Table 7					
Cooling water			As Table 7					
Irrigation			good	good	good	conditional	only light soils, conditional	as Table 7

Possibilities for recovery of materials from sewage

Type of sewage	Recoverable materials
Leachates from brown coal mines	Iron oxide hydrate, $Fe_2O_3 \cdot xH_2O$ (for gas purification, pigments)
Sewage containing fatty acids	Feed protein (through biological protein synthesis)
Household sewage	Vitamin B_{12} (from sludge)
Pickling (metal) plant waste	Metal salts, e. g. iron (II) sulfate
Galvanic wastewater	Copper, nickel, chromium, etc. (by ion exchange)
Cellulose production wastewater	Yeast, alcohol, lactic acid, synthetic fibers
Starch plant wastewater	Protein
Coal refinery wastewater	Phenols, ammonia, sulfur
Petroleum refinery waste	Petroleum and products
Sulfate-containing organic wastes	Sulfur
Potash waste	Magnesium oxide, MgO
Organic sewage and sludges	Methane gas for fuel (through anaerobic digestion)

aeration methods, the wastewater is treated with sulfuric acid, which releases free hydrogen cyanide; this is blown into special traps by air passing through the liquor: $K_2[Zn(CN)_4] + 2\ H_2SO_4 \rightarrow K_2SO_4 + ZnSO_4 + 4\ HCN$; $2\ NaCN + H_2SO_4 \rightarrow Na_2SO_4 + 2\ HCN$. Wastewater containing chromium is treated with $FeSO_4$ solutions, generating chromium(III) salts: $2\ CrO_4^{2-} + 6\ Fe^{2+} + 16\ H^+ \rightarrow 6\ Fe^{3+} + 2\ Cr^{3+} + 8\ H_2O$. Ion exchangers (see) can also be used to remove metals from galvanic plants.

c) Purification of water containing radioactive wastes. Radioactive wastewater is generated mainly by nuclear processes in reactors and by research in laboratories. The concentrations of radioactive substances in such waters are relatively low, but their radiation intensity can still be high, well above the limit of 10^{-4} Ci l^{-1} required for drainage into a body of water. The amount of wastewater can be reduced by evaporation, which simplifies handling. Under some conditions, chemical precipitation with silicates, phosphates and oxide hydrates of the alkaline earth metals, aluminum or iron can adsorb the radioactive materials in the resulting sludges. The radioactively enriched sludges are packaged in prescribed containers and buried in specified radioactive waste dumps. Ion exchangers are also used to purify radioactive wastewater.

The reduction in the amount of phosphates in water is becoming more important, since phosphate is normally the limiting factor for eutrophication of surface bodies of water. It is generally removed by precipitation with lime, iron or aluminum salts.

2) Purification of sewage with mainly organic components. This group includes household wastes and the phenol-containing waste waters from the coal-refining industry. Although chemical methods are used here, biochemical processes are the most important. The sewage from gas, coking and coal-tar plants, oil refineries, hydrogenation plants, generators and plastic plants contain large amounts of phenols, and may also contain fatty acids with various chain lengths, ammonia, amines, hydrogen sulfide, cyanides, alcohols, ketones, aldehydes, etc. Phenols are highly toxic to fish and discolor the water. They may produce foam and they give the water an unpleasant odor. Contamination of the ground water by traces of phenol can greatly decrease its potability, especially after chlorination. Purification of phenol-containing sewage can aim either to destroy the phenols or to recover them. However, the latter type of process only partially purifies the sewage. One of the best pre-purification methods is extraction with benzene or a mixture of aliphatic esters, such as butyl acetate (*phenol solvane process*). In these processes, the

saponification of the butyl acetate is not insignificant, so in many plants diisopropyl ether, which cannot be saponified, is used instead. Before extraction, air or carbon dioxide is blown through the sewage to remove sulfides. In the *vaporization process*, the phenol-containing water is brought to a boil and steam is blown through it; monophenols and other volatile substances are carried off and are trapped in sodium hydroxide as phenolate lye. In the *Koppers process* the sewage is freed of carbon dioxide and hydrogen sulfide before vaporization by passing steam through it for a short time; this prevents the consumption of the sodium hydroxide by undesired materials. In this process, the oligophenols and polyphenols remain in the water. Adsorption on activated charcoal is also used to purify phenol-containing water.

Because of their content of ammonia and ammonium salts, phenol-containing sewage, e.g. from gas plants, can be used under some conditions as fertilizers. The highly active multiclone ash from Winkler generators run on brown coal is suitable for adsorbing phenols from sewage. Some other purification possibilities for this type of sewage, which have so far not been widely applied, are catalytic and anodic oxidation, chlorination, ozonization or chloride dioxide processes, and burning of the sewage.

Since the phenols are not completely removed by a number of methods, the water must usually undergo another step of treatment, biological treatment.

3) Purification of sewage with mixed inorganic and organic components. This groups includes city sewage, in which both household and industrial wastes are combined. Primary sewage treatment consists of physical and chemical treatments, and is normally followed by a biological secondary treatment. Tertiary treatment is sometimes also needed to remove phosphorus and nitrogen compounds.

The sewage entering a treatment plant is first passed through grates to remove large solid objects, and then through a grit channel to remove sand and other large particles. If necessary, the water then passes through a skimmer (Fig. 1). The water leaving this stage is treated with sodium hydroxide or sulfuric acid to neutralize it to a pH of about 7. This initiates precipitation of heavy metal ions, phosphates and arsenates; the process is completed in a settling tank by addition of precipitating and flocculating agents, usually iron(II) sulfate, aluminum sulfate and lime. Flocculation and precipitation can also be carried out in separate tanks, in which case calcium hydroxide, $Ca(OH)_2$, is used to precipitate phosphates, arsenates, fluorides, fatty acids and transition metal ions.

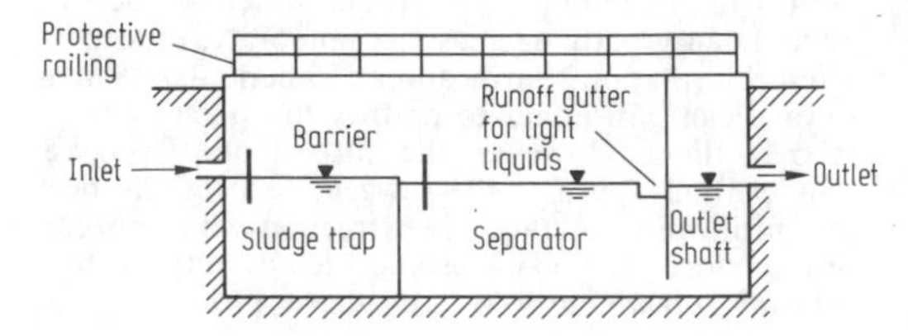

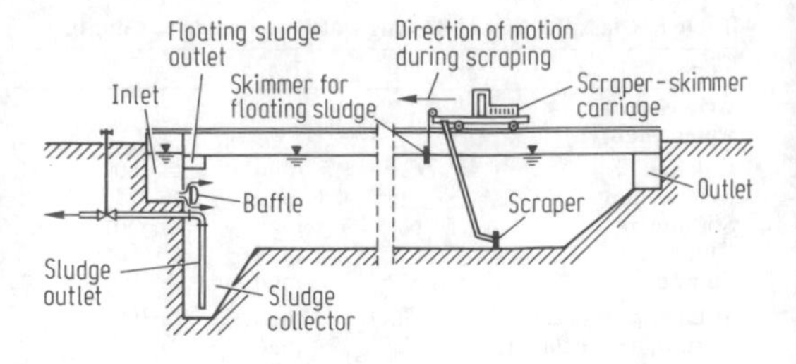

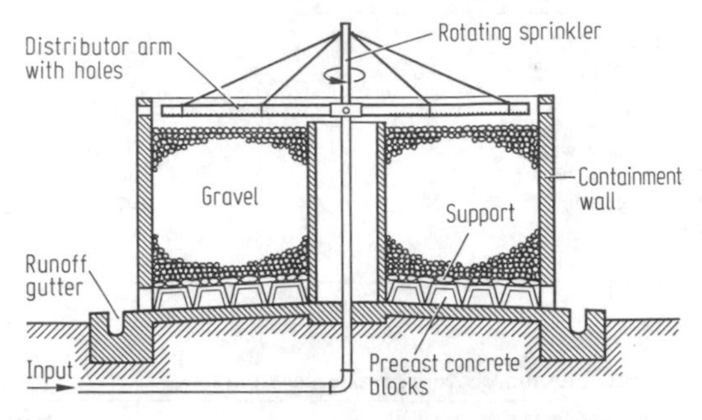

Fig. 1. Rectangular, open skimmer for high flow volumes.

In the next step, the solids formed by precipitation or flocculation are allowed to settle out. Then the cleared water passes through a biological purification stage, in which aerobic bacteria decompose the organic impurities still present in the water. This is done in sprinkling towers, trickle beds or aerated tanks. In the *trickle bed* method (Fig. 3), well aerated round tanks are filled with slag, gravel or other porous materials. The water is sprinkled onto the top of the filler by a rotating sprinkler. A mat of microorganisms forms on the surfaces of the filler and degrades the organic components of the sewage. The tank is usually 3 to 4.5 m high, although towers as high as 20 m have been built. In the *activated sludge* method, the water is added to a tank containing gelatinous sludge particles. The sludge consists of heterotrophic microorganisms and bacteria, colloidal substances and suspenders, and is thus a type of "living flocculant". The microorganisms degrade the organic substances in the water. In plants using either trickle beds or activated sludge, the water must pass through a secondary clarification basin, in which the microorganisms are removed from the water; some are returned to the process, and the rest can be used as fertilizer.

The sewage components to be removed by the microorganisms in either of the above two methods are either adsorbed or absorbed (consumed) by the microorganisms. In all biological treatment methods, the supply of oxygen must be adequate to maintain the microorganisms' aerobic metabolism.

Treated sewage at this stage is already very clean, but may still contain enough phosphorus and nitrogen compounds to cause eutrophication of the surface water into which it is discharged. On the other hand, organic industrial sewage may contain very little phosphorus and nitrogen compounds, and the efficiency of its purification by biological methods can be increased by adding phosphates or ammonium salts. A good example of this type of industrial wastewater is that from cellulose factories. The amount of sulfite liquor from the wood treatment stage of such a factory is 10 m^3 per ton of cellulose. In addition, the factory produces about 1000 m^3 wash water per ton bleached cellulose. In the 10 to 12% sulfite liquor from processing spruce wood, the dry matter is about 20% inorganic (Ca^{2+} and H_2SO_3) and about 80% organic; of the organic matter, about 70% is lignin sulfonate, 22 to 24% is various types of wood sugars, and 5 to 6% is resins. The sugars can be fermented to alcohol by yeast, or the entire amount of carbohydrate can be converted to a protein feed by special types of yeast or mycellar fungi if phosphate and ammonium salts are added. An economical method for large-scale recovery of the calcium hydrogensulfite liquor has not yet been developed. The fibers in the wastewater are retained in sedimentation tanks and recovered by filtration. The lignin substances cannot be degraded by aerobic organisms, but they can be used as raw materials for certain types of plastic. The concentrated sulfite liquors are used, for example, as glues, dust binders for streets, binders for coal briquets and in tanning.

Experience has shown that the most economical treatment of highly contaminated industrial wastewater can be done directly on the site. When such sewage is mixed with household sewage, it nearly always creates problems which require more complicated technological solutions.

The treated sewage can be sprinkled onto large surfaces or drained into city sewers.

Sludge can be digested in closed tanks at temperatures between 30 and 35 °C, at slightly alkaline pH. The anaerobic bacteria produce large amounts of methane gas. Septic tanks for household sewage operate on this principle.

Stages in sewage treatment

Stage	Apparatus	Average time (min)	Comments
Screening of large objects	Coarse and fine screens	1	Required for any treatment plant
Removal of sand and stones	Grit trap	5	Required for any treatment plant
Skimming of substances and liquids which are lighter than water	Skimmer	5- 10	Only some types of sewages
Adjustment of pH	Neutralization tank	5	Only some types (e. g. industrial)
Precipitation of harmful ions, colloids	Precipitation/ flocculation basin	10- 20	Only as needed
Retention of precipitates	Sedimentation I (primary clarification)	60-120	Required for all types
Biological degradation of organic substances	Sprinkling tower, activated sludge, trickle beds	60-120	Necessity; amount of waste depends on matter in water and type of disposal of treated sewage
Retention of precipitates created by biological and chemical treatments	Sedimentation II	60-120	Needed only if biological secondary treatment is used
Elimination of phosphorus and nitrogen compounds (tertiary treatment)	Mixing and flocculation, sedimentation	10- 20	Necessity depends on condition of water into which sewage is discharged
Drainage of water	Drainage pits	60-120	Water can also be reused or further used, e. g. for irrigation, cooling, depending on circumstances

Sex hormones: a group of steroid hormones formed in greatest quantity in the gonads (testes and ovaries). The group includes derivatives of the naturally occurring compounds which have similar effects. The ***male S.*** are called Androgens (see); the ***female S.*** are called Estrogens (see) and Gestagens (see). Female S. are used as Contraceptives (see). Some partly synthetic S. can act as Antiandrogens (see). S. from the other sex are used to treat S.-dependent tumors.

Shaking funnel: same as Separatory funnel (see).

Shale black, a clay, $AlH(SiO_3)$, which contains up to 30% carbon. S. is mined, ground, dried and used as a pigment for coloring cement and artificial stone, and also as a pigment in oil or water-based paints.

Shale oil sulfonates: water-soluble, black-brown products with a high sulfur content and a syrupy consistency used in dermatology because of their antiseptic and antipyretic properties. They are made from shale oils, which are obtained by heating oil shales to 400 °C, followed by distillation. The distilled oil is then sulfonated and converted to the ammonium salts. The best known product is *ichthammol (Ichthyol®)*.

Shell: see Atom, models of.

Shellac: a natural resin, and not a uniform substance. It is a polyester of various alcohols with hydroxycarboxylic acids, such as aleuritic acid (trihydroxypalmitic acid). S. also contains 4 to 10% wax and colored components. It is readily soluble in alcohol, but is only slightly soluble in ether, benzene or light petroleum, and is insoluble in water.

S. is obtained from various trees in India, Thailand and other East Asian countries. It is generated by the feeding puncture made by the female of the shellac insect, *Tachardia lacca*, in the twigs. The sap is partially converted to resin by the wounds, and forms a crust 3 to 8 mm thick. The crusts, called ***stick lac***, are scratched off and sold in granulated form as ***granulated shellac***. This product is colored red by the insects within it, and is therefore called ***ruby lacquer***. The lighter types of S. are obtained by washing the crude product with water and melting it. It is sold in the form of thin plates or leaves. The lightest and for this reason most expensive S. is ***lemon shellac***.

S. can be purified by selective extraction with organic solvents. It can also be dissolved in weak bases, such as soda, and then bleached with chlorine.

Because of its ability to take a polish, S. is the most valuable starting material for production of spirit lacquers. It is also used in the production of phonograph records, to seal porous earthenware objects and in a number of types of lacquer. It can be saponified with borax to give aqueous borax solutions which are used as polishes and sizing for cloth.

Shellac wax: an animal Wax (see). The crude wax is dark brown, but can be bleached to a light yellow. It contains up to 50% free alcohols, mainly myricyl alcohol, and free acids, such as melissic acid, as well as esters of various alcohols with oleic and palmitic acids. S. makes up 4 to 6% of crude shellac, and is isolated during refining of shellac by heat treatment with alcohol or an alkaline solution. Because it is hard, S. is used as a substitute for carnauba wax, e.g. in floor waxes and shoe polish.

Shell model: same as Central field model (see).

Shift polarization: see Polarization.

Shift reagents: see NMR spectroscopy.

Shikimic acid, ***3,4,5-trihydroxycyclohex-1-enecarboxylic acid***: an intermediate in the biosynthesis of phenylpropane derivatives, the amino acids L-phenylalanine, L-tyrosine and L-tryptophan, *p*-aminobenzoic acid and many secondary natural products from plants. It is biosynthesized from phosphoenolpyruvate and erythrose 4-phosphate, both products of carbohydrate metabolism, via 5-dehydroquinoic acid. This pathway for biosynthesis of aromatic compounds is called the *shikimic acid pathway*.

Shoe polish: see Leather care products.

Shore hardness: see Hardness.

Short-range order: see States of matter.

Shot lead: an alloy of lead, antimony and arsenic (4% antimony, 1.5% arsenic, the remainder lead) used to make shot (for shotguns).

Showdomycin: 3-β-D-ribofuranosylmaleimide, a nucleoside antibiotic isolated from culture filtrates of *Streptomyces showdoensis*. The structure of S. is similar to that of uridine, but it is a C-glycoside. Its antibiotic effect is due to the alkylating effect of the maleimide part on SH groups in proteins.

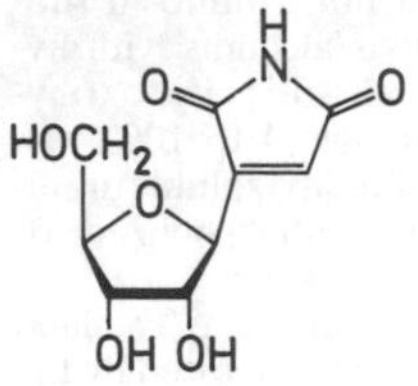

In bacteria, S. inhibits the transport of carbohydrates and amino acids through the cell membrane.

Shrinkage cavity: a cavity in a metal casting caused by rapid solidification of the outer layer of the piece and shrinkage of the more slowly cooling liquid metal in the interior.

Si: 1) symbol for silicon; 2) abb. for silicones (see Polysiloxanes).

Sia: abb. for sialic acid.

Sialic acids, ***N-acylneuraminic acids***, abb. ***Sia***: colorless crystalline substances which are readily soluble in water, but barely soluble in most alcohols. They rotate the plane of polarized light to the left. S. form violet pigments when heated with hydrochloric acid and resorcinol or orcinol in the presence of Fe^{3+} ions. This bial reaction can be used in quantitative analysis.

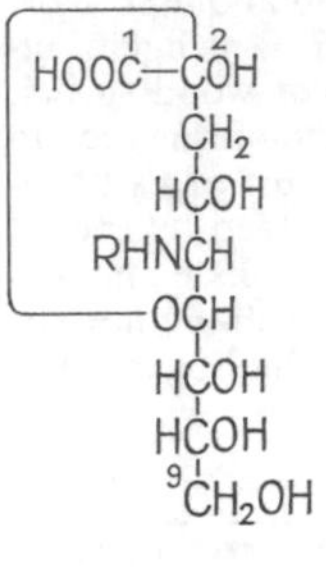

Neu: R = H

NeuAc: R = CH_3CO

NeuGc: R = $HOCH_2CO$

S. are components of glycolipids and glycoproteins in animals, but are rare in other life forms. They are found at the termini of the carbohydrate chains, from which they are removed by neuraminidases. The most common S. is N-acetylneuraminic acid (NeuAc); the hydroxyl groups in the 4, 8 and 9 positions can also be acetylated. Another major S. is N-glycoloylneuraminic acid (NeuGc), which is found, e.g. in extraneural organs of animals. The terminal S. of glycolipids and glycoproteins are thought to have major biological roles. Glycoproteins rich in S. are important mucous substances. The S. residues on the outer cell membrane are involved in essential membrane functions, such as cell-cell, cell-virus and cell-effector interactions (see Gangliosides).

Siccative: a substance which shortens the drying time of oil-based paints by accelerating the polymerization of the oils induced by atmospheric oxygen. S. are metal soaps which become evenly distributed in the varnish or lacquer substance and have a strong catalytic effect. Cobalt, manganese and lead soaps (see Metal soaps) are most commonly used. They are either added to the drying and semidrying oils, or are formed by boiling the varnish after addition of the corresponding metal oxides or salts.

Side-by-side model: see Nucleic acids.

Sident®: a silver alloy containing neither gold nor platinum, used for dental work. S. contains about 87% silver with additions of cadmium, copper, zinc and tin.

Siderin yellow: see Iron(III) chromate.

Sifting, ***air sifting***: a process for separating particles of equal density according to their grain size, or particles of different density according to their density. The separation makes use of gravitational or centrifugal force.

a) In ***gravity sifters***, the material falls freely through a sifting space and the finer particles are carried away from the vertical by a horizontal flow of air. In some cases, the material is forced through a sieve before it falls.

b) In *centrifugal sifters*, the material is sorted by application of centrifugal force.

S. processes are used in the preparation of chemicals, cement and ores. They are ordinarily used when the material is so finely powdered that sieving cannot produce the desired results. The grain sizes for S. are between about 2 μm and a few mm; the low-grain-size range can be handled only by specially constructed centrifugal sifters.

Siemens-Martin process: see Steel.

Siemens ozonizer: see Ozone.

Sigmatropic reaction: see Woodward-Hoffman rules.

SIH: see Statins.

Silanes, ***silicon hydrides***: binary silicon-hydrogen compounds with the general formula Si_nH_{2n+2}. The simplest member of this series, monosilane, SiH_4, is obtained by reduction of silicon tetrachloride with lithium alanate: $SiCl_4 + LiAlH_4 \rightarrow SiH_4 + LiAlCl_4$. Disilane, Si_2H_6, can be made similarly from hexachlorodisilane, Si_2Cl_6 and $LiAlH_4$. The reaction of magnesium silicide with acids produces hydrogen and a mixture of S. which contains mono-, di-, tri-, tetra- and hexasilane.

Table. Physical properties of some silanes

		M_r	m. p. [°C]	b. p. [°C]
Monosilane	SiH_4	32.12	–185	–111.8
Disilane	Si_2H_6	62.22	–132.5	–14.5
Trisilane	Si_3H_8	92.32	–117.4	52.9
Tetrasilane	Si_4H_{10}	122.42	–108	84.3

The tendency to form long chains held together by Si-Si bonds is much lower than the tendency to form long carbon chains, and to date the longest Si chains known contain about eight Si atoms. The S. are colorless gases or liquids (table), and differ from the homologous hydrocarbons the alkanes, in being much more reactive. The combustion of alkanes to carbon dioxide and water requires a considerable activation energy, but the S. are pyrophoric and burn explosively in air to form silicon dioxide and water: $SiH_4 + 2\ O_2 \rightarrow SiO_2 + 2\ H_2O$. Because silicon is less electronegative than hydrogen, the silane bond is polarized in the sense $Si^{\delta+}$-$^{H}\delta$-. The hydrogen has hydridic character. Therefore, S. react with water or other acids to generate hydrogen:

$$SiH_4 + 4\ H_2O \xrightarrow{(OH^-)} Si\ (OH)_4 + 4\ H_2.$$

The substitution products of S. are also considered S.; here, for example, one or more of the hydrogen atoms of monosilane are replaced by organic groups (see Organosilicon compounds) or halogens (see Trichlorosilane).

Silanol: a compound with the structure R_3SiOH (see Polysiloxanes).

Silica gels: highly condensed, X-ray amorphous polysilicic acids with a connective structure which includes widely varying amounts of water. The similarly constituted flocculates are also called S., not quite correctly. Industrially, S. or silica flocculates are made by precipitation of silicic acids from water glass solutions using mineral acids, or by hydrolysis of silicon compounds, e.g. silicon tetrachloride, $SiCl_4$ or silicon disulfide, SiS_2. The porous ***silica xerogels*** are obtained by drying; these have surface areas up to 1000 $m^2\ g^{-1}$. They are used as drying agents for many purposes, including industrial gases and in air conditioners; they are also used as carriers for catalysts and as adsorbents in chromatography. Silica xerogel exists in nature as kieselgur (diatomaceous or infusorium earth). The ***blue gel*** used in the laboratory for drying gas is impregnated with cobalt(II) nitrate as a moisture indicator; when the S. is saturated with moisture, it turns pink.

Silica removal: the removal of silicic acid or silicon dioxide from the feedwater for steam generators. Silicic acid or silicon dioxide forms a hard scale with the calcium or magnesium carbonate in the feedwater: $CaCO_3 + SiO_2 \rightarrow CaSiO_3 + CO_2$. Silica can be removed chemically by precipitation or by ion exchangers. Precipitation can reduce the silicon dioxide concentration to 0.5 to 0.3 mg/l, but with strongly basic ion exchangers, it is possible to reach levels $<$0.02 mg/l, which are required for very high pressure steam generators.

Silicate: a salt of silicic acid. S. are the main component of the earth's crust, and there is immense variety in their forms. They can be made synthetically from silicon dioxide, SiO_2, and metal compounds (see Silicate syntheses). All S. are made up of SiO_4 tetrahedrons, usually linked to one another via their vertices (bridge oxygen atoms). The negative charges on oxygen atoms at non-linked vertices are compensated by metal cations. The great stability of the S. is due to the strongly polar (54% ionic) σ-bonds which are fortified by $p_\pi d_\pi$-double-bond character (Si-O-Si bond angle, 140±5°). Crystalline S. are classified according to structure:

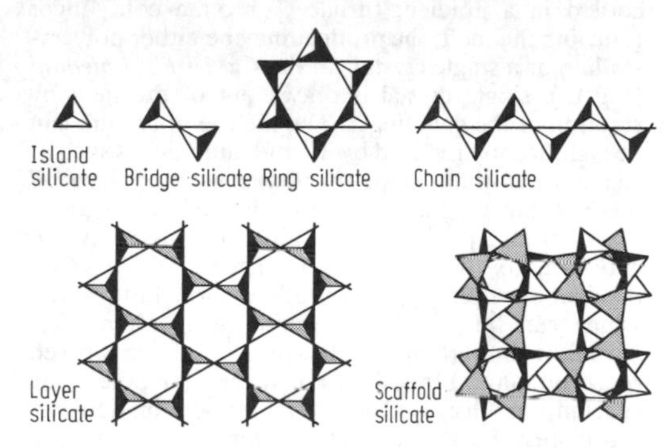

Anion complexes in the silicates. The fourth vertex of each SiO_4 tetrahedron lies above or below the plane of the drawing.

Fig.	Type of silicate	Anion complex	Example
a	Island	$[SiO_4]^{4-}$ Mg_2SiO_4	Forsterite
b	Group	$[Si_2O_7]^{6-}$ $Sc_2Si_2O_7$	Thortveitite
c	Ring n = 3,4,6	$[SiO_3]_n^{2n-}$ $Be_3Al_2Si_6O_{18}$	Beryll
d	Chain	$[SiO_3^{2-}]_x$ $CaSiO_3$	Wollastonite
e	Layer	$[Si_2O_5^{2-}]_{xy}$ $Al_4(OH)_8Si_4O_{10}$	Kaolinite
f	Scaffold	$[AlSi_3O_8^{-}]_{xyz}$ $KAlSi_3O_8$	Orthoclase

There is great structural variety among the S., due to the many possible linkages of the SiO_4 tetrahedrons in the polymeric forms (various identity periods, double rings, chains and layers), and also through substitution of other elements (e.g. aluminum; see Aluminosilicates) for Si. The S. are classified according to their degree of dispersity as coarse (>0.1 μm), fine (<0.1 μm) and molecular (silicate solutions) disperse S.

Silicate analysis: analysis of natural silicate minerals and rocks; some special methods of solubilization and analysis are used in this area.

Silicate concrete: see Binders (building materials).

Silicate pigments, ***water-glass pigments***: suspensions of alkali-stable, inorganic pigments with or without added fillers (e.g. talcum, chalk) in solutions of potassium water glass with a high silicate content. In the suspension, the colored ions of the pigments form water-insoluble colored silicates of varying compositions, with simultaneous precipitation of water-containing silicon dioxide, $SiO_2 \cdot xH_2O$. S. are mixed immediately before use and are applied directly to the base (e.g. metal, glass or wood), which is then etched to achieve very good binding. S. are very resistant to weathering and make wood resistant to burning.

Silicate syntheses: laboratory or industrial methods for synthesis of silicates which either are rare or do not exist in nature, or which are required in very high purity. Four techniques are used. 1) In ***crystallization from the melt*** of oxide components of the desired silicates, the silicate crystals form as the melt is cooled in a gradient furnace, or on a cold finger touching the melt; the product may be either polycrystalline or a single crystal. In the *Czochralski method* (Fig.), a single crystal is drawn out of the melt by rotation of the cold finger. The melt temperature can be significantly reduced by use of halide fluxes such as calcium chloride. 2) In the ***flame melt*** or ***Verneuil method*** (Fig.), a finely divided oxide mixture is transported from a perforated container into an oxyhydrogen flame by the oxygen flow. The silicate reaction product precipitates in crystalline form onto a rod which can be raised or lowered by rotation. This method is important for the synthesis of gems based on corundum. 3) In ***hydrothermal synthesis***, the oxide reactants are dissolved in hot water under high hydrostatic pressure; at a somewhat lower temperature, the

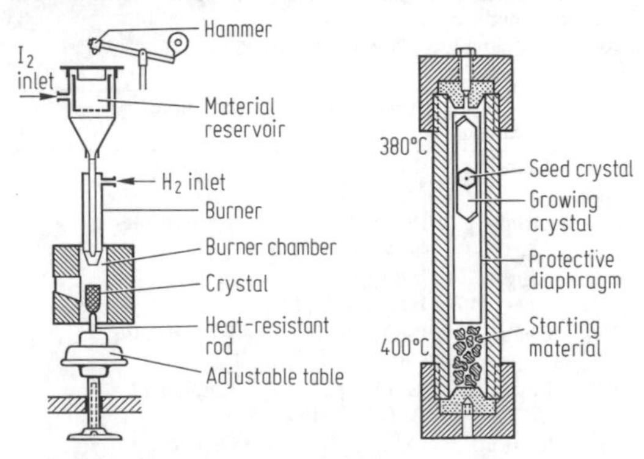

Flame melt method   Hydrothermal synthesis

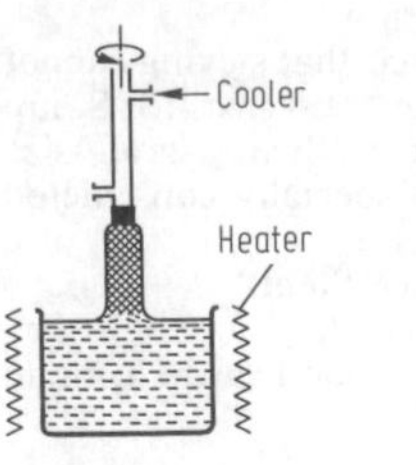

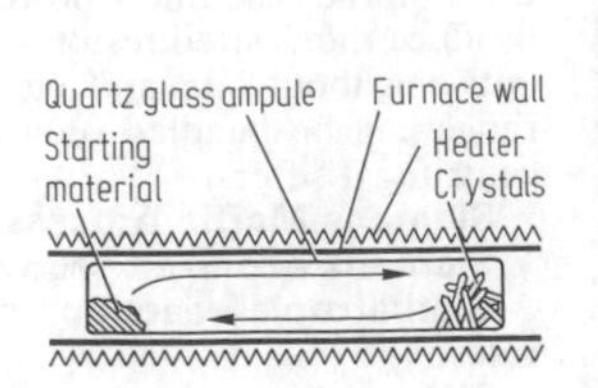

Crystal growing from the melt   Chemical transport

product crystallizes as a single crystal on a seed crystal.

The synthesis of pure quartz (Fig.) starts from impure silicon dioxide, which dissolves at 400 °C and 100 MPa in slightly alkaline water. The solution is transported through a diaphragm, and in a slightly cooler part of the pressure vessel, quartz crystals are formed; these may reach a mass of several kilograms. 4) In ***crystal formation by chemical transport reactions***, the starting material reacts with a transport gas to form a gaseous intermediate. At a site with higher or lower temperature, the reverse reaction occurs with formation of single crystals (see Chemical transport reactions). This method is used in purification of quartz (Fig.); the SiO_2 components of the starting material react with hydrogen at 1400 °C: $SiO_{2(s)} + H_{2(g)} \rightleftharpoons SiO_{(g)} + H_2O_{(g)}$; the back reaction to pure quartz crystals occurs at 1250 °C.

Silicate technology: includes the technologies for production of ceramics, glass, enamel, construction binding materials and silicate adsorbants and for processing silicate waste and byproducts (slag and ashes). The most important steps in the production of industrial silicates are preparation of the raw materials, shaping, temperature treatment and post-heating treatment. Shaping and temperature treatment can occur in the reverse order as well. The nature of the products is determined by the high-temperature process, in which the silicate materials undergo chemical change and acquire their characteristic properties. Subsequent treatment is much more important for the production of glasses than for ceramics. Because the process steps are similar (shaping, firing), the production of non-silicate materials, such as Oxide ceramics (see), Ferrites (see) and Cermets (see) are considered part of S.

Silica xerogels: see Silica gels.

Silicic acid: an oxygen acid of silicon. The simplest S. is ***orthosilicic acid*** (***monosilicic acid***), H_4SiO_4; it forms primarily in the hydrolysis of silicon tetrahalides, e.g. SiF_4 and $SiCl_4$, silicon disulfide, SiS_2, or tetraalkoxysilanes, $Si(OR)_4$. The weak acid (pK_{a1} = 9.8; pK_{a2} = 12.4) is short-lived in a dilute solution at pH 3.2 and rapidly condenses to ***disilicic acid***, $H_6Si_2O_7$, trisilicic acid, $H_8Si_3O_{10}$, ***cyclotrisilicic acid***, $H_6Si_3O_9$, ***cyclotetrasilicic acid***, $H_8Si_4O_{12}$, etc. The two-dimensionally cross-linked ***phyllosilicic acids*** $(H_2Si_2O_5)_{xy}$, which can be produced from layered silicates by acidification, are stable. Condensation of low-molecular weight S. produces colloids (see Silica gels). The Silicates (see), which have structures similar to the S., have been better characterized. There is no monomeric S. corresponding to carbonic acid, H_2SiO_3, because the silicon atom is incapable of forming a true double bond. Compounds with the composition $(H_2SiO_3)_n$ have cyclic or chain molecules.

Silicic acid gel: same as Silica gel (see).

Silicides: binary silicon-metal compounds with variable, often unusual stoichiometric compositions and complicated structures. The S. are obtained by fusion of the elements. Alkali and alkaline earth silicides form silicates and hydrogen when they react with water, and Silanes (see) on reaction with acids, e.g. $Mg_2Si + 2\,H_2SO_4 \rightarrow 2\,MgSO_4 + SiH_4$. The S. of the transition metals have alloy-like properties.

Silicochloroform: same as Trichlorosilane (see).

Silicon, symbol ***Si***: an element from group IVa of the periodic system, the Carbon-silicon group (see), a semimetal, *Z* 14, with natural isotopes with mass numbers 28 (92.21%), 29 (4.70%) and 30 (3.09%), atomic mass 28.0855, valence IV, very rarely II, Mohs hardness 7, density 2.32 to 2.34, m.p. 1410 °C, b.p. 2355 °C.

Properties. The gray, cubic crystals of S. have a metallic luster; their structure is that of diamond. Brown, powdery S. differs only in its degree of dividedness, and sometimes in its impurities. S. is a semiconductor.

Because of its electron configuration and position in the periodic system, S. tends to form four covalent bonds. The resulting tetrahedral configuration of its compounds can be described by sp^3 hybridization of S. The bonds between S. and elements such as oxygen, chlorine and especially fluorine are very stable, and their formation is often the driving force for the reactions of S. and its compounds (e.g. the reaction of silicon dioxide, SiO_2, with hydrofluoric acid, HF). The high bond energy suggests the existence of double bond components, in the sense of a $d_\pi p_\pi$ interaction between the S. and the other element. The availability of energetically favorable d levels also makes possible the expansion of the electron octet to 10 or 12 electrons, and thus the formation of trigonal bipyramidal or octahedral derivatives of five- and six-fold coordinated S. The distinctly electrophilic character of SiX_4 compounds and the high rates of nucleophilic substitutions on S. can also be understood in terms of d-orbital participation.

Finely divided, brown S. is much more reactive than large crystals, due to its large surface area. All the reactions mentioned below require much less activation energy with brown than with gray S. For example, the brown, powdered S. burns in fluorine even at room temperature, forming silicon tetrafluoride, SiF_4. The reaction with the other halogens occurs when the reactants are heated. At very high temperatures, S. burns in air to form silicon dioxide, and around 1400 °C, it forms silicon nitride, Si_3N_4, with nitrogen. S. reacts when hot with many metals, forming Silicides (see). It is insoluble in acids other than hydrofluoric acid, but it dissolves in alkali hydroxide solutions with generation of hydrogen, e.g. $Si + 2\,NaOH + H_2O \rightarrow Na_2SiO_3 + 2\,H_2$. Divalent S. compounds, like the comparable carbon derivatives, are unstable at room temperature. At very high temperatures, compounds such as silicon monoxide, SiO, silicon monosulfide, SiS, or silicon dichloride, $SiCl_2$, can be detected, and in some cases, isolated.

Analysis. S. makes up 27.7% of the earth's crust by mass, and is thus the second most abundant element there (after oxygen). It therefore occurs as widely as all other elements together. It is found in nature as silicon dioxide, and in an extreme variety of Silicates (see) and aluminosilicates.

Production. For industrial production of S., silicon dioxide is reduced in an electric furnace with carbon or calcium carbide. In the laboratory, aluminum or magnesium can be used as reducing agent. When magnesium is used, the S. is obtained in the finely divided brown form. Crude S. to be made into semiconductors is converted either to silicon tetrachloride, $SiCl_4$, or trichlorosilane, $SiHCl_3$. After careful purification by several distillation steps, $SiCl_4$ is again reduced to elemental S. with hydrogen or (this is now more common) the $SiHCl_3$ is decomposed thermally: $SiHCl_3 \rightarrow Si + HCl + Cl_2$. The resulting S. is then purified by zone melting, yielding a product in which the impurity concentration is lower than 10^{-9}%.

Applications. S. is the starting material for synthesis of polysiloxanes, and an important component of alloys (e.g. in alloys of copper, aluminum and steel). It is used in metallurgy as a deoxidizing agent. It is used in very pure form, usually doped, for semiconductors.

Historical. S. was first prepared in 1823 by Berzelius, by reduction of silicon tetrafluoride with potassium.

Silicon bronze: copper-silicon alloy with 1.5 to 3% silicon. They are used for highly stressed, conducting parts of electrical switches.

Silicon carbide: SiC, when pure, forms colorless, hexagonal or cubic crystals; M_r 40.10, density 3.217, subl. and dec. around 2700 °C. Technical S. is known as ***carborundum***, and is usually dark in color, due to the presence of impurities. Its structure is comparable to that of diamond. This explains its hardness (9.6 on the Mohs scale), which is close to that of diamond. S. is unusually inert, thermally and chemically. Even at high temperatures, it is not attacked by oxygen, sulfur or strong acids (even hydrofluoric). In the presence of air, alkali melts convert S. to silicates and carbonates: $SiC + 4\,NaOH + 2\,O_2 \rightarrow Na_2SiO_3 + Na_2CO_3 + 2\,H_2O$. S. is produced from silicon dioxide and carbon in an electric furnace, by the Acheson process (see Graphite). It can also be made directly from the elements at 1500 °C. It is used mainly in abrasives (powders, pastes, sandpaper), in fire-resistant coatings, and as a material for heavily stressed bearings. S. are also used as heating rods for electric resistance furnaces.

Silicon dioxide: SiO_2, M_r 60.08, exists in three modifications at normal pressure. The conversion temperatures are shown in the following diagram:

$$\begin{array}{ccccccc}
\beta\text{-Quartz} & \xrightleftharpoons{870^\circ C} & \beta\text{-Tridymite} & \xrightleftharpoons{1470^\circ C} & \beta\text{-Cristobalite} & \xrightleftharpoons{1732\pm5^\circ C} & \text{Melt} \\
\Big\Updownarrow 573^\circ C & & \Big\Updownarrow 140^\circ C & & \Big\Updownarrow 250^\circ C & & \\
\alpha\text{-Quartz} & & \alpha\text{-Tridymite} & & \alpha\text{-Cristobalite} & &
\end{array}$$

All these crystalline forms contain SiO_4 tetrahedra, with each oxygen atom acting as a bridge between the two tetrahedra to which it belongs; this yields the stoichiometry SiO_2. The tetrahedra are arranged differently in the three forms, so that interconversion requires breaking of stable Si-O bonds. This explains why the interconversions of quartz, tridymite and cristobalite require very high activation energies and occur so slowly that all three forms are found in nature, although only α-quartz is thermodynamically stable at room temperature. The conversion of the high-temperature (β-) to the low temperature (α-) forms, on the other hand, requires only a slight reorientation of the tetrahedra, and occurs rapidly even at low temperatures. In addition to the above normal-pressure modifications, there are two known high-pressure modifications of S., cesite and stishovite, in which each Si atom is surrounded octahedrally by six oxygen atoms.

S. is chemically very inert. It is degraded to SiF_4 and water by hydrofluoric acid, but does not react with other acids or chlorine, bromine or hydrogen. Alkali melts convert S. to silicates, e.g. $SiO_2 + 2\,NaOH \rightarrow Na_2SiO_3 + H_2O$. Reactive metals, such as magnesium or aluminum, reduce S. to silicon at high temperatures.

S. occurs ubiquitously in nature. The most important form is quartz, which occasionally is found in somewhat impure varieties as gems and semiprecious stones. Examples are clear rock crystal, smoky quartz (brown), amethyst (violet), rose quartz (pink), citrine (yellow) and morion (black). Particularly in the form of sand, quartz is widely used to make concrete and glass, silicon and certain silicates, including water glass.

If molten S. is cooled, an amorphous, glass-like product is obtained (quartz glass) which, because of its low thermal expansion coefficient, its transparence to ultraviolet radiation and its chemical inertness, is a valuable material for laboratory apparatus and UV radiation sources. Amorphous forms of S. are also found in nature, e.g. the minerals agate, opal, carneol, onyx, jasper, chrysopas and Kieselguhr (see).

Silicone, ***polysiloxane***: a macromolecular organosilicon compound containing the basic unit -Si(R)$_2$O-. The R groups are most often CH_3. In their structure, the P. resemble natural silicates. The main chain is formed of Si and O atoms, not C atoms. The basic reaction for synthesis of the P. is the acid- and base-catalysed condensation of silanols to siloxanes:

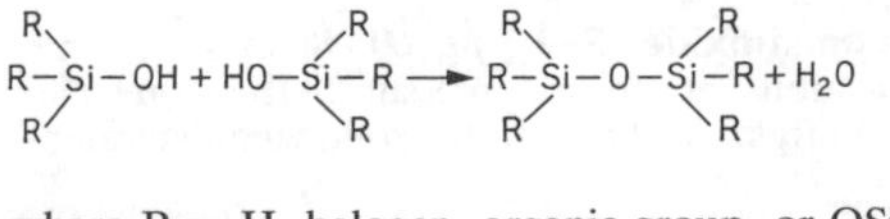

$$R_3Si{-}OH + HO{-}SiR_3 \longrightarrow R_3Si{-}O{-}SiR_3 + H_2O$$

where R = H, halogen, organic group, or $OSiR_3$. If silanediols are used, the condensation leads to chain-like polymers, and if silanetriols are used, the formation of two- or three-dimensional networks is possible.

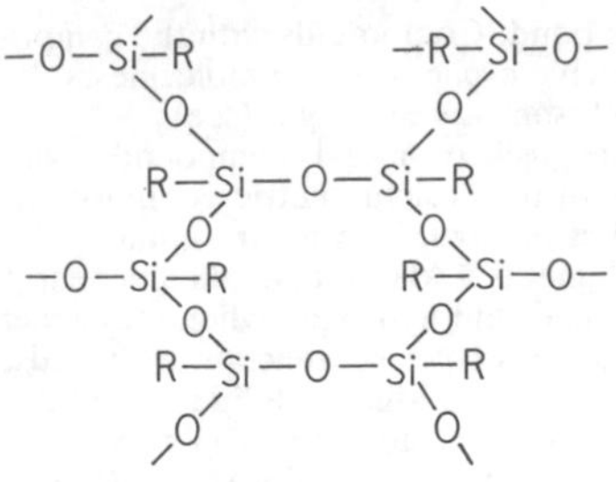

Cross-linked polysiloxanes.

The silanols are obtained by hydrolysis of the corresponding alkylcholorosilanes, and are usually polymerized without isolation, e.g.

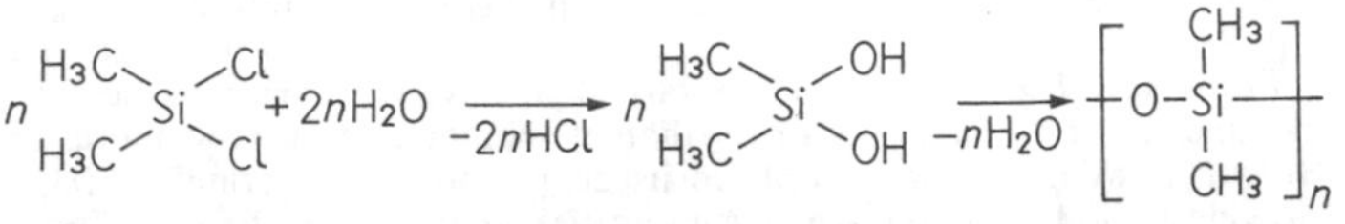

$$n\,(H_3C)_2SiCl_2 + 2n\,H_2O \xrightarrow{-2n\,HCl} n\,(H_3C)_2Si(OH)_2 \xrightarrow{-n\,H_2O} \left[{-}O{-}Si(CH_3)_2{-}\right]_n$$

Depending on the hydrolysis conditions, the product mixture may contain a certain fraction of cyclic polysiloxanes in addition to the long-chain compounds; chiefly $[(CH_3)_2SiO]_3$ and octamethylcyclotetrasiloxane $[(CH_3)_2SiO]_4$. Trimethylchlorosilane, $(CH_3)_3SiCl$, causes chain termination if present in the chlorosilane mixture. The concentration ratio of dimethyldichlorosilane/trimethylchlorosilane determines the average length of the polysiloxane chains and thus the properties of the polymers. If the chlorosilane mixture includes methyltrichlorosilane, CH_3SiCl_3 or even silicon tetrachloride, $SiCl_4$, cross-linked P. are formed. If other chlorosilanes, e.g. phenyl- or H-chlorosilane, are included in the mixture with the methylchlorosilanes, their incorporation into the polymer greatly modifies its properties.

Production and work-up. Monomeric methylchlorosilane is produced industrially by a direct synthesis, the *Müller-Rochow process* in which methylchloride reacts with silicon in the presence of a copper catalyst at 250-300 °C. The product mixture depends on the exact working conditions; it is separated by a painstaking fractionation necessitated by the very similar boiling points of the components. On the average, the mixture contains 80% dimethyldichlorosilane, $(CH_3)_2SiCl_2$ (b.p. 70.2 °C); 8% methyltrichlorosilane, CH_3SiCl_3 (b.p. 66.1 °C); 3% trimethylchlorosilane, $(CH_3)_3SiCl$ (b.p. 57.6 °C); 3% methyldichlorosilane, CH_3SiHCl_2 (b.p. 40.4 °C); 1% silicon tetrachloride, $SiCl_4$ (b.p. 57.6 °C) and about 5% assorted disilanes. For the subsequent hydrolysis, pure dimethyldichlorosilane is usually used, and a polycondensate containing OH endgroups is obtained. This is often done in a heterogeneous phase (e.g. water/toluene), and the polysiloxane dissolves in the organic phase. This is separated, the solvent is removed, and the crude product is either directly processed further to silicone oil or rubber, or the cyclopolysiloxanes are first removed by distillation and purified. The hydrochloric acid produced in the hydrolysis is used to make methyl chloride from methanol, and is thus returned to the process.

Depending on the starting materials and intermediates, and on the type of further processing, the products are liquid (oily), solid (resin-like) or elastic (rubbery) P. All have in common their water-repelling behavior, their resistance to many chemicals, their good electrical insulating qualities, and the constancy of their properties over wide temperature ranges. The temperature resistance of the P. decreases as the length of their organic chains increases. Their relatively great resistance to acids, alkali hydroxides, bromine and other caustic substances is due to the sheathing of the siloxane chains by the organic substituents. Since the substituents in most cases are alkane-like groups, the sheaths they form around the chains also explains the water-repelling behavior of the P.

Silicone oils (M_r 1000 to 15,000) are clear, colorless and odorless liquids which are soluble in benzene, gasoline, carbon tetrachloride and other organic solvents, and in pure form are stable in the range between about -70 and +250 °C. They are chemically inert, neutral and water-repellent. The methyl and methylphenyl silicone oils are of interest. The viscosity of *methyl silicone oils* is very little affected by temperature, and they have very good compressibility. For this reason they are good brake and hydraulic liquids. In the plastic and rubber industry they are used in solutions and aqueous emulsions as separation and unmolding media. They are also added to polishes, paints and varnishes, and are used in medicine, pharmacy and cosmetics. They are used as impregnating agents for textiles, leather and paper. The impregnation of textiles is usually accompanied by an anti-wrinkle treatment.

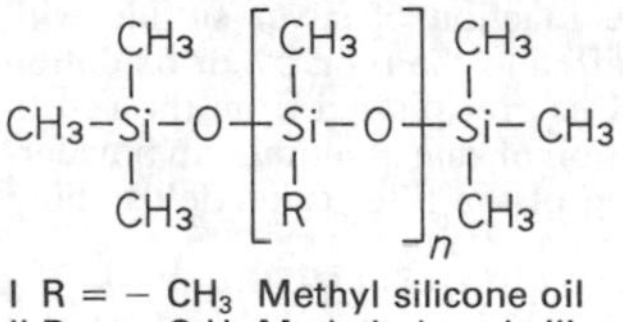

$$CH_3-\underset{CH_3}{\overset{CH_3}{Si}}-O-\left[\underset{R}{\overset{CH_3}{Si}}-O\right]_n-\underset{CH_3}{\overset{CH_3}{Si}}-CH_3$$

I R = – CH_3 Methyl silicone oil
II R = – C_6H_5 Methyl phenyl silicone oil

Methylphenyl silicone oils provide high- and low-temperature lubricants and, after filling with lithium stearate, a basis for silicone grease. Silicone oils are used in transformers and switches, as the working fluid in diffusion pumps, as antifoam and flotation agents, as a dirt-repelling additive to polishes and window-cleaning fluids, and as a heat-transfer liquid. ***Silicone greases*** are obtained by thickening the oils with components such as finely powdered silicate, talcum, graphite or metal soaps; these are used mainly as special sealers and lubricants resistant to high temperatures. Their lubricating qualities can be further improved by addition of molybdenum(IV) sulfide. ***Silicone resins*** are solidified by baking at about 250 °C. Because of their good insulating properties, they are used for encapsulation of electrical and electronic components. Some products, such as vinyltrichlorosilicone and vinyltriethoxysilicone, are used in production of fiber-glass reinforced plastics, as binders between the glass fibers and the plastics. Because of their structure, they are chemically related to both the glass and the plastic, and can form tight bonds with each. Precursors of the silicone resins, the alkali metal alkylsiliconates, are used to protect buildings. Concentrated solutions of these resins are used to impregnate bricks, concrete and mortar to prevent penetration of water, "blooms" of salts on masonry, and also frost damage. The ***silicone resin bake-on paints*** (solutions of silicone resins in toluene, xylene and other solvents) are used as insulating coatings (e.g. in electric motors, transformers and the ceramic industry), as binders for pigments in paints, and to produce heat-resistant insulation with glass fibers. Vulcanization of linear dimethylpolysiloxanes with chain lengths of 6000 to 7000 siloxane units (M_r up to 500,000) produces ***silicone rubber*** (see Rubber).

Covering a surface with a thin film of silicone is called ***siliconization***

Silicone fats: see Silicone.

Silicone oils: see Silicone.

Silicone resin bake-on enamels: see Silicone.

Silicone resin lacquers: see Lacquers.

Silicone rubber: see Rubber.

Siliconization: see Silicone.

Silicon monoxide: SiO, cubic or amorphous, glassy or fibrous, and usually black compound, M_r 44.09, density 2.13, m.p. > 1707 °C, b.p. 1880 °C. S. is oxidized in air even at room temperature to silicon dioxide, SiO_2, and in finely divided form it is pyrophoric. It is made by reduction of SiO_2 with silicon. It is used mainly as a vapor for vapor deposited coatings on optical glasses, for surface treatment of ceramics and ornaments, etc.

Silicon nitride: Si_3N_4, gray amorphic powder; M_r 140.28, density 3.44, m.p. (under pressure) 1900 °C, Mohs hardness 9. It does not react with acids, except for hydrofluoric acid. Alkaline solutions and melts convert it to silicates and ammonia. S. is made by heating silicon to 1250 to 1550 °C in a nitrogen atmosphere. Powdered S. can be sintered at high temperatures to parts which are highly resistant to thermal shock and chemicals. These are used as components for turbines, motors, laboratory equipment, insulation, etc.

Silicon oxide: see Silicon monoxide, see Silicon dioxide.

Silicon tetrachloride, ***tetrachlorosilane***: $SiCl_4$, colorless, mobile liquid which fumes in moist air and has a suffocating odor; M_r 169.90, density 1.483, m.p. -70 °C, b.p. 57.57 °C. The chlorine atoms of S. can be easily exchanged for many other nucleophiles. For example, water decomposes S. in a highly exothermal reaction to hydrochloric and orthosilicic acids, which condenses to SiO_2 via intermediate compounds: $SiCl_4 + 4\,H_2O \rightarrow Si(OH)_4 + 4\,HCl$. Alcohols react similarly, forming tetraalkoxysilates (silicic acid tetraalkyl esters) $Si(OR)_4$. S. reacts with Grignard reagents to form alkyl- and arylchlorosilanes: $SiCl_4 + nRMgCl \rightarrow R_nSiCl_{4-n} + n\,MgCl_2$. Silicon reacts with S. at high temperatures to form hexachlorodisilane and higher chlorosilanes. S. is made by heating silicon or a mixture of silicon dioxide and carbon in a stream of chlorine. It is an important synthetic reagent in organosilicon chemistry. It is used in industry to make pure silicon and silicon dioxide (kiesel gels), and certain silicones.

Silicon tetrafluoride, ***tetrafluorosilane***: SiF_4, a colorless, poisonous gas which fumes in air and has a suffocating odor; M_r 104.08, m.p. -90.2 °C, b.p. -86 °C. S. is decomposed to orthosilicic acid and hydrofluoric acid by water: $SiF_4 + 4 H_2O \rightarrow Si(OH)_4 + 4 HF$. The latter then reacts with unreacted S. to form hexafluorosilicic acid: $SiF_4 + 2 HF \rightarrow H_2SiF_6$. S. is obtained by the reaction of sulfuric acid with a mixture of silicon dioxide and calcium fluoride, CaF_2. The HF released from the latter reacts with SiO_2 ($SiO_2 + 4 HF \rightarrow SiF_4 + 2 H_2O$), and the water formed simultaneously is bound by the sulfuric acid, so that the equilibrium is displaced in favor of SiF_4. This reaction, which is also utilized in the Water drop test (see) is used generally in analytical chemistry to separate SiO_2 and silicates.

Silit®: an electrical resistor material consisting mainly of silicon carbide, SiC, and used to make resistors (up to 1450 °C), heating rods (***silit rods***) and heaters.

Silk fibroin: see Keratins.

Silon®: see Synthetic fibers.

Siloxane: a compound with the structure R_3Si-O-SiR_3 (see Silicone).

Siloxene: $[Si_6O_3H_6]_n$, colorless, solid, polymeric compound which is insoluble in ordinary solvents and sensitive to oxidation. The lattice conssits of buckled sheets of Si_6 rings, which are linked by oxygen atoms. The fourth valence on the Si is saturated by hydrogen. S. is obtained by treatment of calcium silicide, $CaSi_2$, with dilute hydrochloric acid.

Silumin®: an Aluminum alloy.

Silver, symbol ***Ag***: an element from Group Ib of the periodic system, the Copper group (see). S. is a noble metal, *Z* 47. The masses of the natural isotopes are 107 (51.83%) and 109 (48.17%). The atomic mass is 107.868, the valence is I, or, more rarely, II or III. S. is a soft metal, with Mohs hardness 2.7, a density of 10.491, m.p. 960.8 °C, b.p. 2212 °C, electrical conductivity 61.4 S mm^{-1}, standard electrode potential (Ag/Ag^+) + 0.7991 V.

Properties. S. forms cubic face-centered crystals with octahedral coordination. It is very soft, has a white sheen and can be polished. After gold, S. is the most ductile metal; it can be drawn into wire so fine that a 2 km length weighs only one gram. S. can also be hammered into fine foils, blue-green in transmitted light, of 0.00025 mm thickness. S. has the highest electrical and thermal conductivities of any metal. Oxygen dissolves readily in molten S., and as the melt cools, it escapes again, causing the surface to break ("sputtering"). S. forms alloys easily; the very important alloy with copper is harder than pure S., but still has the silvery sheen.

S. has a very high affinity for sulfur; even traces of hydrogen sulfide lead to tarnishing. Non-oxidizing acids, such as hydrochloric, do not attack S., but it is readily soluble in oxidizing acids, such as nitric. At room temperature, dry chlorine forms a protective layer of silver chloride, AgCl, on the surface; however, moist chlorine attacks S. above about 80 °C. In the presence of air, S. will dissolve in alkali cyanide solutions; the reaction is due to a large shift in the oxidation potential of S.. This in turn is a result of the high thermodynamic stability of the dicyanoargentate(I) ions which form. S. is very resistent to caustic alkalies.

S. is monovalent in most of its compounds. With the exception of silver(II) fluoride, the higher oxidation levels, +2 and +3, are observed only in the presence of stabilizing complexes.

Analytical. The characteristic reaction of S. ions is their precipitation with chloride ions as silver chloride. This is insoluble in nitric acid, but dissolves readily in ammonia. Silver chloride is also used for gravimetric analysis of S.; in volumetric analysis, S. is titrated with chloride according to Gay-Lussac or Fajans, or with thiocyanate according to Volhard.

Occurrence. S. makes up about 10^{-5}% of the earth's crust. As it is a noble metal, it is found in the elemental state; most native S. is associated with gold and copper, and often with other elements. S. is also found in the bound state, in the form of silver ores such as argentite, Ag_2S, stromeyerite, CuAgS, pyrargyrite (antimony-silver blende), Ag_3SbS_3, proustite, Ag_3AsS_3, margyrite, $AgSbS_2$, fahlerz, $(Cu,Ag)_3(Sb,As)S_3$ and the rather rare chlorargyrite (horn silver), AgCl. S. also produced from the ores of lead and copper; the S. is enriched in the crude lead or copper. S. is also a byproduct of processing of copper shale.

Production of S. from its ores is now achieved mainly by *cyanide extraction*. The material is ground to a fine sludge, and extracted with well aerated, 0.1 to 0.2% sodium cyanide solution. Both S. and its sulfide and chloride go into solution in the form of sodium dicyanoargentate(I): $2 Ag + H_2O + 1/2 O_2 + 4 NaCN \rightarrow 2 Na[Ag(CN)_2] + 2 NaOH$; $Ag_2S + 4 NaCN \rightleftharpoons 2 Na[Ag(CN)_2] + Na_2S$; $2 AgCl + 4 NaCN \rightarrow 2 Na[Ag(CN)_2] + 2 NaCl$. The equilibrium which is established in the reaction of silver sulfide with sodium cyanide is shifted to the right by air oxidation of the sulfide. The S. is precipitated from the resulting solutions by addition of zinc or aluminum powder. Filtration and melting of the filter cake yields about 95% crude silver.

S. is present in the lead obtained from lead ores at a concentration of 0.01 to 0.03%. This lead is enriched in S., usually by the *Parkes process*. When 1 to 2% metallic zinc is stirred into molten, S.-containing lead below 400 °C, S. is absorbed by the zinc, which forms a floating layer above the lead. After liquation to remove the lead, the zinc is distilled off, and the remaining lead solution contains 8 to 12% S. The *Pattinson process*, which is now declining in use, is based on the fact that as a lead melt containing noble metal cools, pure lead precipitates until the silver content of the remaining melt has risen to 2.5%. Therefore the precipitating lead crystals are continuously removed until an enriched lead is formed.

The crude silver is isolated from the enriched lead by *cupellation*. The metal is melted in a furnace and a stream of air is passed over it, oxidizing the lead to PbO. The lead oxide is removed in liquid form through gutters until crude S.remains (S. blick). This is then purified electrolytically (*Möbius process*): the crude S. is formed into anode plates about 1 cm thick and connected to pure silver cathodes in an electrolysis cell with nitric acid/silver nitrate as electrolyte. Gold and platinum collect in the anode sludge, but the S. precipitates on the cathode as loose, branched crystals (dendrites). These are stripped off and collect on the bottom of the cell. The resulting electrolytic silver is 99.9% pure.

Applications. S. is only rarely worked in the pure state, because it is too soft for most purposes. The widespread use of Silver alloys (see) is a result of their chemical inertness, appearance and good electrical and thermal conductivity. Objects are often coated with galvanic S. Large amounts of S. are used for production of mirrors, thermos bottles and Christmas decorations, and especially for the light-sensitive silver-halide layers in photographic films.

Historical. S. was already in use in the predynastic period in Egypt, in the Sumerian city Ur, in the Early Minoan period on Crete and by the Hethites in Asia Minor. S. ornaments and weapons were present in Western Europe in the early Bronze Age; in antiquity, Spain and Greece were centers of S. mining. The Aztecs in Mexico also mined S. intensively.

Silver alloys: alloys of silver with copper, nickel, zinc, cadmium, tin, palladium and indium. 1) ***Silver-copper alloys*** are harder and more wear-resistant than pure silver. In silver coins, the copper content is between 10 and 50% (coinage silver). Jewelry and household articles usually contain 20% copper (see Ornamental silver). Silver solders contain zinc or cadmium in addition to copper. 2) ***Silver-tin alloys*** with 10 to 50% tin (dental silver) are shredded and mixed with mercury (see Dental alloys) and used to fill teeth. 3) ***Silver-palladium alloys*** containing 30 to 50% palladium and up to 5% copper, or with 5 to 15% cadmium are used as contact material for medium-capacity switches. 4) Other S. To prevent recrystallization of silver laboratory apparatus, pure silver is alloyed with 0.1 to 0.2% nickel. In pressurized water reactors, the absorber material hafnium has been replaced by a silver-indium-cadmium alloy (15% indium, 5% cadmium, remainder silver); this has a similar neutron absorption spectrum and is resistant to radiation damage.

Silver azide: AgN_3, a white salt which turns dark violet in light and is scarcely soluble in water; M_r 149.89. S. is obtained by precipitation from silver nitrate solution with sodium azide. It explodes when heated, and is used as an ignition explosive which is somewhat stronger than lead azide.

Silver bromide: AgBr, yellowish white, cubic crystals; M_r 187.78, density 6.473, m.p. 432 °C. S. is scarcely soluble in water ($S_{AgBr} = 5 \cdot 10^{-13}$), and is obtained by combining aqueous solutions of silver nitrate and potassium bromide. It is slightly soluble in ammonia, and somewhat more soluble in sodium thiosulfate or alkali cyanide solution, forming complexes. S. is more easily reduced and more sensitive to light than silver chloride, so it is used in large amounts in the light-sensitive layers of photographic plates, films and papers. S. is found in nature as bromargyrite.

Silver-cadmium battery: an Electrochemical current source (see) and a secondary cell. Silver oxides, Ag_2O and Ag_2O_2, serve as the cathode. The anode is cadmium, and the electrolyte is 30 to 40% potassium hydroxide. The advantages of the S. are its high cycle life-time (up to about 3000 loading cycles are possible) and its low self-discharge. However, because of its high price, it is used only for special purposes in space and military hardware.

Silver carbonate: Ag_2CO_3, a bright yellow substance; M_r 275.75, density 6.077. When heated to 200 °C, S. splits off carbon dioxide. It is precipitated from aqueous silver nitrate solution to which sodium carbonate has been added, and can also form by the reaction of carbon dioxide with disilver oxide.

Silver chloride: AgCl, white, cubic crystals which turn black in the light; M_r 134.32, density 5.56, m.p. 455 °C, b.p. 1550 °C. S. is scarcely soluble in water (S_{AgCl} $1.7 \cdot 10^{-10}$). It dissolves readily in ammonia, sodium sulfate or alkali cyanide solutions by forming complexes ($[Ag(NH_3)_2]^+$, $[Ag(S_2O_3)_2]^{3-}$, $[Ag(CN)_2]^-$). S. is formed as a "cheesy" white precipitate when silver nitrate solutions are added to chloride solutions. This reaction is used for both qualitative and quantitative determination of chloride or silver ions (see Argentometry). The light sensitivity of S. is of great practical significance; when exposed to light, S. turns violet or black due to precipitation of finely divided silver. A short exposure does not appear to change the S., but the silver crystallization seeds which are formed facilitate local reduction in the subsequent development of photographic material. S. is found in nature as chlorargyrite (horn silver).

Silver complexes: coordination compounds of silver (see Coordination chemistry) in which the coordination number 2 is most common, giving the S. a linear AgL_2 type structure. Complex formation can also stabilize the rare oxidation states +III and +IV of silver, as seen in the orange $[Ag(C_5H_5N)_4]S_2O_8$ or the yellow $Cs[AgF_4]$. S. are used, for example, in cyanide leaching of silver, in galvanic silvering (see Silver cyanide) and in photographic fixing (see Silver thiosulfate).

Silver cyanide: white, hexagonal, rhombohedral crystals; M_r 133.84, density 3.95, m.p. 320 °C (dec.). S. is a linear coordination polymer: ...-Ag-C≡N-Ag-C≡N-... It is formed as a relatively insoluble compound when aqueous silver nitrate and alkali cyanide solutions are mixed. S. reacts with excess alkali cyanide to form water-soluble alkali dicyanoargentate, $M[Ag(CN)_2]$. The dicyanoargentate anion, like the homologous dicyanoaurate ion, is linear. Its high thermodynamic stability is utilized in cyanide leaching of silver. Alkali dicyanoargentate solutions are also used for galvanic silver plating, since the metal precipitates out of these solutions as a dense, tightly adhering layer.

Silver(I) dicyanamide: see Dicyanamides.

Silver fluorides: ***Silver(I) fluoride***, AgF, forms white, cubic crystals; M_r 126.87, density 5.852, m.p. 435 °C, b.p. about 1159 °C. AgF is obtained as a water-soluble compound by reaction of disilver oxide with hydrofluoric acid. ***Silver(II) fluoride***, AgF_2, is colorless when pure; M_r 145.87, density 4.57, m.p. 690 °C. AgF_2 is obtained by the reaction of fluorine with finely divided silver. This compound has considerable thermal stability, and can be used as a fluorinating reagent. ***Disilver fluoride***, Ag_2F, is a bronze-colored, hexagonal compound; M_r 234.74, density 8.57. Ag_2F conducts electric current, and its lattice contains double layers of silver alternating with fluoride layers.

Silver fulminate, ***detonating silver***: AgCNO, a white, light-sensitive and highly explosive salt; K_m 149.88. S. exists in two modifications, an orthorhombic and a trigonal. Both are coordination polymers in which the silver atoms are joined via the fulminate carbon. S. tends to explode violently even upon a light touch or when warmed. Like mercury fulminate, it can be used as an ignition explosive, but it is much more sensitive to impact, even when moist. S. is obtained by reaction of nitric acid with alcohol in the presence of alcohol.

Silver halide photography: see Photographic process.

Silvering: the use of oligodynamically active silver (see Oligodynamism) in water treatment. S. is used to prevent bacterial growth in water reserves and products (drinks, ice, preserves, vinegar), to disinfect drinking water and plumbing, water tanks, filling and measuring machinery, for bottle sterilization and to prevent growth of algae in air conditioning units and pools. The cumasina and catadyne processes (see Disinfection) are based on S.

Silver iodide: AgI, yellow crystals; M_r 234.77, density 6.010, m.p. 558 °C, b.p. 1506 °C. S. exists in three modifications: α-AgI is cubic face-centered, and is stable to 137 °C. Above 137 °C it is converted to β-AgI, which crystallizes in a hexagonal wurtzite lattice. Above 146 °C, this is converted to γ-AgI, which crystallizes in a sodium chloride lattice. S. has the lowest solubility in water of any of the silver halides (S_{AgI}: $8.5 \cdot 10^{-17}$). It is insoluble in ammonia, barely soluble in sodium thiosulfate solution, and readily soluble in alkali cyanide solution. It is obtained by combining aqueous solutions of silver nitrate and iodide. S. was formerly used in photography, and is now used to seed clouds and to prevent hailstorms.

Silver nitrate: $AgNO_3$, colorless, rhombic crystals; M_r 169.87, density 4.352, m.p. 212 °C. S. is dimorphic, and is converted to the hexagonal-rhombohedral form at 159.8 °C. It is caustic and antiseptic, has a bitter, metallic taste, and dissolves in water very readily, but is only slightly soluble in alcohol. S. is obtained by the reaction of silver with nitric acid. It is by far the most important salt of silver, being used to synthesize other silver salts, especially the halides used in photography, for galvanic silver plating, and in the production of silver mirrors. S. is used to make indelible inks and hair dyes. It is an important reagent in the chemical laboratory, used for qualitative and quantitative analysis of halides and pseudohalides.

Silver nitrite: $AgNO_2$, pale yellow, crystalline needles; M_r 153.88, density 4.453, m.p. 140 °C (dec.). It decomposes in light to silver and nitrogen dioxide. S. is obtained by the reaction of silver nitrate with sodium nitrite. It is used, for example, in the synthesis of aliphatic nitro compounds and nitrous acid esters.

Silver(I) oxide: Ag_2O, a black or brown compound, M_r 231.73, density 7.143, m.p. 300 °C (dec.). S. absorbs carbon dioxide from the air, forming silver carbonate, Ag_2CO_3. Above 160 °C, it decomposes into the elements and can easily be reduced with hydrogen. S. is obtained as a very basic precipitate which is difficult to separate from alkali metal ions when alkali hydroxide solutions are added to silver salt solutions. Suspensions of S. are used in preparative chemistry for halide-oxide exchange reactions. It is also used to make ceramic pigments and galvanic elements.

Silver plating: the creation of thin layers of silver on metallic or nonmetallic objects for decoration, corrosion protection or to make mirrors. S. is done by the same methods used for Gilding (see). The most important use of S. is to make mirrors out of glass plates by pouring onto them a silver nitrate solution containing Seignette salt or formaldehyde as a reducing agent.

Silver polish: see Household cleansers.

Silver rhodanide: same as Silver thiocyanate (see).

Silver/silver chloride electrode: a type-two electrode used as a reference electrode. Its construction is similar to that of the Calomel electrode (see). If saturated potassium chloride solution is used as electrolyte, the electrode potential is +0.179 V.

Silver sulfate: Ag_2SO_4, colorless, rhombic crystals, M_r 311.80, m.p. 652 °C, b.p. 1085 °C (dec.). S. is only slightly soluble in water, but dissolves more readily in dilute sulfuric acid and crystallizes out of it as hydrogensulfate, $AgHSO_4$. S. is prepared by dissolving silver powder in concentrated sulfuric acid.

Silver sulfide: Ag_2S, black, rhombic crystals; M_r 247.80, density 7.317, n.p. about 840 °C. S. is the silver compound with the lowest solubility in water, and is obtained by precipitation from silver salt solutions with hydrogen sulfide. It precipitates in crystalline form by reaction of sulfur dioxide with silver at high temperatures. S. is found in nature as argentite (silver glance).

Silver thiocyanate, ***silver rhodanide***: AgSCN, white, coordination polymer with a zig-zag chain structure; M_r 165.99. S. is formed as an insoluble precipitate when equimolar amounts of silver nitrate and alkalithiocyanate solutions in water are mixed. In the presence of excess thiocyanate, water-soluble thiocyanatoargentates(I), $[Ag(SCN)_2]^-$ and $[Ag(SCN)_4]^{2-}$, are formed.

Silver thiosulfate: $Ag_2S_2O_3$, an unstable compound formed as a white, flocculant precipitate when sodium thiosulfate is added to an aqueous silver nitrate solution; M_r 327.87. S. dissolves in excess thiosulfate, forming the complex anion $[Ag(S_2O_3)_2]^{3-}$. Formation of this anion is the basis of the fixing process for photography; the image is fixed by dissolving the non-reduced silver bromide: $AgBr + 2\,S_2O_3^{2-} \rightarrow [Ag(S_2O_3)_2]^{3-} + Br^-$.

Silver-zinc battery: an Electrochemical current source (see) and a secondary cell. Silver oxides, Ag_2O and Ag_2O_2, serve as the cathode. The anode is produced as zinc powder. The electrolyte is a 30 to 40% potassium hydroxide solution saturated with zinc oxide. The advantages of the S. are its high energy and power densities. The disadvantages are its short lifespan and its relatively high price. Therefore it has only limited applications, primarily in space and military technology.

S_E mechanism: see Substitution 2.

Simazine: see Triazine herbicides.

Simmons-Smith reaction: a synthesis reaction for making cyclopropanes from alkenes and an organozinc compound formed from methylene iodide (diiodomethane) and a copper-zinc alloy. The cyclopropane ring is formed by a synchronous *cis*-addition which has the following mechanism (3-center transition state):

Because the reaction occurs smoothly and is highly stereoselective, The S. is often used for the synthesis of complicated alkenes, for example steroids.

SIMS: see Mass spectroscopy.

Simultaneous equilibrium: see Reversibility.

Simultaneous reaction: another term for parallel reaction (see Complex reaction).

Sinalbin: see Glucosinolates.

Sinapic acid: see Hydroxycinnamic acids.

Sinapyl alcohol: see Cinnamyl alcohols.

Single-beam apparatus: see Spectroscopes.

Singlet: see Multiplet structure.

Singlet oxygen: $^1\Delta_g$, an excited state of oxygen with 94 kJ mol^{-1} more energy than the triplet ground state. S. is obtained by triplet-triplet energy transfer with triplet generators (dyes such as methylene blue, eosine, etc.) from triplet oxygen, by reaction of alkaline hydrogen peroxide solution with hypohalites or *N*-halogenamides, thermal decomposition of phosphite-ozone adducts, decomposition of organic per acids, or decomposition of *endo*-peroxides. The lifespan of O. depends on the type of solvent and the presence of quenchers (conjugated polyalkenes, amines).

S. undergoes specific reactions, such as the "ene" reaction to form allylhydroxoperoxides, [2+2] cycloaddition to electron-rich double bonds to form 1,2-dioxetanes (four-membered rings with two oxygen atoms), and [2+2] cycloaddition to conjugated dienes (formation of cyclic peroxides). The anthelminthic ascaridol, synthesized from α-terpenes, was obtained from the first industrial synthesis with S. (Schenck).

Sinigrin: see Glucosinolates.

Sintered glass: see Foam glass.

Sintered metals: in the narrow sense, products made by forming and sintering metal powders (see Powder metallurgy). The best known S. are sintered iron, steel, bronze, aluminum, hard metals, contact and magnetic alloys and composite materials (e.g. fiber-reinforced materials). S. produced simply by pressing and sintering are still more or less porous, depending on the conditions (metal filters are about 50%, self-lubricating bearings between 5 and 30%, and sintered steel for machine parts, about 10%). In the self-lubricating sliding bearings from sintered iron or bronze, the open pores are filled with lubricant. S. with few or no pores are obtained by hot pressing (pressure sintering) or soaking up a low-melting metal.

In the broad sense, the S. also include materials which are made from mixtures of metal powders and nonmetallic components, such as composite materials consisting of iron or bronze with graphite fibers for sliding and friction materials, current carrying parts or cermets.

Sintering: the partial fusion of finely powdered material to form larger pieces by heat treatment at a temperature below the melting temperature, sometimes with application of pressure. The S. process requires good grain contact and the temperature must exceed a certain threshold to permit particle diffusion. Diffusion first occurs at the structurally highly distorted surfaces (surface or grain-boundary diffusion); at higher temperatures, diffusion occurs within the grain (lattice or volume diffusion). The self-diffusion of interlattice particles crosses the grain boundaries, and thus leads to solidification of the product, a process which is promoted by subsequent collective crystallization (Ostwald maturation). However if the crystals which form are too large, the material loses strength. A distinction is made between wet and dry S., depending on the appearance of a melt phase.

Sintering processes are important in the firing of ceramics, in solidification of raw material grains during semi-dry processes of making building materials and in firing of cement clinkers, in iron metallurgy for production of a suitable grain size for furnace feed, in powder metallurgy (see Sintered metals), and in the production of sintered glass and sintered glass ceramics.

Sipal®, ***Palliag®***: a silver-palladium alloy containing about 65% silver, 20 to 25% palladium, 0 to 10% gold and small amounts of copper, zinc and tin. It is used for dental work.

Sirenin: a sesquiterpene which is secreted by the flagellated fungus *Allomyces* as a sexual pheromone to attract male gametes.

Sirius light blue: a dioxazine which is very important as a dye for cotton.

Sitals: same as Vitroceramics.

Sitosterol: a mixture of plant sterols, consisting of stigmasterol, dihydrositosterol and campesterol. It is found in the unsaponifiable fraction of many fatty oils. The most widely occuring is **β-*S.* (*stigmast-5-en-3β-ol*)**. S. is the starting material for partial syntheses of steroids and is used in the therapy of hypercholesterolemia, because it reduces the blood level of cholesterol.

Six-membered heterocycles: carbon compounds with six-membered rings containing one to four heteroatoms (see Heterocycles). The standard form for showing the formula is with the vertex occupied by the highest-priority heteroatom pointing upward (Position 1; see Nomenclature). However, it is fairly common to show the ring in other positions, corresponding to the emphasis on its properties and reactions; for example, the priority heteroatom might point down.

Aromatic S. are π-deficiency heterocycles, that is, they react much more readily with nucleophilic reagents (pyridine, pyrimidine, pyrylium salts) than with electrophilic reagents. Therefore, amino and hydro derivatives can be synthesized from the parent compounds. The conversion of chlorine compounds by nucleophilic substitutions are especially elegant (substitution of -OH, -SH or $-NH_2$ for the -Cl). On the other hand, electrophilic substitution reactions (bromination or nitration) can be carried out on the parent compounds only under extreme conditions. However, these reactions are greatly facilitated by the presence of an electron donor, such as an amino group. For example, 2-aminopyridine can be nitrated in the 3 and 5 positions under normal laboratory conditions. As the number of nitrogen atoms in the ring increases, the π-deficiency character of the compound also increases. This is shown, for example, by the decreasing basicity in the series of di-, tri- and tetrazine. The reason for this is the electron-acceptor effect of the tertiary N-atoms, which is comparable to the effect of a nitro group with its (-)M effect in a benzene ring.

Aromatic S. can be hydrogenated completely or partially; for example, pyridine can be converted to dihydropyridine and especially to piperidine. In addition, S. are very often present in fused benzene and heterocyclic systems.

S. are very common in nature. Some examples are pyridine, quinoline and isoquinoline as components of alkaloids, pyrimidine as a component of the nucleic acids and a variety of purine derivatives. Pyridine 3-carboxamide is nicotinamide, one of the B vitamins; other B vitamins are also derivatives of pyridine and pyrimidine. S. are also components of the active centers of enzymes and natural pigments. They are used in industry as solvents, for example pyridine or 1,4-dioxane, and for synthesis of drugs and dyes.

Sizings: a group of textile conditioners which protect the threads from fraying and wear during processing and make weaving easier. The choice of S. depends on the type of fiber from which the thread is spun. Natural or synthetic colloids are used, usually water-soluble, such as starch (usually solubilized by oxidation), starch ethers, cellulose ethers, polyvinylalcohol, acrylic acid or its salts. Linseed oil is used for acetate silks. In most cases, the sizing must be removed before the cloth is dyed, for which ***desizing agents*** are used. These are based on diastase for starch sizings and proteases for collagen S.; a waterbath is sufficient for water-soluble S. In some cases, however, the S. remains on the cloth, giving it a certain amount of body.

Skatole, ***3-methylindole***: a biogenic amine formed by decarboxylation of tryptophan; m.p. 95 °C, colorless leaflets. It is present in human excrement and is largely responsible for its odor.

Skeleton: see Dendrites.

Skew conformation: see Stereoisomerism, Fig. 41.

Skin preventers: non-volatile additives (0.2 to 2%) which prevent formation of a skin on paints, lacquers, etc. during storage. The active ingredients of S. include guaiacol, 4-*tert.*-butylpyrocatechol, hydroquinone, phenol derivatives, oximes and antioxidants.

Skleroscope hardness: see Hardness.

Skraup synthesis: synthesis of quinolines from a primary aromatic amine (such as aniline), glycerol, concentrated sulfuric acid and an oxidant such as nitrobenzene and iron(III) compounds. In the first step of the reaction, glycerol is converted to acrolein; then the amine adds to its double bond. After ring closure (electrophilic substitution by the O-protonated formyl group) water is eliminated to form the dihydroquinoline, and this is oxidatively dehydrogenated to yield the product:

Aniline + Acrolein → → Dihydroquinoline $\xrightarrow[-H_2O]{+O}$ Quinoline

Variants of this reaction can be universally applied to the synthesis of substituted quinolines. For example, the reaction of aniline with methyl vinyl ketone leads to lepidine (4-methylquinoline).

Slack wax: a mixture of higher paraffin hydrocarbons of chain length about $C_{18}H_{28}$ to $C_{28}H_{58}$. It is of a salve-like consistency. S. is produced, for example, in the processing of soft coal tar to hard paraffin, in the Fischer-Tropsch synthesis, and in paraffination of lubricating oil. Oxidation of S. produces saturated fatty acids which are used, for example, in the production of laundry products (see Paraffin oxidation).

Slag: byproduct of metallurgical processes. S. from different processes can have widely varying compositions, but consist mostly of silicates and other oxides. There are two basic types. 1) ***Blast furnace S.*** are formed by metallurgical processing of ores and metals. ***Thomas S.*** is produced by the Thomas method of steel production; when finely ground, it is a valuable phosphate fertilizer. 2) ***Firing S.*** are residues of burning coal (***coal S.***) or garbage (***garbage S.***). Only some of the S. produced is utilized, as an additive to cements, for fibers or glass-crystalline products.

Slag fibers, ***mineral wool***: fibrous materials prepared from blast furnace slags, which should contain the highest possible concentration of silicon dioxide. In spite of increasing applications, only about 1% of metallurgical slags are used to make S.

The slag is taken directly from the melt in a cupola furnace, which is between 1100 and 1500 °C. It is either blown with steam or compressed air at 0.8 to 1.0 MPa or spun centrifugally into fibers between 5 and 20 mm in length and 0.5 and 15 μm diameter. The main components are 30 to 55% silicon dioxide, 4 to 20% aluminum oxide, 25 to 40% calcium oxide, 3 to 16% magnesium oxide and 1 to 7% iron(II) oxide. S. are less chemically resistant than rock wool, but because they have a lower heat conductivity, they are used loose or in the form of mats and felts. They can replace asbestos as heat or electrical insulation in buildings and machines.

Slaked lime: same as Calcium hydroxide (see). See also Binders (building materials).

Slater orbitals: see Screening.

Slater rules: see Screening.

Sliding materials: materials used for sliding bearings and other parts of machines which must slide past each other, such as hollow cylinders which take up rotating cones (in automatic transmissions in cars, for example). S. must be able to withstand friction, have little tendency to wear, be able to be embedded in other parts, and have adequate hardness, strength and heat conductivity. White metals, copper alloys, aluminum alloys, sintered bearings (metallic S. are often still called ***bearing metals***), thermoset polymers (pressed materials), plastomers (polyamides, polyurethanes, polyoxymethylene, polytetrafluoroethylene), cast iron, soft rubber, graphite, glass and ceramics, gemstones and semiprecious stones are all used as S.

Slip: an aqueous suspension of finely divided raw materials used in making ceramics, cement (wet process) and enamels. By addition of extenders (soda, borax, potssium chloride) or phosphates (alkali metal polyphosphates, phosphoric acid), the consistency of the S. can be adjusted. The viscosity of a S. with a high proportion of solids can be made less viscous by addition of alkalies (alkali metal carbonates or hydroxides, waterglass), which change the flow properties of the clay components.

Slow-reacting substances: see Leukotrienes.

Sludge activation: see Sewage treatment.

Sm: symbol for samarium.

Small-load hardness: the hardness of metals measured with test loads of 2 to 30 N. The S. is used to measure hardness of very thin sheets and small samples which cannot be tested by macro- or microhardness methods. The test for S. is essentially the same as the Vickers method (see Vickers hardness); because they impressions are very small, they must be measured under a microscope.

Smalt: 1) ***cobalt glass***: a cobalt pigment, potassium-cobalt(II) silicate. S. is made by roasting cobalt ores, such as smaltite, and melting the product (zaffer) with potasisum carbonate and quartz powder. The product is a blue glass (see Frit) which is quenched in water, ground and sorted. S. is fairly resistant to high temperatures, reducing kiln atmospheres, alkalies and cold acids, and is used to color ceramics, glass and enamel. Its use has been largely supplanted by the use of cobalt oxide or carbonate, however.

2) Pigments made by mixing colored oxide and alumina-rich clay in a ratio of 2:1. S. are applied to stoneware in the form of a paste and are protected by a salt glaze; because they are soluble and require a high firing temperature, the available colors are limited to the following: green to bluegreen (Cr_2O_3), blue (CoO), violet to brown (manganese oxide), brown (Fe_2O_3), and black ($Fe_2O_3 \cdot Cr_2O_3 \cdot CO$).

S_E Mechanism: see Substitution, 2.

S_N Mechanism: see Substitution, 1.

S_R Mechanism: see Substitution, 3.

Smectic: see Liquid crystals.

Smog: an extreme form of air pollution which occurs during weather inversions in regions with high populations and industrial activity. There are two types. In ***London smog***, soot and dust particles act as condensation nuclei and initiate formation of heavy fogs when the moisture content of the air is high. Sulfur dioxide and sulfur trioxide dissolve in the water droplets and become concentrated near the ground. In ***Los Angeles smog***, nitrogen oxides and unsaturated hydrocarbons released in automobile exhaust react under the influence of intense sunlight (UV) to form ozone and other photochemical products (e.g. peroxyacetyl nitrate, PAN) which are irritating to mucous membranes, phytotoxic and cause damage to plastics and other materials.

Smoke damage: see Emission damage.

Smokeless powder: a group of explosives consisting chiefly of cellulose nitrate. They have almost completely replaced the old black powder as gunpowder. The following products are used. ***Cellulose nitrate powder (nitrocellulose powder)*** is made by swelling and partially dissolving a cellulose nitrate mixture with a nitrogen content of 12.5 to 13.5%. Stabilizers are used, e.g. diphenylamine or urea derivatives. The finished powder is graphitized, to reduce the danger of accumulating static charge. ***Glycerol trinitrate powder***, also known as ***nitroglycerin powder***, is made by mixing cellulose nitrate with glycerol trinitrate under water. Because this powder has a low burning temperature, it does less damage to the barrel of the gun. Diglycol dinitrate with a significantly higher capacity for gelatinizing and a significantly lower sensitivity to impact is used to make ***diglycol powder***. Even lower burning temperatures, with a stronger explosion, due to a higher gas volume, are obtained with ***nitroguanidine powder*** ("cold" powder). It is diglycol powder with a certain content of nitroguanidine. The higher-energy S. are used as rocket fuels.

SM process: see Steel.

Sn: symbol for tin.

SNG: abb. for substitute natural gas, a synthetic natural gas consisting mainly of methane. SNG is formed by reaction of carbon monoxide and hydrogen, and is used mainly as a fuel gas.

SNIA viscose process: see ε-Caprolactam.

Snake toxins: highly toxic mixtures of substances produced by poisonous snakes, usually for attack and capture of prey. They are delivered by a well developed apparatus (highly specialized teeth and poison glands). As a rule, the poison is injected into the victim by means of fangs which are hollow or grooved. Exceptions to this rule are African cobras (*Haemachatus haemachatus* and *Naja nigricollis*), which are able to spray their poison into the eyes of the victim from a distance (up to 2 m).

The main components of S. are peptide toxins and

enzymes which serve both for rapid killing or paralysis and for preliminary digestion. S. are complex mixtures which have the following effects: a) neurotoxicity, i.e. they block nervous impulses (paralyse) and thus impair nervous and muscular activity, including breathing (toxins of the cobras and most adders); b) circulatory effects (loss of blood pressure and circulatory collapse), sometimes leading to cardiac arrest (many adders, vipers, pit vipers); c) local effects, i.e. coagulation and necrosis. The lethal effects can be the result of a peripheral circulatory collapse, or more rarely, due to bleeding in the brain (vipers, pit vipers).

The polypeptides and enzymes mentioned above (hydrolases, proteases, enzymes which affect blood coagulation, glycoside-cleaving enzymes such as hyaluronidase), both initiate digestion and promote absorption of the toxic peptides (synergism). In addition, S. contain unusually high concentrations of zinc ions, probably as a protection of the poison glands against self destruction. (Zinc ions block some enzymes, e.g. phosphatase).

The most poisonous snake known is found in the eastern part of central Australia: *Parademansia microlepidatus*, which is related to *Oxyuranus scutellatus*, also found in Australia. These adders produce almost identical toxins, but the toxin concentration in the first is higher (LD_{50}, mouse, subcutaneously, is 20 or 65 μg kg^{-1}.

The most important snake toxins

Family	Chemical nature of the poisons	Specific toxins and their toxicity (LD_{50} in μg kg^{-1} mouse, subcutaneous)
Adders (*Elapidae*)	Mainly esterases, basic polypeptides	β-Bungarotoxin 40 μg kg^{-1} Cobra toxin, 90 μg kg^{-1} α-Bungarotoxin 210 μg kg^{-1}
Vipers (*Viperidae*)	Esterases, proteinases, rarely specific toxins	Viperotoxin (basic polypeptide)
Pit vipers	Esterases, proteinases, including basic polypeptides	Crotoxin 500 μg kg^{-1} Basic phospholipase A + acidic polypeptide crotapotin.
Sea snakes	Low in enzymes, basic polypeptides	

Soap: a water-soluble potassium or sodium salt of saturated and unsaturated higher fatty acid, a salt of resin acids of colophony or naphthenic acids, or an ammonium or amine salt of a fatty acid. All types of S. are used as Laundry products (see) or for personal hygiene. In some cases, the alkali salts of other organic acids, e.g. the bile acids, are also called S. (bile soaps) and are used as cleansers. Other metal salts of the same acids are generally called Metal soaps (see).

Properties. S. form a clear or slightly opalescent colloidal solution with water. They are sometimes hydrolysed into fatty acid and free alkali. The cleansing effect of S. depends only to a small extent on the solubilization of fats by saponification by the free alkali; most of the effect is due to adsorption of the soap molecules to the fat and dirt particles, which become emulsified, wetted and dispersed, and can thus be washed away. The physical properties are discussed under Washing (see). The solubility of S. in water depends on the type of hydrocarbon residue and the method of production. S. molecules form colloidal micelles in aqueous media; the size of these depends on the length of the fatty acid molecule. S. from short-chain fatty acids form small micelles, while those of long-chain fatty acids form large ones. The larger the micelles, the better the cleansing effect of the S. Two disadvantages of the S. are their sensitivity to water hardness and their instability in the presence of acids. The S. form sticky and greasy or curdy precipitates with calcium and magnesium ions, and also with other metal ions, such as iron and aluminum. These have no cleansing effect; they are also called ***lime*** or ***mineral S.***. The precipitation can be avoided by addition of softeners, such as meta- and polyphosphates, or organic complexing agents. In the presence of small amounts of mineral acid, such as hydrochloric acid, S. are broken down: R-COONa + HCl → R-COOH + NaCl. They are very resistant to alkalies, however. Only at high alkali concentrations is a S. solution salted out. The wetting effect of S. depends largely on the chain length of the hydrophobic part of the molecule; it is usually lower than that of synthetic cleansers with sulfonic acid or sulfate groups. In S. with longer carbon chains, such as stearates, the wetting effect is not achieved except at high temperatures. The dispersive effect of S. is good only at higher concentrations of the S. solution. S. are good emulsifiers of fats and oils, including mineral fats and oils.

Production. The raw materials are the higher fatty acids present in animal and plant fats as triacylglycerols. These are obtained in free form by saponification. The most important natural fatty acids used in making S. are lauric, myristic, palmitic, stearic, oleic and ricinoleic acids. Polyunsaturated fatty acids must be hydrogenated before they can be used in making S. The raw materials are whale oil, fish oils, tallow, bone fat, coconut oil, palm heart fat, palm oil, olive oil, peanut oil, linseed oil, poppyseed oil, hemp oil, sunflower oil, corn oil, soy oil, ricinus oil, etc. Synthetic fatty acids from paraffin oxidation are now used in increasing amounts.

S. are made directly from the fats, or from the free fatty acids obtained by carbonate saponification. The free fatty acids are neutralized with sodium or potassium lye, or with soda, and after a boiling process, a soap mass is produced. In direct production of the S. from the fats, the melted fat is heated with the lye. This produces a heavy foam, to which salt (NaCl) or potassium acetate is added to cause precipitation of a solid S.

Types of S. ***Hard S.*** are the products salted out of the S. foam; they contain up to 33% water. ***S. powder*** is dried and pulverized hard S. to which soda and sodium silicate or sodium metasilicate have been added. ***Toilet S.***: is hard S. made from the purest, most odorless fats, along with pigments and perfumes. Sometimes extra lanolin and fatty alcohols are added to toilet S., yielding ***cosmetic*** or ***baby S. Medicinal S.*** also contain antiseptic substances; for example, sulfur, tar, phenol and other disinfectants are used for prevention of skin diseases. ***Soft S.*** are made of less expensive plant oils by saponification with potassium lye.

Soap chromatography: see Ion-pair chromatography.

Soda, ***sodium carbonate***: a white salt which is readily soluble in water. The anhydrous form is a powder known as ***calcined S.***, Na_2CO_3, density 2.533, m.p. 853 °C. ***Crystal soda***, $Na_2CO_3.10H_2O$, is the most important hydrate. It crystallizes out of aqueous solutions below 32 °C in the form of large, completely clear, monoclinic crystals, density 1.45. These melt in their own water of crystallization at 32 °C, producing the rhombic heptahydrate, $Na_2CO_3.7H_2O$. At 34.5 °C, the rhombic monohydrate, $Na_2CO_3.H_2O$ forms; above 107 °C this loses its water of crystallization and becomes the anhydrous form. ***Caustic S.*** is Sodium hydroxide (see), which is obtained by caustification of S. by the lime-soda process. Aqueous solutions of S. are alkaline, due to hydrolysis. When carbon dioxide is passed through cold-saturated aqueous S. solutions, sodium hydrogencarbonate is formed: $Na_2CO_3 + H_2O + CO_2 \rightarrow 2\ NaHCO_3$. If fatty acids are boiled in a concentrated S. solution, they form soaps.

Occurrence. S. is found in nature in alkaline lakes, for example, in Armenia, Hungary, China, Egypt, East Africa and America, and in alkaline springs. As the lakes dry up, large deposits are formed in which the S. crystallizes as $Na_2CO_3 \cdot NaHCO_3 \cdot 2H_2O$. S. is also found in the ashes of plants which grow near in salty soils, especially near ocean beaches.

cium carbonate, $CaCO_3$, and magnesium hydroxide ($Mg(OH)_2$ precipitate in flocculant form (see Water softening). In the next step, the brine is clarified by settling. The pure brine passes into the washer, where it absorbs ammonia, NH_3, and some of the carbon dioxide, CO_2, from the exhaust gases from the carbonator (precipitation tower). In the absorber, it is enriched with ammonia to a concentration of 85 g $NH_3\ l^{-1}$. The ammoniacal brine is then supersaturated with carbon dioxide in the carbonator, leading to precipitation of crystalline sodium hydrogencarbonate: $NaCl + NH_3 + CO_2 + H_2O \rightarrow NaHCO_3 + NH_4Cl$. The carbon dioxide required for carbonation is obtained from roasting lime and calcination: $CaCO_3 \rightarrow CaO + CO_2$. Calcium hydroxide solution, $Ca(OH)_2$, is made by slaking the burned lime, CaO, with water, and is used to recover the ammonia; part of it is also used to purify the crude brine. The precipitated sodium hydrogencarbonate is separated from the mother liquor (ammonium chloride and sodium chloride solution) by centrifugation or rotating filters. It is calcined (heated in a rotating furnace) to 170-200 °C to convert it to S.: $2\ NaHCO_3 \rightarrow Na_2CO_3 + H_2O + CO_2$. Ammonia is recovered from the filtered liquor (as NH_4Cl, partly as NH_4HCO_3) and returned to the process. Ammonium hydrogencarbonate is cleaved thermally: $NH_4HCO_3 \rightarrow NH_3 + H_2O + CO_2$. After predistillation, the ammonia can be completely recovered by addition of calcium hydroxide to the ammonium chloride solution: $2\ NH_4Cl + Ca(OH)_2 \rightarrow CaCl_2 + 2\ H_2O + 2\ NH_3$. The ammonia is driven off by steam blown through a column and is led back to

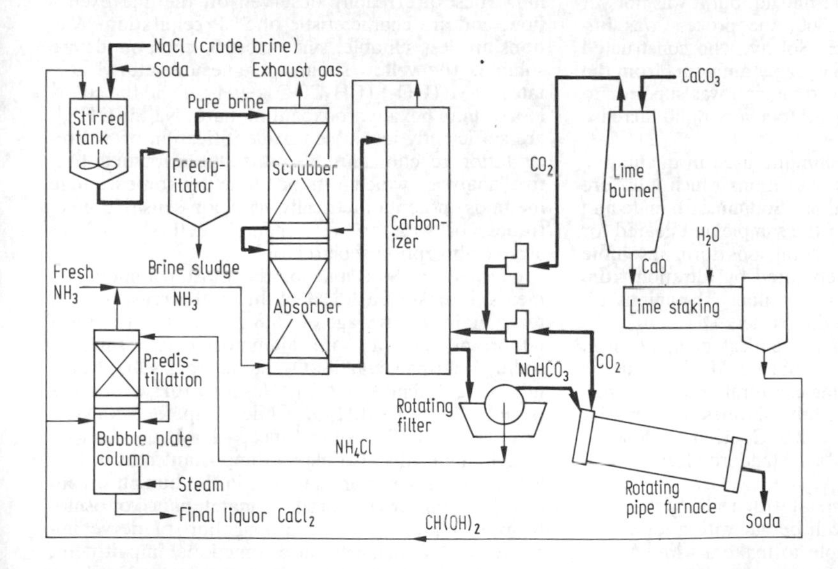

Production. S. is made almost exclusively from sodium chloride and calcium carbonate (using ammonia, coal and water), mostly by the *Solvay process (ammonia-soda process)*: $2\ NaCl + CaCO_3 \rightarrow Na_2CO_3 + CaCl_2$. This overall reaction can only be achieved in stages. The crude NaCl brine is treated to remove calcium and magnesium salts by adding calcium hydroxide, $Ca(OH)_2$, and soda solution. Cal-

the absorber. The solids (mainly calcium chloride) are removed from the remaining liquor, which is then discarded and causes considerable environmental problems.

The *Leblanc process* is no longer used because of its high energy requirements; in it sodium chloride and concentrated sulfuric acid were used. The *cryolite process* is based on the reaction of cryolite (sodium fluoroaluminate) with limestone (calcium carbonate); it is only used now in countries with natural deposits of cryolite.

Applications. S. is used mainly as a flux in making glass and enamel, in textile processing, to make soaps and detergents, as a raw material for production of drugs, sodium hydroxide, nitrates, phosphates, etc., in the production of paper and cellulose, glues and resins. It is also used in ore processing, to remove sulfur from crude and cast iron, to soften water, and in bleaching and tanning. The world production is about 20 million tons per year, making S. a classic mass product of the chemical industry. However, the production has not increased much in recent years, because sodium hydroxide is often used instead.

Historical. S. was discovered by the Egyptians thousands of years ago, as weathering product along the banks of lakes between the Nile and the Libyan coast. They used it in the production of glass and soap, and for washing and bleaching. Until the 18th century, the demand for S. could be met from the ashes of lake and beach plants. At the end of the 18th century, the French Academy of Science established a prize for an industrial method for producing S. from sodium chloride; the prize was won by the French physician Leblanc, who developed the method which bears his name in 1791. Because this method requires rotating furnaces, evaporation and drying pans, its development gave impetus to the construction of apparatus and the rest of the chemical industry. The Leblanc process was the only one in use until 1870. As early as 1838, Dyar and Hemming developed the ammonia-soda process in England, but it was not yet technically feasible. In 1860, the process was improved by the Belgian, E. Solvay, who constructed better apparatus and used cheaper ammonia from the city gas plant; the improved version was superior to the Leblanc process and lead to a very rapid increase in the production of S.

Soda extraction: a technique used in qualitative inorganic analysis to remove cations which interfere with the analysis of the anions. Sodium carbonate and water are added to part of the sample and heated for a time. Most polyvalent metal ions form insoluble carbonates and can be separated by filtration from the soluble sodium salts of the anions. The anions are then determined in the soda extract.

Sodium, symbol ***Na***: a chemical element from group Ia of the periodic system (see Alkali metals); a light metal, *Z* 11, with a single natural isotope; atomic mass 22.9897, valence I, Mohs hardness 0.4, density 0.968, m.p. 97.81 °C, b.p. 882.9 °C, electrical conductivity 23.8 Sm mm^{-2} (at 0 °C), standard electrode potential (Na/Na^+) -2.7109 V.

Properties. S. is a soft metal with a silvery sheen on a freshly cut surface; it can be cut with a knife and pressed through a fine hole to make a wire. Above -163 °C, it crystallizes in a cubic body-centered lattice. Sodium vapor is primarily monoatomic, with only about 1% present as Na_2 molecules. S. dissolves in liquid ammonia or amines, giving a dark blue color. It forms alloys with other alkali metals, which, like sodium amalgam, are liquid at room temperature if the composition is suitable.

The chemical behavior of S. is determined by its strong tendency to form Na^+ cations. S. compounds are typical salts; the element has little tendency to form covalent bonds. The reactivity of S. compared to its homologs in group Ia are discussed under Alkali metals (see).

> In work with sodium, water must be strictly excluded. Only dry apparatus can be used. The eyes should be protected by glasses and the hands by rubber gloves. Leftover sodium should be disposed of by dissolving small pieces in large amounts of alcohol.

S. is a strong reducing agent. Freshly cut surfaces exposed to moist air are rapidly covered with a mixture of sodium oxide and hydroxide. Therefore, S. must always be stored under an inert hydrocarbon, such as light petroleum. It reacts very exothermally with water, and in large amounts, it may explode. The products are elemental hydrogen and sodium hydroxide, NaOH. A similar reaction occurs with alcohol to form sodium alkoxides and hydrogen, but this reaction is much less violent. S. burns in air to form sodium peroxide, Na_2O_2, and it reacts with chlorine, with flames, to give sodium chloride, NaCl. When heated with ammonia, it forms sodium amide, $NaNH_2$, and hydrogen. Elemental hydrogen reduces S. at high temperature to sodium hydride, NaH.

Analytical. The simplest qualitative test for S. is the yellow color it gives to flames, or spectroscopic observation of the yellow double lines at 589.59 and 588.99 nm. These are readily observed on thermal excitation, and are characteristic of S. Precipitation reactions are less suitable, since most S. compounds are soluble. The yellow sodium magnesium uranyl acetate, $NaMg(UO_2)_3(CH_3COO)_9 \cdot 9H_2O$, and the colorless sodium hexahydrooxoantimonate, $Na[Sb(OH)_6]$, are sufficiently insoluble for identification of S. Precipitation reactions are also unsuitable for quantitative analysis, which can be done by ion-exchange methods, potentiometrically with ion-sensitive electrodes, or with an instrumental method, such as atomic absorption sepctroscopy.

Occurrence. S. is one of the most common elements. It makes up 2.63% of the earth's crust; seawater contains an average of 26.8 g NaCl l^{-1}. The most important minerals are aluminosilicates, such as sodium feldspar, $NaAlSi_3O_8$, halite (rock salt), NaCl, mirabilite (Glauber's salt), $Na_2SO_4 \cdot 10H_2O$, natrite (soda), $Na_2CO_3 \cdot 10H_2O$, Chile saltpeter, $NaNO_3$, cryolite, Na_3AlF_6, borax (tincal), $Na_2B_4O_7 \cdot 10H_2O$, etc. Sodium salts also play an important part in biological processes. For example, in the human organism, Na^+ ions are required for maintenance of osmotic pressure and for the propagation of nerve impulses. Na^+ deficiency causes functional impairment, which can be relieved by administration of NaCl.

Production. S. is produced electrolytically; because of its high negative standard potential, water must be excluded. Therefore a melt of sodium hydroxide (*Castner process*) or sodium chloride (*Downs process*) is electrolysed. The latter process is now used almost exclusively. A melt of NaCl containing up to 70% calcium chloride, $CaCl_2$, to reduce its melting point, is electrolysed around 600 °C in a Downs cell. The central graphite anode enters from the bottom; the chlorine generated on it is collected in a hood which dips into the melt. A wire net affixed to the hood hangs down into the melt, separating the anode and cathode spaces, and preventing mixing of the products. The anode is surrounded by a circular iron cathode, where metallic S. rises to the top and is collected (in the absence of air) into a vessel. Very pure S. can be made on a laboratory scale by thermal decomposition of sodium azide, NaN_3.

Applications. S. is the starting material for production of sodium amide, hydride, peroxide and cyanide. Liquid S. is used as a coolant in nuclear reactors, because of its good thermal conductivity. S. wire is used in the laboratory to dry solvents, such as various ethers and hydrocarbons. S. is used in both inorganic and organic syntheses as a reducing agent; to increase its reactivity, Na dispersions are often used. It is used in the Claisen and Wurtz reactions. In metallurgy, S. is used, for example, in the production of titanium. S. is used to make sodium vapor lamps and as an alloy component. The most important industrial use of S. is as a lead alloy for synthesis of tetraethyllead.

Historical. Sodium carbonate (soda) and potassium carbonate (potash) were known in antiquity. Arabian alchemists knew how to prepare sodium and potassium lyes by treating the carbonates with caustic lime. The compounds were not distinguished until the Middle Ages; they are mentioned under the name "nitron" or "nitrum" by Aristotle, Dioscorides and Pliny. The name was changed to "natron" by the Arabian alchemists. The term "alkali" as applied to these compounds is found in writings from the 14th/15th centuries which are ascribed to the alchemist Geber. It was shown experimentally in 1736 by Duhamel de Monceau that soda and potash are different compounds, as had been predicted in 1702 by Stahl. Elemental S. was first obtained in 1807 by Davy, using melt electrolysis of sodium hydroxide. In 1807 and 1808, Gay-Lussac and Thenard prepared S. and potassium by roasting their oxides with iron filings and charcoal.

Sodium acetate: CH_3COONa, colorless, monoclinic crystals which are readily soluble in water and ethanol; M_r 82.03, m.p. 324 °C. S. crystallizes out of water as the trihydrate, $CH_3COONa \cdot 3H_2O$, which melts at 75 °C in its own water of crystallization. Above 120 °C it is converted to the anhydrous form. Aqueous solutions of S. are basic. The compound is made by neutralizing dilute acetic acid with sodium carbonate or sodium hydroxide, and the solution is evaporated. S. is used in the laboratory to make buffer solutions; anhydrous S. is used, among other things, as a weak base and water-withdrawing reagent in organic synthesis. S. is used in the textile, leather and pharmaceutical industries.

Sodium alcoholate: see Sodium alkoxides; see Sodium ethoxide.

Sodium alkoxides, formerly, ***sodium alcoholates***: in a broad sense, compounds in which the hydrogen of an alcoholic OH group has formally been replaced by sodium, for example, sodium methoxide, $NaOCH_3$ (formerly sodium methylate), sodium ethoxide, $NaOC_2H_5$ (formerly sodium ethylate), etc. The S. react with water to form the alcohol and sodium hydroxide. They are made by adding metallic sodium to the alcohol. As strong bases, they have many applications in organic syntheses; they are used as catalysts for rearrangements, as drying accelerators for oil paints, etc.

Sodium alloys: sodium forms alloys with many metals. Those of industrial interest are sodium amalgam (see Mercury alloys), potassium-sodium alloys as coolants in nuclear reactors, and a sodium-lead alloy containing about 10% Na, which is used in the synthesis of tetraethyl lead.

Sodium aluminates: compounds with the composition $NaAlO_2$ and Na_3AlO_3 (anhydrous) or $Na[Al(OH)_4)]$ and $Na_3[Al(OH)_6]$ (hydrated, also called ***sodium hydroxoaluminates***). The anhydrous S. are obtained by fusion of aluminum oxide with soda or sodium hydroxide, and the hydrated S. by dissolving aluminum hydroxide in sodium hydroxide, e.g. $Al(OH)_3 + NaOH \rightarrow Na[Al(OH)_4]$. S. are used in the production of soaps, paper, textiles, paints and glass, and to soften water.

Sodium amide: $NaNH_2$, colorless crystals; M_r 39.02, m.p. 210 °C, sublimable in vacuum, decomposing above 500 °C to sodium imide, Na_2NH, and then to sodium nitride, Na_3N and ammonia, NH_3. S. reacts very violently with water, forming ammonia, NH_3, and sodium hydroxide, NaOH. S. is produced by the reaction of ammonia and molten sodium or by allowing the blue solution of sodium in liquid ammonia to stand. This slowly loses color and develops hydrogen (the reaction is more rapid in the presence of catalytic amounts of iron): $2\,Na + 2\,NH_3 \rightarrow 2\,NaNH_2 + H_2$. S. is used in inorganic and organic synthesis as a strong base and to introduce amine groups. It is also used in the industrial production of sodium azide and sodium cyanide.

Sodium azide: NaN_3, colorless, hexagonal crystals; M_r 65.01. Upon cautious heating, N. decomposes into elemental sodium and nitrogen, and thus serves as a means of preparing very pure sodium. It is synthesized by the reaction of dinitrogen oxide, N_2O, and molten sodium amide. N. is the starting material for synthesis of lead azide and hydrazoic acid.

Sodium bi-: see Sodium hydrogen-.

Sodium boronate, ***sodium hydridoborate, sodium borohydride***: $NaBH_4$, colorless, cubic crystals which are soluble in water, ammonia and dimethylformamide; M_r 37.83, density 1.074. S. reacts slowly with water, forming elemental hydrogen. It is synthesized industrially by heating sodium hydride with trimethyl borate: $B(OCH_3)_3 + 4\,NaOH \rightarrow NaBH_4 + 3\,NaOCH_3$, or from sodium metaborate, aluminum and hydrogen. S. is used as a reducing agent in organic synthesis, for example, to convert carbonyl to hydroxyl groups; an advantage is that the reactions can be done in water.

Sodium borohydride: same as Sodium boranate (see).

Sodium bromate: $NaBrO_3$, colorless, cubic crystals; M_r 150.91, density 3.34, m.p. 381 °C. The water solublity of this compound depends heavily on the temperature: 100 g H_2O dissolves 34.5 g $NaBrO_3$ at 20 °C, and 90.8 g at 100 °C. When heated, S. decomposes to sodium bromide and oxygen. It is poisonous (see Bromates). S. is synthesized by disproportionation of bromine in hot sodium hydroxide: $6\,NaOH + 3\,Br_2 \rightarrow NaBrO_3 + 5\,NaBr + 3\,H_2O$. S. is a strong oxidizing agent, especially in acidic solution, and is used as such in bromatometry. It is also used in electrooptics and as a piezoelectric material. Mixed with sodium bromide, S. is used to extract gold from ores.

Sodium bromide: NaBr, colorless, hygroscopic, cubic crystals soluble in water and ethanol; M_r 102.91, density 3.203, m.p. 747 °C. S. crystallizes out of water below 50.7 °C as the dihydrate, $NaBr \cdot 2H_2O$; above this temperature the crystals are anhydrous. The compound is made by reaction of soda with iron(II,III) bromide or as a byproduct of the disproportionation of bromine (see Sodium bromate). S. is used to make photographic films and papers, and is used in medicine as a sedative.

Sodium carbonate: same as Soda (see).

Sodium chlorate: $NaClO_3$, colorless, hygroscopic, cubic crystals which are deliquescent in moist air; M_r 106.45, density 2.490, m.p. 248-261 °C. S. is very soluble in water: at 20 °C, 100 g water dissolves 101 g $NaClO_3$. At room temperature, S. is stable, but when heated, it disproportionates into sodium chloride and sodium perchlorate: $4\ NaClO_3 \rightarrow NaCl + 3\ NaClO_4$. The latter decomposes at high temperatures to NaCl and oxygen. Because of this formation of oxygen, mixtures of S. with readily oxidized substances, such as sulfur, phosphorus or organic compounds, are unstable and explode when rubbed or heated, or on impact. S. is poisonous. It is made in the laboratory by passing chlorine through a hot, aqueous solution of sodium hydroxide or sodium carbonate. It is made industrially by anodic oxidation of NaCl. S. is used mainly as a herbicide, as an oxidizing reagent in dyeing and printing, and to make sodium perchlorate.

Sodium chloride, ***table salt***: NaCl, colorless, cubic crystals; M_r 58.45, density 2.165, m.p. 801 °C, b.p. 1413 °C. Its solubility in water is nearly independent of temperature; at 20 °C, 35.8 g NaCl dissolves in 100 g water. Pure S. is not hygroscopic.

Occurrence. Seawater contains an average of 2.7% S., while some salt lakes contain more than 20%. S. is also found as halite (rock salt), in deposits which may be more than 1000 m thick. These were formed by evaporation of prehistoric seas and salt lakes. S. is also present in organisms; the adult human body contains between 150 and 300 g S.

Production. Rock salt is generally mined, and can be used directly for many industrial applications; in some cases it is pure enough to be used as table salt as well. S. is also obtained by dissolving rock salt underground and pumping the brine to the surface. Artificial or natural brines are evaporated in flat pans to obtain crystalline S.; if possible, however, the brines are used directly. In areas where the climate is suitable, seawater is let into lagoons to evaporate, and the S. is collected as the residue.

Application. Dietary salt is essential, and it is also used to preserve meats, fish and vegetables. Cattle are provided with cakes of salt made unfit for human consumption with 0.1% iron oxide. S. is used in industry as a raw material for production of nearly all compounds of sodium and chlorine. Chloralkali electrolysis is used to produce chlorine, hydrogen and sodium hydroxide, while the Downs process is used to obtain sodium and chlorine. S. is the basis of production of hydrochloric acid and sodium carbonate. It is used in making soaps, tanning, chlorinating roasting in metallurgy and glazes for ceramics. In the laboratory, S. is used to make cooling mixtures. NaCl solutions isotonic with body fluids are used in medicine as physiological saline. S. is also used to thaw ice on roads - like cattle salt, this salt is mixed with iron oxide, or it may be used in a mixture with magnesium chloride.

Sodium chloride structure: same as Sodium chloride type (see).

Sodium chloride type, ***sodium chloride structure***: a common Structure type (see) for compounds with the composition AB. The crystal lattice of the prototype sodium chloride consists of two cubic face-centered lattices of sodium ions Na^+ and chloride ions Cl^-, which are displaced by 1/2 unit cell with respect to each other. Each ion is octahedrally surrounded by 6 ions of the other type (Fig. 1); this is a coordination lattice in which there are no separate Na^+Cl^- ion pairs or NaCl molecules. To a good approximation, the structure can be described as a cubic close packing of the Cl^- ions in which the octahedral holes are occupied by Na^+ ions. This model takes into account the relative sizes of the ions and the actual space filling (Fig. 2).

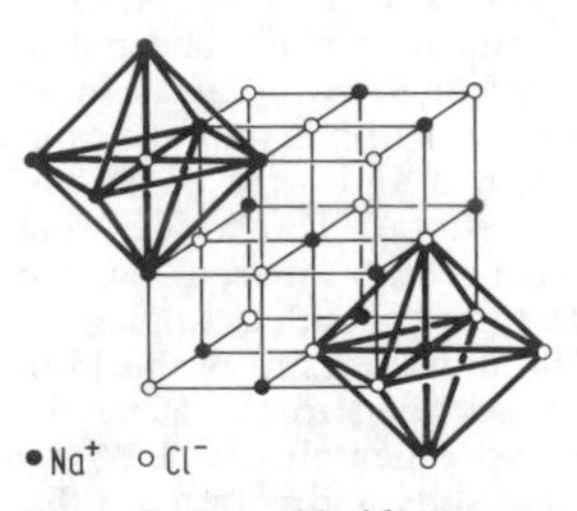

Fig. 1. Sodium chloride structure (schematic; the coordination octahedra are emphasized).

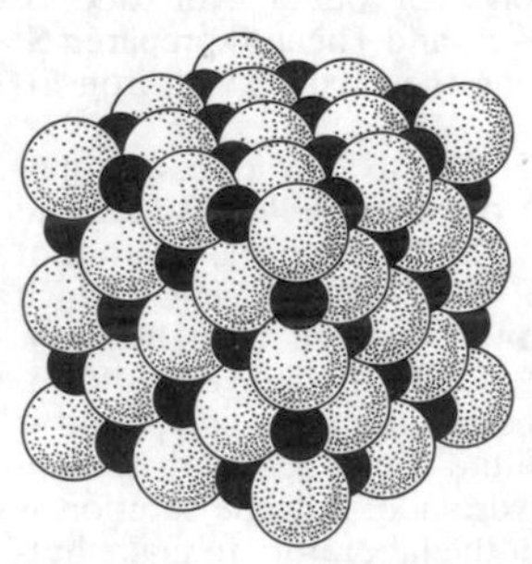

Fig. 2. Sodium chloride structure. The spheres (black, Na^+, white, Cl^-) correctly indicate the relative sizes and packing.

Over 200 compounds crystallize in the N., especially ionic substances, but also compounds with a strong covalent contribution to their bonding. The limiting geometric criterion for formation of a S. structure is that the ratio of the radii of the ions r_A/r_B or atoms is in the range 0.414 to 0.732, or close to it. Most alkali halides (except for CsCl, CsBr and CsI), the alkali hydrides, the alkaline earth chalcogenides (with the exception of the Be salts), various nitrides and carbides (e.g. TiN, TiC, VC), PbS, FeO, AgF, AgCl, AgBr, etc. crystallize in the N.

Sodium chlorite: $NaClO_2$, colorless crystals, soluble in water; M_r 90.45. S. decomposes between 180 and 200 °C, part of it forming sodium chlorate and part oxygen. In the presence of combustible substances, this decomposition occurs explosively. S. is poisonous. It is made by passing chlorine dioxide through a mixture of sodium hydroxide solution and hydrogen peroxide: $2\ ClO_2 + H_2O_2 + 2\ NaOH \rightarrow 2\,NaClO_2 + O_2 + 2\ H_2O$. In acid solution, S. is a strong oxidizing agent (see Chlorites). It is therefore used as a gentle bleach in the textile and paper industries. It is also used to disinfect drinking water.

Sodium chromate: Na_2CrO_4, yellow, rhombic crystals; M_r 161.97, density 2.73. At 413 °C it converts to a yellow, hexagonal modification. S. crystallizes out of water (in which it is very soluble) below 62.8 °C in forms which contain 4 to 6 moles water of crystallization, depending on the temperature. Above 62.8 °C, the anhydrous form is obtained. S. is very poisonous; for its toxicity and synthesis, see Chromium. In aqueous solution, S. is in a pH-dependent equilibrium with sodium dichromate. It is used to make pigments, in dyeing, in corrosion prevention and as an oxidizing agent.

Sodium citrate: $Na_3C_6H_5O_7 \cdot 5.5H_2O$, colorless, very water-soluble, rhombic crystals; M_r 357.16, density 1.857. S. is used in medicine as a coagulation-preventing additive to stored blood and blood products; it is also used in the food industry.

Sodium cyanate: see Cyanates.

Sodium cyanide: NaCN, colorless, hygroscopic, cubic crystals (low temperature form < 6 °C: rhombic), M_r 49.015, density 1.59, m.p. 561.7 °C, b.p. 1500 °C. S. is readily soluble in water, and is stable in the absence of air, carbon dioxide and moisture. However, it is hydrolysed in moist air. Aqueous solutions of S. are very alkaline, due to the strong basicity of the cyanide ion. In aqueous solution, atmospheric oxygen oxidizes S. partially to sodium cyanate, NaNCO. S. is a technical product made by neutralization of sodium hydroxide with hydrogen cyanide. It is used in large amounts in mining to leach gold and silver ores (see Cyanide leaching). It is also used as a flotation aid in the enrichment of sulfidic zinc, lead and copper ores. In galvanic technology, it is used to make anionic cyano complexes of various metals. It is also used to make cyanide pigments, such as Prussian blue. S. is an important raw material for the production of organic intermediates.

Sodium dichromate: $Na_2Cr_2O_7 \cdot 2H_2O$, red-orange, columnar, monoclinic crystals; M_r 298.00, D_4^{13} 2.52, at 100 °C, anhydrous; m.p. 356.7 °C. S. is extremely soluble in water: 100 g water dissolves 433 g $Na_2Cr_2O_7$ at 100 °C. S. is also soluble in alcohol. It is very poisonous; on its toxicity and synthesis, see Chromium. S. is a strong oxidizing agent, especially in acidic solution. It is the starting material for production of many other chromium compounds, and is used to make tannins, pigments, oxidizing agents for the dye industry, in photography, in galvanic technology, and so on.

Sodium dicyanamide: see Dicyanamides.

Sodium disulfite: see Disulfites.

Sodium disulfitoaurate(III): $Na_3[Au(SO_3)_2]$, a compound which is relatively unstable in crystalline form; M_r 426.06. Aqueous solutions of S. are made by reaction of fulminating gold or gold(III) hydroxide with sodium hydroxide and sodium sulfite; the solutions are used for gold electroplating, for making ductile and non-wearing layers, and for white and rose gold electroplating.

Sodium dithionite, formerly ***sodium hydrodisulfite*** or ***sodium hydrosulfite***: $Na_2S_2O_4$, a white, crystalline powder; M_r 174.11, m.p. 52 °C (dec.). It crystallizes out of aqueous solution as the dihydrate, $Na_2S_2O_4 \cdot H_2O$. S. is readily soluble in water; in aqueous solution it is rapidly decomposed, mainly to sodium thiosulfate and sodium disulfate: $2\ Na_2S_2O_4 \rightarrow Na_2S_2O_3 + Na_2S_2O_5$; in the presence of atmospheric oxygen, the main decomposition product is sodium hydrogensulfate. S. is a strong reducing agent, e.g. it precipitates the metals out of gold, silver and copper salt solutions. It is synthesized by reduction of sodium hydrogensulfite in sulfurous acid solution with zinc according to $2\ NaHSO_3 + H_2SO_3 + Zn \rightarrow ZnSO_3 + Na_2S_2O_4 + 2\ H_2O$. It may also be produced by cathodic reduction of sodium hydrogensulfite. S. is used as a reducing agent in vat dyeing and to bleach wool, paper, soaps, etc. In gas analysis, it is used to absorb oxygen.

Sodium dodecylsulfate: same as Sodium laurylsulfate (see).

Sodium ethoxide, formerly ***sodium ethylate*** or ***sodium alcoholate***: $NaOC_2H_5$, colorless, hygroscopic powder which turns dark in the air. S. crystallizes with alcohol of crystallization: $NaOC_2H_5 \cdot 2C_2H_5OH$; crystals of this type are colorless. S. is made by putting sodium in ethanol. Alcoholic solutions of S. gradually turn red in air, due to oxidation. S. is immediately decomposed to ethanol and sodium hydroxide by water. Like all other sodium alkoxides, it is used in organic synthesis as a strong base.

Sodium ethylate: see Sodium ethoxide.

Sodium fluoride: NaF, colorless cube-shaped or octahedron-shaped, cubic crystals; M_r 41.99, D_4^{41} 2.558, m.p. 993 °C, b.p. 1695 °C. S. is only slightly soluble in water and insoluble in alcohol. Its aqueous solutions are weakly basic. ***Sodium hydrogenfluoride***, $NaHF_2$, crystallizes out of a solution containing hydrogen fluoride and S. as colorless rhombohedrons (the nature of the hydrogenfluoride ion $[HF_2^-]$ is discussed in the entry on Fluorides (see)). S. is made by the reaction of hydrogen fluoride with sodium hydroxide or carbonate solution, or by fusion of sodium hexafluoroaluminate (synthetic cryolite) with sodium hydroxide: $Na_3AlF_6 + 6\ NaOH \rightarrow Na_3[Al(OH)_6] + 6\,NaF$. S. is used mainly to protect wood, as an opacifying agent for glass and ceramics, as a flux in metallurgy and to fluoridate drinking water.

Sodium fluoroaluminate: same as Sodium hexafluoroaluminate (see).

Sodium fluorosilicate: same as Sodium hexafluorosilicate.

Sodium formate: HCOONa, colorless, hygroscopic, monoclinic crystals; M_r 68.02, density 1.92, m.p. 253 °C. S. dissolves very readily in water and crystallizes out of it in the form of various hydrates. Above 440 °C it loses hydrogen and is converted to sodium oxalate (COONa). S. is synthesized by neutralization of aqueous formic acid with sodium hydroxide or carbonate, or, on an industrial scale, by the reaction of carbon monoxide with sodium hydroxide under pressure. S. is used to make oxalic acid, in tanning and dyeing, and as a component of preservatives.

Sodium fulminate: see Fulminates.

Sodium hexafluoroaluminate, ***sodium fluoroaluminate***: Na_3AlF_6, colorless, monoclinic crystals; M_r 209.94, density 2.90, m.p. 1000 °C. S. is found in nature as cryolite, and in countries with natural deposits, it is used to make sodium carbonate. S. can be synthesized by dissolving aluminum oxide hydrate and sodium carbonate in aqueous hydrofluoric acid.

Sodium hexafluorosilicate, ***sodium fluorosilicate***, formerly ***sodium silicofluoride***: Na_2SiF_6, colorless, hexagonal crystals which are barely soluble in water; M_r 188.06, density 2.697. S. is made by neutralization of hexafluorosilicic acid, H_2SiF_6, with sodium hydroxide or carbonate. In large amounts, S. is toxic. It is used to fluoridate drinking water, as an opacifying agent in glasses and enamels, and as a pesticide.

Sodium hexahydroxostannate(IV): $Na_2[Sn(OH)_6]$, colorless, hexagonal crystals; M_r 266.75. Above 140 °C, it is converted to the anhydrous form, Na_2SnO_3. S. dissolves in water, and its solublity decreases slightly as the temperature increases. It is made by dissolving a melt of tin(IV) oxide and sodium hydroxide in water. S. is commercially available as ***preparing salt***; it is used in the textile industry, galvanic technology and as a stabilizer for hydrogen peroxide.

Sodium hexanitrocobaltite: $Na_3[Co(NO_2)_6]$, yellow crystalline powder which dissolves in water to give a yellow-brown solution; M_r 403.98. It is made by passing air into a solution of sodium nitrite and cobalt(II) nitrite in acetic acid. S. is used in qualitative analysis as a reagent for detection of K^+ ions, with which it forms the insoluble potassium hexanitrocobaltate, $K_3[Co(NO_2)_6]$.

Sodium hydride: NaH, colorless crystals; M_r 24.01, density 0.92. S. cannot be melted without decomposing into the elements. It is made by the reaction of elemental hydrogen with molten sodium at 300 to 400 °C. It has all the properties of a typical ionic hydride, which indicates the presence of the hydride ion. For example, it immediately (often explosively) forms hydrogen with any proton donor (acid). This is true, for example, of its reactions with water, alcohols, ammonia or hydrogen halides. The basicity of the hydride ion is utilized in the Claisen condensation with S. In addition, S. is a strong reducing agent. When heated in air, it ignites and burns to sodium oxide and water. It combines with elemental sulfur to form sodium sulfide and hydrogen sulfide. S. is also used to reduce organic compounds, to make sodium boranate and to remove the last traces of water from solvents.

Sodium hydridoborate: same as Sodium boranate (see).

Sodium hydrodisulfite: see Sodium dithionite.

Sodium hydrogencarbonate, ***primary*** or ***acidic sodium carbonate, baking soda***, formerly ***sodium bicarbonate***: $NaHCO_3$, colorless monoclinic crystalline powder which is stable in dry air; M_r 84.00, density 2.159. Above 270 °C, S. is converted to sodium carbonate (soda): $2\ NaHCO_3 \rightarrow Na_2CO_3 + CO_2 + H_2O$. In aqueous solution, this reaction begins even at room temperature, and becomes rapid at 65 °C. The moderate solubility of S. in water gives S. its place in the industrial synthesis of Soda (see). Aqueous solutions are weakly basic. S. can be synthesized on a laboratory scale by passing carbon dioxide through an aqueous soda solution: $Na_2CO_3 + CO_2 + H_2O \rightarrow 2\ NaHCO_3$. In industry, S. is an intermediate in soda production by the Solvay process. It is used to make baking powder, as a stomach antacid, in fire extinguishers and in buffer solutions.

Sodium hydrogensulfate, ***primary*** or ***acid sodium sulfate***, formerly ***sodium bisulfate***: $NaHSO_4$, colorless, triclinic crystals which are readily soluble in water; M_r 120.09, D_4^{13} 2.435. Aqueous solutions are acid. When heated, S. is dehydrated to sodium disulfate: $2\ NaHSO_4 \rightarrow Na_2S_2O_7 + H_2O$. When heated still further, it decomposes to sodium sulfate, Na_2SO_4, and sulfur trioxide, SO_3. S. is synthesized by the reaction of moderately warm concentrated sulfuric acid with sodium chloride: $H_2SO_4 + NaCl \rightarrow NaHSO_4 + HCl$. The acidity of S. is utilized in many areas of the chemical, paper and textile industries. It is used in the laboratory, for example, to solubilize insoluble compounds and to clean platinum crucibles.

Sodium hydrogensulfite: ***primary*** or ***acid sodium sulfite***, formerly ***sodium bisulfite***: $NaHSO_3$, colorless crystals which are very soluble in water; M_r 104.07, density 1.48. When heated, S. eliminates water and is converted to sodium disulfite, $Na_2S_2O_5$. It is obtained by passing sulfur dioxide into a cold saturated solution of sodium carbonate: $Na_2CO_3 + 2\ SO_2 + H_2O \rightarrow 2\ NaHSO_3 + CO_2$. S. is used as a reducing agent in dyeing and printing, as a bleach for textiles and paper, in the fermentation industry and to make sodium dithionite.

Sodium hydrosulfite: see Sodium dithionate.

Sodium hydroxide, ***caustic soda***: NaOH, white, opaque, fibrous crystalline, brittle and highly hygroscopic solid; it is sold in the form of scales or wafers. M_r 40.00; density, 2.130, m.p. 318.4 °C, b.p. 1390 °C. NaOH dissolves in water in a strongly exothermal process; at 0 °C, 100 ml water will dissolve 42 g, and at 100 °C, 347 g NaOH. The aqueous solution is called *lye*. The concentration of the solution can be determined from its density:

5	10	20	30	40	50	%NaOH
1.0538	1.1089	1.2192	1.3277	1.4299	1.524	density

NaOH should always be stored in closed containers, because it is rapidly converted to sodium carbonate by the CO_2 in the air. It is synthesized mainly through Chloralkali electrolysis (see); the aqueous solution resulting from that process is evaporated to yield the solid. The reaction of soda with slaked lime (caustification) to produce lye or solid NaOH is now of historical interest only. $Na_2CO_3 + Ca(OH)_2 \rightarrow 2\,NaOH + CaCO_3$. Solid NaOH is used for oxidizing fusions in the dye industry, and NaOH fusion is used in analytical chemistry to dissolve refractive substances. NaOH solutions (lye) are used widely in producing soaps, purifying fats, oils and petroleum, in digesting wood for cellulose production (for non-acid papers), in metallurgy and in the production of many chemicals.

Sodium hydroxide is very irritating to the skin and mucous membranes. Safety glasses and protective clothing should be worn by anyone working with concentrated solutions. If it is splashed onto the body, the affected area should be washed immediately with a large amount of water, followed by 1% acetic acid.

Sodium hypochlorite, NaOCl, M_r 74.44, is unknown in anhydrous form. The isolated salts are hydrates with varying water contents, usually $NaOCl \cdot 2.5\,H_2O$, m.p. 57.5 °C. The aqueous solutions are also known as ***eau de Labarraque***; they are unstable and strongly oxidizing. They are obtained by electrolysis of aqueous sodium chloride solutions without a diaphragm; the chlorine and sodium hydroxide formed by electrolysis immediately react to form S.: $2\,NaOH + Cl_2 \rightarrow NaOCl + H_2O$. Aqueous solutions of S. are used as bleaches for paper, straw, cotton, soaps, etc, and as disinfectants.

Sodium iodate: $NaIO_3$, colorless, rhombic crystals which are readily soluble in water; M_r 197.12, density 4.277, dec. when heated. S. is a strong oxidizing agent. As a component of Chile saltpeter, from which it is isolated, it is the starting material for production of iodine and its compounds.

Sodium iodide: NaI, colorless, cubic crystals which dissolve easily in water, ethanol and acetone; M_r 149.89, density 3.667, m.p. 661 °C, b.p. 1304 °C. S. crystallizes out of water at room temperature as the dihydrate, $NaI \cdot 2H_2O$. It is synthesized by reaction of sodium carbonate with iron(II,III) iodide. S. is used to iodinate table salt, in photography and in medicine.

Sodium laurylsulfate, ***sodium dodecylsulfate***: $C_{12}H_{25}$-O-SO_3Na, M_r 288.38, a valuable surfactant used to condition textiles and in biochemical research for solubilizing proteins (see Alkyl sulfates).

Sodium malonate: see Malonate syntheses.

Sodium metaborate: $NaBO_2$, colorless, water-soluble hexagonal crystals; density 2.464, m.p. 966 °C, b.p. 1434 °C. S. forms various hydrates, for example $NaBO_2 \cdot 4H_2O$, which is converted to the anhydrous form above 120 °C. The basic structure of S. is a cyclic, trimeric metaborate anion, so that its formula is more accurately written as $Na_3[B_3O_6]$; M_r 197.46. The structure of S. is discussed under Borates (see). Aqueous solutions of S. are basic. It is synthesized by fusing borax with sodium carbonate or hydroxide. S. is used as an additive to photo developers, glues, galvanic baths and herbicides; it appears in borax beads.

Sodium metaborate-hydrogen peroxide dihydrate: see Sodium perborate.

Sodium metasilicates: compounds with the general formula Na_2SiO_3 formed by fusing quartz with sodium carbonate or hydroxide; they are commercially available as the water-soluble pentahydrate, $Na_2SiO_3 \cdot 5H_2O$ and the nonahydrate, $Na_2SiO_3 \cdot 9H_2O$. The structures of these salts are based on long-chain or cyclic condensed polysilicate anions, $[SiO_3^{2-}]_n$. Aqueous solutions are basic. S. act as water softeners, emulsifiers and buffering agents, and are added to laundry and cleaning products for the sake of these activities. They are also used to degrease metals.

Sodium nitrate: $NaNO_3$, colorless, hygroscopic, rhombic crystals; M_r 84.99, density 2.261, m.p. 306.8 °C. S. dissolves easily in water, in an endothermic reaction. It is found in nature in Chile saltpeter, but it is now produced by neutralization of nitric acid with sodium carbonate or hydroxide: $Na_2CO_3 + 2\,HNO_3 \rightarrow 2\,NaNO_3 + CO_2 + H_2O$. S. is used as a fertilizer, as an oxidizing agent in fireworks, in the synthesis of potassium nitrate and sodium nitrite, and as a preservative in meat products.

Sodium nitrite: $NaNO_2$, yellowish white, somewhat hygroscopic, rhombic crystals; M_r 69.00, density 2.168, m.p. 271 °C, dec. above 320 °C. S. is often commercially available in small yellowish-white cylinders. It is poisonous. It oxidizes many substances, but is itself oxidized by strong oxidizing agents such as sodium nitrate. S. is made by passing a mixture of nitrogen oxide and nitrogen dioxide into a sodium hydroxide solution. It is converted to nitrous acid or a mixture of nitrogen oxides by even weak acids, and this is the basis for its wide use as a diazotizing reagent with dyes. S. is also used to make hydroxylamine and as a corrosion inhibitor.

Sodium nitroprusside: see Cyanoferrates.

Sodium oleate: $C_{17}H_{33}COONa$, a colorless powder soluble in water and alcohol; M_r 304.44. Aqueous solutions are basic. S. is used as an emulsifier and flotation agent.

Sodium oxalate: $(COONa)_2$, colorless, crystalline needles which are moderately soluble in water; M_r 134.01, density 2.34, dec. at 250-270 °C. S. is present in many plant cells. It is synthesized by heating sodium formate: $2\,HCOONa \rightarrow (COONa)_2 + H_2$. S. is used to treat textiles and in pyrotechnics, and as a titrating substance in manganometry.

Sodium oxide: Na_2O, colorless powder; M_r 61.99, density 2.27, subl.p. 1275 °C. S. dissolves in water with a highly exothermic reaction, forming sodium hydroxide. It is made from sodium peroxide, Na_2O_2, by treatment with sodium, or by reaction of sodium hydroxide with sodium: $2\,NaOH + 2\,Na \rightarrow Na_2O + H_2$. It is used in synthesis as a very basic condensation reagent, and also to dehydrate solvents.

Sodium palmitate: $C_{15}H_{31}COONa$, white, wax-like mass; M_r 278.42. S. is used as an emulsifier in cosmetics and cleansers.

Sodium perborate: the common name for 1) ***sodium metaborate-hydrogen peroxide trihydrate***, $NaBO_2 \cdot H_2O_2 \cdot 3H_2O$, and for 2) ***sodium tetraborate-hydrogen peroxide nonahydrate, perborax***, $Na_2B_4O_7 \cdot H_2O_2 \cdot 9H_2O$.

1) Colorless, monoclinic crystals which decompose above 60 °C, forming water and oxygen. Formula mass 153.88. In aqueous solution, S. releases hydrogen peroxide, slowly when cold and rapidly when warm. This gives it its bleaching and disinfecting activity. S. is synthesized by allowing it to crystallize out of an H_2O_2-containing solution of sodium metaborate, or by the direct reaction of sodium peroxide with boric acid or borates. S. is a component of many bleaches, laundry products, cleansers, pharmaceuticals and cosmetics, such as freckle creams.

2) Colorless crystals; M_r 397.36. S. is soluble in water, releasing H_2O_2, and slowly converting to 1):

$Na_2B_4O_7 \cdot H_2O_2 \cdot 9H_2O \rightarrow NaBO_2 \cdot H_2O_2 \cdot 3H_2O + NaH_2BO_3 + H_3BO_3 + 2\ H_2O$. S. is made by dissolving stoichiometric amounts of boric acid and sodium peroxide in water: $4\ H_3BO_3 + Na_2O_2 + 4\ H_2O \rightarrow Na_2B_4O_7 \cdot H_2O_2 \cdot 9H_2O$. It is used in the same way as 1).

Sodium percarbonate: sodium carbonate containing hydrogen peroxide of crystallization, with the average composition $2\ Na_2CO_3 \cdot 3H_2O_2$; a colorless, unstable powder which decomposes with the evolution of hydrogen peroxide or oxygen. S. is made from sodium carbonate, sodium hydrogencarbonate or carbon dioxide and hydrogen peroxide or sodium peroxide. Like sodium perborate, it is used as a component of laundry products and bleaches.

Sodium perchlorate: $NaClO_4 \cdot H_2O$, colorless, rhombic crystals which are readily soluble in water; M_r 140.47, melts at 482 °C with decomposition into sodium chloride and oxygen. S. is made industrially by electrolysis of an aqueous solution of sodium chlorate. S. is used as a component of explosives.

Sodium peroxide: Na_2O_2, yellowish white powder; M_r 77.99, density 2.805, m.p. 460 °C (dec.). S. dissolves in water in a highly exothermal reaction, forming primarily sodium hydroxide and oxygen. If S. is added to water with very efficient cooling, the octahydrate, $Na_2O_2 \cdot 8H_2O$, can be isolated. In cold sulfuric acid, S. is converted to hydrogen peroxide, H_2O_2. S. is a very strong oxidizing agent. When pure, it is stable and non-explosive, but mixtures with organic substances (straw, paper, etc.) ignite spontaneously, and when an organic solvent, such as ether or benzene, is poured over it, it often ignites. It is therefore kept in lead cans. S. is synthesized by burning sodium in air. It is used as a bleach for wool, cotton and paper, and as a strongly oxidizing solubilizing reagent. Its ability to bind CO_2 from the air and to evolve oxygen ($2\ Na_2O_2 + 2\ CO_2 \rightarrow 2\ Na_2CO_3 + O_2$) is the basis for its use in respiration units for divers and firefighters.

Sodium peroxodisulfate: $Na_2S_2O_8$, is a colorless, water-soluble crystalline powder; M_r 238.10. The aqueous, oxidizing solution produces oxygen when heated: $Na_2S_2O_8 + 2\ H_2O \rightarrow Na_2SO_4 + H_2SO_4 + O_2$. S. is synthesized by anodic oxidation of sodium sulfate. It is used as an oxidizing agent in vat dyeing and as a catalyst in emulsion polymerization.

Sodium phosphates: 1) in the narrow sense, sodium salts of orthophosphoric acid, H_3PO_4.

a) ***Sodium dihydrogenphosphate, primary sodium phosphate***, NaH_2PO_4, colorless, water-soluble, rhombic crystals; M_r 139,000, density 2.040. Primary S. also forms a rhombic dihydrate, $NaH_2PO_4 \cdot 2H_2O$; density 1.91, m.p. 60 °C. Above 100 °C, the water of crystallization is released, and at 200 °C, it condenses to ***disodium dihydrogen phosphate***, $Na_2H_2P_2O_7$. Upon stronger heating, sodium polyphosphates are formed. The aqueous solution is acidic. The primary phosphate is obtained by reaction of equimolar amounts of orthophosphoric acid and sodium hydroxide. It is used in the synthesis of polyphosphates, for softening water and as an additive to feed lime.

b) ***Disodium hydrogenphosphate, secondary sodium phosphate***, Na_2HPO_4, M_r 141.98, forms a rhombic dihydrate, a monoclinic heptahydrate and a rhombic dodecahydrate. Above 100 °C, the hydrates lose their water, and upon stronger heating they are converted to sodium diphosphate, $Na_2P_2O_7$. The secondary phosphate is formed by neutralization of phosphoric acid with stoichiometric amounts of sodium hydroxide or soda. It is used to synthesize fire-retardant cloth impregnation materials, as an additive to feeds, as an additive to bacterial nutrient media, in silk production, in textile dyeing, in ceramic glazes and to clean metal surfaces for soldering and welding.

c) ***Trisodium phosphate, tertiary sodium phosphate***, Na_3PO_4, M_r 163.94, crystallizes as the decahydrate, density 2.536, m.p. 100 °C, and as the dodecahydrate, density 1.62, m.p. 73.3 to 76.7 °C. The tertiary phosphate is soluble in water, and the solution is very alkaline. It is made industrially by neutralization of phosphoric acid or Na_2HPO_4 with sodium hydroxide. It is used in water softening and is a component of cleansers, paint removers and rust protectants. It supports the emulsifying action of detergents and, because of its basic reaction, accelerates the saponification of fats and oils. Chlorinated trisodium phosphate with the approximate composition 92% $Na_3PO_4 \cdot 12H_2O$, 3.25% NaOCl and 5% NaOH is used as a disinfecting, deodorizing and bleaching cleanser.

2) in the broad sense, ***sodium polyphosphates***.

a) ***Sodium diphosphate, sodium pyrophosphate***, $Na_4P_2O_7$, colorless, water-soluble crystals; M_r 265.90, m.p. 880 °C. The salt forms when $Na_2HPO_4 \cdot 12H_2O$ is heated to about 200 °C. There is also a decahydrate, $Na_4P_2O_7 \cdot 10H_2O$. The ability of the diphosphate to form stable complexes with Ca^{2+} and Mg^{2+} ions is the basis of its use as a water softener. It is also a component of laundry products and other cleansers.

b) ***Sodium triphosphate, pentasodium triphosphate***, $Na_5P_3O_{10}$, is a colorless, water-soluble salt, M_r 367.86. It also exists as a hexahydrate. It is synthesized by mixing phosphoric acid with sodium hydroxide solution or soda in the stoichiometric ratio, and the solution is concentrated by spraying. Because of its water-softening effect and its ability to accelerate saponification of fats, and because it acts synergistically with detergents, the triphosphate is widely used in laundry products, but also in cosmetics, cleansers, etc. In aqueous solution, it is gradually degraded, via the diphosphate, to orthophosphate, and is a major contributor to eutrophication of bodies of water.

c) ***Sodium metaphosphates***, $(NaPO_3)_n$, sodium salts of trimeric, tetrameric and in some cases more highly condensed cyclopolyphosphoric acids, which is obtained by heating sodium hydrogenphosphate above 630 °C. The structure of the metaphosphate anion is given under Phosphates (see). In water they are gradually degraded to orthophosphates; when they are heated, the process occurs more rapidly. The metaphosphates are widely used for water softening, in the textile, paper and photographic industries, and as a component of laundry products and cleansers.

d) ***Sodium polyphosphates*** in the narrow sense are the sodium salts of linear, condensed, high-molecular weight polyphosphoric acids. Their anions consist of long chains of (PO_3^-) anions bridged by oxygen atoms. This group includes ***Graham's salt***, which is formed by quenching a melt obtained by heating sodium dihydrogenphosphate above 620 °C. It is a colorless, hygroscopic, water-soluble, glassy mass. This salt is about 90% a mixture of high-molecular weight, unbranched polyphosphates with 30 to 90 PO_3^- and 10% cyclophosphates. It is used as a water softener, as a component of laundry products and cleansers, as a flotation agent and in various ways in tanning, dyeing and paper production. Heating sodium dihydrogenphosphate to 250-500 °C forms ***Maddrell's salt***, which is similar in composition to Graham's salt, but differs in the spatial arrangement of the polyphosphate chains. The result is that when it is boiled in water, it is converted mainly to the tricyclophosphate. ***Kurrol's salt***, which can be obtained from a sodium polyphosphate melt by following defined heating and cooling procedures, has a similar structure. It exists in two different forms, which when heated to 400 °C produce sodium trimetaphosphate $(NaPO_3)_3$ or Maddrell's salt.

Sodium silicofluoride: see Sodium hexafluorosilicate.

Sodium sulfate: Na_2SO_4, colorless, water-soluble orthorhombic crystals; M_r 142.04, density 2.68. Above 32.38 °C, N. crystallizes out of water in anhydrous form; below this temperature it crystallizes as the decahydrate, $Na_2SO_4 \cdot 10H_2O$, which is also known as ***Glauber's salt***. Glauber's salt forms large, monoclinic prisms. The anhydrous form of N. is found in nature as thenardite; various double salts of N. are also known. Large amounts of N. were formerly produced as a byproduct of hydrochloric acid production from sodium chloride and sulfuric acid. It is now obtained mainly from the residues from potassium chloride production. N. is used in the production of glass and paper, in dyeing, to free cellulose from wood, to treat cotton fabrics, to make ultramarine, sodium thiosulfate, sodium sulfide, water glass, etc. It is used medicinally as a laxative.

Sodium sulfide: Na_2S, is colorless when pure, but its crystals are usually yellow because traces of polysulfides are present; M_r 78.05, density 1.856, m.p. 1180 °C. S. dissolves readily in water and crystallizes as the nonahydrate, $Na_2S \cdot 9H_2O$, below 48 °C. This form exists as hygroscopic prisms; density 1.427. Aqueous solutions are basic. S. is gradually converted to sodium thiosulfate by oxygen: $2\ Na_2S + 2\ O_2 + H_2O \rightarrow Na_2S_2O_3 + 2\ NaOH$. Elemental sulfur forms ***sodium polysulfides***, Na_2S_2, with S. The compound is synthesized by reduction of sodium sulfate with carbon: $Na_2SO_4 + 4\ C \rightarrow Na_2S + 4\ CO$. S. is used to make sulfur pigments, to remove hair from hides to be tanned, to reduce nitro compounds, in the production of synthetic fibers and to precipitate metal ions in analytical chemistry.

Sodium sulfite: Na_2SO_3, colorless, water-soluble, hexagonal crystals; M_r 126.06, density 2.633. Below 37 °C, it crystallizes out of water as the monoclinic heptahydrate $Na_2SO_3 \cdot 7H_2O$. The aqueous solution is weakly basic and strongly reducing. S. is made by passing sulfur dioxide into an aqueous sodium carbonate solution. This forms sodium hydrogensulfite, which is converted to S. by neutralization with more sodium carbonate (or hydroxide): $2\ NaHSO_3 + Na_2SO_3 \rightarrow 2\ Na_2SO_3 + CO_2 + H_2O$. S. is used as a solubilizing reagent for cellulose, wood and straw, as a corrosion protection, to prevent oxidation of photographic developers, to pretreat feedwater for steam boilers and as a preservative.

Sodium-sulfur battery: an Electrochemical current source (see), and a secondary cell.

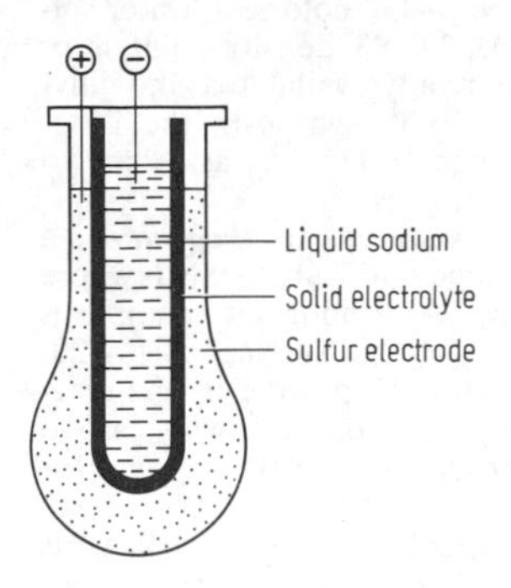

Diagram of the sodium-sulfur battery.

Liquid sodium forms the negative electrode. The discharge process can be formulated as follows: $Na \rightarrow Na^+ + e^-$. A sulfur electrode serves as cathode; since sulfur conducts neither electrons nor ions, the cathode is a mixture of sulfur with molten polysulfide on a graphite-felt matrix. The polysulfide participates in the electrode reaction according to the equation: $(x-1)Na_2S_x + 2\ Na^+ + 2\ e^- \rightarrow x\ Na_2S_{x-1}$. The electrolyte is a solid, nearly always β-aluminum oxide. The N. operates at a temperature of about 350 °C.

Sodium tetraborate decahydrate: same as Borax (see).

Sodium tetraborate-hydrogen peroxide nonahydrate: see Sodium perborates.

Sodium tetrafluoroborate: $NaBF_4$, colorless, water-soluble rhombic crystals; M_r 109.79, density 2.47, m.p. 384 °C. S. is made by neutralizing tetrafluoroboric acid, HBF_4, with sodium carbonate or hydroxide. It is used as a fluorination reagent and, in the laboratory, to make boron fluoride.

Sodium tetraphenylboranate: $Na[B(C_6H_5)_4]$, colorless, water-soluble crystals; M_r 342.24. It is used in quantitative analysis to precipitate K^+, Rb^+, Cs^+ and NH_4^+ ions.

Sodium thioantimonate, ***Schlippe's salt***: $Na_3SbS_4 \cdot 9H_2O$, bright yellow, water-soluble, cubic crystals; M_r 481.11, density 1.806, m.p. 87 °C. S. is formed by the reaction of sodium hydroxide and sulfur with antimony(III) sulfide, and is used to make antimony(V) sulfide.

Sodium thiocyanate: see Thiocyanates.

Sodium thiosulfate: $Na_2S_2O_3 \cdot 5H_2O$, colorless, water-soluble, monoclinic crystals; M_r 248.20, density 1.729, m.p. 40 to 45 °C. Above 100 °C, it is converted to the anhydrous form, which is also monoclinic. Addition of strong acid to an aqueous solution of S. forms thiosulfuric acid, which immediately decomposes to sulfur and sulfur dioxide: $Na_2S_2O_3 + 2\ HCl \rightarrow 2\ NaCl + H_2O + S + SO_2$. S. is oxidized to sodium hydrogensulfate by elemental chlorine: $Na_2S_2O_3 + 4\ Cl_2 + 5\ H_2O \rightarrow 2\ NaHSO_4 + 8\ HCl$. It is therefore used to remove chlorine from bleached fibers or textiles. S. is synthesized by reaction of sodium sulfite solution with elemental sulfur or sodium polysulfides.

The ability of S. to form very stable thiosulfate complexes makes it useful as a fixing salt for photographic films. It is also used to prepare silver ores, and in analytical chemistry (iodometry).

Sodium tungstate: Na_2WO_4, colorless, water-soluble, rhombic crystals; M_r 293.83, density 4.179, m.p. 698 °C; it is often commercially available as the dihydrate, $Na_2WO_4 \cdot 2H_2O$. For its synthesis, see Tungsten. S. is used to impregnate textiles, as an aid to dyeing textiles, and in ceramics.

Sodium xanthate, ***sodium xanthogenate***: a sodium salt of a xanthogenic acid ROC(S)SNa (where R is an aliphatic group). An important example is ***sodium ethyl xanthate***, $C_2H_5OC(S)SNa$, M_r 144.20, which is obtained as a yellowish powder by the reaction of ethanol and sodium hydroxide with carbon disulfide. It is used primarily as a flotation agent for sulfide ores.

Softening rinses: special cationic surfactants which adhere to textile fibers, giving them a soft, plush feel and also acting as an antistatic. S. must usually be added to the textiles after washing, but in combination with non-ionogenic surfactants, they can be used in the washing step. Distearyldimethylammonium chloride and long-chain alkylimidazolinium methylsulfate are examples of S.

Softening point, ***softening temperature***: the temperature at which a substance loses its rigidity and begins to collapse into itself. Crystalline solids have a defined conversion point, the Melting point (see), at which they pass from the solid to the liquid state. Amorphous substances, including glasses and most polymeric substances, however, soften before they enter the liquid or liquid-like state (if they are capable of liquidity). The S. of plastics can be changed by addition of Softeners (see). At low concentrations of softener, under certain conditions, behavior analogous to Freezing point lowering (see) can be observed. The lowering of the S. is proportional to the concentration of softener and independent of the structure of the softener molecule.

Softener, ***plastifier***: a liquid or solid, non-volatile organic compound which can be worked into polymeric substances and give them long-lasting, desirable physical properties, such as plasticity, elasticity, less hardness and lower hardening temperature. A S. should be chemically stable, physiologically harmless and compatible with high polymers. There are two main explanations of the action of S., a thermodynamic theory and a "theory of lubricating S." The second of these explains the action of S. as a lubrication between the macromolecules, while the thermodynamic theory interprets the action in terms of intermolecular forces. According to this theory, there is a thermodynamic equilibrium between solvated and unsolvated softener molecules and polymer molecules. The constants of this equilibrium depend on the temperature and on the relative molar masses and concentrations of the S. and the polymer. A strong polarity of the S. and the polymer leads to a high equilibrium constant, and thus to higher solvation. In structural terms, it is thought that the S. molecules penetrate between the individual macromolecules, weaken their non-covalent bonding, and occupy these active sites of the macromolecule themselves. This causes the structure of the polymer to loosen, makes it possible for the macromolecules to slide past each other and increases the plasticity of the mixture. The effects of S. and the deformation mechanics of softened plastics can be described by the temperature dependence of the dynamic shear and elasticity moduli. The lowering of the glass temperature of a given polymer by a given concentration of S. is a measure of the softening capacity of that S.

There are two basic types of softening, internal and external. *Internal softening* is due to a change in the structure of the macromolecule. For example, partial conversion of free hydroxyl groups into ethers or esters leads to a certain loosening of the rigid cellulose lattice, so that the material becomes softer. Similarly, postchlorination of polyvinyl chloride, PVC, and various types of mixed polymerization which lead to irregular polymers, can be considered internal S. *External softening* is the improvement of properties of a polymer by addition of S.

Classification. A distinction is made between gelatinizing S., which dissolve the polymer, and those which neither dissolve the polymer nor cause it to swell, but are to some degree miscible with it. S. can also be divided into lower and higher molecular weight categories; the high-molecular weight S. are generally polycondensates with molar masses between 600 and 8000. However, by far the larger number of S. have low molecular weights.

1) ***Esters*** are used to soften polar polymers. a) *Phosphoric acid esters*, especially the triesters of aliphatic alcohols and phenols are used, e.g. tributyl phosphate for vinyl polymers, cellulose derivatives, natural and synthetic rubbers; trimethylglycol phosphate and tributylglycol phosphate for cellulose nitrate; triphenyl phosphate for cellulose acetate; and tricresyl phosphate for many cellulose derivatives and for polyalkenes, polyisoalkenes and polyvinyls, especially polyvinyl chloride. b) ***Carboxylic acid esters***. Glycerol triacetate (triacetin), glycerol tripropionate (tripropionin), glycerol triethylbutyrate and triglycol ethylbutyrate are used to soften polyvinylbutyral; diglycoldicaprylate for synthetic rubber, and butyl stearate as a non-dissolving S. for cellulose nitrate. This category includes esters of low-boiling fatty acids (from paraffin oxidation) with monoalcohols such as tetrahydrofurfuryl alcohol or 2-ethylhexanol, or with polyols, such as hexanetriol, pentaerythritol or trimethylpropane. Among the most important S. are the esters of straight-chain, saturated dicarboxylic acids, such as adipic, azelaic, sebacic and succinic acids (e.g. the very cold-resistant dioctyladipate, dioctyl azelainate and dibutyl sebacinate). The esters of glycolic, lactic and ricinolic acids, as well as esters of benzene dicarboxylic acids, especially the phthalates, are also used.

2) ***Hydrocarbons*** are nonpolar S. They can be used for polar polymers only in a mixture with polar S. (esters), and otherwise are limited to use with nonpolar polymers. The aromatic and hydroaromatic hydrocarbons (e.g. higher alkylbenzenes, bisphenyl and its derivatives), alkyl naphthalenes and mixtures containing aromatics (anthracenol and naphthenic heavy petroleum distillates) can be used as S.

3) In a few special cases, ***alcohols***, especially polyols such as glycerol, triethyleneglycol or hexanetriol, or ***phenols***, especially polynuclear condensation prod-

ucts of ketones with phenols, are used as S. An example of the last type is diphenylolpropane.

4) ***Halogen-containing S.*** include long-chain, chlorinated paraffins, chlorohydrin esters, halogen derivatives of bisphenyl and its homologs, and chlorinated polynuclear aromatics. These are often used to reduce flammability.

5) The technically important ***sulfur-containing S.*** include a few thioethers, such as the dibenzyl ether of thiodiglycol (this is a mixture of 80% ditolylsulfide and 20% thianthrene), the esters of thioethers with mono- and dicarboxylic acids, and the alkylarylsulfones, especially the esters of aliphatic sulfonic acids with phenols and *N*-alkylated arylsulfonamides.

6) ***Nitrogen-containing S.*** include amides, nitriles and aliphatic and aromatic nitro compounds.

7) ***Epoxidized fatty acid esters***, such as epoxidated linoleates and soy oil esters, improve the both stability of soft PVC to heat and light and its behavior when cold.

8) Camphor, the oldest of all S., is still the best known of the S. with carbonyl groups, $-C{=}O$.

9) ***High-molecular-weight S.*** include polyolefins and polyesters with low degrees of polymerization.

Processing. 1) The S. is mixed directly with the polymer, which can be powdered or granulated, if necessary at an elevated temperature. 2) The polymer is dissolved in a solution of the S., as in the processing of cellulose nitrate in alcoholic camphor solution to make celluloid, and in all processes of making lacquer. The solvent is evaporated after the mixing. 3) The plastic absorbs the S. from an emulsion or solution. 4) The S. can be worked into the polymerization solution or dispersion, a method which is especially common when crude polymerization solutions are used, e.g. in the paint industry. 5) The S. is added to the monomers before polymerization. The S. used are of many different types, depending on the properties desired in the finished products. In some cases, up to 100% S. can be worked in.

Abbreviation	Compound
ABG	Adipic acid butyleneglycol polyester
ASE	Alkylsulfonic acid ester of phenol or cresols
DBP	Dibutyl phthalate
DBS	Dibutyl sebacinate
DIBP	Di-*iso*-butyl phthalate
DIDA	Di-*iso*-dodecyl phthalate
DIOP	Di-*iso*-octyl phthalate
DOA	Di-2-ethylhexyl adipate
DOP	Di-2-ethylhexyl phthalate
DOS	Di-2-ethylhexyl sebacinate
DOZ	Di-2-ethylhexyl acelainate
ELO	Epoxidized linoleate
ESO	Epoxidized soy oil ester
TCF	Tricresyl phosphate
TOF	Tri-2-ethylhexyl phosphate
TPF	Triphenyl phosphate

Sol: a colloidal distribution of a solid. Depending on the aggregate state of the dispersion medium, it can be an aerosol (gas), lyosol (liquid) or xerosol (solid).

Solanaceae alkaloids: see Tropane alkaloids.

Solanidanine: see Steroid alkaloids.

Solanum alkaloids: see Steroid alkaloids.

Solar distillation: see desalinization.

Solders: alloys, or more rarely, pure metals and glasses, which are used to join metals, glass, ceramic or carbon (soldering). An S. must have a lower melting point than the materials it is to bind. It must also wet the surfaces and be sucked into the finest pores by capillary action. With metallic S., the bonding occurs mainly through diffusion of the two metals into each other; zones of alloy form at the interface. If the diffusion of the solder components occurs mainly along the grain boundaries of the other metal, it leads to brittleness (see Metal corrosion, chemical). The working temperature is always above the solidus temperature, and may be the same as the liquidus temperature.

S. are classified, according to the working temperature T into soft ($T < 450\,°C$) and hard ($T > 450\,°C$) S. 1) ***Soft solder***: alloys based on lead and tin are used to solder heavy metals (including silver and gold), steel, platinum alloys and hard metals. The lead content is usually greater than 38.1%. If the adjacent material must not be damaged by the heat, S. with solidus temperatures below 183 °C are used: Wood's metal (see), Rose's metal (see) and Newton's metal (see). Zinc-tin-cadmium, cadmium-zinc or zinc-tin alloys are used for soft S. for aluminum and its alloys; these S. contain 35-70% tin, 17 to 65% zinc and 24 to 83% cadmium.

2) ***Hard solders***: copper-based alloys with nickel, cadmium and zinc, and silver- or aluminum-based alloys are used as hard S. for heavy metals and aluminum. Heavy metals are usually soldered with brass alloys. To prevent volatilization of the zinc, brass alloys contain a few tenths percent silicon or up to 5% silver. If greater strength is required, new silver is used. Silver alloys with at least 12% silver are used for hard soldering of noble metals. Aluminum-silicon alloys with about 12% silicon are used to hard solder aluminum materials.

S. for binding different types of materials. If the S. does not adequately wet the non-metallic material (glass, ceramic or graphite), a suitable metal coating must be applied before soldering. Steel can be soldered to hard metal with silver S. (copper-silver alloys) and copper S. (pure copper). Silver and lead-tin S. are used to join metals to ceramics. Metals can be soldered to glass with indium, bismuth-tin-cadmium and zinc-lead S. Metals can be soldered to graphite with titanium-zirconium-boron and molybdenum S. Glass solders (see) are increasingly used to join glass and ceramics.

Solidus curve: see Melting diagram.

Solid spiritus: Denatured alcohol (see) which has been made into a solid gel by addition of soap or cellulose acetate.

Solid bath heaters: see Heating baths.

Solid petroleum: a solid mass of about 90 parts petroleum, 10 parts sodium salts of fatty acids and 7.5 parts sodium chloride.

Solid-phase synthesis, ***Merrifield synthesis***: a method of synthesizing polypeptides on insoluble, polymeric carriers. The first amino acid is bound via its carboxyl group to an insoluble, easily filtered polymer; and the peptide is synthesized from its C-terminal end (Fig.). A suitable carrier is a copolymer of polystyrene and 1 to 2% 1,4-divinylbenzene. Beads (20 to 100 μm in diameter) of the polymer are allowed to swell in the organic solvent used for the synthesis, and thus become permeable to the reagents. The classic anchor group is the chloromethyl group, which can be introduced into the polymer by a Friedel-Crafts reaction in the presence of tin(IV) chloride.

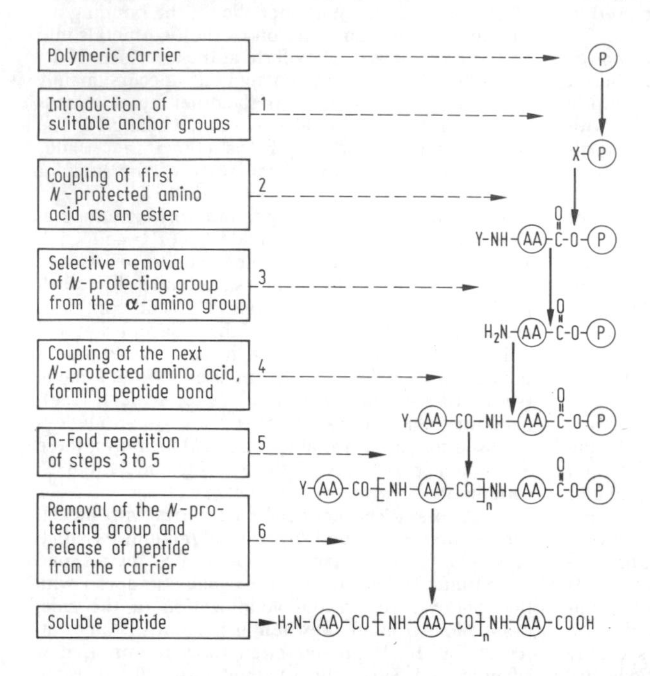

The chloromethyl polymer is esterified with the N-protected amino acid (*tert.*-butyloxycarbonyl- or fluorenyl-9-methoxycarbonyl-amino acids are generally used). After the first *N*-protected amino acid has been coupled, all operations are carried out in a glass reactor provided with a filter and supply and drain lines. The process can be carried out either manually or automatically in commercial synthesizers. The N^{α}-protecting group is removed from the covalently anchored amino acid, and the next *N*-protected amino acid is coupled, usually by the dicyclohexylcarbodiimide method. In this way, the polypeptide chain is lengthened, one amino acid at a time, in the interior of the polymer bead. After the last amino acid has been added, the covalent bond between the C-terminal amino acid and the anchor group of the polymeric carrier is cleaved. The insoluble carrier can be separated from the peptide, which is now in solution, by filtration. In principle, S. eliminates the complicated and time-consuming purification of intermediates. The desired product remains bound to the polymer, while excess reagents and byproducts of the reaction are removed by filtration. However, uniform peptide products are only obtained if the repeated coupling and deblocking reactions (Fig., steps 3 and 4) are practically quantitative. Since this is not entirely possible in practice, partial sequences and wrong sequences accumulate in the course of S., and make purification of the final product a difficult problem. However, modern separation techniques, such as preparative high-pressure liquid chromatography, make an adequate final purification possible in principle.

Various biologically active peptides and proteins have successfully been synthesized using S., even though the products were not always pure. In spite of its limitations, S. is a valuable aid in the synthesis of polypeptides and small proteins, and with further improvements, especially in the procedures for separation and purification of products, it will make possible the chemical synthesis of more proteins.

The principle of the Merrifield synthesis can also be applied to the synthesis of oligo- and polynucleotides. Synthesis automats are commercially available.

In 1963, a S. in which the peptide is lengthened from the *N*-terminal end was described by Letsinger. In spite of further improvements, however, this ***Letsinger synthesis*** is not widely used.

Solid state chemistry: that area of chemistry dealing with the reactions of substances in the solid state (see Solid state reactions). In the broad sense, it is Crystal chemistry.

Solid-state reactions: chemical reactions in which at least one of the reactants is a solid. In S. in the narrower sense, two or more solids react with each other, but in many cases liquid or gaseous phases are involved. S. also include conversion processes of a solid substance in which there are no other reactants.

S. have a number of peculiarities not shared with other chemical reactions. Most of these are due to the fact that solids are usually much less reactive than liquids, dissolved substances or gases. In order for the reaction rate to become significant, the components of the solid which are involved must be highly mobile within the lattice. This is usually achieved only at high temperatures, and depends on the presence of Crystal lattice defects (see).

The most important requirement for many S. is *oxygen transport* through the lattice to the site of reaction. This process occurs at a rate in solids which is many orders of magnitude lower than in liquids or gases, and there are different types of diffusion (Fig.).

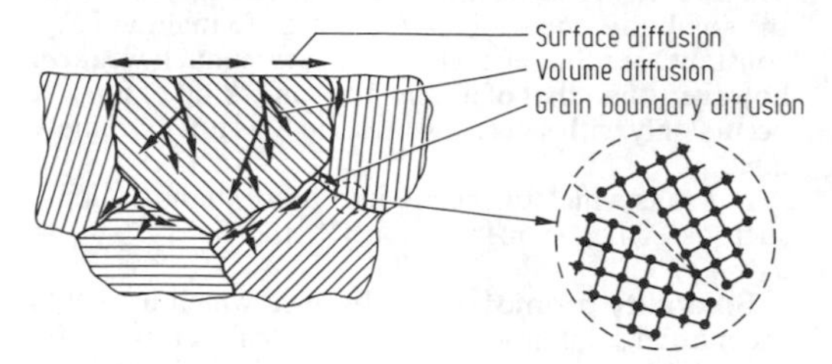

Two-dimensional scheme of the possible forms of diffusion in a polycrystalline solid.

Since a direct exchange of position in the interior of a solid requires high activation energies, and is therefore very improbable, crystal lattice defects play an essential role in *volume diffusion*. Point defects in particular make possible matter transport through vacancies and interlattice positions, but displacements also permit increased particle mobility along the defect. *Grain-boundary diffusion* depends on the loosening of the lattice structure in the region of grain boundaries. If crystal components or adsorbed particles migrate along the surface of the solid, it is called *surface diffusion*. The diffusion coefficients for volume, grain-boundary and surface diffusion increase markedly in the order listed, often by a few orders of magnitude. In real solids, all these types of diffusion occur. In addition to diffusion, transport of matter via the gas phase is important in S. (see Transport reactions).

S. can be classified according to the number and the physical state of the reactants. S. which have only one solid starting phase include *modification changes* of solids (see Polymorphism). These are often inhibited, leading to the existence of metastable phases. Another important group are the *decomposition reactions*, which usually require the input of energy and follow the scheme: $A_{solid} \rightarrow B_{solid} + C_{gas}$. They can be followed using thermo-gravimetric methods. Some examples which are applied on an industrial scale are the thermal decomposition of calcium carbonate at temperatures between 1200 and 1300 K ("lime burning") and the preparation of calcined soda from sodium hydrogencarbonate. Another such reaction is the photochemical decomposition of silver halides according to the equation: $AgX \rightarrow Ag + 1/2\ X$. The S. with more than one starting phase include *solid-gas* reactions with a solid product. Those with the greatest practical significance are the reduction of solids, such as oxides, sulfides, etc. with hydrogen and the reaction of metals with various gases, such as oxygen, halogens, etc. The solid products of these reactions can form a layer coating the starting phase; a thin layer is called *tarnish*, and thicker ones are called *scale*. Both are important in corrosion processes.

If the coating layer is porous, the reaction can continue rapidly, due to diffusion of the gas through the pores, eventually consuming the entire solid. However, if the layer is coherent, the reaction is inhibited (e.g. formation of an effective protective layer of Al_2O_3 on metallic aluminum; see Anodic oxidation). *Solid-liquid* reactions include, for example, corrosion processes in liquid media, in which solid coating layers can also form. The dissolution of crystals in a solvent are usually not included among the S., if there are no solid products. In principle, this group also includes electrochemical processes used to produce or refine metals.

Solid-solid reactions do not occur rapidly without a) a suitable pretreatment of the reactants, in order to convert them into a reactive state (see Tribochemistry), b) thorough mixing of the components, to provide the largest possible area of contact (in addition to mechanical mixing, there are other methods of homogenization, such as mixed precipitation, spray drying and soaking one of the components in a solution of the other) and c) elevated temperatures. High reaction rates can also be achieved when the S. is carried out in the temperature range at which the modification of the solid phase changes (*Hedvall effect*). One group of solid-solid reactions are the *additive S.*, in which complex compounds are formed from simple oxides, sulfides, halides, etc. An example which is well characterized is the formation of spinell, $MgAl_2O_4$, from magnesium and aluminum oxides: $MgO + Al_2O_3 \rightarrow MgAl_2O_4$. It has been shown that the reaction mechanism is a counterdiffusion of cations. Other examples are reactions between solid halides (e.g. $2\ AgI + HgI_2 \rightarrow Ag_2HgI_4$) and silicate formation: $2\ CaO + SiO_2 \rightarrow Ca_2SiO_4$. *Double reactions* or *exchange reactions* are S. of the type $AX + BY \rightarrow AY + BX$ (A and B are cations, X and Y are anions). An example is the reaction $BaO + MgCO_3 \rightarrow BaCO_3 + MgO$. *Displacement reactions*, e.g. $Pb + 2\ AgCl \rightarrow PbCl_2 + 2\ Ag$, follow a similar reaction scheme.

Transport reactions (see) are also S. in the sense of the definition given at the beginning.

The main applications of S. are in inorganic chemistry. S. of crystalline organic compounds have been known for a long time, but they have so far been relatively little studied and have no great practical significance at present. However, lattice-controlled reactions are used in certain areas of polymer chemistry; the bifunctional monomers are packed in the crystal in such a way that polymerization can occur without matter transport or destruction of the latter.

S. have great practical significance and are used whenever the reactants are too resistant to temperature or for other reasons cannot be made to react in a liquid form. Some examples are the previously mentioned lime burning and calcination of soda; in addition S. are involved in the production of cement, many silica-containing ceramic products (chamotte, clinker and porcelain) and non-silicate ceramics with special properties, such as barium titanate, $BaTiO_3$, and ferrites with spinell structure. S. are also involved in the production of industrial materials by powder metallurgy. Recently composites of metallic and nonmetallic components, the cermets, have become important.

Solubility, ***miscibility***: the ability of two substances to form a homogeneous solution or mixture with one another. There is no reliable method of predicting S. at present, although there are thermodynamic criteria. For liquids, the rule is that similar substances are mutually soluble (see Solvent). The decisive factor is the nature and magnitude of intermolecular interactions; dipole moments, dielectric constants, polarizability and the ability to form hydrogen bridges are crucial considerations. Lattice type and similarity of ionic radii are important factors in the formation of mixed crystals in solid solutions. S. is dependent on temperature (see Solution) and can be affected by Solubility promoters (see).

S. is reported quantitatively by indicating the saturation concentration, often in grams solute per 100 g solvent, and for gases, by indication of the Henry absorption coefficient (see Henry Dalton law).

Solubility curve: see Melting diagram.

Solubility product, abb. S: an equation which is derived from the mass action law as applied to electrolytic dissociation of relatively insoluble salts. For a binary salt AB, the dissociation equilibrium is $AB \rightleftharpoons A^+ + B^-$, and by the mass action law,

$$K = \frac{a_{A^+} + a_{B^-}}{a_{AB}}$$

Here a_{A^+}, a_{B^-} and a_{AB} are the activities of the ions A^+ and B^- and the undissociated salt AB. In a saturated solution with a precipitate of AB, a_{AB} is constant, and can be combined with K to a new constant: $Ka_{AB} = S_{AB}$. The resulting equation $S_{AB} = a_{A}+a_{B^-}$ is called the S. The constant S_{AB} has a characteristic value for each salt, which depends on the temperature (table). Since $a = fc$ (see Activity), the S. can also be formulated in terms of concentrations:

$$S_{AB} = c_{A^+}c_{B^-}\, f_{A^+}f_{B^-}.$$

For salts which are very sparingly soluble, the ion concentrations are so low that the activity coefficients f_{A^+} and f_{B^-} approach 1 and $S = c_{A}+c_{B^-}$.

Solubility products of some sparingly soluble salts at 293 K

Salt	S in mol² l⁻²	Salt	S in mol² l⁻²
AgCl	$1.10 \cdot 10^{-10}$	CuS	$1.34 \cdot 10^{-11}$
AgBr	$4.80 \cdot 10^{-13}$	CdS	$9.41 \cdot 10^{-11}$
AgI	$9.60 \cdot 10^{-17}$	HgS	$3.14 \cdot 10^{-15}$
AgSCN	$2.10 \cdot 10^{-12}$	$CaCO_3$	$2.07 \cdot 10^{-8}$
$PbSO_4$	$1.90 \cdot 10^{-8}$	$CaSO_4$	$2.16 \cdot 10^{-4}$
$PbCrO_4$	$9.54 \cdot 10^{-14}$	$SrSO_4$	$4.29 \cdot 10^{-7}$
		$BaSO_4$	$1.06 \cdot 10^{-10}$

The following results can be derived from the S.: 1) If a solution of a salt AB is saturated, $c_{A^+} = c_{B^-} \approx c_s$, where c_s is the saturation concentration of the salt. It follows that $c_s = \sqrt{S_{AB}}$, for example for AgCl: $c_s = \sqrt{1 \cdot 10^{10}\ \text{mol}^2\ \text{l}^{-2}} = 1.05 \cdot 10^{-5}$ mol l⁻¹, or 1.5 mg per liter. 2) If the concentration of one of the two types of ion is increased by addition of more of the same ion (e.g. addition of KCl to a saturated AgCl solution), S_{AB} is initially exceeded. A sufficient amount of AB precipitates to restore the S. Thus addition of one of the ions reduces the solubility of AB. This is why an excess of precipitant is used in gravimetric analysis.

3) Addition of foreign ions, i.e. ions which do not arise by dissociation of the salt AB (e.g. addition of KNO_3 to AgCl) has in the first approximation no effect on the S. However, these ions increase the ionic strength of the solution, and thus reduce the activity coefficients f_{A^+} and f_{B^-} (see Debye-Hückel theory). Therefore, the concentrations c_{A^+} and c_{B^-}, and thus the solubility, must increase in order to maintain S_{AB} constant (see Salt effect). This effect is much weaker, however, than that of adding the same kind of ions. It occurs only with salts for which the S. is not extremely low.

For salts which dissociate into more than two ions, such as $A_nB_m \rightarrow nA^{m+} + mB^{n-}$, the S. is $S_{A_nB_m} = a_A{}^n a_B{}^m$.

Solubility promoter: a substance which increases the mutual solubility of two substances. For example, the solubility of methanol and water in hydrocarbons or chlorinated hydrocarbons is increased by addition of alcohols or ethers with medium-length carbon chains. The S. must have structural features which permit strong molecular interactions with both of the components to be mixed. Addition of S. to a binary system with a mixing gap forms a ternary system with a smaller mixing gap. Salts can also increase solubility (see Salt effect).

Solubilization: 1) dissolving a substance in a solvent in which it is normally not soluble. In the case of S. in water, surfactant solutions above the critical concentration for micelle formation (see Colloids) can be used to bring slightly soluble or insoluble materials into solution. Below the critical micellar concentration (CMC), the solubility of the substance remains constant, and above it, it increases rapidly. There is always a distinct limiting value for the uptake of the substance to be solubilized, and this distinguishes the process from emulsification. The physical-chemical mechanism of S. corresponds to a distribution equilibrium of a substance between two immiscible phases. S. is widely used in practice. There are many pharmaceuticals which have limited solubility in water, and are therefore administered in solubilized form, e.g. steroids, antibiotics, sulfonamides. Other applications are found in cosmetics, the production and applications of herbicides, fungicides and insecticides. S. is also important in laundry washing and for fat absorption and membrane transport processes in organisms.

2) a preparatory step for chemical analysis which is required only when the usual treatments with acids, bases or other solvents can neither decompose nor dissolve the sample. In general, it is used to convert insoluble substances into substances which dissolve readily in water or acid. Many methods are known for the most diverse substances. A. with gaseous reagents are usually carried out in sealed, pressure-resistant apparatus to prevent loss of the products, which are also gaseous or volatile. *Oxygen* is used mainly for oxidative decomposition of organic substances. The ***Schöninger S.*** was introduced for this purpose in 1955; in it the organic substance is burned in pure oxygen in a closed glass vessel, and the products are absorbed by a solution present in the vessel. This method is used mainly to determine the halogen, sulfur and phosphorus contents of organic substances. *Chlorine* can be used for S. of metals and ores; simul-

taneously the volatile chlorides, e.g. of antimony, arsenic and tin, are separated from the nonvolatile ones.

S. with liquid reagents have become more important since the introduction of pressure vessels coated with polytetrafluoroethylene. Nearly all silicates can be rapidly dissolved in conc. hydrofluoric acid in such vessels at temperatures of 110-200 °C.

Such vessels can also be used for S. of organic substances, such as blood or foods, in conc. nitric acid at temperatures of 150-170 °C.

S. with solid reagents are simple in execution. The sample is mixed with an excess of reagent, and the mixture is heated in a crucible to the point of sintering or melting, then left for a time at this temperature. Sodium carbonate alone or mixed with potassium carbonate can be used for S. of silicates, tungstates, fluorides and sulfates in a platinum crucible. The products are the corresponding sodium compounds and the acid-soluble metal carbonates or oxides.

Acid-resistant steels, chromium ores and fire-resistant ceramics can be solubilized with sodium peroxide in crucibles of metallic zirconium. All the elements in the compounds are converted into compounds with the highest possible oxidation states. Organic substances can also be solubilized in closed stainless-steel or nickel vessels with sodium peroxide (***bomb solubilization***. Sodium disulfate is used for S. of insoluble metal oxides, e.g. Fe_2O_3, TiO_2, Nb_2O_5 and Cr_2O_3 in a platinum crucible; the corresponding metal sulfates are formed.

In the ***Freiberger S.***, the melting of arsenic, antimony or tin ores in a porcelain crucible with a mixture of sodium carbonate and sulfur, these elements form water-soluble thio salts, and can thus easily be separated from other metals, which are present as insoluble sulfides under these conditions.

Smith's S. with a mixture of calcium carbonate and ammonium chloride has become obsolete; it was used for determination of alkali metals in silicates.

Solution: a term for a liquid or solid mixture of several substances when one of the components, the Solvent (see), is present in large excess. The other components are called solutes. S. are always homogeneous, and the components are molecularly disperse. Colloidal S. (see Colloid) are not true S. in this sense; they are two-phase microheterogeneous systems. In S. of electrolytes (acids, bases, salts), the solutes are partially or completely dissociated into ions.

Before mixing, solutes can be solid, liquid or gaseous. The solubilization process is often associated with a chemical reaction. For example, when gaseous hydrogen chloride is dissolved in water, it dissociates into ions. Many salts display Solvolysis (see) in S. In addition, the ions are surrounded by a solvate shell (see Solvation), the extent of which depends on the ionic radius, charge and polarity of the solvent.

Solid S. can exist either in the amorphous state as glasses, or as mixed crystals (see Melting point diagram). Dissolved gases are usually incorporated into interlattice positions in the host crystal. Hydrogen is very soluble in many metals, especially in palladium and platinum. It is incorporated into the metal lattice in the atomic state. Molten salts and metals have high capacities for dissolving many substances; gases in the supercritical state can dissolve solids. The chemical compositions of S. are given by various Concentration measures (see). Many substances have a maximum Solubility (see) in a given solvent. A S. which contains this maximum amount of solute is called ***saturated***, and the concentration is the saturation concentration. The concentration is lower in an ***unsaturated S.*** and higher in a ***supersaturated S.***. A supersaturated S. is not stable: after formation of seeds of the new phase, the excess dissolved solute leaves the S. until the saturation limit is reached.

The *saturation concentration* c_s is a constant, dependent on temperature and pressure, for each combination of solvent and solute. The temperature dependence is given by the equation $d \ln c_s/dT = \Delta_S H/RT$, where R is the gas constant, and $\Delta_S H$ the molar enthalpy of solution of the dissolved substance when an infinitely dilute S. is formed (first heat of solution). For solids with positive heats of solution (e.g. sodium chloride or potassium iodide in water), the solubility increases with temperature; if the heat of solution is negative, solubility decreases with temperature (e.g. sodium iodide/water). The solubility can be increased or decreased by addition of a third substance (see Solubility product, Solubility promoter, Salt effect). The differences in solubility of a substance in two mutually non-miscible phases are utilized in extraction methods.

S. of solids in liquid solvents have a lower vapor pressure and melting point, and a higher boiling point, than the solvent.

As with mixtures, a distinction must be made between ***ideal*** and real S. Deviations from the ideal behavior are taken into account by introducing Activities (see). In an ideal dilute S., the dilution is so great that although there are interactions between the solute and solvent, there are no interactions between the molecules of solute. In solutions of nonelectrolytes, this condition is generally achieved at concentrations of 0.1 to 1 mol %. In electrolytes, dilution to an ideal state is extremely difficult, due to the coulomb interactions between the ions.

Solution enthalpy: same as Mixing heat (see).

Solvation: the interaction between molecules of solvents and solutes such as ions, molecules and colloids. A ***solvate*** is a compound formed between the solute and solvent; some can be detected only in solution, but some can be isolated. The number of solvent molecules bound to a dissolved particle is called the ***solvation number***. Some important solvates are hydrates, ammoniates, alcoholates and etherates. The most important special case of S. is Hydration (see).

The solvation capacity of a substance is essentially determined by its donor and acceptor numbers. Most solvent molecules capable of S. have high dipole moments. Cations are generally more highly solvated than anions. A salt is soluble in a solvent if the energy released by S. of cations and anions, the ***solvation energy***, is greater than its lattice energy.

S. also plays an essential role in the stabilization of lyophilic colloids.

Solvatochromicity: dependence of light absorption by a substance on the solvent. Positive S. is defined as a shift of the absorption to longer wavelengths when the solute is moved from a less polar to a more polar solvent (e.g. in the series hexane, chloroform, acetone, methanol, water). Negative S. is a shift in the opposite direction. S. arises when the ground state and/or the excited state of the substance is affected differently by interaction with the solvent, so that the distance between them is not the same in different solvents. Positive S. is observed, for example, with absorption bands based on $\pi \rightarrow \pi^*$ transitions (see UV-VIS spectroscopy), and negative S. with $n \rightarrow \pi^*$ transitions. In the latter, the free electron pair which is responsible for the light absorption is occupied by interactions (H-bonds) with the polar solvent, so that more energy is required to excite it.

Solvay process: see Soda.

Solvent: a liquid compound which can dissolve gases, liquids or solids without reacting with them. The S. can be separated from the solute by physical methods, e.g. distillation or adsorption. In a solution consisting of several components, the component present in excess is considered the S. The chemical structure of the S. is related to its capacity to dissolve substances: it will dissolve those substances with characteristics similar to its own.

The classes of organic compounds often used as S. are the aliphatic and aromatic hydrocarbons and heterocycles, their halogen and nitro derivatives, alcohols, phenols and amines, carboxylic acids and their derivatives (esters, amides, nitriles), ethers, ketones and sulfoxides. For special applications (e.g. solid, transparent solutions for photochemistry), mixtures are also used.

S. are often classified according to their physical properties, especially when the intended application places certain limitations on them. Low, medium and high-boiling S. are those with boiling points (at $1.01 \cdot 10^5$ Pa) below 100 °C, between 100 and 150 °C, and above 150 °C, respectively. Since there is no direct relationship between boiling point and volatility, the latter is also used as a basis of classification. The *volatility* depends on the heat of vaporization, and is related to the evaporation number of ether = 1 at 20 °C and 65% + 5% relative humidity of the atmosphere. Volatile S. have an evaporation number < 10; somewhat volatile S. have numbers from 10 to 35, and non-volatile S., above 35. The *viscosity* (in centipoise, cP) is also a basis for classification: low viscosity is less than 2 cP, medium viscosity is from 2 to 10 cP, and high viscosity is greater than 10 cP. The *polarity* of a S. is very important because it is responsible for certain properties, such as the dissociation effect, ionization and energetic stabilization. The concept of polarity has not been clearly defined, however. One of the criteria is the dipolarity of the S., which is due to a permanent dipole moment in the molecule and causes the polarity. Thus S. are more or less dipolar (depending on the size of the dipole moment) or nonpolar, if the S. molecule has no dipole moment. Although there is no direct relationship between dipole moment and dielectric constant, the latter is often taken as a criterion for the polarity of the S. A different measure of the polarity of the S. is the sum of all specific and nonspecific interactions between S. and solute molecules: the solvation capacity. These interactions include coulomb (ion-ion, ion-dipole, dipole-dipole) and chemical (electron-donor-acceptor complexes, hydrogen bonds) interactions. Single physical parameters such as dipole moment, dielectric constant or index of refraction do not give a satisfactory comparison of polarity, and a number of empirical parameters have been proposed for this purpose. These are based on the LFE relation (see) or on compounds especially chosen for their Solvatochromicity (see). One example is the desmotropic constant L, which uses the enolization capacity of a S. based on acetoacetic ethyl ester as the criterion of polarity. Another is the Y value of Winstein and Grunwald, which is the difference in the logarithms of the rate constants for solvolysis of t-butylchloride at 25 °C in a S. to be characterized and in a standard S.; this is taken as a quantitative measure of ionization capacity. An analogous relationship between logarithms of rate constants for the reaction of bromine with tetramethyltin (S_E2 mechanism) is given by X values; the reference value is X = 0 for glacial acetic acid. Still another parameter is derived from the product ratio of the kinetically controlled Diels-Alder reaction of cyclopentadiene with acrylic methyl ester: $\Omega = \log k_{endo}k_{exo}$. The spectroscopically determined parameters are based on the solvatochromaticity of various types of compounds. The negative solvatochromic behavior of the intramolecular charge-transfer transition of complexes of the S. with 1-ethyl-4-methoxycarbonylpyridinium iodide or pyridine-1-oxide is the basis for the Z value; S. effects on the IR absorption of X=O or X-H...B serves to define G values. The E_T values have been widely accepted as a measure of S. polarity. The E_T scale is based on the solvatochromaticity of a tetraphenyl-substituted *N*-phenolpyridinium betaine. The advantage of this reference compound is that the polarity can be determined visually from the color of the solution. E_T values for numerous S. and S. mixtures have been published in tables, and these values are equal to the electron excitation energies calculated from the equation: $E_T = h \cdot c \cdot \bar{\nu} \cdot N_A = 2.859 \cdot 10^{-3} \cdot \text{cm}^{-1}$. Here h is Planck's constant, c the velocity of light, $\bar{\nu}$ the absorption wavenumber, N_A Avogardro's number. All the criteria of polarity, such as rate constants, equilibrium constants or spectral shifts S can be represented in the form of an LFE relation: $\log k_S/k_E = SR$, where k_S is the measured parameter for a solvent, k_E is the corresponding parameter for ethanol, S is the solvent constant, and R is a susceptibility parameter. An example of such a parameter is the retention index I from gas chromatography, which compares the interaction between the solute and liquid (stationary phase) with the interaction of a paraffin of corresponding length. All these S. parameters are highly correlated with one another.

For certain applications, the acid-base properties of the S. are important. Depending on whether the S. ionizes spontaneously, it is classified as a ***protic S.*** or ***aprotic S.*** Within these two groups, the S. are further classified as acidic, basic or neutral. The acidic protic S. include mineral acids, carboxylic acids and phenols; basic protic S. include ammonia, amines, diamines and carboxamides; neutral protic. S. include water, alcohols, diols and polyols. Nitroalkanes are acidic aprotic S., pyridine, *N,N*-dimethylformamide and dimethylsulfoxide are basic aprotic S., and hydrocarbons, chlorinated hydrocarbons, ethers, ketones, nitriles, esters and nitrobenzene are neutral aprotic S.

The distinction between nucleophiles and electrophiles is important in the choice of S. as media for organic chemical reactions. In this sense, protic S. have both nucleophilic and electrophilic properties, while aprotic S. are primarily nucleophilic. S. with electrophilic properties are Lewis acids, e.g. sulfur dioxide, boron trihalides (in ether), aluminum chloride in chlorinated hydrocarbons or in sulfur dioxide. Finally, a few S., e.g. saturated hydrocarbons, are neither electrophilic nor nucleophilic, and have little tendency to interact. These properties are quantitatively expressed by the Acceptor number (see) and donor number (see Donicity).

Physical constants such as boiling and melting points, index of refraction and density can be used to determined the purity of a S. These data, along with the flash point, dipole moment, dielectric constant and E_T value for a number of common S. are shown in the table on the inside front cover. For many applications, pure S. are required. The purification of a S., and especially drying it, depends on its chemical structure. The available methods are distillation or azeotrope distillation, distillation over sodium, potassium or one of their alloys; and standing or boiling over or with phosphorus(V) oxide, magnesium, calcium or barium oxide or perchlorate, sodium or potassium hydroxide, anhydrous copper sulfate, sodium or potassium carbonate, calcium chloride or carbide, or aluminum oxide followed by filtration and distillation.

The special characteristics of a S. which may be required for a given application include its dissolving capacity, evaporation time, water-solubility, flammability, boiling limits, dilution capacity, purity, price, toxicity, explosiveness and ability to be recovered. S. are used industrially to produce lacquers, paints, films, foils, plastics, synthetic fibers, printing inks, glues, intermediate products, dyes, pigments and pharmaceuticals. They are also important as the media for chemical reactions, in which case suitable melting and boiling ranges, dissolving capacities, chemical inertness or a specific solvation of reactants may be required. In the following, S. suitable for some common organic chemical reactions are listed. *Reduction with hydrogen*: alcohols, glacial acetic acid, hydrocarbons, dioxane. *Oxidation*: glacial acetic acid, pyridine, nitrobenzene. *Halogenation*: carbon tetrachloride, tetrachloroethane, di- and trichlorobenzene, glacial acetic acid. *Esterification*: benzene, toluene, xylene, butyl ether. *Nitration*: glacial acetic acid, dichlorobenzene, nitrobenzene. *Diazotization*: ethanol, glacial acetic acid, benzene, dimethylformamide. *Grignard reactions*: diethyl ether, higher ethers, tetrahydrofuran. *Azo coupling*: ethanol, methanol, glacial acetic acid, pyridine. *Friedel-Crafts reaction*: nitrobenzene, benzene, carbon disulfide, carbon tetrachloride, tetrachloroethane, ethylene chloride. *Dehydration*: benzene, toluene, xylene. *Sulfonation*: nitrobenzene, dioxane. *Dehydrohalogenation*: quinoline.

Recrystallization is a broad area of application for S. The substance to be purified must have the greatest possible difference in solubility at low and high temperatures in a suitable S., while the impurities must be either very soluble or completely insoluble over the entire temperature range. To remove residual S., it must have an adequate volatility. Extraction and distribution, e.g. in chromatography, require S. systems with mixing gaps. Suitable systems have advantageous distribution coefficients, high capacity and different densities of the phases. They cannot form emulsions and must permit easy work-up after the phases have been separated. In addition, the S. must have an elution capacity (polarity) and viscosity suitable for the separation; they must be very pure and the mixtures to be separated must be highly soluble in them. S. for spectral measurements in the IR, UV and visible ranges are chosen with regard for their transparency and stability in the spectral range of interest, their dissolving capacity, their purity and their compatibility with the cuvette material (IR spectroscopy).

In the choice of a suitable S., considerations of safety are essential. Combustible S. are divided into two groups on the basis of their miscibility of water: group A is insoluble or only partially miscible with water, and B is miscible in any proportion with water. Within each of these classes, there are three hazard classes: I, flash point < 21 °C; II, flash point between 21 and 55 °C, and III, flash point between 55 and 100 °C. Some S. form explosive mixtures with air, e.g. carbon disulfide, ether, methanol, ethanol, acetone, benzene and petroleum ether. All S. (except water) are more or less hazardous to health; there is no such thing as an absolutely harmless S. vapor! The hazard comes mainly from their capacity to dissolve lipids. There are three hazard groups: I (very hazardous to health): benzene, carbon disulfide, tetrachloroethane, dichloroethane, carbon tetrachloride, trichloroethane, trichloroethylene, tetrahydrofuran, dioxane and methanol. II (medium hazard): xylene, cyclohexanone, methylcyclohexanone, ethyl acetate. III (slight hazard or none): all other S. For storage, transport and work with S., proper precautions for workplace safety must be taken.

Solvent concept: see Acid-base concepts.

Solvent deparaffinization: a method for removal of paraffin from lubricating oils, high-boiling petroleum fractions and diesel fuels (see Paraffin).

Solvent effect: see Ingold rule.

Solvent naphtha: a mixture of hydrocarbons from coal tar which contains a high proportion of aromatics; it boils between about 150 and 200 °C and is used as a paint and cleaning solvent.

Solvolysis: substitution reactions in which one of the reactants is the solvent. Reaction with water is called *Hydrolysis* (see); with alcohol, *alcoholysis* (see Ester); with ammonia, *ammonolysis* (see Aminolysis).

Soman: see Chemical weapons.

Somatoliberin: see Liberins.

Somatostatin: see Statins.

Somatotropin, ***somatotropic hormone***, abb. ***STH***; ***growth hormone***, abb. ***GH***: a single-chain, polypeptide hormone consisting of 191 amino acids and two intrachain disulfide bridges. Together with other hormones (insulin, thyroid hormone, etc.), S. influences growth, differentiation and continuous renewal of the body. In particular, it stimulates the growth of the epiphysial cartilage, and thus the lengthening of the bones. S. is synthesized in the anterior pituitary under the control of the hypothalamic hormones somatoliberin and somatostatin. Human S. was first isolated in 1956. Its amino acid sequence was published and then repeatedly corrected in the following years. The sequences of the GH from other species differ both with respect to chain length and to degree of homology. Only S. from primates are active in human beings. The sequence is very similar to those of prolactin and the placenta hormone human choriosomatomammotropin (HCS), which is known to have effects similar to those of S. and prolactin. In 1979, the first genetically engineered S. was prepared; this success opens the way to therapy of hypophysial dwarfism, muscular dystrophy, osteoporosis (insufficient calcification of the bones), bleeding stomach ulcers, and so on.

Somatotropin release inhibiting hormone: see Statins.

Somatotropin releasing hormone: see Liberins.

Sommelet reaction: a synthesis reaction for aldehydes in which alkyl or aralkyl halides are heated with urotropin (hexamethylene tetramine) in aqueous alcoholic solution:

$$R\text{–}CH_2\text{–}Cl \xrightarrow{+C_6H_{12}N_4} R\text{–}CH_2\text{–}C_6H_{12}\overset{+}{N_4}Cl \longrightarrow R\text{–}CH_2\text{–}N{=}CH_2$$

$$\longrightarrow R\text{–}CH_2\text{–}NH_2 \xrightarrow[-2H]{} R\text{–}CH{=}NH \xrightarrow[-NH_3]{+\,H_2O/H^+} R\text{–}CHO.$$

The chloromethylarenes, many of which are easily made by chloromethylation (yields 50 to 80%), react especially well. Alkyl iodides are required for aliphatic aldehydes. Halogenophenols tend to give side reactions, and aromatic aldehydes with occupied 2-positions cannot be made, due to steric hindrance. It is assumed that the S. proceeds by formation of the quarternary ammonium salt and then the methylenamine; this is hydrolysed, dehydrogenated and again hydrolysed. The CH_2-NH molecule which is formed from the urotropin serves as the hydrogen acceptor and is converted to methylamine.

Sommelet rearrangement: rearrangement of quaternary ammonium salts with one or two benzyl groups in the presence of sodium amide in liquid ammonia or of phenyllithium, forming tertiary amines:

$$C_6H_5\text{–}CH_2\text{–}\overset{\oplus}{N}(CH_3)_2(CH_2C_6H_5)\ Br^- \xrightarrow[-NaBr,\,-NH_3]{+NaNH_2} C_6H_5\text{–}\overset{\ominus}{C}H\text{–}\overset{\oplus}{N}(CH_3)_2(CH_2C_6H_5) \longrightarrow C_6H_5\text{–}CH(N(CH_3)_2)\text{–}C_6H_4\text{–}CH_3\ (ortho)$$

The reaction mechanism is thought to involve formation of an N-ylide, which then undergoes a cleavage and recombination process at low temperature; either an ion or a radical pair is formed. The new bond is formed at the *ortho* position of the migrating benzyl ion, in contrast to the Stevens rearrangement (see).

Sonnenschein's reagent: see Alkaloids.

Soot: microcyrstalline graphite. S. is formed by combustion of gaseous carbon compounds under conditions of oxygen deficiency (gas, oil or acetylene soots, for example). It is used as a black pigment, and as a filler in rubber.

Sorbic acid, ***(E),(E)-Hexa-2,4-dienoic acid***: CH_3-CH=CH-CH=CH-COOH, a doubly unsaturated monocarboxylic acid. It crystallizes in colorless needles or leaflets; m.p. 134.5 °C, b.p. 228 °C (dec.). It is slightly soluble in water and somewhat more soluble in alcohol, acetone or glacial acetic acid. S. undergoes the reactions typical of carboxylic acids. Its salts and esters are called *sorbates*. Because of its conjugated C=C double bond system, it can also undergo many addition and polymerization reactions. S. occurs in nature mainly in the form of parasorbic acid, in unripe service berries and in wine. It can be synthesized by condensation of crotonic aldehyde and malonic acid or ketene, or by oxidation of sorbic aldehyde. Because it is nontoxic, S. is used as a preservative in foods, pharmaceutical preparations and cosmetics.

Sorbite: in metallography, a term for a perlite structure of steel in which the cementite lamellae are so close together that they can be barely resolved by a light microscope (see Troostite). S. is formed at cooling rates lower than 200 °C/s or by tempering at temperatures between 400 °C and almost 700 °C.

Sorbitol: same as D-Glucitol (see).

L-Sorbose, ***xylo-hexulose***: a ketohexose monosaccharide, colorless crystals, m.p. 165 °C. It is very soluble in water, slightly soluble in ethanol and practically insoluble in ether. S. is an intermediate in the synthesis of ascorbic acid.

Sorel cement: see Magnesium chloride.

Sorption: the uptake of a substance either on the surface (see Adsorption) or the interior (see Absorption, Extraction) of a sorbant material. S. is based on the distribution of the sorbed substance between different phases, without a thermal phase change (evaporation, condensation or crystallization).

Sorting, ***enrichment***: in the processing of ores, the separation of the mine run into its component minerals. Removal of the blank rock from the concentrate enriches it in the desired mineral. Sorting methods are based in differences in the physical properties of the substances being separated, and those of the valuable minerals must be different from the others. For each method, there is a suitable range of grain size. If the last two requirements are fulfilled by grinding the ore, the S. can be based on magnetic separation, flotation, heavy-liquid processing, electrostatic precipitation, settling or other process.

Sowiden®: see Synthetic fibers.

Soxhlet extractor: see Extraction.

Soybean inhibitor: see Trypsin.

Soy fiber: a Protein fiber (see).

Space group: see Crystal.

Spacer: see Affinity chromatography; see Liquid crystals.

Spanish green: same as Verdigris (see).

Spark spectrum: see Atomic spectroscopy.

Sparteine: an alkaloid found in scotch broom (*Cytisus scoparius = Sarothamnus scoparius*) and in the seeds of *lupinus luteus*; M_r 234.4. It is a colorless oil which is only slightly soluble in water. S. is usually sold as the sulfate, which forms colorless, water-soluble crystals. It is practically insoluble in chloroform and ethers. S. is a quinolizidine alkaloid, and is used therapeutically for various cardiac diseases (tachycardia), and also occasionally in obstetrics.

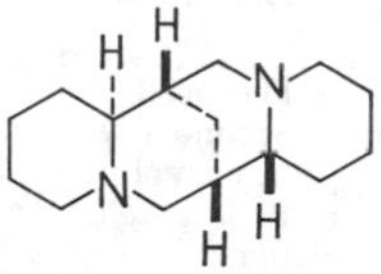

Spasmolytics: compounds which relieve cramps in the smooth muscles. This applies especially to spasms in the gastrointestinal tract, the bronchii and the urinary tract. *Neurotorpic S.*, such as atropine, act as parasympathicolytics, while *musculotropic S.* act directly on the muscle. An example is papaverine. Most synthetic S. have both musculotropic and neutrotropic effects. Some examples are *demelverine*, a tertiary araliphatic amine, *dipiproverine*, *Denaverine* and *propiverine*. The last two are basic esters of benzilic acid ethers. Propiverine is used for spasms of the urinary tract.

Spatial isomerism: same as Stereoisomerism (see).

Spearmint oil: a light yellow to light green essential oil with a penetrating, characteristic odor. The main component is (-)-carvone (up to 60%); it also contains dihydrocouminyl acetate (the main source of the aroma), limonene, α-pinene, phellandrene and esters of dihydrocarveol, linalool, cineol and pulegone. C. is used to make candies and spice mixtures, in toothpaste and chewing gum. It is used in lavender and jasmine perfumes, especially those for soaps.

Specific conductivity: see Electrical conductivity.

Specific free surface energy: see Surface tension.

Specific heat: see Molar heat.

Specific parameters: see State, parameters of.

Specific rotation: see Polarimetry.

Spectral analysis: The goal of S. is to determine the qualitative or quantitatve composition of the sample from its spectrum. Either the *emission spectrum* (see Spectrum) or the *absorption spectrum* of a substance can be analysed. The emission spectrum is obtained from a gaseous substance which has been excited to the point of glowing, while the absorption spectrum is used in the case of substances which cannot be vaporized conveniently – organic substances, for example. In qualitative S., the presence of certain elements or compounds is indicated by the presence of certain lines in the spectrum. Quantitative S. makes use of the intensity of one emission or absorption line to determine the concentration of the substance of interest. The term S. is now used in a somewhat narrower sense to mean the detection and determination of elements.

S. was founded in 1859 by Bunsen and Kirchhoff, who discovered that metal salts, when vaporized in a flame, give the flame characteristic colors. Observation of the flame with a Spectroscope (see) showed that each metal produced a characteristic spectrum, which permitted identification. The great sensitivity of the method made it possible to discover new elements, because these produced new lines which could not be assigned to previously known elements. Bunsen and Kirchhoff themselves discovered the elements rubidium and cesium, but later many others were found in this way, including tellurium, indium, gallium and thallium, many rare-earth metals and noble gases.

S. plays an essential role in astrophysics, because with it, the light from celestial bodies can be studied. In 1868, during an eclipse of the sun, Janssen and Lockyer observed lines in the solar atmosphere which could not be assigned to any element then known on earth. This new element was therefore named helium. In 1814, Fraunhofer discovered the lines named for him in the solar spectrum. These were later assigned to elements present on earth and known to science at the time.

Some important Fraunhofer lines

Line	Element
H	Calcium
h	Hydrogen
G	Iron/Calcium
F	Hydrogen
b	Magnesium
E	Iron
D	Sodium
C	Hydrogen
B	Oxygen
A	Oxygen

Spectral sensitization: the extension of photographic sensitivity beyond the spectral range of the light-sensitive material. The sensitizing pigment must absorb the light in a certain part of the spectrum and transfer it to the silver halide grains in a form in which it can be used to form the latent image. *Desensitization* is the reduction of the sensitivity of the photographic material; an increase in the sensitivity of the material itself by addition of special pigments is called the *Capri-blue effect*. The superadditive increase in sensitivity of photographic emulsions by mixtures of pigments is called *supersensitization*.

Only polymethynyl pigments (cyanin and merocyanins) are of practical significance as sensitizing pigments. The requirements for an effective pigment are that it be solid, that it be adsorbed in sufficient amounts to the silver halide grains, that the lowest unoccupied molecular orbital of the pigment (LUMO) be lower than the lower boundary of the silver halide conduction band, that the highest occupied pigment orbital (HOMO) be above the upper boundary of the silver halide valence band, that the pigment be planar and rigid, and that it have a strong tendency to aggregate. Cyanine pigment aggregates (reversible molecular complexes) absorb in spectral regions where the individual molecules do not, and thus can serve to modify the sensitizing range. In addition, they function in the conduction of the absorbed light energy and in the mechanism of supersensitization.

Two different mechanisms have been proposed for S.: 1) *energy transfer*: the energy absorbed by the pigment is transferred to the silver halide grain and acts to raise electrons out of the electron states which lie between the valence and conduction bands. 2) *Electron transfer*: the excited electron in the pigment LUMO is transferred to the silver halide conduction band. The pigment is regenerated by an electron from the valence band of the silver halide. The above requirements for the positions of the pigment LUMO and HOMO relative to the conduction and valence bands of the silver halide apply to electron transfer; they find expression in certain values for the redox potentials of the pigments which delimit the range of effective S. The sensitizing properties of the pigments also depend on conditions of the environment, such as air oxygen and moisture, pAg and pH values in the emulsion, and crystal habit of the silver halide. The relative quantum yield, i.e. the ratio of the quantum yields of monochromatic light in emulsions with and without sensitizer, is used as a measure for the S. A spectrosensitogram permits estimation of the sensitizing capacities and ranges of pigments. Emulsions which are specifically sensitive to green light ($\lambda \leqq 600$ nm) are called orthochromatic; while red-sensitive emulsions ($\lambda \leqq 700$ nm) are called panchromatic.

Spectrochemical series: see Ligand field theory.

Spectrograph: see Spectrophotometer.

Spectrometry: same as Spectroscopy (see).

Spectrophotometer: an apparatus for measurement of spectra. All S. serve to measure the intensity of radiation absorbed or emitted by the sample as a function of wavelength or frequency. However, since these measurements are made in very different regions of the electromagnetic spectrum and are based on different kinds of interaction between the sample and the electromagnetic radiation, the components of S. vary widely, depending on their application. In the following, S. for measurements in the optical range (ultraviolet, visible and infrared) are described; these depend on optical components such as lenses, prisms, mirrors and diffraction gratings. For S. used with other spectral ranges, see Spectroscopy.

The main components of an S. are a source of radiation (light), a monochromator for spectral resolution and detector, which may be coupled to an amplifier and recording device (Fig. 1). For measurement of absorption spectra, light sources which emit light at all wavelengths over a wide range are used. The sample is placed into a cuvette or sample holder which is located either between the light source and the monochromator or between the monochromator and the detector.

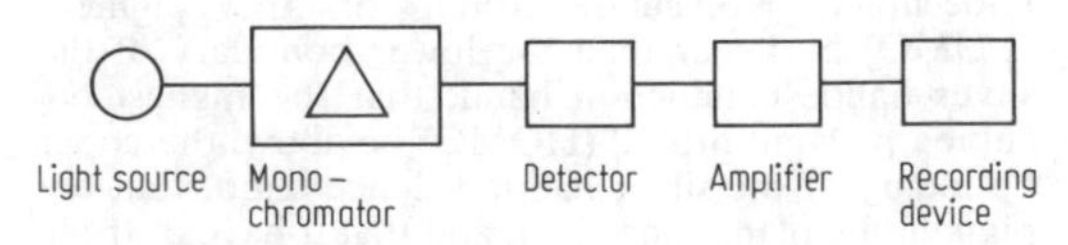

Fig. 1. Main components of spectrophotometers.

No cuvette or sample holder is needed for measurement of emission spectra, for which the sample is excited by absorption of some form of energy (flame, electric arc, electric discharge) and emits light; it occupies the position of the light source shown in Fig. 1. The table gives a summary of the components which can be used in the optical range; different ones must be used with different spectral ranges. For example, there is no material which can be used to make a prism for all wavelengths, because glass is not transparent to ultraviolet or infrared radiation. The detectors respond only to light in limited ranges, and no source emits radiation over the entire range.

Components for spectrometers in the optical range

Range	Light source	Dispersive system	Detector
UV	Deuterium or xenon lamp	Quartz, NaCl or CaF_2 prism, Diffraction grating	Photocell Secondary electron amplifier Photographic film
Visible	Incandescent light	Glass prism	Photocell
	Tungsten ribbon lamp	Diffraction grating	Photographic film
IR	Silicon carbide or Nernst rod, Ceramic rods	CsBe, KBr, NaCl or LiF prism Diffraction grating	Thermoelement Bolometer Golay cell

Classification 1) Depending on the means of dispersing the light, S. are classified as prism, grating or interference S.

2) S. are classified as spectroscopes, spectrographs, spectrometers and spectral photometers, depending on the type of detector used. ***Spectroscopes*** are S. used for visual observation of a spectrum in the visible range. ***Spectrographs*** record the spectrum on a photographic plate, while ***spectrometers*** register the spectrum using physical detectors such as photocells or thermoelements. ***Spectrophotometers*** (often called spectrometers) permit both spectral dispersion of the light and photometric measurement of its intensity.

3) Single- and dual-beam instruments. ***Single-beam instruments*** measure first the intensity of the light at a given wavelength in the absence of the sample, then the intensity with the sample inserted into the light path. The ratio of the two intensities I_T/I_0 (I_0 is the intensity without the sample, I_T with the sample) is the Transparency (see) of the sample at the wavelength in question. In order to obtain the entire spectrum, the transmission is measured at certain intervals over the entire spectral range. The wavelength is usually changed by rotation of the prism or grating, or of a mirror in the monochromator. This may be done manually or automatically.

Most modern S., especially those used to measure absorption spectra, are automatically recording ***dual-beam instruments***, for which the completely automatic infrared apparatus described below may serve as an example (Fig. 2). The light source is a silicon carbide rod. The light emitted by it is divided into two pathways of equal length, one passing through the sample cuvette, and the other through a cuvette containing the blank. (The blank consists of a solution similar to the sample solution but lacking only the substance whose spectrum is to be measured.) In this way the light losses caused by reflection, scattering and absorption by the solvent and cuvette are the same in the two light paths, so that the difference in intensity in the two beams is due solely to absorption by the sample.

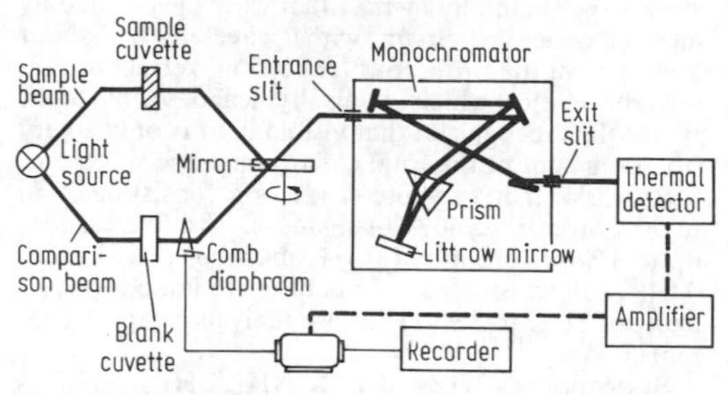

Fig. 2. Schematic view of a completely automatic infrared spectrophotometer.

The two beams then impinge on different sides of a rotating mirror with two transparent and two reflective 90° sections. In this way the reference beam and the sample beam are reflected alternately into the entrance slit of the monochromator. In the monochromator the light is dispersed by a prism, then reflected toward the exit slit by the Littrow mirror, which permits light of a single wavelength to fall on the exit slit. Rotation of the Littrow mirror causes the entire spectrum to pass through the slit in sequence; the rotation of the mirror is coupled with the advance of the paper in the recorder and is regulated so that the abcissa of the recorded spectrum is linear with respect to the wavelengths. The detector is a thermoelement. If the absorption in the two light paths is the same, the current generated in the thermoelement is a direct current, which is not picked up by the amplifier and the recording system. If the intensity of the sample beam is decreased by absorption, an alternating current is generated which, after amplification, is used to drive a motor. This moves a screen into the reference beam until the absorption by the screen exactly matches that by the sample. At this point the alternating current is replaced by a direct current. The distance the screen has moved is a measure for the absorption by the sample; if the motion of the screen is combined with that of a pen, the latter produces the absorption spectrum of the sample.

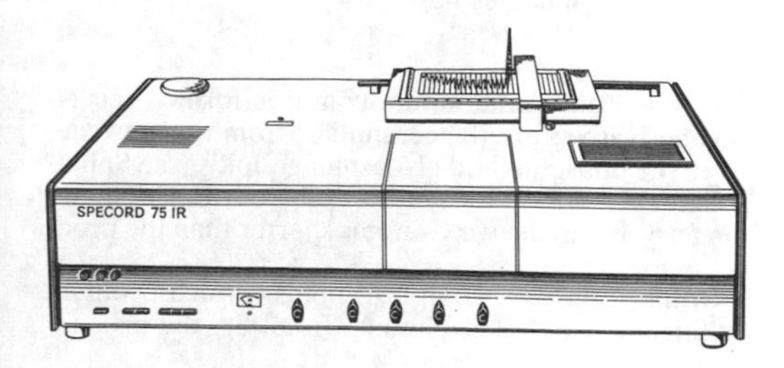

Spectroscopic moment: a parameter which describes the effect of substituents on the electronic excitation spectrum of benzene; it is related to the electron withdrawing or contributing effect of the substituent.

Spectroscopic shift rule: see Atomic spectroscopy.

Spectroscopy, ***spectrometry***: that branch of science dealing with the interaction between Electromagnetic radiation (see) and matter as a function of wavelength of the radiation. The intensity of electromagnetic radiation as a function of the frequency, wavelength or wavenumber is called the Spectrum (see). Devices for spectroscopic studies are discussed under Spectral instruments (see).

Forms of S. can be classified in various ways: 1) The sample can either absorb (absorption S.) or emit (emission S., see Spectrum) the radiation, or reflected (see Reflection spectroscopy) or scattered (see Raman spectroscopy) light may be examined. 2) The nature of the sample determines whether a method is Atomic S. (see), Molecular S. (see) or solid-state S.; according to the region of the electromagnetic spectrum examined, methods are categorized as X-ray S. (see), Gamma ray S. (see), UV-VIS S. (see), Infrared S. (see), Microwave S. (see) or High-frequency S. (see). 4) Electronic, vibrational, rotational, Electron-spin resonance S. (see) NMR (see) and Nuclear quadrupole resonance S. (see) are named according to the type of interaction with the radiation. 5) Certain types of S. are named for their discoverers and/or founders, e.g. Raman S., Rounier S.

In the broader sense, S. includes techniques in which particles are emitted as a result of absorption of electromagnetic radiation (see Mössbauer S., Photoemission S., Auger electron S.) or in which ions or secondary electrons are formed as a result of bombarding the sample with electrons (see Mass S.)

The significance of S. for chemistry is that it is used not only to study light per se, but light which has previously interacted with a substance. The appearance of discrete frequencies in the spectrum indicates that the interaction is directly related to properties of the substance which are determined by quantum mechanical conditions. If E_1 and E_2 are quantized energy states in the substance under study, the equation $\Delta E = E_2 - E_1 = h\nu$ relates the frequency ν of electromagnetic radiation emitted or absorbed to the energy released when the system passes from state 1 to state 2. (h is Planck's constant.) Thus it is understandable that S. provides important information on the structure of atoms, molecules and solids. In the wide range of electromagnetic radiation, the energy quanta correspond to the energy changes associated with a variety of physical phenomena, such as electronic, vibrational, electron spin or nuclear spin states. S. is therefore used to study a number of different properties of samples.

Spectrum: in the broad sense, the arrangement of an intensity distribution as a fuction of certain properties, e.g. the abundances of mass particles as a function of their mass in the mass S.. In the narrower sense, it is the intensity distribution of electromagnetic radiation as a function of its frequency, wavelength or wavenumber (see Electromagnetic S.). S. are often represented graphically, with the intensity being given in various units (see Transparency, Absorption, Extinction, Lambert-Beer law). Depending on the range of frequency and the interactions with the sample, S. vary widely. They can be classified similarly to the various types of Spectroscopy (see), e.g. as reflection, infrared, Mössbauer S.

Originally, the band of colors which appears when visible light is passed through a prism was called a S. If light, e.g. from an arc lamp, is passed through a collecting lens, then a slit at the focal point of the lens, then a prism and a second lens which focuses it on a screen, it does not form a white image of the slit, but rather a broad colored band in which the colors lie in the series red, orange, yellow, green, indigo and violet (Fig.).

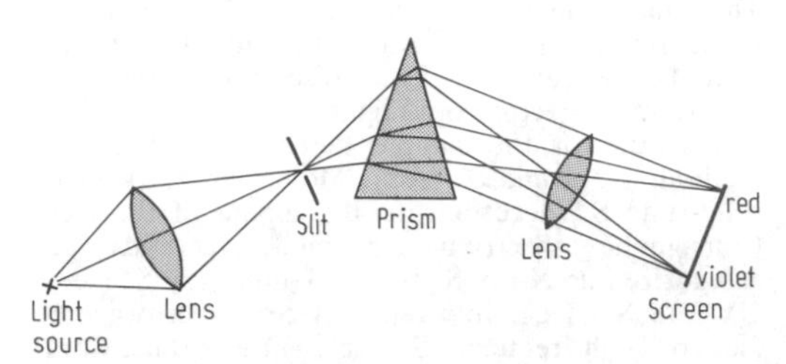

Spectral dispersion of visible light.

The path of the light is deflected from its original direction, and the amount of deflection increases from red to violet, thus forming the S. of the light source. The S. is a series of colored images of the slit which arise because violet light is more strongly deflected than red. The spectral colors named above differ in their wavelengths (and frequencies); the red limit of the visible S. is at 770 nm and the violet is at 400 nm.

The distinction between *emission* and *absorption* S. is an important one. Emission S. are formed by release of energy, while absorption S. indicate uptake of energy in the form of electromagnetic radiation. Glowing solids and liquids produce continuous emission S. which contain all wavelengths over a large spectral range. Glowing gases emit line S. (***atomic S.***) if atoms are the source of the light, or band S. ***molecular S.*** if molecules are emitting. ***Line S.*** consist of a number of individual spectral lines. ***Band S.*** consist of individual lines, but these are close together and form distinct groups (bands).

If, for example, a piece of sodium is vaporized in a Bunsen burner flame, it emits an emission S. which is dominated by the "yellow sodium line" at 589 nm. With a spectroscope (see Spectral instruments), this can be seen as a glowing yellow line on the dark background. On the other hand, if the white light from an arc lamp shines through the glowing sodium vapor, and this light is then observed in the spectroscope, it appears as a continuous band of color in which there is a black line at the location of the yellow sodium line. The sodium vapor has absorbed the light at the wavelength which is characteristic for it, generating an absorption S. An absorption S. can be considered the "negative" of the emission S. The best known absorption S. is the solar S., which contains many dark lines named for the optician Fraunhofer. The glowing solar nucleus emits a continuous S., from which the substances in the solar atmosphere absorb the wavelengths which are characteristic for them.

As early as 1800, F.W. Herschel realized that there is an invisible continuation of the red end of the S. which can be recognized by its heat effects (infrared). J.W. Ritter and W.H. Wollaston discovered in 1801 the invisible radiation extending from the violet end of the S. (ultraviolet). Finally H. Herz was able to show experimentally in 1887 that waves generated by an open oscillating circuit, with a wavelength of about 1 m, had all the properties (reflection, refraction, interference, etc.) which are also typical of visible light. It was thus recognized that visible light is only a very small segment of the total electromagnetic spectrum.

Specular iron: a white crude iron consisting of 46 to 90% iron, 6 to 40% manganese, 4 to 5% carbon, up to 1% silicon, about 0.1% phosphorus and up to 0.04% sulfur. Small amounts of it are added to cast-iron and steel melts to increase their manganese contents.

Spermidine: $H_2N\text{-}(CH_2)_3\text{-}NH\text{-}(CH_2)_4\text{-}NH_2$, a strongly basic polyamine which, together with ***spermine***, $H_2N\text{-}(CH_2)_3\text{-}NH\text{-}(CH_2)_4\text{-}NH(CH_2)_3\text{-}NH_2$, is present in human sperm and forms complexes with the DNA.

Spermine: see Spermidine.

Spherical functions: see Atom, models of.

Spherical micelles: see Colloids.

Sphinganine: see Sphingoids.

Sphingoids, abb. ***Spd***: derivatives of the long-chain amino alcohol ***sphinganine*** (D-erythro- or (2S,3R)-2-aminooctadecane-1,3-diol). S. are the skeletons of the Sphingolipids (see); at present more than 20 different S. are known. The homologs of sphinganine are indicated by a prefix which indicates the number of carbon atoms, e.g. icosa- and hexadecasphinganine for the C_{20} and C_{16} compounds, respectively. The most common S. in animals is ***sphingosine*** (4-sphingenine, abb. Sph, (2S,3R,4E)-2-amino-octadec-4-ene-1,3-diol). Plants contain derivatives of 4-hydroxysphinganine formerly called ***phytosphingosine***.

Sphinganine: CH_2OH–$HCNH_2$–$HCOH$–$(CH_2)_{14}$–CH_3

4-Sphingenine (Sphingosine): CH_2OH–$HCNH_2$–$HCOH$–CH=HC–$(CH_2)_{12}$–CH_3

4-D-Hydroxy sphinganine: CH_2OH–$HCNH_2$–$HCOH$–$HCOH$–$(CH_2)_{13}$–CH_3

The S. are not found naturally in free form. Their N-acyl derivatives are the ceramides; from them are derived the phospho- and glycosphingolipids (see Sphingolipids). An S. is biosynthesized from L-serine and the fatty aldehyde two C-atoms shorter than the product S.

Sphingolipids: complex lipids which contain Sphingoids (see) as skeletons. In all S., the primary amino group of the sphingoid is acylated with a fatty acid, so they are derivatives of the N-acylsphingoids (ceramides). Depending on the residue on the primary hydroxyl group of the ceramide, the S. are subdivided into the ***phosphosphingolipids***, or sphingomyelins, and ***glycosphingolipids***. The sphingomyelins are the most important group of phosphosphingolipids. In contrast to the glycerophospholipids, S. are nearly insoluble in acetone and ether. They can be extracted with hot pyridine, after other lipids have been removed with ether and acetone. The phosphosphingolipids crystallize out of the pyridine solution on cooling, while the glycosphingolipids remain in solution.

The mono- and oligoglycosylceramides (see Cerebrosides) and mono- and oligoglycosylsphingoids are neutral glycosphingolipids. The latter group does not occur free in nature, but are produced by alkaline hydrolysis of cerebrosides. The acidic glycosphingolipids contain carboxyl groups from sialic acid residues or sulfate esters. Sialoglycosylsphingolipids are called Gangliosides (see). Unlike other S., the gangliosides are soluble in water. Sulfoglycosphingolipids with sulfate groups on the hydroxyls of the glycoside were formerly called sulfatides. Sulfoglycosylsphingolipids are stronger acids than the sialoglycosylsphingolipids. There are essentially 7 main groups of glycosphingolipids with different basic oligosaccharide structures linked glycosidically to the ceramide; further monosaccharides can be linked to the basic olgicosaccharide:

Globo-series (Gb): Gal(α1→4)Gal(β→4)GlcCer
Isoglobo-series (iGb): Gal(α1→3)Gal(β1→4)GlcCer
Muco-series (Mc): Gal(β1→4)Gal(β→4)GlcCer
Lacto-series (Lc): GlcAc(β1→3)Gal(β1→4)GlcCer
Neolacto-series (nLc): Gal(β1→4)GlcNAc(β1→3) Gal(β1→4)GlcCer
Ganglio-series (Gg): GalNAc(β1→4)GlcNAc(β1→4) GlcCer
Gala-series (Ga): Gal(α1→4)GalCer.

These oligosaccharide groups have been given trivial names ending in -biose, -triose, -tetraose, etc. to indicate the number of monosaccharide units. Differences in the bonding (1→4 or 1→3) are indicated by the prefixes iso- and neo-, as in isoglobo- and neolacto-. Examples: Galabiose ($GaOse_2$), Globotriose ($GbOse_3$), Globotetraose ($GbOse_4$). Only in the ganglio- and neolacto-series are there terminal sialic acid groups (see Gangliosides). Only the glycosphingolipids in the outer half of the cell membrane are antigenic.

Glycosphingolipids are degraded in steps by enzymes. In certain genetic diseases (*sphingolipidoses*) the lack of certain glycosphingolipid degrading enzymes (sphingolipid hydrolases) leads to accumulation of specific S., mainly in the brain and kidney, which severely impairs the function of these organs. Examples of such pathologically stored S. are GM2 (Tay-Sachs disease), GM1 (generalized gangliosidosis) and glucose cerebroside (Gaucher's disease).

Sphingomyelins: ceramide-1-phosphorylcholine, an important group of phosphosphingolipids (see Sphingolipids). S. can be isolated from bovine brain as colorless crystals which are soluble in chloroform/ methanol (1:1) and in hot ethanol, but insoluble in ether or acetone. Cerebrosides, if present, can be removed by column chromatography on aluminum oxide; glycerophospholipids, by mild alkaline hydrolysis.

$CH_2OP(=O)(O^-)OCH_2CH_2N^+(CH_3)_3$
|
HCNH – Fatty acid group
|
HCOH
|
CH
||
HC
|
$(CH_2)_{12}$
|
CH_3

Sphingosine: see Sphingoids.

Sphondin: see Furocoumarins.

Spider toxins: toxins which spiders inject into their prey by means of a special venom apparatus. Only a few S. are dangerous to human beings. The main components of S. are free amino acids, biogenic amines, enzymes and neurotoxins. The S. are toxicologically similar to the Snake toxins (see) and Scorpion toxins (see). The strongest S. is the venom of *Phoneutria* species. Their bite is extremely painful, and death occurs after 2 to 5 hours. The effective toxin is a nerve poison with central and peripheral effects; death is caused by respiratory paralysis.

Spiegeleisen: see Specular iron.

Spike oil, ***spike lavender oil***: a greenish-yellow essential oil which smells and tastes strongly of turpentine. The main components of S. are (-)-linalyl acetate and linalool; borneol, camphor, cineol, geraniol, thymol and small amounts of coumarin are also present. S. is obtained by steam distillation of various types of lavender, and can also be made synthetically. It is used as a substitute for the more valuable true lavender oil in perfumes.

Spin crossover: see Coordination chemistry.

Spin decoupling: see NMR spectroscopy.

Spindle oil: a light, thin lubricating oil for rapidly running, low-load machines.

Spinel type, ***spinel structure***: a common Structure type (see) for ternary compounds with the general

R^1	R^2	R^3	Sphingolipids
Alkyl, alkenyl	H	H	Spingoids
Alkyl, alkenyl	Fatty acid	H	Ceramides
Alkyl, alkenyl	Fatty acid	Phosphorylcholine	Sphingomyelins
Alkyl, alkenyl	Fatty acid	Mono- or oligoglycosyl	Glycosylceramide (Cerebrosides)
Alkyl, alkenyl	H	Mono- or oligoglycosyl	Glycosylsphingoid
Alkyl, alkenyl	Fatty acid	Sialoglycosyl	Ganglioside

composition AB_2O_4. In the crystal lattice of the mineral spinel, $MgAl_2O_4$, the O^{2-} ions are nearly cubic close packed, while the Al^{3+} ions occupy half the octahedral holes, and the Mg^{2+} ions, one eighth of the tetrahedral holes. In order to crystallize in the S., the cations of the compound must have certain size ratios. The representatives of the S. can be divided into two series with the compositions $A^{2+}B^{3+}{}_2O_4$ (e.g. $MgAl_2O_4$, $MnFe_2O_4$, Fe_3O_4 = $Fe^{2+}Fe^{3+}{}_2O_4$) and $A^{4+}B^{2+}{}_2O_4$ (e.g. $TiZn_2O_4$, $SnCo_2O_4$, Pb_3O_4 = $Pb^{4+}Pb^{2+}{}_2O_4$).

Spin labeling: a special technique in electron spin resonance spectroscopy used to study biomolecules, such as enzymes, nucleic acids and proteins, which are diamagnetic (see Magnetochemistry) and thus cannot be studied directly by ESR. Certain stable radicals linked to or incorporated into the biomolecules are used as ESR probes, or spin labels. The most common spin labels are nitroxide radicals, which contain the ESR-active -N-O- group. Since the free electron interacts primarily with the ^{14}N nucleus (I = 1), a triplet hyperfine structure appears in the ESR spectrum. The effect of the molecular environment of the spin label on the line shape of the triplet makes it possible to draw inferences concerning the local structure and conformational changes in the molecule.

Spin-lattice relaxation: see NMR spectroscopy.

Spin-orbit coupling: see Multiplet structure.

Spin overlap: see Coordination chemistry.

Spin-tickling technique: special double resonance technique used in NMR spectroscopy (see); among other things, it permits identification of mutually coupled nuclei in complex spectra. In this technique, during an NMR experiment, a second alternating magnetic field is applied to the sample; this field has the same frequency as one of the transitions. The strength of the alternating field is adjusted so that it interferes with the energy levels involved in the transition. One then observes that the lines of all transitions which begin or end on those two levels are split into two equal components, or they undergo changes in intensity.

Spirane: see Nomenclature, sect. III E 2.

Spirit: a term for technical ethanol, which is usually produced by fermentation. *Crude spirit* contains fusel oils, aldehydes and acids, and is purified by distillation in columns and filtration over active charcoal. Commercial ethanol contains 4.7% water, which cannot be removed by distillation (azeotropic mixture).

Fermentation of the sulfite liquors from cellulose production yields *sulfite spirit*, and *wood alcohol* is obtained by fermentation of the carbohydrates resulting from saccharification of wood. Other technical processes also yield alcohol, for example, catalytic high-pressure hydrogenation of acetaldehyde produces *carbide spirit*.

Spirit black: see Nigrosine.

Spironolactone: an aldosterone antagonist used as a diuretic in cases of hyperproduction of aldosterone. S. is a partially synthetic steroid compound made by splitting off thioacetic acid to form a double bond which is conjugated with the other double bonds.

Spirostane: see Saponins.

Spirosystems: see Nomenclature, sect. III E 2.

Splicing: see Gene technology.

Spodium: see Charcoal.

Sporidesmolides: nonsymmetric cyclic depsipeptides with antibiotic activity. ***Sporidesmolide I***, *cyclo*-(Hyv-D-Val-D-Leu-Hyv-Val-MeLeu-), is a metabolic product of the fungus *Sporidesmium bakeri*. ***Sporidesmolide II*** has a similar structure in which the D-Val residue is replaced by D-Ile, while ***sporidesmolide III*** has Leu instead of MeLeu (methylleucine). In all S., L-α-hydroxyisovaleric acid (Hyv) acts as the hydroxyacid.

Spray drying, ***atomization drying***: a method of Drying (see) by Atomization (see). It is used to make powdered, paste or crystalline products from solutions, emulsions or suspensions. It is used, for example, to make dried milk, yeast powder, wood sugar, laundry detergents, dry glue, powdered eggs, etc.

For *cold atomization*, a crystallizing component must be present. In *hot atomization*, the substance is sprayed into warm air, superheated steam or fuel gas, and the solvent evaporates, at the same time cooling the gas. The dissolved substance precipitates in solid form, and the solvent vapors are drawn off. Atomization is often done through a nozzle or on a rotating disk (Fig.).

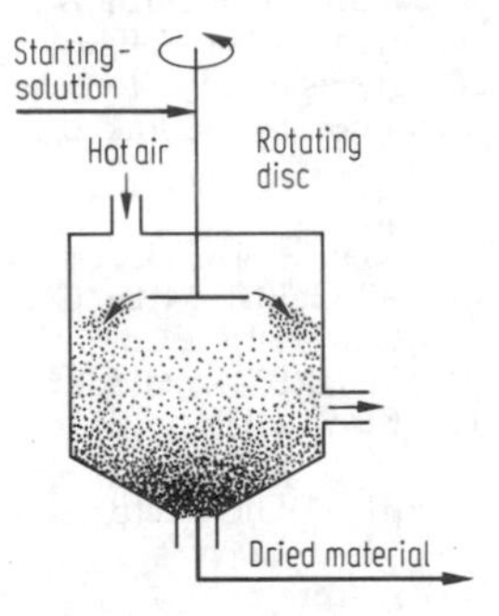

Spray drier

Spray metallizing: same as Metal spraying (see).

Spreading: the formation of a uniform film of an insoluble or slightly soluble substance on the surface of a liquid or solid. S. of a surface-active substance forms a monomolecular film. The S. of liquids on liquids or solids (complete wetting) forms multimolecular layers. The S. of surface-active substances on water is due to the strong solvation of the polar groups, and the resulting repulsion between them, and the relatively weak attractive forces between the non-polar chains. The S. of liquids on liquids or solids is always associated with a change in the sign of the dispersive force (see Van der Waals forces). This change in sign is only possible in a macroscopic phase.

Spruce needle oil: an oil obtained from young shoots and fresh needles of the spruce or red fir native to northern and central Europe at a maximum 0.25% yield. The main components of S. are bornyl acetate (up to 40%), α-pinene, dipentene, cadinene and camphene. S. is bacteriocidal and neurotonizing. It is used medically as a spray and for inhalation for treatment of respiratory diseases, as a bath additive for neuro-vegetative exhaustion and as sprays for room deodorizing. It is also used in the perfume and cosmetic industries.

Spruce oil: same as Tall oil (see).

Spruce shaving reaction: a qualitative detection reaction for various heterocyclic compounds. The vapors of these compounds give characteristic colors to a shaving of spruce wood which has been moistened with hydrochloric acid. For example, furan gives a green color, while pyrrole and indole give red.

Sputtering: sudden escape of dissolved gases from a hardening metallic melt, just above the melting point. The phenomenon, which is especially destructive of silver castings, can be avoided by alloying with other metals.

Squalene: an acyclic triterpene hydrocarbon first isolated from the liver oil of sharks (*Squalus*); it is also contained in the non-saponifiable fraction of many other fats and fatty oils, for example, plant oils and animal skin fats. The Steroids (see) are biosynthesized from S.

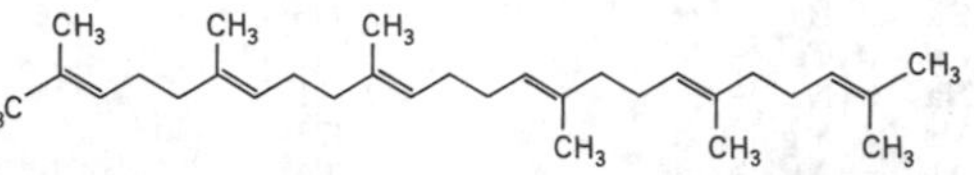

CH_3 CH_3 CH_3 CH_3 H_3C CH_3 CH_3 CH_3

Sr: symbol for strontium.

SR: see Photophysical processes.

SRC process: see Coal hydrogenation.

S_E reaction: see Substitution 2.

S_N reaction: see Substitution 1.

S_R reaction: see Substitution 3.

SRH: see Liberins.

SRS: see Leukotrienes.

Stabilizers: substances added to reactive substances to inhibit premature reaction or decomposition. For example, urea derivatives, diphenylamine or urethanes are added to cellulose nitrate explosives to bind nitrous oxides as they are released. Small amounts of phosphoric acid, sodium diphosphate, sodium, magnesium or aluminum silicate, urea or dilute sulfuric acid are added to hydrogen peroxide to prevent its decomposition to water and oxygen. Pure benzaldehyde undergoes autooxidation, but is stabilized by hydroquinone. The S. are extremely important for the plastics industry. To prevent spontaneous polymerization of the monomers in storage, hydroquinone can be added to styrene and acrylonitrile; *p-tert*-butylpyrocatechol to styrene and vinyl chloride; diphenylamine, copper arsenate or sulfur to vinyl acetate; and methylene blue to acrylonitrile and methacrylates; and pyrogallol to methacrylates. Concentrations of the S. are 10^{-5} to 10^{-1}%. They are removed by distillation or ion exchangers before the monomers are processed.

Most polymers must also be stabilized against light, heat, atmospheric oxygen, moisture, high-energy radiation and microbes. S. are added to plastics, elastics or synthetic fibers in amounts between 0.1 and 5%, or are bonded chemically into the polymer. The mechanisms of stabilization vary. For example, stabilization to light can be achieved by adding UV absorbers, which inhibit the primary photochemical reactions. Some important light S. and antioxidants are: *sterically hindered phenols*, e.g. 2,6-di-*tert.*-butyl-4-methylphenol (Fig. 1); *peroxide decomposers*, e.g. dilauryl β,β-thiodipropionate (Fig. 2), *metal deactivators*, e.g. amides (Fig. 3); *UV absorbers*, e.g. resorcinyl monobenzoate (Fig. 4) and *quenchers*, which return the polymer to the ground state by energy transfer, and themselves release the energy as heat (Fig. 5).

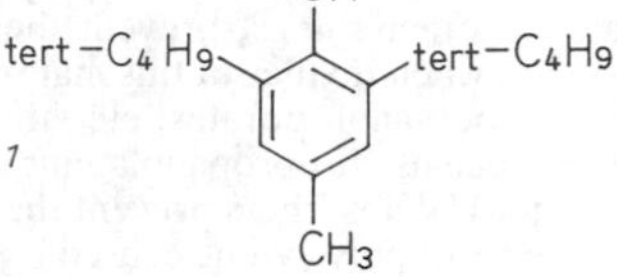

OH; tert$-C_4H_9$; tert$-C_4H_9$; CH_3

1

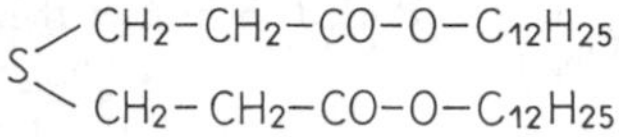

S$<$ $CH_2-CH_2-CO-O-C_{12}H_{25}$ / $CH_2-CH_2-CO-O-C_{12}H_{25}$

2

$[HO-C_6H_2(tert\text{-Bu})_2-CH_2-CH_2-CO-NH-]_2$

3

OH; $-CO-O-$

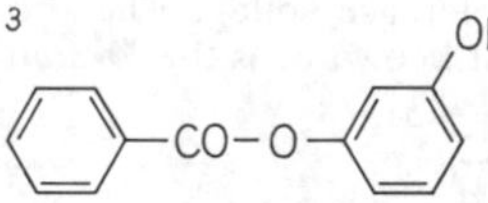

4

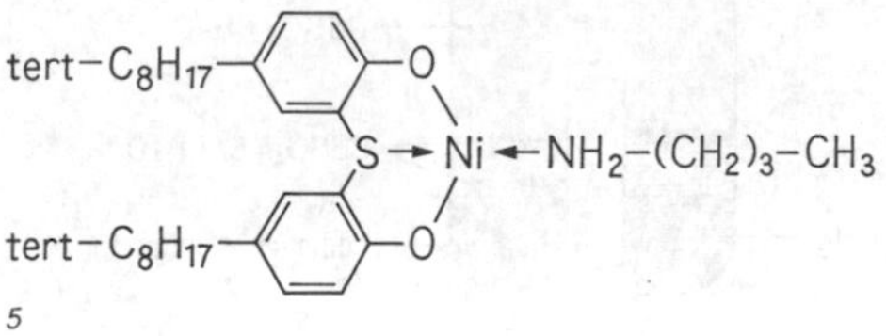

tert$-C_8H_{17}$; O; S → Ni ← $NH_2-(CH_2)_3-CH_3$; O; tert$-C_8H_{17}$

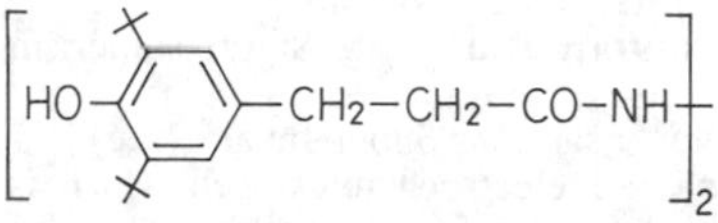

5

$n\ C_4H_9$, $n\ C_4H_9$; Sn; $O-CO-C_{11}H_{23}$, $O-CO-C_{11}H_{23}$

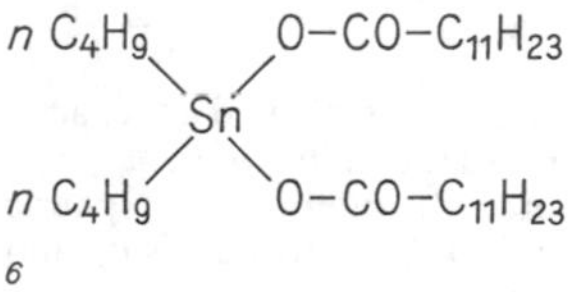

6

Some important stabilizers

S. for polyvinyl chloride are generally compounds which can bind hydrogen chloride. This is very important, because hydrogen chloride released by decomposition processes catalyses further degradation of the PVC. A particularly simple stabilization is achieved by adding about 0.2% sodium carbonate. Nitrogen-containing S. (e.g. diphenylthiourea), basic lead compounds or metal soaps containing unsaturated fatty acids (especially barium cadmium laurate) can also be used. The metal soaps are generally used in combination with antioxidants. The most effective PVC stabilizers are organotin compounds, such as dibutyltin dilaurate (Fig. 6). For heat and oxidation stabilization of polyolefins, e.g. polyethylene and polypropylene, one can use alkylphenols, alkylene bisphenols, aminophenols, thiobisphenols, polyphenols and organic sulfur and phosphorus compounds. Oxidation stabilization of polypropylene used as electrical insulation is very important. The oxidation of the plastic is accelerated by metallic copper, but this can be prevented by metal deactivators. Rubber is stabilized against atmospheric oxidation by antioxidants, e.g. 2-mercaptobenzimidazole and organic phosphites. Other S. for rubber are *N,N'*-diphenyl-*p*-phenylenediamine and condensation products of diphenylamine and acetone. S. for polyacrylonitrile must, among other things, prevent the discoloration which occurs when textiles of this material are heated. 2-Mercaptoethanol, borates, etc. are used for this purpose. S. such as diisopropanolamine, various aminoalcohols and UV absorbers prevent the discoloration and brittleness of polystyrene caused by oxidation and light. S. for polyamides, especially salts of copper and manganese, are used to protect the fibers from heat and light.

Stable: see State.

Stacking defects: see Crystal lattice defects.

Stacking heights: see Polytypism.

Staggered conformation: see Stereoisomerism 2.1.

Stainless Invar®: same as Super Invar® (see).

Standard cell: an electrochemical cell which is used to create a defined reference voltage when no current is flowing. The best known S. is the ***Weston standard cell***.

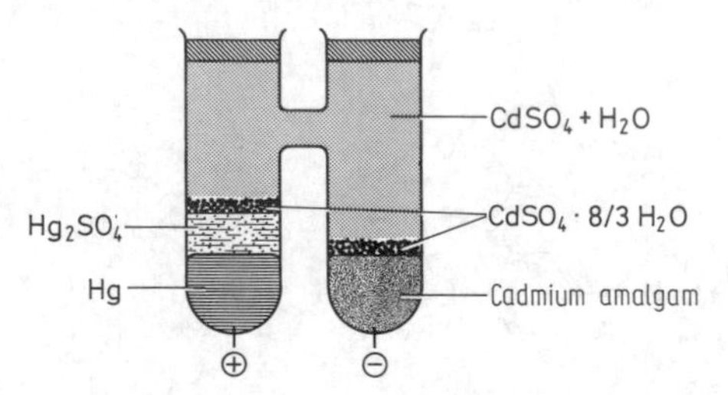

Weston standard cell

The voltage of this cell was established at exactly 1.01830 V at 20 °C by international agreement.

Standard conditions: standard atmospheric pressure (101.3 kPa) and 273 K. At this pressure and temperature, a gas is in its Standard state (see).

Standard electrode potentials, ***electromotive force series***: tables in which corresponding redox pairs are arranged in order of increasing electrode potentials under standard conditions (Table 1 to 3). There are usually separate tables for different types of electrodes (metal/metal ion electrodes, redox and anion electrodes). The values are relative to the standard hydrogen electrode, in aqeuous solution at 25 °C and 101.325 kPa pressure. The activities of the components of the corresponding redox pairs are set at $a = 1$. The position of a redox pair in the table is an indication of its propensity to accept or donate electrons. For example, noble metals have high S., while active metals have low S. If the concentrations of the components are comparable, corresponding redox pairs with relatively low S. act as electron donors to pairs with more positive S.; in other words, the pair with the lower S. acts as the reducing agent see Redox reaction). For example, Cu^{2+} ions are reduced by elemental iron, and Hg_2^{2+} ions by elemental copper (see Cementation). The different oxidizing strengths of the halogens can also be seen in the table. It becomes clear that an iron(III) salt solution oxidizes copper to Cu^{2+} ions, and iodide to iodine, but not bromide to bromine. All metals with negative S. (in the absence of passivation phenomena) dissolve in aqueous solutions with sufficiently high H_3O^+ concentrations. The electromotive force of galvanic cells (with the same ion activities) is the difference in the S. of the two elements, e.g. in the Daniell cell (see). It must be remembered that the tabulated values are for standard conditions, and that the electrode potentials for redox reactions with arbitrary concentrations must be calculated using the Nernst equation (see).

Table 1. Standard electrode potentials of metals

Corresponding redox pair	$E°$ in V
Li ⇌ Li^+ + e	−3.045
K ⇌ K^+ + e	−2.925
Rb ⇌ Rb^+ + e	−2.925
Ba ⇌ Ba^{2+} + 2e	−2.906
Ca ⇌ Ca^{2+} + 2e	−2.766
Na ⇌ Na^+ + e	−2.711
Mg ⇌ Mg^{2+} + 2e	−2.375
Al ⇌ Al^{3+} + 3e	−1.706
Zn ⇌ Zn^{2+} + 2e	−0.763
Cr ⇌ Cr^{3+} + 3e	−0.744
Fe ⇌ Fe^2 + 2e	−0.440
Co ⇌ Co^{2+} + 2e	−0.277
Ni ⇌ Ni^{2+} + 2e	−0.250
Sn ⇌ Sn^{2+} + 2e	−0.136
Pb ⇌ Pb^{2+} + 2e	−0.126
Cu ⇌ Cu^{2+} + 2e	0.337
2 Hg ⇌ Hg_2^{2+} + 2e	0.850
Ag ⇌ Ag^+ + e	0.799
Au ⇌ Au^{3+} + 3e	1.498

Table 2. Electrode potentials of pH-independent redox systems

Corresponding redox pair	$E°$ in V
Cr^{2+} ⇌ Cr^{3+} + e	−0.41
Ti^{2+} ⇌ Ti^{3+} + e	−0.37
Sn^{2+} ⇌ Sn^{4+} + 2e	0.154
Cu^+ ⇌ Cu^{2x} + e	0.167
I^- ⇌ ½ I_2 (solid) + e	0.535
Fe^{2+} ⇌ Fe^{3+} + e	0.771
Hg_2^{2+} ⇌ 2 Hg^{2+} + 2e	0.905
Br^- ⇌ ½ Br_2 (liq.) + e	1.065
Cl^- ⇌ ½ Cl_2 (gas) + e)	1.358
Pb^{2+} ⇌ Pb^{4+} + 2e	1.69
F^- ⇌ ½ F_2 + e	2.85

Table 3. Electrode potentials of pH-dependent redox systems

Corresponding	redox pair	$E°$ in V
½ H_2 (gas)	⇌ H^+ + e	0.000
4 OH^-	⇌ O_2 (gas) + 2 H_2O + 4e	0.401
H_3AsO_3 + H_2O	⇌ H_3AsO_4 + 2 H^+ + 2e	0.559
MnO_2 (solid) + 4	⇌ MnO_4^- + 2 H_2O + 3e	0.587
OH^-	⇌ IO_3^- + 6 H^+ + 6e	1.085
I^- + 3 H_2O	⇌ MnO_2 (solid) + 4 H^+ +	1.236
Mn^{2+} + 2 H_2O	2e	1.34
Cl^- + 4 H_2O	⇌ ClO_4^- + 8 H^+ + 8e	1.36
2 Cr^{3+} + 7 H_2O	⇌ $Cr_2O_7^{2-}$ + 14 H^+ + 6e	1.45
Cl^- + 3 H_2O	⇌ ClO_3^- + 6 H^+ + 6e	1.49
Cl^- + H_2O	⇌ HOCl + H^+ + 2e	1.52
Mn^{2+} + 4 H_2O	⇌ MnO_4^- + 8 H^+ + 5e	1.78
2 H_2O	⇌ H_2O_2 + 2 H^+ + 2e	

Standard galvanic voltage: see Equilibrium galvanic voltage.

Standard hydrogen electrode: see Hydrogen electrode.

Standard state: in physical chemistry, a set of conditions established by convention as those under which data for use in tables is gathered. Such a convention is necessary because the interactions between particles of the different components in Mixtures (see) cause their parameters to differ from those of the pure substances. Furthermore, the parameters of state based on the second law of thermodynamics, e.g. entropy or free energy, depend on the composition of the mixture even when the additional interactions between particles are negligible, that is, in ideal solutions.

The S. is defined as the state of the pure substance at normal atmospheric pressure (101.3 kPa) and a temperature of 298.15 K (25 °C). For gases, ideal behavior is assumed (fugacity equal to 1). The S. is indicated as a superscript to the symbols for the state parameters (^{o}), e.g. the Gibbs free energy of formation, $\Delta_F G^o$, is the molar free energy associated with production of 1 mole of the substance from the elements; the calculation is based on the values for both the starting materials and the compound in the pure state. Since $\Delta_F G^o$ depends on the temperature, the relevant temperature must be indicated; for example, 298 K. The temperature is often appended to the symbol as a subscript: $\Delta_F G^o_{298}$. In some cases, the S. is defined differently. For example, in solutions, the S. of the solute (2) is defined as the pure substance (x_2 = 1); the interactions between the particles are taken to be the same as in an infinitely dilute solution (activity coefficient of 1). Such a S. is naturally a fiction, but it is convenient because, for example, the chemical potential, $\mu_i = \mu_1^o + RT\ln x_i F_i$, in the standard state becomes $\mu_i = \mu_1^o$ (see Activities, Chemical potential).

Stannane, ***tin hydride***: SnH_4, a colorless, poisonous gas; M_r 122.72, m.p. -150 °C, b.p. -52 °C. Of the series of stannanes which might be expected, only S. (actually ***monostannane***) and ***distannane***, Sn_2H_6, are known. This is due to the decreasing stability of element-element bonds in group IVa as one goes from carbon to silicon to tin. S. does not react with dilute acids or bases, but it tends to decay thermally into tin and hydrogen. It is formed by reduction of tin tetrachloride with lithium alanate or of aqueous tin(II) salt solutions with sodium boranates.

Stannates: salts of the oxoacids of tin, which are not known in the free state. ***Stannates(IV)***, $M^I_2SnO_3$ or $M^{II}SnO_3$, can be made by alkaline solubilization of tinstone, or by fusion of the corresponding metal oxides with SnO_2, e.g. $SnO_2 + 2\,NaOH \rightarrow Na_2SnO_3 + H_2O$. The alkali stannates(IV) crystallize out of aqueous solution in a more hydrated form, as ***hexahydroxostannates***, e.g. $Na_2[Sn(OH)_6]$. Addition of tin(II) oxide to alkali hydroxide solutions or treatment of tin(II) salt solutions with bases yields ***hydroxostannates(II), hydroxostannites***, which contain the $[Sn(OH)_3]^-$ or $[Sn(OH)_4]^{2-}$ anions.

Stannic acids: compounds which would best be described as tin dioxide hydrates with undefined water contents, $SnO_2 \cdot xH_2O$. They differ only with respect to their degree of aggregation and particle size. There is no evidence for defined S., such as an H_4SnO_4. When acids are added to aqueous solutions of alkali stannates (see Stannates), or when Sn(IV) compounds are hydrolysed, a voluminous, colorless precipitate is obtained which is soluble in acids and bases. This was formerly called ***α-stannic acid***. Long boiling converts this precipitate to unreactive products which are relatively insoluble in acids and bases, and are identical to the products obtained by the reaction of concentrated nitric acid with tin; they were formerly called β-stannic acid.

Stannum: Latin name for Tin (see).

Stanozolol: see Anabolics.

Starch: a mixture of plant reserve polysaccharides consisting exclusively of glucose units. S. is a white, non-crystalline powder which is insoluble in cold water. In hot water, the S. grains are broken, allowing formation of a colloid which becomes a gel on cooling.

S. is found in the seeds and storage organs of plants in the form of starch grains. Wheat, maize, and rye seeds contain 50 to 70% S., rice seeds, 70 to 80%, and potatoes, 17 to 24%. S. can be obtained from plant parts by milling and flotation with water. The S. grains display concentric or eccentric layer lines; their form and size is characteristic of the plant of their origin. The grains of rice S. are small, 4 to 6 µm, while those of potato S. are much larger (5 to 100 µm) (Fig.).

Grains of potato starch.

S. generally consists of 15 to 25% amylose and 75 to 85% amylopectin. ***Amylose*** is essentially an unbranched α-D-glucan with an average of 10^3 glucose residues in a chain

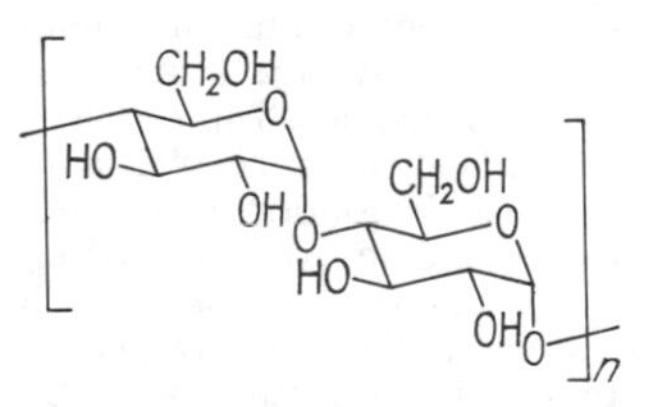

Amylose (conformation).

The amylose in grain seeds is more highly polymerized than that in potato starch. The glucan chain forms a helix. Amylose is a white powder which is soluble in hot water; it forms a dark blue inclusion complex with iodine. ***Amylopectin*** is a branched glucan, in which the average chain length in the tree-like branches is 20 to 25 glucose residues.
Here the average molecule consists of 10^4 to 10^5 units, and is thus much larger than an amylose molecule. The branches are made by α-(1→6) linkages. Amylopectin is a white powder which gives a colloid in hot water. Its inclusion compound with iodine is red-violet.

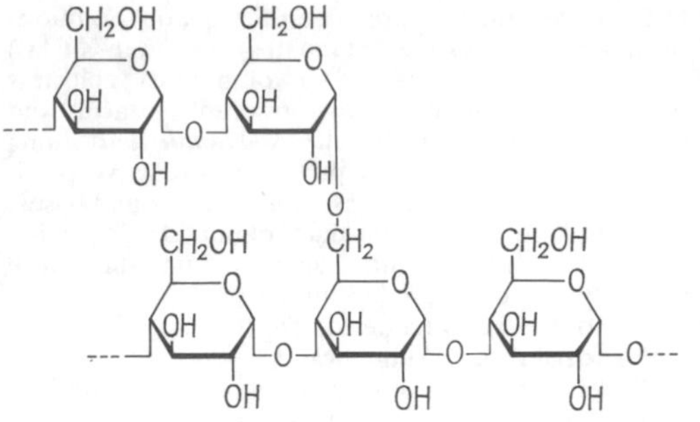

Partial structure of amylopectin (branching point).

Degradation. Acid catalyses hydrolysis to glucose, via the disaccharide maltose. S. partially degraded by dilute acid is called *soluble S.*. S. is degraded by enzymes called amylases; these are found in the saliva and pancreatic juice of animals, in plants and microorganisms. The α-amylases completely degrade amylose, and amylopectins by about 90%, mainly to maltose and to α-dextrins (see Dextrin) rich in α(1→6) links, which are not attacked by the enzyme.

Applications. S. are the major source of energy in the diets of nearly all human beings. Purified S. are used in the food industry to produce pudding powders, baby foods, sauces, cremes, etc. They are used in the pharmaceutical industry and in the textile industry. Soluble S. is used as an indicator for iodine in iodometry.

Stark effect: Splitting of spectral lines by an electric field acting upon the absorbing or emitting particles. The S. was discovered in 1913 by Johannes Stark. There are a linear and a quadratic S.: in the linear S. the splitting is proportional to the electric field strength, and in the quadratic S., to the square of the field strength. The linear S. is observed in the H atom and hydrogen-like atoms in which electronic states with the same principle quantum number but different orbital angular momentum quantum numbers are degenerate, i.e. they have the same or at least approximately the same energies. This degeneracy is lifted by the electric field. In atoms which are not hydrogen-like, and which do not display such degeneracy, the S. arises because an external field induces a dipole moment proportional to itself, which can only assume certain orientations with respect to the external field (directional quantization); this causes a splitting of the energy terms. Since the splitting depends on the product of the field strength and the dipole moment, and the latter is itself dependent on the field strength, the degree of splitting is proportional to the square of the field strength. The splitting due to the S. is generally quite complicated. Like the Zeeman effect (see), the S. has made important contributions to spectral analysis.

The S. is also observed in molecules and is applied, for example, to Rotational spectra (see) to determine dipole moments spectroscopically. The interatomic S. is the spreading of spectral lines which result from randomly oriented electric fields of neighboring electrons and ions.

State: in thermodynamics, a term for the condition of a macroscopic material system. The ***external S.*** applies to the system as a whole, e.g. its position in the gravitational field of the earth or a uniform motion; the ***internal S.*** is determined by the relationships among the components of the system. A S. is exactly characterized by the parameters of state (See State, parameters of). S. can be *stable* or *unstable*. If a system which has been slightly disturbed returns spontaneously to its original S., it is stable. If the conversion of an unstable S. to a stable one is inhibited by internal or external influences, the system is in a ***metastable S.***. An open system may achieve a ***stationary S.*** or ***steady S.*** (see Equilibrium).

The transition of a system from one S. to another is called a ***change of state*** or ***process***. It can be Reversible (see) or irreversible. Often one of the variables of S. is held constant during a change of S. Depending whether this is the temperature, the pressure or the volume, the change is referred to as isothermal, isobaric or isochoric. An adiabatic process is a change of S. without heat exchange with the environment. If a partial heat exchange occurs, the change is polytropic. If a series of changes of S. leads back to the starting S., the process is cyclic.

State, diagram of: same as Phase diagram (see).

State, equation of: an equation showing the relationship between thermodynamic parameters of state.

1) The ***thermal equation of state*** for a homogeneous substance gives the relationship between its volume v, pressure p and temperature T: $v = f(T,p)$. As a rule, the volume increases with increasing temperature and decreasing pressure. However, an important exception is water, the volume of which decreases with increasing temperature between 273 and 277 K and does not begin to increase until a temperature above 277 K is reached. The volumes of gases are much more strongly dependent on pressure and temperature than condensed substances. The coefficients of expansivity and compressibility reflect the changes in one state parameter when another is held constant: the *thermal expansivity* $\alpha = (\delta v/\delta T)_p/v$ and the *compressibility* $\gamma = -(\delta v/\delta p)_T/v$. Since at constant volume, $(\delta v/\delta T)_p\, dT + (\delta v/\delta p)_T\, dp = 0$,

$$\left(\frac{\delta p}{\delta T}\right)_v = \frac{-(\delta V/\delta T)_p}{(\delta v/\delta p)_T} = \frac{\alpha}{\beta}$$

1.1) The thermal equation of state for ***ideal gases*** is $pv = nRT$, where n is the number of moles and R is the general gas constant. This equation applies to all ideal gases, and is therefore also called the ***ideal gas law***. Special cases of the thermal equation of state for ideal gases are *Gay-Lussac's law*, $v = v_0(1 + \alpha\vartheta)$ for isobaric processes (p a constant) and $p = p_0(1 + \alpha\vartheta)$ for isochoric processes (v a constant), and *Boyle's law* pv = constant for isothermal processes (T a constant). Here v_0 and p_0 are the volume and pressure at 273.15 K, i.e. 0 °C, and ϑ is the Celsius temperature. For ideal gases, $(\delta v/\delta T)_p/v = (\delta v/\delta p)_T/p = 1/273.15$ K^{-1}. This is *Charles' law*. The ideal gas law can be derived from kinetic gas theory (see Gas theory, kinetic).

1.2) Thermal ***equations of state of real gases*** are empirical or semiempirical equations which are based on the equation of state of ideal gases; the deviations from ideal behavior are taken into account by introduction of correction factors. The most useful of them are discussed below:

1.2.1) The *virial equation of Clausius* is a series expansion of p: $pV = RT + Bp + Cp^2 + Dp^3 + ...$ The constants B, C, D, etc. are called the second, third, fourth, etc. virial coefficients. They depend on the temperature and type of gas, and have to be determined experimentally. If the forces between molecules (particles) are known, an approximate statistical calculation is also possible. For nonpolar gases and at low pressure, the series can be interrupted after the second term; this approximation is the *Callendar equation*.

1.2.2) the *van der Waals equation* (see) $pV = (p + a/V^2)(V - b)$, where V is the molar volume, and a and b are substance-specific constants.

1.2.3) The *Beattie-Bridgman equation*, $pV^2 = RT(1-\varepsilon)(V + B - A/V^2)$, with the substance-specific parameters ε, A and B is used for high pressures. All equations of state of real gases are shown here in the form for 1 mol. There have been efforts to develop a general equation of state for real gases by introduction of Reduced parameters (see; see also Corresponding states).

1.3) ***Thermal equations of state of liquids and solids.*** Just as there is no general equation of state for real gases, there is none for liquids and solids. The equations for real gases, especially the van der Waals and virial equations, are used for approximate descriptions of their thermal properties. Liquids consisting of simple molecules obey *Eucken's equation* over a wide pressure range: $p + A/V^3 = BT/V^3 + C/V^6$; A, B and C are substance parameters. For high pressures, Tamman's equation (see) is suitable. Ideal solids, which have ideal lattice structures, would have the property that the thermal coefficients α and β would be zero. It follows from the general equation of state, $\alpha v \mathrm{d}T = \beta v \mathrm{d}p$, that their volume would be independent of pressure and temperature. Such an ideal solid does not exist, even in the immediate neighborhood of absolute zero.

2) The ***caloric equation of state*** describes the dependence of internal energy and enthalpy on temperature, pressure and volume: $e = f(T,v)$ or $h = f(T,p)$ (see Thermodynamics, first law). Since for ideal gases, $(\delta e/\delta v)_{T,p} = (\delta h/\delta p)_{T,v} = 0$, only the molar heat capacities are needed to describe the temperature dependence. For real substances, the two partial differentials are not equal to zero, and are thus required in caloric equations of state. The Joule-Thomson effect provides a possible point of entry.

3) ***Canonical equations of state*** (after Planck) describe the dependence of entropy s on the internal energy e and the volume v: $s = f(e,v)$ or $\mathrm{d}s = (\delta s/\delta e)_v \mathrm{d}e + (\delta s/\delta v)_e \mathrm{d}v$. From the definition of *Entropy* (see) and the first law of thermodynamics, of obtains $\mathrm{d}s = \mathrm{d}q_{rev}/T = (\mathrm{d}e + p\mathrm{d}v)/T$ and thus $(\delta s/\delta e)_v = 1/T$ and $(\delta s/\delta v)_u = p/T$. These two partial differentials immediately yield $T = f(e,v)$ and $p = f(e,v)$, which are new relationships among thermodynamic paramaters, if the canonical equation $s = f(e,v)$ is known.

State function: see State, parameters of.

State, parameters of: physical parameters which serve to describe the State (see) of a material system. S. are all physical parameters which have a definite value for a given state of the system, independent of the manner in which the system has reached this state. For example, the state of a system consisting of a pure, homogeneous and isotropic substance can be indicated by the mass m or molar amount n, temperature T, pressure p and volume v. Only three of these are needed for an unequivocal characterization of the state, because the fourth is fixed by the thermal equation of state $V = f(n, T, p)$ (see State, equation of). In systems consisting of several phases or substances, additional parameters are needed to indicate the phase fractions and the chemical composition (see Composition parameters).

A distinction is made between extensive and intensive S. ***Extensive parameters*** are those which depend on the size of the system. If the value of a parameter is not affected by the size, it is an ***intensive parameter***. The extensive parameters include, e.g. the molar amount of substance n, the volume v, the internal energy e, the Helmholz free energy f or the Gibbs free energy g. The intensive parameters include, e.g. density, temperature and pressure. If an extensive parameter is divided by the molar amount n or the mass m, the result is a ***molar*** or ***specific parameter***: molar parameter = extensive parameter/n; specific parameter = extensive parameter/m. To distinguish the three types, extensive parameters are denoted by lower-case letters, intensive molar parameters by capital letters, and specific parameters by the subscript "spec".

Examples: $E = e/n$ = molar internal energy (units J mol^{-1}); $E_{spec} = e/m$ = specific internal energy (units J kg^{-1} or J g^{-1}); $V = v/n$ = molar volume (units m^3 mol^{-1} or cm^3 g^{-1}). The reciprocal of the specific volume is the density, i.e. $d = m/v$ (units kg m^{-3} or g cm^{-3}). Since for a pure substance the ratio of mass m/ number of moles n = molar mass M (see Mole), it follows from the above equations that molar parameter = specific parameter x times molar mass, e.g. $V = V_{spec}M$.

A ***function of state*** is the mathematical form in which the dependence of a S. depends on other parameters, the ***state variables***, which by changing can lead to a change in state. For example, the volume of a system which contains a fixed amount n of a pure gas depends only on the pressure p and temperature T: $v = f(p,T)$. The pressure and temperature are the state variables in the function of state of the volume. Since by definition a S. is independent of the path by which the state is reached, a change in volume $\mathrm{d}v$ due to small changes in pressure and temperature is given by the total differential $\mathrm{d}v = (\delta v/\delta p)_T \mathrm{d}p + (\delta v/\delta T)_P \mathrm{d}T$. If the gas is ideal, the thermal equation of state $v = nRT/p$ applies to $v = f(p,T)$; here R is the general gas constant. It follows for the total differential: $\mathrm{d}v = -(nRT/p^2)\mathrm{d}p + (nR/p)\mathrm{d}T = -(v/p)\mathrm{d}p + (v/T)\mathrm{d}T$.

State sum, abb. $\boldsymbol{Z}$: a central parameter in statistical thermodynamics which indicates how the total energy of a molecule can be distributed among its different Degrees of freedom (see). It is defined by the equation

$$Z = \sum_j g_i e^{-\varepsilon_j / k_B T}$$

where k_B is the Boltzmann constant, T is the absolute temperature, ε_i is a defined energy state i, and g_i is the number of states with the same energy (energy degeneracy). The sum is taken over all energy states. It is possible with the S. to calculate macroscopic thermodynamic parameters from molecular data. For example, $E = RT^2(\mathrm{d}\ln Z/\mathrm{d}T)$; $S = R\ln Z + RT(\mathrm{d}\ln Z/\mathrm{d}T)$ and $F = -RT\ln Z$. On the condition that the total energy ε_i of a molecule is the sum of the energy contributions of the individual degrees of freedom, $\varepsilon_i = \varepsilon_{trans} + \varepsilon_{rot} + \varepsilon_{vib} + \varepsilon_{el} + \varepsilon_{nucl}$, the e function of Z is the product of e functions and can therefore be split into partial state sums for these various degrees of freedom: $Z = Z_{trans} \cdot Z_{rot} \cdot Z_{vib} \cdot Z_{el} \cdot Z_{nucl}$. The following equations are valid for these partial state sums: for 3 translational degrees of freedom, $Z_{trans} = h^{-3}(2\pi m k_B T)^{3/2} v$ with h = Planck's constant, m = molecular mass and v = volume of the gas; for the rotation of the molecule, $Z_{rot} = h^{-2}\sigma^{-1}(8\pi^2 I k_B T)$ for linear molecules and $Z_{rot} = h^{-3}\sigma^{-1}\pi^{1/2}(8\pi^2 k_B T)^{3/2}(I_a I_b I_c)^{1/2}$ for non-linear molecules, with σ = symmetry number (number of identical positions in a 360° rotation), I is the principle moment of inertia of the linear molecule, I_a, I_b and I_c are the moments of inertia with respect to the three axes of rotation of the molecule; for the state sum for vibration, Z_{vib}

$$Z_{vib} = \prod_{j=1}^{3N-6} \tfrac{1}{2}\,[\sin h(e_j/2)]^{-1}$$

with j = index of the $3N$-6 vibrational degrees of freedom and $e_j = h\nu_j/k_B T$, where ν_j is the vibrational frequency of the jth mode of vibration. Linear molecules have $3N$-5 degrees of vibrational freedom, so that for them j runs from 1 to $3N$-5.

Chemical reactions occur without change in the nuclear state, and thermal processes are generally in the electronic ground state, so that Z_{nucl} and Z_{el} have no effect on the total S. and are therefore set equal to 1. Thus if the molecular data m, I and all ν_j are known, the S. and thus the thermodynamic parameters and equilibria can be calculated.

State variable: see State, parameters of.

Statins, ***release inhibiting hormones***: neurohormones formed in the small-cell regions of the hypothalamus and transported from there by the blood to the anterior pituitary. There they inhibit the secretion of somatotropin, prolactin and melanotropin. Together with the three corresponding liberins, the S. regulate the levels of these three pituitary hormones, which are not subject to negative feedback from the periphery.

Melanostatin, melanotropin release inhibiting hormone, abb. ***MIH***. The tripeptide amide Pro-Leu-Gly-NH_2 is contained in the sequence of oxytocin, and it has been proposed that it is released from oxytocin by a membrane-bound peptidase in the eminentia mediana of the hypothalamus (R. Walter, 1971).

Prolactostatin, prolactin release inhibiting hormone, abb. ***PIH***, inhibits prolactin release.

Somatostatin, somatotropin release inhibiting hormone, abb. ***SIH***, is a heterodetic cyclic peptide of 14 amino acid residues (Fig.). It inhibits secretion not only of somatotropin, but of almost all anterior pituitary hormones, and of the tissue hormones of the gastrointestinal tract. It also inhibits release of insulin and glucagon from the pancreas.

Ala–Gly–Cys–Lys–Asn–Phe–Phe–Trp–Lys–Thr–Phe–Thr–Ser–Cys

Somatostatin

Somatostatin has been found in various endocrine glands, the gastrointestinal mucosa, the islets of Langerhans and in the central and peripheral nervous system; outside the hypothalamus it acts as a neurotransmitter, and in the hypothalamus-adenohypophysis system, as a neurohormone. The variety of its biological effects limits the applications of synthetic SIH, so that there is interest in developing synthetic analogs with limited effects, which might also be applied orally. Cyclic analogs of SIH with shorter sequences have proven very active. SIH was the first peptide hormone to be synthesized by a genetically engineered system, in 1977. Its biosynthesis begins with translation of ***pre-prosomatostatin***, a polypeptide of 121 amino acids with the SIH sequence at its C-terminus.

Stationary phase: see Chromatography.

Stationary state: see Equilibrium, see State.

Stationary state principle, ***Bodenstein principle***: an approximation used in the kinetic treatment of Complex reactions (see) with short-lived intermediates. The S. assumes that within a short induction period at the beginning of the reaction, a low concentration of the intermediate builds up. It then remains approximately constant (quasi-stationary) for most of the period of the reaction. As an approximation, $\mathrm{d}c_i/\mathrm{d}t$ can be set = 0, where c_i is the stationary concentration of the intermediate. The S. greatly simplifies the systems of differential equations for kinetic calculations, but its applicability must be carefully tested in each case.

Stauffer grease: a salve-like lubricant for machines. It is made from animal or plant oils or fats, mixed with a little mineral oil and saponified by heating with lime or soda.

Stauff-Klevens equation: a linear relationship between the logarithm of the critical micellar concentration c_M of a surfactant and the number n of carbon atoms in its lipophilic group: $\lg c_M = A - Bn$. A and B are constants which differ from one series of surfactants to the next.

Steady-state equilibrium: see Equilibrium.

Steam reforming: see Synthesis gas.

Stearic acid, ***octadecanoic acid***: CH_3-$(CH_2)_{16}$-COOH, a higher monocarboxylic acid. S. forms colorless and odorless leaflets of a waxy consistency; m.p. 69.4 °C, b.p. 232 °C at 2 kPa. S. is insoluble in water, but is soluble in alcohol, ether, benzene and chloroform. Its salts and esters are called *stearates*. S. is found in free form in grain fusel oil, various fungi, hops and ox gall. The glycerol esters of S. are important components of many plant and animal fats and oils, such as linseed oil, butter, tallow and cod liver oil. S. can be prepared by saponification of fats or catalytic hydrogenation of oleic and elaidic acids. It is used in the manufacture of candles, soaps, wetting agents, foaming agents, cosmetic and pharmaceutical preparations and rubber.

Stearin: a white to yellowish, water-insoluble mass consisting mainly of palmitic and stearic acids; it is obtained industrially by saponification of fats (animal and plant tallow, bone fat, palm fat), separation of the oily component by pressing, and purification by steam or high-vacuum distillation. S. is used to make candles, soaps, and cosmetics, and as a conditioner for textiles, rubber and leather.

Stearin pitch is a firm to hard, black mass which remains after fatty acids are distilled. It is used as a paint, a substitute for linseed oil varnish, etc.

Stearyl alcohol, ***octadecan-1-ol***: $CH_3(CH_2)_{16}CH_2OH$, a colorless, wax-like substance which is insoluble in water; m.p. 59 °C. S. is soluble in ethanol and ether. It is made industrially by catalytic hydrogenation of butyl stearate. It is used to introduce long-chain alkyl groups (fat residues) into organic molecules (color couplers for color film, sodium alkylsulfates as surfactants and emulsifiers).

Steel: an Iron-carbon alloy (see) containing up to 2.1% carbon. S. is the most important industrial material; it can be cast, pressed, hammered and rolled. Its properties, e.g. strength, hardness, resistance to corrosion and magnetic susceptibility, can be varied within broad limits by alloying the melt or tempering and surface treatments of the shaped parts. S. are classified according to their composition, method of production, structural types and applications.

Plain carbon S. are simple S. with carbon contents of 0.05 to 1.5%, phosphorus contents up to 0.09% (Thomas S., 0.3%) and sulfur contents up to 0.07%; they have no other alloy components, or only small amounts, and account for about 95% of all S. produced. They are subdivided on the basis of their carbon contents; *low-carbon S.* contain up to 0.2% carbon, *medium-carbon S.* contain 0.2 to 0.5% carbon, and *high-carbon S.* contain more than 0.5% carbon. Another system of terminology refers to S. with less than 0.83% carbon as *hypoeutectoid*, those with 0.83% carbon as *eutectoid*, and those with more than 0.83% C as *hypereutectoid*.

Alloy S. contain defined fractions of alloy components which give them certain properties. The most important alloy components are manganese, silicon, chromium, nickel, cobalt, molybdenum, tungsten, titanium, vanadium, copper, boron, niobium, aluminum and rare earth metals. These metals are added to the S. melt as iron alloys.

S. production is the refining of liquid crude iron and scrap iron (see Scrap metal) by selective oxidation at 1600 °C of the larger part of the carbon, silicon, phosphorus and manganese impurities. For example, the carbon content of crude iron is reduced from 3.5 to 4.5% to 0.2 to 2.1%. The metallic impurities are converted to a liquid, oxidized slag or are gasified. Oxygen is introduced by blowing air or oxygen into the iron melt. Some oxygen is taken up from the furnace atmosphere, and some is provided by addition of iron ore (Fe_2O_3). There are two basic procedures for making liquid S., the open hearth and blast furnace (Fig.).

I) In the open hearth process, which is used mainly for refining scrap, the heat is supplied by fuel gas or electric energy.

A) There are several varieties of the ***Siemens-Martin process*** (SM process), which was named for the furnace builders: crude iron-scrap, scrap-coal and crude iron-ore processes. The S. is melted in a regeneratively heated, pit-shaped hearth which is lined with dolomite or magnesite bricks. The regenerative chambers are lined with chromite fire bricks and are heated by the exhaust gases, or they are heated with oil, coking or natural gas, using preheated air or oxygen. Siemens-Martin slag is used as an additive to concrete and as a material for road construction.

B) In the ***electric furnace process***, an electric arc or induction furnace is used; the temperature can be precisely controlled and this permits a high melting capacity. The furnace atmosphere can be controlled to provide either an oxidizing or a reducing environment. With the aid of lime-containing additives, this

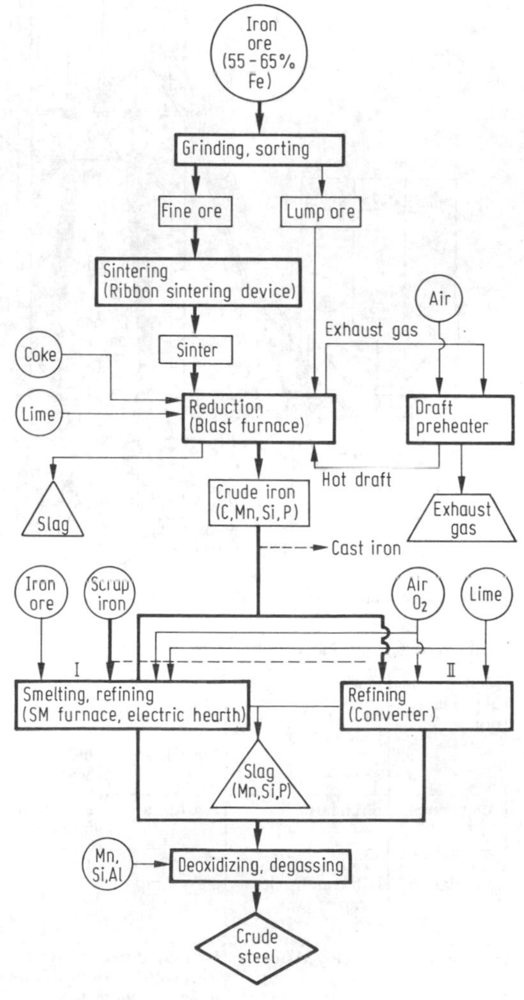

Diagram of crude iron and steel production. *I*, Open Hearth process; *II*, Blast furnace process.

permits removal of nearly all sulfur. This technique is used mainly for scrap, from which high-quality alloyed S. can be made.

II) In the blast furnace technique, liquid crude iron and scrap are converted to S.

A) ***Base blasting***. 1) In the *Bessemer process*, the oxidation process takes place in a converter (Bessemer converter) which is lined with acidic firebricks and can be tipped. The liquid crude iron is poured into the horizontal converter, which is then rotated into the vertical position and air or oxygen-enriched air is blasted into it. This rapidly oxidizes the impurities. Because the converter is lined with acidic silicate material, low-phosphorus crude iron must be used. Bessemer slag is used in the same way as Siemens-Martin slag.

The ***Thomas process*** differs from the Bessemer process in that the converter (Thomas converter) is lined with basic firebrick (dolomite bricks), so that high-phosphorus crude iron can be processed in it. Lime-containing additives yield a phosphate-containing Thomas slag, which can be used as fertilizer (Thomas phosphate).

A serious disadvantage of the base-blasting techniques is that the air jets become corroded, and therefore only a limited oxygen enrichment of the blast air (30%) is possible. With the development of oxygen blasting techniques, the base-blasting methods have

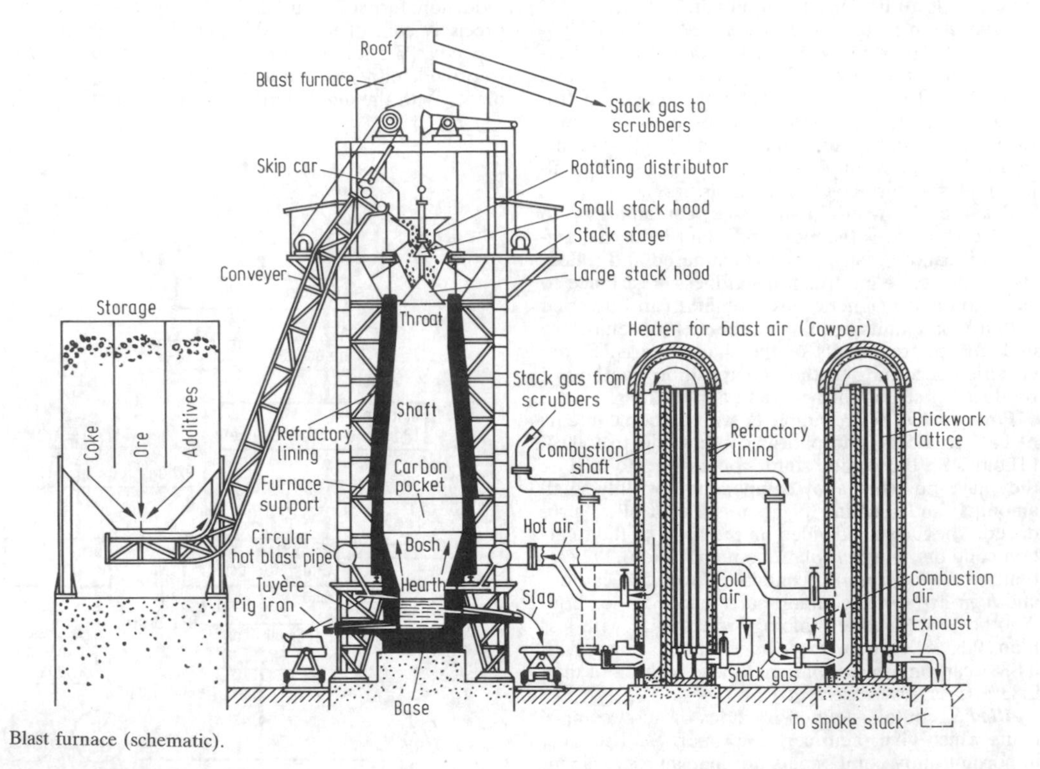

Blast furnace (schematic).

Methods of steel production

Method of production	Hearth processes			Blast furnace processes			
	Siemens-Martin (open-hearth)	Electric melting		Base blasting methods (Converter process)		Oxygen blasting methods LD-, LDAC-, OLP-, LDP-, Caldo-, rotor methods	
		Electric arc	Induction heating	Bessemer process	Thomas process		
Aggregates	SM hearth	Arc furnace	Induction furnace	Base-blasting converter (Bessemer, Thomas)		Blast converter	
Raw materials	Crude iron, liquid and solid scrap ore	Crude iron, solid alloy scrap		Liquid crude iron low phosphorus	high phosphorus	Liquid crude iron low phosphorus	high phosphorus
				scrap		scrap	
Products	Unalloyed steel and alloys with low contents of other metals	Unalloyed steel; alloyed steel with low or high contents of other metals		Unalloyed steel		Unalloyed steel; steel alloys with low contents of other metals	

become much less important. Recently, however, a base-blasting oxygen converter has been developed in which the jets are cooled with a hydrocarbon gas. This greatly reduces corrosion.

B) The ***basic oxygen process*** was developed by the Austrian firm Linz and Donawitz (LD process), and is now the most important method of processing crude iron. The liquid iron is refined in a mixture with scrap and lime in a blast converter lined with basic firebricks. The capacity of these converters can be as high as 400 t, and up to 40% scrap can be used. A number of other oxygen blasting techniques have been developed since the LD process; they differ only in technical details (*LDAC, OLP, LDP and Kaldo and rotor techniques*).

In all methods of S. production, the oxidation is followed by deoxidation, sulfur removal, degassing and alloying. The melt is homogenized by pouring and stirring, and placed under vacuum to remove hydrogen gas and to reduce the contents of oxygen and carbon. Powder injection methods are used for deoxidation with carbon, silicon, manganese and aluminum, and for removal of sulfur with calcium silicide, CaSi. The melt is poured into a form and cooled. It may be subjected later to a heat treatment to improve its quality. High-quality S. and S. alloys can be made by remelting with the alloy materials in a vacuum (see Vacuum metallurgy).

Stellite®: hard, cast, high-carbon alloys in which cobalt is the main element; they also contain chromium, tungsten and molybdenum carbides. The most common S. have the following composition: 50-66% cobalt, 26-33% chromium, 6-13% tungsten, 1-2.5% carbon. Because they are stable at high temperatures and resist abrasion and corrosion, they can resist nearly all kinds of wear. They are used mainly in the production of cutting tools and for surface treatment of steel. S. are prepared by melt sintering and melting in an electric arc.

Stephen reduction: a reaction for synthesis of aldehydes by reduction of nitriles with anhydrous tin(II) chloride in ether, with the passage of hydrogen chloride through the mixture:

$$R{-}C{\equiv}N + SnCl_2 + 2HCl \longrightarrow [R{-}C{\equiv}\overset{+}{N}H]\ HSnCl_4^- \xrightarrow{HCl} [R{-}CH{=}NH_2]_2^{2+}\ SnCl_6^{2-} \xrightarrow{H_2O} R{-}CHO$$

As to the mechanism of the reaction, it is believed that first the tetrachlorotin(II) acid is formed, which then gives an adduct with the nitrile. Reduction to the aldimine and its hydrolysis are the subsequent steps. In a few cases, the hexachlorostannate(IV) of the aldimine can be isolated in crystalline form. The S. has the advantage that the reaction stops at the aldehyde level. In contrast, the more frequently used lithium aluminum hydride can easily reduce further to the alcohol, especially when it is present in excess.

Sterane: see Steroid.

Stercobilin: see Bile pigments.

Stercobilinogen: see Bile pigments.

Stereochemistry: the area of chemistry dealing with the three-dimensional aspects of structures and reactions. Modern S. began with the work of van't Hoff and Le Bel on the tetrahedral shape of tetravalent carbon. In 1874 they explained the optical activity of organic molecules by the existence of stereoisomers (see Stereoisomerism), which are possible when an asymmetric carbon atom is present.

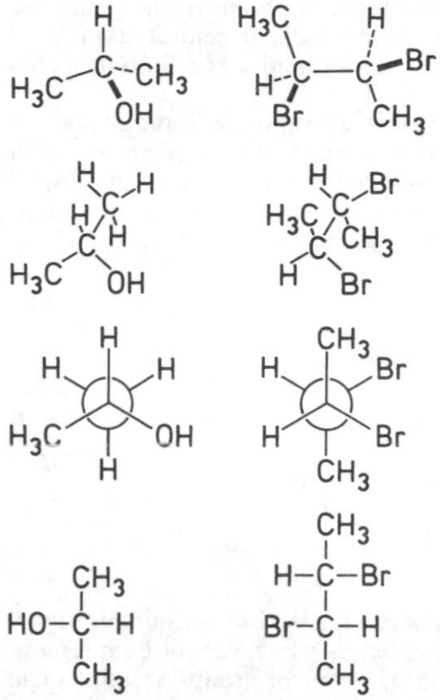

wedge and dash formulas

perspective formulas

Newman projection

Fischer projection

Until the middle of this century, S. was mainly ***static S.***, concerned with the description of spatial structures of molecules in terms of bond lengths and angles, classification of stereoisomers, optical activity, separation of Racemates (see) and relative and absolute configurations. The analysis of spatial structures was greatly simplified by the advent of physical-chemical techniques such as X-ray crystal analysis, NMR spectroscopy and chirooptic methods. These also led to the development of ***dynamic S.***, the study of relationships between spatial structure and reactivity, and of reaction mechanisms. This branch of S. treats problems such as steric hindrance in reactions, stereospecificity and stereoselectivity, inversion and retention of conformation in substitution and sigmotropic reactions, and stereoregulated polymerization.

The classic structural formulas are not able to show the S. of molecules. The best representations of spatial structure are given by *molecular models*, of which there are several kinds. 1) *Space-filling models*, such as those of Stuart and Briegleb, show atomic distances, bond angles and effective atomic radii. 2) *Stereomodels*, e.g. of Dreiding, show only the distances and angles exactly, and 3) *ball and stick* models show only the bond angles. However, these simpler models are sufficient for many purposes, especially for determining symmetry relationships.

There are several conventions for representing three-dimensional formulas on a two-dimensional surface (Fig.): 1) *wedge and dash* stereoformulas, in which the bonds projecting above the plane are indicated by heavy or wedge-shaped marks, and those below the plane by dashed lines; bonds in the plane are indicated by dashes in regular type; 2) *perspective formulas*; 3) *Newman projections*, in which the molecule is observed along one bond, so that the atom in front (represented as a circle) hides the one behind it. All other bonds are projected in the plane. 4) In *Fischer projections*, the molecule is projected onto the plane in such a way that the substituents lying below the plane are shown above or below the central atom, while those lying above the plane are shown to the left and right of the central atom.

Stereoformula: see Formula; see Stereochemistry.

Stereoheterotropic groups: see Topic groups.

Stereoisomerism: a type of isomerism in which the isomers (***stereoisomers***) have the same structures, but the arrangements of the atoms in three dimensions are different (Scheme 1).

Scheme 1.

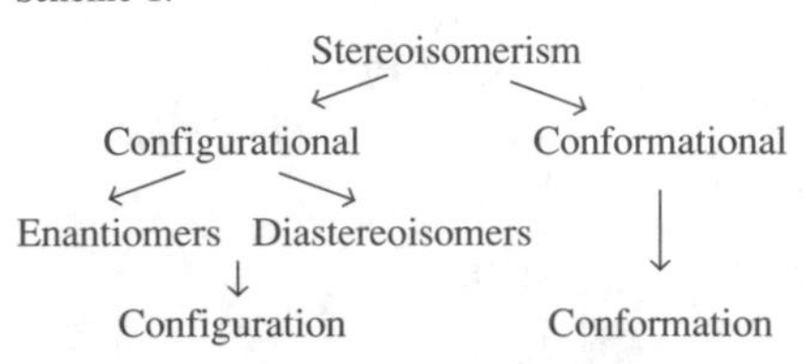

1. ***Configurational isomerism***. Configurational isomers are stereoisomers which cannot be interconverted by rotation of atoms or groups around single bonds. Unlike conformational isomers (below), most configurational isomers can be isolated.

1.1. ***Enantiomers*** are configurational isomers which are mirror images of each other. They are optically active regardless of their physical state, that is, they rotate the plane of linearly polarized light passing through them. The two enantiomers rotate polarized light through the same angle, but in opposite directions. The one which rotates light clockwise (to the right) is denoted by a (+) prefixed to its name; the counterclockwise-rotating isomer is denoted by a (-) (see Polarimetry). In addition to the direction in which they rotate polarized light, enantiomers differ with regard to their crystal habit, their physiological effects and their reactions with enantiomers of other compounds. In all other physical properties and in their reactions with non-optically active compounds, they are identical. Pure enantiomers (of tartaric acid) were first isolated by Pasteur.

There are two enantiomers of a chemical compound if its molecules cannot be made to coincide with their mirror images. This occurs when the molecule has neither a plane of symmetry (σ), a center of inversional symmetry (i) nor an axis of rotational symmetry (S_n) (an axis of rotation, c_n, may be present). Such molecules have the topological property of handedness (*chirality*), which is thus a necessary and sufficient criterion for the existence of enantiomers and optical activity (Scheme 2).

Scheme 2.

Symmetry	C_n	σ, i, S_n	Optical activity
Achiral symmetry	+	+	–
Chiral, axial symmetry	+	–	+
Chiral, asymmetrical	–	–	+

The chirality of a compound is generated by a *chiral element*. The great majority of all optically active compounds have ***asymmetric carbon atoms***, i.e. a carbon atoms with four different substituents, as their *chirality centers* (Fig. 1). Glyceraldehyde, lactic acid and tartaric acid, and in fact, nearly all natural organic substances, contain asymmetric carbon atoms. Other asymmetrically substituted central atoms, such as silicon in silanes, nitrogen in quaternary ammonium salts and amino oxides, sulfur in sulfoxides and metal atoms (M) in complex compounds (Fig. 2), can also be centers of chirality. *Chirality axes* are present in some substituted allenes, spiranes and diphenyl derivatives (Fig. 3). *trans*-Cyclooctene (Fig. 4) has a *plane of chirality*.

Helicity is a special type of chirality in which the enantiomers differ in the handedness of the pitch of the helix. This type is found in macromolecules and angular condensed aromatics (*helicenes*) (Fig. 5).

D-(-)lactic acid
(R)-(-)lactic acid

(2R,3R)-tartaric

Fig. 1. Asymmetric carbon atom (*) as chirality center.

Fig. 2. Si, N, S and M as chirality centers.

Pseudoasymmetry is observed when there are two different achiral ligands and two structurally identical, chiral ligands on the central atom.

Geometric enantiomers are a special case in which the C atom at one end of a double bond has two different achiral groups, and the C atom at the other end has two enantiomeric groups with opposite configurations (Fig. 6).

Fig. 3. Molecules with chirality axis.

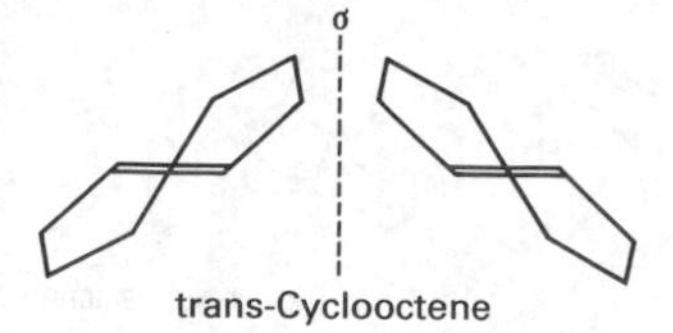

Fig. 4. Molecules with planes of chirality.

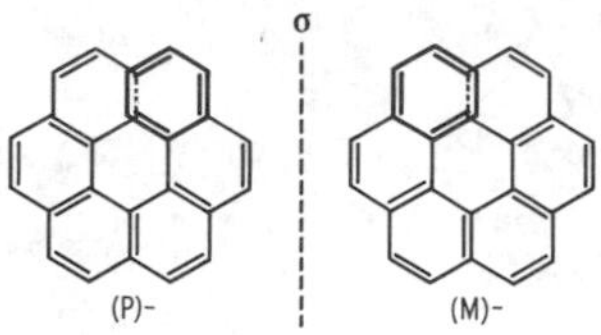

(P)- (M)-

Fig. 5. Hexahelicene. (P)-, abb. of Plus; (M)-, abb. of Minus.

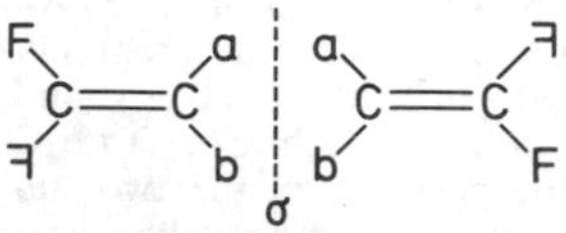

Fig. 6. Geometric enantiomers.

The two enantiomers of a chiral compound differ in their *absolute configuration*, which was first determined in 1951 by Bijvoet by X-ray diffraction of the rubidium salt of tartaric acid. Enantiomers are designated by their *relative configuration*, which has nothing to do with the direction in which they rotate polarized light. For example, the sodium salt of dextrorotatory lactic acid rotates light to the left. The direction in which light is rotated is thus not suitable for designating configuration.

In the configurational notation developed by E. Fischer, the formulae of enantiomers are drawn in the Fisher projection (see Stereochemistry): the longest carbon chain is written vertically, and the highest oxidation state is at the top. The substituents are written horizontally, and the enantiomer in which the substituent is on the right of the vertical axis has D-configuration; the other has L- (Fig. 7). This system is still used with natural products such as carbohydrates and amino acids, but it fails if it is not possible to determine which is the longest carbon chain, the highest oxidation state or which atom should be regarded as substituent, for example with CFClBrI.

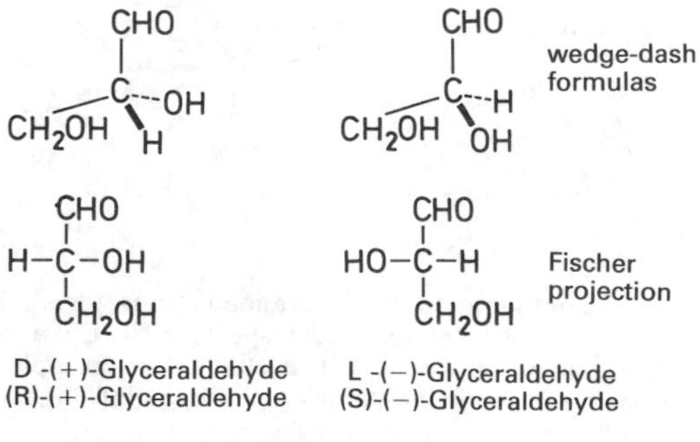

wedge-dash formulas

Fischer projection

D-(+)-Glyceraldehyde
(R)-(+)-Glyceraldehyde

L-(−)-Glyceraldehyde
(S)-(−)-Glyceraldehyde

The Cahn-Ingold-Prelog convention (Fig. 8) is unequivocal, logical and can be used with all chiral compounds. First the ligands on a chirality center are arranged according to decreasing atomic number (in the sequence A > B > C > D). Double bonds are resolved as two single bonds with the same substituent. If the priority of substituents cannot be determined from the atoms directly bonded to the chiral center, those farther from it must be considered. Looking at the molecule from the side opposite substituent d, the series of substituents a → b → c will be seen to progress either clockwise (***R-configuration***) or counterclockwise (***S-configuration***). In compounds with pseudosymmetry centers, the configurations are denoted by r or s.

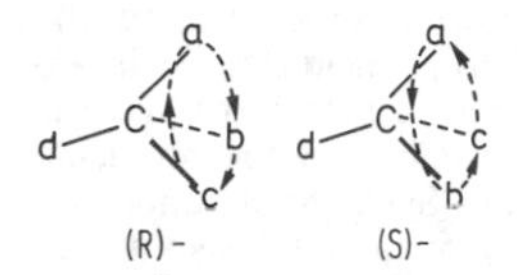

Fig. 8. Configuration in the Cahn-Ingold-Prelog system.

Equimolar mixtures of two enantiomers (optical antipodes) of a compound are called Racemates (see).

1.2 ***Diastereoisomers*** are configurational isomers which are not enantiomers. They occur in compounds with more than one chirality element (σ-diastereoisomers), in alicyclic compounds (σ-diastereoisomers), compounds with double bonds (π-diastereoisomers) and in complexes.

1.2.1. *Diastereoisomerism due to the presence of several chirality elements*. If a compound has *n* chirality elements (usually asymmetrical carbon atoms), the maximum number of configurational isomers is 2^n. Those which have opposite configurations at each asymmetric carbon atom are enantiomers (there are 2^{n-1} enantiomeric pairs). All others, which have the same configuration about at least one asymmetric carbon atom, are diastereoisomers. Compounds which have two structurally non-equivalent asymmetric carbon atoms (e.g. the aldotetroses) thus exist in 4 con-

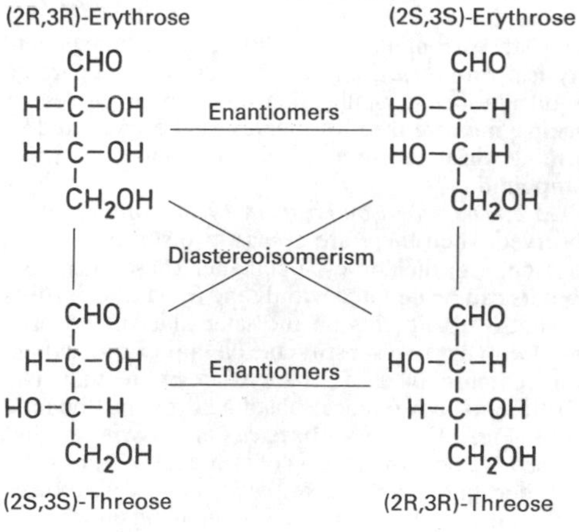

(2R,3R)-Erythrose — Enantiomers — (2S,3S)-Erythrose

Diastereoisomerism

(2S,3S)-Threose — Enantiomers — (2R,3R)-Threose

Fig. 9.

figurational isomers. Of these, (2R,3R)- and (2S,3S)-erythrose and (2R,3S)- and (2S,3R)-threose are enantiomers; the remaining 4 combinations are diastereoisomers (Fig. 9).

Diastereoisomer pairs with neighboring chirality centers are indicated by the prefix *erythro-*, if the two substituents point in the same direction in the Fisher projection, or by *threo-* if they point in opposite directions.

In the Newman projection, an *erythro-* pair of substituents can be superimposed by a suitable rotation, while a *threo-* pair cannot. The number of configurational isomers is reduced from 2^n if any of the chirality centers are structurally equivalent, that is, if two asymmetric carbon atoms have identical sets of 4 substituents. For example, in tartaric acid (Fig. 10), the (2R,3R)- and (2S,3S)-forms are enantiomers, but the diastereoisomeric (2R,3S)-tartaric acid is optically inactive, because it has a plane of symmetry and the optical rotations of the two halves of the molecule compensate for each other. Such compounds are called *internal racemates* (they cannot be split into enantiomers) and are denoted by the prefix *meso-*.

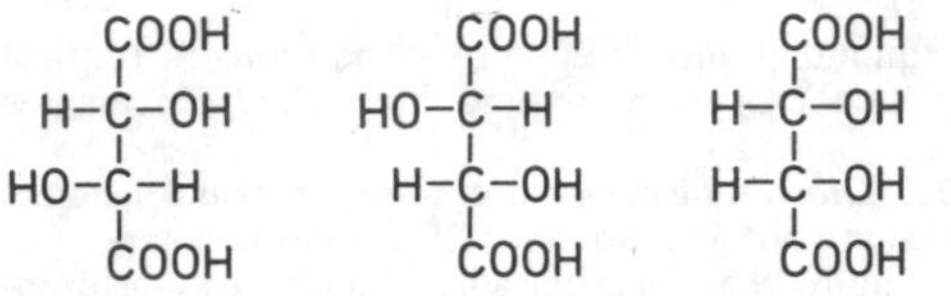

```
   COOH            COOH            COOH
    |               |               |
 H-C-OH         HO-C-H          H-C-OH
    |               |               |
HO-C-H           H-C-OH         H-C-OH
    |               |               |
   COOH            COOH            COOH
```

(2R,3R)-Tartaric acid (2S,3S)-Tartaric acid (2R,3S)-Tartaric acid *meso*-Tartaric acid

Fig. 10

Epimers are diastereoisomers in compounds with several asymmetric carbon atoms, only one of which has a different configuration in the two members of the pair; for example, D-glucose and D-mannose are epimers. ***Epimerization*** is thus the change in configuration at only one of several asymmetric carbon atoms. If the carbon in question is the C1 of a monosaccharide, which affects the glycosidic OH group in the cyclic semiacetal form, one speaks of *anomers*, which are called the α- and β-forms (see Mutarotation).

Diastereosiomers have different chemical and physical properties, and can therefore be separated by distillation, crystallization or other processes. A racemic mixture (see Racemate) can be separated by forming diastereoisomers with an optically active compound.

1.2.2. *Diastereoisomerism in cyclic compounds* is observed when there are at least two substituents on the ring, e.g. dichlorocyclopentane. These diastereoisomers can be denoted with R and S, but the prefixes *cis* (both substituents on the same side of the ring) and *trans* (the two substituents on opposite sides) are more commonly used. If there are more than two substituents, a reference substituent (r) must be indicated (Fig. 11). The prefixes *cis-* and *trans-* can also be used to show the joining of two rings, e.g. *cis-* and *trans-*decaline. The prefixes *endo-* and *exo-* are used in naming bicyclic compounds, depending on whether the substituent is *cis* or *trans* to the bridge (Fig. 12.)

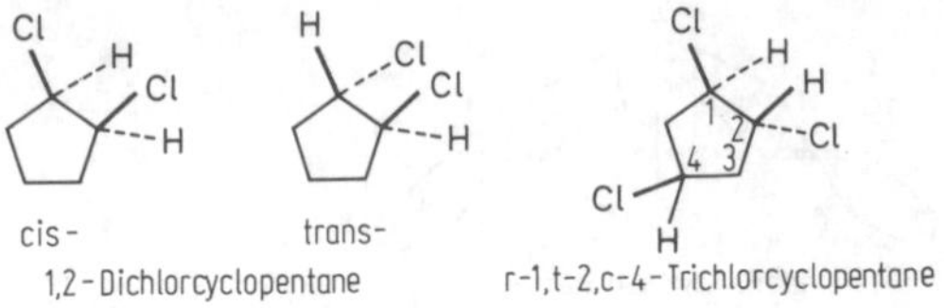

cis- trans- 1,2-Dichlorcyclopentane

r-1,t-2,c-4-Trichlorcyclopentane

Fig. 11.

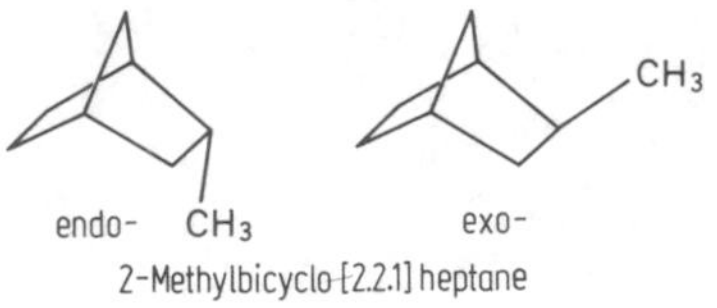

endo- exo- 2-Methylbicyclo-[2.2.1] heptane

Fig. 12.

1.2.3. *Diastereoisomerism in compounds with double bonds* is called *cis-trans*-isomerism or *(E,Z)-isomerism*. It occurs when there are two different substituents at each end of the double bond: (a,b)C=C(c,d). For (a,b)C=C(a,b) or (a,b)C-=C(a,c) compounds, the positions of the two substituents relative to each other are indicated by *cis* or *trans*. In the general case (which may also be applied to the two special cases above), the Cahn-Ingold-Prelog priority sequence is used to establish which substituent on each end of the double bond has priority. If the two higher-priority substituents are on the same side of the bond, the prefix *Z* is used; if they are on opposite sides, the prefix *E* is used. This convention also applies to double bonds on nitrogen atoms, where the terms *syn* and *anti* may also be used (Fig. 13). The terms *cis*, *syn* and *Z* (or *trans*, *anti* and *E*) are not always synonymous.

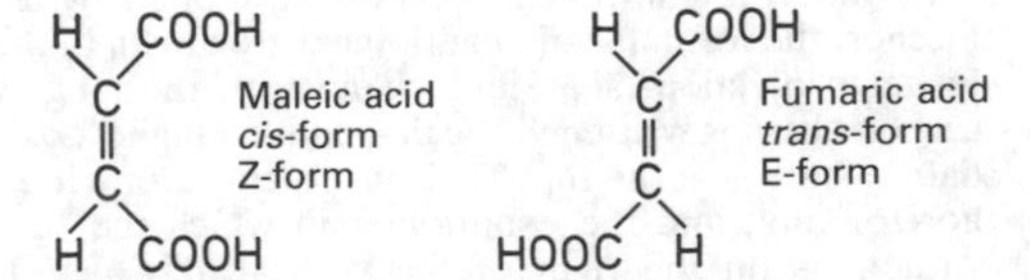

Maleic acid *cis*-form Z-form

Fumaric acid *trans*-form E-form

syn-Benzaldoxime (*E*)-Benzaldoxime

anti-Benzaldoxime (*Z*)-Benzaldoxime

E-form

Z-form

Fig. 13.

The configuration is determined from physical properties, spectroscopic data (especially IR, UV and NMR spectra) and chemical reactions. For example, the *cis* form usually has a lower melting point, higher solubility and larger dipole moment than the *trans* form. The *trans* form is usually more stable, due to electronic and steric interactions, although there are exceptions, such as *cis*-1,2-dichloroethene and *cis*-cyclooctane. *cis,trans*-Isomerization also occurs in coordination compounds.

2. ***Conformational isomers*** are stereoisomers which can be brought into the same configuration by rotation around C-C single bonds. They can only be isolated if the rotation is hindered (see Atropisomerism).

2.1. *Conformations of acyclic compounds*. Even in ethane, the rotation around the C-C double bond is not completely free. In a rotation through 360°, there are three energy barriers (rotation barriers) of 11.7 kJ mol^{-1}; these represent the energy differences between two conformations, *staggered* and *ecliptic* (Fig. 14). The staggered form has the lower energy; the higher energy of the ecliptic form is due to repulsion between the electron clouds around the H atoms (*Pitzer strain, torsional strain*). In the rotation around the middle C-C bond on n-butane, there are a number of conformations, which are shown in Fig. 15 along with their energy differences and terminology.

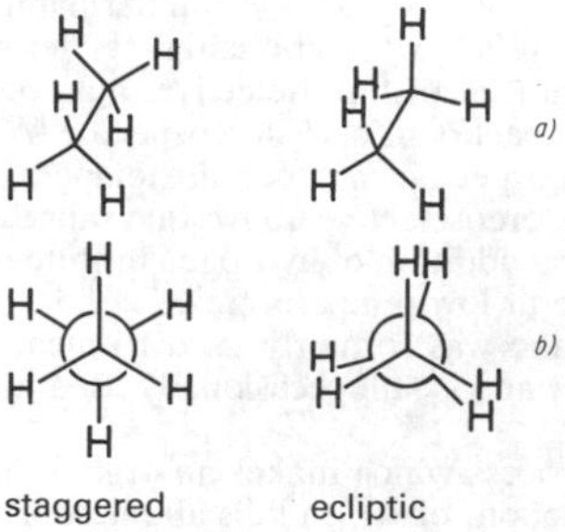

staggered ecliptic

Fig. 14. Conformation of ethane: *a* in the perspective stereo formula, *b*, in the Newman projection.

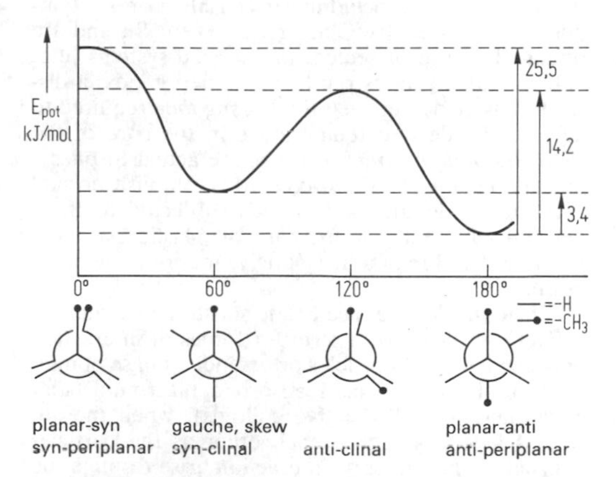

Fig. 15. Conformations of n-butane.

Consideration of the conformations of acyclic compounds leads to the following conclusions: a) Rotation around the C-C single bond is possible, but it is not completely free, as there are rotational barriers. b) At *conformational equilibrium*, the conformations with the lowest conformation energies (sum of all energies which affect the stability of conformations) are the most stable. The largest substituents assume the anti-periplanar conformation; this leads to the zig-zag arrangement of the n-alkanes. c) The presence of oppositely charged groups or hydrogen bridges can lead to a preference for the gauche conformation. d) Only in a few cases, e.g. in 2,2'-disubstituted diphenyl derivatives (see Atropisomerism), are the rotation barriers are so high that rotation is impossible under normal conditions. Cyclic linkage of the groups to each other also prevents rotation.

Investigation of such conformational equilibria and the estimation of stabilities of individual conformational isomers from physical properties (especially dipole moments) and spectroscopy (especially NMR) is called *conformational analysis*.

2.2 *Conformation of cyclohexanes*. Conformational analysis is of particular importance for cyclic compounds. In contrast to Bayer's original assumption (1885; see Bayer's tension hypothesis) the stable conformations of cycloalkanes are not planar; this was suggested as early as 1890, by Sachse. According to Sachse and Mohr, cyclohexane exists in two nonplanar conformations which are free of bond-angle strain: the chair and the boat forms (Fig. 16).

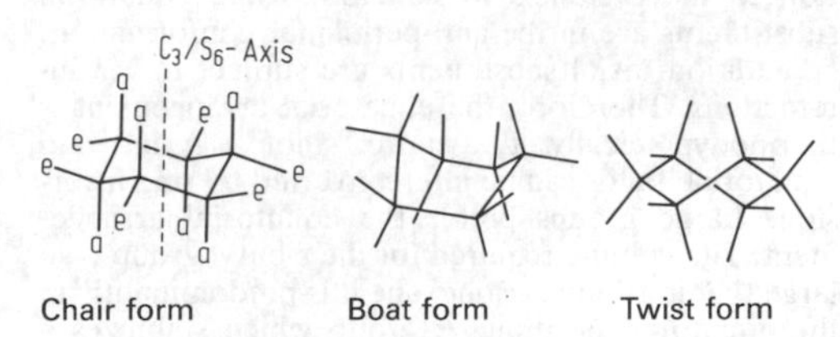

Fig. 16. Forms of cyclohexane.

The chair form has a three-fold axis of rotation and a six-fold axis of rotational symmetry; it is rigid and 23 kJ mol^{-1} more stable than the boat form. It contains only gauche conformations around the C-C bonds. The *boat* form has higher energy than the chair form, because it contains two ecliptic, syn-periplanar conformations in addition to four gauche conformations. In addition, there are repulsive forces between the H atoms in the 1 and 4 positions. The boat form is flexible, passing by *pseudorotation* into alternative boat forms. In the course of pseudorotation it passes through a *twist boat form* which is about 3 kJ mol^{-1} more stable (Fig. 16). Because of the energy difference between the chair and boat forms, cyclohexane at room temperature is 99.9% in the chair form. However, the boat form can be stabilized in bicyclenes or by attractive forces between the substituents on the 1 and 4 positions (e.g. H-bonds) (Fig. 17).

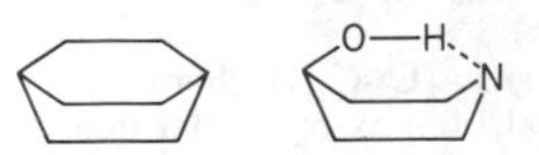

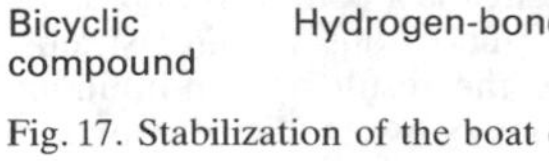

Bicyclic compound Hydrogen-bonded compound

Fig. 17. Stabilization of the boat conformation.

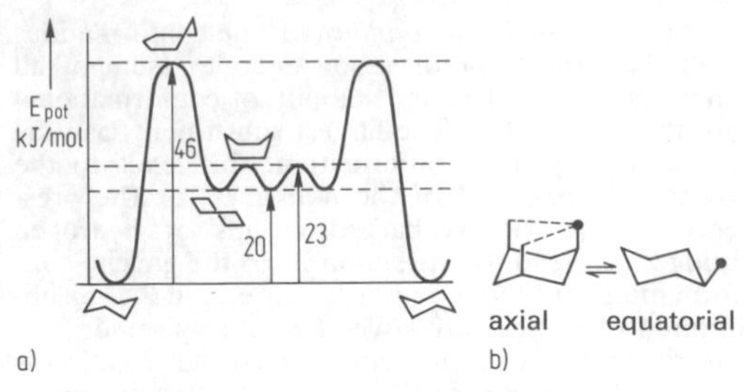

Fig. 18. Ring inversion of cyclohexane (a) and substituted cyclohexane (b)

The chair form can convert to the twist or boat form, or to an alternative chair form, by *ring inversion* (Fig. 18). The inversion barrier, which corresponds to an *envelope form*, is 46 kJ mol^{-1}. Each carbon atom in cyclohexane can bond to two substituents; in the chair form, six of these available bonds are axial (symbol, a), that is, parallel to the symmetry axis, and pointing either up or down in alternation. The remaining six are equatorial (symbol, e) and form an angle of 70° with the axis; they also alternate between pointing up and down (Fig. 15). Axial substituents are in the gauche conformation to the cyclohexane skeleton, while equatorial substituents are in the anti-periplanar conformation. In addition, axial substituents are subject to 1,3 interactions. Therefore, the equatorial arrangement is thermodynamically the more stable. Axial and equatorial bonds can be interconverted by ring inversion. Large groups prefer the equatorial arrangement. The volume required for the t-butyl group is so large that in t-butylcyclohexane it is predominantly in the equatorial position. A group which stabilizes a certain conformation in this way is called a *conformation anchor*.

2.3 *Conformation of other ring systems* (Fig. 19). Cyclobutane is puckered, which eliminates the ecliptic positioning of the H atoms. Alternative conformations are reached through pseudorotation. The energy content is determined, as in cyclopropane, essentially by bond angle strain. Cyclopentane and cycloheptane have conformations, such as the envelope or half-chair form and the twist chair form, which are compromises between bond-angle and torsional strain. In these two cases, pseudorotation leads to rapid interconversion of the alternative conformations.

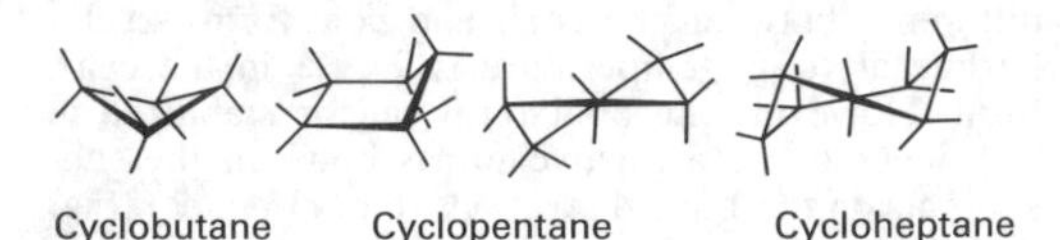

Fig. 19. Conformations of various ring systems.

In rings of medium size (C_8-C_{11}), there are a number of conformational isomers of similar energy available in each case. Each is a compromise among bond-angle, torsional and transannular strains. Transannular strains are the result of non-bonding, van-der-Waals interactions between the H atoms in the interior of the ring. Large rings (> C_{11}) exist as double, zig-zag chains.

Stereoselective synthesis: see Asymmetric synthesis.

Stereoselectivity: see Stereospecificity.

Stereospecificity: the property of a reaction which produces stereochemically differentiated products from stereochemically differentiated reactants. For example, the electrophilic addition of bromine to but-2-ene is a stereospecific *trans*-addition: *trans*-but-2-ene is converted to *meso*-2,3-dibromobutane, while *cis*-but-2-ene gives rise to (D,L)-2,3-dibromobutane. The reverse reaction, iodide-catalysed bromine elimination, is also stereospecific: the *meso*-product produces *trans*-but-2-ene, and the (D,L)-compound gives the *cis* isomer.
In stereospecific reactions, the stereochemistry of the product can be predicted from the stereochemistry of the reactant.
Reactions which yield a greater proportion of one of several possible stereoisomeric products are called *stereoselective*. For example, the elimination of HCl from 1-chloro-1,2-diphenylethane yields an excess of *trans*-stilbene. Stereoselectivity can be given in percent, e.g. 70% stereospecific or 100% stereoselective. From the definitions given by H.E. Zimmermann (1959) and E. L. Eliel (1966), it can be said that every stereospecific reaction is also stereoselective, but not every stereoselective reaction is also stereospecific. A reaction can be stereospecific in a certain temperature range, but only stereoselective above that range, for example the radical addition of hydrogen bromide to 2-bromo-but-2-ene at low temperatures.
The term stereospecific was formerly used to mean 100% stereoselective, and is still occasionally used in that sense.

Sterilization: a process which makes an organism incapable of reproduction; or which kills all microorganisms in a product. In food chemistry, this is achieved by heat treatment; a combination of temperature and time are chosen to kill all microorganisms in the product, including bacterial spores. Temperatures above 110 °C are required for S., and are obtained by use of sealed, pressurized systems (e.g. autoclaves). In industrial S. of canned goods, a distinction is made between the *heating time* required to achieve the desired temperature in the core of the food, the *holding time* which is the actual S. phase, and the *cooling time* required to reestablish normal pressure inside the cans. Foods subjected to S. in hermetically sealed cans, jars or plastic packages aquire a shelf life of at least one year at room temperature.

Steric hindrance: see Steric substituent effects.

Steric inhibition: a term for forces of interaction between colloidal particles or extended phase boundaries in the presence of adsorbed macromolecular compounds (see Protective colloids), when the adsorbed layers penetrate each other as the particles approach. The forces of interaction arise through the increase in concentration in the transition region (osmotic component) and changes in configuration (entropic component).

Steric substituent effect: the size-related effects of substituents on the reactivity of a reaction center. *Steric resonance hindrance* is a prevention of conjugation which occurs when the size of the substituents prevents the molecule from becoming planar. For example, 3,5-dimethyl-4-nitrophenol is much less acidic than 2,6-dimethyl-4-nitrophenol, because of interaction between the nitro and methyl groups in the first compound. *Steric hindrance* is the screening of the reaction center of a molecule by adjacent substituents, so that the reaction is inhibited. For example, trimethylacetates are more slowly saponified than fluoroacetates, and these more slowly than acetates. *Steric substituent constants* serve as a measure of the degree of steric hindrance.

Stern potential: see Electrochemical double layer.

Stern-Volmer equations: equations which indicate the changes in photophysical processes or photochemical reactions when the concentration of a reagent (substrate or quencher) is changed. In the simplest case of a monomolecular reaction, a plot of the ratio Ψ^0/Ψ or I^0/I (quantum yield Ψ or light intensity I in the absence and presence of a quencher Q) versus the quencher concentration is linear: Ψ^0/Ψ $(I^0/I) = 1 + k^q\tau[Q]$, where k^q is the absolute rate constant (quenching constant) and τ is the lifetime of the excited state. For bimolecular reactions, there is a double reciprocal relationship. The constant k^q indicates how quickly excited species are quenched. The S. is used to elucidate the kinetics of photochemical reactions and photophysical processes.

Steroid: a compound derived from the tetracyclic hydrocarbon *perhydro-1H-cyclopenta[a]phenanthrene*. The trivial name for this hydrocarbon is *sterane* if the stereochemistry is unknown, or *gonane*, if the rings B/C and C/D are *trans* to one another. Some important groups of naturally occurring S. include the Sterols (see), Bile acids (see), Steroid hormones (see), cardenolides (see Cardiac glycosides) and various N-containing derivatives (see Steroid alkaloids). Many sapogenins (see Saponins) are also S.

Structure, nomenclature. Rings A/B, B/C and C/D can be joined in either the *cis* or the *trans* orientation. In natural S., rings B/C are always *trans*. Most S. arc derived from gonane. In the cardenolides and bufadienolides, rings C/D are *cis*. However, many S. contain double bonds in ring A or B, and are therefore planar or nearly so. The CH_3 group at position 13 serves as a reference point for stereochemical nomenclature; it is always located above the plane of the ring, that is, in the β position. The 5α and 5β series of S. are distinguished on the basis of the orientation of the H atom at position 5 (α or β). In the 5α-series, rings A/B are *trans*, while in the 5-β-series they are *cis*. Most naturally occurring S. have CH_3 groups in positions 13 (e.g. estrane) or in positions 10 and 13 (e.g. androstane), as well as an oxygen function (either a hydroxy or an oxo group) in position 3. They usually contain an alkyl group at position 17. Ring contraction is indicated by the prefix *nor-* placed before the letter of the affected ring (A, B, C or D), while ring expansion is indicated by the prefix *seco*. In this case, the numbers of the C atoms between which the new atom has been inserted are indicated, as in the prefix 9,10-*seco-*, applied to the calciferols.

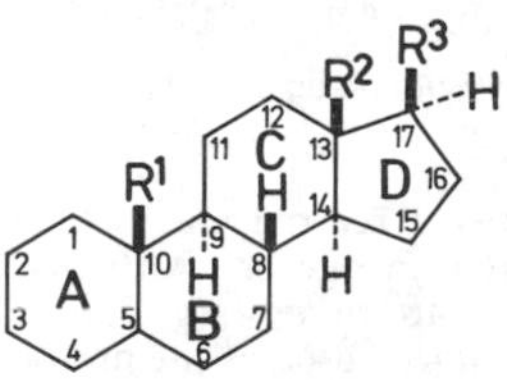

The biosynthesis of S. begins with squalene and proceeds via squalene oxide; in mammals it continues via lanosterol to cholesterol. In both animals and fungi, the first stable cyclic product of sterol biosynthesis is lanosterol, while in algae and higher plants, it is cycloartenol. Lanosterol and cycloartenol are called methylsterols, because of the three additional methyl groups (in positions 4, 4 and 14) they contain; they are considered to be triterpenes.

Occurrence. S. are found in animals, plants and microorganisms, especially fungi. Fungi contain the sterols and steroid carboxylic acids of the ergastane and stigmastane series, as well as methylsterols. The S. of animals are synthesized from cholesterol. Complete removal of the side chain at C17 produces the mammalian steroid hormones (pregnane, androstane and estrane series). In invertebrates, the ecdysteroids act as molting hormones. Bile acids are degradation products of cholesterol; these are steroid carboxylic acids. Plant S. display a much greater variety of structure. The alkyl chain at C17 may be expanded (stigmastane series), additional rings containing O or N may be added (saponins, steroid alkaloids) or an unsaturated lactone ring (cardenolides, bufadienolides) may be present. Most plant S. are present as glycosides; the sugar groups are usually bound to the 3β-hydroxy group. S. carboxylic acids are found in petroleum.

Steroid syntheses are very important economically, as they are used for preparation of drugs and hormones for human beings and domestic animals. Both partial and total syntheses are used. Partial syntheses start from the cheapest possible S., especially diosgenin, cholesterol, stigmasterol, bile acids and S. alkaloids. The conversion to the desired S. can be carried out by microorganisms or chemically. Microbial conversions are used when the available chemical methods are not adequate, are too expensive or their

Name	R^1	R^2	R^3	Derivatives
Gonane	H	H	H	
Estrane	H	CH_3	H	Estrogens
Androstane	CH_3	CH_3	H	Androgens
Pregnane	CH_3	CH_3	C_2H_5	
Cholane	CH_3	CH_3	$CH(CH_3)CH_2CH_2CH_3$	Bile acids
Cholestane	CH_3	CH_3	$CH(CH_3)CH_2CH_2CH(CH_3)_2$	Cholesterol
Ergostane	CH_3	CH_3	$CH(CH_3)CH_2CH_2CH(CH_3)CH(CH_3)_2$	Ergosterol
Stigmastane	CH_3	CH_3	$CH(CH_3)CH_2CH_2CH(C_2H_5)CH(CH_3)_2$	Stigmasterol and other plant sterols

yields are too low; this is often the case when a high level of selectivity is required. Hydroxylations at positions 11, 16, 17α or 21, introduction of Δ^1 double bonds (in the synthesis of corticosteroids) and removal of the side chain at C17 are carried out microbiologically. In such conversions, the S. is dissolved in a solvent which is miscible with water and added to the culture solution of the microorganisms. After the reaction is complete, the mixture of S. is extracted with a nonpolar solvent and the solution is worked up. Chemical methods of partial synthesis are used most often for oxidation, dehydrogenation, halogenation, methylation, fusion of rings (see Robinson fusion) or ethynylation. In the past few years, total syntheses have become economically significant due to increases in the prices of the starting materials for partial syntheses and to progress in the development of stereoselective syntheses. This is especially true for the syntheses of estrogens and 19-*nor*-S.

Steroid alkaloids: nitrogen-containing steroids occurring mainly in higher plants of the *Solanaceae, Liliaceae, Apocyanaceae* and *Buxaceae* families. The S. of plants are derived from the skeletons tomatanine, solanidanine, cevanine, veratranine and jervanine. ***Cevanine, veratranine*** and ***jervanine*** are D-homo-C-nor-steroids. The names of the saturated compounds end in "-anine", and those of unsaturated S. in "-enine", "adienine", etc.

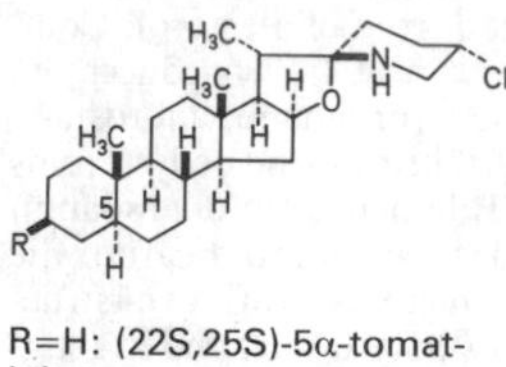

R=H: (22S,25S)-5α-tomatinine
R=OH: tomatidine

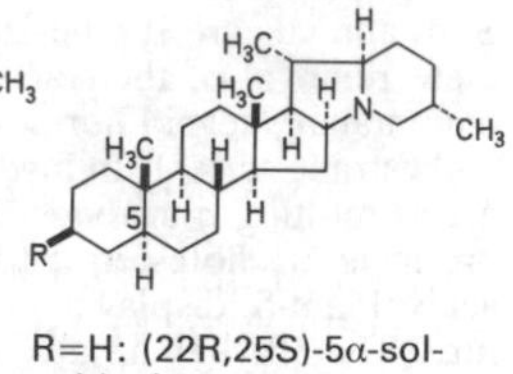

R=H: (22R,25S)-5α-solanidanine
Solanidane: 5

Tomatinine is an aza analog of spirostane (see Saponins). The tomatanine derivatives are therefore also considered spirosolane alkaloids, because they contain the spirane structure formed by the iminoketal. The carbon skeleton is the same as that of cholesterol. The S. in plants are mostly 3-glycosides, usually with branched oligosaccharides. The alkaloids of the *Solanum* genus (***solanum alkaloids***) are derived from tomatanine and solanidanine. These alkaloids include ***tomatine***, which is present in tomato leaves; it consists of the steroid tomatidine bound to two glucose, a galactose and a xylose group. Derivatives of ***solanidanine*** have been isolated from potatoes, including the veratrum alkaloid rubijervin. Other veratrum alkaloids belong to the veratranin (e.g. veratramine) and jervanine series. The S. of plants act like saponins. Some are pharmacologically important (e.g. veratrum alkaloids), while others serve as starting materials for partial syntheses of steroids. Some S. are found in animals, including Batrachotoxin (see) and the ***salamander alkaloids***. ***Salamandrine***, the main salamander alkaloid, is an A-homo-steroid.

Steroid carboxylic acids: same as Bile acids (see).

Steroid hormones: a group of hormones which have the chemical structure of steroids, including Sex hormones (see), Adrenal corticosteroids (see), Ecdysone (see) and similar molting hormones.

Sterols: (name from the Greek "steros" = "solid") steroids with a 3β-hydroxyl group and a cholestane, ergostane or stigmastane skeleton. S. occur naturally in free form or as fatty acid esters (see Waxes). Plants and microorganisms contain S. derived from ergostane (see Ergosterol) or stigmastane (see Stigmasterol). The most important vertebrate S. is Cholesterol (see). S. can be extracted from the nonsaponifiable fraction of fats and fatty oils; they form colorless crystals. They form insoluble precipitates with saponins; precipitation with digitonin is used for their identification.

Stetter synthesis: a synthesis for carboxylic acids with longer chains, starting from cyclohexane-1,3-dione and reactive alkyl halides. The reaction is a C-alkylation in the 2 position of the cyclohexane-1,3-dione, followed by acid cleavage to the ketocarboxylic acid and reduction of the carbonyl group by a Clemmensen or Wolff-Kishner reduction. The product is a carboxylic acid with a chain lengthened by six C atoms; for example, suberic acid is obtained if bromoacetate is used as the halogen reactant. Variants of the reaction consist of reaction of cyclohexa-1,3-dione with aldehydes or with α,β-unsaturated ketones, carboxylic acid esters or nitriles in the sense of a Michael addition (see).

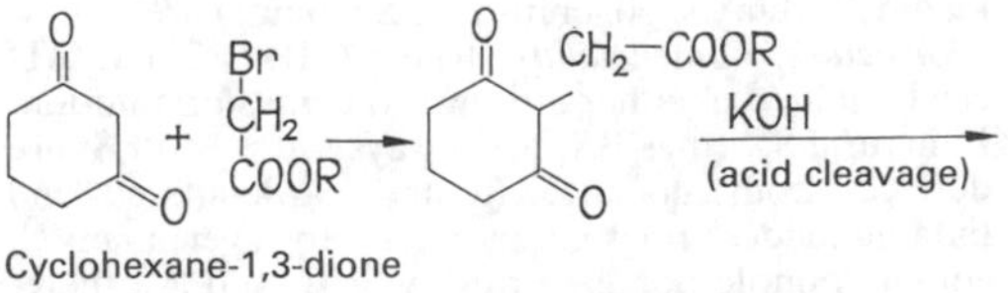

Cyclohexane-1,3-dione + $BrCH_2COOR$ → 2-(CH_2–COOR)-cyclohexane-1,3-dione $\xrightarrow[\text{(acid cleavage)}]{KOH}$ $HOOC-(CH_2)_2-CO-(CH_2)_2-COOH$ $\xrightarrow[-H_2O]{+4H}$ $HOOC-(CH_2)_6-COOH$

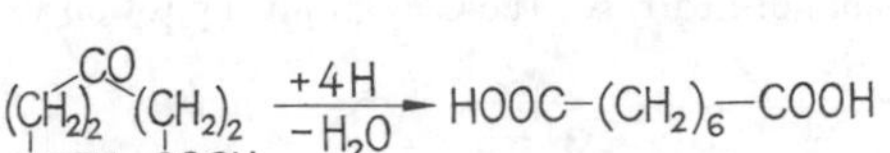

Suberic acid

Stevens rearrangement: a rearrangement of a quaternary ammonium salt with one or two benzyl groups, or aryl group(s), to a tertiary amine. Phenyllithium or sodium amide is used as the proton acceptor. For example, dibenzyldimethylammonium bromide is converted to an ylide, which rearranges to dimethylaminodiphenylethane (Fig.).

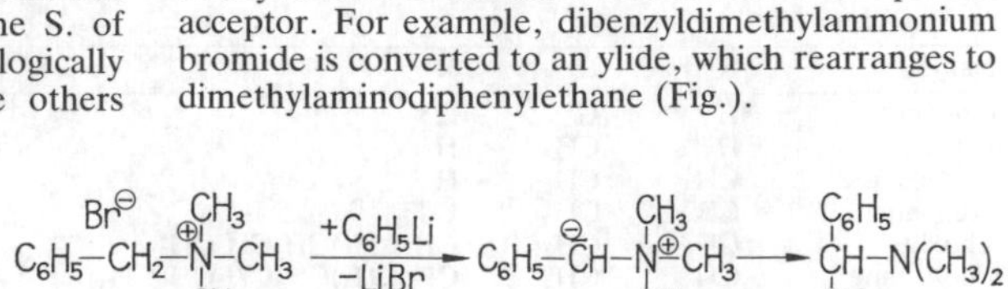

$$C_6H_5-CH_2-\overset{\oplus}{N}(CH_3)_2-CH_2-C_6H_5\ Br^{\ominus} \xrightarrow[-LiBr,\ -C_6H_6]{+C_6H_5Li} C_6H_5-\overset{\ominus}{C}H-\overset{\oplus}{N}(CH_3)_2-CH_2-C_6H_5 \longrightarrow C_6H_5-CH(N(CH_3)_2)-CH_2-C_6H_5$$

The mechanism is thought to be an ionic or radical cleavage and recombination. The bond is made via the CH_2-group of the migrating benzyl residue, in contrast to the similar Sommelet rearrangement.

STH: see Somatotropin.

Stibane, ***stibine, antimony hydride***: SbH_3, a colorless, very poisonous gas with an unpleasant odor; M_r 124.77 °C, density^{-25} 2.26; m.p. -88 °C, b.p. -17.1 °C. A. is formed by the decomposition of magnesium antimonide, Mg_3Sb_2, in sulfuric acid, or by the reaction of nascent hydrogen with Sb compounds. It burns in air to antimony or antimony(III) oxide, Sb_2O_3, and water, depending on the conditions. It decomposes when heated into antimony and hydrogen. This last reaction is applied in the Marsh test (see).

Stiffeners: products for stiffening fabrics after washing, usually a polymer such as potato, corn or rice starch or partially saponified polyvinylacetate.

Stigmastane: see Steroids.

Stigmast-5-en-3β-ol: see Sitosterol.

Stigmasterol:, stigmasta-5,22-dien-3β-ol, a plant sterol which forms colorless crystals with m.p. 170 °C, $[\alpha]_D$ -51°. S. is practically insoluble in water, but is soluble in organic solvents. It makes up 12 to 25% of the unsaponifiable fraction of soybean oil and is an important starting material for partial syntheses of steroids.

Stigmasterol

Stilbene, ***1,2-diphenylethene***: C_6H_5-CH=CH-C_6H_5, an unsaturated, aliphatic-aromatic hydrocarbon which occurs as E and Z isomers. The more stable compound is (E)-S. (*trans*-S.); m.p. 124.5 °C, b.p. 305 °C at 96 kPa; n_D^{17} 1.6264. The liquid ***(Z)-stilbene (cis-stilbene)*** is formed by UV irradiation of *trans*-S.; m.p. 5 °C, b.p. 141 °C, n_D^{20} 1.6130. S. is insoluble in water, but is soluble in ethanol and benzene. ***(E)-S.*** is steam volatile. S. is synthesized by a Grignard reaction of benzaldehyde with benzylmagnesium chloride and dehydration of the resulting 1,2-diphenylethanol. Bromine addition and dehydrobromination with potassium hydroxide yields tolane, and catalytic hydrogenation yields 1,2-diphenylethane. S. is the parent compound of the ***stilbene dyes***.

Stilbene dyes: a group of substantive diazo dyes which are derived from stilbene. The starting material for all S. is 4-nitrotoluene-2-sulfonic acid, which on heating in alkaline medium forms 4,4'-dinitrostilbene-2,2'-disulfonic acid. Further reaction leads to linear, yellow stilbene azo compounds. Commercial products usually contain several compounds, as a result of the synthetic conditions. The S. dye cellulose, animal fibers and polyamides yellow, orange or brownish red colors.

Stilon®: see Synthetic fibers.

Stirred reactor: a container provided with a stirrer and used for carrying out chemical reactions. S. are usually provided with heating or cooling coils as well, although for highly exothermal reactions, an external cooling circuit is needed (Fig. 1). S. are built with volumes up to about 300 m³.

1) Discontinuous S. The S. is filled with the reactants and brought to the required operating temperature. When the reaction is complete, it is cooled and emptied. Discontinuous operation has the disadvantage that the time required for filling, heating, cooling and emptying the reactor is substantial, and reduces the possible output. In addition, the amount of labor required is relatively high. The reaction time for a first-order reaction, A → B + C, is given by the following equation: $r = dc_A/dt = -k \cdot c_A$. By integrating, one obtains $\lg c_A^0/c_A = 0.434\ kt$ and $t = (2.30/k) \lg c_A^0/c_A$, where r is the reaction rate, k is the rate constant, c_A^0 is the initial concentration of A and c_A is the final concentration of A.

For a 90% turnover of A (i.e. $c_A^0 = 1$ and $c_A = 0.1$), $t = 2.30/k$.

For highly exothermal reactions, discontinuous operation is not possible, because the high heat of reaction cannot be removed.

2) S. with semicontinuous operation (*semibatch reactors*). One of the reactants is put into the S. and it is heated to the necessary temperature. The second component is then added at a rate which can maintain the reaction temperature. After a stoichiometric amount of the second component has been added, the reaction is allowed to continue for a time, and then the S. is cooled and emptied. Semicontinuous operation is used for highly exothermal reactions.

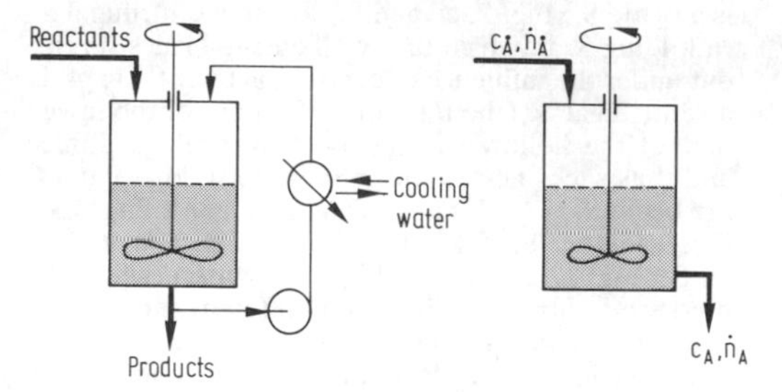

Fig. 1. (*left*) Stirred reactor with external cooling. Fig. 2. (*right*) Stirred reactor with continuous operation.

3) S. with continuous operation. The reactants are continuously fed into the S., and the products are continuously drawn off (Fig. 2). In an ideal S., all concentrations and thus the reaction rate are constant in space and time. The amount of manual maintenance required is less than with discontinuous or semicontinous operation, and there is no lost time for filling and emptying. For a first order reaction A → B + C, $r = dc_A/dc = -k \cdot c_A = dn_A/dt \cdot V_R$ and $\dot{n}_A^0 - kc_A \cdot V_R = \dot{n}_A$. Here V_R is the volume of the reactants, $\dot{n}_A^0$ is the molar flow of A into the reactor, $\dot{n}_A$ is the molar flow of A out of the reactor, and $\dot{v}$ is the amount of material injected or withdrawn per unit of time. The mean residence time $t_m = V_R/\dot{v}$.

For 90% turnover of A ($c_A^0 = 1$ and $c_A = 0.1$), $tm = 9.00/k$.

Comparison with the reaction time under discontinous operation reveals that in a continuous operation, the material must spend more time in the reactor than is required to achieve the same turnover in a discontinuous operation.

Use of a reactor cascade can greatly reduce the time required for a desired degree of reaction (Fig. 3).

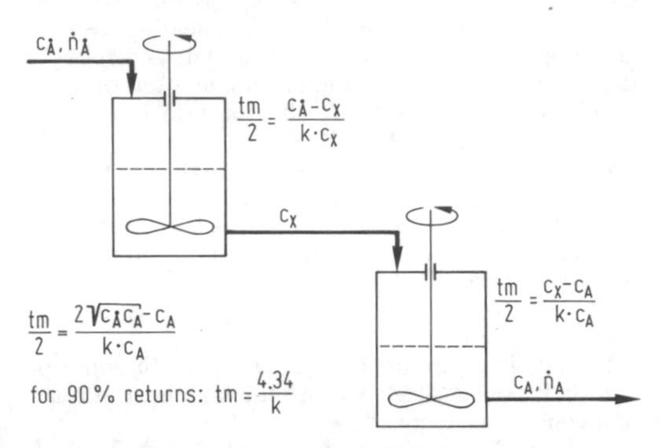

Fig. 3. Scheme of a simple reactor cascade.

Stirrer: a device for stirring, mixing or kneading substances, usually a rotating device. S. are designed to give different types of flow: tangential (blade and anchor S.), axial (e.g. propeller S.) or radial (e.g. turbines). Various types of stirring are achieved with rotating, pendulum, jet, bubble or vibrating S. Some common shapes for S. are shown in the figures. Fig. 1 is a blade S., Fig. 2, a vane S., Fig. 3, a centrifugal S. (a folding S. in which the two hanging blades spread out under the influence of centrifugal force). Fig. 4. is a centrifugal S. (the liquid is pulled out of the lower part of the hollow stirring rod by centrifugal force, and thus sucks new liquid into the top). Fig. 5 and 6 are beam S., Fig. 7 is an anchor S., Fig. 8 a finger S., Fig. 9 a lattice S., Fig. 10 a spoon S., Fig. 11 a propeller S., Fig. 12 a spiral S., Fig. 13 a Wittscher S. (a S. used especially to mix liquids of different densities. It consists of a glass bulb which is open underneath and has holes in its sides. When it rotates in a liquid, the liquid is expelled through the side holes by centrifugal force, and more liquid is sucked up from the bottom). Fig. 14 is a turbine S. It has 6 to 12 blades on a vertical axis. Double turbine S. have two separate sets of blades, one of which sucks the liquid down from the top, while the other brings it up from the bottom. Fig. 15 and 16 are magnetic S. used to stir systems under pressure and in vacuum. Fig. 17 is a vibrating S., which vibrates in time with the frequency of alternating current. A planetary S. (not pictured) has several axes; the axes rotate with respect to each other and simultaneously move around the main axis.

Stirring techniques. The S. shown in Fig. 1 to 14 are linked directly to an electric motor by means of a vertical axis. The motor can be regulated by an internal governor or by means of gears. Flexible axes can be used to transfer power for a distance up to about 1 m. In some cases, water turbines are used instead of electric motors. In the rotating S. (Fig. 18), the vessel is rotated in an inclined position. Flasks of this type for laboratory use are provided with ground glass fittings and fit onto a rotating, ground glass rod. These flasks can also be provided with cooling jackets. They are sealed with stopcock grease, silicone, glycerol or conc. sulfuric acid, depending on the material being stirred. In the mammoth S. (Fig. 19), pneumatic S. (Fig. 20), cyclone S. (Fig. 21) and mixing valve (Fig. 22), the kinetic energy of the liquid or gas is utilized for stirring.

Stirring: a process of combining materials in a liquid phase. S. leads to an even distribution of liquids, gases or solids in a liquid, and is a form of Mixing (see). S. can also be used to promote heat transfer and other goals. Depending on the consistency of the liquid and the nature of the process, different types of Stirrers (see) are used.

Stishovite: see Rutile type.

Stobbe condensation: the reaction of succinate esters with aldehyde or ketones in the presence of sodium methylate or potassium *tert.*-butylate; the products are alkylidene monosuccinates. These can be easily converted to β,γ-unsaturated carboxylic acids by hydrolysis and decarboxylation. The first step of the S. is an aldol reaction, which is followed by an intramolecular transesterification to a γ-lactone. The cleavage of the rings occurs by electron shifting. The resulting monoesters can be obtained in very good yields.

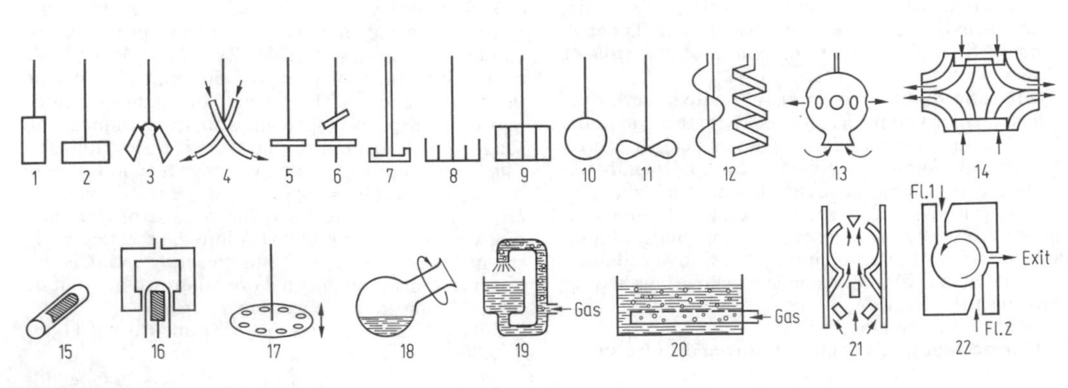

$$R^1\text{-}CHO + {}^{\ominus}CH(COOR)\text{-}CH_2\text{-}COOR \rightarrow R^1\text{-}CH(O^{\ominus})\text{-}CH_2\text{-}CH(COOR)\text{-}C(=O)OR \xrightarrow{-RO^{\ominus}}$$

$$\text{lactone } R^1\text{-}CH\text{-}O\text{-}C(=O)\text{-}CH(COOR)\text{-}CH_2 \xrightarrow[-ROH]{+RO^{\ominus}} R^1\text{-}CH\text{-}O\text{-}C(=O)\text{-}C^{\ominus}(COOR)\text{-}CH_2 \rightarrow R^1\text{-}CH{=}C(COOR)\text{-}CH_2\text{-}COO^{\ominus}$$

$$\xrightarrow[-ROH]{+H_2O/H^+} \cdots \xrightarrow{-CO_2} R^1\text{-}CH{=}CH\text{-}CH_2\text{-}COOH$$

β,γ-unsaturated carboxylic acid

Stock nomenclature: see Nomenclature, sect. II B 2b.

Stoichiometric coefficients. 1) ***Particle stoichiometric coefficient***: a number ν in a chemical formula which gives the molar fraction of an element in the compound. This number is placed as a subscript to the element symbol, and the number 1 is usually omitted, as in NH_3. S. are assigned from the experimentally determined stoichiometric ratios (see Stoichiometry). They express ratios, and are usually integral, but not always. For example, for NH_3, ν_N = 1 and ν_H = 3, while for $Ca(NO_3)_2$, ν_{Ca} = 1 and ν_{NO_3} = 2.

2) ***Reaction coefficient***: a number which expresses the molar ratio of each participant in a chemical reaction relative to the other participants, assuming that chemically equivalent amounts of the substances react. This S. is given a sign; reactants which are consumed have negative signs, while products have positive signs. The absolute values of these numbers (*stoichiometric factors*) are part of the reaction equation itself. The smallest possible integers are supposed to be used as stoichiometric factors, but this is not an absolute requirement. In the equation $N_2 + 3 H_2 \rightarrow 2 NH_3$, with $\nu_{N_2} = -1$, $\nu_{H_2} = -3$ and $\nu_{NH_3} = 2$, the reaction S. indicate that 1 molecule N_2 and 3 molecules H_2 react to form 2 molecules of NH_3 (*elemental formula turnover*), or that 1 mol N_2 and 3 mol H_2 react to form 2 mol NH_3 (*molar formula turnover*) The correctness of the S. is checked by balancing the equation: the sum of each type of atom on the left side must be equal to the sum of that type on the right.

Stoichiometric compounds: same as Daltonides.

Stoichiometric valence: see Valence.

Stoichiometric proportions, laws of: the basic rules of stoichiometry, which predict the mass ratios of reactants and products in chemical reactions and of the elements which make up a compound.

The law of equivalent proportions (J.B. Richter, 1791) states that elements combine to form compounds in the ratio of their equivalent masses, or small integral multiples of their equivalent masses.

According to the ***law of constant proportions*** (J.L. Proust, 1799, and J. Dalton, 1808), each chemical compound contains its constituent elements in a definite, constant ratio of masses. For example, 100 g of pure, natural water always contains 88.81 g. oxygen and 11.19 g hydrogen. If two elements form several different compounds with each other, then according to the ***law of multiple proportions*** (J. Dalton, 1808), the masses of one element, relative to a constant mass of the other, stand in ratios of small integers. For example, in the five nitrogen oxides N_2O, NO, N_2O_3, NO_2 and N_2O_5, for each gram of nitrogen there are 0.571, 1.142, 1.714, 2.284 and 2.856 g oxygen, respectively.

If matter consists of atoms, and the atoms of each element have a characteristic mass, the law of equivalent proportions implies the other two. All three laws are considered the basis of Stoichiometry (see). They were the result of efforts to bring theory into agreement with experimental observation, and are the basis on which Dalton proposed his atomic theory. Compounds which obey these laws are called Daltonides (see); there is a group of compounds, called Berthollides (see), the composition of which can vary, depending on the conditions under which they are synthesized.

Stoichiometry: the basis for calculations of composition of chemical compounds and mixtures, and of mass, volume and charge ratios in reactions. The energies of reactions are not part of the field of S. The development of S. was an important step in the growth of chemistry as an exact science; the founder is considered to be J.B. Richter (1762-1807).

S. is based on the repeated observation that chemical elements are completely conserved during chemical reactions (*law of conservation of elements*) and that the total mass of the substances is constant (*law of conservation of mass*). The actual laws of S. include the law of stoichiometric proportions, Gay-Lussac's law and Avogadro's law (see Stoichiometric proportions, law of; Gay-Lussac's law of chemical volumes; Avogadro's law). All were developed around the turn of the 19th century. Faraday's laws (see) are the basis of the S. of electrolytic processes, and the equations of state of gases (see State, equation of) are needed for reactions in which gases are involved. The application of S. to mixtures or solutions includes the use of Composition parameters (see). Stoichiometric calculations are needed in nearly all branches of chemistry. The following applications are the most common ones.

1) The ***S. of the composition of compounds*** is based on the parameters mass m, molar mass M and molar amount n. These are related by the defining equation $M = m/n$. The stoichiometric parameters ν_i in a formula give molar ratios, e.g. for Fe_3O_4, $n_{Fe}:n_O = \nu_{Fe}:\nu_O = 3:4$. Some typical applications are the calculation of the mass proportions of the elements in a compound from the chemical formula, or determination of the formula from the experimentally determined mass ratios of the elements (see Elemental analysis). The mass fraction w_{Fe} of iron in Fe_3O_4 is $w_{Fe} = \nu_{Fe}M_{Fe}/M_{Fe_3O_4}$; with $\nu_{Fe} = 3$, $M_{Fe} = 55.85$ g mol^{-1} and $M_{Fe_3O_4} = 231.54$ g mol^{-1}, $w_{Fe} = 0.7236 = 72.36\%$.

2) ***S. of chemical reactions***. Calculation of the substance turnover in chemical reactions starts from the reaction equation, which gives the stoichiometric ratios of the reactants and products. The equation $N_2 + 3\ H_2 \rightarrow 2\ NH_3$ shows the *elemental formula turnover* (reaction of one N_2 molecule with three H_2 molecules produces two NH_3 molecules) and at the same time expresses the *molar formula turnover* (1 mol = 28.0 g N_2 reacts with 3 mol = 6.0 g H_2 to produce 2 mol = 34.0 g NH_3). With the equation $m = nM$, the molar amounts can be converted to masses. If the reactants and/or products are gases, the reaction equation also predicts the volume changes, because molar amounts and volumes are directly proportional. In this example, 1 volume N_2 and 3 volumes H_2 form 2 volumes NH_3.

For various reasons, the actual amount of products is smaller than predicted by the chemical equation for a complete reaction. This is expressed quantitatively as the *yield*, which is the amount of product actually formed divided by the amount which is theoretically possible. Yield calculations are very important in practice, especially for optimizing of synthetic processes.

3) ***Production, dilution and mixing of solutions***. Solutions of definite composition and amount can be made either by dissolving the substance(s) in pure solvent, by addition of solvent to a concentrated solution (dilution) or by mixing solutions of higher and lower concentration. The stoichiometric calculations for these three methods depend on the equations defining the composition parameter used, and in many cases, also on the equation $m = nM$. When a solution is formed by direct dissolving, the amount of solute is usually weighed. When concentrated solutions are diluted, the molar amount and mass of the solute are unchanged. If the subscript 1 indicates the state before dilution, and 2 the state afterwards, $n_1 = n_2$ and $m_1 = m_2$. For the molar concentration c or the mass fraction w, the equations $c_1v_1 = c_2v_2$ and $w_1m_1 = w_2m_2$ apply for all dilution calculations. Example: how many ml conc. HCl (11.7 M) is needed to produce 250 ml 0.2M HCl? Solution: $v_1 = c_2v_2/c_1 = 0.2$ mol l$^{-1}\cdot$0.250 l/11.7 mol l^{-1} = 4.27 ml.

When several solutions (subscript i) of the same substance are mixed, the basic rule is that the total amount of dissolved substance in the mixture (subscript M) is $n_M = \sum_i n_i$. The equations derived from this, $c_Mv_M = \sum_i c_iv_i$ and $w_Mm_M = \sum_i w_imK_{m_i}$, are called the *mixing equations*. When they are used, it must be remembered that the volume of the mixture is equal to the sum of the component volumes only in the ideal case; for real solutions, $v_M \neq \sum_i v_i$. Mixtures of very dilute solutions or of concentrated solutions when the difference in concentration is not great, display approximately ideal behavior. Sample application: the following solutions are mixed: 250 g with $w = 0.2$, 450 g with $w = 0.3$ and 500 g with $w = 0.8$. The total amount of solution is m_M = 250 g + 450 g + 500 g = 1200 g. The mass fraction is w_M = (0.2·250 g + 0.3·450 g + 0.8·250 g)/1200 g = 0.488 or 48.8 mass %.

4) ***Application of S. in quantitative chemical analysis***. In Gravimetric analysis (see), the fraction to be determined is converted by suitable reactions and operations into a stoichiometrically well defined, weighable form $A_{\nu_A}B_{\nu_B}$, and its mass $m(A_{\nu_A}B_{\nu_B})$ is determined by weighing. The molar amount of A is $n(A) = \nu_A n(A_{\nu_A}B_{\nu_B})$. Substituting $n = m/M$ for the molar amounts and solving for m(A), i.e. the mass of A, one obtains

$$m(A) = \nu_A \frac{M(A)}{M\,(A_{\nu_A}B_{\nu_B})}\, m\,(A_{\nu_A}B_{\nu_B})$$

For a given given gravimetric method for determination of A, ν_A and the two molar masses are constants which can be collected into the *gravimetric factor* f_{grav} for this method. For example, for gravimetric determination of iron with Fe_2O_3 as the weighed form, the gravimetric factor is derived from $\nu_{Fe} = 2$, M(Fe) = 55.85 g mol^{-1} and $M(Fe_2O_3)$ = 159.69 g mol^{-1}: $f_{grav}(Fe/Fe_2O_3) = 0.6994$. Multiplication of the weight of Fe_2O_3 by this factor gives the amount of iron for any analogous iron determination. Knowledge of gravimetric factors simplifies routine analytical work, and there are published tables for this purpose. (However, gravimetric analysis has been replaced in most routine applications by other methods of determination which are more readily automated.)

In *Volumetric analysis* (see), the stoichiometric analysis of a titration depends on the fact that when the equivalence point has been reached, equivalent molar amounts $n_{eq} = nz$ of titrand (subscript 1) and titrator (subscript 2) have reacted. Here z is the stoichiometric valence. Since the definition of molar concentration is $c = n/v$, knowledge of c_2 and v_2 of the standard solution consumed permits calculation of the titrand content in the unknown solution: $c_1 = (c_2z_2v_2)/(z_1v_1)$. For the mass of the titrand, this yields $m_1 = (M_1c_2z_2v_2)/z_1$. Usually the titrator is used in the form of a normal solution, of which the equivalent concentration $c_{eq,2} = c_2z_2$. For a given titrimetric determination and a given normality of the standard solution, M_1, z_1 and c_2 are constants, which can be collected in the *titrimetric factor* $f_{titr} = M_1c_2/z_1$. Titrimetric factors are useful in routine analytical work, because multiplication of the volume of the standard solution used by the titrimetric factor gives the mass of titrand directly. They are therefore available in tables. If the standard solution is different from the standard normality (usually 0.1 N or 0.01 N), this fact is taken into account in the stoichiometric analysis by multiplying v_2 by a correction factor, the *titer* F of the standard solution, which is determined experimentally. If $F < 1$, the precise normality is less than the standard; for $F > 1$, it is larger. The mass of the titrand is then $m_1 = f_{titr}v_2F$. In both gravimetric and volumetric analysis, it must be remembered that parallel determinations are usually carried out on aliquots of the sample solution, so that the total mass of unknown in the sample must be obtained by multiplication of the mass in the aliquot by the number of aliquots.

Stokes' line: see Raman spectroscopy.

Stokes' law: see Sedimentation.

Stokes' rule: a rule proposed in 1852 by Stokes, which says that the wavelength of light emitted as luminescence is longer than that of the exciting radiation, or the corresponding absorption. There are exceptions to the rule, when the sample was already in an excited state before interacting with the radiation (for example, in an excited vibrational state), and releases this energy in the light emission. In this case, when the emitted light is shifted to a shorter wavelength, one speaks of anti-Stokes lines (see Raman spectroscopy).

Storage battery: see Secondary cell.

Straight-run gasoline: see Gasoline.

Streaming potential: see Zeta potential.

Strecker synthesis: synthesis for α-amino acids by reaction of aldehydes with hydrocyanic acid or sodium cyanide and ammonia. The first products are α-amino nitriles, which are hydrolysed:

$$R{-}CHO + NH_3 \longrightarrow RCH(OH){-}NH_2 \xrightarrow[-H_2O]{+H^+,\,+CN^-} R{-}CH(NH_2){-}CN \xrightarrow[-NH_3]{+2\,H_2O\,/H^+} R{-}CH(NH_2){-}COOH$$

A disadvantage of the method is that the cyanide components are poisonous; on the other hand, a wide variety of end products can be obtained in yields up to 75%. The S. is very important in the industrial production of methionine, lysine and glutamic acid.

Street octane rating: see Octane rating.

Streptocyanins: see Cyanin pigments.

Streptokinase: see Plasmin.

Streptomycin, ***streptomycin A***: an aminoglycoside antibiotic isolated from *Streptomyces griseus* by Waksman in 1944. S. consists of the aglycon streptidine and the disaccharide streptobiosamine, which in turn consists of the furanose streptose and *N*-methyl-L-2-glucosamine (Fig.). S. is a triacidic base, and is commonly administered as the neutral sulfate. Catalytic hydrogenation of the aldehyde group yields dihydrostreptomycin, which, however, is no longer used in therapy. S. is active against gram-negative bacteria, and is used to treat tuberculosis. However, it is somewhat toxic to nerves and auditory organs.

Strontium, symbol ***Sr***: chemical element of group IIa of the periodic system, the Alkaline earth metals (see), a light metal, *Z* 38, with natural isotopes with mass numbers 84 (0.56%), 86 (9.86%), 87 (7.02%) and 88 (82.56%), atomic mass 87.62, valency II, Mohs hardness 1.8, density 2.6, m.p. 769 °C, b.p. 1384 °C, electric conductivity 3.3 Sm mm^{-2} (at 0 °C), standard electrode potential (Sr/Sr^{2+}) -2.89 V.

In addition to the 4 natural isotopes, there are 12 known synthetic isotopes, of which the most famous is ^{90}Sr, a product of nuclear weapon explosions. ^{90}Sr from fallout is incorporated into the bones in the calcium phosphate lattice, and because it is a strong β-emitter with a half-life of 28.1 years, it can be a serious threat to health.

Properties. S. is a soft, lustrous white metal, but when exposed to air, it acquires a gray tarnish. There are three allotropic modifications, with conversion points at 235 and 540 °C. The reactivity of S. is greater than that of its lighter homologs (see Alkaline earth metals). Its chemistry is very similar to that of calcium. Many strontium compounds are isomorphic with the corresponding calcium derivatives. Volatile S. salts give a flame a carmine red color.

S. is a strong reducing agent and tends to donate both its valence electrons, i.e. to form Sr^{2+} cations. Its compounds are generally ionic. At high temperatures it reacts directly with halogens, sulfur, phosphorus and carbon to form the corresponding binary compounds. Finely divided S. is pyrophoric in air, and even the compact metal is ignited when rubbed with a hard object; it burns to strontium oxide, SrO, and strontium nitride, Sr_3N_2. S. is therefore stored under petroleum or a protective gas. It reacts violently with water, releasing elemental hydrogen: $Sr + 2\,H_2O \rightarrow Sr(OH)_2 + H_2$. It also generates hydrogen as it dissolves in alcohol and most acids. However, due to passivation, it is only slightly dissolved in sulfuric or fuming nitric acid. It forms alloys with many metals, including Ni, Sn, Hg and As.

Analysis. In the quantitative analysis scheme, S. is precipitated with the ammonium carbonate group and is identified as strontium sulfate, $SrSO_4$, strontium chromate, $SrCrO_4$ or strontium iodate, $Sr(IO_3)_2$. It can also be readily identified by spectral analysis, due to its characteristic red and blue emission lines.

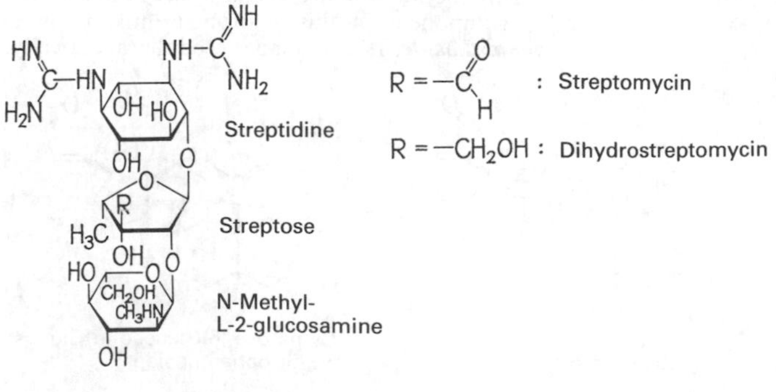

Streptidine

Streptose

N-Methyl-L-2-glucosamine

R = –CHO : Streptomycin

R = $-CH_2OH$: Dihydrostreptomycin

Suitable methods for quantitative determination are precipitation as strontium sulfate, $SrSO_4$, strontium oxalate, SrC_2O_4, or strontium carbonate, $SrCO_3$, complexometric titration with EDTA, or – especially for low concentrations – atomic absorption spectroscopy.

Occurrence. S. makes up 0.03% of the earth's crust. Its most important minerals are coelestine, $SrSO_4$, and strontianite, $SrCO_3$.

Production. S. is obtained by melt electrolysis of strontium chloride, to which potassium chloride is added to reduce its melting point. The metal can also be produced by heating strontium oxide and aluminum powder above the boiling temperature of S.

Applications. S. has little industrial importance. It is used as a getter metal and as a deoxidizing agent in metallurgy.

Historical. In 1793, Klaproth discovered a new element in the mineral strontianite, and named it for the site of discovery of the mineral (Strontian, Scotland). The free metal was first prepared by Davy in 1808.

Strontium bromide: $SrBr_2$, colorless, hexagonal crystals which are readily soluble in alcohol and water; M_r 247.44, density 4.216, m.p. 643 °C. S. crystallizes out of water as the hexahydrate, $SrBr_2 \cdot 6H_2O$. It is obtained by dissolving strontium carbonate in hydrobromic acid. S. is used as a tranquilizer, to prevent excessive stomach acidity, and as an x-ray contrast medium.

Strontium carbonate: $SrCO_3$, a colorless powder or rhombic crystals; M_r 147.63, density 3.70, m.p. at a CO_2 pressure of about 7 MPa, 1497 °C. Above 1340 °C, it is completely decomposed into strontium oxide and carbon dioxide: $SrCO_3 \rightarrow SrO + CO_2$. S. is slightly soluble in water ($K_S = 1.6 \cdot 10^{-9}$ at 25 °C). Acids convert S. to the corresponding salts and carbon dioxide. S. dissolves in water which contains carbon dioxide, forming strontium hydrogencarbonate: $SrCO_3 + CO_2 + H_2O \rightarrow Sr(HCO_3)_2$. It is found in nature as strontianite. The pure compound is obtained by precipitation of strontium salt solutions with soluble carbonates, or by melting strontium sulfate with soda. S. is used mainly to make strontium oxide and strontium hydroxide, fireworks and glass.

Strontium chloride: $SrCl_2$, colorless, cubic crystals soluble in alcohol and water; M_r 158.53, density 3.052, m.p. 875 °C, b.p. 1250 °C. S. crystallizes out of water as the hexahydrate, which is converted at 60 °C to a dihydrate and becomes anhydrous at 100 °C. S. is made by dissolving strontium carbonate in hydrochloric acid. It is the starting material for preparation of elemental strontium and other strontium compounds. It is used in pyrotechnics to make red fire.

Strontium hydroxide: $Sr(OH)_2$, colorless, amorphic mass; M_r 121.63, density 3.625, m.p. 375 °C. At 710 °C, S. is dehydrated to strontium oxide: $Sr(OH)_2 \rightarrow SrO + H_2O$. With a small amount of water, it forms an octahydrate, $Sr(OH)_2 \cdot 8H_2O$, which loses its water of crystallization at 100 °C. S. is only moderately soluble in water, and its solutions are strongly basic. Carbon dioxide causes strontium carbonate to precipitate out of these solutions: $Sr(OH)_2 + CO_2 \rightarrow SrCO_3 + H_2O$. S. is made by dissolving strontium oxide in water. It was formerly used to precipitate sugar from molasses, but has since been replaced by calcium hydroxide, which is cheaper.

Strontium nitrate: $Sr(NO_3)_2$, colorless cubic crystals which are readily soluble in water; M_r 211.63, density 2.986, m.p. 570 °C. S. crystallizes out of water as the tetrahydrate, which becomes anhydrous at 100 °C. When heated to a higher temperature, it first forms strontium nitrite, $Sr(NO_2)_2$, then strontium oxide, SrO. S. is obtained by the reaction of nitric acid with strontium carbonate. It is used mainly to make red flame in fireworks.

Strontium oxide: SrO, colorless, powdery mass or cubic crystals; M_r 103.62, density 4.7, m.p. 2430 °C, b.p. about 3000 °C. S. reacts exothermally with water, forming strontium hydroxide. It is obtained by roasting strontium carbonate or strontium nitrate. It is a starting material for production of strontium hydroxide and oxide cathodes.

Strontium sulfate: $SrSO_4$, colorless, rhombohedral crystals, M_r 183.68, density 3.96, m.p. 1605 °C; at higher temperatures it eliminates sulfur trioxide. S. is only slightly soluble in water ($K_S = 3.8 \cdot 10^{-7}$ at 18 °C, which corresponds to a solubility of 11.3 mg $SrSO_4$ in 100 g water) and precipitates when solutions containing Sr^{2+} and SO_4^{2-} ions are mixed. In nature, S. is found as celestine. It is the starting material for production of strontium, and is used in fireworks and as a white pigment (strontium white).

Strophanthidine: see Strophanthus glycosides.

Strophanthine: see Strophanthus glycosides.

Strophanthus glycosides: cardiac glycosides found in the seeds of *Strophanthus* species. The seeds of *S. gratus* contain 3.5 to 8% ***g-strophanthin (ouabain)***, which has L-rhamnose bound α-glycosidically to the aglycon ***strophanthidin***. The large number of hydroxy functions on the compound makes it very polar, and causes it to be poorly absorbed after oral administration (3%). In addition, the daily elimination rate is about 50%, which makes it suitable only for injection. The seeds of *S. kombé* contain 8 to 10% of a cardenolide mixture called ***k-strophanthin***; the main component of this amorphous mixture is ***k-strophanthoside***. It is no longer used therapeutically.

Rhamnose
g-Strophanthin

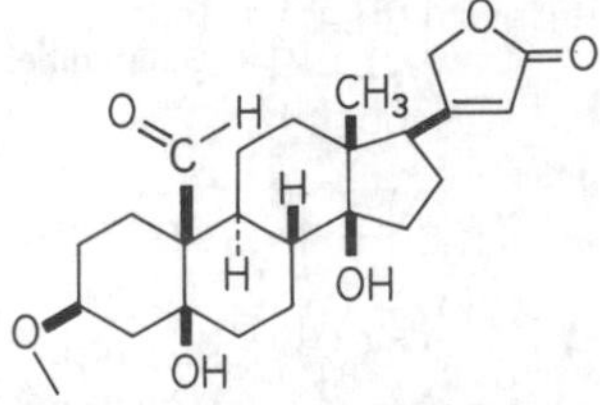

Cymarose-(β)Glucose-(α)Glucose
k-Strophanthoside

Stropeside: see Purpurea glycosides.

Structural analysis: see Analysis.

Structural formula: see Formula.

Structural isomerism: see Isomerism.

Structural viscosity, ***Bingham flow***: a decrease in viscosity with increasing shear stress. It occurs in gels, macromolecular solutions and colloidal dispersions. The viscosity is not constant, but decreases with increasing shear gradients (or increasing shear stress). At low shear stress, the material may not flow (flow limit). S. is always associated with a decrease of structure in the suspension; this is caused by shear.

Structure: by the structure of a chemical compound is meant the spatial arrangement of its constituent atoms in molecules or crystals and the nature of the bonds between them. The most rational description of the S. is still the classical structural formula (see Formula), to which projection formulas can be added to show steric arrangements and resonance canonical formulas to show resonance structures.

Structure elucidation, ***structure analysis***: determination of the structure of a molecule or solid.

Until about 1950, S. was done entirely by chemical methods. For organic carbon compounds, general structural characteristics can be determined by Preliminary tests (see). Subsequent quantitative Elemental analysis (see) yields the percent contents of the elements. At the same time, a ratio formula, such as $(C_7H_4O_2)_x$ can be determined. Mass determination, for which freezing point lowering (cryoscopy) was most often used in classical analysis, yields the value of x, e.g. $x = 2$. Thus the stoichiometric formula is $C_{14}H_8O_4$. For this formula there are many possible structural isomers. Further chemical reactions are required to discover the carbon skeleton (aliphatic, alicyclic, heterocyclic or aromatic) and the presence of functional groups. There are many classical reagents for detection of functional groups, such as Fehling's solution or Tollens' reagent for detection of reducing functions, dinitrophenylhydrazine to detect aldehydes and ketones, etc. If the substance is a compound which has already been described in the literature, one attempts to demonstrate its identity by comparison of the physical constants, and by converting the compound to two crystalline derivatives and comparing the melting points of these derivatives with those reported in the literature. If the compound is absolutely new and its probable constitution has not been suggested by its synthesis (for example, if it were a natural product), the S. can be very difficult, especially with classical chemical methods (oxidative degradation reactions, etc.).

Information about the carbon skeleton and its stereochemistry, and about the presence of functional groups, is much more rapidly obtained by use of spectroscopic methods. The advantages of physical methods are that they require much less time and smaller amounts of substance, and, except for mass spectroscopy, they are nondestructive of the sample. The application of modern methods of S. (table) has made the process much more efficient. For example, the S. of cholesterol, which was carried out primarily by chemical methods, required about 30 years. Today the S. of complicated natural products can often be completed within a few weeks. However, the apparatus is very expensive.

Important physical methods for S

UV-VIS spectroscopy	Recognition of substances with double bonds (e. g. polyenes, aromatics, polymethynes)
Infrared spectroscopy	Recognition of structure elements and functional groups, and indications of fine structure (H-bonds, configuration, conformation)
Raman spectroscopy	Different selection rules; same general type of information as infrared spectroscopy
NMR spectroscopy	Recognition of structural elements and functional groups containing magnetically active nuclei (e. g. ^1H, ^{13}C), recognition of neigh boring groups, indication of fine structure (configuration, conformation)
Mass spectroscopy	Molecular mass, empirical formula, recognition of heteroatoms and structure-specific fragments
Electron spin resonance spectroscopy	Electron distribution in paramagnetic substances (e. g. free radicals, complex compounds)
X-ray crystallography	Determination of complete molecular structure including bond lengths and angles

Except for x-ray crystallographic analysis, the above methods do not provide complete information on a structure, but only indicate partial structures, which must be combined to ascertain the total structure. Since the various methods complement each other to some extent, the use of several is desirable. The application of computers for data collection and processing, especially for searching empirical tables of structural elements or spectral catalogs, offers further improvements.

S. must be preceded by purification operations, usually chromatographic, which assure the purity of the sample.

Structure type, ***crystal structure type, lattice type***: a generalized description of a crystal structure which describes only the geometrical arrangement of the components, not the components themselves. There are many structural analogies among crystalline substances; S. permit these to be comprehended and classified. Isotypism (see) of substances, usually with similar chemical properties but sometimes with quite dissimilar ones, means that they belong to the same S.

In general, the individual S. are named after a type compound which represents the group. In the older literature, especially for organic compounds, the S. were often indicated by capital letters and numerals, e.g. B1 for the sodium chloride type. However, in light of the large number and the amazing diversity of crystal structures (by 1981 structural data for 35,000 organic and 7000 inorganic compounds had been stored in data banks), this approach to classification is no longer practical.

The most important and common simple S. include the following:

Elements: see Copper type, Tungsten type and Magnesium type (about 75% of the 100+ metallic element modifications and the noble gases crystallize in these three types), see also Diamond type and Graphite structure.

AB compounds: see Sodium chloride type, Cesium chloride type, Zinc blende type, Wurtzite type and Nickel arsenide type.

AB_2 compounds: see Fluorite type, Rutile type, Crystobalite type and Cadmium iodide type.

AB_3 compounds: see Rhenium trioxide type.

A_2B_2 compounds: see Corundum type.

Simple ternary compounds: see Perovskite type and Spinel type.

Intermetallic compounds: S. of the Laves, Hume-Rothery and Zintl phases (see Intermetallic compounds).

The structures of compounds with simple complex ions such as NH_4^+, O_2^{2-}, CO_3^{2-}, NO_3^-, SO_4^{2-}, $[PtCl_6]^{2-}$, etc. can be assigned to known S. in many cases, if one ignores small changes in the lattice caused by the fact that complex ions are less symmetric than spherical ions. For example, urotropin, $C_6N_4H_{12}$ crystallizes in the tungsten type, and adamantane, $(CH)_4(CH_2)_6$, in the copper type. Long molecules, such as those of the paraffins, tend to form chain structures in which the anisotropy of the molecules is reflected by the dimensions of the unit cell. This also is true of planar molecules such as anthracene, $C_{14}H_{10}$, which usually form layered structures.

Strychnine: a monoterpene indole alkaloid found together with brucine in the seeds and other parts of the Southeast Asian tree *Strychnos nus-vomica*. The crystals of S. are very bitter-tasting prisms, which are barely soluble in water and readily soluble in alcohol; m.p. 286-288 °C, $[\alpha]_D^{20}$ -139° (in chloroform). The compound is usually applied as the mononitrate; m.p. 268 °C, $[\alpha]_D^{20}$ -31° (in water). S. is biosynthesized from tryptophan and a monoterpene. The seeds (semen strychni, nux vomica, poison nuts) are used medicinally; their total alkaloid content is between 2 and 5%, and 50% of this is S.

S. is an analeptic which, in therapeutic doses, stimulates the circulation and respiration and increases the muscle tone. In toxic doses, it causes convulsions. It is one of the best known convulsion poisons; death occurs by respiratory paralysis. Mainly because of its bitter taste, alcoholic extracts of the seeds are used as tonics.

S. was discovered in 1818 by Pelletier and Caventou. Around 1950, Robinson and Prelog established the correct structural formula. In 1954, Woodward discovered the absolute configuration and synthesized the compound.

Brucine (2,3-dimethoxystrychnine) forms prisms containing 4 mol water of crystallization; m.p. 105 °C. The anhydrous form melts at 178 °C. Brucine has a bitter taste and only about 1/50 the analeptic effect of strychnine. It was discovered in 1818 by Pelletier and Caventou.

R = H : Strychnine
R = OCH_3 : Brucine

Strychnos alkaloids: see Alkaloids.

Stuart-Briegleb ball model: see Stereochemistry.

Styphnic acid, ***2,4,5-trinitroresorcinol***: a yellow, crystalline compound which is slightly soluble in water and readily soluble in ethanol and ether; m.p. 175 °C. S. forms readily crystallized salts with sharp melting points, the ***styphnates***, with organic bases, and is used to characterize the bases. The styphnates explode when heated, but most are not used as explosives. The exception is the lead salt, lead trinitroresorcinate.

Styrene, ***phenylethene***: C_6H_5-CH=CH_2, an unsaturated aliphatic aromatic hydrocarbon; a colorless, combustible liquid which smells like benzene; m.p. -30.6 °C, b.p. 145.2 °C, n_D^{20} 1.5468. It is insoluble in water, but is soluble in ethanol and ether. S. is present in coal tar, petroleum and balsam of styrax. It is produced industrially by reaction of benzene with ethene in the presence of sulfuric acid. The immediate product, ethylbenzene, is then dehydrogenated in the gas phase with iron, chromium and potassium oxides as catalysts. Other syntheses are the tetramerization of ethyne and the dimerization of butadiene. In the laboratory, one can make S. by pyrolysis of cinnamic acid. S. polymerizes very easily to polystyrene, an important industrial plastic. S. is copolymerized with butadiene to make synthetic rubber.

Styryl-: a term for the atomic group C_6H_5-CH=CH- in a molecule, the radical C_6H_5-CH=CH· or the cation C_6H_5-CH=CH^+.

Suberic acid, ***octanedioic acid***: HOOC-$(CH_2)_6$-COOH, a saturated dicarboxylic acid. S. forms colorless crystals; m.p. 144 °C, b.p. 219.5 °C at 1.33 kPa. S. is slightly soluble in water, and more soluble in alcohol and ether. It is found naturally in toad poison. It can be synthesized by oxidation of cork or ricinus oil with nitric acid, or by carbonylation of hexane-1,6-diol. S. is used in the synthesis of polyamides and esters.

Suberin: see Cutin.

Suberinite: see Macerals.

Suberone: same as Cycloheptanone (see).

Subglaze pigments: see Ceramic pigments.

Sublimate: 1) see Mercury chloride; 2) see Sublimation.

Sublimation: the conversion of a substance from the solid to the gaseous state. It may then condense again as a solid ***sublimate***, without ever having passed through the liquid state. S. is possible when the vapor pressure of the solid is greater than that of the liquid at a given temperature (***sublimation temperature*** or ***sublimation point***; see the figure to Phase diagrams). The plot of sublimation pressures vs. temperature is called the ***sublimation pressure curve***, and is described by the Clausius-Clapeyron equation. To reach this curve, it is often necessary to apply vacuum or high vacuum techniques. The ***rate of sublimation*** is determined by the difference between the vapor pressure in the sublimator and condenser. S. can be accelerated by the presence of an inert carrier gas. The ***heat of sublimation*** of a substance is the sum of its heats of evaporation and melting.

S. is applied to purify compounds which are sensitive to temperature and oxygen. It is used on an industrial scale in freeze-drying.

Sublimation enthalpy: see Phase change heats.

Sublimation heat: see Sublimation.

Sublimation pressure: see Vapor pressure.

Subliquidus separation: the separation of a glass into two or more amorphous phases below its liquidus temperature (its theoretical crystallization temperature). In separated glasses, the phases with different compositions can interpenetrate (interpenetration structure), or droplets can separate out (droplet structure). The tendency to separate is greatest for those glass compositions which are intermediate between stable compounds.

Substance A: see Neocarcinostatin.

Substance P: Arg-Pro-Lys-Pro-Gln-Gln-Phe-Phe-Gly-Leu-Met-NH_2, a linear peptide with a broad spectrum of action. S. has been isolated from the intenstinal tracts of various mammals, and also from the brains of human beings, mammals, birds, reptiles and fish. S. promotes contraction of the smooth muscles of the intestinal tract and reduces blood pressure. It stimulates secretion of pancreatic juice and saliva. S. has been identified in the central and peripheral nervous systems by radioimmunological techniques. It is probably released in response to stimulus of primary sensory, afferent nerves, so that in this part of the nervous system, it can act as a neurotransmitter. S. also appears to suppress the effects of morphine and endogenous opiates (see Endorphins), and to ameliorate the effects of abstinence in morphine addicts. It is possible that S. has a general function in resistance to stress-induced disorders.

S. was detected in 1931 by von Euler in the intestine and brain of the horse. The sequence was determined in 1970/1971 by Leeman. The dried extract from the first isolation was indicated by P (from powder), and this is the origin of the name.

Substantive dyes, ***direct dyes***: a group of synthetic dyes which can be used directly on untreated plant fibers (cotton, rayon), and in some cases on untreated animal fibers. Most S. are sodium salts of organic acids. Cotton, which consists mainly of cellulose and has no acidic or basic groups, has no affinity to most dyes which combine directly with animal fibers (wool and silk). The S., however, are bound to the cotton by association or adsorptive forces, such as hydrogen bond formation. The first synthetic S. was Congo red.

Substituent: a heteroatom or atomic group introduced into a compound by substitution. A distinction is made between Characteristic groups (see) and Radicals (see). S. on a benzene ring are classified as first or second order according to their effects on the location of a second substitution.

1) ***First-order S.*** increase the electron density on the benzene ring and direct the new S. in an electrophilic second substitution into the *ortho-* and *para-*positions. First order S. generally do not contain double bonds; examples are -OH, -OR, -NH_2, -NR_2, -SH, -O^-, -R, -F, -Cl, -Br and -I. While the halogens direct the second substitution to the *o-* and *p-*positions, their strong -I-effect makes the reaction difficult.

2) ***Second order S.*** reduce the electron density in the benzene ring and direct the entering S. to the *meta-*position. They have a -I and/or -M-effect (but no +M-effect). These S. include -NO_2, -SO_3H, -CHO, -COR, -COOH, -COOR, -CN, -CCl_3, -$^+NH_3$ and -$^+NR_3$.

Substituent constant: a measure for the influence of electronic and steric effects of a substituent on the reaction center of a molecule. S. are obtained by measurement of equilibrium constants or rate constants and substitution of these numerical values in an LFE equation. They are positive if the substituent withdraws electrons from the reaction center (-I- and -M-substituents) and negative if the substituent increases the negative charge at the reaction center (+I- and +M substituents). A distinction is made between S. which describe the effects of substituents on aromatic compounds, mainly benzene derivatives, and S. which characterize purely inductive effects of substituents. 1) The ***Hammett S.*** describe the substituent effects in benzene derivatives; they are indicated by σ_p and σ_m. σ_p expresses the effect of a substituent *para* to the reaction center, while σ_m applies to the *meta-*substituent. The Hammett S. are determined using the Hammett equation. 2) ***Inductive S.*** are a measure of the inductive effects of the substituents, and there are several types. a) The ***Taft S.*** σ^* characterizes the effect of α-substituents on the reactivity of aliphatic compounds, and are determined using the Taft equation (see). They are proportional to the Hammett σ_m values (Table). b) The σ_I values are determined by the Ingold-Taft method (see), and are related to the Hammett S. by the equation $\sigma_I = (3\,\sigma_m - \sigma_p)/2$. c) The σ' values of Roberts and Moreland can be determined from the dissociation constants of 4-substituted bicyclo[2.2.2]octane-1-carboxylic acids in 50% ethanol at 25 °C, relative to the dissociation constants of the unsubstituted carboxylic acids. σ' and σ_I values are nearly identical. 3) The ***steric S.***, E_s, describe the steric effects of a substituent on the reaction center, and, like the σ_I values, they are determined by the Ingold-Taft method.

Substituent constants

Substituent	σ_m	σ_p	σ^*
$N(CH_3)_2$	–0.21	–0.84	
NH_2	–0.16	–0.66	
OH	0.12	–0.37	1.55
CH_3	0.12	–0.27	1.45
$C(CH_3)_3$	–0.10	–0.20	–0.30
C_2H_5	0.07	–0.15	0.10
CH_3	–0.07	–0.17	0.00
C_6H_5	0.06	–0.01	0.60
COO^-	–0.10	0.00	–1.45
H	0.00	0.00	0.49
F	0.34	0.06	3.1
Cl	0.37	0.23	2.9
Br	0.39	0.23	2.8
I	0.35	0.18	2.4
$COOC_2H_5$	0.37	0.45	1.9
$COCH_3$	0.38	0.50	1.7
CN	0.56	0.66	3.6
NO_2	0.71	0.78	3.9
$N(CH_3)_3$	0.88	0.82	5.3

Substituent effect: see Resonance effect.

Substitution: in a chemical compound, the replacement of an atom or group of atoms by another atom or group. S. are classified according to their mechanisms as nucleophilic, electrophilic and radical S.

1) ***Nucleophilic S. (S_N reaction)***. S_N reactions occur between nucleophilic reagents, such as OH^-, Cl^-, CN^-, H_2O or NH_3, and electrophilic substrates, such as saturated C atoms and activated aromatic compounds. The net reaction is the displacement of an atom or group (nucleofuge) by the nucleophilic reagent. Some examples are the formation of esters of inorganic acids and their saponification, ether formation, the Williamson synthesis, alkylation and quaternization of amines, the Finkelstein reaction, the Koble nitrile synthesis, ring opening of epoxides, synthesis of thiols and thioethers, and synthesis of nitroalkanes. The following types of reactions can occur on saturated C atoms:

1.1 *Monomolecular nucleophilic S.* (S_N1). The rate-determining step is the cleavage (heterolysis) of a C-X bond to form a carbenium ion. This ion, which can be detected as an intermediate, reacts with the nucleophilic reagent to give the product. Eliminations or rearrangements can occur as side reactions leading to more stable products. S_N1 reactions obey first order kinetics. The carbenium ion intermediate can be trapped by reaction with active nucleophilic reagents (N_3^-, SCN^-, H^-, halide):

$$R{-}\overset{|}{\underset{|}{C}}{-}X \xrightarrow{-X^-} R{-}\overset{|}{\underset{|}{C}}{}^+ \xrightarrow{+Y^-} R{-}\overset{|}{\underset{|}{C}}{-}Y.$$

The attack of the nucleophile from either side of the planar carbenium ion leads to racemization of optically active starting materials. Steric and electronic effects (e.g. those in the series of primary, secondary and tertiary alkyl compounds) which stabilize the carbenium ion increase the rate of the S_N1 reaction. The degree of solvation has a large effect on the reaction rate. Since the steric requirements for formation of the intermediate carbenium ion are not great, the activation entropies are close to 0 J K^{-1} mol^{-1}.

1.2 *Bimolecular nucleophilic S.* (S_N2) occur by a synchronous mechanism; both substrate and nucleophilic reagent are involved in the rate-determining step. Reagent Y^- approaches from the side opposite the C-X bond which is to break, and simultaneously leaving group X exits. Formation of the bond with the nucleophilic reagent occurs simultaneously with breaking of the C-X bond.

$$Y| + C{-}X \longrightarrow Y{\cdots}C{\cdots}X \longrightarrow Y{-}C + X|$$

The rate-determining step obeys a second order rate law, and the reaction rate increases with the concentration of the nucleophilic reagent. The optical activity of enantiomers is retained, but the configuration is inverted (inversion, Walden inversion). Although the enthalpies of activation are lower than in S_N1 reactions, the activation entropies are negative, due to the relatively strict steric requirements on the transition state of S_N2 reactions.

Nucleophilic S. depend on the conformation of the substrate. The S. of atoms or groups at the bridgeheads of bicyclic systems can occur only by an S_N1 mechanism, and even then only with difficulty, because the carbenium ion cannot be fully planar, due to the ring tension. In cyclohexyl compounds, axial atoms or groups can be more readily replaced than equatorial atoms, in S_N1 or S_N2 reactions, because the attack of the nucleophile is less sterically hindered. Stereochemical considerations are also observed to play a role in the reactions of primary and secondary alcohols with thionyl chloride. In the first reaction step, an alkyl chlorosulfite is formed, and from it, an internal ion pair forms. The product retains the configuration. The mechanism here is called *"internal" nucleophilic S.* (S_Ni). Nucleophilic reagents with ambidence react in accordance with the Kornblum rule (see).

The mechanisms of bond cleavage of the leaving group are classified as the A1 mechanism (see) and A2 mechanism (see).

In certain nucleophilic substitution reactions, e.g. in certain solvolysis reactions, the rate is increased by the presence of certain β-substituents. In reactions on an optically active atom, configuration may be retained due to neighboring group effects (e.g. in nucleophilic S. on α-halocarboxylic acids). Nucleophilic S. of allyl halides can involve allyl rearrangement (*S_N1' or S_N2' substitution*).

$$R{-}CH{=}CH{-}CH_2{-}X \xrightarrow{-X^-} R{-}CH{=}CH{-}\overset{+}{C}H_2 \longleftrightarrow R{-}\overset{+}{C}H{-}CH{=}CH_2$$

$$\xrightarrow{+Y^-} R{-}CH{=}CH{-}CH_2{-}Y \ (S_N1) \qquad \xrightarrow{+Y^-} R{-}CH(Y){-}CH{=}CH_2 \ (S_N1')$$

$$Ph{-}\overline{\underline{S}}|^- + CH_2{=}CH{-}\overset{|}{\underset{|}{C}}{-}Cl \longrightarrow [Ph{-}S{\cdots}CH_2{\cdots}CH{\cdots}C{\cdots}Cl] \xrightarrow{-Cl^-} Ph{-}S{-}CH_2{-}CH{=}C\langle \qquad S_N2'$$

1.3. S. in aromatic and heterocyclic compounds occur by one of the following three mechanisms.

1.3.1. *S_N1 mechanism*, e.g. when arenediazonium salts are heated.

1.3.2. *Addition-elimination mechanism*. This applies to S_N2 S., where the attacking nucleophilic reagent forms a partial bond with the reaction center before the bond to the leaving group is broken. Some examples are the S. of the CH_3 group in 2,4,6-trinitroanisol by -OR, the alkali melting of arenesulfonic acids (S. of $-SO_3H$ by -OH) and nucleophilic S. on pyridine. This mechanism applies in most esterifications, the acid hydrolysis of carboxylic acid esters and nucleophilic S. of vinyl halides and β-halogen vinylcarbonyl compounds with nucleophiles which are not too strong.

1.3.3. *Elimination-addition mechanism*. Nucleophilic S. on benzoid compounds (see cine-Substitution), vinyl halides, β-halogen carbonyl compounds and β-halogen carboxylic acids with strongly basic nucleophiles occur by this mechanism. First the group to be replaced and an atom or group from a neighboring C atom leave the molecule, forming a multiple bond. A nucleophile then adds to the unsaturated intermediate.

2. ***Electrophilic S. (S_E reaction)***. Electrophilic reagents react with nucleophilic substrates (mainly aromatic compounds) by the *S_E mechanism*. S_E reactions obey second order kinetics; the reaction occurs in two stages. First the benzoid ring is attacked by an electrophilic reagent, which can be a cation or dipolar

molecule, and a π-complex forms. In a second reaction step, the electrophilic reagent forms a σ-bond with a certain ring C atom, forming a carbenium ion. The positive charge of this ion is delocalized over the rest of the conjugated system (σ-complex). The σ-complex stabilizes itself by cleavage of an H atom as a proton (normal electrophilic S.) in the presence of a base, or by cleavage of another atom or group (see ipso-Substitution); a new aromatic system is formed. The S_E mechanism is analogous to electrophilic addition to alkenes, but in the latter reaction, the σ-complex stabilizes by addition of the base. In general, the base is the anion resulting from formation of the electrophilic reagent.

$$+ X^+ \rightleftharpoons [\pi\text{-Complex}] \rightleftharpoons [\sigma\text{-Complex}] + IB \xrightarrow{-HB} C_6H_5X$$

π-Complex σ-Complex

In most cases, the electrophilic reagent is formed in an equilibrium established prior to the S_E reaction. For example, in nitration with nitric acid, the equilibrium is $HONO_2 + HONO_2 \rightleftharpoons H_2O + NO_2^+ + NO_3^-$; with nitrating acid, it is $HNO_2 + 2\,H_2SO_4 \rightleftharpoons NO_2^+ + H_3O^+ + 2\,HSO_4^-$. The equilibrium for sulfonating with free SO_3 or HSO_3^+ cations is $H_2SO_4 + SO_3 \rightleftharpoons H_2S_2O_7 \rightleftharpoons H_2SO_4 + HO\text{-}SO_2^+$. Other examples of S_E reactions are halogenation, Friedel-Crafts alkylations and acylation, Gattermann and Gattermann-Koch syntheses, the Houben-Hoesch reaction, the Vilsmeier reaction, hydroxymethylation, chloromethylation (Blanc reaction), aminomethylation, the Kolbe-Schmitt reaction, nitrosylation, azo coupling, mercurization, oxidative coupling to form indamines and indophenols, and acid-catalysed reaction of benzene derivatives with aldehydes or ketones (synthesis of triphenylmethane pigments). First and second order substituents influence electrophilic substitution reactions and affect the ease of the reaction.

3. ***Radical S. (S_R reaction)***. Substitution reactions in which radicals are formed take place by the *S_R mechanism*, which in many cases is a radical chain mechanism (see Radical reaction). Usually C-H bonds are attacked, and H atoms replaced. S_R reactions occur as chain reactions only when the enthalpy of the total reaction is negative (exothermal), although stages of the reaction may have positive enthalpies. These reactions can be directed by choice of specific conditions (see Halogenization). The reactivity of the C-H bond in radical S. increases in the order primary < secondary < tertiary C atom (decrease in bond dissociation energy and increasing radical stability). For the regioselectivity of these substitution reactions, it is a general rule that higher reactivity is associated with lower selectivity, and conversely. Thus S_R reactions can be classified as:

3.1 exothermal reactions with high-energy radicals X·. The transition state is very similar to the starting material, and electronic and steric effects are very small (low kinetic H/D isotope exchange effect).

3.2. thermoneutral reactions with low-energy radicals X·. The structure of the transition state is similar to that of the product, and electronic and steric influences are important (large kinetic H/D effect). Radicals are reagents with differing electrophilic properties. Halogen and oxygen radicals are strongly electrophilic ("hard" in the sense of the HSAB principle), while alkyl and phenyl radicals are less electrophilic ("soft" in the sense of the HSAB principle; see Acid-Base concepts):

Electron deficiency due to adjacent – I or –M substituent	Electron excess due to adjacent + I or +M substituent

$$R{-}CH_2{-}C({=}O){-}O{-}CH_2{-}R'$$

Alkyl radical Phenyl radical	Oxygen radical Halogen radical
soft	hard

Industrially important S. occuring by the S_R mechanism are halogenation, sulfochlorination, sulfoxidation, nitration and autooxidation (peroxygenation).

Substitution defect: see Crystal lattice defects.

Substitution mixed crystal: see Mixed crystal.

Substrate: a substance reacting with a reagent, or the substance being acted on by a catalyst (especially an enzyme). The distinction between a S. and a reagent is somewhat arbitrary, but usually the more complex reactant is the S. In an enzyme-catalysed reaction, all the reactants may be considered S.

Substrate pigments, ***carrier pigments***: pigments which are precipitated from solution onto a substrate, or which are mixed with a filler. The substrate is quite often a mixture of aluminum hydroxide, $Al(OH)_3$, and barium sulfate, $BaSO_4$, made in sequence from aluminum sulfate, sodium carbonate and barium chloride: $Al_2(SO_4)_3 + 3\ Na_2CO_3 + 3\ H_2O \rightarrow 2\,Al(OH)_3 + 3\,Na_2SO_4 + 3\,CO_2$; $Na_2SO_4 + BaCl_2 \rightarrow BaSO_4 + 2\,NaCl$. In the second case, blanc fixe, barite, gypsum, kaolin, etc. can be used as the filler. S. are often sold in the form of a paste, and are used to make colored paper and wallpaper.

Subtilisin: a single-chain serine protease formed by *Bacillus subtilis* and other microorganisms. S. is a relatively unspecific endopeptidase, and is used in large quantities as an additive to biologically active laundry products.

Succinaldehyde, ***butanedial, succinic dialdehyde***: $OHC\text{-}CH_2\text{-}CH_2\text{-}CHO$, a colorless, non-viscous liquid with a sweetish, pungent odor; b.p. 169-170 °C. S. is irritating to the skin. It is readily soluble in water and most organic solvents. When it stands for a long time, or in the presence of a small amount of water, it polymerizes to a tough, glassy mass, which can be reconverted to the monomeric form by heating. When S. is oxidized with nitrous or nitric acid, the product is succinic or oxalic acid. Like all Dialdehydes (see), S. is very reactive, undergoing the usual aldehyde addition and condensation reactions at each of its two aldehyde groups. It can be made by ozonide cleavage of diallyl, by oxosynthesis from acrolein diacetate, by Rosenmund-Saizew reduction of succinoyl dichloride, or by ring opening of pyrrole. It is used mainly for the synthesis of five-membered heterocyclic rings, such as thiophene or pyrrole, and as a starting material for production of a number of alkaloids, such as tropane alkaloids.

Succinate: a salt or ester of Succinic acid (see) with the general formula $R^1O-CO-CH_2-CH_2-CO-OR^2$, where R^1 and R^2 are monovalent metal ions, ammonium ions, or alkyl or aryl groups. R^1 and R^2 can be the same or different. As a dibasic acid, succinic acid can also form acid S., that is, one group R can also be an H atom. Some important salts are listed below: ***sodium succinate***, $(CH_2-CO-O)_2Na \cdot 6H_2O$, monoclinic crystalline prisms which are readily soluble in water and not soluble in alcohol. ***Calcium succinate***, $(CH_2-CO-O)_2Ca$, colorless crystals which are nearly insoluble in water. Calcium succinate is the most common salt of succinic acid found in nature. It is present in unripe fruits, algae, fungi and lichens. It is used in medicine in combination with salicylic acid to treat fevers.

Various esters of succinic acid with polyalcohols are used as softeners and solvents for waxes and synthetic resins. Some other industrially important S. are: ***dibenzyl succinate***, $(CH_2-CO-OC_4H_9)_2$, a colorless, crystalline compound which is nearly insoluble in water, but is readily soluble in alcohol and ether; m.p. 49-50 °C, b.p. 245 °C at 2 kPa, n_D^{20} 1.596. Dibenzylsuccinate is used as a softener and solvent for cellulose paints. ***Di-n-butylsuccinate***, $(CH_2-CO-OC_4H_9)_2$, is a colorless liquid, insoluble in water but miscible with most organic solvents; m.p. -29.2 °C, b.p. 274.5 °C, n_D^{20} 1.4299. This ester is used as an insecticide. ***Bissuccinylcholine dichloride***, $[CH_2-CO-OCH_2-CH_2-N^+(CH_3)_3]_2 2Cl^-$, a colorless, crystalline compound which is readily soluble in water and scarcely soluble in alcohol and ether; m.p. 160-164 °C (as the dihydrate). It is used in medicine as a muscle relaxant.

Succinate dehydrogenase: a flavin enzyme which catalyses the dehydrogenation of succinic to fumaric acid. S. has a relative molecular mass of 97,000 and consists of two subunits, one of which (M_r 70,000) contains covalently bound flavin adenine dinucleotide (FAD), and the other of which is an iron-sulfur protein (M_r 27,000).

Succinic acid, ***butanedioic acid***: $HOOC-CH_2-CH_2-COOH$, forms colorless, triclinic or monoclinic prisms; m.p. 188 °C, b.p. 235 °C with formation of the anhydride. S. is readily soluble in boiling water, alcohol and acetone, and slightly soluble in ether. It undergoes the reactions typical of dicarboxylic acids. The salts and esters are called Succinates (see). S. is present in free or bound form, especially as potassium succinate, in unripe fruits, amber and other fossil resins, and in algae, fungi and lichens. Free S. is an intermediate in the citric acid cycle. S. can be synthesized chemically by catalytic hydrogenation of maleic or fumaric acid, by oxidation of furan derivatives or carbonylation of acetylene. It is used in the preparation of pigments, drugs, polyester and alkyde resins, and in flavorings. The esters of S. with monohydroxy alcohols are used as softeners and solvents, and as synthesis components in the Stobbe condensation (see).

Succinic acid bischoline chloride ester: see Succinates.

Succcinic acid dibenzyl ester: see Succinates.

Succinic acid dibutyl ester: see Succinates.

Succinic acid imide: see Succinimide.

Succinic anhydride, ***tetrahydrofuran-2,5-dione***: colorless prisms; m.p. 119 °C, b.p. 261 °C. It is soluble in chloroform, carbon tetrachloride and alcohol. S. is made by heating succinic acid with dehydrating reagents such as acetic anhydride or phosphorus oxygen chloride.

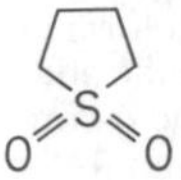

S
O O

Succinic dialdehyde: same as Succinaldehyde (see).

Succinimide, ***pyrrolidine-2,5-dione***: colorless platelets, m.p. 126 °C, b.p. 287-289 °C (dec.). S. is readily soluble in water, alcohol and acetone, and insoluble in ether and chloroform. It is synthesized by heating succinic acid in a stream of ammonia, or by heating succinic acid and urea to 175 °C. The hydrogen atom on the NH group can be replaced by a metal or halogen atom. *N*-chloro- and *N*-bromosuccinimide are especially important in organic syntheses. S. is converted to pyrrole by zinc powder distillation.

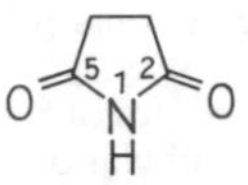

O 5 1 2 O
N
H

Succinite: same as Amber (see).

Sucrose, ***cane sugar, beet sugar***: β-D-fructofuranosyl-α-D-glucopyranoside. A very common disaccharide found in plants; it is "sugar" in the everyday sense. S. forms white crystals, m.p. 185-188 °C; above this temperature it decomposes to caramel. $[\alpha]_D^{20}$ +66.5°; there is no mutarotation. In S., both glycosidic hydroxyl groups are substituted, so that it does not give typical monosaccharide reactions such as reduction or oxazone formation.

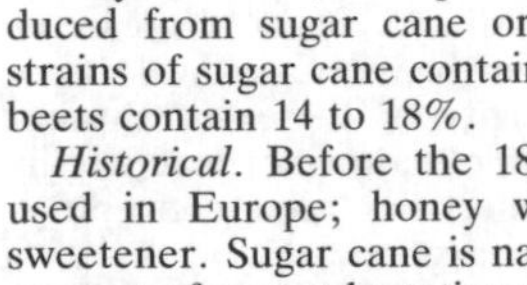

CH_2OH $HOCH_2$
O O
OH HO
HO O CH_2OH
OH OH

Acid-catalysed hydrolysis produces a mixture of glucose and fructose called invert sugar. S. dissolves readily in water, but sparingly in ethanol. It is produced from sugar cane or sugar beets; cultivated strains of sugar cane contain 8 to 17% S., and sugar beets contain 14 to 18%.

Historical. Before the 18th century, little S. was used in Europe; honey was essentially the only sweetener. Sugar cane is native to India. The development of sugar plantations in the Caribbean and the triangular trade in slaves (from Africa), rum and molasses (from the Caribbean) and cloth (Europe) was a major economic force in the 18th and 19th centuries. In 1747, Marggraf observed that beetroots contain S., and this led to the development of higher-yielding strains. Practical production of S. from sugar beets was introduced in 1800 by Achard, and this led to both a great decrease in price and an increase in consumption. S. now supplies up to 20% of the calories in the diets of some individuals in industrialized countries.

Sudan pigments: an assortment of synthetic pigments. Some examples are ***sudan I***, 1-phenylazo-2-naphthol (yellow crystals, m.p. 131-133 °C) and ***sudan orange G*** (4-phenylazo)resorcinol, m.p. 143-146 °C. S. are soluble in fats, oils, waxes, resins, hydrocarbons, chlorinated hydrocarbons and esters, and insoluble in water. They are used to color fats, oils, resins, waxes and mixtures of them, e.g. in shoepolish, floor wax, candles, wax flowers, etc., and as vital stains for microscopy.

Sugar: in the narrow sense, common cane or beet sugar (see Sucrose); in the broad sense, a term for any mono- or oligosaccharide (see Carbohydrates).

Sugar acid: see Aldaric acid.

Sugar alcohol: see Alditol.

Sugar anhydrides: intramolecular acetals of monosaccharides. S. do not give the typical sugar reactions based on the potential carbonyl group. The best known S. is *levoglucosan* (1,6-anhydro-D-glucopyranose), which is formed by dry distillation of starch or cellulose. S. can be hydrolysed to the free sugars; they are used as starting materials for syntheses of glycosides.

Sugar dicarboxylic acid: same as Aldaric acid (see).

Sugar esters: mono- or oligosaccharides with esterified hydroxyl groups. *Sugar phosphates* (esters of phosphoric acid) are intermediates in carbohydrate metabolism; in them, the end hydroxyl groups are esterified, as in glucose 1-phosphate, glucose 6-phosphate or fructose 1,6-bisphosphate. These compounds are highly acidic, stable to alkalies, but sensitive to acids. The *monosulfates* are also highly acidic; they are found in acidic mucopolysaccharides and other polysaccharides, and in glycolipids. The Sugar nucleotides (see) are derived from the aldose 1-phosphates.

Various S. of organic acids also occur in nature, e.g. *acetates* in the lanata glycosides, *fatty acyl esters* in lipopolysaccharides or *gallates* in hydrolysable tannins. Acetates, benzoates and various sulfonates are very important in carbohydrate chemistry; the acyl residues are used as protective groups and this group of S. provides starting materials for numerous exchange reactions. Synthetic fatty acyl esters of sugars are used as emulsifying agents in foods and other products.

Sugar nucleotides: aldose or alditol 1-esters of nucleoside diphosphates which supply glycosyl residues in the biosynthesis of oligo- and polysaccharides, glycosides, glycolipids, glycoproteins, teichoic acids, etc. One of the most important is *uridine diphosphoglucose*, which donates glucose units, e.g. in the synthesis of glycogen.

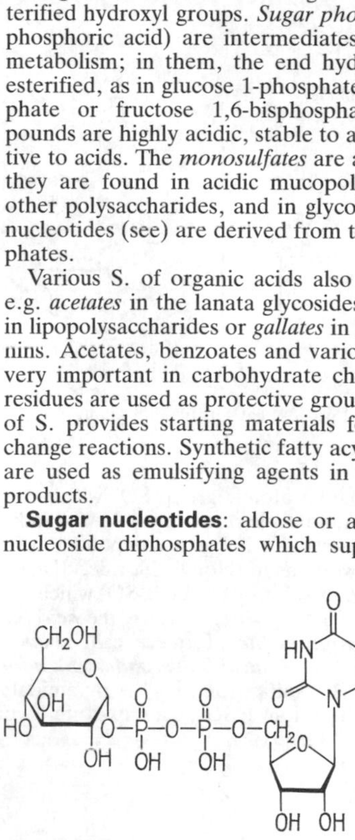

Uridine diphosphoglucose

Sugar substitutes: see Sweeteners.

Sulbentin: an Antimycotic (see).

Sulfacetamide: see Sulfonamides.

Sulfadimethoxin: see Sulfonamides.

Sulfamerazine: see Sulfonamides.

Sulfamethoxazole: see Sulfonamides.

Sulfamethoxypyrazine: see Sulfonamides.

Sulfamidochrosoidin: see Sulfonamides, historical section.

Sulfaminic acid: see Amidosulfonic acid.

Sulfane: a hydrogen compound of sulfur; see Hydrogen sulfide, see Polysulfanes.

Sulfane disulfonic acids: same as Polythionic acids.

Sulfane sulfonic acids: oxoacids of sulfur derived from hydrogen sulfide or polysulfanes by substitution of an -SO_3H group for one or both protons. The most important of the ***sulfane monosulfuric acids***, HS_n-SO_3H is monosulfane monosulfonic acid, HS-SO_3H, known as Thiosulfuric acid (see). The ***sulfane disulfonic acids*** are the Polythionic acids (see), HO_3S-S_n-SO_3H.

Sulfanilic acid, ***4-aminobenzenesulfonic acid***: colorless crystals which are usually hydrated. Anhydrous S. is obtained by heating to 100 °C; m.p. 288 °C. S. is slightly soluble in boiling water, and insoluble in ethanol and ether. It forms salts with bases, but does not react with acids, because it exists as an internal salt. It is prepared by brief heating of aniline with concentrated sulfuric acid to 200 °C. S. is used in the production of azo dyes. The pharmaceutical Sulfonamides (see) are derivatives of S.

Sulfatases: see Esterases.

Sulfate: 1) a *salt* of sulfuric acid, H_2SO_4. As a dibasic acid, sulfuric acid forms two series of salts, the ***acid S., primary S.*** or ***hydrogensulfates***, formerly known as ***bisulfates***, M^IHSO_4, and the ***neutral S., secondary S.*** or ***normal S.***, $M^I_2SO_4$. The sulfate ion is tetrahedral. The hydrogensulfates are readily soluble, and only those of the alkali metals are known in the solid state. When heated, they split off water and form disulfates. With the exception of the calcium, strontium, barium and lead S., the neutral S. are soluble in water. They are thermally stable, and are split into the metal oxide and sulfur trioxide only at very high temperatures. They often crystallize with five or seven moles of water (see Vitriols). Double sulfates of the type $M^IM^{III}(SO_4)_2 \cdot 12H_2O$ are known as Alums (see). The S. are obtained by dissolving the metals, metal oxides, hydroxides or carbonates in the stoichiometric amount of sulfuric acid. They can also be formed by the reaction of sulfuric acid with salts of volatile acids, such as nitrates and halides. S. are found widely in nature, for example as anhydrite, $CaSO_4$, gypsum, $CaSO_4 \cdot 2H_2O$ or barite, $BaSO_4$.

2) An *ester* of sulfuric acid, that is, the diesters $(RO)_2SO_2$, and the monoesters, RO-SO_3H, and their salts, RO-SO_3M^I. Some important examples of this class of compound are the Alkyl sulfates (see) and Dimethylsulfate (see).

Sulfate lead white: see Lead(II) sulfate.

Sulfate process: see Lignin.

Sulfathiourea: see Sulfonamides.

Sulfation: an industrial method of preparing organic sulfates, that is, esters of sulfuric acid, R-O-SO_3H. These compounds are made by reaction of alcohols or unsaturated compounds with sulfuric acid.

Sulfenic acids: compounds with the general formula R-SOH, e.g. ***methane sulfenic acid***, CH_3SOH or ***benzene sulfenic acid***, C_6H_5-SOH. The best known is ***propene-1-sulfenic acid***, which is the lacrimator in freshly cut onions. The acyl chlorides of these acids, the Sulfenyl chlorides (see) are important in organic synthesis; they are more stable than the free acids, but are also very reactive.

Sulfenyl chlorides: derivatives of sulfenic acids which can be used in many organic syntheses. They are made by reaction of thiols or disulfides with chlorine: R-SH + $Cl_2 \rightarrow$ R-SCl + HCl, or R-S-S-R + $Cl_2 \rightarrow$ 2 R-SCl. S. add to alkenes, forming β-chloroalkyl thioethers, or to alkynes, forming β-chlorovinyl thioethers. The chlorine atom can be readily replaced by nucleophilic reagents. For example, the reaction of an S. and potassium cyanide leads to a thiocyanate (rhodanide), and the reaction of an S. and sodium sulfite gives a Bunte salt: R-SCl + $Na_2SO_3 \rightarrow$ R-S-$SO_3^- Na^+$ + NaCl.

Sulfides: 1) binary compounds of sulfur in which the sulfur is in the - 2 oxidation state. Metal sulfides can be considered salts of the dibasic acid hydrogen sulfide, H_2S. Correspondingly, two series of salts are formed, the ***primary S., acid S., hydrogensulfides***, M^IHS, and the ***secondary S., neutral S., normal S.***, M^I_2S. The electropositive elements form S. which are largely ionic, e.g. Na_2S, CaS, Al_2S_3, but bonds in the S. of transition metals have a significant covalent component. Non-metal S. are covalent, e.g. H_2S, CS_2, P_4S_{10}. The sulfide ion, S^{2-}, is a strong base (pK_B = 1.1). Therefore, ionic S. are protolysed in aqueous solution, and these solutions are basic; for example, $Na_2S + H_2O \rightleftharpoons 2\,Na^+ + HS^- + OH^-$. While the S. of the alkali and alkaline earth metals are colorless, other metals form insoluble S. which are often characteristically colored. Their precipitation is used for analytical separation of cations in the H_2S separation step (see Hydrogen sulfide). A few S. react with sulfur in solution or in the melt to form Polysulfides (see), while others combine with excess sulfide ions to form complex thio salts, e.g.: $Sb_2S_5 + 3\,Na_2S \rightarrow 2\,Na_3SbS_4$. Most S. are converted by heating in the air (roasting) to the corresponding metal oxides and sulfur dioxide. The S. are formed by melting the elements together (e.g. Fe + S $\rightarrow$ FeS), by reduction of sulfates ($CaSO_4$ + 4 C $\rightarrow$ CaS + 4 CO) or precipitation with H_2S from the respective metal salt solution. The alkali metal S. containing water of crystallization are made by neutralization of the hydroxide solutions with H_2S, while the anhydrous salts are obtained by reaction of sulfur with the alkali metal in liquid ammonia. S. are found very widely in nature, and these minerals are called blendes, glances or pyrites, for example, copper pyrite, $CuFeS_2$, lead glance (galenite), PbS, and zinc blende (sphalerite), ZnS.

2) *Organic S.* see Thioethers.

Sulfinic acids: compounds with the general formula R-SO_2H. They are viscous oils or crystalline solids which are oxidized in air to sulfonic acids. Reduction with zinc and hydrochloric acid produces thiols. S. are medium strong acids and form stable salts, the ***sulfinates.*** Reaction with thionyl chloride produces ***sulfinyl chlorides***; these are reactive. For example, they react with triethylamine to form the ***sulfines*** R-CH=S=O, which are analogs of the ketenes and are very reactive. S. are synthesized by reduction of sulfonyl chlorides with zinc powder or reaction of Grignard compounds with sulfur dioxide.

Sulfinol process: a method of removing acidic components (carbon dioxide, hydrogen sulfide and carbon oxygen sulfide) from natural, city and synthetic gases. The acidic components are washed out of the pressurized crude gas by a countercurrent of solvent in an absorption column; the solvent is a mixture sulfolane and diisopropanolamine (Fig.). The purified gas is sucked out of the top of the absorption column, and further processed. The acidic components are removed from the solvent by reducing the pressure to atmospheric pressure, which causes part of the dissolved gas to escape. The larger part of the acidic gases, however, is removed in a degassing column, after which the solvent mixture is led back to the absorption column. The acidic gases released in the two steps can be converted to sulfur dioxide in a Claus furnace.

The main advantage of the S. is that carbon oxygen sulfide is also removed from the crude gas, and no conversion is needed as in the Sulfosolvane process (see).

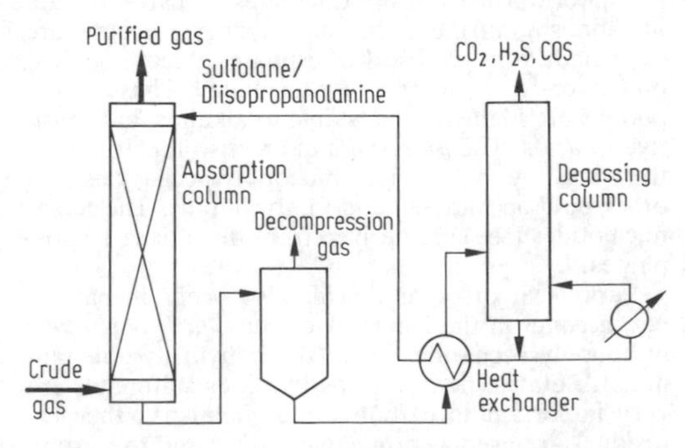

Diagram of a sulfinol process gas purification plant.

Sulfinyl group: the atomic group SO, which can exist as a part of a molecule or an unstable free radical. If the S. is part of a halogen compound, it is called a ***thionyl group***, as in thionyl chloride, $SOCl_2$.

Sulfite: 1) a *salt* of sulfurous acid, H_2SO_3 which, as a dibasic acid, forms two series of salts, the ***acid S., primary S.*** or ***hydrogensulfites***, formerly called ***bisulfites***, M^IHSO_3, and the ***neutral S., secondary S., normal S.***, $M^I_2SO_3$. The sulfite ion SO_3^{2-} is pyramidal; the hydrogen sulfite ion HSO_3^- can exist in two tautomeric forms (Fig.).

The hydrogen sulfite is in equilibrium with disulfite in concentrated aqueous solutions: 2 HSO_3^- ⇌ $S_2O_5^{2-}$ + H_2O. It is therefore not possible to isolate solid hydrogen sulfites. S. are decomposed by strong acids, releasing sulfur dioxide. They react with sulfur to form thiosulfates. S. are decidedly redox amphoteric, although the reducing effects are stronger. They are converted by strong oxidizing agents (e.g. Cl_2, MnO_4^-) to sulfates; manganese(IV) oxide converts S. to dithionates, $M^I_2S_2O_6$. SO_2-containing sulfite solutions are reduced to dithionites, $M^I_2S_2O_4$ by zinc, and strong reducing agents reduce it to sulfur or hydrogen sulfide. The S. are obtained by absorption of SO_2-containing gases into metal hydroxide solutions; Sodium sulfite (see) and Calcium hydrogensulfite (see) (solutions) are important in industry.

2) An *ester* of sulfurous acid with the general formula $(RO)_2SO$. They are made from thionyl chloride and alcohols or phenols in the presence of a reagent to bind HCl, and are used as alkylation and arylation reagents in organic syntheses.

Sulfite liquor: the liquor in which pulp has been cooked in the production of cellulose. It contains about 10% solids; 3 to 5% is carbohydrate, mainly lignin sulfonic acids, but also pentoses and hexoses. The S. contains about 50% of the solid matter of the pulped wood, and therefore an economical use must be found for it.

Applications. 1) Combustion. After evaporation in vacuum, the S. yields a dry substance with a heating value of 16,000 kJ/kg; this amount is theoretically sufficient to provide the steam required for the cooking of the pulp. 2) Fermentation. The sulfur dioxide is removed, and the S. is neutralized with quicklime or limestone at 90 °C. After cooling to 34 °C, it is fermented to ethanol by yeast. The yeast is separated and recycled to the fermenter. For each ton of cellulose, about 100 l 96% alcohol can be obtained from the waste liquor. 3) Yeast production. The cooled, neutralized S. is supplemented with potassium chloride and ammonium phosphate and used as a nutrient medium for yeast. With good aeration, the S. remains for about 4 1/2 hours at 32 °C in a 200-300 m^3 fermenter, from which it is then harvested in a continuous operation. The harvested medium contains 1.2 to 1.5% solid matter, which is separated, condensed in an evaporator, and dried to yield yeast with a water content of 8%. Each m^3 S. yields 12 to 15 kg yeast, which is used mainly as a high-protein feed for animals. 4) S. is also used to produce tannins and vanillin, and as a soil conditioner. The lignin sulfonic acids and their salts (lignin sulfonates) in the S. can be used as binders in the production of plywood, as sources of furfural, as wetting and dispersion agents in the construction industry, as flotation agents in fluorspar production, and as components of rubber and fiber coatings.

Sulfite process: see Lignin.

o-Sulfobenzimide: same as Saccharin (see).

Sulfochloride: same as Sulfonyl chloride (see).

Sulfo group: the very acidic atomic group $-SO_3H$ found in sulfonic acids.

Sulfolane, ***thiolane dioxide, tetrahydrothiophene-1,1-dioxide***: colorless crystals or a liquid; m.p. 27 °C, b.p. 285 °C, n_D^{20} 1.4840. It is soluble in water and organic solvents. S. is made by hydrogenation of sulfolene, which can be obtained by addition of sulfur trioxide to butadiene. S. is a versatile solvent, and is used to extract sulfur compounds from industrial gases, and aromatic compounds from pyrolysis fractions (sulfolane process).

O O O

Sulfolipids: see Glycerolipids.

Sulfonal, ***2,2-bis(ethylsulfonyl)propane***: C_2H_5-SO_2-$C(CH_3)_2$-SO_2-C_2H_5, colorless crystals; m.p. 124-126 °C, b.p. 300 °C. S. is barely soluble in water, but is soluble in alcohol, benzene and chloroform. It is obtained by condensation of acetone with ethanethiol and subsequent oxidation of the thioether with potassium permanganate. S. has a hypnotic effect and was formerly one of the most widely used sleep inducers. However, because it is slowly absorbed and accumulates in the body, leading to side effects, it has been replaced by the barbiturates.

Sulfonamide: an amide of sulfonic acid containing the group $-SO_2NRR'$. S. are formed by reaction of sulfonylchlorides R-SO_2-Cl with ammonia or primary or secondary amines. The sulfonamide bond is more stable to hydrolysis than the carboxylic amide bond. S. which have an H atom on the N atom form salts with alkalies. In the following, the S. used as drugs (see the table, p. 1048) are discussed.

S. which are not acylated on the aromatic N atom are amphoteric compounds. Because the aromatic amino group is slightly basic, they form salts with strong acids, which are hydrolysed in water. These salts are not suitable for use as drugs. The acidity of the S. is due to the electron acceptor effect of the SO_2 group, and is affected by the group bound to the sulfonamide nitrogen. Groups which are strongly electron-withdrawing, e.g. acetyl, make the compounds relatively strongly acidic (e.g. sulfacetamide).

The S. are generally synthesized from acetanilide, CH_3CO-NH-C_6H_5, which reacts with chlorosulfonic acid to form 4-acetamidobenzenesulfonyl chloride. This is usually made to react with an aminoheteroaromatic compound. Finally, the acetyl group is split off.

Derivatives of 4-aminobenzenesulfonamide (sulfanilamide) are used as bacteriostats and oral antidiabetics. The primary amino group in position 4 is essential for the bacteriostatic effect. The strength and duration of the effect is greatly affected by substituents on the sulfonamide group. A heteroaromatic ring at this position increases effectiveness. S. are classified according to the duration of their effect. Short-term S. are eliminated from the body with half-lives less than 8 hours, medium-term S. between 8 and 20 hours, and long-term S., with half-lives greater than 20 hours. *Sulfacetimide* has a relatively weak effect; it is used as a solution of a weakly basic sodium salt in eye drops. S. which are not readily absorbed are acylated on the aromatic amino group; they acquire antibacterial activity only after enzymatic deacylation in the intestine. *Mafenide* deviates from the classic structure of the S. by having a primary aliphatic amino group on the benzene ring. Its salt with sulfathiourea is often used in treatment of burns.

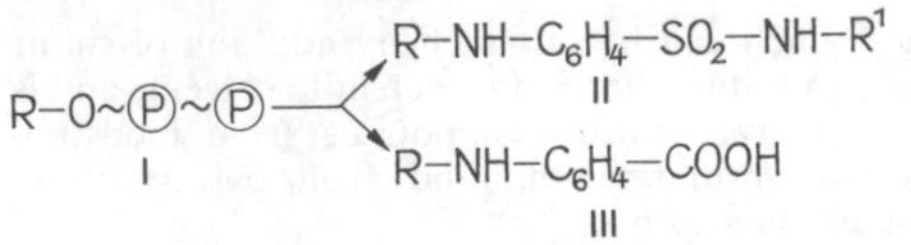

Folate synthetase (see), an enzyme required by bacteria for synthesis of folic acid, has a higher affinity for S. than its natural substrate, 4-aminobenzoic acid. Thus the S. act as competitive inhibitors of the reaction forming 7,8-dihydropteroic acid (III) from its precursor (I); and they are incorporated into homologous compounds (II) which in turn inhibit formation of dihydrofolic acid from III. The bacteriostatic action of S. is due to this inhibition; since folic acid is a vitamin for human beings, the S. are innocuous to most of us.

Dihydrofolic acid is further metabolized by hydrogenation of the bond between the N5 and C6 atoms by dihydrofolate reductase. The product, tetrahydrofolic acid, is a coenzyme which binds and transfers C_1 units, and is essential for the biosynthesis of pyrimidine and purine nucleotides.

The S. are sometimes combined with an inhibitor of dihydrofolate reductase, such as trimethoprim. Because its components act at different points in metabolism, such a mixture has a greater bacteriostatic effect. Sulfamethoxazole and sulfamerazine are used in such mixtures.

Long-term use of S. can lead to resistance.

Historical. The first S. used as an antibacterial drug was the azo dye sulfamidochrysoidine (Prontosil); introduced in 1935 by Domagk, Mietzsch and Klarer. Shortly thereafter, Fréfouel, Nitti and Bovet recognized that the active form of this compound is sulfanilamide. In 1940, Woods observed that *p*-aminobenzoic acid counteracts the effect of the S., and Fildes (1940) suggested that S. act as antimetabolites of *p*-aminobenzoic acid in bacterial metabolism.

Sulfonamide resins: synthetic resins made by condensation of sulfonamides (e.g. 4-toluenesulfonamide or sulfonamide phenol ether) with aldehydes, usually formaldehyde; other components are sometimes added to harden the resin. S. are not macromolecular polymers of *N*-methylenearylsulfonamides; they are supercooled melts of trimeric *N*-

Table. Typical sulfonamides

Category	Name	Formula
Short-term sulfonamides	Sulfacetimide	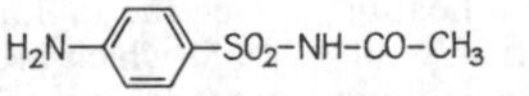
	Sulfathiourea	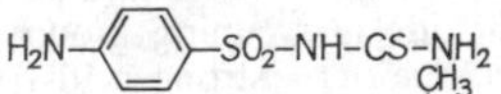
	Sulfisomidine	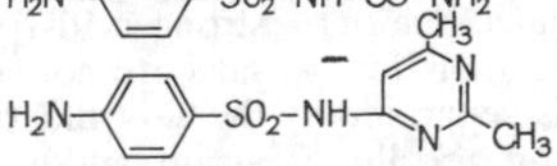
Medium-term sulfonamides	Sulfamethoxazole	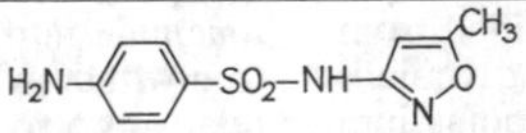
Long-term sulfonamides	Sulfamerazine	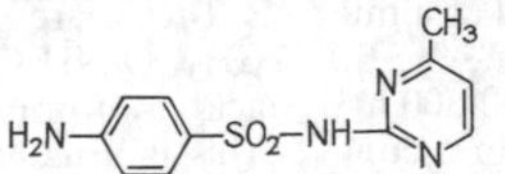
	Sulfaclomid®	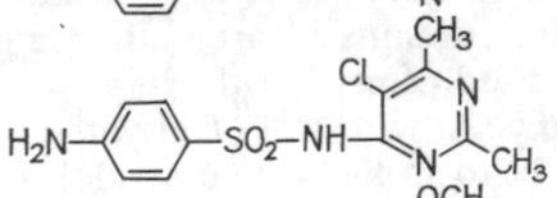
	Sulfadimethoxine	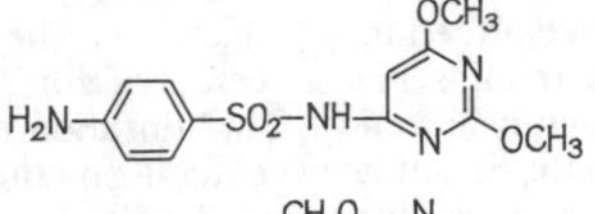
	Sulfamethoxypyrazine	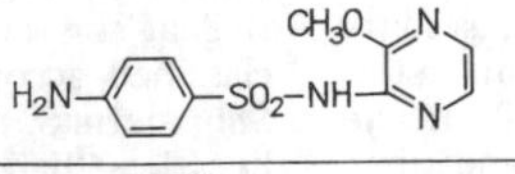
Not readily absorbed	Phthalylsulfathiazole	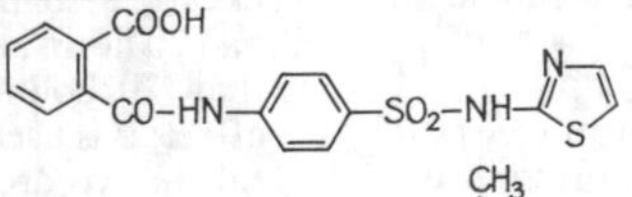
	Formylsulfisomidine	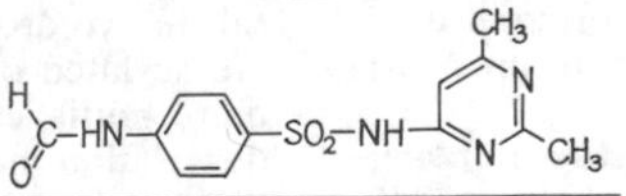
Other	Mafenide	

methylenearylsulfonamides mixed with unreacted sulfonamide. S. are used in mixtures with other components, such as cellulose acetate or polyvinyl acetate, to make paints with excellent color fastness to light.

Sulfonate: see Sulfonic acid.

Sulfonation, ***sulfation***: direct introduction of the sulfo group, $-SO_3H$, into an organic compound to synthesize a sulfonic acid. S. of aromatic compounds is achieved by an S_E reaction with concentrated or fuming sulfuric acid, sulfur trioxide dissolved in pyridine or dioxane, chlorosulfonic acid or a mixture of sulfuric acid and thionyl chloride. S. is reversible, and therefore often leads to product isomerization. Aliphatic sulfonic acids are obtained by Sulfoxidation (see), Sulfochlorination (see) and subsequent hydrolysis of alkanes, reaction of alkenes with sodium hydrogensulfite, reaction of haloalkanes with sodium sulfite, or oxidation of thiols.

Sulfones: organic sulfur compounds with the general formula $R-SO_2-R'$. S. are oxidation products of the sulfides (thioethers), and can be either symmetrical (R = R') or unsymmetrical (R ≠ R'). According to the IUPAC rules, S. are to be considered alkyl- or arylsulfonyl derivatives of hydrocarbons, and to be named accordingly, e.g. methylsulfonylmethane (see Dimethylsulfone). S. contain two S=O double bonds, and are formed by the oxidation of sulfides by an excess of hydrogen peroxide in acetic acid, or by conc. nitric acid. S. are generally crystalline, very stable compounds which as a rule can be reduced only under extreme conditions. The sulfonyl group exerts a strong I effect, which is approximately equivalent to that of a cyano group; it has the effect that S. with α-H atoms are CH acidic. S. are often used to characterize aliphatic and even cyclic sulfides. Alkylarylsulfones are also used as softeners.

Sulfonic acids: organic compounds with the general formula $R-SO_3H$. R can be either an alkyl (***alkane S.***) or aryl (***arene S.***) group. Most aliphatic A. are viscous liquids, while the aromatic S. form hygroscopic crystals which dissolve readily in water. Some S. form stable, crystalline hydrates. S. are strong acids; their salts with alkali and alkaline earth metals and lead are readily soluble in water; these salts are known as ***sulfonates*** and can be precipitated from aqueous solutions with sodium chloride. Alkane sulfonic acids can be synthesized by sulfoxidation or sulfochlorination of alkanes, reaction of halogen alkanes with sodium sulfite, or oxidation of thiols. Arene sulfonic acids can be made by sulfonation of arenes. S. are very reactive and can be converted to Sulfonyl chlorides (see), sulfonamides (see) and esters: $R-SO_3H + PCl_5 \rightarrow R-SO_2Cl + POCl_3 + HCl$. The sulfonic acid group, $-SO_3H$, is a good leaving group and can readily be replaced, e.g. by a nitro group, or by a hydrogen atom in a reaction with water and acid at a high temperature. The alkali salts of arene sulfonic acids react with many nucleophilic reagents at higher temperatures, forming amines, phenols, thiols, carboxylic acids or nitriles. S. are also important intermediates for the production of pigments, pharmaceuticals, washing, wetting, emulsifying and flotation compounds (see Alkane sulfonates).

Sulfonium salts: organic sulfur compounds with the general formula $R_3S^+X^-$, where R are alkyl groups. The crystalline salts are formed by reaction of haloalkanes with thioethers:

$$R_2S + R{-}X \longrightarrow R_2\overset{+}{S}{-}RX^-$$

When the S. are heated, they revert to the starting materials. They react with silver oxide and water to form the very basic sulfonium hydroxides, which decompose into dialkyl sulfides and alkenes when heated:

$$(C_2H_5)_3S^+X^- + AgOH \rightarrow (C_2H_5)_3S^+OH^- \xrightarrow[-H_2O]{\Delta} (C_2H_5)_2S + CH_2{=}CH_2.$$

Loss of an α-H atom produces an S-ylide from a trialkyl sulfonium halide and methyllithium, e.g.:

$$(CH_3)_3S^+Br^- + CH_3Li \xrightarrow[-CH_4/-LiBr]{} (CH_3)_2\overset{+}{S}{-}\overset{-}{C}H_2.$$

S-Ylides are very reactive intermediates in the synthesis of cyclopropanes and oxiranes.

S. with three different alkyl groups have a center of chirality (trivalent sulfur in a pyramidal structure) and can be separated into optical antipodes.

Sulfonyl chlorides: derivatives of sulfonic acids with the general forumula $R-SO_2Cl$. As in carboxylic acyl chlorides, an OH group in the sulfo group is replaced by a chlorine. S. react as readily as carboxylic acyl chlorides with ammonia, amines and alcohols, forming analogous sulfonic acid derivatives. The aliphatic S. are solids or liquids with relatively low melting points; they are synthesized mainly by sulfochlorination of alkanes. The aromatic S. are obtained by sulfonation of arenes and subsequent treatment with phosphorus pentachloride, or directly from arenes and chlorine sulfonic acid. The S. are relatively stable to water, and are only slowly hydrolysed. In the presence of alkali hydroxide, the reaction occurs more rapidly, yielding ***alkali sulfonates***. Reduction of these compounds with zinc under different conditions yields thiols or sulfinic acids. S. react with many nucleophilic reagents to give sulfonic acid derivatives such as ***sulfonic acid esters, sulfonamides, sulfonic acid azides***, etc. In the presence of Friedel-Crafts catalysts ($AlCl_3$), S. react with arenes to form sulfones: $R-SO_2Cl + C_6H_6 \rightarrow R-SO_2-C_6H_5 + HCl$. Mesyl chloride (methanesulfonyl chloride), benzene sulfonyl chloride and tosyl chloride (toluene-4-sulfonyl chloride) are well known S. used widely as organic intermediates or reagents.

Sulfonyl group: the group SO_2, which can occur as a fragment of a molecule or as an unstable free radical. If the S. is part of a halogen compound, it is called the ***sulfuryl group***, as in sulfuryl chloride. The term "sulfonyl..." is also used for derivatives of sulfonic acids, that is for the group $R-SO_2-$, e.g. sulfonyl chloride.

Sulfonylurea herbicides: see Urea herbicides.

Sulfosalicylic acid, ***3-carboxy-4-hydroxybenzenesulfonic acid***, a sulfonic acid derived from salicylic acid. S. forms white crystals, m.p. 120 °C, which are readily soluble in water, ethanol and ether. S. is an antiseptic, and is used in medicines. It is also used for colorimetric determination of iron.

Sulfosolvane process, ***alkazide process***: a method for removing hydrogen sulfide from synthesis gases. In it, the synthesis gas is scrubbed by flowing counter to the sulfosolvane solution, a 30 to 35% aqueous solution of sodium sarconisate or other amino acid salt (see Gas scrubbing). The sulfosolvane solution takes up the hydrogen sulfide at 20 to 30 °C to form salts, and releases it again when heated to 100 to 110 °C:

$$CH_3\text{–}NH\text{–}CH_2\text{–}COONa + H_2S \underset{100\text{-}110°C}{\overset{20\text{-}30°C}{\rightleftharpoons}}$$

$$CH_3\text{–}NH\text{–}CH_2\text{–}COOH + NaSH$$

Carbon dioxide is also removed, but incompletely. Carbon oxygen sulfide is not absorbed by the sulfosolvane solution, so the COS must be removed by heating the gas in the presence of an acidic catalyst to 150 to 200 °C; this converts it to CO_2 and H_2S (see Conversion): $COS + H_2O \rightarrow CO_2 + H_2S$. This stage of the process can be avoided when the scrubbing solvent is a mixture of sulfolane and diisopropanolamine (see Sulfinol process).

Sulfoxidation: preparation of an aliphatic sulfonic acid, especially those with long chains, by simultaneous reaction of sulfur dioxide and oxygen with a long-chain alkane or cycloalkane, in the presence of UV radiation and an initiator (such as ozone or a peroxy acid). The reaction occurs by a radical mechanism. Saturated alcohols, ethers, nitriles, carboxylic acids and alkyl chlorides can undergo this reaction, but alkenes and aromatics cannot be sulfoxidized by this method.

Sulfoxide: an organic sulfur compound with the general formula R-SO-R'. S. are oxidation products of the sulfides (thioethers) and can therefore be either symmetrical (R = R') or unsymmetrical (R ≠ R'). According to the IUPAC rules, they are to be named as alkyl or aryl sulfinyl derivatives of hydrocarbons, e.g. methyl sulfinylmethane (see Dimethylsulfoxide). S. contain an S=O double bond, and are not very stable. They are soluble in water and, as weak bases, form salts with hydrogen chloride: $(CH_3)_2S{=}O + HCl \rightarrow (CH_3)_2S\text{-}OHCl$.

S. can be reduced to sulfides with lithium aluminum hydride or zinc and hydrochloric acid. Stable sulfones are obtained with oxidizing agents. Unsymmetrical S. can be separated into optical antipodes (trivalent sulfur in a pyramidal structure as the center of chirality). S. are formed from thioethers by oxidation with periodic acid at 0 °C, with dilute nitric acid or hydrogen peroxide in acetic acid. The most important S. is Dimethylsulfoxide (see).

Sulfoxylic acid, formerly ***hyposulfurous acid***: H_2SO_2, a short-lived compound which exists only in aqueous solution. Three tautomeric forms of its derivatives are known:

HO–S–OH (1), H–S(=O)–OH (2), H–S(=O)(=O)–H (3).

S. can be made by hydrolysis of sulfur dichloride according to the equation: $SCl_2 + 2\,H_2O \rightarrow H_2SO_2 + 2\,HCl$. It is an oxidizing agent, and is capable, for example, of converting iodide to iodine and Fe^{2+} to Fe^{3+}. It is reduced to sulfur in the process. With acidic hydrogen sulfite solutions, it forms trithionic acid, $H_2S_3O_3$, and with thiosulfate, it forms pentathionic acid, $H_2S_5O_6$. The Sulfinic acids (see) R-C(O)OH are derived from form (2) of S., and the Sulfones (see), R_2SO_2, from form (3).

Sulfur, symbol ***S***: an element of the 6th main group of the periodic system (see Oxygen-sulfur group). S is a nonmetal with atomic number 16. The masses of the natural isotopes are 32 (95%), 33 (0.76%), 34 (4.22%) and 36 (0.014%). The atomic mass is 32.064; the valences are II, IV and VI. Solid α-S has a hardness of 2 on the Mohs scale, D. 2.07, m.p. 119.0 °C, b.p. 444.674 °C, standard electrode potential (S^{2-}/S) = -0.508 V.

Properties. S. exists in several enantiotropic modifications at different temperatures. The stable form at room temperature is the yellow α-S, which is nearly colorless at low temperatures. It consists of crown-shaped S_8 molecules, ***cyclooctasulfur*** (Fig. 1), which form a rhombic crystal lattice. At 95.4 °C, α-S is converted to the light-yellow, monoclinic β-S, which also consists of S_8 molecules. β-S melts at 119.6 °C.

Cyclooctasulfur, S_8

There are various modifications present in a temperature-dependent equilibrium in liquid S. Immediately above the melting point, the light yellow, mobile melt consists mainly of S_8 molecules (λ-S) in equilibrium with about 5% π-S; this equilibrium is reached after about 12 hours tempering at 120 °C. π-S is a mixture of S_6, S_7 and S_x rings ($x > 8$), and its formation is responsible for the fact that this melt freezes about 5 °C below the melting point of pure β-S. As the temperature increases, S_8 molecules break up and join to form macromolecular chains; this modification is μ-S. The increase in molar mass is accompanied with an increase in the viscosity of the melt which reaches a maximum about 187 °C. As the temperature rises above this point, the viscosity decreases, and the color changes to dark reddish-brown. If the 200 °C melt is quenched, for example by pouring it into cold water, ***plastic S*** is produced. This consists of S_8 molecules and μ-S, and slowly reverts to the rhombic form. Immediately above the boiling point, sulfur vapor is dark red, and consists mainly of S_8 molecules, in addition to S_6, S_4 and S_2 molecules. As the temperature increases, the proportion of S_2 molecules also increases; above 1800 °C they decompose into atoms. In addition to the modifications described above, others can be synthesized which consist of rings containing 6 to 20 S atoms.

α-S is only slightly conductive of heat and electric current. It acquires a large negative charge when rubbed. The solubility of S in carbon disulfide depends on the molecular structure. Although α- and β-

S (cyclooctasulfur) dissolve readily (100 g CS_2 dissolves 24.0 g S at 0 °C, and 181.3 g at 55 °C), μ-S is essentially insoluble. S. is practically insoluble in water, and is only slightly soluble in benzene, ethanol, ether and other organic solvents.

Both the valence (-2, +4, +6) and the coordination numbers (1 through 6) of S. vary considerably. Its tendency to form long, unbranched sulfur chains, for example in polysulfanes and polysulfane disulfonic acids, is notable. S bonding is the result of its $[Ne]s^2p^4$ configuration, which explains its tendency to take up two electrons and form sulfide anions S^{2-}. The two unpaired p-electrons can form two covalent bonds, with which a noble gas configuration is achieved. This covalent bonding exists, e.g., in elemental S, hydrogen sulfide, mercaptans and sulfur dichloride; the molecules are puckered. S also has energetically favorable d levels; the formation of hybrid orbitals involving them readily explains the formation of tetracovalent (e.g. SF_4) and hexacovalent (SF_6) bonding. S double bonds, for example those to oxygen in thionyl or sulfuryl chloride, sulfites, sulfates, etc., must be interpreted on the basis of d-orbital participation in $d_\pi p_\pi$ interactions. Since S is in the second period, its ability to form $p_\pi p_\pi$ bonds puts it in a somewhat unusual position. Its preference for two σ bonds instead of a double bond becomes clear when the molecular structures of oxygen and S are compared. On the other hand, there are many compounds with structures best interpreted as the result of sp^2 hybridization and p_π participation of the p_z orbital of the S. This applies to derivatives in which electronegative substituents bound to the S (particularly oxygen) withdraw negative charge and lead to orbital contraction around the S atom; this stabilizes the double bond (e.g. in SO_2, gaseous SO_3, R-N=S=O, R-N=S=N-R, etc.). It also applies to compounds in which the S atom is a component of a resonance-stabilized system (thiourea, thiocarboxylic acid derivatives, thioketones, numerous S-heterocycles, etc.).

At room temperature, S is relatively unreactive. Around 260 °C, it ignites in air and burns with a weak, blue flame to a mixture of sulfur dioxide, SO_2, and small amounts of sulfur trioxide, SO_3. S. combines with hydrogen around 600 °C to form hydrogen sulfide, H_2S. It reacts with fluorine in the cold to form sulfur hexafluoride, SF_6. Chlorine reacts with molten S to form disulfur dichloride, S_2Cl_2. S forms sulfides with most metals; the reactions are often exothermic. For example, $2\,Al + 3\,S \rightarrow Al_2S_3$. Concentrated sulfuric acid is reduced around 200 °C by S to sulfur dioxide: $2\,H_2SO_4 + S \rightarrow 3\,SO_2 + 2\,H_2O$. S reacts with boiling aqueous ammonia or alkali metal hydroxide solutions to form polysulfides and thiosulfates.

S is not poisonous to human beings. After long exposure to S, the skin becomes somewhat irritated. S. is also not toxic to lower animals and plants, so long as it is not reduced to hydrogen sulfide or oxidized to sulfur dioxide by the organism. The more finely divided the sulfur is, the more readily it is reduced or oxidized. For example, solutions of colloidal S are used to treat fungal diseases and spider mites in vineyards and gardens.

Analytical. Elemental S. can be recognized by its yellow color and by its combustion product, SO_2, which is analysed as sulfite. The Hepar test is used for a general qualitative analysis for sulfur, or the sample is melted with sodium and the S is detected as sulfide. Organically bound S is determined as sulfate (see Elemental analysis).

Occurrence. S. makes up 0.048% of the earth's crust. It occurs naturally in elemental form, as sulfides and as sulfates. Some important minerals are iron pyrite, FeS_2, chalcopyrite, $CuFeS_2$, galenite, PbS, sphalerite (zinc blende), ZnS, anhydrite, $CaSO_4$, gypsum, $CaSO_4 \cdot 2H_2O$, kieserite, $MgSO_4 \cdot H_2O$, barite (heavy spar), $BaSO_4$, mirabilite (Glauber's salt) and $Na_2SO_4 \cdot 10H_2O$. Both bituminous and anthracite coal usually contain 1 to 1.5% S, and petroleum, oil shales, etc. contain varying amounts of S, depending on their origin. S. is a component of the essential amino acids cysteine and methionine and other essential biological compounds, such as pantothenic acid; some plants produce mustard oils.

Production. S is obtained both from deposits of elemental S and from the hydrogen sulfide in petroleum and synthesis gases. The S-containing rock is mined and the S, in some cases after enrichment by flotation, is melted out by heating in a furnace or with superheated steam.

In the early development of the S industry in Italy, round furnaces called *calcaroni* were developed; in them, some of the S was burned to produce the heat needed to smelt the rest. Later, multiple-chamber furnaces, *forni*, were developed in which the chambers could be used either for melting or burning. The first process could extract 50 to 70% of the S in the ore; the second achieved about 80% yield.

Modern mining is based on the *Frasch process*, which can be used in suitable geological formations to extract the elemental S without mining ore. In this process, a triple coaxial pipe is driven into a well drilled to the bottom of the seam (Fig. 2). Superheated steam at about 170 °C is pumped into the well through the outer section of the pipe, which has openings in the upper part of its base to permit the steam to escape into the S seam and melt it.

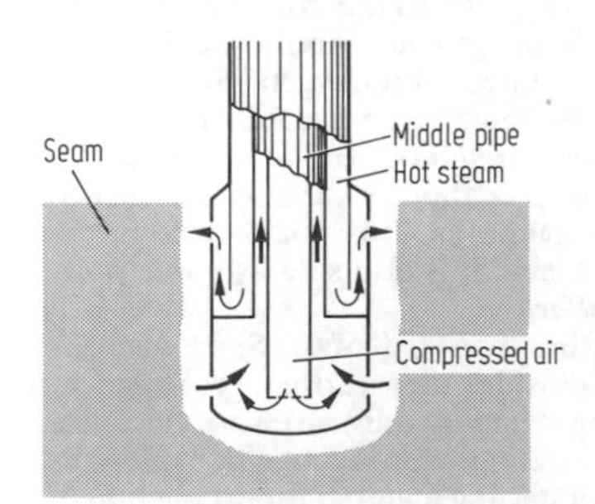

Fig. 2. Base section of a Frasch sulfur pump.

The liquid S collects at the lowest point, and is forced into the middle section of the pipe by compressed air which is pumped into the well through the innermost section of the pipe. Because the middle section of the pipe is jacketed by the steam pipe, the S remains liquid during its passage to the top. It is then collected in large containers, where it solidifies. The product is generally more than 99.5% pure.

The hydrogen sulfide from synthesis and natural gas is removed not only for its economic value, but because it is necessary to prevent air polution when the gas is burned. The H_2S is usually first enriched by an adsorption process, or by scrubbing, and then is converted to sulfur in one of the numerous variants of the *Claus process* (Fig. 3).

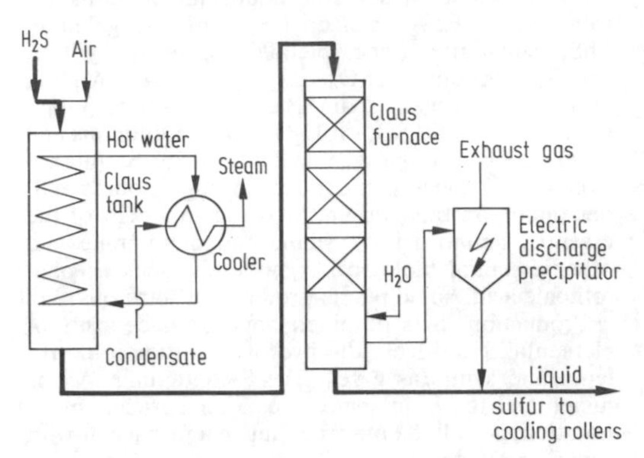

Fig. 3. Schematic diagram of a Claus plant.

Oxygen or air is added to the reaction gas and passed over a bauxite catalyst: $6\,H_2S + 3\,O_2 \rightarrow 6\,S + 6\,H_2O$; ΔH = -1331 kJ mol^{-1}. In the original Claus process, this reaction was carried out in a single step. Because of the large amount of heat released at the catalyst, it was difficult to control, and for that reason, the more recent versions of the process first remove about one third of the H_2S and burn it to SO_2 in the absence of catalyst: $2\,H_2S + 3\,O_2 \rightarrow 2\,SO_2 + 2\,H_2O$; ΔH = -1308 kJ mol^{-1}. The SO_2 is then mixed with the remaining two-thirds of the H_2S and allowed to react in the presence of catalyst: $2\,SO_2 + 4\,H_2S \rightarrow 6\,S + 4H_2O$; ΔH = -293 kJ mol^{-1}. Iron oxide hydrate is also used to separate the H_2S from synthetic or natural gas; it is converted to iron sulfide: $Fe_2O_3 + 3\,H_2S \rightarrow Fe_2S_3 + 3\,H_2O$. When the iron oxide hydrate has been exhausted, the product is removed from the gas filter and regenerated with oxygen: $Fe_2S_3 + 1.5\,O_2 \rightarrow Fe_2O_3 + 3\,S$. The S. is extracted with carbon disulfide or ammonium sulfide solution. The S. obtained by this method is about 99.5% pure, which is sufficient for most purposes. It is usually sold in the form of rods. Sublimed S. is also sold as a fine powder, ***flowers of sulfur***.

Applications. More than 90% of the S. produced is burned to sulfur dioxide, then further processed to sulfuric acid. Large amounts are used to vulcanize rubber. S. is also used in the production of explosives and matches. It is a starting material for production of carbon disulfide, dyes and pharmaceuticals, and is used as a pesticide, especially against mildew in vineyards. In medicine, colloidal S. is used in salves for various skin diseases. The sulfur dioxide generated by burning S is used as a disinfectant for wine barrels, beer bottles, etc. S. is used as a binder in sulfur concrete.

Historical. S was known in antiquity. The Egyptians used the smoke of burning S to bleach cloth. With the introduction of gunpowder, S became very important. It has been used since the end of the 18th century for production of sulfuric acid.

Sulfuration: same as Sulfonization (see).

Sulfur bromides. ***Disulfur dibromide***: S_2Br_2, an unstable, dark red liquid, M_r 223.95, density 2.63, m.p. -40 °C. It readily dissolves sulfur, forming ***dibromosulfanes, dibromopolysulfanes*** or ***bromosulfanes***, Br-S_x-Br (x = 2-24). These are red, oily liquids which decompose easily. S. react with water to form hydrogen bromide, sulfur dioxide and sulfur.

Sulfur chlorides: ***Disulfur dichloride***, S_2Cl_2, yellow-orange liquid with a repulsive odor. It fumes in moist air and is a lacrimator; M_r 135.03, density 1.678, m.p. -80 °C, b.p. 135.6 °C. S_2Cl_2 reacts slowly with water to form hydrogen chloride, sulfur dioxide and sulfur. It is made by chlorination of sulfur: $2\,S + Cl_2 \rightarrow S_2Cl_2$. It is used in cold vulcanization of rubber, as a chlorination reagent and in the production of factice (a rubber substitute). S_2Cl_2 dissolves in elemental sulfur, forming ***dichlorosulfanes, dichloropolysulfanes, chlorosulfanes***, Cl-S_x-Cl with chain lengths varying from x = 2 to 100. The reaction of chlorine with S_2Cl_2 yields ***sulfur dichloride***, SCl_2, a dark red liquid, M_r 102.97, density 1.621, m.p. -78 °C, b.p. 59 °C. At room temperature, this slowly decomposes to S_2Cl_2 and Cl_2. Further chlorination of SCl_2 at -80 °C yields the yellow, solid ***sulfur tetrachloride***, SCl_4, M_r 173.88, which is stable only at low temperatures. It probably has the structure $[SCl_3]^+Cl^-$; it melts at - 30 °C and then decomposes to SCl_2 and Cl_2.

Sulfur dyes: a term for a large class of synthetic dyes obtained by treating aromatic amines, phenols and aminophenols with sulfur and/or sodium polysulfide in the absence (baked dyes) or presence of solvent, such as water, ethanol, etc. (boiled dyes). The most common structural element in the baked dyes is the benzthiazole group (Fig. 1). Most of the baked dyes are yellow, orange or brown, for example ***immedial yellow GG*** (a mixture of x = 2, y = 0 or 2, n = 0). The boiled dyes are blue, green, violet and black, and most are derivatives of phenylthiazones, phenazones and phenoxazones, e.g. ***pyrogen indigo*** (Fig. 2; R^1 = phenylamine, R^2 = H, n = 0 to 3), ***immedial brilliant blue CL*** (R^1 = dimethylamine, R^2 = H, n = 0 to 3) and ***immedial black V*** (R^1 = R^2 = NH_2, n = 0). The structure, color and fastness of the S. depend greatly on the reaction conditions. The S. are very fast to light and washing, but not to chlorine. They are used mainly to dye cotton and other plant fibers in a sodium sulfide bath. A subsequent treatment with metal salts can improve the quality of the dyeing. The sodium sulfide solution (which is made alkaline by addition of soda) reduces the disulfide bridges in the S. to sulfhydryl groups, making it water-soluble. When the dyed cloth is hung in air, or oxidized with potassium dichromate or sodium perborate, the disulfide bonds are reformed, and the dye is fixed to the fibers. Under special precautions, the S. can also be used to dye wool and silk.

The most important of the S. and the synthetic dye produced in largest amounts is ***sulfur black***, which is made from dinitrophenol and polysulfides. Other S. are ***sulfur blue*** (from chlorodinitrohydroxydiphenylamine), ***sulfur indigo*** (from 4-aminophenol and phenol), ***thione green B*** (from 4-hydroxyphenylthiourea), ***thione violet*** (from chlorodinitrobenzene,

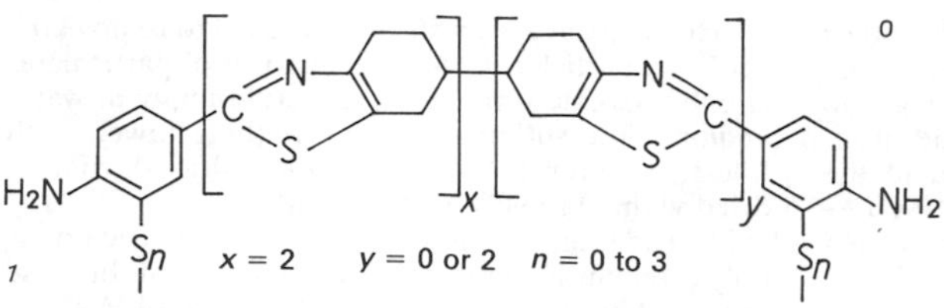

H_2N … S_n $x = 2$ $y = 0$ or 2 $n = 0$ to 3

Fig. 1. Benzthiazole group of the baked dyes.

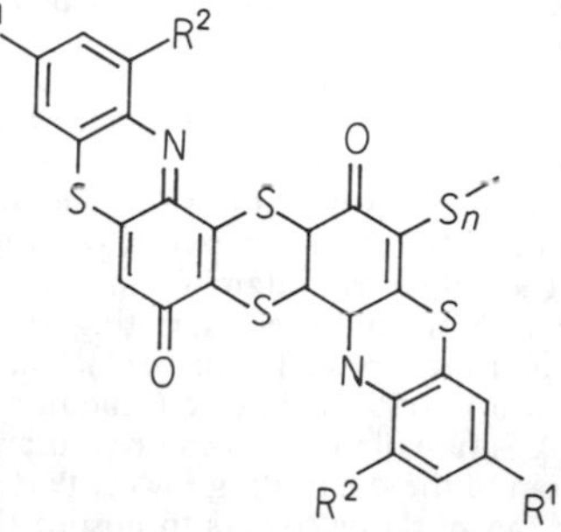

Fig. 2. Basic structure of the boiled dyes.

4-aminophenol, 4-nitrosodimethylaniline) and ***thiogen dark red*** (from 2,4-dinitro-2-hydroxydiphenylamine). The hydrone and immedial dyes are also S.

Sulfur fluorides: the most important S. is ***sulfur hexafluoride***, SF_6, a colorless and odorless gas which is slightly soluble in water; M_r 146.05, m.p. -51 °C (at 2.3 MPa), subl.p. -63.8 °C. SF_6 is both chemically and thermally very stable; its unreactivity is due to steric hindrance around the sulfur and the nonpolarity of the octahedral molecule. It is obtained by an exothermal reaction of fluorine with sulfur: $S + 3 F_2 \rightarrow SF_6$. Because it resists spark conduction, it is used as an insulating gas in high-voltage equipment. ***Sulfur tetrafluoride***, SF_4, is a colorless, reactive gas, M_r 108.06, m.p. -124 °C, b.p. -40 °C. In water it is immediately hydrolysed to sulfur dioxide and hydrogen fluoride. It is made by the reaction of sulfur dichloride with sodium fluoride: $3 SCl_2 + 4 NaF \rightarrow SF_4 + S_2Cl_2 + 4 NaCl$. It is used as a fluorination agent. There are two known isomers of ***sulfur difluoride***, S_2F_2, M_r 102.12. The first is $S{=}SF_2$, m.p. -165 °C, b.p. -10.6 °C, which can be obtained from S_2Cl_2 and KF, and the second is ***disulfur difluoride***, F-S-S-F, m.p. -133 °C, b.p. -1 °C; it can be obtained from silver difluoride and sulfur. ***Disulfur decafluoride***, S_2F_{10}, is a colorless, very poisonous liquid, M_r 254.11, density 2.08, m.p. -92 °C, b.p. 20 °C. Like SF_4, it is a by-product of SF_6 production from sulfur and fluorine, and can be isolated from the product mixture by fractional distillation. ***Sulfur difluoride***, SF_2, is an unstable blue compound.

Sulfuric acid: H_2SO_4, one of the most important and strongest inorganic acids. Pure S. is an oily, colorless and odorless liquid, density 1.83213, m.p. 10.36 °C. Pure, that is 100%, S. is not usually commercially available. Commercial S. comes in three concentrations: concentrated, 98% H_2SO_4, density 1.841, m.p. 3 °C, b.p. 338 °C; dilute S. (about 1 M), about 10% H_2SO_4, density 1.06 to 1.111; and normal S., 4.904% H_2SO_4, density 1.031. Fuming S. is also called Oleum (see). The H_2SO_4 content of a solution is usually determined by measuring the density and locating the corresponding concentration from a table. However, for concentrations over 90%, determination with a hydrometer is imprecise.

Density (at 20° C)	%	Density (at 20° C)	%
1.000	0.2609	1.425	53.01
1.025	4.000	1.450	55.45
1.050	7.704	1.475	57.84
1.075	11.26	1.500	60.17
1.100	14.73	1.525	62.45
1.125	18.09	1.550	64.71
1.150	21.38	1.575	66.91
1.175	24.58	1.600	69.09
1.200	27.72	1.625	71.25
1.225	30.79	1.650	73.37
1.250	33.82	1.675	75.49
1.275	36.78	1.700	77.63
1.300	39.68	1.725	79.81
1.325	42.51	1.750	82.09
1.350	45.26	1.775	84.61
1.375	47.92	1.800	87.69
1.400	50.50	1.825	92.25

Pure S. is obtained by dissolving the calculated amount of sulfur trioxide, SO_3, in conc. S. 100% S. cannot be distilled; when heated, it produces a vapor with a high content of sulfur trioxide, until a constant-boiling mixture of 98.3% S. and 1.7% water is obtained at 338 °C. This mixture is also obtained if less concentrated solutions of S. are evaporated. If S. vapor is heated above 338 °C, it begins to decompose to sulfur trioxide and water vapor. At 450 °C, the dissociation equilibrium is almost completely on the right side: $H_2SO_4 \rightleftharpoons SO_3 + H_2O$. If excess sulfur trioxide is added, disulfuric acid is formed: $H_2SO_4 + SO_3 \rightarrow H_2S_2O_7$. S. forms several hydrates with water, releasing considerable amounts of heat in the process: $H_2SO_4 \cdot H_2O$ (m.p. 8.5 °C), $H_2SO_4 \cdot 2H_2O$ (m.p. -38 °C), $H_2SO_4 \cdot 4H_2O$ (m.p. -27 °C), $H_2SO_4 \cdot 6H_2O$ (m.p. -54 °C) and $H_2SO_4 \cdot 8H_2O$ (m.p. -62 °C). The affinity of S. for water is so great that S. splits water out of many organic substances, such as sugars, and thus carbonizes and destroys them. The carbonization due to the water-withdrawing effect is supported by its oxidative capacity. S. can thus dissolve metals which have higher positive normal potentials than hydrogen, such as copper, silver and mercury. Silver, for example, is first oxidized to silver oxide, and then dissolved as silver sulfate: $2 Ag + H_2SO_4 \rightarrow Ag_2O + H_2O + SO_2$; $Ag_2O + H_2SO_4 \rightarrow Ag_2SO_4 + H_2O$. Nonmetals such as sulfur, phosphorus and carbon are also oxidized, sulfur, for example, to sulfur dioxide: S + 2

$H_2SO_4 \rightarrow 2\ H_2O + 3\ SO_2$. Dilute S., on the other hand, is not a strong oxidizing agent, and can only dissolve those metals which have a more negative standard potential than hydrogen, such as lead, iron, zinc and aluminum. The metal is converted to the sulfate with generation of hydrogen: $Zn + H_2SO_4 \rightarrow ZnSO_4 + H_2$. The salts and esters of S. are called Sulfates (see).

> Because of the large amount of heat released when sulfuric acid and water are mixed, the acid should always be added to the water, with stirring, never water to acid. If water is poured into conc. sulfuric acid, it is immediately vaporized by the heat of solution, and causes the very caustic acid to splatter.

Analytical. With barium ions, S. produces a white precipitate of barium sulfate which is insoluble in water and acids: $Ba^{2+} + H_2SO_4 \rightarrow BaSO_4 + 2\ H^+$. Quantitative determination is done gravimetrically, using barium sulfate which is heated to 600 °C and then weighed, or volumetrically, by titration with sodium hydroxide solution using methyl orange as indicator.

Production. All processes involve the following steps: 1) production of gases containing sulfur dioxide; 2) purification of the gas, 3) oxidation of the sulfur dioxide to sulfur trioxide, and absorption of the sulfur trioxide.

1) Production of sulfur dioxide. a) From elemental sulfur. Solid or premelted sulfur is burned with atmospheric oxygen in a rotating furnace: $S + O_2 \rightarrow SO_2$, $\Delta_R H = -239$ kJ mol^{-1}. The gas then passes into a cooling vessel where the high heat of reaction is used to generate steam. b) From sulfide ores. Roasting of sulfidic ores, especially iron pyrite, FeS_2, yields metal oxides and sulfur dioxide. Depending on the amount of air allowed into the reaction chamber, the roasting of pyrite follows one of the following equations: $2\ FeS_2 + 5.5\ O_2 \rightarrow Fe_2O_3 + 4\ SO_2$, $\Delta_R H = -1720$ kJ mol^{-1}; or $3\ FeS_2 + 8\ O_2 \rightarrow Fe_3O_4 + 6\ SO_2$, $\Delta_R H = -2553$ kJ mol^{-1}. The filter masses used to remove hydrogen sulfide from heating gases or to purify synthesis gas with zinc oxide can be roasted in the same way as the sulfidic ores. c) From anhydrite (anhydrous gypsum). The gypsum is treated by the *gypsum-sulfuric acid process (Müller-Kühne process)*, in which it is heated to 1400 °C in a rotating furnace with coal, sand and clay: $CaSO_4 + C \rightarrow CaO + SO_2 + CO$; $\Delta_R H = +390$ kJ mol^{-1}. The resulting calcium oxide, sand and clay are ground finely to yield a high-quality Portland cement. For each ton of sulfur dioxide, about 1.2 ton Portland cement is formed. d) From natural gas, coal or petroleum refining. The hydrogen sulfide is removed from the fuels by gas scrubbing and subsequent Claus process (see Sulfur). Depending on the process, a gas consisting of 7 to 12% sulfur dioxide is obtained.

2) Gas purification. The gas from combustion of sulfur is very pure and can be oxidized directly, but the gases from the roasting or gypsum-sulfuric acid process must first be purified. Particulates are first removed in cyclones and settling chambers, after which electrical gas purification is used to remove the very fine particles. Complete removal of particulates is only possible through a wet scrubbing with water and/or dilute sulfuric acid in scrubbing towers with fillers. To prevent corrosion, the sulfur dioxide is dried with 70 to 90% sulfuric acid.

3) Oxidation of sulfur dioxide to sulfur trioxide is only economical in the presence of a catalyst, because the equilibrium of $SO_2 + 1/2\ O_2 \rightleftharpoons SO_3$, $\Delta_R H = -95$ kJ mol^{-1} is approached rapidly only at high temperatures. At such temperatures, however, it lies far to the left. At the low temperatures which favor the products at equilibrium, a catalyst must be used to make the reaction sufficiently rapid to be economical. Two processes are used industrially, the tower or nitrosyl process and the contact process.

In the *tower process*, a mixture of nitrogen monoxide and nitrogen dioxide acts as an oxygen-transfer catalyst (Fig. 1). The hot gases containing sulfur dioxide flow together with air into the denitrating tower (1) from below, and are brought into contact here with the nitrosylsulfuric acid coming from the nitrating towers (5,6). A S. free of nitrogen oxides is drawn off from the bottom of the denitrating tower. Part of the S. is led into the nitrating towers to absorb the nitrosyl hydrogensulfide gases. The sulfur dioxide-containing gas and the nitrogen oxides driven off in the denitrating tower enter the production towers (2,3,4). About 70% of the sulfur dioxide is oxidized in tower 2, and the remainder in towers 3 and 4. Towers 3 and 4 also serve as buffers against variations in the production process. Water is sprayed into the production towers and nitric acid is added to replenish the nitrogen oxides which are lost. In the nitrating towers (5,6), the nitrogen oxides are oxidized with atmospheric oxygen and react with the S. brought in from the denitrating tower (1) to form nitrosyl hydrogensulfate. This is led into the denitrating tower and production tower 2. The following reaction equations are significant in the production of S. by the tower process: $2\ NO + O_2 \rightleftharpoons 2\ NO_2$; $SO_2 + H_2O \rightleftharpoons H_2SO_3$; $NO + NO_2 + H_2O \rightleftharpoons 2\ HNO_2$; $NO + NO_2 + 2\ H_2SO_4 \rightleftharpoons 2\ NOHSO_4 + H_2O$; $H_2SO_3 + 2\ HNO_2 \rightleftharpoons H_2SO_4 + H_2O + 2\ NO$. Since a certain amount of water is required for the tower process, it can generate at most 80% S., which is also not very pure. It is used mainly to solubilize phosphates to make superphosphate, because here the water content is desirable, and for production of ammonium sulfate. The tower process can be used to work up gases containing low or variable concentrations of sulfur dioxide, including those obtained as byproducts of metallurgical processes. About 20% of the world production of S. is currently produced by the tower process. The tower process is an improvement on the *lead chamber process*, which is no longer used; in it too, nitrogen oxides were used as oxygen carriers. A special form of the tower process is the *Kachkaroff process*, in which S. and nitric acid are produced simultaneously in a ratio of 2:1.

In the *contact process*, gases containing sulfur dioxide are preheated to about 350 °C, and led into the contact furnace. The catalyst here is usually vanadium(V) oxide, V_2O_5, with a certain amount of pumice, silica gel or zeolite as a carrier. Less frequently, finely divided platinum or iron(III) oxide,

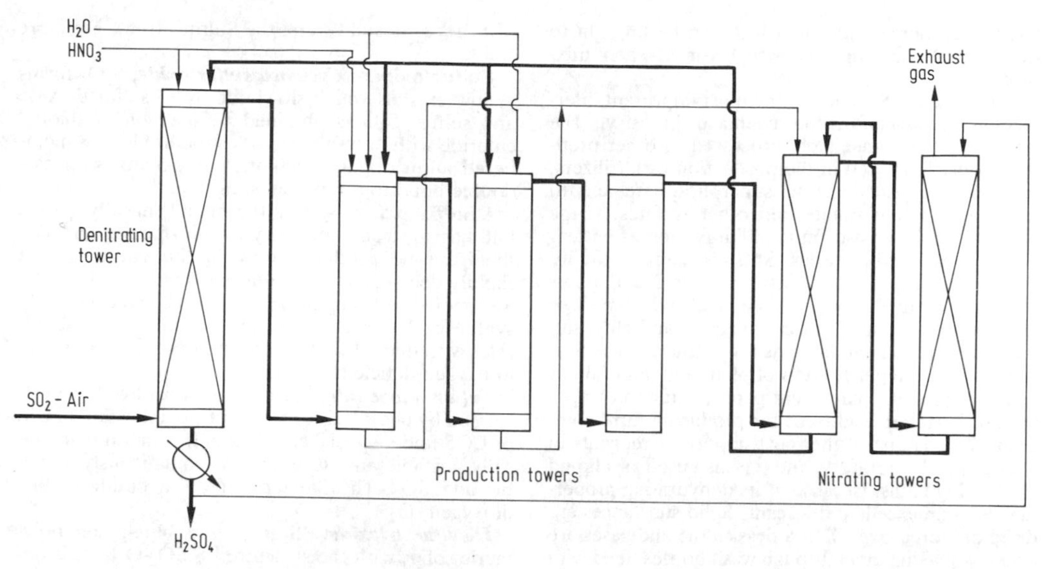

Fig. 1. Production of sulfuric acid by the tower process.

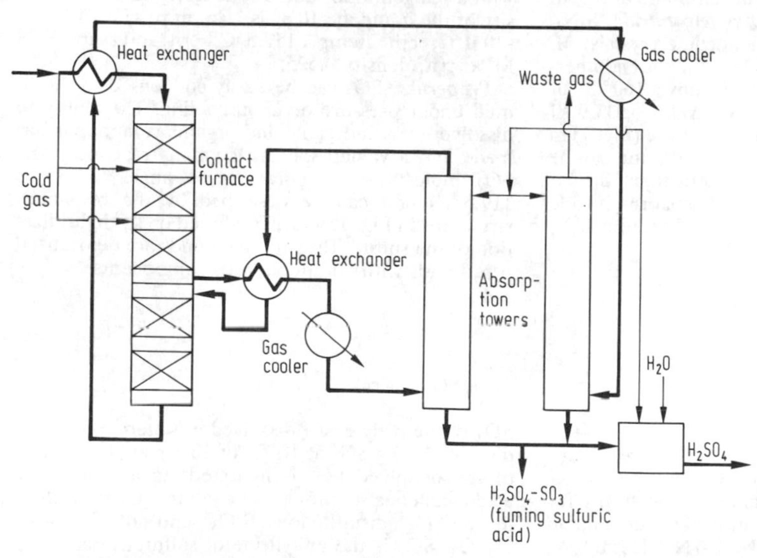

Fig. 2. Production of sulfuric acid by the double contact process.

Fe_2O_3, is used as catalyst. The contact mass is held in vertical pipes or spread on screens; in modern plants, the two arrangements are combined. The gas containing sulfur dioxide flows through the contact furnace, and the sulfur dioxide is oxidized at 425 to 600 °C, depending on the type of catalyst: $2\,SO_2 + O_2 \rightarrow 2\,SO_3$; $\Delta_R H = -190$ kJ mol^{-1}. The hot gas leaving the contact furnace preheats the cold gas in heat exchangers, and is then further cooled in pipe coolers with air or water. It then enters trickle towers, where it is absorbed first with oleum (oleum absorber), then with 98% S. (S. absorber). Some of the SO_3 is bound as disulfuric acid. The high-percent oleum formed in the oleum absorber is drawn off as such, or, like the lower percent oleum formed in the S. absorber, it can be converted by a measured inflow of 70-78% washing acid or by water to 98% S.

In the *double contact process* (Fig. 2), there is an intermediate absorption of the sulfur trioxide. This removes the sulfur trioxide from the reaction mixture as it passes through several contact layers, and permits higher percent conversion in subsequent layers. This method also contributes significantly to reduced emissions to the atmosphere. The contact process yields high-percent oleum, which can be brought to the desired concentration with S. or S.-water mixtures.

Applications. S. is one of the most important intermediate products of the chemical industry. The largest part is immediately processed to other products. Most S. is used in the production of fertilizers, to solubilize phosphate to superphosphate and to make ammonium sulfate and other sulfates. Large amounts of S. are used in metallurgy, and as battery acid for lead-acid batteries. S. is also used in the industrial production of hydrofluoric acid and phosphoric acid. In the organic chemical industry, S. is used in the production of caprolactam and alkyl sulfonates for precipitation baths for synthetic silks. S. is also used in the purification of plant and mineral oils and fats, to make parchment paper, and as a component of Nitrating acid (see) for producing nitro compounds. S. is one of the most important reagents in the chemical laboratory, and it is also used as a liquid for heating baths. Because of its dehydrating properties, S. is an excellent dessicant. Solid substances are dried or stored over S. in a dessicator, and gases are dried by passing them through wash bottles filled with S.

Sulfur indigo: same as Thioindigo (see).

Sulfur nitrides: compounds of sulfur and nitrogen. The most important is ***tetrasulfur tetranitride***, S_4N_4, which forms yellow-orange, monoclinic crystals; M_r 184.28, density 2.22, subl.p. 179 °C. S_4N_4 is endothermal and decomposes explosively above 160 °C, or upon impact: $S_4N_4 \rightarrow 4\ S + 2\ N_2$, ΔH = -533.9 kJ mol^{-1}. The molecules have a cage structure (Fig. 1) in which all S-N bonds are of equal length, and can be described in terms of S-N-S $d_\pi p_\pi$ interactions; the situation is similar to that in the phosphorus nitride dichlorides. Tin(II) chloride in ethanol reduces S_4N_4 to tetrasulfur tetraimide $(\text{-NH-S-})_4$.

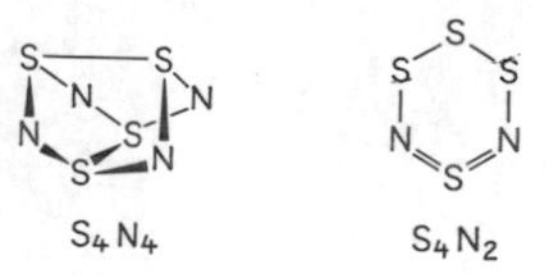

Hydrolysis produces ammonia and a mixture of various thio- and oxoacids of sulfur. S_4N_4 can be synthesized either by a reaction of ammonia with disulfur dichloride, or by disproportionation of sulfur in liquid ammonia: $10\ S + 10\ NH_3 \rightarrow S_4N_4 + 6\ NH_4HS$. If S_4N_4 vapor is passed through a tube filled with fine silver wire at low pressure and 300 °C, it forms the extremely unstable ***disulfur dinitride***, S_2N_2. This compound forms colorless crystals; its molecules consist of four-membered, planar rings with alternating S and N atoms. If S_4N_4 is heated with carbon disulfide and sulfur in an autoclave to 120 °C, one of the products is ***tetrasulfur dinitride***, S_4N_2, which forms dark red crystals; M_r 156.27, density 1.901, m.p. 23 °C. Its molecular structure is probably that shown in Fig. 2.

Sulfurous acid: H_2SO_3, a dibasic acid, pK^+ = 1.8, K_2 = 7.0. S. can be detected only in aqueous solution, as part of the equilibrium $SO_2 + H_2O \rightleftharpoons H_2SO_3$, which, however, lies far to the left. The rapid establishment of the above equilibrium prevents isolation of S. Its salts and esters, the Sulfites (see), are more stable.

Sulfur oxides: ***Cyclooctasulfur oxide***, S_8O, forms yellow crystals which slowly decay to sulfur dioxide and sulfur. S_8O is obtained by reaction of thionyl chloride with polysulfanes. Its structure is an 8-membered sulfur ring in which one of the S atoms forms a double bond to an oxygen substituent.

Disulfur oxide: S_2O, is thermodynamically stable, but has a strong tendency to polymerize and can therefore be produced only in the gas phase. It rapidly decomposes to sulfur and sulfur dioxide. The S_2O molecule is angular, like the SO_2 molecule. It is synthesized, for example, by reaction of metal sulfides with thionyl chloride, or by combustion of sulfur in oxygen-deficient conditions.

Sulfur monoxide, SO, is a short-lived species formed by photolysis of SO_2, or by combustion of CS_2 or COS under special conditions. It is thermodynamically unstable and decomposes spontaneously to sulfur and SO_2. The SO molecule is a biradical, like dioxygen, O_2.

Disulfur trioxide, $(S_2O_3)_x$, is a blue-green polymeric solid, with the structure $[\text{-S-SO}_2\text{-O-}]_x$. It is obtained by mixing sulfur with liquid SO_3.

Sulfur dioxide, SO_2, is a colorless, poisonous gas with a pungent odor and a sour taste; M_r 64.06, density of the liquid at -10 °C is 1.46, m.p. -75.5 °C, b.p. -10.0 °C, crit. temp. 157.5 °C, crit. pressure 7.88 MPa, crit. density 0.525.

Properties. SO_2 can be easily condensed to the liquid under pressure or upon cooling. The ability to dissolve many inorganic and organic compounds has made SO_2 a valued solvent for many reactions. The SO_2 molecule is angular (bond angle O-S-O is 119.5 °C) and can be described by the resonance structures in Fig. 1, which are based on sp^2 hybridization of the sulfur. The multiple bonds are delocalized $p_\pi p_\pi$ bonds with additional $p_\pi d_\pi$ components.

Fig. 1. SO_2 molecule.

SO_2 is relatively easily dissolved in water: 100 g H_2O dissolve 15.4 g SO_2 at 10 °C, or 10.5 g at 20 °C. Most of the dissolved SO_2 is hydrated. In addition, the acidic aqueous solutions also contain hydrogensulfite ions, HSO_3^-, disulfite ions, $S_2O_5^{2-}$ and sulfurous acid, H_2SO_3. SO_2 is the anhydride of sulfurous acid, and when it dissolves in metal hydroxide solutions, it forms the corresponding Sulfites (see). As a derivative of sulfur in the +IV oxidation state, it is redox amphoteric. However, it is able to oxidize only very strong reducing agents. For example, it reacts with hydrogen sulfide to form sulfur ($SO_2 + 2\ H_2S \rightarrow 3\ S + 2\ H_2O$), and magnesium burns in SO_2 ($2\ Mg + SO_2 \rightarrow 2\ MgO + S$). The reducing effect of SO_2 is stronger. For example, it reduces iodine to iodide ($SO_2 + I_2 + 2\ H_2O \rightarrow H_2SO_4 + 2\ HI$), chlorate to chloride ($ClO_3^- + 3\ SO_2 + 3\ H_2O \rightarrow Cl^- + 3\ H_2SO_4$) or permanganate in acidic solution to Mn^{2+} ($2\ MnO_4^- + 5\ SO_2 + 2\ H_2O \rightarrow 2\ Mn^{2+} + 5\ SO_4^{2-} + 4\ H^+$). The thermodynamically favored oxidation of SO_2 to sulfur trioxide by oxygen ($2\ SO_2 + O_2 \rightarrow 2\ SO_3$, ΔH = -198 kJ mol^{-1}) occurs extremely slowly at room temperature, but can be greatly accelerated by catalysts (Pt, V_2O_5); this is the basis for the contact process for production of sulfuric acid. Because of its strong reducing effect, SO_2 bleaches many organic pigments. Lower organisms are prevented from growing or are killed by SO_2. The reaction of SO_2 with phosphorus pentachloride yields thionyl chloride ($PCl_5 + SO_2 \rightarrow POCl_3 + SOCl_2$), and it combines with chlorine to form sulfuryl chloride ($SO_2 + Cl_2 \rightarrow SO_2Cl_2$). With sufficiently active organometal compounds, SO_2 reacts by insertion into the metal-carbon bond. Due to the free electron pair on the sulfur, SO_2 can act as a ligand in transition metal complexes.

Sulfur dioxide is irritating to the mucous membranes, especially those of the respiratory tract. At a concentration of 400 to 500 ppm in air, SO_2 can produce life-threatening conditions within a few minutes. An antidote is inhalation of water vapor or finely atomized 0.5% sodium hydrogencarbonate solution.

Analysis. SO_2 can be recognized qualitatively by its characteristic odor and its ability to bleach iodine or permanganate solution. For quantitative measurements, the gas is passed through an iodine solution and the excess iodine is then back-titrated. Determination of SO_2 in roasting exhaust, stack gases, etc. is usually done continuously with automatic equipment which measures its characteristic absorption in the UV and IR ranges.

Occurrence. SO_2 occurs naturally in volcanic gases and natural gas. Enormous amounts of it enter the atmosphere as a result of burning fossil fuels and various industrial processes. Either as the gas, or dissolved in water as "acid rain", SO_2 causes serious damage to the biosphere. The death of forests in recent years is ascribed mainly to SO_2, along with other factors. The facades of buildings are also attacked by SO_2 in the atmosphere. Overall, SO_2 is probably the most significant environmental pollutant.

Production. In the laboratory, SO_2 can be made by burning sulfur in a stream of oxygen ($S + O_2 \rightarrow SO_2$), by treating sulfite solutions with sulfuric acid (e.g., $Na_2SO_3 + 2\,H_2SO_4 \rightarrow 2\,NaHSO_4 + SO_2 + H_2O$), or by reduction of sulfuric acid with copper ($2\,H_2SO_4 + Cu \rightarrow CuSO_4 + SO_2 + 2\,H_2O$. On an industrial scale, SO_2 is an intermediate in the production of Sulfuric acid (see). The SO_2 used in these syntheses is obtained by combustion of sulfur or hydrogen sulfide-containing gases, from roasting of sulfide ores, or by reductive cleavage of calcium sulfate. Pure SO_2 is obtained by condensation of gas mixtures at -70 °C and 5 MPa pressure, or by absorption into alkaline aqueous solutions, from which SO_2 is released again by increasing the temperature.

Applications. SO_2 is used mainly in the production of sulfuric acid. It is also a starting material in the production of some industrially important salts, such as sulfites, disulfites, dithionites and thiosulfates. It is used in organic synthesis for sulfoxidation and sulfochlorination. Liquid SO_2 is used as a coolant and in the Edeleanu process of refining petroleum. It is used to fumigate wine and beer barrels and kegs, bottles and even entire rooms, to control vermin and to preserve foods.

Sulfur trioxide, SO_3, M_r 80.06, exists in several solid modifications. Two well-defined modifications are the orthorhombic, cyclic trimeric, *ice-like* γ-SO_3, density 1.920, m.p. 16.8 °C, b.p. 44.8 °C, and the polymeric *asbestos-like* SO_3, which forms infinite chains. Two forms of asbestos-like SO_3 are known: α-SO_3, m.p. 62.2 °C, and β-SO_3, m.p. 30.5 °C. The ends of these molecules are -OH groups, so that α- and β-SO_3 are not, in the strict sense, SO_3 modifications, but mixtures of long-chain polysulfuric acids. In both forms (Fig. 2), the sulfur is in a tetrahedral configuration (sp^3 hybridized).

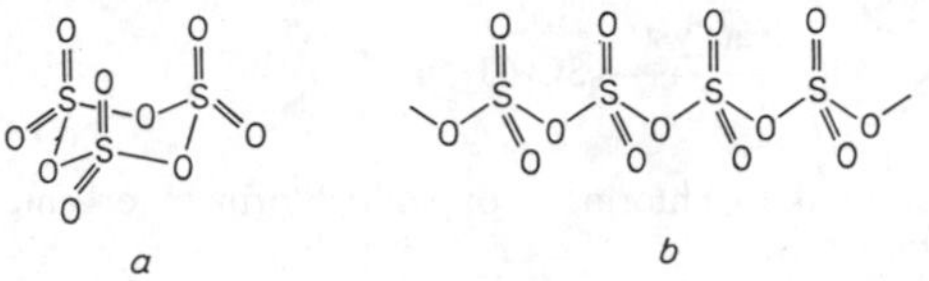

Fig. 2. Molecular structure of ice-like (a) and asbestos-like (b) SO_3.

In the gas phase, SO_3 is monomeric and forms trigonal planar molecules, which can be described as resonance hybrids with $p_\pi p_\pi$ and $p_\pi d_\pi$ bonds between the sulfur and oxygen; some possible limiting formulas are shown in Fig. 3.

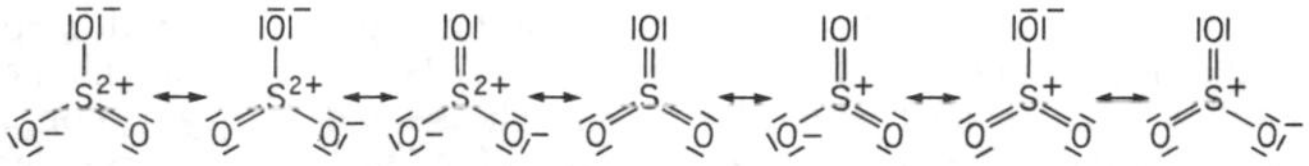

Fig. 3. Molecular structure of gaseous SO_3.

SO_3 is the anhydride of sulfuric acid, and combines with water in a vigorous, highly exothermal reaction to form the acid ($SO_3 + H_2O \rightarrow H_2SO_4$, ΔH = -95.6kJ mol^{-1}). Many organic compounds, e.g. carbohydrates, are dehydrated by SO_3. As a strong Lewis acid, it forms crystalline adducts with various electron-pair donors (e.g. pyridine, aliphatic amines, dioxane). It has a pronounced oxidizing effect, converting SCl_2 to $SOCl_2$, P_4 to P_4O_{10} and PCl_3 to $POCl_3$. It reacts with sulfur to form sulfur dioxide ($2\,SO_3 + S \rightarrow 2\,SO_2$). SO_3 is usually made in the laboratory by heating fuming sulfuric acid, by dehydrating sulfuric acid with P_4O_{10}, or by heating disulfates (e.g., $Na_2S_2O_7 \rightarrow Na_2SO_4 + SO_3$). Catalytic oxidation of sulfur dioxide to sulfur trioxide with oxygen is an important step in the industrial production of Sulfuric acid (see). SO_3 is used mainly as a sulfonating reagent.

Sulfur tetroxide, $(SO_4)_x$ and ***disulfur heptoxide***, $(S_2O_7)_x$ are peroxo compounds of sulfur. They are solid, polymeric substances formed by the action of electric discharges on a mixture of SO_2 or SO_3 and oxygen. They are derived structurally from asbestos-like SO_3 in that the S-O-S bridges are replaced by S-OO-S bridges; in SO_4 all the bridges are peroxide bridges, and in S_2O_7, half of them are peroxides. Both peroxides are strong oxidizing agents. They react with water, forming among other things peroxomonosulfuric acid, H_2SO_5, and peroxodisulfuric acid, $H_2S_2O_8$.

Sulfuryl chloride: SO_2Cl_2, the dichloride of sulfuric acid; a colorless liquid which fumes in moist air and has a suffocating odor; the molecular structure is tetrahedral; M_r 134.97, density 1.6674, m.p. -54.1 °C, b.p. 69.1 °C. When it stands for a long time, S. decomposes partially to sulfur dioxide and chlorine. The dissolved Cl_2 gives old S. a yellow color. S. reacts with an equimolar amount of water to form chlorosulfuric acid and hydrogen chloride: $SO_2Cl_2 + H_2O \rightarrow HSO_3Cl + HCl$. Excess water decomposes it to sulfuric acid and hydrogen chloride. S. is synthesized by combination of sulfur dioxide with chlorine in sunlight, or in the presence of camphor or activated charcoal as catalyst:

$$SO_2 + Cl_2 \xrightarrow{\text{(catalyst)}} SO_2Cl_2.$$

It is used to chlorinate or sulfochlorinate organic compounds.

Sulfuryl group: see Sulfonyl group.

Sump phase: see Coal hydrogenation.

Superacid: an acid with an extremely high proton activity (see Acid-base concept, section on Brönsted definition) which is able to convert even an extremely weak base into its protonated form. This definition is based on the natural assumption that the reaction medium has an extremely low basicity. Some superacidic media are the hydrogen fluoride/antimony pentafluoride and the fluorosulfuric acid/antimony pentafluoride systems. In the first of these, the H_2F cation produced by the reaction $SbF_5 + 2\,HF \rightleftharpoons H_2F^+$ acts as an acid with $pK_a = -17$. The second system is also called ***magic acid***; the cation $H_2SO_3F^+$ produced in the equilibrium $SbF_5 + 2\,HSO_3F \rightleftharpoons H_2SO_3F^+ + F_5SbOSO_2F^-$ ($pK_a = -20$) is such a strong acid that it is able to protonate even saturated hydrocarbons, such as methane, to generate carbonium ions.

Superalloys: multicomponent alloys based on nickel or cobalt, and containing a large amount of chromium. Smaller amounts of molybdenum, tungsten, aluminum and titanium are added, the latter two to make the metals hardenable. The nickel-based alloys with 50 to 80% nickel contain various carbide and intermetallic phases, which are present in a finely divided form in a nickel-rich nickel-cobalt-chromium crystal matrix. The Nimonic® (see) type alloys also contain nickel as the major component. The thermally resistant S. are made as casting and kneading alloys, and are used as blades in gas turbines, as components in the combustion chambers of rocket engines, in petroleum cracking plants and in other types of chemical installation.

Supercooled liquids: see Amorphous.

Superfiltration: a process based on the principle of Reverse osmosis (see), used mainly to desalinate water. The water is pressed through a specially prepared membrane, e.g. of cellulose acetate or other plastic, which is impermeable to the dissolved salt. The applied pressure must be greater than the osmotic pressure of the salt solution ("reverse osmosis").

S. has so far been used only to desalinate water with lower salt concentrations than are present in seawater (brackish water); the desalination of seawater by S. will require development of more robust membranes and is now only sporadically in use.

Superfluidity: see Helium.

Super-Invar®, ***stainless Invar®***: a cobalt alloy with high thermal dimensional stability. The composition is 54% cobalt, 36.5% iron and 9.5% chromium.

Supermalloy®: same as Permalloy® (see).

Supernucleophilicity: see Nucleophilicity.

Superoxides: compounds containing the O_2^- ion, usually with alkali metal counterions. The structure of the superoxide ion, O_2^-, can be understood in terms of the MO scheme for the O_2 molecule (see Molecular orbital theory, Fig. 4); an unpaired electron is present in a π^* molecular orbital. This is responsible for the paramagnetism of the S. The yellow to orange S. of potassium, rubidium or cesium can be obtained by burning the metals in a stream of oxygen; sodium superoxide is formed when sodium peroxide is heated to 500 °C in an oxygen atmosphere at 30 MPa. A lithium superoxide is unknown. The S. are strong oxidizing agents; they react with water to release oxygen: $2\,O_2^- + H_2O \rightarrow O_2 + HO_2^-, 2\,HO_2^- \rightarrow 2\,OH^- + O_2$. The reaction of the S. with carbon dioxide is of particular interest, e.g. $4\,KO_2 + 2\,CO_2 \rightarrow K_2CO_3 + 3\,O_2$; it is used to absorb carbon dioxide and regenerate oxygen in a closed space.

Supersensitizing: see Spectral sensitization.

Superstructure: an ordered distribution of the components of a substitution Mixed crystal (see). Normally, the substituted atoms in such a crystal are distributed randomly among the lattice positions (unordered mixed crystal), but under certain conditions of stoichiometry and crystallization, the distribution may be nonrandom (ordered mixed crystal). The system copper/gold provides a good example. In an alloy with the composition CuAu, above about 693 K and after rapid cooling to room temperature (quenching), the Cu and Au atoms are randomly distributed among the positions in the cubic face-centered lattice. If the alloy is cooled very slowly below 673 K, or if quenched samples are tempered, the Au atoms occupy the vertices and the Cu atoms the face centers of the unit cell; the result is a S. with a primitive unit cell with cubic symmetry (Fig.).

At a composition Cu_3Au, another S. forms under the same conditions, in which the lattice consists of alternating layers of the two types of atoms and has only tetragonal symmetry (Fig.). In general, formation of

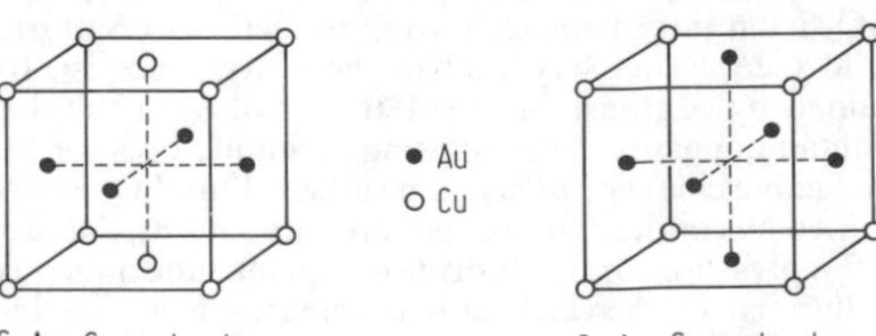

CuAu - Superstructure Cu_3Au - Superstructure

a S. leads to a lower symmetry than displayed by lattices of the pure components, or to a change of the Bravais lattice (see Crystal) or to an increase in the unit cell.

The appearance of an ordering process in the course of mixed crystal formation can be demonstrated by the observation of additional "superstructure lines" in the X-ray diffraction diagram.

Many alloy properties such as hardness, electrical conductivity and magnetic behavior are significantly changed by formation of a S.

Suppressor: see Ion chromatography.

Suprafacial: see Woodward-Hoffmann rules.

Suramin: a synthetic compound with a complicated structure. It was the first successful chemotherapeutic for sleeping sickness caused by trypanosomes.

Surface charge density: σ_0, the number of charges per unit area of surface, a parameter for characterizing macroscopic surfaces and colloidal particles. It is related to the potential at the surface, Ψ_0, by the following equation: $\sigma_0 = -\varepsilon_0\varepsilon(\mathrm{d}\Psi_0/\mathrm{d}x)_{x=0}$, where ε_0 is an influence constant, ε is the dielectric constant and $\mathrm{d}\Psi/\mathrm{d}x$ is the potential change at distance x from the surface.

Surface defect: see Crystal lattice defect.

Surface diffusion: see Solid state reactions.

Surface pretreatment: removal of moisture, electrolytes, fats, oils, soot, scale and rust from a metallic surface (usually steel) before a protective layer is applied, and generating a degree of roughness suitable for the intended protective layer. The bonding of protective layers is reduced by the undesired substances, and this leads to rusting under the protective layer. Rust adsorbs salts and sulfuric acid, which promote rusting. If the surface produced by milling or rolling steel is cracked, the corrosion medium has access to the non-passivated steel surface and causes it to corrode. As a result, the surface splits off, taking the protective layer with it.

S. can be done mechanically or chemically. Mechanically, it is done by sanding or brushing, either manually or by machine. Mechanical cleaning can be done by directing a high-pressure stream of slag, corundum, or steel particles against the surface to be cleaned, or by sudden heating of the surface with a flame, which causes the rust and scale to split off due to the rapid expansion of the metal below. Chemical cleaning includes removal of fats with organic solvents or detergents, and pickling, with or without alternating or direct current, in a bath of acid, lye, salt or salt melt (see Pickling of metals). Pickling is done in baths, after removal of fats, and thus cannot be done on steel parts of buildings. After pickling, the metal is carefully rinsed and neutralized. Since pickled steel surfaces are chemically very active and are rather smooth, the bonding surface for paint can be improved by Phosphatizing (see) in a bath process.

Due to the high cost of traditional methods of removing rust, and their interference with normal use of buildings, new ways of removing scale from steel building components by Rust converters (see) have been sought. Such products have not proved useful in protecting parts from atmospheric corrosion in the chemical industry. However, hand treatment with Penetrating agents (see) and Passivation agents (see) does prolong the lifespan of rust-preventing paints on the treated surfaces.

Surface symbol: see Crystal.

Surface tension, ***specific surface free energy***: σ, the reversible work of enlarging a surface by 1 m^2. Defined phenomenologically, it corresponds to the force acting on an imaginary line, 1 m long, on the surface. Since the two parameters have the same dimension, the unit of G. is either the newton meter^{-1} or the joule meter^{-2}. When the surface of a liquid is increased, the composition of the phase boundary does not change, but with solids, two S. must be distinguished: the S. at constant composition, which is found in formation of solids, e.g. during crystallization; and the S. under changing composition, which is observed when solids are deformed. The numerical value of the S. depends on the intermolecular forces (or atoms or ions) of the substance. In the interior of a phase, all the molecules are subject to the same interaction forces in all three directions (Fig. a). At the phase boundary, however, molecules are exposed to interactions with others in the same phase on one side, and with molecules in the other phase on the other side (Fig. b).

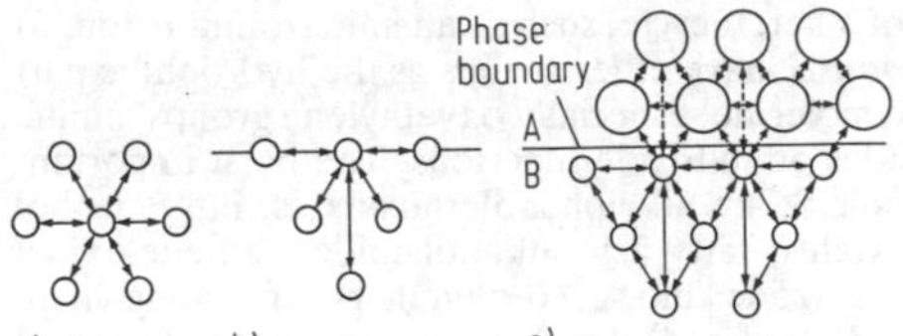

Phase boundary
A
B
a) b) c)

Substances in which only London forces act between the atoms (molecules) have a low S., while ionic crystals and metals have high S. The interaction between molecules in such substances is shown schematically in Fig. c. The force on the surface molecules of substance A is the difference between the interaction energy V_{AA} of molecules in A, and the energy V_{AB} of interactions with molecules of B. The boundary molecule B experiences the analogous force $V_{BB} - V_{AB}$. The total interaction energy V at the surface of two immiscible liquids is thus $V = V_{AA} + V_{BB} - 2\,V_{AB}$. This equation would be easy to solve if all the energy terms could be derived from the dispersion forces, but this is not the case.

The S. of water exposed to air or to immiscible liquids can be reduced by surfactants, and this is often utilized, for example to wet solids and in washing. The working of solids or grinding processes are also made more effective by reduction of the S. (wet milling, deformation in the presence of liquids). The S. of liquids can be measured, for example using the Du Noüy ring method or the Wilhelmy plate method. In the *ring method*, the greatest force on the perimeter of a ring is measured as the ring is drawn through the phase boundary. In the *plate method*, a completely wettable plate is brought into contact with the surface; thereafter it is lifted and the opposing force of the liquid is determined.

Surface viscosity: 1) the viscosity of adsorption layers of surface-active compounds at fluid phase boundaries, measured under tangential shear. 2) The dissipative parameter of adsorption layers of surface active substances at fluid phase boundaries, measured under compression and dilation. Both parameters are important for the stability of foams and emulsions.

Surfactants: in a broad sense, all compounds which are enriched on phase boundaries and which reduce surface tension. These compounds are bipolar, with a lyophilic and a lyophobic part of the molecule. The compounds can consist of small molecules or macromolecules (see Protective colloids). In the narrow sense, S. are small to medium sized molecules with at least one hydrocarbon group consisting of 8 to 20 carbon atoms; the hydrophilic part can be a charged or uncharged polar group. 1) ***Anionic S.*** have negatively charged hydrophilic groups such as carboxyl, sulfonate, sulfate, phosphonate or phosphate. The most important anionic S. are soaps, alkylbenzenesulfonates, alkyl ether sulfates, alkane sulfonates, alkyl sulfates, lignin and petroleum sulfonates. 2) ***Cationic S.*** have positively charged hydrophilic groups such as ammonium, phosphonium or protonated amine *N*-oxide groups. Ammonium salts with one long-chain alkyl groups are important as disinfectants; those with two long-chain alkyl groups are used as softeners. The ammonium group can also be part of a heterocycle, such as an imidazolinium ion. 3) ***Nonionic S.*** have ether groups as the hydrophilic part of the molecule, especially oxyethylene groups, amine *N*-oxides or sulfoxide functions. The most important nonionic S. are alkylphenolethoxylates, fatty alcohol ethoxylates, fatty acid alkanolamides, ethene-oxide/propene-oxide block co-oligomers and long-chain amine *N*-oxides. 4) ***Amphoteric S.*** contain at least one zwitterionic group in the molecule. Some important ones are carbo- and sulfobetaines with one ammonium and one carboxyl or sulfonic acid group, and with a hydrophobic part of the moleule. 5) ***Special S.***. *Fluorine S.* contain a (per)fluorinated hydrocarbon group, and are important because they can reduce the surface tension in aqueous solutions to a very low value; *silicon S.* are also important.

Properties. S. are adsorbed to surfaces in oriented fashion, and in this way they reduce the surface tension. Above the critical micellar concentration, they form association colloids (micelles) which can solubilize water-insoluble substances in aqueous solutions. As the concentration of the S. increases, it leads to anisotropic structures and non-newtonian flow behavior. Rod-shaped or cylindrical micelles are intermediates in the formation of laminar and lyotropic liquid crystalline phases; the washing properties of S. have recently been ascribed to these phases.

Applications. The reduction of surface tension at the phase boundaries between water and oil, or between air and wetting of solids such as fibers and hard surfaces, and the solubilization capacities of S. are the basis of their use in products for washing and cleaning, cosmetics, preparation of textiles and leather, emulsion polymerization, lacquers and paints, pesticides, pharmaceuticals, construction, fire extinguishers and enhanced recovery of petroleum (see Surfactant-polymer flooding). The world production of S. is about 13 million tons per year, of which 8 million tons are soaps. The production is expected to climb as S. are increasingly used in production of petroleum.

To some extent, the terms "detergent" and "surfactant" are used interchangeably, but they are not identical. All detergents are S., but some S. (notably soaps) are not detergents.

Surfactant-polymer flooding: a combined technology for enhanced recovery of petroleum from deposits which have been depleted with respect to primary (e.g. the pressure of the oil field itself) and secondary (e.g. flooding with water) extraction methods. An aqueous solution of surfactant, or preformed microemulsion (micellar flooding) is used to reduce the capillary forces in the pores of the rock, and the oil/water surface tension, so that the oil is locally mobilized, and partly solubilized. It can then flow into oil chambers. This oil is pushed out to the collection pipe by water to which small amounts of special polymers, such as polyacrylamides or polysaccharides, have been added to increase the viscosity of the solution.

Susceptibility: see Magnetochemistry.

Suspended zone process: see Crystal growing.

Sutherland model: see Gas theory, kinetic.

Suxethonium: see Muscle relaxants.

Suzukacillin: see Alamethicin.

Swain-Scott relationship: see Nucleophilicity.

Swamp gas: see Methane.

Swarts reaction: a reaction of chloro- or bromoalkanes with inorganic fluorides, such as antimony(III) fluoride, silver(I) fluoride or cobalt(III) fluoride, to form fluoroalkanes. The S. is used most often for the partial exchange of chlorine or bromine atoms for fluorine in polyhaloalkanes or polyhaloalkenes, for example, $ClCH{=}C(Cl){-}CCl_3 + SbF_3 \rightarrow ClCH{=}C(Cl){-}CF_3 + SbCl_3$. The ease of the halogen-fluorine exchange depends on the fluorination reagent and the structure and bonding in the halogen compound. Iodine and bromine are easier to replace than chlorine. Antimony(III) fluoride reacts only with haloalkanes which have at least two halogen atoms on the same C atom.

Swedish green: see Copper(II) arsenite.

Sweeteners: organic compounds with a sweet taste used in foods. Those which supply no energy are used as substitutes for glucose, sucrose or glucose-containing oligosaccharides in foods for the overweight. Cyclamate (see) and Saccharin (see) are the most widely used of this category. S. are also used in lipsticks, toothpastes, mouth washes and drugs. ***Sugar substitutes*** are used by diabetics; these include fructose and alditols, especially glucitol (sorbitol) and xylitol.

Compounds with widely varying chemical structures (Table) have sweet tastes, but not all of them are completely harmless. Mono- and oligosaccharides and alditols taste sweet. Chlorination considerably increases the sweetness of sucrose. Although L-amino acids have a bitter taste, other amino acids, such as glycine and certain D-amino acids, are sweet. Many synthetic peptides and their derivatives are also sweet, for example *aspartame* (N-α-aspartylphenylalanine 1-methyl ester). A few proteins from African plants are intensely sweet: *thaumatin* from *Thaumatococcus danielli* and *monellin* from *Dioscoreophyllum comminsii*. However, these proteins are not very stable. Dihydrochalcones synthesized from naringin or neohesperidin, from citrus fruits, are

Table. Sweetening power of some substances, relative to sucrose = 1

Compound	Sweetening power
Sucrose	1
Glucose	0.7
Fructose	1.4
Lactose	0.2
4,1,6,6-Tetrachlorosucrose	100
1,6,6-Trichlorosucrose	2000
Glucitol	0.5
Xylitol	1.0-1.25
Glycine	1.5
D-Phenylalanine	7.3
D-Tryptophan	35
Aspartame	150-200
Thaumatin	1600
Monellin	1500
Cyclamate	40
Dulcin (4-ethoxyphenylurea)	250
Saccharin	400
Dihydrochalcones	300-7000
1-Propoxy-2-amino-4-nitrobenzene	4000

also intensely sweet, as are many synthetic compounds.

Studies of the relationships between structure and taste have suggested the following structural model of a sweet substance: the essential structural elements are an H-bond donor (AH), an H-bond acceptor (B) and a hydrophobic group (X) to intensify the sweetness (AH, B, X hypothesis of Shallenberger, Acree and Kier).

Structural model of a sweet substance

The AH/B systems are OH/O in sugars and dihydrochalcones, NH/SO in cyclamate and saccharin, or NH_3^+/COO^- in amino acids and peptides. The spatial arrangement of AH and B is also essential (Fig.).

Sydnones: crystalline, mesoionic, heterocyclic, aromatic compounds. Various zwitterionic resonance structures can be formulated for the S. (Fig.).

The structure shown as c has been shown by ESCA measurements to be the most probable. S. can be obtained by dehydration of *N*-substituted *N*-nitrosoglycines; for example, the parent S., 3-phenyl-1,2,3-oxadiazol-5-one, is formed by the reaction of nitrous acid with anilinoacetic acid, via the nitrosoamine. S. are very reactive and can be converted to many other heterocycles.

symm-: see Nomenclature, section III D.

Symbol: 1) ***chemical S.***: an international abbreviation for a chemical element. The S. for known elements consist of one or two letters; examples are Fe for iron (Latin ferrum), Cu for copper (Latin cuprum), S for sulfur, I for iodine, V for vanadium, etc. For elements which have not yet been given systematic names, the S. consists of the two or three initial letters of the Latin name for its atomic number, e.g. Unp for element 105 (unnilpentium) (see Nomenclature, sect. II.A.1). For computer data processing, where there is danger of confusion of the letters I and V with Roman numerals I and V, iodine and vanadium can be indicated by the symbols Id and Va. The S. for all the elements are given in alphabetical order in the table on the inside back cover of Vol. 1.

These S. are uniform in chemical literature in all languages, although there are variations found in older literature: A for argon (Ar) in English and French, Az (azote) for nitrogen (N) in French, Cb (columbium) for francium (Fr) in English and French, Gl (glucinium) for beryllium (Be) in French, J (Jod) for iodine (I) in German, Lw for lawrencium (Lr), Nt (niton) for radon (Rn) in English, Tu (tuli) for thulium (Tm) in Russian, X for xenon (Xe).

The mass number, atomic number, number of atoms and ionic charge of an element are indicated by sub- and superscripts appended to its S. as follows: the mass number is a superscript to the left, the atomic number is a subscript to the left, the number of atoms is a subscript to the right, and the ionic charge is a superscript to the right. For example ${}^{32}_{16}S_2^{2+}$ means a molecular ion of two sulfur atoms with a double positive charge; the sulfur has mass number 32 and atomic number 16. The indication of the atomic number is redundant, since all atoms of the element denoted by the S. have the same atomic number. The mass number, however, distinguishes different isotopes of the element. Individual symbols for the isotopes exist only for hydrogen (${}^1H$ is H, ${}^2H$ is D, and ${}^3H$ is T) and these are not widely used.

2) In the nomenclature, an abbreviation for a group of atoms (e.g. amino acid, polymer, organic group or ligand).

3) A combination of letters denoting a concept of physical chemistry, analysis, biochemistry, etc. established by the IUPAC.

Symmetric: see Nomenclature, sect. III D.

Symmetry: in the broadest sense, the property of a spatial or mathematical structure of being converted into itself by certain operations. S. of geometrical objects is most easily comprehended; it is the repetition of a structural or morphological motif according to certain rules. Symmetrical objects which are important in chemistry are, for example, molecules, coordination polyhedra and crystals. In physics and mathematics, the concept of S. is also applied to abstract relationships and laws, if these are invariant under certain transformations.

An object is *symmetrical* if it can be transferred into a position which is indistinguishable from its starting position by a geometrical operation, the *symmetry operation*. For mathematical reasons, the *identity operation*, which leaves the object in its original position, is also considered a symmetry operation, although it is actually only a pseudooperation. The indistinguishable positions of the object can thus be either symmetry equivalent or identical. A prerequisite for carrying out a symmetry operation is the existence of a geometric location, i.e. a point, line or surface, to which the operation refers: the *symmetry element*. Symmetry operations and elements require each other: the symmetry operation is defined with respect to a certain symmetry element, and can only be applied to it, while the existence of a certain symmetry element can only be demonstrated by carrying out the corresponding symmetry operation. Usually the corresponding symmetry elements and operations are denoted by the same symbol. For description of the S. of molecules (see Symmetry, molecular), the symbolism of Schoenflies is generally used; for the S. of a crystal, however, the international symbolism of Hermann-Mauguin is preferred.

The following symmetry operations or elements can be defined for finite spatial objects (Fig. 1) (the Schoenflies symbols are indicated):

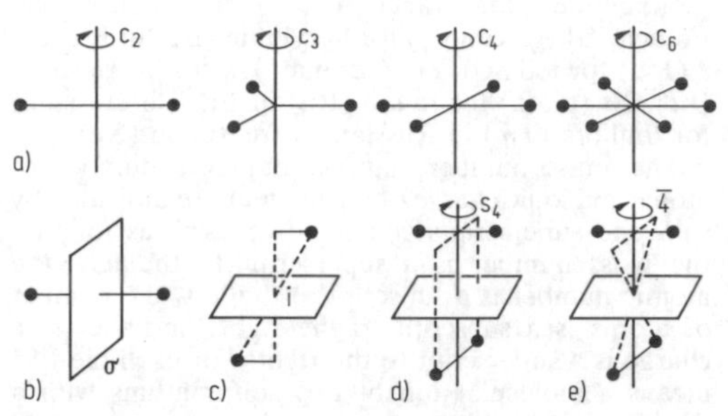

Fig. 1. Point symmetry elements: a) axis of rotation, b) mirror plane, c) center of symmetry, d) rotary reflection axis, e) rotary inversion axis.

1) *Identity operation* (symbol *E*).

2) *Rotation* around a *rotary axis* (symbol C_n). If a rotation through 360° produces *n* indistinguishable positions, or if the smallest angle of rotation required to bring the object into an indistinguishable position is 360°/*n*, the axis has *multiplicity n* (*n*-fold rotary axis). If there are several axes of rotation, the one with the highest multiplicity is the *principle axis of rotation*. The principle axis of rotation of an axially symmetric object (rotation through any angle produces an indistinguishable position) has infinite multiplicity (symbol $C_{;\mathrm{inf}}$).

3) *Reflection* in a *mirror plane* (plane of symmetry, symbol σ). A mirror plane divides an object into two halves which are related as object and reflection. If the position of a mirror plane is given in reference to a principle axis of rotation, a subscript is added to its symbol (Fig. 2): σ_h, the mirror plane is perpendicular

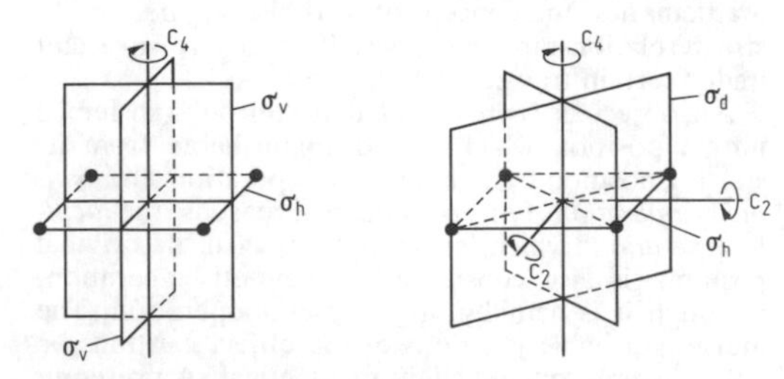

Fig. 2. Types of symmetry planes.

to the principle axis; σ_v, the mirror plane contains the principle axis; δ_d, the mirror plane contains the principle axis and bisects the angle between two C_2 axes which are perpendicular to the principle axis.

4) *Inversion* through a *center of symmetry* (inversion center, symbol *i*). For every point, there is a symmetrically equivalent counterpart at the same distance from the center of symmetry, but in the opposite direction from it.

5) *Rotational reflection* on a *rotational reflection axis* (symbol S_n). This is a composite symmetry operation and consists of a rotation c_n around a rotational axis and subsequent reflection σ_h in a plane perpendicular to the axis. A two-fold rotational reflection axis S_2 has the same effect as a symmetry center *i*.

6) *Rotational inversion* at a *rotational inversion axis* (no Schoenflies symbol; Hermann-Mauguin symbol $\bar{n}$). This is another composite symmetry operation consisting of a rotation C_n and subsequent inversion *i*. Each rotational inversion axis is equivalent to a rotational reflection axis, although the multiplicities can be different; for example, $\bar{1} = S_2$, $\bar{2} = S_1$, $\bar{3} = S_6$, $\bar{4} = S_4$. The use of rotational reflection and rotational inversion are alternatives; in chemistry the former is used, and in crystallography, the latter.

Aside from the identity element, geometric objects can have one or more symmetry elements, or they may have none. The set of all symmetry operations on an object constitutes a mathematical group. Since during any possible symmetry operation on a finite object at least one point retains its position, the corresponding group of symmetry operations are called a *symmetry point group* or simply *point group*. For the determination and symbolism of the point groups of molecules, see Symmetry, molecular.

All possible S. which can be displayed by the external form of a crystal (macroscopic S.) can be described by one of 32 point groups, which are also called the 32 *Crystal classes* (see Crystal). For the description of microscopic S., *translation* must also be included as a symmetry operation; this is a parallel shift over a given distance which can be repeated arbitrarily often; it is the basis of the infinitely extensive lattice structure of the crystal. Coupling of translation with the two point symmetry operations rotation and reflection yields two new microsymmetry operations (Fig. 3):

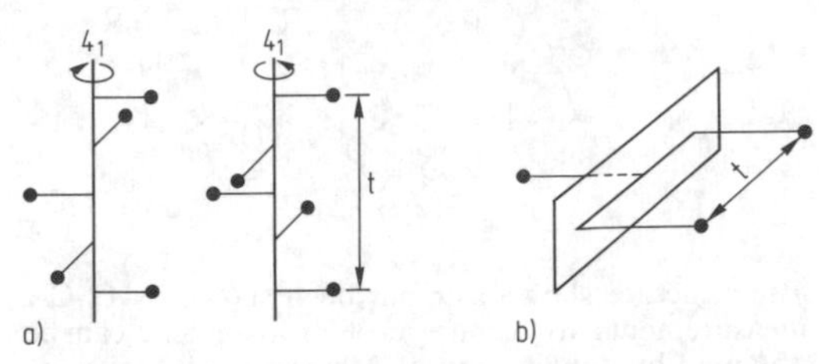

Fig. 3. Symmetry elements with translation components: a) spiral axis (with right- or left-handed twist), b) translational mirror plane.

1) *Twist* along a *spiral axis* (Hermann-Mauguin symbol n_p) involves a rotation through 360°/*n* and a simultaneous translation parallel to the rotation axis over a distance

$\frac{p}{n} t$

where t = translation period and $p = 1, 2, \ldots (n\text{-}1)$. For example on a 4_1 axis, four rotations through 90°, each coupled with a translation over $t/4$, will bring a point into a translationally equivalent position (Fig. 3a).

2) *Translational reflection* on a *translational mirror plane*. Here a point is shifted by half of the translation period and reflected in a plane parallel to the direction of translation (Fig. 3b). The symbol for this operation is determined by the direction of translation (a, b, c, n or d).

The systematic combination of all possibilities for crystallographic structural symmetry produces the 230 *space groups*.

The concept of symmetry is used in the sciences to make the application of theory to experimental results easier. The advantage is that symmetry arguments are easily comprehended, and on the other hand, S. is amenable to an exact mathematical treatment through group theory. The chemical and physical properties of many chemical species are determined by their S. or the S. of their geometric linkages. The elucidation and understanding of chemical structures depends greatly on knowledge of their symmetry properties. Some important areas of application of S. are quantum chemistry, spectroscopy, the interpretation of dipole moments and optical activity and determination of structure using X-ray, neutron and electron diffraction.

Symmetry center: see Symmetry.

Symmetry element: see Symmetry.

Symmetry, molecular: the Symmetry (see) of the spatial arrangement of the atoms or groups of atoms in a molecule. The S. is described in terms of the following symmetry elements: axis of rotation C_n, mirror plane σ, symmetry center i and rotational reflection axis S_n. These are illustrated in several examples in Fig. 1.

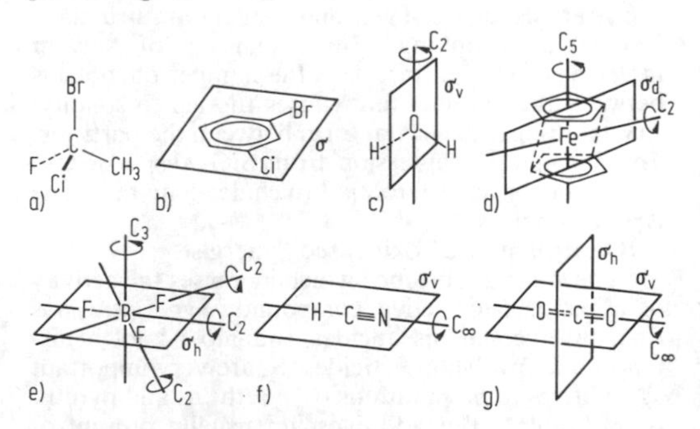

Fig. 1. Molecules with symmetry elements (only those relevant to determination of the symmetry point groups are illustrated).

The symmetry of a molecule is the result of interactions of all symmetry elements present, and can be uniquely characterized by indicating the corresponding molecular symmetry point group or its symbol (according to Schoenflies). Since molecular symmetry is expressed in a number of physical and chemical properties, e.g. the dipole moment, optical activity, spectroscopic behavior or the optical and magnetic properties of complexes (see Ligand field theory), and is essential for their interpretation, its determination has great practical significance. The assignment of a molecule with known geometry to the appropriate symmetry point group is easy if a certain algorithm is applied (Fig. 2).

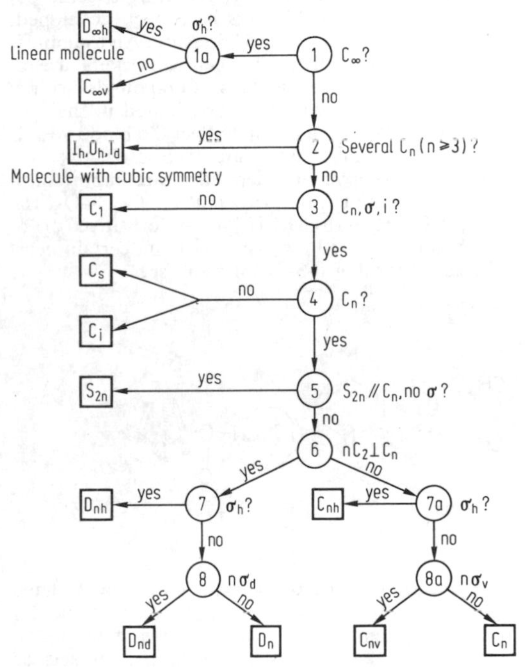

Fig. 2. Algorithm for determination of the symmetry point group of a molecule.

First the rotation axis (or axes) with the highest multiplicity is located, if there is one, and the presence or absence of other symmetry elements corresponding to the branching of the algorithm are noted. In this way, the molecules a to g shown in Fig. 1 can be assigned to the following point groups (the symmetry elements relevant to the assignment are given in parentheses in the order of their appearance in the algorithm): a, C_1; b, C_s (σ); c, C_{2v} (C_2, σ_v); d, D_{5d} (C_5, $C_2 \perp C_5$, σ_d); e D_{3h} (C_3, $C_2 \perp C_3$, σ_h); f $C_{\infty v}$ (C_∞); g $D_{\infty h}$ (C_∞, σ_h).

S. has a decisive role in the phenomenon of optical activity. A carbon atom is considered *asymmetric* if it is linked to four different atoms or atomic groups. Optically active compounds do not have to be asymmetric, which means "without symmetry" and corresponds only to point group C_1. The necessary and sufficient condition for optical activity is that the molecule belong to point group C_1, C_n or D_n. Such molecules, which have no symmetry center, no mirror plane and no rotary reflection axis, are called *dissymmetric* or *chiral* (see Stereoisomerism 1).

Symmetry operation: see Symmetry.

Symmetry point group: see Symmetry.

Sympathicolytics: compounds which weaken or extinguish excitation of the sympathetic nervous system, and are thus antagonists of the Sympathicomimetics (see) such as noradrenalin. S. are also called *adrenolytics*, and are classified as α_1, α_2, β_1 and β_2 adrenolytics. All older S. are unspecific α-adrenolytics, used mainly for their vasodilatory effects. The most important representatives of the group are the Ergot alkaloids (see) with peptide structures and Tolazolin (see). More recently, relatively specific peripheral α_1-receptor blockers have been developed; these are used to lower blood pressure. An example is prazosin. β-Adrenolytics (β-receptor blockers) are relatively recent developments. Compounds with a specific β_1-adrenolytic effect are limited in their effects to the heart, and do not affect the blood vessels or the bronchial musculature (cardioselective S.). The aryloxypropylamine derivatives are important β-S.; an example is *propranolol* ($C_{10}H_7$-O-CH_2-CH(OH)-CH_2-NH-CH(CH_3)$_2$), a derivative of α-naphthol. It is used for hypertonia and certain heart diseases. Another β-S. is Talinolol (see).

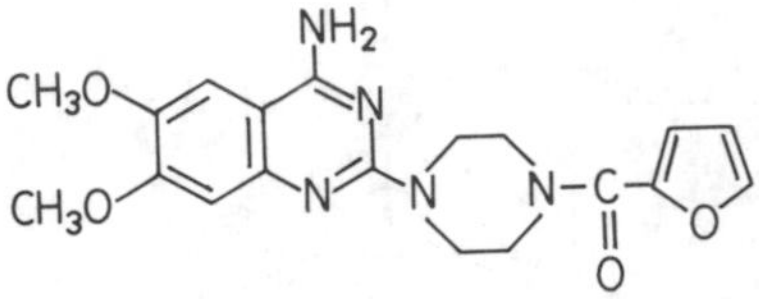

Prazosin

Sympathicomimetica: compounds which have the same effects on the peripheral (autonomic) nervous system as an electrical stimulation of the sympathetic nervous system. Noradrenalin acts as a neurotransmitter in postganglionic sympathetic nerves; it is released from presynaptic vesicles, crosses the synaptic cleft, and interacts with specific receptors on the postsynaptic membranes. Thereafter, it is rapidly degraded. The term *adrenergics*, which is derived from the name of the neurotransmitter, is also used for S. There are three types of postsynaptic receptors for noradrenalin and structurally similar compounds (e.g. adrenalin and isoproterenol): α-, β_1 and β_2 receptors. These are distributed differently in the various adrenergically innervated organs, and the strengths of their interactions with the S. also vary. The α-adrenergics lead to constriction of the blood vessels, and thus higher blood pressure; β_1-adrenergics stimulate of the heart; and β_2-adrenergic effects lead to dilation of the bronchial musculature and the peripheral blood vessels. Noradrenalin has primarily α-adrenergic effects, while its *N*-methyl derivative, adrenalin, has α- and β-effects. Isoproterenol activates β_1 and β_2 receptors, while terbutaline and clenbuterol are relatively specific for β_2 receptors. S. can also act indirectly, by inducing release of the natural transmitters from the presynaptic vesicles. Ephedrin, for example, acts as a weak S. by this mechanism. The biologically active forms of the S. are the (-)-forms with *R* configuration of the carbon atom adjacent to the benzene ring. Since the *S* enantiomers are not antagonists, the racemic mixtures can be applied therapeutically.

S. are β-phenylethylamine derivatives with varying substitution patterns on the benzene ring, in the intermediate chain and on the nitrogen atom. The most important S. are Noradrenalin (see), Pholedrine (see), Isoproterenol (see), Terbutaline (see), Dopamine (see) and Ephedrin (see). Comparison of structures and activities indicates that substitution on the N atom with a bulky residue shifts the α-effect present in the unsubstituted amino group (noradrenalin) to the β-effect (isoproterenol, terbutaline). The most active S. have a 3,4-dihydroxybenzene structure. 3,5-Dihydroxy compounds are somewhat less effective, but they are more stable. The monohydroxybenzene compounds (pholedrine) are less effective than the dihydroxy compounds; the meta compounds are, as a rule, somewhat more effective than the para compounds. The least active are compounds with no substituents on the benzene ring. Replacement of the alcoholic OH group by a hydrogen group weakens activity. Compounds with an alkyl group adjacent to the nitrogen group have weaker effects, but are less rapidly metabolized.

syn: see Stereoisomerism 1.2.3.

Synchrotron radiation: polarized electromagnetic radiation emitted by electrons in circular accelerators (synchrotrons) when they are deflected by magnetic fields perpendicular to their direction of travel. The phenomenon is explained by Maxwell's theory, which says that any acceleration of an electric charge leads to emission of radiation. The angular and spectral distribution of S. can be calculated. At the electron energies which can be achieved today, the spectrum is continuous into the x-ray range. In the short-wave ultraviolet range, S. is more intense than any other source of radiation, and is used for spectroscopic studies in that range.

Syn-clinal conformation: see Stereoisomerism, Fig. 14.

Syndiotactic: see Polymers.

Syneresis: aging of gels and coagulation structures leading to shrinking. The exclusion of solvent molecules during S. increases the number of contacts between the particles and causes the gel to solidify. Crystal bridges may even form between the particles. The geological progression from SiO_2 dispersion to coagulation structure to opal to chalcedony to quartz is an example of S.

Synergism: see Toxins; see Synergist.

Synergist: a substance which increases the activity of a biologically active compound. Synergism was first observed in insecticides and most known instances still involve insecticides; S. are very important as additives to preparations of pyrethrins and pyrethroids (Table). The additives are usually present at concentrations several times that of the biologically active compound. For example, piperonyl butoxide is mixed with pyrethrins at a ratio of 6:1 to 10:1. S. may be either natural substances (e.g. sesame oil) or synthetic compounds, which are often characterized by the presence of a methylene dioxy group. The mechanism of their action is usually postulated to be inhibition of enzymatic degradation, and thus detoxication, in the insect. They thus are functionally analogous to Resistance breakers (see).

Name	Formula	b. p.
Piperonyl butoxide		180°C at 133 Pa
Sesamex		137-141°C at 10 Pa
Sulfoxide		cannot be distilled
Piperin		m. p. 129°C
S 421®		144-150°C at 133 Pa

Synol process: a process for producing low-boiling alcohols from carbon monoxide and hydrogen. As in the Fischer-Tropsch synthesis, the synthesis gas is passed over a catalyst consisting of 97% iron oxide, 2.5% aluminum oxide and 0.5% potassium oxide. The gas is at a pressure of 2 to 3 MPa and 190 to 200 °C. The product is a mixture known as *synol*, consisting of 30 to 70% primary, unbranched alcohols with C_2 to C_{20} chains.

Syn-periplanar conformation: see Stereoisomerism, Fig. 14.

Synproportionation: same as Coproportionation (see).

Syntanes: see Tannins.

Synthases: a main group of Enzymes (see).

Synthesis: the production of chemical compounds. S. from the elements generally yields only simple compounds, such as metal oxides, sulfides or halides. The S. of more complicated compounds is accomplished in steps, from simpler compounds; the intermediates must often be isolated and converted to a reactive form. The S. of organic-chemical compounds which occurs in organisms is called ***biosynthesis***, while the preparation of a natural product in the laboratory, from simple reagents, is called ***total synthesis***. Production of a pure optically active compound is ***asymmetric S.***. Many S. are named after their discoverers (see Name reactions).

Synthesis gas: a mixture of carbon dioxide and hydrogen or of nitrogen and hydrogen used to produce methanol, ammonia and other basic chemicals. There are several types of S., obtained by different processes: ***water gas*** is obtained from coal and steam ($C + H_2O \rightleftharpoons CO + H_2$), ***generator gas*** is obtained from coal and air ($C + 1/2O_2 \rightleftharpoons CO$), and ***cracking gas*** is obtained by cracking of natural gas and petroleum ($CH_4 + H_2O \rightleftharpoons CO + 3\,H_2$). S. for specific purposes are made: ***methanol S.*** ($CO + 2\,H_2$), ***ammonia S.*** ($N_2 + 3\,H_2$) and ***oxo S.*** ($CO + H_2$) are used, respectively, in the syntheses of methanol and ammonia, and for hydroformylation of olefins (oxo synthesis). Until the beginning of the 1950s, S. was produced almost exclusively by gasification of anthracite or bituminous coal coke. Thereafter, methods were developed which use gaseous or liquid fuels. The advantage of these methods is that they produce a higher hydrogen content. For example, the hydrogen:carbon ratio of coal is about 1:1, that of petroleum is 2:1, that of gasoline is 2.4:1, and that of methane-rich natural gas is about 4:1. Another disadvantage of the classical methods is that the gasification occurs at atmospheric pressure, and therefore there is considerable additional expense for compression of the S. to the pressure required for the synthesis. Gasification of gaseous or liquid fuels under pressure is technologically much simpler than pressurized gasification of solids. However, with the rising prices and increasing scarcity of petroleum products, the gasification of solids has again become important, and the new developments are in the direction of pressurized gasification.

Production of a S. from *coal gasification*. The actual reactions in gasification with steam and oxygen are the exothermal partial combustion of carbon and the endothermal formation of water gas:

$2\,C + O_2 \rightleftharpoons 2\,CO$; $\Delta_R H = -218$ kJ mol^{-1}
$C + H_2O \rightleftharpoons CO + H_2$ $\Delta_R H = +130$ kJ mol^{-1}.

There are other reactions as well, e.g. the *water gas equilibrium*: $CO + H_2O \rightleftharpoons CO_2 + H_2$; $\Delta_R H = -42$ kJ mol^{-1}; the *Boudouard equilibrium*: $C + CO_2 \rightleftharpoons 2CO$; $\Delta_R H = +172$ kJ mol^{-1}, the *methane formation equilibrium*: $C + 2\,H_2 \rightleftharpoons CH_4$, $\Delta_R H = -75$ kJ mol^{-1}, and the *methanization reaction*: $CO + 3H_2 \rightleftharpoons CH_4 + H_2O$; $\Delta_R H = -205$ kJ mol^{-1}. A common feature of all coal gasification processes is that the endothermal partial reaction must be carried out at a high temperature (900 to 1000 °C). In the discontinous classical water gas process, the heat is supplied in the heating phase by burning part of the coke in the generator. The water gas process can be run continuously if oxygen and steam are added simultaneously. *Fluidized bed, solid bed pressure* and *fly ash gasification* methods are used. In the *Winkler process*, finely ground lignite coal is gasified at atmospheric pressure with air and steam at a temperature of 800 to 1100 °C. The H_2/CO ratio of the S. made this way is about 1.4:1. A disadvantage of this method, introduced in 1926, is that about 20% of the coal is not burned and remains behind in the ash. A widely used solid-bed method is *Lurgi pressure gasification*. Here lumps of anthracite coal or briquets of lignite coal are first de-gassed at a pressure of 2 to 3 MPa and a temperature of 600-750 °C. The subsequent gasification with oxygen and steam is done at 1200 °C. When anthracite is used, a crude gas consisting of 9 to 11% methane, 15 to 18% carbon monoxide, 30 to 32% carbon dioxide and 38 to 40% (vol. %) hydrogen is obtained. Byproducts are benzene, phenols and tar. In the *Koppers-Totzek process*, finely powdered coal is gasified at 1500 °C with oxygen and steam in a dust cloud flame. The *Saarberg-Otto process* is a pressurized dust process in which nearly any kind of coal can be used, especially coals with high ash contents (up to 40%). In the former USSR, USA, Japan and Germany, *plasma gasification* is under development. The process heat is generated electrically. Coal gasification (see) using the high-temperature process heat from nuclear reactors is being developed in the former USSR, USA and Germany.

b) S. from *natural gas* and *petroleum cracking*. The generation of S. from liquid fuels in the presence of steam depends on coupling of exothermal and endothermal reactions:

$$\text{-CH}_2\text{-} + 1/2\ O_2 \rightarrow CO + H_2;\ \Delta_R H = -92\,\text{kJ mol}^{-1}$$
$$\text{-CH}_2\text{-} + H_2O \rightarrow CO + 2\ H_2;\ \Delta_R H = +152\,\text{kJ mol}^{-1}.$$

As with coal gasification, the Boudouard, water-gas and methane-formation equilibria are established.

There are two methodological principles used in S. generation from natural gas and petroleum, steam reforming and oxygen pressure gasification.

1) In *steam reforming* (Fig. 1, 2) gasification occurs in the presence of steam and a catalyst. The necessary process heat is added from outside (allothermal catalytic gasification). The starting material – petroleum, refinery gas or gasoline – and steam are passed over a $Ni/K_2O/Al_2O_3$ catalyst at 700 to 900 °C and 2 to 3 MPa. Because the cracking catalyst is sensitive to sulfur, the starting material must be subjected to hydrogen pressure refining before it enters the gasification reactor; the hydrogen sulfide formed in this step is absorbed on zinc oxide in an absorption tower. The gasification occurs in a gas or oil-heated furnace (primary reformer), which is followed by a brick-lined shaft reactor (secondary reformer).

If ammonia S. is being generated, the primary reformer is run at 800 °C and the endothermal gasification is incomplete. The starting material not converted in the primary reformer is then burned in air in the secondary reformer, in which the temperature rises to about 1200 °C. The air flow is adjusted so that after conversion of the carbon monoxide ($CO + H_2O \rightleftharpoons CO_2 + H_2$) and removal of the carbon dioxide, the S. has the composition needed for ammonia synthesis ($N_2 + 3\ H_2$). The same catalyst is used in both primary and secondary reformers. The gas leaving the secondary reformer is cooled to 350 °C and then led into *high-temperature conversion*. Here, at a temperature of 350 to 400 °C, the carbon monoxide is converted to carbon dioxide on a Cr_2O_3/Fe_2O_3 catalyst. After further cooling of the S. to 250 °C, a *low-temperature conversion* is carried out on a $CuO/ZnO/Cr_2O_3$ catalyst to remove the remaining carbon monoxide. The carbon dioxide and steam still remaining in the gas are removed in a hot potash scrub: $K_2CO_3 + H_2O + CO_2 \rightleftharpoons 2\ KHCO_3$. After the low-temperature conversion, the gas still contains up to 0.5% carbon monoxide. Since ammonia S. must not contain any CO, the gas is removed by methanization on a Ni/Cr_2O_3 catalyst at 100 to 120 °C: $CO + 3\ H_2 \rightleftharpoons CH_4 + H_2O$. Methane does not interfere with ammonia synthesis.

In the generation of methanol S., the temperature in the primary reformer is raised (900 °C) so that all the hydrocarbon is gasified, and a secondary reformer is not used. Conversion of the S. is not needed if gasoline is used, because the S. generated from it has the composition $CO + 2\ H_2$.

2) In the autothermal *oxygen pressure gasification*

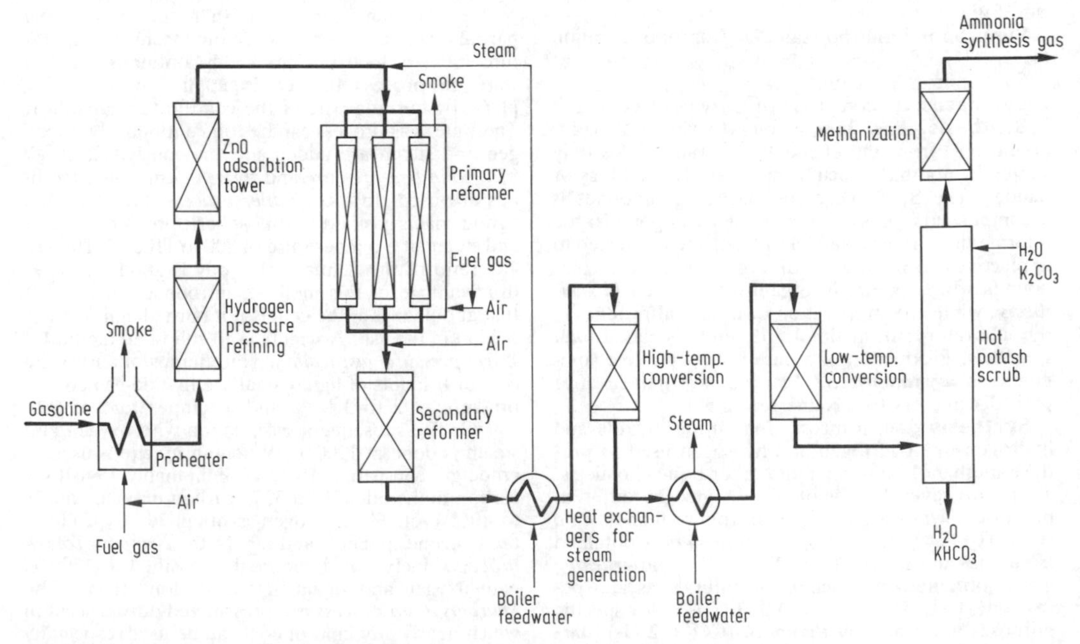

Fig. 1. Scheme of a steam reforming plant for gasoline used to generate ammonia synthesis gas.

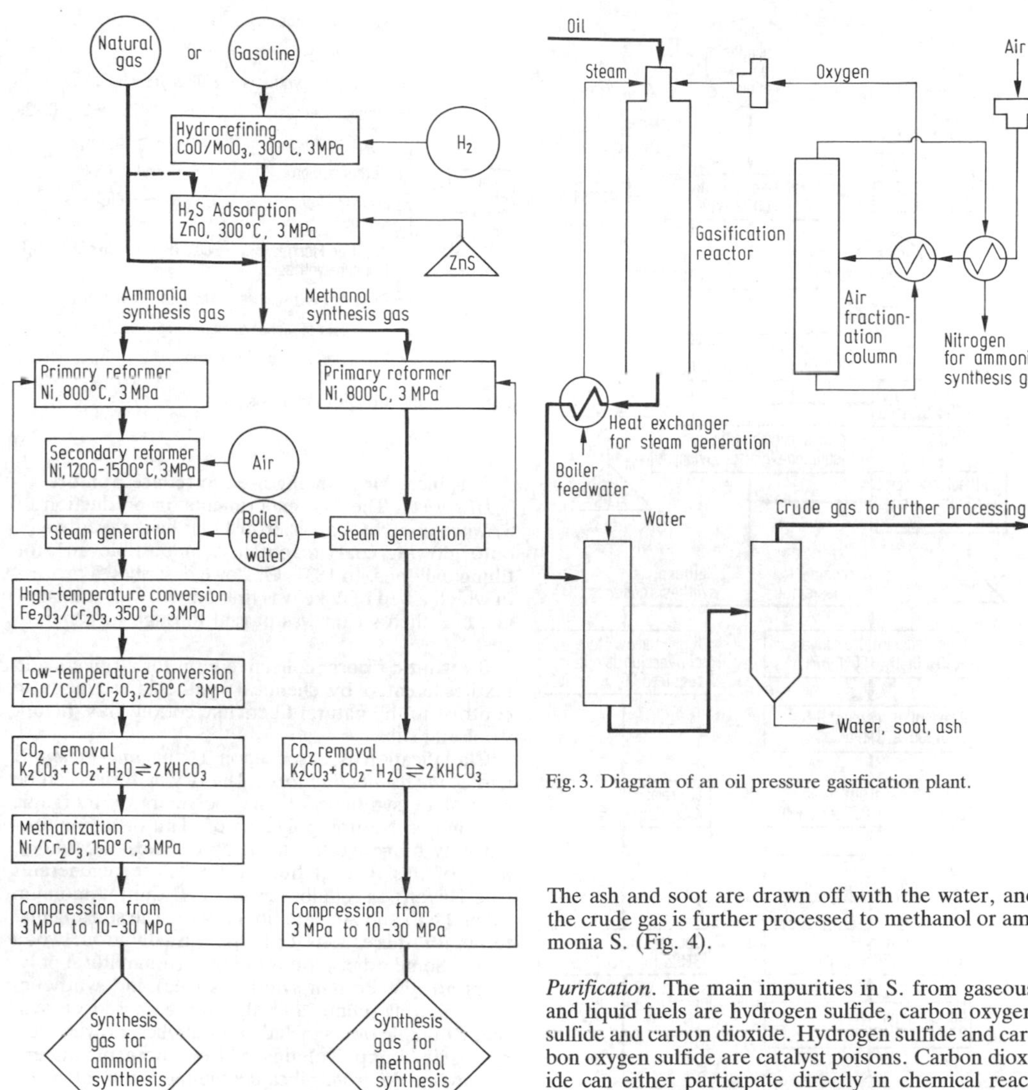

Fig. 2. Generation of ammonia and methanol synthesis gases by steam reforming of natural gas or gasoline.

Fig. 3. Diagram of an oil pressure gasification plant.

process, petroleum and petroleum fractions, especially refinery residues, are gasified with pure oxygen (*oil pressure gasification*, Fig. 3). The starting material is led into a nozzle burner and partially burned at 3 to 6 MPa. The flame zone has a temperature of 1200 to 1600 °C; here the exothermal and endothermal gasification reactions occur. The gas leaving the reactor contains some soot and ash. The gas is cooled to 250-300 °C; the heat is used to generate steam. Then water is sprayed into the gas (quenching) to reduce the temperature below 100 °C.

The ash and soot are drawn off with the water, and the crude gas is further processed to methanol or ammonia S. (Fig. 4).

Purification. The main impurities in S. from gaseous and liquid fuels are hydrogen sulfide, carbon oxygen sulfide and carbon dioxide. Hydrogen sulfide and carbon oxygen sulfide are catalyst poisons. Carbon dioxide can either participate directly in chemical reactions, or may lead to formation of an inert gas cushion. There is a variety of gas scrubbing methods (see Absorption), such as the *Rectisol process* (see) (methanol), the *sulfinol process*) (see) (sulfolane/diisopropanolamine), the *Purisol process* (see) (Methylpyrolidone) and the *sulfosolvane process* (see) (alkali salts of aminocarboxylic acids), or pressurized scrubbing with aqueous potassium carbonate solution. In older ammonia synthesizing plants, copper hydroxide solution is still used to remove residual carbon monoxide from ammonia S.

Applications (Fig. 5). The main consumers of S. are ammonia and methanol syntheses. S. is also used for hydroformylation of olefins (see Oxo synthesis), synthesis of hydrocarbons by the Fischer-Tropsch synthesis (see) and for production of carbon monoxide and hydrogen by low temperature pressure distillation or absorption methods. Synthetic natural gas (SNG) for fuel can be produced by methanization.

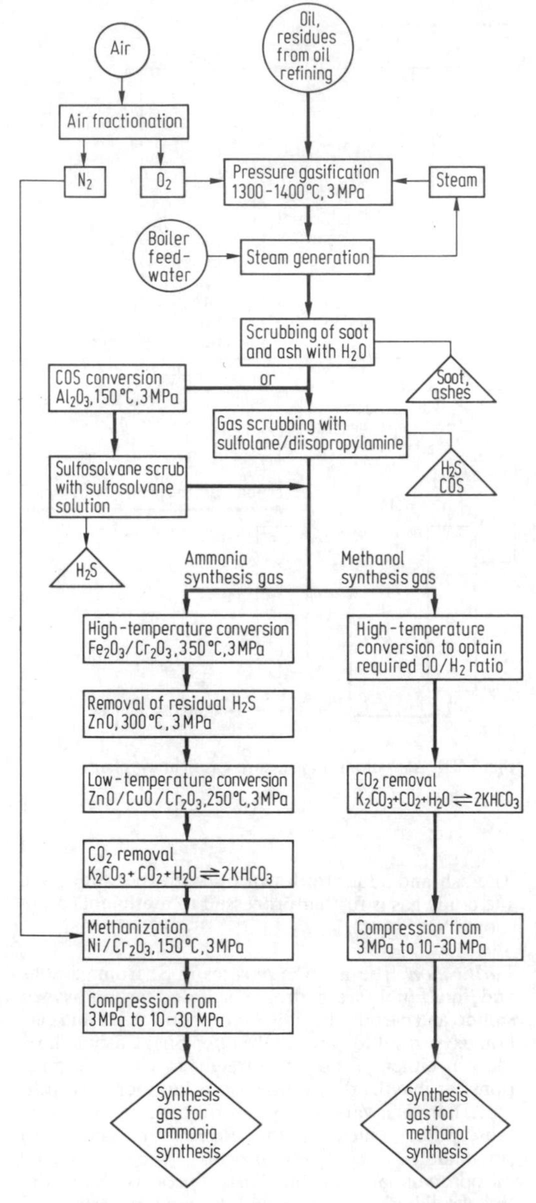

Fig. 4. Generation of ammonia and methanol synthesis gases by oil pressure gasification.

An energy transfer process based on methanization has been developed at the Jülich nuclear research institute (Germany): the "Adam and Eve" circulation. Here methane is converted to a mixture of carbon monoxide and hydrogen with steam in an endothermal reaction ("Eve") utilizing the heat from a nuclear reactor. It is transported by pipeline to consumers, where it is exothermally converted to methane and water ("Adam"). The methane is returned to the reactor for reuse.

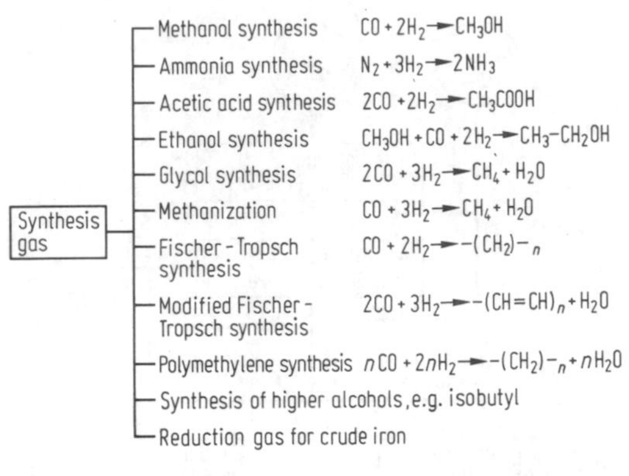

Fig. 5. Applications of synthesis gas.

S. is increasingly being used to reduce iron ore.

Historical. The first experiments on production of S. were done by Lavoisier in 1783; he sprayed water onto glowing coals to obtain "combustible air" for filling balloons. In 1831, G. Lowe described a process in which a bed of coke was first burned in a stream of air, and then steam was passed through to generate gas.

Synthetic fibers: collective term for all fibers and textiles created by chemical synthesis; it stands in contrast to the natural fibers like cotton, flax, hemp, wool and silk.

Classification is based upon 1) the *raw materials* and 2) *the production form*. The raw materials can be natural or synthetic organic polymers or inorganic substances. Natural polymers of plant or animal origin may be converted to S. by chemical treatment; most of this type of fiber are cellulose regenerates (see Rayon) or cellulose esters, e.g. Cuprammonium rayon (see) and Acetate fibers (see). The starting material for these S. is cellulose (usually from wood pulp). Some other groups of fibers from natural polymers are the Protein synthetics (see) and synthetics from natural rubber and alginate. S. based on synthetic organic fibers include polyacrylics, polyamides, polyesters and polyolefins. S. from inorganic materials are made from silica-containing melts of glass-forming materials, which are obtained from rocks, minerals or slag, and contain 30 to 100% silicon dioxide. Although they are "glasses", these fibers are included among the mineral fibers. Inorganic S. also include metal and carbon fibers.

2) The S. are manufactured in the form of ***staple, tow*** and ***infinite fibers***. The staple fibers are of finite lengths suitable for spinning, between 2 and 12 cm, similar to those of cotton or wool.

Manufacture. *S. from natural polymers* are obtained by melting or chemically treating a natural, macromolecular raw material to make it soluble (e.g. cellulose as cellulose xanthogenate or acetate), forcing the solution through the opening of a spinneret, and then removing the solvent with a counter-current of hot air (dry spinning process); alternatively, the fibers are extruded into a bath in which the S. precipitates out of solution as an insoluble thread (e.g. rayon). The fibers made by this wet spinning process can be stretched immediately or after drying, then processed further.

S. from synthetic polymers are either produced from a solution by a wet or dry spinning process, e.g. polyacrylic fibers, or they are melted and extruded through fine holes; the fibers formed in this way cool and harden in the air (melt spinning), e.g. polyamides and polyesters. Extrusion processes are very important in the production of C., e.g. for polypropylene fibers. The essential process in preparing S. for textiles is stretching them to many times their original length; this can be done at room temperature or an elevated temperature, in a wet or dry state. Stretching orients the macromolecules parallel to the axis of the fiber, where they can then form non-covalent bonds between neighboring molecules. Two-component fibers are made by extruding two different S. through separate holes in the spinneret; a specific cross-sectional profile of the fiber is achieved by the arrangement of the holes and gives it specific properties, such as curl.

The macromolecules in the fiber are packed in a partially crystalline arrangement, which becomes oriented parallel to the fiber axis during stretching. The tensile strength of S. is a result of the orientation of the macromolecules, the strong non-covalent bonding between them and their high packing density within the fibers.

The great advantage of synthetic fibers over natural fibers is their greater resistance to tearing and abrasion and good elasticity. Some resist "pilling", many are resistant to water, light, weather, chemicals, rotting, insects, are nonflammable or not easily flammable, and some are wooly and therefore good heat insulators. Like natural fibers, they can be refined, dyed and made into textiles and industrial fabrics. Since it is possible to vary the properties of S. and to mix different S. in composite fibers, they can be designed for special purposes. The basic properties of the different S. which are taken into account in such designs are the hygroscopic quality of rayon fibers, the curl of polyacrylics, the strength of the polyamides and the resistance of polyesters to stretching or shrinking.

Historical and look ahead. The first S. was developed in 1931 from polyvinyl chloride. The work of W.H. Carothers between 1929 and 1936 led to nylon. In 1937/38, P. Schlack developed perlon L (polycaprolactam), and O. Bayer developed perlon U (polyurethane). In 1941, Whinfield and Dickson developed polyesters, and about the same time Rein and Houtz, working independently, obtained S. from polyacrylonitrile. Many new S. based on polyvinyl alcohol, polyvinyl chloride and vinyl mixed polymers were developed about the same time.

In 1979 the world production of S. was more than 14 million tons. In 1981 the first closings of plants for S. production were publicized. Since natural fibers, the four basic types of S. (polyamide, polyacrylonitrile, polyethylene and polyvinylidene chloride) and a few special fibers can meet the needs of the textile industry, it can be expected that future developments will tend to be specific modification of S. rather than the creation of new basic types. For example, the flammability of S. based on cellulose can be reduced by addition of compounds containing phosphorus or halogen; fibers of one substance can be coated with other compounds to create fibers with bacteriocidal or ion-exchange properties. A considerable increase in the production of carbon fibers can be expected. Polybenzimidazole fibers, which do not burn and resist temperatures up to 560 °C, are being used in increasing measure to replace asbestos in the production of protective clothing with good wearing qualities.

Overview of the most important synthetic materials

Material	Properties and main uses	Trade names
Fibers of natural polymers		
1) Rayon	Swells, little water resistance. Outerwear, upholstery, draperies, industrial fabrics.	Argona (Poland) Arylana (Czech.) Avril Avtex (USA) Cefofibra (Rumania) Fibro (UK) Flox (Germany) Macosa (Hungary) Macrolana (Czech.) Merona (Poland) Regan (Germany) Spolana (Czech.) Super 1, 2 or 3 (Russia) Textra (Poland) Viscon (Poland) Wistom (Poland) Zantrel
2) Modified rayons (VIM), HWM fibers, polynosic fibers	Retain shape, high wet strength; easy care; used in light outer wear, winter and rain garments; often mixed with cotton or polyester	Erlan (UK) Hipolan (Japan) Kopolon (Italy) Modulon (Russia) Swilosa (Bulgaria) Viscona (Poland)
3) Cuprammonium rayon	Properties similar to rayons; used for linings, blouses, underwear, light dresses with silk-like properties	Bemberg (Germany, Italy, USA) Cuprama (Germany) Regan (Germany)

Synthetic fibers

Overview of the most important synthetic materials

Material	Properties and main uses	Trade names
4) Acetate fibers Cellulose diacetates	For non-stretch fabrics, similar to natural silk; for raincoats and linings	Acetosilan (Russia) Amcel (Belgium) Carolan (Japan) Dicel (UK) Rhodia (Germany)
Triacetates	Similar to synthetics; thermoplastic; used for non-stretch textiles, tablecloths, decorator fabrics, knits, wire insulation	Arnel (Belgium, USA) Rhonel (France) Soalon (Japan) Tri-a-faser (Germany) Tricel (UK)
5) Protein fibers	Not water-resistant. Only casein fibers are used, in mixtures with other fibers, for hats, felts and upholstery	Merinova (Italy) Wipolan (Poland)
6) Rubber fibers	Highly elastic; used for support garments, swimwear	Gummitex (Poland) Lactron (Italy) Polyisopren SK3 (Russia)
7) Alginate fibers	Novelty threads for stockings and lace	
Synthetic polymer fibers 1) Polyesters	Retain strength when wet; high loft, wrinkle resistant; for suits dresses, shirts, decorator fabrics, sewing thread, sails, filter fabrics, conveyor belts, belts in machines, ropes, electrical insulation, curtains	Avlin Dacron (USA) Diolen (Germany) Encron Elana (Poland) Fortrel Grisbuten (Germany) Jambolen (Bulgaria) Kodel (USA) Lavsan (Russia) Monsanto Terylene (UK, Canada) Tesil (Czech.) Torlen (Poland) Trevira (Germany)
2) Polyamides	Little swelling, good abrasion resistance; for stockings, underwear, blouses and dresses, knits, decorator fabrics, industrial fabrics, fishing nets, ropes, electrical insulation, surgical thread, bristles for brushes	Adimin (Russia) Anid (Russia) Caprolan (USA) Chemlon (Czech.) Danamid (Hungary) Dederon (Germany) Nylon 6 (USA, Germany, UK, France) Nylon 6,6 Perlon (Germany) Polana (Poland) Quiana Redon (Rumania) Rilsan (France, Italy) Silon (Czech.) Stilon (Poland) Vidlon (Bulgaria)
3) Polyurethanes	Elastic, chemical resistant; used only as silks for support garments and swimwear	Ameliotex Dorlastan (Germany) Elaston (Poland) Glospan Lycra (USA) Numa Vyrene (USA)
4) Polyacrylonitriles Acrylics, including mixed polymers with less than 15% other components, e. g. vinyl chloride, vinyl acetate or methyl acrylate	Elastic, chemical resistant; for outer garments, knits, decorator and upholstery fabrics, tents, uniforms, rainwear, tropical clothing, filler for quilts, industrial fabrics, e. g. sails, filters, ropes and fish nets	Acrilan (USA, Canada) Ahilana (Poland) Bulana (Bulgaria) Cashmilon (Japan) Courtelle (UK) Creslan (USA, Canada) Crylor (France) Dolan (Germany) Dralon (Germany) Dunakril (Hungary) Kanekalon (Japan) Krumeron (Hungary)

Material	Properties and main uses	Trade names
		Melana (Rumania) Nitron (former USSR) Orlon (Canada) Redon (Germany) Relana (Rumania) Wolpryla (Germany) Zefran (USA)
5) Polyvinyl chloride fibers (including some post-chlorinated polyvinyl chloride	Protective clothing against acids; medicinal underwear for rheumatism, carpets, filters ropes, insulation, brushes, fishnets	Chlorin (former USSR) Envilon (Japan) Movil (Italy) Piviacid (Germany) Rhovyl (France) Tevilon (Japan) Thermovyl (France)
Vinyl chloride and Vinyl acetate		Vinyon (USA)
Vinyl chloride and acrylonitrile		Dynel (USA)
Post-chlorinated Vinyl chloride and cellulose nitrate		Vinitron (former USSR)
6) Polyvinylidene chloride fibers	Uniforms, mosquito netting	
Vinylidene chloride		
Vinylidene chloride and vinyl chloride	Drapery and upholstery fabrics, filter cloth, conveyor belts, acid-resistant cables	Dawbarn DLP, Permalon (USA) Saran (Germany, UK, Japan, Netherlands, Sweden, USA, Canada) Soviden (former USSR) Velon (USA, Canada) Szaniv (former USSR)
7) Polyolefin fibers	Outerwear, underwear for sports, fish nets, insulation	Bolathene (USA) Courlene (UK) Daplen (Austria) Istrona (Czech.) Meraklon (Italy) Nobelex (Czech.) Northylen (Germany) Pevlen (Czech.) Popril (Hungary) Reevon (USA) Trofil (Germany) Velon LP (USA)
8) Polyvinyl alcohol fibers	Good resistance to chemicals; suits, dresses, linings, bandages, ropes	Alprona (Poland) Evanit (former USSR) Kuralon (Japan)
9) Polyfluoroethylene fibers	Protective clothing against heat and chemicals; filters and insulation	Ftorlon (former USSR) Gore-Tex Politen (former USSR) Teflon (USA)
10) Polycarbonate fibers	Protective clothing against heat; insulating material in space technology	
11) Polycarbamide fibers	Ropes	
12) Mod. acrylics fibers	Mixes polymers containing 35 to 85% acrylic Machine textiles, swimwear, wigs	Akricel (former USSR) Crylor (France) Kanekalon (Japan) Monsanto Verel

Synthetic horn: a water-insoluble, easily dyed plastic similar to natural horn. It is made by condensation of casein, the protein of milk, with a 4 to 10% formaldehyde solution. The casein is solidified by cross-linking with methylene groups, $=CH_2$. However, S. is not very resistant to water. Its advantages are its low production cost, the ease of working it, its excellent surface gloss and its ready absorption of color. It is resistant to organic solvents.

The uses for S. are similar to those of natural horn, turtle shell and ivory. However, it has been largely replaced by other condensation resins and polymers which are more resistant to water. Commercially, S. is available under the international name "galalith".

Synthoil process: see Coal hydrogenation.

Synthol process: see Fischer-Tropsch synthesis.

Synzyme: a synthetic, non-protein Enzyme (see).

Syringa aldehyde: see Vanillin.

Syringin: see Cinnamyl alcohols.

Syringinin: see Cinnamyl alcohols.

System, ***thermodynamic S.***: a macroscopic object which is separated from its environment by idealized walls and can be regarded as an object of thermodynamic inquiry. If the material content of the S. is in a single phase, it is homogeneous. Heterogeneous S. consist of more than one phase. ***Open S.*** are those which exchange matter and energy with the environment, while ***closed S.*** do not. The state of a S. is determined by parameters of state (see State, parameters of), such as temperature, pressure, volume, amount of substance and chemical composition. Changes in state, and the associated chemical and energetic processes can be treated in terms of the state functions of thermodynamics.

Systemic pesticide: a pesticide absorbed by the roots or leaves of a plant and distributed throughout it by the sap. Insecticides, nematocides, fungicides or bacteriocides may be S.i. The advantage over a surface treatment and action is that the active substance is not exposed to the atmosphere and the new tissues generated by growth are also protected.

Szilard-Chalmers effect: see Radionuclides.

T

t: symbol for triton; see Tritium.

T: 1) symbol for tritium; 2) abb. for thymidine; 3) T, symbol for temperature.

2,4,5-T: see Growth hormone herbicides.

Ta: symbol for tantalum.

Tabun: see Chemical weapons.

Tachykinins: a group of peptides which have a rapid stimulating effect on the smooth muscles; this is in contrast to the slowly acting kinins. The family of T. is summarized below. Substance P (see) is of particular interest with regard to its occurrence and action spectrum.

	1	2	3	4	5	6	7	8	9	10	11
Physalaemin	Glu	Ala	Asp	Pro	Asn	Lys	Phe	Tyr	Gly	Leu	Met-NH_2
Uperolein	Glu	Pro	Asp	Pro	Asn	Ala	Phe	Tyr	Gly	Leu	Met-NH_2
Eledoisin	Glu	Pro	Ser	Lys	Asp	Ala	Phe	Ile	Gly	Leu	Met-NH_2
Phyllomedusin	Glu	Asn	Pro	Asn	Arg	Phe	Ile	Gly	Leu	Met-NH_2	
Substanz P	Arg	Pro	Lys	Pro	Gln	Gln	Phe	Phe	Gly	Leu	Met-NH_2
Kassinin	Asp-Val	Pro	Lys	Ser	Asp	Gln	Phe	Val	Gly	Leu	Met-NH_2

Eledoisin (formerly called moschatin) was discovered in 1949 in the salivary glands of gastropods (*Eledone moschata* and *Eledone aldrovandi*) by Erspamer. It was isolated and its sequence elucidated in 1962. Boisonnas confirmed the structure by total synthesis in the same year. Eledoisin excites the extravasal smooth musculature and reduces arterial blood pressure in human beings; the hypotensive effect lasts longer than that of the plasma kinin, bradykinin. Subcutaneous application stimulates the salivary glands and gastrointestinal secretion. The C-terminal hexapeptide has full biological activity.

Physalaemin has less effect on the smooth muscles than eledoisin, but is three to four times more effective in reducing the blood pressure. It was isolated in 1964 from skin extracts from the South American frog *Physalaemus fuscumaculatus*. ***Phyllomedusin*** from the skin of the South American frog *Phyllomedusa bicolor* and ***uperoleon*** from the skin of the Australian frog *Uperoleia rugosa* have similar effects. About 12,000 frogs were required for the isolation of ***kassinin*** from the skin of the African frog *Kassina senegalesis*. Its total synthesis was achieved in 1977 by Yajima.

Tachysterol: see Vitamin D.

Tacot: a short name for tetranitrobenzotetraazapentalene, an explosive which can be stored at relatively high temperatures. T. has a detonation velocity of 7250 m s^{-1}, and is still functional after 4 weeks of storage at 200 °C.

Tetranitrobenzo-1,3a,4, 6a-tetraazapentalene

Tacticity: the relative arrangement of chirality centers (see Stereoisomerism 1.1) in diastereomeric macromolecules with several chirality centers. In *mono-* or *isotactic polymers* (one chirality center per monomer), all the chirality centers can have the same configuration. *Syndiotactic polymers* are those in which the sequence of chirality centers alternates between R and S configurations. Iso- and syndiotactic polymers are *stereoregular polymers*. In *atactic polymers*, the R and S centers are arranged in random sequence. The first synthesis of a stereoregular polymer was achieved in 1954 by G. Natta, using styrene and propene.

TAED: see Bleach activators.

Taft equation: an LFE equation (see) introduced in 1952 by R.W. Taft; it is analogous to the Hammett equation (see). The T. describes the effect of α-substituents on the reactivity of aliphatic compounds: $\lg (k_i/k_0) = \varrho^*\sigma_i^*$. Here k_0 and k_i are the rate constants for the unsubstituted compound and the compound with substituent i in the α position relative to the reaction center, σ_i^* is the inductive Substituent constant (see), and ϱ^* is the Taft reaction constant. The numerical values of σ_i^* and ϱ^* are not identical to the corresponding Hammett values. σ_i^* is determined using the hydrolysis rates for substituted and unsubstituted esters of acetic acid. It is somewhat more difficult to measure than the Hammett constant, because steric interactions must be eliminated. There is therefore a modified form of the T. $\lg (k_i/k_0) = \varrho^*\sigma_i^* + \delta E_s$. Here δ is a steric reaction constant and E_s is the steric substituent constant. The values of E_s are proportional to the Van der Waals radii of the substituents.

Talastin: see Antihistamines.

Talinolol: 4-(C_6H_{11}-NH-CO-NH)-C_6H_4-O-CH_2-CH(OH)-CH_2-NH-C(CH_3)$_3$, a phenyloxypropylamine derivative with an additional urea group. It is a relatively cardioselective β-receptor blocker.

Tall oil: a natural mixture of resin acids, saturated and unsaturated fatty acids, fatty acid esters and higher alcohols. It is a byproduct of cellulose production by the sulfate process. T. is a dark brown liquid, which can be separated by distillation into 60 to 85% fatty acids and 15 to 35% resin acids. The resin component (***tall resin***) contains abietic acid as its main component. T. is used as a paint, a drying oil for oil paints, an emulsion additive and flotation agent, and in soaps and plastics.

Talmi gold: a Brass (see) or silver alloy containing gold.

Talose: see Monosaccharides.

Tamman equation: a thermal equation of state for gases and liquids which is suitable for high pressures: $(p + \Pi(V - V_\infty) = cT$, where p is pressure,

V is the molar volume, T is Kelvin temperature and Π, V_∞ and c are substance-specific constants.

Tamoxifen: an antiestrogen used to treat metastasizing mammary carcinoma.

Tanabe-Sugano diagram: see Ligand field theory.

Tannic acid: see Tannin.

Tannin: a substance which is capable of converting an animal hide to leather (see Tanning). There are vegetable, mineral and synthetic organic T. ***Vegetable T.*** (tannins, tannic acids) are weakly acidic polyphenols which are soluble in water, and sometimes in ethanol. They form insoluble precipitates with proteins, alkaloids and heavy metal ions, and they form blue to green complexes with iron(III) ions. The vegetable T. are classified in two groups on the basis of their structures, the Gallotannins (see) and the Catechol tannins (see). T. are found widely in seed plants, especially in bark and leaves. Turkish and Chinese sumach, oak bark, hamamelis (witch hazel) leaves and rhatany root are especially rich in T. Coffee and black tea also contain large amounts of T. The amount of T. in a solution is usually determined by the powdered skin method, in which the T. is bound by the powdered skin.

The most important ***mineral T.*** are basic chromium(III) salts, but zirconium salts, polymeric phosphates, etc. are also used.

Synthetic organic T. are aldehydes, dialdehydes, sulfochlorides and condensation products of phenols and formaldehyde; their water-soluble sulfonic acids, the syntans, are similar to vegetable T. in their tanning properties. The syntans are sometimes produced in a process utilizing the sulfite liquors from treatment of wood for papermaking; the lignin sulfonic acids in these liquors also have a tanning effect.

Because of their astringent effects, which are due to precipitation of protein, the T. are used to treat various skin diseases or hemorrhoids. They are also used to make inks and as protective colloids in industry.

Tanning: the production of leather from animal hides or pelts. Leather is made from a wide variety of skins, especially cattle, horse, pig, sheep and goat, but also snake, lizard, etc. In general only the cutis is used for leather; the epidermis with the hair and the subcutis are removed. However, pelts of fur-bearing animals are tanned with the fur on.

The raw skins are conserved by salting and drying; they are first softened in water, freed of dirt and salt, and treated with calcium hydroxide and/or sodium sulfide to loosen the hair, or are allowed to decay slightly.

The loosened hairs are mechanically removed along with the epidermis, as is the connective tissue from the inside of the skin. The hairs are valuable for production of felt, and in some cases they are spun. The subcuticular connective tissue is used to make glue (woodworking glue) or it may be processed to make animal feed. The cleaned hide is then ***tanned***, that is, it is treated with Tannins (see) which form chemical bonds with the proteins in the skin. The leather which results is no longer as susceptible to rotting as the original protein, and can no longer be made into glue.

To obtain thin leather from thick hides, the cleaned hide or the leather may be split. Soft leather is made by treating the hide before tanning with an enzyme preparation, and the excess alkalies are neutralized with a weak organic acid.

There are various methods of T. In liquor T., plant tannins such as ground oak or spruce bark are spread over the skins in a pit (see below). *Tanning extracts*, e.g. from chestnut wood, various fruits, roots and leaves, are used to prepare or strengthen the liquor; synthetic tannins (see Tannins) are also used. The hides are first treated in a dilute liquor, which gives them the brown color of leather. In ***pit T.***, they are then spread out flat and stacked, with a layer of plant tannins between each hide and the next, in a T. pit. They are left for several months in the pit. With higher concentrations of tannins and mechanical rotation of the bath, the time required can be considerably shortened (***vat T.***). 2) In ***chrome T.***, basic salts of chromium(III) are used, giving the leather a gray-green color. 3) ***White T. (alum T.)*** makes use of aluminum compounds (e.g. alum), which give the leather a white color. 4) In chamoising, the hides are treated with oxidizable (unsaturated) oils. For many types of leather, various methods of T. are combined.

The raw leather is further processed after T.: It is shaved to an even thickness, dyed, treated with fat, stretched and pounded to smooth it, dried, bleached on the inside, softened, pebbled to remove natural scars and further soften it, and rolled to set the fibers. Colored leathers are often given another coat of surface color; sometimes artificial scars are applied. Smooth leathers are ironed or shined. Rough or velour leathers are brushed to given them a soft surface.

Tantalates: see Tantalum oxide.

Tantalic acid: see Tantalum oxide hydrate.

Tantalum, symbol ***Ta***: an element from group Vb of the periodic system, the Vanadium group (see), a heavy metal, Z 73, with natural isotopes with mass numbers 181 (99.9877%) and 180 (0.0123%, weakly radioactive, half-life $2.0 \cdot 10^{13}$ years), atomic mass 180.9479, valency mainly V, also IV, III, density 16.677, m.p. 2996 °C, b.p. 5425 °C, electrical conductivity 7.6 Sm mm^{-2}, superconductivity below 4.48 K, standard electrode potential (Ta/Ta^{5+}) -0.812 V.

Properties. Z. is a lustrous gray, strong metal which crystallizes in a cubic body-centered lattice. It is more ductile in pure form than niobium, can be cold worked, rolled, hammered and polished. Like niobium, it undergoes passivation and therefore is chemically very resistant to most acids and bases; it is only slightly solubilized by alkali hydroxide melts, but dissolves in hydrofluoric acid and in hot concentrated sulfuric acid. When heated, T. is oxidized by atmospheric oxygen to tantalum(V) oxide. T. powder burns, when heated in air, with a bright flame. Fluorine reacts with T. even when cold, while chlorine reacts with it only at higher temperatures, forming tantalum(V) chloride. Like vanadium and niobium, T. can absorb hydrogen even at room temperature; nitrogen reacts at higher temperatures to form tantalum nitride. In its compounds, T. is usually in the +V oxidation state; tantalum(V) oxide, tantalum(V) chloride and the fluorotantalates(V) are particularly important.

Analysis. Because its atomic and ionic radii are very similar to those of niobium, T. is chemically very similar to that element. The two elements are separated by ion-exchange and paper chromatography of the fluorometallates(V) (see Niobium. X-ray fluorescence analysis and atomic emission spectroscopy are important methods of determination. Gravimetric analysis can be done by precipitating the T. as tantalum(V) oxide hydrate and roasting it to tantalum(V) oxide.

Occurrence. T. makes up $2.9 \cdot 10^{-4}$% of the earth's crust, and is always found associated with niobium. Its most important mineral is columbite, (Fe, Mn)(Nb, $TaO_3)_2$, which is called tantalite when it contains a larger fraction of the tantalum component. T. is also found in association with the elements of group IIIb, as in fergusonite, Y(Ta, NbO_4), yttrotantalite, $(Y,U,FeZr)(Ta,NbO_3)_2$ and samarskite, (Ln, Fe^{II}, U, Ca)(Nb, Ta, $Fe^{III}O_3)_2$. Microlite is essentially a calcium ditantalate, $Ca_2Ta_2O_7$. Pyrochlor, $(Ca, Na)_2(Nb, Ta, TiO_3)_2$ (OH, F) is important as an ore for T.

Production. Niobium-T. ores are used as raw materials. The processing of these ores and separation of the two elements is discussed under Niobium (see). Metallic T. is obtained chiefly by reaction of potassium heptafluorotantalate(V) with sodium, reduction of tantalum(V) chloride or fluoride in a hydrogen plasma, or by electrolysis, e.g. of a tantalum(V) oxide/potassium heptafluorotantalate(V)/alkali chloride/alkali fluoride melt, and then refined in a way similar to that described for the homologous Niobium (see).

Applications. T. is used in many areas because of its chemical resistance and strength, including the construction of chemical equipment, tantalum rectifiers and capacitors, cathodes for electron and x-ray tubes, as a getter in electron tubes, to make medical equipment and prostheses, such as pins for bones, tweezer tips, cannulas, needles, clamps and screws for bone replacements and for dental drills. Tips for fountain pens are also made of T. In the past, it was used in light bulb filaments. Tantalum alloys (see) are also very versatile.

Historical. The discovery of T. was closely linked to that of Niobium (see). T. was first prepared in elemental form in 1903 by Von Bolten.

Tantalum alloys: alloys in which tantalum is the main element. T. containing 93% tantalum and 7% tungsten have good heat resistance and the ability to trap gas; they are used in the construction of electronic tubes. T. are used as carbide formers in hard alloys and in rust- and acid-resistant chrome-nickel steels. They are also used in making high-temperature materials for space and rocket technology. The tantalum is added to steel in the form of ***ferrotantalum***, which consists essentially of iron and tantalum.

Tantalum carbide: TaC, brass-yellow, cubic crystals; M_r 192.69, density 13.9, m.p. 3880 °C, b.p. around 5500 °C, Mohs hardness 9 to 10. T. is chemically very inert, dissolving only in hydrofluoric or sulfuric acid. It is made by reaction of tantalum powder with flame soot. T. is an important hard metal and is used to make cutting tools.

Tantalum chlorides:. ***Tantalum(V) chloride, tantalum pentachloride***, $TaCl_5$, colorless crystals; M_r 358.21, density 3.68, m.p. 215.9 °C, b.p. 232.9 °C. In the gas phase, $TaCl_5$ has a trigonal bipyramidal form; in the solid and in solutions, it has a dimeric structure similar to that of $NbCl_5$ (see Niobium chloride). It is a strong Lewis acid, and as such is an effective Friedel-Crafts catalyst; it forms octahedral complexes with donor molecules: $TaCl_5D$. $TaCl_5$ is obtained by heating tantalum, tantalum carbide or tantalum nitride in a stream of chlorine. Lower T. are obtained by reduction of $TaCl_5$, e.g. with aluminum in the presence of aluminum chloride around 230 °C.

Tantalum(III) chloride, tantalum trichloride, has a narrow range of homogeneity between $TaCl_{2.9}$ and $TaCl_{3.1}$. The lowest T. which have been confirmed are $TaCl_{2.5}$ and $TaCl_{2.33}$, which contain cluster structures: $[Ta_6Cl_{12}]Cl_n$ (n-2, 3). In the cationic clusters, the metal atoms occupy octahedral positions, while the 12 octahedral edges are occupied by chlorine atoms.

Tantalum(V) fluoride: ***tantalum pentafluoride***: TaF_5, colorless, prismatic crystals; M_r 275.94, density 4.74, m.p. 96.8 °C, b.p. 229.5 °C. T. in the gas phase exists as monomeric trigonal bipyramidal molecules, but in the solid it forms a tetrameric ring structure. It reacts with Lewis bases to form complexes; for example, with alkali fluorides it forms the important alkali fluorotantalates(V), $M[TaF_6]$, $M_2[TaF_7]$ and $M_3[TaF_8]$, of which the potassium heptafluorotantalate(V) is used in the separation of niobium and tantalum because of its insolubility in water. T. is obtained from the reaction of tantalum(V) chloride with hydrogen fluoride, while alkali fluorotantalates are obtained by reaction of alkali fluorides with tantalum(V) oxide dissolved in aqueous hydrofluoric acid.

Tantalum nitride: TaN, bronze, hexagonal crystals; M_r 194.95, density 16.30, m.p. 3360 °C. T. is obtained by heating tantalum in a nitrogen atmosphere.

Tantalum oxide, ***tantalum(V) oxide, ditantalum pentoxide***: Ta_2O_5, white, rhombic crystal powder; M_r 441.89, density 8.2, m.p. 1800 °C. T. is chemically inert and, because of its acidity, is insoluble in any acid except hydrofluoric. In an alkali hydroxide melt, it forms ***tantalates***, which in composition and chemistry are very similar to the niobates. T. is obtained by heating tantalum in a stream of oxygen or by roasting tantalum(V) oxide hydrate.

Tantalum(V) oxide hydrate: $Ta_2O_5 \cdot xH_2O$, colorless, jelly-like substance with a varying water content; it is often called ***tantalic acid***. Freshly precipitated T. is soluble in strong acids and bases. Heating converts it to tantalum(V) oxide. T. is an intermediate in the production of tantalum (see Niobium).

Tapping: the withdrawing of liquid metals, slags or carbides (calcium or boron carbide) from metallurgical or electrothermal furnaces by opening the tap hole.

Tar: a liquid to semisolid, dark brown to black byproduct of thermal treatment of petroleum, coal, wood, peat, etc., or a product of polymerization and condensation processes.

1) ***Coal tar*** is a brown to black, viscous to semisolid mass which, in the early days of the coke and city gas industry, was an undesired byproduct and a considerable nuisance. It became important only when it began to be used as an impregnating agent and with the discovery of coal tar dyes. Today it is a versatile starting material for the chemical industry, because dyes,

pharmaceuticals, plastics, explosives, perfumes, pesticides, disinfectants, fuels, solvents, etc. can be made from it. Coal T. is formed in dry distillation (coking and carbonization) of coal. Its composition depends mainly on the temperature of the coking process, and only to a small degree on the type of coal used.

Low-temperature T. are formed during the carbonization of coal at temperatures up to 600 °C. In a thin layer, this T. is a dark brown, transparent oil which smells like carbon disulfide. The different types of carbonization retorts produce different grades of T.: ***heating surface T.*** are formed in retorts which are heated from the outside, while ***rinse gas T.*** is formed in furnaces in which the heat is generated in the furnace. Heating surface T. are liquids which become stabilized on the retort walls through overheating; they have a pitch content of about 30%. Rinse gas T. are viscous and only partially stabilized, because the temperatures are not high enough to crack unstable components; polymerization of the unstable components leads to pitch contents of 50 to 60%. Distillation of these T. yields heating and fuel oils; by hydrogenation, they can be converted to high-quality diesel oils and carburetor fuels. Light oil, with a boiling point up to 200 °C, can be obtained in up to 10% yield (relative to the amount of T.) by distillation with superheated steam and subsequent chemical treatment. The low-temperature T. resemble lignite T. in their composition (see below).

Medium-temperature T. are formed by coking at 800 °C. They are fairly well cracked, and therefore are similar to the high-temperature T., but they differ from the latter in having larger contents of benzene, phenol and homologs, and a lower pitch content of about 50%.

High-temperature T. are the most important group for industrial processing. They are oily to viscous, dark brown to black, shiny liquids with a creosote-like odor. High temperature T. are formed by coking of coal at temperatures above 1000 °C, and their formation involves a high degree of secondary decomposition of the original T. The more thorough this secondary process is, the more the T. is divided into two types of product: the high-molecular weight substances (e.g. pitch) and the low-molecular-weight substances (e.g. benzene, naphthalene and gas). The amount of T. produced by coking varies between 3 and 4% of the dry coal, depending on the type of coal used. The T. yield also increases with the oxygen content of the coal. High temperature T. is processed by distillation, and there are now only a few coking plants where this is done. Instead, the T. is taken to special processing plants. About 90% of the high temperature T. is used as a liquid or solid mixture for various purposes. It is estimated that high temperature T. contains about 10,000 compounds, but only a few hundred of these have been characterized. Only a relatively few of them are isolated in pure form; they are used mainly in the dye and pharmaceutical industries.

The crude T. contains 2 to 5% water, which is either removed before the actual distillation, or it is driven off with the first fraction, the light oil. Ammonia or ammonium salts can be obtained from the aqueous solution. The T. is generally distilled in a continuous, vacuum process from retorts with 50 to 70 t capacity into columns. The distillate is separated into a first light oil and four main fractions.

a) Up to 80 °C, a small amount of light oil is obtained; cyclopentadiene and carbon disulfide are obtained from it.

b) *Light oil* is obtained at temperatures between 80 and 180 °C (about 3% of the T.). It is a yellow to brown, mobile liquid with a characteristic odor. It contains 45 to 50% benzene, 8 to 17% alkyl aromatics (such as toluene and xylene), 12 to 15% naphthalene, 8 to 12% phenols, 3 to 5% alkenes, 0.5 to 1% alkanes, 1 to 1.5% unsaturated and saturated cyclic compounds, 1 to 3% pyridine bases, 0.1% carbon disulfide, thiophene and other sulfur compounds, 0.2 to 0.3% nitriles, 1 to 1.5% acetone and cumarone. There are a number of methods for processing light oil; the one used depends on its composition. The pyridine bases are extracted by washing with dilute acids, and the phenols by extraction with alkalies. The neutral substances are separated by distillation. The fractions obtained in this way are used, after purification, as additives for fuels, as gasoline, as solvents, etc. Cumarone resin remains as a distillation residue. Some of the compounds isolated pure from the light oil fraction are benzene, toluene, o-, m- and p-xylene, pyridine, α-picoline, α,α'-lutidine, collidine, cyclopentadiene, pseudocumol, dicyclopentadiene and indene.

c) At temperatures between 180 and 230 °C, *medium oil* is distilled over (about 10% of the T.). It is a yellow to brown, mobil to viscous oil, and contains mainly 30 to 40% naphthalene, 15 to 25% phenols, and up to 5% pyridine, quinoline and quinaldine. Part of the naphthalene is allowed to crystallize out by cooling the oil in pans; it is isolated by centrifugation and pressing. The remaining liquid is distilled into two fractions, *carbolated oil* (190 to 220 °C) and *naphthalene oil* (220 to 240 °C). The carbolated oil contains most of the phenols and pyridine bases. The phenols are first extracted with alkalies, then released from the extract by carbon dioxide, and finally purified by distillation. The pyridine bases of the carbolated oil are then extracted with an acidic wash, released by ammonia, and also fractionated. Naphthalene is isolated from the naphthalene oil. The neutral oils remaining after isolation of the phenols, pyridine bases and naphthalene are used partly as such, and partly in mixtures with other T. oils as heating oils, impregnating oils and so on. Compounds isolated pure from carbolated oil include phenol, o-, m- and p-cresol.

d) *Heavy oil* (6 to 10%) is obtained at temperatures between 230 and 270 °C. This is a yellow, liquid to viscous mass containing 14 to 20% naphthalene and phenanthrene, 8 to 10% cresols and xylenols, 6% pyridine and quinoline bases, and hydrocarbons of unknown structure. Further distillation of heavy oil yields *naphthalene oil I*, which corresponds to medium oil, *naphthalene oil II* and a residue, which is added to the anthracene oil. Heavy oil is used as a heating and fuel oil (for heavy-oil motors). After removal of the valuable components, it is used as an impregnating agent for wood (carbolineum). Other products isolated from the heavy oil fraction are quinoline, isoquinoline, quinaldine, lepidine, indole, diphenyl, α- and β-methylnaphthalene, fluorene and diphenylene oxide.

e) From 270 °C to the beginning of decomposition, *anthracene oil* (15 to 20%) distills over. It is a somewhat greenish, very viscous oil. When it is cooled, 10 to 15% crystal mass (crude anthracene) precipitates, and this, when freed of the oil, is called the anthracene residue. The oily fractions are used as heating oil, and to make impregnating oil, carbolineum, T. fat oils, gasometer oils, etc.

f) The distillation residue is *Pitch* (see) (50 to 55%). This is allowed to cool in flat pans or special forms, and is further processed, either as granulated block pitch, or while still hot, as liquid pitch. Coal T. pitch is most often used together with T. oils to produced ***prepared T.*** These include the various kinds of T. for painting and insulation, parquet glue, roofing, streets, etc.

2) ***Lignite T.*** is a brown to brown-black solid. It is an important product of carbonization of lignite coal. The amount and composition of lignite T. depends on the coal used and the nature of the carbonization process. Externally heated carbonization produces less T., and specifically heavier T., while rinse gas carbonization produces a much larger yield of T. which has a high alkane content. ***Lignite carbonization T.*** is produced at temperatures of 550 to 650 °C, while ***lignite high temperature T.*** is produced at coking temperatures of 1000 to 1200 °C. The main product at these high temperatures is lignite high temperature coke (see Coking). Unlike coal T., which consists mainly of aromatic compounds, lignite coal T. consists mainly of alkanes of C_7 to C_{32}, but it also contains large amounts of unsaturated hydrocarbons up to C_{10}. Oxygen-containing compounds are also present; most of them are acidic and soluble in alkalies. Lignite coal T. also contains small amounts of nitrogen compounds in the form of pyridine bases and sulfur-containing compounds, mainly thiophene homologs. The processing of lignite coal T. depends on the nature of the T. Most methods depend on distillation, cracking and hydrogenation. The crude T. is first freed of dust, then of water, before further processing. Distillation yields a head product, light oil, and medium oil and hard paraffin as side-stream products. The distillation residue is further processed to make electrode coke or pitch. The medium oil is freed of the creosotes with sodium hydroxide solution, and of resinous components with sulfuric acid. It is then distilled again; after removal of the fraction boiling below 200 °C, the main products are diesel fuel and heating oil. The hard paraffins are used, after purification, mainly in the production of candles and waxes. The yields of individual products depend on the quality of the T. and the way it has been processed. There is about 10% loss of the original T. From 3 to 5% carburetor fuel, 25 to 45% diesel fuel, 15 to 35% heating oil, 10 to 11% paraffin, 10 to 15% pitch and 2 to 4% electrode coke can be obtained.

Hydrogenation of lignite coal T. can yield, depending on the process used, gasoline, medium oil, diesel fuel, paraffin or lubricating oil (see Coal hydrogenation). Lignite coal T. can also be cracked: it is heated at 4 to 6 MPa to about 500 °C, and then expanded into columns, producing mainly gasoline and diesel oil.

3) ***Wood T.*** is a dark brown to black, viscous, oily mass with a characteristic, sharp odor. It is produced by dry distillation of wood.

The compositions of various wood T. depends greatly on the type of wood used, and the method of carbonization. Deciduous trees give the best yields of T.; the T. from beech wood is especially valuable. Distillation produces 2% acetic acid, 0.5% wood alcohol, 5% light oil, 10% heavy oil, 65% pitch (soft pitch) and 17.5% water. The composition of conifer T. varies widely, depending mainly on the turpentine and resin oil content of the wood. The turpentine oil produced by distillation is used in paints.

Wood T. is used as a protective paint for wood and steel parts, to impregnate ropes, as a flotation substance and as heating oil; as a dermatological preparation and as a medicament for the respiratory passages.

4) ***Peat T.*** is a highly viscous, black liquid with a penetrating, sharp odor. It contains phenols, saturated and unsaturated aliphatic and aromatic hydrocarbons, pyridine bases, sulfur compounds and fatty acids. The yield of T. depends on the type of peat and the method of coking, and is in general between 3 and 8%. Crude gasoline, diesel oil, soft paraffin and so on can be obtained from peat T. by distillation and processing of the distillates. The residue is peat T. pitch.

5) ***Shale T.*** is a dark brown liquid with a density of 0.757 to 1, formed by carbonization of oil shale. It is processed mainly to lubricating and diesel oils.

6) ***Oil T.*** is formed by thermal decomposition of petroleum to oil gas, and in the production of water gas. Its properties and composition are similar to those of coal T., but its density and viscosity are lower. In addition, the amounts of phenols and basic substances in it are very low. Oil T. is often used as a heating or diesel fuel.

7) ***Water gas T.*** is a dark brown, oily liquid mass with a high water content (up to 36%). It is formed in the generation of water gas or generator gas, and contains mainly alkanes and aromatic decomposition products.

8) ***Fat T.*** is a brown, viscous mass formed during fractional distillation of fats and fatty oils. It is redistilled into various fatty acids. The residue is tough and, after cooling, rather hard. It is called stearic or fat pitch, and is used to insulate electrical cables.

9) ***Bone T.*** (***oil of hartshorn, animal oil***) is formed during carbonization of defatted, and often chipped bones. It is a blackish brown, thick liquid with an unpleasant smell from which Dippel's oil is obtained by distillation. The distillation residue is bone T. pitch. The yield of T. is 1.5 to 2%.

Targesin: same as Diacetyltannin protein silver (see).

Tarichatoxin: see Tetrodotoxin.

Tar pigments: see Pigments.

Tar sands: same as Oil sands (see).

Tartar: same as Potassium hydrogentartrate.

Tartrazine, ***hydrazine yellow O, fast light yellow, flavazine T***: a golden yellow, acidic pyrazolone dye which is readily soluble in water. It is made industrially by the reaction of phenylhydrazine-4-sulfonic acid with dihydroxytartaric acid or oxaloacetate. T. is used to dye animal fibers, to make inks for graphics, in the paper industry, to make light filters for photography and orthochromatic film, and in some countries, as a food coloring. T. was first synthesized in 1884 by H. Ziegler.

Tartronic acid, ***hydroxypropanedioic acid, hydroxymalonic acid***: HOOC-CH(OH)-COOH, the simplest hydroxydicarboxylic acid. T. forms colorless crystals which sublime above 110 °C; m.p. 156-158 °C (dec.). It is readily soluble in water and alcohol, and is slightly soluble in ether. T. is formed by oxidation of glycerol with potassium permanganate.

Taurine: see Cysteine.

Tautomerism: chemical equilibrium between structural isomers of a compound which are formed by intramolecular migration of a proton and simultaneous rearrangement of the bonding electrons. There are several classes of T. involving migration of the proton between different elements; some important ones are: ***three-carbon T.*** (in β,γ-unsaturated carboxylic acids), ***three-nitrogen T.*** (in diazoamino compounds), ***oxo-cyclo T.*** (in sugars), ***keto-enol T.*** (e.g. in 1,3-dicarbonyl compounds), ***ketol-endiol T.*** (in reductions), ***transannular T.*** (e.g. anthrone $\rightleftharpoons$ anthranol) and ***amide-iminol T.*** (in amides).

Taxin: an alkaloid present in the needles, twigs and wood, but not in the red seed coat, of the yew (*Taxus baccata*). It is only slightly soluble in water, but is soluble in organic solvents. T. is the toxic substance in yew needles and is responsible for the health-impairing effects of working yew wood (skin irritation, headache, general malaise).

Tb: symbol for terbium.

TBA: see Herbicides.

TBTO: see Fungicides.

Tc: symbol for technetium.

TCA: abb. for trichloroacetic acid.

TCC process: see Cracking.

TCDD: abb. for 2,3,7,8-tetrachlorodibenzo-1,4-dioxin.

tce: abb. for Tons of coal equivalent (see).

TCNB: see Fungicides.

TD nickel: a pseudoalloy of sintered nickel or nickel-chromium matrix and mechanically embedded thorium or yttrium oxide particles (1 to 2%). When an oxidation-resistant nickel-chromium matrix with 80% nickel and 20% chromium is used, the high heat resistance of the material is due to dispersion and mixed crystal reinforcement. These materials are used in space vehicles.

Te: symbol for tellurium.

Tear gas: see Chemical weapons.

Technetium, symbol ***Tc***: a radioactive element not found in nature; it belongs in group VIIb of the periodic system, the Manganese group (see). T. is a heavy metal, *Z* 43; isotopes with mass numbers 91 to 110 and half-lives between 0.83 s and 4.2 · 10 a have been prepared. The most stable isotopes (half-life, decay type in parentheses) are: ^{97}Tc (2.6 · 10^6 a, K-capture), ^{98}Tc (4.2 · 10^6 a, β^-), ^{99}Tc (2.12 · 10^5 a, β^-). The most stable isotope has a mass of 98; the valencies are IV to VII, 0; density 11.49, m.p. 2550 °C, b.p. 4877 °C.

T. is a silvery white metal crystallizing in a hexagonal close packed lattice; its chemistry is more similar to that of rhenium than to that of manganese. It is insoluble in hydrochloric acid and alkaline peroxide solutions, but is soluble in nitric acid. It burns above 400 °C in an oxygen stream to technetium(VII) oxide, Tc_2O_7. T. is prepared by reduction of ammonium pertechnetiate with hydrogen, or by cathodic precipitation from acidic solution. ^{99}Tc is a uranium fission product produced in 6% yield. It is isolated by methyl pyridine extraction of an aqueous pertechnetiate solution obtained by oxidation and subsequent removal of uranium and plutonium. T. is superconducting below 11 K.

Historical. In 1925, the discoverers of rhenium, I. Tacke and W. Noddack, also reported x-ray spectroscopic evidence for the still unknown element 43 in an enrichment fraction of columbite; they called it masurium. The report could not be confirmed, however, so that T. was actually discovered in 1937 by Perrier and Segrè, who obtained it as the first synthetic element by bombarding molybdenum with deuterium.

Technetium compounds: In most of its compounds, technetium is in the +7 oxidation state, but the +4 to +6 states are also known. It this, technetium is closer in its chemistry to rhenium than to manganese. When technetium is heated in a stream of oxygen to a temperature above 400 °C, it forms bright yellow, crystalline ***technetium(VII) oxide***, Tc_2O_7, m.p. 119.5 °C, b.p. 310.6 °C. Tc_2O_7 is hygroscopic, and dissolves in water with formation of ***pertechnetic acid***, $HTcO_4$. $HTcO_4$ is a strong acid, which can be obtained from solution in the form of dark red crystals; it forms pale yellow, stable salts, $M[TcO_4]$. Dark brown ***technetium(VII) sulfide***, Tc_2S_7, is precipitated out of hydrochloric acid solutions of pertechnetiates by hydrogen sulfide; this compound is isomorphic with Re_2S_7. In alkaline media, Tc_2S_7 is easily converted by hydrogen peroxide to ***pertechnetiates***. These salts are much weaker oxidizing agents than the corresponding permanganates. With fluorine, technetium forms yellow ***technetium(VI) fluoride***, TcF_6, m.p. 37.4 °C, b.p. 55.3 °C, while chlorination leads to green ***technetium(VI) chloride***, $TcCl_6$. The chloride is converted by thermal decomposition into a red octahedral polymer with chlorine bridges, ***technetium(IV) chloride***, $TcCl_4$. Technetium forms complexes, such as ***potassium nonahydridotechnetiate***, $K_2[TcH_9]$, which is stable up to 200 °C, or the colorless, crystalline ***technetium carbonyl, ditechnetium decacarbonyl***, $Tc_2(CO)_{10}$, m.p. 159 °C.

Technical diagnostics: see Materials testing.

Technical gases: combustible and non-combustible gases synthesized and used on an industrial scale (see separate entries on Fuel gases, Liquid gases, Generator gas, City gas, Water gas, Mixed gas). Before use, T. must be purified (see Gas treatment).

Technofthalam: see Bacteriocide.

Technological microbiology: see Biotechnology.

Technological diagram: a diagram of the apparatus and machinery in a plant. Standardized *symbols* (Fig. 1-4) are used. The basis of the T. is a *flow chart* (Fig. 3). The symbols are connected to indicate the flow of material. Devices for measurement, direction and control of the flow, containment and safety devices are also indicated symbolically. The *montage diagram* shows all the pipes in a plant; the diameters, nominal pressures and materials of the pipes are indicated. Unlike the T., the montage diagram shows all machines and apparatus to scale. In the montage of new plants, it is now common practice to work with scale models. *Material* and *energy balance graphics* are often used to show the flow of materials and energy (Fig. 4).

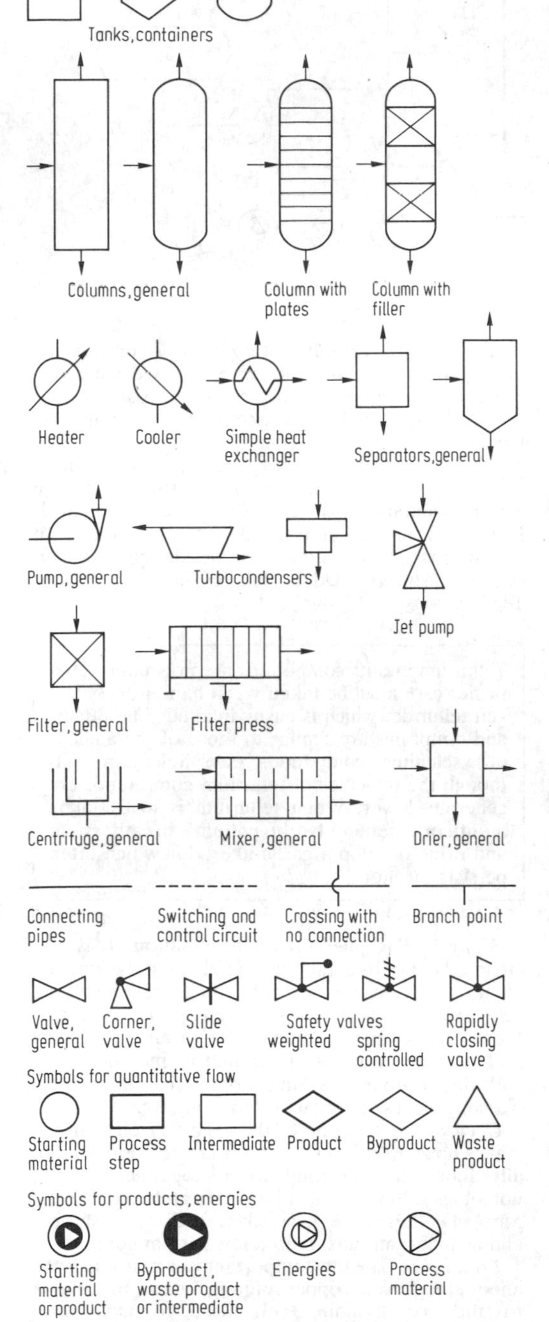

Fig. 1. Symbols for chemical technology.

Teichmann's crystals: see Hemin.

Teichoic acids: components of the cell walls of gram-positive bacteria. T. consist of glycerol (*glycerol T.*), ribitol (*ribitol T.*) or *N*-acetylglucosamine units linked by phosphate groups. Free hydroxyl groups can be bound in ester linkages to D-alanine or in glycosidic linkages to monosaccharides. The T. are bound to the murein by the terminal phosphate group. T. play a role in the regulation of ion permeability and enzyme activity.

Teichoic acid from *Micrococcus sp.*

Glycerol-teichoic acid from *Bacillus stearothermophilus* (R^1 = α-D-glucopyranosyl; R^2 = D-alanine).

TEL: see Anti-knock compounds.

Telinite: see Macerals.

Tellurates: the salts of Telluric acid (see).

Telluric acid, more precisely, ***orthotelluric acid***: H_6TeO_6, colorless, monoclinic crystals; M_r 229.64, density 3.158, m.p. 136 °C. An acid with a composition corresponding to sulfuric or selenic acids, H_2TeO_4, is not known. T. is a weak hexabasic acid (pK_1 = 7.70), and forms several series of salts, the ***hydrogentellurates***, $M^I_nH_{6-n}TeO_6$, and the ***neutral tellurates***, $M^I_6TeO_6$. In addition, tellurates are known in which the anions are derivatives of condensed polytelluric acids. T. and tellurates are poisonous (see Tellurium).

Tellurides: the salts of Hydrogen telluride (see).

Tellurites: the salts of Tellurious acid (see).

Tellurium, symbol ***Te***: a chemical element from group VIa of the periodic system, the Oxygen-sulfur group (see). It is a semimetal, *Z* 52, with natural isotopes with mass numbers 120 (0.089%), 122 (2.46%), 123 (0.78%), 124 (4.61%), 125 (6.99%), 126 (18.71%), 128 (31.79%), 130 (34.48%), atomic mass 127.60, valency IV, II, VI, Mohs hardness 2.5, density 6.25, m.p. 452 °C, b.p. 1390 °C, standard electrode potential (Te^{2-}/Te) -0.92 V.

Properties. T. has an amorphous brown form and a single crystalline modification, the hexagonal, lustrous white, brittle, metallic T. The crystal structure of this modification is the same as that of gray metallic selenium. The golden-yellow vapor of T. consists mainly of Te_2 molecules. As a semimetal, T. has a low electrical conductivity, which increases when light falls on it, but less markedly than that of selenium. The preferred oxidation states of T. are -II, +IV and +VI; the increasing stability of derivatives in the +IV state compared to the +VI state, which is already noticeable on going from sulfur to selenium, is con-

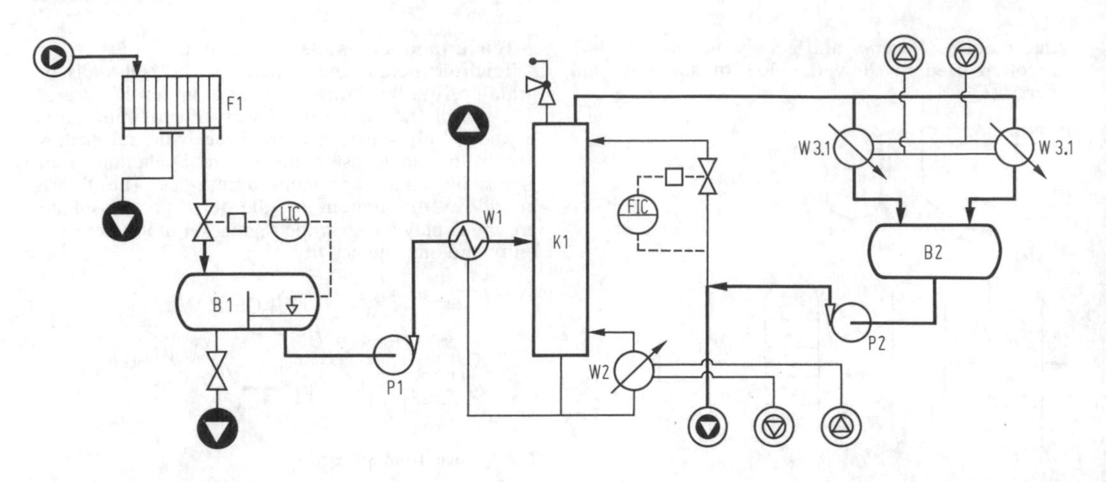

Fig. 2. Technological diagram of a chemical plant.

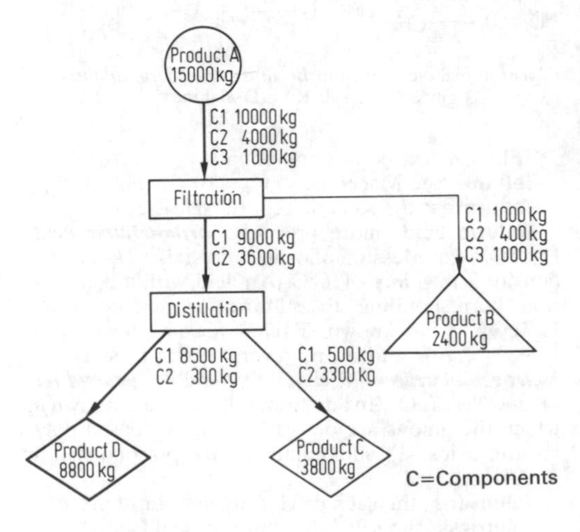

Fig. 3. Flow chart.

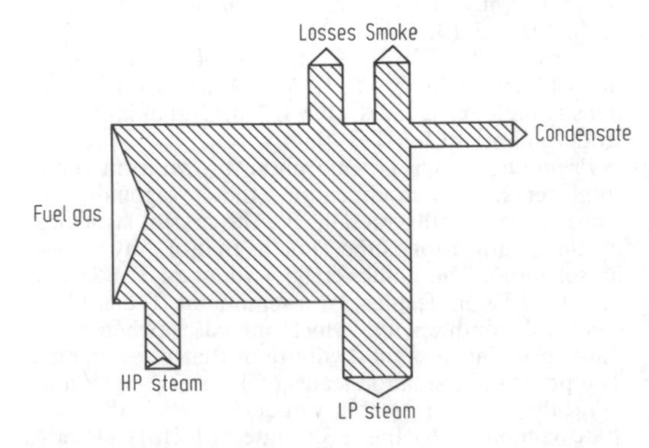

Fig. 4. Energy balance graphics.

tinued here. As a result, tellurates are strong oxidizing agents. Under the influence of an electric discharge, T. reacts with water to form hydrogen telluride, TeH_2. The halogens convert T. to tellurium halides with the compositions TeX_2, TeX_4 and TeX_6. The element burns in the air to form tellurium dioxide, TeO_2. Strong oxidizing agents, such as chloric acid or sodium peroxide, convert T. to telluric acid, H_6TeO_6, or into tellurates, $M^I_nH_{6-n}TeO_6$. The intense red color which is observed when T. dissolves in concentrated sulfuric acid is due to the formation of Te_4^{2+} cations.

> Tellurium and its compounds are poisonous. Particular care must be taken when handling Hydrogen telluride, which is easily inhaled. The effects and symptoms are similar to those of the analogous selenium compounds (see Selenium), although the toxicities of tellurium compounds are generally lower. When tellurium is ingested by mouth or parenterally, the patient's breath, sweat and urine develop a garlic-like odor which often persists for months.

Analysis. For qualitative determination, T. is precipitated out of its compounds in elemental form, and identified by treatment with concentrated sulfuric acid, with which it forms an intense red solution. Iodometry (oxidizing effect of Te^{4+}) and, especially at low concentrations, photometric methods (e.g. with dithiocarbamates) and atomic absorption spectroscopy are used for quantitative analysis.

Occurrence. T. is one of the very rare elements; it makes up $2 \cdot 10^{-7}\%$ of the earth's crust. It is occasionally found in elemental form, together with its homologs sulfur and selenium. It is also found in the form of tellurides, e.g. as silver and lead telluride, and as tellurium dioxide in a few rare minerals.

Production. The most important raw material is the anode sludge from copper refining, which, in addition to the corresponding selenides, contains small amounts of copper, silver and gold telluride. This material is solubilized by fusion with soda/sodium nitrate, or by treatment with nitric acid. After it has been dissolved in water, T. is separated from selenium by precipitation of tellurium dioxide out of acidic solution or by fractional reduction with sulfur dioxide. The precipitated tellurium dioxide, TeO_2, is reduced electrolytically to the element, which, like directly precipitated T., can be purified by electrolysis or distillation.

Applications. T. is used as a component of copper and lead alloys. Some tellurides, e.g. bismuth and lead telluride, are of some importance as semiconductors.

Historical. T. (from the Latin "tellus" = "earth") was first discovered in 1782 in Rumanian gold tellurides by Müller von Reichenstein, and was recognized as a chemical element in 1798 by Klaproth.

Tellurium oxides: ***Tellurium(IV) oxide, tellurium dioxide***, TeO_2, colorless, tetragonal crystals which are barely soluble in water; M_r 159.60, density 5.67, m.p. 733 °C, b.p. 1245 °C. TeO_2 is amphoteric, dissolving in alkali metal hydroxide solutions to form the corresponding tellurites ($TeO_2 + 2\,OH^- \rightarrow TeO_3^{2-} + H_2O$), and in strong acids to form salts: $TeO_2 + 4\,H^+ \rightarrow Te^{4+} + 2\,H_2O$. TeO_2 is made by dissolving tellurium in nitric acid, then diluting the solution with water.

Tellurium(VI) oxide, tellurium trioxide, TeO_3, is an orange, amorphous powder; M_r 175.60, density 5.067. When heated, TeO_3 releases oxygen to form TeO_2. It is obtained by dehydration of telluric acid, H_6TeO_6. The back reaction of the acid anhydride, TeO_3, with water to form telluric acid, is very slow. The T. are poisonous (see Tellurium).

Tellurocyanate: $[NCTe]^-$, an anion which is a Pseudohalide (see).

Tellurious acid: H_2TeO_3, a weak acid not known in isolated form ($pK_1 = 2.48$, $pK_2 = 7.70$). Its salts, the ***tellurites***, $M^I{}_2TeO_3$, are formed by dissolving tellurium(IV) oxide, TeO_2, in alkali metal hydroxide solutions, or by fusion with the coresponding metal oxides or carbonates. T. and tellurites are poisonous (see Tellurium).

Telogens: see Telomerization.

Telomerization: a special form of polymerization which generally occurs by a radical mechanism; the initial reaction is the light- or heat-induced decomposition of a ***telogen***, e.g. a halogen hydrocarbon (carbon tetrachloride, chloroform or dichloromethane), acyl chloride or silicon chloride, into two radicals, which then add to a vinyl monomer such as ethene or tetrafluoroethylene. T. is usually carried out as a solution polymerization, in which the solvent acts as telogen. For example: T. of ethene in carbon tetrachloride:

$$CCl_4 \rightarrow CCl_3\cdot + Cl\cdot \quad CCl_3\cdot + Cl\cdot$$
$$+\, nCH_2{=}CH_2 \rightarrow CCl_3[CH_2\text{-}CH_2]_nCl.$$

The product ***telomers*** have low relative molar masses. They are used primarily as lubricants, softeners, polishes, textile conditioners and synthetic waxes.

Tempera paints: dry pigments, usually inorganic, used as artist's colors. They are emulsified in water, and the emulsions also contain an oil such as linseed or poppyseed oil. To prevent separation, a protective colloid, such as eggwhite, casein (casein tempera) gum arabic (gum tempera), wax solution or soaps, is added. The paints are preserved by phenol, salicylic acid or a similar compound. They do not lose their solubility in water after they dry.

Temperature, symbol T: a parameter which characterizes systems in thermodynamic equilibrium. In the narrow sense, T. applies only to equilibrium systems, but in the wider sense, it is ordinarily applied also to systems which are not in equilibrium, provided that in small, but still macroscopic regions, there is a local equilibrium. The concept of T. must be clearly distinguished from that of heat, which is a form of energy. The T. is an intensive parameter, which has the same value for all systems in thermodynamic equilibrium with each other. All parts of a thermodynamically closed, physical system exchange heat with one another until all have the same T. This is utilized to measure the T. with a thermometer. The *empirical T.* is the assignment of a numerical value to all bodies which are found to be equally warm with reference to a particular body, while the *absolute T. (thermodynamic T.)* is defined in terms of the efficiency of a Carnot cycle (see). The amounts of heat exchanged by two heat reservoirs, $Q_{1,2}$, are in the same ratio as the absolute T. of the two reservoirs: $Q_1/Q_2 = T_1/T_2$.

The SI unit of thermodynamic T. is the kelvin (K). It is defined as one 273.16th of the thermodynamic T. of the triple point of water: $T = 273.16$ K. The kelvin is a unit of both point T. and T. ranges (differences). The difference between the T. of a system and the ice point, $t = T - T_0$, where $T_0 = 273.15$ K, is defined as the Celsius temperature (t) and is given in degrees (°C).

Critical points (see) play a special role.

Low T. can be reached using liquid nitrogen (down to 77 K), liquid hydrogen (to 22 K) or liquid helium (1 K). Still lower T., down to 10^{-3} or 10^{-6} K can be reached by adiabatic demagnetization of paramagnetic electrons or nuclear spin systems. According to the 3rd law of thermodynamics (Nernst's heat theorem), it is impossible in principle to achieve absolute zero, $T = 0$. The highest T. are observed in plasmas. For short times, nuclear fusion plasmas can be generated in a laboratory with T. of 10^7 to 10^8 K.

The ***measurement*** of T. with thermometers or thermoelements is due to transfer of heat to or from the thermometer, which is in direct contact with the system. The amount of energy absorbed or released by the thermometer must be as low as possible, in order to prevent significant change in the T. of the system. With radiation measurements, it is possible to measure T. above 1000 K with pyrometers. T. up to 3300 K can also be measured with carbon thermocouples. The Seger cones (see) which are widely used for firing ceramics give only an approximate reading of T.

Tempering: heat treatment of materials. 1) In metallurgy, it is a long roasting of graphite-free pieces of cast iron to destroy cementite (Fe_3C) (see Iron, cast).

2) In silicate technology, the temperature treatment of silicate materials (see Glass, Enamel, Glass ceramics) to improve their mechanical, thermal and chemical properties, e.g. through reduction of stresses and control of demixing processes.

Template syntheses: syntheses in which metal

ions are used as reaction-directing matrices. T. are used in many ways, particularly for the synthesis of macrocycles containing nitrogen (e.g. via Schiff's base reactions) or for production of phthalocyanins from phthalodinitriles.

Tepa: a Chemical sterilant (see).

Terbium, symbol ***Tb***: a chemical element from group IIIb of the periodic system, and one of the Lanthanoids. It is a heavy, rare-earth metal and has only one isotope, *Z* 65, atomic mass 158.9254, valency III, rarely IV, m.p. 1356 °C, b.p. 2480 °C, standard electrode potential (Tb/Tb^{3+}) -2.391 V.

T. is a silvery-gray, ductile metal which is hexagonal (density 8.272) below 1315 °C and cubic-body-centered above that temperature. It makes up $8.5 \cdot 10^{-5}$% of the earth's crust, and is always found in association with the other rare-earth metals, especially yttrium and the heavier lanthanoids. It is relatively enriched in gadolinite (ytterbite), xenotim, fergusonite and yttrotantalite; it is present at about 0.03% concentration in monacite. Further properties, analysis, production and history are discussed under Lanthanoids (see). T. is used in the form of Cerium mixed metal (see).

Terbium compounds: compounds in which terbium is generally in the +III state, less commonly in the +IV state. Terbium(IV) oxide, TbO_2, has a fluorite structure; it is made by oxidation of Tb_2O_3 with atomic oxygen at 450 °C. Fluorination of terbium(III) fluoride with fluorine at 300 to 400 °C produces the colorless terbium(IV) fluoride, TbF_4, which has the same structure as CeF_4 and ThF_4. The general properties and production of the terbium(III) compounds are discussed under Lanthanoid compounds (see). Some important terbium(III) compounds are: ***terbium(III) chloride hydrate***, $TbCl_3 \cdot 6H_2O$, colorless prisms, M_r 373.38, density (anhydrous), 4.35, m.p. (anhydrous), 588 °C; ***terbium(III) fluoride***, TbF_3, M_r 215.92, m.p. 1172 °C; ***terbium(III) oxide***, Tb_2O_3, white, M_r 365.85, density 7.90; it is used as an activator for green luminophores. ***Terbium(III) nitrate***, $Tb(NO_3) \cdot 6H_2O$, forms colorless, monoclinic crystals, M_r 453.03, and ***terbium(III) oxalate***, $Tb(C_2O_4)_3 \cdot 10H_2O$, forms white crystals, M_r 762.06, density 2.60.

Terbumetone: see Triazine herbicides.

Terbutaline: $3,5(OH)_2C_6H_3$-CHOH-CH_2-NH=$C(CH_3)_3$, a phenylethylamine derivative which acts as a sympathicomimetic. It is used medicinally for asthma; it has no significant effect on the heart.

Terbutryn: see Triazine herbicides.

Terephthalic acid, ***benzene-1,4-dicarboxylic acid***: HOOC-C_6H_4-COOH; colorless and odorless crystalline prisms or needles, m.p. 425 °C (in a closed tube), sublimation above 300 °C. T. is insoluble in water, glacial acetic acid, ether, chloroform and cold alcohol; it is somewhat soluble in boiling alcohol and dissolves readily in dimethylacetamide. T. undergoes the reactions typical of dicarboxylic acids. It is synthesized industrially by air oxidation of *p*-xylene in glacial acetic acid in the presence of heavy metal catalysts and bromine compounds, by oxidation of 4-methylbenzaldehyde, or by rearrangement of potassium phthalate in the presence of zinc or cadmium salts. T. is used mainly for production of polyester fibers.

Term: see Multiplet structure.

Terminal voltage: E_t, the voltage between the poles of an electrochemical cell.

Terpenes: groups of natural products which can be regarded formally as polymerization products of the hydrocarbon isoprene. Depending on the number of isoprene (C_5) units they contain, T. are classified as mono- (2), sesqui- (3), di- (4), sester- (5), tri- (6), tetra- (8) or polyterpenes. T. may be acyclic or contain one or more rings; they also may be classified according to their functional groups as hydrocarbons, alcohols, aldehydes, ketones, etc.

The biosynthesis of T. starts from mevalonic acid (actually its pyrophosphate ester), which is formed from three molecules of acetyl-coenzyme A after reductive cleavage of the coenzyme A and phosphorylation. Removal of water and decarboxylation forms isopent-3-enylpyrophosphate ("active isoprene"), which is converted by an isomerase into the more stable pyrophosphate of 3,3-dimethylallyl alcohol (prenol). With pyrophosphate as a leaving group, this compound is a strong electrophile, and attacks the double bond of isopentenyl pyrophosphate; the resulting addition product is a C_{10} compound (see Monoterpenes), geranyl pyrophosphate. This type of linkage is called *head-to-tail condensation*; T. synthesized in this way have the methyl groups in the 1,5-positions. As an allyl ester, geranyl pyrophosphate in turn can attack another molecule of isopentenyl pyrophosphate, forming a C_{15} compound (see Sesquiterpenes), farnesyl pyrophosphate. Further extensions by C_5 units (prenyl groups) lead to Diterpenes (see) and Sesterterpenes (see). Dimerization of farnesyl or geranylgeranyl pyrophosphate leads to Triterpenes (see) and Tetraterpenes (see). The dimerizations occur by *tail-to-tail condensation*, and leave methyl groups in the middle part of the molecule in 1,6-positions. In addition to such intermolecular substitution-elimination processes, which lead to acyclic compounds, there are intramolecular cycloadditions. In mono- and sesquiterpenes, a double bond makes a nucleophilic attack on the C atom with the terminal pyrophosphate. This produces a 6- or 11-membered carbocation, from which the various basic structures of the cyclic mono- and sesquiterpenes are formed. Monoterpenes tend to form cyclohexane derivatives, while sesquiterpenes can form rings with up to 10 members. Diterpenes tend to undergo another type of intramolecular cyclization, in which the starting materials are not phosphorylated. The reaction starts from the double bond in the terminal isoprene unit, which is oxidized to an epoxide and thus forms a cationic center on the C2 atom; this attacks the nearest double bond of the polyene. This initiates a sequence of electrophilic, stereospecific cycloadditions to tri-, tetra- or pentacyclic rings. A typical example is the intramolecular cycloaddition of squalene via squalene oxide to the methylsterols. The distribution of the methyl groups in the T. corresponds to that in isoprene (*isoprene rule*). Deviations from the isoprene rule arise by simultaneous methyl and hydride shifts, for example in the methylsterols derived from triterpenes. (These are precursors of the steroids, which have 3 C atoms fewer). The multitude of naturally occuring structures arise from alkyl and hydride shifts, dehydrogenations, oxidations, eliminations and other secondary changes.

Occurrence. The T. are typical secondary plant metabolites, but they are also produced by microorganisms (e.g. carotenoids and sesterterpenes). The biosynthesis of squalene, as the starting material for steroids, is very common in animals; however, animals can also synthesize carotenoids and other T. (e.g. cantharidine, insect hormones, pheromones). In plants, T. are present in high concentrations in the volatile oils, resins, balsams and latexes. The T. also include bitter substances (e.g. that of hops), pigments (e.g. carotenoids) and saponins. Chlorophyll, various vitamins and alkaloids also contain isoprenoid components.

Applications. T. are widely used in the form of volatile oils or mixtures of synthetic compounds as perfumes, spices, aromas and essences. Components of readily available volatile oils, such as pinene from terpentine oil, are used as starting products for industrial partial syntheses. Rubber is of great industrial importance. Many T. are important because of their biological activity, for example insecticides (pyrethrins), antiseptics (thymol-containing volatile oils), anthelminthics (such as santonine, ascaridol), analeptics (e.g. picrotoxin) or skin irritants (e.g. cantharidine). Some T. are phytohormones.

1,8-Terpin, ***trans-terpin***: 1-methyl-4-isopropylcyclohexane-1,8-diol, m.p. 104-105 °C, b.p. 258 °C, m.p. (hydrate) 120-121 °C. 1,8-T. is poorly soluble in cold water, but is soluble in alcohol. It is probably not a primary component of essential oils, but is formed in the presence of aqueous acids from turpentine, α- and β-pinene, α- and β-terpineol or acyclic monoterpene alcohols such as linalool, geraniol or limonene. 1,8-T. is used to make terpineol.

CH_3 OH OH H_3C CH_3

Terpinenes: monocyclic, monoterpene hydrocarbons. The T. are found in numerous essential oils, such as cardamom and coriander oils. The main component is α-T., which is accompanied by γ-T. β-T. is rare. α-T. is a colorless oil with a lemon-like odor; b.p. 180-182 °C. When treated with hydrochloric acid, it forms terpinene dihydrochloride. The T. are used to make artificial essential oils and to improve the odors of technological products.

CH_3 CH_2 CH_3
α-Terpinene β-Terpinene γ-Terpinene
H_3C CH_3 H_3C CH_3 H_3C CH_3

Terpineols: monocyclic monoterpene alcohols. T. are formed when terpine hydrate is treated with dehydrating reagents. Commercial T. is obtained from turpentine, either directly from pinene or via terpine hydrate. It is a liquid which consists chiefly of α-T. with a little β-T. and probably some γ-T. α-T., 1-methyl-4-isopropylcyclohex-1-ene-8-ol, exists in a d- and an l-form, and is found free and esterified in many essential oils. It forms colorless crystals which smell like lilac; m.p.$_d$ 38-40 °C, m.p.$_l$ 35-38 °C, m.p.$_{dl}$ 35 °C, b.p.$_{dl}$ 217-219 °C. α-T. is not stable in the presence of acids or bases. It is often used in perfumes and soaps. β-T., 1-methyl-4-isopropenylcyclohexan-1-ol, exists in a *cis* and a *trans* form. γ-T. is 1-methyl-4-isopropylidenecyclohexan-1-ol.

CH_3
α-Terpineol
OH
H_3C CH_3

Terra cotta: a ceramic product with a porous body, usually red and unglazed; it is made of clay, sand, brick and shamotte powder and fired at 900 to 1050 °C. T. has been used since antiquity, mainly for decoration. The Roman ***terra sigillata*** is a refined T. made from a red-firing clay with stamped decorations. It has a clearer firing color, finer structures and an unglazed but dense appearing, semimatte surface with a slip of finely ground clay.

Tertiary structure: see Biopolymers; see Proteins.

Terylene®: see Synthetic fibers.

Test gasoline: see Gasoline.

Testosterone: see Androgens.

Testosterone enanthate: see Androgens.

Tetrabenzotetraazoporphins: same as Phthalocyanins (see).

Tetrabromofluorescein: see Eosin pigments.

Tetracain: see Local anesthetics.

Tetracene: see Acenes.

Tetrachloroauric(II) acid: see Gold chlorides.

2,3,7,8-Tetrachlorodibenzo-1,4-dioxin, abb. ***TCDD*** (commonly called "***dioxin***": a highly mutagenic, teratogenic and carcinogenic derivative of 1,4-dioxin. It is the most toxic chemical compound so far synthesized (LD_{50}: 10 μg kg^{-1} guinea pig, p.o.). 2,3,7,8-T. is a byproduct in the synthesis of polychlorinated phenols formed by thermal dimerization and dehydrohalogenation of polychloropyrocatechols. 2,3,7,8-T. has been detected in the smoke from garbage incineration plants and from combustion of polychlorinated plastics.

In 1976, about 135 g 2,3,7,8-T. was released in an accident at the chlorophenol plant owned by the firm ICMESA in Seveso, Italy, and contaminated an area of more than 100 ha. Enormous damage was done in Vietnam, where between 1961 and 1971 the US Army released 110 kg 2,3,7,8-T. as a contaminant of the defoliant Agent Orange (see).

1,1,2,2-Tetrachloroethane, ***sym. tetrachloroethane, acetylene tetrachloride***: $Cl_2CH-CHCl_2$, a colorless, highly refractive liquid with a smell like chloroform; m.p. -36 °C, b.p. 146.2 °C, n_D^{20} 1.4940. 1,1,2,2-T. is slightly soluble in water and readily soluble in most organic solvents. In the absence of light and air, it can be stored for a long period without decomposition. It reacts with alkalies under mild conditions to produce trichloroethylene.

> 1,1,2,2-Tetrachloroethane is probably the most poisonous of all chlorinated hydrocarbons. Its toxicity is about 10 times higher than that of carbon tetrachloride. 1,1,2,2-tetrachloroethane can be absorbed through the lungs or skin. Intoxication can damage the nervous system, gastrointestinal tract, kidneys and liver. Work with this compound should always be done under a hood with a good draft.

1,1,2,2-T. is made by addition of chlorine to acetylene in the presence of iron(III) or antimony(V) chloride, or from chlorine and ethylene-1,2-dichloroethane mixtures. It is used mainly to produce trichloroethylene and as a solvent. The isomeric 1,1,1,2-tetrachloroethane, which is also formed in the reaction of chlorine and ethene, has no major industrial applications.

Tetrachloroethene, ***perchloroethene***: $Cl_2C{=}CCl_2$, a colorless liquid which smells like ether; m.p. -19 °C (-22.18° has also been reported), b.p. 121 °C, n_D^{20} 1.5053. T. is insoluble in water, but dissolves well in ether, benzene, chloroform and alcohol. It is relatively stable, and can be stored for long periods in the presence of small amounts of stabilizer. It is synthesized by high-temperature chlorination of lower alkanes or their chlorine derivatives; in addition to T., carbon tetrachloride is formed as a major byproduct. T. is used as a solvent for fats, oils and waxes, and as an intermediate for the synthesis of dichloroacetic acid and fluorohydrocarbons.

Tetrachloromethane: same as Carbon tetrachloride (see).

Tetrachlorosilane: same as Silicon tetrachloride (see).

3′,4′,5′,6′-Tetrachloro-2,4,5,7-tetraiodofluorescein: same as Rose bengal (see).

Tetracyanoethene: same as Tetracyanoethylene (see).

Tetracyanoethylene, ***tetracyanoethene, ethene tetracarbonitrile***: $(NC)_2C{=}C(CN)_2$, colorless, poisonous crystals; m.p. 201-202 °C (sealed capillary), b.p. 223 °C. T. is soluble in acetone. In the presence of moisture, it is slowly decomposed with loss of hydrogen cyanide. T. is an alkene with an electron-deficient C=C double bond, and as a result, it reacts with many aromatic compounds to form intensely colored charge-transfer complexes. With suitable alkenes or dienes, it undergoes cycloaddition to form four- and six-membered rings. T. can be obtained by the reaction of chlorine with malonic dinitrile, or of dibromomalonic dinitrile with copper powder. It is used for organic syntheses.

Tetracyanomethane: see Cyanides.

Tetracyclines: a group of antibiotics with a skeleton consisting of four linearly fused rings. The T. are synthesized by various species of *Streptomyces*; the structures of some are altered by partial synthesis. The biosynthetic pathway starts with acetate units. The first T. was ***chlorotetracycline (Aureomycin®)***, which was discovered in 1948 by Dugar in the culture media of *Streptomyces aureofaciens*; m.p. 169 °C, $[\alpha]_D^{23}$ -275° (in methanol). Shortly thereafter, ***oxytetracycline (Terramycin®)*** was discovered; it is synthesized by *Streptomyces rimosus*; for the dihydrate, m.p. 185 °C (dec.), $[\alpha]_D^{25}$ -197° (in 0.1 M HCl). ***Tetracycline*** was first obtained by reductive dehalogenation of chlorotetracycline, but now obtained by fermentation technology, using low-chloride nutrient media. ***Doxycycline*** is obtained by partial synthesis from oxytetracycline, and ***rolitetracycline*** is made by aminomethylation (Mannich reaction) of tetracycline. The T. are amphoteric compounds with a basic dimethylamino group (pK_a 9.7) and acidic hydroxyl groups on C3 (pK_a 3.3, as a vinyl carbonyl group) C10 and C12 (pK_a 7.7). The isoelectric point is at pH 4.8. Thus the compounds are zwitterions in the weakly acidic range. The T. form complexes with polyvalent metal ions, such as Ca^{2+}. They are not stable in aqueous solutions, particularly in acidic or basic solutions, undergoing epimerization, isomerization and dehydrations. They are broad-spectrum antibiotics which are effective against many gram-positive and gram-negative pathogens. The individual T. have nearly identical action spectra, but they are absorbed and eliminated with differing degrees of efficiency.

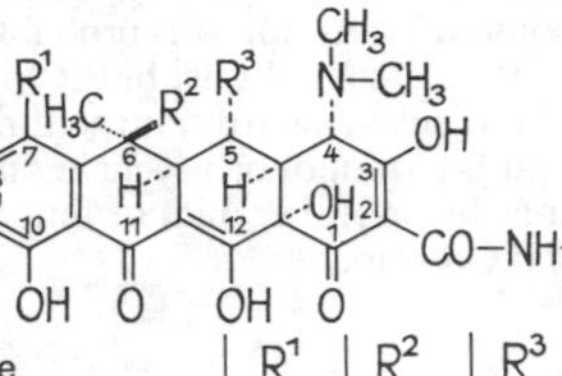

Name	R^1	R^2	R^3	R^4
Tetracycline	H	OH	H	H
Oxytetracycline	H	OH	OH	H
Chlortetracycline	Cl	OH	H	H
Doxycycline	H	H	OH	H
Rolitetracycline	H	OH	H	$-CH_2-N$(pyrrolidine)

Tetradecanoic acid: same as Myristic acid (see).

Tetradifon: an Acaricide (see).

Tetraethyllead: see Antiknock compounds, Organolead compounds.

Tetrafluoroethene, ***perfluoroethene, tetrafluoroethylene***: $F_2C = CF_2$, a colorless and odorless gas; m.p. -142.5 °C, b.p. -76.3 °C. T. is very reactive and forms highly explosive peroxides with oxygen. The addition of halogens or hydrogen halides leads to the corresponding fluoroethane derivatives. T. dimerizes in the gas phase at high temperatures, forming octafluorocyclobutane, a valuable coolant. T. is obtained by pyrolysis of chlorodifluoromethane, with loss of hydrogen chloride, or by thermal cleavage of trifluoroacetic acid: $2\ CF_3C\text{-}COOH \rightarrow F_2C{=}CF_2 + 2\ HF + 2\ CO_2$. T. is used mainly for the synthesis of polytetrafluoroethylene.

Tetrafluoroethylene: same as Tetrafluoroethene (see).

Tetrafluorosilane: same as Silicon tetrafluoride (see).

Tetragonal: see Crystal.

Tetrahedral model of carbon: a model explaining the isomerism of tetravalent carbon. In 1874, J.H. van't Hoff and J.A. LeBel independently postulated the tetrahedral arrangement of the carbon valences to explain an isomerism (Stereoisomerism, see) which is different from structural isomerism. For an asymmetric C atom (one with four different ligands), two isomeric forms (enantiomers) which cannot be superimposed are possible. With this model, it was possible correctly to predict the isomers of the 13 optically active compounds known in 1874. The T. has since been given a theoretical basis in molecular orbital theory, in the form of sp^3 hybridization.

Tetrahydro-1,4-dioxin: see Dioxane.

Tetrahydrofolic acid: see Folic acid.

Tetrahydrofuran, abb. ***THF, oxolan***: a colorless, water-clear, easily burned liquid which smells like ether; m.p. -108.5 °C, b.p. 66 °C, n_D^{20} 1.4070. It is readily soluble in water and organic solvents. The azeotrope with water boils at 63.5 °C and contains 94.6% T. Like ether, it forms explosive peroxides upon standing in air, and these must be destroyed before the T. is used. The vapors of T. are poisonous, irritating the mucous membranes and having a narcotic effect in larger amounts. If the T. contains peroxides, it can also cause damage to liver and kidneys.

T. is made industrially by the *Reppe synthesis* from acetylene and formaldehyde, via butyne diol and butane diol, which is dehydrated with sulfuric acid. Other syntheses start from 1,2-dichlorobut-2-ene. Furfural, which can be obtained from oat bran, can be decarbonylated and dehydrated to form T. It is an important industrial solvent for many polymeric materials, such as polyvinyl chloride, polyvinyl acetate, polyvinyl ether, polyacrylates, natural and synthetic resins. It is also used in the manufacture of paints, printing inks, succinic acid, adipic acid and glues. Reaction of T. with ammonia leads to pyrrolidine, and T. is also an intermediate in an industrial synthesis of butadiene.

Tetrahydrofuran-2,5-dione: same as Succinic anhydride (see).

Tetrahydrofuran-2-one: same as γ-Butyrolactone (see).

Tetrahydrofurfurylalcohol: a hydrogenation product of furfuryl alcohol, a colorless liquid which is miscible with water, ethanol and ether; b.p. 177-178 °C, n_D^{20} 1.4517. It is produced industrially by catalytic hydrogenation of furfuryl alcohol, and is used as a solvent for alkyde resins, chlorinated rubber, dyes, waxes, lubricants, cellulose esters and ethers, and as an antifreeze, hydraulic fluid, and raw material for the synthesis of softeners.

Tetrahydro-4-methyl-2-(2-methylpropenyl)-2H-pyran: same as Rose oxide (see).

Tetrahydro-1,4-oxazine: same as Morpholine (see).

Tetrahydropteroylglutamic acid: see Coenzymes.

Tetrahydropyran-2,6-dione: same as Glutaric anhydride.

1,2,3,6-Tetrahydropyridazine-3,6-dione: same as maleic anhydride.

Tetrahydropyrrole: same as Pyrrolidine.

Tetrahydrothiophene, abb. ***THT, thiolan***: a colorless, poisonous liquid which smells like carbonizing gas; m.p. -96.2 °C, b.p. 121.1 °C, n_D^{20} 1.5048. It is soluble in alcohol, ether, acetone, benzene and chloroform, and insoluble in water. The vapors irritate the skin and eyes, and cause vascular congestion in the lungs. T. is made by heating 1,4-dichlorobutane with sodium sulfide in aqueous dimethylformamide. It is used as a gas odorizer for natural gas, and as a solvent and intermediate for organic syntheses.

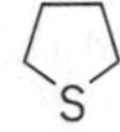

Tetrahydrothiophene-1,1-dioxide: same as Sulfolan (see).

Tetralin, ***1,2,3,4-tetrahydronaphthalene***: a partly hydrogenated aromatic hydrocarbon. T. is a colorless liquid with a characteristic odor; m.p. -35.8 °C, b.p. 207.6 °C, n_D^{20} 1.5414. It is insoluble in water and readily soluble in ethanol, ether and benzene. T. is obtained by pressure hydrogenation of naphthalene in the presence of nickel catalysts.

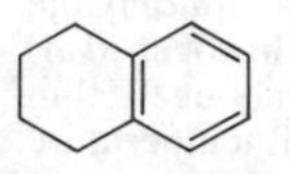

Friedel-Crafts reactions occur exclusively in the 2 position of the aromatic ring. Because such substituted T. can be dehydrogenated with sulfur, selenium or palladium on activated charcoal, they can be used to make naphthalenes uniformly substituted in the 2-position (β-position). T. is used as a solvent for rubber, resins, lacquers, iodine and sulfur.

Tetralone: a ketone derived from tetralin. ***α-T.*** is a colorless liquid; m.p. 8 °C, b.p. 255-257 °C, n_D^{20} 1.5672. It is synthesized by an intramolecular Friedel-Crafts reaction (see) of γ-phenylbutyryl chloride in the presence of aluminum chloride. ***β-T.*** is a colorless liquid, m.p. 18 °C, b.p. 234-240 °C, n_D^{20} 1.5598. It is scarcely soluble in water, but dissolves readily in ether and benzene. It can be synthesized by reduction of β-naphthol. T. are used as intermediates in organic chemistry.

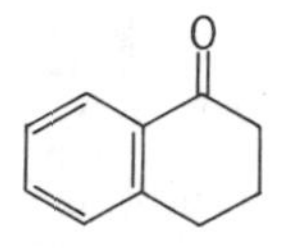

α-Tetralone

Tetramethrin: see Pyrethroids.

Tetramethylene: same as Cyclobutane.

Tetramethylenediamine: same as Putrescine (see).

Tetramethylenimine: same as Pyrrolidine (see).

Tetramethyllead: see Antiknock compounds, Organolead compounds.

Tetranitrobenzo-1,3a,4,6a-tetraazapentalene: see Tacot.

Tetranitromethane: $C(NO_2)_4$, a colorless, mobile liquid; m.p. 14.2°C, b.p. 126°C, n_D^{20} 1.4384. T. is slightly soluble in water, and readily soluble in alcohol and ether. It forms highly explosive mixtures with oxidizable substances, such as hydrocarbons. With unsaturated compounds, e.g. alkenes, dienes and aromatics, it gives a yellow to red-orange color reaction which serves as a qualitative test for these compounds. Alkynes do not react under these conditions. T. is a very reactive nitrating reagent, which donates one nitro group and is converted to nitroform (trinitromethane): $R\text{-}H + C(NO_2)_4 \rightarrow R\text{-}NO_2 + CH(NO_2)_3$. T. is obtained by the reaction of fuming nitric acid with acetic anhydride, or by nitration of nitroform. It is used as a nitrating reagent, an oxidant for rocket fuels, and, in a mixture with toluene, as an explosive.

Tetranitro-N-methylaniline: same as 2,4,6-Trinitrophenylmethylnitramide (see).

Tetranitropentaerythritol: same as Pentaerythritol tetranitrate (see).

1,3,5,7-Tetranitro-1,3,5,7-tetraazaoctane, ***octogen***: a white, crystalline powder, m.p. 280°C. T. is a byproduct of hexogen synthesis, or it can be made by treating 1,5-methylene-3,7-dinitro-1,3,5,7-tetraazacyclooctane with ammonium nitrate, nitric acid and acetic acid. It exists in 4 polymorphic forms, but only the β-form is of practical significance. Like hexogen, T. is a powerful explosive (detonation rate about 9100 m s^{-1}), and it is used as such for high-temperature work.

Tetrapeptides: see Peptides.

Tetraphenylhydrazine: $(C_6H_5)_2N\text{-}N(C_6H_5)_2$, a colorless, crystalline compound; m.p. 144°C. In nonionizing solvents, e.g. benzene, T. dissolves to give a green color which is due to homolytic cleavage of the N-N bond and formation of resonance-stabilized diphenylnitrogen radicals: $(C_6H_5)_2N\text{-}N\text{-}(C_6H_5)_2 \rightarrow 2\,(C_6H_5)_2N$.

In concentrated sulfuric acid, T. undergoes a benzidine rearrangement to *N,N'*-diphenylbenzidine, which is oxidized to a dark blue diphenoquinone diimine derivative. T. can be synthesized by oxidation of diphenylamine with potassium permanganate.

Tetrapyrroles: compounds with 4 pyrrole rings, some of which are partially hydrogenated, linked by methine or methylene bridges. The Porphyrins (see) are ***cyclic T.***, and the Bile pigments (see) are ***acyclic T.***.

Tetraterpenes: terpenes constructed from 8 isoprene units, which thus contain 40 C atoms. T. display less variety in structure than other groups of terpenes. A tail-to-tail coupling in the middle of the molecule is typical for T., nearly all of which are Carotenoids (see).

Tetrazines: six-membered heterocyclic compounds with four nitrogen atoms in the ring. The T. are classified according to the positions of the nitrogen atoms as ***1,2,4,5-T. (symm. T.)***, ***1,2,3,4-T. (vic. T.)*** and ***1,2,3,5-T. (as. T.)***.

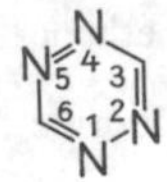

1,2,4,5-Tetrazine

Of these three theoretically possible isomers, only 1,2,4,5-T. is known as such; the others exist only in the form of condensed ring systems. 1,2,4,5-T. forms red-purple crystals; m.p. 99°C (subl.). It is soluble in water, alcohol and ether. It is formed by ring closure from two molecules of diazoacetate, followed by oxidation, saponification and decarboxylation.

Tetrazole, ***IH-tetrazole***: a five-membered, heterocyclic compound with four nitrogen atoms in the ring. T. crystallizes in colorless leaflets; m.p. 156°C (subl.). It is readily soluble in water and alcohol, and relatively insoluble in benzene and ether. The aqueous solution is weakly acidic; the acid has the same strength as acetic acid. T. is formed by addition of hydrocyanic acid to hydrazoic acid. Tetrazolium salts (see) are used in chemical analysis, and Pentetrazole (see) is an analeptic used in medicine.

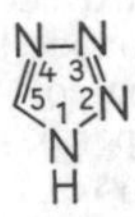

Tetrazolium salts: the salts of 2,3,5-trisubstituted tetrazole. T. are colorless to light yellow, ionic compounds formed by dehydrogenating cyclization of the highly colored formazanes with mercury oxide or lead(IV) acetate in acidic solution. They are converted back to formazanes by reduction. 2,3,5-Triphenyltetrazolium chloride is an important example; it is used in the assay of hydrogenase systems.

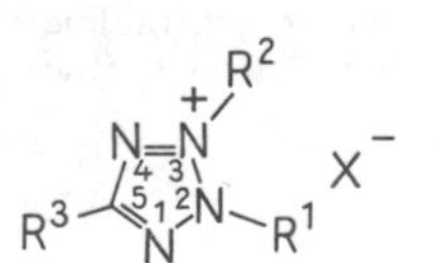

Tetrels: see Nomenclature, sect. II A 2.

Tetrodotoxin: a Fish toxin (see) which is best known because it is present in the fugu fish, a member of the family *Tetraodontidae* or puffer fish, which is prized as a culinary delicacy. The toxin is found in other species of the *Tetraodontidae* as well as in members of the *Diodontidae*, or porcupine fish, many of which are used for food. In all, about 80 species produce T. It is probably synthesized in the ovaries, but other organs also contain it. Only the blood and muscle tissue of the fish are not poisonous, so that preparation as food requires a great deal of knowledge and care. The symptoms of poisoning include faintness, weakness, nausea, muscle pains, respiratory difficulty, blood pressure decrease and paralysis. Death occurs through respiratory arrest.

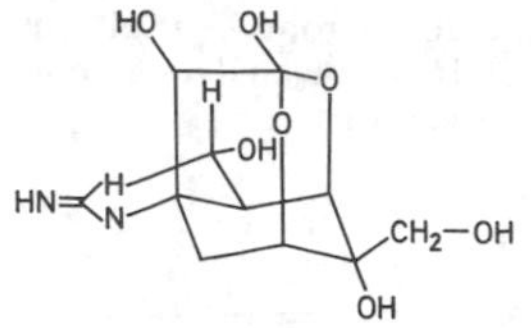

The toxin is related to Saxitonin (see); its structure is known (Fig.). The lethal dose is estimated to be 1 mg for an adult; T. is thus one of the strongest non-peptide poisons.

T. is identical to ***tarichatoxin***, which was isolated from the Californian salamander *Taricha torosa*.

Tetryl: see 2,4,6-Trinitrophenylmethylnitramide.

Textile stiffening rinses: laundry treatments containing copolymers of vinyl acetate and unsaturated organic acids, along with small amounts of surfactants and polywax. F. are used to stiffen textiles and can be used in a washing machine.

Textinite: see Macerals.

Textite: see Coal.

Texture: 1) the frequency distribution of the crystallites of a Polycrystalline material (see) with respect to their crystallographic orientation. Measurement of T. (*texture analysis*) is done by X-ray and neutron diffraction, using texture goniometers. T. is created by plastic deformation (rolling or tensile T.), recrystallization, or growth by precipitation from a melt, solution or a gas. It is the cause of the Anisotropy (see) of physical properties in polycrystalline materials; examples are the anisotropy of mechanical properties in construction materials shaped by rolling or the magnetic porperties of transformer parts.

2) The appearance of Liquid crystals (see) in a polarizing microscope.

TG: abb. for thermogravimetry.

Thalidomide, ***Contergan***®: a compound introduced as a hypnotic. However, because of its neurotoxicity and teratogenic side effects, its use had to be forbidden. A high incidence of birth defects occurred in children whose mothers had used T. in the early months of pregnancy, and it was shown that T. was the probable cause of these malformations.

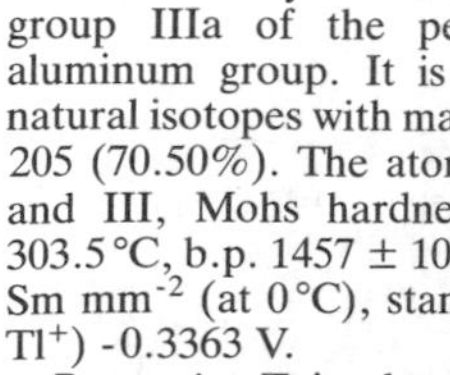

Thallium: symbol ***Tl***: a chemical element from group IIIa of the periodic system, the Boron-aluminum group. It is a heavy metal, *Z* 81, with natural isotopes with mass numbers 203 (29.50%) and 205 (70.50%). The atomic mass is 204.37, valence I and III, Mohs hardness 1.3, density 11.85, m.p. 303.5 °C, b.p. 1457 ± 10 °C, electrical conductivity 6.2 Sm mm^{-2} (at 0 °C), standard electrode potential (Tl/Tl$^+$) -0.3363 V.

Properties. T. is a lustrous, soft metal which rapidly acquires a lead-like matte surface in the air. It can be cut with a knife. The modification stable at room temperature, α-thallium, has a tetragonal lattice.

In most of its compounds, T. is in the +I oxidation state (see Boron-aluminum group). Thallium(I) compounds are similar in their properties both to the derivatives of the alkali metals (water-solubility and basic reaction of TlOH, solubility of Tl_2CO_3) and to silver compounds (color and insolubility of the halides, chromate and oxide, solubility of Tl_2SO_4). Thallium(III) compounds are strong oxidizing agents and are readily reduced to Tl^+ compounds.

T. is stable in dry air and oxygen-free water. In the presence of air, it dissolves in water to form thallium(I) hydroxide, TlOH. It can be oxidized with oxygen at high temperatures to form thallium(I) oxide, Tl_2O, and thallium(III) oxide, Tl_2O_3. In strong acids, e.g. sulfuric or nitric, it dissolves with evolution of hydrogen to give thallium(I) salt solutions. Because of the low solubility of TlCl, it is unreactive with hydrochloric acid. Halogens oxidize T. even at room temperature, while it must be heated to react with sulfur, selenium or tellurium, with which it forms derivatives of Tl^+ and Tl^{3+}.

Analysis. The simplest method of qualitative determination of T. is by spectroscopy (green light at 535.1 nm). Depending on the expected concentration range, quantitative analysis is done by complexometry with EDTA or atomic absorption spectroscopy.

Thallium and all its compounds are very poisonous. They cause severe damage to the nervous system, digestive tract, kidneys and skin. Thallium salts dissolved in water can also be rapidly absorbed through the skin. Intoxication can be recognized by nausea, nerve pains, loss of hair, etc. Antidotes for oral intake: induce vomiting with a warm solution of table salt; rinse the stomach with 1% sodium iodide or 3% sodium thiosulfate solution, then administer activated charcoal.

T. makes up about 10^{-5}% of the earth's crust. It occurs widely, but always in low concentrations, as a companion of zinc, copper, iron and lead in their sulfide ores. The manganese nodules in the Pacific also contain T.

Production. The raw material is the T.-containing fly ash from pyrite roasting for sulfuric acid production. The ash contains T. as the sulfate, and is extracted with water. The T. is separated as thallium chloride, then precipitated electrolytically from sulfuric acid solution.

Applications. T. is not widely used, due to its toxicity. The conductivity of thallium sulfide changes when it is subjected to IR irradiation, and it is therefore used in photocells and IR detectors.

Historical. T. was discovered in 1861 by Crookes by spectroscopy (Greek "thallos" = "green twig") and was prepared about simultaneously in 1862 by Crookes and Lamy.

Thallium(I) carbonate: Tl_2CO_3, colorless, monoclinic crystals; M_r 468.75, density 7.11, m.p. 273 °C.

T. is poisonous (see Thallium). It is the only heavy metal carbonate which dissolves readily in water. It is made by passing carbon dioxide through a solution of thallium(I) hydroxide.

Thallium chlorides: ***Thallium(I) chloride***, TlCl, colorless, cube-shaped crystals; M_r 239.82, density 7.00, m.p. 430 °C, b.p. 720 °C. TlCl is barely soluble in water and therefore precipitates when chloride ions are added to thallium(I) salt solutions. With dry chlorine, it forms the double compound $TlCl \cdot TlCl_3$. It is occasionally used as a chlorination catalyst.

Thallium(III) chloride: $TlCl_3$, colorless, hygroscopic, water-soluble, hexagonal leaflets; M_r 310.73, m.p. 25 °C. $TlCl_3$ is obtained by passing chlorine through a slurry of TlCl in water. It crystallizes out of the aqueous solution as the tetrahydrate, $TlCl_3 \cdot 4H_2O$, which can be readily converted to the monohydrate and anhydride by dehydrating agents. $TlCl_3$ forms chlorothallates, $M^I_3[TlCl_6]$ with various chlorides.

The T. are poisonous (see Thallium).

Thallium(I) hydroxide: TlOH, colorless, water-soluble needles, M_r 221.38, dehydration above 140 °C to thallium(I) oxide. The aqueous solution is strongly basic and absorbs carbon dioxide, forming thallium(I) carbonate, Tl_2CO_3. T. is poisonous (see Thallium). It is obtained by reaction of thallium(I) sulfate with barium hydroxide.

Thallium oxides: ***Thallium(I) oxide***: Tl_2O, black, hygroscopic, rhombohedral crystal powder; M_r 424.74, density 9.52, m.p. 300 °C. Tl_2O reacts with water to form thallium(I) hydroxide. It is obtained by heating thallium(I) hydroxide or thallium(I) carbonate. ***Thallium(III) oxide***, Tl_2O_3, is a dark brown powder or hexagonal crystals; M_r 456.74, density 10.19, m.p. 715±5 °C. Above 875 °C, T. is converted to Tl_2O by releasing oxygen. It is made by thermal decomposition of thallium(III) nitrate, and is used to make synthetic gems and special optical glasses.

The T. are poisonous (see Thallium).

Thallium(I) sulfate: Tl_2SO_4, colorless rhombic crystals which are only slightly soluble in water; M_r 504.80, density 6.77, m.p. 632 °C. T. is poisonous (see Thallium). With aluminum sulfate, it forms an alum, $TlAl(SO_4)_2 \cdot 12H_2O$. T. is obtained by dissolving thallium in dilute sulfuric acid, and is used as a rodenticide.

Thaumatin: see Sweeteners.

Thebaine: an opium alkaloid present at concentrations between 0.5 and 7.4% in opium; m.p. 193 °C, $[\alpha]_D^{20}$ -219° (in ethanol). It is the only alkaloid in the roots of a *Papaverbracteatum* strain, where it is present in relatively large amounts. Since T. can be converted to codeine, but not morphine, by partial synthesis, this plant is a reasonable starting material for production of codeine which cannot be diverted to the production of morphine and heroin. T. itself has no therapeutic significance. It is not analgesic, but, like strychnine, it causes convulsions.

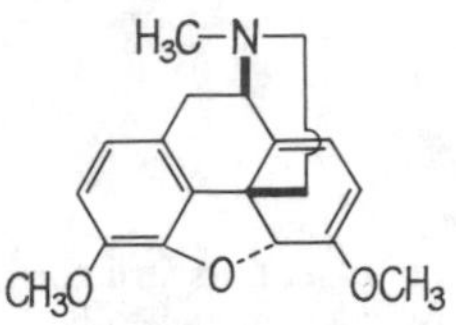

Thenard's blue: same as Cobalt blue 1) (see).

Theobromine: 3,7-dimethylxanthine, an important purine alkaloid. Crystals of T. are colorless and odorless, and have a bitter taste; m.p. 351 °C with sublimation. T. is slightly soluble in ethanol and water. It is an amphoteric compound with an acidic H atom on the N1 atom and displays lactam-lactim tautomerism. T. is found in the cocoa bean, where it comprises 1 to 4% of the mass, and in smaller amounts in Chinese tea and the cola seed. It is extracted from cocoa beans or synthesized. T. has very little effect on the central nervous system, but is a diuretic.

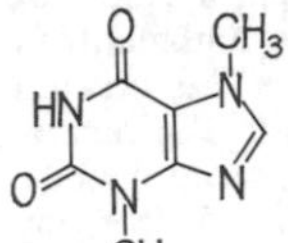

Theophyllin: 1,3-dimethylxanthine, a therapeutically important purine alkaloid. The colorless crystals have a slightly bitter taste and contain 1 mol water of crystallization; m.p. 268 °C. T. is soluble in hot water, and nearly insoluble in ether. It is an amphoteric compound. To improve its solubility, salts with aliphatic amines are prepared. The most important product is *aminophyllin*, which consists of equimolar amounts of T. and ethylenediamine. T. is found in small amounts in tea leaves and coffee beans. For commercial purposes, it is synthesized chemically. *N,N'*-Dimethylurea is condensed with cyanoacetic acid or cyanoacetate to form 6-amino-1,3-dimethyluracil. A nitroso group is then introduced in position 5 and reduced to an animo group. The resulting 5,6-diamino compound is cyclized to T. with formic acid or formamide. T. has only a weak stimulating effect on the central nervous system, but it is a strong diuretic. It relaxes spasms of blood vessels and bronchi, and improves heart function. Various derivatives substituted on N7, e.g. with hydroxyalkyl groups, are used as drugs. *Etofyllin* (oxyethyltheophyllin) has a β-hydroxyethyl group on N7.

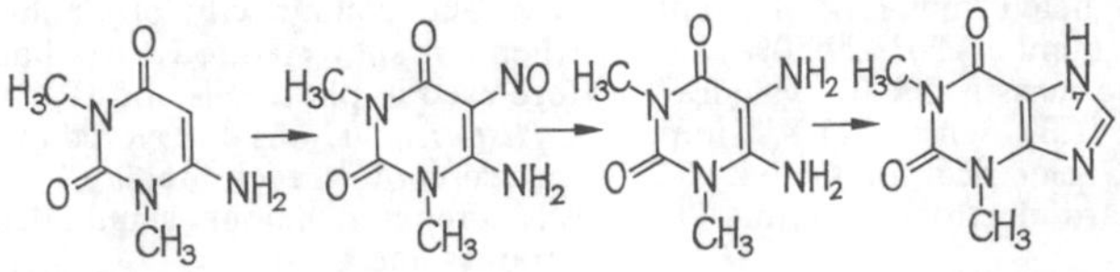

6-Amino-1,3-dimethyluracil

Theophyllin

Synthesis of theophyllin

Theory of the activated complex: see Kinetics of reactions (theory).

Theory of the transition state: see Kinetics of reactions (theory).

Thermal analysis: a technique for determination of changes, especially changes of aggregate state. It is an important method for establishing diagrams of state, particularly Melting diagrams (see) for systems of one or more substances. The basis of T. is Newton's cooling law. In practice, a sample is completely melted and then allowed to cool slowly; the temperature is recorded as a function of time and the values are plotted on a graph. Plateaus or breaks in such cooling curves indicate changes in aggregate state. In pure substances with no changes of state, the result is an exponential curve (K_1 in Fig. 1), which obeys Newton's cooling law. If there is a first-order change in the pure substance, there is a plateau (Fig. 1, K_2) in the curve at the temperature of the change, that is, the temperature remains constant until the entire amount of substance has changed (melted, solidified, evaporated, sublimed or converted to another modification).

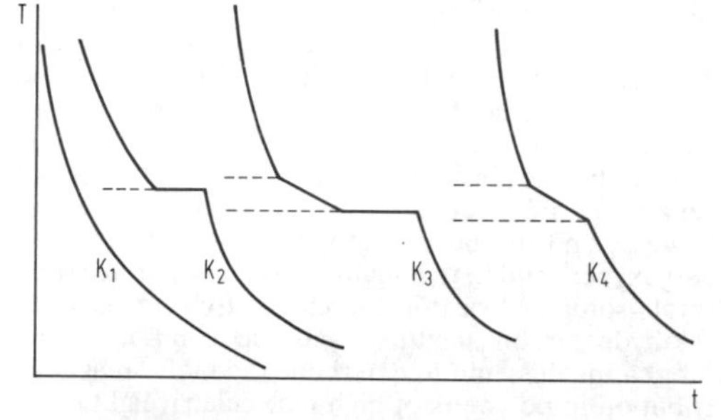

Fig. 1. Thermal analysis. Cooling curves: K_1 has no plateau or break, K_2 has a plateau, K_3 has a break and a plateau, and K_4 has two breaks.

Since the conversion temperatures are changed by the presence of impurities, T. is a sensitive method for testing the purity of a substance. In binary systems, a plateau will only be seen if both substances are present in concentrations such that they are completely miscible in the liquid phase, but are either immiscible in the solid (form a eutectic mixture) or precipitate by forming a compound. Otherwise, for example, if only one component solidifies out of the mixture, the cooling curve will display a break followed section of curve which is less steep. This is followed either by a plateau (Fig. 1, K_3) or another break point (Fig. 1, K_4). Evaluation of the cooling curves is shown in Fig. 2. Samples of known concentration are allowed to cool. The temperatures of the break points on the curves are plotted against the concentration, and in this way, the phase limit lines of the phase diagrams are obtained point by point.

Higher-order changes in pure substances, such as the conversion of iron from the ferromagnetic to the paramagnetic state, are usually not characterized by a certain temperature, but rather by a temperature interval. Thus the cooling curves generally do not contain plateaus; instead, there are break regions which are not as distinct as break points.

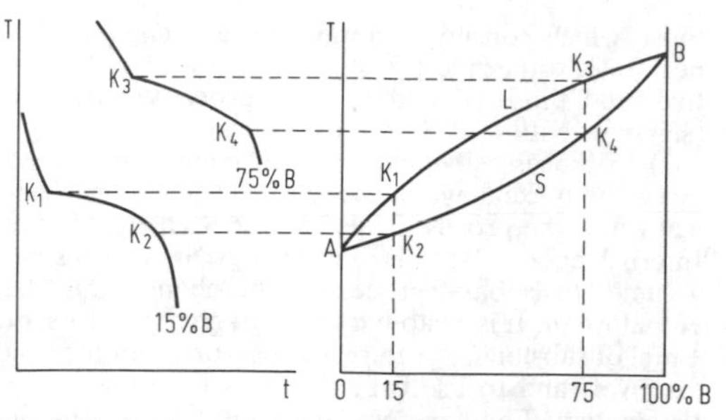

Fig. 2. Evaluation of a thermal analysis: cooling curves (*left*) and melting diagrams of a system in which the components are completely miscible in the solid and liquid phase (*right*). K_1 to K_4, break points; L, liquid line; S, solid line.

Occasionally warming curves are used instead of cooling curves for T. These are similar to cooling curves, but the breaks occur in reverse order. In some cases, especially with metallic systems, the temperatures of the plateaus and break points in the cooling curves are not the same as in the warming curves (thermal hysteresis).

Differential thermoanalysis (see) is an especially powerful form of T.

Thermite: a mixture of iron oxides and aluminum powder which, when they react (for example, $Fe_2O_3 + 2\ Al \rightarrow 2\ Fe + Al_2O_3$) generate very high temperatures. T. are used to make high-melting metals (thermite process; see Aluminothermal process), for welding (aluminothermal welding) and as pyrophorics. The hard aluminum oxide slag is used as a polishing material.

Thermite process: same as Aluminothermal process (see).

Thermoanalytical methods: a collective term for those analytical methods which measure changes in sample properties, e.g. mass, volume, electrochemical potentials, dipole moments or magnetic susceptibility, which occur when the sample is heated or cooled. Under defined experimental conditions, these changes are characteristic for the samples and can be used for qualitative and quantitative analysis. Of this group of methods, the most important are Thermogravimetry (see), Thermometry (see), Differential thermoanalysis (see) and thermometric titration. With modern apparatus (e.g. derivatographs), several thermoanalytical methods can often be carried out at once.

Thermobimetal: same as Bimetal (see).

Thermochemistry: an area of chemical thermodynamics dealing with heat exchanges during phase changes and chemical reactions. The main activities in this field are experimental determination of heats of reactions, phase changes and mixing, and determinations their relations to the state parameters of the first law of thermodynamics. The data are collected in tables and data banks, and used to calculate caloric effects. T. is based on a series of observations and thermodynamic laws.

1) Phase changes always involve changes in energy, and mixing processes usually do. In chemical reactions, energy exchange occurs simultaneously with the matter transformation. Processes which release energy to the environment are called *exothermic*;

those which consume energy are *endothermic*. The heats of exothermic processes are considered negative, and those of endothermic processes, positive (sign convention of thermodynamics).

2) In processes which are irreversible, i.e. carried out without exchange of useable electrical or mechanical work, the process heats represent changes in the Internal energy (see) Δe, if they occur at constant volume. In isobaric processes, the change is in the enthalpy Δh. It is customary in T. to give the heats for 1 mol of substance, or in reactions, for 1 mol formula turnover, and to use capital letters for these molar parameters. For example, the molar Phase change heats (see) are $\Delta_p H$; molar Mixing heats (see) are $\Delta_M H$; and molar reaction enthalpies, $\Delta_R H$, and energies, $\Delta_R E$ (see Reaction heats).

3) Since the absolute values of the internal energy and enthalpy are not known, the Standard state (see) is established as a convention. This is the state of the pure substance at 101.325 kPa (1 atm) and 298.15 K (25 °C). The molar energies and enthalpies of the elements in their stable form (physical state, modification) are set equal to zero. The reaction heat of formation of 1 mol of a chemical compound from the elements under standard conditions is called the molar standard energy of formation, $\Delta_F E^0_{298}$ or standard enthalpy of formation, $\Delta_F H^0_{298}$. Standard enthalpies and energies can be used instead of the unknown absolute quantities in all thermodynamic calculations, and these are collected in data banks. The quantities can be converted to other temperatures using Kirchhoff's law (see), and to other pressures and volumes using the caloric State equation (see). The calculation of reaction heats in real mixtures or solutions from the standard parameters is more difficult, because it requires knowledge of the mixing or excess parameters, or of the partial molar energies or enthalpies. Because of the multitude of possible mixtures, these data are seldom available in collections.

4) As a result of the conservation of energy, when the direction of a process is changed, the sign of the associated energy change is changed, but not the magnitude. For example, for ice at 273.14 K, the molar heat of melting, $\Delta_S H_{H_2O} = 6.0$ kJ mol^{-1}, and the molar heat of freezing for liquid water, $\Delta_L = -6.0$ kJ/mol^{-1}. For the reaction $CO_2 + C \rightarrow 2\ CO$, $\Delta_R H_{298} = 2\ \Delta_F H_{CO} - \Delta_F H_{CO_2} - \Delta_F H_C = 172.2$ kJ mol^{-1}. For the reverse reaction, $2\ CO \rightarrow C + CO_2$, $\Delta_R H_{298} = 2\,\Delta_F H_{CO_2} + \Delta_F H_C - \Delta_F H_{CO} = -172.2$ kJ mol^{-1}.

Thus the reaction equation must be specified when the molar energy or enthalpy of reaction is given, since the direction of the reaction determines the sign, and the choice of stoichiometric coefficients determines the magnitude of 1 mol formula turnover and thus the amount of energy.

For reactions which involve several steps, Hess' law (see) applies.

5) Experimental determination of thermochemical parameters is done by Calorimetry (see). Standard enthalpies of formation are usually calculated from enthalpies of combustion and fluorination.

Thermochromism: in the broadest sense, a shift in light absorption by a substance when the temperature changes. Such shifts can be due to a temperature-dependent association or dissociation equilibrium, such as occurs during aggregate formation of certain pigments. T. in the narrower sense is due to a reversible, temperature-dependent equilibrium between two monomeric molecular forms A and B; the thermochrome B which forms at the higher temperature absorbs light at a longer wavelength. The molecular forms A and B can be tautomers or structural or conformational isomers.

Thermodiffusion: transport process in gases and liquids (also called the ***Ludwig-Soret effect*** in liquids) caused by a temperature difference; it leads to partial separations. If there is a temperature difference in a mixture of gases, T. causes the lighter components to become enriched at the warmer location, and the heavier ones at the cooler location, thus forming a concentration gradient. This effect is used together with a convection current in the ***diffusion method*** of isotope separation developed by Clusius and Dickel (1938). A vertical separation tube a few centimeters in diameter and up to 20 m long contains an axial, electrically heated wire and is filled with a gaseous isotope mixture. The lighter isotopes are enriched along the hot wire, and are transported upward by thermal convection; the heavier isotopes collect along the cold wall and are carried downward. The gas at the top is thus enriched in the lighter isotope, and that at the bottom, in the beavier isotope. The theory of T. is relatively complicated. In 1917 Chapman and Enskog predicted the effect for gases and demonstrated it in the same year.

T. in the gas phase is used, among other things, for separation of hydrocarbons; to enrich noble gas isotopes and the heavy isotopes of carbon, nitrogen and oxygen; and in the liquid phase, to separate structural isotopes, paraffins, cycloparaffins, aromatics, thiohydrocarbon mixtures and the components of lubricating oils; and to determine the molar mass distribution in polymers of high molecular weight.

The reverse of T. is the ***diffusion thermoeffect***. A diffusion flow based on a concentration gradient (see Diffusion) leads simultaneously to a heat flow, and thus causes a temperature gradient to form.

Thermodynamics: an area of physics and physical chemistry concerned with the properties of thermodynamic systems. The State (see) of such a material system is uniquely described by sets of parameters (State parameters, see). Changes of state are associated with an energy exchange and with conversion of one form of energy into another, e.g. of mechanical or electrical work into heat.

I) Phenomenological or ***equilibrium T.*** is a purely macroscopic discipline for which it is unimportant that substances have molecular structure. Material systems are studied at thermodynamic Equilibrium (see). Experimental data such as temperature, pressure, volume, chemical composition and heat capacities are used to characterize the equilibrium state. These state parameters are not all independent; some are related by state functions. The laws of thermodynamics (see below) permit a closed, generally valid description of these systems in terms of a set of state functions. T. provides quantitative information about the changes in energy associated with changes in state (see Reaction heats, Phase-change heats), the direction and driving forces of processes (see Affinity), the maximum available useful work which can be obtained from them and the position of equilibria.

Only qualitative predictions are possible concerning non-equilibrium states. The rate at which equilibrium is reached is not an object of thermodynamic studies; time is not included as a variable in the mathematical formalism of T. For chemical reactions, the time dependence is treated by Kinetics of reactions (see) and the T. of irreversible processes (see below). Historically, equilibrium T. was developed first. It is based on the laws of T., which are derived from experience and lead to the definition of the state parameters temperature, internal energy and entropy.

Zeroth law of T.: If two bodies are in thermal equilibrium with a third, they are also in thermal equilibrium with each other. It follows that for macroscopic systems, there is a state parameter which is the same in all systems which are in equilibrium with each other; this parameter is the temperature T.

First law of T.: This is an extension of the law of conservation of energy from classical mechanics to include heat as a form of energy. The first law could not be formulated until it was recognized that heat and mechanical energy are related by a fixed ratio, the mechanical heat equivalent, and that they can be interconverted (J.R. Mayer, 1842). The same is true for electrical energy (J.P. Joule, 1841), and for this reason, all energy forms are now measured in one SI unit, the Joule (see). H. von Helmholtz formulated the law of conservation of energy in 1847; this can be considered the general form of the first law of T.: energy can neither be created nor destroyed. It can only be converted from one form into another.

From this, it follows that a perpetual motion machine of the first type, i.e. a machine which continuously produces more energy (e.g. mechanical or electrical work) than is supplied to it (e.g. as heat), cannot exist. Another formulation of the first law is: in a closed system, the sum of all energies is constant.

The totality of the energy within a system is called its internal energy e. Changes in the internal energy are equivalent to changes in the state of the system. They can occur by exchange of heat Δq and/or work Δw with the environment according to $\Delta e = \Delta w + \Delta q$, or in very small, differential increments, according to $de = \delta w + \delta q$. There is a sign convention in chemical thermodynamics: parameters added to the system are given a positive sign, and those removed from the system are given a negative sign. Unlike the internal energy, work and heat are not state functions, but depend on the way in which a change in state occurs. This is why their changes are indicated by δw and δq, in contrast to de, in differential notation.

Applications of the first law.

a) Pure homogeneous substances: in a closed system (mass m or molar amount n is constant), the state is uniquely determined by the state variables p, v and T. These three parameters are linked in the thermal equation of state, so it is enough to represent the internal energy as a function of two of these variables.

Let $e = e(T,v)$. Another state parameter is the *enthalpy* $h = e + pv$. For reasons of simplicity in the following demonstration, let $h = h(T,p)$.

Changes in state parameters are independent of the path, thus total differentials: $de = (\delta p/\delta T)_v dT + (\delta e/\delta v)_T dv$ and $dh = (\delta h/\delta T)_p dT$ are used. Since pure substances can exchange work only in the form of volume work, $dw = -pdv$ and $dh = de + pdv + vdp = vdp + \delta q$. It follows that for isochoric processes (v = constant), $de = \delta q$, and for isobaric processes (p = constant), $dh = \delta q$.

If a closed system exchanges heat with its environment, this is equal in isochoric processes to the change in the internal energy; in isobaric processes, it is equal to the change in enthalpy. Since chemical processes are usually carried out isobarically, the heats which appear are changes in enthalpy. This is the reason that the definition of h is so useful.

The following equations can be derived for the partial differentials in the two total differential equations above:

$$(\partial e/\partial T)v = c_v = nC_v,\ (\partial H/\partial T)_p = c_p = nC_p,\ (\partial e/\partial T)_v = T(\partial p/\partial T)v - p,\ (\partial h/\partial p)_T = v - T(\partial v/\partial T)_p.$$

Here n is the number of moles, c_v and C_v are heat capacity and molar heat at constant volume, and c_p and C_p are heat capacity and molar heat at constant pressure.

The differentials $(\delta p/\delta T)_v$ and $(\delta v/\delta T)_p$ can be obtained from the thermal equation of state of the substance which make up the system. For example, it follows from the ideal gas equation that $(\delta e/\delta v)_T = 0$ and $(\delta h/\delta p)_T = 0$ (second Gay-Lussac law). In other words, the internal energy and the enthalpy of ideal gases are independent of volume and pressure. For real gases, $(\delta e/\delta v)_T \leqq 0$, and $(\delta h/\delta p)_T \leqq 0$ (see Joule-Thomson effect).

b) Pure substances in two-phase systems: Here an additional variable is the phase composition. If n indicates the molar amount in one phase, then the functions $e = e(T,v,n)$ and $h = h(T,p,n)$ apply, and for h the total differential is $dh = (\delta h/\delta T)_{p,n}dT + (\delta h/\delta p)_{T,n}dp + (\delta h/\delta n)_{T,p}dn$. The differential $(\delta h/\delta n)_{T,p}$ gives the change in enthalpy if one mole of the substance is transferred from one phase into another at constant temperature and pressure: $(\delta h/\delta n)_{T,p} = \Delta_p H$ (molar phase change enthalpy; see Phase change heat).

c) Chemical reactions: for a chemical reaction to occur, the thermodynamic system must contain several substances. Since the internal energy and the enthalpy are the sums of all partial energies of a system, in the ideal case $e = e_1 + e_2 + ... = \Sigma e_i$, and $h = h_1 + h_2 + ... - \Sigma h_i$, where i is the index denoting the ith component. A mixture for which this additivity is valid is called an ideal mixture. If there are additional interactions between the particles of the various components, the additivity is not valid; this is true of real mixtures.

For example, if ammonia is formed in a mixture of the ideal gases hydrogen and nitrogen: $N_2 + 3\,H_2 \rightarrow 3\,NH_3$, a turnover of 1 formula mole will consume 1 mole N_2 and 3 moles H_2 and produce 2 moles NH_3. From the additivity relation, the change in the enthalpy is $2H_{NH_3} - H_{N_2} - 3H_{H_2} = \Delta_R H$, where H_{NH_3}, H_{N_2} and H_{H_2} are the molar enthalpies of the product and reactants. At 298 K and 0.1 MPa, $\Delta_R H = -92.2$ kJ mol^{-1}, that is, the reaction enthalpy for formation of 2 mol NH_3.

For the general chemical reaction $|\nu_H|A + |\nu_B|B + ... \rightleftharpoons \nu_p P + \nu_q Q + ...$ it follows by analogy that $\Sigma \nu_i E_i =$

$\Delta_R E$ and $\Sigma \nu_i H_i = \Delta_R H$. Here $\Delta_R E$ and $\Delta_R H$ are the molar reaction energy and enthalpy; they are also called the molar reaction heats at constant volume and pressure, respectively. The experimental determination and calculation of reaction energies and enthalpies is the object of Thermochemistry (see).

2) Second law of T. (entropy law): The second law predicts the direction of natural processes, and defines Entropy (see) s as a new state parameter which defines this directional dependence. There are various formulations of the second law:

1) Heat can never be spontaneously transferred (i.e. without input of work from without) from a colder to a warmer body (Clausius).

2) It is impossible to construct a periodically working machine which does nothing more than to generate mechanical work by cooling a heat reservoir (Planck and Thomson). This type of impossible machine is called a perpetual motion machine of the second type. It would be a heat machine which in a series of Cyclic processes (see) would continuously (periodically) convert heat at a certain temperature level into work; it would obey the first law.

3) The entropy s is a state parameter. In a closed system, the entropy can never decrease, but can only increase (in irreversible processes) or remain constant (in reversible processes).

Mathematically formulated, the second law is $ds > 0$ for spontaneous, that is, irreversible processes, and $ds = 0$ at equilibrium and in reversible processes. This applies only to closed systems. However, systems which exchange energy with the environment are very important; in this case, the entropy change of the environment must also be taken into account.

For example, in the reaction $2\,H_2 + O_2 \rightarrow 2\,H_2O_{liq}$ in the standard state, the entropy change in the reacting system per formula mole of turnover, ΔS, is $\Delta S_{internal} = -327\,J\,K^{-1}\,mol^{-1}$, and the reaction enthalpy $\Delta_R H = -572\,kJ\,mol^{-1}$. Under isothermal conditions (298 K), $\Delta_R H$ is completely transferred to the environment, and increases its entropy by

$$\Delta S_{external} = -\Delta_R H/T = \frac{572\times10^3\,J\,mol^{-1}}{298\,K} = 1919\,J\,K^{-1}\,mol^{-1}.$$

The total change ΔS_{total} of the closed system, consisting of the reacting system and the environment, is $\Delta S_{total} = \Delta S_{internal} + \Delta S_{external} = (-327 + 1919)\,J\,K^{-1}\,mol^{-1} = 1592\,J\,K^{-1}\,mol^{-1}$. Since $\Delta S_{total} > 0$, the process occurs spontaneously.

Two other state parameters which have been defined are the Helmholtz Free energy (see) $f = e - Ts$ and the Gibbs free enthalpy $g = h - Ts$. These parameters have the advantage that they can be applied to closed systems without taking into account the changes occurring in the environment as a result of the energy exchange.

Using the functions s and g or f, a complete formalism for the second law of T. can be developed.

Application of the second law.

a) Pure substances: Like the state parameters of the first law, entropy is a function of p, v and T: $s = f(T,v)$, or $s = f(T,p)$. For example, for an ideal gas, $ds = nC_v d\ln T + nR\,d\ln v$, or $ds = nC_p\,d\ln T + nR\,d\ln p$.

If the system consists of two phases, the state function s also depends on the phase fractions: $S = f(T,p,n)$, where n is the molar amount of the pure components in one of the two phases. The total differential then contains in addition the partial differential $(\delta s/\delta n)_{T,p} = \Delta_p S$. The molar phase change entropy $\Delta_p S$ indicates how much the entropy of the substance (statistically speaking, the "order") changes when 1 mole goes from one phase into the other. $\Delta_p S = \Delta_p H/T_p$, where $\Delta_p H$ is the molar phase change enthalpy, and T_p is the temperature of the phase change.

b) Mixtures. Mixing processes, like all spontaneous natural processes, are irreversible. Therefore even when an ideal mixture is made, $\Delta s_M/\Sigma n_i = \Delta S^M = -R\,\Sigma\,x_i \ln x_i$. Here n_i is the molar amount and s_i the molar fraction of the ith component, and R is the general gas constant. ΔS^M is the mean molar entropy of mixing. Since x_i is always < 1, it follows that $\Delta S^M > 0$. The mean molar mixing functions of the internal energy e and the enthalpy h are zero for ideal mixtures (first law).

In real mixtures, the interaction forces lead to additional changes in the entropy which are taken into account by introduction of the activities $a_i = f_i x_i$ (f_i = activity coefficient):

$$\Delta S^M_{real} = -R\,\Sigma\,x_i \ln a_i.$$

There are similar equations for the free energy and the free enthalpy.

c) Reversible work and equilibrium conditions. If a process is associated with the change dh in enthalpy, the maximum fraction of this energy difference which can be obtained as work is dw_{rev}, when the process is carried out reversibly: $dh = dw_{rev} + dq_{rev}$. Any irreversible step (such as frictional or heat conduction losses) shifts the distribution of dh away from the work. The reversible work dw_{rev} corresponds to the change dg in the Gibbs free energy $g \equiv h - Ts = e + pv - Ts$; in isochoric processes, dw corresponds to the change df in the free Helmholtz energy $f \equiv e - ts$.

Processes occur spontaneously if they are able to produce work when conducted reversibly ($dg < 0$). A reverse in direction means the sign is reversed. If work is done on the system ($dg > 0$), a process can be forced which would not occur spontaneously. Some examples are the transfer of heat from a reservoir of lower temperature to one of higher temperature as work is done in a refrigerator, or the charging of a capacitor. If the reversible work capacity is zero, the process comes to a stop; thermodynamic equilibrium has been reached. In summary, the following criteria apply:

$dq < 0$, $df < 0$, $ds > 0$:	spontaneous process
$dq > 0$, $df > 0$, $ds < 0$:	process forced by expenditure of work
$dq = 0$, $df = 0$, $d\mu_i = 0$, $ds = 0$:	thermodynamic equilibrium

d) Chemical reactions. For the reaction $|\nu_A|A + |\nu_B|B + \ldots \rightleftharpoons \nu_P P + \nu_Q Q + \ldots$, the changes in g and s per formula mole turnover are $\Delta g \equiv \Delta_R G = \Sigma\,\nu_i \overline{G}_i = \Sigma \nu_i \mu_i$ and $\Delta s \equiv \Delta_R S = \Sigma \nu_i\,\overline{S}_i$, where $\overline{G}$ and $\overline{S}$ are the partial molar free enthalpies and entropies in the reaction mixture (see Partial molar parameters). $\Delta_R G$ is the molar free reaction enthalpy and $\Delta_R S$ is the molar reaction entropy. Based on the definition of g, the two are related by the Gibbs-Helmholtz equation: $\Delta_R G = \Delta_R H - T\Delta_R S$. $\Delta_R G$ is the part of the reaction enthalpy which can be obtained as work when the reaction is carried out reversibly at constant temperature and pressure. A chemical reaction occurs spontaneously in the direction which enables the system to do work ($\Delta_R G < 0$), so that $-\Delta_R G$ is the thermodynamic measure of the affinity of chemical reactions.

If the chemical potentials $\mu_i = \mu_i^\circ + RT \ln a_i$ are introduced, $\Delta_R G = \Sigma\nu_i\mu_i^\circ + \Sigma\nu_i RT \ln a_i = \Delta_R G^\circ = -RT\Sigma\nu_i \ln a_i$ (van't Hoff's reaction isotherm), where μ_i° is the chemical potential of substance i under standard conditions (i.e. in the pure state, the Standard state (see)), a_i is the activity of the substance i in the mixed phase, and $\Delta_R G^\circ = \Sigma\nu_i\mu_i^\circ$ is the standard free reaction enthalpy.

At equilibrium, $\Delta_R G = 0$. From this follows $\Delta_R G^\circ = -RT\,\Sigma\nu_i \ln a_i$, or $\pi a_i^{n_i} = \exp(-\Delta_R G^\circ/RT) = K$. The last equation is the thermodynamic formulation of the mass action law. The equation $\Delta_R G^\circ + RT \ln K$ makes possible the thermodynamic calculation of equilibrium constants K from standard free enthalpies of reactions. The latter can be taken directly from thermodynamic tables, or calculated using the additivity equations $\Delta_R H^\circ = \Sigma\nu_i\Delta_R H_i^\circ$ and $\Delta_R S^\circ = \Sigma\nu_i s_i^\circ$ and the conventional standard entropies ΔS_i° in the Gibbs-Helmholtz equation $\Delta_R G^\circ = \Delta_R H^\circ - T\Delta_R S^\circ$.

Third law of T. (Nernst's heat theorem): From numerous calorimetric measurments at low temperatures, Nernst concluded in 1906 that as absolute zero is approached, the change Δs in entropy of a pure substance in internal equilibrium goes to zero: $\lim_{T\to 0} \Delta s = 0$. The entropy asymtotically approaches a constant value s_0, the zero point entropy, which by convention is set equal to zero (Planck): $\lim_{T\to 0} s = s_0 = 0$. From this it also follows that:

$$\lim_{T\to 0} C_v = \lim_{T\to 0}\left(\frac{\partial h}{\partial T}\right)_p = 0.$$

For the statistical interpretation of entropy, $s = k \ln P$, $s = 0$ means that $P = 1$, i.e. that the system can be arranged in only one way, because all its components are in the ground state. However, it must be remembered that at very low temperatures, quantum effects become important and an exact treatment must be based on quantum statistics.

It follows from the Nernst heat theorem that absolute zero cannot be reached. Since for $T \to 0$, $c_p \to 0$ also, the smallest amount of heat induces a finite temperature change. However, it is not possible in the real world completely to prevent heat flow, and thus it is also impossible to cool a system to $T = 0$ K. In addition, the third law permits calculation of molar standard entropies of pure substances $S_T = \int_{T=0}^{T} dS = \int_{T=0}^{T} (C_p/T)dT$, because $S_0 = 0$. If phase changes occur in the temperature interval 0 to T, the phase-change entropies must be added to the integral: $S_T = \int_{T=0}^{T} (C_p/T)dT + \Sigma(\Delta_p H/T_p)$. Molar standard entropies are published in tables.

II) T. of irreversible processes. Like phenomenological T., this is a theory of macroscopic systems, but it treats processes in systems which are not at equilibrium. It gives a quantitative description of the course of transport reactions, equilibration processes and chemical reactions. The treatment of coupled processes of the thermodiffusion type is of great significance. The T. of irreversible processes can be given a theoretical basis in nonequilibrium statistics (kinetic theory). There are two types of T. of irreversible process, linear and nonlinear.

1) ***Linear T. of irreversible processes*** applies near equilibrium. It provides little insight into reaction kinetics, because chemical reactions generally begin far from equilibrium and the rate equations are highly nonlinear.

Linear T. characterizes a process by a *flux J*, e.g. a flow of particles in diffusion or of heat in heat conduction. The flux is caused by *forces X*, that is, deviation of certain potentials, or parameters proportional to them, from thermodynamic equilibrium. Irreversible T. is based on three essential postulates:

a) The fluxes J_i are linearly dependent on all forces X_f which cause them: $J_i = \sum_f L_{if} X_f$. The proportionality coefficients L_{if} are the transport coefficients. Some examples are Fick's first law of diffusion and the law of heat conductivity.

b) In nonequilibrium states, the entropy is replaced by its rate of change, and entropy production $P = ds/dt > 0$. The system develops until it has reached equilibrium, at which point entropy has reached a maximum, and entropy production has gone to zero. From this, it follows that an important condition for evolution in a closed system is $dP \leq 0$. Entropy production is related to forces and fluxes by the equation $P = \sum_i X_i J_i$.

c) When two or more processes occur simultaneously, the *Onsager reciprocity equation* applies to the transport coefficients: $L_{if} = L_{fi}$. For example, the coefficients of thermodiffusion and the diffusion thermal effect are equal. The equation was extended to forces in a magnetic field and rotating systems by Casimir.

2) ***The nonlinear T. of irreversible processes*** is still in the development stage. Since the relationship between forces and fluxes is no longer linear, there is a possibility for formation of structure in open, irreversible systems (see Dissipative structures). Because of their significance for biochemistry and reaction kinetics, the questions of structure formation, stability, periodic processes and evolution in such systems are being intensively studied.

III) Statistical T.. The goal is calculation of macroscopic thermodynamic substance parameters and state parameters from molecular data, the molecular motions and interactions. The field is limited to systems which are in thermodynamic equilibrium (equilibrium statistics).

Because of the large number of particles in a macroscopic system (for example, 1 dm^3 air under standard conditions contains $2.7 \cdot 10^{22}$ molecules), the parameters cannot be obtained by applying methods of classical or quantum mechanics to each individual particle. Instead, methods of probability calculation and mathematical statistics are used with a few assumptions about the properties of molecules. The properties of the particles, e.g. location, velocity or energy, are not described by giving a discrete value for each particle, but by distribution functions. A distribution function gives the probability of observing the property of interest in the system. The parameter which is measured by macroscopic, physical methods is the statistical average obtained by summation or integration of the distribution function.

The *thermodynamic probability* P plays an important role. P is the number of microscopic states which can result in a given macroscopic state. A macroscopic state is the state of a macroscopic system, such as a gas, which can be characterized by pressure, temperature and internal energy. The same macroscopic state can be the result of a very large number of different microscopic states, such as different distributions of energy among the individual molecules of the system. Each of these different distributions is called a microscopic state. P, in contrast to ordinary probability, is an integer and is usually much larger than 1. The system tends toward the macroscopic state with the highest thermodynamic probability; this is the equilibrium state. The probability is related to entropy by the equation $s = k \ln P$, where k is Boltzmann's constant. According to the laws of phenomenological T., entropy is also at a maximum at equilibrium.

There is a problem in counting the number of possible microscopic states, because it is determined by the nature of the distribution function. In classical mechanics, the individual particles can be distinguished from each other, and any arbitrary number of them can occupy the same energy state. The exchange of two particles produces a new microscopic state. This type of counting leads to the Boltzmann distribution. In quantum mechanics, the individual particles cannot be distinguished, due to the uncertainty relation. An exchange does not lead to a new microscopic state. Furthermore, particles with half-integral spin (fermions), including electrons, are subject to the Pauli principle, which states that only one particle can occupy a given energy state. This leads to Fermi-Dirac statistics (see). Particles with integral spin (bosons), like classical particles, can occupy the same energy state; this leads to Bose-Einstein statistics.

The State sums (see) are derived from the distribution function and make possible calculation of the thermodynamic potentials, other thermodynamic parameters and the equations of state.

Thermodynamic control: the experimental observation that the result of a process is determined by the laws of thermodynamics. In synthetic chemistry, a reaction is said to be under T. if the relative concentrations of products and reactants correspond to thermodynamic equilibrium concentrations. T. occurs when the reactions of formation of all products are reversible and the reaction time is long enough to permit equilibrium to be reached. The opposite of T. is Kinetic control (see).

Thermodynamic equilibrium: see Equilibrium.

Thermodynamic potentials: the thermodynamic parameters of state internal energy e, enthalpy h, Helmholtz energy f and Gibbs free enthalpy g, represented as state functions of the characteristic state variables temperature T, entropy s, pressure p, volume v and mole number n: $e = f(s, v, n_i)$, $h = f(s, p, n_i)$, $f = f(T, v, n_i)$ and $g = f(T, p, n_i)$. In this system, changes in the parameters of state are given by:

$$de = Tds - pdv + \Sigma\mu_i dn_i,$$
$$dh = Tds - vdp + \Sigma\mu_i dn_i,$$
$$df = - sdT - pdv + \Sigma\mu_i dn_i,$$
$$dg = - sdT + vdp + \Sigma\mu_i dn_i,$$

where μ_i is the Chemical potential (see) of substance i. These four equations are called ***Gibbs' fundamental equations***; they show the relationship between the various thermodynamic parameters of state in a very clear fashion.

The T. in a narrow sense often means the free enthalpy g.

Thermodynamic probability: see Thermodynamics III.

Thermodynamic system: see System.

Thermogravimetry, abb. ***TG***: a thermoanalytical method for determining the change in mass in a substance as a function of temperature as it is heated. T. can be used to study the evaporation, sublimation, desorption, dehydration, decomposition and oxidation of solids. It can also be used to solve analytical problems, such as determining the water content and suitable drying or roasting temperatures for precipitates obtained by Gravimetry (see). An important area of applications is the study of polymers. On the one hand, information can be obtained on decomposition temperatures and mechanisms, and on the other, thermograms for different polymers can be made which permit their identification.

An apparatus for thermogravimetric analyses consists of the following components: 1) a sensitive *thermobalance* with which the mass changes in a sample in an oven can be followed continuously; 2) a *programmable heat source*, which delivers a linear increase in temperature with at a preselected rate (usually between 0.5 and 25 °C/min); 3) a *thermoelement* for measurement of the temperature which is as close to the sample as possible; 4) a *recording device* which draws the thermogram, in which the mass of the sample is given as a function of temperature.

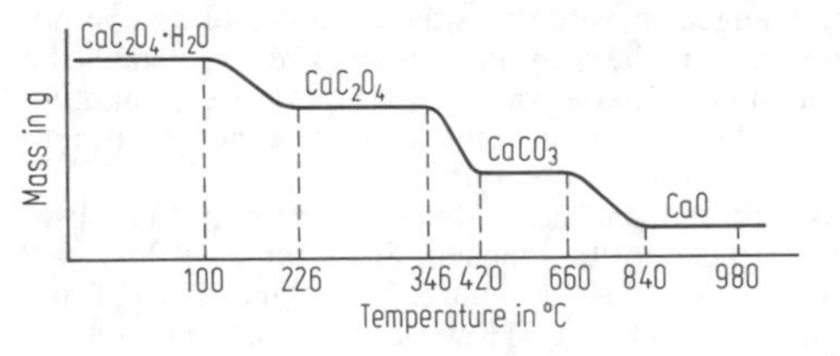

Fig. 1. Thermogram for decomposistion of calcium oxalate monohydrate.

Fig. 1 shows the thermogram obtained for calcium oxalate monohydrate at a temperature increase rate of 5 °C/min. The horizontal sections of the curve correspond to temperature regions in which the indicated compounds are stable. In between the plateaus, there is defined decomposition of the sample. The thermogram indicates the thermal conditions under which a given compound can be gravimetrically determined. Often the first derivative of the thermogram is shown as well, which makes certain conversion points clearer (***differential thermogravimetry***). T. can also be used for quantitative analysis of mixtures, as is shown in Fig. 2 in the thermogram of calcium, strontium and barium oxalate monohydrate. While the transition from the hydrates to the anhydrides occurs at approximately the same temperature, further increase in the temperature leads to decomposition first of calcium carbonate, then strontium and finally barium carbonate, which permits quantitative determination of the elements.

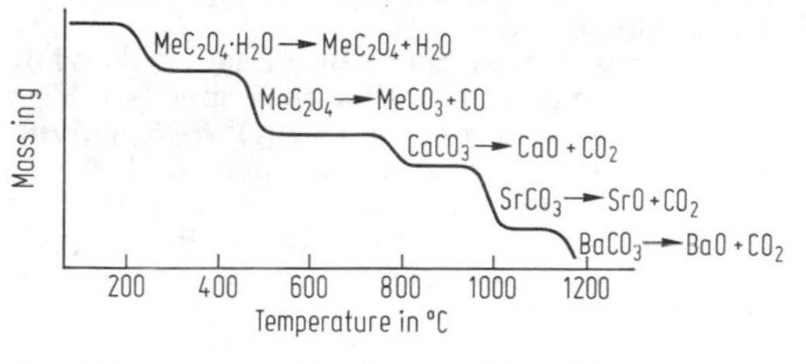

Fig. 2. Thermogram of the decomposition of $CaC_2O_4 \cdot H_2O$, $SrC_2O_4 \cdot H_2O$ and $BaC_2O_4 \cdot H_2O$.

A disadvantage of T. is that overlapping thermal processes are usually difficult to separate.

Thermobalances are often equipped with a device which permits registration of the rate of mass change. This procedure is called ***derivative T.***, and the resulting curves are *DTG curves*. They display sharp minima at temperatures at which the mass changes occur. The methods of T. are often coupled with differential thermoanalysis in a single apparatus, the *derivatograph*. In this way, a maximum of information on the thermal behavior of the substance is obtained.

Thermolysin: a zinc-containing endopeptidase (see Proteases) which catalyses hydrolysis of proteins and peptides at the amino ends of leucine, isoleucine, valine and phenylalanine. T. is formed by *Bacillus thermoproteolyticus* (M_r 37,500) and, because it contains a large proportion of hydrophobic regions, it is biologically active up to 80 °C. Its thermostability can be increased by addition of calcium ions; these stabilize the protein molecule through complex binding to asparagine and glutamic acid residues.

Thermometric analysis: same as Thermometry (see).

Thermometry, ***thermometric analysis***, abb. ***TA***: thermoanalytical methods. If a sample is heated in an oven at a constant rate, a diagram can be made showing the sample temperature as a function of time. This is normally a straight line with a slope which depends on the rate of heating. However, if a reaction occurs during this time, then heat is either absorbed or released, and the curve deviates from linearity. If the reaction is exothermal, it goes through a maximum, and if endothermal, a minimum. After the reaction ends, the normal linear progression is re-established. Because less equipment is required to establish a cooling curve, this is often preferred in practice to a heating curve.

T. can be used to measure simple reactions or phase changes which occur relatively rapidly in a narrow temperature interval. However, for many reactions it is too insensitive, and other thermoanalytical methods (see, e.g., Differential thermoanalysis) are used instead.

Thermoset plastic: see Plastics.

Thermotropic: see Liquid crystals.

THF: abb. for tetrahydrofuran.

Thia-: a prefix often used to designate a sulfur-containing compound. With a few exceptions, this prefix indicates formal exchange of a sulfur atom for one of the carbon atoms in the parent compound.

Thiabendazol: an Anthelminthic (see) used in veterinary medicine; see also Benzimidazole fungicides.

Thialene: see Pseudoazulenes.

Thiamazol: see Thyreostatics.

Thiamin: same as Vitamin B_1 (see).

Thiamin pyrophosphate: see Coenzymes.

Thiazepins: seven-membered, heterocyclic compounds containing one sulfur and one nitrogen atom in the ring. These compounds are designated 1,2-T., 1,3-T. and 1,4-T., depending on the positions of the heteroatoms relative to one another.

Thiazides: see Diuretics.

Thiazine: a six-membered, heterocyclic compound with a sulfur and a nitrogen atom in the ring. There are eight possible isomers. ***1,4-Thiazine*** is a colorless liquid, b.p. 77 °C.

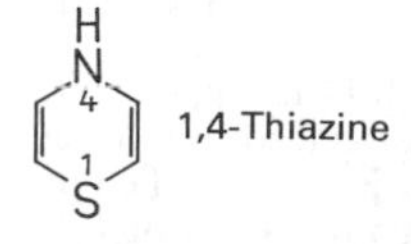

It can be synthesized from thiodiglycolamide by reduction with aluminum. The Thiazine pigments (see) are derived from ***1,4-phenothiazine***.

Thiazine pigments: a group of synthetic pigments which contain thiazine skeletons (see Triarylmethane pigments). The T. include several sulfur pigments and methylene blue.

Thiazole: a colorless, highly hygroscopic liquid which smells like pyridine; b.p. 116.8 °C, n_D^{20} 1.5969. It is soluble in water, alcohol and ether.

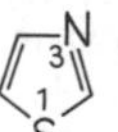

It is an isomer of ***isothiazole***. Its physical and chemical properties are very similar to those of pyridine. T. is a weak base, and is quaternized at the nitrogen atom by haloalkanes. Nucleophilic substitutions take place in position 2, and electrophilic substitutions in position 5, with more difficulty. T. is difficult to reduce and hydrogenate. It is obtained by the reaction of chloroacetaldehyde and thioformamide. Derivatives of T. can be made from α-haloketones by condensation with thioamides. The dihydro derivatives of T. are called ***thiazolines***, and the completely hydrogenated derivatives, ***thiazolidines***. These can be made by special condensation reactions. The most important derivatives of T. are ***2-aminothiazole***, which is important as an intermediate in the synthesis of sulfonamides and as a vulcanization accelerator. Some examples of natural T. derivatives are thiamin and the penicillins which contain thiazolidine rings.

Thiazolidine: see Thiazole.

Thiazoline: see Thiazole.

Thick oils: 1) residues from the distillation of fatty alcohols from oxosynthesis. D. are good defoaming agents. 2) In paints, all oils with artificially increased viscosities.

Thienamycin: see β-Lactam antibiotics.

Thieno[2,3]thiophene: same as Thiophthene (see).

Thieno[3,2]thiophene: see Thiophthene.

Thiepin: a triply unsaturated, seven-membered heterocyclic compound with a sulfur atom in the ring. The parent compound T. is not known, and even substituted T. are very unstable. However, thiepin-1,1-dioxide is stable.

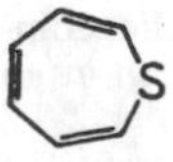

Thiiran: same as Ethylene sulfide (see).

Thiirans: same as Episulfides (see).

Thin acid: dilute sulfuric acid (about 10 to 20%) with dissolved iron sulfate, a byproduct of the chemical industry and metallurgy, especially from ore solubilization, titanium dioxide production and metal treatment. Reprocessing of T is complicated; it involves evaporation, causing iron sulfate to crystallize out of the concentrated solution. It is also possible to precipitate the iron sulfate by adding a solvent which reduces its solubility. The solvent is then distilled off to concentrate the acid. The ocean transport of T. in special ships (North Sea and Atlantic) is very dangerous to the environment.

Thin-layer chromatography, abb. ***TLC***: a micromethod of Adsorption chromatography carried out on thin layers of silica gel or aluminum oxide, with or without calcium sulfate. The adsorbent is poured onto a glass plate or aluminum sheet, or it is spread with a spreader, then activated. The layer is 250 to 300 μm thick. 1 to 3 mg sample is applied, the chromatograms are developed and the spots are made visible as in Paper chromatography (see). However, the separation times are shorter than in paper chromatography: 30 to 60 minutes.

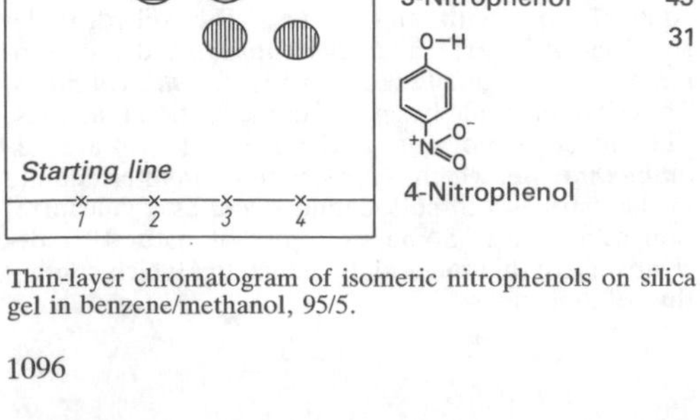

Solvent front

hR$_F$-value

2-Nitrophenol 92

3-Nitrophenol 43

4-Nitrophenol 31

Starting line

1 2 3 4

Thin-layer chromatogram of isomeric nitrophenols on silica gel in benzene/methanol, 95/5.

Methods. T. is usually done with ascending eluent, in one or two dimensions. The chromatograms are developed in normal (> 3 mm gas space) or narrow (< 3 mm gas space) chambers. It is possible to work with layer or solvent gradients, and the separated components can be subjected to a reaction before they are separated in the second direction (***SRS method***, for separate-react-separate). In ***preparative T.***, layers 1 to 2 mm thick are used with samples in the range of 5 to 50 mg.

Evaluation of T. is done as in paper chromatography. In addition, the spots can be made visible by heating or treating them with iodine vapor or concentrated sulfuric acid.

The column height can be significantly reduced by using adsorbents with smaller grain sizes (see High performance liquid chromatography). ***High performance thin-layer chromatography***, abb. ***HPTLC*** is a horizontal method which can be done in a linear or circular (U chamber) fashion. This method reduces the chromatographic parameters (column height, analysis time) by about a factor of 10.

Historical. T. was developed in 1938 by Ismailov and Shraiber, and was introduced as a standardized laboratory method by Stahl in 1958.

Thin-layer reactor: a reactor for liquid-gas reactions in which a film of liquid flows down the walls of a pipe and reacts with the gas in the pipe. The T. can be run with the gas flowing either in the same direction as the liquid or in the opposite direction. It is especially useful for reactions in which the temperature must be precisely controlled.

Thio-: (from the Greek "theion" = "sulfur") the most commonly used prefix used to indicate the presence of sulfur in a compound. This prefix is almost always used to indicate the formal substitution of a sulfur atom for an oxygen atom.

Thioacetals: sulfur analogs of the acetals with the general formula $R^1R^2C(OR^3)(SR^4)$ and $R^1R^2C(SR^3)_2$, which, formally, are derived by stepwise replacement of oxygen atoms by sulfur. T. are made by reaction of ketones or aldehydes with thioalcohols in the presence of hydrogen chloride or zinc chloride:

$$R^1R^2C{=}O + 2\,HS{-}R^3 \xrightarrow{(H)^+} R^1R^2C(SR^3)_2 + H_2O.$$

Thioacetamide: CH_3-CS-NH_2, colorless, monoclinic prisms; m.p. 115-116 °C. T. is readily soluble in water and alcohol, but is only slightly soluble in ether. In the presence of catalytic amounts of acids or bases, it is hydrolyzed to hydrogen sulfide and ammonium acetate. T. is therefore often used in analytical chemistry in the form of a 2% solution, in place of hydrogen sulfide. It can also be used to combat molds. T. is synthesized by the reaction of phosphorus(V) sulfide and acetamide.

Thioacetic acid: CH_3-CO-SH, a bright yellow liquid with an unpleasant, pungent odor; m.p. - 17 °C, b.p. 87 °C (93 °C also reported), n_D^{20} 1.4648. T. is soluble in most organic solvents. It is hydrolysed in water to hydrogen sulfide and acetic acid. T. is a strong acid, which exists in a tautomeric equilibrium between the thiol and thione forms: CH_3-CO-SH $\rightleftharpoons$ SH_3-CS-OH. Like all other Thiocarboxylic acids (see)., T. reacts primarily from the thiol form. It is a strong acetylating reagent. T. is made by the reaction of phosphorus(V) sulfide with glacial acetic acid, or from hydrogen sulfide and acetyl chloride in the presence of pyridine. It is often used in analytical chemistry instead of hydrogen sulfide, and is also used in the synthesis of thiols.

Thioalcohols, older name, ***mercaptans***: organic sulfur compounds with the general formula R-SH, in which the thiol group is linked to an alkyl group R. T. are the sulfur analogs of alcohols. In systematic nomenclature, the name of the T. is formed by adding the suffix "-thiol" to the root of the hydrocarbon, if the thiol group is the principal group. If the thiol group is not the principal group, the prefix "mercapto-" is added to the name of the parent compound to indicate an unsubstituted -SH group. Examples: methanethiol, SH_3SH, ethanethiol, C_2H_5SH, *p*-mercaptobenzoic acid, HS-C_6H_4-COOH. With the exception of the gaseous methanethiol, the T. are liquid or solid compounds. Their boiling points are lower than those of the corresponding alcohols, because they have less tendency to form hydrogen bonds. They are also less soluble in water than alcohols. T. are toxic and have a repulsive odor. The threshhold for perception of the odor of ethanethiol is $4.6 \cdot 10^{-8}$ mg, that is, even traces are noticeable. For example, (*E*)-but-2-ene-1-thiol is the main component of skunk oil.

T. are stronger acids than the corresponding alcohols, and therefore, in contrast to alcohols, they form salts with alkali hydroxides in aqueous solutions; these are called ***thiolates***: R-SH + NaOH → R-S^-Na^+ + H_2O. Salts are also formed with heavy metal ions; these are rather insoluble covalent compounds, and one reaction for detection of T. is based of the formation of lead(II) and mercury(II) thiolates. In fact, the toxic effect of these heavy metal ions is due to their tight binding to thiol groups on biological molecules (see Cysteine), which blocks their activity. The alkali thiolates are strongly nucleophilic reagents which form thioethers by reaction with haloalkanes: $R\text{-}S^- + R'\text{-}Br \rightarrow R\text{-}S\text{-}R' + Br^-$.

The behavior of T. with respect to oxidizing agents is also distinctly different from that of alcohols. Instead of thiocarbonyl compounds, disulfides, R-S-S-R, are formed in presence of mild oxidizing agents, such as air or hydrogen peroxide. Strong oxidizing agents, such as potassium permanganate or nitric acid, produce sulfonic acids, R-SO_3H. Other reactions of the T. include formation of thiol esters, R-COSR', with carboxylic acids (this reaction is catalysed by strong acids), formation of mercaptals and loss of sulfur to form hydrocarbons. This last reaction is very important for removal of T. from petroleum fractions; in hydrorefining, it takes place in the presence of hydrogen with molybdenum(IV) or tungsten(IV) sulfide catalysts at 300-400 °C and pressures of 2-4 MPa.

The formation of yellow lead(II) and colorless mercury(II) thiolates can be used in qualitative analysis to detect T. The 2,4-dinitrophenyl sulfides, formed by reaction of the T. with chloro-2,4-dinitrobenzene, are derivatives with characteristic melting points. The IR spectra of T. have typical bands in the regions of 2550-2600 cm^{-1} (S-H valence vibrations) and 570-705 cm^{-1} (C-S valence vibrations).

Synthesis 1) From haloalkanes and alkali hydrogensulfides: R-Br + NaSH → R-SH + NaBr. Dialkyl sulfides (thioethers) can form as byproducts as the T. react further in basic solution with excess haloalkane. 2) From haloalkanes and thiourea via S-alkylisothiuronium salts, which are hydrolysed with alkali hydroxide. No thioethers are formed:

$$R-Br+S=C(NH_2)_2 \rightarrow RS-C(NH_2)=NH_2^+Br^- \xrightarrow{OH^-} R-SH+O=C(NH_2)_2.$$

3) From alcohols and hydrogen sulfide at temperatures of 300-400 °C on aluminum oxide catalysts by replacement of oxygen by sulfur. 4) By addition of hydrogen sulfide to alkenes at 150-300 °C in the presence of aluminum oxide and nickel sulfide catalysts: $R-CH=CH_2+H_2S \rightarrow R-CH(CH)-CH_3$.

Applications. T. are used for organic syntheses of pesticides and indigoid dyes, in the rubber industry as vulcanization accelerators and ageing retardants.

Thioaldehydes: organic compounds which contain a thioaldehyde group, $-CH=S$, in the molecule. T. are thiocarbonyl compounds, and are significantly more reactive than the corresponding oxygen analogs, the aldehydes. It is difficult to keep T. in free, monomeric form, because they readily react to form oligomers and polymers. They can be obtained by pyrolysis of allyl sulfides:

$$R-CH(H)(-S-CH_2-CH=CH_2) \xrightarrow{T} R-CHS + CH_3-CH=CH_2$$

This method can also be used to make simple T., such as thioacrylaldehyde and thiobenzaldehyde. Reaction of the T. with nucleophiles is similar to that of the corresponding Aldehydes (see).

Thioamides: amides of thiocarboxylic acids with the general formula R-CS-NH_2. These compounds are named analogously to the carboxamides, using the prefix thio- (e.g. thioacetamide). T. can be obtained from carboxamides by thionization with phosphorus(V) sulfide. They are important as synthetic precursors for the Hantzsch thiazole synthesis (see).

Thiobarbitals: see Barbitals.

Thiobarbiturates: see Barbiturates.

Thiocarbamide: same as Thiourea (see).

Thiocarbamates: same as Thiourethanes (see).

Thiocarbamic hydrazide: see Thiosemicarbazide.

Thiocarbanilide: same as *N,N'*-diphenylthiourea.

Thiocarbonic acid: a sulfur analog of carbonic acid in which one, two or three oxygen atoms can be replaced by sulfur. ***Monothiocarbonic acid*** is HO-$-$CO$-$SH ⇌ HO$-$CS$-$OH, ***dithiocarbonic acid*** is HO$-$CS$-$SH ⇌ HS$-$CO$-$SH and ***trithiocarbonic acid*** is HS$-$CS$-$SH; only the last of these can be isolated in free form. The mono- and dithiocarbonic acids are known only in the form of a few derivatives, e.g. thiourea and xanthogenic acid esters. Trithiocarbonic acid is a red, oily compound with a pungent odor; m.p. -30 °C, b.p. 57 °C (dec.). It is soluble in ether, chloroform and toluene, and decomposes readily, forming carbon disulfide and hydrogen sulfide. It can be made by reaction of alkali sulfides with carbon disulfide, which forms alkali trithiocarbonates, and subsequent acidification with concentrated hydrochloric acid:

$$K_2S + CS_2 \rightarrow S{=}C(SK)_2 \xrightarrow{(H)^+} HS{-}CS{-}SH.$$

Stable diesters of trithiocarbonic acid can be obtained by reaction of alkyl halides with its alkali salts; these compounds are used, for example, as pesticides.

Thiocarbonyl compounds: organic compounds containing a thiocarbonyl group, >C = S The T. include, for example, Thioaldehydes (see), Thioketones (see), Thiocarboxylic acids (see), Thiocarbonic acid (see) and several acyl derivatives. T. are often very unstable compounds with unpleasant odors. On the other hand, there are very stable T., including, for example, the substituted thioureas.

Thiocarbonyl group: the functional group >C = S, which is found in thiocarbonyl compounds. If the C atom of the T. is linked to two C atoms of aliphatic, aromatic or heterocyclic groups, the group is called the ***thioketo*** or ***S-heteroanalogous keto group***. The T. is more reactive than the Carbonyl group (see), but its reactions form essentially the same type of products.

Thiocarboxylic acids: compounds related to carboxylic acids, but with one or both oxygen atoms replaced by sulfur atoms. In the ***monothiocarboxylic acids***, one oxygen atom is replaced by a sulfur atom. These are tautomers, but the equilibrium is usually completely on the thiolic acid side.

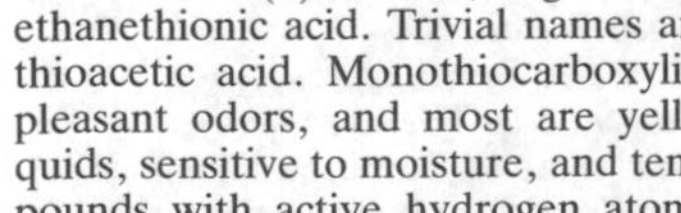

$$R{-}C(=O){-}SH \rightleftharpoons R{-}C(=S){-}OH$$

Thiolic acid Thionic acid

The names of the acids are derived from those of the related carboxylic acids, with "-thiol-" or "-thione-" inserted between the root name of the group R and the suffix "-(o)ic acid", e.g. ethanethiolic acid or ethanethionic acid. Trivial names are also used, e.g. thioacetic acid. Monothiocarboxylic acids have unpleasant odors, and most are yellow. They are liquids, sensitive to moisture, and tend to acylate compounds with active hydrogen atoms. For example, they form carboxylic acid esters with alcohols. Monothiocarboxylic acids are formed by reaction of carboxylic acids with phosphorus(V) sulfide, or by the reaction of carboxylic acid chlorides or anhydrides with H_2S. They are used in analytical chemistry to precipitate heavy metal sulfides. Thioesters are derivatives of the T.; the thiolates, R^1-CO-SR^2, are derivatives of thiolic acids, and the thionates, R^1-CS-OR^2, of thionic acids. ***Thionethiolic acids (dithiocarboxylic acids)*** have the structures R-CS-SH. These are very unstable, yellowish red liquids which have very bad odors. In air, they are rapidly oxidized to thioacyldisulfides, R-CS-S-S-CS-R. The free acids are formed by addition of Grignard compounds to CS_2, and the esters by alkylation of the acids. They are used for thioacylations. The Thioamides (see) are derivatives of the T. T. are in general more reactive than the corresponding carboxylic acids.

Thiochrome: a compound derived from Vitamin B_1. It crystallizes as yellow prisms; m.p. 227-228 °C (other reports, 278 °C). It is soluble in water and methanol, and the solutions have an intense blue fluorescence. T. are formed by oxidative ring closure of thiamin (vitamin B_1) in alkaline media.

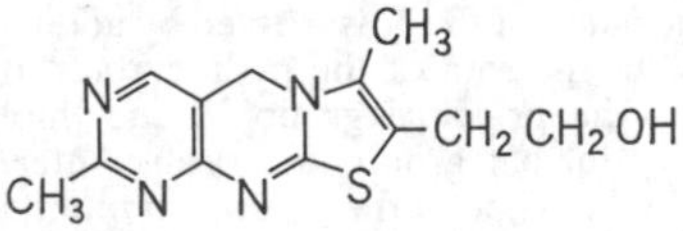

Thiocyanates, ***rhodanides***: derivatives of thiocyanic acid (more correctly, isothiocyanic acid), H-NCS.

1) Inorganic T. The important *ionic T.* are based on the linear, resonance-stabilized thiocyanate anion, $[NCS]^-$, which is one of the Pseudohalides (see). Alkali thiocyanates, such as ***sodium*** or ***potassium thiocyanate***, are typical, colorless salts, which are made by fusion of alkali cyanides with sulfur. Oxidation of these T., for example with manganese dioxide, yields the pseudohalogen ***dithiocyanogen*** (***dirhodane***), NCS-SCN, a compound which is stable only at low temperatures; m.p. -3 °C. Protonation of the thiocyanate anion yields ***hydrogen isothiocyanate***, H-NCS, m.p. +5 °C, which is stable only at low temperatures; in water, it forms the medium strong ***isothiocyanic acid*** (p*K* $1.1 \cdot 10^{-1}$).

The *covalent T.* of nonmetals are generally of the isothiocyanate type, e.g. ***boron triisothiocyanate***, $B(NCS)_3$, is a colorless liquid, b.p. (at 13 Pa) 92 °C; ***silicon tetraisothiocyanate***, $Si(NCS)_4$, colorless prisms, m.p. 146 °C, b.p. (at 9 Pa) 87-88 °C.

The thiocyanate anion is an excellent complex ligand, forming coordinate bonds both via the nitrogen (***isothiocyanate complexes,*** M-NCS) and via the sulfur (***thiocyanate complexes***, M-SCN). The type of coordination is affected by various factors. In bridge positions, the anion can also act as a polydentate ligand; if it is bidentate, it has the coordination type M-NCS-M'.

The T. are used in analysis as reagents for iron, and the decided tendency of the anion to form complexes is also utilized in extraction and purification of metals.

2) The organic T. with the general formula R-SCN can also be considered ***thiocyanic acid esters***. They are colorless oils or crystalline compounds, depending on R., and smell like leeks. They are insoluble in water, but are soluble in alcohol and ether. Some T. can be rearranged in the presence of catalysts or thermally to the isomeric Isothiocyanates (see). Reduction of T. produces thiols:

$$R{-}S{-}C{=}N \xrightarrow{(Zn/H^+)} R{-}SH + HCN$$

Aliphatic T. can be synthesized by reaction of alkali thiocyanates with alkyl halides or dialkyl sulfates: NaSCN + C_2H_5-Cl → C_2H_5-SCN + NaCl. Aromatic T. can be synthesized by the Sandmeyer reaction (see) from arene diazonium salts and alkali thiocyanates, or by reaction of thiols with cyanogen halides: R-SH + Cl-N → R-SCN + HCl. In the presence of activating substituents, aromatic halogen compounds can be converted to aromatic T. with alkali rhodanides:

O_2N–C₆H₃(NO_2)–Cl + NaSCN → O_2N–C₆H₃(NO_2)–SCN + NaCl

T. are used as pesticides and in organic syntheses.

Thioesters: esters of thioacids. Two series of T. are known, corresponding to the two tautomeric structures of the thioacids: the ***thiol esters*** (***S-esters***) and the ***thione esters*** (***O-esters***). The thiol esters are formed either by direct acid-catalysed esterification of carboxylic acids with mercaptans, or by reaction of activated carboxylic derivatives (such as acyl chlorides) with mercaptans. The thione esters are formed from imidoesters and hydrogen sulfide. The T. are more reactive with nucleophiles than are the carboxylic acid esters, due to the greater tendency of the SR^1 group to split off. Acetyl coenzyme A. is a T. of great biochemical importance.

$R{-}C({=}O){-}SR^1$ Thiol ester; $R{-}C({=}S){-}OR^1$ Thione ester; $R{-}C({=}S){-}SR^1$ Dithio ester

Dithio esters are derived from dithiocarboxylic acids; they are formed by addition of a mercaptan to a nitrile and subsequent reaction of the imidothioester with hydrogen sulfide. Three types of T. are derived from thiophosphoric and dithiophosphoric acids; these are important as pesticides.

Thioethanolamine: same as Cysteamine (see).

Thioether, ***organic sulfides***: compounds with the general formula R-S-R', in which R and R' can be any alkyl or aryl groups. T. are the sulfur analogs of ethers, and they can be symmetric or unsymmetric. According to the IUPAC rules, they are to be named as dialkyl, alkylaryl or diaryl sulfides, e.g. diethylsulfide, CH_3-CH_2-S-CH_2-CH_3 and diphenylsulfide, C_6H_5-S-C_6H_5. T. are unpleasant-smelling compounds which are insoluble in water. They can be synthesized by: 1) reaction of alkali thiolates with haloalkanes, $R\text{-}S^- + R'\text{-}X \rightarrow R\text{-}S\text{-}R' + X^-$, or of haloalkanes with alkali sulfides, which always yield symmetric T: $2\ R\text{-}X + Na_2S \rightarrow R\text{-}S\text{-}R + 2\ NaX$. 2) Addition of thiols to alkenes in the presence of peroxides or under UV irradiation: $C_2H_5\text{-}SH + CH_2{=}CH_2 \rightarrow C_2H_5\text{-}S\text{-}C_2H_5$. 3) Reaction of sulfenyl halides with arenes in the presence of Lewis acids:

$$Ar{-}SCl + Ar'{-}H \xrightarrow{AlCl_3} Ar{-}S{-}Ar' + HCl$$

Diarylsulfides can also be made in this way.

T. can be oxidized to sulfoxides or sulfones, depending on the reaction conditions, and they form sulfonium salts. Thus T. are important intermediates for organic synthesis. Some T. are also used as softeners.

Thioformamide: H_2N-CH=S, the sulfur analog of formamide. T. forms colorless prisms with a characteristic, unpleasant odor; m.p. 29 °C. It decomposes slowly in the presence of air or when heated. It is readily soluble in alcohol and ether. T. can be made by reaction of hydrocyanic acid and hydrogen sulfide, or by the reaction of ammonia with dithioformic acid: HS-CH=S + H_2N-CH=S + H_2S. Direct sulfuration of formamide with phosphorus pentasulfide gives only small yields. T. is used for quantitative separation of arsenic, antimony and tin, and for synthesis of heterocyclic compounds.

Thiogen dark red: see Sulfur pigments.

Thioglycolic acid, ***mercaptoacetic acid***: HS-CH_2-COOH, a colorless, skin-irritating liquid with an unpleasant odor; m.p. -16.5 °C, b.p. 101 °C at 1.33 kPa, n_D^{20} 1.5030. T. is soluble in water, alcohol and ether. It is easily oxidized in air, forming dithiodiacetic acid (HOOC-CH_2-S-S-CH_2-COOH); heavy metals, such as iron or copper, accelerate the oxidation. T. undergoes the reactions typical of carboxylic acids and thiols. Its salts and esters are called *thioglycolates*. It is synthesized by reaction of monochloroacetic acid with sodium thiosulfate, followed by hydrolysis of the thiosulfate ester.

$$Cl{-}CH_2{-}COONa + NaS{-}SO_3Na \xrightarrow[-NaCl]{}$$

$$NaOOC{-}CH_2{-}S{-}SO_3Na \xrightarrow[-NaHSO_4]{+H_2O} HS{-}CH_2{-}COONa$$

$$\xrightarrow[-Na^+]{+H^+} HS{-}CH_2{-}COOH.$$

T. is used mainly for the synthesis of thioglycolates, which are needed for the production of permanent waves, depilatories, sulfur-containing pigments, heterocycles, antioxidants for rubber and softeners. T. is also used in inorganic analysis as a special reagent for various metals, including iron, cobalt, copper and chromium. Thionalide is a thioglycolamide derivative used in analysis.

Thiohydantion, ***2-thioxoimidazolidin-4-one***: the cyclic thiourea compound of glycine. 2-T. forms yellow prisms; m.p. 229-231 °C. It is soluble in water, alcohol and ether. 2-T. is formed when glycine ethyl ester hydrochloride is heated with potassium rhodanide, or when acetyl thiourea is heated with chloroacetic acid in aqueous solution. 2-T. plays an important role in the synthesis of amino acids and in the structure elucidation of peptides.

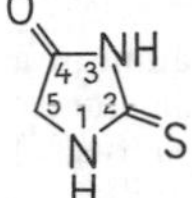

Thioindigo, ***sulfur indigo, anthrared B, vat red B***: a synthetic Indigo pigment (see), and a pure red vat dye (see Fig. to Indigo pigments). The crystals of T. are flat, red-brown needles which are insoluble in water, but dissolve in xylene, giving a yellow color, and in concentrated sulfuric acid, giving a green color. The yellow vat dye formed by reduction with sodium dithionite is a substantive dye, and is bound to wool and cotton. When the cloth or yarn is hung in the air, the dye reoxidizes to T. The colors are more stable to air, light, water and chemicals than indigo itself.

Some important dyes derived from T. are algol orange (diethoxythioindigo) and indanthrene brilliant pink (dichlorodimethylthioindigo). The parent compound of all thioindigo pigments is thionaphthene.

T. was first prepared in 1905 by Friedländer, and was the first red vat dye.

Thioindoxyl: see Benzothiophene.

Thioketals: a term, no longer accepted by the IUPAC Nomenclature Commission, for the sulfur analogs of ketals.

Thioketo group: see Thiocarbonyl group.

Thioketones: organic compounds containing a thioketo group, $>C=S$, in the molecule. T. are thiocarbonyl compounds. All are colored. Aliphatic substituted T. trimerize readily to form colorless trithianes. Aromatic and heterocyclic substituted T. are primarily crystalline compounds which have no tendency to trimerize. T. can be made by the reaction of hydrogen sulfide and hydrogen chloride with ketones, or by pyrolysis of geminal dithiols or allyl sulfides. Because of their extremely unpleasant, penetrating odor, work with the unstable, red aliphatic T. is possible only with good ventilation (under a hood). T. are used mainly for organic syntheses.

Thiokol: see Rubber.

Thiol: an organic sulfur compound which contains the ***thiol group*** -SH. The T. are sulfur analogs of hydroxyl compounds, and may be subdivided into the Thioalcohols (see), in which the -SH is bound to an aliphatic group, and Thiophenols (see), in which it is bound to an aromatic group.

Thiolan: see Tetrahydrothiophene.

Thiolan dioxide: same as Sulfolan (see).

Thiolates, obsolete term, ***mercaptides***: salts of thioalcohols. They are more stable in water than alcoholates are. The alkali thiolates are made by reaction of thioalcohols with alkali hydroxides, and the insoluble lead and mercury salts can be precipitated by addition of the heavy metal ions (see Thioalcohols).

Thiol ester: see Thioester.

Thiolesterases: see Esterases.

Thiol group: see Thiol.

Thiomersal: a mercury organic compound used, among other things, as a preservative for eye medications.

Thiometon: see Organophosphate insecticides.

Thionaphthene: see Benzothiophene.

Thione ester: see Thioester.

Thione green B: see Sulfur pigments.

Thionation: replacement of an oxygen atom in a molecule by a sulfur atom. For example, thioamides can be obtained by T. of amides with phosphorus(V) sulfide.

Thionethiolic acids: see Thiocarboxylic acids.

Thionyl chloride: $SOCl_2$, the dichloride of sulfurous acid, a colorless liquid with a pungent odor which fumes in moist air; M_r 118.97, density 1.655, m.p. -105 °C, b.p. 78.8 °C. The $SOCl_2$ molecule can be described in terms of an sp^3 hybridization of the sulfur and a $d_\pi p_\pi$ bond between sulfur and oxygen (Fig.). Above its boiling point, it decomposes into sulfur dioxide, disulfur dichloride and chlorine. T. is converted to sulfur dioxide and hydrogen chloride by water: $SOCl_2 + 2\,H_2O \rightarrow SO_2 + 2\,HCl$. It is made by the reaction of sulfur dioxide with phosphorus(V) chloride ($SO_2 + PCl_5 \rightarrow POCl_3 + SOCl_2$), or by combining sulfur dichloride with sulfur trioxide ($SCl_2 + SO_3 \rightarrow SOCl_2 + SO_2$). T. is frequently used as a reagent for converting alcohols to alkyl chlorides, and carboxylic acids to acyl chlorides. It is also used to dehydrate metal halides which contain water of crystallization.

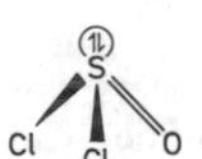

Thionyl group: see Sulfinyl group.

Thiophanate: see Benzimidazole fungicides.

Thiophene: a five-membered, heterocyclic compound with one sulfur atom in the ring. T. is a colorless, mobile liquid which smells like benzene; m.p. -38.2 °C, b.p. 84.2 °C, n_D^{20} 1.5289. It is readily soluble in alcohol, ether, chloroform and benzene, but is insoluble in water. T. is a cyclic conjugated system which is most likely to undergo electrophilic substitution reactions. Bromination with *N*-bromosuccinimide yields ***2-bromothiophene***, nitration leads to ***2-nitrothiophene***, and concentrated sulfuric acid sulfonates it to ***thiophene-2-sulfonic acid***. T. has almost no unsaturated characteristics; palladium-catalysed hydrogenation yields ***tetrahydrothiophene***. T. is desulfurized by hydrogen in the presence of Raney nickel. T. can undergo a Diels-Alder reaction with especially reactive alkynes, such as dicyanoacetylene. The *indophenin reaction* is used to detect T., e.g. as an impurity in benzene; in this reaction, T. turns blue when heated with isatin and concentrated sulfuric acid. T. is a very toxic compound, and can damage the liver, kidneys and heart.

T. is found in the carbonization oils of bituminous shales and in the light oil fraction of coal tars. It is synthesized industrially by passing acetylene over pyrite at 300 °C, or by reaction of butane with sulfur in the vapor phase. T. and its derivatives can also be obtained by distillation of disodium succinate with phosphorus(III) sulfide, or by reaction of 1,4-dicarbonyl compounds with phosphorus(V) sulfide. T. is used to synthesize drugs, pesticides and pigments, and as a solvent.

Thiophenol, ***benzene thiol***: the simplest aromatic compound with a thiol group. It is derived formally from phenol by substitution of a sulfur atom for the oxygen. T. is a colorless, poisonous liquid with a repulsive odor; m.p. -14.8 °C, b.p. 168.7 °C, n_D^{20} 1.5893.

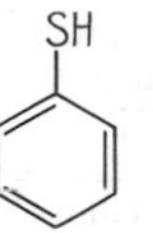

T. is insoluble in water, but soluble in ethanol, ether and benzene. It is distinctly acidic and can be titrated in alcohol solution with alkali hydroxides and phenolphthalein. Neutralization replaces the hydrogen of the thiol group with a metal atom, forming a ***thiophenolate***. T. is very reactive. It is oxidized to diphenyldisulfide by standing in air, and upon reaction with chlorine it forms the very reactive phenyl sulfuryl chloride, C_6H_5SCl. T. is synthesized by reduction of benzene sulfonyl chloride with tin or zinc powder and sulfuric acid, and isolated by steam distillation. T. is used in organic syntheses as the starting material for many sulfur compounds and for polymerization regulators.

Thiophenols, ***arene thiols***: organic sulfur compounds in which the thiol group, -SH, is bound directly to an arene ring. The simplest T. is Thiophenol (see). The properties of the T. are similar to those of Thioalcohols (see). They are toxic and have an extremely unpleasant odor. The thiol group can react in many ways, such as oxidation to disulfides or sulfonic acids, formation of salts with alkali hydroxides or heavy metal ions, reaction of the alkali salts with alkylating reagents to form thioethers (sulfides), addition to activated alkenes to form special thioethers, and so on.

Thiophthene, ***thieno[2,3-b]thiophene***: a colorless oil; m.p. 6.5 °C, b.p. 106 °C at 2.13 kPa. It is soluble in ether.

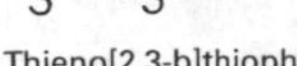
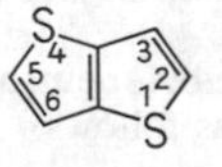

Thieno[2,3-b]thiophene Thieno[3,2-b]thiophene

T. is formed by heating citric acid with phosphorus(V) sulfide. It is also formed from acetylene and sulfur at 440 °C, in a mixture with its isomer, ***thieno[3,2-b]thiophene*** This compound forms colorless crystals, m.p. 56 °C, b.p. 221-224 °C. The two isomers can be separated by distillation.

Thiopyrans: a group of six-membered heterocyclic compounds with a sulfur atom in the ring. ***2H-T.*** and ***4H-T.*** differ in the position of the two double bonds in the ring.

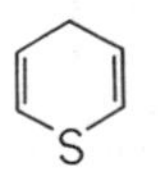

4H-thiopyran

Both isomeric parent compounds have been synthesized. Thiopyrylium salts can be formed by abstraction of a hydride ion from T. by a hydride-ion acceptor such as trityl perchlorate. The T. and thiopyrylium salts are important sensitizers for non-conventional imaging processes.

Thiosulfuric acid, ***monosulfane monosulfonic acid***: $H_2S_2O_3$, a dibasic acid derived formally from sulfuric acid by substitution of a sulfur atom for one of the oxygens. It is stable only at low temperatures or dissolved in organic solvents. In aqueous solution, it decomposes in a reaction which is not understood, forming, among other products, sulfur, sulfur dioxide, hydrogen sulfide and sulfuric acid. The salts of T., the Thiosulfates (see) are stable. T. is obtained as the etherate by the reaction of sulfur trioxide with hydrogen sulfide at -78 °C:

$$SO_3 + H_2S \xrightarrow{\text{(ether)}} H_2S_3O_3$$

or, in pure form, by the reaction of hydrogen sulfide with chlorosulfuric acid: $H_2S + HO_3SCl \rightarrow H_2S_2O_3 + HCl$.

Thiosemicarbazide, ***thiocarbamic acid hydrazide***: $H_2N\text{-}CS\text{-}NH\text{-}NH_2$, a colorless, crystalline compound; m.p. 183 °C. T. is readily soluble in hot water, and only slightly soluble in organic solvents. It forms salts with acids, and metal complexes with a few metal salts. When T. reacts with aldehydes and ketones, the products are ***thiosemicarbazones***, which are usually easily crystallized: $R\text{-}CHO + H_2N\text{-}NH\text{-}CS\text{-}NH_2 \rightarrow R\text{-}CH{=}N\text{-}NH\text{-}CS\text{-}NH_2 + H_2O$. T. is synthesized by the reaction of ammonium rhodanide with hydrazine sulfate: $HS\text{-}C{\equiv}N + H_2N\text{-}NH_2 \rightarrow H_2N\text{-}CS\text{-}NH\text{-}NH_2$. T. is used as a reagent for identifying aldehydes and ketones. Thiosemicarbazones formed in this type of reaction are also used in medicine to treat tuberculosis, and as pesticides.

Thiosemicarbazones: see Thiosemicarbazides.

Thiosulfates: the salts of thiosulfuric acid, $H_2S_2O_3$; their general formula is $M^I{}_2S_2O_3$. T. crystallize readily, and with the exception of barium, silver, lead and thallium(I) thiosulfate, they are soluble in water. The tetrahedral thiosulfate ion is derived formally from the sulfate ion by substitution of a sulfur atom for an oxygen. It is a good complex ligand and forms thiosulfato complexes with many transition metal cations, e.g. $[Ag(S_2O_3)_2]^{3-}$. The T. are obtained by boiling the corresponding sulfite solutions with finely divided sulfur. The most important representative is Sodium thiosulfate (see).

Thiotepa: a Chemical sterilant (see).

2-Thiouracil, ***4-hydroxy-2-mercaptopyrimidine***: an antagonist of uracil. It forms colorless prisms; m.p. 340 °C. 2-T. is slightly soluble in water and alcohol, but dissolves in alkalies. It is synthesized by reaction of sodium ethyl formylacetate with thiourea. Derivatives of 2-T. are used as thyreostatics in medicine.

Thiourea, ***thiocarbamide***: $H_2N\text{-}CS\text{-}NH_2$, the diamide of thiocarbonic acid, which cannot be isolated in the free state. T. crystallizes in colorless, rhombic prisms; m.p. 181-182 °C. It is soluble in water and hot alcohol, but is insoluble in ether. Like urea, T. forms crystal-lattice inclusion compounds with many branched, unbranched and cyclic aliphatic compounds. It forms addition and complex compounds with metal oxides and inorganic salts. T. forms salts with strong acids. It is known in only one tautomeric form, but many of its reaction products indicate the existence of another tautomeric form:

$$S{=}C\begin{smallmatrix}NH_2\\NH_2\end{smallmatrix} \rightleftharpoons HS{-}C\begin{smallmatrix}NH_2\\NH\end{smallmatrix}$$

Thiourea Isothiourea

In reactions with alkyl or acyl halides, the electrophilic attack occurs on the S atom, e.g. T. forms stable *S*-alkyl isothiouronium salts with alkyl bromides: $H_2N\text{-}CS\text{-}NH_2 + R\text{-}Br \rightarrow R\text{-}S\text{-}C(NH_2){=}NH_2Br$. It reacts with 1,2- and 1,3-dicarbonyl compounds to form 5 or 6-membered heterocycles, e.g. thiobarbituric acid is formed from diethyl malonate and T. T. can be obtained by heating ammonium rhodanide, in analogy to the Wöhler urea synthesis. It is prepared industrially from calcium cyanamide and hydrogen sulfide: $N{\equiv}C\text{-}NCa + 2\ H_2S \rightarrow H_2N\text{-}CS\text{-}NH_2 + CaS$. T. is used in the synthesis of aminoplastics, pharmaceuticals, blueprint paper and vulcanization accelerators, and as an additive to color developers.

Thiourethanes, ***thiocarbamic acid esters***: the esters of mono- and dithiocarbamic acids. The ***monothiocarbamates*** include the O esters, $R_2N\text{-}CS\text{-}OR$, and the S esters, $R_2N\text{-}CS\text{-}SR$. There is a uniform type of ***dithiocarbamate***, $R_2N\text{-}CS\text{-}SR$. *N*-Substituted T. in the three different series can be synthesized as follows: monothiocarbamic acid O esters from isothiocyanates and alcohols, monothiocarbamic acid S esters from isothiocyanates and thiols, and dithiocarbamic acid esters from isothiocyanates and thiols.

2-Thioxoimidazolidin-4-one: same as 2-Thiohydantoin (see).

Thiram, ***TMTD***: 1) vulcanization accelerator; 2) fungicide (see Dithiocarbamate fungicides).

Thixotropy: a phenomenon observed in gels of macromolecular compounds and coagulation structures of colloidal particles. When shaken, stirred or exposed to supersonic vibrations, thixotropic gels flow, but a short time after the mechanical stress stops, the sol resumes its gel structure. Solutions of gelatin or polyvinyl alcohol display T., as do dispersions of oxide hydrates or clays. The reverse phenomenon is called Dilatance (see).

Thomas process: see Steel.

Thomson's equation: an equation derived by W. Thomson, later Lord Kelvin of Largs. It says that very small droplets of liquids have higher vapor pressures than large surfaces of liquids. The equation is $\Delta p/p_s = 2\sigma V/rRT$, where Δp is the increase in vapor pressure of a sphere of liquid with the radius r compared to that of a compact liquid. p_s is the saturation vapor pressure of the compact phase, σ is the surface tension, V is the molar volume of the liquid, R is the gas constant and T is the temperature. For water droplets of 0.1 µm diameter, at 300 K, the vapor pressure is increased by 1%.

Thorium, symbol ***Th***: a radioactive member of the Actinoids (see) of the periodic system; a heavy metal, *Z* 90, with natural isotopes with the following mass numbers (decay type; half-life in parentheses): 232 (α; $1.405 \cdot 10^{10}$ a); 227, radioactinium (α; 18.72 d); 228, radiothorium (α; 1.913 a); 230, ionium (α; $7.52 \cdot 10^4$ a), 231, uranium Y (α; β^-; 25.64 h); 234, uranium X_1 (β^-, 24.10 d). There are also synthetic isotopes with mass numbers 223 (α; 0.66 s); 224 (α; 1.03 s); 225 (α; K-capture; 8 min); 226 (α; 3.9 min); 229 (α; $7.34 \cdot 10^3$ a); 233 (β^-; 22.3 min); 235 (β^-; 6.9 min); 236 (β^-; 37.5 min); 215 (α; 1.2 s); and 9 very short-lived isotopes with mass numbers 213, 214, and 216 to 222. The atomic mass of Th is 232.0381, valence IV, density 11.724, m.p. 1755 °C, b.p. 4800 °C.

Properties. T. is a silvery white, very soft, ductile metal which displays two modifications, a cubic face-centered and a cubic body-centered; the conversion temperature is 1345 °C. Pure T. is stable in air for months, but in the presence of oxide impurities, it becomes tarnished in air. Finely divided Th is pyrophoric, and in compact form it is oxidized by atmospheric oxygen around 250 °C. Dilute mineral acids, such as hydrofluoric, nitric or sulfuric acids, and concentrated hydrochloric or phosphoric acid, dissolve T. very slowly. Concentrated nitric acid passivates it. At 500 to 1000 °C, T. reacts with nitrogen to form thorium nitride, Th_3N_4; it forms the hydrides ThH_2 and Th_4H_{15} with hydrogen at high temperatures. Th reacts with graphite below 1200 °C to form thorium carbide, ThC, and above 1300 °C, it forms ThC_2.

The most important isotope, ^{232}Th, is the first member of the *thorium decay series*; it decays by α-emission to ^{228}Ra (mesothorium I). The fourth member of this decay series is also an α-emitting isotope of T., ^{228}Th (radiothorium). The final product is the stable lead isotope ^{208}Pb (thorium D). The two other natural decay series, the actinium and uranium decay series, also contain two T. isotopes each. The primary product of uranium 235 decay (actinium series) is the β-emitter ^{231}Th (radiothorium). The fifth member of the same decay series is another β-emitter, ^{227}Th (radioactinium). The most abundant natural isotope of uranium, ^{238}U, emits an α-particle to form ^{234}Th (uranium X_1), and the fifth member of the uranium decay series is ^{230}Th (ionium), an α-emitter.

In a breeder reactor, neutron bombardment of ^{232}Th produces ^{233}U as follows:

$$^{232}Th \xrightarrow{+\,^1n} {}^{233}Th \xrightarrow{-\beta^-} {}^{233}Pa \xrightarrow{-\beta^-} {}^{233}U.$$

Like ^{235} and ^{239}Pu, ^{233}U is capable of fission, so that ^{232}Th is potentially a nuclear fuel.

Analysis. In the reactions used for analysis, T. is very similar to the lanthanoids and to the metals zirconium and hafnium from group IVb. It can be precipitated from aqueous solution as thorium(IV) oxygen hydroxide, oxalate, iodide or fluoride. In the classical separation series, T. appears as thorium(IV) oxygen hydroxide in the urotropic group. The precipitation of thorium(IV) iodate from strong nitric acid solution is especially specific. T. can also be separated from the rare earth metals by utilizing the solubility of thorium(IV) oxalate in excess ammonium oxalate solution, where it forms the water-soluble complex $(NH_4)_4[Th(C_2O_4)_4]$. For gravimetric determination, T. is precipitated as the oxygen hydrate and heated to form thorium(IV) oxide. It can be determined volumetrically by complexometric titration with EDTA using pyrocatechol violet or xylene orange as indicator. Traces of T. (20 to 200 µg in 100 ml solution) can be determined spectrophotometrically by formation of a red complex with 2-(2-hydroxy-3,6-disulfo-1-naphthylazo)phenylarsonic acid (Thoron).

Occurrence. T. makes up about $1 \cdot 10^{-3}$% of the earth's crust, and is thus about four times as abundant as uranium. T. is usually found associated with rare-earth metals, for example in monacite, (Ce, Th)(P, Si)O_4. Some other, very rare T. minerals are thorite, $ThSiO_4$, its variant orangite, and thorianite, (Th, U)O_2.

Production. The most important raw material is monacite sand, which is solubilized in conc. sulfuric acid or sodium hydroxide solution. The classical separation method, i.e. precipitation of T. along with the rare-earth metals in the form of their oxalates and conversion of thorium(IV) oxalate to the water-soluble ammonium oxalatothorate(IV), $(NH_4)_4[Th(C_2O_4)_4]$, has now been largely replaced by extraction processes which use methyl isobutyl ketone, tributylphosphate or another phosphate ester. The metal is produced on a large scale by reduction of thorium (IV) fluoride with calcium in the presence of zinc(II) chloride and subsequent remelting in an electric arc furnace. Other methods utilize the reduction of thorium(IV) oxide with calcium or electrochemical precipitation of T. by electrolysis of $K[ThF_5]$ or $ThCl_4$ in KCl/NaCl melts; very pure T. is made by Vacuum deposition (see).

Applications. T. is used as the oxide or carbide (ThO_2 or ThC_2) along with uranium as a breeding substance in high-temperature reactors. It is also used as an alloy component for making heating wires, as an additive to magnesium alloys, to make wires for amplifier and transmitter tubes, as a getter metal in high vacuum systems, and in silver alloys for contacts.

A few nuclides of other elements were also called "thorium", because they are part of the thorium decay scheme: thorium A (^{216}Po), thorium B (^{212}Pb), thorium B' (^{216}At), thorium C (^{212}Bi), thorium C' (^{212}Po), thorium C'' (^{208}Tl), thorium D or thorium lead (^{208}Pb) and thorium X (^{224}Ra).

Historical. T. was discovered in the form of its oxide in 1828 by Berzelius in a Norwegian mineral (thorite) and was named for the god Thor. Berzelius also made the first attempts to obtain the metal by reduction of its fluoro or chloro complexes with potassium or sodium. After the periodic system had been established, T. was first assigned to group IVb, until the discovery of the transuranium elements led to formulation of the actinoid series.

Thorium carbide: ThC_2, yellow, monoclinic crystals; M_r 256.06, density 8.96, m.p. ≈ 2655 °C, b.p. ≈ 500 °C. Other forms of T. stable at higher temperatures crystallize in tetragonal and cubic lattices. T. becomes superconducting around 9 K. It is hydrolysed by water, forming ethane, hydrogen and unsaturated hydrocarbons. T. is obtained by reaction of thorium with graphite above 1300 °C. $(Th,U)C_2$ is used as a fuel in gas-cooled, high-temperature reactors. The mixed carbide is formed by reaction of thorium and uranium oxides with carbon at 1600 to 2000 °C.

Thorium(IV) chloride: $ThCl_4$, colorless, crystalline needles; M_r 373.85, density 4.59, m.p. 770 °C, b.p. 928 °C (dec.); it sublimes at 750 °C in a vacuum. T. is hydrolysed by water, and forms chlorocomplexes of the types $M[ThCl_5]$, $M_2[ThCl_6]$ and $M_3[ThCl_7]$ with alkali chlorides. It forms adducts with amines and oxygen-containing Lewis bases. T. is obtained by conducting tetrachloromethane or a mixture of Cl_2 and CO over thorium(IV) oxide at red heat: $ThO_2 + 2\,Cl_2 + 2\,CO \rightarrow ThCl_4 + 2\,CO_2$.

Thorium emanation: see Radon.

Thorium(IV) nitrate: $Th(NO_3)_4$, a colorless compound which is readily soluble in alcohol; M_r 480.06; it crystallizes out of water with 4 or 12 mol H_2O. T. is an important intermediate in the production of thorium(IV) oxide and thorium.

Thorium(IV) oxalate: $Th(C_2O_4)_2 \cdot 6H_2O$, a white compound which is insoluble in water and dilute acids, but dissolves readily in excess alkali or ammonium oxalate solution; M_r 516.17. T. is an intermediate in the separation of thorium from the rare-earth metals in the course of processing monacite sand.

Thorium(IV) oxide, ***thorium dioxide***: ThO_2, white, heavy, cubic crystals; M_r 264.04, density 9.87, m.p. 3200 °C, b.p. 4400 °C. T. is found in nature as the mineral thorianite. It is obtained in powdered form by roasting thorium(IV) oxalate at 800 to 1200 °C. The resulting product, which is soluble in acid, can be converted to the crystalline form, which is nearly insoluble in acids, by melting it with a suitable flux. T., which is present in small amounts in rare earths, gives a very bright light in a bunsen flame. It is therefore used, together with about 10% cerium(IV) oxide, to make gas lighting tubes. It is also used in the production of catalyst for the Fischer-Tropsch synthesis, for cracking processes, in the production of oxide ceramic products and, mixed with lanthanum oxide, in the preparation of optical glasses for cameras and optical instruments. In the form of (Th, U)O_2, T. is used as a nuclear fuel in gas-cooled, high-temperature reactors.

Thorium(IV) sulfate: $Th(SO_4)_2$, a colorless, very water-soluble compound; M_r 424.16, density 4.255. T. crystallizes out of aqueous solution in the form of hydrates with water contents depending on the method of preparation: the tetrahydrate, $Th(SO_4)_2 \cdot 4H_2O$, M_r 496.22, crystallizes in white needles. With alkali sulfates in aqueous solution, T. forms relatively insoluble sulfatothorates. Anhydrous T. is obtained by treating thorium(IV) oxide with concentrated sulfuric acid.

Thoron: see Radon.

Thorpe reaction: addition reaction of CH-acidic nitriles with each other; it is initiated by sodium amide or sodium ethylate, and at low temperatures (in boiling ether) it leads to β-iminonitriles by dimerization:

$$R{-}CH_2{-}C{\equiv}N \xrightarrow[-ROH]{+RO^-} R{-}\bar{C}H{-}C{\equiv}N \xrightarrow{R-CH_2-CN} R{-}CH_2{-}C({=}N{-}R){-}CH{-}C{\equiv}N \xrightarrow{+H^+} R{-}CH_2{-}C({=}NHR){-}CH{-}C{\equiv}N \rightleftharpoons R{-}CH_2{-}C(NH_2){=}C(R){-}C{\equiv}N.$$

At higher temperatures, e.g. in boiling dibutyl ether, there is a trimerization to 4-aminopyrimidines or 2,4-diaminopyridines. When dinitriles are used, cyclization occurs; the yields depend on the reaction conditions and the length of the carbon chain between the two cyano groups. Five- and six-membered cycloalkanones can readily be obtained in 90% yields, and for larger rings, the Ruggli-Ziegler dilution principle (see) is applied.

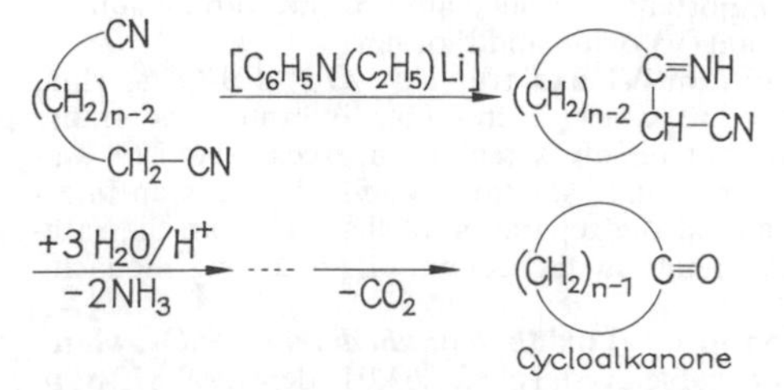

The lithium salt of *N*-ethylaniline can be used instead of sodium ethylate, and this makes possible the synthesis of 8- and 14-33-membered rings in good (85%) yields; the intermediate ring sizes cannot be obtained.

Thr: abb. for Threonine.

Three-center bond: in molecular orbital theory, a partially localized form in which three atomic orbitals X_1, X_2 and X_3 from different atoms are combined into three molecular orbitals. The superposition of the orbitals leads to a bonding (Ψ_1), a non-bonding (Ψ_2) and an antibonding (Ψ_3) orbital. If only two electrons are available from the three atoms (centers), these occupy the binding molecular orbital, forming a ***two-electron T.*** (Fig. a). This type of bond explains, for example, the bonding in diborane (B_2H_6). Starting from sp^3-hybridized boron atoms, four $sp^3_B1s_H$ two-center bonds are formed with the terminal hydrogen atoms, and two $sp^3_B1s_Hsp^3_B$T. bonds with the bridging hydrogen atoms. Since only two electrons are available for each T., B_2H_6 is an *electron-deficiency compound*.

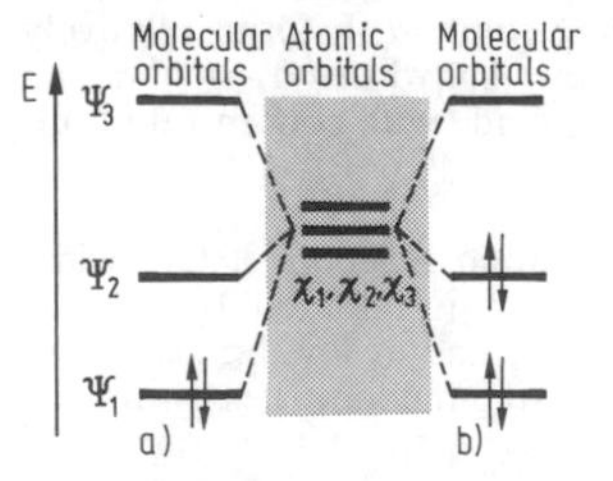

Molecular orbital scheme of a three-center bond: *a*, two-electron, *b*, four-electron

In the ***four-electron T.***, the four electrons occupy the bonding and the nonbonding molecular orbitals (Fig. b). This model applies to the bonding in noble gas compounds, such as xenon difluoride. The combination of two singly occupied p_z orbitals of the fluorine atom with the doubly occupied p_z orbital of the xenon atom leads to a stable bonding state with a linear structure. If the number of electrons is larger than the number of centers, as in xenon difluoride, the molecule is also called an *electron-excess compound*.

Three-dimensional lattice: see Crystal.

threo-: prefix indicating a specific configuration (see Monosaccharides, see Stereoisomerism, 1.2.1).

Threonine, abb. ***Thr***: L-*threo*-α-amino-β-hydroxybutyric acid, an essential amino acid with two asymmetric C atoms (formula and physical properties, see Amino acids, Table 1). T. exists in a *threo*- and an *allo*-form, and each of these forms a racemate. Of the four stereoisomers produced by chemical synthesis, only the L-*threo*-form found in proteins is biologically active. T. is now synthesized by the *Akabori process*, in which the reactive methylene groups of the glycine-copper complex with acetaldehyde are converted to the T.-copper complex. The reaction at 30 to 80 °C in the presence of base gives 80% yield, with a *threo:allo* ratio of about 2:1. After separation of the mixture of diastereomers (renewed formation of copper complex and aldehyde reaction), *N*-chloroacetyl-DL-threonine is converted to L-T. by enzymatic hydrolysis. T. is a limiting essential amino acid, and is used to supplement nutritional proteins and animal feeds.

Threshhold: that concentration of a gaseous or particulate pollutant at which it becomes damaging to human beings, animals or plants; see Workplace concentration, see Emission.

Threshhold energy: see Kinetics of reactions (theory), see Potential surface.

Thrombin: a highly specific serine protease which catalyses the cleavage of peptide and ester bonds in which the carboxyl group is supplied by L-arginine. T. (M_r 39,000) circulates in the blood in the form of its inactive precursor ***prothrombin*** (M_r 72,000). T. is released by proteolysis of prothrombin, and in turn converts fibrinogen into fibrin, which forms a clot. The catalytically active amino acids histidine, aspartic acid and serine are found on the *B-chain* of T. (M_r 33,000, 260 amino acids), which is bound to the *A chain* (M_r 6000, 49 amino acids) by a disulfide bridge.

Thromboxane: see Prostaglandins.

Throughput: the amount of material flowing through an apparatus in a given time interval. The T. can be measured in kg/s, kg/min or kg/h.

THT: same as Tetrahydrothiophene (see).

Thujane: see Monoterpenes.

Thulium, symbol ***Tm***: a chemical element from group IIIb of the periodic system, a Lanthanoid (see) rare-earth metal. T. has only one isotope, *Z* 69, atomic mass 168.9342, valence III, less often II, density 9.318, m.p. 1545 °C, b.p. 1725 °C, standard electrode potential (Tm/Tm^{3+}) -2.278 V.

T. is a lustrous, white metal which crystallizes in a hexagonal lattice. It makes up only about $1.9 \cdot 10^{-5}$% of the earth's crust, and is thus a rare lanthanoid. In nature, it is always found in association with other rare-earth metals, especially yttrium and the heavier lanthanoids. It is relatively enriched (0.25%) in gadolinite (ytterbite); in monacite its concentration is 0.007%. Other properties, analysis, production and history are discussed under Lanthanoids (see). Because of its rarity, T. has so far not found many practical applications. The synthetic isotope ^{170}Tm (γ-emitter, $t_{1/2}$ 129 d) is used in industry to measure the thickness of metallic and non-metallic materials, and in portable "x-ray" machines which do not require an external source of electricity.

Thulium compounds: compounds in which thulium is usually in the +III oxidation state, or less often, in the +II state. Violet-red thulium(II) compounds can be made by reduction of the corresponding thulium(III) compounds. For example, thulium(II) halides are formed by reaction of thulium(III) halides with hydrogen or thulium at elevated temperatures. The general properties and synthesis of T. are discussed under Lanthanoid compounds (see). Some important thulium(III) compounds: ***thulium(III) chloride***, $TmCl_3 \cdot 7H_2O$, green crystals, M_r 401.40; ***thulium(III) fluoride***, TmF_3, M_r 225.93, m.p. 1158 °C, b.p. > 2200 °C; ***thulium(III) oxalate***, $Tm_2(C_2O_4)_3 \cdot 6H_2O$, M_r 710.02, greenish white; ***thulium(III) oxide***, Tm_2O_3, greenish white, M_r 385.87.

Thy: abb. for thymine.

Thylox process: a process for removing hydrogen sulfide from gas mixtures; it is similar to the sulfosolvane process.

Thymeretics: see Psychopharmaceuticals.

Thyme oil: a colorless to yellowish, sometimes also reddish essential oil. The main components are thymol and carvacrol; smaller amounts of cymol, borneol, linalool and pinene are also present. T. is obtained by steam distillation of wild thyme; the yield is about 1% or less. T. is used in cosmetics, in medicine and as an additive to disinfecting cleansers.

Thymidine, abb. ***dThyd*** or ***T***: 1-β-D-2'-deoxyribofuranosylthymine, a nucleoside. It forms crystalline needles, M_r 242.2, m.p. 186 °C, $[\alpha]_D^{25}$ + 30.6° (in 1 N NaOH); soluble in water, methanol and hot ethanol and pyridine. It is practically insoluble in cold ethanol and chloroform. T. is a component of deoxyribonucleic acids. It is biosynthesized as thymidine 5'-monophosphate by methylation of 2'-deoxyuridine 5'-monophosphate.

Thymine, abb. ***Thy***: 5-methyluracil, 2,5-dihydroxy-5-methylpyrimidine, a nucleic acid base. It forms colorless crystals; M_r 126.1, m.p. 326 °C (dec.), sublimes. It is scarcely soluble in water or alcohol, slightly soluble in ether, but dissolves in alkali hydroxide solutions by forming salts. T. is part of thymidine, and as such is a component of deoxyribonucleic acids.

Thymol, ***2-isopropyl-5-methylphenol***: colorless, bitter-tasting crystals which smell like thyme; m.p. 51 °C, b.p. 233 °C. T. is slightly soluble in water, and soluble in ethanol, ether and other organic solvents. It is found along with its isomer carvacrol in essential oils from herbs used as seasonings, especially thyme, oregano and marjoram.

Thymol

It can be synthesized by a Friedel-Crafts reaction of *m*-cresol and isopropyl chloride. T. is less poisonous, but more strongly antiseptic than phenol. It does not hurt intact skin, and even mucous membranes tolerate solutions of T. well. It is therefore used in medicine as a disinfectant with bacteriocidal and also fungicidal activity. It is a component of salves, mouth rinses, toothpastes and cough syrups, and is effective as a treatment for digestive upsets and worms. It is also used to preserve anatomical specimens and as a raw material for production of menthol by hydrogenation of the benzene ring and subsequent racemate separation.

Thymol blue: see Bromphenol blue.

Thymoleptics: see Psychopharmaceuticals.

Thymopoietin-5: see Thymopoietin II.

Thymopoietin II: a polypeptide hormone from the thymus which consists of 49 amino acids. The hormone causes T-cells to differentiate, and the active center of this activity is the segment 32-35, Arg-Lys-Asp-Val-Tyr. This is known as ***thymopoietin-5 (TP-5)***, and is easily synthesized; it is administered to strengthen the unspecific immune system. T. was isolated in 1975 from calf thymus by Schlesinger and Goldstein.

Thymosin α_1: Ac-Ser-Asp-Ala-Ala-Val-Asp-Thr-Ser-Ser-Glu-Ile-Thr-Thr-Lys-Asp-Leu-Lys-Glu-Lys-Lys-Glu-Val-Val-Glu-Glu-Ala-Glu-Asn, a 28-amino-acid peptide from the thymus which stimulates the immune system. T. is a component of the peptide mixture known as thymosin fraction 5 from calf thymus. The primary structure was elucidated in 1977 by Goldstein, and was confirmed in 1979 by two independent total syntheses. T. and the standardized thymosin fraction 5 have some promise as treatments for congenital immune deficiencies, leukemia and other types of caner. In vivo, T. seems to have an important regulatory function in the late stages of T-cell differentiation.

Thyreostatics, ***antithyroid drugs***: compounds which inhibit biosynthesis of the thyroid hormones; they are used to treat thyroid hyperfunction. In cases of pronounced hyperfunction of the thyroid, for example in Basedow's disease, the body metabolism is accelerated and the patient loses weight, is excitable and displays tachycardia and exophthalamus. The use of T. requires simultaneous administration of thyroid hormones, to prevent excessive production of thyroid-stimulating hormone and a resulting development of goiter and thyroid hypofunction. T. include the salts of inorganic acids, such as potassium perchlorate, $KClO_4$, and sulfur-containing heterocyclic compounds. The anions of the inorganic salts competitively inhibit uptake of iodine into the thyroid, while the heterocyclic compounds inhibit oxidation of iodide to iodine, and thus prevent its incorporation into thyroxin and triiodothyronin. The active heterocyclic compounds include *6-methylthiouracil (MTU), 6-propylthiouracil* and *thiamazole* (1-methyl-2-mercaptoimidazole).

Thyreotropin: same as Thyrotropin (see).

Thyrocalcitonin: same as Calcitonin (see).

Thyroid hormones: compounds formed in the thyroid, stored in the form of the glycoprotein thyreoglobulin and released into the blood in free form. S. are derivatives of *thyronin*: *levothyroxin* or *thyroxin*, 3,3',5,5'-tetraiodo-L-thyronine [m.p. 228-233 °C (dec.), $[\alpha]_D^{20}$ + 20° (in 0.3N ethanolic hydrochloric acid)], and *liothyronin*, 3,5,5'-triiodo-L-thyronine [m.p. 236-237 °C (dec.), $[\alpha]_D^{20}$ +21° (in 0.3N ethanolic hydrochloric acid). Levothyroxin (T4) is considerably more acidic (pK 6.7) than liothyronin (T3, pK 9.2). T4 is present in the serum mainly as a phenolate, in contrast to T3.

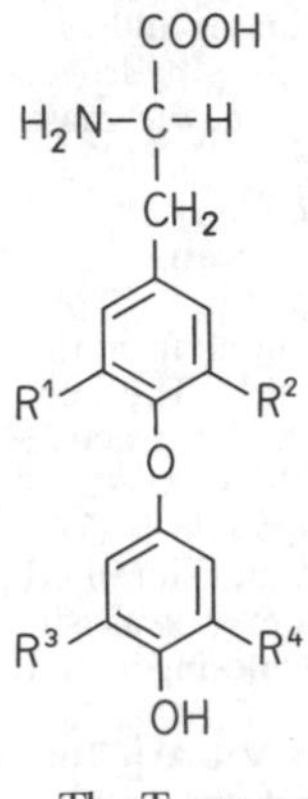

$R^1, R^2, R^3, R^4 = H$: L-Thyronin

$R^1, R^2, R^3 = I; R^4 = H$: Liothyronin

$R^1, R^2, R^3, R^4 = I$: Levothyroxin

The T. promote mental and physical development, increase the energy turnover in the organism, promote protein biosynthesis and accelerate the degradation of carbohydrates and fats. The S. are used for replacement therapy in hypothyroses and after surgical removal of the thyroid. Insufficient thyroid function, for example in myxedema, is characterized by reduced metabolism and by physical and mental lethargy. A congenital insufficiency produces *cretinism*. The effects of T4 and T3 are qualitatively the same, but they are more rapidly induced by T3 and also disappear more rapidly. The majority of T3 is formed from T4 by reductive deiodization in the tissues.

Historical. As early as 1890, myxedema was treated with thyroid extracts. In 1915, Kendall isolated T4, and in 1920 its structure was elucidated by Harington. In 1927, Harrington and Barger confirmed the structure by total synthesis. T3 was not isolated until relatively late, in 1952, by Gross and Pitt-Rivers.

Thyroliberin: see Liberins.

Thyronin: see Thyroid hormones.

Thyrotropin, ***thyroid stimulating hormone***, abb. ***TSH***: a glycoprotein hormone from the anterior pituitary. T. consists of two subunits; the α-subunit of bovine T. contains 96 amino acids, and the β-subunit, 113 amino acids. The β-subunit of the human hormone contains 112 amino acids and differs from the bovine sequence in 12 positions. T. is formed in the basophilic cells of the anterior pituitary, in response to thyroliberin. It stimulates the formation and secretion of the thyroid hormones thyroxin and triiodothyronin.

Thyrotropin releasing hormone: see Liberins.

Thyroxin: see Thyroid hormones.

Ti: symbol for titanium.

Tillman's reagent: same as sodium 2,6-dichlorophenolindophenol.

Tin, symbol ***Sn***: a chemical element from group IVa of the periodic system, the Carbon-silicon group (see), and a heavy metal; *Z* 50, with natural isotopes with mass numbers 112 (0.96%), 114 (0.66%), 115 (0.35%), 116 (14.30%), 117 (7.61%), 118 (24.03%), 119 (8.58%), 120 (32.85%), 122 (4.72%) and 124 (5.94%), atomic mass 118.69, valence II and IV.

Properties. T. exists in three modifications. The form stable at room temperature is metallic, β-T., which is silvery white in color and forms a tetragonal lattice; density 7.28, m.p. 231.88 °C, b.p. 2260 °C, electrical conductivity 8.96 Sm mm^{-2} at 0 °C, standard electrode potential (Sn/Sn^{2+}) -0.1364 V. β-T. is relatively soft and very ductile. It can be rolled into thin foils (tinfoil). At 162 °C, it is converted to brittle, gray rhombic γ-T., which can easily be powdered. Below 13.2 °C, the gray, powdery cubic α-T. is stable; it crystallizes in a diamond lattice. It normally forms very slowly from β-T., but if crystallization nuclei have formed during a period of prolonged cold, the conversion to a microcrystalline powder leads to destruction of the entire object (tin plague or tin pest, so called because it led to destruction of medieval organ pipes). This conversion can be accelerated by treating the surface of the T. with ammonium hexachlorostannate or by the presence of various metals (aluminum, magnesium, manganese, zinc), but in tin alloys with small amounts of arsenic, bismuth or lead, it does not occur at all. If a tin rod is bent, it makes a grating sound (***tin cry***) which is due to rubbing of the crystals past one another.

In its compounds, T. is di- or tetravalent; the divalent state is more stable and more important than in the lighter homologs, carbon, silicon and germanium. Sn^{2+} can be converted to Sn^{4+} by strong oxidizing agents. The four-fold coordinated, tetrahedral derivatives contain sp^3-hybridized Sn atoms, and the bonds are largely covalent. In tin(II) compounds, the bonds are more ionic. The Lewis acidity of Sn(IV) compounds is stronger than that of the comparable silicon derivatives, and they have a pronounced tendency towards five- and six-fold coordination, e.g. $(CH_3)_3SnCl \cdot$ pyridine, $[SnF_6]^{2-}$, $[SnCl_6]^{2-}$. Sn(II) compounds also interact with electron pair donors and form derivatives of triply coordinated T., e.g. solid tin(II) chloride, $SnCl \cdot 2H_2O$ or the $[SnCl_3]^-$ anion. T. is stable towards air, water and dilute acids and bases. Its standard electrode potential is very close to zero, and T. therefore dissolves only in concentrated HCl or bases, with evolution of hydrogen, e.g. $Sn + 2\,NaOH + 4\,H_2O \rightarrow 2\,H_2 + Na_2[Sn(OH)_6]$. When heated, T. burns to tin(IV) oxide, SnO_2 (***tin ash***). Concentrated nitric acid also oxidizes T. to SnO_2, while more dilute HNO_3 dissolves it, forming tin(II) nitrate, $Sn(NO_3)_2$. T. reacts with sulfur to form tin(II) sulfide, SnS, or tin(IV) sulfide, SnS_2, depending on the stoichiometric ratio of the reactants. T. reacts with chlorine even at room temperature to form tin tetrachloride, $SnCl_4$.

Analysis. T. is qualitatively determined in the H_2S step of the analysis scheme, by the reducing action of Sn^{2+}, e.g. on molybdatophosphoric acid, mercury(II) chloride or gold(III) chloride (see Cassius' gold purple). The Flame test (see) is also a suitable test. For quantitative gravimetric determination, stannic acid is precipitated and weighed as tin dioxide. Low concentrations can be determined by photometric methods (diphenylcarbazone, morin, phenylfluoron), atomic absorption spectroscopy and x-ray fluorescence analysis.

Occurrence. T. makes up $3.5 \cdot 10^{-3}$% of Earth's crust. The most important mineral is cassiterite (tinstone), SnO_2; stannite (tin pyrite), $Cu_2S \cdot FeS \cdot SnS_2$, is less common.

Production. The ore is first enriched in SnO_2 by wet mechanical (based on the different densities or ore and gangue) or chemical pretreatment (removal of sulfur and arsenic by roasting). It is then reduced in a furnace with carbon. This produces a metal which is about 97% pure and a slag which still contains large amouts of T. More T. can be obtained from it by reduction, i.e. melting with coal and lime ($SnSiO_3 + CaO + C \rightarrow Sn + CaSiO_3 + CO$), or by precipitation, i.e. melting with scrap iron and coal ($SnSiO_3 + Fe \rightarrow Sn + FeSiO_3$).

There are a number of methods used to purify the crude T. For example, in seiving, the crude metal is heated to about the melting temperature of the T. on a grate, so that the pure element can flow out and the iron and higher-melting Fe-Sn alloy remains behind. It can be oxidized by blasting air or water into the crude T. melt, or submerging fresh wood into the melt (*poling*). The oxidized impurities then float to the surface as ***T. dross***. Electrolytic refining methods are also in use.

The recovery of T. from tin plate is significant. The method used is *chlorine detinning*; the T. is first converted to tin tetrachloride, $SnCl_4$, by liquid chlorine, then distilled off, while the iron does not react. Electrolytic methods are also used in which the scrap tin plate is enclosed in an iron wire cage and used as the anode, and the T. is precipitated on a pure T. cathode.

Applications. T. was formerly an important metal for production of household ware, but is now used mainly for production of tin plate and as an alloy metal (see Tin alloys). Organotin compounds are being used increasingly as stabilizers for various plastics, as biocides in agriculture, as wood preservatives, etc.

Historical. Pure T. was probably first obtained around 1800 B.C. in China and Japan, long after the production of bronze by reduction of a mixture of T. and copper ores was practiced.

Tin alloys: alloys of tin with lead, antimony and copper, often with small amounts of iron. Tin solders contain 30 to 70% tin along with lead and antimony. These are used as soft solders to join various materials (see Solders). The White metals (see) used for sliding parts contain 4.5 to 81% tin, together with lead, antimony and copper. Tin is a component of a few alloys with exceptionally low melting temperatures (see Rose's metal, Wood's metal, Lipowitz's metal). Tin is very significant as a component of the Tin bronzes (see). Tin organ pipes contain about 30% lead, and tin for figurines, about 40% lead. Household articles (trays, plates, etc.) are made of *britannia metal*, which contains 88 to 90% tin, 8-10% antimony and 2% copper.

Tin ash: see Tin oxide.

Tin bronze: copper-tin alloy with a maximum of 9% tin in kneaded alloys and up to 14% tin in cast alloys. Because the α-mixed crystals separate extremely slowly at room temperature, the T. exist as metastablc, homogeneous alloys. The kneaded alloys have great compressional and tensile strength, and are used for screws, bands for metal tubes, springs, wire screens, slide bearings, pipes for pressure-measuring instruments and gear wheels. They are also chemically resistant, and are used for equipment in the chemical industry. Cast T. with 20 to 22% tin are used in clocks because of their hardness and damping qualities. Cast alloys with 11 to 13% tin are used for sliding bearings, gears, armatures, housings for pumps and turbines used in mines, and in the chemical industry.

Historical. The Sumerians began to make objects from T. about 4000 B.C. In the Indus culture (Mohenyo-Daro), T. was introduced about 2800 B.C., in Babylon about 2400 B.C. Because of the widespread use of the material, this period is called the Bronze Age.

Tin butter: see Tin chlorides.

Tin chlorides. ***Tin(II) chloride, tin dichloride***: $SnCl_2$, colorless, rhombic crystals soluble in water, alcohol, ether and other organic solvents; M_r 189.60, density 3.95, m.p. 246 °C, b.p. 652 °C. $SnCl_2$ crystallizes out of aqueous solution as a monoclinic dihydrate, $SnCl_2 \cdot 2H_2O$ (***tin salt***). The $SnCl_2$ molecule (Fig.) has an electron sextet and in the gas state is angular, as would be expected (a). In the solid salt, the electron deficiency situation is counteracted by formation of polymeric $[SnCl_2]$ chains (b). The ability of the compound to add more chloride ions to form pyramidal $[SnCl_3]^-$ anions (c) is another manifestation of the electron deficiency. The structure of the dihydrate, in which one H_2O is directly bound to the Sn (d), while the second H_2O is a true water of crystallization and can be readily removed, is in agreement with the notion that three-fold coordination of Sn(II) is preferred.

$SnCl_2$ molecule

$SnCl_2$ is a strong reducing agent, which is able, for example, to reduce Fe^{3+} to Fe^{2+}, chromate to Cr^{2+}, permanganate to Mn^{2+}, arsenate to As^{3+} and Au^{3+}, Ag^{+} or Hg^{2+} to the metals. In dilute aqueous solutions it is gradually hydrolysed to the basic chloride. $SnCl_2$ is obtained by heating tin in a stream of HCl, and the dihydrate is obtained by dissolving tin in concentrated hydrochloric acid. It is used in industry for reducing organic nitro-, azo- and diazonium compounds, and in printing textiles. It is also a versatile laboratory reducing agent.

Tin(IV) chloride, tin tetrachloride: $SnCl_4$, colorless liquid which fumes in air; M_r 260.50, density 2.226, m.p. - 33 °C, b.p. 114.1 °C. $SnCl_4$ is much less sensitive to hydrolysis than silicon tetrachloride. It dissolves in water and crystallizes out of the solution as the pentahydrate, $SnCl_4 \cdot 5H_2O$ (tin butter). In hydrochloric acid solution, it adds two molecules of HCl to form hexachlorostannic acid, H_2SnCl_6, which can be isolated as the hexahydrate; it forms salts, the hexachlorostannates, with the composition $M^{I}_2[SnCl_6]$ (e.g. Pink salt (see)). In dilute aqueous solution, the salt is hydrolysed to tin dioxide hydrate, $SnO_2 \cdot xH_2O$. $SnCl_4$ is made by the reaction of chlorine with tin, a reaction which is also used for recovery of tin from scrap tin plate. $SnCl_4$ is used mainly as a catalyst for chlorination and condensation reactions. It is also a starting material for synthesis of organotin compounds.

Tin foil: a thin foil, 0.007 to 0.13 mm thick, consisting approximately of 96% tin, 2% lead, 1% copper, 0.3% nickel and 0.09% iron. It was widely used as a packaging material for cigarettes, chocolate and the like, but has now been largely replaced by aluminum foil, which is cheaper.

Tin oxides. ***Tin(II) oxide, tin monoxide***: SnO, blue-black, cubic srystals which are only slightly soluble in water; M_r 134.69, density 6.446. At 1080 °C, SnO melts with partial disproportionation to Sn and SnO_2. When heated, it burns in air to SnO_2. SnO is amphoteric and dissolves in aqueous acids to form Sn(II) salt solutions; and in alkali hydroxide solutions, it forms hydroxostannates(II) (hydroxostannites), such as $Na[Sn(OH)_3]$ or $Na_2[Sn(OH)_4]$. It is obtained by heating tin(II) oxide hydrate, $SnO \cdot xH_2O$ in the absence of air; the tin(II) oxide hydrate can be precipitated out of tin(II) salt solutions with bases.

Tin(IV) oxide, tin dioxide, tin ash: SnO_2 exists as an amorphous powder or trigonal, hexagonal or rhombic crystals; M_r 150.69, density 6.95, m.p. 1630 °C, subl.p. 1800 to 1900 °C. SnO_2 is insoluble in aqueous acids and bases. It can be solublized with sulfur and soda (*Freiberg solublization*), which convert it to soluble sodium thiostannate: $2\ SnO_2 + 2\ Na_2CO_3 + 9\ S \rightarrow 2\ Na_2[SnS_3] + 3\ SO_2 + 2\ CO_2$. It can also be converted to soluble stannates by fusion with alkali hydroxides, as in the reaction $SnO_2 + 2\ NaOH \rightarrow Na_2SnO_3 + H_2O$. It is reduced to tin by carbon or hydrogen. SnO_2 occurs in nature as cassiterite (tinstone); it is formed either by combustion of tin or by roasting of stannic acid. The mineral is the starting material for tin production. SnO_2 is also used as an opacifier for white glazes and enamels.

Tin plate: tin-coated iron plate, made on a large scale by dipping pickled iron plate into molten tin, or by electrolytic tin plating. T. is generally 1 to 1.5% tin, and is more resistant to chemicals than galvanized (zinc-coated) iron. It is used to make tin cans (e.g. for foods), kitchen equipment and other objects.

Tin plating: the coating of metal objects with a layer of tin to protect them from corrosion (primarily for objects which come into contact with foods), for soldering and to reduce friction of cylinders in internal combustion engines. The object can be dipped in molten tin, galvanized by electrolysis in alkaline or acid baths or electrochemically plated, either by dipping or by contact. The protective layer is 0.025 to 0.125 mm thick.

Tin salt: see Tin chlorides.

Tin sulfides: ***Tin(II) sulfide***, SnS, brown powder or cubic or monoclinic crystals; M_r 150.75, density 5.22, m.p. 882 °C, b.p. 1230 °C. SnS is insoluble in water, dilute acids or colorless ammonium hydrogensulfide solution (solubility product $K_s = 10^{-28}$), and it therefore precipitates when hydrogen sulfide is passed through a tin(II) salt solution. It dissolves in concentrated hydrochloric acid. Polysulfide solution oxidizes SnS to soluble thiostannates(IV).

Tin(IV) sulfide, SnS_2, is a yellow powder or hexagonal crystals; M_r 182.82, density 4.5, dec. 600 °C. SnS_2 is only slightly soluble in water and dilute acids, and is made by passing H_2S through acidic solutions of Sn(IV) compounds. Ammonium sulfide dissolves Sn_2S to thiostannates: $SnS_2 + S^{2-} \rightarrow SnS_3^{2-}$. SnS_2 is also soluble in alkali hydroxide solutions and concentrated hydrochloric acid. The T. are used to separate tin in the H_2S step in qualitative analysis. SnS_2 is also used as a gold-colored pigment (see Mosaic gold).

Titanates, ***metal oxotitanates(IV)***: compounds which may be obtained by reaction of titanium(IV) oxide with the corresponding metal oxides. In addition to alkali titanates of the types M_4TiO_4, M_2TiO_3, $M_2Ti_2O_5$ and $M_2Ti_3O_7$, there are alkaline earth and lead titanates $MTiO_3$ (M = Ca, Sr, Pb), which are significant ferroelectrics. Lead titanate is used as a yellow pigment. Ilmenite, $FeTiO_3$, and perovskite, $CaTiO_3$, are naturally occurring T.

Titanic acid esters, ***titanium(IV) alkoxides*** or ***aryl oxides***, $Ti(OR)_4$, covalent, oligomeric liquids or low-melting solids which can be hydrolysed. T. are obtained by reaction of titanium(IV) chloride with alcohols or phenols, sometimes in the presence of hydrogen chloride acceptors, such as ammonia or amines. They are used widely as wetting agents and hardeners, e.g. for alkyde, epoxide and terephthalate resins, to impregnate textiles to make them water-repellent, as glues for films, to provide surface protection for glassware and as catalyst components in olefin polymerization.

Titanium, symbol ***Ti***: a chemical element from group IVb of the periodic system, the Titanium group (see). It is a light metal, *Z* 22, with natural isotopes with mass numbers 48 (73.94%), 46 (7.93%), 47 (7.28%), 49 (5.51%) and 50 (5.34%); atomic mass 47.90, valence IV, III, occasionally II, density 4.506, m.p. 1667 °C, b.p. 3287 °C, electrical conductivity 2.3 Sm mm^{-2}, standard electrode potential (Ti/TiO_2^{2+}) -0.882 V.

Properties. Pure T. is a lustrous, white, ductile and easily hammered metal. At room temperature it exists in a hexagonal close-packed modification, and above 882.5 °C, in a cubic modification. Although its oxidation potential indicates that it is a reactive metal, it forms a tight oxide layer on its surface and as a result, in compact form, it is unreactive. It does not react with oxygen until heated red hot, but reacts with chlorine above 300 °C to form the tetrachloride. T. dissolves in hot hydrochloric acid to form titanium(III) chloride, $TiCl_3$, and reacts with hydrofluoric acid to give hexafluorotitanic acid, H_2TiF_6. Finely divided T. burns in oxygen or nitrogen to form titanium(IV) oxide, TiO_2, or titanium nitride, TiN. Oxygen reacts with a freshly exposed surface of compact T. under excess oxygen pressure (2.5 MPa), even at room temperature. In general, the uptake of even small amounts of oxygen or nitrogen makes T. brittle. Hydrogen is reversibly absorbed, forming hydrides. Nonmetals, such as carbon, nitrogen and boron react at high temperatures to form hard and high-melting inclusion compounds: titanium carbide, TiC, titanium nitride, TiN, and titanium boride, TiB_2. T. forms alloys with a number of metals; those with aluminum, molybdenum, manganese, chromium and iron are technologically important. In its compounds, T. is only rarely found in the +II oxidation state, and titanium(II) compounds are stable only in the solid phase. The titanium(III) compounds have characteristic colors, and are strong reducing agents, while titanium(IV) compounds are similar in many respects to the homologous tin(IV) derivatives.

Analysis. The reaction of titanyl sulfate, $TiOSO_4$ with hydrogen peroxide to give orange titanium peroxosulfate, TiO_2SO_4, is used for both detection and colorimetric determination of T. In a reducing flame, T. colors phosphorus salt beads violet, due to formation of titanium(III) salts. For gravimetric determination, titanium oxide hydrate is precipitated, then roasted to the dioxide, TiO_2. After reduction with zinc amalgam to titanium(III), T. can be determined by titration with $NH_4Fe(SO_4)_2$ solution (an oxidizing agent) with potassium thiocyanate as indicator.

Occurrence. T. makes up 0.43% of the earth's crust, and is thus a relatively abundant element; in fact, it is the 10th most abundant element. However, it is very widely distributed; the most important ore is ilmenite (titanium iron ore), $FeTiO_3$; other titanium minerals, such as perovskite, $CaTiO_3$, and titanite, $CaTi[SiO_4]O$, and the titanium(IV) oxide modifications rutile, anatas and brookite, are less important. T. is found widely in soils, and plants contain about 1 ppm. So far there is no evidence that it has a biological function.

Production. T. cannot be obtained by reduction of titanium(IV) oxide, TiO_2, with coal, because the reaction forms titanium carbide, TiC, and, in the presence of atmospheric nitrogen, titanium nitride, TiN; in some cases copper-colored mixed crystals of $TiC \cdot 4TiN$ are formed. T. is obtained by reduction of titanium(IV) chloride or fluoride with sodium in powder form. In the industrial *Kroll process*, titanium(IV) chloride is reduced with magnesium at 800 to 900 °C under a protective gas: $TiCl_4 + Mg \rightarrow Ti + 2 MgCl_2$. The resulting finely divided, pyrophoric T. (titanium sponge) is separated from magnesium chloride and unreacted magnesium by vacuum distillation and remelted to form the compact metal. In a process similar to Zone melting (see), titanium(IV) iodide can be formed and thermally degraded as a further purification step. Ferrotitanium (see) is produced for industrial purposes.

Applications. In addition to its resistance to corrosion, T. has other excellent properties, such as high mechanical strength at low mass, high melting point and low thermal expansion coefficient. It is thus a versatile material which combines the qualities of stainless steel and aluminum alloys. It is therefore used, often in the form of Titanium alloys (see), in the construction of ships, aircraft, rockets, reactors and desalination and chemical plants. Titanium electrodes, coated with a noble metal or noble metal oxide, are used in chloralkali electrolysis, the production of perchlorates, electrodialysis and galvanic technology. T. is used in light bulbs as a getter, an application which utilizes its reactivity with oxygen and nitrogen. It is used to make bone pins, prostheses and needles for medical purposes.

Historical. T. was discovered independently by Gregor in 1791 and Klaproth in 1795 as a component of ilmenite and rutile, respectively. In 1825, it was prepared by Berzelius by reduction of potassium hexafluorotitanate with sodium. The production and applications of T. on an industrial scale are a relatively recent phenomenon.

Titanium alloys: alloys of titanium with aluminum, vanadium, tin, molybdenum, zirconium, iron and a few platinum metals. The T. containing aluminum and vanadium have the strength of tempered steels. Under reducing conditions, they have the same or better resistance to corrosion than pure titanium. Because of their high strength and low densities, such alloys are used to build air and space craft. "Pure" titanium of various grades of strength is used as the body or cladding on chemical apparatus. Such "pure" titanium types can be seen as T. with differing trace amounts of oxygen, nitrogen and iron. Addition of about 0.2% of cathode-acting noble metal (platinum, palladium, rhodium or ruthenium) increases the resistance to corrosion under reducing conditions. Various types of Ferrotitanium (see) are used as prealloys.

Titanium boride: TiB_2, hexagonal crystals with a metallic luster; M_r 69.52, density 4.52, Mohs hardness 9, m.p. 2870 °C. T. is a good electrical conductor and becomes superconducting below 1.26 K. It is made by reaction of titanium with boron or of titanium(IV) oxide with boron carbide at high temperature. T. is used as a chemically inert, thermally stable, hard material, especially in the temperature range 1100 to 1700 °C; it is harder than any other metal boride. T. is used as an electrode material, and in combination with molybdenum disilicide and graphite to make jets. Combined with titanium carbide and nitride, it is used to make cutting tools.

Titanium carbide: TiC, gray-black, lustrous compound, M_r 59.91, denstiy 4.93, m.p. 3140 °C, b.p. 4820 °C. T. is similar in appearance and reactivity to titanium, and is a good electrical conductor, but is even more resistant to acids than the element. It is made by Zone melting (see) or by a sintering process in which titanium or titanium(IV) oxide is heated with carbon to 2000 °C; the resulting T. is pressed into rods at a pressure of 200 MPa, and again heated to

2500 to 3000 °C. After grinding to a powder and repeated pressing, it is heated close to its melting point, at which temperature any remaining impurities vaporize. T. is a component of many hard and cutting metals, due to its hardness and resistance. In the production of hard metal plates, it is often applied as highly wear-resistant layers, 5 to 10 μm thick, by gas-phase precipitation. Sometimes such layers consist of a combination of T. with titanium nitride. T. is also a component of titanium-containing cast iron.

Titanium chlorides: ***Titanium(II) chloride, titanium dichloride***: $TiCl_2$, black, microcrystalline compound formed by heating titanium(III) chloride to 500 °C, at which temperature it disproportionates; M_r 118.81, density 3.13.

Titanium (III) chloride, titanium trichloride, $TiCl_3$, dark violet crystals; M_r 154.26, density 2.64, m.p. 440 °C (dec.). $TiCl_3$ is formed by passing $TiCl_4$ vapor and hydrogen through a tube heated to 500 °C. Reactions of $TiCl_3$ with suitable ligands (e.g. Cl^-, THF) produce octahedral complexes ($TiCl_6^{3-}$, $TiCl_3(THF)_3$). With water it yields green or violet coordination compounds, e.g. $[TiCl_2(OH_2)_4]Cl \cdot 2H_2O$ and $[Ti(H_2O)_6]Cl_3$ (hydrate isomerism). Aqueous solutions of $TiCl_3$ are used in volumetric analysis (see Titanometry).

Titanium (IV) chloride, titanium tetrachloride, $TiCl_4$, colorless liquid which fumes in moist air and has a pungent odor; M_r 189.71, density 1.726, m.p. -25 °C, b.p. 136.4 °C. The covalent, tetrahedral compound adds chloride ions and other Lewis base ligands to form complexes ($TiCl_6^{2-}$, $TiCl_4L_2$). It is hydrolysed by water ($TiCl_4 + 2\,H_2O \rightarrow TiO_2 + 4\,HCl$), and reacts with alcohols to form titanate esters. $TiCl_4$ is made by passing chlorine over a mixture of titanium(IV) oxide and carbon at 900 °C. It is an important intermediate in the production of titanium and titanium compounds. It is also used to treat textiles and leather, and to improve lenses.

$TiCl_3$ and $TiCl_4$ have particular technological significance as catalyst components for low-pressure ethene polymerization.

Titanium (IV) fluoride, ***titanium tetrafluoride***: TiF_4, a colorless compound; M_r 123.89, density 2.798, subl.p. 284 °C. T. is obtained by reaction of titanium(IV) chloride with hydrofluoric acid. It reacts with alkali fluorides to form alkali hexafluorotitanates, $M_2[TiF_6]$; the potassium, rubidium and cesium salts are suitable for detection of titanium because of their characteristic crystal forms. T. is used in the production of synthetic rubies and sapphires, and as emory for grinding.

Titanium group: Group IVb of the periodic system, including the metals titanium (Ti), zirconium (Zr), hafnium (Hf) and unnilquadrium (Unh; also called kurčatovium, Ku, or rutherfordium, Rf. See Transactinoides). These elements are similar to the metals of Group IIIb, but there are also similarities to Group IVa. Because their atomic and ionic radii are very similar, the metals zirconium and hafnium are closely related; element 104 is synthetic and its physical and chemical properties are not well known. The naturally occurring elements of the T. have high melting and boiling points; their electropositive character increases with increasing nuclear charge, and they tend to form compounds in the +4 oxidation state. The tendency to form compounds with lower oxidation states decreases sharply from titanium to hafnium. The T. elements form volatile, readily hydrolysed tetrachlorides, MCl_4. The basicity of their dioxides increases from the rather acidic titanium(IV) oxide to the basic hafnium(IV) oxide.

Titanium hydride: $TiH_{1\text{-}2}$, a gray-black, metallic compound obtained by the reaction of hydrogen with titanium powder, or by reduction of titanium(IV) oxide with calcium hydride in a hydrogen atmosphere at a temperature above 600 °C. T. is used to join glass and metal, as a getter, to deoxidize metals and as a driver in the production of metal foams.

Titanium nitride: TiN, golden yellow, cubic crystals; M_r 61.91, density 5.22, Mohs hardness 8.9, m.p. 2930 °C. T. is very stable, and is not affected by cold hydrochloric, sulfuric, nitric or hydrofluoric acids, or sodium or potassium hydroxide, steam at 100 °C or chlorine at 270 °C. It does dissolve in hydrofluoric acid in the presence of strong oxidizing agents, and hot potassium hydroxide and superheated steam decompose it with formation of ammonia. T. is a semiconductor. It is made by reaction of titanium(IV) chloride with nitrogen in a hydrogen plasma, by nitration of titanium, or by the Zone melting (see) process. It is used to make protective surfaces on machine parts and reactor vessels.

Titanium oxides: ***Titanium(II) oxide, titanium monoxide***, TiO, golden yellow, hard compound which crystallizes in a sodium chloride lattice; M_r 63.90, density 4.93, m.p. 1750 °C. TiO is obtained by reduction of TiO_2 with titanium in vacuum at temperatures above 1500 °C.

Titanium(III) oxide, Ti_2O_3, is an amethyst-colored compound isomorphic with aluminum oxide, Al_2O_3; M_r 143.80, density 4.6, m.p. 2130 °C (dec.). Ti_2O_3 is formed by reduction of TiO_2 with hydrogen around 1000 °C.

Properties of the Group IVb elements

	Ti	Zr	Hf
Atomic number	22	40	72
Electron configuration	[Ar] $3d^2\,4s^2$	[Kr] $4d^2\,5s^2$	[Xe] $4f^{14}\,5d^2\,6s^2$
Atomic mass	47.90	91.22	178.49
Atomic radius [pm]	132.4	145.4	144.2
Ionic radius (M^{4+}) [pm]	68	74	75
Electronegativity	1.32	1.22	1.23
Standard electrode potential (M/MO^{2+}) [V]	–0.882	–1.43	–1.57
Density [g cm^{-3}]	4.506	6.508	13.31
m. p. [°C]	1677	1857	2150
b. p. [°C]	3262	4377	5400

Titanium(IV) oxide, titanium dioxide, TiO_2, is a white compound which turns yellow to orange when heated. It is trimorphic, forming rutile (tetragonal), anatas (tetragonal) and brookite (rhombic); M_r 79.90, density 4.26 (rutile), 3.84 (anatas) and 4.17 (brookite), m.p. 1825 °C, b.p. 2500-3000 °C. TiO_2 is insoluble in water, dilute bases and acids, but reacts with concentrated sulfuric acid to form water-soluble titanium sulfate, $Ti(SO_4)_2$. Its fusion with alkali hydroxides or carbonates yields titanates. In nature, TiO_2 is usually found as a companion of silicon, primarily as rutile, and more rarely as anatas or brookite. It is usually obtained from ilmenite by solubilization with sulfuric acid, subsequent hydrolysis to titanium(IV) oxide hydrate, and calcination. T. is a very good white pigment, which is sold in a mixture with barium sulfate as Titanium white (see) and is also used to make Nickel titanium yellow (see). The rutile form is preferred as a white pigment, because it has a higher index of refraction, and therefore also a higher covering ability than anatas. It is also less likely to flake off. T. is used to make printer's inks, to dye plastics, to make synthetic fibers matte and for decorative enamels and ceramic products. Because of its nontoxicity, T. is also often used in skin creams, make up, lipsticks, sun-protection creams and powders.

Titanium(IV) oxide hydrate, ***titanium dioxide hydrate***: $TiO_2 \cdot nH_2O$, slimy, white precipitate also called titanic acid. T. is obtained by hydrolysis of alkali titanates or titanium salts. Defined titanic acids such as H_4TiO_4 or H_2TiO_3 apparently do not exist.

Titanium sulfates: ***Titanium(III) sulfate***, $Ti_2(SO_4)_3$, a green crystalline powder which is insoluble in water but dissolves in dilute sulfuric or hydrochloric acid to give a violet color in solution; M_r 338.98. It is obtained by cathodic reduction of a solution of titanium(IV) in sulfuric acid solution. ***Titanium(IV) sulfate***, $Ti(SO_4)_2$, is formed when titanium(IV) oxide is treated with concentrated sulfuric acid. It is hydrolysed by water to ***titanium oxygen sulfate***, $TiOSO_4$, which upon reaction with hydrogen peroxide gives orange ***titanium peroxosulfate***, TiO_2SO_4. This reaction is used to indicate the presence of titanium or hydrogen peroxide.

Titanium(IV) sulfide, ***titanium disulfide***: TiS_2, brass-yellow leaflets with a metallic luster; M_r 112.03, density 3.22. At room temperature, it is stable in air; and it is stable in boiling water, dilute sulfuric acid, hydrochloric acid and ammonia. However, it decomposes in nitric acid or hot sulfuric acid, leading to precipitation of sulfur. It is formed by passing a mixture if titanium(IV) chloride and hydrogen sulfide through a glowing porcelain tube: $TiCl_4 + 2\,H_2S \rightarrow TiS_2 + 4\,HCl$. T. is used as a heat-stable lubricant.

Titanium white: a white pigment consisting of titanium(IV) oxide, barium sulfate and sometimes zinc oxide. T. has excellent covering ability, good coloring ability, is very fast to light, and is stable to bases and acids. It is obtained from ilmenite, which is solubilized with 90% sulfuric acid. The reaction product is dissolved in water, and after it cools, precipitated iron(III) sulfate is centrifuged out. Any remaining iron(III) is reduced to iron(II) by addition of scrap iron, and the titanium(IV) in the solution is precipitated by hydrolysis to titanium dioxide hydrate. This is finally converted to T. by roasting and grinding. The addition of barium sulfate, and sometimes also of zinc oxide, reduces its tendency to flake off. T. is used mainly for paints, but also in printing textiles.

Titanometry: a method of redox analysis in which titanium(III) ion is used as titrator. This is a strong reducing agent with a redox potential of -0.1 V. It is one of the few cases in which a reducing agent is used in redox analysis. The standard solutions are usually made by dilution of commercial titanium(III) salt solutions, and must be stored under an inert gas. In spite of this, their titer changes, so they must be calibrated daily. T. is therefore not suited to single determinations, although it is used widely in serial iron determinations in a number of industrial products. The endpoint of this titration can easily be recognized by the bleaching of the iron(III) thiocyanate complex. Furthermore, T. is important in the determination of nitro-, nitroso- and azo-compounds, which are reduced to the corresponding amines by an excess of titanium(III) solution. This excess is then determined by back titration with iron(III).

Titer: a correction factor for the exact concentration of a standard solution (see Composition parameters) which was not prepared by weighing of Primary titer substances (see).

Titration: the method used in Volumetric analysis (see).

Titrimetry: same as Volumetric analysis (see).

Titrimetric factor: see Stoichiometry.

Tl: symbol for thallium.

Tm: symbol for thulium.

T-metal: see Metal.

TML: see Antiknock substances.

TMTD: same as Thiram (see).

TNRS: abb for lead tetranitroresorcinate.

TNT: abb. for 2,4,6-trinitrotoluene.

TNV: see Flue gas scrubbing.

TLC: abb. for Thin-layer chromatography.

Toad toxins: see Amphibian toxins.

Tobias acid: an Alphabet acid (see).

Tocoquinone: see Vitamin E.

Tocol: see Vitamin E.

Tocopherol: same as Vitamin E.

Tolane, ***Diphenylethyne, diphenylacetylene***: $C_6H_5\text{-}C{\equiv}C\text{-}C_6H_5$, an arene-substituted ethyne derivative. T. forms colorless leaflets, m.p. 63.5 °C. It is synthesized by addition of bromine to stilbene, and heating the resulting 1,2-dibromo-1,2-diphenylethane in alcoholic potassium hydroxide solution. T. is used for organic syntheses.

Tolazoline: a derivative of imidazoline, with an α-sympathicomimetic (blood-vessel dilating) effect.

Tolbutamide: see Antidiabetics.

Tolidines, ***dimethylbenzidine***: symmetrical dimethyl derivatives of benzidine. ***o-T. (4,4'-diamino-3,3'-dimethylbiphenyl)*** forms colorless leaflets which are barely soluble in water, but dissolve easily in alcohol and ether; m.p. 131 °C-132 °C. Its reactions with oxidizing agents produce various colors. o-T. is synthesized by Benzidine rearrangement (see) of 2,2'-dimethylhydrazobenzene. It is an important intermediate in the production of azo pigments. It is also used in analytical chemistry to determine chlorine and oxygen in water, and for colorimetric determination of gold, manganese and cerium. ***m-T. (4,4'-***

diamino-2,2'-dimethylbiphenyl) forms prismatic crystals; m.p. 108-109 °C. It is somewhat soluble in water and readily soluble in alcohol and ether. m-T. is also produced by benzidine rearrangement. It is used mainly for production of pigments.

H_2N NH_2 CH_3 CH_3

o-Tolidine

Tollens test: see Monosaccharides (Table 1).

Tolnaftate: an Antimycotic (see).

Total hardness: see Hardness 2.

Tolubalsam: a brown, viscous mass with an aromatic odor which dissolves readily in alkali hydroxide solutions, chloroform and hot alcohol. T. contains mainly cinnamic acid esters of resin alcohols as well as cinnamic and benzoic acids, essential oils and vanillin. T. is obtained from the tolubalsam tree *Myroxylum balsumum*, which grows in tropical America. T. is used in medicine against bronchitis and as an antiseptic for treating wounds; it is also used as a perfume in cosmetics and in chewing gum.

Toluene, ***methylbenzene***: an aliphatic-aromatic hydrocarbon; a colorless, combustible liquid with a typical aromatic odor; m.p. -95 °C, b.p. 110.6 °C, n_D^{20} 1.4961. T. is slightly soluble in water, but is miscible with ethanol, ether, acetone, chloroform and carbon disulfide. It forms an azeotrope with water which boils at 84 °C; this contains 81.4% T. Like benzene, T. is toxic, but it is less so; however, it causes comparable damage. T. is present in petroleum coking gases and the light oil fraction from coal tar distillation; it is separated from these by fractional distillation. T. can be synthesized by aromatization of the heptane fraction of cracking gasoline or petroleum on chromium oxide-aluminum oxide contacts at about 500 °C. In the laboratory, it can be synthesized by chloromethylation of benzene and hydrogenation of the resulting benzyl chloride. T. is an important solvent and is used as a thinner for paints and chlorinated rubber. It is also the starting material for the synthesis of benzaldehyde, benzoic acid, trinitrotoluene, saccharin and other organic compounds.

CH_3

Toluene-benzoic acid process: see Phenol.

Toluene-4-sulfonamide: CH_3-C_6H_4-SO_2NH_2, colorless crystalline leaflets from water or ethanol; m.p. 137-138 °C. T. is made by reaction of toluene-5-sulfonyl chloride with ammonia. It is used as a softener for synthetic resins and as an intermediate in the synthesis of the antiseptics chloramine T and dichloramine T.

Toluene-4-sulfonic acid: CH_3-C_6H_4-SO_3H, colorless, hygroscopic crystals, m.p. 104 °C (anhydrous). T. is readily soluble in water, soluble in ethanol and ether, but insoluble in benzene and toluene. It is synthesized by sulfonation of toluene with concentrated sulfuric acid at 140 to 150 °C. T. irritates the skin and mucous membranes; in aqueous solution its acidity is equal to that of sulfuric acid. Its salts and esters, the toluenesulfonates, are also called ***tosylates***. T. is used in the laboratory in place of other strong acids to esterify, etherify and dehydrate, as a hardener, and as an intermediate in the synthesis of antiseptics, pigments, etc. Its esters are good alkylation reagent, and are used to make ethers, nitriles, thiols and sulfides.

Toluidine, ***aminotoluene***: the three isomeric amino derivatives of toluene. ***o-Toluidine (2-aminotoluene)*** is an oily liquid; m.p. -14.7 °C, b.p. 202 °C, n_D^{20} 1.5725; ***m-T. (3-aminotoluidine)*** is an oily liquid, m.p. -30.4 °C, b.p. 203.3 °C, n_D^{20} 1.5681; and ***p-T. (4-aminotoluene)*** forms shiny platelets or leaflets; m.p. 44-45 °C, b.p. 200.5 °C.

T. are colorless compounds which turn reddish to brown in the presence of air and light. They are scarcely soluble in water, but dissolve more readily in alcohol and ether. T. undergo all the reactions typical of aromatic primary amines. They dissolve in dilute mineral acids, forming salts. They are made by reduction of the corresponding nitrotoluenes, and are used mainly in the synthesis of dyes, but also as intermediates in the production of vulcanization accelerators and treatments for textiles.

H_3C NH_2 (Position 2, 3 or 4)

Toluol: same as Toluene (see).

Toluylene red: same as Neutral red.

Tolylaldehdye: one of the methyl derivatives of benzaldehyde. ***4-T.*** is the most important of this series; it is a colorless liquid which smells like peppermint, b.p. 204 °C. 4-T. is scarcely soluble in water, but dissolves readily in most organic solvents. It undergoes the addition and condensation reactions typical of the aromatic Aldehydes (see). 4-T. is generally made by the Gattermann-Koch synthesis (see) from toluene. It is used as an aroma in the production of perfumes and soaps.

Tomatanin: see Steroid alkaloids.

Tomatin: see Steroid alkaloids.

Tombac, ***tombak***: see Brass.

Toner: see Electrophotography.

Tons of coal equivalent, abb. ***tce***: a unit for comparison of fuels in energy economics. 1 tce = $29310 \cdot 10^3$kJ = 29310 MJ = 29.31 GJ. 1 tce = 0.7 t petroleum = 830 m^3 natural gas under standard conditions.

Topical magnetic resonance: see NMR spectroscopy.

Topic groups: groups (atoms, atomic groups or molecules) in a compound can differ either in their constitution or in their chemical environment (*topography*). The relationships among identical groups are summarized in the diagram. Groups which are both structurally and topographically identical are called *homotopic (equivalent) groups*; all others are called *heterotopic groups*. To distinguish among homotopic, constitutopic and diastereotopic groups either the symmetry test or the substitution test can be applied. In the *substitution test*, first one and then the other of the groups in question is replaced by a different group and the two resulting compounds are compared.

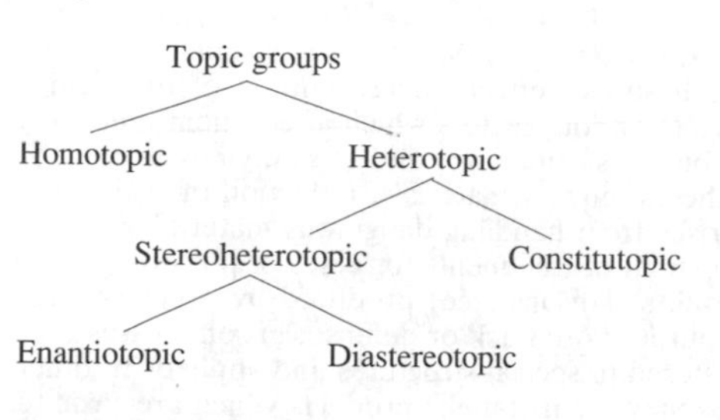

Homotopic groups can be converted into one another by a rotation around an axis (C_n), and identical molecules are formed in the substitution test. Each of the molecules a and b in Figure 1. has a C_2 axis. Using bromine in the substitution test, molecule a gives rise to the identical molecules c and d. If the substitution test produces constitutional isomers, the groups are *constitutopic*; if stereoisomers are formed, they are *stereoheterotopic*. *Enantiotopic* groups are those which can be interconverted by reflection in a mirror and which give rise to enantiomers in the substitution test. Molecules with enantiotopic groups are *prochiral* (Fig. 2). If two groups would be identical outside the molecule, and can be interconverted in the molecule by a symmetry operation, and if the substitution test produces diastereomers, the groups are *diastereotopic* (Fig. 3). The topographic relationships of the sides above and below the trigonal plane of unsaturated compounds are determined in the same way as for substituents. There are homotopic (e.g. in formaldehyde), enantiotopic (e.g. in acetaldehyde) and diastereotopic sides (e.g. in 3-phenylbutan-2-one).

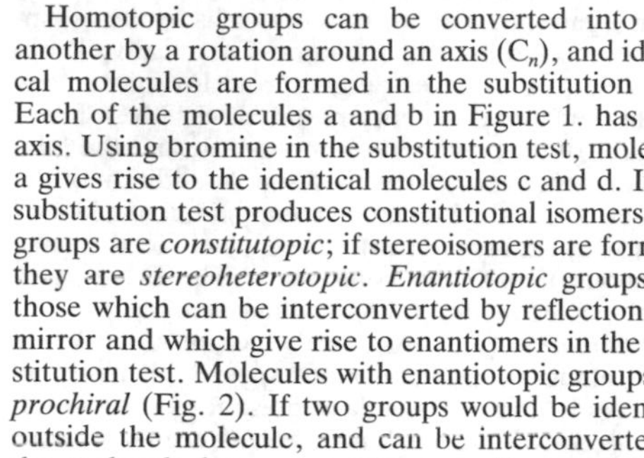

Fig. 1. Homotopic H and C atoms.

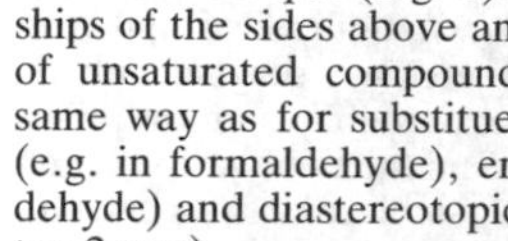

Fig. 2. Enantiotopic H atoms.

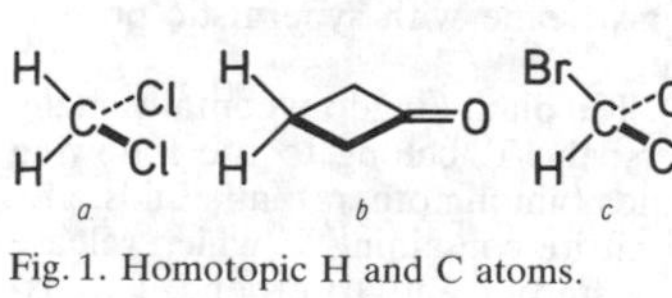

Fig. 3. Diastereotopic H atoms.

Enantiotopic groups and sides are indicated by means of the Cahn-Ingold-Prelog convention (see Stereoisomerism 1.1). An enantiotopic group or side is indicated by R if, when observed from the perspective of that group or side, the remaining three groups in the carbon atom are in the order a → b → c when read clockwise; but if this order corresponds to a counterclockwise progression, the group or side is S (Fig. 4).

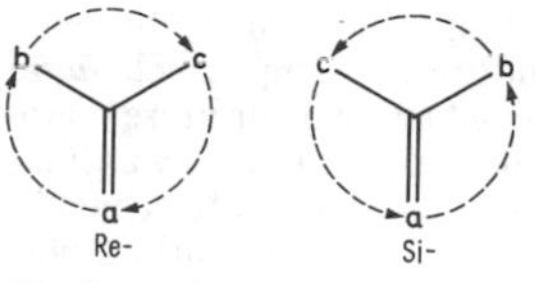

Configuration

Fig. 4. Designation of enantiotopic sides.

This classification of the topography of groups is useful for predicting their chemical and physical properties. Homotopic groups and sides cannot be distinguished. Diastereotopic groups are chemically and physically (e.g. in NMR) different. Enantiotopic groups can only be distinguished by reaction with chiral reagents (such as enzymes) or by measurements in a chiral medium.

Topochemistry: a special field of chemistry concerned with chemical reactions occurring on surfaces or in the interiors of solids, where the reactants are locally bound. In such *topochemical reactions*, both the course of the reaction and the properties of the products are affected to a greater or lesser degree by the surface structure or the internal structures of the solid reactants. The terms "T." and "topochemical reactions" were invented by V. Kohlschütter (1919), who made a systematic study of solid-state reactions. There are three types of topochemical reactions:

1) Reactions in which the nature of the solid reactant affects the surface structure and degree of dispersity of the products, and lattice and other defects (see Crystal lattice defects) in the crystalline products. An example is the observation that in thermal decomposition of various calcium compounds at the same temperature of 1200 °C, powdered calcium oxide, CaO, with very different characteristics is obtained, although the structures of the products are identical.

2) Reactions in which the crystal structure of the product is affected by the solid starting material. For example, a different type of zinc hydroxide (α-$Zn(OH)_2$) is obtained from leaching certain solid, basic zinc salts with layered lattice structures than from hydrolysis of zinc salt solutions in which amorphous precipitates form (most of which are converted to ε-$Zn(OH)_2$).

3) Reactions in which the solid starting material determines the chemical composition of the product. These often lead to compounds which either cannot be obtained in another way, or at least cannot be obtained by analogous reactions in the liquid or gas phase. Such topochemical reactions are of considerable preparative interest. For example, they are the basis for the production of graphite compounds such as graphite oxide or fluoride. In this case, the layered lattice of the Graphite structure (see) represents the reaction matrix, the special atomic arrangement of which is necessary for the formation of the product.

The area of T. also includes problems of corrosion, passivation and protection of metal surfaces, and heterogeneous catalysis, e.g. the influence of the synthesis conditions on the activity of a catalyst. For example, the exchange processes which occur on permutites and zeolites are topochemical reactions.

Topological isomerism: isomerism which occurs when chemically independent subunits are differently linked by mechanical bonds. For example, ***catenanes*** are cycloalkane systems in which two (or more) saturated rings are held together in the same way as links in a chain ("catena" is the Latin word for "chain"). ***Rotaxanes*** consist of a saturated hydrocarbon ring with a hydrocarbon chain passing through its middle and held in place by bulky end groups.

$(CH_2)_m$ $(CH_2)_n$ $R-(CH_2)_n-R$ $(CH_2)_m$

Catenanes Rotaxanes

Topotaxis: the phenomenon in which the crystallographic orientation of a product of a solid state reaction is related to that of the starting material. It is characteristic of topotactic reactions that certain structural elements, such as densely occupied lattice planes or lines, are conserved during the reaction. The dehydration of magnesium hydroxide (brucite) is an example. The reaction equation is $Mg(OH)_2 \rightarrow MgO + H_2O$; it occurs at temperatures between 525 and 635 K, and there is a clear relationship between the structures of the two solids, which is manifest in the retention of the most densely occupied oxygen positions. However, here as in nearly all other cases of T., the product is not obtained as a single crystal; it is polycrystalline.

Top residue: the residue from atmospheric Petroleum distillation (see).

Torlen®: see Synthetic fibers.

Torsion tension: see Stereoisomerism 2.1.

Tosyl-: CH_3-C_6H_4-SO_2-, an atomic group derived from toluene-4-sulfonic acid.

Tosylation: introduction of the tosyl group into a molecule, e.g. by reaction of toluene-4-sulfonyl chloride with alcohols or phenols.

Tosylates: salts and esters of Toluene-4-sulfonic acid (see).

Tosyl chloride: same as Toluene-4-sulfonyl chloride.

Tower process: see Sulfuric acid.

Toxicology: the study of disorders which are the result of interactions of chemical substances with organisms. These disorders can be either immediate damage after intake of the substance (once, several times or chronically), or changes in the genes which are manifest either in the affected individual or its progeny. The field of T. is subdivided into human, veterinary and phytotoxicology. ***Medical T.*** deals mostly with the effects of pollutants and the side effects of pharmaceuticals, and prevention or cure of such effects. ***Chemical T.*** is concerned with the chemical and biochemical conversion of pollutants in the organism and the environment, and with their effects on the estimates of toxicological risks from these pollutants. ***Ecotoxicology*** studies the complex behavior of pollutants in ecosystems and subsystems. ***Military T.*** is concerned with the behavior and effects of Chemical weapons (see). ***Industrial T.*** is concerned with problems of workplace chemicals. The center of interest of T. is the prevention of poisonings and chemical damage to genes, from elucidation of the mechanisms of effects to recognition of the significance of various factors which affect them (chemical and biotransformations) and to safety engineering of syntheses for new materials, reduction of toxicological risks from handling dangerous materials and development of treatments for cases of poisoning.

Toxins: Poisons (see) produced by organisms for the purpose of attack or defense; in some cases they are stored in special structures and applied. In other cases they are metabolic products which are toxic to other organisms, and are stored in cells of the organism or secreted into the medium. Chemically, T. are very diverse substances ranging from low-molecular-weight compounds to enzyme mixtures. Most T. are mixtures of different toxic components, and the mixtures are often more toxic than simple addition of the toxic effects of the individual components would lead one to expect (synergism). The T. include the most poisonous of all known compounds. The T. are traditionally classified according to their sources.

1) ***Animal T.*** (see). Animals which produce T. are found alongside non-poisonous species in nearly all phyla; only a few classes lack poisonous representatives (bivalves, cephalopods, crustaceans, birds, mammals). Toxicologically interesting T. are produced by the coelenterates (jellyfish), molluscs (some snails), worms, some insects (ants, bees, beetles), arachnids (scorpions, spiders), millipedes and centipedes, echinoderms (sea urchins, sea cucumbers, starfish), vertebrates (some amphibians and reptiles, and fish). Most of the T. are mixtures of various components; in addition to low-molecular-weight compounds, toxic proteins and polypeptides, enzymes and enzyme systems (some with synergistic properties) are important.

2) ***Plant T.*** (see). The plant kingdom contains many poisonous members; the T. belong to the following groups of compounds (among others): alkaloids, glycosides (including nitrite-containing T. which release cyanide, coumarin, steroid and triterpene T.), oxalates, proteins and polypeptides, formic acid derivatives and amines. There are other substances whose structures are not yet known. See also Mushroom poisons.

3) ***Microbial T.*** Bacteria and lower fungi (see Mycotoxins) produce T. which are important contaminants of food and water; the long-term effects of some have recently been recognized (see Aflatoxins). Some bacterial T. have extremely high acute toxicity.

Toxicity: a measure of the poisonous effect of a substance. The ***acute T.*** is the average amount (dose) of a poison required to achieve a certain toxic effect, given either as the amount per kg body weight or as a concentration-time product ("Haber product") in mg min^{-1} m^{-3}. The ***chronic T.*** describes the effect of a substance under long-term, regular application of a particular dose. The ***late T.*** includes toxic effects which appear after a long incubation period, usually mutagenicity, carcinogenicity or teratogenicity. In addition, there may be organ-specific or neurological late and chronic effects.

The absolute value of T. (*toxic dose*) depends on a number of factors, of which the most important are the type and site of application (inhalation, oral, subcutaneous, percutaneous, intramuscular, intravenous, intraperitoneal, etc.), the application vehicle (e.g. the solvent), characteristics of experimental animals (species, age and sex, individual differences between animals, predisposition, health and living conditions) and chronobiological factors (time of application). In order that toxicity reports be comparable, these parameters should always be included in the report.

T. is usually determined in animal experiments, i.e. in vivo, although in certain cases, or for characterization of certain toxic properties, in vitro models may be useful.

Common terms for indication of acute toxic doses

Term	Abb.	Dimensions	Explanation
Median lethal dose	LD_{50}	mg/kg	Dose at which 50% of the experimental animals die
Minimum lethal dose	LD_{LO}	mg/kg	Minimum dose at which individual experimental animals die
Absolute lethal dose	LD_{100}	mg/kg	Dose at which all experimental animals die
Median effective dose	ED_{50}	mg/kg	Median dose required to produce a certain effect
Median lethal concentration-time product	LCt_{50}	$mg\ min^{-1}\ m^3$	Product of the exposure time and concentration in breathing air at which 50% of the animals die

TP-5: see Thymopoietin II.

TPN: see Nicotinamide nucleotides.

Trace analysis: analysis of components constituting less than 1% of the sample. The indication of ppm (parts per million) is common in T.; 1 ppm = 10^{-4} % or 1 g/t. The main chemical methods of T. are UV-VIS spectroscopy (see) and electrochemical methods, such as Polarography (see) and Potentiometry (see). Purely physical methods, such as Atomic absorption spectroscopy (see), Atomic emission spectroscopy (see), X-ray fluorescence analysis (see) and Activation analysis (see), are becoming more and more significant.

Trace elements: elements found in natural or synthetic materials in very low concentrations. 1) In descriptions of the biosphere, T. is used for the ***microelements***, those chemical elements which are involved in biological processes in trace amounts. A lack of T. leads to characteristic deficiency diseases in animals, human beings and plants. The T. essential for all organisms include iron, manganese, copper and zinc; other T., including cobalt, molybdenum, vanadium, iodine, chlorine, fluorine and boron, are required only by certain classes or species of plants or animals. The transition elements (see Periodic system) are found in the biosphere as components of metalloproteins, including metalloenzymes, oxygen and electron transport proteins and metal-storage proteins. For example, the transition metals iron, molybdenum and copper take part in the catalysis of redox processes, manganese activates certain enzymes and plays a role in photosynthesis, zinc catalyses hydrolysis and dehydrogenation processes, cobalt is a component of vitamin B_{12}, molybdenum plays an important role in the nitrogen metabolism of plants and microorganisms, and iron takes part in the transport and storage of oxygen. Iodine is a component of the thyroid hormone thyroxin, and fluorine is incorporated into bones and teeth. The human requirements for T. are provided by diet. In some cases, supplementation is needed; examples are iodine or iron preparations, or fluoridation of drinking water. T. deficiency diseases in plants, such as the chlorosis resulting from manganese deficiency, can be cured by provision of special fertilizers or by mobilizing T. present in the soil.

2) ***Dopants*** or ***activators*** are elements added in small amounts to certain materials, such as alloys, semiconductors or luminophor bases, to give them desired properties.

Trace metals: collective term for the metals gallium, germanium, indium and thallium, which are usually found in traces as isomorphic constituents of other minerals. Minerals in which the T. are main components are rare. In the processing of the minerals in which they occur, the T. are automatically enriched in the intermediate products and are isolated as byproducts.

Tracer: a substance (atoms, molecules or macroscopic particles) with which chemical, physical, biological or industrial processes can be followed. Substances used as T. must be labelled by addition (to elements) or incorporation (into molecules) of radionuclides (*radiolabels*) or stable isotopes which differ markedly in their relative abundances from the natural isotopic composition of the element. Usually it is the less abundant isotope which is enriched. The labelled substances behave identically to the unlabelled ones in the process being studied. Radionuclides are detected with radiation detectors or autoradiography of the sample or organism. Stable isotopes usually cannot be detected directly in the object under study, but must be measured, e.g. by mass spectroscopy.

Applications. Radionuclides are used where studies are carried out directly on the object. A disadvantage is the expense and inconvenience of handling and disposing of radioactivity. Stable isotopes of an element are used when there are no radioisotopes of this element, or only very short-lived ones, or when radiation damage must be prevented, e.g. in studies on living organisms. In biology and medicine metabolic and transport processes can be followed by T. labelled with deuterium (D), tritium (T), carbon-13 (^{13}C), carbon-14 (^{14}C), nitrogen-15 (^{15}N), oxygen-18 (^{18}O), phosphorus-32 (^{32}P) or iodine-131 (^{131}I). For example, the iodine uptake of the thyroid can be measured using the radioactive iodine isotope ^{131}I. Metabolic processes can be followed using the radioactive carbon isotope ^{14}C incorporated, for example, into glucose. The uptake of the sugar into the blood stream can be followed, and the ^{14}C in the exhaled carbon dioxide can be detected. The radiation load on the living organism must be kept as low as possible in such measurements. In plants, metabolism can also be followed using radioactive phosphorus ^{32}P. In chemistry, T. are used to elucidate reaction mechanism and exchange processes, to study kinetics, to monitor the completion of reactions, etc. In industry, radionuclides are used, e.g. to study wear. Measurement of the radioactivity permits even small amounts of the abraded material (e.g. in lubricants) to be detected. In addition, T. permit flow and distribution measurements to be made.

Traganth: see Gum resins.

Trajectory: see Kinetics of reactions (theory).

Trakephon®: see Herbicides.

Tranexamic acid: see Antifibrinolytics.

Tranquilizer: see Psychopharmaceuticals.

trans: see Stereoisomerism 1.2.3.

Transactinoids: the short-lived 6d elements following lawrencium (atomic number 103) in the periodic table: kurčatovium or rutherfordium (104), bohrium or hahnium (105), eka-tungsten (106), eka-rhenium (107), eka-osmium (108) and eka-iridium (109). Partly to remove competition among discoverers of new elements for the honor of naming them, the IUPAC has recommended the use of numerical names (and three-letter symbols) for the T. Thus element 104 is unnilquadrium (Unq), 105 is unnilpentium (Unp), etc. However, this recommendation has not been universally accepted.

The T. belong in groups IVb to VIIIb of the periodic system. They are obtained in extremely tiny amounts by bombardment of plutonium or americium targets with neon nuclei (Unq, Unp), by bombardment of californium targets with oxygen nuclei or lead targets with chromium nuclei (Unh), bombardment of bismuth foil with chromium (eka-Re) or iron (eka-Ir), or of lead foil with very fast iron nuclei (eka-Os).

It is thought that more stable T. with atomic numbers around 112 might exist, but they have not yet been confirmed experimentally.

Transaminases: same as Aminotransferases (see).

Transamination: the reversible transfer of an amino group, $-NH_2$, from an α-amino acid to an α-keto acid: $R^1\text{-}CH(NH_2)\text{-}COOH + R^2\text{-}CO\text{-}COOH \rightleftharpoons R^1\text{-}CO\text{-}COOH + R^2\text{-}C(NH_2)\text{-}COOH$. This type of reaction is catalysed by enzymes (see Aminotransferases) and is very important for the metabolism of α-amino acids. T. is the link between carbohydrate and protein metabolism and the mechanism by which amino acids can be synthesized from sugars or degraded to sugars.

Transcriptase: same as RNA polymerase (see).

Transcription: see Nucleic acids.

Trans-effect: see Coordination chemistry.

Transesterification: exchange of alcohol groups in an ester through alcoholysis (reaction with an alcohol) or exchange of the carboxylic acid group of an ester through acidolysis (reaction with a carboxylic acid).

Tranferases: a major group of Enzymes (see). Some important examples are the Kinases (see) and the Aminotransferases (see).

Transferrin: a protein in vertebrate blood which serves as a storage and transport protein for Fe^{3+} ions (M_r 90,000). It is a glycoprotein with two binding sites for Fe ions, which it transports to the red bone marrow for incorporation into red blood cells. The serum levels of T. are elevated toward the end of pregnancy and in acute hepatitis.

Transfer RNA: see Nucleic acids.

Transformation: 1) see Gene technology; 2) see Polymorphism.

Transformation range: see Glass state.

Transition: the change of a system from one energy state to another; the process involves emission or absorption of Electromagnetic radiation (see). There are Selection rules (see) which determine whether a T. is allowed or forbidden.

Transition elements: see Periodic system.

Transition metals: see Periodic system, see Metals.

Transition state: see Kinetics of reactions (theory).

Translational reflection: see Symmetry.

Transmitter: see Sympathicomimetics.

Transparency: the quotient I_T/I_0 obtained when light passes through a sample. I_T = intensity of the transmitted radiation; I_0 = intensity of the radiation falling on the sample. $I_T/I_0 \cdot 100$ gives the percent transmission.

Transport coefficient: see Kinetic gas theory.

Transport number: the contribution of the ith ion to the total specific conductivity of an electrolyte:

$$t_i = \frac{|z_i| F u_i c_i}{\Sigma |z_j| F u_j c_j}.$$

Here t_i is the T. of the ith ion, z is the charge number, F is the Faraday constant, u is the ionic mobility and c is the concentration.

For a single electrolyte solution,

$$t_+ = \frac{u_+}{u_+ + u_-} = \frac{l_+}{l_+ + l_-}.$$

and

$$t_- = \frac{u_-}{u_+ + u_-} = \frac{l_-}{l_+ + l_-}.$$

$t_+ + t_- = 1$. (l is the ionic limit conductivity.)

T. can be measured by various methods. The most useful is *Hittorf's method*, which makes use of an electrolysis cell divided into anode, cathode and intermediate spaces. The entire cell is filled with an electrolyte of known concentration, and current is passed through it. As a result of electrolysis, the concentrations of the electrolyte in the anode and cathode spaces change. ***Hittorf's T.*** is calculated from these changes in concentration. Due to solvation, the migration of the ions causes the transport of different amounts of water, so that the method does not measure the ***true T.***. This can be obtained by correcting for the systematic error.

Transport numbers t of anions and cations of some aqueous electrolyte solutions at 25°C and infinite dilution

Electrolyte	t_+	t_-
KCl	0.491	0.509
HCl	0.821	0.179
KOH	0.274	0.726
NaCl	0.396	0.604
$CuSO_4$	0.375	0.625

Transport reaction: a chemical reaction in which a solid or liquid substance A reacts with a gas B (or with several gases) to form a gaseous product C (or several gaseous products); at another site in the system, the reverse reaction occurs with precipitation of A. The result of the T. corresponds to a sublimation or distillation of substance A, but in fact A, which has no appreciable vapor pressure at the temperature used, is chemically transported through the gas phase. The requirements for a T. to occur are the reversibility of the heterogeneous reaction and the presence of a concentration gradient between the two reaction sites. Each T. can be subdivided into three steps: 1) reaction of the solid or liquid with the gaseous transport reagent, 2) matter transport in the gas phase and 3) back reaction with precipitation of the starting material and release of the transport reagent. The gas motion needed for the T. can be produced by gas flow, diffusion or convection. In *flow methods*, the gas flows over the starting material and reacts with it at temperature T_1, and the back reaction at temperature T_2 occurs downstream. In *diffusion methods*, the closed transport tube is heated to create a temperature gradient, and in *convection methods* the transport tube slopes, so that thermal convection is added to the diffusion process and may greatly increase the rate of transport.

A long-known example of a T. is the migration of Fe_2O_3 in an HCl current according to the transport equation $Fe_2O_3 + 6\,HCl \rightleftharpoons 2\,FeCl_3 + 3\,H_2O$; Fe_2O_3 is consumed at about 1300 K and precipitated again about 1100 K (the temperature gradient of a T. is given in the form 1300 → 1100 K).

The field of T. has been systematically developed principally by H. Schäfer (since about 1950); a series of new applications have been added to those known previously. T. are used mainly 1) for purification of solids, e.g. preparation of pure nickel by the Mond-Langer process ($Ni + 4\,CO \rightleftharpoons Ni(CO)_4$, 350 → 475 K); the iodide method of van Arkel and de Boer for purification of zirconium ($Zr + 2\,I_2 \rightleftharpoons ZrI_4$, 550 → 1725 K) and many other metals, such as Fe, Cr, Ti, V, Nb, Ta etc., 2) for preparation of single crystals (see Crystal growing), and 3) as an aid in preparative chemistry, in which a T. is added to the synthetic reaction, e.g. the formation of a covering layer in Solid state reactions (see) can be avoided by constant removal of the product by a T.

Transuranium elements: the elements in the periodic system with atomic numbers >92. These include the Actinoides (see) with atomic numbers 93 to 103, neptunium (Np), plutonium (Pu), americium (Am), curium (Cm), berkelium (Bk), californium (Cf), einsteinium (Es), fermium (Fm), mendelevium (Md), nobelium (No), and lawrencium (Lw), and the Group IVb to VIIIb elements 104 to 109 (see Transactinoids). All T. are radioactive; a number of nucleotides with varying half-lives are known for each of the elements 93 to 103. T. with atomic numbers > 100 have very short half-lives.

Of the T., only neptunium and plutonium are found in nature, and these only in minute amounts in uranium minerals as a result of neutron interactions with ^{238}U. The actinoids are synthesized by nuclear reactions, directly or indirectly from uranium. Plutonium (^{239}Pu) is now made on an industrial scale, neptunium (^{237}Np) is available in kg amounts, americium (^{241}Am, ^{243}Am) and curium (^{244}Cm) on the 100 g scale, berkelium (^{249}Bk), californium (^{252}Cf) and einsteinium (^{253}Es, ^{254}Es) in mg amounts, and fermium (^{257}Fm) is available in µg quantity. Mendelevium, nobelium and lawrencium have not yet been produced in weighable amounts. The 6d elements 105 to 109 have so far been made only in minute traces, generally a few atoms only.

Trapidil: see Coronary pharmaceuticals.

Trapping reaction: a reaction of an unstable intermediate with a suitable reagent to make a stable product which can be identified.

Traube purine synthesis: a series of reactions described in 1904 by W. Traube. Starting from urea and cyanoacetate, it leads in good yield to uric acid, from which many purine derivatives can be obtained. The starting materials condense to cyanoacetylurea, which is cyclized to 4-aminouracil with sodium hydroxide. Nitrosylation and reduction produce 4,5-diaminouracil, which is converted to uric acid by melting with urea or reaction with chloroformate. Derivatives of purine are obtained mainly via 2,6,8-trichloropurine, which is easily made from uric acid.

+NaOH

+HNO₂ / −H₂O

+4H / −H₂O

4-Aminouracil

4,5-Diaminouracil

a) Cl−CO−OR (−HCl/−ROH)
b) H₂N−CO−NH₂ (−2NH₃)

Uric acid (lactam form)

Trehalose: α-D-glucopyranosyl-α-D-glucopyranoside, a non-reducing disaccharide. T. forms colorless, sweet-tasting crystals; m.p. 97°C (anhydrous). It is soluble in water and hot ethanol, but insoluble in ether. Acid-catalysed hydrolysis of T. produces two molecules of D-glucose. T. is found in insects, young mushrooms and other lower plants, and can be obtained from cake yeast.

Tretamine, *TEM*: triethylenemelamine, a Chemosterilant (see).

Trevira®: see Synthetic fibers.

TRH: see Liberins.

Tri: 1) abb. for trichloroethylene; 2) abb. for 2,4,6-trinitrotoluene.

Triacetate: see Cellulose acetate.

Triacetonamine: see Diacetonamine.

Triad theory: see Pigment theory.

Triadimefon: see Azole fungicides.

Triadimenol: see Azole fungicides.

Tri-a-faser®: see Synthetic fibers.

Triamcinolone: see Adrenal cortex hormones.

2,4,6-Triamino-1,3,5-triazine: same as Melamine.

Triamphos: see Fungicides.

Triampur®: see Diuretics.

Triamterene: see Diuretics.

Triarimol: see Pyrimidine fungicides.

Triarylmethane pigments: in the narrow sense, pigments in which a central carbon atom is bound to three aryl groups (usually phenyl groups); in the broad sense, all arylogenic methine and polymethine pigments and their aza analogs. They are classified as follows:

1) ***Diphenyl*** and ***triphenylmethane pigments*** (Fig. 1). Some important examples are the *malachite green type* (see Malachite green), *crystal violet type* (see Crystal violet) and *phenolphthalein type* (see Phenolphthalein).

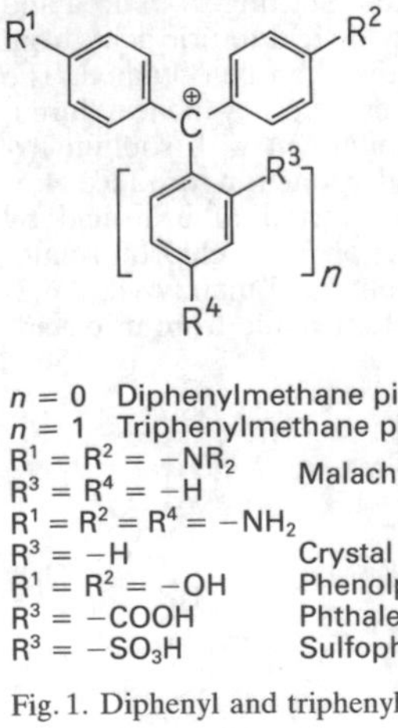

$n = 0$ Diphenylmethane pigment
$n = 1$ Triphenylmethane pigment
$R^1 = R^2 = -NR_2$, $R^3 = R^4 = -H$ Malachite green type
$R^1 = R^2 = R^4 = -NH_2$, $R^3 = -H$ Crystal violet type
$R^1 = R^2 = -OH$ Phenolphthalein type
$R^3 = -COOH$ Phthalein
$R^3 = -SO_3H$ Sulfophthalein

Fig. 1. Diphenyl and triphenylmethane pigments.

If a carboxyl group is in the 2-position relative to the central C atom, the compound is a Phthalein (see); if a sulfonyl group (-SO_3H) is in this position, it is a sulphophthalein.

The diphenyl and triphenylmethane pigments can be classified according to their dye characteristics as acid, basic, chrome and substantive dyes; they are used for silk, wool, treated cotton, paper, food coloring, cosmetics, paints, inks and microbiological stains. However, because they are not very fast to light, acids or bases, the use of di- and triphenylmethane dyes is declining.

2) ***Acridines*** and ***xanthenes (thioxanthenes)*** (Fig. 2). See Acridine pigments, Xanthene pigments.

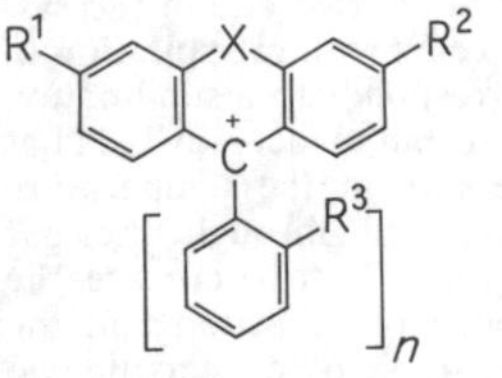

$n = 0$ or 1, X = >N-R Acridines
$n = 0$ or 1, X = O or S Xanthenes
$R^3 = -H$
$R^3 = -COCH$ Phthaleins
$R^3 = -SO_3H$ Sulfophthaleins

Fig. 2. Acridines and xanthenes.

3) ***Quinonimines*** can be further subdivided into the *indamine type, indaniline type* and *indophenol type* (Fig. 3). See Quinonimine pigments.

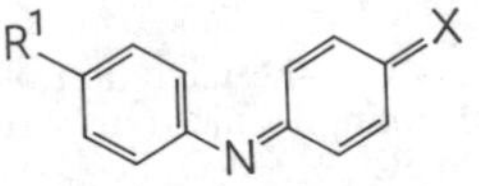

$R^1 = -NR_2$, $X = -NR_2$ Indamine type
$R^1 = -NR_2$, X = O Indaniline type
$R^1 = -OH$, X = O Indophenol type

Fig. 3. Quinonimines.

4) ***Azines, oxazines*** and ***thiazines*** (Fig. 4), see Azine pigments, Oxazine pigments, Thiazine pigments.

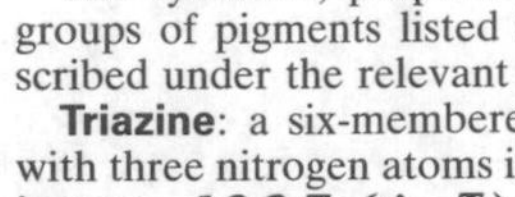

Z = >N-R Azines
Z = O Oxazines
Z = S Thiazines

Fig. 4. Azines, oxazines and thiazines.

The synthesis, properties and applications of the groups of pigments listed in sections 2 to 4 are described under the relevant entries.

Triazine: a six-membered heterocyclic compound with three nitrogen atoms in the ring. There are three isomers, ***1,2,3-T. (vic. T.)***, ***1,2,4-T. (unsym. T.)*** and ***1,3,5-T. (sym. T.)***

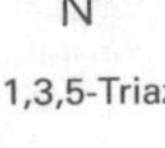

1,3,5-Triazine

The most important of these are 1,3,5-T. and its derivatives. 1,2,4-T. is a pale yellow oil; m.p. 17 °C, b.p. 156 °C, n_D^{25} 1.5149. It is synthesized from 1,2,4-triazine 3-carboxylic acid by decarboxylation, or from 3-hydrazino 1,2,4-triazine by oxidation with manganese dioxide. 1,3,5-T. forms colorless crystals; m.p. 86 °C, b.p. 114 °C. It is soluble in alcohol and ether, and in water it decomposes to formamidine. This decomposition is accelerated by dilute acids. 1,3,5-T. has a considerable resistance to electrophilic substitution. It cannot be sulfonated or nitrated, because the acids immediately hydrolyse the T. The reaction with chlorine at 140 °C yields 2,4-dichlorotriazine, and the reaction with bromine at 120 °C yields 2,4-dibromo-1,3,5-triazine hydrobromide. 1,3,5-T. can be synthesized from formimidoethyl ester, from formamidine or formamide by treatment with anhydrous organic bases. The most important of the 1,3,5-T. derivatives are Cyanuric acid (see), Cyanuric chloride (see) and Melamine (see). These compounds are used in the production of Triazine herbicides (see), as textile conditioners and in the syntheses of pigments and plastics.

Triazine herbicides: 1,3,5-triazine derivatives used as Herbicides (see). After the growth-hormone herbicides, the T. are now economically the most important group of herbicides. The starting material for their synthesis is cyanuric chloride. The stepwise exchange of two of the three chlorine atoms for two alkylamino residues, which may be either the same or different, leads to the ***triazines*** in the narrower sense. Exchange of the third chlorine atom for a methoxy residue leads to a ***triatone***; exchange for a methylthio residue porduces a ***triatryn***. Like the urea herbicides, the T. act chiefly by inhibiting photosynthesis. However, they also affect other biochemical processes in the plant. There are considerable differences in the action spectra, length of time they remain active, selectivity and persistence in the soil of the T., so that they are used in different areas of agriculture. In general, their persistence decreases in the order triazines, triatones, triatryns (Table).

Table. Triazine herbicides

R1, R2, R3 on the triazine ring: R¹ at C-2; R³–HN– and –NH–R² at the other positions.

	Name	R¹	R²	R³	F. in °C
Triazine	Simazin	Cl	C_2H_5	C_2H_5	225···227
	Atrazin	Cl	C_2H_5	$(CH_3)_2CH$	173···175
	Propazin	Cl	$(CH_3)_2CH$	$(CH_3)_2CH$	212···214
Triatone	Atraton	CH_3O	C_2H_5	$(CH_3)_2CH$	94···96
	Prometon	CH_3O	$(CH_3)_2CH$	$(CH_3)_2CH$	91···92
	Terbumeton	CH_3O	C_2H_5	$(CH_3)_3C$	123···124
Triatryne	Desmetryn	CH_3S	CH_3	$(CH_3)_2CH$	84···86
	Prometryn	CH_3S	$(CH_3)_2CH$	$(CH_3)_2CH$	118···120
	Terbutryn	CH_3S	C_2H_5	$(CH_3)_3C$	104···105

Simazin and *atrazin* are used mainly as selective soil herbicides in maize fields. *Desmetryn* is used as a leaf herbicide to control weeds in members of the cabbage family, *prometryn* for weed control in potatoes, vegetables and grains. Some T. are used in combination with other herbicides to control grassy weeds in grains.

Triazine polymers, ***polytriazines***: polycondensates of bisphenols (e.g. dian) and cyanuric chloride. Cyclotrimerization of the biscyanamide formed as an intermediate forms a highly cross-linked polycondensate which is stable to about 260 °C. T. are used mainly as resins in printed circuits.

Triazine ring with RO– substituents at C-2, C-4, C-6; R = $-C_6H_4-C(CH_3)_2-C_6H_4-O-$

Triazine pigments: see Azo pigments.

Triazoles: five-membered heterocyclic compounds with three nitrogen atoms in the ring. The isomeric forms are 2H-1,2,3-T. and 1H-1,2,4-T. The tautomeric forms indicated are favored in the equilibrium. ***2H-1,2,3-T.*** forms colorless hygroscopic crystals; m.p. 23 °C, b.p. 204 °C, n_D^{25} 1.4854. It is soluble in water, alcohol, acetone and ether. 1,2,3-T. is obtained by 1,3-dipolar cycloaddition of hydrazoic acid to acetylene. ***1H-1,2,4-T.*** forms colorless prisms or needles; m.p. 120-121 °C, b.p. 260 °C.

1H-1,2,4-Triazole

It is soluble in water and alcohol. 1,2,4-T. is readily synthesized by condensation of hydrazine with formic acid followed by reductive deamination of the resulting 4-amino-1,2,4-triazole.

T. are used as optical brighteners, herbicides, parent compounds for azo pigments and as organic reagents.

Triballoy®: a cobalt alloy which is extremely resistant to abrasive wear. In it, a hard intermetallic matrix is embedded in a softer matrix. T. has the composition 52 to 62% cobalt, 0 to 50% nickel, 28 to 35% molybdenum, 0 to 17% chromium and 2 to 10% silicon. It has high thermal and dimensional stability in addition to resistance to wear and corrosion. It is synthesized from the powdered components by fusion in a plasma beam or by isostatic compression and hot sintering.

Tribochemistry: an area of physical chemistry dealing with the chemical reactions induced by mechanical energy in solids. T. is a sub-field of *mechanochemistry*, which is the study of mechanical effects on any system, including liquids and gases.

The transfer of mechanical energy can occur by an impulse effect, e.g. by an impact of solid particles, gas or liquid streams, or by tension and pressure forces, as in rubbing, grinding or scratching. These cause numerous physical-chemical processes on both micro and macro scales, which change the structure, physical properties and reactivity of the solid. According to the "tribo-plasma" model developed in the 1960's by

P.A. Thiessen, at the moment of mechanical impact, energy accumulates on the sub-microscopic scale, causing local temperatures of 10^5 to 10^6 K; these create a very short-lived (10^{-7} s) plasma-like state. This highly excited energy state is characterized by extensive dissociation of crystal components and high reactivity; it can be studied, e.g. by electron and light emission (triboluminescence) and morphological studies. A long-lived form of excitation (10^{-3} to 10^6 s) remaining in the solid is Crystal lattice defects (see), which also contribute to increased reactivity. The mechanically induced defects can either extend over the entire volume or they may be concentrated on the surface, depending on the type of mechanical stress applied. The increased chemical reactivity can be utilized in two different ways: either the solid is made to react with other substances after the mechanical treatment (*static activation*), or the other reactants are present during the activation (*dynamic activation*). In dynamic activation, which is much more effective, not only is the surface area increased and its structure changed, but the high excitation states elevate the free energy, and thus the activity, of the solid.

The kinetics of tribochemical reactions are somewhat different from those of thermal reactions. For example, their rates are largely independent of temperature, but they are directly proportional to the intensity of mechanical activation. An example of the acceleration of a reaction by tribomechanical activation is the production of nickeltetracarbonyl $Ni(CO)_4$ from nickel powder and carbon monoxide by the Mond-Langer process. In a vibrating mill, the reaction will occur even at room temperature, and at a much higher rate than the thermal process (320 to 350 K). Under impact working, reactions which would otherwise require high temperatures or pressures can be carried out under normal conditions. Examples are formation of methane from the elements and formation of molybdenumhexacarbonyl $Mo(CO)_6$ from molybdenum and carbon monoxide. In addition, reactions with positive free enthalpies, which do not occur spontaneously, can be driven by addition of free energy in the mechanical working process.

T. is of considerable practical and industrial significance. Mechanical activation of solids has long been used to reduce the reaction and sinter temperatures, e.g. in the cement and ceramic industries, in roasting lime and reduction of metal oxides on an industrial scale. The effects of heterogeneous catalysts and the rates of solution of many substances can be considerably elevated by mechanical pretreatment. Tribochemical reactions have an essential role in friction and lubrication processes. In dry rubbing, frictional oxidation causes corrosion of the mechanically activated surfaces. The addition of a lubricant can either reduce wear (by reaction of the lubricant with the metal surface, forming a stable intermediate layer with high mechanical stability) or promote it, by generation of reaction products which are removed or can themselves cause corrosion.

Tribocorrosion: see Wear corrosion.

Tribromomethane: same as Bromoform (see).

Tributylphosphate: see Phosphates 2).

Tricarboxylic acid cycle: same as Citric acid cycle (see).

Tricel®: see Synthetic fibers.

Trichlorfon: O,O-dimethyl 2,2,2-trichloro-1-hydroxyethylphosphonate, an Organophosphate insecticide (see) (phosphonate ester), which acts as a contact and oral poison. T. is a colorless crystalline substance, D_4^{20} 1.73, m.p. 83-84 °C, b.p. 100 °C at 13 Pa, vapor pressure at 20 °C, $1.0 \cdot 10^{-3}$ Pa. T. is soluble in water at 15.4 g/100 ml, and in organic solvents. In alkaline solution, hydrogen chloride is cleaved off, converting T. to Dichlorvos (see).

The industrial synthesis starts from phosphorus trichloride, which is reacted with methanol in the absence of base. Chloromethane is split off, leading to dimethylphosphite. The reaction of the latter with chloral leads directly to T. If the 1-hydroxy group is esterified with butyric acid, the product is *butonate*, which also has insecticidal effects, but its oral acute toxicity (LD_{50} = 1070-1310 mg/kg rat) is still lower than that of the relatively nontoxic T. (435 mg/kg rat).

Due to its relatively low toxicity, T. is preferred as a hygiene insecticide in the house and barn; in veterinary practice it is used against grubs, bots and screwworms. It is also used to control insects in grain, vegetables, fruits and ornamental plants.

Trichloroacetic acid, abb. ***TCA***: CCl_3-COOH, one of the strongest organic acids; $pK_1 = 0.66$. T. forms colorless, hygroscopic crystals which are very soluble in water; m.p. 57.5 °C. When aqueous solutions of T. are boiled, it decomposes to chloroform, $CHCl_3$, and CO_2. T. is caustic and precipitates proteins. It is made by oxidation of Chloralhydrate (see) with concentrated nitric acid. The sodium salt is used as a Herbicide (see).

Trichlorobenzene: one of three structural isomers with the general formula $C_6H_3Cl_3$. ***1,2,3-T. (vic.-T.)*** forms colorless, flaky crystals; m.p. 53-54 °C, b.p. 218-219 °C. It is insoluble in water, but soluble in ether, benzene and carbon disulfide. 1,2,3-T. is formed in low yields, along with other chloro derivatives, by direct chlorination of benzene. It can be synthesized in pure form by diazotization of 3,4,5-trichloroaniline and subsequent reduction with alcohol. ***1,2,4-T. (asym.-T.)*** is a colorless liquid; m.p. 17 °C, b.p. 213.5 °C, n_D^{20} 1.5715. It is insoluble in water, and soluble in ether, benzene and carbon disulfide. 1,2,4-T. can also be formed by direct chlorination of benzene, along with other chlorobenzenes, or in pure form by Diazotization (see) and the Sandmeyer reaction (see) of 2,4-, 2,5- or 3,4-dichloroaniline and copper(I) chloride. ***1,3,5-T. (sym.-T.)*** forms colorless, crystalline needles; m.p. 63-64 °C, b.p. 208 °C. It is insoluble in water, but soluble in ether, benzene and carbon disulfide. It is synthesized by diazotization of 2,4,6-trichloroaniline and subsequent reduction with alcohol.

The three T. are steam-volatile. They are used as pesticides, e.g. against termites, as solvents and as intermediates in the production of pigments.

Trichloroethene: same as Trichloroethylene (see).

Trichloroethylene, ***trichloroethene***, abb. ***Tri***: $CHCl{=}CCl_2$, a colorless liquid which smells like chloroform; m.p. -73 °C (-81.8 °C also reported), b.p. 87 °C, n_D^{20} 1.4773. T. is scarcely soluble in water, but dissolves readily in most organic solvents. It forms azeotropic mixtures with many alcohols, water and glacial acetic acid. In the presence of stabilizers, e.g. amines, T. can be stored for a long period. Pure, unstabilized T. decomposes in air, forming hydrogen chloride, phosgene, carbon monoxide and dichloroacetyl chloride. Under normal conditions, the chlorine atoms cannot be replaced by nucleophilic reagents.

T. is made by cleavage of hydrogen chloride out of 1,1,2,2-tetrachloroethane in the presence of an alkali or by catalysed pyrolysis:

$$Cl_2CH\text{–}CHCl_2 \xrightarrow[-\,HCl]{} CHCl{=}CCl_2.$$

Newer methods of synthesizing T. are based on chlorination of 1,2-dichloroethane. Because it does not burn, T. is a useful solvent for fats, oils, waxes, resins, rubber, paraffin, cellulose acetate, phosphorus and sulfur. It is also used to dry alcohol, separate methanol from crude spirits, as an emulsifying agent in the textile industry, as an inhalation anaesthetic and for organic syntheses.

> Trichloroethylene is poisonous. Inhalation of high concentrations can cause damage to the central nervous system, the lungs, liver and kidneys. Acute intoxication can lead to death.
> Antidote: artificial respiration with oxygen.

Trichlorofluoromethane: CCl_3F, a colorless, noncombustible gas, m.p. -111 °C, b.p. 23.7 °C. T. can be readily liquefied and is soluble in most organic solvents, but is insoluble in water. It is synthesized by chlorine exchange from carbon tetrachloride and hydrogen fluoride in the presence of antimony(III) chloride; the product is mixed with other chlorofluoromethanes. These can be separated by fractional distillation. T. is used as a propellant for aerosols and foams, and as a coolant (R 11).

Trichloromethane: same as Chloroform (see).

Trichloromethanesulfenyl chloride: same as Perchloromethyl mercaptan (see).

Trichloromethylbenzene: same as Benzene trichloride (see).

Trichloronitromethane, ***chloropicrin, nitrochloroform***: $Cl_3C\text{-}NO_2$, a colorless, very poisonous and highly refractive liquid with a pungent odor; m.p. -64.5 °C, b.p. 111.8 °C, n_D^{20} 1.4622. T. is barely soluble in cold water, but dissolves readily in alcohol, acetone, glacial acetic acid and benzene. In the presence of light, it slowly discolors, going from yellow to red-brown. When heated or exposed to UV radiation, it decomposes to phosgene and nitrosyl chloride. T. is highly volatile, and its vapors irritate the eyes and damage the lungs. Poisoning also leads to liver and kidney damage. T. is formed by chlorination of nitromethane. It is used as an insecticide. It was formerly used as a chemical weapon.

Trichlorophenol: one of six isomeric phenol derivatives, of which 2,4,5-T. and 2,4,6-T. are important.

2,4,5-T. forms colorless needles with a phenol-like odor; m.p. 68-70 °C (subl.), m.p. 253 °C. The substance is steam volatile, scarcely soluble in water, and readily soluble in polar organic solvents. It is synthesized by alkaline hydrolysis of 1,2,4,5-tetrachlorobenzene. Nearly all of it is treated with chloroacetic acid to form 2,4,5-trichlorophenoxyacetic acid, a plant growth hormone used as an herbicide.

2,4,5-Trichlorophenol   2,4,6-Trichlorophenol

A byproduct is Dioxin (see), which is a very toxic compound.

2,4,6-Trichlorophenol, colorless, steam-volatile crystals which smell like phenol; M_r 69.5 °C, b.p. 246 °C. The substance is slightly soluble in water, and soluble in ethanol, ether and acetone. It is easily synthesized by chlorination of phenol, and is a relatively strong acid, so that it dissolves even in sodium hydrogencarbonate solution. This T. is a strong antiseptic and is used as a disinfectant and as a preservative for leather, wood and textiles.

2,6,8-Trichloropurine: crystalline leaflets; m.p. 187-189 °C (dec.). It is synthesized by reaction of uric acid with phosphorus oxygen chloride. 2,6,8-T. is an important intermediate in organic syntheses of purine derivatives. The individual chlorine atoms have different reactivities, e.g. in the presence of sodium hydroxide, nucleophilic substitution can occur only in the 6 position. This forms 2,8-dichlorohypoxanthine, which is reduced to hypoxanthine. When 2,6,8-T. is reduced with hydrogen iodide, the product is purine.

Trichlorosilane, ***silicochloroform***: $SiHCl_3$, a colorless liquid which fumes in air and has a suffocating odor; M_r 135.45, density 1.34, m.p. -126.5 °C, b.p. 33 °C. T. is rapidly and quantitatively hydrolysed by water. Mixtures of T. with oxygen or air can explode violently when heated or upon impact. T. is synthesized by passing a hydrogen chloride-chlorine mixture over heated silicon or ferrosilicon. It is used mainly to produce pure silicon, but it also plays a role in production of special silicones. In the laboratory, it is an important starting material for synthesis of organosilicon compounds.

α,α,α-Trichlorotoluene: same as Benzene trichloride (see).

2,4,6-Trichloro-1,3,5-triazine: same as Cyanuric chloride (see).

Trichotoxin: see Alamethicin.

Triclinic: see Crystal.

Trickle column: a highly effective distillation column in which the condensed vapor moves down the walls as a counterflow to the rising vapors. It is distinguished by a very low loss of pressure of the vapors rising in the column.

Trickle phase reactor: a stirred or pipe reactor in which a three-phase system is used. The reactants are generally in the gaseous and liquid phases, and the catalyst is the solid phase. The T. is used for low-pressure polymerization of ethene and propene and for Fischer-Tropsch syntheses.

Trickle process: see Hydrorefining.

Tricuran®: see Muscle relaxants.

Tricresyl phosphates: $(CH_3\text{-}C_6H_4\text{-}O)_3PO$. The significant members of this class are ***tri-o-cresyl phosphate***, a pale yellow liquid (m.p. 11 °C, b.p. 410 °C) and ***tri-p-cresyl phosphate***, colorless and odorless needles (m.p. 77-78 °C, b.p. 224 °C at 0.47 kPa). They are insoluble in water, but soluble in most organic solvents. T. are very poisonous, and are easily taken up through the skin. They are synthesized by reaction of phosphorus oxygen chloride with *o*- or *p*-cresol, or technically with a crude cresol, in xylene as solvent. T. were used as softeners for plastics, as additives to leaded gasoline to prevent corrosion of the spark plugs (formation of lead phosphate) and as additives to lubricants. However, because of their toxicity, they are no longer used for these purposes. They are still used to absorb phenols in benzene scrubbers. They are suitable for use as the stationary phase in gas chromatography for separation of aromatic and aliphatic hydrocarbons, ketones and esters.

Tricyanomethanide: $[C(CN)_3]^-$, a Pseudohalide (see), a planar, resonance-stabilized anion. ***Alkali tricyanomethanides***, such as ***potassium*** and ***sodium tricyanomethanides***, $KCN(CN)_3$ and $NaC(CN)_3$, are typical salts which can be obtained by reaction of bromomalonic dinitrile, $(NC)_2CHBr$ with alkali cyanide. The transition metal tricyanomethanides are only slightly soluble in water; they are coordination polymers. Some examples are the brown ***copper(II) tricyanomethanide***, $Cu[C(CN)_3]_2$, the bright blue ***nickel(II) tricyanomethanide***, $Ni[C(CN)_3]_2$ or the colorless ***silver(I) tricyanomethanide***, $AgC(CN)_3$. There are also covalent T. derivatives of nonmetals, such as ***dicyanoketenimine (cyanoform)***, $NHCC(CN)_2$; this is a very strong acid in water (pK-5.1). Other examples are the colorless ***halotricyanomethanes***, $ClC(CN)_3$ (crystalline needles, m.p. 47 °C) and $BrC(CN)_3$. T. is similar in its complexing behavior to the homologous pseudohalides dicyanamide and cyanate. Like them, it preferentially forms coordinate bonds via the cyano nitrogen, but is able to act as a multidentate ligand in bridge functions.

Tricyanophosphane: see Cyanides.

Tricyclic: a term for carbon compounds in which three carbon rings are joined together.

Tridemorph: see Morpholine and piperazine fungicides.

Tridymite type, ***tridymite structure***: a Structure type (see) for compounds with the general composition AB_2, in which the ratio of radii is small: $r_A/r_B \leq 0.414$. In the crystal lattice of the SiO_2 modification tridymite, the Si atoms form what would be a wurtzite structure (see Wurtzite type) among themselves, and are surrounded tetrahedrally by 4 O atoms, each of which is centered between two Si atoms. Ordinary ice crystallizes in the T.; the oxygen atoms occupy the Si places, and the hydrogen atoms are arranged (though asymmetrically) on the lines connecting them.

Triels: see Nomenclature, sect. II A 2.

Triethylaluminum: see Organoaluminum compounds.

Triethylamine: see Ethylamines.

Triethyleneglycol, ***triglycol***: $HO\text{-}CH_2\text{-}CH_2\text{-}O\text{-}CH_2\text{-}CH_2\text{-}O\text{-}CH_2\text{-}CH_2\text{-}OH$, a colorless, hygroscopic liquid; m.p. -5 °C, b.p. 278.3 °C, n_D^{20} 1.4531. T. is formed as a byproduct of ethylene glycol synthesis, and is separated from it by fractional distillation. The viscous liquid is used as a solvent, e.g. for dehydrohalogenation reactions, as a disinfectant and as a heating bath liquid.

Triethyl orthoformate: $HC(OC_2H_5)_3$, the triethyl ester of the hypothetical orthoformic acid, $HC(OH)_3$. It is the best known of the Ortho esters (see). T. can be distilled without decomposition; it is a pleasant-smelling liquid, m.p. -76 °C, b.p. 146 °C, n_D^{20} 1.3922. It is readily soluble in most organic solvents. T. is hydrolysed by water and acids, but is stable to bases. It can be synthesized by reaction of sodium ethanolate with chloroform: $3\ C_2H_5ONa + HCCl_3 \rightarrow HC(OC_2H_5)_3 + 3\ NaCl$; or by alcoholysis of ethyl formimide hydrochloride: $HC(=NH_2)\text{-}OC_2H_5Cl^- + 2\,C_2H_5OH \rightarrow HC(OC_2H_5)_4 + NH_4Cl$. T. is used in synthetic chemistry to make acetals, aldehydes, hydroxymethylketones and cyanin pigments.

Triethyl phosphate: see Phosphates 2).

Triethylphosphite: see Phosphites 2).

Trifenmorph: a Molluscicide (see).

Triflate: see Trifluoromethanesulfonic acid.

Trifluoroacetic acid: $CF_3\text{-}COOH$, a colorless, highly hygroscopic liquid which fumes in air; m.p. -15.2 °C, b.p. 72.4 °C. T. is soluble in water, alcohol, ether and acetone. In aqueous solution, it is nearly completely dissociated, and is thus a very strong organic acid; its pK_a value is 0.23, which is in the same range as that of hydrochloric acid. The fluorine atoms of T. are very resistant to hydrolysis, and cannot be removed, for example, by boiling with alkalies. T. is very caustic and causes burns if it comes in contact with the skin. It is synthesized by oxidation of 4-(trifluoromethyl)aniline with potassium permanganate or chromic acid, or by electrolysis of acetic anhydride in anhydrous hydrogen fluoride. T. and its anhydride are used as esterification catalysts, condensation reagents, aids to peptide syntheses and as intermediates for organic syntheses.

Trifluoromethanesulfonic acid: $F_3C\text{-}SO_3H$, one of the strongest organic acids (in the anhydrous state); a colorless liquid, b.p. 162 °C. T. is miscible with ether, but is decomposed in the presence of ethanol. It forms stable salts and esters (***triflates***). T. is used as a solvent for polymers, to remove peptide protecting groups, and to catalyse isomerizations and polymerizations. Its esters are excellent alkylation reagents. The anhydride, $(F_3C\text{-}SO_2)_2O$, is a colorless, hygroscopic, caustic, toxic liquid (b.p. 84 °C), and is used in preparative chemistry to introduce the trifluoromethanesulfonyl group (triflate group).

Trifluoromethylbenzene: same as Benzene trifluoride (see).

α,α,α-Trifluorotoluene: same as Benzene trifluoride (see).

Trifluraline: see Herbicides.

Triforin: see Morpholine and piperazine fungicides.

Triglycol: same as Triethylene glycol.

Trihexyphenidyl: see Antiparkinson's compounds.

Trihydroxybenzene: see Benzene triol.

3,4,5-Trihydrobenzoic acid: same as Gallic acid (see).

3,4,5-Trihydroxycyclohex-1-enecarboxylic acid: same as Shikimic acid (see).

2,6,8-Trihydroxypurine: same as Uric acid (see).

Triiodomethane: same as Iodoform (see).

Triiodothyronine: see Hormones.

Trimer: see Polymers.

Trimethoprim: a diaminopyrimidine derivative which acts as an inhibitor of dihydrofolic acid reductase, and is used as an antibiotic, usually in combination with sulfonamides. T. strongly inhibits the bacterial enzymes, but the mammalian liver enzyme is not so strongly affected. Other diaminopyrimidine derivatives used as antimalaria agents, e.g. pyrimethamine (Tindurin®) have the same mechanism of action.

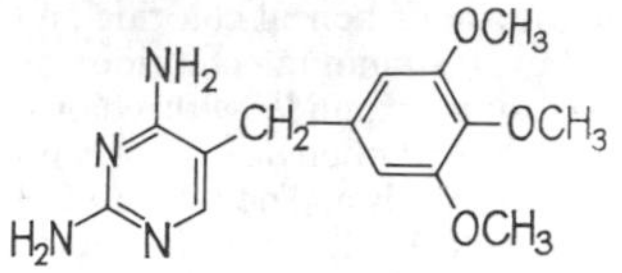

Trimethylacetic acid: see Valeric acids.

2,4,6-Trimethylbenzaldehyde: same as Mesitylaldehyde.

Trimethylbenzenes: aliphatic-aromatic hydrocarbons found in coal tar. There are three isomers: mesitylene, pseudocumene and hemellitene.

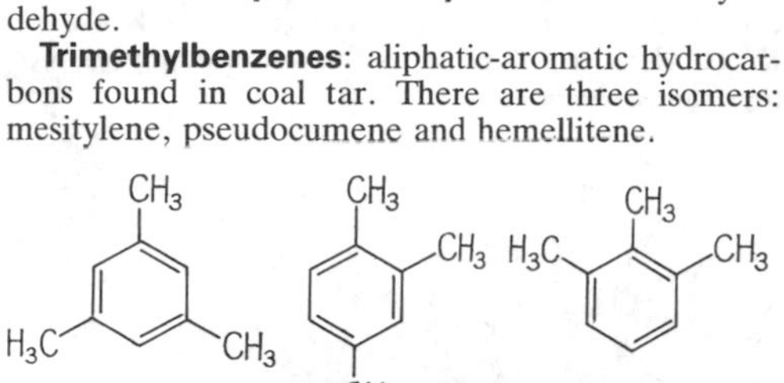

Mesistylene Pseudocumene Hemellitene

Mesitylene (1,3,5-trimethylbenzene) is a colorless liquid with a green fluorescence; m.p. -44.7°C, b.p. 164.7°C n_D^{20} 1.4994. It is insoluble in water, but is readily soluble in ethanol and ether. It can be synthesized by reaction of toluene or *m*-xylene with methyl chloride in a Friedel-Crafts reaction, by disproportionation of certain xylene fractions on an aluminum chloride contact, or, in the laboratory, from acetone and sulfuric acid.

Pseudocumene (1,2,4-trimethylbenzene) is a colorless liquid with the same solubility properties as mesitylene; m.p. -43.8°C, b.p. 169.4°C, n_D^{20} 1.5048. Pseudocumene is an intermediate in the dye and perfume industries.

Hemellitene (1,2,3-trimethylbenzene) is a colorless liquid with the same solublity properties as mesitylene; m.p. -25.4°C, b.p. 176.1°C, n_D^{20} 1.5139. All T. are combustible and can be used as solvents.

Trimethylborate, ***boric acid trimethyl ester***: $B(OCH_3)_3$, the ester of methanol and boric acid, a colorless liquid which is unstable in the presence of moisture. It burns with a green flame. m.p. -29.3°C, b.p. 67-69°C, n_D^{20} 1.3568. B. is soluble in organic solvents and is saponified by water to boric acid and methanol: $B(OCH_3)_3 + 3\,H_2O \rightarrow H_3BO_3 + 3\,CH_3OH$. It is made by the reaction of boric acid with methanol in the presence of conc. sulfuric acid. T. can be determined qualitatively from the green flame, and quantitatively by hydrolysis with calcium hydroxide suspension, evaporation, roasting and weighing of the residue.

Trimethylenimine: same as Azetidine (see).

Trimethylene oxide: same as Oxetane (see).

Trimethylpyridine: see Collidines.

Trimipramin: see Dibenzodihydroazepines.

Trimolecular reaction: an Elementary reaction (see) among three reactants.

Trimorphamide: see Morpholine and piperazine fungicides.

Trimorphism: see Polymorphism.

2,4,6-Trinitroaniline: same as Picramide (see).

Trinitrobenzenes: three structural isomers: ***1,2,3-T.*** forms colorless to greenish prisms or needles; m.p. 127.5°C. ***1,2,4-T.*** forms colorless prisms; m.p. 61.2°C; ***1,3,5-T.***, colorless, crystalline leaflets; m.p. 121-122°C, b.p. 315°C. The T. are poisonous (see Nitrobenzene) and are scarcely soluble in water. Only 1,3,5-T. can be obtained by direct nitration, and only in low yields. The three T. can be obtained in good yields by oxidation of 2,4-, 2,6- or 3,5-dinitroaniline with peroxotrifluoroacetic acid. 1,3,5-Trinitrobenzene can also be synthesized by oxidation of 2,4,6-trinitrotoluene to 2,4,6-trinitrobenzoic acid, which is then decarboxylated.

Trinitromethane: same as Nitroform (see).

2,4,6-Trinitrophenol: same as Picric acid (see).

2,4,6-Trinitrophenylmethylnitramide, ***tetranitro-N-methylaniline***, abb. ***Tetryl***: $(NO_2)_3C_6H_2$-N$(CH_3)NO_2$, poisonous, colorless crystals which turn yellowish in light; the technical product is intense yellow; m.p. 129.4°C. 2,4,6-T. is scarcely soluble in water or cold alcohol, but dissolves readily in benzene and acetone. It is sensitive to acids and bases, and burns with a bright flame. The compound is made by nitration of *N,N*-dimethylaniline with simultaneous oxidation of a methyl group to a carboxyl group, decarboxylation and *N*-nitration. The industrial synthesis usually begins with nitration of dinitromethylaniline, which is made by reaction of dinitrobenzene with methylamine. 2,4,6-T. is a powerful explosive, which is used mainly to make ignition caps. Its detonation velocity is 7200 m s^{-1}. It has been replaced as a general explosive by the cheaper and more powerful explosives nitropenta, hexogen and octogen. 2,4,6-T. is also used as a pH indicator.

2,4,6-Trinitroresorcinol: same as Styphnic acid (see).

Trinitroresorcinyllead:, abb. ***TNRS, lead styphnate***: $C_6H(NO_3)_2O_2Pb \cdot H_2O$, the lead salt of styphnic acid (trinitroresorcinol). It forms red-brown, heavy crystals or a yellow basic salt; density 2.9. T. is made in batches or continuously by precipitation from a magnesium trinitroresorcinate solution at a pH of about 5.4 with a lead nitrate solution, $Pb(NO_3)_2$. It explodes violently on impact or ignition (detonation rate 5200 m s^{-1}). T. is used in a mixture with lead azide as an ignition explosive which is weaker than lead azide, but is still extremely sensitive to flames or sparks. Because of its strong tendency to accumulate electric charge, and its ability to be detonated even by small discharge sparks, T. is very dangerous to handle. It is used as an additive for detonators, and as the main component of the non-rusting priming for percussion caps.

2,4,6-Trinitrotoluene, abb. ***TNT, Tri***: bright yellow, crystalline needles; m.p. 80.8 °C. 2,4,6-T. can be distilled in vacuum without decomposition. It is practically insoluble in water, but dissolves readily in ether, acetone, benzene and pyridine. It can be produced by direct nitration of toluene with nitrating acid. Mono- and dinitrotoluenes are often used as starting materials. 2,4,6-T. is the most important conventional explosive, with a detonation velocity of 6900 m s^{-1}. It is used for both military and commercial purposes as a relatively safe explosive which is not sensitive to impacts. It explodes only upon ignition. Because of its low melting point, 2,4,6-T. can be safely melted in steam and poured into molds.

The explosive effects of fusion and fission bombs are expressed in kilotons and megatons TNT.

1,3,5-Trinitro-1,3,5-triazacyclohexane: trivial names ***hexogen, cyclotrimethylene trinitramine***: a white, finely crystalline powder which has no taste or odor. T. is nearly insoluble in water, is slightly soluble in alcohol and ether, and dissolves readily in acetone. It melts unchanged around 200 °C, and decomposes at higher temperatures without exploding, but with production of smoke. T. is synthesized by nitration of hexamethylenetetraamine with conc. nitric acid, or by reaction of acetic anhydride with paraformaldehyde. After nitropenta and octogen, hexogen is the most powerful explosive and is used mainly by the military. Pressed hexogen has a detonation velocity of 9,000 m s^{-1}. To reduce its sensitivity to impact and to make it easier to compress, T. is usually phlegmatized by adding 5 to 10% montane wax, which does not significantly reduce its explosiveness. Pressed hollow charges are used as filling for tank-destroying grenades. Pourable mixtures of T. and 2,4,6-trinitrotoluene are used to fill bombs, and for special purposes such as petroleum exploration.

Triolefin process: a method of converting propene to ethene and butene by disproportionation of the propene at 150 °C and 3 to 4 MPa pressure on a molybdenum, tungsten or rhenium carrier catalyst.

1,3,5-Trioxane: the cyclic trimer of Formaldehyde (see). It forms colorless needles; m.p. 64 °C, b.p. 114.5 °C, and is soluble in water, alcohol and ether. When heated to 150 to 200 °C, it depolymerizes to monomeric formaldehyde.

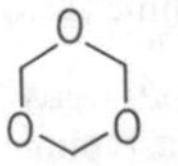

Tripelenamine: see Antihistamines.

Tripeptide: see Peptides.

Triperidene: see Antiparkinsons's compounds.

Triphenylcarbinol: ***triphenylmethanol***: $(C_6H_5)_3$ COH, a tertiary alcohol derived from triphenylmethane. It forms colorless crystals, m.p. 164 °C, and is insoluble in water, but dissolves readily in alcohol, ether and benzene. It is synthesized by reaction of phenylmagnesium bromide with benzophenone or ethyl benzoate (Grignard reaction), or by oxidation of triphenylmethane with chromic acid. The OH-group of T. cannot be esterified with organic acyl derivatives. However, trityl chloride or trityl perchlorate can be very easily made by the reaction of hydrochloric or perchloric acid with it.

Triphenylformazane: see Formazane pigments.

Triphenylmethane, ***tritane***: an aliphatic-aromatic hydrocarbon which forms colorless crystals; m.p. 83 °C, b.p. 359 °C. It is insoluble in water and scarcely soluble in cold ethanol, but is readily soluble in ether and hot ethanol. T. can be made in various ways, e.g. from benzene and chloroform or benzal chloride and benzene in the presence of aluminum chloride, by reaction of phenylmagnesium bromide with chloroform, or by condensation of diphenylcarbinol with benzene and sulfuric acid as catalyst. The tertiary CH group of T. is very reactive. T. is oxidized to triphenylcarbinol by chromium(VI) compounds. Reaction with potassium amide in liquid ammonia leads to replacement of the H atom by potassium, forming a red salt, tritylpotassium: $(C_6H_5)_3CH + KNH_2 \rightarrow K^+[(C_6H_5)_3C^-] + NH_3$. T. is the parent compound of the ***triphenylmethane pigments***.

Triphenylmethane pigments: see Triaryl pigments.

Triphenylmethanol: same as Triphenylcarbinol (see).

Triphenylmethyl-, abb. ***trityl-***: a term for the atomic group $(C_6H_5)_3C$- in a molecule, the $(C_6H_5)_3C\cdot$ radical or the $(C_6H_5)_3C^+$ cation. The ***triphenyl radical*** was the first free organic radical discovered. Its lifetime is long compared to that of the very short-lived alkyl radicals, due to resonance stabilization. That is, the free electron is not localized on the central C-atom (delocalization is demonstrated by the EPR spectrum). The triphenylmethyl radical is formed by the reaction of finely divided silver with triphenylchloromethane in benzene, in the absence of air, and gives the solution a yellow color. A colorless dimer can also form; it was formerly thought to be hexaphenylethane. More recent studies have shown, however, that its structure is the following:

Colorless dimer

In solution, the triphenylmethyl radical is in equilibrium with its dimer, with the position of the equilibrium depending on the concentration, temperature and nature of the solvent. When a benzene solution is evaporated, or acetone is added, the dimer can be precipitated (m.p. 146-147 °C). The radical is very reactive, and forms the colorless triphenylmethyl peroxide with atmospheric oxygen, and triphenyliodomethane with iodine. The ***triphenylmethyl cation*** exists in salt form in triphenylmethyl perchlorate (see Trityl perchlorate), and has become very interesting as a hydride ion acceptor.

2,3,5-Triphenyltetrazolium chloride: an important tetrazolium salt. It crystallizes in colorless, shiny needles; m.p. 243 °C (dec.). It is soluble in water, alcohol and acetone, and insoluble in ether. In light it turns yellow. 2,3,5-T. is synthesized by coupling benzaldehyde phenylhydrazone with benzenediazonium chloride to form triphenylformazane, which is then oxidatively cyclized. 2,3,5-T. is readily reduced to the dark red triphenylformazane, which has a metallic sheen; it is therefore used as a redox indicator, especially for staining cells to locate reducing enzyme systems. It is used to test the germination capacity of seeds, for detection of reducing sugars on paper chromatograms, and as a rapid test for the effectiveness of antibiotics and disinfectants.

2,3,5-triphenyltetrazolium chloride — Triphenylformazane

Triphosphopyridine nucleotide: see Nicotinamide nucleotides.

Triphosphoric acid: see Polyphosphoric acids.

Triple point: a point in the Phase diagram (see) of a pure substance at which three different phases are in thermodynamic equilibrium; see Gibbs phase law.

Triplet: 1) see Multiplet structure; 2) see Code, genetic.

Triplet generator: a species which is capable of highly effective triplet-triplet transfer can induce triplet states for photochemical reactions in molecules in which it is difficult to generate a triplet state directly. T. are compounds in which the Intercombination process (see Photophysical processes) is very effective and in which the triplet state is relatively long-lived and energetic (e.g. arylketones). T. are used, among other things, to decide whether a photochemical reaction occurs via a triplet state by observing its rate in the absence and presence of a T.

Tris, ***tromethamol, tris(hydroxymethyl)methylamine***: $(HOCH_2)_3C\text{-}NH_2$, a water-soluble aminoalcohol, m.p. 170-174 °C. T. is used to make buffer solutions and as an aqueous infusion in cases of blood acidosis.

Tritane: same as Triphenylmethane (see).

Triterpenes: terpenes constructed from 6 isoprene units, which thus contain 30 C atoms. They include the acyclic hydrocarbon Squalene (see), but most are polycyclic compounds. The skeleton of the tetracyclic T. is perhydrocyclopentanophenanthrene, from which the Methylsterols (see) are derived. Most T. are pentacyclic and are derived from the skeleton compounds *1* (perhydropicene) and *2*.

Basic structures of the triterpenes.

Compound *1* is the parent of *oleanane* (CH_3 groups in positions 4,4, 10, 14, 17, 20,20), *ursane* (CH_3 groups in positions 4,4, 10, 14, 17, 19, 20) and *friedelane* (CH_3 groups in positions 4, 5, 9, 13, 14, 17, 20,20); compound *2* is the parent of *hopane* (CH_3 groups in positions 4,4, 8, 10, 14, 18 and isopropyl group at 22) and *lupane* (CH_3 groups in positions 4,4, 8, 10, 14, 17; isopropyl group at 19). The Δ_{12}-derivatives of oleanane and ursane are β-amyrane and α-amyrane, respectively; their 3-hydroxy drivatives are β-*amyrin* and α-*amyrin*. T. are solid, nonvolatile compounds which are not present in volatile oils, but are found free, esterified or as ethers in plant extracts, resins (see Resin acids) and saponins. Methylsterols and their derivatives, the steroids, are also synthesized by microorganisms and animals.

Trithiones: a class of five-membered, heterocyclic compounds derived from 3*H*-1,2-dithiol-3-thione. T. are stable in air and acids, crystallize readily, and are dark red-orange to yellow-orange in color. They are decomposed by alkalies. The parent compound, ***trithione*** (R = H), forms orange prisms; m.p. 79-81 °C. T. are found in nature in cabbage and brussels sprouts. They can be synthesized by reaction of sulfur with activated alkenes or by dehydrogenation of 1,3-dimercaptopropanes with excess sulfur. Because of their affinity for metal surfaces, they are used as inhibitors, e.g. in the treatment of iron with hot hydrochloric acid.

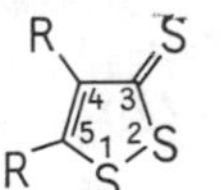

Tritium, ***T***, ^{3}***H***: Hydrogen isotope with mass number 3; the nucleus (*triton*, symbol t) contains one proton and two neutrons; atomic mass 3.01605. T. is a weak β^--emitter, and is converted with a half-life of 12.26 years to the helium isotope ^3He. In the natural isotope mixture of hydrogen, it is present at a concentration of about $10^{-16}\%$. It is formed in the upper layers of the atmosphere as a result of cosmic rays, which produce fast neutrons. These collide with nitrogen: $^{14}N(^1n; {}^4He)^3H$. T. is used mainly for radioactive labelling of organic compounds, which in turn are used to elucidate biochemical processes. The fusion of tritium and deuterium nuclei to form helium, which releases enormous amounts of energy, is the basis of the hydrogen bomb. If the reaction can be controlled, it might make a decisive contribution to the solution of the world energy problem. T. was synthesized in 1934 by Rutherford, who obtained it by bombarding deuterium compounds with deuterons.

Triton: see Tritium.

Trityl-: abb. for triphenylmethyl.

Trityl perchlorate: $(C_6H_5)_3C^+ClO_4^-$, a compound which contains the triphenylmethyl cation. T. is a suitable dehydrogenation reagent for removal of hydride ions, e.g. from cyclohepta-1,3,5-triene, to form the tropylium cation. In such reactions, the perchlorate anion is available as a negative counterion, and triphenylmethane is formed as a byproduct.

Trivial name: see Nomenclature, sect. II A.

Tröger base: an organic compound with two asymmetric tertiary N atoms incorporated into a configurationally stable ring system. The racemic T. crystallizes in colorless needles; m.p. 136-137 °C. Because pyramidal inversion at the N atoms is prevented by the ring structure, the two stereoisomers of T. can be separated.

Troostite: a metallographic term for a pearlite structure in Iron-carbon alloys (see). The distance between the cementite lamellae is so small that they can no longer be resolved by light microscopy. T. is formed at cooling rates between 300 and 500 °C/s, or when the metal is tempered at a temperature below 400 °C.

Tropane: 8-methyl-8-azabicyclo[3.2.1]octane, the skeleton of the Tropane alkaloids (see). ***Nortropane*** has a hydrogen atom instead of the CH_3 group. Tropine and Ψ-tropine are cyclic amino alcohols derived from T. ***Tropine*** (tropan-3α-ol) has the OH-group in the axial position, while Ψ-tropine (tropan-3β-ol) has it in the equatorial position. ***Egonine*** (2β-carboxytropan-3β-ol) is a derivative of Ψ-tropine with an additional axial carboxyl group. T. is a liquid with b.p. 167 °C.

Tropane Tropine Ψ-Tropine Ecgonine

Tropine forms hygroscopic platelets; m.p. 63 °C, b.p. 229 °C. Ψ-Tropine crystallizes in prisms; m.p. 198 °C, $[\alpha]_D^{20}$ -45.4°.

Tropane alkaloids, ***tropa alkaloids***: ester alkaloids in which the Tropane (see) derivatives tropine, Ψ-tropine and ecgonine are esterified to various organic acids. They are biosynthesized from L-ornithine. The ***Solanaceae alkaloids*** are derived from tropine; some important examples are hyoscyamine, atropine and scopolamine. ***Coca alkaloids*** are derived from ecgonine or Ψ-tropine. Cocaine and tropacocaine are important examples. The Solanaceae alkaloids are also called tropine-type alkaloids, while the coca alkaloids are called ecgonine-type alkaloids.

Tropeolin, ***orange IV***: the sodium salt of 4'-anilinoazobenzene-4-sulfonic acid, an azo dye. T. is sold as a yellow-orange, water-soluble powder and is used to dye wool and silk; in inorganic analysis it is used as an indicator for the pH range 1.3-3.2 (color change from red to yellow) and as a reagent for magnesium.

Tropilidene, ***cycloheptatriene***, ***cyclohepta-1,3,5-triene***: an unsaturated hydrocarbon; a colorless liquid; m.p. -79.5 °C, b.p. 117 °C, n_D^{20} 1.5543. T. is insoluble in water, but soluble in ether and ethanol. T. was first synthesized by degradation reactions of the alkaloids atropine and cocaine. It can be synthesized by reaction of benzene with diazomethane. In this reaction, methylene is first added to the benzene ring, forming norcaradiene (bicyclo[4,1,0]hepta-2,4-diene) and then the valence-isomeric cycloheptatriene is immediately obtained. T. can be converted to the tropylium bromide by reaction with bromine; this compound contains the aromatic tropylium cation. T. can be converted to tropolone, which is also aromatic, by treatment with potassium permanganate.

Tropine: see Tropane.

α-Tropolone, ***2-hydroxycyclohepta-2,4,6-trien-one***: a colorless, crystalline compound; m.p. 51-52 °C, sublimation at 40 °C (533 Pa). α-T. is soluble in water and most organic solvents.

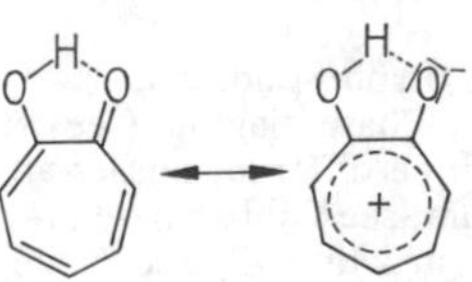

It is aromatic, and does not react with typical carbonyl reagents to form the corresponding ketone derivatives. The cause of this behavior can be understood by considering the resonance structures. α-T. is thus a non-benzoid aromatic compound. It can be nitrated, brominated and coupled with diazonium salts in the same way as benzoid aromatics. α-T. is found in nature only in the form of derivatives, such as colchicine, thujaplicins and purpurogallin. In addition to α-T., the β- (1,3-) and γ- (1,4-)T. isomers are known, but they are less important.

Tropomycin: see Actin.

Tropone, ***cyclohepta-2,4,6-trienone***: a colorless, viscous liquid; m.p. -7 °C, b.p. 104-105 °C at 1.33 kPa, n_D^{22} 1.6172. T. is soluble in water. Unlike other ketones, it does not react with phenylhydrazine and barely reacts with semicarbazide or hydroxylamine. The reason for this behavior is the electron distribution in T., which can give rise to a stable π-electron sextet. The structure of T. is best described by resonance structures. This formulation makes it clear why T. reacts with hydrogen chloride to form an ionic hydroxytropylium chloride.

T. can be synthesized by reaction of bromobenzene with diazomethane under UV irradiation, or by partial hydrogenation of anisol and subsequent reaction with dibromocarbene.

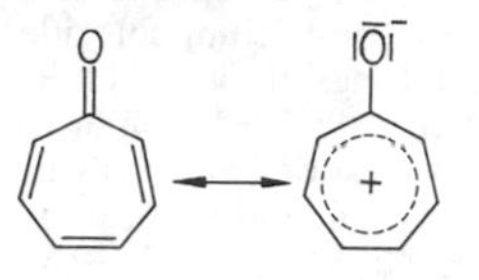

Troponin: see Actin.

Tropylium-: a name for the aromatic carbenium ion $C_7H_7^+$, which is stabilized by a π-electron sextet and exists in the form of salts, e.g. in tropylium bromide. Its existence was predicted in 1931 by Hückel and demonstrated in 1954 by Doering and Knox. Tropylium bromide is obtained by reaction of cyclohepta-1,3,5-triene with bromine, or, still better, from tropylium perchlorate by reaction of the same hydrocarbon with trityl perchlorate, which extracts a hydride ion. The perchlorate is a very sensitive explosive in dry form.

Trotyl: an abb. for 2,4,6-Trinitrotoluene.

Trp: abb. for tryptophan.

TRT technique: see Thin-layer chromatography.

Truxillic acids, ***2,4-diphenylcyclobutane-1,3-dicarboxylic acids***: structural isomers of Truxic acids (see).

Truxic acids, ***3,4-diphenylcyclobutane-1,2-dicarboxylic acids***: alicyclic dicarboxylic acids formed along with Truxillic acids (see) by hydrolysis of coca secondary alkaloids, or by photochemical dimerization of Cinnamic acid (see).

Trypsin: a serine protease formed in the pancreas and secreted into the duodenum. T. consists of 223 amino acid residues (M_r 23,300, optimum pH 7.5 to 8.5); it is released from its inactive precursor, ***trypsinogen***, by removal of an *N*-terminal hexapeptide. The enzyme catalyses the hydrolysis of peptide bonds in which the carboxyl group is provided by lysine or arginine. The serine group at position 183 and the histidine group at position 46 are involved in catalysis. There are a number of trypsin inhibitors, including *soybean trypsin inhibitor*, a plant protein of 181 amino acids (M_r 21,000) which forms an enzymatically inactive complex with T., and organic serine protease inhibitors (see Proteases). In soybean products intended for food, the inhibitor is inactivated by a baking process. Carrier-bound T. is used in sequence analysis of peptide chains and for preparative isolation of proteinase inhibitors.

Tryptamine: 3-(2-aminoethyl)indole, a biogenic amine formed by decarboxylation of tryptophan; m.p. 145-146 °C (dec.). It forms colorless crystals which are slightly soluble in water, and readily soluble in ether and chloroform.

Tryptophan, abb. ***Trp***: α-amino-β-indolylpropionic acid, an essential proteogenic amino acid found in small amounts in proteins. It is completely destroyed by acid hydrolysis of the proteins (formula and physical properties, see Amino acids, Table 1). L-T. is now made from indole and serine, using the enzyme tryptophan synthetase; yields of 80% are obtained, relative to the DL-serine starting material. T. is gaining importance as an additive to nutritional and feed proteins, and in medicine in the treatment of sleep disorders.

Tschitschibabin reaction: see Čičibabin reaction.

Tschugajev's reagent: see Dimethylglyoxime.

TSH: see Thyrotropin.

TTC process: see Cracking.

TTH process: see Hydrorefining.

Tuberculostearic acid, ***10-methylstearic acid:*** $H_3C(CH_2)_7CH(CH_3)(CH_2)_8COOH$, a branched fatty acid which was isolated from the wax of the pathogen of tuberculosis, *Mycobacterium tuberculosis*.

Tuberculostatics, ***antituberculosis compounds***: compounds used to treat tuberculosis caused by *Mycobacterium tuberculosis*. Because the therapy must be continued over a long period, and high doses are sometimes required, the compound must be relatively well tolerated. Because of the danger that a resistant strain will be generated, a mixture of several T. with different types of structure is generally administered. Effective T. were first discovered after World War II, e.g. *p*-aminosalicylic acid in 1946, isoniazide and pyrazinamide (pyrazine carboxamide) in 1952, and ethambutol in 1961. Aminoglycoside type antibiotics, such as streptomycin, and ansamycin type antibiotics such as rifampicin are also used.

Tubocurarine hydrochloride: see Muscle relaxants.

Tuftsin: (named for the site of its discovery, Tufts University in Boston), Thr-Lys-Pro-Arg, a peptide released from a γ-globulin fraction by enzymatic action; it stimulates phagocytosis. Synthetic T. has all the same properties as the natural compound. T. and synthetic analogs are of interest as potential drugs for use against certain infective diseases and tumors. T.

was first isolated in 1970 by Najjar and Nishioka from leukokinin.

Tungstates: in the narrower sense, ***tungstates(VI)***, salts of tungstic(VI) acids, of which the monotungstates, M_2SO_4, are of particular importance. T. are formed by dissolving tungsten(VI) oxide in alkali, ammonia or carbonate solutions. The tetrahedral monotungstate ion $[WO_4]^{2-}$ is present only in very alkaline solutions. If the H^+ concentration is increased, at pH > 5 there is reversible formation of hexatungstate(VI) ions, $[HW_6O_{21}]^{5-}$, which are present in hydrated form; an example is the "paratungstate ion A", $[H_7W_6O_{23}]^{5-}$. These in turn are slowly equilibrated with dodecatungstate ions $[W_{12}O_{41}]^{10-}$ (hydrated as "paratungstate ions Z", e.g. $[H_2W_{12}O_{43}]^{10-}$ and $[H_8W_{12}O_{46}]^{10-}$). In the pH range <5, metatungstate ions $[W_{12}O_{39}]^{6-}$ are formed, while tungstamide hydrates precipitate at pH < 1.5. By combination of polytungstates with suitable reactants, such as non-metal acids, heteropolytungstates can be obtained.

In the broad sense, the T. also include anionic acido complexes of tungsten, e.g. the 8-fold coordinated octacyanotungstates(VI), $M_4[W(CN)_8]$.

Tungsten, symbol ***W***: a chemical element from group VIb of the periodic system, the Chromium group (see); a heavy metal, *Z* 74, with natural isotopes with mass numbers 184 (30.64%), 186 (28.41%), 182 (26.41%), 183 (14.40%), 180 (0.135%); atomic mass, 183.5, valence usually VI, but also V, IV, III, II, 0, density 19.26, m.p. 3410°C, b.p. 5660°C, electrical conductivity 17.8 sm mm^{-2}, standard electrode potential (W/W^{3+}) -0.11 V.

Properties. T. is a shiny, white metal which, when very pure, is ductile and easily shaped; it crystallizes in a cubic body-centered lattice. Even small amounts of carbon or oxygen makes the metal very hard and brittle. At room temperature, it is completely stable in the presence of dry air. At red heat, it burns to tungsten(VI) oxide, WO_3. Steam will oxidize glowing T. to tungsten(IV) oxide, WO_2. T. does not react easily with nitrogen, but fluorine oxidizes it even at room temperature to tungsten hexafluoride, WF_6. The other halogens react with T. only at higher temperatures, forming tungsten hexachloride, pentabromide or triiodide, respectively. T. powder reacts with ammonia at high temperatures, forming the nitride. Acids, even aqua regia and hydrofluoric acid, react only slowly with T., due to passivation. However, T. is soluble in a mixture of hydrofluoric and concentrated nitric acids, and also in alkali hydroxide melts.

In T. compounds, the most common state is +6, but +5, +4, +3 and +2 are also important; they are represented by tungsten(V) bromide, octacyanotungstate(IV), and the cluster compounds W_6Cl_{18} and W_6Cl_{12}. The oxidation states 0 and -2 are present in the carbonyl compounds $W(CO)_6$ and $[W(CO)_5]^{2-}$. In T. complexes, the most common coordination numbers are 6 to 8.

Analysis. An important indicator reaction for T. is based on the reduction of tungsten(VI) compounds to Tungsten blue (see). For colorimetric determination of T., tungstates(VI) are reduced with tin(II) chloride in the presence of potassium thiocyanate, forming the yellow thiocyanatotungstates(V). For gravimetric determination of T., tungsten(VI) oxide, mercury(I) tungstate and tungsten oximates are used.

Occurrence. T. makes up $1.3 \cdot 10^{-4}$% of the earth's crust. The most important ores are tungstates, e.g. wolframite (Mn,Fe)(WO_4), an isomorphic mixture of berberite, $FeWO_4$ and hubnerite, $MnWO_4$, scheelite, $CaWO_4$, and stolzite, $PbWO_4$. Tungstite (tungsten ocher) is a weathering product of wolframite, $WO_3 \cdot xH_2O$.

Production. Tungsten ores, usually wolframite or scheelite, are first enriched by flotation or an electromagnetic process. They are solubilized by melting them with sodium carbonate in a furnace at 800°C, and the sodium tungstate formed in this step is then leached with water, or pressure leached with aqueous sodium hydroxide solution. Tungstic acid is either precipitated directly by addition of hydrochloric acid to the sodium tungstate solution, or calcium chloride is first added to obtain calcium tungstate as an intermediate, which is then converted by reaction with hydrochloric acid to tungsten(VI) oxide hydrate. This is filtered off, dehydrated by heating, and reduced by hydrogen to T. at 1000 to 1200°C: $WO_3 + 3\,H_2 \rightarrow W + 3\,H_2O$. In this reaction, T. is produced as a powder with a purity of about 98%; the metal can be obtained in compact form by sintering in a hydrogen atmosphere and hammering. T. can also be obtained by reaction of tungsten(VI) sulfide with calcium oxide in an electric arc. The metal produced in this way is not completely pure, and is therefore hard and brittle. It can be made ductile by hammering, and then drawn into thin wires. Pressed threads, when heated to about 2500°C in the presence of thorium dioxide, can form meter-long single crystal fibers which are very flexible. The T. used to make T. steel is made in the form of ***ferrotungsten*** by reduction of T. and iron mixtures with carbon in an electric furnace.

Applications. About 90% of the world production of T. is utilized in the form of ferrotungsten to make T. steels. Of the metals, T. has the highest melting and boiling points; its low vapor pressure and high melting point are reasons for its use to make lightbulb filaments and cathodes and anticathodes for electronic tubes. T. is used for thermoelements, rocket valves and heat shields for space vehicles. Highly stressed electric contacts are usually made of T., to which copper and silver are added before sintering. T. wires are used for production of group IVb and Vb metals by the vacuum deposition method. The use of T. to make hard metals and cutting tools is of special importance. In its alloys (see Tungsten alloys), T. transfers its properties to these systems, to a greater or lesser extent.

Historical. In 1781, Scheele detected tungsten(VI) oxide as a component of Scheelite. The Spanish scientists Fausto and Joseph d'Elhujar discovered the oxide in wolframite as well, and in 1783, prepared the element. The name "wolfram" was coined by Berzelius, while the name "tungsten" was supplied by the Swede Gronstedt, who in 1768 named a heavy T. mineral "heavy stone" (in Swedish, "tungstein").

Tungsten alloys: Multicomponent systems in which tungsten is the main component. High-tungsten alloys with silver, copper or nickel (see Tungsten bronzes) are used for electrical contacts, and because of their high densities, are used for counterweights, flywheels or damping members in physics apparatus. They were used as drive masses in automatic wristwatches. These alloys are produced by sintering powdered mixtures or impregnation of a previously sintered tungsten matrix with metal. To prevent recrystallization of tungsten wires in light bulbs, small amounts of oxides (thorium, silicon, aluminum oxides) are added. Tungsten is used as an alloy element in steels; it is added in the form of ***ferrotungsten***, which consists of 81 to 83% tungsten.

Tungsten blue: a collective term for bright blue tungsten(IV)-tungsten(VI) mixed oxides, $WO_{3-x}(OH)_x$ (for example, $x = 0.35$ or 0.5), obtained by reduction of freshly precipitated tungsten(VI) oxide hydrate with tin(II) chloride or zinc and hydrochloric acid. A very sensitive test for tungsten is based on the formation of T. T. is used as an aquarell pigment which is stable to light and air.

Tungsten bronzes: 1) intensely colored, shiny, semimetallic mixed compounds with the general formula Na_xWO_3 (x = 0-1). If x is 0.3, the compound is blue-violet; if $x = 0.9$, the compound is golden yellow. T. crystallizes very readily and combines a metallic appearance with high chemical inertness. It is made by reduction of molten sodium tungstates with hydrogen, zinc or tungsten, or electrochemically. T. are used as coating paints.

2) ***Tungsten special bronze***, a tungsten-copper alloy consisting of 33.3 to 66.6% tungsten and the rest copper; it is used for welding electrodes. It is produced by powder metallurgical techniques.

Tungsten carbides: ***Tungsten carbide***: WC, gray crystals with a metallic luster; M_r 195.86, density 15.6, m.p. about 2870 °C, b.p. about 6000 °C. WC is about as hard as diamond, and is the most important carbide in the production of hard metals. ***Ditungsten carbide***, W_2C, forms gray, very hard hexagonal crystals; M_r 379.71, density 17.15, m.p. about 2860 °C, b.p. about 6000 °C. In the T., the tungsten atoms are in a close-packed array, with either half (W_2C) or all (WC) of the octahedral holes occupied by C atoms. Both T. are very inert, but W_2C, unlike WC, is soluble in a mixture of hydrofluoric and nitric acids. The T. are made by reaction of tungsten or tungsten(VI) oxide with carbon at high temperatures, and are used as hardeners for hard metals.

Tungsten carbonyl, ***tungsten hexacarbonyl***: $W(CO)_6$, colorless, volatile, rhombic crystals; M_r 351.91, density 2.65, m.p. 180 °C (dec.); octahedral molecules. The compound is obtained by reaction of carbon monoxide with tungsten(VI) chloride under pressure, using aluminum as a reducing agent.

Tungsten chlorides: ***Tungsten(VI) chloride, tungsten hexachloride***, WCl_6, is a dark violet, crystalline compound soluble in organic solvents, and susceptible to hydrolysis; M_r 369.57, density 3.52, m.p. 275 °C, b.p. 346.7 °C. WCl_6 is obtained by passing chlorine over hot tungsten powder. Above its boiling point, it decomposes into ***tungsten(V) chloride, tungsten pentachloride***, WCl_5, which forms dark green, crystalline needles; M_r 361.12, density 3.875, m.p. 248 °C, b.p. 275.6 °C. Like Niobium(V) chloride (see), WCl_5 is probably dimeric in the solid state. It is obtained by repeated distillation of WCl_6 in a stream of hydrogen. A byproduct is ***tungsten(IV) chloride, tungsten tetrachloride***, WCl_4, a black, hygroscopic compound; M_r 325.66, density 4.624. WCl_4 disproportionates when heated ($3\,WCl_4 \rightarrow WCl_2 + 2\,WCl_5$). The resulting gray, hexameric ***tungsten(II) chloride***, density 5.436, is a strongly reducing cluster compound, $[W_6Cl_8]Cl_4$, in the cations of which the tungsten atoms are arranged octahedrally, and the chlorine atoms occupy the corners of a cube. Reduction of WCl_6 with aluminum in a temperature gradient from 475 to 200 °C yields ***tungsten(III) chloride***, a black, hexameric compound which is a cluster, $[W_6Cl_{12}]Cl_6$, with tungsten in the octahedral positions and chlorine on the octahedral edges.

Tungsten oxides: the most important W. is ***tungsten(VI) oxide, tungsten trioxide***, WO_3, a crystalline powder which is intensely yellow at room temperature and turns orange when heated; M_r 231.85, density 7.16, m.p. 1473 °C. WO_3 is volatile above 1750 °C. The crystal lattice of rhombic WO_3 consists of WO_6 octahedra which are connected in all three dimensions by shared vertices. WO_3 is completely insoluble in water and acids, but it reacts with bases to form tungstates. When the tungstate solutions are acidified, tungsten(VI) oxide hydrate gels precipitate out. WO_3 is obtained by roasting tungsten or tungstate compounds in the presence of air. It is used as a contact and as a yellow pigment for ceramics. If WO_3 is reduced with hydrogen at high temperatures, it forms $W_{10}O_{29}$ (blue-violet) and W_5O_{14} (red-violet) as intermediates, and is finally converted to brown ***tungsten(IV) oxide, tungsten dioxide***, WO_2, M_r 215.85, which is insoluble in water, acids and bases; it crystallizes in a rutile lattice. WO_2 sublimes around 800 °C, and is relatively easily oxidized to WO_3 or reduced to tungsten.

Tungsten(VI) oxide hydrate: $WO_3 \cdot xH_2O$, often called ***tungstic acid***, although this is not strictly correct. It forms as a yellow or white precipitate when tungstate solutions are acidified. Precipitation at low temperatures frequently first produces white T., which is then converted on standing or heating into the defined yellow hydrates, $WO_3 \cdot 2H_2O$ and $WO_3 \cdot H_2O$. These hydrates do not contain discrete H_2WO_4 molecules.

Tungsten sulfides: ***Tungsten(IV) sulfide, tungsten disulfide***, WS_2, gray-black, hexagonal crystals; M_r 247.98, density 7.5, m.p. 1250 °C (dec.). WS_2 is chemically inert and is soluble only in a mixture of nitric and hydrofluoric acids. It readily burns in air to form tungsten(VI) oxide, WO_3. When it is heated in the absence of air to very high temperatures, the sulfur splits off and tungsten is formed. WS_2 is obtained by heating WS_3 in the absence of air, or by melting a mixture of WO_3, potassium carbonate and sulfur. ***Tungsten(VI) sulfide, tungsten sulfide***, WS_3, is a dark brown powder which readily forms colloidal solutions; M_r 280.04. When heated in the air, WS_3 is converted to WO_3. It is obtained by acidifying thiotungstate solutions, which in turn are made by passing hydrogen sulfide through tungstate solutions.

Tungsten type, ***tungsten structure***: a structure type in which the atoms are in a cubic body-centered lattice (see Crystal). The coordination number is 8, if only nearest neighbors are considered; if the next-nearest neighbors, which are only 16% farther away, are included, it is 14. The packing density, assuming spherical atoms (see Packing of spheres), is 0.68. In addition to the prototype tungsten, more than 20% of all known metallic element modifications crystallize in the T., e.g. the alkali metals, the heavier alkaline earths calcium, strontium and barium, the transition metals vanadium, niobium, tantalum, molybdenum and iron (up to 1197 K as α-Fe, and above 1677 K as δ-Fe).

Tungstic acid: see Tungsten(VI) oxide hydrate.

Turbidity measurement: see Nephelometry.

Turbidity point: a characteristic temperature for non-ionic surfactants in aqueous solution at which desolvation of the ether oxygen leads to separation into a water-poor, surfactant-rich phase and a surfactant-poor, water-rich phase. As the number of oxyethylene groups increases, the T. is shifted to higher temperatures.

Turkish red: an aluminum-calcium color lake of alizarin. It is formed when cotton which has been treated with turkish red oil and alumina is boiled with lime and alizarin. The alizarin gives a pink color with a tin mordant, a violet color with an iron mordant, and a brown with chromium mordant.

Turkish red oil is a dispersive consisting of ricinus oil, ricinolic acid, ricinolic anhydrides, lactones and sulfates, polyricinol and dihydroxystearic acid derivatives.

Turpentine: a colorless to yellowish, thin essential oil with a pungent odor; it burns readily. The main components are α- and β-pinene. T. is obtained by steam distillation of balsams, that is, the sticky exudates from wounded bark of conifers, especially pine or larch, but also fir, spruce or pistachio. The colophony, which is the other main component of the balsam is left as a residue. Some special T. are pine oil, wood T. (obtained by steam distillation of stumps) and sulfate or sulfite T. (byproducts of the sulfite or sulfate methods of cellulose production).

T. is rapidly changed in the presence of air, as it takes up oxygen and becomes viscous, then forming a brittle mass of resin. Its main components are α- and β-pinene and cavene. When T. vapor is passed over a glowing platinum wire, isoprene is formed; T. is also a cheap raw material for isolation of camphor. It is used mainly to dissolve resins, oils and rubber, to prepare lacquers, varnishes and oil paints, for products such as shoe polish and floor wax, and for pharmaceutical purposes.

Turnbull's blue: a coordination polymeric compound which is identical to Prussian blue (see); $Fe^{III}_4[Fe^{II}(CN)_6]_3 \cdot nH_2O$. It was first believed to have a structure different from that of Prussian blue. T. is formed by reaction of iron(II) salt solutions with potassium ferrocyanate, $K_3[Fe(CN)_6]$. The first step of the reaction is an electron transfer (Fe^{2+} + $[Fe(CN)_6]^{3-} \rightarrow Fe^{3+} + [Fe(CN)_6]^{4-}$), which is followed by combination of Fe^{3+} and $[Fe(CN)_6]^{4-}$ to form Prussian blue.

Turquois green: same as Cobalt green (see).

Twinning plane: see Crystal lattice defects.

Twins: see Crystal.

Twistane, ***tricyclo[4.4.0.0³,8]decane***: a twisted hydrocarbon; colorless crystals, m.p. 163-165 °C. Its synthesis is relatively complicated. It is especially interesting that T. is optically active, because of its molecular dissymmetry. The formula shows the (+)-enantiomer.

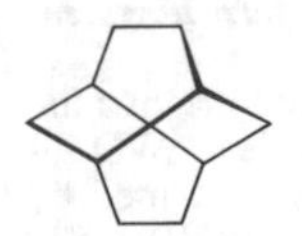

Twist displacement: see Crystal lattice defect.

Twist form: see Stereoisomerism 2.2.

Tyndall effect, ***Faraday-Tyndall effect***: In 1857, M. Faraday discovered that a beam of light is scattered when it passes through a colloid or true solution. J. Tyndall determined in 1863 that the scattered light is polarized. If an electromagnetic wave propagates through a medium, the molecules or atoms of the medium are polarized, and the positive and negative charge centers vibrate with respect to each other. This constitutes an oscillating electric dipole which emits light of the same wavelength as the exciting light. In an isotropic crystal, in which all the atoms are in their rest positions, scattered light is not emitted, because for each volume element which emits light, there is another which emits light 180° out of phase with the first; the scattered light is thus quenched by interference. However, if the polarization of the light emitted from microregions (such as colloidal particles) varies, scattering is observed. Scattered light is measured by Nephelometry (see).

Tyndallometry: see Nephelometry.

Type metal: a term for lead-antimony-tin alloys with added copper and nickel; they are used mainly to make type for printing. ***Letter metal*** is used to make complete type casts (e.g. with a composition of 28% antimony, 5% tin, 0.3% copper and nickel, the remainder lead). ***Linometal*** is used for linotype and other typesetting machines (e.g. 12% antimony, 5% tin, 0.05% copper and nickel). ***Note metal*** is used to make engraved plates for musical scores (e.g. 16% tin, 4.5% antimony, 0.3% copper and nickel, the rest lead).

Tyr: abb. for tyrosine.

Tyramine, ***4-hydroxyphenylethylamine***: a biogenic amine formed by decarboxylation of tyrosine; m.p. 164-165 °C. It forms colorless crystals and is only slightly soluble in water.

Tyrocidins: peptide antibiotics produced by *Bacillus brevis*. T. are homodetic, homeomeric cyclic decapeptides. They contain the same pentapeptide sequence which is found in gramicidin S but differ in the second pentapeptide. The primary structures of T. A through E are:

	1	2	3	4	5	6	7	8	9	10	
A cyclo-(	–Val–	Orn–	Leu–	D–Phe–	Pro–	Phe–	D–Phe–	Asn–	Gln–	Tyr–	)
B cyclo-(	–	–	–	–	–	Trp	–	–	–	–	)
C cyclo-(	–	–	–	–	–	Trp–	D–Trp	–	–	–	)
D cyclo-(	–	–	–	–	–	–	–	–	–	Phe–	)
E cyclo-(	–	–	–	–	–	–	–	Asp	–	Phe–	)

T. mixtures, often combined with about 20% gramicidin, are used to treat skin infections and infections of the oral cavity (tyrothricin). Open-chain analogs of the T. are biologically inactive.

Tyrosine, abb. ***Tyr***: α-amino-β-(*p*-hydroxyphenyl)propionic acid. As a proteogenic amino acid, Tyr occurs widely in nature (formula and physical properties, see Amino acids). T. is now obtained entirely from protein hydrolysates, from which it is easily isolated as the least soluble amino acid. L-T. is a starting material for synthesis of ***L-dopa (3,4-dihydroxyphenylalanine)***, a drug used to treat Parkinson's disease (see Antiparkinson compounds).

Tyrothricin: see Tyrocidins.

U

u: unit symbol for atomic mass units.

U: 1) symbol for uranium; 2) abb. for uridine.

Ubiquinone, ***coenzyme Q***, abb. ***Q***: a 2,3-dimethoxy-5-methylbenzo-1,4-quinone with an isoprenoid side chain in position 6. Individual U. differ in the number *n* of isoprene units in their side chains (U.-*n*, *n* = 6-10). These compounds were formerly also indicated by the number of C atoms in the side chain. U.-10 is found in higher animals and plants; the shorter-chain U. are found in lower animals and plants. U.-6, for example, is present in yeast. Many bacteria have ***menaquinone*** instead of U. The U. are part of the electron transport chain of the cell.

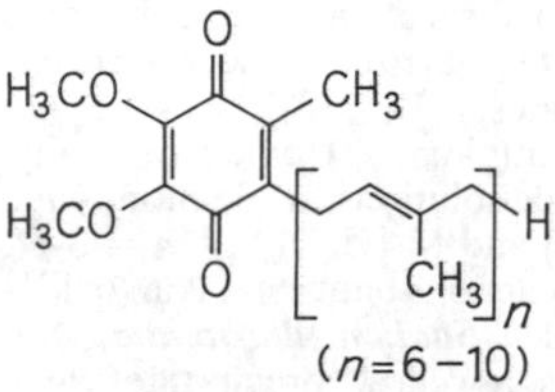

(*n*=6−10)

UDP: see Uridine.

Ugi reaction: a reaction for synthesis of α-aminocarboxamides in which an aldehyde and an amine react with an acid to form a carbenium-immonium salt; this then reacts with an isonitrile. Hydrolysis of the unstable intermediate forms the product:

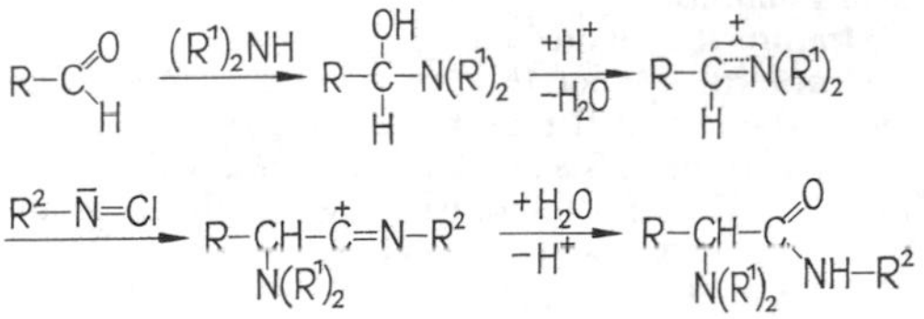

$$\text{R-CHO} \xrightarrow{(\text{R}^1)_2\text{NH}} \text{R-CH(OH)-N(R}^1)_2 \xrightarrow[-\text{H}_2\text{O}]{+\text{H}^+} \text{R-CH=}\overset{+}{\text{N}}(\text{R}^1)_2$$

$$\xrightarrow{\text{R}^2\text{-}\bar{\text{N}}\text{=C|}} \text{R-CH(N(R}^1)_2)\text{-}\overset{+}{\text{C}}\text{=N-R}^2 \xrightarrow[-\text{H}^+]{+\text{H}_2\text{O}} \text{R-CH(N(R}^1)_2)\text{-C(=O)-NH-R}^2$$

Ketones or α-ketocarboxylic acid derivatives can be used in place of aldehydes, while ammonia, primary or secondary amines, hydrazine or Schiff's bases can be used as the amine component.

Ulich approximations: equations for approximate thermodynamic calculations of equilibrium constants (see Mass action law). For reactions at constant pressure, the ***First U.*** follows from the equation $\Delta_R G^o = -RT \ln K_e$ and the Gibbs-Helmholtz equation (see): $\ln K_e = -\Delta_R H^o/RT + \Delta_R S^o/R$. Here K_e is the equilibrium constant at temperature T, R is the gas constant and $\Delta_R H^o$ and $\Delta_R S^o$ are the molar standard enthalpy and entropy. The latter two parameters are taken from tables of thermodynamic data. The first U. is based on the assumption that $\Delta_R H^o$ and $\Delta_R S^o$ are independent of temperature, which is a rather rough approximation. It applies best to gas reactions in which the number of moles does not change, and to these at high temperatures; thus the tabulated values should be chosen for the temperature which is closest to the temperature T of interest.

In the ***second U.***, the temperature dependencies of $\Delta_R H^o$ and $\Delta_R S^o$ are taken into account by using the Kirchhoff law (see) for $\Delta_R H^o$ and the equation $\delta(\Delta_R S^o)/\delta T)_p = \Delta C^o/T$, so that the first U. becomes:

$$\ln K_a = -\frac{\Delta_R H^0}{RT} + \frac{\Delta_R S^0}{R} + \frac{\Delta C_p^0}{R}\left(\ln\frac{T}{T_0} + \frac{T_0}{T} - 1\right).$$

Here $\Delta C_p^o = \sum_i \vartheta_i C_{pi}$, the sum of the molar heats of all reactants at temperature T_0, is treated as independent of temperature.

In the ***third U.***, the temperature dependence of ΔC_p^o is taken into account. It can be used only with data tables if these indicate the temperature dependence of the molar heats of all components, e.g. in the form of exponential series (see Molar heats).

A much less accurate approximation is the ***Nernst approximation***: $\ln K_e = -\Delta_R H^o/RT + 1.75\, \Delta\nu_i \ln T + 2.303\, \Sigma C_i$. Here $\Delta\nu_i$ is the difference in numbers of moles of products and reactants, and C_i are chemical constants which are derived from well examined equilibria and are available in tables.

Ullmann reaction: a means of synthesizing biphenyl and its derivatives by heating iodoarenes, such as iodobenzene, with copper powder in a high-boiling solvent, such as toluene, dimethylformamide, nitrobenzene or naphthalene. The less reactive bromo-

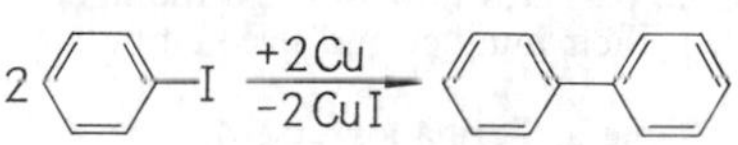

$$2\ \text{C}_6\text{H}_5\text{-I} \xrightarrow[-2\text{CuI}]{+2\text{Cu}} \text{C}_6\text{H}_5\text{-C}_6\text{H}_5$$

Iodobenzene Biphenyl

and chloroarenes can only be used if they contain activating groups, predominantly nitro groups. For example, it is possible to react picryl chloride (1-hydro-2,4,6-trinitrobenzene) with iodobenzene to make a correspondingly substituted biphenyl. The reaction can be inhibited by steric hindrance or by electron donor groups. Reactive functional groups (OH, NH_2, COOH, etc.) also interfere with the reaction and lead to side reactions.

Ulminite: see Macerals.

Ultracentrifuge: a special type of centrifuge with rotation speeds up to 10^6 min^{-1}. This generates a centrifugal force one million times stronger than that of the earth's gravitational field. There are ***preparative*** and ***analytical U.***. With the latter, molecular masses and molecular mass distributions of dissolved molecules and particle masses of colloids can be determined. Recently, the interactions between colloidal particles have been studied in U.

Ultrafiltration: 1) A method for separation of colloids from solutions, using filters with pore sizes smaller than the size of the colloidal particles. These are usually membranes of nitrocellulose, acetylcellulose or polyvinylalcohol. U. may be combined with dialysis or electrodialysis.

2) a method based on reverse osmosis (see Osmosis) used for desalinating water. The water is pressed through specially prepared membranes, e.g. of cellulose acetate, which are impermeable to the dissolved salts. The pressure applied must be greater than the osmotic pressure of the salt solution. This method has so far been successful only with brackish water; the desalination of seawater using U. must await the development of stronger membranes and at present is only occasionaly used.

Ultraforming process: see Reforming.

Ultramarine: an important pigment, a sulfur-containing sodium aluminum silicate, $Na_8[Al_6Si_6O_{24}]S_{2-4}$, with variable composition. U. is usually blue, but can also be green, red or violet; it is nontoxic and is fast to light, air and soaps. It is insoluble in any solvent, except in acids. It is found in nature as azure (lapis lazuli), and was formerly obtained from this source. Now it is synthesized by heating a mixture of white clay, quartz, soda, sulfur and charcoal. The mass is ground, boiled in water, and made into a slurry. U. is used as a pigment for oil and water colors, colored paper, wallpaper, rubber products and synthetic resins, to blue laundry (bluing), feathers, starch and sugar, and as a sun coating on glass windows.

Ultramicroscope: a device built in 1903 by Siedentopf and Zsigmondy for study of colloids (e.g. emulsions, sols and aerosols), which are not visible in the light microscope because of the small size of their particles. The light scattered by the particles is measured, which makes them appear spherical regardless of their actual shape. In the older versions, the sample is observed perpendicular to the incident light. In newer versions, the forward separation at small scattering angles is measured, so the primary beam must be blocked. If the particles flow through the lighted volume (***flow U.***), their number and size can be analysed.

Ultraperm®: same as Permalloy® (see).

Ultraviolet spectroscopy: see UV-VIS spectroscopy.

Umbelliferone: see Coumarin.

Umber: a natural brown clay pigment formed by weathering of iron and manganese ores. It contains about 25 to 35% iron(III) oxide, Fe_2O_3, 7 to 14% manganese(III) oxide, Mn_2O_3, 7 to 14% aluminum oxide, Al_2O_3, 20 to 30% silicon dioxide, SiO_2, 4 to 8% calcium carbonate and 10 to 17% water. U. is used in artist's and ordinary paints. If the paint has a linseed oil base, the manganese oxides act as catalysts for drying.

UMP: see Uridine.

Undecane, ***hendecane***: $CH_3(CH_2)_9CH_3$, an alkane hydrocarbon and a colorless, combustible liquid; m.p. -25 °C, b.p. 195 °C, n_D^{20} 1.4172. It is very soluble in ethanol and ether, and is insoluble in water. *n*-U. is found in a few essential oils, in American petroleum and in coal tar. It can be synthesized by Clemmensen reduction of undecanone.

Unifining process: see Hydrorefining.

Unit cell: see Crystal.

Unit operations: see Process technology.

Unit processes: see Process technology.

Uperolein: see Tachykinins.

Ura: abb. for uracil.

Uracil, abb. ***Ura***: 2,4-dihydroxypyrimidine, a nucleic acid base; colorless crystals, m.p. 335 °C (dec.). It is soluble in hot water and in alkali hydroxide solutions, with salt formation; it is scarcely soluble in cold water, and is practically insoluble in alcohol and ether. U. can be synthesized from alkylisothiourea and formylacetic acid esters via alkylmercaptopyrimidine. It occurs naturally as the phosphate ester of uridine. U. is a component of the ribonucleic acids.

Uracil 6-carboxylic acid: see Orotic acid.

Uramil, ***dialuramide, 5-aminobarbituric acid, 5-aminopyrimidine-2,4,6-triol***: colorless crystals which turn red on standing in air; m.p. 400 °C. U. is insoluble in cold water and ether, but is soluble in acids and bases.

Uraniates: in the broad sense, anionic complexes of uranium, and in the narrow sense, ***oxouraniates(VI)***. The diuraniates, $M_2[U_2O_7]$, can be obtained by reaction of uranium(VI) oxide or uranyl salts with alkali hydroxide solutions, while mono- and polyuraniates, $M_2[UO_4]$ and $M_2[U_nO_{3n+1}]$ (n = 3-6), can be made by fusion reactions between alkali oxides and uranium(VI) oxide. ***Sodium diuraniate***, $Na_2[U_2O_7].6H_2O$, forms as a yellow precipitate when sodium hydroxide is added to uranyl salt solutions, and is used as a pigment for glass and porcelain. ***Ammonium diuraniate*** is usually represented by the formula $(NH_4)_2U_2O_7$, although recent studies have shown that it is hydrated uranyl hydroxide containing ammonium ions. It is formed as a yellow, powdery precipitate by addition of ammonia to uranyl salt solutions, and is readily soluble in ammonium carbonate solutions.

Uranin: see Fluorescein.

Uranium, symbol ***U***: a radioactive chemical element from group IIIb of the periodic system and an Actinoid (see); it is a heavy metal, *Z* 92, with natural isotopes with the following mass numbers (abundance, decay type and half-life in parentheses): 238 or U. I (99.2739%, α, $4.51 \cdot 10^9$ a); 235 or actinouranium (0.7205%, α, $6.96 \cdot 10^8$ a); and 234 or U. II (0.0056%, α, $2.44 \cdot 10^5$ a). Synthetic isotopes have mass numbers 226 (α, 0.5 s), 227 (α, 1.1 min); 228 (α, K-capture, 9.2 min); 229 (K, α, 58 min); 230 (α, 20.8 d); 231 (K, α, 4.2 d) 232 (α, 71.7 a); 233 (α, $1.59 \cdot 10^5$ a); 236 (α, $2.342 \cdot 10^7$ a); 237 (β^-, 6.75 d); 239 (β^-, 23.54 min); 240 (β^-, 14.1 h). The atomic mass is 238.029, valency is usually VI or IV, less often V or III; density 18.97, m.p. 1132 °C, b.p. 3900 °C, standard electrode potential U/U^{3+}) -1.789 V.

Properties. U. is a silvery white, lustrous metal which rapidly becomes coated in air; it is soft and dense, and exists in three modifications, rhombic, tetragonal and cubic body-centered. The conversion temperatures are 667 °C and 772 °C. Finely divided U. appears gray to black; it is very reactive and pyrophoric. U. reacts with boiling water with evolution of hydrogen: $U + 2\,H_2O \rightarrow UO_2 + 2\,H_2$; dilute acids also react with it to form hydrogen and uranium(IV) salts, while alkali hydroxides do not react with it. Hydrogen reacts at 250 to 300 °C to form uranium hydride, UH_3, which is an important intermediate in the synthesis of other U. compounds. Thermal decomposition of the hydride at higher temperatures produces a particularly reactive form of U. U. reacts with nitrogen above 450 °C to form uranium nitrides, UN and U_2N_2. Even at moderate temperatures, U. burns in air to uranium(IV,VI) oxide, U_3O_8, sending off a shower of sparks. Reaction with halogens leads to uranium halides, e.g. UF_4 and UF_6, UCl_5 and UCl_4, UBr_4 and UI_3.

In its compounds, U. prefers the +IV and +VI oxidation states; uranium(III) and uranium(V) compounds are rapidly oxidized in air. The chemstry of U. is rather similar to that of the actinoid elements which follow it, neptunium, plutonium and americium; like these, it forms linear dioxometal cations in the +V and +VI oxidation states. These are characterized by a pale lavender (UO_2^+) or yellow (UO_2^{2+}) color. U. and its compounds are very toxic.

All the naturally occurring U. isotopes are radioactive. The most abundant of them, ^{238}U (U. I) is the first member of the uranium-radium decay series. It is an α-emitter and decays via ^{234}Th (also known as U. X_1), ^{234}Pa (also called U. X_2 or U. Z) ^{234}U (U. II) and a number of further intermediates to the stable nuclide ^{208}Pb (uranium lead). ^{235}U is the first member of the uranium-actinium decay series, and as an α-emitter, it decays via ^{231}Th (also called U. Y) and a number of other intermediates to the lead nuclide ^{207}Pb (actinium lead). Unlike ^{238}U, ^{235}U can be induced to split by slow neutrons. The intermediate nucleus ^{236}U formed by capture of a neutron fissions into two nuclei with mass numbers around 95 and 138, releasing energy and more neutrons. The following equations represent possible fission events: $^{236}U \rightarrow {}^{92}Kr + {}^{142}Ba + 2\,{}^{1}n$ and $^{236}U \rightarrow {}^{90}Sr + {}^{143}Xe + 3\,{}^{1}n$. Each fission event releases two or three neutrons, which permit the process to continue in a chain reaction. When ^{238}U captures a neutron, it forms ^{239}Pu according to the equation:

$$^{238}U \xrightarrow{+\,^{1}n} {}^{239}U \xrightarrow[23.54\ min]{-\beta^-} {}^{239}Np \xrightarrow[2.355\ d]{-\beta^-} {}^{239}Pu.$$

This reaction is utilized in breeder reactors to generate plutonium.

Analysis. In the classical cation separation scheme, U. is found in the urotropin group as ammonium diuraniate, $(NH_4)_2U_2O_7$, which is roasted to U_3O_8 prior to gravimetric determination. Some characteristic diagnostic reactions are the rather sensitive precipitation of brown uranyl hexacyanoferrate(II) $(UO_2)_2[Fe(CN)_6]$, with potassium hexacyanoferrate(II), or the formation of yellow-orange peroxouraniate, $M_4[UO_2(O_2)_3]$ in alkaline solution. Thiocyanate complexes are used for spectrophotometric determination. With the aid of fluorimetry, the limit of detection is about 10^{-10} g U.

Occurrence. U. makes up $3.2 \cdot 10^{-4}$% of the earth's crust. It is always found in the form of compounds, but of the 150 U. minerals known, only a few are economically significant. The most important is uraninite (uranium pitchblende), UO_2. Other U. minerals are torbernite (uranium mica), $Cu(UO_2)_2(PO_4)_2 \cdot 8H_2O$, zeunerite, $Cu(UO_2)_2(AsO_4)_2 \cdot 8H_2O$ and autunite (uranite), $Ca(UO_2)_2(PO_4)_2 \cdot 8H_2O$. U. is also found in euxenite, (U, Th, Ce, Ca, Y)(Nb, Ta, Ti)$_2$O$_6$ and carnotite, $KUO_2[VO_4] \cdot 1.5H_2O$. Brown coal, some gold ores, phosphates and oil shales can also contain U. Seawater contains about 0.002 ppm U. Ores with a concentration higher than 0.1% are now economically useful.

Production: U. is extracted from the ore by an oxidizing acid or alkaline leaching. After acid leaching with sulfuric acid, the companion metals are removed by absorption of the uranylsulfate complexes (e.g. of the type $[UO_2(SO_4)_2]^{2-}$) on anion-exchange resins. Alkaline leaching with a sodium carbonate/sodium hydrogencarbonate mixture is sometimes carried out under pressure and at elevated temperature: $UO_2 + 1/2\,O_2 + Na_2CO_3 + 2\,NaHCO_3 \rightarrow Na_4[UO_2(CO_3)_3] + H_2O$. The formation of the anionic carbonate complex is rather selective; it is followed by addition of NaOH to precipitate sodium diuraniate, $Na_2U_2O_7$. For further purification, the intermediate products are converted to uranyl nitrate, which is extracted from aqueous solution with tributylphosphate. Uranium(IV,VI) oxide, U_3O_8, is obtained from the solution and reduced with hydrogen to uranium(IV) oxide, UO_2. This is converted to uranium(IV) fluoride, UF_4, with gaseous HF, and the fluoride is then reduced to the metal with calcium, magnesium or sodium.

Applications. Since the discovery of uranium fission, U. has become extremely important as a fissionable material for nuclear reactors (^{235}U) and as the starting material for production of plutonium (^{238}U). A significant fraction of the world's electricity is now supplied by uranium reactors. In modern light-water reactors, the nuclear fuel is enriched to 4% ^{235}U. The enrichment of ^{235}U, which is also important for reprocessing spent fuel rods, is done by separation of uranium(VI) fluoride in gas-diffusion, gas centrifugation or separatory jet plants.

Historical. U. was discovered in 1789 by Klaproth, and was named for the planet Uranus, which was discovered in the same decade. Metallic U. was first prepared in 1841 by Péligot. In 1896, Becquerel discovered that uranium preparations emitted radiation which was similar in certain respects to the recently discovered x-rays. He was thus the discoverer of radioactivity. Hahn, Strassmann and Meitner discovered neutron-induced fission of U. nuclei in 1939, and on the basis of this discovery, the first atomic weapons were developed during World War II in the USA (a ^{235}U bomb was dropped on Hiroshima in 1945). In subsequent years, the technology of nuclear power plants was developed. The enormous destructive potential of modern nuclear arsenals make urgently necessary an international agreement banning their use (freeze in development, dismantling and prohibition of nuclear weapons).

Uranium carbides: uranium monocarbide and dicarbide are of industrial importance. ***Uranium monocarbide***, UC, gray, cubic crystal latice; M_r 250.04, density 13.63, m.p. 2550 °C, formed by reaction of uranium(IV) oxide with graphite at 1400 to 2000 °C in an inert gas atmosphere. ***Uranium dicarbide***, UC_2, dark brown, tetragonal crystals; M_r 262.05, density 11.2, m.p. 2350 to 2400 °C. UC_2 is obtained by reaction of uranium with carbon, or of uranium powder with hydrocarbons at higher temperatures. Both U. are hydrolysed by water, primarily to methane, hydrogen and ethane (UC) or methane, ethane and higher alkanes (UC_2). They are used as nuclear fuels in the form of the mixed carbides (U,Pu)C and (U, Th)C_2.

Uranium chlorides. ***Uranium(III) chloride***, UCl_3, dark red, very hygroscopic crystalline needles; M_r 344.39, density 5.44, m.p. 842 °C. UCl_3 is obtained by heating UCl_4 in a stream of hydrogen, or by reaction of uranium hydride with hydrochloric acid. ***Uranium(IV) chloride***, UCl_4, dark green, octahedral crystals; M_r 379.84, density 4.87, m.p. 590 °C, dissolves very easily in water to give a green solution in which it is partially hydrolysed. With alkali chlorides it forms chlorouranates(IV), $M_2[UCl_6]$. Sodium hexachlorouranate(IV) can be used for electrochemical uranium production. UCl_4 is made by liquid-phase chlorination of uranium(VI) oxide (heating under reflux with hexachloropropane) or by reaction of uranium hydride with chlorine. It is an important starting material for production of other uranium(IV) compounds. ***Uranium(V) chloride*** is made by chlorination of UCl_4; it forms red-brown crystalline needles which are composed of dimeric molecules, U_2Cl_{10}, M_r 830.60, m.p. 327 °C; these are similar in structure to tantalum(V) or molybdenum(V) chloride molecules. ***Uranium(VI) chloride***, UCl_6, forms dark green, hexagonal crystals which are very hygroscopic, and can be distilled in vacuum; the compound is made by chlorination of a mixture of U_3C_8, a uranium carbide, and carbon at 380 °C.

Uranium fluorides: The most important U., which is made on an industrial scale, is ***uranium(VI) fluoride***, UF_6, colorless orthorhombic crystals. It is a volatile compound which is very reactive and is hydrolysed by water; M_r 352.02, m.p. 64.05 °C, subl.p. 56.54 °C. In the gas state, UF_6 has a regular octahedral configuration which is only slightly distorted in the crystal. It is a strong oxidizing fluorination agent, and adds fluoride ions to give fluorouranates(VI), $[UF_7]^-$ and $[UF_8]^{2-}$. In the laboratory, UF_6 can be made by fluorination of uranium oxides with fluorine or halogen fluorides. On a large technical scale, it is made by fluorination of uranium(IV) fluoride with fluorine around 500 °C in a fluidized bed or flame reactor. UF_6 is the most volatile of the known uranium compounds (vapor pressure at 20 °C: 10.63 kPa), and is therefore used in nuclear technology for isotope separation of uranium by gas diffusion, gas centrifugation or separatory jet processes. ***Uranium(IV) fluoride***, UF_4, forms green, non-volatile, triclinic crystalline needles, M_r 314.02, density 6.70, m.p. 1036 °C. It is scarcely soluble in water, and is made by fluorination of uranium(IV) oxide with hydrofluoric acid or tetrachlorodifluoroethane at elevated temperatures. The dihydrate, $UF_4 \cdot 2H_2O$, is fairly insoluble in water, and precipitates when hydrofluoric acid is added to uranium(IV) salt solutions. In the UF_4 lattice, the uranium is eight-fold coordinated. It reacts with fluorides to yield fluorocomplexes of the types $[UF_5]^-$, $[UF_6]^{2-}$, $[UF_7]^{3-}$ and $[UF_8]^{4-}$. ***Uranium(V) fluoride***, UF_5, M_r 333.02, is a hygroscopic, colorless compound which exists in two modifications; its structure is polymeric. It is obtained by photochemical reduction of UF_6 with carbon monoxide or by fluorination of UF_4 with fluorine at 240 °C. UF_5 adds fluoride ions to form fluorouranates(V), $[UF_6]^-$, $[UF_7]^{2-}$ and $[UF_8]^{3-}$. It disproportionates when heated: 3 $UF_5 \rightarrow U_2F_9 + UF_6$. ***Diuraniumnonafluoride***, U_2F_9, is ***uranium(IV,VI) fluoride***, $UF_4 \cdot UF_5$. ***Uranium(III) fluoride***, UF_3, forms red-violet, high-melting, nonvolatile crystals, M_r 295.03; it is made by reduction of UF_4 with aluminum or uranium at 900 °C.

Uranium hydride: UH_3, is a dark brown compound, M_r 241.05, density 10.95, which is formed by heating uranium with hydrogen at 250 to 300 °C. U. is an important raw material for synthesis of other uranium compounds. It reacts with chlorine and bromine to give the uranium(IV) halides UCl_4 and UBr_4, with hydrogen chloride to give uranium(III) chloride, and with ammonia or phosphane to give uranium nitrides or phosphides.

Uranium nitrides: ***Uranium nitride***, UN, forms brown, cubic crystals; M_r 252.04, density 14.31, m.p. 2850 °C. UN is formed by reaction of uranium powder or uranium hydride with nitrogen or ammonia, and is used as a reactor fuel. ***Diuranium trinitride***, U_2N_3, forms cubic crystals of the Se_2O_3 type.

Uranium oxides: ***Uranium(VI) oxide***, UO_3, is a yellow-orange, water-insoluble, amphoteric compound which exists in six modifications; M_r 286.03, density 7.29. UO_3 reacts with acids to give uranyl salts (UO_3 + 2 HX $\rightarrow UO_2S_2 + H_2O$), and with alkali hydroxides it forms diuraniates (2 UO_3 + 2 MOH $\rightarrow$ $M_2[U_2O_7] + H_2O$). It is made by thermal decomposition of uranyl nitrate, $UO_2(NO_3)_2 \cdot 6H_2O$, or ammonium diuraniate, $(NH_4)_2U_2O_7$, at approximately 350 °C.

Triuranium octaoxide, uranium(IV,VI) oxide, U_3O_8, is a dark olive-green compound, M_r 842.00, density 8.30, which crystallizes in several modifications and is stable to about 1300 °C. It is formed by heating UO_3 or UO_4 in the air to about 700 °C, which leads to oxygen uptake or release.

Uranium(VI) oxide, UO_2, is a dark brown, basic compound which crystallizes in rhombic or cubic form; M_r 270.03, density 10.96, m.p. 2500 °C. It is obtained by heating UO_3 in a stream of hydrogen or carbon monoxide to about 700 °C. UO_2 has a relatively broad phase width (UO_2 to $UO_{2.25}$). It is found in nature as uranium pitchblende. In the form of ceramics or the mixed oxide (U, Pu)O_2, UO_2 is an important nuclear reactor fuel.

Uranyl salts, ***dioxouranium(VI) salts***: UO_2X_2, yellow compounds which fluoresce in UV light; they contain the UO_2^{2+} cation and are obtained by reaction of uranium(VI) oxide with the corresponding acids. The most important U. are: ***Uranyl acetate***, $UO_2(OOCCH_3)_2 \cdot 2H_2O$, yellow, rhombic crystals which are moderately soluble in water, M_r 422.19, density 2.893; when heated to 110 °C they lose their water of crystallization, and at 275 °C they are converted to uranium(VI) oxide. With alkali acetates they form easily crystallized acetate complexes of the type $M[UO_2(OOCCH_3)_3]$. ***Sodium-magnesium uranyl acetate***, $NaMg(UO_2)_3(OOCCH_3)_9 \cdot 9H_2O$, is used for sodium determination. Uranyl acetate is used for amplification and toning in photochemistry and as a contrast reagent in electron microscopy. ***Uranyl carbonate*** or ***uranyl carbonate complexes***, e.g. of the type $M_2[UO_2(CO_3)_2]$, are important intermediates in the processing of uranium ores. ***Uranyl nitrate***, $UO_2(NO_3)_2 \cdot 6H_2O$, is a lemon-yellow compound which gives a yellow-green fluorescence. It dissolves readily in water, alcohol, ether and esters, and forms deliquescent, columnar crystals; M_r 502.12, density 2.81, m.p. 59.5 °C. The crystals display triboluminescence when ground. Uranyl nitrate is used as an amplifier and toner in photochemistry. It is an important intermediate in uranium production, where fine purification is achieved by extracting uranyl nitrate solutions with tributylphosphate. ***Uranyl sulfate***, $UO_2(SO_4) \cdot 3H_2O$, forms yellow-green crystals which are readily soluble in water; M_r 420.14, density 3.28. Uranyl sulfate and uranyl sulfate complexes are the products of leaching uranium ores with sulfuric acid.

Urates: see Uric acid.

Urd: abb. for uridine.

U. can be synthesized from cyanamide and water: $H_2N\text{-}C{\equiv}N + H_2O \rightarrow O{=}C(NH_2)_2$. It can also be synthesized from ammonia and carbonic esters or phosgene, or from ammonium isocyanate. Industrially, U. is synthesized from ammonia and carbon dioxide under pressure (10 to 20 MPa) and at 150 to 200 °C; ammonium carbamate is formed as an intermediate: $2\,NH_3 + CO_2 \rightleftharpoons H_2N\text{-}CO\text{-}O^- \cdot NH_4^+$ ($\Delta_R H$ = -126 kJ mol^{-1}) $\rightleftharpoons H_2N\text{-}CO\text{-}NH_2 + H_2O$ ($\Delta_r H$ = +33 kJ mol^{-1}). Because the starting material does not react completely, and because the ammonium carbamate decomposes partially to ammonia and carbon dioxide when the pressure is reduced, it is essential to return the ammonia and carbon dioxide to the reactor. The product is decompressed in two stages. The resulting equimolar mixture of U. and water is evaporated to a U. melt, which is sprayed into a beading tower. The rising air cools the melt and it solidifies. Depending on the method, ammonium hydrogencarbonate, ammonium sulfate or ammonium nitrate can be obtained as byproducts of U. synthesis.

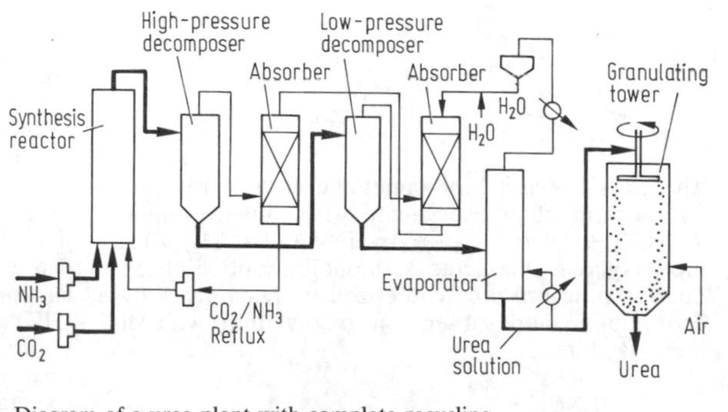

Diagram of a urea plant with complete recycling.

Urea, ***carbamide, carbonic acid diamide***: $H_2N\text{-}CO\text{-}NH_2$, colorless and odorless, prismatic crystals; m.p. 132.7 °C. U. is soluble in water and ethanol, and insoluble in ether and chloroform. Aqueous solutions are neutral. U. forms insoluble salts with some acids, for example, urea nitrate and oxalate. These reactions are also used for qualitative analysis of U. U. forms crystal-lattice inclusion compounds with many substances, such as alkanes with more than 5 C atoms or substituted alkanes. When aqueous U. solutions are heated, ammonium carbonate is formed; in the presence of acids or bases, U. is hydrolysed to ammonia and carbon dioxide. This cleavage can also take place under physiological conditions. When slowly heated above its melting point, U. is converted to Biuret (see). The reaction of U. with alcohols produces Urethanes (see); reaction with carboxylic acid derivatives yields Ureides (see), and Semicarbazides (see) are formed from hydrazine and U.

U. is the most important excreted product of protein metabolism in human and animal bodies. An adult human being excretes 29 to 30 g U. daily. U. is found in the plant kingdom, but only in small amounts in some higher plants and in fungi.

More than 85% of the world production of U. is used as a high-quality nitrogen fertilizer. U. is being used increasingly as a nitrogen source for ruminants and for production of urea-formaldehyde resins. It is also used as an intermediate in the production of drugs, dyes, glues, cosmetics, textile conditioners and stabilizers.

Historical. U. was discovered in urine in 1773 by Rouelle, and was isolated and characterized in 1797 by Vauquelin, Fourcroy and Pront. In 1828, Wöhler synthesized U. from potassium cyanate and ammonium sulfate via ammonium cyanate: $[N{\equiv}C\text{-}O]^- NH_4^+ \rightarrow H_2N\text{-}CO\text{-}NH_2$. This was the first laboratory synthesis of an organic compound.

Urea chloride: same as Carbamoyl chloride (see).

Urea-formaldehyde plastics: same as Urea resins (see).

Urea herbicides: Herbicides (see) which contain urea as a structural element. After the growth-hormone and triazine herbicides, U. are economically the most important group of herbicides.

Table. Urea herbicides Ar–HN–C(=O)–N(CH_3)(R)

Name	Ar	R	F. in °C
Fenuron	C_6H_5–	CH_3	136
Monuron	Cl–C_6H_4–	CH_3	174···175
Diuron	Cl–C_6H_3(Cl)–	CH_3	158···159
Metoxuron	CH_3O–C_6H_3(Cl)–	CH_3	126···127
Chlortoluron	CH_3–C_6H_3(Cl)–	CH_3	147···148
Isoproturon	$(H_3C)_2CH$–C_6H_4–	CH_3	151···153
Chloroxuron	Cl–C_6H_4–O–C_6H_4–	CH_3	151···152
Benzthiazuron	benzothiazol-2-yl	H	287
Monolinuron	Cl–C_6H_4–	OCH_3	79···80
Linuron	Cl–C_6H_3(Cl)–	OCH_3	93···94
Metobromuron	Br–C_6H_4–	OCH_3	95···96

The classic *N*-aryl-*N,N'*-dimethylurea is synthesized by reaction of arylisocyanate with dimethylamine: $ArNCO + HN(CH_3)_2 \rightarrow Ar\text{-}HN\text{-}CO\text{-}N(CH_3)_2$. The methoxyurea herbicides (monolinuron, linuron, metobromuruon) are synthesized by reaction of hydroxylamine and subsequent methylation with dimethylsulfate:

$$ArNCO \xrightarrow{H_2NOH} Ar\text{–}HN\text{–}CO\text{–}NHOH \longrightarrow Ar\text{–}HN\text{–}CO\text{–}N(CH_3)OCH_3$$

The H. act mainly as photosynthesis inhibitors and are mostly used as soil herbicides. Linuron and monolinuron are used primarily in potato fields, metobromuron is also used on a number of other crops, and methabenzthiazuron (*N*-benzthiazolyl-*N,N'*-dimethylurea) for control of annual weed grasses in grains. The ***sulfonylurea*** herbicides have a special position; the main example is chlorosulfuron, which is used in very low amounts for control of weeds in wheat, oats and barley.

$$2\text{-}Cl\text{-}C_6H_4\text{–}SO_2\text{–}NH\text{–}C(=O)\text{–}HN\text{–}(\text{triazinyl with } OCH_3, CH_3)$$

Chlorosulfuron

Urea resins, ***carbamide resins, urea-formaldehyde plastics***: a group of thermoset plastics made by polycondensation of urea and formaldehyde. The formaldehyde is used in the form of a 30 to 40% aqueous solution, as paraformaldehyde or as hexamethylenetetraamine. When the urea is partially replaced by thiourea, the condensation process is delayed, so that the reaction can be more easily controlled and condensates with exceptionally high mechanical strength and water resistance are formed. The primary reaction in the synthesis of U. is the formation of hydroxymethylurea compounds. Depending on the conditions, the reaction is: $H_2N\text{-}CO\text{-}NH_2 + CH_2O \rightarrow H_2N\text{-}CO\text{-}NH\text{-}CH_2OH$ (monohydroxymethylurea) or $H_2N\text{–}CO\text{-}NH_2 + 2\ CH_2O \rightarrow HOCH_2\text{-}NH\text{-}CO\text{-}NH\text{-}CH_2OH$ (dihydroxymethylurea), or sometimes various ring compounds are formed. Further condensation of monohydroxymethylurea, which is formed when the concentration of formaldehyde is limiting, leads to linear molecules with the approximate structure $H_2N\text{-}CO\text{-}NH\text{-}(CH_2\text{-}NH\text{-}CO\text{-}NH)_n\text{-}CH_2\text{-}NH\text{-}CO\text{-}NH_2$. The chains can be cross-linked by more formaldehyde molecules. Dihydroxymethylureas, which are formed when the formaldehyde is in excess, condense in the presence of acids to high-molecular-weight, cross-linked molecules. If water is split off the methylhydroxymethylurea, the product is methyleneurea, $H_2N\text{-}CO\text{-}N{=}CH_2$, which can polymerize into chains.

The U. are generally used with fillers, such as sawdust, short-fiber α-cellulose and textile fibers, to form white plastics which are resistant to light, have no odor or taste, and are not attacked by ether, alcohol, light petroleum, aromatic hydrocabons, esters or chlorohydrocarbons. These plastics are used mainly for household appliances and furniture. However, they are not as resistant to water and heat as products made from phenoplastics or melamine resins; in addition, the loss of formaldehyde from the finished products is a problem. The condensation products of urea and formaldehyde are important components as glues (HF glue) for plywood. Non-elastic foams made from U. are widely used for heat insulation; a solution of urea and formaldehyde in water and emulsifying agent are beaten into a foam by vigorous stirring, and polycondensed in this form. U. are also used as textile treatments to make wrinkle-free cloth, and to impregnate paper and textiles used in making household articles. Condensation of urea with formaldehyde in the presence of an alcohol, such as butanol, produces excellent paint resins with a high sheen and good resistance to scratching and water. These resins mix well with cellulose nitrate and alkyde resins, and are used especially for solvent-resistant, non-yellowing bake-on paints. Condensation products of urea and formaldehyde are also used to improve soil, especially for high-priced crops in greenhouses.

Urease: an enzyme which catalyses the hydrolysis of urea into ammonia, carbon dioxide and water: $CO(NH_2)_2 + 2\ H_2O \rightarrow 2\ NH_3 + CO_2 + H_2O$. U. occurs widely in plants and microorganisms. It was the first enzyme to be obtained in crystalline form (Sumner, 1926). The enzyme isolated from soy beans consists of 8 subunits, and has a relative molecular mass of 489,000. It has a pH optimum of 7.0 and is remarkably stable to the denaturing action of its substrate.

Urea separation, ***extractive crystallation***: a deparaffinization process in which urea is used to precipitate alkanes from hydrocarbon mixtures. In 1949, F. Bengen observed that urea forms inclusion compounds with normal alkanes, but not with branched alkanes, naphthenes or aromatics. The alkane-urea compounds can be separated from the remaining oil by filtration or centrifugation, washed with suitable solvents and cleaved in hot water. The normal alkanes then separate as an oil layer, and can be separated into the individual compounds by distillation. The aqueous urea solution is evaporated and the urea is returned to the process. In most installations for paraffin production by U., methanolic urea solutions are used.

U. is used mainly to remove straight-chain hydrocarbons from diesel fuels and lubricants, to improve their Stock points; it is also used to obtain the alkanes and to improve the octane numbers of fuels by removing the alkanes, which tend to cause knocking.

Thiourea forms inclusion compounds with isoalkenes, but no inclusion compounds of alkanes with thiourea are known.

Today U. is used only to obtain alkanes > C_{18}; for C_{10} to C_{18} alkanes, molecular sieve methods are more economical.

Ureide: a *N*-acyl derivative of urea with the general formula H_2N-CO-NH-CO-R. U. are stable, crystalline compounds formed by reaction of carboxylic acyl halides, esters or anhydrides with urea; for example, acetyl chloride and urea form acetylurea (acetic ureide): CH_3-CO-Cl + H_2N-CO-NH → CH_3-CO-NH-CO-NH_2 + HCl. In the reaction of dicarboxylic acid derivatives with urea, both NH_2 groups react to form cyclic ureides, e.g. diethyl malonate and urea react in the presence of sodium methylate to form barbituric acid (malonylurea). U. can be readily cleaved into the starting components under basic conditions. Many U. are used in medicine as sedatives and hypnotics.

5-Ureidohydantoin: same as Allantoin (see).

Urethanes: the esters of carbamic acid with the general formula H_2N-CO-OR. They are also called ***carbamates***. Unlike the unstable carbamic acid, U. are stable, crystalline compounds. They are formed by reaction of chlorocarbonate esters or carbonate esters with ammonia:

$$O{=}C\langle^{OR}_{X} + NH_3 \longrightarrow O{=}C\langle^{OR}_{NH_2} + HX$$

(X=Cl or OR). The U. can also be synthesized from isocyanic acid or its esters and alcohols:

$$O{=}C{=}N{-}R^1 + R^2{-}OH \rightarrow O{=}C\langle^{NHR^1}_{OR^2}$$

(R^1 = alkyl, aryl).

N-Phenylurethanes are crystalline compounds often used to characterize unknown alcohols. When *N*-methylurethane is nitrosylated, it yields ***N-methyl-N-nitrosourethane***, which can be used to synthesize diazomethane. Polyurethanes (see) are made from diisocyanates and glycols. U. are also used as pesticides and drugs.

Uric acid, ***2,6,8-trihydroxypurine***: colorless crystals which decompose, without melting, when heated above 400 °C; they release hydrogen cyanide. H. can exist in two tautomeric forms; the lactam form is more prevalent. It is insoluble in alcohol and ether, and barely soluble in water. However, it dissolves in glycerol and in solutions of alkali hydroxides, carbonates, acetates and phosphates. It is a weak acid and forms two series of salts, the water-soluble, neutral ***urates*** and the acid urates, which are much less soluble in water. U. is detected through the murexid reaction.

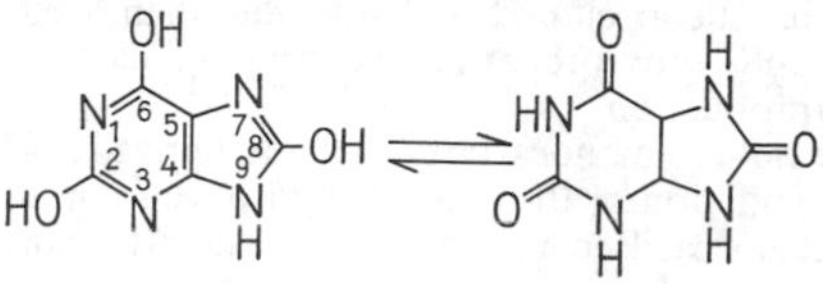

Lactam form Lactim form

U. was discovered in 1776 by Scheele as a component of kidney stones. It was also isolated from the tissue cells of patients with gout, and from the excrements of snakes and birds. Human beings and other primates excrete small amounts of U. in their urine as the endproduct of purine metabolism, while in other mammals, U. is converted to allantoin by uricases. In adult human beings, 1 to 3% of the total nitrogen content of the urine is in the form of U., which amounts to a daily excretion of 0.4 to 1.3 g.

The hormones ACTH and cortisone and certain diseases cause increased excretion of U. The deposition of U. in the joints is an important factor in the pathology of gout (arthritis urica). Kidney stones consisting of U. can often be removed by drug treatment.

Many syntheses for U. have been developed, most based on the Traube purine synthesis (see). U. is readily converted to trichloropurine and can react further by nucleophilic substitution of the chlorine atoms. In this way it is possible to obtain purine derivatives such as hypoxanthine, xanthine, guanine and adenine.

Uridine, abb. ***Urd*** or ***U***: 1-β-D-ribofuranosyluracil, a nucleoside. It forms colorless crystals, M_r 244.2, m.p. 165-168 °C. It is soluble in water, but practically insoluble in ether. U. is a component of ribonucleic acids. The 5'-nucleotides ***uridine monophosphate (UMP), uridine diphosphate (UDP)*** and ***uridine triphosphate (UTP)*** play an essential role in intermediary metabolism. UMP is biosynthesized by decarboxylation of orotidine monophosphate. UTP is the starting material for biosynthesis of thymidine and cytosine derivatives. UDP serves as a carrier for sugars (e.g. UDP-glucose) in the synthesis of polysaccharides and glycoproteins.

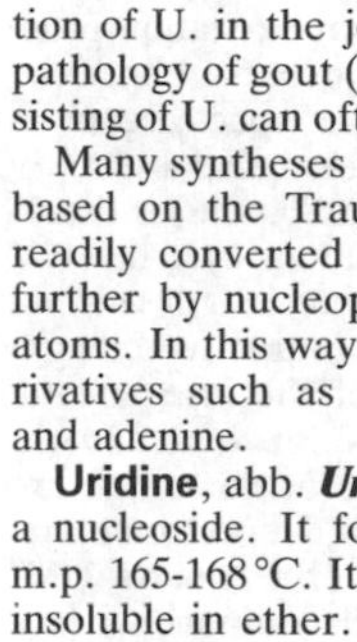

Uridine diphosphoglucose: see Sugar nucleotides.

Urokinase: a trypsin-like enzyme which catalyses the conversion of plasminogen to plasmin in the blood plasma. Plasmin is an enzyme which dissolves blood clots, and is essential for prevention of thromboses. It is possible, for example, to use U. to reopen arteries blocked by heart infarction. U. can be isolated from human urine or kidney cell cultures; the U. gene has been cloned and introduced into *Escherichia coli*, thus preparing the way for biotechnological production.

Uronic acids: monocarboxylic acids derived formally by oxidation of the primary hydroxyl group of monosaccharides. Because they retain the glycosidic hydroxyl group, U. give the typical monosaccharide reactions, including formation of glycosides. U. are usually isolated as the lactones. They cannot be synthesized from the free monosaccharides; the glycosides, acetates, acetals or ketals are used as starting materials for the oxidation, for example with dinitrogen tetroxide, N_2O_4. U. are found in Mucopolysaccharides (see), Hemicelluloses (see) and numerous vegetable mucuses and gums, such as tragacantha, gum arabic, pectin substances and alginic acids.

Uroporphyrin: see Porphyrins.

Uroporphyrinogen: see Porphyrins.

Ursane: see Triterpenes.

UTP: see Uridine.

UV-VIS spectroscopy, ***UVS spectroscopy***: study of the absorption or emission of ultraviolet and visible light by a sample. The same type of excitation process, namely electronic transitions, occurs in both the ultraviolet and visible ranges, so that spectroscopy in the two ranges can be treated as a single subject.

The range of *visible light* (400 to 800 nm) is established by the sensitivity of the human eye; the UV range is divided into the shorter-wavelength *vacuum* or *far UV* (10 to 200 nm) and the longer-wavelength *near* or *quartz UV* on the basis of the instrumental techniques used in the two regions. Since air absorbs UV below 200 nm, measurements in this range are possible only in special, evacuated spectrometers, and are rarely undertaken. The term "quartz UV" arose because most of the optical components (see Spectral instruments) for use in the near UV are made of quartz.

Both absorption and emission spectra of atoms in the UV-VIS are studied (see Atomic spectroscopy); but molecules are studied almost entirely by means of absorption spectra.

Absorption spectra of molecules. Molecular spectra can be recorded from samples in any of the 3 physical states of matter, but they are most commonly taken with the sample in solution in a non-absorbing solvent such as water, alcohol or alkanes. The spectra are displayed graphically as a plot of the intensity (% absorption, % transmission, extinction ε, or molar extinction coefficient) vs. the wavelength in nm, or, more rarely, against wavenumber in cm^{-1}. The amount of substance required is very small; solutions in the concentration range 10^{-2} to 10^{-5} mol l^{-1} are generally used.

If a sample absorbs more intensely than a reference substance at an absorption band in the UV-VIS range, the change is called a "hyperchromic shift"; if it absorbs less intensely, a "hypochromic shift". If the absorption maximum is shifted towards longer wavelengths, it is called a "bathochromic shift"; a shift towards shorter wavelengths is a "hypsochromic shift".

Theory. Long before the theory of atomic structure had provided a physical explanation of light absorption processes, chemists had noted empirical relationships between chemical structure and absorption of UV-VIS light (see Color theory). It had been recognized that multiple bonds (see Chromophores) had to be present for longer-wavelength absorption to occur. With the exception of the azo group, atomic groups containing multiple bonds absorb in the UV unless they are conjugated with several other chromophoric groups or with Auxochromic groups (see); in this case the absorption is shifted to the visible.

Quantum theory of light absorption. Electron transitions between orbitals with different energies are associated with absorption of definite frequencies of light: $E_2 - E_1 = h\nu$. In principle, the energies of the orbitals can be calculated using quantum mechanics, but for molecules with several atoms, exact calculations are not possible. There are several methods of approximate calculation (see Molecular orbital theory); the one most commonly used at present is the LCAO-MO method, which describes molecular orbitals as linear combinations of atomic orbitals. For example, combination of the 1*s* orbitals of two hydrogen atoms creates two molecular hydrogen orbitals, one of which causes bonding (σ*s*) and the other (the anti-bonding, σ**s* orbital) prevents it (Fig. 1.).

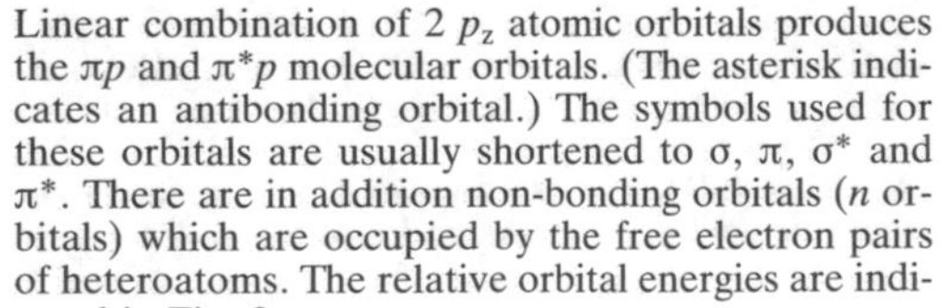

Fig. 1. Bonding and anti-bonding molecular orbitals and their formation from atomic orbitals.

Linear combination of 2 p_z atomic orbitals produces the π*p* and π**p* molecular orbitals. (The asterisk indicates an antibonding orbital.) The symbols used for these orbitals are usually shortened to σ, π, σ* and π*. There are in addition non-bonding orbitals (*n* orbitals) which are occupied by the free electron pairs of heteroatoms. The relative orbital energies are indicated in Fig. 2.

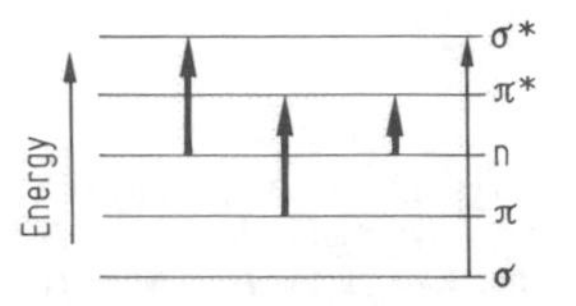

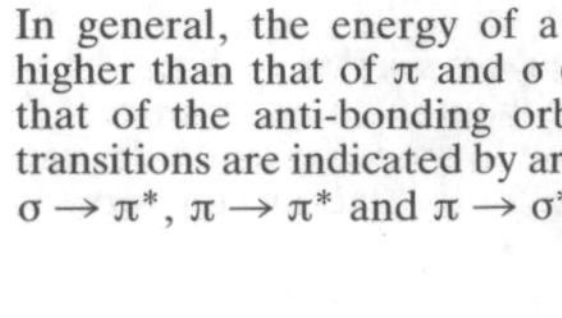

Fig. 2. Relative energies of the molecular orbitals and possible transitions between them.

In general, the energy of a non-bonding orbital is higher than that of π and σ orbitals, but lower than that of the anti-bonding orbitals. Possible electron transitions are indicated by arrows. It can be seen that $\sigma \rightarrow \pi^*$, $\pi \rightarrow \pi^*$ and $\pi \rightarrow \sigma^*$ transitions require less energy than $\sigma \rightarrow \sigma^*$ transitions. The latter are therefore observed only in the vacuum UV, while the other transitions can appear in the near UV and sometimes even in the visible. The relative energies of these molecular orbitals can be affected by various factors, the most significant of which is conjugation of double bonds. An isolated double bond gives rise to a $\pi \rightarrow \pi^*$ transition around 180 nm. In dienes, combination of the π orbitals of the separate double bonds forms new orbitals, two bonding, π_1 and π_2, and two anti-bonding, π_3 and π_4. It can be seen in Fig. 3 that this leads to a new $\pi \rightarrow \pi^*$ transition with significantly lower energy, and dienes therefore absorb at longer wavelengths than alkenes. The bathochromic shifting of absorption continues as further double bonds enter the conjugated system, and there is a simultaneous hyperchromic shift as well (Table 1). As a rule of thumb, the longer the conjugated system in a compound, the longer the wavelength of its absorption maximum.

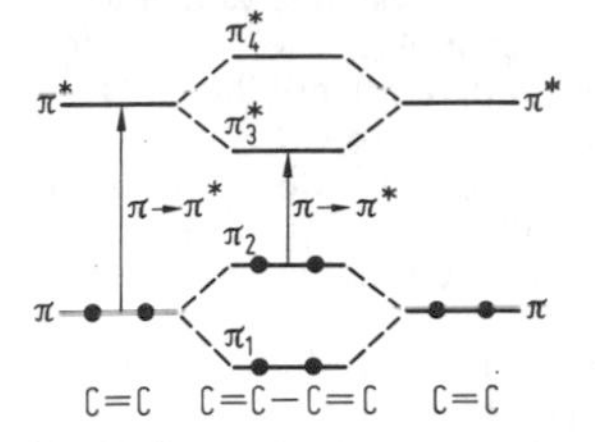

Fig. 3. Change in the energy of molecular orbitals upon conjugation.

Table 1. Longest wavelength absorption maximum of some polyaldehydes CH_3–(CH = CH)$_n$–CHO

n	λ_{max} in nm	ε_{max} in $l \cdot mol^{-1} \cdot cm^{-1}$
1	220	15 000
2	270	27 000
3	312	40 000
4	343	45 000
5	370	57 000
6	393	65 000
7	415	63 000

Electronic transitions are accompanied by vibrational and rotational transitions. Each electronic absorption line includes contributions from several vibrational bands and rotational fine structure. Usually intermolecular interactions smear out the fine structure, so that only a broad absorption maximum is seen, although in some cases, e.g. in the gas phase or in nonpolar solvents, the vibrational structure appears. The vibrational structure can provide information on the bond strengths and interatomic distances in the molecule in its excited state.

Selection rules for electronic transitions involve symmetry and spin. Transitions are possible only between electronic states with certain symmetries. It can be calculated from the symmetry of the wavefunction and the states which combine with one another whether a given electronic transition is allowed or forbidden. In addition, transitions between states with different multiplicities are forbidden (see Multiplet structure).

An electron in an excited state can return to the ground state by re-emission of radiation (see Fluorescence, Phosphorescence), through photochemical reactions, or through conversion of the energy of excitation into heat energy.

Applications. 1) ***Qualitative U.*** *Absorption by organic substances*. Conjugated π electron systems are most often studied. For example, the number of conjugated double bonds n in a polyene R^1-(C=C-)$_n$-R_2 can be determined; for the longest-wavelength absorption maximum $\lambda_{max} = A\sqrt{n} + B$, where A and B are constants which depend on the substituents R^1 and R^2. Saturated ketones can be distinguished from α,β-unsaturated ketones on the basis of their UV absorption; the saturated ketones have their longest-wavelength maxima around 270 nm, but this transition is shifted by conjugation with a double bond to about 320 nm. The absorption by substituted dienes, trienes and α,β-unsaturated ketones can be calculated using empirical rules (*Woodward absorption rules*).

Benzene and its derivatives have a characteristic UV absorption (Fig. 4). As shown in Table 2, the spectrum of benzene is shifted in a regular way towards longer wavelengths when substituents are present in the molecule. The π-electron system of condensed aromatics also has a characteristic absorption spectrum which is very useful for identification of these compounds (Fig. 5).

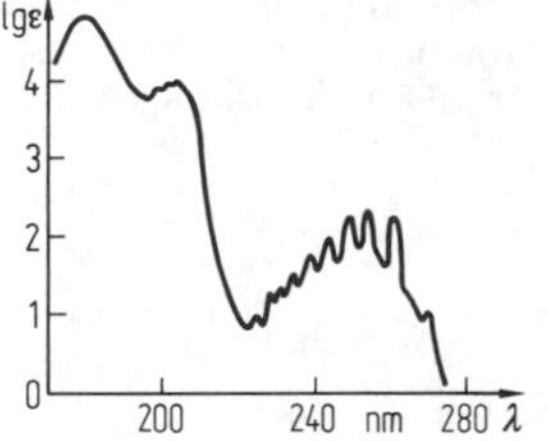

Fig. 4. Absorption spectrum of benzene.

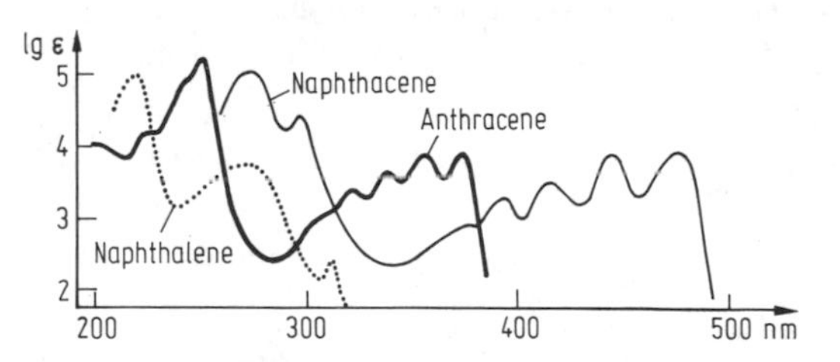

Fig. 5. Spectra of condensed aromatics.

Table 2. Absorption maxima of some monosubstituted benzenes

Substituent	λ_{max} in nm	ε_{max} in $l \cdot mol^{-1} \cdot cm^{-1}$	λ_{max} in nm	ε_{max} in $l \cdot mol^{-1} \cdot cm^{-1}$
H	203	7400	254	204
NH_3^+	203	7500	254	160
CH_3	206	7000	261	225
Cl	209	7400	263	190
Br	210	7900	261	192
OH	210	6200	270	1450
COOH	230	11600	273	970
NH_2	230	8600	280	1430

The longest-wavelength absorption maxima in polymethylenes depend on the number of conjugated double bonds in similar fashion to polyenes: $\lambda_{max} = A_n + B$. Here, however, there is a linear relationship between λ and n. Each additional double bond shifts the absorption maximum by about 100 nm towards longer wavelengths. Steric hindrance which occurs in conjugated systems can also be detected by U. For example, the absorption spectrum of biphenyl is shifted towards shorter wavelengths after introduction of 2 methyl substituents. The cause is the steric hindrance between the methyl groups, which tends to force the two benzene groups out of a single plane and thus reduces the amount of conjugation between them.

Absorption of inorganic compounds. The transition metal ions and their complexes tend to be highly colored, due to transitions of the d electrons (or f electrons in the actinoids and lanthanoids). Ligand field theory (see) deals with d-d transitions, which, however, are usually not very intense. Complexes often also undergo charge-transfer transitions, which have much higher intensities and fall in the UV. Here the electron moves from an oxidizable ligand to a central atom which is highly oxidized. The energy required for this transition depends on the relative redox potentials of the ligand and central cation. Thus transition metal complexes are usually characterized by intense absorption in the UV (charge-transfer transitions) and much weaker absorption in the visible (d-d transitions (Fig. 6). Absorption in the UV may also occur because of $\pi \rightarrow \pi^*$ or $n \rightarrow \pi^*$ transitions within the ligands.

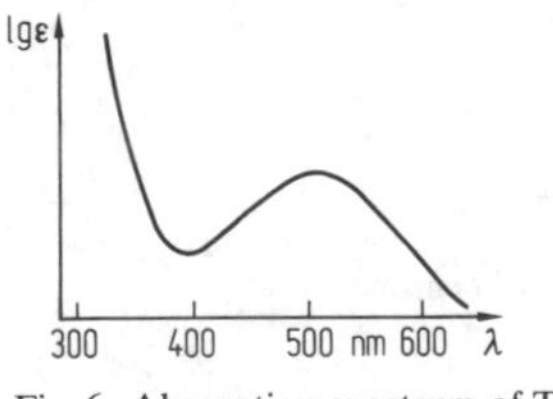

Fig. 6. Absorption spectrum of $Ti(H_2O)_6^{3+}$

Various elements in the periodic system, e.g. the halogens, form colored molecules. The color is due to excitation of free electron pairs. Molecules which are composed of related elements of this type can also be colored (e.g. SCl_2). Certain elements, e.g. carbon, silicon and boron, are black in the elemental state. This is related to the fact that they are semiconductors.

2) ***Quantitative U.*** The basis for quantitative analyses is the Lambert-Beer law (see): $c = A/(\varepsilon d)$. If the molar extinction coefficient ε and layer thickness d are known, the concentration c can be calculated from the measured absorption A. Multicomponent analyses can also be carried out using this law. Quantitative U. can be used to study concentration- and temperature-dependent equilibria (e.g. tautomerism, molecular association, complex formation, thermochromism, photochromism, solvatochromism, dissociation constants of acids and bases). For U. to be applicable, the species in the equilibrium must have differences in their π-electron systems, and thus in their absorptions. For example, the absorption spectra of the tautomers 4-nitrosophenol (benzoid nature) and 1,4-benzoquinone monooxime (quinoid nature) are distinctly different. Quantitative analysis indicates that, e.g. in ether solution, there is an equilibrium between 70% 1,4-benzoquinone monooxime and 30% 4-nitrosophenol.

U. is also very important for quantitative determinations by Photometry (see).

Time-resolved U. With an ordinary spectrometer, a UV-VIS spectrum can be measured in a few minutes. For the study of short-lived excitation states and intermediates, and in kinetic studies, however, methods with much better time resolution are needed. In the 1960s, rapid spectrometers were developed which gave resolutions between 10^{-2} and 1^{-5} s. Using Flash spectroscopy (see), spectra can be made in 10^{-5} to 10^{-7} s. Introduction of laser technology made possible nanosecond and picosecond spectroscopy. With these modifications, U. has made great contributions to the development of photophysics, photochemistry and reaction kinetics.

V

v: symbol for volume.

V: 1) symbol for vanadium; 2) V, symbol for molar volume.

Vacancies: see Crystal lattice defects.

Vacuum: a state in a gas-filled space in which the pressure is lower than atmospheric pressure. Using the equation of state for an ideal gas ($PV = nRT$), the number of gas molecules in a given volume can be calculated if the pressure and temperature are known. Under standard conditions (P = 0.1 MPa, T = 273 K), there are $2.68 \cdot 10^{19}$ molecules per cm^3 of gas; at the lowest pressure which has been achieved in a laboratory, about $1.3 \cdot 10^{-12}$ Pa, there are still approximately 350 molecules per cm^3 gas. It is customary to divide the entire range of pressure below 0.1 Mpa into four ranges according to the special physical phenomena observed and the techniques used to achieve them (Table).

Pressure ranges in vacuum technology and their characteristics. (Numbers have been rounded to powers of ten and apply at room temperature)

	Partial vacuum	Vacuum	High vacuum	Ultra-high vacuum
Pressure in pascals	$10^5 - 10^2$	$10^2 - 10^{-1}$	$10^{-1} - 10^{-5}$	$< 10^{-5}$
Particles/cm^3	$10^{19} - 10^{16}$	$10^{16} - 10^{13}$	$10^{13} - 10^{9}$	$< 10^{9}$
Mean free path in cm	$10^{-5} - 10^{-2}$ less than container dimensions	$10^{-2} - 10^{1}$ less than or equal to container dimensions	$10^{1} - 10^{5}$ greater than container dimensions	$> 10^{5}$ much greater than container dimensions

The characteristic parameters of these ranges are the number of molecules per unit volume and the mean free pathlength of the molecules. As the V. is increased, the number of molecules decreases and the mean free pathlength increases. In high V., the mean free path is longer than the linear dimensions of the receptacle, and there are more molecules adsorbed to the walls of the receptacle than are moving freely in the space between the walls. In ultra-high V., only a relatively small number of molecules move.

V. technology includes the production (*evacuation*) and measurement of V., and the technology of processes which occur in V. Low pressures are achieved by V. pumps, such as gas-ballast and oil-sealed rotary and diffusion pumps; for ultra-high V., getters, ion pumps and cryopumps are used. The V. is measured by spring-tube, membrane, heat-conduction, ionization and compression V. meters.

Applications. 1) Partial V. and V. are used in chemistry for distillation, calcination, melting, sublimation, impregnation, filtration, drying, crystallization and cooling. High V. are used in the production of electronic components, photocells and secondary electron amplifiers. They are also used in the production of light bulbs, V. evaporation and vapor deposition. Many types of thin layers are applied by vapor deposition in high V., for example, aluminum layers on mirrors and other reflectors, fluoride coatings on glass lenses, interference layers for optical filters, and protective layers on metals, plastics and textiles. High V. are used in the purification of rare metals by reduction and distillation, and in sintering of metals and metal carbides . Titanium, molybdenum, zirconium, stainless steels and special metals are melted in V. electric arc furnaces. High V. are also needed in mass spectroscopes, V. spectrographs, electron microscopes, particle accelerators and high-voltage devices.

Vacuum deposition method, ***van Arkel-de Boer method, hot wire method***: a method for obtaining very pure metals and semimetals (titanium, zirconium, hafnium, thorium, vanadium, tantalum, rhenium, boron, silicon) and certain compounds (e.g. titanium carbide, TiC, titanium nitride, TiN, zirconium nitride, ZrN) by thermal decomposition of a volatile metal compound (metal halide). This compound is either present in the reaction chamber or is generated by a chemical reaction. The chamber is evacuated and heated to 500 to 600 °C. It contains a tungsten wire, 40 µm in diameter, which is heated to 1200 to 1800 °C (Fig.).

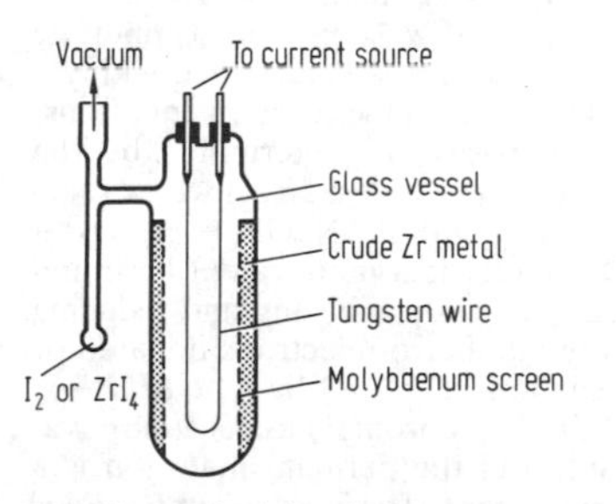

Hot wire method for zirconium transport.

The gaseous metal halide decomposes on the hot wire and deposits the pure metal, leaving the elemental halogen free for further Transport reaction (see) with the starting material. If compounds are to be made, a steady supply of a gaseous reactant (e.g. N_2) must be provided. The method was developed in 1924 by van Arkel and de Boer.

Vacuum distillation: see Distillation.

Vacuum drying: see Drying.

Vacuum metallurgy: a collective term for all metallurgical processes (production from ores, refining and casting) which are carried out under reduced pressure (as low as 10^{-6} Pa). Vacuum techniques are used to obtain high-melting metals (steel, molybdenum, tungsten, titanium, etc.) in very pure form by remelting, degassing and addition of alkaline earth metals with high affinities to oxygen (e.g. magnesium).

Val: abb. for valine.

Valency: As originally defined by Frankland (1852), the number of bonds a given atom is able to form with other atoms. The concept is closely related to the Stoichiometry (see) of chemical bonds and reactions, and has been modified in the course of time.

1) The *stoichiometric V. of an element* is a number n which indicates how many monovalent atoms or groups an atom of this element can bind or replace in a compound. In a binary compound, the total V. of the two elements are equal. The total V. is the V. of each element times the number of atoms ν of that element in the stoichiometric formula: $n \cdot \nu$ In other words, there can be no "free" V. in a stable compound. Therefore, if the composition of a binary compound and the V. of one of the elements are known, the V. of the other can be calculated. Hydrogen and fluorine always have V. of 1. Therefore, the V. of the elements Na, O, B, Si and S in the compounds NaH, H_2O, BF_3, SF_4 and SF_6 are 1, 2, 3, 4 and 6, respectively. For most elements, the stoichiometric V. is not a constant, but can differ from one compound to the next. For example, the V. of nitrogen in the oxides N_2O, NO, N_2O_3, NO_2 and N_2O_5 are 1, 2, 3, 4 and 5. The V. is always integral. However, when the formal V. for iron is calculated for the compound Fe_3O_4, a value of $n_{Fe} = 8/3$ is obtained. This is interpreted to mean that one Fe atom is divalent, while the other two are trivalent. Values of n can be computed for groups as well as for single atoms. For example, the formulas NaOH, H_2SO_4, H_3PO_4 and NH_4Cl yield $n_{OH} = 1$, $n_{SO_4} = 2$, $n_{PO_4} = 3$ and $n_{NH_4} = 1$.

The *stoichiometric V. of a compound* cannot be derived from its formula, but is determined with respect to certain reactions from the reaction equations. The V. of acids and bases are determined by the number of protons added or lost in an acid-base reaction. According to the equation $2\ NaOH + H_2SO_4 \rightarrow Na_2SO_4 + 2\ H_2O$, NaOH is a monovalent base and H_2SO_4 is a divalent acid. For oxidizing and reducing agents, the V. is the number of electrons donated or accepted. In the reaction $5\ Fe^{2+} + MnO_4^- + 8H^+ \rightarrow 5\ Fe^{3+} + Mn^{2+} + 4\ H_2O$, the iron(II) ion is a monovalent reducing agent, and the permanganate ion is a pentavalent oxidizing agent. The V. of a compound is by no means a constant; it often depends on the reaction conditions. The V. determines the size of the corresponding Equivalent (see).

2) The *ionic V. (ionic charge, charge number)* is the number of charges on an ion. It is indicated by a superscript numeral and plus or minus sign (e.g. Ag^+, NH_4^+, Cl^-, SO_4^{2-}).

3) The *bonding valency* gives the number of bonds in which an atom in a compound participates (see Chemical bond).

4) The *oxidation state* (see) is the ionic charge an element in a compound would have if this compound were made up of atomic ions. It is symbolized by a number written above the formula (e.g. $\overset{+1}{K}\overset{+7}{Mn}\overset{-2}{O_4}$).

5) The *formal charge* of an atom in a compound is the difference between the number of valence electrons in the neutral atom and the number of valence electrons which it would have after homolytic cleavage of all the bonds in which it participates.

6) The *coordination number* of an atom in a molecule is the number of other atoms to which it is bound. If it participates only in single bonds, the coordination number is equal to the bonding valency, but if multiple bonds are present, the coordination number is smaller than the valency. The coordination number is particularly important in Coordination chemistry (see).

Valence band: see Energy band model.

Valence bond method: A quantum mechanical approximation method used to describe bonding in molecules. In contrast to the Molecular orbital theory (see), the V. is based on the assumption that atomic orbitals are largely retained in the molecule, and is usually applied only to the valence electrons. Different possible distributions of these electrons among the atomic orbitals of a molecule produce different valence structures. These determinant wavefunctions (see Antisymmetry) can be constructed mathematically from the occupied atomic orbitals, and represented symbolically by valence dash formulas. The total wavefunction Ψ of the molecule is approximated in the V. by a linear combination of valence structures: $\Psi = \Sigma C_I \varphi_I$, where φ_I is the wavefunction of valence structure I. The coefficients C_I are determined using the variation method to minimize the total energy. Valence structures do not actually exist, but are only a device used to represent the true bonding state. For the hydrogen molecule, the linear combination of the covalent valence structure H_A-H_B with the two ionic valence structures $\overset{\oplus}{H_A}\overset{\ominus}{H_B}$ and $\overset{\ominus}{H_A}\overset{\oplus}{H_B}$ gives a good description of the bonding state; in it, the covalent valence structure makes the greatest contribution. The bonding in benzene can be represented in simplified form as a combination of the Kekulé formulas K_1 and K_2 plus the Dewar formulas D_1, D_2 and D_3 (see Resonance). It has been calculated that each Kekulé structure contributes about 39% of the total wavefunction, and each Dewar structure, about 7.3%. If only one valence structure makes a significant contribution to the energy of the ground state, one speaks of approximately localized valency; otherwise, it is nonlocalized valency. The calculations for larger molecules using the V. are complicated by the large numbers of valence structures which have to be taken into acount. The method is therefore most important for qualitative interpretation of molecular bonding.

The V. was developed in large part through the work of Heitler, London, Slater and Pauling (1927 to 1931).

Valence-dash formula: see Formula.

Valence electrons: electrons in the outermost (valence) shell of an atom (see Chemical bond).

Valence electron concentration: see Intermetallic compounds.

Valence isomerization: synchronous isomerization reactions with cyclic transition states in which the reactant and product are structural isomers. Electrocyclic and sigmatropic reactions are examples (see Woodward-Hoffmann rules).

Valence shell electron pair repulsion model, *VSEPR model*: a model for qualitative prediction of molecular structure which assumes that the shape of a molecule AX_mE_{n-m} is determined largely by the repulsive interactions between electron pairs of the valence shell of the central atom A. Here m is the number of ligand atoms X, E are lone electron pairs, and n is the number of valence electron pairs on the central atom. The first models were suggested by Sidgewick and Powell (1940). The VSEPR models were then further developed by Gillespie and Nyholm (1967) for predicting molecular structures, mainly of inorganic compounds of main group elements. The basic assumption of the VSEPR method is that, due to the Pauli principle, the electron pairs in a valence shell are oriented in such a way that the distances between them are maximized. The energetically favored orientations for various numbers of valence electron pairs on the central atom A in molecules of the type AX_mE_{n-m} are shown in the figure; the molecular structures follow from these orientations.

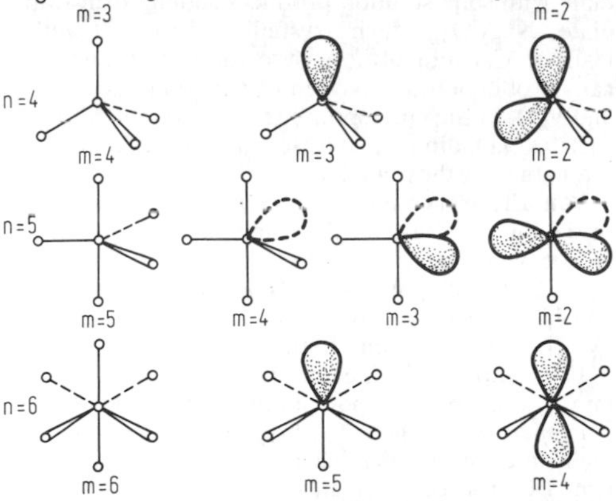

n=2 m=2
n=3 m=3 m=2
n=4 m=4 m=3 m=2
n=5 m=5 m=4 m=3 m=2
n=6 m=6 m=5 m=4

Energetically favored orientations of n electron pairs in molecules AX_mE_{n-m}.

The bond angles in molecules AX_mE_{n-m} differ from those in AX_m molecules ($m = n$). For example, the bond angles in CH_4 (109.5°), NH_3 (107.3°) and H_2O (104.5°) decrease as the number of non-bonding electron pairs increases. This reduction in bond angle is interpreted as a requirement for more room for lone pairs than for bonded pairs. Thus whenever possible, lone electron pairs also assume orientations which reduce interaction with other electron pairs, such as the equatorial position in a trigonal bipyramid (Fig.). In molecules with multiple bonds, such as $COCl_2$ and HCN, the two or three electron pairs in the double or triple bond are treated as electron pairs with respect to space requirements. Their requirements are between those of a bonding electron pair and a lone pair. The table gives the molecular geometries predicted by the VSEPR for several molecules of the type AX_mE_{n-m}.

Structures of the molecules AX_mE_{n-m}

n	Type	Geometry	Examples
2	AX_2	Linear	BeH_2, CO_2
3	AX_3	Trigonal planar	BF_3
	AX_2E	V-shaped	SO_2, $SnCl_2$
4	AX_4	Tetrahedral	CH_4, BF_4^-
	AX_3E	Trigonal pyramidal	NH_3, SO_3^{2-}
	AX_2E_2	V-shaped	H_2O, H_2S
5	AX_5	Trigonal bipyramidal	PF_5, PCl_5
	AX_4E	Distorted tetrahedral	SF_4
	AX_3E_2	T-shaped	ClF_3, BrF_3
	AX_2E_3	Linear	ICl_2^-, I_3^-
6	AX_6	Octahedral	SF_6
	AX_5E	Square pyramidal	BrF_5, IF_5
	AX_4E_2	Square planar	XeF_4, ICl_4

Valence state: see Hybridization.

Valence vibrations, ν *vibrations*: Vibrations of a molecule in line with the valence direction of the bound atoms, leading to changes in the interatomic distances. For example, "$\nu_{C=O}$" denotes the V. of the carbonyl group. There is a difference between symmetric (ν_s) and asymmetric (ν_{as}) V. (Fig.); the asymmetric V. have somewhat higher energies than the symmetric V.

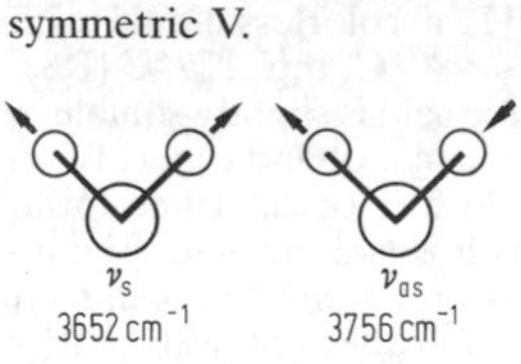

ν_s 3652 cm^{-1} ν_{as} 3756 cm^{-1}

Symmetric and asymmetric valence vibrations in the H_2O molecule.

V. require more energy than Deformation vibrations (see).

Valepotriates: triesters of unsaturated monoterpene alcohols derived from dimethylcyclopentapyran. The most important examples are found in the valerian root, ***valtrate*** and ***didrovaltrate***. The V. are believed to be the pharmacologically active components of valerian.

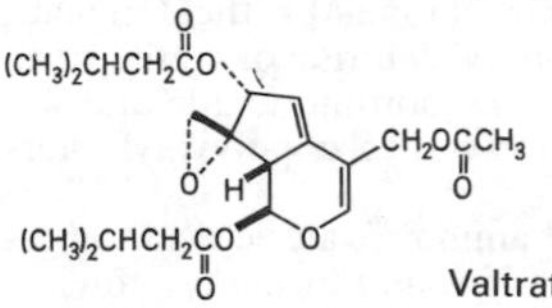

$(CH_3)_2CHCH_2COO$ CH_2OCCH_3 O H $(CH_3)_2CHCH_2COO$

Valtrate

Valeraldehyde, ***pentanal***: CH_3-$(CH_2)_3$-CHO, a colorless liquid with a pungent odor; m.p. - 91 °C, b.p. 104 °C, n_D^{20} 1.3947. *n*-V. is slightly soluble in water, but dissolves readily in most organic solvents. It is very reactive and undergoes the addition and condensation reactions typical of Aldehydes (see). V. is synthesized by catalytic dehydrogenation of amyl alcohol, or by reduction of valeric acid with formic acid on a manganese(II) oxide contact. It is used in the cosmetic industry and as a vulcanization accelerator.

Valeric acids: the four isomeric saturated monocarboxylic acids with the general formula C_4H_9-COOH.

1) ***Valeric acid, pentanoic acid***: CH_3,-$(CH_2)_3$-COOH, a colorless liquid with an unpleasant odor; m.p. 34.5 °C, b.p. 186 °C. V. is slightly soluble in water, and readily soluble in alcohol and ether. It occurs in free form in wood vinegar and in the foul water from low-temperature carbonization of soft coal. V. is synthesized industrially by oxidation of pentan-1-ol. It is used as an intermediate in the production of aromas and flavorings; its esters with lower alcohols are of particular significance. Its esters with higher alcohols are used as softeners for plastics.

2) ***Isovaleric acid, 3-methylbutanoic acid***: $(CH_3)_2$-CH-CH_2-COOH, a colorless liquid with an unpleasant odor reminiscent of valerian; m.p. -29.3 °C, b.p. 176.7 °C, n_D^{20} 1.4033. Isovaleric acid is slightly soluble in water, but dissolves readily in alcohol, ether and chloroform. It occurs naturally in free and esterified form in essential oils from medicinal herbs, tea leaves and especially in valerian roots, and can be isolated from the latter by steam distillation. In industry, isovaleric acid is synthesized from isobutene, carbon monoxide and water in the presence of nickel(II) iodide and nickel tetracarbonyl. Isovaleric acid is used in the syntheis of perfumes and medicaments, includings sedatives and hypnotics.

3) ***2-Methylbutanoic acid, ethylmethylacetic acid***: CH_3-CH_2-$CH(CH_3)$-COOH, a colorless liquid with an unpleasant odor; m.p. $<$ -80 °C, b.p. 177 °C (RS), n_D^{20} 1.4051. 2-Methylbutanoic acid is slightly soluble in water and readily soluble in alcohol and ether. It occurs in many plant juices, in free or esterified form, and sometimes together with isovaleric acid. The industrial synthesis of 2-methylbutanoic acid can start from *n*-butenes, carbon monoxide and water. It is also formed by the Grignard reaction of *sec*-butylmagnesium chloride with carbon dioxide. 2-Methylbutanoic acid can be used to synthesize aromas and flavorings.

4) ***Pivalic acid, trimethylacetic acid, 2,3-dimethylpropanoic acid***: $(CH_3)_3$C-COOH, colorless needles; m.p. 35 °C, b.p. 164 °C. Pivalic acid is slightly soluble in water, and readily soluble in alcohol and ether. Because it is highly branched, pivalic acid has some properties which differ from those of the other V. For example, ester formation and the hydrolysis of pivalate esters is much more difficult. P. can be synthesized by oxidation of pinacolone, by the Grignard reaction (see) from *tert.*-butyl chloride or by the Koch synthesis from isobutene, carbon monoxide and water. Pivalic acid is used to synthesize polyvinyl esters and pharmaceuticals.

Valine, abb. ***Val***: L-α-aminoisovaleric acid, an essential amino acid which is found in many proteins. (Formula and physical properties, see Amino acids). It is synthesized from isobutyraldehyde according to Strecker and Bucher. The racemate is separated by enzymatic hydrolysis of acetyl-DL-valine. V. is a component of penicillin, and D-valine is the starting material for synthesis of the contact insecticide fluvalinate.

Valinomycin: cyclo-(-D-Val-Lac-Val-D-Hyv-)$_3$, a cyclic depsipeptide with an ion-selective, antibiotic effect. V. forms colorless crystals; m.p. 187-190 °C. It dissolves readily in ether, chloroform and acetone, but is nearly insoluble in water. The compound was named for its high valine content; it also contains L-lactic acid (Lac) and D-α-hydroxyisovaleric acid (D-Hyv), which are not amino acids. The hydrophobic side chains of V. make the K^+-V. complex dissolve readily in the nonpolar hydrocarbon layer of the membrane, and in this way allows it to transport K^+ ions through biological and artificial membranes. V. is a classic example of an ionophore, and was the first peptide antibiotic for which the spatial structure was elucidated. V. was isolated from the mycelium of *Streptomyces fulvissimus* by Brockmann et al.

Valium®: see Benzodiazepines.

Valtrate: see Valepotriates.

Vanadiates: salts of vanadic acids. Aqueous alkaline vanadiate solutions contain a number of vanadiate ions. Above pH 13, the most abundant are orthovanadiate, $[VO_4]^{3-}$, while at lower pH values, protonated orthovanadiate, divanadiate, $[V_2O_7]^{4-}$, metavanadiate $[VO_3]^-$, as well as protonated divanadiate and metavanadiate are more abundant. In the pH range 7 to 0, the orange decavanadiates $[V_{10}O_{28}]^{6-}$, $[HV_{10}O_{28}]^{5-}$ and $[H_2V_{10}O_{28}]^{4-}$ are formed. The addition of ammonium compounds to aqueous vanadiate solutions leads to precipitation of ammonium metavanadiate, NH_4VO_3, which crystallizes as insoluble, light yellow, rhombic platelets. This salt is an important intermediate in the technical production of vanadium. Reaction of ammonium metavanadiate with soda solution produces sodium orthovanadiate, Na_3VO_4, which crystallizes in bright yellow leaflets. Calcium ortho-, pyro- and metavanadiates can be obtained by reaction of the components CaO and V_2O_5 in the corresponding molar ratios. Minerals such as vanadinite and roscoelite (vanadium mica) are natural orthovanadiates.

Vanadium, symbol ***V***: an element from group Vb of the periodic system, the Vanadium group (see). It is a heavy metal, *Z* 23, with isotopes with mass numbers 51 (99.76%) and 50 (0.24%, weakly radioactive, half-life $6 \cdot 10^{15}$ a), atomic mass 50.9414, valency V, IV, III, II, less often, I, 0, -I, density 6.092, m.p. 1919 °C, b.p. 3400 °C, electrical conductivity 5.02 Sm mm^{-2}, standard electrode potential (V/V^{3+}) -0.876 V.

Properties. Pure V. is a bright, white, ductile and malleable metal which has a cubic body-centered lattice. Because of passivation, compact V. is not dissolved by non-oxidizing acids (except hydrofluoric acid) or alkalies at room temperature. However, it dissolves in oxidizing acids, such as nitric and sulfuric acids. Above 660 °C, V. is oxidized by oxygen to vanadium(V) oxide, while at high temperatures the reaction with chlorine yields vanadium(IV) chloride. With nonmetals, such as carbon and nitrogen, V. reacts at white heat to form vanadium carbide, VC, or vanadium nitride, VN. It readily forms alloys with metals such as iron, nickel, copper, cobalt, aluminum or tin. In its compounds, most of which are colored, V. is most often found in the +V oxidation state, but the +IV, +III and +II states are also represented. In addition, suitable complex ligands stabilize it in the +I, 0 and -1 states.

V. is a biologically significant trace element; in the mammal, it has a special role in fat metabolism, and it is present in the blood pigments of certain marine animals in the form of hemovanadium. It promotes nitrogen fixation by bacteria.

> In higher concentrations, vanadium compounds are highly toxic for human beings. Chronic exposure leads to poisoning, and high doses paralyse the respiratory center.

Analysis. V. is characterized by formation of brown to red tetrathiovanadiate(V), $[VS_4]^{3-}$, with ammonium sulfide in aqueous solution. This can be converted to vanadium(V) sulfide, V_2S_5, by addition of acid. Vanadic acid reacts with hydrogen peroxide to form red-brown peroxovanadiate(V), $[V(O_2)_4]^{3-}$. There are also organic reagents which are suitable for detection of V., e.g. oxime, 1,10-phenanthroline or *N*-benzoyl-*N*-phenylhydroxylamine. For quantitative determination, V. is first reduced, then titrated with sulfurous acid with manganate as an indicator.

Occurrence. V. makes up $9.0 \cdot 10^{-3}$% of the earth's crust, and is thus a relatively abundant element, but, like titanium, it occurs very widely in low concentrations. Typical V. ores are rare. Iron ores, bauxite and phosphates are important sources; certain types of petroleum also contain V. The most important V. minerals are: vanadinite, $Pb_5(VO_4)_3Cl$, patronite, V_2S_5, roscoelite (vanadium mica), $K(Al,V)_2[AlSi_3O_{10}](OH,F)_2$ and carnotite, $KUO_2VO_4 \cdot 3/2\ H_2O$. Granites, clays, coals and petroleum can also contain V. The concentration of V. in soils is about 200 ppm, while plants contain about 1 ppm V. (dry weight).

Production. In addition to V. ores, iron and chromium ores, phosphates and petroleum ash are used as sources of V. The ores are roasted with added sodium chloride, which converts V. to water-soluble sodium metavanadiate. When V.-containing iron ores are processed, the V. is dissolved in the crude iron, and can be driven into the slag by stepwise blasting. The slag is then roasted in the presence of sodium chloride, Glauber's salt, soda or sodium hydroxide to form sodium metavanadiate. This is leached out and precipitated as vanadic acid by addition of sulfuric acid, or converted to ammonium metavanadiate by addition of ammonium salts. These vanadium(V) compounds are then heated to convert them to vanadium(V) oxide, V_2O_5, which is reduced by heating with aluminum. As an intermediate, a vanadium-aluminum alloy is obtained; the aluminum is distilled out in vacuum at 1700 °C. In a frequently used alternative method, calcium is used as a reducing agent at 950 °C: $V_2O_5 + 5\ Ca \rightarrow 2\ V + 5\ CaO$. Final purification of V. can be done by Vacuum deposition (see). Most of the V. used today is obtained as Ferrovanadium (see).

Applications. V. is used mainly in the form of ferrovanadium, as an alloy component for steels. V. is also a component of high-temperature alloys. Because of its low neutron-capture cross section, pure V. is suitable as a cladding for nuclear fuel rods. V. compounds are used as heterogeneous catalysts for oxygen transfer reactions, e.g. in SO_2 or naphthalene oxidation, or as homogeneous catalyst components for low-pressure ethene polymerization.

Historical. After the new element was first discovered by del Rio in 1801 in a Mexican lead ore, the results were doubted and the element was rediscovered in 1830 in a Swedish iron ore by Sefström, who named it for Vanadis (another name for the goddess Freya). Roscoe perpared the metal in 1867 by reduction of vanadium(III) chloride with hydrogen.

Vanadium carbide: VC, hard, alloy-like, cubic compound; M_r 62.95, density 5.77, m.p. 2810 °C, b.p. 3900 °C. V. conducts electric current. It is made from vanadium(V) oxide and carbon at 1100 to 1200 °C. It is used to make hard metals and as an alloy additive for steels.

Vanadium chlorides: ***Vanadium(II) chloride, vanadium dichloride***, VCl_2, green leaflets with a mica-like sheen; M_r 121.85, density 3.23. VCl_2 is readily soluble in water, forming the gray-violet hexaaqua complex $[V(H_2O)_6]Cl_2$. It is obtained by reducing VCl_3 with hydrogen. ***Vanadium(III) chloride, vanadium trichloride***, VCl_3, is a hygroscopic compound the color of peach blossoms; M_r 157.30, density 3.0. It can be conveniently prepared by heating vanadium(V) oxide with disulfur dichloride under reflux: $2\ V_2O_5 + 6\ S_2Cl_2 \rightarrow 4\ VCl_3 + 5\ SO_2 + 7\ S$. ***Vanadium(IV) chloride, vanadium tetrachloride***, VCl_4, is a red-brown, oily liquid which is very sensitive to moisture; M_r 192.75, density 1.816, m.p. -28 °C, b.p. 148.5 °C. VCl_4 is obtained by chlorination of vanadium, vanadium nitride or vanadium silicide.

Vanadium complexes: coordination compounds of vanadium. The most common type are octahedral, e.g. the hexaaqua complexes of vanadium(II) and (III), $[V(H_2O)_6]^{2+}$ (violet) and $[V(H_2O)_6]^{3+}$ (green); the oxopentaaquavanadium(IV) ion $[VO(H_2O)_5]^{2+}$ (blue), cyano complexes of the type $[V(CN)_6]^{n-}$ (n = 3-5) and $[VO(CN)_5]^{3-}$. There are also carbonyl compounds $[V(CO)_6]$ (blue-black, paramagnetic) and $[V(CO)_6]^-$, dipyridyl complexes $[V dipy]^{n-}$ (n = 0-2) and $[VF_6]^-$. Thus the V. can have the coordination number 6 in any of its oxidation states from -I to +V. Tetrahedral structures are found, e.g. in orthovanadiate(V), $[VO_4]^{3-}$, and in adducts of vanadium(III) chloride, $[VCl_3D]$. The pentagonal bipyramidal coordination type is unusual for vanadium, but it is found in the ruby-red potassium heptacyanovanadiate(III), $K_4[V(CN)_7]$.

Vanadium fluorides: ***Vanadium(III) fluoride, vanadium trifluoride***, $VF_3 \cdot 3H_2O$, forms dark green, rhombohedral crystals; it can be made by reaction of vanadium(III) hydroxide with hydrofluoric acid; M_r 191.98. ***Vanadium(IV) fluoride, vanadium tetrafluoride***, VF_4, is a brownish yellow, very hygroscopic powder which can be made from vanadium(IV) chloride and anhydrous hydrofluoric acid; M_r 126.94, density 2.975, m.p. 325 °C (dec.). In aqueous solution, VF_4 forms oxofluorovanadiates, $M_2[VOF_4]$, with fluorides. ***Vanadium(V) fluoride, vanadium penta-***

lignin used). Under laboratory conditions, yields up to 25% can be obtained by alkaline nitrobenzene oxidation. If lignins from broad-leaf trees are used, 2 to 3 times as much *syringaaldehyde* (3,5-dimethoxy-4-hydroxybenzaldehyde) is formed as V. For example, up to 30% syringaaldehyde and 12% V. can be obtained from aspen lignin.

V. is used as a flavoring for foods and as a component of synthetic perfumes. It can be oxidized to ***vanillic acid***, the derivatives and polycondensates of which (polyesters) have interesting material properties.

Van't hoff equation: see Osmosis.

Van't Hoff reaction isobar: see Mass action law.

Van't Hoff reaction isochore: see Mass action law.

Van't Hoff reaction isotherm: see Mass action law.

Van Urk reaction: see Ergot alkaloids.

Vapor: The gas form of a substance in the neighborhood of its condensation point. If a substance evaporates at normal or slightly less than normal pressure, the V. also contains air. If it is in thermodynamic equilibrium with the liquid, it is said to be ***saturated***. In the absence of the liquid phase, the partial pressure can be lower than the saturation vapor pressure (***unsaturated*** or ***superheated V.***), or, within certain limits, it can be higher (***supersaturated V.***). The supersaturated state of a V. is unstable. It exists because the formation of a new phase in a very clean system can be inhibited (see Thomson's equation). If nuclei of the liquid phase have formed (e.g. on dust particles or ions), the supersaturated V. is very rapidly converted to the saturated state by partial condensation. For unsaturated V., the degree of saturation is given by the ratio of the partial pressure p to the saturation vapor pressure p_s. In meteorology, it is conventional to give the percent humidity, which is $100p/p_s$.

Vapor compression: a distillation process based on the principle of a heat pump. Distillations requiring a large amount of energy can be run more economically if the vapors leaving the top of the column are compressed by a compressor and the heat applied to the liquid in the distillation kettle. The method saves on heat for the column and cooling water for the distillation cooler; only the energy to run the compressor must be applied.

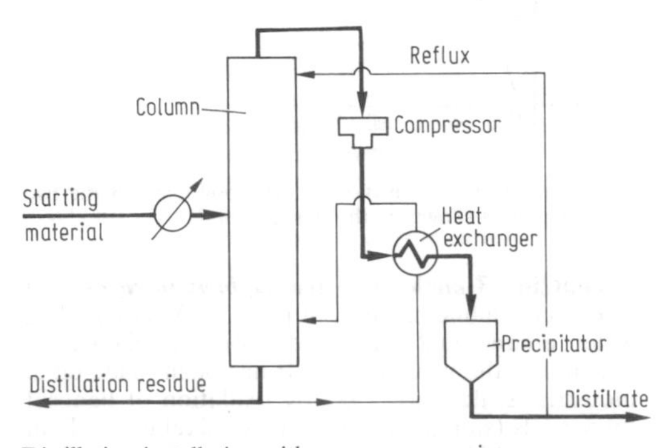

Distillation installation with vapor compression.

Vaporization process: see Sewage treatment.

Vapor phase inhibitor, abb. ***VPI***: long-chain aliphatic or alicyclic aminonitrites or aminocarbonates which are used for Corrosion protection (see). One of the most effective D. is dicyclohexylammonium nitrate (Dichan®, Leukorrosion C®). The D. evaporates and forms an adsorption film on the surface of the metal which protects it from water and oxygen. Some also attract NO_2 ions to the metallic surface and thus passivate it. D. have the advantage over fats and waxes used for corrosion protection that they are simple to apply, and do not have to be removed before the object is used. A disadvantage is that they cause corrosion of some nonferrous metals and discolor some plastics.

VPI are used to protect objects made of steel or other metals during storage, transport or mothballing. The object is packed in paper impregnated with the VPI or a powdered VPI is placed in the container along with a drying agent, such as silica gel. The method is most successful when the packaging is sealed (e.g. in plastic).

Vapor pressure: the pressure which is established over a liquid in a closed vessel in the absence of other gases. Solid substances also have a V., although it is more correct to call this the ***sublimation pressure***.

For a pure substance, the V. (also called ***saturation vapor pressure***) depends only on the substance and the temperature. It increases with increasing temperature; a plot of the V. versus the temperature is represented graphically as its ***vapor pressure curve*** (see Phase diagram). This curve can be described by the Clausius-Clapeyron equation (see). The V. at the melting point is equal to the sublimation pressure of the solid at that temperature. A liquid boils when its V. is equal to the external pressure; atmospheric pressure is 101.3 kPa. If the pressure over the liquid is decreased by pumping off the gas phase, the boiling point is reduced. This is used in vacuum distillation to separate substances under gentle conditions.

If additional gases are present in the closed vessel, the saturation vapor pressure is scarcely affected, so long as these gases are not soluble in the liquid. The total V. over two immiscible liquid phases is equal to the sum of the V. of the two phases. However, for homogeneous mixtures, the V. of the components depend on the ratio of components (see Vapor pressure diagram, see Raoult's law.

Strictly speaking, the saturation V. apply to plane liquid surfaces. Small droplets have an elevated V. (see Thomson's equation).

There are many ways to measure the V. In *statistical methods*, the liquid is placed in an evacuated vessel and the V. is measured as a function of temperature, using a closed manometer. In *dynamic methods*, the boiling temperatures are determined as a function of the external pressure. In *transport methods*, the liquid is allowed to flow into an insoluble gas so slowly that it becomes saturated with the vapor. The amount of liquid carried off by a given volume of gas is then measured. Very low vapor and sublimation pressures are measured in a *Knudsen cell*. This is a closed container in which the sample is placed. It is connected by a very fine valve to a high vacuum system. The amount of substance flowing through the valve by Knudsen diffusion is measured, e.g. by mass spectroscopy. It is proportional to the V. of the sample.

Vapor pressure curve: the graph of the vapor pressure of a pure substance versus the temperature (see Phase diagram). It begins at the melting point and ends at the critical point.

Vapor pressure diagram: the representation of the vapor pressure over a mixture at a given constant temperature as a function of the composition of the liquid phase. The composition is generally given as the mole fraction (or mole percent), but can be given as the mass fraction (or percent) (see Composition parameters). If a mixture contains two volatile components A and B, then the composition can be described by the mole fraction, because $x_A + x_B = 1$, or $x_B = 1 - x_A$. If x_B is plotted on the abcissa, pure substance A is present at $x_B = 0$, and pure B at $x_B = 1$.

The total vapor pressure p over the mixture is, according to Dalton's law (see), the sum of the partial pressures $p_A + p_B$. In the ideal case, Raoult's law (see) holds for both partial pressures: $p_A = p_{oA}x_A = p_{oA}(1-x_B)$ and $p_B = p_{oB}x_B$, so that $p = p_{oA} + (p_{oB} - p_{oA})x_B$, where p_{oA} and p_{oB} are the vapor pressures of the pure components. The total vapor pressure increases or decreases linearly with x_B, depending on whether $p_{oB} - p_{oA}$ is less or greater than 0 (Fig. 1).

The gas phase does not have the same composition as the liquid phase, but is richer in the more volatile component (i.e. the one with the higher vapor pressure). This is the basis of separation of a mixture by distillation. In the figure a second curve g is shown, which gives the mole fractions in the coexisting gas phase. These are obtained by moving horizontally from the vapor pressure curve p to curve g, and reading from the abcissa the mole fraction corresponding to this point. In Fig. 1, point 1 corresponds to the total pressure of a liquid mixture with the mole fraction $x_B = x_{liq}$, and point 2 gives the composition of the gas phase in equilibrium with it ($x_B = x_g$).

Real mixtures deviate from Raoult's law. If the interactions between the different kinds of molecule are less than in the pure substances ($f > 1$), the partial pressure curves and the total pressure curve are convex, i.e. the volatility is increased. In the opposite case, the curves are concave. Figs. 2 and 3 show examples where the effects are so strong that the total pressure has a maximum or a minimum. At the maximum or minimum, the liquid and gas have the same composition; the mixture is an azeotrope which cannot be separated by distillation.

If a mixture contains more than two components (n components), n-1 molar fractions are required to establish the composition. The D. in this case cannot be represented in a plane. For ternary systems, a triangular representation is possible. The vertices correspond to the three components, and the vapor pressures are indicated as isobars similar to the elevation lines on a map.

Diagrams in which the boiling temperature at constant pressure is plotted as a function of composition are called Boiling diagrams (see).

Vapor pressure reduction: see Raoult's law.

Variamine blue: 4-amino-4'-methoxydiphenylamine hydrochloride, a water-soluble, greenish-blue powder used as an indicator for iron determination and as a redox indicator. As diazo components with naphthol AS, ***variamine blue base B*** (4-amino-4'-methoxydiphenylamine) and ***variamine blue FG*** (4-amino-3-methoxydiphenylamine) give blue development dyes which are extremely fast.

Varnish: a transparent coating applied to wood surfaces to seal and protect them. The V. is applied as a liquid or paste which dries to form a thin, tightly bound layer. It consists of a solution of suitable organic film-forming compounds in a solvent; in some cases drying agents and/or pigments are also present.

The film-forming compounds are usually macromolecular substances, but sometimes small molecules are used which polymerize during the drying process. The most important film-forming compounds are plant oils, cellulose nitrate, vinyl polymers, alkyde resins, polyesters, polyurethanes, chlorinated latex, and urea, melamine, phenol, epoxide and natural resins. The mechanical properties of the V., e.g. shine, hardness and ability to be sanded and polished, are improved to a certain degree by increasing the relative molecular mass of the film-forming compound.

Softeners are added to improve the elasticity of the film. The most common of these are various phthalic acid esters, but esters of phosphoric, adipic and long-chain fatty acids are also used. Oil V. contain drying agents (siccatives) to accelerate the polymerization which occurs during drying. The most common siccatives are heavy metal salts of fatty acids. Pigments can be either inorganic or organic; barite, hematite or graphite are typical mineral pigments, while some ex-

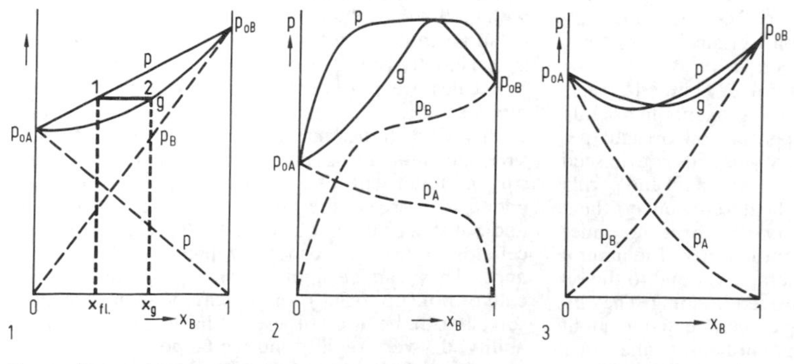

Types of vapor pressure diagram.

amples of organic pigments are bordeaux, permanent red and heliogen blue. Ether, acetone, benzene, toluene, petroleum distillates, amyl acetate, butyl acetate and terpentine oil are used as solvents.

The properties of a V. are determined by its nonvolatile components, which are usually a mixture of substances with complementary properties. The final consistency of the V. film depends not only on the product applied, but also on the conditions under which it applied and the surface to which it is applied. There are essentially two types of drying, physical and chemical. 1) With *physically drying V.*, such as cellulose nitrate, cellulose ether, chlorinated latex and polyvinyl V., the film is formed by evaporation of the solvent. The process is reversible, that is, the film can be removed by the same solvent. For industrial purposes, the significance of this type of V. is the speed with which they dry. 2) With *chemically drying V.*, the films form as a result of chemical reactions. In principle, polymerization, polycondensation and polyaddition can lead to film formation. a) In oxidatively drying V., the film is formed after the solvent has evaporated, by uptake of oxygen and subsequent polymerization (e.g. oil and alkyde V.) b) Other V. form films by polycondensation (e.g. phenol resin and urea-melamine-alkyde resin V.) or by catalysts (e.g. acid-hardening V.) Chemical drying is irreversible, that is, the film is insoluble in solvents. In some V., physical and chemical drying are combined; such combination V. can contain both cellulose nitrate or polyvinyl compounds and drying oils or alkyde resins.

In many cases, several coats of V. are applied. The simplest system consists of a primer and a finish coat. The primer gives good adhesion to the surface, and the finish coat protects the surface from mechanical, chemical and weathering effects, and gives the desired appearance.

V. can be produced by hot or cold processes. The cold process is preferred when the raw materials are readily soluble and the solvents have low flame points. These V. are clarified by filtration, centrifugal sieving or centrifugation. The hot process of V. production is called cooking; the method leads to transesterification, condensation or polymerization, and is used with oil and synthetic resin V.

There are many kinds of V., including the following important types. ***Spirit V.*** are solutions of natural (e.g. shellack, copal, colophony) or synthetic resins (e.g. alkyde resins and polyvinyl compounds) in volatile organic solvents. The most commonly used solvent is 90 to 96% ethanol. Spirit V. is used for shellack furniture polishes and spirit enamel V. for toys, musical instruments, baskets, etc.

Oil V. are solutions of natural or synthetic resins and drying oils with added drying agents in volatile organic solvents such as terpentine oil or light petroleum. The most common drying oils are linseed and wood oil; tall oil, ricinus oil, oiticia oil and perilla oil are also used. The oils are heated to convert them to stand oils. In this process, two to three molecules of oil generally condense to an oligomer. The natural or synthetic resins used as fillers are bound to the oil molecules by covalent or hydrogen bonding. They include natural copals which have been heated to about 360 °C, colophony-modified phenol resins and other synthetic resins. In general, the higher the ratio of resin to oil, the shorter the drying time, and the harder the V.; however, the pliability and weather-resistance of the V. also decreases at higher resin to oil ratios. Oil V. are used for many purposes where the coating can be given enough time to dry.

Industrial production of consumer items led to a demand for V. which would dry more quickly than the oil V. The following synthetic V. dry rapidly and provide tough, shiny coatings.

Alkyde resin V. contain mixed esters of polyvalent alcohols, such as glycerol, pentaerythritol, hexitol and sorbitol, with polybasic carboxylic acids, such as phthalic, isophthalic, adipic and sebacic acids. Free hydroxyl groups are subsequently esterified with unsaturated fatty acids. Modified alkyde resins are becoming more and more important for V. These include reaction products of styrene, acrylic acid and polyamides with the alkyde resins. The polyamide-modified alkyde resins are highly thixotropic, so they do not run down vertical surfaces to which they are applied. The drying of alkyde resin films can be significantly accelerated by addition of cobalt, lead and manganese oleates or naphthenates. Strictly speaking, alkyde resin V. are oil V., but they are superior to the classical oil V. with respect to shine, drying and color-fastness; they are commercially available as air-drying, heat-drying and bake-on V. Heat drying is only an accelerated form of air drying which occurs at 60 to 80 °C, while baking is done at temperatures above 80 °C, for times of 10 to 60 min.

Silicon resin V.: the properties of a V., such as surface hardness, covering and pigment compatibility are increased by reaction of organosilicon compounds with monomers or polymers, e.g. epoxides, melamine, phenol, alkyde and polyster resins. For example, silicon alkyde resins produce V. coats with greater resistance to heat and chemicals, and with more durable shine and color. Silicon resin bake-on V. are very important (see Polysiloxanes).

Polyester V. are solutions of unsaturated polyesters (e.g. glycol maleate) in monomeric liquids (e.g. styrene) to which organic peroxides are added as reaction accelerators. Polyester V. are used without solvents, or with relatively little solvent. The resulting films are highly resistant to weathering and chemicals. The main use of polyester V. is for production of wood V. and of spackel for wood and iron.

Cellulose nitrate V. are also known as ***zapon V.***; they are solutions of collodium in acetone and acetate esters. The modern cellulose nitrate V. can be diluted with methanol, toluene or other organic solvents. The cellulose to be used for these V. is nitrated and then subjected to pressure cooking, so that the macromolecules are partially degraded. This makes them more soluble and makes the solution less viscous. Nearly all cellulose nitrate V. contain softeners. Resins, such as shellack, colophony, copal or alkyde resins modified with phenol, urea, ketones or oil, are added to increase the adhesion, weather resistance and polishing quality of the V. The alkyde resins give cellulose nitrate V. especially high weather resistance. They can be applied with spray guns, either cold or hot (up to 80 °). In this way, V. with very little solvent can be used; these V. have good covering ability, dry very rapidly and can be polished to give a high shine. Cellulose nitrate V. are used on furniture, leather, paper and metals (automobiles), and especially for industrial assembly line production.

V. from vinyl polymers are highly resistant to light and very elastic. The raw materials are polyvinyl chloride, post-chlorinated polyvinyl chloride, polyvinyl acetate, polyvinyl alcohols, polyacrylamide and polymethacrylic acid derivatives, polystyrene and polyvinyl ethers. The individual polyvinyl compounds have special properties, and the V. made from them have special applications. For example, V. from post-chlorinated polyvinyl chloride are used on aircraft, and V. made from polyacrylic acid derivatives are used for light metals and for elastic spackels.

Epoxide resins give chemically resistant films with excellent adhesion, especially to metals. They are used mainly for industrial coatings, e.g. of tanks, machine parts, plate metal for construction, pipes, etc.

Polyurethane V. are both hard and elastic, and are very resistant to moisture, chemicals and ageing; they also have good electrical properties. They are used as insulating V., and also as metal coatings and underwater paints.

V. from urea, melamine and phenol resins are hardened either by temperature (baking) or by addition of acid (acid-hardening V.). Urea and melamine resin V. are used mainly for decorative purposes, while phenol resin V. are used mainly for protective coatings, because they are intrinsically yellow to brown in color.

In ***chlorinated latex V.***, chlorinated natural rubber is the film-producing compound. These very rapidly drying V. are highly resistant to water and chemicals, and are used mainly in mines and chemical plants.

Asphalt, bitumen and tar V. are made by melting the bituminous binder and addition of solvent to it, in many cases, a nonpolar hydrocarbon. Both bitumen and tar can be emulsified in water by suitable emulsifying agents. The thoroughly dried coatings are water repellent and have considerable resistance to chemical and mechanical stress.

Japan lacquer is made from the Japanese lacquer tree *Rhus vernicifera*. It is used for arts and crafts, but is not used for industrial purposes.

Corrosion-preventing V. include all V., but the term is generally applied to those V. which protect a metal surface from aggressive liquids, gases and vapors, including moist air (which causes rusting).

Drying processes and times for various film-forming compounds

Compound	Body of V. in %	Type of drying	Time of drying in h at indicated temperature	
Linseed oil	95-100	Chemical	20°C	36
Oil	70- 80	Chemical	20°C	24
Alkyde resin	50- 60	Chemical	20°C	12
			80°C	1
			130°C	0.5
Cellulose nitrate/ alkyde resin (1:2)	40- 45	Physical and chemical	20°C	2-4
			60°C	1
Cellulose nitrate	20- 25	Physical	20°C	0.5
Polyester/ polyurethane	20- 25	Chemical	180°C	0.5

The components of all V. can be combined, within limits of compatibility, and there are many modern ***combination V.*** with special properties for special applications.

Varnish remover: a mixture of organic solvents (e.g. acetone, toluene, halogenated hydrocarbons) used to remove varnish.

Varnish resins: natural or synthetic resins used to make varnishes.

Vaseline: a natural or synthetic mixture of solid and liquid petroleum hydrocarbons with melting points between 35 and 60 °C. ***Natural vaseline*** consists of paraffin hydrocarbons, viscous mineral oils and small amounts of *n*-paraffins. It is homogeneous, with a salve-like consistency, and can be drawn into a thread. It is transparent in a thin layer. It is slightly soluble in alcohol, readily soluble in ether, chloroform, benzene and petroleum ether, and insoluble in water. Natural V. is obtained from distillation residues of paraffin-based petroleum, or from the residues left after de-paraffinization of heavy petroleum distillates (petrolatum). ***Synthetic V.*** is a cheaper substitute, is made by dissolving paraffin, usually with added ceresin, in refined mineral oils. Because the paraffin tends to crystallize, synthetic V. loses its salve-like consistency after a time. It also tears off rather than drawing a thread, is granular and not transparent.

V. is used mainly in the pharmaceutical and cosmetic industries; the technical grades are used as rust protection and ball-bearing grease, and to make leather and textiles water-repellent.

Vaseline oil: see Paraffins.

Vaska complex: *trans*-chlorocarbonylbis(triphenylphosphane) iridium(I), *trans*-$[IrCl(CO)(P(C_6H_5)_3)_2]$, the most thoroughly studied of the iridium(I) complexes. V. forms yellow crystals, and can be obtained by reaction of iridium(III) chloride with triphenylphosphine in 2-methoxyethanol. It undergoes oxidative addition reactions with many compounds, such as H_2, Cl_2, HX and CH_3I. The products are octahedral iridium(III) complexes. Molecular oxygen is reversibly added (see Dioxygen complex).

Vasoactive intestinal peptide, abb. ***VIP***: a peptide synthesized in the duodenum and also present in the cerebral cortex. It has about 1/3 the hyperglycemic effect of glucagon, and stimulates the smooth musculature. It also affects the central nervous system. V.I. consists of 28 amino acids; the *N*-terminal sequence is very similar to the sequences of glucagon and secretin, and is also homologous to gastrin inhibiting peptide. *C*-Terminal partial sequences also have biological activity.

Vasopressin, ***antidiuretic hormone***, abb. ***ADH, antidiuretin, adiuretin***: a neurohypophysial peptide hormone. V. is a heterodetic cyclic peptide with an antidiuretic effect.

Cys–Tyr–Phe–Gln–Asn–Cys–Pro–Arg–Gly–NH_2

In higher doses, it also increases blood pressure. Because of structural similarity to oxytocin, it also has a slight oxytocin effect. Porcine pituitary contains $[Lys^8]$V., but bovine and other mammalian pituitaries have [Arg^8]V. When V. is deficient, diabetes insipidus occurs; in this disorder, because of inadequate resorption by the kidneys, up to 20 l urine per day is passed. The biosynthesis of V. goes via the vasopressin-neurophysin II precursor, the amino acid sequence of which was elucidated in 1982. This glycoprotein contains 166 amino acid residues; after the *N*-terminal signal sequence of 19 amino acids comes the [Arg^8]V. sequence. This is followed by a glycine residue, which provides the terminal amide group of V. after hydrolysis. After a pair of basic amino acids comes the sequence of neurophysin II (95 amino acids), and after another Arg residue is a 39-amino acid glycopolypeptide.

V. is synthesized in the nucleus supraopticus of the hypothalamus. It is transported through the supraoptico-hypophysial tract to the posterior pituitary, which acts as a storage and release facility for V. and oxytocin. Very active analogs of V. are known which have the advantage that they can conveniently be applied through the nose. The commercially synthesized 1-deamino-[D-Arg^8]V. (Abb. DDAVP) has 400 times the effect of V. on the kidneys, but its effect on blood pressure is practically negligible. There are other useful analogs of V., such as [De-Gly-NH_2^9] V., which has CNS activity.

The sequence of V. was determined in 1953/54 by du Vigneaud, who was also the first to synthesize it.

Vat dyes: water-insoluble dyes which can be applied to the yarn only in the reduced state; they are very fast to light and washing. The most important V. are derivatives of anthraquinone, indigo, naphthalene or perylene. Before application, V. must be reduced with sodium dithionite, $Na_2S_2O_4$, zinc powder, iron(II) salts or an enzyme (see Indigo), which makes them soluble in an alkaline solution and usually also bleaches them or changes their color markedly. The yarn or cloth is dipped into the reduced alkaline solution and colored, then hung in the air, where the atmospheric oxygen reoxidizes the dye to its original, water-insoluble state. In modern dyeing procedures, the dye can be applied as the pigment to the yarn and then reduced in situ.

Some important V. are indigo, indigo dyes, hydrone dyes, indanthrene dyes and helindone dyes.

VB method: same as Valence structure theory (see).

Vegard's rule: a relationship established in 1921 by L. Vegard for substitution Mixed crystals (see) with continuous miscibility and a random distribution of components. According to V., there is a linear relationship between the lattice constants and the composition of such mixed crystals. In the process of substitution, a host lattice with larger components is compressed and one with smaller components is expanded. If the composition of the mixed crystal and the lattice constants of the pure components are known, V. can be used to calculate the lattice constants of the mixed crystal. V. is obeyed mainly by metallic systems (Fig.), but often even these obey only approximately. Often instead of the straight lines predicted by the V. (and actually observed, e.g. for the gold/platinum system), a slightly curved line is observed, which may lie either below (e.g. gold/silver) or above (e.g. gold/copper) the straight line.

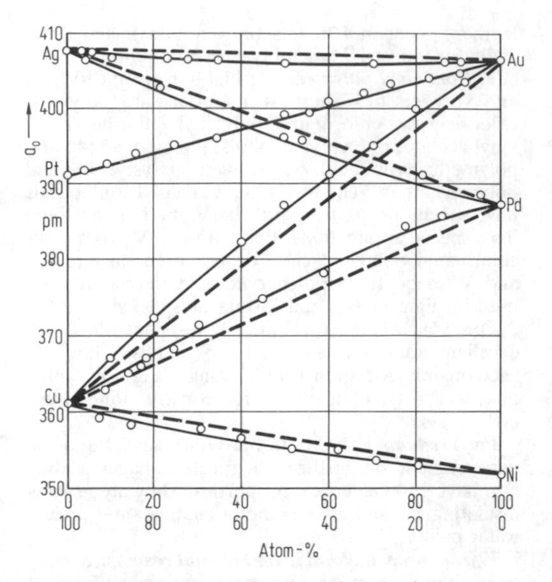

Lattice constants of metallic mixed crystals of two components as a function of their composition. The dashed lines correspond to Vegard's rule.

Vehicle exhaust: the gaseous products of combustion in internal combustion engines. The main components of V. are nitrogen, carbon dioxide and water; it also contains nitrogen oxides and unburned hydrocarbons. The nitrogen oxides form as a result of reactions at high temperatures: $N_2 + O_2 \rightarrow 2\,NO$; $2\,NO + O_2 \rightarrow 2\,NO_2$. The unburned products include carbon monoxide, hydrocarbons, aldehydes and similar compounds. V. also contain aerosols of unchanged fuel and lubricating oil. Badly maintained diesel engines also produce considerable quantities of solids, which make up *diesel smoke*. Gasoline may contain lead compounds which raise the octane rating (see Knocking). Arranged in order of quantity, the most important air pollutants are carbon monoxide, nitrogen oxides and hydrocarbons. The oxygen-containing organic compounds, mostly aldehydes (and especially formaldehyde), irritate mucous membranes. In the presence of UV radiation, V. is converted to photochemical smog.

The amounts of air pollutants in V. can be reduced by improved construction of engines and carburetors, proper maintenance of the vehicle, measures to minimize unfavorable conditions during idling, acceleration and delays (microelectronic regulatory systems in the engine) use of unleaded Fuels (see) and catalytic combustion of the V.

Velinvar®: a cobalt alloy consisting of 56-63% cobalt, 29-34% iron and 7-10% vanadium. It is used similarly to Co-Elinvar® (see).

Velocity distribution: see Maxwell-Boltzmann velocity distribution.

Venturi washer: a device for removing particulates from stack gases, e.g. from blast furnaces, converters or garbage burning plants. The scrubbing liquid is sprayed into the gas and the wetted particles are precipitated downstream.

Venzar®: see Herbicides.

Verapamil: see Coronary pharmaceuticals.

Veratranin: see Steroid alkaloids.

Veratrol, ***1,2-dimethoxybenzene, pyrocatechol methyl ether***: a phenol ether, colorless crystals; m.p. 22.5 °C, b.p. 207 °C, n_D^{21}. V. is slightly soluble in water, and soluble in ethanol and ether. It is easily synthesized by methylation of pyrocatechol or guaiacol with dimethylsulfate. It is also a degradation product of various natural products, e.g. lignin or the alkaloid papaverine. V. is used as an intermediate in the synthesis of veratrum alkaloids and for various analytical detection reactions.

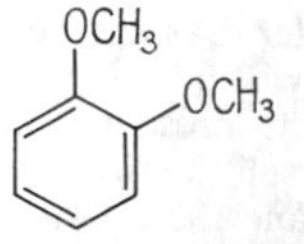

OCH_3 OCH_3

Veratrum aldehyde, ***3,4-dimethoxybenzaldehyde***: crystalline needles with a vanillin-like odor; m.p. 44 °C (58 °C also reported), b.p. 258 °C. V. is slightly soluble in water, and readily soluble in alcohol and ether. It undergoes the reactions typical of aromatic aldehydes. V. is found in nature in a few essential oils, such as cevadilla seed oil. V. is biosynthesized by decarboxylation of opianic acid; chemically, it can be synthesized by methylation of vanillin. It is used to make flavors and aromas, and in drugs.

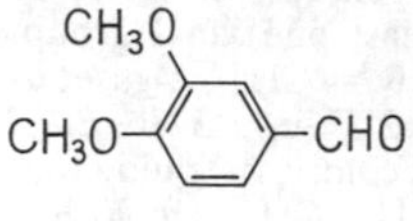

CH_3O CH_3O CHO

Verdigris: a mixture of basic copper(II) acetates obtained by the reaction of acetic acid vapor with copper or brass in the presence of air. V. can also be formed by acidic foods in copper containers. It is used as a pigment in paints and in the preparation of gilder's wax for fire gilding.

Verel: see Synthetic fibers.

Verneuil process: 1) a general method of Crystal growing (see); 2) a special method for Silicate synthesis (see).

Verodoxin: see Purpurea glycosides.

Veronal®: see Barbitals

Vesicular process: a method of reproduction in which the image is formed by scattering of light. The light-sensitive compound (e.g. diazonium salt) is embedded in a thermoplastic material and splits off a gas (in the case of diazonium salts, nitrogen) where the light strikes it. A heat treatment enlarges the gas bubbles, which act as scattering centers. A special system is based on the decomposition of hydrogen peroxide with photolytically produced silver. The image is fixed by slow decomposition of the diazonium salt or hydrogen peroxide, during which the gas can diffuse away. The V. has a high resolution and sensitivity, and is used mainly for reproduction of microfilms and black-and-white movie films.

Vesuvin, ***Bismark brown R, leather brown, excelsior brown***: a diazo dye. V. forms a black-brown powder which dissolves in water to give a brown color; it is also readily soluble in alcohol. It consists mainly of the dihydrochloride of benzene-1,3-bis(azo-3-phenylenediamine), and is obtained by diazotization of 3-phenylenediamine; in a strongly acidic solution, the diazotized 3-phenylenediamine couples with unreacted starting material. In addition, small amounts of triaminoazobenzene and other bases are formed. V. is now only sporadically used to dye wool or tannin-treated cotton, but it is still important in the dyeing of leather and paper. V. is also used as a vital stain, especially for bacteria.

Vetivazulene, ***4,8-dimethyl-2-isopropylazulene***: a red-violet, crystalline hydrocarbon (see Azulene); m.p. 31-32 °C, b.p. 140-160 °C at 270 Pa. V. occurs naturally in vetiver oil and is used as a mild antiseptic.

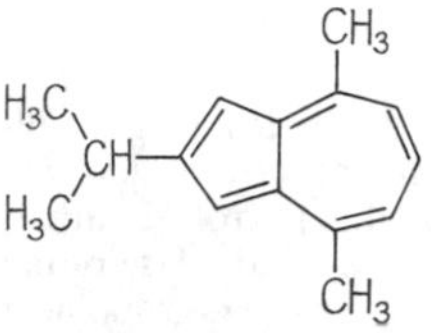

CH_3 H_3C CH H_3C CH_3

Vetiver oil: a colorless, transparent essential oil of varying consistency. It has an unusual, sweet, woody odor which persists for a long time. The main components of V. are α- and β-vetivone, which comprise 60%, and several sesquiterpene alcohols, which account for the odor. V. is obtained from the dried roots of vetiver grass (*Andropogon squarrosus*) by steam distillation and subsequent petroleum-ether extraction of the distillate. It is used in cosmetics and as a source of vetivenols (the sum of the sesquiterpene alcohols).

VF: abb. for volcanic fiber.

VI: abb. for viscose fiber.

vic-: see Nomenclature, sect. III D.

Vicilin: see Lectins.

Vicinal: see Nomenclature, sect. III D.

Vickers hardness: the hardness of a metal, determined by a method developed by the English firm Vickers. A diamond four-sided pyramid with an angle of 136° between opposing sides is pressed into the metal at a pressure of 294 N, and the ratio of pressure to the surface of the resulting impression is the V. Up to 300 degrees hardness, this measure agrees with the Brinell hardness (see). The V. can be readily measured in many situations, and is especially suitable for hardness determination of very small samples and thin sheets (small-load hardness).

Victoria green: see Chromium(III) oxide hydrate.

Victoria yellow: same as Metanil yellow (see).

Vidal black: the first industrially produced sulfur pigment. It is made by heating 3,5-dinitrophenol with sodium sulfide and sulfur, and is used to dye cotton. V. was first made in 1893 by Vidal.

Vilsmeier-Haack reaction: a reaction of organic compounds with formylation reagents suitable for synthesis of aldehydes. Tertiary aromatic amines, phenol ethers, activated arenes and π-excess heteroaromatics can be used as substrates; in these compounds an H atom is replaced by a formyl group -CHO in the course of the reaction. The formylation reagent, called the Vilsmeier reagent (V.-R.) is produced from dimethylformamide (DMF) and phosphorus oxide chloride, $POCl_3$, in solvents such as 1,2-

dichloroethene, or without a solvent in an excess of $POCl_3$.

$$[(CH_3)_2\overset{+}{N}{=}CH{-}Cl]\,PO_2Cl_2^-$$

Vilsmeier reagent

The attack of the Vilsmeier reagent on an electron-rich position of the substrate leads to an intermediate which can be isolated in exceptional cases, but which is usually immediately hydrolysed:

$$(CH_3)_2N{-}C_6H_5 + V.R. \longrightarrow$$
$$[(CH_3)_2N{-}C_6H_4{-}CH = N(CH_3)_2]^+Cl^- + H_2O \longrightarrow$$
$$(CH_3)_2N{-}C_6H_4{-}CHO + (CH_3)_2N^+H_2Cl^-.$$

Amino-substituted aromatic aldehydes, e.g. 4-dimethylaminobenzaldehyde, phenol ether aldehyde and heterocyclic aldehydes of the pyrrole-2-carbaldehyde type, can be made in good yields. Use of the reaction for ketones with a $-CH_2-$ or $-CH_3$ group next to the carbonyl group, which was discovered in 1959 by Z. Arnold, leads to substituted β-chlorovinyl aldehydes, e.g. 3-chlorocinnamaldehyde:

$$C_6H_5{-}CO{-}CH_3 + DMF/POCl \longrightarrow$$
$$C_6H_5{-}C(Cl){=}CH{-}N^+(CH_3)_2{-}Cl^- \longrightarrow$$
$$C_6H_5{-}C(Cl){=}CH{-}CHO + (CH_3)_2N^+H_2Cl^-.$$

Another formylation reaction with a comparable mechanism is the introduction of formyl groups into exocyclic methylene groups or in amides and imides.

VIM: see Modal fibers.

Vinblastin: see Alkaloids.

Vinca alkaloids: see Alkaloids.

Vinclozoline: see Carboximide fungicides.

Vincristine: see Alkaloids.

Vinitron®: see Synthetic fibers.

Vinyl: a term for the group $CH_2{=}CH-$ in a molecule, the radical $CH_2{=}CH\cdot$ or the cation $CH_2{=}CH^+$.

Vinyl acetate: $CH_3{-}CO{-}O{-}CH{=}CH_2$, a colorless liquid; m.p. -93.2 °C, b.p. 72.2 °C, n_D^{20} 1.3959. V. is slightly soluble in water, but dissolves readily in most organic solvents. In the presence of light or peroxides, it readily polymerizes, forming polyvinylacetate. V. undergoes the reactions typical of alkenes and esters. It can be hydrolysed to acetic acid and acetaldehyde. It is synthesized by addition of acetic acid to acetyls in the presence of mercury(II) or zinc salts, or by reaction of ethylene, glacial acetic acid and oxygen on palladium catalysts: $CH_2{-}CH_2 + CH_3{-}COOH + 1/3\ O_2 \rightarrow CH_3{-}CO{-}O{-}CH{=}CH_2 + H_2O$. V. is used mainly to produce polyvinylacetate or copolymers.

Vinylacetylene, ***monovinylacetylene***, abb. ***Mova*** ***(divinylacetylene***, abb. ***diva***), ***vinylethyne, but-1-en-3-yne***: $HC{\equiv}C{-}CH{=}CH_2$, the dimer of ethyne, an easily liquefied, gaseous hydrocarbon; b.p. 5.5 °C. It is synthesized industrially by passing ethyne through a solution of copper(I) chloride and ammonium chloride. The products also include trimers, such as 1,2-divinylethyne, $CH_2{=}CH{-}C{\equiv}C{-}CH{=}CH_2$, b.p. 85 °C, and hexa-1,3-dien-5-yne, $CH_2{=}CH{-}CH{=}CH{-}C{\equiv}CH$, b.p. 83.4 °C, which can be separated by fractional distillation and condensation. Some important reactions of V. are the addition of water to yield methylvinyl ketone and the addition of hydrogen chloride to form chloroprene (2-chlorobuta-1,3-diene).

Vinyl carboxylic acid: same as Acrylic acid.

Vinyl chloride, ***chloroethene***: $CH_2{=}CH{-}Cl$, a colorless, combustible gas with a sweetish odor and a narcotic effect; m.p. -153.8 °C, b.p. -13.4 °C. V. is nearly insoluble in water, and almost infinitely soluble in most organic solvents. In the presence of oxygen and light, it polymerizes to form polyvinyl chloride. V. undergoes the addition reactions typical of Alkenes (see). The chlorine atom is very tightly bound, and can be replaced only under drastic reaction conditions. V. is made industrially by addition of hydrogen chloride to acetylene in the presence of mercury(II) chloride catalysts, or by splitting hydrogen chloride out of 1,2-dichloroethane. The latter process can be done thermally, catalytically or by using alkalies. V. is used mainly for the production of polyvinyl chloride.

Vinyl esters: the esters of vinyl alcohol, $CH_2{=}CH{-}OH$, which is nearly impossible to isolate, with carboxylic acids. The general formula is $R{-}CO{-}O{-}CH{=}CH_2$. V. are obtained industrially by vinylation of carboxylic acids and acetylene: $HC{=}CH + R{-}COOH \rightarrow R{-}CO{-}O{-}CH{=}CH_2$. V. are used to make polyvinyl esters; an important example is Vinyl acetate (see).

Vinyl ethers: ethers with the general formula $R{-}O{-}CH{=}CH_2$, which contain at least one vinyl group, $-CH{=}CH_2$. ***Methyl vinyl ether***, R = CH_3-, a gaseous compound, m.p. - 122 °C, b.p. 12 °C, n_D 1.3730. ***Ethyl vinyl ether***, R = $CH_3{-}CH_2-$, a colorless liquid, b.p. 35.5 °C. ***Divinyl ether***, R = $CH_2{=}CH-$, a colorless liquid, m.p. - 101 °C, b.p. 28-31 °C. Isobutyl vinyl and octadecyl vinyl ethers are also industrially important. The V. are made industrially by the Reppe synthesis, that is, by addition of alcohols to ethyne at elevated temperature and under pressure; the reaction conditions can be varied depending on the alcohol used. Divinyl ether is made from bis(2-chloroethyl) ether by a double cleavage of hydrogen chloride. The V. are used in large amounts to make polyvinyl ethers; these are thermoplastic and are used mainly as textile sizings and paint ingredients. Divinyl ether is also used as an inhaled anaesthetic.

Vinylethyne: same as Vinylacetylene (see).

Vinylidene-: a term for the atomic group $CH_2{=}C\langle$ in a molecule.

Vinylidene chloride, ***1,1-dichloroethene, 1,1-dichloroethylene***: $CH_2{=}CCl_2$, a colorless liquid with a pleasant, sweetish odor; m.p. -122.1 °C, b.p. 37 °C, n_D^{20} 1.4249. V. is only slightly soluble in water, but dissolves readily in most organic solvents. It reacts with oxygen to form peroxides, which catalyse polymerization. In the presence of stabilizers, such as hydroquinone, V. can be stored for a long time. It is obtained by the reaction of alkalies with 1,1,2-trichloroethane, with loss of hydrogen chloride:

$$Cl{-}CH_2{-}CHCl_2 \xrightarrow[-HCl]{} CH_2{=}CCl_2.$$

V. is most often used in the form of a mixed polymer with vinyl chloride, acrylonitrile, etc.

Vinylidene cyanide, ***1,1-dicyanoethene***: CH_2-$C(CN)_2$, a colorless liquid; m.p. 9.7 °C, b.p. 40 °C at $6.65 \cdot 10^2$ Pa, n_D^{20} 1.4411. V. is soluble in benzene and trichloroethylene. In the presence of water, alcohols or amines, V. polymerizes spontaneously to form polyvinylidene cyanide. It is obtained by pyrolytic cleavage of 1,1,3,3-tetracyanopropane.

Vinylogic principle: a partial transfer of the properties of a C atom, for example a C atom adjacent to a carbonyl group, to the terminal C atom of a conjugated system. The reactivity of the terminal C atom becomes similar to that of the atom from which the transfer arose; in the example used here, its CH acidity is increased.

1-Vinylpyrrolidin-2-one: same as *N*-vinylpyrrolidone.

N-Vinylpyrrolidone, ***1-vinylpyrrolidin-2-one***: a colorless liquid; m.p. 13 °C, b.p. 215 °C, n_D^{20} 1.5120. It is soluble in water and most organic solvents. The first step in the synthesis of *N*-V. is oxidation of butane-1,4-diol with oxygen at 200 °C on a copper contact; this forms γ-butyrolactone, which is treated with ammonia at 200 °C to form pyrrolidin-2-one. This compound is vinylated with acetylene to yield *N*-V. It is used mainly in the synthesis of polyvinylpyrrolidone.

Vinyon®: see Synthetic fibers.

Violanin: see Delphinidin.

Violuric acid, ***alloxan 5-oxime, 5-(hydroxyimino)-barbituric acid***: The compound forms colorless to light yellow crystals, m.p. 242 °C. It dissolves readily in water and gives it a violet color; it is also readily soluble in alcohol. V. gives a blue color with iron(III) chloride solution. It is synthesized by reaction of barbituric acid with nitrous acid, or by reaction of alloxan with hydroxylamine.

V. forms characteristic, intensely colored chelates with alkali metal, alkaline earth metal and heavy metal ions; these can be identified by paper chromatography. In clinical chemistry, V. is used to detect metal ions in the urine.

VIP: see Vasoactive intestinal peptide.

Virial equation: see State, equations of.

Viridogrisein: same as Etamycin.

Virostatics: compounds which inhibit the replication of viruses in an organism, without significantly affecting normal cell function. Since viruses do not have an independent metabolism, and only a limited number of their own enzymes, they require the resources of a living host cell for replication. This makes it difficult to discover suitable V. for therapy, and the results which have so far been achieved are not satisfactory. There are a few V. available for therapy, but they are very limited in the range of viruses they can inhibit, the routes of application and the time of application. Theoretically, there are various points at which viruses can be attacked: penetration of the host cell, replication in the cell and exit from the cell. Examples of current V. are idoxuridine, cytarabin and amantadin.

VIS: abb. for visible; applied to spectra in the visible range (see UV-VIS spectroscopy).

Visbreaking: a thermal cracking process in which residual oils from petroleum refining are converted to medium distillates under mild conditions (see Cracking). V. was originally used to reduce the viscosity of heating oils, but as petroleum becomes less abundant, processing of the residues becomes more important, and residual oils are less used for heating oils.

Because of its flexibility and simple technology, V. is an important method for processing heavy residues. It is done in a pipe furnace (Fig.). To stop the cracking reactions, the products leaving the furnace are quenched with the cold stream of starting material. In a subsequent step, they are distilled.

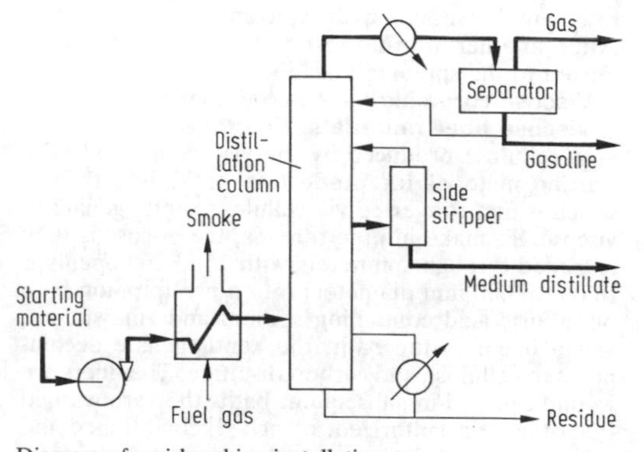

Diagram of a visbreaking installation.

The most important process parameters of V. are the reaction temperature and the time that the reactants remain at that temperature. In the original method for reducing the viscosity of residues, the temperatures were between 450 and 470 °C, and the reaction times were several minutes. To obtain higher yields of medium distillates, higher reaction temperatures (480-490 °C) are used, and the reaction times are shortened (to a few minutes). The following table compares the yields under these conditions:

Reaction time Temperature °C	Several minutes 450-470	A few minutes 480-490
Yields (vol. %)		
Gas	1.7	1.3
Gasoline	8.2	4.3
Medium distillate	7.5	16.9
Residue	84.6	77.5

When the conditions of V. are too severe, and especially when the starting materials have high asphalt contents, the colloidal solution of asphaltenes is disturbed and coke formation produces solids in the V. residues. This leads to precipitation of solids on storage and phase separation on mixing with other products. One way to suppress coking during V. is to work in the presence of hydrogen (***hydrovisbreaking***).

Catalytic methods of refining residual oils (both hydrocatalytic and catalytic processes) have not been widely adopted. Although efforts are being made to develop catalysts for direct processing of materials with metal contents >100 ppm, at present such residues cannot be processed without pretreatment.

Viscose: a viscous solution of cellulose xanthogenate used to make Viscose fiber materials (see), viscose cellophane and viscose sponges.

Production. The bleached cellulose is treated with 18 to 22% sodium hydroxide solution to remove hemicelluloses. This leads to formation of sodium cellulose; after a ripening period, this is reacted with carbon disulfide and sodium hydroxide to make sodium cellulose xanthogenate. The sticky, orange cellulose xanthogenate is then dissolved in dilute sodium hydroxide solution and filtered. The filtered solution is allowed to ripen for 15 to 20 hours, during which time the cellulose xanthogenate is gradually hydrolysed. The resulting V. contains 6 to 9% cellulose, 6 to 7% sodium hydroxide and about 2% sulfur. After another filtration and deaeration, the V. is moved to the spinning machines.

Viscose cord silk: see Viscose fiber materials.

Viscose fiber materials: fibers made of regenerated cellulose produced by the viscose method. The starting material for production of V. is cellulose, which is first converted via cellulose xanthogenate to viscose. To make infinite threads, the viscose is then extruded through spinnerets with 25 to 160 openings (0.06 to 0.08 mm diameter) into a precipitation bath of sulfuric acid containing sodium and zinc sulfate. When it enters the bath, the xanthogenate decomposes to cellulose and carbon disulfide. The fibers are wound up, and in subsequent baths they are washed free of acid, desulfurized, bleached, conditioned and dried.

To make ***viscose staple***, spinnerets with 1000 to 15,000 openings are used. The very fine fibers are drawn out of the precipitation bath with rotating rollers. Unlike the silks, they are not rolled up as an infinite thread, but instead are collected into a thick bundle. This runs via a roller into the subsequent baths as an endless ribbon. The ribbon is later cut to the desired staple length of 3 to 20 cm. It is dried to a moisture content suitable for work-up, 10 to 12%. Better quality viscose fibers are obtained when the fiber cable is cut immediately after it leaves the spinning machine (acid cutting). In this case, the fibers are worked up as a mat.

The V. swell and stretch easily when moist. They are easily bleached, dried and cleaned. They can be made with a smooth or roughened surface, and fibers can be made with any degree of shine from very glossy to totally matte. V. can be dyed either during production or afterwards. They are not very stable to water, acids, lyes or chemicals. The high wet modulus materials (see Rayon) are V. with a high stability in water. *Viscose silk* is similar to natural silk, and is used to make clothing and decoration textiles. *Viscose cord silk* consists of coarser threads, and is used to make industrial fabrics, such as cords for tires, conveyor belts and cables. *Viscose fibers* are fleeces with silk-like, matte appearance and great absorption capacities. They can be made very similar to wool in feel, appearance, insulating and wearing quality by curling and crimping the fibers.

V. are often mixed with natural fibers, such as cotton or wool, to make textiles, etc. They can be conditioned with synthetic resins to make non-wrinkling fabrics.

The most important trade names are listed under Synthetic fibers (see).

Viscose silk: see Viscose fiber materials.

Viscosity: the result of internal friction between molecules of a liquid or gas. Deformable solids, such as lacquer, ice and most metals, display V. In gases, the V. increases with increasing temperature, while in liquids, it decreases with increasing temperature.

Physically, V. is defined by the frictional resistance to relative displacement of parallel layers of the substance. *Dynamic V.*, η, is expressed in terms of the force required to move two layers A and B, of equal size and separated by the substance in question, past each other. The appropriate SI unit is the pascal second (Pa s). A substance has a dynamic V. of 1 Pa s when the distance between the layers is 1 m, and a force p of 1 N in the plane of the layers causes the layers to move past each other with a velocity v of 1 m s^{-1}. This applies only for laminar flow between the layers and for Newtonian fluids. The *kinematic V.* ν is equal to the dynamic V. divided by the density d: $\nu = \eta/d$. Its SI unit is the square meter per second (m^2/s).

V. is measured (***viscosimetry***) with ***viscosimeters***. In these devices, the time required for a certain volume of liquid or gas to pass through a capillary of a certain diameter (capillary viscosimeter) is measured, or the rate of fall of a sphere in the liquid (falling body viscosimeter) is determined. In gases, the damping of a freely swinging disk (disk viscosimeter) can be used. The Engler viscosimeter is a flow viscosimeter used to determine a relative V., which is given in Engler degrees.

Viscosity, structural, ***Bingham's flow***: the decrease in viscosity with increasing shear stress. It is observed in gels, macromolecular solutions and colloidal dispersions. The viscosity is not constant, but decreases with increasing shear gradient or stress. At low shear stress, the colloid may not flow at all (flow limit). The flow is related to loss of structures in the dispersion caused by the shear stress.

Visual pigment: see Vitamin A.

Vitallium: cobalt alloy consisting of 64% cobalt, 30% chromium and 3% molybdenum. Since V. is inert to body fluids, and causes no damage to tissues, it is used as a replacement for bones and teeth.

Vital stain: an organic pigment used to stain living tissues or organisms (especially microorganisms) for microscopy. The V. are either taken up with food by animals and transported into the tissue of interest, or they are injected subcutaneously or intravenously. When the staining is completed, the animal is killed and disected, and the stained tissue can be studied.

The most important V. are methylene blue, methyl violet, janus green, vesuvin, neutral red, eisone, rhodamine pigments, Nile blue, fat-soluble Sudan pigments and red carotenoids.

Vitamin: [Latin "vita" = "life" + "amine"] an organic compound required for life, growth or reproduction by the human or animal body, which it cannot synthesize and must ingest with the diet. A V. is required only in small amounts, as a biocatalyst; the definition does not include proteins, carbohydrates or fats, and in particular does not include the essential fatty acids, which do not act as catalysts. Most V. are synthesized by plants; some are also synthesized by intestinal bacteria. Some V. are present in the diet as ***provitamins***, that is, as precursors which the organism can convert to V.

The V. are termed *accessory nutrients*. Many of them act as coenzymes or as the prosthetic groups of enzymes: nicotinamide and riboflavin serve as hydrogen-transfer catalysts, while thiamin, pyridoxine, pantothenic acid and folic acid act as group-transfer agents. Vitamin D (which is only a V. in the absence of sufficient ultraviolet irradiation of the skin) acts as a hormone; vitamin A may also have regulatory effects in addition to its well known role in the visual process. Vitamins C and E act as antioxidants.

The complete absence of a V. leads to an *avitaminosis*, and an inadequate intake, to a *hypovitaminosis*. Vitamins A and D can be stored in the liver (unlike the water-soluble V., which are not stored), so that excessive intake of these two V. leads to *hypervitaminoses*. Avitaminoses are now rare in the industrial countries, but scurvy (lack of V. C) was formerly a major cause of death in Europe. Hypovitaminoses occur in the industrial countries due to improper diet or increased requirements, as in certain diseases and pregnancy. In developing countries, hypovitaminosis A is very common, especially among young children, and causes hundreds of thousands of cases of blindness each year (the affected children rarely survive). Hypervitaminoses do not result from normal diets; they can be induced only by improper administration of synthetic V. The V. contents of foods vary; fresh vegetables contain most V., while other types of food are rich only in certain types of V. Refined sugar and non-enriched, white flour or polished rice contain almost no V. Poor cooking and storage methods can lead to significant losses of V.

There are twelve groups of related compounds which are V. for human beings (the requirements of other animals vary somewhat). These are commonly classified in two groups, the ***fat-soluble*** and ***water-soluble*** V., based on the ability of ether or water to extract them from foods. The V. were originally named for the symptoms which they are able to cure, e.g. antirachitic or antiscorbutic V. Early in the period of V. research, however, the isolated compounds were indicated by letters of the alphabet. When it was later discovered that several different V. were contained in what had originally been thought a single compound, subscripts were added (e.g. V. B_2, V. B_6). It is now preferred in medicine and pharmacy to use internationally recognized trivial names which indicate the effect (e.g. retinol, ergocalciferol) or the structure (thiamin, riboflavin) of the molecules.

The effectiveness of V. preparations was originally measured in arbitrary units. After the structures were elucidated, the effect of a certain amount of a pure V. was established as an international unit (IU).

Historical. The beginning of scientific V. research was the observation of Eijkmann (1897) that chickens fed exclusively on polished rice developed a condition very similar to human beri-beri. He deduced that the condition was caused by a lack of some dietary factor. Later, Stepp and Hopkins showed that animals fed pure protein, carbohydrate and fats did not survive. Animal experiments were of great significance for further V. research. Thiamin (V. B_1) was the first V. to be isolated in pure form (1926). Most of the V. were purified between 1926 and 1940. The name "vitamin" was coined by Funk around 1910; he was working with thiamin, which is an amine. The name was universally adopted, although many V. are not amines. The discovery that riboflavin is a component of flavin enzymes (1933) was the beginning of the elucidation of the coenzyme and cofactor character of V.

Vitamin A: a group of fat-soluble vitamins formed from carotenes with β-ionone rings by oxidative cleavage of the middle double bond. V. A_1 is found in most animals. V. A_2, the corresponding 3-dehydro compound, is the major form in fresh-water fish. Both forms can be isolated from the non-saponifiable fractions of fish liver oils as alcohols (retinol or 3-dehydroretinol). ***Vitamin A₁ (retinol, axerophthol)*** is usually isolated as a yellow oil; in pure form, m.p. 63-64 °C, λ_{max} of the palmitate, 325 nm (isopropanol). It is practically insoluble in water, but is soluble in organic solvents. It is very unstable in the presence of atmospheric oxygen and light, and its decomposition is catalysed by heavy metal ions. Antioxidants are added to solutions of the vitamin to stabilize it. Esters such as the acetate or palmitate are more stable than the free alcohol. Retinol gives a color reaction with antimony(III) chloride (*Carr-Price reaction*) which can be used for its quantitative determination. Acids can remove water from retinol, producing the biologically inactive ***anhydroretinol***. Retinol generally has the *all-trans*-configuration. Natural retinol also contains about 50% 13-*cis*-retinol, which is also called ***neovitamin A***.

The industrial synthesis of retinol starts from citral, and goes via β-ionone. A Wittig synthesis is also used. The aldehyde of V. A, ***retinal***, is the active form of the molecule in the visual process. It is bound covalently to proteins, the *opsins*, by reaction of the aldehyde group with the ε-amino group of a lysine group. The resulting chromoproteins, the *visual pigments*, have absorption maxima which depend both on the configuration of the double bonds of the retinal and on the structure of the protein. The pigment in the rod cells is called *rhodopsin*.

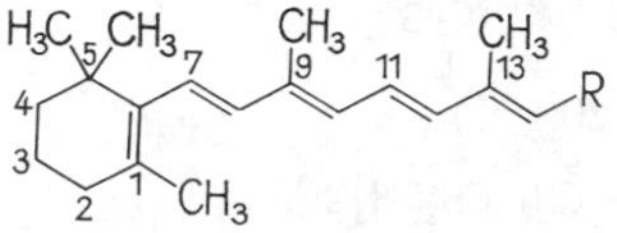

H_3C CH_3 CH_3 CH_3 5 7 9 11 13 R 4 3 2 1 CH_3

Vitamin A_1

R = CH_2OH : Retinol
R = CHO : Retinal
R = COOH : Retinoic acid

The thermodynamically most stable form of retinal and its derivatives is the all-*trans*-form. In the visual process, the 11-*cis*- derivative is converted to the all-

trans-form. This induces a change in the conformation of the protein, and initiates a series of reactions which lead to a nerve impulse being generated by the cell.

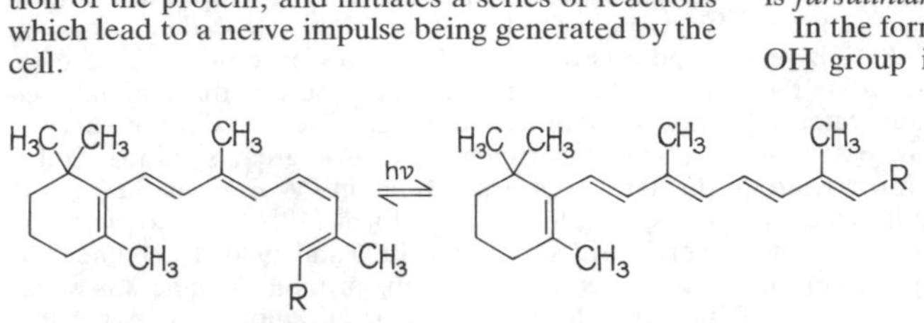

R = CHO : 11-*cis*,12-S-*cis*-retinal — all-*trans*-retinal
R = CH=N-opsin : rhodopsin — meta-rhodopsin

The retina of the human eye contains only about 0.005% of the body reserve of V. A. However, night blindness is an early symptom of V. A deficiency. V. A is also required for fertility, growth and cell development, especially of epithelial cells. ***Retinoic acid*** also has these effects, although it is not active in the visual process. Overdoses of V. A have toxic effects.

One international unit corresponds to the activity of 0.344 μg retinol acetate in 0.1 mg cottonseed oil.

Retinal-protein pigments serve not only as light sensors, but are part of the light-energy conversion system in the halobacteria. In the purple membranes of these bacteria, a 13-*cis*-retinal molecule is bound to bacteriorhodopsin.

Vitamin B_1, *thiamin, aneurin, antineuritic vitamin*: a water-soluble vitamin found widely in nature. The chloride hydrochloride forms white needles with a bitter taste and yeast-like odor; m.p. 245-248 °C (dec.). It is soluble in water and slightly soluble in ethanol. It decomposes in alkaline media and is sensitive to heat and oxidation; the oxidation product, *thiochrome*, has a strong blue fluorescence. V. B_1 is relatively abundant in rice bran, yeast, the germs of grains, fresh vegetables and liver. The structure of V. B_1 consists of a pyrimidine and a thiazole ring and contains a quaternary N atom. Commercial V. B_1 is entirely synthetic; it is made by reaction of the pyrimidine derivative (I) with the thiazole derivative (II) followed by exchange of the bromide for chloride (Fig.). Because V. B_1 absorption after oral application is limited, derivatives are often administered. These are derived from the tautomeric thiol form of the vitamin in which the thiazole ring is open and contains free mercapto and aldehyde groups; an example is *fursulthiamin*.

In the form of thiamin pyrophosphate, in which the OH group is esterified to pyrophosphate, V. B_1 is part of the coenzymes of decarboxylases, such as pyruvate decarboxylase, and of aldehyde transferases.

V. B_1 deficiency disrupts carbohydrate metabolism. The typical deficiency syndrome is beri-beri, which is a result of a diet consisting only of polished rice. In this disease and other V. B_1 hypovitaminoses, there is impairment of the central and peripheral nervous systems (polyneuritis). V. B_1 and its derivatives are used to treat hypovitaminoses and their complications, such as neuralgias and neuritis.

Historical. V. B_1 was isolated in 1926 from rice bran by Jansen and Donath; it was the first vitamin to be purified. It was isolated from yeast in 1932 by Windaus, and in 1935 the structure was elucidated by Williams and Windaus. A little later, the compound was synthesized by Grewe, Williams, Andersag and Westphal.

Vitamin B_2, *riboflavin, lactoflavin*: 7,8-dimethyl-10-(D-1'-ribityl)isoalloxazine, a yellow flavin derivative which is found in nature almost exclusively in the form of flavin nucleotides or flavin enzymes. V. B_2 is an orange-yellow, odorless substance with a bitter taste; m.p. 280 °C (dec.), $[\alpha]_D^{25}$ -114° (in 0.5 M NaOH). V. B_2 is insoluble in ether and petroleum ether, slightly soluble in water, and soluble in ethanol. Its UV spectrum depends strongly on the pH. In an acetic acid/sodium acetate buffer, the maxima are at 267, 375 and 444 nm. In aqueous solution, V. B_2 has a yellowish green fluorescence, which disappears when hydrochloric acid or sodium hydroxide is added. V. B_2 occurs widely in nature. It is found in yeast, milk, liver and legumes. Milk contains free riboflavin.

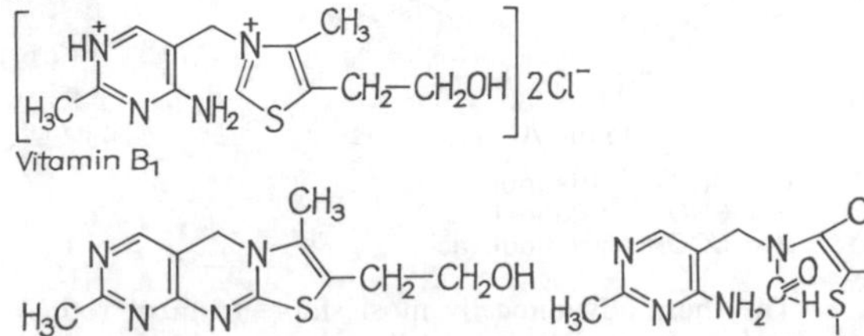

I + II →

Vitamin B_1

Thiochrome

Fursulthiamin

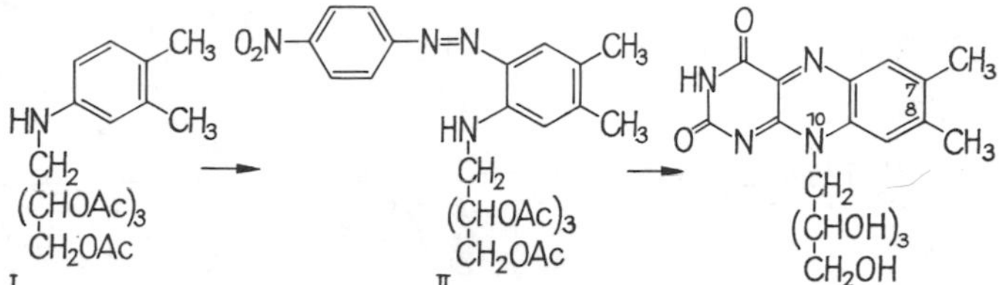

I II Vitamin B_2

V. B_2 is made by total chemical synthesis or by fermentation using certain ascomycetes. The chemical synthesis starts with 3,4-dimethylaniline. Reaction with acetylated D-ribose and hydrogenation yields the D-ribamine derivative (I). The reaction of (I) with 4-nitrophenyldiazonium chloride leads to (II), which is converted to V. B_2 by reaction with barbituric acid and deacetylation.

V. B_2 is a component of flavin mononucleotide (FMN) and flavin adenine dinucleotide (FAD), which are coenzymes of oxidoreductases.

A dietary deficiency of V. B_2 is ordinarily associated with a deficiency of nicotinic acid, and the resulting avitaminosis is called pellagra. Deficiency of V. B_2 alone leads to lesions around the corners of the mouth, seborrhoic changes in the nose and ears, and keratitis of the eyes.

Historical. V. B_2 was isolated in 1934 by Kuhn, Szent-Györgyi and Wagner-Jauregg from whey. Its structure was elucidated in 1935 by Kuhn and Karrer, and was confirmed by synthesis.

Vitamin B_2 complex: a term for a group of water-soluble vitamins which includes riboflavin, folic acid, pantothenic acid, and nicotinic acid/nicotinamide.

Vitamin B_6: ***pyridoxine, pyridoxol, adermine*** (R = CH_2OH), 2-methyl-3-hydroxy-4,5-di(hydroxymethyl)pyridine; ***pyridoxal*** (R=CHO) and ***pyridoxamine*** (R=CH_2NH_2) are forms of the vitamin found ubiquitously in plants and animals. It is especially plentiful in liver, vegetables and yeast. Pyridoxine is readily soluble in water, soluble in ethanol and acetone, and slightly soluble in ether. It forms colorless needles; m.p. 160 °C (dec.). Pyridoxine hydrochloride is used in vitamin preparations; m.p. 210 °C (dec.). The three forms of V. B_6 are readily interconverted in the body; the active form is pyridoxal phosphate, in which the hydroxymethyl group is esterified with phosphoric acid. This compound is a coenzyme in many enzymes, such as aminotransferases, C_1 transferases and decarboxylases. V. B_6 deficiency symptoms are extremely rare in human beings. The vitamin is administered in therapy of nervous disorders, anemia and motion sickness (kinetosis). It is also used to ameliorate the effects of long-term treatment with certain drugs, such as isoniazide, which are B_6 antagonists.

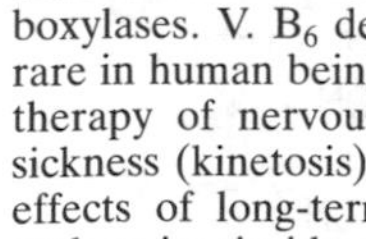

Vitamin B_{12}: a group of water-soluble compounds derived from the cyclic tetrapyrrole Corrin (see). The corrin derivative with acetate, propionate and methyl groups, and with Co(III) as the central atom, is called ***cobyric acid***, and its hexamide is called ***cobyrinic acid***. All the derivatives found in animals are derived from ***cobalamin***, which have a 5,6-dimethylbenzimidazole ribofuranoside group as the α-ligand of the central atom. This is linked to the cobyric acid via a phosphopropanolamine bridge.

V. B_{12} is produced by industrial fermentation, sometimes as a byproduct of microbiological antibiotic production. It is isolated as ***cyanocobalamin***, Coα-[α-5,6-dimethylbenzimidazolyl)]-Coβ-cyanocobamide, abb. CN-Cbl, because this form is more stable and more readily isolated than the physiological forms. Cyanocobalamin is a dark red, crystalline powder, which is moderately soluble in water, and only slightly soluble in ethanol.

The human body contains 3 to 6 mg V. B_{12}; the reserves are stored in the liver. The daily requirement for V. B_{12} is about 1 to 2 μg daily, and it is normally provided by the diet and the microorganisms in the intestinal tract. Absorption of V. B_{12} from the intestines is mediated by *intrinsic factor*, a glycoprotein formed in the gastric mucosa. When the gastric mucosa atrophies, the absorption of V. B_{12} is impaired, leading to pernicious anemia.

About 65% of the V. B_{12} present in human blood is in the form of methylcobalamin. In the liver, about 60% is adenosylcobalamin and 1% is methylcobalamin. The remainder is aquacobalamin. In ***aquacobalamin***, V. B_{12a}, Coβ-aquacobalamin, a water molecule is the 6th ligand of the central atom. ***Methylcobalamin***, Coβ-methylcobalamin, abb. Me-Cbl, and ***adenosylcobalamin***, Coβ-5'-deoxyadenosylcobalamin, abb. Ado-Cbl, are ***B_{12} coenzymes***. These are the active forms of V. B_{12}. In human beings, the corrinoid-dependent enzymes are homocysteine methyltransferase and methylmalonyl-CoA mutase; in microorganisms, there are many corrinoid-dependent mutases and transferases. The coenzymes are organometal compounds, in which there is a direct C-Co bond in the β-position. The B_{12} coenzymes are extremely sensitive to light, which converts the β-ligand to a radical. V is biosynthesized only by microorganisms, mainly the streptomycetes and nocardias. However, animals are capable of partial synthesis to the coenzyme forms. Fig., see p. 1162.

Historical. Cyanocobalamin was first isolated and crystallized in 1948, and the structure was completely elucidated in 1955 by x-ray analysis (Hodgkin, Todd, et al.). The total synthesis was achieved in 1976 by Woodward and Eschenmoser.

Vitamin C, ***ascorbic acid***: a water-soluble vitamin, 3-oxo-L-gulono-γ-lactone. The compound has two chiral C-atoms; only the L(+)-*xylo* form is biologically active. V. C is colorless and dissolves readily in

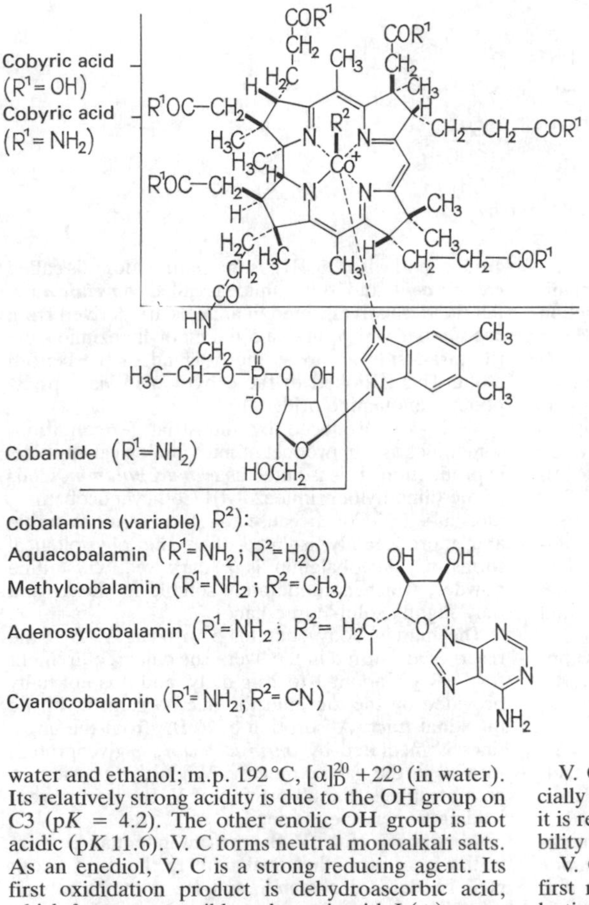

Cobyric acid (R^1= OH)
Cobyric acid (R^1= NH_2)
Cobamide (R^1=NH_2)
Cobalamins (variable) R^2):
Aquacobalamin (R^1=NH_2 ; R^2=H_2O)
Methylcobalamin (R^1=NH_2 ; R^2=CH_3)
Adenosylcobalamin (R^1=NH_2 ; R^2= H_2C–)
Cyanocobalamin (R^1= NH_2; R^2= CN)

water and ethanol; m.p. 192 °C, $[\alpha]_D^{20}$ +22° (in water). Its relatively strong acidity is due to the OH group on C3 (pK = 4.2). The other enolic OH group is not acidic (pK 11.6). V. C forms neutral monoalkali salts. As an enediol, V. C is a strong reducing agent. Its first oxididation product is dehydroascorbic acid, which forms a reversible redox pair with L(+) ascorbic acid (Fig. 1). Several reactions for detection of the vitamin depend on its reducing properties, e.g. the hydrogenation of Tillman's reagent, a dark blue compound which becomes colorless on reduction, and the reduction of silver nitrate or iodine.

L(+)-Ascorbic acid   Dehydroascorbic acid

V. C is stable in solid form, but in solution, especially in the presence of copper(II) and iron(III) ions, it is readily oxidized by atmospheric oxygen. The stability optimum is at pH 5.6.

V. C is now synthesized from D-glucose, which is first reduced to D-sorbitol (D-glucitol). This is selectively oxidized to L-sorbose by *Acetobacter suboxydans*. The hydroxyl groups on C2, C3, C4 and C6 are then protected by formation of acetals, after which the C1 atom is oxidized to a carboxyl group with $KMnO_4$. The protective groups are then removed in an acid-catalysed step, and the compound is lactonized to make V. C (Fig. 2).

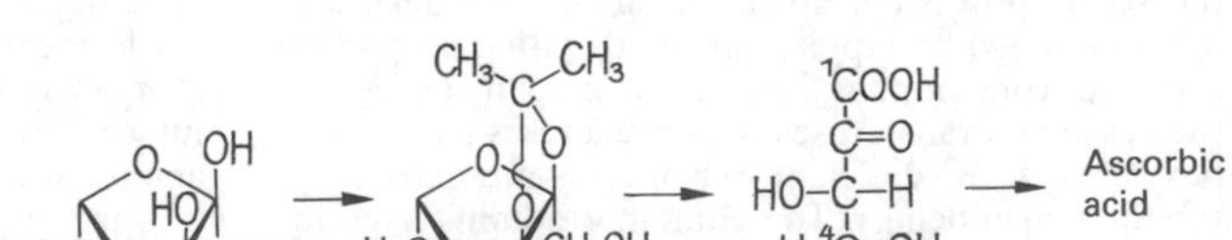

L-Sorbose → → 2-oxo-L-gulonic acid → Ascorbic acid

V. C occurs widely in nature, and is particularly abundant in citrus and other fruits, rose hips and fresh vegetables. V. C is a vitamin only for human beings, monkeys and guinea pigs. Most animals are able to synthesize it. It is not a coenzyme, but takes part in hydroxylation reactions and acts as a protective agent against oxidation by oxygen radicals. The recommended daily intake is 75 mg per day for an adult. An inadequate intake leads to scurvy, the symptoms of which are bleeding of the mucous membranes and painful swelling of the joints. The gums recede and the teeth rot and fall out. The disease is fatal within a few months. Subclinical deficiency leads to tiredness, susceptibility to infectious diseases and bleeding of the gums. Whether high doses of V. C prevent infections and cancer is debated. There have been reports of successful treatment of cancers with V. C.

Historical. V. C was first isolated from adrenal glands in 1928 by Szent-Györgyi. It was shown in 1932 that this compound was identical to the active agent in antiscorbutic plant extracts. The structure was elucidated and the compound synthesized in 1933 by Haworth, Reichstein, Karrer and Micheel.

Vitamin D, *calciferol*: a group of fat-soluble, antirachitic vitamins with steroid structures. They are formed from provitamins with sterol structures by UV irradiation. The most important are V.D_2 and D_3. V.D_1, a molecular complex of lumisterol$_2$ and V.D_2, and V.D_4 (22-dihydroergocalciferol), which has a saturated side chain on C17, are not effective.

V.D$_2$ (ergocalciferol) forms colorless and tasteless needles which are insoluble in water and soluble in ether, acetone, ethanol and chlorform; m.p. 121 °C, $[\alpha]_D^{20}$ + 83° (in acetone). ***V.D$_3$ (cholecalciferol)*** has similar properties; m.p. 87-89 °C, $[\alpha]_D^{20}$ +112° (in ethanol). Both compounds are easily decomposed by acids, light and oxygen.

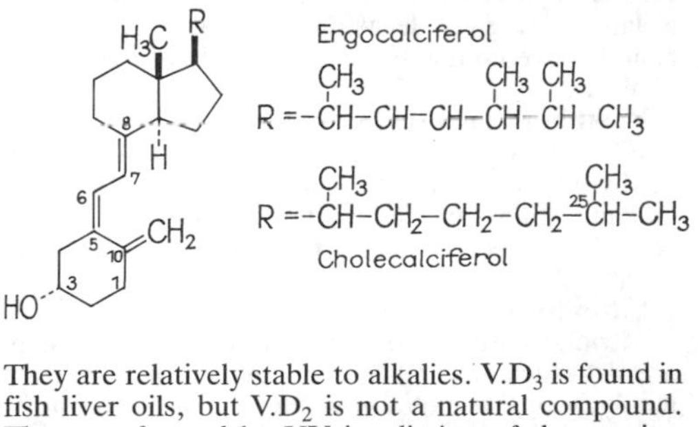

Ergocalciferol: R = –CH(CH$_3$)–CH=CH–CH(CH$_3$)–CH(CH$_3$)–CH$_3$

Cholecalciferol: R = –CH(CH$_3$)–CH$_2$–CH$_2$–CH$_2$–^{25}CH(CH$_3$)–CH$_3$

They are relatively stable to alkalies. V.D_3 is found in fish liver oils, but V.D_2 is not a natural compound. They are formed by UV irradiation of the provitamins ergosterol (D_2) and 7-dehydrocholesterol (D_3); the double bonds between C5 and C6 and between C7 and C8 are necessary for this conversion. UV light causes ring B of the sterane skeleton to open between C9 and C10, forming pre-ergocalciferol or precholecalciferol; these trienes are converted by thermal equilibrium processes to V.D_2 or V. D_3. Both compounds have a *Z*-configuration at the double bond on C5 and an *E*-configuration at the C7 double bond. The double bond in the side chain of V.D_2 has *E*-configuration. Recyclization leads to *lumisterols* as byproducts; *tachysterols* arise by isomerization. V.D_2 is isolated from the irradiation solution, after removal of byproducts, by formation of the 3,5-dinitrobenzoate ester.

V.D_2 and V.D_3 give an orange color with antimony(III) chloride in chloroform (*Brockmann-Chen reaction*), which can be used for photometric determination at 500 nm. Direct spectroscopic determination is also possible, at 265 nm. The V.D content of fish liver oils is occasionally still given in international units (IU); 1 IU = 0.025 µg V.D_2.

V.D promotes absorption of calcium and phosphate ions from the intestine and inhibits phosphate excretion from the kidneys. It regulates calcium uptake into or resorption from bones. Provitamin D_3 is synthesized from cholesterol in the liver and stored in the skin, where it is converted to V.D_3 by UV light.

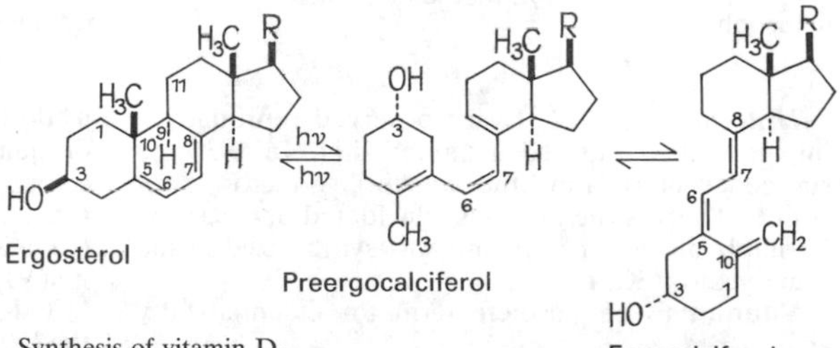

Synthesis of vitamin D_2.

Hypovitaminoses occur when there is inadequate exposure of the skin to light. In infants and small children, V.D deficiency leads to rickets; in adults, to osteomalacia (softening of the bones). V.D_2 and V.D_3 are used for therapy and prophylaxis of rickets. The V.D are hydroxylated at C25 and C1-α to form the active compounds. V.D are stored in the liver; massive doses can be toxic.

Historical. In 1920, rickets was recognized as a vitamin deficiency disease. In 1926, Windaus and Hess discovered that ergosterol could be converted to an antirachitic product by UV irradiation. In 1932, Windaus purified V.D_2; in 1936, Brockmann isolated V.D_3 from cod liver oil. In 1948, Crowfoot-Hodgin elucidated the stereochemistry by X-ray analysis. Inhoffen reported the total synthesis in 1960.

Vitamin E, *tocopherol*: a group of fat-soluble vitamins derived from *tocol*, 6-hydroxy-2-methyl-2-(4',8',12'-trimethyltridecyl)chromane. Tocol is a chromane derivative with an isoprenoid side chain. The individual tocopherols differ with respect to the number and positions of the methyl groups on their aromatic rings. The most important is ***α-tocopherol*** (5,7,8-trimethyltocol), which also has the strongest antisterility activity in rats. It has 3 chiral C atoms. Natural α-tocopherol has the 2R, 4'R and 8'R configuration and is dextrorotatory. It can be synthesized by condensation of trimethylhydroquinone with phytol. If natural phytol is used, an epimeric mixture is obtained with the configuration 2 RS, 4'R, 8'R. If synthetic (±)-phytol is used, the resulting tocopherol is totally racemic. α-Tocopherol is usually used in the form of the more stable acetate, in which the OH group is esterified with acetic acid. It is a yellowish oil. V. E is easily oxidized to tocopherol-*p*-quinones (*tocoquinones*), and can therefore be used as antioxidants.

V. E is found widely in nature. Abundant sources are wheat germ, peanut and soy oil. It is also present in milk, milk products and vegetables. V. E hypovitaminoses are not known in human beings. In rats, V. E deficiency leads to reduced fertility, and in some other animals it causes changes in the musculature. The biological function of V. E may be to protect natural lipids from oxidation by peroxides. V. E is used mainly as an antioxidant for fats and fatty oils.

Microorganisms form ***V. K₂ (menaquinone-n)***, in which the side chain consists of a variable number (n) of isoprene groups, each with an *E* configuration of the double bond. Menaquinone-7, V. $K_{2(35)}$, and menaquinone-6, V. $K_{2(30)}$, also known as *farnoquinone*, are important forms of V. K_2. V. K_2 is a light yellow powder, which is similar in its other properties to V. K_1.

V. K_3 (menadione, menaquinone) is the parent compound without a substituent group on the C3 atom. It forms lemon-yellow crystals which are slightly soluble in water and have a characteristic odor; m.p. 108 °C. It has an antihemorrhagic effect,

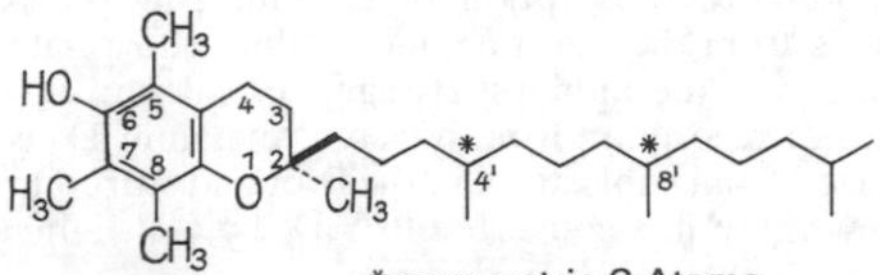

*asymmetric C-Atoms

α-Tocopherol

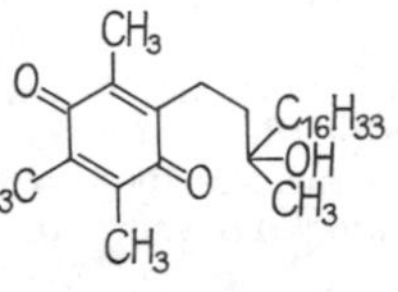

α-Tocoquinone

Historical. In 1922, Evans observed reproductive impairment in rats fed a limited diet. In 1936, he succeeded in isolating the antisterility factor, α-tocopherol. Its structure was elucidated in 1938 by Fernholz and the compound was synthesized in the same year by Karrer.

Vitamin F: an obsolete term for essential fatty acids.

Vitamin H: same as Biotin (see).

Vitamin H': 4-aminobenzoic acid, a component of Folic acid (see).

Vitamin K: a group of fat-soluble compounds with antihemorrhagic effects. Their common skeleton is 2-methyl-1,4-naphthoquinone. The naturally occurring V. K have isoprenoid side chains on C3. The form of the vitamin found in plants, ***V. K_1 (phytomenadione, phylloquinone)*** has a phytyl group with 20 C atoms, an *E*-configured double bond and two chiral C atoms with *R* configuration. It is a viscous, yellow oil, which is readily soluble in nonpolar solvents, slightly soluble in ethanol and insoluble in water. It is sensitive to UV light, alkalies and strong acids, but is relatively stable to heat and air; m.p. -20 °C. V. K_1 can be synthesized by reaction of V. K_3 (see below) with phytyl bromide in the presence of zinc chloride or boron trifluoride.

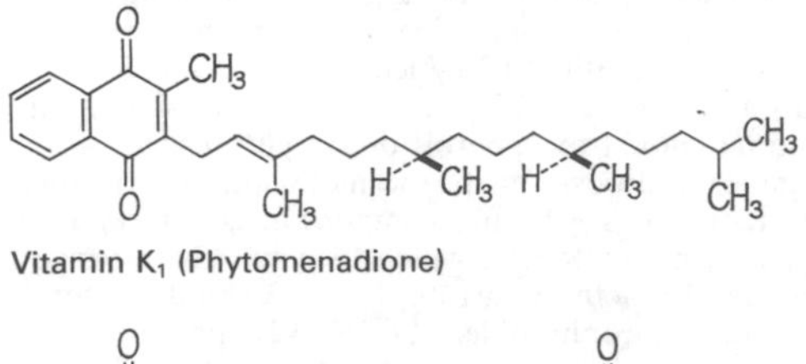

Vitamin K_1 (Phytomenadione)

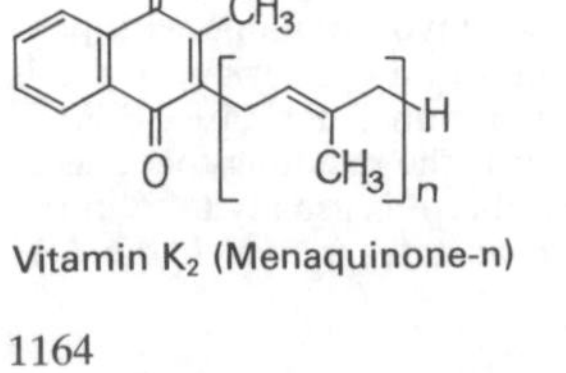

Vitamin K_2 (Menaquinone-n)

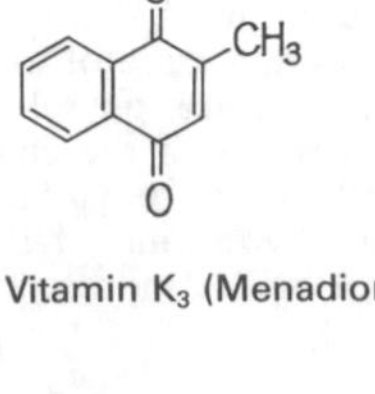

Vitamin K_3 (Menadione)

as do its reduction product 2-methyl-1,4-naphthohydroquinone (***V. K_4, menadiol***) and derivatives of this compound in which, for example, one OH group is replaced by an NH_2 group (***V. K_5***). V. K_3 is made by oxidation of 2-methylnaphthalene with chromium(VI) oxide.

V. K_1 and K_3 are most commonly used in therapy, the latter as the more soluble sodium hydrogensulfite addition product. V. K is required for the synthesis of γ-carboxyglutamic acid residues (Gla) in coagulation factors II, VIII, IX and X (which are enzymatically inactive in the absence of their Gla residues). Anticoagulants such as coumarin and Warfarin inhibit Gla formation; this is the basis for their activity. V. K is administered to counteract overdoses of anticoagulants or K hypovitaminoses induced by other drugs.

Historical. In 1929, Dam observed that animals were susceptible to impaired blood coagulation when a factor called V. K was absent from their diets. In 1939, Karrer idolated V. K_1 from alfalfa, and Doisy isolated V. K_2 from decaying fish meal. In 1959, Isler et al. discovered that there are several different forms of V. K_2.

Vitamin P: same as Rutin (see).

Vitamin U: *S*-methylmethionine, L-methionine methylsulfonium chloride. It is not one of the classical vitamins, although methionine is an essential amino acid. It is used to some extent as a treatment for ulcers.

Vitrinite: see Macerals.

Vitriols: an obsolete term for sulfates of divalent metals which crystallize with seven, or less commonly, with five moles of water. The lattice of the V. is formed of hexaaqua- or tetraaquametal cations and SO_4^{2-} anions, with one water molecule bonded by hydrogen bonds to an oxygen atom on the sulfate and a water coordinated to the metal. This structure is given the general formula $[M^{II}(H_2O)_6]SO_4 \cdot H_2O$ or $[M^{II}(H_2O)_4]SO_4 \cdot H_2O$. Some examples of V. are nickel vitriol, $NiSO_4 \cdot 7H_2O$, cobalt vitriol, $CoSO_4 \cdot 7H_2O$ and copper vitriol, $CuSO_4 \cdot 5H_2O$.

Vitrocerams, ***sitals, pyrocerams***: glass crystalline materials (see Glass ceramics) based on oxide glasses. They are partially crystalline; the degree of crystallinity is determined by the presence of crystallization nuclei and the time and temperature at which they are tempered. They are stronger and more resistant to temperature changes than ordinary glass. The crystallization nuclei can be noble metals, oxides of transition group elements or fluorides which are only slightly soluble in the glass. Certain metallurgical slags (copper slag, blast furnace slag) can be crystallized to ***slag vitrocerams***.

Photovitrocerams are light-sensitive glasses in which silver crystals precipitate upon exposure to light. These serve as crystallization centers for the formation of silicate crystalline phases above the glass transition temperature; these phases have a solubility in hydrofluoric acid which is different than that of the glass matrix. These materials can be made into very complicated shapes by a process of covering with an opaque pattern, irradiation, tempering and dissolving (etching) the exposed areas.

Voltammetry: a collective term for all electrochemical methods of analysis in which current-voltage curves are utilized, e.g. Cyclic triangular-wave voltammetry (see), Inverse voltammetry (see) or Polarography (see).

Voltaic column: see Electrochemical current source.

Volume contraction: a reduction in the volume in a process. A V. can occur when two substances are mixed, e.g. a salt or ethanol with water. It is caused by strong interactions in the mixed phase. For example, when a salt dissolves, tightly packed solvation shells develop around each ion. Gas reactions which reduce the number of moles also lead to V. In a few special cases, melting of a solid is also coupled to a V., for example melting of ice, antimony or bismuth. In such substances, the melting point decreases as the pressure increases (see Clausius-Clapeyron equation).

Volume diffusion: see Solid state reactions.

Volume fraction: see Composition parameters.

Volume percent, abb. ***vol.%***: see Composition parameters.

Volume work: see Thermodynamics, first law.

Volumetric analysis: a form of quantitative analysis in which the amount of a dissolved substance is calculated from the volume of a standard reagent solution with which it reacts. Reactions used for V. must be strictly stoichiometric, and occur very rapidly; in addition, the equilibrium must lie very far on the side of the products. The sample solution contains the titrand to be determined. A reagent solution is added slowly from a buret; this is the ***standard solution*** of the titrator, the concentration of which is often given as the equivalent concentration (see Composition parameters). Such normal solutions have the advantage that the same volumes of different titrators with the same normality are equivalent to each other. The addition of the standard solution is called ***titration***. It ends when the amount of titrator added is equivalent to the amount of titrand; this point is called the *equivalence point* of the titration. The equivalence point usually cannot be recognized without an Indicator (see) which changes color at the endpoint of the titration. The indicator should be chosen so that its change, and thus the endpoint, is as close as possible to the equivalence point of the titration. If this is not the case, systematic errors can occur. In electrochemical methods, especially Potentiometry (see), the apparatus records the entire titration curve, from which the equivalence point can be determined.

The methods of V. are classified according to the underlying reaction type: Neutralization analysis (see), Redox analysis (see), Precipitation analysis (see), Complexometry (see). These methods can be subdivided according to the titrator used.

All these methods are usually carried out in aqueous solution. The use of other solvents can be advantageous in some cases.

Volumetric equivalent: see Stoichiometry.

VPI: see Vapor phase inhibitor.

VSEPR model: same as Electron-pair repulsion model (see).

Vulcanization: a process of making rubber and other polymers more elastic and stable by incorporating chain-linking sulfur bridges. See Rubber.

Vulcanization pigments, ***vulcanosine pigments***: a group of synthetic pigments used to color vulcanized rubber; red, wine, black, orange and green types are available.

Vulcanized fiber, abb. ***VF***: chemically, cellulose hydrate. It is one of the oldest synthetic cellulose products. V. is a tough material which can be worked in a number of ways. The products can be hard, like horn, or soft and flexible with a leather-like quality. Horn-like V. is a pure cellulose hydrate, while flexible products have higher moisture contents; this is achieved by addition of hygroscopic materials. The products can withstand heat up to 70 °C, or at most, 90 °C. V. is very resistant to pressure and impact, oils, fats and solvents. When it is coated with paraffin, lacquer or an impregnating agent, it can be made water-resistant.

V. is used to make luggage, textiles, automobile parts and electrical devices; it is available as films, plates or foams.

VX: see chemical weapons.

Vycor glasses: see Silica glass.

W

w: a symbol for mass fraction; see Composition parameters.

W: symbol for tungsten.

Wackenroder liquid: see Polythionic acids.

Wacker process: see Acetic anhydride.

Wagner-Meerwein rearrangement: a nucleophilic 1,2-rearrangement in which R = H, alkyl or aryl groups, and the two C atoms can be primary, secondary or tertiary. X is an atom (or group) which can leave the molecule with its bonding electron pair, causing R, as an anion, to approach the positively charged C atom. The electron hole resulting from this maneuver can be filled by readdition of X or by reaction with the solvent Y (see Pinacol-pinacolone rearrangement).

$$\mathrm{R{-}C(X){-}C{-}} \xrightarrow{-X^-} [\text{nonclassical ion}] \longrightarrow \mathrm{{-}C^+{-}C(R){-}}$$

nonclassical ion

In the terpene series, for example, the W. can be used to synthesize camphor from α-pinene. Addition of hydrogen chloride to α-pinene forms the relatively unstable 2-chloropinane, which rearranges into 2-chlorobornane (bornyl chloride). The subsequent reaction sequence bornyl chloride, camphene, isobornylformate involves two W. Finally, isoborneol is formed by saponification and then oxidized to camphor.

The rearrangement occurs in polar solvents, e.g. nitromethane or sulfur dioxide, and can be catalysed by Lewis acids such as tin(IV) chloride or iron(III) chloride. The reaction is reversible. The Retropinacolin rearrangement (see) is similar to the W.

Wagner's reagent: see Alkaloids.

Walden inversion: inversion of the configuration at an asymmetric C atom during an S_N2 reaction; it is generally associated with a reversal of the optical rotation. For example, the reaction of (S)-(-)-malic acid with phosphorus(V) chloride yields (R)-(+)-chlorosuccinic acid:

$$\mathrm{HOOC{-}CH(OH){-}CH_2{-}COOH} \xrightarrow[-HCl,\,-POCl_3]{+PCl_5} \mathrm{HOOC{-}CH(Cl){-}CH_2{-}COOH}$$

L-Malic acid D-Chlorosuccinic acid

The inversion in configuration is due to the fact that the nucleophilic reagent approaches from the side opposite the leaving substituent, for electrostatic reasons. After the system passes through the typical S_N2 transition state, all the bonds on the asymmetric C atom "invert"; a process which can be visualized as similar to the inversion of an umbrella in a high wind.

Wallach rearrangement: the conversion of azoxybenzene to 4-hydroxyazobenzene in concentrated acids. The reaction can be used as a means of detecting azoxy compounds.

$$\mathrm{C_6H_5{-}\bar{N}{=}\overset{+}{N}(O^-){-}C_6H_5} \xrightarrow{H^+} \mathrm{C_6H_5{-}N{=}N{-}C_6H_4{-}OH}$$

The reaction mechanism is thought to be a nucleophilic attack of water molecules on the 4-position. A side reaction is an *ortho*-rearrangement in which 2-hydroxyazobenzene is formed. This is the major product when the reaction is initiated photochemically. The reaction is assumed to be an intramolecular rearrangement.

Warburg's respiratory enzyme: same as Cytochrome oxidase (see).

Warfarin: see Rodenticides.

Wash-active substances: the analytically determined surfactant content in various technical forms of surfactants and in surfactant-containing finished products.

Washing: the removal of dirt from the body or textiles using water and a Laundry product (see), mechanical and heat energy. The dirt to be removed consists of water-soluble and insoluble impurities and deposits (salts, fats, proteins, carbohydrates, pigments, discoloring impurities); this is *primary washing activity*. Stabilization of the dirt in the wash water to prevent its readsorption to the textiles is *secondary washing activity*. The physical chemistry of W. can be broken down into the following processes: 1) wetting of the cloth by reducing the surface tension of the water; 2) removal of fats and oils, most of which are liquid at the wash temperature, that is, the phase boundary between the cloth and the fat is replaced by the phase boundary between the cloth and the surfactant solution; and 3) increasing the negative surface charge of fibers and pigments by increasing the pH value and adsorption of surfactants and builders to them. This leads to mutual repulsion between the cloth and the pigment. The ability of the wash water to carry the dirt is improved by stable dispersion of solid particles and by solubilization of water-insoluble compounds in the micelles of the surfactant.

The calcium and magnesium ions which cause water hardness are bound by the builders; calcium ions are extracted from the dirt and the cloth fibers by ion exchange; colored impurities are oxidized, and protein dirt is degraded by enzymes in some products.

Washout: a process in which rain drops adsorb gaseous or particulate air pollutants during their free fall to earth. W. contributes to cleaning the atmosphere in polluted regions.

Wasp venom: see Bee venom.

Waste: solid, liquid and gaseous substances which accumulate as byproducts of industrial production or consumption. W. include industrial and community garbage, sewage and smoke and other exhaust gases (see Air pollution). These must either be disposed of in such a way that they do not damage the environment, or they can be reutilized; the latter approach is becoming more widely accepted. It consists of utilizing substances present in the W. (*secondary raw materials*) or their reuse in production (*recycling*), for example, of scrap metal, old textiles, paper, tires and oil. The main methods of disposal of W. are burning, composting and dumping (landfill). Biotechnological processing is now being developed.

The most rational means of handling W. is to include plans for it in the planning of new industrial plants or communities; investment should be made in waste-free technology which permits the most rational utilization of natural resources, including energy, and protection of the environment, while meeting the needs of society.

Waste gases: the unusable gases formed by combustion or production processes. W. contain air pollutants like carbon oxides, sulfur oxides, nitrogen oxides, ash and soot. W. from combustion of solid or liquid fuels are also called Flue gases (see). The W. from motor vehicles contain carbon dioxide, nitrogen, water vapor, carbon monoxide, nitrogen oxides, aldehydes and hydrocarbons.

If the W. are hot enough, a fraction of their heat content can be used to generate steam and warm water, or to preheat the combustion air. That fraction of the heat content needed to make the W. rise in the chimney is not available for other work. The W. from internal combustion engines can be used to drive a turbine which precondenses the carburetor air (turbocharger).

High concentrations of W. in the atmosphere can lead to smog. To protect the environment, Waste gas treatment (see) is necessary.

Waste gas treatment: the separation of Air pollutants (see) from waste gases. Solid particles are removed by methods of Ash removal (see). There are essentially three methods of removing gases and liquid aerosols:

1) *Absorption*. Water is the preferred absorbant, due to its cheapness. If the absorptivity of water is not adequate, other substances must be added to it to react with the gases (chemisorption). For example, acidic components are removed with NaOH. Absorbers (scrubbers) operated as spray jets, turbulent wet scrubbers or Venturi scrubbers, achieve high levels of absorption. Oils (oil scrubbing) are used as absorbants for organic substances. The disadvantage of the absorption methods is that the scrubbing water must be disposed of. Flue gas desulfurizing (see) is a special case.

2) *Adsorption*. The adsorbers are usually filled with activated charcoal, and are regenerated with steam or by heating. Since some time is spent on regeneration, several adsorbers must be available for alternating use. In the very smallest installations, the adsorbers are thrown away.

3) *Afterburning*. There is a difference between thermal and catalytic afternurning. In thermal processes, temperatures of 900 °C are required and the gas must remain in the combustion space for a sufficient period of time. In the simplest case, thermal afterburning can be done in other parts of the same plant. Otherwise, the required temperature must be provided by burning fuels. The process then becomes very cumbersome and creates more air pollution. However, since the air pollutants are reduced to carbon dioxide and water (aside from the heteroelements nitrogen, sulfur and halogens present in the organic pollutants), this method is optimum from the point of view of air hygiene. In catalytic afterburning, the combustion temperature is reduced to 350 to 400 °C by use of catalysts. This process is very important for the treatment of exhaust from motor vehicles.

Waste liquors: liquid wastes from industrial sources, most of which contain high concentrations of substances damaging to the natural environment. W. must always be treated in some fashion (see Sewage treatment), unless they can be used elsewhere.

Water: H_2O, hydrogen oxide; K_m 18.0153, density of the solid at 0 °C, 0.9186, density of the liquid at 0 °C, 0.99987, at 4 °C, 1.0000, at 20 °C, 0.99823; m.p. (at 101.3 kPa), 0.00 °C, b.p. (at 101.3 kPa) 1.00 °C, n_D^{20}, 1.3330, crit. temp. 373.9 °C, crit. pressure 22.05 MPa, crit. density 0.3155, melting enthalpy 6.010 kJ mol^{-1}, enthalpy of evaporation at 100 °C, 40.651 kJ mol^{-1}, enthalpy of formation 285.89 kJ mol^{-1}, specific heat at 15 °C, 4.1868 J g-1 K^{-1}, viscosity at 20 °C, $72.75 \cdot 10^{-3}$N m^{-1}, dielectric constant at 18 °C, 80.84, electrical conductivity at 0 °C, $6.35 \cdot 10^{-12}$ Sm mm^{-2}.

Properties. W. is a colorless, odorless and tasteless liquid, which freezes to ice when cooled. The melting point of ice, 273.15 K, is the zero point of the Celcius temperature scale. The water molecule is angular (bond angle H-O-H 104.5 °C, bond length 96 pm). The bond in H_2O can be described in terms of a hybrid state of the oxygen atom, in which the bonding orbitals have a somewhat higher p contribution, and the two free electron pairs have a somewhat lower p contribution, than in pure sp^3 hybridization. The difference in electronegativity of oxygen and hydrogen and the availability of free electron pairs on the oxy-

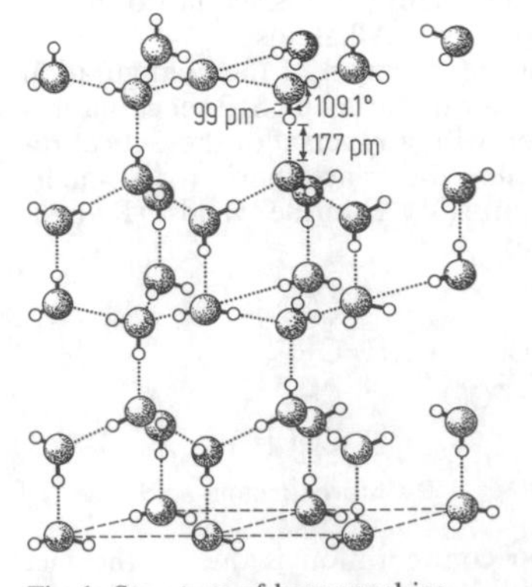

Fig. 1. Structure of hexagonal ice.

gen atom make the W. molecule polar, and are the basis for formation of stable hydrogen bonds in the condensed phases. These intermolecular interactions determine the structure of solid and liquid W.

Under normal pressure, W. crystallizes in a hexagonal lattice comparable to that of tridymite (Fig. 1). (A total of eight polymorphic forms of ice are known).

At the melting point of ice, this relatively bulky structure collapses, and the packing density of the H_2O molecules increases, leading to an increase in the density of the liquid phase. However, fragments of the ice lattice are preserved above the melting point, and it is their gradual degradation, together with the increasing mobility of the H_2O molecules, which causes liquid water to have its density maximum at 4 °C. Even in liquid W., a certain structure can be detected; the H_2O molecules are bound by hydrogen bonds into large, fluctuating aggregates which are tetrahedral around the oxygen atoms (Fig. 2).

Fig. 2. Two-dimensional representation of the interactions in liquid water.

This ***density anomaly*** of W. has important consequences for the planet Earth. As natural bodies of water cool, the denser, colder W. sinks to the bottom, until a temperature of 4 °C (in fresh water) is reached. However, as the water cools further, the colder W. remains on the surface and freezes. Because it has a lower density, ice floats, and because it acts as a heat insulater, it prevents freezing of the deeper layers, even when the temperature is very low. This is essential for organisms living in lakes and rivers. If ice did not float, the oceans of the planet would presumably consist of thick layers of ice on the bottom with relatively shallow layers of melted surface water, and this would cause the climates of the planet to be entirely different. Because the volume of water increases as it freezes, freezing water pipes burst, and rocks are cracked apart because of water which has seeped into them. The high degree of cross-linking of H_2O molecules in solid and liquid W. is the reason for its high enthalpies of melting and vaporization, which in turn have an important effect on the climate. The aggregation also accounts for the relatively low vapor pressures of solid and liquid W., and for the high melting and boiling points (compared to those of the heavier hydrogen chalcogenides). The high surface tension and specific heats are also due to the aggregation.

Its bond polarity and angular structure are the reason that the W. molecule has a dipole moment. This gives W. an unusually high dielectric constant and the ability to dissolve polar and ionic compounds. This effect is amplified by the solvation capacity of W., which is also due to its polar nature. It can act as either a donor or acceptor of electrons, and can thus solvate either anions or cations (see Hydration). As a result, many compounds tend to dissociate into ions in aqueous solution. Small, highly charged ions in particular acquire a relatively stable hydration shell (see Aqua complexes) which is often incorporated into the crystals of the salt (see Crystal water).

W. has both acidic and basic properties, and is the standard for determination of acid and base strength (see Acid-base concepts, Brønsted definition). It undergoes autoprotolysis: $H_2O + H_2O \rightleftharpoons H_3O^+ + OH^-$; $a_{H_3O^+} \cdot a_{OH^-} = K_w = 1 \cdot 10^{-14}$ at 22 °C (see Ion product).

W. is thermodynamically a very stable compound. Its high enthalpy of formation is the driving force for many reactions which release it. Strong reducing agents, such as glowing iron powder, can release hydrogen from W: $H_2O + Fe \rightarrow FeO + H_2$. Strong oxidizing agents, such as fluorine, can release oxygen from W: $2\,H_2O + 2\,F_2 \rightarrow 4\,HF + O_2$. W. is also split into oxygen and hydrogen by electrolysis. W. reacts with many nonmetal oxides to give acids, e.g. $SO_3 + H_2O \rightarrow H_2SO_4$. Many metal oxides are converted to metal hydroxides by W., e.g. $CaO + H_2O \rightarrow Ca(OH)_2$.

Analysis. Water may be detected by salts which form colored hydrates (for example, colorless $CuSO_4$ reacts to form blue $CuSO_4 \cdot 5H_2O$, and blue $CoCl_2$ forms pink $CoCl_2 \cdot 6H_2O$), or by salts which are converted to colored products (e.g. colorless $K[PbI]_3$ is converted to yellow PbI_2 and KI). The water content of a solution can be measured quantitatively through its reaction with ionic or complex hydrides, e.g. calcium hydride, CaH_2, or lithium alanate, $LiAlH_4$, which release hydrogen. This can be measured in a gas buret. A widely used method for water determination is based on the Karl-Fischer solution (see). There are also automatic methods in which the water content of a stream of gas or a solution is determined from its IR absorption, or by ^1H NMR spectroscopy, gas chromatography or electrolysis.

Occurrence. W. is found in all three aggregate states on Earth: as solid polar ice, sea ice and glaciers, as surface and ground water (oceans, rivers, lakes, underground reservoirs and streams), and as water vapor, which makes up a varying proportion of the lower atmosphere. About 71% of the Earth's surface is covered with W. The total volume of the oceans is estimated as $1348 \cdot 10^6$ km^3, making up 97.4% of the total amount of W. on Earth. Only about 2.6% of the W. reserves are fresh water, and of this amount, more than three quarters are polar and sea ice. Surface and ground waters always contain dissolved salts, the composition and concentrations of which depend on the source and cause Hardness (see) of water. The salt content of sea water is an average of 3.5%, of which 2.7% is sodium chloride. W. with an evaporation residue less than 0.1% is called ***fresh water***. Many minerals also contain W. as crystal water. In addition to H_2O, natural W. always contains small amounts of ***heavy W.***, D_2O (see Deuterium oxide) and HDO.

Production. W. is formed when hydrogen is burned, and as a byproduct of many other chemical reactions, such as combustion of hydrocarbons and carbohydrates, condensations, etc. W. is usually obtained by purifying natural W.; the applicable process of Water treatment (see) depends on the impurities in the W. and its intended purpose. Especially pure ***conductivity W.*** is obtained for physical chemical measurements by repeated distillation in quartz or noble metal stills under special conditions. For pharmacy and general chemical laboratory use, ***distilled water*** is ordinarily sufficient. ***Doubly distilled W.*** is sometimes used for medical purposes.

Significance, applications. The presence of W. is a requirement for life. The human body consists of 60 to 70% W., and plants contain up to 95%. W. is the raw material for photosynthesis and the solvent for nutrients in both organisms and soil. The body temperatures of many organisms are regulated by controlled evaporation of W.

The applications of W. are extraordinarily broad, and different degrees of purity are required for different applications (see Water treatment, Water provision). ***Drinking water*** is consumed for actual drinking, cooking and cleaning, and toilet flushing, at a rate of about 150 l per person and day in developed industrial countries. W. of varying quality is used as a coolant, solvent, raw material and reagent in many industrial processes. It is used as an energy carrier and, in agriculture, for irrigation.

Water analysis: methods for analysis of drinking, process and waste water. In addition to the usual methods, which in some cases are standardized and normalized for W., there are special methods for determining water hardness. (See Water hardness, determination of). W. is important for environmental protection.

Water baths: see Heating baths.

Water chemistry, ***hydrochemistry***: an area of applied chemistry concerned with the physical and chemical properties of water and substances dissolved in it. W. also includes water analysis and the technological bases of Water treatment (see), Sewage treatment (see) and surface water chemistry.

Water drop test: a test for qualitative analysis of silicon, silicates or silicon dioxide, SiO_2, or fluoride. The reaction is based on formation of volatile silicon tetrafluoride, SiF_4, from hydrogen fluoride, HF, and SiO_2 or silicates. SiF_4 is then hydrolysed to orthosilic acid, which condenses to polysilicic acids (see Silicic acids). The substance to be analysed for silicate is mixed with calcium fluoride, CaF_2 and sulfuric acid, H_2SO_4, in a lead or platinum crucible and heated. For detection of fluoride, the sample is mixed with SiO_2 and S_2SO_4, then heated. The crucible is covered with a lid which has a hole in it; above the hole one holds a glass rod with a drop of water hanging on it. The SiF_4 flowing out of the hole is hydrolysed in this drop, and the resulting polysilic acids can be recognized as turbidity or a gel in the drop.

Water gas: a synthetic gas consisting mainly of hydrogen and carbon monoxide. ***Coke water gas*** (obsolete term, ***blue water gas***) is about 50% hydrogen, 40% carbon monoxide and 5% each of carbon dioxide and nitrogen. Its heating value is about 11,000 kJ m^{-3}. The values for ***coal water gas (double gas)***, the values are about 1500 kJ higher. For more information, see Synthesis gas.

Water gas tar: see Tar.

Water glass: a mixture of various sodium and potassium silicates which hardens to a transparent, glassy solid. It is colorless when pure, but is often greenish or yellowish, due to the presence of iron. W. is obtained by melting silicon dioxide with sodium carbonate (***soda W.***) or potassium carbonate (***potash W.***) at 1400 to 1500 °C. The resulting stoichiometric ratio of SiO_2/Na_2O or K_2O determines its properties, and also the nature of the silicate anions (see Silicates) it contains, e.g. $SiO_2 + Na_2CO_3 \rightarrow Na_2SiO_3 + CO_2$ or $2\,SiO_2 + Na_2CO_3 \rightarrow Na_2Si_2O_5 + CO_2$. The hardened melts are barely soluble in cold water, but under pressure they dissolve in hot water to form clear, very alkaline solutions, which are also commercially available as W. The addition of strong acids to these solutions precipitates silicic acids of varying degrees of condensation; they can be separated and processed to silica gels.

W. is used to obtain silicate materials and for synthesis of special aluminosilicates (zeolite); it is a component of cleansers and is used to impregnate paper and textiles to retard flame. W. solutions are used as a mineral glue for joining silicate materials (glass, porcelain, etc.), for impregnating paper and for preserving eggs.

Water glass pigments: same as Silicate pigments (see).

Water hardness, determination of: quantitative determination of the Hardness (see) of water.

1) The total hardness is the sum of alkaline earth cations present as the carbonates, sulfates, chlorides, nitrates and phosphates. It is now determined by Complexometry (see), which is much simpler and more precise than previous methods, e.g. those of Blacher, Boutron/Boudet and Splittgerber/Mehr. The procedure is further simplified by the availability of tablets of disodium EDTA which, for a given volume of sample, correspond to 1° or 5° hardness. One only has to determine the number of tablets which must be added before an indicator in the sample turns color. The indicator is also supplied in the form of indicator-buffer tablets, which insure that the solution has the appropriate pH.

If the water contains traces of iron or other heavy metal ions, these must be masked before determination (see Masking reagents), because they would otherwise block the indicator and prevent recognition of the endpoint.

2) Carbonate hardness is determined by titration with 0.1 M hydrochloric acid, using methyl orange as indicator. If the value determined for the carbonate hardness is greater than that for the total hardness, the latter is given as the value for carbonate hardness.

3) Non-carbonate hardness is determined from the difference between the total and carbonate hardness.

Water pollutants: anthropogenic solid, liquid or gaseous materials present in water which degrade its quality. W. can be present as suspensions, emulsions colloids or in true solution. They are removed from drinking and process water by Water treatment (see), and from sewage by Sewage treatment (see). See table, p. 1171.

Water quality: see Sewage.

Water softening: removal of calcium and magnesium salts dissolved in water, e.g. $Ca(HCO_3)_2$, $Mg(CHO_3)_2$, $MgSO_4$, $MgCl_2$ and $CaSO_4$.

1) ***Chemical water softening***: Carbonate, or temporary hardness (see Hardness, def. 2) can be par-

Table. A selection of highest allowable concentrations of important pollutants (from the Water Pollutant Catalog of 1974, Liebmann and Stammer, 1960, and from WHO)

Pollutant	Concentration g m^{-3}
Acrylonitrile	2.0
Arsenic	0.05
Benzene	0.5
Cadmium	0.005
Chlorobenzene	0.02
DDT	0.008
Dichloroethane	2.0
Dimethylformamide	10.0
Formaldehyde	0.5
Caprolactam	1.0
Nitrochlorobenzene	0.05
Phenols, total	0.003
Pyridine	0.2
Carbon tetrachloride	5.0
Crude oil	0.3

tially removed (down to about 0.72 mval = 20 mg CaO/l) as follows:

a) ***Thermal W.***. Heating converts the dissolved calcium and magnesium hydrogencarbonate in the water to calcium carbonate, carbon dioxide and water, according to equation (1):

$$Ca(HCO_3)_2 \rightleftharpoons CaCO_3+CO_2+H_2O \quad (1)$$

This process depends on time and temperature.

b) ***(Quick)lime*** and ***caustic soda*** process. The amount of calcium hydroxide corresponding to the amount of carbonate hardness is calculated (equation 2), and the equivalent amount of calcium oxide (quicklime) is added. (The lime reacts with water to form calcium hydroxide.) Alternatively, the amount of sodium hydroxide (caustic soda) required in equation (3) is added; the amount of hardness can be reduced to 0.72 mval in this way:

$$Ca(HCO_3)_2+Ca(OH)_2 \rightarrow 2\,CaCO_3+2\,H_2O \quad (2)$$

$$Ca(HCO_3)_2+NaOH \rightarrow CaCO_3+NaHCO_3+H_2O \quad (3)$$

Magnesium carbonate, $MgCO_3$, forms in reactions corresponding to (2) and (3); some of it (up to 94.4 mg/l) remains in solution. A further addition of lime or caustic soda causes the less soluble magnesium hydroxide to form (solubility 9 mg/l):

$$MgCO_3+Ca(OH)_2 \rightarrow Mg(OH)_2+CaCO_3 \quad (4)$$

Inocculation method: the carbonate hardness in the water is reduced to 0.36 mval by adding a calculated amount of acid (sulfuric or hydrochloric):

$$Ca(HCO_3)_2+2\,HCl \rightarrow CaCl_2+2\,CO_2+2\,H_2O \quad (5)$$

This method increases the non-carbonate hardness, and produces a high carbon dioxide concentration.

d) ***Lime excess method***. Lime is first added to the untreated water in excess, leading to reaction (2). The water is now very alkaline, which promotes removal of iron, manganese and, to some degree, of other substances. The water is subsequently almost neutralized by addition of acid.

In chemical removal of the total hardness (see Hardness), reactions (1)-(5) also occur. The following processes are used:

a) ***Lime-soda process***. Addition of lime first removes the carbonate hardness portion of the calcium and magnesium according to equations (2) and (4). Then soda is added to precipitate the non-carbonate hardness (e.g. $CaSO_4$):

$$CaSO_4+Na_2CO_3 \rightarrow CaCO_3+Na_2SO_4 \quad (6)$$

The hardness of the water can be reduced to 0.36-0.72 mval in this way.

b) ***Soda-caustic soda process***. The process is analogous to the lime-soda process, but the soda and sodium hydroxide can be added simultaneously instead of sequentially.

c) ***Trisodium phosphate***. Although it is relatively expensive, trisodium phosphate is capable of reducing the concentration of hardness to 0.036-0.108 mval by precipitating magnesium and calcium phosphates as a flocculent sludge:

$$3\,Ca(HCO_3)_2+2\,Na_3PO_4 \rightarrow Ca_3(PO_4)_2+6NaHCO_3 \quad (7)$$

$$3\,CaSO_4+2\,Na_3PO_4 \rightarrow Ca_3(PO_4)_2+3\,Na_2SO_4 \quad (8)$$

d) ***Balcke process***: First the hardness is reduced to 0.36 mval using the cheaper lime, soda or caustic soda as described above; then the remaining hardness is reduced to 0.036 mval using Na_3PO_4.

e) ***Neckar process***: the total hardness is removed with soda alone. First the non-carbonate hardness reacts with the soda according to equation (6); the sodium hydroxide formed in these reactions then precipitates the carbonate hardness according to equation (3).

f) ***Barite process***. Barium-containing compounds, especially barium carbonate, $BaCO_3$, are used. This process is expensive and is only rarely used. The residual hardness at 20 °C is 1.44 to 1.8 mval.

g) Modern water softeners work with ion exchangers and exchange adsorbants, which can produce water equivalent in quality to distilled water.

2) Softening by physical-chemical and purely physical methods.

a) ***Electrolytic process***. A constant-current voltage of 6 to 10 V applied to the inner surface of steam boiler pipes causes the hardness to precipitate as a very finely divided sludge.

b) ***Electroosmotic process***. see Desalination.

c) ***Distillation process***. see Desalination.

d) ***Static process***. Evacuated glass balls filled with neon and droplets of mercury are kept in motion in the water. Electric charges form as the mercury drops are broken apart and reunite, and these charges cause the hardness to precipitate as a sludge.

e) ***Freezing out***. see Desalination.

f) ***Magnetic process***. The water to be treated is passed through a magnetic field, with the direction of flow perpendicular to the magnetic field lines. This does not prevent precipitation of the carbonate hardness, but the treatment causes it to precipitate as an amorphous sludge which can be more readily removed. It thus can be prevented from forming incrustations within the equipment; in some cases the method can even be used to remove old incrustations.

Water, provision of: water must be provided for domestic consumption, industry, agriculture, transportation and other purposes in sufficient amounts and adequate quality (see Water).

For hygienic reasons, the favored source of drinking water is groundwater. In areas where there is little ground water, it must often be taken by filtration through riverbanks or from artificial reservoirs, and in some cases, the water must be taken from rivers. Process water is generally taken from surface sources, usually rivers. When the water is pure enough, it can be delivered directly to the consumers. In a few cases (high-altitude reservoirs and springs) it can be delivered by gravity flow, but usually it must be pumped. Intermediate reservoirs (such as water towers) are used to regulate the pressure. Transport is through underground pipes.

Water quality. drinking water and process water for the food industry must meet certain hygienic requirements. It may contain no pathogenic bacteria, and the number of other bacteria must be low. The number of bacteria (coliform bacteria) is given by the Colititer (see). Process water must have properties to match the requirements of its consumers, which are generally very speific (see Water treatment).

Water treatment: production of water of a quality suitable for use from surface or ground water. In general, W. is divided into treatment for drinking and industrial quality.

According to the World Health Organization, drinking water must meet the following qualifications: it must be free of pathogens and substances which can impair health; it must have a low bacterial content and be appetizing; it must be colorless, cool and free of foreign odors and tastes; it must not be too hard (see Hardness 2); and it must not attack materials or lead to deposits or incrustations, that is, it must be in Calcium-carbon-dioxide equilibrium (see).

Substance	mg/l
Iron	≦ 0.1
Manganese	≦ 0.05
Potassium permanganate consumption	≦ 12.0
Ammonium	not detectable
Nitrite	not detectable
Lead	not detectable
Phenols	not detectable
Chloride	≦ 250
Nitrate	≦ 20
Sulfate	≦ 250
Fluoride	≦ 1.0

Water for industrial purposes must meet different specifications, depending on the use. For example, steam generators require water with no hardness, while textile plants, bakeries and breweries require soft water with no iron. Manganese must also be present in no more than traces. Cooling water must not produce deposits in the cooling system. The specifications for industrial water are sometimes very demanding (see, for example, Feedwater), and the problems of W. are so numerous that independent disciplines have arisen to treat them (e.g. feedwater chemistry). The qualities of crude water vary greatly, as do the requirements for treated water, and as a result many methods of treatment have been developed and tested on an industrial scale.

To achieve the desired quality of treated water, several processes are combined at each stage of treatment, and in each case, protection against corrosion, high technological and economic efficiency and optimum purification must be considered. The processes in turn often consist of several steps. The most important of the processes are: a) Gas exchange (see); b) removal of suspended particles and colloids by sieving, Sedimentation (see) and Filtration (see); c) removal of dissolved substances by Deferrization (see), Demanganization (see), Water softening (see), Desalination (see), chemical stabilization (see Protective layer formation), Deacidification (see) and Clarification (see) in the form of adsorption (see Sorption) or Oxidation processes (see); d) Groundwater enrichment (see) and e) Disinfection (see).

There are a number of other processes for certain substances, for extreme demands for quality or for extreme climatic or other conditions. These include Degasification (see), Evaporation (see), Oil removal (see), Silica removal (see), sea water Desalination (see), deactivation (see Sewage treatment) and Fluoridation (see).

Watson-Crick model: see Nucleic acids.

Wave function: see Atom, models of.

Waxes: originally a term for diverse mixtures of water-insoluble biological products in which the main components are esters of long-chain fatty acids with long-chain primary alcohols. W. often also contain a great variety of free acids (corresponding to the esterified acids), ketones, alcohols and limit hydrocarbons.

The modern concept arises from industrial use of W. and the extensive substitution of mineral and synthetic materials for natural products. W. are now considered to be substances and mixtures with certain technological properties: at 20 °C they are malleable to brittle solids, with large to small crystals, are transparent to opaque, but not glassy, melt above 40 °C without decomposition, and have relatively low viscosity a few degrees above their melting points. Their consistencies and solubilities are highly temperature dependent. If a substance displays these properties, it is considered a W. (in borderline cases, one of the above properties may not be completely present). It is a characteristic of W. that they not only consist of a large number of similar chemical compounds, but that compounds representing a number of chemical classes are present in them.

Economically, the most important W. is solid paraffin from petroleum (it accounts for more than 90% of the world production); in some countries, paraffin is still extracted from soft coal in significant amounts. For special applications or for production of certain compositions, natural products are chemically modified (mineral wax, fatty acid mixtures). W. may be completely synthetic (polyethylene, copolymers) or may be obtained from natural sources (e.g. candelilla, carnauba, and palm waxes, beeswax, shellac and wool waxes).

W. are used mainly to produce candles, protective coatings for floors, automobiles and leather; to impregnate paper, cardboard and fiberboard, and as bases for cosmetic and pharmaceutical products.

Wegscheider principle: a rule of reaction kinetics: in parallel reactions (see Reaction, complex), the ratio of the amounts of the various products at any time is the same as the ratio of the rate constants of the steps which lead to their formation. The W. is not valid unless the parallel steps have the same form of rate equation, and the initial concentrations of all products are zero.

Weissenberg process: see X-ray structure analysis.

Weston standard cell: a Standard cell (see).

Wet metallurgy: same as Hydrometallurgy (see).

Wet grinding: the grinding of particles in a paste or liquid suspension in a funnel, ball or colloid mill.

Wetting: spreading of one liquid over another liquid which is not miscible with it, or over a solid surface. If a drop spreads over the entire surface as a film, one speaks of ***complete W.***. Complete W. is required of a glue. In ***incomplete W.***, a Contact angle (see) forms which corresponds to a thermodynamic equilibrium. When a liquid spreads over either another liquid or a solid, there are two possible states. Either the spreading liquid is in equilibrium with the surface of the substrate (Fig., a + b), or it is in equilibrium with a spread film (Fig., c + d). The equilibrium states are defined as follows:

a) $\sigma_{13} = \sigma_{12}\cos\Theta_{12} + \sigma_{23}\cos\Theta_{23}$,
b) $\sigma_{13} = \sigma_{12}\cos\Theta_{12} + \sigma_{23}$,
c) $\sigma_F = \sigma_{12}\cos\Theta_{12} + \sigma_{23}\cos\Theta_{23}$ and
$\sigma_F = \sigma_{12}\cos\Theta_{12} + \sigma_{23}$,

where $\sigma_F < \sigma_{13}$. σ is the surface tension and Θ is the contact angle.

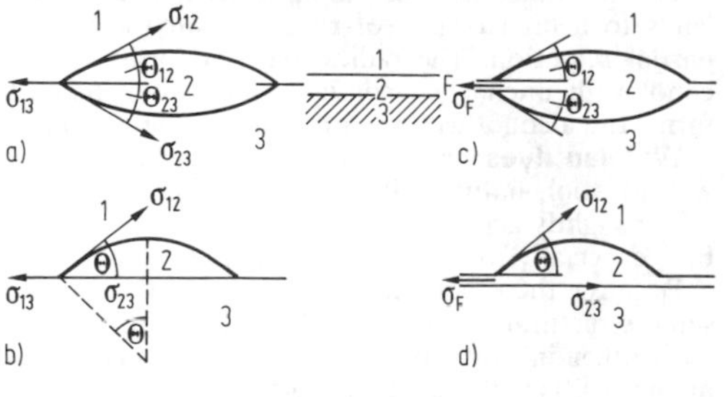

Changes in wetting are important industrially in flotation, washing and layering processes.

Wetting agent: a natural or synthetic compound which reduces the surface tension of water or another liquid (see Surface active substances). A W. makes possible better penetration of liquids into solid surfaces, such as those of wood, metal or textile fibers.

Whisker, ***hair crystal***: a needle or hair-shaped growth form of a crystal which has special physical properties. W. are very thin (about 10^{-4} mm in diameter) and may be up to several centimeters long. They are formed mainly by metals, but under certain conditions can also be formed by semiconductors or ionic substances. Because their lattices have extremely few defects, they can be very strong, for example, they may be about 1000 times stronger than a normal single crystal. They are used in combination with plastics or glasses to make new composite materials.

White gold: see Gold alloys.

White metals: alloys of tin or lead with antimony, copper, cadmium, arsenic and nickel. The W. are used in sliding bearings. They are cast onto steel or cast iron bases. The W. consists of a soft, eutectic ground substance in which the hard and brittle crystals of intermetallic phases (in high-tin W., Cu_6Sn_5 and SbSn) are embedded. To save tin, lead-alkali bearing metals have been developed with additives of calcium, sodium, lithium, magnesium and barium; these are used especially in railroad cars (***railroad metal***).

White oils: refined mineral oils used for lubrication and production of cosmetics and pharmaceuticals.

Wiesner reaction: a reaction used to detect lignin in woody material. The substance to be examined, such as paper, is sprayed with 12% hydrochloric acid, and then a 5 to 10% alcoholic solution of phloroglucin is applied. The presence of lignin is indicated by a purple-red to violet color which becomes stronger with time.

Willgerodt-Kindler reaction: a synthesis of amides found in the original methods of Willgerodt, published 1887. Aliphatic-aromatic ketones are heated with ammoniumpolysulfide in an autoclave to 150-200 °C and undergo redox amidination, e.g. phenylacetamide is formed from acetophenone:

$$C_6H_5\text{–CO–}CH_3 \xrightarrow{(NH_4)_2S_x} C_6H_5\text{–}CH_2\text{–}CONH_2$$

Subsequent hydrolysis produces the carboxylic acid with the same number of carbon atoms as the original ketone, e.g. phenylacetic acid is formed from acetophenone. It was found in 1923 by Kindler that the same reaction with sulfur and primary or secondary amines, especially morpholine, is possible in an open flask at 130 °C. The reaction mechanism is rather complicated and passes through many intermediates, including enamines and thionamides:

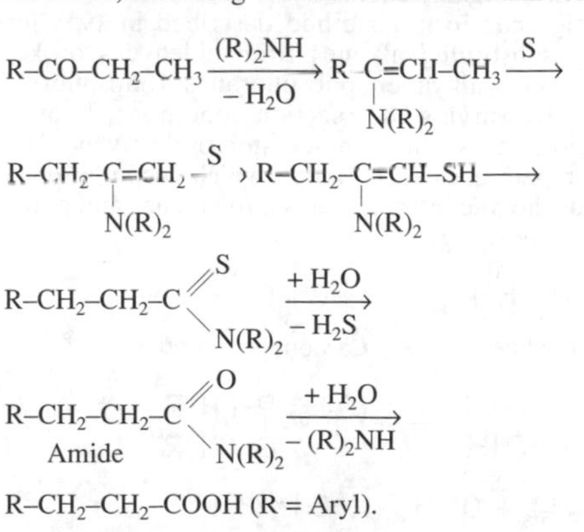

$$R\text{–CO–}CH_2\text{–}CH_3 \xrightarrow[-H_2O]{(R)_2NH} R\text{–}\underset{N(R)_2}{C}\text{=CH–}CH_3 \xrightarrow{S}$$

$$R\text{–}CH_2\text{–}\underset{N(R)_2}{C}\text{=}CH_2 \xrightarrow{S} R\text{–}CH_2\text{–}\underset{N(R)_2}{C}\text{=CH–SH} \longrightarrow$$

$$R\text{–}CH_2\text{–}CH_2\text{–}C(\text{=S})\text{–}N(R)_2 \xrightarrow[-H_2S]{+H_2O}$$

$$\underset{\text{Amide}}{R\text{–}CH_2\text{–}CH_2\text{–}C(\text{=O})\text{–}N(R)_2} \xrightarrow[-(R)_2NH]{+H_2O}$$

$R\text{–}CH_2\text{–}CH_2\text{–COOH}$ (R = Aryl).

Williamson synthesis: a method for making symmetric and unsymmetric dialkyl ethers and alkylaryl ethers by an S_N2 reaction of alcoholates or phenolates with halogen alkanes: $R\text{-}O^-Na^+ + R'\text{-}X \rightarrow R\text{-}O\text{-}R' + NaX$. Dimethylsulfoxide is a very suitable solvent. Alkene formation can occur as a secondary reaction, especially when secondary and tertiary haloakanes are used. For phenol ethers, dimethyl or diethyl sulfate is useful instead of the haloalkane, especially for production of methoxy- and ethoxyarenes: $C_6H_5O^-Na^+ + (CH_3)_2SO_4 \rightarrow C_6H_5OCH_3 + CH_3OSO_3Na$; or $C_6H_5O^-Na^+ + (C_2H_5)_2SO_4 \rightarrow C_6H_5OC_2H_5 + C_2H_5OCO_3Na$.

Winkler process: see Fluidized bed process; see Generator; see Synthesis gas.

Wintergreen oil: an essential oil with a pleasant aromatic odor; when pure, it is colorless, but it is often yellowish to reddish. The main component is methyl salicylate, at 95 to 99% concentration. W. is obtained by steam distillation from the chopped blossoms, leaves and stems of wintergreen; the yield is up to 0.8%. W. is used mainly in perfumes and soaps, but also in medicine. Methyl salicylate is often called ***synthetic W.***

Wipolan®: see Synthetic fibers.

Wittig-Horner reaction: a special case of the Wittig reaction in which the alkene is supplied by a phosphine oxide or phosphonate (*PO-activated olefination* of Horner):

$$R{-}P(R)(=O){-}\bar{C}^{\ominus}(R^2){-}R^1 + R^3{-}C(=O){-}R^4 \rightarrow R{-}P(R)(=O){-}C(R^1)(R^2){-}C(R^3)(R^4){-}\bar{O}^{\ominus}$$

$$\rightarrow R^1R^2C{=}CR^3R^4 + R{-}P(R)(=O){-}O^{\ominus}$$

Alkene

Like the Wittig reaction, it is used to make alkenes from aldehydes or ketones. The W. is used widely in preparative organic chemistry.

Wittig reaction: a method described in 1953 for making substituted alkenes from aldehydes or ketones, using alkylidenephosphoranes (phosphorus-ylides). By varying the reaction conditions, E or Z diasterioisomers can be made stereoselectively. The reaction path goes via betaines, which decompose via oxaphosphetane intermediates into alkenes and phosphane oxides, e.g.:

$$(C_6H_5)_3\overset{+}{P}{-}\bar{C}H{-}R + R'{-}CH{=}O \longrightarrow$$

Phosphorylide — Carbonyl compound

$$(C_6H_5)_3\overset{+}{P}{-}CH(R){-}CH(R'){-}O^- \longrightarrow (C_6H_5)_3P{-}CH(R){-}CH(R'){-}O \text{ (ring)}$$

$$\longrightarrow (C_6H_5)_3P{=}O + R{-}CH{=}CH{-}R'$$

Alkene

In this *carbonylolefination* with P-ylides, there is thus formally an exchange of the oxygen of the carbonyl compound for the alkylide group of the P-ylide, a reaction which has found a wide range of applications in organic chemistry, especially for production of unsaturated compounds in the natural products field (e.g. carotinoids, vitamin A and vitamin D compounds). The reaction can often be carried out as a "one-flask" reaction, i.e. the quarternary phosphonium halide used to form the p-ylide is mixed with a strong base (e.g. sodium amide, sodium methylate or butyllithium) in ether or tetrahydrofuran, and then the carbonyl compound is added immediately.

Wittig rearrangement: conversion of benzyl ethers, dibenzyl ethers, fluorene methyl ethers or the like into substituted alcohols with phenyllithium:

$$C_6H_5{-}CH_2{-}O{-}CH_3 \xrightarrow{C_6H_5Li} C_6H_5CHLi{-}O{-}CH_3 \rightarrow$$

$$C_6H_5{-}CH(CH_3){-}OLi \xrightarrow[-\,LiOH]{+H_2O} C_6H_5{-}CH(CH_3){-}OH.$$

The reaction probably occurs by a two-step cleavage-recombination mechanism following the metallization of the CH_2 group:

$$C_6H_5{-}\overset{\ominus}{C}H{-}O{-}C(H)(CH_3)(C_2H_5)\;Li^{\oplus}\;(1) \longrightarrow \left(C_6H_5{-}\dot{C}H{-}OLi\;\;\dot{C}(H)(H_5C_2){-}CH_3\right) \rightleftharpoons \left(C_6H_5{-}\dot{C}H{-}OLi\;\;\dot{C}(H)(H_5C_2){-}CH_3\right)$$

$$2a \rightarrow C_6H_5{-}CH(OLi){-}C(H)(CH_3)(C_2H_5) \qquad 2b \rightarrow C_6H_5{-}CH(OLi){-}C(H_5C_2)(CH_3)(H)$$

The rearrangement of the optically active ether *1* leads to a high degree of racemization (80%) and partial retention. The radical pair *2a* forms the alcohol with unchanged configuration, while pair *2b* forms the alcohol with the opposite configuration.

Wofalan dyes: metal complex dyes which adhere well to wool, natural silk and polyamide fibers from either slightly acidic or neutral solutions. W. are very fast, especially to light, and give even coloration.

Because their adhering qualities are due to the same structural features, the W. can be mixed with each other in any proportions without concern that nuance shifts will change the colors as they are taken up by the yarn.

Wofanol fast pigments: alcohol- and ester-soluble pigments which are very fast to light and can withstand heats up to 180°C (in some cases 240°C or higher). W. are used for transparent coloring of lacquers and plastics.

Wofatite®: a synthetic resin used as an Ion-exchanger (see).

Wofatox®: a methylparathion preparation.

Wöhler's urea synthesis: the first published synthesis of an organic compound (urea) from an inorganic substance (ammonium cyanate) outside a living cell; he used it to disprove the idea that a "vital force" was needed for such reactions: $[N{\equiv}C{-}O]^-NH_4^+ \rightarrow O{=}C(NH_2)_2$.

Wohl-Ziegler reaction: the method for allyl bromination of unsaturated compounds first described by A. Wohl in 1919 and further studied and expanded by K. Ziegler in 1942. The bromination was originally done with *N*-bromoacetamide, until the superior properties of *N*-bromosuccinimide were recognized:

3-Bromocyclohexene

In this reaction, the double bond is not attacked; instead, substitution takes place on the neighboring, sp^3-hybridized C atom. Methylene groups react more rapidly than methyl groups, and tertiary C atoms in general do not react. The reaction is promoted by irradiation with UV light and addition of peroxides. Radical traps, such as phenols, inhibit the reaction, which probably occurs via a radical chain mechanism. The applications extend to simple alkenes, cycloalkenes, isoprene derivatives, including steroids, and polyenes. *N*-Bromosuccinimide also reacts with ketones in the α-position; with arenes it reacts at the ring and with methyl substituted aromatics and heteroaromatics, it reacts at the methyl group.

Wolff-Kishner reduction: method for converting an aldehyde or ketone into a hydrocarbon with the same number of C atoms. The carbonyl compound is converted by reaction with hydrazine or semicarbazide into the corresponding hydrazone or semicarbazone, which is then cleaved into the hydrocarbon and nitrogen by potassium hydroxide in triethylene glycol at 200 °C or by potassium *tert*.-butylate in dimethylsulfoxide at 188 °C:

$$R_2C{=}N{-}NH_2 \xrightarrow[-H_2O]{+OH^-} R_2C{=}N{-}\bar{N}H^{\ominus} \longrightarrow R_2CH{-}N{=}N^{\ominus}$$

Hydrazone

$$\xrightarrow{-N_2} R_2CH^{\ominus} \xrightarrow[-OH^-]{+H_2O} R_2CH_2$$

Hydrocarbon

This reduction is especially suitable with higher-boiling ketones, e.g. aliphatic-aromatic ketones, cycloketones and keto-carboxylic acids.

Wolff rearrangement: a rearrangement of α-diazoketones to ketenes, which are available for further reactions, e.g. with water to form carboxylic acids, with alcohols to form esters, or with ammonia to form carboxylic acid amides. The rearrangement occurs on heating in the presence of silver oxide; ketocarbenes are assumed to be formed as intermediates. These form ketenes by anionotropic migration of the R substituent. In some cases, ketenes have been isolated in the absence of water or alcohol. In the photochemical rearrangement, an oxirene intermediate is probably involved:

$$R{-}CO{-}CR'N_2 \xrightarrow[-N_2]{h\nu} R{-}CO{-}\ddot{C}R'$$

Amide

$$R{-}\ddot{C}{-}CO{-}R' \rightleftharpoons R{-}C{=}C{-}R' \text{ (oxirene)}$$

$$\longrightarrow RR'C{=}C{=}O \quad \text{Ketene}$$

Since α-diazoketones are very readily prepared from acyl chlorides and diazomethane, the W. is a versatile method for producing carboxylic acids and their derivatives from starting materials which are one CH_2 group shorter than the products.

Wolpryla®: see Synthetic fibers.

Wood: the supportive, non-living tissue in the stems of plants. Its dry material consists of a little less than 50% cellulose, lignin (up to about 14% in broadleaf W., or up to 20% in conifer W.) and hemicelluloses. The cellulose in W. consists mainly of glucose, but contains some mannose and xylose units as well. W. also contains resins (conifer W. only), fats, oils, tannins and pigments (see Colored woods), proteins, soluble polysaccharides, alkaloids, glucosides, latex and other organic compounds, and minerals. The chemical composition of all W. is approximately the same: 50% carbon, 6% hydrogen, 43% oxygen, 0.3% nitrogen, and the rest minerals which form ash when the W. is burned. The substances present in W. can be crudely separated by treatment with steam, ether, ethanol and hot water, and the fractions can be further separated by chromatographic methods.

W. is used to construct buildings, furniture, utensils, etc., and to make pulp (see Wood pulp) and Cellulose (see). It is also the raw material (biomass) for gasification (see Wood, gasification of), Charcoal (see) and Wood hydrolysis (see). W. with high resin contents yield terpentine, colophony and similar products.

Wood preservatives (see) and Flame retardants (see) are applied to protect W.

Wood alcohol: see Methanol.

Wood ash: The noncombustible inorganic residue remaining after wood is burned. The ash content of the wood depends on the species, the growth conditions, the season in which the wood is cut, and so on. On the average, it is 0.3 to 0.6%; bark, needles and leaves have higher ash contents than the wood. The main components of W. are alkali and alkaline earth carbonates (14 to 19% K_2O, Na_2O; 30-40% CaO; 5-11% MgO). In addition, ash contains iron (1-3% Fe_2O_3), phosphates (0.4-5% P_2O_5), silicates (1.8-3% SiO_2), and sulfates (3-5% SO_3). W. was originally the source of potassium carbonate (potash).

Wood, gasification of: dry conversion of wood in the presence of limited amounts of an oxidizing agent, e.g. air, pure oxygen, water or carbon dioxide, or a mixture of these. The desired product is a gas consisting of carbon monoxide, hydrogen, methane and other hydrocarbons as combustibles, as well as carbon dioxide and nitrogen. Depending on the conditions, lean gas (see Generator gas), Synthesis gas (see) or a heating gas with a high methane content is obtained.

Wood hydrolysis: cleavage of glycosidic bonds in the polysaccharides of wood with acid catalysis. The reaction decreases the average length of the polysaccharides, eventually to the level of monosaccharides. Other reactions occur simultaneously, including dehydration, condensation and other reactions, which degrade the sugars. The temperature, acid concentration and reaction time are decisive factors. At temperatures above 180 °C (up to 240 °C), the rate of sugar formation is much higher than that of decomposition.

W. is important for production of xylose and furfural from the hemicelluloses, and for production of glucose (and its subsequent biotechnological conversion to yeast for feeds and ethanol), cellulose for textiles, microcrystalline cellulose, cellulose acetate and nitrate from the cellulose component of the wood.

W. is carried out by the *Scholler-Tornesch process* (percolation of wood chips at 180-190 °C and 1.2-1.4 MPa with 0.5 to 1% sulfuric acid) and the *Bergius-Rheinau process* (hydrolysis with conc. hydrochloric acid). The Scholler-Tornesch process yields about 50 kg sugar in solution from 100 kg dry conifer wood residues; about 18 to 24 kg yeast can be obtained from this sugar. Solid hydrolysis residues (lignin) are obtained in 35 to 40% yield; it consists of 50 to 70% polysaccharides, 7 to 19% resins and fats, 1.5 to 2% sulfuric acid and 1 to 10% ash. The Bergius-Rheinau process yields, from 100 kg dry conifer wood residues, 66 kg sugar in solution, 2 kg acetic acid and 33 kg lignin. From the sugar solution, one can obtain 31 kg crystalline glucose and 34 l ethanol or 25 kg yeast.

The advantages of the Bergius-Rheinau process over the Scholler-Tornesch process are hydrolysis at normal pressure and temperature, the production of relatively concentrated sugar solutions, and a relatively reactive hydrolysis lignin; however, the disadvantages are the expense of regenerating the hydrochloric acid and corrosion problems which limit the choice of materials.

Wood pulp, mechanical: a fibrous material made by pulping wood (usually conifer or poplar wood) or sawdust and used to produce paper, cardboard and fiberboard. As a product of mechanical treatment of wood, it contains all the components of wood, mainly cellulose, hemicelluloses and lignin. Before the mechanical pulping process, the wood structure may be loosened by treatment with neutral alkali sulfite solution buffered with alkali hydrogencarbonate (*chemical pulping*). The size of the fibers and degree of grinding are important factors in the end uses of W. W. can be made not only from cordwood, by pressing the outer surface of the wood against a rotating grindstone (stone-ground wood, abb. SGW), but from wood chips (refiner mechanical pulp, RMP). This process has been further developed by thermal pretreatment at an exactly regulated temperature, which softens the lignins (thermomechanical pulp, TMP), and thus reduces the amount of energy required for pulping. The purpose of this treatment is to make W. with technological properties similar to cellulose (in particular, high resistance to tearing); a similar result is obtained by chemical pretreatment of the wood chips, yielding chemothermomechanical cellulose (CTMP).

Wood preservatives: compounds or mixtures of substances for treatment of wood. Protection may be needed against fungi (rot), insects (especially termites) and fire (see Fire retardants). ***Wood care products*** prevent cracking and oxidative discoloration. There are *water-soluble* W. (magnesium silicofluoride, zinc silicofluoride, sodium fluoride, potassium dichromate, potassium hydrogen fluoride, sodium arsenate, sodium pentachlorophenol, borax, etc.) *oily* H. (tar oils, pentachlorophenol, organotin compounds, etc.) and *gaseous* H. (such as phosphorus hydride).

Protective paints contain pentachlorophenol or organotin compounds for protection against rot organisms, and alkali silicates or organic foaming agents for fire protection. ***Wood care finishes***, such as polyvinylacetate dispersions, are also available.

W. can be applied by brushing or spraying, dipping, pressure treatment, diffusion, injection, soaking or gas treatment of the wood.

Wood, pyrolysis of: see Charcoal.

Wood's metal: an alloy of 50% bismuth, 25% lead, 12.5% cadmium and 12.5% tin named for its inventor. W. melts about 60 °C and is therefore used for melt fuses, heating-bath fluid and soft solder for heat-sensitive materials.

Wood stains: There are two types of stain, one which creates the color by a reaction of a caustic material with the substances in the wood itself, usually a tannin, and the other a pigment which is applied to the wood. A variant of the first method is a two-step process: in the first step, the wood surface is impregnated with a tannin or similar material (e.g. pyrogallol, pyrocatechol, *p*-phenylenediamine), and in the second, the color is developed by application of a heavy metal salt (copper, chromium, nickel, manganese or iron). In some cases, the salt is dissolved in a solution of ammonia.

In ***wax staining***, the metal salts are emulsified with crude montane wax, colophony or some other substance. In this way, the wood acquires both color and gloss. Wax stains cover the wood texture to some degree.

The pigments used for wood staining are divided into two classes. ***H-stains*** are water-soluble, acidic aniline pigments with supplements. As a rule, spring wood takes up more aqueous color than summer wood, so that the color intensities of these two components are reversed. Wood surfaces treated with H-stains are sensitive to water. ***S-stains*** are basic aniline pigments in organic solvents, usually ethanol, which penetrate relatively deeply into the wood. They are not removed by water, but because the solvent evaporates rapidly, it is difficult to distribute them evenly over the wood surface.

It is also possible to inject stains into the living tree and allow them to be distributed by the movement of liquids in it. However, even when penetration promoters (e.g. dimethylsulfoxide) are used, it is not possible to achieve a homogeneous coloration of the wood in this way.

Wood tar: see Tar.

Wood vinegar: the foulwater from pyrolysis and carbonification of wood. It is an aqueous solution of acetic acid, methanol, acetone and "soluble tar" (tar components which are either water-soluble or are solubilized by the other components of the solution).

The methanol/acetone fraction can be separated by distillation. The acetic acid can be removed from the residual solution as calcium acetate by adding calcium carbonate, or by extraction with ethyl acetate. 30 to 50 kg acetic acid and 13 to 16 kg of a methanol/acetone mixture can be obtained as byproducts from the production of charcoal from 1 m^3 wood.

Woodward-Hoffman rules: a system developed by R.B. Woodward and R. Hoffmann (1965) on the basis of molecular orbital theory for predicting the courses of reactions. The basic concept is that the course of a synchronous reaction is essentially determined by conservation of orbital symmetry properties of reactants and products. It is sufficient to consider qualitatively π- and semilocalized σ-molecular orbitals, which can be characterized by the signs of their wavefunctions and the number of nodal surfaces. Molecular orbitals must be either symmetric or antisymmetric to the symmetry elements of the molecular system. If this is not the case, symmetry-satisfying orbitals can be generated by suitable linear combinations of orbitals. When the symmetry properties of the orbitals involved in the reaction are characterized, a correlation diagram can be constructed by drawing lines between the energy levels of orbitals with the same symmetry. From the positions of the bonding and anti-bonding energy levels of the initial and final states, the energy of the transition state can be estimated, and thus the reaction course may be predicted. In the following, the application of the W. to selected reactions is discussed.

1) In the reaction of two ethene molecules to form cyclobutane (cycloaddition), there are two planes of symmetry, m_1 and m_2 (Fig. 1).

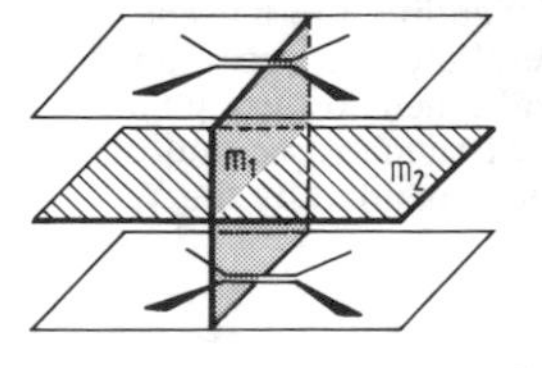

Fig. 1. Approach of two parallel ethene molecules.

During the reaction, two π-bonds are converted into two σ-bonds of the sp^3-sp^3 type (see Hybridization). Since the corresponding molecular orbitals π_1, π_2, π_1^*, π_2^* or σ_1, σ_2^*, σ_1^* and σ_2^* do not completely fulfill the symmetry requirements of the system, suitable linear combinations of them are formed. The symmetry-satisfying orbitals for the initial and final states and their symmetry with respect to m_1 and m_2 are indicated in the correlation diagram (Fig. 2).

In the transition from the initial to the final state, a high energy of activation is required, because the energy levels of the bonding and anti-bonding orbitals cross. Thus a synchronous reaction of two ethene molecules in the ground state to form cyclobutane ([2+2]-cycloaddition) cannot occur with conservation of orbital symmetry. After photochemical excitation of an electron from a π- to a π^* orbital, the energy balance for the reaction is more favorable, and

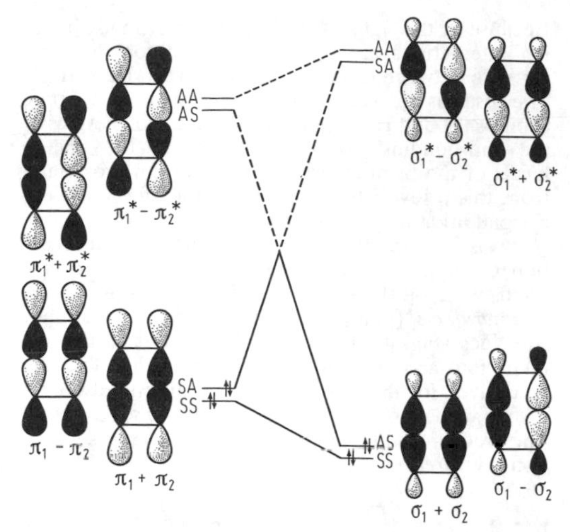

Fig. 2. Correlation diagram for formation of cyclobutane from two ethene molecules. SA, symmetric with respect to m_1 and antisymmetric with respect to m_2; AS, the reverse; SS, symmetric with respect to both m_1 and m_2; AA, antisymmetric with respect to m_1 and m_2.

ethene dimerizes to cyclobutane. If one considers the [2+2]-cycloaddition as thermally forbidden and photochemically allowed, analogous considerations for the reaction of butadiene and ethene (Diels-Alder reaction) indicate that the [4+2] cycloaddition is thermally allowed and photochemically forbidden. The activation energy, and thus the course of cycloaddition, can also be predicted using the *FO method* (frontier orbital method) developed by Fukui (1966). Frontier orbitals are the highest occupied molecular orbital (*HOMO*) and the lowest unoccupied molecular orbital (*LUMO*) of a molecule. The HOMO-LUMO interaction is predictive of the course of a reaction because this orbital interaction essentially determines whether the energy levels of bonding and non-bonding orbitals cross. The frontier orbitals of ethene (H1, L1) and butadiene (H2, L2) are shown in Fig. 3.

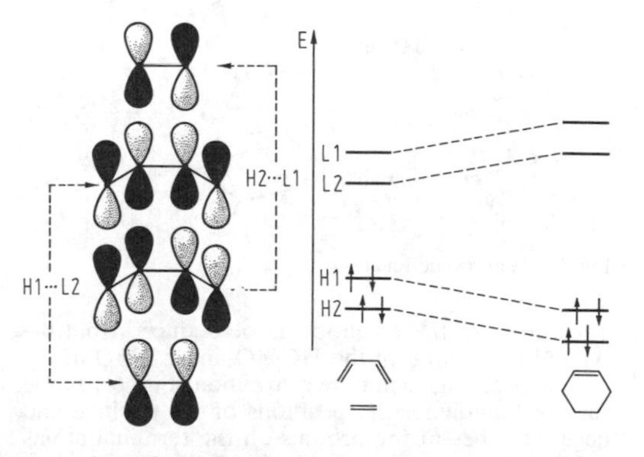

Fig. 3. [4 + 2] cycloaddition.

Because of the signs of the molecular orbitals, bonding interactions in a [4+2] cycloaddition are only possible between H1 and L2 and between H2 and L1. The result is a decrease in the energy of the bonding orbitals H1 and H2, and an increase in the energies of the anti-bonding orbitals L1 and L2. The energy levels of the frontier orbitals thus do not cross, and from this a low activation energy for thermal [4+2] cycloaddition may be inferred.

It was formerly thought that the attack of the reaction partners was *suprafacial* (s), i.e. that the new bonds were on the same side of the reacting system. In *antarafacial* (a) attack, the new bonds are on opposite sides (Fig. 4). If these different possibilities are taken into account, the rules shown in the table can be derived for the course of cycloaddition between a system with m π-electrons and one with n π-electrons. The predictions are equally valid for the reverse reaction (*cycloreversion*).

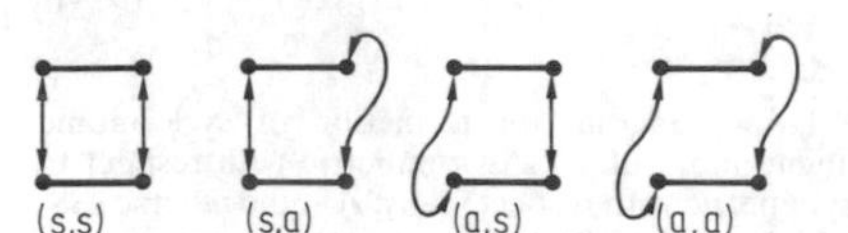

Fig. 4. Possible modes of attack in cycloadditions. s = suprafacial, a = antarafacial.

Table. Rules for $[m + n]$ cycloadditions

$[m + n]$	Thermally allowed Photochemically forbidden	Photochemically allowed Thermally forbidden
$4q$	$[m_s + n_a]$	$[m_s + n_s]$
	$[m_a + n_s]$	$[m_a + n_a]$
	$[m_s + n_s]$	$[m_s + n_a]$
$4q + 2$	$[m_a + n_a]$	$[m_a + n_s]$

$q = 1, 2, \ldots$, s = suprafacial, a = antarafacial

2) Electrocyclic reactions are defined as those which form a single bond between the ends of a polyene system. In order for the ring to close, the p_π-orbitals must be rotated so that a σ-bond can form. The reaction may be either *conrotatory* or *disrotatory* (Fig. 5). The two reaction courses can be distinguished because they lead to different sterospecific products in labelling experiments.

Fig. 5. Electrocyclic reactions.

The course of the reaction can be deduced from the type of interaction of the HOMO and LUMO of the π-system during each type of rotation (Fig. 6.) In the case of butadiene, the positions of the positive and negative lobes of the orbitals on the terminal atoms are such that conrotatory rotation leads to a bonding interaction for the HOMO and an anti-bonding interaction for the LUMO. This explains the fact that thermal cyclization of butadiene is conrotatory. Generalization of this concept leads to the rule that thermal electrocyclic reactions of systems with k π-electrons are disrotatory if $k = 4q+2$, and conrotatory if $k = 4q$ ($q = 0, 1, 2\ldots$). For photochemical electrocyclic reactions, the converse rule applies.

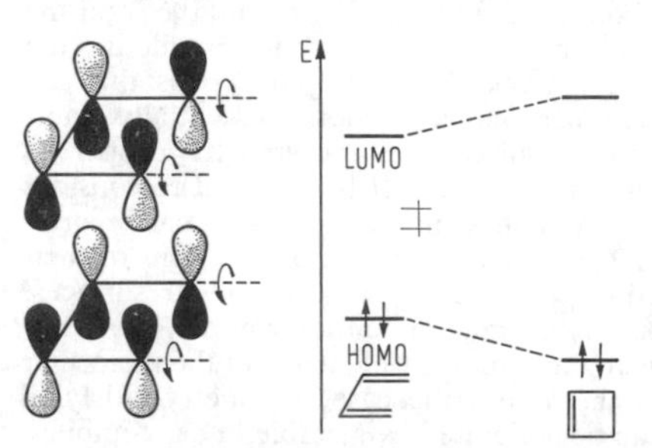

Fig. 6. Frontier orbitals in the conrotatory cyclization of butadiene.

3) A sigmatropic reaction is the intramolecular migration of a σ-bond which is flanked by one or more π-electron systems. The sigmatropic migration, e.g. of a hydrogen atom, can follow one of two topologically distinct pathways. In the suprafacial process, the migrating hydrogen atom remains on one side of the π-system, while in antarafacial migration, it moves from one side to the other. A cheletropic reaction is a process in which an odd-numbered ring is formed by insertion of an atom between the ends of an open-chain, even-numbered polyene chain, or conversely, the formation of an open chain by addition to a cyclic compound. An example is the photolytic decomposition of 3,5-heptadione into hexatriene and carbon monoxide (Fig. 7). The courses of sigmatropic and chelotropic reactions can also be explained by the W.

Fig. 7. CO loss from 3,5-cycloheptadienone.

In addition to correlation diagrams and the frontier orbital method, there is another method for estimating the energy of the transition states of the above reactions, the *Evans principle* introduced in 1939. It is based on consideration of the aromaticity properties of cyclic transition states in *pericyclic reactions*, in which the transition from the initial to the final structure can be formally represented by a shift in the bonding electron pair, as in the Cope rearrangement or the Diels-Alder reaction. Thermal pericyclic reactions tend to involve aromatic transition states, while photochemical pericyclic reactions lead most often to products which would have to be formed thermally via antiaromatic transition states. Because of the high degree of stereospecificity in synchronous reactions, the W. are of great importance in the chemistry of natural products.

Wool grease: an animal wax which melts between 36 and 42 °C; it is obtained from wool by washing or extraction with ether, benzene, petroleum ether, carbon disulfide or trichloroethane, and subsequent purification. It is a mixture of various esters of fatty acids and free alcohols, especially cholesterols. W. is used as a source of wool grease alcohols (cholesterol and related compounds, including lanolin) and as a salve base.

Wulff process: see Ethyne.

Wurster's red: see Semiquinones.

Wurtz-Fittig synthesis: a method for synthesizing alkylbenzenes first described by A. Wurtz in 1855 and expanded by R. Fittig starting in 1864. It consists of heating a mixture of benzylhalide and an alkylhalide with metallic sodium in a low-boiling solvent such as ether: C_6H_5-X + X-R + 2 Na → C_6H_5-R + 2 NaX. The R in this equation is an alkyl group, X is bromine or iodine, more rarely chlorine. The symmetric compounds biphenyl and the alkanes R-R are formed as byproducts, but these can easily be separated. The mechanism is thought to involve the formation of metallo-organic intermediates (see Wurtz reaction).

Wurtz-Grignard synthesis: a variant of the Wurtz-Fittig synthesis (see) for production of alkylbenzenes from bromobenzenes, magnesium and halogenoalkanes via phenylmagnesium bromide: C_6H_5-Br + Mg → C_6H_5-MgBr; C_6H_5-MgBr + RX → C_6H_5-R + MgBrX. Dimethyl or diethyl sulfate can also be used instead of the halogenoalkane to introduce a methyl or ethyl group into an arene ring by this method.

Wurtzite type, ***Wurtzite structure***: a common Structure type (see) for compounds with the general composition AB with small ratios of atomic radii, in the ideal case in the range of 0.225 to 0.414. The mineral wurtzite (zinc sulfide, ZnS) forms a crystal lattice in which the sulfur atoms are in hexagonal close packing, and the zinc atoms occupy alternate tetrahedral holes. Each atom is surrounded tetrahedrally by 4 atoms of the other type (Fig.). Thus the coordination is the same as in the Zinc blende type (see). Differences appear only when the next-nearest neighbors are considered: in the zinc blende type, the atomic arrangements are stacked, while in the W. there is an ecliptic atomic arrangement in the direction of the ZnS bond.

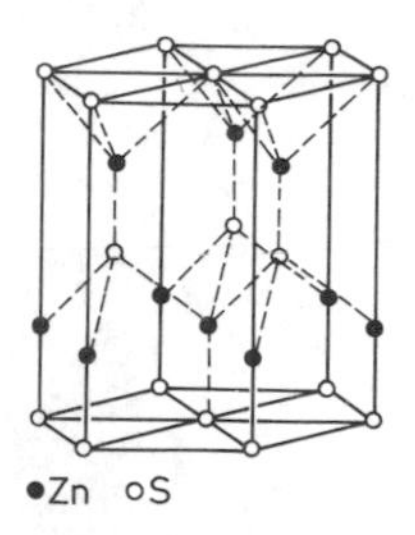

Wurtzite structure.

Crystallization of compounds in the W. has the same requirements as for the zinc blende type, and besides ZnS there are a number of compounds which crystallize in either structure (see Polymorphism).

Wurtz reaction: a reaction of halogenated alkanes with metallic sodium, discovered by A. Wurtz in 1855, used mainly to produce higher homologs of alkanes. Bromoalkanes are usually used, because they are easily synthesized and sufficiently reactive. It is not wise to use starting materials with different alkyl residues, because the products are mixtures which are very difficult to separate. The W. is usually carried out at relatively low temperatures; metallo-organic intermediates have been detected, e.g. CH_3-CH_2-Br + 2 Na → CH_3-CH_2-Na + NaBr; CH_3-CH_2-Na + Br-CH_2-CH_3 → CH_3-CH_2-CH_2-CH_3 + NaBr. These observations do not support a radical mechanism which was originally proposed, nor does the fact that optically active α-chloroethyl benzene, 2-chlorobutane and 2-chloroctane yield optically active hydrocarbons in the reaction; in other words there is no racemization. Racemization observed in other cases in the reaction of alkyl bromides are ascribed to a reversible halogen-metal exchange. The W. of α,ω-dihalogen alkanes leads to cyclic hydrocarbons and makes possible the synthesis of cycloalkanes of ring size C_3 to C_6.

Wurzschmitt bomb: see Solubilization.

Wurzschmitt method: a qualitative method for detection of phosphorus in organic compounds. The substance to be tested is heated with sodium peroxide and ethylene glycol; the reaction yields phosphate ions.

Wustite phase: see Crystal lattice defects.

X

x: symbol for amount of substance (mole fraction).

X: 1) abb. for xanthosine; 2) in French, symbol for xenon (Xe).

Xan: abb. for xanthine.

XANES: see X-ray spectroscopy.

Xanthate: see Xanthogenate.

Xanthene, ***dibenzo-γ-pyran***: colorless leaflets, m.p. 100.5 °C, b.p. 310-312 °C. It is readily soluble in ether, chloroform and benzene, slightly soluble in alcohol and insoluble in water. Oxidation produces Xanthone (see). X. is synthesized by pressure hydrogenation of xanthone, or by distillation over zinc powder. X. is the parent compound of the Xanthene pigments (see).

Xanthen-9-ol: same as Xanthydrol (see).

Xanthen-9-one: same as Xanthone (see).

Xanthene pigments: a large group of pigments derived structurally from xanthene. X. include fluorescein, the eosine dyes, rhodamine dyes (which were formerly classified as a subgroup of the aminofluorines), rosamines and some mordant dyes. Of the latter, a few chrome dyes are important, such as chromoxan brilliant red BL and eriochrome blue GGK, which are very colorfast and washfast.

Some important xanthene pigments

Pigment	R^3	R^4	R^5	R^6	$R^{2'}$	$R^{4'}$	$R^{5'}$	$R^{7'}$
Fluorescein	H	H	H	H	H	H	H	H
Gallein	H	H	H	H	H	OH	OH	H
Eosine yellow	H	H	H	H	Br	Br	Br	Br
Eosine blue	H	H	H	H	Br	NO_2	NO_2	Br
Phloxine	Cl	H	H	Cl	Br	Br	Br	Br
Erythrosine	H	H	H	H	I	I	I	I
Rose bengal	Cl	Cl	Cl	Cl	I	I	I	I

Xanthine, abb. ***Xan***: 2,6-dihydroxypurine. X. forms colorless crystals which sublime and decompose above 280 °C. It is slightly soluble in hot water, but is practically insoluble in cold water. It dissolves in alkali hydroxide solutions or mineral acids by forming salts. In the form of xanthosine 5'-monophosphate, X. is an intermediate in the biosynthesis of guanosine monophosphate. X. is also an intermediate in the degradation of the purines, which are converted via hypoxanthine and X. to uric acid in the presence of the enzyme xanthine oxidase. The methylated X. caffeine, theophylline and theobromine are used as drugs and are present in coffee, tea and chocolate. The name X. was derived from the Greek "xanthos" = "yellow"; X. gives a yellow color with nitric acid.

Xanthocillins: antibiotics with an isonitrile structure; they are made by *Penicillium notatum*. The preparation Brevicid® contains X. X as its main component. However, because of its toxicity, it can only be used externally.

Xanthogenate, ***xanthate***: a salt or ester of xanthogenic acid. In systematic nomenclature, X. are also named as metal salts or esters of dithiocarboxylic acid O esters. The X. of monovalent metals have the general formula RO-CS-S$^-$ Me$^+$, e.g. potassium ethyl xanthogenate, $C_2H_5O\text{-}CS\text{-}S^-K^+$. They are formed by the reaction of alkali alcoholates with carbon disulfide: $RO^-Na^+ + S{=}C{=}S \rightarrow RO\text{-}CS\text{-}S^-Na^+$.

Alkali xanthogenates are water-soluble, yellow, crystalline compounds which, when acidified, release unstable xanthogenic acids. Anhydrous alkali xanthogenates are stable compounds which melt only at higher temperatures, with decomposition. Alkali xanthogenates with different O ester groups are used mainly as flotation agents in the preparation of sulfidic, nonferrous metal ores, as vulcanization accelerators and as pesticides. ***Cellulose xanthogenates*** are very important intermediates in the production of viscose fibers.

The esters of xanthogenic acids with the general formula R^1O-CS-SR^2 can be made by alkylation of alkali metal X. with alkyl halides: R^1O-CS-SNa + R^2-X → R^1O-CS-SR^2 + NaX. The esters are used only to a limited extent in industry, e.g. as additives for high-pressure lubricants. They are used mainly for special organic syntheses, e.g. for synthesis of olefins by the Čugaev reaction or thiols by the Leuckart reaction.

Xanthogenic acids: the O esters of dithiocarboxylic acids with the general formula RO-CS-SH. The free X. are very unstable, oily liquids, which are scarcely soluble in water. The aqueous solutions are weakly acidic. In the presence of aqueous acids, X. decompose to carbon disulfide and the corresponding alcohols. They can be synthesized from their esters and salts, which are more stable (see Xanthogenates).

Xanthone, *xanthen-9-one, dibenzo-γ-pyrone*: colorless, readily sublimable needles; m.p. 174 °C, b.p. 349°-350 °C. It is soluble in alcohol, ether, chloroform and benzene, and insoluble in water. X. dissolves in conc. sulfuric acid, giving a yellow color and an intense, bright blue fluorescence. It can be reduced to xanthene and xanthydrol. Reaction with Grignard compounds gives rise to xanthylium salts. X. is synthesized by distillation of salicylic acid phenyl ester. X. itself does not occur in nature, but it is the parent compound of a series of plant pigments, such as euxanthone, corymbiferin and gentisin.

Xanthoommatin: see Ommochromes.

Xanthophylls: oxygen-containing carotenoids. The oxygen is present as an hydroxy or carbonyl group, or more rarely, as an oxirane. The X. include numerous lipophilic pigments of fruits, blossoms and autumn leaves. *Capsanthin* from peppers (*Capsicum*) and *Zeaxanthin* from maize kernels are examples of X.

Xanthoprotein reaction: see Proteins.

Xanthopterin: see Pterin.

Xanthosine: abb. ***Xao*** or ***X***: 9-β-D-ribofuranosylxanthine, a nucleoside which is formed from xanthosine 5'-monophosphate (XMP) as an intermediate in the biosynthesis of guanosine monophosphate.

Xanthotoxal: see Furocoumarins.

Xanthotoxin: same as methoxsalene; see Furocoumarins.

Xanthohydrol, *xanthen-9-ol*: colorless, crystalline needles; m.p. 123 °C (dec.). X. is soluble in alcohol and chloroform, but insoluble in water. It forms salts with mineral acids, and is easily oxidized to xanthone. It is synthesized by reduction of xanthone with sodium amalgam in alcoholic solution. X. is used for determination of urea in blood, for phenol determination and as a reagent for determination of indole and indole derivatives on thin-layer chromatograms.

Xanthylium perchlorate: see Xanthylium salts.

Xanthylium salts: the oxonium salts of xanthene. They are formed, for example, when xanthydrol is dissolved in concentrated mineral acids. X. are intensely colored compounds and fluoresce strongly in solution. ***Xanthylium perchlorate*** form bronze-colored crystals; m.p. 225-226 °C. It can be synthesized from xanthene by heating with trityl perchlorate in glacial acetic acid.

Xao: abb. for xanthosine.

Xe: symbol for xenon.

Xenobiotics: organic compounds of non-biological origin which enter an organism. The most important group of X. with which the human organism must deal are synthetic drugs. Most X. are altered by enzymatic reactions (biotransformation). The term also refers to anthropogenic organic compounds found in the environment; see Ecological chemistry.

Xenon: symbol ***Xe***: a chemical element from group 0 or VIII or the periodic system, the Noble gases (see); *Z* 54, with natural isotopes with mass numbers 124 (0.096%), 126 (0.090%), 128 (1.92%), 129 (26.44%), 130 (4.08%), 131 (21.18%), 132 (26.89%), 134 (10.44%) and 136 (8.87%). Its atomic mass is 131.30, valence 0 (with respect to fluorine and oxygen, II, IV, VI and VIII), density 5.887 g l^{-1} at 0 °C, m.p. -111.9 °C, b.p. -107.1 °C, crit. temp. 16.6 °C, crit pressure 5.8 MPa.

Properties. The element is a colorless, odorless, tasteless, monoatomic noble gas, and is chemically very unreactive. It forms stable compounds only with the highly electronegative elements, such as fluorine and oxygen (see Xenon compounds). Hydrates and a few other clathrates of X. are also known.

Analysis. X. is identified spectroscopically after gas-chromatographic separation.

Occurrence. X. is one of the rarest elements, making up $2.4 \cdot 10^{-9}$% of the earth's crust; it makes up $8 \cdot 10^{-6}$ vol. % of the air.

Production. The relatively high-boiling X. is washed out of a very large volume of purified and precooled air with liquid nitrogen or oxygen; krypton is also present in the extract and the two noble gases are separated by distillation. They can then be purified by fractional adsorption and desorption on activated charcoal.

Applications. The wide use of X. as a filler gas in light bulbs, which would be very advantageous, is precluded by its high price. High-pressure discharge lamps filled with X. emit a very bright, white light similar to daylight, and are used for special purposes (lighthouse lights, stage lighting, etc.). Mixed with oxygen, X. can be used in medicine as a narcotic.

Historical. X. was discovered in 1898 by Ramsey, as a solid residue remaining after cautious evaporation of krypton from crude krypton cooled to the temperature of liquid air. The spectrum of the vaporized solid revealed that it was a new element, which Ramsey named after the Greek word "xenos" = "foreign".

Xenon compounds: Three xenon fluorides are formed by reaction of elemental fluorine and xenon: ***xenon(II) fluoride***, XeF_2, ***xenon(IV) fluoride***, XeF_4 and ***xenon(VI) fluoride***, XeF_6. With a suitable molar ratio of the reactants, XeF_2 or XeF_6 can be obtained directly in pure form. XeF_4 must be purified by separation from the other reaction products, e.g. by complex formation with sodium fluoride, NaF, or antimony fluoride, SbF_5. The three xenon fluorides are colorless, crystalline compounds which can be kept for an unlimited time in absolutely dry air (table).

The structure of the xenon fluorides agrees completely with the predictions of the VSEPR model. For the 10e system of XeF_2 (2 bonding, 3 non-bonding electron pairs), a linear structure is predicted; for XeF_4 (12e system, 4 bonding and 2 non-bonding electron pairs), a square planar structure is predicted; and

Some physical properties of the xenon fluorides

	XeF_2	XeF_4	XeF_6
M. p. [°C]	129.0	114	49.47
B. p. [°C]			75.57
Enthalpy of formation of the solid at 298K, $\Delta_F H_{(S)}$ [kJ mol^{-1}]	−164	−227.4	−361.1
Molecular structure	linear	planar	distorted octahedral

for XeF_6 (14e system, 6 bonding and one non-bonding electron pairs) a distorted octahedral configuration is expected. All these predictions have been confirmed experimentally.

Under suitable conditions, the fluoride ligands in the xenon fluorides can be exchanged for oxygen or oxygen-containing groups. For example, XeF_2 + $HOSO_2F \rightarrow F\text{-}Xe\text{-}OSO_2F + HF$. With an exactly stoichiometric amount of water, XeF_4 reacts to form $XeOF_2$, and XeF_6 to form $XeOF_4$. Complete hydrolysis of XeF_6 produces ***xenon(VI) oxide***, XeO_3, a highly explosive gas which dissolves in water. It then forms a certain amount of ***xenic acid***, H_2XeO_4, which gives a weakly acidic solution and which forms ***xenate(VI) salts***, M^+HXeO_4 with bases. In strongly alkaline solution or at elevated temperatures, the xenate(VI) salts disproportionate to xenon and ***perxenates(VIII)***: $2\,HXeO_4^- + 2\,OH^- \rightarrow XeO_6^{4-} + Xe + O_2 + 2\,H_2O$. If a sodium or barium perxenate solution is acidified with sulfuric acid, the light-yellow, solid and highly explosive ***xenon(VIII) oxide*** is obtained: $Ba_2XeO_4 + 2\,H_2SO_4 \rightarrow XeO_4 + 2\,BaSO_4 + 2\,H_2O$.

The colorless, crystalline, endothermic ***xenon(II) chloride*** $XeCl_2$ is formed by high-frequency excitation of a mixture of xenon, fluorine and silicon tetrachloride or carbon tetrachloride. Above 80 °C, it decomposes into the elements.

Xenopsin: see Neurotensin.

Xerogel: see Gel.

Xerography: see Electrophotography.

X-ray amorphous: see Amorphous.

X-ray contrast media: compounds used to make organs opaque to x-rays for diagnostic purposes. Compounds of elements with medium atomic numbers (50 to 60) are used because they absorb very well in the wavelength range of 0.01 to 0.05 nm used for diagnostic x-rays. The most commonly used X. are iodine-containing organic compounds which are applied orally or parenterally; some are accumulated by certain organs, such as the gall bladder or kidneys. Some iodine-containing R. are *iodipamide (adipiodone)*, *amidotrizoate* (3,5-diacetylamino-2,4,6-triiodobenzoic acid) and *iomeglamic acid* (Falignost®). Sodium or amine salts of these compounds are used for injection.

COOH, I, I, NH–CO–(CH$_2$)$_4$–CO–HN, COOH, I, I, I, I

Adipiodone

H_3C–N–C(=O)–CH$_2$–CH$_2$–CH$_2$–COOH, I, I, NH$_2$, I

Iomeglamic acid

The gastrointestinal tract is made visible by administration of finely divided barium sulfate, $BaSO_4$; it is so insoluble that it is not poisonous.

X-ray fluorescence analysis: an X-ray spectroscopic method used for qualitative and especially for quantitative determination of elements. X. can be used with liquid, powdered or compact solid samples; it is dependable and largely independent of the bonding state of the elements. The measurement time is very short, about 1 minute.

In X., the sample is irradiated with a primary X-ray beam from an X-ray tube. As a result of its interaction with the X-rays, the sample emits secondary X-rays (proper radiation) which are analysed in an X-ray spectrometer (see X-ray spectroscopy). Because the emitted X-rays are induced by other X-rays, they are a type of Fluorescence (see). The basis for qualitative analysis is Moseley's law, according to which the wavelengths of proper radiation are specific for the element. Using X., all elements with atomic numbers greater than 9 (fluorine) can be determined; in special cases, even those between 6 (carbon) and 9 can be included. X-ray spectra have relatively few lines, so that superpositions of lines from different elements occur only rarely. Quantitative analysis is based on the relationship between the intensity of the emission and the concentration of the element in the sample. However, the intensity also depends on the composition and physical state of the sample (see Matrix effects), which must be taken into account. For example, if the fluorescence from one element is sufficiently energetic to excite another element, the proper radiation of the first element will be weakened. The absorption of the excitation X-rays by the matrix also affects the fluorescence intensities. The parameters of the apparatus may also prevent a linear relationship between intensity and concentration. It is therefore necessary to work with calibrated standards. The range of concentrations which can be determined is between 0.1 and 100%. Under favorable conditions (e.g. a heavy element in a lighter matrix), traces down to 10^{-4}% can also be detected. Precision $\Delta c/c = 0.1$ to 0.5% is achieved. The sample for X. is analysed to a depth of 0.1 to 0.5 mm from the surface, and an area of with a diameter of about 40 mm is included.

Apparatus for X. is constructed on the same principles as X-ray spectrometers and consists of the same structural elements (see X-ray spectroscopy). There are two types of apparatus: sequence and multi-channel. *Sequence devices* are equipped with tunable monochromators with which it is possible to analyse desired wavelengths in sequence (one-channel devices). Sequence devices are most suitable for laboratories which analyse samples with varying elemental compositions. *Multichannel devices* contain several monochromators, each set at a fixed wavelength. Each monochromator is coupled to an independent electronic detection unit. The number of monochromators must be at least as great as the number of elements to be analysed in the samples. Multichannel devices are used where the samples to be analysed all have the same elemental compositions, and only the relative concentrations must be determined. In modern instruments, the entire measurement process is automated and is controlled by a computer, which also analyses the data, identifies the elements present in the sample and calculates the concentration of each from the intensities of the emissions. The relationship between intensity and concentration can be approximated by a system of linear equations, the coefficients of which are determined from calibrated standards.

The main application of X. is in the control of manufacturing processes. It is used in the chemical industry, in the analysis of steel and other metallurgical products, of reactor metals, reactor fuels, cements and other refractory products, petroleum products, etc. X. is the most frequently used method of X-ray spectroscopic analysis.

X-ray spectral analysis: same as X-ray spectroscopy (see).

X-ray spectroscopy: a form of spectroscopy applied to the interaction of X-rays (see Electromagnetic spectrum) and material systems. X. is used mainly for qualitative and quantitative determination of chemical elements. There are three main areas, X-ray absorption and X-ray emission spectroscopy, and X-ray fluorescence analysis (see). *X-ray absorption spectroscopy* measures the decrease in intensity of X-rays on passage through a sample. *X-ray emission spectroscopy* and *X-ray fluorescence analysis* are used to examine the X-rays emitted from an excited sample; in the first case, the excitation is usually caused by electron collision, and in the second case, by photons of appropriate energy.

Bremsstrahlung and proper radiation. X-rays are emitted when high-velocity electrons are suddenly slowed down by collision with a solid substance (e.g. the anticathode of the X-ray tube). The maximum energy E_{max} of the X-ray released by an electron corresponds to complete conversion of the electron's kinetic energy into the energy of the radiation quantum: $E_{max} = eV = h\nu = hc/\lambda$, where e is the charge of the electron, V the applied voltage, h is Planck's constant, ν the frequency and λ the wavelength of the radiation, and c is the velocity of light. By simple rearrangement of the above equation, one obtains the Duane-Hunt law, $\lambda_{min} = hc/eV$, which represents the short-wave limit of the X-rays emitted. This limit is determined by the voltage V and is, to a first approximation, independent of the anticathode material. Since the electrons generally do not release their total energy in a single braking event, most of the X-rays emitted are at longer wavelengths than the Duane-Hunt minimum. *Bremsstrahlung* thus consists of a continuum which has a sharp boundary at the short-wave end (Fig. 1).

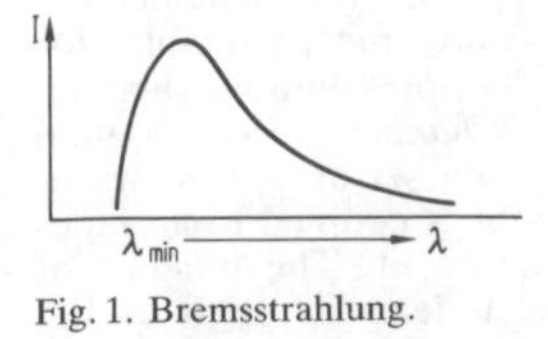

Fig. 1. Bremsstrahlung.

If the voltage is gradually increased, the boundary is shifted to shorter and shorter wavelengths, and very strong radiation is observed at certain positions in the spectrum; these positions depend on the anticathode material (Fig. 2).

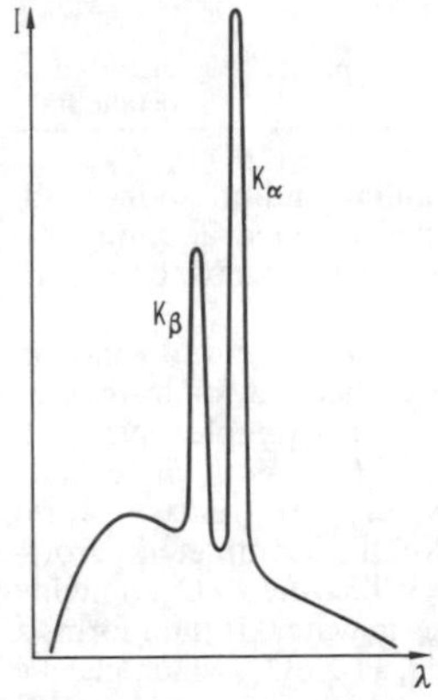

Fig. 2. Superposition of bremsstrahlung and proper radiation.

This radiation is called *proper* or *characteristic X-radiation* i.e. the wavelength of this radiation is determined by the material of the anticathode. Proper radiation is the basis for X-ray emission spectroscopy and X-ray fluorescence analysis.

Source of the X-ray spectrum. If sufficient amounts of energy are added to atoms, for example by bombarding them with high-energy electrons, electrons can be knocked out of the inner orbitals of the atoms. The resulting vacancies in the inner shells are filled again by electrons from outer shells which jump into these vacancies, and the energy released by this process is emitted in the form of X-rays according to the equation $E_2\text{-}E_1 = h\nu$. The wavelengths of these proper X-rays are determined by the distances between the inner-electron energy levels of the atom, and are thus characteristic of the type of atom. If an electron is knocked out of the innermost, or K shell, one observes the K series, which arises by electron transitions from the L, M or higher shells into the K shell. As Fig. 3 shows, the corresponding lines are called K_α, K_β, K_λ, etc. If an electron is removed from the L shell by excitation, the L series is generated by electron transitions from higher shells into the L shell. The M, N, O and P series are also possible for atoms with high atomic numbers. Production of the K series leads to vacancies in the L, M and N shells, which are filled by electrons from other outer shells. Thus the appearance of the K_α line necessarily implies the appearance not only of the other lines of the K series, but also of the L, M and N series – depending on the atomic number of the element.

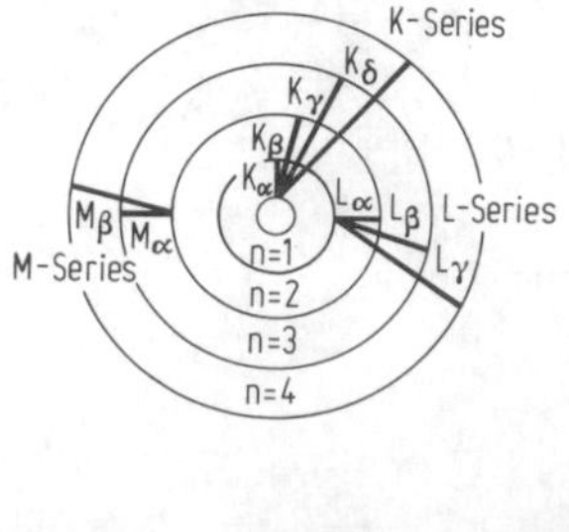

Fig. 3. Formation of the X-ray spectrum.

The existence of the X-ray bremsstrahlung continuum means that the X-ray spectra arising from electron excitation always consist of a continuous background with superposed lines of the proper radiation. The frequencies of the lines in the X-ray spectrum are determined by *Moseley's Law*, according to which the frequencies of analogous lines for different elements (e.g. the K_α line) are proportional to the square of the difference between the atomic number and a constant: $\nu \approx (Z-a)^2$; ν is the frequency of the proper radiation of an element with atomic number Z and a is the shielding constant. In the case of K_α radiation, a is 1, because the effective nuclear charge for a K-shell electron is reduced by 1 by the negative charge of the other K-shell electron.

More precise studies have shown that the transitions pictured in Fig. 3 are further split. This is due to the fact that every energy state with the primary quantum number n is actually a group of $2n-1$ states with similar energies. The K shell consists of only a single state, the L shell of 3 (called L_I, L_{II} and L_{III}), the M shell of 5, and so on. As a result, a variety of transitions can occur between the L and K shells. However, not all of these theoretically possible transitions actually occur, because there are certain Selection rules (see) which apply to them. Furthermore, radiationless transitions between inner states can also occur, leading to emission of a further electron (see Auger electron spectroscopy). This effect reduces the radiation yield, especially with light elements.

Apparatus. The structure of an X-ray spectrometer is in principle the same as that of other spectrometers (see Spectrometers). There are 3 main elements: 1) excitation source, 2) monochromator and 3) detector. In the case of X-ray emission spectra, there are primary spectra, usually caused by fast electrons, but also by protons or ions, and secondary spectra, which are generated by excitation with X-rays. In order to measure the X-ray spectrum of a substance excited by electrons, the sample is placed on the anode of an X-ray tube which can be disassembled for this purpose. The disadvantage of this method is that the sample is strongly heated, and may undergo chemical reactions. Measurement of secondary spectra is less harsh; here the braking spectrum of a tungsten anode is commonly used, but K_α or L_α radiation may also be used. Because there is no background of bremsstrahlung, secondary spectra usually have greater contrast than primary spectra. The intensive continuous X-ray spectrum needed for absorption spectra is usually obtained from an X-ray tube with a tungsten anode.

The spectral resolution of X-rays in a monochromator is accomplished by single crystals or ruled lattices corresponding to Bragg's relation. For studies in the range $\lambda < 2$ nm crystals are usually used; for longer wavelengths ($\lambda > 2$ nm), lattices are used. Monochromatic radiation can be obtained by rotation of the crystal or lattice, which changes the angle of incidence. These methods of spectral resolution of X-rays are called *wavelength dispersive processes*; there are also *energy dispersive processes* in use. In this case, the radiation is not passed through a monochromator, but goes directly into the detector (a special semiconductor detector), which not only detects X-ray quanta but indicates their energy. The detector in this case also serves as the dispersive element. Energy dispersive methods have lower resolution than wave-length dispersive methods, and are not useful for study of spectral fine structure.

Recently, quantum detectors have been increasingly used as detectors with the wavelength dispersive methods. In both cases the detector must be tuned to the spectral range of interest. Some examples of quantum detectors are ionization and scintillation counters, secondary electron amplifiers and semiconductor counters. The construction of a spectrometer depends on the wavelength range for which it is to be used. There are open, or short-wave spectrometers ($\lambda < 0.2$ nm), vacuum or long-wave spectrometers ($\lambda \approx 0.2$ to 1 nm) and high vacuum or ultra-long-wave spectrometers ($\lambda > 2$ nm).

X-ray emission spectra. These are obtained when the radiation emitted by a sample excited with electrons or X-rays is spectrally resolved. X-ray emission spectra contain information needed for elemental analysis. In qualitative analysis, the wavelengths of the spectral lines emitted by the sample are recorded and assigned to elements using tables. An element can be definitely identified if several lines of a spectral series of this element are shown to be present. For quantitative analysis, the intensities of the spectral lines are also required. As a rule, the K_α or L_α lines are used for this, and their intensities are compared to those of a calibrated standard. The energy differences between inner electron levels can be calculated from the wavelengths of the main X-ray lines, K_α, L_α, etc. (Fig. 3), and the shortest wavelength line of a series gives the energy difference between the corresponding electron level and the optical level. These differences depend on the bonding state of the emitting atom so that structural analysis may be done using X-ray emission spectra.

The methods of X-ray fluorescence analysis (see) and Electron beam microanalysis (see) are special cases of X. which are of practical importance.

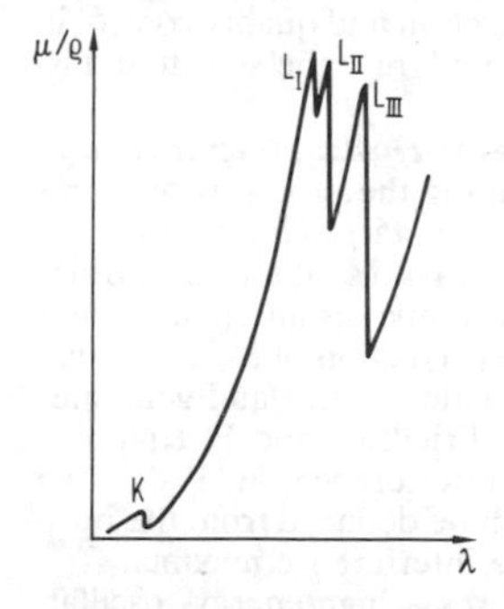

X-ray absorption spectra. These are measured as the decrease in intensity of an X-ray beam as it passes through a layer of a substance. The absorption obeys Beer's law (see). The degree of absorption is determined by the nature and concentration of the absorbing material. If scattering is ignored, it can be said that the absorbed X-rays cause electrons to be emitted from the sample or raised into higher, excited states. Absorption of the K_α or K_β lines is apparently not possible, because the L and M shell are completely occupied, so that the K electrons cannot jump into them. When a K electron absorbs an X-ray, it can only enter an unoccupied optical level or enter the continuous range above the ionization threshold. In practice, this means that a K electron can only absorb from the series continuum of the K series. This extends from the ionization threshold of the corresponding level towards shorter wavelengths, with decreasing wavelengths. X-ray absorption spectra therefore do not consist of lines. If the absorption is plotted as a function of wavelength (Fig. 4), the absorption coefficient rises sharply when the ionization threshold is reached. This sudden increase in the absorption coefficient is called the K absorption edge. X-ray quanta with with energy equal to or greater than the K absorption edge can knock electrons out of the K shell, but those quanta with lower energy cannot. The absorption coefficient does not fall to 0 on the long-wavelength side of the edge because here the extension of the series limit continuum of the L shell is active. The position of the absorption edge is characteristic for each element and can serve to identify it (qualitative analysis). Quantitative information can be obtained by the height of the absorption edge. At higher resolution, one often observes a fine structure of the absorption edges (*XANES*, abb. for *x*-ray *a*bsorption *n*ear *e*dge *s*tructure). This is due in part to the splitting of the L, M and N levels into sub-states, and to the possibility of transition of an electron into one of the unoccupied optical levels. In addition, the exact position of the absorption edge depends on the bonding state of the atom.

Of the X-ray methods, emission spectroscopy is more widely applied than absorption spectroscopy. In addition to theoretical information on atomic structure and bonding, X. gives qualitative and especially quantitative evidence of elements, even in the trace range. It is therefore important in chemical analysis, environmental protection, forensic chemistry and similar areas. Because of their versatility, their high sensitivity and high reliability, X-ray spectroscopic methods are used in production and quality control in the chemical industry, metallurgy, the silicate industry and other areas.

X-ray structure analysis, ***crystal structure analysis***: a method of determining the positions of atoms or ions in the unit cell (see Crystal) by diffraction of x-rays. The wavelengths of x-rays are of the same order of magnitude as the interatomic distances in crystals (10^{-8} cm). According to a prediction of M. von Laue, the crystal acts as a diffraction lattice and generates interference patterns. W. Friedrich and P. Knipping first demonstrated the interference in 1912. The structure of the crystal can be deduced from the position and intensities of the interference maxima.

Theoretical. X-rays posses a high-energy, oscillating electric field, which interacts with electrons. An electron struck by an x-ray begins to vibrate in resonance with it, and thus itself emits x-rays. The energy required for emission is extracted from the incident, primary x-ray, which loses a corresponding amount of energy. The secondary radiation emitted by the electrons spreads out in all directions (*coherent scattering*). The scattering capacity of an atom is directly proportional to the number of electrons it contains; it is expressed by the atomic form factor *f*. Since the sizes of atoms are appoximately the same as the wavelengths of x-rays, interference patterns are observed as a function of the scattering angle Θ. The cooperative scattering by atoms in a crystal is called *diffraction*. Diffraction maxima are observed only when the beam of x-rays falls on the crystal from certain directions. The angle of incidence of the primary beam and the spatial directions of the diffraction maxima can be calculated using Bragg's law. The generation of diffraction maxima is interpreted geometrically as reflection of the incident waves on the crystal by the lattice planes hkl of the crystal.

Constructive interference occurs only when the difference in pathlengths Δl travelled by the incident and reflected beams is an integral multiple of the wavelength λ ($\Delta l = n\lambda$; $n = 1, 2...$). If the angle of incidence is Θ, the difference in pathlengths Δl for a given distance between lattice planes d is given by $\Delta l = 2d \sin \Theta$ (Fig. 1).

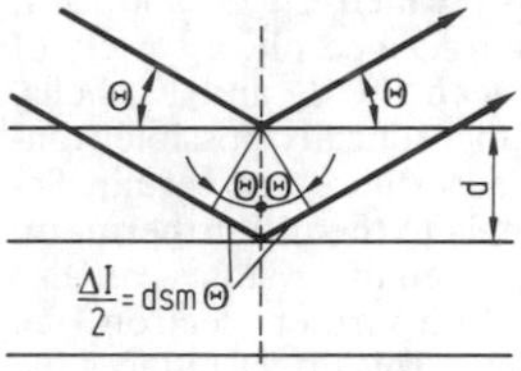

Fig. 1. Reflection of the X-rays from a lattice plane.

Thus a condition for reflection is *Bragg's law*, $2d \sin \Theta = n\lambda$ ($n = 1, 2, 3...$). Based on this interpretation, the diffraction maxima are called *reflexes*. Information on the size and shape of the unit cell is obtained from the diffraction angles of the reflexes. The intensity I (hkl) of the reflex gives information on the arrangement of the atoms in the crystal and its symmetry. The intensity of the diffracted x-rays is proportional to the square of the absolute value of the structure factor:

$$I(\mathrm{hkl}) \propto |F(\mathrm{hkl})|^2. \quad (1)$$

$|F(\mathrm{hkl})|$ corresponds to the amplitude of the diffracted wave, and is called the *structure amplitude* The *structure factor* is a complex number, and can be calculated as follows:

$$F(\mathrm{hkl}) = \sum_{j=1}^{N} f_j \exp 2\pi i\,(\mathrm{h}x_j + \mathrm{k}y_j + \mathrm{l}z_j). \quad (2)$$

Here N is the number of atoms in the unit cell, fj are the atomic form factors, and x_j, y_j and z_j are the coordinates of the atoms in the unit cell. The structure factor can also be expressed in terms of the structure amplitude:

$$F(\mathrm{hkl}) = |F(\mathrm{hkl})| \exp(i\alpha). \quad (3)$$

Here α is the phase of the structure factor. The phase angle cannot yet be determined experimentally; this is referred to as the *phase problem of X*. If the structure amplitudes and phases of the reflexes are known, a *Fourier synthesis* according to eq. 4 can be used to calculate the electron density $\varrho(x, y, z)$ in the unit cell:

$$\rho(x,y,z) = \frac{1}{V} \sum_h \sum_k \sum_l F(hkl) \exp - [2\pi i(hx + ky + lz)] \quad (4)$$

The maxima of the electron densities from eq. 4 then indicate the equilibrium positions of the atoms in the unit cell. The most important methods for calculating the phase angle are the heavy atom and direct methods.

In the *heavy atom method*, the compound must contain a few atoms with high electron numbers. The experimentally determined $|F(hkl)|^2$ values are used as Fourier coefficients in eq. 4. The result is the interatomic vectors of the "heavy atoms", and thus their coordinates. Using eq. 2, this approximate structure can be used to calculate phase angles. With eq. 4, this phase model yields the coordinates of more atoms. *Direct methods* attempt a computational derivation of the phase angle from the scattering amplitudes $|F(hkl)|$ The possible relationships among individual phases are limited by the requirement that the Fourier synthesis can only give positive values for the function. Thus the sum of the phases of three reflexes whose h,k,l indices add up to the triplet 0,0,0 is, with a certain probability, equal to zero:

$$\alpha h_1k_1l_1 + \alpha h_2k_2l_2 + \alpha h_3k_3l_3 \approx 0. \quad (5)$$

(with $h_1 + h_2 + h_3 = 0$, $k_1 + k_2 + k_3 = 0$; $l_1 + l_2 + l_3 = 0$). The greater the structure amplitudes of the reflexes, the more probable the solution of eq. 5. Using combinatorial methods, a set of phases is calculated in which the largest possible number of equations satisfies eq. 5. The resulting set of phases is then used to calculate the electron densities by eq. 4, and from these, the coordinates of the atoms are calculated.

Using the method of the least squares, the squares of the differences between observed structure factors F_0 (determined experimentally using eq. 1) and the calculated structure factors F_c (from eq. 3) of all reflexes are minimized, and thus the atomic coordinates are refined. The accuracy of an X. is indicated by the *R-value*.

$$R = \frac{\Sigma \, ||F_0| - |F_c||}{\Sigma \, |F_0|} \quad (6)$$

R-values of 0.05 are achieved in very precise structure determinations. Fig. 2 shows the structure of ascorbic acid determined by X. as an example.

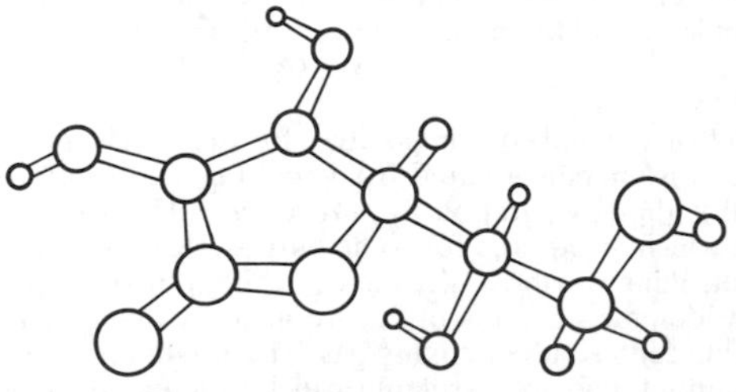

Fig. 2. Structure of ascorbic acid.

Experimental procedure. The systematic procedure in an X. consists of determination of the lattice constants, the symmetry of the crystals and the reflex intensities I(hkl). In the single-crystal method, crystals with edge lengths of 0.1 mm are used.

In the *rotating crystal* and *tilting methods*, the crystal is arranged coaxially to a cylindrical film camera (Fig. 3a) with a film on its wall. The crystal is rotated through 360° (rotating crystal) or smaller angles (tilting). The reflexes are arranged along layer lines (Fig. 3b), and the distances between the layer lines are inversely proprotional to the lattice constant of the axis of rotation. These methods are used to determine the lattice constants and to adjust the crystal.

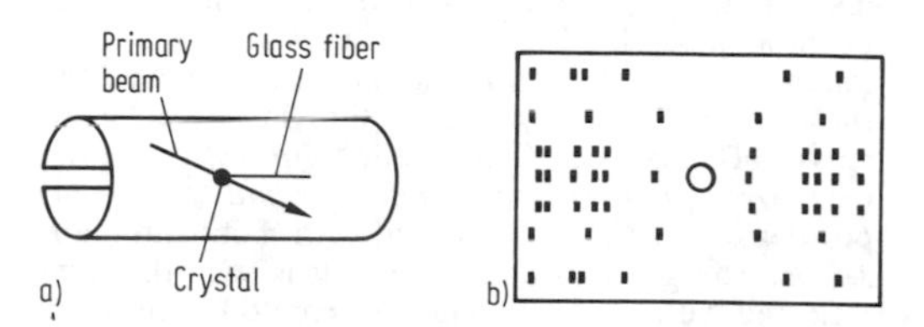

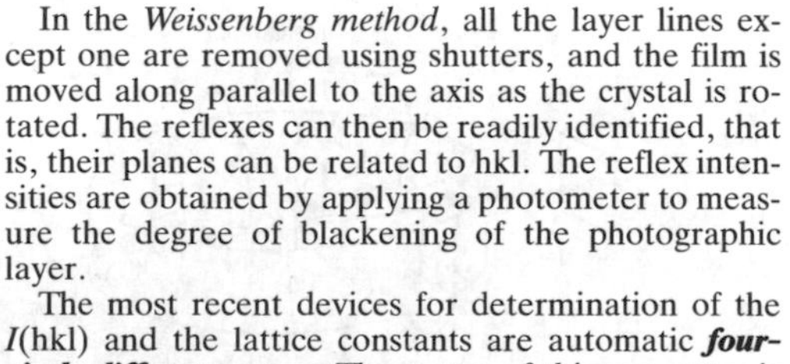

Fig. 3. Set up for a rotating or Weissenberg camera (a); rotating crystal exposure (b).

In the *Weissenberg method*, all the layer lines except one are removed using shutters, and the film is moved along parallel to the axis as the crystal is rotated. The reflexes can then be readily identified, that is, their planes can be related to hkl. The reflex intensities are obtained by applying a photometer to measure the degree of blackening of the photographic layer.

The most recent devices for determination of the I(hkl) and the lattice constants are automatic ***four-circle diffractometers***. The center of this apparatus is an Euler cradle (Fig. 4). The goniometer head to which the crystal is attached is set on the φ-axis of the cradle, which is in the plane of the X-circle and can rotate around the normal to this circle, the X-axis. The X-circle can rotate around the ω-axis which is perpendicular to the equatorial plane of the diffractometer and coaxial with the axis of rotation (2Θ-axis) of the radiation detector (proportional counter, scintillation counter). The four angles are calculated

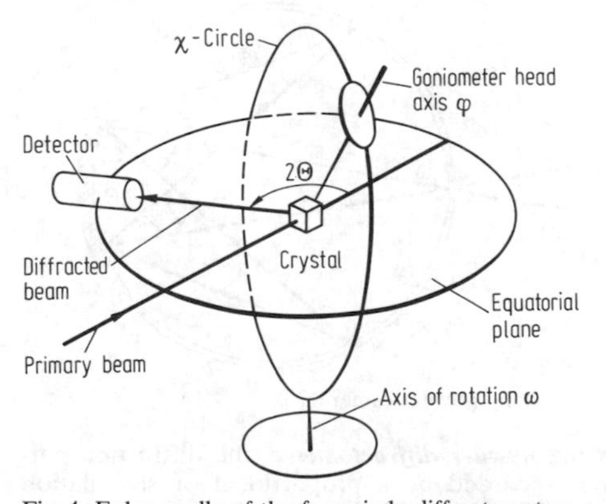

Fig. 4. Euler cradle of the four-circle diffractometer.

from the indices hkl and the lattice constants, and are set on the diffractometer. Then the intensity is measured and the values are stored in a data system (magnetic tape).

Powder methods are used with microcrystalline samples (particle sizes 10^{-2} to 10^{-3} mm). In the *Debye-Scherrer method* (Fig. 5), the powder sample is set in a cylindrical camera (the most common radii are 90 mm/π = 28.6 mm and 180 mm/π = 57.4 mm). The primary beam is perpendicular to the axis of the film cylinder. The Bragg equation is fulfilled by the random orientation of the crystallites. The preparation is rotated during the exposure, in order to make the distribution of orientations of the crystallites as uniform as possible. The reflected rays make up a system of cones with the primary beam as the axis, and these intersect the cylindrical film in characteristic curves. The exposed film is called a *powder diagram* or *Debyeogram*. In the *Guinier method*, a focusing powder camera (Fig. 6) is used with monochromatic x-rays. The geometric arrangement is such that rays reflected with the same angle Θ from different parts of the sample are focused on a point. This reduces the width of the interference lines. Single crystals of quartz or graphite are used as monochromators.

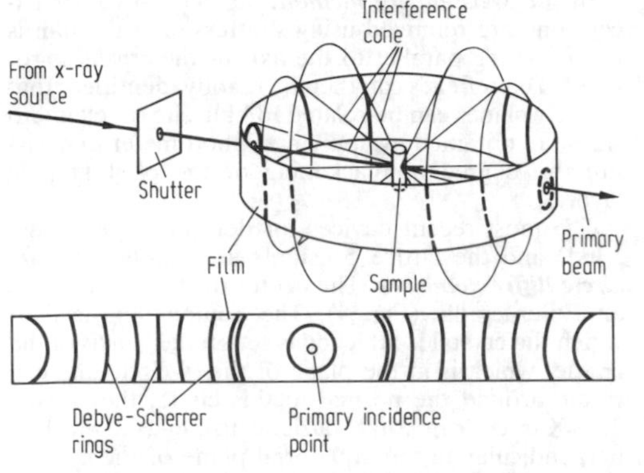

Fig. 5. Debye-Scherrer method: *a*, experimental arrangement; *b*, Debyeogram of sodium chloride.

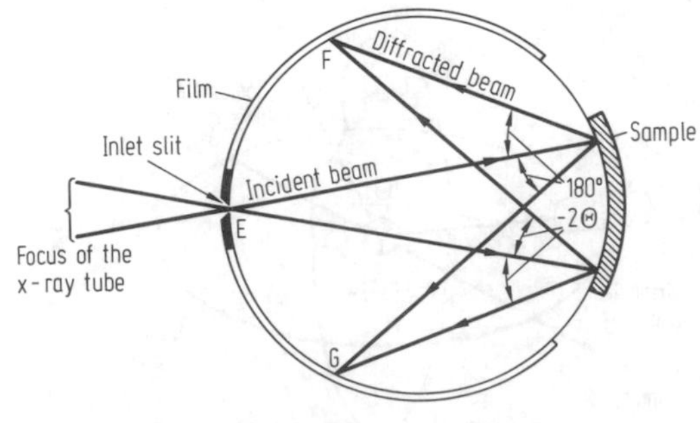

Fig. 6. Focusing powder camera.

In the ***powder diffractometer***, the diffraction pattern is recorded by a proportional or scintillation counter. The Bragg-Bretano focusing system is used: the focal point of the tube, the input slit of the detector system and the sample are on the focus circle (Fig. 7). The sample is in the shape of a disc, and is rotated at half the rate of rotation of the detector.

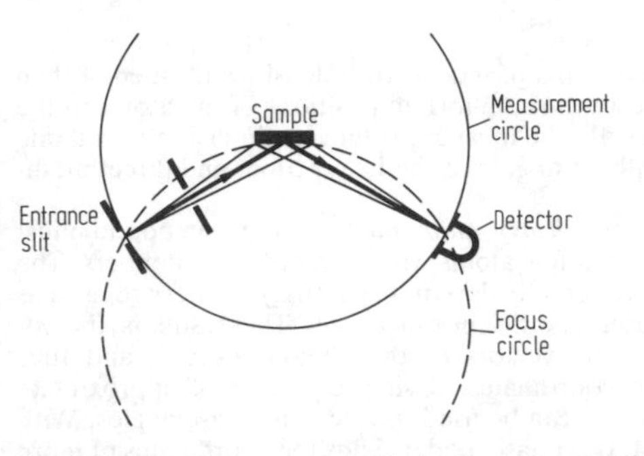

Fig. 7. Powder diffractometer with the Bragg-Bretano focusing circle.

Xylan: a homopolysaccharide which yields xylose when subjected to acid-catalysed hydrolysis. α(1→4)-D-X. are components of the hemicelluloses of plant cell walls. β(1→3)-D-X. are found in the cell walls of various algae in place of cellulose.

Xylene, ***xylol, dimethylbenzene***: one of a group of three isomeric aliphatic-aromatic hydrocarbons. ***o-X. (1,2-dimethylbenzene)*** is a colorless, combustible liquid with a typical aromatic odor;

CH_3 CH_3 CH_3 CH_3 CH_3 CH_3

o-Xylene *m*-Xylene *p*-Xylene

m.p. -25.2 °C, b.p. 144.4 °C, n_D^{20} 1.5055. *o*-X., like the other isomers, is only barely soluble in water, but is soluble in alcohol, ether and other organic solvents. The vapors form explosive mixtures with air. They are also toxic, and their effects are similar to those of Benzene (see). Liquid X. cause inflammation of the skin.

m-X. (1,3-dimethylbenzene) is a colorless liquid with the same general properties as the *ortho* isomer; m.p. -47.8 °C, b.p. 139.1 °C, n_D^{20} 1.4972.

p-X. (1,4-dimethylbenzene) is also a colorless liquid with the same general properties as the other isomers. However, it is more toxic than the *o*- and *m*- compounds. In addition, it forms colorless, monoclinic crystals when cooled; m.p. 13.3 °C, b.p. 138.4 °C, n_D^{20} 1.4958.

Petroleum and coal are sources for the X. The mixture of isomers is isolated from coal tar, and is often used without separation as *pure xylene*. The separation which is required for some purposes can be done by distilling off the *o*-X., then freezing out the *p*-X. Considerable amounts of X. are also obtained from the benzene scrub of coking gas. The relatively small amounts found in petroleum can be increased by a hydroforming process.

X. are used in large amounts as solvents for fats, oils, waxes, synthetic resins, rubber, cellulose ethers and esters. Because of their high octane ratings, they are used as additives to aviation and automobile gasolines. They are important thinners for lacquers and paints, because they are less volatile than benzene and toluene, and thus make possible the composition of solvent mixtures with different rates of evaporation. The X. are starting materials for many important organic products, including phthalic acid by oxidation of *o*-X. and terephthalic acid by oxidation of *p*-X. The reaction with nitrating acid to form nitroxylenes and their reduction to aminoxylenes is important in the pigment industry. Finally, *m*-X. is used to synthesize isophthalic acid and ***xylene musk*** (1,3-dimethyl-5-isopropyl-2,4,6-trinitrobenzene), a synthetic aroma substance with a musk-like odor.

Xylene-formaldehyde resin: see Hydrocarbon resins.

Xylene musk: see Xylene.

Xylenol, ***dimethylphenol***: one of six structural isomers in which the hydroxyl group is in position 1, and the methyl groups are in different positions relative to it. The general formula is $HO-C_6H_3-(CH_3)_2$. In the systematic names of the isomers, the position of the methyl groups is given before the term "xylenol": ***2,3-X.***, m.p. 75 °C, b.p. 218 °C; ***2,4-X.***, m.p. 26 °C, b.p. 211 °C; ***2,5-X.***, m.p. 75 °C, b.p. 213 °C; ***2,6-X.***, m.p. 49 °C, b.p. 212 °C; ***3,4-X.***, m.p. 66 °C, b.p. 225 °C; and ***3,5-X.***, m.p. 68 °C, b.p. 219 °C. The X. are colorless, crystalline compounds which are only slightly soluble in water, but dissolve readily in ethanol and ether. They are present in coal and wood tars, and are obtained from those sources by fractional distillation. In industry, mixtures of X. are usually used to make pesticides, indicator pigments, drugs, aromas and antioxidants. X. have a bacteriocidal effect which is 15 times stronger than that of phenol, and they are therefore used for disinfection, for synthesis of antiseptic soaps and antiparasitic cleansers. 3,5-X. is generally used for synthesis of phenol plastics, because in it the reactive *ortho-* and *para*-positions of the benzene ring are free of substituents.

Xylenol blue: see Bromphenol blue.

Xylenol orange: see Bromphenol blue.

Xylidines, ***aminoxylenes, aminodimethylbenzenes, dimethylanilines***: six isomeric amino derivatives of xylene.

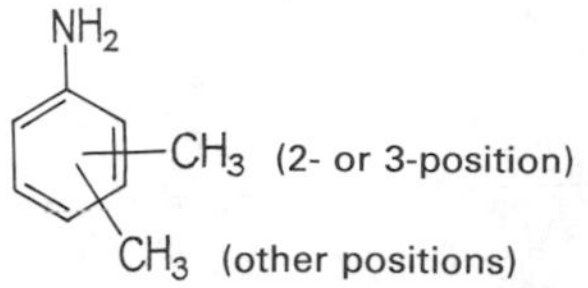

NH_2 ... CH_3 (2- or 3-position) ... CH_3 (other positions)

With the exception of the 3,4-dimethyl derivative, X. are colorless liquids which rapidly develop a dark color in the presence of light and air. This discoloration can be prevented by stabilizers. The reactions of X. are typical of primary aromatic amines. They give basic reactions and form salts with acids. ***2,4-dimethylaniline*** (2,4-xylidine) is industrially important; m.p. -14.3 °C, b.p. 214 °C, n_D^{20} 1.5569. The X. are barely soluble in water, but are soluble in alcohol and ether. Like aniline, the X. are blood poisons. They are synthesized mainly by reduction of the corresponding nitroxylenes. They are used as intermediates in the syntheses of dyes.

Xylites: see Coal.

Xylitol: see Alditols.

xylo- a prefix indicating a certain configuration (see Monosaccharides).

xylo-Hexulose: same as L-Sorbose (see).

Xylol: same as Xylene.

Xylose: a pentose. Commercial α-D-xylopyranose is a white, crystalline powder which has a slightly sweet taste. It is readily soluble in water, but only slightly soluble in ethanol; $[\alpha]_D^{20}$ at mutarotational equilibrium, +19-20°. D-X. can be obtained by acid hydrolysis of corn cobs or wheat straw, or from the wastes from cellulose production. It is used in medicine to test absorption in the small intestine (xylose test) and to recognize absorption disorders.

Xylyl bromides, ***methylbenzyl bromide, α-bromoxylene***: the three structural isomers of xylene with a bromine atom in the side chain. The X. are insoluble in water, but are soluble in alcohol, ether and chloroform. They can be made by bromination of the corresponding xylenes. X. are strong lacrimators. ***1,2-(o)-Xylylbromide***, colorless crystals; m.p. 21 °C, b.p. 223-224 °C. It was used as a tear gas.

CH_2-Br ... CH_3 position 2, 3 or 4

Y

Y: symbol for yttrium.

Yara-yara: same as 2-Methoxynaphthalene (see).

Yb: symbol for ytterbium.

Yeast adenylic acid: see Adenosine phosphates.

Yellow enzymes: same as Flavin enzymes (see).

Yield: for a chemical reaction, the molar amount of the desired product per mole of one of the reactants. Y. is often given in percent. The theoretically possible Y. can be calculated from the stoichiometry and the equilibrium constant, but the practical Y. is always less than the theoretical. It is affected by parallel and subsequent reactions, losses in isolation and purification of the desired product and other factors.

Ylang-ylang oil: an essential oil from the blossoms of *Canaga odorata*. It has a fruity, sweet odor and is used in expensive perfumes.

Ylenes: see Ylides.

Ylides: zwitterionic, very reactive N, P or S compounds which contain a negatively charged CH_2 group, the *methylide group* (more generally, *alkylide groups*). The positive charge is taken on by the N, P or S atom.

These compounds are synthesized from the tetraalkylammonium or phosphonium salts, or from trialkylsulfonium salts, by reaction with butyllithium or phenyllithium, e.g.

$$H_3C-\overset{\oplus}{N}(CH_3)_2-CH_2\overset{\delta-}{-}\overset{\delta+}{H} + R-Li \xrightarrow{-RH/Li^+} H_3C-\overset{\oplus}{N}(CH_3)_2-\overset{\ominus}{\bar{C}}H_2$$

Trimethylammonium methylide

In the trimethylammonium methylide, the covalent bond between the nitrogen and carbon (yl) has superimposed on it an ionic bond (ide). The N ylides are not stabilized, in contrast to the P and S ylides, and are therefore much more reactive than the latter. In P and S ylides, the possibility of octet expansion contributes to resonance stabilization, which is indicated in the formula by a P=C or S=C double bond:

$$(R)_3\overset{\oplus}{P}-\overset{\ominus}{C}H_2 \longleftrightarrow (R)_3P{=}CH_2,\ (R)_2\overset{\oplus}{S}-\overset{\ominus}{C}H_2 \longleftrightarrow (R)_2S{=}CH_2$$

The double bonds are very polar and have a strong tendency to undergo addition (***ylenes***).

Addition of aldehydes or ketones leads via various intermediates to substituted alkenes, in the case of the P-ylides (see Wittig reaction); the reaction is stereoselective. In the case of S-ylides, the products are oxiranes. N-Ylides are intermediates in many isomerizations and rearrangements, e.g. in the Stevens rearrangement (see) or Sommelet rearrangement (see).

$$(R)_3\overset{\oplus}{N}-\overset{\ominus}{\bar{C}}H_2 \quad (R)_3\overset{\oplus}{P}-\overset{\ominus}{\bar{C}}H_2 \quad (R)_3\overset{\oplus}{S}-\overset{\ominus}{\bar{C}}H_2$$

N–ylide P–ylide S–ylide

Yohimbine: the principle alkaloid in the rind of *Pausinystalia yohimbe*. It is a monoterpene indole alkaloid, with a structure similar in some respects to that of the rauwolfia alkaloids. It is biosynthesized from tryptophan and a monoterpene, and may be considered one of the β-carboline alkaloids. m.p. 241 °C, $[\alpha]_D^{20}$ +108° (in pyridine). ***Yohimbine hydrochloride***, m.p. 302 °C, $[\alpha]_D^{20}$ +103° (in water). Y. is a sympathicolytic with a dilating effect on blood vessels, and is sometimes used for hypertonia.

Yperite: see Chemical weapons.

Ytterbium, symbol ***Yb***: a chemical element from group IIIb of the periodic system, and a Lanthanoid (see) rare-earth, heavy metal, *Z* 70, with natural isotopes with mass numbers 174 (31.8%), 172 (21.9%), 173 (16.1%), 171 (14.3%), 176 (12.7%) 170 (3.06%) and 168 (0.14%); atomic mass 173.04, valency III, rarely II, density 6.972, m.p. 816 °C, b.p. 1193 °C, standard electrode potential (Yb/Yb^{3+}) -2.267 V.

Y. is a ductile metal with a silvery sheen. It crystallizes in two modifications: below 798 °C, in a cubic face-centered lattice, and above 798 °C, in a cubic body-centered lattice. Under high pressure (4000 MPa), the α-modification, which is a metallic conductor at room temperature, becomes a semiconductor. Y. makes up $2.5 \cdot 10^{-4}$% of the earth's crust, and is always found accompanied by the other rare-earth metals, especially yttrium and the heavier lanthanoids. It is relatively highly enriched in gadolinite (ytterbite), xenotim, fergusonite, samarskite, yttrotantalite and euxenite. Further properties, analysis, production and history are discussed under Lanthanoids (see). Y. is used to dope lasers.

Ytter earths: oxides of the rare-earth metals yttrium and scandium and the lanthanides europium to lutetium. Because their ionic radii are very similar, the Y. are found in association in nature; yttrium is always the main component of the Y. minerals. Some examples are xenotin, YPO_4, gadolinite,

$Y_2M^{II}{}_3Si_2O_{10}$ (M^{II} = Fe, Be) and euxenite, (Y) (Nb,Ta)$TiO_6 \cdot xH_2O$. There are three subgroups of Y. The ***terbine earths*** include europium, gadolinium and terbium oxides, the ***erbine earths***, dysprosium, holmium, erbium and thulium oxides, and the ***ytterbine earths***, ytterbium and lutetium oxides.

Yttrium, symbol ***Y***: a chemical element from group IIIb of the periodic system, and a rare-earth metal belonging to the Scandium group (see). There is only one isotope of this light metal, *Z* 39, atomic mass 88.9059, valency III, m.p. 1552 °C, b.p. 3337 °C, electrical conductivity 15.4 Sm mm^{-2}, standard electrode potential (Y/Y^{3+}) -2.3772 V.

Properties. Y. is a ductile, silvery metal which occurs in two modifications. Below 1490 °C it crystallizes in a hexagonal lattice (density 4.472), and above this temperature, in a cubic body-centered lattice. At room temperature, Y. is relatively stable in air, but at higher temperatures it reacts with oxygen to form yttrium(III) oxide, Y_2O_3, or with chlorine to form yttrium(III) chloride, YCl_3. It is always present in the +III oxidation state in its compounds.

Analysis. Y., like all other members of the rare-earth group, can be precipitated as the oxalate, $Y_2(C_2O_4)_3 \cdot 9H_2O$, from weakly acidic solution. Y. is identified and its purity tested by x-ray spectroscopy.

Occurrence. Y. makes up $3.0 \cdot 10^{-3}$% of the earth's crust, and is always accompanied by other rare-earth metals. It is the major component in minerals such as gadolinite (ytterbite), $Y_2M^{II}{}_3Si_2O_{10}$ (M^{II} = Fe, Be), thalenite, $Y_2[Si_2O_7]$ and xenotim, YPO_4. It is now produced on an industrial scale mainly from monacite sand and bastnesite, which contain 3 and 0.2% Y., respectively.

Production. The ores are solubilized with sulfuric acid; this is followed by a complicated series of separations, usually including extraction, ion exchange and complex formation, to separate Y. from the lanthanoids and lanthanum with which it is associated (see Lanthanoid separation). It is finally precipitated as the oxalate, roasted to the oxide, and converted by reaction with hydrofluoric acid to yttrium(III) fluoride. The fluoride is reduced to the metal by reaction with calcium/magnesium, and the alkaline earth metals are distilled out of the resulting yttrium-magnesium-calcium alloy at high vacuum at 1000 to 1200 °C. The residual yttrium sponge is then converted to the compact metal by melting in an electric arc.

Applications. Y. is an important industrial metal, and its broad use – also in the form of certain of its compounds – has led to a remarkable surge in the rare-earth metal industry. Because of its low absorption cross section for thermal neutrons, Y. is used in nuclear reactors for the pipes which hold the fuel rods in the reactor core. Its high melting point, low density, resistance to corrosion and inability to fuse with uranium are also important properties in this application. Y. is also used as a deoxidant for vanadium and other nonferrous metals. When added to aluminum and magnesium alloys, it makes them stronger. Y.-cobalt alloys of the types YCo_5 and and Y_2Co_{17} are used in industry as ferromagnets.

Historical. Y. was discovered in 1843 by Mosander, as part of the mineral gadolinite (ytterbite) which had first been studied by Gadolin in 1797; the mineral was obtained near the town of Ytterby in Sweden, and hence the name.

Yttrium compounds: compounds of yttrium, usually of the type YX_3. Most are colorless and diamagnetic. Y. are intermediate in their properties between the corresponding derivatives of aluminum and the alkaline earth metals. ***Yttrium(III) chloride***, YCl_3, colorless, translucent leaflets, M_r 195.26, density 2.67, m.p. 709 °C, b.p. 1507 °C; it is made by reaction of yttrium with hydrogen chloride at 700 °C, by heating a mixture of yttrium(III) oxide and coal in a stream of chlorine, or by dehydration of the hexahydrate, $YCl_3 \cdot 6H_2O$ in a stream of hydrogen chloride. ***Yttrium(III) fluoride***, YF_3, is a white compound, M_r 145.90, density 4.01, m.p. 1152 °C, is made on a technical scale by reaction of yttrium(III) oxide and coal with hydrogen fluoride in a rotating furnace. Yttrium reacts at high temperatures with hydrogen to form the black ***yttrium hydride***, which has the approximate composition YH_3. Precipitation of yttrium(III) salt solutions with alkali hydroxide, carbonate or oxalate produces ***yttrium(III) hydroxide***, $Y(OH)_3$, a white powder, M_r 139.93, density 4.01, ***yttrium(III) carbonate***, $Y_2(CO_3)_3 \cdot 3H_2O$, a reddish-white powder, M_r 411.88, and ***yttrium(III) oxalate***, $Y_2(C_2O_4)_3 \cdot 9H_2O$, a white powder, M_r 604.01. These compounds are converted by roasting to ***yttrium(III) oxide***, Y_2O_3, a colorless to yellow compound, M_r 225.81, density 5.01, m.p. 2680 °C. Yttrium(III) hydroxide is intermediate in basicity between the weak base scandium(III) hydroxide and the stronger base lanthanum(III) hydroxide. Like yttrium(III) oxide, it dissolves in nitric acid, forming ***yttrium(III) nitrate***, $Y(NO_3)_3 \cdot 4H_2O$, reddish-white prisms, M_r 396.98, density 2.682.

Y. are now used on a large scale. Yttrium vanadate, oxide and oxygen sulfide, activated by europium, are used as red components for color television screens. Yttrium iron oxide and yttrium aluminum oxides of the types $Y_3Fe_5O_{12}$ and $Y_3Al_5O_{12}$, are used in electronics and data processing as storage units and modulators, and in lasers. A transparent ceramic material based on 90% yttrium(III) oxide and 10% thorium(IV) oxide (yttrilox) is resistant to high temperatures and transparent to UV and IR radiation; it is used for high-power lights and gas-discharge lamps. Combined with zirconium(IV) oxide, yttrium(III) oxide forms the radiation source in the Nernst lamp.

Z

z: symbol for stoichiometric valence.

Z: 1) abb. for atomic number (nuclear charge number); 2) configuration symbol, see Stereoisomerism 1.2.3.

Zaffer, ***zaffre***: a reddish to bluish mixture of cobalt oxides and arsenates, containing nickel compounds as impurities. It is an intermediate in the processing (roasting) of cobalt ores and is sometimes used as a pigment.

Zajcev product: see Elimination.

Zantrel: see Synthetic fibers.

Zapon: a cellulose ester varnish; see Lacquer.

Zeatin: see Cytokinins.

Zeaxanthin: see Xanthophylls.

Zeeman effect: splitting of spectral lines when the emission or absorption of the light occurs in a magnetic field. The cause of the Z. is the quantized orientation of the atoms in an external magnetic field. The ***normal Z.*** occurs in singlet systems in which the total spin moment S is zero, and only the orbital magnetic moment can interact with the external magnetic field. It consist of splitting of the lines into triplets, the middle component of which has the frequency of the original line, while the other two lines are symmetrically spaced on either side of it. The shift of the components is proportional to the magnetic field strength. When observed perpendicular to the field direction, all 3 components display linear polarization. The unshifted component oscillates parallel to the magnetic field lines, while the shifted components oscillate perpendicular to the field lines (*transverse Z.*). Observation in the direction of the magnetic field reveals only the two shifted components, which are circularly polarized in opposite directions from one another (*longitudinal Z.*).

In the ***anomalous Z.***, which occurs in non-singlet atoms, the splitting of terms is more complicated. There are more lines, and the individual components are spaced unevenly. The cause for this is the *magnetomechanical anomaly*: the magnetic moment of the electron due to its spin is twice as large as its orbital magnetic moment. The splittings observed in the anomalous Z. are seen only in weak external magnetic fields in which the Zeeman splitting is smaller than the distance between components of the natural multiplet. In strong magnetic fields, in which the Zeeman splitting is greater than the distance between components of the natural multiplet, the splitting is simplified, and the normal Zeeman triplets are again observed (*Paschen-Back effect*). In the region of transition between the anomalous Z. and the Paschen-Back effect, the splitting is very confusing and is also difficult to treat theoretically.

Like the Stark effect (see), the Z. is one of the best methods for empirical determination of the quantum numbers of an atomic state, and is therefore very important for term analysis.

Zeeman splitting is the basis of Electron spin resonance spectroscopy (see), and the nuclear Z. is the basis of NMR spectroscopy (see).

Zefran: see Synthetic fibers.

Zein: see Prolamins.

Zein fibers: Protein fibers (see).

Zeisel method: a method for quantitative determination of methoxy or ethoxy groups in carbon compounds by means of ether cleavage. The sample is treated with boiling hydroiodic acid, and the alkyl iodide resulting from ether cleavage is distilled into an alcoholic solution of silver nitrate. Silver iodide is precipitated and determined gravimetrically: C_2H_5-O-C_2H_5 + 2 HI $\rightarrow$ 2 C_2H_5I + H_2O; C_6H_5-O-CH_3 + HI $\rightarrow$ CH_3I + C_6H_5-OH; CH_3I + $AgNO_3$ + H_2O $\rightarrow$ AgI + CH_3OH + HNO_3. Sulfur-containing substances interfere with the reaction by precipitating silver sulfide. Therefore, the alkyl iodide in sulfur-containing samples is oxidized with bromine in glacial acetic acid/potassium acetate, the resulting iodate is treated with potassium iodide, and the precipitated iodine is titrated with sodium thiosulfate.

Zeise's salt: see Organoplatinum compounds.

Zeolites: a group of crystalline minerals with the general formula $(M^I_2, M^{II})O \cdot Al_2O_3 \cdot nSiO_2 \cdot mH_2O$; M^I = sodium or, more rarely, potassium; M^{II} = calcium, or more rarely, barium or strontium. The X. are used as adsorbants (molecular sieves). According to a rule proposed by Loewenstein, the AlO_4 tetrahedra in the aluminosilicate network of Z. are linked exclusively to SiO_4 tetrahedra. The Z. have a loose, three-dimensional network structure (tectosilicates) with regular channels which hold the water molecules and alkali or alkaline earth metal cations. The Z. are classified in several types on the basis of the size of the pores and the ratio n of SiO_4 to AlO_4 tetrahedra. A-type Z. have $n = 2$, X-type have $n = 2.2$, and Y-type have $n = 3$ to 6; these are formed by combinations of 4-, 6-, 8- and 12-membered rings which form cubic octahedra. Y-type Z. have large pores with diameters of 8 to $9 \cdot 10^{-10}$ m, and small pores with $2.4 \cdot 10^{-10}$m in which adsorption and exchange processes can occur. The water can be reversibly released, or ion-exchange processes can occur without changes in the basic structure of the Z.

Synthetic Z. can maintain their structures up to 800 °C. There are more than 30 known natural Z., which arose by the action of basic mineral salt solutions on aluminosilicates over very long periods at 100 to 300 °C (see geochemical Hydrothermal synthesis). Because natural Z. are difficult to separate and purify by extraction of impurities or recrystallization, many natural types of Z. and some new ones are chemically synthesized from alkali metal or alkaline earth metal aluminates and silicates. The various types of synthetic Z. can be produced by variation of the starting materials (e.g. water glass, SiO_2 sol, alkali and alkaline earth metal chlorides and hydroxides), the time and temperature of crystallization, the nature and amounts of additives, and technological parameters such as stirring and degree of washing out.

Z. are applied in separation processes (drying and purification of gases and liquids, selective separations of gas mixtures), ion exchange processes, catalysis (adsorption, molecular sieve and carrier functions), vacuum technology (adsorption pumps) and dosing of radioactive and toxic catalysts, accelerators and stabilizers.

Zero charge potential: see Electrocapillarity.

Zero point energy: see Infrared spectroscopy.

Zeta potential, *ζ-potential, electrokinetic potential*: the potential which arises at the shear plane between a mobile and a stationary phase when an electrical field is applied to the system. The Z. is smaller than the surface potential, and smaller than or equal to the Stern potential (see Electrochemical double layer). Knowledge of the Z. is important for determining the stability of colloidal dispersions. The Z. of particles can be determined by microscopic observation of the migration velocity in the electric field. The Z. which arises at the shear plane of colloidal particles and solutions during sedimentation is also called the ***sedimentation potential***; and the Z. in aqueous solutions at the shear plane between a solid and the liquid is called the ***flow potential***.

Ziegler-Natta catalysts: see Catalysis, sect. II, 4.

Zineb: see Dithiocarbamate fungicides.

Zinc: symbol ***Zn***: a chemical element from group IIb of the periodic system, the Zinc group (see); a heavy metal, *Z* 30, with natural isotopes with the following mass numbers: 64 (48.49%), 66 (27.81%), 67 (4.11%) and 70 (0.62%); atomic mass 63.38, valency II, Mohs hardness 2.5, density 7.133, m.p. 419.58 °C, b.p. 908.5 °C, electrical conductivity 16.5 S m mm^{-2}, standard electrode potential (Zn/Zn^{2+}) 0.7268 V.

Properties. Z. is a bluish white metal with an unusual lattice structure. It forms a hexagonal close packing which is expanded in the direction of the six-fold lattice axis. Z. is rather brittle at room temperature, but at 150 °C it becomes soft and ductile, so that in this temperature range, it can be rolled to thin sheet or drawn into wire. Above 200 °C, it then becomes so brittle that it can be pulverized. Z. has relatively low melting and boiling points. The vapor consists mainly of atoms, as well as a small proportion of weakly bound Zn_2 molecules. Z. is stable in air, because it becomes covered with a tight protective layer of zinc oxide, ZnO, or zinc carbonate, $ZnCO_3$. When heated in air to the boiling point, it burns with a bright, greenish-blue flame to zinc oxide. At red heat, Z. is also oxidized by steam or carbon dioxide. It forms alloys with many metals (see Zinc alloys). Non-oxidizing acids, such as hydrochloric or dilute sulfuric acid, dissolve Z. with evolution of hydrogen. The discharge of the protons on pure Z. is inhibited, but can be accelerated by formation of local elements, for example by addition of copper sulfate to the solution. Z. reacts with water to form a protective, insoluble layer of zinc hydroxide, $Zn(OH)_2$; because this is readily soluble in bases, with formation of zincates, Z. readily reacts with bases with evolution of hydrogen.

The electronic configuration of Z., $3d^{10}4s^2$ is responsible for the strong tendency of zinc compounds to be in the +II oxidation state, and for the fact that they are colorless. Only the elevated volatility of Z. in the presence of zinc(II) chloride, $ZnCl_2$, at 285-350 °C, indicates the existence of dimeric zinc(I) chloride, Zn_2Cl_2, in this temperature range. Zinc(II) readily forms complexes; tetrahedral and octahedral complexes with coordination numbers of 4 and 6, respectively, are most common.

Z. is a trace element required by organisms. It is involved in regulation of oxidation and reduction processes, carbohydrate and protein metabolism, and synthesis of chlorophyll. Because of its high affinity for nitrogen and sulfur ligands, Z. is generally bound to amino acids, proteins or nucleic acids in the cell. Over 25 Z.-containing enzymes are known, including dehydrogenases, phosphatases, carboxypeptidases and carbonic acid anhydrase. The human body contains 2 to 3 g Z., most of which is localized in the cells. A lack of Z. in soils can lead to low crop yields, so addition of Z. in fertilizers is necessary in such cases.

> In spite of the life-supporting functions of zinc, the metal and its compounds can be toxic in high concentrations. Finely divided zinc powder or zinc oxide can cause diseases of the respiratory passages. First aid for poisonings with zinc compounds: emetic followed by egg in milk.

Analysis. There is no specific detection reagent for Z. in the presence of other metals. In the classical separation scheme, Z. is precipitated with the ammonium sulfide group, together with nickel, cobalt and manganese. Zinc sulfide differs from the other sulfides in this group in its color and in the fact that it can be precipitated out of a weakly acidic solution. After the associated metals have been removed, Z. can be identified by precipitation reactions, e.g. with $K_4[Fe(CN)_6]$ or $K_2[Hg(SCN)_4]$ to form $Zn_2[Fe(CN)_6]$ or $Zn[Hg(SCN)_4]$, or by formation of characteristically colored zinc-cobalt mixed oxides (see Cobalt green). For gravimetric determination, Z. may be precipitated as zinc sulfide, ZnS, zinc carbonate, $ZnCO_3$, or ammonium zinc phosphate (NH_4ZnPO_4). The most commonly used method today is complexometric titration with EDTA under alkaline conditions. As little as 2 ppm Z. can be detected by atomic absorption spectroscopy.

Occurrence. Z. makes up $5.8 \cdot 10^{-3}$% of Earth's crust. Zinc sulfide, ZnS, is an important form for extraction; it is found as the cubic sphalerite (zinc blende) or the hexagonal wurtzite. Other Z. ores are: smithsonite (zinc spar), $ZnCO_3$, hemimorphite, $Zn_4Si_2O_7(OH)_2 \cdot H_2O$, and willemite, Zn_2SiO_4. Z. is present in soils as a trace element, at about 50 ppm, and in plants at about 3 ppm (dry matter). Z. deficiency causes dwarf growth, chlorophyll defects (mosaic disease in the leaves) and considerable disturbance of the phosphoric acid economy in plants.

Production. Z. is obtained industrially either by a dry process, reduction of zinc oxide with coal, or by a wet process, electrolysis of zinc sulfate solution. The *reduction process* is a discontinuous process. Zinc oxide is obtained from sphalerite by roasting, or from smithsonite by combustion; this is heated with excess powdered coal in a horizontal, closed distillation retort to 1100-1300 °C. The resulting Z. escapes as a vapor and is condensed in a firebrick receptacle. Residues of the vapor precipitate as zinc powder in steel cans beyond the receptacle. The crude Z. obtained in this way is 97 to 98% pure; the impurities are lead and small amounts of iron, cadmium and arsenic. It is purified by remelting and fractional distillation. 99.99% Z. can be obtained by repeated distillation, while the more volatile cadmium settles out with some Z. as "cadmium powder" (about 40% Cd). Together with the zinc powder from the reduction step, this is processed to obtain the cadmium.

A more rational process is the continuous *New Jersey process*, in which vertical silicon carbide retorts are used. Pellets of pressed zinc oxide and coal are fed in from the top and are heated from the outside with generator gas to 1200 to 1400 °C. The crude Z. which condenses in the condensors is purified by fractional condensation of the vapors in a column.

In *wet extraction*, roasted smithsonite or burned sphalerite is extracted with sulfuric acid, and the resulting zinc sulfate solution is electrolysed between a lead anode and an aluminum cathode. The Z. deposited on the aluminum cathode is 99.99% pure after remelting. The application of this process depends on a high degree of purity of the zinc sulfate solution. The purification of the solution is not necessary with the *Kuss process (amalgam process)*, in which mercury is used as the cathode material. With this process, 99.999% pure zinc can be obtained.

Applications. Because of its stability in air, Z. is used for roofing and gutters, and to plate sheet iron and iron wire. Large amounts of zinc plate are used to make dry cells (batteries). Metal can be coated with Z. by dipping in molten metal, metal spraying or electrolytically. Very fine Z. powder used to paint iron effectively protects it from rust. Z. powder is used in industry and the laboratory as a reducing agent. In metallurgy, it is used, for example, in the Parkes process of removing silver from lead, and in cyanide leaching to precipitate silver and gold. Zinc alloys (see) are used widely, especially those with copper (see Brass), copper and nickel (see New silver).

Historical. As a pure metal, Z. was first discovered in ancient India and in China. In European antiquity, Z. was known in the form of its copper alloy, brass, which was obtained by melting copper with an ore known in Greece as "cadmeia", and in Rome (Pliny) as "cadmia" (this is the source of the name of cadmium, which is found with Z.). A.S. Marggraf is given credit for the discovery of Z. He obtained it by reduction of zinc oxide with coal. Industrial production of Z. began in England.

Zinc acetate: $Zn(OOCH_3)_2$, colorless octahedra, which can be sublimed without decomposition under reduced pressure; M_r 183.46, density 1.84, m.p. 242 °C (dec.). Z. is readily soluble in water and ethanol, and is obtained by reaction of zinc nitrate, $Zn(NO_3)_2 \cdot 6H_2O$ with acetic anhydride. If zinc oxide, ZnO, is dissolved in acetic acid, it precipitates as the dihydrate, $Zn(OOCCH_3)_2 \cdot 2H_2O$, in the form of shinyl, colorless, monoclinic leaflets; M_r 219.49, density 1.835. Z. is used to make textiles resistant to flame or water. It is also used in solutions for gargling and to bathe the skin in cases of skin diseases.

Zinc alloys: alloys of zinc with aluminum, copper, magnesium, manganese, lithium or titanium. Aluminum and copper greatly increase the strength of zinc. Very small amounts of lead, cadmium or tin make the zinc-aluminum-copper alloys susceptible to intercrystalline corrosion, but this tendency can be reduced by addition of magnesium or lithium, and by limiting the amounts of lead, cadmium and tin. The most common alloys contain 3.5 to 4.3% aluminum and 0.6 to 1.0% copper; these are used as pressure casting materials. As these alloys solidify, supersaturated, zinc-rich mixed crystals form; at room temperature, aluminum- and copper-rich phases precipitate out of these crystals. This causes the alloys to shrink and lose strength as they age. The zinc-aluminum-copper alloys are used in machines, motor vehicles, household and kitchen machines and for fine tools. Most zinc is consumed in the production of Brass (see) and New silver (see).

Zincates, ***hydrozincates***: anionic zinc complexes with hydroxide ligands which are formed by dissolving zinc hydroxide, $Zn(OH)_2$, in alkali hydroxide solutions. The most important Z. are the alkali tetrahydroxozincates, $M_2[Zn(OH)_4]$.

Zinc blende type, ***zinc blende structure***: a common Structure type (see) for compounds with the general composition AB. The mineral zinc blende (zinc sulfide, ZnS) has a crystal lattice in which the two types of atoms form cubic face-centered partial lattices displaced with respect to one another along the direction of the diagonal of the unit cube and by 1/4 of the distance. The structure can also be seen as a cubic close-packed arrangement of sulfur atoms with alternate occupation of the tetrahedral holes by the zinc atoms. Each atom is surrounded tetrahedrally by 4 atoms of the other kind (Fig.).

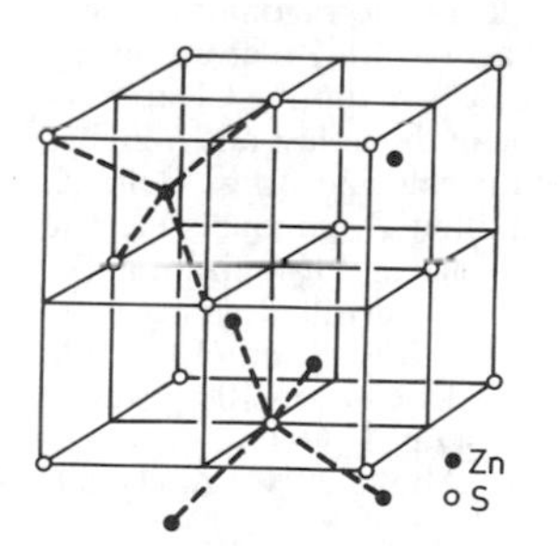

Zinc blende structure.

In Z. compounds with a small ratio of atomic radii (in the ideal case, between 0.225 and 0.414), the qualitative composition obeys the Grimm-Sommerfeld rule (see). The bonding forces in the crystal lattice are a transition type between ionic and covalent bonding, and many of the compounds are semiconductors. The phenomenon of Polytypism (see) occurs frequently in the Z. Compounds with complicated compositions can also crystallize in the Z., in that the Zn atoms are replaced in a regular or even random fashion by other atoms. For example, the *chalcopyrite type* (named after the mineral chalcopyrite, $CuFeS_2$) is a derivative of the Z. in which the metal atom positions are alternately occupied by two different types of atoms, such as Cu/Fe, Zn/Ge, Ag/Al. If all the lattice positions in the Z. are occupied by a single type of atom, the crystal is the Diamond type (see). The Z. is closely related to the Wurtzite type (see), in which zinc sulfide can also crystallize (see Polymorphism).

Zinc bloom: a mixture of zinc hydroxide and basic zinc salts which forms as a crust on zinc or zinc-rich alloys in the presence of condensation water or aqueous solutions.

Zinc bromide: $ZnBr_2$, white, dimorphic, hexagonal or tetragonal crystals; M_r 225.19, density 4.20, m.p. 394 °C, b.p. 697 °C. Z. is very readily soluble in water, and crystallizes out of aqueous solutions as hydrates, $ZnBr_2 \cdot 2H_2O$ and $ZnBr_2 \cdot 3H_2O$. Z. is made by reaction of zinc with bromine at about 550 °C, or by reaction of zinc with hydrobromic acid, followed by sublimation of the product into HBr/N_2-bromine. Z. is used to make the light-sensitive layers for photographic film, and in medicine to treat epilepsy, paralysis and St. Vitus' dance.

Zinc carbonate: $ZnCO_3$, a white compound; M_r 125.39, density 4.398. Z. is scarcely soluble in water, and when heated to 150 °C, it splits off carbon dioxide. It is found in nature as smithsonite (zinc spar). It reacts with water to form basic zinc carbonates, which have the approximate composition 2 $ZnCO_3 \cdot 3Zn(OH)_2$. A basic zinc carbonate with this composition is found in nature as hydrozincite (zinc bloom). Z. is obtained by reaction of carbon dioxide with a freshly precipitated slurry of zinc oxide in water, or by precipitation from a zinc salt solution to which a CO_2-saturated alkali hydrogencarbonate solution is added. It is used as an activator and filler for rubber, as a pigment and in the production of ceramics.

Zinc chloride: $ZnCl_2$, a very hygroscopic, white compound; M_r 136.28, density 2.91, m.p. 275 °C, b.p. 756 °C. Z. exists in three forms: tetragonal α-Z. has Cl atoms in close-packed array with Zn atoms in the tetrahedral holes; monoclinic β-Z. has the Cl atoms in a hexagonal array with zinc in the holes; and γ-Z. has a layered arrangement of the chlorine atoms. Solid Z. is a nonelectrolyte, but molten Z. conducts electric current, probably due to partial self-ionization. Z. has a metallic taste and can be distilled without decomposition; it is very caustic. It is extremely soluble in water: at 25 °C, 432 g Z. dissolves in 100 g water. Due to hydrolysis to form $H_2[ZnCl_2(OH)_2]$, the aqueous solutions are very acid. Above 28 °C, Z. crystallizes out of aqueous solution in anhydrous form, and below this temperature, it forms various hydrates. Z. is made by heating zinc in a stream of chlorine or by reaction of zinc oxide, hydroxide, carbonate or sulfide with hydrochloric acid. It is used in the laboratory as a water-removing reagent. Because of the toxicity of zinc(II) compounds to microorganisms, Z. is used to impregnate wood. It is also used in textile printing, soldering and as a caustic.

Zinc cyanide: $Zn(CN)_2$, white, cubic crystals; M_r 117.41, density 1.852, m.p. 800 °C (dec.). Z. is poisonous (see Hydrocyanic acid), and is insoluble in water or ethanol; however, it dissolves readily in alkali cyanide solutions, forming tetracyanozincate $[Zn(CN)_4]^{2-}$. It is obtained by combining equivalent amounts of alkali cyanide and zinc salt in aqueous solution. Z. is used, for example, to make non-inflammable paints. Solutions of alkali tetracyanozincate and tetracyanocuprate(I) are used in galvanizing to give brass coatings, while zinc plating is done with solutions which contain $[Zn(CN)_4]^{2-}$ and hydroxozincate.

Zinc fluoride: ZnF_2, white, monoclinic crystals; M_r 103.37, density 4.95, m.p. 872 °C, b.p. 1502 °C. Z. is poisonous (see Zinc). It is scarcely soluble in water, and unlike the other zinc halides, it is largely ionic. It is made by heating zinc oxide with hydrofluoric acid to red heat, or by dehydration of the tetrahydrate, $ZnF_2 \cdot 4H_2O$, which is obtained from the reaction of zinc carbonate with aqueous hydrofluoric acid. Z. is a mild fluorination reagent, and is also used for preserving wood, glazes, enamels and galvanic zinc plating.

Zinc gray: a pigment obtained as a byproduct of roasting zinc ores to make zinc oxide. It contains about 10% zinc oxide, and other metals or semimetals such as lead, arsenic and antimony.

Zinc green: a mixed pigment consisting of zinc yellow and Prussian blue. Z. is used as a light-fast mineral pigment for paints.

Zinc group: Group IIb of the periodic system, consisting of zinc (Zn), cadmium (Cd) and mercury (Hg). Strictly speaking, these metals are not transition elements, as their outer d and s shells are filled: nd^{10}, $(n+1)s^2$. They have relatively low melting and boiling points, which decrease with increasing atomic mass; mercury is the only metal which is liquid at room temperature. Zinc and cadmium are very similar in their chemistry, occurring only in the +2 oxidation state. Mercury compounds of both the +1 and +2 oxidation states are known; mercury(I) compounds are always dimolecular and contain Hg-Hg bonds.

Properties of the elements of the zinc group

	Zn	Cd	Hg
Nuclear charge	30	48	80
Electron configuration	[Ar] $3d^{10}$ $4s^2$	[Kr] $4d^{10}$ $5s^2$	[Xe] $4f^{14}$ $5d^{10}$ $6s^2$
Atomic mass	65.38	112.40	200.59
Atomic radius	124.9	141.3	144.0
Ionic radius (M^{2+}) [pm]	74	97	110
Electronegativity	1.66	1.46	1.44
Standard electrode potential (M/M^{2+}) in V	–0.762 8	–0.402 9	+0.850 0
Density [g/cm³]	7.133	8.642	13.534
m.p. [°C]	419.58	320.9	–38.84
b.b. [°C]	908.5	767.3	356.95

The heavy metals of the Z. have higher ionization potentials, smaller atomic and ionic radii, and lower energies of formation than the alkaline earth metals, and their compounds have more covalent character than the corresponding alkaline earth derivatives. The Z. metals become less reactive with increasing atomic number.

Zinc hexafluorosilicate, formerly ***zinc silicofluoride***: $ZnSiF_6 \cdot 6H_2O$, colorless, hexagonal prisms; M_r 315.54, density 2.104, m.p. 100 °C (dec.). Z. is extremely soluble in water (769.6 g in 100 g water at 10 °C). It is made by dissolving zinc oxide in dilute hexafluorosilicic acid, and is used to protect buildings.

Zinc hydroxide: $Zn(OH)_2$, a white substance; M_r 99.38, density 3.053, decomposition begins above 120 °C. Z. exists in six different modifications, α-, β_1-, β_2-, γ-, δ- and ε-$Zn(OH)_2$. Only rhombic ε-$Zn(OH)_2$ is stable in water below 39 °C; above this temperature, zinc oxide, ZnO, forms the phase which is stable in contact with water. Z. is soluble in bases, with formation of zincates. When it reacts with ammonia, it forms tetra- or hexaammine complexes. Z. is formed by the reaction of zinc salt solutions with alkali hydroxide solutions as a white, gelatinous precipitate.

Zinc iodide: ZnI_2, yellow, dimorphic crystals; M_r 319.18, density 4.74, m.p. 446 °C, b.p. 625 °C (dec.). Z. is obtained by direct reaction of the elements, or by heating a solution of iodine in ether with zinc under reflux.

Zincke aldehyde: trivial name for 5-(*N*-methylanilino)penta-2,4-dienal, an *N*-methylaniline derivative of glutaconic dialdehyde. Z. is a starting material for the synthesis of azulene.

Zinc loss: a form of corrosion of brass, which can occur either in spots or layers. In non-aerated, chloride-containing solutions, both copper and zinc ions dissolve at first. Later the copper forms a spongy precipitate on the more active brass surface, which appears as a zinc-depleted surface.

Zinc-mercury oxide cell: an Electrochemical current source (see), and a primary cell. The cathode is about 90% mercury oxide (HgO) and 10% graphite as the conductor.

⊖ Cover (negative pole of battery)
Seal and insulation
Zinc powder anode
Electrolyte-soaked paper
Separator combination
Metal support
Mercury oxide cathode
⊕ Battery case (positive pole of battery)

Cross section of a zinc-mercury oxide cell.

Zinc powder is used as anode, and the electrolyte is a potassium hydroxide solution saturated with zinc oxide. The overall electrode reactions can be described by the following equations:

Cathode: $HgO + H_2O + 2e^- \rightarrow Hg + 2\,OH^-$
Anode: $Zn + 4\,OH^- \rightarrow Zn(OH)_4^{2-} + 2e^-$.

The Z. has a high energy density, 241 Wh/kg, and can be stored 5 to 8 years. Because of its relatively high price, however, its use is limited to a few special applications, e.g. quartz watches and pocket calculators.

Zinc molybdate: see Molybdate pigments.

Zinc nitrate: $Zn(NO_3)_2 \cdot 6H_2O$, colorless, clear, four-sided columnar prisms; M_r 297.47, density 2.065, m.p. 36.4 °C. Z. is very hygroscopic and dissolves readily in water; it crystallizes out of aqueous solution in the temperature range between -17.6 and +36.4 °C as the hexahydrate. Other hydrates with 2, 4 and 9 mol H_2O are known. Z. is obtained by reaction of zinc with nitric acid. Anhydrous Z., $Zn(NO_3)_2$, is obtained by reaction of zinc with dinitrogen tetroxide, which yields a primary adduct, $Zn(NO_3)_2 \cdot 2N_2O_4$. The adduct is heated to 100 °C in vacuum to convert it to Z.

Zinc nitride: Zn_3N_2, dark gray, cubic crystals; M_r 224.12, density 6.22. Z. is hydrolysed by water. It is obtained by the reaction of ammonia with zinc powder around 600 °C, or by thermal decomposition of zinc amide at 350 °C.

Zinc oxide: ZnO, colorless, hexagonal crystals or white powder which is insoluble in water; M_r 81.37, density 5.606, m.p. 200 °C (at about 5 MPa pressure), subl.p. 1975 °C (dec.). When Z. is heated, it turns yellow (thermochromism), due to a small excess of zinc which is formed. Yellow, green or red, semiconducting preparations can also be made by heating Z. with zinc vapor. Z. is found in nature as zincite (red zinc ore). It is produced by burning zinc vapor, which is made by reduction of oxidic zinc minerals with coke. To make pure Z., one starts with precipitates such as zinc carbonate or hydroxide. Z. is used mainly as a paint pigment (see Zinc white), a pigment lightener, in soap fabrication, in printing paper and wallpaper, as an additive to glass and filler for rubber, in scouring powders and to make other zinc compounds. Z. is also used by itself or mixed with other oxides as a catalyst for methanol synthesis from carbon monoxide and hydrogen. Pure Z. is used to make powders and pastes for cosmetic purposes, for dusting powder and as a component of medicinal salves.

Zinc peroxide: $ZnO_2 \cdot 1/2O_2$, a bright yellow powder; M_r 106.38, density 3.00. Z. is scarcely soluble in water. It splits off oxygen above 150 °C; in the presence of organic substances or transition metal ions, such as Cu^{2+} or Mn^{2+}, it decomposes rapidly above room temperature. Z. is used, e.g. as an antiseptic for treatment of wounds.

Zinc phosphides: ***Trizinc diphosphide***, Zn_3P_2, dark gray, tetragonal needles or leaflets; M_r 258.06, denisty 4.55, m.p. >420 °C, subl.p. (under H_2) 1100 °C. Zn_3P_2 is insoluble in water and is poisonous (see Zinc). It is obtained by heating the components to about 750 °C, and is used as a rodenticide.

Zinc diphosphide, ZnP_2, orange, tetragonal crystals; M_r 137.67, obtained from the elements when an excess of phosphorus is present.

Zinc pigments: inorganic zinc compounds which are used as pigments. Some important examples (see separate entries for each) are: Zinc white, Zinc yellow, Zinc green and Lithopone.

Zinc plating: the plating of metal objects, especially those made of steel, with a zinc layer as a corrosion protection. Because zinc is highly electronegative, it provides effective protection against rusting. The object may be dipped into molten zinc, the molten zinc can be sprayed onto the object with a spray gun (see Metal spraying), or it may be ***galvanized***. In this process, the zinc is applied by electrolysis in an acid or basic bath, with an anode of electrolysis zinc. The amount of zinc applied is 300 to 400 g m^{-2} for dipped or sprayed coatings ($\geqq$= 0.5 mm thick), and 100 g m^{-2} for galvanic coatings. In sherardization the zinc is diffused into the surface of the object being protected.

Zinc powder distillation: process for converting an oxygen-containing ring system into the corresponding aromatic hydrocarbon. The substance is distilled with zinc powder in a stream of hydrogen. The method was first used by A. von Baeyer in 1867 for converting oxindole to indole, and then became very important in the elucidation of structures of natural substances. For example, anthracene was obtained from alizarin, and phenanthrene from morphine.

Zinc selenide: ZnSe, dimorphous, reddish yellow crystals; M_r 144.33, density 5.42, m.p. $\rightleftharpoons$ 1100 °C, subl.p. about 2000 °C. Z. is obtained in cubic form by heating zinc and selenium, or by reaction of zinc oxide and sulfide with selenium at 800 °C: $2\,ZnO + ZnS + 3\,Se \rightarrow 3\,ZnSe + SO_2$. Reaction of zinc and selenium at about 1350 °C produces the hexagonal modification. Z. is often used in a mixture with other zinc or cadmium chalcogenides as a luminophore.

Zinc-silver cell: an Electrochemical current source (see) and a primary cell. It is constructed similarly to the Zinc-mercury cell (see), but the cathode material is silver oxides (Ag_2O, Ag_2O_2).

Zinc silicate: Zn_2SiO_4, white, microcrystalline compound; M_r 222.84, density 4.10, m.p. 1512 °C. Z. is obtained by heating the components zinc oxide and silicon dioxide in a molar ratio 2:1. Pure Z. is activated with manganese and used as a luminophore for UV or cathode ray excitation.

Zinc sulfate: $ZnSO_4$, rhombic crystals, M_r 161.43, density 3.54, m.p. 600 °C (dec.). Z. is readily soluble in water and decomposes when strongly heated, especially in the presence of reducing agents, to form zinc oxide, sulfur dioxide, sulfur trioxide and oxygen. It crystallizes out of aqueous solution as the heptahydrate, $ZnSO_{42} \cdot 7H_2O$, which is also called ***zinc vitriol***. This compound forms large, colorless, columnar rhombic crystals; M_r 287.54, density 1.957, m.p. 100 °C. When heated cautiously, zinc vitriol first releases six molecules of crystal water, and is then converted to the anhydrous form at about 280 °C. When Z. is crystallized out of aqueous solutions above 39 °C, a hexahydrate is obtained, $ZnSO_4 \cdot 6H_2O$, and above 60 °C, the monohydrate, $ZnSO_4 \cdot H_2O$. Z. is found sporadically in nature both in anhydrous form (zincosite) and as the heptahydrate (goslarite, natural zinc vitriol, white vitriol). It is made in large amounts by cautious sufating roasting of sphalerite (zinc blende): $ZnS + 2\,O_2 \rightarrow ZnSO_4$, by leaching roasted zinc sulfide or burned smithsonite (zinc spar) with sulfuric acid, or by reaction of zinc with sulfuric acid. Z. is an intermediate in the production of further zinc compounds. It is used to obtain electrolyte zinc, in printing textiles and to make lithopone, to impregnate wood and hides, as an additive to spinning baths for production of synthetic silks, and in electroplating. Z. is occasionally used in medicine as an emetic, and in very dilute solution, as an agent to inhibit secretions and as an astringent for washing.

Zinc sulfide: ZnS, a white powder; M_r 97.43, density 4.102 (sphalerite), 3.98 (wurtzite), m.p. (under 15 MPa) 1750 °C, subl.p. 1185 °C. Z. is found in nature as cubic sphalerite (zinc blende) and hexagonal wurtzite. Sphalerite crystallizes in a diamond lattice; wurtzite differs from this lattice in having the opposite orientation of individual ZnS_4 and SZn_4 tetrahedra. Z. is obtained by quantitative precipitation of buffered zinc salt solutions adjusted to pH 3 with hydrogen sulfide; it forms a white precipitate which is converted to the cubic crystal form by heating to 650 °C. The conversion temperature sphalerite/wurtzite is around 1020 °C. Z. is a semiconductur and has remarkable optical properties which are widely utilized. Metal-doped preparations (activation e.g. by copper or silver) fluoresce when irradiated with UV, cathode, X- or γ-rays. It also displays electroluminescence. This property is the basis for use of Z. as a luminophore, e.g. for television screens. Mixed with barium sulfate, it is used as a white paint (see Lithopone). Z. is also used as a filler for rubber and other polymers.

Zinc vitriol: see Zinc sulfate.

Zinc white: Zinc oxide (see) used as a white pigment. Z. is very fast to light and is weather-resistant. As a basic pigment, it can neutralize acids, and forms water-resistant zinc soaps with oil-containing binders. It fluoresces in UV light. Z. has a photochemical stabilizing effect, which contributes to increased light fastness of mixtures of Z. with organic pigments.

Zintl phases: see Intermetallic compounds.

Ziram: see Dithiocarbamate fungicides.

Zirconates(IV): esters $Zr(OR)_4$ and salts of the (hypothetical) ortho- and metazirconic acids, M_2ZrO_3 or M_4ZrO_4. Alkalizirconates(IV), M_2ZrO_3 or M_4ZrO_4, are formed by fusion of zirconium(IV) oxide with alkali hydroxides. Lead and barium zirconate, $PbZrO_3$ or $BaZrO_3$, are ferroelectrics, like the homologous titantates. Lead zirconate titanates are used with added lanthanum in the construction of optical data storage devices.

Zirconic acid: see Zirconium(IV) oxide.

Zirconium, symbol ***Zr***: an element of group IVb of the periodic system, the Titanium group (see); a heavy metal, *Z* 40, with natural isotopes with mass numbers 90 (51.46%), 94 (17.40%), 92 (17.11%), 91 (11.23%) and 96 (2.80%); atomic mass 91.22, valency usually IV, less often III, II, I, density 6.508, m.p. 1857 °C, b.p. 4377 °C, electrical conductivity 2.3 Sm mm^{-2}, standard electrode potential (Zr/ZrO^{2+}) -1.43 V.

Properties. Pure Z. is a shiny metal which looks similar to steel. It is relatively soft and ductile, with a hexagonal lattice at room temperature. At 867 °C, it converts to a cubic body-centered β-form. In contrast to hafnium, Z. has a very small neutron absorption cross section. The compact metal is passivated by formation of a thin, tight oxide layer, so that it is not oxidized by oxygen until heated white hot. It does not react with water, hydrochloric, nitric or sulfuric acid or aqueous alkalies, even when warm (or reacts slightly). However, it dissolves easily in aqua regia, hydrofluoric acid or melted alkalies. At dark red heat, Z. reacts with chlorine to form zirconium(IV) chloride. Z. powder burns in air to form the zirconium oxides ZrO and ZrO_2, zirconium nitride, ZrN, and zirconium oxygen nitride, Zr_2ON_2. In an oxygen atmosphere, finely divided Z. burns with the highest known temperature for a metal flame, 4660 °C. Burning Z. cannot be extinguished with water, carbon dioxide or carbon tetrachloride, but has to be covered with dry sand. Powdered Z. tends to take up oxygen, nitrogen and hydrogen. Z. forms Zirconium alloys (see) with various metals. When added to light metals, Z. makes the microcrystalline structure finer, and thus hardens the alloys.

In its comopunds, Z. is generally in the +IV oxidation state. Zirconium(IV) complexes are found to have coordination numbers of 6 (e.g. $[ZrF_6]^{2-}$), 7 ($[ZrF_7]^{3-}$) and 8 ($[Zr(C_2O_4)_4]^{4-}$). Compounds in which Z. is in a lower oxidation state, III, II or I, are less common than zirconium(IV) compounds.

Analysis. Organic reagents are often used to detect Z., e.g. tannin, kupferon, phenylarsonic acid, oxime, alizarin S or xylenol orange. Conversion to zirconium oxygen chloride, $ZrOCl_2 \cdot 8H_2O$, which crystallizes in characteristic thin needles, is also recommended. For gravimetric determination, Z. is converted to zirconium oxide hydrate by precipitation with ammonia, and this is roasted to form ZrO_2.

Occurrence. Z. makes up $1.4 \cdot 10^{-2}$% of Earth's crust. It is found as zircon, $ZrSiO_4$, and Baddeleyite (brazilite, zirconium earth), ZrO_2. Z. is also found in minerals of other elements, such as rutile, ilmenite, apatite and magnetite, in amounts up to a few tenths of a percent. On the other hand, zirconium minerals generally contain 1 to 5% hafnium oxide.

Production. The older techniques for producing powdered Z. make use of reduction of potassium hexafluorozirconate or zirconium tetrachloride with sodium. The present day industrial process begins with zircon sands, which are solubilized with sodium hydroxide solution and converted to zirconium(IV) oxide. This is converted to zirconium carbonitride with coal in an electric arc furnace; the carbonitride then reacts with chlorine to form zirconium(IV) chloride. Reduction with magnesium in the *Kroll process* produces zirconium sponge ($ZrCl_4 + 2\,Mg \rightarrow Zr + 2\,MgCl_2$), from which magnesium chloride and unreacted magnesium are removed by distillation in vacuum. The metal is finally remelted in an electrode ray furnace. Further purification can be done by Vapor deposition (see). The Z. for many industrial applications must be free of hafnium. To produce it, the first wet chemical steps include a separation of hafnium, usually by liquid-liquid extraction (see Hafnium).

Applications. Because of its low neutron-capture cross section, hafnium-free Z. is used mainly as a material for reactor fuel rod sheaths, in the form of alloys with tin, iron, chromium or nickel. The high corrosion resistance of Z. has also led to its use in construction of chemical plant equipment, especially for high-performance parts such as valves, pumps, stirrers, spinnerets and heat exchangers. It is used as a getter because of its ability to absorb gases such as nitrogen or oxygen. In medicine, Z. is used in the form of surgical instruments, clamps and screws. Z. powder is used in fireworks. It is also used to make smokeless flash bulb powder. Z. is used in metallurgy as ferrosilicon zircon for steel production, and as a deoxidizing agent for metal casting. It is used as a welding foil and solder wire for welding of molybdenum and tungsten. Z. is a component of many alloys (see Zirconium alloys).

Historical. Z. was first isolated in 1789 by Klaproth in the form of zirconium dioxide, which he obtained from a zircon from Ceylon. In 1824, Berzelius succeeded in preparing the element in powder form by reduction of potassium hexafluorozirconate with potassium. Industrial methods of producing pure Z. were only developed in the last few decades in response to the development of nuclear reactors.

Zirconium alloys: alloys in which zirconium is the main element. Because of its low capture cross section for slow neutrons, Z. with niobium, tin, iron, nickel and chromium (to give stability to heat and corrosion) are used to make fuel rod sheaths for pressurized water reactors. These alloys must not contain any hafnium, which has a large capture cross section. In the former USSR, an alloy of 1% niobium with 99% zirconium is used. The USA and Germany use Zircaloy 2® and Zircaloy 4®, which contain 1.5% tin, 0.15 to 0.2% iron, 0.1% chromium, 0.05% nickel (only in Zircaloy 2) and the rest zirconium. In chemical equipment, zirconium alloyed with 2 to 2.5% hafnium is used to clad pipes and other parts exposed to alkalies, acids and urea.

Zirconium carbide: ZrC, a metal-like, very hard compound which crystallizes in a cubic lattice; M_r 103.23, density 6.73, m.p. 3540 °C, b.p. 5100 °C. Z. does not react with alkalies or acids, not even concentrated hydrochloric or nitric acid, but it does dissolve in aqua regia or hot, concentrated sulfuric acid. It is also oxidized by halogens. It is obtained by reduction of zirconium compounds with carbon. Z. is used as a hard metal in cutting tools and as an extremely fire-resistant material.

Zirconium chlorides: ***Zirconium(IV) chloride, zirconium tetrachloride***, $ZrCl_4$, white crystals, M_r 233.03, density 2.803, subl.p. 331 °C. $ZrCl_4$ is a coordination polymer in the solid state ($ZrCl_6$ octahedra with two shared edges), and in the gas phase, consists of tetrahedral monomers. It fumes in air and is hydrolysed in water to ***zirconium oxide dichloride***, which crystallizes out of aqueous solution as the octahydrate, $ZrOCl_2 \cdot 8H_2O$. $ZrCl_4$ forms hydrolysis-resistant alkali hexachlorozirconates, $M_3[ZrCl_6]$, with alkali chlorides. It is made by reaction of zirconium or zirconium carbide with chlorine at high temperatures. It is an important industrial intermediate in the production of metallic zirconium. It is also used as a catalyst in Friedel-Crafts synthesis, and can be used as a catalyst component in low-pressure olefin polymerization. The reduction of $ZrCl_4$ leads to the chlorides of divalent and trivalent zirconium; these are stable only in the solid phase, and react with water with evolution of hydrogen. ***Zirconium(III) chloride, zirconium trichloride***, $ZrCl_3$, forms dark red-brown hexagonal crystals, M_r 197.58, density 3.00, m.p. 350 °C (dec.), and ***zirconium(II) chloride, zirconium dichloride***, $ZrCl_2$, is a black compound, M_r 162.13, density 3.6.

Zirconium diboride: ZrB_2, hexagonal crystals with a metallic appearance; M_r 112.84, density 6.085,

Mohs hardness 8, m.p. about 3040 °C. Z. does not react with hydrochloric or nitric acid, chlorine or oxygen. It is made by heating zirconium powder with boron, boron(III) oxide or boron carbide. Z. is used as a fire-resistant material for making such items as crucibles and cladding of thermoelements.

Zirconium(IV) fluoride, ***zirconium tetrafluoride***: ZrF_4, white, highly refractive, monoclinic compound which is scarcely soluble in water; M_r 167.21, density 4.43, subl.p. 903 °C. Z. reacts with fluorides to form fluorozirconates(IV), $[ZrF_6]^{2-}$, $[ZrF_7]^{3-}$ and $[ZrF_8]^{4-}$. It is obtained as the hydrate from a solution of zirconium dioxide hydrate in hydrofluoric acid. Z. can be used as a catalyst for Friedel-Crafts syntheses.

Zirconium hydride: ZrH_2, greenish black powder, stable in air; M_r 93.24, density 5.6. Z. is not decomposed by water, but is very easily oxidized. It is made by reducing zirconium(IV) oxide with calcium hydride in a hydrogen atmosphere above 660 °C. Z. is used as a getter; it is also used in pyrotechnics, powder metallurgy, the production of foam metals and for neutron diffraction.

Zirconium nitride: ZrN, a very hard, lustrous compound which looks somewhat like silver, but is yellowish; M_r 105.23, density 7.09, m.p. 2980 °C. Z. is a good electrical conductor which can be made by reaction of zirconium with nitrogen at high temperatures. Very pure Z. is made by Vapor deposition (see). It is used to make very high-temperature materials.

Zirconium(IV) oxide, ***zirconium dioxide, zirconium earth***: ZrO_2, white compound, very stable to alkalies and acids; M_r 123.22, density 5.89, m.p. about 2700 °C, b.p. about 5000 °C. Naturally occurring Z., baddeleyite (brazilite) forms monoclinic crystals. Above 1000 °C, it is reversibly converted to a second, tetragonal modification. Although slightly warmed Z. is still soluble in mineral acids, highly mineralized Z. can be dissolved only by concentrated sulfuric or hydrofluoric acid. Alkali hydroxides or carbonates convert Z. to Zirconates (see), generally of the type M_2ZrO_3. If Z. is heated to very high temperatures, it radiates a blinding white light, a property which was formerly utilized in the "Nernst rod" - a rod of 85% ZrO_2 and 15% Y_2O_3 heated above 1000 °C by electrical resistance. Z. is made from aqueous zirconium salt solutions by precipitation with ammonia. The precipitate, ***zirconium(IV) oxide hydrate***, formerly also called ***zirconic acid***, is then roasted to Z. Z. is very resistant to thermal, mechanical and chemical resistance stress, and is used to make crucibles and other chemical apparatus, to clad high-temperature furnaces, as an insulator material and catalyst carrier. Mixed with graphite, it is used as an electrical heating resistor. It is used as an abrasive, white pigment (zirconium white), opacifier for enamels and protective coating for metallic materials exposed to high temperatures. It is used to make casting outlets for cast iron and in pharmacy for pastes to treat skin diseases.

Zirconium silicate: $ZrSiO_4$, colorless, tetragonal crystals; M_r 183.30, density 4.56, m.p. 2550 °C. Z. is resistant to mineral acids, aqua regia and alkalies. It is found in nature as zircon. Transparent crystals of Z. are used as semiprecious stones for jewelry. Z. is used as a raw material for making fire-resistant materials, as a filler for plastics, to make coudy enamels and glazes and as casting sand for steel.

Zirconium sulfate: $Zr(SO_4)_2$, white compound which dissolves in water in a highly exothermal reaction; M_r 283.35, density 3.22, m.p. 410 °C (dec.). Z. crystallizes out of water as the tetrahydrate, $Zr(SO_4)_2 \cdot 4H_2O$; M_r 355.41, density 3.22. It is obtained by treating zirconium(IV) oxide with sulfuric acid, and is hydrolysed in water to basic Z., which is used in tanning to make a white, high-quality leather. Z. is also used to impregnate textiles to make them resistant to flame and water-repellant.

Zn: symbol for zinc.

Zonal enrichment: see Zonal melting.

Zonal fractionation: see Zonal melting.

Zonal melting: the migration of a melt zone along a rod, used for high purification of substances; the impurities or secondary components become enriched in the melt zone and are transported to the end of the rod (Fig.). If the liquid zone is made to pass through the rod several times in the same direction, the purity of the material at the end where the passes begin can be increased still more. Z. is used for purification (***zonal purification***), separation of binary mixtures (***zonal fractionation***) and enrichment of traces (***zonal enrichment***). The greatest significance of Z. is for the purification of semiconductors (see Zonal melt process) and production of single crystals (see Crystal growing).

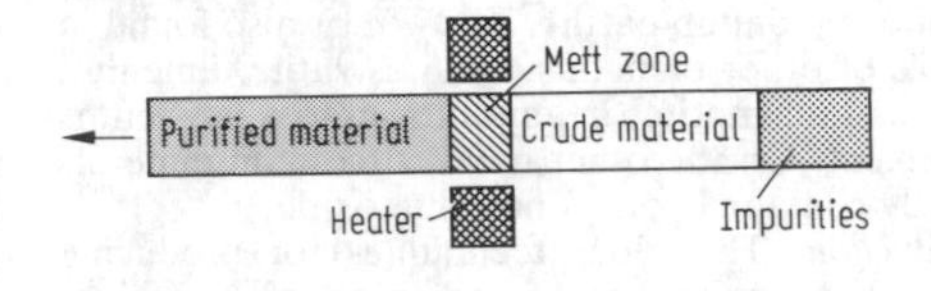

Zonal melting process: a method for purifying metals, semiconductors, inorganic and organic compounds. Z. is used in industry for refining element semiconductors, such as silicon and germanium, and semiconducting compounds, such as GaAs and InSb. Single crystals can also be produced by the Z. The purification effect is due to the differential distribution of the impurities in the solid and liquid phases of the material. Induction heating is used to melt a narrow zone in a rod of the material, and this zone is

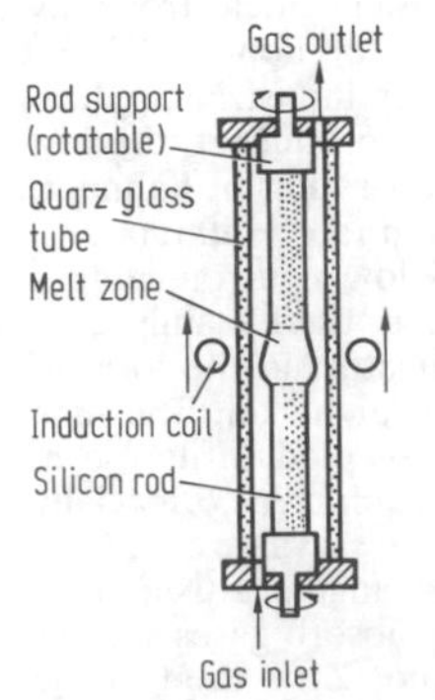

Production of extremely pure silicon by the zonal melt process.

moved along the material by motion of the induction coil. The impurities tend to become concentrated in the melt zone, but if they are stably incorporated into the solid lattice, they may become enriched in it. The more often the melt zone travels through the material, the more highly purified it becomes. The part of the material in which the impurities have been concentrated is removed and either reprocessed or thrown away.

The process can be either horizontal, in which case the material to be melted is placed in a long crucible or an ampule, or vertical, as shown in the figure. In this case, the melt zone is held in place between the solid regions by the surface tension of the liquid.

To prevent chemical reactions of the material, Z. is carried out in a high vacuum or under an inert gas (argon, helium or hydrogen). For purification of very high-melting materials, an electron beam furnace is used instead of inductive heating. Z. was developed in 1952 by Pfann for purification of germanium, and in 1957 it was first applied to organic substances by Schildknecht.

Zonal purification: see Zonal melting.

Zr: symbol for zirconium.

Zwitterions: compounds in which the molecules contain both positive and negative centers of charge. They thus have the structure of "internal salts". Common positively charged groups are ammonium and sulfonium, while negative groups arise by loss of protons from carboxyl, sulfonyl, phosphate, phenolic or enolic hydroxyl groups, etc. The amino acids, phosphatides (lecithins), amine oxides and oxides of various heterocycles (e.g. pyridine-*N*-oxide), diazoalkanes and diazoketones, ylides, and so on, are well-known Z.

Zymogens: inactive precursors of enzymes, especially proteolytic enzymes of digestion or blood coagulation. At the site of action, they are converted to the active form by limited proteolysis. Z. are often the storage forms of the enzymes.

SI Basis Units

Quantity Name	Quantity Symbol	Unit Name	Unit Symbol
length	l	meter	m
time	t	second	s
mass	m	kilogram	kg
amount of substance	n	mole	mol
electric current	I	ampere	A
thermodynamic temperature	T	kelvin	K
luminous intensity	I_v	candela	cd

SI Prefixes

Factor	Prefix	Symbol	Factor	Prefix	Symbol
10^{18}	exa	E	10^{-1}	deci	d
10^{15}	peta	P	10^{-2}	centi	c
10^{12}	tera	T	10^{-3}	milli	m
10^{9}	giga	G	10^{-6}	micro	μ
10^{6}	mega	M	10^{-9}	nano	n
10^{3}	kilo	k	10^{-12}	pico	p
10^{2}	hecto	h	10^{-15}	femto	f
10^{1}	deka	da	10^{-18}	atto	a

Important Conversion Factors

Quantity	SI Unit	Conversion Factor
length	meter m	1 inch = 25,40 mm; 1 foot = 30,478 cm 1 yard = 0,9144 m; 1 mile = 1,6093 km 1 Ångström = 10^{-10} m
mass	kilogram kg	1 ounce = 28,35 g; 1 pound = 0,4536 kg 1 atomic mass unit = $1{,}66057 \cdot 10^{-27}$ kg
temperature	kelvin K	$t \hat{=} T - 273{,}15$ K (Celsius temperature t in °C, thermodynamic temperature T in K)
force	newton N	1 N = 10^{5} dyn = 0,10197 kp
pressure	pascal Pa	1 Pa = 10^{-5} bar = $0{,}987 \cdot 10^{-5}$ atm = 0,0075 Torr; 1 atm = 1,01325 bar
work (energy)	joule J	1 J = 0,2390 cal = $6{,}242 \cdot 10^{18}$ eV = $2{,}778 \cdot 10^{-7}$ kWh; 1 cal = 4,187 J
power	watt W	1 W = 859,8 cal h^{-1} = $1{,}35962 \cdot 10^{-3}$ PS
radioactivity	bequerel Bq	1 Curie (Ci) = $3{,}700 \cdot 10^{10}$ Bq
energy dose	gray Gy	1 Rad (rd) = 0,01 Gy

Number of Protons (Atomic Number) and

Element Name	Element Symbol	Atomic Number	Relative Atomic Mass
Actinium*	Ac	89	(227)
Aluminium	Al	13	26.981539
Americium*	Am	95	(243)
Antimony (Stibium)	Sb	51	121.757
Argon	Ar	18	39.948
Arsenic	As	33	74.92159
Astatine*	At	85	(210)
Barium	Ba	56	137.327
Berkelium*	Bk	97	(247)
Beryllium	Be	4	9.012182
Bismuth	Bi	83	208.98037
Boron	B	5	10.811
Bromine	Br	35	79.904
Cadmium	Cd	48	112.411
Caesium	Cs	55	132.90543
Calcium	Ca	20	40.078
Californium*	Cf	98	(251)
Carbon	C	6	12.011
Cerium	Ce	58	140.115
Chlorine	Cl	17	35.4527
Chromium	Cr	24	51.9961
Cobalt	Co	27	58.93320
Copper	Cu	29	63.546
Curium*	Cm	96	(247)
Dysprosium	Dy	66	162.50
Einsteinium*	Es	99	(252)
Erbium	Er	68	167.26
Europium	Eu	63	151.965
Fermium*	Fm	100	(257)
Fluorine	F	9	18.998403
Francium*	Fr	87	(223)
Gadolinium	Gd	64	157.25
Gallium	Ga	31	69.723
Germanium	Ge	32	72.61
Gold	Au	79	196.96654
Hafnium	Hf	72	178.49
Helium	He	2	4.002602
Holmium	Ho	67	164.93032
Hydrogen	H	1	1.00794
Indium	In	49	114.818
Iodine	I	53	126.90447
Iridium	Ir	77	192.22
Iron	Fe	26	55.847
Krypton	Kr	36	83.80
Lanthanum	La	57	138.9055
Lawrencium*	Lr	103	(260)
Lead	Pb	82	207.2
Lithium	Li	3	6.941
Lutetium	Lu	71	174.967
Magnesium	Mg	12	24.3050
Manganese	Mn	25	54.93805
Mendelevium*	Md	101	(258)
Mercury	Hg	80	200.59
Molybdenum	Mo	42	95.94
Neodymium	Nd	60	144.24
Neon	Ne	10	20.1797
Neptunium*	Np	93	(237)
Nickel	Ni	28	58.6934
Niobium	Nb	41	92.90638
Nitrogen	N	7	14.00674
Nobelium*	No	102	(259)
Osmium	Os	76	190.23